CHILTON'S GUIDE TO
FUEL INJECTION & ELECTRONIC ENGINE CONTROLS—1988-90
GENERAL MOTORS

President	Gary R. Ingersoll
Senior Vice President	Ronald A. Hoxter
Publisher	Kerry A. Freeman, S.A.E.
Editor-In-Chief	Dean F. Morgantini, S.A.E.
Managing Editor	David H. Lee, A.S.E., S.A.E.
Manager of Manufacturing	John J. Cantwell
Production Manager	W. Calvin Settle, Jr., S.A.E.
Senior Editor	Richard J. Rivele, S.A.E.
Senior Editor	Nick D'Andrea
Senior Editor	Ron Webb

CHILTON BOOK COMPANY
ONE OF THE ABC PUBLISHING COMPANIES,
A PART OF CAPITAL CITIES/ABC, INC.

Manufactured in USA
© 1989 Chilton Book Company
Chilton Way, Radnor, PA 19089
ISBN 0–8019–7954–4
ISSN 1052–9144
3456789012   0987654321

HOW TO USE THIS MANUAL

For ease of use, this manual is divided into sections as follows:

SECTION 1 Basic Electricity
SECTION 2 Troubleshooting and Diagnosis
SECTION 3 Self-Diagnostic Systems
SECTION 4 Electronic Ignition Systems
SECTION 5 Fuel Injection Systems

The **CONTENTS** summarize the subjects covered in each section.

To quickly locate the proper service section, use the application chart on the following pages. It references applicable **CAR AND TRUCK MODELS** and **SERVICE SECTIONS** for major electronic engine control systems.

It is recommended that the user be familiar with the applicable **GENERAL INFORMATION, SERVICE PRECAUTIONS** and **TROUBLESHOOTING AND DIAGNOSIS TECHNIQUES** before testing or servicing any engine control system.

Major service sections are grouped by vehicle manufacturer, with each engine control system subsection containing:

- **GENERAL INFORMATION** pertaining to the operation of the system, individual components and the overall logic by which components work together.

- **SERVICE PRECAUTIONS** (if any) of which the user should be aware to prevent injury or damage to the vehicle or components.

- **FAULT DIAGNOSIS** in the form of diagnostic charts or test procedures which lead the user through the various system circuit tests and explain the trouble codes stored in the computer memory.

SAFETY NOTICE

Proper service and repair procedures are vital to the safe, reliable operation of all motor vehicles, as well as the personal safety of those performing service or repairs. This manual outlines procedures for servicing and repairing vehicles using safe, effective methods. The procedures contain many NOTES and CAUTIONS which should be followed along with standard safety procedures to eliminate the possibility of personal injury or improper service which could damage the vehicle or compromise its safety.

It is important to note that repair procedures and techniques, tools and parts for servicing motor vehicles, as well as the skill and experience of the individual performing the work vary widely. It is not possible to anticipate all of the hazards that may result. Standard and accepted safety precautions and equipment should be used when handling toxic or flammable fluids and safety goggles or other protection should be used during cutting, grinding, chiseling, prying or any other process that that can cause material removal or projectiles. Similar protection against the high voltages generated in all electronic ignition systems should be employed during service procedures.

Some procedures require the use of tools or test equipment specially designed for a specific purpose. Before substituting another tool or procedure, you must be completely satisfied that neither your personal safety, nor the performance of the vehicle will be endangered.

PART NUMBERS

Part numbers listed in this reference are not recommendations by Chilton for any product by brand name. They are references that can be used with interchange manuals and aftermarket supplier catalogs to locate each brand supplier's discrete part number.

Although information in this manual is based on industry sources and is complete as possible at the time of publication, the possibilty exists that some car manufacturers made later changes which could not be included here. While striving for total accuracy, Chilton Book Company cannot assume responsibility for any errors, changes or omissions that may occur in the compilation of this data.

Contents

SECTION 1 BASIC ELECTRICITY

Fundamentals of Electricity	1-2
Units of Electrical Measurement	1-2
OHM's Law	1-3
Electrical Circuits	1-3
Magnetism and Electromagnets	1-4
Basic Solid State	1-6
Microprocessors, Computers and Logic Systems	1-7

2 TROUBLESHOOTING AND DIAGNOSIS

Diagnostic Epuipment and Special Tools	2-2
Safety Precautions	2-2
Organized Troubleshooting	2-2
Jumper Wires	2-3
12 Volt Test Light	2-3
Voltmeter	2-4
Ohmmeter	2-5
Ammeter	2-6
Multimeters	2-7
Special Test Equipment	2-7
Reading Wiring Diagrams	2-8
Wiring Repairs	2-13
Mechanical Test Equipment	2-17

SECTION 3 SELF-DIAGNOSTIC SYSTEMS

All Models Except E, K and Cadillac C-Body	**3-2**
Entering Diagnostic Mode	3-2
Diagnostic Trouble Code Chart	3-4
E, K, and Cadillac C-Body	**3-6**
1988-90 Buick Riviera and Reatta	3-6
1988-90 Cadillac DeVille and Fleetwood	3-14
1988-90 Cadillac Eldorado and Seville	3-17

SECTION 4 ELECTRONIC IGNITION SYSTEMS

High Energy Ignition (HEI) System	4-2
Electronic Spark Timing (EST) System	4-10
Electronic Spark Control (ESC) System	4-13/4-15
Computer Controlled Coil Ignition (C^3I)	4-15
Direct Ignition System (DIS)	4-15
Integrated Direct Ignition (IDI) System	4-15

SECTION 5 FUEL INJECTION SYSTEMS

Cadillac Digital Fuel Injection (DFI) System	**5-4**
DeVille and Fleetwood Diagnostic Charts	5-12
Eldorado and Seville Diagnostic Charts	5-56
Multi-Port (MPI) and Tuned Port Injection (TPI)	**5-107**
2.0L (VIN M) Diagnostic Charts	5-128
2.3L (VIN D and A) Diagnostic Charts	5-153
2.8L (VIN S) Diagnostic Charts	5-184
2.8L (VIN W) Diagnostic Charts	5-210
3.0L (VIN L) Diagnostic Charts	5-333
3.1L (VIN T) Diagnostic Charts	5-358
3.3L (VIN N) Diagnostic Charts	5-386
5.0L (VIN F) Diagnostic Charts	5-436
5.7L (VIN 8) Diagnostic Charts	5-460
Throttle Body Injection (TBI) System	**5-498**
2.0L (VIN 1) Diagnostic Charts	5-522
2.0L (VIN K) Diagnostic Charts	5-566
2.5L (VIN R) Diagnostic Charts	5-590
2.5L (VIN U) Diagnostic Charts	5-612
5.0L (VIN E) Diagnostic Charts	5-632
4.3L (VIN Z) Diagnostic Charts	5-658
5.0L (VIN E) and 5.7L (VIN 7) Diagnostic Charts	5-660
2.5L (VIN E) Diagnostic Charts	5-680
2.8L (VIN R) Diagnostic Charts	5-682
4.3L (VIN Z) Diagnostic Charts	5-683
5.0L (VIN H) and 5.7L (VIN K) Diagnostic Charts	5-684
7.4L (VIN N) Diagnostic Charts	5-686
Sequential Fuel Injection (SFI) System	**5-732**
3800 (VIN C) Diagnostic Charts-All Except E-Body	5-744
3.8L (VIN 3) Diagnostic Charts	5-776
3800 (VIN C) Diagnostic Charts-E-Body	5-814

BUICK

Models	Series VIN	Engine	Fuel System	Engine VIN Code	Models	Series VIN	Engine	Fuel System	Engine VIN Code
1988					**1989-90**				
Century	A	2.8L	MPI	W	Century	A	2.8L	MPI	W
Century	A	2.5L	TBI	R	Century	A	3.3L	MPI	N
Century	A	3.8L	SFI	3	Century	A	2.5L	TBI	R
Electra	C	3.8L	SFI	3	Electra	C	3.8L	SFI	C
Electra	C	3.8L	SFI	C	LeSabre	H	3.8L	SFI	C
LeSabre	H	3.8L	SFI	3	LeSabre Wagon	B	5.0L	Carb.	Y
LeSabre	H	3.8L	SFI	C	Reatta	E	3.8L	SFI	C
LeSabre Wagon	B	5.0L	Carb.	Y	Regal	W	2.8L	MPI	W
Reatta	E	3.8L	SFI	C	Regal	W	3.1L	MPI	T
Regal	W	2.8L	MPI	W	Riviera	E	3.8L	SFI	C
Riviera	E	3.8L	SFI	C	Skyhawk	J	2.0L	TBI	K
Skyhawk	J	2.0L	TBI	K	Skyhawk	J	2.0L	TBI	1
Skyhawk	J	2.0L	TBI	1	Skylark	N	2.5L	TBI	U
Skylark	N	2.5L	TBI	U	Skylark	N	2.3L	MPI	D
Skylark	N	2.3L	MPI	D	Skylark	N	3.3L	MPI	N
Skylark	N	3.0L	MPI	L					

CADILLAC

Models	Series VIN	Engine	Fuel System	Engine VIN Code	Models	Series VIN	Engine	Fuel System	Engine VIN Code
1988					**1989-90**				
Brougham	D	5.0L	Carb.	Y	Brougham	D	5.0L	Carb.	Y
Brougham	D	5.0L	Carb.	9	Brougham	D	5.0L	Carb.	9
Cimmaron	J	2.8L	MPI	W	Cimmaron	J	2.8L	MPI	W
Deville	C	4.5L	DFI	5	Deville	C	4.5L	DFI	5
Eldorado	E	4.5L	DFI	5	Eldorado	E	4.5L	DFI	5
Fleetwood	C	4.5L	DFI	5	Fleetwood	C	4.5L	DFI	5
Seville	K	4.5L	DFI	5	Seville	K	4.5L	DFI	5

Carb. Carburetor
DFI Digital Fuel Injection
MPI Multi-Port Fuel Injection
SFI Sequential Fuel Injection
TBI Throttle Body Injection
TPI Tuned Port Injection

CHEVROLET

Models	Series VIN	Engine	Fuel System	Engine VIN Code	Models	Series VIN	Engine	Fuel System	Engine VIN Code
1988					1989-90				
Beretta	L	2.0L	TBI	1	Beretta	L	2.0L	TBI	1
Beretta	L	2.8L	MPI	W	Beretta	L	2.8L	MPI	W
Camaro	F	2.8L	MPI	S	Camaro	F	2.8L	MPI	S
Camaro	F	5.0L	MPI	F	Camaro	F	5.0L	MPI	F
Camaro	F	5.7L	MPI	8	Camaro	F	5.7L	MPI	8
Camaro	F	5.0L	TBI	E	Camaro	F	5.0L	TBI	E
Caprice	B	4.3L	TBI	Z	Caprice	B	4.3L	TBI	Z
Caprice	B	5.0L	TBI	E	Caprice	B	5.0L	TBI	E
Caprice	B	5.0L	Carb.	H	Caprice	B	5.0L	Carb.	Y
Caprice	B	5.7L	Carb.	6	Cavalier	J	2.0L	TBI	1
Cavalier	J	2.0L	TBI	1	Cavalier	J	2.8L	MPI	W
Cavalier	J	2.8L	MPI	W	Celebrity	A	2.5L	TBI	R
Celebrity	A	2.5L	TBI	R	Celebrity	A	2.8L	MPI	W
Celebrity	A	2.8L	MPI	W	Corsica	L	2.0L	TBI	1
Corsica	L	2.0L	TBI	1	Corsica	L	2.8L	MPI	W
Corsica	L	2.8L	MPI	W	Corvette	Y	5.7L	TPI	8
Corvette	Y	5.7L	TPI	8	Lumina	W	2.5L	TBI	R
Lumina	W	2.5L	TBI	R	Lumina	W	3.1L	MPI	T
Lumina	W	3.1L	MPI	T	Monte Carlo	G	4.3L	TBI	Z
Monte Carlo	G	4.3L	TBI	Z	Astro Van	M	2.5L	TBI	E
Monte Carlo	G	4.3L	Carb.	Z	Light Truck	C/K	2.5L	TBI	E
Astro Van	M	2.5L	TBI	E	S-10 and S-15	S/T	2.5L	TBI	E
Light Truck	C/K	2.5L	TBI	E	Van	G	2.5L	TBI	E
S-10 and S-15	S/T	2.5L	TBI	E	Astro Van	M	2.8L	TBI	R
Van	G	2.5L	TBI	E	Light Truck	C/K	2.8L	TBI	R
Astro Van	M	2.8L	TBI	R	S-10 and S-15	S/T	2.8L	TBI	R
Light Truck	C/K	2.8L	TBI	R	Van	G	2.8L	TBI	R
S-10 and S-15	S/T	2.8L	TBI	R	Light Truck	C/K	4.3L	TBI	Z
Van	G	2.8L	TBI	R	S-10 and S-15	S/T	4.3L	TBI	Z
Light Truck	C/K	4.3L	TBI	Z	Van	G	4.3L	TBI	Z
S-10 and S-15	S/T	4.3L	TBI	Z	Light Truck	C/K	5.0L	TBI	H
Van	G	4.3L	TBI	Z	Van	G	5.0L	TBI	H
Light Truck	C/K	5.0L	TBI	H	Van	G	5.7L	TBI	K
Van	G	5.0L	TBI	H	Light Truck	C/K	7.4L	TBI	N
Light Truck	C/K	5.7L	TBI	K					
Van	G	5.7L	TBI	K					
Light Truck	C/K	7.4L	TBI	N					

Carb.   Carburetor
DFI   Digital Fuel Injection
MPI   Multi-Port Fuel Injection
SFI   Sequential Fuel Injection
TBI   Throttle Body Injection
TPI   Tuned Port Injection

OLDSMOBILE

Models	Series VIN	Engine	Fuel System	Engine VIN Code	Models	Series VIN	Engine	Fuel System	Engine VIN Code
1988					**1989-90**				
Calais	N	2.3L	MPI	D	Calais	N	2.3L	MPI	D
Calais	N	2.5L	TBI	U	Calais	N	2.3L	MPI	A
Calais	N	3.0L	MPI	L	Calais	N	2.5L	TBI	U
Ciera	A	2.5L	TBI	R	Calais	N	3.3L	MPI	N
Ciera	A	2.8L	MPI	W	Ciera	A	2.5L	TBI	R
Ciera	A	3.8L	SFI	3	Ciera	A	2.8L	MPI	W
Cutlass Supreme	W	2.8L	MPI	W	Ciera	A	3.3L	MPI	N
Cutlass Cruiser	A	2.5L	TBI	R	Cutlass Supreme	W	2.8L	MPI	W
Cutlass Cruiser	A	2.8L	MPI	W	Cutlass Supreme	W	3.1L	MPI	T
Custom Cruiser	B	5.0L	Carb.	Y	Cutlass Cruiser	A	2.5L	TBI	R
Delta 88	H	3.8L	SFI	3	Cutlass Cruiser	A	2.8L	MPI	W
Delta 88	H	3.8L	SFI	C	Cutlass Cruiser	A	3.3L	MPI	N
Firenza	J	2.0L	TBI	1	Custom Cruiser	B	5.0L	Carb.	Y
Firenza	J	2.0L	TBI	K	Delta 88	H	3.8L	SFI	C
Ninety-Eight	C	3.8L	SFI	3	Firenza	J	2.0L	TBI	1
Ninety-Eight	C	3.8L	SFI	C	Firenza	J	2.0L	TBI	K
Toronado	E	3.8L	SFI	C	Ninety-Eight	C	3.8L	SFI	C
					Toronado	E	3.8L	SFI	C

PONTIAC

Models	Series VIN	Engine	Fuel System	Engine VIN Code	Models	Series VIN	Engine	Fuel System	Engine VIN Code
1988					**1989-90**				
Bonneville	H	3.8L	SFI	3	Bonneville	H	3.8L	SFI	C
Bonneville	H	3.8L	SFI	C	Firebird	F	2.8L	MPI	S
Fiero	P	2.5L	TBI	R	Firebird	F	5.0L	MPI	F
Fiero	P	2.8L	MPI	9	Firebird	F	5.7L	MPI	8
Firebird	F	2.8L	MPI	S	Firebird	F	5.0L	TBI	E
Firebird	F	5.0L	MPI	F	Grand Am	N	2.0L	MPI	M
Firebird	F	5.7L	MPI	8	Grand Am	N	2.3L	MPI	D
Firebird	F	5.0L	TBI	E	Grand Am	N	2.3L	MPI	A
Grand Am	N	2.0L	MPI	M	Grand Am	N	2.5L	TBI	U
Grand Am	N	2.3L	MPI	D	Grand Prix	W	2.8L	MPI	W
Grand Am	N	2.5L	TBI	U	Grand Prix	W	3.1L	MPI	T
Grand Prix	W	2.8L	MPI	W	Safari	B	5.0L	Carb.	Y
Safari	B	5.0L	Carb.	Y	Sunbird	J	2.0L	MPI	M
Sunbird	J	2.0L	MPI	M	Sunbird	J	2.0L	TBI	K
Sunbird	J	2.0L	TBI	K	Transport	U	3.1L	TBI	D
Transport	U	3.1L	TBI	D	6000	A	2.5L	TBI	R
6000	A	2.5L	TBI	R	6000	A	2.8L	MPI	W
6000	A	2.8L	MPI	W	6000	A	3.1L	MPI	T

Carb. Carburetor	MPI Multi-Port Fuel Injection	TBI Throttle Body Injection
DFI Digital Fuel Injection	SFI Sequential Fuel Injection	TPI Tuned Port Injection

Basic Electricity

INDEX

FUNDAMENTALS OF ELECTRICITY **1-2**
    Units of Electrical Measurement 1-2
    Ohm's Law 1-3

ELECTRICAL CIRCUITS **1-3**
    Circuit Breakers 1-4
    Series Circuits 1-4
    Parallel Circuits 1-4
    Voltage Drop 1-4

MAGNETISM AND ELECTROMAGNETS **1-4**
    Relays 1-5
    Buzzers 1-6
    Solenoids 1-6

BASIC SOLID STATE **1-6**
    Diodes 1-6
    LED's 1-6
    Transistors 1-7
    Integrated Circuits 1-7

MICROPROCESSORS, COMPUTERS AND LOGIC SYSTEMS **1-7**
    Basic Logic Functions 1-8
    Input Devices 1-8
    Output Devices 1-8
    Logic Circuits 1-8
    Programs 1-9
    Computer Memory 1-9
    CalPak 1-10
    Mem-Cal 1-10

FUNDAMENTALS OF ELECTRICITY

A good understanding of basic electrical theory and how circuits work is necessary to successfully perform the service and testing outlined in this manual. Therefore, this section should be read before attempting any diagnosis and repair.

All matter is made up of tiny particles called molecules. Each molecule is made up of two or more atoms. Atoms may be divided into even smaller particles called protons, neutrons and electrons. These particles are the same in all matter and differences in materials (hard or soft, conductive or non-conductive) occur only because of the number and arrangement of these particles. In other words, the protons, neutrons and electrons in a drop of water are the same as those in an ounce of lead, there are just more of them (arranged differently) in a lead molecule than in a water molecule. Protons and neutrons packed together form the nucleus of the atom, while electrons orbit around the nucleus much the same way as the planets of the solar system orbit around the sun.

The proton is a small positive natural charge of electricity, while the neutron has no electrical charge. The electron carries a negative charge equal to the positive charge of the proton. Every electrically neutral atom contains the same number of protons and electrons, the exact number of which determines the element. The only difference between a conductor and an insulator is that a conductor possesses free electrons in large quantities, while an insulator has only a few. An element must have very few free electrons to be a good insulator and vice-versa. When we speak of electricity, we're talking about these free electrons.

In a conductor, the movement of the free electrons is hindered by collisions with the adjoining atoms of the element (matter). This hindrance to movement is called **RESISTANCE** and it varies with different materials and temperatures. As temperature increases, the movement of the free electrons increases, causing more frequent collisions and therefore increasing resistance to the movement of the electrons. The number of collisions (resistance) also increases with the number of electrons flowing (current). Current is defined as the movement of electrons through a conductor such as a wire. In a conductor (such as copper) electrons can be caused to leave their atoms and move to other atoms. This flow is continuous in that every time an atom gives up an electron, it collects another one to take its place. This movement of electrons is called electric current and is measured in amperes. When 6.28 billion, billion electrons pass a certain point in the circuit in one second, the amount of current flow is called 1 ampere.

The force or pressure which causes electrons to flow in any conductor (such as a wire) is called **VOLTAGE**. It is measured in volts and is similar to the pressure that causes water to flow in a pipe. Voltage is the difference in electrical pressure measured between 2 different points in a circuit. In a 12 volt system, for example, the force measured between the two battery posts is 12 volts. Two important concepts are voltage potential and polarity. Voltage potential is the amount of voltage or electrical pressure at a certain point in the circuit with respect to another point. For example, if the voltage potential at one post of the 12 volt battery is 0, the voltage potential at the other post is 12 volts with respect to the first post. One post of the battery is said to be positive (+); the other post is negative (−) and the conventional direction of current flow is from positive to negative in an electrical circuit. It should be noted that the electron flow in the wire is opposite the current flow. In other words, when the circuit is energized, the current flows from positive to negative, but the electrons actually move from negative to positive. The voltage or pressure needed to produce a current flow in a circuit must be greater than the resistance present in the circuit. In other words, if the voltage drop across the resistance is greater than or equal to the voltage input, the voltage potential will be

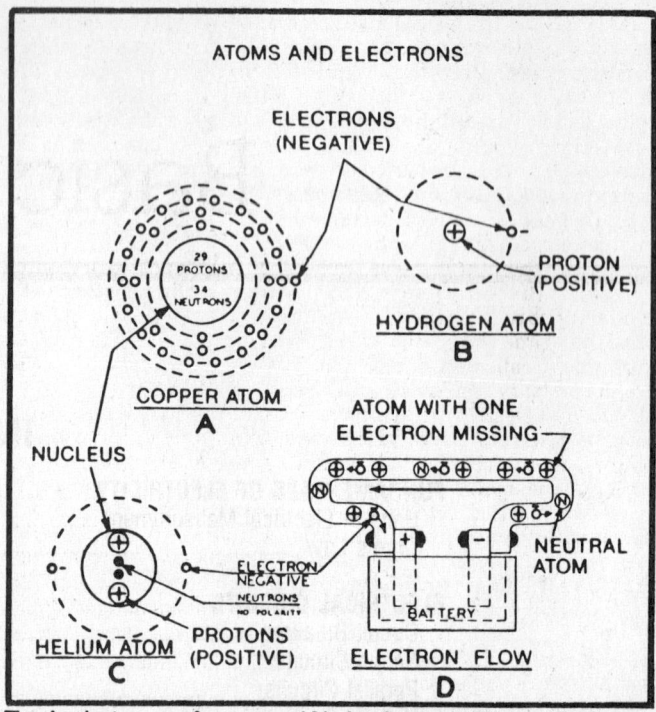

Typical atoms of copper (A), hydrogen (B) and helium (C). Electron flow in battery circuit (D)

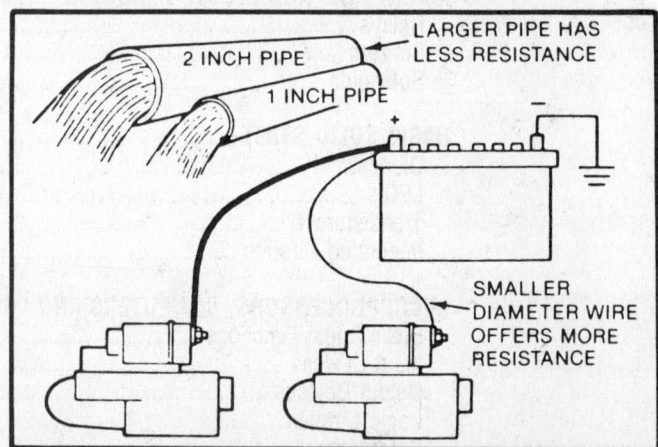

Electrical resistance can be compared to water flow through a pipe. The smaller the wire (pipe), the more resistance to the flow of electrons (water)

zero — no voltage will flow through the circuit. Resistance to the flow of electrons is measured in ohms. One volt will cause 1 ampere to flow through a resistance of 1 ohm.

Units Of Electrical Measurement

There are 3 fundamental characteristics of a direct-current electrical circuit: volts, amperes and ohms.

VOLTAGE in a circuit controls the intensity with which the loads in the circuit operate. The brightness of a lamp, the heat of an electrical defroster, the speed of a motor are all directly proportional to the voltage, if the resistance in the circuit and/or mechanical load on electric motors remains constant. Voltage available from the battery is constant (normally 12 volts), but as it operates the various loads in the circuit, voltage decreases (drops).

AMPERE is the unit of measurement of current in an electrical circuit. One ampere is the quantity of current that will flow through a resistance of 1 ohm at a pressure of 1 volt. The amount of current that flows in a circuit is controlled by the voltage and the resistance in the circuit. Current flow is directly proportional to resistance. Thus, as voltage is increased or decreased, current is increased or decreased accordingly. Current is decreased as resistance is increased. However, current is also increased as resistance is decreased. With little or no resistance in a circuit, current is high.

OHM is the unit of measurement of resistance, represented by the Greek letter Omega (Ω). One ohm is the resistance of a conductor through which a current of one ampere will flow at a pressure of one volt. Electrical resistance can be measured on an instrument called an ohmmeter. The loads (electrical devices) are the primary resistances in a circuit. Loads such as lamps, solenoids and electric heaters have a resistance that is essentially fixed; at a normal fixed voltage, they will draw a fixed current. Motors, on the other hand, do not have a fixed resistance. Increasing the mechanical load on a motor (such as might be caused by a misadjusted track in a power window system) will decrease the motor speed. The drop in motor rpm has the effect of reducing the internal resistance of the motor because the current draw of the motor varies directly with the mechanical load on the motor, although its actual resistance is unchanged. Thus, as the motor load increases, the current draw of the motor increases, and may increase up to the point where the motor stalls (cannot move the mechanical load).

Circuits are designed with the total resistance of the circuit taken into account. Troubles can arise when unwanted resistances enter into a circuit. If corrosion, dirt, grease, or any other contaminant occurs in places like switches, connectors and grounds, or if loose connections occur, resistances will develop in these areas. These resistances act like additional loads in the circuit and cause problems.

OHM'S LAW

Ohm's law is a statement of the relationship between the 3 fundamental characteristics of an electrical circuit. These rules apply to direct current (DC) only.

Ohm's law provides a means to make an accurate circuit analysis without actually seeing the circuit. If, for example, one wanted to check the condition of the rotor winding in a alternator whose specifications indicate that the field (rotor) current draw is normally 2.5 amperes at 12 volts, simply connect the rotor to a 12 volt battery and measure the current with an ammeter. If it measures about 2.5 amperes, the rotor winding can be assumed good.

An ohmmeter can be used to test components that have been removed from the vehicle in much the same manner as an ammeter. Since the voltage and the current of the rotor windings used as an earlier example are known, the resistance can be calculated using Ohms law. The formula would be ohms equals volts divided by amperes.

If the rotor resistance measures about 4.8 ohms when checked with an ohmmeter, the winding can be assumed good. By plugging in different specifications, additional circuit information can be determined such as current draw, etc.

Electrical Circuits

An electrical circuit must start from a source of electrical supply and return to that source through a continuous path. Circuits are designed to handle a certain maximum current flow. The

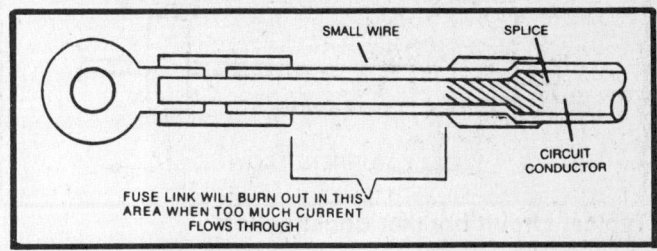

$$I = \frac{E}{R} \quad \text{or} \quad AMPERES = \frac{VOLTS}{OHMS}$$

$$R = \frac{E}{I} \quad \text{or} \quad OHMS = \frac{VOLTS}{AMPERES}$$

$$E = I \times R \quad \text{or} \quad VOLTS = AMPERES \times OHMS$$

Ohms Law is the basis for all electrical measurements. By simply plugging in two values, the third can be calculated using the illustrated formula.

$$R = \frac{E}{I} \qquad \text{Where:} \quad E = 12 \text{ volts}$$
$$I = 2.5 \text{ amperes}$$
$$R = \frac{12 \text{ volts}}{2.5 \text{ amps}} = 4.8 \text{ ohms}$$

An example of calculating resistance (R) when the voltage (E) and amperage (I) is known.

SMALL WIRE   SPLICE

CIRCUIT CONDUCTOR

FUSE LINK WILL BURN OUT IN THIS AREA WHEN TOO MUCH CURRENT FLOWS THROUGH

Typical fusible link wire

maximum allowable current flow is designed higher than the normal current requirements of all the loads in the circuit. Wire size, connections, insulation, etc., are designed to prevent undesirable voltage drop, overheating of conductors, arcing of contacts and other adverse effects. If the safe maximum current flow level is exceeded, damage to the circuit components will result; it is this condition that circuit protection devices are designed to prevent.

Protection devices are fuses, fusible links or circuit breakers designed to open or break the circuit quickly whenever an overload, such as a short circuit, occurs. By opening the circuit quickly, the circuit protection device prevents damage to the wiring, battery and other circuit components. Fuses and fusible links are designed to carry a preset maximum amount of current and to melt when that maximum is exceeded, while circuit breakers merely break the connection and may be manually reset. The maximum amperage rating of each fuse is marked on the fuse body and all contain a see-through portion that shows the break in the fuse element when blown. Fusible link maximum amperage rating is indicated by gauge or thickness of the wire. Never replace a blown fuse or fusible link with one of a higher amperage rating.

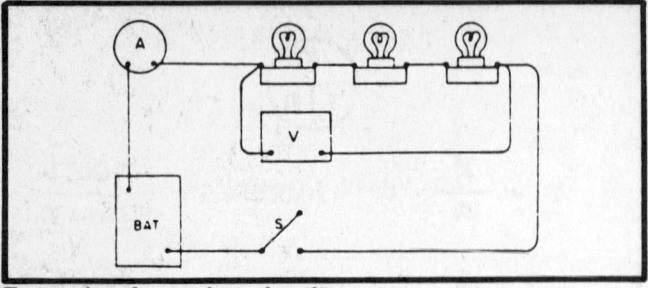

Example of a series circuit

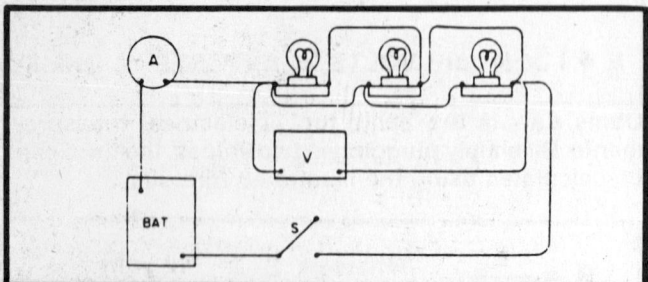

Example of a parallel circuit

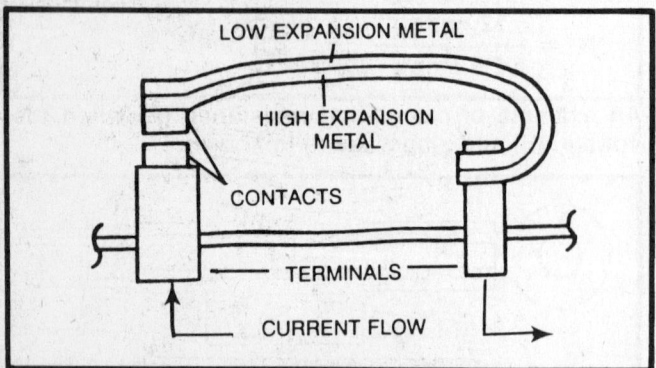

Typical circuit breaker construction

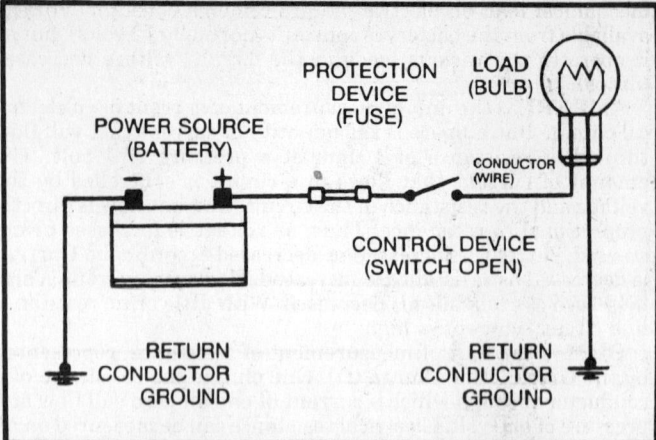

Typical circuit with all essential components

CAUTION

Resistance wires, like fusible links, are also spliced into conductors in some areas. Do not make the mistake of replacing a fusible link with a resistance wire. Resistance wires are longer than fusible links and are stamped "RESISTOR-DO NOT CUT OR SPLICE."

Circuit breakers consist of 2 strips of metal which have different coefficients of expansion. As an overload or current flows through the bimetallic strip, the high-expansion metal will elongate due to heat and break the contact. With the circuit open, the bimetal strip cools and shrinks, drawing the strip down until contact is re-established and current flows once again. In actual operation, the contact is broken very quickly if the overload is continuous and the circuit will be repeatedly broken and remade until the source of the overload is corrected.

The self-resetting type of circuit breaker is the one most generally used in automotive electrical systems. On manually reset circuit breakers, a button will pop up on the circuit breaker case. This button must be pushed in to reset the circuit breaker and restore power to the circuit. Always repair the source of the overload before resetting a circuit breaker or replacing a fuse or fusible link. When searching for overloads, keep in mind that the circuit protection devices protect only against overloads between the protection device and ground.

There are 2 basic types of circuit; Series and Parallel. In a series circuit, all of the elements are connected in chain fashion with the same amount of current passing through each element or load. No matter where an ammeter is connected in a series circuit, it will always read the same. The most important fact to remember about a series circuit is that the sum of the voltages across each element equals the source voltage. The total resistance of a series circuit is equal to the sum of the individual resistances within each element of the circuit. Using ohms law, one can determine the voltage drop across each element in the circuit. If the total resistance and source voltage is known, the amount of current can be calculated. Once the amount of current (amperes) is known, values can be substituted in the Ohms law formula to calculate the voltage drop across each individual element in the series circuit. The individual voltage drops must add up to the same value as the source voltage.

A parallel circuit, unlike a series circuit, contains 2 or more branches, each branch a separate path independent of the others. The total current draw from the voltage source is the sum of all the currents drawn by each branch. Each branch of a parallel circuit can be analyzed separately. The individual branches can be either simple circuits, series circuits or combinations of series-parallel circuits. Ohms law applies to parallel circuits just as it applies to series circuits, by considering each branch independently of the others. The most important thing to remember is that the voltage across each branch is the same as the source voltage. The current in any branch is that voltage divided by the resistance of the branch. A practical method of determining the resistance of a parallel circuit is to divide the product of the 2 resistances by the sum of 2 resistances at a time. Amperes through a parallel circuit is the sum of the amperes through the separate branches. Voltage across a parallel circuit is the same as the voltage across each branch.

By measuring the voltage drops the resistance of each element within the circuit is being measured. The greater the voltage drop, the greater the resistance. Voltage drop measurements are a common way of checking circuit resistances in automotive electrical systems. When part of a circuit develops excessive resistance (due to a bad connection) the element will show a higher than normal voltage drop. Normally, automotive wiring is selected to limit voltage drops to a few tenths of a volt. In parallel circuits, the total resistance is less than the sum of the individual resistances; because the current has 2 paths to take, the total resistance is lower.

Magnetism and Electromagnets

Electricity and magnetism are very closely associated because when electric current passes through a wire, a magnetic field is created around the wire. When a wire carrying electric current

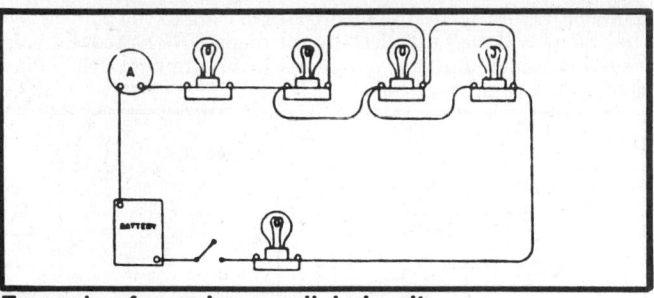

Example of a series-parallel circuit

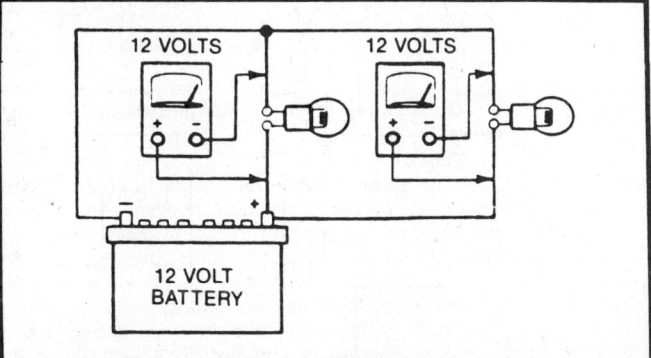

Voltage drop in a parallel circuit. Voltage drop across each lamp is 12 volts

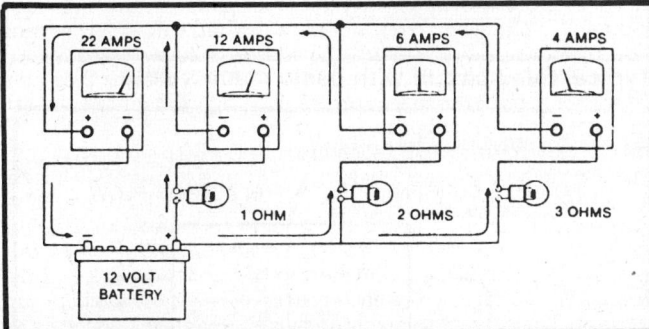

Total current in parallel circuit: 4 + 6 + 12 = 22 amps

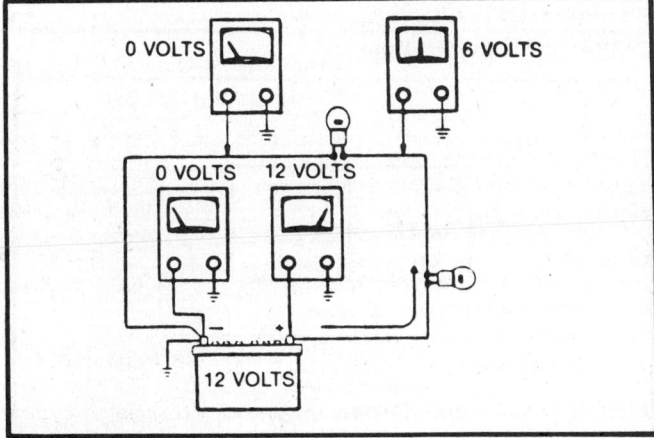

Voltage drop in a series circuit

ELECTRO-MAGNETS

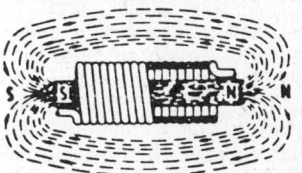

FORCE FIELD SURROUNDING A CURRENT CARRYING COIL
(WITHOUT IRON CORE)
ALL FORCE LINES ARE COMPLETE LOOPS

FORCE FIELD WITH SOFT IRON CORE
NOTE CONCENTRATION OF LINES IN IRON CORE

Magnetic field surrounding an electromagnet

MAGNETISM & PERMANENT MAGNETS

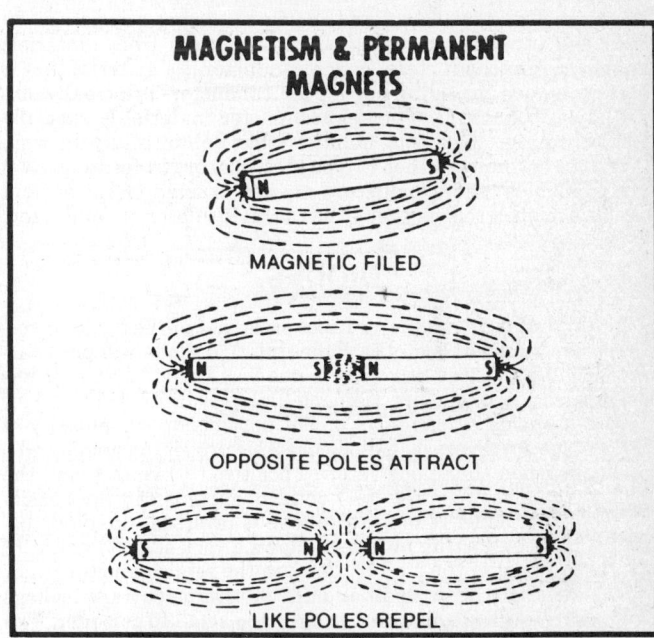

MAGNETIC FILED

OPPOSITE POLES ATTRACT

LIKE POLES REPEL

Magnetic field surrounding a bar magnet

is wound into a coil, a magnetic field with North and South poles is created just like in a bar magnet. If an iron core is placed within the coil, the magnetic field becomes stronger because iron conducts magnetic lines much easier than air. This arrangement is called an electromagnet and is the basic principle behind the operation of such components as relays, buzzers and solenoids.

A relay is basically just a remote-controlled switch that uses a small amount of current to control the flow of a large amount of current. The simplest relay contains an electromagnetic coil in series with a voltage source (battery) and a switch. A movable armature made of some magnetic material pivots at one end and is held a small distance away from the electromagnet by a spring or the spring steel of the armature itself. A contact point, made of a good conductor, is attached to the free end of the armature with another contact point a small distance away. When the relay is switched on (energized), the magnetic field created by the current flow attracts the armature, bending it until the contact points meet, closing a circuit and allowing current to flow in the second circuit through the relay to the load the circuit operates. When the relay is switched off (de-energized), the armature springs back and opens the contact points, cutting off the current flow in the secondary, or controlled, circuit. Relays can be designed to be either open or closed when energized, depending on the type of circuit control a manufacturer requires.

A buzzer is similar to a relay, but its internal connections are different. When the switch is closed, the current flows through the normally closed contacts and energizes the coil. When the coil core becomes magnetized, it bends the armature down and breaks the circuit. As soon as the circuit is broken, the spring-loaded armature remakes the circuit and again energizes the coil. This cycle repeats rapidly to cause the buzzing sound.

A solenoid is constructed like a relay, except that its core is allowed to move, providing mechanical motion that can be used to actuate mechanical linkage to operate a door or trunk lock or control any other mechanical function. When the switch is closed, the coil is energized and the movable core is drawn into the coil. When the switch is opened, the coil is de-energized and spring pressure returns the core to its original position.

Basic Solid State

The term "solid state" refers to devices utilizing transistors, diodes and other components which are made from materials known as semiconductors. A semiconductor is a material that is neither a good insulator nor a good conductor; principally silicon and germanium. The semiconductor material is specially treated to give it certain qualities that enhance its function, therefore becoming either P-type (positive) or N-type (negative) material. Most semiconductors are constructed of silicon and can be designed to function either as an insulator or conductor.

DIODES

The simplest semiconductor function is that of the diode or rectifier (the 2 terms mean the same thing). A diode will pass current in one direction only, like a one-way valve, because it has low resistance in one direction and high resistance on the other. Whether the diode conducts or not depends on the polarity of the voltage applied to it. A diode has 2 electrodes, an anode and a cathode. When the anode receives positive (+) voltage and the cathode receives negative (−) voltage, current can flow easily through the diode. When the voltage is reversed, the diode becomes non-conducting and only allows a very slight amount of current to flow in the circuit. Because the semiconductor is not a perfect insulator, a small amount of reverse current leakage will occur, but the amount is usually too small to consider. The application of voltage to maintain the current flow described is called "forward bias."

A light-emitting diode (LED) is made of a particular type of crystal that glows when current is passed through it. LED's are used in display faces of many digital or electronic instrument clusters. LED's are usually arranged to display numbers (digital readout), but can be used to illuminate a variety of electronic graphic displays.

Like any other electrical device, diodes have certain ratings that must be observed and should not be exceeded. The forward current rating (or bias) indicates how much current can safely pass through the diode without causing damage or destroying it. Forward current rating is usually given in either amperes or milliamperes. The voltage drop across a diode remains constant regardless of the current flowing through it. Small diodes designed to carry low amounts of current need no special provision for dissipating the heat generated in any electrical device, but large current carrying diodes are usually mounted on heat sinks

to keep the internal temperature from rising to the point where the silicon will melt and destroy the diode. When diodes are operated in a high ambient temperature environment, they must be de-rated to prevent failure.

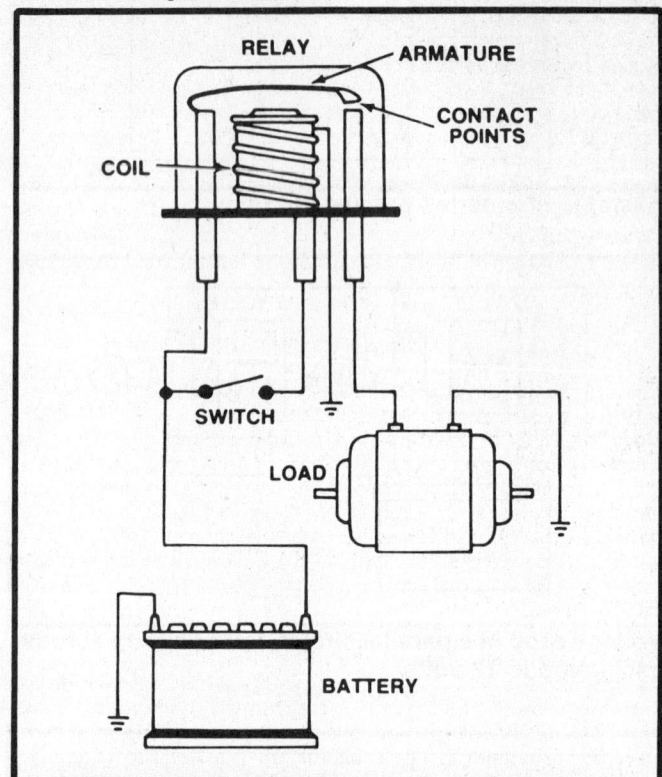

Typical relay circuit with basic components

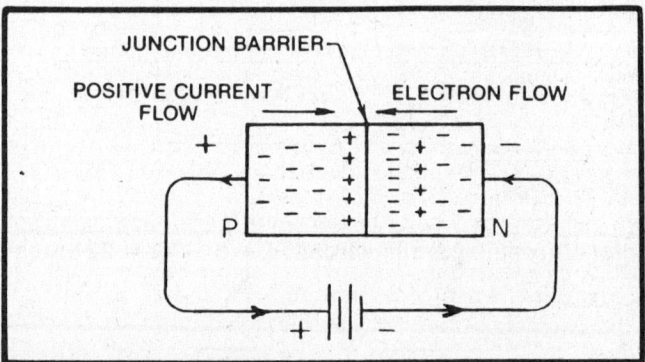

Diode with forward bias

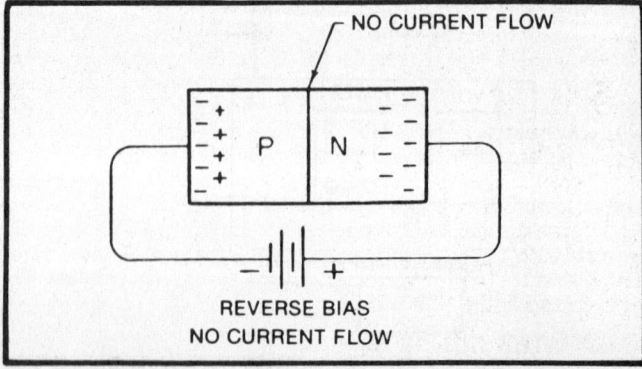

Diode with reverse bias

Another diode specification is its peak inverse voltage rating. This value is the maximum amount of voltage the diode can safely handle when operating in the blocking mode. This value can be anywhere from 50–1000 volts, depending on the diode. If voltage amount is exceeded, it will damage the diode just as too much forward current will. Most semiconductor failures are caused by excessive voltage or internal heat.

One can test a diode with a small battery and a lamp with the same voltage rating. With this arrangement one can find a bad diode and determine the polarity of a good one. A diode can fail and cause either a short or open circuit, but in either case it fails to function as a diode. Testing is simply a matter of connecting the test bulb first in one direction and then the other and making sure that current flows in one direction only. If the diode is shorted, the test bulb will remain on no matter how the light is connected.

TRANSISTORS

The transistor is an electrical device used to control voltage within a circuit. A transistor can be considered a "controllable diode" in that, in addition to passing or blocking current, the transistor can control the amount of current passing through it. Simple transistors are composed of 3 pieces of semiconductor material, P and N type, joined together and enclosed in a container. If 2 sections of P material and 1 section of N material are used, it is known as a PNP transistor; if the reverse is true, then it is known as an NPN transistor. The 2 types cannot be interchanged.

Most modern transistors are made from silicon (earlier transistors were made from germanium) and contain 3 elements; the emitter, the collector and the base. In addition to passing or blocking current, the transistor can control the amount of current passing through it and because of this can function as an amplifier or a switch. The collector and emitter form the main current-carrying circuit of the transistor. The amount of current that flows through the collector-emitter junction is controlled by the amount of current in the base circuit. Only a small amount of base-emitter current is necessary to control a large amount of collector-emitter current (the amplifier effect). In automotive applications, however, the transistor is used primarily as a switch.

When no current flows in the base-emitter junction, the collector-emitter circuit has a high resistance, like to open contacts of a relay. Almost no current flows through the circuit and transistor is considered OFF. By bypassing a small amount of current into the base circuit, the resistance is low, allowing current to flow through the circuit and turning the transistor ON. This condition is known as "saturation" and is reached when the base current reaches the maximum value designed into the transistor that allows current to flow. Depending on various factors, the transistor can turn on and off (go from cutoff to saturation) in less than one millionth of a second.

Much of what was said about ratings for diodes applies to transistors, since they are constructed of the same materials. When transistors are required to handle relatively high currents, such as in voltage regulators or ignition systems, they are generally mounted on heat sinks in the same manner as diodes. They can be damaged or destroyed in the same manner if their voltage ratings are exceeded. A transistor can be checked for proper operation by measuring the resistance with an ohmmeter between the base-emitter terminals and then between the base-collector terminals. The forward resistance should be small, while the reverse resistance should be large. Compare the readings with those from a known good transistor. As a final check, measure the forward and reverse resistance between the collector and emitter terminals.

INTEGRATED CIRCUITS

The integrated circuit (IC) is an extremely sophisticated solid

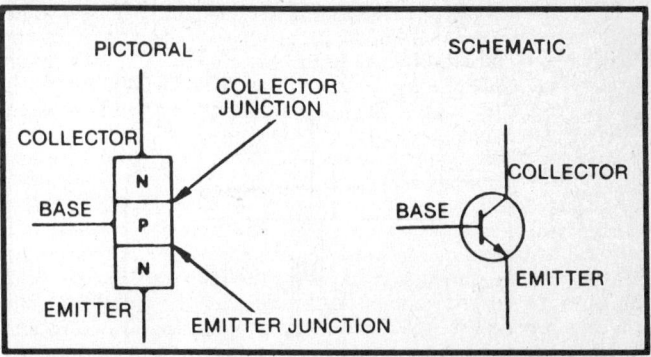

NPN transistor illustrations (pictorial and schematic)

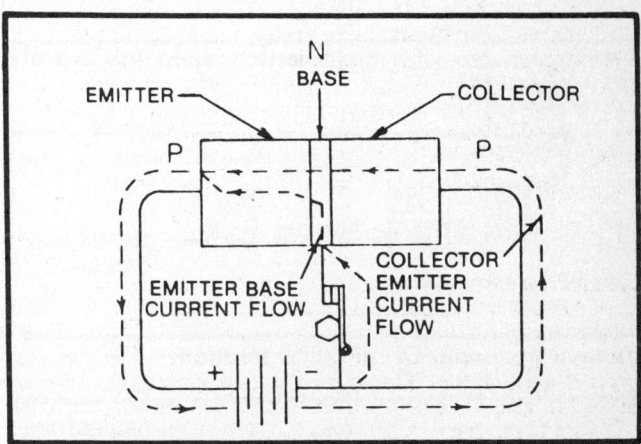

PNP transistor with base switch closed (base emitter and collector emitter current flow)

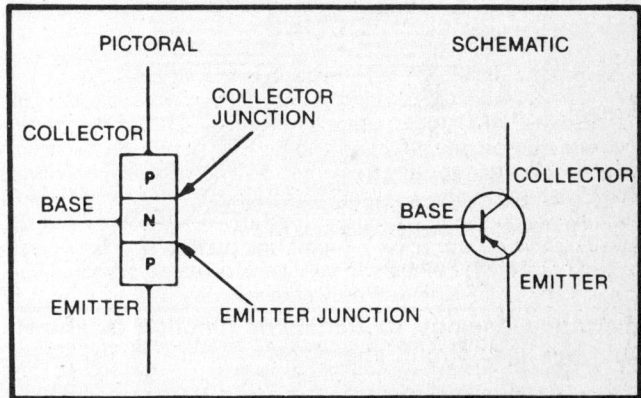

PNP transistor illustrations (pictorial and schematic)

state device that consists of a silicone wafer (or chip) which has been doped, insulated and etched many times so that it contains an entire electrical circuit with transistors, diodes, conductors and capacitors miniaturized within each tiny chip. Integrated circuits are often referred to as "computers on a chip" and are largely responsible for the current boom in electronic control technology.

Microprocessors, Computers and Logic Systems

Mechanical or electromechanical control devices lack the precision necessary to meet the requirements of modern control standards. They do not have the ability to respond to a variety of

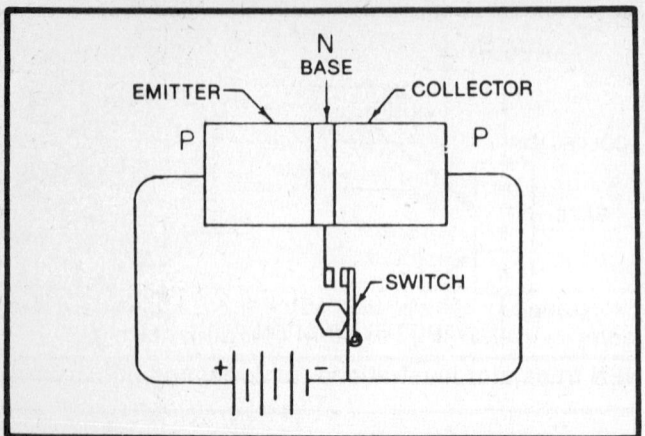

PNP transistor with base switch open (no current flow)

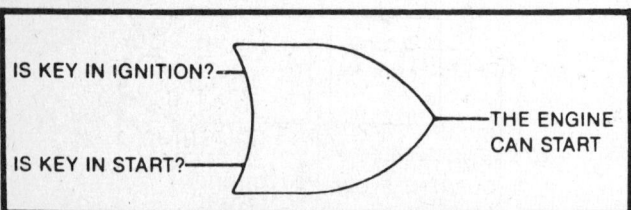

Typical two-input OR circuit operation

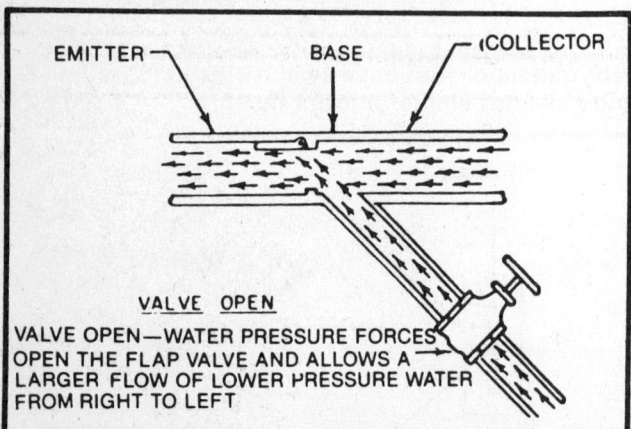

Hydraulic analogy to transistor function is shown with the base circuit energized

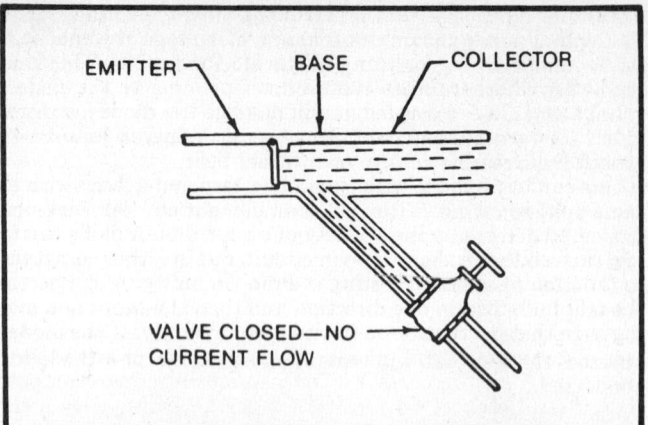

Hydraulic analogy to transistor function is shown with the base circuit shut off

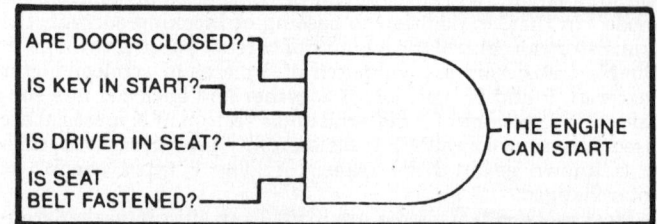

Multiple input AND operation in a typical automotive starting circuit

input conditions common to antilock brakes, climate control and electronic suspension operation. To meet these requirements, manufacturers have gone to solid state logic systems and microprocessors to control the basic functions of suspension, brake and temperature control, as well as other systems and accessories.

One of the more vital roles of microprocessor-based systems is their ability to perform logic functions and make decisions. Logic designers use a shorthand notation to indicate whether a voltage is present in a circuit (the number 1) or not present (the number 0). Their systems are designed to respond in different ways depending on the output signal (or the lack of it) from various control devices.

There are 3 basic logic functions or "gates" used to construct a microprocessor control system: the AND gate, the OR gate or the NOT gate. Stated simply, the AND gate works when voltage is present in 2 or more circuits which then energize a third (A and B energize C). The OR gate works when voltage is present at either circuit A or circuit B which then energizes circuit C. The NOT function is performed by a solid state device called an "inverter" which reverses the input from a circuit so that, if voltage is going in, no voltage comes out and vice versa. With these three basic building blocks, a logic designer can create complex systems easily. In actual use, a logic or decision making system may employ many logic gates and receive inputs from a number of sources (sensors), but for the most part, all utilize the basic logic gates discussed above.

Stripped to its bare essentials, a computerized decision-making system is made up of three subsystems:

    a. Input devices (sensors or switches)
    b. Logic circuits (computer control unit)
    c. Output devices (actuators or controls)

The input devices are usually nothing more than switches or sensors that provide a voltage signal to the control unit logic circuits that is read as a 1 or 0 (on or off) by the logic circuits. The output devices are anything from a warning light to solenoid-operated valves, motors, linkage, etc. In most cases, the logic circuits themselves lack sufficient output power to operate these devices directly. Instead, they operate some intermediate device such as a relay or power transistor which in turn operates the appropriate device or control. Many problems diagnosed as computer failures are really the result of a malfunctioning intermediate device like a relay. This must be kept in mind whenever troubleshooting any microprocessor-based control system.

The logic systems discussed above are called "hardware" systems, because they consist only of the physical electronic components (gates, resistors, transistors, etc.). Hardware systems do not contain a program and are designed to perform specific or "dedicated" functions which cannot readily be changed. For many simple automotive control requirements, such dedicated logic systems are perfectly adequate. When more complex logic functions are required, or where it may be desirable to alter these functions (e.g. from one model vehicle to another) a true computer system is used. A computer can be programmed through its software to perform many different functions and, if that program is stored on a separate integrated circuit chip called a ROM (Read Only Memory), it can be easily changed simply by plugging in a different ROM with the desired program. Most on-board automotive computers are designed with this capability. The on-board computer method of engine control offers the manufacturer a flexible method of responding to data from a variety of input devices and of controlling an equally large variety of output controls. The computer response can be changed quickly and easily by simply modifying its software program.

The microprocessor is the heart of the microcomputer. It is the thinking part of the computer system through which all the data from the various sensors passes. Within the microprocessor, data is acted upon, compared, manipulated or stored for future use. A microprocessor is not necessarily a microcomputer, but the differences between the 2 are becoming very minor. Originally, a microprocessor was a major part of a microcomputer, but nowadays microprocessors are being called "single-chip microcomputers". They contain all the essential elements to make them behave as a computer, including the most important ingredient–the program.

All computers require a program. In a general purpose computer, the program can be easily changed to allow different tasks to be performed. In a "dedicated" computer, such as most on-board automotive computers, the program isn't quite so easily altered. These automotive computers are designed to perform one or several specific tasks, such as maintaining the passenger compartment temperature at a specific, predetermined level. A program is what makes a computer smart; without a program a computer can do absolutely nothing. The term "software" refers to the computer's program that makes the hardware preform the function needed.

The software program is simply a listing in sequential order of the steps or commands necessary to make a computer perform the desired task. Before the computer can do anything at all, the program must be fed into it by one of several possible methods. A computer can never be "smarter" than the person programming it, but it is a lot faster. Although it cannot perform any calculation or operation that the programmer himself cannot perform, its processing time is measured in millionths of a second.

Because a computer is limited to performing only those operations (instructions) programmed into its memory, the program must be broken down into a large number of very simple steps. Two different programmers can come up with 2 different programs, since there is usually more than one way to perform any task or solve a problem. In any computer, however, there is only so much memory space available, so an overly long or inefficient program may not fit into the memory. In addition to performing arithmetic functions (such as with a trip computer), a computer can also store data, look up data in a table and perform the logic functions previously discussed. A Random Access Memory (RAM) allows the computer to store bits of data temporarily while waiting to be acted upon by the program. It may also be used to store output data that is to be sent to an output device. Whatever data is stored in a RAM is lost when power is removed from the system by turning **OFF** the ignition key, for example.

Computers have another type of memory called a Read Only Memory (ROM) which is permanent. This memory is not lost when the power is removed from the system. Most programs for automotive computers are stored on a ROM memory chip. Data is usually in the form of a look-up table that saves computing time and program steps. For example, a computer designed to control the amount of distributor advance can have this information stored in a table. The information that determines distributor advance (engine rpm, manifold vacuum and temperature) is coded to produce the correct amount of distributor advance over a wide range of engine operating conditions. Instead of the computer computing the required advance, it simply looks it up in a pre-programmed table. However, not all electronic control functions can be handled in this manner; some must be

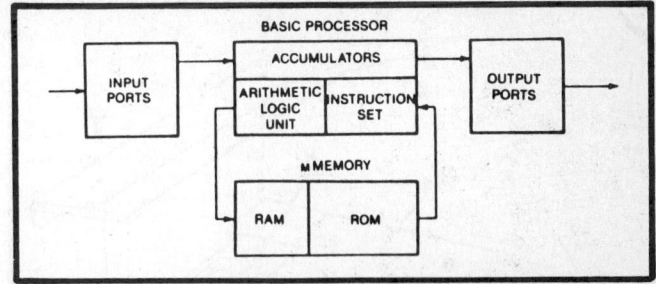

Schematic of typical microprocessor based on-board computer showing essential components

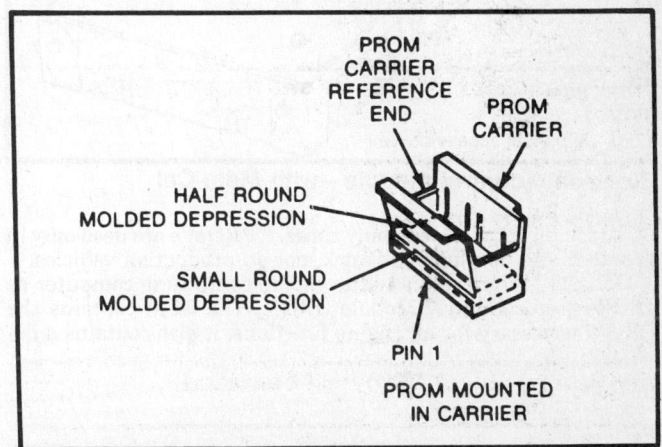

Typical PROM showing carrier refernce markings

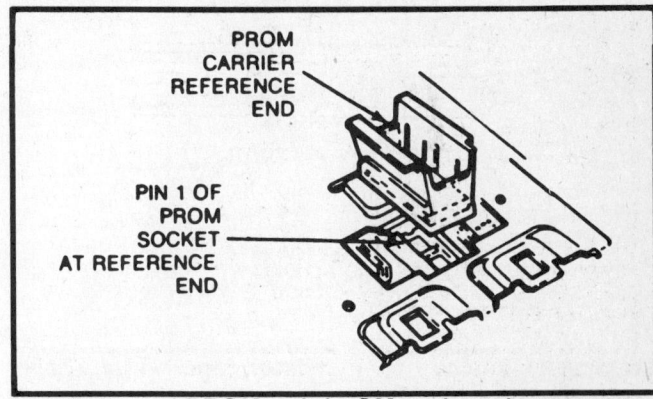

Installation of PROM unit in GM on-board computer

computed. On an antilock brake system, for example, the computer must measure the rotation of each separate wheel and then calculate how much brake pressure to apply in order to prevent one wheel from locking up and causing a loss of control.

There are several ways of programming a ROM, but once programmed the ROM cannot be changed. If the ROM is made on the same chip that contains the microprocessor, the whole computer must be altered if a program change is needed. For this reason, a ROM is usually placed on a separate chip. Another type of memory is the Programmable Read Only Memory (PROM) that has the program "burned in" with the appropriate programming machine. Like the ROM, once a PROM has been programmed, it cannot be changed. The advantage of the PROM is that it can be produced in small quantities economically, since it is manufactured with a blank memory. Program changes for various vehicles can be made readily. There is still another type of memory called an EPROM (Erasable PROM) which can be

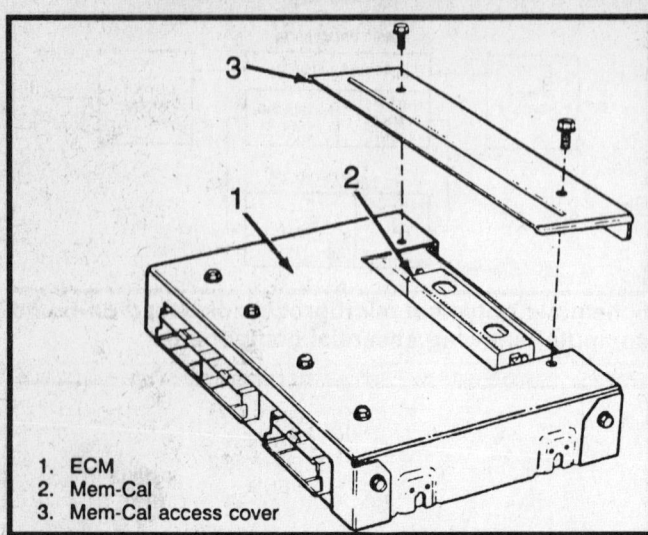

1. ECM
2. Mem-Cal
3. Mem-Cal access cover

Electronic control module — with Mem-Cal

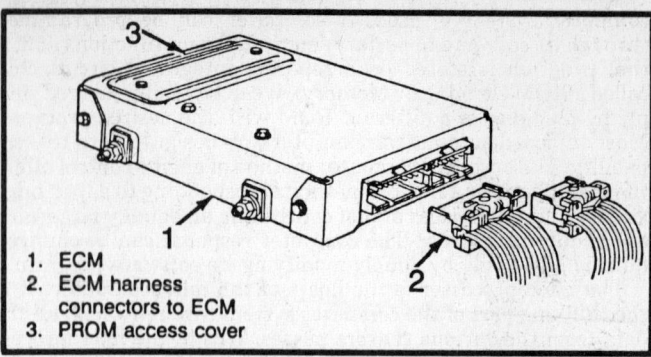

1. ECM
2. ECM harness
   connectors to ECM
3. PROM access cover

Electronic control module — with PROM and CalPak

erased and programmed many times. EPROM's are used only in research and development work, not on production vehicles.

General Motors refers to the engine controlling computer as an Electronic Control Module (ECM). The ECM contains the PROM necessary for all engine functions, it also contains a de-vice called a CalPak. This allows the fuel delivery function should other parts of the ECM become damaged. It has an access door in the ECM, like the PROM has. There is a third type control module used in some ECMs called a Mem-Cal. The Mem-Cal contains the function of PROM, CalPak and Electronic Spark Control (EST) module. Like the PROM, it contains the calibrations needed for a specific vehicle, as well as the back-up fuel control circuitry required if the rest of the ECM should become damaged and the spark control. An ECM containing a PROM and CalPak can be identified by the 2 connector harnesses, while the ECM containing the Mem-Cal has 3 connector harnesses attached to it.

Troubleshooting and Diagnosis

2 SECTION

INDEX

Diagnostic Equipment and Special Tools 2-2
Safety Precautions 2-2
Organized Troubleshooting 2-2
Jumper Wires 2-3

12 VOLT TEST LIGHT 2-3
Self-Powered Test Light 2-4
Open Circuit Testing 2-4
Short Circuit Testing 2-4

VOLTMETER 2-4
Available Voltage Measurement 2-4
Voltage Drop 2-5
Indirect Computation of Voltage Drop 2-5
Direct Measurement of Voltage Drop 2-5
High Resistance Testing 2-5

OHMMETER 2-5
Ohmmeter Calibration 2-6
Continuity Testing 2-6
Resistance Measurement 2-6

AMMETERS 2-6
Battery Current Drain Test 2-6
Ammeter Connection 2-6

MULTIMETERS 2-7
Volt-Ammeter 2-7
Tach-Dwell Meter 2-7

SPECIAL TEST EQUIPMENT 2-7
Hand-Held Testers 2-7
Adapters 2-7

WIRING DIAGRAMS 2-8
Electrical Symbols 2-8
Circuit Conductors 2-11
Wire Gauge 2-11

WIRING REPAIRS 2-13
Solderng Techniques 2-14
Wire Harness and Connectors 2-15
Repair Procedure 2-15
Connector Types 2-15
Weatherpak Connectors 2-16
Repairing Hard Shell Connectors 2-17

MECHANICAL TEST EQUIPMENT 2-17
Vacuum Gauge 2-17
Hand Vacuum Pump 2-17
Compression Gauge 2-18
Fuel Pressure Gauge 2-18
Fuel Injector Cleaning 2-18
Fuel Line Tools 2-19

TROUBLESHOOTING AND DIAGNOSIS

Diagnostic Equipment and Special Tools

While we may think that with no moving parts, electronic components should never wear out, in the real world malfunctions do occur. The problem is that any computer-based system is extremely sensitive to electrical voltages and cannot tolerate careless or haphazard testing or service procedures. An inexperienced individual can literally do major damage looking for a minor problem by using the wrong kind of test equipment or connecting test leads or connectors with the ignition switch **ON**. Therefore, when selecting test equipment, make sure the manufacturers instructions state that the tester is compatible with whatever type of electronic control system is being serviced. Read all instructions carefully and double check all test points before installing probes or making any connections.

The following section outlines basic diagnosis techniques for dealing with computerized engine control systems. Along with a general explanation of the various types of test equipment available to aid in servicing modern electronic automotive systems, basic repair techniques for wiring harnesses and connectors is given. Read the basic information before attempting any repairs or testing on any computerized system, to provide the background of information necessary to avoid the most common and obvious mistakes that can cost both time and money. Likewise, the individual system sections for engine controls, fuel injection and feedback carburetors should be read from the beginning to the end before any repairs or diagnosis is attempted. Although the replacement and testing procedures are simple in themselves, the systems are not, and unless one has a thorough understanding of all components and their function within a particular fuel injection system (for example), the logical test sequence these systems demand cannot be followed. Minor malfunctions can make a big difference, so it is important to know how each component affects the operation of the overall electronic system to find the ultimate cause of a problem without replacing good components unnecessarily. It is not enough to use the correct test equipment; the test equipment must be used correctly.

Safety Precautions

—————————— CAUTION ——————————
Whenever working on or around any computer-based microprocessor control system, always observe these general precautions to prevent the possibility of personal injury or damage to electronic components:

- Never install or remove battery cables with the ignition key **ON** or the engine running. Jumper cables should be connected with the ignition key **OFF** to avoid power surges that can damage electronic control units. Engines equipped with computer controlled systems should avoid both giving and getting jump starts due to the possibility of serious damage to components from arcing in the engine compartment when connections are made with the ignition **ON**.
- Always remove the battery cables before charging the battery. Never use a high-output charger on an installed battery or attempt to use any type of "hot shot" (24 volt) starting aid.
- Exercise care when inserting test probes into connectors to insure good connections without damaging the connector or spreading the pins. Always probe connectors from the rear (wire) side, never the pin side, to avoid accidental shorting of terminals during test procedures.
- Never remove or attach wiring harness connectors with the ignition switch **ON**, especially to an electronic control unit.

- Do not drop any components during service procedures and never apply 12 volts directly to any component (like a solenoid or relay) unless instructed specifically to do so. Some component electrical windings are designed to safely handle only 4–5 volts and can be destroyed in seconds if 12 volts are applied directly to the connector.
- Remove the electronic control unit if the vehicle is to be placed in an environment where temperatures exceed approximately 176°F (80°C), such as a paint spray booth or when arc-welding or gas-welding near the control unit location in the car. Always disconnect the battery, and when possible remove the electronic control unit, when arc-welding.

Organized Troubleshooting

When diagnosing a specific problem, organized troubleshooting is a must. The complexity of a modern automobile demands the approach to any problem be in a logical, organized manner. There are certain troubleshooting techniques that are standard:

1. Establish when the problem occurs. Does the problem appear only under certain conditions? Were there any noises, odors, or other unusual symptoms? Make notes on any symptoms found, including warning lights and trouble codes, if applicable.

2. Isolate the problem area. To do this, make some simple tests and observations; then eliminate the systems that are working properly. Check for obvious problems such as broken wires or split or disconnected vacuum hoses. Always check the obvious before assuming something complicated is the cause.

3. Test for problems systematically to determine the cause once the problem area is isolated. Are all the components functioning properly? Is there power going to electrical switches and motors? Is there vacuum at vacuum switches and/or actuators? Is there a mechanical problem such as bent linkage or loose mounting screws? Doing careful, systematic checks will often turn up most causes on the first inspection without wasting time checking components that have little or no relationship to the problem.

4. Test all repairs after the work is done to make sure that the problem is fixed. Some causes can be traced to more than one component, so a careful verification of repair work is important to pick up additional malfunctions that may cause a problem to reappear or a different problem to arise. A blown fuse, for example, is a simple problem that may require more than just replacing a fuse. If the problem that caused a fuse to blow isn't searched for, a shorted wire may go undetected.

The diagnostic tree charts are designed to help solve problems by leading the user through closely defined conditions and tests so that only the most likely components, vacuum and electrical circuits are checked for proper operation when troubleshooting a particular malfunction. By using the trouble trees to eliminate those systems and components which normally will not cause the condition described, a problem can be isolated within one or more systems or circuits without wasting time on unnecessary testing. Experience has shown that most problems tend to be the result of a fairly simple and obvious cause, such as loose or corroded connectors or air leaks in the intake system. A careful inspection of components during testing is essential to quick and accurate troubleshooting. Frequent references to special test equipment will be found in the text and in the diagnosis charts. These devices or compatible equivalents are necessary to perform some of the more complicated test procedures listed, but many components can be functionally tested with the quick checks outlined in the "On-Car Service" procedures. Aftermarket testers are available from a variety of sources, as well as from the vehicle manufacturer, but care should be taken that any test equipment being used is designed to diagnose that particular system accurately without damaging the control unit (ECU) or components being tested.

NOTE: Pinpointing the exact cause of trouble in an electrical system can sometimes only be done using special test equipment. The following describes commonly used test equipment and explains how to put it to best use in diagnosis. In addition to the information covered below, the manufacturer's instructions booklet provided with the tester should be read and clearly understood before attempting any test procedures.

Jumper Wires

Jumper wires are simple, yet extremely valuable pieces of test equipment. Jumper wires are merely wires that are used to bypass sections of a circuit. The simplest type of jumper wire is merely a length of multistrand wire with an alligator clip at each end. Jumper wires are usually fabricated from lengths of standard automotive wire and whatever type of connector (alligator clip, spade connector or pin connector) that is required for the particular vehicle being tested. The well-equipped tool box will have several different styles of jumper wires in several different lengths. Some jumper wires are made with 3 or more terminals coming from a common splice for special-purpose testing. In cramped, hard-to-reach areas it is advisable to have insulated boots over the jumper wire terminals in order to prevent accidental grounding, sparks, and possible fire, especially when testing fuel system components.

Jumper wires are used primarily to locate open electrical circuits, on either the ground (–) side of the circuit or on the hot (+) side. If an electrical component fails to operate, connect the jumper wire between the component and a good ground. If the component operates only with the jumper installed, the ground circuit is open. If the ground circuit is good, but the component does not operate, the circuit between the power feed and component is open. Sometimes a jumper wire connected directly from the battery to the hot terminal of the component can be used, but first make sure the component uses 12 volts in operation. Some electrical components, such as fuel injectors, are designed to operate on about 4 volts and running 12 volts directly to the injector terminals can burn out the wiring. By inserting an in-line fuseholder between a set of test leads, a fused jumper wire can be used for bypassing open circuits. Use a 5 amp fuse to provide protection against voltage spikes. When in doubt, use a voltmeter to check the voltage input to the component and measure how much voltage is being applied normally. By moving the jumper wire successively back from the lamp toward the power

source, the area of the circuit where the open is located can be isolated. When the component stops functioning, or the power is cut off, the open is in the segment of wire between the jumper and the point previously tested.

─────── **CAUTION** ───────

Never use jumpers made from wire that is of lighter gauge than used in the circuit under test. If the jumper wire is of too small gauge, it may overheat and possibly melt. Never use jumpers to bypass high-resistance loads (such as motors) in a circuit. Bypassing resistances, in effect, creates a short circuit which may, in turn, cause damage and fire. Never use a jumper for anything other than temporary bypassing of components in a circuit.

12 Volt Test Light

The 12 volt test light is used to check circuits and components while electrical current is flowing through them. It is used for voltage and ground tests. Twelve volt test lights come in different styles but all have 3 main parts; a ground clip, a probe, and a light. The most commonly used 12 volt test lights have pick-type probes. To use a 12 volt test light, connect the ground clip to a good ground and probe wherever necessary with the pick. The pick should be sharp so that it can penetrate wire insulation to make contact with the wire, without making a large hole in the insulation. The wrap-around light is handy in hard to reach areas or where it is difficult to support a wire to push a probe pick into it. To use the wrap around light, hook the wire to be probed with the hook and pull the trigger. A small pick will be forced through the wire insulation into the wire core.

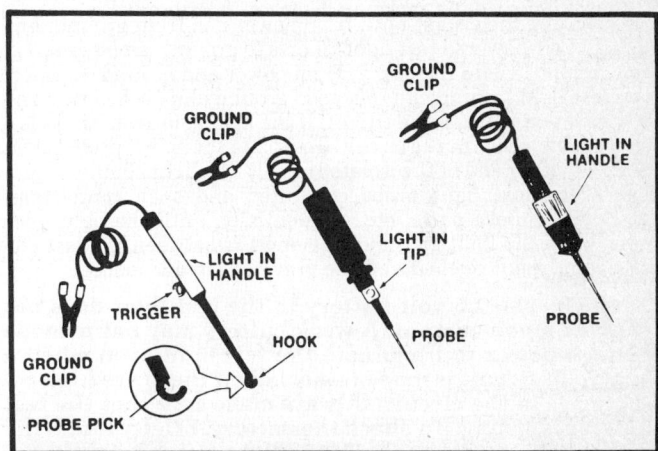

Examples of various types of 12 volt test lights

─────── **CAUTION** ───────

Do not use a test light to probe electronic ignition spark plug or coil wires. Never use a pick-type test light to probe wiring on computer controlled systems unless specifically instructed to do so. Any wire insulation that is pierced by the test light probe should be taped and sealed with silicone after testing to weatherproof it.

Like the jumper wire, the 12 volt test light is used to isolate opens in circuits. But, whereas the jumper wire is used to bypass the open to operate the load, the 12 volt test light is used to locate the presence of voltage in a circuit. If the test light glows, there is power up to that point; if the 12 volt test light does not glow when its probe is inserted into the wire or connector, there is an open circuit (no power). Move the test light in successive steps back toward the power source until the light in the handle does glow. When it does glow, the open is between the probe and point previously probed.

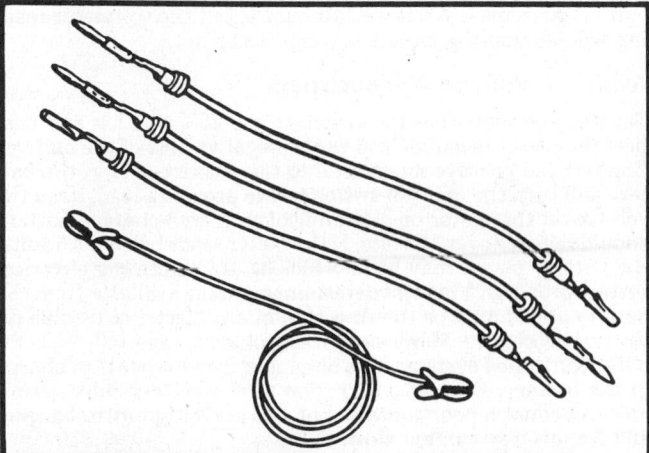

Typical jumper wires with various terminal ends

NOTE: The test light does not detect that 12 volts (or any particular amount of voltage) is present; it only detects that some voltage is present. It is advisable before using the test light to touch its terminals across the battery posts to make sure the light is operating properly.

Self-Powered Test Light

The self-powered test light usually contains a 1.5 volt penlight battery. One type of self-powered test light is similar in design to the 12 volt test light. This type has both the battery and the light in the handle and pick-type probe tip. The second type has the light toward the open tip, so that the light illuminates the contact point. The self-powered test light is dual-purpose piece of test equipment. It can be used to test for either open or short circuits when power is isolated from the circuit (continuity test). A powered test light should not be used on any computer controlled system or component unless specifically instructed to do so. Many engine sensors can be destroyed by even this small amount of voltage applied directly to the terminals.

Open Circuit Testing

To use the self-powered test light to check for open circuits, first isolate the circuit from the vehicle's 12 volt power source by disconnecting the battery or wiring harness connector. Connect the test light ground clip to a good ground and probe sections of the circuit sequentially with the test light. (start from either end of the circuit). If the light is out, the open is between the probe and the circuit ground. If the light is on, the open is between the probe and end of the circuit toward the power source.

Short Circuit Testing

By isolating the circuit both from power and from ground, and using a self-powered test light, shorts to ground can be found in the circuit. Isolate the circuit from power and ground. Connect the test light ground clip to a good ground and probe any easy-to-reach test point in the circuit. If the light comes on, there is a short somewhere in the circuit. To isolate the short, probe a test point at either end of the isolated circuit (the light should be on). Leave the test light probe connected and open connectors, switches, remove parts, etc., sequentially, until the light goes out. When the light goes out, the short is between the last circuit component opened and the previous circuit opened.

NOTE: The 1.5 volt battery in the test light does not provide much current. A weak battery may not provide enough power to illuminate the test light even when a complete circuit is made (especially if there are high resistances in the circuit). Always make sure that the test battery is strong. To check the battery, briefly touch the ground clip to the probe; if the light glows brightly the battery is strong enough for testing. Never use a self-powered test light to perform checks for opens or shorts when power is applied to the electrical system under test. The 12-volt vehicle power will quickly burn out the 1.5 volt light bulb in the test light.

Voltmeter

A voltmeter is used to measure voltage at any point in a circuit, or to measure the voltage drop across any part of a circuit. It can also be used to check continuity in a wire or circuit by indicating current flow from one end to the other. Voltmeters usually have various scales on the meter dial and a selector switch to allow the selection of different voltages. The voltmeter has a positive and a negative lead. To avoid damage to the meter, always connect the negative lead to the negative (–) side of circuit (to ground or nearest the ground side of the circuit) and connect the positive lead to the positive (+) side of the circuit (to the power source or the nearest power source). Note that the negative voltmeter lead will always be black and that the positive voltmeter

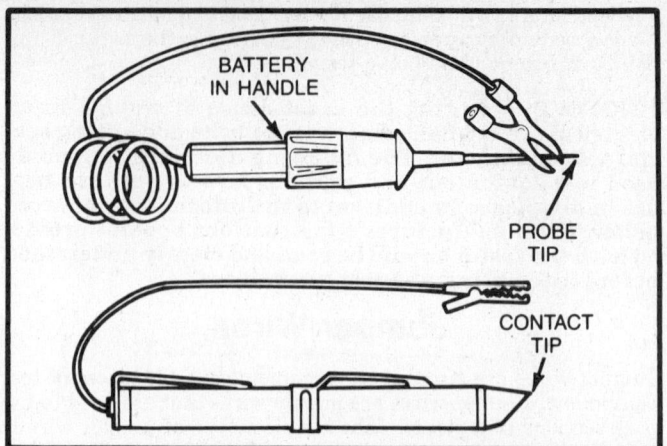

Two types of self-powered test lights

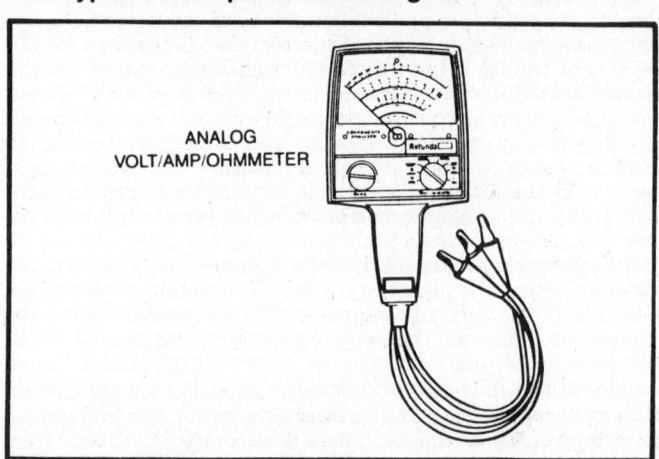

Typical analog-type voltmeter

will always be some color other than black (usually red). Depending on how the voltmeter is connected into the circuit, it has several uses.

A voltmeter can be connected either in parallel or in series with a circuit and it has a very high resistance to current flow. When connected in parallel, only a small amount of current will flow through the voltmeter current path; the rest will flow through the normal circuit current path and the circuit will work normally. When the voltmeter is connected in series with a circuit, only a small amount of current can flow through the circuit. The circuit will not work properly, but the voltmeter reading will show if the circuit is complete or not.

Available Voltage Measurement

Set the voltmeter selector switch to the 20V position and connect the meter negative lead to the negative post of the battery. Connect the positive meter lead to the positive post of the battery and turn the ignition switch ON to provide a load. Read the voltage on the meter or digital display. A well-charged battery should register over 12 volts. If the meter reads below 11.5 volts, the battery power may be insufficient to operate the electrical system properly. This test determines voltage available from the battery and should be the first step in any electrical trouble diagnosis procedure. Many electrical problems, especially on computer controlled systems, can be caused by a low state of charge in the battery. Excessive corrosion at the battery cable terminals can cause a poor contact that will prevent proper charging and full battery current flow.

Normal battery voltage is 12 volts when fully charged. When the battery is supplying current to 1 or more circuits it is said to

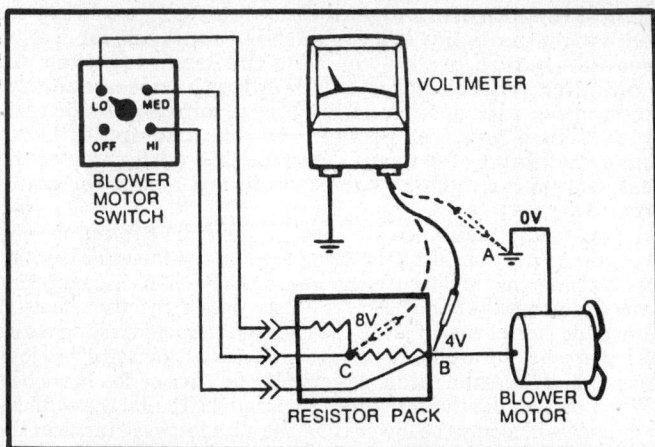

Measuring available voltage in a blower circuit

be "under load". When everything is **OFF** the electrical system is under a "no-load" condition. A fully charged battery showing about 12.5 volts at no load; may drop to 12 volts under medium load; and will drop even lower under heavy load. If the battery is partially discharged the voltage decrease under heavy load may be excessive, even though the battery shows 12 volts or more at no load. When allowed to discharge further, the battery's available voltage under load will decrease more severely. For this reason, it is important that the battery be fully charged during all testing procedures to avoid errors in diagnosis and incorrect test results.

VOLTAGE DROP

When current flows through a resistance, the voltage beyond the resistance is reduced (the larger the current, the greater the reduction in voltage). When no current is flowing, there is no voltage drop because there is no current flow. All points in the circuit which are connected to the power source are at the same voltage as the power source. The total voltage drop always equals the total source voltage. In a long circuit with many connectors, a series of small, unwanted voltage drops due to corrosion at the connectors can add up to a total loss of voltage which impairs the operation of the normal loads in the circuit.

Indirect Computation of Voltage Drops

1. Set the voltmeter selector switch to the 20 volt position.
2. Connect the meter negative lead to a good ground.
3. Probe all resistances in the circuit with the positive meter lead.
4. Operate the circuit in all modes and observe the voltage readings.

Direct Measurement of Voltage Drops

1. Set the voltmeter switch to the 20 volt position.
2. Connect the voltmeter negative lead to the ground side of the resistance load to be measured.
3. Connect the positive lead to the positive side of the resistance or load to be measured.
4. Read the voltage drop directly on the 20 volt scale.
Too high a voltage indicates too high a resistance. If, for example, a blower motor runs too slowly, if there is too high a resistance in the resistor pack can be determine. By taking voltage drop readings in all parts of the circuit, the problem can be isolated. Too low a voltage drop indicates too low a resistance. If, for example, a blower motor runs too fast in the **MED** and/or **LOW** position, the problem can be isolated in the resistor pack by taking voltage drop readings in all parts of the circuit to locate a possibly shorted resistor. The maximum allowable volt-

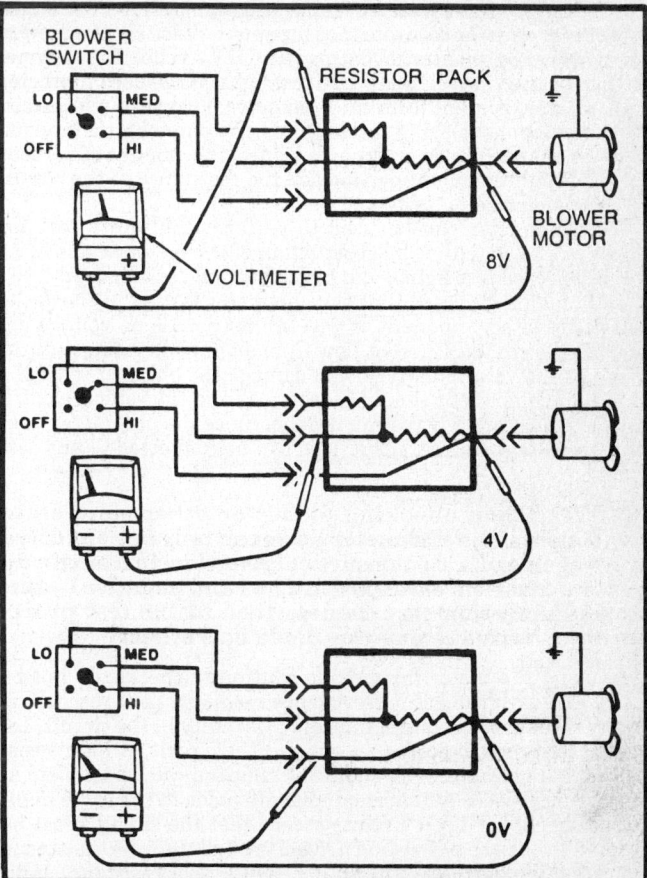

Direct measurement of voltage drops in a circuit

age drop under load is critical, especially if there is more than one high resistance problem in a circuit because all voltage drops are cumulative. A small drop is normal due to the resistance of the conductors.

High Resistance Testing

1. Set the voltmeter selector switch to the 2 volt position.
2. Connect the voltmeter positive lead to the positive post of the battery.
3. Turn **ON** the headlights and heater blower to provide a load.
4. Probe various points in the circuit with the negative voltmeter lead.
5. Read the voltage drop. Some average maximum allowable voltage drops are:
Fuse panel—0.7 volts
Ignition switch—0.5volts
Headlight switch—0.7 volts
Ignition coil (+)—0.5 volts
Any other load—0.5–1.3 volts

NOTE: Voltage drops are all measured while a load is operating; without current flow, there will be no voltage drop.

Ohmmeter

The ohmmeter is designed to read resistance (ohms) in a circuit or component. Although there are several different styles of ohmmeters, all will usually have a selector switch which permits the measurement of different ranges of resistance (usually the selector switch allows the multiplication of the meter reading by

10, 100, 1000, 10,000, etc.). A calibration knob allows the meter to be set at zero for accurate measurement. Since all ohmmeters are powered by an internal battery (usually 9 volts), the ohmmeter can be used as a self-powered test light. When the ohmmeter is connected, current from the ohmmeter flows through the circuit or component being tested. Since the ohmmeter's internal resistance and voltage are known values, the amount of current flow through the meter depends on the resistance of the circuit or component being tested.

The ohmmeter can be used to perform continuity test for opens or shorts (either by observation of the meter needle or as a self-powered test light), and to read actual resistance in a circuit. It should be noted that the ohmmeter is used to check the resistance of a component or wire while there is no voltage applied to the circuit. Current flow from an outside voltage source (such as the vehicle battery) can damage the ohmmeter, so the circuit or component should be isolated from the vehicle electrical system before any testing is done. Since the ohmmeter uses its own voltage source, either lead can be connected to any test point.

NOTE: When checking diodes or other solid state components, the ohmmeter leads can only be connected one way in order to measure current flow in a single direction. Make sure the positive (+) and negative (-) terminal connections are as described in the test procedures to verify the one-way diode operation.

In using the meter for making continuity checks, do not be concerned with the actual resistance readings. Zero resistance, or any resistance readings, indicate continuity in the circuit. Infinite resistance indicates an open in the circuit. A high resistance reading where there should be none indicates a problem in the circuit. Checks for short circuits are made in the same manner as checks for open circuits except that the circuit must be isolated from both power and normal ground. Infinite resistance indicates no continuity to ground, while zero resistance indicates a dead short to ground.

Resistance Measurement

The batteries in an ohmmeter will weaken with age and temperature, so the ohmmeter must be calibrated or "zeroed" before taking measurements. To zero the meter, place the selector switch in its lowest range and touch the 2 ohmmeter leads together. Turn the calibration knob until the meter needle is exactly on zero.

NOTE: All analog (needle) type ohmmeters must be zeroed before use, but some digital ohmmeter models are automatically calibrated when the switch is turned

ON. Self-calibrating digital ohmmeters do not have an adjusting knob, but it's a good idea to check for a zero readout before use by touching the leads together. All computer controlled systems require the use of a digital ohmmeter with at least 10 megohms impedance for testing. Before any test procedures are attempted, make sure the ohmmeter used is compatible with the electrical system, or damage to the on-board computer could result.

To measure resistance, first isolate the circuit from the vehicle power source by disconnecting the battery cables or the harness connector. Make sure the key is **OFF** when disconnecting any components or the battery. Where necessary, also isolate at least one side of the circuit to be checked to avoid reading parallel resistances. Parallel circuit resistances will always give a lower reading than the actual resistance of either of the branches. When measuring the resistance of parallel circuits, the total resistance will always be lower than the smallest resistance in the circuit. Connect the meter leads to both sides of the circuit (wire or component) and read the actual measured ohms on the meter scale. Make sure the selector switch is set to the proper ohm scale for the circuit being tested to avoid misreading the ohmmeter test value.

CAUTION

Never use an ohmmeter with power applied to the circuit. Like the self-powered test light, the ohmmeter is designed to operate on its own power supply. The normal 12 volt automotive electrical system current could damage the meter.

Ammeters

An ammeter measures the amount of current flowing through a circuit in units called amperes or amps. Amperes are units of electron flow which indicate how fast the electrons are flowing through the circuit. Since Ohm's Law dictates that current flow in a circuit is equal to the circuit voltage divided by the total circuit resistance, increasing voltage also increases the current level (amps). Likewise, any decrease in resistance will increase the amount of amps in a circuit. At normal operating voltage, most circuits have a characteristic amount of amperes, called "current draw" which can be measured using an ammeter. By referring to a specified current draw rating, measuring the amperes, and comparing the 2 values, one can determine what is happening within the circuit to aid in diagnosis. An open circuit, for example, will not allow any current to flow so the ammeter reading will be zero. More current flows through a heavily loaded circuit or when the charging system is operating.

An ammeter is always connected in series with the circuit being tested. All of the current that normally flows through the circuit must also flow through the ammeter; if there is any other

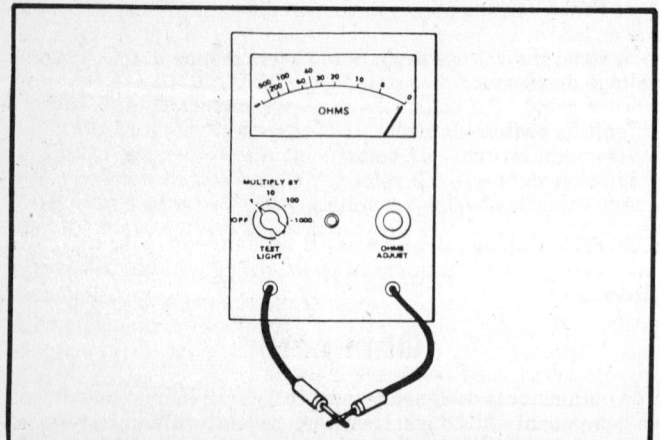
Analog ohmmeters must be calibrated before use by touching the probes together and adjusting the knob

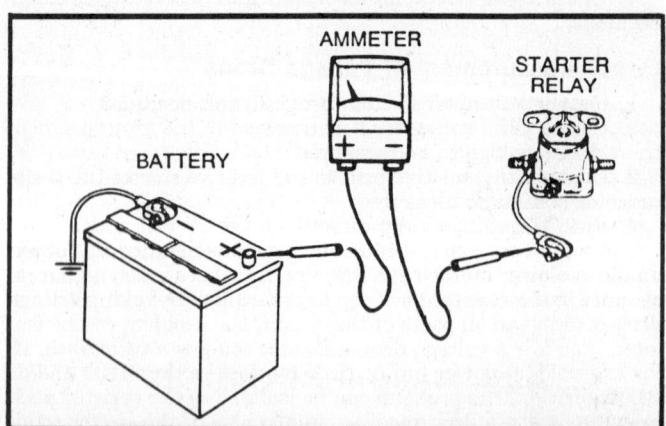
Battery current drain test

path for the current to follow, the ammeter reading will not be accurate. The ammeter itself has very little resistance to current flow and therefore will not affect the circuit, but it will measure current draw only when the circuit is closed and electricity is flowing. Excessive current draw can blow fuses and drain the battery, while a reduced current draw can cause motors to run slowly, lights to dim and other components not to operate properly. The ammeter can help diagnose these conditions by locating the cause of the high or low reading.

Multimeters

Different combinations of test meters can be built into a single unit designed for specific tests. Some of the more common combination test devices are known as Volt-Amp testers, Tach-Dwell meters, or Digital Multimeters. The Volt-Amp tester is

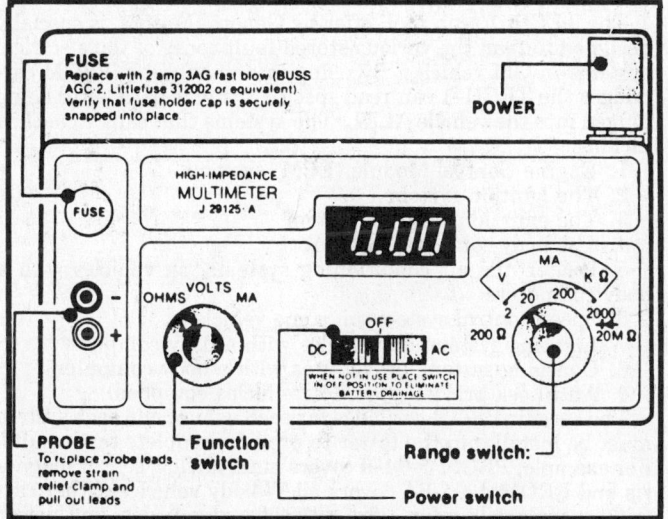
Typical multimeter used to test GM systems

Hand-held aftermarket tester used to diagnosis electronic engine control systems

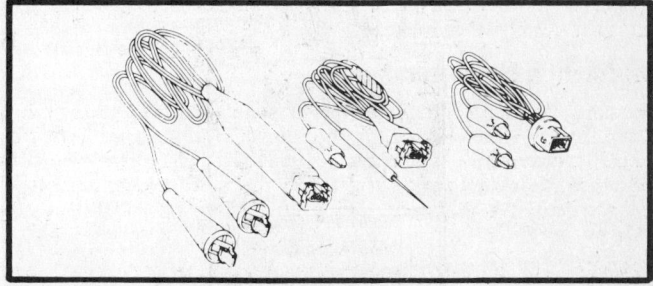
Special purpose test connections for use on some systems made up from factory connectors and jumper wires

used for charging system, starting system or battery tests and consists of a voltmeter, an ammeter and a variable resistance carbon pile. The voltmeter will usually have at least 2 ranges for use with 6, 12 and 24 volt systems. The ammeter also has more than 1 range for testing various levels of battery loads and starter current draw and the carbon pile can be adjusted to offer different amounts of resistance. The Volt-Amp tester has heavy leads to carry large amounts of current and many later models have an inductive ammeter pickup that clamps around the wire to simplify test connections. On some models, the ammeter also has a zero-center scale to allow testing of charging and starting systems without switching leads or polarity. A digital multimeter is a voltmeter, ammeter and ohmmeter combined in an instrument which gives a digital readout. These are often used when testing solid state circuits because of their high input impedance (usually 10 megohms or more).

The tach-dwell meter combines a tachometer and a dwell (cam angle) meter and is a specialized kind of voltmeter. The tachometer scale is marked to show engine speed in rpm and the dwell scale is marked to show degrees of distributor shaft rotation. In most electronic ignition systems, dwell is determined by the control unit, but the dwell meter can also be used to check the duty cycle (operation) of some electronic engine control systems.

Special Test Equipment

A variety of diagnostic tools are available to help troubleshoot and repair computerized engine control systems. The most so-

phisticated of these devices are the console-type engine analyzers that usually occupy a garage service bay, but there are several types of aftermarket electronic testers available that will allow quick circuit tests of the engine control system by plugging directly into a special connector located in the engine compartment or under the dashboard. Several tool and equipment manufacturers offer simple, hand-held testers that measure various circuit voltage levels on command to check all system components for proper operation. Although these testers usually cost about $300–500, consider that the average computer control unit (or ECM) can cost just as much and the money saved by not replacing perfectly good sensors or components in an attempt to correct a problem could justify the purchase price of a special diagnostic tester the first time it's used.

These computerized testers can allow quick and easy test measurements while the engine is operating or while the vehicle is being driven. In addition, the on-board computer memory can be read to access any stored trouble codes; in effect allowing the computer to tell where it hurts and aid trouble diagnosis by pin-

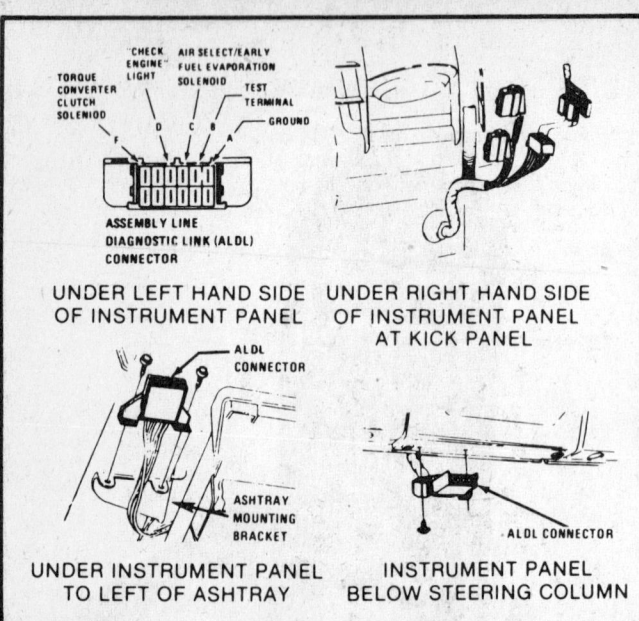

Typical diagnostic terminal locations on GM models. The diagnosis terminals are usually mounted under the dash or in the engine compartment

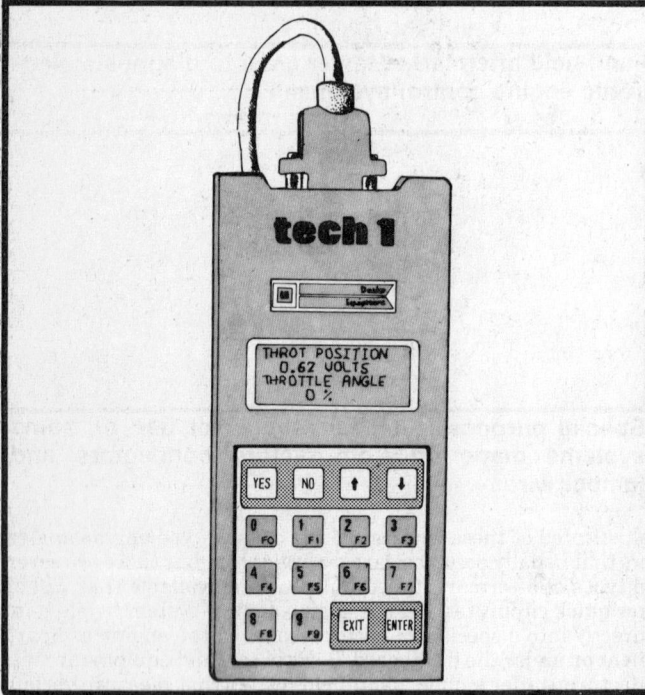

General Motor TECH–1 tester

pointing exactly which circuit or component is malfunctioning. In the same manner, repairs can be tested to make sure the problem has been corrected. The biggest advantage these special testers have is their relatively easy hookups that minimize or eliminate the chances of making the wrong connections and getting false voltage readings or damaging the computer accidentally.

NOTE: It should be remembered that these testers check voltage levels in circuits; they don't detect mechanical problems or failed components if the circuit

voltage falls within the preprogrammed limits stored in the tester PROM unit. Also, most of the hand-held testers are designed to work only on 1 or 2 systems made by a specific manufacturer.

A variety of aftermarket testers are available to help diagnose different computerized control systems. Owatonna Tool Company (OTC), for example, markets a device called the OTC Monitor which plugs directly into the Assembly Line Diagnostic Link (ALDL). The OTC tester makes diagnosis a simple matter of pressing the correct buttons and, by changing the internal PROM or inserting a different diagnosis cartridge, it will work on any model from full size to subcompact, over a wide range of years. An adapter is supplied with the tester to allow connection to all types of ALDL links, regardless of the number of pin terminals used. By inserting an updated PROM into the OTC tester, it can be easily updated to diagnose any new modifications of computerized control systems.

The TECH–1 scan tool, offer by General Motors, is specially designed to read the various stored fault codes of the electrical systems in GM vehicles. By using interchangeable PROM cartridges the TECH–1 can read specific systems data, after being pluged into the vehicle ALDL. The systems that can be checked include:
1. Engine Control Module (ECM) systems
2. The ignition system
3. The emission controls system
4. The body computer system
5. Heater and air conditioning systems, on vehicles with a body computers
6. Speed control system, on some vehicles
7. Lighting system, on vehicles with a body computer
8. Charging system, on vehicles with a body computer
9. Anti-Lock brake system (on vehicles equipped)

The specific PROM cartridge for each vehicle and each system must be install into the tester to provide complete test results. For example, PROM J–94–9 covers mass airflow sensor diagnosis and PROM J–94–8A covers all W body vehicles. Other cartridges are availble for different test and vehicles, and newer cartridges will be made as models and options change. Because of the many changeable cartridges, the TECH–1 displays faults in words rather than code numbers.

Wiring Diagrams

The average automobile contains about ½ mile of wiring, with hundreds of individual connections. To protect the many wires from damage and to keep them from becoming a confusing tangle, they are organized into bundles, enclosed in plastic or taped together and called wire harnesses. Different wiring harnesses serve different parts of the vehicle. Individual wires are color-

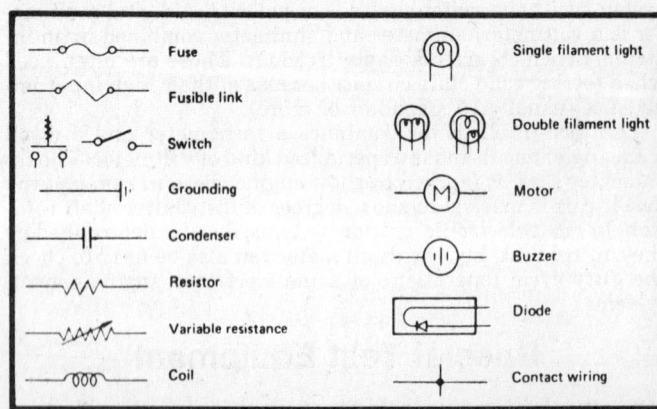

Typical electrical symbols found on wiring diagrams

COMMON SYMBOLS FOR AUTOMOTIVE COMPONENTS USED IN SCHEMATIC DIAGRAMS

Automotive service manuals use schematic diagrams to show how electrical and other types of components work, and how such components are connected to make circuits. Components that are shown whole are represented in full lines in a rectangular shape, and are identified by name; where only a part of a component is shown in a schematic diagram, the rectangular shape is outlined with a dashed line.

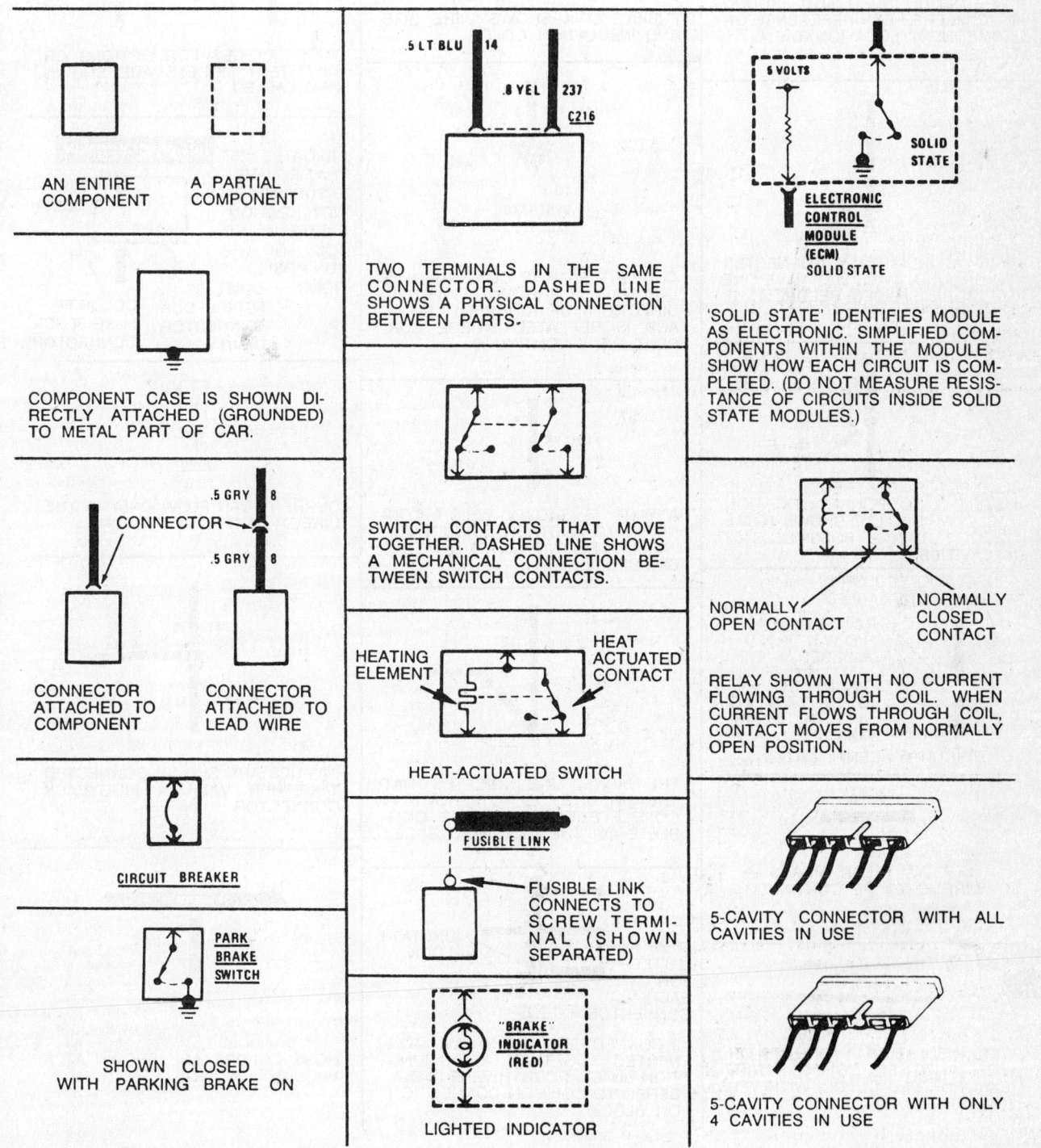

AN ENTIRE COMPONENT

A PARTIAL COMPONENT

COMPONENT CASE IS SHOWN DIRECTLY ATTACHED (GROUNDED) TO METAL PART OF CAR.

CONNECTOR

CONNECTOR ATTACHED TO COMPONENT

CONNECTOR ATTACHED TO LEAD WIRE

CIRCUIT BREAKER

PARK BRAKE SWITCH

SHOWN CLOSED WITH PARKING BRAKE ON

.5 LT BLU 14

.8 YEL 237

C216

TWO TERMINALS IN THE SAME CONNECTOR. DASHED LINE SHOWS A PHYSICAL CONNECTION BETWEEN PARTS.

SWITCH CONTACTS THAT MOVE TOGETHER. DASHED LINE SHOWS A MECHANICAL CONNECTION BETWEEN SWITCH CONTACTS.

HEATING ELEMENT

HEAT ACTUATED CONTACT

HEAT-ACTUATED SWITCH

FUSIBLE LINK

FUSIBLE LINK CONNECTS TO SCREW TERMINAL (SHOWN SEPARATED)

"BRAKE" INDICATOR (RED)

LIGHTED INDICATOR

5 VOLTS

SOLID STATE

ELECTRONIC CONTROL MODULE (ECM) SOLID STATE

'SOLID STATE' IDENTIFIES MODULE AS ELECTRONIC. SIMPLIFIED COMPONENTS WITHIN THE MODULE SHOW HOW EACH CIRCUIT IS COMPLETED. (DO NOT MEASURE RESISTANCE OF CIRCUITS INSIDE SOLID STATE MODULES.)

NORMALLY OPEN CONTACT

NORMALLY CLOSED CONTACT

RELAY SHOWN WITH NO CURRENT FLOWING THROUGH COIL. WHEN CURRENT FLOWS THROUGH COIL, CONTACT MOVES FROM NORMALLY OPEN POSITION.

5-CAVITY CONNECTOR WITH ALL CAVITIES IN USE

5-CAVITY CONNECTOR WITH ONLY 4 CAVITIES IN USE

WIRE IS GROUNDED, AND GROUND IS NUMBERED FOR REFERENCE ON COMPONENT LOCATION TABLE.

WIRE IS INDIRECTLY CONNECTED TO GROUND. (WIRE MAY HAVE ONE OR MORE SPLICES BEFORE IT IS GROUNDED.)

FEMALE TERMINAL

MALE TERMINAL

CONNECTOR REFERENCE NO. IS LISTED IN COMPONENT LOCATION TABLE, WHICH ALSO SHOWS TOTAL NO. OF TERMINALS POSSIBLE: C103 (6 CAVITIES).

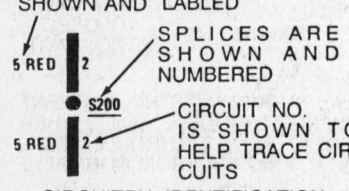

INSULATION COLOR IS SHOWN AND LABLED

SPLICES ARE SHOWN AND NUMBERED

CIRCUIT NO. IS SHOWN TO HELP TRACE CIRCUITS

CIRCUITRY IDENTIFICATION

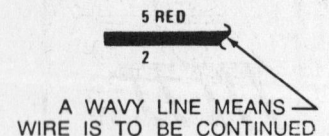

A WAVY LINE MEANS WIRE IS TO BE CONTINUED

2 RED/YEL

79

WIRE INSULATION IS ONE COLOR, WITH ANOTHER COLOR STRIPE (EXAMPLE: RED COLOR, WITH YELLOW STRIPE).

FUSIBLE LINK SHOWS WIRE SIZE AND INSULATION COLOR.

1 YEL 5

A

TO GENERATOR PAGE 109

CURRENT PATH IS CONTINUED AS LABLED. THE ARROW SHOWS THE DIRECTION OF CURRENT FLOW, AND IS REPEATED WHERE CURRENT PATH CONTINUES.

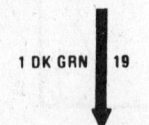

1 DK GRN 19

A WIRE IS SHOWN WHICH CONNECTS TO ANOTHER CIRCUIT. THE WIRE IS SHOWN AGAIN ON THAT CIRCUIT.

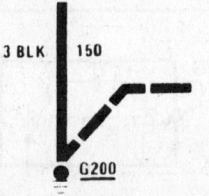

3 BLK 150

G200

THE DASHED LINE INDICATES THAT THE CIRCUITRY IS NOT SHOWN IN COMPLETE DETAIL BUT IS COMPLETE ON THE INDICATED PAGE.

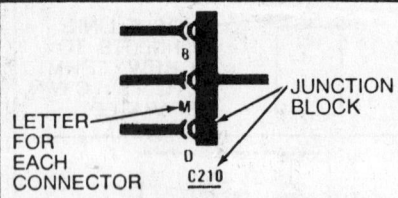
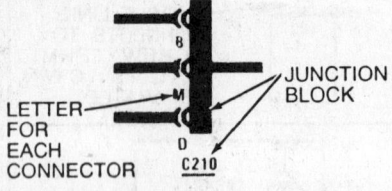

JUNCTION BLOCK

LETTER FOR EACH CONNECTOR

C210

3 CONNECTORS ARE SHOWN CONNECTED TOGETHER AT A JUNCTION BLOCK. FOURTH WIRE IS SOLDERED TO COMMON CONNECTION ON BLOCK.

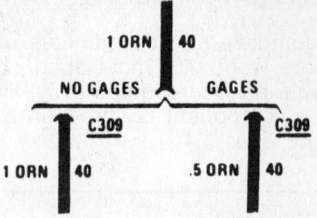

1 ORN 40

NO GAGES **GAGES**

C309 **C309**

1 ORN 40 **.5 ORN 40**

WIRE CHOICES FOR OPTIONS OR DIFFERENT MODELS ARE SHOWN AND LABLED.

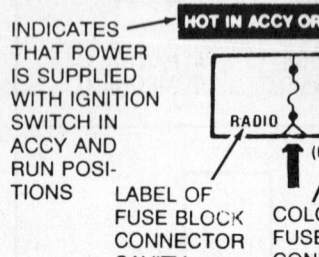

INDICATES THAT POWER IS SUPPLIED WITH IGNITION SWITCH IN ACCY AND RUN POSITIONS

HOT IN ACCY OR RUN

FUSE BLOCK

RADIO

(GRN)

LABEL OF FUSE BLOCK CONNECTOR CAVITY

COLOR OF FUSE BLOCK CONNECTOR

DIODE

CURRENT CAN FLOW ONLY IN THE DIRECTION OF THE ARROW

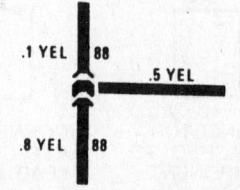

.1 YEL 88

.5 YEL

.8 YEL 88

3 WIRES ARE SHOWN CONNECTED TOGETHER WITH A PIGGYBACK CONNECTOR

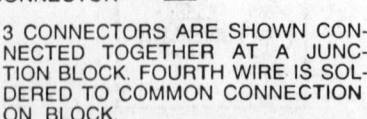

BLUE **BROWN**

RED

HOSE COLORS ARE SHOWN AT A VACUUM JUNCTION.

coded to help trace them through a harness where sections are hidden from view.

A loose or corroded connection or a replacement wire that is too small for the circuit will add extra resistance and an additional voltage drop to the circuit. A 10% voltage drop can result in slow or erratic motor operation, for example, even though the circuit is complete. Automotive wiring or circuit conductors can be in any 1 of 3 forms:

1. Single strand wire
2. Multistrand wire
3. Printed circuitry

Single strand wire has a solid metal core and is usually used inside such components as alternators, motors, relays and other devices. Multistrand wire has a core made of many small strands of wire twisted together into a single conductor. Most of the wiring in an automotive electrical system is made up of multistrand wire, either as a single conductor or grouped together in a harness. All wiring is color-coded on the insulator, either as a solid color or as a colored wire with an identification stripe. A printed circuit is a thin film of copper or other conductor that is printed on an insulator backing. Occasionally, a printed circuit is sandwiched between 2 sheets of plastic for more protection and flexibility. A complete printed circuit, consisting of conductors, insulating material and connectors for lamps or other components is called a printed circuit board. Printed circuitry is used in place of individual wires or harnesses in places where space is limited, such as behind instrument panels.

Wire Gauge

Since computer-controlled automotive electrical systems are very sensitive to changes in resistance, the selection of properly sized wires is critical when systems are repaired. The wire gauge number is an expression of the cross section area of the conductor. The most common system for expressing wire size is the American Wire Gauge (AWG) system.

Wire cross section area is measured in circular mils. A mil is $\frac{1}{1000}$ of an in. (0.001); a circular mil is the area of a circle one mil in diameter. For example, a conductor ¼ in. in diameter is 0.250 in. or 250 mils. The circular mil cross section area of the wire is 250 squared or 62,500 circular mils.

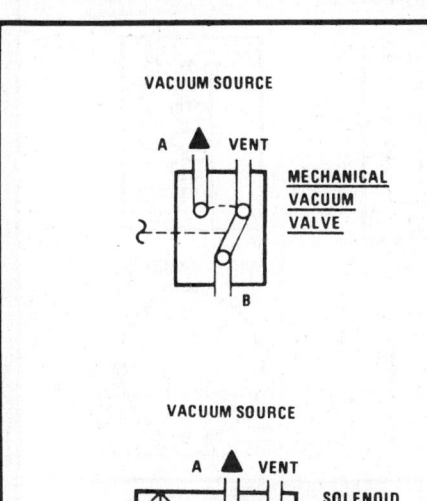

2-POSITION VACUUM MOTORS

IN THE 'AT REST' POSITION SHOWN, THE VALVE SEALS PORT 'A' AND VENTS PORT 'B' TO THE ATMOSPHERE. WHEN THE VALVE IS MOVED TO THE 'OPERATED' POSITION, VACUUM FROM PORT 'A' IS CONNECTED TO PORT 'B'. THE SOLENOID VACUUM VALVE USES THE SOLENOID TO MOVE THE VALVE.

VACUUM MOTORS OPERATE LIKE ELECTRICAL SOLENOIDS, MECHANICALLY PUSHING OR PULLING A SHAFT BETWEEN TWO FIXED POSITIONS. WHEN VACUUM IS APPLIED, THE SHAFT IS PULLED IN. WHEN NO VACUUM IS APPLIED, THE SHAFT IS PUSHED ALL THE WAY OUT BY A SPRING.

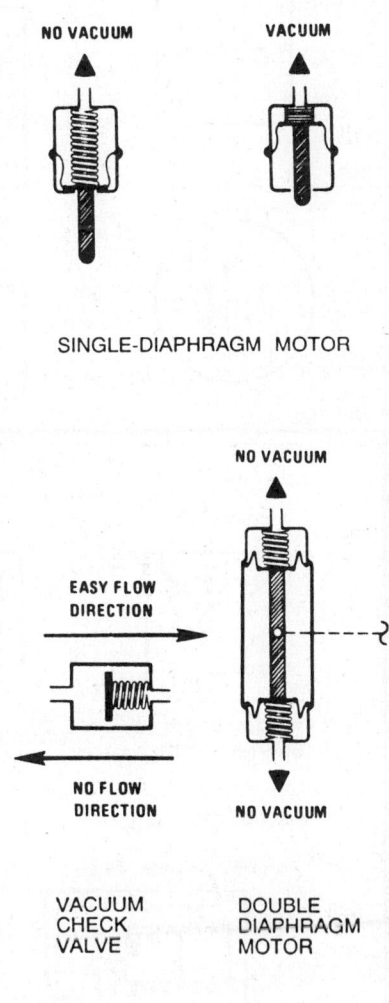

SINGLE-DIAPHRAGM MOTOR

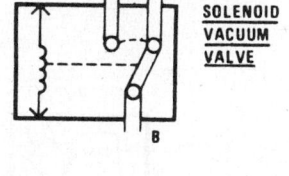

VACUUM CHECK VALVE

DOUBLE DIAPHRAGM MOTOR

DOUBLE-DIAPHRAGM MOTORS CAN BE OPERATED BY VACUUM IN TWO DIRECTIONS. WHEN THERE IS NO VACUUM, THE MOTOR IS IN THE CENTER 'AT REST' POSITION.

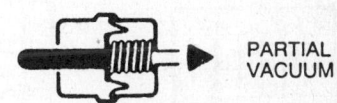

SERVO MOTOR

SOME VACUUM MOTORS, SUCH AS THE SERVO MOTOR IN THE CRUISE CONTROL, CAN POSITION THE ACTUATING ARM AT ANY POSITION BETWEEN FULLY EXTENDED AND FULLY RETRACTED. THE SERVO IS OPERATED BY A CONTROL VALVE THAT APPLIES VARYING AMOUNTS OF VACUUM TO THE MOTOR. THE HIGHER THE VACUUM LEVEL, THE GREATER THE RETRACTION OF THE MOTOR ARM. SERVO MOTORS WORK LIKE THE TWO-POSITION MOTORS; THE ONLY DIFFERENCE IS IN THE WAY THE VACUUM IS APPLIED. SERVO MOTORS ARE GENERALLY LARGER AND PROVIDE A CALIBRATED CONTROL.

METRIC SIZE	AWG SIZES
.22	24
.35	22
.5	20
.8	18
1.0	16
2.0	14
3.0	12
5.0	10
8.0	8
13.0	6
19.0	4
32.0	2

Wire Size Conversion Table

Gauge numbers are assigned to conductors of various cross section areas. As gauge number increases, area decreases and the conductor becomes smaller. A 5 gauge conductor is smaller than a 1 gauge conductor and a 10 gauge is smaller than a 5 gauge. As the cross section area of a conductor decreases, resistance increases and so does the gauge number. A conductor with a higher gauge number will carry less current than a conductor with a lower gauge number.

NOTE: Gauge wire size refers to the size of the conductor, not the size of the complete wire. It is possible to have 2 wires of the same gauge with different diameters because one may have thicker insulation than the other.

Twelve volt automotive electrical systems generally use 10, 12, 14, 16 and 18 gauge wire. Main power distribution circuits and larger accessories usually use 10 and 12 gauge wire. Battery cables are usually 4 or 6 gauge, although 1 and 2 gauge wires are occasionally used. Wire length must also be considered when making repairs to a circuit. As conductor length increases, so does resistance. An 18 gauge wire, for example, can carry a 10 amp load for 10 feet without excessive voltage drop; however if a 15 foot wire is required for the same 10 amp load, a 16 gauge wire must be used.

An electrical schematic shows the electrical current paths when a circuit is operating properly. It is essential to understand how a circuit works before trying to figure out why it doesn't. Schematics break the entire electrical system down into individual circuits and show only one particular circuit. In a schematic, no attempt is made to represent wiring and components as they physically appear on the vehicle; switches and other components are shown as simply as possible. Face views of harness connectors show the cavity or terminal locations in all multi-pin connectors to help locate test points.

When backprobing a connector while it is on the component, the order of the terminals must be mentally reversed. The wire

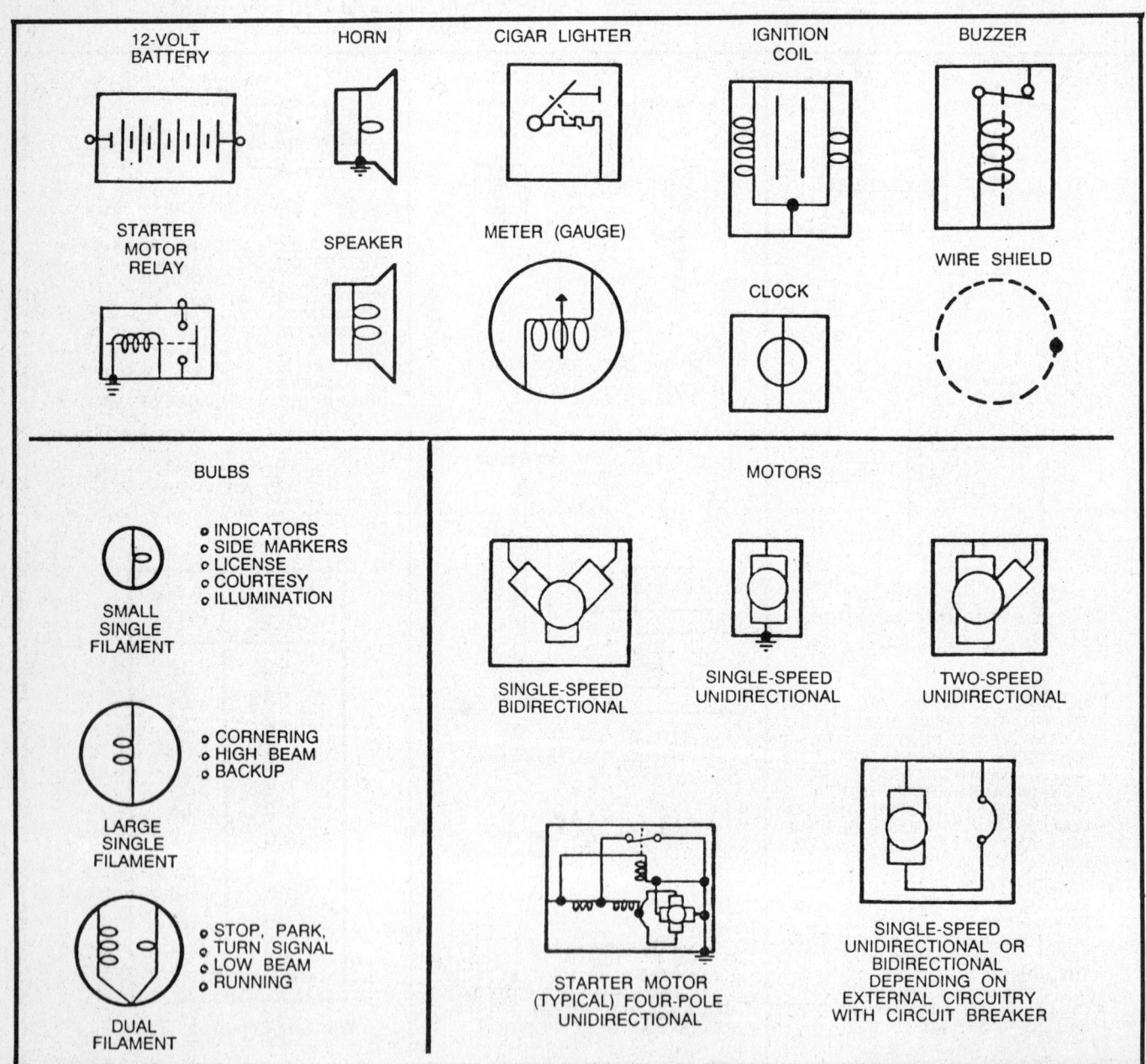

color code can help in this situation, as well as a keyway, lock tab or other reference mark.

Wiring Repairs

Soldering is a quick, efficient method of joining metals permanently. Everyone who has to make wiring repairs should know how to solder. Electrical connections that are soldered are far less likely to come apart and will conduct electricity much better than connections that are only "pig-tailed" together. The most popular (and preferred) method of soldering is with an electrical soldering gun. Soldering irons are available in many sizes and wattage ratings. Irons with higher wattage ratings deliver higher temperatures and recover lost heat faster. A small soldering iron rated for no more than 50 watts is recommended, especially on electrical systems where excess heat can damage the components being soldered.

There are 3 ingredients necessary for successful soldering; proper flux, good solder and sufficient heat. A soldering flux is necessary to clean the metal of tarnish, prepare it for soldering and to enable the solder to spread into tiny crevices. When soldering, always use a resin flux or resin core solder which is non-corrosive and will not attract moisture once the job is finished. Other types of flux (acid core) will leave a residue that will attract moisture and cause the wires to corrode. Tin is a unique metal with a low melting point. In a molten state, it dissolves and alloys easily with many metals. Solder is made by mixing tin with lead. The most common proportions are 40/60, 50/50 and 60/40, with the percentage of tin listed first. Low priced solders usually contain less tin, making them very difficult for a begin-

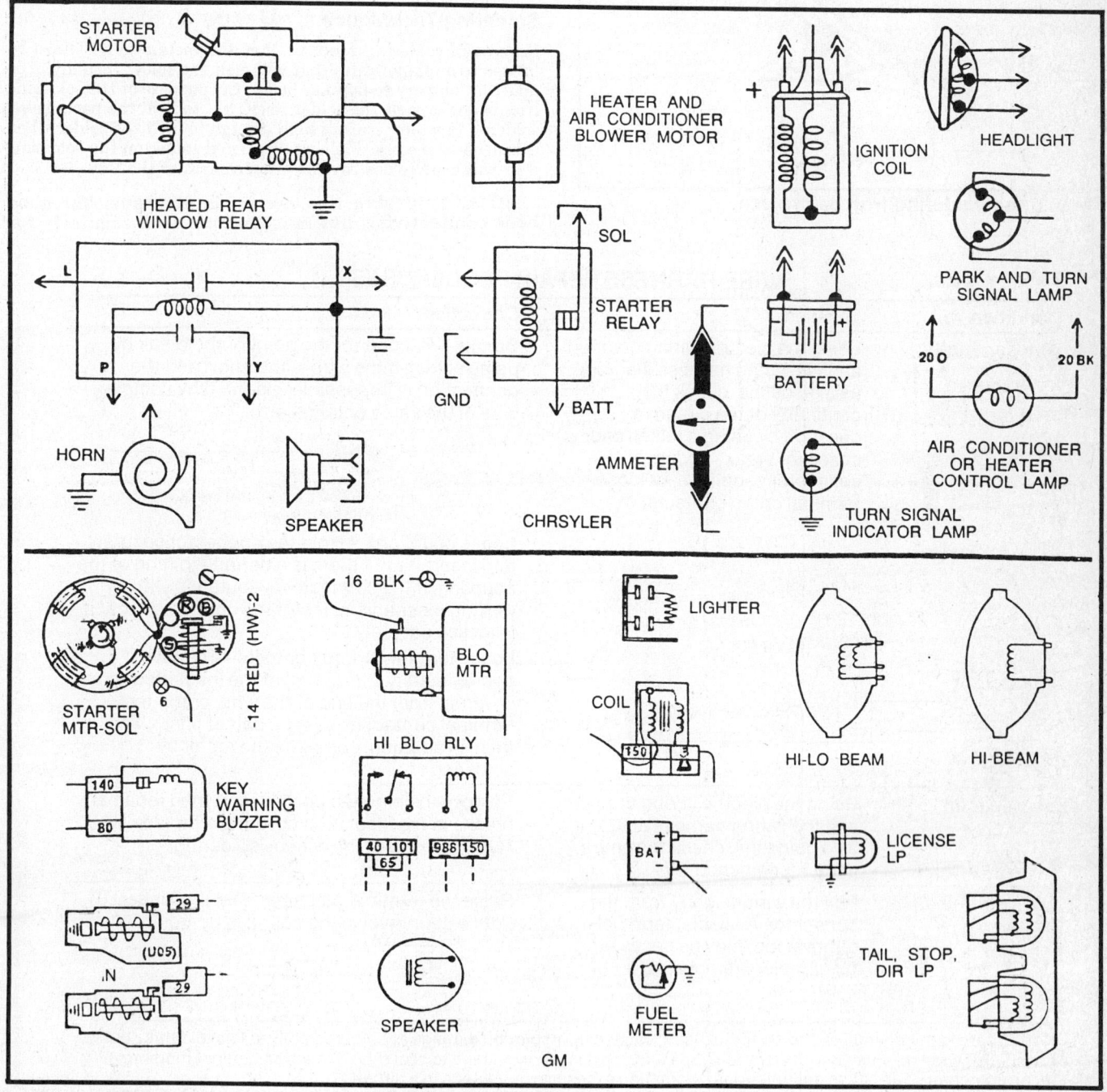

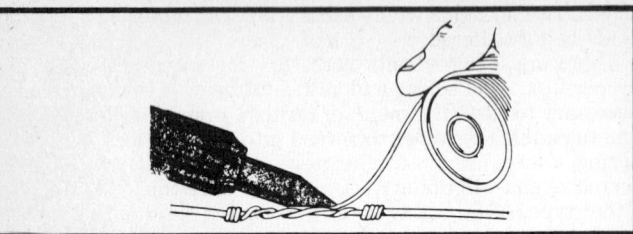

Proper soldering method. Allow the soldering iron to heat the wire first, then apply the solder as shown

Tinning the soldering iron before use

Various types of soldering guns

ner to use because more heat is required to melt the solder. A common solder is 40/60 which is well suited for general use, but 60/40 melts easier, has more tin for a better joint and is preferred for electrical work.

Soldering Techniques

Successful soldering requires that the metals to be joined be heated to a temperature that will melt the solder (usually 360–460°F). Contrary to popular belief, the purpose of the soldering iron is not to melt the solder itself, but to heat the parts being soldered to a temperature high enough to melt the solder when it touches the work. Melting flux-cored solder on the soldering iron will usually destroy the effectiveness of the flux.

NOTE: Soldering tips are made of copper for good heat conductivity, but must be "tinned" regularly for

WIRE HARNESS REPAIR PROCEDURES

Condition	Location	Correction
Non-continuity	Using the electric wiring diagram and the wiring harness diagram as a guideline, check the continuity of the circuit in question by using a tester, and check for breaks, loose connector couplings, or loose terminal crimp contacts.	**Breaks**—Reconnect the point of the break by using solder. If the wire is too short and the connection is impossible, extend it by using a wire of the same or larger size. Solder Be careful concerning the size of wire used for the extension **Loose couplings**—Hold the connector securely, and insert it until there is a definite joining of the coupling. If the connector is equipped with a locking mechanism, insert the connector until it is locked securely. **Loose terminal crimp contacts**—Remove approximately 2 in. (5mm) of the insulation covering from the end of the wire, crimp the terminal contact by using a pair of pliers, and then, in addition, complete the repair by soldering. Crimp by using pliers / Solder
Short-circuit	Using the electric wiring diagram and the wiring harness diagram as a guideline, check the entire circuit for pinched wires.	Remove the pinched portion, and then repair any breaks in the insulation covering with tape. Repair breaks of the wire by soldering.
Loose terminal	Pull the wiring lightly from the connector. A special terminal removal tool may be necessary for complete removal.	Raise the terminal catch pin, and then insert it until a definite clicking sound is heard. Catch pin

Note: There is the chance of short circuits being caused by insulation damage at soldered points. To avoid this possibility, wrap all splices with electrical tape and use a layer of silicone to seal the connection against moisture. Incorrect repairs can cause malfunctions by creating excessive resistance in a circuit.

TWISTED/SHIELDED CABLE

DRAIN WIRE

OUTER JACKET

MYLAR→

1. REMOVE OUTER JACKET.
2. UNWRAP ALUMINUM/MYLAR TAPE. DO NOT REMOVE MYLAR.

3. UNTWIST CONDUCTORS. STRIP INSULATION AS NECESSARY.

DRAIN WIRE

4. SPLICE WIRES USING SPLICE CLIPS AND ROSIN CORE SOLDER. WRAP EACH SPLICE TO INSULATE.
5. WRAP WITH MYLAR AND DRAIN (UNINSULATED) WIRE.

6. TAPE OVER WHOLE BUNDLE TO SECURE AS BEFORE

TWISTED LEADS

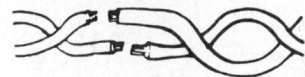

1. LOCATE DAMAGED WIRE.
2. REMOVE INSULATION AS REQUIRED.

SPLICE & SOLDER

3. SPLICE TWO WIRES TOGETHER USING SPLICE CLIPS AND ROSIN CORE SOLDER.

4. COVER SPLICE WITH TAPE TO INSULATE FROM OTHER WIRES.
5. RETWIST AS BEFORE AND TAPE WITH ELECTRICAL TAPE AND HOLD IN PLACE.

quick transfer of heat to the project and to prevent the solder from sticking to the iron. To "tin" the iron, simply heat it and touch the flux-cored solder to the tip; the solder will flow over the hot tip. Wipe the excess off with a clean rag, but be careful as the iron will be hot.

After some use, the tip may become pitted. If so, simply dress the tip smooth with a smooth file and "tin" the tip again. An old saying holds that "metals well cleaned are half soldered." Flux-cored solder will remove oxides but rust, bits of insulation and oil or grease must be removed with a wire brush or emery cloth. For maximum strength in soldered parts, the joint must start off clean and tight. Weak joints will result if there are gaps too wide for the solder to bridge.

If a separate soldering flux is used, it should be brushed or swabbed only on areas that are to be soldered. Most solders contain a core of flux and separate fluxing is unnecessary. Hold the work to be soldered firmly. It is best to solder on a wooden board, because a metal vise will only rob the piece to be soldered of heat and make it difficult to melt the solder. Hold the soldering tip with the broadest face against the work to be soldered. Apply solder under the tip close to the work, using enough solder to give a heavy film between the iron and the piece being soldered, while moving slowly and making sure the solder melts properly. Keep the work level or the solder will run to the lowest part and favor the thicker parts, because these require more heat to melt the solder. If the soldering tip overheats (the solder coating on the face of the tip burns up), it should be retinned. Once the soldering is completed, let the soldered joint stand until cool. Tape and seal all soldered wire splices after the repair has cooled.

Wire Harness and Connectors

The on-board computer (ECM) wire harness electrically connects the control unit to the various solenoids, switches and sensors used by the control system. Most connectors in the engine compartment or otherwise exposed to the elements are protected against moisture and dirt which could create oxidation and deposits on the terminals. This protection is important because

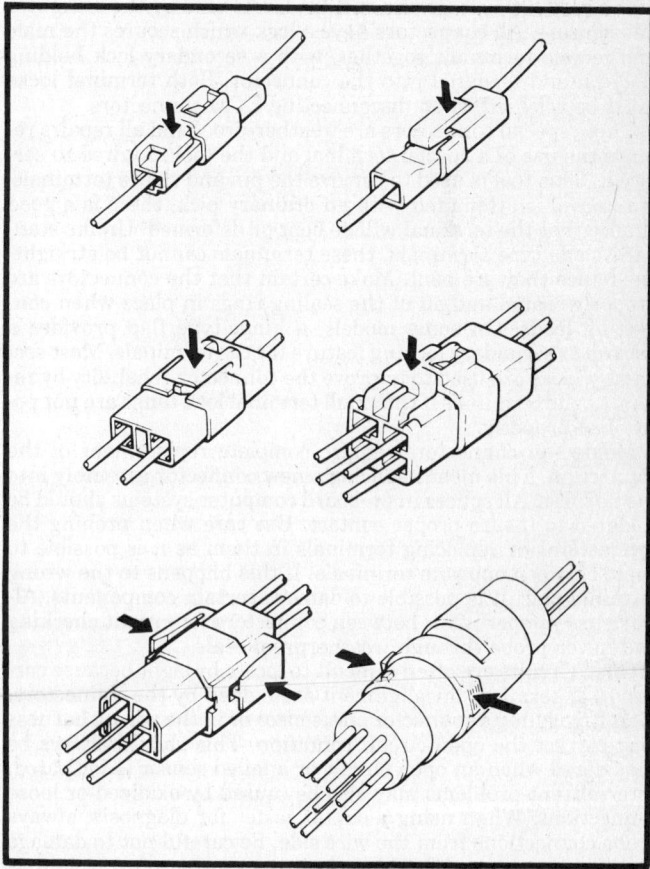

Various types of locking harness connectors. Depress the locks at the arrows to separate the connectors

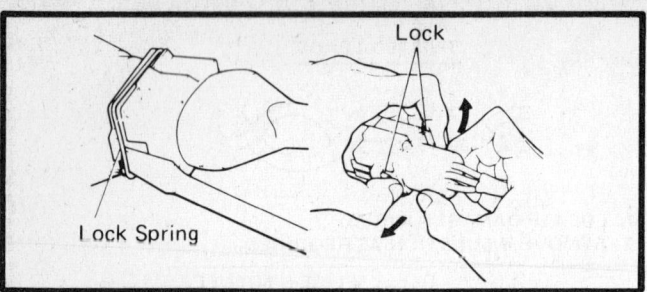

Some electrical connectors use a lock spring instead of the molded locking tabs

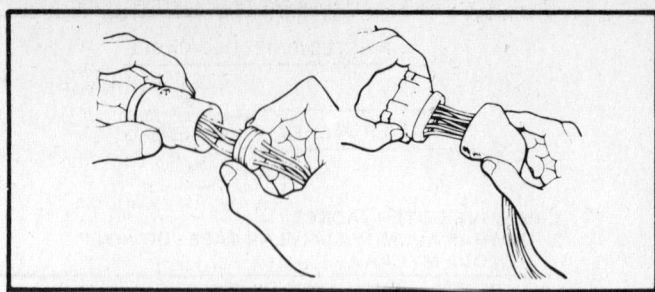

Slide back the weatherproof seals or boots on sealed terminals for testing

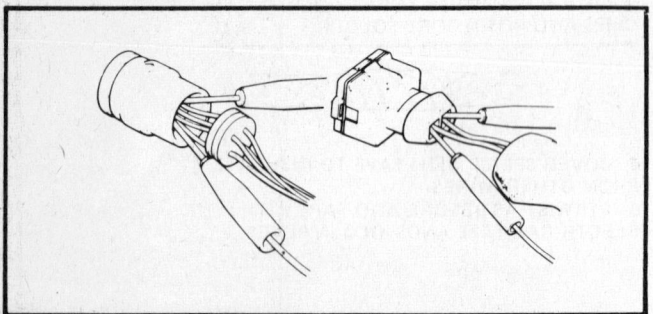

Correct method of testing weatherproof connectors. Do not pierce connector seals with test probes

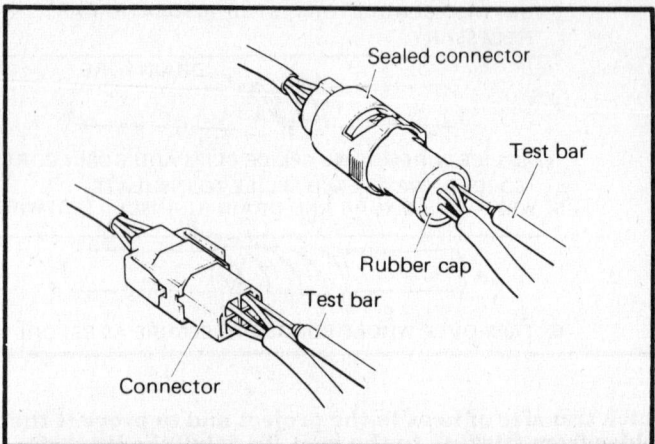

Probe all connectors from the wire side when testing

of the very low voltage and current levels used by the computer and sensors. All connectors have a lock which secures the male and female terminals together, with a secondary lock holding the seal and terminal into the connector. Both terminal locks must be released when disconnecting ECM connectors.

These special connectors are weather-proof and all repairs require the use of a special terminal and the tool required to service it. This tool is used to remove the pin and sleeve terminals. If removal is attempted with an ordinary pick, there is a good chance that the terminal will be bent or deformed. Unlike standard blade type terminals, these terminals cannot be straightened once they are bent. Make certain that the connectors are properly seated and all of the sealing rings in place when connecting leads. On some models, a hinge-type flap provides a backup or secondary locking feature for the terminals. Most secondary locks are used to improve the connector reliability by retaining the terminals if the small terminal lock tangs are not positioned properly.

Molded-on connectors require complete replacement of the connection. This means splicing a new connector assembly into the harness. All splices in on-board computer systems should be soldered to insure proper contact. Use care when probing the connections or replacing terminals in them as it is possible to short between opposite terminals. If this happens to the wrong terminal pair, it is possible to damage certain components. Always use jumper wires between connectors for circuit checking and never probe through weatherproof seals.

Open circuits are often difficult to locate by sight because corrosion or terminal misalignment are hidden by the connectors. Merely wiggling a connector on a sensor or in the wiring harness may correct the open circuit condition. This should always be considered when an open circuit or a failed sensor is indicated. Intermittent problems may also be caused by oxidized or loose connections. When using a circuit tester for diagnosis, always probe connections from the wire side. Be careful not to damage sealed connectors with test probes.

All wiring harnesses should be replaced with identical parts, using the same gauge wire and connectors. When signal wires are spliced into a harness, use wire with high temperature insu-

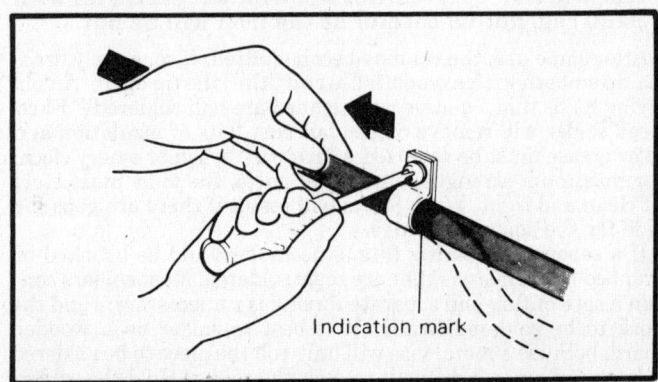

Secure the wiring harness at the indication marks, if used, to prevent vibrations from causing wear and a possible short

lation only. With the low voltage and current levels found in the system, it is important that the best possible connection at all wire splices be made by soldering the splices together. It is seldom necessary to replace a complete harness. If replacement is necessary, pay close attention to insure proper harness routing. Secure the harness with suitable plastic wire clamps to prevent vibrations from causing the harness to wear in spots or contact any hot components.

NOTE: Weatherproof connectors cannot be replaced with standard connectors. Instructions are provided with replacement connector and terminal packages. Some wire harnesses have mounting indicators (usually pieces of colored tape) to mark where the harness is to be secured.

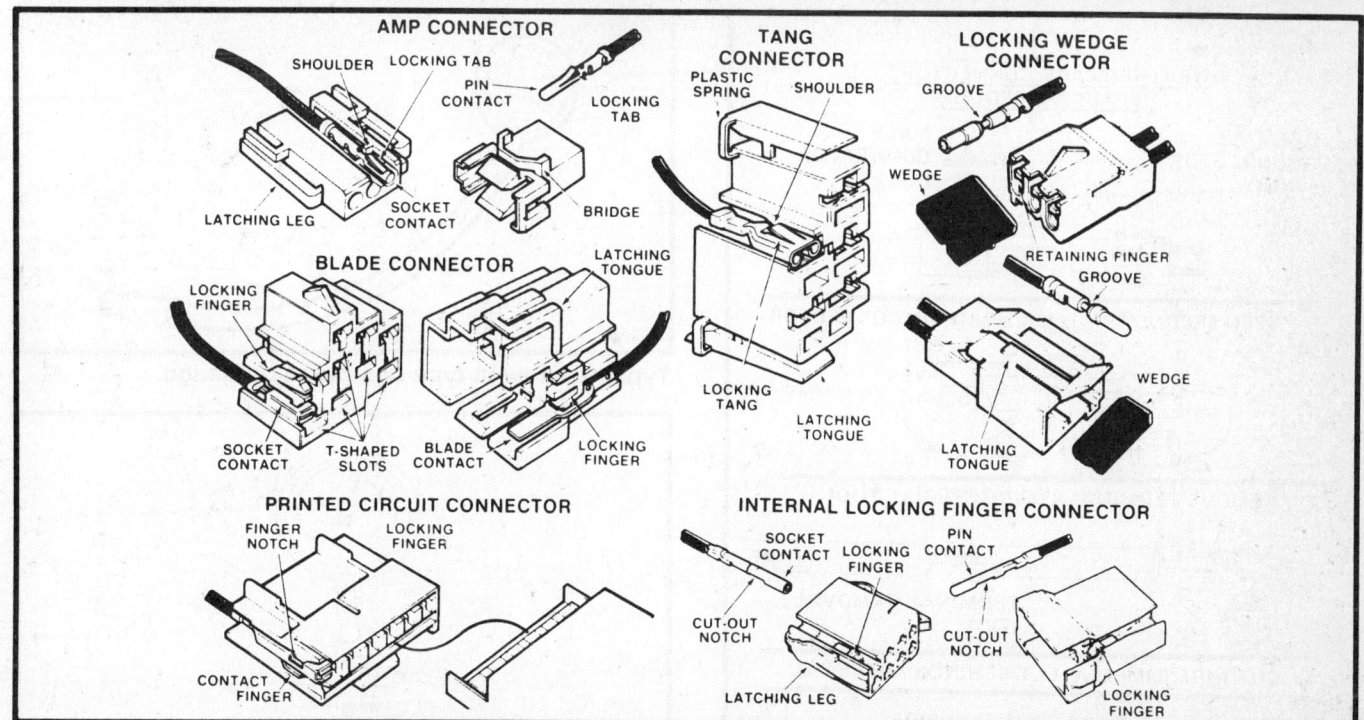

In making wiring repairs, it's important to always replace damaged wires with wires that are the same gauge as the wire being replaced. The heavier the wire, the smaller the gauge number. Wires are color-coded to aid in identification and whenever possible the same color coded wire should be used for replacement. A wire stripping and crimping tool is necessary to install solderless terminal connectors. Test all crimps by pulling on the wires; it should not be possible to pull the wires out of a good crimp.

Wires which are open, exposed or otherwise damaged are repaired by simple splicing. Where possible, if the wiring harness is accessible and the damaged place in the wire can be located, it is best to open the harness and check for all possible damage. In an inaccessible harness, the wire must be bypassed with a new insert, usually taped to the outside of the old harness.

When replacing fusible links, be sure to use fusible link wire, NOT ordinary automotive wire. Make sure the fusible segment is of the same gauge and construction as the one being replaced and double the stripped end when crimping the terminal connector for a good contact. The melted (open) fusible link segment of the wiring harness should be cut off as close to the harness as possible, then a new segment spliced in as described. In the case of a damaged fusible link that feeds 2 harness wires, the harness connections should be replaced with 2 fusible link wires so that each circuit will have its own separate protection.

NOTE: Most of the problems caused in the wiring harness are due to bad ground connections. Always check all vehicle ground connections for corrosion or looseness before performing any power feed checks to eliminate the chance of a bad ground affecting the circuit.

Repairing Hard Shell Connectors

Unlike molded connectors, the terminal contacts in hard shell connectors can be replaced. Weatherproof hard-shell connectors with the leads molded into the shell have non-replaceable terminal ends. Replacement usually involves the use of a special terminal removal tool that depress the locking tangs (barbs) on the connector terminal and allow the connector to be removed from the rear of the shell. The connector shell should be replaced if it shows any evidence of burning, melting, cracks, or breaks. Replace individual terminals that are burnt, corroded, distorted or loose.

NOTE: The insulation crimp must be tight to prevent the insulation from sliding back on the wire when the wire is pulled. The insulation must be visibly compressed under the crimp tabs, and the ends of the crimp should be turned in for a firm grip on the insulation.

The wire crimp must be made with all wire strands inside the crimp. The terminal must be fully compressed on the wire strands with the ends of the crimp tabs turned in to make a firm grip on the wire. Check all connections with an ohmmeter to insure a good contact. There should be no measurable resistance between the wire and the terminal when connected.

Mechanical Test Equipment

VACUUM GAUGE

Most gauges are graduated in inches of Mercury (in. Hg), although a device called a manometer reads vacuum in inches of water (in. H_2O). The normal vacuum reading usually varies between 18 and 22 in. Hg at sea level. To test engine vacuum, the vacuum gauge must be connected to a source of manifold vacuum. Many engines have a plug in the intake manifold which can be removed and replaced with an adapter fitting. Connect the vacuum gauge to the fitting with a suitable rubber hose or, if no manifold plug is available, connect the vacuum gauge to any device using manifold vacuum, such as EGR valves, etc. The vacuum gauge can be used to determine if enough vacuum is reaching a component to allow its actuation.

HAND VACUUM PUMP

Small, hand-held vacuum pumps come in a variety of designs. Most have a built-in vacuum gauge and allow the component to

2-17

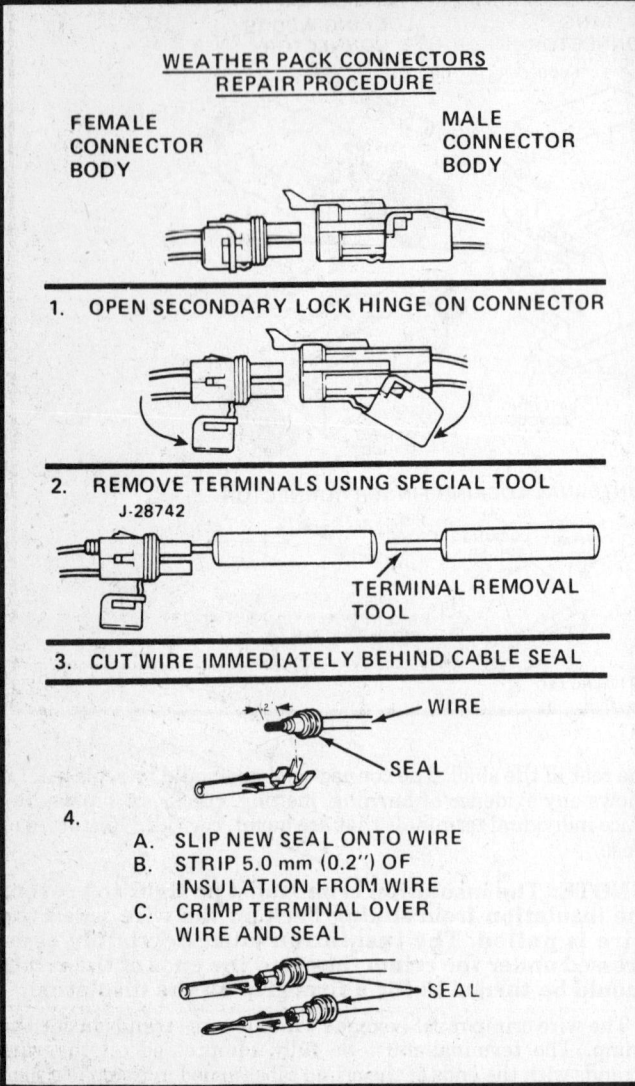

Repairing GM Weatherpak connectors. Note special terminal removal tools

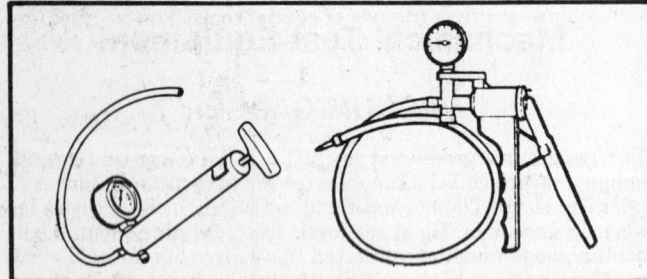

Typical hand vacuum pumps

be tested without removing it from the vehicle. Operate the pump lever or plunger to apply the correct amount of vacuum required for the test specified in the diagnosis routines. The level of vacuum in inches of Mercury (in. Hg) is indicated on the pump gauge. For some testing, an additional vacuum gauge may be necessary.

Intake manifold vacuum is used to operate various systems and devices on late model cars. To correctly diagnose and solve problems in vacuum control systems, a vacuum source is neces-

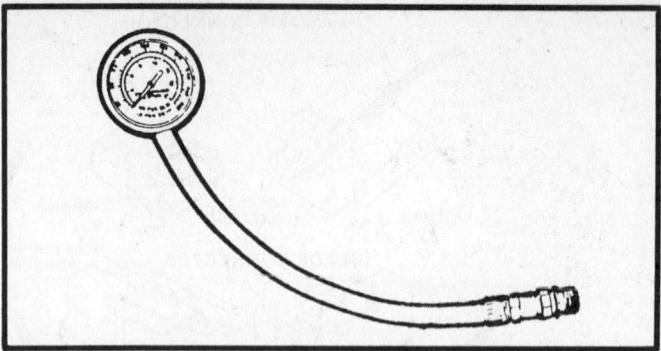

Typical screw-in type compression gauge

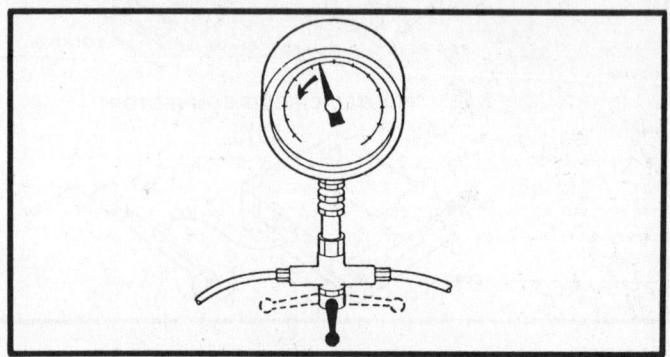

Fuel pressure gauge with 3-way adapter

sary for testing. In some cases, vacuum can be taken from the intake manifold when the engine is running, but vacuum is normally provided by a hand vacuum pump. These hand vacuum pumps have a built-in vacuum gauge that allow testing while the device is still attached to the car. For some tests, an additional vacuum gauge may be necessary.

COMPRESSION GAUGE

A compression gauge is designed to measure the amount of pressure in psi that a cylinder is capable of producing. Some gauges have a hose that screws into the spark plug hole while others have a tapered rubber tip which is held in the spark plug hole. Engine compression depends on the sealing ability of the rings, valves, head gasket and spark plug gaskets. If any of these parts are not sealing properly, compression will be lost and the power output of the engine will be reduced. The compression in each cylinder should be measured and the variation between cylinders should be noted. The engine should be cranked through 5 or 6 compression strokes while warm, with all plugs removed, ignition disabled and throttle valves wide open.

FUEL PRESSURE GAUGE

A fuel pressure gauge is required to test the operation of the fuel delivery and injection systems. Some systems also need a 3-way valve to check the fuel pressure in various modes of operation. Gauges may require special adapters for making fuel connections. Always observe the cautions outlined in the individual fuel system service section, when working around any pressurized fuel system.

FUEL INJECTOR CLEANING

Restricted fuel injectors can be caused by a build-up of deposits on the pintle of the injector, resulting in lean air/fuel mixtures, incorrect spray pattern and poor engine operation.

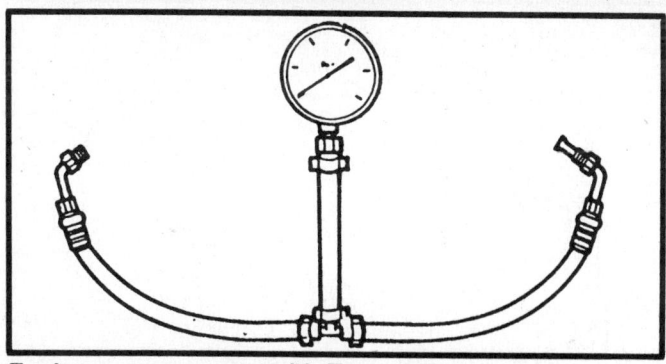

Fuel pressure gauge with Tee adapter

Various types of injector cleaning equipment are available to clean injectors. To use fuel injector cleaning equipment the fuel system must be in good working order and must be able to hold pressure. If the system will not hold pressure, it must be serviced first.

CAUTION

When using injector cleaner, always wear protective clothing, eye and face protection. Injector cleaner is under pressure and is flammable, so all precautions used with gasoline and fuel system service must be followed.

Injector cleaner is available in single application cans as well as concentrate. Follow the directions with cleaner for information on the proper ratio of cleaner to gasoline. Most cleaner kits will the following or similar procedures for use:
1. Disable fuel pump and release fuel system pressure.
2. Attach adapter to fuel rail or line and connect pressure test gauge.
3. Attach fuel injector cleaner to fuel gauge T-connector or to fuel rail.
4. Raise pressure of injector cleaner tank to 25 psi.
5. Open fuel injector cleaner valve and start engine.
6. Run engine at 2000 rpm for 10 minutes.
7. Stop engine and close fuel injector cleaner valve.
8. Release fuel pressure if necessary to recap fuel rail or disconnect adapter from fuel line.
9. Reconnect vehicle fuel pump circuit.
10. Start engine, allow to run at 2000 rpm for at least 5 minutes to flush out injector cleaner. Check for fuel leaks.

NOTE: Never add injector cleaner to the fuel tank, it will damage pump and other fuel system components.

FUEL LINE TOOLS

CAUTION

Do not attempt to repair nylon fuel lines.

Always replace nylon fuel lines with the same type and size removed. Follow normal safety precautions before disconnecting fuel lines. Joining the steel pipes to the nylon fuel lines is accom-

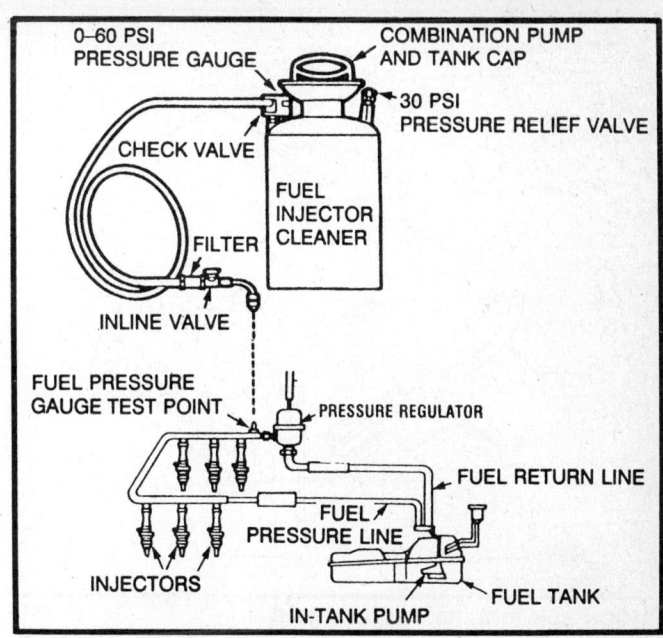

Fuel injector cleaning set-up

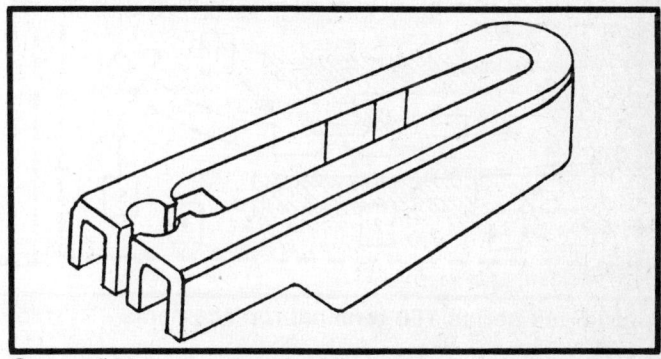

General Motors nylon fuel line removal tool, 2 sizes available

plished through the use of quick-connect fittings. These fittings snap onto the pipe/lines without the use of tools. To disconnect the fuel lines requires the use of special tools. The 5/16 fuel line requires tool J–37088–1 and the 3/8 fuel line requires tool J–37088–2.

1. Release fuel pressure and disconnect the negative battery cable.
2. Insert the fuel line separator tool into the quick-connect fitting.
3. Squeeze the tool evenly until the connector releases. It help to push on the connection slightly while squeezing the tool.
4. Remove the old fuel line.

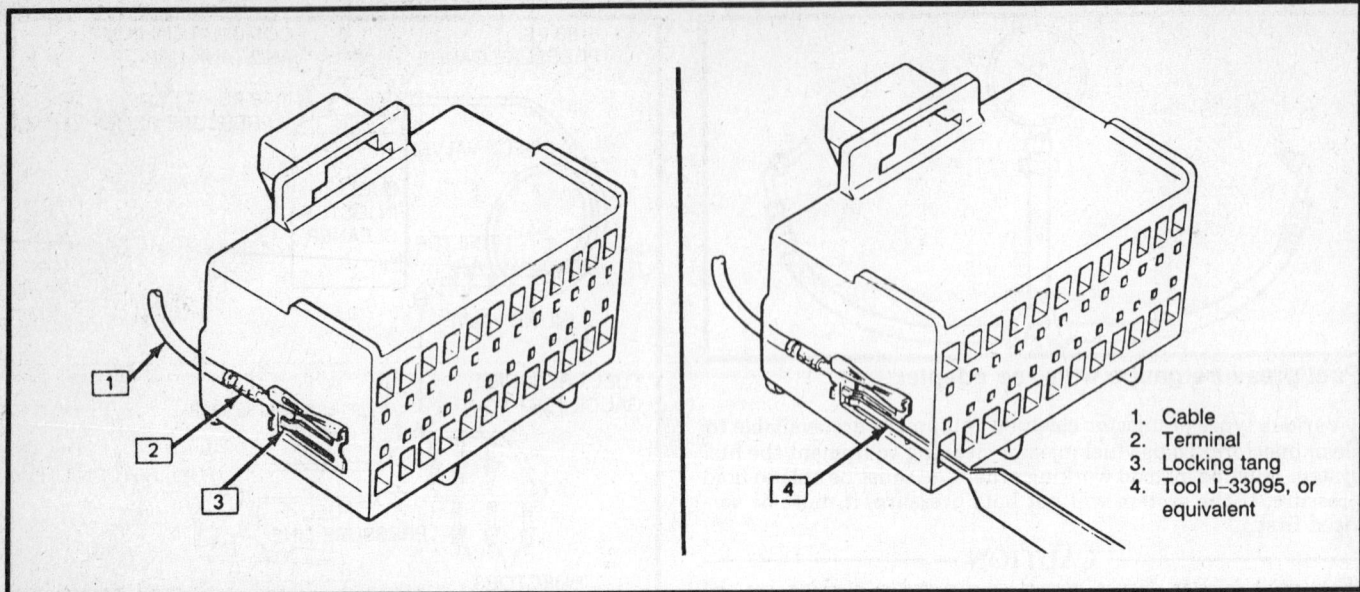

1. Cable
2. Terminal
3. Locking tang
4. Tool J–33095, or
   equivalent

Mirco-Pack terminal replacement

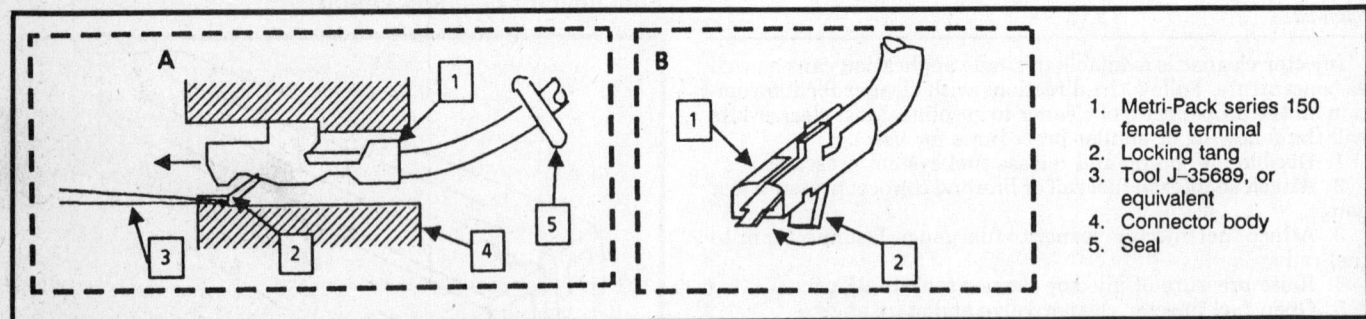

1. Metri-Pack series 150
   female terminal
2. Locking tang
3. Tool J–35689, or
   equivalent
4. Connector body
5. Seal

Metri-Pack series 150 terminal replacements

Self-Diagnostic Systems

3 SECTION

INDEX

ALL MODELS EXCEPT E, K AND CADILLAC C-BODY...... **3-2**
  Intermittent or Hard Trouble Codes........................... 3-2
  Assembly Line Communication Link (ALCL) 3-2
  Entering Diagnostic Mode 3-2
  Field Service Mode... 3-2
  Clearing the Trouble Codes..................................... 3-2
  ALDL Scan Tools 3-3
  Diagnostic Circuit Check 3-3
  Diagnostic Trouble Code Chart 3-4

E, K AND CADILLAC C-BODY................................. **3-6**
  1988–90 Buick Riviera and Reatta 3-6
  Graphic Control Center... 3-6
  How to Enter the Diagnostic Service Mode.............. 3-7
  Trouble Code Display 3-7
  Diagnostic Codes List 3-7
  Climate Control in Service Mode 3-8
  Operating the Service Mode..................................... 3-9
  Data Displays 3-10
  Input Displays... 3-11
  Output Displays 3-12
  Override Displays 3-12
  Troubleshooting Hints.. 3-13
  System Diagnosis... 3-13
  1988–90 Cadillac DeVille and Fleetwood 3-14
  Trouble Codes 3-14
  Intermittent and Hard Failure Codes...................... 3-14
  Intermittent Problem Diagnosis 3-14
  Entering Diagnostic Mode 3-15

Trouble Code Display... 3-15
Diagnostic Trouble Codes List.................................. 3-15
Clearing Trouble Codes.. 3-16
Exiting Diagnostic Mode... 3-16
EEC Program Override 3-16
1988–90 Cadillac Eldorado and Seville...................... 3-17
  System Diagnosis... 3-17
  Self-Diagnostic Trouble Codes............................... 3-17
  Entering the Diagnostic Mode 3-17
  Diagnostic Trouble Codes List................................ 3-18
  Self-Diagnostic System Flowchart........................... 3-19
  Trouble Code Display 3-19
  Service Diagnostics Operations 3-20
  Data Displays 3-21
  Input Displays... 3-22
  Output Displays 3-22
  Override Displays 3-23
1988–90 Oldsmobile Toronado 3-23
  Troubleshooting 3-24
  Self-Diagnostic Trouble Codes............................... 3-24
  Entering the Diagnostic Mode 3-24
  Diagnostic Trouble Codes List................................ 3-25
  Trouble Code Display 3-25
  Service Diagnostic Operation 3-25
  Self-Diagnostic System Flowchart (non-CRT) 3-26
  Self-Diagnostic System Flowchart (CRT).................. 3-27
  Data Displays 3-27
  Input Displays... 3-29
  Output Displays 3-29
  Override Display 3-30

ALL MODELS EXCEPT E, K AND CADILLAC C-BODY

The ECM is equipped with a self-diagnostic capability which can detect system failures and aids the technician by identifying the fault via a trouble code system and a dash mounted indicator light, marked either "SERVICE ENGINE SOON" or "CHECK ENGINE". The light is mounted on the instrument panel and has 2 functions:

1. It is used to inform the operator that a problem has occurred and the vehicle should be taken in for service as soon as reasonably possible.

2. It is used by the technician to read out stored trouble codes in order to localize malfunction areas during the diagnosis and repair phases.

As a bulb and system check, the light will come on with the ignition key in the **ON** position and the engine not operating. When the engine is started, the light will turn off. If the light does not turn off, the self diagnostic system has detected a problem in the system. If the problem goes away, the light will go out, in most cases after 10 second, but a trouble code will be set in the ECM's memory.

Intermittent or Hard Trouble Codes

An intermittent code is one which does not reset itself and is not present when initiating the trouble codes. It is often be caused by a loose connection which, with vehicle movement, can possibly cure its self but intermittently reappear. A hard code is an operational malfunction which remains in the ECM memory and will be presented when calling for the trouble code display.

The Electronic Control Module (ECM) is actually a computer. It uses numerous sensors to look at many engine operating conditions. It has been programmed to know what certain sensor readings should be under most all operating conditions and if the sensor readings are not what the ECM thinks it should be, the ECM will turn on the "SERVICE ENGINE SOON" or "CHECK ENGINE" indicator light and will store a trouble code in its memory. When called up, the trouble code directs the technician to examine a particular circuit in order to locate and repair the trouble code setting defect.

Assembly Line Communication Link (ALCL)

In order to access the ECM to provide the trouble codes stored in its memory, the Assembly Line Communication Link (also known as the Assembly Line Diagnostic Link or ALDL) is used.

NOTE: This connector is utilized at the assembly plant to insure the engine is operating properly before the vehicle is shipped.

Terminal **B** of the diagnostic connnector is the diagnostic terminal and it can be connected to terminal **A**, or ground, to enter the diagnostic mode, or the field service mode on fuel injection models.

ENTERING DIAGNOSTIC MODE

If the diagnostic terminal is grounded with the ignition in the **ON** position and the engine stopped, the system will enter the diagnostic mode. In this mode, the ECM will accomplish the following:

1. The ECM will display a Code 12 by flashing the "SERVICE ENGINE SOON" or "CHECK ENGINE" light, which indicates the system is working. A Code 12 consists of one flash, followed by a short pause, then 2 flashes in quick succession.

a. This code will be flashed 3 times. If no other codes are stored, Code 12 will continue to flash until the diagnostic terminal is disconnected from the ground circuit.

b. On a carbureted engine, the engine should not be started with the diagnostic terminal grounded, because it may contin-

ue to flash a Code 12 with the engine running. Also, if the test terminal is grounded after the engine is running any stored codes will flash, but Code 12 will flash only if there is a problem with the distributor reference signal.

c. On fuel injected engines, codes can only be obtained with the engine stopped. Grounding the diagnostic terminal with the engine running activates the Field Service Mode.

2. The ECM will display any stored codes by flashing the "SERVICE ENGINE SOON" or "CHECK ENGINE" light. Each code will be flashed 3 times, then Code 12 will be flashed again.

a. On carbureted engines, if a trouble code is displayed, the memory is cleared, then the engine is operated to see if the code is a hard or intermittent failure.

b. If the code represents a hard failure, a diagnostic code chart is used to locate the area of the failure.

c. If an intermittent failure is determined, the problem circuits can be examined physically for reasons of failure.

d. On fuel injected engines, if a trouble code is displayed, a diagnostic code chart is used to locate the area of failure.

3. The ECM will energize all controlled relays and solenoids that are involved in the current engine operation.

a. On carbureted engines, the ISC motor, if equipped, will move back and forth and the mixture control solenoid will be pulsed for 25 seconds or until the engine is started, which ever occurs first.

b. On fuel injected engines, the IAC valve is moved back and forth or is fully extended, depending upon the engine family.

Field Service Mode
FUEL INJECTION MODELS

If the diagnostic terminal is grounded with the engine operating, the system will enter the Field Service Mode. In this mode, the "SERVICE ENGINE SOON" or "CHECK ENGINE" indicator light will show whether the system is in open or closed loop operation.

When in the open loop mode, the indicator light will flash 2½ times per second.

When in the closed loop mode, the indicator light will flash once every second. Also, in closed loop, the light will stay out most of the time if the system is too lean. The light will stay on most of the time is the system is too rich. In either case, the Field Service Mode Check, which is part of the diagnostic circuit check, will direct the technician to the fault area.

While in the Field Service Mode, the ECM will be in the following mode:

1. The distributor will have a fixed spark advance.
2. New trouble codes cannot be stored in the ECM.
3. The closed loop timer is bypassed.

Clearing the Trouble Codes

When the ECM sets a trouble code, the "SERVICE ENGINE SOON" or "CHECK ENGINE" light will be illuminated and a trouble code will be stored in the ECM's memory. If the problem is intermittent, the light will go out after 10 seconds when the fault goes away, however, the trouble code will stay in the ECM memory until the battery voltage to the ECM is removed. Removing the battery voltage for 10 seconds will clear all stored trouble codes.

To prevent damage to the ECM, the ignition key must be in the **OFF** position when disconnecting or reconnecting the power to the ECM through the battery cable, ECM pigtail, ECM fuse, jumper cables, etc.

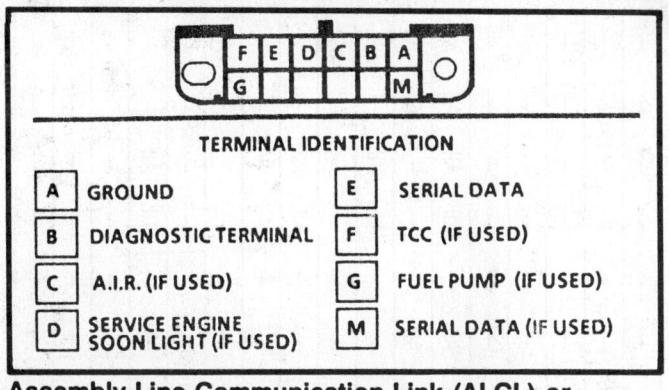

TERMINAL IDENTIFICATION

A	GROUND	E	SERIAL DATA
B	DIAGNOSTIC TERMINAL	F	TCC (IF USED)
C	A.I.R. (IF USED)	G	FUEL PUMP (IF USED)
D	SERVICE ENGINE SOON LIGHT (IF USED)	M	SERIAL DATA (IF USED)

Assembly Line Communication Link (ALCL) or Assembly Line Data Link (ALDL) – typical

NOTE: If posiable clear codes by disconnecting the pigtail or ECM fuse. Disconnecting the negative battery cable will clear the memory from other on-board computers systems, such as pre-set radio, seat control, etc.

All trouble codes should be cleared after repairs have been accomplished. In some cases, such as through a diagnoistic routine, the codes may have to be cleared first to allow the ECM to set a trouble code during the test, should a malfunction be present.

NOTE: The ECM has a learning ability to perform after the battery power has been disconnected to it. A change may be noted in the vehicle's performance. To teach the vehicle, make sure the engine is at normal operating temperature and drive it at part throttle, at moderate acceleration and idle conditions, until normal performance returns.

ALDL Scan Tools

The ALDL connector, located under the dash, has a variety of information available on terminals **E** and **M** (depending upon the engine used). There are several tools on the market, called Scan units for reading the available information.

The use of the Scan tools do not make the diagnostics unnecessary. They do not tell exactly where a problem is in a given circuit. However, with an understanding of what each position on the instrument measures and the knowledge of the circuit involved, the tool can be very useful in getting information which could be more time consuming to get with outer test equipment. It must be emphasized that each type scanner instrument must be used in accordance with the manufacturers instructions.

If a problem seems to be related to certain parameters, they can be checked with the Scan tool while driving the vehicles. If there does not seem to be any correlation between the problem and any specific circuit, the Scan tool can be checked on each position, watching for a period of time to see if there is any change in the readings that indicate intermittent operations.

Diagnostic Circuit Check

After a physical and visual underhood inspection, the diagnostic circuit check it the starting point for all diagnostic procedures. If the on-vehicle diagnostics aren't working, the diagonstic circuit check will lead to a diagnostic chart, to correct the problem. If no fault is found in the diagnostic check circuit and no trouble codes are stored by the ECM, then the use of a scan tool is necessary to obtain serial data from the ECM. The data readings from the SCAN tool should be compared to specified parameters for each circuit to help isolate the system malfunction.

Trouble Codes

The Diagnostic Trouble Code chart contains all the trouble

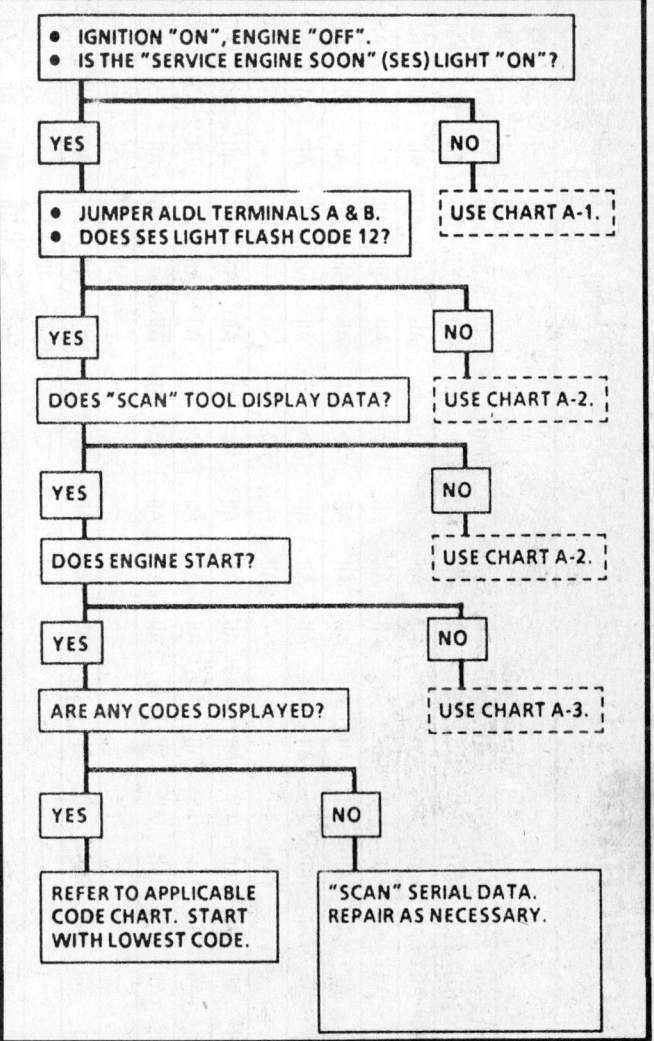

Diagnostic flow chart – typical

codes for all General Motors car and light truck engines from 1988–90 except E-body series.

The chart is labeled with the engine (in liters) and the engine VIN designation at the top of the chart. Along the side of the chart are the trouble codes and the abbreviated function. Within the chart are all the applicable body series letter designation. If a particular code applies the all the body series that is equipped with the listed engine, then the word "ALL" will appear in block.

The abbreviation for the chart functions are as follows:
A/C – Air Conditioning
CTS – Coolant Temperature Sensor
ECM – Electronic Control Module
EGR – Exhaust Gas Recirculation
ESC – Electronic Spark Control
EST – Electronic Spark Timing
IAC – Idle Air Control
M/C SOL – Mixture Control Solenoid
MAF – Mass Air Flow sensor
MAP – Manifold Absolute Pressure
MAT – Manifold Air Temperature sensor
TCC – Torque Converter Clutch
TPS – Throttle Position Sensor
VAC – Vacuum
VATS – Vehicle Anti-Theft System
VSS – Vehicle Speed Sensor

1988–90 GENERAL MOTORS CORPORATION DIAGNOSTIC TROUBLE CODES

	2.0L (M)	2.0L (K)	2.3L (A)	2.3L (D)	2.5L (R)	2.5L (U)	2.8L (9)	2.8L (S)	2.8L (W)	3.0L (L)	3.1L (D)	3.1L (T)	3300 (N)	3.8L (3)	3800 (C)	4.3L (Z)	5.0L (E)	5.0L (Y)	5.0L (8)	5.7L (F)	5.7L (7)	All Light Truck/Van
12—Speed reference pulse	All	All	All	All	All	All	All	All	All	All	All	All	All	All	⑫	All	All	All	All	All	All	All
13—Oxygen sensor	All	All	All	All	All	All	All	All	All	All	All	All	All	All	⑫	All	All	All	All	All	All	All
14—CTS (Temp. HI)	All	All	All	All	All	All	All	All	All	All	All	All	All	All	⑫	All	All	All	All	All	All	All
15—CTS (Temp. LO)	All	All	All	All	All	All	All	All	All	All	All	All	All	All	⑫	All	All	All	All	All	All	All
16—System volts HI	—	—	—	—	—	—	All	All	—	—	—	—	All	—	⑫	—	—	—	—	—	—	—
21—TPS (Volts HI)	All	All	All	All	All	All	All	All	All	All	All	All	All	All	⑫	All	All	All	All	All	All	All
22—TPS (Volts LO)	All	All	All	All	All	All	All	All	All	All	All	All	All	All	⑫	All	All	All	All	All	All	All
23—MAT (Temp LO)	All	All	All	All	All	All	All	All	All	—	All	All	—	All	⑫	All	All	—	All	All	All	All
23—M/C solenoid (open or ground)	—	—	—	—	—	—	—	—	—	—	—	—	—	—	—	—	—	All	—	—	—	—
24—VSS	All	All	All	All	All	All	All	All	All	All	All	All	All	All	⑫	All	All	All	All	All	All	All
24B—Park/neutral switch	—	—	—	—	—	—	—	—	—	—	—	—	—	—	—	—	—	All	—	—	—	—
25—MAT (Temp. HI)	All	All	All	All	All	All	All	All	All	All	All	All	All	All	⑫	All	All	—	All	All	All	All
26—Quad-driver	—	—	All	All	—	—	—	—	W	—	—	①	—	—	⑫	—	—	—	—	—	—	—
27/28—Gear switch	—	—	—	—	—	—	—	—	—	—	—	—	⑨	—	—	—	—	—	—	—	—	—
28/29—Gear switch	—	—	—	—	—	—	—	—	②	—	—	—	⑩	—	⑫	—	—	—	—	—	—	—
31—Park/neutral switch	—	—	—	—	—	—	—	—	—	—	—	—	All	—	⑫	—	—	—	—	—	—	—
31—Wastegate overboost	All	—	—	—	—	—	—	—	—	—	—	—	—	—	—	—	—	—	—	—	—	—
31—Canister purge (Volts HI)	—	—	—	—	—	—	—	—	—	—	—	—	—	—	—	—	—	All	—	—	—	—
32—EGR failure	All	All	⑧	⑧	⑧	All	All	All	②	All	All	All	—	All	⑫	All	All	All	All	All	All	All
33—MAP (Volts HI/Vac LO)	All	All	⑦	⑦	All	All	All	—	②	All	All	All	—	—	—	All	All	—	—	All	All	All
33—MAF (gm/sec HI)	—	—	—	—	—	—	—	All	④	All	—	—	All	All	⑫	—	—	All	All	—	—	—
34—MAP (Volts LO/Vac HI)	All	All	⑧	⑧	All	All	All	—	②	All	All	All	—	—	—	All	All	—	—	All	All	All
34—MAF (gm/sec HI)	—	—	—	—	—	—	All	All	④	All	—	—	All	All	⑫	—	—	All	All	—	—	—
35—Idle speed error	All	All	⑧	⑧	All	All	All	All	All	All	All	All	All	All	⑫	All	All	All	All	All	All	All
35—IAC	—	—	—	—	—	—	—	—	—	—	—	—	—	—	—	—	—	—	—	—	—	All
36—Close throttle air flow HI	—	—	⑦	⑦	—	—	—	—	—	—	—	—	—	—	—	—	—	—	—	—	—	—
36—MAF burn-off	—	—	—	—	—	—	—	—	—	—	—	—	—	—	—	—	All	—	All	All	—	—
38—Brake switch	—	—	—	—	—	—	—	—	—	—	—	—	—	—	⑫	—	—	—	—	—	—	—
39—TCC	—	—	⑧	⑧	—	—	—	—	—	—	All	All	All	All	⑫	—	All	All	—	All	All	—
41—1 X reference	—	—	⑧	⑧	—	—	—	—	—	—	—	—	—	—	—	—	—	—	—	—	⑥	—

Engine (VIN)

Trouble Codes	2.0L (M)	2.0L (K)	2.0L (1)	2.3L (A)	2.3L (D)	2.5L (R)	2.5L (U)	2.8L (9)	2.8L (S)	2.8L (W)	3.0L (L)	3.1L (D)	3.1L (T)	3300 (N)	3.8L (3)	3800 (C)	4.3L (Z)	5.0L (E)	5.0L (Y)	5.0L (8)	5.7L (F)	5.7L (7)	All Light Truck/Van
41—Cylinder select error/MEM-CAL	—	—	—	—	—	—	—	—	All	All	—	—	All	—	—	—	—	—	—	—	—	—	—
41—Cam sensor	—	—	—	—	—	—	—	—	—	—	—	—	—	—	All	⑫	—	—	—	—	—	—	—
42—EST	All	All	All	All	All	All	All	All	All	All	All	All	All	All	All	⑫	All	All	All	All	All	All	All
43—ESC	All	—	All	All	All	All	All	All	—	—	—	All	All	All	All	⑫	All	All	All	All	All	All	All
44—Oxygen sensor (lean)	All	All	All	All	All	All	All	All	All	All	All	All	All	All	All	⑫	All	All	All	All	All	All	All
45—Oxygen sensor (rich)	All	All	All	All	All	All	All	All	All	All	All	All	All	All	All	⑫	All	All	All	All	All	All	All
46—Power steering press switch	—	—	—	—	—	—	—	—	—	—	—	—	—	All	—	⑫	—	—	—	—	—	—	—
46—VATS	—	—	—	—	—	—	—	—	—	—	—	—	—	—	—	—	—	—	—	All	All	—	—
48—Misfire diagnosis	—	—	—	—	—	—	—	—	—	—	—	—	—	All	—	⑫	—	—	—	—	—	—	—
51—PROM or MEM-CAL error	All	All	All	All	All	All	All	All	All	All	All	All	All	All	All	⑫	All	All	All	All	All	All	All
52—CALPAK error	—	—	—	—	—	—	—	All	All	⑤	All	—	A	All	—	—	—	—	—	—	—	—	—
53—System overvoltage	—	All	—	All	—	—	All	All	All	All	All	All	All	—	All	—	All	All	—	—	—	—	—
53—EGR (Improper VAC)	—	—	—	—	—	—	—	—	—	—	—	—	—	—	—	—	—	—	⑪	—	—	—	—
54—Fuel pump (Volts LO)	—	—	—	—	—	—	—	—	—	—	—	All	All	—	All	All	All	All	—	—	All	All	All
54—M/C solenoid (Volts HI)	—	—	—	—	—	—	—	—	—	—	—	—	—	—	—	—	—	—	All	—	—	—	—
55—ECM error	—	—	—	—	—	—	—	All	All	All	All	①	①	All	All	All	All	All	All	All	All	All	All
61—Degraded oxygen sensor	—	—	—	—	—	—	—	All	All	All	—	All	All	All	All	—	All	All	—	—	All	All	All
62—Trans. gear switch	—	—	—	All	All	—	—	—	—	—	—	—	—	—	—	—	—	—	—	—	—	—	—
63—MAP (Volts HI/Vac LO)	—	—	—	—	—	—	—	—	—	④	—	—	—	—	—	—	—	—	—	—	—	—	—
63—EGR flow check	—	—	—	—	—	—	—	—	—	—	—	—	—	—	—	⑫	—	—	—	—	—	—	—
64—MAP (Volts LO/Vac HI)	—	—	—	—	—	—	—	—	—	④	—	—	—	—	—	—	—	—	—	—	—	—	—
64—EGR flow check	—	—	—	—	—	—	—	—	—	—	—	—	—	—	—	⑫	—	—	—	—	—	—	—
65—Fuel injector (LO amps)	—	—	—	All	All	—	—	—	—	—	—	—	—	—	—	—	—	—	—	—	—	—	—
65—EGR flow check	—	—	—	—	—	—	—	—	—	—	—	—	—	—	—	⑫	—	—	—	—	—	—	—
66—A/C press switch	—	—	—	All	All	—	—	—	—	—	—	—	—	N	—	—	—	—	—	—	—	—	—

① 1989–90 W-Body vehicles
② All except 1988 A and L-Body vehicles
③ Except J-Body vehicles
④ 1988 A and L-Body vehicles
⑤ Except 1988 W-Body vehicles
⑥ No distributor reference pulse (all models)
⑦ 1988 2.3L (VIN D) only (all models)
⑧ All 1989–90 models only
⑨ Vehicles equipped with THM 125-C (3T40) transaxle (all models)
⑩ Vehicles equipped with THM 440-T4 (4T60) transaxle (all models)
⑪ All California models only
⑫ All models except E-Body vehicles

E, K AND CADILLAC C-BODY

1988–90 Buick Riviera and Reatta

Numerous electronic components are located on the vehicle as a part of an electrical network, designed to control various engine and body subsystems.

At the heart of the computer system is the Body Computer Module (BCM). The BCM is located on the front driver's side of the floor console and has an internal microprocessor, which is the center for communication with all other components in the system. All sensors and switches are monitored by the BCM or other major components that complete the computer system. The components are as follows:

1. Electronic Control Module (ECM)
2. Instrument Panel Cluster (IPC)
3. Cathode Ray Tube Controller (CRTC)
4. Programmer/Heating/Ventilation/Air Conditioning

A combination of inputs from these major components and the other sensors and switches communicate together to the BCM, either as individual inputs, or on the common communication link called the serial data line. The various inputs to the BCM combine with program instructions within the system memory to provide accurate control over the many subsystems involved. When a subsystem circuit exceeds pre-programmed limits, a system malfunction is indicated and may provide certain back-up functions. Providing control over the many subsystems from the BCM is done by controlling system outputs. This can be either direct or transmitted along the serial data line to 1 of the other major components. The process of receiving, storing, testing and controlling information is continuous. The data communication gives the BCM control over the ECM's self-diagnostic capacities in addition to its own.

Between the BCM and the other 4 major components of the computer system, a communication process has been incorporated which allows the devices to share information and thereby provide for additional control capabilities. In a method similar to that used by a telegraph system, the BCM's internal circuitry rapidly switches to a circuit between 0 and 5 volts like a telegraph key. This process is used to convert information into a series of pulses which represents coded data messages understood by other components. Also, much like a telegraph system, each major component has its own recognition code. When a message is sent out on the serial data line, only the component or station that matches the assigned recognition code will pay attention and the rest of the components or stations will ignore it. This data transfer of information is most important in understanding how the system operates and how to diagnosis possible malfunctions within the system.

In order to access and control the BCM self-diagnostic features, additional electronic components are necessary, which are the Cathode Ray Tube Controller (CRTC) and the Graphic Control Center (GCC) picture tube. As part of the GCC's SERVICE MODE page, a 22 character display area is used to display diagnostic information. When a malfunction is sensed by the computer system, a driver warning messages is displayed on the GCC under the DIAGNOSTIC category. When the Service Mode is entered, the various BCM, ECM, or IPC parameters, fault codes, inputs, outputs as well as override commands and clearing code capability are displayed when commanded through the GCC.

The GCC becomes the device to enter the diagnostics and access the Service Diagnostic routines. The CRTC is the device which controls the display on the GCC and interprets the switches touched on the GCC and passes this information along to the BCM. This communication process allows the BCM to transfer any of its available diagnostic information to the GCC for display during SERVICE MODE.

By touching the appropriate pads on the GCC, data messages can be sent to the BCM from the CRTC over the data line, requesting the specific diagnostic feature required.

Graphic Control Center

GENERAL OPERATION

The GCC system provides the operator with fingertip electronic control of the heating/AC system and the radio/tape player. It also provides clock and trip information. This control and information is entered and displayed by the use of a cathode ray tube mounted in the dash. The GCC has adjustments for viewing ease. The are the GCC dimming and face plate light. The dimming is a pulse width modulated signal which is regulated by the BCM. Incandescent face plate dimming is controlled by a pulse width modulator circuit in the instrument panel cluster.

The Graphic Control Center system consists of a Cathode Ray Tube (CRT) monitor, a Cathode Ray Tube Controller (CRTC) and a body computer module (BCM). The information is sent to and from the BCM, the CRTC and the electronic A/C controller with the serial data link.

Data is sent between the CRTC and the radio on a separate data link.

GENERATOR INDICATOR

The BCM senses generator (alternator) current. When the BCM detects that the battery is not charging, data line information is sent to the instrument panel cluster to turn on the warning red generator (alternator) indicator lamp.

COMPUTER SYSTEM SERVICE PRECAUTIONS

The computer system is designed to withstand normal current draws associated with vehicle operation. However, care must be taken to avoid overloading any of these circuits. In testing for open or short circuits, do not ground or apply voltage to any of the circuits unless instructed to do so by the diagnosis procedures. These circuits should only be tested by using a High Impedance Mulimeter (Kent Moore J–29125A or its equivalent), if the tester remains connected to any of the computers. Power should never be applied or removed to any of the computers with the key in the **ON** position. Before removing or connecting battery cables, fuses or connectors, always turn the ignition switch to the **OFF** position.

SELF-DIAGNOSTIC TROUBLE CODES

In the process of controlling the various subsystems, the ECM and the BCM continually monitor operating conditions for possible malfunctions. By comparing systems conditions against standard operating limits, certain circuit and component malfunctions can be detected. A 3 digit numerical trouble code is stored in the computer memory when a problem is detected by this self-diagnostic system. These trouble codes can later be displayed by the service technician as an aid in system repair.

The occurrence of certain system malfunctions require that the vehicle operator be alerted to the problem so as to avoid prolonged operation of the vehicle under degrading system operations. The computer controlled diagnostic messages and/or telltales will appear under these conditions which indicate that service or repairs are required.

If a particular malfunction would result in unacceptable system operation, the self-diagnostics will attempt to minimize the effect by taking a FAILSOFT action. FAILSOFT action refers to any specific attempt by the computer system to compensate for the detected problem. A typical FAILSOFT action would be the substitution of a fixed input value when a sensor is detected to have an open or shorted circuit.

HOW TO ENTER THE DIAGNOSTIC SERVICE MODE

To enter the diagnostic service mode, proceed as follows:
1. Turn the ignition switch to the **ON** position.

2. Touch the OFF and the WARM pads on the CRT's climate control page, simultaneously and hold until a double BEEP is heard or a page entitled SERVICE MODE appears on the CRT.

NOTE: Operating the vehicle in the SERVICE MODE for an extended time period without the engine operating or without a trickle battery charger attached to the battery, can cause the battery to become discharged and possibly relate false diagnostic information or cause an engine no-start. Avoid lengthy (over ½ hour) SERVICE MODE operation.

TROUBLE CODE DISPLAY

After the SERVICE MODE is entered, any trouble codes stored

ECM DIAGNOSTIC CODES

CODE	DESCRIPTION	COMMENTS	CODE	DESCRIPTION	COMMENTS
E013	Oxygen Sensor Circuit [Canister Purge]	Ⓐ/Ⓑ	E041	Cam Sensor Circuit	Ⓐ/Ⓑ
E014	Coolant Temperature Sensor (Too High)	Ⓐ/Ⓑ/Ⓕ	E042	Electronic Spark Timing (EST)	Ⓐ/Ⓑ/Ⓙ
E015	Coolant Temperature Sensor (Too Low)	Ⓐ/Ⓑ/Ⓕ	E043	Electronic Spark Control (ESC)	Ⓐ/Ⓑ
E016	Battery Voltage Too High (All Solenoide)	Ⓐ/Ⓑ	E044	Lean Exhaust Signal	Ⓐ/Ⓑ/Ⓘ
			E045	Rich Exhaust Signal	Ⓐ/Ⓑ/Ⓘ
E021	Throttle Position Sensor Circuit (Volt High) [TCC]	Ⓐ/Ⓑ	E046	Power Steering Pressure Switch Circuit [A/C Clutch]	Ⓒ
E022	Throttle Position Sensor Circuit (Volt Low) [TCC]	Ⓐ/Ⓑ			
E023	MAT Sensor Circuit (Low Temp)	Ⓐ/Ⓑ	E047	BCM–ECM Communication [A/C Clutch & Cruise]	Ⓐ/Ⓑ Ⓒ
E024	Vehicle Speed Sensor Circuit [TCC]	Ⓐ/Ⓑ	E048	Misfire	
E025	MAT Sensor Circuit (HighTemp)	Ⓐ/Ⓑ			
E026	QDM Circuit	Ⓐ/Ⓑ			
E027	Second Gear Circuit	Ⓒ	E051	PROM Error	Ⓐ/Ⓑ/Ⓙ/Ⓚ
E028	Third Gear Circuit	Ⓒ			
E029	Fourth Gear Circuit	Ⓒ	E063	Small EGR Fault	Ⓐ/Ⓑ
			E064	Medium EGR Fault	Ⓐ/Ⓑ
E031	P/N Switch Circuit	Ⓒ	E065	Large EGR Fault	Ⓐ/Ⓑ
E034	MAF Sensor Circuit	Ⓐ/Ⓒ			
E038	Brake Switch Circuit	Ⓒ			
E039	TCC Circuit	Ⓒ			

DIAGNOSTIC CODE COMMENTS

Ⓐ	"Service Engine Soon" telltale in IPC "ON".
Ⓑ	Displays Diagnostic Message on IPC.
Ⓒ	No telltale light or message.
Ⓓ	Displays "ERROR" in season odometer.
Ⓕ	Forces Cooling Fans on.

Ⓖ	Displays "Electrical Problem" Message on IPC.
Ⓘ	Forces OL Operation.
Ⓙ	Causes system to operate on bypass spark.
Ⓚ	Causes system to operate on back-up fuel.
[ ]	Functions within bracket are disengaged while specified malfunction remains current.

BCM AND CRTC DIAGNOSTIC CODES

Code	Circuit Affected	Code	Circuit Affected
B110	Outside Temp Sensor	B420	Relays
B111	A/C Hi Side Temp Sensor	B440	Air Mix Door
B112	A/C Lo Side Temp Sensor	B446	Low Refrigerant Warning
B113	In-Car Temp Sensor	B447	Very Low Refrigerant Problem
B115	Sunload Temp Sensor	B448	Low Refrigerant Pressure
B119	Twilight Photocell	B449	A/C HI Temp
B120	Twilight Delay Pot	B450	Coolant HI Temp – A/C
B121	Twilight Enable Switch	B482	Anti Lock Brake Pressure
B122	Panel Lamp Dimming Pot	B552	BCM Memory Reset
B123	Courtesy Light Switch	B556	E² Prom
B124	VSS	B660	Cruise – Not In Drive
B132	Oil Pressure Sensor	B663	Cruise – Speed Difference
B140	Phone	B664	Cruise – Acceleration
B332	Compass	B667	Cruise – Switch Shorted
B334	ECM Data	B671	Cruise – Position Sensor
B335	CRTC Data	B672	Cruise – Vent Solenoid
B336	IPC Data	B673	Cruise – Vacuum Solenoid
B337	Programmer Data	C331	Tape Deck
B410	Charging System	C339	Radio Data
B411	Battery Low	C553	CRTC Memory Reset
B412	Battery High	C710	CRT Switching

Diagnostic trouble codes – 1988–90 Buick Riviera and Reatta

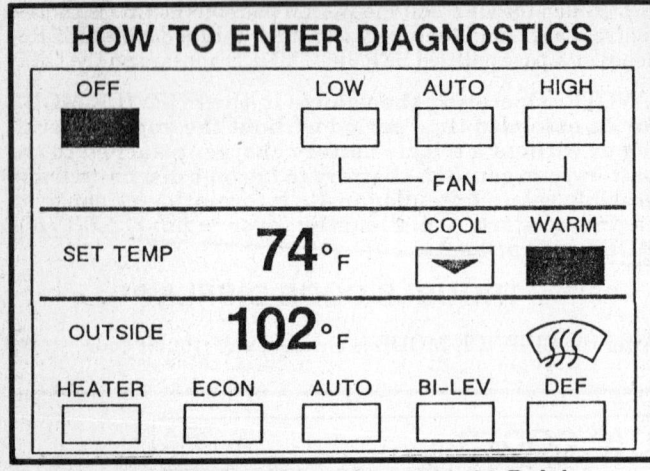

HOW TO ENTER DIAGNOSTICS

How to enter diagnostic mode — 1988–90 Buick Riviera and Reatta

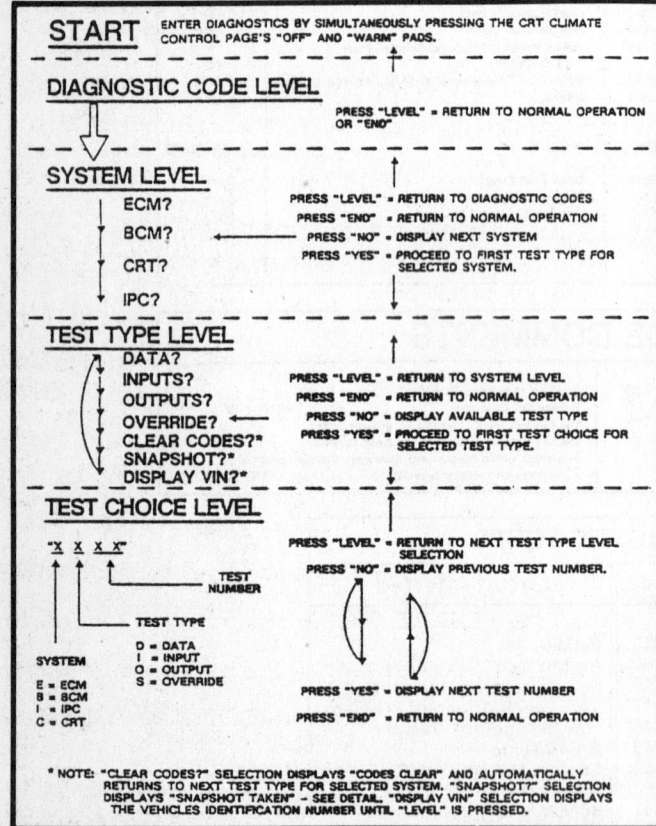

Self-diagnostic system flowchart — 1988–90 Buick Riviera and Reatta

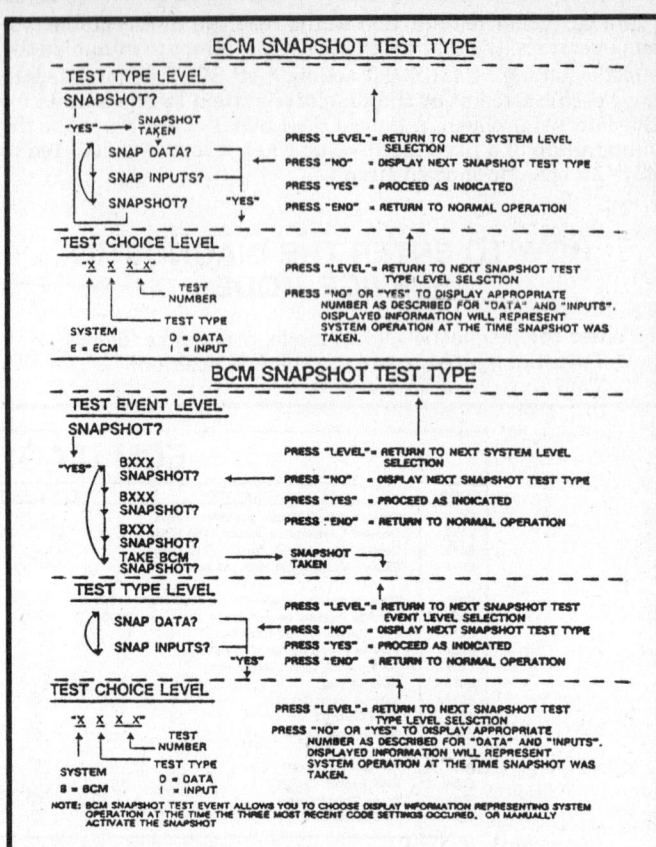

Selecting snapshot — 1988–90 Buick Riviera and Reatta

in the computer memory will be displayed. ECM codes will be displayed first. If no ECM codes are stored, the CRT will display a NO ECM CODES message for approximately 2 seconds. All ECM codes will be prefixed with an E (Example-EO13, EO23 and etc.)

The lowest numbered ECM code will be displayed first, followed by progressively higher numbered codes present. Codes will be displayed consecutively for 2 second intervals until the highest code present has been displayed. When all ECM codes have been displayed, BCM codes will be displayed. Following the highest ECM code present or the NO ECM CODES message, the lowest numbered BCM code will be displayed for approximately 2 seconds. BCM codes displayed will also be accompanied by CURRENT or HISTORY. HISTORY indicates the failure was not present the last time the code was tested and CURRENT indicates the fault still exists. Since the ECM is not capable of making this determination, these messages do not appear when the ECM codes are being displayed. All BCM codes will be prefixed with the letter B (Example-B110 and etc.). Progressively higher numbered BCM codes, if present, will be displayed for approximately 2 second intervals until the highest code present has been displayed. If no BCM trouble codes are stored, NO BCM CODES message will be displayed. At any time during the display of the ECM or BCM codes, if the NO pad is touched, the display will bypass the codes. At any time during the display of trouble codes, it the EXIT pad is touched, the CRT will exit the SERVICE MODE and go back to normal vehicle operation.

CLIMATE CONTROL IN SERVICE MODE

Upon entering the SERVICE MODE, the climate control will operate in whatever setting was being commanded just prior to depressing the OFF and WARM pads. Even though the display may change just as the pads are touched, the prior operating setting is remembered and will resume after the SERVICE MODE is entered.

During the SERVICE MODE, the climate control can be operated the same as normally by touching the climate control border pad and calling up climate control page. To get back to SERVICE MODE page, simply touch the border pad marked DIAGNOSTICS. This will take the system back to the exact same spot in the SERVICE MODE as what it was before the climate control border pad was touched. The climate control and the diagnostic border pads are the only 2 border pads that will operate while in the SERVICE MODE.

OPERATING THE SERVICE MODE

After trouble codes have been displayed, the SERVICE MODE can be used to perform several tests on different systems at the time. Upon completion of code display, a specific system may be selected for testing or a segment check can be performed.

Selecting the System

Following the display of trouble codes, the first available system will be displayed (i.e. ECM?). When selecting a system to test, any of the following actions may be taken to control the display.

1. Touching the EXIT pad will stop the system selection process and return the display to the beginning of the trouble code sequence.

2. Touching the NO pad will display the next available system selection. This allows the display to be stepped through all systems choices. This list of systems can be repeated following the display of the last system.

3. Touching the YES pad will select the displayed system for testing. At this point, the first available test type will appear with the selected name above it.

Selecting the Test Type

Having selected a system, the first available test type will be displayed (i.e. ECM DATA?). While selecting a specific test type, any of the following actions may be taken to control the display.

1. Touching the EXIT pad will stop the test type selection process and return the display to the next available system selection.

2. Touching the NO pad will display the next available test type for the selected system. This allows the display to be stepped through all available test type choices. The list of test types can be repeated following the display of the last test type.

3. Touching the YES pad will select the displayed test type. At this point, the display will either indicate the at the selected test type is in progress or the first of several specific tests will appear. If NO DEVICES is displayed, no tests are available.

Selecting the Test

Selection of the DATA?, INPUTS?, OUTPUTS? or OVER-RIDE? test types will result in the first available test being displayed.

If a SELECT ERR message ever appears, this test is not allowed with the engine operating. Turn the engine OFF and try again.

Four characters of the display will contain a test code to identify the selection. The first 2 characters are letters which identify the system and the test type (i.e. ED01 for throttle position). While selecting a specific test, any of the following actions may be taken to control the display.

1. Touching the EXIT pad will stop the test selection process and return the display to the next available test type for the selected system.

2. Touching the NO pad will display the next smaller test number for the selected test type. If this pad is touched with the lowest test number displayed, the highest test number will then appear.

3. Touching the YES pad will display the next larger test number for the selected test type. If this pad is touched with the highest test number displayed, the lowest test number will then appear.

Selecting the CODE RESET? Test

Selection of CODE RESET? test type will result in the message CODES CLEAR being displayed with the selected system name above it after the YES pad has been touched. This message will

appear for 3 seconds to indicate all stored trouble codes have been erased from that system's memory. After 3 seconds, the display will automatically return to the next available test type for the selected system

IPC Segment Check

Whenever the key is **ON** and the vehicle is in the **PARK** position, pressing the TEST button on the instrument panel cluster, will cause the IPC to sequentially illuminate and blank all segments and telltales in the cluster. This is helpful in determining if any bulbs or segments or the cluster's vacuum fluorescent display are out or always on. To provide more time to study the various segments whenever in the SERVICE MODE and not in the middle of running a test, such as when the CRT displays ECM?, if the TEST button on the ICP is depressed, the segment check will operate 10 times slower than when not in the SERVICE MODE.

Exiting Service Mode

To exit the service mode, repeatedly touch the EXIT pad until the SERVICE MODE page disappears or turn the ignition switch to the **OFF** position. Trouble codes are not erased when this is done.

DATA DISPLAYS

When troubleshooting a malfunction, the ECM and BCM data display can be used to compare the vehicle with problems to a vehicle which is functioning properly. A brief summary of each parameter is provided as follows:

ELECTRONIC CONTROL MODULE (ECM)

ED01 – The THROTTLE POSITION (TPS) is displayed in millivolts (mv) from 0–5100.

ED04 – The COOLANT TEMPERATURE is displayed in degrees (C.) from minus 40–151.

ED06 – The INJECTOR PULSE WIDTH is displayed in milliseconds (MS), from 0–1002.

ED07 – The OXYGEN SENSOR VOLTAGE is displayed in Millivolts (mv) from 0–1128 mv.

ED08 – The SPARK ADVANCE value is displayed in degrees, from 0–70.

ED10 – The BATTERY VOLTAGE is read in volts, from 0–25.5.

ED11 – The ENGINE SPEED is displayed in rpm from 0–6375.

ED12 – The VEHICLE SPEED is in miles per hour (mph) from 0–159.

ED16 – The KNOCK RETARD (ESC) is displayed in degrees of timing pulled out, from 0–45.

ED17 – The KNOCK SENSOR ACTIVITY is displayed as counts. This counter changes whenever the ESC line is LO, which indicates sensor activity (detonation).

ED18 – The OXYGEN SENSOR CROSS COUNTS is displayed as the number of times the oxygen sensor crossed the reference line each second.

ED19 – The FUEL INTEGRATOR displayed in counts, from 0–255.

ED20 – The BLOCK LEARN MULTIPLIER (BLM) is displayed in counts, from 0–255.

ED21 – The MASS AIR FLOW (MAF) is displayed in grams per second of air flow from 0–255.

ED22 – The IDLE AIR CONTROL (IAC) is displayed in steps of travel from 0–255.

ED23 – The MANIFOLD AIR TEMPERATURE (MAT) also called air temperature sensor is displayed in Celsius (°C) from minus 40–151.

ED98 – The IGNITION CYCLE COUNTER value is the number of times the ignition has been cycled to **OFF** since an ECM Trouble Code was last detected. After 50 ignition cycles without any malfunction being detected, all stored ECM codes are cleared.

ED99—The ECM PROM I.D. is displayed as a number up to 3 digits long which can be used to verify that the proper PROM is installed in the ECM.

BODY CONTROL MODULE (BCM)

BD20—The COMMANDED BLOWER VOLTAGE is read in volts, from 0 to system voltage.

BD21—The COOLANT TEMPERATURE is displayed in degrees (°C) from minus 40–151. This value is sent from the ECM to the BCM. If this circuit malfunctions as determined by the ECM, the ECM will send the BCM a Failsoft value for display.

BD22—The COMMANDER AIR MIX DOOR POSITION is displayed in percent (%). A value close to 0% represents a cold air mix and a value close to 100% represents a warm air mix.

BD23—The ACTUAL AIR MIX DOOR POSITION is displayed in percent (%). This value should follow the commanded air mix door position (BD22) except when the door is commanded beyond its mechanical limits of travel.

BD24—The AIR DELIVERY MODE is displayed as a number from 0–10. Each number is a code which represents the following air delivery modes:

**0—Auto-Recirc/MAX A/C
**1—Auto-A/C (vents)
**2—Auto-Bi-Level
**3—Auto-Heater-Def
**4—Auto-Heater
**5—OFF
**6—Normal Purge
**7—Cold (DEF) Purge
**8—Forced/Def
**9—Forced/Heater
**10—Forced/Bi-Level

BD25—The IN-CAR TEMPERATURE is displayed in degrees (°C), from minus 40–102.

BD26—The ACTUAL OUTSIDE TEMPERATURE is displayed in degrees (°C), from minus 40–58. This valve represents actual sensor temperatures and is not restricted by the features used to minimize engine heat affects on the customer display valve.

BD27—The HIGH SIDE TEMPERATURE (condenser output) is displayed in degrees (°C) from minus 40–215.

BD28—The LOW SIDE TEMPERATURE (evaporator input) is displayed in degrees (°C) from minus 40–93.

BD32—The SUNLOAD TEMPERATURE SENSOR is displayed in degrees (°C) from minus 23–102.

BD40—The ACTUAL FUEL LEVEL is read in gallons between 0–25.5. This value represents actual sensor position and is not restricted by the features used to eliminate fuel slosh effects on the customer display value.

BD42—The DIMMING POT is displayed in percent (%). A value close to 0% represents maximum dimming and a value close to 100% represents maximum brightness.

BD43—The TWILIGHT DELAY POT is displayed in percent (%). A value close to 0% represents minimum delay time and a value close to 100% represents maximum delay time.

BD44—The TWILIGHT PHOTOCELL is displayed in percent (%). A value close to 0% represents daylight and value close to 100% represents darkness.

BD45—The PHONE MODE is displayed as a number from 0–7. Each number is a code which represents the following phone modes:

**0—No phone in system
**1—OFF
**2—ON
**3—ON-no service
**4—ON-ROAM
**5—ON-Call in progress
**6—ON-Call received
**7—System Problem

BD50—The BATTERY VOLTAGE is read in volts between 0 and 25.5.

BD51—The GENERATOR FIELD is displayed in percent (%). A value close to 0% represents minimum regulator on time and a value close to 100% represents maximum regulator on time.

BD60—The VEHICLE SPEED is displayed in miles per hours (mph)from 0–159.

BD61—The ENGINE SPEED is displayed in rpm from 0–6375.

BD70—The CRUISE SERVO POSITION is displayed in percent (%) from 0–100. A value close to 0% represents reset position and a value close to 100% represents WOT

BD71—The OIL PRESSURE SENSOR is displayed in pressure (psi) from 0–255.

BD90—The OPTION CONTENT No. 1 is displayed as a number from 0–255. Each number is a code which represents the following option content and only the following numbers are valid:

Riveria
**128—Illuminated Entry
**130—Twilight Sentinel and Illuminated Entry
**132—No options (assumes U.S., Domestic, Unleaded fuel
****and GM 30)
**134—Twilight Sentinel
**136—Illuminated Entry and Coolant Level
**138—Twilight Sentinel, Illuminated Entry and Coolant
****Level
**140—Coolant Level
**142—Twilight Sentintel and Coolant Level

Reatta
**144—Illuminated Entry
**148—No options (assumes U.S., Domestic, Unleaded fuel
****and GM 30/33)
**152—Coolant Level and Illuminated Entry
**156—Coolant Level

BD91—The OPTION CONTENT No. 2 is displayed as a number from 0–255. Each number is a code with only the following numbers valid:

**0—No options
**4—Oil Level
**8—Tape Deck
**12—Tape Deck and Oil Level
**16—Compass
**20—Compass and Oil Level
**24—Compass and Tape Deck
**28—Compass, Tape Deck and Oil Level
**32—Anti-skid Brakes
**36—Anti-skid Brakes and Oil Level
**40—Anti-skid Brakes and Tape Deck
**44—Anti-skid Brakes, Tape Deck and Oil Level
**48—Anti-skid Brakes and Compass
**52—Anti-skid Brakes, Compass and Oil Level
**56—Anti-skid Brakes, Compass and Tape Deck
**60—Anti-skid Brakes, Compass, Tape Deck and Oil Level
**64—Theft Deterrent
**68—Theft Deterrent and Oil Level
**72—Theft Deterrent and Tape Deck
**76—Theft Deterrent, Tape Deck and Oil Level
**80—Theft Deterrent and Compass
**84—Theft Deterrent, Compass and Oil Level
**88—Theft Deterrent, Compass and Tape Deck
**92—Theft Deterrent, Compass, Tape Deck and Oil Level
**96—Theft Deterrent and Anti-skid Brakes
**100—Theft Deterrent, Anti-skid Brakes and Oil Level
**104—Theft Deterrent, Anti-skid Brakes and Tape Deck
**108—Theft Deterrent, Anti-skid Brakes, Tape Deck and Oil Level
**112—Theft Deterrent, Anti-skid Brakes and Compass
**116—Theft Deterrent, Anti-skid Brakes, Compass and Oil Level

120—Theft Deterrent, Anti-skid Brakes, Compass and Tape Deck

124—Theft Deterrent, Anti-skid Brakes, Compass, Tape Deck and Oil Level

128—Washer Fluid Level

132—Washer Fluid Level and Oil Level

136—Washer Fluid Level and Tape Deck

140—Washer Fluid Level, Tape Deck and Oil Level

144—Washer Fluid Level and Compass

148—Washer Fluid Level, Compass and Oil Level

152—Washer Fluid Level, Compass and Tape Deck

156—Washer Fluid Level, Compass, Tape Deck and Oil Level

160—Washer Fluid Level and Anti-skid Brakes

164—Washer Fluid Level, Anti-skid Brakes and Oil Level

168—Washer Fluid Level, Anti-skid Brakes and Tape Deck

172—Washer Fluid Level, Anti-skid Brakes, Tape Deck and Oil Level

176—Washer Fluid Level, Anti-skid Brakes and Compass

180—Washer Fluid Level, Anti-skid Brakes, Compass and Oil Level

184—Washer Fluid Level, Anti-skid Brakes, Compass and Tape Deck

188—Washer Fluid Level, Anti-skid Brakes, Compass, Tape Deck and Oil Level

192—Washer Fluid Level and Theft Deterrent

196—Washer Fluid Level, Theft Deterrent and Oil Level

200—Washer Flyuid Level, Theft Deterrent and Tape Deck

204—Washer Flyuid Level, Theft Deterrent, Tape Deck and Oil Level

208—Washer Fluid Level, Theft Deterrent and Compass

212—Washer Fluid Level, Theft Deterrent, Compass and Oil Level

216—Washer Fluid Level, Theft Deterrent, Compass and Tape Deck

220—Washer Fluid Level, Theft Deterrent, Compass, Tape Deck and Oil Level

224—Washer Fluid Level, Theft Deterrent and Anti-skid Brakes

228—Washer Fluid Level, Theft Deterrent, Anti-skid Brakes and Oil Level

232—Washer Fluid Level, Theft Deterrent, Anti-skid Brakes and Tape Deck

236—Washer Fluid Level, Theft Deterrent, Anti-skid Brakes, Tape Deck and Oil Level

240—Washer Fluid Level, Theft Deterrent, Anti-skid Brakes and Compass

244—Washer Fluid Level, Theft Deterrent, Anti-skid Brakes, Compass and Oil Level

248—Washer Fluid Level, Theft Deterrent, Anti-skid Brakes, Compass and Tape Deck

252—Washer Fluid Level, Theft Deterrent, Anti-skid Brakes, Compass, Tape Deck and Oil Level

BD92—The OPTION CONTENT No. 3 is displayed as a number. An 87 indicates the EEPROM was programmed by the assembly plant. This value will show up as some other number, other than the 87, if the odometer is ever re-programmed.

BD98—The IGNITION CYCLE value is the number of times that the ignition has been cycled to **OFF** since a BCM Trouble Code was last detected. After 50 ignition cycles without any malfunction being detected, all BCM codes are cleared.

BD99—The BCM PROM I.D. is displayed as a number, up to 4 digits long, which can be used to verify that the proper PROM was installed in the BCM.

CATHODE RAY TUBE (CRT)

CD99—The CRT PROM I.D. (PROM ID) is displayed as a number up to 4 digit long used to verify that the proper PROM was installed in the CRTC.

INPUT DISPLAYS

Input displays are operated as outlined under the heading of HOW TO OPERATE A SERVICE MODE. When troubleshooting a malfunction, the ECM, BCM or IPC input display can be used to determine if the switched inputs can be used to determine if the switched inputs can be properly interpreted. When 1 of the various input tests is selected, the state of that device is displayed as HI or LO. In general, the HI and LO refer to the input terminal voltage for that circuit. The display also indicates if the input changed state so that the technician could activate or deactivate any listed device and return to the display to see if it changed state. If a change of state occurred, an X will appear next to the HI/LO indicator, otherwise, an O will remain displayed. The X will only appear once per selected input, although the HI/LO indication will continue to change as the input changes. Some tests are momentary and the X can be used as and indication of a change. The following is a list of the ECM, BCM and IPC inputs:

ELECTRONIC CONTROL MODULE (ECM)

EI71—The BRAKE SWITCH display is LO when the brake pedal is depressed.

EI74—The PARK–NEUTRAL (P/N) display is LO when the vehicle is in the **PARK** or **NEUTRAL** position.

EI78—The POWER STEERING PRESSURE SWITCH display is LO when the power steering pressure (effort) is high, such as when the wheel is in a turned position.

EI79—The 2ND GEAR SWITCH is HI when the vehicle is in **2nd, 3rd or 4th** gears.

EI80—The 3RD GEAR SWITCH is HI when the vehicle is in **3rd or 4th** gears.

EI82—The 4TH GEAR SWITCH is HI when the vehicle is in **4th** gear.

BODY CONTROL MODULE (BCM)

BI01—The COURTESY LAMP PANEL SWITCH display is LO when the courtesy lights are illuminated by the switch.

BI02—The PARK LAMP SWITCH display is LO when the parklamp switch is in the **OFF** position

BI03—The DRIVER DOOR AJAR SWITCH display is LO when the driver's door is ajar.

BI04—The PASSENGER DOOR AJAR SWITCH display is LO when the passenger's door is ajar.

BI05—The DOOR JAMB SWITCH display is LO when the driver's or passenger's door is open.

BI06—The DOOR HANDLE SWITCH display is momentarily LO when either outside door handle button is depressed.

BI08—The LOW REFRIGERANT PRESSURE SWITCH display is LO when the system is low on refrigerant.

BI09—The WASHER FLUID LEVEL SWITCH display is LO when the vehicle is low on washer fluid.

BI10—The LOW COOLANT LEVEL display is LO when the engine is low on coolant.

BI16—The KEY IN THE IGNITION display is LO only when the key is in the **LOCK** position. (Cannot be seen, as diagnostics do not work with key **OFF**. Therefore, should always read HI.)

BI18—The BRAKE PRESSURE SIGNAL is LO when the brake controller detects a pressure problem.

BI21—The LOW BRAKE FLUID display is LO when brake fluid level is low.

BI22—The PARK BRAKE SWITCH display is LO when the parking brake is engaged.

BI24—The REVERSE GEAR SWITCH DISPLAY is HI only when the vehicle is in **REVERSE**.

BI25—The SEAT BELT SWITCH display is HI when the driver's seatbelt is fastened.

54, BI51—The GENERATOR FEEDBACK display is LO when there is a generator (alternator) problem (or engine not **RUNNING**).

BI71—The CRUISE CONTROL BRAKE SWITCH display is HI when the cruise ON/OFF switch is engaged and the brake pedal is not depressed (free state).

BI75—The CRUISE CONTROL ON/OFF SWITCH display is HI when the switch is **ON**.

BI76—The CRUISE CONTROL SET/COAST SWITCH display is HI when the cruise ON/OFF switch is **ON** and the set/coast switch is depressed.

BI77—The CRUISE CONTROL RESUME/ACCEL SWITCH display is HI when the cruise ON/OFF switch is **ON** and the Resume/Accel switch is depressed.

BI78—The HEADLAMP SWITCH display is HI whenever the headlamps are illuminated.

BI79—The HIGH BEAM SWITCH display is LO as long as the lever is pulled out.

BI82—The TWILIGHT ENABLE SWITCH is LO whenever the twilight sentinel is in operation.

BI83—The FOG LAMP SWITCH displays HI whenever the fog lamps are illuminated.

BI88—The LOW OIL LEVEL SWITCH (LOW OIL) displays LO whenever the oil level is low.

INSTRUMENT PANEL CLUSTER (IPC)

II10—The REAR DEFOG SWITCH display is LO as long as the switch in the right switch assembly is held depressed.

II11—The FRONT DEFROST SWITCH display is LO as long as the switch on the right switch assembly is held depressed.

II15 (1988 only)—The TAILLAMP OUT display is HI whenever a taillamp is out.

II16 (1988 only)—The HEADLAMP OUT display is HI whenever a headlamp is out.

II17 (1988 only)—The PARKLAMP OUT display is HI whenever a parklamp is out.

OUTPUT DISPLAYS

Output displays are operated as previously outlined under the heading HOW TO OPERATE THE SERVICE MODE. When troubleshooting a malfunction, the ECM and BCM output cycling can be used to determine if the output tests can be actuated regardless of the inputs and normal program instructions. Once a test in outputs has been selected, except for ECM IAC, the test will display HI and LO for 3 seconds in each state to indicated the command and output terminal voltage. A summary of each output is as follows:

ELECTRONIC CONTROL MODULE (ECM)

EO00—The NO OUTPUTS display will not display HI or LO as this is a resting spot where no outputs can be cycled.

EO01—The TORQUE CONVERTER CLUTCH (TCC) display will be LO when the TCC is energized.

EO04—The EGR SOLENOID No. 1 display will be LO when the EGR solenoid No. 1 is energized.

EO05—The EGR SOLENOID No. 2 display will be LO when the EGR solenoid No. 2 is energized.

EO06—The EGR SOLENOID No. 3 display will be LO when the EGR solenoid No.3 is energized.

EO07—The CANISTER PURGE SOLENOID display will be LO when the solenoid is ON (energized).

EO08—The A/C CLUTCH display will be HI when the clutch is energized.

EO09—The COOLANT FAN RELAY display is LO when the coolant fan is energized.

EO10—The COOLANT FAN RELAY No. 2 display will be LO when the pusher fan is energized.

EO11—The IAC MOTOR display will be LO when the pintel is extended and HI during is retraction.

BODY CONTROL MODULE (BCM)

BO00—The NO OUTPUTS display will not display HI or LO as this is a resting spot where no outputs will be cycled.

BO01—The CRUISE CONTROL VENT SOLENOID display is HI when the vent solenoid is energized. The cruise ON/OFF switch must be **ON** for this output to cycle.

BO02—The CRUISE CONTROL VACUUM SOLENOID display is HI when the vacuum solenoid is energized. The cruise ON/OFF switch must be **ON** for this output to cycle.

BO03—The RETAINED ACCESSORY POWER (RAP) relays display LO when the relays are energized.

BO04—The COURTESY RELAY display is LO when the relay is energized.

BO05—The TWILIGHT HEADLAMP RELAYS display is LO when the relays are energized and lights are on.

BO06—The HI/LO-BEAM RELAYS display is LO when the relays are operating (energized with hi beams on).

BO10—The CHIME 1 display is LO when the intermediate chime is operating (sounding).

BO11—The CHIME 2 display is LO when the intermediate chime is operating (sounding).

BO12—The FOGLAMP display is LO when the relay is energized (Reatta only).

BO13—The PARKLAMP display is LO when the relay is energized and the parklamps are illuminated.

OVERRIDE DISPLAYS

Override displays are operated as outlined under the heading HOW TO OPERATE THE SERVICE MODE. When troubleshooting a malfunction, the BCM override feature allows testing of certain system functions regardless of normal program instructions.

Upon selecting a test, that selections current operation will be represented as a percentage of it full range and this value will be displayed on the CRT below the exit pad.

Touching the pads above or below OVERRIDE on the GCC begins the override. Touching the pad above the OVERRIDE increases the value while the pad below OVERRIDE decreases the value. Normal program control can be resumed in the following ways.

Regain Normal Program

1. Selection of another override test will cancel the current override.
2. Selection of another system will cancel the current override.
3. Overriding the value beyond either extreme (0–99) will display – – – – momentarily and then jump to the normal program control.

The override test type is unique in that any other test type within the selected system may be active at the same time. After selecting an override test, touching the EXIT button will allow selection of an other test type, however, the CRT will continue to display the selected override. By selecting another test type and test, then touching the pads above or below OVERRIDE, it is possible to monitor the effect of the override on different vehicle parameters.

Selecting Override Test

ELECTRONIC CONTROL MODULE (ECM)

ES00—The NO OVERRIDES display will not display any value as this is a resting spot where no overrides will be controlled. This is a handy spot to go to stop the override without having to go back to the level where the system is selected.

ES01—The TCC SOLENOID display will show the state of the command (00 not engaged; 99 engaged). Pressing the up arrow will engage the TCC and pressing the down arrow will disengage the TCC. The vehicle must be in **3rd** or **4th** gear for this override to operate.

ES02—The EGR No. 1 SOLENOID can be turned OFF 00 (OFF) by pressing the down arrow, or ON 99 (EGR FLOW) by pressing the up arrow.

ES03—The EGR No. 2 SOLENOID can be turned OFF 00 (OFF) by pressing the down arrow, or ON 99 (EGR FLOW) by pressing the up arrow.

ES04—The EGR No. 3 SOLENOID can be turned OFF 00 (OFF) by pressing the down arrow, or ON 99 (EGR FLOW) by pressing the up arrow.

ES05—The CANISTER PURGE SOLENOID can be turned OFF 00 by pressing the down arrow, or ON 99 (purging) by pressing the up arrow.

ES06—The A/C CLUTCH RELAY can be turned OFF 00 (OFF) by pressing the down arrow, or ON 99 (engaged) by pressing the up arrow.

ES07—The COOLING FAN RELAY NO. 1 (puller) can be turned OFF 00 (OFF) by pressing the down arrow, or ON 99 (engaged) by pressing the up arrow.

ES08—The COOLING FAN RELAY NO. 2 (pusher) can be turned OFF 00 (OFF) by pressing the down arrow, or ON 99 (engaged) by pressing the up arrow.

ES09—The IAC MOTOR display will show the position the IAC is being commanded to in a percentage of travel. In order for this override to work the vehicle must be in **P** or **N**. The down arrow will retract the motor and the up arrow will extend the motor.

ES10—The INJECTOR CONTROL OVERRIDE will work only when the vehicle is in **P** or **N**. Upon entering injector override the display will show 0, all injectors working under normal program control. The arrow up button will increment the display by 1, up to 6 and wrap around to 0. This display number indicates the cylinder number of the injector that is being held OFF (e.g., 4 = injector in cylinder No. 4 OFF). The corresponding injector to the number displayed will be held OFF for as long as the number is displayed. The arrow down button is used to turn the injector back ON and then OFF (toggle effect); i.e., if display shows 4, injector 4 is OFF. Touching arrow down will turn injector 4 ON, and the display will show 0. Touching arrow down again will turn injector 4 back OFF and display will show a 4. Arrow down toggles the injector last displayed ON/OFF.

BODY COMPUTER MODULE (BCM)

BS00—The NO OVERRIDES display will not display any value as this is a resting spot where no overrides will be controlled. This is a handy spot to go to stop the override without having to go back to the level where the system is selected.

BS01—PROGRAM NUMBER. As operating conditions change, this number will automatically change in response. The automatic calculation of program number can be bypassed by a manual override feature, using the pads above and below the word OVERRIDE on the CRT. Touching the pad above override will force the program, number to increase at a controlled rate until the value of 99 is reached. 99 represents the MAX HEAT mode of climate control operation. Touching the pad below override will force the program number to decrease until the value of 0 is reached which represents the MAX A/C mode. This manual override of the automatic program number calculation will continue until the override is stopped. This allows the technician to control the program number to any number from 0–99 and simultaneously observe the reaction of any of the BCM data parameters.

BS02—VACUUM FLUORESCENT (VF) DIMMING. As the dim pot is moved, the VF dimming number is will automatically change in response (if outside lights are illuminated). The automatic control of dimming can be bypassed by a manual override feature using the pads above or below the word OVERRIDE. 99 represents the maximum brightness of VF displays and the 0 represents the maximum dimming of the VF displays (outside lights do not have to be illuminated to override the VF dim-

ming). This allows the technician to control the dimming number to any number from 0–99.

BS03—INCANDESCENMT DIMMING. As the dim pot is moved, the incandescent dimming number will automatically change in response (if outside lights are on). The automatic control of the dimming can be bypassed by a manual override feature, using the pads above and below the word OVERRIDE. 99 represents the maximum brightness of the incandescent displays and 0 represents the maximum dimming of the incandescent lighting. (Outside lighting does not have to be operating to override incandescent dimming). This allows the technician to control the dimming number from 0–99.

BS05—CRUISE CONTROL. The cruise control ON/OFF switch must be **ON** for this test to work. As operating conditions change, the cruise control position number will automatically change in response. The automatic calculation of cruise control position number can be bypassed by a manual override feature using the pads above and below the word OVERRIDE on the CRT. 99 represents the max cruise position (WOT) and 0 represents the minimum cruise position (closed throttle). This manual override of automatic cruise control position number calculation will continue until the override is stopped. This allows the technician to control the cruise control position number to any number from 0–99 and simultaneously observe the reaction of the data value for cruise servo position (BD70). Both should agree with each other within ± 3% except when the servo is commanded beyond its mechanical limits of travel. Since this test is not allowed with engine **RUNNING**, remember that there is usually only enough vacuum in the storage tank to stroke the servo 3 times.

BS06—OPTION CONTENT No. 1.

BS07—OPTION CONTENT No. 2.

System Check

1. Place the ignition switch in the **RUN** position, wait a few seconds for the GCC to warm up.
   a. The summary page will be displayed on the GCC.
   b. The summary hardkey L.E.D. lights.
2. Operate all hard and soft switches on the GCC face.
   a. Audible sound (beep-beep) should be heard for each page that appears.
   b. Hardswitch the L.E.D. lights for the switches pressed.
   c. The system reacts to each soft switch pressed on the GCC screen.
3. Turn the parking lamps **ON**.
   a. The GCC monitor dims according to the panel light setting.

TROUBLESHOOTING HINTS

1. Check the Cranking Circuit fuse (16) by turning ignition switch to the **START** position and verify that the instrument panel cluster goes blank.
2. If any diagnostic codes are displayed or the diagnostics cannot be entered, refer to the accompanying charts and diagnostic procedures.
3. If the CRT has erratic display during engine cranking, check the No. 805 circuit (YEL/BLK) wire for an open circuit and repair as required. If the circuit is good, replace the CRTC.

System Diagnosis

VISUAL INSPECTION

The most important check that must be done before any diagnostics are attempted is a careful visual inspection of the suspected wiring and components. Inspecting all vacuum hoses for pinches, cuts or disconnects is another check that must be done. These simple steps can often lead to fixing a problem without further steps.

This visual inspection is very critical and must be done carefully and thoroughly. Inspect the hoses that are difficult to see, such as under the air cleaner assembly, the air compressor, alternator and etc. Inspect all the related wiring for disconnects, burned or chaffed spots, pinched wires or contact with sharp edges or hot exhaust manifolds.

NOTE: Should a problem exist in a vehicle that has a history of body repair work, the area of repairs should be scrutinzed for damages to the wiring, connectors, vacuum hoses or other sub-components that could contribute to component problems. After being satisfied that the concerned area appear trouble free, expand the diagnosis as required.

1988–90 Cadillac DeVille and Fleetwood

At the center of the self-diagnostic system is the Body Computer Module (BCM), located behind the glove compartment opening. An internal microprocessor is used to control various vehicle function, based on monitored sensor and switch inputs. The ECM is located on the right side of the instrument panel and is the major factor in providing self-diagnostic capabilities for those subsection which it controls.

When both the BCM and ECM are used, a communication process has been incorporated which allows the 2 modules to share information and thereby provide for additional control capability. In a method similar to a telegraph key operator, each module's internal circuitry rapidly switches a circuit between 0 and 5 volts. This process is used to convert information into a series of pulses which represent coded data messages understood by the other components.

One of the data messages transferred from the BCM is a request for specific ECM diagnostic action. This action may affect the ECM controlled output or require the ECM to transfer some information back to the BCM. This communication gives the BCM control over the ECM's self-diagnostic capabilities in addition to its own.

In order to access and control the self-diagnostic features, available to the BCM, 2 additional electronic components are utilized by the service technician. Located to the right of the steering column is the Climate Control Panel (CCP) and located to the left of the steering column is either the Fuel Data Center (FDC), used with DFI equipped vehicles, or the Diesel Data Center (DDC), used with diesel engines.

These devices provide displays and keyboard switches used with several BCM controlled subsystems. This display and keyboard information is transferred over the single wire data circuits which carry coded data back and forth between the BCM and the display panels. This communication process allows the BCM to transfer any of its available diagnostic information to the instrument panel for display during service. By depressing the appropriate buttons on the CCP, data messages can be sent to the BCM, requesting the specific diagnostic features required.

The ECM/BCMs are designed to withstand normal current draws associated with the vehicle operation. However, care must be exercised to avoid overloading any of these circuits. In testing for opens or shorts, do not ground or apply voltage to any of the ECM/BCM circuits, unless instructed to do so by a diagnostic procedure. These circuits should only be tested using a high impedance multimeter (10 megohms minimum), should they remain connected to the ECM/BCM.

Power should never be applied to the ECM with the ignition in the **ON** position. Before removing or connecting battery cables, ECM/BCM fuses or ECM/BCM connectors, always turn the ignition to the **OFF** position.

Trouble Codes

In the process of controlling its various subsystems, the ECM and BCM continually monitor the operating conditions for possible system malfunctions. By comparing system conditions against standard operating limits, certain circuit and component malfunctions can be detected. A 2 digit numerical trouble code is stored in the computer's memory when a problem is detected by this self-diagnostic system. These trouble codes can later be displayed by the service technician as an aid in system repair.

If a particular malfunction would result in unacceptable system operation, the self-diagnostics will attempt to minimize the effect by taking FAIL–SAFE action. FAIL–SAFE action refers to any specific attempt by the computer system to compensate for the detected problem. A typical FAIL–SAFE action would be the substitution of a fixed input value when a sensor/circuit is detected to be open or shorted.

INTERMITTENT AND HARD FAILURE CODES

For Codes 12 through 51, the "SERVICE SOON" or "SERVICE NOW" indicator will go out automatically if the malfunction clears. However, the ECM stores the trouble code associated with the detected failure until the diagnostic system is cleared or until 50 ignition cycles have occurred without any fault reappearing. This condition is known as an intermediate failure.

Therefore, the ECM may have 2 types of trouble codes stored in its memory. These 2 codes types are:

1. A code for malfunction which is a hard failure. A hard failure turns on the appropriate service indicator lamp and keeps it on as long as the malfunction is present.

2. A code for intermediate malfunction which has occurred within the last 50 ignition cycles. An intermediate failure is one that was previously present, but was not detected the last time the ECM tested the circuit. The service indicator lamp turns out after the ECM tests the circuit without the defect being detected.

The first pass of the diagnostic codes, preceded by "..E", will contain all history codes, both hard and intermittent. The second pass contains only the hard codes that are present and will be preceded by ".E.E".

For Codes 52 through 67, the service indicator lamp will never come on. These codes indicate that a specific condition occurred of which the technician should be aware. Since these codes can be operator induced, a judgement must be made whether or not the code requires investigation. These codes will also be stored until the diagnostic system is cleared or until 50 ignition cycles have occurred without any faults reappearing.

INTERMITTENT PROBLEM DIAGNOSIS

It should be noted that diagnostic charts cannot be used to diagnose intermittent failures. The testing required at various points of the chart depends upon the fault to be present in order to locate its source in order to correct it.

If the fault is intermittent, an unnecessary ECM replacement could be indicated and the problem could remain.

Since many of the intermittent problems are caused at electrical connections, diagnosis of intermittent problems should start with a visual and physical inspection of the connectors involved in the circuit. Disconnect the connectors, examine and reconnect before replacing any component of the system.

Some causes of connector problems are:
1. Improperly formed terminals or connector bodies.
2. Damaged terminals or connector bodies.
3. Corrosion, body sealer, or other foreign matter on the terminal mating surfaces which could insulate the terminals.
4. Incomplete mating of the connector halves.
5. Connectors not fully seated in the connector body.
6. Terminals not tightly crimped to the wire.

If an affected circuit is one that may be checked by the status light on the ECC, the switch tests, the output cycling tests, or the engine data displays, make the check on the appropriate circuit. Some of the trouble codes include a "Note On Intermittents" describing a suggested procedure for isolating the location of intermittent malfunctions.

ENTERING DIAGNOSTIC MODE

To enter the diagnostic mode, proceed as follows:
1. Turn the ignition switch to the **ON** position.
2. Depress the OFF and the WARMER buttons on the CCP, simultaneously and hold them in until all display segments illuminate, which indicated the beginning of the diagnostic readout.

NOTE: If any of the segments are inoperative, the diagnosis should not be attempted, as this could lead to misdiagnosis. The display in question would have to be replaced before the diagnosis procedure is initiated.

Trouble Code Display

After the display segment check is completed, any trouble codes stored in the computer memory will be displayed on the Data Center panel as follows:
1. Display of the trouble codes will begin with an "8.8.8" on the data center panel for approximately one second. "..E" will then be displayed which indicates the beginning of the ECM stored trouble codes.
2. This first pass of ECM codes includes all detected malfunctions whether they are currently present or not. If no ECM trouble codes are stored, the "..E" display will be bypassed.
3. Following the display of "..E", the lowest numbered ECM trouble code will be displayed for approximately 2 seconds. All ECM trouble codes will be prefixed with the letter "E". (i.e. E12, E13, etc.).

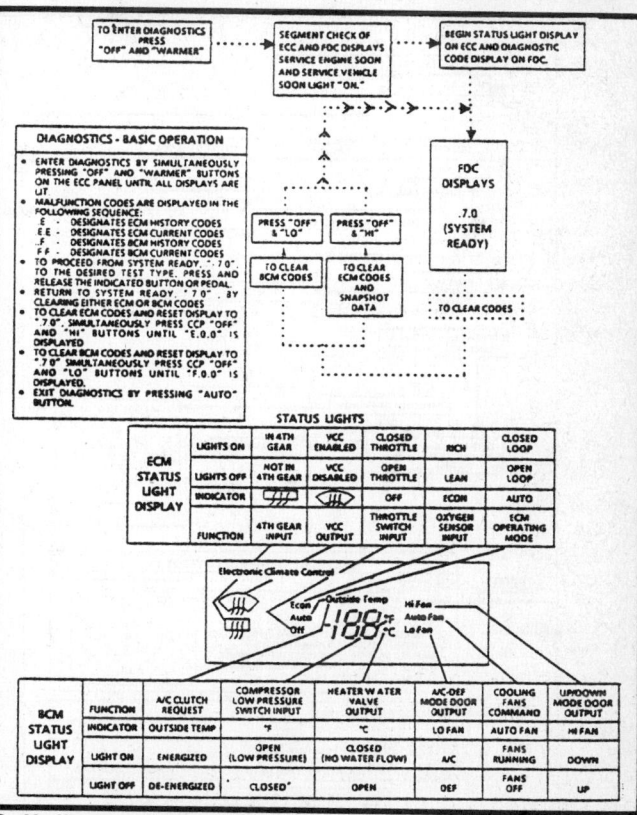

Self-diagnostic system flowchart—1988–90 Cadillac DeVille and Fleetwood

ECM DIAGNOSTIC CODES

CODE	DESCRIPTION	COMMENTS	CODE	DESCRIPTION	COMMENTS
E12	No Distributor Signal	Ⓐ	E38	Open MAT Sensor Circuit [AIR]	Ⓐ
E13	Oxygen Sensor Not Ready [AIR & CL]	Ⓐ	E39	VCC Engagement Problem	ⒶⒻ
E14	Shorted Coolant Sensor Circuit [AIR]	ⒶⒷ			
E15	Open Coolant Sensor Circuit [AIR]	ⒶⒷ	E40	Open Power Steering Pressure Switch Circuit	Ⓐ
E16	Generator Voltage Out Of Range [All Solenoids]	ⒷⒸ	E44	Lean Exhaust Signal [AIR & CL]	ⒶⒸ
E18	Open Crank Signal Circuit	Ⓑ	E45	Rich Exhaust Signal [AIR & CL]	ⒶⒸ
E19	Shorted Fuel Pump Circuit	Ⓑ	E47	BCM—ECM Data Problem [A/C Clutch]	Ⓐ
			E48	EGR System Fault [EGR]	Ⓐ
E20	Open Fuel Pump Circuit	Ⓑ			
E21	Shorted Throttle Position Sensor Circuit	ⒶⒷ	E52	ECM Memory Reset Indicator	Ⓒ
E22	Open Throttle Position Sensor Circuit	ⒶⒷ	E53	Distributor Signal Interrupt	Ⓒ
E23	EST/Bypass Circuit Problem [AIR & EGR]	ⒶⒷⒺ	E55	TPS Misadjusted	Ⓒ
E24	Speed Sensor Circuit Problem [VCC & Cruise]	ⒶⒹⒻ	E59	VCC Temperature Sensor Circuit Problem	ⒶⒻ
E26	Shorted Throttle Switch Circuit	Ⓐ	E60	Cruise—Transmission Not In Drive [Cruise]	Ⓒ
E27	Open Throttle Switch Circuit	Ⓐ			
E28	Open Third Or Fourth Gear Circuit	Ⓐ	E63	Cruise—Car Speed And Set Speed Difference Too High [Cruise]	Ⓒ
E30	RPM Error Too Great	Ⓐ	E64	Cruise—Car Acceleration Too High [Cruise]	Ⓒ
E31	Shorted MAP Sensor Circuit [AIR]	ⒶⒷ	E65	Cruise—Coolant Temperature Too High [Cruise]	Ⓒ
E32	Open MAP Sensor Circuit [AIR]	ⒶⒷ	E66	Cruise—Engine RPM Too High [Cruise]	Ⓒ
E34	MAP Sensor Signal Too High [AIR]	ⒶⒻ	E67	Cruise—Cruise Switch Shorted During Enable [Cruise]	Ⓒ
E37	Shorted MAT Sensor Circuit [AIR]	Ⓐ			

BCM DIAGNOSTIC CODES

CODE	DESCRIPTION	COMMENTS		DESCRIPTION	COMMENTS
F10	Outside Air Temperature Circuit Problem	Ⓒ	F40	Air Mix Door Problem	Ⓒ
F11	A/C High Side Temperature Circuit Problem	ⒸⓀ	F43	Heated Windshield Failure	Ⓒ
F12	A/C Low Side Temperature Circuit Problem [A/C Clutch]	Ⓒ	F46	Low A/C Refrigerant Condition Warning	Ⓒ
F13	In-Car Temperature Circuit Problem	Ⓒ	F47	Very Low A/C Refrigerant Condition Warning [A/C Clutch]	Ⓜ
			F48	Very Low A/C Refrigerant Pressure Condition [A/C Clutch]	Ⓜ
F30	CCP To BCM Data Problem	ⒸⒾⓁ	F49	High Temperature Clutch Disengage [A/C Clutch]	Ⓒ
F31	FDC To BCM Data Problem	ⒸⒾ	F61	BCM PROM Error	ⒸⓃ
F32	ECM—BCM Data Problem	ⒸⒾ			

DIAGNOSTIC CODE COMMENTS

Ⓐ	Turns On "SERVICE ENGINE SOON" Light.		Ⓘ	Displays "c" For Clock Problem Or "d" For Data Problem.
Ⓑ	Turns On "SERVICE VEHICLE SOON" Light.		Ⓙ	Turns On Front Defog AT 75° F.
Ⓒ	Does Not Turn On Any Telltale Light.		Ⓚ	Turns On "SERVICE AIR COND" Light For A Period Of Time.
Ⓓ	Disables Cruise For Entire Ignition Cycle.		Ⓜ	Turns On "SERVICE AIR COND" Light For A Period Of Time, & Switches ECC Mode To ECON.
Ⓔ	Causes System To Operate On Bypass Spark.		Ⓝ	Displays "-151" On CCP And Turns On Front Defog.
Ⓕ	Disengages VCC For Entire Ignition Cycle.		Ⓟ	Enable Canister Purge.
Ⓖ	Forces Cooling Fans On Full Speed.		[ ]	Functions Within Bracket Are Disengaged While Specified Malfunction Remains Current.
Ⓗ	Turns On Cooling Fans Whenever A/C Clutch Is Engaged.			

Diagnostic trouble codes—1988–90 Cadillac DeVille and Fleetwood

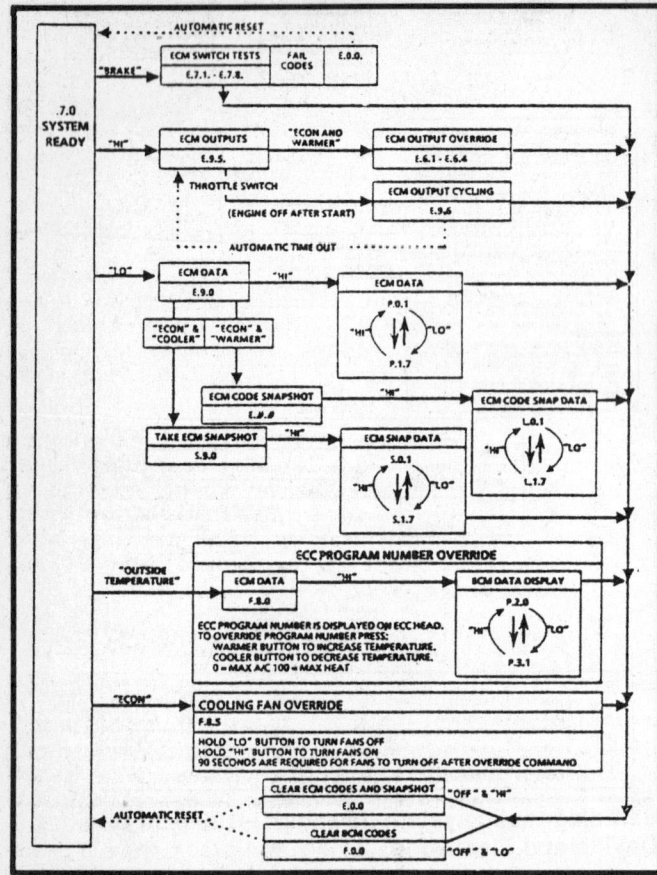

Self-diagnostic system flowchart (cont.) — 1988–90 Cadillac DeVille and Fleetwood

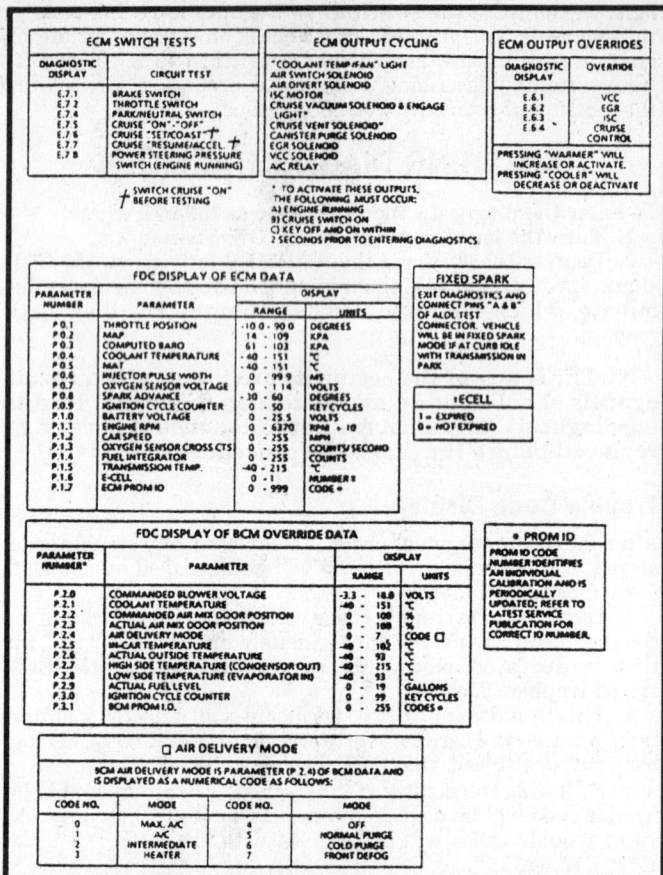

Self-diagnostic system flowchart (cont.) — 1988–90 Cadillac DeVille and Fleetwood

4. Progressively higher numbered trouble codes will be displayed, until the highest code present has been displayed.

5. ".E.E" will then be displayed, which indicates the beginning of the second pass of the ECM trouble codes.

6. On the second pass, only "HARD" trouble codes will be displayed. These are the codes which indicate a currently present malfunction.

7. Codes which are displayed during the first pass, but not on the second, are classified as "intermittent" trouble codes. If all the ECM codes are considered "intermittent", the ".E.E" display will be bypassed.

8. When all the ECM trouble codes have been displayed, the BCM codes will then displayed in a similar fashion. The only exceptions during the BCM code display are as follows:

a. "..F" precedes the first display pass.
b. The BCM codes are prefixed by an "F".
c. ".F.F" precedes the second display pass.

9. After all the ECM and BCM trouble codes have been displayed or if no codes are present, Code ".7.0" will be displayed. This code indicates that the system is ready for the next diagnostic feature to be selected.

NOTE: If a Code E51 is detected, it will be displayed continuously until the diagnostic mode is exited. During this display of Code E51, none of the other diagnostic features will be possible.

Clearing Trouble Codes

Trouble codes stored in the ECM's memory may be cleared (erased) by entering the diagnostic mode and then depressing the OFF and HI buttons on the CCP simultaneously. Hold the buttons in until "E.0.0"appears on the display. Trouble codes stored in the BCM's memory may be cleared by depressing the OFF and LO buttons simultaneously until "F.0.0" appears. After "E.0.0" or "F.0.0" is displayed, ".7.0." will appear. With the ".7.0"displayed, turn the ignition **OFF** for at least 10 seconds before re-entering the diagnostic mode.

Exiting Diagnostic Mode

To get out of the diagnostic mode, depress the AUTO button or turn the ignition switch **OFF** for 10 seconds. Trouble codes are not erased when this is done. The temperature setting will reappear in the display panel.

NOTE: The climate control system will operate in whatever mode was commanded prior to depressing the necessary buttons to enter the diagnostic system. The prior operating mode will be remembered and will resume after the diagnostic mode is entered.

Status Display

While in the diagnostic mode, the mode indicators on the CCP are used to indicate the status of certain operating systems. These different modes of operation are indicated by the status light either being turned off or on.

EEC Program Override

During the BCM display of data, a manual override of the system can be accomplished by the technician, for different levels of heating and cooling effort. Since this is not in the realm of this diagnostic coverage, no procedural explanation is given.

1988–90 Cadillac Eldorado and Seville

Numerous electronic components are located on the vehicle as a part of an electrical network, designed to control various engine and body subsystems.

At the heart of the computer system is the Body Computer Module (BCM), located in the center of the instrument panel, behind the Climate Control Driver Information Center (CCDIC) display. This BCM has an internal microprocessor, which is the center for communication with all other components in the system. All sensors and switches are monitored by the BCM or by other major components that complete the computer system. The major components are as follows:

1. Electronic Control Module (ECM)
2. Instrument Panel Cluster (IPC)
3. Programmer/Heating/Ventilation/Air Conditioning
4. Climate Control/Driver Information Center

A combination of inputs from these major components, the other sensors and switches communicate together to the BCM, either as individual inputs, or on the common communication link called the serial data line. The various inputs to the BCM combine with program instructions within the system memory to provide accurate control over the many subsystems involved. When a subsystem circuit exceeds pre-programmed limits, a system malfunction is indicated and may provide certain back-up functions. Providing control over the many subsystems from the BCM is done by controlling system outputs. This can be either direct or transmitted along the serial data line to 1 of the other major components. The process of receiving, storing, testing and controlling information is continuous. The data communication gives the BCM control over the ECM's self-diagnostic capacities in addition to its own.

Between the BCM and the other 4 major components of the computer system, a communication process has been incorporated which allows the devices to share information and thereby provide for additional control capabilities. In a method similar to that used by a telegraph system, the BCM's internal circuitry rapidly switches to a circuit between 0 and 5 volts like a telegraph key. This process is used to convert information into a series of pulses which represents coded data messages understood by other components. Also, much like a telegraph system, each major component has its own recognition code. When a message is sent out on the serial data line, only the component or station that matches the assigned recognition code will pay attention and the rest of the components or stations will ignore it. This data transfer of information is most important in understanding how the system operates and how to diagnosis possible malfunctions within the system.

System Diagnosis

A systematic approach is needed to begin the vehicle's self-diagnostic capabilities along with an understanding of the basic operation and procedures, necessary to determine external or internal malfunctions of the computer operated circuits and systems. A systematic beginning is to determine if the SERVICE ENGINE SOON telltale lamp is illuminated when the ignition key is in the **ON** position and the engine not operating. If the lamp is not illuminated, a problem could be in the power supply circuits of the systems.

If the lamp is illuminated, can the SERVICE MODE be accessed? If the display is not operating, the self diagnostics can not be used.

Is there a trouble code displayed? If a trouble code is identified, using the self diagnostics mode, a malfunction or problem has been detected by the system.

Code charts are included to assist the technician in determining in what area malfunctions or problems have been detected. However, if no malfunction or problem is recorded, a SELF-DIAGNOSTIC SYSTEM CHECK chart will guide the technician to appropriate symptom reference charts for those symptoms not detected by the Self-Diagnosis charts.

VISUAL INSPECTION

The most important checks that must be done before any diagnostics are attempted is a careful visual inspection of the suspected wiring and components. Inspecting all vacuum hoses for pinches, cuts or disconnects is another check that must be done. These simple steps can often lead to fixing a problem without further steps.

This visual inspection is very critical and must be done carefully and thoroughly. Inspect the hoses that are difficult to see, such as under the air cleaner assembly, the air compressor, alternator and etc. Inspect all the related wiring for disconnects, burned or chaffed spots, pinched wires or contact with sharp edges or hot exhaust manifolds.

NOTE: Should a problem exist in a vehicle that has a history of body repair work, the area of repairs should be scrutinzed for damages to the wiring, connectors, vacuum hoses or other sub-components that could contribute to component problems. After being satisfied that the concerned area appears trouble free, expand the diagnosis as required.

COMPUTER SYSTEM SERVICE PRECAUTIONS

The computer control system is designed to withstand normal current draws associated with vehicle operation. However, care must be taken to avoid overloading any of these circuits. In testing for open or short circuits, do not ground or apply voltage to any of the circuits, unless instructed to do so by diagnostic procedures. These circuits should only be tested using a high impedance Multimeter, such as Kent Moore tool No. J-29125A or its equivalent, if they are to remain connected to any of the computers. Power should never be remove or applied to any of the computers with the key switch in the **ON** position. Before removing or connecting battery cables, fuses or connectors, always turn the ignition to the **LOCK** position.

SELF-DIAGNOSTIC TROUBLE CODES

In the process of controlling the various subsystems, the ECM and BCM continually monitor operating conditions for possible system malfunctions. By comparing system conditions against standard operating limits, certain circuit and component malfunctions can be detected. A 3 digit numerical TROUBLE CODE is stored in the computer memory when a problem is detected by this self diagnostic system. These TROUBLE CODES can be displayed by the technician as an aid in the system repairs.

The occurrence of certain system malfunctions require that the vehicle operator be alerted to the problem so as to avoid prolonged vehicle operation under the downgraded system operation, which could affect other systems and components. Computer controlled diagnostic messages and/or telltales will appear under these conditions which indicate that service is required.

If a particular malfunction would result in unacceptable system operation, the self diagnostics will attempt to minimize the effect by taking FAILSOFT action. FAILSOFT action refers to any specific attempt by the computer system to compensate for the detected problem. A typical FAILSAFE action would be the substitution of a fixed input value when a sensor is detected to be open or shorted.

ENTERING THE DIAGNOSTIC MODE

To enter the diagnostic mode, proceed as follows:

ECM DIAGNOSTIC CODES

CODE	DESCRIPTION	COMMENTS	CODE	DESCRIPTION	COMMENTS
E012	No Distributor Signal	Ⓐ	E038	Open MAT Sensor Circuit [AIR]	Ⓐ
E013	Oxygen Sensor Not Ready [AIR & CL]	Ⓐ Ⓔ	E039	VCC Engagement Problem	Ⓐ Ⓕ
E014	Shorted Coolant Sensor Circuit [AIR]	Ⓐ Ⓔ	E040	Open Power Steering Pressure Switch Circuit	Ⓐ
E015	Open Coolant Sensor Circuit [AIR]	Ⓐ Ⓔ	E044	Lean Exhaust Signal [AIR & CL]	Ⓐ Ⓕ
E016	Generator Voltage Out Of Range [All Solenoids]	Ⓑ Ⓒ	E045	Rich Exhaust Signal [AIR & CL]	Ⓐ Ⓕ
E018	Open Crank Signal Circuit	Ⓑ	E047	BCM—ECM Data Problem [A/C Clutch]	Ⓐ
E019	Shorted Fuel Pump Circuit	Ⓑ	E048	EGR System Fault [EGR]	Ⓐ
E020	Open Fuel Pump Circuit	Ⓑ	E052	ECM Memory Reset Indicator	Ⓐ
E021	Shorted Throttle Position Sensor Circuit	Ⓐ Ⓔ	E053	Distributor Signal Interrupt	Ⓐ
E022	Open Throttle Position Sensor Circuit	Ⓐ Ⓔ	E055	TPS Misadjusted	Ⓐ
E023	EST/Bypass Circuit Problem [AIR & EGR]	Ⓐ Ⓔ	E058	Pass Key Fuel Enable	Ⓐ
E024	Speed Sensor Circuit Problem [VCC & Cruise]	Ⓐ Ⓕ	E059	VCC Temperature Sensor Circuit Problem	Ⓐ Ⓕ
E026	Shorted Throttle Switch Circuit	Ⓐ	E060	Cruise—Transmission Not In Drive [Cruise]	Ⓘ
E027	Open Throttle Switch Circuit	Ⓐ	E063	Cruise-Car Speed And Set Speed Difference Too High [Cruise]	Ⓘ
E028	Open Third Or Fourth Gear Circuit	Ⓐ	E064	Cruise-Car Acceleration Too High [Cruise]	Ⓘ
E030	RPM Error Too Great	Ⓐ	E065	Cruise-Coolant Temperature Too High [Cruise]	Ⓘ
E031	Shorted MAP Sensor Circuit [AIR]	Ⓐ Ⓕ	E066	Cruise-Engine RPM Too High [Cruise]	Ⓘ
E032	Open MAP Sensor Circuit [AIR]	Ⓐ Ⓕ	E067	Cruise-Cruise Switch Shorted During Enable [Cruise]	Ⓘ
E034	MAP Sensor Signal Too High [AIR]	Ⓐ Ⓕ			
E037	Shorted MAT Sensor Circuit [AIR]	Ⓐ			

BCM DIAGNOSTIC CODES

CODE	DESCRIPTION	COMMENTS	CODE	DESCRIPTION	COMMENTS
B110	Outside Air Temperature Circuit Problem	Ⓒ Ⓙ	B410	Charging System Problem	Ⓒ
B111	A/C High Side Temperature Circuit Problem		B411	Battery Volts Too Low	Ⓒ
B112	A/C Low Side Temperature Circuit Problem [A/C Clutch]	Ⓒ	B412	Battery Volts Too High	Ⓒ
B113	In-Car Temperature Circuit Problem	Ⓒ	B420	Relay Circuit Problem	Ⓒ
B115	Sunload Temperature Circuit Problem	Ⓒ	B440	Air Mix Door Problem	Ⓒ
B119	Twilight Sentinel Photosensor Circuit Problem	Ⓒ	B441	Cooling Fans Problem	Ⓒ Ⓓ
B120	Twilight Sentinel Delay Pot Circuit Problem	Ⓒ	B446	Low A/C Refrigerant Condition Warning	Ⓒ
B121	Twilight Sentinel Enable Pot Circuit Problem	Ⓒ	B447	Very Low A/C Refrigerant Condition Warning [A/C Clutch]	Ⓒ Ⓒ
B122	Panel Lamp Dimming Pot Circuit Problem	Ⓒ	B448	Very Low A/C Refrigerant Pressure Condition [A/C Clutch]	Ⓒ Ⓒ
B123	Panel Lamp Enable Circuit Problem	Ⓒ	B449	A/C High Side Temperature Too High [A/C Clutch]	Ⓒ
B124	Speed Sensor Circuit Problem	Ⓒ Ⓒ	B450	Coolant Temperature Too High [A/C Clutch]	Ⓒ
B127	PRND321 Sensor Circuit Problem	Ⓒ	B552	BCM Memory Reset Indicator	Ⓒ
B334	Loss Of ECM Data [A/C Clutch]	Ⓒ Ⓒ	B556	BCM EEPROM Error	Ⓒ
B335	Loss Of CCDIC Data	Ⓒ			
B336	Loss Of IPC Data	Ⓒ			
B337	Loss Of Programmer Data	Ⓒ			

DIAGNOSTIC CODE COMMENTS

Ⓐ	Turns On "SERVICE ENGINE SOON" Light.	Ⓖ	Displays Square Box Around Each PRND321 Position On IPC.
Ⓑ	Displays "SERVICE CAR SOON" On DIC.	Ⓗ	Displays 'Error' In Season Odometer.
Ⓒ	Displays Status Message On DIC.	Ⓘ	Switches ECC Mode To ECON.
Ⓓ	Does Not Turn On Any Telltale Light Or Display Any Message.	Ⓙ	Forces Cooling Fans On Full Speed.
Ⓔ	Causes System To Operate On Bypass Spark.	Ⓟ	Enables Canister Purge.
Ⓕ	Disengages VCC For Entire Ignition Cycle.	Ⓒ	Disables Cruise Control for Ignition Cycle.
		[ ]	Functions Within Bracket Are Disengaged While Specified Malfunction Remains Current.

Diagnostic trouble codes—1988–90 Cadillac Eldorado and Seville

1. Turn the ignition switch to the **ON** position.
2. Touch the OFF and the WARMER buttons on the Climate Control panel simultaneously and hold until a segment check is displayed on the Instrument Panel Cluster (IPC) and Climate Control Driver Information Center, usually around 3 seconds.

—— CAUTION ——
Operating the vehicle in the SERVICE MODE for extended time periods (exceeding ½ hour) without the engine operating or without a trickle type charger connected to the battery, can cause the the battery to discharge, resulting in possible relaying of false diagnostic information or causing a no-start condition.

SEGMENT CHECK

The purpose of illuminating the Instrument Panel Cluster (IPC) and the Climate Control Driver Information Center (CCDIC), is to check that all segments of the vacuum fluorescent displays are working. On the IPC, however, the turnsignal indicators do not light during this check. Diagnosis should not be attempted unless all CCDIC segments appear, as this could lead to misdiagnosis, If any portions or segments of the CCDIC displays are inoperative, it must be replaced.

STATUS LAMPS

While in the diagnostic service mode, the mode indicator lamps on the Climate Control Panel of the CCDIC are used to indicate

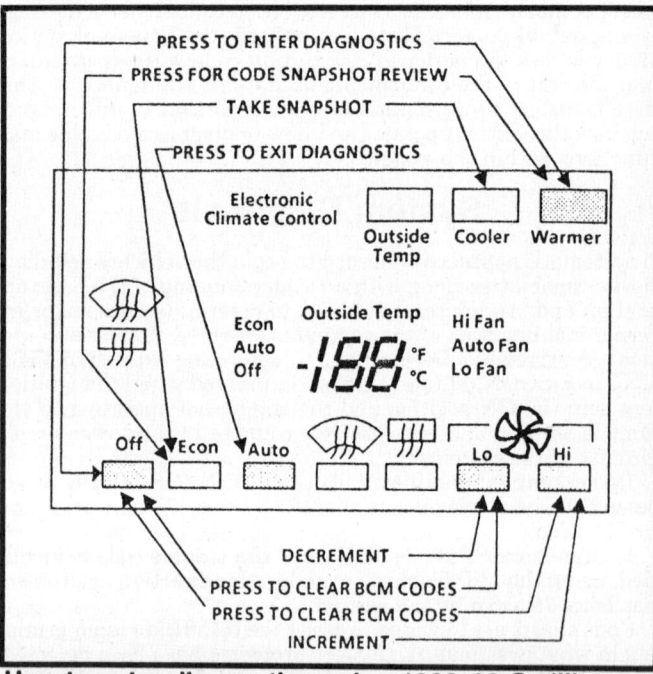

How to enter diagnostic mode—1988–90 Cadillac Eldorado and Seville

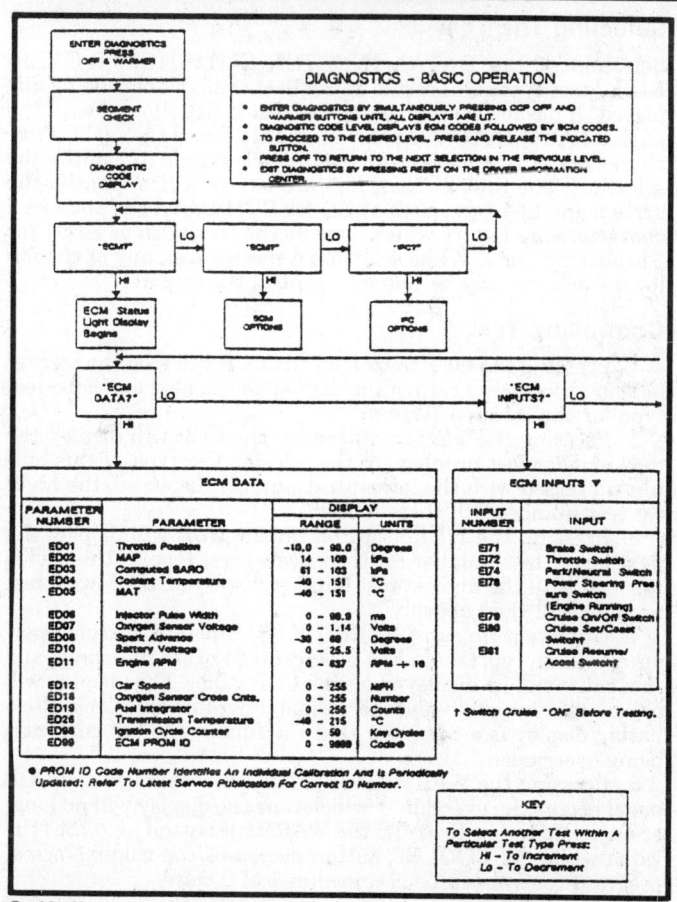

Self-diagnostic system flowchart — 1988–90 Cadillac Eldorado and Seville

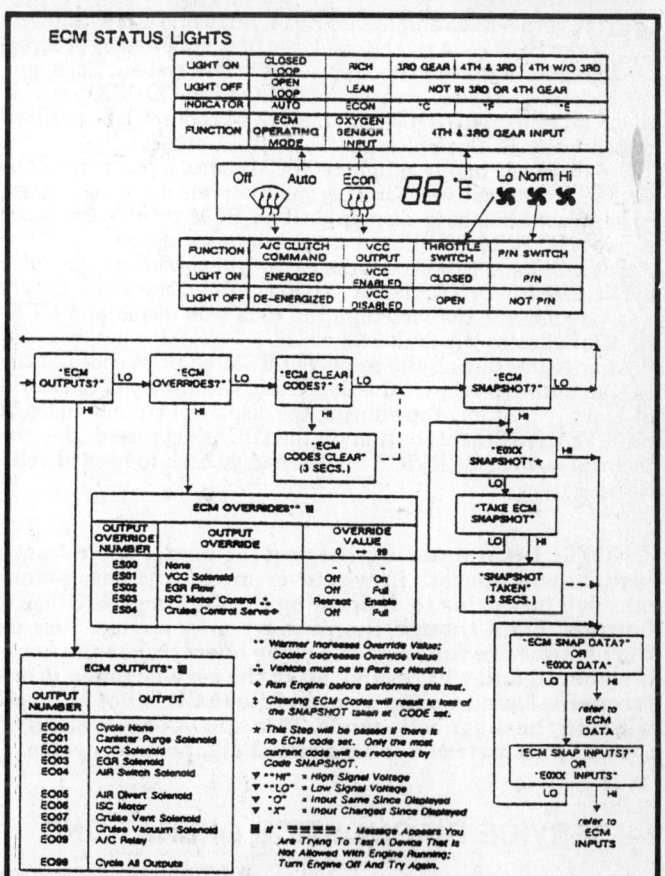

Self-diagnostic system flowchart (cont.) — 1988–90 Cadillac Eldorado and Seville

the status of the certain operating modes. The different modes of operation are indicated by the status lamp being turned **ON** or **OFF** in the following manner of operation;

1. The OFF status indicator lamp is illuminated whenever the ECM is commanding the Viscous Converter Clutch (VCC) to engage. The lamp only indicates whether the VCC is enabled or disabled by the ECM. Actual operation depends upon the integrity of the VCC system.

2. The AUTO status indicator lamp is turned **ON** whenever the ECM is operating in the CLOSED LOOP fuel control mode. The lamp should illuminate after the coolant and oxygen sensors have reached normal operating temperatures.

3. The ECON status indicator lamp is turned **ON** whenever the oxygen sensor signal to the ECM indicates a RICH exhaust condition. This lamp should switch between RICH and LEAN (flash **ON** and **OFF** during warm, steady throttle operation).

4. The C status indicator lamp is turned **ON** whenever ECM senses that the 3rd and 4th gear pressure switches are open. The lamp should be **ON** while in the 4th gear operation and the ECM had received a 3rd gear input signal.

5. The F status indicator lamp is turned **ON** whenever the ECM senses that the 3rd and 4th gear pressure switches are open. The lamp should be **ON** while in the 4th gear operation and the ECM had received a 3rd gear input signal.

6. The E status indicator lamp is turned **ON** whenever the ECM senses that the 4th gear pressure switch is open, but not the 3rd gear switch. The lamp should be **ON** while in 4th gear operation and the ECM had not received a 3rd gear input signal.

7. The FRONT DEFOG status indicator lamp is turned **ON** whenever the BCM is commanding the ECC compressor clutch to engage. The lamp indicates whether the clutch is enabled or disabled by the BCM and actual operation depends on the integrity of the compressor clutch system.

8. The REAR DEFOG status indicator lamp is turned **ON** whenever the BCM senses that the low refrigerant pressure switch is open. The lamp should remain **OFF** if the refrigerant system is fully charged and being properly controlled. However, when the ambient temperature drops below approximately –5°F (–21°C.), the lamp will come **ON** due to the pressure-temperature relationship of refrigerant.

9. The LO FAN status indicator lamp is turned **ON** whenever the BCM is commanding the A/C-DEF mode door to divert air flow to the A/C outlets, as in the A/C or normal purge modes. This lamp will be **OFF** whenever the ECC system is in the heater, intermediate, defrost and cold purge modes.

10. The NORM FAN SYMBOL status indicator lamp is turned **ON** whenever the BCM is commanding the heater water valve to block coolant flow through the heater core. The lamp should remain **OFF** except when the air mix door is being commanded to the MAX A/C position.

11. The HI FAN SYMBOL status indicator lamp is turned **ON** whenever the BCM is commanding the UP-DOWN mode door to divert air flow up, away from the heater outlet. The lamp will be **OFF** whenever the ECC system is in the heater or normal purge mode.

TROUBLE CODE DISPLAY

After the SERVICE MODE is entered, any trouble codes stored in the computer memory will be displayed. ECM codes will be displayed first. If no ECM trouble codes are stored, a NO ECM CODES message will be displayed. All ECM codes will be prefixed with a E. Examples are E013, E014 and etc. The lowest numbered ECM code will be displayed first, followed by progressively higher numbered codes present in the system. Following the highest ECM code present or the NO ECM CODES message, the BCM codes will be displayed. All BCM codes will be prefixed with a letter B. Examples are B110, B111 and etc.

Progressively higher numbered BCM codes, if present, will be displayed consecutively for 2 second intervals until the highest code present has been displayed. If no BCM trouble codes are stored, NO BCM CODES message will be displayed.

Any BCM and ECM codes displayed will also be accompanied by CURRENT or HISTORY. HISTORY indicates the failure was not present the last time the code was tested and CURRENT indicates the fault still exists.

At any time during the display of ECM or BCM codes, if the LO fan button on the ECC is depressed, the display of codes will be bypassed. At any time during the display of trouble codes, if the RESET/RECALL button on the DIC is depressed, the system will exit the SERVICE MODE and go back to normal vehicle operation.

NOTE: Upon entering the service mode, the climate control will operate in whatever mode was being commanded just prior to depressing the OFF and WARMER buttons. Even though the displays may change just as the buttons are touched, the prior operating mode is remembered and will resume after the service mode is entered. Extended Compressor at Idle (ECI) is not allowed while in the diagnostic mode. This allows observation of system parameters during normal compressor cycles.

SERVICE DIAGNOSTICS OPERATIONS

After the trouble codes have been displayed, the SERVICE MODE can be used to perform several tests on the different systems, 1 at a time. Upon completion of code display, a specific system may be selected for testing.

Selecting The System

Following the display of the trouble codes, the first available system will be displayed. As an example, ECM.

While selecting the system to test, any of the following actions may be taken to control the display.

1. Pressing the OFF button, the Climate Control Panel (CCP) will stop the system selection process and return the display to the beginning of the trouble code sequence.

2. Pressing the LO fan button on the CCP will display the next available system selection. This allows the display to be stepped through all system choices. The list of systems can be repeated following the display of the last system.

3. Pressing the HI fan button on the CCP will select the system for testing.

Selecting The Test Type

Having selected a system, the first available test type will be displayed, such as ECM DATA?. While selecting a specific test type, any of the following actions may be taken to control the display.

1. Pressing the OFF button on the CCP will stop the test type selection process and return the display to the next available system selection.

2. Pressing the LO fan button on the CCP will display the next available test type for the selected system. This allows the display to be stepped through all available test type choices. The list of test types can be repeated following the display of the last test type.

3. Pressing the HI fan button on the CCP will select the displayed test type. At this point, the first of several specific tests will appear.

Selecting The Test

Selection of the DATA?, INPUTS?, OUTPUTS? or OVERRIDE? test types will result in the first available test being displayed. If dashes ever appears, this test is not allowed with the engine **RUNNING**. Turn the engine **OFF** and try again. Four characters of the display will contain a test code to identify the selection. The first 2 characters are letters which identify the system and test type, such as ED for ECM DATA and the last 2 characters are letters which identify the test, such as ED01 for Throttle Position. While selecting a specific test, any of the following actions may be taken to control the display.

Controlling Test

1. Pressing the OFF button on the CCP will stop the test selection process and return the display to the next available test type for the selected system.

2. Pressing the LO fan button on the CCP will display the next smaller test number for the selected test type. If this button is pressed with the lowest test number displayed, the highest test number will then appear.

3. Pressing the HI fan button on the CCP will display the next larger test number for the selected test type. If this pad is touched with the highest test number displayed, the lowest test number will then appear.

4. Upon selecting an OVERRIDE test function, current operation will be represented as a percentage of its full range and this value will be displayed on the CCP panel. This display will alternate between – – and the normal program value. This alternating display is a reminder that the function is not currently being overridden.

5. Pressing the WARMER or COOLER buttons on the CCP panel begins the override at which time the display will no longer alternate to – –. Pressing the WARMER button increases the value while the COOLER button decreases the value. Normal program control can be resumed in 1 of 3 ways.

Regaining Normal Program Control

1. Selection of another override test will cancel the current override.

2. Selection of another system (ECM, BCM or IPC) will cancel the current overide.

3. Overriding the value beyond either extreme (0 or 99) will display – – momentarily and then jump to the opposite extreme. If the button is released while – – is displayed, normal program control will resume and the display will again alternate.

The override test type is unique in that any other test type within the selected system may be active at the same time. After selecting an override test, pressing the OFF button will allow selection of another test type, DATA, INPUTS or OUTPUTS. The CCP panel will continue to display the selected override. By selecting another test type and test, while at the same time pressing the WARMER or COOLER buttons, it is possible to monitor the effects of the override on different vehicle parameters.

Selecting CLEAR CODES?

Selection of the CLEAR CODES? test type will result in the message CODES CLEAR being displayed along with the selected system name. This message will appear for 3 seconds to indicate that all stored trouble codes have been erased from that system's memory. After 3 seconds, the display will automatically return to the next available test type for the selected system

Selecting SNAPSHOT?

Selection of SNAPSHOT? test type will result in the message SNAPSHOT TAKEN being displayed with the selected system name preceeding it. This message will appear for 3 seconds to indicate that all system data and inputs have been stored in memory. After 3 seconds, the display will automatically proceed to the first available snapshot test type, for example SNAP DATA. While selecting a snapshot test type, any of the following actions can be taken to control the display;

1. Pressing the OFF button on the CCP will stop the test type selection process and return the display to the next available system selection.

2. Pressing the LO button on the CCP will display the next available snapshot test type. This allows the display to be stepped through all available choices. This list of snapshot test types can be repeated following the display of the last choice.

3. Pressing the HI button with SNAP DATA? or SNAP INPUT? displayed, will select that test type. At this point, the display is controlled as it would be for non-snapshot data and inputs displays. However, all values and status information represents memorized vehicle conditions.

4. Pressing the HI button on the CCP with SNAPSHOT? displayed will again display the SNAPSHOT TAKEN message to indicate that new information has been stored in the memory. Access to this information is obtained the same as previously outlined.

Exiting Service Mode

To exit the service mode, press the RESET/RECALL button on the DIC or turn the ignition switch to the **OFF** position. Trouble codes are not erased when this is done.

DATA DISPLAYS

When troubleshooting a malfunction, the ECM and BCM data display can be used to compare the vehicle with problems to a vehicle which is functioning properly. A brief summary of each parameter is provided as follows:

ELECTRONIC CONTROL MODULE (ECM)

ED01 – The THROTTLE POSITION (TPS) is displayed in millivolts (mv) from 0–5100 (1988–89 displayed in degree form, 10–90 degrees).

ED02 – The MANIFOLD AIR PRESSURE (MAP) is displayed in kilopascals (kPa) from 14–109.

ED03 – The COMPUTED BAROMETRIC PRESSURE (BARO) is displayed in kilopascals (kPa) from 61–103.

ED04 – The COOLANT TEMPERATURE is displayed in degrees (C.) from minus 40–151, or degrees (F.) from 40–306.

ED05 – The MANIFOLD AIR TEMPERATURE (MAT) also called air temperature sensor, is displayed on degrees (C.) from minus 40–151, or degrees (F.) from 40–306.

ED06 – The INJECTOR PULSE WIDTH is displayed in milliseconds (MS), from 0–99.9.

ED07 – The OXYGEN SENSOR VOLTAGE is displayed in volts from 0–1.14.

ED08 – The SPARK ADVANCE value is displayed in degrees, from minus 30–60.

ED10 – The BATTERY VOLTAGE is read in volts, from 0 – 25.5.

ED11 – The ENGINE SPEED is displayed in RPM from 0 – 6370.

ED12 – The VEHICLE SPEED is in miles per hour (mph) from 0–159.

ED18 – The OXYGEN SENSOR CROSS COUNTS is displayed as the number of times the O_2 sensor crossed the reference line each second.

ED19 – The FUEL INTEGRATOR displayed in counts, from 88–160.

ED26 – The VCC TEMPERATURE is displayed as a voltage level at the ECM. The normal range is from 0–5.12 volts (1988–89 display is in degrees C. from –40–215.

ED70 – The CRUISE SERVO position is displayed in percentages (%) from 0–100. A value close to 0% represents at rest position and a value close to 100% represents wide open throttle.

ED98 – The IGNITION CYCLE COUNTER value is the number of times the ignition has been cycled to **OFF** since an ECM trouble code was last detected. After 50 ignition cycles without any malfunctions being detected, all stored ECM codes are cleared.

ED99 – The ECM PROM I.D. is displayed as a number up to 3 digits long, which can be used to verify that the proper PROM is installed in the ECM.

BODY CONTROL MODULE (BCM)

BD20 – The COMMANDED BLOWER VOLTAGE is read in volts from 3.3–18.0.

BD21 – The COOLANT TEMPERATURE is displayed in degrees (°C) from minus 40–151. This value is sent from the ECM to the BCM. If this circuit malfunctions as determined by the ECM, the ECM will send the BCM a Failsoft value for display.

BD22 – The COMMANDER AIR MIX DOOR POSITION is displayed in percent (%). A value close to 0% represents a cold air mix and a value close to 100% represents a warm air mix.

BD23 – The ACTUAL AIR MIX DOOR POSITION is displayed in percent (%). This value should follow the commanded air mix door position (BD22) except when the door is commanded beyond its mechanical limits of travel.

BD24 – The AIR DELIVERY MODE is displayed as a number from 0–7. Each number is a code which represents the following air delivery modes:

**0 – MAX A/C
**1 – A/C
**2 – Intermediate
**3 – Heater
**4 – Off
**5 – Normal Purge
**6 – Cold Purge
**7 – Front Defog

BD25 – The IN-CAR TEMPERATURE is displayed in degrees (°C), from –40–102.

BD26 – The ACTUAL OUTSIDE TEMPERATURE is displayed in degrees (°C), from –40–93. This valve represents actual sensor temperatures and is not restricted by the features used to minimize engine heat affects on the customer display valve.

BD27 – The HIGH SIDE TEMPERATURE (condenser output) is displayed in degrees (°C) from 40–215.

BD28 – The LOW SIDE TEMPERATURE (evaporator input) is displayed in degrees (°C) from 40–93.

BD32 – The SUNLOAD TEMPERATURE SENSOR is displayed in degrees (°C) from 40–102.

BD40 – The ACTUAL FUEL LEVEL is read in gallons from 0–19.0. The display can read to 25. This value represents actual sensor position and is not restricted by the features used to eliminate fuel slosh effects on the customer display value.

BD41 – The P-R-N-D-3-2-1 is displayed in percent (%). The signal is calculated as a percent of available reference voltage to determine P-R-N-D-3-2-1 position. The value close to 100% represents **PARK** and a value close to 0% represents 1.

BD42 – The DIMMING POT is displayed in percent (%). A value close to 0% represents maximum dimming and a value close to 100% represents maximum brightness.

BD43 – The TWILIGHT DELAY POT is displayed in percent (%). A value close to 0% represents minimum delay time and a value close to 100% represents maximum delay time.

BD44 – The TWILIGHT PHOTOCELL is displayed in percent (%). A value close to 0% represents daylight and value close to 100% represents darkness.

BD50 – The BATTERY VOLTAGE is read in volts between 0–16.3.

BD51 – The GENERATOR (ALTERNATOR) FIELD is displayed in percent (%). A value close to 0% represents minimum regulator on time and a value close to 100% represents maximum regulator on time.

BD52 – The INCANDESCENT BULB REFERENCE is displayed in volts. With the parking lamps/headlamps **OFF**, the value is zero and with either the parking lamps or the head-lamps on, the value is battery voltage.

BD60 – The VEHICLE SPEED is displayed in miles per hours (mph)from 0–159.

BD61 – The ENGINE SPEED is displayed in rpm from 0-6375.

BD70 – The CRUISE SERVO POSITION is displayed in percent (%) from 0–100. A value close to 0% represents reset position and a value close to 100% represents W.O.T.

BD90 – Represents OPTION 1.

BD91 – Represents OPTION 2.

BD98 – The IGNITION CYCLE value is the number of times that the ignition has been cycled to **OFF** since a BCM Trouble Code was last detected. After 100 ignition cycles, without any malfunction being detected, all BCM codes are cleared.

BD99 – The BCM PROM I.D. is displayed as a number, up to 4 digits long, which can be used to verify that the proper PROM was installed in the BCM.

INPUT DISPLAYS

Input displays are operated as outlined under the heading of HOW TO OPERATE A SERVICE MODE. When troubleshooting a malfunction, the ECM, BCM or IPC input display can be used to determine if the switched inputs can be used to determine if the switched inputs can be properly interpreted. When one of the various input tests is selected, the state of that device is displayed as HI or LO. In general, the HI and LO refer to the input terminal voltage for that circuit. The display also indicates if the input changed state so that the technician could activate or deactivate any listed device and return to the display to see if it changed state. If a change of state occurred, an X will appear next to the HI/LO indicator. The X will only appear once per selected input, although the HI/LO indication will continue to change as the input changes. Some tests are momentary and the X can be used as and indication of a change. The following is a list of the ECM, BCM and IPC inputs:

ELECTRONIC CONTROL MODULE (ECM)

EI71 – The BRAKE SWITCH display is LO when the brake pedal is depressed.

EI72 – The THROTTLE SWITCH display is HI when the accelerator pedal is depressed.

EI74 – The PARK/NEUTRAL (P/N) display is LO when the vehicle is in the **PARK** or **NEUTRAL** position.

EI78 – The POWER STEERING PRESSURE SWITCH display is LO when the power steering pressure (effort) is high, such as when the wheel is in a crimped position.

EI79 – The Cruise Control ON/OFF switch display is HI when the switch is engaged.

E180 – The Cruise Control SET/COAST switch display is HI when the cruise ON/OFF switch is engaged and the SET/COAST switch is depressed.

EI81 – The Cruise Control RESUME/ACCEL switch display is HI when the Cruise ON/OFF switch is engaged and the RESUME/ACCEL switch is pushed.

BODY CONTROL MODULE (BCM)

BI01 – The COURTESY LAMP PANEL SWITCH display is LO when the courtesy lights are on from the switch.

BI02 – The PARK LAMP SWITCH display is LO when the parklamp switch is in the **OFF** position.

BI03 – The DRIVER DOOR AJAR SWITCH display is LO when the driver's door is ajar.

BI04 – The PASSENGER DOOR AJAR SWITCH display is LO when the passenger's door is ajar.

BI05 – The DOOR JAMB SWITCH display is LO when any door is open.

BI06 – The DOOR HANDLE SWITCH display is momentarily LO when either outside door handle button is depressed.

BI07 – The TRUNK OPEN SWITCH display is LO when the trunk lid is open.

BI08 – The LOW REFRIGERANT PRESSURE SWITCH display is LO when the system is low on refrigerant.

BI09 – The WASHER FLUID LEVEL SWITCH display is LO when the vehicle is low on washer fluid.

BI10 – LO-COOLANT is not used on 1988 E and K vehicles.

BI30 – The TEMP/TIME SWITCH display is LO when the button is depressed.

BI41 – The COOLING FAN FEEDBACK display is LO when the cooling fans are **RUNNING**.

BI151 – The GENERATOR FEEDBACK display is LO when there is a generator (alternator) problem (or engine not **RUNNING**).

BI71 – The CRUISE CONTROL BRAKE SWITCH display is HI when the cruise ON/OFF switch is **ON** and the brake pedal is not depressed (free state).

BI75 – The CRUISE CONTROL ON/OFF SWITCH display is HI when the switch is **ON**.

BI76 – The CRUISE CONTROL SET/COAST SWITCH display is HI when the cruise ON/OFF switch is **ON** and the set/coast switch is depressed.

BI77 – The CRUISE CONTROL RESUME/ACCEL SWITCH display is HI when the cruise ON/OFF switch is **ON** and the Resume/Accel switch is depressed.

INSTRUMENT PANEL CLUSTER (IPC)

II78 – The HEADLAMP SWITCH display is HI whenever the headlamps are **ON**.

II79 – The HIGH BEAM SWITCH display is LO as long as the lever is pulled in.

II80 – The DIMMING SENTINEL SWITCH display is LO whenever the system is **ON** .

II81 – The DIMMING SENTINEL PHOTOSENSOR DISPLAY IS HI whenever it senses light.

II82 – The TWILIGHT ENABLE SWITCH display is LO whenever the system is **ON**.

OUTPUT DISPLAYS

When troubleshooting a malfunction, the ECM and BCM output cycling can be used to determine if the output tests can be actuated regardless of the inputs and normal program instructions. Once a test in outputs has been selected, the test will display HI and LO for 3 seconds in each state to indicated the command and output terminal voltage. A summary of each output is provided.

ELECTRONIC CONTROL MODULE (ECM)

EO00 – This test displays CYCLE NONE as no outputs are activated at this point.

EO01 – The CANISTER PURGE SOLENOID display will be LO when the solenoid is energized.

EO02 – The VISCOUS CONVERTER CLUTCH (VCC) display will be LO when the solenoid is energized.

EO03 – The EGR SOLENOID display will be LO when the EGR Solenoid is energized.

EO04 – The AIR SWITCH SOLENOID display will be LO when the solenoid is energized.

EO05 – The AIR DIVERT SOLENOID display will be LO when the solenoid is energized.

EO06 – The ISC MOTOR display will be LO when the plunger is retracting and HI during its extension.

EO07 – The Cruise Control vent solenoid display is HI when the vent solenoid is energized. The Cruise ON/OFF switch must be engaged and the engine not **RUNNING** for the output to cycle.

EO08 – The Cruise Control vent solenoid display is HI when the vacuum solenoid is energized. The Cruise Control ON/OFF switch must be engaged and the engine not **RUNNING** for the output to cycle.

EO09 – Air Conditioning relay display will be in HI when the relay is energized.

BODY CONTROL MODULE (BCM)

BO00 – The NO OUTPUTS display will not display HI or LO as this is a resting spot where no outputs will be cycled.

BO01 – The CRUISE CONTROL VENT SOLENOID display is HI when the vent solenoid is energized. The cruise ON/OFF switch must be **ON** for this output to cycle.

BO02 – The CRUISE CONTROL VACUUM SOLENOID display is HI when the vacuum solenoid is energized. The cruise ON/OFF switch must be **ON** and the engine **OFF** for this output to cycle.

BO03 – The RETAINED ACCESSORY POWER (RAP) relays display is LO when the relays are energized.

BO04 – The COURTESY RELAY display is LO when the relay is energized.

BO05 – The TWILIGHT HEADLAMP RELAYS display is LO when the relays are energized and the lights are illuminated.

BO06 – The HI/LO BEAM RELAYS display is LO when the relays are energized, high beams illuminated.

OVERRIDE DISPLAYS

Override displays are operated as outlined under the heading HOW TO OPERATE THE SERVICE MODE. When troubleshooting a malfunction, the BCM override feature allows testing of certain system functions regardless of normal program instructions.

Upon selecting a test, that function's current operation will be represented as a percentage of its full range and this value will be displayed on the CCP panel. The display will alternate between – – for 1 second, followed by the normal program value for 10 seconds. This alternating display is a reminder that the function is not currently being overridden.

Touching the WARMER or COOLER buttons on the CCP panel begins the override at which time the display will no longer alternate to – –. Touching the WARMER button increases the value while the COOLER button decreases the value. Upon release of the button, the display may either remain at the override value or automatically return to normal program control. This depends on which function is being overridden at the time. If the display remains at the override value, normal program control can be resumed in 1 of 3 ways.

1. Selection of another override test will cancel the current override.

2. Selection of another system will cancel the current override.

3. Overriding the value beyond either extreme (0–99) will display – – momentarily and then jump to the opposite extreme. If the button is released while – – is displayed, normal program control will resume and the display will again alternate.

The override test type is unique in that any other test type within a selected system may be active at the same time. After selecting an override test, touching the OFF button will allow selection of another test type. However, the CCP panel will continue to display the selected override. By selecting another test type and test, while at the same time touching the WARMER or COOLER button, it is possible to monitor the effect of the override on the different vehicle parameters.

BODY CONTROL MODULE (BCM)

BS00 – This test will display NONE as no overrides are active at this point.

BS01 – When the PROGRAM NUMBER OVERRIDE is selected, the CCP will display the program number which is currently being used by the climate control system. As operating conditions change, this number will automatically change in response. The automatic calculation of program number can be

overridden with the WARMER and COOLER buttons. A value of 99 represents maximum heating where 0 is maximum cooling.

BS02 – The VACUUM FLUORESCENT (VF) DIMMING override can be controlled from 0 (max dim)–99 (max bright). The display will hold the override value upon release of the buttons.

BS03 – The INCANDESCENT BULB DIMMING override can be controlled from O (max dim)–99 (max bright) if the parking lamps have been turned **ON**. The display will hold the override value upon release of the buttons.

BS04 – The COOLING FANS override will control the 0 (fan **OFF**) or 99 (max fans) as long as the button is held. Normal control will resume upon release of the button.

6 **BS06** – The GENERATOR (ALTERNATOR) DISABLE override will control to O (generator operating) or 99 (generator disabled) as long as the button is held. Normal control will resume upon release of the buttons.

ENGINE CONTROL MODULE (ECM)

ES00 – This test will display NONE as no overrides are active at this point.

ES01 – The VCC solenoid override can be activated by the WARMER or COOLER buttons. Its status is displayed as ON/OFF.

ES02 – The EGR solenoid override status is displayed as ON/OFF.

ES03 – The ISC motor override status is displayed as UP/DN.

ES04 – The Cruise Control servo override status is displayed as UP/DN.

1988–90 Oldsmobile Toronado

Numerous electronic components are located on the vehicle as a part of an electrical network, designed to control various engine and body subsystems.

At the heart of the computer system is the Body Computer Module (BCM), located behind the glove box. This BCM has an internal microprocessor, which is the center for communication with all other components in the system. All sensors and switches are monitored by the BCM or other major components that complete the computer system. The components are as follows:

1. Electronic Control Module (ECM)
2. Instrument Panel Cluster (IPC)
3. Electronic Climate Control Panel (ECC)
4. Programmer/Heating/Ventilation/AC
5. Chime/Voice module

A combination of inputs from these major components and the other sensors and switches communicate together to the BCM, either as individual inputs, or on the common communication link called the serial data line. The various inputs to the BCM combine with program instructions within the system memory to provide accurate control over the many subsystems involved. When a subsystem circuit exceeds pre-programmed limits, a system malfunction is indicated and may provide certain back-up functions. Providing control over the many subsystems from the BCM is done by controlling system outputs. This can be either direct of transmitted along the serial data line to the other major components. The process of receiving, storing, testing and controlling information is continuous. The data communication gives the BCM control over the ECM's self-diagnostic capacities in addition to its own.

Between the BCM and the other 5 major components of the computer system, a communication process has been incorporated which allows the devices to share information and thereby provide for additional control capabilities. In a method similar to that used by a telegraph system, the BCM's internal circuitry rapidly switches to a circuit between 0 and 5 volts like a telegraph key. This process is used to convert information into a series of pulses which represents coded data messages understood by other components. Also, much like a telegraph system, each major component has its own recognition code. When a message is sent out on the serial data line, only the component or station that matches the assigned recognition code will pay attention and the rest of the components or stations will ignore it. This data transfer of information is most important in understanding how the system operates and how to diagnosis possible malfunctions within the system.

In order to access and control the BCM self diagnostic features, additional electronic components are necessary, the Instrument Panel Cluster (IPC) and the Electronic Climate Control panel (ECC). As part of the IPC, a 20 character display area called the Information Center is used. During normal engine operation, this area displays Toronado or is a Tachmeter, displaying the engine rpm. When a malfunction is sensed by the BCM, a driver's warning messages is displayed in this area. When the diagnostic mode is entered, the various BCM or ECM diagnostic codes are displayed. In addition to the codes of the ECM/BCM data parameters, discrete inputs and outputs, as well as output override messages are also displayed when commanded for, through the ECC.

The Electronic Climate Control Panel (ECC) provides the controls for the heating and air conditioning systems. It also becomes the controller to enter the diagnostics and access the BCM self-diagnostics. This communication process allows the BCM to transfer any of its available diagnostic information top the instrument panel for display during service. By pressing the appropriate buttons on the ECC, data messages can be sent to the BCM over the serial data line requesting the specific diagnostic features desired. When in the Override mode of the BCM diagnostics, the amount of Override is displayed at the ECC where the outside and set temperatures are normally displayed.

TROUBLESHOOTING

System Diagnosis

A systematic approach is needed to begin the vehicle's self-diagnostic capabilities along with an understanding of the basic operation and procedures, necessary to determine external or internal malfunctions of the computer operated circuits and systems. A systematic beginning is to determine if the SERVICE ENGINE SOON telltale lamp is illuminated when the ignition key is in the ON position and the engine not operating. If the lamp is OFF, a problem could be in the power supply circuits of the systems.

If the lamp is illuminated, can the SERVICE MODE be accessed? If the Electronic Climate Control panel is not operating, the self diagnostics cannot be used.

Is there a trouble code displayed? If a trouble code is identified, using the self diagnostics mode, a malfunction or problem has been detected by the system.

Code charts are included to assist the technician in determining in what area malfunctions or problems have been detected. However, if no malfunction or problem is recorded, a SELF-DIAGNOSTIC SYSTEM CHECK chart will guide the technician to appropriate symptom reference charts for those symptoms not detected by the Self-Diagnosis charts.

VISUAL INSPECTION

The most important checks that must be done before any diagnostics are attempted is a careful visual inspection of the suspected wiring and components. Inspecting all vacuum hoses for pinches, cuts or disconnects is another check that must be done. These simple steps can often lead to fixing a problem without further steps.

This visual inspection is very critical and must be done carefully and thoroughly. Inspect the hoses that are difficult to see, such as under the air cleaner assembly, the air compressor, al-

ternator and etc. Inspect all the related wiring for disconnects, burned or chaffed spots, pinched wires or contact with sharp edges or hot exhaust manifolds.

NOTE: Should a problem exist in a vehicle that has a history of body repair work, the area of repairs should be scrutinized for damages to the wiring, connectors, vacuum hoses or other areas that could contribute to component problems. After being satisfied that the concerned area appears trouble free, expand the diagnosis as required.

COMPUTER SYSTEM SERVICE PRECAUTIONS

The computer control system is designed to withstand normal current draws associated with vehicle operation. However, care must be taken to avoid overloading any of these circuits. In testing for open or short circuits, do not ground or apply voltage to any of the circuits, unless instructed to do so by diagnostic procedures. These circuits should only be tested using a high impedance Multimeter, such as Kent Moore tool No. J–29125A or its equivalent, if they are to remain connected to any of the computers. Power should never be remove or applied to the computers with the key in the ON position. Before removing or connecting battery cables, fuses or connectors, always turn the ignition to the LOCK position.

SELF-DIAGNOSTIC TROUBLE CODES

In the process of controlling the various subsystems, the ECM and BCM continually monitor operating conditions for possible system malfunctions. By comparing system conditions against standard operating limits, certain circuit and component malfunctions can be detected. A 3 digit numerical TROUBLE CODE is stored in the computer memory when a problem is detected by this self diagnostic system. These TROUBLE CODES can be displayed by the technician as an aid in the system repairs.

The occurance of certain system malfunctions require that the vehicle operator be alerted to the problem so as to avoid prolonged vehicle operation under the downgraded system operation, which could affect other systems and components. Computer Controlled diagnostic messages and/or telltales will appear under these conditions which indicate that service is required.

If a particular malfunction would result in unacceptable system operation, the self diagnostics will attempt to minimize the effect by taking FAILSOFT action. FAILSOFT action refers to any specific attempt by the computer system to compensate for the detected problem. A typical FAILSAFE action would be the substitution of a fixed input value when a sensor is detected to be open or shorted.

ENTERING THE DIAGNOSTIC MODE

To enter the diagnostic mode, proceed as follows:
1. Turn the ignition switch to the ON position.
2. Touch the OFF and the WARM buttons on the Electronic Climate Control (ECC) panel simultaneously and hold until a segment check is displayed on the Instrument Panel Cluster (IPC) and Electronic Climate Control (ECC), usually around 3 seconds.

——————————— **CAUTION** ———————————

Operating the vehicle in the SERVICE MODE for extended time periods (exceeding ½ hour) without the engine operating or without a trickle type charger connected to the battery, can cause the the battery to discharge, resulting in possible relaying of false diagnostic information or causing a no-start condition.

ECM DIAGNOSTIC CODES

CODE	DESCRIPTION	COMMENTS	CODE	DESCRIPTION	COMMENTS
E013	Oxygen Sensor Circuit [Canister Purge]	(A)/(B)	E041	Cam Sensor Circuit	
E014	Coolant Temperature Sensor (Too High)	(A)/(B)/(F)	E042	Electronic Spark Timing (EST)	(A)/(B)/(C)
E015	Coolant Temperature Sensor (Too Low)	(A)/(B)/(F)	E043	Electronic Spark Control (ESC)	(A)/(B)/(C)
E016	Battery Voltage Too High [All Solenoids]	(A)/(B)	E044	Lean Exhaust Signal	(A)/(B)/(C)
			E045	Rich Exhaust Signal	(A)/(B)/(C)
E021	Throttle Position Sensor Circuit (Volt High) [TCC]	(A)/(B)	E046	Power Steering Pressure Switch Circuit [A/C Clutch]	(C)
E022	Throttle Position Sensor Circuit (Volt Low) [TCC]	(A)/(B)	E047	BCM-ECM Communication [A/C Clutch & Cruise]	(A)/(B)
E023	MAT Sensor Circuit (Low Temp)	(A)/(B)	E048	Misfire	(C)
E024	Vehicle Speed Sensor Circuit [TCC]	(A)/(B)			
E025	MAT Sensor Circuit (HighTemp)	(A)/(B)	E051	PROM Error	(A)/(B)/(O)/(X)
E026	QDM Circuit	(A)/(B)			
E027	Second Gear Circuit	(C)	E063	Small EGR Fault	(A)/(B)
E028	Third Gear Circuit	(C)	E064	Medium EGR Fault	(A)/(B)
E029	Fourth Gear Circuit	(C)	E065	Large EGR Fault	(A)/(B)
E031	P/N Switch Circuit	(C)			
E034	MAF Sensor Circuit	(A)/(B)			
E038	Brake Switch Circuit	(C)			
E039	TCC Circuit	(C)			

DIAGNOSTIC CODE COMMENTS

(A)	"Service Engine Soon" telltale in IPC "ON".	(G)	Displays "Electrical Problem" Message on IPC.
(B)	Displays Diagnostic Message on IPC.	(O)	Forces OL Operation.
(C)	No telltale light or message.	(W)	Causes system to operate on bypass spark.
(D)	Displays "ERROR" in season odometer.	(X)	Causes system to operate on back-up fuel.
(F)	Forces Cooling Fans on.	[ ]	Functions within bracket are disengaged while specified malfunction remains current.

BCM DIAGNOSTIC CODES

CODE	DESCRIPTION	COMMENTS	CODE	DESCRIPTION	COMMENTS
B110	Outside Air Temperature Circuit Problem	(B)	B448	Very Low A/C Refrigerant Pressure Condition [A/C Clutch]	(B)
B111	A/C High Side Temperature Circuit Problem	(B)/(F)	B449	A/C High Side Temperature Too High [A/C Clutch]	(B)/(F)
B112	A/C Low Side Temperature Circuit Problem [A/C Clutch]	(D)	B450	Coolant Temperature Too High [A/C Clutch]	(B)/(F)
B113	In-Car Temperature Circuit Problem	(B)	B482	Anti-Lock Brake Pressure Circuit Problem	(B)
B115	Sunload Temperature Circuit Problem	(B)	B552	BCM Memory Reset Indicator	(C)
B119	Twilight Sentinel Photosensor Circuit Problem	(B)	B555	BCM EEPROM Error	(D)
B120	Twilight Sentinel Delay Pot Circuit Problem	(B)	B660	Cruise - Transmission Not In Drive [Cruise]	(C)
B122	Panel Lamp Dimming Pot Circuit Problem	(C)	B663	Cruise - Car Speed and Set Speed Difference Too High [Cruise]	(C)
B123	Panel Lamp Enable Circuit Problem	(C)	B664	Cruise - Car Acceleration Too High [Cruise]	
B124	Speed Sensor Circuit Problem [Cruise]	(A)	B667	Cruise - Cruise Switch Shorted During Enable [Cruise]	(B)
B127	PRNDL Sensor Problem	(B)	B671	Cruise - Servo Position Sensor Circuit Problem [Cruise]	(B)
B132	Oil Pressure Sensor	(B)	B672	Cruise - Vent Solenoid Circuit Problem [Cruise]	(B)
B140	Phone System Problem		B673	Cruise - Vacuum Solenoid Circuit Problem [Cruise]	(B)
B332	Loss of Compass Data				
B334	Loss of ECM Data [Cruise, A/C Clutch and ECM Service Mode]	(A)/(B)			
B335	Loss of ECC or CRT Data	(C)			
B336	Loss of IPC Data [IPC Service Mode]	(C)			
B337	Loss of Programmer Data	(D)			
B410	Charging System Problem	(D)			
B411	Battery Volts Too Low [Cruise]	(D)			
B412	Battery Volts Too High [Cruise]	(D)			
B420	Relay Circuit Problem	(D)			
B440	Air Mix Door Problem	(C)			
B446	Low A/C Refrigerant Condition Warning	(C)			
B447	Very Low A/C Refrigerant Problem [A/C Clutch]	(B)			

CRT DIAGNOSTIC CODES §

CODE	DESCRIPTION	COMMENTS
C553	CRT Keep Alive Memory Reset	(C)
C710	CRT - CRTC Communications Problem	(C)

Diagnostic trouble codes—1988–90 Oldsmobile Toronado

TROUBLE CODE DISPLAY

After the SERVICE MODE is entered, any trouble codes stored in the computer memory will be displayed. ECM codes will be displayed first. If no ECM trouble codes are stored, the IPC will display a NO ECM CODES message for approximately 2 seconds. All ECM codes will be prefixed with a E. Examples are E013, E014 and etc. The lowest numbered ECM code will be displayed first, followed by progressively higher numbered codes present in the system.

The codes will be displayed consecutively for 2 second intervals until the highest code present has been displayed. When all ECM codes have been displayed, the BCM codes will be displayed. The lowest numbered BCM code will be displayed for appropriately 2 seconds. BCM codes accompanied by CURRENT indicates the fault still exits. Since the ECM is not capable of making this determination, this message does not appear when the ECM codes are being displayed. All BCM codes will be prefixed with a letter B. Examples are B110, B111 and etc.

Progressively higher numbered BCM codes, if present, will be displayed consecutively for 2 second intervals until the highest code present has been displayed. If no BCM trouble codes are stored, NO BCM CODES message will be displayed. At any time during the display of ECM or BCM codes, if the LO fan button on the ECC is depressed, the display of codes will be bypassed. At any time during the display of trouble codes, if the BI-LEV

button is depressed, the BCM will exit the SERVICE MODE and go back to normal vehicle operation.

NOTE: Upon entering the service mode, the climate control will operate in whatever mode was being commanded just prior to depressing the OFF and WARM buttons. Even though the displays may change just as the buttons are touched, the prior operating mode is remembered and will resume after the service mode is entered.

SERVICE DIAGNOSTICS OPERATION

After the trouble codes have been displayed, the SERVICE MODE can be used to perform several tests on the different systems, 1 at a time. Upon completion of code display, a specific system may be selected for testing.

Selecting The System

Following the display of the trouble codes, the first available system will be displayed. As an example, ECM.

While selecting the system to test, any of the following actions may be taken to control the display.

1. Pressing the OFF button will stop the system selection process and return the display to the beginning of the trouble code sequence.

2. Pressing the LO fan button will display the next available

system selection. This allows the display to be stepped through all system choices. The list of systems can be repeated following the display of the last system.

3. Pressing the HI fan button will select the displaced system for testing. At this point, the first available test type will appear with the selected system name above it.

4. Pressing the BI–LEV button will exit diagnostics and return to normal IPC and ECC operation.

Selecting The Test Type

Having selected a system, the first available test type will be displayed, such as ECM DATA?. While selecting a specific test type, any of the following actions may be taken to control the display.

1. Pressing the OFF button will stop the test type selection process and return the display to the next available system selection.

2. Pressing the LO fan button will display the next available test type for the selected system. This allows the display to be stepped through all available test type choices. The list of test types can be repeated following the display of the last test type.

3. Pressing the HI fan button will select the displayed test type. At this point, the display will either indicate that the selected test type is in progress or the first of several specific tests will appear.

4. Pressing the BI–LEVEL button will exit the diagnostics.

Selecting The Test

Selection of the DATA?, INPUTS?, or OUTPUTS? test types will result in the first available test being displayed. If a EEEE message ever appears, this test is not allowed with the engine **RUNNING**. Turn the engine **OFF** and try again. The last 4

characters of the display will contain a test code to identify the selection. The first 2 characters are letters which identify the system and test type, such as ED for ECM DATA. and the last 2 characters are letters which identify the test, such as ED01 for Throttle Position. While selecting a specific test, any of the following actions may be taken to control the display.

1. Pressing the OFF button will stop the test selection process and return the display to the next available test type for the selected system.

2. Pressing the LO fan button will display the next smaller test number for the selected test type. If this button is touched with the lowest test number displayed, the highest test number will then appear.

3. Pressing the HI fan button will display the next larger test number for the selected test type. If this pad is touched with the highest test number displayed, the lowest test number will then appear.

Upon selecting an OVERRIDE test function, current operation will be represented as a percentage of its full range and this value will be displayed on the ECC panel. This display will alternate between – – and the normal program value. This alternating display is a reminder that the function is not currently being overridden.

Pressing the WARM or COOL buttons on the ECC panel begins the override at which time the display will no longer alternate to – –. Pressing the WARM button increases the value while the COOL button decreases the value. Normal program control can be resumed in 1 of 3 ways.

1. Selection of another override test will cancel the current override.

2. Selection of another system (ECM, BCM or IPC) will cancel the current overide.

3. Overriding the value beyond either extreme (0 or 99) will display – – momentarily and then jump to the opposite extreme.

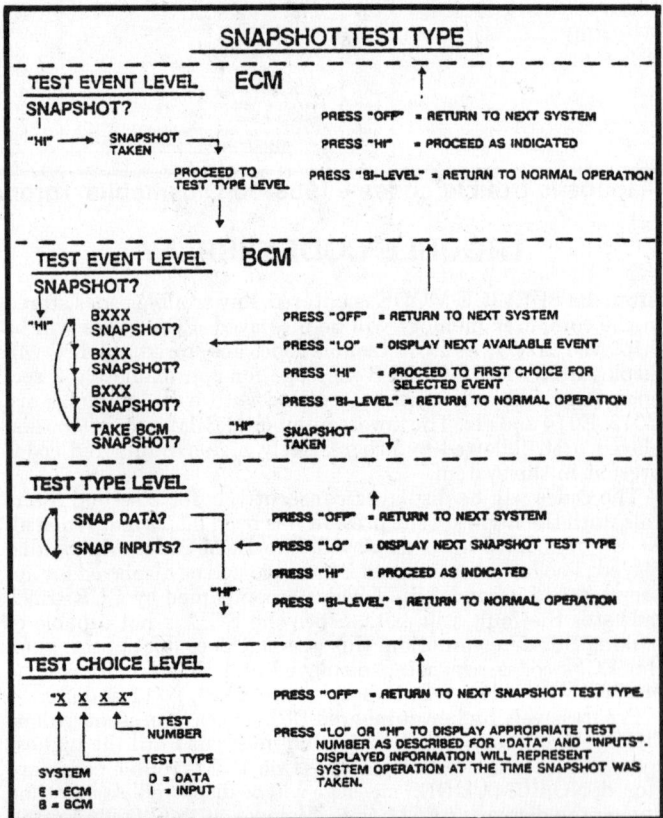

Self-diagnostic system flowchart (non-CRT) – 1988–90 Oldsmobile Toronado

Selecting snapshot (non-CRT) – 1988–90 Oldsmobile Toronado

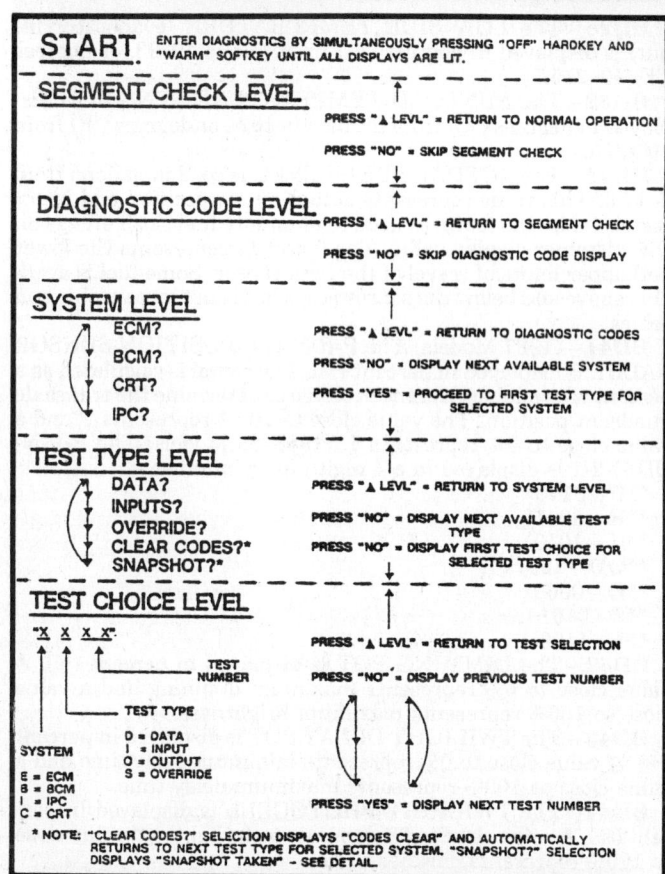

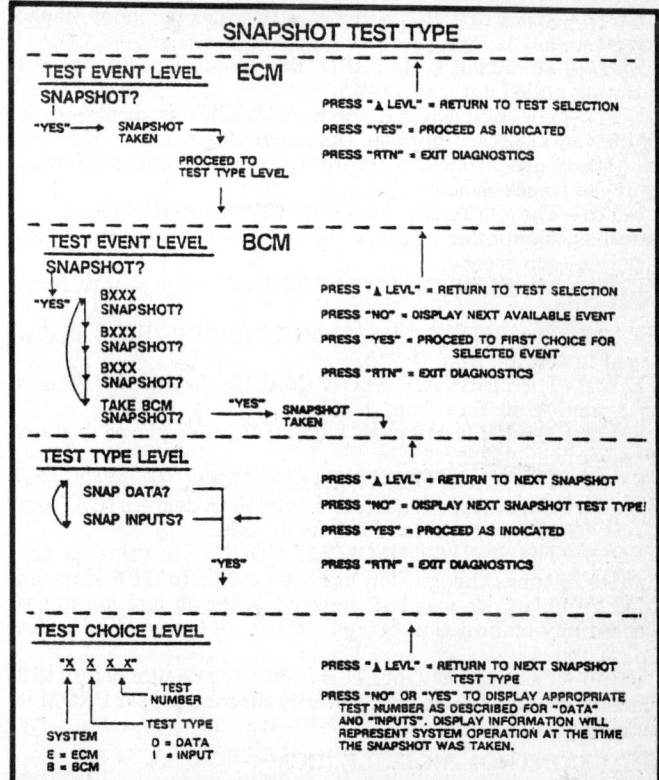

Self-diagnostic system flowchart (CRT) – 1989–90 Oldsmobile Toronado

If the button is released while – – is displayed, normal program control will resume and the display will again alternate.

The override test type is unique in that any other test type within the selected system may be active at the same time. After selecting an override test, pressing the OFF button will allow selection of another test type, DATA, INPUTS or OUTPUTS. The ECC panel will continue to display the selected override. By selecting another test type and test, while at the same time pressing the WARM or COOL button, it is possible to monitor the effects of the override on different vehicle parameters.

Selecting Clear Codes

Selecting reset codes will result in the message CLEAR ECM CODES? or CLEAR BCM CODES? depending which system was being tested. At this point, the following action may be taken:

1. Pressing the OFF button will stop the test selection process and return the display to the next available test type for the selected system.

2. Pressing the LO fan button will display the next test type available.

3. Pressing the HI fan button will select CLEAR CODES. A message ECM CODES CLEARED or BCM CODES CLEARED will appear to indicate those codes have been cleared from memory.

4. Pressing BI-LEV will exit diagnostics.

IPC Segment Check

Whenever the key is **ON**, pressing the SYSTEM MONITOR button on the left switch assembly will cause the IPC and the ECC to sequentially illuminate and darken all segments and

Selecting snapshot (CRT) – 1989–90 Oldsmobile Toronado

telltales in the clusters. This is helpful in determining if any bulbs or segments or the clusters vacuum fluorescent display are out or always on. To provide more time to study the various segments, whenever service diagnostics are entered, a total illumination of all segments and bulbs on the IPC and ECC will also occur.

Exiting Service Mode

To exit the service mode, press the BI-LEV button. Trouble codes are not erased when this is done. Any mode button will exit diagnostics, however, BI-LEV was chosen for procedural consistency.

DATA DISPLAYS

When troubleshooting a malfunction, the ECM and BCM data display can be used to compare the vehicle with problems to a vehicle which is functioning properly. A brief summary of each parameter is provided as follows:

ELECTRONIC CONTROL MODULE (ECM)

ED01 – The THROTTLE POSITION (TPS) is displayed in millivolts (mv) from 0–5100.

ED04 – The COOLANT TEMPERATURE is displayed in degrees (C.) from minus 40–152, or degrees (F.) from 40–306.

ED06 – The INJECTOR PULSE WIDTH is displayed in milliseconds (MS), from 0–1002.

ED07 – The OXYGEN SENSOR VOLTAGE is displayed in Millivolts (mv) from 0–1128 mv.

ED08 – The SPARK ADVANCE value is displayed in degrees, from 0–70.

ED10 – The BATTERY VOLTAGE is read in volts, from 0–25.5.

ED11 – The ENGINE SPEED is displayed in rpms from 0–6375.

ED12 – The VEHICLE SPEED is in miles per hour (mph) from 0–159.

ED16 – The KNOCK RETARD (ESC) is displayed in degrees of timing pulled out, from 0–45.

ED17 – The KNOCK SENSOR ACTIVITY is displayed as counts, an arbitrary number that increases from 0 –255 and then starts over (rollover) according to the amount of activity from the Knock Sensor.

ED18 – The OXYGEN SENSOR CROSS COUNTS is displayed as the number of times the O_2 sensor crossed the reference line each second.

ED19 – The FUEL INTEGRATOR displayed in counts, from 0–255.

ED20 – The BLOCK LEARN MULTIPLIER (BLM) is displayed in counts, from 0–255.

ED21 – The MASS AIR FLOW (MAF) is displayed in grams per second of air flow from 9–255.

ED22 – The IDLE AIR CONTROL (IAC) is displayed in steps of travel from 0–255.

ED23 – The MANIFOLD AIR TEMPERATURE (MAT) also called air temperature sensor, is displayed in degrees (C.) from minus 40–152, or degrees (F.) from 40–306.

ED98 – The IGNITION CYCLE COUNTER value is the number of times the ignition has been cycled to **OFF** since an ECM Trouble Code was last detected. After 50 ignition cycles without any malfunction being detected, all stored ECM codes are cleared.

ED99 – The ECM PROM I.D. is displayed as a number up to 3 digits long which can be used to verify that the proper PROM is installed in the ECM.

BODY CONTROL MODULE (BCM)

BD20 – The COMMANDED BLOWER VOLTAGE is read in volts 0 to system voltage.

BD21 – The COOLANT TEMPERATURE is displayed in degrees (°C) from minus 40–151, or degrees (°F) from 40–306. This value is sent from the ECM to the BCM. If this circuit malfunctions as determined by the ECM the ECM, will send the BCM a Failsoft value for display.

BD22 – The COMMANDER AIR MIX DOOR POSITION is displayed in percent (%). A value close to 0% represents a cold air mix and a value close to 100% represents a warm air mix.

BD23 – The ACTUAL AIR MIX DOOR POSITION is displayed in percent (%). This value should follow the commanded air mix door position (BD22) except when the door is commanded beyond its mechanical limits of travel.

BD24 – The AIR DELIVERY MODE is displayed as a number from 0–11. Each number is a code which represents the following air delivery modes:
- **0 – Auto-Recirculate Upper
- **1 – Auto-Outside Upper
- **2 – Auto-Bi-Level
- **3 – Auto-Heater-Defog
- **4 – Auto-Lower
- **5 – OFF
- **6 – Normal Purge
- **7 – Cold (DEF) Purge
- **8 – Defog
- **9 – Forced Lower
- **10 – Forced Upper
- **11 – Forced Bi-Level

BD25 – The IN-CAR TEMPERATURE is displayed in degrees (°C), from minus 40–102, or degrees (°F) from 40–215.

BD26 – The ACTUAL OUTSIDE TEMPERATURE is displayed in degrees (°C), from minus 40–93, or degrees (°F) from 40–140. This valve represents actual sensor temperatures and is not restricted by the features used to minimize engine heat affects on the customer display valve.

BD27 – The HIGH SIDE TEMPERATURE (condenser output) is displayed in degrees (°C) from minus 40–215, or degrees (°F) from 40–415.

BD28 – The LOW SIDE TEMPERATURE (evaporator input) is displayed in degrees (°C) from minus 40–93, or degrees (°F) 40–215.

BD32 – The SUNLOAD TEMPERATURE SENSOR is displayed in degrees (°C) from minus 40–102 , or degrees (°F) from 40 –215.

BD40 – The ACTUAL FUEL LEVEL is read in gallons from 0–17.0. This value represents actual sensor position and is not restricted by the features used to eliminate fuel slosh effects on the customer display value. The 0 and 17 represents the lower and upper limits of travel of the tank sensor. Some fuel is available above and below the sensor stops but is not measurable and varies.

BD41 – (1987 Models) The P-R-N-D-L POSITION SENSOR VALUE is displayed in percent (%). The signal is calculated as a percent of available reference voltage to determine the transaxle quadrant position. The value close to 100% represents P and a value close to 0% represents 1. (1988-89 Models) The P-R-N-OD-D-2-1 is displayed in a 4 digit binary number.
- **P – 0110
- **R – 0011
- **N – 1010
- **OD – 1001
- **D – 0000
- **2 – 0101
- **1 – 1100

BD42 – The DIMMING POT is displayed in percent (%). A value close to 0% represents maximum dimming and a value close to 100% represents maximum brightness.

BD43 – The TWILIGHT DELAY POT is displayed in percent (%). A value close to 0% represents minimum delay time and a value close to 100% represents maximum delay time.

BD44 – The TWILIGHT PHOTOCELL is displayed in percent (%). A value close to 0% represents daylight and value close to 100% represents darkness.

BD45 – The PHONE MODE is displayed as a number from 0–7. Each number is a code which represents the following phone modes:
- **0 – No phone in system
- **1 – OFF
- **2 – ON
- **3 – ON-no service
- **4 – ON-ROAM
- **5 – ON-Call in progress
- **6 – ON-Call received
- **7 – System Problem

BD50 – The BATTERY VOLTAGE is read in volts between 0 and 25.5.

BD51 – The GENERATOR FIELD is displayed in percent (%). A value close to 0% represents minimum regulator on time and a value close to 100% represents maximum regulator on time.

BD60 – The VEHICLE SPEED is displayed in miles per hours (mph) from 0–159.

BD61 – The ENGINE SPEED is displayed in rpms from 0–6375.

BD70 – The CRUISE SERVO POSITION is displayed in percent (%) from 0–100. A value close to 0% represents reset position and a value close to 100% represents W.O.T.

BD71 – The OIL PRESSURE SENSOR is displayed in pressure (psi) from 0–80.

BD90 – OPTION 1 is displayed a a number between 0–255.

BD91 – OPTION 2 is displayed as a number between 0–255.

NOTE: Options not designated as to numbers.

BD92 – The OPTION CONTENT No. 3 is displayed as a number. An 87 indicates the EEPROM was programmed by the assembly plant. This value will show up as some other number, other than the 87, if the odometer is ever re-programmed.

BD98 – The IGNITION CYCLE value is the number of times that the ignition has been cycled to **OFF** since a BCM Trouble

Code was last detected. After 99 ignition cycles, where each **OFF, CRANK, OFF** cycle counts as 2, without any malfunction being detected, all BCM codes are cleared.

BD99 – The BCM PROM I.D. is displayed as a number, up to 4 digits long, which can be used to verify that the proper PROM was installed in the BCM.

INPUT DISPLAYS

Input displays are operated as outlined under the heading of HOW TO OPERATE A SERVICE MODE. When troubleshooting a malfunction, the ECM, BCM or IPC input display can be used to determine if the switched inputs can be used to determine if the switched inputs can be properly interpreted. When 1 of the various input tests is selected, the state of that device is displayed as HI or LO. In general, the HI and LO refer to the input terminal voltage for that circuit. The display also indicates if the input changed state so that the technician could activate or deactivate any listed device and return to the display to see if it changed state. If a change of state occurred, an X will appear next to the HI/LO indicator. The X will only appear once per selected input, although the HI/LO indication will continue to change as the input changes. Some tests are momentary and the X can be used as and indication of a change. The following is a list of the ECM, BCM and IPC inputs:

ELECTRONIC CONTROL MODULE (ECM)

EI71 – The BRAKE SWITCH display is LO with the service brake depressed.

EI74 – The PARK–NEUTRAL (P/N) display is LO when the vehicle is in the **PARK** or **NEUTRAL** position.

EI78 – The POWER STEERING PRESSURE SWITCH display is LO when the power steering pressure (effort) is high, such as when the wheel is in a crimped position.

EI79 – The 2ND GEAR display is HI when the vehicle is in 2nd gear.

EI80 – The 3RD GEAR display is HI when the vehicle is in 3rd gear.

EI82 – The 4TH GEAR SWITCH is HI when the vehicle is in 4th gear.

BODY CONTROL MODULE (BCM)

BI01 – The COURTESY LAMP PANEL SWITCH display is LO when the courtesy lights are illuminated from the switch.

BI02 – The PARKLAMP SWITCH display is LO when the parklamp switch is in the **OFF** position.

BI03 – The DRIVER DOOR AJAR SWITCH display is LO when the driver's door is ajar.

BI04 – The PASSENGER DOOR AJAR SWITCH display is LO when the passenger's door is ajar.

BI05 – The DOOR JAMB SWITCH display is LO when the driver's or passenger's door is open.

BI06 – The DOOR HANDLE SWITCH display is momentarily LO when either outside door handle button is depressed.

BI08 – The LOW REFRIGERANT PRESSURE SWITCH display is LO when the system is low on refrigerant.

BI09 – The WASHER FLUID LEVEL SWITCH display is LO when the vehicle is low on washer fluid.

BI10 – The LOW COOLANT LEVEL display is LO when the engine is low on coolant.

BI18 – The brake pressure BRK PRESS display is LO when the ABS brake pressure is below a set value. This input has no meaning on non-ABS equipped vehicles.

BI21 – The LOW BRAKE FLUID display is LO when brake fluid level is low.

BI22 – The PARK BRAKE SWITCH display is LO when the parking brake is applied.

BI25 – The SEAT BELT SWITCH display is HI when the driver's seat belt is fastened.

BI51 – The GENERATOR FEEDBACK display is LO when there is a GENERATOR (alternator) problem (or engine not **RUNNING**).

BI71 – The CRUISE CONTROL BRAKE SWITCH display is HI when the cruise ON/OFF switch is **ON** and the brake pedal is not depressed (free state).

BI75 – The CRUISE CONTROL ON/OFF SWITCH display is HI when the switch is **ON**.

BI76 – The CRUISE CONTROL SET/COAST SWITCH display is HI when the cruise ON/OFF switch is **ON** and the set/coast switch is depressed.

BI77 – The CRUISE CONTROL RESUME/ACCEL SWITCH display is HI when the cruise ON/OFF switch is **ON** and the Resume/Accel switch is depressed.

BI78 – The HEADLAMP SWITCH display is HI whenever the headlamps are illuminated.

BI88 – The LOW OIL display will be LO when the engine oil level is low.

INSTRUMENT PANEL CLUSTER (IPC)

II01 – The FUEL RANGE DISPLAY IS LO whenever the fuel range button is pressed.

II02 – The FUEL REST display is LO whenever the fuel rest button is pressed.

II03 – The FUEL ECONOMY display is LO whenever the fuel economy button is pressed.

II04 – The FUEL USED display is LO whenever the fuel used button is pressed.

II05 – The ENGLISH-METRIC display is LO whenever the English-Metric button is pressed.

II06 – The SYSTEM MONITOR display is LO whenever the system monitor button is pressed.

II07 – The TRIP ODOMETER display is LO whenever the trip odometer button is pressed.

II08 – The EXPANDED FUEL GAUGE is LO whenever the fuel gauge button is pressed.

II09 – The TACHOMETER display is LO whenever the tachometer button is pressed.

II10 – The TRIP RESET display is LO whenever the trip reset button is pressed.

II11 – The INS ECON economy display is HI when the instant fuel economy button is depressed.

II12 – The ENG DATA input display is HI when the engine data button is depressed.

OUTPUT DISPLAYS

Output displays are operated as previously outlined under the heading HOW TO OPERATE THE SERVICE MODE. When troubleshooting a malfunction, the ECM and BCM output cycling can be used to determine if the output tests can be actuated regardless of the inputs and normal program instructions. Once a test in outputs has been selected, except for ECM IAC, the test will display HI and LO for 3 seconds in each state to indicated the command and output terminal voltage. A summary of each output is as follows:

ELECTRONIC CONTROL MODULE (ECM)

EO00 – The NO OUTPUTS display will not display HI or LO as this is a resting spot where no outputs can be cycled.

EO01 – The TORQUE CONVERTER CLUTCH (TCC) display will be LO when the TCC is energized.

EO04 – The EGR SOLENOID No. 1 display will be LO when the EGR solenoid No. 1 is energized.

EO05 – The EGR SOLENOID No. 2 display will be LO when the EGR solenoid No. 2 is energized.

EO06 – The EGR SOLENOID No. 3 display will be LO when the EGR solenoid No.3 is energized.

EO07 – The CANISTER PURGE SOLENOID display will be LO when the solenoid is ON (energized).

EO08 – The A/C CLUTCH display will be HI when the clutch is energized.

EO09 — The COOLANT FAN RELAY (FAN REL 1) display is LO when the coolant fan is energized.

EO10 — The COOLANT FAN RELAY (FAN REL 2) display will be LO when the pusher fan is energized.

EO11 — The IAC MOTOR display will be LO when the pintel is extended and HI during is retraction.

NOTE: This output test should not be activated if the IAC has been removed from the throttle body.

BODY CONTROL MODULE (BCM)

BO00 — The NO OUTPUTS display will not display HI or LO as this is a resting spot where no outputs will be cycled.

BO01 — The CRUISE CONTROL VENT SOLENOID display is HI when the vent solenoid is energized. The cruise ON/OFF switch must be **ON** for this output to cycle.

BO02 — The CRUISE CONTROL VACUUM SOLENOID display is HI when the vacuum solenoid is energized. The cruise ON/OFF switch must be **ON** for this output to cycle.

BO04 — The COURTESY RELAY display is LO when the relay is energized.

BO05 — The TWILIGHT HEADLAMP RELAYS display is LO when the relays are energized and the lights illuminated.

BO10 — The CHIME 1 display is LO when the intermediate chime is sounding.

BO12 — The CHIME 2 display is LO when the slow chime is sounding.

OVERRIDE DISPLAY

When troubleshooting a malfunction, the BCM override feature allows testing of certain system functions regardless of normal program instructions.

ELECTRONIC CONTROL MODULE (ECM)

ES00 — The NO OVERRIDES display will not display any value as this is a resting spot where no overrides will be controlled. This is a handy spot to go to stop the override without having to go back to the level where the system is selected.

ES01 — The TCC SOLENOID display will show the state of the command (00 not engaged; 99 engaged). Pressing the up arrow will engage the TCC and pressing the down arrow will disengage the TCC. The vehicle must be in **3rd** or **4th** gear for this override to operate.

ES02 — The EGR No. 1 SOLENOID can be turned OFF 00 (OFF) by pressing the down arrow, or ON 99 (EGR FLOW) by pressing the up arrow.

ES03 — The EGR No. 2 SOLENOID can be turned OFF 00 (OFF) by pressing the down arrow, or ON 99 (EGR FLOW) by pressing the up arrow.

ES04 — The EGR No. 3 SOLENOID can be turned OFF 00 (OFF) by pressing the down arrow, or ON 99 (EGR FLOW) by pressing the up arrow.

ES05 — The CANISTER PURGE SOLENOID can be turned OFF 00 by pressing the down arrow, or ON 99 (purging) by pressing the up arrow.

ES06 — The A/C CLUTCH RELAY can be turned OFF 00 (OFF) by pressing the down arrow, or ON 99 (engaged) by pressing the up arrow.

ES07 — The COOLING FAN RELAY NO. 1 (puller) can be turned OFF 00 (OFF) by pressing the down arrow, or ON 99 (engaged) by pressing the up arrow.

ES08 — The COOLING FAN RELAY NO. 2 (pusher) can be turned OFF 00 (OFF) by pressing the down arrow, or ON 99 (engaged) by pressing the up arrow.

ES09 — The IAC MOTOR display will show the position the IAC is being commanded to in a percentage of travel. In order for this override to work the vehicle must be in **P** or **N**. The down arrow will retract the motor and the up arrow will extend the motor.

ES10 — The INJECTOR CONTROL OVERRIDE will work only when the vehicle is in **P** or **N**. Upon entering injector override the display will show 0, all injectors working under normal program control. The arrow up button will increment the display by 1, up to 6 and wrap around to 0. This display number indicates the cylinder number of the injector that is being held OFF (e.g., 4 = injector in cylinder No. 4 OFF). The corresponding injector to the number displayed will be held OFF for as long as the number is displayed. The arrow down button is used to turn the injector back ON and then OFF (toggle effect); i.e., if display shows 4, injector 4 is OFF. Touching arrow down will turn injector 4 ON, and the display will show 0. Touching arrow down again will turn injector 4 back OFF and display will show a 4. Arrow down toggles the injector last displayed ON/OFF.

BODY COMPUTER MODULE (BCM)

BS00 — The no overrides display will not display any value as this is a resting spot where no overrides will be controlled. This is a handy spot to go to stop the override without having to go back to the level where the system is selected.

BS01 — PROGRAM NUMBER. When program number is first selected, the ECC will display the program number which is currently being used by the climate control system. As operating conditions change, this number will automatically change in response. The automatic calculation of program number can be overridden with the WARM and COOL buttons. A value of 99 represents maximum heating where 0 is maximum cooling. This always control of the program number to any number from 0–99 and while simultaneously observing the reaction of any of the BCM data parameters.

BS03 — PANEL DIMMING. When panel dimming is first selected, the ECC will display the number which is currently being used by the BCM to control panel dimming. As the dim pot is moved this number will automatically change in response. The automatic control of dimming can be bypassed by a manual override feature using the WARM and COOL buttons 99 represents the max brightness of the VF and incandescent displays and 0 represents the max dimming of panel lighting. (Outside lights do not have to be on to override dimming). This manual override of dimming will continue until the override is stopped. This is done in 1 of the ways detailed in selecting the test choice. This way, the technician can control the dimming to any number from 0–99.

BS05 — CRUISE CONTROL. The cruise control ON/OFF switch must be engaged for this test to work. When cruise control override is first selected, the ECC will display the servo position number which is currently being used by the cruise control system. As operating conditions change, this number will automatically change in response. The automatic calculation of cruise control position number can be bypassed by a manual override feature using the WARM and COOL buttons on the ECC panel. 99 represents the max cruise position (WOT) and 0 represents the minimum cruise position (closed throttle). This manual override of automatic cruise control position number calculation will continue until the override is stopped. Selecting the Test choice in Override Test Type. This allows the control of the cruise control position to any number from 0–99 and simultaneously observe the reaction of the data value for cruise servo position (BD70). Both should agree with each other within ± 3%, except when the servo is commanded beyond its mechanical limits of travel. Since this test is not allowed with engine **RUNNING**, remember that there is usually only enough vacuum in the storage tank to stroke the servo 3 times.

BS06 — BLOWER MOTOR. When BS06 is first selected, the ECC will display the number for blower speed that the BCM is currently using. This can be overridden using the WARM and COOL buttons on the ECC panel. 99 indicates maximum blower speed while 0 indicates minimum blower speed. This always control of blower speed over its entire range plus the ability to observe other DATA values while varying blower speed.

BS07 — Option No. 1. Programmed at factory. No codes available.

BS08 — Option No. 2. Programmed at factory. No codes available.

Electronic Ignition Systems

SECTION 4

INDEX

HIGH ENERGY IGNITION (HEI) SYSTEM 4-2
  General Information 4-2
  Troubleshooting 4-2
  Distributor Component Testing 4-3
    Coil in Cap Distributor 4-3
    Distributor with Separate Coil-Type 1 4-4
    Distributor with Separate Coil-Type 2 4-6
    Distributor with Separate Coil-Type 3 4-7
    Distributor with Separate Coil-Type 4-Tang Drive 4-9

ELECTRONIC SPARK TIMING (EST) SYSTEM 4-10
  General Description 4-10
    Hall Effect Switch 4-10
    How Code 42 is Determined 4-10
    Code 12 4-11

ELECTRONIC SPARK CONTROL (ESC) SYSTEM 4-13
  General Description 4-13
    Code 43 4-14
    Diagnosis 4-14
    Basic Ignition Timing 4-14

COMPUTER CONTROLLED COIL IGNITION (C³I) SYSTEM ... 4-15
DIRECT IGNITION SYSTEM (DIS) 4-15
INTEGRATED DIRECT IGNITION (IDI) SYSTEM 4-15
ELECTRONIC SPARK CONTROL (ESC) 4-15
  General Description 4-15
    Secondary System 4-15
    The Primary Control and Triggering System 4-17
  C³I System Components 4-24
    Ignition Coils 4-24
    C³I Module 4-24
    Camshaft Sensor 4-24
    Crankshaft Sensor 4-24
    Combination Sensor 4-25
    Electronic Control Module (ECM) 4-25
    Electronic Spark Control (ESC) 4-25
  C³I Electronic Spark Timing (EST) Circuits 4-27
  Diagnostics and Testing 4-27
  Direct Ignition System Components 4-27
    Crankshaft Sensor 4-27
    Ignition Coils 4-28
    DIS Module 4-28
    Direct Ignition Electronic Spark Timing (EST) Circuits ... 4-28
  Diagnosis and Testing 4-31
    Direct Ignition System 4-31
    Checking EST Performance 4-31
    Crankshaft Sensor R&R 4-32
    Camshaft Position Sensor R&R 4-32
  DIS Modifications 4-34

HIGH ENERGY IGNITION (HEI) SYSTEM

NOTE: This manual has the ability to save time in diagnosis and prevent the replacement of good parts. The key to using this manual successfully for diagnosis, lies in the technician's ability to understand the system he is trying to diagnose as well as an understanding of the manual's layout and limitations. The technician should review this manual to become familiar with the way this manual should be used.

General Information

The High Energy Ignition (HEI) distributor is still being used on some engines. The ignition coil is either mounted to the top of the distributor cap or is externally mounted on the engine, having a secondary circuit high tension wire connecting the coil to the distributor cap and interconnecting primary wiring as part of the engine harness.

The HEI distributor is equipped to aid in spark timing changes, necessary for emissions, economy and performance. This system is called the Electronic Spark Timing Control (EST). The HEI distributors use a magnetic pick-up assembly, located inside the distributor containing a permanent magnet, a pole piece with internal teeth and a pick-up coil. When the teeth of the rotating timer core and pole piece align, an induced voltage in the pick-up coil signals the electronic module to open the coil primary circuit. As the primary current decreases, a high voltage is induced in the secondary windings of the ignition coil, directing a spark through the rotor and high voltage leads to fire the spark plugs. The dwell period is automatically controlled by the electronic module and is increased with increasing engine rpm. The HEI system features a longer spark duration which is instrumental in firing lean and EGR (Exhaust Gas Recirculation) diluted fuel/air mixtures. The condenser (capacitor) located within the HEI distributor is provided for noise (static) suppression purposes only and is not a regularly replaced ignition system component.

All spark timing changes in the HEI (EST) distributors are done electronically by the Electronic Control Module (ECM), which monitors information from the various engine sensors, computes the desired spark timing and signals the distributor to change the timing accordingly. With this distributor, no vacuum or centrifugal advances are used.

Troubleshooting

NOTE: An accurate diagnosis is the first step to problem solution and repair. For several of the following steps, a HEI spark tester, tool ST 125, which has a spring clip to attach it to ground. Use of this tool is recommended, as there is more control of the high energy spark and less chance of being shocked. If a tachometer is connected to the TACH terminal on the distributor, disconnect it before proceeding with this test.

SECONDARY CIRCUIT

Testing

SECONDARY SPARK

1. Check for spark at the spark plugs by attaching the HEI spark tester, tool ST 125, to one of the plug wires, grounding the HEI spark tester on the engine and cranking the starter.
2. If no spark occurs on one wire, check a second. If spark is present, the HEI system is good.
3. Check fuel system, plug wires, and spark plugs.
4. If no spark occurs from EST distributor, disconnect the 4 terminal EST connector and recheck for spark. If spark is present, EST system service check should be performed.

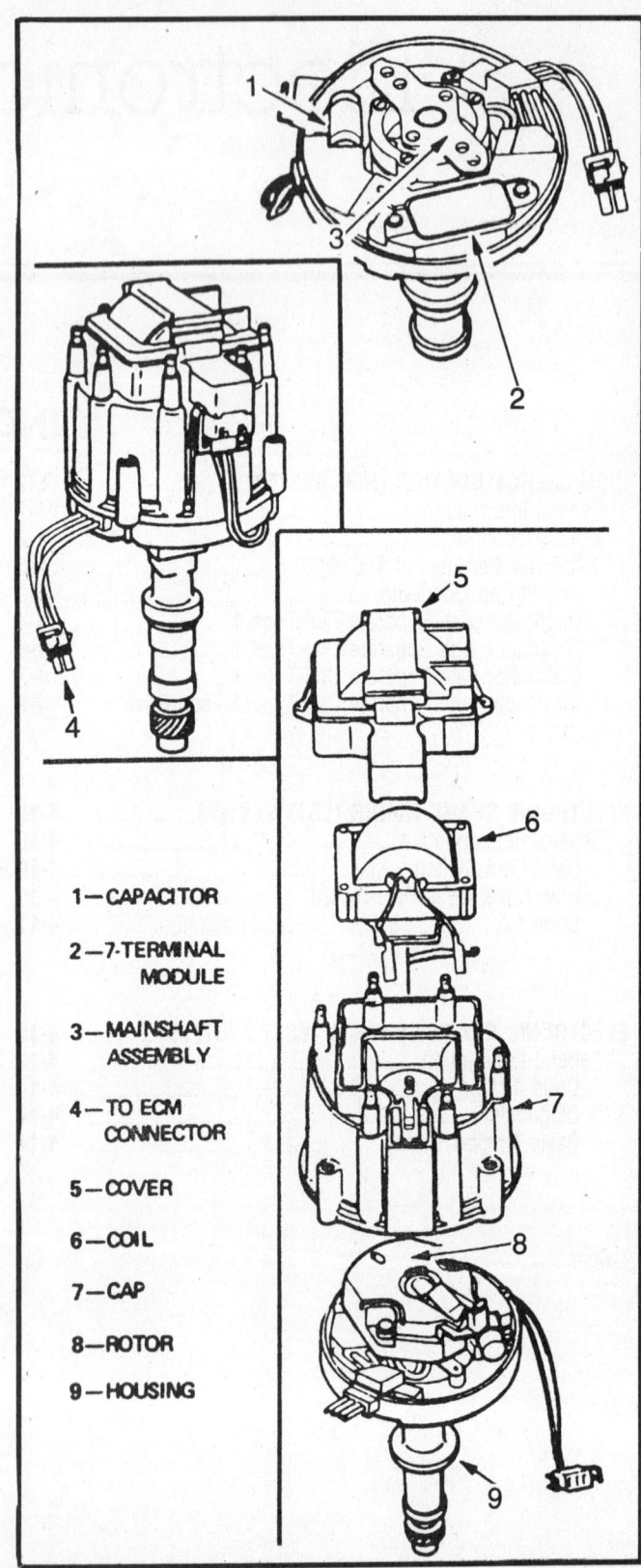

1 — CAPACITOR

2 — 7-TERMINAL MODULE

3 — MAINSHAFT ASSEMBLY

4 — TO ECM CONNECTOR

5 — COVER

6 — COIL

7 — CAP

8 — ROTOR

9 — HOUSING

Typical HEI distributor assembly

NOTE: Before making any circuit checks with test meters, be sure that all primary circuit connectors are properly installed and that spark plug cables are secure at the distributor and at the plugs.

Distributor Component Testing

COIL IN CAP DISTRIBUTOR

Testing

IGNITION COIL

1. Remove the cap from the distributor.
2. With coil attached to cap, connect an ohmmeter to the distributor cap terminals C and ground. The reading should be zero or nearly zero. If not, replace the ignition coil.
3. Position the ohmmeter leads to terminal B+ and the high tension rotor contact in the center of the cap.
4. Using the ohmmeter high scale, measure the resistance. Reverse the ohmmeter leads and again measure the resistance.
5. Replace the coil ONLY if BOTH readings are infinite.

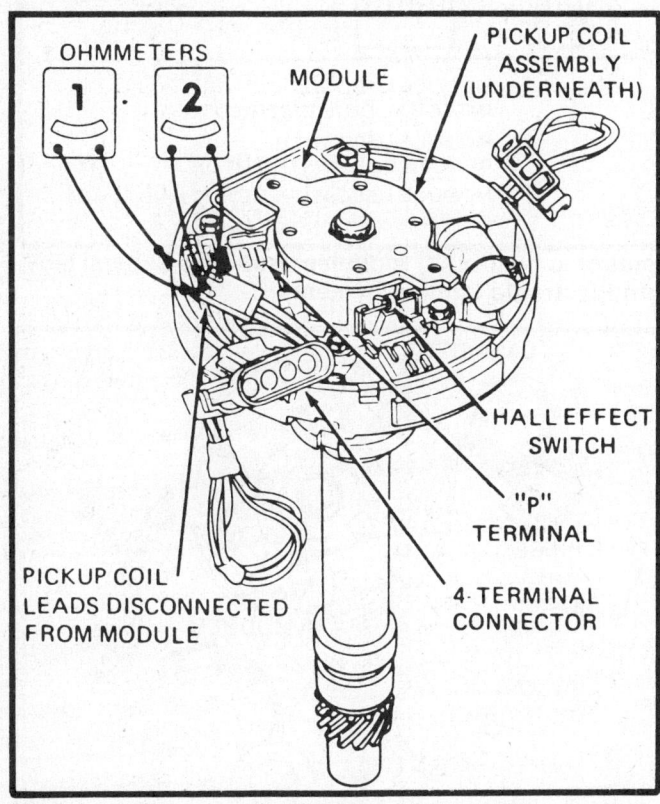

Testing coil-in-cap distributor with Hall Effect switch

PICK-UP COIL

1. Disconnect the rotor and pick-up coil leads from the module.
2. Using an ohmmeter, connect one lead to the distributor housing and the second lead to one of the pick-up terminals in the connector.
3. The reading should be infinite.

NOTE: While testing, flex the leads to determine if wire breaks are present under the wiring insulation.

4. Place the ohmmeter leads into both the pick-up terminals of the connector and measure the resistance.
5. The reading should be steady at one value, between 500–1500 ohms.

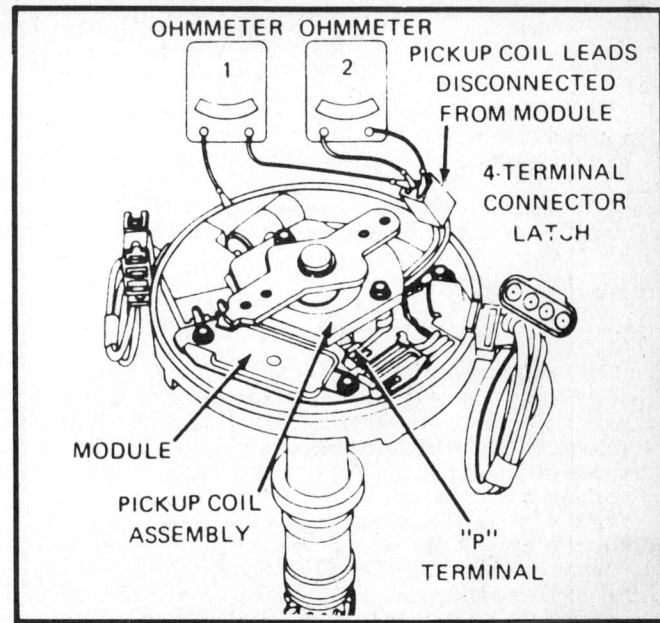

Testing pick-up coil used in coil-in-cap distributor

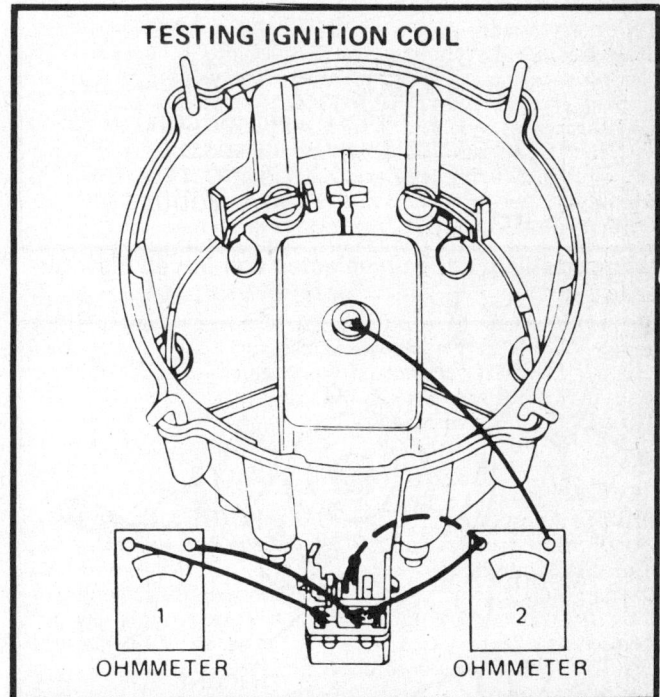

Testing ignition coil used in coil-in-cap distributor

IGNITION MODULE

Because of the complexity of the internal circuitry of the HEI/EST module, it is recommended the module be tested with an accurate module tester.

It is imperative that silicone lubricant be used under the module when it is installed, to prevent module failure due to overheating.

NOTE: The module can be replaced without distributor disassembly. However, the distributor must be disassembled to replace the pick-up coil, magnet and pole piece.

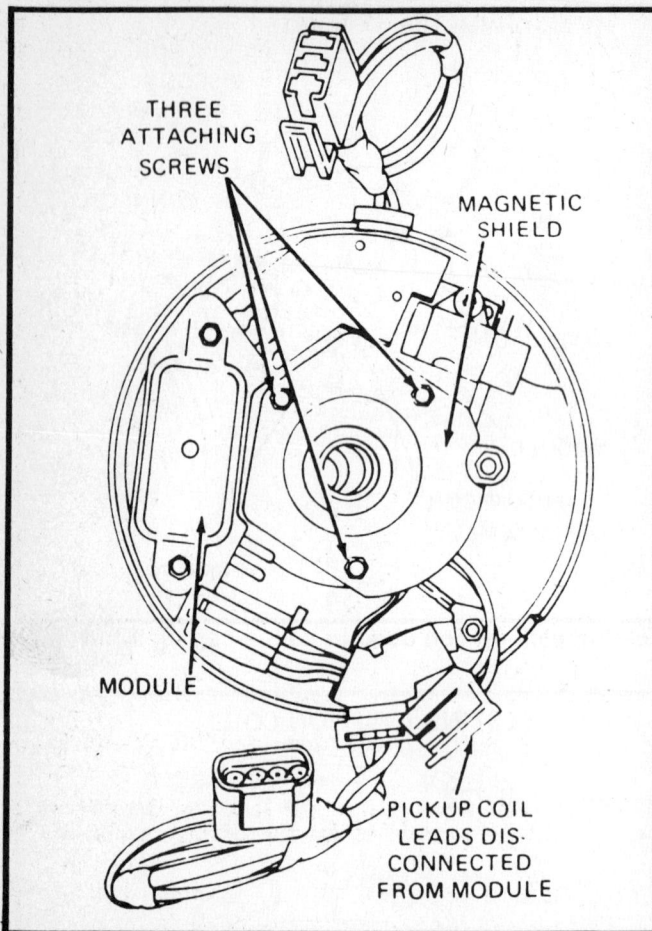

Magnetic shield used on selected coil-in-cap distributors

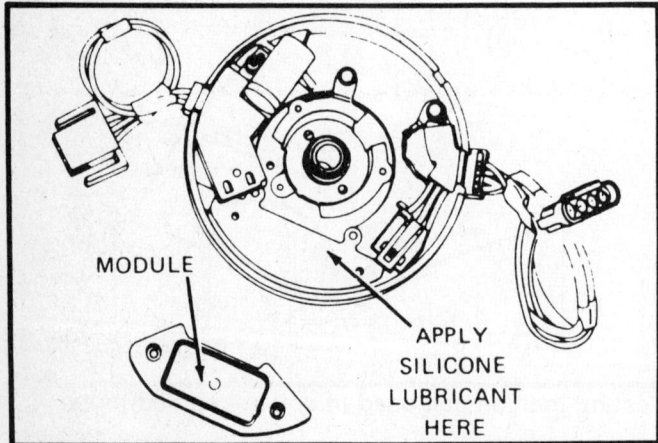

Module replacement and use of silicone lubricant, coil-in-cap distributor

DISTRIBUTOR WITH SEPARATE COIL TYPE 1

NOTE: This type distributor has no vacuum or centrifugal advance mechanism and has the pick-up coil mounted above the module. This distributor is used with the EST system.

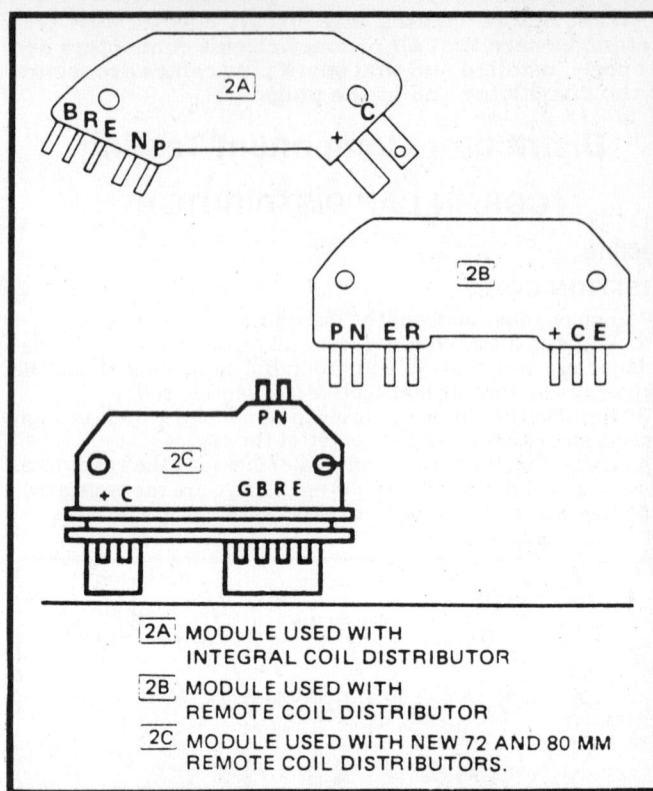

2A MODULE USED WITH INTEGRAL COIL DISTRIBUTOR

2B MODULE USED WITH REMOTE COIL DISTRIBUTOR

2C MODULE USED WITH NEW 72 AND 80 MM REMOTE COIL DISTRIBUTORS.

Electronic distributor modules used with General Motors Electronic Ignition systems

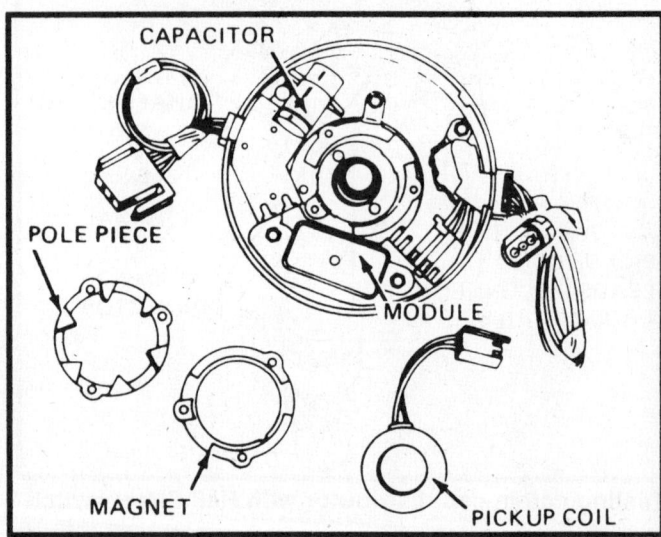

Pick-up coil components, coil-in-cap distributor

Testing

IGNITION COIL

1. Disconnect the primary wiring connectors and secondary coil wire from the ignition coil.
2. Using an ohmmeter on the high scale, connect one lead to a grounding screw and the second lead to one of the primary coil terminals.
3. The reading should be infinite. If not, replace the ignition coil.
4. Using the low scale, place the ohmmeter leads on both the primary coil terminals.

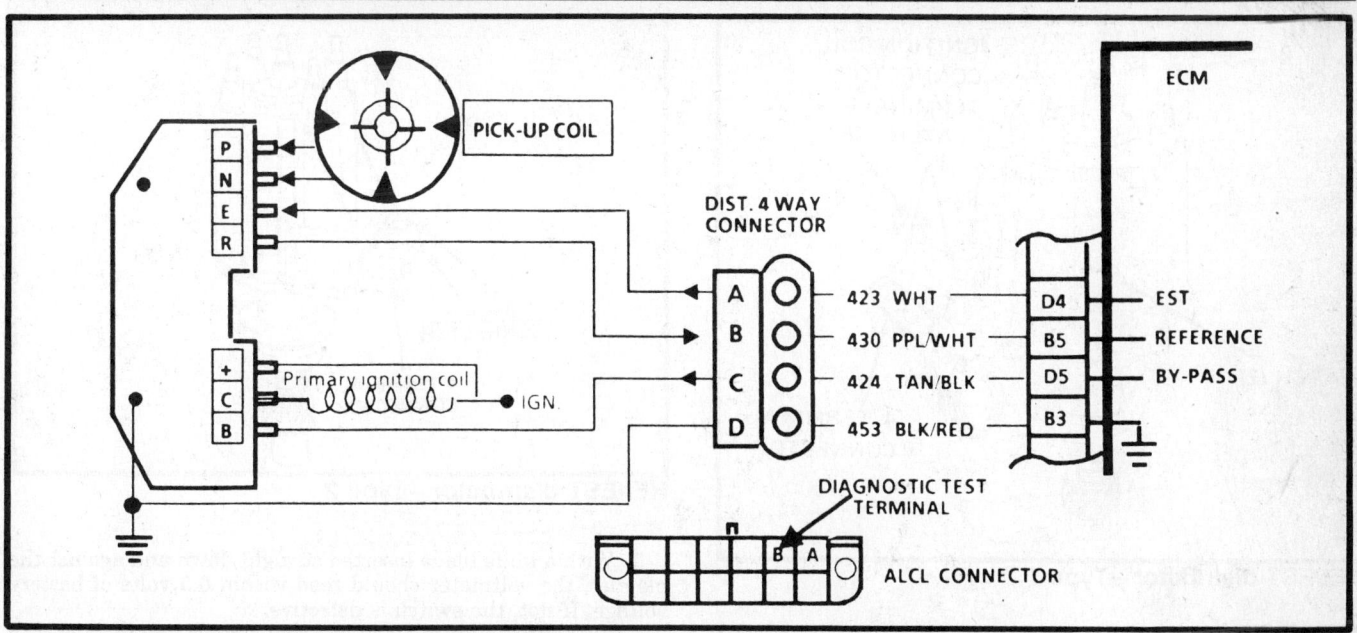

Typical wiring schematic for electronic ignition system using remote ignition coil

5. The reading should be very low or zero. If not, replace the ignition coil.

6. Using the high scale, place one ohmmeter lead on the high tension output terminal and the other lead on a primary coil terminal.

7. The reading should NOT be infinite. If it is, replace the ignition coil.

PICK-UP COIL

1. Remove the rotor and pick-up coil leads from the module.

2. Using an ohmmeter, attach one lead to the distributor base and the second lead to one of the pick-up coil terminals of the connector.

3. The reading should be infinite at all times.

4. Position both leads of the ohmmeter to the pick-up terminal ends of the connector.

5. The reading should be a steady value between 500–1500 ohms.

6. If not within the specification value, the pick-up coil is defective.

NOTE: While testing, flex the leads to determine if wire breaks are present under the wiring insulation.

IGNITION MODULE

Because of the complexity of the internal circuitry of the HEI/EST module, it is recommended the module be tested with an accurate module tester.

It is imperative that silicone lubricant be used under the module when it is installed, to prevent module failure due to overheating.

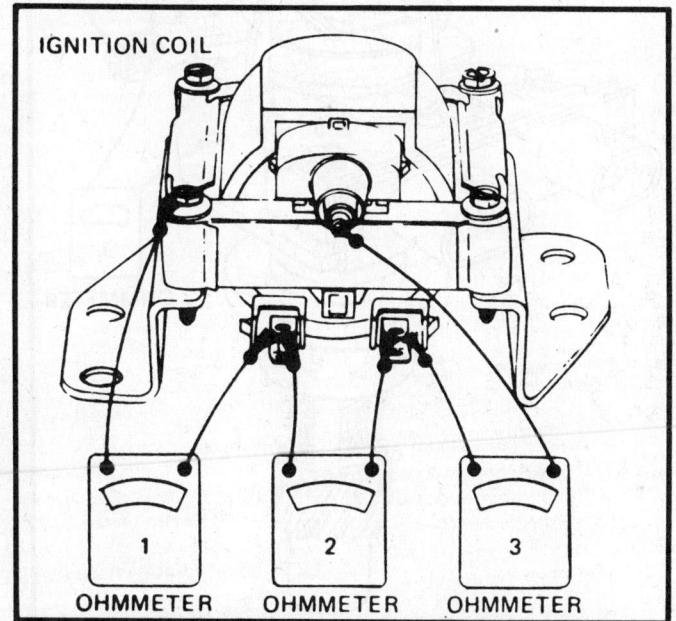

Testing ignition coil—Type 1 distributor

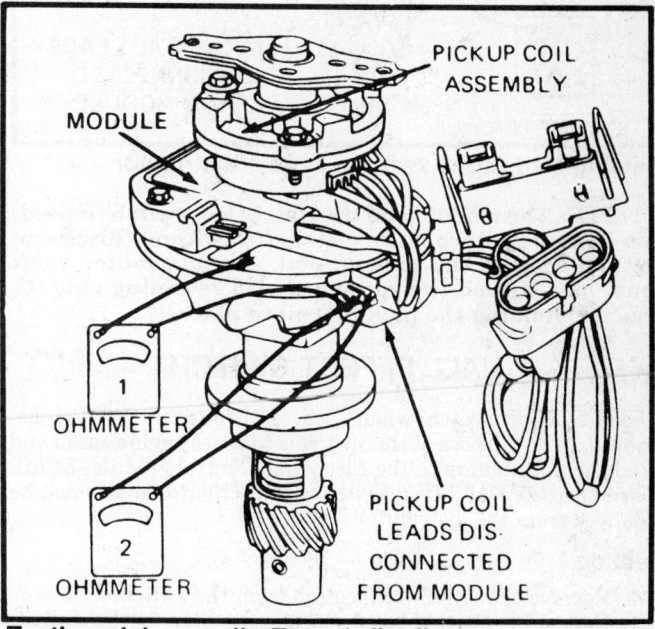

Testing pick-up coil—Type 1 distributor

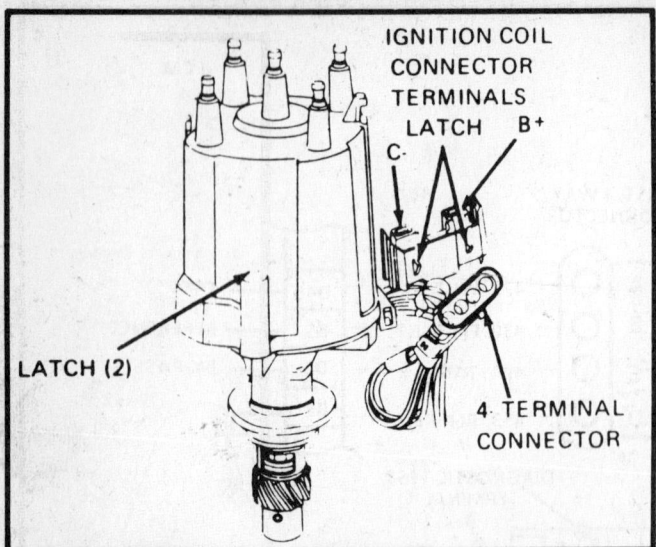

HEI/EST distributor – Type 1

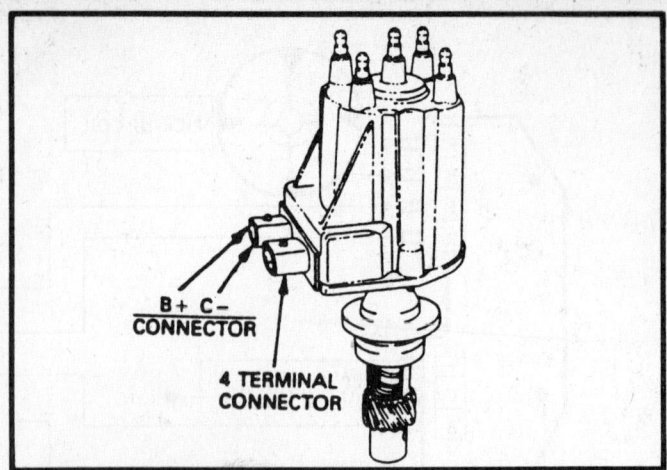

HEI/EST distributor – Type 2

3. With a knife blade inserted straight down and against the magnet, the voltmeter should read within 0.5 volts of battery voltage. If not, the switch is defective.

4. Without the knife blade inserted against the magnet, the voltmeter should read less than 0.5 volts. If not, the switch is defective.

DISTRIBUTOR WITH SEPARATE COIL TYPE 2

NOTE: This type distributor has no vacuum or centrifugal advance mechanisms and the module has 2 outside terminal connections for the wiring harness. This distributor is used with the EST system.

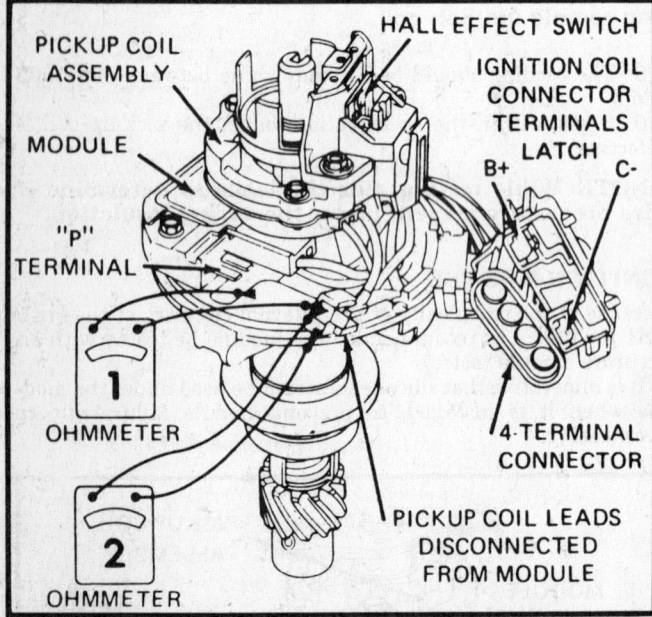

Testing Hall Effect switch – Type 1 distributor

NOTE: The module and the Hall Effect switch (if used) can be removed from the distributor without disassembly. To remove the pick-up coil, the distributor shaft must be removed to expose a waved retaining ring (C-washer) holding the pick-up coil in place.

HALL EFFECT SWITCH

The Hall Effect switch, when used, is installed in the HEI distributor. The purpose of the switch is to sense engine speed and send the information to the Electronic Control Module (ECM). To remove the Hall Effect switch, the distributor shaft must be removed from the distributor.

Testing

1. Remove the switch connectors from the switch.
2. Connect a 12 volt battery and voltmeter to the switch. Note and follow the polarity markings.

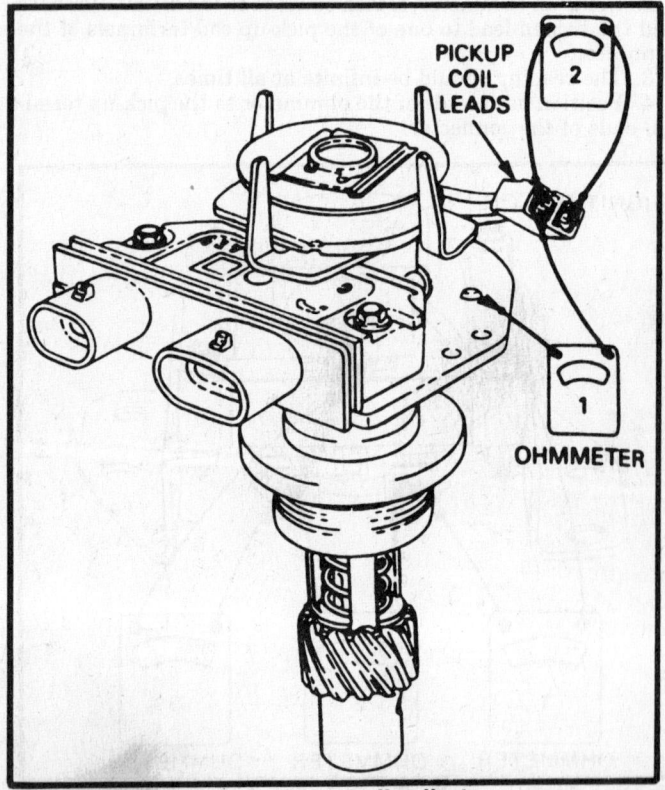

Testing pick-up coil – Type 2 distributor

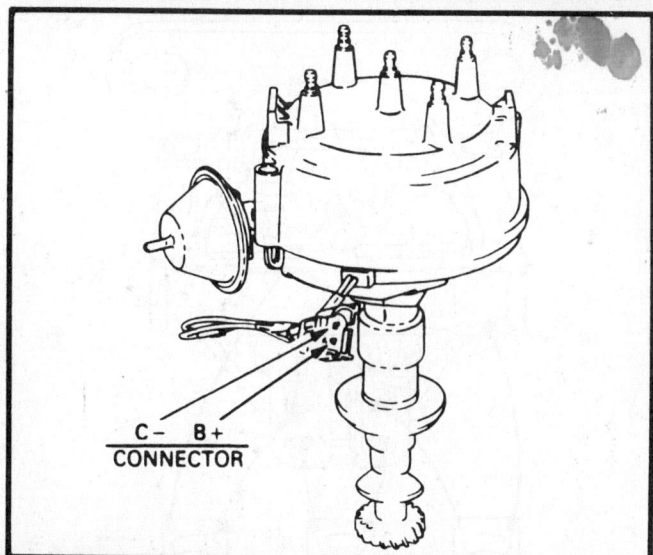

HEI distributor—Type 3

Testing

IGNITION COIL

1. Using an ohmmeter set on the high scale, place one lead on a ground of the ignition coil.
2. Place the second lead into one of the rearward terminals of the ignition coil primary connector.
3. The ohmmeter scale should read infinite. If not, replace the ignition coil.
4. Using the low scale, place the ohmmeter leads into each of the outer terminals of the coil connector.
5. The reading should be zero or very low. If not, replace the ignition coil.
6. Using the high scale, place one ohmmeter lead on the coil secondary terminal and the second lead into the rearward terminal of the ignition coil primary connector.
7. The reading should not be infinite. If so, replace the ignition coil.

PICK-UP COIL

1. Remove the rotor and pick-up leads from the module.
2. Using an ohmmeter, connect one of the leads to the distributor base.
3. Connect the second lead to one of the pick-up coil lead terminals
4. The reading should be infinite. If not, the pick-up coil is defective.

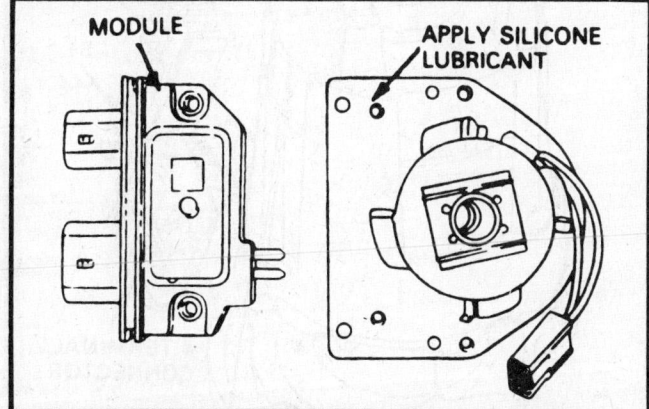

Module replacement and use of silicone lubricant— Type 2 distributor

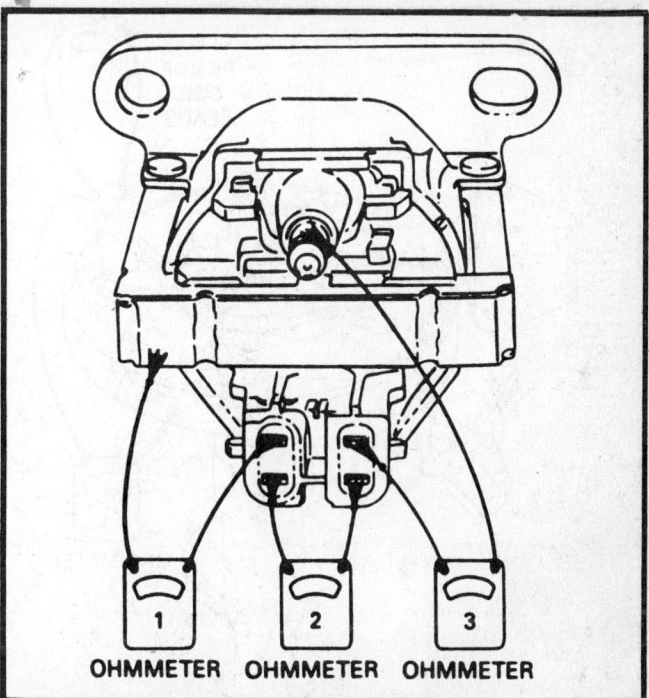

Testing ignition coil—Type 2 distributor

NOTE: During the ohmmeter tests, flex the leads by hand to check for intermediate opens in the wiring.

5. Connect both ohmmeter lead to the pick-up coil terminals at the connector.
6. The reading should be of one steady value, between 500–1500 ohms.
7. If the reading is not within specifications, the pick-up coil must is defective.

IGNITION MODULE

Because of the complexity of the internal circuitry of the HEI/EST module, it is recommended the module be tested with an accurate module tester.

It is imperative that silicone lubricant be used under the module when it is installed, to prevent module failure due to overheating.

NOTE: The module can be removed without distributor disassembly. To remove the pick-up coil, the distributor shaft must be removed. A retainer can then be removed from the top of the pole piece and the pick-up coil removed.

DISTRIBUTOR WITH SEPARATE COIL TYPE 3

NOTE: This distributor uses vacuum and centrifugal advance units and does not have the EST system.

Testing

IGNITION COIL

1. Disconnect the primary wiring connectors and secondary coil wire from the ignition coil.
2. Using an ohmmeter on the high scale, connect one lead to a grounding screw and the second lead to one of the primary coil terminals.
3. The reading should be infinite. If not, replace the ignition coil.

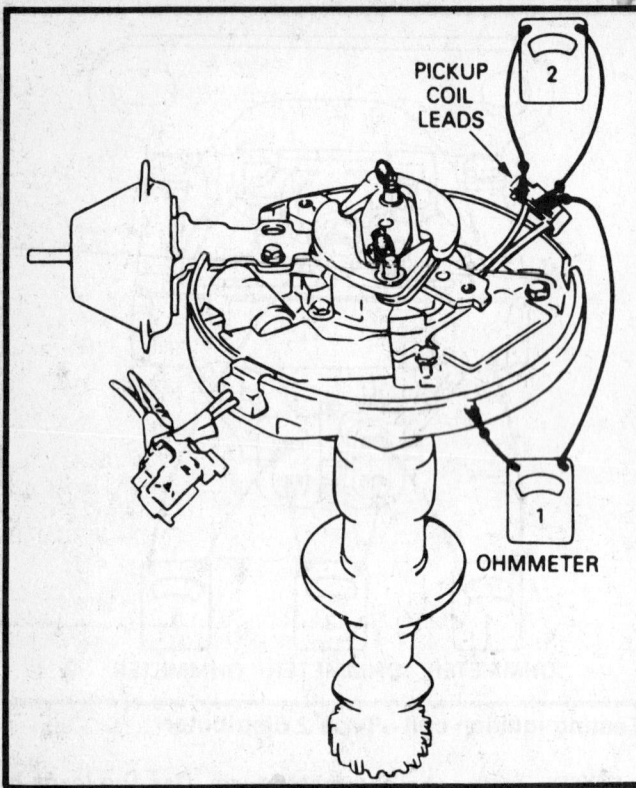

Testing the pick-up coil—Type 3 distributor

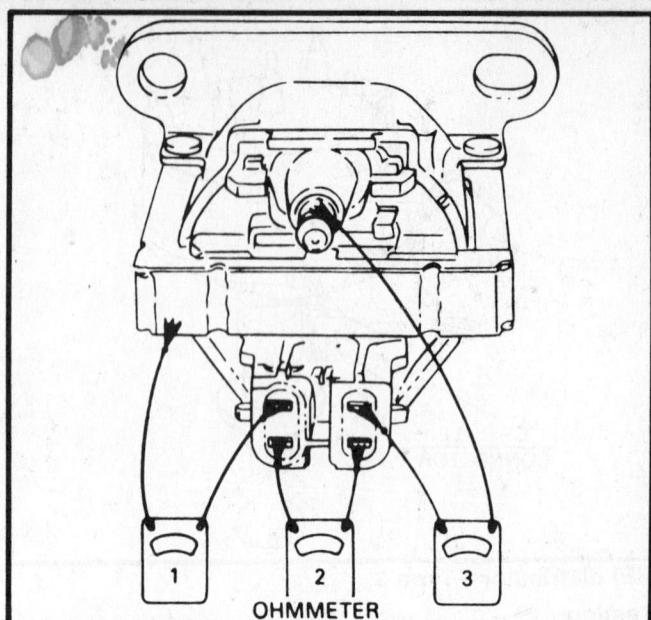

Testing ignition coil—Type 3 distributor

DISTRIBUTOR WITH SEPARATE COIL TYPE 4—TANG DRIVE

NOTE: This distributor is used with the EST system. The unit is mounted horizontally to the valve cover housing and is driven by the camshaft, through a tang on the distributor shaft.

Testing

IGNITION COIL

1. Using an ohmmeter set on the high scale, place one lead on a ground of the ignition coil.
2. Place the second lead into one of the rearward terminals of the ignition coil primary connector.
3. The ohmmeter scale should read infinite. If not, replace the ignition coil.

4. Using the low scale, place the ohmmeter leads on both the primary coil terminals.
5. The reading should be very low or zero. If not, replace the ignition coil.
6. Using the high scale, place one ohmmeter lead on the high tension output terminal and the other lead on a primary coil terminal.
7. The reading should NOT be infinite. If it is, replace the ignition coil.

PICK-UP COIL

1. Remove the rotor and pick-up coil leads from the module.
2. Using an ohmmeter, attach one lead to the distributor base and the second lead to one of the pick-up coil terminals of the connector.
3. The reading should be infinite at all times.
4. Position both leads of the ohmmeter to the pick-up terminal ends of the connector.
5. The reading should be a steady value between 500–1500 ohms.
6. If not within the specification value, the pick-up coil is defective.

NOTE: While testing, flex the leads to determine if wire breaks are present under the wiring insulation.

IGNITION MODULE

Because of the complexity of the internal circuitry of the HEI/EST module, it is recommended the module be tested with an accurate module tester.

It is imperative that silicone lubricant be used under the module when it is installed, to prevent module failure due to overheating.

NOTE: The module can be removed without distributor disassembly. The distributor shaft and C-clip must be removed before the pick-up coil can be removed.

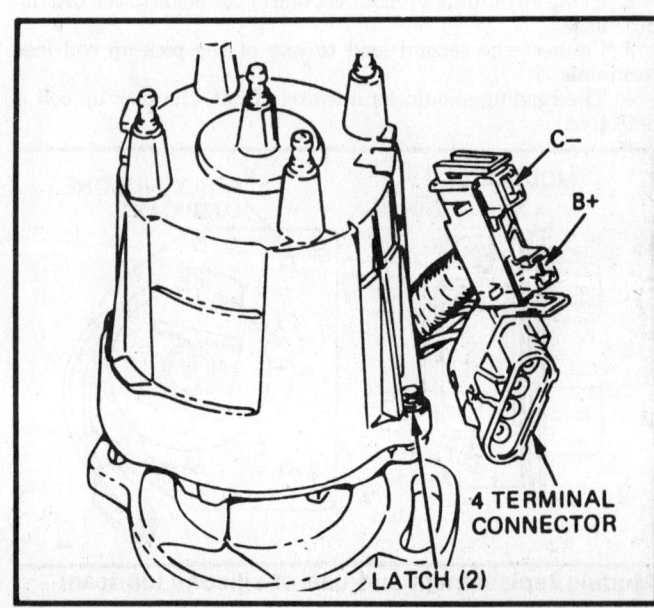

HEI/EST distributor—Type 4 with tang drive

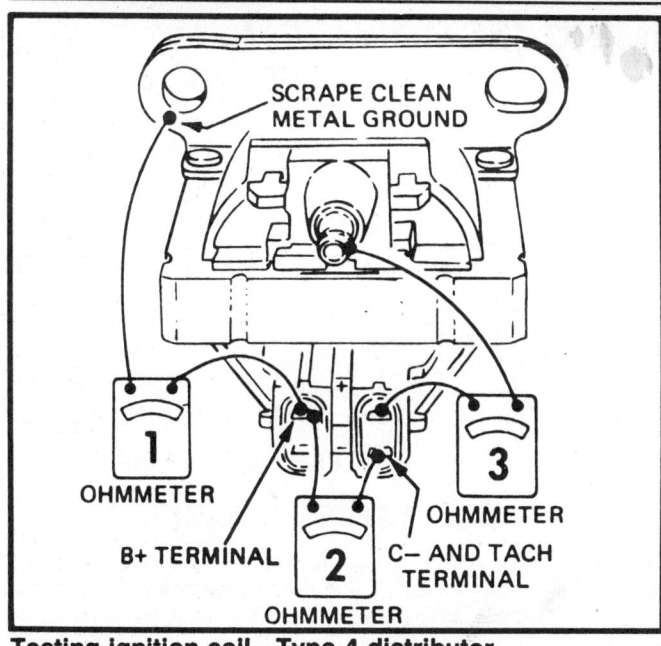

Testing ignition coil—Type 4 distributor

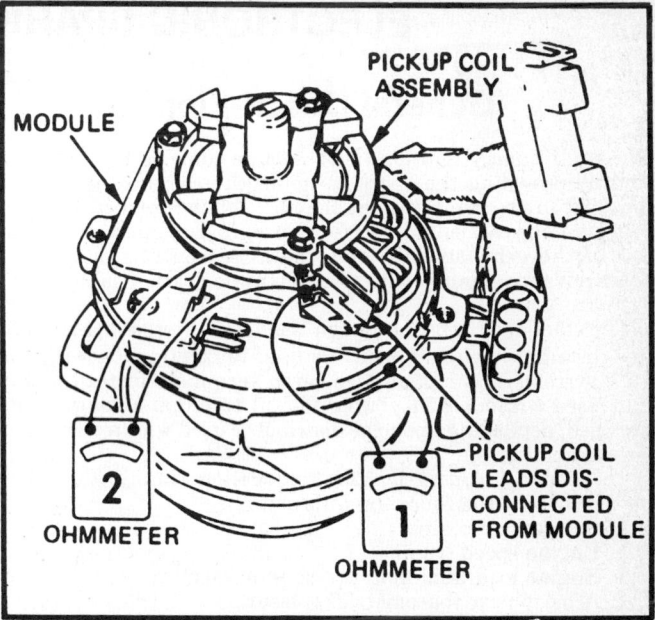

Testing pick-up coil—Type 4 distributor

4. Using the low scale, place the ohmmeter leads into each of the outer terminals of the coil connector.

5. The reading should be zero or very low. If not, replace the ignition coil.

6. Using the high scale, place one ohmmeter lead on the coil secondary terminal and the second lead into the rearward terminal of the ignition coil primary connector.

7. The reading should not be infinite. If so, replace the ignition coil.

PICK-UP COIL

1. Remove the rotor and pick-up leads from the module.

2. Using an ohmmeter, connect one of the leads to the distributor base.

3. Connect the second lead to one of the pick-up coil lead terminals

4. The reading should be infinite. If not, the pick-up coil is defective.

NOTE: During the ohmmeter tests, flex the leads by hand to check for intermediate opens in the wiring.

5. Connect both ohmmeter lead to the pick-up coil terminals at the connector.

6. The reading should be of one steady value, between 500–1500 ohms.

7. If the reading is not within specifications, the pick-up coil must is defective.

IGNITION MODULE

Because of the complexity of the internal circuitry of the HEI/EST module, it is recommended the module be tested with an accurate module tester.

It is imperative that silicone lubricant be used under the module when it is installed, to prevent module failure due to overheating.

NOTE: The module can be removed without distributor disassembly. The distributor shaft and C-clip must be removed before the pick-up coil can be removed. Before removing the roll pin from the distributor tang drive to shaft, a spring must first be removed.

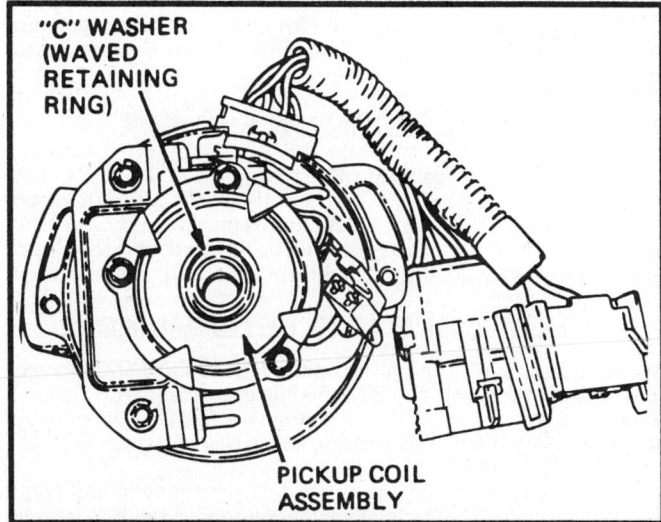

Replacement of pick-up coil with distributor shaft removed—Type 4 distributor

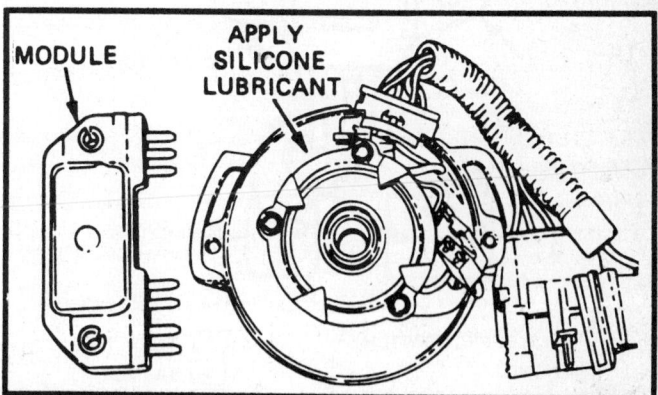

Module replacement and use of silicone lubricant—Type 4 distributor

ELECTRONIC SPARK TIMING (EST) SYSTEM

General Description

The High Energy Ignition (HEI) system controls fuel combustion by providing the spark to ignite the compressed air/fuel mixture, in the combustiuon chamber, at the correct time. To provide improved engine performance, fuel economy and control of the exhaust emissions, the ECM controls distributor spark advance (timing) with the Electronic Spark Timing (EST) system.

The standard High Energy Ignition (HEI) system has a modified distributor module which is used in conjunction with the EST system. The module has seven terminals instead of the four used without EST. Two different terminal arrangements are used, depending upon the distributor used with a particular engine application.

To properly control ignition/combustion timing, the ECM needs to know the following information:
1. Crankshaft position
2. Engine speed (rpm)
3. Engine load (manifold pressure or vacuum)
4. Atmospheric (barometric) pressure
5. Engine temperature
6. Transmission gear position (certain models)

The EST system consists of the distributor module, ECM and its connecting wires. The distributor has 4 wires from the HEI module connected to a 4 terminal connector, which mates with a 4 wire connector from the ECM.

These circuits perform the following functions:
1. Reference ground—Terminal A—This wire is grounded in the distributor and makes sure the ground circuit has no voltage drop, which could affect performance. If this circuit is open, it could cause poor performance.
2. Bypass—Terminal B—At approximately 400 rpm, the ECM applies 5 volts to this circuit to switch the spark timing control from the HEI module to the ECM. An open or grounded bypass circuit will set a Code 42 and the engine will run at base timing, plus a small amount of advance built into the HEI module.
3. Distributor reference—Terminal C—This provides the ECM with rpm and crankshaft position information.
4. EST—Terminal D—This triggers the HEI module. The ECM does not know what the acrtual timing is, but it does know when it gets its reference signal. It then advances or retards the

spark timing from that point. Therefore, if the base timing is set incorrectly, the entire spark curve will be incorrect.

An open circuit in the EST circuit will set a Code 42 and cause the engine to run on the HEI module timing. This will cause poor performance and poor fuel economy. A ground may set a Code 42, but the engine will not run.

The ECM uses information from the MAP or VAC and coolant sensors, in addition to rpm, in order to calulate spark advance as follows:
1. Low MAP output voltage (high VAC sensor output voltage) would require MORE spark advance.
2. Cold engine would require MORE spark advance.
3. High MAP output voltage (low VAC sensor output voltage) would require LESS spark advance.
4. Hot engine would require LESS spark advance.

HALL EFFECT SWITCH

The Hall Effect switch is used on the 2.5L engine for the S/T truck series. This switch is mounted above the pick-up coil in the distributor. It takes place of the reference **R** terminal on the module. The switch an electronic device, which puts out a voltage signal controlled by the presence or absence of a magnectic field on an electronic circuit. This system tells the ECM which cylinder is next to fire.

Hall Effect Test

1. Disconnect the negative battery cable.
2. Disconnect and remove the hall effect switch from the distributor.
3. Noting the polarity marking on the switch, connect a 12 volt battery and voltmeter.
4. The voltmeter should read less than 0.5 volts without the blade against the magnet. Replace the switch if the voltage is above 0.5 volts.
5. With the blade against the magnet, the voltage should be within 0.5 volts of the battery voltage. Replace the switch if there is a low voltage reading.

RESULTS OF INCORRECT EST OPERATION

Detonation could be caused by low MAP output (high VAC sensor output), or high resistance in the coolant sensor circuit.

Poor performance could be caused by high MAP output (low VAC sensor output) or low resistance in the coolant sensor circuit.

EST PERFORMANCE CHECK

The ECM will set a specified value timing when the ALDL diagnostic terminal is grounded. To check the EST operation, record the timing at 2000 rpm with the diagnostic terminal not grounded. Then, ground the diagnostic terminal and the timing should change at 2000 rpm, indicating the EST is operating.

HOW CODE 42 IS DETERMINED

When the systems is operating on the HEI module with no voltage in the bypass line, the HEI module grounds the EST signal. The ECM expects to sense no voltage on the EST line during this condition. If it senses voltage, it sets Code 42 and will not go into the EST mode.

When the rpm for EST is reached (approximately 400 rpm), the ECM applies 5 volts to the bypass line and the EST should no longer be grounded in the HEI module, so the EST voltage should be varying.

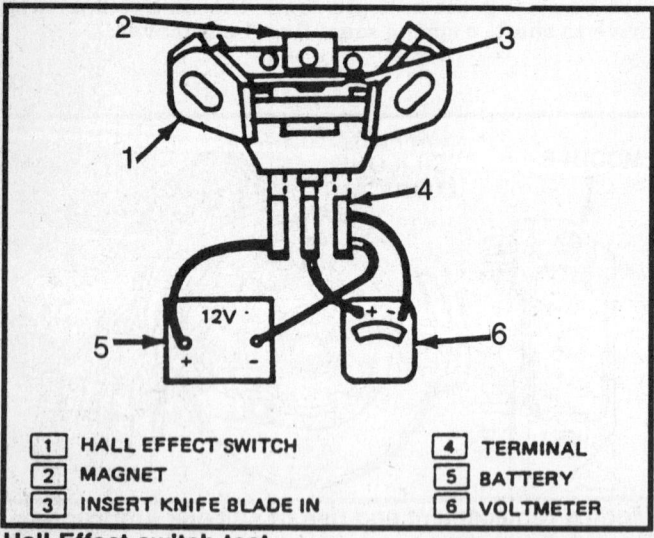

1	HALL EFFECT SWITCH		4	TERMINAL
2	MAGNET		5	BATTERY
3	INSERT KNIFE BLADE IN		6	VOLTMETER

Hall Effect switch test

If the bypass line is open, the HEI module will not switch to the EST mode, so the EST voltage will be low and Code 42 will be set.

If the EST line is grounded, the HEI module will switch to the EST, but because the line is grounded, there will be no EST signal and the engine will not operate. A Code 42 may or may not be set.

CODE 12

Code 12 is used during the Diagnostic Circuit Check procedure to test the code display ability of the ECM. This code indicates that the ECM is not receiving the engine rpm (REFERENCE) signal. This occurs when the ignition key is in the **ON** position and the engine is not running. The reference signal, also, triggers the fuel injection system. Without a reference signal the engine cannot run.

SETTING TIMING

Set timing according to the instructions on the Vehicle Emission Control Information Label.

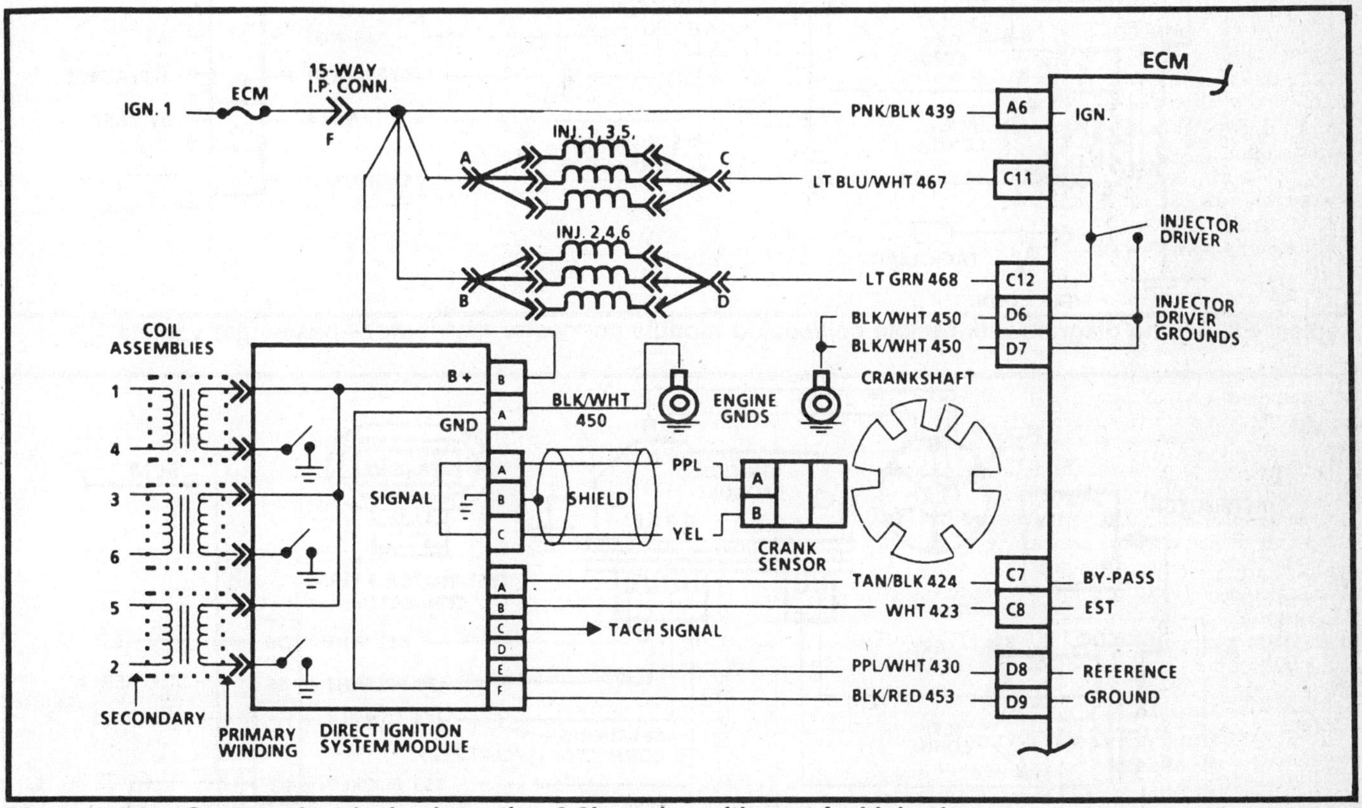

Direct Ignition System electrical schematic—2.8L engine with port fuel injection

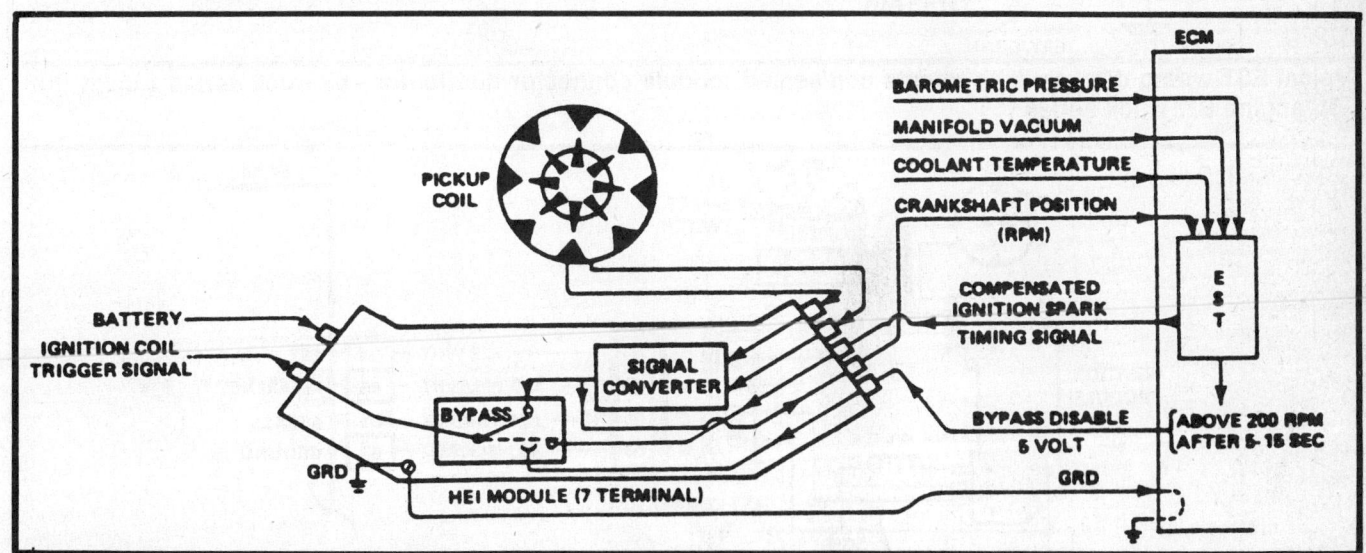

Typical EST control circuit schematic

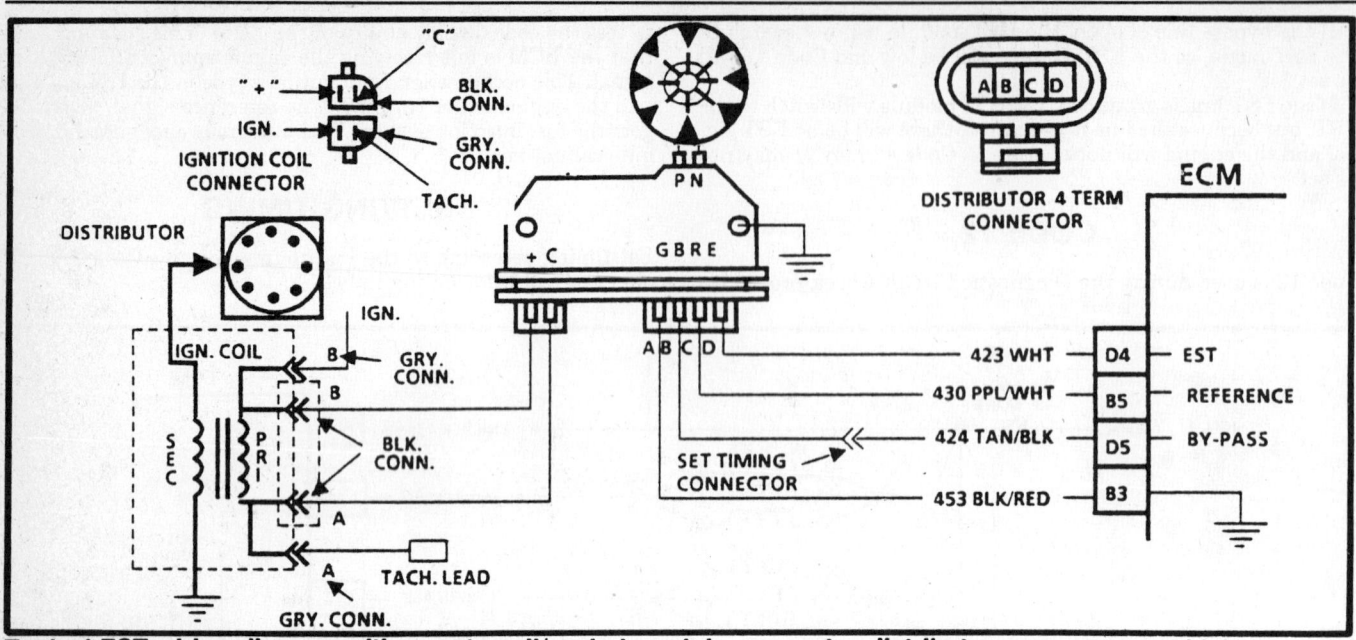

Typical EST wiring diagram with remote coil/sealed module connector distributor—passenger vehicle

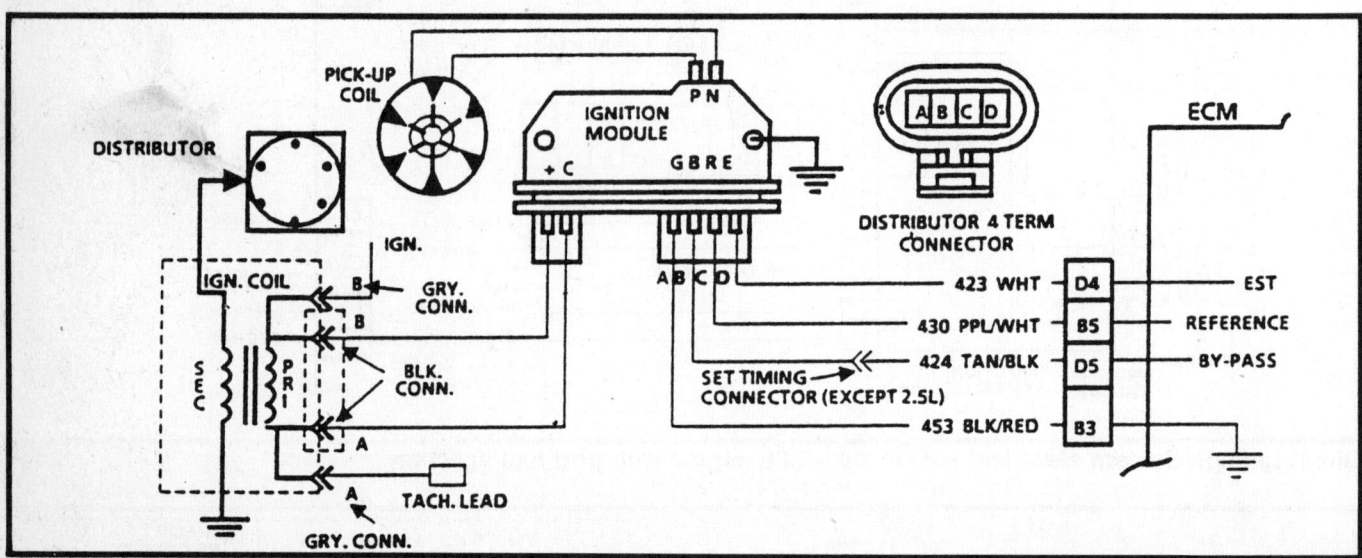

Typical EST wiring diagram with remote coil/sealed module connector distributor—all truck series except the 2.5L engine S/T truck series

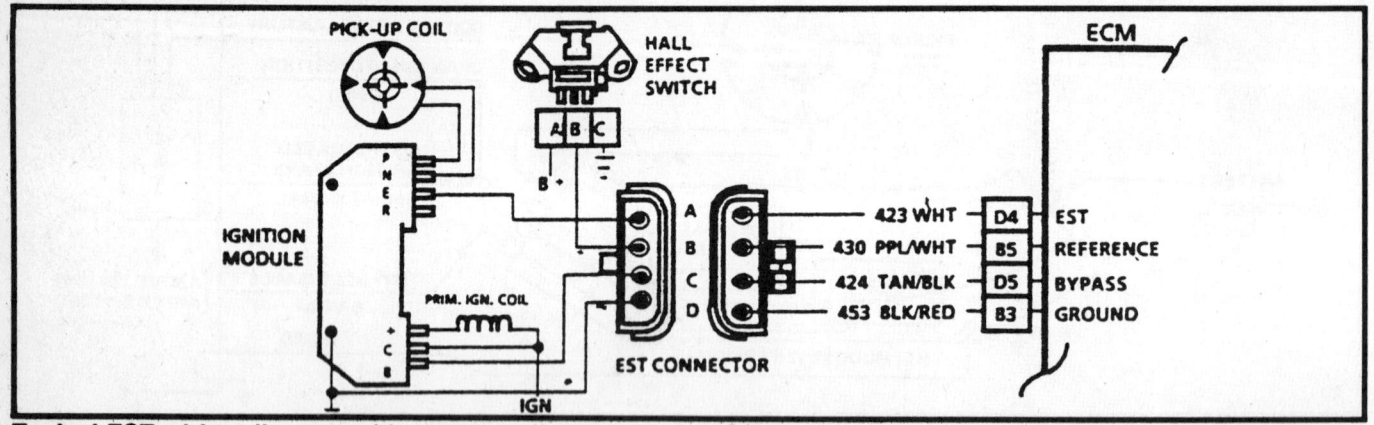

Typical EST wiring diagram with remote coil—2.5L engine S/T truck series

ELECTRONIC SPARK CONTROL (ESC) SYSTEM

General Description

The Electronic Spark Control (ESC) system is designed to retard spark timing up to 20 degrees to reduce spark knock (detonation) in the engine. This allows the engine to use maximum spark advance to improve driveability and fuel economy.

Varying octane level's in today's gasoline can cause detonation in an engine. Detonation is called spark knock.

The ESC knock sensor detects abnormal vibration (spark knocking) in the engine. The sensor is mounted in the engine block near the cylinders. The ESC module receives the knock sensor information and sensds a signal to the ECM. The ECM then adjusts the electronic spark timing to reduce the spark knocking.

The ESC module sends a voltage signal (8–10 volts) to the ECM when no spark knocking is detected by the ESC knock sensor and the ECM provides normal spark advance.

When the knock sensor detects spark knock, the module turns **OFF** the circuit to the ECM. The ECM then retards the EST to reduce spark knock.

There are 3 basic components of the Electronic Spark Control (ESC) system:

DETONATION (KNOCK) SENSOR

The knock sensor detects the presence (or absence) and intensity of the detonation by the vibration characteristics of the engine. The output is an electrical signal that goes to the controller. A sensor failure would allow no spark retard.

DISTRIBUTOR

The distributor is an HEI/EST unit with an electronic module, modified so it can respond to the ESC controller signal. This command is delayed when detonation is occurring, thus providing the level of spark retard required. The amount of spark retard is a function of the degree of detonation.

CONTROLLER

The Electronic Spark Control (ESC) controller processes the sensor signal into a command signal to the distributor, to adjust the spark timing. The process is continuous, so that the presence of detonation is monitored and controlled. The controller is a hard wired signal processor and amplifier which operates from 6-16 volts. Controller failure would be no ignition, no retard or full retard. The controller has no memory storage.

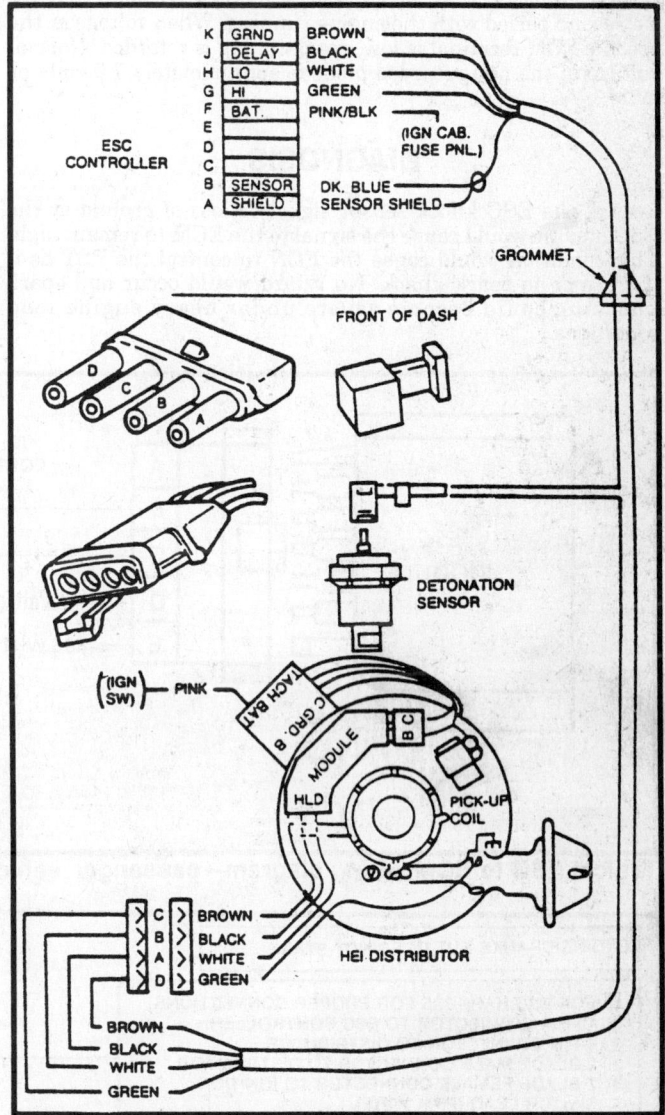

Typical ESC control circuit schematic

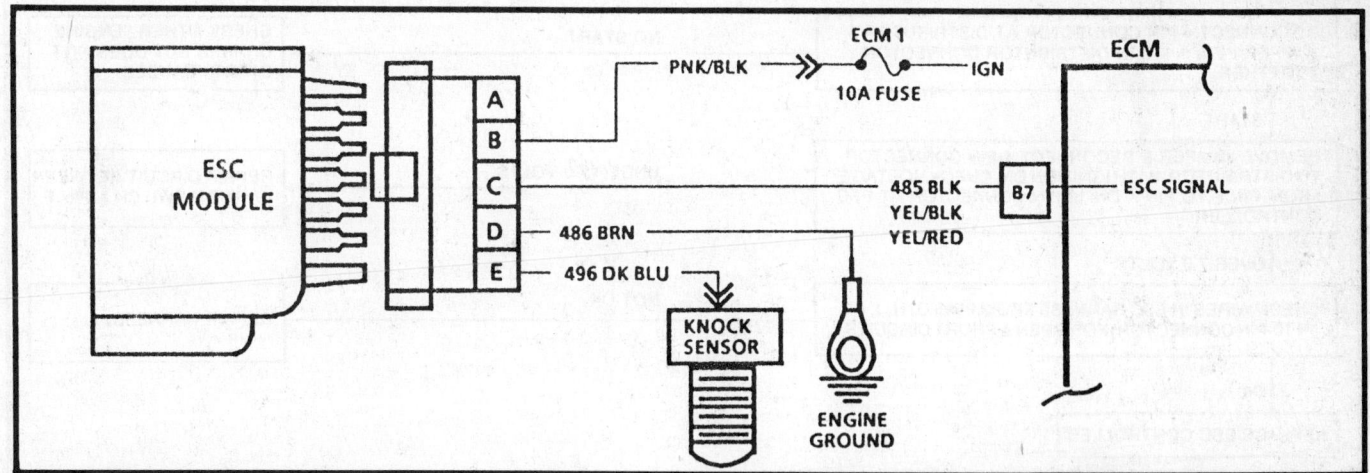

Typical ESC to ECM wiring diagram — all truck series except the 2.5L engines S/T truck series

CODE 43

Code 43 indicates that the ECM is receiving less than 6 volts for a 4 second period with the engine running. When voltage at the specific ECM terminal is low, spark timing is retarded. Normal voltage in the non-retarded mode is approximately 7.5 volts or more.

DIAGNOSIS

Loss of the ESC knock sensor signal or loss of ground at the ESC module would cause the signal to the ECM to remain high. This condition would cause the ECN to control the EST as if there was no spark knock. No retard would occur and spark knocking could become severe under heavy engine load conditions.

Spark retard without the knock sensor connected could indicate a noise signal on the wire to the ECM or a malfunctioning ESC module. Loss of the ESC signal to the ECM would cause the ECM to constantly retard EST. This could result in sluggish performance and cause a Code 43 to be set.

BASIC IGNITION TIMING

Basic ignition timing is critical to the proper operation of the ESC system. Always follow the Vehicle Emission Control Information label procedures when adjusting ignition timing.

Some engines will incorporate a magnetic timing probe hole for use with special electronic timing equipment. Consult the manufacturer's instructions for the use of this electronic timing equipment.

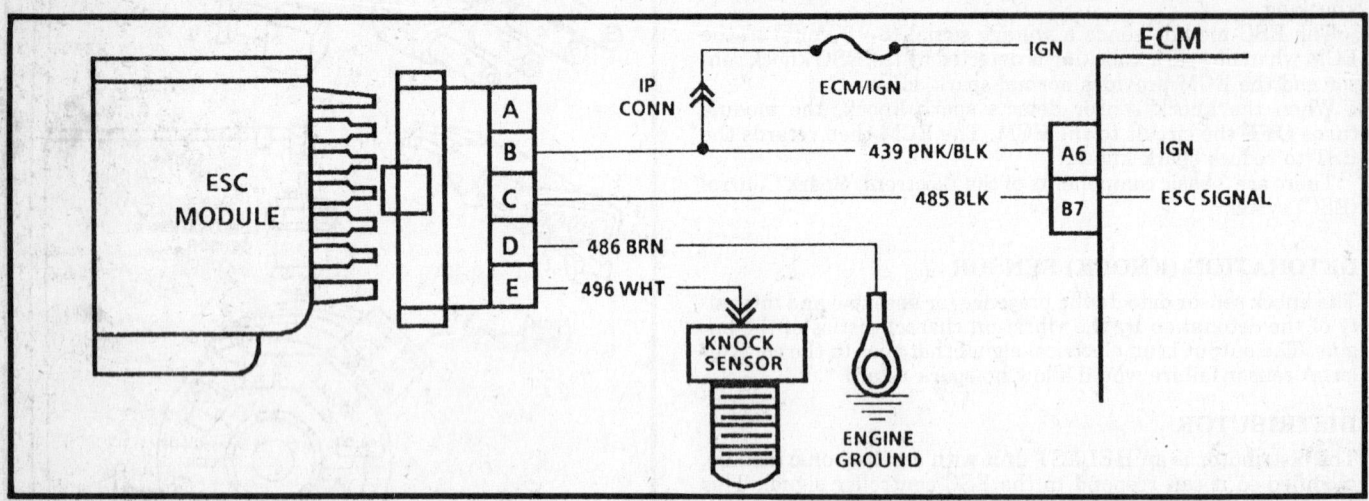

Typical ESC to ECM wiring diagram—passenger vehicle

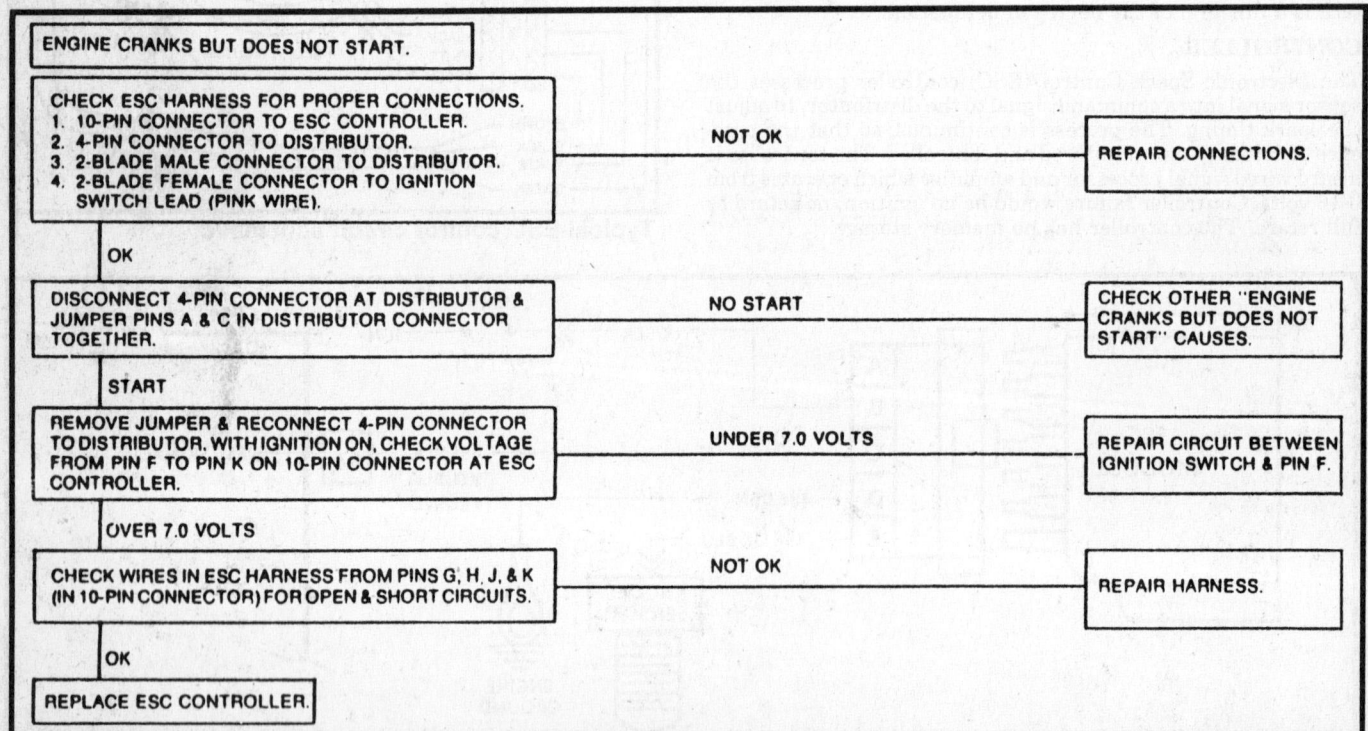

General ESC system diagnosis guide for all light truck series

COMPUTER CONTROLLED COIL IGNITION (C³I) SYSTEM
DIRECT IGNITION SYSTEM (DIS)
INTEGRATED DIRECT IGNITION (IDI) SYSTEM
ELECTRONIC SPARK CONTROL (ESC)

NOTE: To properly diagnosis the ignition systems and their problems, it will be necessary to refer to the diagnostic charts in either the fuel injection section or feedback carburetor section of this manual. Locate the charts that applied the vehicle in question and follow them from start to stop. The proper use of this manual, along with the proper tools available to today's technician, will provide systematic procedures to address such problems as, no start conditions, self diagnosis codes, intermittent problems and operator errors when no trouble codes are stored to aid the technician.

General Description

The Computer Control Coil Ignition System or C³I systems, use a coil pack, an ignition module and various types of Hall Effect crankshaft sensors. There are 3 different types of C³I systems. It is important to determine which system is used in a specific application, since either Type 1 or 2 may be used on the same year, make and model vehicle. The difference is determined by visual examination of the top of the coil pack. Type 1 has all 3 coils molded together while Type 2 has individually serviceable coils.

The third system, known as Type 3 or fast start is similar in appearance to Type 1. In fact the coil packs are interchangeable, however, the module is electronically different and the harness connector plugs are not compatible. There are distinguishable differences in the harness, the crankshaft sensor and the harmonic balancer. The Type 3 system is used only on the 3800 Buick engine.

The primary difference between the C³I system and the Direct Ignition System (DIS), is the method of sensing the crankshaft position. Rather than the Hall Effect switches mounted on the front of the engine, this system uses a magnetic sensor mounted on the side of the engine block.

The Integrated Direct Ignition (IDI) system, is similar to the DIS system in operation, with a few additional advantages. Because the wires are not the resistor type and are very short in length, there is reduced secondary capacitance. This gives a faster rise time to the secondary coil. Secondary current is nearly double that of the HEI distributors because the primary module

Computer controlled coil ignition system components

current limiter is 8.5–10.0 amps versus 3.0–5.5 amps on the HEI. Secondary voltage is tha same as DIS with 3.3 kilovolts available under all load condtions.

The main characteristic of the IDI system is that there are no secondary carbon spark wires. Instead a flat wire circuit is used to distribute secondary voltage to the spark plugs. The flat wire circuit is located in the housing.

SECONDARY SYSTEM

While discussing secondary ignition systems, we will use electron flow instead of conventional current flow. Conventional flow is from positive to negative, electron flow is from negative to positive.

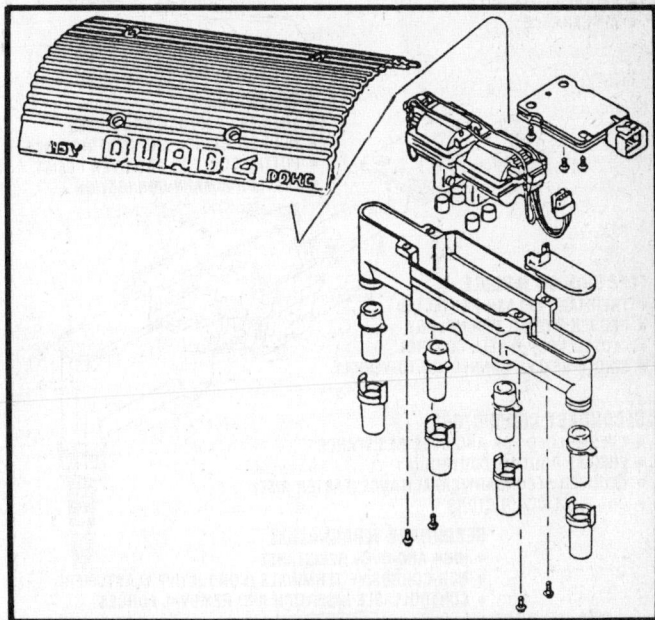

Typical integrated direct ignition system parameters

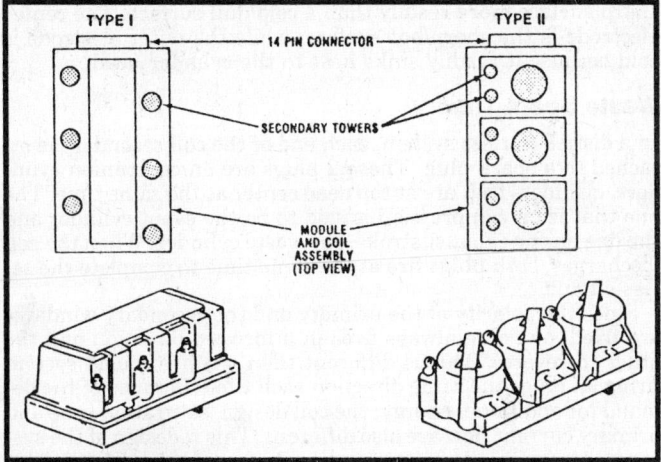

Typical DIS Type 1 and Type 2 system identification

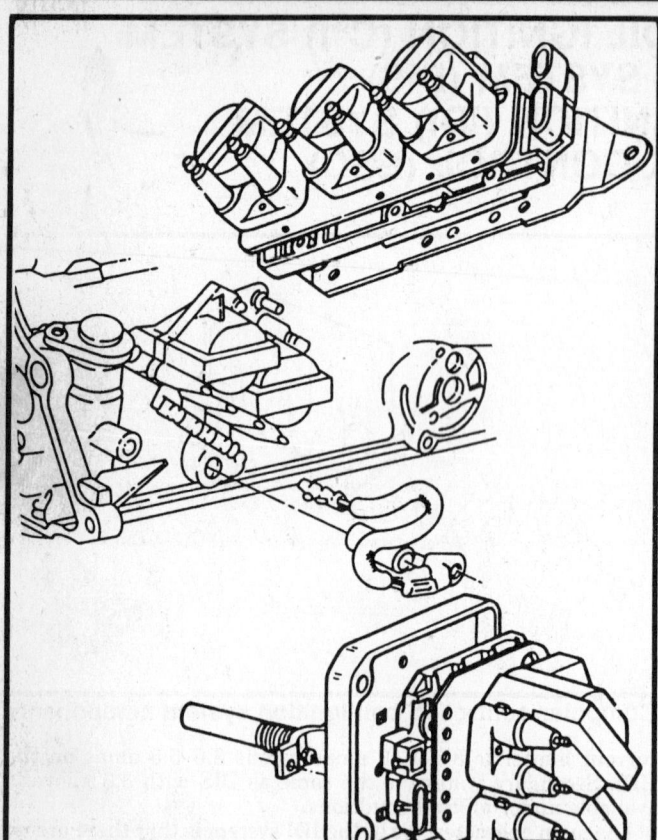

Typical direct ignition systems

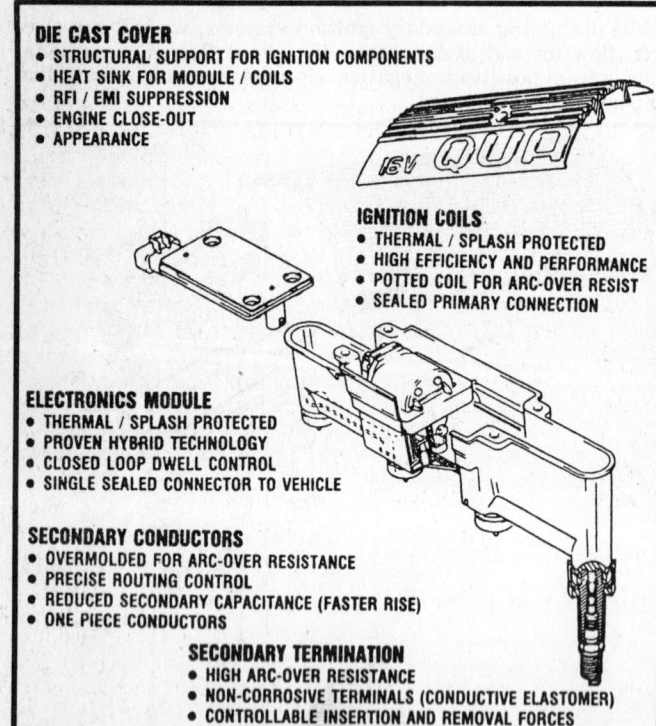

DIE CAST COVER
- STRUCTURAL SUPPORT FOR IGNITION COMPONENTS
- HEAT SINK FOR MODULE / COILS
- RFI / EMI SUPPRESSION
- ENGINE CLOSE-OUT
- APPEARANCE

IGNITION COILS
- THERMAL / SPLASH PROTECTED
- HIGH EFFICIENCY AND PERFORMANCE
- POTTED COIL FOR ARC-OVER RESIST
- SEALED PRIMARY CONNECTION

ELECTRONICS MODULE
- THERMAL / SPLASH PROTECTED
- PROVEN HYBRID TECHNOLOGY
- CLOSED LOOP DWELL CONTROL
- SINGLE SEALED CONNECTOR TO VEHICLE

SECONDARY CONDUCTORS
- OVERMOLDED FOR ARC-OVER RESISTANCE
- PRECISE ROUTING CONTROL
- REDUCED SECONDARY CAPACITANCE (FASTER RISE)
- ONE PIECE CONDUCTORS

SECONDARY TERMINATION
- HIGH ARC-OVER RESISTANCE
- NON-CORROSIVE TERMINALS (CONDUCTIVE ELASTOMER)
- CONTROLLABLE INSERTION AND REMOVAL FORCES

Typical integrated direct ignition system

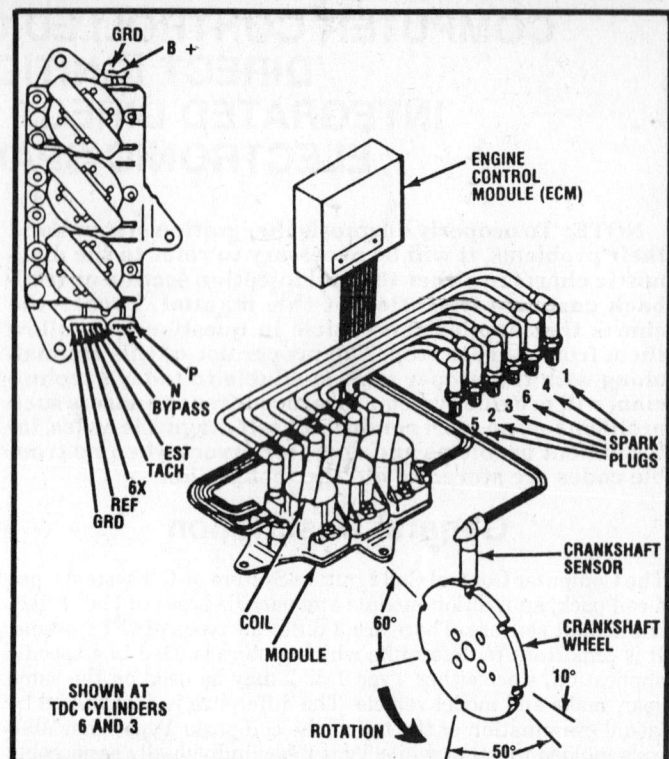

Exploded view of the direct ignition system on the 2.8L engine

On standard ignition systems the center terminal of the plug is always negative. That is, the electrons depart from the center electrode and arrive at the outer electrode, to return through the block and other circuitry to the source, the secondary coil winding. Consider the secondary winding as a source in itself, the energy does not have to be common to the 12 volt source at any point in order to operate properly.

In the past, some primary and secondary windings were connected together for convenience in packaging only. Technicians were warned not to reverse the polarity of the ignition coil or a weak spark and a misfire could result. This was partly die to primary current limitations (3.0–5.0 amps) so the coil could not produce more than 20–35 kilovolts. It requires approximately 30% more energy to fire a spark plug backwards, (from edge or outer to the center electrode) than forwards (from center to outer or edge). This is because the electrons tend to leave a hot/sharp surface more readily than a cold/dull surface. The center electrode is the sharp/hot surface, while the outer electrode is cold because it readily sinks heat to the cylinder head.

Waste Spark Theory

In a distributorless system, each end of the coil secondary is attached to a spark plug. These 2 plugs are on companion cylinders, cylinders that are at top dead center at the same time. The one that is on compression is said to be the event cylinder and the one on the exhaust stroke, the waste cylinder. When the coil discharges, both plugs fire at the same time to complete the series circuit.

Since the polarity of the primary and the secondary windings are fixed, one plug always fires in a forward direction and the other in reverse. This is different than a conventional system firing all plugs the same direction each time. Because of the demand for additional energy; the coil design, saturation time and primary current flow are also different. This redesign of the system allows higher energy to be available from the distributorless coils, greater than 40 kilovolts at all rpm ranges.

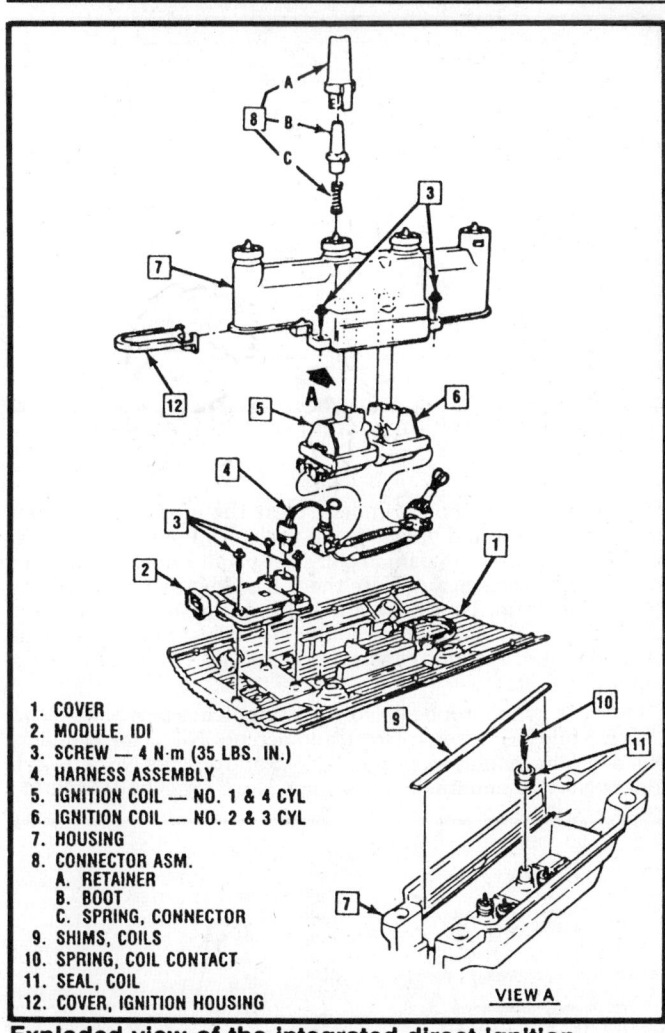

1. COVER
2. MODULE, IDI
3. SCREW — 4 N·m (35 LBS. IN.)
4. HARNESS ASSEMBLY
5. IGNITION COIL — NO. 1 & 4 CYL
6. IGNITION COIL — NO. 2 & 3 CYL
7. HOUSING
8. CONNECTOR ASM.
   A. RETAINER
   B. BOOT
   C. SPRING, CONNECTOR
9. SHIMS, COILS
10. SPRING, COIL CONTACT
11. SEAL, COIL
12. COVER, IGNITION HOUSING

Exploded view of the integrated direct ignition system on the 2.3L engine

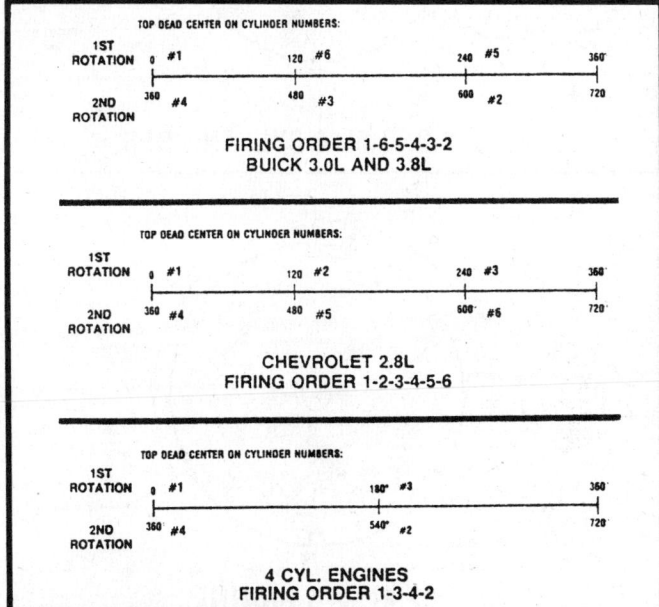

FIRING ORDER 1-6-5-4-3-2
BUICK 3.0L AND 3.8L

CHEVROLET 2.8L
FIRING ORDER 1-2-3-4-5-6

4 CYL. ENGINES
FIRING ORDER 1-3-4-2

Typical DIS firing orders

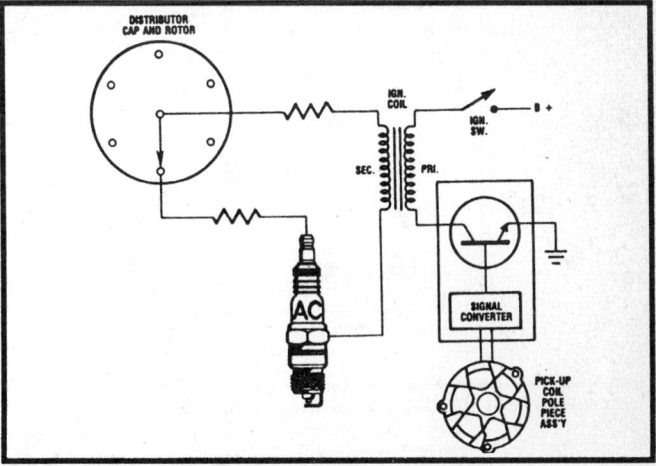

Conventional electronic ignition system

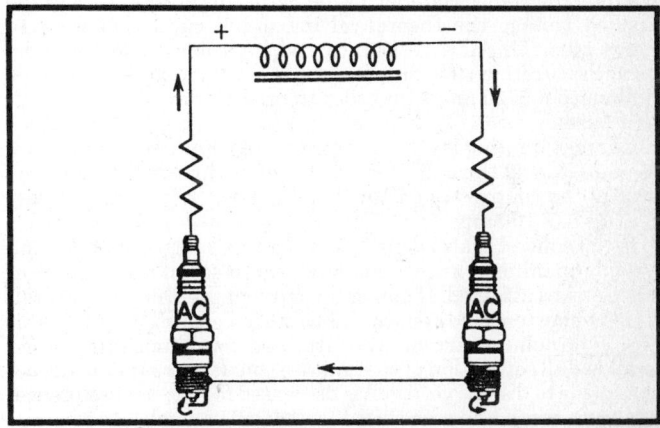

Typical DIS current flow chart

A Buick V6 has a firing order of 1–6–5–4–3–2. To begin with, we will concentrate on cylinders 1 and 4 (companion cylinders). The ignition system is unaware of and has no need of sensing which of these 2 cylinders is on compression. When the coil fires, on plug fires forward and the other fires backward simultaneously. The voltage dropped across each plug is determined by the polarity and the cylinder pressure. The cylinder on compression will require more voltage to create an arc than the one on exhaust.

While the 2.8L Chevrolet V6 has a different firing order (1–2–3–4–5–6) and a different coil sequence order, the companion cylinders are the same. The coil firing order is the same in all V6 systems. With the exception of the Buick 3800, the second coil in the firing order fires first at initial set-up. This firing order of the coils is maintained by the module logic circuitry.

All of the 4 cylinder engines use the same firing order of 1–3–4–2. At initial set-up, the module energizes the 3–2 coil first and then alternates the sequence.

THE PRIMARY CONTROL AND TRIGGERING SYSTEM

CURRENT AND DWELL CONTROL

Primary coil current is controlled by 2 transistors (4 cylinder) or 3 transistors (6 cylinder) in the ignition module. These devices complete the ground circuit of the ignition coil primary. The timing and the sequencing of these transistors is determined by several circuits within the module and the external triggering devices.

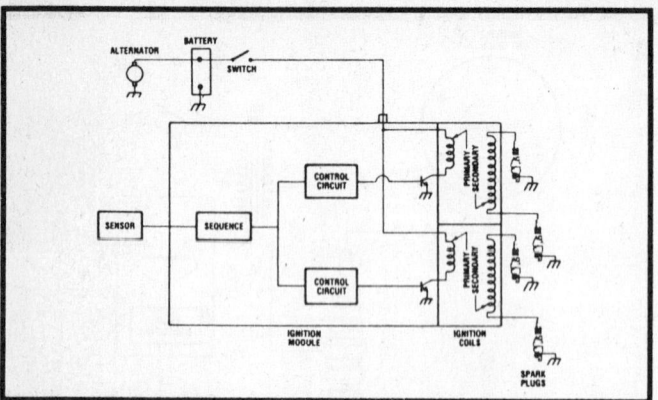

Typical direct ignition system current flow

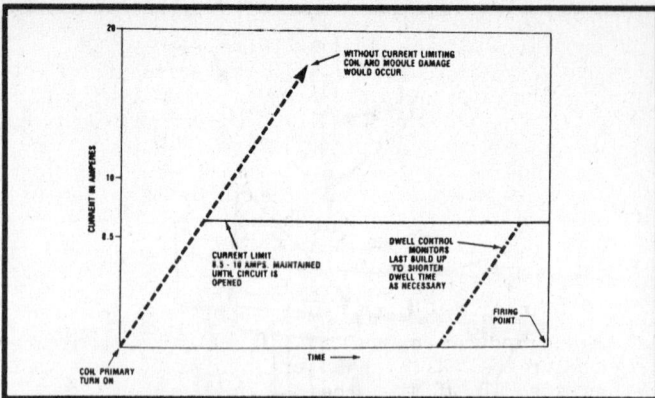

Typical direct ignition system closed loop dwell control

The internal control circuit of the module is responsible for primary current flow and dwell time limiting. The primary coil winding resistance is under 1 ohm, with this and 14 volts as the applied voltage the theoretical maximum current flow is 14 amps plus. This low resistance and high amperage helps decrease saturation time but, the maximum current flow must be limited to 8.5–10 amps, in order to prevent damage to the system parts.

Current limiting is accomplished by the ignition control module monitoring the circuit current and modifying the base current of the control transistor to limit the collector emitter current to 8.5–10 amps.

Dwell control is also done by the ignition control module. The module monitors the last coil build up to see of the maximum current was attained. If maximum current is attained, the dwell time is shortened to reduce the wattage consumed by the system. If minimum current is not attained, dwell time is increased to allow full saturation of the ignition coil. If current limiting occurs prior to discharge, dwell is decreased for the next sequence. This process is know as closed loop dwell control.

TRIGGERING

Magnetic Sensor and Reluctor Type

The various distributorless systems use different types of sensing devices to determine crankshaft and camshaft positions. The DIS and the IDI systems on the 2.0L, 2.3L, 2.5L, and the 2.8L engines use a magnetic sensor and reluctor.

This single sensor is referred to as a magnetic sensor because it has a permanent magnet surrounded by a winding or wire. The sensor is placed 0.050 ± 0.020 from the crankshaft. The magnetic field of the sensor is modulated by the machine surface on the crankshaft called the reluctor of the sensor system.

As the crankshaft rotates, the notches in the reluctor wheel cause the magnetic field to change in intensity because the field travels through metal at times and air at others. This actions induces a small AC or push pull voltage in the winding or wire around the sensor. Minimum peak voltages of the 2.0L, 2.5L, 2.8L sensors are 250 millvolts, AC at cranking speeds. Winding resistances are 1000 ohms ± 200 ohms. The 2.3 IDI system in the Oldsmobile Quad 4 engine has a slightly different resistance, 625–875 ohms and AC, peak voltage is slightly higher at cranking speeds.

The crankshaft has 7 notches which provide 7 signals to the ignition module. Both the 4 cylinder and the 6 cylinder engines use the same 7 notch reluctor. There are 6 equally spaced notches at 60 degree intervals around the machined surface on the crankshaft. The 7th notch is 10 degrees away from one of the 6 evenly spaced notches and is used to sync or synchronize the coil sequence to crankshaft position. The module contains specific programming to trigger events on the different engines.

On the 4 cylinder model, the sync, notch programs the module to skip notch No. 1 and accept notch 2 as the 10 degrees BTDC

for the 2–3 companion cylinders. Next the module would skip notches 3 and 4 and accept No. 5 to determine the 10 degrees BTDC for the 1–4 companion pair. The 6 and 7 notches pass and the process starts again. Note the 2–3 coil is the first to fire on the initial set up. This is the coil that fires the second cylinder in the firing order. With the exception of the Fast Start system, all of the DIS/C³I systems engergize the second cylinder in the firing order first.

On the 6 cylinder model, No. 6 and 7 notches pass the sensor. The module skips notch No. 1 and accepts No. 2 for cylinders No. 2–5. The module skips notch No. 3 and accepts No. 4 for cylinders No. 3–6 and finally uses notch No. 6 for cylinders No. 1–

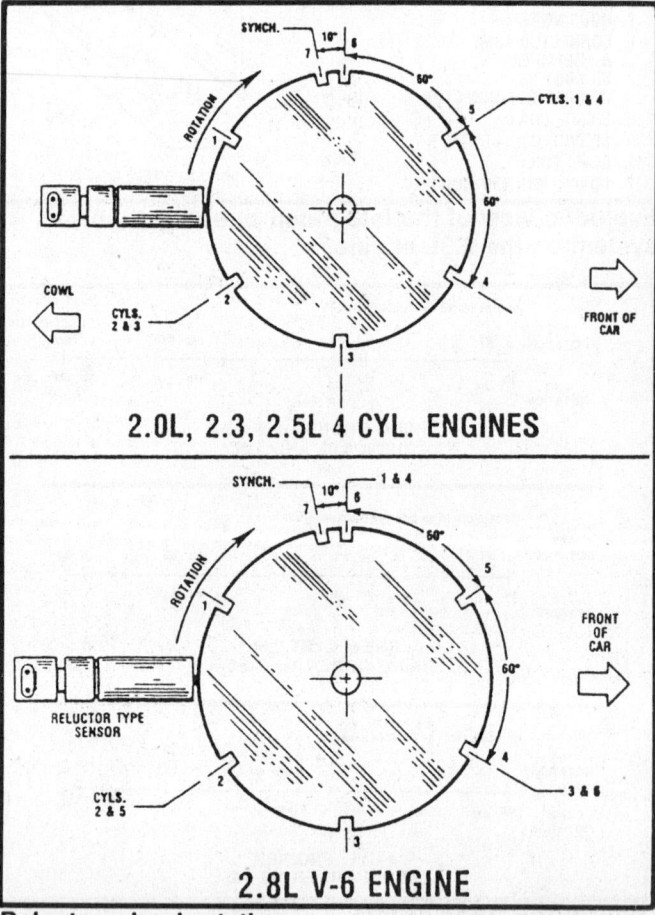

Reluctor wheel rotation

4. The firing order is 123 in the first crankshaft revolution and 456 in the second crankshaft revolution.

Since 2 plugs are fired by a coil at the same time, the system is primarily concerned with coil order and the position of crank rotation in which to fire them. These descriptions apply to the systems when at base timing in the module or bypass mode. The reference pulse (CKT 430 PPL/W) wire is pulled low by the notch that is 60 degrees ahead of the cylinder event notch and high at the cylinder event notch. This pulse is used for Electronic Spark Timing (EST) and Fuel Injection.

C³I Primary (Hall Effect Type)

The Buick produced engines use a different system to synchronize and fire the coils at the proper time. These sensors are referred to as Hall Effect switches.

The word Hall Effect is named after the person that discovered this effect (E.H. Hall). The Hall Effect was discovered over a hundred years ago. Recently, this effect has been applied to modern transistor theory and allows the device to sense the position of the ferrous metal object without the aid of contacting parts. There are various Hall sensors used to detect the crankshaft and the camshaft positions for ignition as well as sequential port fuel injection.

The effect states, when a steady current flows through a semiconductor material no voltage potential can be perceived measuring at right angles to the flow or current. However, when a magnetic field is introduce perpendicular to the current flow, a voltage can be perceived at right angles to the current flow. This voltage is used to switch a transistor on and off as a magnet is exposed to and then shielded from the semiconductor material.

When a magnetic field is exposed to the switch, the hall voltage turns on the base of a transistor. The collector/emitter circuit of the transistor pulls down a reference voltage from the module. This pull down or pull up of the reference voltage allows the module logic to detect the position of a device.

When the magnetic field is shielded from the switch, no hall voltage is generated, the transistor is left in the off position, so the reference voltage is left in the high state.

These sensor pull down a regulated reference voltage supplied by the ignition module, signaling the ignition module of the position of the crankshaft when a ferrous metal object passes between the permanent magnet and the semiconductor material. In the case of the 3.8L SFI cam sensor, a moving magnet passes the hall effect switch to turn it on, pulling down the reference voltage. The 3.8L Buick SFI engine uses both a crank sensor and a cam sensor. The cam sensors are in different locations and appear different; however, they serve the same purpose and have identical electrical signals.

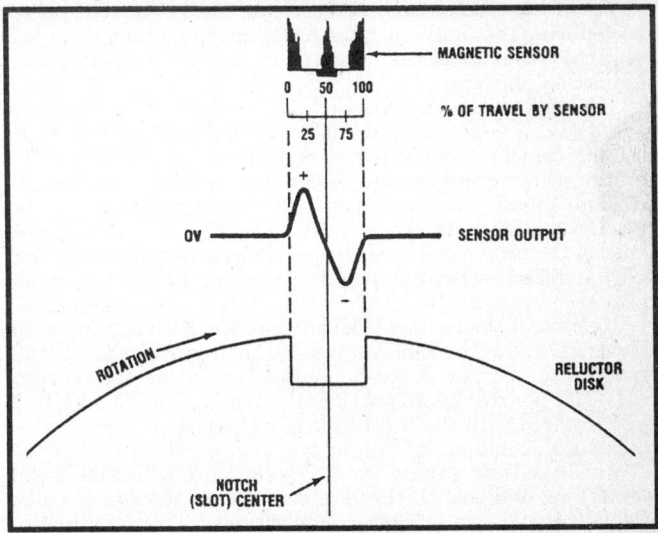

Notch effect on the output signal

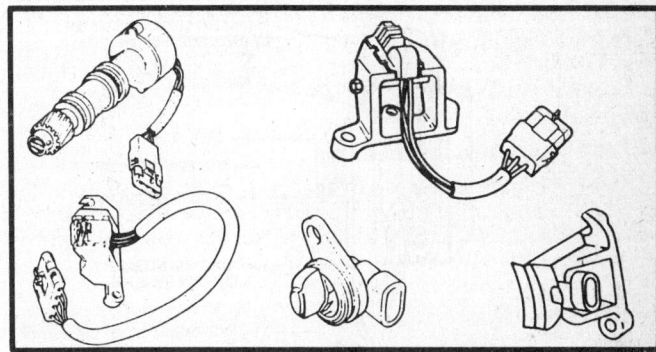

The various hall effect sensors available

As the crankshaft rotates, 3 equally spaced metal interrupters pass between the permanent magnet and the switch. As the transistor in the switch turns on and off it toggles the module reference voltage high (6–8 volts) to low (0–0.5 volts). The three signals are identical in time and amplitude so, the ignition module cannot detect which of these signals it should assign to which ignition coil.

This is where the cam sensor enters the ignition set up routine. As the crank rotates. the cam sprocket also rotates at half speed, the magnet mounted on the cam sprocket toggles the Hall switch mounted in the front cover. The signal from this switch is sensed by the module when the switch pulls down the module's reference voltage.

The cam signal is synchronized with one of the crank signal pull down and identifies which of the 3 identical cranks references is to be assigned to the correct coil. Once this takes place, during cranking, the ignition module is able to remember the se-

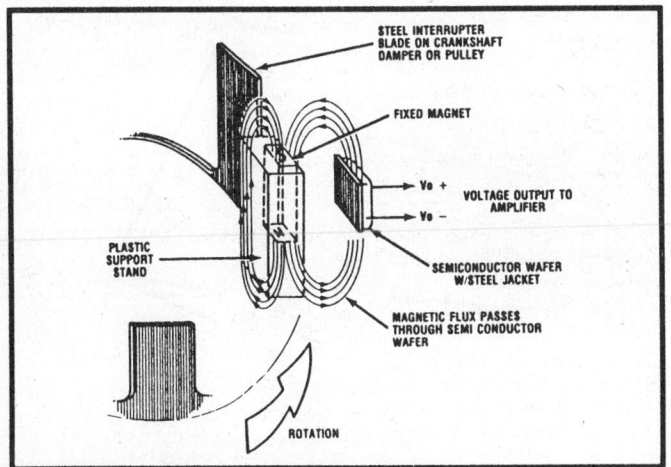

Hall Effect switch in the on position

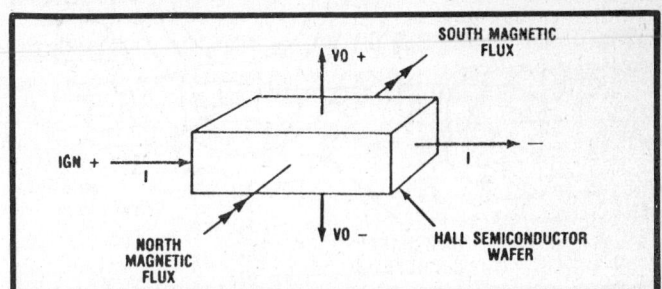

Typical hall effect magnetic field

quence of the crank sensor and does not need to be re-synchronized during that ignition cycle. As far as the ignition module is concerned, the cam sensor serves no purpose after the initial set up task is complete.

The Buick 3.0L with C³I is slightly different than the 3.8L engine. First, it is not a sequentially injected engine, therefore a discrete camshaft signal is not necessary. The ECM provides multiport injection from processing the crankshaft signal only. The crankshaft sensor still has 3 identical Hall switch voltage toggles so, some method of identifying which of these toggles belongs to the proper coil is necessary. This crankshaft has 2o Hall Effect switches, one on either side of the permanent magnet.

The back of the harmonic balancer of the 3.0L has 2 concentric rings, one is the same as the 3.8L and provides the crankshaft sensor toggles. A second one with only one open area or one toggle per revolution provides the sync signal. This toggle is synchronized with the 6–3 toggle of the crank sensor.

The 3.0L combination sensor has 4 wires, the sensor power (10 volt) the logic ground from the ignition module, the crank signal line, and the synchronize signal line (sometimes called the cam signal line in earlier publications). This combination sensor has 2 hall switches, one on either side of the permanent

magnet. One damper ring runs in the inboard slot and the other in the outboard slot. Once again the synchronize or sync signal is only needed during cranking and set up. Both the 3.0L and the 3.8L will run without a cam or a sync signal if the engine is running before the failure occurs, but neither will start without it. Note that some 3.0L engines may start if the sync signal is pulled low and the driver repeatedly tries to start the vehicle. If this circuit is pulled low while the engine is running the module will re-sync and usually stalls the engine.

C³I Type 3 – Fast Start 1

The 1988 Buick 3800 engine has a slightly different system known as the C³I – (Type 3) – FAST START 1. Advantages include: A faster start up Walk home protection for a cam sensor failure More accurate measurement of the crankshaft sensor

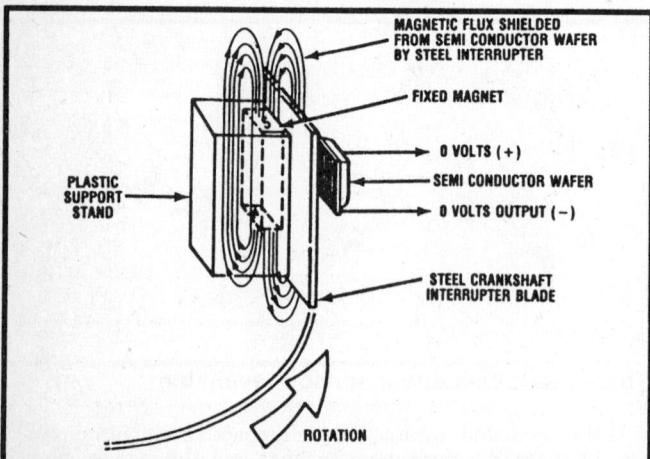

Hall Effect switch in the off position

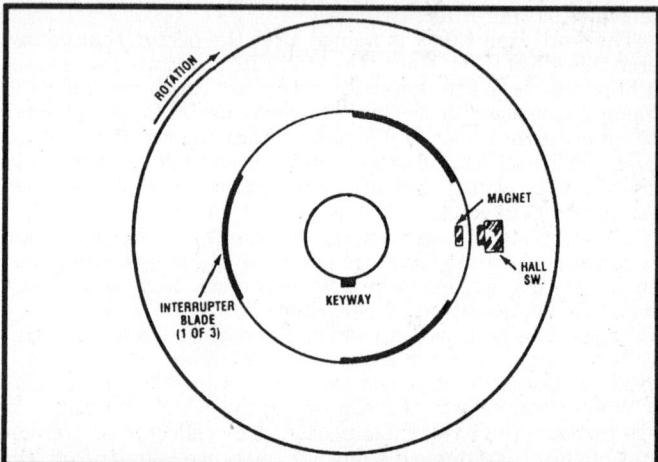

Typical Buick 3.8L SFI crankshaft damper viewed from the front

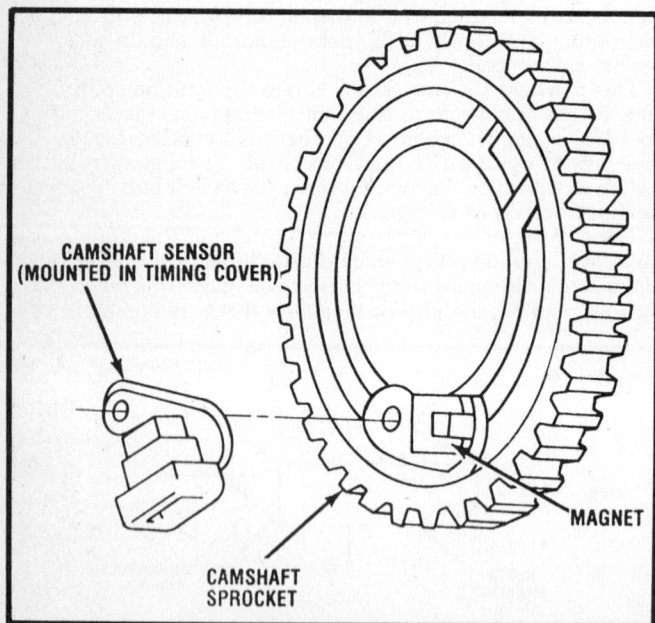

Typical 3.8L camshaft sensor

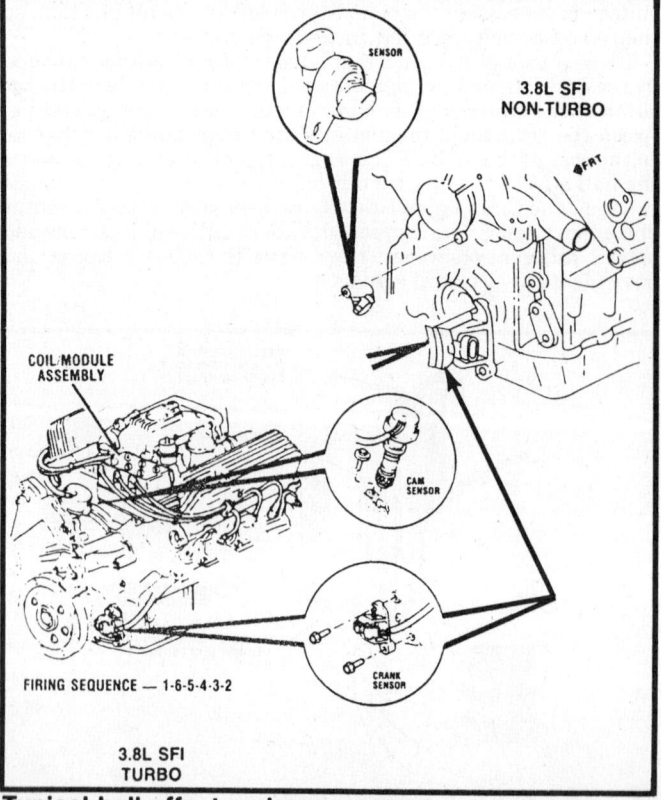

Typical hall effect and cam sensors

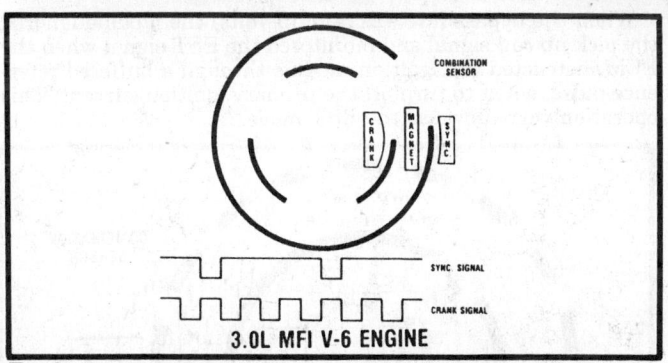

Typical 3.0L MPI engine crankshaft and sync signal

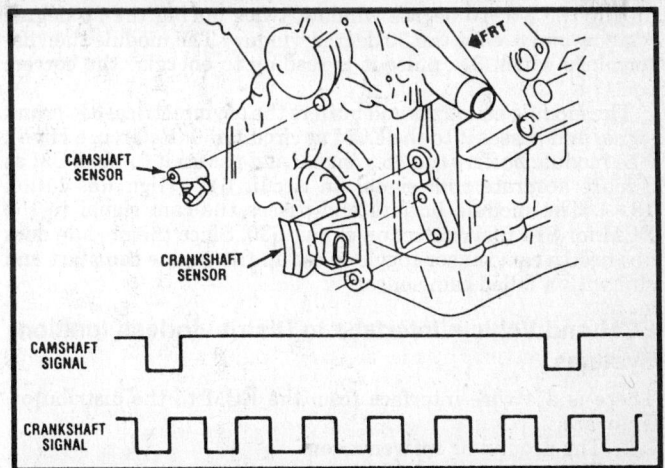

Typical 3.8L SFI engine crankshaft and camshaft signal

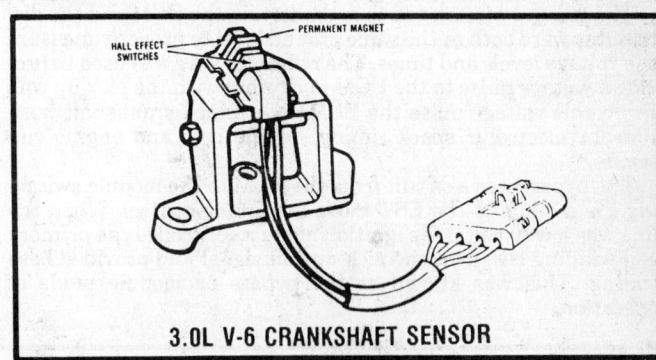

Typical 3.0L V6 crankshaft sensor

signals. The system contains a dual crank sensor similar to the 3.0L and a separate cam sensor in the front cover like the 3.8L. The interruptor vanes on the back side of the balancer differ in configuration from the other systems. First the outside vane, known as the crank 1 signal has 18 evenly spaced interruptors. This signal is known as asymmetrical vanes and is known as the 3x signal. The air gaps of 10, 20 and 30 degrees are spaced 100, 100 and 90 degrees apart respectively.

The system can identify the proper coil in as little as 120 degrees of crankshaft rotation. The system will also start on any coil it identifies first. Because of the inequities of the 3x signal, the module does not need to wait for the cam or the sync signal to energize the correct coil.

TDC No. 1 and No. 4 cylinders occurs 75 degrees after the trailing edge of the 10 degree window TDC No. 2 and No. 5 cylinders occurs 75 degrees after the trailing edge of the 20 degree window. The No. 6 and No. 3 cylinders TDC 75 degrees after the trailing edge of the 30 degree window. The trailing edges of the windows are each 120 degrees apart.

The module uses the $18 \times$ signal as a clock pulse to measure the duration of each $3 \times$ pulse. The $18 \times$ will change state once

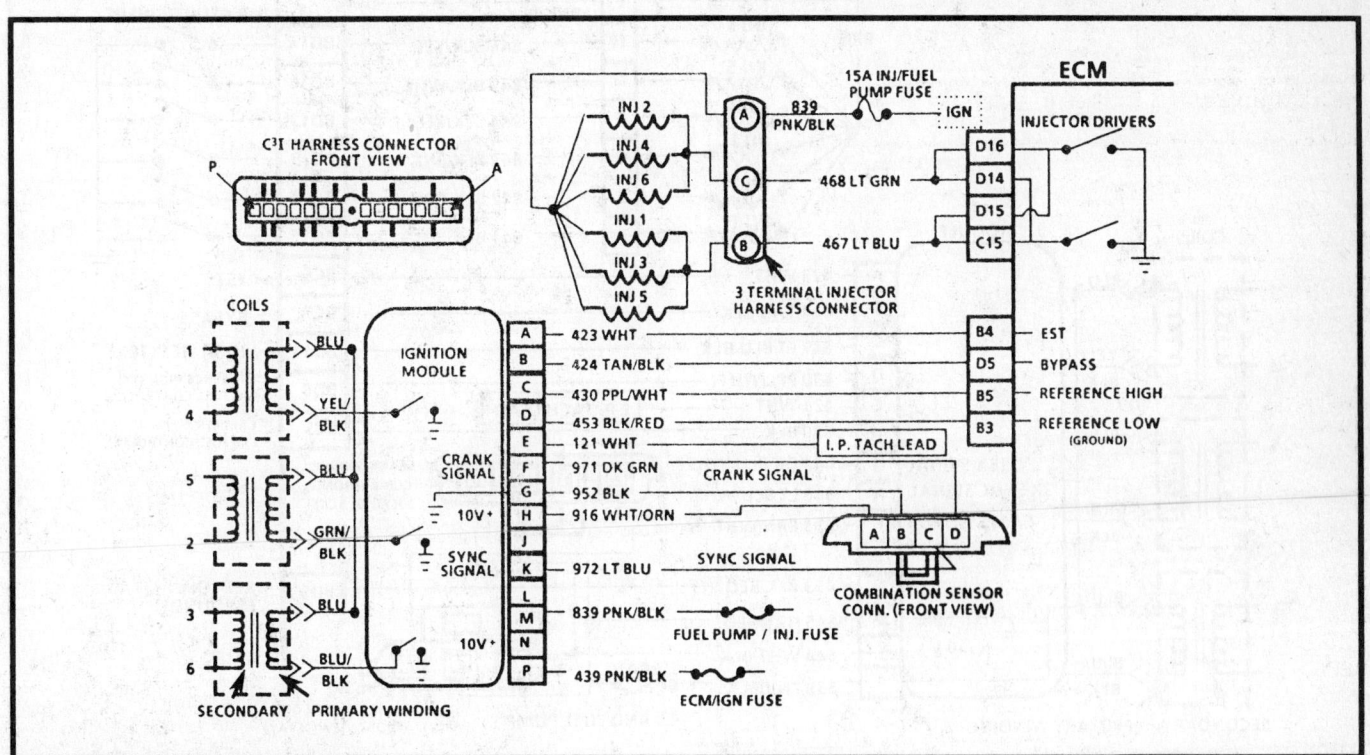

Typical C³I wiring schematic

during the 3× 10 degree window, twice during the 20 degree window and 3× for the 30 degree window. The module then determines which 3× pulse it is reading to energize the correct coil.

The module equalizes and buffers the asymmetrical 3× crank signal and passes it to the ECM as circuit 430 (Reference High). The module buffers the 18× signal and passes it to the ECM as a more accurate rpm signal on circuit 647 (High Resolution 18×). The module buffers and passes the cam signal to the ECM for SFI fuel control on circuit 630. Since the module does not need a cam sensor toggle to set up the engine can start and run with a failed cam sensor.

ECM and Vehicle Interface to Distributorless Ignition Systems

There is a 4 wire interface from the ECM to the distributor. They are:
1. The ground or reference low
2. The reference, or reference high
3. The bypass
4. The electronic spark timing, or EST

The ground wire served to make certain the ECM and the distributor were both at the same ground level to properly measure the voltage levels and times. The reference wire was used to provide a voltage pulse to the ECM in rhythm with the pick up coil. From this voltage pulse the ECM determined crankshaft position for electronic spark timing, engine rpm and engine run status.

The bypass was a circuit from the ECM to the module switching the module to the EST mode for timing control. When the line was low (0 volts) the ignition module controlled the primary coil winding based on the pick up coil signal and provided base timing. This was known as the bypass or module mode of operation.

When the bypass line was high (5 volts) the module ignored the pick up coil signal and monitored the EST signal when the ECM instructed the ignition module through a buffered reference pulse, when to turn off the primary ignition current. This operation was knows as the EST mode.

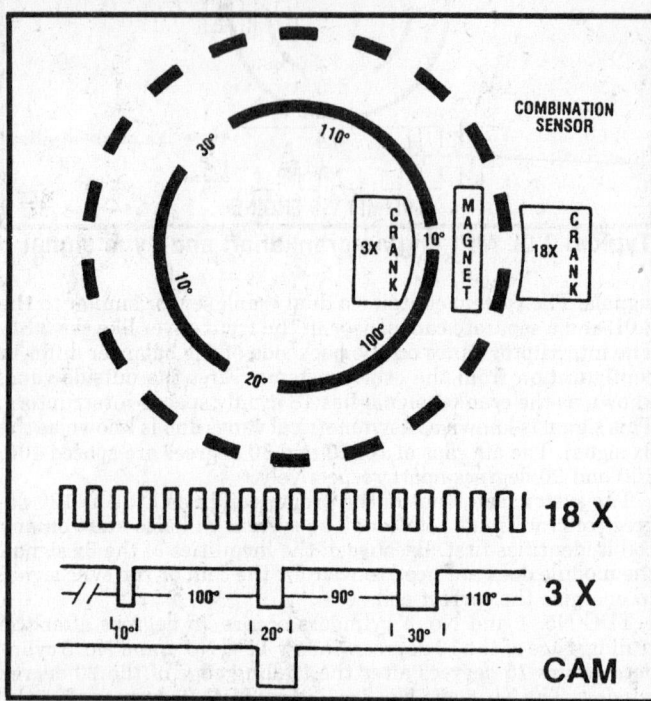

Triggering the C³I Type 3—Fast Start 1 system

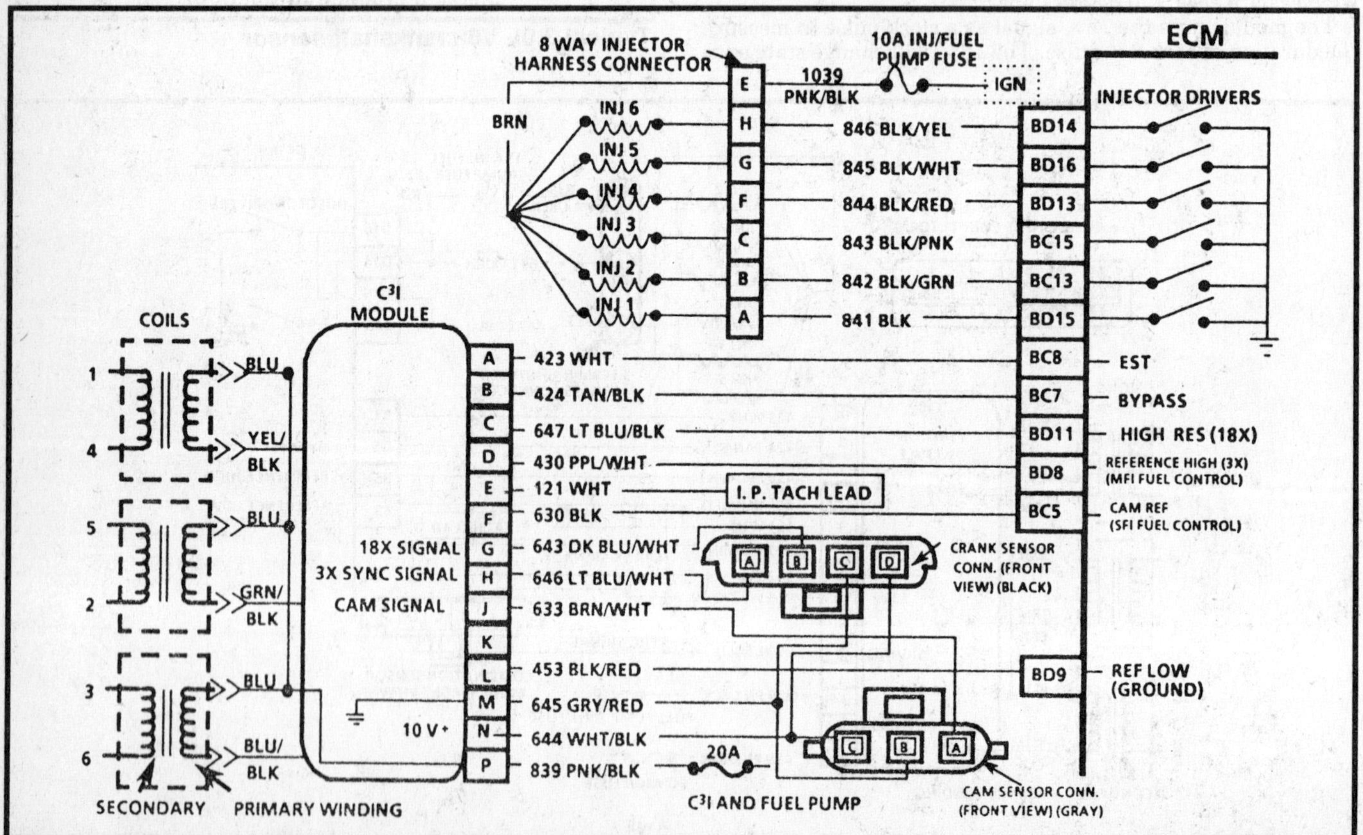

Typical wiring schematic of the C³I Type 3—Fast Start 1 system

The EST wire was a 5 volt high-low toggle to the module and was a buffered reference signal. This signal pulsed the module and fired the coil in advance of the pick up coil base timing pulse. This was accomplished by the ECM referencing the proceeding pulse and then delaying the output so the returning pulse was shifted from the reference. This allows the timing to be advanced above base.

These same 4 wires exist on all distributorless igniton systems and they serve the same purpose, with the exception that the reference pulse is generated by the crankshaft sensor through the ignition module rather than the pick up coil.

These 4 wires are joined by a buffered cam signal toggle on vehicles equipped with sequential fuel injection. The additional cam sensor circuit on the sequentially injected engines provides the ECM with information so the ECM can determine the opening point of the intake valves. The ECM opens on injector at a time rather than all at once to enhance performance and emissions control.

The ignition module processes the camshaft sensor signal and passes it along the ECM for sequential injection. The cam signal at the ECM (circuit 630) coincides with the rise in the Reference High signal (circuit 430). Once the ECM has identified which 430 signal is 70 degrees BTDC No. 1 cylinder firing, it can pulse the injector that is 70 degrees BTDC intake stroke (No. 4 cylinder in this case). In this way, the ECM can time or sequence fuel injection so that fuel is delivered just prior to intake valve opening.

If this circuit is opened or grounded, the ECM will terminate sequential injection in favor of multiport injection (all injectors at once). A Code 41 will be stored with a service engine soon light on. If this problem occurs in circuit 630 it will not cause the

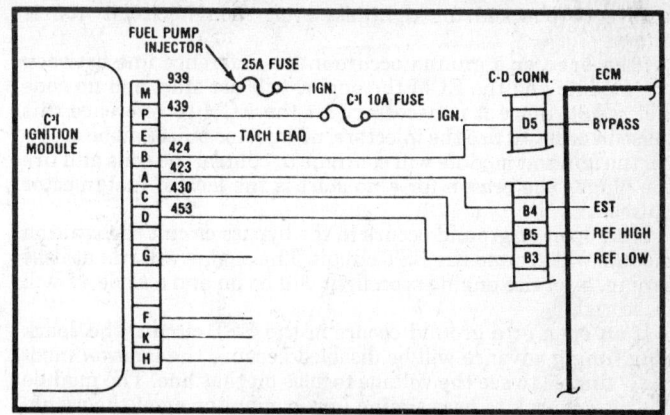

Typical wiring schematic of the C³I ignition module to ECM interface

engine to stall or have difficulty in starting but, some degraded performance and higher emissions may occur.

If an open occurs in the ground line between the ignition module and the ECM it may or may not effect operation, the symptoms can vary from no effect at all to a no start condition. This discrepancy in symptoms results from the integrity of the redundant vehicle ground circuits and the ground level of the ignition module. the mounting studs of the module may effect the EST as well as the primary ignition coil circuit so, they should

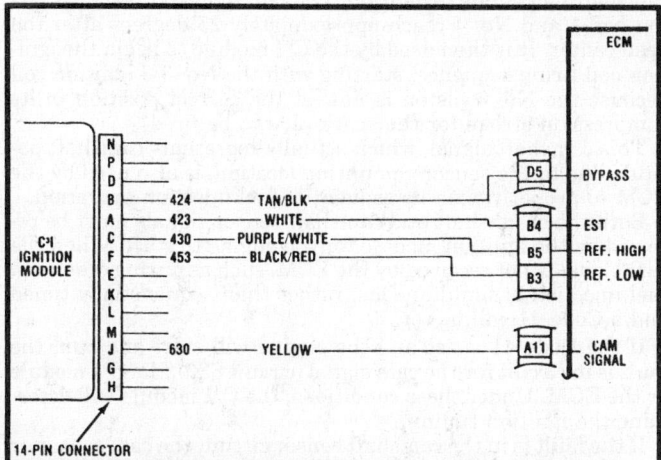

Typical wiring schematic of the C³I (SFI) ignition module to ECM interface

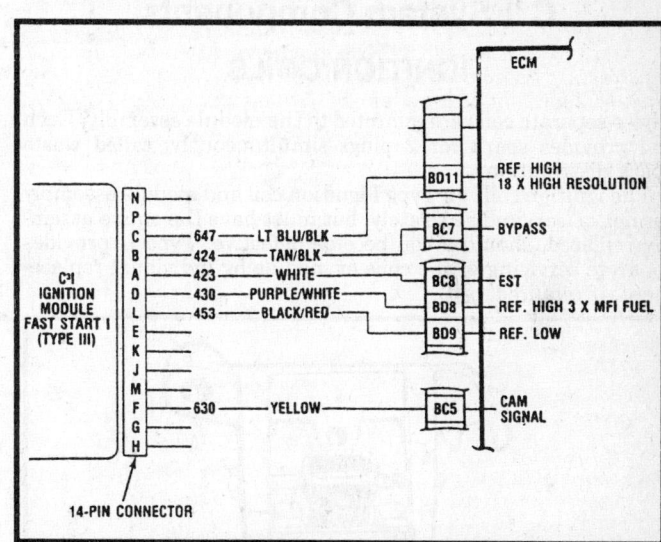

Typical wiring schematic of the C³I ignition module to ECM interface on the Buick 3800

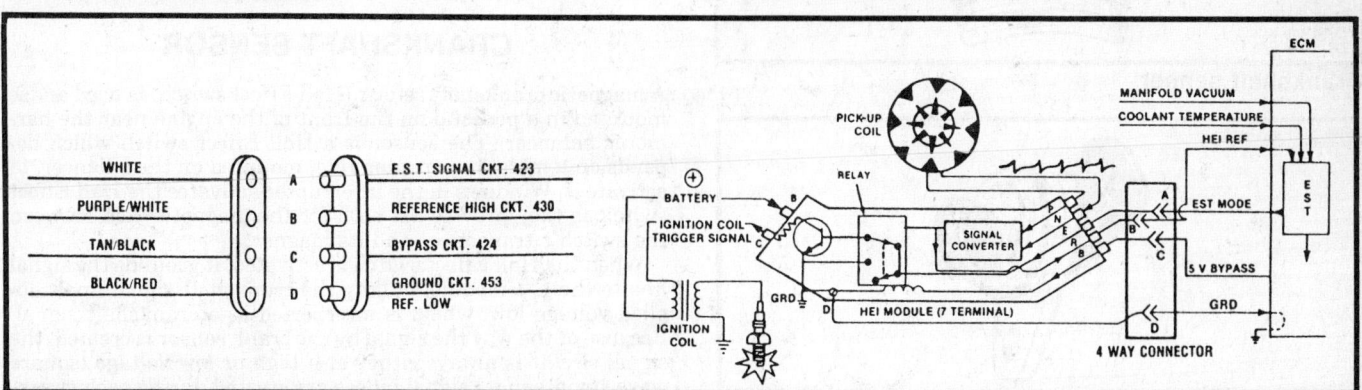

Typical wiring schematic of the standard EST interface

always be checked for tightness even though circuit 453 is intact.

If an open or a ground occurs in the reference line between the module and the ECM the engine will not start and no code will result. Since it is necessary for the ECM to reference this pulse in order to fire the injectors, no injector pulse can be created. the ignition module will continue to control the coils and fire the plugs. The reason for a no start is the lack of fuel injector pulses.

If an open or ground occurs in the bypass circuit the ignition module will ground the EST circuit. The engine will run at base timing, a service engine soon light will be on and a Code 42 will be stored.

If an open or a ground occurs in the EST circuit, the spark plug timing advance will be disabled because the ignition module is unable to see the voltage toggles on that line. The module will revert back to base timing (direct monitoring of the crank/cam sensor) for ignition timing, the service being soon light will be on and a Code 42 will be stored. Note that Code 42 has 4 possible failures-an open or grounded bypass circuit and an open or grounded EST circuit.

The 3800 engine uses an additional wire to the ECM the $18 \times$ signal. This signal (circuit 647) issues 18 pulses per crankshaft revolution to the ECM This signal provides more accuracy than the $3 \times$ (circuit 430) particularly at low engine speeds. This high resolution $18 \times$ signal also offers great versatility for future considerations, such as crankshaft acceleration and deceleration rates.

C^3I System Components

IGNITION COILS

Three separate coils are mounted to the module assembly. Each coil provides spark for 2 plugs simultaneously, called waste spark distribution.

The ignition coils on Type I ignition coil and module assembly cannot be serviced separately, but must have the entire assembly replaced, should a coil become defective. Type II provides separate servicing of the coils or module by individual replacement, if required.

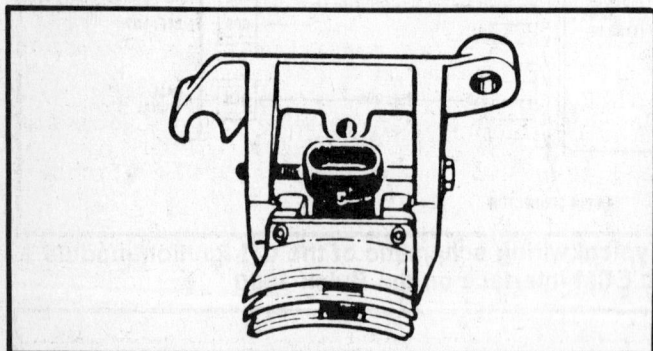

Crankshaft sensor

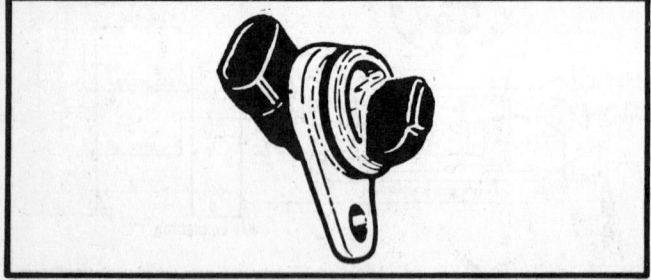

Camshaft sensor

C^3I MODULE

The Ignition module monitors the cam and crankshaft signals. This information is passed on to the ECM so that the correct spark and fuel injector timing can be maintained during all driving conditions. During the cranking mode, it monitors the camshaft signal to begin the ignition firing sequence and the fuel injection timing (SFI).

Below 400 rpm, the module controls the spark advance by triggering each of the 3 coils at a predetermined interval, based on engine speed only. Above 400 rpm, the C^3I module relays the crankshaft signal to the ECM as a reference signal. The ECM then controls the spark timing and compensates for all driving conditions. The C^3I module must receive a camshaft and then a crankshaft signal, in that order, to enable the engine to start.

The C^3I module is not repairable. When a module is replaced, the 3 coils must be transferred to the new module.

CAMSHAFT SENSOR

The camshaft sensor is located on the timing cover, behind the water pump, near the camshaft sprocket. As the camshaft sprocket turns, a magnet mounted on it, activates the Hall Effect switch in the cam sensor. This grounds the signal line to the C^3I module, pulling the crankshaft signal line's applied voltage low. This is interpreted as a cam signal (sync. pulse). Because the way the signal is created by the crankshaft sensor, the signal circuit is always either at a high or a low voltage, known as a square voltage signal. While the camshaft sprocket continues to turn, the Hall Effect switch turns off as the magnetic field passes the cam sensor, resulting in one signal each time the camshaft makes one revolution. The cam signal is created as piston No. 1 and No. 4 reach approximately 25 degrees after top dead center. It is then used by the C^3I module to begin the ignition coil firing sequence, starting with the No. 3/6 ignition coil because the No. 6 piston is now at the correct position in its compression stroke for the spark plug to be fired.

This camshaft signal, which actually represents camshaft position due to the sensors mounting location, is also used by the ECM to properly time its sequential fuel injection operation.

Both the crankshaft and camshaft sensor signals must be received by the ignition module for the engine to start. When the cam signal is not received by the ECM, such as during cranking, fuel injection is simultaneous, rather than sequenstially timed and a Code 41 will be set.

If the Code 41 is set and the engine will start and run, the fault is in circuit for the cam signal (circuit 630), the C^3I module or the ECM. Under these conditions, the C^3I module will determine the ignition timing.

If the fault is in the camshaft sensor circuit, the cam sensor or the cam signal portion of the C^3I module, Code 41 may also be present, but the engine will not start, since the C^3I module can not determine the position of the No. 1 piston.

CRANKSHAFT SENSOR

A magnetic crankshaft sensor (Hall Effect switch) is used and is mounted in a pedestal on the front of the engine near the harmonic balancer. The sensor is a Hall Effect switch which depends on a metal interrupter ring, mounted on the balancer, to activate it. Windows in the interrupter activates the Hall Effect switch as they provide a a path for the magnetic field between the switch's transducer and its magnet.

When the Hall Effect switch is activated, it grounds the signal line to the C^3I module, pulling the crankshaft signal line's applied voltage low, which is interpreted as a crankshaft signal. Because of the way the signal by the crank sensor is created, the signal circuit is always either at a high or low voltage (square wave signal) and 3 signal pulses are created during each crankshaft revolution. This signal is used by the C^3I module to create

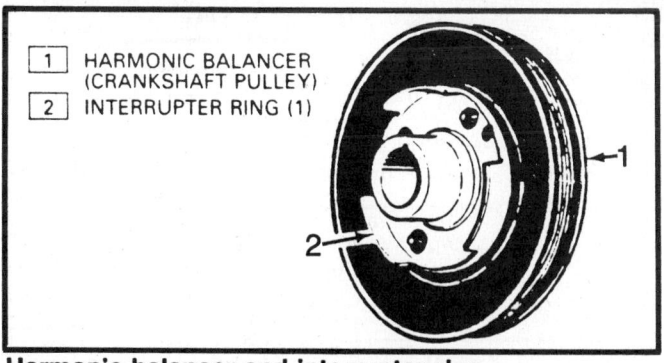

1. HARMONIC BALANCER (CRANKSHAFT PULLEY)
2. INTERRUPTER RING (1)

Harmonic balancer and interrupter ring

a reference signal, which is also a square wave signal, similar to the crank signal. The reference signal is used to calculate engine rpm and crankshaft position by the ECM. A misadjusted sensor or bent interrupter ring could cause rubbing of the sensor resulting in potential driveability problems, such as rough idle, poor engine performance, or a no-start condition.

NOTE: Failure to have the correct clearance between the sensor and the interrupter ring could damage the sensor.

The crankshaft sensor is not adjustable for ignition timing but positioning of the interrupter ring is very important. A clearance of 0.025 in. is required on either side of the interrupter ring.

A crankshaft sensor that is damaged due to mispositioning, or a bent interrupter ring can result in an engine hesitation, sag stumble or dieseling condition. To determine if the crankshaft sensor is at fault, observe the diagnostic display ECM data, ED11 (engine rpm), while driving the vehicle or use a scan tool. An erratic display indicates that a proper reference pulse has not been received by the ECM, which may be the result of a malfunctioning crankshaft sensor.

COMBINATION SENSOR

On certain engine applications, such as the 3.0L engine, the crankshaft and camshaft functions are combined into one dual sensor, called a combination sensor, which is mounted at the harmonic balancer. It functions the same as though both camshaft and crankshaft sensors are used.

The reason this type of sensor can be used is the fact that the 3.0L engine is a simultaneously injected engine, which does not require the actual camshaft signal. Instead, it uses a sync Pulse signal from the combination sensor. at the rate of one per each revolution of the crankshaft. The combination sensor is activated and controls its signal lines in the same way the crankshaft sensor does on the other engines. The only difference is the sync pulse portion of the sensor, which serves the same purpose as the cam sensor on the other engines, relative to ignition operation. That is , it starts the ignition coil firing sequence, starting with the 3–6 ignition coil.

ELECTRONIC CONTROL MODULE (ECM)

The ECM is responsible for maintaining the proper spark and fuel injection timing for all driving conditions.

To provide optimum driveability and emissions, the ECM monitors input signals from the following components in calculating Electronic Spark Timing (EST):
1. Ignition module
2. Coolant temperature
3. Manifold air temperature sensor
4. Mass air flow sensor
5. Park/neutral switch

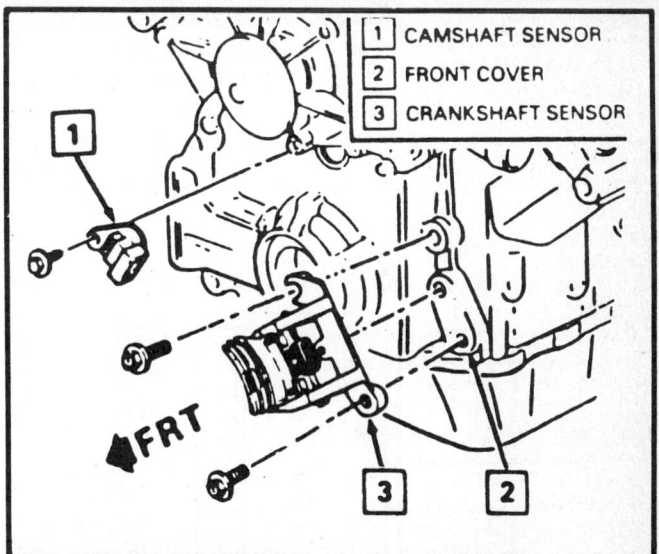

1. CAMSHAFT SENSOR
2. FRONT COVER
3. CRANKSHAFT SENSOR

FRT

Crankshaft and camshaft sensor locations

6. ESC module
7. Throttle position sensor (TPS)
8. Vehicle speed sensor

Under 400 rpm, the ECM will start injector timing (simultaneous) as soon as the C³I module receives a camshaft signal, syncronizes the spark and produces a reference signal for the ECM to calculate the fuel ignition timing sequence. The C³I module controls the spark timing during this period. Over 400 rpm, the ECM controls timing (EST) and also changes the mode of fuel injection to sequential, providing a camshaft signal is received.

ELECTRONIC SPARK CONTROL (ESC)

The ESC systems is comprised of a knock sensor and an ESC module. The ECM monitors the ESC signal to determine when engine detonation occurs.

As long as the ESC module is sending a voltage signal of 8–10 volts to the ECM, indicating that no detonation is detected by the ESC sensor, the ECM provides normal spark advance. When the knock sensor detects detonation, the ESC module turns off the circuit to the ECM and the voltage at the ECM terminal B7 drops to 0 volts. The ECM then retards EST to reduce detonation.

NOTE: Retarded timing can be the result of excessive engine noise, caused by valve lifters, pushrods or other mechanical engine or transmission noise.

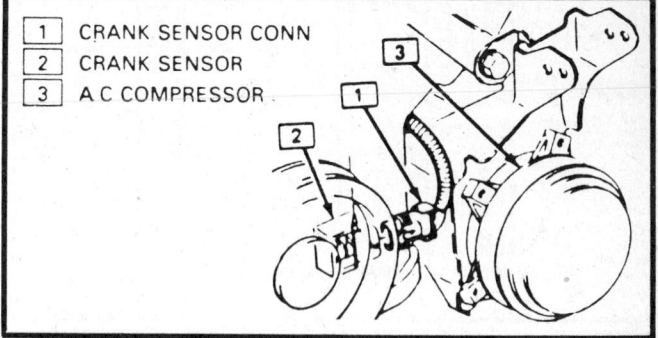

1. CRANK SENSOR CONN
2. CRANK SENSOR
3. A C COMPRESSOR

Typical combination sensor

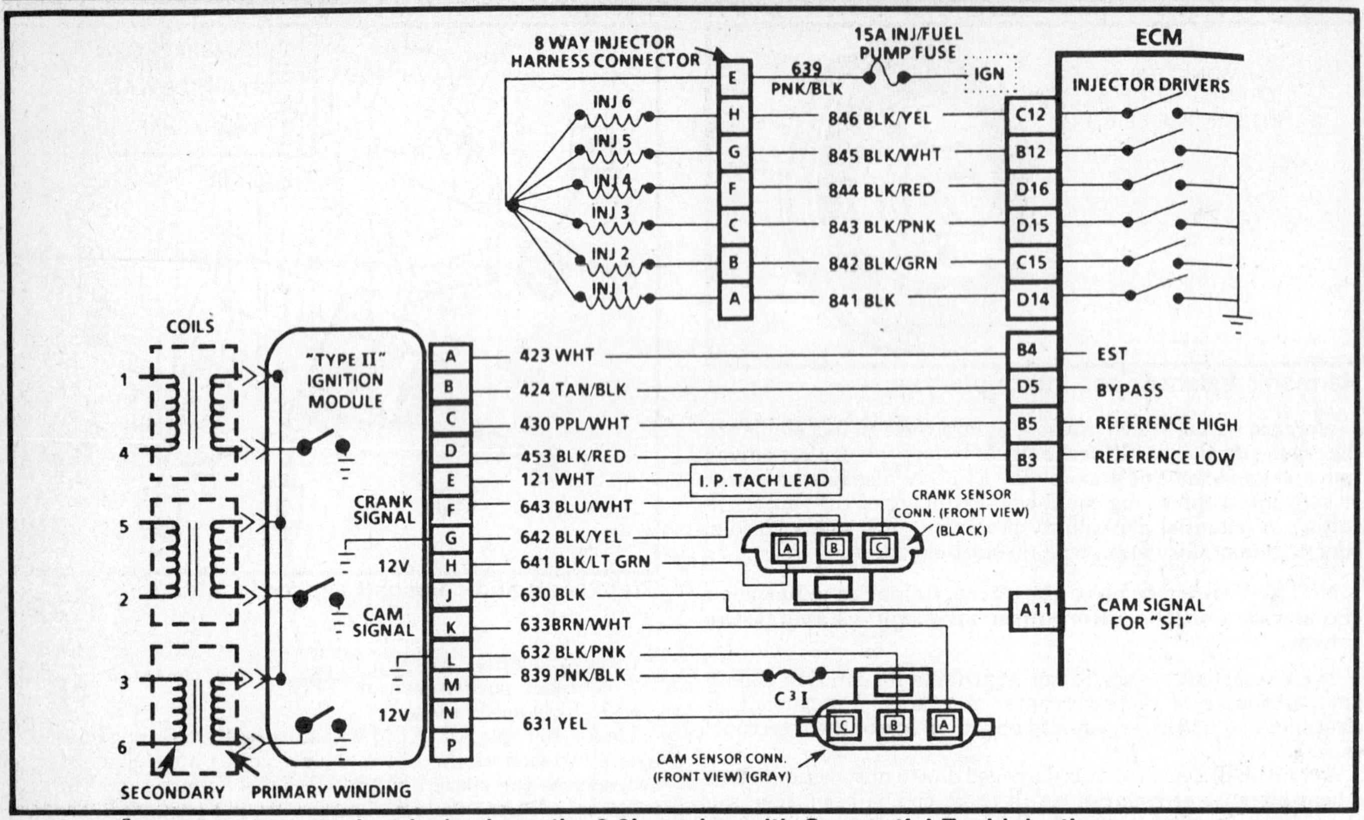

ELECTRONIC IGNITION SYSTEMS
DIRECT IGNITION

8 WAY INJECTOR HARNESS CONNECTOR

15A INJ/FUEL PUMP FUSE

IGN

639 PNK/BLK

ECM

INJECTOR DRIVERS

E	639 PNK/BLK	
H	846 BLK/YEL	C12
G	845 BLK/WHT	B12
F	844 BLK/RED	D16
C	843 BLK/PNK	D15
B	842 BLK/GRN	C15
A	841 BLK	D14

INJ 6, INJ 5, INJ 4, INJ 3, INJ 2, INJ 1

COILS

"TYPE II" IGNITION MODULE

1, 4, 5, 2, 3, 6

A	423 WHT	B4 — EST
B	424 TAN/BLK	D5 — BYPASS
C	430 PPL/WHT	B5 — REFERENCE HIGH
D	453 BLK/RED	B3 — REFERENCE LOW
E	121 WHT	I. P. TACH LEAD
F	643 BLU/WHT	
G	642 BLK/YEL	
H	641 BLK/LT GRN	
J	630 BLK	A11 — CAM SIGNAL FOR "SFI"
K	633 BRN/WHT	
L	632 BLK/PNK	
M	839 PNK/BLK	
N	631 YEL	
P		

CRANK SIGNAL
12V
CAM SIGNAL
12V

CRANK SENSOR CONN. (FRONT VIEW) (BLACK)
A B C

C³I

CAM SENSOR CONN. (FRONT VIEW) (GRAY)
C B A

SECONDARY   PRIMARY WINDING

Type II C³I Ignition system electrical schematic, 3.8L engine with Sequential Fuel Injection

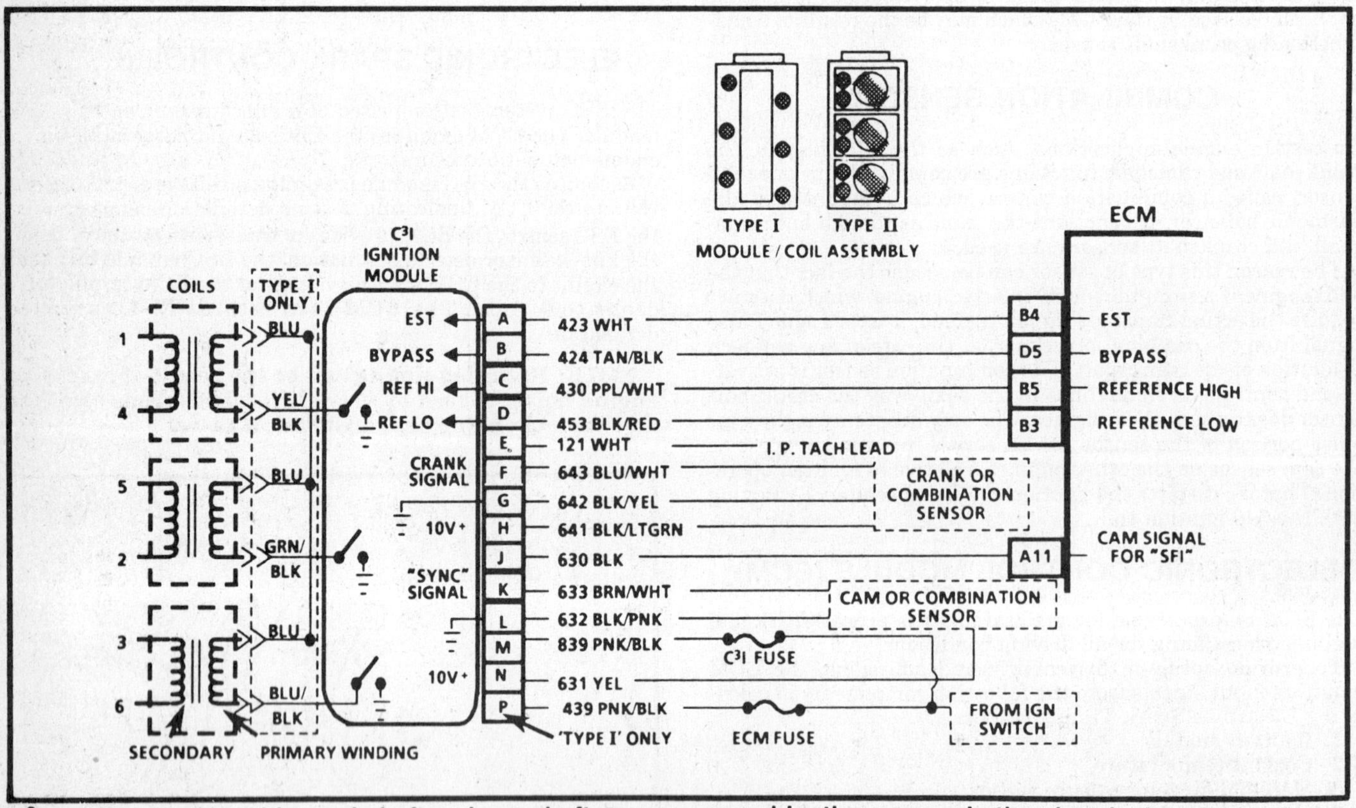

TYPE I   TYPE II
MODULE / COIL ASSEMBLY

ECM

COILS

C³I IGNITION MODULE

TYPE I ONLY

1 — BLU
4 — YEL/ BLK
5 — BLU
2 — GRN/ BLK
3 — BLU
6 — BLU/ BLK

EST ←	A	423 WHT	B4 — EST
BYPASS ←	B	424 TAN/BLK	D5 — BYPASS
REF HI ←	C	430 PPL/WHT	B5 — REFERENCE HIGH
REF LO ←	D	453 BLK/RED	B3 — REFERENCE LOW
	E	121 WHT	I. P. TACH LEAD
CRANK SIGNAL	F	643 BLU/WHT	
	G	642 BLK/YEL	CRANK OR COMBINATION SENSOR
10V+	H	641 BLK/LTGRN	
	J	630 BLK	A11 — CAM SIGNAL FOR "SFI"
"SYNC" SIGNAL	K	633 BRN/WHT	
	L	632 BLK/PNK	CAM OR COMBINATION SENSOR
10V+	M	839 PNK/BLK	C³I FUSE
	N	631 YEL	
	P	439 PNK/BLK	ECM FUSE — FROM IGN SWITCH

'TYPE I' ONLY

SECONDARY   PRIMARY WINDING

C³I Ignition system with crankshaft and camshaft sensor or combination sensor in the electrical system

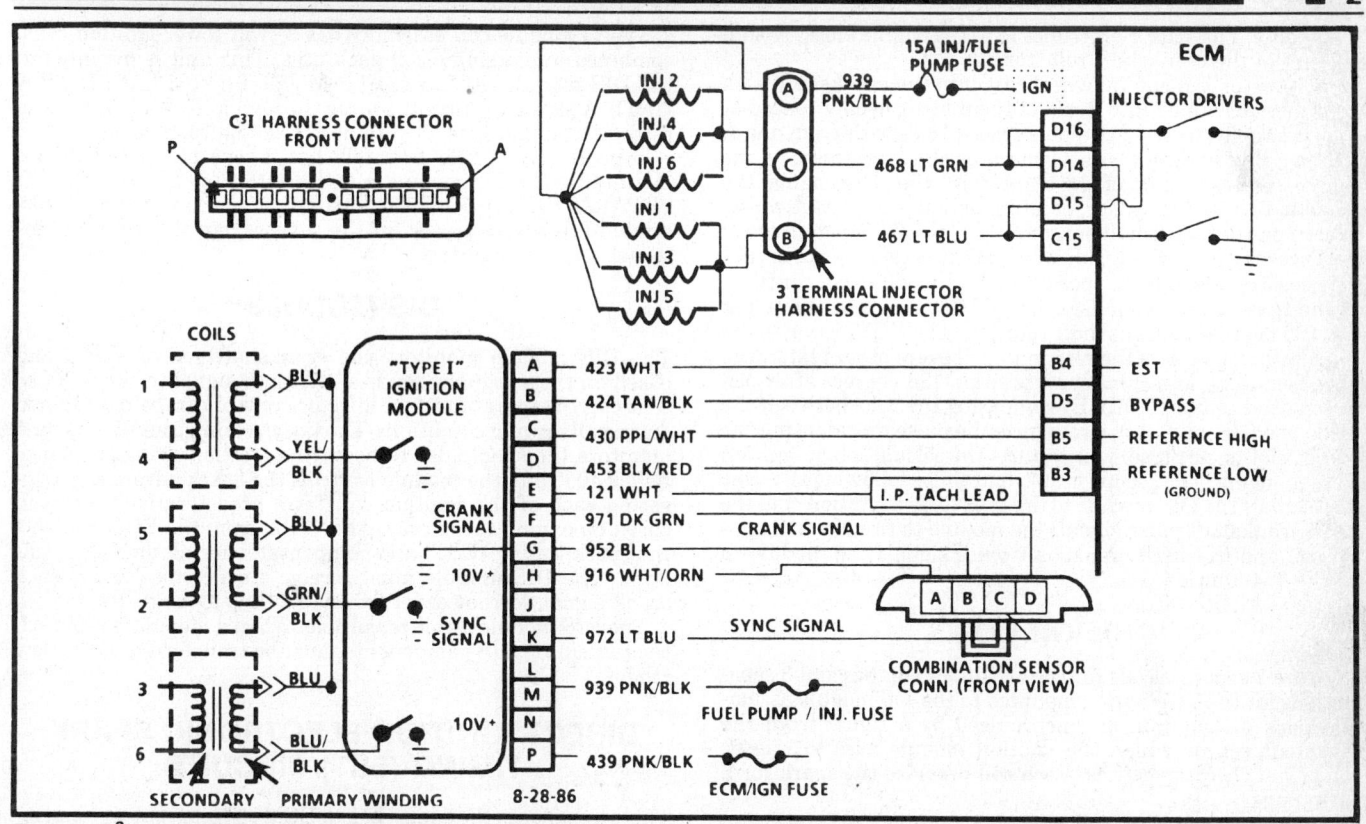

Type I C³I Ignition system with only combination sensor

C³I ELECTRONIC SPARK TIMING (EST) CIRCUITS

This system uses the same EST to ECM circuits that the distributor type systems with EST use. However, a difference does exist between the C³I system and the DIS.

The following is a brief description for the EST circuits and the camshaft signal circuit (circuit 630).

REFERENCE SIGNAL, CIRCUIT 430

This circuit provides the ECM with the rpm and crankshaft position information from the C³I module. The C³I module receives the signal from the crankshaft sensor's Hall Effect switch.

This signal will either be high or low, depending upon the position of the interrupter ring. This high-low signal is used to trigger the C³I module for ignition operation and by the ECM to calculate fuel injection timing. Both the camshaft and crankshaft sensor signals must be received by the C³I module in order for a reference signal to be produced on circuit 430. A loss of the reference signal would prevent the engine from running.

BYPASS SIGNAL, CIRCUIT 424

At approximately 400 rpm, the ECM applies 5 volts to this circuit to switch spark timing control from the C³I module to the ECM.

An open or grounded bypass circuit will set a Code 42 and result in the engine operating in a back-up ignition timing mode (module timing) at a calculated timing value. This may cause poor performance and reduced fuel economy.

EST SIGNAL, CIRCUIT 423

The C³I module sends a reference signal to the ECM when the engine is cranking. While the engine is under 400 rpm, the C³I module controls the ignition timing. When the engine speed exceeds 400 rpm, the ECM applies 5 volts to the bypass line to switch the timing to the ECM control (EST).

An open or ground in the EST circuit will stall the engine and set a Code 42. The engine can be restarted, but will operate in a back-up ignition timing mode (module timing) at a calculated timing value. This may cause poor performance and reduced fuel economy.

CAM SIGNAL, CIRCUIT 630

The ECM uses this signal to determine the position of the No. 1 piston in its compression stroke. This signal is used by the ECM to calculate the sequential fuel injection (SFI) mode of operation. A loss of this signal will set a Code 41. If the cam signal is lost while the engine is running, the fuel injection system will shift to the simultaneous injection mode of operation and the engine will continue to operate. The engine can be re-started, but will continue to run in the simultaneous mode as long as the fault is present.

Diagnostics and Testing
C³I IGNITION SYSTEM/EST

NOTE: Verification of Type I or Type II systems is very important, because the diagnostics are not the same for both types.

If the engine cranks, but will not operate, or starts and immediately stalls, further diagnosis must be made to determine if the failure is in the ignition system or the fuel system.

Direct Ignition System Components

NOTE: The Direct Ignition System/EST is used with TBI and Ported fuel injection systems.

CRANKSHAFT SENSOR

A magnetic crankshaft sensor (Hall Effect switch) is used and is remotely mounted on the opposite side of the engine from the

DIS module. The sensor protrudes in to the engine block, within 0.050 in. of the crankshaft reluctor.

The reluctor is a special wheel cast into the crankshaft with seven slots machined into it, six of them being evenly spaced at 60 degrees apart. A seventh slot is spaced 10 degrees from one of the other slots and serves as a generator of a sync pulse. As the reluctor rotates as part of the crankshaft, the slots change the magnetic field of the sensor, creating an induced voltage pulse.

Based on the crankshaft sensor pulses, the DIS module sends reference signals to the ECM, which are used to indicate crankshaft position and engine speed. The DIS module will continue to send these reference pulses to the ECM at a rate of one per each 120 degrees of crankshaft rotation. The ECM actvates the fuel injectors, based on the recognition of every other reaference pulse, beginning at a crankshaft position 120 degrees after piston top dead center (TDC). By comparing the time between the pulses, the DIS module can recognize the pulse representing the seventh slot (sync pulse) which starts the calculation of ignition coil sequencing. The second crankshaft pulse following the sync pulse signals the DIS module to fire the No. 2–5 ignition coil, the fourth crankshaft pulse signals the module to fire No. 3–6 ignition coil and the sixth crankshaft pulse signals the module to fire the 1–4 ignition coil.

IGNITION COILS

There are 2 separate coils for the 4 cylinder engines and 3 separate coils for the V6 engines, mounted to the coil/module assembly. Spark distribution is synchronized by a signal from the crankshaft sensor which the ignition module uses to trigger each coil at the proper time. Each coil provides the spark for 2 spark plugs.

Two types of ignition coil assemblies are used, Type I and Type II During the diagnosis of the systems, the correct type of ignition coil assembly must be identified and the diagnosis directed to that system.

Type I module/coil assembly has 3 twin tower ignition coils, combined into a single coil pack unit. This unit is mounted to the DIS module. ALL 3 COILS MUST BE REPLACED AS A UNIT. A separate current source through a fused circuit to the module terminal **P** is used to power the ignition coils.

Type II coil/module assembly has 3 separate coils that are mounted to the DIS module. EACH COIL CAN BE REPLACED SEPARATELY. A fused low current source to the module terminal **M**, provides power for the sensors, ignition coils and internal module circuitry.

DIS MODULE

The DIS module monitors the crankshaft sensor signal and based on these signals, sends a reference signal to the ECM so that correct spark and fuel injector control can be maintained during all driving conditions. During cranking, the DIS module monitors the sync-pulse to begin the ignition firing sequence. Below 400 rpm, the module controls the spark advance by triggering each of the ignition coils at a predetermined interval, based on engine speed only. Above 400 rpm, the ECM controls the spark timing (EST) and compensates for all driving conditions. The DIS module must receive a sync-pulse and then a crank signal, in that order, to enable the engine to start.

The DIS module is not repairable. When a module is replaced, the remaining DIS components must be transferred to the new module.

DIRECT IGNITION ELECTRONIC SPARK TIMING (EST) CIRCUITS

This system uses the same EST to ECM circuits that the distributor type systems with EST use. However, a difference does exist between the C³I system and the DIS.

The following is a brief description for the EST circuits:

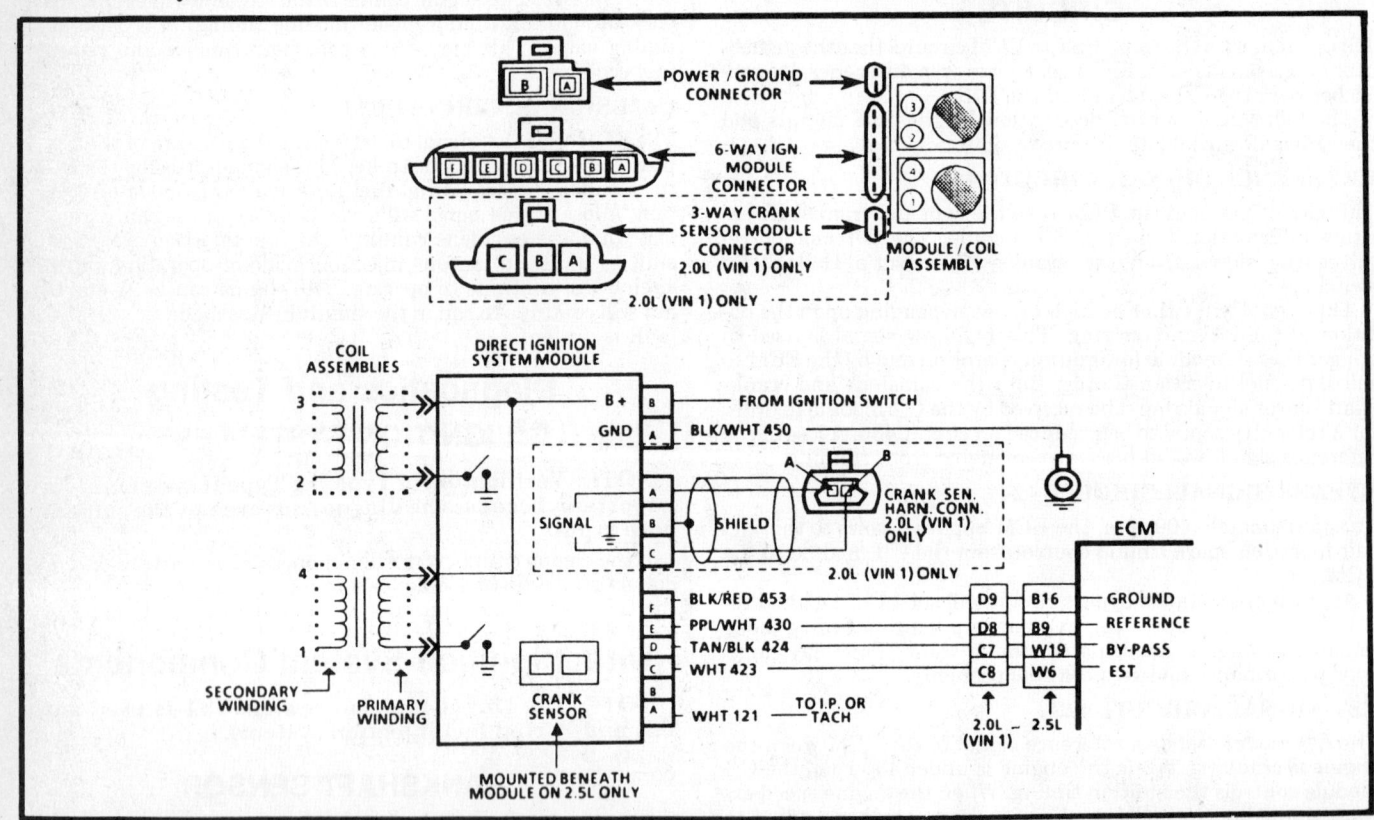

Typical Direct Ignition system electrical schematic

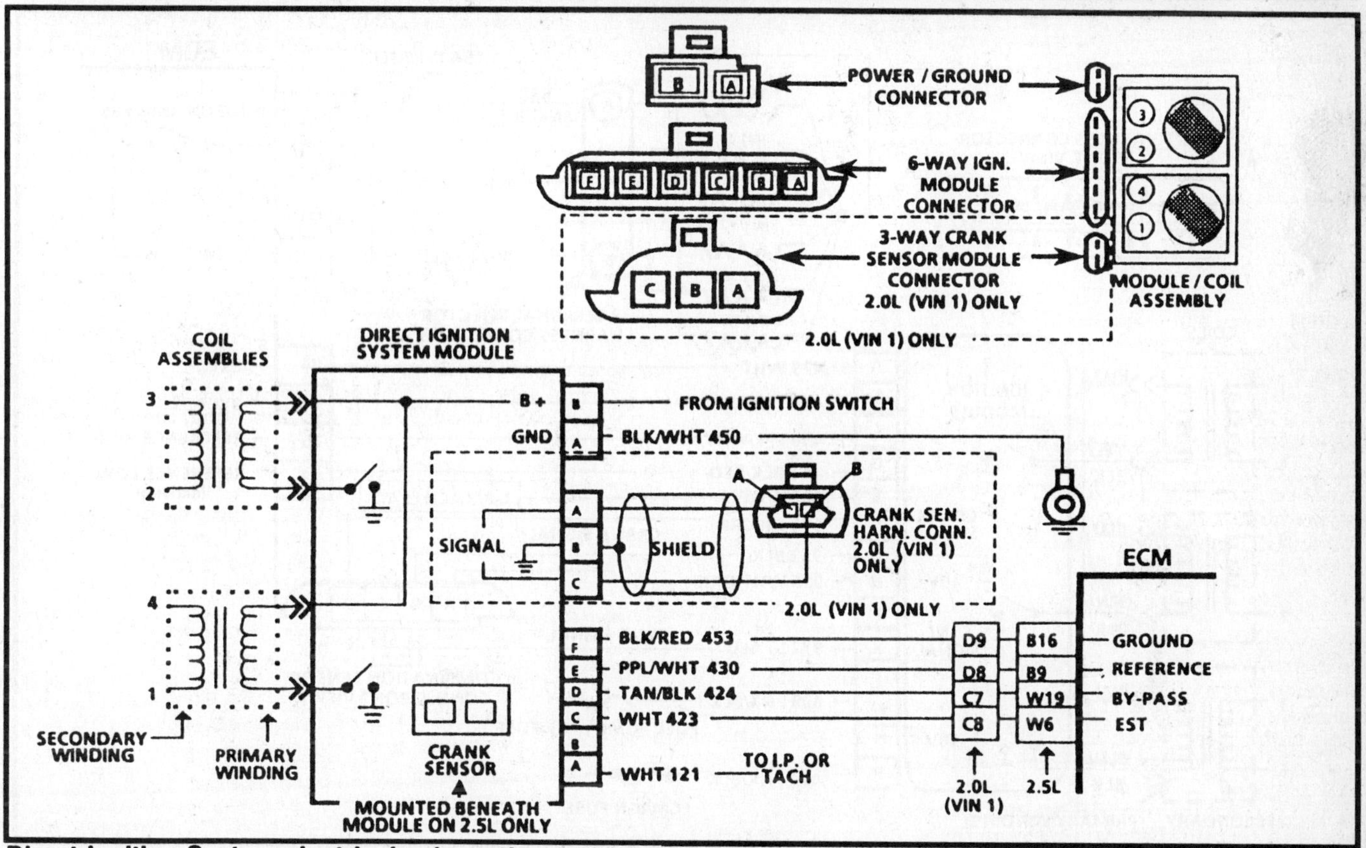

POWER / GROUND CONNECTOR

B A

6-WAY IGN. MODULE CONNECTOR

F E D C B A

3-WAY CRANK SENSOR MODULE CONNECTOR 2.0L (VIN 1) ONLY

C B A

2.0L (VIN 1) ONLY

MODULE / COIL ASSEMBLY

3 2 4 1

COIL ASSEMBLIES

DIRECT IGNITION SYSTEM MODULE

3 2 4 1

SECONDARY WINDING

PRIMARY WINDING

B+ B FROM IGNITION SWITCH
GND A BLK/WHT 450

SIGNAL A B C SHIELD

CRANK SEN. HARN. CONN. 2.0L (VIN 1) ONLY

2.0L (VIN 1) ONLY

A B

CRANK SENSOR MOUNTED BENEATH MODULE ON 2.5L ONLY

F BLK/RED 453 D9 B16 GROUND
E PPL/WHT 430 D8 B9 REFERENCE
D TAN/BLK 424 C7 W19 BY-PASS
C WHT 423 C8 W6 EST
B
A WHT 121 TO I.P. OR TACH

ECM

2.0L (VIN 1) 2.5L

Direct Ignition System electrical schematic—2.0L/ 2.5L engines

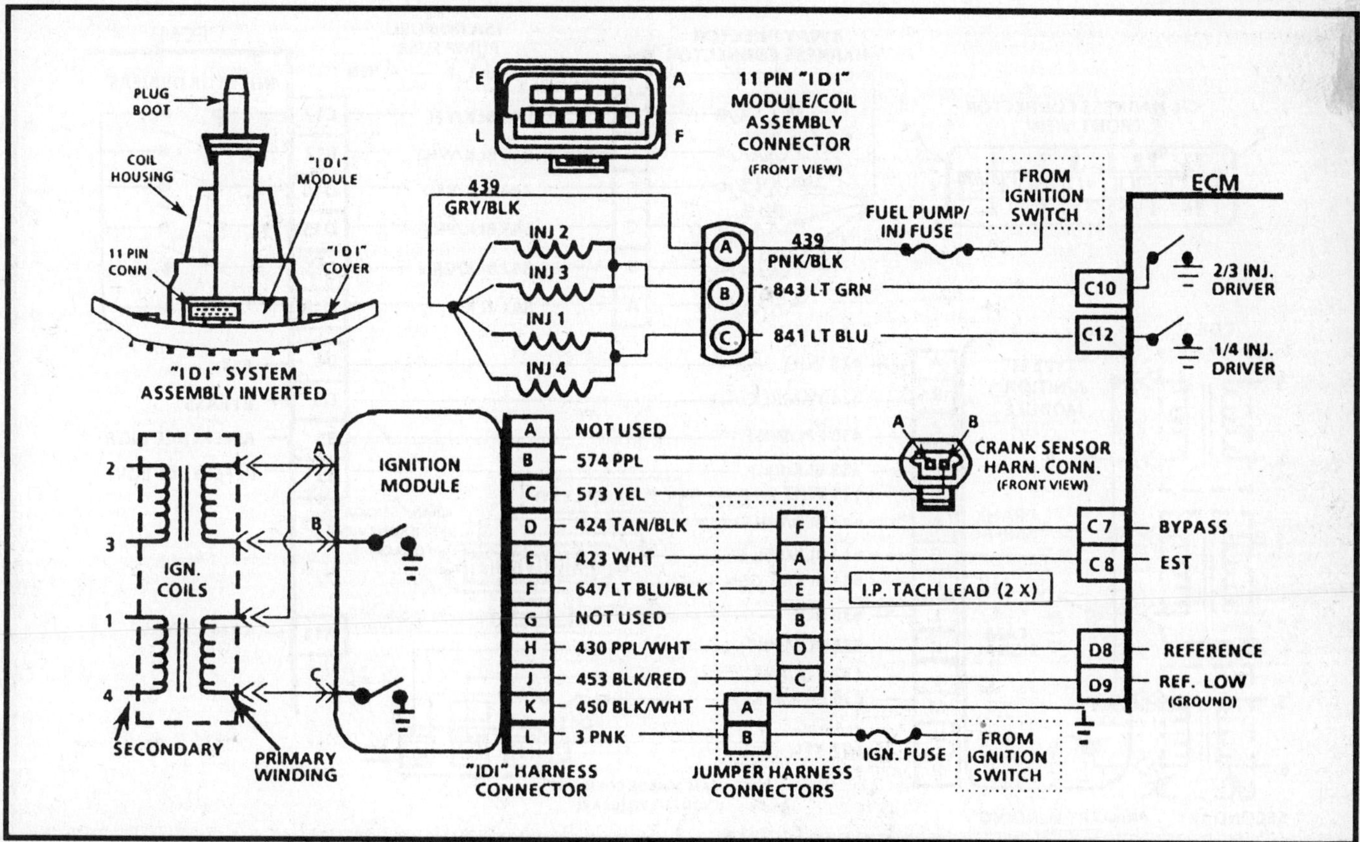

PLUG BOOT

COIL HOUSING

"IDI" MODULE

11 PIN CONN

"IDI" COVER

"IDI" SYSTEM ASSEMBLY INVERTED

11 PIN "IDI" MODULE/COIL ASSEMBLY CONNECTOR (FRONT VIEW)

E   A
L   F

439 GRY/BLK

INJ 2
INJ 3
INJ 1
INJ 4

A 439 PNK/BLK
B 843 LT GRN
C 841 LT BLU

FUEL PUMP/ INJ FUSE

FROM IGNITION SWITCH

ECM

C10 2/3 INJ. DRIVER
C12 1/4 INJ. DRIVER

IGNITION MODULE

2 3 1 4

IGN. COILS

SECONDARY

PRIMARY WINDING

A NOT USED
B 574 PPL
C 573 YEL
D 424 TAN/BLK F
E 423 WHT A
F 647 LT BLU/BLK E
G NOT USED B
H 430 PPL/WHT D
J 453 BLK/RED C
K 450 BLK/WHT A
L 3 PNK B

A B

CRANK SENSOR HARN. CONN. (FRONT VIEW)

C7 BYPASS
C8 EST

I.P. TACH LEAD (2 X)

D8 REFERENCE
D9 REF. LOW (GROUND)

IGN. FUSE

FROM IGNITION SWITCH

"IDI" HARNESS CONNECTOR

JUMPER HARNESS CONNECTORS

Integrated Direct Ignition System electrical schematic—2.3L Quad 4 engine

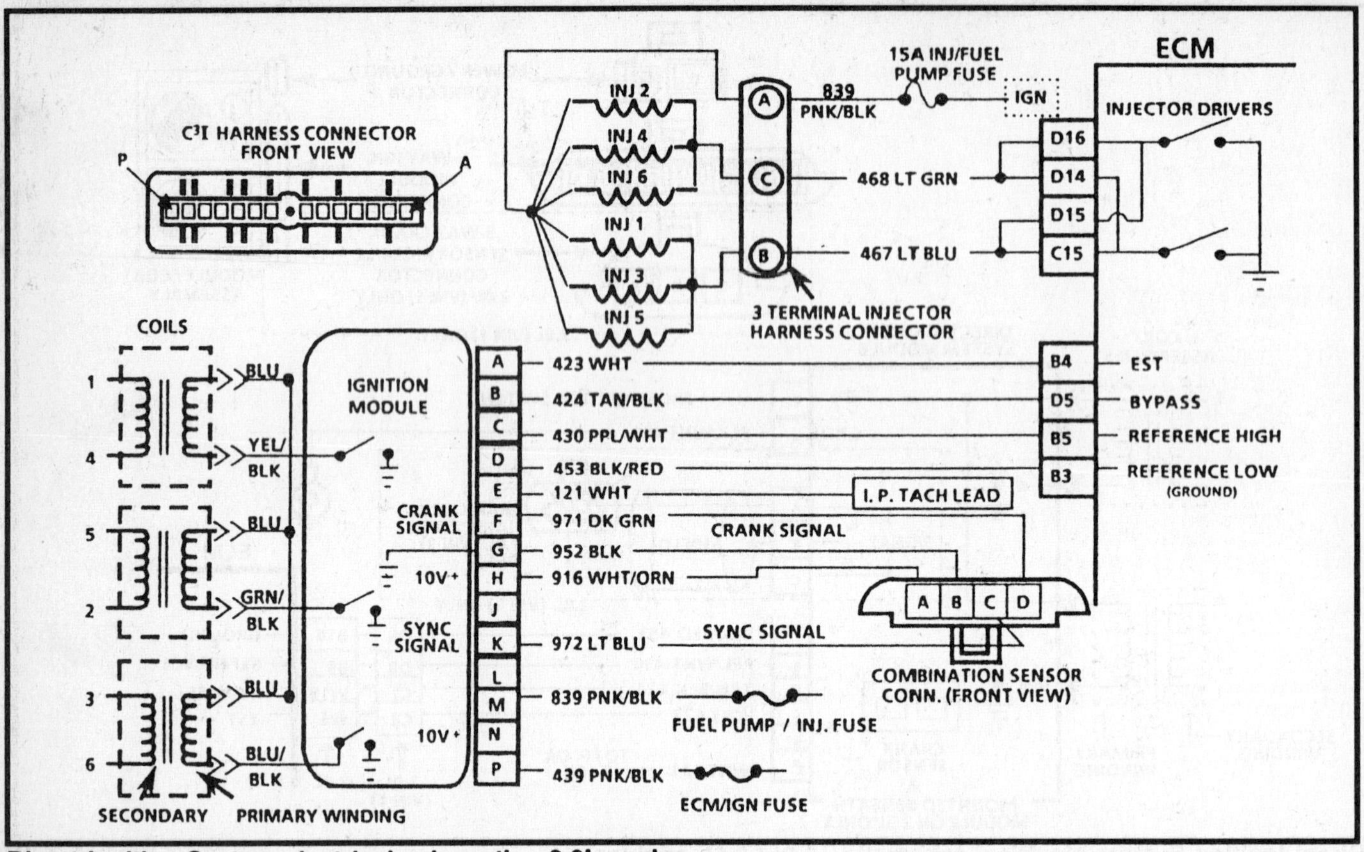

Direct Ignition System electrical schematic—3.0L engine

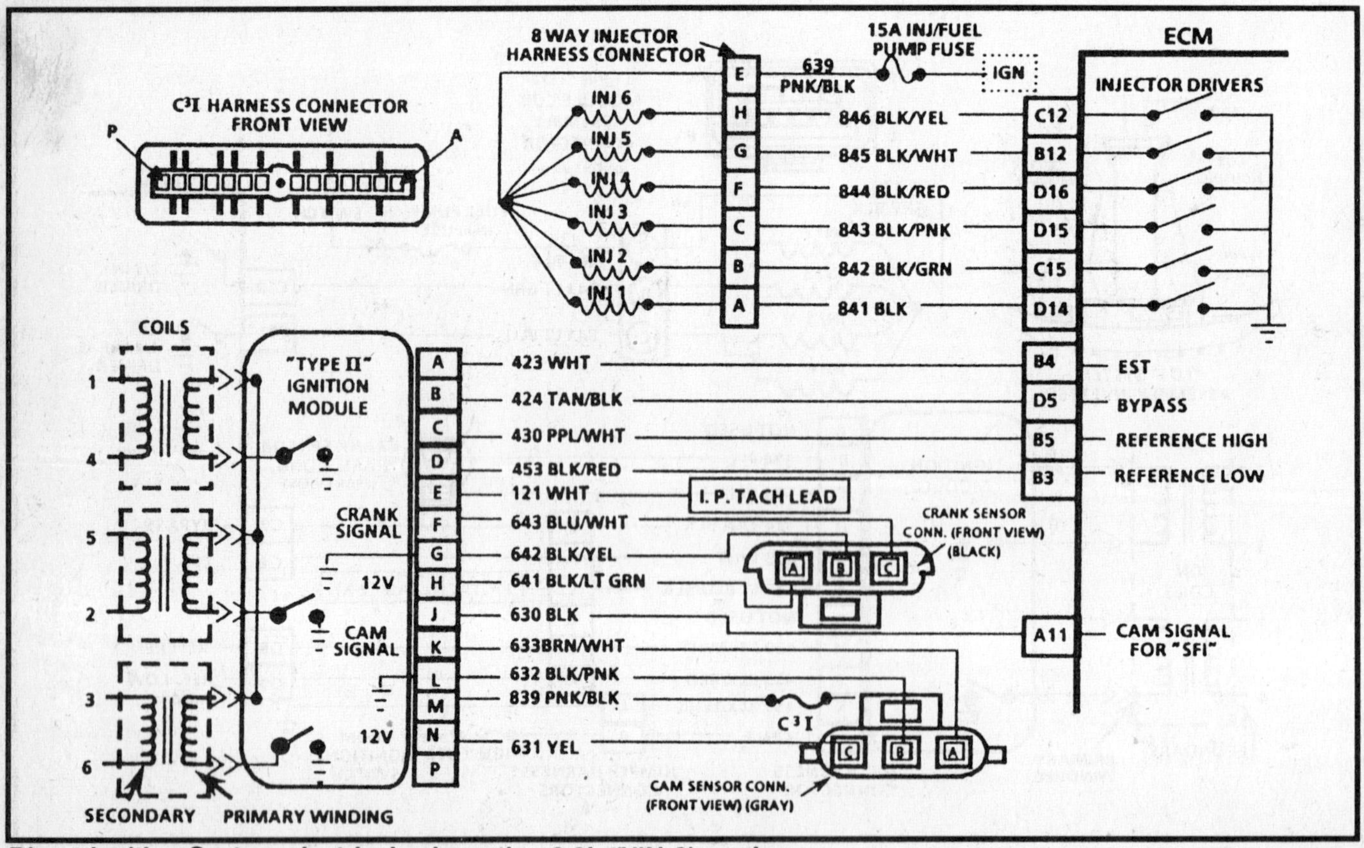

Direct Ignition System electrical schematic—3.8L (VIN 3) engine

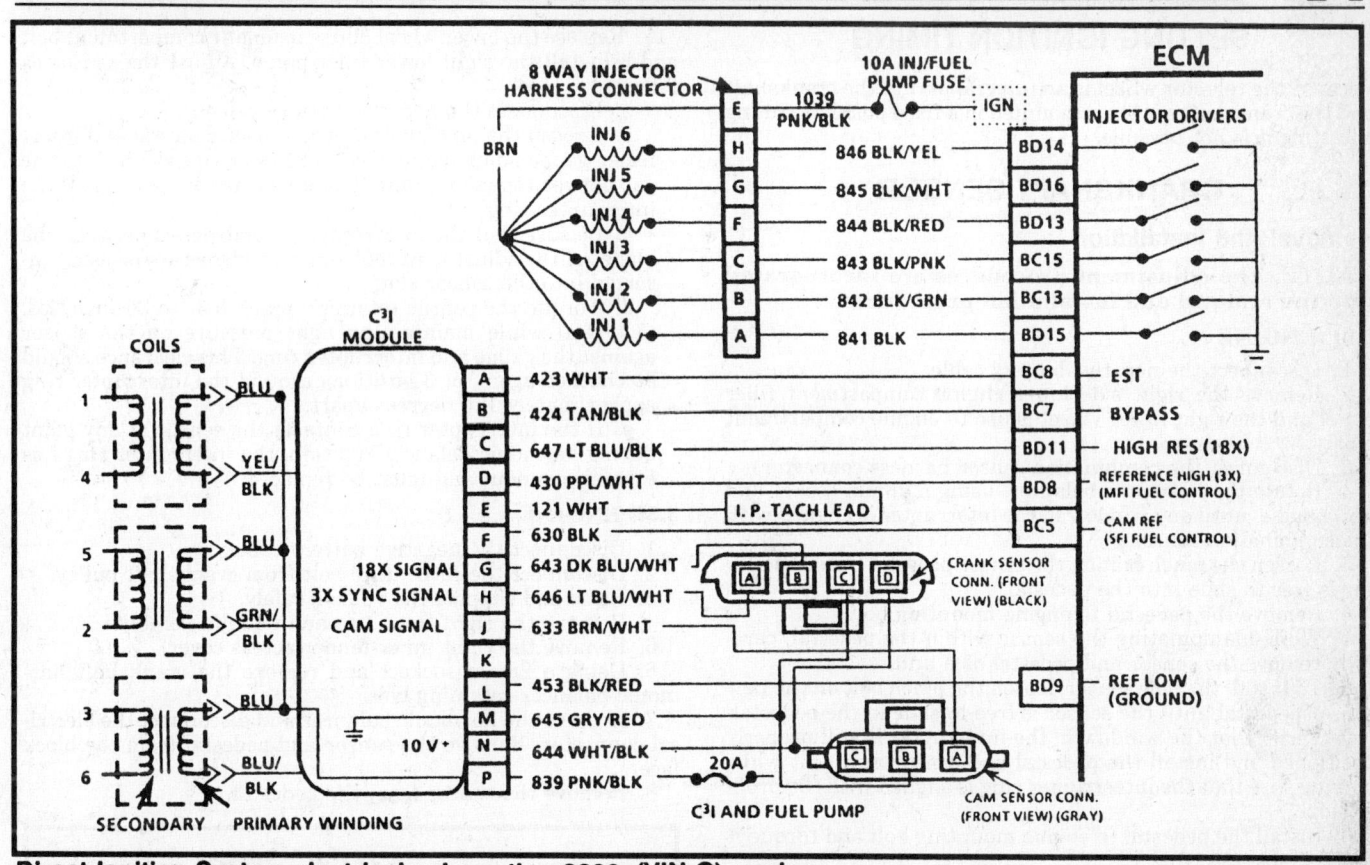

Direct Ignition System electrical schematic—3800 (VIN C) engine

DIS REFERENCE, CIRCUIT 430

The crankshaft sensor generates a signal to the ignition module, which results in a reference pulse being sent to the ECM. The ECM uses this signal to calculate crankshaft position and engine speed for injector pulse width.

NOTE: The crankshaft sensor is mounted to the base of the DIS module on the 2.5L 4 cylinder engines and is mounted directly into the side of the engine block.

REFERENCE GROUND, CIRCUIT 453

This wire is grounded through the module and insures that the ground circuit has no voltage drop between the ignition module and the ECM, which can affect performance.

BYPASS, CIRCUIT 424

At approximately 400 rpm, the ECM applies 5 volts to this circuit to switch spark timing control from the DIS module to the ECM. An open or grounded by pass circuit will set a Code 42 and result in the engine operating in a back-up ignition timing mode (module timing) at a calculated timing value. This may cause poor performance and reduced fuel economy.

ELECTRONIC SPARK TIMING (EST), CIRCUIT 423

The DIS module sends a reference signal to the ECM when the engine is cranking. While the engine is under 400 rpm, the DIS module controls the ignition timing. When the engine speed exceeds 400 rpm, the ECM applies 5 volts to the Bypass line to switch the timing to the ECM control (EST).

An open or ground in the EST circuit will result in the engine continuing to run, but in a back-up ignition timing mode (module timing mode) at a calculated timing value and the "SERVICE ENGINE SOON" light will not be on. If the EST fault is still present, the next time the engine is restarted, a Code 42 will be set and the engine will operate in the module timing mode.

This may cause poor performance and reduced fuel economy.

Diagnosis and Testing

DIRECT IGNITION SYSTEM

The ECM uses information from the MAP and coolant sensors, in addition to rpm to calculate spark advance as follows:
1. Low MAP output voltage—more spark advance
2. Cold engine—more spark advance
3. High MAP output voltage—less spark advance
4. Hot engine—less spark advance

Therefore, detonation could be caused by low MAP output or high resistance in the coolant sensor circuit.

Poor performance could be caused by high MAP output or low resistance in the coolant sensor circuit.

If the engine cranks but will not operate, or starts, then immediately stalls, diagnosis must be accomplished to determine if the failure is in the DIS system or the fuel system.

CHECKING EST PERFORMANCE

The ECM will set timing at a specified value when the diagnostic TEST terminal in the ALDL connector is grounded. To check for EST operation, run the engine at 1500 rpm with the terminal ungrounded. Then ground the TEST terminal. If the EST is operating, there should be a noticeable engine rpm change. A fault in the EST system will set a trouble Code 42.

CODE 12

Code 12 is used during the diagnostic circuit check procedure to test the diagnostic and code display ability of the ECM. This code indicates that the ECM is not receiving the engine rpm (reference) signal. This occurs with the ignition key in the **ON** position and the engine not operating.

SETTING IGNITION TIMING

Because the reluctor wheel is an integral part of the crankshaft and the crankshaft sensor is mounted in a fixed position, timing adjustment is not possible.

CRANKSHAFT SENSOR

Removal And Installation

NOTE: The adjustment procedures are incorporated into the removal and installation procedures.

3.0L ENGINE

1. Disconnect the negative battery cable.
2. Remove the right side lower engine compartment filler panel and the right lower wheel house to engine compartment bolt.
3. Disconnect the combination sensor harness connector.
4. Rotate the harmonic balancer, using a 28mm socket and pull handle, until any window in the interrupter is aligned with the combination sensor.
5. Lossen the pinch bolt on the sensor pedestal until the sensor is free to slide into the pedestal.
6. Remove the pedestal to engine mounting bolt.
7. While manipulating the sensor within the pedestal, carefully remove the sensor and pedestal as a unit.
8. To install the new sensor, loosen the pinch bolt on the new sensor pedestal until the sensor is free to slide in the pedestal.
9. Verify that the window in the interrupter is still properly positioned and install the pedestal and sensor as a unit while making sure that the interrrupter ring is aligned with the proper slot.
10. Install the pedestal to engine mounting bolt and torque it to 22 ft. lbs. (30 Nm).

11. Replace the lower wheel house to engine compartment bolt and reinstall the right lower filler panel. Adjust the sensor as follows:
  a. Disconnect the negative battery cable.
  b. Loosen the pinch bolt on the sensor pedestal and insert feeler gauge adjustment tool J–36179 or equivalent into the gap between the sensor and the interrupter on each side of the interrupter ring.
  c. Be sure that the interrupter is sandwiched between the blades of the adjustment tool and both blades are properly inserted into the sensor slot.
  d. Torque the sensor retaining pinch bolt to 30 inch. lbs. (3.4 Nm) while maintaining light pressure on the sensor against the gauge and interrupter ring. This clearance should be checked again, at 3 positions around the interrupter ring approximately 120 degrees apart.
  e. If the interrupter ring contacts the sensor at any point during harmonic balancer rotation, the interrupter ring has excessive runout and must be replaced.

3.3L ENGINE

1. Disconnect the negative battery cable.
2. Disconnect the serpentine belt from crankshaft pulley.
3. Raise and support the vehicle safely.
4. Remove the right front tire and wheel assembly.
5. Remove the right inner fender access cover.
6. Using a 28mm sockect and remove the crankshaft harmonic balancer retaining bolt.
7. Remove the harmonic balancer and disconnect the electrical connector. Remove the sensor and pedestal from the block face.
8. Remove the sensor from the pedestal.

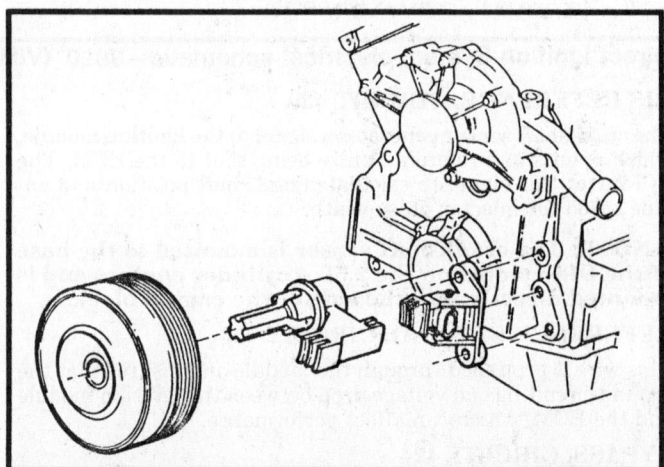

Position special tool J–37089 on the crankshaft

9. To install and adjust us the following procedure:
  a. Loosely install the crankshaft sensor on the pedestal. Position the sensor with the pedestal attached on special crankshaft sensor adjustment tool J–37089.
  b. Position the special tool on the crankshaft. Install the bolts to hold the pedestal to block and torque them to 14–28 ft. lbs. (20–40 Nm).
  c. Torque the pedestal pinch bolt to 30–35 inch. lbs. (3–4 Nm). Remove special tool J–37089.
  d. Place special tool J–37089 onto the harmonic balancer and turn it. If any vane of the harmonic balancer touches the tool, replace the balancer assembly.
  e. Install the balancer onto the crankshaft and torque the bolt to 200–239 ft. lbs. (270–325 Nm).
  f. Install the inner fender shield. Install the tire and wheel assembly. Lower the vehicle and install the serpentine belt and negative battery cable.

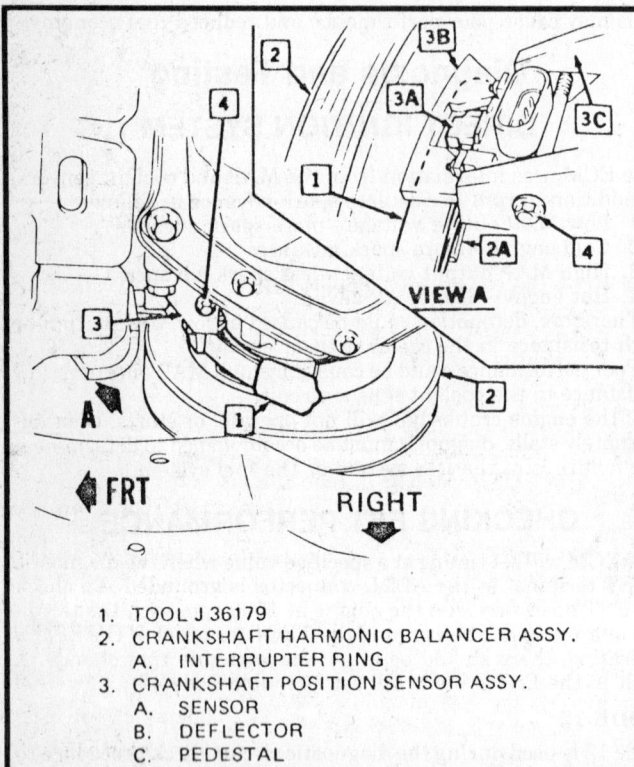

1. TOOL J 36179
2. CRANKSHAFT HARMONIC BALANCER ASSY.
  A. INTERRUPTER RING
3. CRANKSHAFT POSITION SENSOR ASSY.
  A. SENSOR
  B. DEFLECTOR
  C. PEDESTAL
4. PINCH BOLT

Combination sensor adjustment — 3.0L and 3.8L (VIN 3) engine

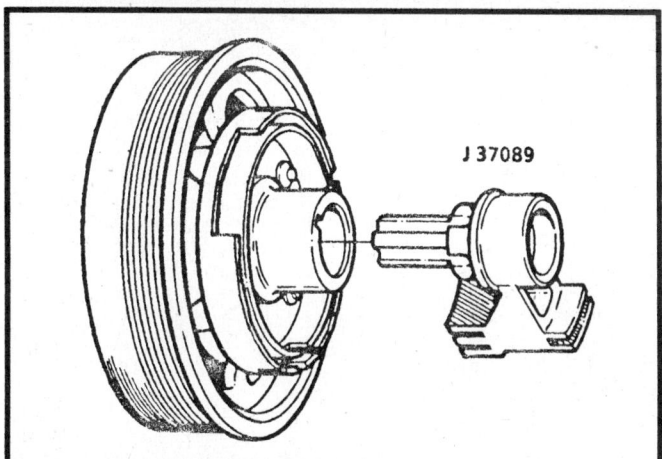

Checking the harmonic balancer vanes

3.8L (VIN 3) ENGINE

1. Disconnect crank sensor harness connector.
2. Using a 28mm socket and pull handle, rotate the harmonic balancer until any window in the interrupter is aligned with the crank sensor.
3. Loosen the pinch bolt on the sensor pedestal until the sensor is free to slide in the pedestal.
4. Remove the pedestal to engine mounting bolts.
5. While manipulating the sensor within the pedestal, carefully remove the sensor and pedestal as a unit.
6. To install, loosen the pinch bolt on the new sensor pedestal until the sensor is free to slide the pedestal.
7. Verify that the window in the interrupter is still properly positioned and install sensor and pedestal as a unit while making sure that the interrupter ring is aligned within the proper slot.
8. Install pedestal to engine mounting bolts and torque to 22 ft. lbs. (30 Nm).
9. To adjust use the following procedure:

   a. Using a 28mm socket and pull handle, rotate the harmonic balancer until the interrupter ring(s) fills the sensor slot(s) and edge of interrupter window is aligned with edge of the deflector on the pedestal.

   b. Insert adjustment tool (J–36179 or equivalent) into the gap between sensor and interrupter on each side of interrupter ring. If gauge will not slide past sensor on either side of interrupter ring, the sensor is out of adjustment or interrupter ring is bent. This clearance should be checked at 3 positions around the outer interrupter ring, approximately 120 degrees apart.

NOTE: If found out of adjustment, the sensor should be removed and inspected for potential damage.

   c. Loosen the pinch bolt on sensor pedestal and insert adjustment tool (J–36179 or equivalent) into the gap between sensor and interrupter on each side of interrupter ring.

   d. Slide the sensor into contact against gauge and interrupter ring.

   e. Torque sensor retaining pinch bolt to 30 inch lbs. (3.4 Nm) while maintaining light pressure on sensor against gauge and interrupter ring. This clearance should be checked again, at 3 positions around the interrupter ring, approximately 120 degrees apart. If interrupter ring contacts sensor at any point during harmonic balancer rotation, the interrupter ring has excessive runout and must be replaced.

3800 (VIN C) ENGINE

1. Disconnect the negative battery cable.
2. Disconnect the serpentine belt from the crankshaft pulley.

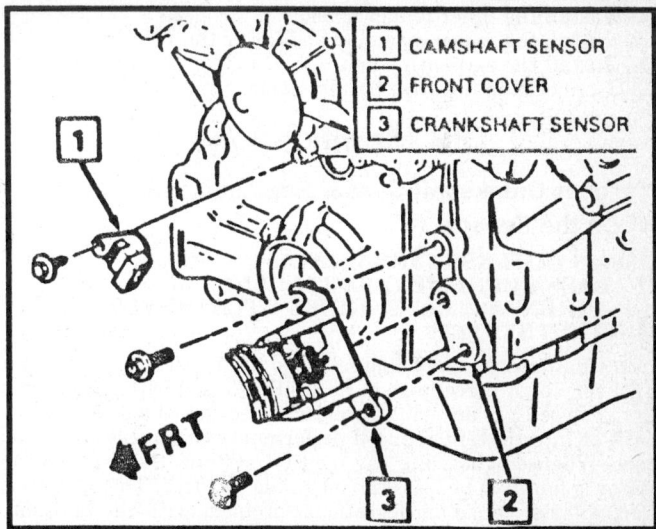

Camshaft and crankshaft removela and installation 3800 VIN C engine

1	CAMSHAFT SENSOR
2	FRONT COVER
3	CRANKSHAFT SENSOR

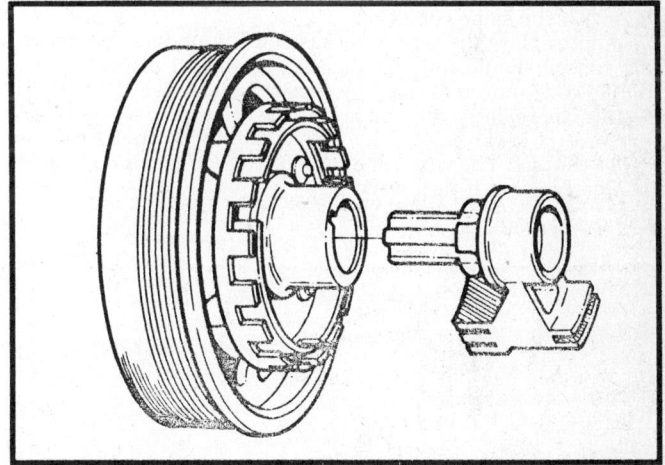

Position special tool J–37089 on the crankshaft 3800 VIN C engine

3. Raise the vehicle and support safely.
4. Remove the right front wheel assembly.
5. Remove the right inner fender access cover.
6. Remove the crankshaft harmonic balancer retaining bolt using a 28mm socket.
7. Remove the crankshaft harmonic balancer.
8. Disconnect the sensor electrical connector.
9. Remove the sensor and pedestal from the block face.
10. Remove the sensor from the pedestal.
11. To install, loosely install the crankshaft sensor on the pedestal.
12. Position the sensor with the pedestal attached on special tool J–37089 or equivalent.
13. Position the special tool on the crankshaft.
14. Install the bolts to hold the pedestal to the block and torque to 14–28 ft. lbs. (20–40 Nm).
15. Torque the pedestal pinch bolt to 30–35 inch lbs. (3–4 Nm).
16. Remove the special tool J–37089 or equivalent.
17. Place the special tool J–37089 or equivalent, on the harmonic balancer and turn. If any vane of the harmonic balancer touches the tool, replace the balancer assembly.
18. Install the balancer on the crankshaft.
19. Install the crankshaft bolt and torque to 200–239 ft. lbs. (270–325 Nm).

20. Install the inner fender shield.
21. Install the wheel assembly and lower the vehicle.
22. Install the serpentine belt.
23. Connect the negative battery cable.

DIS Modifications

Improper Crankshaft Sensor Replacements Due to Oil On the Sensor

1988–89 CUTLASS CIERA (2.8L)
CUTLASS SUPREME (2.8L/3.1L) AND
1988 FIRENZA (2.0L) ENGINES WITH DIRECT IGNITION SYSTEM

Crankshaft sensors are being returned to the parts dealer showing a very high percentage of no trouble found with these sensors. The sensors are usually replaced because of a driver comment of intermittent reduced performance, hesitation or a no start. When diagnosising the ignition system, the crankshaft sensor is found to be wet with oil and is replaced. The oil, however, will not cause an operational problem with the ignition system and the sensor should not be replaced for this reason.

The likely cause for this performance problem is ant intermittent connection to the sensor. Replacing the sensor may cure the condition for a time, but the vehicle could return at a later time with the same condition.

Do not replace the sensor because oil is found on the sensor. The sensor should only be replaced if normal diagnosis or a trouble code indicates that the sensor is defective. The sensor may also be replaced if an oil leak is the original concern (this is very rare). Check the wire harness connector to the sensor for proper contact. The wire terminals, or the connector body itself may, be replaced if necessary.

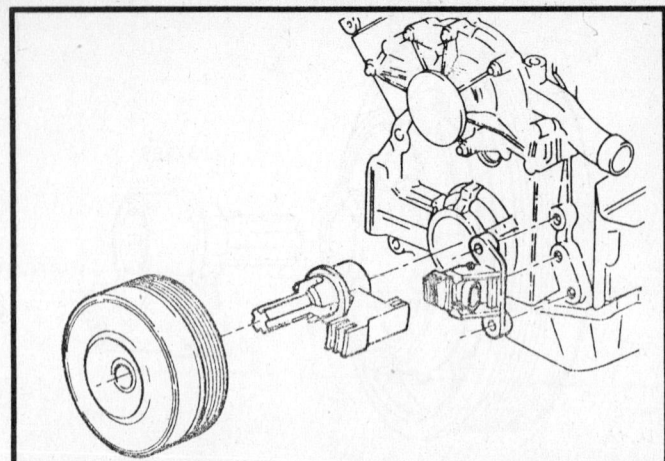

Checking the harmonic balancer vanes 3800 VIN C engine

CAMSHAFT POSITION SENSOR

Removal and Installation
3.8L VIN 3 AND 3800 VIN C ENGINES

1. Disconnect negative battery cable.
2. Remove attaching screw securing sensor.
3. Disconnect 3 terminal sensor connector and remove sensor.
4. Installation is the reverse of removal.

Fuel Injection Systems

INDEX

CADILLAC DIGITAL FUEL INJECTION (DFI)

Application Chart	**5-4**
General Information	**5-4**
Fuel Supply System	5-4
Air Induction System	5-5
Data Sensors	5-5
Electronic Control Module (ECM)	5-7
Body Control Module (BCM)	5-8
Electronic Spark Timing (EST)	5-8
Idle Speed Control System	5-8
Emission Controls	5-9
Closed Loop Fuel Control	5-9
Viscous Converter Clutch (VCC) Control	5-9
Service Precautions	**5-9**
Diagnosis and Testing	**5-10**
Visual Inspection	5-10
Fuel System	5-10
C-Body	**5-12 — 5-55**
System Check Chart	5-12
Charts A–1 through A–8	5-13 — 5-19
Component Locations	5-20
Charts C–1 through C–8	5-19 — 5-25
Diagnostic Charts	5-25 — 5-30
ECM Diagnostic Code Charts	5-31 — 5-55
Testing Conditions	5-55
E/K-Body	**5-56 — 5-99**
System Check Chart	5-56
Charts A–1 through A–8	5-57 — 5-61
Component Locations	5-62
Charts C–1 through C–8	5-63 — 5-67
Diagnostic Charts	5-67 — 5-73
ECM Wiring Diagram	5-73 — 5-74
ECM Diagnostic Code Charts	5-75 — 5-98
BCM Connector Wiring Idenitification	5-99
Component Replacement	**5-100**
Service Precautions	5-100
Throttle Body R&R	5-100
Fuel Meter Cover R&R	5-100
Fuel Injector R&R	5-100
Fuel Meter Body R&R	5-101
Idle Speed Control (ISC) R&R and Adjustment	5-102
Minimum Air, TPS and IAC Motor Adjustment	5-103
Specifications Chart	5-104

MULTI-PORT (MPI) AND TUNED PORT INJECTION (TPI)

Application Chart	**5-107**
General Information	**5-107**
Multi-Port Fuel Injection (MPI) System	5-107
Tuned Port Injection (TPI)	5-108
Fuel Control System	5-110
Synchronized Mode	5-112

Non-synchronized Mode	5-112
ECM Parameters	5-112
ALCL Connector	5-112
MPI Fuel Control Systems Components	**5-112**
Throttle Body Unit	5-112
Fuel Injector	5-113
Fuel Rail	5-113
Pressure Regulator	5-113
Idle Air Control (IAC)	5-114
Fuel Pump	5-114
Fuel Pump Relay Circuit	5-114
In-Line Fuel Filter	5-115
Cold Start Injector	5-115
Data Sensors	**5-116**
Electronic Control Module (ECM)	5-116
Electronic Spark Timing (EST)	5-117
Electronic Spark Control (ESC)	5-117
Computer Controlled Coil Ignition System (C^3I)	5-117
Direct Ignition System (DIS)	5-118
Engine Coolant Temperature	5-118
Oxygen Sensor	5-119
Manifold Absolute Pressure (MAP) Sensor	5-119
Manifold Air Temperature (MAT) Sensor	5-120
Mass Air Flow (MAF) Sensor	5-120
Vehicle Speed Sensor (VSS)	5-120
Throttle Position Sensor (TPS)	5-120
Crankshaft Sensor	5-120
Dual Crank Sensor/Combination Sensor	5-121
Air Conditioning Pressure Sensor	5-121
Non-Air Conditioning Program Input	5-121
Detonation (Knock) Sensor	5-121
Park/Neutral Switch	5-122
Air Conditioning Request Signal	5-122
Torque Converter Clutch Solenoid	5-122
Power Steering Pressure Switch	5-122
Oil Pressure Switch	5-122
Service Precautions	5-122
Diagnosis and Testing	**5-122**
ALCL Connector	5-122
Field Service Mode	5-122
Diagnostic Mode	5-123
ECM Learning Ability	5-123
Reading the Data Stream	5-123
Integrator and Block Learn	5-123
Closed Loop Fuel Control	5-124
Reading Codes	5-124
Clearing the Trouble Codes	5-124
Trouble Code Test Lead	5-124
Engine Performance Diagnosis	5-125
System Diagnostic Circuit Check	5-125
Intermittent Check Engine Light	5-126
Undervoltage to the ECM	5-126
Overvoltage to the ECM	5-126
Electronic Fuel Injection ALCL Scan Tester Info	5-126
Scan Tools Used with Intermittents	5-127
Wire Harness and Connector Service	5-127
Tools Needed to Service the System	5-127

2.0L Turbo (VIN M) Diagnostic Charts	
J and N-Body	**5-128 — 5-152**
Component Locations — Grand Am	5-128
ECM Wiring Diagram Grand Am	5-128
ECM Connector Terminal End View — Grand Am	5-129
Component Locations — Sunbird	5-129
ECM Wiring Diagram — Sunbird	5-129 — 5-130
ECM Connector Terminal End View — Sunbird	5-130
Diagnostic Circuit Check Charts	5-131
Charts A-1 through A-7	5-131 — 5-134
Diagnostic Code Charts	5-135 — 5-143
Section B Symptoms Charts	5-143 — 5-147
Charts C-1A through C-12	5-148 — 5-152
2.3L (VIN D and A) Diagnostic Charts	
N-Body	**5-153 — 5-183**
Component Locations — Grand Am	5-153
Component Locations — Grand Am	5-153
ECM Wiring Diagram — Grand Am	5-153
ECM Connector Terminal End View — Grand Am	5-153
Component Locations — Skylark	5-154
ECM Wiring Diagram — Skylark	5-155
ECM Connector Terminal End View — Skylark	5-155
Component Locations (VIN D) — Calais	5-156
Component Locations (VIN A) — Calais	5-156
ECM Wiring Diagram — Calais	5-156
ECM Connector Terminal End View — Calais	5-157
Diagnostic Circuit Check Charts	5-158
Charts A-1 through A-7	5-158 — 5-162
Diagnostic Code Charts	5-163 — 5-175
Section B Symptoms Charts	5-176 — 5-177
Charts C-1A through C-12	5-178 — 5-183
2.8L (VIN S) Diagnostic Charts	
Chevrolet and Pontiac F-Body	**5-184 — 5-209**
Component Locations	5-184
ECM Wiring Diagram	5-185 — 5-186
ECM Connector Terminal End View	5-186
Diagnostic Circuit Check Charts	5-186
Charts A-1 through A-9	5-188 — 5-191
Diagnostic Code Charts	5-191 — 5-200
Section B Symptoms Charts	5-201 — 5-202
Charts C-1 through C-12	5-202 — 5-209
2.8L (VIN W) Diagnostic Charts	
Chevrolet L-Body	**5-210 — 5-233**
Component Locations	5-210
ECM Wiring Diagram	5-210
ECM Connector Terminal End View	5-211
Diagnostic Circuit Check Charts	5-211
Charts A-1 through A-7	5-212 — 5-215
Diagnostic Code Charts	5-215 — 5-224
Section B Symptoms Charts	5-225 — 5-226
Charts C-1A through C-12	5-227 — 5-233
Chevrolet and Pontiac A-Body	**5-234 — 5-257**
Component Locations	5-234
ECM Wiring Diagram	5-234
ECM Connector Terminal End View	5-235
Diagnostic Circuit Check Charts	5-235
Charts A-1 through A-7	5-236 — 5-239

Diagnostic Code Charts 5-240–5-249
Section B Symptoms Charts 5-249–5-251
Charts C-1A through C-12 5-251–5-257
Buick and Oldsmobile A-Body 5-258–5-282
Component Locations 5-258
ECM Wiring Diagram 5-258
ECM Connector Terminal End View 5-259
Diagnostic Circuit Check Charts 5-259
Charts A-1 through A-7 5-260–5-263
Diagnostic Code Charts 5-264–5-273
Section B Symptoms Charts 5-273–5-275
Charts C-1A through C-12 5-275–5-282
Chevrolet and Cadillac J-Body 5-282–5-306
Component Locations – Cavalier 5-282
ECM Wiring Diagram – Cavalier 5-282
ECM Connector Terminal End View – Cavalier 5-283
Component Locations – Cimarron 5-284
ECM Wiring Diagram – Cimarron 5-284
ECM Connector Terminal End View – Cimarron 5-285
Diagnostic Circuit Check Charts 5-285
Charts A-1 through A-7 5-286–5-289
Diagnostic Code Charts 5-289–5-297
Section B Symptoms Charts 5-298–5-300
Charts C-1A through C-12 5-300–5-306
Buick and Oldsmobile W-Body 5-307–5-333
Component Locations 5-307
ECM Wiring Diagram 5-307
ECM Connector Terminal End View 5-308
Diagnostic Circuit Check Charts 5-308
Charts A-1 through A-7 5-309–5-312
Diagnostic Code Charts 5-312–5-324
Section B Symptoms Charts 5-325–5-326
Charts C-1A through C-12 5-327–5-333
3.0L (VIN L) Diagnostic Charts
Buick and Oldsmobile N-Body 5-333–5-357
Component Locations 5-333
ECM Wiring Diagram 5-334
ECM Connector Terminal End View 5-334
Diagnostic Circuit Check Charts 5-335
Charts A-1 through A-7 5-335–5-339
Diagnostic Code Charts 5-340–5-347
Section B Symptoms Charts 5-347–5-349
Charts C-1A through C-12C 5-349–5-357
W-Body 5-358–5-386
Pontiac and Chevrolet 5-358–5-359
Component Locations 5-358
ECM Wiring Diagram 5-358
ECM Connector Terminal End View 5-358
Buick and Oldsmobile 5-359–5-360
Component Locations 5-359
ECM Wiring Diagram 5-359
ECM Connector Terminal End View 5-360
Diagnostic Circuit Check Charts 5-361
Charts A-1 through A-7 5-361–5-364
Diagnostic Code Charts 5-365–5-376
Section B Symptoms Charts 5-376–5-378
Charts C-1A through C-12 5-378–5-386
3.3L (VIN N) Diagnostic Charts
Oldsmobile A-Body 5-386–5-412
Component Locations 5-386
ECM Wiring Diagram 5-386
ECM Connector Terminal End View 5-387
Diagnostic Circuit Check Charts 5-388
Charts A-1 through A-7 5-388–5-392
Diagnostic Code Charts 5-392–5-403
Section B Symptoms Charts 5-403–5-405
Charts C-2A through C-12C 5-406–5-412
Buick and Oldsmobile N-Body 5-413–5-436
Component Locations 5-413
ECM Wiring Diagram 5-413
ECM Connector Terminal End View 5-414
Diagnostic Circuit Check Charts 5-414
Charts A-1 through A-7 5-415–5-418
Diagnostic Code Charts 5-419–5-428
Section B Symptoms Charts 5-429–5-431
Charts C-2A through C-12C 5-431–5-436
5.0L (VIN F) Diagnostic Charts
Chevrolet and Pontiac F-Body 5-436–5-459
Component Locations 5-436
ECM Wiring Diagram 5-436
ECM Connector Terminal End View 5-437
Diagnostic Circuit Check Charts 5-438
Charts A-1 through A-7 5-438–5-441
Diagnostic Code Charts 5-441–5-451
Section B Symptoms Charts 5-451–5-453
Charts C-2A through C-12 5-453–5-459
5.7L (VIN 8) Diagnostic Charts
Chevrolet Y-Body 5-460–5-483
Component Locations 5-460
ECM Wiring Diagram 5-460
ECM Connector Terminal End View 5-461
Diagnostic Circuit Check Charts 5-461
Charts A-1 through A-7 5-462–5-464
Diagnostic Code Charts 5-465–5-474
Section B Symptoms Charts 5-474–5-476
Charts C-1A through C-12 5-477–5-483
Component Testing 5-484
Cold Start Injector Valve 5-484
Minimum Air Rate Check 5-484
TPS Output Check 5-484
Crankshaft Sensor Inspection 5-484
Fuel System Pressure 5-485
Component Replacement 5-485
Fuel Rail R&R 5-485
Fuel Injectors R&R 5-488
Fuel Pressure Regulator R&R 5-489
Throttle Body Assembly R&R 5-490
Cold Start Tube and Valve Assembly R&R 5-491
Idle Air Control Valve R&R 5-491
Throttle Position Sensor R&R 5-492
Oxygen Sensor R&R 5-493
Crankshaft Sensor or Combination Sensor R&R 5-493
Changes and Corrections 5-494
Coolant Temperature Sensor R&R 5-494
Torque Converter Clutch Solenoid R&R 5-494
Vehicle Speed Sensor R&R 5-495
Manifold Absolute Pressure Sensor R&R 5-495
Power Steering Pressure Switch R&R 5-495
Mass Air Flow Sensor R&R 5-495
Mass Air Temperature Sensor R&R 5-495
Electronic Control Module R&R 5-495
Mem-Cal R&R 5-495
Prom R&R 5-496
Cal-Pak R&R 5-496
Specifications Chart 5-496

THROTTLE BODY INJECTION SYSTEM (TBI)

Application Chart 5-498
General Information 5-498
Synchronized Mode 5-498
Nonsynchronized Mode 5-498
Components and Operation 5-499
Cranking Mode 5-500
Clear Flood Mode 5-500
Run Mode 5-500
Acceleration Enrichment Mode 5-501
Deceleration Leanout Mode 5-502
Deceleration Fuel Cut-Off Mode 5-502
Battery Voltage Correction Mode 5-502
Fuel Cut-Off Mode 5-502
Backup Mode 5-502
ECM Parameters 5-502
ALCL Connector 5-502
Electronic Fuel Injection Subsystems 5-502
Fuel Supply System 5-503
Throttle Body Injector (TBI) Assembly 5-504
Idle Air Control (IAC) 5-504
Idle Speed Control 5-505
Rough Idle/Low Idle Speed 5-505
High Idle Speed/Warm-Up Idle Speed (No Kickdown) 5-505
Data Sensors 5-505
Engine Coolant Temperature 5-505
Oxygen Sensor 5-506
Manifold Absolute Pressure Sensor 5-506
Manifold Air Temperature Sensor 5-506
Vehicle Speed Sensor (VSS) 5-506
Throttle Position Sensor (TPS) 5-506
Crankshaft and Camshaft Sensor 5-506
Park/Neutral Switch 5-507
Air Conditioner Request Signal 5-507
Torque Converter Clutch Solenoid 5-507
Power Steering Pressure Switch 5-507
Oil Pressure Switch 5-507
Electronic Spark Timing (EST) 5-507
Electronic Spark Control (ESC) 5-507
Direct Ignition System (DIS) 5-507
Emission Control 5-509
Evaporative Emission Control (EEC) Systems 5-510
Service Precautions 5-511
Diagnosis and Testing 5-512
ALCL Connector 5-512
Field Service Mode 5-512
Diagnostic Mode 5-512
ECM Learning Ability 5-513
Reading the Data Stream 5-513
Integrator and Block Learn 5-513
Closed Loop Fuel Control 5-514
Reading Codes 5-514
Clearing the Trouble Codes 5-514
Trouble Code Test Lead 5-514
Engine Performance Diagnosis 5-514
System Diagnostic Circuit Check 5-515
Intermittent Check Engine Light 5-515
Trouble Code Identification Chart 5-516
Undervoltage to the ECM 5-518
Overvoltage to the ECM 5-518
EFI – ALCL Scan Tester Information 5-518
Scan Tools Used with Intermittents 5-518
Wire Harness and Connector Service 5-519
Tools Needed to Service the System 5-520
2.0L (VIN 1) Diagnostic Charts
Chevrolet L-Body 5-522–542
Component Locations 5-522
ECM Wiring Diagram 5-523
ECM Connector Terminal End View 5-523
Diagnostic Circuit Check Charts 5-524
Charts A- through A-7 5-524–5-527
Diagnostic Code Charts 5-528–5-534
Section B Symptoms Charts 5-535–5-538
Charts C-1D through C-12 5-538–5-542
J-Body 5-543–566
Component Locations – Cavalier 5-543
ECM Wiring Diagram – Cavalier 5-543
ECM Connector Terminal End View – Cavalier 5-544
Component Locations – Firenza 5-544
ECM Wiring Diagram – Firenza 5-544
ECM Connector Terminal End View – Firenza 5-544
Component Locations – Skyhawk 5-545
ECM Wiring Diagram – Skyhawk 5-545
ECM Connector Terminal End View – Skyhawk 5-546
Diagnostic Circuit Check Charts 5-547
Charts A-1 through A-7 5-547–5-551
Diagnostic Code Charts 5-551–5-558
Section B Symptoms Charts 5-558–5-561
Charts C-1A through C-12 5-561–5-566
2.0L (VIN K, OHC) Diagnostic Charts
J-Body 5-566–5-589
Component Locations 5-566
ECM Wiring Diagram 5-566
ECM Connector Terminal End View 5-567
Pontiac J-Body 5-567–5-568

Component Locations 5-567
ECM Wiring Diagram 5-568
ECM Connector Terminal End View 5-568
Buick and Oldsmobile J-Body .... **5-569—5-570**
Component Locations 5-569
ECM Wiring Diagram 5-569
ECM Connector Terminal End View 5-570
Diagnostic Circuit Check Charts 5-570
Charts A-1 through A-7 .... 5-571—5-574
Diagnostic Code Charts .... 5-574—5-582
Section B Symptoms Charts .... 5-582—5-585
Charts C-1A through C-12 .... 5-586—5-589

2.5L (VIN R) Diagnostic Charts
All Models Except Pontiac P-Body .... **5-590—612**
Component Locations 5-590
ECM Wiring Diagram 5-591
ECM Connector Terminal End View 5-592
Component Locations — Lumina 5-592
ECM Wiring Diagram — Lumina 5-592
ECM Connector Terminal End View — Lumina 5-593
Diagnostic Circuit Check Charts — 5-594
Charts A-1 through A-7 .... 5-594—5-597
Diagnostic Code Charts .... 5-598—5-605
Section B Symptoms Charts .... 5-605—5-608
Charts C-1A through C-12 .... 5-608—5-612

2.5L (VIN U) Diagnostic Charts
All Models .... **5-612—5-631**
Component Locations 5-612
ECM Wiring Diagram 5-613
ECM Connector Terminal End View 5-613
Diagnostic Circuit Check Charts 5-614
Charts A-1 through A-7 .... 5-614—5-617
Diagnostic Code Charts .... 5-618—5-625
Section B Symptoms Charts .... 5-625—5-627
Charts C-1A through C-10 .... 5-628—5-631

5.0L (VIN E) Diagnostic Charts
Chevrolet and Pontiac F-Body .... **5-632—5-657**
ECM Wiring Diagram 5-632
ECM Connector Terminal End View 5-632
Component Locations 5-633
Diagnostic Circuit Check Charts 5-637
Charts A-1 through A-7 .... 5-637—5-640
Diagnostic Code Charts .... 5-640—5-649
Section B Symptoms Charts .... 5-649—5-650
Charts C-1A through C-8B .... 5-651—5-657

4.3L (VIN Z) Diagnostic Charts
B-Body .... **5-658—5-680**
Component Locations 5-658
ECM Wiring Diagram 5-659
ECM Connector Terminal End View 5-659

5.0L (VIN E) and 5.7L (VIN 7) Diagnostic Charts
B-Body .... **5-660—5-680**
Component Locations 5-660
ECM Wiring Diagram 5-660
Connector Terminal End View 5-661
Diagnostic Circuit Check Charts 5-661
Charts A-1 through A-7 .... 5-661—5-664
Diagnostic Code Charts .... 5-664—5-672
Section B Symptoms Charts .... 5-672—5-674
Charts C-1A through C-8 .... 5-674—5-680

LIGHT TRUCK AND VAN
DIAGNOSTIC CHARTS .... **5-680—5-712**
2.5L (VIN E)
Component Locations — M-Series 5-680
Component Locations — ST-Series 5-680
ECM Wiring Diagram 5-681
ECM Connector Terminal End View 5-681

2.8L (VIN R)
Component Locations — ST-Series 5-682
ECM Wiring Diagram 5-682
ECM Connector Terminal End View 5-682

4.3L (VIN Z)
Component Locations — ST-Series 5-683
Component Locations — CK-Series 5-683
Component Locations — G-Series 5-683
Component Locations — M-Series 5-683
Component Locations — RV-Series 5-684

5.0L (VIN H) and 5.7L (VIN K)
Component Locations — CK-Series 5-684
Component Locations — G-Series 5-684
Component Locations — RV-Series 5-684
Component Locations — P-Series 5-685

4.3L (VIN Z) and V8 ECM
Wiring Diagram 5-685
ECM Connector Terminal End View 5-686

7.4L (VIN N)
Component Locations — CK and RV-Series 5-686
Component Locations — G-Series 5-686
Section 2 Symptoms Charts .... 5-687—5-688
Charts A-1 through A-6 .... 5-688—5-693
Diagnostic Code Charts .... 5-693—5-702
Diagnostic Aid Charts .... 5-702—5-712
Component Testing 5-713
TPS Output Check Test 5-713
Controlled Idle Speed/Min. Idle Air Rate Chart 5-714
Component Replacement 5-715
Service Precautions 5-715
Throttle Body Injection Specifications Chart 5-715
Cleaning and Inspection 5-716
Throttle Body R&R 5-716
Fuel Meter Cover R&R 5-716
Fuel Injector R&R 5-717
Fuel Meter Body R&R 5-721
Pressure Regulator R&R 5-722
Idle Air Control Valve R&R 5-722
Throttle Position Sensor R&R 5-723
Direct Ignition System (DIS) Assembly R&R 5-723
Crankshaft Sensor R&R 5-724
Coolant Temperature Sensor R&R 5-724
Torque Converter Clutch Solenoid R&R 5-724
Vehicle Speed Sensor R&R 5-724
Oxygen Sensor R&R 5-724
Manifold Absolute Pressure Sensor R&R 5-725
Power Steering Pressure Switch R&R 5-725
Mass Air Temperature Sensor R&R 5-725
Electronic Control Module R&R 5-725
Mem-Cal R&R 5-725
Prom R&R 5-725
Cal-Pak R&R 5-726
Minimum Idle Speed Adjustment 5-727
Controlled Idle Speed and Minimum Idle Air
 Rate Chart — Light Truck and Van 5-729

SEQUENTIAL FUEL INJECTION SYSTEM (SFI)

Application Chart 5-732
General Information 5-723
Fuel Control System 5-734
Information Data Sensors 5-738
Computer Controlled Coil Ignition System (C³I) 5-740
ALDL Connector 5-740
Field Service Mode 5-740
Clearing the Trouble Codes 5-740
Closed Loop Fuel Control 5-740
ECM Learning Ability 5-740
Service Precaution 5-741
Diagnosis and Testing 5-741
Fuel Injector Balance Test 5-742
Diagnostic Circuit Check 5-743
3800 (VIN C) Diagnostic Charts
All Models Except E-Body .... **5-744—5-776**
Component Locations 5-744
ECM Wiring Diagram 5-744
ECM Connector Terminal End View 5-745
Diagnostic Circuit Check Charts 5-745
Charts A-1 through A-7 .... 5-746—5-753
Diagnostic Code Charts .... 5-754—5-765
Section B Symptoms Charts .... 5-766—5-767
Charts C-2A through C-12C .... 5-767—5-776
E-Body .... **5-814—835**
Component Locations 5-814
ECM Wiring Diagram 5-814
ECM Connector Terminal End View 5-815
Diagnostic Circuit Check Charts 5-815
Diagnostic Code Charts .... 5-820—5-835
3.8L (VIN 3) Diagnostic Charts
All Models .... **5-776—5-813**
Component Locations — A-Body 5-776
Component Locations — C and H-Body 5-776
ECM Wiring Diagram — A-Body 5-777
ECM Wiring Diagram — C and H-Body 5-777
ECM Connector Terminal End View — A-Body 5-778
ECM Connector Terminal End View — C and H-Body ... 5-778
Diagnostic Circuit Check Charts 5-779
Charts A-1 through A-7 .... 5-779—5-788
Diagnostic Code Charts .... 5-789—5-797
Section B Symptoms Charts .... 5-797—5-798
Charts C-1A through C-12K .... 5-799—5-813
1988–90 Riviera and Reatta
 BCM Wiring Diagram .... 5-836—5-838
1988–90 Riviera and Reatta
 BCM Connector View .... 5-838—5-839
1988–90 Riviera and Reatta
 BCM Diagnostic Charts .... 839—5-845
1988–90 Toronado BCM Wiring Diagram .. 5-845—5-847
1988–90 Toronado BCM Connector View 5-847
1988–90 Toronado BCM Diagnostic Charts 5-847—5-855
Component Replacement 5-856
Fuel Injectors 5-856
Fuel Rail 5-856
Fuel Pressure Regulator 5-856
Throttle Body Assembly 5-856
Idle Air Control Valve 5-857
Minimum Idle Speed Adjustment 5-858
Throttle Position Sensor 5-858
Mass Air Flow Sensor 5-859
Crankshaft Sensor 5-859
Camshaft Position Sensor 5-860
Oxygen Sensor 5-860

CADILLAC DIGITAL FUEL INJECTION (DFI)

APPLICATION CHART
1988–90 Digital Fuel Injection Vehicles

Year	Models	Series VIN	Engine	Engine VIN Code
		CADILLAC		
1988-90	DeVille	C	4.5L	5
	Fleetwood	C	4.5L	5
	Eldorado	E	4.5L	5
	Seville	K	4.5L	5

General Information

The Digital Fuel Injection (DFI) system, is a speed density fuel system that accurately controls the air/fuel mixture into the engine in order to achieve desired performance and emission goals. The Manifold Absolute Pressure (MAP) sensor and the Manifold Air Temperature (MAT) are used to determine the density (amount) of air entering the engine.

The High Energy Ignition (HEI) distributor provides the engine with speed (rpm) information. All of this information is then fed to the Electronic Control Module (ECM) and the ECM performs a high speed digital computations to determine the proper amount of fuel necessary to achieve the desired air/fuel mixture.

Once the ECM has calculated how much fuel to deliver, it signals the fuel injectors to meter the fuel into the throttle body. When the combustion process has been completed, some hydrocarbon (HC), carbon monoxide (CO) and nitrous oxides (NOx) result, therefore, each DFI engine has an emission system to reduce the amount of these gases into the exhaust stream.

The dual bed catalytic converter coverts these gases into a more inert gases, however, the conversion process is most efficient (lower emission levels) at an air/fuel mixture of 14.7:1.

Once the engine is warmed up, the ECM uses the input from the oxygen sensor to more precisely control the air/fuel mixture to 14.7:1. This correction process is known a s closed loop operation. Because a vehicle is driven under a wide range of operating conditions, the ECM must provide the correct quantity of fuel under all operating conditions.

Therefore, additional sensors and switches are necessary to determine what operating conditions exist so that the ECM can provide an acceptable level of engine control and driveability under all operating conditions. So the closed loop DFI operation provides the acceptable level of driveability and fuel economy while improving emission levels.

The following subsystems combine to form the DFI closed loop system:
- Fuel Delivery
- Air Induction
- Data Sensors
- Electronic Control Module
- Body Control Module
- Electric Spark Timing
- Idle Speed Control
- Emission Controls
- Closed Loop Fuel Control
- System Diagnosis
- Cruise Control
- Torque Converter Clutch

All models are also equipped with a Body Control Module (BCM) that is used to control various vehicle body functions based upon data sensors and switch inputs. The ECM and BCM exchange information to maintain efficient operation of all vehicle functions. This transfer of information gives the BCM control over the ECM's self-diagnostic capabilities as well as its own.

Both the ECM and the BCM have the capability to diagnose faults with the various inputs and systems they control. When the ECM recognizes a problem, it lights a "Service Soon" tell-tale lamp on the instrument panel to alert the driver that a malfunction has occurred.

The DFI consists of a pair of electronically actuated fuel metering valves, which, when actuated, spray a calculated quantity of fuel into the engine intake manifold. These valves or injectors are mounted on the throttle body above the throttle blades with the metering tip pointed into the throttle throats. The injectors are normally actuated alternately.

Fuel is supplied to the inlet of the injectors through the fuel lines and is maintained at a constant pressure across the injector inlets. When the solenoid-operated valves are energized, the injector ball valve moves to the full open position. Since the pressure differential across the valve is constant, the fuel quantity is changed by varying the time that the injector is held open.

The amount of air entering the engine is measured by monitoring the intake Manifold Absolute Pressure (MAP), the intake Manifold Air Temperature (MAT) and the engine speed (in rpm). This information allows the computer to compute the flow rate of air being inducted into the engine and, consequently, the flow rate of fuel required to achieve the desired air/fuel mixture for the particular engine operating condition.

FUEL SUPPLY SYSTEM

The fuel supply system components provide fuel at the correct pressure for metering into the throttle bores by the injectors. The pressure regulator controls fuel pressure to a nominal 10.5 psi across the injectors. The fuel supply system is made up a fuel tank mounted electric pump, a full-flow fuel filter mounted on the vehicle frame, a fuel pressure regulator integral with the throttle body, fuel supply and fuel return lines and 2 fuel injectors. The timing and amount of fuel supplied is controlled by the computer.

An electric motor-driven twin turbine-type pump is integral with the fuel tank float unit. It provides fuel at a positive pressure to the throttle body and fuel pressure regulator. The pump is specific for DFI application and is not repairable. However, the pump may be serviced separately from the fuel gauge unit.

Fuel pump operation is controlled by the fuel pump relay, located in the relay center. Operation of the relay is controlled by a signal from the computer. The fuel pump circuit is protected by a fuse, located in the mini-fuse block. The computer turns the pump on with the ignition ON or START. However, if the engine is not cranked within 1 second after the ignition is turned ON, the computer signal is removed and the pump turns off.

Fuel is pumped from the fuel tank through the supply line and the filter to the throttle body and pressure regulator. The injectors supply fuel to the engine in precisely timed bursts as a result of electrical signals from the computer. Excess fuel is returned to the fuel tank through the fuel return line.

The fuel tank incorporates a reservoir directly below the sending unit-in-tank pump assembly. The "bathtub" shaped reservoir is used to ensure a constant supply of fuel for the in-tank pump even at low fuel level and severe maneuvering conditions.

FUEL PUMP RELAY

When the key is first turned ON (key in the RUN position with

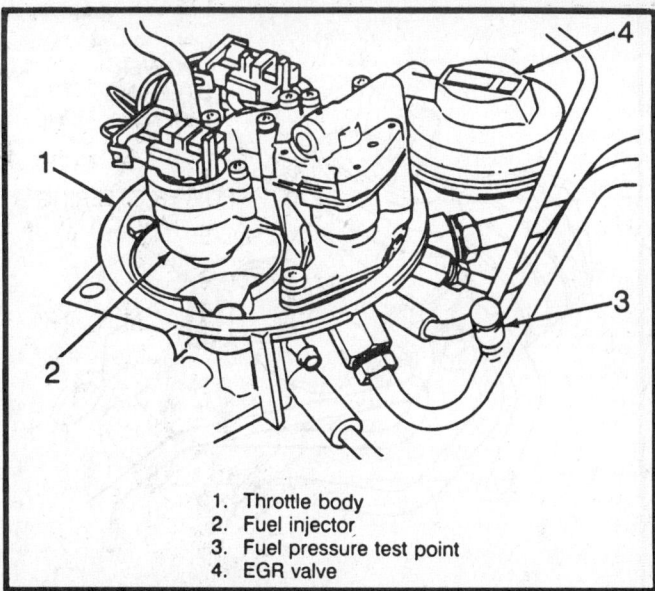

1. Throttle body
2. Fuel injector
3. Fuel pressure test point
4. EGR valve

Throttle body injection unit

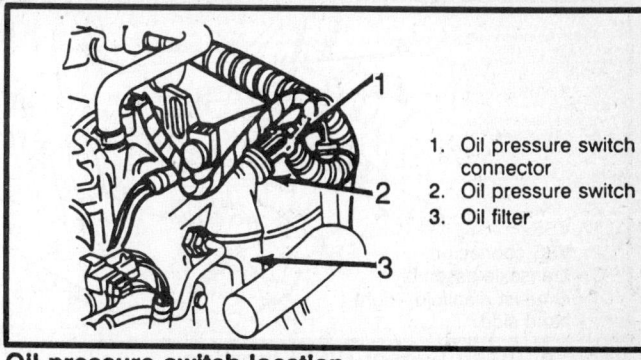

1. Oil pressure switch connector
2. Oil pressure switch
3. Oil filter

Oil pressure switch location

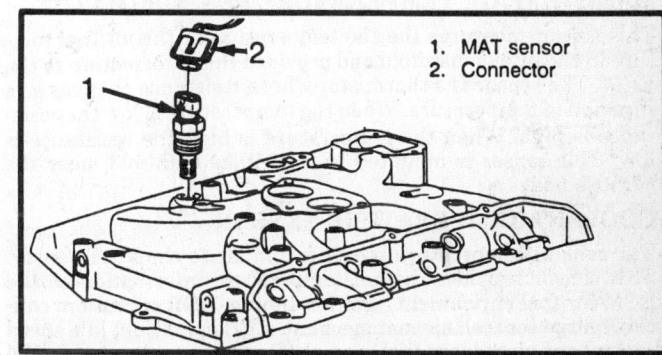

1. MAT sensor
2. Connector

Manifold Air Temperature (MAT) sensor location

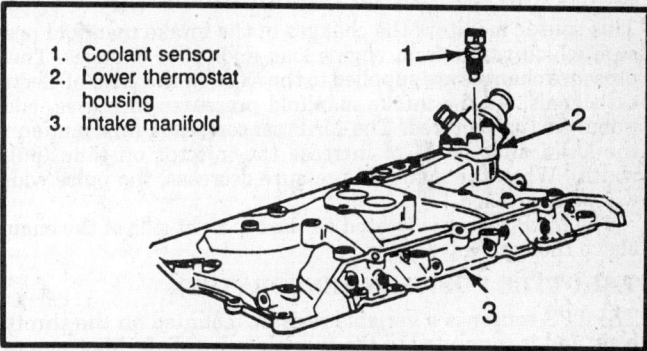

1. Coolant sensor
2. Lower thermostat housing
3. Intake manifold

Coolant temperature sensor location

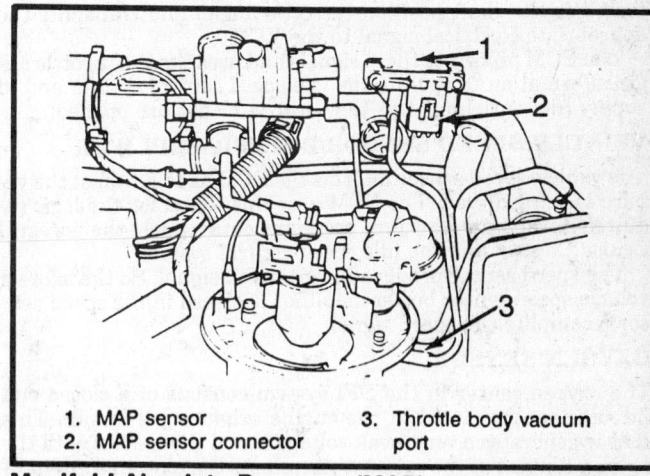

1. MAP sensor
2. MAP sensor connector
3. Throttle body vacuum port

Manifold Absolute Pressure (MAP) sensor location

the engine **OFF**), the ECM will turn the fuel pump relay on for 2 seconds. This builds up the fuel pressure for cranking. If the engine is not started within 2 seconds, the ECM will shut off the fuel pump and wait until the engine starts. As soon as the engine is cranked, the ECM will turn the relay on a run the fuel pump.

OIL PRESSURE SWITCH

As a backup system to the fuel pump relay, the fuel pump can also be turned on by the oil pressure switch. The oil pressure switch has 2 circuits internally. One operates the oil pressure indicator lamp while the other is a normally closed open switch which closes when the oil pressure reaches about 4 psi. If the fuel pump relay fails, the oil pressure switch contacts will close and run the fuel pump.

An inoperative fuel pump relay can result in long cranking times, particularly if the engine is cold. The fuel pump relay inoperative will result in a trouble code. The oil pressure switch acts as a backup to the relay to turn on the fuel pump as soon as the oil pressure reaches 4 psi.

An inoperative fuel pump would cause a no start condition. A fuel pump which does not provide enough pressure can result in poor performance.

AIR INDUCTION SYSTEM

The air induction system consists of a throttle body and an intake manifold. Air for the combustion enters the throttle body and is distributed to each cylinder through the intake manifold. The throttle body contains a special distribution skirt below each injector to improve the fuel distribution.

The air flow rate is controlled by the throttle valves, which are connected to the accelerator linkage. The idle speed is determined by the position of the throttle valves and is controlled by the idle speed control (ISC).

DATA SENSORS

The purpose of the sensors is to supply electronic impulses to the ECM. The ECM then computes the spark timing and fuel delivery rate necessary to maintain the desired air/fuel mixture, thus controlling the amount of fuel delivered to the engine. The data sensors are inter-related.

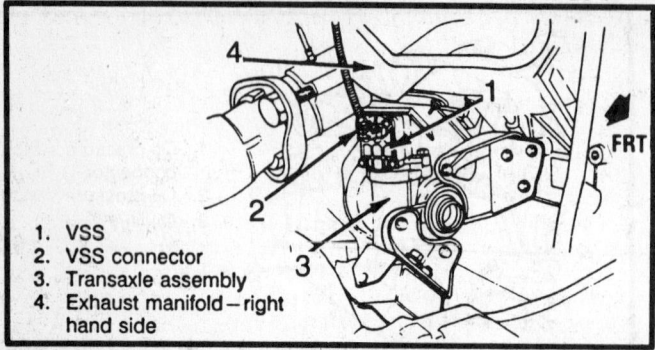

1. VSS
2. VSS connector
3. Transaxle assembly
4. Exhaust manifold—right hand side

Vehicle Speed Sensor (VSS) location

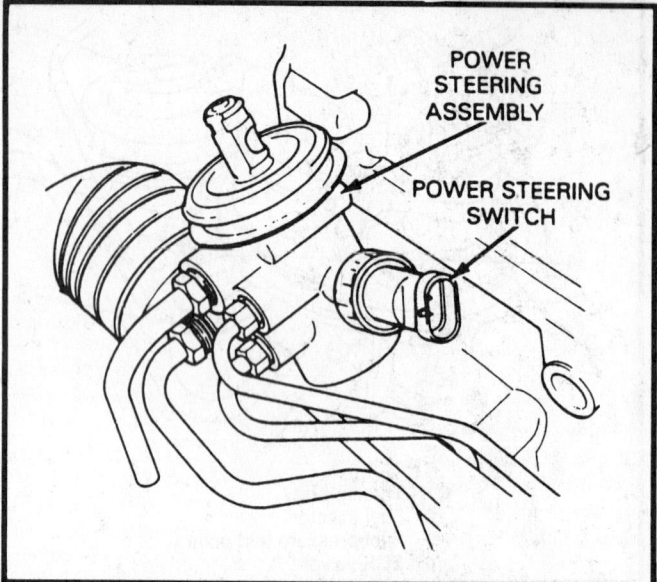

Power steering pressure switch location

MANIFOLD AIR TEMPERATURE SENSOR (MAT)

This sensor measures the the temperature of the air/fuel mixture in the intake manifold and provided this information to the ECM. The sensor is a thermistor whose resistance changes as a function of temperature. When the temperature is low the resistance is high. When the temperature is high the resistance is low. This sensor is mounted in the intake manifold, near the throttle body.

COOLANT TEMPERATURE SENSOR

The coolant temperature sensor is similar to the MAT sensor. This sensor provides coolant temperature information to the ECM for fuel enrichment, ignition timing, EGR operation, canister purge control, air management, EFE operation, idle speed control and closed loop fuel control. This sensor is located on the top left side of the engine near the distributor.

MANIFOLD ABSOLUTE PRESSURE SENSOR (MAP)

This sensor monitors the changes in the intake manifold pressure which result from engine load and speed changes. These pressure changes are supplied to the ECM in the form of electrical signals. As the intake manifold pressures increases, additional fuel is required. The MAP sensor sends information to the ECM and the ECM increase the injector on time (pulse width). When the manifold pressure decrease, the pulse width will be shortened.

The MAP sensor is located on the top right side of the engine above the rear valve cover.

THROTTLE POSITION SENSOR (TPS)

The TPS sensor is a variable resistor mounted on the throttle body and is connected to the throttle valve shaft. Movement of acceleration causes the throttle shaft to rotate and the throttle shaft rotation opens or closes the throttle blades. The sensor determines the shaft position (throttle angle) and transmits the appropriate electrical signal to the ECM.

The ECM processes these signals and uses the the throttle angle information to operate the idle speed control system and to supply fuel enrichment as the throttle blades are opened.

VEHICLE SPEED SENSOR/BUFFER/AMPLIFER

The vehicle speed sensor informs the ECM as to how fast the vehicle is being driven. The ECM uses this signal for the logic required to operate the fuel economy data panel, the integral Cruise Control and the idle speed control system.

The speed sensor produces a very week signal. So therefore a vehicle speed sensor buffer amplifier is placed in the speed sensor to amplify the speed signal.

OXYGEN SENSOR

The oxygen sensor in the DFI system consists of a closed end Zirconia sensor placed in the engine exhaust gas stream This sensor generates a very weak voltage signal that varies with the oxygen content of the exhaust stream. As the oxygen content of the exhaust stream increases relative to the surrounding atmosphere, a lean fuel mixture is indicated by a low voltage output, as the oxygen content decrease, a rich fuel mixture is indicated by a rising voltage output from the sensor.

When the oxygen sensor is warm 392°F (200°C) The output voltage swings between 200 millivolts (lean mixture) and 800 millivolts (rich mixture). However, when the oxygen sensor is cold (below 392°F). the voltage output drops below this range and the response time of the sensor is much slower. The sensor cannot react quickly to rich-lean or lean-rich transitions; therefore, the sensor does not supply accurate information to the ECM. The output voltage may be read using a high impedance digital voltmeter.

NOTE: The high impedance digital voltmeter must be a minimum 5 megohm input impedance. Digital voltmeters with a lower input impedance may cause an inaccurate reading or force the system to behave incorrectly.

ENGINE SPEED SENSOR

The engine speed sensor signal comes from the 7 terminal HEI module in the distributor. Pulse from the distributor are sent to the ECM where the time between these pulses is used to calculate engine speed. The ECM adds the spark advance modifications to the signal and sends the signal back to the distributor.

POWER STEERING SWITCH

The power steering pressure switch is normally closed and opens with power steering pressure. The power steering pressure switch receives 12 volts from a fuse which is located in the relay center.

When high power steering pressures occur, the switch contacts are opened by the power steering oil pressure and the ECM reads 0 volts on the line which goes from the switch to the ECM. The ECM uses the power steering switch input to extend the ISC motor when high power steering loads occur, to help maintain and stable idle. Switch test can be used to check the power steering pressure switch for proper operation.

PARK/NEUTRAL SWITCH

The park/neutral switch is part of the transmission neutral safety-backup switch, pin A of the 6-way weather pack connector on this switch is the park/neutral switch contacts. The park/neutral switch contacts are closed in **PARK** or **NEUTRAL**, shorting the neutral safety-backup switch, pin A to ground. In any other gear range, pin A is open.

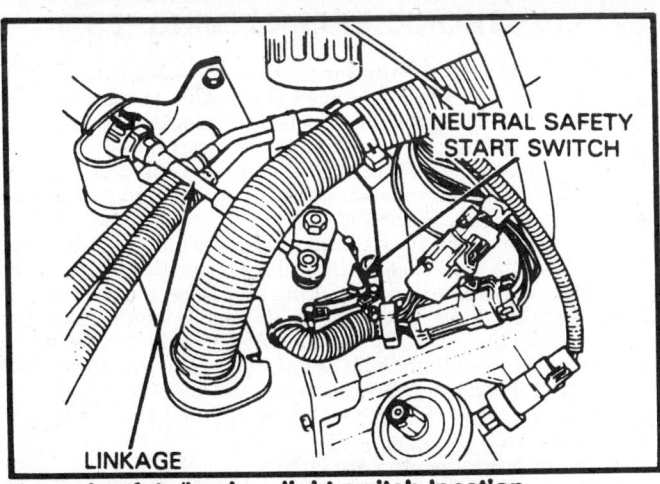

Neutral safety/backup light switch location

The ECM sends 12 volts to pin **A** of the neutral safety-backup switch. When the gear selector is in **PARK** or **NEUTRAL**, the 12 volt signal from the ECM is shorted to ground, resulting in 0 volts at the ECM. In reverse or forward gears, the switch will be opened, resulting in 12 volts at the ECM.

The switch test can be used to check the park/neutral switch for proper operation. An inoperative park/neutral switch could cause improper idle speed, cruise control or transmission converter clutch.

TRANSMISSION TEMPERATURE SENSOR

The transmission temperature sensor is mounted in a transmission cooler line. The sensor measures the temperature of the transmission fluid as it leaves the transaxle and provides this information to the ECM.

The transmission temperature sensor is a thermistor whose resistance changes as a function of temperature. When the temperature is low, the resistance is high. The resistance decreases as the temperature increases. The transmission temperature sensor is a 2-wire sensor with a signal voltage coming from the ECM to the sensor pin B and a sensor reference ground to the sensor or pin A.

CRANK SIGNAL

The ECM looks at the starter solenoid to tell when the engine is cranking. It uses this to tell when the car is in the starting mode. If the crank signal is not received by the ECM, a code will be stored. Under these circumstances, the vehicle may be difficult to start.

ELECTRONIC CONTROL MODULE (ECM)

The Electronic Control Module (ECM) provides all computation and controls for the DFI system. Sensor inputs are fed into the computer from the various sensors. They are processed to produce the appropriate pulse duration for the injectors, the correct idle speed for the particular operating condition and the proper spark advance. Analog inputs from the sensors are converted to digital signals before processing. The computer assembly is mounted under the instrument panel and consists of various printed circuit boards mounted in a protective metal box. The computer receives power from the vehicle battery. When the ignition is set to the **ON** or **CRANK** position, the following information is received from the sensors:
- Engine coolant temperature
- Intake manifold air temperature
- Intake manifold absolute pressure
- Engine speed
- Throttle position

The following commands are transmitted by the ECM:
- Electric fuel pump activation
- Idle speed control
- Spark advance control
- Injection valve activation
- EGR solenoid activation

The desired air/fuel mixture for various driving and atmospheric conditions are programmed into the computer. As signals are received from the sensors, the computer processes the signals and computes the engine's fuel requirements. The computer issues commands to the injection valves to open for a specific time duration. The duration of the command pulses varies as the operating conditions change.

The digital electronic fuel injection system is activated when the ignition switch is turned to the **ON** position. The following events occur at this moment:

1. The computer receives the ignition **ON** signal.
2. The fuel pump is activated by the ECM. The pump will operate for approximately 1 second only, unless the engine is cranking or running.
3. All engine sensors are activated and begin transmitting signals to the computer.
4. The EGR solenoid is activated to block the vacuum signal to the EGR valve at coolant temperatures below 110°F.
5. The "CHECK ENGINE" and "COOLANT" lights are illuminated as a functional check of the bulb and circuit.
6. Operation of the fuel economy lamps begins.

The following events occur when the engine is started:

1. The fuel pump is activated for continuous operation.
2. The idle speed control motor will begin controlling idle speed, including fast idle speed, if the throttle switch is closed.
3. The spark advance shifts from base (bypass) timing to the computer programmed spark curve.
4. The fuel pressure regulator maintains the fuel pressure at 9–12 psi by returning excess fuel to the fuel tank.
5. The following sensor signals are continuously received and processed by the computer:
   a. Engine coolant temperature
   b. Intake manifold air temperature
   c. Intake manifold absolute air pressure
   d. Engine speed
   e. Throttle position changes
6. The computer alternately grounds each injector, precisely controlling the opening and closing time (pulse width) to deliver fuel to the engine.

CENTRAL PROCESSING UNIT (CPU)

The digital signals received by the CPU are used to perform all mathematical computations and logic functions necessary to deliver the proper air/fuel mixture. The CPU also calculates the spark timing and idle speed information. The CPU commands the operation of emission controls, closed loop fuel control, cruise control and diagnostic system.

INPUT AND OUTPUT DEVICES

These integral devices of the ECM, convert the electrical signals received by the data sensors and change them over to digital signals for use by the CPU.

POWER SUPPLY

The main source of power for the ECM is from the battery. The ECM requires battery feed power for memory and ignition feed battery voltage.

ECM MEMORY

There are 3 types of memory in the ECM and they are as follows:

Programmable Read Only Memory (PROM) — The purpose of this memory is to contain the calibration information about each engine, transmission, body and rear axle ratio combination. If the battery voltage is lost for any reason, the PROM information is retained. The PROM chip can be easily changed if necessary.

Read Only Memory (ROM) — The read only memory is programmed information that can only be read by the ECM. The read only memory program cannot be changed. If the battery voltage is lost at any time, the read only memory will be retained.

Random Access Memory (RAM) — Random access memory acts as the scratch pad for the CPU. Information can be read into or out of the RAM memory hence it is called the scratch pad memory. The engine sensor information, diagnostic codes, and the results of calculations are temporarily stored here. If the battery voltage is removed, all the information in the RAM memory will be lost (this is similar to a hand held calculator when the switch is turned **OFF**).

These 3 memory devices are all removable from the ECM unit.

To demonstrate how the ECM operates, the following is a list of events that will occur when the ignition switch is turned **ON**:

1. The ECM receives the ignition **ON** signal.
2. The fuel pump is activated by the ECM. The pump will operate for approximately 1–2 seconds only, unless the engine is being cranked or has started.
3. All engine sensors are activated and begin transmitting signals to the ECM.
4. The EGR solenoid is activated to block the vacuum signal to the EGR valve at a temperatures below 175°F.
5. The coolant light will be illuminated as a functional check of the bulb and the circuit.
6. The HEI bypass line is pulled down to 0 volts.

The following events are what occurs when the engine is being cranked:

1. The 12 volt crank signal is sent to the ECM.
2. The fuel pump is operating.
3. After a short series of prime pulses, injectors alternately deliver a fuel pulse on each distributor reference pulse.
4. The engine sensors continue to transmit signals to the ECM.
5. The "Service Soon" and "Service Now" lights are illuminated as a functional check of the bulb and circuit.
6. The other events are similar to the events which occur when the ignition switch is turned **ON**.

The following events are what occurs when the engine is started:

1. The crank signal is removed from the ECM.
2. The injectors deliver fuel pulses alternately for each distributor reference pulse.
3. The HEI bypass line is pulled up to 5 volts and the HEI module receives spark advance signals from the ECM.
4. The ISC motor begins to control the idle speed if the throttle switch is closed.
5. The fuel pump operates continuously.
6. The pressure regulator maintains fuel pressure at 10.5 psi by returning excess fuel to the fuel tank.
7. The other events are similar to the events which occur when the ignition switch is turned **ON**.

BODY CONTROL MODULE (BCM)

The BCM monitors and controls Electronic Climate Controls (ECC), rear defogger, outside temperature display, cooling fan control, fuel data display center display information, vacuum florescent display dimming, self diagnostics and retained accessory power system functions.

The BCM consists of input and output devices, CPU, power supply and memories that coincide with those of the ECM. The BCM exchanges information with the ECM to provide maximum reliability and improved serviceability of the body related systems.

ELECTRONIC SPARK TIMING (EST)

The EST type HEI distributor receives all spark timing information from the computer when the engine is running. The computer provides spark plug firing pulses based upon the various engine operating parameters. The electronic components for the electronic spark control system are integral with the computer. The 2 basic operating modes are cranking (or bypass) and normal engine operation.

When the engine is in the cranking/bypass mode, ignition timing occurs at a reference setting (distributor timing set point) regardless of other engine operating parameters. Under all other normal operating conditions, basic engine ignition timing is controlled by the computer and modified or added to, depending on particular conditions such as altitude and/or engine loading.

The HEI distributor communicates to the ECM through a 4 terminal connector which contains 4 circuits, these 4 circuits are as follows:

- The distributor reference circuit
- The bypass circuit
- The EST circuit
- The ground circuit

Whenever the pickup coil signals the HEI module to open the primary circuit, it also sends the spark timing signals to the ECM through the reference line.

When the voltage on the HEI bypass line is 0 volts (engine cranking), the HEI module is forced into the bypass mode which means that the HEI module provides spark advance at base timing and disregards the spark advance signal from the ECM. If the voltage on the HEI bypass line is 5 volts (engine running), the HEI module accepts the spark timing signal provided by the ECM.

IDLE SPEED CONTROL SYSTEM

The idle speed control system is controlled by the computer. The system acts to control engine idle speed in 3 ways: as a normal idle (rpm) control, as a fast idle device and as a "dashpot" on decelerations and throttle closing. The normal engine idle speed is programmed into the computer and no adjustments are possible. Under normal engine operating conditions, idle speed is maintained by monitoring idle speed in a closed loop fashion. To accomplish this loop, the computer periodically senses the engine idle speed and issues commands to the idle speed control to move the throttle stop to maintain the correct speed.

For engine starting, the throttle is either held open by the idle speed control for a longer (cold) or a shorter (hot) period to provide adequate engine warm-up prior to normal operation. When the engine is shut off, the throttle is opened by fully extending the idle speed control actuator to get ready for the next start.

Signal inputs for transmission gear, air conditioning compressor clutch (engaged or not engaged) and throttle (open or closed) are used to either increase or decrease throttle angle in response to these particular engine loadings.

Vehicle idle speed is controlled by an electrically driven actuator (idle speed control) which changes the throttle angle by acting as a movable idle stop. Inputs to the ISC actuator motor come from the ECM and are determined by the idle speed required for the particular operating condition. The electronic components for the ISC system are integral with the ECM. An integral part of the ISC is the throttle switch. The position of the switch determines whether the ISC should control idle speed or not. When the switch is closed, as determined by the throttle lever resting upon the end of the ISC actuator, the ECM will issue the appropriate commands to move the idle speed control to provide the programmed idle speed. When the throttle lever moves off the idle speed control actuator from idle, the throttle switch is opened. The computer than extends the actuator and stops sending idle speed commands and the driver controls the engine speed.

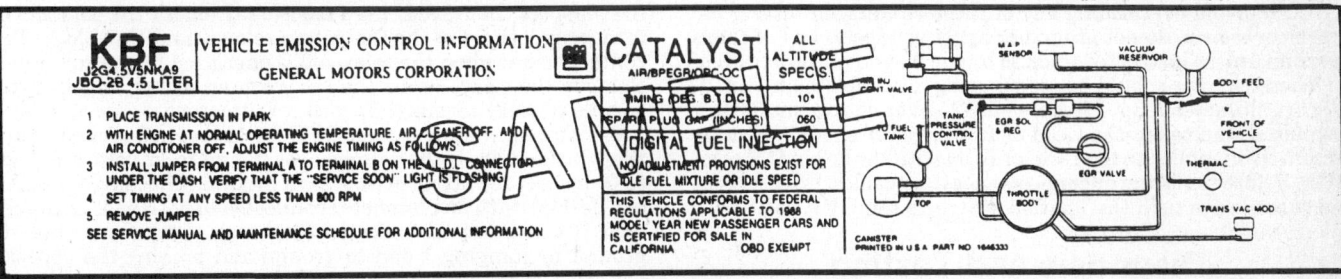

Sample vehicle emission control information label — DeVille and Fleetwood

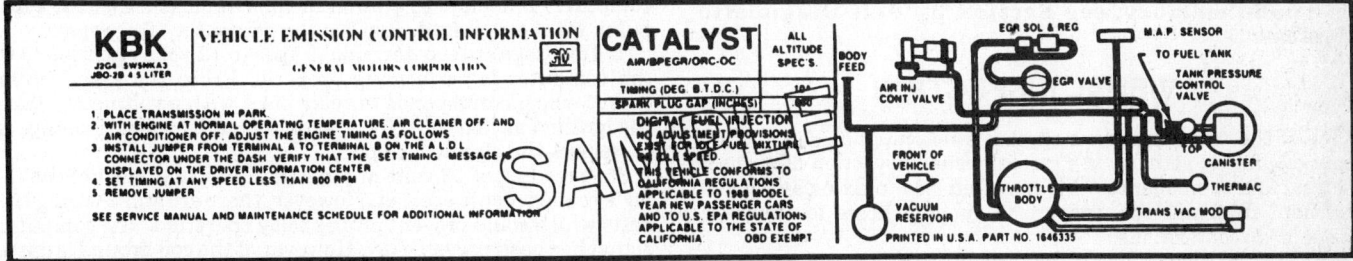

Sample vehicle emission control information label — Eldorado and Seville

EMISSION CONTROLS

EGR Operation

The ECM controls controls the operation of the EGR system. Whenever the EGR solenoid is energized by the ECM, the EGR system is disabled (meaning no exhaust gas will be recirculated through the intake manifold). When the EGR solenoid is deenergized, the EGR system is enabled and exhaust gas will be recirculated through the intake manifold.

Air Management Operation

The ECM controls the operations of the air management system. The air pump delivers the air to the divert (control) valve which sends the air to the air cleaner and the air switching valve. The switching valve sends the air either to the catalytic converter or the exhaust ports of the engine.

When the switching valve is energized by the ECM, it direct the air to the exhaust ports to aid in quickly heating the oxygen sensor to 600°F. When the engine is warm or in closed loop operation, the ECM will deenergize the air switching valve and direct the air to the catalytic converter to assist in oxidation of the hydrocarbon (HC) and the carbon monoxide (CO).

If the air control valve detects a rapid increase in intake manifold (deceleration), if certain operating modes exist or if the ECM detects any failure in the system, the air is diverted to the air cleaner or dumped into the atmosphere.

Canister Purge Control Operation

The ECM controls the operations of the canister purge control system. The ECM energizes the canister purge control solenoid by supplying a ground signal to the solenoid. The solenoid is energized by the ECM during cold engine operation. When the solenoid is energized, the vacuum to the canister line is blocked.

When the engine is at normal operating temperature, the ECM deenergizes the canister purge control solenoid by removing the ground signal. When the solenoid is deenergized, the vacuum is supplied to the canister.

CLOSED LOOP FUEL CONTROL

The purpose of closed loop fuel control is to precisely maintain an air/fuel mixture 14.7:1. When the air/fuel mixture is main-tained at 14.7:1, the catalytic converter is able to operate at maximum efficiency which results in lower emission levels.

Since the ECM controls the air/fuel mixture, it needs to check its output and correct the fuel mixture for deviations from the "ideal" ratio. The oxygen sensor feeds this output information back to the ECM.

A catalytic converter is used on all vehicles to reduce the emissions levels of the 3 major pollutants; hydrocarbon (HC), carbon monoxide (CO) and oxides of nitrogen (NOx). There are 2 converter designs being used, a pellet type on all California models and a monolithic honeycomb design on all others.

The converter, whether pellets or ceramic monolith, is coated with platinum, palladium and rhodium. When exhaust gases come into contact with these metals, the 3 pollutants are oxidized (burned off), further reducing the total engine emission levels.

CRUISE CONTROL

The ECM receives input signals from the cruise control engagement switches, instrument panel switch, brake release switch, drive switch and speed sensor. The ECM processes the cruise control inputs together with the DFI engine control inputs and transmits command signals to the vacuum control solenoid valve and power unit solenoid valve to control the vehicle speed.

VISCOUS CONVERTER CLUTCH (VCC) CONTROL

The ECM controls an electrical solenoid mounted in the automatic transmission. When the vehicle reaches a specified speed, the ECM energizes the solenoid and allows the viscous converter to mechanically couple the engine to the transmission.

When operating conditions indicate the transmission should operate as a normal fluid-coupled transmission (deceleration, passing, etc.) the solenoid is de-energized. The transmission also returns to normal (fluid-coupled) automatic operation when the brake pedal is depressed.

SERVICE PRECAUTIONS

The ECM and BCM are designed to withstand normal current draws associated with vehicle operation, however care must be taken to avoid overloading any of these circuits. In testing for opens or shorts, do not ground or apply voltage to any of these circuits unless instructed to do so by the diagnostic procedures.

These circuits should be tested using a suitable high impedance multimeter J–34029–A or J–29125–A or equivalent, if they remain connected to the ECM or BCM. Power should never be removed or applied to the ECM or BCM with the key in ON position. Before removing or connecting battery cables, fuses or connectors always turn the ignition switch to the OFF position.

Diagnosis and Testing

NOTE: For self-diagnostic system and accessing trouble code memory, see Section 3 "Self-Diagnostic Systems".

VISUAL INSPECTION

One of the most important checks, which must be done before any diagnostic activity, is a careful visual inspection of suspect wiring and components. This can often lead to fixing a problem without further steps. Inspect all vacuum hoses for pinches, cuts or disconnections.

Be sure to inspect hoses that are difficult to see beneath the air cleaner. Inspect all the related wiring for disconnects, for example, burned or chaffed spots, pinched wires, or contact with sharp edges or hot exhaust manifolds. This visual inspection is very important. It must be done carefully and thoroughly.

A visual inspection consists of checking the following:
   a. The "Engine Control System" telltale is working.
   b. "Service Mode" can be accessed.
   c. No trouble codes are stored.
   d. A careful visual check found no problems.

FUEL SYSTEM

1. Low or no fuel pressure diagnosis should begin by trying to determine if the fuel pump is operating or not. This is most easily accomplished by turning the ignition ON and listening for the 1 second run of the fuel pump and the associated relay clicks. Since this may not be possible in some shops, the best test is to probe both sides of the fuel pump fuse in the mini-fuse block with a voltmeter. Observe the meter as the ignition is turned ON. It should go to battery voltage (12 volts) and then, after 1 second, to 0 volts.

2. If the fuel pump circuit is operating properly, the computer signal and the relay are okay. The last connector in the fuel pump circuit is the connector at the tail panel. The voltage actions seen at the fuse should be repeated. If not, there is an open in the circuit. Check the connectors and repair wiring as required. Individual sections of the wiring can be tested with an ohmmeter.

3. If the fuel pump signal is correct at the tail panel connector, a fuel delivery system situation exists. If the pump cannot be hears to run during the 1 second on period, the pump should be replaced. This observation is more easily made if helper turns the ignition ON as the technician listens at the fuel tank or filler neck area. If the fuel pump can be heard to run, disconnect the fuel return line at the throttle body and install a plug in the throttle body opening. This will effectively "dead head" the fuel pump and eliminate the pressure regulator. If the pump is able to produce above 9 psi under these conditions (with the ignition ON), replace the fuel pressure regulator as it is controlling at too low pressure.

4. If the fuel pressure remains below 9 psi with the return line plugged, a restriction in the fuel supply line may exist. A blocked fuel line or fuel filter can be determined by visually inspecting the filter element and the fuel line routing for kinks, damage, etc. If lines and filter are okay, replace fuel pump.

5. If the voltage at the fuel pump fuse does not go to 12 volts,

first inspect the fuse. If the fuse is okay, check the contacts at terminals 1 and 4 of the fuel pump relay connecting block. The contacts close when the relay coil is energized by the computer. Remove the relay. Probe the relay center socket, which corresponds to relay terminal 5, with a voltmeter.

6. After ignition has been OFF for at least 10 seconds, turn the ignition ON. If the voltage does not go to 12 volts and back to 0, inspect for opens or shorts to ground. Checking for opens can be done with an ohmmeter connected to both ends of the circuit. If continuity is indicated (0 ohms), check for a short to ground by jumping 1 end to ground and probing the opposite end with an ohmmeter to ground. An infinite reading indicates the wire is not grounded and the circuit is okay. Replace the computer. If a short to ground in the circuit is found the computer will be damaged. Replace the computer after repair.

7. If voltage at relay terminal 5 goes to 12 volts and then to 0, the computer is performing properly. Probe the relay center socket which corresponds to relay pin 4 with a voltmeter. Turn the ignition switch to RUN, it should read 12 volts. If voltage is 0, inspect the circuit for open.

8. A reading of 12 volts at pin 4 indicates that most of the relay's requirements are met. However, there are still 2 wiring circuits which could prevent proper relay operation. The computer signal has been proven okay. However, if the coil ground is open, the coil will not be energized. To check, probe the relay center socket which corresponds to relay pin 2, with an ohmmeter connected to ground. Continuity (0 ohms) indicates a good circuit. An infinite reading indicates an open. Repair wire as required.

9. The second circuit in which an open could occur, preventing the proper voltage at the fuse, is between the fuel pump relay terminal 1 and the mini-fuse block. This circuit can be checked by probing both ends with an ohmmeter. If the circuit has continuity as indicated by a 0 ohms reading, replace fuel pump relay.

10. High fuel pressure is caused by either a malfunction of the pressure regulator or a restriction in the fuel return line. To determine which of these problems exists, disconnect the fuel return line at the throttle body and connect a suitable fitting to the throttle body to accept a length of flexible rubber fuel hose. Insert the open end of the hose into a suitable fuel container. Observe the fuel pressure as the ignition switch is turned on. If the fuel pressure remains above 12 psi, replace the pressure regulator as it is unable to regulate with no return restriction.

11. If the fuel pressure now falls into the correct pressure range of 9–12 psi, the restriction has been eliminated by bypassing the return system. A restricted fuel return line can be located by visually inspecting the line routing for kinks, damage, etc.

Diagnosing Poor Performance

1. Unsatisfactory engine performance complaints which are related to the digital electronic fuel injection system are caused either by improper fuel delivery or improper fuel delivery or improper ignition advance (controlled by the computer). To isolate the problem to one of these systems, remove the air cleaner and observe the injector spray pattern of both injectors at idle. The spray pattern should be compared to a proper injector spray pattern of a known good car.

2. Improper fuel delivery which affects both injectors is most likely a fuel delivery system problem. By switching injector connectors, it can be determined if the problem is the injector assembly or the signal to the injector. If the problem remains with the original injector, it is most likely an injector problem. Replace the injector. If the problem moves with the injector connector, an improper signal circuit is indicated. The injectors are powered through the 3 amp fuses in the mini-fuse block. The fuse block receives battery voltage from the starter solenoid battery terminal when the fuel pump relay is energized by the computer during CRANK or RUN. With the relay closed, the circuits apply 12 volts to the injector. The computer provides the ground to energize the solenoid.

── CAUTION ──

Do not apply direct battery voltage to the injector, 12 volts will destroy the fuel injector coil within ½ second.

3. Check the injector fuses by visually inspecting the fuse filament. If the fuses are okay there is a harness problem in the voltage feed or ground. The troubled circuit should be investigated. Check for opens with an ohmmeter. If harness checks okay, replace computer.

4. A blown injector fuse should be replaced. If it blows again, a short to ground is indicated. To check for this condition, connect ohmmeter between the red or white wire at the injector connector and ground (ignition **OFF** and computer disconnected). A low reading indicates a short circuit. Repair as required. If harness is okay, replace the computer.

5. A proper injector spray pattern indicates that improper ignition timing is the main DFI component which can cause poor performance. Ground terminal B of the ALDL for the "set timing" mode and check timing. It should be at the base (or initial) value of 10 degrees BTDC (900 rpm or less). If not, reset to 10 degrees.

6. Once it has been established that the ignition timing is using the proper reference signal, the system's ability to advance the spark must also be determined. Since the actual ignition advance produced is the result of other variables besides engine rpm and manifold vacuum, it is not possible to establish checkpoints. However, if the system does advance, the shape of the advance curve can be assumed to be correct, since it is determined by the electronic circuitry which was able to recognize that some advance was required and did respond to this information. Disconnect the "set timing" jumper and check ignition timing. At normal idle, this should be approximately 20–30 degrees BTDC.

7. If the ignition timing does not advance as a result of disconnecting the "set timing" jumper, a problem with the advance system is indicated. Since the computer selects and determines the spark advance curve, it is necessary to determine if the computer is operating properly or not. During cranking, no spark advance is desired. The computer limits the advance to base timing by turning off the voltage signal. The pick-up coil pulse is used directly to turn the HEI module on. When the engine starts, the computer turns on the voltage in the circuit and the pick-up coil pulse is sent to the computer, modified by the computer and sent back to the module. Whether the computer advances the timing or not depends upon whether it applied a voltage to the circuit or not (no voltage = base timing; voltage = electric spark timing). To check the circuit, disconnect the 4-way connector at the distributor while the engine is idling. This will stop the engine. Probe the harness side pin C (not distributor side harness) with a voltmeter while the ignition switch remains **ON**. If voltage is greater than 4 volts, refer to HEI diagnosis because the computer has signaled that EST should be used, but timing did not advance when checked.

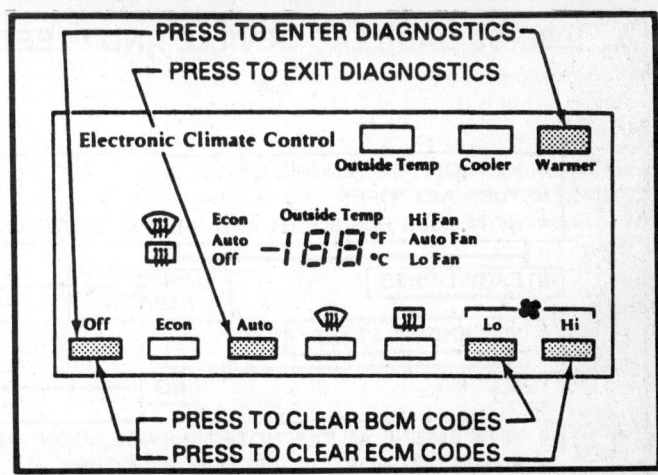

Climate Control Panel

8. Voltage less than 1 volt indicates that either the computer signal is not being produced or the circuit is open or shorted to the ground. To check for shorts, disconnect the black/green computer connector and probe harness pin D with an ohmmeter to ground (ignition **OFF** and distributor connector disconnected). A 0 ohm reading indicates a short. Repair as required.

9. If the ohmmeter reads infinity (∞), check for opens by jumping distributor connector pin C to ground. If the ohmmeter reading remains infinite, circuit is open. repair as required. If the ohmmeter reads zero ohms, the circuit is okay. Check MAT and coolant sensor circuits for an open between the splice and the computer. Attention should be focused on the bulk head and computer connectors. If the circuit is okay, substitute a new computer and observe performance.

10. The EGR system utilizes various controls in order to provide EGR gases only when they are needed for emission control. One of these controls is the EGR solenoid, with power feed from the ignition switch through a fuse. Ground for the solenoid is provided by the computer. The computer provides this ground whenever the coolant temperature signal from the coolant sensor says the temperature is below 43°C (110°F). This energizes the solenoid and blocks the flow of vacuum to the EGR transducer thus preventing EGR operation at cold engine temperatures. Above 71°C (160°F), the solenoid ground is removed and the solenoid opens, allowing vacuum to the EGR valve. This vacuum signal is a ported vacuum which exists only off idle. This means that even on a warm engine, there is no EGR vacuum signal and no EGR flow at idle.

11. If EGR operation is okay, check to make sure that throttle valves open to wide open throttle when the accelerator pedal is wide open. If this is okay, the performance problem is not related to the DFI system. If EGR problems are found, check hoses, etc.

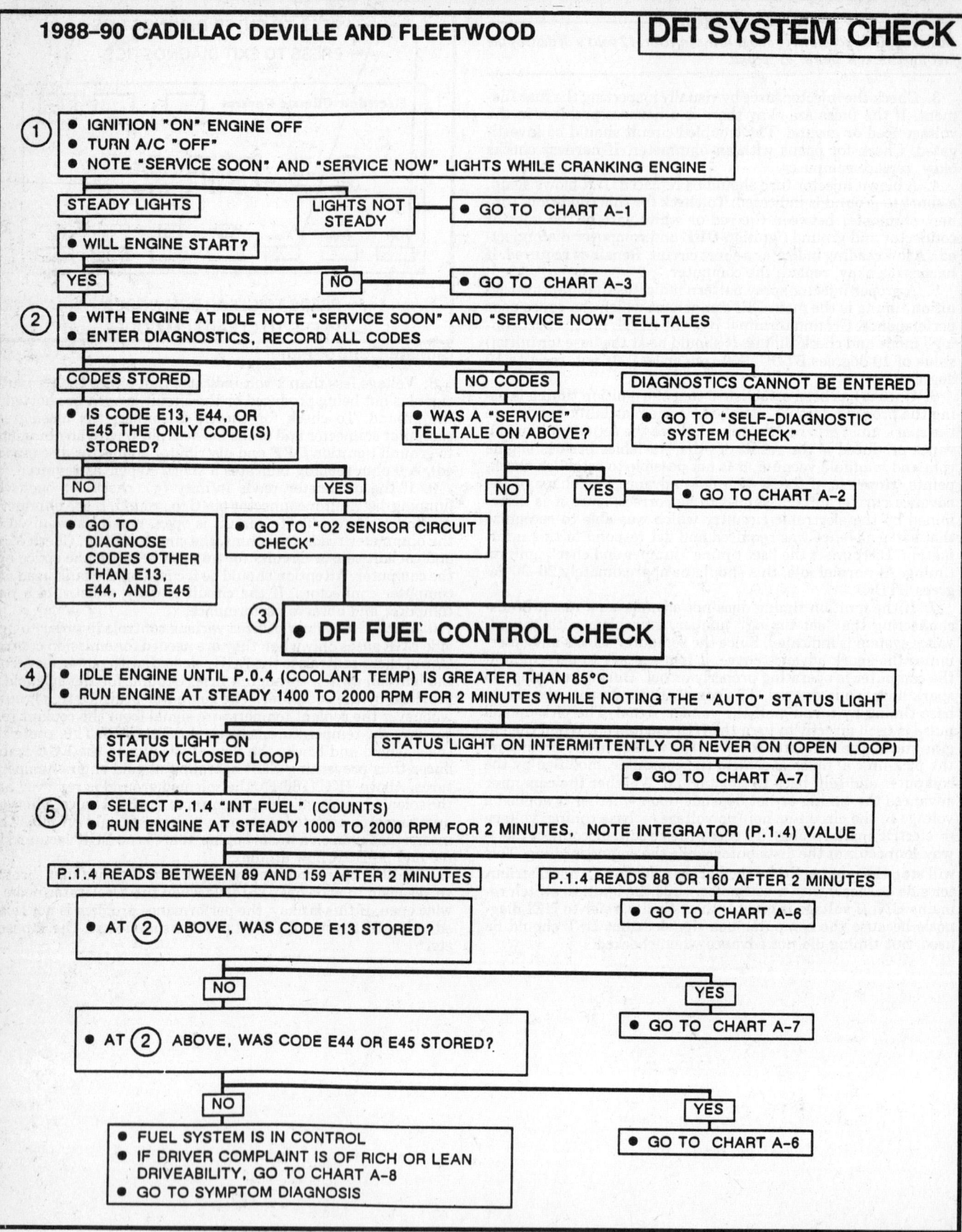

1988–90 CADILLAC DEVILLE AND FLEETWOOD

DFI SYSTEM CHECK

(1)
- IGNITION "ON" ENGINE OFF
- TURN A/C "OFF"
- NOTE "SERVICE SOON" AND "SERVICE NOW" LIGHTS WHILE CRANKING ENGINE

STEADY LIGHTS

LIGHTS NOT STEADY → • GO TO CHART A-1

- WILL ENGINE START?

YES

NO → • GO TO CHART A-3

(2)
- WITH ENGINE AT IDLE NOTE "SERVICE SOON" AND "SERVICE NOW" TELLTALES
- ENTER DIAGNOSTICS, RECORD ALL CODES

CODES STORED

NO CODES

DIAGNOSTICS CANNOT BE ENTERED

- IS CODE E13, E44, OR E45 THE ONLY CODE(S) STORED?

- WAS A "SERVICE" TELLTALE ON ABOVE?

- GO TO "SELF-DIAGNOSTIC SYSTEM CHECK"

NO

YES

NO

YES → • GO TO CHART A-2

- GO TO DIAGNOSE CODES OTHER THAN E13, E44, AND E45

- GO TO "O2 SENSOR CIRCUIT CHECK"

(3) • **DFI FUEL CONTROL CHECK**

(4)
- IDLE ENGINE UNTIL P.0.4 (COOLANT TEMP) IS GREATER THAN 85°C
- RUN ENGINE AT STEADY 1400 TO 2000 RPM FOR 2 MINUTES WHILE NOTING THE "AUTO" STATUS LIGHT

STATUS LIGHT ON STEADY (CLOSED LOOP)

STATUS LIGHT ON INTERMITTENTLY OR NEVER ON (OPEN LOOP)

- GO TO CHART A-7

(5)
- SELECT P.1.4 "INT FUEL" (COUNTS)
- RUN ENGINE AT STEADY 1000 TO 2000 RPM FOR 2 MINUTES, NOTE INTEGRATOR (P.1.4) VALUE

P.1.4 READS BETWEEN 89 AND 159 AFTER 2 MINUTES

P.1.4 READS 88 OR 160 AFTER 2 MINUTES

- GO TO CHART A-6

- AT (2) ABOVE, WAS CODE E13 STORED?

NO

YES

- GO TO CHART A-7

- AT (2) ABOVE, WAS CODE E44 OR E45 STORED?

NO

YES

- GO TO CHART A-6

- FUEL SYSTEM IS IN CONTROL
- IF DRIVER COMPLAINT IS OF RICH OR LEAN DRIVEABILITY, GO TO CHART A-8
- GO TO SYMPTOM DIAGNOSIS

1988–90 CADILLAC DEVILLE AND FLEETWOOD

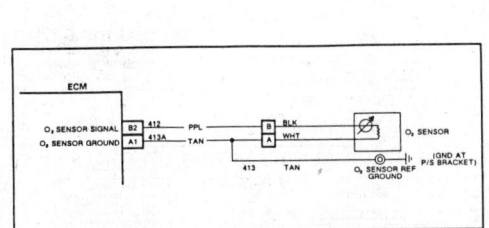

OXYGEN SENSOR CIRCUIT CHECK

Description:

The ECM provides a 0.45 volt reference to the oxygen sensor on CKT 412. When warm, a properly operating oxygen sensor will drive the 0.45 volt reference lower (below 0.45 volts) to indicate lean mixtures and higher (above 0.45 volts) to indicate a rich mixture. When the oxygen sensor is cold (below 200°C), the output voltage will be around 0.45 volts and the ECM will keep the system in open loop operation.

When the ECM sees that the oxygen sensor is varying from the cold voltage of 0.45 volt and the coolant sensor value is above 85°C, it will send the system into closed loop operation. In closed loop operation, the ECM will adjust the fuel delivery rate based on the oxygen sensor readings.

Notes On Fault Tree:

1. The ECM compares voltage on ckt 412 to the ground voltage on ckt 413. It is essential that the oxygen sensor ground and ECM ground on pin A–1 show good continuity.

2. At key on with oxygen sensor disconnected, the Parameter P.0.7 should read 0.45 volts (Nominal). Lower or higher voltage indicates circuit problems.

3. At this point the Oxygen Sensor circuit between the 2–way connector (at the sensor) and the ECM, all the Oxygen Sensor grounds and the ECM are all functioning correctly and have good contact.

4. This series of tests will determine if a faulty reading is due to poor ground connections, circuit faults or problem with the ECM connection or ECM.

1988–90 CADILLAC DEVILLE AND FLEETWOOD

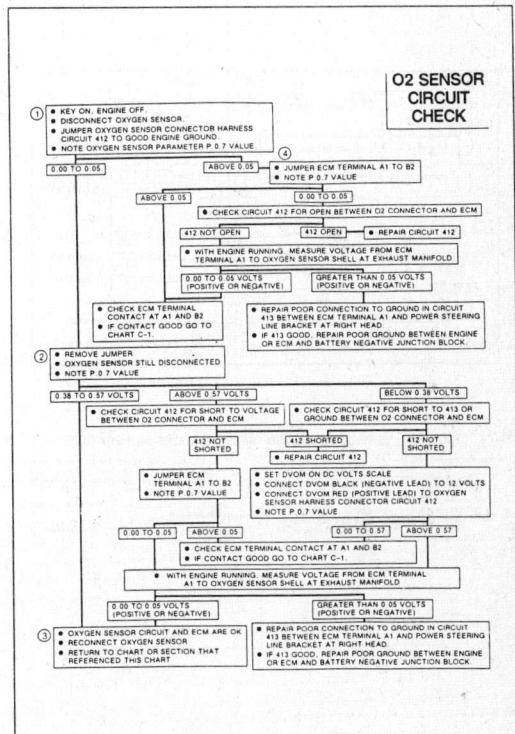

1988–90 CADILLAC DEVILLE AND FLEETWOOD

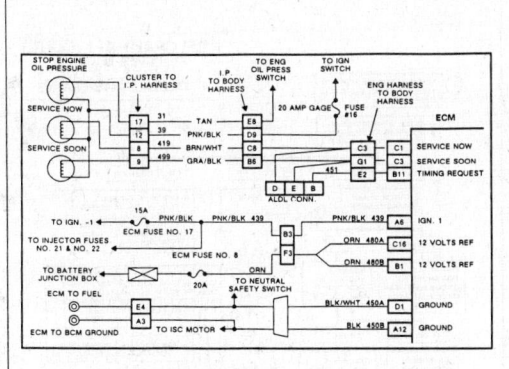

CHART A–1
"SERVICE NOW" AND/OR "SERVICE SOON" INOPERATIVE

Description:

The "Service Now" and "Service Soon" lamps are powered through the 20 AMP "gage" fuse #3 on the fuse panel. The ECM supplies a ground to ckt 419 (terminal C1) to illuminate "Service Now" and the ckt 499 (terminal C3) to illuminate "Service Soon".

Notes On Fault Tree:

1. The "Service Soon" and "Service Now" lamps both should illuminate (bulb check) during cranking.

2. Service Soon will flicker in ALDL mode or set timing. ALDL mode can be induced by a resistance path to ground from ckt 451, ALDL request.

3. Checking for power to "Service Now" in ckt 419 (terminal C1) to illuminate "Service Now", ground open to bulb or burned out bulb.

4. Checking for power to "Service Soon" bulb, ground open to bulb or burned out bulb.

5. If the bulb will illuminate with a ground at the ECM connector, carefully perform "ECM Replacement Check".

1988–90 CADILLAC DEVILLE AND FLEETWOOD

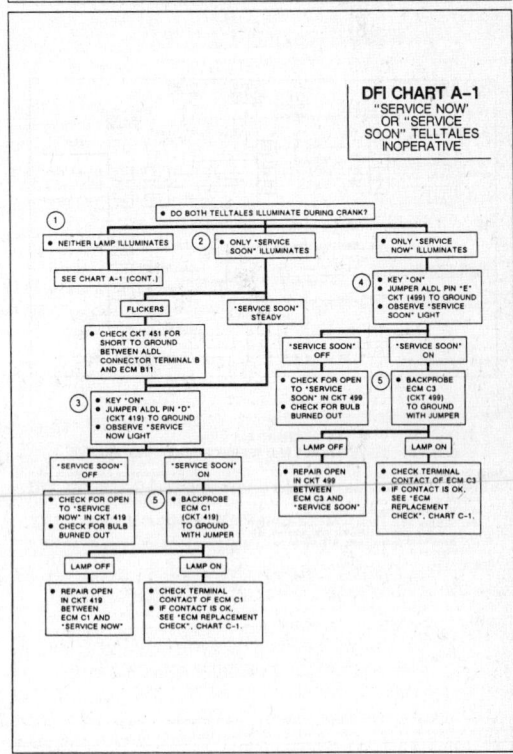

1988–90 CADILLAC DEVILLE AND FLEETWOOD

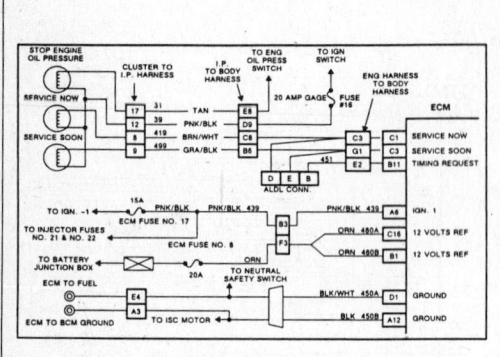

CHART A-1
(CONT'D) "SERVICE NOW" AND "SERVICE SOON" INOPERATIVE

Notes On Fault Tree:

1. Three possible causes for both "Service" lights inoperative are:
   A. No power to bulbs from "gage" fuse
   B. No crank input to ECM — the crank input must be present for bulbs to check. With no crank input present, code E18 should be stored hard.
   C. ECM not powering up due to loss of power or ground. If ECM loses power, engine should not start or run, code F32 should be stored as a hard code.

2. Checking for "key on" power and both grounds present at ECM.

3. Checking for full time memory power present at ECM.

1988–90 CADILLAC DEVILLE AND FLEETWOOD

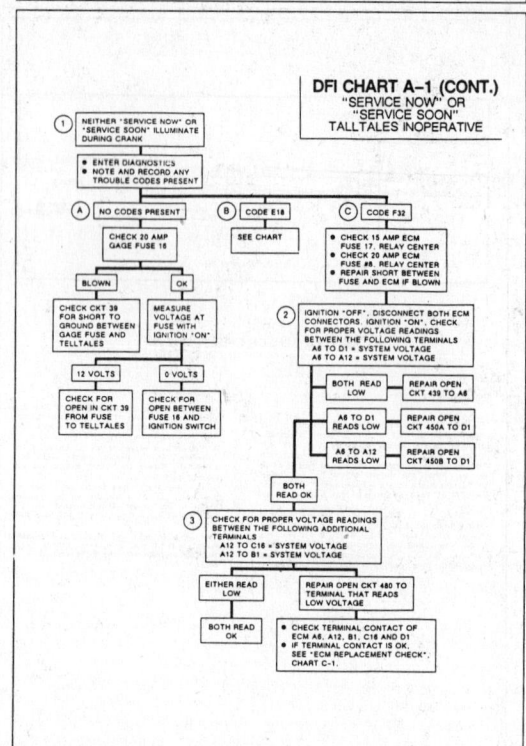

DFI CHART A-1 (CONT.)
"SERVICE NOW" OR "SERVICE SOON" TALLTALES INOPERATIVE

1988–90 CADILLAC DEVILLE AND FLEETWOOD

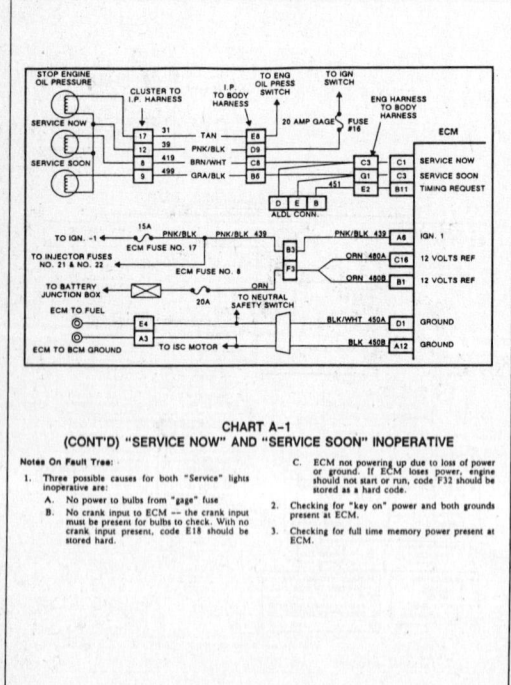

CHART A-1
(CONT'D) "SERVICE NOW" AND "SERVICE SOON" INOPERATIVE

Notes On Fault Tree:

1. Three possible causes for both "Service" lights inoperative are:
   A. No power to bulbs from "gage" fuse
   B. No crank input to ECM — the crank input must be present for bulbs to check. With no crank input present, code E18 should be stored hard.
   C. ECM not powering up due to loss of power or ground. If ECM loses power, engine should not start or run, code F32 should be stored as a hard code.

2. Checking for "key on" power and both grounds present at ECM.

3. Checking for full time memory power present at ECM.

1988–90 CADILLAC DEVILLE AND FLEETWOOD

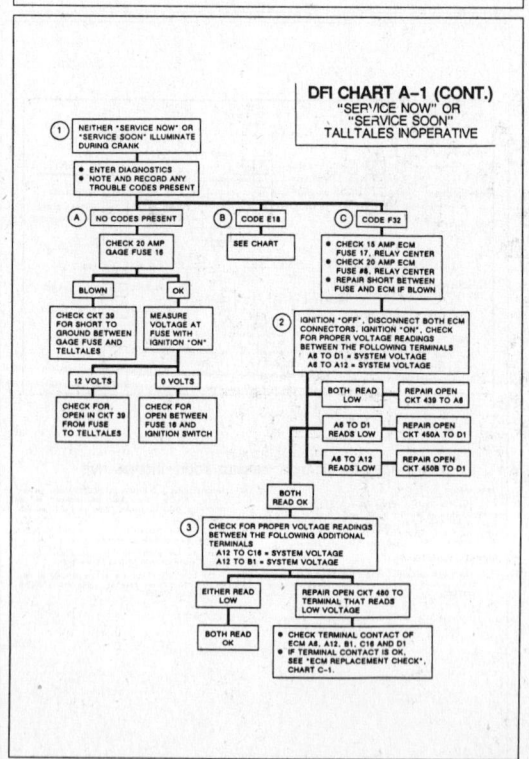

DFI CHART A-1 (CONT.)
"SERVICE NOW" OR "SERVICE SOON" TALLTALES INOPERATIVE

1988–90 CADILLAC DEVILLE AND FLEETWOOD

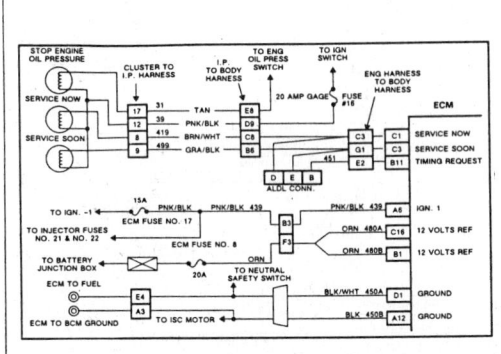

CHART A-2
DFI "SERVICE NOW" AND/OR "SERVICE SOON" LIGHTS ON
NO HARD CODES

Description:

The "Service Soon" and "Service Now" lights are powered from the ignition switch through the 20 amp gage fuse #3 on the fuse panel. The ECM supplies ground to ckt 419 (terminal C1) to illuminate "Service Now" and ckt 499 (terminal C3) to illuminate "Service Soon". The "Service Soon" and "Service Now" lamps illuminate when most ECM trouble codes set.

Notes On Fault Tree:

1. The ECM supplies ground to the "Service Now" light through circuit #419. If this wire is shorted to ground, the "Service Now" light will be on whenever the ignition switch is on. No vehicle performance problems are caused by a shorted circuit #419.

2. The ECM supplies ground to the "Service Soon" light through circuit #499. If this wire is shorted to ground, the "Service Soon" light will be on whenever the ignition switch is on. No vehicle performance problems are caused by a shorted circuit #499.

   A. If circuit 451 is shorted to ground, the ECM will send Serial Data over the Service Now lamp circuit, Service Soon will flicker.

3. Both the "Service Soon" and "Service Now" lights will be on with no codes set if a short to voltage is present on the crank signal input to the ECM (circuit #806).

4. If all the wiring circuits are OK, then the PROM and ECM must be investigated for operating in hard backup. See Chart C-1, "ECM Replacement Check".

1988–90 CADILLAC DEVILLE AND FLEETWOOD

DFI CHART A-2
"SERVICE NOW" AND/OR "SERVICE SOON" TELLTALES NO CODES PRESENT

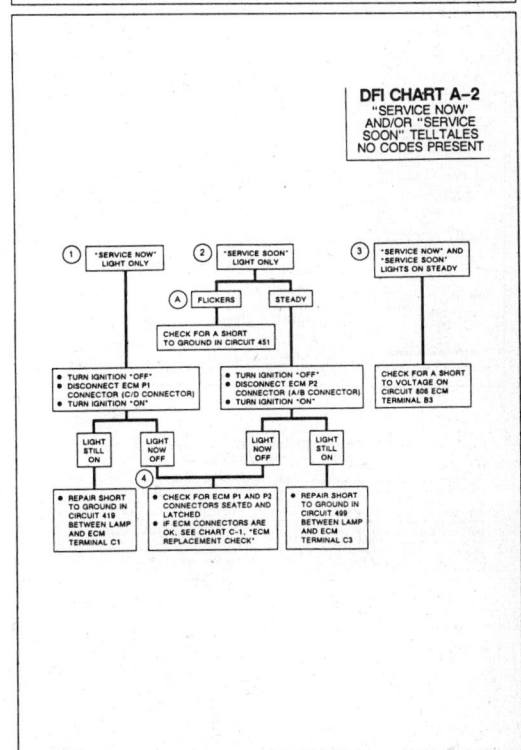

1988–90 CADILLAC DEVILLE AND FLEETWOOD

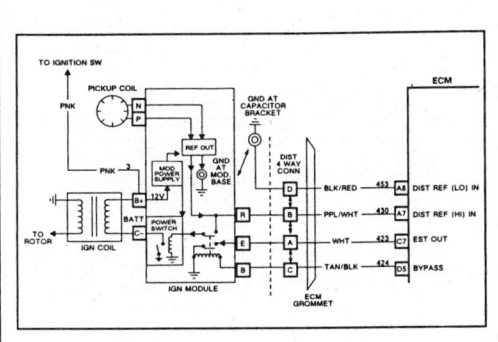

CHART A-3
NO START OR STALL AFTER START

Description:

All internal combustion engines require spark, fuel, air, and proper timing to operate. The DFI system is no different. If the battery is at the proper charge level, the first step should be to determine which of these elements is missing.

Notes On Fault Tree:

1. Checking for codes stored and for proper "Service Now" "Service Soon" operation. Repair stored codes and improper telltale operation before proceding with chart A-3.

2. Injectors should spray only when the engine is cranking or running. Look for spray, drips or leaks at key on/engine off.

3. Checking for both injectors to spray fuel while cranking.

4. Fuel is OK, checking for spark. Note: Use an ST 125 to test. A spark plug with a wide gap or allowing a plug wire to arc to ground may not test the HEI for sufficient output and may damage the coil, cap or rotor.

5. This step bypasses the EST system. If the vehicle will start and run with the EST system disabled (Bypass open), then the "No Start" condition may be due to an EST system fault. If the ECM has a poor ground to the engine or if the distributor has a poor ground connection to the ECM, the distributor may not be able to recognize EST pulses and the engine will stall as the ECM tries to enable EST (use EST to control spark timing.)

6. If the vehicle will not start, then the fuel system must be checked next. Connect the fuel pressure gage J-25400-300 and observe the fuel pressure while cranking. The gage should be installed in the fuel inlet line at the service fitting. A fuel pressure reading of between 9-12 PSI during cranking indicates that the fuel system is operating properly. Improper fuel pressure indicates a fuel problem, refer to Chart A-4 "Fuel System Diagnosis".

   Since we now have spark, and fuel spray from both injectors at the correct fuel pressure, all DFI functions for starting are operating normally. The cause of the no start condition is a mechanical problem (spark plugs, valves, valve timing, etc.).

1988–90 CADILLAC DEVILLE AND FLEETWOOD

DFI CHART A-3
NO START OR STALL AFTER START

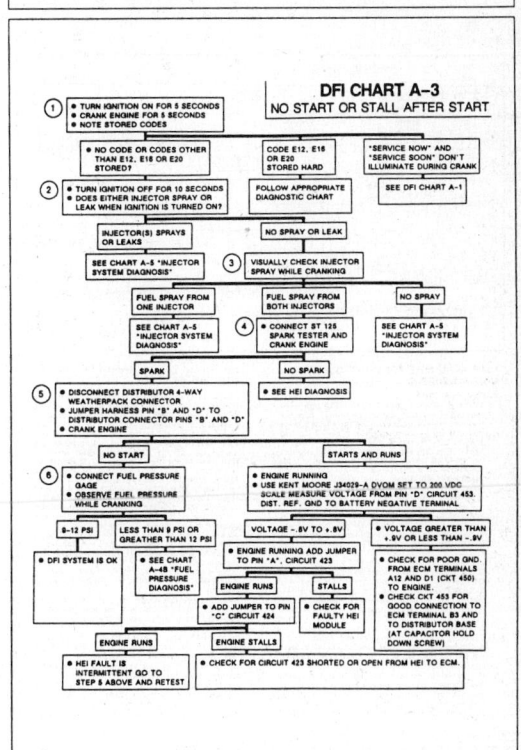

1988–90 CADILLAC DEVILLE AND FLEETWOOD

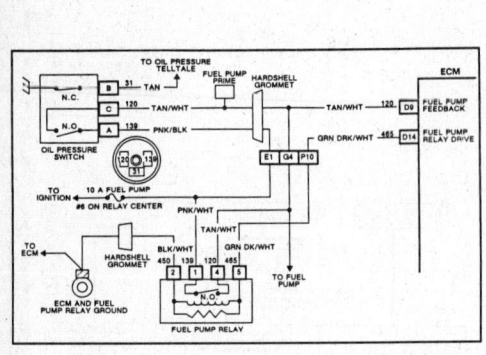

CHART A-4A
FUEL SYSTEM DIAGNOSIS

The DFI system requires that fuel pressure be 9.0–12.0 PSI and steady under all driving conditions. The J25400-300 (0–60 PSI) pressure gage may not show small variations in fuel pressure. For a more precise and more responsive gage reading, use the 0–15 PSI gage head (from the essential tool gage number J29658/BT8205) attached to the hose with schrader valve fitting from the essential tool gage J34730-1.

Notes On Fault Tree:

1. Fuel pressure should be 9.0–12.0 PSI and steady. If the fuel pressure is not in this range, go to chart A-4B to diagnose fuel pressure out of range.
2. If the pressure drops off (leaks down) at key off, either the fuel pressure regulator cannot hold pressure or the fuel pump check ball is not seating. This branch of the trouble tree will determine which is causing the pressure leak down.

1988–90 CADILLAC DEVILLE AND FLEETWOOD

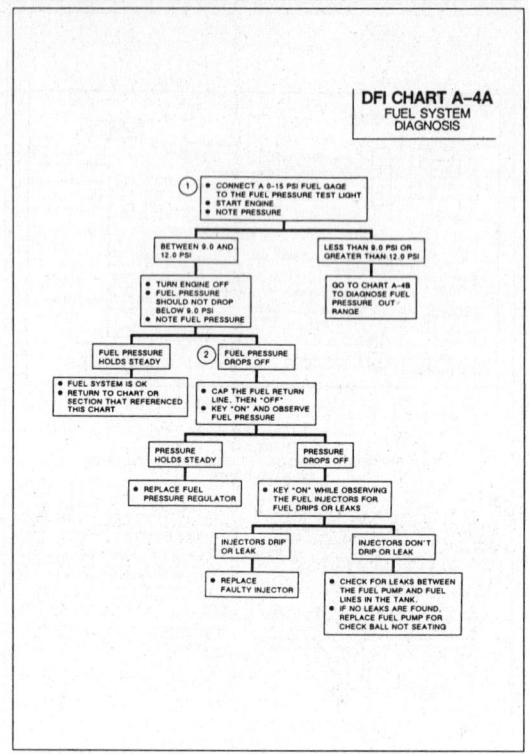

DFI CHART A-4A
FUEL SYSTEM DIAGNOSIS

1988–90 CADILLAC DEVILLE AND FLEETWOOD

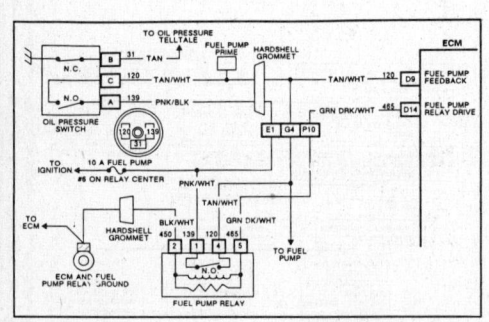

CHART A-4B, FUEL PRESSURE OUT OF RANGE DIAGNOSIS

Notes On Fault Tree:

1. The fuel pressure gage J-25400-300 should be installed at the fuel line service fitting. Measure the fuel pressure while cranking the engine. If the fuel pressure is between 9.0 and 12.0 PSI, then refer to the fuel system diagnosis chart A-4A.
NOTE: The DFI system requires that fuel pressure be 9.0–12.0 PSI and steady under all driving conditions. The J25400-300 0-60 PSI pressure gage may not show small variations in fuel pressure. For a more precise and more responsive gage reading, use the 0–15 PSI gage head (from the essential tool gage number J29658/BT8205) attached to the hose with schrader valve fitting from the essential tool gage J34730-1.
2. If the fuel pump relay or ECM were the cause of a low fuel pressure, there would be an ECM code E20 set. This step is to check for voltage supply to the fuel tank five-way connector.
3. If the voltage signal to the fuel tank connector is OK, then an open may exist between the five-way fuel tank connector and the fuel pump. If the fuel pump runs with an alternative power source connected, then the fuel tank unit is OK, check throttle body fuel metering assembly for cause of low pressure.
4. Checking for fuel supply system (tank, filter, pump, sender, supply line) able to deliver at least 9.0 PSI pressure or for throttle body fuel pressure regulator fault.
5. If pressure is low with fuel return line plugged, then throttle body fuel metering assembly is not at fault. A restriction or blockage may exist in the fuel supply system. The fuel supply line should be checked visually for kinks, damage, etc.; the fuel

filter element can also restrict flow. Check for proper fuel line routing.

Check sender tubes for restrictions, check for rubber coupler between pump and sender leaking or cracked, check for fuel strainer in tank collapsed, mispositioned or restricted. If all of the above are OK, replace the fuel pump.
6. Fuel pressure above 12.0 PSI is caused either by a malfunction of the pressure regulator or by a restriction in the fuel return line. It should be noted that a secondary condition of spark plug fouling, code E45 or oxygen sensor contamination resulting in code E13 accompanied by E45 may result from the too rich fuel flow. To isolate the cause of the high fuel pressure, disconnect the return line at the throttle body and connect a suitable fuel container and insert a length of flexible rubber fuel hose. Insert the other end of the hose into a suitable fuel container and observe the fuel pressure as the ignition switch is turned on. If the fuel pressure remains above 12.0 PSI, replace the fuel metering assembly—it is unable to control pressure properly.
If the fuel pressure drops into the correct pressure range of 9.0–12.0 PSI, with the fuel return line bypassed, then the fuel return line is restricted. A restricted fuel return line can be diagnosed by visually inspecting the line for kinks, damage, etc. A kink on the Teflon fuel line (braided stainless steel clad) may not be visually obvious.
7. If the fuel pump will not run with externally applied power, the fault is an electrical open in the fuel sender unit wiring to the pump, an open in the RFI suppression connector inside the tank on the pump or a faulty fuel pump. The fuel sending unit must be removed from the vehicle to check.

1988–90 CADILLAC DEVILLE AND FLEETWOOD

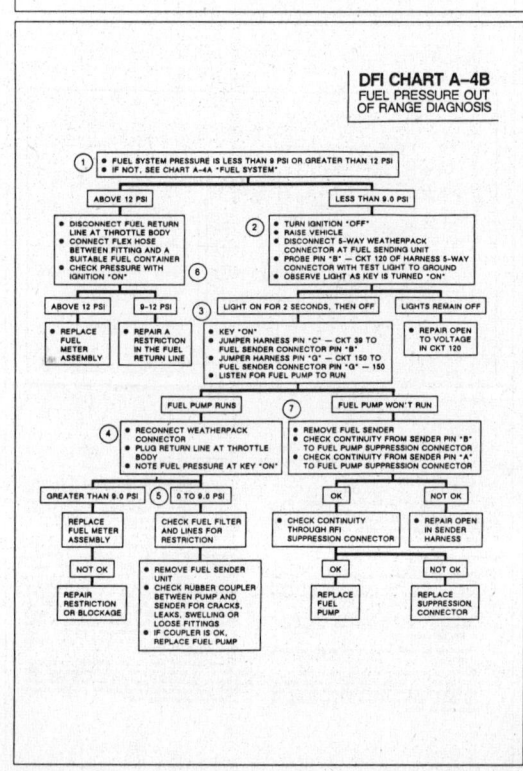

DFI CHART A-4B
FUEL PRESSURE OUT OF RANGE DIAGNOSIS

1988–90 CADILLAC DEVILLE AND FLEETWOOD

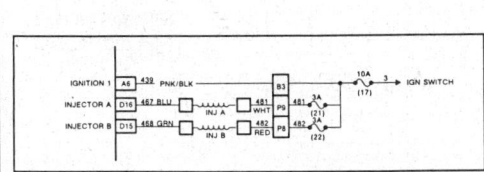

CHART A-5
INJECTOR SYSTEM DIAGNOSIS

Notes On Fault Tree:

1. If this procedure is being followed for a "no-start" condition, crank the engine for 5 seconds to check the distributor reference signal. If a Code E12 does not set, turn the ignition off for 10 seconds and observe the injectors as the ignition is turned back on. If there is no spray, then they are not stuck open. Observe the injectors while cranking the engine.

2. To determine if the injector is being activated electrically, repeat the above procedure with the electrical connector removed. If the injector continues to spray, it is defective and must be replaced. If the injector no longer sprays, the drive circuit of the affected injector must either be shorted to ground or the ECM is grounding internally.

3. If both injectors spray or if neither injector sprays, it must be determined if the fuel system is operating properly. The fuel pressure gage J-25400-300 should be installed at the fuel line service fitting. Measure the fuel pressure while cranking the engine. If the fuel pressure is not between 9-12 PSI, then refer to the performance diagnosis chart A-4A, "Fuel System Diagnosis".

4. If the fuel pressure is between 9-12 psi and there was no spray from either injector while cranking, the injector circuit must be checked for proper voltage. If there is voltage at the injector fuses, then the ECM must be faulty

because it is not grounding both injector circuits. If there is no voltage at the fuse, check for voltage at the 10A amp ECM fuse which feeds the injector fuses. If there is voltage here then an open must exist between the fuses on circuit #439. If there is no voltage at the ECM fuse then circuit #3 must be repaired for an open or short to ground.

5. If the fuel pressure is between 9-12 psi and there was fuel spray from both injectors while cranking, then check the injector system to determine if the injectors leak. To check for injector leakage, proceed as follows:

   A. Start and run the engine for 10 seconds.
   B. Turn the engine off for at least 10 seconds.
   C. Turn the ignition on to pressurize the injectors.
   D. Visually check for dripping fuel from the bottom of the injector.

   If the fuel is dripping, check for damaged "O" rings and if OK, replace the injector. If the fuel does not drip, then the fuel system is OK.

6. If there is spray from only one injector, then there is a malfunction in the injector assembly or in the signal to the injector assembly. The malfunction can be isolated by switching the injector connectors. If the problem remains with the original injector after switching the connector, then the injector is defective. Replace the injector.

1988–90 CADILLAC DEVILLE AND FLEETWOOD

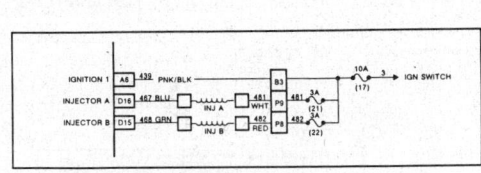

CHART A-5 (CONTINUED)
INJECTOR CIRCUIT DIAGNOSIS

Important

The injector circuit test should only be performed when the driveability condition is present or when one injector is visually verified to be inoperative. For the injector circuit test to result in a positive diagnosis of an injector circuit fault, the voltage measurements requested by Chart A-5 must be taken with the injector fault present.

Description:

The fuel injectors are powered through the 3 amp injector fuses. The ECM turns on the injectors by applying a ground to ECM connector pins D15 and D16, injector drive circuits. The ECM grounds the injector drive circuits to turn on the fuel injector to supply fuel to the engine. ECM Data Parameter P.0.6, Injector Pulse Width, reflects the time in milli-seconds that the ECM turns on the ground to the injector for each injector pulse.

Notes On Fault Tree:

When measuring voltages between ECM connector pins D15 and D16, the voltmeter should read 0.0 volts if the ECM is able to control both fuel injectors to provide equal amounts of fuel. If one injector loses power from the fuse or loses ground from the ECM, the voltage measured between pin D15 and D16 will not be zero, reflecting unequal voltage drops across the injector coils. The voltage reading obtained when an injector fault is present can be used to diagnose injector open to

voltage, injector open to ECM, injector coil open or ECM unable to ground the injector drive circuits.

To obtain an accurate and usable voltage reading, the essential tool J34029-A DVOM must be used, the DVOM must be connected as directed by CHART A-5 (cont'd) and must be set on the 200 VDC scale.

1. **DVOM Reads 00.0 Volts:** If the DVOM reads 00.0 volts, then both injectors are being powered through the fuses and grounded by the ECM an equal amount; no fault exists at this time.

2. **DVOM Reads 00.2 and 02.0 Volts:** If the voltage polarity is negative, then injector B is inoperative, pin D15 of the ECM is not grounding injector drive circuit 468. The cause is an ECM connector fault or an ECM fault.

   If the voltage polarity is positive then injector A is inoperative, pin D16 of the ECM is not grounding injector drive circuit 467. The cause is an ECM connector fault or an ECM fault.

3. **DVOM Reads Above 2.0 Volts:** If the DVOM reads above 2.0 volts, then the fault is an open from one of the 3A injector fuses to the ECM. A negative voltage polarity means that an open exists between injector A voltage supply and the ECM—check circuit 481, check injector coil of the injector connected to Blue and Red wires, check circuit 467 for opens or shorts to ground.

   A positive voltage polarity means that an open exists between injector B voltage supply and the ECM—check circuit 482, check injector coil of the injector connected to Green and White wires, check circuit 468 for opens or shorts to ground.

1988–90 CADILLAC DEVILLE AND FLEETWOOD

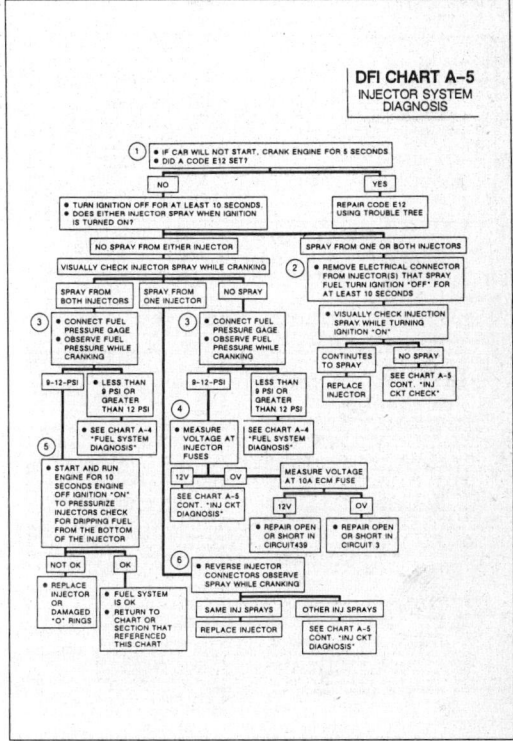

DFI CHART A-5
INJECTOR SYSTEM DIAGNOSIS

1988–90 CADILLAC DEVILLE AND FLEETWOOD

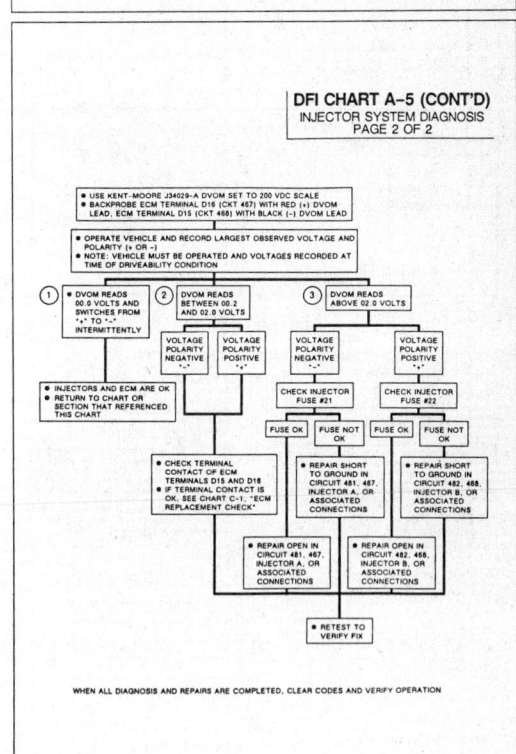

DFI CHART A-5 (CONT'D)
INJECTOR SYSTEM DIAGNOSIS
PAGE 2 OF 2

WHEN ALL DIAGNOSIS AND REPAIRS ARE COMPLETED, CLEAR CODES AND VERIFY OPERATION

1988–90 CADILLAC DEVILLE AND FLEETWOOD

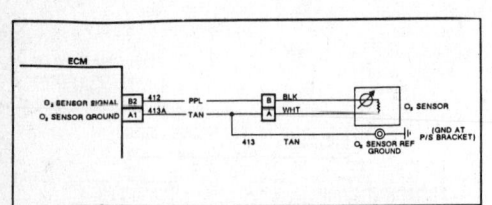

DFI CHART A-6
Rich or Lean Exhaust Code (E44 or E45) Diagnosis

Description:

The ECM provides a 0.45 volt reference signal to the oxygen sensor on CKT 412. When the oxygen sensor is cold (below 200°C), the output voltage will be around 0.45 volts and the ECM will keep the system in open loop operation. When warm, a properly operating oxygen sensor will drive the 0.45 volt reference lower (below 0.45 volts) to indicate a lean mixture and higher (above 0.45 volts) to indicate a rich mixture. The oxygen sensor signal voltage will swing from rich to lean rapidly, at least one swing every two seconds - if the ECM is in good control of the air fuel mixture.

Low oxygen sensor voltage readings are normally evidence that the air-fuel mixture is lean and the closed loop system is unable to compensate sufficiently due to a failure in some part of engine emission or fuel system.

Fixed high oxygen sensor voltage readings are normally evidence that the air-fuel mixture is rich and the closed loop system is unable to compensate sufficiently due to a failure in some part of the engine emission or fuel system.

Less likely is the possibility that the oxygen sensor has failed and is giving an incorrect reading. If the oxygen sensor is giving false rich readings, the closed loop fuel system will be overcompensating and causing lean operation while the O₂ sensor is indicating rich. Likewise, a lean indication will cause rich operation (i.e. black smoke, fouled spark plugs, poor fuel economy, high HC and CO, etc.) while the O₂ sensor is indicating lean.

Notes On Fault Tree:

1. Steps contained in this section of the diagnostic tree could cause both an indicated rich and lean operation as follows:
   - A fuel delivery system which is not functioning properly may cause an incorrect fuel mixture. This malfunction can be caused by fuel pressure which is greater than 12 PSI or less than 9 PSI at the injectors, by defective injectors, etc. Refer to Chart A-4A "Fuel System Diagnosis" for

additional information.
   - Injector dripping or not injecting can cause rich or lean exhaust (depending on failure). To check injectors, see Chart A-5.
   - Excessive EGR flow displaces oxygen and causes a rich exhaust indication. A loss of EGR will cause a lean exhaust indication. Refer to EGR diagnosis in Chart C-7, for additional information.

2. If a Code was not set, the "DFI System Check" did not refer to the "Oxygen Sensor Circuit Check" which will be required before further diagnosis.

3. Steps contained in this section could cause only an indicated rich operation as follows:
   - Carbon canister loaded with fuel can cause rich operation, see chart C-3.
   - Vacuum leak to the MAP hose can cause a false high MAP reading. High MAP readings cause the ECM to deliver too much fuel for current driving conditions.
   - A restricted air cleaner could cause a rich fuel mixture. Inspect the cleaner and replace if necessary.
   - If none of the above have indicated a problem, the code should be cleared and a road test should be performed. This code can be the result of extreme purge rates under high temperature conditions, in such a case no further diagnosis is required and no repairs should be attempted.

4. Steps contained in this section could cause only an indicated lean operation as follows:
   - If the AIR management system were to send air to the exhaust ports at all times, this would give a lean indication at the oxygen sensor. Refer to Chart C-6, for additional information.
   - Check for vacuum leaks as hoses, intake manifold and throttle body gaskets.
   - Check for fuel contamination.
   - If no problems found, replace the Oxygen Sensor.

1988–90 CADILLAC DEVILLE AND FLEETWOOD

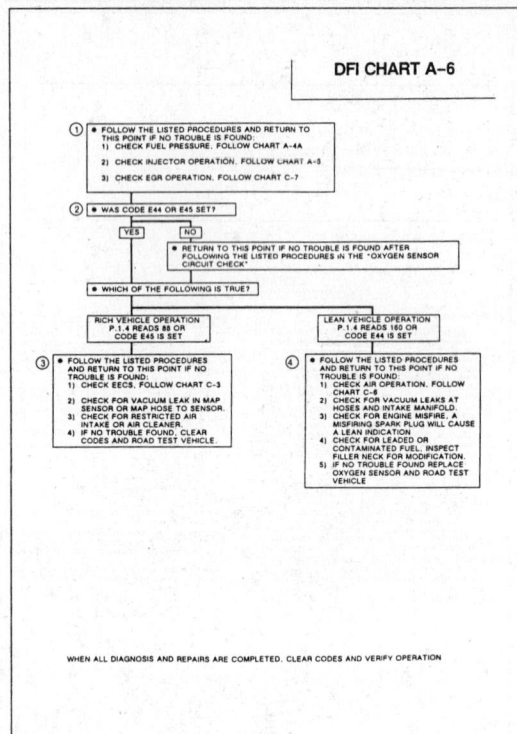

DFI CHART A-6

WHEN ALL DIAGNOSIS AND REPAIRS ARE COMPLETED, CLEAR CODES AND VERIFY OPERATION

1988–90 CADILLAC DEVILLE AND FLEETWOOD

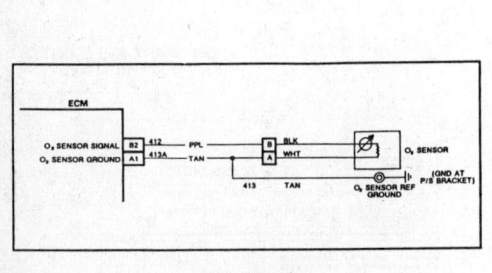

DFI CHART A-7

Open Loop Diagnosis (Code E13)

Description:

The ECM provides a 0.45 volt reference signal to the oxygen sensor on CKT 412. When the oxygen sensor is cold (below 200°C), the output voltage will be around 0.45 volts and the ECM will keep the system in open loop operation. When warm, a properly operating oxygen sensor will drive the 0.45 volt reference lower (below 0.45 volts) to indicate a lean mixture and higher (above 0.45 volts) to indicate a rich mixture. The oxygen sensor signal voltage will swing from rich to lean rapidly, at least one swing every two seconds - if the ECM is in good control of the air fuel mixture.

Notes On Fault Tree:

1. If a Code was not set, the "DFI System Check"

did not refer to the "Oxygen Sensor Circuit Check" which will be required before further diagnosis. If the "Oxygen Sensor Circuit Check" indicates a problem, clear codes and road test the vehicle.

2. The Code E13 is intermittent and the remainder of the Oxygen sensor circuit has passed the "Oxygen Sensor Circuit Check". If none of the above have indicated a problem, the code should be cleared and a road test should be performed.

3. Check for fuel contamination which would result in the failure of the Oxygen Sensor replace the Oxygen Sensor and road test the vehicle.

1988–90 CADILLAC DEVILLE AND FLEETWOOD

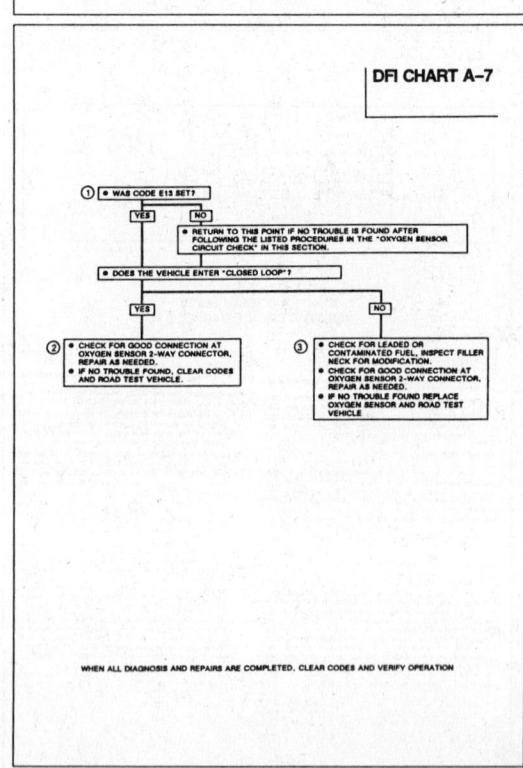

DFI CHART A-7

WHEN ALL DIAGNOSIS AND REPAIRS ARE COMPLETED, CLEAR CODES AND VERIFY OPERATION

1988–90 CADILLAC DEVILLE AND FLEETWOOD

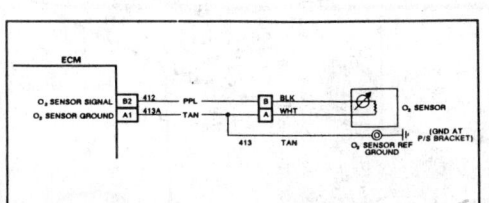

DFI CHART A–8

Rich or Lean Driveability Diagnosis

The ECM provides a 0.45 volt reference signal to the oxygen sensor on CKT 412. When the oxygen sensor is cold (below 200°C), the output voltage will be around 0.45 volts and the ECM will keep the system in open loop operation. When warm, a properly operating oxygen sensor will drive the 0.45 volt reference lower (below 0.45 volts) to indicate a lean mixture and higher (above 0.45 volts) to indicate a rich mixture. The oxygen sensor signal voltage will swing from rich to lean rapidly, at least one swing every two seconds – if the ECM is in good control of the air fuel mixture.

Low oxygen sensor voltage readings are normally evidence that the air-fuel mixture is lean and the closed loop system is unable to compensate sufficiently due to a failure in some part of engine emission or fuel system.

Fixed high oxygen sensor voltage readings are normally evidence that the air-fuel mixture is rich and the closed loop system is unable to compensate sufficiently due to a failure in some part of the engine emission or fuel system.

Less likely is the possibility that the oxygen sensor has failed and is giving an incorrect reading. If the oxygen sensor is giving false rich readings, the closed loop fuel system will be overcompensating and causing lean operation while the O_2 sensor is indicating rich. Likewise, a lean indication will cause rich operation (i.e. black smoke, fouled spark plugs, poor fuel economy, high HC and CO, etc.) while the O_2 sensor is indicating lean.

Notes On Fault Tree:

1. Steps contained in this section of the diagnostic tree could cause both an indicated rich or lean operation as follows:
   –The "DFI System Check" did not refer to the "Oxygen Sensor Circuit Check" which will be required before further diagnosis.
   – A fuel delivery system which is not functioning properly may cause an incorrect fuel mixture.

This malfunction can be caused by fuel pressure which is greater than 12 PSI or less than 9 PSI at the injectors, by defective injectors, etc. Refer to Chart A-4A "Fuel System Diagnosis" for additional information.
– Injector dripping or not injecting can cause rich or lean exhaust (depending on failure). To check injectors, see Chart A-5.
– Excessive EGR flow displaces oxygen and causes a rich exhaust indication. A loss of EGR will cause a lean exhaust indication. Refer to EGR diagnosis Chart C-7, for additional information.

2. Steps contained in this section could cause rich operation as follows:
   – A restricted air cleaner could cause a rich mixture. Inspect the cleaner and replace if necessary.
   – Carbon canister loaded with fuel can cause rich operation, see chart C-3.
   – If the AIR management system were to send air to the exhaust ports at all times, this would give a lean indication at the oxygen sensor. Refer to Air Management Diagnosis Chart C-6, for additional information.
   – Vacuum leak on the MAP hose can cause a false high MAP reading. High MAP readings cause the ECM to deliver too much fuel for current driving conditions.
   – If none of the above have indicated a problem, symptom diagnosis is required.

3. Steps contained in this section could cause only an indicated lean operation as follows:
   – Check for vacuum leaks as hoses, intake manifold and throttle body gaskets.
   – Check for fuel contamination.
   – If none of the above have indicated a problem, symptom diagnosis is required.

1988–90 CADILLAC DEVILLE AND FLEETWOOD

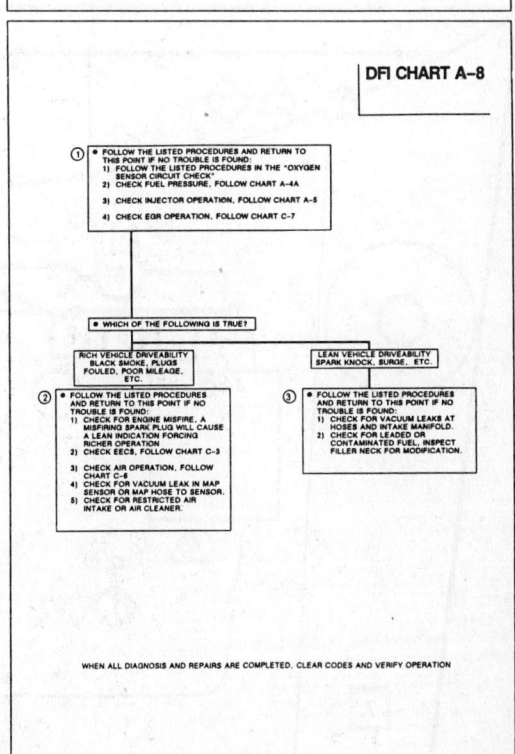

1988–90 CADILLAC DEVILLE AND FLEETWOOD

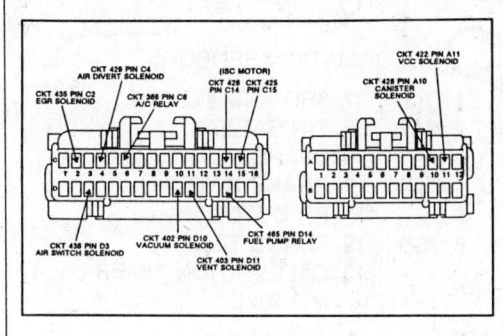

DFI CHART C–1

ECM REPLACEMENT CHECK

Description:

Prior to replacing an ECM, the vehicle must be tested for:

1. Poor connector terminal to ECM contact.
2. Direct short to battery voltage on an ECM ground circuit.
3. Shorted solenoids or relays. If a short is found, the circuit must be repaired prior to replacing the ECM to prevent repeat ECM failures.

Notes on Fault Tree:

1. Check for poor terminal contact due to weak or dirty ECM terminals. Remove suspected terminals to inspect, replace if broken or dirty. If coolant is present at the ECM connector, replace coolant sensor, coolant sensor connector, and the coolant sensor signal and ground wires. Also, replace ECM connector terminal and blow coolant out of harness. Clean ECM connector with alcohol or spray contact cleaner and replace ECM.

2. Checking for a short to ignition or shorted solenoid or relay. All terminals must be tested since several are connected together internally in the ECM. A short in one circuit may cause another circuit in the ECM to be inoperative. An example might be that the VCC solenoid shorted can cause both VCC to be inoperative and the EGR to be enabled at all times. Any circuit testing below 20 ohms is shorted and should be diagnosed for the cause of the short.

3. Check ISC circuit (EXTEND/RETRACT) for shorted to ground or shorted together. Normal resistance for an ISC motor circuit is 4–100 ohms.

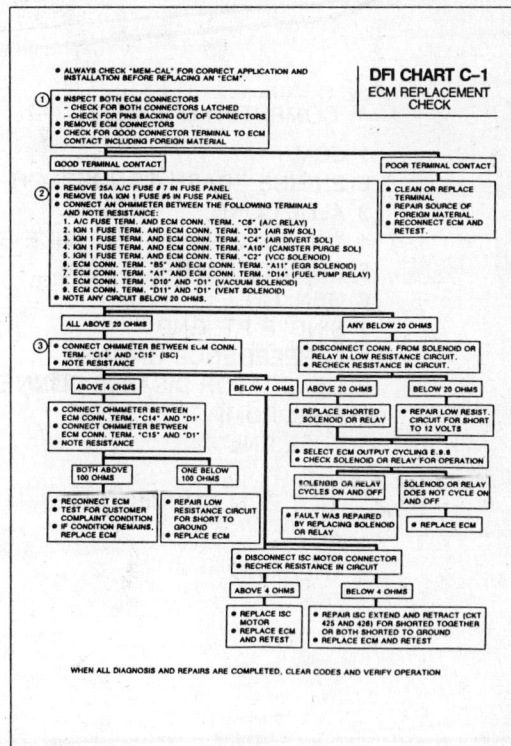

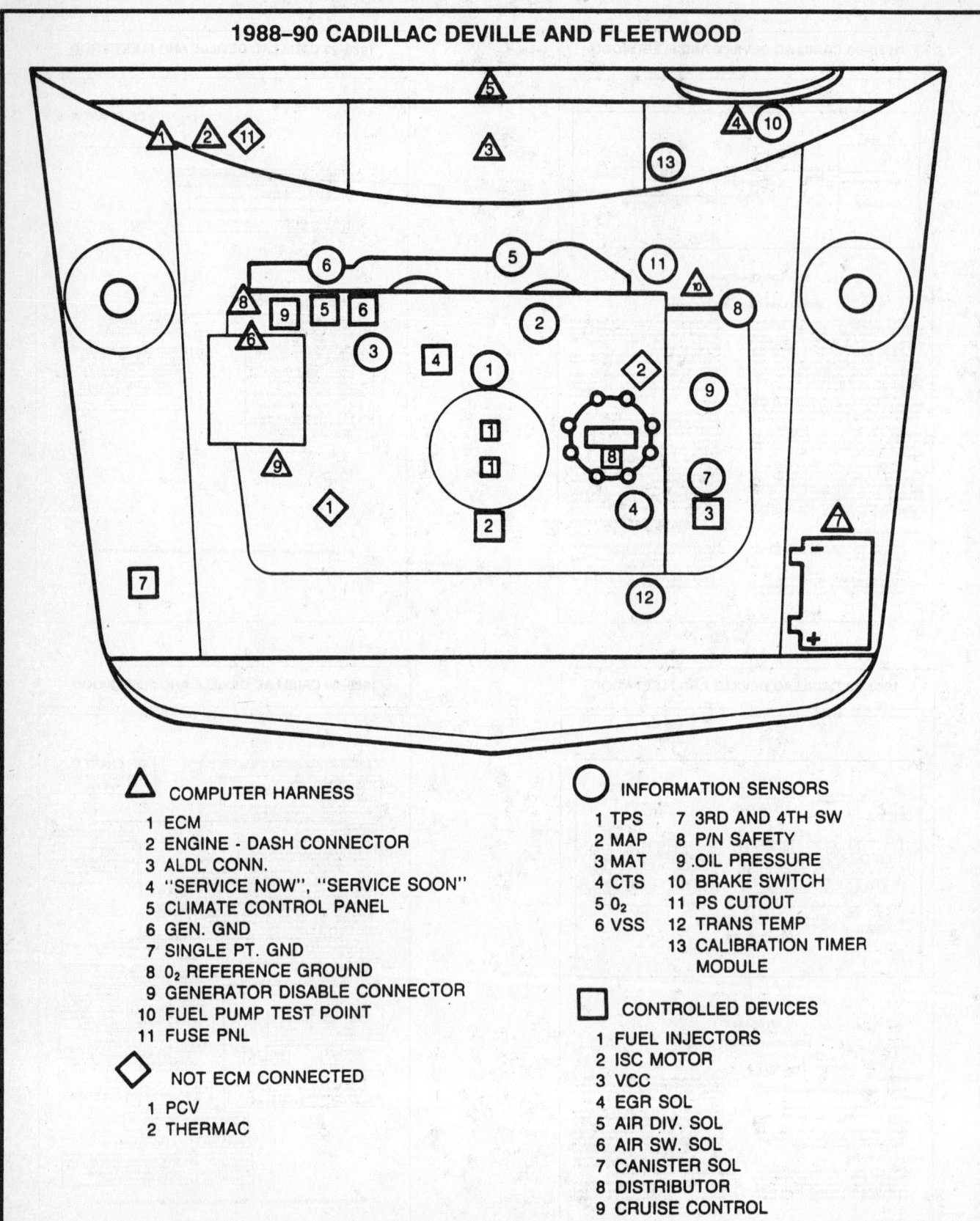

1988–90 CADILLAC DEVILLE AND FLEETWOOD

△ COMPUTER HARNESS

1 ECM
2 ENGINE - DASH CONNECTOR
3 ALDL CONN.
4 ''SERVICE NOW'', ''SERVICE SOON''
5 CLIMATE CONTROL PANEL
6 GEN. GND
7 SINGLE PT. GND
8 O_2 REFERENCE GROUND
9 GENERATOR DISABLE CONNECTOR
10 FUEL PUMP TEST POINT
11 FUSE PNL

◇ NOT ECM CONNECTED

1 PCV
2 THERMAC

◯ INFORMATION SENSORS

1 TPS   7 3RD AND 4TH SW
2 MAP   8 P/N SAFETY
3 MAT   9 .OIL PRESSURE
4 CTS   10 BRAKE SWITCH
5 O_2 11 PS CUTOUT
6 VSS   12 TRANS TEMP
             13 CALIBRATION TIMER
                  MODULE

▢ CONTROLLED DEVICES

1 FUEL INJECTORS
2 ISC MOTOR
3 VCC
4 EGR SOL
5 AIR DIV. SOL
6 AIR SW. SOL
7 CANISTER SOL
8 DISTRIBUTOR
9 CRUISE CONTROL

1988–90 CADILLAC DEVILLE AND FLEETWOOD

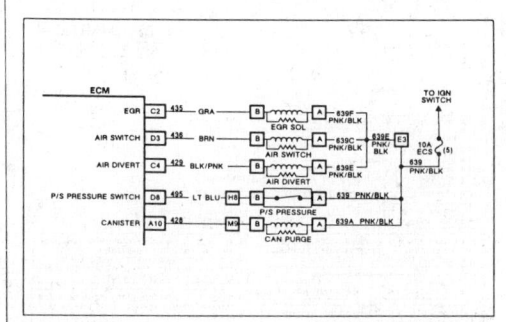

DFI CHART C-3

CANISTER PURGE CONTROL DIAGNOSIS

Description

The canister purge solenoid receives 12 volts from the 10 amp number 5 fuse on the underhood relay panel through circuit 639. The ECM energizes the canister purge solenoid by grounding pin A10 (circuit 428). When the solenoid is energized it allows the canister to purge.

The canister is commanded to purge when:

- Coolant is above 70°C.
- Closed loop has been achieved for at least 9 seconds.
- Throttle switch is open.
- OR
- Code E13, E44, or E45 is present.

The ECM will de-energize the solenoid (not allow purge) when codes E16 is set or the ECM is running in the backup mode (no normal program control).

Notes on Fault Tree:

1. Checking for the canister purge solenoid able to hold vacuum. If the solenoid can't hold vacuum, it should be replaced.
2. Vacuum should release when the solenoid is cycled on in output cycling.

3. Checking for proper vacuum signals from the throttle body.
4. Checking for proper electrical signals to the canister purge solenoid.

FUNCTIONAL TEST OF CHARCOAL CANISTER

Disconnect the hose leading from the TPCV to the lower port of the charcoal canister at the canister end. Attach a hose to the fuel tank port of the TPCV and attempt to blow through it. Little or no air should pass through the TPCV.

With a hand vacuum pump, apply vacuum (15" Hg. or 51 kPa) through the TPCV upper tube (manifold vacuum supply port). If the diaphragm does not hold vacuum for at least 20 seconds, the diaphragm is leaking, and the TPCV must be replaced.

If the diaphragm holds vacuum, try to blow through the hose connected to the lower tube while vacuum is still being applied. An increased flow of air should be observed. If not, replace the TPCV.

Attach a hose to the lower port of the canister and attempt to blow through it. Air **should** pass into the canister. If not, replace the canister.

Disconnect the canister purge solenoid electrical connector, attach a hose to the upper port of the canister, and attempt to blow through it. Air **should not** pass into the canister. If air passes, replace the purge solenoid.

Measure the resistance of the purge solenoid. Resistance should be greater than 20 ohms and less than 100 ohms. If the solenoid resistance is not in this range, replace the solenoid.

1988–90 CADILLAC DEVILLE AND FLEETWOOD

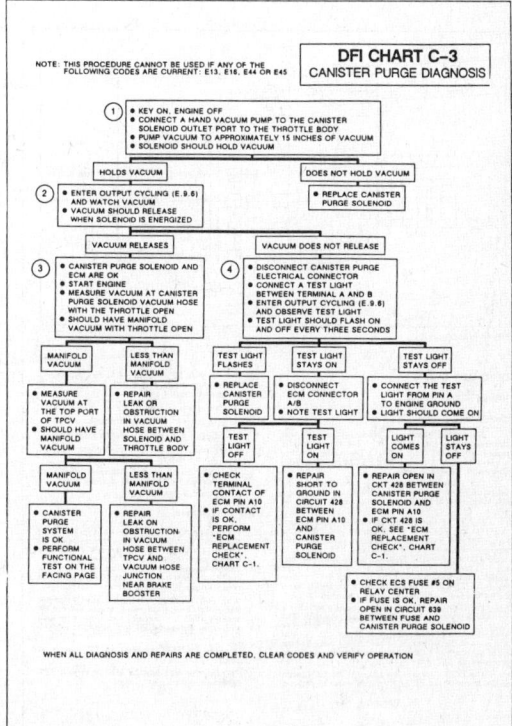

1988–90 CADILLAC DEVILLE AND FLEETWOOD

ON-CAR SERVICE

SETTING TIMING

The initial base timing is set by jumpering pins A&B together at the ALDL connector while not in diagnostic display. The ALDL connector is located above a hatch in the hush panel at the center of the dash. Then set the timing to the specification shown on the Emission Control Information label located in the engine compartment. Jumpering pin A to B of the ALDL connector will cause the "Service Soon" light to flicker which indicates that the ECM is in set timing mode.

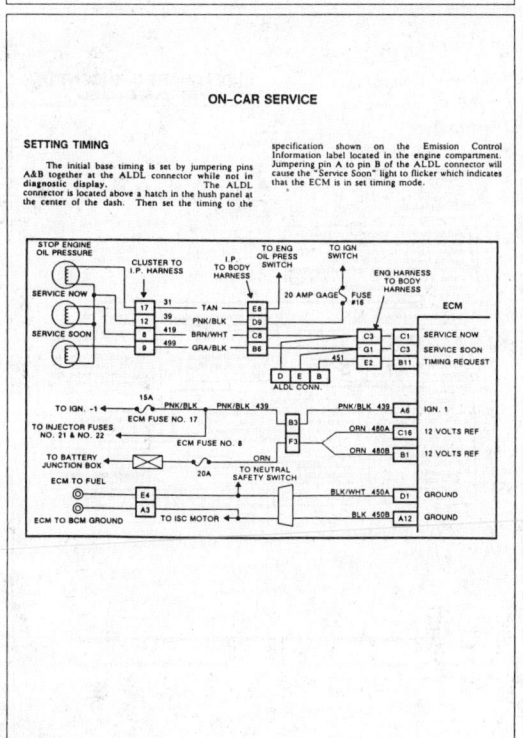

1988–90 CADILLAC DEVILLE AND FLEETWOOD

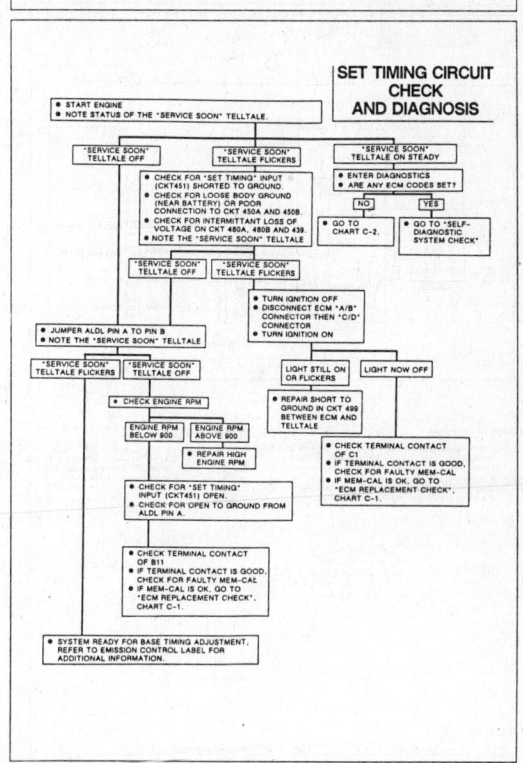

1988–90 CADILLAC DEVILLE AND FLEETWOOD

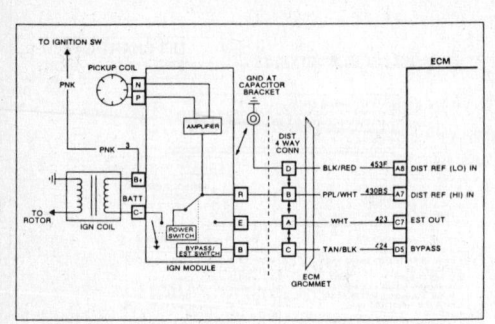

DFI CHART C-4
HEI SYSTEM CHECK

Description

The HEI distributor produces one 5 volt Distributor Reference Pulse each time that a cylinder reaches 10° BTDC (base timing). The reference pulses are sent from the distributor to the ECM on Circuit 430, Distributor Reference Signal.

The ECM adds spark timing information to the reference pulses received from the HEI and sends out 5 volt Electronic Spark Timing pulses (EST) on Circuit 423.

The ECM can choose to control spark timing. The HEI Bypass Circuit 424 is turned on by the ECM (approximately 5 volts) when the ECM wishes to control timing. When the ECM wants the HEI module to control timing, the HEI Bypass circuit is turned off and the HEI module grounds Circuit 423, EST. The ECM always sends EST pulses to the HEI; the HEI module shorts the EST pulses to ground when the bypass signal from the ECM to the module is low (less than .5 volts).

The Distributor Reference Ground (Circuit 453) is a common ground between the ECM and the HEI. The ECM and HEI compare Bypass, EST, and Reference voltages to the ground voltage on Circuit 453. Circuit 453 open to the HEI or ECM can cause a failure of the ECM to recognize a reference pulse or of the module to recognize an EST pulse.

The ECM compares the status of the EST circuit to the status of the Bypass circuit in order to detect a

fault and set code E23. If the ECM has Bypass at 5 volts, the module should accept EST information, EST voltage should be high (.5 – 2.8 volts).

Possible causes of code E23 are:

A. Circuit 423 (EST) open/ground.
B. Circuit 424 (Bypass) open/grounded.
C. Circuit 453 (Reference Ground) open to ground at HEI or ECM.
D. ECM with a poor connection to the engine through the Circuit 430 grounds.
E. ECM or HEI module faults.

Notes on Fault Tree:

1. Four jumper wires must be obtained for use in the diagnostic procedure. The jumpers should be about 12 inches long with a male and female weatherpack connector on either end (P/N 12014836 and 12014837). With the distributor connector disconnected, use the jumpers to reconnect the terminals per instructions provided.

2. With only Distributor Reference and Distributor Reference Ground jumpered, engine will run on backup spark. Checking for proper ground connection between ECM and the engine and the ECM and HEI.

3. Checking for ECM providing an EST output on Circuit 423.

4. Checking for HEI module able to ground EST signal with open Bypass Circuit.

1988–90 CADILLAC DEVILLE AND FLEETWOOD

DFI CHART C-4
EST SYSTEM CHECK

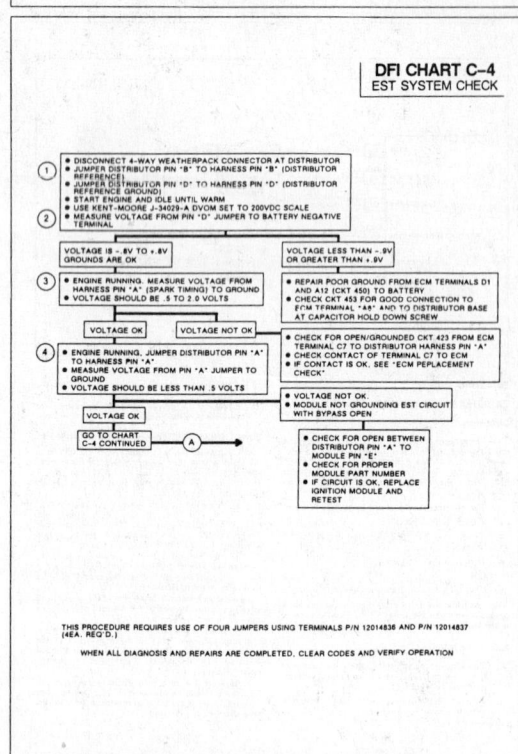

THIS PROCEDURE REQUIRES USE OF FOUR JUMPERS USING TERMINALS P/N 12014836 AND P/N 12014837 (4EA. REQ'D.)

WHEN ALL DIAGNOSIS AND REPAIRS ARE COMPLETED, CLEAR CODES AND VERIFY OPERATION

1988–90 CADILLAC DEVILLE AND FLEETWOOD

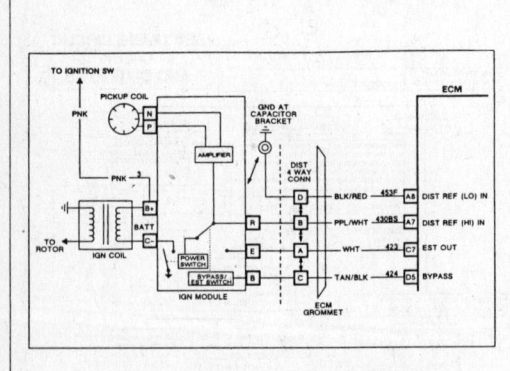

Notes on Fault Tree:

5. Checking for HEI module able to recognize a voltage on the Bypass Circuit and to stop grounding EST (ECM controlling timing).

6. Checking for Bypass signal to the module. If Bypass is being sent by the ECM to the HEI and if the module is interpreting Bypass voltage correctly,

then the module will switch off the ground to the EST.

7. If the chart leads to "EST circuit is OK", then a fault may exist in the four-way weatherpack. Check for proper connector mating and for pins backing out of the weatherpack. If the connector is no trouble found, reconnect the four way, clear codes, and retest.

1988–90 CADILLAC DEVILLE AND FLEETWOOD

DFI CHART C-4 (CONT'D)
EST SYSTEM CHECK

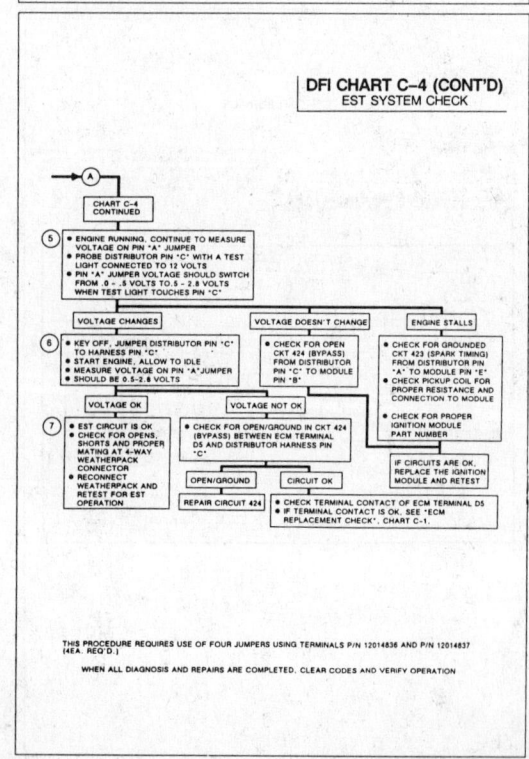

THIS PROCEDURE REQUIRES USE OF FOUR JUMPERS USING TERMINALS P/N 12014836 AND P/N 12014837 (4EA. REQ'D.)

WHEN ALL DIAGNOSIS AND REPAIRS ARE COMPLETED, CLEAR CODES AND VERIFY OPERATION

1988–90 CADILLAC DEVILLE AND FLEETWOOD

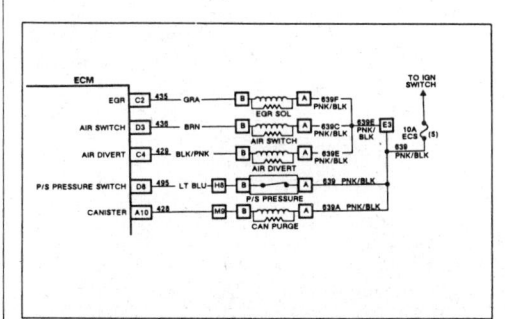

DFI CHART C-6
AIR MANAGEMENT DIAGNOSIS

Description

The AIR divert and switch solenoids are supplied with 12 volts on circuit 639 from the 10 amp number 5 solenoid fuse on the underhood relay panel. The ECM controls air management by grounding AIR switch solenoid circuit 436 and AIR divert solenoid circuit 429.

To divert air to the air cleaner, the divert solenoid is de-energized. When pin C4 at the ECM is grounded, the divert solenoid energizes and diverts air to the switching valve. When the switch solenoid is de-energized and when pin D3 at the ECM is grounded the air is routed to the exhaust ports.

Notes on Fault Tree:

1. With a cold engine, the air management system should switch air to the exhaust ports (the heads)

to help oxidize the rich mixture that exists during cold starts and warm-up.

2. With a warm engine, the air should be switched to the converter to oxidize the exhaust gas flowing to the second bed of the catalytic converter.

3. Checking for the air system to "Divert" air away from the converter during acceleration and deceleration. The divert function protects the converter from the rich conditions that occur A.) during acceleration B.) during power enrichment and C.) when the throttle is closed to start deceleration.

4. Checking for vacuum supply to the air management valves.

5. Checking for air pump operating.

6. This branch of the tree will check for ECM ability to control air switch and air divert solenoids.

7. Checking for open to voltage to solenoid.

8. Checking for a low resistance solenoid.. If resistance is less than 20 ohms, replace the air management solenoid.

1988–90 CADILLAC DEVILLE AND FLEETWOOD

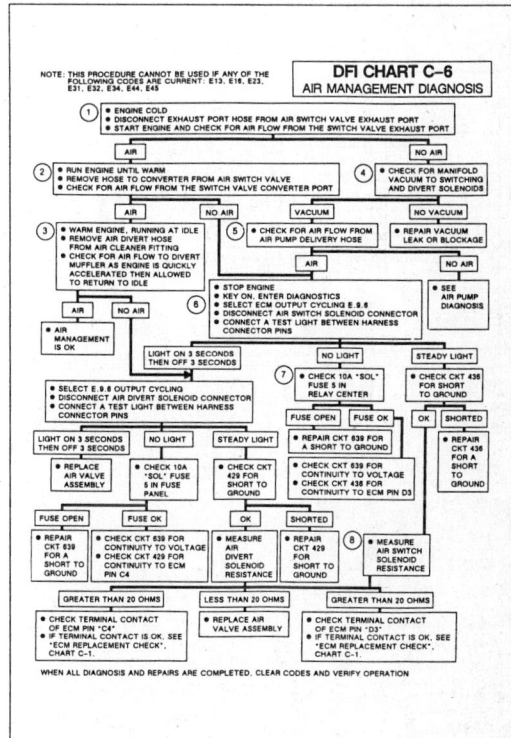

1988–90 CADILLAC DEVILLE AND FLEETWOOD

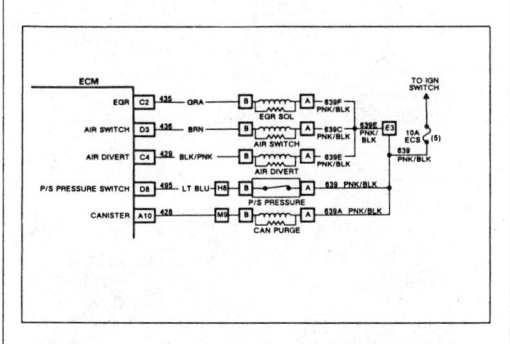

DFI CHART C-7
EGR DIAGNOSIS

Before Starting:

- Check 10 amp ECS fuse (# 5) on fuse panel.
- Measure EGR solenoid resistance - should be 20-100 ohms.

The 4.5L DFI engine uses a positive backpressure EGR valve which limits EGR flow with low exhaust backpressure (idle, decel, etc.). It is very important that exhaust tubes which decrease backpressure (pull air through) are **not** hooked to the vehicle when diagnosing the EGR system.

VACUUM TEST

1. Connect a vacuum gage to the source side of the EGR solenoid. Start the engine, manifold vacuum should be present. If it is not, repair leaks or obstruction between the EGR solenoid and throttle body.

2. Connect a vacuum gage to the EGR valve vacuum supply. There should be no vacuum with

the engine idling. If there is, follow Chart C-7

3. With the gage hooked to the EGR valve vacuum supply, disconnect the EGR solenoid connector. There should be more than 8 inches of vacuum available. If not, repair leak or obstruction in EGR valve vacuum hose.

Notes on Fault Tree:

1. Checking for EGR valve's ability to flow sufficient exhaust gas to engine. RPM should drop and idle should become rough or stall. Compare the RPM drop to a known good car.
2. Checking for blocked EGR passages in the intake manifold.
3. Checking for EGR solenoid able to pass vacuum.

Diagnostic Aids:

1. Check vacuum supply hoses for leaks, restrictions, and obstructions. Clean out or replace as necessary.
2. Check intake manifold passages for any obstructions that would prevent exhaust gases from passing through.

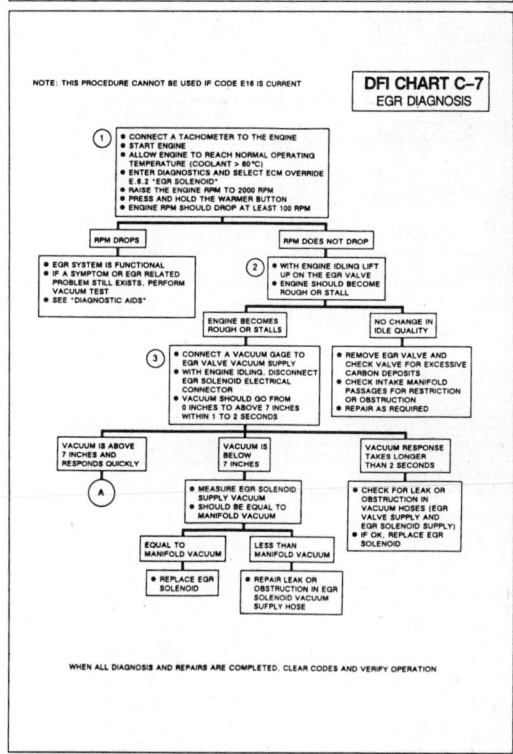

1988–90 CADILLAC DEVILLE AND FLEETWOOD

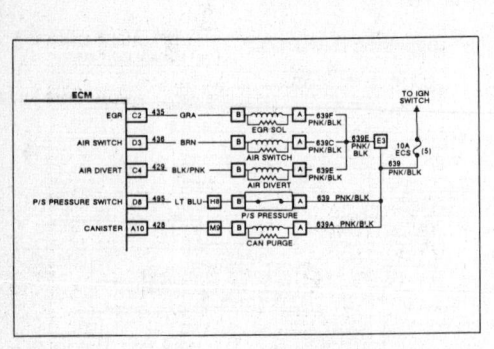

Notes on Fault Tree:

1. Checking for voltage to EGR solenoid and ground signal from the ECM. If either of these is missing, the EGR solenoid will command EGR on at all times.

2. Checking for ECM ability to turn solenoid off. If the light goes off, the electrical portion of the EGR system is OK.

1988–90 CADILLAC DEVILLE AND FLEETWOOD

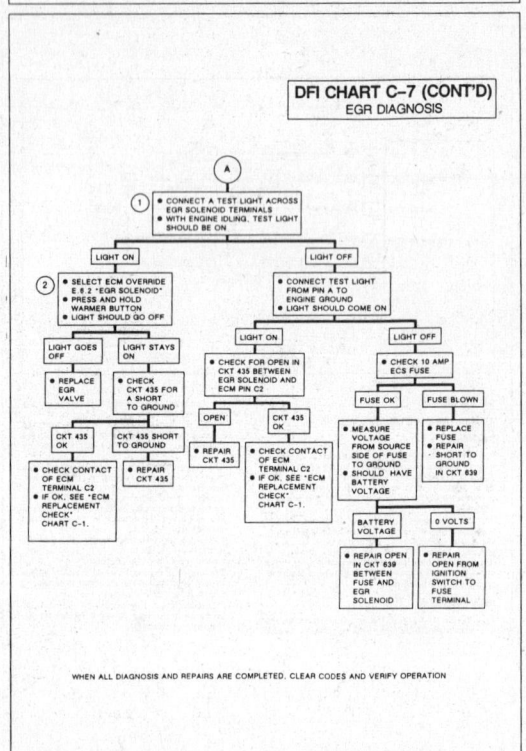

DFI CHART C-7 (CONT'D)
EGR DIAGNOSIS

WHEN ALL DIAGNOSIS AND REPAIRS ARE COMPLETED, CLEAR CODES AND VERIFY OPERATION

1988–90 CADILLAC DEVILLE AND FLEETWOOD

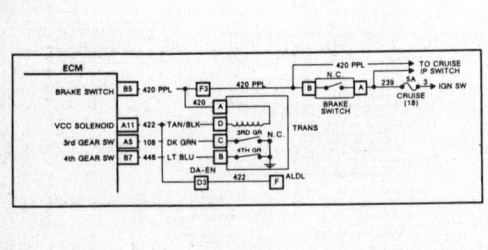

DFI CHART C-8

VCC DIAGNOSIS

Description

The viscous converter clutch solenoid is turned on when the ECM applies ground to CKT 422, "VCC Solenoid Circuit". The power for the VCC solenoid comes from the 5A fuse #18 in the fuse block, flows through the VCC brake switch to pin A of five-way connector on the transmission. At pin A, current flows into the transmission to the VCC solenoid and then returns to ground through the ECM.

Notes on Fault Tree:

1. Checking for continuity from brake switch through solenoid coil to CKT 422.

2. Checking for ECM to ground CKT 422.

3. Checking for VCC to apply using the ECM override. If VCC does not apply, refer to Section 7A for hydraulic diagnosis of the VCC.

4. VCC electrical system is operating properly.

1988–90 CADILLAC DEVILLE AND FLEETWOOD

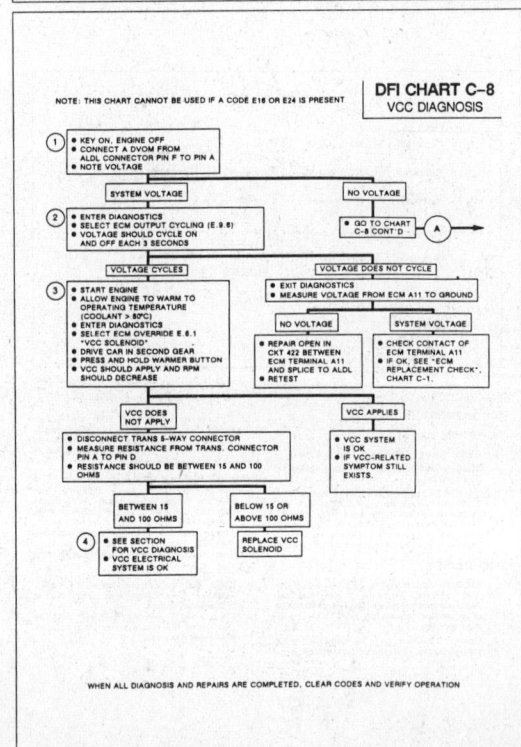

NOTE: THIS CHART CANNOT BE USED IF A CODE E16 OR E24 IS PRESENT

DFI CHART C-8
VCC DIAGNOSIS

WHEN ALL DIAGNOSIS AND REPAIRS ARE COMPLETED, CLEAR CODES AND VERIFY OPERATION

1988–90 CADILLAC DEVILLE AND FLEETWOOD

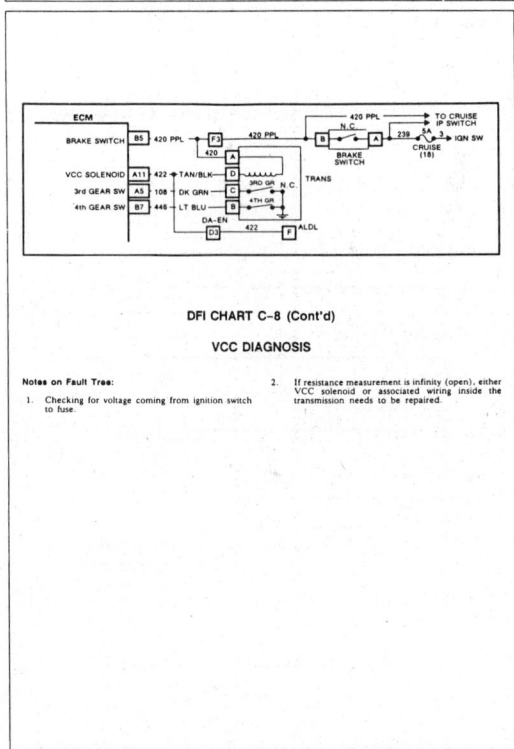

DFI CHART C–8 (Cont'd)

VCC DIAGNOSIS

Notes on Fault Tree:

1. Checking for voltage coming from ignition switch to fuse.

2. If resistance measurement is infinity (open), either VCC solenoid or associated wiring inside the transmission needs to be repaired.

1988–90 CADILLAC DEVILLE AND FLEETWOOD

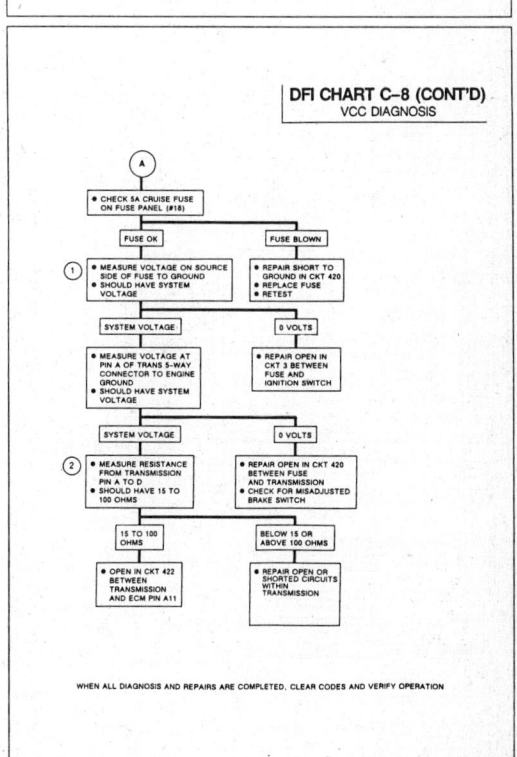

1988–90 CADILLAC DEVILLE AND FLEETWOOD

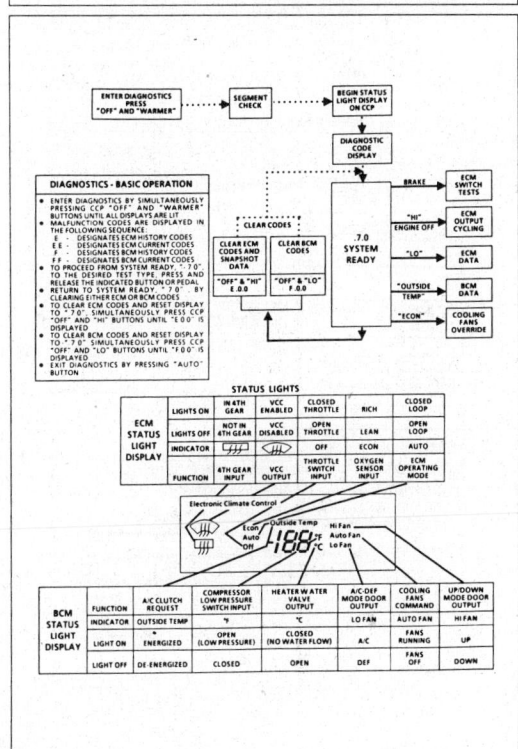

1988–90 CADILLAC DEVILLE AND FLEETWOOD

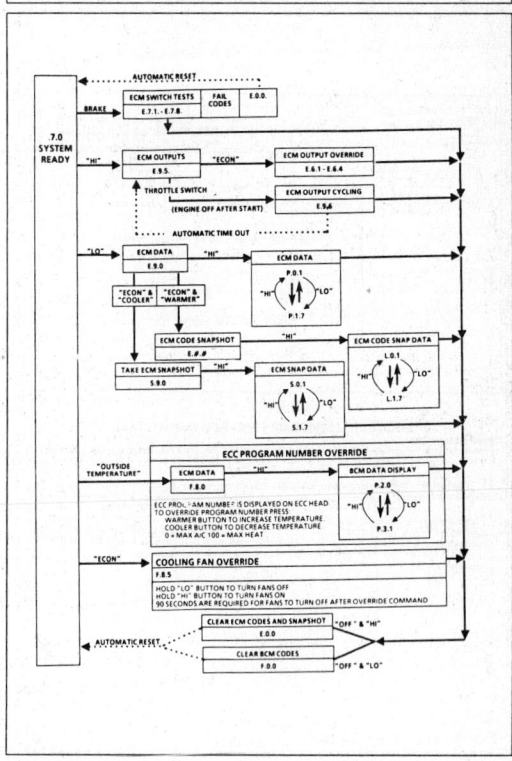

1988–90 CADILLAC DEVILLE AND FLEETWOOD

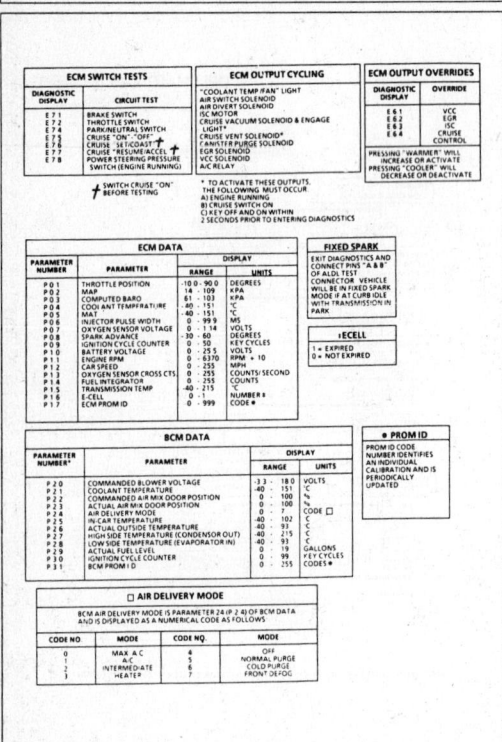

ECM SWITCH TESTS

DIAGNOSTIC DISPLAY	CIRCUIT TEST
E 71	BRAKE SWITCH
E 72	THROTTLE SWITCH
E 74	PARK/NEUTRAL SWITCH
E 75	CRUISE "ON" - "OFF"
E 76	CRUISE "SET/COAST" †
E 77	CRUISE "RESUME/ACCEL"
E 78	POWER STEERING PRESSURE SWITCH (ENGINE RUNNING)

† SWITCH CRUISE "ON" BEFORE TESTING

ECM OUTPUT CYCLING

"COOLANT TEMP /FAN" LIGHT
AIR SWITCH SOLENOID
AIR DIVERT SOLENOID
ISC MOTOR
CRUISE VACUUM SOLENOID & ENGAGE LIGHT *
CRUISE VENT SOLENOID*
CANISTER PURGE SOLENOID
EGR SOLENOID
VCC SOLENOID
A/C RELAY

* TO ACTIVATE THESE OUTPUTS, THE FOLLOWING MUST OCCUR.
A) ENGINE RUNNING
B) CRUISE SWITCH ON
C) KEY OFF AND ON WITHIN 2 SECONDS PRIOR TO ENTERING DIAGNOSTICS

ECM OUTPUT OVERRIDES

DIAGNOSTIC DISPLAY	OVERRIDE
E 61	VCC
E 62	EGR
E 63	ISC
E 64	CRUISE CONTROL

PRESSING "WARMER" WILL INCREASE OR ACTIVATE
PRESSING "COOLER" WILL DECREASE OR DEACTIVATE

ECM DATA

PARAMETER NUMBER	PARAMETER	RANGE	UNITS
P 01	THROTTLE POSITION	10.0 - 90.0	DEGREES
P 02	MAP	14 - 109	KPA
P 03	COMPUTED BARO	61 - 103	KPA
P 04	COOLANT TEMPERATURE	-40 - 151	°C
P 05	MAT	-40 - 151	°C
P 06	INJECTOR PULSE WIDTH	0 - 99.9	MS
P 07	OXYGEN SENSOR VOLTAGE	0 - 1.14	VOLTS
P 08	SPARK ADVANCE	-30 - 60	DEGREES
P 09	IGNITION CYCLE COUNTER	0 - 50	KEY CYCLES
P 10	BATTERY VOLTAGE	0 - 25.5	VOLTS
P 11	ENGINE RPM	0 - 6370	RPM ÷ 10
P 12	CAR SPEED	0 - 255	MPH
P 13	OXYGEN SENSOR CROSS CTS.	0 - 255	COUNTS/ SECOND
P 14	FUEL INTEGRATOR	0 - 255	COUNTS
P 15	TRANSMISSION TEMP	-40 - 215	°C
P 16	E-CELL	0 - 1	NUMBER #
P 17	ECM PROM ID	0 - 999	CODE #

FIXED SPARK

EXIT DIAGNOSTICS AND CONNECT PINS "A & B" OF ALDL TEST CONNECTOR. VEHICLE WILL BE IN FIXED SPARK MODE IF AT CURB IDLE WITH TRANSMISSION IN PARK

‡ECELL

1 = EXPIRED
0 = NOT EXPIRED

BCM DATA

PARAMETER NUMBER*	PARAMETER	RANGE	UNITS
P 20	COMMANDED BLOWER VOLTAGE	-3.3 - 18.0	VOLTS
P 21	COOLANT TEMPERATURE	-40 - 151	°C
P 22	COMMANDED AIR MIX DOOR POSITION	0 - 100	%
P 23	ACTUAL AIR MIX DOOR POSITION	0 - 100	%
P 24	AIR DELIVERY MODE	0 - 7	CODE □
P 25	IN-CAR TEMPERATURE	-40 - 102	C
P 26	ACTUAL OUTSIDE TEMPERATURE	-40 - 93	C
P 27	HIGH SIDE TEMPERATURE (CONDENSOR OUT)	-40 - 215	C
P 28	LOW SIDE TEMPERATURE (EVAPORATOR IN)	-40 - 93	C
P 29	ACTUAL FUEL LEVEL	0 - 19	GALLONS
P 30	IGNITION CYCLE COUNTER	0 - 50	KEY CYCLES
P 31	BCM PROM ID	0 - 255	CODE #

● PROM ID

PROM ID CODE NUMBER IDENTIFIES INDIVIDUAL CALIBRATION AND IS PERIODICALLY UPDATED

□ AIR DELIVERY MODE

BCM AIR DELIVERY MODE IS PARAMETER 24 (P 2.4) OF BCM DATA AND IS DISPLAYED AS A NUMERICAL CODE AS FOLLOWS:

CODE NO.	MODE	CODE NO.	MODE
0	MAX A C	4	OFF
1	A C	5	NORMAL PURGE
2	INTERMEDIATE	6	COLD PURGE
3	HEATER	7	FRONT DEFOG

1988–90 CADILLAC DEVILLE AND FLEETWOOD

SELF-DIAGNOSTIC SYSTEM CHECK

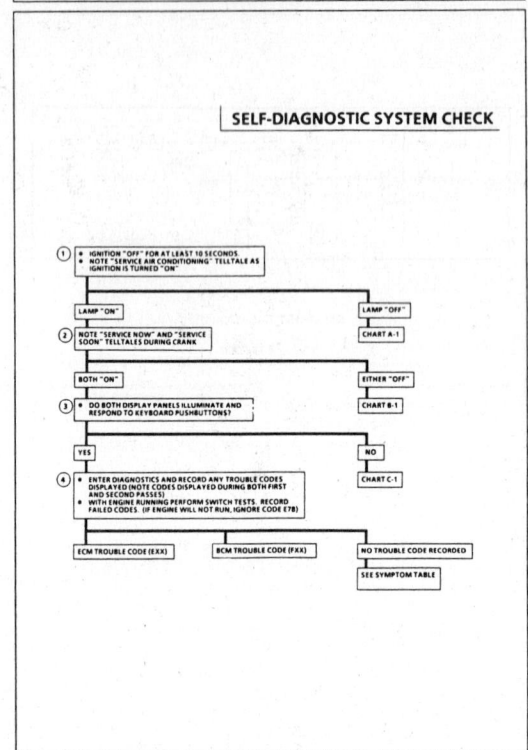

1988–90 CADILLAC DEVILLE AND FLEETWOOD

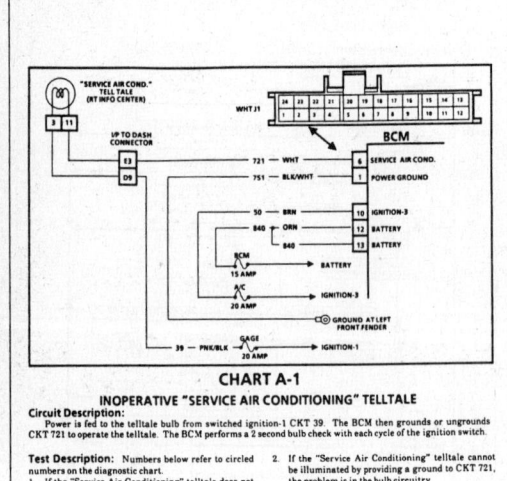

CHART A-1
INOPERATIVE "SERVICE AIR CONDITIONING" TELLTALE

Circuit Description:
Power is fed to the telltale bulb from switched ignition-1 CKT 39. The BCM then grounds or ungrounds CKT 721 to operate the telltale. The BCM performs a 2 second bulb check with each cycle of the ignition switch.

Test Description: Numbers below refer to circled numbers on the diagnostic chart.

1. If the "Service Air Conditioning" telltale does not illuminate at key-on and a "c", "d" or "-151" appears on one of the displays, the BCM's microprocessor is not functioning properly which could be due to improper prom insertion or faulty components in the BCM.

2. If the "Service Air Conditioning" telltale cannot be illuminated by providing a ground to CKT 721, the problem is in the bulb circuitry.

3. If the bulb can be illuminated, then battery ignition and ground integrity must be checked to the BCM. If circuits check OK, the BCM connector or BCM is faulty.

1988–90 CADILLAC DEVILLE AND FLEETWOOD

CHART A-1
INOPERATIVE "SERVICE AIR CONDITIONING" TELLTALE

(REFER TO "SERVICE AIR CONDITIONING" TELLTALE CIRCUIT)

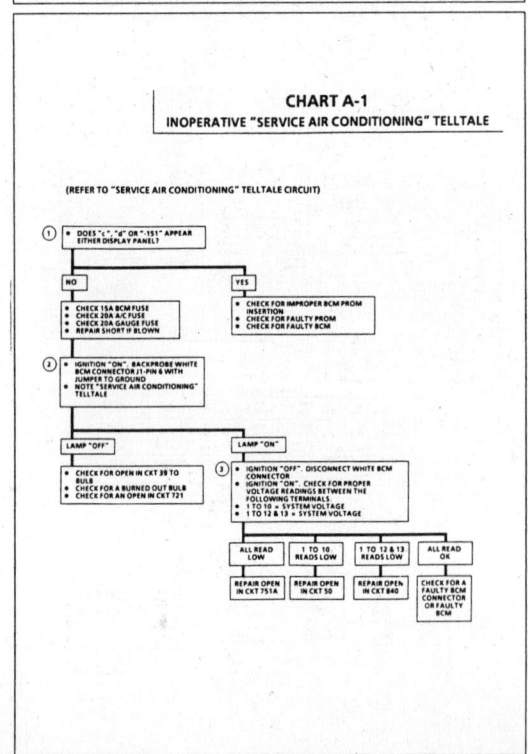

1988–90 CADILLAC DEVILLE AND FLEETWOOD

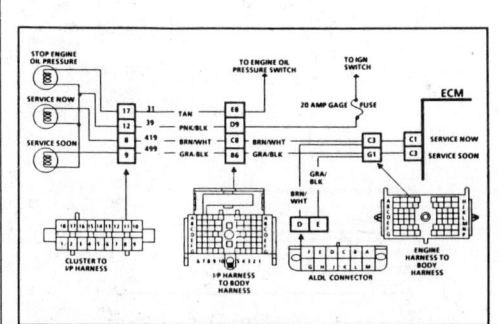

CHART B-1
INOPERATIVE "SERVICE NOW" AND/OR "SERVICE SOON" TELLTALE

Circuit Description:

The "Service Now" and "Service Soon" lamps are powered through the 20 amp "gage" fuse #16 on the fuse panel.

The ECM supplies a ground to CKT 419 (terminal "C1") to illuminate the "Service Now" light and to CKT 499 (terminal "C3") to illuminate the "Service Soon" light.

Test Description: Numbers below refer to circled numbers on the diagnostic chart.

1 If the "Service Now" and "Service Soon" telltales do not illuminate while cranking, go to CHART B-2

2 If the "Service Soon" telltale operates normally, the power circuit (39) is OK and the cause of the problem is the "Service Now" bulb or the ground side of the bulb. The ECM is responsible for turning "ON" the "Service Now" light by grounding CKT 419. The ECM's activity can be simulated by jumpering the ALDL connector pin "D" to ground. If the light comes "ON", check CKT 419 for an open to the ECM. If CKT 419 is OK, check for a faulty ECM connector or a faulty ECM. If the bulb does not light the cause is a burned out bulb or an open in CKT 419

3 "Service Soon" will flicker if ALDL pin "B" is grounded. The ECM sends ALDL Data to the "Service Soon" lamp and ALDL connector pin "B", when in ALDL or set timing modes.

4 If the "Service Now" telltale operates normally, the power circuit (39) is OK and the cause of the problem is the "Service Soon" bulb or the ground side of the bulb. The ECM is responsible for turning "ON" the "Service Soon" light by grounding CKT 499. The ECM's activity can be simulated by jumpering the ALDL connector pin "E" to ground. If the light comes "ON", check CKT 499 for an open to the ECM. If CKT 499 is OK, check for a faulty ECM connector or a faulty ECM. If the bulb does not light, the cause is a burned out bulb or an open in CKT 499

1988–90 CADILLAC DEVILLE AND FLEETWOOD

CHART B-1
(Page 1 of 2)
INOPERATIVE "SERVICE NOW" AND/OR "SERVICE SOON" TELLTALE

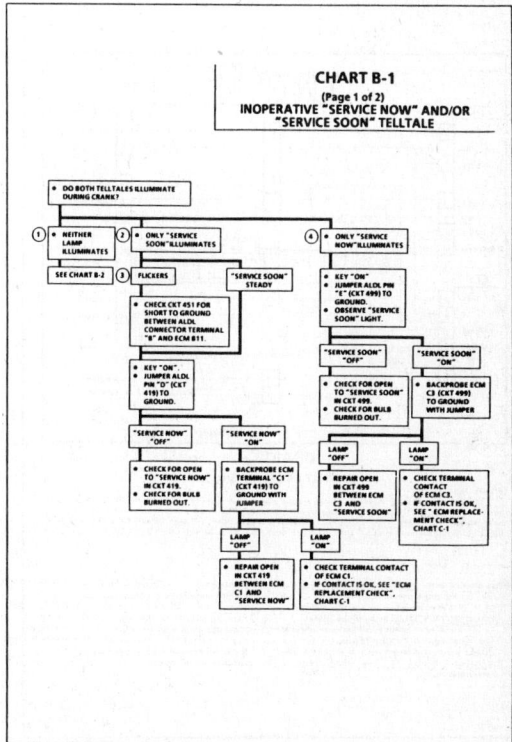

1988–90 CADILLAC DEVILLE AND FLEETWOOD

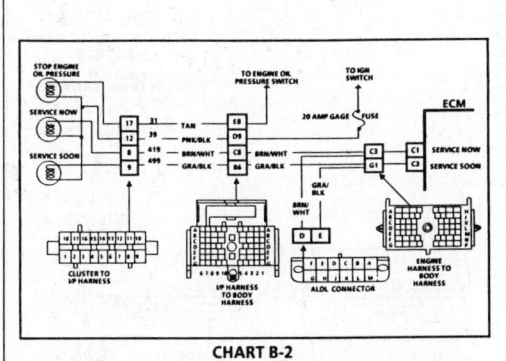

CHART B-2
INOPERATIVE "SERVICE NOW" AND/OR "SERVICE SOON" TELLTALE

Test Description: Numbers below refer to circled numbers on the diagnostic chart.

1 If neither lamp illuminates, cause is a) no power to the bulbs, b) ECM does not receive crank input (Code E18) or c) ECM inoperative due to lack of power, lack of ground or blown ECM.

2 Checking for cause of no power to bulbs.

3 Checking for blown fuse on ECM memory power or ECM ignition power.

4 Checking for power and grounds present at the ECM connector.

1988–90 CADILLAC DEVILLE AND FLEETWOOD

CHART B-2
(Page 2 of 2)
INOPERATIVE "SERVICE NOW" AND/OR "SERVICE SOON" TELLTALE

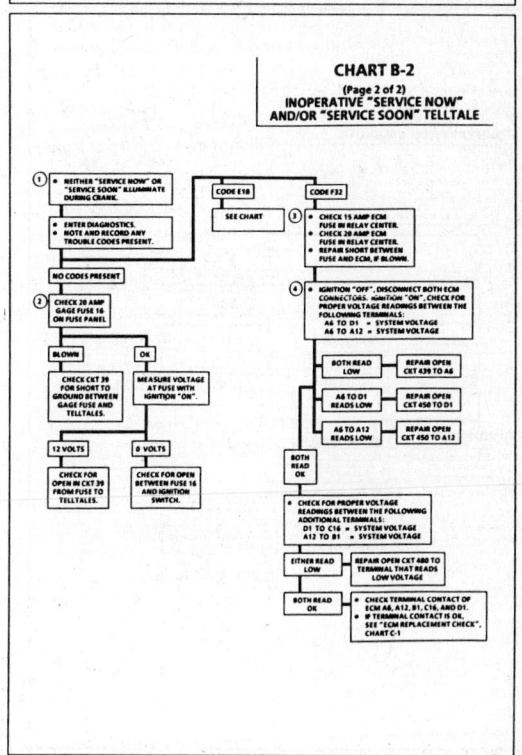

1988–90 CADILLAC DEVILLE AND FLEETWOOD

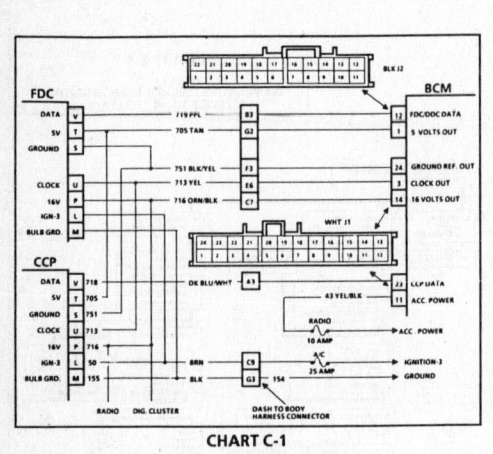

CHART C-1
DISPLAY PANEL DIAGNOSIS

Circuit Description:
The BCM supplies the FDC and CCP with certain information. Through CKT 719, the BCM sends data to the FDC. Through CKT 705, the BCM supplies the FDC and CCP with a 5 volt reference. Through CKT 751, the BCM supplies a reference ground. A clock signal is provided to the FDC and CCP over CKT 713. The BCM supplies a 16 volt source to the CCP and the FDC through CKT 716. Through CKT 718, the BCM sends data to the CCP.

Test Description: Numbers below refer to circled numbers on the diagnostic chart.
1 A "c" displayed on one or both panels indicates a loss of the "clock" signal to the affected panel.
2 A "d" displayed on either panel indicates a loss of the "data" signal to the affected panel.
3 If one panel is affected, only branches of the critical circuits to that panel require investigation.

4 If both panels are affected, several circuits then require investigation. If depressing the "OFF" button on the CCP does not result in the blower turning off, then the panels are not in communication with the BCM. This could be caused by a loss of 5 volts or ground to the panels. If the blower turns off, all remaining critical circuits must be checked.

1988–90 CADILLAC DEVILLE AND FLEETWOOD

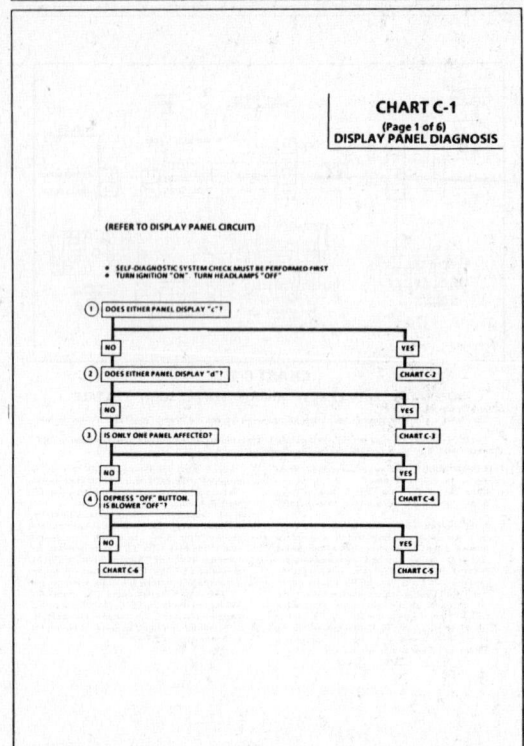

CHART C-1
(Page 1 of 6)
DISPLAY PANEL DIAGNOSIS

1988–90 CADILLAC DEVILLE AND FLEETWOOD

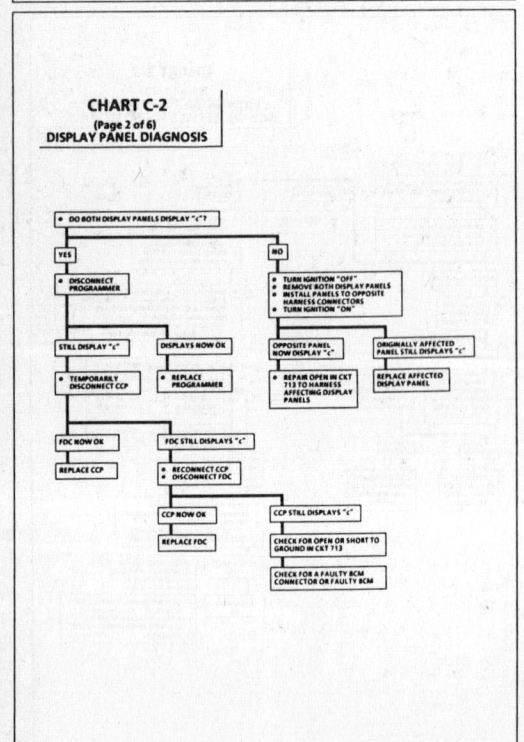

CHART C-2
(Page 2 of 6)
DISPLAY PANEL DIAGNOSIS

1988–90 CADILLAC DEVILLE AND FLEETWOOD

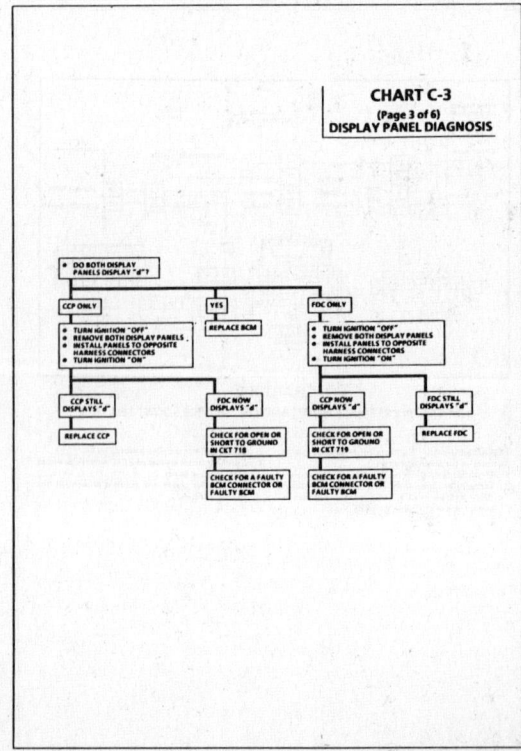

CHART C-3
(Page 3 of 6)
DISPLAY PANEL DIAGNOSIS

1988–90 CADILLAC DEVILLE AND FLEETWOOD

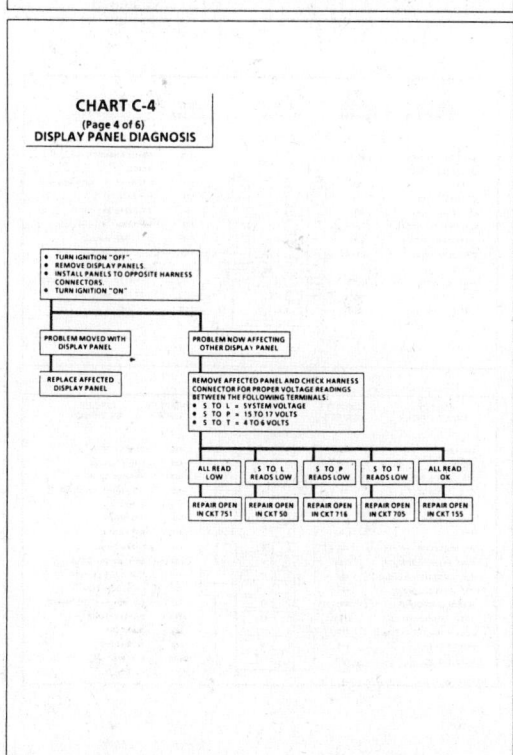

1988–90 CADILLAC DEVILLE AND FLEETWOOD

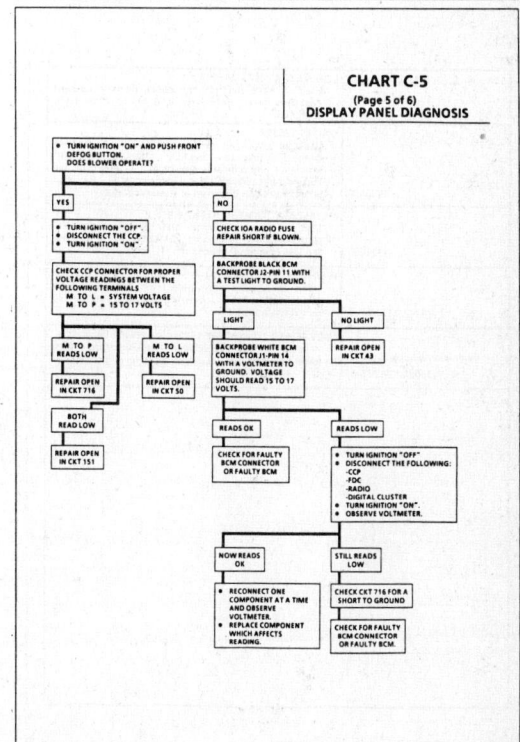

1988–90 CADILLAC DEVILLE AND FLEETWOOD

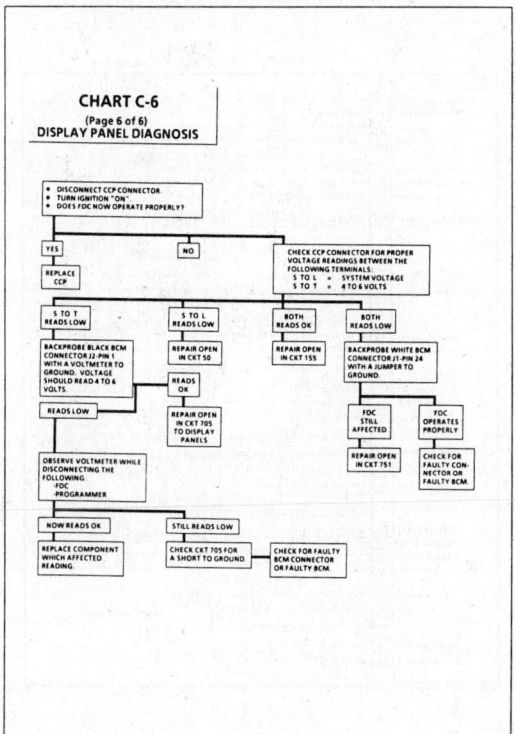

1988–90 CADILLAC DEVILLE AND FLEETWOOD

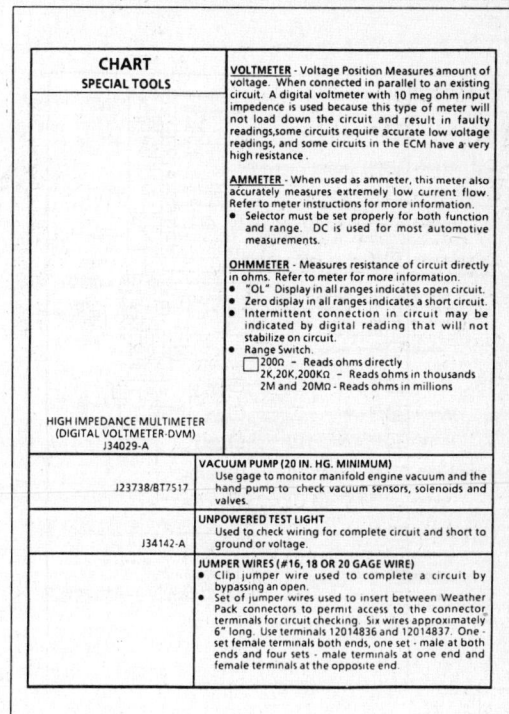

1988–90 CADILLAC DEVILLE AND FLEETWOOD

J35616	**CONNECTOR TEST ADAPTER KIT** Used to make electrical test connections in current Weather Pack, Metri-Pack and Micro-Pack style terminals.
J34636	**CIRCUIT TESTER** Used to check all relays and solenoids before connecting them to a new ECM. Measures the circuit resistance and indicates pass or fail via green or red LED. Amber LED indicates current polarity. Can also be used as a non-powered continuity checker.
J35689	**METRI-PACK TERMINAL REMOVER** Used to remove 150 series Metri-Pack "pull-to-seat" terminals from connectors. Refer to wiring harness service for removal procedure.
J28742/BT8234-A	**WEATHER PACK TERMINAL REMOVER** Used to remove terminals from Weather Pack connectors. Refer to wiring harness service for removal procedure.
J33095/BT8234-A	**EDGE BOARD CONNECTOR TERMINAL REMOVER** Used to remove terminal from Micro-Pack connectors. Refer to wiring harness service for removal procedure.

1988–90 CADILLAC DEVILLE AND FLEETWOOD

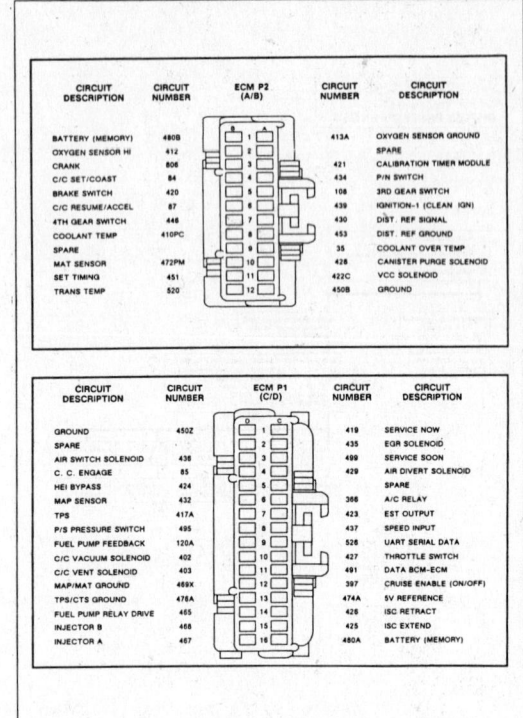

1988–90 CADILLAC DEVILLE AND FLEETWOOD

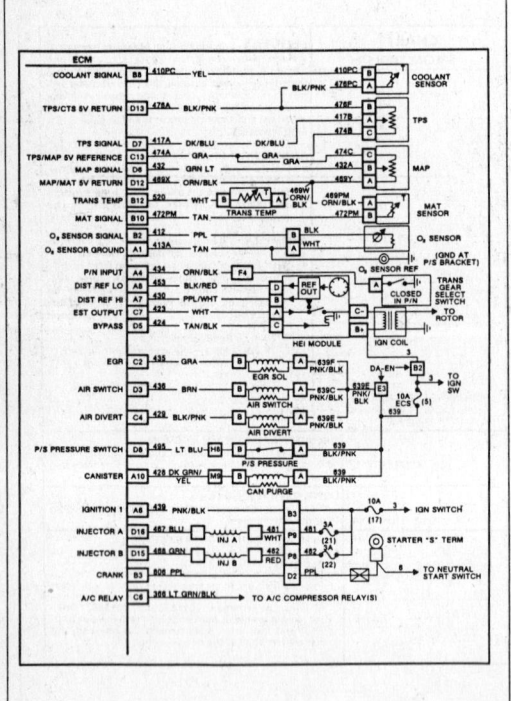

1988–90 CADILLAC DEVILLE AND FLEETWOOD

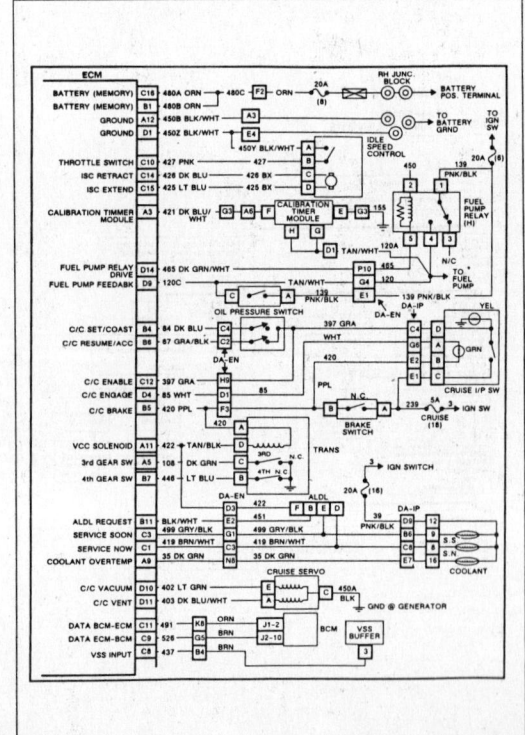

1988–90 CADILLAC DEVILLE AND FLEETWOOD

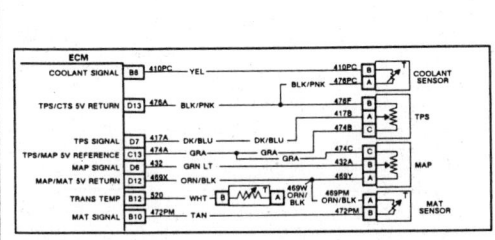

CHART A—MULTIPLE CODES STORED HARD

The conditions diagnosed in ECM Chart A, are caused by a single circuit failure yet result in multiple diagnostic codes. If any of these conditions are met, follow the appropriate correction procedure before using the procedures for the individual codes.

Chart A-1

If Codes E22 and E32 are both stored hard, the cause is probably the loss of 5 volts on circuit #474 (5 volt reference to MAP and TPS). To verify this condition, probe the following harness terminals with a voltmeter to ground:

- MAP sensor harness connector Pin C

If the voltage is 0 for the two sensors, circuit #474 must be investigated for an open or short to ground. If a defect is not found in the wiring, either the ECM connector or ECM itself is faulty. If the proper 5 volt signal is observed on both of the sensors' terminals, then the diagnostic procedures for each individual code must be followed. Diagnose E22 first, then Code E32.

Chart A-2

If a code E15 is stored hard along with a hard Code E21, or E26 the cause is probably an open in the sensor ground circuit #476 for the TPS and coolant

sensors. To verify this condition, probe the following harness terminals with a voltmeter to 12 volts:

- Coolant Temp sensor Pin A, Blk/Pnk wire

If the voltage is 0 at both sensors, circuit #476 must be investigated for an open. If a defect is not found in the wiring, either the ECM connector or ECM itself is faulty. If 12 volts is observed at either of the two sensors, then the diagnostic procedures for each individual code must be followed. Diagnose code E15 then E21 then E26.

Chart A-3

If a Code E38 is stored hard along with a hard Codes E34 or E59 this is probably caused by an open in the sensor ground circuit #469 for the TPS, MAT and TRANS TEMP. sensors. To verify this condition, probe the following harness terminals with a voltmeter to 12 volts:

- MAT sensor Pin A
- MAP sensor harness Pin A
- Trans Temp sensor Pin A

If the voltage is 0 for the sensors, circuit #469 must be investigated for an open. If a defect is not found in the wiring, either the ECM connector or ECM itself is faulty. If 12 volts is observed at the sensors, then the diagnostic procedures for each individual code must be followed.

1988–90 CADILLAC DEVILLE AND FLEETWOOD

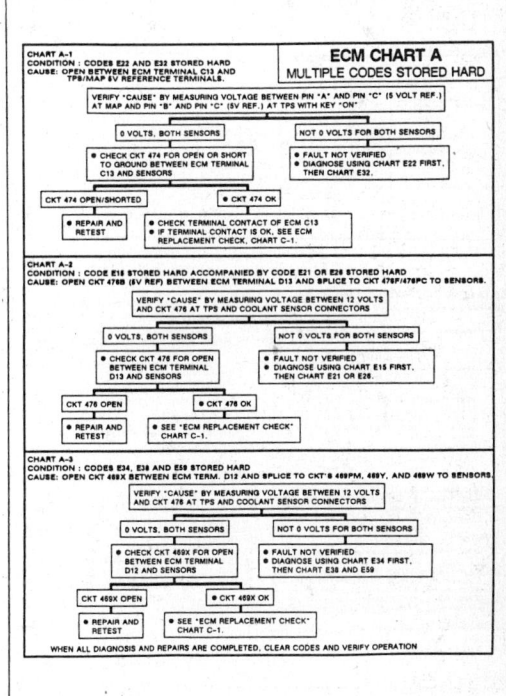

1988–90 CADILLAC DEVILLE AND FLEETWOOD

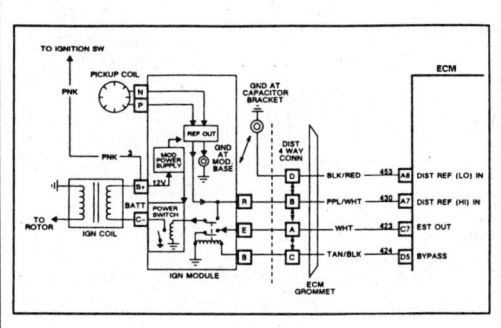

ECM CODE E12, NO DISTRIBUTOR (TACH) SIGNAL

TEST CONDITIONS: Code E12 is tested for during engine cranking (crank input to the ECM at system voltage and no Code E18 pending).

FAILURE CONDITIONS: If ECM does not see distributor reference pulses for 2.1 seconds with crank input to the ECM at system voltage, current Code E12 will be set.

ACTION: 1. ECM turns on "Service Now" light.

Description:

Possible causes of "no distributor reference" are: 1. Open pick-up coil – the pick-up coil produces an A.C. voltage that is used by the HEI module to create distributor reference pulses; 2. HEI module unable to process pick-up coil signals; 3. Open or short on CKT 430 and 453 from module to 4-way weatherpack to ECM; 4. Loss of 12 volts to "BAT" terminal on HEI.

Notes On Fault Tree:

1. Checking for proper voltage output of HEI system. If voltmeter shows .5–2.8 volts, HEI is producing reference pulses.

2. Checking for proper ground connection between ECM and the engine and the ECM and HEI.
3. Checking for proper voltage through CKT 430 from HEI to ECM. If ECM terminal A7 sees .5–2.8 volts, then ECM is receiving reference pulses.
4. If the HEI will produce spark then fault is not pick-up coil or module. Check CKT 430 for opens or short to ground from module terminal "R" to ECM terminal "R" of 4-way weatherpack.

Note On Intermittents

If Code E12 is stored as a history code, start engine and allow to idle while manipulating CKT's 430 and 453. An intermittent open will cause the engine to stumble or quit when the ECM loses distributor reference.

If wiring and ECM connectors are no trouble found, check HEI pick-up coil leads for intermittent open circuit.

Check for short to voltage on CKT 806. If pin B3 has voltage at key on – engine off, the ECM will record a false Code E12. If found, repair short to voltage on CKT 806.

1988–90 CADILLAC DEVILLE AND FLEETWOOD

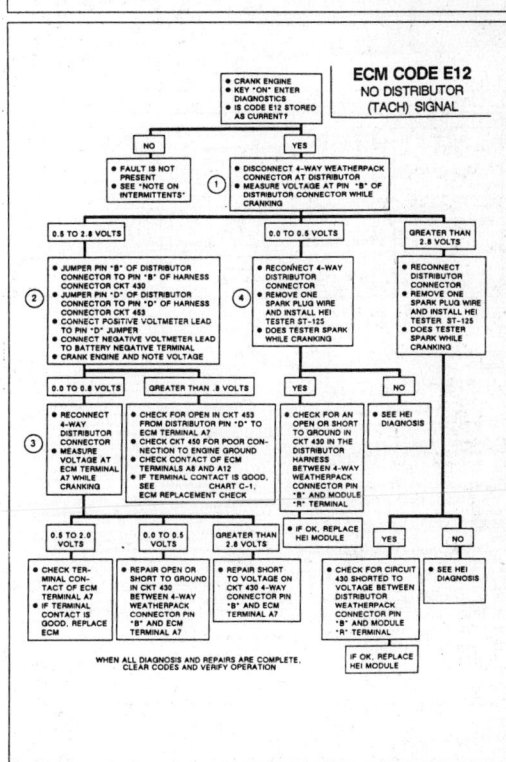

1988–90 CADILLAC DEVILLE AND FLEETWOOD

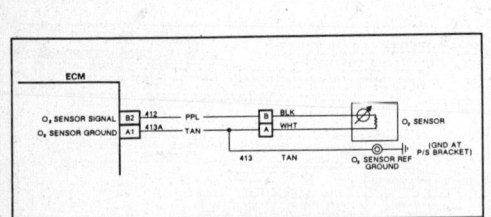

ECM CODE E13, OXYGEN SENSOR NOT READY

Important

If Code E13 is accompanied by current or History Code E45, use the Code E45 fault tree to diagnose. If Code E13 is accompanied by current or history Code E44, use the Code E44 fault tree to diagnose.

TEST CONDITIONS: 1. Codes E14, E15, E21, E22, E26, E27 clear; 2. Coolant temp greater than or equal to 80°C; 3. TPS value between 6 and 29 degrees; 4. Throttle switch open; 5. RPM greater than or equal to 800.

FAILURE CONDITIONS: Oxygen sensor voltage stays between .275 volts and .630 volts for more than 40 seconds (oxygen sensor voltage not toggling).

ACTION TAKEN: 1. ECM turns on "Service Soon" light; 2. ECM turns off canister purge, switches AIR system to air cleaner; 3. Disables Closed Loop.

Description:

The ECM provides a .45 volt reference to the oxygen sensor on CKT 412. When warm, a properly operating oxygen sensor will drive the .45 volt reference lower (below .45 volts) to indicate lean mixtures and higher (above .45 volts) to indicate a rich mixture. If under the above test conditions the oxygen sensor doesn't vary from the "cold" or "not ready" voltage,

the ECM assumes that the sensor can't respond to air-fuel mixtures rich or lean and Code E13 is set.

Notes On Fault Tree:

1. At key on with oxygen sensor disconnected, the parameter P.0.7 should read about .45 volts. Lower or higher voltage indicates circuit problems.

2. The ECM compares voltage on CKT 412 to the ground voltage on CKT 413. It is essential that the oxygen sensor ground and ECM ground on pin A1 show good continuity (no voltage difference) with engine running.

3. Fault is most likely at ECM connector or ECM. Before ECM is replaced, perform "ECM Replacement Check", Chart C-1.

Notes On Intermittents:

If Code E13 is a history code, start engine, enter diagnostics. Operate engine at fast idle until "Auto" closed loop status light is turned on. Observe parameter while manipulating CKT's 412 and 413 wiring. If parameter P.0.7 changes from fluctuating to fixed at about .45 volts, repair intermittent open circuit.

Check CKT 413 ground to engine for proper installation (clean, tight, star washer installed.) Ground for CKT 413 is attached to the front of the engine at the right or rearmost cylinder head.

Review the snapshot data parameters stored with the code to verify failure condition. Refer to "Symptoms" for complete details on using the "Snapshot on code set" feature. If the failure condition is verified, check and repair intermittent wiring connection or sensor.

1988–90 CADILLAC DEVILLE AND FLEETWOOD

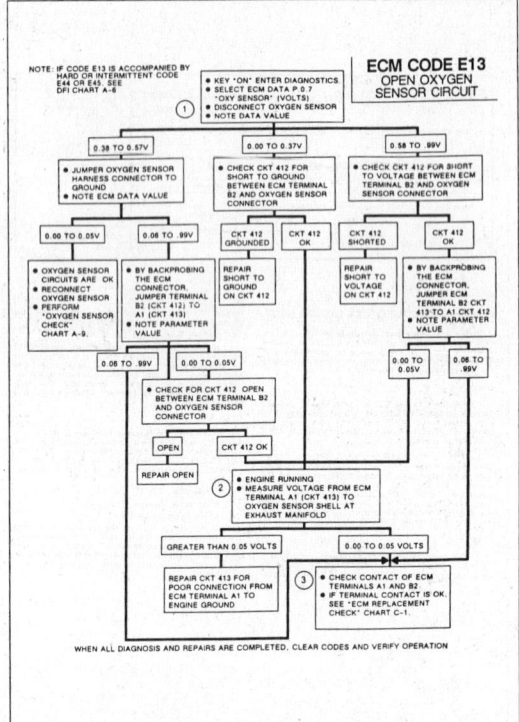

ECM CODE E13
OPEN OXYGEN SENSOR CIRCUIT

WHEN ALL DIAGNOSIS AND REPAIRS ARE COMPLETED, CLEAR CODES AND VERIFY OPERATION

1988–90 CADILLAC DEVILLE AND FLEETWOOD

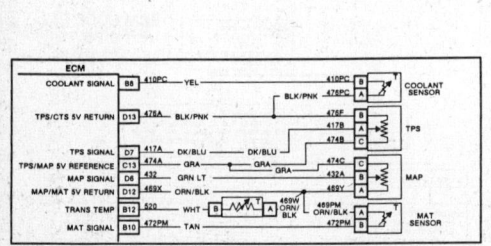

ECM CODE E14, SHORTED COOLANT SENSOR CIRCUIT

TEST CONDITIONS: 1. Code E37 and E38 not set; 2. MAT Sensor Value less than or equal to 100°C.

FAILURE CONDITIONS: Coolant Sensor Value greater than or equal to 144°C. (Sensor Signal shorted to ground.)

ACTION TAKEN: 1. ECM turns on "Service Soon" telltale; ECM turns off both AIR solenoids; ECM uses MAT sensor value in place of coolant sensor value for all calculations.

Description:

The coolant sensor is a thermistor or variable resistor. The coolant sensor is a "Two wire" sensor with a sensor signal voltage coming from the ECM on Pin B, CKT 410 and a sensor reference ground on Pin A, CKT 476.

As the temperature of the sensor increases, sensor resistance decreases. The signal voltage from the ECM to Pin A decreases as sensor temperature increases and current flows through the sensor element to Pin A sensor ground.

Code E14 "Shorted Coolant Sensor" sets because the ECM assumes that the coolant temperature can't be 144°C (291°F) or greater when MAT is 100°C or less. (Sensor signal shorted to ground).

Notes On Fault Tree:

1. With sensor shorted, parameter P.0.4 should read 144°C or greater. If not, then the sensor is not shorted, see "Note on Intermittents."

2. Checking for sensor shorted on CKT 410 shorted. If the parameter stays at 144° to 151° with the sensor unplugged, then the short is in CKT 410, between the sensor and ECM.

3. Fault is most likely at ECM connector or ECM. Before ECM is replaced, perform "ECM Replacement Check", chart C-1.

Notes On Intermittents:

Manipulate CKT 410 wiring, the coolant sensor and ECM connector while observing ECM Parameter P.0.4. If a failure is induced, the coolant temperature will jump from its normal value to the shorted reading of 144° to 151°C. Remove and replace both the COOLANT and ECM connectors and ensure that they are latched. If wiring and connector check out OK, substitute a known good coolant sensor and retest.

Review the snapshot data parameters stored with the code to verify failure condition. Refer to "Symptoms" for complete details on using the "Snapshot on code set" feature. If the failure condition is verified, check and repair intermittent wiring connection or sensor.

1988–90 CADILLAC DEVILLE AND FLEETWOOD

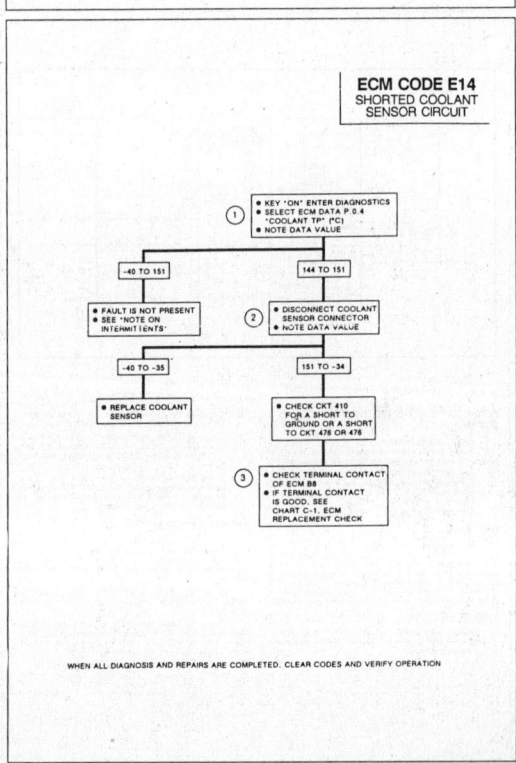

ECM CODE E14
SHORTED COOLANT SENSOR CIRCUIT

WHEN ALL DIAGNOSIS AND REPAIRS ARE COMPLETED, CLEAR CODES AND VERIFY OPERATION

1988–90 CADILLAC DEVILLE AND FLEETWOOD

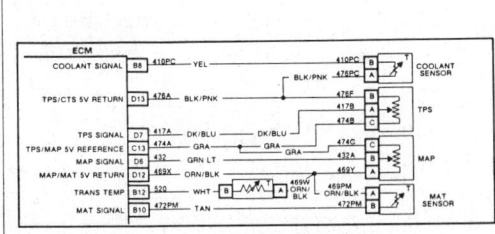

ECM CODE E15, OPEN COOLANT SENSOR CIRCUIT

TEST CONDITIONS: 1. No problems detected on the MAT sensor circuit; 2. MAT Sensor Value greater than or equal to 2.5°C.

FAILURE CONDITIONS: Coolant Sensor Value less than or equal to –32°C (Sensor Signal Open).

ACTION TAKEN: 1. ECM turns on "Service Soon" light. 2. ECM turns off both AIR solenoids. 3. ECM uses MAT sensor value in place of coolant for all calculations.

Description:

The coolant sensor is a thermistor or variable resistor. The coolant sensor is a "two wire" sensor with a sensor signal or reference voltage coming from the ECM to sensor pin B on CKT 410 and sensor reference ground CKT 476 connecting sensor pin A to the ECM sensor ground. As the temperature of the sensor decreases, sensor resistance is higher. The signal voltage from the ECM to sensor pin B increases as sensor temperature decreases and less current flows through the sensor element to pin A sensor ground.

Code E15 "Open Coolant Sensor" sets because the ECM assumes that coolant temperature can't be –32°C or less when the MAT is –2.5°C or greater; sensor signal is open to sensor.

Notes On Fault Tree:

1. If sensor is open, parameter P.0.4 should read –32°C or less. If not then sensor signal is not open, see "Note on Intermittents."

2. Checking for open sensor signal CKT 410 from ECM to sensor connector. If parameter P.0.4 reads 144°–151°C with connector shorted then CKT 410 is OK.

3. Checking for open sensor ground CKT 476 from ECM to sensor. If pin "B" to ground will jump from 476 reading of –32°C to the open circuit reading of –32°C or less. Disconnect and reconnect both the coolant sensor and ECM connectors and ensure that they are latched before replacing components. If wiring and connectors check out OK, substitute a known good coolant sensor and retest.

4. Checking for ECM's ability to recognize a short to ground or low voltage on B8, coolant sensor input.

5. Fault is most likely at ECM connector or ECM. Before ECM is replaced, perform "ECM Replacement Check", Chart C-1.

Notes On Intermittents:

Manipulate CKT 410 and 476 wiring, the coolant sensor and ECM connector while observing ECM parameter P.0.4. If the failure is induced, the coolant temperature will jump from normal value to the open circuit reading of –32°C or less. Disconnect and reconnect both the coolant sensor and ECM connectors and ensure that they are latched before replacing components. If wiring and connectors check out OK, substitute a known good coolant sensor and retest.

Review the snapshot data parameters stored with the code to verify failure condition. Refer to "Symptoms" for complete details on using the "Snapshot on code set" feature. If the failure condition is verified, check and repair intermittent wiring connection or sensor.

1988–90 CADILLAC DEVILLE AND FLEETWOOD

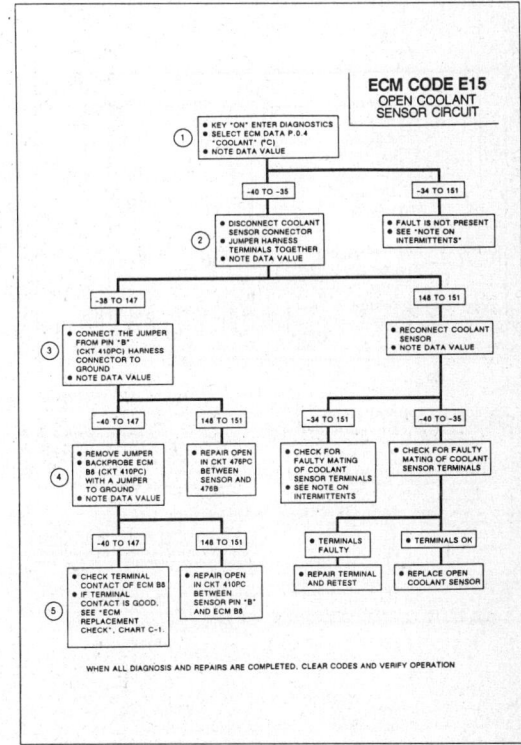

ECM CODE E15
OPEN COOLANT SENSOR CIRCUIT

WHEN ALL DIAGNOSIS AND REPAIRS ARE COMPLETED, CLEAR CODES AND VERIFY OPERATION

1988–90 CADILLAC DEVILLE AND FLEETWOOD

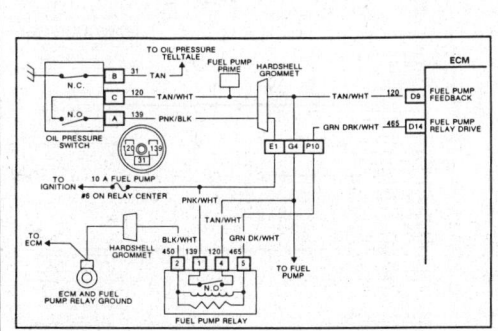

ECM CODE E16, GENERATOR VOLTAGE OUT OF RANGE

TEST CONDITIONS: 1. Code E18 and E20 Clear; 2. RPM greater than or equal to 800; 3. Not Cranking (Crank input to ECM at 0 volts).

FAILURE CONDITIONS: Voltage on Fuel Pump Feedback to ECM less than 10.0 volts or greater than 16.0 volts for 5 seconds or more.

ACTION TAKEN: 1. ECM turns on "Service Now" light; 2. ECM turns off EGR solenoid, canister purge solenoid and both AIR solenoids. 3. ECM disables Cruise Control 4. ECM disables VCC for entire key cycle.

Description:

The ECM monitors vehicle electrical system voltage or battery voltage indirectly by monitoring the voltage on the fuel pump feedback CKT 120.

The voltage readings taken by the ECM are used to detect fuel pump voltage faults and to modify injector pulse width as system voltage varies; the ECM increases injector pulse width as system voltage decreases.

Code E16 is set when the ECM sees low or high system voltage with an engine RPM high enough to allow almost full generator output.

Notes On Fault Tree:

1. Checking for generator output high or low at idle.

2. Checking for insufficient generator output when at high electrical loads.

3. Checking for generator output voltage too high.

4. Checking for proper voltage at fuel pump test point on fuel pump power circuit.

5. Checking for open between ECM pin D9 and fuel pump feed circuit.

Note On Intermittents:

If history Code E16 is found, start engine, enter diagnostics and display ECM parameter P.1.0. Manipulate CKT 120 wiring from ECM pin D9 to the open at CKT 120. Watch for the parameter to drop to less than 10.0 volts. Many intermittent Code E16's can be traced to intermittent charging or battery system faults such as insufficient charging system output at high accessory loads or generator voltage out of regulation high.

Check for occurrence of "No Charge" telltale ON, "Service Electrical System" telltale ON, and for low or overcharged battery.

Review the snapshot data parameters stored with the code to verify failure condition. Refer to "Symptoms" for complete details on using the "Snapshot on code set" feature. If the failure condition is verified, check and repair intermittent wiring connection or sensor.

1988–90 CADILLAC DEVILLE AND FLEETWOOD

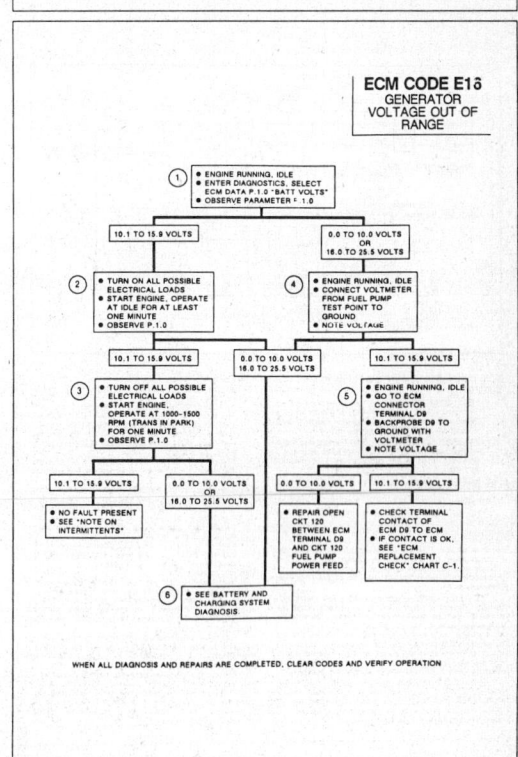

ECM CODE E16
GENERATOR VOLTAGE OUT OF RANGE

WHEN ALL DIAGNOSIS AND REPAIRS ARE COMPLETED, CLEAR CODES AND VERIFY OPERATION

1988–90 CADILLAC DEVILLE AND FLEETWOOD

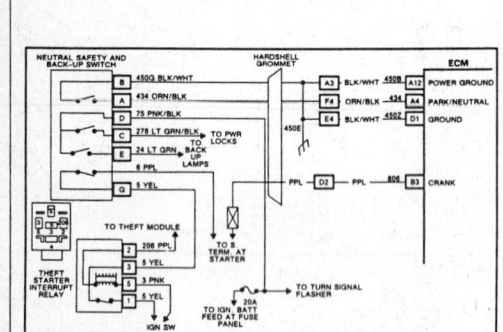

ECM CODE E18, OPEN CRANK SIGNAL CIRCUIT

TEST CONDITIONS: 1. Engine RPM changed from 0 to greater than or equal to 800 while in park or neutral.
FAILURE CONDITIONS: No crank signal has been received by the ECM since the last time that RPM was 0 RPM. (Engine started, ECM didn't register a crank signal.)
ACTION TAKEN: 1. ECM turns on "Service Soon" light; the light stays on for the entire key cycle.

Description:

The crank input to the ECM is an ON/OFF signal that comes from the starter solenoid, through the neutral safety backup switch through a fuse link then to pin B3 of the ECM. During crank, the signal is battery voltage, at all other times the signal is at 0 volts.

The code E18 is set if the ECM detects the engine running at 800 RPM or greater without having the crank input go to system voltage.

Notes On Fault Tree:

1. The fuse link is to prevent the starter solenoid from pulling in due to a short to voltage on CKT 806. If 806 becomes shorted to voltage or is grounded when the key is turned to "crank", the fuse link will blow.

2. Checking for voltage to the fuse link while cranking.

Note On Intermittents

Back probe ECM pin B3—CKT 806 with a voltmeter, you should get a reading of 12 volts. Check for intermittent opens while manipulating wiring. Check the engine – dash connector for opens, pin D2 for good contact.

If code E18 doesn't set immediately after cranking, check for plug wires arcing, loose engine grounds or other sources of Electromagnetic Interference (EMI). Large transient voltage surges can "Erase" the crank signal from the ECM's memory and cause code E18 to set while driving.

1988–90 CADILLAC DEVILLE AND FLEETWOOD

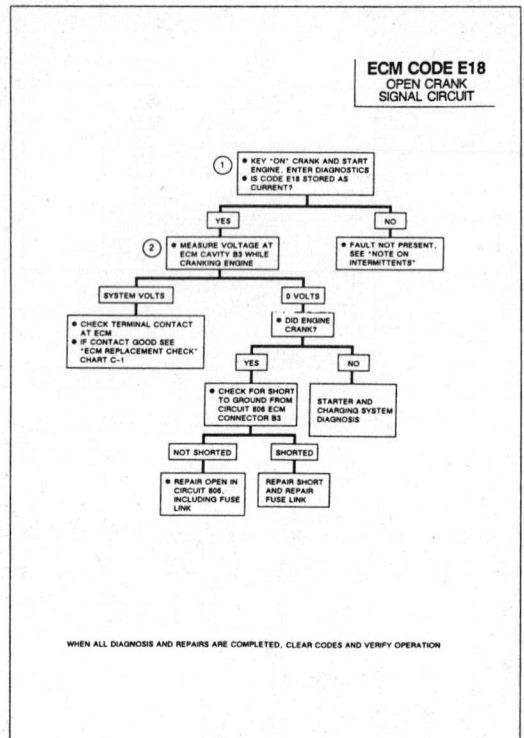

ECM CODE E18
OPEN CRANK SIGNAL CIRCUIT

WHEN ALL DIAGNOSIS AND REPAIRS ARE COMPLETED, CLEAR CODES AND VERIFY OPERATION

1988–90 CADILLAC DEVILLE AND FLEETWOOD

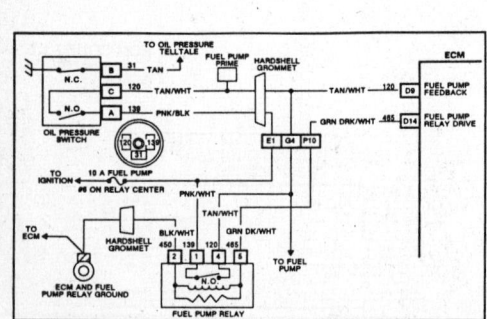

ECM CODE E19, SHORTED FUEL PUMP CIRCUIT

TEST CONDITIONS: 1. No problems detected on the crank circuit. 2. RPM = 0 3. Crank input to ECM at 0 volts (not cranking) 4. "Coolant" light in bulb check (key "on" engine off).
FAILURE CONDITIONS: Fuel pump feedback voltage greater than or equal to 11.0 volts for 7 seconds. (Fuel pump powered with engine not running.)
ACTION TAKEN: 1. ECM turns on "Service Soon" light.

Description:

The fuel pump receives power from CKT 120. Ckt 120 is fed by two alternative sources, the fuel pump relay and the oil pressure switch. The relay is energized by the ECM. The relay is energized for 2 seconds at key "ON" to provide fuel pressure for starting. After the 2 second fuel pump run, the ECM will not power the relay again until distributor reference pulses are received indicating that the engine is turning (crank or run).

The oil pressure switch is provided as a "backup". Normal operating engine oil pressure will close the connection between CKT 139 and 120. If a relay fault occurs, the engine will continue to run as the oil pressure switch powers the fuel pump. A vehicle with the relay inoperative will crank until oil pressure builds (oil telltale light off) then start and run as the oil pressure switch provides power to the fuel pump.

The ECM monitors voltage on the fuel pump power CKT 120 to detect fuel system supply faults. Code E19 is set when the fuel pump feedback (ECM pin D9) is at 11.0 volts or greater when the fuel pump feed voltage should be 0 volts.

Note On Fault Tree:

1. At key on engine off, the fuel pump should not be running and the fuel pump feedback (ECM parameter P.1.0) should show 0 volts. (Note: The fuel pump runs for 2 seconds then turns off at the start of each key "on" cycle. After the initial 2 second run, the fuel pump will not be powered until the ECM sees distributor reference pulses from the HEI.)

2. This branch of the tree checks for oil pressure switch contacts shorted, powering the fuel pump at all times.

3. Checking for fuel pump relay providing fuel pump power due to a relay or relay drive circuit fault.

4. Checking CKT 120 for a short to a 12 volt source.

5. Checking ECM relay drive circuit for a fault that allows fuel pump drive coil to be energized continuously at key on.

6. Fault is most likely at ECM connector or ECM. Before ECM is replaced, perform "ECM Replacement Check", Chart C-1.

Note On Intermittents:

Backprobe ECM pin D9 with a voltmeter to ground. Key on and observe voltmeter – it should read battery voltage for 2 seconds then drop to 0. If the voltmeter stays at battery voltage for longer than 2 seconds, check for fuel pump relay contacts sticking. Repeat the test several times.

Continue backprobing D9 to ground and crank and start engine then key off – voltage should drop immediately to 0. If not, check oil pressure switch for sticking contacts.

Review the snapshot data parameters stored with the code to verify failure condition. Refer to "Symptoms" for complete details on using the "Snapshot on code set" feature. If the failure condition is verified, check and repair intermittent wiring connection or sensor.

1988–90 CADILLAC DEVILLE AND FLEETWOOD

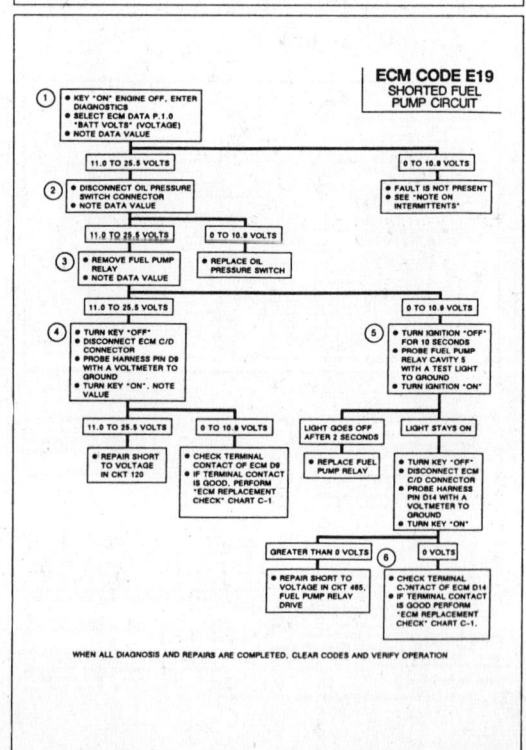

ECM CODE E19
SHORTED FUEL PUMP CIRCUIT

WHEN ALL DIAGNOSIS AND REPAIRS ARE COMPLETED, CLEAR CODES AND VERIFY OPERATION

1988–90 CADILLAC DEVILLE AND FLEETWOOD

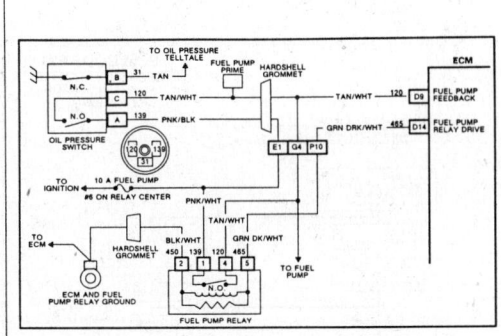

ECM CODE E20, OPEN FUEL PUMP CIRCUIT

TEST CONDITIONS: 1. Engine RPM greater than or equal to 24 RPM.

FAILURE CONDITIONS: Fuel pump feedback voltage less than or equal to 2 volts for 3 seconds or more.

ACTION TAKEN: 1. Store code as hard until engine is turned off; 2. ECM turns on "Service Now" telltale.

Description:

The ECM monitors voltage on the fuel pump power CKT 120 to detect fuel pump voltage supply faults. Code E20 is set when the ECM sees the fuel pump not energized (0 volts on the feedback) with the engine cranking or running. The code is designed to detect a fuel pump relay fault; relay not powering the fuel pump.

Notes On Fault Tree:

1. Engine starts with hard E20 indicates that the engine started with fuel pump powered through the oil pressure switch – the relay circuit may be at fault.

2. 0 volts with engine running indicates an open from fuel pump power circuit to ECM – the fuel pump relay is not at fault.

3. Testing for ECM's ability to energize relay coil.
4. Testing for open from 20A fuel pump fuse to relay pin "1".
5. Testing for open from relay pin "2" to ground.
6. If light turns on then all relay circuitry is OK. Fault causing the code must be the relay.
7. If relay or relay drive circuit are grounded, the ECM may be damaged by excessive current draw. Repair the ground or replace the relay and check for proper operation.
8. Fault is most likely at ECM connector or ECM. Before ECM is replaced, perform "ECM Replacement Check", Chart C-1.

Notes On Intermittents:

If intermittent Code E20 is stored, unplug the oil pressure switch, start engine and allow to idle. Manipulate affected wiring, ensure that the ECM P1 (C/D) connector is latched, check the relay for proper installation into relay center socket. If the fault is induced, the engine will stall, Code E20 will set.

Review the snapshot data parameters stored with the code to verify failure condition. Refer to "Symptoms" for complete details on using the "Snapshot on code set" feature. If the failure condition is verified, check and repair intermittent wiring connection or sensor.

1988–90 CADILLAC DEVILLE AND FLEETWOOD

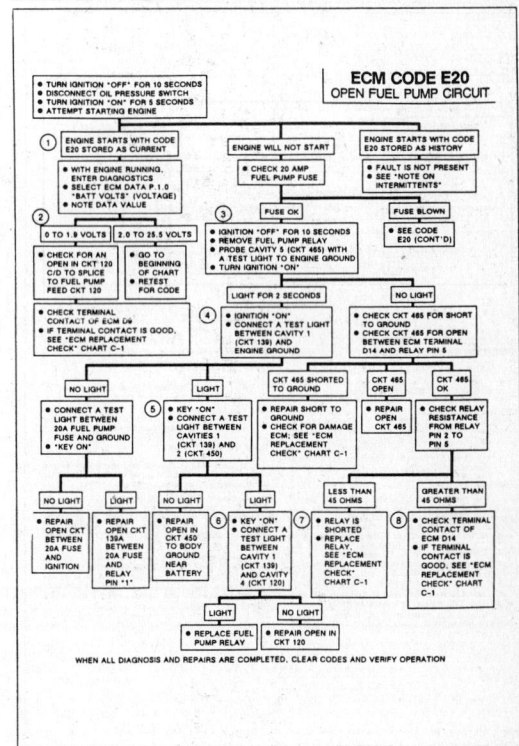

ECM CODE E20
OPEN FUEL PUMP CIRCUIT

WHEN ALL DIAGNOSIS AND REPAIRS ARE COMPLETED, CLEAR CODES AND VERIFY OPERATION

1988–90 CADILLAC DEVILLE AND FLEETWOOD

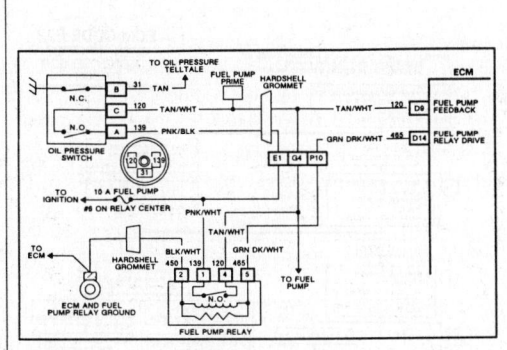

ECM CODE E20 (CONT'D) FUEL PUMP FUSE BLOWN

Notes On Fault Tree:

1. Checking for an open or short to ground on CKT 139.
2. Checking for short to ground on CKT 120.
3. Checking for ECM shorting 120 to ground.
4. Measuring resistance of fuel pump CKT inside the fuel tank.
5. Check for pinched wires inside the fuel tank.
6. Checking for a short to ground in the wire harness leading to the fuel tank.
7. Wire harness is shorted to ground. Check for pinched or grounded wires at the fuel prime test point, oil pressure switch, and connectors leading to the fuel pump.

8. CKT 139 is being grounded all the time. Checking for faulty component.

Notes On Intermittents:

Replace fuse, start engine and allow to idle. Manipulate CKT 120 and CKT 139 wiring – look for shorts to ground. If fault is induced, the fuse will blow and code E20 will be stored as a hard code.

Review the snapshot data parameters stored with the code to verify failure condition. Refer to "Symptoms" for complete details on using the "Snapshot on code set" feature. If the failure condition is verified, check and repair intermittent wiring connection or sensor.

1988–90 CADILLAC DEVILLE AND FLEETWOOD

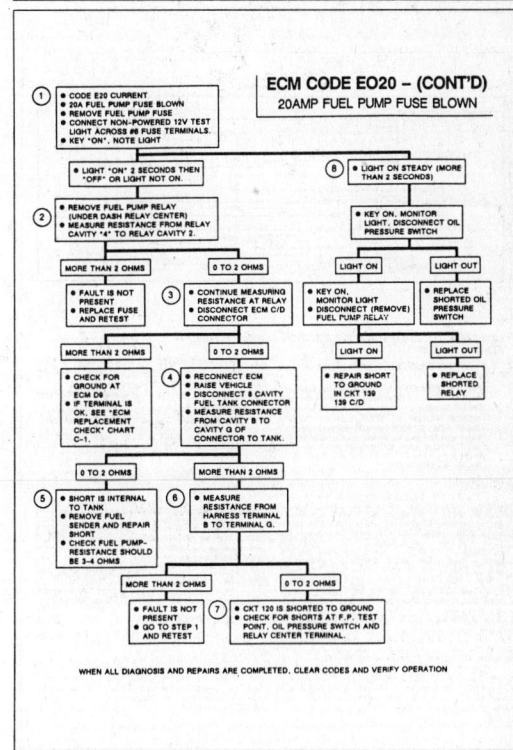

ECM CODE EO20 – (CONT'D)
20AMP FUEL PUMP FUSE BLOWN

WHEN ALL DIAGNOSIS AND REPAIRS ARE COMPLETED, CLEAR CODES AND VERIFY OPERATION

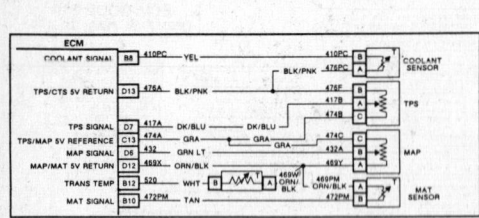

ECM CODE E21, SHORTED THROTTLE POSITION SENSOR CIRCUIT

TEST CONDITIONS: 1. RPM between 25 and 1,000.

FAILURE CONDITIONS: TPS value greater than or equal to 83 degrees for 1.5 seconds.

ACTION TAKEN: 1. ECM turns on "Service Soon" light. 2. ECM uses 12 degrees for TPS value with ISC throttle switch open and 6 degrees for TPS value when ISC throttle switch closes; 3. ECM uses RPM for all ISC control calculations (won't use TPS reading for coast down throttle control.

Description:

The Throttle Position Sensor (TPS) is a "3 wire" sensor or potentiometer with a 5 volt reference input from the ECM to the sensor CKT 474, a reference ground from the ECM to the sensor CKT 476 and a sensor output signal (CKT 417) from the sensor to the ECM. The sensor output signal is a D.C. voltage that varies with throttle angle. At low throttle angle, the TPS signal voltage is low (about .5 volts at minimum air setting) and at high throttle angles, the TPS signal voltage is high (about 4.5 volts at wide open throttle).

Code E21 is set when the ECM sees a high throttle angle (high TPS voltage) at low engine RPM. The code

is designed to detect a TPS signal (CKT 417) shorted to voltage.

Notes On Fault Tree:

1. To determine location of short to voltage –– TPS or CKT 417.

2. Checking for an open ground on CKT 474. A ground open will result in high TPS values whenever the TPS is plugged in.

Notes On Intermittents:

If Code E21 is intermittent, manipulate related wiring while observing ECM parameter P.0.1. Check the TPS connectors for shorts to voltage.

Cycle the TPS through its travel and tap on the TPS with a pencil or pocket screwdriver to test for intermittent TPS. If the fault is induced, the parameter will skip to a high throttle angle. If wiring and connectors check out OK, substitute a known good TPS sensor and retest.

Review the snapshot data parameters stored with the code to verify failure condition. Refer to "Symptoms" for complete details on using the "Snapshot on code set" feature. If the failure condition is verified, check and repair intermittent wiring connection or sensor.

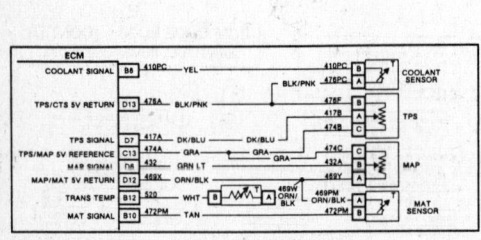

ECM CODE E22, OPEN THROTTLE POSITION SENSOR CIRCUIT

TEST CONDITIONS: 1. RPM greater than or equal to 600.

FAILURE CONDITIONS: TPS value less than –6° degrees (TPS parameter reads –10 to –6 degrees) for 1.5 seconds.

ACTION TAKEN: 1. ECM turns on "Service Soon" light. 2. ECM uses 12 degrees for TPS value with ISC throttle switch open, and 6 degrees for TPS value when ISC throttle switch closes 3. ECM stops using TPS to control ISC – uses RPM for all ISC calculations.

Description:

The Throttle Position Sensor (TPS) is a "3 wire" sensor or potentiometer with a 5 volt reference input from the ECM to the sensor (CKT 474), a reference ground from the ECM to the sensor (CKT 476) and a sensor output signal (CKT 417) from the sensor to the ECM. The sensor output signal is a D.C. voltage that varies with throttle angle. At low throttle angle, the TPS signal voltage is low (about .5 volts at minimum air setting) and at high throttle angles, the TPS signal voltage is high (about 4.5 volts at wide open throttle).

Code E22 is set when the ECM sees a throttle angle that is out of limits low (low TPS voltage) with engine running at idle or faster. The code is designed to detect a TPS signal (CKT 417) open to ECM pin D7.

Notes On Fault Tree:

1. Jumpering 5 volts to TPS signal to see if the ECM parameter will respond to high signal voltage on TPS inputs (CKT 417).

2. Checking for TPS reference (CKT 474) open/shorted.

3. Checking for TPS signal (CKT 417) grounded.

4. Fault is most likely at ECM connector or ECM. Before ECM is replaced, perform "ECM Replacement Check", Chart C-1.

Notes On Intermittents:

If a Code E22 is intermittent, manipulate related wiring while observing ECM parameter P.0.1. Cycle the TPS through its travel and tap on the TPS with a pencil or pocket screwdriver to test for intermittent TPS. If the fault is induced, the parameter will jump to –6 or less. If wiring and connectors check out OK, substitute a good TPS sensor and retest.

Review the snapshot data parameters stored with the code to verify failure condition. Refer to "Symptoms" for complete details on using the "Snapshot on code set" feature. If the failure condition is verified, check and repair intermittent wiring connection or sensor.

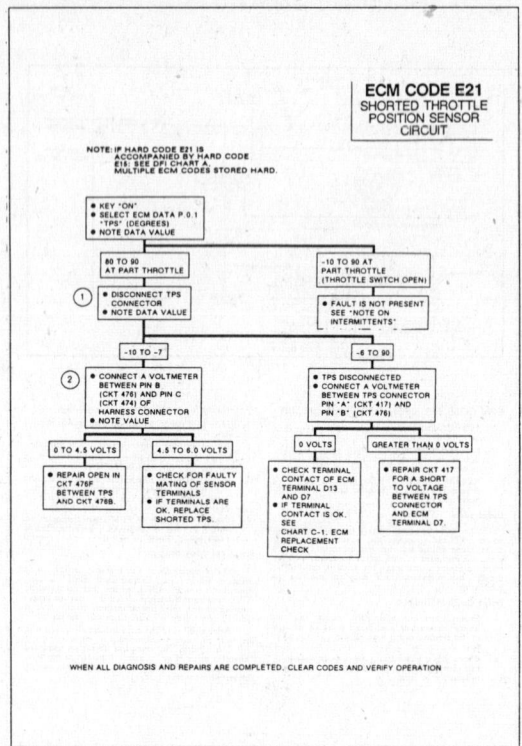

ECM CODE E21
SHORTED THROTTLE POSITION SENSOR CIRCUIT

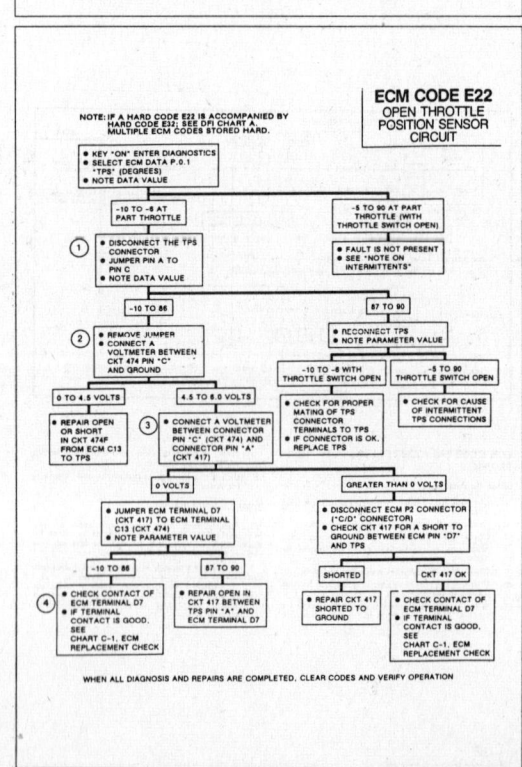

ECM CODE E22
OPEN THROTTLE POSITION SENSOR CIRCUIT

1988–90 CADILLAC DEVILLE AND FLEETWOOD

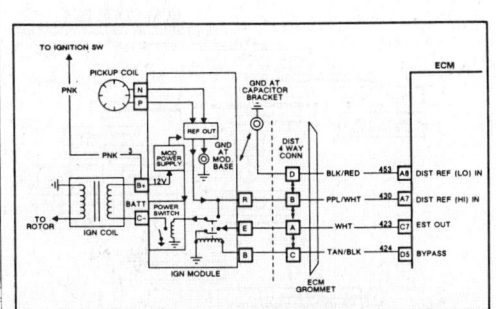

ECM CODE E23 EST CIRCUIT PROBLEM

TEST CONDITIONS: 1. Key "on"; 2. RPM greater than or equal to 648.

FAILURE CONDITIONS: EST pulses are seen by the ECM before bypass circuit is turned "on" (5 volts) or EST signal remains grounded after the bypass circuit is turned "on" (5 volts).

ACTION TAKEN: 1. ECM turns on the "Service Soon" lamp. 2. If ECM sees 5 volt EST pulses with bypass voltage at 0 volts (caused by EST circuit open or bypass circuit shorted to voltage) the ECM will not enable spark – the engine will start and run on base timing. 3. If the EST circuit is shorted to ground, the vehicle will start and stall when the engine reaches 648 RPM (when EST is enabled). When restarted, the ECM will not try to turn bypass on, code E23 will set, vehicle will start and run on base timing.

Description:

The EST distributor produces one 5 volt Distributor Reference Pulse each time that a cylinder reaches 10° BTDC (base timing). The reference pulses are sent from the distributor to the ECM on circuit 430, Distributor Reference Signal.

The ECM adds spark timing information to the reference pulses received from the HEI and sends out 5 volt Electronic Spark Timing pulses (EST) on circuit 423.

The ECM can choose to control spark timing. The HEI Bypass circuit 424 is turned on by the ECM (approximately 5 volts) when the ECM wishes to control timing. When the ECM wants the HEI module to control timing, the HEI Bypass circuit is turned off and the HEI module grounds circuit 423, EST. The ECM always sends EST pulses to the HEI; the HEI module shorts the EST pulses to ground when the bypass signal from the ECM to the module is low (less than .5 volts).

Notes On Fault Tree:

1. Four jumper wires must be obtained for use in the diagnostic procedure. The jumpers should be about 12 inches long with a male and female weatherpack connector on either end (P/N 12014836 and 12014837). With the distributor connector disconnected, use the jumpers to reconnect the terminals per instructions provided on the code E23 chart.

2. With only Distributor Reference and Distributor Reference Ground jumpered, engine will run at backup spark. Checking for proper ground connection between ECM and the engine and the ECM and HEI.

3. Checking for ECM providing an EST output on circuit 423.

4. Checking for HEI module able to ground EST signal with open Bypass Circuit.

1988–90 CADILLAC DEVILLE AND FLEETWOOD

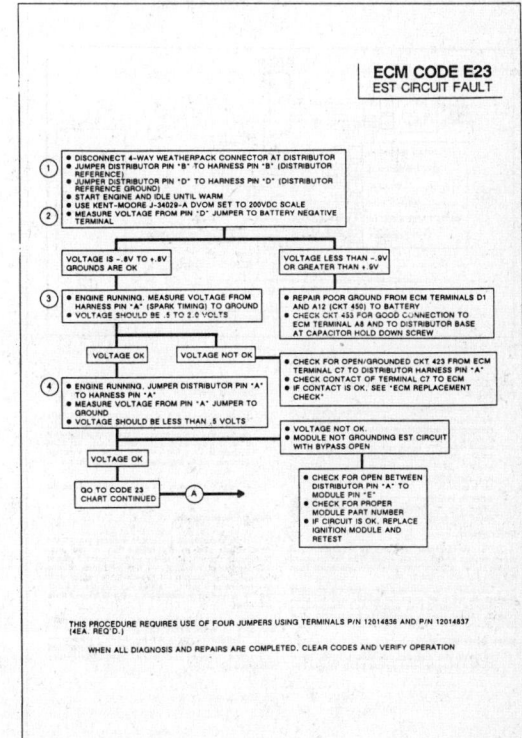

ECM CODE E23
EST CIRCUIT FAULT

GO TO CODE E23 CHART CONTINUED → A

THIS PROCEDURE REQUIRES USE OF FOUR JUMPERS USING TERMINALS P/N 12014836 AND P/N 12014837 (4EA. REQ'D.)

WHEN ALL DIAGNOSIS AND REPAIRS ARE COMPLETED, CLEAR CODES AND VERIFY OPERATION

1988–90 CADILLAC DEVILLE AND FLEETWOOD

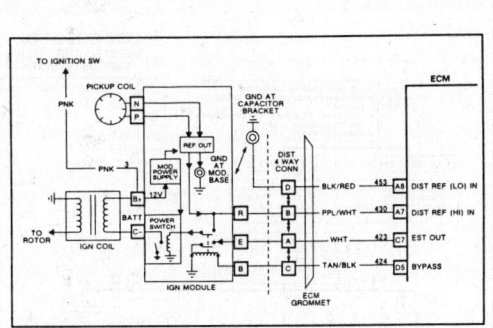

ECM CODE E23 EST CIRCUIT PROBLEM (Cont'd)

Description (Cont'd) (From Page 8D1–32):

The Distributor Reference Ground (Circuit 453) is a common ground between the ECM and the HEI. The ECM and HEI compare Bypass, EST and Reference voltages to the ground voltage on circuit 453. Circuit 453 open to ground or ECM can cause a failure of the ECM to recognize a reference pulse or of the module to recognize an EST pulse.

The ECM compares the status of the EST circuit to the status of the Bypass circuit in order to detect a fault and set code E23. If the ECM has Bypass at 5 volts, the module should accept EST information. EST voltage should be high (.5–2.8 volts). If bypass is at 0 volts, the module should ground the EST circuit and the EST voltage should be low (less than .5 volts).

Possible causes of code E23 are:

A. Circuit 423 (EST) open/ground.

B. Circuit 424 (Bypass) open/grounded.

C. Circuit 453 (Reference Ground) open to ground at HEI or ECM.

D. ECM with a poor connection to the engine through the circuit 450 grounds.

E. ECM or HEI module faults.

Notes On Fault Tree (cont'd):

5. Checking for HEI module able to recognize a voltage on the Bypass circuit and to stop grounding EST (ECM controlling timing).

6. Checking for Bypass signal to the module. If Bypass is being sent by the ECM to the HEI and if the module is interpreting Bypass voltage correctly, then the module will switch off the ground to the EST.

7. If the chart leads to "EST circuit is OK" then a fault may exist in the 4 way weatherpack. Check for proper connector mating and for pins backing out of the weatherpack. If the connector is no trouble found, reconnect the four way, clear codes and retest.

1988–90 CADILLAC DEVILLE AND FLEETWOOD

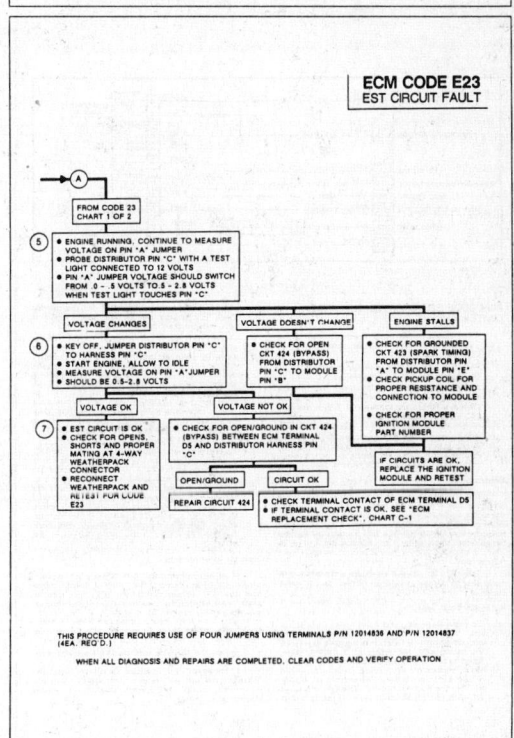

ECM CODE E23
EST CIRCUIT FAULT

THIS PROCEDURE REQUIRES USE OF FOUR JUMPERS USING TERMINALS P/N 12014836 AND P/N 12014837 (4EA. REQ'D)

WHEN ALL DIAGNOSIS AND REPAIRS ARE COMPLETED, CLEAR CODES AND VERIFY OPERATION

1988–90 CADILLAC DEVILLE AND FLEETWOOD

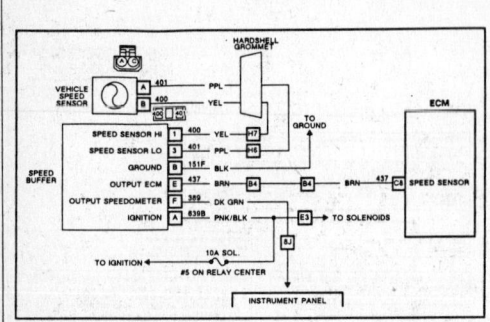

ECM CODE E24, SPEED SENSOR CIRCUIT PROBLEM

TEST CONDITIONS: 1. No problems detected on the throttle switch, fourth gear, or MAP sensor circuits; 2. Gear shift in drive or reverse; 3. Not braking (brake switch closed) 4. ISC throttle switch open; 5. MAP less than or equal to 85 KPA; 6. Fourth gear switch open (Trans in fourth gear); 7. RPM greater than or equal to 1400.

FAILURE CONDITION: Vehicle Speed = 0 MPH for 3 seconds.

ACTION TAKEN: 1. ECM turns on "Service Soon" light. 2. Disable VCC and cruise for the entire key cycle.

Description:

The speed sensor generates an electrical signal representative of the vehicle speed. The vehicle speed sensor buffer amplifies and conditions the signal from the speed sensor to the ECM.

Code E24 indicates that a speed signal is not being received by the ECM when the vehicle is in 4th gear. This can be observed in engine data display as a "vehicle speed" reading (parameter P.1.2) of 0 MPH while driving.

Notes On Fault Tree:

1. To begin the diagnosis, determine if the speed signal is being received by the ECM. This can be observed by noting the vehicle speed sensor reading (parameter P.1.2) with the drive wheels turning.

2. If the voltage is 6.5 volts or greater, check for a faulty vehicle speed sensor buffer connector or a faulty speed sensor buffer. If the voltage present is 5 to 6.5 volts, fault is most likely at ECM connector or ECM. Before ECM is replaced, perform "ECM Replacement Check", Chart C-1.

3. If output voltage is OK at the buffer, check the output at ECM terminal C8.

Note On Intermittents:

If an intermittent Code E24 is being set, lift the drive wheels and let the vehicle idle in low gear. Manipulate the related wiring while observing engine data parameter P.1.2. If the failure is induced, the "vehicle speed" reading will drop from its normal value to a reading of 0 MPH. If wiring and connectors check out OK, substitute a known good vehicle speed sensor and retest.

Review the snapshot data parameters stored with the code to verify failure condition. Refer to "Symptoms" for complete details on using the "Snapshot on code set" feature. If the failure condition is verified, check and repair intermittent wiring connection or sensor.

1988–90 CADILLAC DEVILLE AND FLEETWOOD

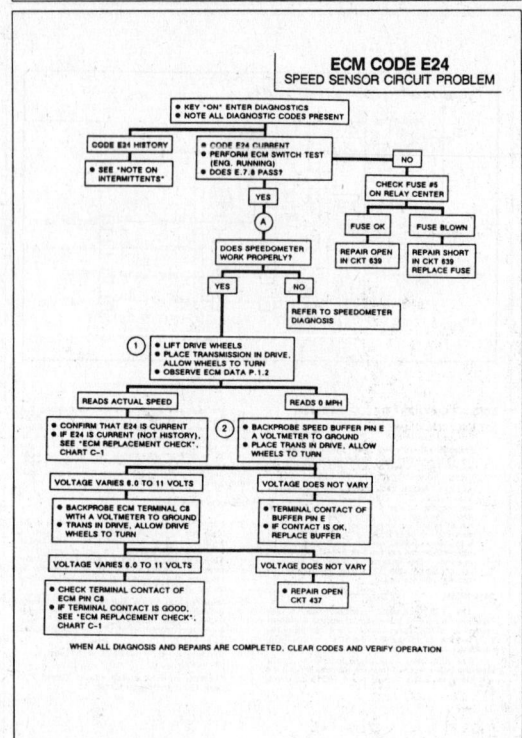

ECM CODE E24
SPEED SENSOR CIRCUIT PROBLEM

WHEN ALL DIAGNOSIS AND REPAIRS ARE COMPLETED, CLEAR CODES AND VERIFY OPERATION

1988–90 CADILLAC DEVILLE AND FLEETWOOD

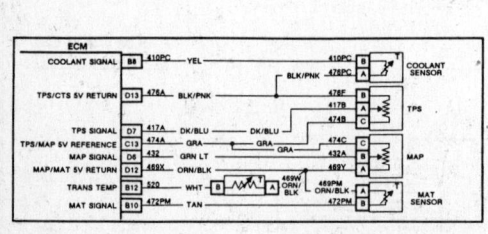

ECM CODE E26, SHORTED THROTTLE SWITCH CIRCUIT

TEST CONDITIONS: No problems detected on the TPS circuit; 2. TPS value between 15 and 83 degrees.

FAILURE CONDITION: Throttle switch input to ECM grounded for 1.5 seconds.

ACTION TAKEN: 1. ECM turns on "Service Soon" light. 2. ECM assumes throttle switch is closed if brakes are applied and TPS is less than 13°. 3. ECM assumes throttle switch is open if brakes are not applied or TPS is greater than 13°.

NOTE: With Code E26 stored hard, the throttle switch status light ("Off") displays the failsoft throttle switch status. To display the actual switch status, enter diagnostics and clear codes.

Description:

The throttle switch is part of the ISC motor assembly. Pin B of the ISC motor 4 way weather pack connector is the throttle switch input to the ECM. The ECM provides a 5 volt signal to pin B of the ISC. When the throttle lever contacts the ISC plunger, the throttle switch closes, shorting the 5 volt signal at pin B to pin A which is a ground (closed throttle = throttle switch input voltage low).

Code E26 is designed to detect a throttle switch input that is always grounded, throttle switch never opens.

Notes On Fault Tree:

1. To begin diagnosis, clear ECM codes and observe the "throttle switch" status light (the "Off" annunciator on the CCP). When a code is present, the status light will represent a failsoft value. Therefore, it is important that any codes present are cleared before proceeding. Diagnostics should not be exited since the ECM will not set a code while in the diagnostic mode. The light should go off upon opening of the throttle. If the light does not go off, refer to the diagnostic decision tree for Code E.7.2 since the throttle switch signal is remaining in the "closed" mode.

2. If the switch test E.7.2 passes, check the TPS for proper operation. If the sensor is free and correctly adjusted, enter diagnostics and display engine data parameter P.0.1. Under the condition of a Code E26 the "throttle angle" reading will be greater than 13 degrees at closed throttle. If the opposite condition (low TPS signal) can be created, then self-diagnostics can be used to detect this failure by displaying engine data parameter P.0.1. With the opposite condition created, a "throttle angle" reading of –10 to –7 degrees should be displayed.

3. To create a low sensor signal, disconnect the TPS connector. If the "throttle angle" reading is –10 to –7 degrees, circuit #476-A must be checked for an open. If the voltage between circuits #474 and #476-A is 0 volts, circuit #476-A is open. If the voltage is 5 volts, the circuit and the wiring are OK. The fault would be improper mating of sensor terminals or a faulty TPS sensor. If the "throttle angle" reading is –6 degrees or greater, then there is a short to voltage in the wiring harness or the ECM.

Note On Intermittents:

If an intermittent Code E26 is being set, operate the vehicle while observing the "throttle switch" status light (the "Off" annunciator on the CCP) and "throttle angle" parameter P.0.1 simultaneously. If the throttle switch ever fails to open upon opening of the throttle, investigate that circuit for a malfunction. If the "throttle angle" ever jumps above 13 degrees with the throttle closed, investigate the TPS circuits for a malfunction.

Review the snapshot data parameters stored with the code to verify failure condition. Refer to "Symptoms" for complete details on using the "Snapshot on code set" feature. If the failure condition is verified, check and repair intermittent wiring connection or sensor.

1988–90 CADILLAC DEVILLE AND FLEETWOOD

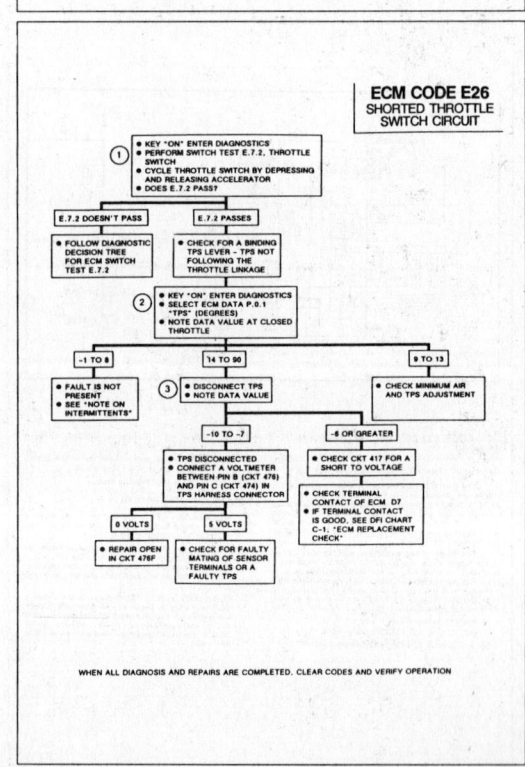

ECM CODE E26
SHORTED THROTTLE SWITCH CIRCUIT

WHEN ALL DIAGNOSIS AND REPAIRS ARE COMPLETED, CLEAR CODES AND VERIFY OPERATION

1988–90 CADILLAC DEVILLE AND FLEETWOOD

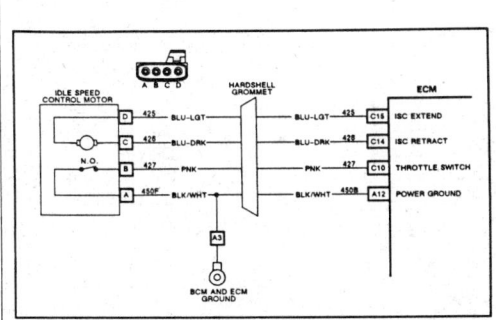

ECM CODE E27, OPEN THROTTLE SWITCH CIRCUIT

TEST CONDITIONS: Always tested.
FAILURE CONDITIONS: 1. Throttle switch was never closed during the last key cycle. 2. Brake switch was applied with engine running during last ignition key cycle. 3. Throttle switch has not closed in the current key cycle.
ACTION TAKEN: 1. ECM assumes throttle switch is closed if TPS is less than 13° and brake switch is open (braking). 2. ECM assumes that the throttle switch is open if TPS is greater than 13° or the brake switch is closed (not braking).
Note: When Code E27 is stored hard, the throttle switch status light ("Off") will display the failsoft or substituted throttle switch status. To display actual throttle switch status, enter diagnostics and clear codes.

Description:

The throttle switch is part of the ISC motor assembly. Pin B of the ISC motor 4-way weatherpack connector is the throttle switch input to the ECM. The ECM provides a 5 volt signal to pin B of the ISC. When the throttle lever contacts the ISC plunger, the throttle switch closes, shorting the 5 volt signal at pin B to pin A which is a ground. (Closed throttle = throttle switch input voltage low).

Notes On Fault Tree:

1. If throttle switch test E.7.2 doesn't pass, use chart E.7.2 to diagnose.
2. If throttle switch E.7.2 cycles, E27 should not remain as a hard code.

Notes On Intermittents:

Check for throttle shaft and throttle blades binding, throttle spring weak or distorted, cruise or TV cable binding throttle linkage, TPS for proper installation.

Intermittent Code E27 can be induced by resting foot on the accelerator during a key "On", key "Off" for 10 seconds, key "On", cycle.

Check for intermittent open on CKT 427 between ECM and ISC.

1988–90 CADILLAC DEVILLE AND FLEETWOOD

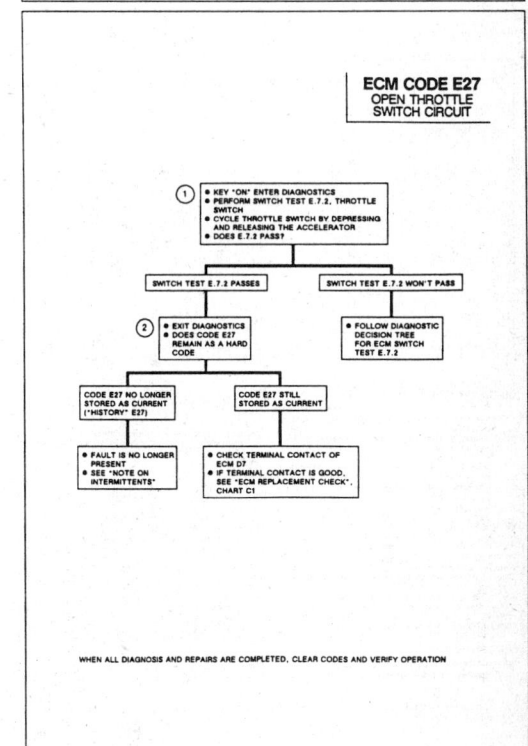

ECM CODE E27
OPEN THROTTLE SWITCH CIRCUIT

WHEN ALL DIAGNOSIS AND REPAIRS ARE COMPLETED, CLEAR CODES AND VERIFY OPERATION

1988–90 CADILLAC DEVILLE AND FLEETWOOD

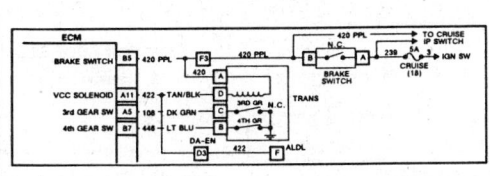

ECM CODE E28, OPEN THIRD OR FOURTH GEAR CIRCUIT

TEST CONDITIONS: 1. Code E24 clear; 2. Transmission in park or neutral; 3. Vehicle speed less than or equal to 4 MPH.
FAILURE CONDITIONS: Third gear input to ECM not grounded or Fourth gear input to ECM not grounded.
ACTION TAKEN: 1. ECM turns on "Service Soon" lamp.

Description:

The third and fourth gear switches in the transmission are normally closed switches that are opened by transmission oil pressure. The ECM sends a 12 volt signal to each of the switches. When not in third or fourth gear, the third and fourth gear inputs are grounded by the normally closed switches. When third gear is achieved, the third gear switch is opened by third gear oil pressure and the third gear input changes from 0 volts to 12 volts. When fourth gear is achieved, the fourth gear switch is opened by fourth gear oil pressure and the fourth gear input changes from 0 to 12 volts. Code E28 is designed to detect a false third or fourth gear indication to ECM.

Notes On Fault Tree:

1. To begin diagnosis, observe the fourth gear status light (the "rear defogger" annunciator on the CCP) to determine if the failure resulted from the third gear circuit or fourth gear circuit. If the 4th gear status light is off, the fault is the 3rd gear switch, proceed to step 6.
2. If the fourth gear status light is on, determine if the ECM can recognize the grounding of circuit #438 at the transmission connector. If the "fourth gear" status light remains on, then there is an open in either circuit #438 or the ECM. If the light remains on with circuit #438 grounded at the ECM then either the ECM connector or ECM is faulty. If the light goes off, then circuit #438 is open.
3. If the status light is on with Pin B grounded, then the ECM is functioning properly and the fault must be in the transmission. If the test light on pin B at transmission does not light, then an open must exist in the wiring of third gear switch. If the test light fails to light on the fourth gear switch terminal, then it is open and should be replaced. If the test light does light, the open is between the transmission connector and fourth gear switch connector.

4. If the test light on pin B lights, then the fourth gear switch is working properly. If the test light goes out when the engine is started, then the switch is receiving oil pressure and should not be.
5. If the test light stays on, then the transmission connector should be reconnected and the "fourth gear" status light should be checked for proper operation. If the light is on, then the fault must be improper mating of circuit #446 terminals at pin B of the transmission connector. If the light is off, then the malfunction is not present at this time.
6. If the fourth gear status light was off without Pin B grounded, determine if the ECM can recognize the grounding of circuit #422 at the transmission connector. If the "Service Soon" telltale lamp is on when out of the diagnostic mode, then there is an open in either circuit #422 or the ECM. If the light remains on with circuit #422 grounded at the ECM then either the ECM connector or the ECM is faulty. If the light goes off, then circuit #422 is open.
7. If the service soon light is off with Pin B and Pin C grounded, then the ECM is functioning properly and the fault must be in the transmission. If the test light on Pin C at the transmission does not light, then an open must exist in the wiring of third gear switch. If the test light fails to light on the third gear switch terminal, then it is open and should be replaced. If the test light does light, the open is between the transmission connector and third gear switch connector.
8. If the test light on Pin C lights, then the third gear switch is working properly. If the test light goes out when the engine is started, then the switch is receiving oil pressure and should not be.
9. If the test light stays on, then the transmission connector should be reconnected and the "Service Soon" light should be checked for proper operation. If the light is on, then the fault must be improper mating of circuit #422 terminals at Pin C of transmission connector. If the light is off, then the malfunction is not present at this time.

Note On Intermittents:

At key "ON", engine "OFF", trans in park, manipulate affected wiring and connectors while observing "Service Soon" light. "Service Soon" flashes on, repair intermittent open in CKT 422 (third gear) or CKT 438 (fourth gear).

1988–90 CADILLAC DEVILLE AND FLEETWOOD

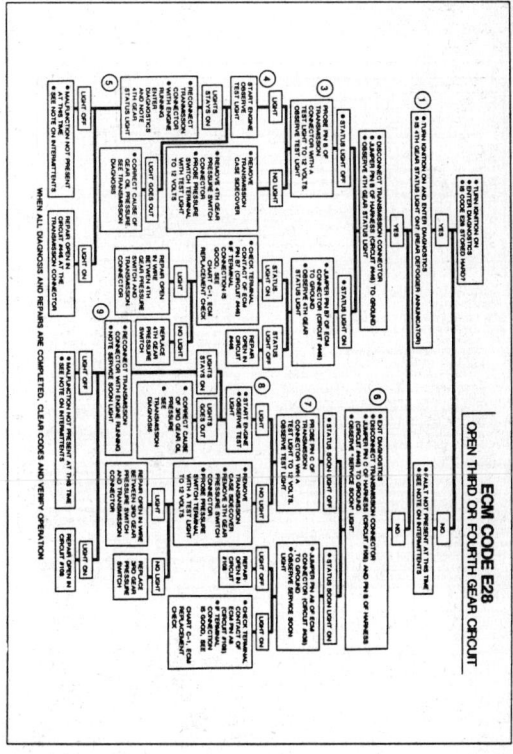

ECM CODE E28
OPEN THIRD OR FOURTH GEAR CIRCUIT

WHEN ALL DIAGNOSIS AND REPAIRS ARE COMPLETED, CLEAR CODES AND VERIFY OPERATION

1988–90 CADILLAC DEVILLE AND FLEETWOOD

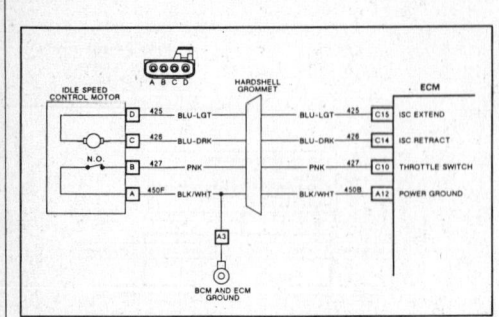

ECM CODE E30, IDLE SPEED CONTROL CIRCUIT PROBLEM

TEST CONDITIONS: 1. No problems detected on the TPS, vehicle speed, or throttle switch circuits; 2. Speed = 0; 3. Throttle switch closed; 4. Battery voltage greater than or equal to 11.0 volts; 5. RPM is greater than 152 RPM different than desired; 6. Not a cold start.

FAILURE CONDITIONS: ISC plunger extending with TPS less than 12.75 or ISC retracting with TPS greater than 1.7 for 15 seconds.

ACTION TAKEN: 1. ECM turns on "Service Soon" light for entire key cycle.

Description:

The ECM controls engine idle by increasing or decreasing the throttle opening using the idle speed control motor (ISC).

The ISC will be active in controlling idle speed at any time the throttle switch is closed. At vehicle speeds less than 6 MPH, the ECM controls idle speed on engine RPM. When vehicle speed is greater than 6 MPH, the ECM controls idle based primarily on throttle opening (TPS) with adjustments to maintain a minimum engine RPM.

Code E30 is designed to detect engine RPM out of limits high or low.

Notes On Fault Tree:

1. Checking for TPS misadjusted.
2. Checking for proper throttle switch, power steering switch, park/neutral switch operation. The ECM must receive accurate switch status information in order to control idle.

3. Checking for proper ISC motor operation.
4. Many engine fuel and emissions system faults may cause unstable idle. If the base engine idle is not steady, the ISC may not be able to control idle to within 150 RPM of commanded.
5. Out of adjustment minimum air rate, ISC or TPS can cause ISC not to control idle or code E30 to set falsely. Check adjustment.

Notes On Intermittents:

Display ECM parameter P.0.1, manipulate TPS wiring and connectors, watch for TPS value to jump or skip.

Enter output cycling and manipulate ISC wiring and connectors, watch for ISC to stop cycling.

Enter diagnostics and observe throttle switch status light – check for binding throttle linkage or throttle return spring weak.

Verify that minimum air RPM, ISC and TPS are set to specification.

If idle is unstable, see "Symptoms" for "Rough, Unstable, Incorrect Idle, Stalling".

A Code E30 may be stored along with a Code E19 – Shorted Fuel Pump Circuit. If Code E30 no longer appears after Code E19 has been cleared, do not investigate any further. The voltage on the fuel pump feedback circuit made the ECM improperly test Code E30.

Review the snapshot data parameters stored with the code to verify failure condition. Refer to "Symptoms" for complete details on using the "Snapshot on code set" feature. If the failure condition is verified, check and repair intermittent wiring connection or sensor.

1988–90 CADILLAC DEVILLE AND FLEETWOOD

ECM CODE E30
IDLE SPEED CONTROL CIRCUIT PROBLEM

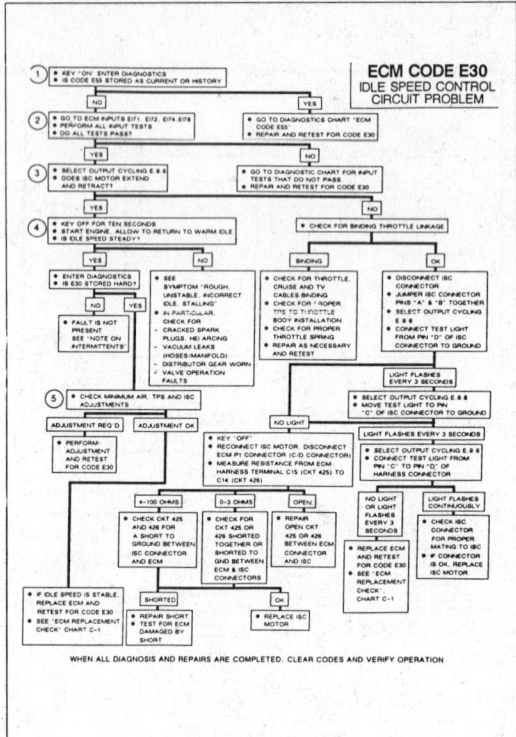

WHEN ALL DIAGNOSIS AND REPAIRS ARE COMPLETED, CLEAR CODES AND VERIFY OPERATION

1988–90 CADILLAC DEVILLE AND FLEETWOOD

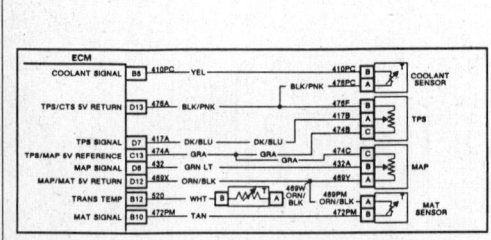

ECM CODE E31, SHORTED MANIFOLD ABSOLUTE PRESSURE SENSOR CIRCUIT

TEST CONDITIONS: Always tested.

FAILURE CONDITIONS: MAP value greater than 107 KPA (106–109 kPa) for at least 2 seconds.

ACTION TAKEN: 1. ECM turns on "Service Now" light; 2. Turn off air management solenoids; 3. ECM uses a substitute MAP sensor value from a table of substitute MAP vs RPM when throttle switch is closed; 4. ECM uses a substitute MAP sensor value of 81.8 kPa with open throttle switch; 5. ECM assumes BARO value is 92.2 kPa.

Description:

The MAP sensor output signal voltage is a DC voltage that varies with manifold absolute pressure. As MAP decreases, voltage decreases (low engine load, high vacuum). As MAP increases, voltage increases (high engine load, low vacuum).

The ECM uses MAP sensor values as an indicator of engine load. A high MAP reading indicates heavy load and low MAP reading indicates low load.

Code E31 is designed to set when the ECM detects a MAP sensor signal out of limits high – MAP signal at 4.85 volts or more, MAP value greater than 105 kPa.

Notes On Fault Tree:

1. If ECM parameter P.0.2 goes to 14–16 kPa with the sensor unplugged, the fault is at the MAP sensor or sensor connector.
2. Checking for an open circuit from Pin A of the sensor connector to ECM Pin D12 (ground). If the ground is open, the sensor can't divide reference voltage to make the signal voltage vary – the signal voltage is always high.

Note On Intermittents:

Code E31 can be set by 1. short to voltage on CKT 432; 2. Open from sensor to ECM on CKT 476;3. Defective MAP sensor.

Manipulate affected wiring and connectors while observing ECM parameter P.0.2. Apply and release vacuum to the MAP sensor vacuum part using a vacuum source. If the ECM parameter P.0.2 displays greater than 105 kPa, the condition has been induced and the cause of the intermittent should be repaired. If wiring and connectors check out OK, substitute a known good MAP sensor and retest.

Review the snapshot data parameters stored with the code to verify failure condition. Refer to "Symptoms" for complete details on using the "Snapshot on code set" feature. If the failure condition is verified, check and repair intermittent wiring connection or sensor.

1988–90 CADILLAC DEVILLE AND FLEETWOOD

NOTE: IF A HARD CODE E31 IS ACCOMPANIED BY A HARD CODE E36, SEE DFI CHART A. MULTIPLE ECM CODES STORED HARD.

ECM CODE E31
SHORTED MAP SENSOR CIRCUIT

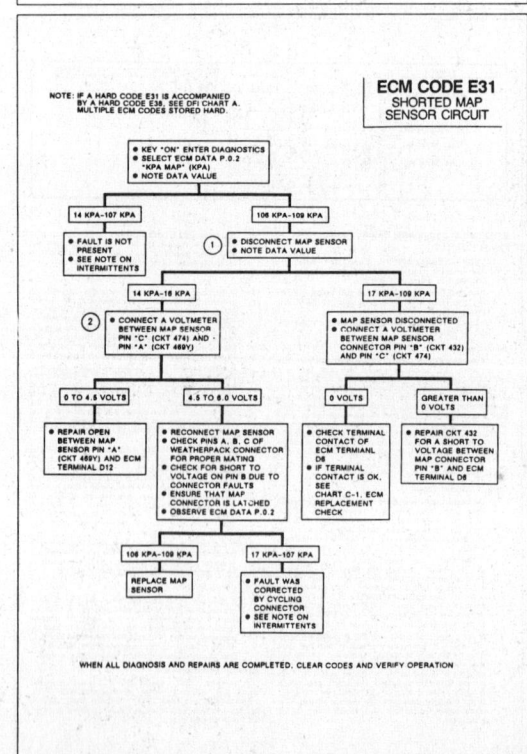

WHEN ALL DIAGNOSIS AND REPAIRS ARE COMPLETED, CLEAR CODES AND VERIFY OPERATION

1988–90 CADILLAC DEVILLE AND FLEETWOOD

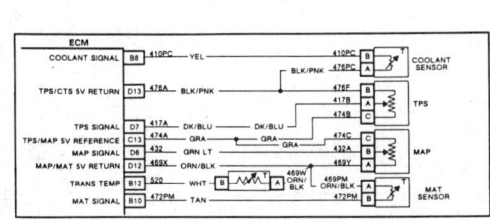

ECM CODE E32, OPEN MANIFOLD ABSOLUTE PRESSURE SENSOR CIRCUIT

TEST CONDITIONS: 1. With ISC throttle switch closed, RPM less than or equal to 700, TPS less than or equal to 13°; 2. With ISC throttle switch open, TPS greater than or equal to 12°.
FAILURE CONDITIONS: MAP value less than 16 kPa.
ACTION TAKEN: 1. ECM turns on "Service Now" light. 2. Turn off both air solenoids; 3. ECM uses a substitute MAP sensor value of 81.8 kPa with open throttle switch; 4. ECM uses a substitute MAP value from a table of substitute MAP vs RPM when throttle switch is closed; 5. ECM uses a substitute BARO value of 92.2 kPa.

Description:

The MAP sensor output voltage is a DC voltage that varies with Manifold Absolute Pressure (MAP). As MAP decreases, voltage on CKT 432 decreases (low engine load, high engine vacuum). As MAP increases, voltage on CKT 432 increases (high engine load, low engine vacuum).

Code E32 is designed to set when the ECM detects a MAP sensor signal out of limits low – MAP signal at 16 kPa or less, MAP voltage at .08 volts or less.

Notes On Fault Tree:

1. Checking for circuitry from sensor to ECM able to respond to a 5 volt signal on MAP input. A

reading of 106–109 means wiring and ECM are OK.
2. Checking for 5 volt reference present at sensor connector.
3. Checking for CKT 432 shorted to ground.
4. Checking for ECM able to respond to a 5 volt signal voltage on MAP input.
5. Fault is most likely at ECM connector or ECM. Before ECM is replaced, perform "ECM Replacement Check", Chart C-1.

Note On Intermittents:

Code E32 can be set by 1. Open 5 volt reference between ECM and sensor; 2. Open in MAP signal between sensor and ECM; 3. Defective MAP sensor.

Manipulate affected wiring and connectors while observing ECM parameter P.0.2. Apply and release vacuum to the MAP sensor vacuum port using a vacuum source. If the ECM parameter P.0.2 displays less that 17 kPa, the condition has been induced and the cause of the intermittent can be repaired. If wiring and connectors check out OK, substitute a known good MAP sensor and retest.

Review the snapshot data parameters stored with the code to verify failure condition. Refer to "Symptoms" for complete details on using the "Snapshot on code set" feature. If the failure condition is verified, check and repair intermittent wiring connection or sensor.

1988–90 CADILLAC DEVILLE AND FLEETWOOD

NOTE: IF A HARD CODE E32 IS ACCOMPANIED BY A HARD CODE E22, SEE DFI CHART A. MULTIPLE ECM CODES STORED HARD.

ECM CODE E32
OPEN MAP SENSOR CIRCUIT

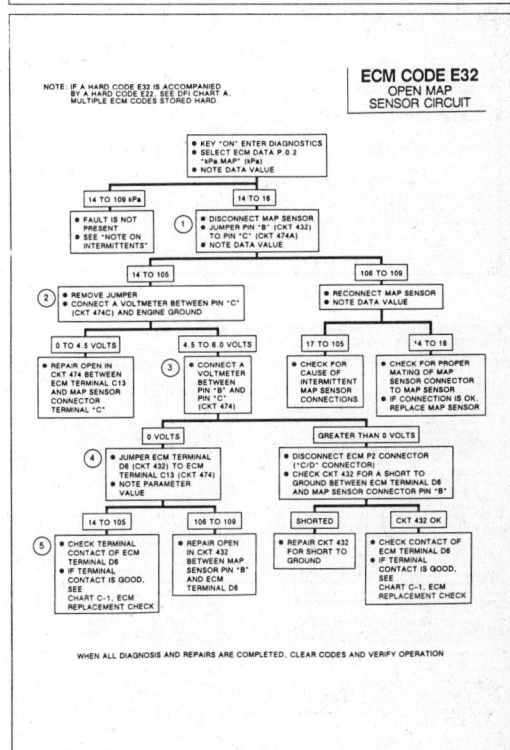

WHEN ALL DIAGNOSIS AND REPAIRS ARE COMPLETED, CLEAR CODES AND VERIFY OPERATION

1988–90 CADILLAC DEVILLE AND FLEETWOOD

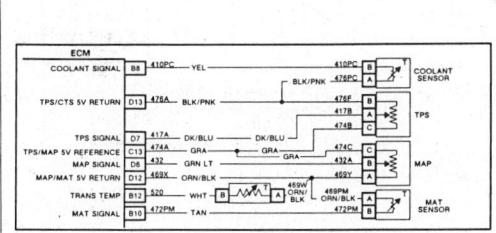

ECM CODE E34, MAP SIGNAL TOO HIGH

TEST CONDITIONS: 1. Codes E21 E22, E26, E27, E31 and E32 clear; 2. Key on; 3. RPM greater than or equal to 400; 4. Throttle switch closed; 5. TPS less than or equal to 12 degrees; 6. BARO greater than or equal to 68 kPa.
FAILURE CONDITIONS: MAP value within 3.3 kPa of the BARO value for 2 seconds.
ACTION TAKEN: 1. The ECM turns on the "Service Now" light. 2. ECM turns off both air management solenoids; 3. ECM uses a substitute MAP sensor value of 81.8 kPa with throttle switch open; 4. ECM looks up a table of "Substitute MAP" value from a table of "Substitute MAP" vs. RPM when the throttle switch is closed.

Description:

The MAP sensor output voltage is a DC voltage that varies with Manifold Absolute Pressure (MAP). As MAP decreases, voltage on CKT 432 decreases (low engine load, high engine vacuum). As MAP increases, voltage on CKT 432 increases (high engine load, low engine vacuum).

The ECM uses the MAP sensor reading as an indicator of engine load. Low MAP readings mean low engine load, high MAP readings mean high engine load.

Code E34 is designed to set when the ECM detects a MAP sensor signal that is much too high for the closed throttle test conditions. Code E34 usually indicates that there is a fault in the vacuum supply to the MAP sensor.

Notes On Fault Tree:

1. MAP at idle should be 30–60 kPa, depending on engine load. BARO should be 85–105, kPa, depending upon altitude.

2. Check for vacuum at the MAP sensor hose with a vacuum gage. At idle, typical vacuum readings are 15"–20" Hg. depending on engine load.
3. Vacuum supply to sensor is OK, checking for MAP sensor or MAP circuitry fault.
4. Checking for sensor ground open from sensor to ECM.
5. Checking for short to voltage on sensor signal, CKT 432.
6. Fault is most likely at ECM connector or ECM. Before ECM is replaced, perform "ECM Replacement Check", Chart C-1.

Code E34 is usually set by a vacuum supply problem to the MAP sensor. Check for proper vacuum routing (see DFI vacuum routing diagrams) for MAP hose connected to the proper throttle body port, MAP hose chafed, pinched, or cut.

Apply vacuum to the MAP hose at the throttle body and look for vacuum leaks in the MAP hose or MAP sensor.

Manipulate affected wiring and connections while observing ECM parameter P.0.2. Apply and release vacuum to the MAP hose at the throttle body using a vacuum source. If the ECM parameter P.0.2 jumps or skips high with vacuum applied, the condition has been induced and the cause of the intermittent can be repaired. If wiring and connectors check out OK, substitute a known good MAP sensor and retest.

Review the snapshot data parameters stored with the code to verify failure condition. Refer to "Symptoms" for complete details on using the "Snapshot on code set" feature. If the failure condition is verified, check and repair intermittent wiring connection or sensor.

1988–90 CADILLAC DEVILLE AND FLEETWOOD

ECM CODE E34
MAP SIGNAL TOO HIGH

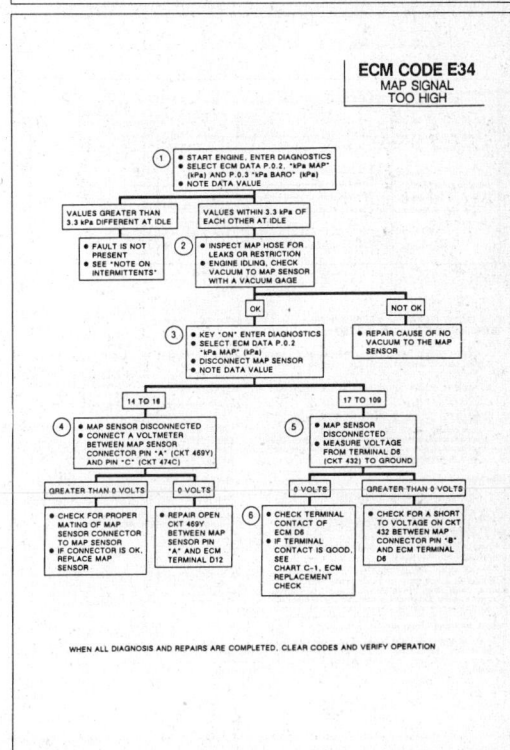

WHEN ALL DIAGNOSIS AND REPAIRS ARE COMPLETED, CLEAR CODES AND VERIFY OPERATION

1988–90 CADILLAC DEVILLE AND FLEETWOOD

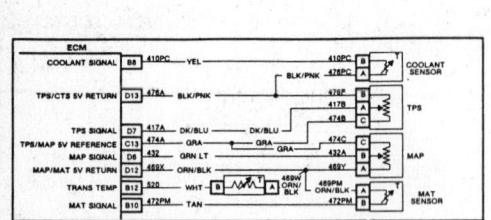

ECM CODE E37, SHORTED MANIFOLD AIR TEMPERATURE SENSOR CIRCUIT

TEST CONDITIONS: 1. No problems detected on the coolant sensor circuit; 2. Coolant sensor less than or equal to 100°C.

FAILURE CONDITIONS: MAT sensor value greater than or equal to 144°C.

ACTION TAKEN: 1. ECM turns on "Service Soon" light. 2. ECM turns off both air management solenoids; 3. ECM substitutes 40°C for MAT when coolant is greater than or equal to 40°C; 5. ECM uses coolant temp for MAT when coolant temp is less than or equal to 40°C.

Description:

The MAT sensor is a thermistor or variable resistor that varies in resistance with temperature. The MAT sensor is a "two wire" sensor with a signal voltage coming from the ECM to sensor pin B, CKT 472 and a sensor reference ground on pin A, CKT 469pm.

As the temperature of the sensor increases, sensor resistance is lower. The signal voltage from the ECM to pin B decreases as the sensor resistance decreases as more current flows from pin B through the sensor element to pin A, sensor ground (voltage is dropped across the sensor element). High temperature means low signal voltage on CKT 472 and low temperature means high signal voltage on CKT 472.

Code E37, shorted MAT sensor sets because the ECM assumes that manifold air temperature can't be 144°C or greater with a coolant of less than 190°C.

Notes On Fault Tree:

1. With a shorted sensor, ECM parameter P.0.5 should read 144°C or greater, if not, then the sensor is not shorted, see note on intermittents.

2. Checking for sensor shorted or CKT 472 shorted. If ECM parameter P.0.5 stays at 144-151°C with the sensor unplugged, then the short is in CKT 472.

3. Fault is most likely at ECM connector or ECM. Before ECM is replaced, perform "ECM Replacement Check", Chart C-1.

4. MAT sensors can be damaged by a backfire in the intake. If this vehicle has had more than one MAT sensor replaced, check for signs of backfire or high intake manifold temperatures due to improper valve train operation.

Notes On Intermittents:

Manipulate CKT 472 wiring, the MAT sensor and ECM connector while observing ECM parameter P.0.5. If the failure is induced, the manifold air temperature will jump from its normal value to the shorted reading of 144-151°C. If wiring and connectors check out OK, substitute a known good MAT sensor and retest.

Review the snapshot data parameters stored with the code to verify failure condition. Refer to "Symptoms" for complete details on using the "Snapshot on code set" feature. If the failure condition is verified, check and repair intermittent wiring connection or sensor.

1988–90 CADILLAC DEVILLE AND FLEETWOOD

ECM CODE E37
SHORTED MAT SENSOR CIRCUIT

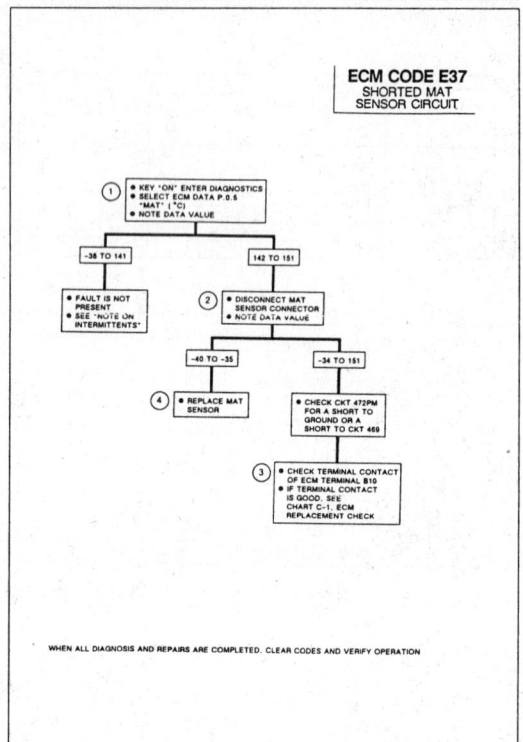

WHEN ALL DIAGNOSIS AND REPAIRS ARE COMPLETED, CLEAR CODES AND VERIFY OPERATION

1988–90 CADILLAC DEVILLE AND FLEETWOOD

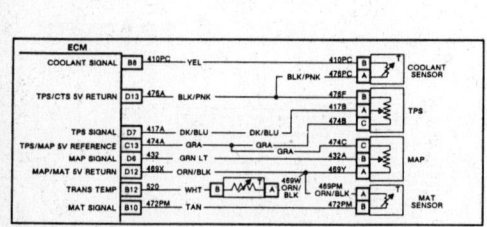

ECM CODE E38, OPEN MANIFOLD AIR TEMPERATURE SENSOR CIRCUIT

TEST CONDITIONS: 1. No problems detected on the coolant sensor circuit; 2. Coolant sensor greater than or equal to -2.5°C.

FAILURE CONDITIONS: MAT sensor value less than -32°C.

ACTION TAKEN: 1. ECM turns on "Service Soon" 2. ECM turns off both air management solenoids; 3. ECM uses 40°C for a MAT value when coolant temp is greater than 40°C; 4. ECM substitutes coolant temp for MAT when coolant is less than 40°C.

Description:

The MAT sensor is a thermistor or variable resistor that varies in resistance with temperature. The MAT sensor is a "two wire" sensor with a reference or signal voltage coming from the ECM to sensor pin B, CKT 472 and a sensor reference ground on pin A, CKT 469pm. As the temperature of the sensor decreases, sensor resistance increases.

The signal voltage from the ECM to pin B increases as sensor temperature decreases because less current flows from sensor pin B through the sensor element to ground (less of the signal voltage is dropped across the sensor element). Low temperature means high signal voltage on CKT 472 and high temperature means low signal voltage on CKT 472.

Code E38, open MAT sensor, sets because the ECM assumes that the manifold air temperature can't be -32 to -40°C with a coolant temp of -2.5°C or greater.

Notes On Fault Tree:

1. If the sensor is open, ECM parameter P.0.5 should read -32°C or less. If not, then sensor signal is not open at this time, see "Notes on Intermittents".

2. Checking ECM and sensor circuitry from ECM to sensor connector. If ECM parameter P.0.5 reads 148°-151°C with connector pin B shorted to pin A then the sensor circuits and ECM are OK.

3. Checking for open sensor ground.

4. Checking ECM's ability to recognize short to ground on MAT input.

5. Fault is most likely at ECM connector of ECM. Before ECM is replaced, perform "ECM Replacement Check", Chart C-1.

6. MAT sensor can be damaged by backfire in the intake or by excessive intake heat due to valve train faults. If the vehicle has had multiple MAT sensor replacements, check for signs of backfire or high intake manifold air temperature due to improper valve train operation.

Notes On Intermittents:

Manipulate CKT 472 and 469 wiring, the MAT connector and ECM connector while observing ECM parameter P.0.5. If failure is induced, the MAT will jump from its normal value to the open signal circuit reading of -32 to -40°C. If wiring and connectors check out OK, substitute a known good MAT sensor and retest.

Review the snapshot data parameters stored with the code to verify failure condition. Refer to "Symptoms" for complete details on using the "Snapshot on code set". If the failure condition is verified, check and repair intermittent wiring connection or sensor.

1988–90 CADILLAC DEVILLE AND FLEETWOOD

ECM CODE E38
OPEN MAT SENSOR CIRCUIT

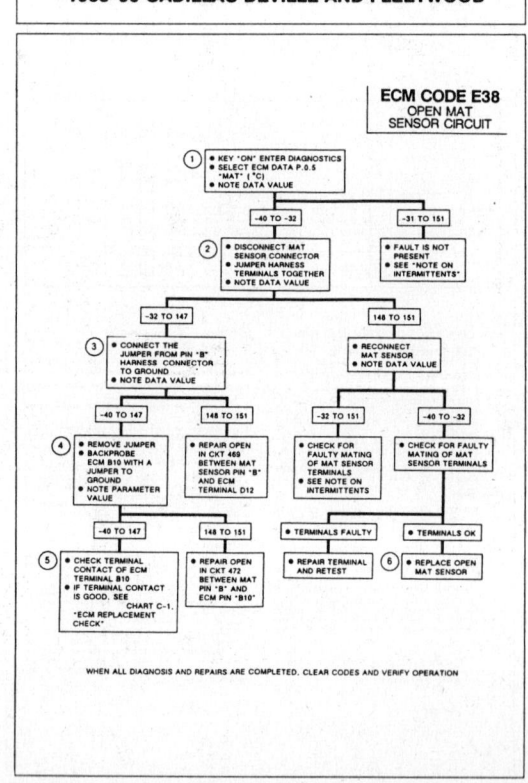

WHEN ALL DIAGNOSIS AND REPAIRS ARE COMPLETED, CLEAR CODES AND VERIFY OPERATION

1988–90 CADILLAC DEVILLE AND FLEETWOOD

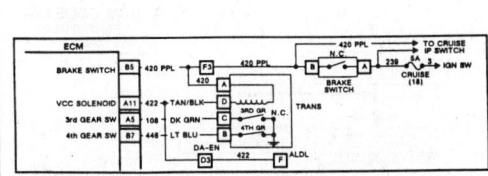

ECM CODE E39, VCC ENGAGEMENT PROBLEM

TEST CONDITIONS: 1. No problems detected on the fourth gear circuit; 2. Trans in fourth gear; 3. Throttle switch open; 4. Closed loop; 5. Not accelerating or decelerating; 6. Trans not in park or neutral; 7. Not braking; 8. VCC engaged. 9. MAP ≤ 80 kPa.

FAILURE CONDITIONS: Engine RPM too high for vehicle speed — ECM has a 9 place table of speed vs RPM. If RPM exceeds the RPM for the speed listed in table for 8 seconds, Code E39 will set.

ACTION TAKEN: 1. ECM turns on the "Service Soon" lamp.

Description:

The viscous converter clutch solenoid is turned on when the ECM applies ground to circuit 422, "VCC Solenoid Circuit". The power for the VCC solenoid comes from the 5A "Cruise" fuse in the fuse block, flows through the VCC brake switch to pin A of five-way connector on the transmission. At pin A, current flows into the transmission to the VCC solenoid and then to the transmission overtemp switch.

The transmission overtemp switch is normally closed and opens when the transmission sump temperature is 211 °C or greater, to disable VCC and allow the transmission to cool. From the overtemp switch, current returns to ground through the ECM.

Code E39 is set when the engine RPM exceeds the normal value for a given speed with VCC applied. With VCC applied, the engine is coupled directly to the transmission through the VCC — almost no slippage should occur, so the RPM/speed ratio is nearly constant. If RPM is too high at a given speed, the ECM diagnoses the condition as the VCC is slipping or not applied and code E39 logged.

Notes On Fault Tree:

1. Checking for continuity from the fuse through the solenoid on CKT 422.
2. Checking ECM's ability to ground the solenoid.
3. Circuit is OK, checking for transmission hydraulic system faults.

Notes On Intermittents:

The vehicle must be driven for 5 seconds in the test conditions for code E39 to occur. This may necessitate an extended test drive, at least 15 minutes of operation at above 45 MPH (50 MPH if BARO ≤ 92 kPa).

Review the snapshot data parameters stored with the code to verify failure condition. Refer to "Symptoms" for complete details on using the "Snapshot on code set" feature. If the failure condition is verified, check and repair intermittent wiring connection or sensor.

1988–90 CADILLAC DEVILLE AND FLEETWOOD

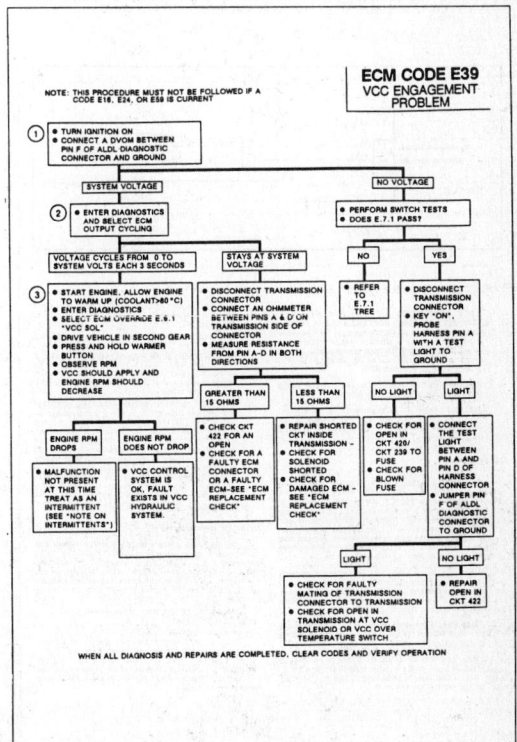

ECM CODE E39 VCC ENGAGEMENT PROBLEM

NOTE: THIS PROCEDURE MUST NOT BE FOLLOWED IF A CODE E16, E24, OR E59 IS CURRENT

WHEN ALL DIAGNOSIS AND REPAIRS ARE COMPLETED, CLEAR CODES AND VERIFY OPERATION

1988–90 CADILLAC DEVILLE AND FLEETWOOD

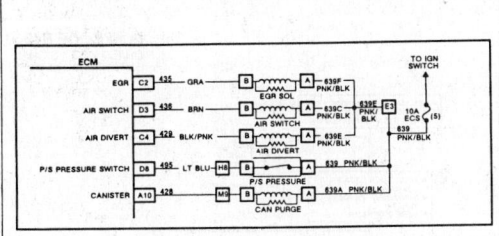

ECM CODE E40 OPEN POWER STEERING PRESSURE CIRCUIT

TEST CONDITIONS: 1. Vehicle speed greater than or equal to 45 MPH.

FAILURE CONDITIONS: Power steering switch opens (ECM sees 0 volts on pin D8, steering input) for ten seconds.

ACTION TAKEN: 1. ECM turns on "Service Soon" light.

Description:

The power steering pressure switch is normally closed and opens with power steering pressure. The power steering pressure switch receives 12 volts from the 10 AMP solenoid fuse on CKT 639 and sends the 12 volts signal to the ECM on CKT 495.

When high power steering pressures occur, the switch contacts are opened by power steering oil pressure and CKT 495 voltage is read by the ECM as 0 volts.

Note On Intermittents:

Key "ON", backprobe ECM terminal D8 to ground with a 12 volt test light. The light should remain "ON" unless steering is turned to full lock. Manipulate power steering pressure switch connector, CKT 495 wiring and ECM D8 connector while observing the test light. If the light goes out, repair intermittent open or short to ground. If wiring and connectors check out OK, substitute a known good power steering switch and retest.

Review the snapshot data parameters stored with the code to verify failure condition. Refer to "Symptoms" for complete details on using the "Snapshot on code set" feature. If the failure condition is verified, check and repair intermittent wiring connection or sensor.

1988–90 CADILLAC DEVILLE AND FLEETWOOD

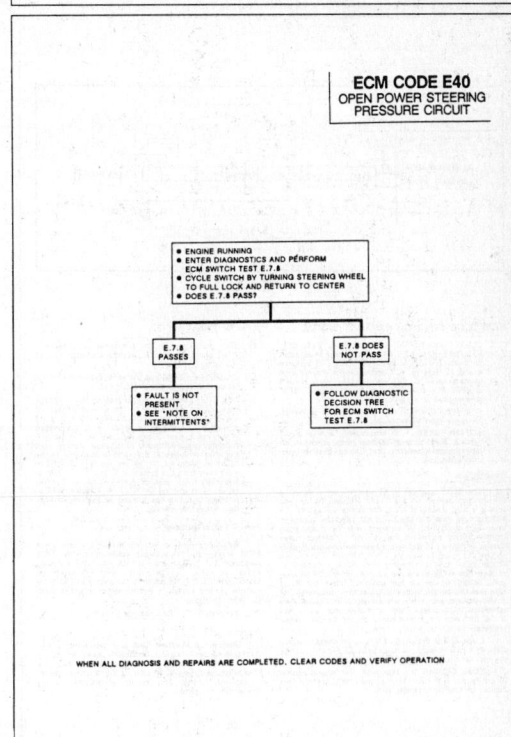

ECM CODE E40 OPEN POWER STEERING PRESSURE CIRCUIT

- ENGINE RUNNING
- ENTER DIAGNOSTICS AND PERFORM ECM SWITCH TEST E.7.8
- CYCLE SWITCH BY TURNING STEERING WHEEL TO FULL LOCK AND RETURN TO CENTER
- DOES E.7.8 PASS?

E.7.8 PASSES	E.7.8 DOES NOT PASS
• FAULT IS NOT PRESENT • SEE "NOTE ON INTERMITTENTS"	• FOLLOW DIAGNOSTIC DECISION TREE FOR ECM SWITCH TEST E.7.8

WHEN ALL DIAGNOSIS AND REPAIRS ARE COMPLETED, CLEAR CODES AND VERIFY OPERATION

1988–90 CADILLAC DEVILLE AND FLEETWOOD

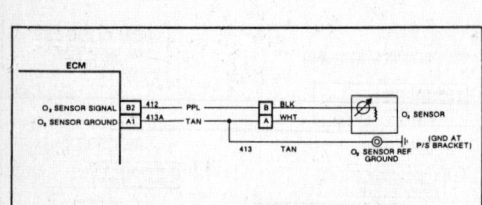

ECM Code E44, LEAN EXHAUST SIGNAL

TEST CONDITIONS: 1. Codes E14, E15, E16, E21, E22, E26, E27, E31, E32 and E34 all must be clear; 2. Throttle switch open; 3. TPS between 6 and 29 degrees; 4. Coolant sensor greater than or equal to 80°C; 5. Oxygen sensor ready (closed loop); 6. Not accelerating or decelerating; 7. RPM greater than or equal to 800.

FAILURE CONDITIONS: Oxygen sensor status stays lean for more than 51 seconds.

ACTION TAKEN: 1. ECM turns on "Service Now" light. 2. ECM turns on canister purge solenoid, air management solenoids; 3. ECM switches to open loop operation.

Description:

The ECM provides a .45 volt reference signal to the oxygen sensor on CKT 412. When the oxygen sensor is cold (below 200°C), the oxygen sensor signal voltage will be around .45 volts and the ECM will keep the system in open loop operation. When the oxygen sensor is warm (above 200°C), the oxygen sensor signal voltage will swing from rich to lean rapidly, at least one swing every two seconds, if the ECM is in good control of the air fuel mixture.

When the ECM sees that the oxygen is varying from the cold voltage of .45 volts, it will send the system into closed loop operation. In closed loop operation, the ECM will adjust the fuel delivery rate to the engine based on the oxygen sensor readings.

Code E44, is designed so that if the oxygen sensor stays at a lean voltage for more than 51 seconds during the test conditions, Code E44 will set.

Code E44 will set when: 1. There is an oxygen sensor circuit fault giving a false lean indication or 2. When the air fuel ratio is actually lean due to a vacuum leak or fuel control system fault.

Notes On Fault Tree:

1. With the oxygen sensor disconnected, parameter P.0.7 should remain at reference voltage (.38 to .63 volts).

2. Checking for sensor circuitry able to record rich readings. The DVOM set on volts will provide a few billionths of an amp to drive CKT 412 to above .64 volts (rich). Similar results may be obtained by placing one finger on battery positive terminal and another finger on oxygen sensor CKT 412 harness terminal.

3. The ECM compares oxygen sensor signal voltage received on circuit 412 to the ground voltage on circuit 413. If the ECM doesn't have a good ground to the engine on circuit 413, the oxygen sensor can appear falsely high or low. With engine running, use a voltmeter to measure voltage from the oxygen sensor at the exhaust manifold to the ECM terminal A1. If the voltage is -.05 volts to +.05 volts then the ground is OK. If the voltage is less than -.05 volts or greater than +.05 volts, repair poor ground on CKT 413 between ECM terminal A1 and the ground at front of engine, right (rearmost) head.

Notes On Intermittents

Engine running, manipulate the oxygen sensor and ECM wiring and connectors while observing ECM parameter P.0.7. If the fault is induced, P.0.7 will jump below .37 volts and the "ECON" status light will go off. Manipulate CKT 412 ground to the engine and look for a loose ground eyelet or ground eyelet installed at wrong location.

If lean engine operation is suspected, perform "DFI System Check."

Review the snapshot data parameters stored with the code to verify failure condition. Refer to "Symptoms" for complete details on using the "Snapshot on code set" feature. If the failure condition is verified, check and repair intermittent wiring connection or sensor.

1988–90 CADILLAC DEVILLE AND FLEETWOOD

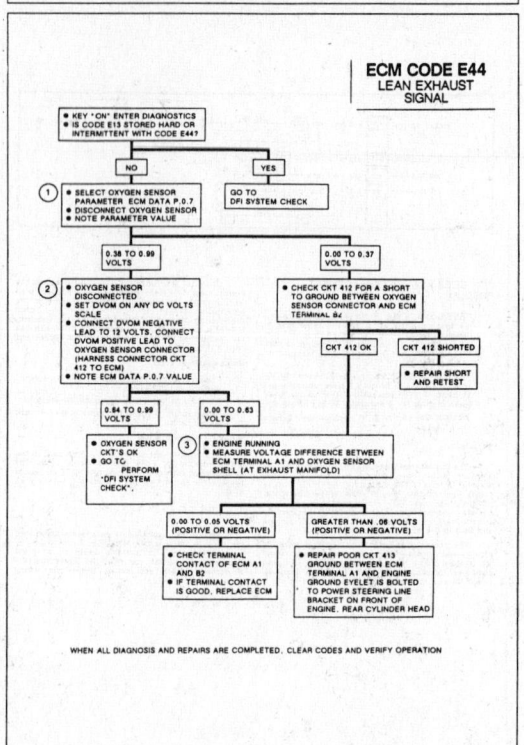

ECM CODE E44
LEAN EXHAUST SIGNAL

WHEN ALL DIAGNOSIS AND REPAIRS ARE COMPLETED, CLEAR CODES AND VERIFY OPERATION

1988–90 CADILLAC DEVILLE AND FLEETWOOD

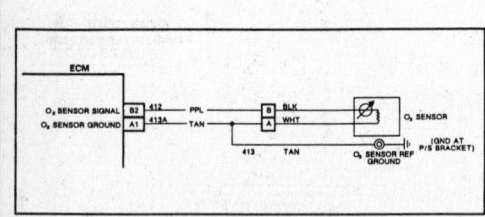

ECM Code E45, RICH EXHAUST SIGNAL

TEST CONDITIONS: 1. Codes E14, E15, E16, E21, E22, E26, E27, E31, E32, and E34 all must be clear; 2. Throttle switch open; 3. TPS between 6 and 29 degrees; 4. Coolant sensor greater than or equal to 80°C; 5. Oxygen sensor ready (closed loop); 6. Not accelerating or decelerating; 7. RPM greater than or equal to 800.

FAILURE CONDITIONS: Oxygen sensor stays rich for more than 51 seconds.

ACTION TAKEN: 1. ECM turns on "Service Now" light. 2. ECM turn off canister purge solenoid, air management solenoids; 3. ECM switches to open loop operation.

Description:

The ECM provides a .45 volt reference signal to the oxygen sensor on CKT 412. When the oxygen sensor is cold (below 200°C), the output voltage will be around 0.45 volts and the ECM will keep the system in open loop operation. When warm, a properly operating oxygen sensor will drive the .45 volt reference lower (below .45 volts) to indicate a lean mixture and higher (above .45 volts) to indicate a rich mixture. The oxygen sensor signal voltage will swing from rich to lean rapidly, at least one swing every two seconds, if the ECM is in good control of the air fuel mixture. When the ECM sees that the oxygen sensor is varying from the cold voltage of 0.45 volts, it will send the system into closed loop operation. In closed loop operation, the ECM will meter fuel into the engine based on the oxygen sensor readings.

Code E45 is designed so that if the oxygen sensor stays at a rich voltage for more than 51 seconds during the test conditions, Code E45 will set.

Code E45 can be caused by: 1. Oxygen sensor circuit faults; 2. Air fuel ratio actually rich due to a fuel control or emissions system fault.

Notes On Fault Tree:

1. With the oxygen sensor disconnected, parameter P.0.7 should remain at reference voltage (.38 to .63 volts).

2. Checking for ECM's ability to recognize lean input on oxygen sensor signal CKT 412.

3. The ECM compares oxygen sensor signal voltage received on circuit 412 to the ground voltage on circuit 413. If the ECM doesn't have a good ground to the engine on circuit 413, the oxygen sensor can appear falsely high or low. With engine running, use a voltmeter to measure voltage from the oxygen sensor at the exhaust manifold to the ECM terminal A1. If the voltage is -.05 volts to +.05 volts then the ground is OK. If the voltage is less than -.05 volts or greater than +.05 volts, repair poor ground on CKT 413 between ECM terminal A1 and the ground at front of engine, right (rearmost) head.

Note On Intermittents

Engine running, manipulate the oxygen sensor and ECM wiring and connectors while observing ECM parameter P.0.7. If the fault is induced, P.0.7 will jump above .63 volts. "ECON" status light will come on. Manipulate CKT 413 ground to the engine and look for a loose ground eyelet or ground eyelet installed in an improper location.

Perform "DFI System Check".

Review the snapshot data parameters stored with the code to verify failure condition. Refer to "Symptoms" for complete details on using the "Snapshot on code set" feature. If the failure condition is verified, check and repair intermittent wiring connection or sensor.

1988–90 CADILLAC DEVILLE AND FLEETWOOD

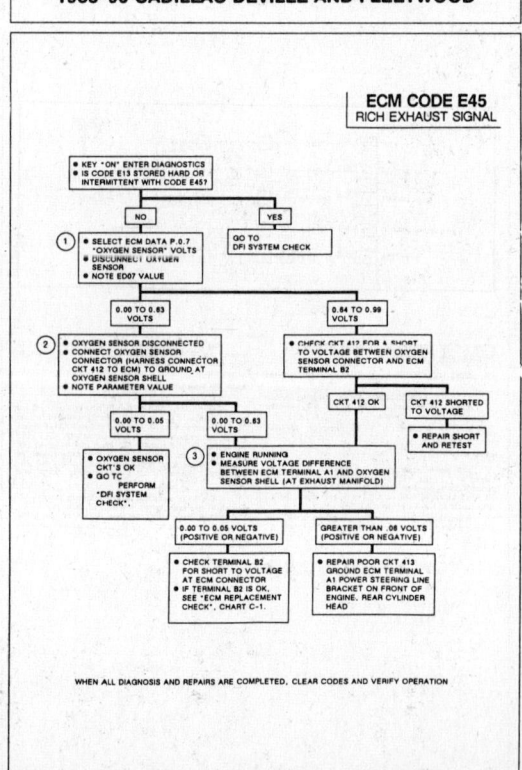

ECM CODE E45
RICH EXHAUST SIGNAL

WHEN ALL DIAGNOSIS AND REPAIRS ARE COMPLETED, CLEAR CODES AND VERIFY OPERATION

1988–90 CADILLAC DEVILLE AND FLEETWOOD

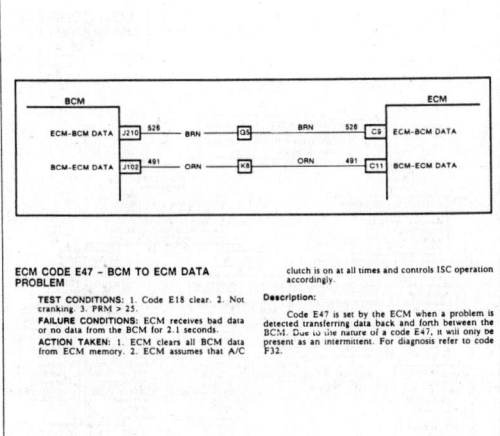

ECM CODE E47 – BCM TO ECM DATA PROBLEM

TEST CONDITIONS: 1. Code E18 clear. 2. Not cranking. 3. PRM > 25.

FAILURE CONDITIONS: ECM receives bad data or no data from the BCM for 2.1 seconds.

ACTION TAKEN: 1. ECM clears all BCM data from ECM memory. 2. ECM assumes that A/C clutch is on at all times and controls ISC operation accordingly.

Description:

Code E47 is set by the ECM when a problem is detected transferring data back and forth between the BCM. Due to the nature of a code E47, it will only be present as an intermittent. For diagnosis refer to code F32.

1988–90 CADILLAC DEVILLE AND FLEETWOOD

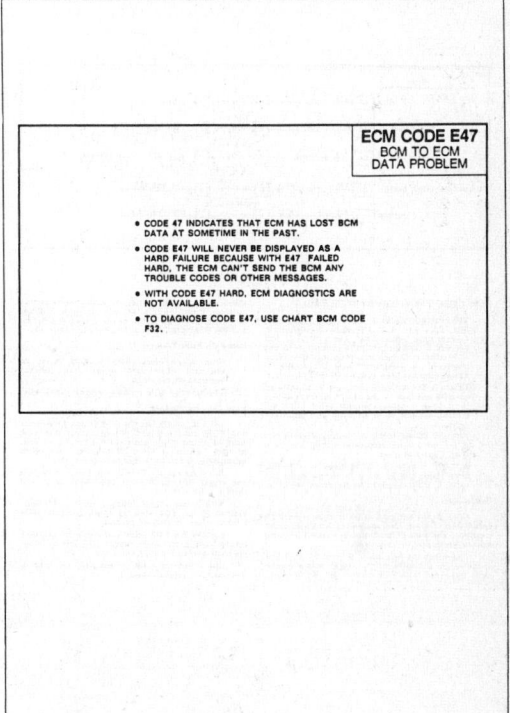

ECM CODE E47
BCM TO ECM
DATA PROBLEM

- CODE 47 INDICATES THAT ECM HAS LOST BCM DATA AT SOMETIME IN THE PAST.
- CODE E47 WILL NEVER BE DISPLAYED AS A HARD FAILURE BECAUSE WITH E47 FAILED HARD, THE ECM CAN'T SEND THE BCM ANY TROUBLE CODES OR OTHER MESSAGES.
- WITH CODE E47 HARD, ECM DIAGNOSTICS ARE NOT AVAILABLE.
- TO DIAGNOSE CODE E47, USE CHART BCM CODE F32.

1988–90 CADILLAC DEVILLE AND FLEETWOOD

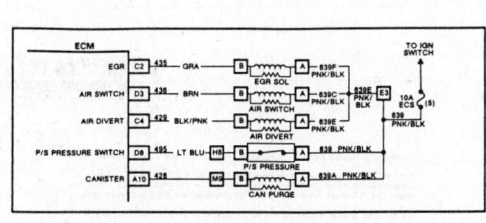

(1 of 2)

ECM CODE E48 – EGR SYSTEM FAULT

TEST CONDITIONS: 1. Codes E13, E14, E15, E21, E22, E31, E32, E34, E44, E45 clear; 2. Coolant temp between 85° and 110°C; 3. TPS between 7° and 14°; 4. RPM between 1450 and 1650 RPM 5. Oxygen sensor swinging (closed loop operation); 6. MPH > 35; 7. 10 minute timer after startup expired; 8. Steady throttle.

TEST: In the test conditions, the ECM will turn off EGR and look for a leaner mixture to indicate a leaner mixture. The ECM will perform the test up to six times in a given key cycle.

FAILURE CONDITIONS: If the oxygen sensor fails to indicate a leaner mixture in at least 3 of the 5 tests, Code E48 is set.

ACTION TAKEN: 1. ECM turns on "Service Soon" lamp and latches it on for the entire key cycle; 2. EGR is disabled for the entire key cycle.

Description:

Code E48 is designed to set if there is an EGR system fault. The test for Code E48 is performed under conditions where the EGR is normally enabled (allowing exhaust gas to flow into the intake).

To perform the test, the ECM turns off EGR flow to the engine and monitors the oxygen sensor (closed loop) integrator. With EGR turned off, the integrator should swing to a higher value reflecting leaner air/fuel mixtures. If not, the ECM assumes that either EGR was turned off before the test started or that EGR is flowing and the ECM doesn't have the ability to turn it off.

Notes On Fault Tree:

1. Checking for EGR operation using ECM override.

2. Checking for EGR gases to enter the intake manifold by raising the EGR valve off of its seat.
3. Checking for EGR solenoid able to pass vacuum.

Note On Intermittents:

At idle, manipulate the EGR solenoid connector and ECM connector and related wiring. The EGR solenoid should remain energized (CKT 435 grounded) to block vacuum to the EGR valve/vent vacuum to atmosphere. Listen for a change in idle quality.

Drive the car at TPS between 8 and 15 degrees and engine RPM at 1450 to 1650 RPM to try and duplicate the code.

With a hot engine (idling), apply and release vacuum to the EGR valve to check for EGR valve binding in the up or down position.

Remove the EGR valve and check for excessive carbon build up that would restrict EGR flow or foreign materials holding the EGR valve open.

Check vacuum hoses for pinched, cut, kinked, misrouted, or hoses blocked.

VACUUM TEST

1. Connect a vacuum gage to the source side of the EGR solenoid. Start the engine, manifold vacuum should be present. If it is not, repair leaks or obstruction between the EGR solenoid and throttle body.
2. Connect a vacuum gage to the EGR valve vacuum supply. There should be no vacuum with the engine idling. If there is, follow Chart on the following page.
3. With the gage hooked to the EGR valve vacuum supply, disconnect the EGR solenoid connector. There should be more than 8 inches of vacuum available. If not, repair leak or obstruction in EGR valve vacuum hose.

1988–90 CADILLAC DEVILLE AND FLEETWOOD

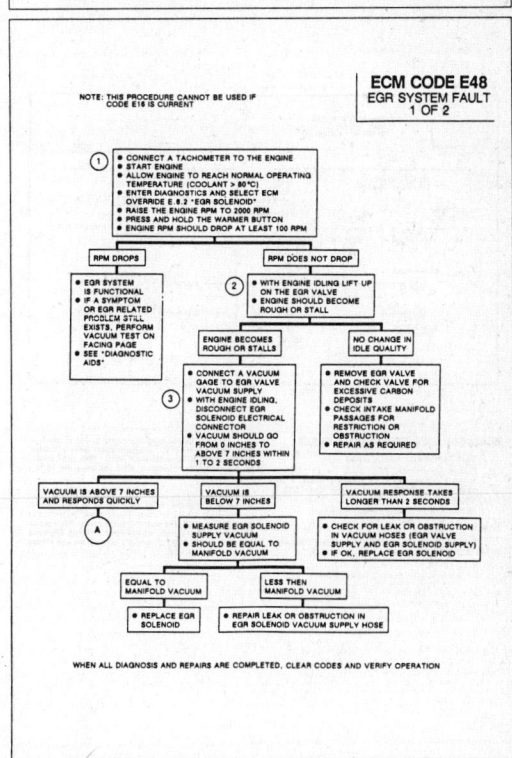

ECM CODE E48
EGR SYSTEM FAULT
1 OF 2

1988–90 CADILLAC DEVILLE AND FLEETWOOD

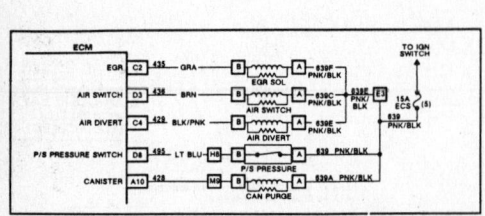

(2 of 2)

ECM CODE E48 – EGR SYSTEM FAULT

TEST CONDITIONS: 1. Codes E13, E14, E15, E21, E22, E31, E32, E34, E44, E45 clear; 2. Coolant temp between 85° and 110°C; 3. TPS between 7° and 14°; 4. RPM between 1450 and 1650 RPM. 5. Oxygen sensor toggling (closed loop operation). 6. Integrator below 144 counts. 7. Closed loop has been enabled for 600 seconds.

TEST: In the test conditions, the ECM will turn off EGR and look for the oxygen sensor integrator to indicate a leaner mixture. The ECM will perform the test up to six times in a given key cycle.

FAILURE CONDITIONS: If the oxygen sensor fails to indicate a leaner mixture in at least 3 of the 5 tests, Code E48 is set.

ACTION TAKEN: 1. ECM turns on "Service Soon" lamp and latches it on for the entire key cycle; 2. EGR is disabled for the entire key cycle.

Description:

Code E48 is designed to set if there is an EGR system fault. The test for Code E48 is performed under conditions where the EGR is normally enabled (allowing exhaust gas to flow into the intake).

To perform the test, the ECM turns off EGR flow to the engine and monitors the oxygen sensor (closed loop) integrator. With EGR turned off, the integrator should swing to a higher value reflecting leaner air/fuel mixtures. If not, the ECM assumes that either EGR was turned off before the test started or that EGR is flowing and the ECM doesn't have the ability to turn it off.

Notes On Fault Tree:

1. With engine idling, EGR solenoid should be energized. Test light across EGR solenoid terminals should be lit.
2. Checking for ECM's ability to turn solenoid off.

Note On Intermittents:

At idle, manipulate the EGR solenoid connector and ECM connector and related wiring. The EGR solenoid should remain energized (CKT 435 grounded) to block vacuum to the EGR valve/vent vacuum to atmosphere. Listen for a change in idle quality.

Drive the car at TPS between 8 and 15 degrees and engine RPM at 1450 to 1650 RPM to try and duplicate the code.

With a hot engine (idling), apply and release vacuum to the EGR valve to check for EGR valve binding in the up or down position.

Remove the EGR valve and check for excessive carbon build up that would restrict EGR flow or foreign materials holding the EGR valve open.

Check vacuum hoses for pinched, cut, kinked, misrouted, or hoses blocked.

1988–90 CADILLAC DEVILLE AND FLEETWOOD

ECM CODE E48
EGR SYSTEM FAULT
2 OF 2

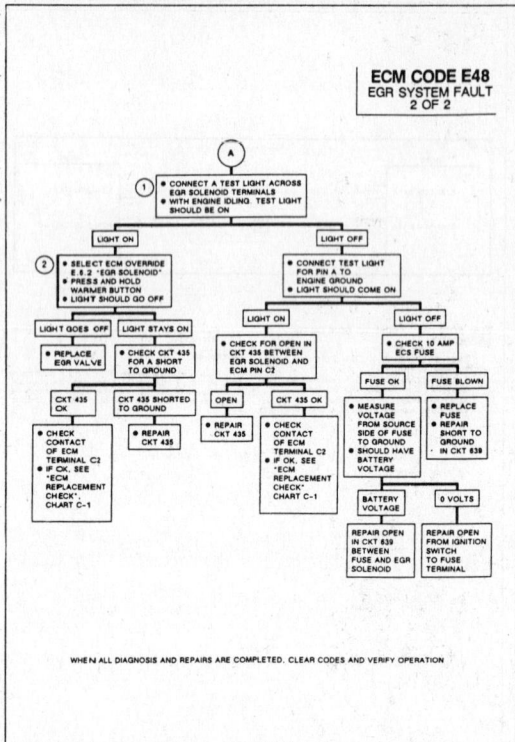

WHEN ALL DIAGNOSIS AND REPAIRS ARE COMPLETED. CLEAR CODES AND VERIFY OPERATION

1988–90 CADILLAC DEVILLE AND FLEETWOOD

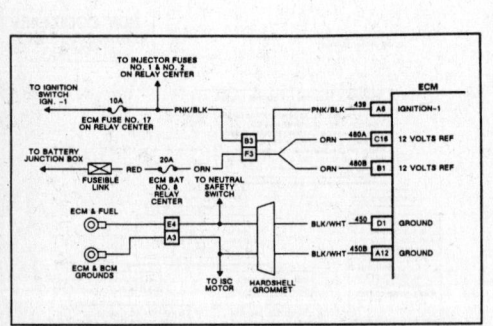

ECM CODE E52 – ECM MEMORY RESET

Test Conditions: Always tested.

Failure Conditions: ECM looks for all full time memory information to be erased or reset. If full time memory is completely erased, code E52 is logged.

Action Taken: E52 is stored in memory.

Description:

Code E52 indicates that the "long term" memory in the ECM has been reset.

This will be the case whenever power is removed from the ECM (i.e., disconnecting battery cables, disconnecting ECM connector, etc.). This code should be "cleared" from memory after restoring the ECM's power supply.

If Code E52 is seen accompanied by complaint of engine quit, stumble, telltale's flashing or other stored codes, check for intermittent loss of power or ground to the ECM on CKT 480 and 450 respectively. Manipulate wiring and connections with engine running. Remove and replace the ECM P1 and P2 connectors and ensure that they are latched.

1988–90 CADILLAC DEVILLE AND FLEETWOOD

ECM CODE E52
ECM MEMORY RESET

Code E52 indicates that the "long term" memory in the ECM has been reset.

This will be the case whenever power is removed from the ECM (i.e., disconnecting battery cables, disconnecting ECM connector, etc.) This code should be "cleared" from memory after restoring the ECM's power supply.

If Code E52 is seen accompanied by complaint of engine quit, stumble, telltale's flashing or other stored codes, check for intermittent loss of power or ground to the ECM on CKT 480 and 450 respectively. Manipulate wiring and connections with engine running. Remove and replace the ECM J1 and J2 connectors and ensure that they are latched.

1988–90 CADILLAC DEVILLE AND FLEETWOOD

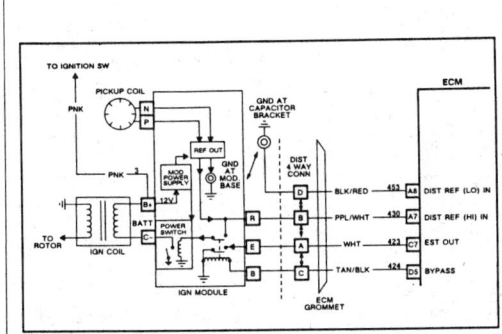

ECM CODE E53 – DISTRIBUTOR SIGNAL INTERRUPT

TEST CONDITIONS: 1. RPM above 448. 2. Fuel pump voltage > 10 volts.

FAILURE CONDITIONS: No distributor reference pulses for .7 seconds.

ACTION TAKEN: 1. Code E53 is stored as a history code. (No teltales are illuminated).

Description:

Code E53 is set if the ECM does not receive distributor reference pulses from the HEI for more than .7 seconds. Since the DFI system requires HEI pulses in order to fire the injectors, most occurrences of Code E53 will be accompanied by a stall.

Code E53 can be set by ignition switch contact timing – when turning the key off, if the Ignition 3 contacts that feed the HEI open before the Ignition 1 contacts to the ECM, a Code E53 may be set. Code E53 does not turn on any teltales or "service" messages. Do not attempt to diagnose a history Code E53 unless the customer complains of stumble, miss, stall or other driveability condition that could be caused by loss of spark or fuel injection.

Notes On Fault Tree:

1. Checking for hard failure of distributor reference which will cause code E12 to set.
2. Do not diagnose Code E53 unless the customer complains of stumble, miss, stall or other driveability condition that could be caused by loss of spark or fuel.
3. Checking for Code E53 set due to ignition key cycling.
4. See "Description", above.
5. See "Note On Intermittents".

Note On Intermittents:

Code E53 can be caused by:
- Loss of ground on CKT 453
- Loss of distributor reference signal on CKT 430
- Loss of battery power to "B+" terminal of the distributor.
- Ignition switch contact timing.

Do not attempt to diagnose Code E53 unless the customer symptom is stumble, stall, miss or other driveability condition that could be caused by loss of spark or fuel.

1988–90 CADILLAC DEVILLE AND FLEETWOOD

ECM CODE E53
DISTRIBUTOR SIGNAL INTERRUPT

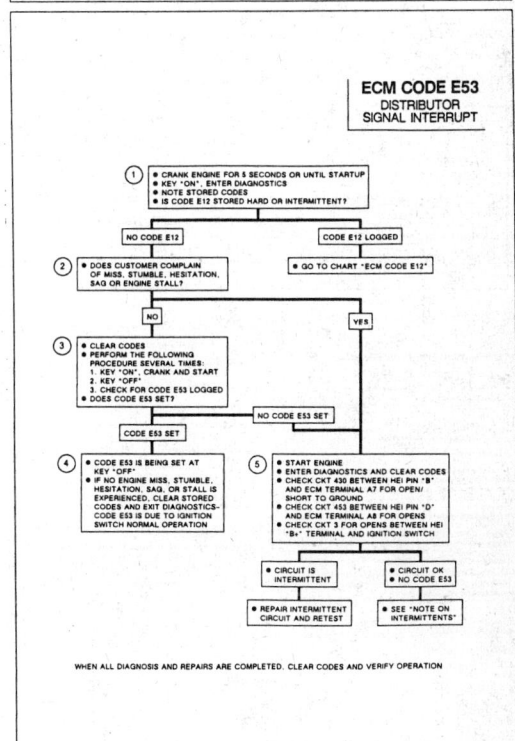

WHEN ALL DIAGNOSIS AND REPAIRS ARE COMPLETED, CLEAR CODES AND VERIFY OPERATION

1988–90 CADILLAC DEVILLE AND FLEETWOOD

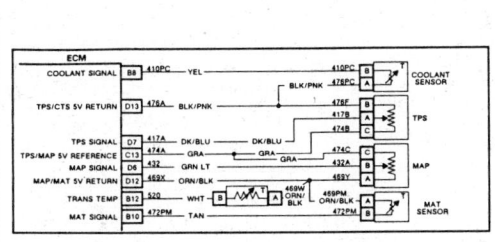

ECM CODE E55 – TPS OUT OF ADJUSTMENT

TEST CONDITIONS: 1. Codes E21, E22, E26, E27 clear. 2. ECM tests for Code E55 at key off. At key off the ISC will retract until the throttle switch opens. The TPS value is read by the ECM and a correction factor is stored. If the correction factor value is the same on two consecutive key off cycles, the TPS setting is "relearned".

FAILURE CONDITION: If TPS correction needed is from –2.9° to +3° "TPS out of adjustment" flag is set.

ACTION TAKEN: At the next key on, the ECM will see the "TPS out of adjustment" flag and log Code E55 as current. No teltale or "Service" message will appear.

Description:

The TPS on DFI is "self adjusting". At key-off, the ECM executes a TPS learning routine. After key off, the ECM will retract the ISC until the throttle switch opens and the throttle linkage is resting on the minimum air screw. At that time, the ECM stores the TPS value and calculates a correction. If the same correction factor occurs on two consecutive key "off" cycles, the TPS is then corrected to 0 degrees using the correction factor "learned". If the value needs correction by more than –2.9 degrees or +3.0 degrees, Code E55 will be stored in memory at the next key

"ON" cycle. **Note:** Parameter P.0.1, TPS, displays uncorrected TPS values.

Notes On Fault Tree:

1. Checking TPS adjustment. ECM parameter P.0.1, TPS, displays uncorrected TPS so that it can be used to check TPS adjustment.
2. TPS adjustment is OK.
3. If TPS adjustment is OK, ISC and throttle switch operation need to be thoroughly checked. Check for TPS return spring for proper operation — throttle, cruise and TV cables not binding, proper throttle return spring operation throttle shaft and blades free to move.

Notes On Intermittents:

Enter diagnostics. Manipulate ISC wiring while observing "OFF" throttle switch status light and while observing ISC operation during ECM output cycling.

Manipulate TPS wiring and connector while observing ECM parameter P.0.1 for jumps, skips, intermittent behavior. Check for TPS secured to throttle body (both TORX screws tight). Cycle the TPS through its full travel while observing P.0.1. Check for proper part number TPS installed on vehicle.

Remove and replace the TPS, ISC and ECM connectors and ensure that they are latched. If wiring and connectors check out OK, substitute a known good TPS sensor and retest.

1988–90 CADILLAC DEVILLE AND FLEETWOOD

ECM CODE E55
TPS MISADJUSTED

NOTE: IF CODES E21, E26, E27 OR E30 ARE LOGGED AS CURRENT OR HISTORY, DO NOT USE THIS PROCEDURE. SEE THE CODE E21, E22, E26, E27 AND E30 PROCEDURES TO DIAGNOSE

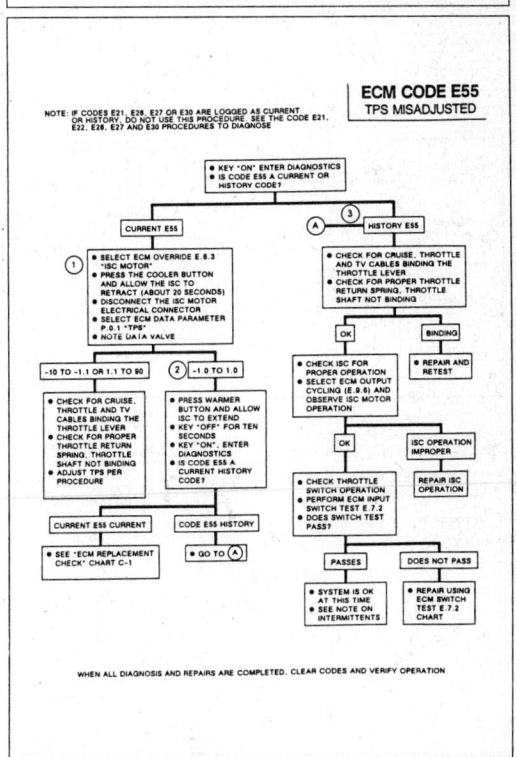

WHEN ALL DIAGNOSIS AND REPAIRS ARE COMPLETED, CLEAR CODES AND VERIFY OPERATION

1988–90 CADILLAC DEVILLE AND FLEETWOOD

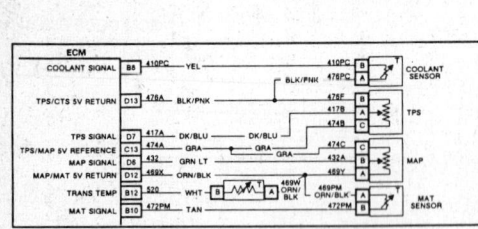

**ECM CODE E59, TRANSMISSION
TEMPERATURE SENSOR CIRCUIT**

TEST CONDITIONS: Always tested.

FAILURE CONDITIONS: Transmission temperature greater than or equal to 211°C or transmission temperature less than or equal to −35°C.

ACTION TAKEN: 1. VCC is engaged at a higher than normal vehicle speed.

Description:

The transmission temperature sensor is a two wire sensor that provides the temperature of the transmission oil as it enters the transmission oil cooler. The transmission temperature sensor receives a 5 volt reference from the ECM on CKT 520. As the temperature of the transmission oil increases, the sensor resistance decreases, and the signal voltage on CKT 520 increases. The ECM interprets low voltage as a high temperature and a high voltage as a low temperature. The transmission temperature sensor is a thermistor (like the Coolant Temp Sensor).

The ECM uses the transmission temperature input to delay VCC engagement in cases of very high transmission oil temperature.

Code E59 is designed to detect a transmission temperature sensor open or shorted. Code E59 will not set due to transmission overheating or very cold ambient conditions.

Note On Fault Tree:

1. The transmission temp sensor CKT 520 is shorted to ground.

2. CKT 520 is open to the sensor or the sensor is open to ground.

Note On Intermittents:

Manipulate the transmission temperature sensor and transmission connector and the CKT 520 wiring while observing parameter P.1.5.

Remove and replace the transmission temperature sensor and ECM connectors and ensure that they are latched. If the wiring and connectors check out OK, substitute a known good transmission temperature sensor and retest.

Review the snapshot data parameters stored with the code to verify failure condition. Refer to "Symptoms" for complete details on using the "Snapshot on code set" feature. If the failure condition is verified, check and repair intermittent wiring connection or sensor.

1988–90 CADILLAC DEVILLE AND FLEETWOOD

ECM CODE E59
VCC TEMP SENSOR CIRCUIT

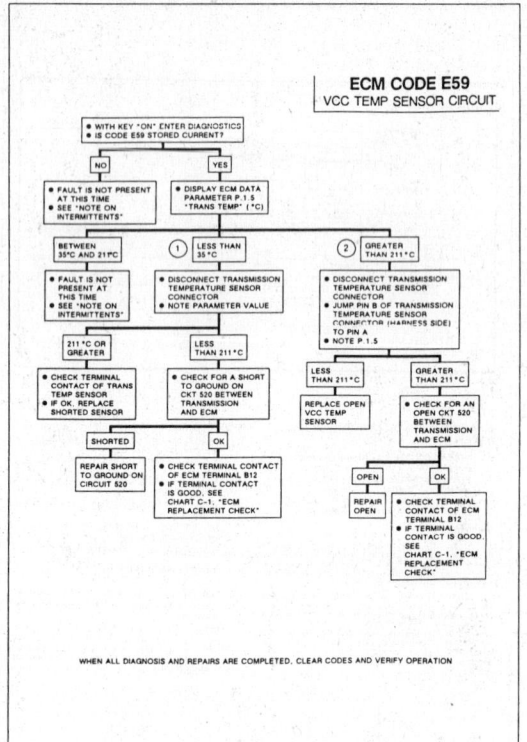

WHEN ALL DIAGNOSIS AND REPAIRS ARE COMPLETED, CLEAR CODES AND VERIFY OPERATION

1988–90 CADILLAC DEVILLE AND FLEETWOOD

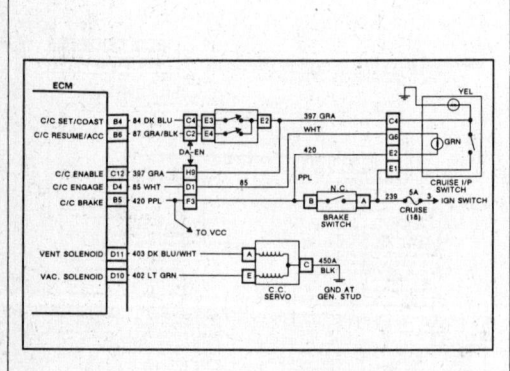

ECM CODE E60, CRUISE – NOT IN DRIVE

TEST CONDITIONS: 1. Cruise control enabled; 2. Cruise control engaged.

FAILURE CONDITIONS: Transmission in Park/Neutral.

ACTION TAKEN: 1. Disengage cruise control.

Description:

Code E60 will set if the cruise control is engaged and the Park/Neutral switch is closed, indicating the transmission is in Park/Neutral.

Notes On Intermittents:

If Code E60 is stored as an intermittent code, check the operation of the Park/Neutral Input test E.7.4. If E.7.4 does not pass, repair using trouble tree for switch test E.7.4.

If no trouble is found, explain to the customer that this code can be induced if the transmission is inadvertently put into neutral while the cruise control is engaged.

1988–90 CADILLAC DEVILLE AND FLEETWOOD

ECM CODE E60
*CRUISE ENGAGED WITH TRANS
IN PARK OR NEUTRAL*

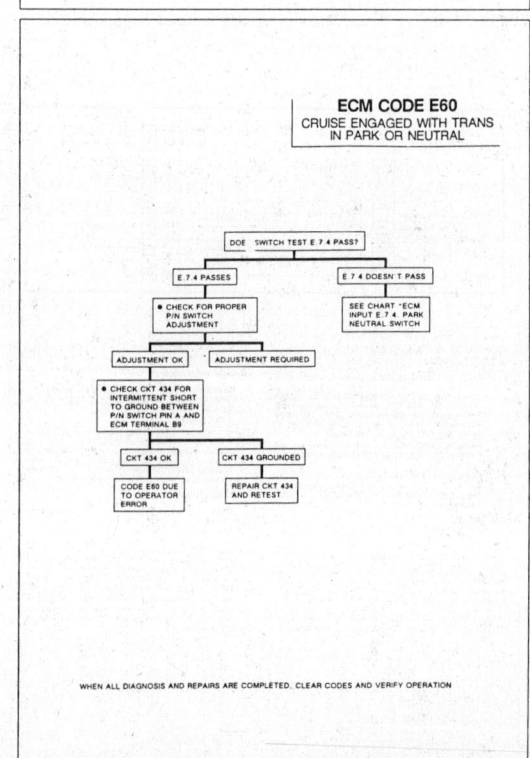

WHEN ALL DIAGNOSIS AND REPAIRS ARE COMPLETED, CLEAR CODES AND VERIFY OPERATION

1988–90 CADILLAC DEVILLE AND FLEETWOOD

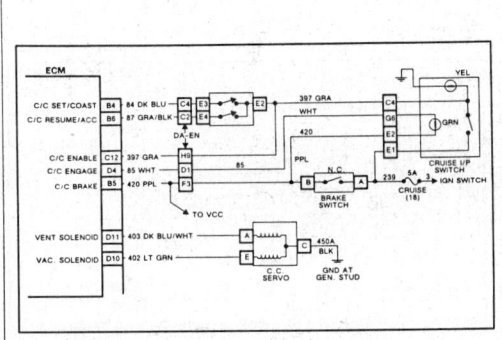

ECM Code E63, SET SPEED/VEHICLE SPEED DIFFERENCE TOO HIGH

TEST CONDITIONS: 1. Cruise control enabled; 2. Cruise control engaged; 3. Cruise control not in the resume mode.

FAILURE CONDITIONS: Vehicle speed 20 MPH higher than set speed for .5 seconds.

ACTION TAKEN: 1. Cruise control disengaged.

Description:

Code E63 will set and disengage the cruise control if vehicle speed is 20 MPH higher than the set speed. This code is used as a type of safety valve to the cruise control system, and can be set under normal conditions if the operator accelerates using the gas pedal and goes 20 MPH over the cruise set speed.

Notes On Fault Tree:

1. Checking for the ability of the servo to pull in.
2. This step checks for the ability of the servo to be released.
3. This step checks the vent and vacuum outputs of the ECM for proper operation.

Notes On Intermittents:

If Code E63 is setting intermittently make sure owner is aware that overrunning the cruise set speed by more than 20 MPH will cause this code and disengage the cruise system.

This code will not be caused by vacuum leaks or vacuum supply problems. It can only be set by the vehicle speed exceeding the set speed by 20 MPH.

1988–90 CADILLAC DEVILLE AND FLEETWOOD

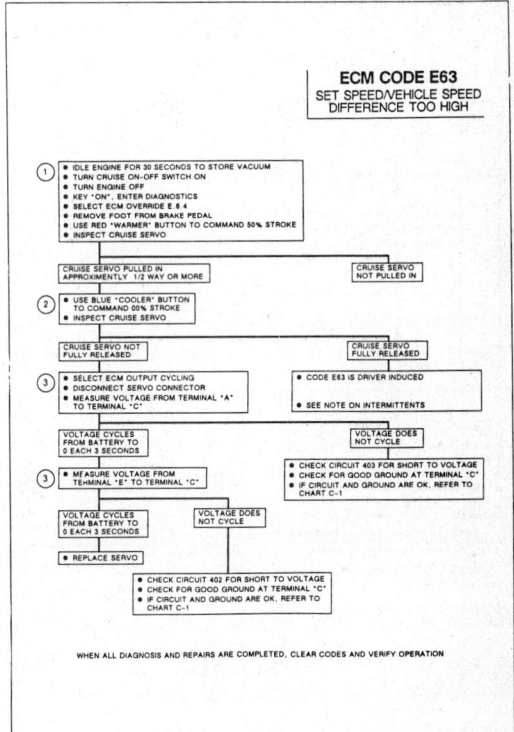

ECM CODE E63
SET SPEED/VEHICLE SPEED DIFFERENCE TOO HIGH

WHEN ALL DIAGNOSIS AND REPAIRS ARE COMPLETED, CLEAR CODES AND VERIFY OPERATION

1988–90 CADILLAC DEVILLE AND FLEETWOOD

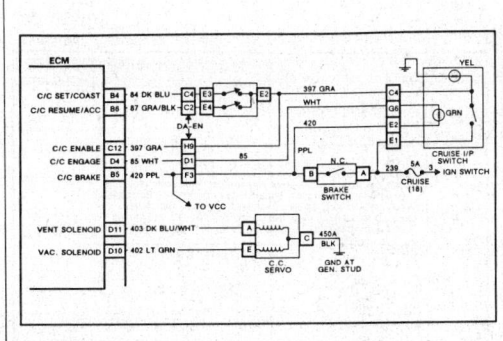

ECM CODE E64, ACCELERATION TOO HIGH CRUISE ENGAGED

TEST CONDITIONS: 1. Cruise control enabled; 2. Cruise control engaged.

FAILURE CONDITIONS: Vehicle speed increases more than 5 MPH in 1/3 second.

ACTION TAKEN: 1. Disengaged cruise control.

Description:

Code E64 is designed to set when the vehicle speed is increasing at an extremely rapid rate

(wheelspin). It is a protective measure so that the wheels will not be under cruise control when on ice.

Notes On Fault Tree:

If Code E64 is stored as an intermittent code and the customer complains of frequent loss of cruise control, drive the car while observing ECM data ED12, vehicle speed. If the speed displayed is erratic, check the operation and integrity of the speed sensor circuit.

1988–90 CADILLAC DEVILLE AND FLEETWOOD

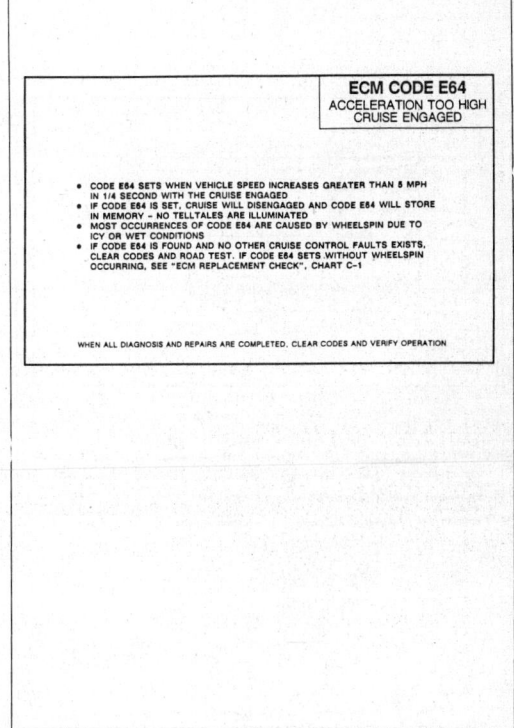

ECM CODE E64
ACCELERATION TOO HIGH
CRUISE ENGAGED

- CODE E64 SETS WHEN VEHICLE SPEED INCREASES GREATER THAN 5 MPH IN 1/4 SECOND WITH THE CRUISE ENGAGED
- IF CODE E64 IS SET, CRUISE WILL DISENGAGED AND CODE E84 WILL STORE IN MEMORY – NO TELLTALES ARE ILLUMINATED
- MOST OCCURRENCES OF CODE E64 ARE CAUSED BY WHEELSPIN DUE TO ICY OR WET CONDITIONS
- IF CODE E64 IS FOUND AND NO OTHER CRUISE CONTROL FAULTS EXISTS, CLEAR CODES AND ROAD TEST. IF CODE E64 SETS WITHOUT WHEELSPIN OCCURRING, SEE "ECM REPLACEMENT CHECK", CHART C-1

WHEN ALL DIAGNOSIS AND REPAIRS ARE COMPLETED, CLEAR CODES AND VERIFY OPERATION

1988–90 CADILLAC DEVILLE AND FLEETWOOD

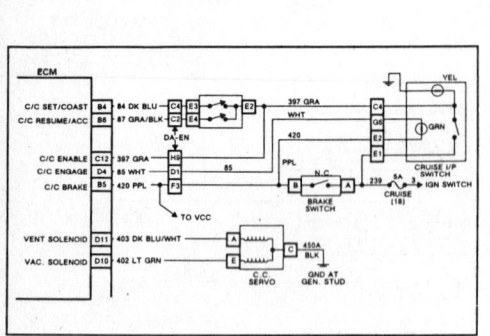

ECM CODE E65, CRUISE – COOLANT TEMP TOO HIGH

TEST CONDITIONS: 1. Cruise engaged.
FAILURE CONDITIONS: 1. Coolant > 126°C.
ACTION TAKEN: 1. Disable cruise control.

Description:

Code E65 is designed to protect the engine from overheating. The ECM monitors coolant temperature and will disable cruise if the coolant temperature rises above 126°C.

A Code E65 will be stored by the ECM to alert the technician to the reason the cruise was disabled.

Notes On Intermittents:

If an intermittent code E65 is stored, manipulate the coolant sensor wiring cable observing ECM data P0.4 "Coolant Temp". If the parameter jumps to a high reading (> 126°C) repair the intermittent wiring or connector.

Review the snapshot data parameters stored with this code to verify failure condition. Refer to "Symptoms" for complete details on using the "Snapshot on code set" feature.

1988–90 CADILLAC DEVILLE AND FLEETWOOD

ECM CODE E65
CRUISE COOLANT TEMP
TOO HIGH

- CODE E65 SETS WHEN COOLANT TEMPERATURE GOES ABOVE 126 °C WITH THE CRUISE ENGAGED
- IF CODE E65 IS SET, CRUISE WILL DISENGAGE AND CODE E65 WILL STORE IN MEMORY—NO TELLTALES ARE ILLUMINATED
- IF CODE E65 IS ACCOMPANIED BY HARD OR INTERMITTENT CODE E14, GO TO CHART "CODE E14" IN 8D1
- IF CODE E65 IS FOUND, DIAGNOSE FOR ENGINE OVERHEATING

WHEN ALL DIAGNOSIS AND REPAIRS ARE COMPLETED, CLEAR CODES AND VERIFY OPERATION

1988–90 CADILLAC DEVILLE AND FLEETWOOD

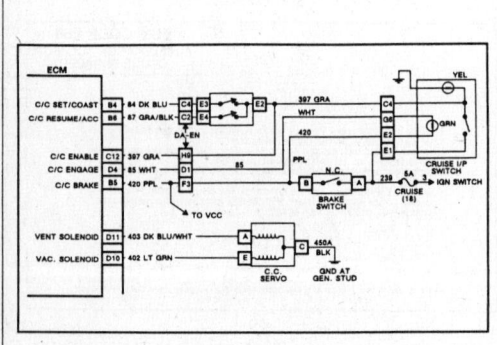

ECM CODE E66, CRUISE – ENGINE RPM TOO HIGH

TEST CONDITIONS: 1. Cruise control enabled; 2. Cruise control engaged.

FAILURE CONDITIONS: Engine RPM ≥ 4800 for .25 seconds.

ACTION TAKEN: 1. Disengage cruise control.

Description:

This code will set if engine RPM exceeds 4800 RPM. This may occur on slippery pavement, extended wide open throttle acceleration; or for some mechanical problems (such as transmission slippage). Under these conditions, Code E66 will set and should be considered normal. The driver should be advised why the cruise deenergized. Clear the code and road test vehicle to verify normal operation.

1988–90 CADILLAC DEVILLE AND FLEETWOOD

ECM CODE E66
CRUISE–ENGINE RPM
TOO HIGH

THIS CODE WILL SET IF ENGINE RPM EXCEEDS 4800 RPM. THIS MAY OCCUR ON SLIPPERY PAVEMENT, EXTENDED WIDE OPEN THROTTLE ACCELERATION, OR FOR SOME MECHANICAL PROBLEMS (SUCH AS TRANSMISSION SLIPPAGE). UNDER THESE CONDITIONS, CODE E66 WILL SET AND SHOULD BE CONSIDERED NORMAL. THE DRIVER SHOULD BE ADVISED WHY THE CRUISE DISENGAGED. CLEAR THE CODE AND ROAD TEST VEHICLE TO VERIFY NORMAL OPERATION

WHEN ALL DIAGNOSIS AND REPAIRS ARE COMPLETED, CLEAR CODES AND VERIFY OPERATION

1988–90 CADILLAC DEVILLE AND FLEETWOOD

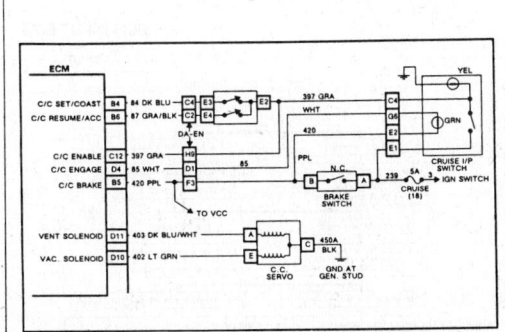

ECM CODE E67, CRUISE – SET/COAST OR RESUME/ACCEL SWITCH SHORTED

TEST CONDITIONS: 1. Always tested.

FAILURE CONDITIONS: Set/Coast or Resume/Accel signal present when the cruise ON/OFF switch is turned "ON" or Set/Coast and Resume/Accel signal present at the same time.

ACTION TAKEN:
If PROM ID (P.1.7) is 701 or 702, disable cruise for complete ignition cycle.

If PROM ID (P.1.7) is 851 or 852, disable cruise until failure conditions are not true.

Description:

When the cruise control ON/OFF switch is "ON", system voltage is available at one side of the normally open contacts on the Set/Coast and Resume/Accel switches. If the cruise control ON/OFF switch is in the "ON" position this system voltage is available to the Set/Coast and Resume/Accel switches at key on.

If the Set/Coast or the Resume/Accel switches were stuck or their signal wires to the ECM were shorted to voltage, the vehicle could begin cruise operation. To prevent this, Code E67 will set and disable cruise control if signal voltage from the Set/Coast (CKT 84) or Resume/Accel (CKT 87) is HI when the cruise control ON/OFF switch is turned "OFF" to "ON" or when the ignition key is turned on and the cruise control ON/OFF switch was left "ON".

Notes On Fault Tree:

1. Checking for proper operation of the cruise control switches.

Notes On Intermittents:

This code can be set by: 1. The operator if the cruise ON/OFF switch is turned "ON" and the operator was depressing either the Set/Coast or Resume/Accel switches or; 2. The ON/OFF switch was on when the ignition key is turned on and the operator was depressing either the Set/Coast or Resume/Accel switches or; 3. Both the Set/Coast and Resume/Accel switches were pressed at the same time with the dash switch on.

This code will not be caused by vacuum leaks, vacuum servo, or vacuum supply problems.

1988–90 CADILLAC DEVILLE AND FLEETWOOD

ECM CODE E67
CRUISE–SET/COAST OR RESUME/ACCEL SWITCH SHORTED

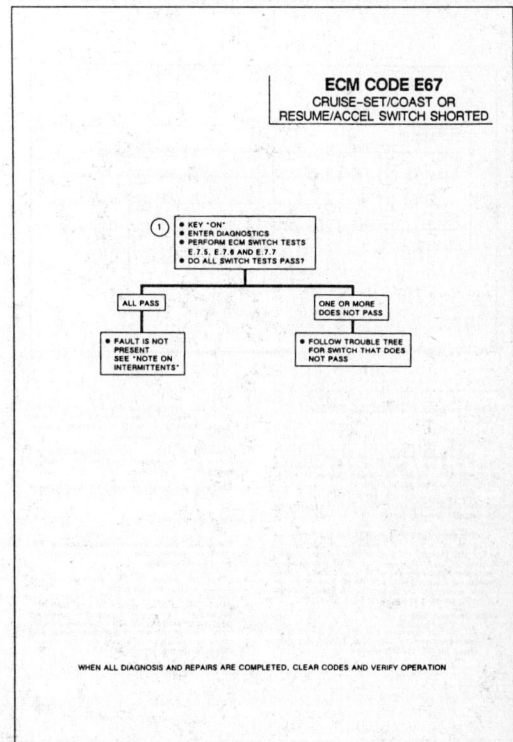

WHEN ALL DIAGNOSIS AND REPAIRS ARE COMPLETED, CLEAR CODES AND VERIFY OPERATION

1988–90 CADILLAC DEVILLE AND FLEETWOOD

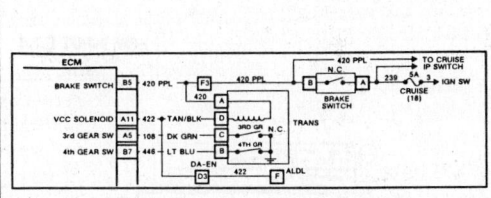

ECM INPUT E.7.1 CRUISE CONTROL/VCC BRAKE CIRCUIT

Description:

The VCC brake switch is normally closed and opens when the brake pedal is depressed. The VCC brake switch serves three functions. The VCC brake switch supplies power to the VCC solenoid and the cruise vacuum valve and a brake ON/OFF indication to the ECM.

The VCC brake switch receives 12 volts from the 5 AMP cruise fuse #18 on circuit 239 and sends the 12 volt signal; A.) To the ECM and cruise control on CKT 420, and B.) to the VCC solenoid on CKT 420.

When the VCC brake switch opens, power is off (0 volts) to the cruise vacuum valve and to the VCC solenoid.

The ECM uses VCC brake input as an aid in providing failsoft values for codes and for braking status for cruise and VCC operation.

Notes On Fault Tree:

1. At key on, not braking, should have 12 volts on both sides of VCC brake switch. Test light to ground should light if either pin on brake switch connector is backprobed.

2. Light on one side means that brake switch is open as if the brake pedal were depressed.

3. No light indicates open between fuse and the brake switch.

4. Checking for proper brake switch operation. If switch test doesn't pass and there is power at the switch, fault is switch, circuit or ECM.

Note On Intermittents:

- Check brake switch for proper adjustment.
- Check for intermittent open between fuse and brake switch and ECM.
- If cruise fuse blows intermittently, check for short to ground from fuse to brake switch, VCC, cruise or ECM.

1988–90 CADILLAC DEVILLE AND FLEETWOOD

ECM INPUT E.7.1
VCC BRAKE SWITCH

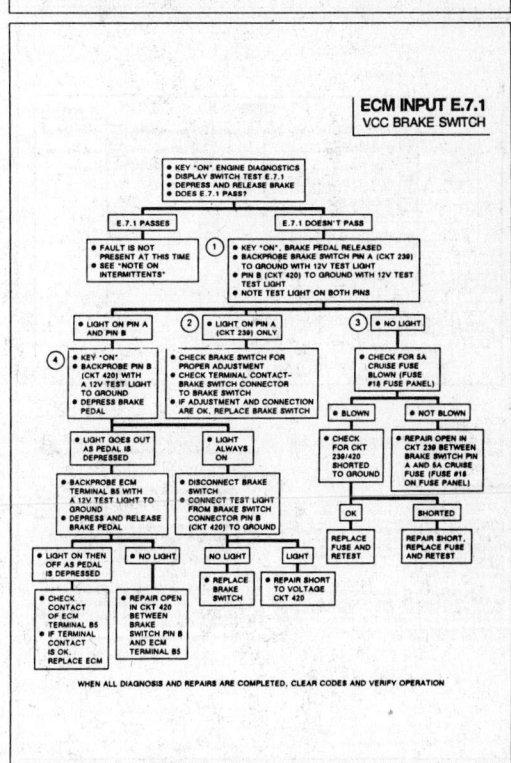

WHEN ALL DIAGNOSIS AND REPAIRS ARE COMPLETED, CLEAR CODES AND VERIFY OPERATION

5–51

1988-90 CADILLAC DEVILLE AND FLEETWOOD

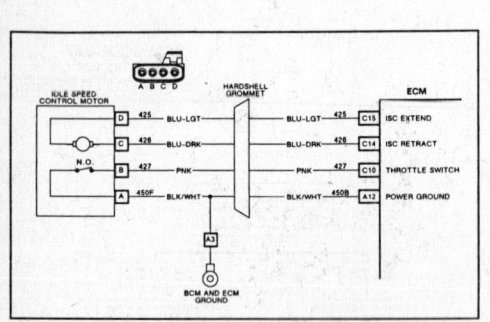

ECM INPUT E.7.2, THROTTLE SWITCH CIRCUIT

Description:

The Throttle Switch is part of the Idle Speed Control motor (ISC). Pin "B" of the four-way weather pack connector on the ISC is the throttle switch or closed throttle input to the ECM.

The ISC throttle switch contacts are normally open and are closed when the throttle linkage contacts the ISC plunger (closed throttle, ISC in control of idle speed).

The ECM sends a 5 volts signal to pin "B" of the ISC motor on CKT 427. When the throttle linkage rests on the ISC plunger, the throttle switch contacts close, shorting ISC pin "B" to pin "A" (CKT 450 ground). The 5 volt signal from the ECM is grounded at closed throttle, resulting in 0 volts at ECM pin C10. When the throttle is opened, the throttle switch opens pin "B" CKT 427, resulting in 5 volts at ECM pin C10.

Notes On Fault Tree:

1. ISC plunger depressed and released should allow the switch test to pass. If E.7.2 doesn't pass, check for ISC/circuit problems.

2. When the ISC motor is completely retracted, the throttle switch will open due to ISC internal design. Make sure the ISC is extended partially before continuing diagnosis.

3. Checking for ISC or circuit problem. If E.7.2 passes then the problem is at the ISC motor or connector.

4. Checking for continuity on CKT 427 between ECM and ISC motor.

Note On Intermittents:

- Check for binding throttle linkage due to TV, cruise or throttle cables, TPS miss-installed or throttle shaft binding.

- Check for proper throttle return spring and throttle return spring installation.

- Probe ECM C10 to ground with a voltmeter. Manipulate wiring and connectors at closed throttle and watch for 5.0 volts, indicating an open from C10 to ground. Manipulate wiring and connectors at open throttle and watch for 0 volts, indicating short to ground on 427.

1988-90 CADILLAC DEVILLE AND FLEETWOOD

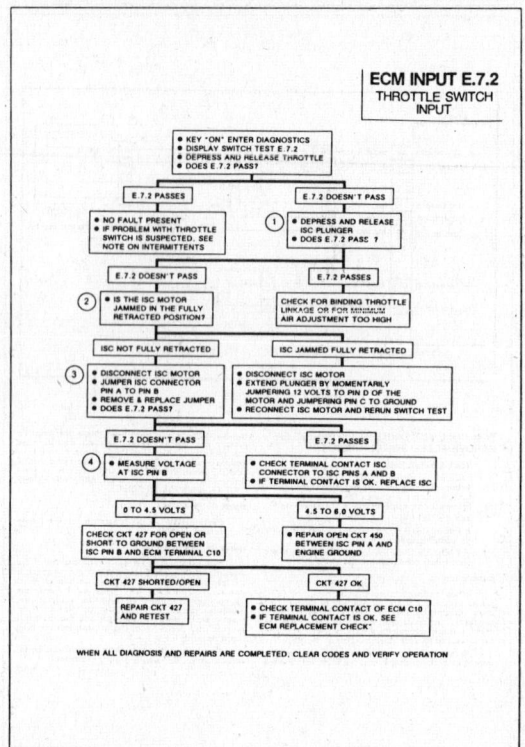

1988-90 CADILLAC DEVILLE AND FLEETWOOD

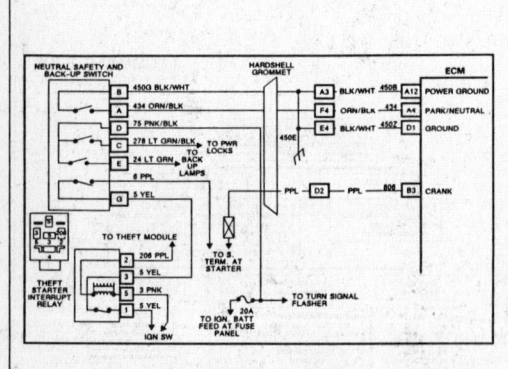

ECM INPUT E.7.4, PARK/NEUTRAL (NSBU) SWITCH CIRCUIT

Description:

The Park Neutral switch is a part of the transmission Neutral Safety – Back-up switch (NSBU). Pin "A" of the six-way weather pack connector on the NSBU switch is the park/neutral input for the ECM. The NSBU park/neutral switch contacts are closed in park or neutral, shorting NSBU switch pin "A" to ground. In any other gear range, pin "A" is open to ground.

The ECM sends 12 volts to pin "A" of the NSBU on CKT 434. When the gear selector is in park or neutral, the 12 volt signal from the ECM is shorted to ground, resulting in 0 volts at ECM pin A4. In reverse or forward gears, CKT 434 is open to ground, resulting in 12 volts at ECM pin A4.

The ECM uses park/neutral status for fuel and idle speed control and as a test condition for many trouble codes.

Notes On Fault Tree:

1. Checking for hard or intermittent fault.
2. Checking for fault in wiring and switch.
3. If E.7.4 passes by jumpering, the fault is at the switch or switch contactor.
4. Checking for open to ground at the switch.
5. Checking for ECM ability to recognize a ground at terminal A4.

Notes On Intermittents:

- Check for NSBU switch out of adjustment.
- Check for trans shift indicator out of adjustment.
- Check for open or short to ground in CKT 434 between ECM and switch.
- Check for open to ground on switch pin B.

1988-90 CADILLAC DEVILLE AND FLEETWOOD

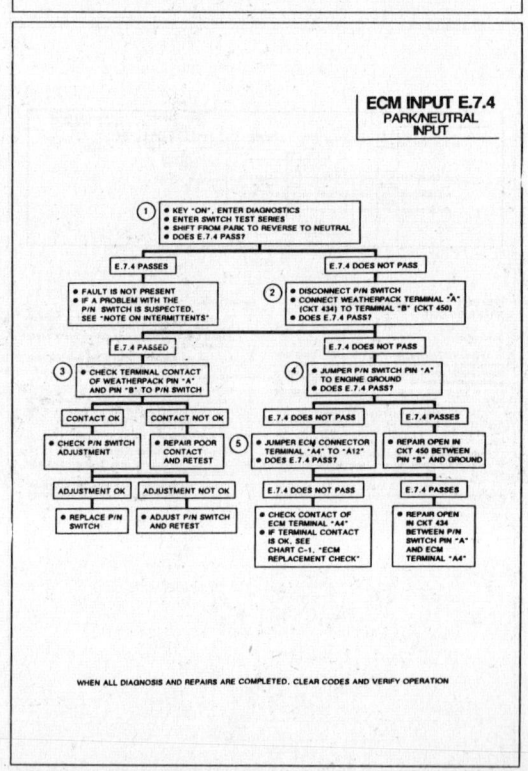

1988–90 CADILLAC DEVILLE AND FLEETWOOD

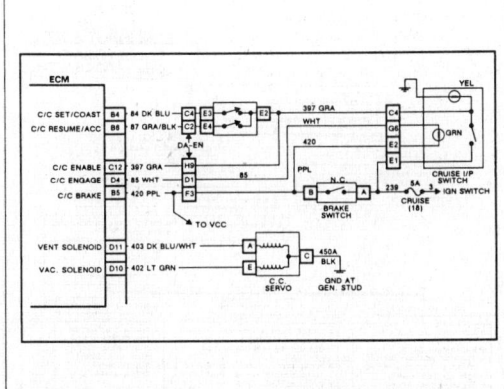

ECM INPUT E.7.5, CRUISE CONTROL ON/OFF CIRCUIT

Description:

The cruise enable or cruise dash switch provides power to the cruise enable input of the ECM. The ECM can't allow cruise operation until the enable or instrument panel switch is turned to the "ON" position, which passes 12 volts to ECM terminal C12.

Note On Intermittents:

- The amber cruise "ON" indicator can be used to monitor the cruise dash switch status. Turn the dash switch "ON" and manipulate wiring and connectors. If the switch or circuit from the fuse to the switch are open, the amber light will blink out.
- If the cruise 5A fuse blows intermittently, turn the switch on, manipulate wiring, connectors, cruise steering column switch. If the circuit becomes grounded, the fuse will blow.

1988–90 CADILLAC DEVILLE AND FLEETWOOD

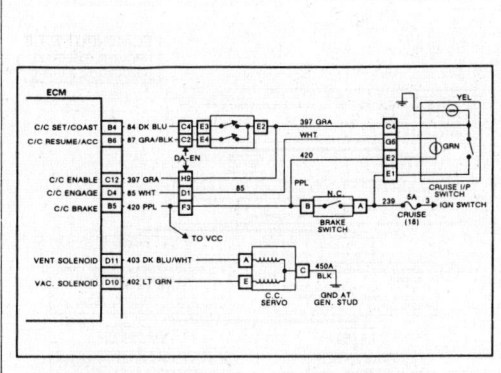

ECM INPUT E.7.6, CRUISE CONTROL "SET/COAST" CIRCUIT

Description:

The ECM Cruise "Set/Coast" circuit is a normally open switch that provides a 12 volt "Set/Coast" signal to the ECM on terminal B4 when depressed. When the cruise dash switch is in the "ON" position, the cruise dash switch provides 12 volts to the cruise steering column switch on pin E2, CKT 397. When the Set/Coast button is depressed, the steering column switch pin E2 is connected to pin E3, CKT 84, providing the ECM with a 12 volt Set/Coast input on terminal B4.

Notes On Intermittents:

- To diagnose an intermittent Set/Coast function, turn the cruise dash switch to "ON" and backprobe ECM B4 with a voltmeter to ground. B4 should show 12 volts with Set/Coast depressed and 0 volts with Set/Coast released. Cycle the switch while observing the meter; manipulate wiring and connectors.

1988–90 CADILLAC DEVILLE AND FLEETWOOD

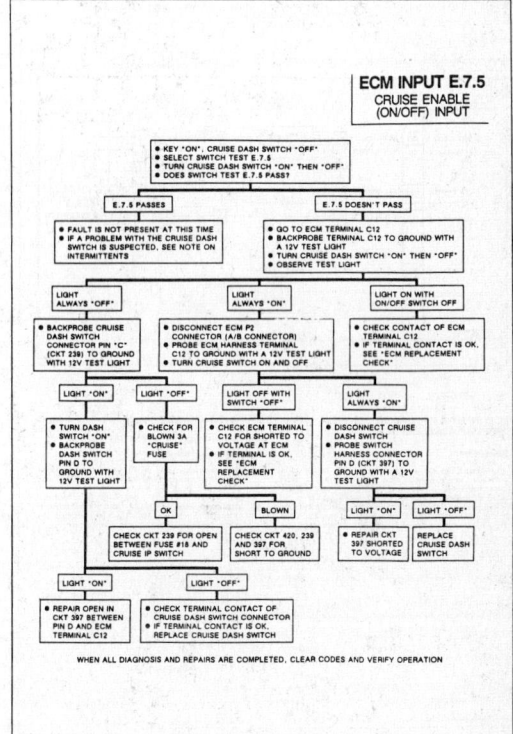

ECM INPUT E.7.5
CRUISE ENABLE (ON/OFF) INPUT

1988–90 CADILLAC DEVILLE AND FLEETWOOD

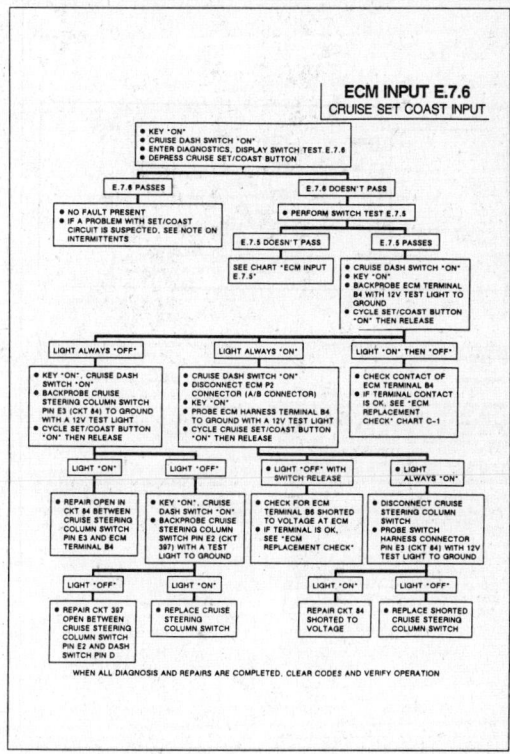

ECM INPUT E.7.6
CRUISE SET COAST INPUT

1988–90 CADILLAC DEVILLE AND FLEETWOOD

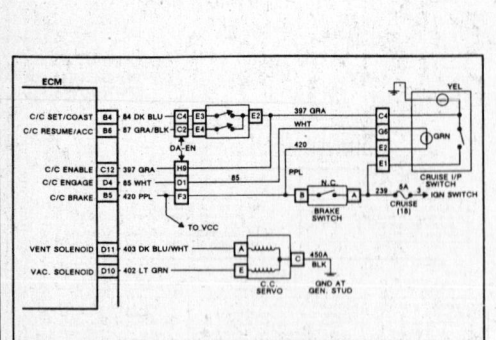

ECM INPUT E.7.7, CRUISE CONTROL "RESUME/ACCELERATION" CIRCUIT

Description:

The ECM cruise "Resume/Accel" switch is a normally open switch that provides a 12 volt "Resume/Accel" signal to the ECM on terminal B6 when depressed. When the cruise dash switch is in the "ON" position, the cruise dash switch provides 12 volts to the cruise steering column switch on pin E2, CKT 397. When the Resume/Accel button is depressed, the

steering column switch pin E2 is connected to pin C, CKT 87, providing the ECM with a 12 volt Resume/Accel input on terminal B6.

Notes On Intermittents:

- To diagnose an intermittent Resume/Accel function, turn the cruise dash switch to "ON" and backprobe ECM B6 with a voltmeter to ground. B6 should show 12 volts with Resume/Accel depressed and 0 volts with Resume/Accel released. Cycle the switch while observing the meter, manipulate wiring and connectors.

1988–90 CADILLAC DEVILLE AND FLEETWOOD

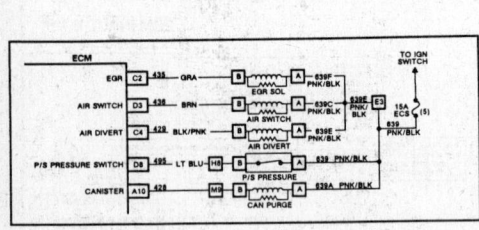

ECM INPUT E.7.8, POWER STEERING PRESSURE CIRCUIT

Description:

The Power Steering switch is normally closed and opens with high power steering pressure. During normal driving, the power steering switch receives 12 volts from the 15A AMP solenoid fuse on CKT 639 and sends the 12 volt signal to the ECM on CKT 495. When high power steering pressures occur, the switch opens and CKT 495 voltage is read by the ECM as 0 volts.

The ECM uses the power steering input to extend the ISC motor when high power steering pressures occur at low speeds, such as parking maneuvers.

Notes On Fault Tree:

1. Checking ECM and circuits ability to respond to cycling CKT 495.

2. Checking for open power steering switch or switch connector. If E.7.8 passes, the ECM and wiring are OK.

3. Checking for 12 present on CKT 639.

4. Checking for ECM able to recognize a signal on terminal D8.

5. If 15A fuse is blown, check for CKT 495 or 639 shorted to ground.

Note On Intermittents:

At key on, probe ECM terminal D8 with a voltmeter to ground. Manipulate wiring and connectors while observing voltmeter. If voltage drops to 0 volts at key "ON" engine "OFF" repair open or short to ground.

1988–90 CADILLAC DEVILLE AND FLEETWOOD

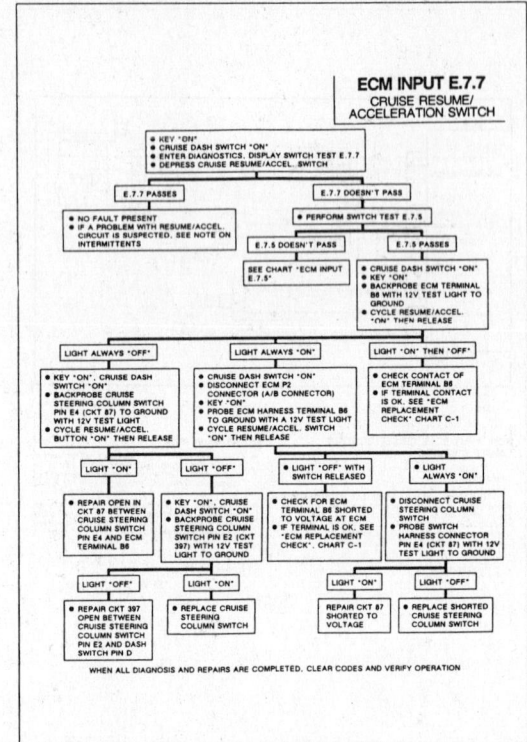

1988–90 CADILLAC DEVILLE AND FLEETWOOD

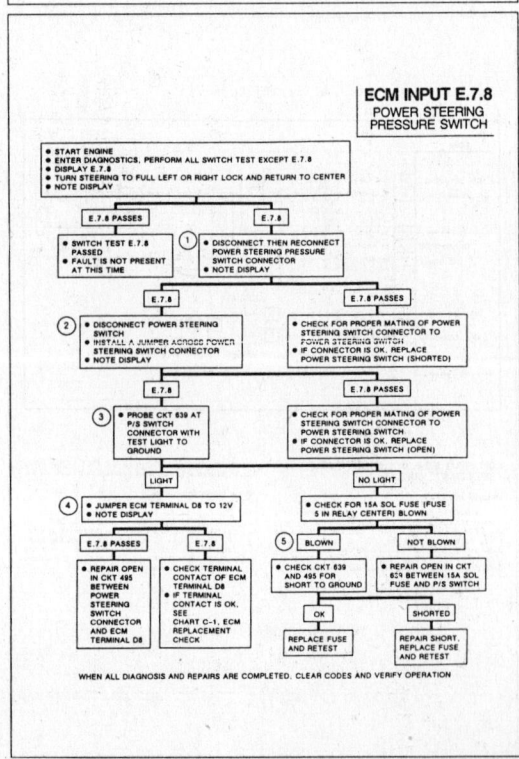

1988–90 CADILLAC DEVILLE AND FLEETWOOD

ADDITIONAL DIAGNOSTIC INFORMATION

The following charts define the testing and failure conditions associated with each trouble code. This information is presented here to help explain how the ECM determines that a malfunction has occurred. The information contained in the charts is detailed below:

1. **Trouble Codes** – This is the code which is displayed in diagnostic display if the testing and failure requirements are both met. Codes E20, E23, E32 and E39 have more than one set of requirements which will cause them to set. These different conditions are identified by a letter following the trouble code (i.e. 20–A and 20–B).

2. **Testing Requirements** – These are the conditions which must be met before the ECM will test for the failure requirements.

3. **Failure Requirements** – This is the input which the ECM identifies as abnormal under the conditions the testing requirements. If a time is included in the requirements, this is how long all of the testing and failure requirements must be present in order to satisfy setting the trouble code.

4. **Failsoft Action** – After the ECM has identified a system malfunction, it may take some action to keep the vehicle operational. The following failsoft actions are taken by the ECM and are identified on the chart:

   N – No action taken
   1 – Coolant temp = MAT
   2 – Throttle angle set at 13° with throttle switch open, 6° with throttle switch closed
   3 – HEI held in bypass spark
   4 – Brake on and TPS less than 13° = closed switch. Brake off or TPS more than 13° = open switch.

5. Code is kept hard for entire ignition cycle in which malfunction occurs and automatically stored as an intermittent upon starting of new ignition cycle

6. MAP is set at:
   - Table lookup if throttle switch is closed
   - 81.8 kPa if throttle switch is open

7. BARO set at 92.2 kPa

8. MAT is set at:
   - Coolant temp if coolant temp is less than 40°C
   - 40°C if coolant temp is greater than 40°C

9. ECM uses CALPAK for backup spark and fuel

5. **Disabled Functions** – Depending on the detected malfunction, certain functions will be disabled. These functions are disabled since their proper operation is dependent upon the input which has malfunctioned. The following functions are disabled and are identified on the chart:

   N – No Action Taken
   A – Cruise Control
   B – Air management (de-energize both solenoids)
   C – EGR (de-energize solenoid)
   D – Canister Purge (de-energize solenoid)
   E – VCC (de-energize solenoid)
   F – Cruise Control disabled for entire ignition cycle in which malfunction occurred.
   G – Closed Loop Control
   H – ECM diagnostics disabled

1988–90 CADILLAC DEVILLE AND FLEETWOOD

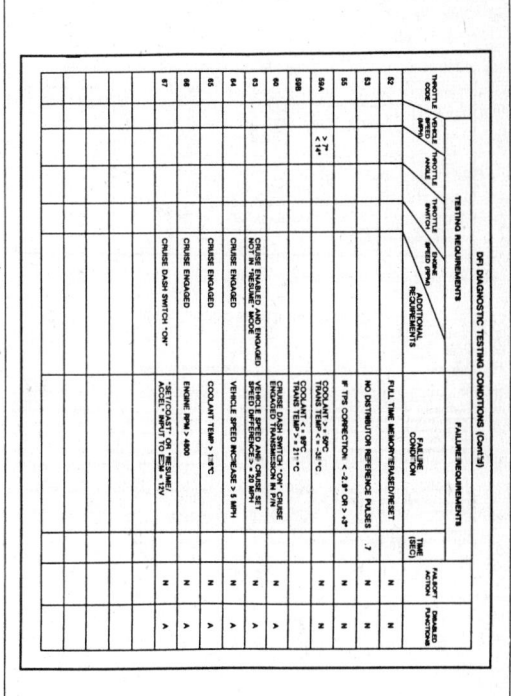

1988–90 CADILLAC DEVILLE AND FLEETWOOD

1988–90 CADILLAC DEVILLE AND FLEETWOOD

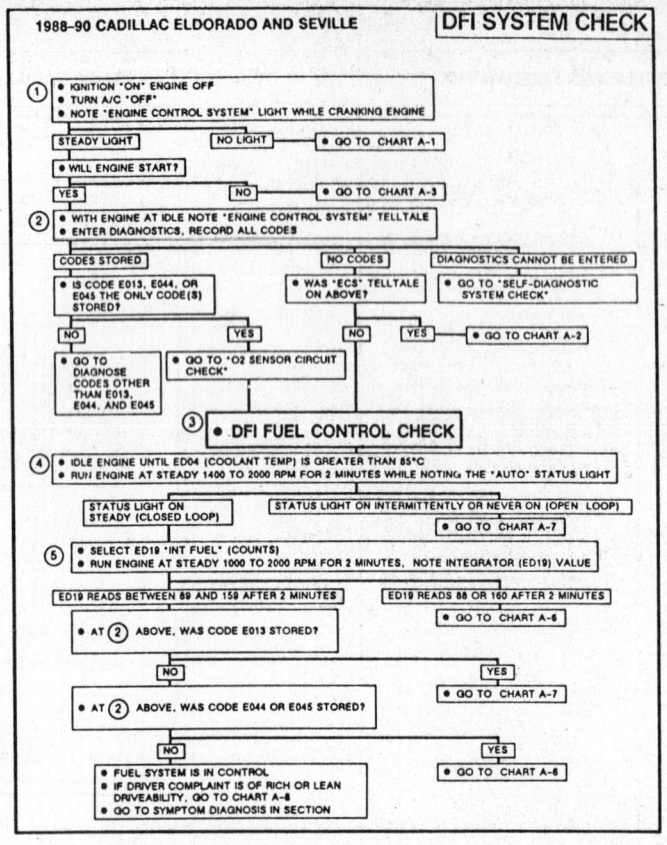

1988-90 CADILLAC ELDORADO AND SEVILLE — DFI SYSTEM CHECK

1988-90 CADILLAC ELDORADO AND SEVILLE

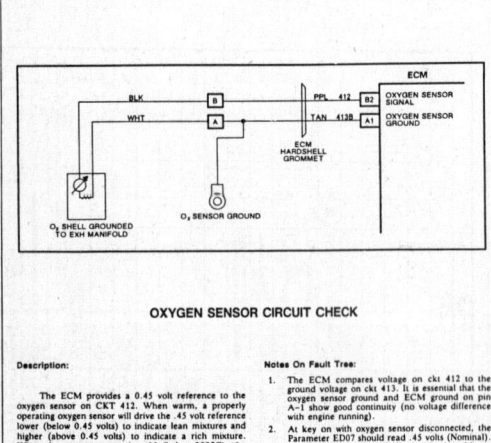

OXYGEN SENSOR CIRCUIT CHECK

Description:

The ECM provides a 0.45 volt reference to the oxygen sensor on CKT 412. When warm, a properly operating oxygen sensor will drive the .45 volt reference lower (below 0.45 volts) to indicate lean mixtures and higher (above 0.45 volts) to indicate a rich mixture. When the oxygen sensor is cold (below 200°C), the output voltage will be around 0.45 volts and the ECM will keep the system in open loop operation.

When the ECM sees that the oxygen sensor is varying from the cold voltage of 0.45 volt and the coolant sensor value is above 85°C, it will send the system into closed loop operation. In closed loop operation, the ECM will adjust the fuel delivery rate based on the oxygen sensor readings.

Notes On Fault Tree:

1. The ECM compares voltage on ckt 412 to the ground voltage on ckt 413. It is essential that the oxygen sensor ground and ECM ground on pin A-1 show good continuity (no voltage difference with engine running).

2. At key on with oxygen sensor disconnected, the Parameter ED07 should read .45 volts (Nominal). Lower or higher voltage indicates circuit problems.

3. At this point the Oxygen Sensor circuit between the 2-way connector (at the sensor) and the ECM, all the Oxygen Sensor grounds and the ECM are all functioning correctly and have good contact.

4. This series of tests will determine if a faulty reading is due to poor ground connections, circuit faults or problem with the ECM connection or ECM.

1988-90 CADILLAC ELDORADO AND SEVILLE — O2 SENSOR CIRCUIT CHECK

1988-90 CADILLAC ELDORADO AND SEVILLE

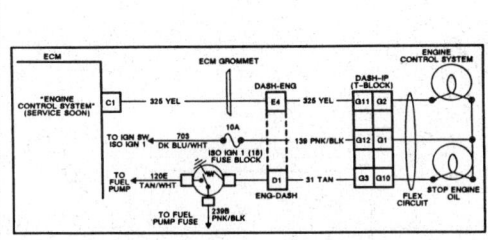

CHART A-1
"ENGINE CONTROL SYSTEM" TELLTALE INOPERATIVE

Description:

The "Engine Control System" lamp is powered through the 10 AMP "IGN-1" fuse 18 in the fuse block and grounded by the ECM Pin "C1" to illuminate. The ECS telltale will bulb check at key "ON" is illuminated steadily during cranking and is illuminated whenever a trouble code with service message is set. If a service trouble code is set, the ECM will send a service status to the BCM over the UART data link. At the same time, the ECM grounds Pin "C1" to turn "ON" the "Engine Control System" telltale.

Notes On Chart A-1

1. If Code E018 is present, diagnose Code E018 before proceeding. The ECS lamp illuminates on cranking only if the crank input is present to the ECM.
2. If Code B334 is current "No ECM Data" is displayed) then go to Chart A-1 Continued to check for loss of power or ground to the ECM.
3. If "Stop Engine/Oil" lamp is displayed at key "ON", engine "OFF", then power to the cluster telltales is ok. Need to diagnose for location of open ground circuit to the ECS telltale or for blown ECS telltale bulb.
4. Starting at the ECM, this branch of the chart follows CKT 325, grounding CKT 325 as the ECM would do to turn the bulb "ON". If a ground at a given point on CKT 325 brings the telltale "ON", then the circuit from the grounded point to the telltale is OK.

1988-90 CADILLAC ELDORADO AND SEVILLE

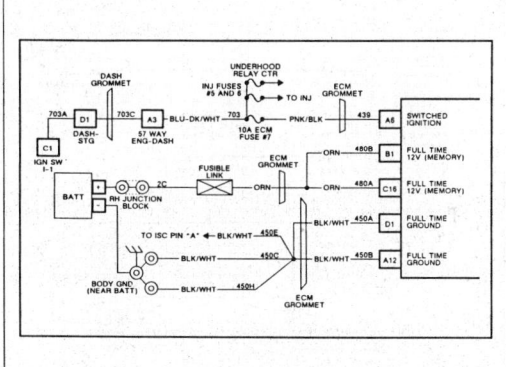

CHART A-1 (CONT'D)
"ENGINE CONTROL SYSTEM" TELLTALE INOP/CODE B334 CURRENT

If the "Engine Control System" telltale is inoperative and Code B334 is current, then the ECM may not be "powering up" at key ON, may not be receiving full time power from battery or may be damaged due to electrical overload or water intrusion.

Notes On Chart A-1 (Cont'd)

1. Checking for switched ignition power to ECM.
2. Checking for ground to ECM. Pin A12 and Pin D1 are redundant grounds – if either ground is OK, the ECM should be able to operate normally.

If both grounds are open, the ECM will not power up.

3. Checking for full time memory voltage supply to ECM. Pin C16 and B1 are redundant power supplies to the ECM. If either power supply is OK, the ECM can operate normally.
4. If power, grounds and connections are OK, then be sure to perform Chart C-1 before replacing ECM. The ECM may have been damaged by electrical overload due to a low resistance component.

1988-90 CADILLAC ELDORADO AND SEVILLE

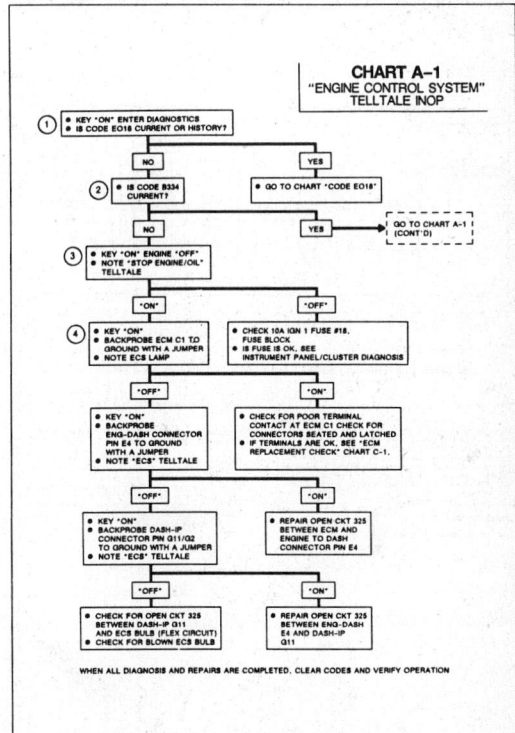

CHART A-1
"ENGINE CONTROL SYSTEM" TELLTALE INOP

WHEN ALL DIAGNOSIS AND REPAIRS ARE COMPLETED, CLEAR CODES AND VERIFY OPERATION

1988-90 CADILLAC ELDORADO AND SEVILLE

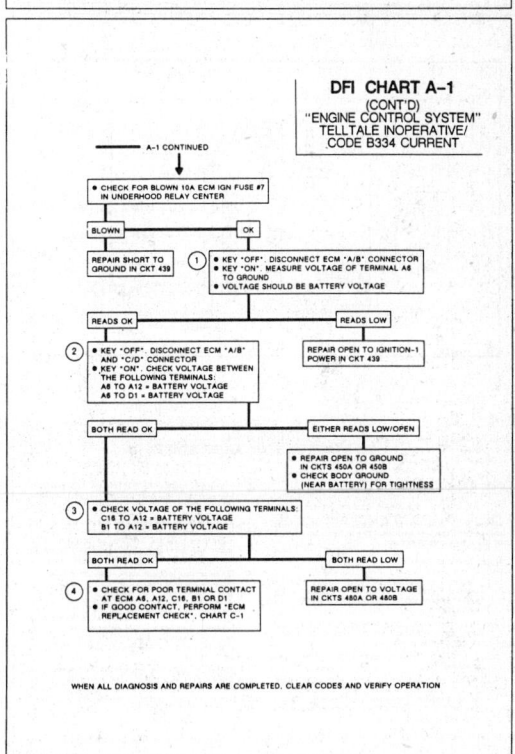

DFI CHART A-1
(CONT'D)
"ENGINE CONTROL SYSTEM" TELLTALE INOPERATIVE/ CODE B334 CURRENT

WHEN ALL DIAGNOSIS AND REPAIRS ARE COMPLETED, CLEAR CODES AND VERIFY OPERATION

1988–90 CADILLAC ELDORADO AND SEVILLE

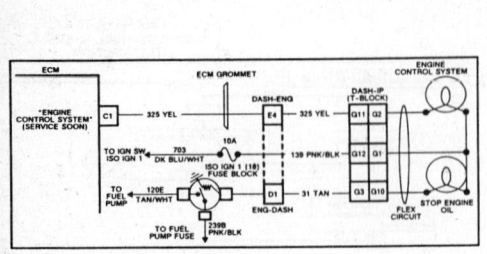

CHART A-2
"ENGINE CONTROL SYSTEM" TELLTALE ON, NO CODES PRESENT

Description:

The "Engine Control System" lamp is powered through the 10 AMP "Ign-1" Fuse 18 in the fuse block and grounded by the ECM Pin "C1" to illuminate. The ECS telltale will bulb check at key on, is illuminated steadily during cranking and is illuminated whenever a trouble code is set, the ECM will send a Service status to the BCM over the UART data link. At the same time, the ECM grounds Pin "C1" to turn "ON" the "Engine Control System" telltale.

Notes On Chart A-2

1. A flickering "Engine Control System" lamp may be caused by an incorrect "Set Timing Request".

2. Checking for bulb grounded by ECM through Pin "C1" or through a circuit fault.

3. Remove and reconnect both ECM connectors and ensure that they are latched. Check the Mem-Cal for proper orientation and fully seated in the Mem-Cal socket. Substitute a test Mem-Cal to see if the Mem-Cal is causing the light to flicker. If the test Mem-Cal does not correct the condition, replace the ECM. Be sure to perform "ECM Replacement Check", Chart C-1.

1988–90 CADILLAC ELDORADO AND SEVILLE

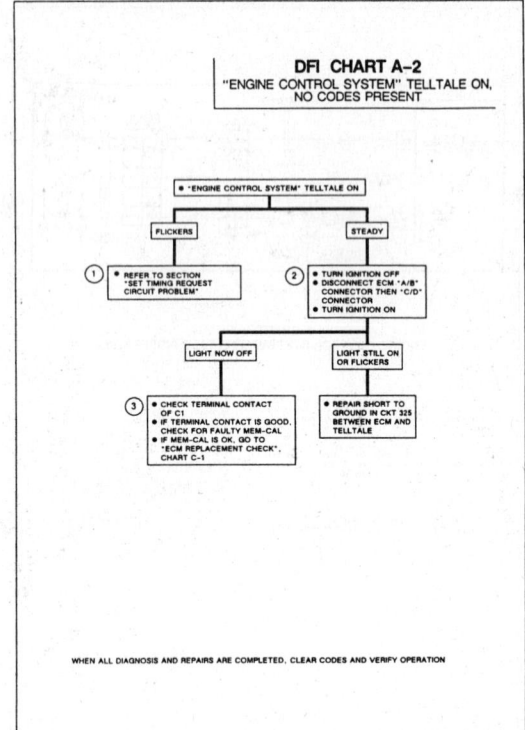

1988–90 CADILLAC ELDORADO AND SEVILLE

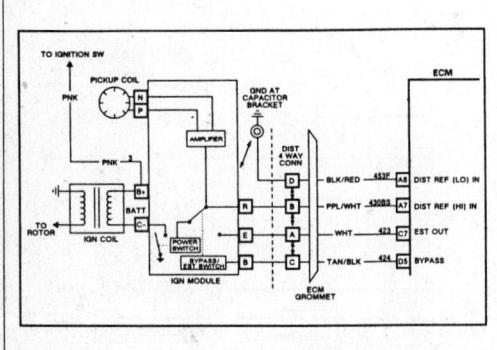

CHART A-3
NO START OR STALL AFTER START

Description:

All internal combustion engines require spark, fuel, air, and proper timing to operate. The DFI system is no different. If the battery is at the proper charge level, the first step should be to determine which of these elements is missing.

Notes On Fault Tree:

1. Checking for codes stored and for proper "Engine Control System" operation. Repair stored codes and improper telltale operation before preceding with chart A-3.

2. Injectors should spray only when the engine is cranking or running. Look for spray, drips or leaks at key on/engine off.

3. Checking for both injectors to spray fuel while cranking.

4. Fuel is OK, checking for spark. Note: Use an ST 125 to test. A spark plug with a wide gap or allowing a plug wire to arc to ground may not test the HEI for sufficient output and may damage the coil, cap or rotor.

5. This step bypasses the EST system. If the vehicle will start and run with the EST system disabled (Bypass open), then the "No Start" condition may be due to an EST system fault. If the ECM has a poor ground connection to the engine or if the distributor has a poor ground connection to the ECM, the distributor may not be able to recognize EST pulses and the engine will stall as the ECM tries to enable EST (use EST to control spark timing).

6. If the vehicle will not start, then the fuel system must be checked next. Connect the fuel pressure gage J-25400-300 and observe the fuel pressure while cranking. The gage should be installed in the fuel inlet line at the service fitting. A fuel pressure reading of between 9–12 PSI during cranking indicates that the fuel system is operating properly. Improper fuel pressure indicates a fuel problem, refer to Chart A-4B "Fuel Pressure Out Of Range".

Since we now have spark, and fuel spray from both injectors at the correct fuel pressure, all DFI functions for starting are operating normally. The cause of the no start condition is a mechanical problem (spark plugs, valves, valve timing, etc.).

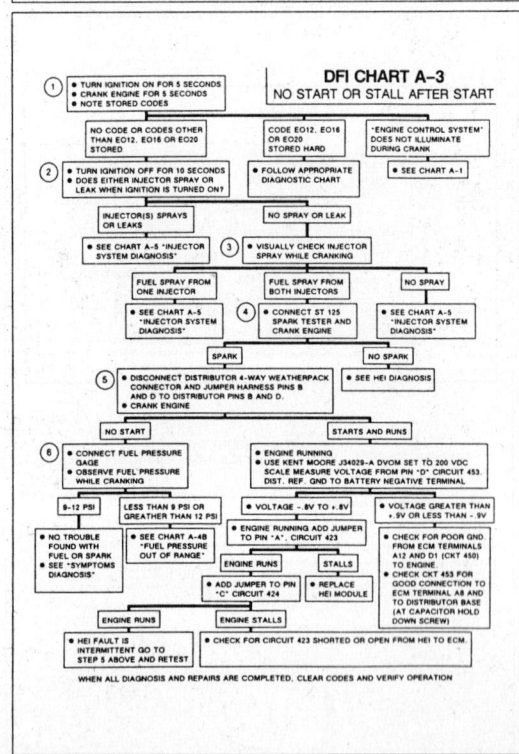

1988–90 CADILLAC ELDORADO AND SEVILLE

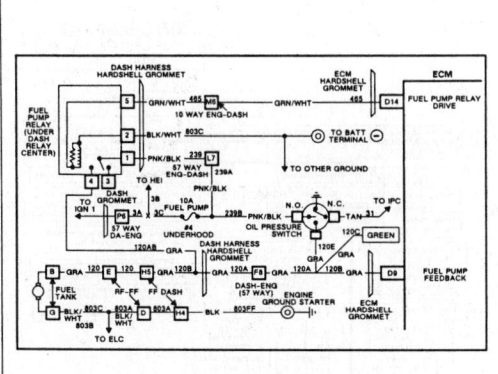

CHART A-4A
FUEL SYSTEM DIAGNOSIS

The DFI system requires that fuel pressure be 9.0–12.0 PSI and steady under all driving conditions. The J25400-300 (0–60 PSI) pressure gage may not show small variations in fuel pressure. For a more precise and more responsive gage reading, use the 0–15 PSI gage head (from the essential tool gage number J29658/BT8205) attached to the hose with schrader valve fitting from the essential tool gage J34730-1.

Notes On Fault Tree:
1. Fuel pressure should be 9.0–12.0 PSI and steady. If the fuel pressure is not in this range, go to chart A-4B to diagnose fuel pressure out of range.
2. If the pressure drops off (leaks down) at key off, either the fuel pressure regulator cannot hold pressure or the fuel pump check ball is not seating. This branch of the trouble tree will determine which is causing the pressure leak down.

1988–90 CADILLAC ELDORADO AND SEVILLE

DFI CHART A-4A
FUEL SYSTEM DIAGNOSIS

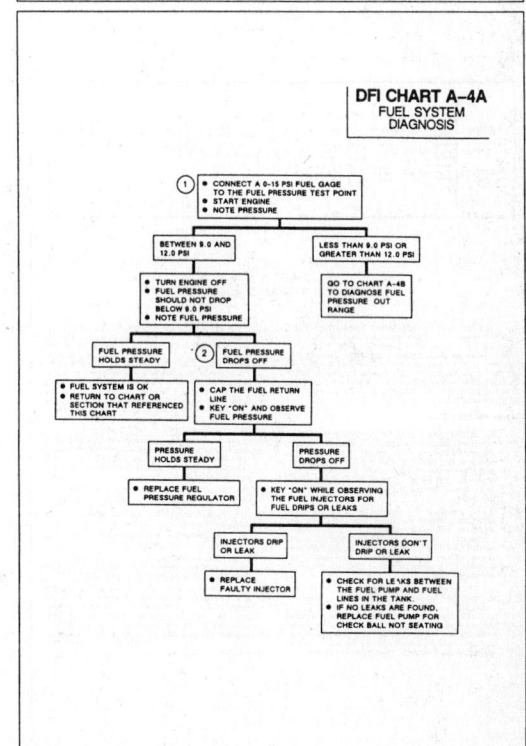

1988–90 CADILLAC ELDORADO AND SEVILLE

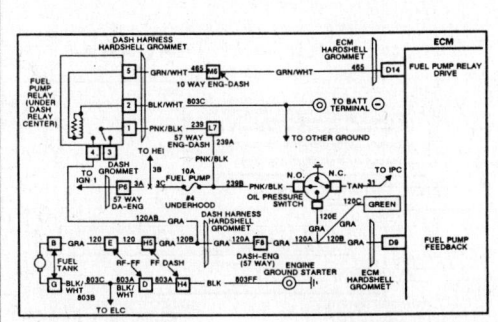

CHART A-4B, PRESSURE OUT OF RANGE DIAGNOSIS

This procedure tests for fuel supply system problems that can cause incorrect fuel pressure or incorrect fuel pump operation.

Notes On Fault Tree:
1. The fuel pressure gage J-25400-300 should be installed at the fuel line service fitting. Measure the fuel pressure while cranking the engine. If the fuel pressure is between 9.0 and 12.0 PSI, then refer to the performance diagnosis chart A-4A "Fuel System Diagnosis".
   NOTE: The DFI system requires that fuel pressure be 9.0–12.0 PSI and steady under all driving conditions. The J25400-300 0–60 PSI pressure gage may not show small variations in fuel pressure. For a more precise and more responsive gage reading, use the 0–15 PSI gage head (from the essential tool gage number J29658/BT8205) attached to the hose with schrader valve fitting from the essential tool gage J34730-1.
2. If the fuel pump relay or ECM were the cause of a low fuel pressure, there would be an ECM code E020 set. This step is to check for voltage supply to the fuel tank five-way connector.
3. If the voltage signal to the fuel tank connector is OK, then an open may exist between the five-way fuel tank connector and the fuel pump. If fuel pump runs with an alternative power source connected, the fuel tank unit is OK; check throttle body fuel metering assembly for cause of low pressure.
4. Checking for fuel supply system (tank, filter, pump, sender, supply line) able to deliver at least 9.0 PSI pressure or for throttle body fuel pressure regulator fault.
5. If pressure is low with fuel return line plugged, then throttle body fuel metering assembly is not at fault. A restriction or blockage may exist in the

fuel supply system. The fuel supply line should be checked visually for kinks, damage, etc.; the fuel filter element can also restrict flow. Check for proper fuel line routing.

Check sender tubes for restrictions, check for rubber coupler between pump and sender leaking or restricted, check for fuel strainer in tank collapsed, mispositioned or restricted. If all of the above are OK, replace the fuel pump.
6. Fuel pressure above 12.0 PSI is caused either by a malfunction of the pressure regulator or by a restriction in the fuel return line. It should be noted that a secondary condition of spark plug fouling, code E045 or oxygen sensor contamination resulting in code E013 accompanied by E045 may result from the too rich fuel flow. To isolate the cause of the high fuel pressure, disconnect the return line at the throttle body and connect a suitable fitting to the throttle body which will accept a length of flexible rubber fuel hose. Insert the other end of the hose into a suitable fuel container and observe the fuel pressure as the ignition switch is turned on. If the fuel pressure remains above 12.0 PSI, replace the fuel metering assembly—it is unable to control pressure properly.
If the fuel pressure drops into the correct pressure range of 9.0–12.0 PSI, with the fuel return line bypassed, then the fuel return line is restricted. A restricted fuel return line can be diagnosed by visually inspecting the line for kinks, damage, etc. A kink in the Teflon fuel line (braided stainless steel clad) may not be visually obvious.
7. If the fuel pump will not run with externally applied power, the fault is an electrical open in the fuel sender unit wiring to the pump, an open at the RFI suppression connector inside the tank on the pump or a faulty fuel pump. The fuel sending unit must be removed from the vehicle to check.

1988–90 CADILLAC ELDORADO AND SEVILLE

DFI CHART A-4B
PRESSURE OUT OF RANGE DIAGNOSIS

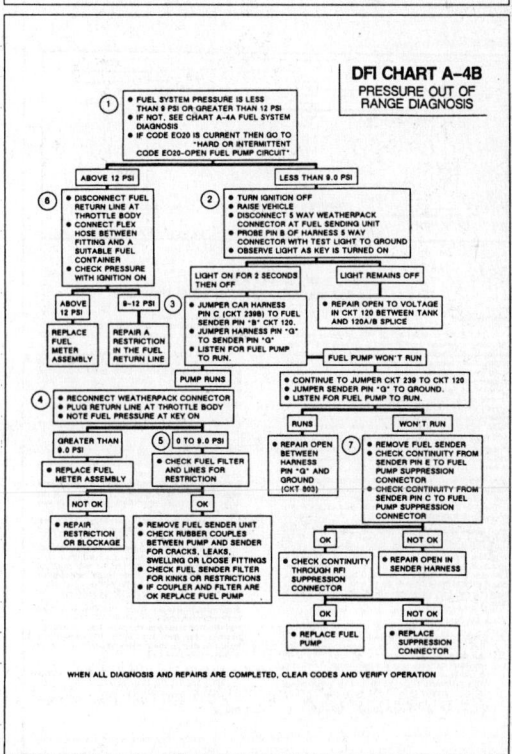

WHEN ALL DIAGNOSIS AND REPAIRS ARE COMPLETED, CLEAR CODES AND VERIFY OPERATION

1988–90 CADILLAC ELDORADO AND SEVILLE

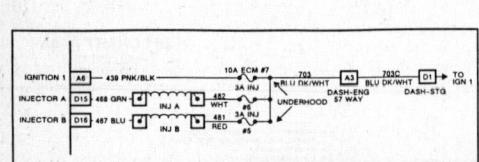

CHART A-5
INJECTOR SYSTEM DIAGNOSIS

The fuel injectors are powered through the 3 AMP injector fuses. The ECM turns on the injectors by applying a ground to ECM connector pins "D15" and "D16", injector drive circuits. The ECM grounds the injector drive circuits to turn on the fuel injector to supply fuel to the engine. ECM Data Parameter ED06 Injector Pulse width, reflects the time in milli-seconds that the ECM turns on the ground to the injectors for each injector pulse.

Notes On Fault Tree:

1. If this procedure is being followed for a "no-start" condition, crank the engine for 5 seconds to check the distributor reference signal. If a Code E012 does not set, turn the ignition off for 10 seconds and observe the injectors as the ignition is turned back on. If there is no spray, then they are not stuck open. Observe the injectors while cranking the engine.

2. To determine if the injector is being activated electrically, repeat the above procedure with the electrical connector removed. If the injector continues to spray, it is defective and must be replaced. If the injector no longer sprays, the drive circuit of the affected injector must either be shorted to ground or the ECM is grounding internally.

3. If both injectors spray or if neither injector sprays, it must be determined if the fuel system is operating properly. The fuel pressure gage J-25400-300 should be installed at the fuel line service fitting. Measure the fuel pressure while cranking the engine. If the fuel pressure is not between 9-12 PSI, then refer to the performance diagnosis chart A-4A, "Fuel System Diagnosis".

4. If the fuel pressure is between 9-12 psi and there was no spray from either injector while cranking, the injector circuit must be checked for proper voltage.
   If there is voltage at the injector fuses, then the ECM must be faulty because it is not grounding both injector circuits. If there is no voltage at the fuse, check for voltage at the 10 amp ECM fuse which feeds the injector fuses. If there is voltage here then an open must exist between the fuses on circuit #439. If there is no voltage at the ECM fuse then circuit #3 must be repaired for an open or short to ground.

5. If the fuel pressure is between 9-12 psi and there was fuel spray from both injectors while cranking, then check the injector system to determine if the injectors leak. To check for injector leakage, proceed as follows:
   A. Start and run the engine for 10 seconds.
   B. Turn the engine off for at least 10 seconds.
   C. Turn the ignition on to pressurize the injectors.
   D. Visually check for dripping fuel from the bottom of the injector.
   If the fuel is dripping, check for damaged "O" rings and if OK, replace the injector. If the fuel does not drip, then the fuel system is OK.

6. If there is spray from only one injector, then there is a malfunction in the injector assembly or in the signal to the injector assembly. The malfunction can be isolated by switching the injector connectors. If the problem remains with the original injector after switching the connector, then the injector is defective. Replace the injector.

1988–90 CADILLAC ELDORADO AND SEVILLE

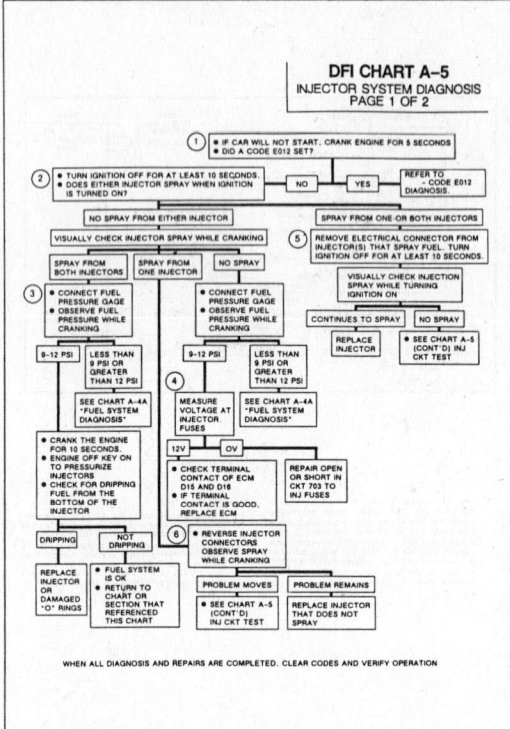

DFI CHART A-5
INJECTOR SYSTEM DIAGNOSIS
PAGE 1 OF 2

WHEN ALL DIAGNOSIS AND REPAIRS ARE COMPLETED, CLEAR CODES AND VERIFY OPERATION

1988–90 CADILLAC ELDORADO AND SEVILLE

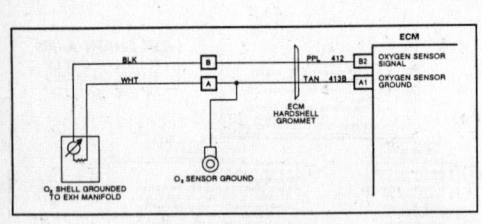

DFI CHART A-6
Rich or Lean Exhaust Code (E044 or E045) Diagnosis

Description:

The ECM provides a 0.45 volt reference signal to the oxygen sensor on CKT 412. When the oxygen sensor is cold (below 200°C), the output voltage will be around 0.45 volts and the ECM will keep the system in open loop operation. When warm, a properly operating oxygen sensor will drive the 0.45 volt reference lower (below 0.45 volts) to indicate a lean mixture and higher (above 0.45 volts) to indicate a rich mixture. The oxygen sensor signal voltage will swing rich and lean rapidly, at least one swing every two seconds – if the ECM is in good control of the air fuel mixture.

Low oxygen sensor voltage readings are normally evidence that the air-fuel mixture is lean and the closed loop system is unable to compensate sufficiently due to a failure in some part of engine emission or fuel system.

Fixed high oxygen sensor voltage readings are normally evidence that the air-fuel mixture is rich and the closed loop system is unable to compensate sufficiently due to a failure in some part of the engine emission or fuel system.

Less likely is the possibility that the oxygen sensor has failed and is giving an incorrect reading. If the oxygen sensor is giving false rich readings, the closed loop fuel system will be overcompensating and causing lean operation while the O_2 sensor is indicating rich. Likewise, a lean indication will cause rich operation (i.e. black smoke, fouled spark plugs, poor fuel economy, high HC and CO, etc.) while the O_2 sensor is indicating lean.

Notes On Fault Tree:

1. Steps contained in this section of the diagnostic tree could cause both an indicated rich and lean operation as follows:
   – A fuel delivery system which is not functioning properly may cause an incorrect fuel mixture. This malfunction can be caused by fuel pressure which is greater than 12 PSI or less than 9 PSI at the injectors, by defective injectors, etc. Refer to Chart A-4A "Fuel System Diagnosis" for

additional information.
   – Injector dripping or not injecting can cause rich or lean exhaust (depending on failure). To check injectors, see Chart A-5.
   – Excessive EGR flow displaces oxygen and causes a rich exhaust indication. A loss of EGR will cause a lean exhaust indication. Refer to EGR diagnosis in Chart C-7, for additional information.

2. If a Code was not set, the "DFI System Check" did not refer to the "Oxygen Sensor Circuit Check" which will be required before further diagnosis.

3. Steps contained in this section could cause only an indicated rich operation as follows:
   – Carbon canister loaded with fuel can cause rich operation, see Chart C-3.
   – Vacuum leak to the MAP hose can cause a false high MAP reading. High MAP readings cause the ECM to deliver too much fuel for current driving conditions.
   – A restricted air cleaner could cause a rich fuel mixture. Inspect the cleaner and replace if necessary.
   – If none of the above have indicated a problem, the code should be cleared and a road test should be performed. This code can be the result of extreme purge rates under high temperature conditions, in such a case no further diagnosis is required and no repairs should be attempted.

4. Steps contained in this section could cause only an indicated lean operation as follows:
   – If the AIR management system were to send air to the exhaust ports at all times, this would give a lean indication at the oxygen sensor. Refer to Air Management Diagnosis
     – Check for vacuum leaks at hoses, intake manifold and throttle body gaskets.
     – Check for fuel contamination.
     – If no problems found, replace the Oxygen Sensor.

1988–90 CADILLAC ELDORADO AND SEVILLE

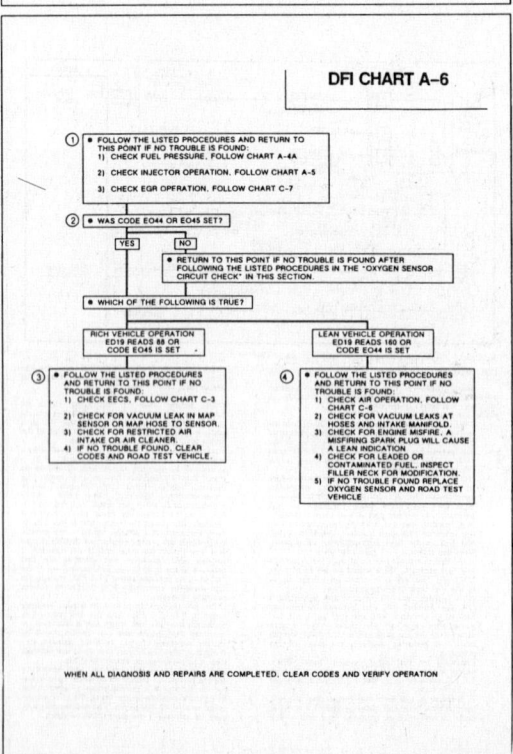

DFI CHART A-6

WHEN ALL DIAGNOSIS AND REPAIRS ARE COMPLETED, CLEAR CODES AND VERIFY OPERATION

1988–90 CADILLAC ELDORADO AND SEVILLE

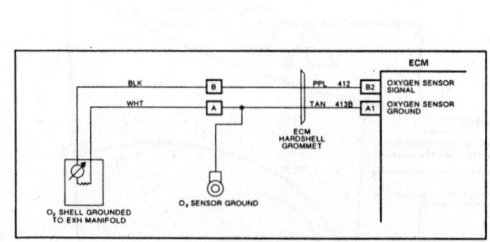

DFI CHART A-7

Open Loop Diagnosis (Code EO13)

Description:

The ECM provides a 0.45 volt reference signal to the oxygen sensor on CKT 412. When the oxygen sensor is cold (below 200°C), the output voltage will be around 0.45 volts and the ECM will keep the system in open loop operation. When warm, a properly operating oxygen sensor will drive the 0.45 volt reference lower (below 0.45 volts) to indicate a lean mixture and higher (above 0.45 volts) to indicate a rich mixture. The oxygen sensor signal voltage will swing from rich to lean rapidly, at least one swing every two seconds – if the ECM is in good control of the air fuel mixture.

Notes On Fault Tree:

1. If a Code was not set, the "DFI System Check"

did not refer to the "Oxygen Sensor Circuit Check" which will be required before further diagnosis. If the "Oxygen Sensor Circuit Check" indicates a problem, clear codes and road test the vehicle.

2. The Code EO13 is intermittent and the remainder of the Oxygen sensor circuit has passed the "Oxygen Sensor Circuit Check". If none of the above have indicated a problem, the code should be cleared and a road test should be performed.

3. Check for fuel contamination which would result in the failure of the Oxygen Sensor replace the Oxygen Sensor and road test the vehicle.

1988–90 CADILLAC ELDORADO AND SEVILLE

DFI CHART A-7

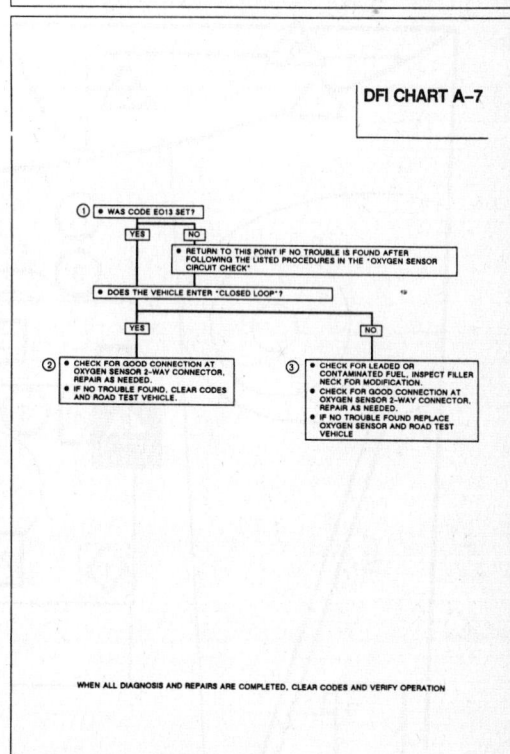

WHEN ALL DIAGNOSIS AND REPAIRS ARE COMPLETED, CLEAR CODES AND VERIFY OPERATION

1988–90 CADILLAC ELDORADO AND SEVILLE

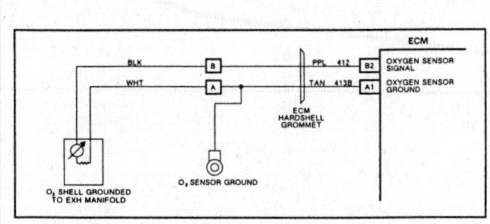

DFI CHART A-8

Rich or Lean Driveability Diagnosis

The ECM provides a 0.45 volt reference signal to the oxygen sensor on CKT 412. When the oxygen sensor is cold (below 200°C), the output voltage will be around 0.45 volts and the ECM will keep the system in open loop operation. When warm, a properly operating oxygen sensor will drive the 0.45 volt reference lower (below 0.45 volts) to indicate a lean mixture and higher (above 0.45 volts) to indicate a rich mixture. The oxygen sensor signal voltage will swing from rich to lean rapidly, at least one swing every two seconds – if the ECM is in good control of the air fuel mixture.

Low oxygen sensor voltage readings are normally evidence that the air-fuel mixture is lean and the closed loop system is unable to compensate sufficiently due to a failure in some part of engine emission or fuel system.

Fixed high oxygen sensor voltage readings are normally evidence that the air-fuel mixture is rich and the closed loop system is unable to compensate sufficiently due to a failure in some part of the engine emission or fuel system.

Less likely is the possibility that the oxygen sensor has failed and is giving an incorrect reading. If the oxygen sensor is giving false rich readings, the closed loop fuel system will be overcompensating and causing lean operation while the O₂ sensor is indicating rich. Likewise, a lean indication will cause rich operation (i.e. black smoke, fouled spark plugs, poor fuel economy, high HC and CO, etc.) while the O₂ sensor is indicating lean.

Notes On Fault Tree:

1. Steps contained in this section of the diagnostic tree could cause both an indicated rich or lean operation as follows:
   – The "DFI System Check" did not refer to the "Oxygen Sensor Circuit Check" which will be required before further diagnosis.
   – A fuel delivery system which is not functioning properly may cause an incorrect fuel mixture.

This malfunction can be caused by fuel pressure which is greater than 12 PSI or less than 9 PSI at the injectors, by defective injectors, etc. Refer to Chart A-4A "Fuel System Diagnosis" for additional information.
   – Injector dripping or not injecting can cause rich or lean exhaust (depending on failure). To check injectors, see Chart A-5.
   – Excessive EGR flow displaces oxygen and causes a rich exhaust indication. A loss of EGR will cause a lean exhaust indication.

2. Steps contained in this section could cause rich operation as follows:
   – A restricted air cleaner could cause a rich fuel mixture. Inspect the cleaner and replace if necessary.
   – Carbon canister loaded with fuel can cause rich operation, see chart C-3.
   – If the AIR management system were to send air to the exhaust ports at all times, this would give a lean indication at the oxygen sensor. Refer to Air Management Diagnosis

   – Vacuum leak to the MAP sensor can cause a false high MAP reading. High MAP readings cause the ECM to deliver too much fuel for current driving conditions.
   – If none of the above have indicated a problem, symptom diagnosis is required.

3. Steps contained in this section could cause only an indicated lean operation as follows:
   – Check for vacuum leaks at hoses, intake manifold and throttle body gaskets.
   – Check for fuel contamination.
   – If none of the above have indicated a problem, symptom diagnosis is required.

1988–90 CADILLAC ELDORADO AND SEVILLE

DFI CHART A-8

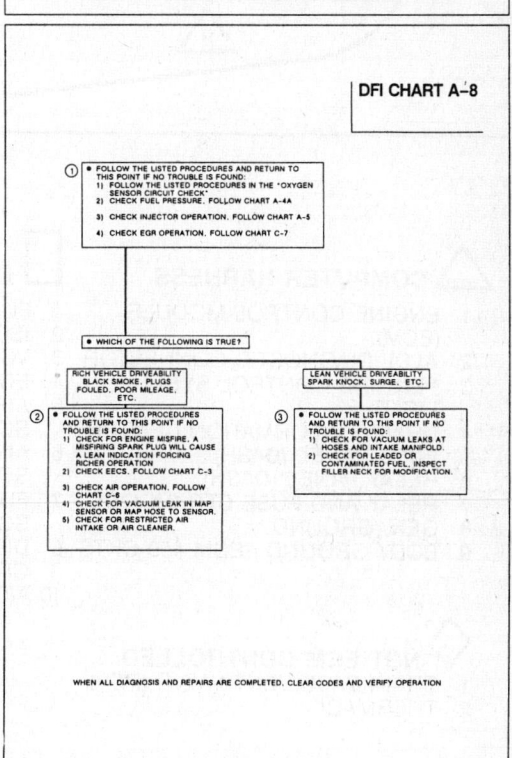

WHEN ALL DIAGNOSIS AND REPAIRS ARE COMPLETED, CLEAR CODES AND VERIFY OPERATION

1988–90 CADILLAC ELDORADO AND SEVILLE

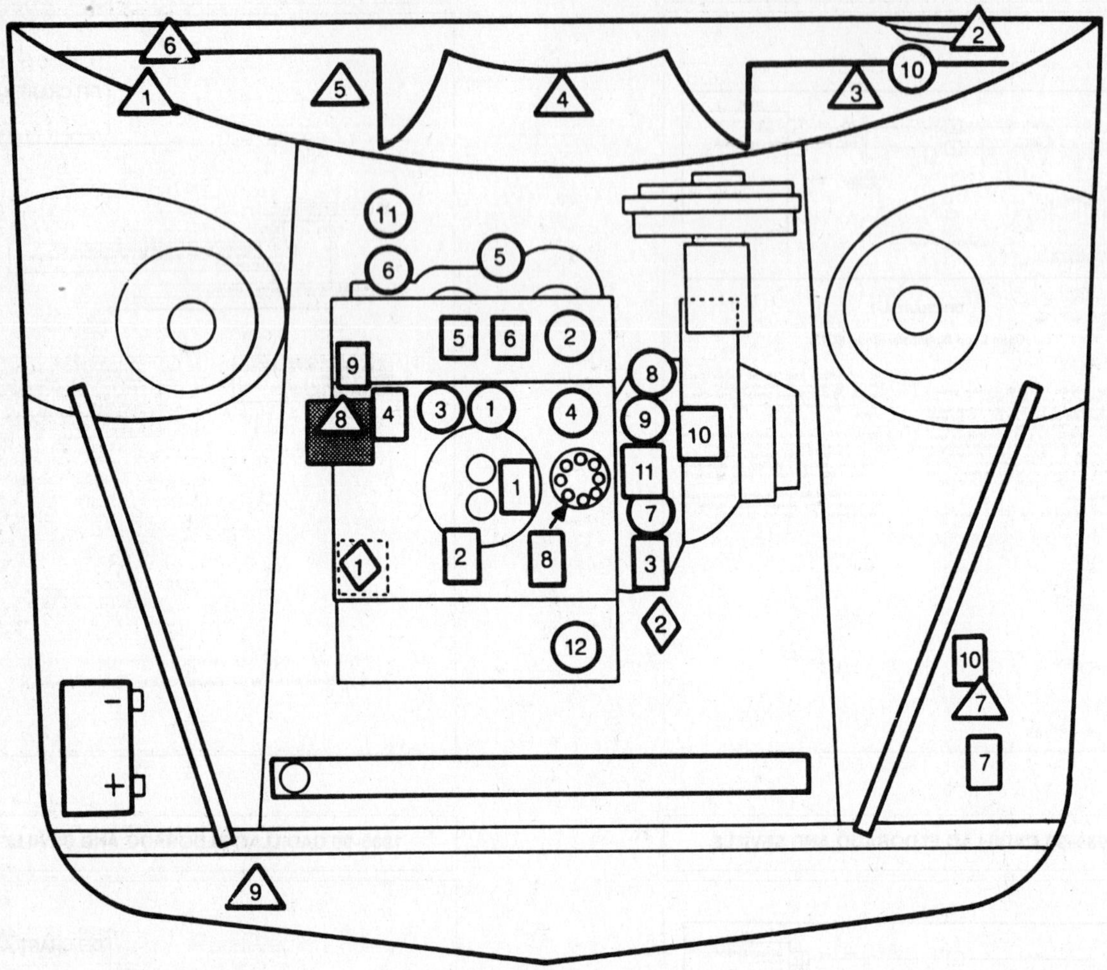

COMPUTER HARNESS

1 ENGINE CONTROL MODULE (ECM)
2 ALDL DIAGNOSTIC CONNECTOR
3 "ENGINE CONTROL SYSTEM" LIGHT
4 DRIVER INFORMATION CENTER
5 FUSE PANEL (DASH)
6 RELAY PANEL (DASH)
7 RELAY AND FUSE CTR (HOOD)
8 GEN. GROUND
9 BODY GROUND - ECM 450 CKTS

CONTROLLED DEVICES

1 FUEL INJECTORS
2 ISC MOTOR
3 VCC
4 EGR SOLENOID
5 AIR INJECTION DIVERT SOLENOID
6 AIR INJECTION SWITCHING SOLENOID
7 FUEL VAPOR CANISTER SOLENOID
8 DISTRIBUTOR
9 CRUISE CONTROL
10 A/C RELAY

INFORMATION SENSORS

1 TPS
2 MAP
3 MAT
4 CTS
5 O_2
6 VSS
7 3RD AND 4TH SWITCH
8 P/N SAFETY SWITCH
9 OIL PRESSURE SWITCH
10 BRAKE SWITCH
11 P/S SWITCH
12 TRANS. TEMP. SENSOR

NOT ECM CONTROLLED

1 CRANKCASE VENT VALVE (PCV)
2 THERMAC

1988–90 CADILLAC ELDORADO AND SEVILLE

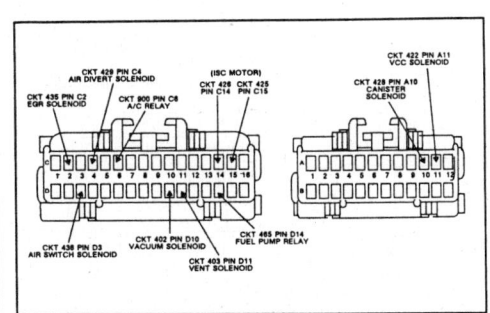

DFI CHART C-1

ECM REPLACEMENT CHECK

Description:

Prior to replacing an ECM, the vehicle must be tested for :
1. Poor connector terminal to ECM contact.
2. Direct short to battery voltage on an ECM ground circuit.
3. Shorted solenoids or relays. If a short is found, the circuit must be repaired prior to replacing the ECM to prevent repeat ECM failures.

Notes on Fault Tree:

1. Check for poor terminal contact due to weak or dirty ECM terminals. Remove suspected terminals to inspect, replace if broken or dirty. If coolant is present at the ECM connector, replace coolant sensor, coolant sensor connector, and the coolant sensor signal and ground wires. Also, replace ECM connector terminal and blow coolant out of harness. Clean ECM connector with alcohol or spray contact cleaner and replace ECM.

2. Checking for a short to ignition or shorted solenoid or relay. All terminals must be tested since several are connected together internally in the ECM. A short in one circuit may cause another circuit in the ECM to be inoperative. An example might be that the VCC solenoid shorted can cause both VCC to be inoperative and the EGR to be enabled at all times. Any circuit testing below 20 ohms is shorted and should be diagnosed for the cause of the short.

3. Check ISC circuit (EXTEND/RETRACT) for shorted to ground or shorted together. Normal resistance for an ISC motor circuit is 4–100 ohms.

1988–90 CADILLAC ELDORADO AND SEVILLE

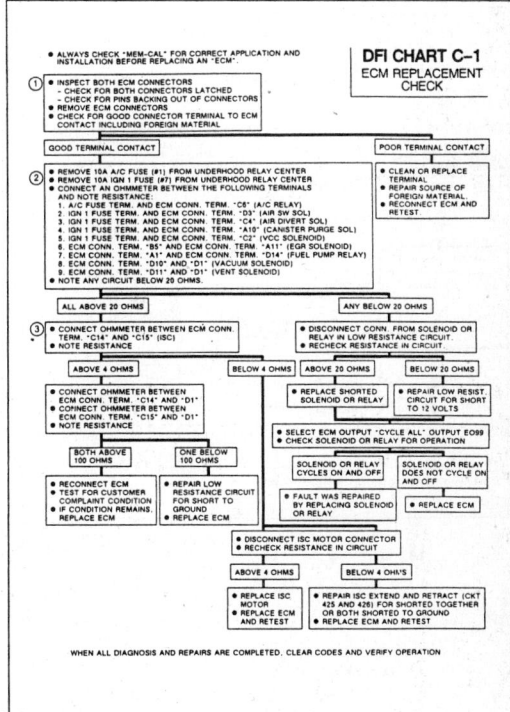

1988–90 CADILLAC ELDORADO AND SEVILLE

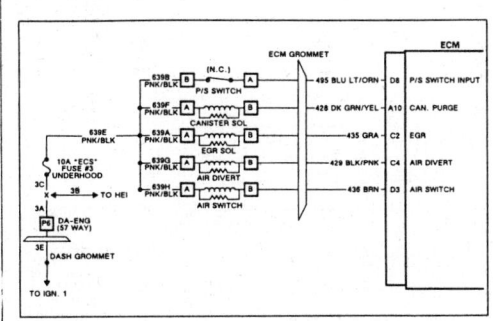

DFI CHART C-3

CANISTER PURGE CONTROL DIAGNOSIS

Description

The canister purge solenoid receives 12 volts from the 10 amp number 3 fuse on the underhood relay panel through circuit 639. The ECM energizes the canister purge solenoid by grounding pin A10 (circuit 428). When the solenoid is energized it allows the canister to purge.

The canister is commanded to purge when:

- Coolant is above 70°C.
- Closed loop has been achieved for at least 9 seconds.
- Throttle switch is open.
OR
- Code E013, E044, or E045 is present.

The ECM will de-energize the solenoid (not allow purge) when codes E016 is set or the ECM is running in the backup mode (no normal program control).

Notes on Fault Tree:

1. Checking for the canister purge solenoid able to hold vacuum. If the solenoid can't hold vacuum, it should be replaced.
2. Vacuum should release when the solenoid is cycled on in output cycling.

3. Checking for proper vacuum signals from the throttle body.
4. Checking for proper electrical signals to the canister purge solenoid.

FUNCTIONAL TEST OF CHARCOAL CANISTER

Disconnect the hose leading from the TPCV to the lower port of the charcoal canister at the canister end. Attach a hose to the fuel tank port of the TPCV and attempt to blow through it. Little or no air should pass through the TPCV.

With a hand vacuum pump, apply vacuum (15" Hg. or 51 kPa) through the TPCV upper tube (manifold vacuum supply port). If the diaphragm does not hold vacuum for at least 20 seconds, the diaphragm is leaking, and the TPCV must be replaced.

If the diaphragm holds vacuum, try to blow through the hose connected to the lower tube while vacuum is still being applied. An increased flow of air should be observed. If not, replace the TPCV.

Attach a hose to the lower port of the canister and attempt to blow through it. Air **should** pass into the canister. If not, replace the canister.

Disconnect the canister purge solenoid electrical connector, attach a hose to the upper port of the canister, and attempt to blow through it. Air **should not** pass into the canister. If air passes, replace the purge solenoid.

Measure the resistance of the purge solenoid. Resistance should be greater than 20 ohms and less than 100 ohms. If the solenoid resistance is not in this range, replace the solenoid.

1988–90 CADILLAC ELDORADO AND SEVILLE

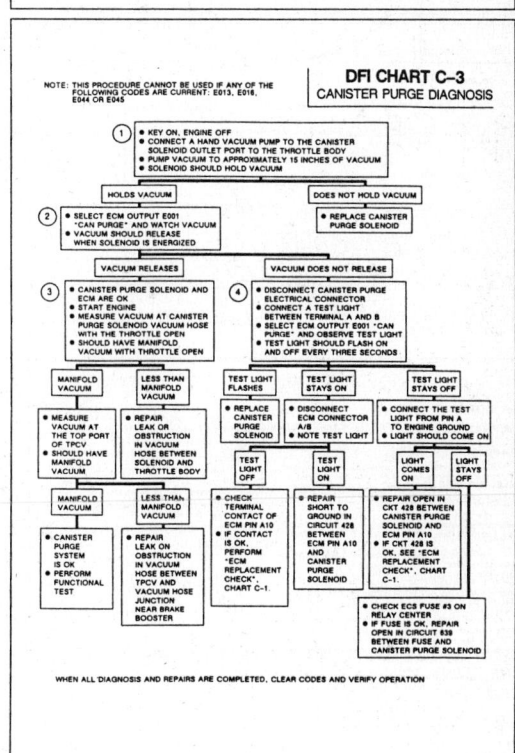

1988–90 CADILLAC ELDORADO AND SEVILLE

ON–CAR SERVICE

SETTING TIMING

The initial base timing is set by jumpering pins A&B together at the ALDL connector while not in **diagnostic display.** The ALDL connector is located near the parking brake pedal just below the dash. Then set the timing to the specification shown on the Emission Control Information label located in the engine compartment. Jumpering pin A to pin B of the ALDL connector will cause the "SET TIMING" message to appear on the CCDIC which indicates that the ECM is in set timing mode.

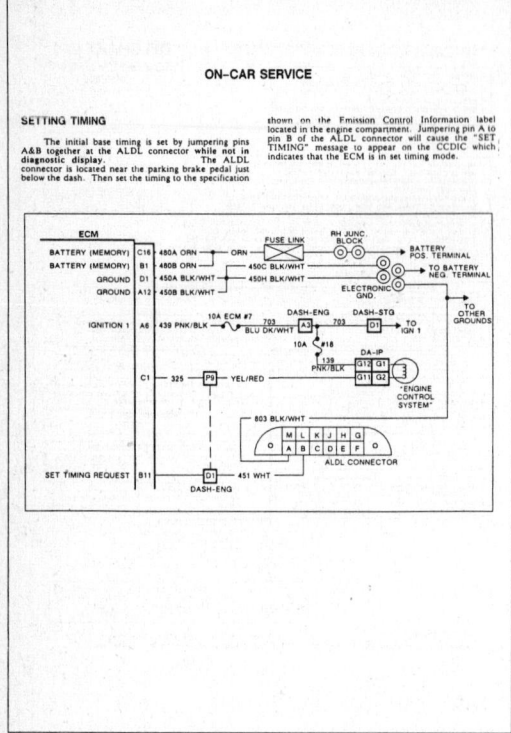

1988–90 CADILLAC ELDORADO AND SEVILLE

SET TIMING CIRCUIT CHECK AND DIAGNOSIS

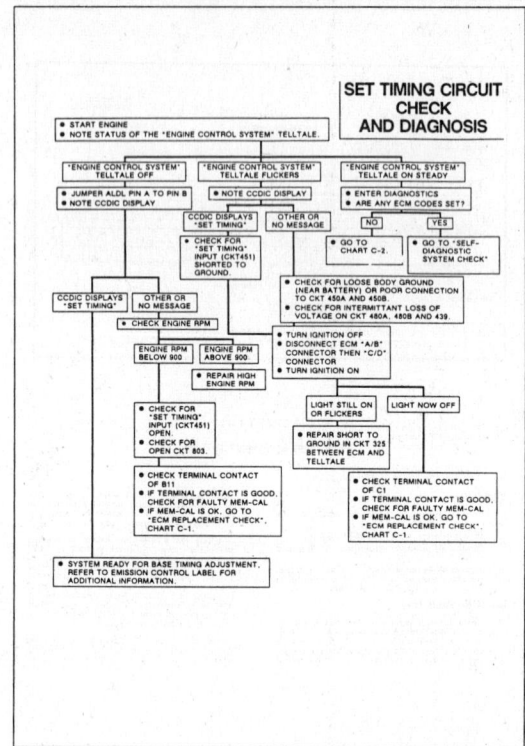

1988–90 CADILLAC ELDORADO AND SEVILLE

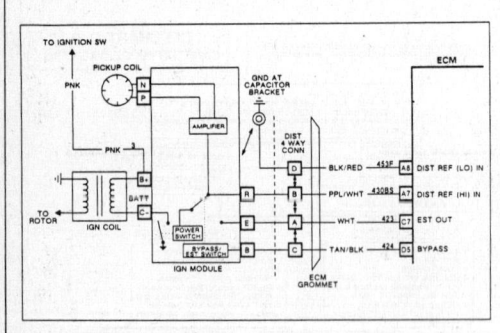

DFI CHART C-4

DFI SYSTEM CHECK

Description

The HEI distributor produces one 5 volt Distributor Reference Pulse each time that a cylinder reaches 10° BTDC (base timing). The reference pulses are sent from the distributor to the ECM on Circuit 430, Distributor Reference Signal.

The ECM adds spark timing information to the reference pulses received from the HEI and sends out 5 volt Electronic Spark Timing pulses (EST) on Circuit 423.

The ECM can choose to control spark timing. The HEI Bypass Circuit 424 is turned on by the ECM (approximately 5 volts) when the ECM wishes to control timing. When the ECM wants the HEI module to control timing, the HEI Bypass circuit is turned off and the HEI module grounds Circuit 423, EST. The ECM always sends EST pulses to the HEI; the HEI module shorts the EST pulses to ground when the bypass signal from the ECM to the module is low (less than .5 volts).

The Distributor Reference Ground (Circuit 453) is a common ground between the ECM and the HEI. The ECM and HEI compare Bypass, EST, and Reference voltages to the ground voltage on Circuit 453. Circuit 453 open to the ground voltage can cause a failure of the ECM to recognize a reference pulse or of the module to recognize an EST pulse.

The ECM compares the status of the EST circuit to the status of the Bypass circuit in order to detect a

fault and set code E023. If the ECM has Bypass at 5 volts, the module should accept EST and EST voltage should be high (.5 – 2.8 volts).

Possible causes of code E023 are:

A.   Circuit 423 (EST) open/ground.

B.   Circuit 424 (Bypass) open/grounded.

C.   Circuit 453 (Reference Ground) open to ground at HEI or ECM.

D.   ECM with a poor connection to the engine through the Circuit 430 grounds.

E.   ECM or HEI module faults.

Notes on Fault Tree:

1.   Four jumper wires must be obtained for use in the diagnostic procedure. The jumpers should be about 12 inches long with a male and female weatherpack connector on either end (P/N 12014836 and 12014837). With the distributor connector disconnected, use the jumpers to reconnect the terminals per instructions provided.

2.   With only Distributor Reference and Distributor Reference Ground jumpered, engine will run on backup spark. Checking for proper ground connection between ECM and the engine and the ECM and HEI.

3.   Checking for ECM providing an EST output on Circuit 423.

4.   Checking for HEI module able to ground EST signal with open Bypass Circuit.

1988–90 CADILLAC ELDORADO AND SEVILLE

DFI CHART C–4
EST SYSTEM CHECK

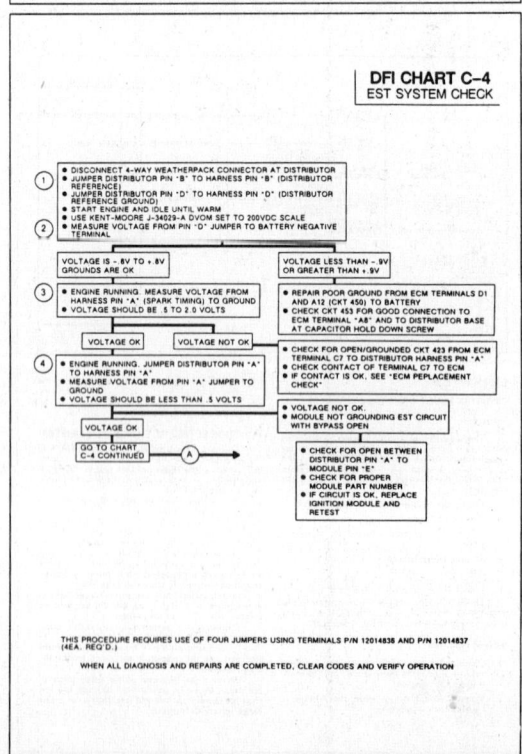

THIS PROCEDURE REQUIRES USE OF FOUR JUMPERS USING TERMINALS P/N 12014838 AND P/N 12014837 (4EA. REQ'D.).

WHEN ALL DIAGNOSIS AND REPAIRS ARE COMPLETED, CLEAR CODES AND VERIFY OPERATION

1988–90 CADILLAC ELDORADO AND SEVILLE

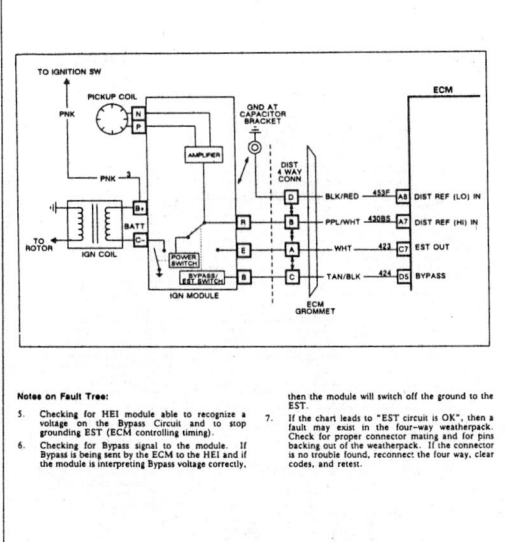

Notes on Fault Tree:

5. Checking for HEI module able to recognize a voltage on the Bypass Circuit and to stop grounding EST (ECM controlling timing).

6. Checking for Bypass signal to the module. If Bypass is being sent by the ECM to the HEI and if the module is interpreting Bypass voltage correctly,

then the module will switch off the ground to the EST.

7. If the chart leads to "EST circuit is OK", then a fault may exist in the four-way weatherpack. Check for proper connector mating and for pins backing out of the weatherpack. If the connector is no trouble found, reconnect the four way, clear codes, and retest.

1988–90 CADILLAC ELDORADO AND SEVILLE

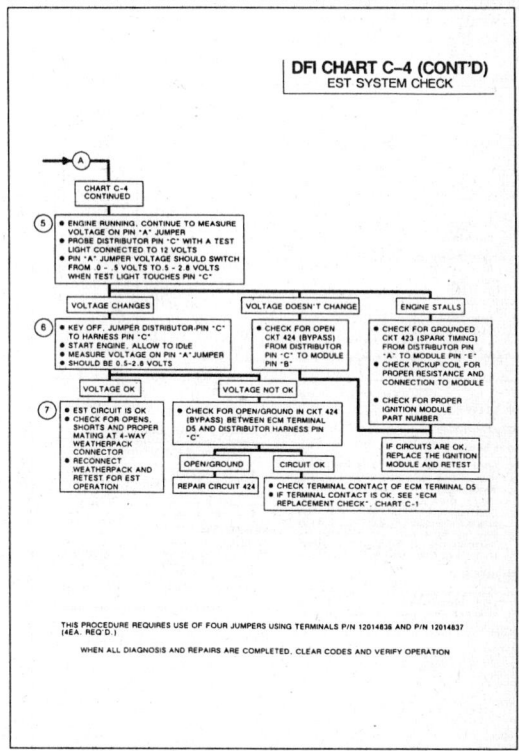

DFI CHART C–4 (CONT'D)
EST SYSTEM CHECK

THIS PROCEDURE REQUIRES USE OF FOUR JUMPERS USING TERMINALS P/N 12014836 AND P/N 12014837 (4EA. REQ'D.)

WHEN ALL DIAGNOSIS AND REPAIRS ARE COMPLETED, CLEAR CODES AND VERIFY OPERATION

1988–90 CADILLAC ELDORADO AND SEVILLE

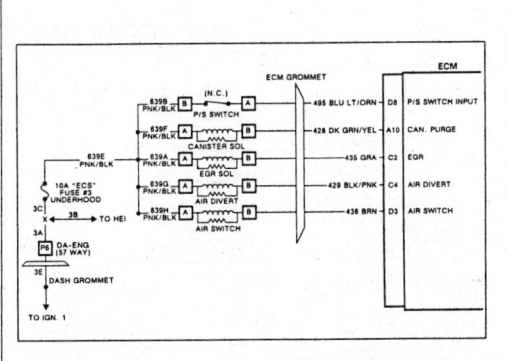

DFI CHART C-6

AIR MANAGEMENT DIAGNOSIS

Description

The AIR divert and switch solenoids are supplied with 12 volts on circuit 639 from the 10 amp number 3 solenoid fuse on the underhood relay panel. The ECM controls air management by grounding AIR switch solenoid circuit 436 and AIR divert solenoid circuit 429.

To divert air to the air cleaner, the divert solenoid is de-energized. When pin C4 at the ECM is grounded, the divert solenoid energizes and diverts air to the switching valve. When the switch solenoid is de-energized the air is routed to the catalytic converter and when pin D3 at the ECM is grounded the air is routed to the exhaust ports.

Notes on Fault Tree:

1. With a cold engine, the air management system should switch air to the exhaust ports (the heads)

to help oxidize the rich mixture that exists during cold starts and warm-up.

2. With a warm engine, the air should be switched to the converter to oxidize the exhaust gas flowing to the second bed of the catalytic converter.

3. Checking for the air system to "Divert" air away from the converter during acceleration and deceleration. The divert function protects the converter from the rich conditions that occur A.) during acceleration B.) during power enrichment and C.) when the throttle is closed to start deceleration.

4. Checking for vacuum supply to the air management valves.

5. Checking for air pump operating.

6. This branch of the tree will check for ECM ability to control air switch and air divert solenoids.

7. Checking for open to voltage to solenoid.

8. Checking for a low resistance solenoid. If resistance is less than 20 ohms, replace the air management solenoid.

1988–90 CADILLAC ELDORADO AND SEVILLE

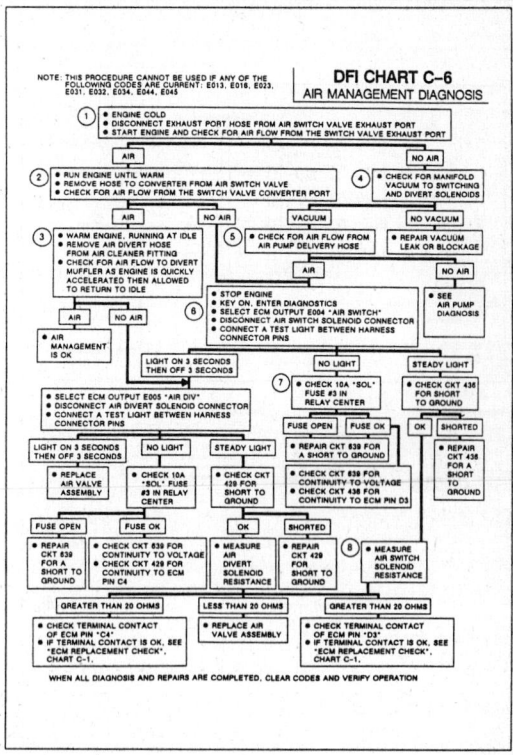

NOTE: THIS PROCEDURE CANNOT BE USED IF ANY OF THE FOLLOWING CODES ARE CURRENT: E013, E018, E023, E031, E032, E034, E044, E045

DFI CHART C-6
AIR MANAGEMENT DIAGNOSIS

WHEN ALL DIAGNOSIS AND REPAIRS ARE COMPLETED, CLEAR CODES AND VERIFY OPERATION

1988–90 CADILLAC ELDORADO AND SEVILLE

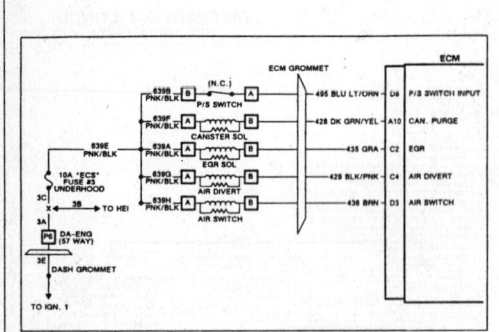

DFI CHART C-7
EGR DIAGNOSIS

Before Starting:

- Check 10 amp ECS fuse (# 3) on fuse panel.
- Measure EGR solenoid resistance - should be 20-100 ohms.

  The 4.5L DFI engine uses a positive backpressure EGR valve which limits EGR flow with low exhaust backpressure (idle, decel, etc.). It is very important that exhaust tubes which decrease backpressure (pull air through) are not hooked to the vehicle when diagnosing the EGR system.

VACUUM TEST

1. Connect a vacuum gage to the source side of the EGR solenoid. Start the engine, manifold vacuum should be present. If it is not, repair leaks

or obstruction between the EGR solenoid and throttle body.

2. Connect a vacuum gage to the EGR valve vacuum supply. There should be no vacuum with the engine idling. If there is, follow Chart C-7.

3. With the gage hooked to the EGR valve vacuum supply, disconnect the EGR solenoid connector. There should be more than 8 inches of vacuum available. If not, repair leak or obstruction in EGR valve vacuum hose.

Notes on Fault Tree:

1. Checking for EGR valve's ability to flow sufficient exhaust gas to engine. RPM should drop and idle should become rough or stall. Compare the RPM drop to a known good car.

2. Checking for blocked EGR passages in the intake manifold.

3. Checking for EGR solenoid able to pass vacuum.

1988–90 CADILLAC ELDORADO AND SEVILLE

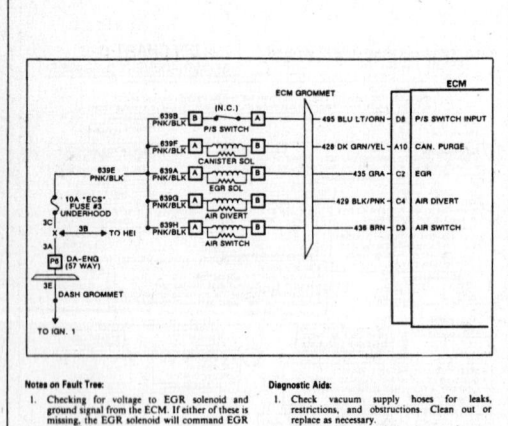

Notes on Fault Tree:

1. Checking for voltage to EGR solenoid and ground signal from the ECM. If either of these is missing, the EGR solenoid will command EGR on at all times.

2. Checking for ECM ability to turn solenoid off. If the light goes off, the electrical portion of the EGR system is OK.

Diagnostic Aids:

1. Check vacuum supply hoses for leaks, restrictions, and obstructions. Clean out or replace as necessary.

2. Check intake manifold passages for any obstructions that would prevent exhaust gases from passing through.

1988–90 CADILLAC ELDORADO AND SEVILLE

NOTE: THIS PROCEDURE CANNOT BE USED IF CODE E16 IS CURRENT

DFI CHART C-7
EGR DIAGNOSIS

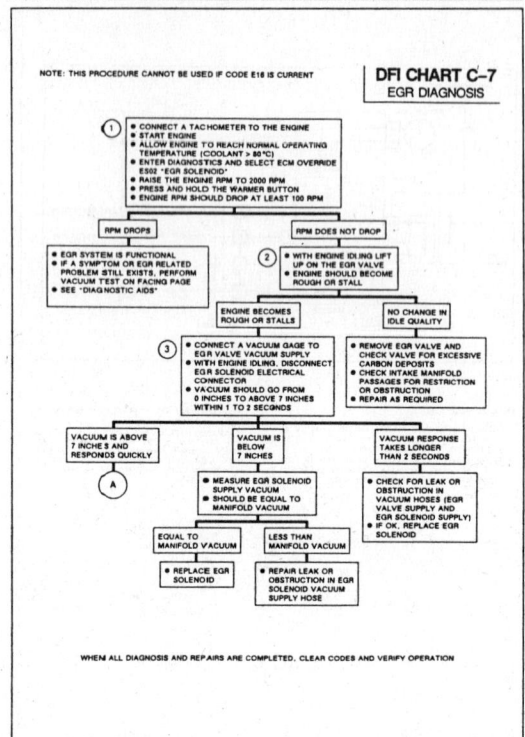

WHEN ALL DIAGNOSIS AND REPAIRS ARE COMPLETED, CLEAR CODES AND VERIFY OPERATION

1988–90 CADILLAC ELDORADO AND SEVILLE

DFI CHART C-7 (CONT'D)
EGR DIAGNOSIS

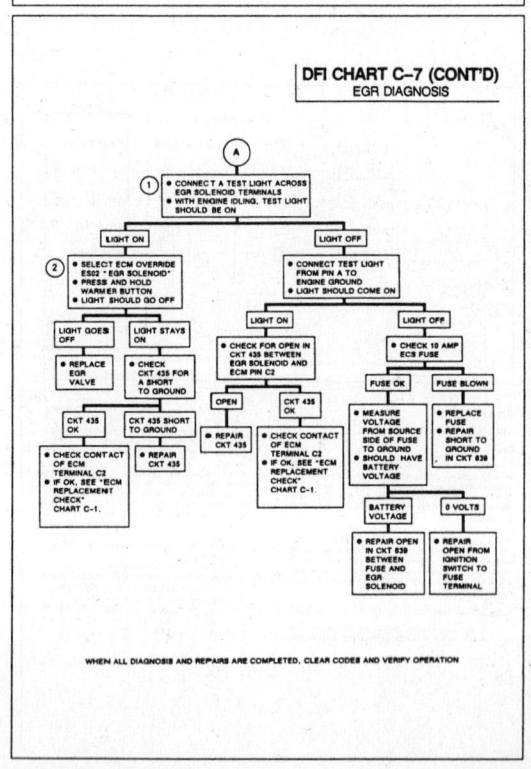

WHEN ALL DIAGNOSIS AND REPAIRS ARE COMPLETED, CLEAR CODES AND VERIFY OPERATION

1988–90 CADILLAC ELDORADO AND SEVILLE

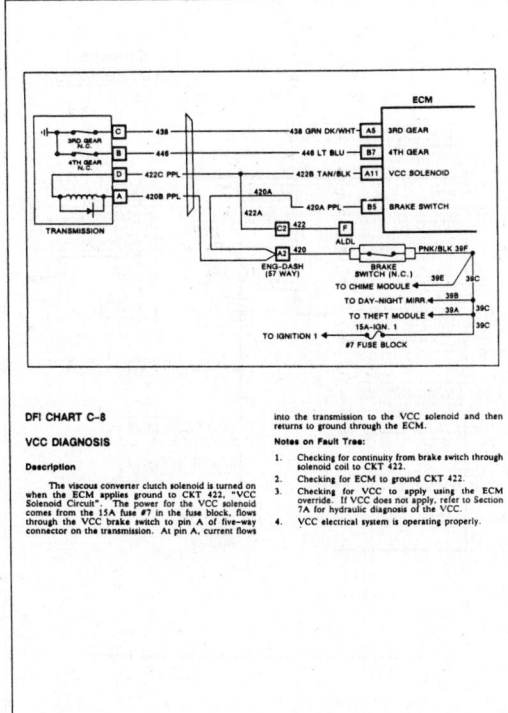

DFI CHART C-8

VCC DIAGNOSIS

Description

The viscous converter clutch solenoid is turned on when the ECM applies ground to CKT 422, "VCC Solenoid Circuit". The power for the VCC solenoid comes from the 15A fuse #7 in the fuse block, flows through the VCC brake switch to pin A of five-way connector on the transmission. At pin A, current flows into the transmission to the VCC solenoid and then returns to ground through the ECM.

Notes on Fault Tree:

1. Checking for continuity from brake switch through solenoid coil to CKT 422.
2. Checking for ECM to ground CKT 422.
3. Checking for VCC to apply using the ECM override. If VCC does not apply, refer to Section 7A for hydraulic diagnosis of the VCC.
4. VCC electrical system is operating properly.

1988–90 CADILLAC ELDORADO AND SEVILLE

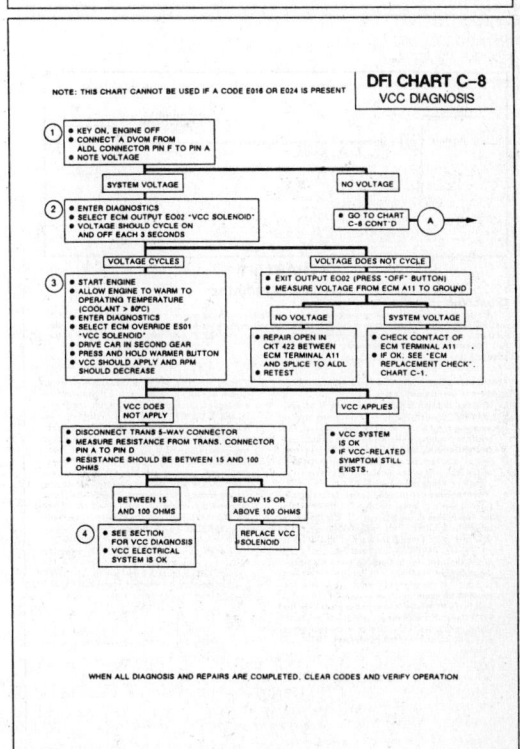

WHEN ALL DIAGNOSIS AND REPAIRS ARE COMPLETED, CLEAR CODES AND VERIFY OPERATION

1988–90 CADILLAC ELDORADO AND SEVILLE

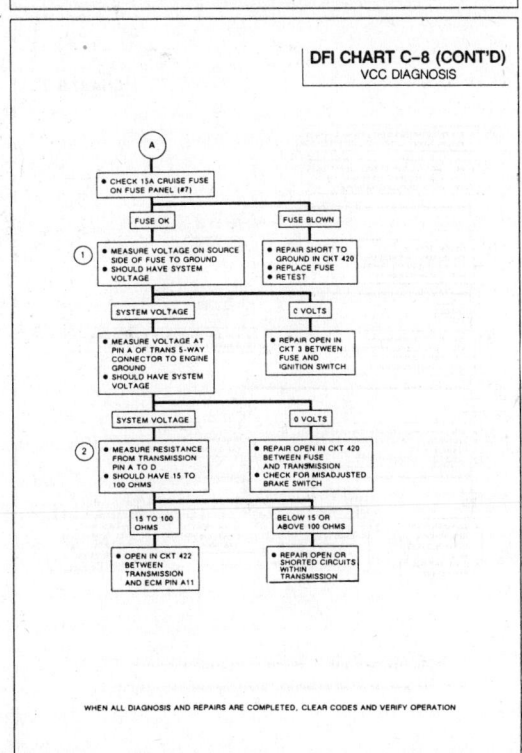

WHEN ALL DIAGNOSIS AND REPAIRS ARE COMPLETED, CLEAR CODES AND VERIFY OPERATION

1988–90 CADILLAC ELDORADO AND SEVILLE

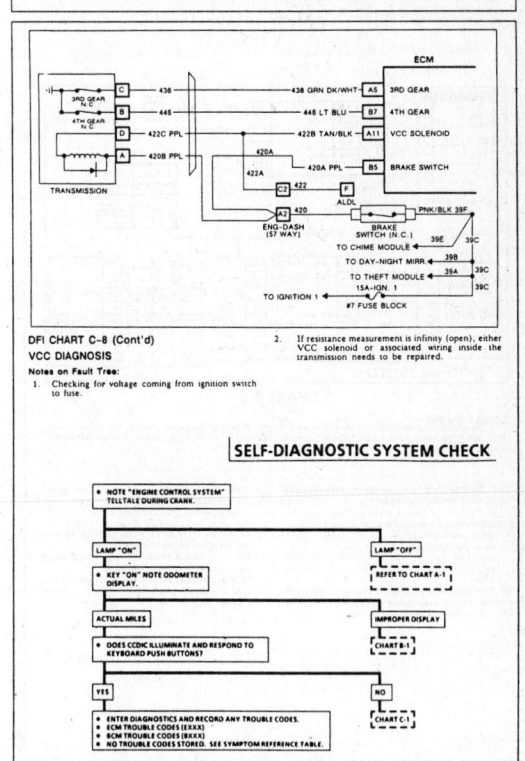

DFI CHART C-8 (Cont'd)
VCC DIAGNOSIS
Notes on Fault Tree:
1. Checking for voltage coming from ignition switch to fuse.

2. If resistance measurement is infinity (open), either VCC solenoid or associated wiring inside the transmission needs to be repaired.

SELF-DIAGNOSTIC SYSTEM CHECK

1988–90 CADILLAC ELDORADO AND SEVILLE

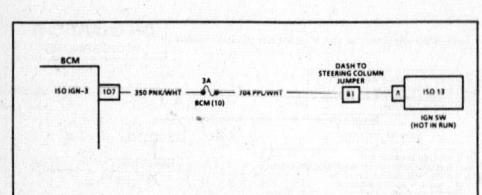

CHART B-1
ODOMETER PROBLEM

Circuit Description:
An incorrect odometer reading can be caused by problems relating to power and ground circuits, (not only the IPC, but those of other components within the computer network) computer system wake-up signals, the vacuum fluorescent power supply, the odometer mileage memory chip (EE PROM) contained in the BCM, or the data line between computer components.

Test Description: Numbers below refer to circled numbers on the diagnostic chart.
1. In order to begin to isolate the cause of an IPC odometer problem, if no codes are set in the diagnostic memory, observe the state of the odometer reading with the headlamp switch and twilight sentinel switch in the "OFF" position. this is important because these signals can wake-up the BCM and otherwise cause a misdiagnosis of a malfunctioning primary circuit.
2. When the odometer reads all zeroes, actuating the interior courtesy lamps switch will help to isolate the exact cause of the problem. This symptom is caused by a failure in the computer system other than the IPC. Further isolation of this fault is analyzed in CHARTS B-2, B-3 and B-4.
3. When the odometer is blank or reads all eights, actuating the headlamp will help to isolate the problem. If the odometer reads all zeroes, the cause of the problem is due to a faulty ignition-3 feed to the BCM (open or shorted to ground) or the BCM itself. If the odometer is blank or reads all eights, the cause is due to one of several other problems within the computer network.

4. Cycling the high beam, or headlamp dimmer switch, will help to further isolate the cause of the problem. If the high beams cycle, the problem is related to the IPC, IPC ground circuit, vacuum fluorescent power supply or the components related to these circuits. This analysis is contained in CHART B-5.
5. If the high beams did not cycle when the dimmer switch was actuated, turning "ON" the courtesy lamps switch will help to further isolate the system problem. If the courtesy lights turn "ON", the problem is related to the IPC itself, the CPS to IPC logic circuits ground, the 12volt power circuits from the CPS to the IPC or the CPS itself. Further problem isolation is contained in CHART B-6.
6. If the courtesy lamps did not turn "ON" when the switch was actuated, several other causes for the problem can be isolated by following CHARTS 7 and 8, depending on whether the "LIGHTS ON" chime sounds when the test is run.

1988–90 CADILLAC ELDORADO AND SEVILLE

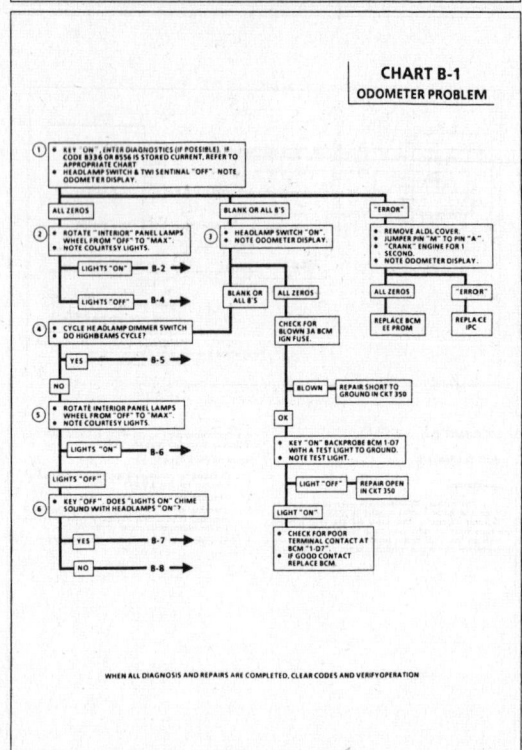

CHART B-1
ODOMETER PROBLEM

WHEN ALL DIAGNOSIS AND REPAIRS ARE COMPLETED, CLEAR CODES AND VERIFY OPERATION

1988–90 CADILLAC ELDORADO AND SEVILLE

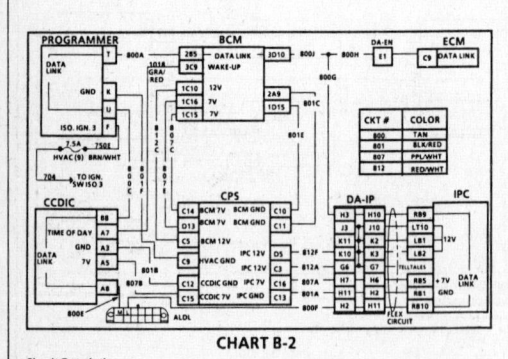

CHART B-2

Circuit Description:
You are on this chart because the odometer, at key "ON" read all zeroes and the courtesy lamps turned "ON" when the switch was actuated. These symptoms indicate a loss of BCM data communication to the IPC, which is the result of one of several different possible problems.

Test Description: Numbers below refer to circled numbers on the diagnostic chart.
1. If the ALDL connector cover is off or loose, this could cause a break in the data circuit, cutting off BCM communication to the IPC under certain electrical conditions. Should this be the case, securely connect the ALDL cover and return to the beginning of the "Self-Diagnostic System Check".
2. This step isolates the ALDL connector, the ECM and the programmer from the data circuit in order to determine the location of the fault.

If the odometer reads actual accumulated miles, then the fault lies within the network between the ALDL connector, the ECM and the programmer.
3. This step checks to see if the fault is due to the ECM and its related data circuits or the programmer and its related circuits.
4. This step will determine if the condition is associated with the IPC or the BCM.
5. This step determines if the fault is due to the programmer, the programmer's logic or, ground CKT 801 from the CPS.

1988–90 CADILLAC ELDORADO AND SEVILLE

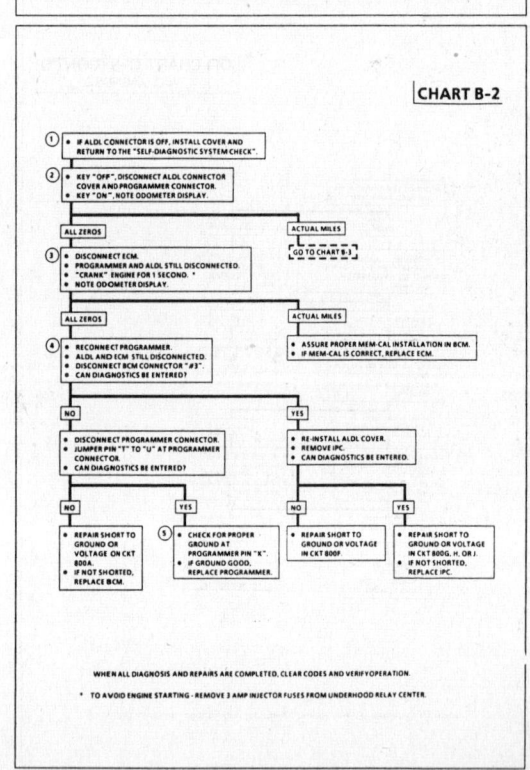

CHART B-2

WHEN ALL DIAGNOSIS AND REPAIRS ARE COMPLETED, CLEAR CODES AND VERIFY OPERATION

* TO AVOID ENGINE STARTING - REMOVE 3 AMP INJECTOR FUSES FROM UNDERHOOD RELAY CENTER.

1988–90 CADILLAC ELDORADO AND SEVILLE

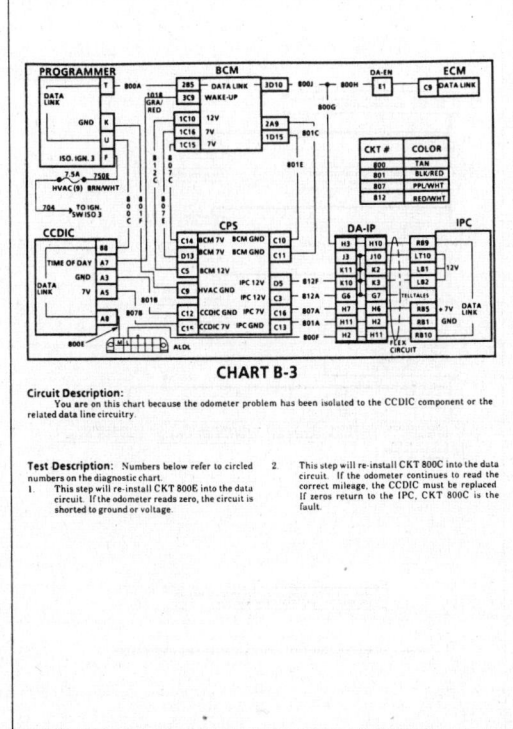

CHART B-3

Circuit Description:
You are on this chart because the odometer problem has been isolated to the CCDIC component or the related data line circuitry.

Test Description: Numbers below refer to circled numbers on the diagnostic chart.
1. This step will re-install CKT 800E into the data circuit. If the odometer reads zero, the circuit is shorted to ground or voltage.

2. This step will re-install CKT 800C into the data circuit. If the odometer continues to read the correct mileage, the CCDIC must be replaced. If zeros return to the IPC, CKT 800C is the fault.

1988–90 CADILLAC ELDORADO AND SEVILLE

CHART B-3

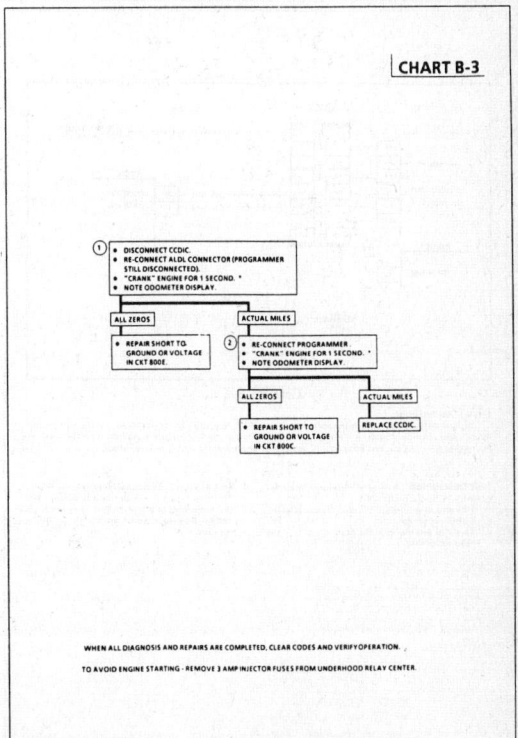

WHEN ALL DIAGNOSIS AND REPAIRS ARE COMPLETED, CLEAR CODES AND VERIFY OPERATION.

TO AVOID ENGINE STARTING - REMOVE 3 AMP INJECTOR FUSES FROM UNDERHOOD RELAY CENTER.

1988–90 CADILLAC ELDORADO AND SEVILLE

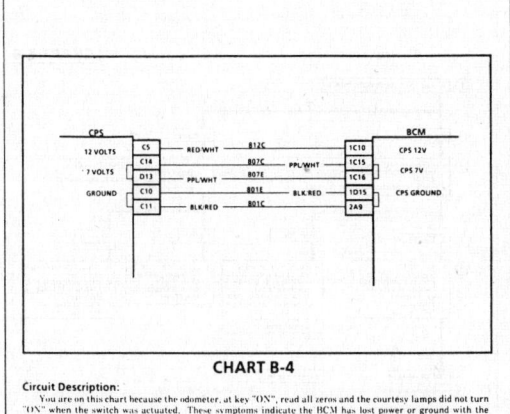

CHART B-4

Circuit Description:
You are on this chart because the odometer, at key "ON", read all zeros and the courtesy lamps did not turn "ON" when the switch was actuated. These symptoms indicate the BCM has lost power or ground with the central power supply (CPS).

Test Description: Numbers below refer to circled numbers on the diagnostic chart.
1. Checking for maximum blower speed with key "ON" will indicate whether the BCM has lost logic ground (CKT 801E) with the CPS.
2. This step determines if the loss of logic ground is due to the BCM, the CPS, or the 801E wire and terminals.

3. This step checks for 12 volts to the BCM. If 12 volts is not present at the BCM, then the fault is due to the CPS or the wiring.
4. This step checks for 7 volts to the BCM, which has two individual inputs fed from the CPS. Loss of both signals would result in a system fault, otherwise the fault is due to a bad CPS or bad BCM.

1988–90 CADILLAC ELDORADO AND SEVILLE

CHART B-4

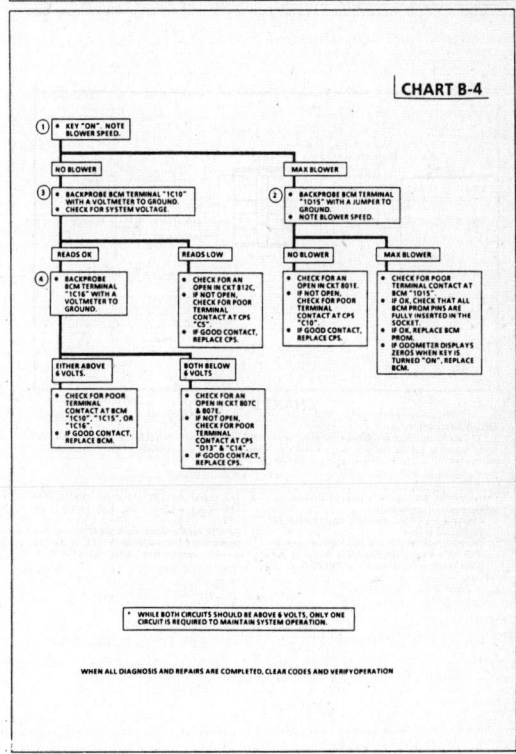

* WHILE BOTH CIRCUITS SHOULD BE ABOVE 6 VOLTS, ONLY ONE CIRCUIT IS REQUIRED TO MAINTAIN SYSTEM OPERATION.

WHEN ALL DIAGNOSIS AND REPAIRS ARE COMPLETED, CLEAR CODES AND VERIFY OPERATION

1988–90 CADILLAC ELDORADO AND SEVILLE

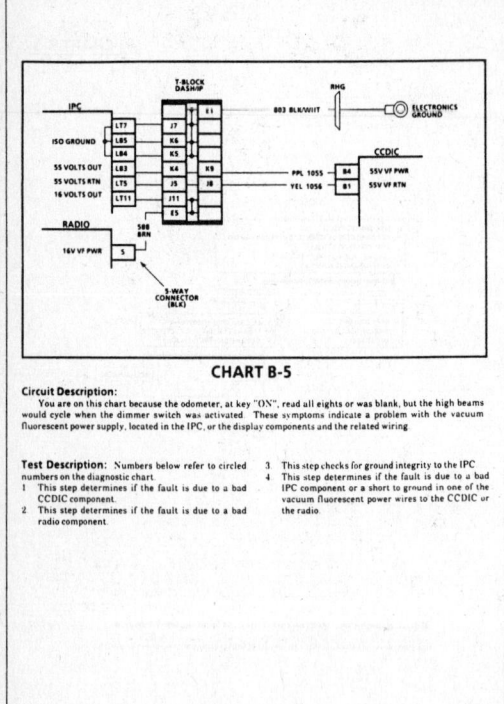

CHART B-5

Circuit Description:
You are on this chart because the odometer, at key "ON", read all eights or was blank, but the high beams would cycle when the dimmer switch was activated. These symptoms indicate a problem with the vacuum fluorescent power supply, located in the IPC, or the display components and the related wiring.

Test Description: Numbers below refer to circled numbers on the diagnostic chart.
1. This step determines if the fault is due to a bad CCDIC component.
2. This step determines if the fault is due to a bad radio component.
3. This step checks for ground integrity to the IPC.
4. This step determines if the fault is due to a bad IPC component or a short to ground in one of the vacuum fluorescent power wires to the CCDIC or the radio.

1988–90 CADILLAC ELDORADO AND SEVILLE

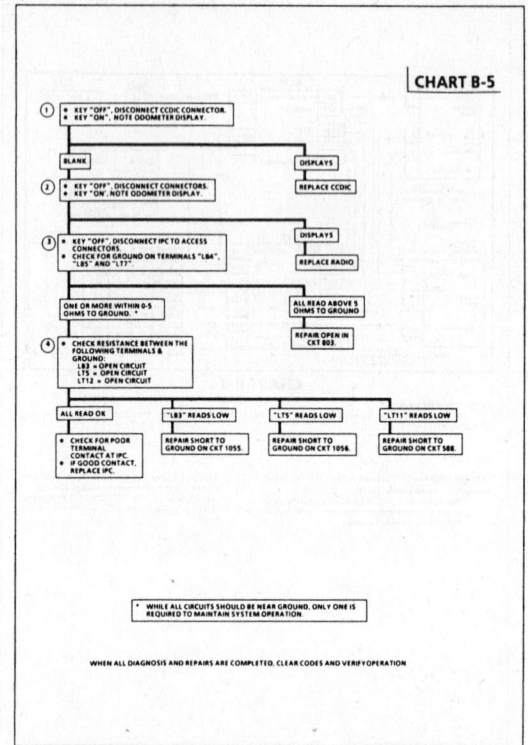

CHART B-5

1988–90 CADILLAC ELDORADO AND SEVILLE

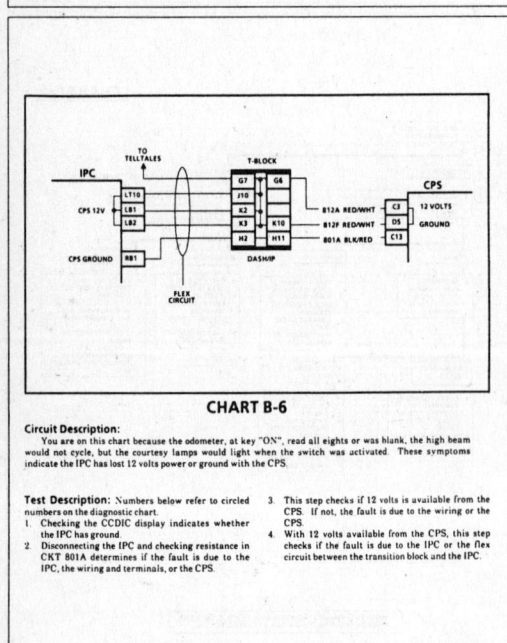

CHART B-6

Circuit Description:
You are on this chart because the odometer, at key "ON", read all eights or was blank, the high beam would not cycle, but the courtesy lamps would light when the switch was activated. These symptoms indicate the IPC has lost 12 volts power or ground with the CPS.

Test Description: Numbers below refer to circled numbers on the diagnostic chart.
1. Checking the CCDIC display indicates whether the IPC has ground.
2. Disconnecting the IPC and checking resistance in CKT 801A determines if the fault is due to the IPC, the wiring and terminals, or the CPS.
3. This step checks if 12 volts is available from the CPS. If not, the fault is due to the wiring or the CPS.
4. With 12 volts available from the CPS, this step checks if the fault is due to the IPC or the flex circuit between the transition block and the IPC.

1988–90 CADILLAC ELDORADO AND SEVILLE

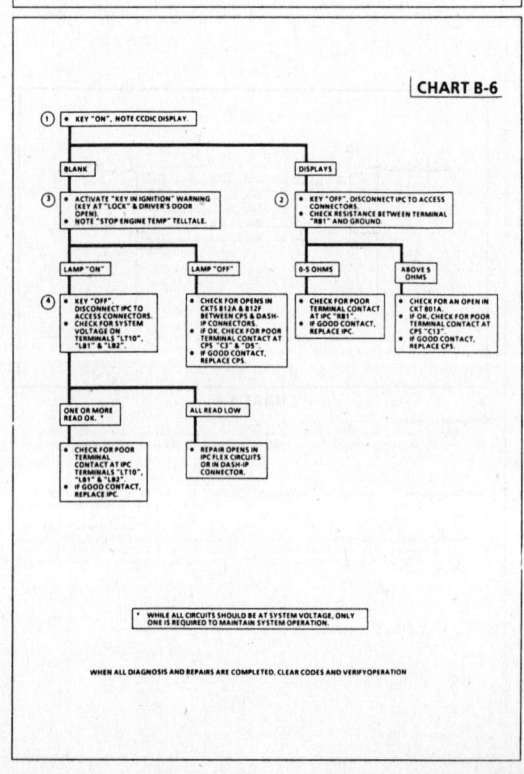

CHART B-6

1988–90 CADILLAC ELDORADO AND SEVILLE

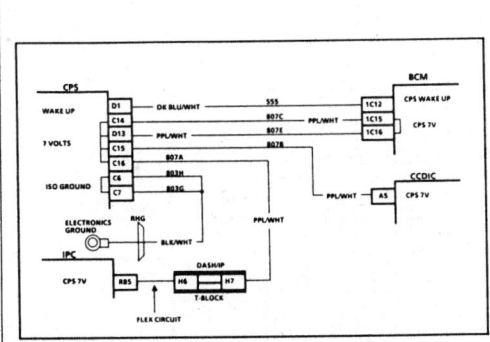

CHART B-7

Circuit Description:
You are on this chart because the odometer, at key "ON", read all eights or was blank, the high beams would not cycle, the courtesy lamps would not turn "ON", but the "lights on" chime would sound when activated. These symptoms indicate a loss of 7 volts from the CPS to the data network components, used to power the microprocessor chip. Or, the CPS has lost the system wake-up signal from the BCM, which is used to turn "ON" the 7 volt power supply.

Test Description: Numbers below refer to circled numbers on the diagnostic chart.
1. This step checks CKT 555, the CPS wake-up signal, for a short to ground, or a fault in the BCM or the BCM connector, or a bad CPS.
2. This step checks for an open in CKT 555.
3. This step checks for an open in the CPS ground, which will disable the 7 volt power supply.
4. This step checks for a short to ground in each 7 volt circuit from the CPS, or for a faulty component in the network, or a faulty CPS.

1988–90 CADILLAC ELDORADO AND SEVILLE

CHART B-7

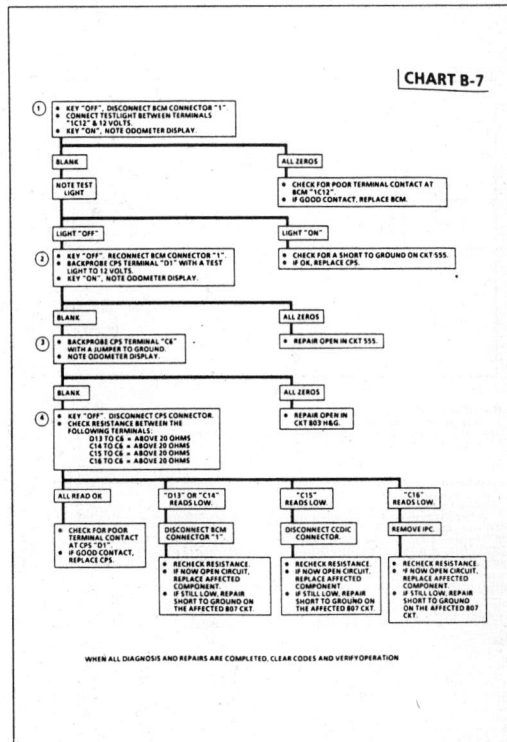

1988–90 CADILLAC ELDORADO AND SEVILLE

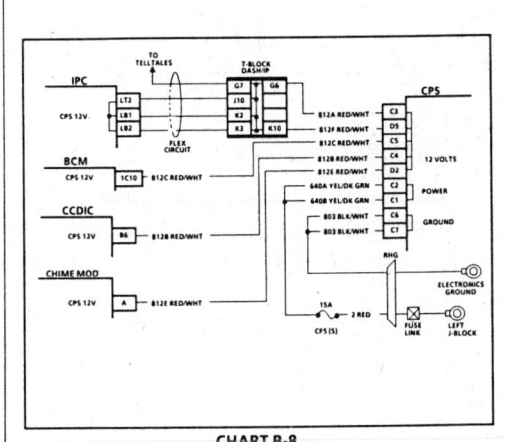

CHART B-8

Circuit Description:
You are on this chart because the odometer, at key "ON", read all eights or was blank, the high beams would not cycle, the courtesy lamps would not turn "ON", and the "LIGHTS ON" chime would not sound when the test was attempted. These symptoms indicate a loss of 12 volts, from the CPS to the other components in the system.

Test Description: Numbers below refer to circled numbers on the diagnostic chart.
1. This step check whether battery voltage is feeding the CPS.
2. This step checks for an open in CKT 640, the CPS battery feed, poor terminal contact or a faulty CPS.
3. This step checks for an intermittent short to ground condition on any of the 12 volt circuits from the CPS, or in the 640 feed circuit.
4. This step checks for a short to ground in CKT 640.
5. This step checks each individual 12 volt circuit from the CPS for a short to ground, a faulty component or a faulty CPS.

1988–90 CADILLAC ELDORADO AND SEVILLE

CHART B-8

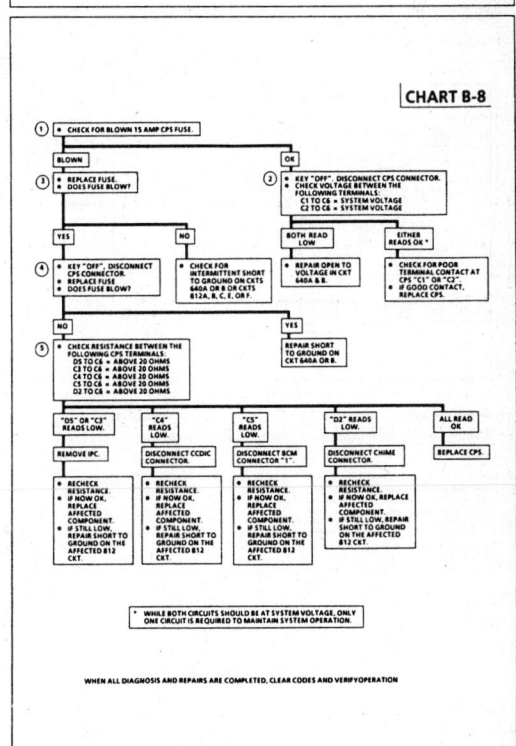

1988–90 CADILLAC ELDORADO AND SEVILLE

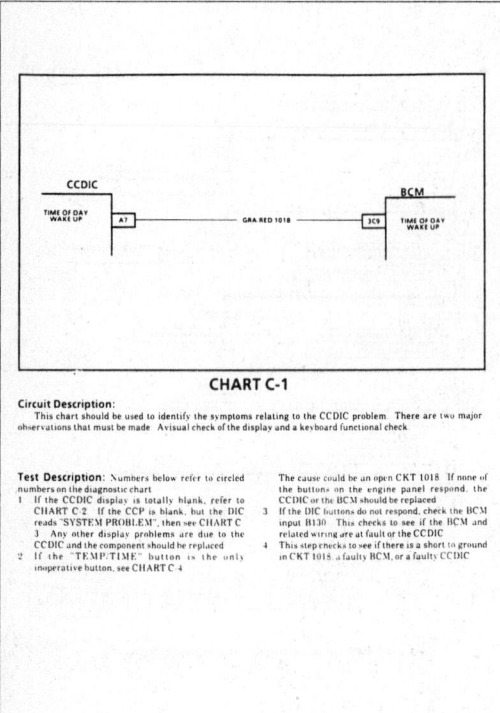

CHART C-1

Circuit Description:
 This chart should be used to identify the symptoms relating to the CCDIC problem. There are two major observations that must be made. A visual check of the display and a keyboard functional check.

Test Description: Numbers below refer to circled numbers on the diagnostic chart.
1 If the CCDIC display is totally blank, refer to CHART C-2. If the CCP is blank, but the DIC reads "SYSTEM PROBLEM", then see CHART C 3. Any other display problems are due to the CCDIC and the component should be replaced.
2 If the "TEMP/TIME" button is the only inoperative button, see CHART C-4.

The cause could be an open CKT 1018. If none of the buttons on the engine panel respond, the CCDIC or the BCM should be replaced
3 If the DIC buttons do not respond, check the BCM input B130. This checks to see if the BCM and related wiring are at fault or the CCDIC
4 This step checks to see if there is a short to ground in CKT 1018, a faulty BCM, or a faulty CCDIC

1988–90 CADILLAC ELDORADO AND SEVILLE

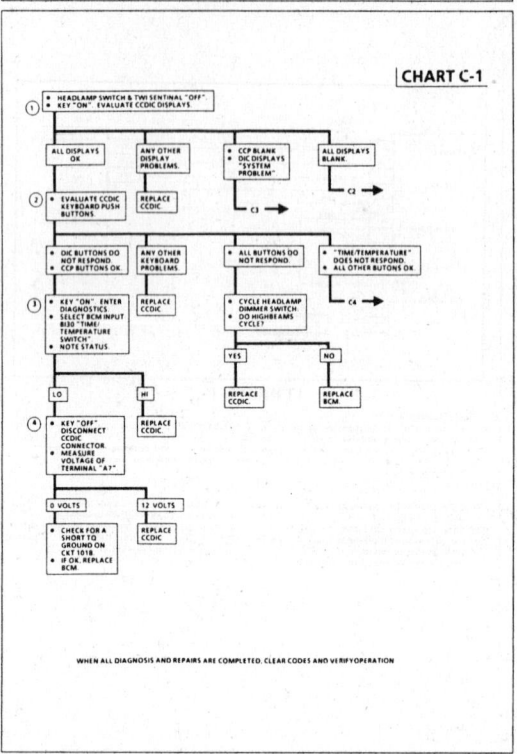

CHART C-1

WHEN ALL DIAGNOSIS AND REPAIRS ARE COMPLETED, CLEAR CODES AND VERIFY OPERATION

1988–90 CADILLAC ELDORADO AND SEVILLE

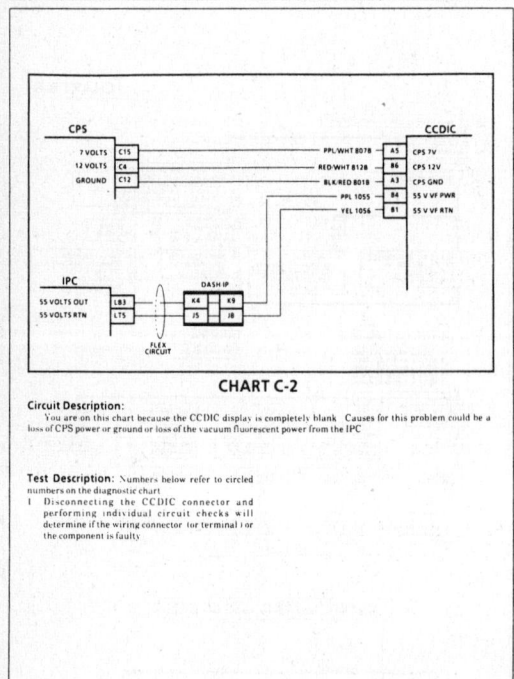

CHART C-2

Circuit Description:
 You are on this chart because the CCDIC display is completely blank. Causes for this problem could be a loss of CPS power or ground or loss of the vacuum fluorescent power from the IPC

Test Description: Numbers below refer to circled numbers on the diagnostic chart
1 Disconnecting the CCDIC connector and performing individual circuit checks will determine if the wiring connector for terminal 1 or the component is faulty

1988–90 CADILLAC ELDORADO AND SEVILLE

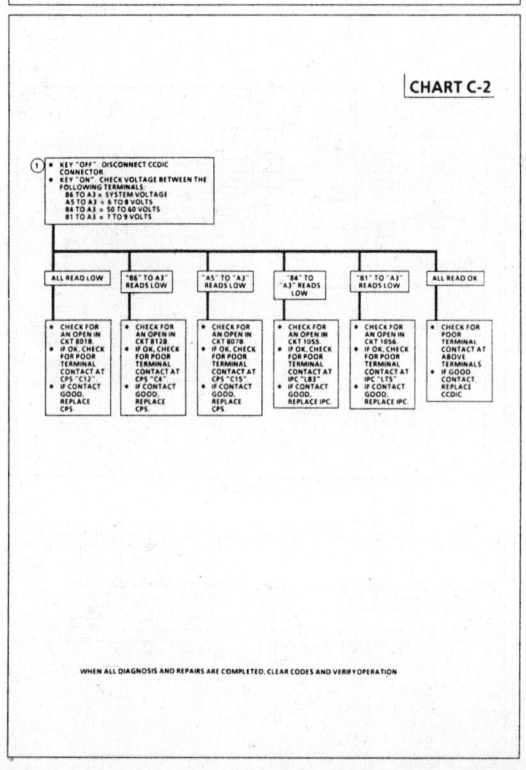

CHART C-2

WHEN ALL DIAGNOSIS AND REPAIRS ARE COMPLETED, CLEAR CODES AND VERIFY OPERATION

1988–90 CADILLAC ELDORADO AND SEVILLE

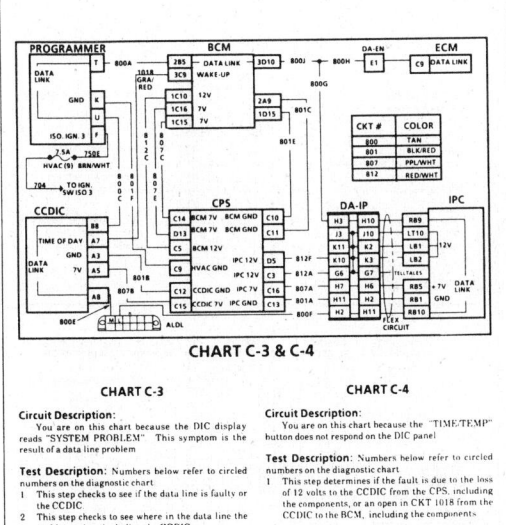

CHART C-3 & C-4

CHART C-3

Circuit Description:
You are on this chart because the DIC display reads "SYSTEM PROBLEM." This symptom is the result of a data line problem.

Test Description: Numbers below refer to circled numbers on the diagnostic chart.
1. This step checks to see if the data line is faulty or the CCDIC.
2. This step checks to see where in the data line the problem exists, including the CCDIC.

CHART C-4

Circuit Description:
You are on this chart because the "TIME/TEMP" button does not respond on the DIC panel.

Test Description: Numbers below refer to circled numbers on the diagnostic chart.
1. This step determines if the fault is due to the loss of 12 volts to the CCDIC from the CPS, including the components, or an open in CKT 1018 from the CCDIC to the BCM, including the components.

1988–90 CADILLAC ELDORADO AND SEVILLE

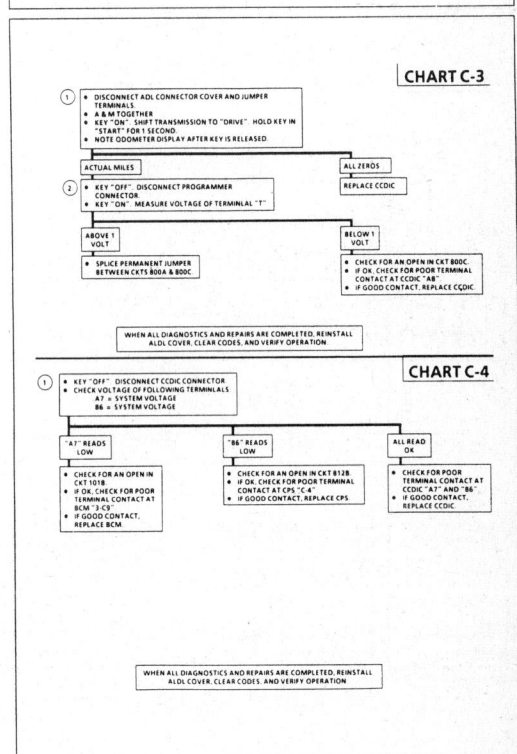

1988–90 CADILLAC ELDORADO AND SEVILLE

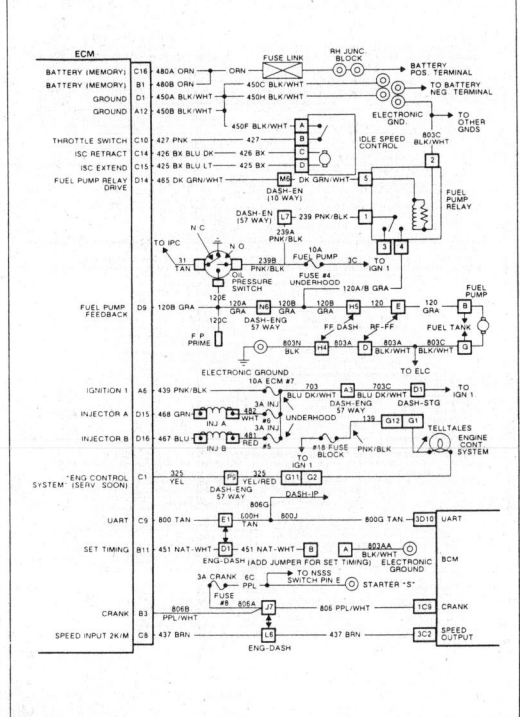

1988–90 CADILLAC ELDORADO AND SEVILLE

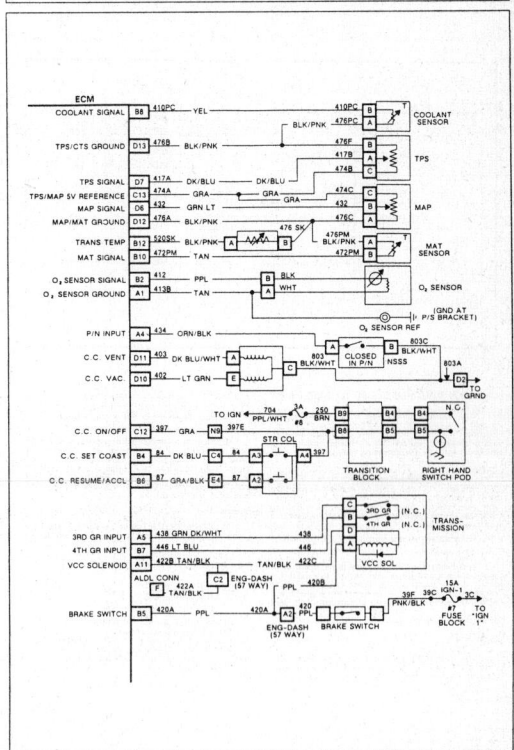

1988–90 CADILLAC ELDORADO AND SEVILLE

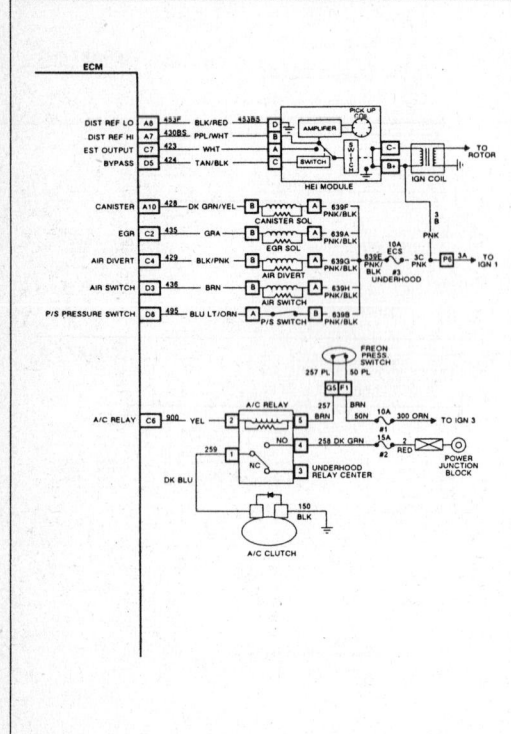

1988–90 CADILLAC ELDORADO AND SEVILLE

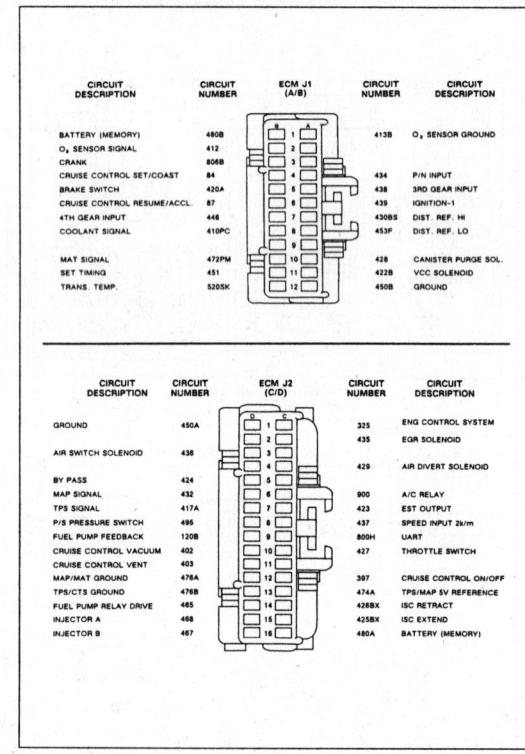

1988–90 CADILLAC ELDORADO AND SEVILLE

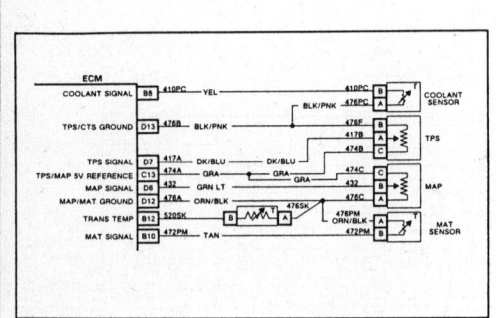

CHART A—MULTIPLE CODES STORED HARD

The conditions diagnosed in ECM Chart A, are caused by a single circuit failure yet result in multiple diagnostic codes. If any of these conditions are met, follow the appropriate correction procedure before using the procedures for the individual codes.

Chart A–1

If Codes E022 and E032 are both stored hard, the cause is probably the loss of 5 volts on circuit #474 (5 volt reference to MAP and TPS). To verify this condition, probe the following harness terminals with a voltmeter to ground:

• MAP sensor harness connector Pin C
• TPS connector Pin C

If the voltage is 0 for the two sensors, circuit #474 must be investigated for an open or short to ground. If a defect is not found in the wiring, either the ECM connector or ECM itself is faulty. If the proper 5 volt signal is observed on both of the sensors' terminals, then the diagnostic procedures for each individual code must be followed. Diagnose E022 first, then Code E032.

Chart A–2

If a code E015 is stored hard along with a hard Code E021, or E026 the cause is probably an open in the sensor ground circuit #476 for the TPS and coolant

sensors. To verify this condition, probe the following harness terminals with a voltmeter to 12 volts:

• Coolant Temp sensor Pin A, blk/pnk wire
• TPS connector Pin B, blk/pnk wire

If the voltage is 0 at both sensors, circuit #476 must be investigated for an open. If a defect is not found in the wiring, either the ECM connector or ECM itself is faulty. If 12 volts is observed at either of the two sensors, then the diagnostic procedures for each individual code must be followed. Diagnose code E015 then E021 then E026.

Chart A–3

If a Code E031 is stored hard along with hard Codes E038 and E059 this is probably caused by an open in the sensor ground circuit #476 for the MAP, MAT and TRANS TEMP sensors. To verify this condition, probe the following harness terminals with a voltmeter to 12 volts:

• MAT sensor Pin A
• MAP sensor harness Pin A
• TRANS TEMP sensor Pin A

If the voltage is 0 for the sensors, circuit #476A must be investigated for an open. If a defect is not found in the wiring, either the ECM connector or ECM itself is faulty. If 12 volts is observed at the sensors, then the diagnostic procedures for each individual code must be followed. Diagnose code E031, then E038 and then E059.

1988–90 CADILLAC ELDORADO AND SEVILLE

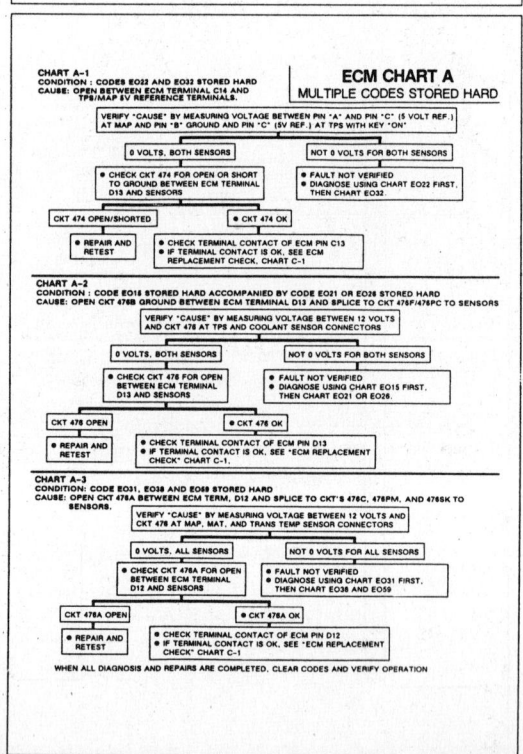

1988–90 CADILLAC ELDORADO AND SEVILLE

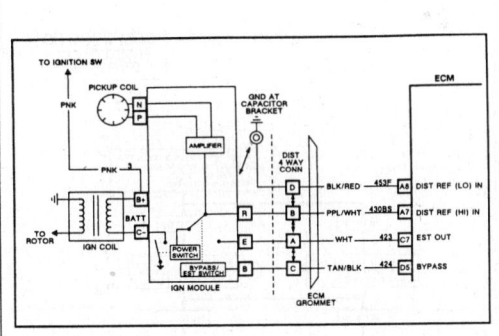

ECM CODE E012, NO DISTRIBUTOR (TACH) SIGNAL

TEST CONDITIONS: Code E012 is tested for during engine cranking (crank input to the ECM at system voltage and no Code e018 pending).

FAILURE CONDITIONS: If ECM does not see distributor reference pulses for 2.1 seconds with crank input to the ECM at system voltage, current Code E012 will be set.

ACTION TAKEN: 1. ECM turns on "Engine Control System" telltale; 2. ECM commands BCM to display "Service Now" message on CCDIC.

Description:

Possible causes of "no distributor reference" are: 1. Open pick-up coil – the pick-up coil produces an A.C. voltage that is used by the HEI module to create distributor reference pulses; 2. HEI module unable to process pick-up coil signals; 3. Open or shorts on CKT 430 and 453 from module to 4-way weatherpack to ECM; 4. Loss of 12 volts to "BAT" terminal on HEI.

Notes On Fault Tree:

1. Checking for proper voltage output of HEI system. If voltmeter shows .5–2.8 volts, HEI is producing reference pulses.

2. Checking for proper ground connection between ECM and the engine and the ECM and HEI.

3. Checking for proper voltage through CKT 430BS from HEI to ECM. If ECM terminal A7 sees .5–2.8 volts, then ECM is receiving reference pulses.

4. If the HEI will produce spark then fault is not pick-up coil or module. Check CKT 430BS for opens or short to ground from module terminal "R" to pin "B" of 4-way weatherpack.

Note On Intermittents:

If Code E012 is stored as a history code, start engine and allow to idle while manipulating CKT's 430 and 453. An intermittent open will cause the engine to stumble or quit when the ECM loses distributor reference.

If wiring and ECM connectors are no trouble found, check HEI pick-up coil leads for intermittent open circuit.

Check for short to voltage on CKT 806, crank input to ECM pin B3. If pin B3 has voltage at key on – engine off, the ECM will record a false Code E012. If found, repair short to voltage on CKT 806.

1988–90 CADILLAC ELDORADO AND SEVILLE

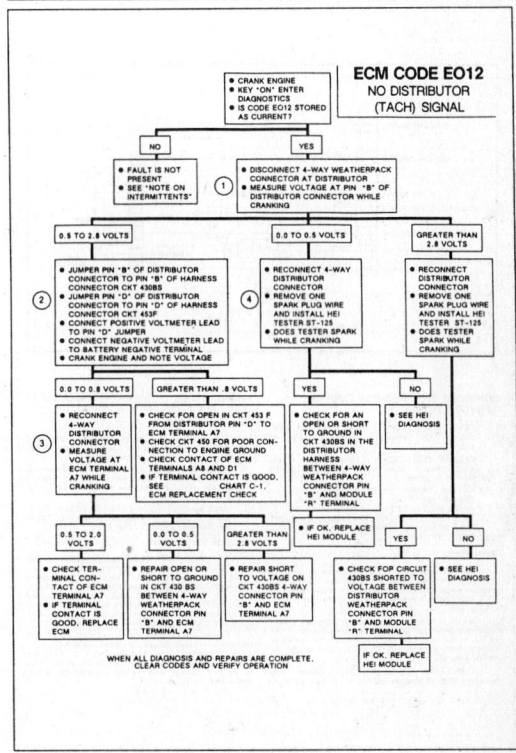

ECM CODE E012 NO DISTRIBUTOR (TACH) SIGNAL

WHEN ALL DIAGNOSIS AND REPAIRS ARE COMPLETE. CLEAR CODES AND VERIFY OPERATION

1988–90 CADILLAC ELDORADO AND SEVILLE

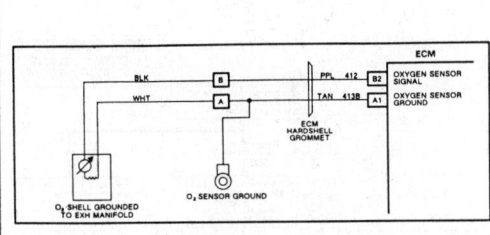

ECM CODE E013, OXYGEN SENSOR NOT READY

Important

If Code E013 is accompanied by current or History Code E045, use the E045 fault tree to diagnose. If Code E013 is accompanied by current or history Code E044, use the Code E044 fault tree to diagnose.

TEST CONDITIONS: 1. Codes E014, E015, E021, E022, E026, E027 clear; 2. Coolant temp greater than or equal to 80°C; 3. TPS value between 6 and 29 degrees; 4. Throttle switch open; 5. RPM greater than or equal to 800.

FAILURE CONDITIONS: Oxygen sensor voltage stays between .275 volts and .630 volts for more than 30 seconds (oxygen sensor voltage not toggling).

ACTION TAKEN: 1. ECM turns on "Engine Control System" telltale; 2. ECM commands BCM to display "Service Soon" message on CCDIC; 3. ECM turns on canister purge, diverts AIR system to air cleaner; 4. Closed Loop is disabled.

Description:

The ECM provides a .45 volt reference to the oxygen sensor on CKT 412. When warm, a properly operating oxygen sensor will drive the .45 volt reference lower (below .45 volts) to indicate lean mixtures and higher (above .45 volts) to indicate a rich mixture. If under the above test conditions the oxygen sensor doesn't vary from the "cold" or "not ready" voltage,

the ECM assumes that the sensor can't respond to air-fuel mixtures rich or lean and Code E013 is set.

Notes On Fault Tree:

1. At key on with oxygen sensor disconnected, the parameter ED07 should read about .45 volts. Lower or higher voltage indicates circuit problems.
2. The ECM compares voltage on CKT 412 to the ground voltage on CKT 413. It is essential that the oxygen sensor ground and ECM ground on pin A1 show good continuity (no voltage difference) with engine running.
3. Fault is most likely at ECM connector or ECM. Before ECM is replaced, perform "ECM Replacement Check", Chart C-1.

Notes On Intermittents:

If Code E013 is a history code, start engine, enter diagnostics. Operate engine at fast idle until "Auto" closed loop status light is turned on. Observe parameter ED07 while manipulating CKT's 412 and 413 wiring. If parameter ED07 changes from fluctuating to fixed at about .45 volts, repair intermittent open circuit.

Check CKT 413 ground to engine for proper installation (clean, tight, star washer installed.) Ground for CKT 413 is attached to the front of the engine at the right or rearmost cylinder head.

Review the snapshot data parameters stored with this code to verify failure condition. Refer to "Symptoms" for complete details on using the "Snapshot on code set" feature. If the failure condition is verified, check and repair intermittent wiring connection, or sensor.

1988–90 CADILLAC ELDORADO AND SEVILLE

NOTE: IF CODE E013 IS ACCOMPANIED BY HARD OR INTERMITTENT CODE E044 OR E045, SEE DFI CHART A-8

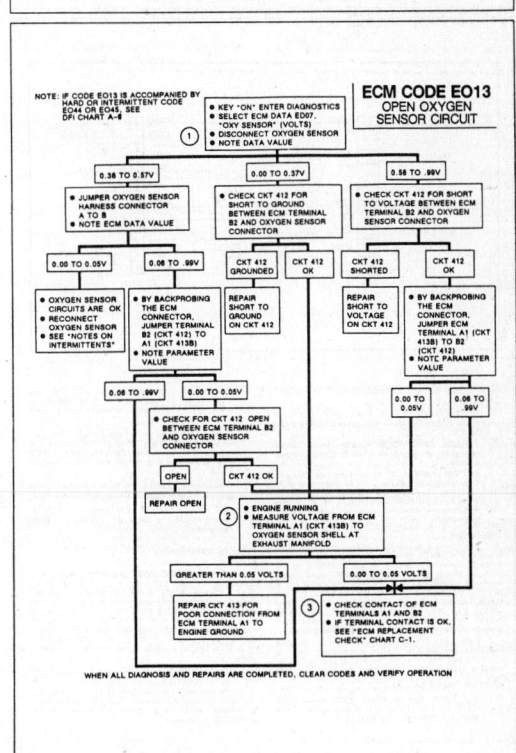

ECM CODE E013 OPEN OXYGEN SENSOR CIRCUIT

WHEN ALL DIAGNOSIS AND REPAIRS ARE COMPLETED, CLEAR CODES AND VERIFY OPERATION

1988–90 CADILLAC ELDORADO AND SEVILLE

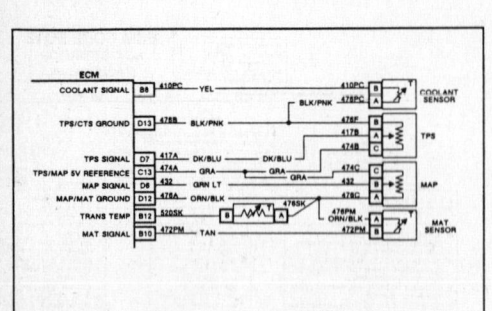

ECM CODE E014, SHORTED COOLANT SENSOR CIRCUIT

TEST CONDITIONS: 1. Codes E037 and E038 clear; 2. MAT Sensor Value less than or equal to 105°C.

FAILURE CONDITIONS: Coolant Sensor Value greater than or equal to 142°C. (Sensor Signal shorted to ground.)

ACTION TAKEN: 1. ECM turns on "Engine Control System" telltale; 2. ECM commands BCM to display "Service Soon" message on CCDIC; 3. ECM turns off both AIR solenoids; 4. ECM uses MAT sensor value in place of coolant sensor value for all calculations.

Description:

The coolant sensor is a thermistor or variable resistor. The coolant sensor is a "Two wire" sensor with a sensor signal voltage coming from the ECM on Pin B, CKT 410PC and a sensor reference ground on Pin A, CKT 476PC.

As the temperature of the sensor increases, sensor resistance decreases. The signal voltage from the ECM to Pin A decreases as sensor temperature increases and current flows through the sensor element to Pin A, sensor ground.

Code E014 "Shorted Coolant Sensor" sets because the ECM assumes that the coolant temperature

can't be 142°C (287°F) or greater when MAT is 105°C or less. (Sensor signal shorted to ground).

Notes On Fault Tree:

1. With sensor shorted, parameter ED04 should read 142°C or greater. If not, then the sensor is not shorted, see "Note on Intermittents."

2. Checking for sensor shorted on CKT 410 shorted. If the parameter stays at 142° to 151° with the sensor unplugged, then the short is in CKT 410, between the sensor and ECM.

3. Fault is most likely at ECM connector or ECM. Before ECM is replaced, perform "ECM Replacement Check", chart C-1.

Notes On Intermittents:

Manipulate CKT 410 wiring, the coolant sensor and CKT 410 connector while observing ECM Parameter ED04. If the failure is induced, the coolant temperature will jump from its normal value to the shorted reading of 142° to 151°C. Remove and replace both the COOLANT and ECM connectors and ensure that they are latched. If wiring and connector check out OK, substitute a known good coolant sensor and retest.

Review the snapshot data parameters stored with this code to verify failure condition. Refer to "Symptoms" for complete details on using the "Snapshot on code set" feature. If the failure condition is verified, check and repair intermittent wiring connection, or sensor.

1988–90 CADILLAC ELDORADO AND SEVILLE

ECM CODE E014
SHORTED COOLANT SENSOR CIRCUIT

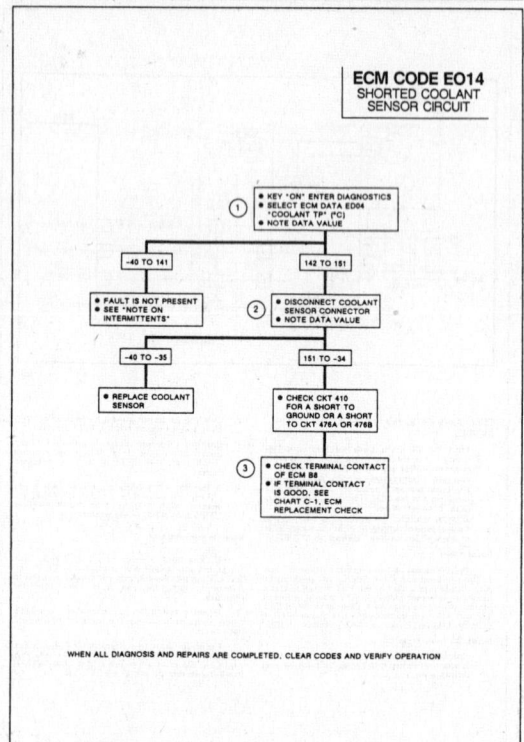

WHEN ALL DIAGNOSIS AND REPAIRS ARE COMPLETED, CLEAR CODES AND VERIFY OPERATION

1988–90 CADILLAC ELDORADO AND SEVILLE

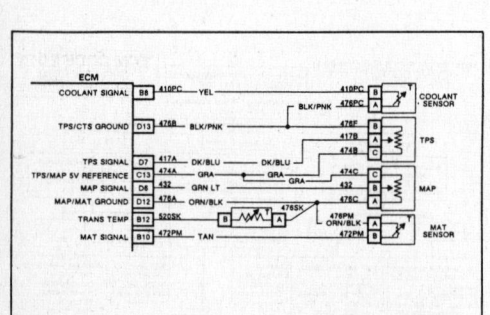

ECM CODE E015, OPEN COOLANT SENSOR CIRCUIT

TEST CONDITIONS: 1. Code, E037 and E038 clear; 2. MAT Sensor Value greater than or equal to 5°C.

FAILURE CONDITIONS: Coolant Sensor Value less than or equal to −35°C. (Sensor Signal Open).

ACTION TAKEN: 1. ECM turns on "Engine Control System" telltale; 2. ECM commands BCM to display "Service Soon" message on CCDIC; 3. ECM turns off both AIR solenoids; 4. ECM uses MAT sensor value in place of coolant for all calculations.

Description:

The coolant sensor is a thermistor or variable resistor. The coolant sensor is a "two wire" sensor with a sensor signal or reference voltage coming from the ECM to sensor pin B on CKT 410PC and sensor reference ground CKT 476PC connecting sensor pin A to the ECM sensor ground. As the temperature of the sensor decreases, sensor resistance is higher. The signal voltage from the ECM to sensor pin B increases as sensor temperature decreases and less current flows through the sensor element to pin A sensor ground.

Code E015 "Open Coolant Sensor" sets because the ECM assumes that coolant temperature can't be −35°C or less when the MAT is 5°C or greater; sensor signal is open to sensor.

Notes On Fault Tree:

1. If sensor is open, parameter ED04 should read −35°C or less. If not then sensor signal is not open, see "Note on Intermittents."

2. Checking for open sensor signal CKT 410 from ECM to sensor connector. If parameter ED04 reads 148°–151°C with connector shorted then CKT 410 is OK.

3. Checking for open sensor ground CKT 476 from ECM to sensor. If pin "B" to ground causes ED04 to read 148°–151°C then there is an open in CKT 476 between sensor pin A and ECM terminal D13.

4. Checking for ECM's ability to recognize a short to ground or low voltage on B8, coolant sensor input.

5. Fault is most likely at ECM connector or ECM. Before ECM is replaced, perform "ECM Replacement Check", Chart C-1.

Notes On Intermittents:

Manipulate CKT 410PC and 476PC wiring, the coolant sensor and ECM connector while observing ECM parameter ED04. If the failure is induced, the coolant temperature will jump from its normal value to the open circuit reading of −35°C or less. Disconnect and reconnect both the coolant sensor and ECM connectors and ensure that they are latched before replacing components. If wiring and connectors check out OK, substitute a known good coolant sensor and retest.

Review the snapshot data parameters stored with this code to verify failure condition. Refer to "Symptoms" for complete details on using the "Snapshot on code set" feature. If the failure condition is verified, check and repair intermittent wiring connection, or sensor.

1988–90 CADILLAC ELDORADO AND SEVILLE

ECM CODE E015
OPEN COOLANT SENSOR CIRCUIT

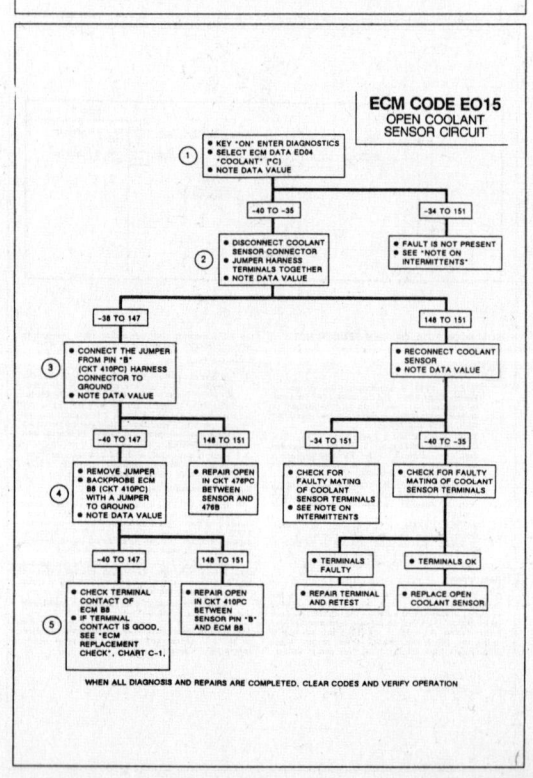

WHEN ALL DIAGNOSIS AND REPAIRS ARE COMPLETED, CLEAR CODES AND VERIFY OPERATION

1988–90 CADILLAC ELDORADO AND SEVILLE

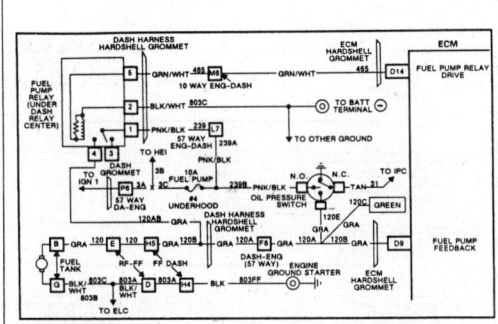

ECM CODE E016, GENERATOR VOLTAGE OUT OF RANGE

TEST CONDITIONS: 1. Code E018 and E020 Clear; 2. RPM greater than or equal to 800; 3. Not Cranking (Crank input to ECM at 0 volts).

FAILURE CONDITIONS: Voltage on Fuel Pump Feedback to 10.0 volts less than 10.0 volts or greater than 16.0 volts for 5 seconds or more.

ACTION TAKEN: 1. ECM turns on "Engine Control System" telltale; 2. ECM commands BCM to display "Service Now" message on CCDIC. 3. ECM turns off EGR solenoid, canister purge solenoid and both AIR solenoids. 4. ECM disables Cruise Control 5. ECM disables VCC for entire key cycle.

Description:

The ECM monitors vehicle electrical system voltage or battery voltage indirectly by monitoring the voltage on the fuel pump feedback CKT 120B.

The voltage readings taken by the ECM are used to detect fuel pump voltage faults and to modify injector pulse width as system voltage varies; the ECM increases injector pulse width as system voltage decreases.

Code E016 is set when the ECM sees low or high system voltage with an engine RPM high enough to allow almost full generator output.

Notes On Fault Tree:

1. Checking for generator output high or low at idle.

2. Checking for insufficient generator output when at high electrical loads.

3. Checking for generator output voltage too high.

4. Checking for proper voltage at fuel pump test point (on fuel pump power circuit).

5. Checking for open between ECM pin D9 and fuel pump feed circuit.

Note On Intermittents:

If history Code E016 is found, start engine, enter diagnostics and display ECM parameter ED10. Manipulate CKT 120A wiring from ECM pin D9 to the splice at CKT 120E. Watch for the parameter to drop to less than 10.0 volts. Many intermittent Code E016's can be traced to intermittent charging or battery system faults such as insufficient charging system output at high accessory loads or generator voltage out of regulation high.

Check for occurrence of "No Charge" telltale ON. If BCM Code B411 or B412 accompany history E016, diagnosis using B411 or B412 procedure.

Check for low or overcharged battery.

Review the snapshot data parameters stored with this code to verify failure condition. Refer to "Symptoms" for complete details on using the "Snapshot on code set" feature. If the failure condition is verified, check and repair intermittent wiring connection, or sensor.

1988–90 CADILLAC ELDORADO AND SEVILLE

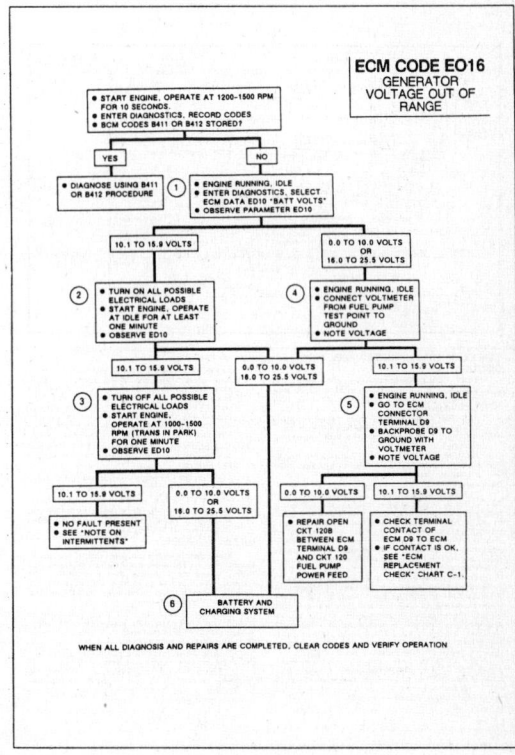

ECM CODE E016
GENERATOR VOLTAGE OUT OF RANGE

WHEN ALL DIAGNOSIS AND REPAIRS ARE COMPLETED, CLEAR CODES AND VERIFY OPERATION

1988–90 CADILLAC ELDORADO AND SEVILLE

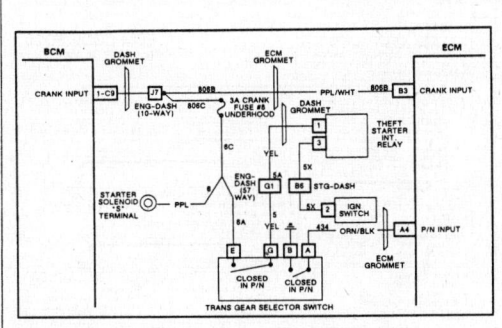

ECM CODE E018, OPEN CRANK SIGNAL CIRCUIT

TEST CONDITIONS: 1. Engine RPM changed from 0 to greater than or equal to 800 while in park or neutral.

FAILURE CONDITIONS: No crank signal has been received by the ECM since the last time that RPM was 0 RPM. (Engine started, ECM didn't register a crank signal.)

ACTION TAKEN: 1. ECM turns on "Engine Control System" telltale 2. ECM commands BCM to display "Service Soon" message on CCDIC. 3. Code is stored as current for entire key cycle.

Description:

The crank input to the ECM is an ON/OFF signal that comes from the starter solenoid, through the 3A crank fuse then to pin B3 of the ECM. During crank, the signal is battery voltage, at all other times the signal is at 0 volts.

Code E018 is set if the ECM detects the engine running at 800 RPM or greater without having the crank input go to system voltage.

Notes On Fault Tree:

1. Do not use this procedure if E018 is a history code. See Notes On Intermittents.

2. If 806 becomes grounded, the crank fuse will blow when the engine is cranked.

3. This branch of the fault tree is to test for ECM or BCM shorting crank input to ground.

4. Voltage signal to crank fuse comes from the ignition switch to the theft starter interrupt relay to the P/N switch to the crank fuse. This branch checks for opens between the crank fuse and the starter solenoid.

Note On Intermittents:

Remove the crank fuse and use an ohmmeter to check continuity on CKT 806 from the crank fuse to ECM pin B3. Check for intermittent opens while manipulating wiring.

If code E018 doesn't set immediately after cranking, check for plug wires arcing, loose engine grounds or other sources of Electromagnetic Interference (EMI). Large transient voltage surges can "Erase" the crank signal from the ECM's memory and cause code E018 to set while driving.

1988–90 CADILLAC ELDORADO AND SEVILLE

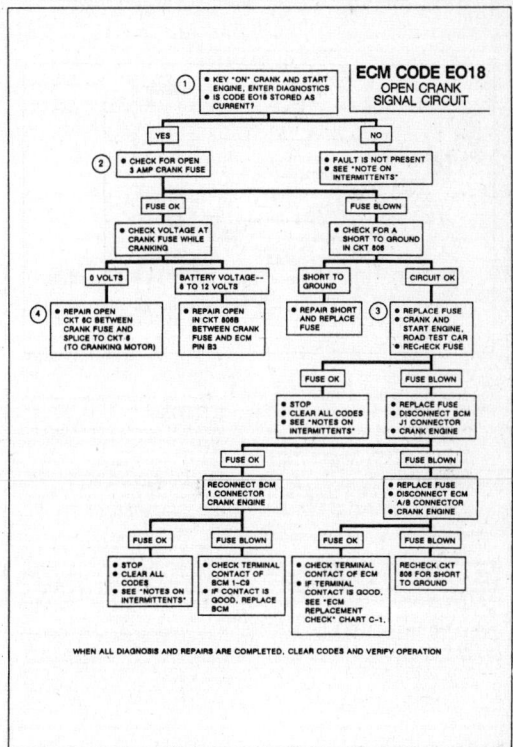

ECM CODE E018
OPEN CRANK SIGNAL CIRCUIT

WHEN ALL DIAGNOSIS AND REPAIRS ARE COMPLETED, CLEAR CODES AND VERIFY OPERATION

1988–90 CADILLAC ELDORADO AND SEVILLE

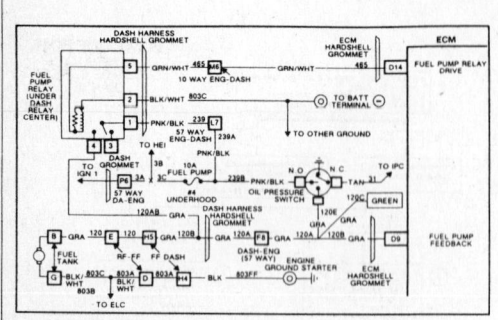

ECM CODE E019, SHORTED FUEL PUMP CIRCUIT

TEST CONDITIONS: 1. Code E018 clear. 2. RPM = 0 ; 3. Crank input to ECM at 0 volts (not cranking) 4. "Coolant" light in bulb check (key "on" engine off).

FAILURE CONDITIONS: Fuel pump feedback voltage greater than or equal to 11.0 volts for 7 seconds. (Fuel pump powered with engine not running.)

ACTION TAKEN: 1. ECM turns on "Engine Control System" telltale. 2. ECM commands BCM to display "Service Soon" message on CCDIC.

Description:

The fuel pump receives power from CKT 120 Ckt 120 is fed by two alternative sources; the fuel pump relay and the oil pressure switch. The relay is controlled by the ECM. The relay is energized by the ECM for 2 seconds at key "ON" to provide fuel pressure for starting. After the 2 second fuel pump run, the ECM will not power the relay again until distributor reference pulses are received indicating that the engine is turning (crank or run).

The oil pressure switch is provided as a "backup". Normal operating engine oil pressure will close the normally open oil pressure switch contacts, providing a connection between CKT 239 and 120. If a relay fault occurs, the engine will continue to run as the oil pressure switch powers the fuel pump. A vehicle with the relay inoperative will crank until oil pressure builds (oil telltale light off) then start and run as the oil pressure switch provides power to the fuel pump.

The ECM monitors voltage on the fuel pump power CKT 120 to detect fuel pump voltage supply faults. Code E019 is set when the fuel pump feedback (ECM pin D9) is at 11.0 volts or greater when the fuel pump feed voltage should be 0 volts.

Note On Fault Tree:

1. At key on engine off, the fuel pump should not be running and the fuel pump feedback (ECM parameter ED10) should show 0 volts. (Note: The fuel pump runs for 2 seconds then turns off at the start of each key "on" cycle. After the initial 2 second run, the fuel pump will not be powered until the ECM sees distributor reference pulses from the HEI.)
2. This branch of the tree checks for oil pressure switch contacts shorted, powering the fuel pump at all times.
3. Checking for fuel pump relay providing fuel pump power due to a relay or relay drive circuit fault.
4. Checking CKT 120 for a short to a 12 volt source.
5. Checking ECM relay drive circuit for a fault that allows fuel pump relay coil to be energized continuously at key on.

Note On Intermittents:

Probe the fuel pump test point with a voltmeter to ground. Key on and observe voltmeter – it should read battery voltage for 2 seconds then drop to 0. If the voltmeter stays at battery voltage for longer than 2 seconds, check for fuel pump relay contacts sticking. Repeat the test several times.

Continue probing the fuel pump test point to ground and crank and start engine then key off - voltage should drop immediately to 0. If not, check oil pressure switch for sticking contacts.

Review the snapshot data parameters stored with this code to verify failure condition. Refer to "Symptoms" for complete details on using the "Snapshot on code set" feature. If the failure condition is verified, check and repair intermittent wiring connection, or sensor.

1988–90 CADILLAC ELDORADO AND SEVILLE

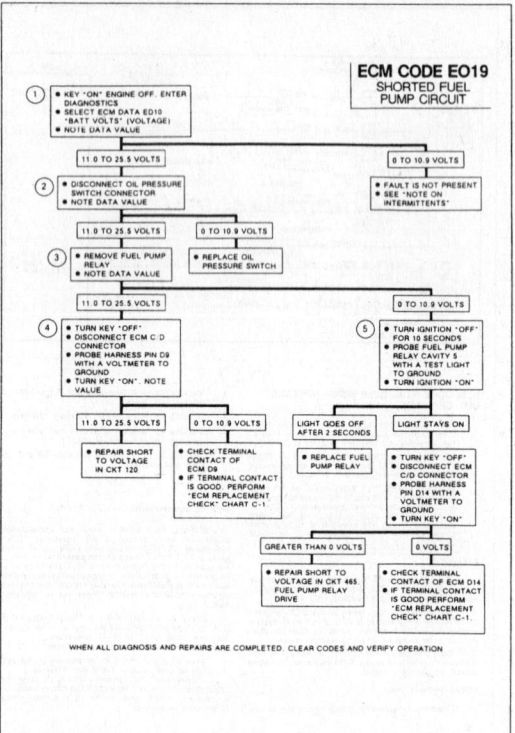

ECM CODE E019
SHORTED FUEL PUMP CIRCUIT

1. • KEY "ON" ENGINE OFF. ENTER DIAGNOSTICS
   • SELECT ECM DATA ED10 ("BATT VOLTS" (VOLTAGE)
   • NOTE DATA VALUE

   11.0 TO 25.5 VOLTS → **0 TO 10.9 VOLTS**

2. • DISCONNECT OIL PRESSURE SWITCH CONNECTOR
   • NOTE DATA VALUE

   [0 TO 10.9 VOLTS branch] • FAULT IS NOT PRESENT
   • SEE "NOTE ON INTERMITTENTS"

   11.0 TO 25.5 VOLTS → **0 TO 10.9 VOLTS**

3. • REMOVE FUEL PUMP RELAY
   • NOTE DATA VALUE

   [0 TO 10.9 VOLTS] • REPLACE OIL PRESSURE SWITCH

   11.0 TO 25.5 VOLTS → **0 TO 10.9 VOLTS**

4. • TURN KEY "OFF"
   • DISCONNECT ECM C/D CONNECTOR
   • PROBE HARNESS PIN D9 WITH A VOLTMETER TO GROUND
   • TURN KEY "ON". NOTE VALUE

5. • TURN IGNITION "OFF" FOR 10 SECONDS
   • PROBE FUEL PUMP RELAY CAVITY 5 WITH A TEST LIGHT TO GROUND
   • TURN IGNITION "ON"

   11.0 TO 25.5 VOLTS → **0 TO 10.9 VOLTS** | **LIGHT GOES OFF AFTER 2 SECONDS** | **LIGHT STAYS ON**

   • REPAIR SHORT TO VOLTAGE IN CKT 120

   • CHECK TERMINAL CONTACT OF ECM D9
   • IF TERMINAL CONTACT IS GOOD, PERFORM "ECM REPLACEMENT CHECK" CHART C-1.

   • REPLACE FUEL PUMP RELAY

   • TURN KEY "OFF"
   • DISCONNECT ECM C/D CONNECTOR
   • PROBE HARNESS PIN D14 WITH A VOLTMETER TO GROUND
   • TURN KEY "ON"

   GREATER THAN 0 VOLTS → **0 VOLTS**

   • REPAIR SHORT TO VOLTAGE IN CKT 465 FUEL PUMP RELAY DRIVE

   • CHECK TERMINAL CONTACT OF ECM D14
   • IF TERMINAL CONTACT IS GOOD PERFORM "ECM REPLACEMENT CHECK" CHART C-1.

WHEN ALL DIAGNOSIS AND REPAIRS ARE COMPLETED, CLEAR CODES AND VERIFY OPERATION

1988–90 CADILLAC ELDORADO AND SEVILLE

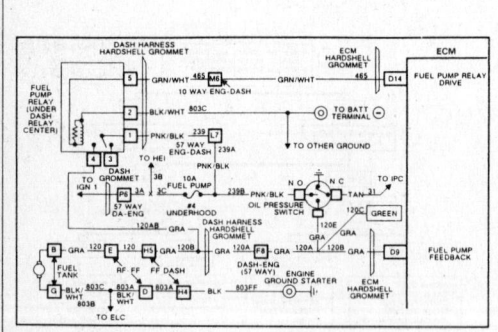

ECM CODE E020, OPEN FUEL PUMP CIRCUIT

TEST CONDITIONS: 1. Engine RPM greater than or equal to 24 RPM

FAILURE CONDITIONS: Fuel pump feedback voltage less than or equal to 2 volts for 3 seconds or more.

ACTION TAKEN: 1. Store code as current until engine is turned off; 2. ECM turns on "Engine Control System" telltale. 3. ECM commands BCM to display "Service Now" message on CCDIC.

Description:

The ECM monitors voltage on the fuel pump power CKT 120 to detect fuel pump voltage supply faults. Code E020 is set when the ECM sees the fuel pump not energized (0 volts on the feedback) with the engine cranking or running. This code is designed to detect a fuel pump relay fault, relay not powering the fuel pump.

Notes On Fault Tree:

1. Engine starts with a current E020 indicates that the engine started with fuel pump powered through the oil pressure switch - the relay circuit may be at fault.

2. 0 volts with engine running indicates an open from fuel pump power circuit to ECM - the fuel pump relay is not at fault.
3. Testing for ECM's ability to energize relay coil.
4. Testing for open from 10A fuel pump fuse to relay pin "1".
5. Testing for open from relay pin "2" to ground.
6. If light turns on then all relay circuitry is OK. Fault causing the code must be the relay.
7. Fault is most likely an ECM connector or ECM. Before ECM is replaced, perform "ECM Replacement Check", Chart C-1.

Notes On Intermittents:

If intermittent Code E020 is stored, unplug the oil pressure switch, start engine and allow to idle. Manipulate affected wiring, ensure that the ECM J2 (C/D) connector is latched, check the relay for proper installation into relay center socket. If the fault is induced, the relay will set. Code E020 will set. If Code E020 sets without an engine stall, the cause is an intermittent open CKT 120A between ECM D9 and splice to 120B.

Review the snapshot data parameters stored with this code to verify failure condition. Refer to "Symptoms" for complete details on using the "Snapshot on code set" feature. If the failure condition is verified, check and repair intermittent wiring connection, or sensor.

1988–90 CADILLAC ELDORADO AND SEVILLE

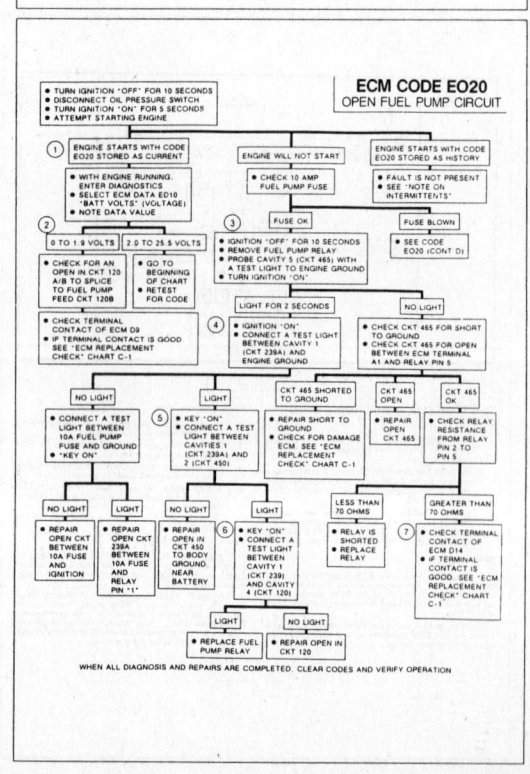

ECM CODE E020
OPEN FUEL PUMP CIRCUIT

• TURN IGNITION "OFF" FOR 10 SECONDS
• DISCONNECT OIL PRESSURE SWITCH
• TURN IGNITION "ON" FOR 5 SECONDS
• ATTEMPT STARTING ENGINE

1. ENGINE STARTS WITH CODE E020 STORED AS CURRENT | ENGINE WILL NOT START | ENGINE STARTS WITH CODE E020 STORED AS HISTORY

   • WITH ENGINE RUNNING, ENTER DIAGNOSTICS
   • SELECT ECM DATA ED10 "BATT VOLTS" (VOLTAGE)
   • NOTE DATA VALUE

   • CHECK 10 AMP FUEL PUMP FUSE

   • FAULT IS NOT PRESENT
   • SEE "NOTE ON INTERMITTENTS"

2. **0 TO 1.9 VOLTS** | **2.0 TO 25.5 VOLTS**   3. **FUSE OK** | **FUSE BLOWN**

   • CHECK FOR AN OPEN IN CKT 120 A/B TO SPLICE TO FUEL PUMP FEED CKT 120B

   • GO TO BEGINNING OF CHART & RETEST FOR CODE

   • IGNITION "OFF" FOR 10 SECONDS
   • REMOVE FUEL PUMP RELAY
   • PROBE CAVITY 5 (CKT 465) WITH A TEST LIGHT TO ENGINE GROUND
   • TURN IGNITION "ON"

   • SEE CODE E020 (CONT D)

   • CHECK TERMINAL CONTACT OF ECM D9
   • IF TERMINAL CONTACT IS GOOD SEE "ECM REPLACEMENT CHECK" CHART C-1

4. **LIGHT FOR 2 SECONDS** | **NO LIGHT**

   • IGNITION "ON"
   • CONNECT A TEST LIGHT BETWEEN CAVITY 1 (CKT 239A) AND ENGINE GROUND

   • CHECK CKT 465 FOR SHORT TO GROUND
   • CHECK CKT 465 FOR OPEN BETWEEN ECM TERMINAL A1 AND RELAY PIN 5

   NO LIGHT | **LIGHT** | **CKT 465 SHORTED TO GROUND** | **CKT 465 OPEN** | **CKT 465 OK**

   • CONNECT A TEST LIGHT BETWEEN 10A FUEL PUMP FUSE AND GROUND
   • "KEY ON"

5. • KEY "ON"
   • CONNECT A TEST LIGHT BETWEEN (CKT 239A) AND 2 (CKT 450)

   • REPAIR SHORT TO GROUND
   • CHECK FOR DAMAGE ECM. SEE "ECM REPLACEMENT CHECK" CHART C-1

   • REPAIR OPEN CKT 465

   • CHECK RELAY RESISTANCE FROM RELAY PIN 2 TO PIN 5

   NO LIGHT | **LIGHT** | **NO LIGHT** | **LIGHT** | **LESS THAN 70 OHMS** | **GREATER THAN 70 OHMS**

   • REPAIR OPEN CKT BETWEEN 10A FUSE AND IGNITION

   • REPAIR OPEN CKT 239A BETWEEN 10A FUSE AND IGNITION

   • REPAIR OPEN IN CKT 450 TO BODY GROUND NEAR BATTERY

6. • KEY "ON"
   • CONNECT A TEST LIGHT BETWEEN CAVITY 1 (CKT 239) AND CAVITY 4 (CKT 120)

   • RELAY IS SHORTED
   • REPLACE RELAY

7. • CHECK TERMINAL CONTACT OF ECM D14
   • IF TERMINAL CONTACT IS GOOD, SEE "ECM REPLACEMENT CHECK" CHART C-1

   LIGHT | **NO LIGHT**

   • REPLACE FUEL PUMP RELAY

   • REPAIR OPEN IN CKT 120

WHEN ALL DIAGNOSIS AND REPAIRS ARE COMPLETED, CLEAR CODES AND VERIFY OPERATION

1988–90 CADILLAC ELDORADO AND SEVILLE

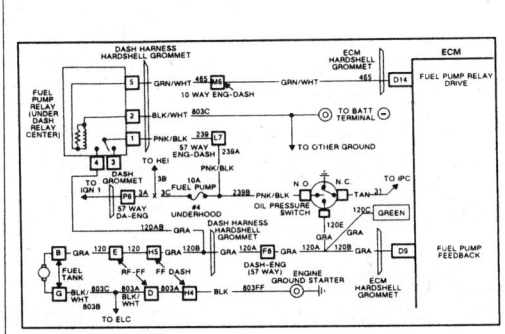

ECM CODE E020 (CONT'D) FUEL FUSE BLOWN

Notes On Fault Tree:

1. At key "ON" light should be "OFF" or glow for 2 seconds as the relay powers up the fuel pump.
2. If a short is present, it is in CKT 120 the fuel pump or ECM. Normal CKT resistance is 3–4 ohms.
3. Checking for ECM shorting 120A to ground.
4. Checking for short internal to fuel tank.
5. Check fuel pump for shorts, check sender harness and check RFI supression connector.
6. Checking for 120 CKT shorted to ground.
7. Check for grounded fuel pump test point, then connection at relay center, oil pressure switch, then diagnose for harness shorted to ground.

8. If the light is on steadily then the non switched portion of the fuel pump circuit is shorted-checking circuit 239.

Notes On Intermittents:

- If the fuel pump test point is grounded (for example if the test point is mistakenly grounded instead of the generator disable connector) the fuse will blow and E020 will set.
- Replace fuse, start engine and allow to idle. Manipulate affected wiring and connectors. If fault occurs, the fuse will blow and the engine will stall.

Review the snapshot data parameters stored with this code to verify failure condition. Refer to "Symptoms" for complete details on using the "Snapshot on code set" feature. If the failure condition is verified, check and repair intermittent wiring, connection, or sensor.

1988–90 CADILLAC ELDORADO AND SEVILLE

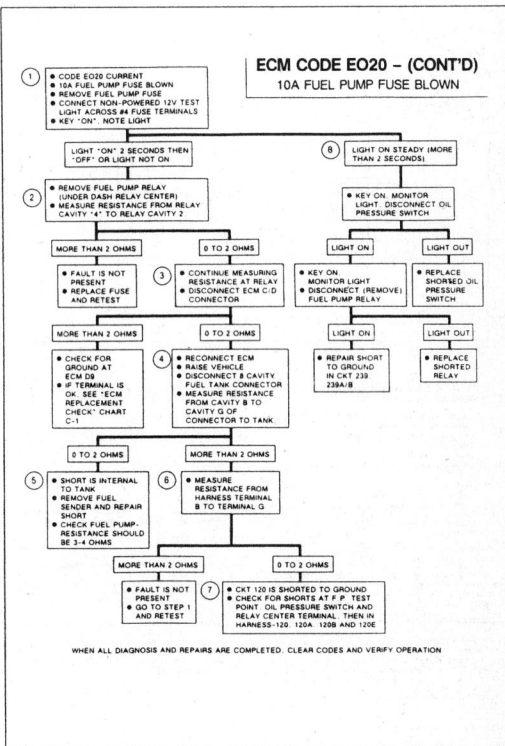

ECM CODE EO20 – (CONT'D)
10A FUEL PUMP FUSE BLOWN

WHEN ALL DIAGNOSIS AND REPAIRS ARE COMPLETED. CLEAR CODES AND VERIFY OPERATION

1988–90 CADILLAC ELDORADO AND SEVILLE

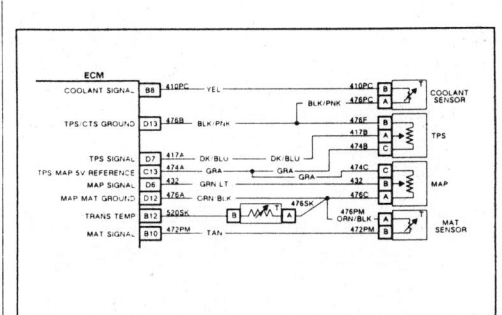

ECM CODE E021, SHORTED THROTTLE POSITION SENSOR CIRCUIT

TEST CONDITIONS: 1 RPM between 25 and 1000

FAILURE CONDITIONS: TPS value greater than or equal to 80 degrees for 3 seconds

ACTION TAKEN: 1. ECM turns on "Engine Control System" telltale 2. ECM commands BCM to display "Service Soon" message on CCDIC 3. ECM uses 12 degrees for TPS value with ISC throttle switch open and 6 degrees for TPS value when ISC throttle switch closes. 4. ECM uses RPM for all ISC control calculations (won't use TPS reading for coast down throttle control).

Description:

The Throttle Position Sensor (TPS) is a "3 wire" sensor or potentiometer with a 5 volt reference input from the ECM to the sensor (CKT 474), a reference ground from the ECM to the sensor (CKT 476) and a sensor output signal (CKT 417) from the sensor to the ECM. The sensor output signal is a DC voltage that varies with throttle angle, the TPS signal voltage is low (about 5 volts at minimum air setting) and at high throttle angles, the TPS signal voltage is high (about 4.5 volts at wide open throttle).

Code E021 is set when the ECM sees a high throttle angle (high TPS voltage) at low engine RPM. The code is designed to detect a TPS signal (CKT 417) shorted to voltage.

Notes On Fault Tree:

1. To determine location of short to voltage — TPS or CKT 417
2. Checking for an open ground on CKT 474. A ground open will result in high TPS values whenever the TPS is plugged in.

Notes On Intermittents:

If Code E021 is intermittent, manipulate related wiring while observing ECM parameter ED01. Check the TPS connector for shorts to voltage.

Cycle the TPS through its travel and tap on the TPS with a pencil or pocket screwdriver to test for intermittent TPS. If the fault is induced, the parameter will skip to a high throttle angle. If wiring and connectors check out OK, substitute a known good TPS sensor and retest.

Review the snapshot data parameters stored with this code to verify failure condition. Refer to "Symptoms" for complete details on using the "Snapshot on code set" feature. If the failure condition is verified, check and repair intermittent wiring, connection, or sensor.

1988–90 CADILLAC ELDORADO AND SEVILLE

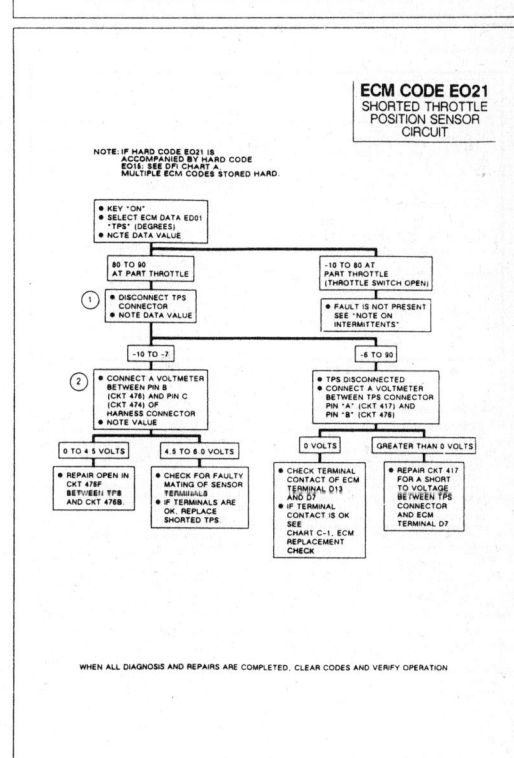

ECM CODE EO21
SHORTED THROTTLE POSITION SENSOR CIRCUIT

NOTE: IF HARD CODE EO21 IS ACCOMPANIED BY HARD CODE EO16, SEE DFI CHART A MULTIPLE ECM CODES STORED HARD.

WHEN ALL DIAGNOSIS AND REPAIRS ARE COMPLETED. CLEAR CODES AND VERIFY OPERATION

1988–90 CADILLAC ELDORADO AND SEVILLE

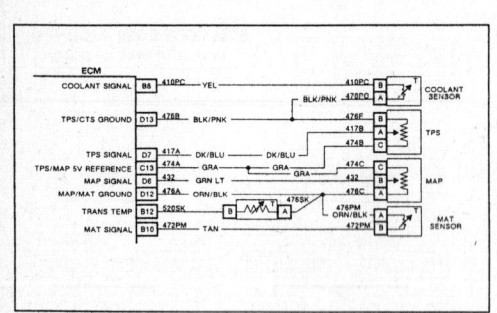

ECM CODE E022, OPEN THROTTLE POSITION SENSOR CIRCUIT

TEST CONDITIONS: 1. RPM greater than or equal to 600.

FAILURE CONDITIONS: TPS value less than –5 degrees (TPS parameter reads –10 to –6 degrees) for 1.5 seconds.

ACTION TAKEN: 1. ECM turns on "Engine Control System" telltale. 2. ECM commands BCM to display "Service Soon" message on CCDIC. 3. ECM uses 12 degrees for TPS value with ISC throttle switch open, and 6 degrees for TPS value when ISC throttle switch closes. 4. ECM stops using TPS to control ISC – uses RPM for all ISC calculations.

Description:

The Throttle Position Sensor (TPS) is a "3 wire" sensor or potentiometer with a 5 volt reference input from the ECM to the sensor (CKT 474), a reference ground from the ECM to the sensor (CKT 476) and a sensor output signal (CKT 417) from the sensor to the ECM. The sensor output signal is a D.C. voltage that varies with throttle angle. At low throttle angle, the TPS signal voltage is low (about .5 volts at minimum air setting) and at high throttle angles, the TPS signal voltage is high (about 4.5 volts at wide open throttle).

Code E022 is set when the ECM sees a throttle angle that is out of limits low (low TPS voltage) with engine running at idle or faster. The code is designed to detect a TPS signal (CKT 417) open to ECM pin D7.

Notes On Fault Tree:

1. Jumpering 5 volts to TPS signal to see if the ECM parameter will respond to high signal voltage on TPS inputs (CKT 417).

2. Checking for TPS reference (CKT 474) open/shorted.

3. Checking for TPS signal (CKT 417) grounded.

4. Fault is most likely at ECM connector or ECM. Before ECM is replaced, perform "ECM Replacement Check", Chart C-1.

Notes On Intermittents:

If a Code E022 is intermittent, manipulate related wiring while observing ECM parameter ED01. Cycle the TPS through its travel and tap on the TPS with a pencil or pocket screwdriver to test for intermittent TPS. If the fault is induced, the parameter will jump to –6 or less. If wiring and connectors check out OK, substitute a known good TPS sensor and retest.

Review the snapshot data parameters stored with this code to verify failure condition. Refer to "Symptoms" for complete details on using the "Snapshot on code set" feature. If the failure condition is verified, check and repair intermittent wiring connection, or sensor.

1988–90 CADILLAC ELDORADO AND SEVILLE

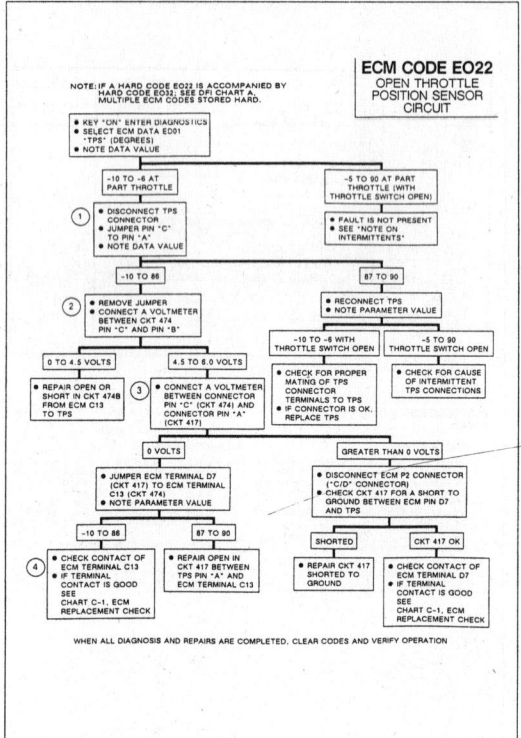

1988–90 CADILLAC ELDORADO AND SEVILLE

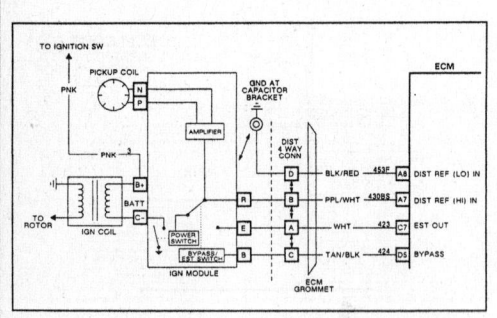

ECM CODE E023 EST CIRCUIT PROBLEM

TEST CONDITIONS: 1. Key "on"; 2. RPM greater than or equal to 648.

FAILURE CONDITIONS: EST pulses are seen by the ECM before bypass circuit is turned "on" (5 volts) or EST signal remains grounded after the bypass circuit is turned "on" (3 volts).

ACTION TAKEN: 1. ECM turns on the "Engine Control System" telltale. 2. ECM commands BCM to display "Service Soon" message on CCDIC. 3. If ECM sees 5 volt EST pulses with bypass voltage at 0 volts (caused by EST circuit open or bypass circuit shorted to voltage) the ECM will not enable spark – the engine will start and run on base timing. 4. If the EST circuit is shorted to ground, the vehicle will start and stall when the engine reaches 648 RPM (when EST is enabled). When restarted, the ECM will not try to turn bypass on, code E023 will set, vehicle will start and run on base timing.

Description:

The EST distributor produces one 5 volt Distributor Reference Pulse each time that a cylinder reaches 10° BTDC (base timing). The reference pulses are sent from the distributor to the ECM on circuit 430, Distributor Reference Signal.

The ECM adds spark timing information to the reference pulses received from the HEI and sends out 5

volt Electronic Spark Timing pulses (EST) on circuit 423.

The ECM can choose to control spark timing. The HEI Bypass circuit 424 is turned on by the ECM (approximately 5 volts) when the ECM wishes to control timing. When the ECM wants the HEI module to control timing, the HEI Bypass circuit is turned off and the HEI module grounds circuit 423, EST. The ECM always sends EST pulses to the HEI; the HEI module shorts the EST pulses to ground when the bypass signal from the ECM to the module is low (less than .5 volts).

Notes On Fault Tree:

1. Four jumper wires must be obtained for use in the diagnostic procedure. The jumpers should be about 12 inches long with a male and female weatherpack connector on either end (P/N 12014836 and 12014837). With the distributor connector disconnected, use the jumpers to reconnect the terminals per instructions provided on the code E023 chart.

2. With only Distributor Reference and Distributor Reference Ground jumpered, engine will run at backup spark. Checking for proper ground connection between ECM and the engine and the ECM and HEI.

3. Checking for ECM providing an EST output on circuit 423.

4. Checking for HEI module able to ground EST signal with open Bypass Circuit.

1988–90 CADILLAC ELDORADO AND SEVILLE

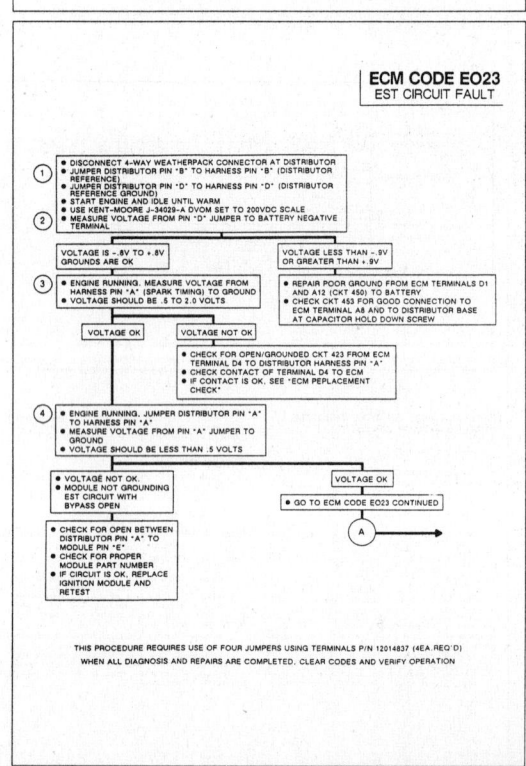

1988–90 CADILLAC ELDORADO AND SEVILLE

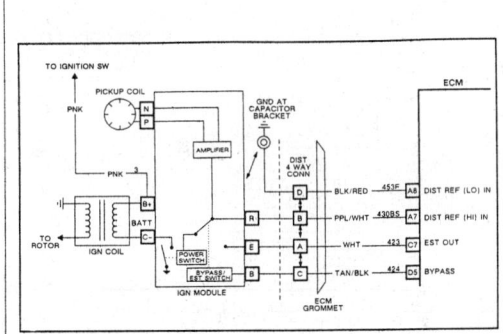

ECM CODE E023 EST CIRCUIT PROBLEM (CONT'D)

Description (Cont'd):

The Distributor Reference Ground (Circuit 453) is a common ground between the ECM and the HEI. The ECM and HEI compare Bypass, EST and Reference voltages to the ground voltage on circuit 453. Circuit 453 open to the HEI or ECM can cause a failure of the ECM to recognize a reference pulse or of the module to recognize an EST pulse.

The ECM compares the status of the EST circuit to the status of the Bypass circuit in order to detect a fault and set code E023. If the ECM has Bypass at 5 volts, the module should accept EST information. EST voltage should be high (.5–2.8 volts). If bypass is at 0 volts, the module should ground the EST circuit and the EST voltage should be low (less than .5 volts).

Possible causes of code E023 are:

A. Circuit 423 (EST) open/ground.
B. Circuit 424 (Bypass) open/grounded.

C. Circuit 453 (Reference Ground) open to ground at HEI or ECM.
D. ECM with a poor connection to the engine through the circuit 450 grounds.
E. ECM or HEI module faults.

Notes On Fault Tree (Cont'd):

5. Checking for HEI module able to recognize a voltage on the Bypass circuit and to stop grounding the EST (ECM controlling timing).

6. Checking for Bypass signal to the module. If Bypass is being sent by the ECM to the HEI and if the module is interpreting Bypass voltage correctly, then the module will switch off the ground to the EST.

7. If the chart leads to "EST circuit is OK" then a fault may exist in the 4 way weatherpack. Check for proper connector mating and for pins backing out of the weatherpack. If the connector is no trouble found, reconnect the four way, clear codes and retest.

1988–90 CADILLAC ELDORADO AND SEVILLE

ECM CODE E023–(CONT'D)
EST CIRCUIT FAULT

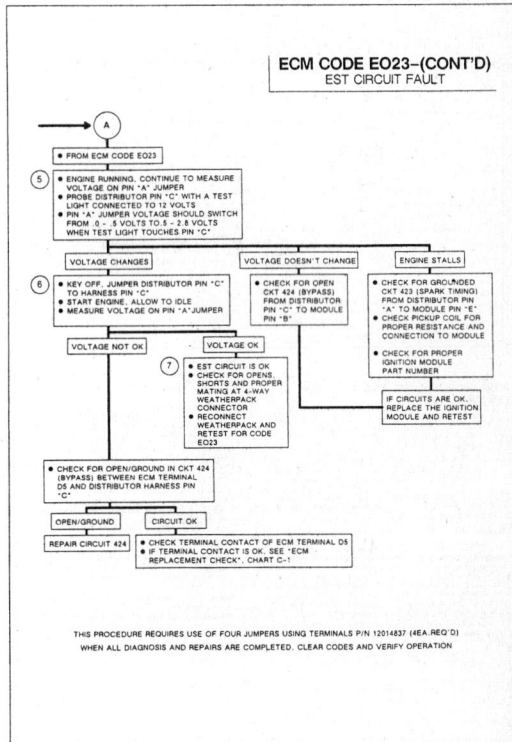

THIS PROCEDURE REQUIRES USE OF FOUR JUMPERS USING TERMINALS P/N 12014837 (4EA. REQ'D)
WHEN ALL DIAGNOSIS AND REPAIRS ARE COMPLETED, CLEAR CODES AND VERIFY OPERATION

1988–90 CADILLAC ELDORADO AND SEVILLE

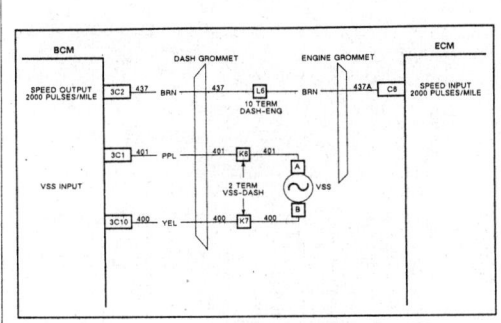

ECM CODE E024, SPEED SENSOR CIRCUIT PROBLEM

TEST CONDITIONS: 1. Codes E026, E027, E028, E031, E032, and E034 clear; 2. Gear shift in drive or reverse; 3. Not braking (Brake switch closed) 4. ISC (throttle switch open); 5. MAP less than or equal to 85 KPA; 6. Fourth gear switch open (Trans in fourth gear); 7. RPM greater than or equal to 1400.

FAILURE CONDITIONS: Vehicle Speed = 0 MPH for 3 seconds.

ACTION TAKEN: 1. ECM turns on "Engine Control System" telltale. 2. ECM commands BCM to display "Service Soon" message on CCDIC. 3. Disable cruise for the entire key cycle. 4. Disable VCC. 5. Calculate vehicle speed based on engine RPM and gear status.

Description:

The Vehicle Speed Sensor (VSS) is a permanent magnet pulse generator that is geared to the transmission output to create 4000 voltage pulses/mile. The BCM receives the speed sensor from the VSS, changes the signal to 2000 pulses per mile then sends the 2000 pulse/mile signal to the ECM or CKT 437.

The ECM uses VSS input for VCC apply and release determinations, to select between RPM and throttle angle control of ISC and as a test condition for many codes.

Notes On Fault Tree:

1. Checking for proper voltage from BCM at ECM connector.

2. Checking for voltage sent from BCM.

3. Before replacing ECM or BCM, remove and replace the BCM "J3" connector and the ECM "J2" connector and ensure that they are latched.

Note On Intermittents:

Backprobe ECM Pin C8 with a voltmeter to ground. Lift both drive wheels of the vehicle, start engine and allow to idle in drive. Manipulate affected wiring and watch for sudden loss of voltage on ECM Pin C8.

Review the snapshot data parameters stored with this code to verify failure condition. Refer to "Symptoms" for complete details on using the "Snapshot on code set" feature. If the failure condition is verified, check and repair intermittent wiring connection, or sensor.

1988–90 CADILLAC ELDORADO AND SEVILLE

ECM CODE E024
SPEED SENSOR CIRCUIT PROBLEM

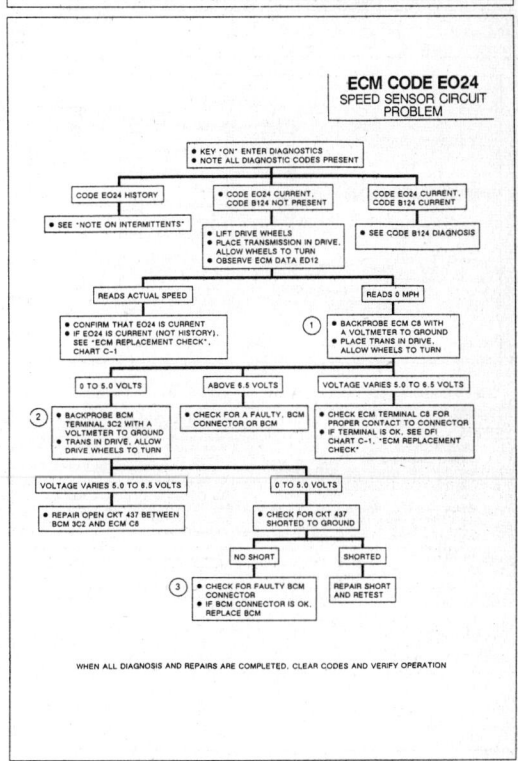

WHEN ALL DIAGNOSIS AND REPAIRS ARE COMPLETED, CLEAR CODES AND VERIFY OPERATION

1988–90 CADILLAC ELDORADO AND SEVILLE

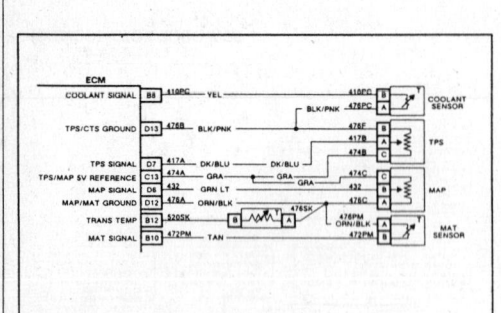

ECM CODE E026, SHORTED THROTTLE SWITCH CIRCUIT

TEST CONDITIONS: 1. Code E021 and E022 clear; 2. TPS greater than 14 degrees; 3. TPS voltage less than 4.5V (about 80°).

FAILURE CONDITION: Throttle switch input to ECM grounded (throttle switch closed).

ACTION TAKEN: 1. ECM turns on "Engine Control System" light. 2. ECM commands BCM to display "Service Soon" message on CCDIC; 3. ECM assumes closed throttle if brake is applied and TPS ≤ 12°. ECM assumes open throttle when brake is off or TPS > 12°.

Description:

The throttle switch is part of the ISC motor assembly. Pin B of the ISC motor 4 way weather pack connector is the throttle switch input to the ECM. The ECM provides a 4.5 to 6.0 volt signal to pin B of the ISC. When the throttle lever contacts the ISC plunger, the throttle switch closes, shorting the 4.5 to 6.0 volt signal at pin B to pin A which is a ground (closed throttle = throttle switch input voltage low).

Code E026 is designed to detect a throttle switch input that is always grounded; throttle switch never opens.

Notes On Fault Tree:

1. Code E026 due to throttle switch input shorted to ground or the ISC throttle switch inoperative should result in E172 not cycling–E172 will be "LO" at all times.

2. Checking for TPS reading too high–TPS fault or misadjustment that results in a reading above 14 Degrees at closed throttle will cause a false Code E026.

3. Checking for TPS too high due to TPS or TPS circuit.

4. If TPS and circuit are both OK, the ECM should be checked per Chart C-1.

Note On Intermittents:

Select ECM input E172. Manipulate affected wiring and connectors and watch the throttle switch status to change. Check for intermittent ground on CKT 427.

Select ECM data ED01. Manipulate affected wiring and connectors and watch for TPS value to go above 14 degrees at closed throttle. Tap lightly on the TPS with a pencil or pocket screwdriver. (Check for binding TPS lever, intermittent open on CKT 476 or intermittent short to 5 volts on CKT 417 at TPS.

1988–90 CADILLAC ELDORADO AND SEVILLE

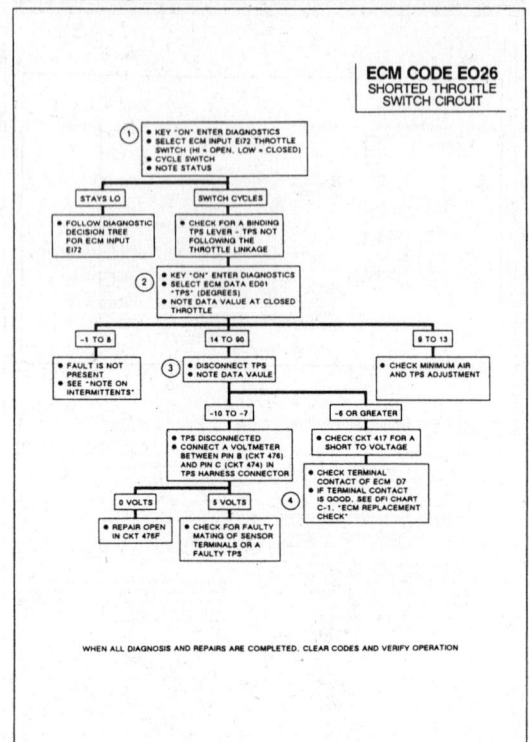

ECM CODE EO26
SHORTED THROTTLE SWITCH CIRCUIT

WHEN ALL DIAGNOSIS AND REPAIRS ARE COMPLETED, CLEAR CODES AND VERIFY OPERATION

1988–90 CADILLAC ELDORADO AND SEVILLE

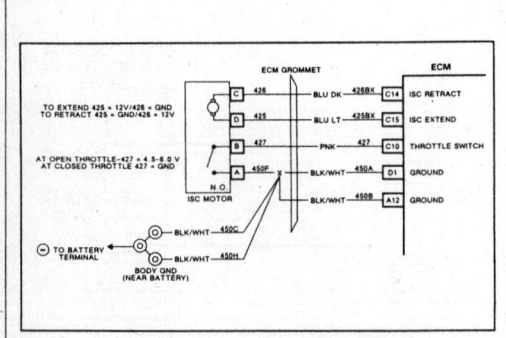

ECM CODE E027, OPEN THROTTLE SWITCH CIRCUIT

TEST CONDITIONS: Always tested.

FAILURE CONDITIONS: 1. Throttle switch was never closed during the last key cycle; 2. Brake switch was applied during the last key cycle; 3. Throttle switch has not closed in the current key cycle.

ACTION TAKEN: 1. ECM turns on "Engine Control System" telltale; 2. ECM commands BCM to display "Service Soon" message on CCDIC; 3. ECM assumes throttle switch is close if TPS is ≤ 12° and the brake switch is closed (not braking).

Description:

The throttle switch is part of the ISC motor assembly. Pin B of the ISC motor 4-way weatherpack connector is the throttle switch input to the ECM. The

ECM provides a 4.5–6.0 volt signal to pin B of the ISC. When the throttle lever contacts the ISC plunger, the throttle switch closes, shorting the 4.5–6.0 volt signal at pin B to pin A which is a ground. (Closed throttle = throttle switch input voltage low). Code E027 is designed to detect a throttle switch that is always open.

Notes On Fault Tree:

1. Code E027 due to throttle switch input open or the ISC throttle switch inoperative should result in E172 not cycling – E172 will be "HI" at all times.

Notes On Intermittents:

Check for throttle shaft and throttle blades binding, throttle spring weak or distorted, cruise or TV cable binding throttle linkage, TPS for proper installation.

Intermittent Code E027 can be induced by resting foot on the accelerator during a key "On" key "Off" for 10 seconds, key "On" cycle.

1988–90 CADILLAC ELDORADO AND SEVILLE

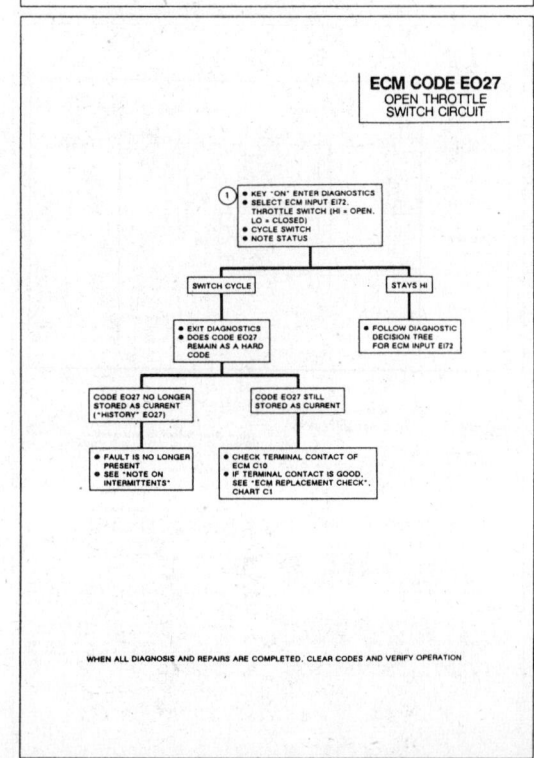

ECM CODE EO27
OPEN THROTTLE SWITCH CIRCUIT

WHEN ALL DIAGNOSIS AND REPAIRS ARE COMPLETED, CLEAR CODES AND VERIFY OPERATION

1988–90 CADILLAC ELDORADO AND SEVILLE

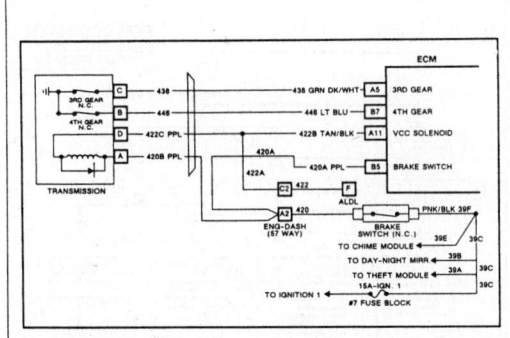

ECM CODE E028, OPEN THIRD OR FOURTH GEAR CIRCUIT

TEST CONDITIONS: 1. Code E024 clear; 2. Transmission in park or neutral; 3. Vehicle speed less than or equal to 4 MPH.

FAILURE CONDITION: Third gear input to ECM not grounded or Fourth gear input to ECM not grounded.

ACTION TAKEN: 1. ECM turns on "Engine Control System" telltale; 2. ECM commands BCM to display "Service Soon" message on CCDIC.

Description:

The third and fourth gear switches in the transmission are normally closed switches that are opened by transmission oil pressure. The ECM sends a 12 volt signal to each of the switches. When not in third or fourth gear, the third and fourth gear inputs are grounded by the normally closed switches. When third gear is achieved, the third gear switch is opened by third gear oil pressure and the third gear input changes from 0 volts to 12 volts. When fourth gear is achieved, the fourth gear switch is opened by fourth gear oil pressure and the fourth gear input changes from 0 to 12 volts.

Code E028 is designed to detect a false third or fourth gear indication to ECM.

Notes On Intermittents:

At key "ON", engine "OFF", trans is park, manipulate affected wiring and connectors while observing the third/fourth gear status light. If third or fourth gear status indicator flashes on repair intermittent open in CKT 438 (third gear) or CKT 446 (fourth gear).

Review the snapshot data parameters stored with this code to verify failure condition. Refer to "Symptoms" for complete details on using the "Snapshot on code set" feature. If the failure condition is verified, check and repair intermittent wiring connection, or sensor.

1988–90 CADILLAC ELDORADO AND SEVILLE

ECM CODE E028
OPEN 3RD OR 4TH GEAR CIRCUIT

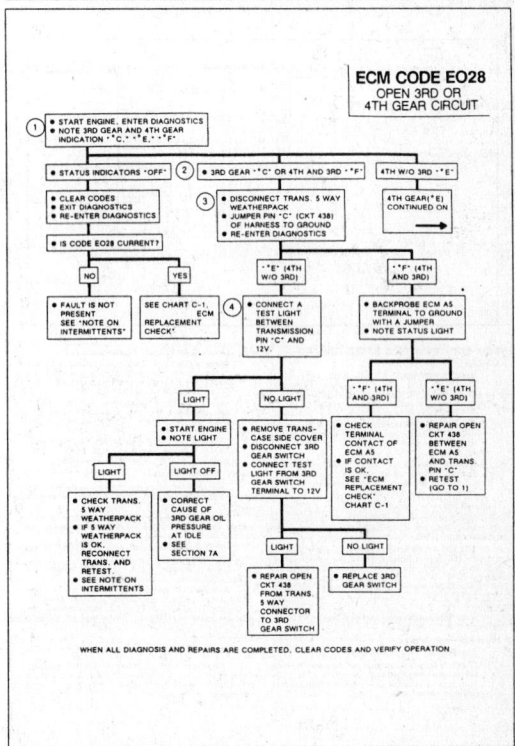

WHEN ALL DIAGNOSIS AND REPAIRS ARE COMPLETED, CLEAR CODES AND VERIFY OPERATION

1988–90 CADILLAC ELDORADO AND SEVILLE

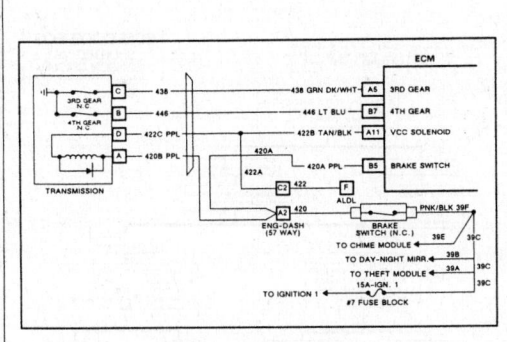

ECM INPUT E028 (CONT'D)

OPEN THIRD OR FOURTH GEAR CIRCUIT (4TH W/O 3RD)

TEST CONDITIONS: 1. Code E024 clear; 2. Transmission in park or neutral; 3. Vehicle speed ≤ 4 MPH.

FAILURE CONDITION: Third gear input to ECM not grounded or Fourth gear input to ECM not grounded.

ACTION TAKEN: 1. ECM turns on "Engine Control System" light; 2. ECM commands BCM to display "Service Soon" message on CCDIC.

Description

The third and fourth gear switches in the transmission are normally closed switches that are opened by transmission oil pressure. The ECM sends a 12 volt signal to each of the switches. When not in third or fourth gear, the third and fourth gear inputs are grounded by the normally closed switches. When third gear is achieved, the third gear switch is opened by third gear oil pressure and the third gear input changes from 0 volts to 12 volts. When fourth gear is achieved, the fourth gear switch is opened by fourth gear oil pressure and the fourth gear input changes from 0 to 12 volts.

The Code E028 is designed to detect a false third or fourth gear input.

Notes On Intermittents:

At key "ON", engine "OFF", manipulate affected wiring and connectors while observing the third/fourth gear status light. If third or fourth gear status indicator flashes on, repair intermittent open in CKT 438 (third gear) or CKT 446 (fourth gear).

Review the snapshot data parameters stored with this code to verify failure condition. Refer to "Symptoms" for complete details on using the "Snapshot on code set" feature. If the failure condition is verified, check and repair intermittent wiring connection, or sensor.

1988–90 CADILLAC ELDORADO AND SEVILLE

ECM CODE E028 – (CONT'D)
(4TH W/O 3RD)
OPEN 3RD OR 4TH GEAR CIRCUIT

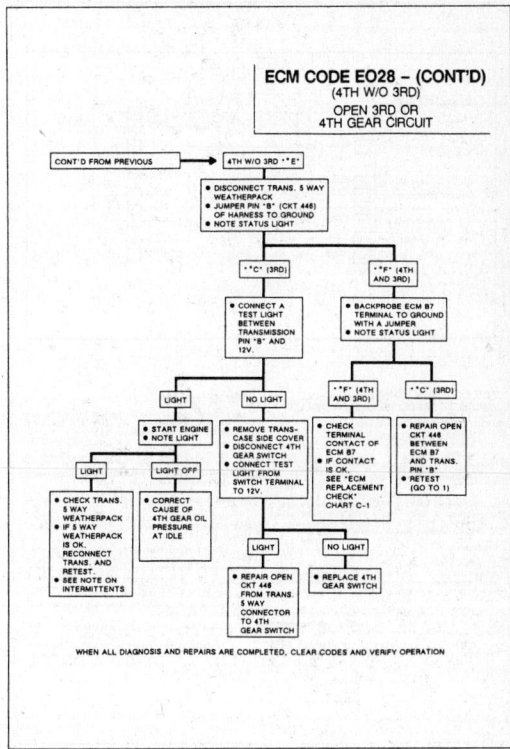

WHEN ALL DIAGNOSIS AND REPAIRS ARE COMPLETED, CLEAR CODES AND VERIFY OPERATION

1988–90 CADILLAC ELDORADO AND SEVILLE

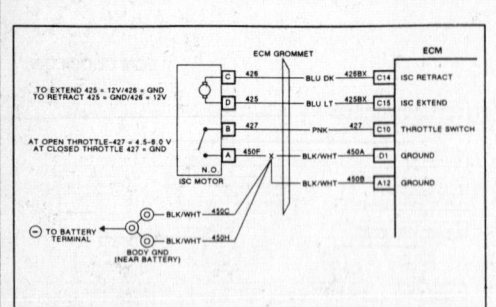

ECM CODE E030, IDLE SPEED CONTROL CIRCUIT PROBLEM

TEST CONDITIONS: 1. Codes E021, E022, E024, E026, and E027 clear; 2. Speed = 0; 3. Throttle switch closed; 4. Battery voltage greater than or equal to 11.0 volts; 5. Minimum air rate or engine RPM greater than 152 RPM different than desired; 6. Not a cold start.

FAILURE CONDITIONS: ISC plunger extending with TPS less than or equal to 9° or ISC retracting with TPS greater or equal to 2.5° for 15 seconds.

ACTION TAKEN: 1. ECM turns on "Engine Control System" telltale light for entire key cycle. 2. ECM commands BCM to display "Service Soon" message on CCDIC.

Description:

The ECM controls engine idle by increasing or decreasing the throttle opening using the idle speed control motor (ISC).

The ISC will be active in controlling idle speed at any time the throttle switch is closed. At vehicle speeds less than 6 MPH, the ECM controls idle based on engine RPM. When vehicle speed is greater than 6 MPH, the ECM controls idle based primarily on throttle opening (TPS) with adjustments to maintain a minimum engine RPM.

Code E030 is designed to detect engine RPM out of limits high or low.

Notes On Fault Tree:

1. Checking for TPS misadjusted.
2. Checking for proper throttle switch, power steering switch, park/neutral switch operation. The ECM must receive accurate switch status information in order to control idle.

3. Checking for proper ISC motor operation.
4. Many engine fuel and emissions system faults may cause unstable idle. If the base engine idle is not steady, the ISC may not be able to control idle to within 150 RPM of commanded.
5. Out of adjustment minimum air rate, ISC or TPS can cause ISC not to control idle or code E030 to set falsely. Check adjustment.
6. Checking ISC motor circuits for proper operation: NOTE: Throttle switch input to ECM must be jumpered in order to allow ISC to cycle.

Notes On Intermittents:

Display ECM parameter ED01, manipulate TPS wiring and connectors, watch for TPS value to jump or skip.

Enter ECM inputs EI72 and observe throttle switch status – check for binding throttle linkage or throttle return spring weak.

Verify that minimum air rate, RPM, ISC and TPS are set to specification.

If idle is unstable, see "Symptoms" for "Rough, Unstable, Incorrect Idle, Stalling".

Review the snapshot data parameters stored with this code to verify failure condition. Refer to "Symptoms" for complete details on using the "Snapshot on code set" feature. If the failure condition is verified, check and repair intermittent wiring connection, or sensor.

1988–90 CADILLAC ELDORADO AND SEVILLE

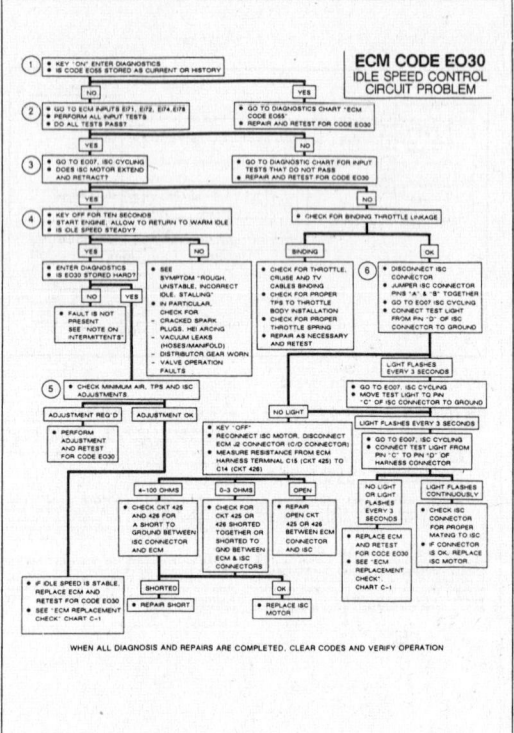

ECM CODE E030
IDLE SPEED CONTROL CIRCUIT PROBLEM

WHEN ALL DIAGNOSIS AND REPAIRS ARE COMPLETED, CLEAR CODES AND VERIFY OPERATION

1988–90 CADILLAC ELDORADO AND SEVILLE

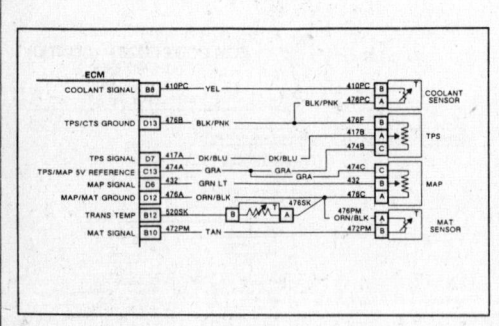

ECM CODE E031, SHORTED MANIFOLD ABSOLUTE PRESSURE SENSOR CIRCUIT

TEST CONDITIONS: Always tested.

FAILURE CONDITIONS: MAP value greater than 105 KPA (106–109 kPa) for at least .2 seconds.

ACTION TAKEN: 1. ECM turns on "Engine Control System" telltale; 2. ECM commands BCM to display "Service Now" on the CCDIC; 3. Turn off air management solenoids; 4. ECM uses a substitute MAP sensor value from a table of substitute MAP vs RPM when throttle switch is closed; 5. ECM uses a substitute MAP sensor value of 81.8 kPa with open throttle switch; 6. ECM assumes BARO value is 92.2 kPa.

Description:

The MAP sensor output signal voltage is a DC voltage that varies with manifold absolute pressure. As MAP decreases, voltage decreases (low engine load, high vacuum). As MAP increases, voltage increases (high engine load, low vacuum).

The ECM uses MAP sensor values as an indicator of engine load. A high MAP reading indicates heavy load and low MAP reading indicates low load.

Code E031 is designed to set when the ECM detects a MAP sensor signal out of limits high – MAP signal at 4.85 volts or more, MAP value greater than 105 kPa.

Notes On Fault Tree:

1. If ECM parameter ED02 goes to 14–16 kPa with the sensor unplugged, the fault is at the MAP sensor or sensor connector.
2. Checking for an open circuit from Pin A of the sensor connector to ECM-Pin D12 (ground). If the ground is open, the sensor can't divide the reference voltage to make the signal voltage vary – the signal voltage is always high.

Note On Intermittents:

Code E031 can be set by 1. short to voltage on CKT 432; 2. Open from sensor to ECM on CKT 476; 3. Defective MAP sensor.

Manipulate affected wiring and connectors while observing ECM parameter ED02. Apply and release vacuum to the MAP sensor vacuum port using a vacuum source. If the ECM parameter ED02 displays greater than 105 kPa, the condition has been induced and the cause of the intermittent should be repaired. If wiring and connectors check out OK, substitute a known good MAP sensor and retest.

Review the snapshot data parameters stored with this code to verify failure condition. Refer to "Symptoms" for complete details on using the "Snapshot on code set" feature. If the failure condition is verified, check and repair intermittent wiring connection, or sensor.

1988–90 CADILLAC ELDORADO AND SEVILLE

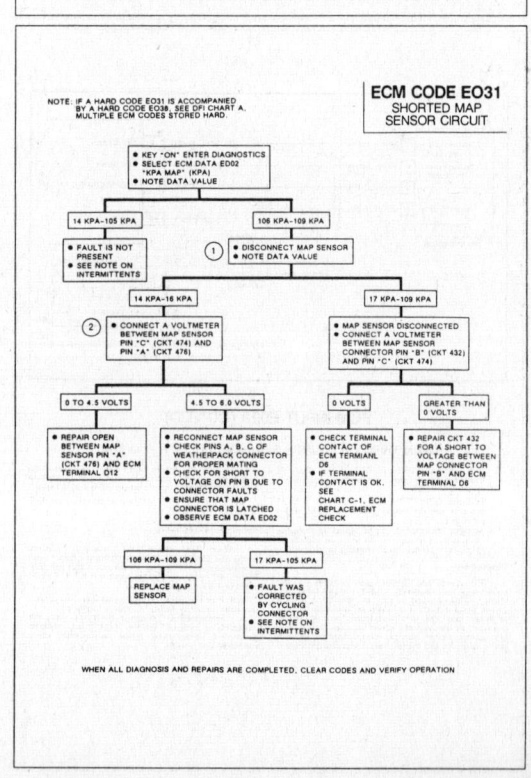

NOTE: IF A HARD CODE E031 IS ACCOMPANIED BY A HARD CODE E038, SEE DFI CHART A. MULTIPLE ECM CODES STORED HARD.

ECM CODE E031
SHORTED MAP SENSOR CIRCUIT

WHEN ALL DIAGNOSIS AND REPAIRS ARE COMPLETED, CLEAR CODES AND VERIFY OPERATION

1988–90 CADILLAC ELDORADO AND SEVILLE

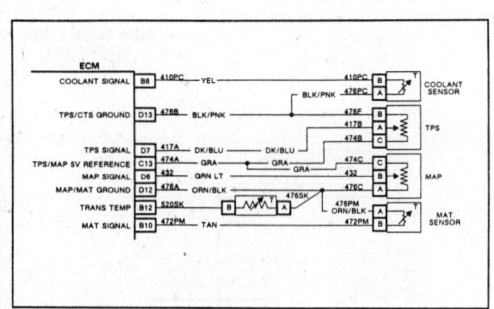

ECM CODE E032, OPEN MANIFOLD ABSOLUTE PRESSURE SENSOR CIRCUIT

TEST CONDITIONS: 1. With ISC throttle switch closed, RPM less than or equal to 700, TPS less than or equal to 12°; 2. With ISC throttle switch open, TPS greater than or equal to 9.6°.

FAILURE CONDITIONS: MAP value less than 16 kPa.

ACTION TAKEN: 1. ECM turns on "Engine Control System" telltale. 2. ECM commands BCM to display "Service Now" message on CCDIC. 3. Turn off both air solenoids; 4. ECM uses a substitute MAP sensor value of 81.8 kPa with open throttle switch; 5. ECM looks up a substitute MAP value from a table of substitute MAP vs RPM when throttle switch is closed; 6. ECM uses a substitute BARO value of 92.2 kPa.

Description:

The MAP sensor output voltage is a DC voltage that varies with Manifold Absolute Pressure (MAP). As MAP decreases, voltage on CKT 432 decreases (low engine load, high engine vacuum). As MAP increases, voltage on CKT 432 increases (high engine load, low engine vacuum).

Code E032 is designed to set when the ECM detects a MAP sensor signal out of limits low – MAP signal at 16 kPa or less, MAP voltage at .08 volts or less.

Notes On Fault Tree:

1. Checking for circuitry from sensor to ECM able to respond to a 5 volt signal on MAP input. A

reading of 106–109 means wiring and ECM are OK.

2. Checking for 5 volt reference present at sensor connector.

3. Checking for CKT 432 shorted to ground.

4. Checking for ECM able to respond to a 5 volt signal voltage on MAP input.

5. Fault is most likely at ECM connector or ECM. Before ECM is replaced, perform "ECM Replacement Check", Chart C-1.

Note On Intermittents:

Code E032 can be set by 1. Open 5 volt reference between ECM and sensor; 2. Open in MAP signal between sensor and ECM; 3. Defective MAP sensor.

Manipulate affected wiring and connectors while observing ECM parameter ED02. Apply and release vacuum to the MAP sensor vacuum port using a vacuum source. If the ECM parameter ED02 displays less that 17 kPa, the condition has been induced and the cause of the intermittent can be repaired. If wiring and connectors check out OK, substitute a known good MAP sensor and retest.

Review the snapshot data parameters stored with this code to verify failure condition. Refer to "Symptoms" for complete details on using the "Snapshot on code set" feature. If the failure condition is verified, check and repair intermittent wiring connection, or sensor.

1988–90 CADILLAC ELDORADO AND SEVILLE

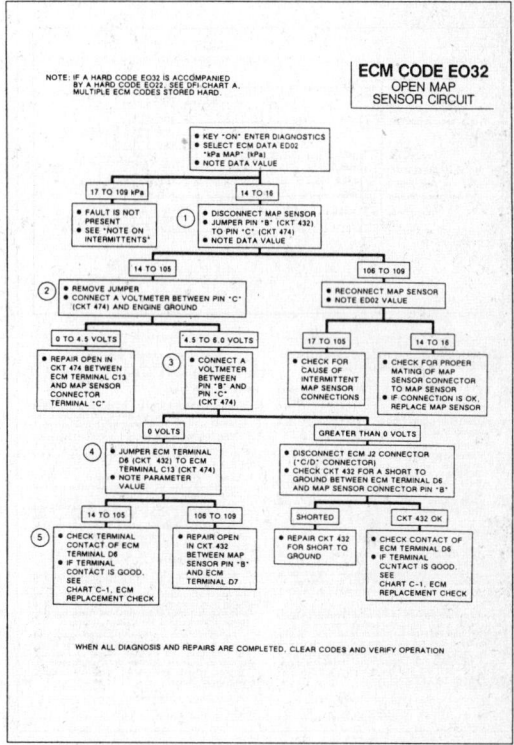

ECM CODE E032
OPEN MAP SENSOR CIRCUIT

NOTE: IF A HARD CODE E032 IS ACCOMPANIED BY A HARD CODE E022, SEE DFI CHART A. MULTIPLE ECM CODES STORED HARD.

WHEN ALL DIAGNOSIS AND REPAIRS ARE COMPLETED, CLEAR CODES AND VERIFY OPERATION

1988–90 CADILLAC ELDORADO AND SEVILLE

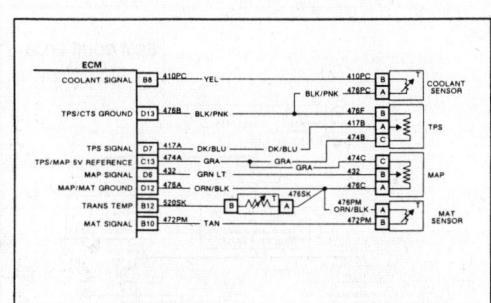

Figure 8D1-51 DFI/Sensor Circuit

ECM CODE E034, MAP SIGNAL TOO HIGH

TEST CONDITIONS: 1. Codes E021, E022, E026, E027, E031, and E032 clear; 2. Key on; 3. RPM greater than or equal to 400; 4. Throttle switch closed; 5. TPS less than or equal to 12 degrees; 6. BARO greater than or equal to 68 kPa.

FAILURE CONDITIONS: MAP value within 6 kPa of the BARO value for 2.1 seconds.

ACTION TAKEN: 1. The ECM turns on the "Engine Control System" telltale. 2. ECM commands BCM to display "Service Now" message on CCDIC. 3. ECM turns off both air management solenoids. 3. ECM uses a substitute MAP sensor value of 81.8 kPa with throttle switch open; 5. ECM looks up a substitute MAP value from a table of "Substitute MAP" vs. RPM when the throttle switch is closed.

Description:

The MAP sensor output voltage is a DC voltage that varies with Manifold Absolute Pressure (MAP). As MAP decreases, voltage on CKT 432 decreases (low engine load, high engine vacuum). As MAP increases, voltage on CKT 432 increases (high engine load, low engine vacuum).

The ECM uses the MAP sensor reading as an indicator of engine load. Low MAP readings mean low engine load, high MAP readings mean high engine load.

Code E034 is designed to set when the ECM detects a MAP sensor signal that is much too high for the closed throttle test conditions. Code E034 usually indicates that there is a fault in the vacuum source to the MAP sensor.

Notes On Fault Tree:

1. MAP at idle should be 30–50 kPa, depending on engine load. BARO should be 85–105, kPa, depending upon altitude.

2. Check for vacuum at the MAP sensor hose with a vacuum gage. At idle, typical vacuum readings are 15″–20″ Hg, depending on engine load.

3. Vacuum supply to sensor is OK, checking for MAP sensor or MAP circuitry fault.

4. Checking for short to voltage on sensor signal, CKT 432.

5. Checking for sensor ground open from sensor to ECM.

6. Fault is most likely at ECM connector or ECM. Before ECM is replaced, perform "ECM Replacement Check", Chart C-1.

Note On Intermittents:

Code E034 is usually set by a vacuum supply problem to the MAP sensor. Check for proper vacuum routing (see emission label underhood), for MAP hose connected to the proper throttle body port. MAP hose chafed, pinched, or cut.

Apply vacuum to the MAP hose on the throttle body and look for vacuum leaks in the MAP hose or MAP sensor.

Manipulate affected wiring and connections while observing ECM parameter ED02. Apply and release vacuum to the MAP hose at the throttle body using a vacuum source. If the ECM parameter ED02 jumps or skips high with vacuum applied, the condition has been induced and the cause of the intermittent can be repaired. If wiring and connectors check out OK, substitute a known good MAP sensor and retest.

Review the snapshot data parameters stored with this code to verify failure condition. Refer to "Symptoms" for complete details on using the "Snapshot on code set" feature. If the failure condition is verified, check and repair intermittent wiring connection, or sensor.

1988–90 CADILLAC ELDORADO AND SEVILLE

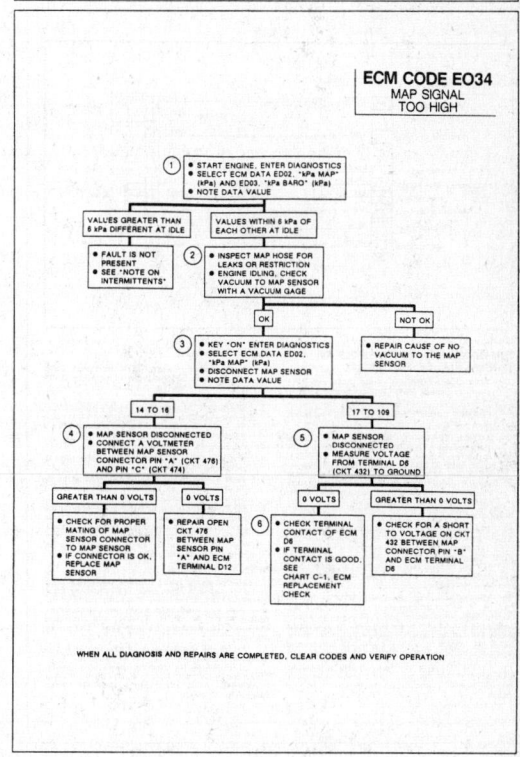

ECM CODE E034
MAP SIGNAL TOO HIGH

WHEN ALL DIAGNOSIS AND REPAIRS ARE COMPLETED, CLEAR CODES AND VERIFY OPERATION

1988–90 CADILLAC ELDORADO AND SEVILLE

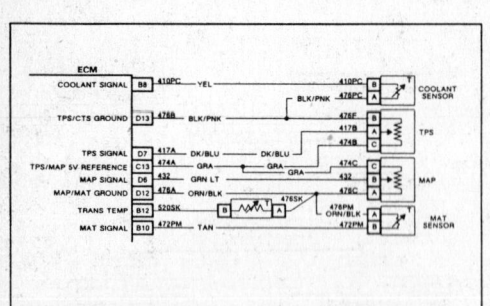

ECM CODE E037, SHORTED MANIFOLD AIR TEMPERATURE SENSOR CIRCUIT

TEST CONDITIONS: 1. Codes E014 and E015 clear; 2. Coolant sensor less than or equal to 105°C.

FAILURE CONDITIONS: MAT sensor value greater than or equal to 142°C.

ACTION TAKEN: 1. ECM turns on "Engine Control System" telltale. 2. ECM commands BCM to display "Service Soon" message on CCDIC. 3. ECM turns off both air management solenoids. 4. ECM substitutes 40°C for MAT when coolant is greater than or equal to 40°C. 5. ECM uses coolant temp for MAT when coolant temp is less than or equal to 40°C.

Description:

The MAT sensor is a thermistor or variable resistor that varies in resistance with temperature. The MAT sensor is a "two wire" sensor with a reference or signal voltage coming from the ECM to sensor pin B, CKT 472 and a sensor reference ground on pin A, CKT 476.

As the temperature of the sensor increases, sensor resistance is lower. The signal voltage from the ECM to pin B decreases as sensor temperature increases because current flows from pin B through the sensor element to pin A, sensor ground (voltage is dropped across the sensor element). High temperature means low signal voltage on CKT 472 and low temperature means high signal voltage on CKT 472.

Code E037, shorted MAT sensor sets because the ECM assumes that manifold air temperature can't be 142°C or greater with a coolant of less than 105°C.

Notes On Fault Tree:

1. With a shorted sensor, ECM parameter ED05 should read 142°C or greater, if not, then the sensor is not shorted, see note on intermittents.

2. Checking for sensor shorted or CKT 472 shorted. If ECM parameter ED05 stays at 142–151°C with the sensor unplugged, then the short is in CKT 472.

3. Fault is most likely at ECM connector or ECM. Before ECM is replaced, perform "ECM Replacement Check", Chart C-1.

4. MAT sensors can be damaged by a backfire in the intake. If this vehicle has had more than one MAT sensor replaced, check for signs of backfire or high intake manifold temperatures due to improper valve train operation.

Notes On Intermittents:

Manipulate CKT 472 wiring, the MAT sensor and ECM connector while observing ECM parameter ED05. If the failure is induced, the manifold air temperature will jump from its normal value to the shorted reading of 142–151°C. If wiring and connectors check out OK, substitute a known good MAT sensor and retest.

Review the snapshot data parameters stored with this code to verify failure condition. Refer to "Symptoms" for complete details on using the "Snapshot on code set" feature. If the failure condition is verified, check and repair intermittent wiring connection, or sensor.

1988–90 CADILLAC ELDORADO AND SEVILLE

ECM CODE EO37
SHORTED MAT SENSOR CIRCUIT

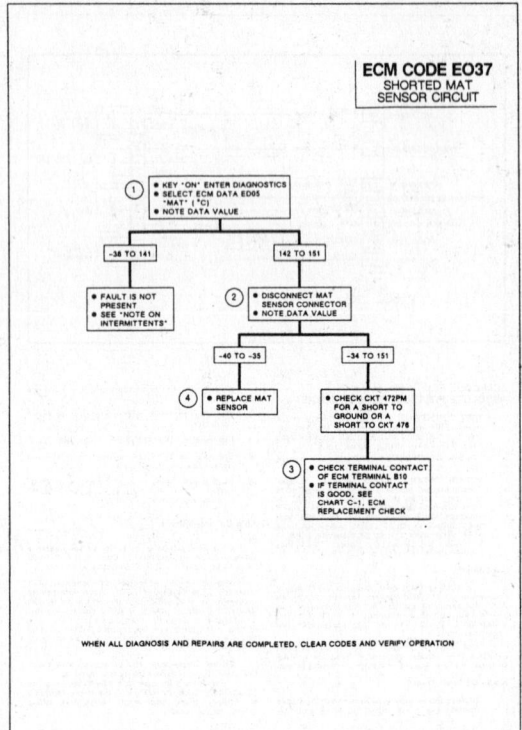

WHEN ALL DIAGNOSIS AND REPAIRS ARE COMPLETED, CLEAR CODES AND VERIFY OPERATION

1988–90 CADILLAC ELDORADO AND SEVILLE

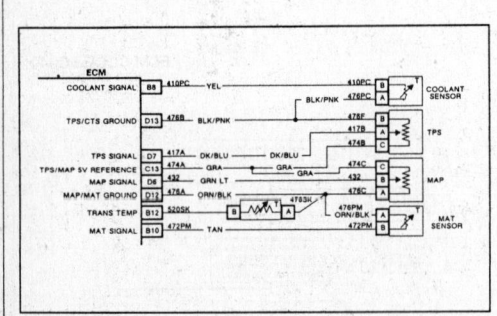

ECM CODE E038, OPEN MANIFOLD AIR TEMPERATURE SENSOR CIRCUIT

TEST CONDITIONS: 1. Codes E014 and E015 clear; 2. Coolant sensor greater than or equal to –5°C.

FAILURE CONDITIONS: MAT sensor value less than –35°C.

ACTION TAKEN: 1. ECM turns on "Engine Control System" telltale. 2. ECM commands BCM to display "Service Soon" message on CCDIC. 3. ECM turns off both air management solenoids. 4. ECM uses 40°C for a MAT value when coolant temp is greater than 40°C; 5. ECM substitutes coolant temp for MAT when coolant temp is less than 40°C.

Description:

The MAT sensor is a thermistor or variable resistor that varies in resistance with temperature. The MAT sensor is a "two wire" sensor with a reference or signal voltage coming from the ECM to sensor pin B, CKT 472 and a sensor reference ground on pin A, CKT 476. As the temperature of the sensor decreases, sensor resistance increases.

The signal voltage from the ECM to pin B increases as sensor temperature decreases because less current flows from sensor pin B through the sensor element to ground (less of the signal voltage is dropped across the sensor element). Low temperature means high signal voltage on CKT 472 and high temperature means low signal voltage on CKT 472.

Code E038, open MAT sensor, sets because the ECM assumes that the manifold air temperature can't be –35 to –40°C with a coolant temp of –5°C or greater.

Notes On Fault Tree:

1. If the sensor is open, ECM parameter ED05 should read –35°C or less. If not, the sensor signal is not open at this time, see "Notes on Intermittents".

2. Checking MAT and sensor circuitry from ECM to sensor connector. If ECM parameter ED05 reads 142°–151°C with connector pin A shorted to pin B then the sensor circuits and ECM are OK.

3. Checking for open sensor ground.

4. Checking for ECM's ability to recognize open to ground on MAT input.

5. Fault is most likely at ECM connector of ECM. Before ECM is replaced, perform "ECM Replacement Check", Chart C-1.

6. MAT sensor can be damaged by backfire in the intake or by excessive intake heat due to valve train faults. If the vehicle has had multiple MAT sensor replacements, check for signs of backfire or high intake manifold air temperature due to improper valve train operation.

Notes On Intermittents:

Manipulate CKT 472 and 476 wiring, the MAT connector and ECM connector while observing ECM parameter ED05. If the failure is induced, the MAT will jump from its normal value to the open signal circuit reading of –35 to –40°C. If wiring and connectors check out OK, substitute a known good MAT sensor and retest.

Review the snapshot data parameters stored with this code to verify failure condition. Refer to "Symptoms" for complete details on using the "Snapshot on code set" feature. If the failure condition is verified, check and repair intermittent wiring connection, or sensor.

1988–90 CADILLAC ELDORADO AND SEVILLE

ECM CODE EO38
OPEN MAT SENSOR CIRCUIT

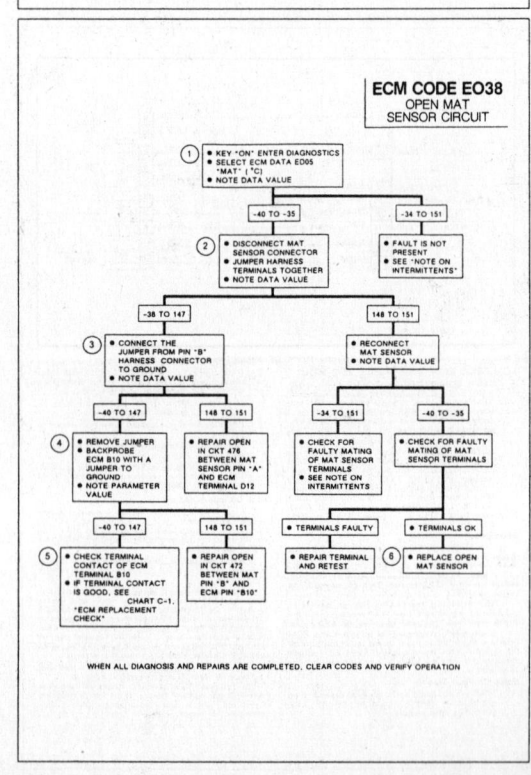

WHEN ALL DIAGNOSIS AND REPAIRS ARE COMPLETED, CLEAR CODES AND VERIFY OPERATION

1988–90 CADILLAC ELDORADO AND SEVILLE

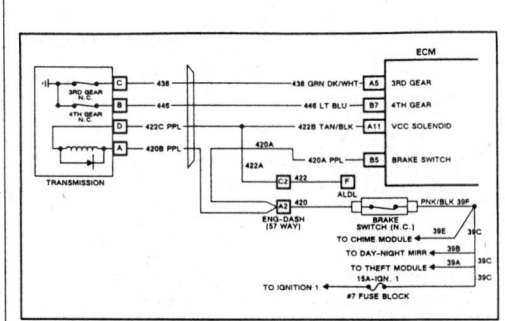

ECM CODE E039, VCC ENGAGEMENT PROBLEM

TEST CONDITIONS: 1. Code E028 clear; 2. Trans in fourth gear; 3. Throttle switch open; 4. Closed loop; 5. Not accelerating or decelerating; 6. Trans not in park or neutral; 7. Not braking 8. VCC engaged. 9. MAP ≤ 80 kPa.

FAILURE CONDITIONS: Engine RPM too high for vehicle speed – ECM has a 9 place table of speed vs RPM. If RPM exceeds the RPM for the speed listed in table for 5 seconds, Code E039 will set.

ACTION TAKEN: 1. ECM turns on the "Engine Control System" telltale. 2. ECM commands BCM to display "Service Soon" message on CCDIC. 3. VCC disengages for ignition cycle.

Description:

The viscous convertor clutch solenoid is turned on when the ECM applies ground to circuit 422, "VCC Solenoid Circuit". The power for the VCC solenoid comes from the 15A IGN-1 fuse in the fuse block, flows through the VCC brake switch to pin A of five-way connector on the transmission. At pin A, current flows into the transmission to the VCC solenoid.

Code E039 is set when the engine RPM exceeds the normal value for a given speed with VCC applied. With VCC applied, the engine is coupled directly to the transmission through the VCC — almost no slippage should occur, so the RPM/speed ratio is nearly constant. If RPM is too high at a given speed, the ECM diagnosis the condition as the VCC is slipping or not applied and code E039 logged.

Notes On Fault Tree:

1. Checking for continuity from the fuse through the solenoid to CKT 422.

2. Checking ECM's ability to ground the solenoid.

3. Circuit is OK, checking for transmission hydraulic system faults.

Notes On Intermittents:

The vehicle must be driven for 5 seconds in the test conditions for code E039 to occur. This may necessitate an extended test drive, at least 15 minutes of operation at above 45 MPH (50 MPH if BARO ≤ 92 kPa).

1988–90 CADILLAC ELDORADO AND SEVILLE

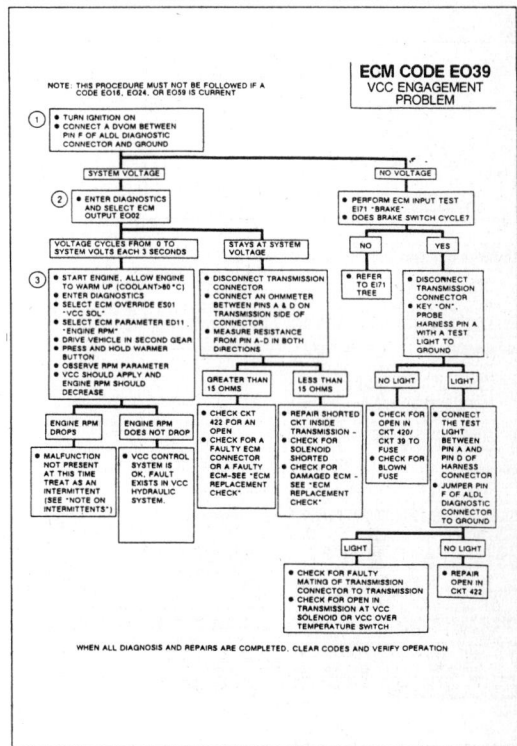

NOTE: THIS PROCEDURE MUST NOT BE FOLLOWED IF A CODE E016, E024, OR E059 IS CURRENT

ECM CODE E039
VCC ENGAGEMENT PROBLEM

WHEN ALL DIAGNOSIS AND REPAIRS ARE COMPLETED, CLEAR CODES AND VERIFY OPERATION

1988–90 CADILLAC ELDORADO AND SEVILLE

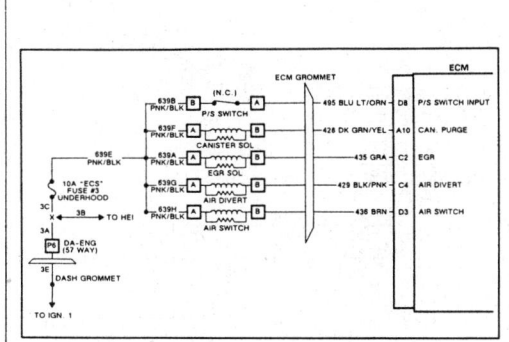

ECM CODE E040 OPEN POWER STEERING PRESSURE CIRCUIT

TEST CONDITIONS: 1. Vehicle speed greater than or equal to 40 MPH.

FAILURE CONDITIONS: Power steering switch opens (ECM sees 0 volts on pin D8, steering input) for ten seconds.

ACTION TAKEN: 1. ECM turns on "Engine Control System" telltale. 2. ECM commands BCM to display "Service Soon" message on CCDIC.

Description:

The power steering pressure switch is normally closed and opens with power steering pressure. The power steering pressure switch receives 12 volts from the 10 AMP solenoid fuse on CKT 639E and sends the 12 volts signal to the ECM on CKT 495.

When high power steering pressures occur, the switch contacts are opened by power steering oil pressure and CKT 495 voltage is read by the ECM as 0 volts.

The ECM uses the power steering switch input to extend the ISC when high power steering loads occur, to help maintain a stable idle speed.

Note On Intermittents:

Test drive the car at above 45 MPH while observing ECM Input E178. Input E178 should be "HI" at all times.

If E178 reads "LO" without turning the steering wheel to full lock, check for intermittent open in CKT 639 to the power steering switch. Check for intermittent open in CKT 495 to the ECM. Manipulate affected wiring and connectors and remove and replace both the power steering switch connector and the J2 ("C/D") ECM connector and ensure that they are latched.

Review the snapshot data parameters stored with this code to verify failure condition. Refer to "Symptoms" for complete details on using the "Snapshot on code set" feature. If the failure condition is verified, check and repair intermittent wiring connection, or sensor.

1988–90 CADILLAC ELDORADO AND SEVILLE

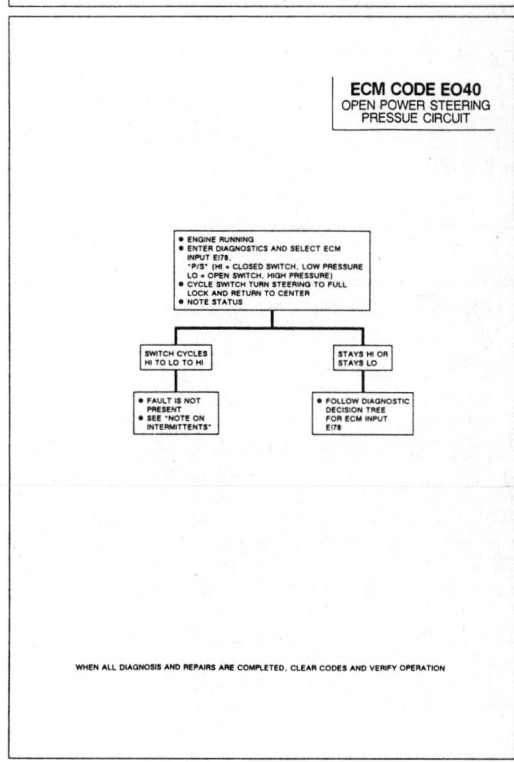

ECM CODE E040
OPEN POWER STEERING PRESSURE CIRCUIT

WHEN ALL DIAGNOSIS AND REPAIRS ARE COMPLETED, CLEAR CODES AND VERIFY OPERATION

1988–90 CADILLAC ELDORADO AND SEVILLE

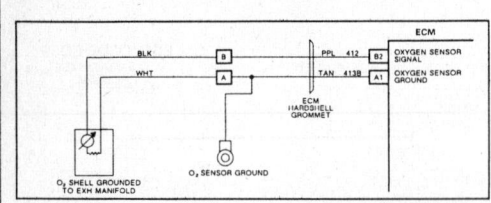

ECM CODE E044, LEAN EXHAUST SIGNAL

TEST CONDITIONS: 1. Codes E014, E015, E016, E021, E022, E026, E027, E031, E032 and E034 all must be clear; 2. Throttle switch open; 3. TPS between 6 and 29 degrees and steady; 4. Coolant sensor greater than or equal to 80°C; 5. Oxygen sensor ready (closed loop); 6. Not accelerating or decelerating; 7. RPM greater than or equal to 800; 8. Vehicle operated at above 13 degrees TPS for 10 minutes; 9. Canister purge enabled.

FAILURE CONDITIONS: Oxygen sensor status stays lean for more than 45 seconds.

ACTION TAKEN: 1. ECM turns on "Engine Control System" telltale. 2. ECM commands BCM to display "Service Now" message on CCDIC. 3. ECM turns on canister purge solenoid and turns off air management solenoids; 4. ECM switches to open loop operation.

Description:

The ECM provides a .45 volt reference signal to the oxygen sensor on CKT 412. When the oxygen sensor is cold (below 200°C), the oxygen sensor signal voltage will be around 0.45 volts and the ECM will keep the system in open loop operation. When the oxygen sensor is warm (above 200°C), the oxygen sensor signal voltage will swing from rich to lean rapidly, at least one swing every two seconds, if the ECM is in good control of the air fuel mixture.

When the ECM sees that the oxygen is varying from the cold voltage of 0.45 volt, it will send the system into closed loop operation. In closed loop operation, the ECM will adjust the fuel delivery rate to the engine based on the oxygen sensor readings.

Code E044, is designed so that if the oxygen sensor stays at a lean voltage for more than 45 seconds during the test conditions, Code E044 will set.

Code E044 will be set when: 1. There is an oxygen sensor circuit fault giving a false lean indication or 2. When the air fuel ratio is actually lean due to a vacuum leak or fuel control system fault.

Notes On Fault Tree:

1. With the oxygen sensor disconnected, parameter ED07 should remain at reference voltage (.38 to .63 volts).

2. Checking for sensor circuitry able to record rich readings. The DVOM set on volts will provide a few billionths of an amp to drive CKT 412 to above .64 volts (rich). Similar results may be obtained by placing one finger on battery positive terminal and another finger on oxygen sensor CKT 412 harness terminal.

3. The ECM compares oxygen sensor signal voltage received on circuit 412 to the ground voltage on circuit 413. If the ECM doesn't have a good ground to the engine on circuit 413, the oxygen sensor can appear falsely high or low. With engine running, use a voltmeter to measure voltage from the oxygen sensor at the exhaust manifold to the ECM terminal A1. If the voltage is –.05 volts to +.05 volts then the ground is OK. If the voltage is less than –.05 volts or greater than +.05 volts, repair poor ground on CKT 413 between ECM terminal A1 and the ground at front of engine, right (rearmost) head.

Notes On Intermittents

Engine running, manipulate the oxygen sensor and ECM wiring and connectors while observing ECM data ED07. If the fault is induced, ED07 will jump below .37 volts and the "ECON" status light will go off. Manipulate CKT 413 ground to the engine and look for a loose ground eyelet or ground eyelet installed in wrong location. If lean engine operation is suspected, perform "DFI System Check." See Symptoms – "Excessive Exhaust Emissions" – NOx. An engine that is misfiring may have a lean oxygen sensor indication because of unburned oxygen in the exhaust stream. See 6E Subsection B to diagnose the misfiring.

If the oxygen sensor circuit appears to be OK, go to DFI Chart A–9, "Oxygen Sensor Diagnosis" to check for faulty or contaminated sensor.

Review the snapshot data parameters stored with this code to verify failure condition. Refer to "Symptoms" for complete details on using the "Snapshot on code set" feature. If the failure condition is verified, check and repair intermittent wiring connection, or sensor.

1988–90 CADILLAC ELDORADO AND SEVILLE

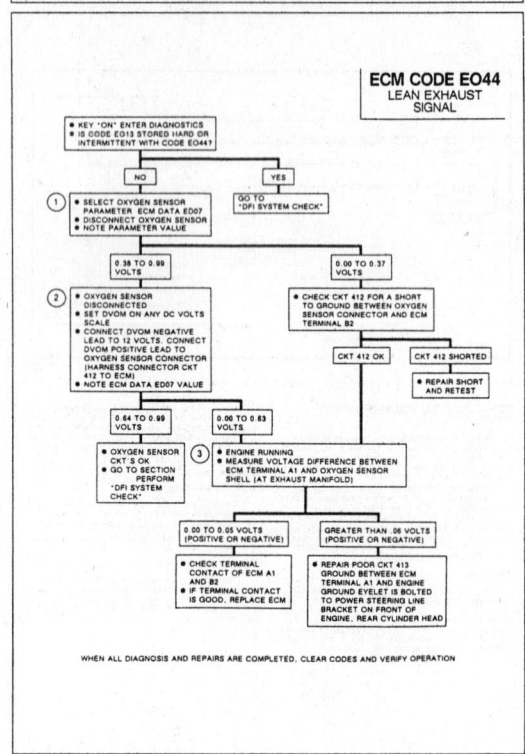

ECM CODE E044
LEAN EXHAUST SIGNAL

WHEN ALL DIAGNOSIS AND REPAIRS ARE COMPLETED, CLEAR CODES AND VERIFY OPERATION

1988–90 CADILLAC ELDORADO AND SEVILLE

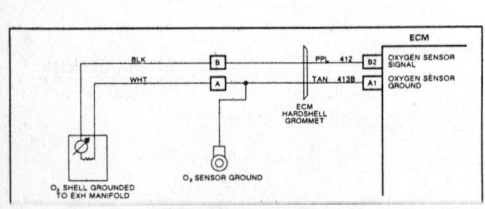

ECM CODE E045, RICH EXHAUST SIGNAL

TEST CONDITIONS: 1. Codes E014, E015, E016, E021, E022, E026, E027, E031, E032, and E034 all must be clear; 2. Throttle switch open; 3. TPS between 6 and 29 degrees and steady; 4. Coolant sensor greater than or equal to 80°C; 5. Oxygen sensor ready (closed loop); 6. Not accelerating or decelerating; 7. RPM greater than or equal to 800; 8. Vehicle operated at above 13 degrees TPS for 10 minutes; 9. Canister purge enabled.

FAILURE CONDITIONS: Oxygen sensor stays rich for more than 45 seconds.

ACTION TAKEN: 1. ECM turns on "Engine Control System" telltale. 2. ECM commands BCM to display "Service Now" message on CCDIC. 3. ECM turn on canister purge solenoid and turns off air management solenoids; 4. ECM switches to open loop operation.

Description:

The ECM provides a .45 volt reference signal to the oxygen sensor on CKT 412. When the oxygen sensor is cold (below 200°C), the output voltage will be around 0.45 volts and the ECM will keep the system in open loop operation. When warm, a properly operating oxygen sensor will drive the .45 volt reference lower (below .45 volts) to indicate a lean mixture and higher (above .45 volts) to indicate a rich mixture. The oxygen sensor signal voltage will swing from rich to lean rapidly, at least one swing every two seconds, if the ECM is in good control of the air fuel mixture. When the ECM sees that the oxygen sensor is not at the cold voltage of 0.45 volts, it will send the system into closed loop operation. In closed loop operation, the ECM will meter fuel into the engine based on the oxygen sensor readings.

Code E045 is designed so that if the oxygen sensor stays at a rich voltage for more than 45 seconds during the test conditions, Code E045 will set.

Code E045 can be caused by: 1. Oxygen sensor circuit faults; 2. Air fuel ratio actually rich due to a fuel control or emissions system fault.

Notes On Fault Tree:

1. With the oxygen sensor disconnected, parameter ED07 should remain at reference voltage (.38 to .63 volts).

2. Checking for ECM's ability to recognize lean input on oxygen sensor CKT 412.

3. The ECM compares oxygen sensor signal voltage received on circuit 412 to the ground voltage on circuit 413. If the ECM doesn't have a good ground to the engine on circuit 413, the oxygen sensor can appear falsely high or low. With engine running, use a voltmeter to measure voltage from the oxygen sensor at the exhaust manifold to the ECM terminal A1. If the voltage is –.05 volts to +.05 volts then the ground is OK. If the voltage is less than –.05 volts or greater than +.05 volts, repair poor ground on CKT 413 between ECM terminal A1 and the ground at front of engine, right (rearmost) head.

Note On Intermittents:

Engine running, manipulate the oxygen sensor and ECM wiring and connectors while observing ECM parameter ED07. If the fault is induced, ED07 will jump above .63 volts, "ECON" status light will come on. Manipulate CKT 413 ground to the engine and look for a loose ground eyelet or ground eyelet installed in an improper location.

Perform – "DFI System Check." See Symptoms – "Poor Fuel Economy" and "Excessive Exhaust Emissions, Odors". If the oxygen sensor circuit appears to be OK, go to DFI Chart A–9, "Oxygen Sensor Diagnosis" to check for a faulty or contaminated sensor.

Review the snapshot data parameters stored with this code to verify failure condition. Refer to "Symptoms" for complete details on using the "Snapshot on code set" feature. If the failure condition is verified, check and repair intermittent wiring connection, or sensor.

1988–90 CADILLAC ELDORADO AND SEVILLE

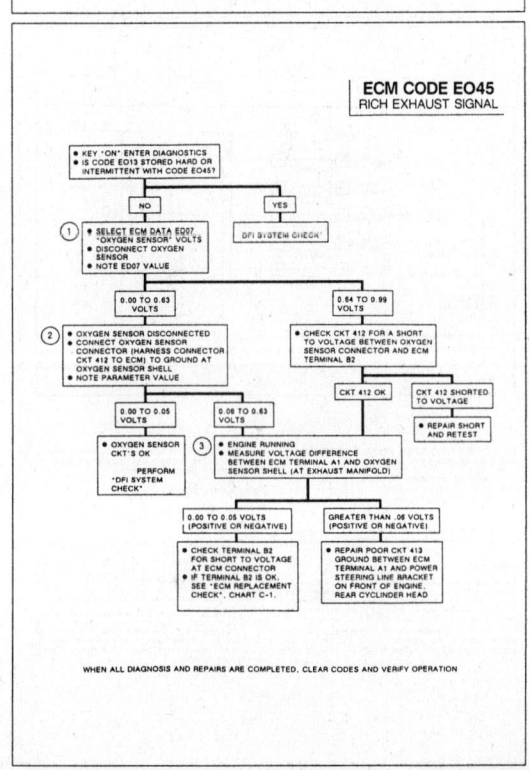

ECM CODE E045
RICH EXHAUST SIGNAL

WHEN ALL DIAGNOSIS AND REPAIRS ARE COMPLETED, CLEAR CODES AND VERIFY OPERATION

1988–90 CADILLAC ELDORADO AND SEVILLE

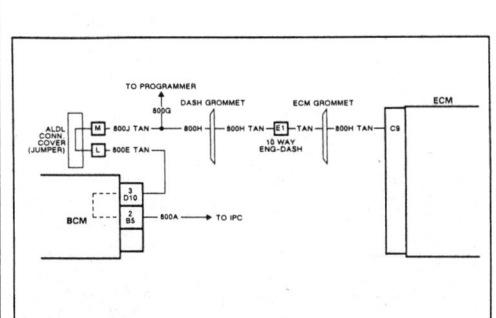

ECM CODE E047 – BCM TO ECM DATA PROBLEM

TEST CONDITIONS: 1. Code E018 clear. 2. Not cranking; 3. RPM > 25.

FAILURE CONDITIONS: ECM receives bad data or no data from the BCM for 2.1 seconds.

ACTION TAKEN: 1. ECM clears all BCM data from ECM memory. 2. ECM assumes that A/C clutch is on at all times and controls ISC operation accordingly. 3. ECM turns on "Engine Control System" telltale. 4. ECM tries to send "Service Soon" message to the BCM over the UART link.

Description:

The ECM and BCM share information information through the UART (Universal Asynchronous Receiver Transmitter) data link. The BCM is the "Master" of the link and data is only transmitted at the BCM's request. Data is sent in eight-character "Words" at the rate of 8192 characters per second.

The data from the ECM to the BCM contains Power Steering Pressure Switch, Park/Neutral Switch and 4th Gear Switch status as well as an engine running and Coolant Temperature Sensor status. The ECM–BCM data also includes Coolant Temperature Sensor, RPM and Injector Pulse width values so the BCM can control cooling fans, display RPM and calculate MPG for display at the driver information center. The BCM sends the ECM air conditioning status to be used for idle speed control and ambient (outside) temperature for use in VCC apply and release decisions.

Code E047 is logged in the ECM in the event of a UART fault. If the fault is a hard failure, the ECM will not be able to communicate with the BCM and Code B334 will be displayed as current. (Code E047 is current in the ECM but can't be sent to the BCM because of the UART fault.) If the UART fault is corrected, both Code E047 and Code B334 will be displayed as history codes.

Code E047 should be diagnosed using the chart for BCM Code B334.

1988–90 CADILLAC ELDORADO AND SEVILLE

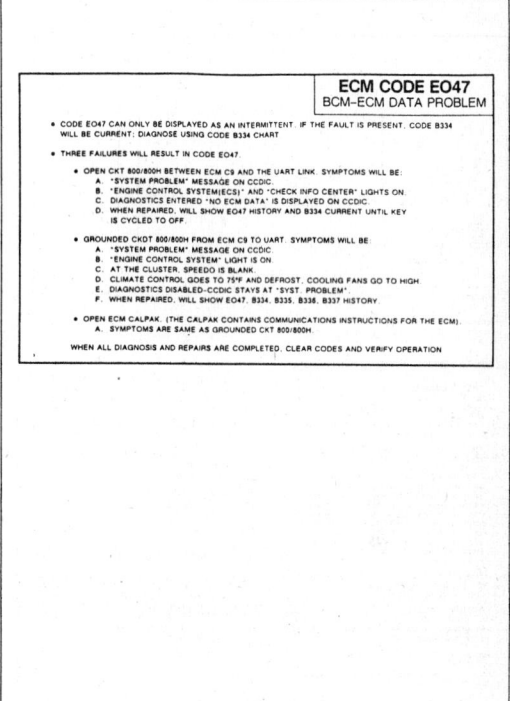

1988–90 CADILLAC ELDORADO AND SEVILLE

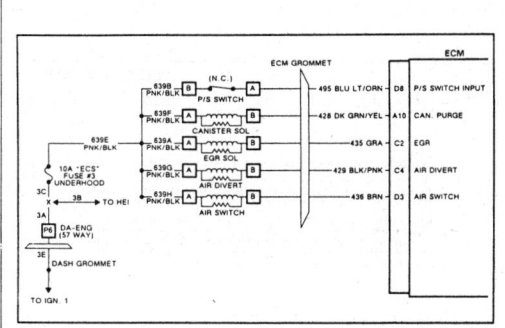

ECM CODE E048 – EGR SYSTEM FAULT

TEST CONDITIONS: 1. Codes E013, E014, E015, E021, E022, E031, E032, E034, E044, E045 clear; 2. Coolant temp between 85° and 110°C; 3. TPS between 7° and 14°; 4. RPM between 1450 and 1650 RPM. 5. Oxygen sensor swinging (closed loop operation). 6. MPH > 35; 7. 10 minute timer after startup expired; 8. Steady throttle.

TEST: During the above test conditions, the EGR system is normally enabled, allowing EGR to flow to the engine. The test that is performed is to turn OFF the EGR flow to the engine for five seconds and to monitor the closed loop integrator. With EGR flow turned OFF, the integrator value should increase by at least 14 counts; for example, the integrator might change from 130 counts to 144 counts or more to indicate that more fuel should be added to compensate for the leaner (less EGR) intake mixture. The ECM will perform the test no more than six times each key cycle; the ECM stops testing after three test failures are recorded.

FAILURE CONDITIONS: If the oxygen sensor fails to indicate a leaner mixture in at least 3 of the 5 tests, Code E048 is set.

ACTION TAKEN: 1. ECM turns on "Engine Control System" telltale. 2. ECM commands BCM to display "Service Soon" message on CCDIC; 3. EGR is disabled for the entire key cycle.

Description:

To perform the test, the ECM turns off EGR flow to the engine and monitors the oxygen sensor (closed loop) integrator. With EGR turned off, the integrator should swing to a higher value reflecting leaner air/fuel mixtures. If not, the ECM assumes that either EGR was turned off before the test started or that EGR is flowing and the ECM doesn't have the ability to turn it off. The ECM tests EGR six times in a given key cycle. If the EGR doesn't respond three or more times in five tests, Code E048 is logged.

Notes On Fault Tree:

1. Many codes disable EGR and inhibit the E048 test. Repair other DFI codes before diagnosis of code E048.
2. At idle, the EGR solenoid is electrically energized, blocking vacuum flow to the EGR valve.

Note On Intermittents:

Drive the car at TPS between 8 and 15 degrees and engine RPM at 1450 to 1650 RPM to try and duplicate the code.

Remove the EGR valve and check for excessive carbon build up that would restrict EGR flow or foreign materials holding the EGR valve open. Check for blocked or restricted EGR passage in the intake manifold reducing EGR flow.

Check vacuum hoses for pinched, cut, kinked, misrouted, or hoses blocked.

1988–90 CADILLAC ELDORADO AND SEVILLE

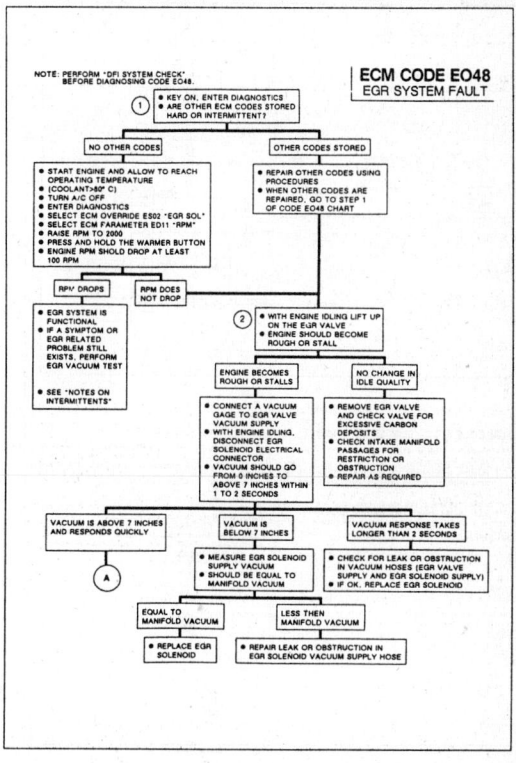

1988–90 CADILLAC ELDORADO AND SEVILLE

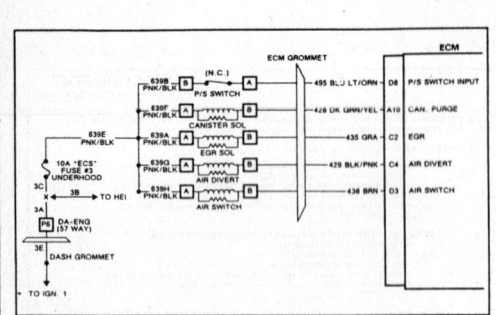

ECM CODE E048 – EGR SYSTEM FAULT

TEST CONDITIONS: 1. Codes E013, E014, E015, E021, E022, E031, E032, E034, E044, E045 clear; 2. Coolant temp between 85° and 110°C; 3. TPS between 8° and 15°; 4. RPM between 1450 and 1650 RPM. 5. Oxygen sensor swinging (closed loop operation).

TEST: During the above test conditions, the EGR system is normally enabled, allowing EGR to flow to the engine. The test that is performed is to turn OFF the EGR flow to the engine for five seconds and to monitor the closed loop integrator. With EGR flow turned OFF, the integrator value should increase by at least 14 counts; for example, the integrator might change from 130 counts to 144 counts or more to indicate that more fuel should be added to compensate for the leaner (less EGR) intake mixture. The ECM will perform the test no more than five times each key cycle; the test stops testing after three test failures are recorded.

FAILURE CONDITIONS: If the oxygen sensor fails to indicate a leaner mixture in at least 3 of the 5 tests, Code E048 is set.

ACTION TAKEN: 1. ECM turns on "Engine Control System" telltale. 2. ECM commands BCM to display "Service Soon" message on CCDIC; 3. EGR is disabled for the entire key cycle.

Description:

To perform the test, the ECM turns off EGR flow to the engine and monitors the oxygen sensor (closed loop) integrator. With EGR turned off, the integrator should swing to a higher value reflecting leaner air/fuel mixtures. If not, the ECM assumes that either EGR was turned off before the test started or that EGR is flowing and the ECM doesn't have the ability to turn it off. The ECM tests EGR five times in a given key cycle. If the EGR doesn't respond three or more times in five tests, Code E048 is logged.

Notes On Fault Tree:

1. Check for power at EGR solenoid.
2. Checking for the ability of the ECM to command EGR on.

Note On Intermittents:

Drive the car at TPS between 8 and 15 degrees and engine RPM at 1450 to 1650 RPM to try and duplicate the code.

Remove the EGR valve and check for excessive carbon build up that would restrict EGR flow or foreign materials holding the EGR valve open. Check for blocked or restricted EGR passage in the intake manifold reducing EGR flow.

Check vacuum hoses for pinched, cut, kinked, misrouted, or hoses blocked.

1988–90 CADILLAC ELDORADO AND SEVILLE

ECM CODE E048
EGR SYSTEM FAULT
(CONT'D)

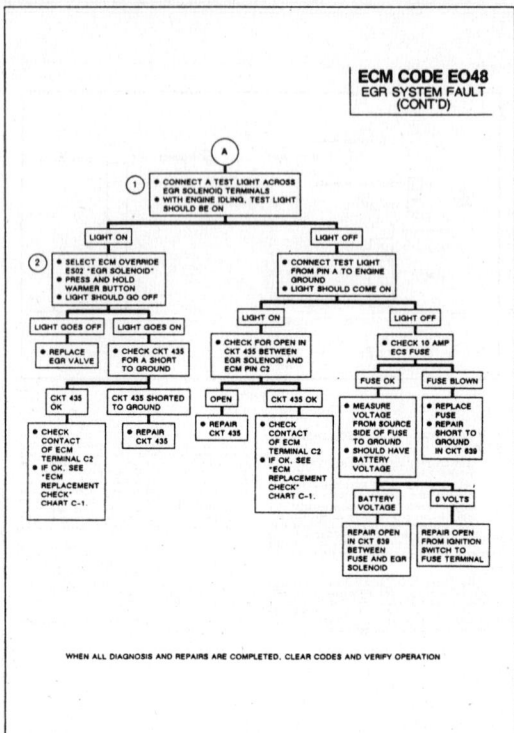

WHEN ALL DIAGNOSIS AND REPAIRS ARE COMPLETED, CLEAR CODES AND VERIFY OPERATION

1988–90 CADILLAC ELDORADO AND SEVILLE

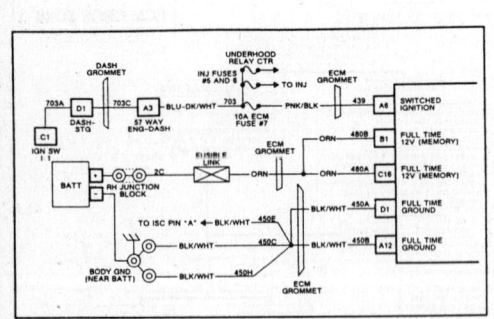

ECM CODE E052 – ECM MEMORY RESET

Test Conditions: Always tested.

Failure Conditions: ECM looks for all time memory information to be erased or reset. If full time memory is completely erased, code E052 is logged.

Action Taken: 1. All block learn value are reset to 128 counts (see ECM learning ability). 2. E052 is stored as a history code.

Description:

Code E052 indicates that the "long term" memory in the ECM has been reset.

This will be the case whenever power is removed from the ECM (i.e., disconnecting battery cables, disconnecting ECM connector, etc.). This code should be "cleared" from memory after restoring the ECM's power supply.

If Code E052 is seen accompanied by complaint of engine quit, stumble, telltale's flashing or other stored codes, check for intermittent loss of power or ground to the ECM on CKT 480 and 450 respectively. Manipulate wiring and connections with engine running. Remove and replace the ECM J1 and J2 connectors and ensure that they are latched.

1988–90 CADILLAC ELDORADO AND SEVILLE

ECM CODE E052
ECM MEMORY RESET

Code E052 indicates that the "long term" memory in the ECM has been reset.

This will be the case whenever power is removed from the ECM (i.e., disconnecting battery cables, disconnecting ECM connector, etc.). This code should be "cleared" from memory after restoring the ECM's power supply.

If Code E052 is seen accompanied by complaint of engine quit, stumble, telltale's flashing or other stored codes, check for intermittent loss of power or ground to the ECM on CKT 480 and 450 respectively. Manipulate wiring and connections with engine running. Remove and replace the ECM J1 and J2 connectors and ensure that they are latched.

1988–90 CADILLAC ELDORADO AND SEVILLE

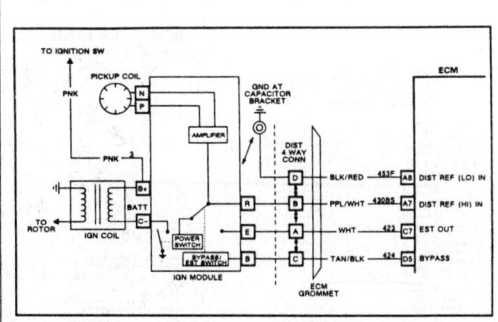

ECM CODE E053 – DISTRIBUTOR SIGNAL INTERRUPT

TEST CONDITIONS: 1. RPM above 448; 2. Fuel pump voltage above 10 volts.

FAILURE CONDITIONS: No distributor reference pulses for .7 seconds.

ACTION TAKEN: 1. Code E053 is stored as a history code. (No telltales are illuminated).

Description:

Code E053 is set if the ECM does not receive distributor reference pulses from the HEI for more than .7 seconds. Since the DFI system requires HEI pulses in order to fire the injectors, most occurrences of Code E053 will be accompanied by a stall.

Code E053 can be set by ignition switch contact timing – when turning the key off, if the Ignition 3 contacts that feed the HEI open before the Ignition 1 contacts to the ECM, a Code E053 may be set. Code E053 does not turn on any telltales or "service" messages. Do not attempt to diagnose a history Code E053 unless the customer complains of stumble, miss, stall or other driveability condition that could be caused by loss of spark or fuel injection.

Notes On Fault Tree:

1. Checking for hard failure of distributor reference which will cause code E053 to set.
2. Do not diagnose Code E053 unless the customer complains of stumble, miss, stall or other driveability condition that could be caused by loss of spark or fuel.
3. Checking for Code E053 set due to ignition key cycling.
4. See "Description", above.
5. See "Note On Intermittents".

Note On Intermittents:

Code E053 can be caused by:
- Loss of ground on CKT 453
- Loss of distributor reference signal on CKT 430
- Loss of battery power to "B+" terminal of the distributor.
- Ignition switch contact timing.

Do not attempt to diagnose Code E053 unless the customer symptom is stumble, stall, miss or other driveability condition that could be caused by loss of spark or fuel.

1988–90 CADILLAC ELDORADO AND SEVILLE

ECM CODE E053
DISTRIBUTOR SIGNAL INTERRUPT

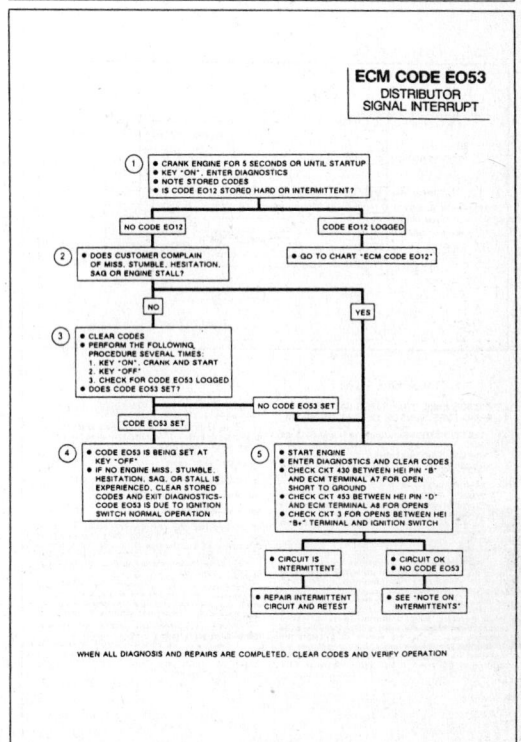

WHEN ALL DIAGNOSIS AND REPAIRS ARE COMPLETED, CLEAR CODES AND VERIFY OPERATION

1988–90 CADILLAC ELDORADO AND SEVILLE

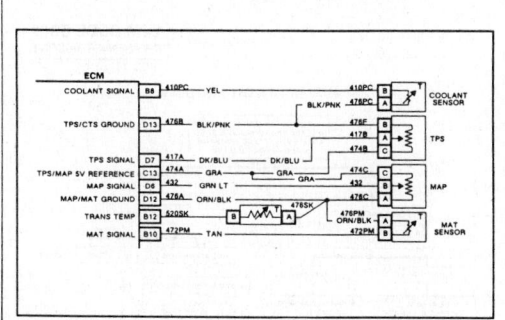

ECM CODE E055 – TPS OUT OF ADJUSTMENT

TEST CONDITIONS: 1. Codes E021, E022, E026, E027 clear; 2. ECM tests for Code E055 at key off. At key off the ECM will retract until the throttle switch opens. The TPS value is read by the ECM and a correction factor is stored. If the correction factor value is the same on two consecutive key off cycles, the TPS setting is "relearned".

FAILURE CONDITION: If TPS correction needed is from -10 to -2.9° or from 3.0 to 90.0° the "TPS out of adjustment" flag is set.

ACTION TAKEN: At the next key on, the ECM will see the "TPS out of adjustment" flag and log Code E055 as current. No telltale or "Service" message will appear.

Description:

The TPS on DFI is "self adjusting". At key-off, the ECM executes a TPS learning routine. After key off, the ECM will retract the ISC until the ISC throttle switch opens and the throttle linkage is resting on the minimum air screw. At that time, the ECM stores the TPS value and calculates a correction. If the same correction factor occurs on two consecutive key "off" cycles, the TPS is then corrected to 0 degrees using the correction factor "learned". If the value needs correction by more than -2.9 degrees and +3.0 degrees,

Code E055 will be stored in memory at the next key "ON" cycle. **Note:** Parameter ED01, TPS, displays uncorrected TPS values.

Notes On Fault Tree:

1. Checking TPS adjustment. ECM parameter ED01, TPS, displays uncorrected TPS so that it can be used to check TPS adjustment.
2. TPS adjustment is OK.
3. If TPS adjustment is OK, ISC and throttle switch operation need to be thoroughly checked. The throttle linkage needs to be checked for proper operation — throttle, cruise and TV cables not binding, proper throttle return spring operation throttle shaft and blades free to move.

Notes On Intermittents:

Manipulate ISC wiring while observing ECM Input E172 and while observing ISC operation during ECM output Cycling E007.

Manipulate TPS wiring and connector while observing ECM parameter ED01 for jumps, skips, intermittent behavior. Check for TPS secured to throttle body (both TORX screws tight). Cycle the TPS through its full travel while observing ED01 for erratic behavior. Check for proper part number TPS installed on vehicle.

Remove and replace the TPS, ISC and ECM connectors and ensure that they are latched.

1988–90 CADILLAC ELDORADO AND SEVILLE

NOTE: IF CODES E021, E026, E027, OR E030 ARE LOGGED AS CURRENT OR HISTORY, DO NOT USE THIS PROCEDURE. SEE THE CODE E021, E022, E026, E027 AND E030 PROCEDURES TO DIAGNOSE

ECM CODE E055
TPS MISADJUSTED

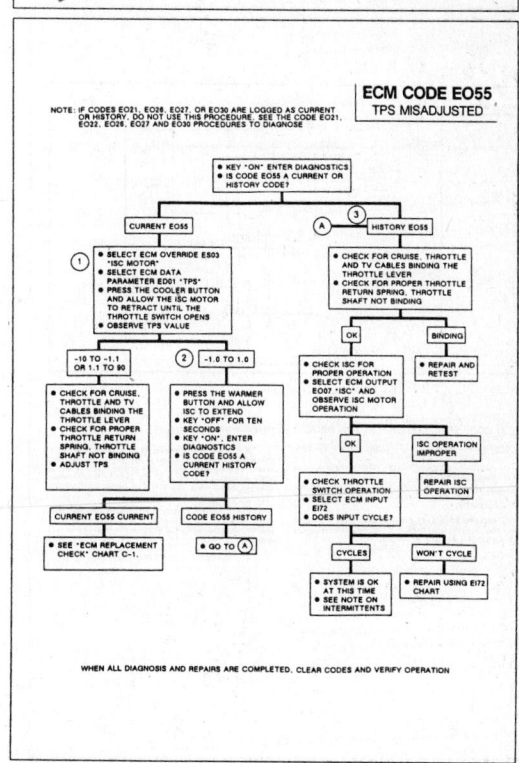

WHEN ALL DIAGNOSIS AND REPAIRS ARE COMPLETED, CLEAR CODES AND VERIFY OPERATION

1988–90 CADILLAC ELDORADO AND SEVILLE

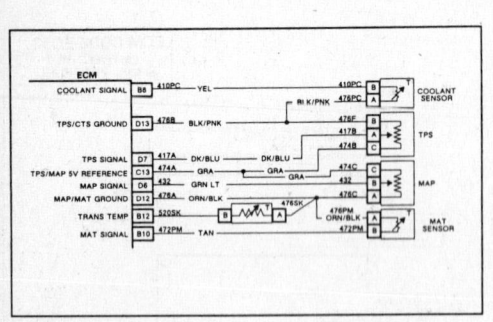

ECM CODE E059, TRANSMISSION FLUID TEMPERATURE SENSOR CIRCUIT PROBLEM

TEST CONDITIONS: 1. Codes E014 and E015 not set; 2. Vehicle speed ≥ 10 MPH for 600 seconds.
FAILURE CONDITIONS: Transmission temperature ≤ -39°C when the coolant temp is ≥ 50°C or Transmission temperature ≥ 148°C when the coolant temp is ≤ 50°C.
ACTION TAKEN: 1. ECM turns on "Engine Control System" telltale; 2. ECM commands BCM to display "Service Soon" message on DIC; 3. ECM substitutes 110°C for transmission temperature.

Description:

The transmission temperature sensor is a thermistor, or variable resistor, that varies in resistance based on temperature. The sensor is a "two-wire" sensor with a reference signal coming from the ECM to the sensor on Pin A (CKT 520) and a sensor reference ground on Pin B (CKT 476). As the temperature of the sensor increases, sensor resistance decreases. As the resistance decreases, the signal voltage from the ECM decreases due to the less voltage drop across the transmission temperature sensor. High temperature will result in low signal voltage on CKT 520 and low temperature will result in high signal voltage on CKT 520.

The ECM uses transmission temperature information to vary VCC apply speeds.

Code E059 will set the ECM sees an excessively low or excessively high VCC temperature when the coolant is hot or cold, respectively.

Notes On Fault Tree:

1. VCC sensor circuit is shorted to ground.
2. VCC sensor circuit is open.

Notes On Intermittents:

Manipulate the transmission connector, ECM connector, and CKT 520 wiring while observing ECM parameter ED26. If the fault is induced, the data value will jump from a normal reading to a low or high reading. Repair the intermittent short to ground or open in CKT 520.

Disconnect and reconnect the ECM and transmission connectors and ensure they are latched. If wiring and connectors check out OK, substitute a known good transmission temperature sensor and retest.

Review the snapshot data parameters stored with this code to verify failure condition. Refer to "Symptoms" for complete details on using the "Snapshot on code set" feature. If the failure condition is verified, check and repair intermittent wiring connection, or sensor.

1988–90 CADILLAC ELDORADO AND SEVILLE

ECM CODE E059
VCC TEMP SENSOR CIRCUIT

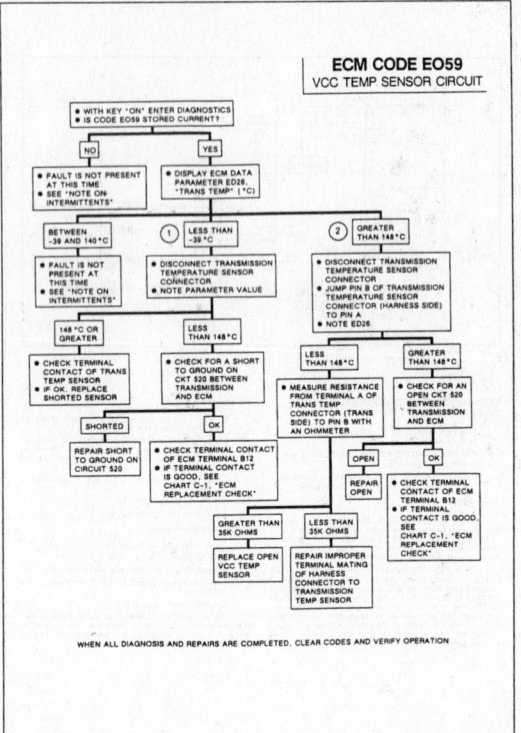

WHEN ALL DIAGNOSIS AND REPAIRS ARE COMPLETED, CLEAR CODES AND VERIFY OPERATION

1988–90 CADILLAC ELDORADO AND SEVILLE

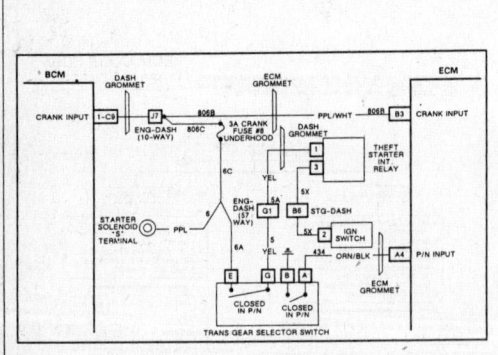

ECM CODE E060, CRUISE – NOT IN DRIVE

TEST CONDITIONS: 1. Cruise control enabled; 2. Cruise control engaged.
FAILURE CONDITIONS: Transmission in Park/Neutral.
ACTION TAKEN: 1. Disengage cruise control.

Description:

Code E060 will set if the cruise control is engaged and the Park/Neutral switch is closed, indicating the transmission is in Park or Neutral.

Notes On Intermittents:

If Code E060 is stored as an intermittent code, check the operation of the Park/Neutral Input test. After selecting input E174, manipulate affected wiring while observing the status of the switch. If the switch changes status while the wiring is being manipulated, repair the intermittent connections.

If no trouble is found, explain to the customer that this code can be induced if the transmission is inadvertently put into neutral while the cruise control is engaged.

1988–90 CADILLAC ELDORADO AND SEVILLE

ECM CODE E060
CRUISE–NOT IN DRIVE

NOTE: THIS PROCEDURE CANNOT BE USED IF BCM CODE B127 IS CURRENT

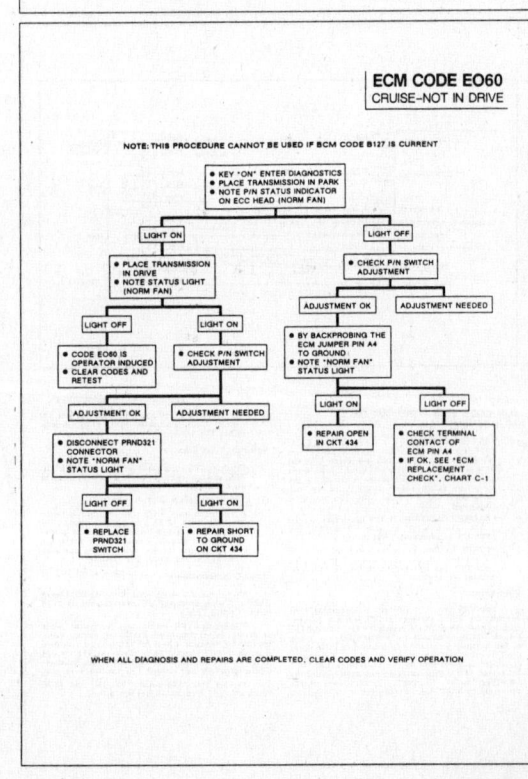

WHEN ALL DIAGNOSIS AND REPAIRS ARE COMPLETED, CLEAR CODES AND VERIFY OPERATION

1988-90 CADILLAC ELDORADO AND SEVILLE

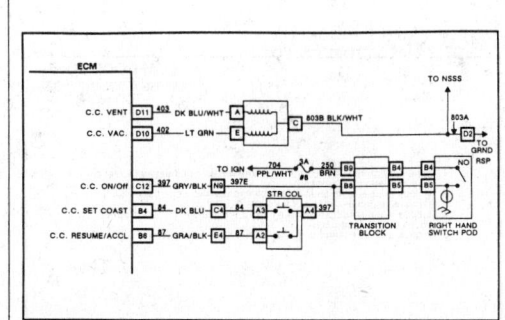

ECM CODE E063, SET SPEED/VEHICLE SPEED DIFFERENCE TOO HIGH

TEST CONDITIONS: 1. Cruise control enabled; 2. Cruise control engaged; 3. Cruise control not in the resume mode.

FAILURE CONDITIONS: Vehicle speed 20 MPH higher than set speed for .5 seconds.

ACTION TAKEN: 1. Cruise control disengaged.

Description:

Code E063 will set and disengage the cruise control if vehicle speed is 20 MPH higher than the set speed. This code is used as a type of safety valve to the cruise control system, and will be set under normal conditions if the operator accelerates using the gas pedal and goes 20 MPH over the cruise set speed.

1. Checking for the ability of the servo to pull in.
2. This step checks for the ability of the servo to be released.
3. This step checks the vent and vacuum outputs of the ECM for proper operation.

Notes On Intermittents:

If Code E063 is setting intermittently make sure owner is aware that overrunning the cruise set speed by more than 20 MPH will cause this code and disengage the cruise system.

This code will not be caused by vacuum leaks or vacuum supply problems. It can only be set by the vehicle speed exceeding the set speed by 20 MPH.

1988-90 CADILLAC ELDORADO AND SEVILLE

ECM CODE E063
SET SPEED/VEHICLE SPEED DIFFERENCE TOO HIGH

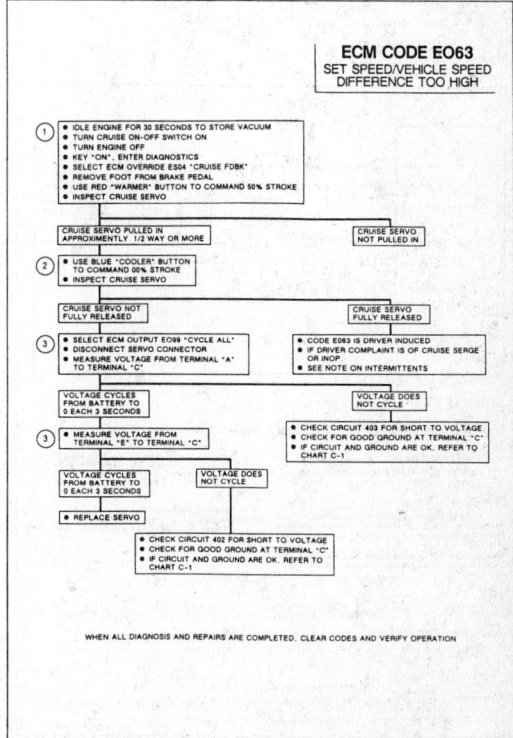

WHEN ALL DIAGNOSIS AND REPAIRS ARE COMPLETED, CLEAR CODES AND VERIFY OPERATION

1988-90 CADILLAC ELDORADO AND SEVILLE

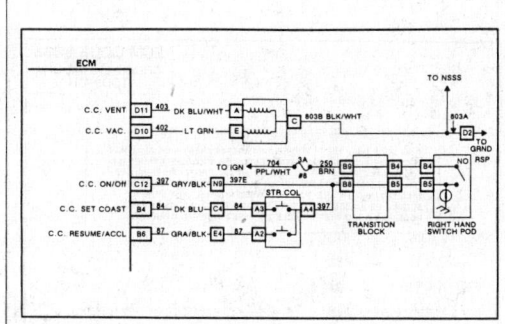

ECM CODE E064, ACCELERATION TOO HIGH CRUISE ENGAGED

TEST CONDITIONS: 1. Cruise control enabled; 2. Cruise control engaged.

FAILURE CONDITIONS: Vehicle speed increases more than 4 MPH in 2/3 second.

ACTION TAKEN: 1. Disengage cruise control.

Description:

Code E064 is designed to set when the vehicle speed is increasing at an extremely rapid rate (wheelspin). It is a protective measure so that the wheels will not be under cruise control when on ice.

Notes On Intermittents:

If Code E064 is stored as an intermittent code and the customer complains of frequent loss of cruise control, drive the car while observing ECM data ED12, vehicle speed. If the speed displayed is erratic, check the operation and integrity of the speed sensor circuit.

1988-90 CADILLAC ELDORADO AND SEVILLE

ECM CODE E064
ACCELERATION TOO HIGH CRUISE ENGAGED

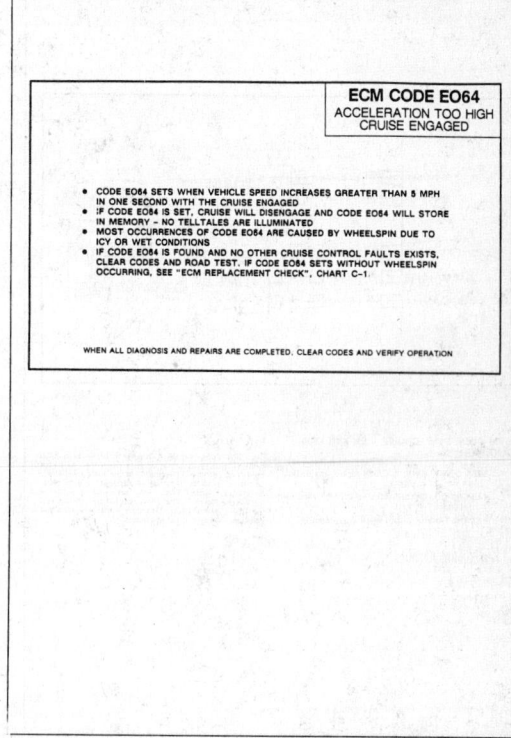

- CODE E064 SETS WHEN VEHICLE SPEED INCREASES GREATER THAN 5 MPH IN ONE SECOND WITH THE CRUISE ENGAGED.
- IF CODE E064 IS SET, CRUISE WILL DISENGAGE AND CODE E064 WILL STORE IN MEMORY – NO TELLTALES ARE ILLUMINATED.
- MOST OCCURRENCES OF CODE E064 ARE CAUSED BY WHEELSPIN DUE TO ICY OR WET CONDITIONS.
- IF CODE E064 IS FOUND AND NO OTHER CRUISE CONTROL FAULTS EXISTS, CLEAR CODES AND ROAD TEST. IF CODE E064 SETS WITHOUT WHEELSPIN OCCURRING, SEE "ECM REPLACEMENT CHECK", CHART C-1.

WHEN ALL DIAGNOSIS AND REPAIRS ARE COMPLETED, CLEAR CODES AND VERIFY OPERATION

1988–90 CADILLAC ELDORADO AND SEVILLE

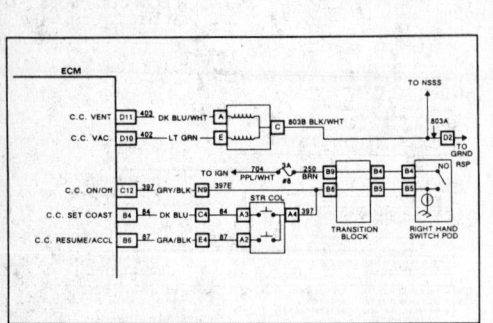

ECM CODE E065, CRUISE - COOLANT TEMP TOO HIGH

TEST CONDITIONS: 1. Cruise engaged.
FAILURE CONDITIONS: 1. Coolant > 126°C.
ACTION TAKEN: 1. Disable cruise control.

Description:

Code E065 is designed to protect the engine from overheating. The ECM monitors coolant temperature and will disable cruise if the coolant temperature rises above 126°C.

A Code E065 will be stored by the ECM to alert the technician to the reason the cruise was disabled.

Notes On Intermittents:

If an intermittent code E065 is stored, manipulate the coolant sensor wiring cable observing ECM data ED04 "Coolant Temp". If the parameter jumps to a high reading (> 126°C) repair the intermittent wiring on connector.

Review the snapshot data parameters stored with this code to verify failure condition. Refer to "Symptoms" for complete details on using the "Snapshot on code set" feature.

1988–90 CADILLAC ELDORADO AND SEVILLE

> **ECM CODE EO65**
> CRUISE COOLANT TEMP TOO HIGH
>
> - CODE EO65 SETS WHEN COOLANT TEMPERATURE GOES BETWEEN 126°C, WITH THE CRUISE ENGAGED
> - IF CODE EO65 IS SET, CRUISE WILL DISENGAGE AND CODE EO65 WILL STORE IN MEMORY—NO TELLTALES ARE ILLUMINATED
> - IF CODE EO65 IS ACCOMPANIED BY HARD OR INTERMITTENT CODE EO14, GO TO CHART "CODE EO14"
> - IF CODE EO65 IS FOUND, DIAGNOSE FOR ENGINE OVERHEATING
>
> WHEN ALL DIAGNOSIS AND REPAIRS ARE COMPLETED, CLEAR CODES AND VERIFY OPERATION

1988–90 CADILLAC ELDORADO AND SEVILLE

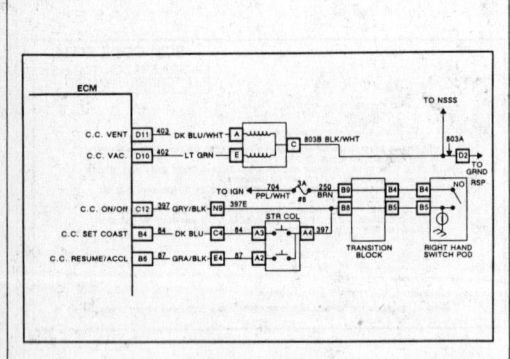

ECM CODE E066, CRUISE - ENGINE RPM TOO HIGH

TEST CONDITIONS: 1. Cruise control enabled; 2. Cruise control engaged.
FAILURE CONDITIONS: 1. Engine RPM ≥ 4800 for 2/3 seconds.
ACTION TAKEN: 1. Disengage cruise control.

Description:

This code will set if engine RPM exceeds 4800 RPM. This may occur on slippery pavement, extended wide open throttle acceleration, or for some mechanical problems (such as transmission slippage). Under these conditions, Code E066 will set and should be considered normal. The driver should be advised why the cruise deenergized. Clear the code and road test vehicle to verify normal operation.

1988–90 CADILLAC ELDORADO AND SEVILLE

> **ECM CODE EO66**
> CRUISE-ENGINE RPM TOO HIGH
>
> THIS CODE WILL SET IF ENGINE RPM EXCEEDS 4800 RPM. THIS MAY OCCUR ON SLIPPERY PAVEMENT, EXTENDED WIDE OPEN THROTTLE ACCELERATION, OR FOR SOME MECHANICAL PROBLEMS (SUCH AS TRANSMISSION SLIPPAGE). UNDER THESE CONDITIONS, CODE EO66 WILL SET AND SHOULD BE CONSIDERED NORMAL. THE DRIVER SHOULD BE ADVISED WHY THE CRUISE DISENGAGED. CLEAR THE CODE AND ROAD TEST VEHICLE TO VERIFY NORMAL OPERATION
>
> WHEN ALL DIAGNOSIS AND REPAIRS ARE COMPLETED, CLEAR CODES AND VERIFY OPERATION

1988–90 CADILLAC ELDORADO AND SEVILLE

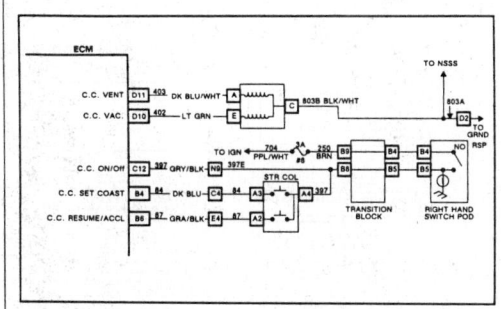

ECM CODE E067, CRUISE SWITCH SHORTED DURING ENABLE

TEST CONDITIONS: 1. Always tested.

FAILURE CONDITIONS: Set/Coast or Resume/Accel signal present when the cruise ON/OFF switch is turned "ON" or if Set/Coast and Resume/Accel switches are on at the same time.

ACTION TAKEN:

If PROM ID (ED99) is 861 or 862, disable cruise for complete ignition cycle.

If PROM ID (ED99) is 1000 or 1001, disable cruise until failure conditions are not true.

Description:

When the cruise control ON/OFF switch is "ON", system voltage is available at one side of the normally open contacts on the Set/Coast and Resume/Accel switches. If the cruise control ON/OFF switch is in the "ON" position this system voltage is available to the Set/Coast and Resume/Accel switches at key on. If the Set/Coast or the Resume/Accel switches were stuck or their signal wires to the ECM were shorted to voltage, the vehicle could begin cruise operation. To prevent this, Code E067 will set and disable cruise control if signal voltage from the Set/Coast (CKT 84) or Resume/Accel (CKT 87) is HI when the cruise control ON/OFF switch is turned from "OFF" to "ON" or when the ignition key is turned on with the cruise control ON/OFF switch "ON".

Notes On Fault Tree:

1. ECM inputs E180 and E181 display Set/Coast and Resume/Accel switch status as HI or LO depending on the voltage state at the ECM. If one of these inputs stays HI when the switches are cycled, that particular switch or signal wire is shorted to voltage.

2. This step checks to see if short circuit reading is due to switch or circuit.

Notes On Intermittents:

If an intermittent Code E067 is being set, manipulate the related wiring while observing ECM inputs E180 and E181. If the failure is induced the reading will jump from LO to HI with the switch off (not depressed). This code can be set by: 1. The operator of the cruise ON/OFF switch is turned "ON" and the operator was depressing either the Set/Coast or Resume/Accel switches or; 2. The ON/OFF switch was on when the ignition key was turned on and the operator was depressing either the Set/Coast or Resume/Accel switches or; 3. Both the Set/Coast and Resume/Accel switches were pressed at the same time with the dash switch on.

Review the snapshot data parameters stored with this code to verify failure condition. Refer to "Symptoms" for complete details on using the "Snapshot on code set" feature. If the failure condition is verified, check and repair intermittent wiring, connection, or sensor.

1988–90 CADILLAC ELDORADO AND SEVILLE

ECM CODE E067
CRUISE SWITCH SHORTED DURING ENABLE

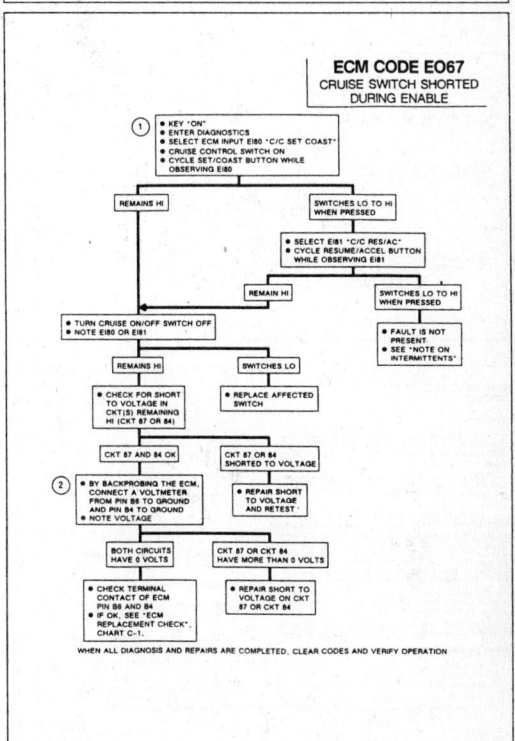

WHEN ALL DIAGNOSIS AND REPAIRS ARE COMPLETED, CLEAR CODES AND VERIFY OPERATION

1988–90 CADILLAC ELDORADO AND SEVILLE

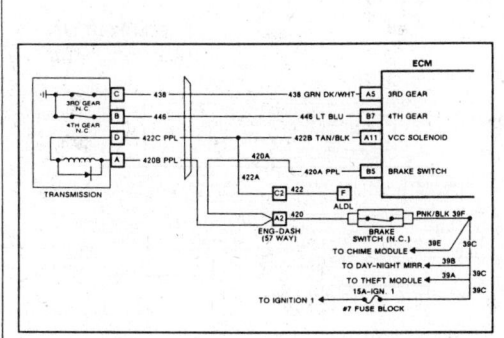

ECM INPUT EI71 VCC BRAKE CIRCUIT

Description:

The VCC brake switch is normally closed and opens when the brake pedal is depressed. The VCC brake switch serves a dual function, the brake switch supplies power to the VCC solenoid and a brake switch receives 12 volts from the 15 AMP "IGN 1" fuse on CKT 39 and sends the 12 volt signal to the ECM and VCC solenoid on CKT 420. When the brake is depressed, the switch opens and CKT 420 voltage is read by the ECM as 0 volts.

On the diagnostic display, "HI" is a closed switch (not braking) while "LO" is an open switch (brake pedal depressed).

The ECM uses VCC brake input to determine braking status for VCC apply and release, as well as a test condition for many codes.

Notes On Fault Tree:

1. At key on/not braking, should have 12 volts on both sides of VCC brake switch. Test light to ground should light if either pin on brake switch connector is backprobed.

2. Light on one side means that brake switch is open as if the brake pedal were depressed.

3. Light on CKT 420 with pedal depressed means that brake switch never opened.

4. Remove and replace ECM "A/B" connector and ensure that it is latched.

Note On Intermittents:

- Check brake switch for proper adjustment
- Check for intermittent open circuit from Fuse #7 (fuse block) to brake switch (CKT 39) then from brake switch to ECM (CKT 420).
- If Fuse #7 is being blown intermittently, check for CKT 39 or CKT 420 for shorts to ground.

1988–90 CADILLAC ELDORADO AND SEVILLE

ECM INPUT EI71
VCC BRAKE CIRCUIT

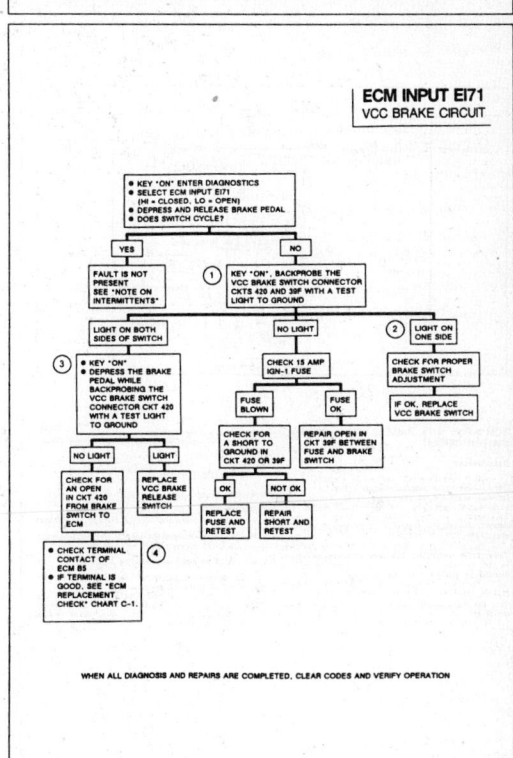

WHEN ALL DIAGNOSIS AND REPAIRS ARE COMPLETED, CLEAR CODES AND VERIFY OPERATION

1988–90 CADILLAC ELDORADO AND SEVILLE

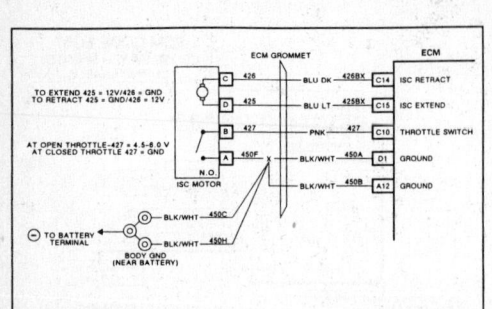

ECM INPUT EI72, THROTTLE SWITCH CIRCUIT

Description:

The Throttle Switch is part of the Idle Speed Control motor (ISC). Pin "B" of the four-way weather pack connector on the ISC is the throttle switch or closed throttle input to the ECM.

The ISC throttle switch contacts are normally open and are closed when the throttle linkage contacts the ISC plunger (closed throttle, ISC in control of idle speed).

The ECM sends a 5 volts signal to pin "B" of the ISC motor on CKT 434. When the throttle linkage rests on the ISC plunger, the throttle switch contacts close, shorting ISC pin "B" to pin "A" (CKT 450 ground). The 5 volt signal from the ECM is grounded at closed throttle, resulting in 0 volts at ECM pin C10. When the throttle is opened, the throttle switch opens pin "B" CKT 427, resulting in 5 volts at ECM pin C10.

On the diagnostic display, "HI" is an open switch, open throttle while "LO" is a closed switch, closed throttle (ISC in control of idle).

The ECM uses throttle switch input to determine when the ISC is in control of idle RPM or throttle angle.

Notes On Fault Tree:

1. With ISC plunger depressed, status should be "LO", closed throttle. When the plunger is released, status should be "HI", open throttle.
2. Check for throttle shaft, throttle plates binding, throttle spring weak or distorted, cruise and TV

cables binding throttle linkage, TPS for proper installation.
3. When the ISC bottoms in the fully retracted position, the throttle switch will open. ISC should be partially extended to control idle.
4. Never connect a voltage source across pins "A" and "B" of ISC. Damage to throttle switch contacts will result.
5. This step simulates the throttle switch shorting Pin "A" to Pin "B". If remove and replacing jumper causes switch to cycle, harness and ECM are OK; fault is in ISC motor connector or ISC motor.
6. Checking for 5V reference signal to ISC.
7. 5V reference from ECM to ISC is open or shorted to ground.
8. 5V reference is OK – check for open CKT 450F from ISC Pin "B" to ground.

Note On Intermittents:

- Check for binding throttle linkage due to TV, cruise or throttle cables, TPS mis-installed or throttle shaft binding.
- Check for proper throttle return spring and throttle return spring installation.
- Probe ECM C10 to ground with a voltmeter. Manipulate wiring and connectors at closed throttle and watch for 5.0 volts, indicating an open from C10 to ground. Manipulate wiring and connectors at open throttle and watch for 0 volts, indicating short to ground on 427.

1988–90 CADILLAC ELDORADO AND SEVILLE

ECM INPUT EI72
THROTTLE SWITCH CIRCUIT

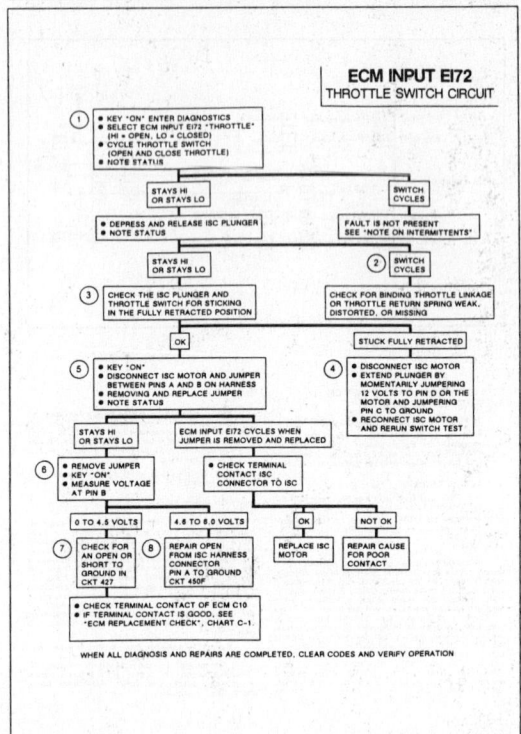

WHEN ALL DIAGNOSIS AND REPAIRS ARE COMPLETED, CLEAR CODES AND VERIFY OPERATION

1988–90 CADILLAC ELDORADO AND SEVILLE

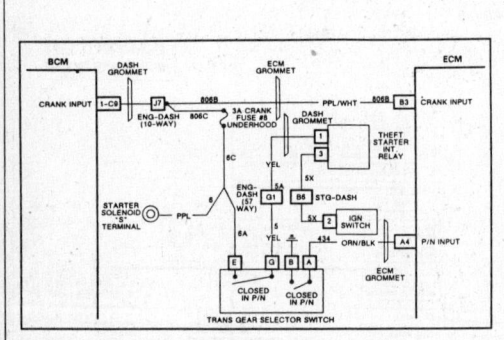

ECM INPUT EI74, PARK/NEUTRAL (NSBU) SWITCH CIRCUIT

Description:

The Park Neutral switch is a part of the transmission Gear Selector Switch. Pin "A" of the six-way weather pack connector on the switch is the park/neutral input for the ECM and BCM. The Park/Neutral Switch contacts are closed in park or neutral, shorting switch pin "A" to ground. In any other gear range, pin "A" is open.

The ECM sends 12 volts to pin "A" of the Gear Selector Switch on CKT 434. When the gear selector is in park or neutral, the 12 volt signal from the ECM is shorted to ground, resulting in 0 volts at ECM pin "A4". In reverse or forward gears, CKT 434 is open to ground, resulting in 12 volts at ECM pin "A4".

On the diagnostic display, "HI" is an open switch indication (not in park or neutral) and "LO" is a closed switch (gear selector in park or neutral).

The ECM uses park/neutral status for fuel and idle speed control and as a test condition for many trouble codes.

Notes On Intermittents:

- Check for trans gear selector switch out of adjustment.
- Check for trans shift indicator out of adjustment.
- Check for intermittent open or short to ground in CKT 434 between ECM and switch.
- Check for open to ground on switch pin B.

Review the snapshot data parameters stored with this code to verify failure condition. Refer to "Symptoms" for complete details on using the "Snapshot on code set" feature. If the failure condition is verified, check and repair intermittent wiring connection, or sensor.

1988–90 CADILLAC ELDORADO AND SEVILLE

ECM INPUT EI74
PARK/NEUTRAL SWITCH

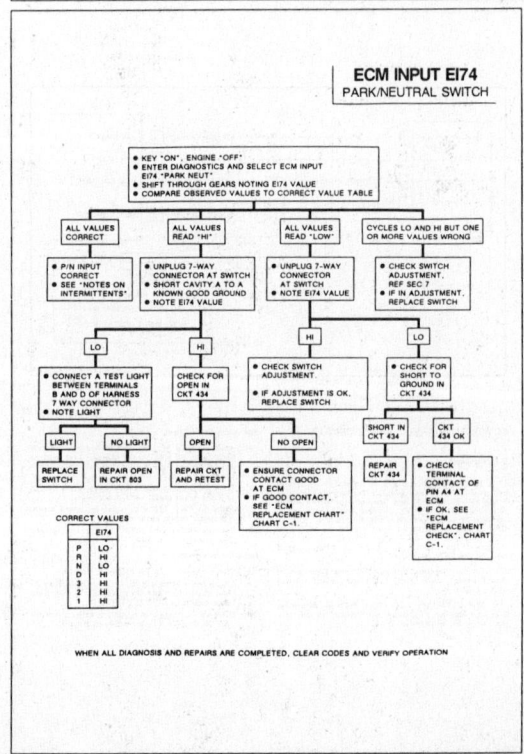

EI74	
P	LO
R	HI
N	LO
D	HI
3	HI
2	HI
1	HI

CORRECT VALUES

WHEN ALL DIAGNOSIS AND REPAIRS ARE COMPLETED, CLEAR CODES AND VERIFY OPERATION

1988–90 CADILLAC ELDORADO AND SEVILLE

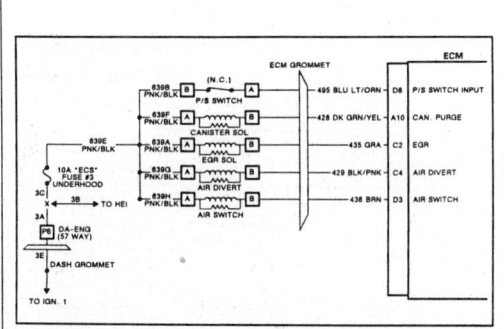

ECM INPUT EI78, POWER STEERING PRESSURE SWITCH

Description:

The Power Steering switch is normally closed and opens with high power steering pressure. During normal driving, the power steering switch receives 12 volts from the 10 AMP ECS Fuse (#3, Underhood Relay Center) on CKT 639B and sends the 12 volt signal to the ECM on CKT 495. When high power steering pressures occur, the switch opens and CKT 495 voltage is read by the ECM as 0 volts.

On the diagnostic display, "HI" is a closed switch, low steering pressure while "LO" is an open switch, high steering pressure.

The ECM uses the power steering input to extend the ISC motor when high power steering pressures occur at low speeds, such as during parking maneuvers.

Notes On Fault Tree:

1. Input Test EI78 must be performed engine running.

2. Status "Stays HI" means that switch is not opening with high pressure or that CKT 495 is shorted to voltage.

3. Status "Stays LO" means that ECM never receives a 12 volt signal on CKT 495.

4. Remove and replace ECM "J2 (C/D)" connector and ensure that it is latched.

Note On Intermittents:

At key on, engine off, manipulate power steering switch connector and related wiring while observing ECM Input "EI78". If EI78 cycles to "LO" at any time, repair source of intermittent open.

Review the snapshot data parameters stored with this code to verify failure condition. Refer to "Symptoms" for complete details on using the "Snapshot on code set" feature. If the failure condition is verified, check and repair intermittent wiring, connection, or sensor.

1988–90 CADILLAC ELDORADO AND SEVILLE

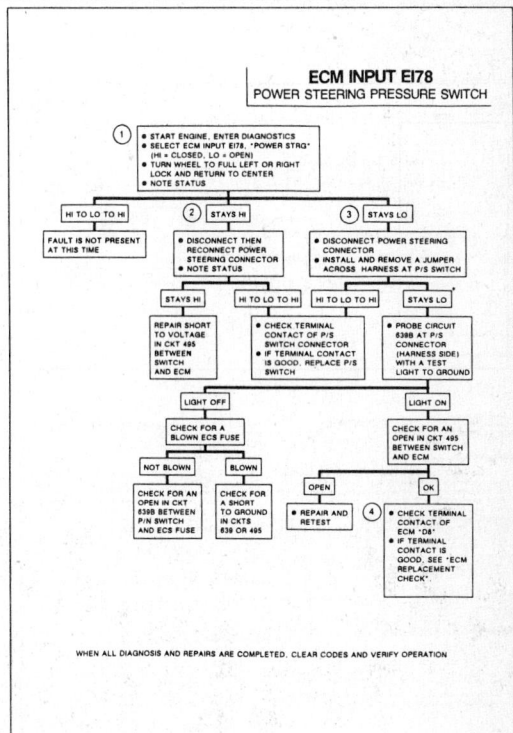

ECM INPUT EI78
POWER STEERING PRESSURE SWITCH

WHEN ALL DIAGNOSIS AND REPAIRS ARE COMPLETED, CLEAR CODES AND VERIFY OPERATION

1988–90 CADILLAC ELDORADO AND SEVILLE

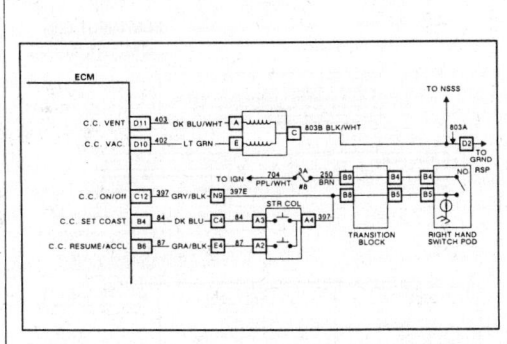

ECM INPUT EI79, CRUISE CONTROL ON/OFF CIRCUIT

Description:

The cruise enable or cruise dash switch provides power to the cruise enable input of the ECM. The ECM can't allow cruise operation until the enable or instrument panel switch is turned to the "ON" position, which passes 12 volts to ECM terminal C12.

Note On Intermittents:

The amber cruise "ON" indicator can be used to monitor the cruise dash switch status. Turn the dash switch "ON" and manipulate wiring and connectors. If the switch or circuit from the fuse to the switch are open, the amber light will blink out.

If the cruise 3A fuse blows intermittently, turn the switch on, manipulate wiring, connectors, cruise steering column switch. If the circuit becomes grounded, the fuse will blow.

1988–90 CADILLAC ELDORADO AND SEVILLE

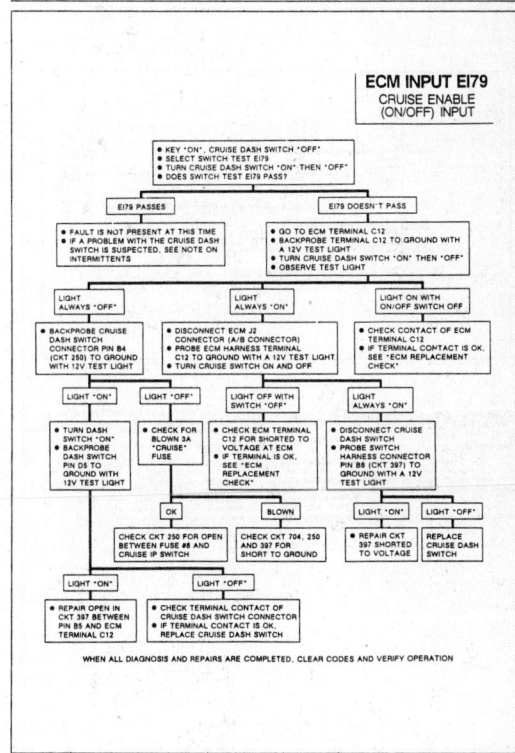

ECM INPUT EI79
CRUISE ENABLE (ON/OFF) INPUT

WHEN ALL DIAGNOSIS AND REPAIRS ARE COMPLETED, CLEAR CODES AND VERIFY OPERATION

1988–90 CADILLAC ELDORADO AND SEVILLE

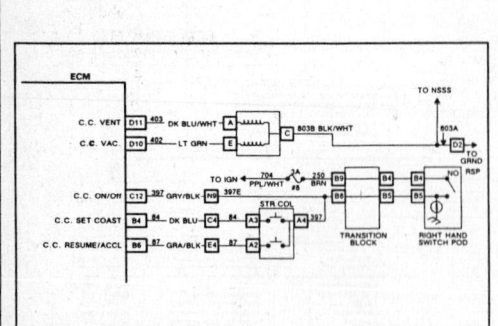

ECM INPUT EI80, CRUISE CONTROL "SET/COAST" CIRCUIT

Description:

The ECM Cruise "Set/Coast" circuit is a normally open switch that provides a 12 volt "Set/Coast" signal to the ECM on terminal B4 when depressed. When the cruise dash switch is in the "ON" position, the cruise dash switch provides 12 volts to the cruise steering column switch on pin A4, CKT 397. When the Set/Coast button is depressed, the steering column switch pin A4 is connected to pin E3, CKT 84, providing the ECM with a 12 volt Set/Coast input on terminal B4.

Notes On Intermittents:

- To diagnose an intermittent Set/Coast function, turn the cruise dash switch to "ON" and backprobe ECM B4 with a voltmeter to ground. B4 should show 12 volts with Set/Coast depressed and 0 volts with Set/Coast released. Cycle the switch while observing the meter; manipulate wiring and connectors.

1988–90 CADILLAC ELDORADO AND SEVILLE

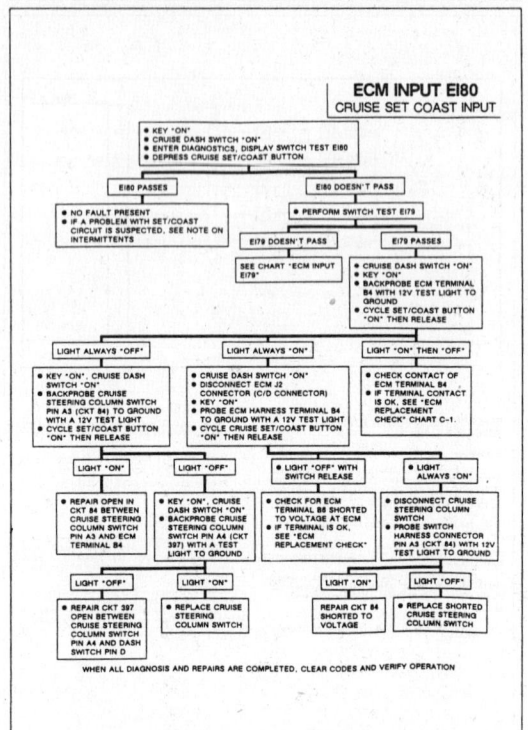

ECM INPUT EI80
CRUISE SET COAST INPUT

WHEN ALL DIAGNOSIS AND REPAIRS ARE COMPLETED, CLEAR CODES AND VERIFY OPERATION

1988–90 CADILLAC ELDORADO AND SEVILLE

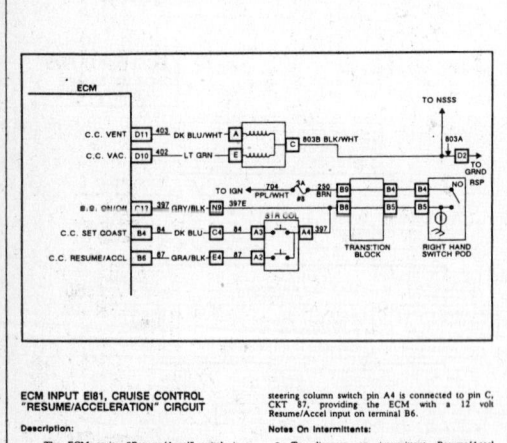

ECM INPUT EI81, CRUISE CONTROL "RESUME/ACCELERATION" CIRCUIT

Description:

The ECM cruise "Resume/Accel" switch is a normally open switch that provides a 12 volt "Resume/Accel" signal to the ECM on terminal B6 when depressed. When the cruise dash switch is in the "ON" position, the cruise dash switch provides 12 volts to the cruise steering column switch on pin A4, CKT 397. When the Resume/Accel button is depressed, the steering column switch pin A4 is connected to pin C, CKT 87, providing the ECM with a 12 volt Resume/Accel input on terminal B6.

Notes On Intermittents:

- To diagnose an intermittent Resume/Accel function, turn the cruise dash switch to "ON" and backprobe ECM B6 with a voltmeter to ground. B6 should show 12 volts with Resume/Accel depressed and 0 volts with Resume/Accel released. Cycle the switch while observing the meter, manipulate wiring and connectors.

1988–90 CADILLAC ELDORADO AND SEVILLE

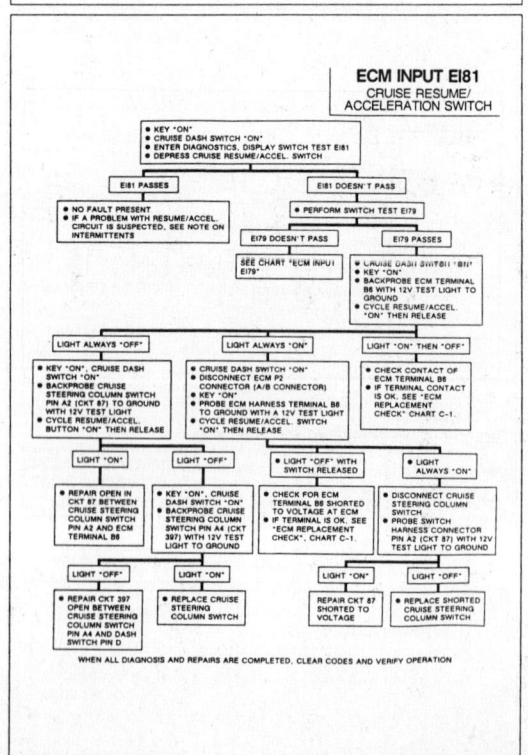

ECM INPUT EI81
CRUISE RESUME/ACCELERATION SWITCH

WHEN ALL DIAGNOSIS AND REPAIRS ARE COMPLETED, CLEAR CODES AND VERIFY OPERATION

1988–90 CADILLAC ELDORADO AND SEVILLE

WIRE SIDE

CIRCUIT	COLOR	CKT #
		SPARE
		SPARE
PRNDL	WHT	776
		SPARE
		SPARE
		SPARE
		SPARE
		SPARE
CRANK INPUT	PPL/WHT	806
CONDITIONED 12 VOLT	RED/WHT	812C
		SPARE
CPS WAKE-UP	DK BLU/WHT	555
		SPARE
		SPARE
CPS 7 VOLT	PPL/WHT	807C
CPS 7 VOLT	PPL/WHT	807E

Connector C1 / D1 / #1 BLACK / C16 / D16, pins 1–16

CIRCUIT	COLOR	CKT #
		SPARE
		SPARE
		SPARE
MIX DR. WIPER	LT BLU	733
		SPARE
		SPARE
ISO IGNITION #3	PNK/WHT	350B
		SPARE
TWILIGHT RELAYS	BLK/PNK	692A
BLOWER CONTROL	PPL/WHT	760
HI/LO BEAM RELAYS	BLK/RED	691A
COURTESY RELAY	GRA/BLK	690
RAP RELAYS	BLK/WHT	707A
5V RETURN (GND)	BLK/PNK	736C
CPS GROUND	BLK/RED	801E
5V REFERENCE	TAN	705B

CIRCUIT	COLOR	CKT #
		SPARE
IN CAR TEMP	DK GRN	734
		SPARE
WASHER FLUID LEVEL	BLK	99
COURTESY ENABLE	BRN/YEL	685
		SPARE
PRNDL DECODER A	BLK/WHT	771
COOLING FAN SIG	BRN	790
CPS GND	BLK/RED	801C
		SPARE
RADIO DIMMING OUTPUT	PPL/YEL	724
INCANDES. DIM OUTPUT	GRA/WHT	717

Connector A1 / B1 / #2 BLACK / A12 / B12, pins 1–12

CIRCUIT	COLOR	CKT #
		SPARE
		SPARE
		SPARE
TWILIGHT DELAY INPUT	PPL/WHT	271
DATA LINE (UART)	BRN	800A
		SPARE
PARKLAMPS ON	GRA/WHT	308B
FUEL LEVER WIPER	PPL	30
		SPARE
PRNDL DECODER B	YEL	772
COOLING FAN FEDBK	DK GRN	791
		SPARE

CIRCUIT	COLOR	CKT #
SPEED SENSOR LOW	PPL	401
		SPARE
PRNDL DECODER	GRA	773
VSS TO ECM	BRN	437
		SPARE
SUNLOAD SENSOR	LT BLU/YEL	590
DOOR HANDLE SW	GRA	157
DOOR JAMB SW	WHT/BLK	156
TIME OF DAY WAKE UP	GRA/RED	1018
SPEED SENSOR HI	YEL	400
ISO IGNITION 1	PNK/BLK	139C
FT/DRIVER DOOR AJAR	GRA/WHT	147
		SPARE
RR/PASS DOOR AJAR	BLK/ORN	158
TRUNK OPEN	DK GRN	146
		SPARE

Connector C1 / D1 / #3 RED / C16 / D16, pins 1–16

CIRCUIT	COLOR	CKT #
		SPARE
		SPARE
		SPARE
A/C LO SIDE TEMP	GRA	731
A/C HI SIDE TEMP	DK BLU	732
		SPARE
		SPARE
TWI PHOTOCELL	WHT/DK GRN	278
		SPARE
DATA	TAN/WHT	800G
GEN FIELD " + "	GRA	23
IP DIM INPUT	TAN	686
		SPARE
OUTSIDE TEMP	LT GRN/BLK	735
GENERATOR ENABLE	RED	225
LO REF PRESSURE	BRN	257A

WIRE SIDE

Component Replacement

SERVICE PRECAUTIONS

When working around any part of the fuel system, take precautionary steps to prevent fire and/or explosion:

   a. Disconnect negative terminal from battery (except when testing with battery voltage is required).

   b. When ever possible, use a flashlight instead of a drop light.

   c. Keep all open flame and smoking material out of the area.

   d. Use a shop cloth or similar to catch fuel when opening a fuel system.

   e. Relieve fuel system pressure before servicing.

   f. Use eye protection.

   g. Always keep a dry chemical (class B) fire extinguisher near the area.

NOTE: Due to the amount of fuel pressure in the fuel lines, before doing any work to the fuel system, the fuel system should be depressurized. To depressurize the fuel system, disconnect the fuel pump electrical connections at the fuel pump and start the vehicle. Let the vehicle run until it burns up the remaining fuel in the fuel lines. This way there will be no pressure left in the fuel system and the repair work can be performed.

THROTTLE BODY

Removal and Installation

1. Remove the air cleaner housing assembly.

2. Disconnect the electrical connectors from the idle air control (IAC) motor, throttle position sensor (TPS) and the fuel injectors.

3. Disconnect the throttle cable, throttle return spring, throttle valve cable and the cruise control cable.

4. Mark and disconnect the vacuum hoses.

5. Disconnect the fuel feed and return lines. Use a 2nd wrench on all fittings.

6. Remove the 3 bolts securing the throttle body to the manifold and remove the throttle body.

7. To install, position the throttle body gasket on the manifold.

8. Install the throttle body gasket and attaching bolts. Torque the bolts alternately to 11 ft. lbs. (15 Nm).

9. Connect the fuel feed and return lines and torque to 17 ft. lbs. (23 Nm).

10. Connect the vacuum lines.

11. Connect the electrical connectors to the ISC motor, TPS and the fuel injectors.

12. Connect the throttle cable, throttle return spring, throttle valve cable and cruise control cable.

13. Install the air cleaner housing.

FUEL METER COVER

Removal and Installation

The fuel meter cover contains the pressure regulator and is only serviced as a complete preset assembly. The fuel pressure regulator is preset and plugged at the factory.

────── **CAUTION** ──────

DO NOT remove the 4 screws securing the pressure regulator to the fuel meter cover. The fuel pressure regulator incorporates a large spring under heavy tension which, if accidentally released, could cause personal injury.

1. Depressurize the fuel system. Remove the air cleaner assembly.

2. Disconnect the electrical connector to the fuel injectors.

3. Remove the 8 screws securing the fuel meter cover to the fuel meter body. Be sure to take note of the location of the 4 short screws.

4. Remove the fuel meter cover from the fuel meter body.

NOTE: DO NOT immerse the fuel meter cover (with pressure regulator) in any type of cleaner. Immersion in cleaner will damage the internal fuel pressure regulator diaphragms and gaskets.

5. Installation is the reverse order of the removal procedure. Be sure to install a new dust seal and all new gaskets.

FUEL INJECTOR

Removal

NOTE: Use care in removing the injector to prevent damage to the electrical connector pins on top of the injector, the injector fuel filter and the nozzle. The fuel injector is serviced as a complete assembly only. The fuel injector is an electrical component and should not be immersed in any type of cleaner.

1. Depressurize the fuel system. Remove the air cleaner assembly.

2. Disconnect the injector electrical connector by squeezing the 2 tabs together and pulling straight up.

3. Remove the fuel meter cover (refer to the fuel meter cover procedure).

4. With the fuel meter cover gasket in place to prevent damage to the casting, use a suitable dowel rod and lay the dowel rod on top of the fuel meter body.

5. Insert a suitable pry tool into the small lip of the injector and pry against the dowel rod lifting the injector straight up. Tool J–26868 or equivalent can also be used.

6. Remove the large O-ring and steel back-up washer.

7. Remove the small O-ring at the bottom of the injector cavity. Be sure to discard both O-rings.

Installation

1. Lubricate the new small O-ring with pertroleum jelly or equivalent. Push the new O-ring on the nozzle end of the injector, pressing the O-ring up against the injector fuel filter.

2. Install a new steel backup washer in the recess of the fuel meter body.

3. Lubricate the new large O-ring with pertroleum jelly or

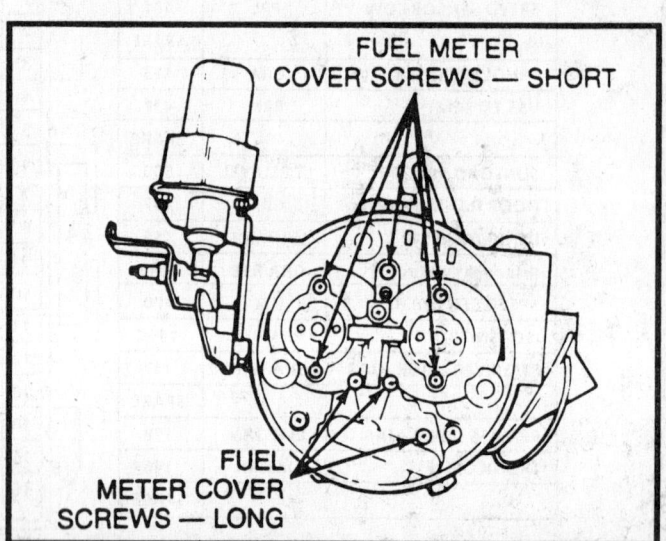

Fuel meter cover removal

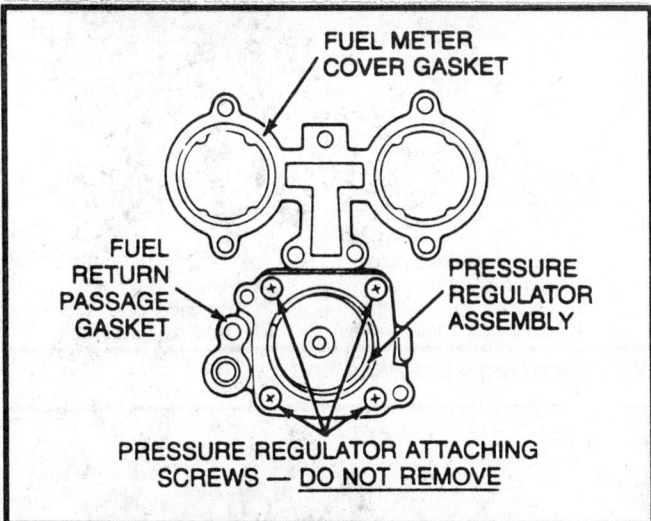

Fuel meter cover installation

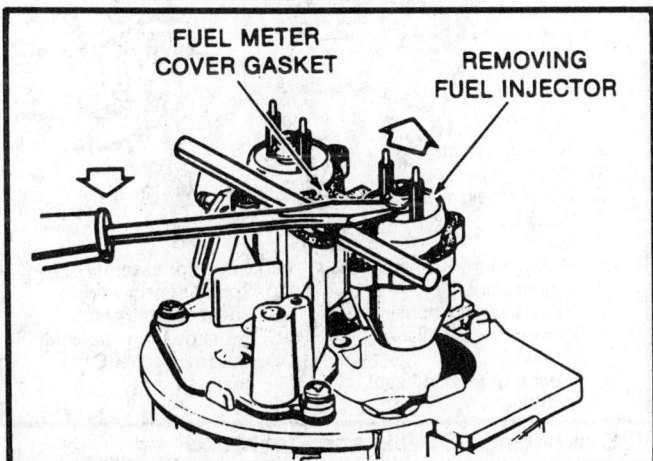

Fuel injector removal

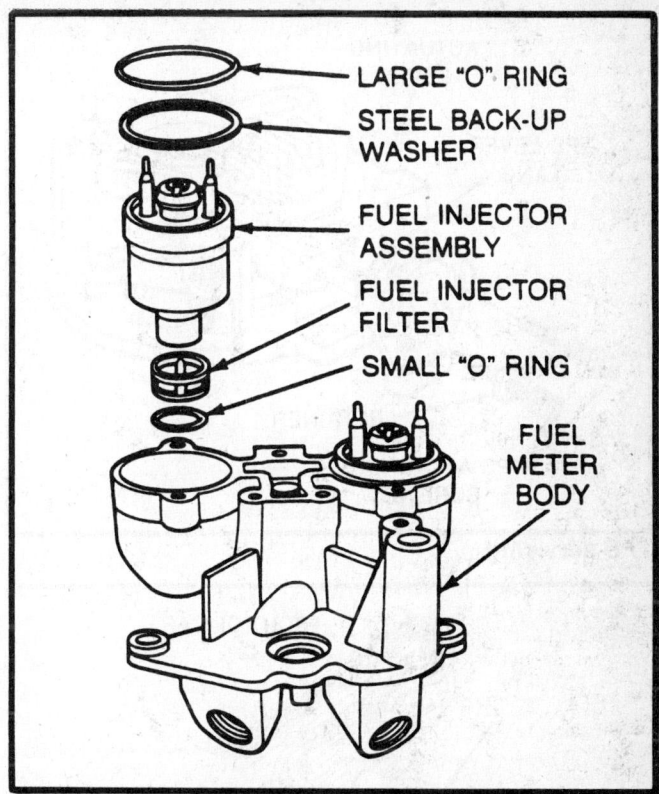

Fuel injector components

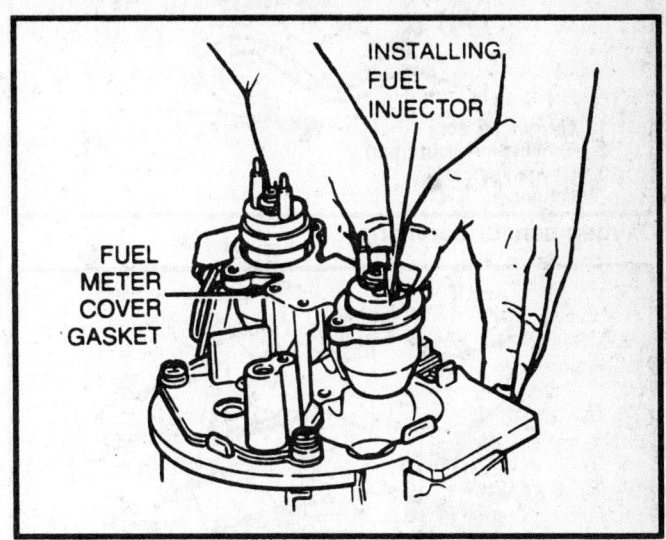

Fuel injector installation

equivalent. Install the new O-ring directly above the backup washer, pressing the O-ring down into the cavity recess. The O-ring is installed properly when it is flush with the fuel meter body casting surface.

NOTE: Do not attempt to reverse the installation of the large O-ring procedure. Install the backup washer and O-ring after the injector is located in the cavity. To do so will prevent the seating of the O-ring in the cavity recess which may result in a fuel leak.

4. Install the injector by using as pushing and twisting motion to center the nozzle O-ring in the bottom of the injector cavity and aligning the raised lug on the injector base with the notch cast into the fuel meter body.
5. Push down on the injector making sure it is fully seated in the cavity. The injector is installed correctly when the lug is seated in the notch and the electrical terminals are parallel to the throttle shaft in the throttle body.
6. Install the fuel meter cover. Install the injector electrical connector and all electrical and vacuum lines. Install the air cleaner assembly.
7. Start the engine and check for leaks and proper injector operation.

FUEL METER BODY

Removal and Installation

1. Depressurize the fuel system. Remove the air cleaner assembly.
2. Remove the fuel meter cover assembly. Remove the fuel meter cover gasket, fuel meter outlet gasket and pressure regulator seal.
3. Remove the fuel injectors. Remove the fuel inlet and fuel outlet nuts and gaskets from the fuel meter body.
4. Remove the 3 screws and lockwashers, then remove the fuel meter body from the throttle assembly.

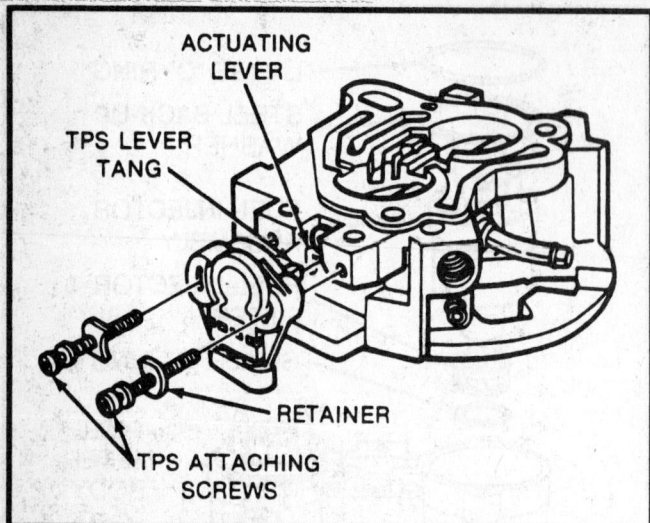

TPS servicing

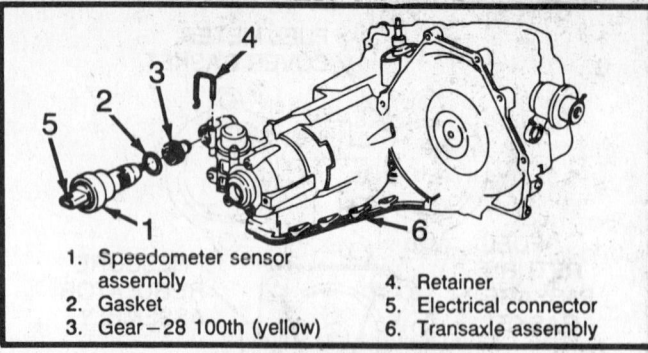

1. Speedometer sensor assembly
2. Gasket
3. Gear—28 100th (yellow)
4. Retainer
5. Electrical connector
6. Transaxle assembly

VSS servicing—Eldorado and Seville

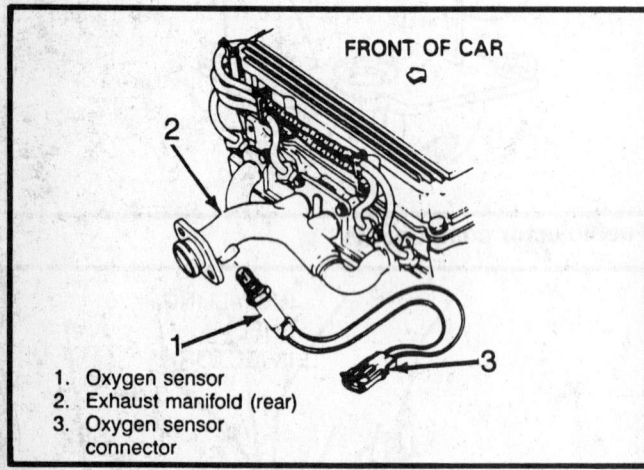

1. Oxygen sensor
2. Exhaust manifold (rear)
3. Oxygen sensor connector

Oxygen sensor servicing

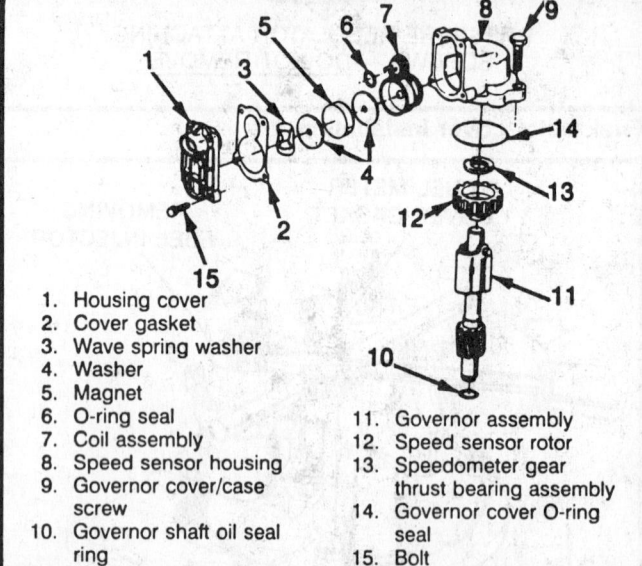

1. Housing cover
2. Cover gasket
3. Wave spring washer
4. Washer
5. Magnet
6. O-ring seal
7. Coil assembly
8. Speed sensor housing
9. Governor cover/case screw
10. Governor shaft oil seal ring

11. Governor assembly
12. Speed sensor rotor
13. Speedometer gear thrust bearing assembly
14. Governor cover O-ring seal
15. Bolt

VSS servicing—DeVille and Fleetwood

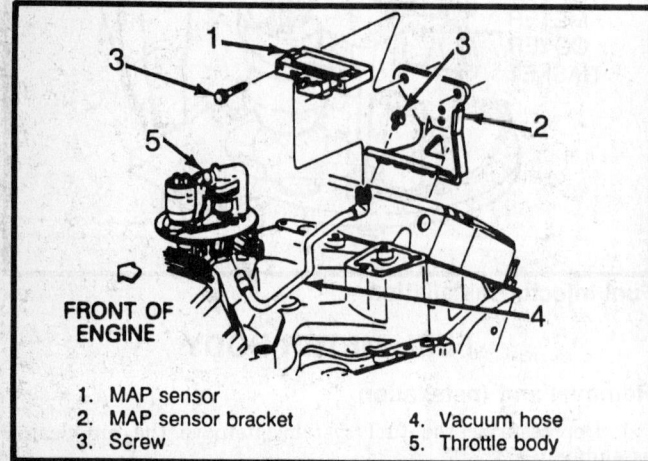

1. MAP sensor
2. MAP sensor bracket
3. Screw
4. Vacuum hose
5. Throttle body

MAP sensor servicing

5. Remove the fuel meter body insulator gasket.
6. To install, position the new meter body insulator gasket on the throttle body assembly.

NOTE: The gasket is installed correctly when the cut-out portions of the gasket match the opening in the throttle body assembly. Be sure to use the appropriate gasket for the 4.5L engine. The correct gasket will have a tab made of gasket material located on the the gasket. This gasket is different from the gasket used on past model 4.1L engines which should not be used on the 4.5L engine.

7. Install the fuel meter body on the insulator gasket. Install the 3 retaining screws and lockwashers. Tighten securely.

8. Install the fuel inlet and outlet nuts and the new gaskets in the fuel meter body. Tighten the inlet nut to 30 ft. lbs. (40 Nm) and the outlet nut to 21 ft. lbs. (29 Nm).

9. Install the injectors with new upper and lower O-rings.
10. Install the fuel meter cover assembly.
11. Start the engine and check for fuel leaks.

IDLE SPEED CONTROL (ISC)

Removal and Installation

1. Disconnect the negative battery cable and remove the air cleaner assembly.
2. Disconnect the electrical connector. Remove the 2 idle speed control mounting screws. Remove the ISC motor from the throttle body.

NOTE: The idle speed control motor is calibrated at the factory and NO attempt should be made to disassemble the unit. Do not immerse the ISC in any type of

cleaner and always remove before the throttle body cleaning and servicing of components. Immersion in cleaner will damage the ISC unit.

3. To install, position the ISC motor to the throttle body and secure it with the retaining screws. Adjust the ISC as necessary. Reconnect the electrical connector and the negative battery cable. Install the air cleaner assembly.

Adjustment

RETRACTING THE ISC PLUNGER

1. Disconnect the ISC connector and connect a jumper harness to the ISC.

2. Connect a jumper wire leading to the ISC terminal **C** to a 12 volt source at the battery or junction block.

3. Apply finger pressure to the ISC plunger (close the throttle switch). Touch the jumper wire connected to the ISC terminal **D** to a ground until the ISC plunger retracts fully and stops. Remove the ground immediately.

NOTE: Do not leave terminal D grounded longer than necessary to fully retract the ISC plunger. If the ISC is stalled retracted for prolonged periods of time, damage to the ISC will result. Apply finger pressure to the ISC plunger before connecting terminal D to ground. Retracting the ISC plunger without pressure on the plunger may cause the internal gears to clash and bind. Never connect a voltage source to the ISC motor terminals A and B as damage to the internal throttle switch contacts will result.

EXTENDING THE ISC PLUNGER

1. Connect the ISC terminal **D** to 12 volts at the battery or the junction block.

2. Connect the ISC terminal **C** to a ground until the ISC plunger is partially or fully extended.

3. If the ISC motor fails to extend, disconnect and reconnect terminal **C** to ground to bump the motor from its stalled fully retracted position.

MINIMUM AIR, THROTTLE POSITION SENSOR (TPS) AND IDLE AIR CONTROL (IAC) MOTOR

Adjustment

1. Preliminary checks:

a. Enter diagnostic and record all the trouble codes displayed. If current or history codes are found, refer to the diagnostic charts. When all the trouble codes are repaired, proceed to the next Step.

b. Perform the ECM input tests (EI71, EI72, EI74 and EI78 for Eldorado and Seville or E.7.1, E.7.2, E.7.4 and E.7.8 for Deville and Fleetwood). Refer to diagnostic charts if inputs do not cycle. If all inputs cycle, proceed to the next Step.

c. Exit diagnostics. While not in diagnostics, enter set timing mode by jumpering pin A to pin B of the ALDL connector. The Climate Control Driver Information Center (CCDIC) will display the "Set Timing" message with the jumper in place. If the "Set Timing" message does not appear, refer to the diagnostic charts to diagnose. Check for timing at 10 degrees BTDC and adjust as necessary. Disconnect the jumper at the ALDL after the timing is verified as correct.

d. Perform DFI System Check in the diagnostic charts. If the chart leads to "Fuel System is in Control," proceed to the next Step.

2. Check minimum idle speed:

a. Remove the air cleaner and plug the thermac vacuum tap. Start the engine.

b. Enter diagnostics; select ECM parameter for coolant

temperature. (P.O.4 for Deville and Fleetwood and ED04 for Eldorado and Seville.) Operate the engine until the engine coolant temperature is greater than 185°F (85°C).

c. Disable the alternator by grounding the green test connector under the hood near the alternator. The alternator will turn off and the "No Charge" telltale light will illuminate.

d. Select the ECM override for ISC Motor. (E.6.3 for Deville and Fleetwood and ES03 for Eldorado and Seville.).

e. Press the "Cooler" button to retract the ISC motor. The ISC motor will slowly retract (about 20 seconds).

f. With the ISC plunger fully retracted, the plunger should not be touching the throttle lever. If contact is noted, adjust the ISC plunger (turn in) with a suitable pair of pliers or tool J–29607 so that it is not touching the throttle lever.

g. Make sure that the throttle lever is being bound by the throttle, cruise or T.V. cables. The throttle lever must be resting on the minimum air screw.

h. Check the minimum idle speed; it should be 475–550 rpm. If the minimum idle speed is outside the limits go on to Step 9. If the minimum idle speed is good, than no further adjustment is necessary.

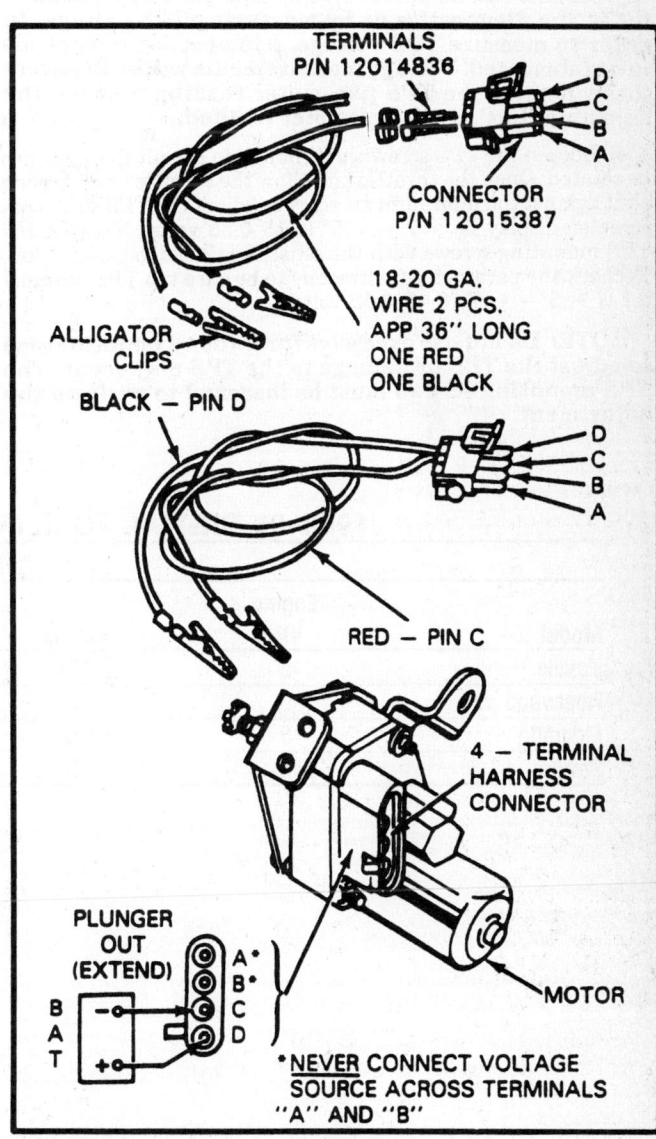

ISC jumper harness procedure

NOTE: If the engine speed in not within specifications, check for a vacuum leak at the throttle body, intake manifold vacuum fittings, tees and hoses. If the minimum air setting is made with a vacuum leak present, fuel control can be adversely affected throughout the driving range.

i. Make sure that the alternator is disabled; the "No Charge" telltale light will be illuminated.

j. If the minimum idle speed is out of specifications, connect a tachometer to the engine and adjust the minimum idle screw to obtain 525 rpm.

3. Check throttle position sensor (TPS) setting by turning the engine **OFF**, key **ON**. Open the throttle and let it snap shut against the minimum air screw. Note the throttle angle displayed in diagnostics on ECM parameter for TPS (ED01 for Eldorado and Seville and P.0.1 for Deville and Fleetwood). The TPS should display −1.0° – +1.0°. If the TPS displays out of tolerance (−10.0° – −1.1° or +1.1° – +90.0°), the TPS requires adjustment.

NOTE: If the throttle angle displayed is outside the specified tolerance, the TPS needs to be adjusted. This adjustment can be performed by using the TPS parameter in the diagnostics or by measuring TPS voltage. In order to measure TPS voltage, a jumper harness needs to be fabricated. Voltage measurements will be in parenthesis next to the TPS parameter reading to make the adjustment using the voltmeter method.

4. Loosen the TPS screws just enough to permit the sensor to be rotated. Open the throttle and allow the throttle lever to snap shut against the minimum air screws. Adjust the TPS so the parameter display is −.5° – +.5° (0.45–0.55 volts). Tighten the TPS mounting screws with the sensor in the adjusted position. Recheck the parameter (voltmeter) to be sure the TPS parameter is −.5° – +.5° (0.45–0.55 volts).

NOTE: Do not use excessive force (don't pound or pry) to adjust the TPS or damage to the TPS may occur. The TPS mounting screws must be loosened to perform the adjustment.

5. Check and adjust the ISC maximum extension:

a. With the ignition **ON**, engine **OFF**, extend the ISC motor by connecting the ISC terminal D to 12 volts at the battery or the junction block.

b. Connect the ISC terminal C to a ground until the ISC plunger is partially or fully extended.

c. If the ISC motor fails to extend, disconnect and reconnect terminal C to ground to bump the motor from its stalled fully retracted position.

d. With the ISC fully extended (so that it ratchets), apply and release power to extend the ISC while watching the nose of the ISC motor.

NOTE: The procedure that should be used is apply voltage, release voltage, note the ISC motor nose with the ISC stopped. This should be done multiple times. With the ISC stopped at full extension, 1 of 2 nose positions will be seen: an extended position at the top of the ratchet and a low position at the bottom, of the ratchet. Look for the highest nose position, which will be seen when the ISC stops at the top of its ratchet. Stop the ISC when the highest nose position is seen.

e. With the ISC at maximum extension the TPS parameter should be 10.5–11.5 (1.05–1.15 volts). If the parameter is not within this range, adjust the ISC plunger in to lower the parameter (voltage) or out to raise the parameter (voltage) as necessary. Recheck the maximum extension TPS parameter (voltage) by again applying power to extend the ISC until a maximum ISC nose position is seen.

6. After all checks and adjustments are completed, disconnect all test equipment and reconnect the ISC motor connector. Turn the key **OFF** for at least 10 seconds. Start the engine and check the ISC motor for proper operation.

7. Remove the ground from the green alternator test lead. Reinstall the air cleaner and reconnect the thermac vacuum line, hot air tube and PCV hose.

8. The above procedure may have turned on a telltale light or set an ECM or BCM code. Enter diagnostics and clear all stored ECM or BCM codes.

1988–90 DIGITAL FUEL INJECTION SPECIFICATIONS

CADILLAC

Model	Engine VIN	Engine	Minimum Idle rpm	TPS Voltage	Fuel Pressure (psi)
DeVille	5	4.5L	475-550	0.45-0.55	9-12
Fleetwood	5	4.5L	475-550	0.45-0.55	9-12
Eldorado	5	4.5L	475-550	0.45-0.55	9-12
Seville	5	4.5L	475-550	0.45-0.55	9-12

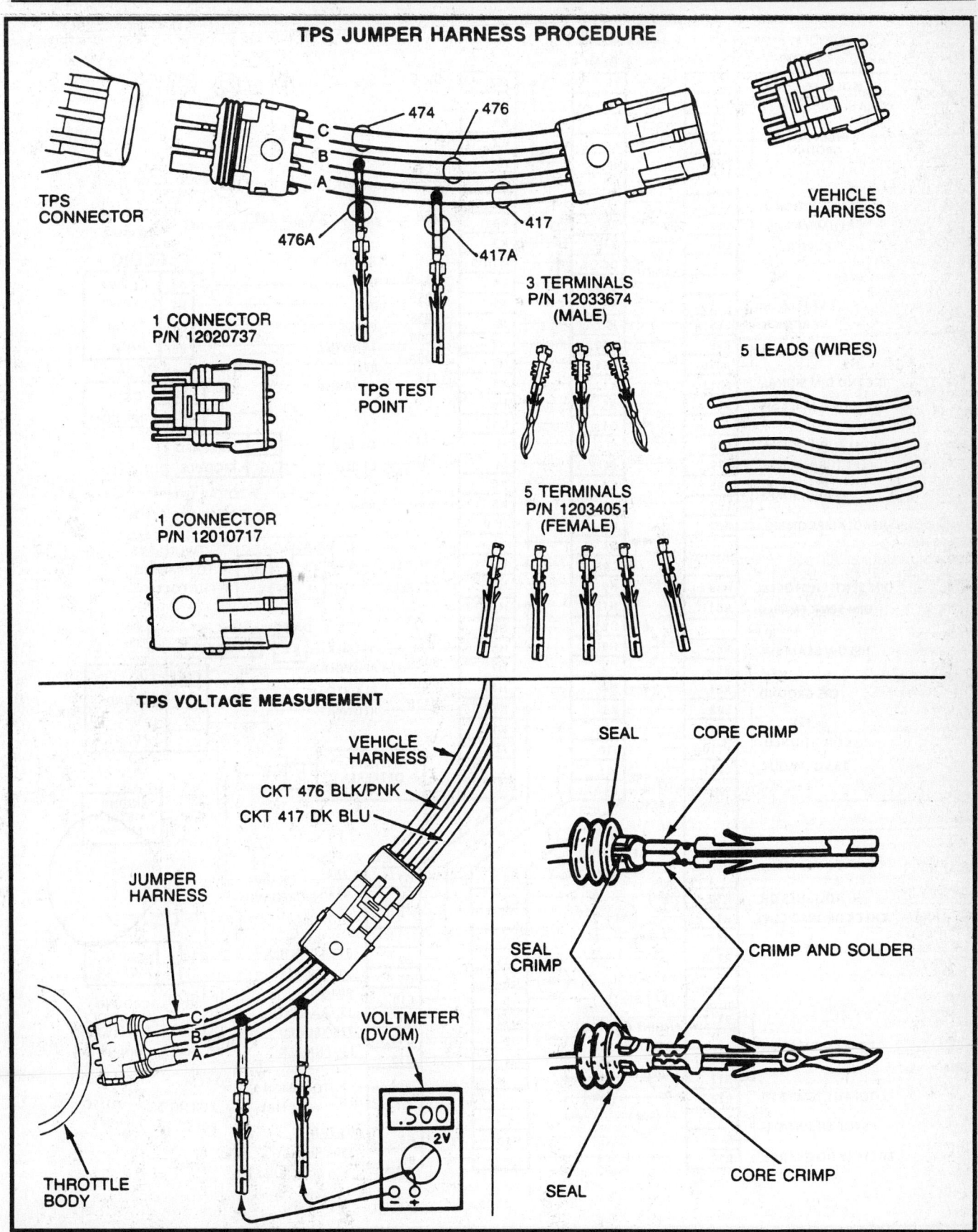

TPS JUMPER HARNESS PROCEDURE

TPS CONNECTOR

474
476
476A
417
417A

C
B
A

VEHICLE HARNESS

1 CONNECTOR P/N 12020737

TPS TEST POINT

3 TERMINALS P/N 12033674 (MALE)

5 LEADS (WIRES)

1 CONNECTOR P/N 12010717

5 TERMINALS P/N 12034051 (FEMALE)

TPS VOLTAGE MEASUREMENT

VEHICLE HARNESS

CKT 476 BLK/PNK
CKT 417 DK BLU

JUMPER HARNESS

C
B
A

VOLTMETER (DVOM)

.500
2V

THROTTLE BODY

SEAL

CORE CRIMP

SEAL CRIMP

CRIMP AND SOLDER

SEAL

CORE CRIMP

TPS jumper harness procedure

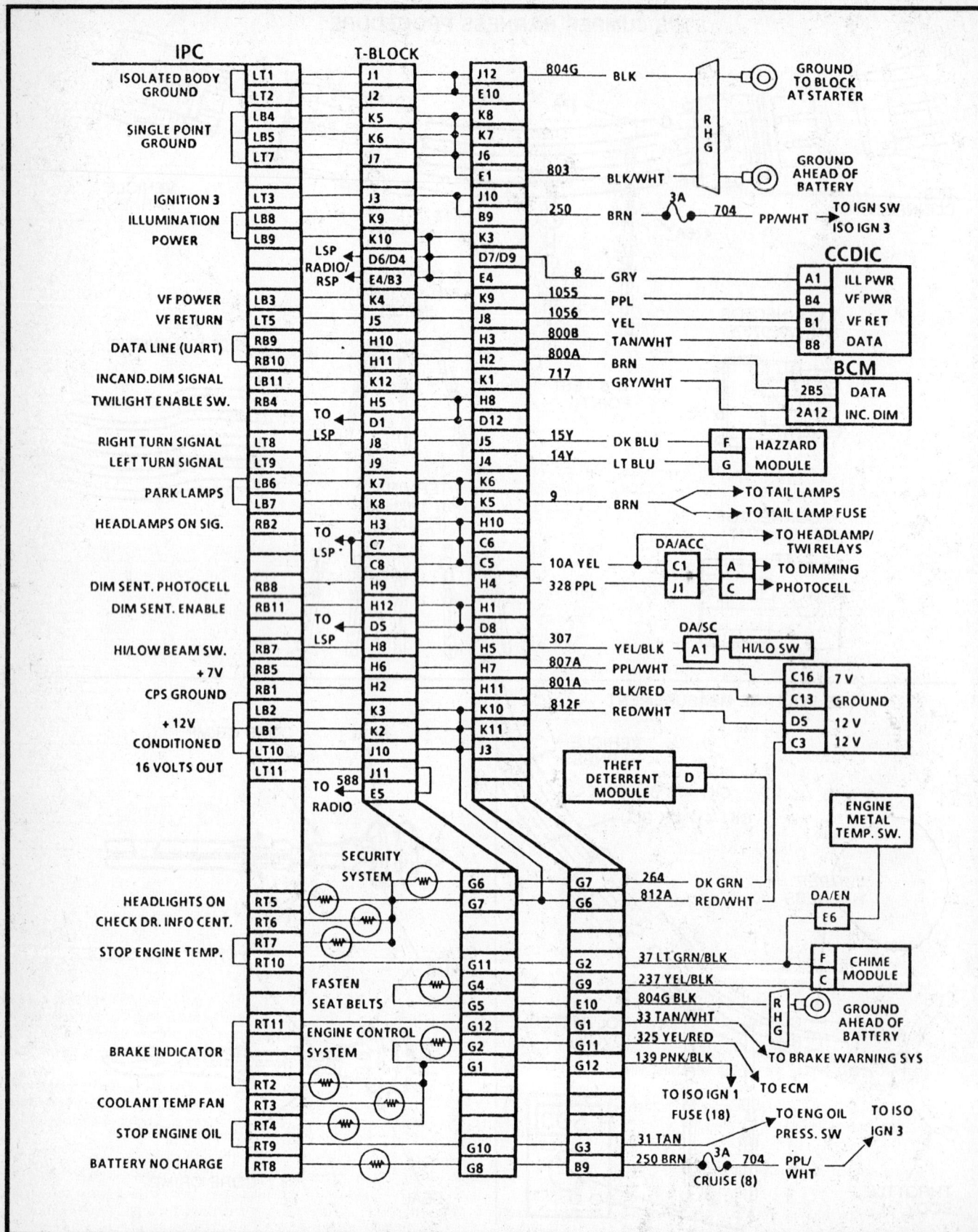

Instrument Panel Cluster (IPC) interface diagram

MULTI-PORT (MPI) AND TUNED PORT INJECTION (TPI)

APPLICATION CHART
1988–90 Multi-Port Injection Vehicles

Year	Models	Series VIN	Engine	Engine VIN Code
		BUICK		
1988	Century	A	2.8L	W
	Regal	W	2.8L	W
	Skylark	N	2.3L	D
	Skylark	N	3.0L	L
1989–90	Century	A	2.8L	W
	Century	A	3.3L	N
	Regal	W	2.8L	W
	Regal	W	3.1L	T
	Skylark	N	2.3L	D
	Skylark	N	3.3L	N
		CADILLAC		
1988	Cimarron	J	2.8L	W
		CHEVROLET		
1988–90	Beretta	L	2.8L	W
	Camaro	F	2.8L	S
	Camaro	F	5.0L	F
	Camaro	F	5.7L	8
	Cavalier	J	2.8L	W
	Celebrity	A	2.8L	W
	Corsica	L	2.8L	W
	Corvette	Y	5.7L	8
		OLDSMOBILE		
1988	Calais	N	2.3L	D
	Calais	N	3.0L	L
	Ciera	A	2.8L	W
	Cutlass	W	2.8L	W

APPLICATION CHART
1988–90 Multi-Port Injection Vehicles

Year	Models	Series VIN	Engine	Engine VIN Code
1988	Cutlass Cruiser	A	2.8L	W
1989–90	Calais	N	2.3L	D
	Calais	N	2.3L	A
	Calais	N	3.3L	N
	Ciera	A	2.8L	W
	Ciera	A	3.3L	N
	Cutlass	W	2.8L	W
	Cutlass	W	3.1L	T
	Cutlass Cruiser	A	2.8L	W
	Cutlass Cruiser	A	3.3L	N
		PONTIAC		
1988	Fiero	P	2.8L	9
	Firebird	F	2.8L	S
	Firebird	F	5.7L	8
	Grand Am	N	2.0L Turbo	M
	Grand Am	N	2.3L	D
	Grand Prix	W	2.8L	W
	Sunbird	J	2.0L Turbo	M
	6000	A	2.8L	W
1989–90	Firebird	F	2.8L	S
	Firebird	F	5.0L	F
	Grand Am	N	2.0L Turbo	M
	Grand Am	N	2.3L	D
	Grand Prix	W	3.1L	T
	Sunbird	J	2.0L Turbo	M
	6000	A	2.8L	W

NOTE: This manual has the ability to save time in diagnosis and prevent the replacement of good parts. The key to using this manual successfully for diagnosis, lies in the technician's ability to understand the system he is trying to diagnose as well as an understanding of the manual's layout and limitations. The technician should review this manual to become familiar with the way this manual should be used.

General Information

The Multi-Port Fuel Injection (MPI) system is controlled by an Electronic Control Module (ECM) which monitors engine operations and generates output signals to provide the correct air/fuel mixture, ignition timing and engine idle speed control. Input to the control unit is provided by an oxygen sensor, coolant temperature sensor, detonation sensor, hot film mass sensor and throttle position sensor. The ECM also receives information concerning engine rpm, road speed, transmission gear position, power steering and air conditioning.

The injectors are located, one at each intake port, rather than the single injector found on the earlier throttle body system. The injectors are mounted on a fuel rail and are activated by a signal from the electronic control module. The injector is a solenoid-operated valve which remains open depending on the width of the electronic pulses (length of the signal) from the ECM; the longer the open time, the more fuel is injected. In this manner, the air/fuel mixture can be precisely controlled for maximum performance with minimum emissions.

Fuel is pumped from the tank by a high pressure fuel pump,

located inside the fuel tank. It is a positive displacement roller vane pump. The impeller serves as a vapor separator and precharges the high pressure assembly. A pressure regulator maintains 28–36 psi (28–50 psi on turbocharged engines) in the fuel line to the injectors and the excess fuel is fed back to the tank. A fuel accumulator is used to dampen the hydraulic line hammer in the system created when all injectors open simultaneously.

The Mass Air Flow (MAF) Sensor is used to measure the mass of air that is drawn into the engine cylinders. It is located just ahead of the air throttle in the intake system and consists of a heated film which measures the mass of air, rather than just the volume. A resistor is used to measure the temperature of the film at 75° above ambient temperature. As the ambient (outside) air temperature rises, more energy is required to maintain the heated film at the higher temperature and the control unit used this difference in required energy to calculate the mass of the incoming air. The control unit uses this information to determine the duration of fuel injection pulse, timing and EGR.

The throttle body incorporates an Idle Air Control (IAC) that provides for a bypass channel through which air can flow. It consists of an orifice and pintle which is controlled by the ECM through a step motor. The IAC provides air flow for idle and allows additional air during cold start until the engine reaches operating temperature. As the engine temperature rises, the opening through which air passes is slowly closed.

The Throttle Position Sensor (TPS) provides the control unit with information on throttle position, in order to determine injector pulse width and hence correct mixture. The TPS is connected to the throttle shaft on the throttle body and consists of as potentiometer with on end connected to a 5 volt source from the ECM and the other to ground. A third wire is connected to the ECM to measure the voltage output from the TPS which changes as the throttle valve angle is changed (accelerator pedal moves). At the closed throttle position, the output is low (approximately 0.4 volts); as the throttle valve opens, the output increases to a maximum 5 volts at Wide Open Throttle (WOT).

The TPS can be misadjusted open, shorted, or loose and if it is out of adjustment, the idle quality or WOT performance may be poor. A loose TPS can cause intermittent bursts of fuel from the injectors and an unstable idle because the ECM thinks the throttle is moving. This should cause a trouble code to be set. Once a trouble code is set, the ECM will use a preset value for TPS and some vehicle performance may return. A small amount of engine coolant is routed through the throttle assembly to prevent freezing inside the throttle bore during cold operation.

TUNE PORT INJECTION (TPI)

This system can be found in the Camaro and Firebird with the

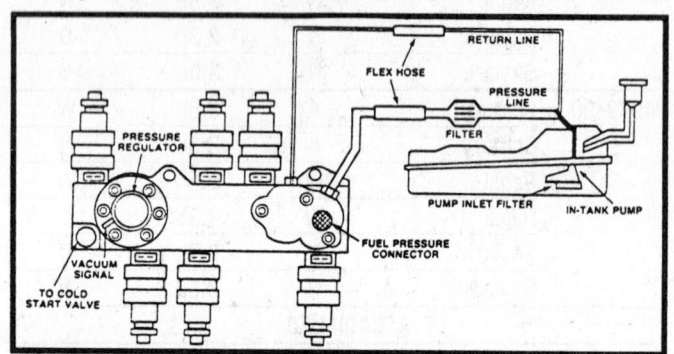

Typical fuel injection diagram

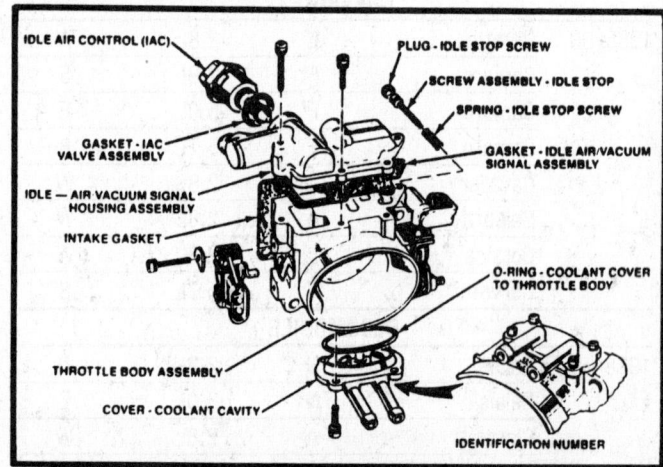

Typical 2.8L throttle body assembly

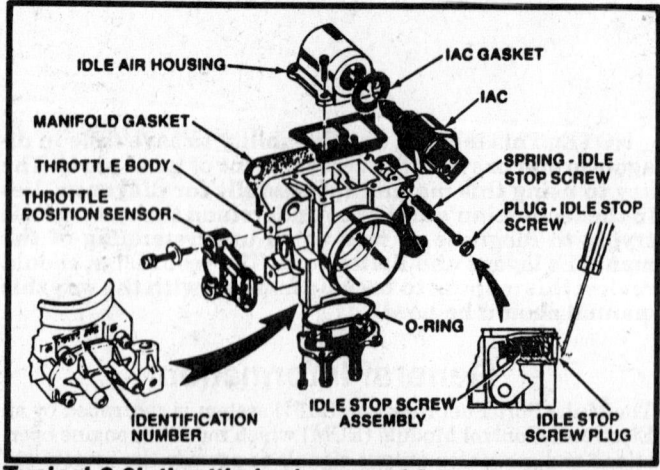

Typical 3.0L throttle body assembly

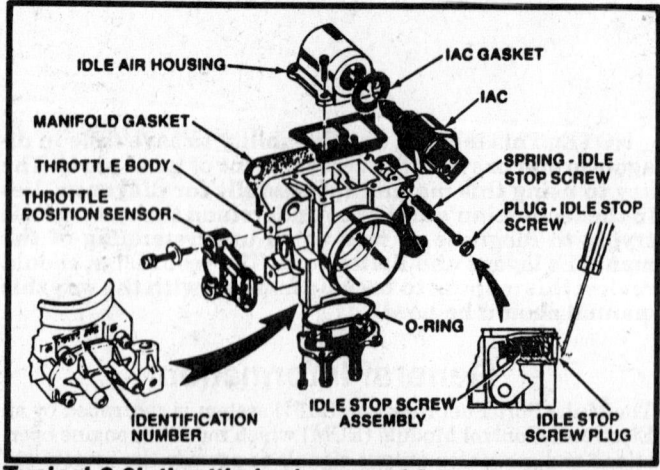

Typical MPFI throttle body assembly

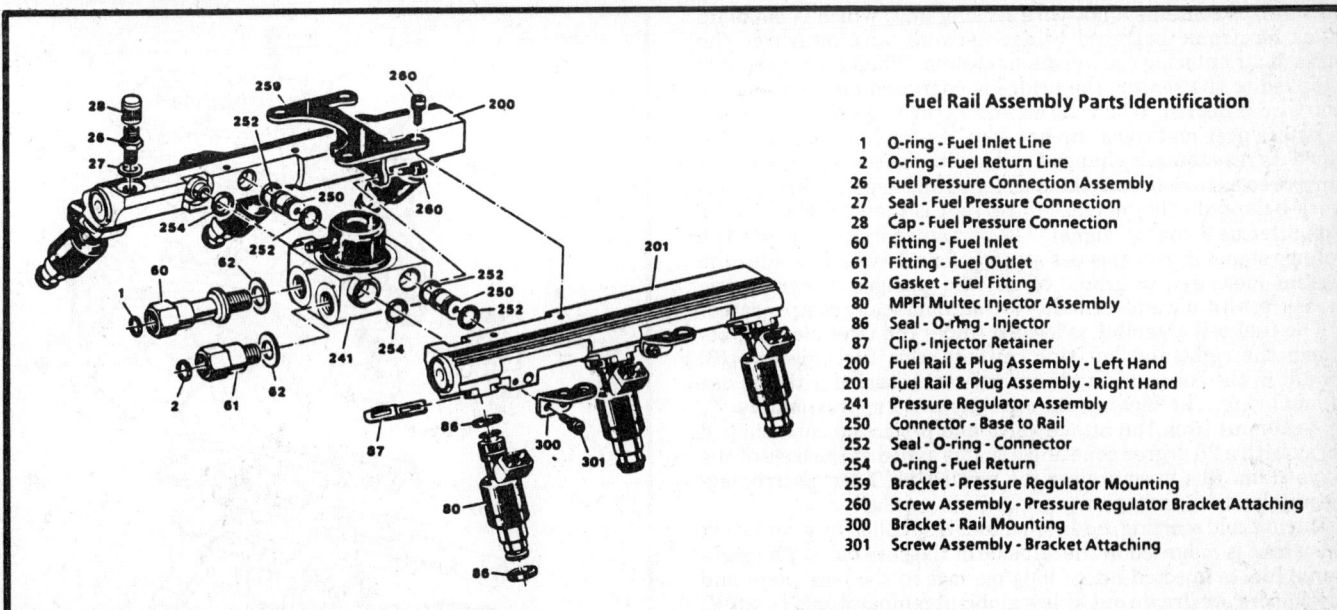

Fuel Rail Assembly Parts Identification

1. O-ring - Fuel Inlet Line
2. O-ring - Fuel Return Line
26. Fuel Pressure Connection Assembly
27. Seal - Fuel Pressure Connection
28. Cap - Fuel Pressure Connection
60. Fitting - Fuel Inlet
61. Fitting - Fuel Outlet
62. Gasket - Fuel Fitting
80. MPFI Multec Injector Assembly
86. Seal - O-ring - Injector
87. Clip - Injector Retainer
200. Fuel Rail & Plug Assembly - Left Hand
201. Fuel Rail & Plug Assembly - Right Hand
241. Pressure Regulator Assembly
250. Connector - Base to Rail
252. Seal - O-ring - Connector
254. O-ring - Fuel Return
259. Bracket - Pressure Regulator Mounting
260. Screw Assembly - Pressure Regulator Bracket Attaching
300. Bracket - Rail Mounting
301. Screw Assembly - Bracket Attaching

F6C model fuel rail assembly

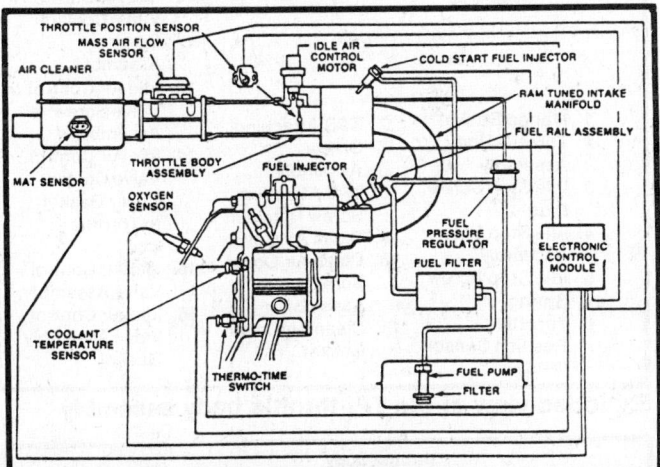

Typical 2.8L fuel system diagram

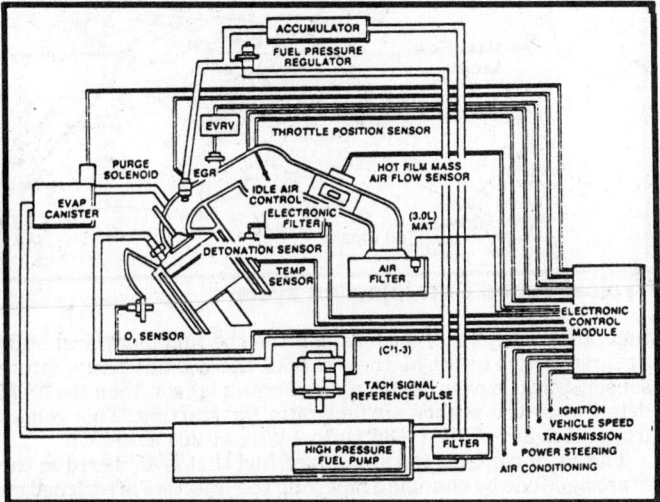

Typical 3.0L fuel system diagram

5.0L and 5.7L V8 engines, and the Corvette with the 5.7L V8 engine. The introduction of this TPI system to these engines has improved the torque and power from both engines. The induction system for the TPI is made up of large forward mounted air cleaners, a new mass airflow sensor, a cast aluminum throttle body assembly with dual throttle blades, a large extended cast aluminum plenum, individual aluminum tuned runners and a protruding dual fuel rail assembly with computer controlled injectors. The base plate is cast aluminum and incorporated the crossover portion of the tuned runners. The base plate also serves as a mounting for the fuel injectors. The individual aluminum runners are designed to provide the best tuning or frequency of air pulses within the runners and for the optimum throttle response throughout the driving range, thus the name Tuned Port Injection. The runners are selected by length and size so as to take advantage of the air pulses set up by the opening and closing of the intake valves. The high pressure pulses result in denser air at each intake valve, and timing the pressure pulses to occur during the valve open period forces more air into the combustion chamber, which results in a more efficient cylinder charging and improved volumetric efficiency.

The 8 fuel injectors fire at the same time, once each crankshaft revolution. During the first injection, fuel is sprayed at the base of the closed intake valve, during the second injection, fuel is sprayed into the air stream entering the combustion chamber. The fuel from the first injection vaporizes from the heat of the intake valve, and the fuel vapors are drawn into the combustion chamber along with the air when the valve opens to charge the cylinder. The regulated pressure of the fuel being injected in the 5.0L V8 engine is 44 psi, and 37 psi in the 5.7L V8 engine. This fuel pressure is regulated constantly, or, as the manifold vacuum changes, the regulator adjusts the fuel pressure to maintain a constant drop in pressure across the injectors.

When the signals are received by the computer from the mass airflow sensor and the engine coolant temperature sensor, the computer will search its pre-programmed information to determine the pulse width of the fuel injectors required to match the input signals. The computer now, based on the engine rpm, signals the injectors to release the required amount of fuel. The computer makes mass air flow sensor readings and fuel requirement calculations every 12.5 milliseconds.

The mass air flow sensor and the individual fuel injectors are made by Bosch and supplied by General Motors. The mass air flow sensor contains a hot-wire sensing unit, which is make up of an electronic balanced bridge network, and measures the mass of air entering the injections system. When ever current is supplied to the sensor, the bridge is energized and the sensing hot-wire is heated. When the air enters the mass air flow sensor is passes over and cools the hot wire. As the hot wire is cooled down its resistance is changed and additional current is then required, so as to maintain the resistance and keep the bridge network balanced. The increase in current is then supplied to the computer as a voltage signal. The computer inturn coverts the voltage signal into grams per second of air flow and because the system measures in grams of air, the changes in barometric pressure, altitude and humidity are automatically compensated.

The fuel rail assembly is located under the inlet plenum, between the right and left side runners. The fuel injectors (8) mount in the base plate and each injector is sealed with the use of an O-ring. The fuel injectors are mounted approximately $7/16$ in. (110mm) from the intake valve and projects a cone-shaped spray with a 20 degree cone angle that is aimed at the base of the valve stem. The injectors are the pintle type, have electromagnetic solenoids, and operate on ignition voltage.

During cold starting, additional fuel is supplied by a cold start valve that is mounted in the left side of the base plate. The additional fuel is injected into a long passage in the base plate and fuel vapors are drawn out at low ambient temperatures ($-20°F$) through orifices to each intake port. During a cold starting situation, the engine requires an extremely rich air-fuel ratio to provide enough fuel vaporization for combustion. Because of the small amount of air drawn into the combustion chamber during cold starting cranking, the cold start valve supplies fuel into the base plate passage where only the vapors are drawn into the combustion chamber. This prevents flooding or fuel fouling the spark plugs that would occur if the fuel needed for a cold start situation was supplied only from the main injectors. This cold start valve is controlled by a type of coolant temperature sensor called the thermo-time switch, and also by the starter cranking circuit. When the ignition key is turned to crank the engine, the thermal-time switch will heat up until it reaches approximately 95°F at which time the switch will open and de-energize the cold start injector. The maximum time for the cold start injector to be energized is 12 seconds, with an engine temperature of $-4°F$ or below. When the engine starts and the ignition key is released, the circuit is also de-energized. Except for the new mass air flow system, the new tuned runners and longer plenum assembly the rest of this TPI system is very similar to the multiport fuel injection system.

FUEL CONTROL SYSTEM

The basic function of the fuel control system is to control the fuel delivery to the engine. The fuel is delivered to the engine by individual fuel injectors mounted on the intake manifold near each cylinder.

The main control sensor is the oxygen sensor which is located in the exhaust manifold. The oxygen sensor tells the ECM how much oxygen is in the exhaust gas and the ECM changes the air/fuel ratio to the engine by controlling the fuel injectors. The best mixture (ratio) to minimize exhaust emissions is 14.7:1 which allows the catalytic converter to operate the most efficiently. Because of the constant measuring and adjusting of the air/fuel ratio, the fuel injection system is called a **CLOSED LOOP** system.

Modes Of Operation

The ECM looks at voltage from several sensors to determine how much fuel to give the engine. The fuel is delivered under one of several conditions, called **MODES**.

STARTING MODE

When the engine is first turned **ON**, the ECM will turn on the

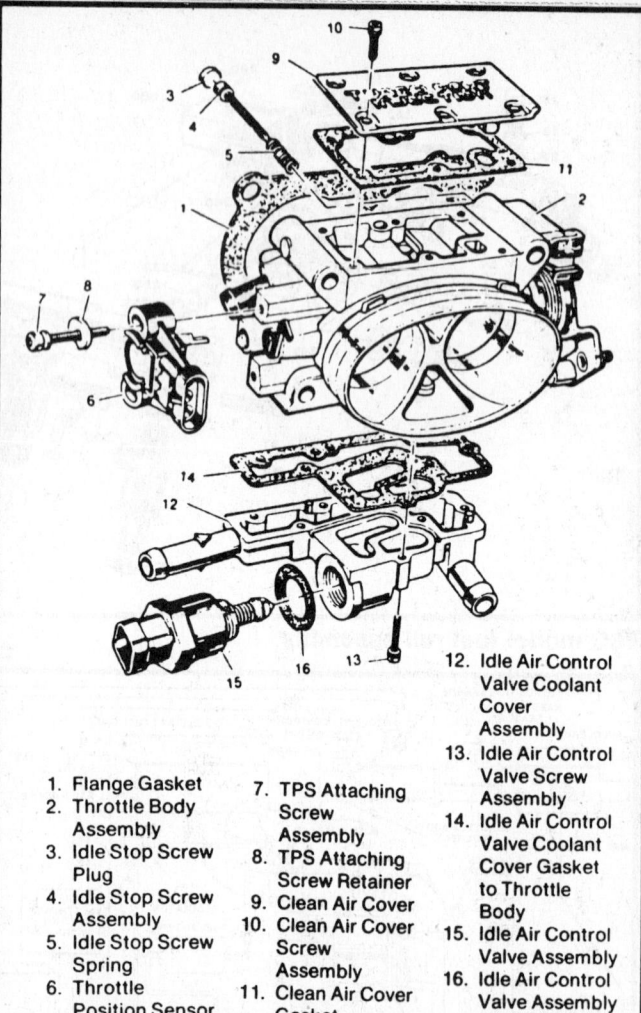

1. Flange Gasket
2. Throttle Body Assembly
3. Idle Stop Screw Plug
4. Idle Stop Screw Assembly
5. Idle Stop Screw Spring
6. Throttle Position Sensor (TPS)
7. TPS Attaching Screw Assembly
8. TPS Attaching Screw Retainer
9. Clean Air Cover
10. Clean Air Cover Screw Assembly
11. Clean Air Cover Gasket
12. Idle Air Control Valve Coolant Cover Assembly
13. Idle Air Control Valve Screw Assembly
14. Idle Air Control Valve Coolant Cover Gasket to Throttle Body
15. Idle Air Control Valve Assembly
16. Idle Air Control Valve Assembly Gasket

Exploded view of the TPI throttle body assembly

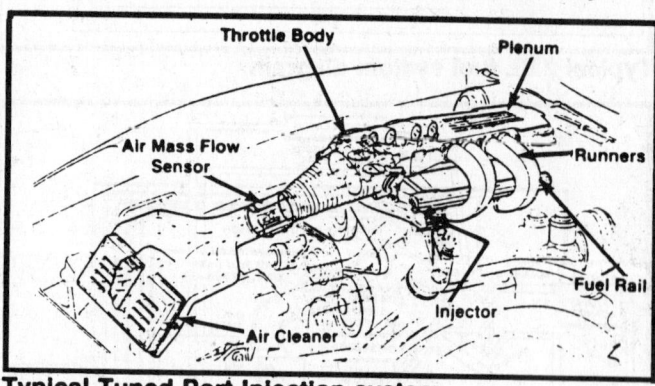

Typical Tuned Port Injection system

fuel pump relay for 2 seconds and the the fuel pump will build up pressure. The ECM then checks the coolant temperature sensor, throttle position sensor and crank sensor, then the ECM determines the proper air/fuel ratio for starting. This ranges from 1.5:1 at $-33°F$ ($-36°C$) to 14.7:1 at 201°F (94°C).

The ECM controls the amount of fuel that is delivered in the Starting Mode by changing how long the injectors are turned on and off. This is done by pulsing the injectors for very short times.

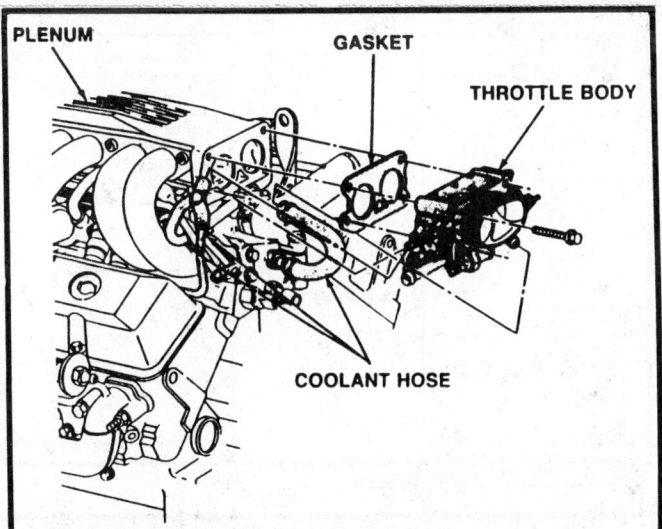

Typical 5.0L and 5.7L TPI throttle body assembly

CLEAR FLOOD MODE

If for some reason the engine should become flooded, provisions have been made to clear this condition. To clear the flood, the driver must depress the accelerator pedal enough to open to wide-open throttle position. The ECM then issues completely turns off the fuel flow. The ECM holds this operational mode as long as the throttle stays in the wide open position and the engine rpm is below 600. If the throttle position becomes less than 62% (2.9mv), the ECM returns to the starting mode. For throttle openings up to 62% (2.9mv) the ECM increases the fuel flow (based on TPS).

RUN MODE

There are 2 different run modes. When the engine is first started and the rpm is above 400, the system goes into open loop operation. In open loop operation, the ECM will ignore the signal from the oxygen (O_2) sensor and calculate the injector on-time based upon inputs from the coolant sensor, MAF sensors and MAT sensors.

During open loop operation, the ECM analyzes the following items to determine when the system is ready to go to the closed loop mode.

1. The oxygen sensor varying voltage output. (This is dependent on temperature).
2. The coolant sensor must be above specified temperature.
3. A specific amount of time must elapse after starting the engine. These values are stored in the PROM.

When these conditions have been met, the system goes into closed loop operation In closed loop operation, the ECM will calculate the air/fuel ratio (injector on time) based upon the signal from the oxygen sensor. The ECM will decrease the on-time if the air/fuel ratio is too rich, and will increase the on-time if the air/fuel ratio is too lean.

ACCELERATION MODE

When the engine is required to accelerate, the opening of the throttle valve(s) causes a rapid increase in Manifold Absolute Pressure (MAP). This rapid increase in MAP causes fuel to condense on the manifold walls. The ECM senses this increase in throttle angle and MAP, and supplies additional fuel for a short period of time. This prevents the engine from stumbling due to too lean a mixture.

DECELERATION MODE

Upon deceleration, a leaner fuel mixture is required to reduce emission of hydrocarbons (HC) and carbon monoxide (CO). To adjust the injection on-time, the ECM uses the decrease in MAP and the decrease in throttle position to calculate a decrease in injector on time. To maintain an idle fuel ratio of 14.7:1, fuel output is momentarily reduced. This is done because of the fuel remaining in the intake manifold. The ECM can cut off the fuel completely for short periods of time.

BATTERY VOLTAGE CORRECTION MODE

The purpose of battery voltage correction is to compensate for variations in battery voltage to fuel pump and injector response. The ECM compensates by increasing the engine idle rpm.

Battery voltage correction takes place in all operating modes. When battery voltage is low, the spark delivered by the distributor may be low. To correct this low battery voltage problem, the ECM can do any or all of the following:

a. Increase injector on time (increase fuel)
b. Increase idle rpm
c. Increase ignition dwell time

FUEL CUT-OFF MODE

When the ignition is **OFF**, no fuel will be delivered by the injectors. Fuel will also be cut off if the ECM does not receive a reference pulse from the distributor. To prevent dieseling, fuel delivery is completely stopped as soon as the engine is stopped. The ECM will not allow any fuel supply until it receives distributor reference pulses which prevents flooding.

CONVERTER PROTECTION MODE

In this mode the ECM estimates the temperature of the catalytic converter and then modifies fuel delivery to protect the converter from high temperatures. When the ECM has determined that the converter may overheat, it will cause open loop operation and will enrichen the fuel delivery. A slightly richer mixture will then cause the converter temperature to be reduced.

FUEL BACKUP MODE

The ECM functions in the fuel backup circuit mode if any one, or any combination, of the following exist:

1. The ECM voltage is lower than 9 volts.
2. The cranking voltage is below 9 volts.
3. The PROM is missing or not functioning.
4. The ECM circuit fails to insure the computer operating pulse. The computer operating pulse (COP) is an internal ECM feature designed to inform the fuel backup circuit that the ECM is able to function.

Some engines run erratically in the fuel backup mode, while others (5.0L and 5.7L) seemed to run very well. Code 52 will be set to indicate a missing CalPak. The fuel backup circuit is ignition fed and senses Throttle Position Sensor (TPS), Coolant Temperature Sensor (CTS) and rpm. The fuel backup circuit controls the fuel pump relay and the pulse width of the injectors.

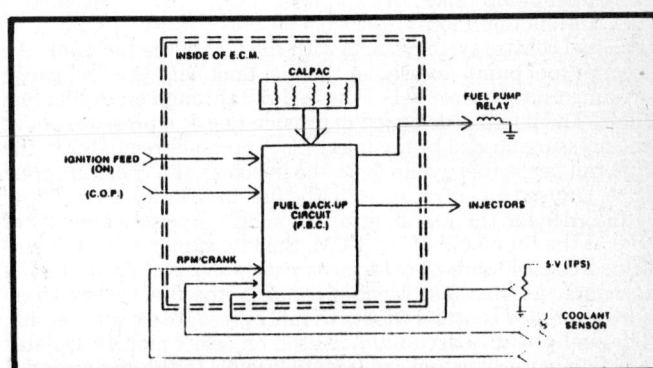

Fuel backup circuit diagram

SYNCHRONIZED MODE

In synchronized mode operation, the injector is pulsed once for each distributor reference pulse.

NONSYNCHRONIZED MODE

In nonsynchronized mode operation, the injector is pulsed once every 12.5 milliseconds or 6.25 milliseconds depending on calibration. This pulse time is totally independent of distributor reference pulses.

Nonsynchronized mode results only under the following conditions:
1. The fuel pulse width is too small to be delivered accurately by the injector (approximately 1.5 milliseconds)
2. During the delivery of prime pulses (prime pulses charge the intake manifold with fuel during or just prior to engine starting)
3. During acceleration enrichment
4. During deceleration leanout

ECM PARAMETERS

The ECM has parameters that it controls and parameters that it senses. The parameters that the ECM senses help control the other parameters.

ALCL CONNECTOR

The Assembly Line Communication Link (ALCL) is a diagnostic connector located in the passenger compartment. It has terminals which are used in the assembly plant to check that the engine is operating properly before it leaves the plant. This connector is a very useful tool in diagnosing EFI engines. Important information from the ECM is available at this terminal and can be read with one of the many popular scanner tools.

NOTE: Some models refer to the ALCL as the Assembly Line Diagnostic Link (ALDL). Either way it is referred to, they both still perform the same function.

MPI Fuel Control Systems Components

The fuel control system is made up of the following components:
1. Fuel supply system
2. Throttle body assembly
3. Fuel injectors
4. Fuel rail
5. Fuel pressure regulator
6. Idle Air Control (IAC)
7. Fuel pump
8. Fuel pump relay
9. In-line fuel filter

The fuel control system starts with the fuel in the fuel tank. An electric fuel pump, located in the fuel tank with the fuel gauge sending unit, pumps fuel to the fuel rail through an in-line fuel filter. The pump is designed to provide fuel at a pressure above the pressure needed by the injectors. A pressure regulator in the fuel rail keeps fuel available to the injectors at a constant pressure. Unused fuel is returned to the fuel tank by a separate line.

In order for the fuel injectors to supply a precise amount of fuel at the command of the ECM, the fuel supply system maintains a constant pressure (approximately 35 psi) drop across the injectors. As manifold vacuum changes, the fuel system pressure regulator controls the fuel supply pressure to compensate. The fuel pressure accumulator used on select models, isolates fuel line noise. The fuel rail is bolted rigidly to the engine and it provides the upper mount for the fuel injectors. It also contains

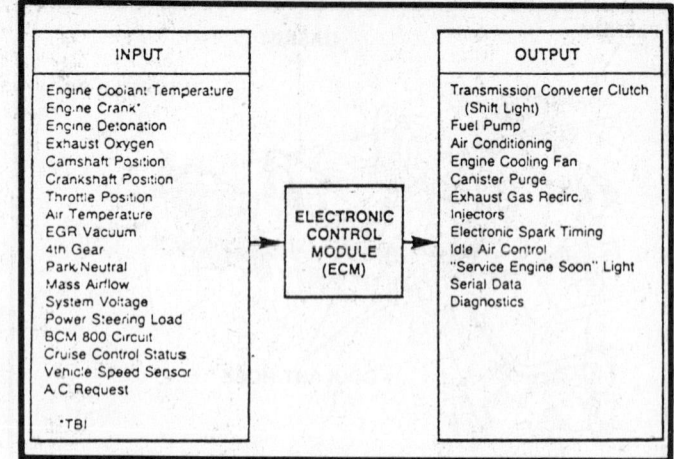

Typical ECM parameters

INPUT	OUTPUT
Engine Coolant Temperature Engine Crank* Engine Detonation Exhaust Oxygen Camshaft Position Crankshaft Position Throttle Position Air Temperature EGR Vacuum 4th Gear Park Neutral Mass Airflow System Voltage Power Steering Load BCM 800 Circuit Cruise Control Status Vehicle Speed Sensor A C Request *TBI	Transmission Converter Clutch (Shift Light) Fuel Pump Air Conditioning Engine Cooling Fan Canister Purge Exhaust Gas Recirc. Injectors Electronic Spark Timing Idle Air Control "Service Engine Soon" Light Serial Data Diagnostics

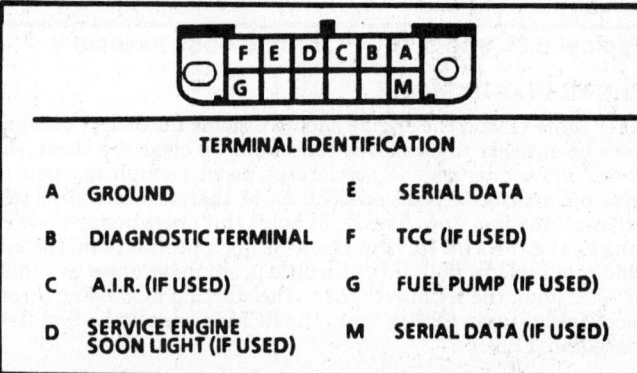

Typical ALCL connector terminal identification

TERMINAL IDENTIFICATION

A	GROUND	E	SERIAL DATA
B	DIAGNOSTIC TERMINAL	F	TCC (IF USED)
C	A.I.R. (IF USED)	G	FUEL PUMP (IF USED)
D	SERVICE ENGINE SOON LIGHT (IF USED)	M	SERIAL DATA (IF USED)

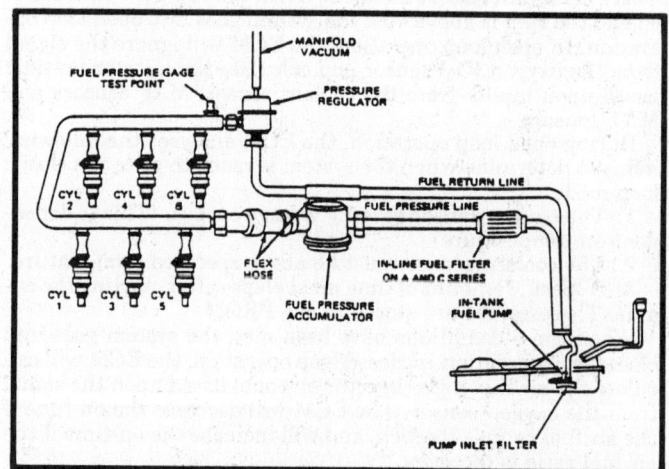

Typical fuel system layout

a spring loaded pressure tap for testing the fuel system or relieving the fuel system pressure.

The injectors are controlled by the ECM. They deliver fuel in one of several modes as previously described. In order to properly control the fuel supply, the fuel pump is operated by the ECM through the fuel pump relay and oil pressure switch.

THROTTLE BODY UNIT

The throttle body unit has a throttle valve to control the amount of air delivered to the engine. The TPS and IAC valve

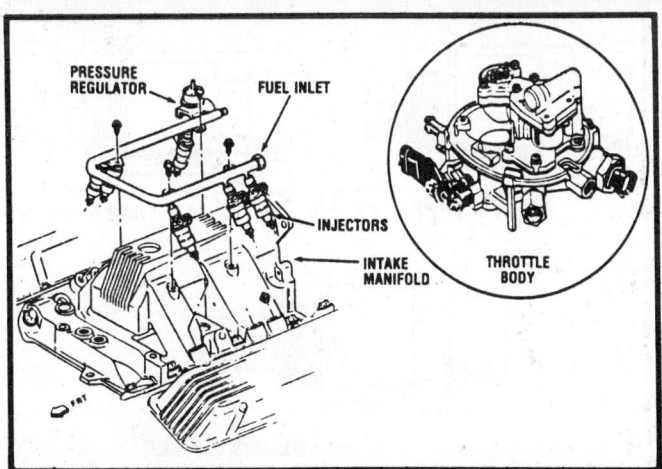

Typical fuel rail and throttle body assemblies

are also mounted onto the throttle body. The throttle body contains vacuum ports located at, above or below the throttle valve. These vacuum ports generate the vacuum signals needed by various components.

On some models, the engine coolant is directed through the coolant cavity at the bottom of the throttle body to warm the throttle valve and prevent icing.

FUEL INJECTOR

A fuel injector is installed in the intake manifold at each cylinder. Mounting is approximately 70–100mm from the center line of the intake valve on V6 and V8 engine applications. The nozzle spray pattern is on a 25 degree angle. The fuel injector is a solenoid operated device controlled by the ECM. The ECM turns on the solenoid, which opens the valve which allows fuel delivery. The fuel, under pressure, is injected in a conical spray pattern at the opening of the intake valve. The fuel, which is not used by the injectors, passes through the pressure regulator before returning to the fuel tank.

An injector that is partly open, will cause loss of fuel pressure after the engine is shut down, so long crank time would be noticed on some engines. Also dieseling could occur because some fuel could be delivered after the ignition is turned to **OFF** position.

There are 2 O-ring seals used. The lower O-ring seals the injector at the intake manifold. The O-rings are lubricated and should be replaced whenever the injector is removed from the intake manifold. The O-rings provide thermal insulation, thus preventing the formation of vapor bubbles and promoting good hot start characteristics. The O-rings also prevent excess injector vibration.

Air leakage at the injector/intake area would create a lean cylinder and a possible driveability problem. A second seal is used to seal the fuel injector at the fuel rail connection. The injectors are identified with an ID number cast on the injector near the top side. Injectors manufactured by Rochester® Products have an **RP** positioned near the top side in addition to the ID number.

NOTE: The most widely used injector for the MPI units are Bosch injectors, but now there will also be a new injector being used. This new injector will be a Multec MPI injector (a Rochester® product). It is classified as a top feed design because the fuel enters the top of the injector and then flows through the entire length of the injector. It is designed to operate with the system fuel pressures ranging from 36–51 psi (250–350 kPa), and uses a high impedance (12.2 ohms) solenoid coil. It is most widely used in the multi-port fuel injection system on some Chevrolet produced 2.8L V6 engines.

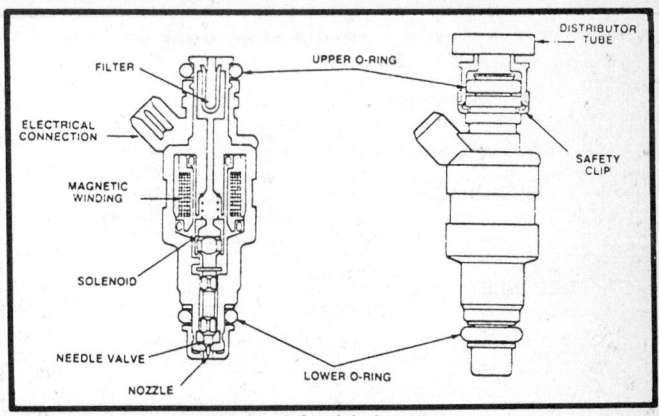

Inside view of a typical fuel injector

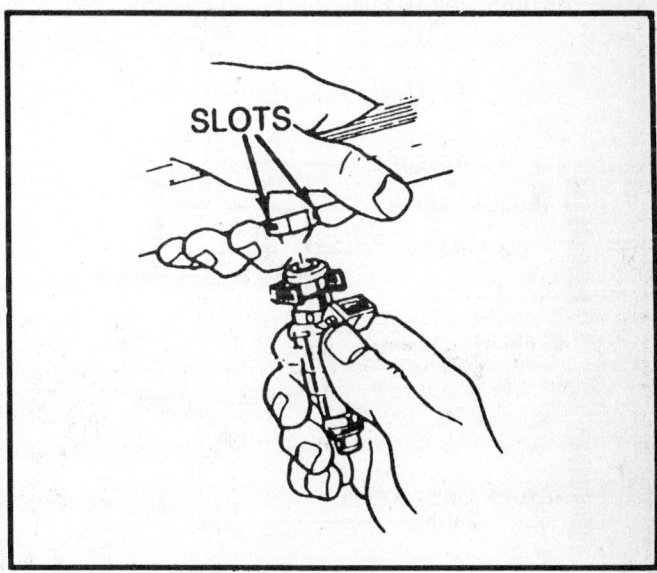

Typical fuel injector installation

FUEL RAIL

The fuel rail is bolted rigidly to the engine and it provides the upper mount for the fuel injectors. It distributes fuel to the individual injectors. Fuel is delivered to the input end of the fuel rail by the fuel lines, goes through the rail, then to the fuel pressure regulator. The regulator keeps the fuel pressure to the injectors at a constant pressure. The remaining fuel is then returned to the fuel tank. The fuel rail also contains a spring loaded pressure tap for testing the fuel system or relieving the fuel system pressure.

PRESSURE REGULATOR

The fuel pressure regulator contains a pressure chamber separated by a diaphragm relief valve assembly with a calibrated spring in the vacuum chamber side. The fuel pressure is regulated when the pump pressure acting on the bottom spring of the diaphragm overcomes the force of the spring action on the top side.

The diaphragm relief valve moves, opening or closing an orifice in the fuel chamber to control the amount of fuel returned to the fuel tank. Vacuum acting on the top side of the diaphragm along with spring pressure controls the fuel pressure. A decrease in vacuum creates an increase in the fuel pressure. An increase in vacuum creates a decrease in fuel pressure.

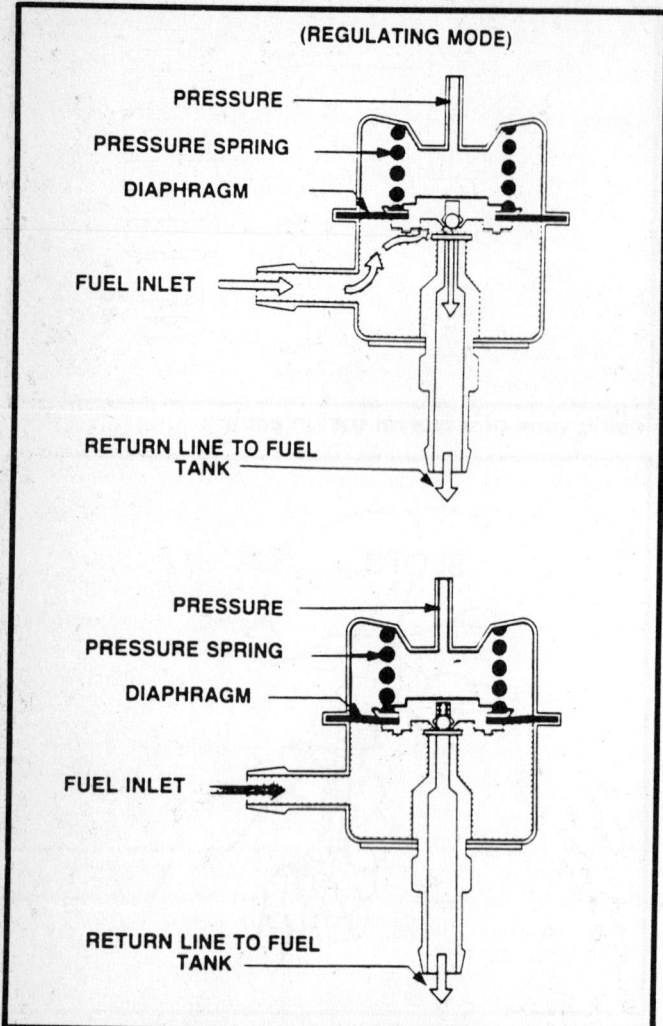

Inside view of the fuel pressure regulator

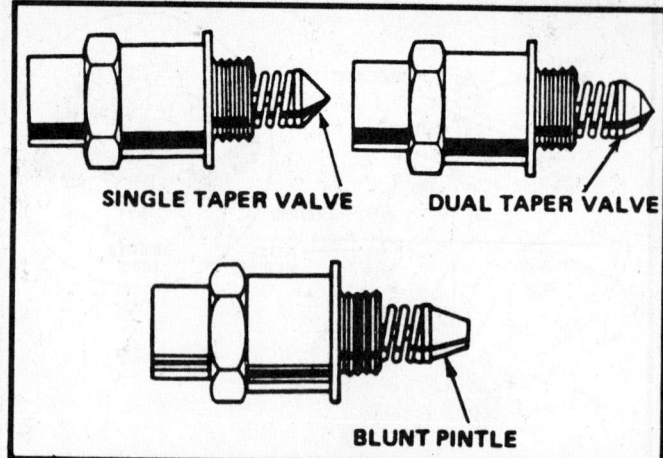

The 3 different designs of the idle air control valve

The first design used is single taper while the second design used is a dual taper. The third design is a blunt valve. Care should be taken to insure use of the correct design when service replacement is required.

The IAC motor has 255 different positions or steps. The zero, or reference position, is the fully extended position at which the pintle is seated in the air bypass seat and no air is allowed to bypass the throttle plate. When the motor is fully retracted, maximum air is allowed to bypass the throttle plate. When the motor is fully retracted, maximum air is allowed to bypass the throttle plate.

The ECM always monitors how many steps it has extended or retracted the pintle from the zero or reference position; thus, it always calculates the exact position of the motor. Once the engine has started and the vehicle has reached approximately 40 mph, the ECM will extend the motor 255 steps from whatever position it is in. This will bottom out the pintle against the seat. The ECM will call this position **0** and thus keep its zero reference updated.

The IAC only affects the engine's idle characteristics. If it is stuck fully open, idle speed is too high (too much air enters the throttle bore) If it is stuck closed, idle speed is too low (not enough air entering). If it is stuck somewhere in the middle, idle may be rough, and the engine won't respond to load changes.

FUEL PUMP

The fuel is supplied to the system from an in-tank positive displacement roller vane pump. The pump supplies fuel through the in-line fuel filter to the fuel rail assembly. The pump is removed for service along with the fuel gauge sending unit. Once they are removed from the fuel tank, they pump and sending unit can be serviced separately.

Fuel pressure is achieved by rotation of the armature driving the roller vane components. The impeller at the inlet end serves as a vapor separator and a precharger for the roller vane assembly. The unit operates at approximately 3500 rpm.

The pressure relief valve in the fuel pump will control the fuel pump maximum psi to 60–90 psi. The fuel pump delivers more fuel than the engine can consume even under the most extreme conditions. Excess fuel flows through the pressure regulator and back to the tank via the return line. The constant flow of fuel means that the fuel system is always supplied with cool fuel, thereby preventing the formation of fuel-vapor bubbles (vapor lock).

FUEL PUMP RELAY CIRCUIT

The fuel pump relay is usually located on the left front inner

An example of this is under heavy load conditions the engine requires more fuel flow. The vacuum decreases under a heavy load condition because of the throttle opening. A decrease in the vacuum allows more fuel pressure to the top side of the pressure relief valve, thus increasing the fuel pressure.

The pressure regulator is mounted on the fuel rail and serviced separately. If the pressure is too low, poor performance could result. If the pressure is too high, excessive odor and a Code 45 may result.

IDLE AIR CONTROL (IAC)

The purpose of the Idle Air Control (IAC) system is to control engine idle speeds while preventing stalls due to changes in engine load. The IAC assembly, mounted on the throttle body, controls bypass air around the throttle plate. By extending or retracting a conical valve, a controlled amount of air can move around the throttle plate. If rpm is too low, more air is diverted around the throttle plate to increase rpm.

During idle, the proper position of the IAC valve is calculated by the ECM based on battery voltage, coolant temperature, engine load, and engine rpm. If the rpm drops below a specified rate, the throttle plate is closed. The ECM will then calculate a new valve position.

Three different designs are used for the IAC conical valve.

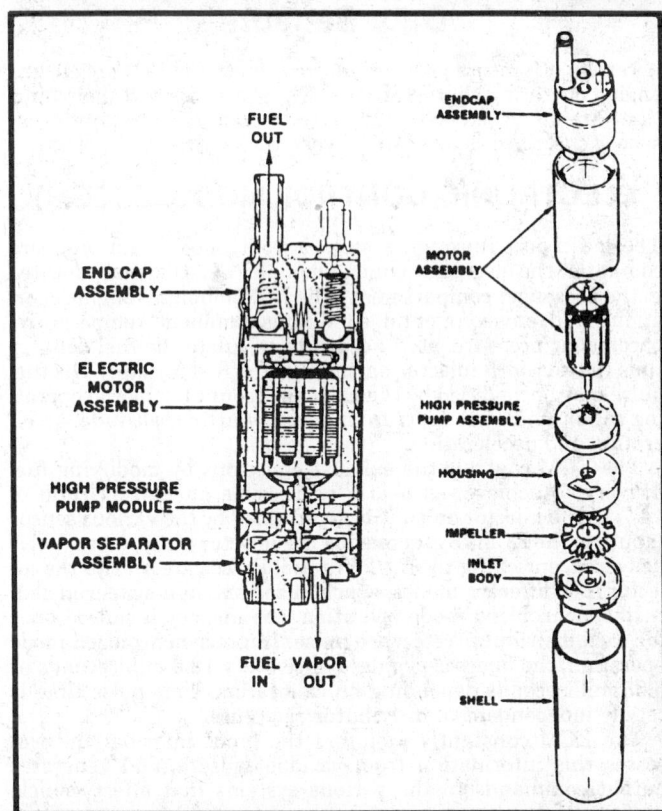

Exploded view of a typical roller vane fuel pump assembly

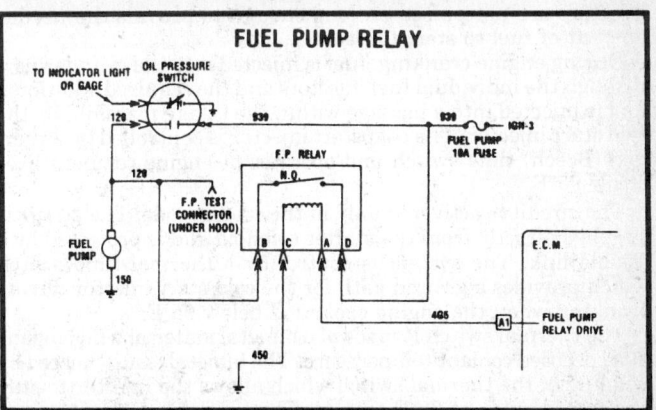

Fuel pump relay schematic in the OFF position

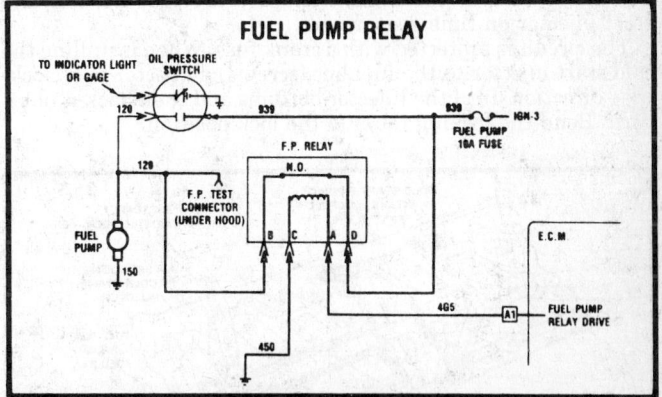

Fuel pump relay schematic in the RUN position

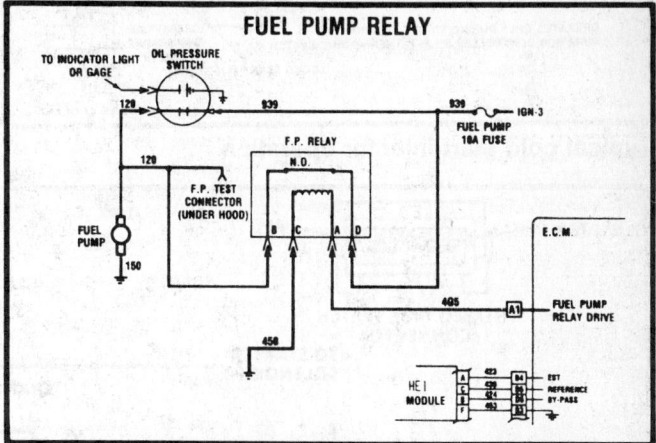

Fuel pump relay schematic in the RUN operation

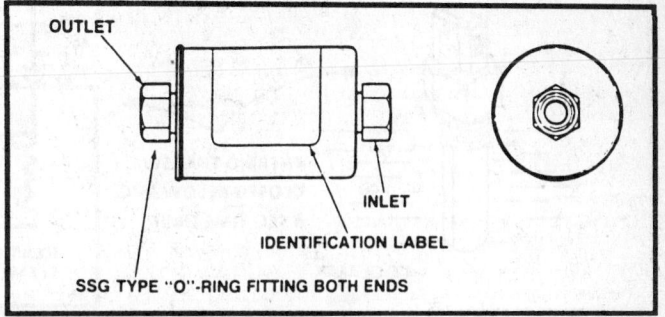

Typical in-line fuel filter

fender (or shock tower) or on the engine side of the firewall (center cowl). The fuel pump electrical system consists of the fuel pump relay, ignition circuit and the ECM circuits are protected by a fuse. The fuel pump relay contact switch is in the normally open (NO) position.

When the ignition is turned **ON** the ECM will for 2 seconds, supply voltage to the fuel pump relay coil, closing the open contact switch. The ignition circuit fuse can now supply ignition voltage to the circuit which feeds the relay contact switch. With the relay contacts closed, ignition voltage is supplied to the fuel pump. The ECM will continue to supply voltage to the relay coil circuit as long as the ECM receives the rpm reference pulses from the ignition module.

The fuel pump control circuit also includes an engine oil pressure switch with a set of normally open contacts. The switch closes at approximately 4 lbs. of oil pressure and provides a secondary battery feed path to the fuel pump. If the relay fails, the pump will continue to run using the battery feed supplied by the closed oil pressure switch. A failed fuel pump relay will result in extended engine crank times in order to build up enough oil pressure to close the switch and turn on the fuel pump.

IN-LINE FUEL FILTER

The fuel filter is a 10–20 micron and is serviced only as a complete unit. The O-rings are used at all threaded connections to prevent fuel leakage. The threaded flex hoses connect the filter to the fuel tank feed line.

COLD START INJECTOR

The cold start injector is used to provide additional fuel during the crank mode to improve cold start-ups. This circuit is important when engine coolant temperature is low because the main injectors are not pulsed on long enough to provide the needed amount of fuel to start the engine.

During engine cranking, fuel is injected into the cylinder port through the individual fuel injectors and the required additional fuel is injected into a passage within the intake manifold by the cold start injector. The cold start injector is controlled by a thermal (Bosch) time switch and operates at engine temperatures below 95°F.

The circuit is activated only in the crank mode. The power is supplied directly from the starter solenoid and is protected by a fusible link. The system is control by a thermal time switch which provides a ground path for the cold start injector during cranking when the engine coolant is below 95°F.

The thermal switch is made of a bimetal material which opens at a specified coolant temperature. The bimetal is also heated by winding in the thermal switch which allows the injector to stay on for 8 seconds at 68°F coolant. The time the thermal switch will stay closed varies inversely with the coolant temperature. In other words, as the coolant temperature goes up, the cold start injector on-time goes down.

The circuit is protected with a crank fuse. When installing the cold start injector to the fuel line, screw the injector in a clockwise direction until the injector bottoms and then back it out 1 turn. Bend the locking tab into the lock position.

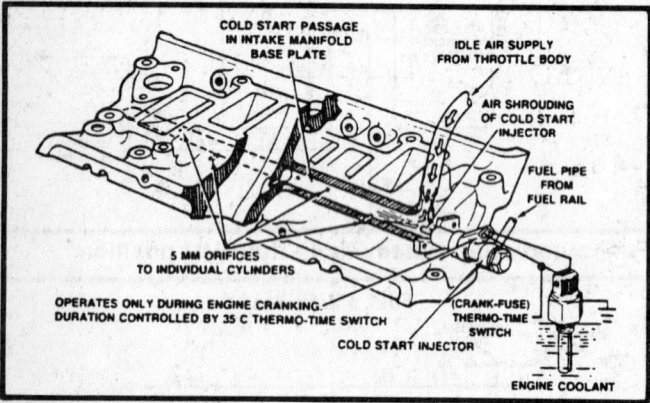

Typical cold start injector operation

Data Sensors

A variety of sensors provide information to the ECM regarding engine operating characteristics. These sensors and their functions are described below. Be sure to take note that not every sensor described is used with every GM engine application.

ELECTRONIC CONTROL MODULE (ECM)

The multi-port injection system is controlled by an on-board computer, the electronic control module (ECM), usually located in the passenger compartment. The ECM monitors engine operations and environmental conditions (ambient temperature, barometric pressure, etc.) needed to calculate the fuel delivery time (pulse width/injector on-time) of the fuel injector. The fuel pulse may be modified by the ECM to account for special operating conditions, such as cranking, cold starting, altitude, acceleration and deceleration.

The ECM controls the exhaust emissions by modifying fuel delivery to achieve, as nearly as possible an air/fuel ratio of 14.7:1. The injector on-time is determined by the various sensor inputs to the ECM. By increasing the injector pulse, more fuel is delivered, enriching the air/fuel ratio. Pulses are sent to the injector in 2 different modes, synchronized and non-synchronized.

In synchronized mode operation, the injector is pulsed once for each distributor reference pulse. In nonsynchronized mode operation, the injector is pulsed once every 12.5 milliseconds or 6.25 milliseconds depending on calibration. This pulse time is totally independent of distributor reference.

The ECM constantly monitors the input information, processes this information from various sensors, and generates output commands to the various systems that affect vehicle performance.

The ability of the ECM to recognize and adjust for vehicle variations (engine transmission, vehicle weight, axle ratio, etc.) is provided by a removable calibration unit (PROM) that is programmed to tailor the ECM for the particular vehicle. There is a specific ECM/PROM combination for each specific vehicle, and the combinations are not interchangeable with those of other vehicles.

The ECM also performs the diagnostic function of the system. It can recognize operational problems, alert the driver through

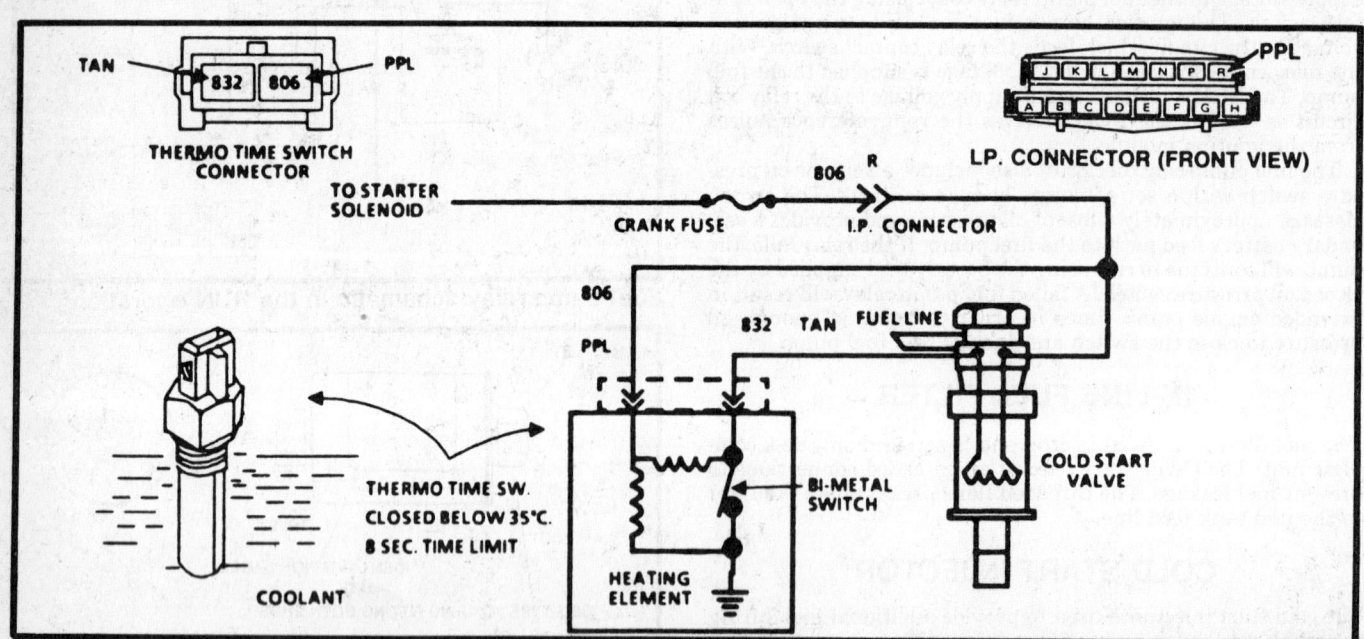

Typical cold start injector circuit

the "CHECK ENGINE" light, and store a code or codes which identify the problem areas to aid the technician in making repairs.

NOTE: Instead of an edgeboard connector, newer ECM's have a header connector which attaches solidly to the ECM case. Like the edgeboard connectors, the header connectors have a different pinout identification for different engine designs.

The 1988 2.8L engine will use a new type of ECM. For service, this ECM only consists of 2 parts a Controller (the ECM without a Mem-Cal) and an assembly called a Mem-Cal (this stands for Memory and Calibration unit).

The other MFI vehicles will use the standard ECM for service. The ECM consists of 3 parts; a Controller (the ECM without a PROM), a Calibrator called a PROM (Programmable Read Only Memory) and a CalPak.

PROM

To allow 1 model of the ECM to be used for many different vehicles, a device called a Calibrator (or PROM) is used. The PROM is located inside the ECM and has information on the vehicle's weight, engine, transmission, axle ratio and other components.

While one ECM part number can be used by many different vehicles, a PROM is very specific and must be used for the right vehicle. For this reason, it is very important to check the latest parts book and or service bulletin information for the correct PROM part number when replacing the PROM.

An ECM used for service (called a controller) comes without a PROM. The PROM from the old ECM must be carefully removed and installed in the new ECM.

CALPAK

A device called a CalPak is added to allow fuel delivery if other parts of the ECM are damaged. It has an access door in the ECM, and removal and replacement procedures are the same as with the PROM. If the CalPak is missing, a Code 52 will be set.

MEM-CAL

This assembly contains the functions of the PROM CalPak and the ESC module used on other GM applications. Like the PROM, it contains the calibrations needed for a specific vehicle as well as the back-up fuel control circuitry required if the rest of the ECM becomes damaged or faulty.

ECM Function

The ECM supplies either 5 or 12 volts to power various sensors and or switches. This is done through resistances in the ECM which are so high in value that a test light will not light when connected to the circuit. In some cases, even an ordinary shop voltmeter will not give an accurate reading because its resistance is too low. Therefore, a 10 megohm input impedance digital voltmeter is required to assure accurate voltage readings.

The ECM controls output circuits such as injectors, IAC, cooling fan relay, etc. by controlling the ground circuit through transistors or a device called a quad driver.

ELECTRONIC SPARK TIMING (EST)

Electronic spark timing (EST) is used on all engines equipped with HEI distributors and direct ignition systems. The EST distributor contains no vacuum or centrifugal advance and uses a 7-terminal distributor module. It also has 4 wires going to a 4-terminal connector in addition to the connectors normally found on HEI distributors. A reference pulse, indicating both engine rpm and crankshaft position, is sent to the ECM. The ECM determines the proper spark advance for the engine operating conditions and sends an **EST** pulse to the distributor.

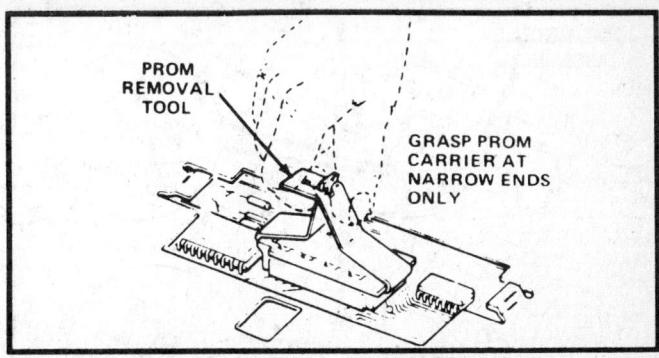

Removing the ECM PROM with PROM removal tool

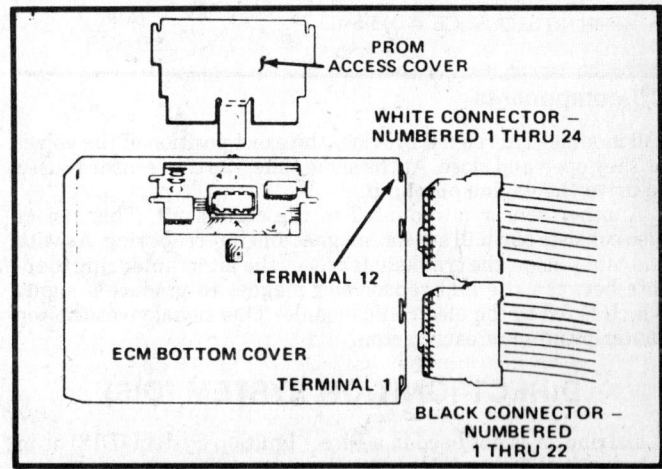

Typical General Motors ECM showing the PROM access cover and harness connectors

The EST system is designed to optimize spark timing for better control of exhaust emissions and for fuel economy improvements. The ECM monitors information from various engine sensors, computes the desired spark timing and changes the timing accordingly. A backup spark advance system is incorporated in the module in case of EST failure.

ELECTRONIC SPARK CONTROL (ESC)

When engines are equipped with ESC in conjunction with EST, ESC is used to reduce spark advance under conditions of detonation. A knock sensor signals a separate ESC controller to retard the timing when it senses knock. The ESC controller signals the ECM which reduces spark advance until no more signals are received from the knock sensor.

COMPUTER CONTROLLED COIL IGNITION SYSTEM (C³I)

The heart of this system is a electronic coil module that replaces the standard distributor and coil. Logic circuits within the module receive and buffer signals from the crankshaft and camshaft and by way of 3 interconnected coils contained in the cover of the module, distribute high voltage current to spark plugs.

The C³I system eliminates the need for a distributor to control the flow or current between the battery and spark plugs. In its place is an electro-magnetic sensor consisting of a hall effect sensor, magnet and interruptor ring. The gear on the shaft of this sensor is connected directly to the camshaft gear.

As the camshaft turns, the interruptor ring, moving at camshaft speed (½ the engine rpm), rotates between the hall sensor and the magnet to produce a signal which is fed to the electronic

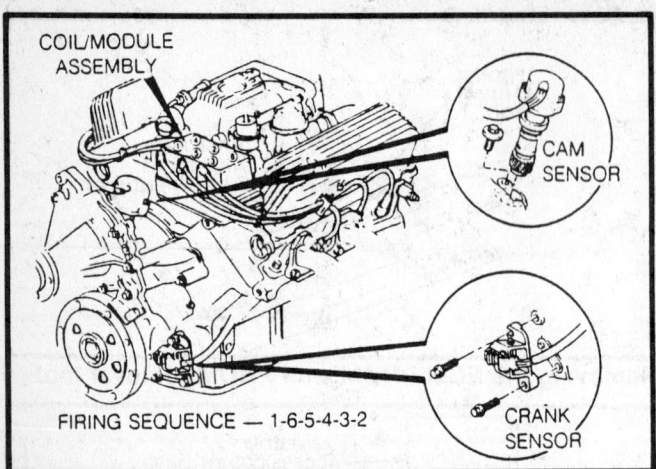

C³I components

coil module. This signal provides the exact position of the valves as they open and close. At the same time, the cam sensor is used to drive the engine oil pump.

Another sensor is mounted to the crankshaft. This sensor also consists of a hall sensor, magnet and interrupt ring. As with the cam sensor, the crankshaft causes the interruptor ring to rotate between the hall sensor and magnet to produce a signal which is fed to the electronic module. This signal gives the top center position of each piston.

DIRECT IGNITION SYSTEM (DIS)

A distributor is not used in a Direct Ignition System (DIS) or in a Integrated Direct Ignition (IDI) system. These 2 systems are basically the same electrically, and where a reference is made only to DIS, the content applies to both DIS and IDI. The components of the DIS are a coil pack, ignition module, crankshaft reluctor ring, magnetic sensor and the ECM. The coil pack consists of 2 or 3 separate, interchangeable, ignition coils. These coils operate in the same manner as previous coils. More than 1 coil is needed because each coil fires for 2 cylinders. The ignition module is located under the coil pack and is connected to the ECM by a 6-pin connector. The ignition module controls the primary circuits to the coils, turning them on and off and controls spark timing below 400 rpm and if the ECM bypass circuit becomes open or grounded.

The magnetic pickup sensor inserts through the engine block, just above the pan rail in proximity to the crankshaft reluctor ring. Notches in the crankshaft reluctor ring trigger the magnetic pickup sensor to provide timing information to the ECM. The magnetic pickup sensor provides a cam signal to identify correct firing sequence and crank signals to trigger each coil at the proper time.

This system uses EST and control wires from the ECM, as with the distributor systems. The ECM controls the timing using crankshaft position, engine rpm, engine temperature and manifold absolute pressure sensing.

When the engine is running at normal operating speeds, the bypass signal from the ECM to the ignition module will effectively connect the base of the transistor to the EST terminal. The EST signal is determined by not only the kind of reference signal the ECM receives, but also by all of the other engine sensors which at the same time are sending voltage signals to the ECM. Under these conditions, the ECM is controlling the timing and is constantly tuning the engine for control of exhaust emissions and good fuel economy.

Under certain conditions, such as during starting, the crankshaft, reference and bypass signals will be different because the engine rpm will be very low. The transistor base effectively will

be connected to the crankshaft input voltage. Thus, the crankshaft sensor will control the timing and duration of the primary current.

This means the engine can start and can run without the ECM if the ECM should not operate properly due to defects in the system. The engine will not run well and a warning light or other signal will tell the driver to obtain service attention at the earliest convenience. This back-up feature allows the operator to continue driving the vehicle until it can be serviced.

In summary, the bypass signal connects the transistor base to the EST signal for complete computer control by the ECM, or to crankshaft sensor for ignition control during cranking and other engine conditions, including some defects. The 2 ends of the coil secondary are connected internally to the 2 high voltage towers of each coil. These towers are then connected with high voltage wiring to 2 spark plugs.

On the 4 cylinder engines with a firing order of 1–3–4–2, one coil secondary would be connected to cylinders No. 1 and 4 and the other coil secondary would be connected to cylinders No. 2 and 3. When high voltage is induced in a secondary, both spark plugs will fire in series. This will occur at the end of the compression stroke on No. 1 cylinder and at the end of the exhaust stroke in the No. 4 cylinder.

Since the exhaust stroke cylinder is at atmosphere pressure, very little voltage is required to fire this plug, leaving most of the secondary voltage available to fire the plug in the cylinder under compression (the beginning of the power stroke). One engine revolution later the same ignition coil will fire No. 4 at the end of compression and No. 1 at the end of exhaust. The other ignition coil fires cylinders No. 2 and 3 in the same manner.

ENGINE COOLANT TEMPERATURE

The coolant sensor is a thermister (a resistor which changes value based on temperature) mounted on the engine coolant stream. As the temperature of the engine coolant changes, the resistance of the coolant sensor changes. Low coolant temperature produces a high resistance (100,000 ohms at −40°C/−40°F), while high temperature causes low resistance (70 ohms at 130°C/266°F).

The ECM supplies a 5 volt signal to the coolant sensor and measures the voltage that returns. By measuring the voltage change, the ECM determines the engine coolant temperature. The voltage will be high when the engine is cold and low when the engine is hot. This information is used to control fuel management, IAC, spark timing, EGR, canister purge and other engine operating conditions.

A failure in the coolant sensor circuit should either set a Code 14 or 15. These codes indicate a failure in the coolant tempera-

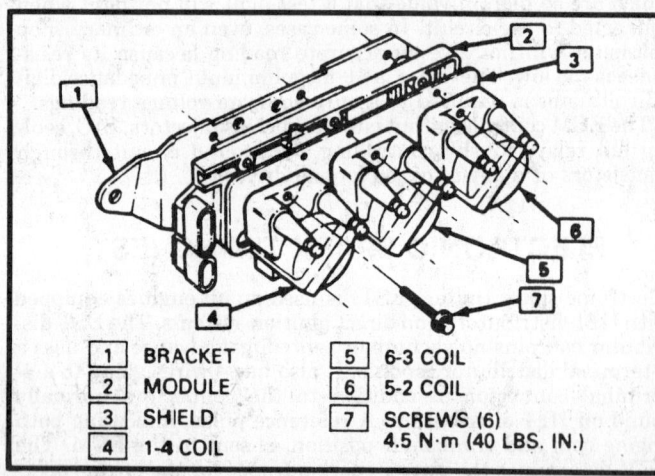

1	BRACKET	5	6-3 COIL
2	MODULE	6	5-2 COIL
3	SHIELD	7	SCREWS (6)
4	1-4 COIL		4.5 N·m (40 LBS. IN.)

Exploded view of a typical Direct Ignition assembly

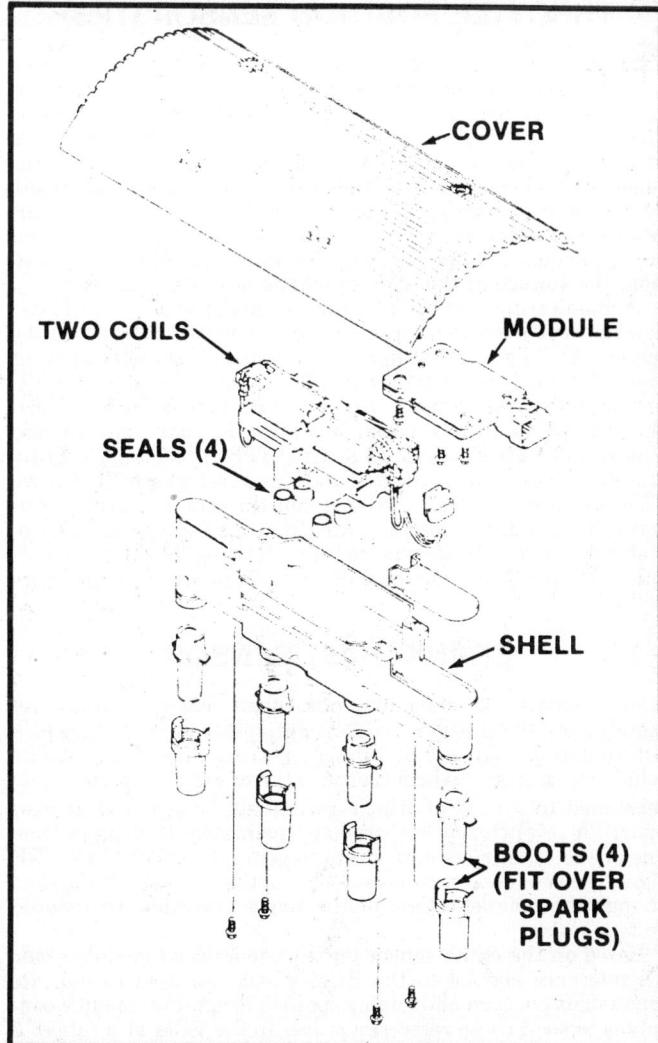

Exploded view of a typical Integrated Direct Ignition assembly

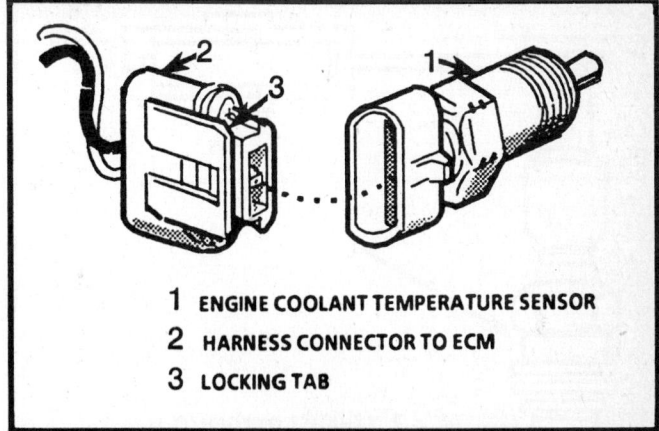

1 ENGINE COOLANT TEMPERATURE SENSOR
2 HARNESS CONNECTOR TO ECM
3 LOCKING TAB

Typical engine coolant temperature sensor

ture sensor circuit. Once the trouble code is set, the ECM will use a default valve for engine coolant temperature.

OXYGEN SENSOR

The exhaust oxygen sensor is mounted in the exhaust system where it can monitor the oxygen content of the exhaust gas stream. The oxygen content in the exhaust reacts with the oxygen sensor to produce a voltage output. This voltage ranges from approximately 100 millivolts (high oxygen — lean mixture) to 900 millivolts (low oxygen — rich mixture).

By monitoring the voltage output of the oxygen sensor, the ECM will determine what fuel mixture command to give to the injector (lean mixture — low voltage — rich command, rich mixture — high voltage — lean command).

Remember that oxygen sensor indicates to the ECM what is happening in the exhaust. It does not cause things to happen. It is a type of gauge: high oxygen content = lean mixture; low oxygen content = rich mixture. The ECM adjust fuel to keep the system working.

The oxygen sensor, if open should set a Code 13. a constant low voltage in the sensor circuit should set a Code 44 while a constant high voltage in the circuit should set a Code 45. Codes 44 and 45 could also be set as a result of fuel system problems.

MANIFOLD ABSOLUTE PRESSURE (MAP) SENSOR

The Manifold Absolute Pressure (MAP) sensor measures the changes in the intake manifold pressure which result from engine load and speed changes. The pressure measured by the MAP sensor is the difference between barometric pressure (outside air) and manifold pressure (vacuum). A closed throttle engine coastdown would produce a relatively low MAP value (approximately 20–35 kPa), while wide-open throttle would produce a high value (100 kPa). This high value is produced when the pressure inside the manifold is the same as outside the manifold, and 100% of outside air (or 100 kPa) is being measured. This MAP output is the opposite of what you would measure on a vacuum gauge. The use of this sensor also allows the ECM to adjust automatically for different altitude.

The ECM sends a 5 volt reference signal to the MAP sensor. As the MAP changes, the electrical resistance of the sensor also changes. By monitoring the sensor output voltage the ECM can determine the manifold pressure. A higher pressure, lower vacuum (high voltage) requires more fuel, while a lower pressure, higher vacuum (low voltage) requires less fuel. The ECM uses the MAP sensor to control fuel delivery and ignition timing. A failure in the MAP sensor circuit should set a Code 33 or Code 34.

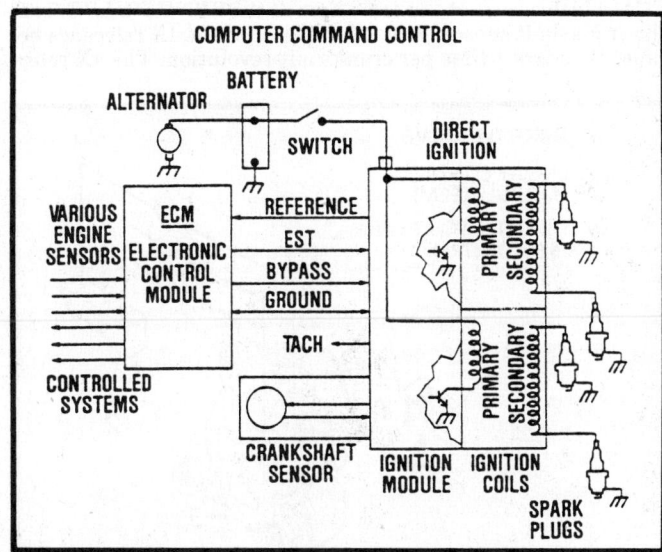

Direct Ignition System schematic — typical

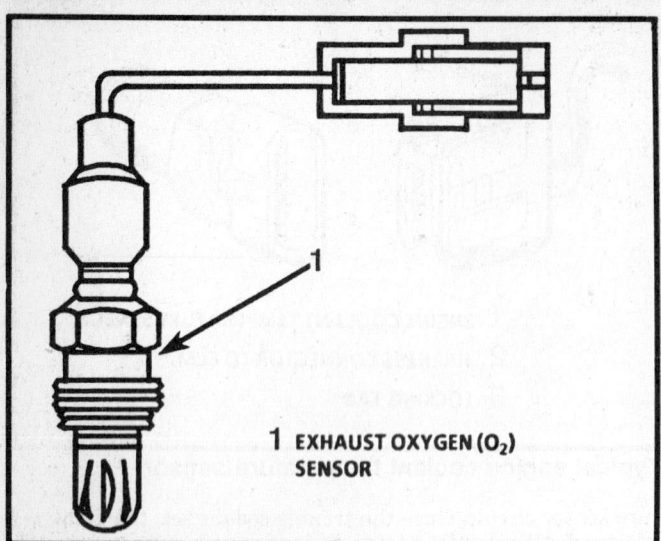

1 EXHAUST OXYGEN (O₂) SENSOR

Typical exhaust oxygen sensor

MANIFOLD AIR TEMPERATURE (MAT) SENSOR

The Manifold Air Temperature (MAT) sensor is a thermistor mounted in the intake manifold. A thermistor is a resistor which changes resistance based on temperature. Low manifold air temperature produces a high resistance (100,000 ohms at −40°F/−40°C), while high temperature cause low resistance (70 ohms at 266°F/130°C).

The ECM supplies a 5 volt signal to the MAT sensor through a resistor in the ECM and monitors the voltage. The voltage will be high when the manifold air is cold and low when the air is hot. By monitoring the voltage, the ECM calculates the air temperature and uses this data to help determine the fuel delivery and spark advance. A failure in the MAT circuit should set either a Code 23 or Code 25. Once the trouble code is set, the ECM will use an artificial default value for the MAT and some vehicle performance will return.

MASS AIR FLOW (MAF) SENSOR

The Mass Air Flow (MAF) sensor measures the amount of air which passes through it. The ECM uses this information to determine the operating condition of the engine, to control fuel delivery. A large quantity of air indicates acceleration, while a small quantity indicates deceleration or idle.

This sensor produces a frequency output between 32 and 150 hertz. A scan tool will display air flow in terms of grams of air per second (gm/sec), with a range from 3gm/sec to 150 gm/sec.

VEHICLE SPEED SENSOR (VSS)

NOTE: A vehicle equipped with a speed sensor, should not be driven without a the speed sensor connected, as idle quality may be affected.

The vehicle speed sensor (VSS) is mounted behind the speedometer in the instrument cluster or on the transmission/speedometer drive gear. It provides electrical pulses to the ECM from the speedometer head. The pulses indicate the road speed. The ECM uses this information to operate the IAC, canister purge, and TCC.

Some vehicles equipped with digital instrument clusters use a Permanent Magnet (PM) generator to provide the VSS signal. The PM generator is located in the transmission and replaces the speedometer cable. The signal from the PM generator drives a stepper motor which drives the odometer. A failure in the VSS circuit should set a Code 24.

THROTTLE POSITION SENSOR (TPS)

The Throttle Position Sensor (TPS) is connected to the throttle shaft and is controlled by the throttle mechanism. A 5 volt reference signal is sent to the TPS from the ECM. As the throttle valve angle is changed (accelerator pedal moved), the resistance of the TPS also changes. At a closed throttle position, the resistance of the TPS is high, so the output voltage to the ECM will be low (approximately 0.5 volt). As the throttle plate opens, the resistance decreases so that, at wide open throttle, the output voltage should be approximately 5 volts. At closed throttle position, the voltage at the TPS should be less than 1.25 volts.

By monitoring the output voltage from the TPS, the ECM can determine fuel delivery based on throttle valve angle (driver demand). The TPS can either be misadjusted, shorted, open or loose. Misadjustment might result in poor idle or poor wide-open throttle performance. An open TPS signals the ECM that the throttle is always closed, resulting in poor performance. This usually sets a Code 22. A shorted TPS gives the ECM a constant wide-open throttle signal and should set a Code 21. A loose TPS indicates to the ECM that the throttle is moving. This causes intermittent bursts of fuel from the injector and an unstable idle. Once the trouble code is set, the ECM will use an artificial default value for the TBI and some vehicle performance will return.

CRANKSHAFT SENSOR

Some systems use a magnetic crankshaft sensor, mounted remotely from the ignition module, which protrudes into the block within approximately 0.050 in. of the crankshaft reluctor. The reluctor is a special wheel cast into the crankshaft with 7 slots machined into it, 6 of which are equally spaced (60 degrees apart). A seventh slot is spaced approximately 10 degrees from one of the other slots and severs to generate a **SYNC PULSE** signal. As the reluctor rotates as part of the crankshaft, the slots change the magnetic field of the sensor, creating an induced voltage pulse.

Based on the crank sensor pulses, the ignition module sends 2X reference signals to the ECM which are used to indicate crankshaft position and engine speed. The ignition module continues to send these reference pulses to the ECM at a rate of 1 per each 180 degrees of the crankshaft rotation. This signal is called the 2X reference because it occurs 2 times per crankshaft revolution.

The ignition also sends a second, 1X reference signal to the ECM which occurs at the same time as the **SYNC PULSE** from the crankshaft sensor. This signal is called the 1X reference because it occurs 1 time per crankshaft revolution. The 1X refer-

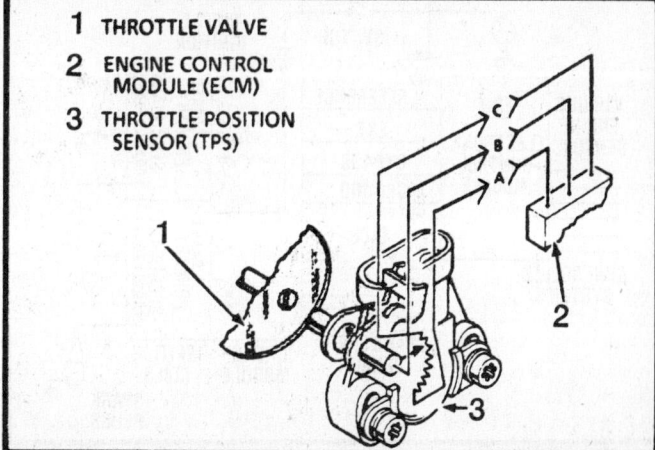

1 THROTTLE VALVE
2 ENGINE CONTROL MODULE (ECM)
3 THROTTLE POSITION SENSOR (TPS)

Typical throttle position sensor

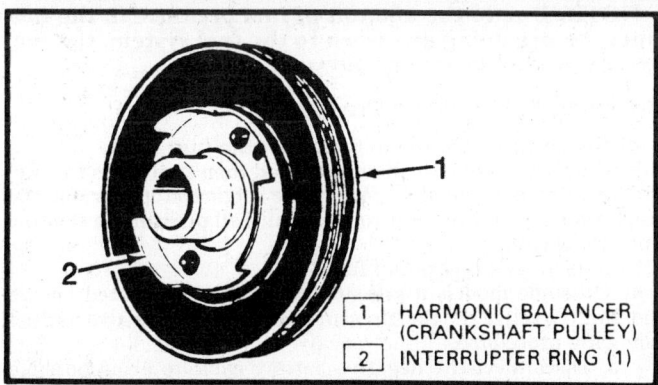

Typical harmonic balancer with interrupter ring

| 1 | HARMONIC BALANCER (CRANKSHAFT PULLEY) |
| 2 | INTERRUPTER RING (1) |

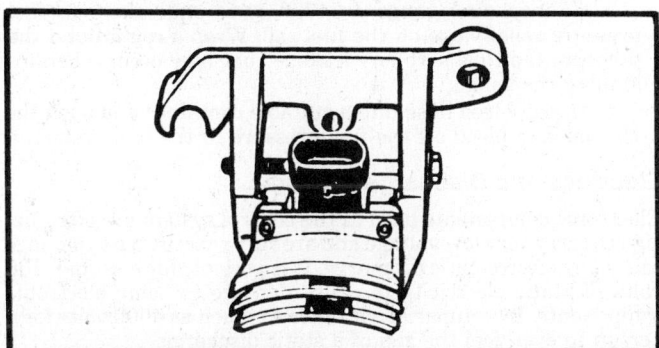

Typical crankshaft sensor

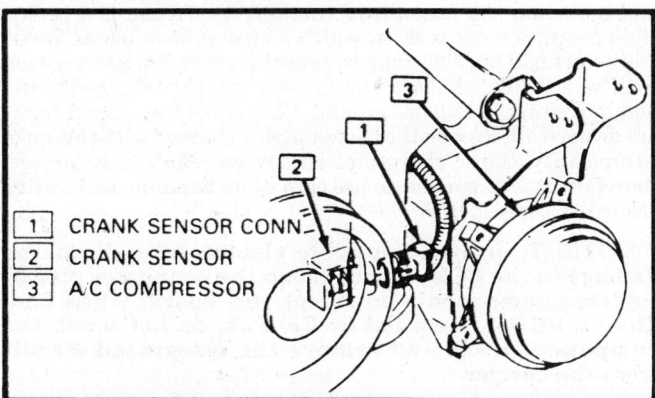

1	CRANK SENSOR CONN.
2	CRANK SENSOR
3	A/C COMPRESSOR

Typical combination sensor

ence and the 2X reference signals are necessary for the ECM to determine when to activate the fuel injectors.

By comparing the time between pulses, the ignition module can recognize the pulse representing the seventh slot (sync pulse) which starts the calculation of the ignition coil sequencing. The second crank pulse following the **SYNC PULSE** signals the ignition module to fire the No. 2–3 ignition coil and the fifth crank pulse signals the module to fire the No. 1–4 ignition coil.

DUAL CRANK SENSOR/ COMBINATION SENSOR

The dual crank sensor is mounted in a pedestal on the front of the engine near the harmonic balancer. The sensor consists of 2 Hall Effect switches, which depend on 2 metal interrupter rings mounted on the balancer to activate them. Windows in the in-

terrupters activate the hall effect switches as they provide a patch for the magnetic field between the switches transducers and magnets. When one of the hall effect switches is activated, it grounds the signal line to the C^3I module, pulling that signal line's (Sync Pulse or Crank) applied voltage low, which is interpreted as a signal.

Because of the way the signal is created by the dual crank sensor, the signal circuit is always either at a high or low voltage (square wave signal). Three crank signal pulses and one **SYNC PULSE** are created during each crankshaft revolution. The crank signal is used by the C^3I module to create a reference signal which is also a square wave signal similar to the crank signal. The reference signal is used to calculate the engine rpm and crankshaft position by the ECM. The **SYNC PULSE** is used by the C^3I module to begin the ignition coil firing sequence starting with No. 3–6 coil. The firing sequence begins with this coil because either piston No. 3 or piston No. 6 is now at the correct position in compression stroke for the spark plugs to be fired. Both the crank sensor and the **SYNC PULSE** signals must be received by the ignition module for the engine to start. A misadjusted sensor or bent interrupter ring could cause rubbing of the sensor resulting in potential driveability problems, such as rough idle, poor performance, or a nor start condition.

NOTE: Failure to have the correct clearance will damage the crankshaft sensor.

The dual crank sensor is not adjustable for ignition timing but positioning of the interrupter ring is very important. A clearance of 0.025 in. is required on either side of the interrupter ring. A dual crank sensor that is damaged, due to mispositioning or a bent interrupter ring, can result in a hesitation, sag stumble or dieseling condition.

To determine if the dual crank sensor could be at fault, scan the engine rpm with a suitable scan tool, while driving the vehicle. An erratic display indicates that a proper reference pulse has not been received by the ECM, which may be the result of a malfunctioning dual crank sensor.

AIR CONDITIONING PRESSURE SENSOR

The air conditioning (A/C) pressure sensor provides a signal to the electronic control module (ECM) which indicates varying high side refrigerant pressure between approximately 0–450 psi. The ECM used this input to the A/C compressor load on the engine to help control the idle speed with the IAC valve.

The A/C pressure sensor electrical circuit consists of a 5 volt reference line and a ground line, both provided by the ECM and a signal line to the ECM. The signal is a voltage that varies approximately 0.1 volt at 0 psi, to 4.9 volts at 450 psi or more. A problem in the A/C pressure circuits or sensor should set a Code 66 and will make the A/C compressor inoperative.

NON—AIR CONDITIONING PROGRAM INPUT

Vehicles not equipped with air conditioning (A/C) have a circuit connecting the ECM terminal **BC3** to ground, to program to operate without A/C related components connected to it. Vehicles with A/C do not have a wire in ECM **BC3** terminal. If this circuit is open on a non-A/C vehicle it may cause false Codes 26 and/or Code 66. If terminal **BC3** is grounded on A/C equipped vehicles, it will cause the compressor relay to be on whenever the ignition is in the **ON** position.

DETONATION (KNOCK) SENSOR

This sensor is a piezoelectric sensor located near the back of the engine (transmission end). It generates electrical impulses which are directly proportional to the frequency of the knock which is detected. A buffer then sorts these signals and eliminates all except for those frequency range of detonation. This information is passed to the ESC module and then to the ECM, so that the ignition timing advance can be retarded until the detonation stops.

PARK/NEUTRAL SWITCH

NOTE: Vehicle should not be driven with the park/neutral switch disconnected as idle quality may be affected in PARK or NEUTRAL and a Code 24 (VSS) may be set.

This switch indicates to the ECM when the transmission is in **P** or **N**. the information is used by the ECM for control on the torque converter clutch, EGR, and the idle air control valve operation.

AIR CONDITIONING REQUEST SIGNAL

This signal indicates to the ECM that an air conditioning mode is selected at the switch and that the A/C low pressure switch is closed. The ECM controls the A/C and adjusts the idle speed in response to this signal.

TORQUE CONVERTER CLUTCH SOLENOID

The purpose of the torque converter clutch system is designed to eliminate power loss by the converter (slippage) to increase fuel economy. By locking the converter clutch, a more effective coupling to the flywheel is achieved. The converter clutch is operated by the ECM controlled torque converter clutch solenoid.

POWER STEERING PRESSURE SWITCH

The power steering pressure switch is used so that the power steering oil pressure pump load will not effect the engine idle. Turning the steering wheel increase the power steering oil pressure and pump load on the engine. The power steering pressure switch will close before the load can cause an idle problem. The ECM will also turn the A/C clutch off when high power steering pressure is detected.

OIL PRESSURE SWITCH

The oil pressure switch is usually mounted on the back of the engine, just below the intake manifold. Some vehicles use the oil pressure switch as a parallel power supply, with the fuel pump relay and will provide voltage to the fuel pump, after approximately 4 psi (28 kPa) of oil pressure is reached. This switch will also help prevent engine seizure by shutting off the power to the fuel pump and causing the engine to stop when the oil pressure is lower than 4 psi.

SERVICE PRECAUTIONS

When working around any part of the fuel system, take precautionary steps to prevent fire and/or explosion:

- Disconnect negative terminal from battery (except when testing with battery voltage is required).
- When ever possible, use a flashlight instead of a drop light.
- Keep all open flame and smoking material out of the area.
- Use a shop cloth or similar to catch fuel when opening a fuel system.
- Relieve fuel system pressure before servicing.
- Use eye protection.
- Always keep a dry chemical (class B) fire extinguisher near the area.

NOTE: Due to the amount of fuel pressure in the fuel lines, before doing any work to the fuel system, the fuel system should be de-pressurized.

Relieving Fuel System Pressure

1. Remove the fuel pump fuse from the fuse block.
2. Start the engine. It should run and then stall when the fuel in the lines is exhausted. When the engine stops, crank the starter for about 3 seconds to make sure all pressure in the fuel lines is released.
3. Replace the fuel pump fuse.
4. On some models a pressure relief valve is located on the fuel rail. To relive the fuel pressure using the relief valve use the following procedure.

   a. Disconnect the negative battery cable to avoid possible fuel discharge if an accidental attempt is made to start the engine. Loosen the fuel filler cap to relieve fuel tank pressure.

   b. Connect fuel gauge J–34730–1, or equivalent, to fuel pressure relief valve on the fuel rail. Wrap a rag around the pressure tap to absorb any leakage that may occur when installing the gauge.

   c. Install bleed hose into a suitable container and open the the valve to bleed off the fuel pressure in the fuel system.

Electrocstatic Discharge Damage

Electronic components used in the control system are often design to carry very low voltage and are very susceptible to damage caused by electrostatic discharge. It is possible for less than 100 volts of static electricity to cause damage to some electronic components. By comparison it takes as much as 4000 volts for a person to even feel the zap of a static discharge.

There are several ways for a person to become statically charged. The most common methods of charging are by friction and induction. An example of charging by friction is a person sliding across a car seat, in which a charge as much as 25000 volts can build up. Charging by induction occurs when a person with well insulated shoes stands near a highly charged object and momentarily touches ground. Charges of the same polarity are drained off, leaving the person highly charged with the opposite polarity. Static charges of either type can cause damage, therefore, it is important to use care when handling and testing electronic components.

NOTE: To prevent possible electrostatic discharge damage to the ECM, do not touch the connector pins or soldered components on the circuit board. When handling a PROM, Mem-Cal or Cal-Pak, do not touch the component leads and remove the integrated circuit from the carrier.

Diagnosis and Testing

ALCL CONNECTOR

The Assembly Line Communication Link (ALCL) (or also known as the Assembly Line Diagnostic Link (ALDL) is a diagnostic connector located in the passenger compartment usually under the instrument panel (except Pontiac Fiero which is located in the console). The assembly plant were the vehicles originate use the connector to check the engine for proper operation before it leaves the plant. Terminal **B** is usually the diagnostic **TEST** terminal (lead) and it can be connected to terminal **A**, or ground, to enter the Diagnostic mode or the Field Service Mode.

FIELD SERVICE MODE

If the **TEST** terminal is grounded with the engine running, the system will enter the the Field Service mode. In this mode, the "CHECK ENGINE" light will show whether the system is in

Typical ALCL connector

Open loop or Closed loop. In Open loop the "CHECK ENGINE" light flashes 2½ times per second. In Closed loop the light flashes once per second. Also in closed loop, the light will stay OUT most of the time if the system is too lean. It will stay on most of the time if the system is too rich. In either case the Field Service mode check, which is part of the Diagnostic circuit check, will lead the technician into choosing the correct diagnostic chart to refer to. While the system is in the Field Service mode, new trouble codes cannot be stored in the ECM and the Closed Loop timer is bypassed.

DIAGNOSTIC MODE

A built-in, self-diagnostic system catches the problems most likely to occur in the Computer Command Control system. The diagnostic system turns on a "CHECK ENGINE" light in the instrument panel when a problem is detected. By grounding a trouble code **TEST** terminal under the dash, (ignition **ON** engine not running) the "CHECK ENGINE" light will flash a trouble code or codes indicating the problem areas.

As a bulb and system check, the "CHECK ENGINE" light will come on with the ignition switch **ON** and the engine not running. If the **TEST** terminal is then grounded, the light will flash a Code 12, which indicates the self-diagnostic system is working. A Code 12 consists of 1 flash, followed by a short pause, then 2 flashes in quick succession. After a longer pause, the code will repeat 2 more times.

When the engine is started, the "CHECK ENGINE" light will turn off. If the "CHECK ENGINE" light remains on, the self-diagnostic system has detected a problem. If the **TEST** terminal is then grounded with the ignition **ON**, engine not running, each trouble code will flash and repeat 3 times. If more than 1 problem has been detected, each trouble code will flash 3 times. Trouble codes will flash in numeric order (lowest number first). The trouble code series will repeat as long as the **TEST** terminal is grounded.

A trouble code indicates a problem in a given circuit (Code 14, for example, indicates a problem in the coolant sensor circuit; this includes the coolant sensor, connector harness, and ECM). The procedure for pinpointing the problem can be found in diagnosis. Similar charts are provided for each code.

Also in this mode all ECM controlled relays and solenoids except the fuel pump relay. This allows checking the circuits which may be difficult to energize without driving the vehicle and being under particular operating conditions. The IAC valve will move to its fully extended position on most models, block the idle air passage. This is useful in checking the minimum idle speed.

ECM LEARNING ABILITY

The ECM has a learning capability. If the battery is disconnect-

ed the learning process has to begin all over again. A change may be noted in the vehicle's performance. To teach the ECM, insure the vehicle is at operating temperature and drive at part throttle, with moderate acceleration and idle conditions, until performance returns.

READING THE DATA STREAM

This information is able to be read by putting the ECM into one of 3 different modes. These modes are entered by inserting a specific amount of resistance between the the ALCL (ALDL) connector terminals **A** and **B**. The modes and resistances needed to enter these modes are as follows:

0 OHMS DIAGNOSTIC MODES

When 0 resistance is between terminals **A** and **B** of the ALCL connector, the diagnostic mode is entered. There are 2 positions to this mode. One with the engine **OFF**, but the ignition **ON**; the other is when the engine is running.

If the diagnostic mode is entered with the engine in the **OFF** position, trouble codes will flash and the idle air control motor will pulsate in and out. Also, the relays and solenoids are energized with the exception of the fuel pump and injector.

In the event the ALCL connector terminal **B** is grounded with the engine running, the ECM goes into the field service mode. In this mode, the "CHECK ENGINE" light light flashes closed or open loop and indicates the rich/lean status of the engine. The ECM runs the engine at a fixed ignition timing advanced above the base.

3.9 KILO-OHMS BACKUP MODE

The backup mode is entered by applying 3.9 kilo-ohms resistance between terminals **A** and **B** of the ALCL connector with the ignition switch in the **ON** position. The ALCL scanner tool can now read 5 of the 20 parameters on the data stream. These parameters are as mode status, oxygen sensor voltage, rpm, block learn and idle air control. There are 2 ways to enter the backup mode. Using a scan tool is one way; putting a 3.9 kilo-ohms resistor across terminals **A** and **B** of the ALCL is another.

10K OHMS SPECIAL MODE

This special mode is entered by applying a 10K ohms resistor across terminals **A** and **B**. When this happens the ECM does the following:

1. Allows all of the serial data to be read
2. Bypasses all timers
3. Add a calibrated spark advance
4. Enables the canister purge solenoid on some engines
5. Idles at 1000 rpm fixed idle air control and fixed base pulse width on the injector
6. Forces the idle air control to reset at part throttle (approximately 2000 rpm)
7. Disables the park/neutral restrict functions

20K OHMS OPEN OR ROAD TEST MODE

The system is in this mode during normal operation.

INTEGRATOR AND BLOCK LEARN

The integrator and block learn functions of the ECM are responsible for making minor adjustments to the air/fuel ratio on the fuel injected GM vehicles. These small adjustments are necessary to compensate for pinpoint air leaks and normal wear.

The integrator and block learn are 2 separate ECM memory functions which control fuel delivery. The integrator makes a temporary change and the block learn makes a more permanent change. Both of these functions apply only while the engine is in **CLOSED LOOP**. They represent the on-time of the injector. Also, integrator and block learn controls fuel delivery on the fuel injected engines as does the MC solenoid dwell on the CCC carbureted engines.

INTEGRATOR

Integrator is the term applied to a means of temporary change in fuel delivery. Integrator is displayed through the ALCL data line and monitored with a scanner as a number between 0 and 255 with an average of 128. The integrator monitors the oxygen sensor output voltage and adds and subtracts fuel depending on the lean or rich condition of the oxygen sensor. When the integrator is displaying 128, it indicates a neutral condition. This means that the oxygen sensor is seeing results of the 14.7:1 air/fuel mixture burned in the cylinders.

NOTE: An air leak in the system (a lean condition) would cause the oxygen sensor voltage to decrease while the integrator would increase (add more fuel) to temporarily correct for the lean condition. If this happened the injector pulse width would increase.

BLOCK LEARN

Although the integrator can correct fuel delivery over a wide range, it is only for a temporary correction. Therefore, another control called block learn was added. Although it cannot make as many corrections as the integrator, it does so for a longer period of time. It gets its name from the fact that the operating range of the engine for any given combinations of rpm and load is divided into 16 cell or blocks.

The computer has a given fuel delivery stored in each block. As the operating range gets into a given block the fuel delivery will be based on what value is stored in the memory in that block. Again, just like the integrator, the number represents the on-time of the injector. Also, just like the integrator, the number 128 represents no correction to the value that is stored in the cell or block. When the integrator increases or decreases, block learn which is also watching the integrator will make corrections in the same direction. As the block learn makes corrections, the integrator correction will be reduced until finally the integrator will return to 128 if the block learn has corrected the fuel delivery.

BLOCK LEARN MEMORY

Block learn operates on 1 of 2 types of memories depending on application, non-volatile and volatile. The non-volatile memories retain the value in the block learn cells even when the ignition switch is turned **OFF**. When the engine is restarted, the fuel delivery for a given block will be based on information stored in memory.

The volatile memories lose the numbers stored in the block learn cells when the ignition is turned to the **OFF** position. Upon restarting, the block learn starts at 128 in every block and corrects from that point as necessary.

INTEGRATOR/BLOCK LEARN LIMITS

Both the integrator and block learn have limits which will vary from engine to engine. If the mixture is off enough so that the block learn reaches the limit of its control and still cannot correct the condition, the integrator would also go to its limit of control in the same direction and the engine would then begin to run poorly. If the integrators and block learn are close to or at their limits of control, the engine hardware should be checked to determine the cause of the limits being reached, vacuum leaks, sticking injectors, etc.

If the integrator is lied to, for example, if the oxygen sensor lead was grounded (lean signal) the integrator and block learn would add fuel to the engine to cause it to run rich. However, with the oxygen sensor lead grounded, the ECM would continue seeing a lean condition eventually setting a Code 44 and the fuel control system would change to open loop operations.

CLOSED LOOP FUEL CONTROL

The purpose of closed loop fuel control is to precisely maintain an air/fuel mixture 14.7:1. When the air/fuel mixture is main-

tained at 14.7:1, the catalytic converter is able to operate at maximum efficiency which results in lower emission levels.

Since the ECM controls the air/fuel mixture, it needs to check its output and correct the fuel mixture for deviations from the ideal ratio. The oxygen sensor feeds this output information back to the ECM.

READING CODES

The codes stores in the ECM's memory can be read either through a hand held diagnostic scanner plugged into the ALCL connector or by counting the number of flashes on the "CHECK ENGINE" light when the diagnostic terminal of the ALCL connector is grounded. The ALCL connector terminal **B** (diagnostic terminal) is usually the second terminal from the right of the top row in the ALCL connector. The terminal is most easily grounded by connecting to terminal **A** (internal ECM ground), the terminal to the right of the terminal **B** on the top row in the ALCL connector.

Once terminals **A** and **B** are grounded, the ignition switch must be turned **ON** with the engine not running. At this point the "CHECK ENGINE" light should flash a Code 12 (3) times consecutively. This would be the following flash sequence; flash, pause, flash-flash, long pause, flash, pause, flash-flash, long pause, flash, pause, flash-flash.

Code 12 indicates that the ECM's diagnostic system is operating. If Code 12 is not indicated, a problem is present within the diagnostic system itself and should by addresses by using the appropriate diagnostic chart.

Following the output of Code 12, the "CHECK ENGINE" light will indicate a diagnostic trouble code 3 times if a code is present or it will simply continue to output Code 12. If more than 1 diagnostic code has been stored in the ECM's memory, the codes will be output from the lowest to the highest, with each code being displayed 3 times.

CLEARING THE TROUBLE CODES

When the ECM finds a problem with the system, the "CHECK ENGINE" light will come on and a trouble code will be recorded in the ECM memory. If the problem is intermittent, the "CHECK ENGINE" light will go out after 10 seconds, when the fault goes away. However the trouble code will stay in the ECM memory until the battery voltage to the ECM is removed. Removing the battery voltage for 10 seconds will clear all trouble codes. Do this by disconnecting the ECM harness from the positive battery terminal pigtail for 10 seconds with the key in the **OFF** position, or by removing the ECM fuse for 10 seconds with the key **OFF**.

NOTE: To prevent ECM damage, the key must be OFF when disconnecting and reconnecting ECM power.

TROUBLE CODE TEST LEAD

The trouble code **TEST** lead terminal is mounted in a multi-terminal connector located under the dash. Grounding this terminal signals the ECM to flash any trouble codes stored in the memory. This is easily done by jumping to the adjacent ground terminal.

If the **TEST** terminal is grounded with the ignition **ON** and the engine stopped, the system will enter the diagnostic mode. In the diagnostic mode the ECM will:
1. Flash a Code 12 (indicating system is operating)
2. Energize all ECM controlled relays

If the **TEST** terminal is grounded with the engine running, the system will enter the field service mode. In this mode, the "CHECK ENGINE" light will indicate whether the system is in open or closed loop. Open loop is indicated by the "CHECK ENGINE" light flashing approximately twice per second. In closed loop, the light flashes approximately once per second.

ENGINE PERFORMANCE DIAGNOSIS

Engine performance diagnosis procedures are guides that will lead to the most probable causes of engine performance complaints. They consider the components of the fuel, ignition, and mechanical systems that could cause a particular complaint, and then outline repairs in a logical sequence.

It is important to determine if the "CHECK ENGINE" light is on or has come on for a short interval while driving. If the "CHECK ENGINE" light has come on, the system should be checked for stored **TROUBLE CODES** which may indicate the cause for the performance complaint.

All of the symptoms can be caused by worn out or defective parts such as spark plugs, ignition wiring, etc. If time and/or mileage indicate that parts should be replaced, it is recommended that it be done.

NOTE: Before checking any system controlled by the Electronic Fuel Injection (EFI) system, the Diagnostic Circuit Check must be performed or misdiagnosis may occur. If the complaint involves the "CHECK ENGINE" light, go directly to the Diagnostic Circuit Check.

Basic Troubleshooting

NOTE: The following explains how to activate the Trouble Code signal light in the instrument cluster and gives an explanation of what each code means. This is not a full CCC System troubleshooting and isolation procedure.

Before suspecting the CCC System or any of its components as faulty, check the ignition system including distributor, timing, spark plugs and wires. Check the engine compression, air cleaner, and emission control components not controlled by the ECM. Also check the intake manifold, vacuum hoses and hose connectors for leaks.

The following symptoms could indicate a possible problem with the system:
1. Detonation
2. Stalls or rough idle-cold
3. Stalls or rough idle-hot
4. Missing
5. Hesitation
6. Surges
7. Poor gasoline mileage
8. Sluggish or spongy performance
9. Hard starting-cold
10. Objectionable exhaust odors (that rotten egg smell)
11. Cuts out
12. Improper idle speed

As a bulb and system check, the "CHECK ENGINE" light will come on when the ignition switch is turned to the **ON** position but the engine is not started. The "CHECK ENGINE" light will also produce the trouble code or codes by a series of flashes which translate as follows. When the diagnostic test terminal under the dash is grounded, with the ignition in the **ON** position and the engine not running, the "CHECK ENGINE" light will flash once, pause, then flash twice in rapid succession. This is a Code 12, which indicates that the diagnostic system is working. After a long pause, the Code 12 will repeat itself 2 more times. The cycle will then repeat itself until the engine is started or the ignition is turned off.

When the engine is started, the "CHECK ENGINE" light will remain on for a few seconds, then turn off. If the "CHECK EN-GINE" light remains on, the self-diagnostic system has detected a problem. If the test terminal is then grounded, the trouble code will flash 3 times. If more than 1 problem is found, each trouble code will flash 3 times. Trouble codes will flash in numerical order (lowest code number to highest). The trouble codes series will repeat as long as the test terminal is grounded.

A trouble code indicates a problem with a given circuit. For

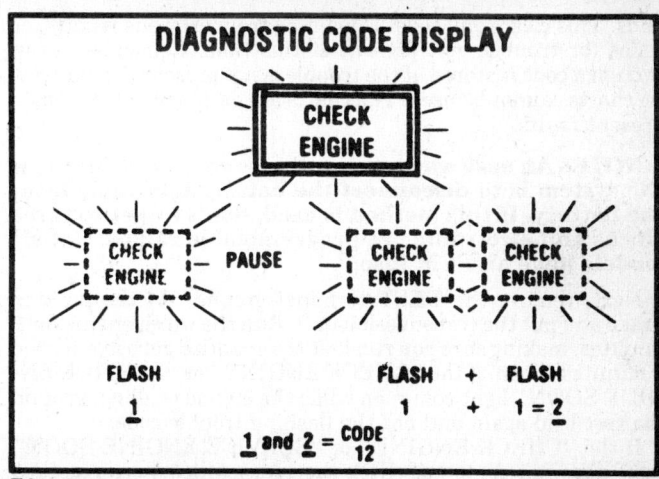

Diagnostic Code 12 display

example, trouble Code 14 indicates a problem in the cooling sensor circuit. This includes the coolant sensor, its electrical harness, and the Electronic Control Module (ECM) Since the self-diagnostic system cannot diagnose every possible fault in the system, the absence of a trouble code does not mean the system is trouble-free. To determine problems within the system which do not activate a trouble code, a system performance check must be made.

In the case of an intermittent fault in the system, the "CHECK ENGINE" light will go out when the fault goes away, but the trouble code will remain in the memory of the ECM. Therefore, it a trouble code can be obtained even though the "CHECK ENGINE" light is not on, the trouble code must be evaluated. It must be determined if the fault is intermittent or if the engine must be at certain operating conditions (under load, etc.) before the "CHECK ENGINE" light will come on. Some trouble codes will not be recorded in the ECM until the engine has been operated at part throttle for about 5–18 minutes. On the CCC System, a trouble code will be stored until terminal **R** of the ECM has been disconnected from the battery for 10 seconds.

An easy way to erase the computer memory on the CCC System is to disconnect the battery terminals from the battery. If this method is used, don't forget to reset clocks and electronic pre-programmable radios. Another method is to remove the fuse marked ECM in the fuse panel. Not all models have such a fuse.

SYSTEM DIAGNOSTIC CIRCUIT CHECK

Begin the Diagnostic Circuit Check by making sure that the diagnostic system itself is working. Turn the ignition to **ON** with the engine stopped. If the "CHECK ENGINE" or "SERVICE ENGINE SOON" light comes on, ground the diagnostic code terminal (test lead) under the dash. If the "CHECK ENGINE" or "SERVICE ENGINE SOON" light flashes a Code 12, the self-diagnostic system is working and can detect a faulty circuit. If there is no Code 12, see the appropriate chart in this section. If any additional codes flash, record them for later use.

If a Code 51 flashes, use chart 51 to diagnose that condition before proceeding with the Diagnostic Circuit Check. A Code 51 means that the "CHECK ENGINE" or "SERVICE ENGINE SOON" light flashes 5 times, pauses, then flashes once. After a longer pause, Code 51 will flash again twice in this same way. To find out what diagnostic step to follow, look up the chart for Code 51 in this section. If there is not a Code 51, follow the NO CODE 51 branch of the chart.

Clear the ECM memory by disconnecting the voltage lead either at the fuse panel or the ECM letter connector for 10 seconds. This clears any codes remaining from previous repairs, or codes for troubles not present at this time. Remember, even though a code is stored, if the trouble is not present the diagnostic charts cannot be used. The charts are designed only to locate present faults.

NOTE: An easy way to erase the computer memory on the system is to disconnect the battery terminals from the battery. If this method is used, don't forget to reset clocks and electronic pre-programmable radios. Not all models have an ECM fuse.

Next, remove the **TEST** terminal ground, set the parking brake and put the transmission in **P**. Run the warm engine for 2 minutes, making sure you run it at the specified curb idle for the 2 minutes. Then, if the "CHECK ENGINE" or "SERVICE ENGINE SOON" light comes on while the engine is idling, ground the test lead again and not the flashing trouble code.

If the "CHECK ENGINE" or "SERVICE ENGINE SOON" light does not come on, check the codes which were recorded earlier. If there were no additional codes, road test the car for the problem being diagnosed to make sure it still exists.

The purpose of the Diagnostic Circuit check is to make sure the "CHECK ENGINE" or "SERVICE ENGINE SOON" light works, that the ECM is operating and can recognize a fault and to determine if any trouble codes are stored in the ECM memory.

If trouble codes are stored, it also checks to see if they indicate an intermittent problem. This is the starting point of any diagnosis. If there are no codes stored, move on to the System Performance Check.

The codes obtained from the "CHECK ENGINE" or "SERVICE ENGINE SOON" light display method indicate which diagnostic charts provide in the section are to be used. For example, Code 23 can be diagnosed by following the step-by-step procedures on chart 23.

NOTE: If more than 1 code is stored in the ECM, the lowest code number must be diagnosed first. Then proceed to the next highest code. The only exception is when a 50 series flashes. Fifty code take precedence over all other trouble codes and must be dealt with first, since they point to a fault in the PROM unit or the ECM.

If the diagnostic procedures call for the ECM to be replaced, the calibration unit (PROM) should be checked first to see if it is functioning correctly. If it is correct, the PROM should be removed from the defective ECM and installed in the new service ECM. THE SERVICE ECM WILL NOT CONTAIN A PROM. Trouble Code 51 indicates the PROM is installed improperly or has malfunctioned. When Code 51 is obtained, the PROM installation should be checked for bent pins or pins not fully seated in the socket. If the PROM is installed correctly and Code 51 still shows, the PROM should be replaced.

NOTE: To prevent internal ECM damage, the ignition switch must be in the OFF position when reconnecting power to the ECM (for example, battery positive cable, ECM pigtail, ECM fuse, jumper cables, etc.).

INTERMITTENT CHECK ENGINE LIGHT

An intermittent open in the ground circuit would cause loss of power through the ECM and intermittent "CHECK ENGINE" light operation. When the ECM loses ground, distributor ignition is lost. An intermittent open in the ground circuit would be described as an engine miss.

Therefore, an intermittent "CHECK ENGINE" light, no code stored and a driveability comment described as similar to a miss will require checking the grounding circuit and the Code 12 circuit as it originates at the ignition coil.

UNDERVOLTAGE TO THE ECM

A circuit breaker will shut off the ECM if the power supply falls below 9 volts. The undervoltage condition will cause the "CHECK ENGINE" light to come on as long as the condition exist.

Therefore, an intermittent "CHECK ENGINE" light, no code stored and a driveability comment described as similar to a miss will require checking the grounding circuit, Code 12 circuit and the ignition feed circuit to terminal C of the ECM. This does nor eliminate the necessity of checking the normal vehicle electrical system for possible cause such as a loose battery cable.

OVERVOLTAGE TO THE ECM

The ECM will also shut off by the circuit breaker when the power supply rises above 16 volts. The overvoltage condition will also cause the "CHECK ENGINE" to come on as long as this condition exist.

A momentary voltage surge in a vehicle's electrical system is a common occurrence. These voltage surges have never presented any problems because the entire electrical system acted as a shock absorber until the surge dissipated. Voltage surges or spikes in the vehicle's electrical system have been known, on occasion, to exceed 100 volts.

The system is a low voltage (between 9 and 16 volts) system and will not tolerate these surges. The ECM will be shut off by any surge in excess of 16 volts and will come back on, only after the surge has dissipated sufficiently to bring the voltage under 16 volts.

A surge will usually occur when an accessory requiring a high voltage supply is turned off or down. The voltage regulator in the vehicle's charging system cannot react to the changes in the voltage demands quickly enough and surge occurs. The driver should be questioned to determine which accessory circuit was turned **OFF** that caused the "CHECK ENGINE" light to come on.

Therefore, intermittent "CHECK ENGINE" light operation, with no trouble code stored, will require installation of a diode in the appropriate accessory circuit.

ELECTRONIC FUEL INJECTION ALCL SCAN TESTER INFORMATION

An ALCL display unit (ALCL tester, scanner, monitor, etc), allows a technician to read the engine control system information from the ALCL connector under the instrument panel. It can provide information faster than a digital voltmeter or ohmmeter can. The scan tool does not diagnose the exact location of the problem. The tool supplies information about the ECM, the information that it is receiving and the commands that it is sending plus special information such as integrator and block learn. To use an ALCL display tool you should understand throughly how an engine control system operates.

An ALCL scanner or monitor puts a fuel injection system into a special test mode. This mode commands an idle speed of 1000 rpm. The idle quality cannot be evaluated with a tester plugged in. Also the test mode commands a fixed spark with no advance. On vehicles with Electronic Spark Control (ESC), there will be a fixed spark, but it will be advanced. On vehicles with ESC, there might be a serious spark knock, this spark knock could be bad enough so as not being able to road test the vehicle in the ALCL test mode. Be sure to check the tool manufacturer for instructions on special test modes which should overcome these limitations.

When a tester is used with a fuel injected engine, it bypasses the timer that keeps the system in open loop for a certain period of time. When all closed loop conditions are met, the engine will go into closed loop as soon as the vehicle is started. This means that the air management system will not function properly and air may go directly to the converter as soon as the engine is started.

These tools cannot diagnose everything. They do not tell the technician where a problem is located in a circuit. The diagnostic charts to pinpoint the problems must still be used. These tester's do not let a technician know if a solenoid or relay has been turned on. They only tell the technician the ECM command. To find out if a solenoid has been turned on, check it with a suitable test light or digital voltmeter, or see if vacuum through the solenoid changes.

SCAN TOOLS USED WITH INTERMITTENTS

In some scan tool applications, the data update rate makes the tool less effective than a voltmeter, such as when trying to detect an intermittent problem which lasts for a very short time. However, the scan tool does allow one to manipulate the wiring harness or components under the hood with the engine not running while observing the scan tool's readout.

The scan tool can be plugged in and observed while driving the vehicle under the condition when the "CHECK ENGINE" light turns on momentarily or when the engine driveability is momentarily poor. If the problem seems to be related to certain parameters that can be checked on the scan tool, they should be checked while driving the vehicle. If there does not seem to be any correlation between the problem and any specific circuit, the scan tool can be checked on each position. Watching for a period of time to see if there is any change in the reading that indicates intermittent operation.

The scan tool is also an easy way to compare the operating parameters of a poorly operating engine with those of a known good one. For example, a sensor may shift in value but not set a trouble code. Comparing the sensor's reading with those of a known good vehicle may uncover the problem.

The scan tool has the ability to save time in diagnosis and prevent the replacement of good parts. The key to using the scan tool successfully for diagnosis lies in the technician's ability to understand the system he is trying to diagnose as well as an understanding of the scan tool's operation and limitations. The technician should read the tool manufacturer's operating manual to become familiar with the tool's operation.

WIRE HARNESS AND CONNECTOR SERVICE

A wire harness should be replaced with the proper part number harnesses. When the signal wires are spliced into a harness, use wire with high temperature insulation only. With low current and voltage levels found in the system, it is important that the best possible bond be made at all wire splices by soldering the splices carefully and correctly.

Use care when probing a connector or replacing connector terminals. It is possible to short between the opposite terminals. If this happens, certain components can be damaged. Always use a jumper wire between connectors for circuit checking. NEVER probe through connector seals, wire insulation, secondary ignition wires, boots, nipples or covers. Even microscopic damage or holes may result in eventual water intrusion, corrosion and or component or circuit failure.

When diagnosing, open circuits are often difficult to locate by sight because oxidation or terminal misalignment are hidden by the sealed connectors. Merely wiggling a connector on a sensor or in the wiring harness may locate the open circuit condition. This should always be considered when an open circuit or failed sensor is indicated. Intermittent problems may also be caused by oxidized or loss connections.

METRI-PACK SERIES 150 TERMINALS

Some ECM harness connectors contain terminals called metri-pack. These may be used at the coolant sensors as well as at the ignition modules. Metri-pack terminal are also called pull to seat terminals, because to install a terminal on a wire, the wire must first be inserted through the seal and connector. The terminal is then crimped on the wire and the terminal pulled back into the connector to seat it in place. To remove one of these metri-pack terminals, proceed as follows:

1. Slide the seal back on the wire.
2. Insert connector tool BT-8518/J-35689 or equivalent, to release the terminal locking tang.
3. Push the wire and terminal out through the connector.
4. If the terminal is being reused, reshape the locking tangs.

WEATHER-PACK CONNECTORS

Special connector tools BT-8234-A/J-28742 or equivalents are needed to service the weather-pack connectors. This tool is used to remove the pin and sleeve terminals. If terminal removal is attempted with an ordinary pick, there is a good chance that the terminal will be bent or deformed an unlike standard blade type terminals, these terminals cannot be straightened once they are bent.

Make certain that the connectors are properly seated and all of the sealing rings in place when connecting the leads. The hinge-type flap provides a secondary locking feature for the connector. It improves the connector reliability by retaining the terminals if the small terminal lock tangs are not positioned properly.

Weather-pack connections cannot be replaced with standard connections. Instructions are provided with the weather-pack connector and terminal packages.

COMPACT 3 CONNECTORS

The compact 3 connector, which looks similar to a weather-pack connector, is not sealed and is used where resistance to the environment is not required. This type connector most likely is used at the AIR control solenoid. Use the standard method when repairing a terminal. Do not use the weather-pack terminal tools on this connector.

MICRO-PACK CONNECTORS

The micro-pack connector terminal replacement requires the use of special connector tool BT-8234-A/J-33095 or equivalent.

TOOLS NEEDED TO SERVICE THE SYSTEM

The system requires a scan tool, tachometer, test light, ohmmeter, digital voltmeter with a 10 megohms impedance, vacuum gauge and jumper wires for diagnosis. A test light or voltmeter must be used when specified in the procedures.

NOTE: Some vehicles will use more sensors than others. Also, a complete general diagnostic section is outlined. The steps and procedures can be altered as required (if necessary) by the technician according to the specific model being diagnosed and the sensors it is equipped with. The wiring diagrams and schematics may not coincide with every GM vehicle in use. If this situation should arise, use the wiring diagram or schematic as a general guide. On some models, the electronic module may have a Learning Ability which allows them to make corrections for minor variations in the fuel system to improve driveability. If the battery power is disconnected for any reason, the volatile memory resets and the learning process begins again. A change may be noted in the performance of the vehicle. To teach the vehicle, ensure that the engine is at normal operating temperature. Then, the vehicle should be driven at part throttle, with moderate acceleration and idle conditions until normal performance returns.

2.0L TURBO (VIN M) COMPONENT LOCATIONS GRAND AM

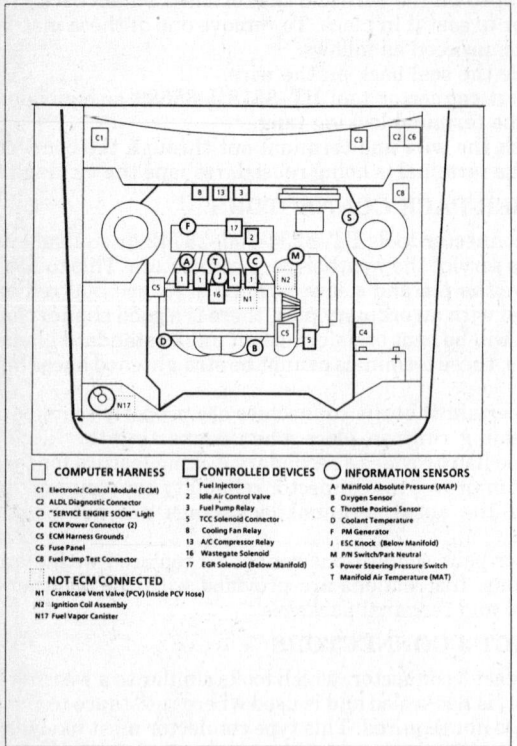

☐ **COMPUTER HARNESS**

C1 Electronic Control Module (ECM)
C2 ALDL Diagnostic Connector
C3 "SERVICE ENGINE SOON" Light
C4 ECM Power Connector (2)
C5 ECM Harness Grounds
C6 Fuse Panel
C7 Fuel Pump Test Connector

☐ **NOT ECM CONNECTED**

N1 Crankcase Vent Valve (PCV) (Inside PCV Hose)
N2 Ignition Coil Assembly
N17 Fuel Vapor Canister

☐ **CONTROLLED DEVICES**

1 Fuel Injectors
2 Idle Air Control Valve
3 Fuel Pump Relay
4 TCC Solenoid Connector
8 Cooling Fan Relay
13 A/C Compressor Relay
16 Wastegate Solenoid
17 EGR Solenoid (Below Manifold)

☐ **INFORMATION SENSORS**

A Manifold Absolute Pressure (MAP)
B Oxygen Sensor
C Throttle Position Sensor
D Coolant Temperature
F PM Generator
J ESC Knock (Below Manifold)
M P/N Switch/Park Neutral
S Power Steering Pressure Switch
T Manifold Air Temperature (MAT)

2.0L TURBO (VIN M) ECM WIRING DIAGRAM GRAND AM

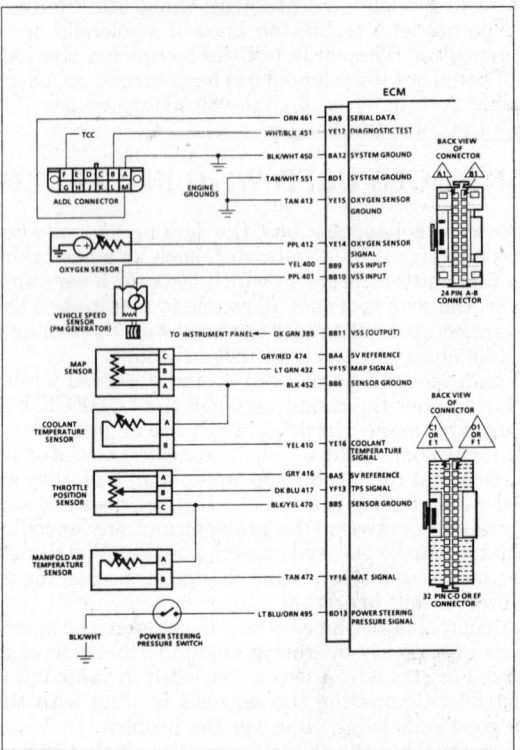

2.0L TURBO (VIN M) ECM WIRING DIAGRAM GRAND AM (CONT.)

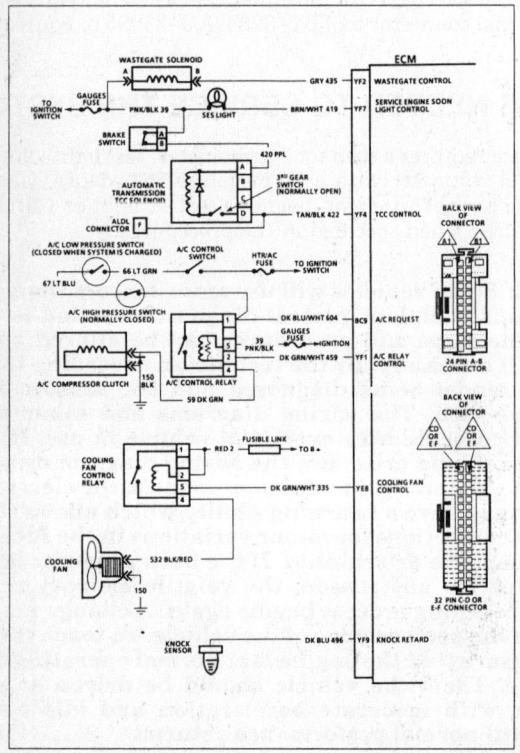

2.0L TURBO (VIN M) ECM WIRING DIAGRAM GRAND AM (CONT.)

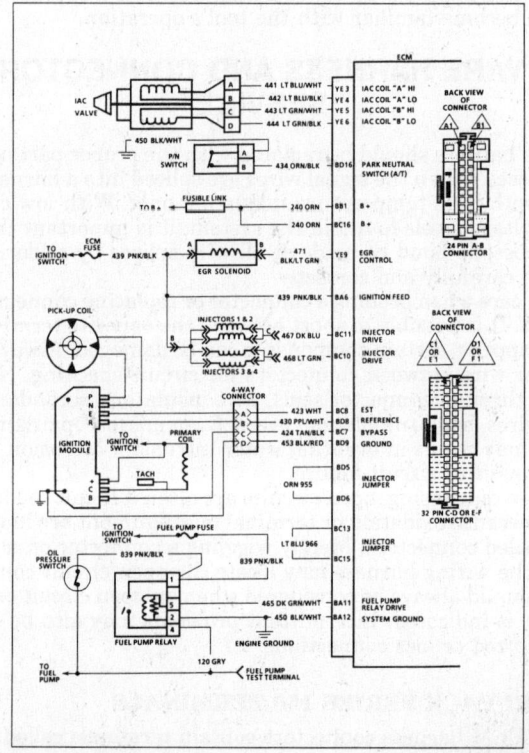

2.0L TURBO (VIN M) ECM CONNECTOR TERMINAL END VIEW—GRAND AM

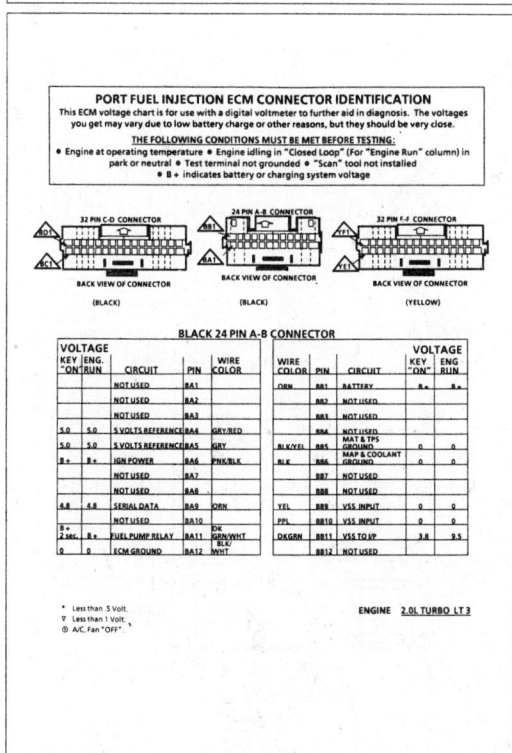

PORT FUEL INJECTION ECM CONNECTOR IDENTIFICATION

This ECM voltage chart is for use with a digital voltmeter to further aid in diagnosis. The voltages you get may vary due to low battery charge or other reasons, but they should be very close.

THE FOLLOWING CONDITIONS MUST BE MET BEFORE TESTING:
• Engine at operating temperature • Engine idling in "Closed Loop" (For "Engine Run" column) in park or neutral • Test terminal not grounded • "Scan" tool not installed
• B+ indicates battery or charging system voltage

BLACK 24 PIN A-B CONNECTOR

VOLTAGE KEY "ON"	ENG. RUN	CIRCUIT	PIN	WIRE COLOR		WIRE COLOR	PIN	CIRCUIT	VOLTAGE KEY "ON"	ENG. RUN
		NOT USED	BA1			ORN	BA2	BATTERY	B+	B+
		NOT USED	BA2				BA2	NOT USED		
		NOT USED	BA3				BA3	NOT USED		
5.0	5.0	5 VOLTS REFERENCE	BA4	GRY/RED			BA4	NOT USED		
5.0	5.0	5 VOLTS REFERENCE	BA5	GRY		BLK/YEL	BA5	MAT & TPS GROUND	0	0
B+	B+	IGN POWER	BA6	PNK/BLK		BLK	BA6	MAP & COOLANT GROUND	0	0
		NOT USED	BA7				BA7	NOT USED		
		NOT USED	BA8				BA8	NOT USED		
4.8	4.8	SERIAL DATA	BA9	ORN		YEL	BA9	VSS INPUT	0	0
B+		NOT USED	BA10			PPL	BA10	VSS INPUT	0	0
2 sec.	B+	FUEL PUMP RELAY	BA11	DK GRN/WHT		DKGRN	BA11	VSS TO V/P	3.8	9.5
0	0	ECM GROUND	BA12	BLK/WHT			BA12	NOT USED		

* Less than 5 Volt.
▽ Less than 1 Volt.
① A/C, Fan "OFF"

ENGINE 2.0L TURBO LT 3

2.0L TURBO (VIN M) ECM CONNECTOR TERMINAL END VIEW—GRAND AM (CONT.)

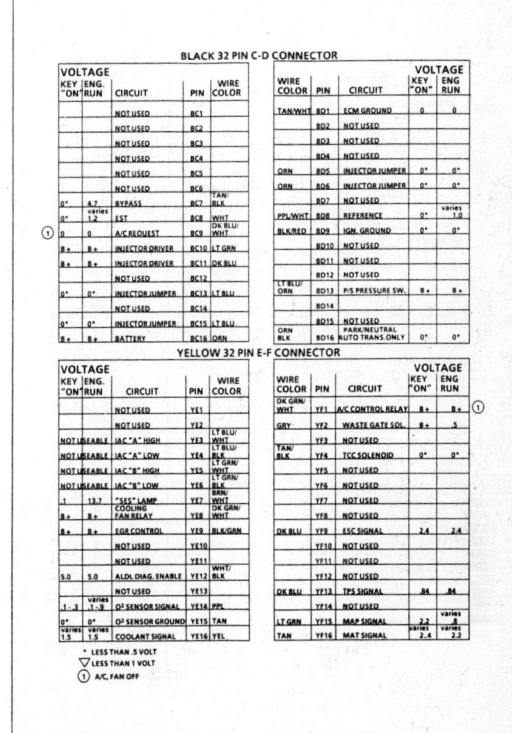

BLACK 32 PIN C-D CONNECTOR

VOLTAGE KEY "ON"	ENG. RUN	CIRCUIT	PIN	WIRE COLOR		WIRE COLOR	PIN	CIRCUIT	VOLTAGE KEY "ON"	ENG. RUN
		NOT USED	BC1			TAN/WHT	BD1	ECM GROUND	0	0
		NOT USED	BC2				BD2	NOT USED		
		NOT USED	BC3				BD3	NOT USED		
		NOT USED	BC4				BD4	NOT USED		
		NOT USED	BC5			ORN	BD5	INJECTOR JUMPER	0*	0*
		NOT USED	BC6			ORN	BD6	INJECTOR JUMPER	0*	0*
0*	4.7	BYPASS	BC7	TAN/BLK			BD7	NOT USED		
0*	varies 1.2	EST	BC8	WHT		PPL/WHT	BD8	REFERENCE	varies	varies 1.0
0	0	A/C REQUEST	BC9	DK BLU/WHT		BLK/RED	BD9	IGN. GROUND	0*	0*
B+	B+	INJECTOR DRIVER	BC10	LT GRN			BD10	NOT USED		
B+	B+	INJECTOR DRIVER	BC11	DK BLU			BD11	NOT USED		
		NOT USED	BC12				BD12	NOT USED		
0*	0*	INJECTOR JUMPER	BC13	LT BLU		LT BLU/ ORN	BD13	P/S PRESSURE SW.	B+	B+
		NOT USED	BC14				BD14			
0*	0*	INJECTOR JUMPER	BC15	LT BLU		ORN BLK	BD15	NOT USED		
B+	B+	BATTERY	BC16	ORN			BD16	PARK/NEUTRAL AUTO TRANS. ONLY	0*	0*

YELLOW 32 PIN E-F CONNECTOR

VOLTAGE KEY "ON"	ENG. RUN	CIRCUIT	PIN	WIRE COLOR		WIRE COLOR	PIN	CIRCUIT	VOLTAGE KEY "ON"	ENG. RUN
		NOT USED	YE1			DK GRN/ WHT	YF1	A/C CONTROL RELAY	B+	B+ ①
		NOT USED	YE2			GRY	YF2	WASTE GATE SOL.	B+	5
NOT USEABLE		IAC "A" HIGH	YE3	LT BLU		TAN/ BLK	YF3	NOT USED		
NOT USEABLE		IAC "A" LOW	YE4	LT BLU/ BLK			YF4	TCC SOLENOID	0*	0*
NOT USEABLE		IAC "B" HIGH	YE5	LT GRN/ BLK			YF5	NOT USED		
NOT USEABLE		IAC "B" LOW	YE6	LT GRN/ BLK			YF6	NOT USED		
.1	13.7	"SES" LAMP	YE7	DK GRN BRN			YF7	NOT USED		
B+	B+	COOLING FAN RELAY	YE8				YF8	NOT USED		
B+	B+	EGR CONTROL	YE9	BLK/GRN		DK BLU	YF9	ESC SIGNAL	2.4	2.4
		NOT USED	YE10				YF10	NOT USED		
		NOT USED	YE11				YF11	NOT USED		
5.0	5.0	ALDL DIAG. ENABLE	YE12	WHT/ BLK			YF12	NOT USED		
		NOT USED	YE13			DK BLU	YF13	TPS SIGNAL	.84	.84
1-.3	varies 1-.8	O² SENSOR SIGNAL	YE14	PPL			YF14	NOT USED		
0*	0*	O² SENSOR GROUND	YE15	TAN		LT GRN	YF15	MAP SIGNAL	2.2 varies	varies .8
varies 1.5	varies 1.5	COOLANT SIGNAL	YE16	YEL		TAN	YF16	MAT SIGNAL	2.4	2.2

* LESS THAN .5 VOLT
▽ LESS THAN 1 VOLT
① A/C, FAN OFF

2.0L TURBO (VIN M) COMPONENT LOCATIONS SUNBIRD

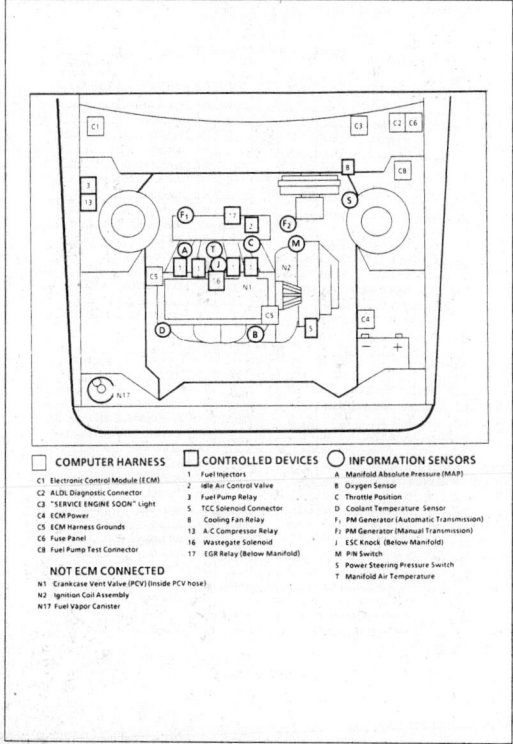

□ COMPUTER HARNESS	■ CONTROLLED DEVICES	○ INFORMATION SENSORS
C1 Electronic Control Module (ECM)	1 Fuel Injectors	A Manifold Absolute Pressure (MAP)
C2 ALDL Diagnostic Connector	2 Idle Air Control Valve	B Oxygen Sensor
C3 "SERVICE ENGINE SOON" Light	3 Fuel Pump Relay	C Throttle Position
C4 ECM Power	7 TCC Solenoid Connector	D Coolant Temperature Sensor
C5 ECM Harness Grounds	8 Cooling Fan Relay	F PM Generator (Automatic Transmission)
C6 Fuse Panel	13 A/C Compressor Relay	F₁ PM Generator (Manual Transmission)
C8 Fuel Pump Test Connector	16 Wastegate Solenoid	J ESC Knock (Below Manifold)
	17 EGR Relay (Below Manifold)	M P/N Switch
NOT ECM CONNECTED		S Power Steering Pressure Switch
N1 Crankcase Vent Valve (PCV) (Inside PCV hose)		T Manifold Air Temperature
N2 Ignition Coil Assembly		
N17 Fuel Vapor Canister		

2.0L TURBO (VIN M) ECM WIRING DIAGRAM SUNBIRD

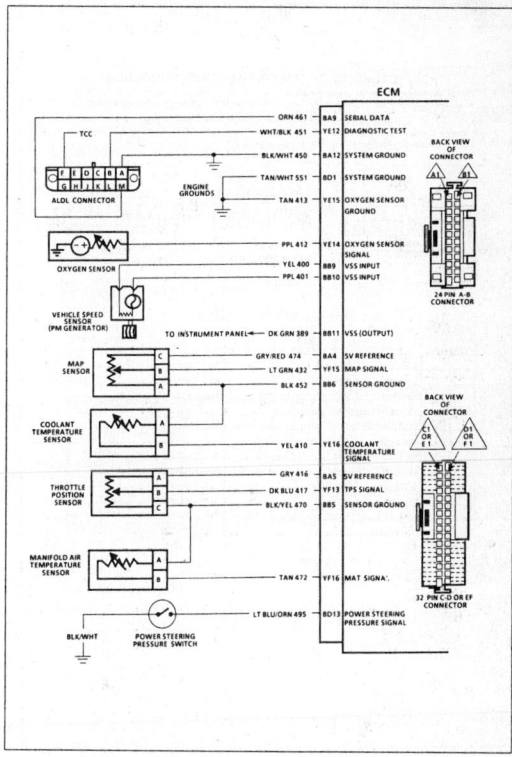

2.0L TURBO (VIN M) ECM WIRING DIAGRAM SUNBIRD (CONT.)

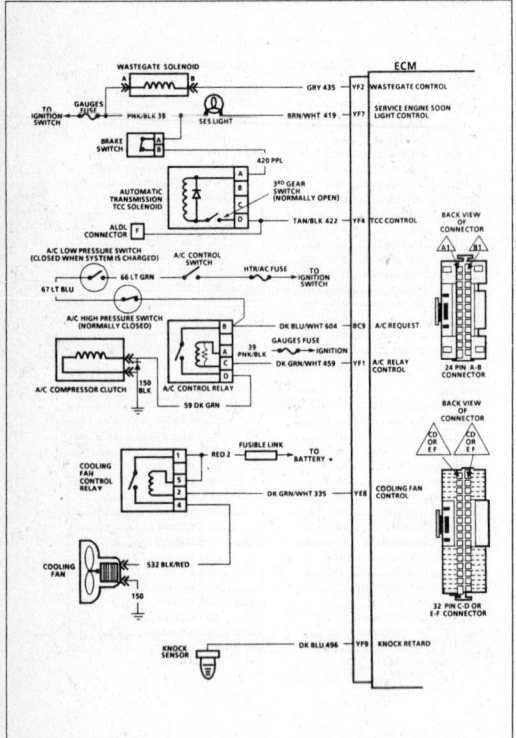

2.0L TURBO (VIN M) ECM WIRING DIAGRAM SUNBIRD (CONT.)

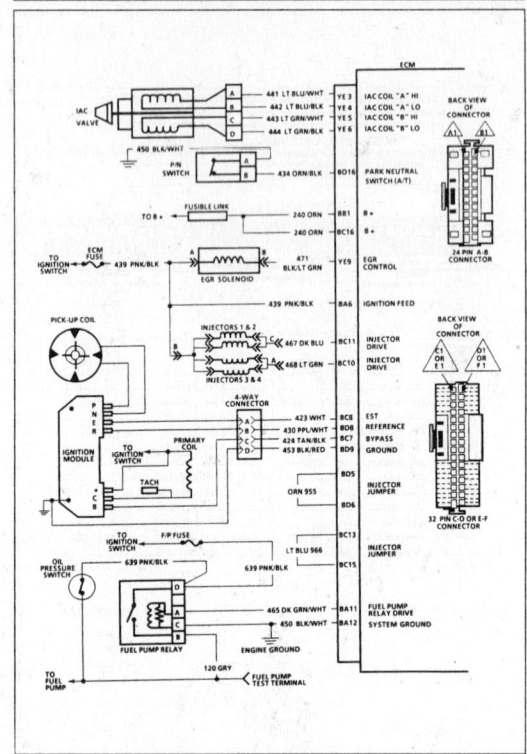

2.0L TURBO (VIN M) ECM CONNECTOR TERMINAL END VIEW—SUNBIRD

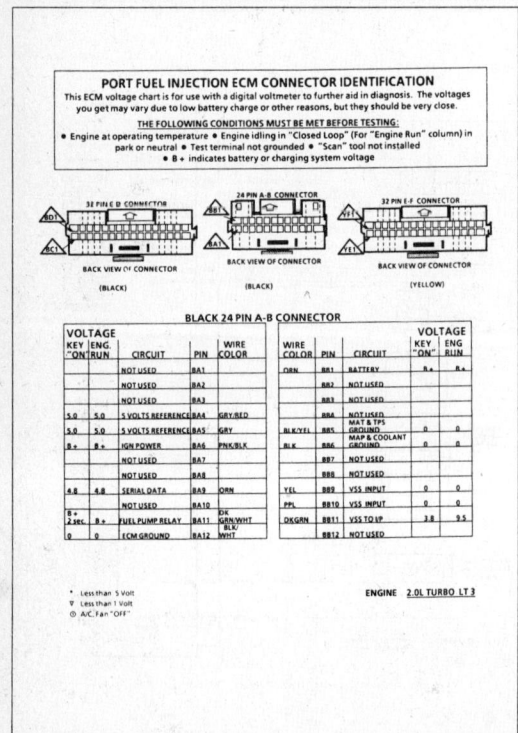

PORT FUEL INJECTION ECM CONNECTOR IDENTIFICATION

This ECM voltage chart is for use with a digital voltmeter to further aid in diagnosis. The voltages you get may vary due to low battery charge or other reasons, but they should be very close.

THE FOLLOWING CONDITIONS MUST BE MET BEFORE TESTING:

- Engine at operating temperature • Engine idling in "Closed Loop" (For "Engine Run" column) in park or neutral • Test terminal not grounded • "Scan" tool not installed
- B+ indicates battery or charging system voltage

BLACK 24 PIN A-B CONNECTOR

VOLTAGE KEY ENG. "ON" "RUN"	CIRCUIT	PIN	WIRE COLOR		WIRE COLOR	PIN	CIRCUIT	VOLTAGE KEY ENG. "ON" "RUN"
	NOT USED	BA1			ORN	BB1	BATTERY	B+ / B+
						BB2	NOT USED	
						BB3	NOT USED	
5.0 / 5.0	5 VOLTS REFERENCE	BA4	GRY/RED			BB4	NOT USED	
5.0 / 5.0	5 VOLTS REFERENCE	BA5	GRY		BLK/YEL	BB5	MAT & TPS GROUND	0 / 0
B+ / B+	IGN POWER	BA6	PNK/BLK		BLK	BB6	MAP & COOLANT GROUND	0 / 0
	NOT USED	BA7				BB7	NOT USED	
	NOT USED	BA8				BB8	NOT USED	
4.8 / 4.8	SERIAL DATA	BA9	ORN		YEL	BB9	VSS INPUT	0 / 0
B+ 2 sec. / B+	FUEL PUMP RELAY	BA10			PPL	BB10	VSS INPUT	0 / 0
		BA11	DK GRN/WHT		DK GRN	BB11	VSS UP	3.8 / 9.5
0 / 0	ECM GROUND	BA12	BLK/WHT			BB12	NOT USED	

* Less than 5 Volt
▽ Less than 1 Volt
① A/C Fan "OFF"

ENGINE 2.0L TURBO LT 3

BLACK 32 PIN C-D CONNECTOR

VOLTAGE KEY ENG. "ON" "RUN"	CIRCUIT	PIN	WIRE COLOR		WIRE COLOR	PIN	CIRCUIT	VOLTAGE KEY ENG. "ON" "RUN"
	NOT USED	BC1			TAN/WHT	BD1	ECM GROUND	0 / 0
	NOT USED	BC2				BD2	NOT USED	
	NOT USED	BC3				BD3	NOT USED	
	NOT USED	BC4				BD4	NOT USED	
	NOT USED	BC5			ORN	BD5	INJECTOR JUMPER	0* / 0*
	NOT USED	BC6			ORN	BD6	INJECTOR JUMPER	0* / 0*
0* / 4.7	BYPASS	BC7	TAN/BLK			BD7	NOT USED	
0 / varies 1.0	EST	BC8	WHT		PPL/WHT	BD8	REFERENCE	0* / varies 1.0
① 0 / 0	A/C REQUEST	BC9	DK BLU/WHT		BLK/RED	BD9	IGN. GROUND	B* / 0*
B+ / B+	INJECTOR DRIVER	BC10	LT GRN			BD10	NOT USED	
B+ / B+	INJECTOR DRIVER	BC11	DK BLU			BD11	NOT USED	
		BC12				BD12	NOT USED	
0* / 0*	INJECTOR JUMPER	BC13	LT BLU		LT BLU/ ORN	BD13	P/S PRESSURE SW.	B+ / B+
	NOT USED	BC14				BD14		
0* / 0*	INJECTOR JUMPER	BC15	LT BLU		ORN BLK	BD15	NOT USED	
	NOT USED	BC16	ORN			BD16	PARK/NEUTRAL AUTO TRANS. ONLY	0* / 0*

YELLOW 32 PIN E-F CONNECTOR

VOLTAGE KEY ENG. "ON" "RUN"	CIRCUIT	PIN	WIRE COLOR		WIRE COLOR	PIN	CIRCUIT	VOLTAGE KEY ENG. "ON" "RUN"
	NOT USED	YE1			DK GRN/ WHT	YF1	A/C CONTROL RELAY	B+ / B+ ①
	NOT USED	YE2			GRY	YF2	WASTE GATE SOL.	B+ / 5
NOT USEABLE	IAC "A" HIGH	YE3	LT BLU/ WHT		TAN/ BLK	YF3	TCC SOLENOID	0* / 0*
NOT USEABLE	IAC "A" LOW	YE4	LT BLU/ BLK			YF4	TCC SOLENOID	0* / 0*
NOT USEABLE	IAC "B" HIGH	YE5	LT GRN/ WHT			YF5	NOT USED	
NOT USEABLE	IAC "B" LOW	YE6	LT GRN/ BRN/			YF6	NOT USED	
.1 / 13.7	"SES" LAMP	YE7	WHT			YF7	NOT USED	
B+ / B+	COOLING FAN RELAY	YE8	DK GRN/ WHT			YF8	NOT USED	
B+ / B+	EGR CONTROL	YE9	BLK/GRN		DK BLU	YF9	ESC SIGNAL	2.4 / 2.4
	NOT USED	YE10				YF10	NOT USED	
	NOT USED	YE11				YF11	NOT USED	
5.0 / 5.0	ALDL DIAG. ENABLE	YE12	BLK/ BLK			YF12	NOT USED	
	NOT USED	YE13			DK BLU	YF13	TPS SIGNAL	.84 / .84
1.3 / 1.9	O² SENSOR SIGNAL	YE14	PPL			YF14	NOT USED	
0* / 0*	O² SENSOR GROUND	YE15			LT GRN	YF15	MAP SIGNAL	varies 2.2 / 2.2
varies 1.5 / 1.5	COOLANT SIGNAL	YE16	YEL		TAN	YF16	MAT SIGNAL	varies 2.4 / 2.2

* LESS THAN 5 VOLT
▽ LESS THAN 1 VOLT
① A/C, FAN OFF

2.0L TURBO (VIN M) ECM CONNECTOR TERMINAL END VIEW—SUNBIRD (CONT.)

1988–90 2.0L TURBO (VIN M) – ALL MODELS

DIAGNOSTIC CIRCUIT CHECK

The Diagnostic Circuit Check is an organized approach to identifying a problem created by an electronic engine control system malfunction. It must be the starting point for any driveability complaint diagnosis because it directs the service technician to the next logical step in diagnosing the complaint.

The "Scan" data listed in the table may be used for comparison after completing the diagnostic circuit check and finding the on-board diagnostics functioning properly with no trouble codes displayed. The "Typical Data Values" are an average of display values recorded from normally operating vehicles and are intended to represent what a normally functioning system would typically display.

A "SCAN" TOOL THAT DISPLAYS FAULTY DATA SHOULD NOT BE USED, AND THE PROBLEM SHOULD BE REPORTED TO THE MANUFACTURER. THE USE OF A FAULTY "SCAN" TOOL CAN RESULT IN MISDIAGNOSIS AND UNNECESSARY PARTS REPLACEMENT.

Only the parameters listed below are used in this manual for diagnosis. If a "Scan" tool reads other parameters, the values are not recommended by General Motors for use in diagnosis. For more description on the values and use of the "Scan" tool to diagnosis ECM inputs, refer to the applicable component diagnosis.

If all values are within the range illustrated, refer to symptoms in Section "B".

"SCAN" TOOL DATA

Test Under Following Conditions: Idle, Upper Radiator Hose Hot, Closed Throttle, Park or Neutral, "Closed Loop," All Accessories "OFF."

"SCAN" Position	Units Displayed	Typical Data Value
Desired RPM	RPM	ECM idle command (varies with temperature)
RPM	RPM	± 50 RPM from desired rpm in drive (AUTO)
		± 100 RPM from desired rpm in neutral (MANUAL)
Coolant Temperature	Degrees Celsius	85 - 105
MAT Temperature	Degrees Celsius	10 - 90 (varies with underhood temperature and sensor location)
MAP	Volts	1 - 2 (varies with manifold and barometric pressures)
BPW (base pulse width)	Milliseconds	.8 - 3.0
O₂	Volts	.1 - 1 (varies continuously)
TPS	Volts	4 - 1.25
Throttle Angle	0 - 100%	0
IAC	Counts (steps)	1 - 50
P/N Switch	P-N and R-D-L	Park/Neutral (P/N)
INT (Integrator)	Counts	110 - 145
BLM (Block Learn Memory)	Counts	118 - 138
Open/Closed Loop	Open/Closed	"Closed Loop" (may enter "Open Loop" with extended idle)
VSS	MPH	0
TCC	ON/OFF	"OFF"
Spark Advance	Degrees	Varies
Battery	Volts	13.5 - 14.5
Fan	ON/OFF	"OFF" (coolant temperature below 102°C)
P/S Switch	Normal/Hi Pressure	Normal
A/C Request	Yes/No	No
A/C Clutch	ON/OFF	"OFF"
Shift Light (M/T)	ON/OFF	"OFF"
Knock Signal	Yes/No	No
Knock Retard	Degrees	0

1988–90 2.0L TURBO (VIN M) – ALL MODELS

DIAGNOSTIC CIRCUIT CHECK
2.0L TURBO (VIN M)

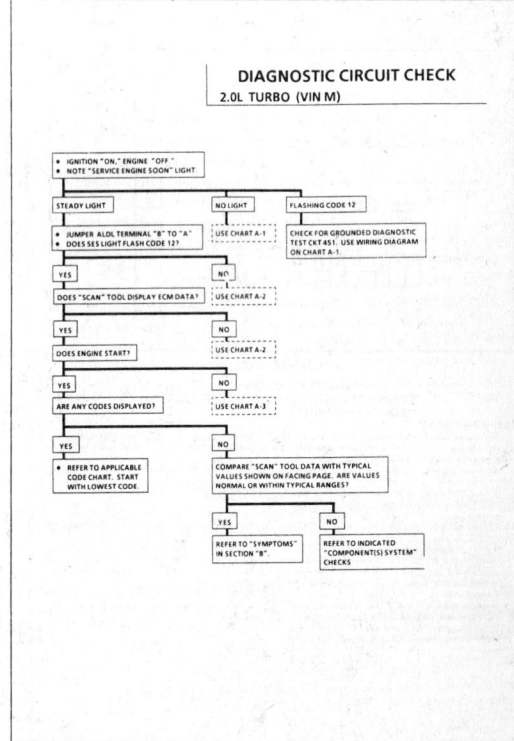

1988–90 2.0L TURBO (VIN M) – ALL MODELS

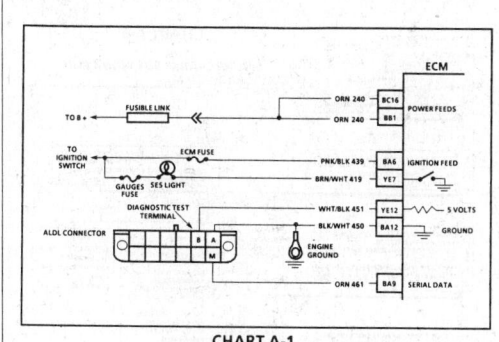

CHART A-1
NO "SERVICE ENGINE SOON" LIGHT
2.0L TURBO (VIN M)

Circuit Description:
There should always be a steady "Service Engine Soon" light when the ignition is "ON" and the engine is "OFF." Battery voltage is supplied directly to the light bulb. The Electronic Control Module (ECM) controls the light and turns it "ON" by providing a ground path through CKT 419 to the ECM.

Test Description: Numbers below refer to circled numbers on the diagnostic chart.
1. Probing CKT 419 to ground creates an alternate ground. If the "Service Engine Soon" light illuminates, this verifies that the trouble is not in the lamp portion of the circuit.

Diagnostic Aids:

If engine runs OK, check the following:
- Faulty light bulb.
- CKT 419 open.
- Gage fuse blown. This will result in no oil or generator lights, seat belt reminder, etc.

If "Engine Cranks But Won't Run," use CHART A-3.

1988–90 2.0L TURBO (VIN M) – ALL MODELS

CHART A-1
NO "SERVICE ENGINE SOON" LIGHT
2.0L TURBO (VIN M)

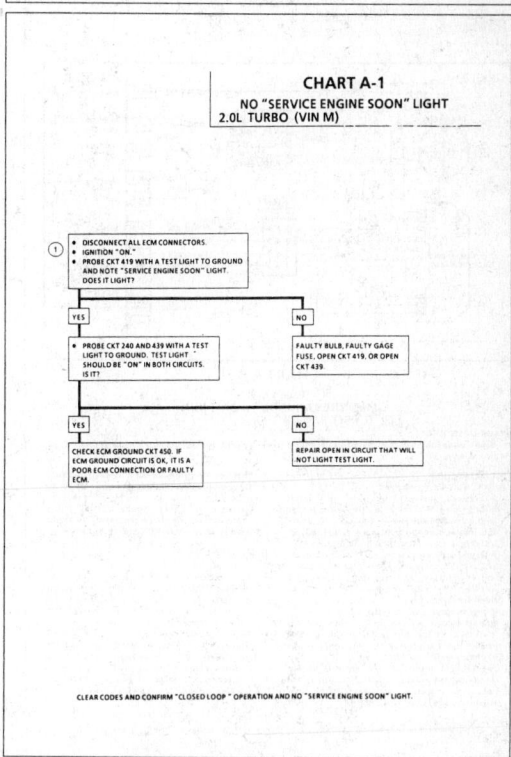

1988–90 2.0L TURBO (VIN M) – ALL MODELS

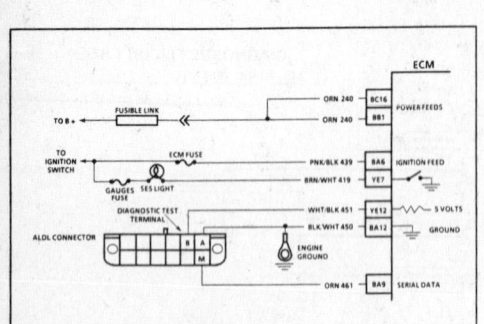

CHART A-2
WON'T FLASH CODE 12
"SERVICE ENGINE SOON" LIGHT "ON" STEADY
2.0L TURBO (VIN M)

Circuit Description:
There should always be a steady "Service Engine Soon" light when the ignition is "ON" and engine stopped. Battery voltage is supplied directly to the light bulb. The electronic control module (ECM) will turn the light "ON" by grounding CKT 419 at the ECM.

With the diagnostic terminal grounded, the light should flash a Code 12 followed by any trouble code(s) stored in memory.

A steady light suggests a short to ground in the light control CKT 419 or an open in diagnostic CKT 451.

Test Description: Numbers below refer to circled numbers on the diagnostic chart.
1. If there is a problem with the ECM that causes a "Scan" tool to not read serial data, then the ECM should not flash a Code 12. If Code 12 does flash, be sure that the "Scan" tool is working properly on another vehicle. If the "Scan" is functioning properly and CKT 461 is OK, the Mem-Cal or ECM may be at fault for the no ALDL symptom.
2. If the light goes "OFF" when the ECM connector is disconnected, then CKT 419 is not shorted to ground.
3. This step will check for an open diagnostic CKT 451.
4. At this point the "Service Engine Soon" light wiring is OK. The problem is a faulty ECM or Mem-Cal. If Code 12 does not flash, the ECM should be replaced using the original Mem-Cal. Replace the Mem-Cal only after trying an ECM, as a defective Mem-Cal is an unlikely cause of the problem.

1988–90 2.0L TURBO (VIN M) – ALL MODELS

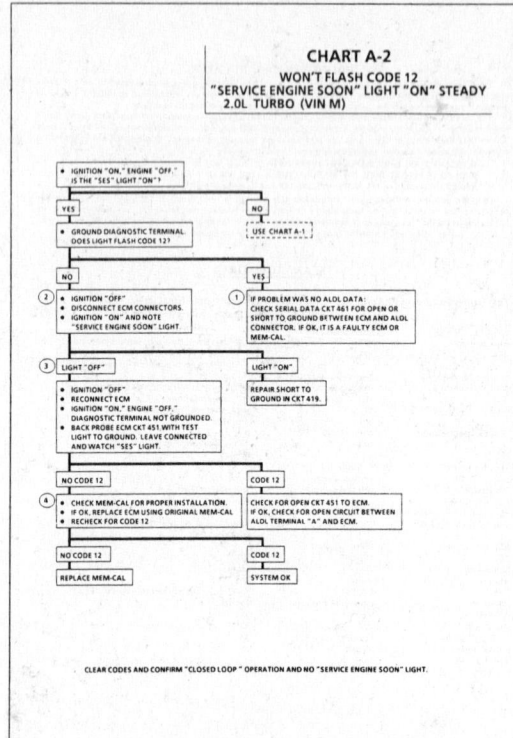

1988–90 2.0L TURBO (VIN M) – ALL MODELS

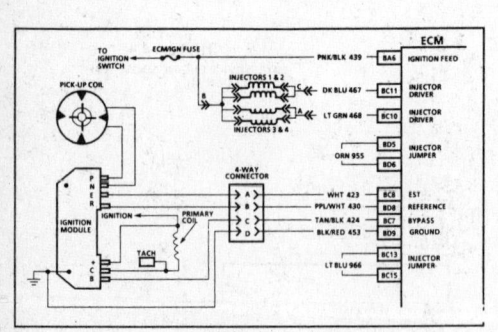

CHART A-3
(Page 1 of 2)
ENGINE CRANKS BUT WON'T RUN
2.0L TURBO (VIN M)

Circuit Description:
Before using this chart, battery condition, engine cranking speed, and fuel quantity should be checked and verified as being OK.

Test Description: Numbers below refer to circled numbers on the diagnostic chart.
1. A "Service Engine Soon" light "ON" is a basic test to determine if there is battery voltage and ignition voltage supplied to the ECM. No ALDL data may be due to an ECM problem. CHART A-2 will diagnose the ECM. If TPS voltage is over 2.5 volts, the engine may be in the "clear flood" mode, which will cause starting problems. The engine will not start without reference pulses and, therefore, the "Scan" tool should indicate engine speed during cranking.
2. If engine speed was indicated during crank, the ignition module is receiving a crank signal, but "no spark" at this test indicates that the ignition module is not triggering the coil or there is a secondary ignition problem.
3. The test light should flash, indicating the ECM is controlling the injectors. The brightness of the light is not important. However, the test light should be a J 34730 or equivalent.
4. This test will determine if the ignition module is not generating the reference pulse, or if the wiring or ECM are at fault. By touching and removing a test light to battery voltage on CKT 430, a reference pulse should be generated. If engine speed is indicated, the ECM and wiring are OK.

Diagnostic Aids:
- Water or foreign material can cause a no start condition during freezing weather. The engine may start after 5 or 6 minutes in a heated shop. The problem may reoccur after an overnight park in freezing temperatures.
- An EGR sticking open can cause a low air/fuel ratio during cranking. Unless the system enters "clear flood" it can result in a flooding condition it can result in a no start.
- Fuel pressure: Low fuel pressure can result in a very lean air/fuel ratio. See CHART A-7.

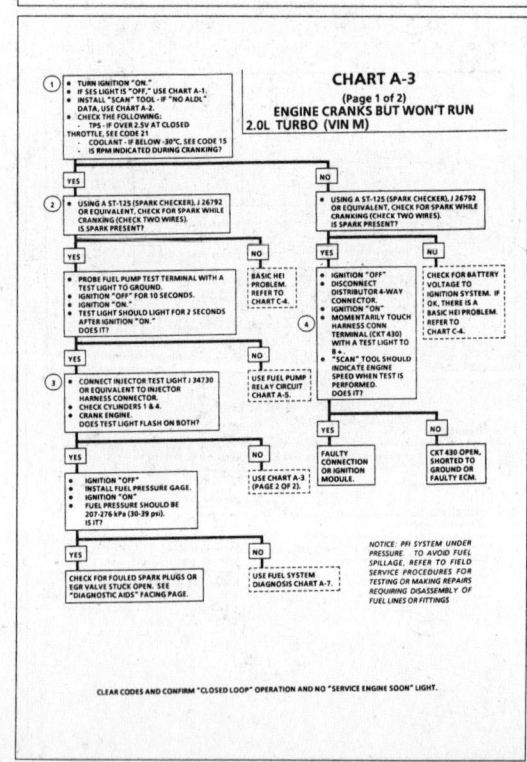

1988–90 2.0L TURBO (VIN M) – ALL MODELS

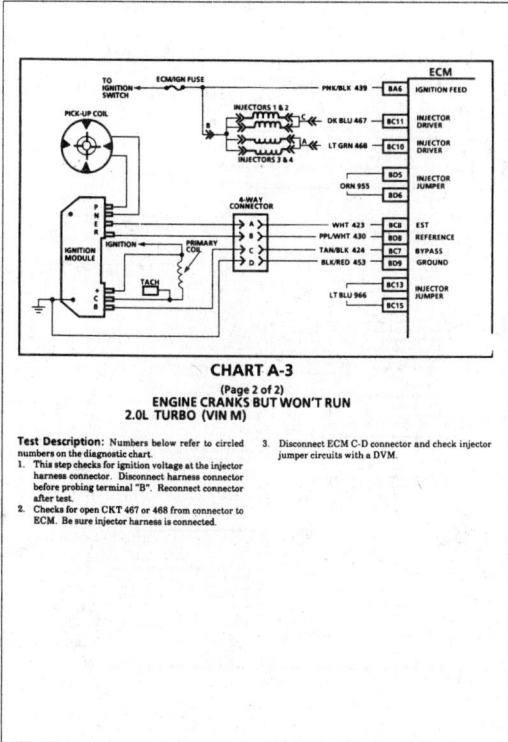

CHART A-3
(Page 2 of 2)
ENGINE CRANKS BUT WON'T RUN
2.0L TURBO (VIN M)

Test Description: Numbers below refer to circled numbers on the diagnostic chart.

1. This step checks for ignition voltage at the injector harness connector. Disconnect harness connector before probing terminal "B". Reconnect connector after test.
2. Checks for open CKT 467 or 468 from connector to ECM. Be sure injector harness is connected.
3. Disconnect ECM C-D connector and check injector jumper circuits with a DVM.

1988–90 2.0L TURBO (VIN M) – ALL MODELS

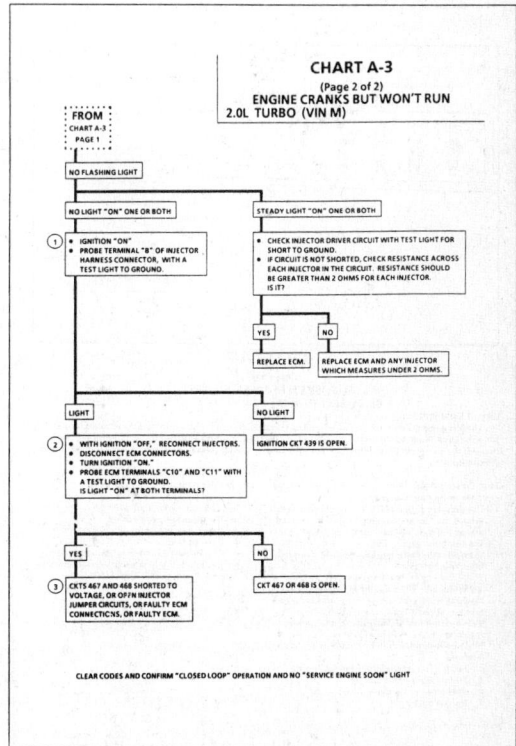

1988–90 2.0L TURBO (VIN M) – ALL MODELS

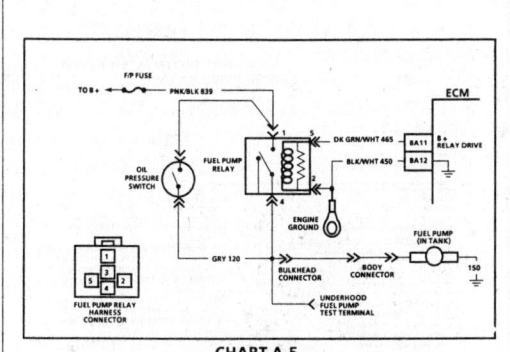

CHART A-5
FUEL PUMP RELAY CIRCUIT
2.0L TURBO (VIN M)

Circuit Description:

When the ignition switch is turned "ON," the Electronic Control Module (ECM) will activate the fuel pump relay with a 12 volt signal and run the in-tank fuel pump. The fuel pump will operate as long as the engine is cranking or running and the ECM is receiving ignition reference pulses. If there are no ignition reference pulses, the ECM will no longer supply the fuel pump relay signal within 2 seconds after the ignition is turned "ON."

Should the fuel pump relay or the 12 volt relay drive from the ECM fail, the fuel pump will receive electrical current through the oil pressure switch back-up circuit.

The fuel pump test terminal is located in the left side of the engine compartment. When the engine is stopped, the pump can be turned "ON" by applying battery voltage to the test terminal.

Diagnostic Aids:

1. This check is to determine if the oil pressure switch is faulty also. Since it is a back-up component, it must be checked without the fuel pump relay connected.
2. This check will determine if the oil pressure switch is shorted internally.

An inoperative fuel pump relay can result in long cranking times. The extended crank period is caused by the time necessary for oil pressure to reach the pressure required to close the oil pressure switch and turn "ON" the fuel pump.

1988–90 2.0L TURBO (VIN M) – ALL MODELS

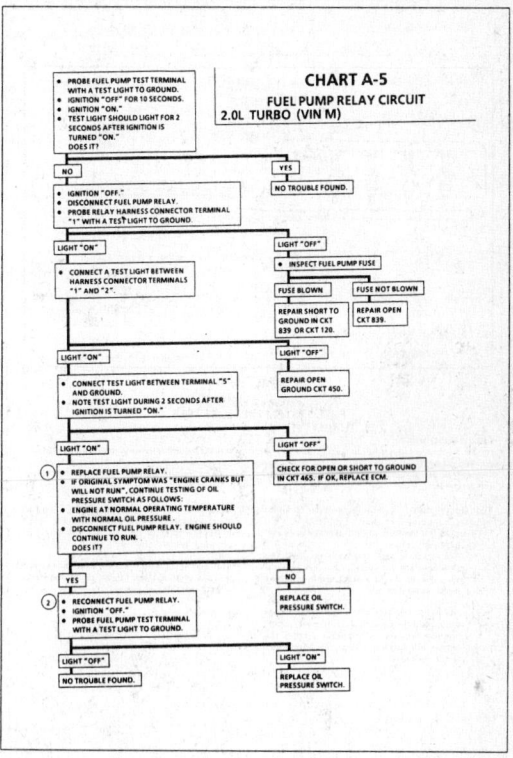

1988–90 2.0L TURBO (VIN M) – ALL MODELS

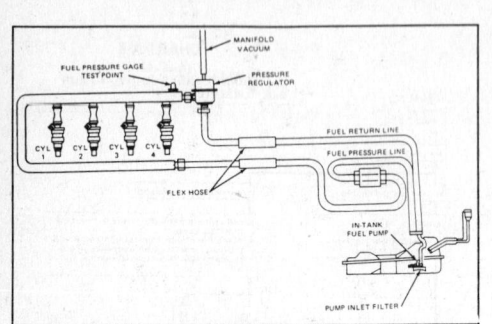

CHART A-7
(Page 1 of 2)
FUEL SYSTEM PRESSURE TEST
2.0L TURBO (VIN M)

Circuit Description:
The fuel pump delivers fuel to the fuel rail and injectors, where the system pressure is controlled from 245 to 256 kPa (35 to 38 psi) by the pressure regulator. Excess fuel is returned to the fuel tank. When the engine is stopped, the pump can be energized by applying battery voltage to the test terminal located in the engine compartment.

Test Description: Numbers below refer to circled numbers on the diagnostic chart.
1. Use pressure gage J 34730-1. Wrap a shop towel around the fuel pressure tap to absorb any small amount of fuel leakage that may occur when installing the gage. (The pressure will not leak down after the fuel pump is stopped on a correctly functioning system.)
2. While the engine is idling, manifold pressure is low (vacuum). When applied to the fuel regulator diaphragm, the pressure will result in a lower fuel pressure at about 190-200 kPa (25-30 psi).
3. The application of vacuum to the pressure regulator should result in a fuel pressure drop.
4. Pressure leak-down may be caused by one of the following:
   - In-tank fuel pump check valve not holding
   - Pump coupling hose leaking
   - Fuel pressure regulator valve leaking
   - Injector sticking open

Diagnostic Aids:

Improper fuel system pressure may contribute to one or all of the following symptoms
- Cranks but won't run
- Code 44 or 45
- Cutting out (May feel like ignition problem)
- Hesitation, loss of power or poor fuel economy
Refer to "Symptoms" in Section "B".

1988–90 2.0L TURBO (VIN M) – ALL MODELS

NOTICE: FUEL SYSTEM IS UNDER PRESSURE. TO AVOID FUEL SPILLAGE, REFER TO FIELD SERVICE PROCEDURES FOR TESTING OR REPAIRS REQUIRING DISASSEMBLY OF FUEL LINES OR FITTINGS.

CHART A-7
(Page 1 of 2)
FUEL SYSTEM PRESSURE TEST
2.0L TURBO (VIN M)

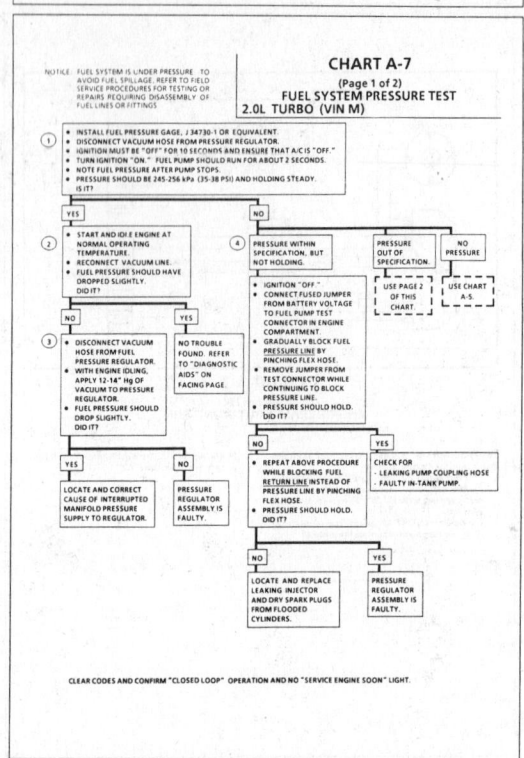

CLEAR CODES AND CONFIRM "CLOSED LOOP" OPERATION AND NO "SERVICE ENGINE SOON" LIGHT.

1988–90 2.0L TURBO (VIN M) – ALL MODELS

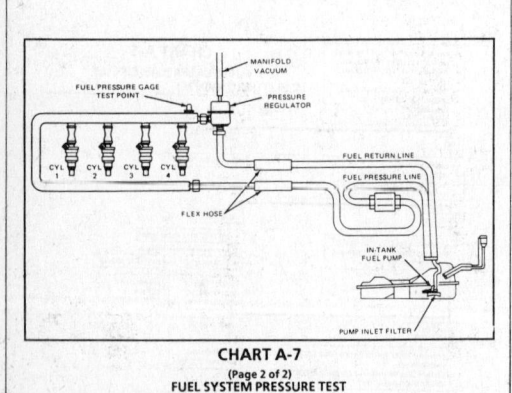

CHART A-7
(Page 2 of 2)
FUEL SYSTEM PRESSURE TEST
2.0L TURBO (VIN M)

Circuit Description:
The fuel pump delivers fuel to the fuel rail and injectors, where the system pressure is controlled from 245 to 256 kPa (35 to 36 psi) by the pressure regulator. Excess fuel is returned to the fuel tank. When the engine is stopped, the pump can be energized by applying battery voltage to the test terminal located in the engine compartment.

Test Description: Numbers below refer to circled numbers on the diagnostic chart.
5. Pressure less than 245 kPa (35 psi) may be caused by one of two problems.
   - The regulated fuel pressure is too low. The system will be running lean and may set Code 44. Also, hard cold starting and overall poor performance is possible.
   - Restricted flow is causing a pressure drop. Normally, a vehicle with a fuel pressure loss at idle will not be driveable. However, if the pressure drop occurs only while driving, the engine will surge and then stop as pressure begins to drop rapidly.
6. Restricting the fuel return line allows the fuel pump to build above regulated pressure. When battery voltage is applied to the pump test terminal, pressure should be above 450 kPa (65 psi).
7. This test determines if the high fuel pressure is due to a restricted fuel return line or a pressure regulator problem.

1988–90 2.0L TURBO (VIN M) – ALL MODELS

NOTICE: FUEL SYSTEM IS UNDER PRESSURE. TO AVOID FUEL SPILLAGE. REFER TO FIELD SERVICE PROCEDURES FOR TESTING OR REPAIRS REQUIRING DISASSEMBLY OF FUEL LINES OR FITTINGS.

CHART A-7
(Page 2 of 2)
FUEL SYSTEM PRESSURE TEST
2.0L TURBO (VIN M)

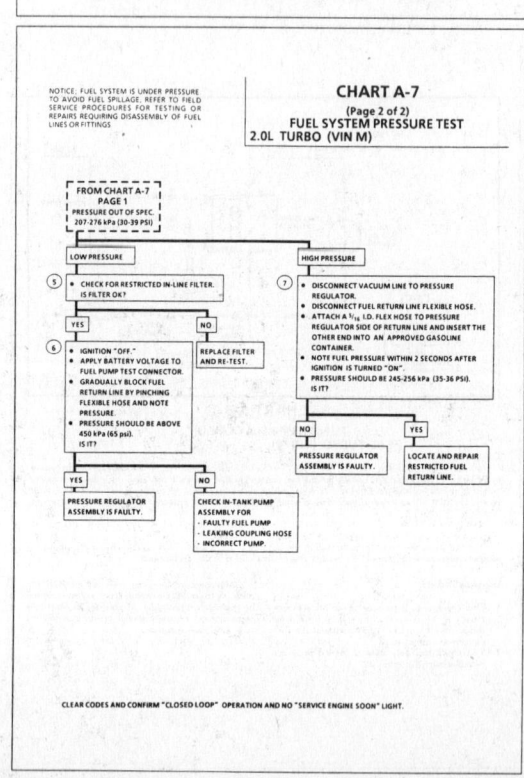

CLEAR CODES AND CONFIRM "CLOSED LOOP" OPERATION AND NO "SERVICE ENGINE SOON" LIGHT.

1988–90 2.0L TURBO (VIN M) – ALL MODELS

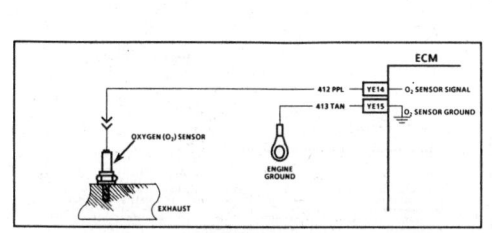

CODE 13
OXYGEN SENSOR CIRCUIT
(OPEN CIRCUIT)
2.0L TURBO (VIN M)

Circuit Description:
The ECM supplies a voltage of about .45 volt between terminals "YE14" and "YE15". (If measured with a 10 megohm digital voltmeter, this may read as low as .32 volt.) The O₂ sensor varies the voltage within a range of about 1 volt if the exhaust is rich, down through about .10 volt if exhaust is lean.

The sensor is like an open circuit and produces no voltage when it is below 360°C (600°F). An open sensor circuit or cold sensor causes "Open Loop" operation.

Test Description: Numbers below refer to circled numbers on the diagnostic chart.
1. Code 13 will set:
   - Engine at normal operating temperature
   - O₂ signal voltage steady between .35 and .55 volt
   - Throttle position sensor signal above 6.5%
   - All conditions must be met for about 60 seconds.

   If the conditions for a Code 13 exist, the system will not go "Closed Loop."
2. This will determine if the sensor is at fault or the wiring or ECM is the cause of the Code 13.
3. In doing this test use only a high impedance digital volt ohmmeter. This test checks the continuity of CKTs 412 and 413 because if CKT 413 is open, the ECM voltage on CKT 412 will be over .6 volt (600 mV).

Diagnostic Aids:
Normal "Scan" voltage varies between 100 mV to 999 mV (.1 and 1.0 volt) while in "Closed Loop." Code 13 will be set in 3 seconds if all criteria have been met and the system will go "Open Loop."
Refer to "Intermittents" in Section "B".

1988–90 2.0L TURBO (VIN M) – ALL MODELS

CODE 13
OXYGEN SENSOR CIRCUIT
(OPEN CIRCUIT)
2.0L TURBO (VIN M)

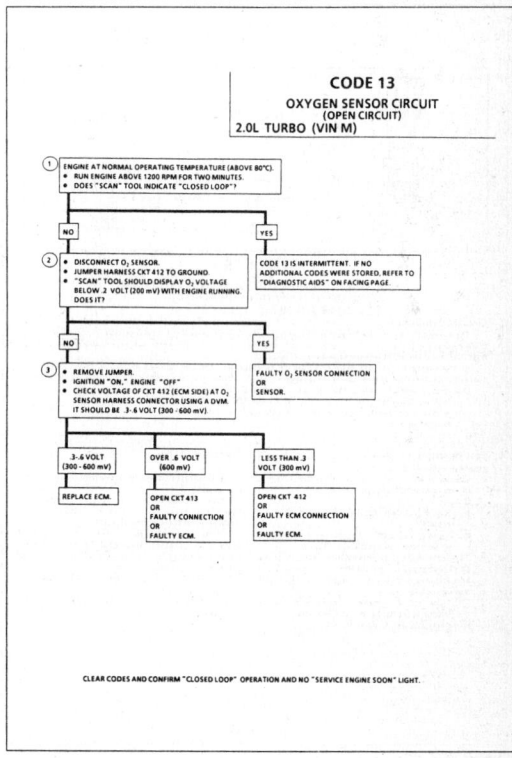

CLEAR CODES AND CONFIRM "CLOSED LOOP" OPERATION AND NO "SERVICE ENGINE SOON" LIGHT.

1988–90 2.0L TURBO (VIN M) – ALL MODELS

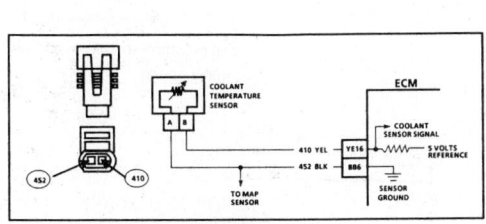

CODE 14
COOLANT TEMPERATURE SENSOR (CTS) CIRCUIT
(HIGH TEMPERATURE INDICATED)
2.0L TURBO (VIN M)

Circuit Description:
The Coolant Temperature Sensor (CTS) uses a thermistor to control the signal voltage to the ECM. The ECM applies a voltage on CKT 410 to the sensor. When the engine is cold, the sensor (thermistor) resistance is high, therefore, the ECM will see high signal voltage.

As the engine warms, the sensor resistance becomes less and the voltage drops. At normal engine operating temperature, the voltage will measure about 1.5 to 2.0 volt at the ECM terminal "YE16".

Coolant temperature is one of the inputs used to control:
- Fuel Delivery
- Electronic Spark Timing (EST)
- Cooling Fan
- Torque Converter Clutch (TCC)
- Idle Air Control (IAC)

Test Description: Numbers below refer to circled numbers on the diagnostic chart.
1. Checks to see if a code was set as a result of hard failure or intermittent condition.
   Code 14 will set if:
   - Engine has been running for more than 10 seconds.
   - Signal voltage indicates a coolant temperature above 135°C (275°F) for 3 seconds.
2. This test simulates conditions for a Code 15. If the ECM recognizes the open circuit (high voltage), and displays a low temperature, the ECM and wiring are OK.

Diagnostic Aids:
A "Scan" tool reads engine temperature in degrees centigrade.
After the engine is started, the temperature should rise steadily to about 90°, then stabilize, when the thermostat opens.
If the engine has been allowed to cool to an ambient temperature (overnight), coolant and MAT temperature may be checked with a "Scan" tool and should read close to each other.
When a Code 14 is set, the ECM will turn "ON" the engine cooling fan.
A Code 14 will result if CKT 410 is shorted to ground.
If Code 14 is intermittent, refer to Section "B".

1988–90 2.0L TURBO (VIN M) – ALL MODELS

CODE 14
COOLANT TEMPERATURE SENSOR CIRCUIT
(HIGH TEMPERATURE INDICATED)
2.0L TURBO (VIN M)

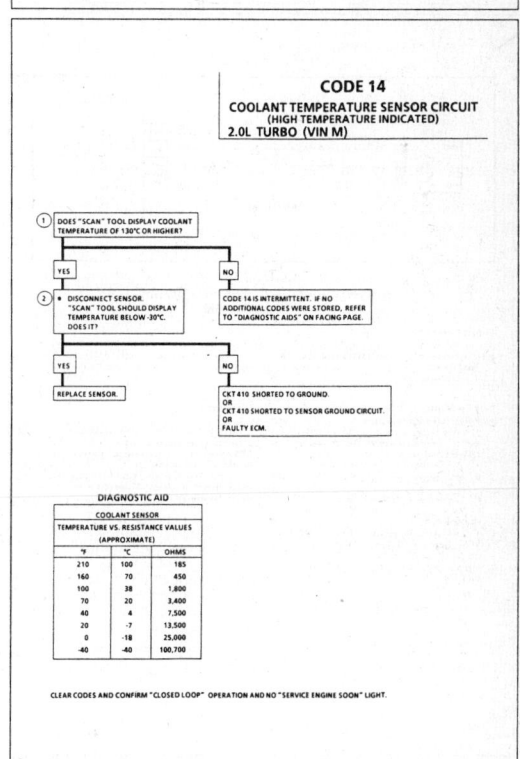

DIAGNOSTIC AID

COOLANT SENSOR TEMPERATURE VS. RESISTANCE VALUES (APPROXIMATE)		
°F	°C	OHMS
210	100	185
160	70	450
100	38	1,800
70	20	3,400
40	4	7,500
20	-7	13,500
0	-18	25,000
-40	-40	100,700

CLEAR CODES AND CONFIRM "CLOSED LOOP" OPERATION AND NO "SERVICE ENGINE SOON" LIGHT.

1988–90 2.0L TURBO (VIN M) – ALL MODELS

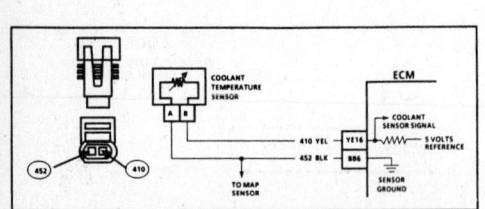

CODE 15
COOLANT TEMPERATURE SENSOR (CTS) CIRCUIT
(LOW TEMPERATURE INDICATED)
2.0L TURBO (VIN M)

Circuit Description:
The Coolant Temperature Sensor (CTS) uses a thermistor to control the signal voltage to the ECM. The ECM applies a voltage on CKT 410 to the sensor. When the engine is cold, the sensor (thermistor) resistance is high, therefore, the ECM will see high signal voltage.
As the engine warms, the sensor resistance becomes less, and the voltage drops. At normal engine operating temperature, the voltage will measure about 1.5 volts to 2.0 volts at the ECM terminal "YE16".
Coolant temperature is one of the inputs used to control
- Fuel Delivery
- Torque Converter Clutch (TCC)
- Electronic Spark Timing (EST)
- Idle Air Control (IAC)
- Cooling Fan

Test Description: Numbers below refer to circled numbers on the diagnostic chart.
1. Checks to see if code was set as result of hard failure or intermittent condition.
   Code 15 will be set if:
   - Engine has been running for more than 50 seconds.
   - Signal voltage indicates a coolant temperature below -30°C (-22°F).
2. This test simulates conditions for a Code 14. If the ECM recognizes the grounded circuit (low voltage) and displays a high temperature, the ECM and wiring are OK.
3. This test will determine if there is a wiring problem or a faulty ECM. If CKT 452 is open, there may also be a Code 21 stored.

Diagnostic Aids:
A "Scan" tool reads engine temperature in degrees centigrade.
After the engine is started, the temperature should rise steadily to about 90°C (194°F), then stabilize when the thermostat opens.
If the engine has been allowed to cool to an ambient temperature (overnight), coolant and MAT temperature may be checked with a "Scan" tool and should read close to each other.
When a Code 15 is set, the ECM will turn "ON" the engine cooling fan.
A Code 15 will result if CKTs 410 or 452 are open.
If Code 15 is intermittent, refer to Section "B".

1988–90 2.0L TURBO (VIN M) – ALL MODELS

CODE 15
COOLANT TEMPERATURE SENSOR CIRCUIT
(LOW TEMPERATURE INDICATED)
2.0L TURBO (VIN M)

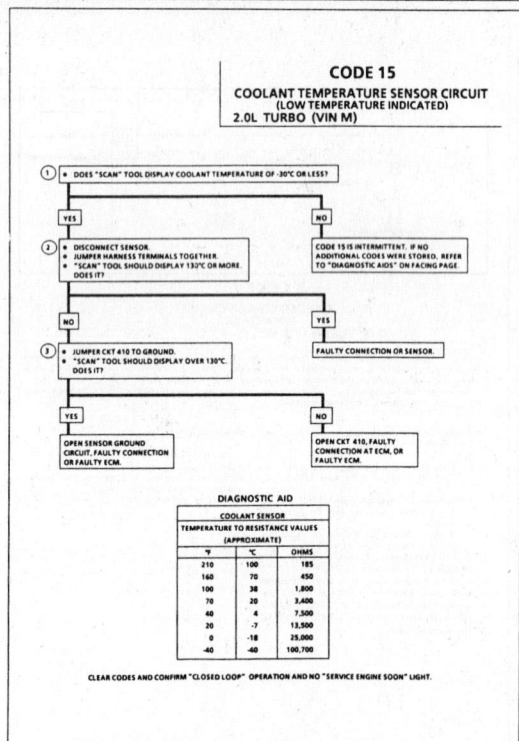

DIAGNOSTIC AID		
COOLANT SENSOR		
TEMPERATURE TO RESISTANCE VALUES		
(APPROXIMATE)		
°F	°C	OHMS
210	100	185
160	70	450
100	38	1,800
70	20	3,400
40	4	7,500
20	-7	13,500
0	-18	25,000
-40	-40	100,700

CLEAR CODES AND CONFIRM "CLOSED LOOP" OPERATION AND NO "SERVICE ENGINE SOON" LIGHT.

1988–90 2.0L TURBO (VIN M) – ALL MODELS

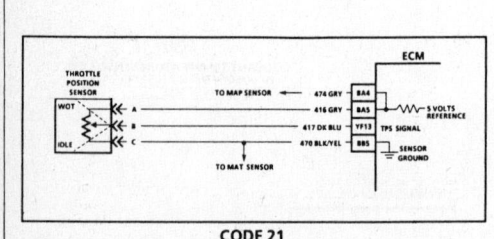

CODE 21
THROTTLE POSITION SENSOR (TPS) CIRCUIT
(SIGNAL VOLTAGE HIGH)
2.0L TURBO (VIN M)

Circuit Description:
The Throttle Position Sensor (TPS) provides a voltage signal that changes relative to the throttle valve. Signal voltage will vary from less than 1.0 volt at idle to about 4.6 volts at wide open throttle (WOT).
The TPS signal is one of the most important inputs used by the ECM for fuel control and for many of the ECM controlled outputs.

Test Description: Numbers below refer to circled numbers on the diagnostic chart.
1. This step checks to see if Code 21 is the result of a hard failure or an intermittent condition.
   A Code 21 will set if:
   - TPS reading above 2.5 volts
   - MAP reading below 70 kPa (M/T) or 81 kPa (A/T)
   - Engine speed less than 1300 rpm
   - All of the above conditions are present for 10 seconds
2. This step simulates conditions for a Code 22. If the ECM recognizes the change of state, the ECM and CKTs 416 and 417 are OK.
3. This step isolates a faulty sensor, ECM, or an open CKT 470.

Diagnostic Aids:
A "Scan" tool displays throttle position in volts. Closed throttle voltage should be less than 1.0 volt. TPS voltage should increase at a steady rate as throttle is moved to WOT.
A Code 21 will result if CKT 470 is open or CKT 417 is shorted to voltage. If Code 21 is intermittent, refer to Section "B".

1988–90 2.0L TURBO (VIN M) – ALL MODELS

CODE 21
THROTTLE POSITION SENSOR (TPS) CIRCUIT
(SIGNAL VOLTAGE HIGH)
2.0L TURBO (VIN M)

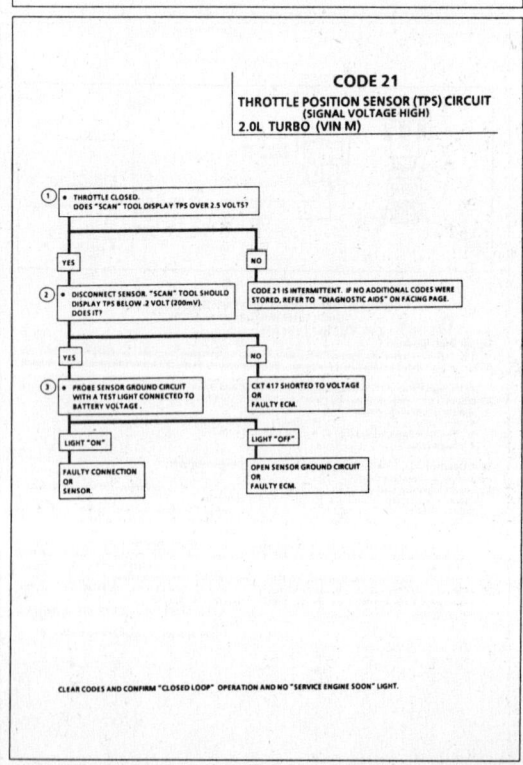

CLEAR CODES AND CONFIRM "CLOSED LOOP" OPERATION AND NO "SERVICE ENGINE SOON" LIGHT.

1988–90 2.0L TURBO (VIN M) – ALL MODELS

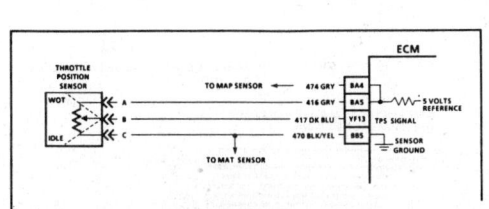

CODE 22
THROTTLE POSITION SENSOR (TPS) CIRCUIT
(SIGNAL VOLTAGE LOW)
2.0L TURBO (VIN M)

Circuit Description:
The Throttle Position Sensor (TPS) provides a voltage signal that changes relative to the throttle valve. Signal voltage will vary from less than 1.0 volt at idle to about 4.5 volts at wide open throttle (WOT).
The TPS signal is one of the most important inputs used by the ECM for fuel control and for many of the ECM controlled outputs.

Test Description: Numbers below refer to circled numbers on the diagnostic chart.
1. Code 22 will set if:
   - Engine is running
   - TPS signal voltage is less than .2 volt for 4 seconds
2. Simulates Code 21: (high voltage). If ECM recognizes the high signal voltage the ECM and wiring are OK.
3. With closed throttle, ignition "ON" or at idle, voltage at "YF13" should be .36–.44 volt. If not, replace the TPS.
4. Simulates a high signal voltage. Checks CKT 417 for an open.

Diagnostic Aids:

A "Scan" tool reads throttle position in volts. Voltage should increase at a steady rate as throttle is moved toward WOT.
Also some "Scan" tools will read throttle angle 0% = closed throttle, 100% = WOT.

An open or short to ground in CKTs 416 or 417 will result in a Code 22.
- Poor Connection or Damaged Harness. Inspect ECM harness connectors for backed out terminal "YF13", improper mating, broken locks, improperly formed or damaged terminals, poor terminal to wire connection, and damaged harness.
- Intermittent Test. If connections and harness check OK, monitor TPS voltage display while moving related connectors and wiring harness. If the failure is induced the display will change. This may help to isolate the location of the malfunction.
- TPS Scaling. Observe TPS voltage display while depressing accelerator pedal with engine stopped and ignition "ON." Display should vary from closed throttle TPS voltage when throttle was closed, to over 4.5 volts (4500 mV) when throttle is held at wide open throttle position.

1988–90 2.0L TURBO (VIN M) – ALL MODELS

CODE 22
THROTTLE POSITION SENSOR (TPS) CIRCUIT
(SIGNAL VOLTAGE LOW)
2.0L TURBO (VIN M)

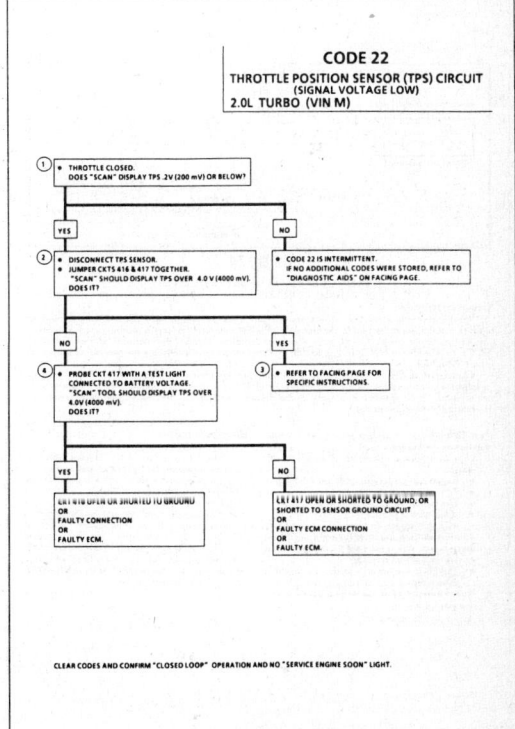

CLEAR CODES AND CONFIRM "CLOSED LOOP" OPERATION AND NO "SERVICE ENGINE SOON" LIGHT.

1988–90 2.0L TURBO (VIN M) – ALL MODELS

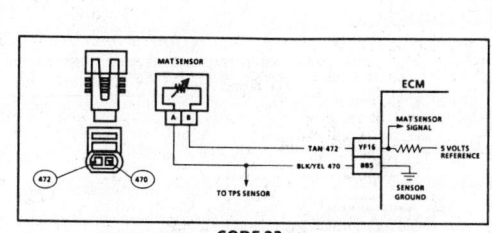

CODE 23
MANIFOLD AIR TEMPERATURE (MAT) SENSOR CIRCUIT
(LOW TEMPERATURE INDICATED)
2.0L TURBO (VIN M)

Circuit Description:
The Manifold Air Temperature (MAT) sensor uses a thermistor to control the signal voltage to the ECM. The ECM applies a voltage (about 5 volts) on CKT 472 to the sensor. When the air is cold the sensor (thermistor) resistance is high, therefore the ECM will see a high signal voltage. If the air is warm, the sensor resistance is low, therefore, the ECM will detect a low voltage.

Test Description: Numbers below refer to circled numbers on the diagnostic chart.
Code 23 will set if:
- A signal voltage indicates a manifold air temperature below −30°C (−22°F).
- Boost conditions have been present for longer than 5 seconds.
Due to the conditions necessary to set a Code 23, the "Service Engine Soon" light will only stay "ON" while the fault is present.

1. A "Scan" tool may not be used to diagnose this fault, due to the ECM transmitting "default" (substitute) values while the fault is present. A Code 23 will set due to an open sensor, wire, or connection. This test will determine if the wiring and ECM are OK.
2. If the resistance is greater than 25,000 ohms, replace the sensor.

1988–90 2.0L TURBO (VIN M) – ALL MODELS

CODE 23
MANIFOLD AIR TEMPERATURE (MAT) SENSOR CIRCUIT
(LOW TEMPERATURE INDICATED)
2.0L TURBO (VIN M)

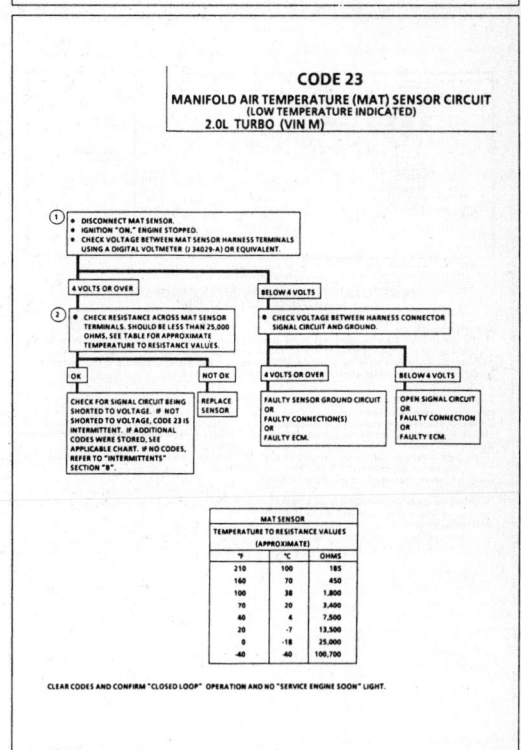

MAT SENSOR		
TEMPERATURE TO RESISTANCE VALUES (APPROXIMATE)		
°F	°C	OHMS
210	100	185
160	70	450
100	38	1,800
70	20	3,400
40	4	7,500
20	-7	13,500
0	-18	25,000
-40	-40	100,700

CLEAR CODES AND CONFIRM "CLOSED LOOP" OPERATION AND NO "SERVICE ENGINE SOON" LIGHT.

1988–90 2.0L TURBO (VIN M) – ALL MODELS

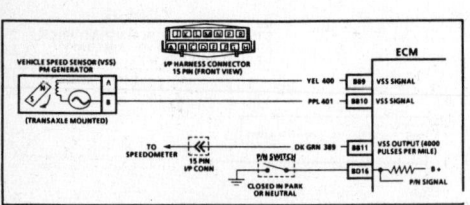

CODE 24
VEHICLE SPEED SENSOR (VSS) CIRCUIT
2.0L TURBO (VIN M)

Circuit Description:
Vehicle speed information is provided to the ECM by the Vehicle Speed Sensor (VSS), which is a permanent magnet (PM) generator, located in the transaxle. The PM generator produces a pulsing voltage whenever vehicle speed is over 3 mph. The AC voltage level and the number of pulses increases with vehicle speed. The ECM then converts the pulsing voltage to mph which is used for calculations, and the mph can be displayed with a "Scan" tool.

The function of VSS buffer used in past model years has been incorporated into the ECM. The ECM then supplies the necessary signal for the instrument panel (4000 pulses per mile) for operating the speedometer and the odometer. If the vehicle is equipped with cruise control the ECM also provides a signal (2000 pulses per mile) to the cruise control module.

Test Description: Numbers below refer to circled numbers on the diagnostic chart.
1. Code 24 will set if vehicle speed equals 0 mph.
   - Engine speed is between 1600 and 4400 rpm
   - TPS indicates closed throttle
   - Low load condition (low air flow)
   - Not in park or neutral
   - All conditions met for 5 seconds
   These conditions are met during a road load deceleration. Disregard Code 24 that sets when drive wheels are not turning.
   - The PM generator only produces a signal if drive wheels are turning greater than 3 mph.
2. Before replacing ECM, check Mem-Cal for proper application.

Diagnostic Aids:
"Scan" should indicate a vehicle speed whenever the drive wheels are turning greater than 3 mph.

A problem in CKT 381 or 389 will not affect the VSS input or the readings on a "Scan."

Check CKTs 400 and 401 to ensure that the connections are clean and tight and that the harness is routed correctly. Refer to "Intermittents" in Section "B".

(A/T) A faulty or misadjusted park/neutral switch can result in a false Code 24. Use a "Scan" and check for proper signal while in drive. Refer to CHART C-1A for P/N switch diagnosis check.

1988–90 2.0L TURBO (VIN M) – ALL MODELS

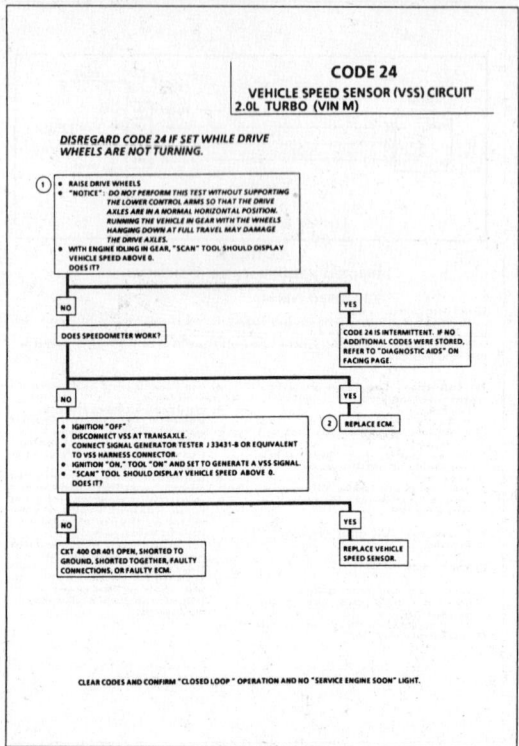

1988–90 2.0L TURBO (VIN M) – ALL MODELS

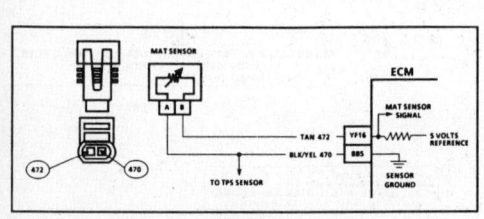

CODE 25
MANIFOLD AIR TEMPERATURE (MAT) SENSOR CIRCUIT
(HIGH TEMPERATURE INDICATED)
2.0L TURBO (VIN M)

Circuit Description:
The Manifold Air Temperature (MAT) sensor uses a thermistor to control the signal voltage to the ECM. The ECM applies a voltage (4-6 volts) on CKT 472 to the sensor. When manifold air is cold, the sensor (thermistor) resistance is high, therefore, the ECM will detect a high signal voltage. If the air warms, the sensor resistance becomes less, and the voltage drops.

Test Description: Numbers below refer to circled numbers on the diagnostic chart.
Code 25 will set if the engine is not experiencing turbocharger boost and the following conditions are met:
- Signal voltage indicates a manifold air temperature greater than 135°C (275°F).
- The above requirement is met for at least 30 seconds.
Due to the conditions necessary to set a Code 25, the "Service Engine Soon" light will only stay "ON" while the fault is present.

1. A "Scan" tool may not be used to diagnose this fault due to the ECM transmitting "default" (substitute) values while the fault is present. If voltage is above 4 volts, the ECM and wiring are OK.
2. If the resistance is less than 100 ohms, replace the sensor.

1988–90 2.0L TURBO (VIN M) – ALL MODELS

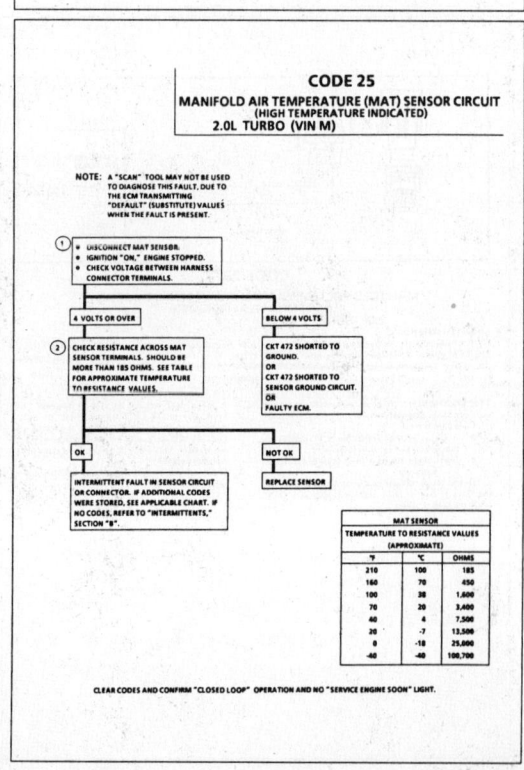

1988–90 2.0L TURBO (VIN M) – ALL MODELS

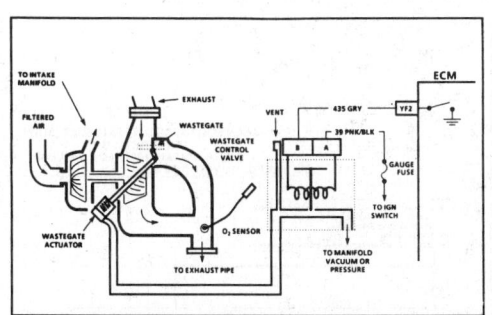

CODE 31
TURBO WASTEGATE OVERBOOST
2.0L TURBO (VIN M)

Circuit Description:

On turbocharged engines, the exhaust gases pass from the exhaust manifold through the turbocharger, turning the turbine blades. The compressor side of the turbocharger also turns, pulling air through the air filter and pushing the air into the intake manifold, pressurizing the intake manifold.

The wastegate is normally closed, but opens to bypass exhaust gas to prevent an overboost condition. The wastegate will open when pressure is applied to the actuator, and is controlled by a wastegate control solenoid valve pulsed "ON" and "OFF" by the ECM. Under normal driving conditions, the control solenoid is energized all the time which closes "OFF" the manifold pressure to the wastegate actuator. This allows for a rapid increase in boost pressure. A boost increase will be detected by the MAP sensor, and the ECM will pulse the wastegate control valve. Manifold pressure will then be allowed to pass to the wastegate actuator, and the actuator will open the wastegate. This will prevent an overboost condition on heavy acceleration. As boost pressure decreases, the ECM closes the control valve and the wastegate actuator pressure bleeds "OFF" through the vent in the control valve. If an overboost does exist as indicated by the MAP sensor, the ECM will reduce fuel delivery to prevent damage to the engine.

Test Description: Numbers below refer to circled numbers on the diagnostic chart.

1. A Code 31 will set when the manifold pressure exceeds 143 kPa of boost for two seconds, and a Code 33 has not previously been set. Code 31 will set, but the "Service Engine Soon" light will stay "ON" only while the overboost exists. The light will stay "ON" for 10 seconds after the condition exists and then go "OUT".
   An overboost condition could be caused by:
   - CKT 435 shorted to ground
   - A sticking wastegate actuator or wastegate
   - A control valve stuck in the closed position
   - A cut or pinched hose
   - A faulty ECM
   - An extremely dirty air filter

With ignition shut "OFF," the control valve solenoid is open.

2. After the 103 kPa (15 psi) is applied to valve and then the pressure source is removed, the actuator should slowly move back and close the wastegate. If the pressure does not bleed "OFF," the vent in the control valve solenoid could be plugged.

3. With the ignition "ON" and the diagnostic terminal grounded, the control valve solenoid should be energized. This closes "OFF" the manifold to the wastegate actuator.

4. Checks the electrical control portion of the system. With key "ON" and engine not running, the solenoid should not be energized.

1988–90 2.0L TURBO (VIN M) – ALL MODELS

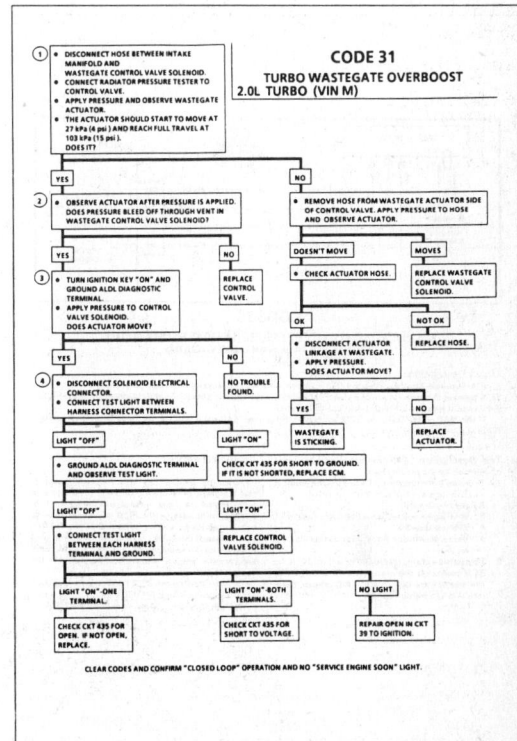

CODE 31
TURBO WASTEGATE OVERBOOST
2.0L TURBO (VIN M)

1988–90 2.0L TURBO (VIN M) – ALL MODELS

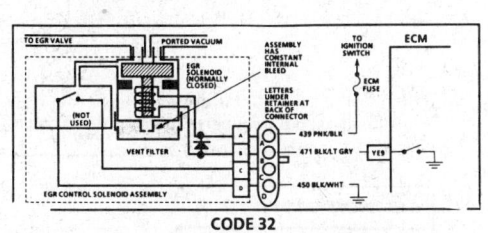

CODE 32
EXHAUST GAS RECIRCULATION (EGR) CIRCUIT
2.0L TURBO (VIN M)

Circuit Description:

The ECM operates a vacuum solenoid to control the Exhaust Gas Recirculation (EGR) valve. This solenoid is normally closed. By providing a ground path, the ECM energizes the solenoid, allowing vacuum to reach the EGR valve.

Under certain conditions, when the EGR valve is normally open, the ECM tests the EGR function by de-energizing the EGR control solenoid, blocking vacuum to the EGR valve diaphragm. Without EGR, the system will sense a lean condition and will increase the fuel integrator rate in response. The ECM monitors the amount of fuel delivery increase. If the increase is below a specified value, the ECM will interpret that the test was failed. The failure indicates that closing the EGR valve when it would normally be open does not make a significant change, indicating a problem in the EGR system.

Test Description: Numbers below refer to circled numbers on the diagnostic chart.

The diagnostic chart covers checks for the entire EGR system. If no trouble is found but Code 32 was set, an intermittent electrical condition or a sticky EGR valve is at fault.

Diagnostic Aids:

The vacuum switch in the EGR solenoid assembly is not used.

An EGR valve stuck open will cause a rough idle.

A plugged EGR solenoid vent filter could cause the EGR valve to remain open or to close slowly.

An inoperative check valve in the ported vacuum line will result in faulty EGR system operation.

1988–90 2.0L TURBO (VIN M) – ALL MODELS

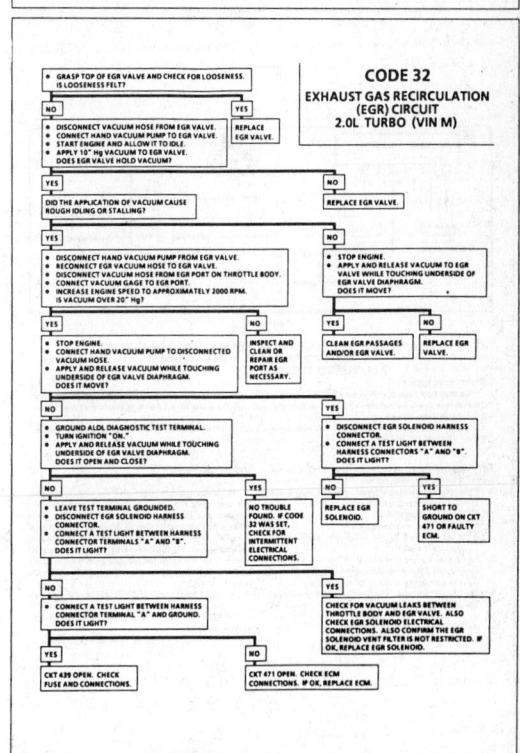

CODE 32
EXHAUST GAS RECIRCULATION (EGR) CIRCUIT
2.0L TURBO (VIN M)

1988–90 2.0L TURBO (VIN M) – ALL MODELS

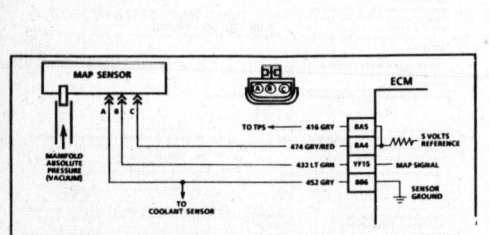

CODE 33
MANIFOLD ABSOLUTE PRESSURE (MAP) SENSOR CIRCUIT
(SIGNAL VOLTAGE HIGH - LOW VACUUM)
2.0L TURBO (VIN M)

Circuit Description:
The Manifold Absolute Pressure (MAP) sensor responds to changes in manifold pressure (vacuum). The ECM receives this information as a signal voltage that will vary from about 1-1.5 volts at closed throttle idle, to 4-4.5 volts at wide open throttle (low vacuum or boost).
If the MAP sensor fails, the ECM will substitute a fixed MAP value and use the throttle position sensor (TPS) to control fuel delivery.

Test Description: Numbers below refer to circled numbers on the diagnostic chart.
1. This step will determine if Code 33 is the result of a hard failure or an intermittent condition.
   A Code 33 will set if:
   • MAP signal voltage is too high (low vacuum).
   • TPS less than 2%.
   • These conditions for a time longer than 5 seconds.
2. This step simulates conditions for a Code 34. If the ECM recognizes the change, the ECM and CKTs 474 and 432 are OK. If CKT 452 is open, there may also be a stored Code 23.

Diagnostic Aids:

With the ignition "ON" and the engine stopped, the manifold pressure is equal to atmospheric pressure and the signal voltage will be high. This information is used by the ECM as an indication of vehicle altitude and is referred to as BARO. Comparison of this BARO reading with a known good vehicle with the same sensor is a good way to check accuracy of a "suspect" sensor. Reading should be the same ± .4 volt.
A Code 33 will result if CKT 452 is open, or if CKT 432 is shorted to voltage or to CKT 474.
If Code 33 is intermittent, refer to Section "B".

1988–90 2.0L TURBO (VIN M) – ALL MODELS

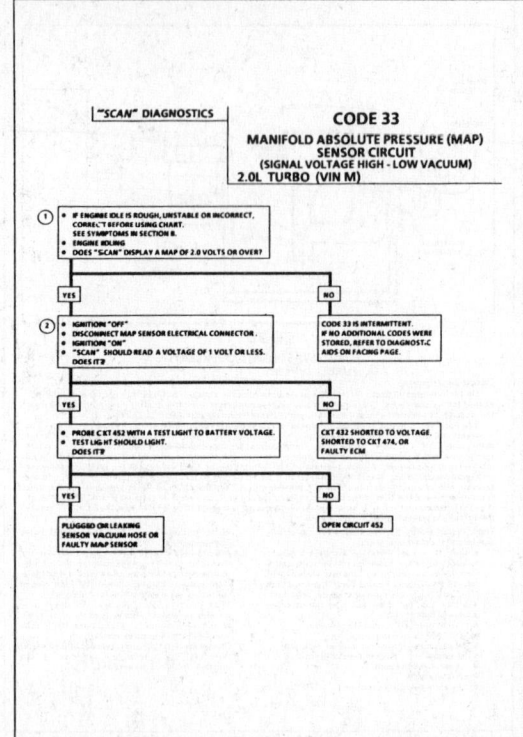

1988–90 2.0L TURBO (VIN M) – ALL MODELS

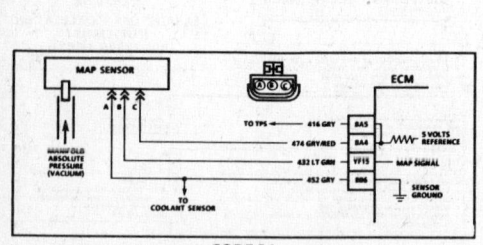

CODE 34
MANIFOLD ABSOLUTE PRESSURE (MAP) SENSOR CIRCUIT
(SIGNAL VOLTAGE LOW - HIGH VACUUM)
2.0L TURBO (VIN M)

Circuit Description:
The Manifold Absolute Pressure (MAP) sensor responds to changes in manifold pressure (vacuum). The ECM receives this information as a signal voltage that will vary from less than 1.0 volt at closed throttle idle, to 4-4.5 volts at wide open throttle.
If the MAP sensor fails, the ECM will substitute a fixed MAP value and use the throttle position sensor (TPS) to control fuel delivery.

Test Description: Numbers below refer to circled numbers on the diagnostic chart.
1. This step determines if Code 34 is the result of a hard failure or an intermittent condition.
   A Code 34 will set when:
   • MAP signal voltage is too low
   • Engine speed below 1200 rpm and/or TPS greater than 20%
2. Jumpering harness terminals "B" to "C", 5 volts to signal, will determine if the sensor is at fault or if there is a problem with the ECM or wiring.
3. The "Scan" tool may not display battery voltage. The important thing is that the ECM recognizes the voltage as more than 4 volts, indicating that the ECM and CKT 432 are OK.

Diagnostic Aids:

With the ignition "ON" and the engine stopped, the manifold pressure is equal to atmospheric pressure and the signal voltage will be high. This information is used by the ECM as an indication of vehicle altitude and is referred to as BARO. Comparison of this BARO reading with a known good vehicle with the same sensor is a good way to check accuracy of a "suspect" sensor. Reading should be the same ± .4 volt.
A Code 34 will result if CKTs 474 or 432 are open or shorted to ground.
If CKT 416 is shorted to ground, there may also be a stored Code 22.
If Code 34 is intermittent, refer to Section "B".

1988–90 2.0L TURBO (VIN M) – ALL MODELS

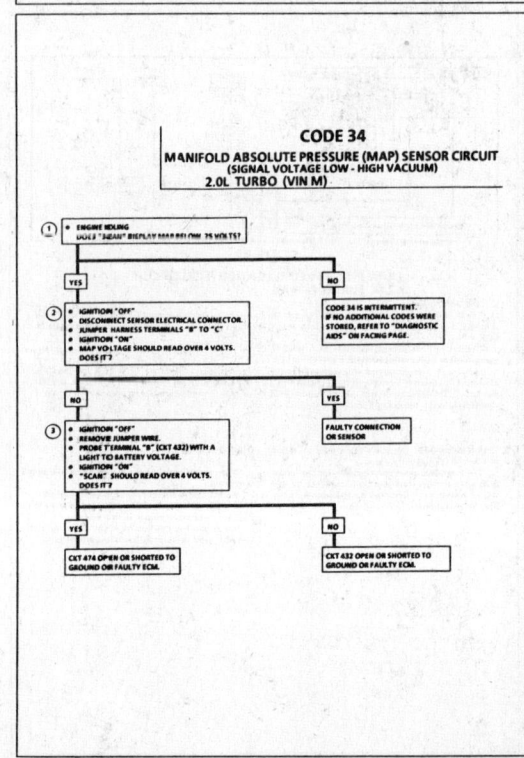

1988–90 2.0L TURBO (VIN M) – ALL MODELS

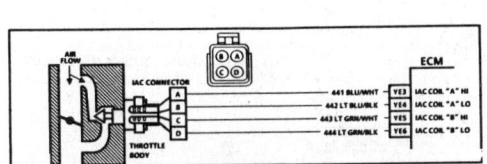

CODE 35
IDLE SPEED ERROR
2.0L TURBO (VIN M)

Circuit Description:
Code 35 will set when the closed throttle engine speed is 225 rpm above or below the desired (commanded) idle speed for 20 seconds.

Test Description: Numbers below refer to circled numbers on the diagnostic chart.

1. Continue with test even if engine will not idle. If idle is too low, "Scan" will display 80 or more counts (steps). If idle is high it will display "0" counts. Occasionally an erratic or unstable idle may occur. Engine speed may vary 200 rpm or more up and down. Engine speed may vary 200 rpm or more up and down. If the condition is unchanged, the IAC is not at fault. There is a system problem. Proceed to diagnostic aids below.

2. When the engine was stopped, the IAC valve retracted (more air) to a fixed "park" position for increased air flow and idle speed during the next engine start. A "Scan" will display 80 or more counts.

3. Be sure to disconnect the IAC valve prior to this test. The test light will confirm the ECM signals by a steady or flashing light on all circuits.

4. There is a remote possibility that one of the circuits is shorted to voltage which would have been indicated by a steady light. Disconnect ECM and turn the ignition "ON" and probe terminals to check for this condition.

Diagnostic Aids:

A slow unstable idle may be caused by a system problem that cannot be overcome by the IAC. "Scan" counts will be above 80 counts if idle is too low and "0" counts if it is too high.

If idle is too high, stop engine. Ignition "ON." Ground diagnostic terminal. Wait at least 30 seconds for IAC to seat, then disconnect IAC. Start engine. If idle speed is above minimum air rate, locate and correct vacuum leak.

- **System too lean (high air/fuel ratio)**
  Idle speed may be too high or too low. Engine speed may vary up and down, disconnecting IAC does not help. This may set Code 44. "Scan" and/or voltmeter will read an oxygen sensor output less than 300 mV (.3 volt). Check for low regulated fuel pressure or water in fuel. A lean exhaust with an oxygen sensor output fixed above 800 mV (.8 volt) will be a contaminated sensor, usually silicone. This may also set a Code 45.
- **System too rich (low air/fuel ratio)**
  Idle speed too low. "Scan" counts usually above 80. System obviously rich and may exhibit black smoke exhaust. "Scan" tool and/or voltmeter will read an oxygen sensor signal fixed above 800 mV (.8 volt).
  Check
  - For fuel in pressure regulator hose
  - High fuel pressure
  - Injector leaking or sticking.
- **Throttle body.** Inspect area beneath throttle plate for a build-up of residue (coking). If coking is present, the throttle body assembly must be cleaned before setting minimum air rate or replacing IAC. Remove IAC and inspect bore for foreign material or evidence of IAC valve dragging the bore.
- **A/C compressor or relay failure.** See CHART C-10 if the A/C control relay drive circuit is shorted to ground or if the relay is faulty an idle problem may exist.
- Refer to "Rough, Unstable, Incorrect Idle or Stalling" in "Symptoms," in Section "B".

1988–90 2.0L TURBO (VIN M) – ALL MODELS

CODE 35
IDLE SPEED ERROR
2.0L TURBO (VIN M)

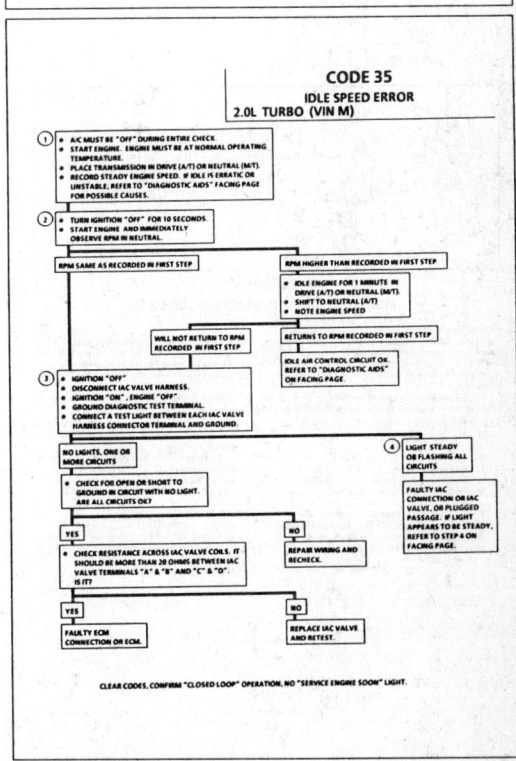

CLEAR CODES, CONFIRM "CLOSED LOOP" OPERATION, NO "SERVICE ENGINE SOON" LIGHT.

1988–90 2.0L TURBO (VIN M) – ALL MODELS

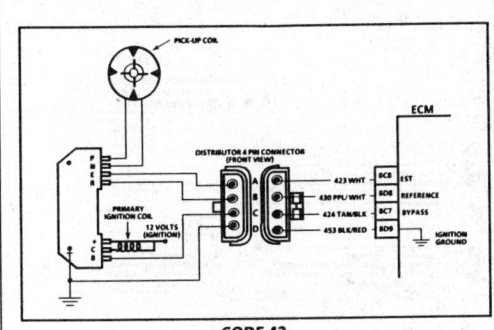

CODE 42
ELECTRONIC SPARK TIMING (EST) CIRCUIT
2.0L TURBO (VIN M)

Circuit Description:
The ignition module sends a reference signal (CKT 430) to the ECM when the engine is cranking. While the engine speed is under 400 rpm, the ignition module will control ignition timing. When the engine speed exceeds 400 rpm, the ECM applies 5 volts to the bypass line (CKT 424) to switch the timing to ECM control (EST CKT 423).

When the system is running "ON" the ignition module, that is, no voltage on the bypass line, the ignition module grounds the EST signal. The ECM expects to see no voltage on the EST line during this condition. If it sees a voltage, it sets Code 42 and will not go into the EST mode.

When the rpm for EST is reached (about 400 rpm), voltage will be applied to the bypass line, the EST should no longer be grounded in the ignition module, so the EST voltage should be varying.

If the bypass line is open or grounded, the ignition module will not switch to EST mode, so the EST voltage will be low and Code 42 will be set.

If the EST line is grounded, the ignition module will switch to EST but, because the line is grounded, there will be no EST signal. A Code 42 will be set.

Test Description: Numbers below refer to circled numbers on the diagnostic chart.

1. Confirms Code 42 and that the fault causing the code is present.

2. Checks for a normal EST ground path through the ignition module. An EST CKT 423 shorted to ground will also read less than 500 ohms, however, this will be checked later.

3. As the test light voltage touches terminal "C7" the module should switch, causing the ohmmeter to "overrange" if the meter is in the 1000-2000 ohms position.

Selecting the 10-20,000 ohms position will indicate above 5000 ohms. The important thing is that the module "switched."

4. The module did not switch and this step checks for
   - EST CKT 423 shorted to ground
   - Bypass CKT 424 open
   - Faulty ignition module connection or module

5. Confirms that Code 42 is a faulty ECM and not an intermittent in CKTs 423 or 424.

1988–90 2.0L TURBO (VIN M) – ALL MODELS

CODE 42
ELECTRONIC SPARK TIMING (EST) CIRCUIT
2.0L TURBO (VIN M)

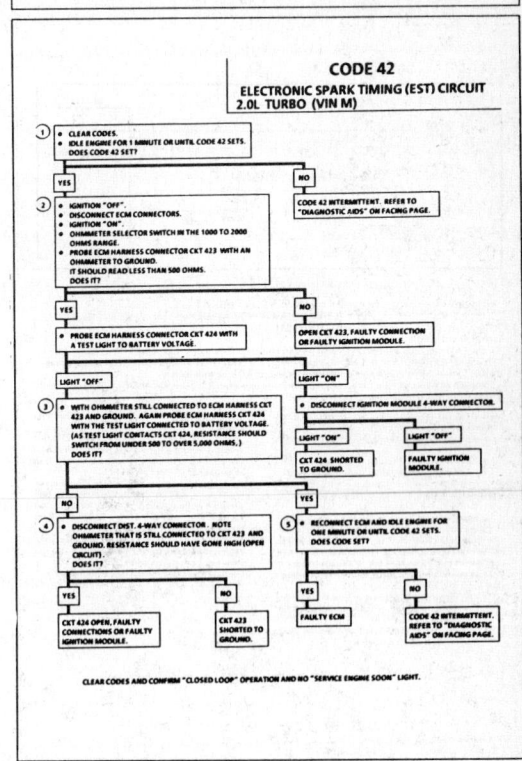

CLEAR CODES AND CONFIRM "CLOSED LOOP" OPERATION AND NO "SERVICE ENGINE SOON" LIGHT.

1988–90 2.0L TURBO (VIN M) – ALL MODELS

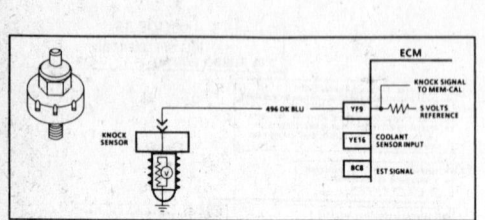

CODE 43
ELECTRONIC SPARK CONTROL (ESC) CIRCUIT
2.0L TURBO (VIN M)

Circuit Description:

The knock sensor is used to detect engine detonation and the ECM will retard the electronic spark timing based on the signal being received. The circuitry within the knock sensor causes the ECM 5 volts to be pulled down so that under a no knock condition, CKT 496 would measure about 2.5 volts. The knock sensor produces an A/C signal which rides on the 2.5 volts, DC voltage. The amplitude and signal frequency is dependent upon the knock level.

If CKT 496 becomes open or shorted to ground the voltage will either go above 3.5 volts or below 1.5 volts. If either of these conditions are met for about 5 seconds, a Code 43 will be stored.

Test Description: Numbers below refer to circled numbers on the diagnostic chart.

1. Code 43 will set when:
   - Coolant temperature is over 90°C
   - MAT temperature is over 0°C
   - High engine load based on MAP and rpm
   - Voltage on CKT 496 goes above 3.5 volts or below 1.5 volts
   - All conditions present for 5 seconds

   If an audible knock is heard from the engine, repair the internal engine problem, as normally no knock should be detected at idle.

2. If tapping on the engine lift hook does not produce a knock signal, try tapping engine closer to sensor before proceeding.
3. The ECM has a 5 volts pull-up resistor, which should be present at the knock sensor terminal.
4. This test determines if the knock sensor is faulty or if the ESC portion of the Mem-Cal is faulty.

Diagnostic Aids:

Check CKT 496 for a potential open or short to ground. Also check for proper installation of Mem-Cal.
Refer to "Intermittents" in Section "B".

1988–90 2.0L TURBO (VIN M) – ALL MODELS

CODE 43
ELECTRONIC SPARK CONTROL (ESC) CIRCUIT
2.0L TURBO (VIN M)

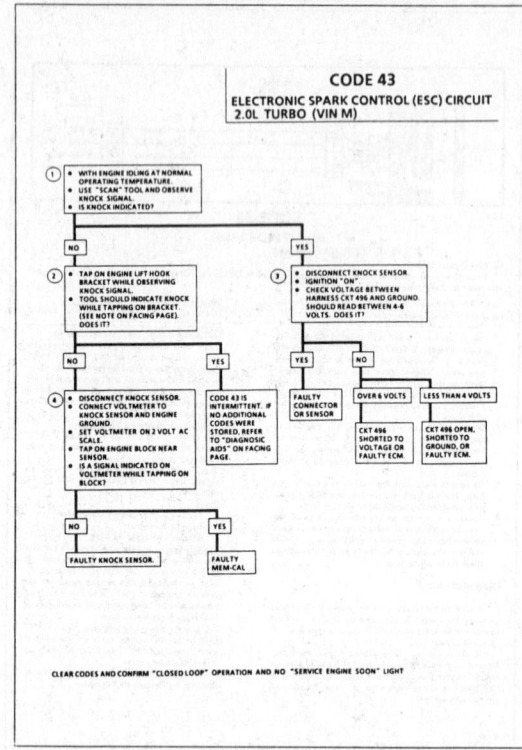

CLEAR CODES AND CONFIRM "CLOSED LOOP" OPERATION AND NO "SERVICE ENGINE SOON" LIGHT

1988–90 2.0L TURBO (VIN M) – ALL MODELS

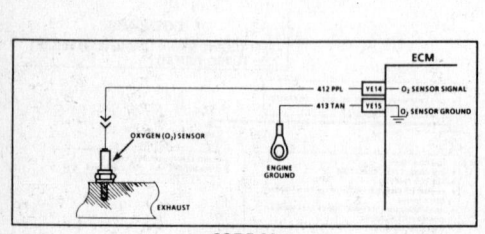

CODE 44
OXYGEN SENSOR CIRCUIT
(LEAN EXHAUST INDICATED)
2.0L TURBO (VIN M)

Circuit Description:

The ECM supplies a voltage of about .45 volt between terminals "YE14" and "YE15". (If measured with a 10 megohm digital voltmeter, this may read as low as .32 volt.) The O2 sensor varies the voltage within a range of about 1 volt if the exhaust is rich, down through about .10 volt if exhaust is lean.

The sensor is like an open circuit and produces no voltage when it is below about 360°C (600°F). An open sensor circuit or cold sensor causes "Open Loop" operation.

Test Description: Numbers below refer to circled numbers on the diagnostic chart.

1. Code 44 is set when the O2 sensor signal voltage on CKT 412:
   - Remains below .2 volt for 60 seconds or more
   - And the system is operating in "Closed Loop"

Diagnostic Aids:

Using the "Scan," observe the block learn values at different rpm and air flow conditions. The "Scan" also displays the block cells, so the block learn values can be checked in each of the cells to determine when the Code 44 may have been set. If the conditions for Code 44 exists, the block learn values will be around 150.

- **O2 Sensor Wire.** Sensor pigtail may be mispositioned and contacting the exhaust manifold.
  Check for intermittent ground in wire between connector and sensor.

- **Lean Injector(s).** Perform injector balance test CHART C-2A.
- **Fuel Contamination.** Water, even in small amounts, near the in-tank fuel pump inlet can be delivered to the injectors. The water causes a lean exhaust and can set a Code 44.
- **Fuel Pressure.** System will be lean if pressure is too low. It may be necessary to monitor fuel pressure while driving the car at various road speeds and/or loads to confirm. See Fuel System diagnosis CHART A-7.
- **Exhaust Leaks.** If there is an exhaust leak, the engine can cause outside air to be pulled into the exhaust and past the sensor. Vacuum or crankcase leaks can cause a lean condition.
- If the above are OK, it is a faulty oxygen sensor.

1988–90 2.0L TURBO (VIN M) – ALL MODELS

CODE 44
OXYGEN SENSOR CIRCUIT
(LEAN EXHAUST INDICATED)
2.0L TURBO (VIN M)

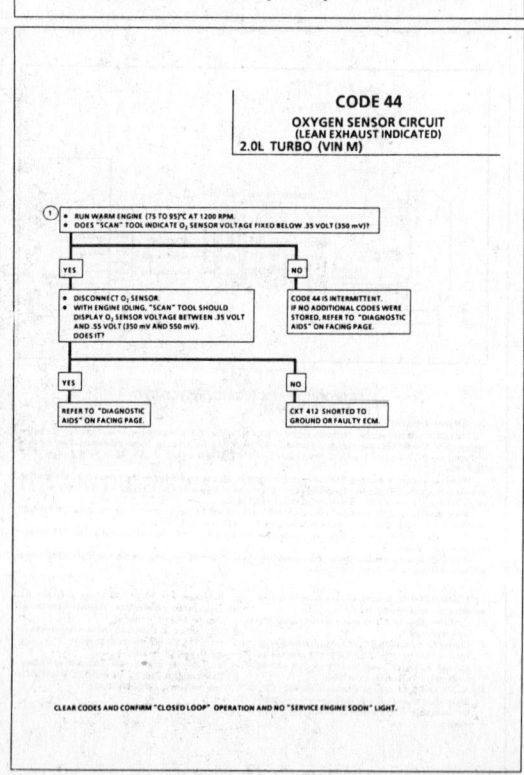

CLEAR CODES AND CONFIRM "CLOSED LOOP" OPERATION AND NO "SERVICE ENGINE SOON" LIGHT.

1988–90 2.0L TURBO (VIN M) – ALL MODELS

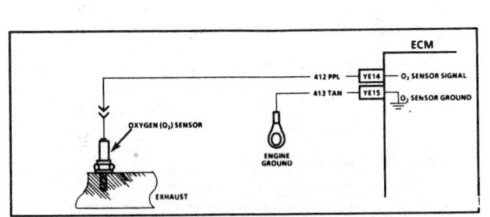

CODE 45
OXYGEN SENSOR CIRCUIT
(RICH EXHAUST INDICATED)
2.0L TURBO (VIN M)

Circuit Description:

The ECM supplies a voltage of about .45 volt between terminals "YE14" and "YE15". (If measured with a 10 megohm digital voltmeter, this may read as low as .32 volt.) The O₂ sensor varies the voltage within a range of about 1 volt if the exhaust is rich, down through about .10 volt if exhaust is lean.

The sensor is like an open circuit and produces no voltage when it is below about 360°C (600°F). An open sensor circuit or cold sensor causes "Open Loop" operation.

Test Description: Numbers below refer to circled numbers on the diagnostic chart.
1. Code 45 is set when the O₂ sensor voltage:
   - Remains above .75 volt for 50 seconds; and the system is in "Closed Loop"

Diagnostic Aids:

Using the "Scan," observe the block learn values at different rpm and air flow conditions. The "Scan" also displays the block cells, so the block learn values can be checked in each of the cells to determine when the Code 45 may have been set. If the conditions for Code 45 exists, the block learn values will be around 115.
- Fuel Pressure. System will go rich if pressure is too high. The ECM can compensate for some increase. However, if it gets too high, a Code 45 may be set. See "Fuel System" diagnosis CHART A-7.
- Rich Injector. Perform injector balance test CHART C-2A.
- Leaking Injector. See CHART A-7.
- Check for fuel contaminated oil.
- HEI Shielding. An open ignition ground CKT 453 may result in EMI, or induced electrical "noise". The ECM looks at this "noise" as reference pulses. The additional pulses result in a higher than actual engine speed signal.

The ECM then delivers too much fuel, causing system to go rich. Engine tachometer will also show higher than actual engine speed, which can help in diagnosing this problem.
- Canister purge. Check canister for fuel saturation. If full of fuel, check canister control and hoses.
- MAP Sensor. An output that causes the ECM to sense a higher than normal manifold pressure can cause the system to go rich. Disconnecting the MAP sensor will allow the ECM to set a fixed value for the sensor. Substitute a different MAP sensor if the rich condition is gone while the sensor is disconnected.
- Check for leaking fuel pressure regulator diaphragm by checking vacuum line to regulator for fuel.
- TPS. An intermittent TPS output will cause the system to go rich, due to a false indication of the engine accelerating.
- EGR. An EGR staying open (especially at idle) will cause the O₂ sensor to indicate a rich exhaust, and this could result in a Code 45.

1988–90 2.0L TURBO (VIN M) – ALL MODELS

CODE 45
OXYGEN SENSOR CIRCUIT
(RICH EXHAUST INDICATED)
2.0L TURBO (VIN M)

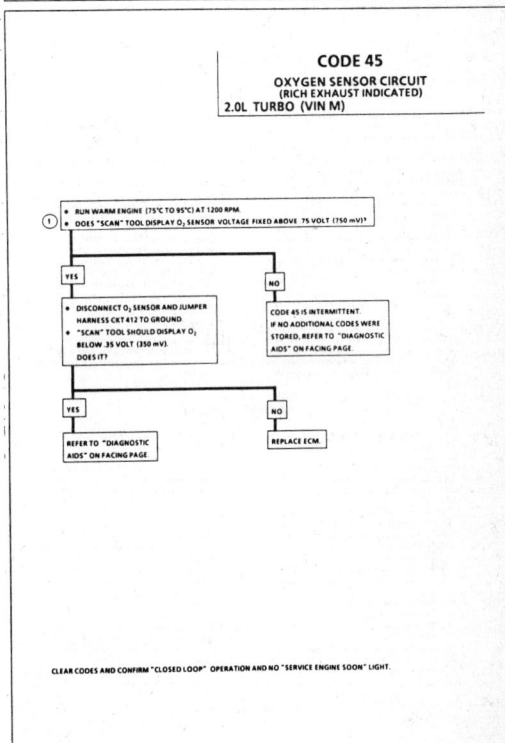

CLEAR CODES AND CONFIRM "CLOSED LOOP" OPERATION AND NO "SERVICE ENGINE SOON" LIGHT.

1988–90 2.0L TURBO (VIN M) – ALL MODELS

CODE 51
PROM ERROR
(FAULTY OR INCORRECT PROM)
2.0L TURBO (VIN M)

CHECK THAT ALL PINS ARE FULLY INSERTED IN THE SOCKET AND THAT PROM IS PROPERLY SEATED. IF OK, REPLACE PROM, CLEAR MEMORY, AND RECHECK. IF CODE 51 REAPPEARS, REPLACE ECM.

CLEAR CODES AND CONFIRM "CLOSED LOOP" OPERATION AND NO "SERVICE ENGINE SOON" LIGHT.

1988–90 2.0L TURBO (VIN M) – ALL MODELS

SECTION B
SYMPTOMS

Performing Symptom Diagnosis
Intermittents
Hard Start
Rough, Unstable, or Incorrect Idle, Stalling
Poor Gas Mileage
Detonation/Spark Knock
Lack of Power, Sluggish, or Spongy
Surges and/or Chuggle
Cuts Out, Misses
Hesitation, Sag, Stumble
Excessive Exhaust Emissions or Odors
Dieseling, Run-On
Backfire

PERFORMING SYMPTOM DIAGNOSIS

The DIAGNOSTIC CIRCUIT CHECK should be performed before using this section. The purpose of this section is to locate the source of a driveability or emissions problem when other diagnostic procedures cannot be used. This may be because of difficulties in locating a suspected sub-system or component.

Many driveability related problems can be eliminated by following the procedures found in Service Bulletins. These bulletins supersede this manual. Be sure to check all bulletins related to the complaint or suspected system.

If the ENGINE CRANKS BUT WILL NOT RUN, use CHART A-3.

The sequence of the checks listed in this section is not intended to be followed as on a step-by-step procedure. The checks are listed such that the less difficult and time consuming operations are performed before more difficult ones.

Most of the symptom procedures call for a careful visual and physical check. *The importance of this step cannot be stressed too strongly.* It can lead to correcting a problem without further checks, and can save valuable time. This procedure includes checking the following.
- Vacuum hoses for splits, kinks, and proper connections, as shown on the Vehicle Emission Control Information label
- Throttle body and intake manifold for leaks
- Ignition wires for cracking, hardness, proper routing, and carbon tracking
- Wiring for proper connections, pinches, and cuts

1988–90 2.0L TURBO (VIN M)–ALL MODELS

INTERMITTENTS

Definition: Problem may or may not activate the "Service Engine Soon" light or store a trouble code.

DO NOT use the trouble code charts in Section "A" for intermittent problems. The fault must be present to locate the problem. If a fault is intermittent, the use of trouble code charts may result in the replacement of good parts.

- Most intermittent problems are caused by faulty electrical connections or wiring. Perform careful checks of suspected circuits for
  - Poor mating of the connector halves and terminals not fully seated in the connector body (backed out)
  - Improperly formed or damaged terminals. All connector terminals in problem circuit should be carefully reformed to increase contact tension
  - Poor terminal to wire connection. This requires removing the terminal from the connector body to check
- If a visual and physical check does not locate the cause of the problem, the car can be driven with a voltmeter connected to a suspected circuit or a "Scan" tool may be used. An abnormal voltage reading while the problem occurs indicates that the problem may be in that circuit.
- Check for loss of trouble code memory. To check, disconnect the TPS and allow the engine to idle until the "Service Engine Soon" light turns "ON." Code 22 should be stored and kept in memory when the ignition is turned "OFF" for at least 10 seconds. If not, the ECM is faulty.
- An intermittent SES light and no trouble codes may be caused by
  - Electrical system interference caused by a defective relay, ECM driven solenoid, or switch. They can cause a sharp electrical surge. Normally, the problem will occur when the faulty component is operated.
  - Improper installation of electrical options, such as lights, 2-way radios, etc
  - EST wires which should be routed away from spark plug wires, ignition system components, and generator. Ground wire from ECM to ignition system which may be faulty.
  - Ignition secondary wire shorted to ground.
  - "Service Engine Soon" light and diagnostic test terminal circuits intermittently shorted to ground.
  - Faulty ECM grounds.

HARD START

Definition: Engine cranks well but does not start for a long time. Engine does eventually start, but may or may not continue to run.

Perform careful visual and physical check as described at the beginning of Section "B".

Perform "Diagnostic Circuit Check."

- **CHECK**
  - For possibility of misfiring, crossfiring, or cutting under load or at idle. Locate misfiring cylinder(s) by performing the following test.
    1. Start engine. Disconnect idle air control valve. Remove one spark plug wire from a spark plug and ground it against the engine.
    2. Note drop in engine speed.
    3. Repeat for all four cylinders.
    4. Stop engine and reconnect idle air control valve.

    If the engine speed dropped equally (within 50 rpm) on all cylinders, refer to "Rough, Unstable, or Incorrect Idle, Stalling" symptom.

    If there was no drop or excessive variation in engine speed on one or more cylinders, check for spark on the respective cylinder(s) with J 26792 (ST-125) spark tester or equivalent.

    If spark is present, remove the spark plugs from the cylinder(s) and check for the following.
    - Cracks
    - Wear
    - Improper gap
    - Burned electrode
    - Heavy deposits
  - For worn distributor shaft
  - For moisture in distributor cap
  - Ignition pickup coil resistance and connections
  - Ignition timing. See Vehicle Emission Control Information label.

1988–90 2.0L TURBO (VIN M)–ALL MODELS

- Fuel for poor quality, "stale" fuel, and water contamination
- Ignition wires for shorts or faulty insulation
- Ignition coil connections
- Fuel pump relay. Connect test light between pump test terminal and ground. Light should be "ON" for 2 seconds following ignition "ON." If not, use CHART A-5.
- Secondary ignition voltage output with ST-125 tester
- Spark plugs. Look for wetness, cracks, improper gap, burned electrodes, and heavy deposits. Visually inspect ignition system for moisture, dust, cracks, burns, etc.
- For faulty ECM and ignition grounds
- Mem-Cal for correct application. (Consult Service Bulletins.) Spray plug wires with fine water mist to check for shorts.
- EGR operation. Use CHART C-7.
- Idle Air Control system. Use Code 35 chart.
- Fuel system for restricted filter or improper pressure. Use CHART A-7.
- Injectors for leakage. Pressurize system by energizing fuel pump through the under hood fuel pump test connector.
- Coolant sensor for a shift in calibration. Use Code 14 or Code 15 chart.
- TPS for sticking or binding. TPS voltage should read less than 1.25 volts on a "Scan" tool.
- Injector balance by performing injector balance test in CHART C-2A
- In-tank fuel pump check valve. A faulty valve would allow the fuel in the lines to drain back to the tank after the engine is stopped. To check for this condition, conduct the following test.
  1. Ignition "OFF."
  2. Disconnect fuel line at the filter.
  3. Remove the tank filler cap.
  4. Connect a radiator test pump to the line and apply 103 kPa (15 psi) pressure. If the pressure will hold for 60 seconds, the check valve is OK.
- For the possibility of an exhaust restriction or improper valve timing by performing the following test.
  1. With engine at normal operating temperature, connect a vacuum gauge to any convenient vacuum port on intake manifold.
  2. Run engine at 1000 rpm and record vacuum reading.
  3. Increase engine speed slowly to 2500 rpm. Note vacuum reading at steady 2500 rpm.
  4. If vacuum at 2500 rpm decreases more than 3" Hg from reading at 1000 rpm, the exhaust system should be inspected for restrictions. Use CHART B-1.
  5. Disconnect exhaust pipe from engine and repeat Steps 3 & 4. If vacuum still drops more than 3" Hg with exhaust disconnected, check valve timing.
- Engine valve timing and compression.

ROUGH, UNSTABLE, OR INCORRECT IDLE, STALLING

Definition: The engine runs unevenly at idle. If severe, the car may shake. Also, the idle speed may vary (called "hunting"). Either condition may be severe enough to cause stalling. Engine idles at incorrect speed.

Perform careful visual and physical check as described at the beginning of Section "B".

Perform "Diagnostic Circuit Check."

- **CHECK**
  - For possibility of misfiring, crossfiring, or cutting under load or at idle. Locate misfiring cylinder(s) by performing the following test.
    1. Start engine. Disconnect idle air control valve. Remove one spark plug wire from a spark plug and ground it against the engine.
    2. Note drop in engine speed.
    3. Repeat for all four cylinders.
    4. Stop engine and reconnect idle air control valve.

    If the engine speed dropped equally (within 50 rpm) on all cylinders, proceed through the causes listed. If there was no drop or excessive variation in engine speed on one or more cylinders, check for spark on the respective cylinder(s) with J 26792 (ST-125) spark tester or equivalent. If there is no spark, see ENGINE ELECTRICAL (SECTION 6D). If spark is present, remove the spark plugs from the cylinder(s) and check for the following.
    - Cracks
    - Wear
    - Improper gap
    - Burned electrode
    - Heavy deposits

1988–90 2.0L TURBO (VIN M)–ALL MODELS

- For worn distributor shaft
- For moisture in distributor cap
- Ignition pickup coil resistance and connections
- Ignition timing. See underhood emission control information label
- MAP sensor. Use CHART C1-D
- Throttle for sticking shaft or binding linkage. This will cause a high TPS voltage (open throttle indication) and the ECM will not control idle. TPS voltage should be less than 1.25 volts with throttle closed
- Battery cables and ground straps for poor contact. Erratic voltage will cause the IAC valve to change its position, resulting in poor idle quality
- Ignition wires for shorts or faulty insulation
- Ignition system for moisture, dust, cracks, burns, etc. Spray plug wires with fine water mist to check for shorts
- Secondary ignition voltage output with ST-1 tester
- Ignition coil connections
- ECM and ignition system for faulty grounds
- Proper operation of EST
- Spark plugs. Look for wetness, cracks, improper gap, burned electrodes, and heavy deposits
- Fuel system for restricted filter or improper pressure. Use CHART A-7
- Injectors for leakage. Pressurize system by energizing fuel pump through the underhood fuel pump test connector
- EGR operation. Use CHART C-7
- For vacuum leaks at intake manifold gasket
- Idle Air Control system. Use Code 35 chart
- Electrical system voltage. IAC valve will not move if voltage is below 9 volts or greater than 17.8 volts. Also check battery cables and ground straps for poor contact. Erratic voltage will cause the IAC valve to change its position, resulting in poor idle quality
- PCV valve for proper operation by placing finger over inlet hole in valve end several times. Valve should snap back. If not, replace valve. Ensure that valve is correct part. Also, check PCV hose
- Canister purge system for proper operation. Use CHART C-3
- Mem-Cal for correct application
- Throttle shaft or TPS for sticking or binding. TPS voltage should read less than 1.25 volts on a "Scan" tool with the throttle closed
- MAP sensor output. Use CHART C1-D and/or check sensor by comparing it to the output on a similar vehicle if possible
- Oxygen sensor for silicone contamination from contaminated fuel or use of improper RTV sealant. The sensor will have a white, powdery coating and will cause a high but false signal voltage (rich exhaust indication). The ECM will reduce the amount of fuel delivered to the engine, causing a severe driveability problem
- Coolant sensor for a shift in calibration. Use Code 14 or Code 15 chart
- A/C refrigerant pressure for high pressure. Check for overcharging or faulty pressure switch
- P/N switch circuit on vehicle with automatic transmission. Use CHART C-1A
- Generator output voltage. Repair if less than 9 volts or more than 16 volts
- Power steering. Use CHART C-1E. The ECM should compensate for power steering loads. Loss of this signal would be most noticeable when steering loads are high such as during parking
- Engine valve timing and compression
- For worn or incorrect basic engine parts such as cam, heads, pistons, etc. Also check for bent pushrods, worn rocker arms, and broken or weak valve springs
- For the possibility of an exhaust restriction or improper valve timing, perform the following test.
  1. With engine at normal operating temperature, connect a vacuum gauge to any convenient vacuum port on intake manifold.
  2. Run engine at 1000 rpm and record vacuum reading.
  3. Increase engine speed slowly to 2500 rpm. Note vacuum reading at steady 2500 rpm.
  4. If vacuum at 2500 rpm decreases more than 3" Hg from reading at 1000 rpm, the exhaust system should be inspected for restrictions.
  5. Disconnect exhaust pipe from engine and repeat Steps 3 & 4. If vacuum still drops more than 3" Hg with exhaust disconnected, check valve timing.
- Injector balance by performing injector balance test in CHART C-2A

1988–90 2.0L TURBO (VIN M)–ALL MODELS

- For overheating and possible causes. Look for the following
  - Low or incorrect coolant solution. It should be a 50/50 mix of GM #1052753 anti-freeze coolant (or equivalent) and water.
  - Loose water pump belt
  - Restricted air flow to radiator, or restricted water flow through radiator
  - Faulty or incorrect thermostat
  - Inoperative electric cooling fan circuit
- If the system is running RICH (block learn less than 118), refer to "Diagnostic Aids" on facing page of Code 45.
- If the system is running LEAN (block learn greater than 138), refer to "Diagnostic Aids" on facing page of Code 44.

POOR GAS MILEAGE

Definition: Gas mileage, as measured by an actual road test, is noticeably lower than expected. Gas mileage is noticeably lower than it was during a previous actual road test.

Perform careful visual and physical check as described at the beginning of Section "B".

Perform "Diagnostic Circuit Check.

- **CHECK**
  - For possibility of misfiring, crossfiring, or cutting under load or at idle. Locate misfiring cylinder(s) by performing the following test.
    1. Start engine. Disconnect idle air control valve. Remove one spark plug wire from a spark plug and ground it against the engine.
    2. Note drop in engine speed.
    3. Repeat for all four cylinders.
    4. Stop engine and reconnect idle air control valve.

    If the engine speed dropped equally (within 50 rpm) on all cylinders, refer to "Rough, Unstable, or Incorrect Idle, Stalling" symptom. If there was no drop or excessive variation in engine speed on one or more cylinders, check for spark on the respective cylinder(s) with J 26792 (ST-125) spark tester or equivalent.

    If spark is present, remove the spark plugs from the cylinder(s) and check for the following.
    - Cracks
    - Wear
    - Improper gap
    - Burned electrode
    - Heavy deposits
  - For worn distributor shaft
  - Ignition timing. See underhood emission control information label.
  - Proper operation of EST.
  - Spark plugs. Look for wetness, cracks, improper gap, burned electrodes, and heavy deposits.
  - Spark plugs for correct heat range
  - Fuel for poor quality, "stale" fuel, and water contamination
  - Fuel system for restricted filter or improper pressure. Use CHART A-7.
  - Injectors for leakage. Pressurize system by energizing fuel pump through the underhood fuel pump test connector.
  - EGR operation. Use CHART C-7.
  - For vacuum leaks at intake manifold gasket
  - Air cleaner element (filter) for dirt or plugging
  - Idle Air Control system. Use Code 35 chart.
  - Canister purge system for proper operation. Use CHART C-3.
  - Mem-Cal for correct application
  - Throttle shaft or TPS for sticking or binding. TPS voltage should read less than 1.25 volts on a "Scan" tool with the throttle closed.
  - MAP sensor output. Use CHART C1-D and/or check sensor by comparing it to the output on a similar vehicle if possible.
  - Oxygen sensor for silicone contamination from contaminated fuel or use of improper RTV sealant. The sensor will have a white, powdery coating and will cause a high but false signal voltage (rich exhaust indication). The ECM will reduce the amount of fuel delivered to the engine, causing a severe driveability problem.
  - Coolant sensor for a shift in calibration. Use Code 14 or Code 15 chart.
  - Vehicle speed sensor (VSS) input with a "Scan" tool to make sure reading of VSS matches that of vehicle speedometer.
  - A/C relay operation. A/C should cut out at wide open throttle. Use CHART C-10.

1988–90 2.0L TURBO (VIN M) – ALL MODELS

- A/C refrigerant pressure for high pressure. Check for overcharging or faulty pressure switch.
- Injector balance by performing injector balance test in CHART C-2A
- Generator output voltage. Repair if less than 9 volts or more than 16 volts.
- Cooling fan operation. Use CHART C-12.
- Power steering. Use CHART C-1E. The ECM should compensate for power steering loads. Loss of this signal would be most noticeable when steering loads are high such as during parking.
- Transmission torque converter operation.
- Transmission for proper shift points.
- Transmission torque converter clutch operation. Use CHART C-8.
- Engine valve timing and compression.
- For worn or incorrect basic engine parts such as cam, heads, pistons, etc. Also check for bent pushrods, worn rocker arms, and broken or weak valve springs.
- For the possibility of an exhaust restriction or improper valve timing by performing the following test.
  1. With engine at normal operating temperature, connect a vacuum gauge to any convenient vacuum port on intake manifold.
  2. Run engine at 1000 rpm and record vacuum reading.
  3. Increase engine speed slowly to 2500 rpm. Note vacuum reading at steady 2500 rpm.
  4. If vacuum at 2500 rpm decreases more than 3" Hg from reading at 1000 rpm, the exhaust system should be inspected for restrictions.
  5. Disconnect exhaust pipe from engine and repeat Steps 3 & 4. If vacuum still drops more than 3" Hg with exhaust disconnected, check valve timing.
- Thermostat for incorrect heat range or being inoperative

- Check driver's driving habits and vehicle conditions which affect gas mileage.
  - Suggest driver read "Important Facts on Fuel Economy" in Owner's Manual.
  - Is A/C "ON" full time (Defroster mode "ON")?
  - Are tires at correct pressure?
  - Are excessively heavy loads being carried?
  - Is acceleration often heavy?
  - Are the wheels aligned correctly?
  - Is the speedometer calibrated correctly?
  - Are the vehicle brakes dragging?
  - Is the brake switch applying excessive force on the brake pedal?
- If the system is running RICH, (block learn less than 118), refer to "Diagnostic Aids" on facing page of Code 45.

DETONATION/SPARK KNOCK

Definition: A mild to severe ping, usually worse under acceleration. The engine makes sharp metallic knocks that change with throttle opening.

Perform careful visual and physical check as described at the beginning of Section "B".

Perform "Diagnostic Circuit Check."

- **CHECK**
  - For possibility of misfiring, crossfiring, or cutting under load or at idle. Locate misfiring cylinder(s) by performing the following test.
    1. Start engine. Disconnect idle air control valve. Remove one spark plug wire from a spark plug and ground it against the engine.
    2. Note drop in engine speed.
    3. Repeat for all four cylinders.
    4. Stop engine and reconnect idle air control valve.

    If the engine speed dropped equally (within 50 rpm) on all cylinders, refer to "Rough, Unstable, or Incorrect Idle, Stalling" symptom. If there was no drop or excessive variation in engine speed on one or more cylinders, check for spark on the respective cylinder(s) with J 26792 (ST-125) spark tester or equivalent. If spark is present, remove the spark plugs from the cylinder(s) and check for the following.
    - Cracks
    - Wear
    - Improper gap
    - Burned electrode
    - Heavy deposits

1988–90 2.0L TURBO (VIN M) – ALL MODELS

- For worn distributor shaft
- Ignition pickup coil resistance and connections
- Turbocharger wastegate control circuit. Follow diagnostic chart for Code 31.
- Ignition timing. See underhood emission control information label.
- Ignition wires for shorts or faulty insulation
- Spark plugs for correct heat range
- Fuel for poor quality, "stale" fuel, and water contamination
- Fuel system for restricted filter or improper pressure. Use CHART A-7.
- For excessive oil entering combustion chamber. Oil will reduce the effective octane of fuel.
- EGR operation. Use CHART C-7.
- For vacuum leaks at intake manifold gasket
- PCV valve for proper operation by placing finger over inlet hole in valve end several times. Valve should snap back. If not, replace valve. Ensure that valve is correct part. Also check PCV hose.
- MAP sensor output. Use CHART C1-D and/or check sensor by comparing it to the output on a Similar vehicle, if possible.
- Coolant sensor for a shift in calibration
- Oxygen sensor for silicone contamination from contaminated fuel or use of improper RTV sealant. The sensor will have a white, powdery coating and will cause a high but false signal voltage (rich exhaust indication). The ECM will reduce the amount of fuel delivered to the engine, causing a severe driveability problem
- Vehicle speed sensor (VSS) input with a "Scan" tool to make sure reading of VSS matches that of vehicle speedometer.
- Transmission torque converter operation.
- Transmission for proper shift points.
- Transmission torque converter clutch operation. Use CHART C-8.
- Vehicle brakes for dragging
- Mem-Cal for correct application
- For overheating and possible causes. Look for the following
  - Low or incorrect coolant solution. It should be a 50/50 mix of GM #1052753 anti-freeze coolant (or equivalent) and water
  - Loose water pump belt
  - Restricted air flow to radiator or restricted water flow through radiator
  - Faulty or incorrect thermostat
  - Inoperative electric cooling fan circuit
- Engine valve timing and compression.
- For worn or incorrect basic engine parts such as cam, heads, pistons, etc. Also check for bent pushrods, worn rocker arms, and broken or weak valve springs.
- For the possibility of an exhaust restriction or improper valve timing by performing the following test.
  1. With engine at normal operating temperature, connect a vacuum gauge to any convenient vacuum port on intake manifold.
  2. Run engine at 1000 rpm and record vacuum reading.
  3. Increase engine speed slowly to 2500 rpm. Note vacuum reading at steady 2500 rpm.
  4. If vacuum at 2500 rpm decreases more than 3" Hg from reading at 1000 rpm, the exhaust system should be inspected for restrictions.
  5. Disconnect exhaust pipe from engine and repeat Steps 3 & 4. If vacuum still drops more than 3" Hg with exhaust disconnected, check valve timing.
- Remove internal engine carbon with top engine cleaner.

  If the system is running LEAN (block learn greater than 138), refer to "Diagnostic Aids" on facing page of Code 44.

1988–90 2.0L TURBO (VIN M) – ALL MODELS

LACK OF POWER, SLUGGISH, OR SPONGY

Definition: Engine delivers less than expected power. There is little or no increase in speed when the accelerator pedal is depressed partially.

Perform careful visual and physical check as described at the beginning of Section "B".

Perform "Diagnostic Circuit Check."

- **CHECK**
  - For possibility of misfiring, crossfiring, or cutting under load or at idle. Locate misfiring cylinder(s) by performing the following test.
    1. Start engine. Disconnect idle air control valve. Remove one spark plug wire from a spark plug and ground it against the engine.
    2. Note drop in engine speed.
    3. Repeat for all four cylinders.
    4. Stop engine and reconnect idle air control valve.

    If the engine speed dropped equally (within 50 rpm) on all cylinders, refer to "Rough, Unstable, or Incorrect Idle, Stalling" symptom. If there was no drop or excessive variation in engine speed on one or more cylinders, check for spark on the respective cylinder(s) with J 26792 (ST-125) spark tester or equivalent. If there is no spark, see ENGINE ELECTRICAL (SECTION 6D). If spark is present, remove the spark plugs from the cylinder(s) and check for the following.
    - Cracks
    - Wear
    - Improper gap
    - Burned electrode
    - Heavy deposits
  - For worn distributor shaft
  - Ignition pickup coil resistance and connections
  - Ignition timing. See underhood emission control information label.
  - Ignition wires for shorts or faulty insulation
  - Ignition system for moisture, dust, cracks, burns, etc. Spray plug wires with fine water mist to check for shorts.
  - Secondary ignition voltage output with ST-125 tester.
  - Ignition coil connections
  - ECM and ignition system for faulty grounds
  - Proper operation of EST.
  - Spark plugs. Look for wetness, cracks, improper gap, burned electrodes, and heavy deposits.
  - Turbocharger wastegate control circuit. Follow diagnostic chart for Code 31.
  - Spark plugs for correct heat range
  - Fuel for poor quality, "stale" fuel, and water contamination
  - Fuel system for restricted filter or improper pressure. Use CHART A-7.
  - EGR operation. Use CHART C-7.
  - For vacuum leaks at intake manifold gasket
  - Air cleaner element (filter) for dirt or plugging
  - Mem-Cal for correct application
  - Throttle shaft or TPS for sticking or binding. TPS voltage should read less than 1.25 volts on a "Scan" tool with the throttle closed.
  - MAP sensor output. Use CHART C1-D and/or check sensor by comparing it to the output on a similar vehicle if possible.
  - Oxygen sensor for silicone contamination from contaminated fuel or use of improper RTV sealant. The sensor will have a white, powdery coating and will cause a high but false signal voltage (rich exhaust indication). The ECM will reduce the amount of fuel delivered to the engine, causing a severe driveability problem.
  - Coolant sensor for a shift in calibration. Use Code 14 or Code 15 chart.
  - Vehicle speed sensor (VSS) input with a "Scan" tool to make sure reading of VSS matches that of vehicle speedometer.
  - Engine for improper or worn camshaft.
  - A/C relay operation. A/C should cut out at wide open throttle. Use CHART C-10.
  - A/C refrigerant pressure for high pressure. Check for overcharging or faulty pressure switch.
  - Generator output voltage. Repair if less than 9 volts or more than 16 volts.
  - Cooling fan operation. Use CHART C-12.
  - Power steering. Use CHART C-1E. The ECM should compensate for power steering loads. Loss of this signal would be most noticeable when steering loads are high such as during parking.
  - Transmission torque converter operation.
  - Transmission for proper shift points.

1988–90 2.0L TURBO (VIN M) – ALL MODELS

- Transmission torque converter clutch operation. Use CHART C-8.
- Vehicle brakes for dragging
- Engine valve timing and compression.
- For worn or incorrect basic engine parts such as cam, heads, pistons, etc. Also check for bent pushrods, worn rocker arms, and broken or weak valve springs.
- For the possibility of an exhaust restriction or improper valve timing by performing the following test.
  1. With engine at normal operating temperature, connect a vacuum gauge to any convenient vacuum port on intake manifold.
  2. Run engine at 1000 rpm and record vacuum reading.
  3. Increase engine speed slowly to 2500 rpm. Note vacuum reading at steady 2500 rpm.
  4. If vacuum at 2500 rpm decreases more than 3" Hg from reading at 1000 rpm, the exhaust system should be inspected for restrictions.
  5. Disconnect exhaust pipe from engine and repeat Steps 3 & 4. If vacuum still drops more than 3" Hg with exhaust disconnected, check valve timing.
- For overheating and possible causes. Look for the following.
  - Low or incorrect coolant solution. It should be a 50/50 mix of GM #1052753 anti-freeze coolant (or equivalent) and water.
  - Loose water pump belt.
  - Restricted air flow to radiator, or restricted water flow through radiator.
  - Faulty or incorrect thermostat.
  - Inoperative electric cooling fan circuit. See CHART C-12.
- If the system is running RICH (block learn less than 118), refer to "Diagnostic Aids" on facing page of Code 45.
- If the system is running LEAN (block learn greater than 138), refer to "Diagnostic Aids" on facing page of Code 44.

SURGES AND/OR CHUGGLE

Definition: Engine power variation under steady throttle or cruise. Feels like the car speeds up and slows down with no change in the accelerator pedal.

Perform careful visual and physical check as described at the beginning of Section "B".

Perform "Diagnostic Circuit Check."

- **CHECK**
  - For possibility of misfiring, crossfiring, or cutting under load or at idle. Locate misfiring cylinder(s) by performing the following test.
  1. Start engine. Disconnect idle air control valve. Remove one spark plug wire from a spark plug and ground it against the engine.
  2. Note drop in engine speed.
  3. Repeat for all four cylinders.
  4. Stop engine and reconnect idle air control valve.

  If the engine speed dropped equally (within 50 rpm) on all cylinders, refer to "Rough, Unstable, or Incorrect Idle, Stalling" symptom. If there was no drop or excessive variation in engine speed on one or more cylinders, check for spark on the respective cylinder(s) with J 26792 (ST-125) spark tester or equivalent. If there is no spark, see ENGINE ELECTRICAL (SECTION 6D). If spark is present, remove the spark plugs from the cylinder(s) and check for the following.
    - Cracks
    - Wear
    - Improper gap
    - Burned electrode
    - Heavy deposits
  - For worn distributor shaft
  - Ignition pickup coil resistance and connections
  - Ignition timing. See Vehicle Emission Control Information label.
  - Ignition wires for shorts or faulty insulation.
  - Ignition system for moisture, dust, cracks, burns, etc. Spray plug wires with fine water mist to check for shorts.
  - Secondary ignition voltage output with ST-125 tester.
  - Ignition coil connections.

1988–90 2.0L TURBO (VIN M) – ALL MODELS

- ECM and ignition system for faulty grounds.
- Proper operation of EST.
- Spark plugs. Look for wetness, cracks, improper gap, burned electrodes, and heavy deposits.
- Spark plugs for correct heat range.
- Fuel for poor quality, "stale" fuel, and water contamination.
- Fuel system for restricted filter or improper pressure. Use CHART A-7.
- Injectors for leakage. Pressurize system by energizing fuel pump through the underhood fuel pump test connector.
- EGR operation. Use CHART C-7.
- Injector balance by performing injector balance test in CHART C-2A
- For vacuum leaks at intake manifold gasket.
- Idle Air Control system. Use Code 35 chart.
- Electrical system voltage. IAC valve will not move if voltage is below 9 volts or greater than 17.8 volts. Also check battery cables and ground straps for poor contact. Erratic voltage will cause the IAC valve to change its position, resulting in poor idle quality.
- PCV valve for proper operation by placing finger over inlet hole in valve end several times. Valve should snap back. If not, replace valve. Ensure that valve is correct part. Also check PCV hose.
- Canister purge system for proper operation. Use CHART C-3.
- Mem-Cal for correct application.
- Throttle shaft or TPS for sticking or binding. TPS voltage should read less than 1.25 volts on a "Scan" tool with the throttle closed.
- MAP sensor output. Use CHART C1-D and/or check sensor by comparing it to the output on a similar vehicle, if possible.
- Oxygen sensor for silicone contamination from contaminated fuel or use of improper RTV sealant. The sensor will have a white, powdery coating and will cause a high but false signal voltage (rich exhaust indication). The ECM will reduce the amount of fuel delivered to the engine, causing a severe driveability problem.
- Coolant sensor for a shift in calibration. Use Code 14 or Code 15 chart.
- Vehicle speed sensor (VSS) input with a "Scan" tool to make sure reading of VSS matches that of vehicle speedometer.
- A/C relay operation. A/C should cut out at wide open throttle. Use CHART C-10.
- P/N switch circuit on vehicle with automatic transmission. Use CHART C-1A.
- Transmission torque converter clutch operation. Use CHART C-8.
- For the possibility of an exhaust restriction or improper valve timing by performing the following test:
  1. With engine at normal operating temperature, connect a vacuum gauge to any convenient vacuum port on intake manifold.
  2. Run engine at 1000 rpm and record vacuum reading.
  3. Increase engine speed slowly to 2500 rpm. Note vacuum reading at steady 2500 rpm.
  4. If vacuum at 2500 rpm decreases more than 3" Hg from reading at 1000 rpm, the exhaust system should be inspected for restrictions.
  5. Disconnect exhaust pipe from engine and repeat Steps 3 & 4. If vacuum still drops more than 3" Hg with exhaust disconnected, check valve timing.
- Engine valve timing and compression.
- For worn or incorrect basic engine parts such as cam, heads, pistons, etc. Also check for bent pushrods, worn rocker arms, and broken or weak valve springs.

- If the system is running RICH (block learn less than 118), refer to "Diagnostic Aids" on facing page of Code 45.
- If the system is running LEAN (block learn greater than 138), refer to "Diagnostic Aids" on facing page of Code 44.

1988–90 2.0L TURBO (VIN M) – ALL MODELS

CUTS OUT, MISSES

Definition: Steady pulsation or jerking that follows engine speed, usually more pronounced as engine load increases. The exhaust has a steady spitting sound at idle or low speed.

Perform careful visual and physical check as described at the beginning of Section "B".

Perform "Diagnostic Circuit Check."

- **CHECK**
  - For worn distributor shaft
  - Ignition pickup coil resistance and connections
  - Ignition timing. See Vehicle Emission Control Information label.
  - Ignition wires for shorts or faulty insulation
  - Ignition system for moisture, dust, cracks, burns, etc. Spray plug wires with fine water mist to check for shorts.
  - Secondary ignition voltage output with ST-125 tester
  - Ignition coil connections
  - ECM and ignition system for faulty grounds
  - Proper operation of EST.
  - Spark plugs. Look for wetness, cracks, improper gap, burned electrodes, and heavy deposits.
  - Spark plugs for correct heat range
  - Fuel for poor quality, "stale" fuel, and water contamination
  - Fuel system for restricted filter or improper pressure. Use CHART A-7.
  - Throttle shaft or TPS for sticking or binding. TPS voltage should read less than 1.25 volts on a "Scan" tool with the throttle closed.
  - Injector balance by performing injector balance test in CHART C-2A
  - For possibility of misfiring, crossfiring, or cutting under load or at idle. Locate misfiring cylinder(s) by performing the following test.
    1. Start engine. Disconnect idle air control valve. Remove one spark plug wire from a spark plug and ground it against the engine.
    2. Note drop in engine speed.
    3. Repeat for all four cylinders.
    4. Stop engine and reconnect idle air control valve.

If the engine speed dropped equally (within 50 rpm) on all cylinders, refer to "Rough, Unstable, or Incorrect Idle, Stalling" symptom. If there was no drop or excessive variation in engine speed on one or more cylinders, check for spark on the respective cylinder(s) with J 26792 (ST-125) spark tester or equivalent.

If spark is present, remove the spark plugs from the cylinder(s) and check for the following.

Cracks
Wear
Improper gap
Burned electrode
Heavy deposits

HESITATION, SAG, STUMBLE

Definition: Momentary lack of response as the accelerator is pushed down. Can occur at all vehicle speeds. Usually most severe when first trying to make the car move, as from a stop sign. May cause the engine to stall if severe enough.

Perform careful visual and physical check as described at the beginning of Section "B".

Perform "Diagnostic Circuit Check."

- **CHECK**

  For possibility of misfiring, crossfiring, or cutting under load or at idle. Locate misfiring cylinder(s) by performing the following test.
  1. Start engine. Disconnect idle air control valve. Remove one spark plug wire from a spark plug and ground it against the engine.
  2. Note drop in engine speed.
  3. Repeat for all four cylinders.
  4. Stop engine and reconnect idle air control valve.

If the engine speed dropped equally (within 50 rpm) on all cylinders, refer to "Rough, Unstable, or Incorrect Idle, Stalling" symptom. If there was no drop or excessive variation in engine speed on one or more cylinders, check for spark on the respective cylinder(s) with J 26792 (ST-125) spark tester or equivalent.

If spark is present, remove the spark plugs from the cylinder(s) and check for the following.

Cracks
Wear
Improper gap
Burned electrode
Heavy deposits

1988–90 2.0L TURBO (VIN M) – ALL MODELS

- For worn distributor shaft
- Ignition pickup coil resistance and connections
- Ignition timing. See Vehicle Emission Control Information label.
- Ignition wires for shorts or faulty insulation
- Ignition system for moisture, dust, cracks, burns, etc. Spray plug wires with fine water mist to check for shorts.
- Secondary ignition voltage output with ST-125 tester
- Ignition coil connections
- ECM and ignition system for faulty grounds
- Proper operation of EST. See Section "C4"
- Spark plugs. Look for wetness, cracks, improper gap, burned electrodes, and heavy deposits
- Spark plugs for correct heat range
- Fuel for poor quality, "stale" fuel, and water contamination
- Fuel system for restricted filter or improper pressure. Use CHART A-7.
- EGR operation. Use CHART C-7.
- For vacuum leaks at intake manifold gasket
- Air cleaner element (filter) for dirt or plugging
- Idle Air Control system. Use Code 35 chart
- Check electrical system voltage. IAC valve will not move if voltage is below 9 volts or greater than 17.8 volts. Also check battery cables and ground straps for poor contact. Erratic voltage will cause the IAC valve to change its position, resulting in poor idle quality.
- PCV valve for proper operation by placing finger over inlet hole in valve end several times. Valve should snap back. If not, replace valve. Ensure that valve is correct part. Also check PCV hose.
- Canister purge system for proper operation. Use CHART C-3
- Mem-Cal for correct application
- Throttle shaft or TPS for sticking or binding. TPS voltage should read less than 1.25 volts on a "Scan" tool with the throttle closed.
- MAP sensor output. Use CHART C1-D and/or check sensor by comparing it to the output on a similar vehicle, if possible.
- Coolant sensor for a shift in calibration. Use Code 14 or Code 15 chart.
- A/C relay operation. A/C should cut out at wide open throttle. Use CHART C-10.
- Injector balance by performing injector balance test in CHART C-2A
- Oxygen sensor for silicone contamination from contaminated fuel or use of improper RTV sealant. The sensor will have a white, powdery coating and will cause a high but false signal voltage (rich exhaust indication). The ECM will reduce the amount of fuel delivered to the engine, causing a severe driveability problem.
- A/C refrigerant pressure for high pressure. Check for overcharging or faulty pressure switch.
- P/N switch circuit on vehicle with automatic transmission. Use CHART C-1A.
- Generator output voltage. Repair if less than 9 volts or more than 16 volts.
- Transmission torque converter operation.
- Transmission for proper shift points.
- Transmission torque converter clutch operation. Use CHART C-8.
- Vehicle brakes for dragging
- Engine valve timing and compression.
- For the possibility of an exhaust restriction or improper valve timing by performing the following test.
  1. With engine at normal operating temperature, connect a vacuum gauge to any convenient vacuum port on intake manifold.
  2. Run engine at 1000 rpm and record vacuum reading.
  3. Increase engine speed slowly to 2500 rpm. Note vacuum reading at steady 200 rpm.
  4. If vacuum at 2500 rpm decreases more than 3" Hg from reading at 1000 rpm, the exhaust system should be inspected for restrictions.
  5. Disconnect exhaust pipe from engine and repeat Steps 3 & 4. If vacuum still drops more than 3" Hg with exhaust disconnected, check valve timing.
- For worn or incorrect basic engine parts such as cam, heads, pistons, etc. Also check for bent pushrods, worn rocker arms, and broken or weak valve springs.
- For overheating and possible causes. Look for the following.
  - Low or incorrect coolant solution. It should be a 50/50 mix of GM #1052753 anti-freeze coolant (or equivalent) and water.

1988–90 2.0L TURBO (VIN M) – ALL MODELS

  - Loose water pump belt
  - Restricted air flow to radiator, or restricted water flow through radiator
  - Faulty or incorrect thermostat
  - Inoperative electric cooling fan circuit. See CHART C-12.
- If the system is running RICH (block learn less than 118), refer to "Diagnostic Aids" on facing page of Code 45.
- If the system is running LEAN (block learn greater than 138), refer to "Diagnostic Aids" on facing page of Code 44.

EXCESSIVE EXHAUST EMISSIONS OR ODORS

Definition: Vehicle fails an emission test or vehicle has excessive "rotten egg" smell. (Excessive odors do not necessarily indicate excessive emissions).

Perform careful visual and physical check as described at the beginning of Section "B".

Perform "Diagnostic Circuit Check."

- **CHECK**
  - EGR valve not opening. Use CHART C-7.
  - Vacuum leaks
  - Faulty coolant system and/or coolant fan operation. Use CHART C-12.
  - Remove carbon with top engine cleaner. Follow instructions on can.
- If the system is running RICH (block learn less than 118), refer to "Diagnostic Aids" on facing page of Code 45.
- If the system is running LEAN (block learn greater than 138), refer to "Diagnostic Aids" on facing page of Code 44.
- If emission test indicates excessive NO_x, check for items which cause car to run lean or too hot.
- If emission test indicates excessive HC and CO or exhaust has excessive odors, check for items which cause car to run RICH.
  - Incorrect fuel pressure. Use CHART A-7.
  - Fuel loading of evaporative vapor canister. Use CHART C-3.
  - PCV valve plugging, sticking, or blocked PCV hose. Check for fuel in crankcase.
  - Catalytic converter lead contamination (Look for removal of fuel filler neck restrictor.)
  - Improper fuel cap installation
  - Faulty spark plugs, plug wires, or ignition components.

DIESELING, RUN-ON

Definition: Engine continues to run after key is turned "OFF," but runs very roughly. (If engine runs smoothly, check ignition switch).

Perform careful visual and physical check as described at the beginning of Section "B".

Perform "Diagnostic Circuit Check."

- **CHECK**

  Injectors for leakage. Pressurize system by energizing fuel pump through the fuel pump test connector.

1988–90 2.0L TURBO (VIN M) – ALL MODELS

BACKFIRE

Definition: Fuel ignites in intake manifold or in exhaust system, making a loud popping sound.

Perform careful visual and physical check as described at the beginning of Section "B".

Perform "Diagnostic Circuit Check."

- **CHECK**
  - For possibility of misfiring, crossfiring, or cutting under load or at idle. Locate misfiring cylinder(s) by performing the following test.
    1. Start engine. Disconnect idle air control valve. Remove one spark plug wire from a spark plug and ground it against the engine.
    2. Note drop in engine speed.
    3. Repeat for all four cylinders.
    4. Stop engine and reconnect idle air control valve.

    If the engine speed dropped equally (within 50 rpm) on all cylinders, refer to "Rough, Unstable, or Incorrect Idle, Stalling" symptom. If there was no drop or excessive variation in engine speed on one or more cylinders, check for spark on the respective cylinder(s) with J 26792 (ST-125) spark tester or equivalent. If spark is present, remove the spark plugs from the cylinder(s) and check for the following.

    Cracks
    Wear
    Improper gap
    Burned electrode
    Heavy deposits

  - For worn distributor shaft
  - Ignition timing. See underhood emission control information label.
  - EGR operation for valve being open all the time. Use CHART C-7.
  - Intake manifold gasket for leaks
  - Spark plugs. Look for wetness, cracks, improper gap, burned electrodes, and heavy deposits.
  - Ignition coil connections
  - Ignition system for moisture, dust, cracks, burns, etc. Spray plug wires with fine water mist to check for shorts.
  - ECM and ignition system for faulty grounds
  - Secondary ignition voltage output with ST-125 tester
  - For vacuum leaks at intake manifold gasket
  - Engine valve timing and compression.

  - For worn or incorrect basic engine parts such as cam, heads, pistons, etc. Also check for bent pushrods, worn rocker arms, and broken or weak valve springs.

1988–90 2.0L TURBO (VIN M) – ALL MODELS

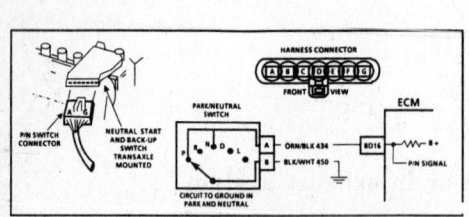

CHART C-1A
PARK/NEUTRAL SWITCH DIAGNOSIS
(AUTO TRANSAXLE ONLY)
2.0L TURBO (VIN M)

Circuit Description:
The Park/Neutral (P/N) switch contacts are a part of the neutral start switch they are closed to ground in park or neutral and open in drive ranges.
The ECM supplies voltage through a current limiting resistor to CKT 434 and senses a closed switch when the voltage on CKT 434 drops to less than one volt.
The ECM uses the P/N signal as one of the inputs to control idle air, VSS diagnostics, and EGR.
If CKT 434 indicates P/N (grounded), while in drive range, the EGR would be inoperative, resulting in possible detonation.
If CKT 434 indicates drive (open) a dip in the idle may exist when the gear selector is moved into drive range.

Test Description: Numbers below refer to circled numbers on the diagnostic chart.
1. Checks for a closed switch to ground in park position. Different makes of "Scan" tools will read P/N differently. Refer to tool operator's manual for type of display used for a specific tool.

2. Checks for an open switch in drive range.
3. Be sure "Scan" indicates drive, even while wiggling shifter, to test for an intermittent or misadjusted switch in drive or overdrive range.

1988–90 2.0L TURBO (VIN M) – ALL MODELS

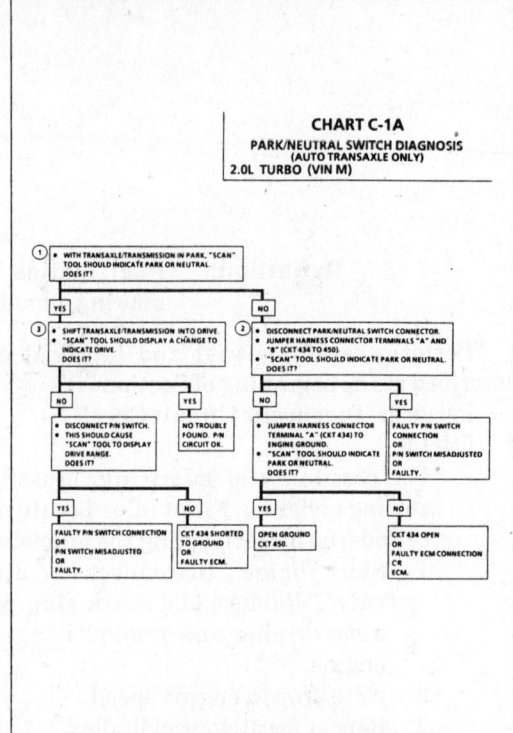

1988–90 2.0L TURBO (VIN M) – ALL MODELS

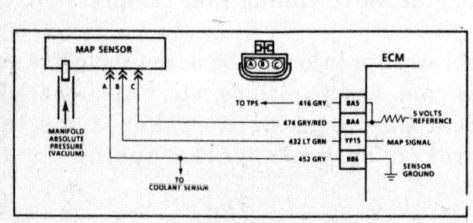

CHART C-1D
MANIFOLD ABSOLUTE PRESSURE (MAP) OUTPUT CHECK
2.0L TURBO (VIN M)

Circuit Description:
The Manifold Absolute Pressure (MAP) sensor measures manifold pressure (vacuum) and sends that signal to the ECM. The MAP sensor is mainly used to calculate engine load, which is a fundamental input for spark and fuel calculations. The MAP sensor is also used to determine barometric pressure.

Test Description: Numbers below refer to circled numbers on the diagnostic chart.
1. Checks MAP sensor output voltage to the ECM. With the ignition "ON" and the engine stopped, the manifold pressure is equal to atmospheric pressure and the signal voltage will be high. This voltage, without engine running, represents a barometer reading to the ECM.
Comparison of this BARO reading with a known good vehicle with the same sensor is a good way to check accuracy of a "suspect" sensor. Readings should be the same ± .4 volt.

2. Applying 34 kPa (10 inches Hg) vacuum to the MAP sensor should cause the voltage to be 1.2-2.3 volts less than the voltage at step 1. Upon applying vacuum to the sensor, the change in voltage should be instantaneous. A slow voltage change indicates a faulty sensor.
3. Check vacuum hose to sensor for leaking or restriction. Be sure no other vacuum devices are connected to the MAP hose.

1988–90 2.0L TURBO (VIN M) – ALL MODELS

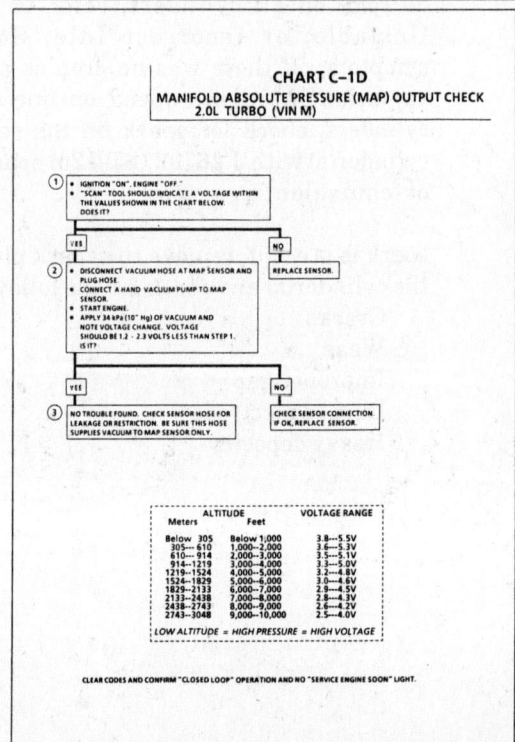

1988–90 2.0L TURBO (VIN M) – ALL MODELS

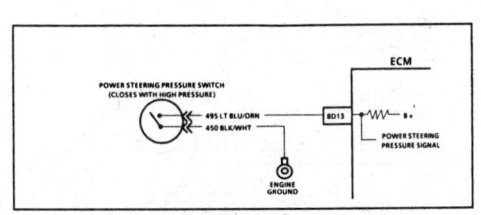

CHART C-1E
POWER STEERING PRESSURE SWITCH (PSPS) DIAGNOSIS
2.0L TURBO (VIN M)

Circuit Description:

The Power Steering Pressure Switch (PSPS) is normally open to ground, and CKT 495 will be near the battery voltage.

Turning the steering wheel increases power steering oil pressure and its load on an idling engine. The pressure switch will close before the load can cause an idle problem.

Closing the switch causes CKT 495 to read less than 1 volt. The ECM will increase the idle air rate and disengage the A/C relay.

- A pressure switch that will not close, or an open CKT 495 or 450, may cause the engine to stop, when power steering loads are high.
- A switch that will not open or a CKT 495 shorted to ground, may affect idle quality and will cause the A/C relay to be de-energized.

Test Description: Numbers below refer to circled numbers on the diagnostic chart.
1. Different makes of "Scan" tools may display the state of this switch in different ways. Refer to "Scan" tool operator's manual to determine how this input is indicated.

2. Checks to determine if CKT 495 is shorted to ground.
3. This should simulate a closed switch.

1988–90 2.0L TURBO (VIN M) – ALL MODELS

CHART C-1E
POWER STEERING PRESSURE SWITCH (PSPS) DIAGNOSIS
2.0L TURBO (VIN M)

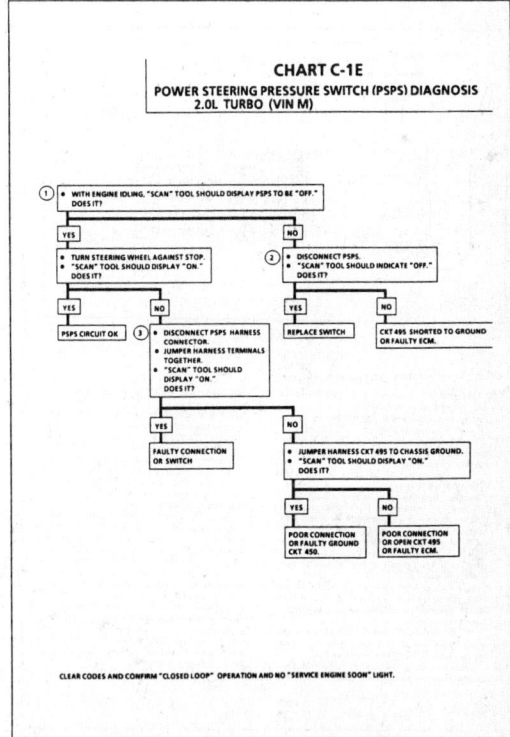

CLEAR CODES AND CONFIRM "CLOSED LOOP" OPERATION AND NO "SERVICE ENGINE SOON" LIGHT.

1988–90 2.0L TURBO (VIN M) – ALL MODELS

CHART C-2A
INJECTOR BALANCE TEST

The injector balance tester is a tool used to turn the injector on for a precise amount of time, thus spraying a measured amount of fuel into the manifold. This causes a drop in fuel rail pressure that we can record and compare between each injector. All injectors should have the same amount of pressure drop (± 10 kPa). Any injector with a pressure drop that is 10 kPa (or more) greater or less than the average drop of the other injectors should be considered faulty and replaced.

STEP 1

Engine "cool down" period (10 minutes) is necessary to avoid irregular readings due to "Hot Soak" fuel boiling. With ignition "OFF" connect fuel gauge J 347301 or equivalent to fuel pressure tap. Wrap a shop towel around fitting while connecting gage to avoid fuel spillage.

Disconnect harness connectors at all injectors, and connect injector tester J 34730-3, or equivalent, to one injector. On Turbo equipped engines, use adaptor harness furnished with injector tester to energize injectors that are not accessible. Follow manufacturers instructions for use of adaptor harness. Ignition must be "OFF" at least 10 seconds to complete ECM shutdown cycle. Fuel pump should run about 2 seconds after ignition is turned "ON." At this point, insert clear tubing attached to vent valve into a suitable container and bleed air from gauge and hose to insure accurate gauge operation. Repeat this step until all air is bled from gauge.

STEP 2

Turn ignition "OFF" for 10 seconds and then "ON" again to get fuel pressure to its maximum. Record this initial pressure reading. Energize tester one time and note pressure drop at its lowest point (Disregard any slight pressure increase after drop hits low point). By subtracting this second pressure reading from the initial pressure, we have the actual amount of injector pressure drop.

STEP 3

Repeat step 2 on each injector and compare the amount of drop. Usually, good injectors will have virtually the same drop. Retest any injector that has a pressure difference of 10kPa, either more or less than the average of the other injectors on the engine. Replace any injector that also fails the retest. If the pressure drop of all injectors is within 10kPa of this average, the injectors appear to be flowing properly. Reconnect them and review "Symptoms," Section "B".

NOTE: *The entire test should not be repeated more than once without running the engine to prevent flooding. (This includes any retest on faulty injectors).*

1988–90 2.0L TURBO (VIN M) – ALL MODELS

CHART C-2A
INJECTOR BALANCE TEST
2.0L TURBO (VIN M)

NOTE: If injectors are suspected of being dirty, they should be cleaned using an approved tool and procedure prior to performing this test. The fuel pressure test in Section "A", Chart A-7, should be completed prior to this test.

Step 1. If engine is at operating temperature, allow a 10 minute "cool down" period then connect fuel pressure gauge and injector tester.
1. Ignition "OFF."
2. Connect fuel pressure gauge and injector tester.
3. Ignition "ON."
4. Bleed off air in gauge. Repeat until all air is bled from gauge.

Step 2. Run test:
1. Ignition "OFF" for 10 seconds.
2. Ignition "ON". Record gauge pressure. (Pressure must hold steady, if not see the Fuel System diagnosis, Chart A-7, in Section "A").
3. Turn injector on, by depressing button on injector tester, and note pressure at the instant the gauge needle stops.

Step 3.
1. Repeat step 2 on all injectors and record pressure drop on each. Retest injectors that appear faulty (Any injectors that have a 10 kPa difference, either more or less, in pressure from the average). If no problem is found, review "Symptoms" Section "B".

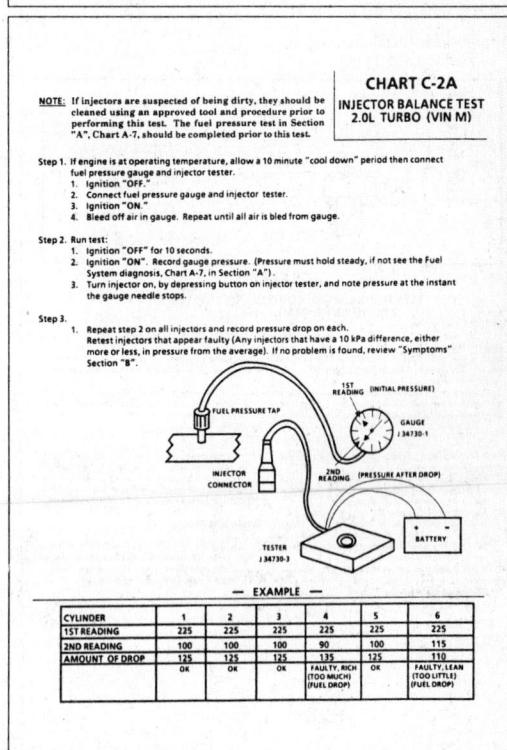

— EXAMPLE —

CYLINDER	1	2	3	4	5	6
1ST READING	225	225	225	90	225	225
2ND READING	100	100	100	135	100	115
AMOUNT OF DROP	125	125	125	135	125	110
	OK	OK	OK	FAULTY, RICH (TOO MUCH) (FUEL DROP)	OK	FAULTY, LEAN (TOO LITTLE) (FUEL DROP)

1988–90 2.0L TURBO (VIN M) – ALL MODELS

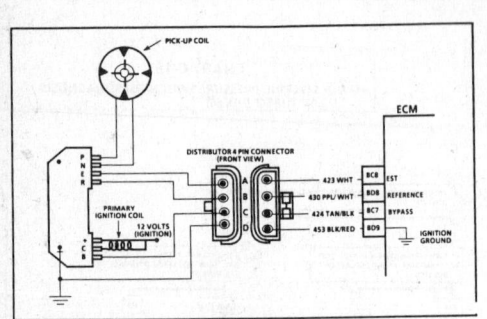

CHART C-4C
IGNITION SYSTEM CHECK
(REMOTE COIL)
2.0L TURBO (VIN M)

Test Description: Numbers below refer to circled numbers on the diagnostic chart.

1. Two wires are checked to ensure that an open is not present in a spark plug wire.

1A. If spark occurs with 4 terminal distributor connector disconnected, pick-up coil output is too low for EST operation.

2. A spark indicates the problem must be the distributor cap or rotor.

3. Normally, there should be battery voltage at the "C" and "+" terminals. Low voltage would indicate an open or a high resistance circuit from the distributor to the coil or ignition switch. If "C" terminal voltage was low, but "+" terminal voltage is 10 volts or more, circuit from "C" terminal to ignition coil or ignition coil primary winding is open.

4. Checks for a shorted module or grounded circuit from the ignition coil to the module. The distributor module should be turned "OFF." Normal voltage should be about 12 volts.
   If the module is turned "ON," the voltage would be low, but above 1 volt. This could cause the ignition coil to fail from excessive heat.
   With an open ignition coil primary winding, a small a mount of voltage will leak through the

module from the "BATT" to the "tach" terminal.

5. Applying voltage (1.5 to 8 volts) to module terminal "P" should turn the module "ON" and the tachometer terminal voltage should drop to about 7-9 volts. This test will determine whether the module or coil is faulty or if the pick-up coil is not generating the proper signal to turn the module "ON." This test can be performed by using a DC battery with a rating of 1.5 to 8 volts. The use of the test light is mainly to allow the "P" terminal to be probed more easily.
   Some digital multi-meters can also be used to trigger the module by selecting ohms, usually the diode position. In this position the meter may have a voltage across its terminals which can be used to trigger the module. The voltage in the ohms position can be checked by using a second meter or by checking the manufacturer's specification of the tool being used.

6. This should turn "OFF" the module and cause a spark. If no spark occurs, the fault is most likely in the ignition coil because most module problems would have been found before this point in the procedure. A module tester (J 24642) could determine which is at fault.

1988–90 2.0L TURBO (VIN M) – ALL MODELS

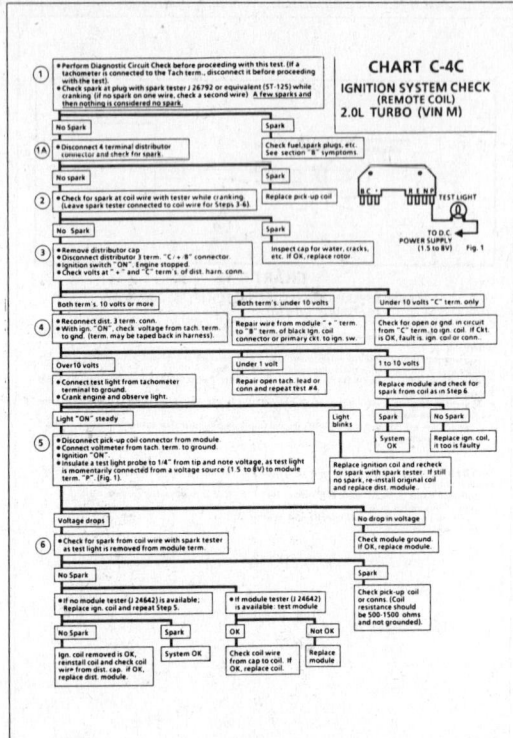

CHART C-4C
IGNITION SYSTEM CHECK
(REMOTE COIL)
2.0L TURBO (VIN M)

1988–90 2.0L TURBO (VIN M) – ALL MODELS

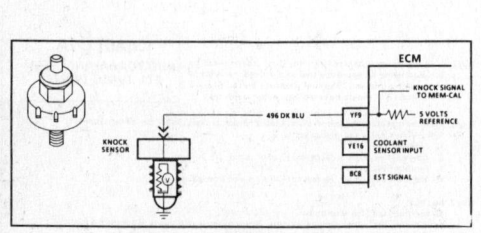

CHART C-5
ELECTRONIC SPARK CONTROL (ESC) SYSTEM CHECK
2.0L TURBO (VIN M)

Circuit Description:
The knock sensor is used to detect engine detonation and the ECM will retard the electronic spark timing based on the signal being received. The circuitry, within the knock sensor, causes the ECM's 5 volts to be pulled down so that under a no knock condition, CKT 496 would measure about 2.5 volts. The knock sensor produces an A/C signal, which rides on the 2.5 volts DC voltage. The amplitude and frequency are dependent upon the knock level.
The Mem-Cal, used with this engine, contains the functions which were part of remotely mounted ESC modules used on other GM vehicles. The ESC portion of the Mem-Cal then sends a signal to other parts of the ECM which adjusts the spark timing to retard the spark and reduce the detonation.

Test Description: Numbers below refer to circled numbers on the diagnostic chart.

1. With engine idling, there should not be a knock signal present at the ECM, because detonation is not likely under a no load condition.

2. Tapping on the engine lift hood bracket should simulate a knock signal to determine if the sensor is capable of detecting detonation. If no knock is detected, try tapping on engine block closer to sensor before replacing sensor.

3. If the engine has an internal problem, which is creating a knock, the knock sensor may be responding to the internal failure.

4. This test determines if the knock sensor is faulty, or, if the ESC portion of the Mem-Cal is faulty. If it is determined that the Mem-Cal is faulty, be sure that it is properly installed and latched into place. If not properly installed, repair and retest.

Diagnostic Aids:
While observing knock signal on the "Scan," there should be an indication that knock is present, when detonation can be heard. Detonation is most likely to occur under high engine load conditions.

1988–90 2.0L TURBO (VIN M) – ALL MODELS

CHART C-5
ELECTRONIC SPARK CONTROL (ESC) SYSTEM CHECK
2.0L TURBO (VIN M)

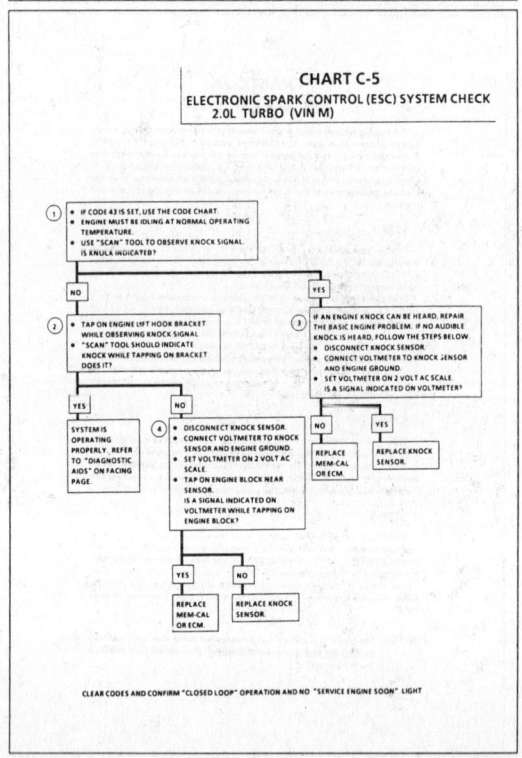

1988–90 2.0L TURBO (VIN M) – ALL MODELS

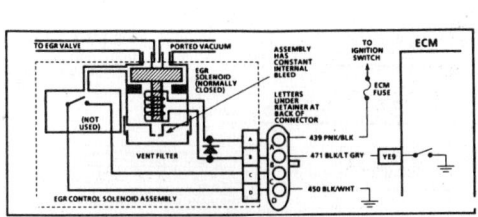

CHART C-7
EXHAUST GAS RECIRCULATION (EGR) CIRCUIT CHECK
2.0L TURBO (VIN M)

Circuit Description:
The ECM operates a vacuum solenoid to control the Exhaust Gas Recirculation (EGR) valve. This solenoid is normally closed. By providing a ground path, the ECM energizes the solenoid, allowing vacuum to reach to the EGR valve.

Under certain conditions, when the EGR valve is normally open, the ECM tests the EGR function by de-energizing the EGR control solenoid, blocking vacuum to the EGR valve diaphragm. Without EGR, the system will sense a lean condition and will increase the fuel integrator rate in response. The ECM monitors the amount of fuel delivery increase. If the increase is below a specified value, the ECM will interpret that the test was failed. The failure indicates that closing the EGR valve when it would normally be open does not make a significant change, indicating a problem in the EGR system.

Test Description: Numbers below refer to circled numbers on the diagnostic chart.

The diagnostic chart covers checks for the entire EGR system. If no trouble is found but Code 32 was set, an intermittent electrical condition or a sticky EGR valve is at fault.

Diagnostic Aids:

The vacuum switch in the EGR solenoid assembly is not used.

An EGR valve stuck open will cause a rough idle.

A plugged EGR solenoid vent filter could cause the EGR valve to remain open or to close slowly.

An inoperative check valve in the ported vacuum line will result in faulty EGR system operation.

1988–90 2.0L TURBO (VIN M) – ALL MODELS

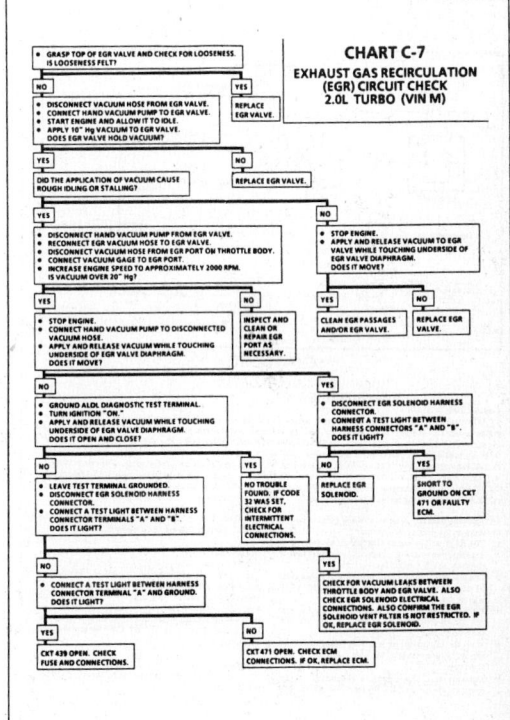

CHART C-7
EXHAUST GAS RECIRCULATION (EGR) CIRCUIT CHECK
2.0L TURBO (VIN M)

1988–90 2.0L TURBO (VIN M) – ALL MODELS

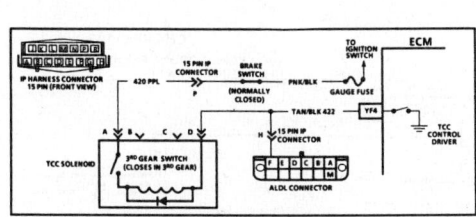

CHART C-8A
TORQUE CONVERTER CLUTCH (TCC) CIRCUIT CHECK
(ELECTRICAL DIAGNOSIS)
2.0L TURBO (VIN M)

Circuit Description:
The purpose of the Torque Converter Clutch (TCC) is to eliminate the power loss of the torque converter when the vehicle is in a cruise condition. This allows the convenience of the automatic transmission and the fuel economy of a manual transmission.

Voltage is supplied to the TCC solenoid through the brake switch and transmission third gear apply switch. The ECM will engage TCC by grounding CKT 422 to energize the solenoid.

TCC will engage under the following conditions:
- Vehicle speed exceeds 30 mph (48 km/h)
- Engine temperature is above 70°C (156°F)
- Throttle position sensor output is not changing faster than a calibrated rate (steady throttle).
- Transaxle third gear switch is closed
- Brake switch is closed

Test Description: Numbers below refer to circled numbers on the diagnostic chart.

1. Light "OFF" confirms that transaxle third gear apply switch is open.
2. At 48 km/h (30 mph), the transmission third gear switch should close. Test light will light and confirm battery supply, and close brake switch.
3. Grounding the diagnostic terminal, with engine "OFF," should energize the TCC solenoid. This test checks the capability of the ECM to control the solenoid.

Diagnostic Aids:

An engine coolant thermostat that is stuck open or opens at too low a temperature may result in an inoperative TCC.

1988–90 2.0L TURBO (VIN M) – ALL MODELS

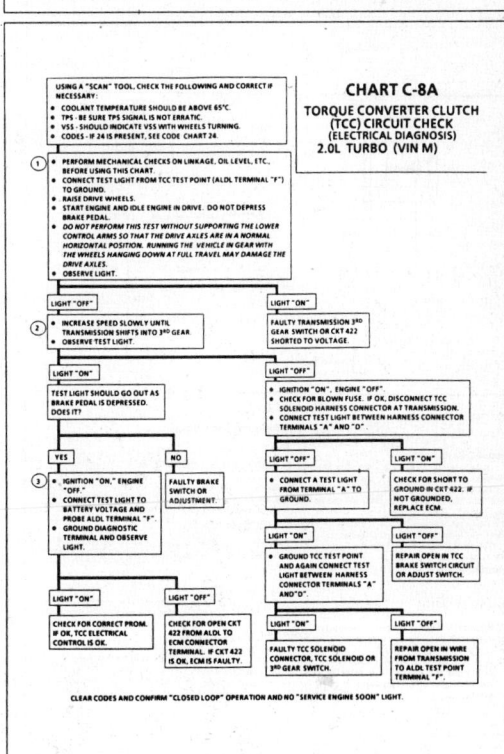

CHART C-8A
TORQUE CONVERTER CLUTCH (TCC) CIRCUIT CHECK
(ELECTRICAL DIAGNOSIS)
2.0L TURBO (VIN M)

1988-90 2.0L TURBO (VIN M) – ALL MODELS

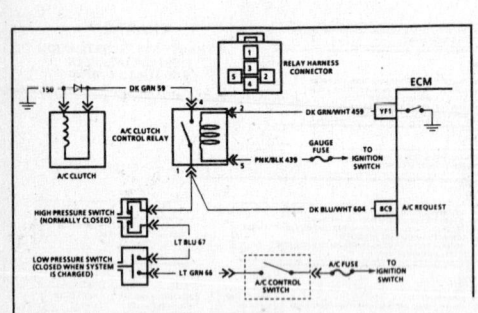

CHART C-10
A/C CLUTCH CONTROL
2.0L TURBO (VIN M)

Circuit Description:

When an A/C mode is selected on the A/C control switch, ignition voltage is supplied to the compressor low pressure switch. If there is sufficient A/C refrigerant pressure, the low pressure switch will be closed and complete CKT 67 to the closed high pressure cut-off switch and to CKT 604. The voltage on CKT 604 to the ECM is shown by the "Scan" tool as A/C request "ON" (voltage present) or "OFF" (no voltage). When a request for A/C is detected by the ECM, the ECM will ground CKT 459 of the A/C clutch control relay. The relay contact will close and current will flow from CKT 604 to CKT 59, causing the A/C compressor clutch to engage. A "Scan" tool will show the grounding of CKT 459 as A/C clutch "ON." Also, when voltage is detected by the ECM on CKT 604, the cooling fan will be turned "ON."

Test Description: Numbers below refer to circled numbers on the diagnostic chart.
1. The ECM will energize the A/C relay only when the engine is running. This test will determine if the relay or CKT 459 is faulty.
2. The low pressure and high pressure switches must be closed so that the A/C request signal (12 volts) will be present at the ECM.
3. A short to ground in any part of the A/C request or A/C clutch control circuits could be the cause of the blown fuse.
4. With the engine idling and A/C "ON," the ECM should be grounding CKT 459, causing the test light to be "ON."
5. Determines if the signal is reaching the low pressure switch on CKT 66 from the A/C control

panel. The signal should be present only when the A/C mode or defrost mode has been selected.

Diagnostic Aids:

• If complaint was insufficient cooling, the problem may be caused by an inoperative cooling fan or A/C pressure fan switch. The engine cooling fan should turn "ON" when A/C pressure exceeds a value to open the switch, which causes the ECM to energize the cooling fan relay. See CHART C-12 for cooling fan diagnosis.

The A/C clutch will be disengaged if a high power steering pressure signal is detected by the ECM. Refer to CHART C1-E.

1988-90 2.0L TURBO (VIN M) – ALL MODELS

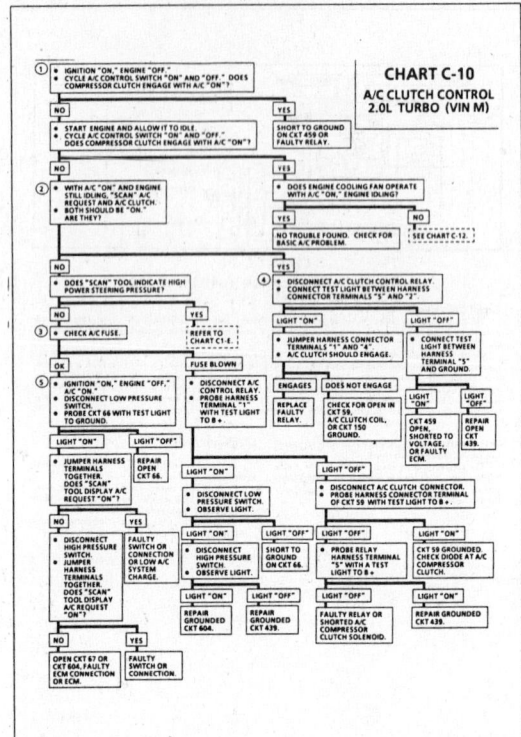

CHART C-10
A/C CLUTCH CONTROL
2.0L TURBO (VIN M)

1988-90 2.0L TURBO (VIN M) – ALL MODELS

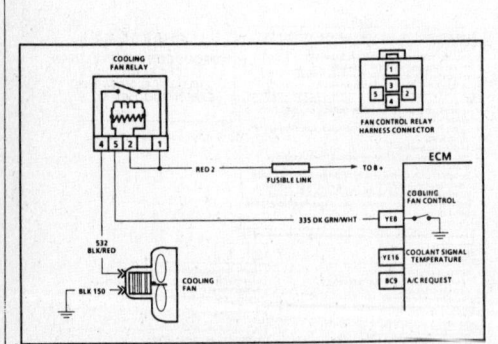

CHART C-12
ENGINE COOLING FAN SYSTEM CHECK
2.0L TURBO (VIN M)

Circuit Description:

Battery voltage to operate the cooling fan motor is supplied to relay terminals "1" & "5". When the ECM grounds CKT 335, the relay is energized and the cooling fan is turned "ON."

The ECM can command the fan to run when the engine is shut down (ignition "OFF"). Under certain conditions (based on MAT, coolant, and engine run time), the ECM will remain "powered up" and ground CKT 335. If the ECM determines a need for cooling after shut down, the fan will run between 20 seconds and 7 minutes after ignition "OFF." This shutdown feature can extend the life of the turbocharger by reducing underhood temperature after shutdown.

The ECM will turn "ON" the fan (ignition "ON") when the coolant temperature is approximately 108°C. The ECM will also command constant fan (ignition "ON") when it senses voltage on A/C request line (A/C "ON") regardless of temperature, or if a Code 14 or 15 is stored.

Diagnostic Aids:

1. Because of the shutdown fan feature (described above), the fan may be running with key "OFF" for up to 7 minutes.
2. If an overheating problem is suspected, it must be determined if it was due to an actual boil over, or the hot light, or temperature gage indicated overheating.

If the gage or light indicates overheating, but no boil over is detected, the gage circuit should be checked. The gage accuracy can be checked by comparing the coolant sensor reading on a "Scan" tool with the gage reading.

If the engine is actually overheating and the gage indicates overheating but the cooling fan is not turning "ON," the coolant sensor has probably shifted out of calibration and should be replaced.

1988-90 2.0L TURBO (VIN M) – ALL MODELS

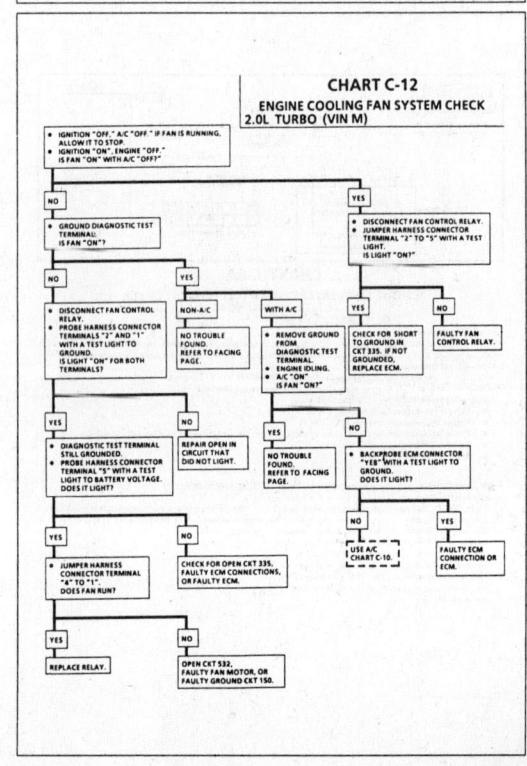

CHART C-12
ENGINE COOLING FAN SYSTEM CHECK
2.0L TURBO (VIN M)

2.3L (VIN D) COMPONENT LOCATIONS—GRAND AM

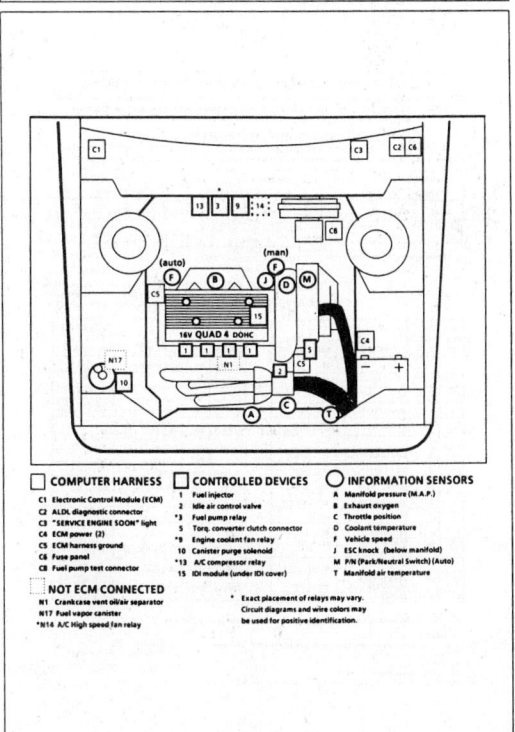

□ COMPUTER HARNESS
C1 Electronic Control Module (ECM)
C2 ALDL diagnostic connector
C3 "SERVICE ENGINE SOON" light
C4 ECM power (2)
C5 ECM harness ground
C6 Fuse panel
C8 Fuel pump test connector

▢ NOT ECM CONNECTED
N1 Crankcase vent oil/air separator
N17 Fuel vapor canister
*N14 A/C High speed fan relay

□ CONTROLLED DEVICES
1 Fuel injector
2 Idle air control valve
*3 Fuel pump relay
5 Torq. converter clutch connector
*9 Engine coolant fan relay
10 Canister purge solenoid
*13 A/C compressor relay
15 IDI module (under IDI cover)

○ INFORMATION SENSORS
A Manifold pressure (M.A.P.)
B Exhaust oxygen
C Throttle position
D Coolant temperature
F Vehicle speed
J ESC knock (below manifold)
M P/N (Park/Neutral Switch) (auto)
T Manifold air temperature

* Exact placement of relays may vary.
Circuit diagrams and wire colors may
be used for positive identification.

2.3L (VIN A) COMPONENT LOCATIONS—GRAND AM

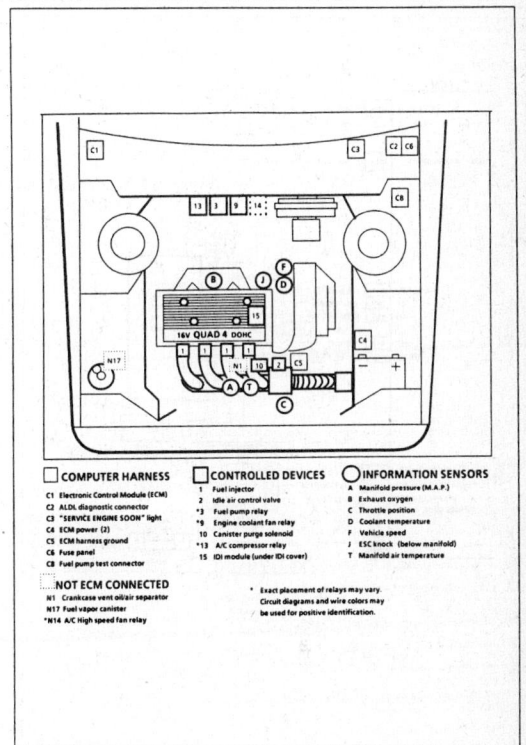

□ COMPUTER HARNESS
C1 Electronic Control Module (ECM)
C2 ALDL diagnostic connector
C3 "SERVICE ENGINE SOON" light
C4 ECM power (2)
C5 ECM harness ground
C6 Fuse panel
C8 Fuel pump test connector

▢ NOT ECM CONNECTED
N1 Crankcase vent oil/air separator
N17 Fuel vapor canister
*N14 A/C High speed fan relay

□ CONTROLLED DEVICES
1 Fuel injector
2 Idle air control valve
*3 Fuel pump relay
*9 Engine coolant fan relay
10 Canister purge solenoid
*13 A/C compressor relay
15 IDI module (under IDI cover)

○ INFORMATION SENSORS
A Manifold pressure (M.A.P.)
B Exhaust oxygen
C Throttle position
D Coolant temperature
F Vehicle speed
J ESC knock (below manifold)
T Manifold air temperature

* Exact placement of relays may vary.
Circuit diagrams and wire colors may
be used for positive identification.

2.3L (VIN D AND A) ECM WIRING DIAGRAM GRAND AM

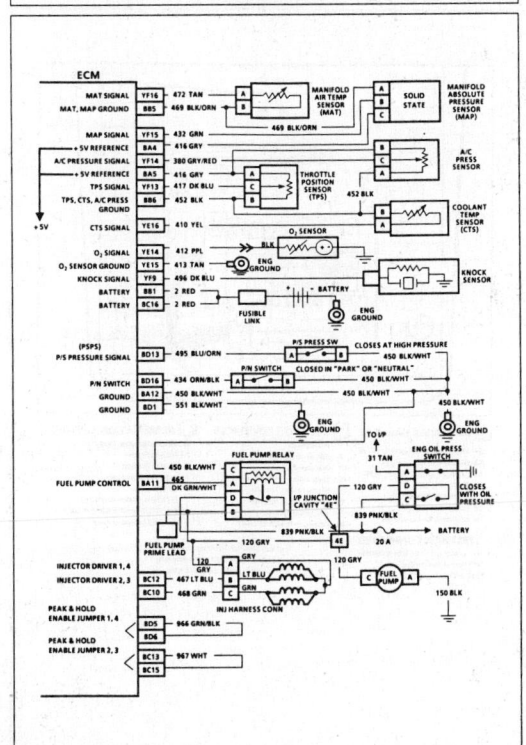

2.3L (VIN D AND A) ECM WIRING DIAGRAM GRAND AM (CONT.)

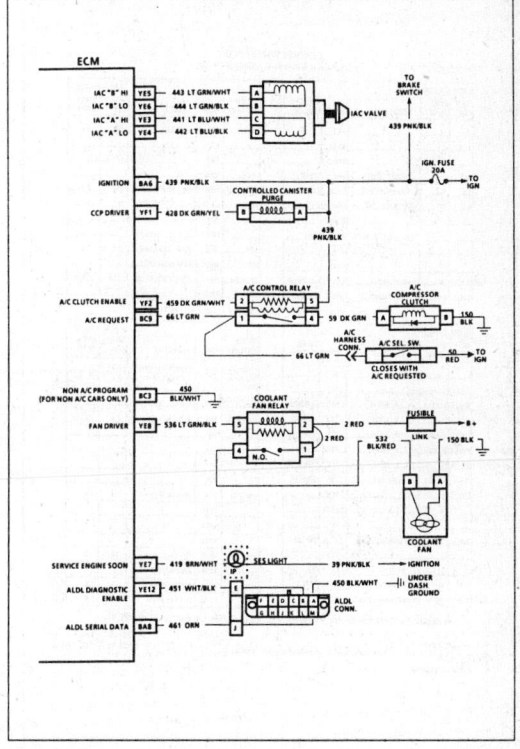

2.3L (VIN D AND A) ECM WIRING DIAGRAM GRAND AM (CONT.)

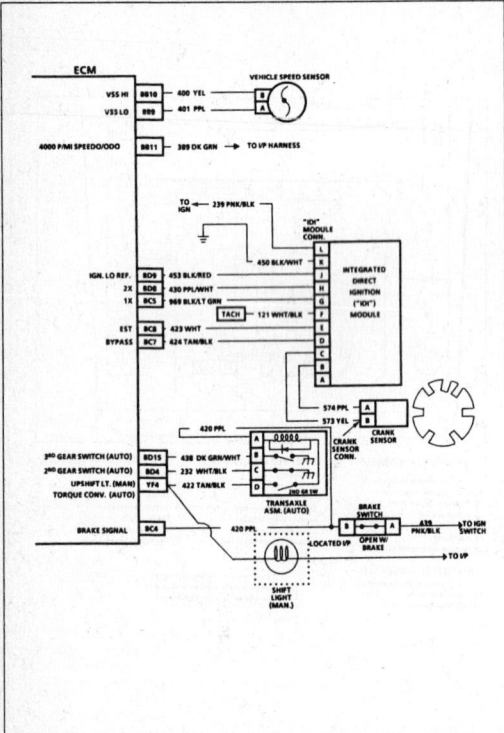

2.3L (VIN D AND A) ECM CONNECTOR TERMINAL END VIEW—GRAND AM

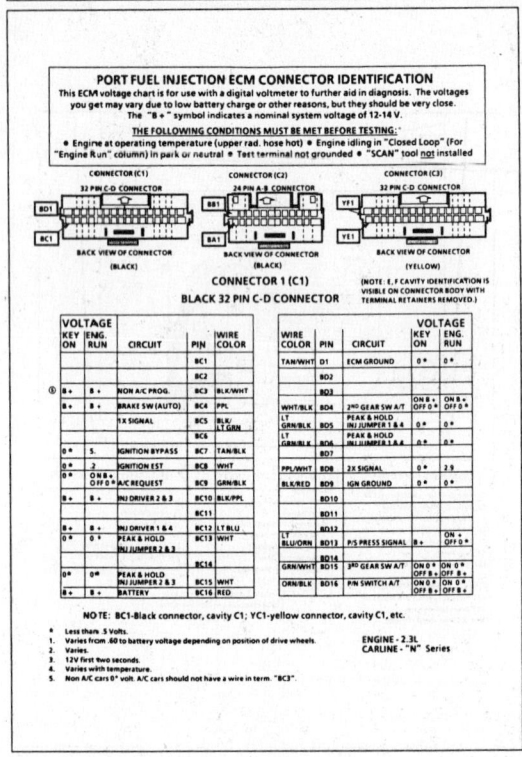

2.3L (VIN D AND A) ECM CONNECTOR TERMINAL END VIEW—GRAND AM (CONT.)

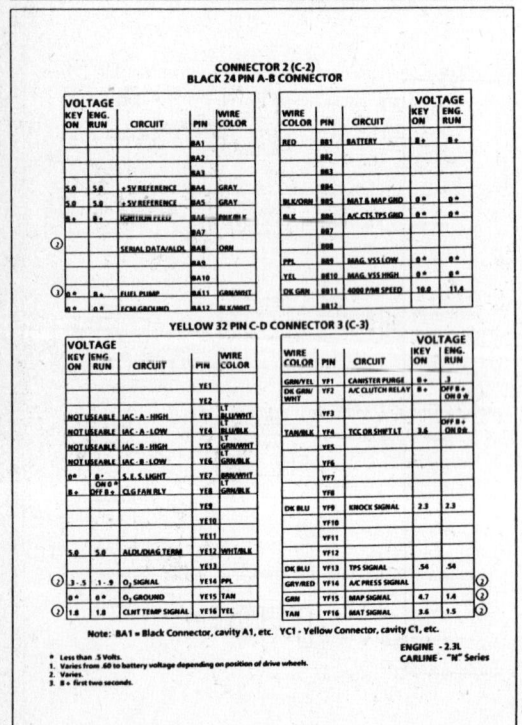

2.3L (VIN D) COMPONENT LOCATIONS—SKYLARK

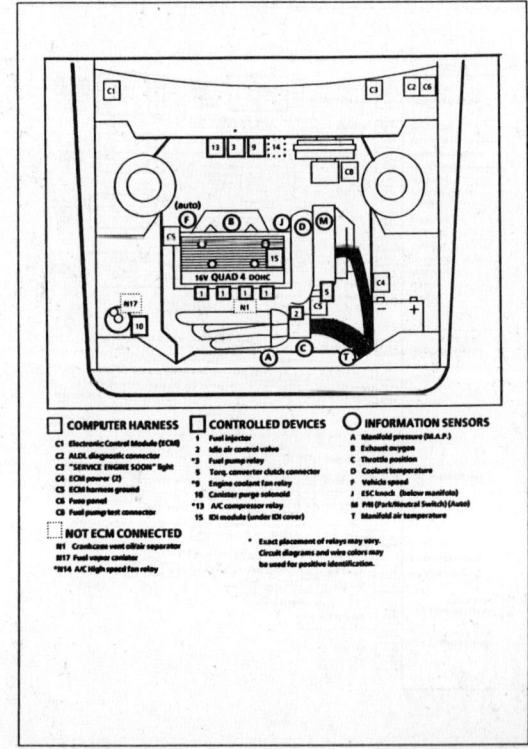

2.3L (VIN D) ECM WIRING DIAGRAM – SKYLARK

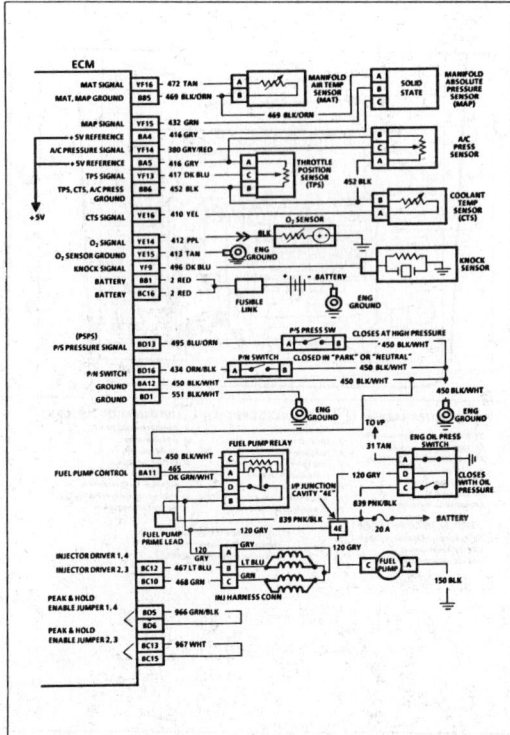

2.3L (VIN D) ECM WIRING DIAGRAM – SKYLARK (CONT.)

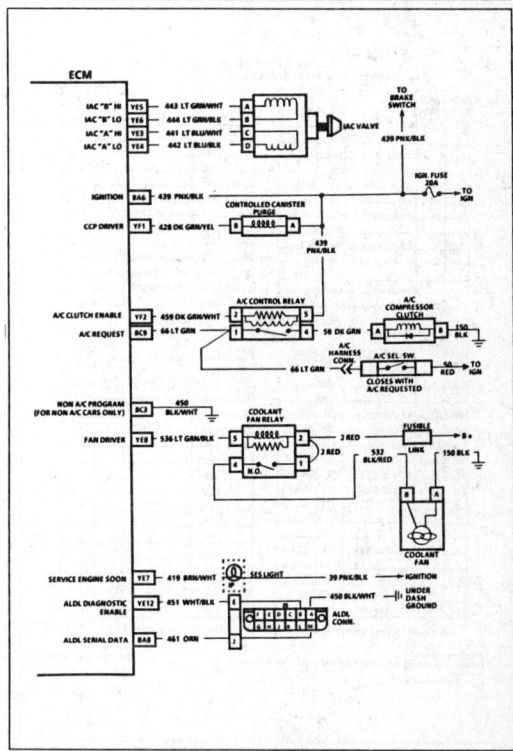

2.3L (VIN D) ECM WIRING DIAGRAM – SKYLARK (CONT.)

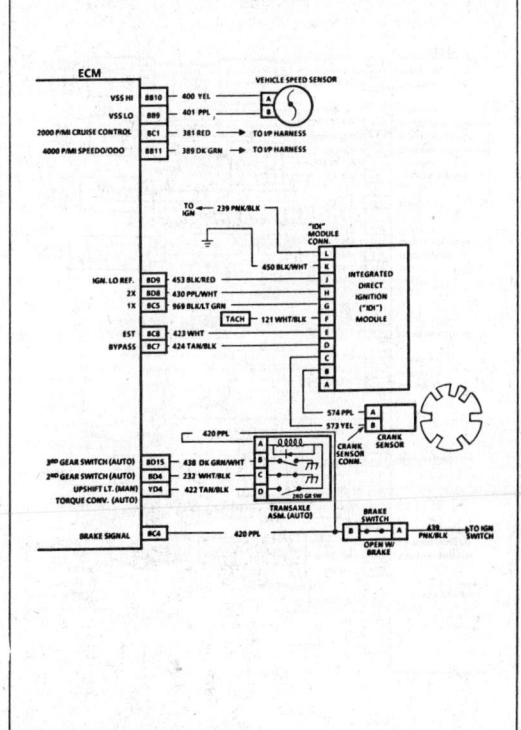

2.3L (VIN D) ECM CONNECTOR TERMINAL END VIEW – SKYLARK

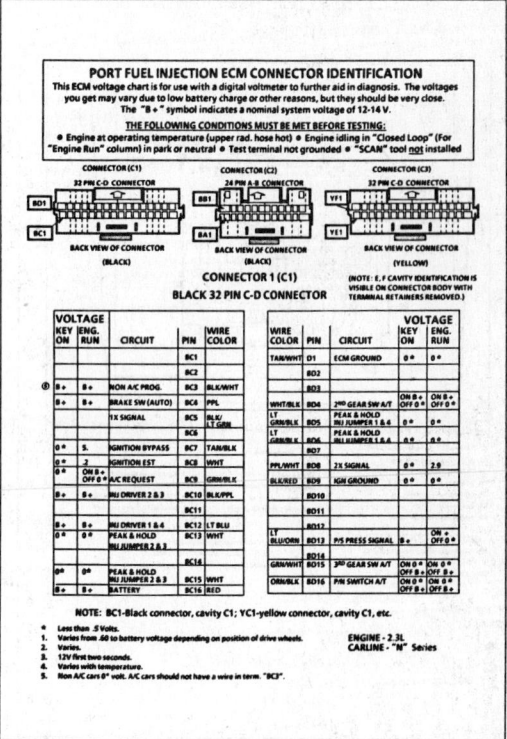

2.3L (VIN D) ECM CONNECTOR TERMINAL END VIEW—SKYLARK (CONT.)

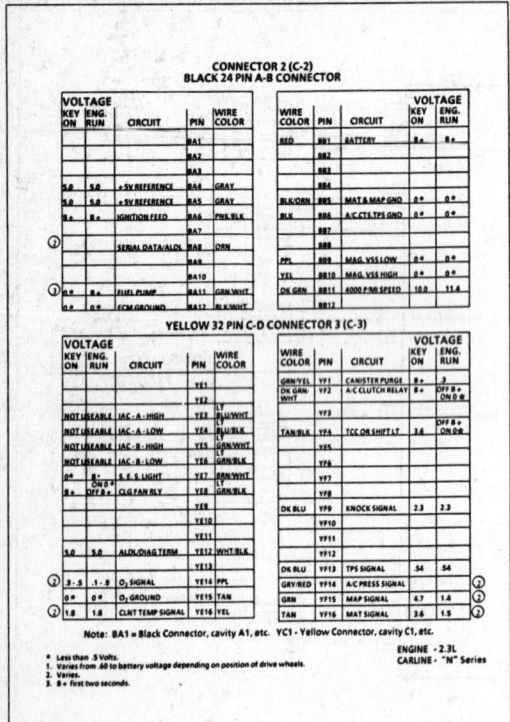

CONNECTOR 2 (C-2)
BLACK 24 PIN A-B CONNECTOR

VOLTAGE KEY ON	ENG. RUN	CIRCUIT	PIN	WIRE COLOR
			BA1	
			BA2	
			BA3	
			BA4	
5.0	5.0	+5V REFERENCE	BA5	GRAY
5.0	5.0	+5V REFERENCE	BA5	GRAY
B+	B+	IGNITION FEED	BA6	PNK/BLK
			BA7	
①		SERIAL DATA/ALDL	BA8	ORN
			BA9	
			BA10	
①	B+	FUEL PUMP	BA11	GRN/WHT
0*	0*	ECM GROUND	BA12	BLK/WHT

WIRE COLOR	PIN	CIRCUIT	KEY ON	ENG. RUN
RED	BB1	BATTERY	B+	B+
	BB2			
	BB3			
	BB4			
BLK/ORN	BB5	MAT & MAP GND	0*	0*
BLK	BB6	A/C,CTS,TPS GND	0*	0*
	BB7			
	BB8			
PPL	BB9	MAG VSS LOW	0*	0*
YEL	BB10	MAG VSS HIGH	0*	0*
DK GRN	BB11	4000 P/MI SPEED	10.0	11.4
	BB12			

YELLOW 32 PIN C-D CONNECTOR 3 (C-3)

VOLTAGE KEY ON	ENG. RUN	CIRCUIT	PIN	WIRE COLOR
			YC1	
			YC2	
NOT USEABLE		IAC - A - HIGH	YC3	LT BLU/WHT
NOT USEABLE		IAC - A - LOW	YC4	BLU/BLK
NOT USEABLE		IAC - B - HIGH	YC5	GRN/WHT
NOT USEABLE		IAC - B - LOW	YC6	GRN/BLK
0*		S.E.S. LIGHT	YC7	BRN/WHT
B+	OFF B+	CLG FAN RLY	YC8	GRN/BLK
			YC9	
			YC10	
			YC11	
5.0	5.0	ALDL/DIAG TERM	YC12	WHT/BLK
			YC13	
①	.3-.5	.1-.9 O₂ SIGNAL	YC14	PPL
①	0*	0* O₂ GROUND	YC15	TAN
②	1.8	1.8 CLNT TEMP SIGNAL	YC16	YEL

WIRE COLOR	PIN	CIRCUIT	KEY ON	ENG. RUN
GRN/YEL	YF1	CANISTER PURGE	B+	3
DK GRN/WHT	YF2	A/C CLUTCH RELAY	B+	OFF B+ ON 0*
	YF3			
TAN/BLK	YF4	TCC OR SHIFT LT	3.6	OFF B+ ON B+
	YF5			
	YF6			
	YF7			
	YF8			
DK BLU	YF9	KNOCK SIGNAL	2.3	2.3
	YF10			
	YF11			
	YF12			
DK BLU	YF13	TPS SIGNAL	.54	.54
GRY/RED	YF14	A/C PRESS SIGNAL		
GRN	YF15	MAP SIGNAL	4.7	1.4
TAN	YF16	MAT SIGNAL	3.6	1.5

Note: BA1 = Black Connector, cavity A1, etc. YC1 - Yellow Connector, cavity C1, etc.

* Less than .5 Volts.
1. Varies from 60 to battery voltage depending on position of drive wheels.
2. Varies.
3. B+ first two seconds.

ENGINE - 2.3L
CARLINE - "N" Series

2.3L (VIN D) COMPONENT LOCATIONS—CALAIS

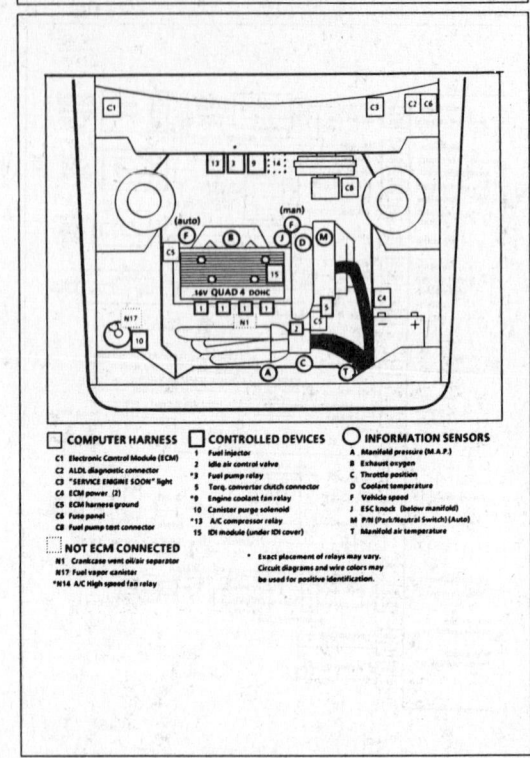

16V QUAD 4 DOHC

COMPUTER HARNESS
C1 Electronic Control Module (ECM)
C2 ALDL diagnostic connector
C3 "SERVICE ENGINE SOON" light
C4 ECM power (2)
C5 ECM harness ground
C6 Fuse panel
C8 Fuel pump test connector

NOT ECM CONNECTED
N1 Crankcase vent oil/air separator
N17 Fuel vapor canister
*N14 A/C High speed fan relay

CONTROLLED DEVICES
1 Fuel injector
2 Idle air control valve
9 Fuel pump relay
6 Torq. converter clutch connector
9 Engine coolant fan relay
10 Canister purge solenoid
*13 A/C compressor relay
15 IDI module (under IDI cover)

INFORMATION SENSORS
A Manifold pressure (M.A.P.)
B Exhaust oxygen
C Throttle position
D Coolant temperature
F Vehicle speed
J ESC knock (below manifold)
M P/N (Park/Neutral Switch) (Auto)
T Manifold air temperature

* Exact placement of relays may vary.
Circuit diagrams and wire colors may
be used for positive identification.

2.3L (VIN A) COMPONENT LOCATIONS—CALAIS

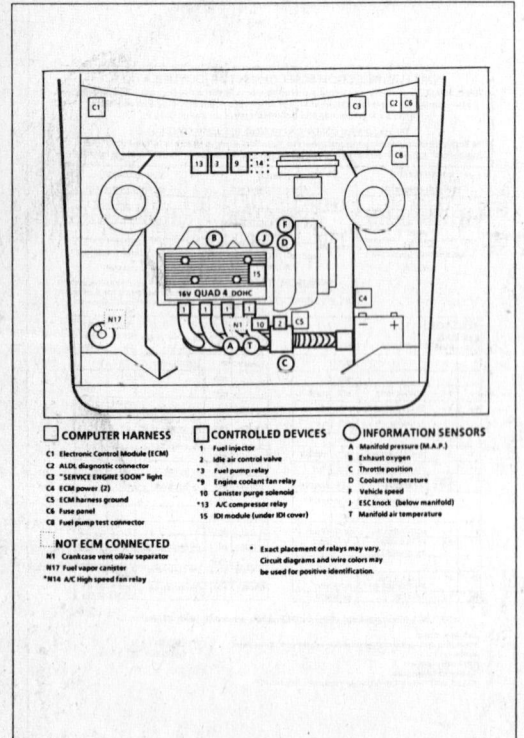

16V QUAD 4 DOHC

COMPUTER HARNESS
C1 Electronic Control Module (ECM)
C2 ALDL diagnostic connector
C3 "SERVICE ENGINE SOON" light
C4 ECM power (2)
C5 ECM harness ground
C6 Fuse panel
C8 Fuel pump test connector

NOT ECM CONNECTED
N1 Crankcase vent oil/air separator
N17 Fuel vapor canister
*N14 A/C High speed fan relay

CONTROLLED DEVICES
1 Fuel injector
2 Idle air control valve
*9 Engine coolant fan relay
6 Torq. converter clutch connector
9 Engine coolant fan relay
10 Canister purge solenoid
*13 A/C compressor relay
15 IDI module (under IDI cover)

INFORMATION SENSORS
A Manifold pressure (M.A.P.)
B Exhaust oxygen
C Throttle position
D Coolant temperature
F Vehicle speed
J ESC knock (below manifold)
T Manifold air temperature

* Exact placement of relays may vary.
Circuit diagrams and wire colors may
be used for positive identification.

2.3L (VIN D AND A) ECM WIRING DIAGRAM—CALAIS

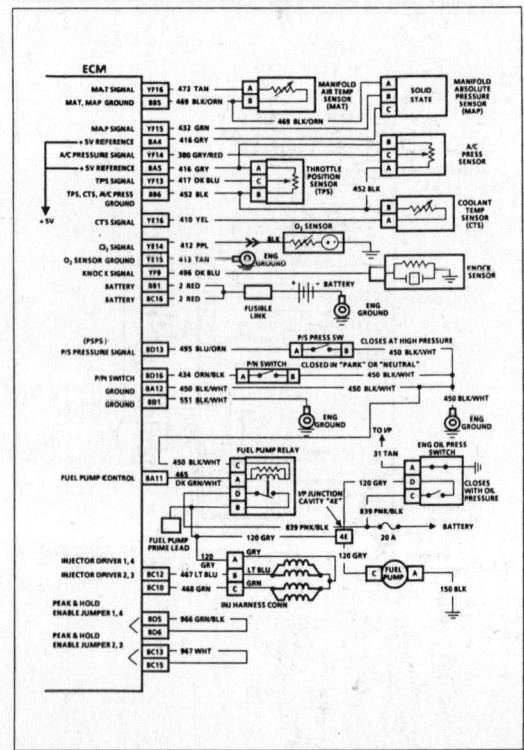

2.3L (VIN D AND A) ECM WIRING DIAGRAM—CALAIS (CONT.)

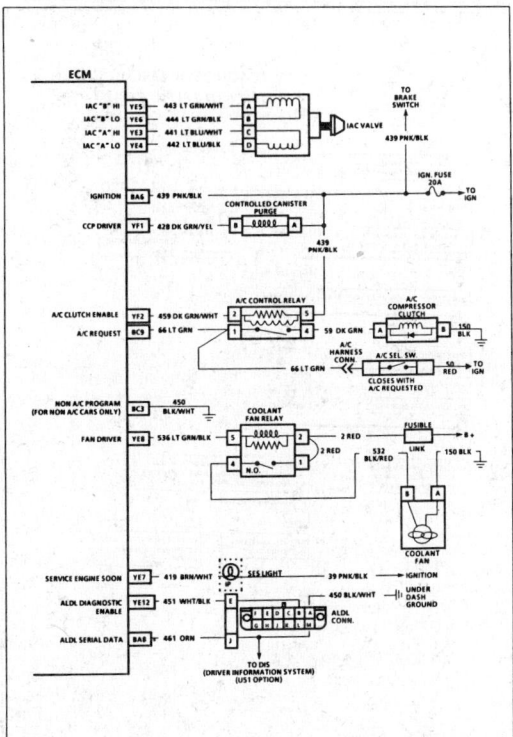

2.3L (VIN D AND A) ECM WIRING DIAGRAM—CALAIS (CONT.)

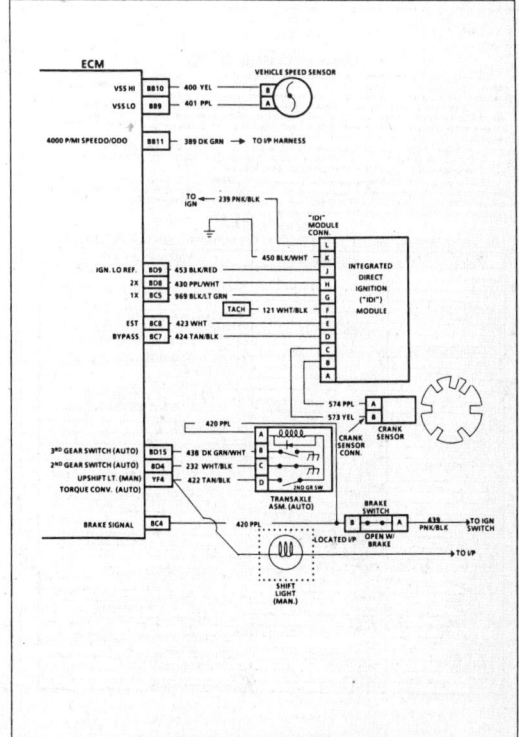

2.3L (VIN D AND A) ECM CONNECTOR TERMINAL END VIEW—CALAIS

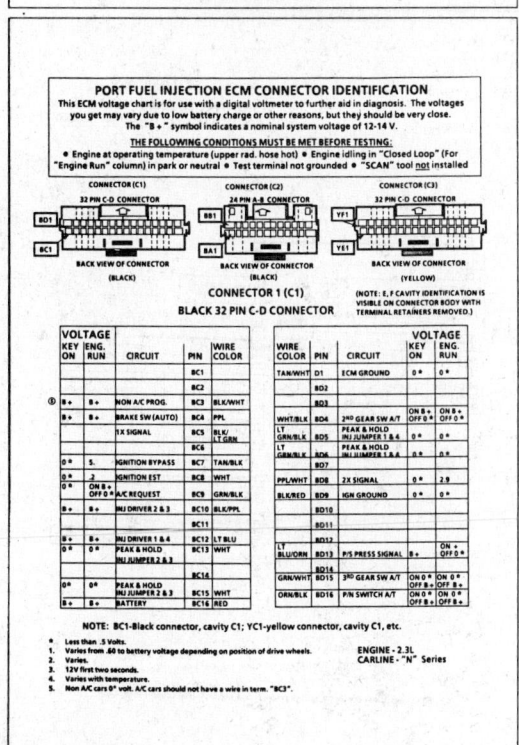

2.3L (VIN D AND A) ECM CONNECTOR TERMINAL END VIEW—CALAIS (CONT.)

CONNECTOR 2 (C-2)
BLACK 24 PIN A-B CONNECTOR

VOLTAGE KEY ON	ENG. RUN	CIRCUIT	PIN	WIRE COLOR	WIRE COLOR	PIN	CIRCUIT	VOLTAGE KEY ON	ENG. RUN
			BA1		RED	BB1	BATTERY	B+	B+
			BA2			BB2			
			BA3			BB3			
5.0	5.0	+5V REFERENCE	BA4	GRAY		BB4			
5.0	5.0	+5V REFERENCE	BA5	GRAY	BLK/ORN	BB5	MAT & MAP GND	0*	0*
5.	B+	IGNITION FEED	BA6		BLK	BB6	A/C CTS,TPS GND	0*	0*
			BA7			BB7			
		SERIAL DATA/ALDL	BA8	ORN		BB8			
②			BA9		PPL	BB9	MAG. VSS LOW	0*	0*
			BA10		YEL	BB10	MAG. VSS HIGH	0*	0*
0*	B..	FUEL PUMP	BA11	GRN/WHT	DK GRN	BB11	4000 P/MI SPEED	10.0	11.4
0*	0*	ECM GROUND	BA12			BB12			

YELLOW 32 PIN C-D CONNECTOR 3 (C-3)

VOLTAGE KEY ON	ENG. RUN	CIRCUIT	PIN	WIRE COLOR	WIRE COLOR	PIN	CIRCUIT	VOLTAGE KEY ON	ENG. RUN
			YE1		GRN/YEL	YF1	CANISTER PURGE	B+	.3
			YE2		DK GRN/ WHT	YF2	A/C CLUTCH RELAY	B+	OFF B+ ON 0*
NOT USEABLE		IAC - A - HIGH	YE3	LT BLU/WHT		YF3			
NOT USEABLE		IAC - A - LOW	YE4	LT BLU/BLK	TAN/BLK	YF4	TCC OR SHIFT LT	.36	OFF B+ ON 0*
NOT USEABLE		IAC - B - HIGH	YE5	GRN/WHT		YF5			
NOT USEABLE		IAC - B - LOW	YE6	GRN/BLK		YF6			
			YE7			YF7			
			YE8		DK BLU	YF8	KNOCK SIGNAL	2.3	2.3
			YE9			YF9			
			YE10			YF10			
			YE11			YF11			
5.0	5.0	ALDL/DIAG TERM	YE12	WHT/BLK		YF12			
			YE13		DK BLU	YF13	TPS SIGNAL	.54	.54
.3-.5	B..	O2 SIGNAL	YE14	PPL	GRY/RED	YF14	A/C PRESS SIGNAL		②
0*	0*	O2 GROUND	YE15	TAN	GRN	YF15	MAP SIGNAL	4.7	1.4
1.8	1.8	CLNT TEMP SIGNAL	YE16	YEL	GRN	YF16	MAT SIGNAL	3.6	1.5

Note: BA1 = Black Connector, cavity A1, etc. YC1 - Yellow Connector, cavity C1, etc.

* Less than .5 Volts.
1. Varies from .60 to battery voltage depending on position of drive wheels.
2. Varies.
3. B+ first two seconds.

ENGINE - 2.3L
CARLINE - "N" Series

1988–90 2.3L (VIN D AND A) – ALL MODELS

DIAGNOSTIC CIRCUIT CHECK

The Diagnostic Circuit Check is an organized approach to identifying a problem created by an electronic engine control system malfunction. It must be the starting point for any driveability complaint diagnosis, because it directs the service technician to the next logical step in diagnosing the complaint.

The "Scan Data" listed in the table may be used for comparison, after completing the diagnostic circuit check and finding the on-board diagnostics functioning properly and no trouble codes displayed. The "Typical Values" are an average of display values recorded from normally operating vehicles and are intended to represent what a normally functioning system would typically display.

A "SCAN" TOOL THAT DISPLAYS FAULTY DATA SHOULD NOT BE USED, AND THE PROBLEM SHOULD BE REPORTED TO THE MANUFACTURER. THE USE OF A FAULTY "SCAN" CAN RESULT IN MISDIAGNOSIS AND UNNECESSARY PARTS REPLACEMENT.

Only the parameters listed below are used in this manual for diagnosing. If a "Scan" reads other parameters, the values are not recommended by General Motors for use in diagnosing. For more description on the values and use of the "Scan" to diagnosis ECM inputs, refer to the applicable diagnosis section in Section "C". If all values are within the range illustrated, refer to symptoms in Section "B".

"SCAN" DATA
Idle / Upper Radiator Hose Hot / Closed Throttle / Park or Neutral / "Closed Loop" / Acc. "OFF"

"SCAN" Position	Units Displayed	Typical Data Value
Desired RPM	RPM	ECM idle command (varies with calibration, temp.)
RPM	RPM	± 100 RPM from desired RPM (± 50 in drive)
Coolant Temp.	C°	85° - 105°C
MAT Temp.	C°	10° - 80°C (depends on underhood temp.)
MAP	Volts/kPa	1 - 2 (depends on Vacuum & Baro pressure)
BARO	Volts/kPa	2.5 - 5.5 (depends on altitude & Baro pressure)
Purge Duty Cycle	%	0-100%
PROM ID	#	Production ECM/PROM ID (not useable)
Injector Pulse Width	m Sec	1 - 4 and varying
O₂	mV	1-1000 and varying
TPS	Volts	400 - 900 (up to 5.0 at wide open throttle)
Throttle Angle	0 - 100%	0% (up to 100% at wide open throttle)
IAC	Counts (steps)	5 - 45
P/N Switch	P/N and RDL	Park/Neutral (P/N)
INT (Integrator)	Counts	Varies
BLM (Block Learn)	Counts	58 - 198 (see Section "C2")
Open/Closed Loop	Open/Closed	Closed Loop (may go open with extended idle)
BLM Cell	Cell Number	21 at idle (depends on Air Flow, RPM, P/N & A/C)
VSS	MPH	0
TCC	On/Off	Off ("ON" with TCC commanded)
Fan	On/Off	Off ("ON" with A/C "ON" or hot eng)
Spark Advance	# of Degrees	Varies
Knock Retard	Degrees of Retard	0
Knock Signal	Yes/No	No
Battery	Volts	13.5 - 14.5
P/S Switch	Normal/Hi Press.	Normal
2nd Gear	Yes/No	No (yes, when in 2nd or 3rd gear)
3rd Gear	Yes/No	No (yes, when in 3rd gear)
A/C Request	Yes/No	No (yes, with A/C requested, ie: selector "ON")
A/C Clutch	On/Off	Off ("ON", with A/C commanded on)
A/C Pressure	psi/volts	0 - 450 psi (depends on high side pressure)
Shift Light (M/T)	On/Off	Off
Time From Start	min/sec	Varies (engine run time since start)
2X Ref Pulse	Yes/No or counts	No (yes, while cranking) or 0 - 255
1X Ref Pulse	Yes/No or counts	Yes (useable for code 41) or 0 - 8
QDM A	High/Low	Low
QDM B	High/Low	High (See Code 26 charts *)

1988–90 2.3L (VIN D AND A) – ALL MODELS

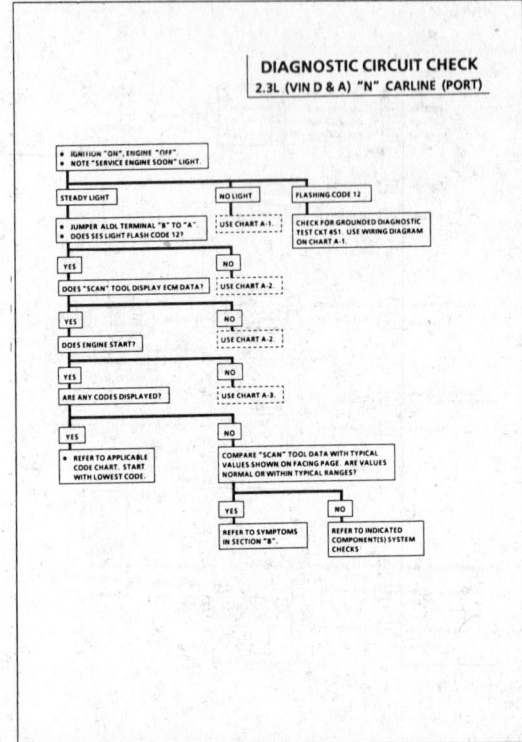

DIAGNOSTIC CIRCUIT CHECK
2.3L (VIN D & A) "N" CARLINE (PORT)

1988–90 2.3L (VIN D AND A) – ALL MODELS

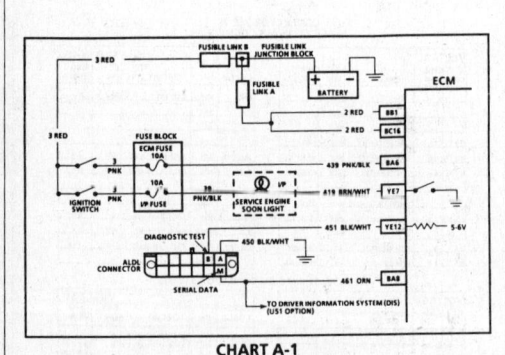

CHART A-1
NO "SERVICE ENGINE SOON" LIGHT
2.3L (VIN D & A) "N" CARLINE (PORT)

Circuit Description:
There should always be a steady "Service Engine Soon" light when the ignition is "ON" and engine stopped. Battery voltage is supplied through the ignition switch directly to the light bulb. The electronic control module (ECM) controls the light and turns it "ON" by providing a ground path through CKT 419 to the ECM.

Test Description: Numbers below refer to circled numbers on the diagnostic chart.
1. If the fusible link is blown, locate and repair short to ground.
2. Using a test light connected to 12V (B+) probe each of the system ground circuits to be sure a good ground is present. See ECM terminal end view in front of this section for ECM pin locations of ground circuits.

Diagnostic Aids:
Engine runs OK, check:
- Faulty light bulb.
- CKT 419 open.
- IP fuse blown; this will result in the loss of "SES" light, oil light, brake light, etc. on IP.
Engine cranks but will not run.
- Continuous battery - fuse or fusible link open.
- ECM ignition fuse open.
- Battery CKT 2 to ECM open.
- Ignition CKT 439 to ECM open.
- Poor connection to ECM.
- Poor ECM ground.

1988–90 2.3L (VIN D AND A) – ALL MODELS

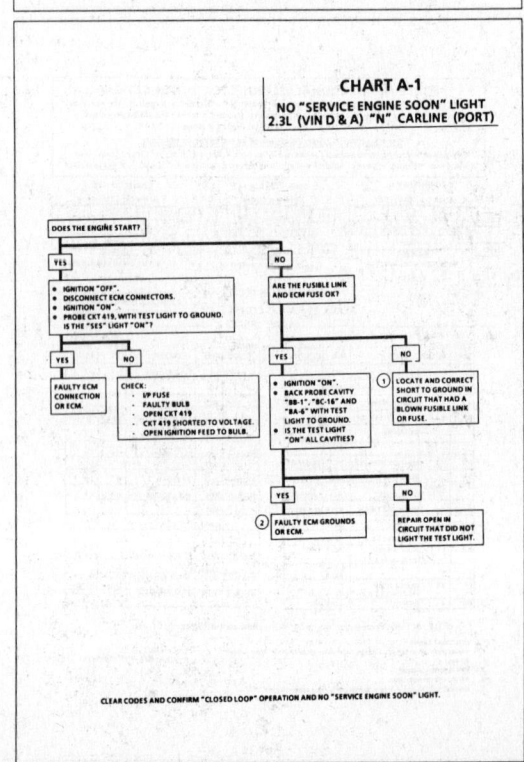

CHART A-1
NO "SERVICE ENGINE SOON" LIGHT
2.3L (VIN D & A) "N" CARLINE (PORT)

CLEAR CODES AND CONFIRM "CLOSED LOOP" OPERATION AND NO "SERVICE ENGINE SOON" LIGHT.

1988–90 2.3L (VIN D AND A) – ALL MODELS

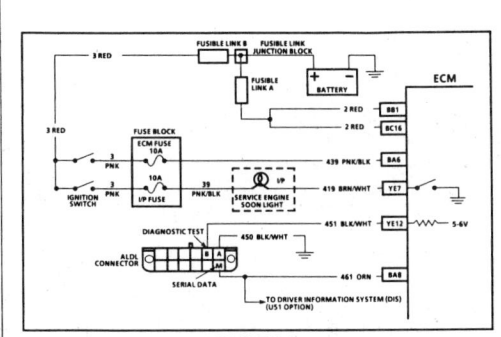

CHART A-2
NO ALDL DATA OR WON'T FLASH CODE 12
"SERVICE ENGINE SOON" LIGHT "ON" STEADY
2.3L (VIN D & A) "N" CARLINE (PORT)

Circuit Description:
There should always be a steady "Service Engine Soon" light when the ignition is "ON" and engine stopped. Battery ignition voltage is supplied to the light bulb. The electronic control module (ECM) turns the light "ON" by grounding CKT 419 at the ECM.

With the diagnostic terminal grounded, the light should flash a Code 12, followed by any trouble code(s) stored in memory.

A steady light suggests a short to ground in the light control CKT 419, or an open in diagnostic CKT 451.

Test Description: Numbers below refer to circled numbers on the diagnostic chart.

1. Light "OFF" with CKT 419 disconnected from ECM indicates that ground circuit was completed through the ECM, not through external short to ground.
2. If there is a problem with the ECM that causes a "Scan" tool to not read Serial data, the ECM should not flash a Code 12. If Code 12 is flashing, check for CKT 451 short to ground. If Code 12 does flash, be sure that the "Scan" tool is working properly on another vehicle. If the "Scan" tool is functioning properly, check CKT 461 for open or short to ground or voltage (including DIS, if equipped). If CKT 461 is OK, the ECM or mem-cal may be the fault for the "NO ALDL" symptom.

3. This step will check for an open diagnostic CKT 451.
4. At this point, the "Service Engine Soon" light wiring is OK. The problem is a faulty ECM or mem-cal. If Code 12 does not flash, the ECM should be replaced using the original mem-cal. Replace the mem-cal only after trying an ECM, as a defective mem-cal is an unlikely cause of the problem.

1988–90 2.3L (VIN D AND A) – ALL MODELS

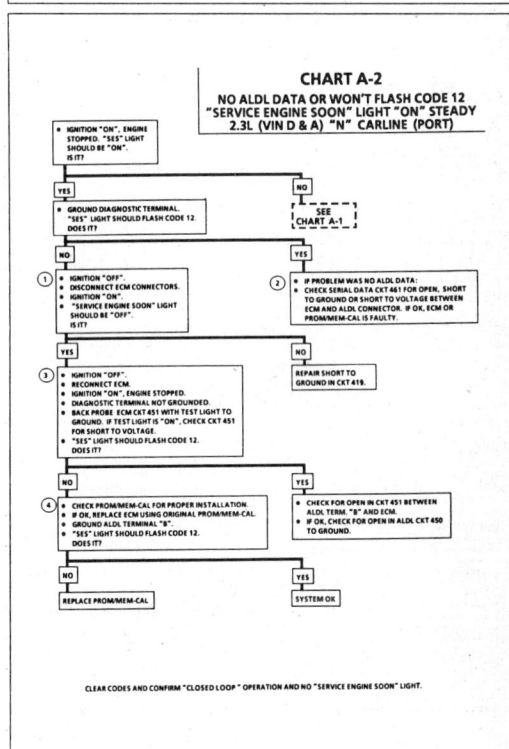

CHART A-2
NO ALDL DATA OR WON'T FLASH CODE 12
"SERVICE ENGINE SOON" LIGHT "ON" STEADY
2.3L (VIN D & A) "N" CARLINE (PORT)

1988–90 2.3L (VIN D AND A) – ALL MODELS

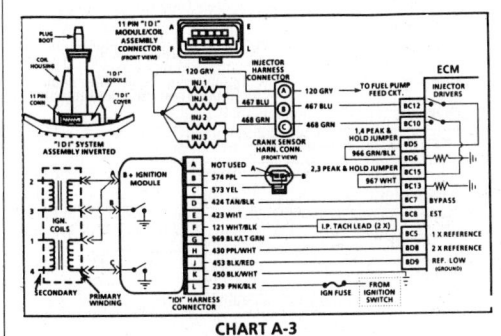

CHART A-3
(Page 1 of 3)
ENGINE CRANKS BUT WON'T RUN
2.3L (VIN D & A) "N" CARLINE (PORT)

Condition:
Engine cranks but won't run, or engine may start, but immediately stops running. Battery condition and engine cranking speed are OK and there is adequate fuel in the tank.

Circuit Description:
This engine is equipped with a distributorless ignition system called the "Integrated Direct Ignition" system (IDI). The primary circuit of the IDI consists of two separate ignition coils, an IDI (ignition) module and crankshaft sensor as well as the related connecting wires and the EST (electronic spark timing) portion of the ECM. Each secondary circuit consists of the secondary winding of the coil, two connecting metal strips molded into the coil housing, spark plug boot/connector assemblies and spark plugs.

Test Description: Numbers below refer to circled numbers on the diagnostic chart.

1. This step verifies that "SES" light operation, on-board diagnostics, cranking rpm, TPS and coolant sensor signals are normal. A blinking test light verifies that the ECM is receiving the IDI reference signal and is attempting to activate the injectors.
2. This step checks injector harness and injectors for opens or shorts. Resistance should measure half that of one injector due to parallel circuit.
3. By installing spark plug jumper leads and testing for spark on two adjacent plug leads (do not use 2 & 3 as they are on same coil), each ignition coil's ability to produce at least 25,000 volts is verified.
4. Checks to see if fuel pump and relay are operating correctly (fuel pump only "ON" 2-3 seconds) and fuel pressure is within proper range.

5. If module can make the test light blink, the fault is coil harness or connections. If not, module or it's connections are faulty.
6. This step determines whether harness or injector is cause of incorrect resistance. Nominal injector resistance is 1.9 to 2.1 ohms at 60°C (140°F). Resistance will increase slightly at higher temperatures.

Diagnostic Aids:

Check For:
- TPS binding or sticking in wide open throttle position or intermittently shorted or open.
- Water or foreign material in fuel.
- Low Compression. (Timing chain failure)
- Verify that only resistor spark plugs are used.

1988–90 2.3L (VIN D AND A) – ALL MODELS

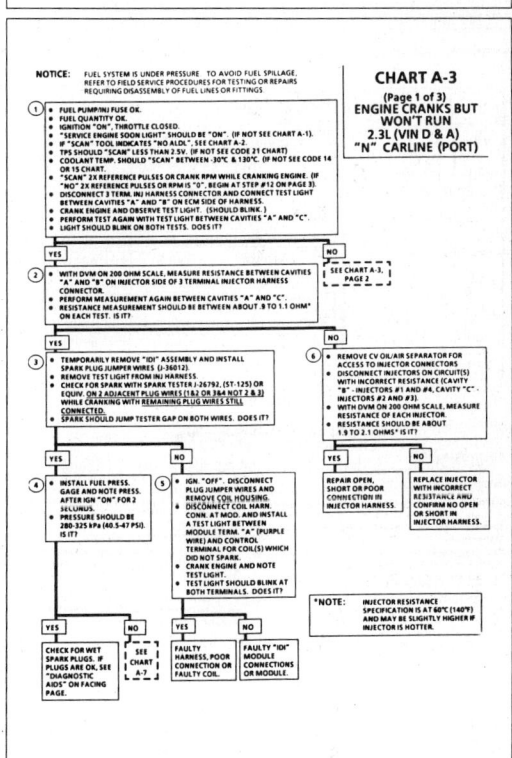

CHART A-3
(Page 1 of 3)
ENGINE CRANKS BUT WON'T RUN
2.3L (VIN D & A) "N" CARLINE (PORT)

1988–90 2.3L (VIN D AND A) – ALL MODELS

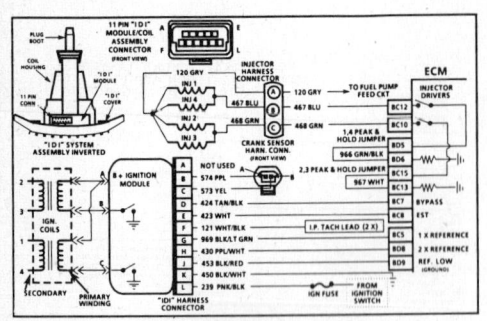

CHART A-3
(Page 2 of 3)
ENGINE CRANKS BUT WON'T RUN
2.3L (VIN D & A) "N" CARLINE (PORT)

Condition:
Engine cranks but won't run, or engine may start, but immediately stops running. Battery condition and engine cranking speed are OK and there is adequate fuel in the tank.

Circuit Description:
This engine is equipped with a distributorless ignition system called the "Integrated Direct Ignition" system (IDI). The primary circuit of the IDI consists of two separate ignition coils, an IDI ignition module and crankshaft sensor as well as the related connecting wires and the EST (electronic spark timing) portion of the ECM. Each secondary circuit consists of the secondary winding of the coil, two connecting metal strips molded into the coil housing, the spark plug boot/connector assemblies and spark plugs.

Test Description: Numbers below refer to circled numbers on the diagnostic chart.
7. Battery voltage should be available at cavity "A" whenever the fuel pump power feed circuit is switched "ON". The ECM should switch the fuel pump "ON" for 2-3 seconds after ignition is turned "ON" (and when ECM is receiving ignition reference pulses, as while cranking or running). The ignition must be turned "OFF" for at least 10 seconds to assure that the ECM powers down and will then switch the fuel pump back "ON" for 2-3 seconds when ignition is turned back "ON".
8. Light "ON" one circuit only indicates power is available at cavity "A", but grounded circuit is not being completed on the other circuit. This could be due to open circuit or ECM not switching the injector driver circuit to ground.
9. Steady light indicates ground circuit is always completed and is not being switched. This could be due to short to ground in circuit, or faulty ECM injector driver.
10. The fuel pump should be switched "ON" by the ECM for 2-3 seconds after ignition is first turned "ON". It is necessary to turn the ignition "OFF" for at least 10 seconds to assure that the ECM powers down and will then switch the fuel pump back "ON". If the fuel pump operates, but power is not available at the injector harness, circuit must be open. If the fuel pump does not operate, CHART A-5 should be used to diagnose the cause.

1988–90 2.3L (VIN D AND A) – ALL MODELS

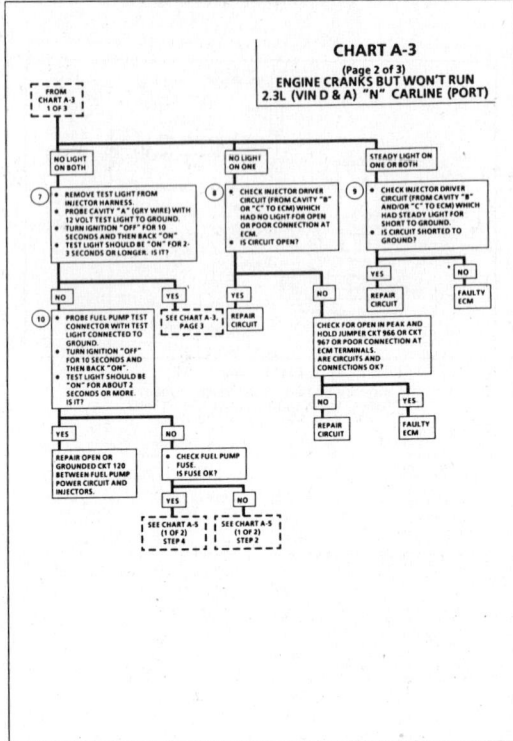

1988–90 2.3L (VIN D AND A) – ALL MODELS

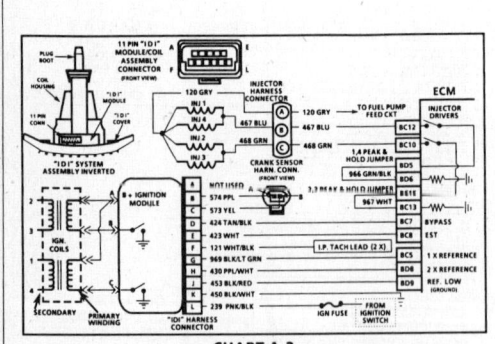

CHART A-3
(Page 3 of 3)
ENGINE CRANKS BUT WON'T RUN
2.3L (VIN D & A) "N" CARLINE (PORT)

Circuit Description:
The "Integrated Direct Ignition" system (IDI) uses a waste spark method of distribution. In this type of system the ignition module triggers the #1-4 coil pair resulting in both #1 and #4 spark plugs firing at the same time. #1 cylinder is on the compression stroke, it the same time #4 is on the exhaust stroke, resulting in a lower energy voltage required to fire #4 spark plug. This leaves the remainder of the high voltage to fire #1 spark plug. On this application, the crank sensor is mounted to, and protrudes through the block to within approximately 0.050" of the crankshaft reluctor. Since the reluctor is a machined portion of the crankshaft and the sensor is mounted in a fixed position on the block, timing adjustments are not possible or necessary.

Test Description: Numbers below refer to circled numbers on the diagnostic chart.
11. Battery voltage should be available at terminal "L" of the IDI 11 pin connector, and terminal "K" should be a good ground.
12. The test light to 12V simulates a reference signal to the ECM which will result in an injector test light blink for every other touch of the test light, if CKT 430 the ECM and the injector driver circuit are all functioning properly.
13. The crankshaft sensor should output a voltage as the crankshaft turns. If no voltage is produced, the indication is a poor sensor connection or faulty sensor.
14. The crank sensors core is a magnet, therefore, it should be magnetized and the resistance should be within a range of 500 to 900 ohms.

CHART A-3
(Page 3 of 3)
ENGINE CRANKS BUT WON'T RUN
2.3L (VIN D & A) "N" CARLINE (PORT)

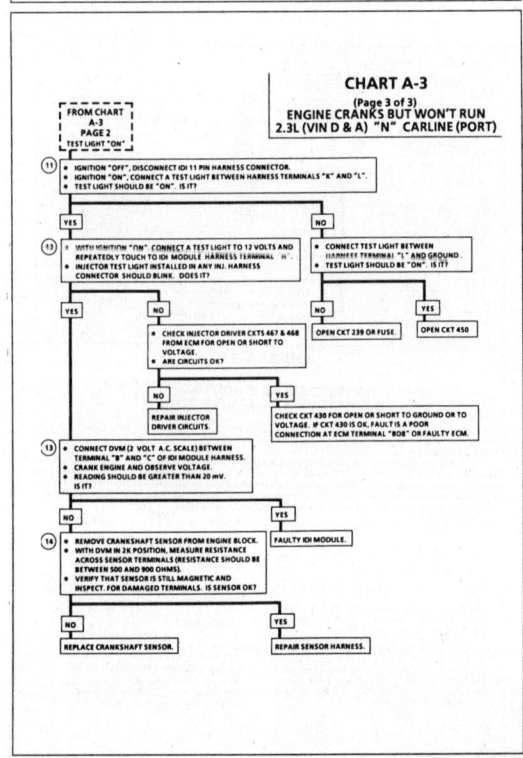

1988–90 2.3L (VIN D AND A) – ALL MODELS

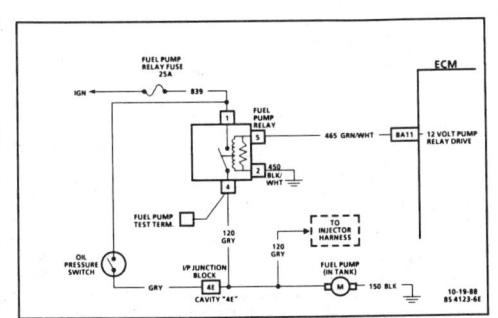

CHART A-5
(Page 1 of 2)
ENGINE CRANKS BUT WILL NOT RUN
(FUEL PUMP CIRCUIT)
2.3L (VIN D & A) "N" CARLINE (PORT)

Circuit Description:
When the ignition switch is turned "ON", the electronic control module (ECM) turns "ON" the in-tank fuel pump. It will remain "ON" as long as the ECM is receiving ignition reference pulses from the Integrated Direct Ignition module (IDI).

If there are no reference pulses, the ECM will shut "OFF" the fuel pump about 2-3 seconds after key "ON", or about 10 seconds after reference pulses stop. If sufficient oil pressure is present to close the oil pressure switch, the fuel pump will remain "ON" during cranking without reference pulses.

The pump delivers fuel to the fuel rail and injectors, then to the pressure regulator, where the system pressure is controlled to 280 - 325 kPa (40.5 - 47 psi) with no manifold vacuum or 211 - 304 kPa (30.5 - 44 psi) at idle. Excess fuel is then returned to the fuel tank.

The fuel pump test terminal is located in the engine compartment. When the engine is stopped, the pump can be turned "ON" by applying battery voltage to the test terminal.

Improper fuel system pressure will result in one or all of the following symptoms:
- Cranks but won't run.
- Cuts out, may feel like ignition problem.
- Code 44
- Poor fuel economy, loss of power.
- Code 45
- Hesitation.

Test Description: Numbers below refer to circled numbers on the diagnostic chart.
1. Determines if the pump circuit is ECM controlled. The ECM will turn "ON" the pump relay. Engine is not cranking or running so the ECM will turn "OFF" the relay within 2 seconds after ignition is turned "ON".

2. If the fuse is blown, this test will confirm a short to ground on CKT 120. To prevent mis-diagnosis, be sure fuel pump is disconnected before test.
3. Turns "ON" the fuel pump if CKT 120 wiring is OK. If the pump runs, it is a basic fuel delivery problem which the following steps will locate.
4. Check for battery voltage at the pump relay.

1988–90 2.3L (VIN D AND A) – ALL MODELS

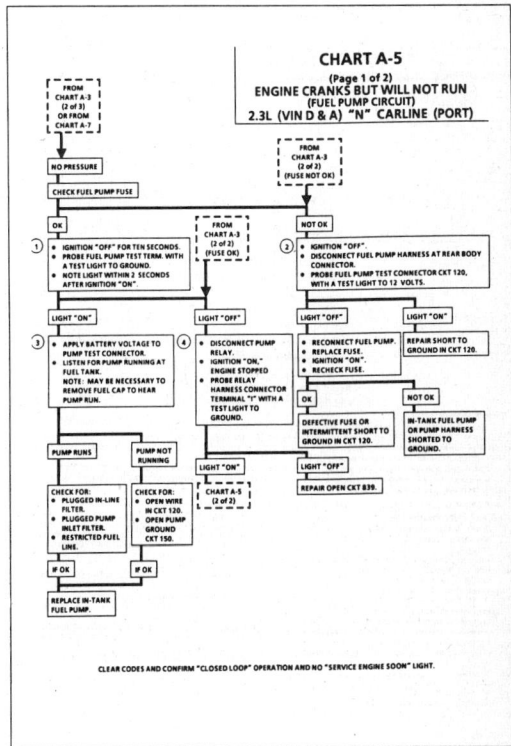

CHART A-5
(Page 1 of 2)
ENGINE CRANKS BUT WILL NOT RUN
(FUEL PUMP CIRCUIT)
2.3L (VIN D & A) "N" CARLINE (PORT)

CLEAR CODES AND CONFIRM "CLOSED LOOP" OPERATION AND NO "SERVICE ENGINE SOON" LIGHT.

1988–90 2.3L (VIN D AND A) – ALL MODELS

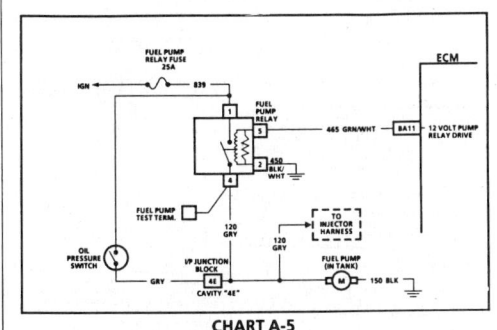

CHART A-5
(Page 2 of 2)
ENGINE CRANKS BUT WILL NOT RUN
(FUEL PUMP CIRCUIT)
2.3L (VIN D & A) "N" CARLINE (PORT)

Circuit Description:
When the ignition switch is turned "ON", the electronic control module (ECM) turns "ON" the in-tank fuel pump. It will remain "ON" as long as the ECM is receiving ignition reference pulses from the Integrated Direct Ignition module (IDI).

If there are no reference pulses, the ECM will shut "OFF" the fuel pump about 2-3 seconds after key "ON", or about 10 seconds after reference pulses stop. If sufficient oil pressure is present to close the oil pressure switch, the fuel pump will remain "ON" during cranking without reference pulses.

The pump delivers fuel to the fuel rail and injectors, then to the pressure regulator, where the system pressure is controlled to 280 - 325 kPa (40.5 - 47 psi) with no manifold vacuum or 211 - 304 kPa (30.5 - 44 psi) at idle. Excess fuel is then returned to the fuel tank.

The fuel pump test terminal is located in the engine compartment. When the engine is stopped, the pump can be turned "ON" by applying battery voltage to the test terminal.

Improper fuel system pressure will result in one or all of the following symptoms:
- Cranks but won't run.
- Cuts out, may feel like ignition problem.
- Code 44
- Poor fuel economy, loss of power.
- Code 45
- Hesitation.

Test Description: Numbers below refer to circled numbers on the diagnostic chart.
5. Check relay ground CKT 450.
6. Check for ECM control of relay through CKT 465.
7. The fuel pump voltage control circuit includes an engine oil pressure switch with a separate set of normally open contacts. The switch closes at about 28 kPa (4 psi) of oil pressure and provides a second battery feed path to the fuel pump. If the relay fails, the pump will continue to run using the battery feed supplied by the closed oil pressure switch.

A failed pump relay will result in extended engine crank time, because of the time required to build enough oil pressure to close the oil pressure switch and turn "ON" the fuel pump. There may be instances when the relay has failed but the engine will not crank fast enough to build enough oil pressure to close the switch. This or a faulty oil pressure switch can result in "Engine Cranks But Will Not Run."

8. Check the oil pressure switch to be sure it provides battery feed to the fuel pump should the pump relay fail.

1988–90 2.3L (VIN D AND A) – ALL MODELS

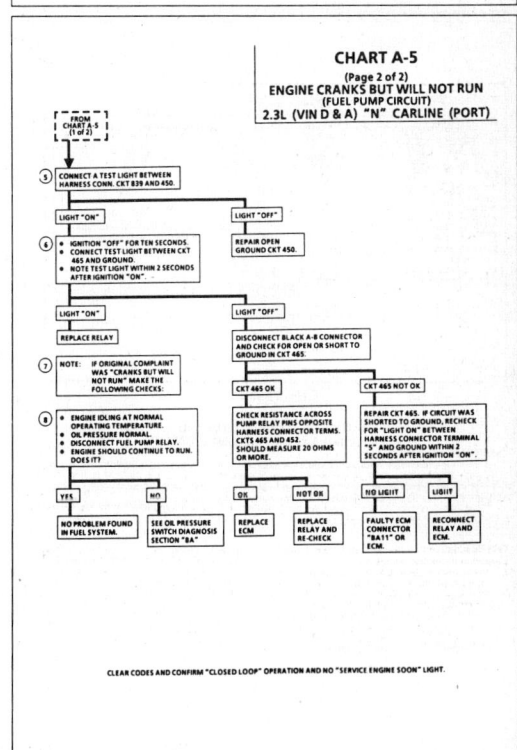

CHART A-5
(Page 2 of 2)
ENGINE CRANKS BUT WILL NOT RUN
(FUEL PUMP CIRCUIT)
2.3L (VIN D & A) "N" CARLINE (PORT)

CLEAR CODES AND CONFIRM "CLOSED LOOP" OPERATION AND NO "SERVICE ENGINE SOON" LIGHT.

1988–90 2.3L (VIN D AND A) – ALL MODELS

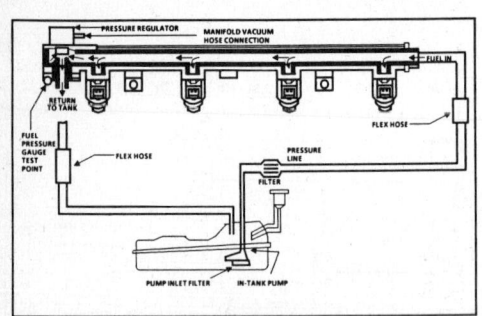

CHART A-7
(Page 1 of 2)
FUEL SYSTEM DIAGNOSIS
2.3L (VIN D & A) "N" CARLINE (PORT)

Circuit Description:
When the ignition switch is turned "ON", the electronic control module (ECM) will turn "ON" the in-tank fuel pump. It will remain "ON" as long as the engine is cranking or running, and the ECM is receiving reference pulses. If there are no reference pulses, the ECM will shut "OFF" the fuel pump in about 2 seconds after ignition "ON" or 10 seconds after reference pulses stop.
The pump delivers fuel to the fuel rail and injectors, where the system pressure is controlled to about 211 - 325 kPa (30.5 - 47 psi) depending on engine operating conditions. Excess fuel is then returned to the fuel tank.

Test Description: Numbers below refer to circled numbers on the diagnostic chart.

1. Wrap a shop towel around the fuel pressure connector to absorb any small amount of fuel leakage that may occur when installing the gage. Ignition "ON" pump pressure should be 280-325 kPa (40-47 psi). This pressure is controlled by spring pressure within the regulator assembly.
2. When the engine is idling, the manifold pressure is low (high vacuum) and is applied to the fuel regulator diaphragm. This will offset the spring and result in a lower fuel pressure. This idle pressure will vary somewhat depending on barometric pressure, however, the pressure idling should be less indicating pressure regulator control.
3. Pressure that continues to fall quickly is caused by one of the following:
   • In-tank fuel pump check valve not holding.
   • Pump coupling hose or pulsator leaking.

• Fuel pressure regulator valve leaking.
• Injector(s) sticking open.
4. An injector sticking open can best be determined by checking for a fouled or saturated spark plug(s). If a leaking injector can not be determined by a fouled or saturated spark plug the following procedure should be used.
   • Remove fuel rail bolts. Follow the procedures in the "Fuel Control" section of this manual, but leave fuel lines connected.
   • Lift fuel rail out just enough to leave injector nozzles in the ports.
CAUTION: *BE SURE INJECTOR(S) ARE NOT ALLOWED TO SPRAY ON ENGINE AND THAT INJECTOR RETAINING CLIPS ARE INTACT. THIS SHOULD BE CAREFULLY FOLLOWED TO PREVENT FUEL SPRAY ON ENGINE WHICH WOULD CAUSE A FIRE HAZARD.*
   • Pressurize the fuel system and observe for injector(s) leaking.

1988–90 2.3L (VIN D AND A) – ALL MODELS

CAUTION: TO REDUCE THE RISK OF FIRE AND PERSONAL INJURY, WRAP A SHOP TOWEL AROUND THE FUEL PRESSURE CONNECTION TO ABSORB ANY FUEL LEAKAGE THAT MAY OCCUR WHEN INSTALLING THE PRESSURE GAGE. PLACE TOWEL IN APPROVED CONTAINER.

CHART A-7
(Page 1 of 2)
FUEL SYSTEM DIAGNOSIS
2.3L (VIN D & A) "N" CARLINE (PORT)

NOTE:
THE IGNITION MAY HAVE TO BE CYCLED "ON" MORE THAN ONCE TO OBTAIN MAXIMUM PRESSURE.
ALSO, IT IS NORMAL FOR THE PRESSURE TO DROP SLIGHTLY WHEN THE PUMP STOPS.

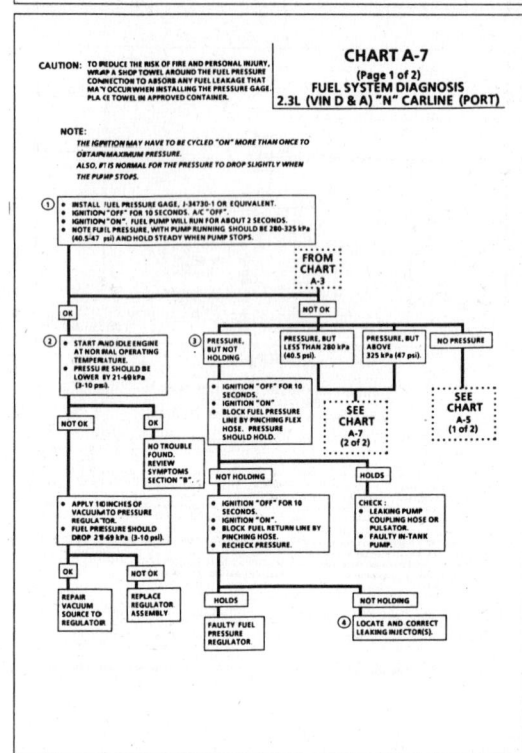

1988–90 2.3L (VIN D AND A) – ALL MODELS

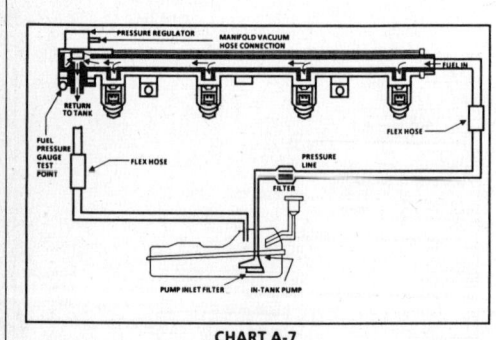

CHART A-7
(Page 2 of 2)
FUEL SYSTEM DIAGNOSIS
2.3L (VIN D & A) "N" CARLINE (PORT)

Circuit Description:
When the ignition switch is turned "ON", the electronic control module (ECM) will turn "ON" the in-tank fuel pump. It will remain "ON" as long as the engine is cranking or running, and the ECM is receiving reference pulses. If there are no reference pulses, the ECM will shut "OFF" the fuel pump in about 2 seconds after ignition "ON" or 10 seconds after reference pulses stop.
The pump delivers fuel to the fuel rail and injectors, then to the pressure regulator, where the system pressure is controlled to about 211 - 325 kPa (30.5 - 47 psi) depending on engine operating conditions. Excess fuel is then returned to the fuel tank.

Test Description: Numbers below refer to circled numbers on the diagnostic chart.

5. Pressure below 280 kPa (40.5 psi) may cause a lean condition and may set a Code 44. It could also cause hard starting cold and poor driveability. Low enough pressure will cause the engine not to run at all. Restricted flow may allow the engine to run at idle, or low speeds, but may cause a surge and stall when more fuel is required, as when accelerating or driving at high speeds.

6. Restricting the fuel return line allows the fuel pump to develop its maximum pressure (dead head pressure). When B+, (about 12 volts) is applied to the pump test terminal, pressure should be above 420 kPa (61 psi).
7. This test determines if the high fuel pressure is due to a restricted fuel return line or a pressure regulator problem. High fuel pressure may cause a rich condition and may set a Code 45 or cause driveability problems.

1988–90 2.3L (VIN D AND A) – ALL MODELS

CHART A-7
(Page 2 of 2)
FUEL SYSTEM DIAGNOSIS
2.3L (VIN D & A) "N" CARLINE
(PORT)

CAUTION: TO REDUCE THE RISK OF FIRE AND PERSONAL INJURY, IT IS NECESSARY TO RELIEVE FUEL SYSTEM PRESSURE BEFORE SERVICING THE FUEL SYSTEM. REFER TO ENGINE FUEL (SECTION 6C) FOR PRESSURE RELIEF PROCEDURE AND FOR FUEL SYSTEM SERVICE INFORMATION.

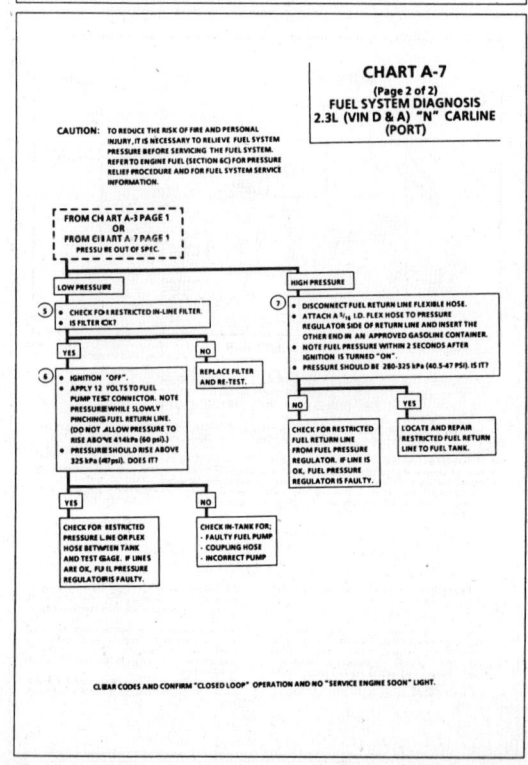

1988–90 2.3L (VIN D AND A) – ALL MODELS

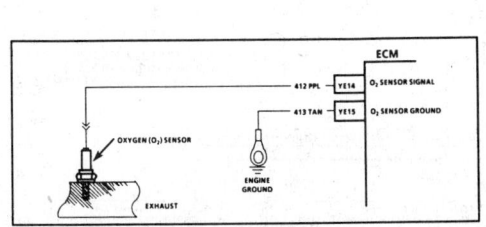

CODE 13
OXYGEN SENSOR CIRCUIT
(OPEN CIRCUIT)
2.3L (VIN D & A) "N" CARLINE (PORT)

Circuit Description:

The ECM supplies a voltage of about .45 volt between terminals "YE14" and "YE15". (If measured with a 10 megohm digital voltmeter, this may read as low as .32 volt.) The O₂ sensor varies the voltage within a range of about 1 volt if the exhaust is rich, down through about .10 volt if exhaust is lean.

The sensor is like an open circuit and produces no voltage when it is below 360°C (600° F). An open sensor circuit or cold sensor causes "Open Loop" operation.

Test Description: Numbers below refer to circled numbers on the diagnostic chart.
1. Code 13 WILL SET under the following conditions:
   - Engine running at least 40 seconds after start.
   - Coolant temperature at least 42.5°C (40.1°F).
   - No Code 21 or 22.
   - O₂ signal voltage steady between .34 and .55 volt.
   - Throttle position sensor signal above 6% for more time than TPS was below 6%. (About .3 volt above closed throttle voltage)
   - All conditions must be met and held for at least 20 seconds.
   If the conditions for a Code 13 exist, the system will not go "Closed Loop".

2. This will determine if the sensor is at fault or the wiring or ECM is the cause of the Code 13.
3. Use only a high impedance digital volt ohmmeter for this test. This test checks the continuity of CKTs 412 and 413; because if CKT 413 is open, the ECM voltage on CKT 412 will be over .6 volt (600 mV).

Diagnostic Aids:

Normal "Scan" voltage varies between 100 mV to 999 mV (.1 volt to 1.0 volt) while in "Closed Loop". Code 13 sets in 20 seconds if voltage remains between .35 volt and .55 volt, but the system will go "Open Loop" in about 15 seconds. Refer to "Intermittents" in Section "B".

1988–90 2.3L (VIN D AND A) – ALL MODELS

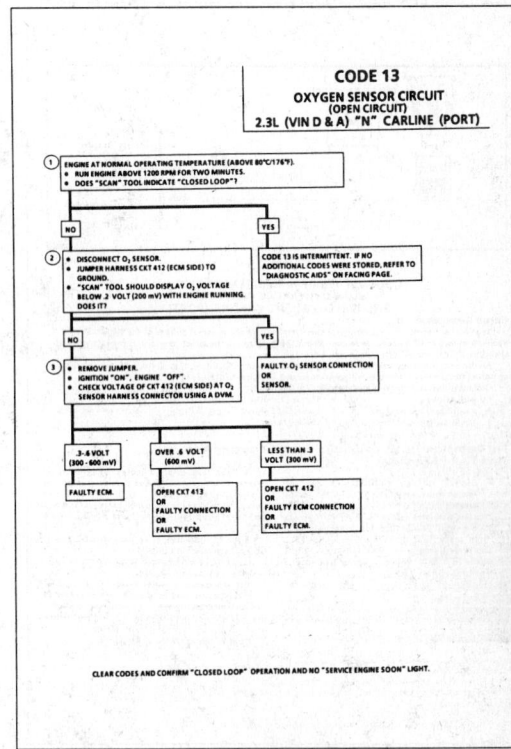

CODE 13
OXYGEN SENSOR CIRCUIT
(OPEN CIRCUIT)
2.3L (VIN D & A) "N" CARLINE (PORT)

1988–90 2.3L (VIN D AND A) – ALL MODELS

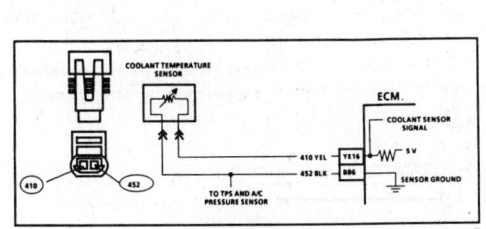

CODE 14
COOLANT TEMPERATURE SENSOR CIRCUIT
(HIGH TEMPERATURE INDICATED)
2.3L (VIN D & A) "N" CARLINE (PORT)

Circuit Description:

The coolant temperature sensor uses a thermistor to control the signal voltage at the ECM. The ECM applies a voltage on CKT 410 to the sensor. When the engine is cold the sensor (thermistor) resistance is high, therefore ECM terminal "YE16" will be high.

As the engine warms, the sensor resistance becomes less, and the voltage drops. At normal engine operating temperature, voltage will measure about 1.5 to 2.0 volts at ECM terminal "YE16".

Coolant temperature is one of the inputs used to control:
- Fuel delivery
- Engine Spark Timing (EST)
- Idle Air Control (IAC)
- Torque Converter Clutch (TCC)
- Controlled Canister Purge (CCP)
- Cooling Fan

Test Description: Numbers below refer to circled numbers on the diagnostic chart.
1. Code 14 will set if:
   - Signal voltage indicates a coolant temperature above 140°C (285°F).
   - Engine running longer than 128 seconds.
2. This test will determine if CKT 410 is shorted to ground which will cause the conditions for Code 14.

Diagnostic Aids:

Check harness routing for a potential short to ground in CKT 410.

"Scan" tool displays engine temperature in degrees celcius. After engine is started, the temperature should rise steadily to about 90°C, and then stabilize when thermostat opens. Refer to "Intermittents" in Section "B".

Verify that engine is not overheating and has not been subjected to conditions which could create an overheating condition (i.e. overload, trailer towing, hilly terrain, heavy stop and go traffic, etc.). The "Temperature To Resistance Value" scale at the right may be used to test the coolant sensor at various temperature levels to evaluate the possibility of a "shifted" (mis-scaled) sensor. A "shifted" sensor could result in poor driveability complaints.

1988–90 2.3L (VIN D AND A) – ALL MODELS

CODE 14
COOLANT TEMPERATURE SENSOR CIRCUIT
(HIGH TEMPERATURE INDICATED)
2.3L (VIN D & A) "N" CARLINE (PORT)

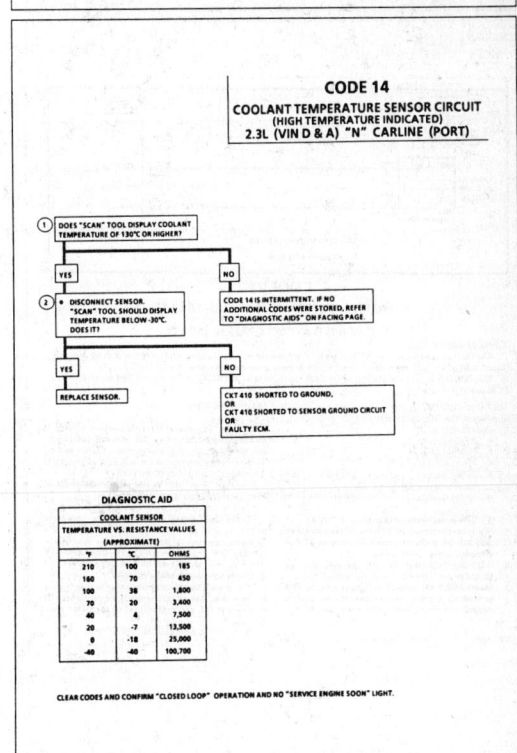

DIAGNOSTIC AID

COOLANT SENSOR TEMPERATURE VS. RESISTANCE VALUES (APPROXIMATE)		
°F	°C	OHMS
210	100	185
160	70	450
100	38	1,800
70	20	3,400
40	4	7,500
20	-7	13,500
0	-18	25,000
-40	-40	100,700

CLEAR CODES AND CONFIRM "CLOSED LOOP" OPERATION AND NO "SERVICE ENGINE SOON" LIGHT.

1988–90 2.3L (VIN D AND A) – ALL MODELS

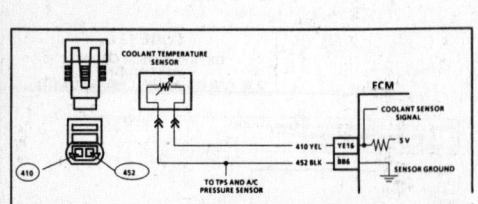

CODE 15
COOLANT TEMPERATURE SENSOR CIRCUIT
(LOW TEMPERATURE INDICATED)
2.3L (VIN D & A) "N" CARLINE (PORT)

Circuit Description:

The coolant temperature sensor uses a thermistor to control the signal voltage at the ECM. The ECM applies a voltage on CKT 410 to the sensor. When the engine is cold, the sensor (thermistor) resistance is high, therefore, ECM terminal "YE16" voltage will be high.

As the engine warms, the sensor resistance becomes less, and the voltage drops. At normal engine operating temperature the voltage will measure about 1.5 to 2.0 volts at ECM terminal "YE16".

Coolant temperature is one of the inputs used to control:
- Fuel delivery
- Torque Convertor Clutch (TCC)
- Engine Spark Timing (EST)
- Controlled Canister Purge (CCP)
- Idle Air Control (IAC)
- Cooling Fan

Test Description: Numbers below refer to circled numbers on the diagnostic chart.
1. Code 15 will be set if:
   - Signal voltage indicates a coolant temperature less than -40°C (-40°F) for 60 seconds.
2. This test simulates a Code 14. If the ECM senses the low signal voltage (high temperature) and the "Scan" reads 130°C, the ECM and wiring are OK.
3. This test will determine if CKT 410 is open. There should be 5 volts present at sensor connector if measured with a DVM.

Diagnostic Aids:

A "Scan" tool displays engine temperature in degrees celcius. After the engine is started the temperature should rise steadily to about 95°C, and then stabilize when the thermostat opens. It is normal for coolant temperature to fluctuate slightly around 95°C.

A faulty connection, or an open in CKT 410 or CKT 452 can result in a Code 15.

Codes 15, 21 and 66 stored at the same time could be the result of an open CKT 452.

The "Temperature to Resistance Value" scale at the right may be used to test the coolant sensor at various temperature levels to evaluate the possibility of a "shifted" (mis-scaled) sensor. A "shifted" sensor could result in poor driveability complaints.

Refer to "Intermittents" in Section "B".

1988–90 2.3L (VIN D AND A) – ALL MODELS

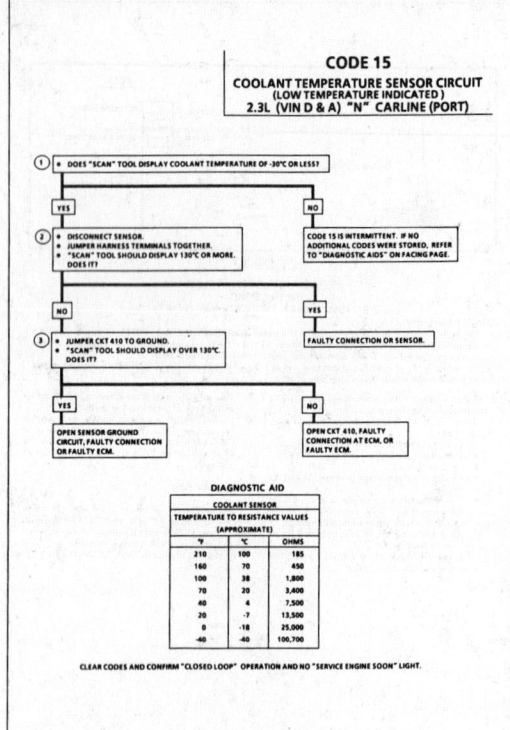

CODE 15
COOLANT TEMPERATURE SENSOR CIRCUIT
(LOW TEMPERATURE INDICATED)
2.3L (VIN D & A) "N" CARLINE (PORT)

DIAGNOSTIC AID

COOLANT SENSOR		
TEMPERATURE TO RESISTANCE VALUES		
(APPROXIMATE)		
°F	°C	OHMS
210	100	185
160	70	450
100	38	1,800
70	20	3,400
40	4	7,500
20	-7	13,500
0	-18	25,000
-40	-40	100,700

CLEAR CODES AND CONFIRM "CLOSED LOOP" OPERATION AND NO "SERVICE ENGINE SOON" LIGHT.

1988–90 2.3L (VIN D AND A) – ALL MODELS

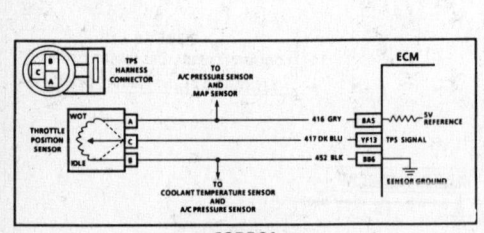

CODE 21
THROTTLE POSITION SENSOR (TPS) CIRCUIT
(SIGNAL VOLTAGE HIGH)
2.3L (VIN D & A) "N" CARLINE (PORT)

Circuit Description:

The throttle position sensor (TPS) provides a voltage signal that changes relative to the throttle opening. Signal voltage will vary from about .5 volt at idle to about 5 volts at wide open throttle (WOT).

The TPS signal is one of the most important inputs used by the ECM for fuel control and for most of the ECM control outputs.

Test Description: Numbers below refer to circled numbers on the diagnostic chart.
1. Code 21 will set if:
   - Engine is running
   - No Code 33 or 34
   - MAP less than 65 kPa
   - TPS signal voltage greater than approximately 4.0 volts (78%)
   - Above conditions exist for over 5 seconds.
   OR
   - TPS voltage greater than about 4.7 volts.
   With throttle closed the TPS should read less than .900 volt. If it doesn't, check for sticking TPS or throttle linkage. If OK, replace TPS.
2. With the TPS disconnected, the TPS voltage should go low if the ECM and wiring are OK.
3. Probing CKT 452 with a test light checks the TPS ground circuit because an open or very high resistance ground circuit will cause a Code 21.

Diagnostic Aids:

A "Scan" tool displays throttle position in volts. It should display .475 volt to .625 (475 mV to 675 mV) volt with throttle closed and ignition "ON" or at idle. Voltage should increase at a steady rate as throttle is moved toward wide open throttle (WOT).

Also some "Scan" tools will display throttle angle %, 0% = closed throttle 100% = WOT.

An open in CKT 452 will result in a Code 21.

Codes 15, 21 and 66 stored at the same time could be the result of an open CKT 452. "Scan" TPS while depressing accelerator pedal with engine stopped and ignition "ON". Display should vary from about .5V (500 mV) when throttle is closed, to over 4500 mV (4.5 volts) when throttle is held wide open.

Check condition of connector and sensor terminals for corrosion, and clean or replace as necessary. If corrosion found, check condition of connector seal, repair and/or replace if necessary.

Refer to "Intermittents" in Section "B".

1988–90 2.3L (VIN D AND A) – ALL MODELS

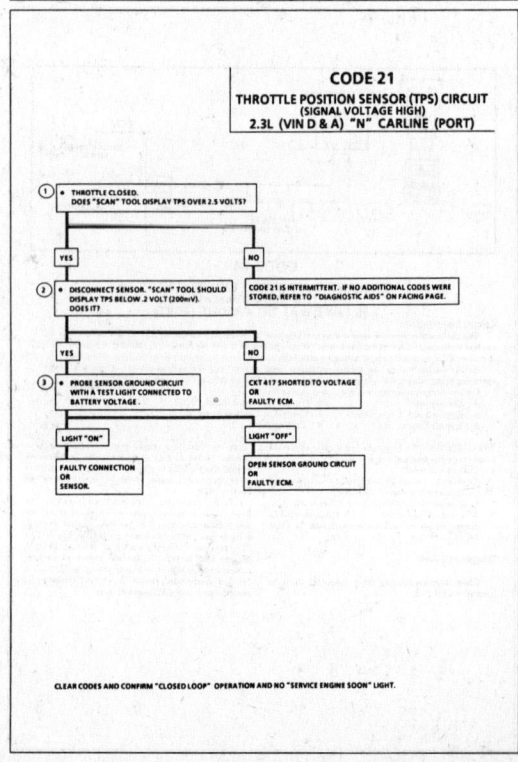

CODE 21
THROTTLE POSITION SENSOR (TPS) CIRCUIT
(SIGNAL VOLTAGE HIGH)
2.3L (VIN D & A) "N" CARLINE (PORT)

CLEAR CODES AND CONFIRM "CLOSED LOOP" OPERATION AND NO "SERVICE ENGINE SOON" LIGHT.

1988–90 2.3L (VIN D AND A) – ALL MODELS

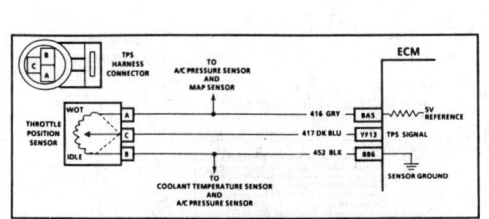

CODE 22
THROTTLE POSITION SENSOR (TPS) CIRCUIT
(SIGNAL VOLTAGE LOW)
2.3L (VIN D & A) "N" CARLINE (PORT)

Circuit Description:
The Throttle Position Sensor (TPS) provides a voltage signal that changes relative to the throttle opening. Signal voltage will vary from about .5 volt at idle to about 5 volts at Wide Open Throttle (WOT).
The TPS signal is one of the most important inputs used by the ECM for fuel control and for most of the ECM control outputs.

Test Description: Numbers below refer to circled numbers on the diagnostic chart.
1. Code 22 will set if:
   - Engine is running
   - TPS signal voltage is less than about .2 volt for 3 seconds.
2. Simulates Code 21 (high voltage). If the ECM recognizes high signal voltage, the ECM and wiring are OK.
3. TPS Check. The TPS has an auto zeroing feature. If the voltage reading is within the range of 400 to .900 volt, the ECM will use that value as closed throttle. The TPS is not adjustable. If the voltage is out of the auto-zeroing range, the TPS must be replaced.
4. This simulates a high signal voltage to check for an open in CKT 417.

Diagnostic Aids:
"Scan" TPS while depressing accelerator pedal with engine stopped and ignition "ON". Display should vary from about 500 mV (.5 volt) when throttle is closed, to over 4500 mV (4.5 volts) when throttle is held wide open.
Also, some "Scan" tools will display throttle angle %: 0% = closed throttle; 100% = WOT.
If Code 22, 33 and/or 66 are set, check CKT 416 for faulty wiring or connections.
Should also check condition of connector and sensor terminals for corrosion, and clean and/or replace as necessary. If corrosion is found, check condition of connector seal and repair or replace if necessary.
" Refer to "Intermittents" in Section "B".

1988–90 2.3L (VIN D AND A) – ALL MODELS

CODE 22
THROTTLE POSITION SENSOR (TPS) CIRCUIT
(SIGNAL VOLTAGE LOW)
2.3L (VIN D & A) "N" CARLINE (PORT)

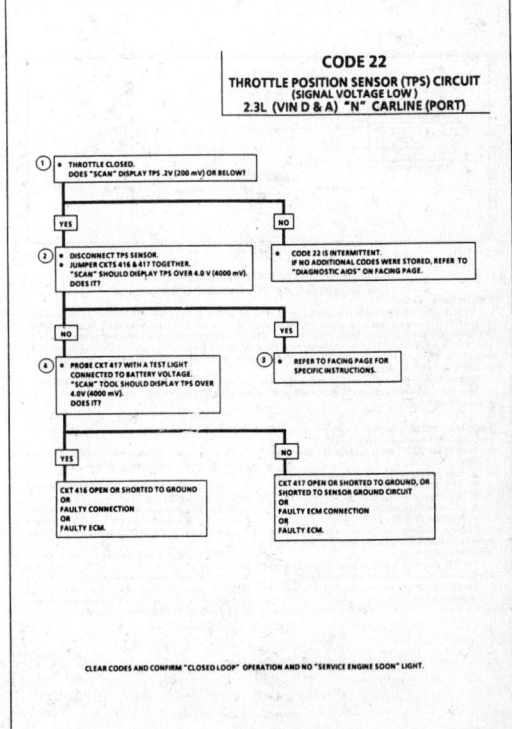

CLEAR CODES AND CONFIRM "CLOSED LOOP" OPERATION AND NO "SERVICE ENGINE SOON" LIGHT.

1988–90 2.3L (VIN D AND A) – ALL MODELS

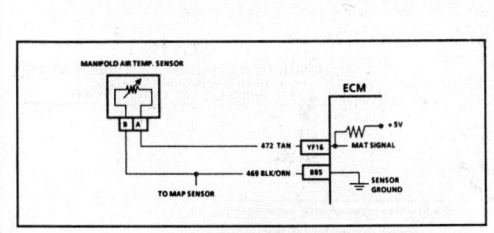

CODE 23
MANIFOLD AIR TEMPERATURE (MAT) SENSOR CIRCUIT
(LOW TEMPERATURE INDICATED)
2.3L (VIN D & A) "N" CARLINE (PORT)

Circuit Description:
The MAT sensor uses a thermistor to control the signal voltage at the ECM. The ECM applies a voltage (about 5 volts) on CKT 472 to the sensor. When the air is cold the sensor (thermistor) resistance is high, therefore the ECM terminal "YF16" voltage will be high. If the air is warm the sensor resistance is low, therefore the ECM terminal "YF16" voltage will be low.

Test Description: Numbers below refer to circled numbers on the diagnostic chart.
1. Code 23 will set if:
   - A signal voltage indicates a manifold air temperature below about -34°C (-29°F).
   - Time since engine start is 320 seconds or longer.
   - Vehicle speed less than 15 mph.
2. A Code 23 will set due to an open sensor, wire, or connection. This test will determine if the wiring and ECM are OK.
3. This will determine if the signal CKT 472 or the 5 volts return CKT 469 is open.

Diagnostic Aids:
A "Scan" tool displays temperature of the air entering the engine, which should be close to ambient air temperature when engine is cold, and rise as underhood temperature increases.
A faulty connection, or an open in CKT 472 or CKT 469 can result in a Code 23.
Codes 23 and 34 stored at the same time, could be the result of an open CKT 469. The "Temperature to Resistance Values" scale at the right may be used to test the MAT sensor at various temperature levels to evaluate the possibility of a "slewed" (mis-scaled) sensor. A "slewed" sensor could result in poor driveability complaints.
Refer to "Intermittents" in Section "B".

1988–90 2.3L (VIN D AND A) – ALL MODELS

CODE 23
MANIFOLD AIR TEMPERATURE (MAT) SENSOR CIRCUIT
(LOW TEMPERATURE INDICATED)
2.3L (VIN D & A) "N" CARLINE (PORT)

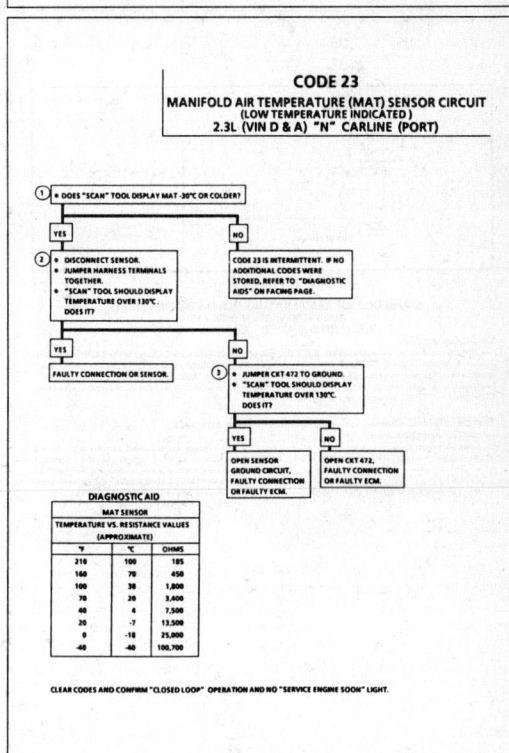

DIAGNOSTIC AID		
MAT SENSOR		
TEMPERATURE VS. RESISTANCE VALUES		
(APPROXIMATE)		
°F	°C	OHMS
210	100	185
160	70	450
100	38	1,800
70	20	3,400
40	4	7,500
20	-7	13,500
0	-18	25,000
-40	-40	100,700

CLEAR CODES AND CONFIRM "CLOSED LOOP" OPERATION AND NO "SERVICE ENGINE SOON" LIGHT.

1988–90 2.3L (VIN D AND A) – ALL MODELS

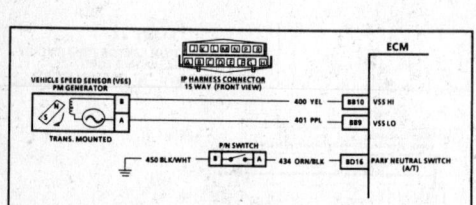

CODE 24
VEHICLE SPEED SENSOR (VSS) CIRCUIT
2.3L (VIN D & A) "N" CARLINE (PORT)

Circuit Description:

Vehicle speed information is provided to the ECM by the vehicle speed sensor (VSS) which is a permanent magnet (PM) generator that is mounted in the transmission. The PM generator produces a pulsing voltage whenever vehicle speed is over about 3 mph, (5 kph). The AC voltage level and the number of pulses increases with vehicle speed. The ECM then converts the pulsing voltage to mph which is used for calculations, and the mph can be displayed with a "Scan" tool. Output of the generator can also be seen by using a digital voltmeter on the AC scale while rotating the generator.

The function of VSS buffer used in past model years has been incorporated into the ECM. The ECM then supplies the necessary signal for the instrument panel for operating the speedometer, the odometer, and for the cruise control module.

Test Description: Numbers below refer to circled numbers on the diagnostic chart.

1. Code 24 will set if vehicle speed is less than 2 mph when:
   - Engine speed is between 1600 and 3600 rpm.
   - TPS is greater than 7%.
   - Not in park or neutral.
   - All conditions met for 20 seconds.
   - No Code 21 or 22.

   These conditions are met during a road load operation. Disregard Code 24 that sets when drive wheels are not turning.

   The PM generator only produces a signal if drive wheels are turning greater than 3 mph (5 kph).

2. Check PROM for correct application before replacing ECM.

Diagnostic Aids:

"Scan" should indicate a vehicle speed whenever the drive wheels are turning greater than 3 mph, (5 kph).

A problem in CKT 434 will not affect the VSS input or the readings on a "Scan".

Check CKT 400 and 401 for proper connections. Be sure they are clean and tight and the harness is routed correctly. Refer to "Intermittents" in Section "B".

(A/T) A faulty or misadjusted park/neutral switch can result in a false Code 24. Use a "Scan" and check for proper signal while in drive (125C). Refer to CHART C-1A for P/N switch diagnosis check.

Code 24 can be falsely set if engine is "brake-torqued" in gear for 20 seconds.

1988–90 2.3L (VIN D AND A) – ALL MODELS

CODE 24
VEHICLE SPEED SENSOR (VSS) CIRCUIT
2.3L (VIN D & A) "N" CARLINE (PORT)

DISREGARD CODE 24 IF SET WHILE DRIVE WHEELS ARE NOT TURNING.

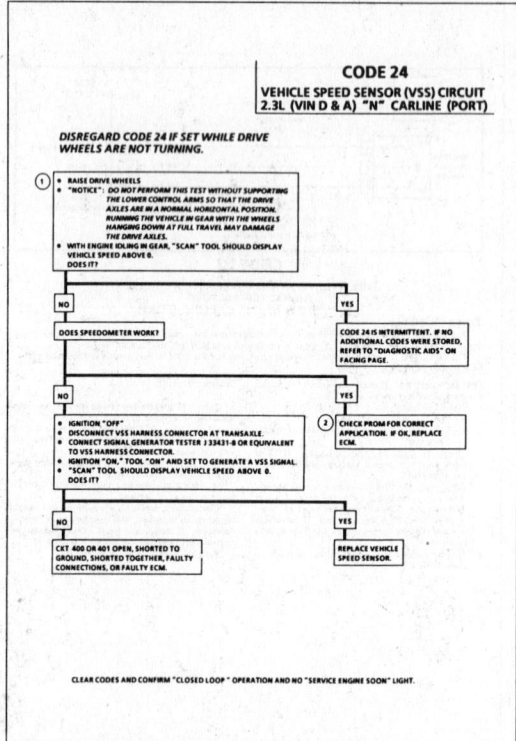

CLEAR CODES AND CONFIRM "CLOSED LOOP" OPERATION AND NO "SERVICE ENGINE SOON" LIGHT.

1988–90 2.3L (VIN D AND A) – ALL MODELS

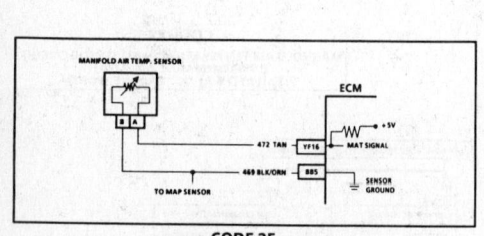

CODE 25
MANIFOLD AIR TEMPERATURE (MAT) SENSOR CIRCUIT
(HIGH TEMPERATURE INDICATED)
2.3L (VIN D & A) "N" CARLINE (PORT)

Circuit Description:

The manifold air temperature sensor uses a thermistor to control the signal voltage to the ECM. The ECM applies a voltage (4-6 volts) on CKT 472 to the sensor. When manifold air is cold, the sensor (thermistor) resistance is high, therefore, the ECM terminal "YF16" voltage is high. As the air warms, the sensor resistance becomes less, and the voltage drops. As the incoming air gets warmer, the sensor resistance decreases, causing ECM terminal "YF16" voltage to decrease.

Test Description: Numbers below refer to circled numbers on the diagnostic chart.

1. Code 25 will set if:
   - Signal voltage indicates a manifold air temperature greater than about 159°C (318°F).
   - Vehicle speed is greater than 15 mph.

Diagnostic Aids:

The "Temperature To Resistance Value" scale at the right may be used to test the MAT sensor at various temperature levels to evaluate the possibility of a "slewed" (mis-scaled) sensor. A "slewed" sensor could result in poor driveability complaints.

Refer to "Intermittents" in Section "B".

1988–90 2.3L (VIN D AND A) – ALL MODELS

CODE 25
MANIFOLD AIR TEMPERATURE (MAT) SENSOR CIRCUIT
(HIGH TEMPERATURE INDICATED)
2.3L (VIN D & A) "N" CARLINE (PORT)

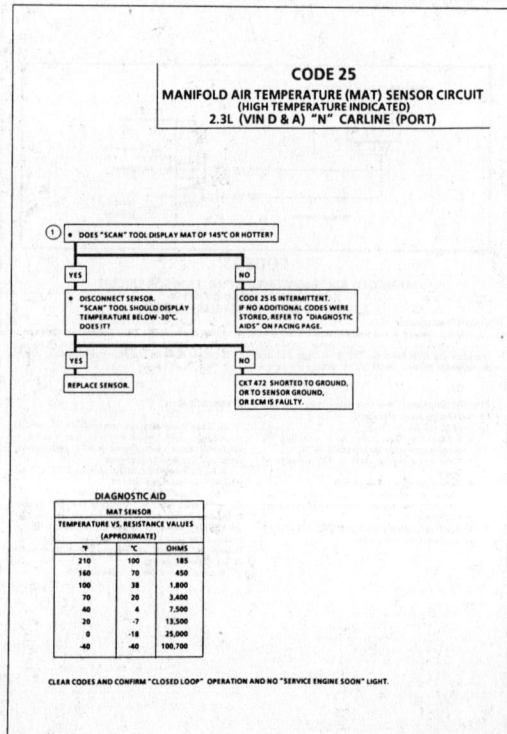

DIAGNOSTIC AID		
MAT SENSOR		
TEMPERATURE VS. RESISTANCE VALUES		
(APPROXIMATE)		
°F	°C	OHMS
210	100	185
160	70	450
100	38	1,800
70	20	3,400
40	4	7,500
20	-7	13,500
0	-18	25,000
-40	-40	100,700

CLEAR CODES AND CONFIRM "CLOSED LOOP" OPERATION AND NO "SERVICE ENGINE SOON" LIGHT.

1988–90 2.3L (VIN D AND A) – ALL MODELS

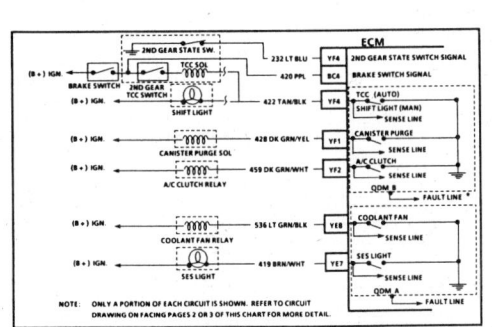

CODE 26
(Page 1 of 3)
QUAD-DRIVER (QDM) CIRCUIT
2.3L (VIN D & A) "N" CARLINE (PORT)

Circuit Description:
The ECM controls most components with electronic switches which complete a ground circuit when turned "ON". These switches are arranged in groups of 4, called quad-driver modules (QDM's) which can independently control up to 4 outputs (ECM terminals), although not all outputs are used. When an output is "ON", the terminal is grounded and its voltage normally will be low. When an output is "OFF", its terminal voltage normally will be high, except for the TCC, as noted below, which depends on the brake and 2nd gear TCC switches.

QDM's are fault protected. If a relay or solenoid coil is shorted, having very low or zero resistance, or if the control side of the circuit is shorted to voltage, it would allow too much current into the QDM. The QDM senses this and the output turns "OFF" or its internal resistance increases to limit current flow and protect the QDM. The result is high output terminal voltage when it should be low. If the circuit from B + or the component is open, or the control side of the circuit is shorted to ground, terminal voltage will be low, even when output is commanded "OFF". Either of these conditions is considered to be a QDM fault. *SEE NOTE BELOW!*

Each QDM has a separate fault line to indicate the presence of a current fault to the ECM's central processor. A "Scan" tool displays the status of each of these fault lines as "Low" = OK, "High" = Fault. Because of the brake and 2nd gear switches in the TCC circuit, Code 26 is set under different conditions for QDM A and QDM B as follows:
- QDM A fault line = "High" for 20 seconds or more.
- QDM B fault line = "High" for 20 seconds or more and
  Brake switch signal indicates brake switch is closed and 2nd gear state switch indicates transaxle is in 2nd or 3rd gear.
  OR
  TCC is commanded "ON".
- *NOTE*: QDM B fault line on an automatic transaxle car will normally be "High" when the car is stopped. The ECM ignores the QDM B fault line except under conditions noted above.

1988–90 2.3L (VIN D AND A) – ALL MODELS

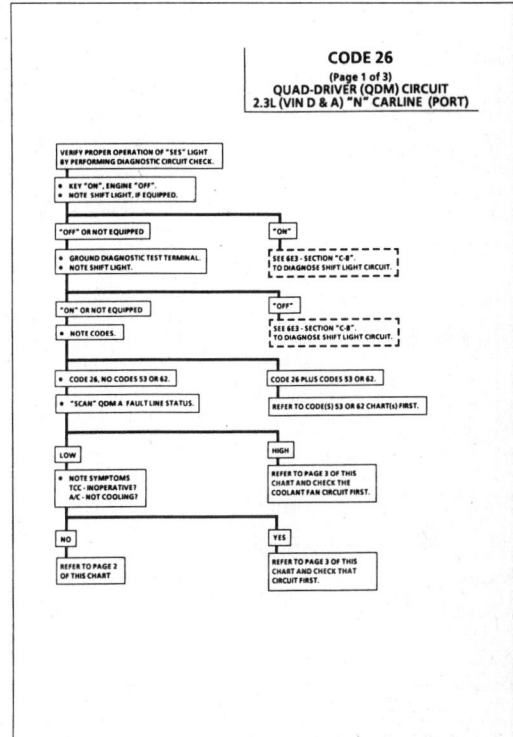

CODE 26
(Page 1 of 3)
QUAD-DRIVER (QDM) CIRCUIT
2.3L (VIN D & A) "N" CARLINE (PORT)

1988–90 2.3L (VIN D AND A) – ALL MODELS

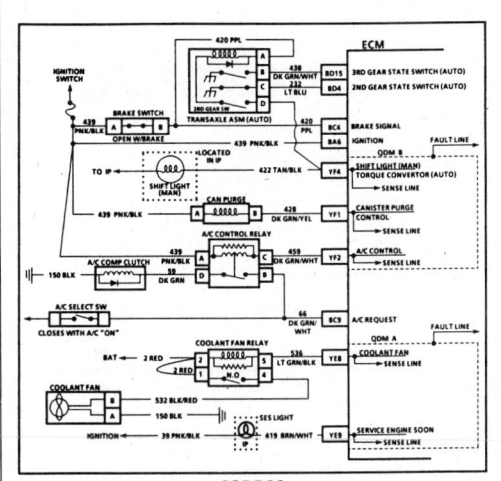

CODE 26
(Page 2 of 3)
QUAD-DRIVER (QDM) CIRCUIT
2.3L (VIN D & A) "N" CARLINE (PORT)

Diagnostic Aids:
Intermittent faults must be continuously present for at least 20 seconds to cause Code 26 to set. QDM controlled circuits should be inspected for poor terminal contact or damaged harnesses. The TCC circuit should be checked with the transaxle at operating temperature if Code 26 sets intermittently and no other cause is found, as a defective TCC solenoid resistance can drop too low (below 20 ohms) at high temperature.

QDM faults can be detected as noted on page 1 of this chart and when outputs are "ON" or "OFF" as follows:
- Open circuit or control circuit short to ground - output commanded "OFF".
- Shorted device or control circuit short to voltage - output commanded "ON".
NOTE: QDM B fault line on an automatic transaxle car will normally be "High" when the car is stopped. The ECM ignores the QDM B fault line except under conditions noted above.

1988–90 2.3L (VIN D AND A) – ALL MODELS

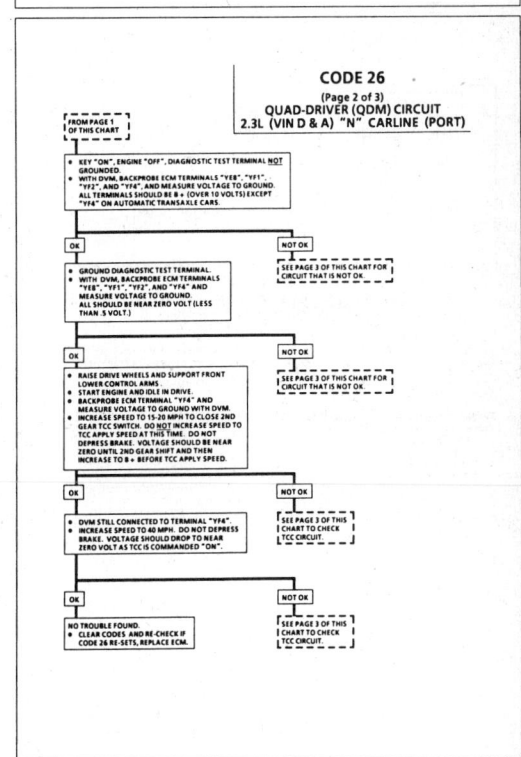

CODE 26
(Page 2 of 3)
QUAD-DRIVER (QDM) CIRCUIT
2.3L (VIN D & A) "N" CARLINE (PORT)

1988–90 2.3L (VIN D AND A) – ALL MODELS

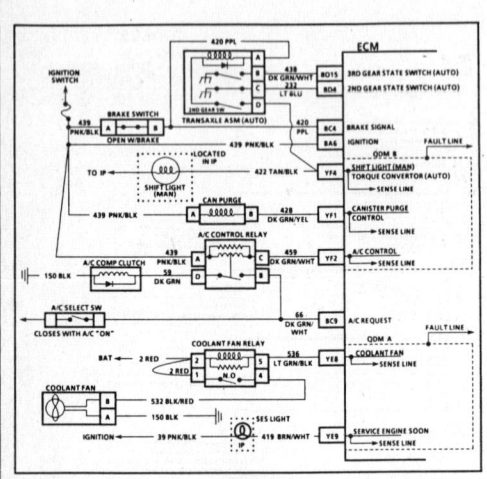

CODE 26
(Page 3 of 3)
QUAD-DRIVER (QDM) CIRCUIT
2.3L (VIN D & A) "N" CARLINE (PORT)

Diagnostic Aids:

Intermittent faults must be continuously present for at least 20 seconds to cause Code 26 to set. QDM controlled circuits should be inspected for poor terminal contact or damaged harnesses. The TCC circuit should be checked with the transaxle at operating temperature if Code 26 sets intermittently and no other cause is found, as a defective TCC solenoid resistance can drop too low (below 20 ohms) at high temperature.

QDM faults can be detected as noted on page 1 of this chart and when outputs are "ON" or "OFF" as follows:
- Open circuit or control circuit short to ground - output commanded "OFF".
- Shorted device or control circuit short to voltage - output commanded "ON"

NOTE   QDM B fault line on an automatic transaxle car will <u>normally be "High"</u> when the car is stopped. The ECM ignores the QDM B fault line <u>except under conditions noted above.</u>

1988–90 2.3L (VIN D AND A) – ALL MODELS

CODE 26
(Page 3 of 3)
QUAD-DRIVER (QDM) CIRCUIT
2.3L (VIN D & A) "N" CARLINE (PORT)

CIRCUIT ISOLATED FROM PRIOR CHARTS

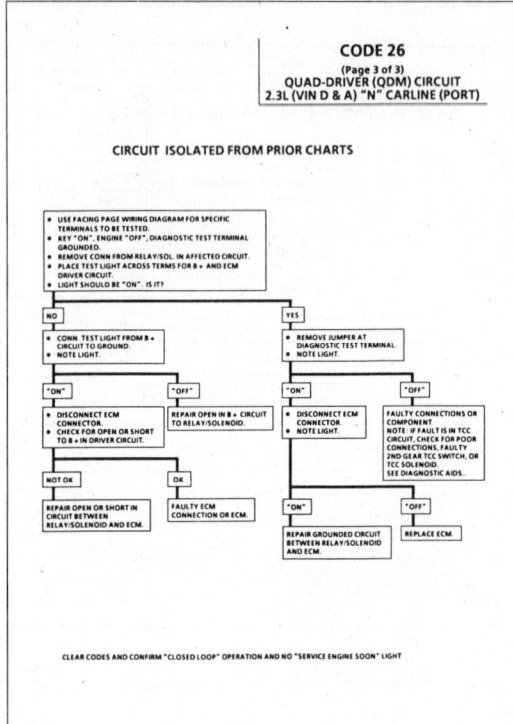

CLEAR CODES AND CONFIRM "CLOSED LOOP" OPERATION AND NO "SERVICE ENGINE SOON" LIGHT.

1988–90 2.3L (VIN D AND A) – ALL MODELS

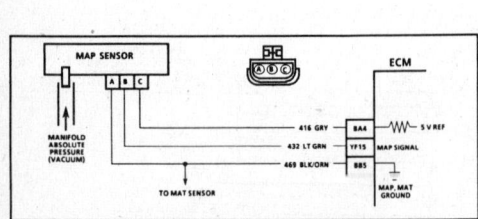

CODE 33
MANIFOLD ABSOLUTE PRESSURE (MAP) SENSOR CIRCUIT
(SIGNAL VOLTAGE HIGH - LOW VACUUM)
2.3L (VIN D & A) "N" CARLINE (PORT)

Circuit Description:

The manifold absolute pressure (MAP) sensor responds to changes in manifold pressure (vacuum). The ECM receives this information as a signal voltage that will vary from about 1 to 1.5 volts at idle, when manifold pressure is low (high vacuum), to 4 - 4.5 volts at wide open throttle (low vacuum or high pressure).

If the MAP sensor fails, the ECM will substitute a fixed MAP value and use the throttle position sensor (TPS) and other sensors to control fuel delivery.

Test Description: Numbers below refer to circled numbers on the diagnostic chart.
1. Code 33 will set when:
   - MAP signal greater than 80 kPa
   - TPS less than 12%
   - VSS less than 1 mph
   - Above conditions met for 5 seconds
2. With the MAP sensor disconnected, the ECM terminal "YF15" voltage should be low if the ECM and wiring are OK.

Diagnostic Aids:

An intermittent open in CKT 469 can cause a Code 33.
Refer to "Intermittents" in Section "B".

1988–90 2.3L (VIN D AND A) – ALL MODELS

CODE 33
MANIFOLD ABSOLUTE PRESSURE (MAP) SENSOR CIRCUIT
(SIGNAL VOLTAGE HIGH - LOW VACUUM)
2.3L (VIN D & A) "N" CARLINE (PORT)

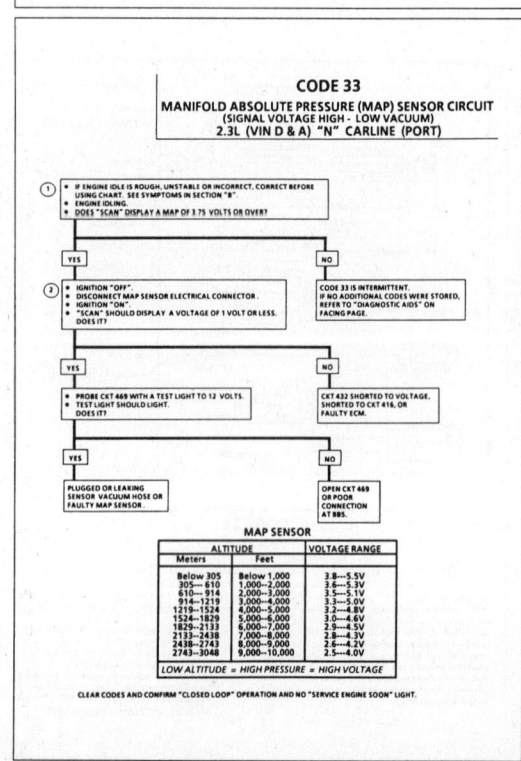

MAP SENSOR

ALTITUDE		VOLTAGE RANGE
Meters	Feet	
Below 305	Below 1,000	3.8—5.5V
305— 610	1,000—2,000	3.6—5.3V
610— 914	2,000—3,000	3.5—5.1V
914—1219	3,000—4,000	3.3—5.0V
1219—1524	4,000—5,000	3.2—4.8V
1524—1829	5,000—6,000	3.0—4.6V
1829—2133	6,000—7,000	2.9—4.5V
2133—2438	7,000—8,000	2.8—4.3V
2438—2743	8,000—9,000	2.6—4.2V
2743—3048	9,000—10,000	2.5—4.0V

LOW ALTITUDE = HIGH PRESSURE = HIGH VOLTAGE

CLEAR CODES AND CONFIRM "CLOSED LOOP" OPERATION AND NO "SERVICE ENGINE SOON" LIGHT.

1988–90 2.3L (VIN D AND A) – ALL MODELS

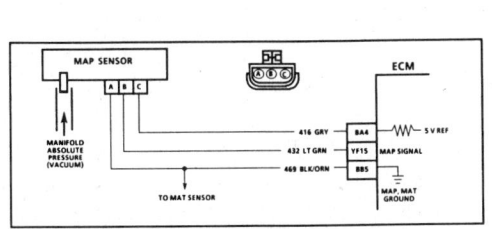

CODE 34
MANIFOLD ABSOLUTE PRESSURE (MAP) SENSOR CIRCUIT
(SIGNAL VOLTAGE LOW - HIGH VACUUM)
2.3L (VIN D & A) "N" CARLINE (PORT)

Circuit Description:

The manifold absolute pressure (MAP) sensor responds to changes in manifold pressure (vacuum). The ECM receives this information as a signal voltage that will vary from about 1 to 1.5 volts at idle, when manifold pressure is low (high vacuum), to 4 - 4.5 volts at wide open throttle (low vacuum or high pressure).

If the MAP sensor fails, the ECM will substitute a fixed MAP value and use the throttle position sensor (TPS) and other sensors to control fuel delivery.

Test Description: Numbers below refer to circled numbers on the diagnostic chart.
1. Code 34 will set when:
   - Engine running
   - No Code 21
   - MAP less than 14 kPa
   - Engine rpm less than 1200 or TPS greater than 15.2%
   - Above conditions met for .2 seconds
2. This tests to see if the sensor is at fault for the low voltage or if there is an ECM or wiring problem.

3. This simulates a high signal voltage to check for an open in CKT 432. If the test light is bright during this test, CKT 432 is probably shorted to ground. If "Scan" reads over 4 volts at this test, CKT 416 can be checked by measuring the voltage at terminal "C" (should be 5 volts).

Diagnostic Aids:

An intermittent on CKT 416 or 469 will result in a Code 34.

Refer to "Intermittents" in Section "B".

1988–90 2.3L (VIN D AND A) – ALL MODELS

CODE 34
MANIFOLD ABSOLUTE PRESSURE (MAP) SENSOR CIRCUIT
(SIGNAL VOLTAGE LOW - HIGH VACUUM)
2.3L (VIN D & A) "N" CARLINE (PORT)

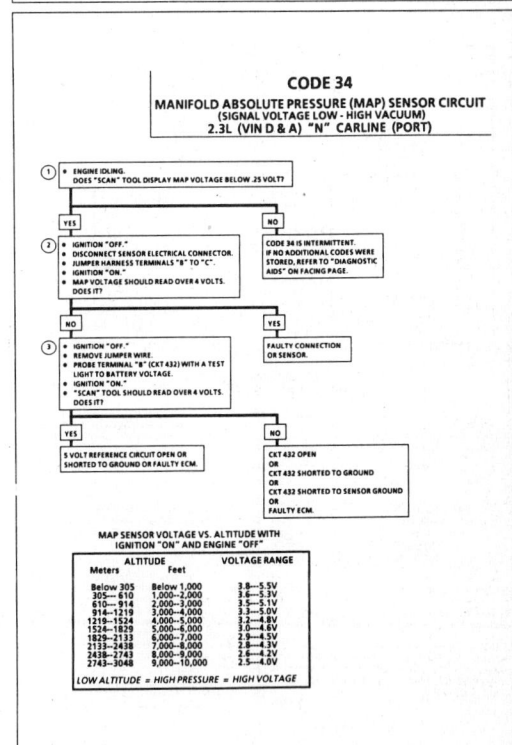

1988–90 2.3L (VIN D AND A) – ALL MODELS

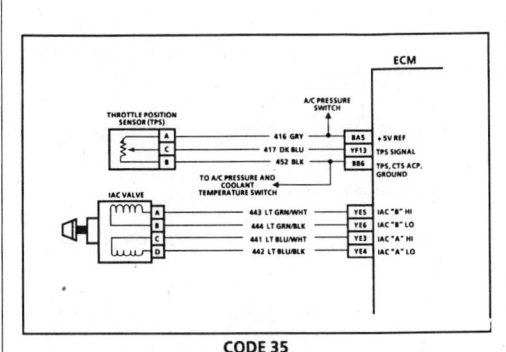

CODE 35
IDLE SPEED ERROR
2.3L (VIN D & A) "N" CARLINE (PORT)

Circuit Description:

The ECM controls idle speed to a calculated, "desired" rpm based on sensor inputs and actual engine rpm, determined by the time between successive 2X ignition reference pulses from the ignition module. The ECM uses 4 circuits to move an idle air control (IAC) valve, which allows varying amounts of air flow into the intake manifold, controlling idle speed.

Code 35 sets when:
   - Engine speed is at least 175 rpm more or less than "desired".
   - TPS voltage indicates throttle is open less than 1 %.
   - VSS indicates vehicle speed is less than 3 mph.
   - All above conditions are continuously met for 5 seconds or more.

Diagnostic Aids:

Check for vacuum leaks, disconnected or brittle vacuum hoses, cuts, etc. Examine manifold and throttle body gaskets for proper seal. Check for cracked intake manifold. Check open, shorts, or poor connections to IAC valve in CKT's 441, 442, 443 and 444.

Check for poor connections at ECM terminals YE3, YE4, YE5 and YE6. An open, short, or poor connection in CKT's 441, 442, 443, or 444 will result in improper idle control and may cause Code 35.

An IAC valve which is stopped and cannot respond to the ECM, a throttle stop screw which has been tampered with, or a damaged throttle body or linkage could cause Code 35. If no problem is found and Code resets, replace ECM.

1988–90 2.3L (VIN D AND A) – ALL MODELS

NOTE: *A VACUUM LEAK IS THE MOST PROBABLE CAUSE OF CODE 35. CHECK THOROUGHLY FOR LEAKS AT VACUUM CONNECTOR BLOCK, THROTTLE BODY AND INTAKE MANIFOLD BEFORE USING CHART.*

CODE 35
IDLE SPEED ERROR
2.3L (VIN D & A) "N" CARLINE (PORT)

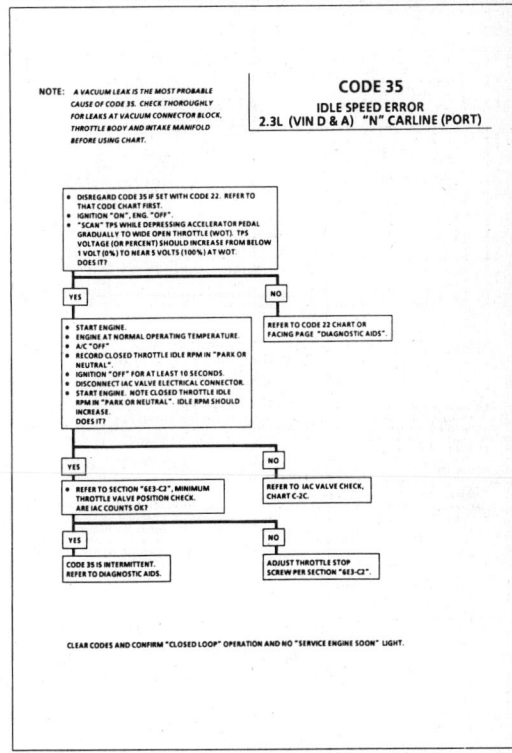

1988–90 2.3L (VIN D AND A) – ALL MODELS

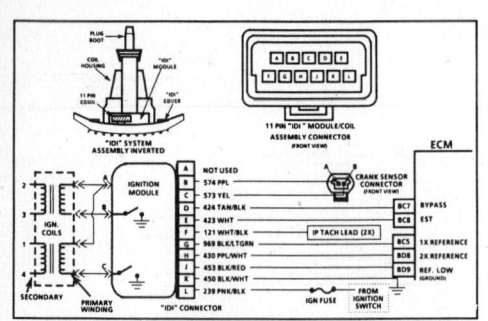

CODE 41
1X REFERENCE CIRCUIT
2.3L (VIN D & A) "N" CARLINE (PORT)

Circuit Description:

The ignition module sends a reference signal to the ECM once per revolution to indicate crankshaft position so that the ECM can determine when to pulse the injectors for cylinders 2 and 3 in the ASDF fuel control mode. This signal may be described as a synchronization signal and is called the 1X reference because it occurs one time per revolution. The ignition module applies 5 volts from terminal "G" through CKT 969 to ECM terminal "BC5" and in effect, switches this circuit to ground for a very short period of time, 125 degrees before TDC of cylinders 2 and 3. Code 41 is set if the ECM receives (8) 2X reference pulses with no 1X reference pulses. When Code 41 is present, the ECM pulses the injectors in the SSDF (simultaneous) mode.

Test Description: Numbers below refer to circled numbers on the diagnostic chart.
1. This determines if the ECM recognizes a problem. If it doesn't set Code 41 at this point, the problem is intermittent and could be due to a loose connection. (See Diagnostic Aids.)
2. This step simulates the 1X signal. The ECM should recognize the drop in voltage as the test light probe is removed, if the circuit and ECM are OK. This step will give accurate results only if the chart sequence is used - ignition "OFF", ignition "ON", "Scan" tool set to 1X reference and terminal "G" touched with test light probe.
3. If the ECM did not recognize the simulation of the 1X signal, CKT 969 may be open or shorted to ground or voltage. If CKT 969 is OK, the ECM is faulty.
4. Step 2 indicated that CKT 969 was OK and the ECM

reference pulse. This indicates either a poor connection at ignition module terminal "G" or a faulty ignition module caused the Code 41.

Diagnostic Aids:

An intermittent may be caused by a poor connection, rubbed through wire insulation, or a wire broken inside the insulation. Inspect ECM harness connector terminal "BC5" and ignition module terminal "G" for improper mating, broken locks, improperly formed or damaged terminals, poor terminal to wire connection and damaged harness.

1988–90 2.3L (VIN D AND A) – ALL MODELS

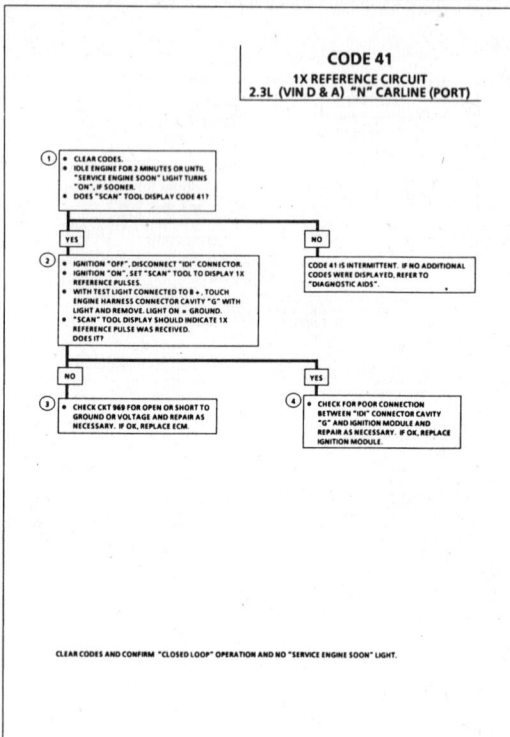

CODE 41
1X REFERENCE CIRCUIT
2.3L (VIN D & A) "N" CARLINE (PORT)

CLEAR CODES AND CONFIRM "CLOSED LOOP" OPERATION AND NO "SERVICE ENGINE SOON" LIGHT.

1988–90 2.3L (VIN D AND A) – ALL MODELS

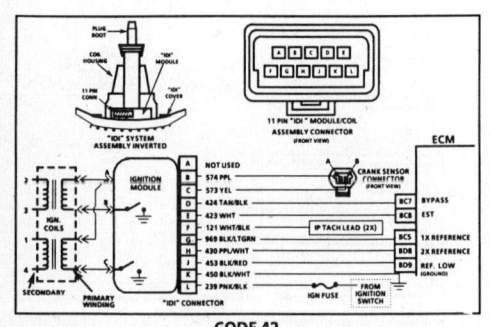

CODE 42
ELECTRONIC SPARK TIMING (EST) CIRCUIT
2.3L (VIN D & A) "N" CARLINE (PORT)

Circuit Description:

The ignition module sends a reference signal to the ECM when the engine is cranking or running. While the engine is under 700 rpm, the ignition module controls the ignition timing. When the engine speed exceeds 700 rpm, the ECM sends a 5 volts signal on the "bypass" CKT 424 to switch the timing to ECM control through the EST CKT 423.

An open or ground in the EST or "bypass" circuit will set a Code 42 and cause the engine to run on module or "bypass" timing. This will result in poor performance and poor fuel economy.

Test Description: Numbers below refer to circled numbers on the diagnostic chart.
1. Checks to see if ECM recognizes a problem. If it doesn't set Code 42 at this point, it is an intermittent problem and could be due to a loose connection.
2. With the ECM disconnected, the ohmmeter should be reading less than 500 ohms, which is the normal resistance of the ignition module. A higher resistance would indicate a fault in CKT 423, a poor ignition module connection or a faulty ignition module.
3. If the test light was "ON" when connected from 12 volts to ECM harness terminal "BC7", either CKT 423 is shorted to ground or the ignition module is faulty.
4. Checks to see if ignition module switches when the bypass circuit is energized by 12V through the test light.

If the ignition module actually switches, the ohmmeter reading should shift to over 8000 ohms.
5. Disconnecting the ignition module should make the ohmmeter read as if it were monitoring an open circuit (infinite reading). If the ohmmeter has a reading other than infinite, CKT 423 is shorted to ground.

Diagnostic Aids:

An intermittent may be caused by a poor connection, rubbed through wire insulation, or a wire broken inside the insulation. Inspect ECM harness connectors for backed out terminals "BC7" or "BC8", improper mating, broken locks, improperly formed or damaged terminals, poor terminal to wire connection, and damaged harness.

1988–90 2.3L (VIN D AND A) – ALL MODELS

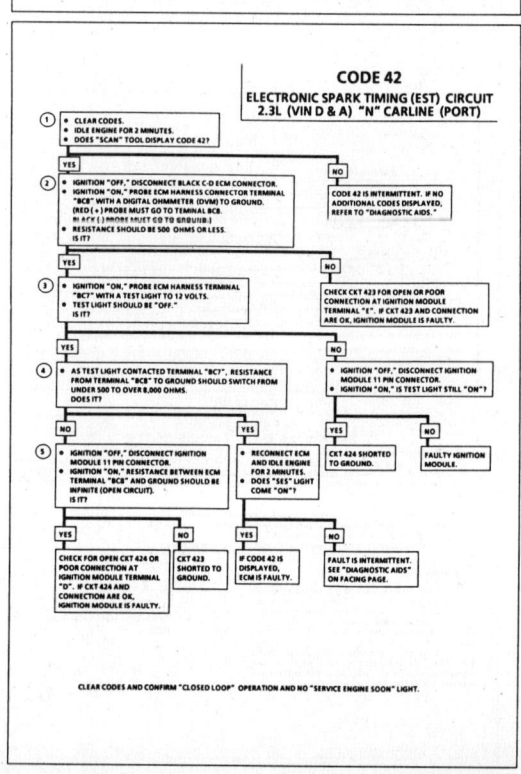

CODE 42
ELECTRONIC SPARK TIMING (EST) CIRCUIT
2.3L (VIN D & A) "N" CARLINE (PORT)

CLEAR CODES AND CONFIRM "CLOSED LOOP" OPERATION AND NO "SERVICE ENGINE SOON" LIGHT.

1988–90 2.3L (VIN D AND A) – ALL MODELS

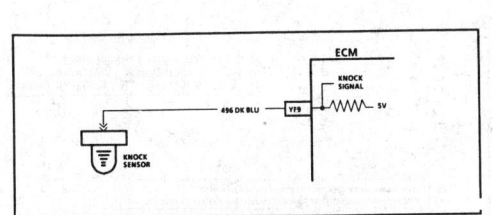

CODE 43
ELECTRONIC SPARK CONTROL (ESC) CIRCUIT
2.3L (VIN D & A) "N" CARLINE (PORT)

Circuit Description:
The knock sensor detects engine detonation and the ECM retards the electronic spark timing based on the signal being received. The circuitry within the knock sensor causes the ECM 5 volts to be pulled down so that, under a no knock condition, CKT 496 would measure about 2.5 volts. The knock sensor produces an AC signal which rides on the 2.5 volts DC voltage. The amplitude and signal frequency are dependent upon the knock level.

The ECM performs two tests on this circuit to determine if it is operating correctly. If either of the tests fail, a Code 43 will be set.

- If there is an indication of knock for 3.67 seconds over a 3.9 second interval with the engine running.
- If ECM terminal "YF9" voltage is either above about 3.75 volts (indicating open CKT 496), or below about 1.25 volts (indicating CKT 496 is shorted to ground) for 5 seconds or more.

Test Description: Numbers below refer to circled numbers on the diagnostic chart.
1. If the conditions for the test, as described above, are being met, the "Scan" tool will always indicate "Yes" when the knock signal position is selected. If an audible knock is heard from the engine, repair the internal engine problem, because normally, no knock should be detected at idle.
2. If tapping on the engine lift hook does not produce a knock signal, try tapping engine closer to sensor before proceeding.
3. The ECM has a 5 volts signal through a pull-up resistor which should be present at the knock sensor terminal.

4. This test determines if the knock sensor is faulty or if the ESC portion of the mem-cal is faulty.

Diagnostic Aids:

Check CKT 496 for a potential open or short to ground.
Also check for proper installation of mem-cal. Refer to "Intermittents" in Section "B".
Mechanical engine knock can cause a knock sensor signal. Abnormal engine noise must be corrected before using this chart.

1988–90 2.3L (VIN D AND A) – ALL MODELS

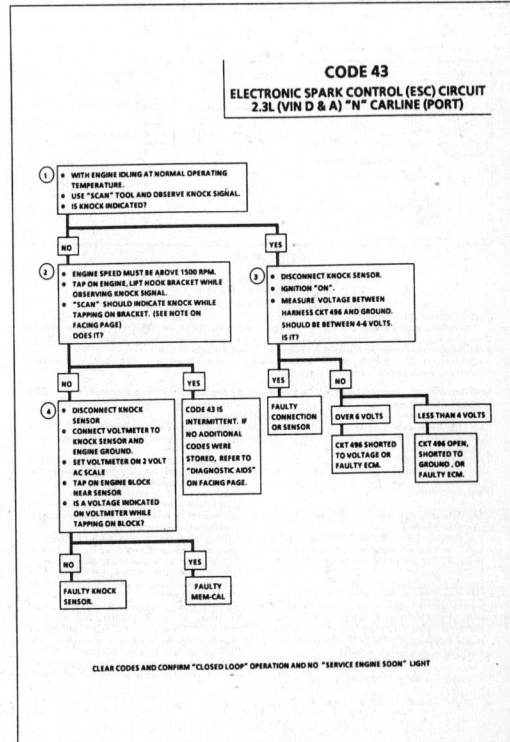

CODE 43
ELECTRONIC SPARK CONTROL (ESC) CIRCUIT
2.3L (VIN D & A) "N" CARLINE (PORT)

1988–90 2.3L (VIN D AND A) – ALL MODELS

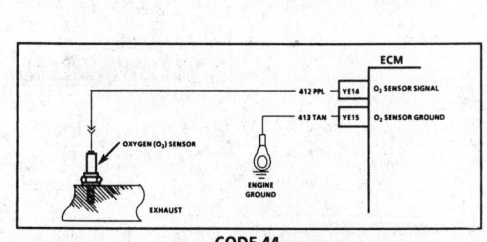

CODE 44
OXYGEN SENSOR CIRCUIT
(LEAN EXHAUST INDICATED)
2.3L (VIN D & A) "N" CARLINE (PORT)

Circuit Description:
The ECM supplies a voltage of about .45 volt between terminals "YE14" and "YE15". (If measured with a 10 megohm digital volt meter, this may read as low as .32 volt.) The O_2 sensor varies the voltage within a range of about 1 volt if the exhaust is rich, down through about .10 volt if exhaust is lean.

The sensor is like an open circuit and produces no voltage when it is below 360° C (600° F). An open sensor circuit or cold sensor causes "Open Loop" operation.

Test Description: Numbers below refer to circled numbers on the diagnostic chart.
1. Code 44 is set when the O_2 sensor signal voltage on CKT 412:
   - Remains below .3 volt for 50 seconds or more.
   - The system is operating in "Closed Loop".
   - No Code 33 or 34.
   - "Closed Loop" integrator active.

Diagnostic Aids:

The Code 44 or lean exhaust is most likely caused by one of the following:
- Fuel Pressure System will be lean if pressure is too low. It may be necessary to monitor fuel pressure while driving the car at various road speeds and/or loads to confirm. Refer to CHART A-7.

- MAP Sensor An output that causes the ECM to sense a lower than normal manifold pressure (high vacuum) can cause the system to go lean. Disconnecting the MAP sensor will allow the ECM to substitute a fixed (default) value for the MAP sensor. If the rich condition is gone when the sensor is disconnected, substitute a known good sensor and recheck.
- Fuel Contamination Water, even in small amounts, near the in-tank fuel pump inlet can be delivered to the injector. The water causes a lean exhaust and can set a Code 44.
- Sensor Harness Sensor pigtail may be mispositioned and contacting the exhaust manifold.
- Engine Misfire A cylinder misfire will result in unburned oxygen in the exhaust, which could cause Code 44. Refer to CHART C4-M and/or "Symptoms" in Section "B".

1988–90 2.3L (VIN D AND A) – ALL MODELS

CODE 44
OXYGEN SENSOR CIRCUIT
(LEAN EXHAUST INDICATED)
2.3L (VIN D & A) "N" CARLINE (PORT)

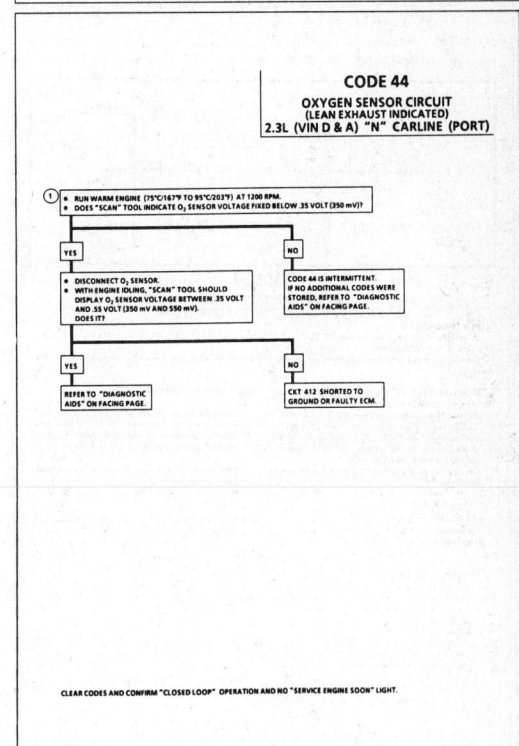

1988–90 2.3L (VIN D AND A) – ALL MODELS

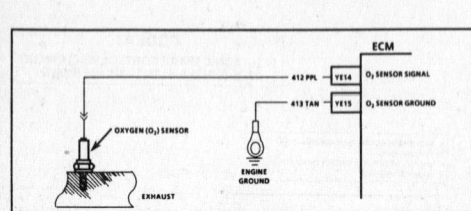

CODE 45
OXYGEN SENSOR CIRCUIT
(RICH EXHAUST INDICATED)
2.3L (VIN D & A) "N" CARLINE (PORT)

Circuit Description:

The ECM supplies a voltage of about .45 volt between terminals "YE14" and "YE15". (If measured with a 10 megohm digital voltmeter, this may read as low as .32 volt.) The O₂ sensor varies the voltage within a range of about 1 volt if the exhaust is rich, down through about .10 volt if exhaust is lean.

The sensor is like an open circuit and produces no voltage when it is below 360° C (600° F). An open sensor circuit or cold sensor causes "Open Loop" operation.

Test Description: Numbers below refer to circled numbers on the diagnostic chart.
1. Code 45 is set when:
   * O₂ voltage is above .75 volt
   * No Code 33 or 34
   * Fuel system in "Closed Loop"
   * TPS above 5%
   * Above conditions met for 50 seconds

Diagnostic Aids:

The Code 45 or rich exhaust is most likely caused by one of the following:
* **Fuel Pressure** System will go rich if pressure is too high. The ECM can compensate for some increase. However, if it gets too high, a Code 45 will be set. See "Fuel System" diagnosis CHART A-7.
* **Leaking Injector** See CHART A-7.
* **HEI Shielding** An open ground CKT 453 may result in EMI or induced electrical noise. The ECM looks at this noise as distributor pulses. The additional pulses result in a higher than actual engine speed signal. The ECM then delivers too much fuel causing system to go rich.

Engine tachometer will also show higher than actual engine speed which can help in diagnosing this problem.
* **Canister Purge** Check for fuel saturation. If full of fuel, check canister control and hoses.
* **MAP Sensor** An output that causes the ECM to sense a higher than normal manifold pressure (low vacuum) can cause the system to go rich. Disconnecting the MAP sensor will allow the ECM to set a fixed value for the MAP sensor. Substitute a different MAP sensor if the rich condition is gone while the sensor is disconnected.
* **Pressure Regulator** Check for leaking fuel pressure regulator diaphragm by checking for the presence of liquid fuel in the vacuum line to the regulator.
* **TPS** An intermittent TPS output will cause the system to go rich due to a false indication of the engine accelerating.
* **O₂ Sensor Contamination** Inspect oxygen sensor for silicone contamination from fuel or use of improper RTV sealant. The sensor may have a white powdery coating and result in a high but false signal voltage (rich exhaust indication). The ECM will then reduce the amount of fuel delivered to the engine causing a severe surge driveability problem.

1988–90 2.3L (VIN D AND A) – ALL MODELS

CODE 45
OXYGEN SENSOR CIRCUIT
(RICH EXHAUST INDICATED)
2.3L (VIN D & A) "N" CARLINE (PORT)

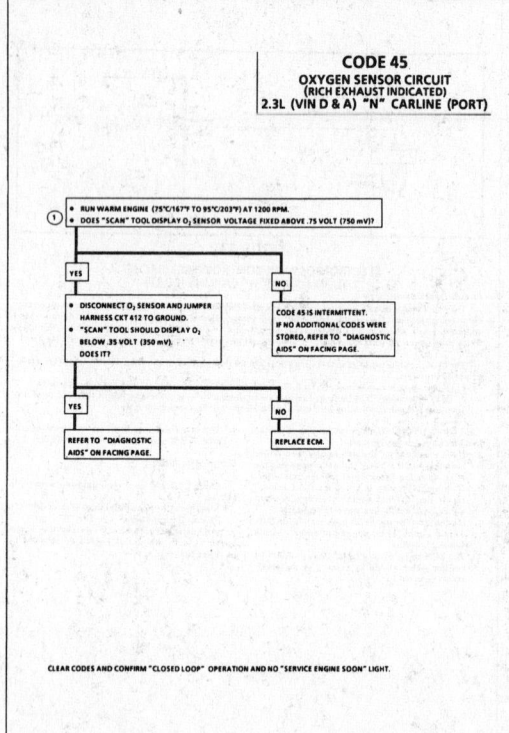

CLEAR CODES AND CONFIRM "CLOSED LOOP" OPERATION AND NO "SERVICE ENGINE SOON" LIGHT.

1988–90 2.3L (VIN D AND A) – ALL MODELS

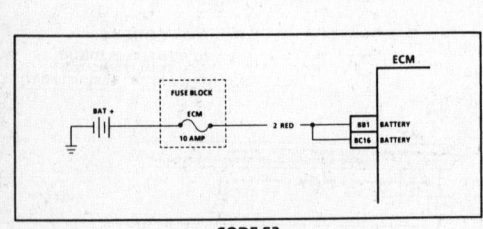

CODE 53
BATTERY VOLTAGE TOO HIGH
2.3L (VIN D & A) "N" CARLINE (PORT)

Circuit Description:

Code 53 will set when the ignition is "ON" and ECM terminal "BB1" and "BC16" voltages are more than 17.1 volts for about 2 seconds.

During the time the failure is present, all ECM outputs will be disengaged. (The setting of additional codes may result.)

Test Description: Numbers below refer to circled numbers on the diagnostic chart.
1. Normal battery output is between 10 - 17.0 volts.
2. Checks to see if the high voltage reading is due to the generator or ECM. With engine running, check voltage at the battery. If the voltage is above 17.1 volts, the ECM is OK.
3. Checks to see if generator is faulty under load condition.

Note On Intermittents:

Charging battery with a battery charger and starting engine, may set Code 53. If code sets when an accessory is operated, check for poor connections or excessive current draw.

Also, check for poor connections at starter solenoid or fusible link junction box.

1988–90 2.3L (VIN D AND A) – ALL MODELS

CODE 53
BATTERY VOLTAGE TOO HIGH
2.3L (VIN D & A) "N" CARLINE (PORT)

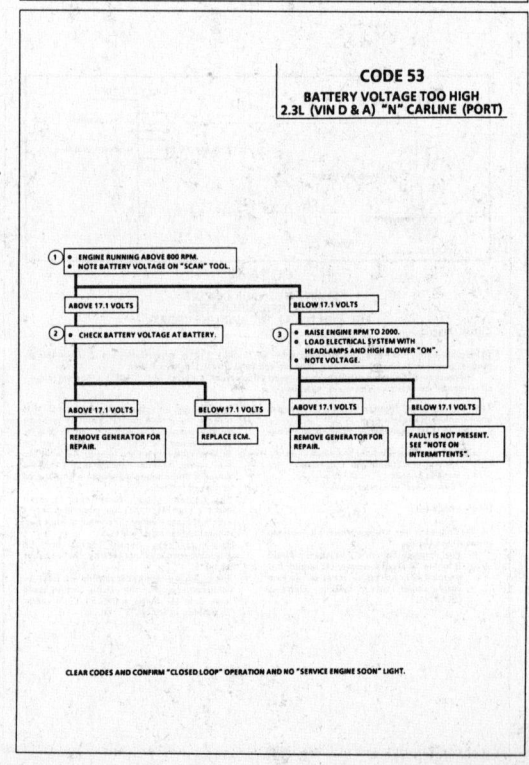

CLEAR CODES AND CONFIRM "CLOSED LOOP" OPERATION AND NO "SERVICE ENGINE SOON" LIGHT.

1988–90 2.3L (VIN D AND A) – ALL MODELS

CODE 51
MEM-CAL ERROR
(FAULTY OR INCORRECT MEM-CAL)
2.3L (VIN D & A) "N" CARLINE (PORT)

CHECK THAT ALL PINS ARE FULLY INSERTED IN THE SOCKET AND THAT MEM-CAL IS PROPERLY LATCHED.
IF OK, REPLACE MEM-CAL, CLEAR MEMORY, AND RECHECK. IF CODE 51 REAPPEARS, REPLACE ECM.

NOTICE: TO PREVENT POSSIBLE ELECTROSTATIC DISCHARGE DAMAGE TO THE ECM OR MEM-CAL, DO NOT TOUCH THE COMPONENT LEADS, AND DO NOT REMOVE THE INTEGRATED CIRCUIT FROM CARRIER.

CLEAR CODES AND CONFIRM "CLOSED LOOP" OPERATION AND NO "SERVICE ENGINE SOON" LIGHT.

1988–90 2.3L (VIN D AND A) – ALL MODELS

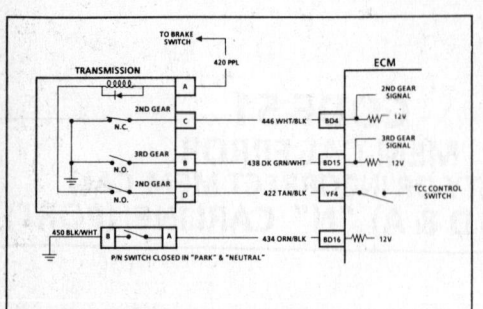

CODE 62
TRANSMISSION GEAR SWITCHES CIRCUIT
2.3L (VIN D & A) "N" CARLINE (PORT)

Circuit Description:
Code 62 diagnostic is used to check the 2nd and 3rd gear switch inputs to the ECM. Transmission gear input signals to the ECM are determined as follows:

1st gear	= 2nd gear switch closed
(also P/N & R)	3rd gear switch open
2nd gear	= 2nd gear switch open
	3rd gear switch open
3rd gear	= 2nd gear switch open
	3rd gear switch closed

Test Description: Numbers below refer to circled numbers on the diagnostic chart.
1. Code 62 will set if any of the following conditions are met for approximately 17 seconds:
   • In "PARK" or "NEUTRAL" and 2nd or 3rd gear indicated.
   • Vehicle speed less than approximately 5 mph and 2nd or 3rd gear indicated.
   • First or second gear indicated, less than approx 3000 rpm, not in P/N, and VSS greater than 50 mph.
   Step one checks for an open or grounded circuit between transmission connector and ECM. Since the ECM terminals are normally high, B+ should be indicated at both terminals with the connector removed.
2. The DVM should measure near 0 volt to terminal "B" because the 3rd gear switch should be open

providing no continuity to ground. Terminal "C" should measure near B+ because the 2nd gear switch should provide continuity to ground.
3. The third step checks to see if the transmission switches function normally. In third gear both switches should have changed from their original 1st gear state.

Note On Intermittents

Clear codes and recheck for Code 62. If code resets, check all harness connections, VSS, rpm, P/N circuits prior to replacing an ECM. Transmission starting in 2nd or 3rd gear can also cause Code 62.

1988–90 2.3L (VIN D AND A) – ALL MODELS

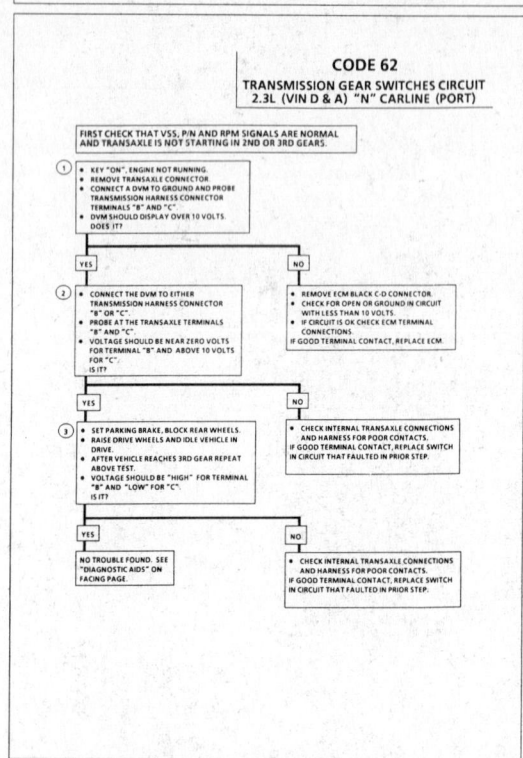

CODE 62
TRANSMISSION GEAR SWITCHES CIRCUIT
2.3L (VIN D & A) "N" CARLINE (PORT)

1988–90 2.3L (VIN D AND A) – ALL MODELS

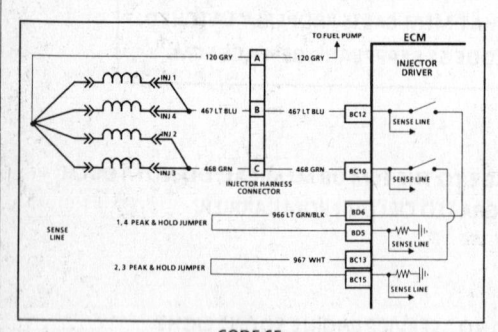

CODE 65
(PAGE 1 OF 2)
FUEL INJECTOR CIRCUIT
(LOW CURRENT)
2.3L (VIN D & A) "N" CARLINE (PORT)

Circuit Description:
The ECM has two injector driver circuits, each of which controls a pair of injectors (1 and 4 or 2 and 3). The ECM monitors the current in each driver circuit by measuring voltage drop through a fixed resistor and is able to control it. The current through each driver is allowed to rise to a "peak" of 4 amps to quickly open the injectors and is then reduced to 1 amp to "hold" them open. This is called "peak and hold". If the current can't reach a 4 amp peak, Code 65 is set as noted below. This code is also set if an injector driver circuit is shorted to voltage.

Test Description: Numbers below refer to circled numbers on the diagnostic chart.
1. Code 65 sets when.
   • 4 amp injector driver current not reached on either circuit
   • Battery voltage greater than 9 volts
   • Injectors commanded "ON" longer than a calibrated pulse width.
   • Above conditions met for 20 seconds
2. Tests ECM and harness wiring to the 3 terminal injector harness connector
3. Tests for open or shorted injector harness or injector. A shorted harness or injector will not cause Code 65
4. Results of step 2 will determine which branch to follow on page 2.
5. Checks remainder of circuit from injectors to ECM as both harnesses were confirmed OK in steps 2 and 3.

6. Determines cause of high resistance found in step 3. (Low resistance or a short will not cause Code 65, but should be corrected if found)
7. Checks for grounded "peak and hold" jumpers. This fault would allow injectors to pulse but would not allow "peak and hold" operation as current would not flow through the resistor in the ECM.

Diagnostic Aids:
Open CKTs 467,468,966 or 967 or CKT 467 or 468 shorted to voltage will cause Code 65 and will also cause a misfire due to an inoperative pair of injectors. CKTs 966 and 967 shorted to ground will cause Code 65 while allowing the injectors to pulse. An intermittent problem would have to be present for at least 20 seconds to set Code 65.

1988–90 2.3L (VIN D AND A) – ALL MODELS

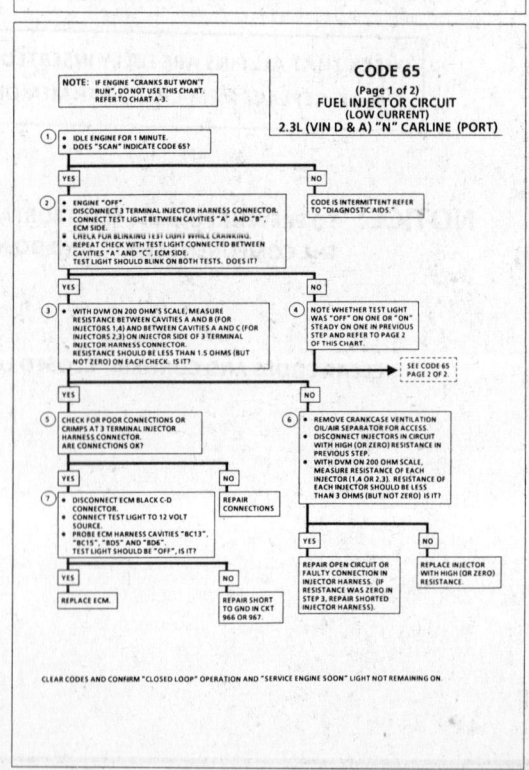

CODE 65
(Page 1 of 2)
FUEL INJECTOR CIRCUIT
(LOW CURRENT)
2.3L (VIN D & A) "N" CARLINE (PORT)

1988–90 2.3L (VIN D AND A) – ALL MODELS

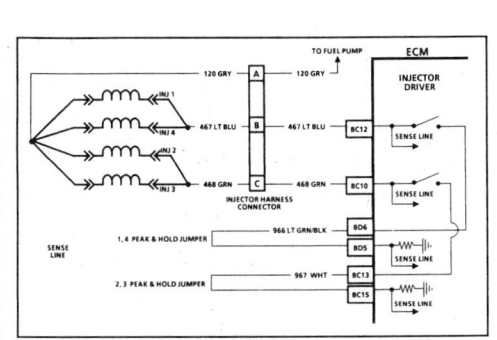

CODE 65
(PAGE 2 OF 2)
FUEL INJECTOR CIRCUIT
(LOW CURRENT)
2.3L (VIN D & A) "N" CARLINE (PORT)

Circuit Description:
The ECM has two injector driver circuits, each of which controls a pair of injectors (1 and 4 or 2 and 3). The ECM monitors the current in each driver circuit by measuring voltage drop through a fixed resistor and is able to control it. The current through each driver is allowed to rise to a "peak" of 4 amps to quickly open the injectors and is then reduced to 1 amp to "hold" them open. This is called "peak and hold". If the current can't reach a 4 amp peak, Code 65 is set as noted below. This code is also set if an injector driver circuit is shorted to voltage.

Test Description: Numbers below refer to circled numbers on the diagnostic chart.
8. This checks for short to voltage in injector driver circuits. It is necessary to crank the engine to assure voltage to CKT 120.
9. Determines whether injector driver CKTs 467 and 468 are shorted to ground.
10. This checks the output at the ECM to determine if CKTs 467 and 468 are OK.
11. Checks for good continuity of "peak and hold" jumpers CKTs 966 and 967.

1988–90 2.3L (VIN D AND A) – ALL MODELS

CODE 65
(Page 2 of 2)
FUEL INJECTOR CIRCUIT
(LOW CURRENT)
2.3L (VIN D & A) "N" CARLINE (PORT)

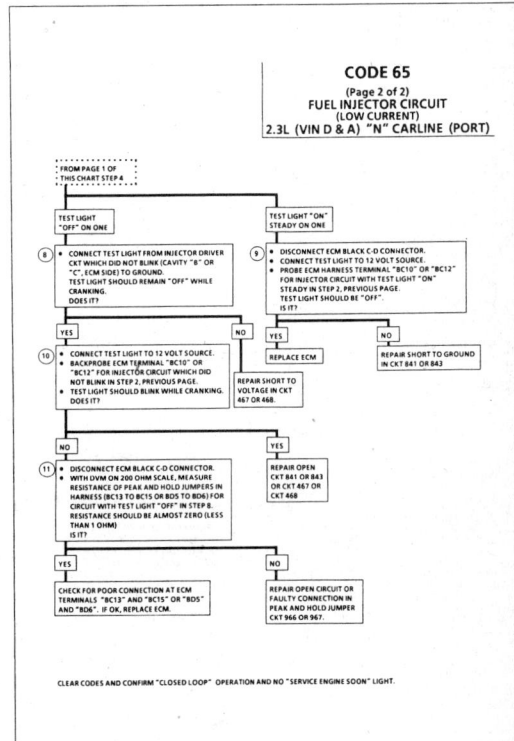

1988–90 2.3L (VIN D AND A) – ALL MODELS

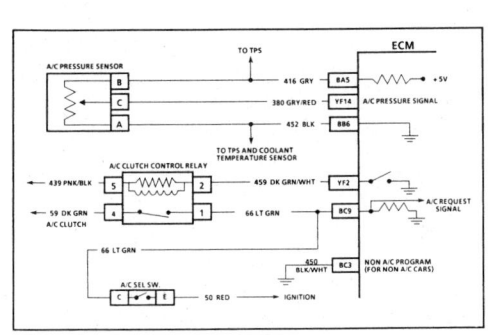

CODE 66
A/C PRESSURE SENSOR CIRCUIT
2.3L (VIN D & A) "N" CARLINE (PORT)

Circuit Description:
The A/C pressure sensor responds to changes in A/C refrigerant system high side pressure. This input indicates how much load the A/C compressor is putting on the engine and is one of the factors used by the ECM to determine IAC valve position for idle speed control. The circuit consists of a 5 volts reference and a ground, both provided by the ECM, and a signal line to the ECM. The signal is a voltage which is proportional to the pressure. The sensor's range of operation is 0 to 450 psi. At 0 psi, the signal will be about .1 volt, varying up to about 4.9 volts at 450 psi or above. Code 66 sets if the voltage is above 4.9 volts or below .1 volt 5 seconds or more. The A/C compressor is disabled by the ECM if Code 66 is present, or if pressure is above or below calibrated values discribed in section "C-10".

Test Description: Numbers below refer to the circled numbers on the diagnostic chart.
1 This step checks the voltage signal being received by the ECM from the A/C pressure sensor. The normal operating range is between .1V and 4.9V.
2 Checks to see if the high voltage signal is from a shorted sensor or a short to voltage in the circuit. Normally, disconnecting the sensor would make a normal circuit go to near zero volt.
3 Checks to see if low voltage signal is from the sensor or the circuit. Jumpering the sensor signal CKT 380 to 5 volts, checks the circuit, connections, and ECM.
4 This step checks to see if the low voltage signal was due to an open in the sensor circuit or the 5 volts reference circuit since the prior step eliminated the pressure sensor.

Diagnostic Aids:

Code 66 sets when signal voltage falls outside the normal possible range of the sensor and is not due to a refrigerant system problem. If problem is intermittent, check for opens or shorts in harness or poor connections. If OK, replace A/C pressure sensor. If Code 66 re-sets, replace ECM.

Non-A/C Program

Code 66 will set on a Non-A/C car if CKT 450 to terminal "BC3" is open or shorted to B +.

1988–90 2.3L (VIN D AND A) – ALL MODELS

CODE 66
A/C PRESSURE SENSOR CIRCUIT
2.3L (VIN D & A) "N" CARLINE (PORT)

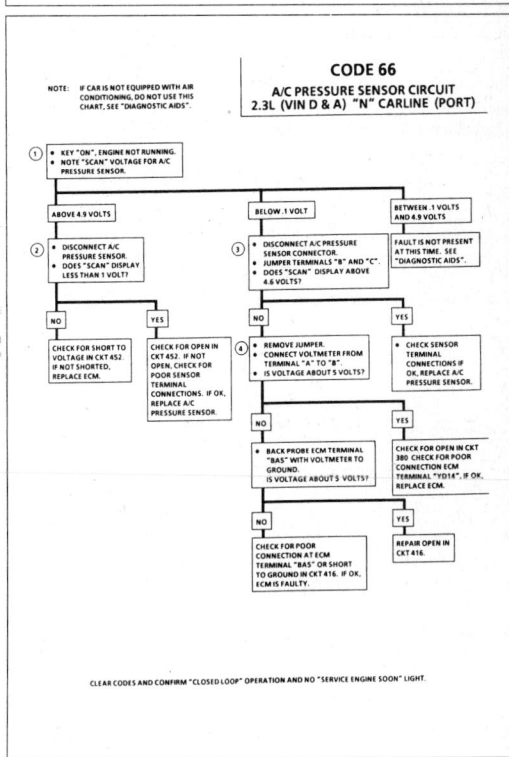

1988–90 2.3L (VIN D AND A) – ALL MODELS

SECTION B
SYMPTOMS

TABLE OF CONTENTS

Before Starting
Intermittents
Hard Start
Hesitation, Sag, Stumble
Surges and/or Chuggle
Lack of Power, Sluggish, or Spongy
Detonation/Spark Knock
Cuts Out, Misses
Backfire
Poor Fuel Economy
Dieseling, Run-On
Rough, Unstable, or Incorrect Idle, Stalling
Excessive Exhaust Emissions Or Odors
Chart B-1 - Restricted Exhaust System Check

BEFORE STARTING

Before using this section you should have performed the DIAGNOSTIC CIRCUIT CHECK and found out that:
1. The ECM and "Service Engine Soon" light are operating.
2. There are no trouble codes.

Verify the customer complaint, and locate the correct SYMPTOM below. Check the items indicated under that symptom.

If the ENGINE CRANKS BUT WILL NOT RUN, see CHART A-3.

Several of the following symptom procedures call for a careful visual check. This check should include:

- ECM grounds for being clean and tight
- Vacuum hoses for splits, kinks, and proper connections, as shown on Emission Control Information label.
- Air leaks at throttle body mounting and intake manifold.
- Wiring for proper connections, pinches, and cuts. The importance of this step cannot be stressed too strongly - it can lead to correcting a problem without further checks and can save valuable time.

1988–90 2.3L (VIN D AND A) – ALL MODELS

INTERMITTENTS
Problem may or may not turn "ON" the "Service Engine Soon" light, or store a code.

DO NOT use the trouble code charts in Section "A" for intermittent problems. The fault must be present to locate the problem. If a fault is intermittent, use of trouble code charts may result in replacement of good parts.

- Most intermittent problems are caused by faulty electrical connections or wiring. Perform careful check as described at start of Section "B". Check for:
  - Poor mating of the connector halves, or terminals not fully seated in the connector body (backed out).
  - Improperly formed or damaged terminals. All connector terminals in problem circuit should be carefully reformed to increase contact tension.
  - Poor terminal to wire connection. This requires removing the terminal from the connector body to check. See Introduction to Section "6E".
- If a visual check does not find the cause of the problem, the car can be driven with a voltmeter connected to a suspected circuit. A "Scan" tool can, also, be used for monitoring input signals to help detect intermittent conditions. An abnormal voltage, or "Scan" reading when the problem occurs, indicates the problem may be in that circuit. If the wiring and connectors check OK and a trouble code was stored for a circuit having a sensor, except for Codes 43, 44, and 45, substitute a known good sensor and recheck.

An intermittent "Service Engine Soon" light with no stored code may be caused by:
- Ignition coil shorted to ground.
- "Service Engine Soon" light wire to ECM shorted to ground. (CKT 419)
- Diagnostic "test" terminal wire to ECM, shorted to ground. (CKT 451)
- ECM power grounds. See ECM wiring diagrams.
- Loss of trouble code memory. To check, disconnect TPS and idle engine until "Service Engine Soon" light comes "ON". Code 22 should be stored, and kept in memory when ignition is turned "OFF". If not, the ECM is faulty.
- Check for an electrical system interference caused by a defective relay, ECM driven solenoid, or switch. They can cause a sharp electrical surge. Normally, the problem will occur when the faulty component is operated.
- Check for improper installation of electrical options, such as lights, 2-way radios, etc.
- EST wires should be kept away from generator. Wire from ECM to ignition system (CKT 453) should be a good connection.
- Check for open diode across A/C compressor clutch, and for other open diodes (see wiring diagrams).

HARD START
Definition: Engine cranks OK, but does not start for a long time. Does eventually run, or may start but immediately dies.

- Make sure driver is using correct starting procedure.
- Perform careful check as described at start of Section "B".
- CHECK:
  - TPS for sticking or binding or a high TPS voltage with the throttle closed (should read less than .7 volt).
  - High resistance in coolant sensor circuit or sensor itself. See CODE 15 chart or, with a "Scan" tool, compare coolant temperature with ambient temperature on a cold engine.
  - Fuel pump relay operation - pump should turn "ON" for 2 seconds when ignition is turned "ON". See CHART A-7 and CHART A-5.
  - Fuel Pressure, CHART A-7.
  - Water contaminated fuel.
  - Ignition system - Check for proper output with spark tester J-26792 or equivalent (ST-125).
  - IAC operation - See CHART C-2C.
- A faulty in-tank fuel pump check valve will allow the fuel in the lines to drain back to the tank after the engine is stopped. To check for this condition:
  - Perform fuel system diagnosis, CHART A-7.
- Remove spark plugs. Check for wet plugs, cracks, wear, improper gap, burned electrodes, or heavy deposits. Repair or replace as necessary.
- Throttle plate stop screw (minimum air rate) adjustment check.

1988–90 2.3L (VIN D AND A) – ALL MODELS

HESITATION, SAG, STUMBLE

Definition: Momentary lack of response as the accelerator is pushed down. Can occur at all car speeds. Usually most severe when first trying to make the car move, as from a stop sign. May cause the engine to stall if severe enough.

- Perform careful visual check as described at start of Section "B".
- CHECK:
  - Fuel Pressure. See CHART A-7. Also check for water contaminated fuel.
  - Spark plugs for being fouled or faulty wiring.
  - Mem-Cal number and Service Bulletins for latest Mem-Cal.
  - TPS for binding or sticking. Voltage should increase at a steady rate as throttle is moved toward WOT.
  - Generator output voltage. Repair, if less than 9 or more than 16 volts.
  - Ignition system ground, CKT 453.
  - Canister purge system for proper operation. See CHART C-3.
  - MAP sensor - See CHART C-10.
- Perform injector balance test, CHART C-2A.

SURGES AND/OR CHUGGLE

Definition: Engine power variation under steady throttle or cruise. Feels like the car speeds up and slows down with no change in the accelerator pedal.

- Be sure driver understands transmission/transaxle converter clutch and A/C compressor operation in owner's manual.
- Perform careful visual inspection as described at start of Section "B".
- To help determine if the condition is caused by a rich or lean system, the car should be driven at the speed of the complaint. Monitoring block learn at the complaint speed will help identify the cause of the problem. If the system is running lean (block learn near 198), refer to "Diagnostic Aids" on facing page of Code 44. If the system is running rich (block learn near 58), refer to "Diagnostic Aids" on facing page of Code 45.
- CHECK:
  - Generator output voltage. Repair if less than 9 or more than 16 volts.
  - Vacuum lines for kinks or leaks.
  - In-line fuel filter. Replace if dirty or plugged.
  - Fuel pressure while condition exists. See CHART A-7.
- Remove spark plugs. Check for cracks, wear, improper gap, burned electrodes, or heavy deposits. Check for proper output voltage using spark tester (ST-125) J-26792 or equivalent.

1988–90 2.3L (VIN D AND A) – ALL MODELS

LACK OF POWER, SLUGGISH, OR SPONGY

Definition: Engine delivers less than expected power. Little or no increase in speed when accelerator pedal is pushed down part way.

- Perform careful visual check as described at start of Section "B".
- Compare customer's car to similar unit. Make sure the customer's car has an actual problem.
- Remove air filter and check air filter for dirt, or for being plugged. Replace as necessary.
- CHECK:
  - Restricted fuel filter, contaminated fuel or improper fuel pressure. See CHART A-7.
  - ECM power grounds, see wiring diagrams.
  - Exhaust system for possible restriction: See CHART B-1.
    - Inspect exhaust system for damaged or collapsed pipes.
    - Inspect muffler for heat distress or possible internal failure.
  - Generator output voltage. Repair if less than 9 or more than 16 volts.
  - Engine valve timing and compression.
  - Engine for proper or worn camshaft.
  - Secondary voltage using a shop oscilloscope or a spark tester J-26792 (ST-125) or equivalent.
  - Check A/C operation. A/C clutch should cut out at WOT. See A/C CHART C-10.

DETONATION /SPARK KNOCK

Definition: A mild to severe ping, usually worse under acceleration. The engine makes sharp metallic knocks that change with throttle opening. Sounds like popcorn popping.

- Check for obvious overheating problems:
  - Low coolant.
  - Loose belt.
  - Restricted air flow to radiator, or restricted water flow through radiator.
  - Inoperative electric cooling fan circuit. See CHART C-12.
- To help determine if the condition is caused by a rich or lean system, the car should be driven at the speed of the complaint. Monitoring block learn, at the complaint speed, will help identify the cause of the problem. If the system is running lean (block learn near 198), refer to "Diagnostic Aids" on facing page of Code 44. If the system is running rich (block learn near 58), refer to facing page of Code 45.
- CHECK:
  - ESC system for no retard - See CHART C-5.
  - Park/Neutral switch. Be sure "Scan" indicates drive with gear selector in drive. See CHART C-1A.
  - TCC operation, TCC applying too soon - see CHART C-8.
  - Fuel system pressure - See CHART A-7.
  - Remove carbon with top engine cleaner. Follow instructions on can.
    - Check for correct Mem-Cal. (See service bulletins)
    - Check for excessive oil in combustion chamber.
    - Check for incorrect basic engine parts such as cam, heads, pistons, etc.
    - Check for poor fuel quality, proper octane rating.

1988-90 2.3L (VIN D AND A) – ALL MODELS

CUTS OUT, MISSES

Definition: Steady pulsation or jerking that follows engine speed, usually more pronounced as engine load increases. Not normally felt above 1500 rpm or 30 mph (48 km/h). The exhaust has a steady spitting sound at idle or low speed.

- Perform careful visual check as described at start of Section "B".
- Check for misfiring cylinder Refer to CHART C-4M.
- Check secondary voltage using a spark tester J-26792 (ST-125) or equivalent.
- Check spark plugs for cracks, broken insulator or closed gap.
- Check for open peak & hold jumper.
- Perform compression check on questionable cylinder(s) found above. If compression is low, repair as necessary.

- Perform the injector balance test. See CHART C-2A.
- Check for restricted fuel filter. Also check fuel tank for water.
- Check for low fuel pressure. See CHART A-7.
- Remove cam housing covers. Check for broken valve springs or worn camshaft lobes. Repair as necessary.
- Check for proper valve timing.

BACKFIRE

Definition: Fuel ignites in intake manifold, or in exhaust system, making a loud popping noise.

- CHECK:
  - Spark plugs.
  - Output voltage of ignition coils, using a spark tester J-26792 (ST-125), or equivalent.
  - Intermittent condition in ignition system.
  - Compression - Look for sticking or leaking valves.
  - Valve timing.
  - Intake manifold gasket vacuum leaks.

1988-90 2.3L (VIN D AND A) – ALL MODELS

POOR FUEL ECONOMY

Definition: Fuel economy, as measured by an actual road test, is noticeably lower than expected. Also, economy is noticeably lower than it was on this car at one time, as previously shown by an actual road test.

- CHECK:
  - Engine thermostat for faulty part (always open) or for wrong heat range. Using a "Scan" tool, monitor engine temperature. A "Scan" displays engine temperature in degrees Celsius. After engine is started, the temperature should rise steadily to about 90°C, then stabilize, when thermostat opens.
  - Fuel pressure. See CHART A-7.
- Check owner's driving habits.
  - Is A/C "ON" full time (Defroster mode "ON")?
  - Are tires at correct pressure?
  - Are excessively heavy loads being carried?
  - Is acceleration too much, too often?
  - Suggest driver read "Important Facts on Fuel Economy" in owner's manual.
- Check air cleaner element (filter) for dirt or being plugged.

- Check for proper calibration of speedometer.
- Visually (physically) check:
  - Vacuum hoses for splits, kinks, and proper connections as shown on Vehicle Emission Control Information label.
- Remove spark plugs. Check for cracks, wear, improper gap, burned electrodes, or heavy deposits. Repair or replace as necessary.
- Check compression.
- Check TCC for proper operation. See CHART C-8. A "Scan" should indicate an rpm drop, when the TCC is commanded "ON".
- Suggest owner fill fuel tank and recheck fuel economy.
- Check for exhaust system restriction. See CHART B-1.

DIESELING, RUN-ON

Definition: Engine continues to run after key is turned "OFF", but runs very roughly. If engine runs smoothly, check ignition switch and adjustment.

- Check injectors for leaking. See CHART A-7.

1988-90 2.3L (VIN D AND A) – ALL MODELS

ROUGH, UNSTABLE, OR INCORRECT IDLE, STALLING

Definition: The engine runs unevenly at idle. If bad enough, the car may shake. Also, the idle may vary in rpm (called "hunting"). Either condition may be bad enough to cause stalling. Engine idles at incorrect speed.

- Perform careful visual check as described at start of Section "B".
- Clean injectors.
- CHECK:
  - Throttle linkage for sticking or binding.
  - TPS for sticking or binding, and be sure "Scan" output is 0% and stable at idle.
  - IAC system. See CHART C-2C.
  - Generator output voltage. Repair if less than 9 or more than 16 volts.
  - P/N switch circuit. See CHART C-1A, or use "Scan" tool, and be sure tool indicates vehicle is in drive with gear selector in drive (125C).
  - Injector balance. See CHART C-2A.
  - Evaporative Emission Control System. CHART C-3.
  - Power steering pressure switch input. The state of the switch should only change when wheels are turned up against the stops. See CHART C-1E.
  - Minimum air rate (minimum throttle valve position check).
  - ECM ground circuits.

- Monitoring block learn values may help identify the cause of the problem. If the system is running lean (block learn near 198) refer to "Diagnostic Aids" on facing page of Code 44. If the system is running rich (block learn values near 58) refer to "Diagnostic Aids" on facing page of Code 45.
- Run a cylinder compression check. See Section "6".
- Check for fuel in pressure regulator hose. If present, replace regulator assembly.
- Check ignition system.
- If problem exists with A/C "ON", check A/C system operation CHART C-10.
- Inspect oxygen sensor for silicon contamination from fuel, or use of improper RTV sealant. The sensor will have a white, powdery coating, and will result in a high but false signal voltage (rich exhaust indication). The ECM will then reduce the amount of fuel delivered to the engine causing a severe driveability problem.

EXCESSIVE EXHAUST EMISSIONS OR ODORS

Definition: Vehicle fails an emission test. Vehicle has excessive "rotten egg" smell. Excessive odors do not necessarily indicate excessive emissions.

- Perform "Diagnostic Circuit Check."
- IF TEST SHOWS EXCESSIVE CO AND HC, (or also has excessive odors):
  - Check items which cause car to run RICH.
    - Make sure engine is at normal operating temperature.
- CHECK:
  - Fuel pressure. See CHART A-7.
  - Canister for fuel loading. See CHART C-3.
  - Injector balance. See CHART C-2A.
  - Spark plugs and ignition components.

  - Check for lead contamination of catalytic converter (look for removal of fuel filler neck restrictor).
  - Check for properly installed fuel cap.

- If the system is running rich, (block learn near 58), refer to "Diagnostic Aids" on facing page of Code 45.
- IF TEST SHOWS EXCESSIVE NOx:
  - Check items which cause car to run LEAN, or to run too hot.
    - Vacuum leaks.
    - Coolant system and coolant fan for proper operation.
    - Remove carbon with top engine cleaner. Follow instructions on can.
- If the system is running lean (block learn near 198), refer to "Diagnostic Aids" on facing page of Code 44.

1988-90 2.3L (VIN D AND A) – ALL MODELS

CHART B-1
RESTRICTED EXHAUST SYSTEM CHECK
2.3L VIN D & A

Proper diagnosis for a restricted exhaust system is essential before any components are replaced. The following procedure may be used for diagnosis.

CHECK AT O2 SENSOR:

1. Carefully remove O2 sensor.
2. Install Borroughs exhaust backpressure tester (BT 8515 or BT 8603) or equivalent in place of O2 sensor (see illustration).
3. After completing test described below, be sure to coat threads of O2 sensor with anti-seize compound P/N 5613695 or equivalent prior to re-installation.

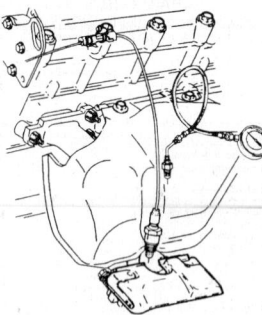

DIAGNOSIS:

1. With the engine idling at normal operating temperature, transaxle in park or neutral, observe the exhaust system backpressure reading on the gauge. Reading should not exceed 3.4 kPa (.5 psi).
2. Increase engine speed to 3000 rpm and observe gauge. Reading should not exceed 5 kPa (.75 psi).
3. If the backpressure at either speed exceeds specification, a restricted exhaust system is indicated.
4. Inspect the entire exhaust system for a collapsed pipe, heat distress, or possible internal muffler failure.
5. If there are no obvious reasons for the excessive backpressure, the catalytic converter is suspected to be restricted and should be replaced using current recommended procedures.

1988–90 2.3L (VIN D AND A) – ALL MODELS

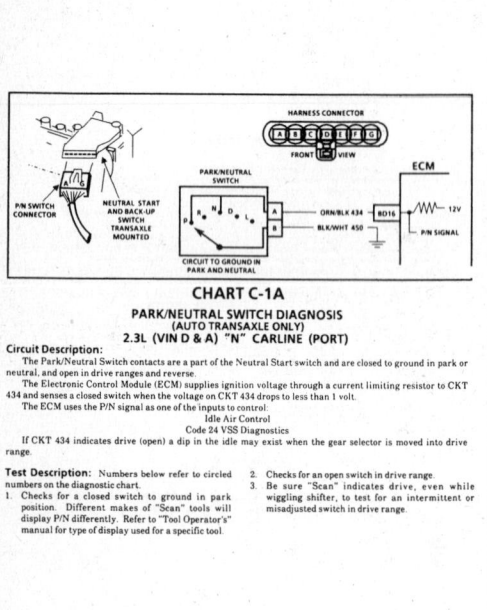

CHART C-1A
PARK/NEUTRAL SWITCH DIAGNOSIS
(AUTO TRANSAXLE ONLY)
2.3L (VIN D & A) "N" CARLINE (PORT)

Circuit Description:
The Park/Neutral Switch contacts are a part of the Neutral Start switch and are closed to ground in park or neutral, and open in drive ranges and reverse.
The Electronic Control Module (ECM) supplies ignition voltage through a current limiting resistor to CKT 434 and senses a closed switch when the voltage on CKT 434 drops to less than 1 volt.
The ECM uses the P/N signal as one of the inputs to control:
    Idle Air Control
    Code 24 VSS Diagnostics
If CKT 434 indicates drive (open) a dip in the idle may exist when the gear selector is moved into drive range.

Test Description: Numbers below refer to circled numbers on the diagnostic chart.
1. Checks for a closed switch to ground in park position. Different makes of "Scan" tools will display P/N differently. Refer to "Tool Operator's" manual for type of display used for a specific tool.
2. Checks for an open switch in drive range.
3. Be sure "Scan" indicates drive, even while wiggling shifter, to test for an intermittent or misadjusted switch in drive range.

1988–90 2.3L (VIN D AND A) – ALL MODELS

CHART C-1A
PARK/NEUTRAL SWITCH DIAGNOSIS
(AUTO TRANSAXLE ONLY)
2.3L (VIN D & A) "N" CARLINE (PORT)

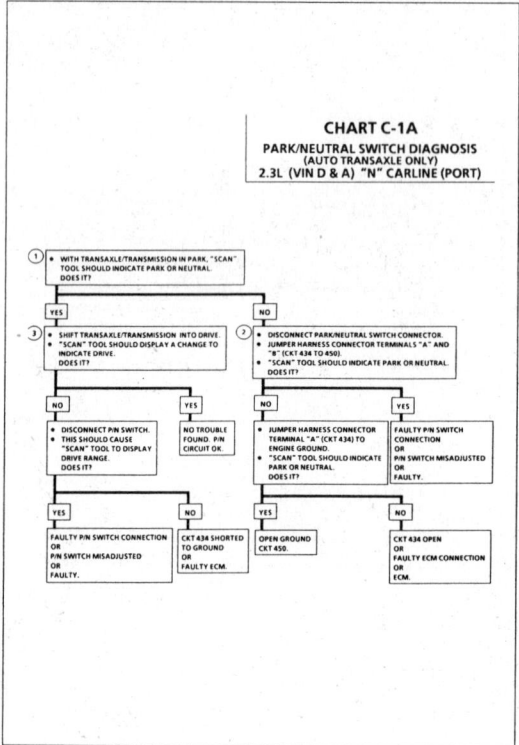

1988–90 2.3L (VIN D AND A) – ALL MODELS

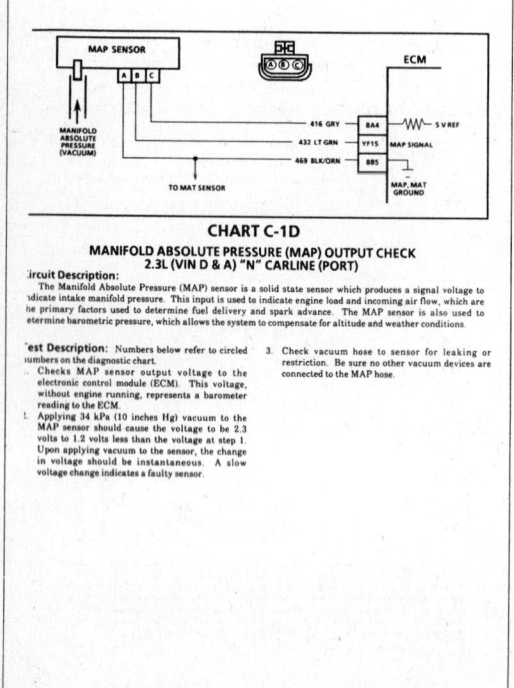

CHART C-1D
MANIFOLD ABSOLUTE PRESSURE (MAP) OUTPUT CHECK
2.3L (VIN D & A) "N" CARLINE (PORT)

Circuit Description:
The Manifold Absolute Pressure (MAP) sensor is a solid state sensor which produces a signal voltage to indicate intake manifold pressure. This input is used to indicate engine load and incoming air flow, which are the primary factors used to determine fuel delivery and spark advance. The MAP sensor is also used to determine barometric pressure, which allows the system to compensate for altitude and weather conditions.

Test Description: Numbers below refer to circled numbers on the diagnostic chart.
1. Checks MAP sensor output voltage to the electronic control module (ECM). This voltage, without engine running, represents a barometer reading to the ECM.
2. Applying 34 kPa (10 inches Hg) vacuum to the MAP sensor should cause the voltage to be 2.3 volts to 1.2 volts less than the voltage at step 1. Upon applying vacuum to the sensor, the change in voltage should be instantaneous. A slow voltage change indicates a faulty sensor.
3. Check vacuum hose to sensor for leaking or restriction. Be sure no other vacuum devices are connected to the MAP hose.

1988–90 2.3L (VIN D AND A) – ALL MODELS

CHART C-1D
MANIFOLD ABSOLUTE PRESSURE (MAP) OUTPUT CHECK
2.3L (VIN D & A) "N" CARLINE (PORT)

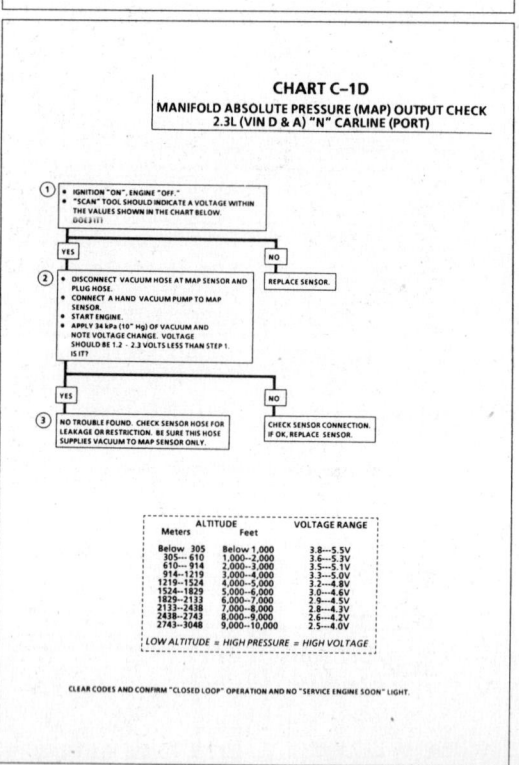

ALTITUDE		VOLTAGE RANGE
Meters	Feet	
Below 305	Below 1,000	3.8—5.5V
305—610	1,000—2,000	3.6—5.3V
610—914	2,000—3,000	3.5—5.1V
914—1219	3,000—4,000	3.3—5.0V
1219—1524	4,000—5,000	3.2—4.8V
1524—1829	5,000—6,000	3.0—4.6V
1829—2133	6,000—7,000	2.9—4.5V
2133—2438	7,000—8,000	2.8—4.3V
2438—2743	8,000—9,000	2.6—4.2V
2743—3048	9,000—10,000	2.5—4.0V

LOW ALTITUDE = HIGH PRESSURE = HIGH VOLTAGE

CLEAR CODES AND CONFIRM "CLOSED LOOP" OPERATION AND NO "SERVICE ENGINE SOON" LIGHT.

1988–90 2.3L (VIN D AND A) – ALL MODELS

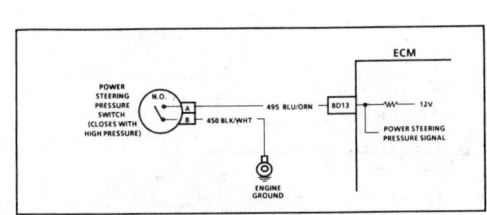

CHART C-1E
POWER STEERING PRESSURE SWITCH (PSPS) DIAGNOSIS
2.3L (VIN D & A) "N" CARLINE (PORT)

Circuit Description:
The Power Steering Pressure Switch (PSPS) is normally open to ground, and CKT 495 will be near battery voltage.
Turning the steering wheel increases power steering oil pressure and its load on an idling engine. The pressure switch will close before the load can cause an idle problem.
Closing the switch causes CKT 495 to read less than 1 volt. The Electronic Control Module (ECM) will increase the idle air rate and disengage the A/C relay.
- A pressure switch that will not close, or an open CKT 495 or 450, may cause the engine to stop when power steering loads are high.
- A switch that will not open, or a CKT 495 shorted to ground, may affect idle quality and will cause the A/C relay to be de-energized.

Test Description: Numbers below refer to circled numbers on the diagnostic chart.
1. Different makes of "Scan" tools may display the state of this switch in different ways. Refer to "Scan" tool operator's manual to determine how this input is indicated.
2. Checks to determine if CKT 495 is shorted to ground.
3. This should simulate a closed switch.

1988–90 2.3L (VIN D AND A) – ALL MODELS

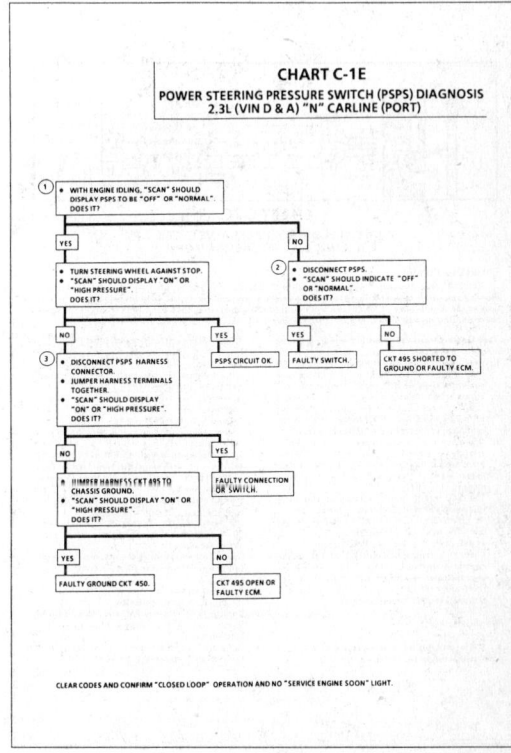

1988–90 2.3L (VIN D AND A) – ALL MODELS

CHART C-2A
INJECTOR BALANCE TEST
2.3L (VIN D & A) "N" CARLINE (PORT)

The injector balance tester is a tool used to turn the injector on for a precise amount of time, thus spraying a measured amount of fuel into the manifold. This causes a drop in fuel rail pressure that we can record and compare between each injector. All injectors should have the same amount of pressure drop (± 10 kpa). Any injector with a pressure drop that is 10 kpa (or more) greater or less than the average drop of the other injectors should be considered faulty and replaced.

STEP 1
Engine "cool down" period (10 minutes) is necessary to avoid irregular readings due to "Hot Soak" fuel boiling. Relieve fuel pressure in the fuel rail using the "fuel pressure relief procedure" described previously in this section. With ignition "OFF" connect fuel gauge J347301 or equivalent to fuel pressure tap.
Disconnect harness connectors at all injectors, and connect injector tester J-34730-3, or equivalent, to one injector. On Turbo equipped engines, use adaptor harness furnished with injector tester to energize injectors that are not accessible. Follow manufacturers instructions for use of adaptor harness. Ignition must be "OFF" at least 10 seconds to complete ECM shutdown cycle. Fuel pump should run about 2 seconds after ignition is turned "ON". At this point, insert clear tubing attached to vent valve into a suitable container and bleed air from gauge and hose to insure accurate gauge operation. Repeat this step until all air is bled from gauge.

STEP 2
Turn ignition "OFF" for 10 seconds and then "ON" again to get fuel pressure to its maximum. Record this initial pressure reading. Energize tester one time and note pressure drop at its lowest point. (Disregard any slight pressure increase after drop hits low point.) By subtracting this second pressure reading from the initial pressure, we have the actual amount of injector pressure drop.

STEP 3
Repeat step 2 on each injector and compare the amount of drop. Usually, good injectors will have virtually the same drop. Retest any injector that has a pressure difference of 10kPa, either more or less than the average of the other injectors on the engine. Replace any injector that also fails the retest. If the pressure drop of all injectors is within 10kPa of this average, the injectors appear to be flowing properly. Reconnect them and review "Symptoms," Section "B".

NOTE: The entire test should not be repeated more than once without running the engine to prevent flooding. (This includes any retest on faulty injectors.)

1988–90 2.3L (VIN D AND A) – ALL MODELS

NOTE: If injectors are suspected of being dirty, they should be cleaned using an approved tool and procedure prior to performing this test. The fuel pressure test in Section "A", CHART A-7, should be completed prior to this test.

CHART C-2A
INJECTOR BALANCE TEST
2.3L (VIN D & A) "N" CARLINE (PORT)

Step 1. If engine is at operating temperature, allow a 10 minute "cool down" period then connect fuel pressure gauge and injector tester.
1. Ignition "OFF".
2. Connect fuel pressure gauge and injector tester.
3. Ignition "ON".
4. Bleed off all air in gauge. Repeat until all air is bled from gauge.

Step 2. Run test:
1. Ignition "OFF" for 10 seconds.
2. Ignition "ON". Record gauge pressure. (Pressure must hold steady, if not see the Fuel System Diagnosis, CHART A-7, in Section "A").
3. Turn injector "ON", by depressing button on injector tester, and note pressure at the instant the gauge needle stops.

Step 3.
1. Repeat step 2 on all injectors and record pressure drop on each. Retest injectors that appear faulty (any injectors that have a 10 kPa difference, either more or less, in pressure from the average. If no problem is found, review Symptoms Section "B".

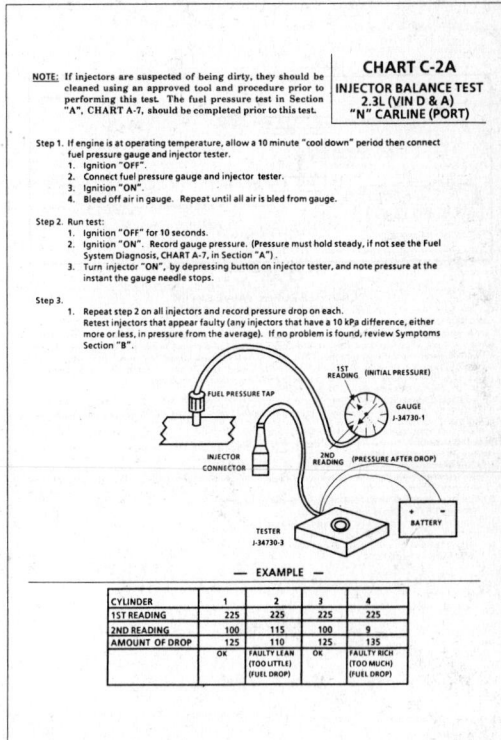

— EXAMPLE —

CYLINDER	1	2	3	4
1ST READING	225	225	225	225
2ND READING	100	115	100	9
AMOUNT OF DROP	125	110	125	135
	OK	FAULTY LEAN (TOO LITTLE) (FUEL DROP)	OK	FAULTY RICH (TOO MUCH) (FUEL DROP)

1988–90 2.3L (VIN D AND A) – ALL MODELS

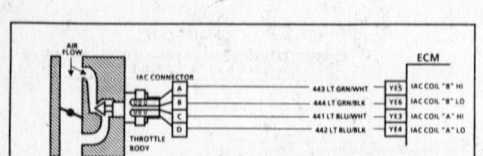

CHART C-2C
IDLE AIR CONTROL (IAC) VALVE CHECK
2.3L (VIN D & A) "N" CARLINE (PORT)

Circuit Description:

The ECM controls idle rpm with the IAC valve. To increase idle rpm, the ECM moves the IAC valve out, allowing more air to bypass the throttle plate. To decrease rpm, it moves the IAC valve in, reducing air flow by-passing the throttle plate. A "Scan" tool will read the ECM commands to the IAC valve in counts. The higher the counts, the more air allowed (higher idle). The lower the counts, the less air allowed (lower idle).

Test Description: Numbers below refer to circled numbers on the diagnostic chart.
1. Continue with test, even if engine will not idle. If idle is to low, "Scan" will display 80 or more counts, or steps. If idle is high, it will display "5" counts or less. Occasionally an erratic or unstable idle may occur. Engine speed may vary 200 rpm or more up and down. Disconnect IAC. If the condition is unchanged, the IAC is not at fault.
2. When the engine was stopped, the IAC Valve retracted (more air) to a fixed "Park" position for increased air flow and idle speed during the next engine start. A "Scan" will display 100 or more counts.
3. The IAC valve should extend as the ECM commands idle speed to decrease to desired rpm.
4. Be sure to disconnect the IAC valve prior to this test. The test light will confirm the ECM signals by a steady or flashing light on all circuits.
5. There is a remote possibility that one of the circuits is shorted to voltage, which would have been indicated by a steady light. Disconnect the ECM and turn the ignition "ON" and probe terminals to check for this condition.

Diagnostic Aids:

A slow unstable idle may be caused by a system problem that cannot be overcome by the IAC.

"Scan" counts will be above 60 counts, if idle is too low, and "5" counts or less, if idle is too high. If idle speed is excessively high, check for and correct any trouble code problem or vacuum leak.
- **System too lean (High Air/Fuel Ratio)**
  Idle speed may be too high or too low. Engine speed may vary up and down, disconnecting IAC does not help. May set Code 44.
  "Scan" and/or voltmeter will read an oxygen sensor output less than 300 mV (.3 volt). Check for low regulated fuel pressure or water in fuel. A lean exhaust, with an oxygen sensor output fixed above 800 mV (.8 volt), will probably be caused by a contaminated sensor (usually silicone). This may also set a Code 45.
- **System too rich (Low Air/Fuel Ratio)**
  Idle speed too low. "Scan" counts usually above 80. System obviously rich and may exhibit black smoke exhaust.
  "Scan" tool and/or voltmeter will read an oxygen sensor signal fixed above 800 mV (.8volt).
  Check:
  - High fuel pressure
  - Injector leaking or sticking
- **Throttle Body** - Remove IAC and inspect bore for foreign material or evidence of IAC valve seat or pintle damage.
- Refer to "Rough, Unstable, Incorrect Idle or Stalling" in "Symptoms" in Section "B".

1988–90 2.3L (VIN D AND A) – ALL MODELS

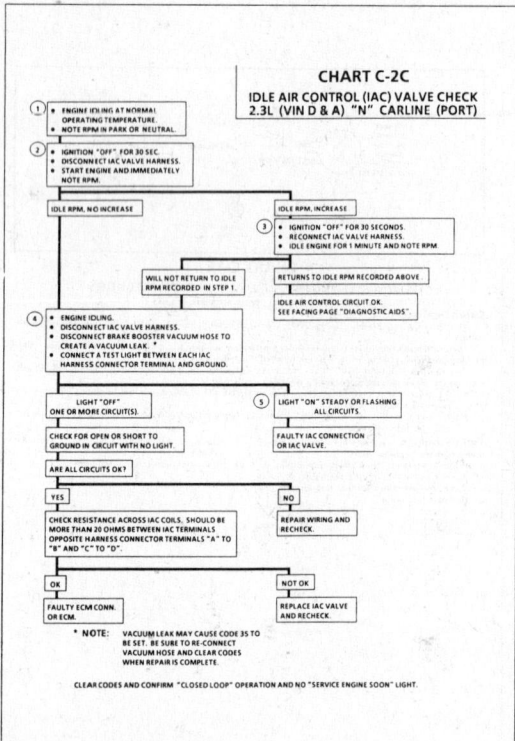

CHART C-2C
IDLE AIR CONTROL (IAC) VALVE CHECK
2.3L (VIN D & A) "N" CARLINE (PORT)

* NOTE: VACUUM LEAK MAY CAUSE CODE 35 TO BE SET. BE SURE TO RE-CONNECT VACUUM HOSE AND CLEAR CODES WHEN REPAIR IS COMPLETE.

CLEAR CODES AND CONFIRM "CLOSED LOOP" OPERATION AND NO "SERVICE ENGINE SOON" LIGHT.

1988–90 2.3L (VIN D AND A) – ALL MODELS

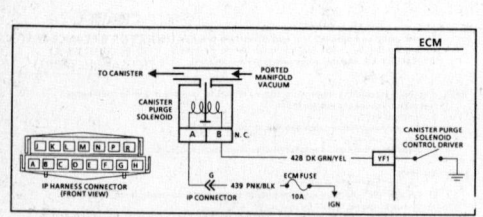

CHART C-3
CANISTER PURGE VALVE CHECK
2.3L (VIN D & A) "N" CARLINE (PORT)

Circuit Description:

Canister purge is controlled by a solenoid that allows manifold and/or ported vacuum to purge the canister when energized. The Electronic Control Module (ECM) supplies a ground to energize the solenoid (purge "ON"). The purge solenoid control by the ECM is pulse width modulated (turned "ON" and "OFF" several times a second). The duty cycle (pulse width) is determined by "Closed Loop" feed back from the O₂ sensor. The duty cycle is calculated by the ECM and the output commanded when the following conditions have been met.
- Engine run time after start more than 65 seconds.
- Coolant temperature above 56°C.

Also, if the diagnostic test terminal is grounded with the engine stopped, the purge solenoid is energized (purge "ON").

Test Description: Numbers below refer to circled numbers on the diagnostic chart.
1. Checks to see if the solenoid is opened or closed. The solenoid is normally de-energized in this step, so it should be closed.
2. Checks to determine if solenoid was open due to electrical CKT problem or defective solenoid.
3. Completes functional check by grounding test terminal. This should normally energize the solenoid opening the valve which should allow the vacuum to drop (purge "ON").

Diagnostic Aids:

Make a visual check of vacuum hose(s). Check throttle body for possible cracked, broken, or plugged vacuum block. Check engine for possible mechanical problem.

1988–90 2.3L (VIN D AND A) – ALL MODELS

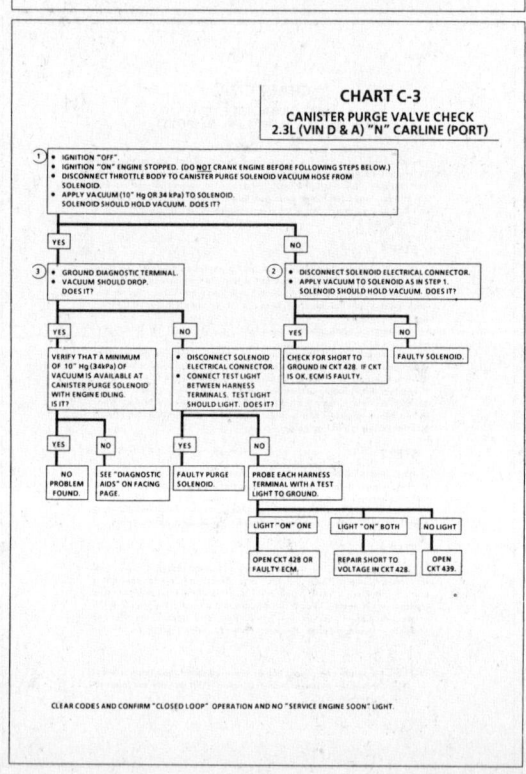

CHART C-3
CANISTER PURGE VALVE CHECK
2.3L (VIN D & A) "N" CARLINE (PORT)

CLEAR CODES AND CONFIRM "CLOSED LOOP" OPERATION AND NO "SERVICE ENGINE SOON" LIGHT.

1988–90 2.3L (VIN D AND A) – ALL MODELS

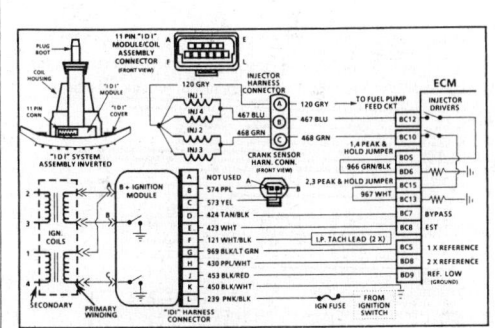

CODE C-4M
INTEGRATED DIRECT IGNITION (IDI) MISFIRE DIAGNOSIS
2.3L (VIN D & A) "N" CARLINE (PORT)

Circuit Description:

The "Integrated Direct Ignition" system (IDI) uses a waste spark method of distribution. In this type of system the ignition module triggers the #1-4 coil pair resulting in both #1 and #4 spark plugs firing at the same time. #1 cylinder is on the compression stroke at the same time #4 is on the exhaust stroke, resulting in a lower energy requirement to fire #4 spark plug. This leaves the remainder of the high voltage to be used to fire #1 spark plug. On this application, the crank sensor is mounted to, and protrudes through the block to within approximately 0.050" of the crankshaft reluctor. Since the reluctor is a machined portion of the crankshaft and the sensor is mounted in a fixed position on the block, timing adjustments are not possible or necessary.

Test Description: Numbers below refer to circled numbers on the diagnostic chart.
1. This checks for equal relative power output between the cylinders. Any injector which when disconnected did not result in an rpm drop approximately equal to the others, is located on the misfiring cylinder.
2. If a plug boot is burned, the other plug on that coil may still fire at idle. This step tests the system's ability to produce at least 25,000 volts at each spark plug.
3. No spark, on one coil, may be caused by an open secondary circuit. Therefore, the coil's secondary resistance should be checked. Resistance readings above 20,000 ohms, but not infinite, will probably not cause a no start but may cause an engine miss under certain conditions.
4. If the no spark condition is caused by coil connections, a coil or a secondary boot assembly, the test light will blink. If the light does not blink, the fault is module connections or the module.
5. Checks for ignition voltage feed to injector and for an open injector driver circuit.
6. An injector driver circuit shorted to ground would result in the test light "ON" steady, and possibly a flooded condition which could damage engine. A shorted injector (less than 2 ohms) could cause incorrect ECM operation.

Diagnostic Aid:

Verify IDI connector terminal "K", CKT 450 resistance to ground is less than .5 ohm. A shorted or low resistance injector may cause a miss in the other injector in that pair (1 & 4 or 2 & 3).

1988–90 2.3L (VIN D AND A) – ALL MODELS

CHART C-4M
INTEGRATED DIRECT IGNITION (IDI) MISFIRE DIAGNOSIS
2.3L (VIN D & A) "N" CARLINE (PORT)

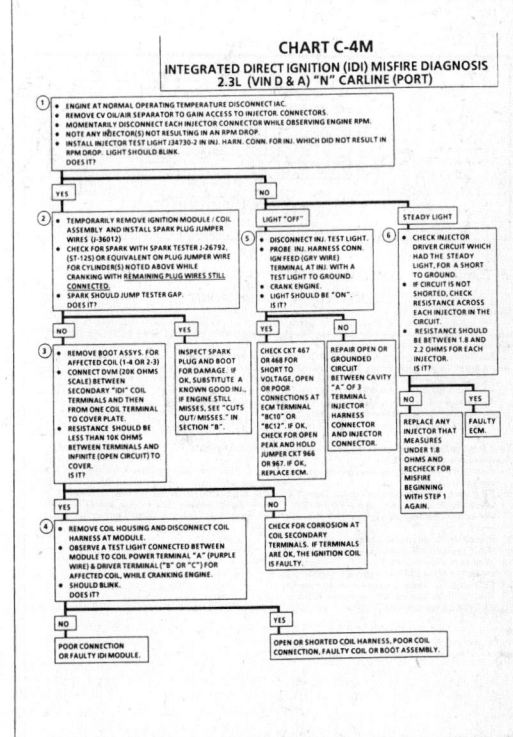

1988–90 2.3L (VIN D AND A) – ALL MODELS

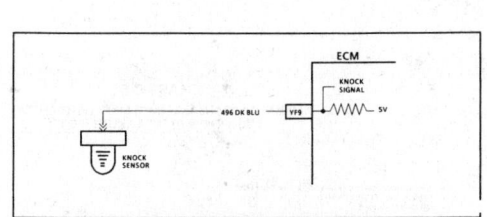

CHART C-5
ELECTRONIC SPARK CONTROL (ESC) SYSTEM CHECK
2.3L (VIN D & A) "N" CARLINE (PORT)

Circuit Description:

The knock sensor is used to detect engine detonation and the ECM will retard the electronic spark timing based on the signal being received. The circuitry within the knock sensor causes the ECM's 5 volts to be pulled down so that CKT 496 would measure about 2.5 volts. The knock sensor produces an AC signal which rides on the 2.5 volts DC signal. The amplitude and frequency are dependent upon the knock level.

The Mem-Cal used with this engine contains the functions which were part of remotely mounted ESC modules used on other GM vehicles. The ESC portion of the Mem-Cal then sends a signal to other parts of the ECM which adjusts the spark timing to retard the spark and reduce the detonation.

Test Description: Numbers below refer to circled numbers on the diagnostic chart.
1. With engine idling, there should not be a knock signal present at the ECM because detonation is not likely under a no load condition.
2. Tapping on the engine lift hood bracket should simulate a knock signal to determine if the sensor is capable of detecting detonation. If no knock is detected, try tapping on engine block closer to sensor before replacing sensor.
3. If the engine has an internal problem which is creating a knock, the knock sensor may be responding to the internal failure.
4. This test determines if the knock sensor is faulty or if the ESC portion of the Mem-Cal is faulty. If it is determined that the Mem-Cal is faulty, be sure that it is properly installed and latched into place. If not properly installed, repair and retest.

Diagnostic Aids:

While observing knock signal on the "Scan," there should be an indication that knock is present when detonation can be heard. Detonation is most likely to occur under high engine load conditions.

CHART C-5
ELECTRONIC SPARK CONTROL (ESC) SYSTEM CHECK
2.3L (VIN D & A) "N" CARLINE (PORT)

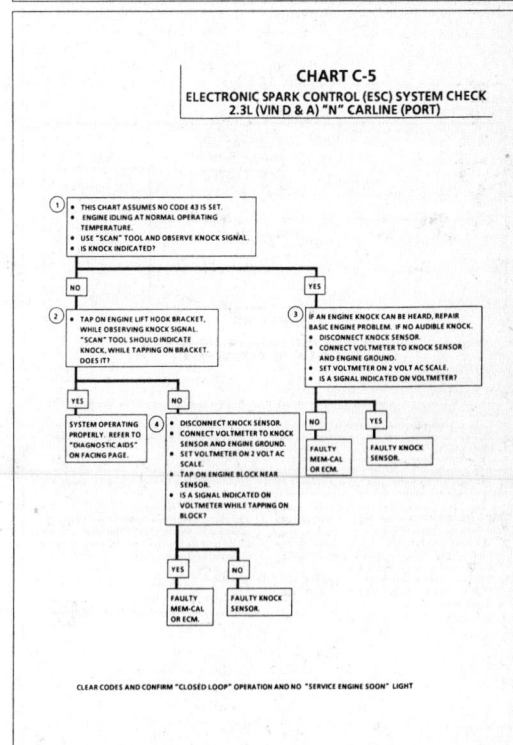

1988–90 2.3L (VIN D AND A) – ALL MODELS

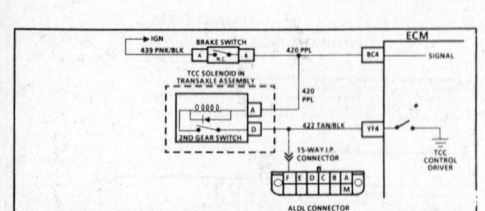

CHART C-8A
125C TORQUE CONVERTER CLUTCH (TCC)
(ELECTRICAL DIAGNOSIS)
2.3L (VIN D) "N" CARLINE (PORT)

Circuit Description:
The purpose of the torque converter clutch feature is to eliminate the power loss of the transaxle converter stage when the vehicle is in a cruise condition. This allows the convenience of the automatic transaxle and the fuel economy of a manual transaxle.

Fused battery ignition is supplied to the TCC solenoid through the brake switch, and transaxle third gear apply switch. The ECM will engage TCC by grounding CKT 422 to energize the solenoid.

TCC will engage when:
- Vehicle speed above a calibrated value (about 34 mph) (55 km/h).
- Throttle position sensor output not changing, indicating a steady road speed.
- Transmission second gear switch closed.
- Brake switch closed.

Test Description: Numbers below refer to circled numbers on the diagnostic chart.
1. Light "OFF" confirms transmission second gear apply switch is open.
2. By 25 mph, the transmission second gear TCC switch should close. Test light will come "ON" and confirm battery supply and closed brake switch.
3. Grounding the diagnostic terminal with ignition "ON," engine "OFF," should energize the TCC solenoid by grounding CKT 422. This test checks the ability of the ECM to supply a ground to the TCC solenoid. The test light connected from 12 volts to ALDL terminal "F" will turn "ON" as CKT 422 is grounded.

Diagnostic Aids:
A "Scan" tool only indicates when the ECM has turned "ON" the TCC driver and this does not confirm that the TCC has engaged. To determine if TCC is functioning properly, engine rpm should decrease when the "Scan" indicates the TCC driver has turned "ON."

1988–90 2.3L (VIN D AND A) – ALL MODELS

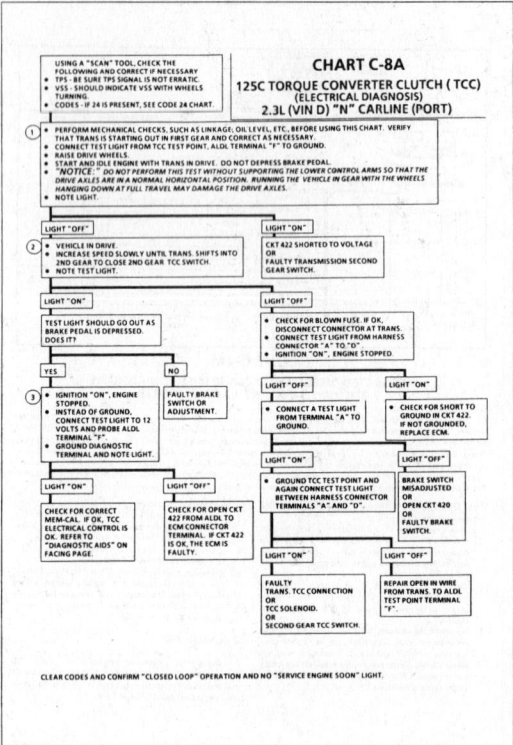

1988–90 2.3L (VIN D AND A) – ALL MODELS

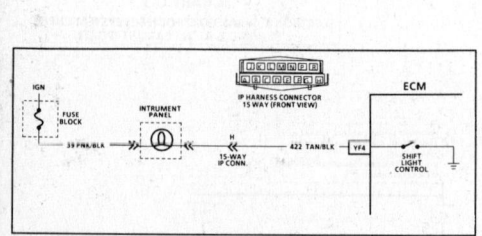

CHART C-8C
MANUAL TRANSAXLE (M/T) SHIFT LIGHT CHECK
2.3L (VIN D & A) "N" CARLINE (PORT)

Circuit Description:
The shift light indicates the best transaxle shift point for maximum fuel economy. The light is controlled by the ECM and is turned "ON" by grounding CKT 422.

The ECM uses inputs from various sensors to calculate when the shift light should be turned "ON" as follows:
- Coolant temperature must be above -10°C (14°F).
- ECM can determine the transaxle has been in a gear for at least 1.2 seconds, by comparison of vehicle speed (from VSS) with engine rpm.
- TPS is between minimum and maximum calibrated values for each gear.
- Rpm is above a calibrated value for each gear (maximum 6500 rpm).

The light will be turned "ON" after a calibrated delay time which is dependent on last gear change or downshift, and will remain on for a maximum of 10 seconds.

Test Description: Numbers below refer to circled numbers on the diagnostic chart.
1. This should not turn "ON" the shift light. If the light is "ON," there is a short to ground in CKT 422 wiring or a fault in the ECM.
2. When the diagnostic terminal is grounded, the ECM should ground CKT 456 and the shift light should come "ON."
3. This checks the shift light circuit up to the ECM connector. If the shift light illuminates, then the ECM connector is faulty or the ECM does not have the ability to ground the circuit.

Diagnostic Aids:
Check for Code 24 (no VSS). Faulty thermostat or incorrect heat range. Incorrect or faulty PROM.

1988–90 2.3L (VIN D AND A) – ALL MODELS

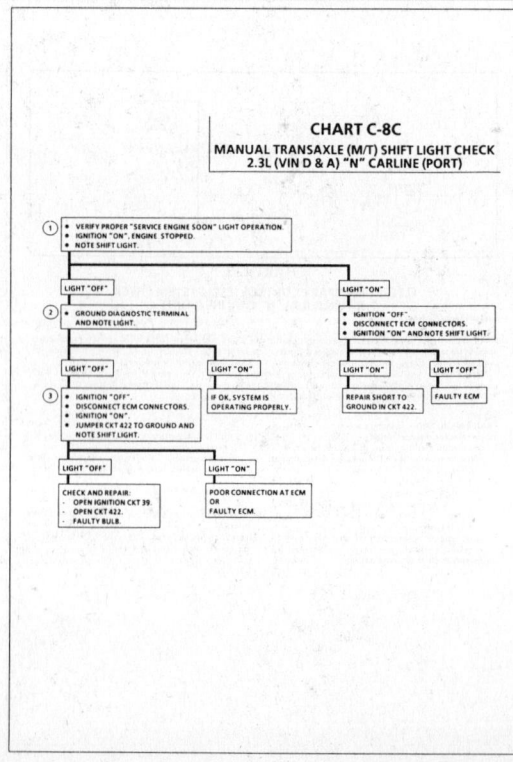

CHART C-8C
MANUAL TRANSAXLE (M/T) SHIFT LIGHT CHECK
2.3L (VIN D & A) "N" CARLINE (PORT)

1988-90 2.3L (VIN D AND A) - ALL MODELS

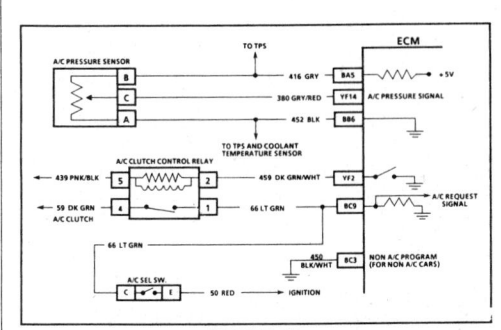

CHART C-10
A/C CLUTCH CONTROL CIRCUIT DIAGNOSIS
2.3L (VIN D & A) "N" CARLINE (PORT)

Circuit Description:
The A/C clutch control relay is energized when the ECM provides a ground path through CKT 459 and A/C is requested. A/C clutch is delayed about .3 seconds after A/C is requested. This will allow the IAC to adjust engine rpm for the additional load.
The ECM will temporarily disengage the A/C clutch relay for calibrated times for one or more of the following:
- Hot engine restart.
- Wide open throttle (TPS over 90%).
- Power steering pressure high (open power steering pressure switch).
- Engine rpm greater than about 6000 rpm.
- During IAC reset.

The A/C clutch relay will remain disengaged when a Code 66 is present, if pressure is out of range described previously in this section, or there is no A/C request signal due to an open A/C select switch or circuit. Refer to Section "1B" for more information on A/C refrigerant systems.

Test Description: Numbers below refer to circled numbers on the diagnostic chart.
1. The ECM will only energize the A/C relay when the engine is running. This test will determine if the relay or CKT 459 is faulty.
2. Determines if the signal is reaching the ECM through CKT 66 from the A/C control panel. Signal should only be present when in A/C mode or defrost mode has been selected.
3. If the ECM is receiving a high power steering pressure signal, the A/C clutch will be disengaged by the ECM.
4. With the engine stopped and diagnostic test terminal grounded, the ECM should be grounding CKT 459, which should cause the test light to be "ON".

Diagnostic Aids:
If complaint is insufficient cooling, the problem may be caused by an inoperative cooling fan. See CHART C-12 for cooling fan diagnosis. If fan operates correctly, see A/C diagnosis in Section "1B". A/C pressure outside of a range of 43 to 428 psi will cause the compressor to be disabled by the ECM. Observe "Scan" A/C pressure for 2 minutes with engine idling and A/C "ON". If pressure goes out of range refer to Section "1B" to measure and diagnose. "Scan" pressure should be within 20 psi of actual. If not, check for a circuit problem using Code 66 chart or replace sensor.

1988-90 2.3L (VIN D AND A) - ALL MODELS

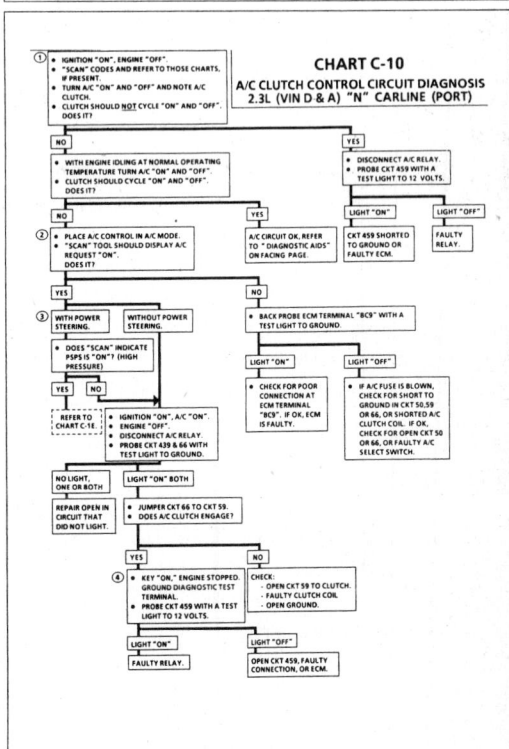

CHART C-10
A/C CLUTCH CONTROL CIRCUIT DIAGNOSIS
2.3L (VIN D & A) "N" CARLINE (PORT)

1988-90 2.3L (VIN D AND A) - ALL MODELS

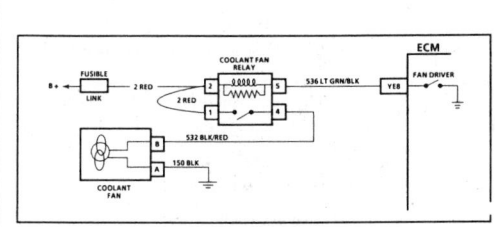

CHART C-12
COOLANT FAN FUNCTIONAL CHECK
2.3L (VIN D & A) "N" CARLINE (PORT)

Circuit Description:
The electric coolant fan is controlled by the ECM through the fan relay based on inputs from the coolant and manifold air temperature sensors, the A/C control switch, A/C pressure sensor and the vehicle speed sensor. The ECM controls the coolant fan by grounding CKT 536 which turns "ON" the fan relay.
The fan relay will be commanded "ON" when.
- Coolant temperature 103°C - 106°C or more.
- A/C clutch requested.
- Vehicle speed is less than 35 mph.
The fan relay will be commanded "ON" regardless of vehicle speed when:
- Code 14 or 15 are set.
- Coolant temperature 115°C - 118°C or more.
- A/C pressure is high.
The coolant fan may be commanded "ON" when the engine is not running under fan "Run-On" conditions described previously in this section.

Test Description: Numbers below refer to circled numbers on the diagnostic chart.
1. With the diagnostic terminal grounded, the coolant fan control driver should close, which should energize the fan control relay.
2. Test to see if fault is in wiring to the fan or fan/fan connection.

Diagnostic Aids:
If the owner complained of an overheating problem, it must be determined if the complaint was due to an actual boil over, or the "hot light," or temperature gage indicated overheating.

If the gage, or light, indicates overheating, but no boil over is detected, the gage or light circuit should be checked. The gage accuracy should also be checked by comparing the coolant sensor reading using a "Scan" tool with the gage reading.
If the engine is actually overheating, and the gage indicates overheating, but the coolant fan is not coming "ON", the coolant sensor has probably shifted out of calibration and should be replaced. See Code 15 chart for a temperature to resistance chart.
If the engine is overheating, and the coolant fan is "ON", the cooling system should be checked.

1988-90 2.3L (VIN D AND A) - ALL MODELS

NOTE: CHECK FOR CODE(S) STARTING WITH A COMPLETE DIAGNOSTIC CIRCUIT CHECK AND REPAIR AS NECESSARY BEFORE USING THIS CHART.

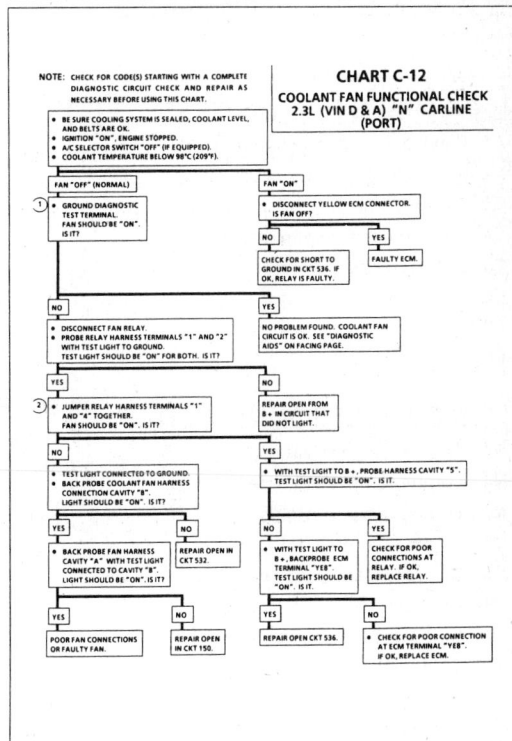

CHART C-12
COOLANT FAN FUNCTIONAL CHECK
2.3L (VIN D & A) "N" CARLINE (PORT)

2.8L (VIN S) COMPONENT LOCATIONS CAMARO AND FIREBIRD

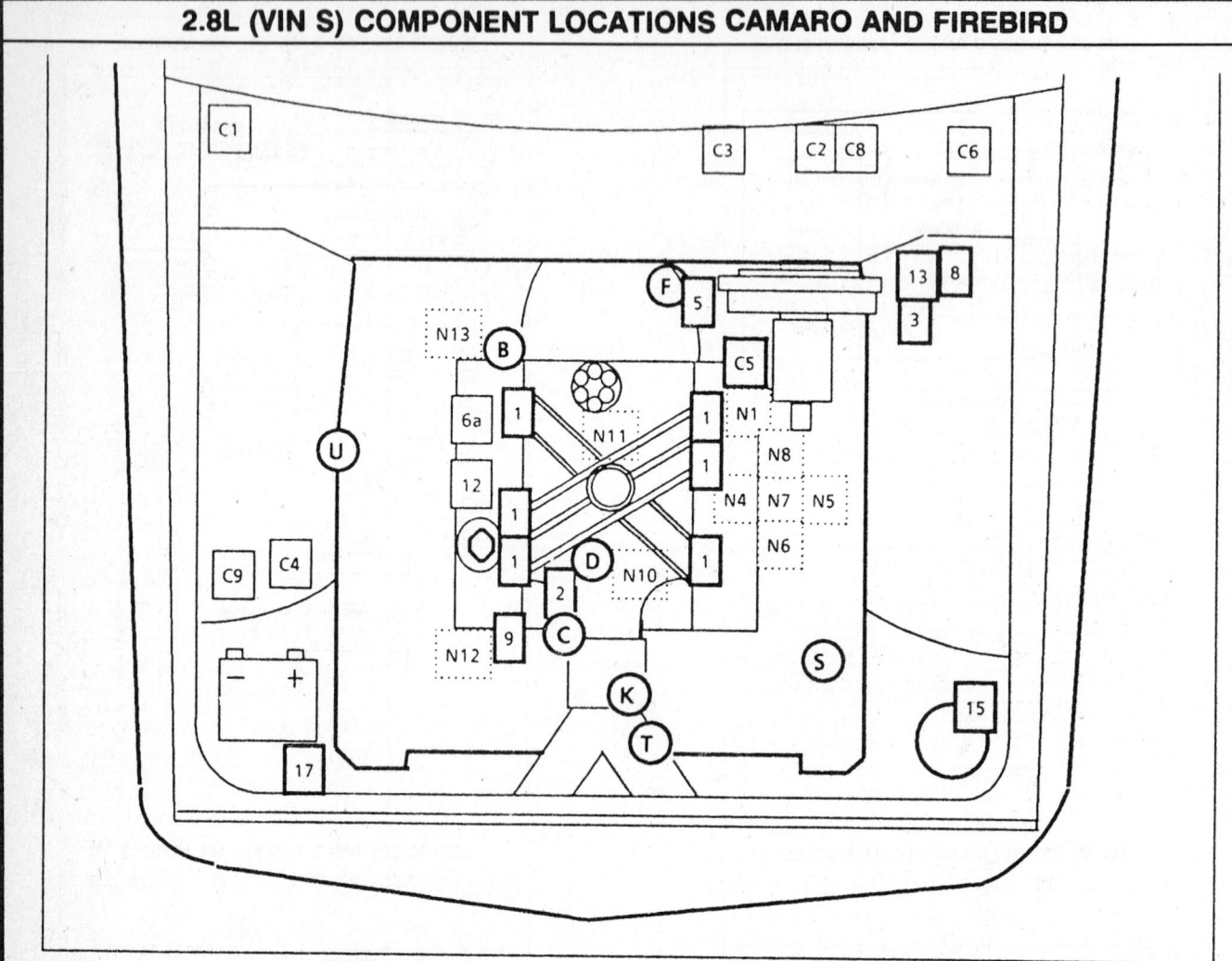

☐ **COMPUTER HARNESS**

C1 Electronic Control Module (ECM)
C2 ALDL diagnostic connector
C3 "SERVICE ENGINE SOON" light
C4 ECM power/fuel pump fuse
C5 ECM harness ground
C6 Fuse panel
C8 Fuel pump test connector (ALDL "G")
C9 MAF fuse

┆ **NOT ECM CONNECTED**

N1 Crankcase vent valve (PCV)
N4 Engine temp. switch (telltale)
N5 Engine temp. sensor (gage)
N6 Oil press. switch (telltale)
N7 Oil press. sensor (gage)
N8 Oil press. swich (fuel pump)
N10 Cold start fuel injection switch
N11 Cold start valve
N12 Deceleration Valve (M/T only)
N13 Fan Override Switch

☐ **CONTROLLED DEVICES**

1   Fuel injector
2   Idle air control motor
3   Fuel pump relay
5   Trans. Conv. Clutch connector
6a Remote ignition coil
8   Engine coolant fan relay
9   Air control solenoid (M.T. only)
12 Exhaust Gas Recirculation solenoid
13 A/C compressor relay
15 Fuel vapor canister solenoid
17 Mass air flow sensor relay

⬡   Exhaust Gas Recirculation valve

◯ **INFORMATION SENSORS**

B   Exhaust oxygen
C   Throttle position
D   Coolant temperature
F   Vehicle speed
K   Mass Air Flow
S   Power steering pressure switch
T   Manifold Air Temperature
U   A/C pressure fan switch

2.8L (VIN S) ECM WIRING DIAGRAM CAMARO AND FIREBIRD

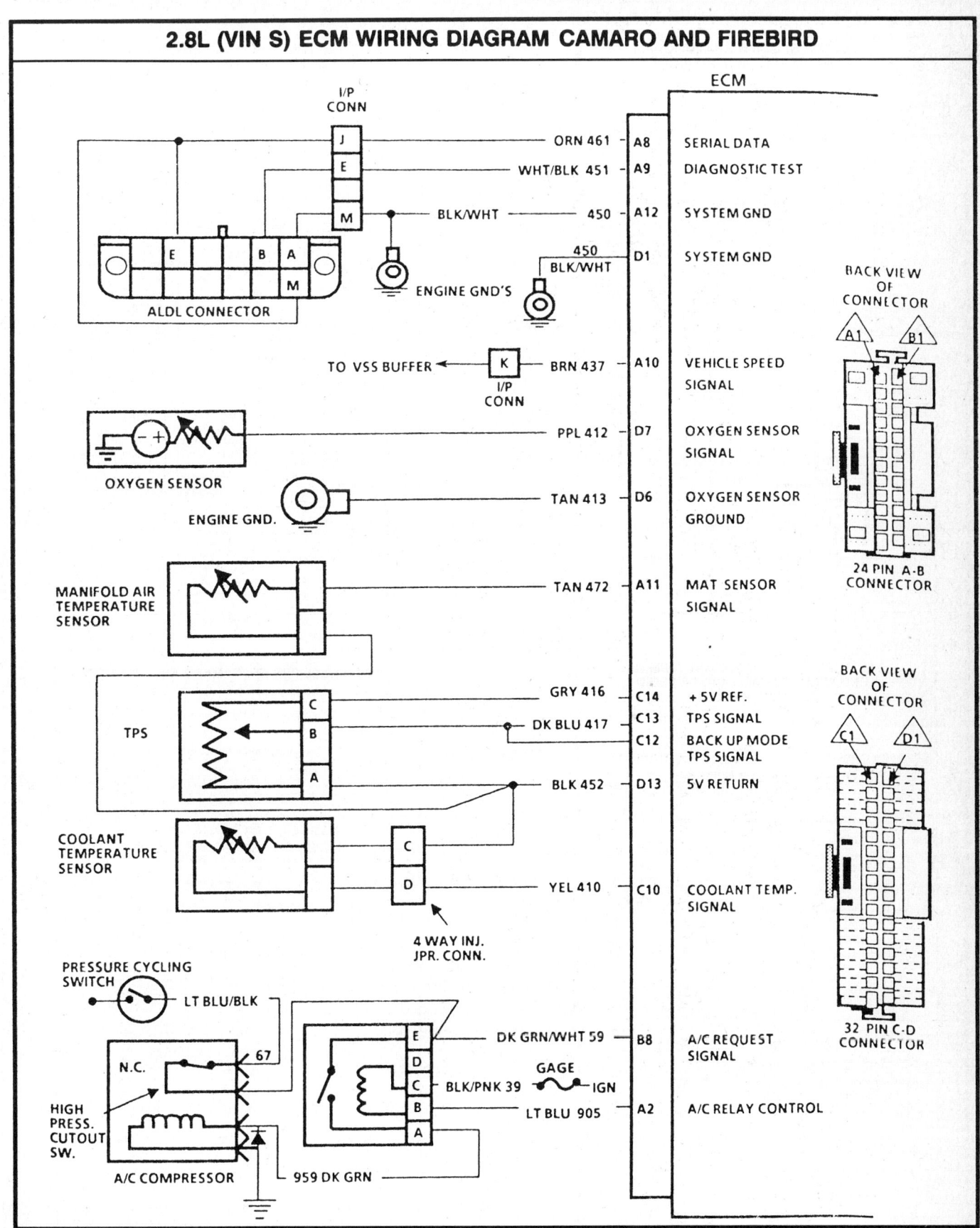

2.8L (VIN S) ECM WIRING DIAGRAM CAMARO AND FIREBIRD (CONT.)

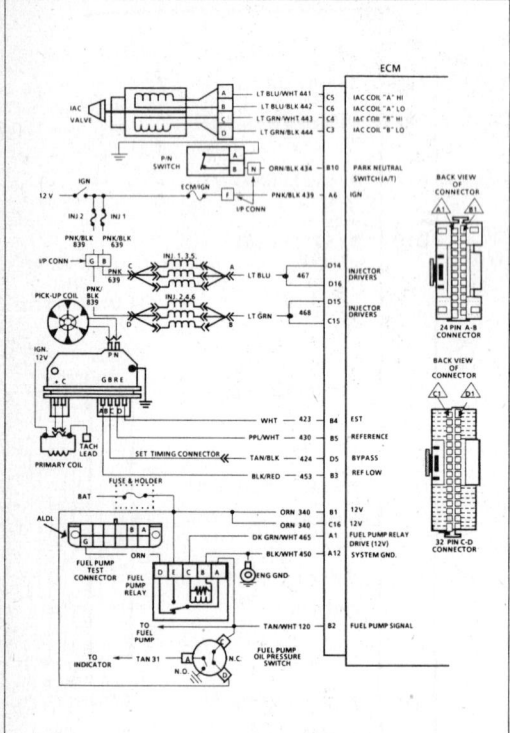

2.8L (VIN S) ECM WIRING DIAGRAM CAMARO AND FIREBIRD (CONT.)

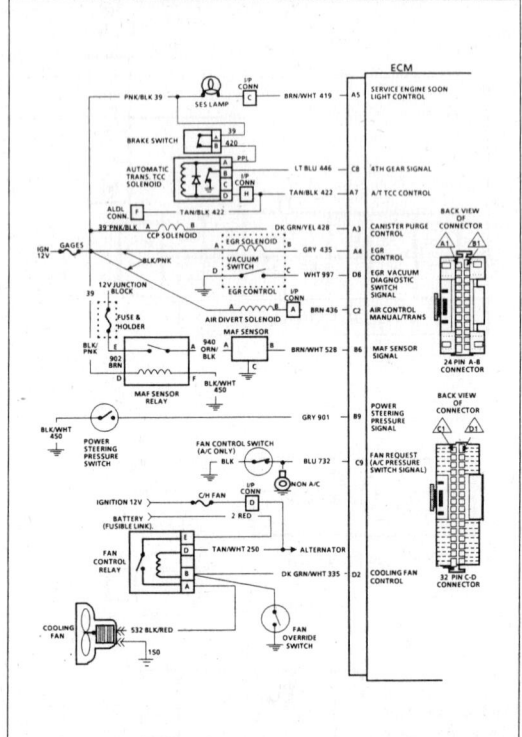

2.8L (VIN S) ECM CONNECTOR TERMINAL END VIEW—CAMARO AND FIREBIRD

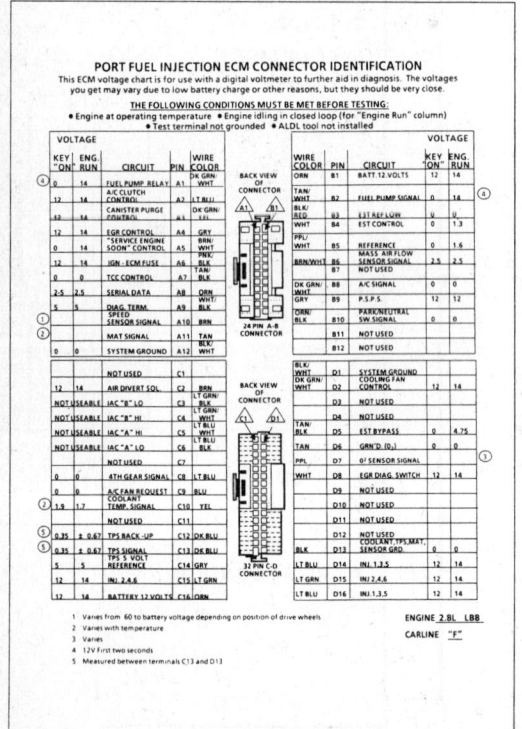

1988–90 2.8L (VIN S)—CAMARO AND FIREBIRD

DIAGNOSTIC CIRCUIT CHECK

The diagnostic circuit check is an organized approach to identifying a problem created by an electronic engine control system malfunction. It must be the starting point for any driveability complaint diagnosis, because it directs the service technician to the next logical step in diagnosing the complaint.

The "Scan Data" listed in the table may be used for comparison, after completing the diagnostic circuit check and finding the on-board diagnostics functioning properly and no trouble codes displayed. The "Typical Values" are an average of display values recorded from normally operating vehicles and are intended to represent what a normally functioning system would typically display.

A "SCAN" TOOL THAT DISPLAYS FAULTY DATA SHOULD NOT BE USED, AND THE PROBLEM SHOULD BE REPORTED TO THE MANUFACTURER. THE USE OF A FAULTY "SCAN" CAN RESULT IN MISDIAGNOSIS AND UNNECESSARY PARTS REPLACEMENT.

Only the parameters listed below are used in this manual for diagnosis. If a "Scan" reads other parameters, the values are not recommended by General Motors for use in diagnosis. For more description on the values and use of the "Scan" to diagnose ECM inputs, refer to the applicable diagnosis section. If all values are within the range illustrated, refer to symptoms in Section "B".

"SCAN" DATA

Idle / Upper Radiator Hose Hot / Closed Throttle / Park or Neutral / Closed Loop / Acc. off

"SCAN" Position	Units Displayed	Typical Data Value
Coolant Temp.	C°	85° - 105°
MAT Temp.	C°	10° - 60° (depends on underhood temp.)
TPS	volts	0.35 - 0.67
MAF	gm/sec	4 - 7
INT (Integrator)	Counts	Varies
BLM (Block Learn)	Counts	118 - 138
IAC	Counts (steps)	5 - 50
rpm	rpm	± 100 rpm from desired rpm (± 50 in drive)
O₂	volts	.1 - 1 and varies
Open/Closed Loop	Open/Closed	Closed Loop (may go open with extended idle)
Spark Advance	# of Degrees	Varies
BPW (base pulse width)	M/Sec	7 - 2.0
EGR Duty Cycle	0-100%	0% (at idle)
A/C Request	Yes/No	No (yes, with A/C requested)
4th gear	Yes/No	No (yes, when in 4th gear)
A/C Clutch	On/Off	OFF (ON, with A/C commanded ON)
P/N Switch	P/N and RDL	Park/Neutral (P/N)
Power Steering Pressure Switch	Normal/Hi pressure	Normal
TCC	ON/OFF	OFF/ (ON, with TCC commanded)
VSS	mph	0

1988–90 2.8L (VIN S) – CAMARO AND FIREBIRD

DIAGNOSTIC CIRCUIT CHECK
2.8L (VIN S) "F" CARLINE (PORT)

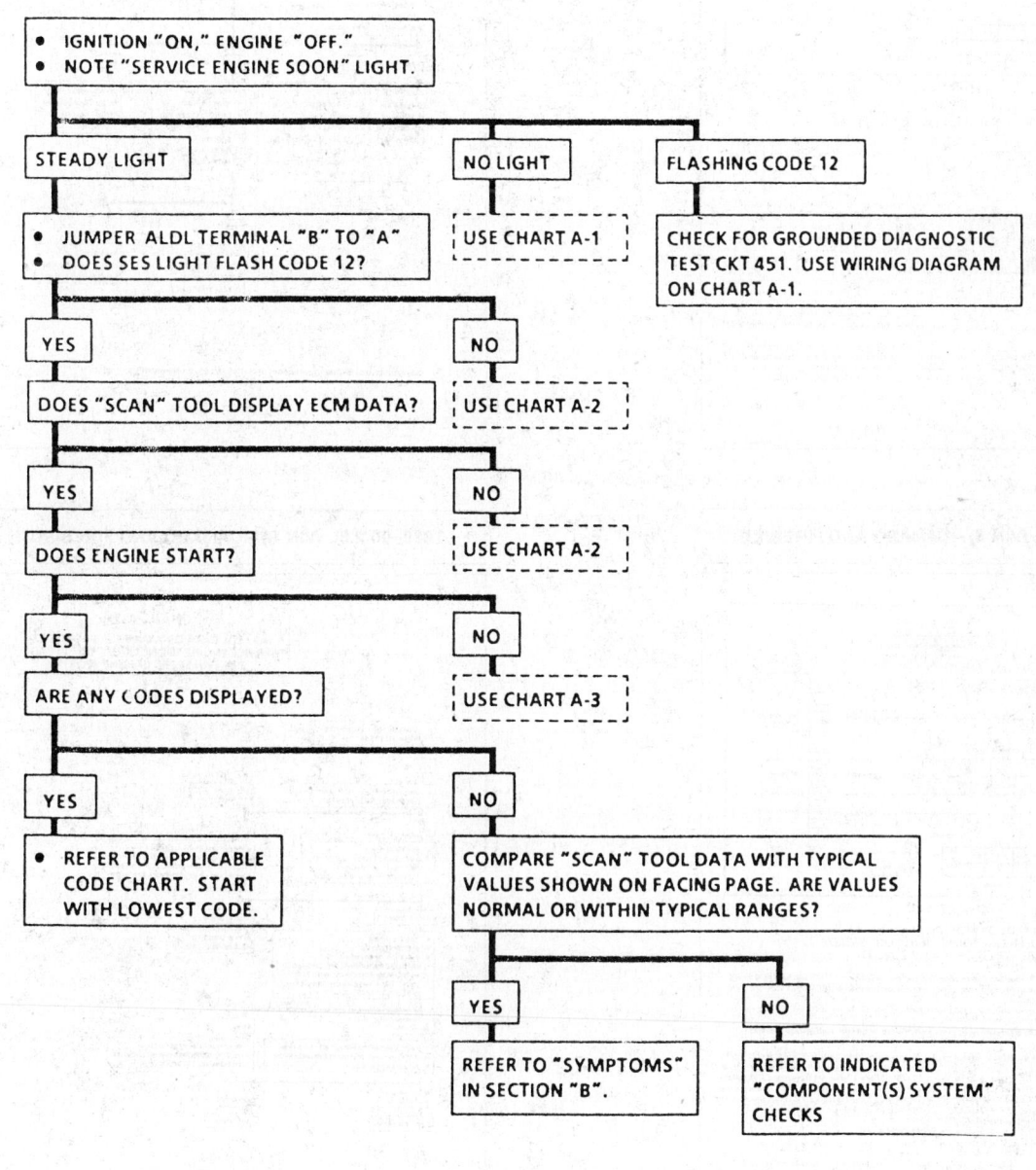

- IGNITION "ON," ENGINE "OFF."
- NOTE "SERVICE ENGINE SOON" LIGHT.

| STEADY LIGHT | NO LIGHT | FLASHING CODE 12 |

- JUMPER ALDL TERMINAL "B" TO "A"
- DOES SES LIGHT FLASH CODE 12?

USE CHART A-1

CHECK FOR GROUNDED DIAGNOSTIC TEST CKT 451. USE WIRING DIAGRAM ON CHART A-1.

YES / NO

DOES "SCAN" TOOL DISPLAY ECM DATA?   USE CHART A-2

YES / NO

DOES ENGINE START?   USE CHART A-2

YES / NO

ARE ANY CODES DISPLAYED?   USE CHART A-3

YES / NO

- REFER TO APPLICABLE CODE CHART. START WITH LOWEST CODE.

COMPARE "SCAN" TOOL DATA WITH TYPICAL VALUES SHOWN ON FACING PAGE. ARE VALUES NORMAL OR WITHIN TYPICAL RANGES?

YES / NO

REFER TO "SYMPTOMS" IN SECTION "B".

REFER TO INDICATED "COMPONENT(S) SYSTEM" CHECKS

1988–90 2.8L (VIN S) – CAMARO AND FIREBIRD

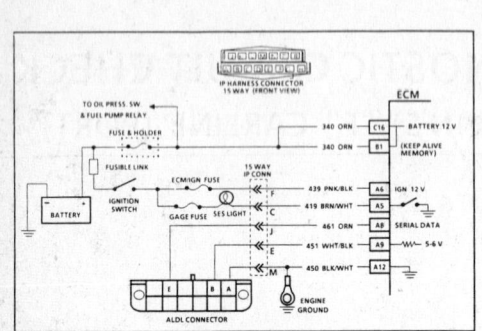

CHART A-1
NO "SERVICE ENGINE SOON" LIGHT
2.8L (VIN S) "F" CARLINE (PORT)

Circuit Description:
   There should always be a steady "Service Engine Soon" light when the ignition is "ON" and engine stopped. Battery is supplied directly to the light bulb. The electronic control module (ECM) will control the light and turn it "ON" by providing a ground path through CKT 419 to the ECM.

Test Description: Numbers below refer to circled numbers on the diagnostic chart.
1. If the fuse in holder is blown refer to facing page of Code 54 for complete circuit.
2. Using a test light connected to 12 volts probe each of the system ground circuits to be sure a good ground is present. See ECM terminal end view in front of this section for ECM pin locations of ground circuits.

Diagnostic Aids:

Engine runs OK, check:
- Faulty light bulb
- CKT 419 open.
- Gage fuse blown. This will result in no oil or generator lights, seat belt reminder, etc.

Engine cranks but will not run.
- Continuous battery - fuse or fusible link open.
- ECM ignition fuse open.
- Battery CKT 340 to ECM open.
- Ignition CKT 439 to ECM open.
- Poor connection to ECM.
   Solenoids and relays are turned "ON" and "OFF" by the ECM, using internal electronic switches called "drivers". Each driver is part of a group of four called "Quad-Drivers". Failure of one driver can damage any other driver in the set. Solenoid and relay coil resistance must measure more than 20 ohms. Less resistance will cause early failure of the ECM "driver".
   Before replacing ECM, be sure to check the coil resistance of all solenoids and relays controlled by the ECM. See ECM wiring diagram for the solenoid(s) and relay(s) and the coil terminal identification.

1988–90 2.8L (VIN S) – CAMARO AND FIREBIRD

CHART A-1
NO "SERVICE ENGINE SOON" LIGHT
2.8L (VIN S) "F" CARLINE (PORT)

CLEAR CODES AND CONFIRM "CLOSED LOOP" OPERATION AND NO "SERVICE ENGINE SOON" LIGHT.

1988–90 2.8L (VIN S) – CAMARO AND FIREBIRD

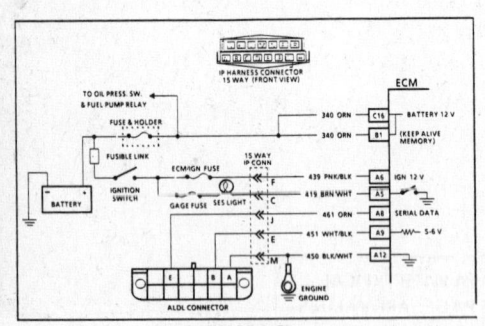

CHART A-2
NO ALDL OR WON'T FLASH CODE 12
"SERVICE ENGINE SOON" LIGHT "ON" STEADY
2.8L (VIN S) "F" CARLINE (PORT)

Circuit Description:
   There should always be a steady "Service Engine Soon" light when the ignition is "ON" and engine stopped. Battery ignition voltage is supplied to the light bulb. The electronic control module (ECM) will turn the light "ON" by grounding CKT 419 at the ECM.
   With the diagnostic terminal grounded, the light should flash a Code 12, followed by any trouble code(s) stored in memory.
   A steady light suggests a short to ground in the light control CKT 419, or an open in diagnostic CKT 451.

Test Description: Numbers below refer to circled numbers on the diagnostic chart.
1. If there is a problem with the ECM that causes a "Scan" tool to not read serial data, the ECM should not flash a Code 12. If Code 12 is flashing check for CKT 451 short to ground. If Code 12 does flash be sure that the "Scan" tool is working properly on another vehicle. If the "Scan" is functioning properly and CKT 461 is OK the PROM or ECM may be at fault for the NO ALDL symptom.
2. If the light goes "OFF" when the ECM connector is disconnected, CKT 419 is not shorted to ground.
3. This step will check for an open diagnostic CKT 451.

4. At this point the "Service Engine Soon" light wiring is OK. The problem is a faulty ECM or PROM. If Code 12 does not flash, the ECM should be replaced using the original PROM. Replace the PROM only after trying an ECM, as a defective PROM is an unlikely cause of the problem.

Diagnostic Aids:
   Solenoids and relays are turned "ON" or "OFF" by the ECM using internal electronic switches called "drivers". Each driver is part of a group of four called "Quad-Drivers". Failure of one driver can damage any other driver in the set.
   Before replacing ECM, be sure to check the coil resistance of all solenoids and relays controlled by the ECM. See ECM wiring diagram for the solenoid(s) and relay(s) and the coil terminal identification.

1988–90 2.8L (VIN S) – CAMARO AND FIREBIRD

CHART A-2
NO ALDL DATA OR WON'T FLASH CODE 12
"SERVICE ENGINE SOON" LIGHT "ON" STEADY
2.8L (VIN S) "F" CARLINE (PORT)

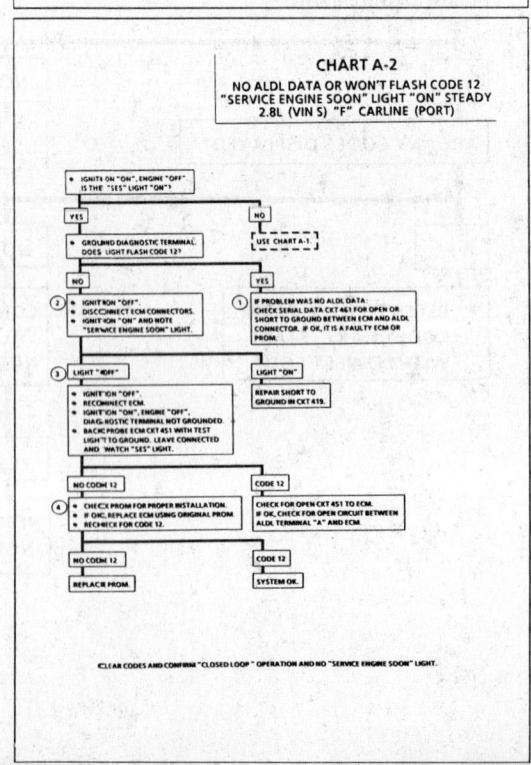

CLEAR CODES AND CONFIRM "CLOSED LOOP" OPERATION AND NO "SERVICE ENGINE SOON" LIGHT.

1988–90 2.8L (VIN S) – CAMARO AND FIREBIRD

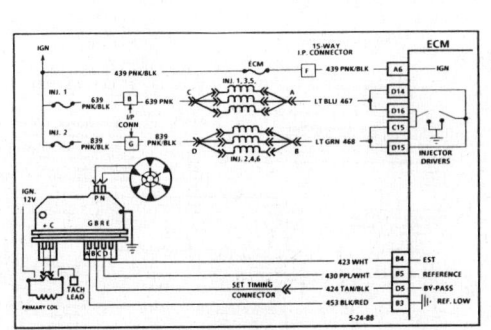

CHART A-3
(Page 1 of 2)
ENGINE CRANKS BUT WON'T RUN
2.8L (VIN S) "F" CARLINE (PORT)

Test Description: Numbers below refer to circled numbers on the diagnostic chart.

1. This chart assumes that battery condition and engine cranking speed are OK, and there is adequate fuel in the tank. If engine starts but immediately stalls, see "Symptoms", Section "B" (Hard Start). A "Service Engine Soon" light "ON" is a basic check for ignition and battery supply to the electronic control module (ECM).
2. No spark indicates a basic HEI problem.
3. This test will determine if the ECM is receiving the reference signal and controlling the injectors. This test could also be performed at the 4-way injector connector by using a test light between terminals "A" and "D".
   If the test light "blinks" while cranking, then ECM control should be considered OK. How bright the test light "blinks" is not important. However, the test light should be a J-34730-2 or equivalent.
4. Use pressure gage J-34730-1. Wrap a shop towel around the fuel pressure tap to absorb any small amount of fuel leakage that may occur when installing the gage.

Diagnostic Aids:

- An EGR valve sticking open can cause a low air/fuel ratio during cranking. Unless engine enters "Clear Flood" at the first indication of a flooding condition, it can result in a no start.
- Check for fouled plugs.
- If the TPS is sticking or binding in the wide open throttle position, the ECM will be in the "Clear Flood" mode.
- A defective cold start circuit or water in fuel line can cause a no start in cold weather. To check cold start circuit: See CHART A-9.
- A defective MAF sensor may cause a no start or a stall after start. To determine if the sensor is causing the problem, disconnect it. The ECM will then use a default value for the sensor, and if the condition is corrected and the connections are OK, replace the sensor.
- Also check that injectors on both sides of engine will cause a test light to "blink". Checking of two injectors on each bank in this manner will locate a shorted injector.
- If above are all OK, refer to "Symptoms" in Section "B" "Hard Start".
- Also check that injectors are not open or shorted. Injector resistance should be greater than ohms.

1988–90 2.8L (VIN S) – CAMARO AND FIREBIRD

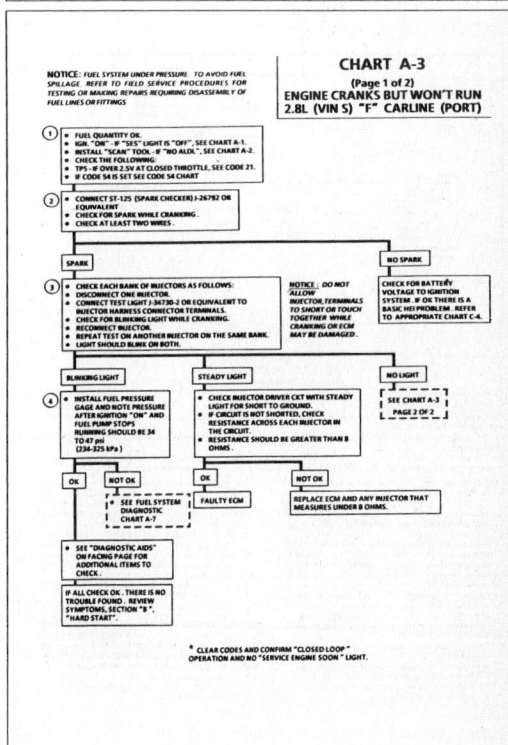

NOTICE: *FUEL SYSTEM UNDER PRESSURE. TO AVOID FUEL SPILLAGE, REFER TO FIELD SERVICE PROCEDURES FOR TESTING OR MAKING REPAIRS REQUIRING DISASSEMBLY OF FUEL LINES OR FITTINGS.*

CHART A-3
(Page 1 of 2)
ENGINE CRANKS BUT WON'T RUN
2.8L (VIN S) "F" CARLINE (PORT)

1988–90 2.8L (VIN S) – CAMARO AND FIREBIRD

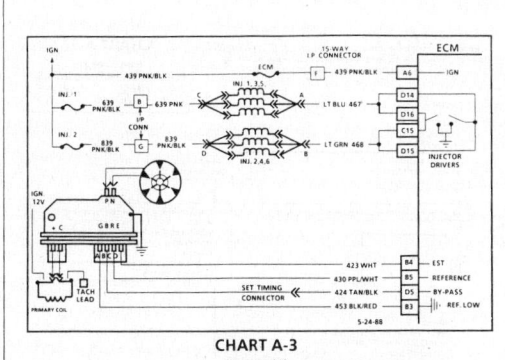

CHART A-3
(Page 2 of 2)
ENGINE CRANKS BUT WON'T RUN
2.8L (VIN S) "F" CARLINE (PORT)

Test Description: Numbers below refer to circled numbers on the diagnostic chart.

5. Checks for 12 volt supply to injectors.
6. This test will determine if the distributor module is not generating the reference pulse or if the wiring or ECM are at fault. By touching CKT 430 with a test light a reference signal is being generated. If the test light (J-34730-2) blinks at the injector, then the ECM and wiring is OK.
7. Each time the test light touches CKT 430, the ECM should turn "ON" the fuel pump for 2 seconds.
8. All checks made to this point would indicate that the ECM is at fault. However, there is a possibility of CKT 467 or 468 being shorted to a voltage source either in the engine harness or in the injector harness.

To test for this condition:
- Disconnect the injector 4-way connector.
- Ignition "ON".
- Probe CKTs 467 and 468 on the ECM side of harness with a test light connected to ground. There should be no light.
- If OK, check the resistance of the injector harness between terminals "A & C", "A & D", "B & D", and "B & C".
- Should be more than 4 ohms.
- If less than 4 ohms check harness for wires shorted together and check each injector resistance.
- Resistance should be more than 10 ohms.
- If all OK, replace ECM.

1988–90 2.8L (VIN S) – CAMARO AND FIREBIRD

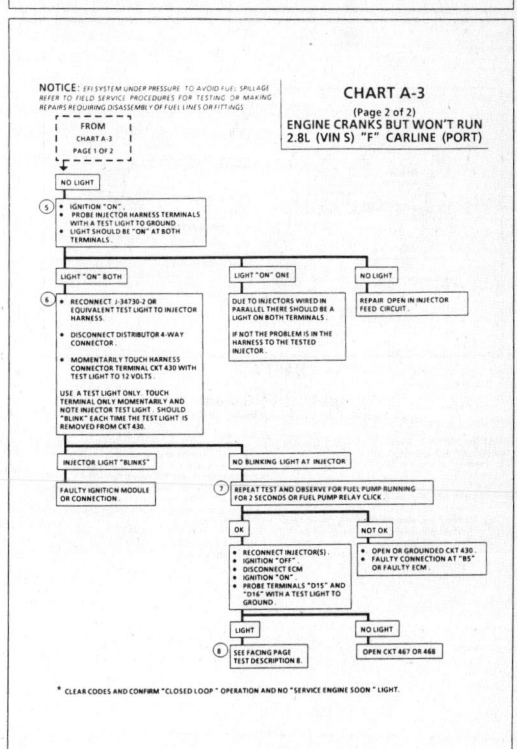

NOTICE: *EFI SYSTEM UNDER PRESSURE. TO AVOID FUEL SPILLAGE, REFER TO FIELD SERVICE PROCEDURES FOR TESTING OR MAKING REPAIRS REQUIRING DISASSEMBLY OF FUEL LINES OR FITTINGS.*

CHART A-3
(Page 2 of 2)
ENGINE CRANKS BUT WON'T RUN
2.8L (VIN S) "F" CARLINE (PORT)

1988–90 2.8L (VIN S) – CAMARO AND FIREBIRD

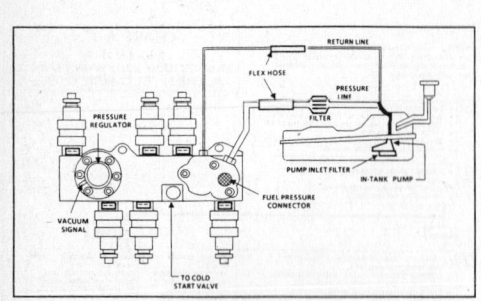

CHART A-7
(Page 1 of 2)
FUEL SYSTEM DIAGNOSIS
2.8L (VIN S) "F" CARLINE (PORT)

Circuit Description:

When the ignition switch is turned "ON", the electronic control module (ECM) will turn "ON" the in-tank fuel pump. It will remain "ON" as long as the engine is cranking or running, and the ECM is receiving HEI distributor reference pulses.

If there are no reference pulses, the ECM will shut "OFF" the fuel pump within 2 seconds after key "ON" or engine stopped.

The pump will deliver fuel to the fuel rail and injectors, then to the pressure regulator, where the system pressure is controlled to about 234 to 317 kPa (34 to 46 psi). Excess fuel is then returned to the fuel tank.

Test Description: Numbers below refer to circled numbers on the diagnostic chart.

1. Use pressure gage J-34730-1. Wrap a shop towel around the fuel pressure tap to absorb any small amount of fuel leakage that may occur when installing the gage.
   Ignition "ON", pump pressure should be 280-325 kPa (40.5-47 psi). This pressure is controlled by spring pressure within the regulator assembly.
2. When the engine is idling, the manifold pressure is low (high vacuum) and is applied to the fuel regulator diaphragm. This will offset the spring and result in a lower fuel pressure. This idle pressure will vary somewhat depending on barometric pressure, however, the pressure idling was less indicating pressure regulator control.
3. Pressure that continues to fall is caused by one of the following:
   • In-tank fuel pump check valve not holding.
   • Pump coupling hose or pulsator leaking.
   • Fuel pressure regulator valve leaking.
   • Injector sticking open.

4. An injector sticking open can best be determined by checking for a fouled or saturated spark plug(s). If a leaking injector can not be determined by a fouled or saturated spark plug the following procedure should be used.
   • Remove Plenum, cold start valve and remove fuel rail bolts. Follow the procedures in the fuel control section of this manual, but leave fuel lines connected.
   • Reconnect cold start valve.
   • Connect a hose to valve nozzle and insert into a gasoline container.
   • Lift fuel rail out just enough to leave injector nozzles in the ports.
   CAUTION: *BE SURE INJECTOR(S) ARE NOT ALLOWED TO SPRAY ON ENGINE AND THAT INJECTOR RETAINING CLIPS ARE INTACT. THIS SHOULD BE CAREFULLY FOLLOWED TO PREVENT FUEL SPRAY ON ENGINE WHICH WOULD CAUSE A FIRE HAZARD.*
   • Pressurize the fuel system.
   • Lift each side of rail up and observe for injector(s) leaking.

1988–90 2.8L (VIN S) – CAMARO AND FIREBIRD

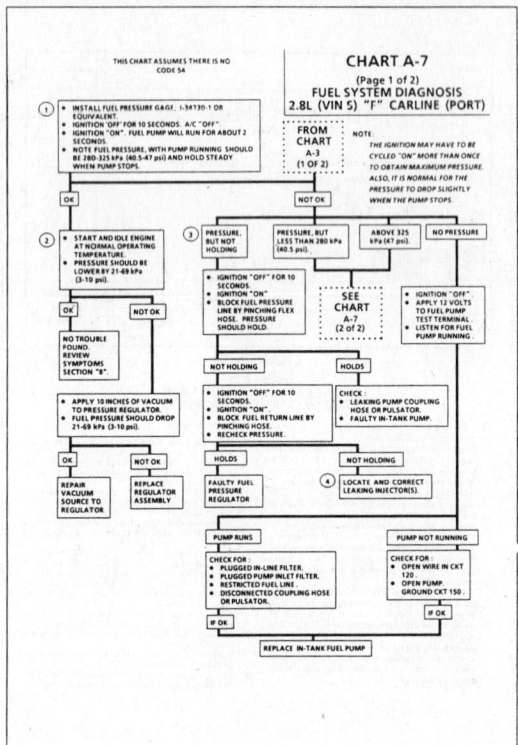

THIS CHART ASSUMES THERE IS NO CODE 54

CHART A-7
(Page 1 of 2)
FUEL SYSTEM DIAGNOSIS
2.8L (VIN S) "F" CARLINE (PORT)

NOTE:
THE IGNITION MAY HAVE TO BE CYCLED "ON" MORE THAN ONCE TO OBTAIN MAXIMUM PRESSURE. ALSO, IT IS NORMAL FOR THE PRESSURE TO DROP SLIGHTLY WHEN THE PUMP STOPS.

1988–90 2.8L (VIN S) – CAMARO AND FIREBIRD

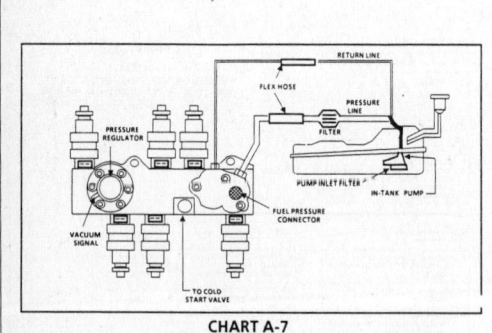

CHART A-7
(Page 2 of 2)
FUEL SYSTEM DIAGNOSIS
2.8L (VIN S) "F" CARLINE (PORT)

Test Description: Numbers below refer to circled numbers on the diagnostic chart.

1. Pressure but less than 280 kPa (40.5 psi) falls into two areas.
   • Regulated pressure, but less than 280 kPa (40.5 psi). Amount of fuel to injectors OK but pressure is too low. System will be lean running and may set Code 44. Also, hard starting cold and overall poor performance.
   • Restricted flow causing pressure drop - Normally, a vehicle with a fuel pressure of less than 165 kPa (24 psi) at idle will not be driveable.

However, if the pressure drop occurs only while driving, the engine will normally surge then stop as pressure begins to drop rapidly.

2. Restricting the the fuel return line allows the fuel pump to develop its maximum pressure (dead head pressure). When battery voltage is applied to the pump test terminal, pressure should be above 414 kPa (60 psi).

3. This test determines if the high fuel pressure is due to a restricted fuel return line or a pressure regulator problem.

1988–90 2.8L (VIN S) – CAMARO AND FIREBIRD

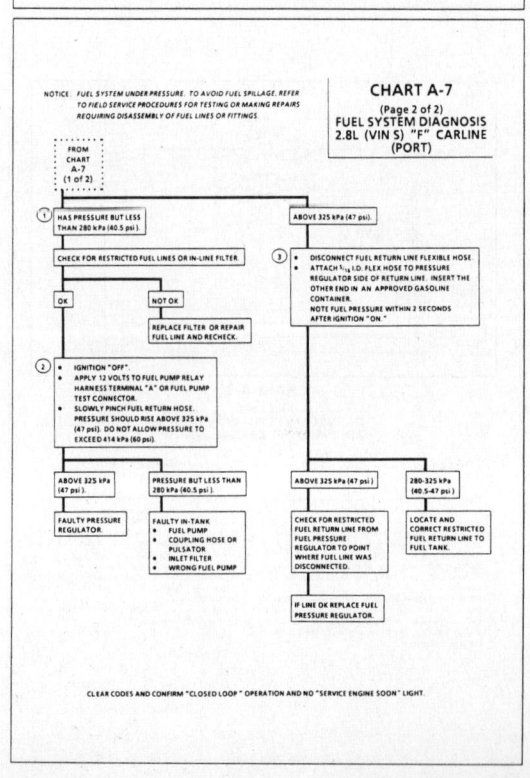

NOTICE: *FUEL SYSTEM UNDER PRESSURE. TO AVOID FUEL SPILLAGE, REFER TO FIELD SERVICE PROCEDURES FOR TESTING OR MAKING REPAIRS REQUIRING DISASSEMBLY OF FUEL LINES OR FITTINGS.*

CHART A-7
(Page 2 of 2)
FUEL SYSTEM DIAGNOSIS
2.8L (VIN S) "F" CARLINE (PORT)

CLEAR CODES AND CONFIRM "CLOSED LOOP" OPERATION AND NO "SERVICE ENGINE SOON" LIGHT.

1988–90 2.8L (VIN S) – CAMARO AND FIREBIRD

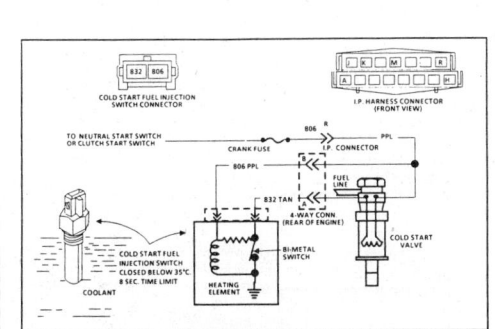

CHART A-9
COLD START VALVE
2.8L (VIN S) "F" CARLINE (PORT)

Circuit Description:
The cold start valve is used to provide additional fuel during the crank mode to improve cold start-ups. This circuit is important when engine coolant temperature is low because the other injectors are not pulsed "ON" long enough to provide the needed amount of fuel to start.

The circuit is activated only in the crank mode. The power is supplied directly from the starter solenoid and is protected by a fuse. The system is controlled by a cold start fuel injection switch which provides a ground path for the valve during cranking when engine coolant is below 95°F (35°C).

The cold start fuel injection switch consists of a bimetal material which opens at a specified coolant temperature. This bimetal is also heated by the winding in the switch which allows the valve to stay "ON" for 8 seconds at -20°F (-5°F) coolant. The time the switch will stay closed varies inversely with coolant temperature. In other words, as the coolant temperature goes up, the cold start valve "ON" time goes down.

Test Description: Numbers below refer to circled numbers on the diagnostic chart.
1. Disconnecting the distributor 4-way connector will disable the other injectors.
2. This test will determine the continuity through the switch to ground.

The amount of pressure drop depends on the temperature of the engine.

1988–90 2.8L (VIN S) – CAMARO AND FIREBIRD

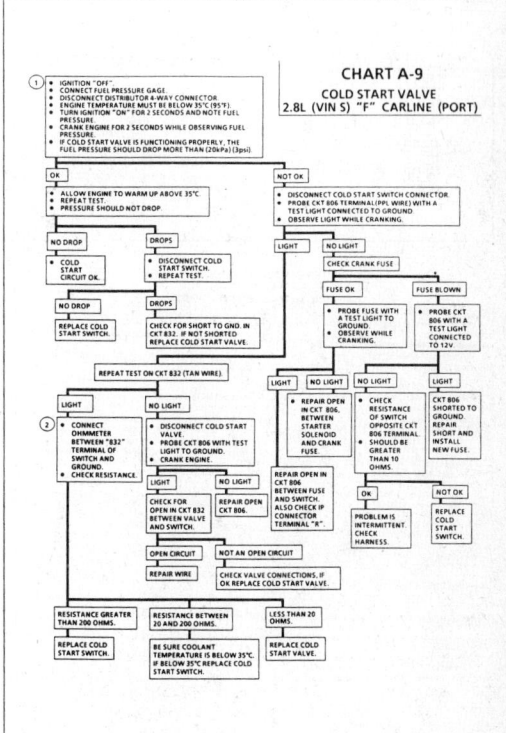

CHART A-9
COLD START VALVE
2.8L (VIN S) "F" CARLINE (PORT)

1988–90 2.8L (VIN S) – CAMARO AND FIREBIRD

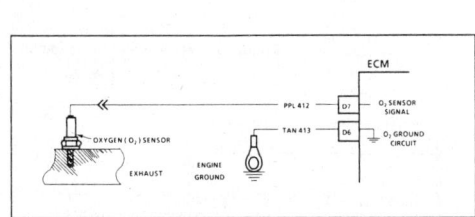

CODE 13
OXYGEN SENSOR CIRCUIT
(OPEN CIRCUIT)
2.8L (VIN S) "F" CARLINE (PORT)

Circuit Description:
The ECM supplies a voltage of about .45 volt between terminals "D6" and "D7". (If measured with a 10 megohm digital voltmeter, this may read as low as 32 volts.) The O₂ sensor varies the voltage within a range of about 1 volt if the exhaust is rich, down through about .10 volt if exhaust is lean.

The sensor is like an open circuit and produces no voltage when it is below 315°C (600°F). An open sensor circuit or cold sensor causes "Open Loop" operation.

Test Description: Numbers below refer to circled numbers on the diagnostic chart.
1. Code 13 will set:
   - Engine at normal operating temperature.
   - At least 2 minutes engine time after start.
   - O₂ signal voltage steady between .35 and .55 volts.
   - Throttle position sensor signal above 4%.
   - All conditions must be met for about 60 seconds.
   If the conditions for a Code 13 exist the system will not go "Closed Loop"
2. This will determine if the sensor is at fault or the wiring or ECM is the cause of the Code 13.

3. In doing this test use only a high impedence digital volt ohmmeter. This test checks the continuity of CKTs 412 and 413 because if CKT 413 is open the ECM voltage on CKT 412 will be over .6 volts (600 mV.)

Diagnostic Aids:

Normal "Scan" voltage varies between 100mV to 999mV (.1 and 1.0 volts) while in "Closed Loop." Code 13 sets in one minute if voltage remains between .35 and .55 volts, but the system will go "Open Loop" in about 15 seconds. Refer to "Intermittents" in Section "B".

1988–90 2.8L (VIN S) – CAMARO AND FIREBIRD

CODE 13
OXYGEN SENSOR CIRCUIT
(OPEN CIRCUIT)
2.8L (VIN S) "F" CARLINE (PORT)

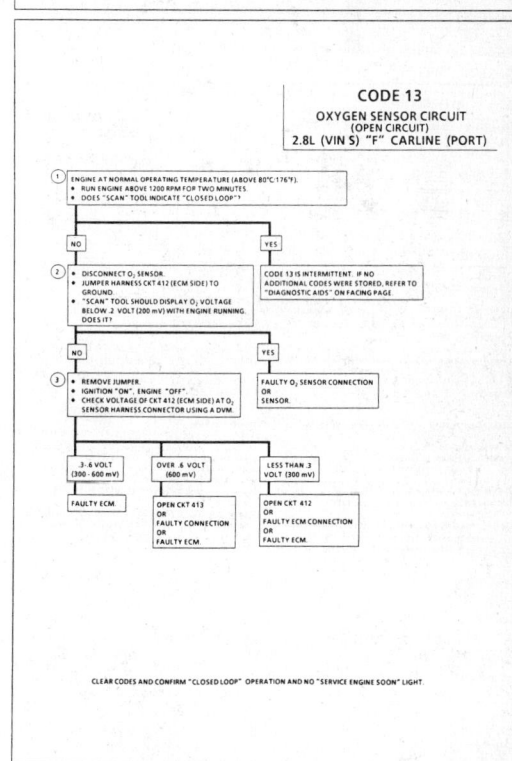

1988–90 2.8L (VIN S) – CAMARO AND FIREBIRD

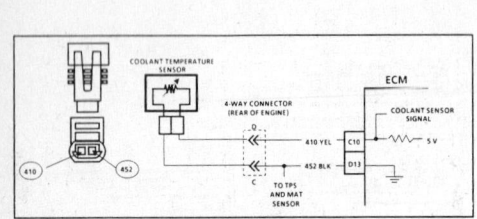

CODE 14
COOLANT TEMPERATURE SENSOR CIRCUIT
(HIGH TEMPERATURE INDICATED)
2.8L (VIN S) "F" CARLINE (PORT)

Circuit Description:

The coolant temperature sensor uses a thermistor to control the signal voltage to the ECM. The ECM applies a voltage on CKT 410 to the sensor. When the engine is cold the sensor (thermistor) resistance is high, therefore the ECM will see high signal voltage.

As the engine warms, the sensor resistance becomes less, and the voltage drops. At normal engine operating temperature (85°C to 95°C) the voltage will measure about 1.5 to 2.0 volts.

Test Description: Numbers below refer to circled numbers on the diagnostic chart.
1. Code 14 will set if:
   - Signal voltage indicates a coolant temperature above 135°C (275°F) for 3 seconds.
2. This test will determine if CKT 410 is shorted to ground, which will cause the conditions for Code 14.

Diagnostic Aids:

Check harness routing for a potential short to ground in CKT 410. "SCAN" tool displays engine temperature in degrees centigrade. After engine is started, the temperature should rise steadily to about 90°C then stabilize when thermostat opens. Refer to "Intermittents" in Section "B".

1988–90 2.8L (VIN S) – CAMARO AND FIREBIRD

CODE 14
COOLANT TEMPERATURE SENSOR CIRCUIT
(HIGH TEMPERATURE INDICATED)
2.8L (VIN S) "F" CARLINE (PORT)

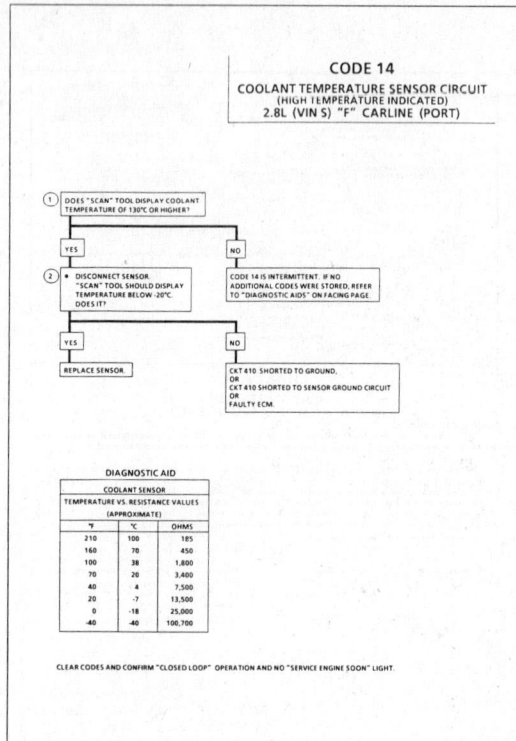

DIAGNOSTIC AID

COOLANT SENSOR		
TEMPERATURE VS. RESISTANCE VALUES		
(APPROXIMATE)		
°F	°C	OHMS
210	100	185
160	70	450
100	38	1,800
70	20	3,400
40	4	7,500
20	-7	13,500
0	-18	25,000
-40	-40	100,700

CLEAR CODES AND CONFIRM "CLOSED LOOP" OPERATION AND NO "SERVICE ENGINE SOON" LIGHT.

1988–90 2.8L (VIN S) – CAMARO AND FIREBIRD

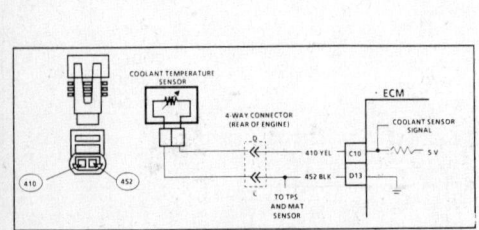

CODE 15
COOLANT TEMPERATURE SENSOR CIRCUIT
(LOW TEMPERATURE INDICATED)
2.8L (VIN S) "F" CARLINE (PORT)

Circuit Description:

The coolant temperature sensor uses a thermistor to control the signal voltage to the ECM. The ECM applies a voltage on CKT 410 to the sensor. When the engine is cold the sensor (thermistor) resistance is high, therefore the ECM will see high signal voltage.

As the engine warms, the sensor resistance becomes less, and the voltage drops. At normal engine operating temperature (85°C to 95°C) the voltage will measure about 1.5 to 2.0 volts at the ECM.

Test Description: Numbers below refer to circled numbers on the diagnostic chart.
1. Code 15 will set if:
   - Signal voltage indicates a coolant temperature less than -44°C (-47°F) for 3 seconds.
2. This test simulates a Code 14. If the ECM recognizes the low signal voltage, (high temperature) and the "Scan" reads 130°C, the ECM and wiring are OK.
3. This test will determine if CKT 410 is open. There should be 5 volts present at sensor connector if measured with a DVM.

Diagnostic Aids:

A "SCAN" tool reads engine temperature in degrees centigrade. After engine is started the temperature should rise steadily to about 90°C then stabilize when thermostat opens.

A faulty connection, or an open in CKT 410 or 452 will result in a Code 15.

If Code 23 or 63 is also set, check CKT 452 for faulty wiring or connections. Check terminals at sensor for good contact. Refer to "Intermittents" in Section "B".

1988–90 2.8L (VIN S) – CAMARO AND FIREBIRD

CODE 15
COOLANT TEMPERATURE SENSOR CIRCUIT
(LOW TEMPERATURE INDICATED)
2.8L (VIN S) "F" CARLINE (PORT)

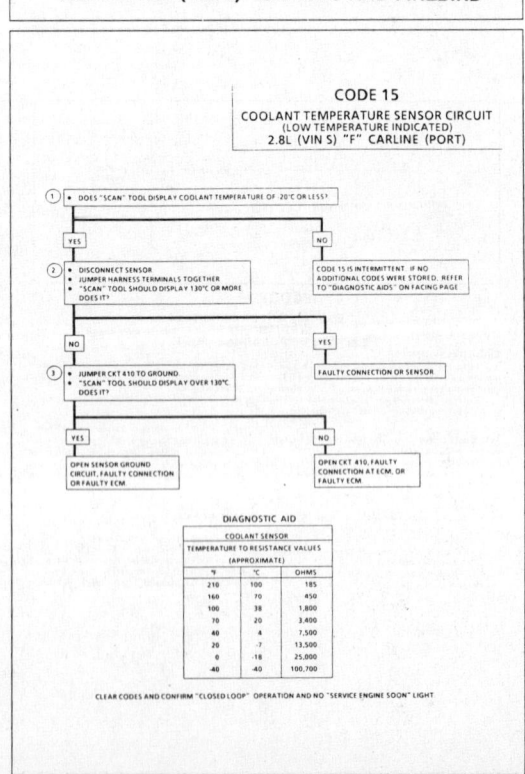

DIAGNOSTIC AID

COOLANT SENSOR		
TEMPERATURE TO RESISTANCE VALUES		
(APPROXIMATE)		
°F	°C	OHMS
210	100	185
160	70	450
100	38	1,800
70	20	3,400
40	4	7,500
20	-7	13,500
0	-18	25,000
-40	-40	100,700

CLEAR CODES AND CONFIRM "CLOSED LOOP" OPERATION AND NO "SERVICE ENGINE SOON" LIGHT.

1988–90 2.8L (VIN S) – CAMARO AND FIREBIRD

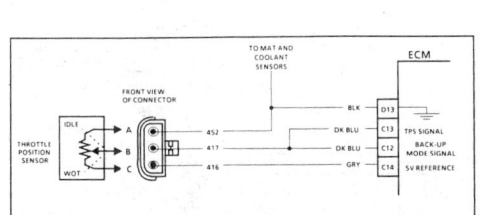

CODE 21
THROTTLE POSITION SENSOR (TPS) CIRCUIT
(SIGNAL VOLTAGE HIGH)
2.8L (VIN S) "F" CARLINE (PORT)

Circuit Description:
The throttle position sensor (TPS) provides a voltage signal that changes relative to the throttle blade. Signal voltage will vary from about .5 at idle to about 5 volts at wide open throttle.
The TPS signal is one of the most important inputs used by the ECM for fuel control and for most of the ECM control outputs.

Test Description: Numbers below refer to circled numbers on the diagnostic chart.
1 Code 21 will set if:
- Engine is running
- TPS signal voltage is greater than 2.5 volts
- Air flow is less than 12 gm/sec
- All conditions met for 5 seconds
OR
- TPS signal voltage over 4.5 volts with ignition "ON"

The TPS has an auto zeroing feature. If the voltage reading is within the range of 0.35 to 0.7 volts, the ECM will use that value as closed throttle. If the voltage reading is out of the auto zero range at closed throttle, refer to "TPS Adjustment" in Section "6E3-C1"

2. With the TPS sensor disconnected, the TPS voltage should go low if the ECM and wiring is OK.
3. Probing CKT 452 with a test light checks the 5 volt return circuit, because a faulty 5 volt return will cause a Code 21.

Diagnostic Aids:

A "SCAN" tool reads throttle position in volts. Voltage should increase at a steady rate as throttle is moved toward WOT.
An open in CKT 452 will result in a Code 21. Refer to "Intermittents" in Section "B".

1988–90 2.8L (VIN S) – CAMARO AND FIREBIRD

CODE 21
THROTTLE POSITION SENSOR (TPS) CIRCUIT
(SIGNAL VOLTAGE HIGH)
2.8L (VIN S) "F" CARLINE (PORT)

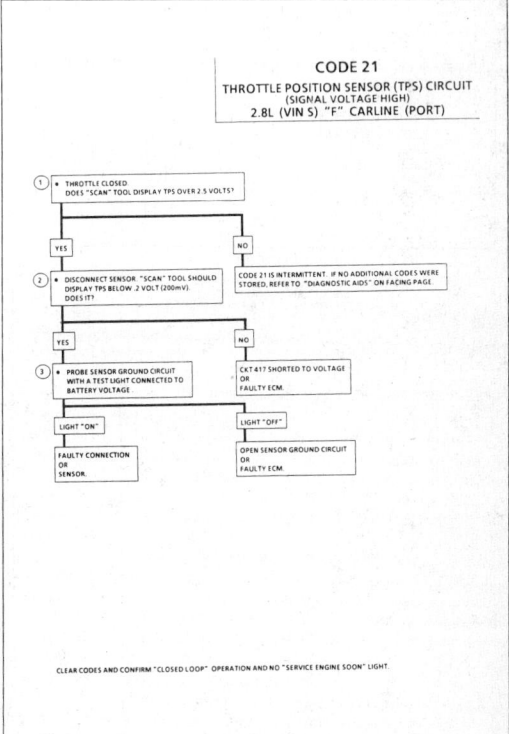

1988–90 2.8L (VIN S) – CAMARO AND FIREBIRD

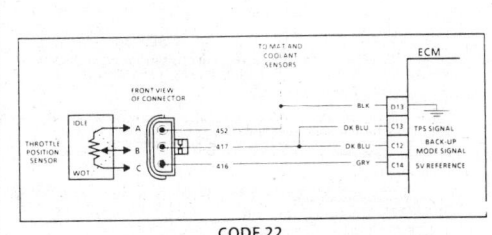

CODE 22
THROTTLE POSITION SENSOR (TPS) CIRCUIT
(SIGNAL VOLTAGE LOW)
2.8L (VIN S) "F" CARLINE (PORT)

Circuit Description:
The throttle position sensor (TPS) provides a voltage signal that changes relative to the throttle blade. Signal voltage will vary from about .5 at idle to about 5 volts at wide open throttle.
The TPS signal is one of the most important inputs used by the ECM for fuel control and for most of the ECM control outputs.

Test Description: Numbers below refer to circled numbers on the diagnostic chart.
1 Code 22 will set if:
- Engine running
- TPS signal voltage is less than .2 volt for 3 seconds
2 Simulates Code 21 (high voltage). If the ECM recognizes the high signal voltage the ECM and wiring are OK.
3 The TPS has an auto zeroing feature. If the voltage reading is within the range of 0.35 to 0.7 volts, the ECM will use that value as closed throttle. If the voltage reading is out of the auto zero range at closed throttle, refer to "TPS Adjustment" in Section "6E3-C1"

4 This simulates a high signal voltage to check for an open in CKT 417.

Diagnostic Aids:

A "Scan" tool reads throttle position in volts. Voltage should increase at a steady rate as throttle is moved toward WOT
An open or short to ground in CKTs 416 or 417 will result in a Code 22.
Refer to "Intermittents" in Section "B".

1988–90 2.8L (VIN S) – CAMARO AND FIREBIRD

CODE 22
THROTTLE POSITION SENSOR (TPS) CIRCUIT
(SIGNAL VOLTAGE LOW)
2.8L (VIN S) "F" CARLINE (PORT)

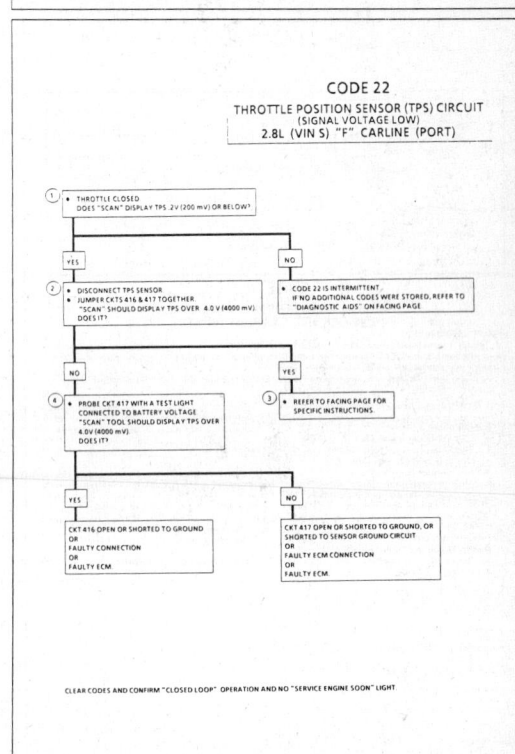

1988–90 2.8L (VIN S) – CAMARO AND FIREBIRD

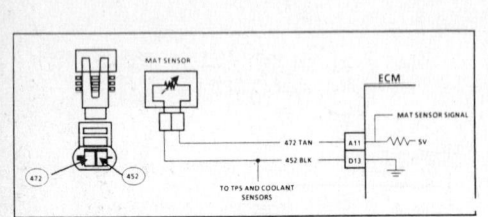

CODE 23
MANIFOLD AIR TEMPERATURE (MAT) SENSOR CIRCUIT
(LOW TEMPERATURE INDICATED)
2.8L (VIN S) "F" CARLINE (PORT)

Circuit Description:

The MAT sensor uses a thermistor to control the signal voltage to the ECM. The ECM applies a voltage (about 5 volts) on CKT 472 to the sensor. When the air is cold the sensor (thermistor) resistance is high, therefore the ECM will see a high signal voltage. If the air is warm the sensor resistance is low, therefore, the ECM will see a low voltage.

Test Description: Numbers below refer to circled numbers on the diagnostic chart.
1. Code 23 will set if:
   - A signal voltage indicates a manifold air temperature below −35°C (−31°F) for 3 seconds.
   - Time since engine start is 8 minutes or longer.
   - No VSS.
2. A Code 23 will set, due to an open sensor, wire, or connection. This test will determine if the wiring and ECM are OK.
3. This will determine if the signal CKT 472 or the 5V return CKT 452 is open.

Diagnostic Aids:

A "SCAN" tool reads temperature of the air entering the engine and should read close to ambient air temperature when engine is cold, and rises as underhood temperature increases.

A faulty connection, or an open in CKT 472 or 452 will result in a Code 23.

Refer to "Intermittents" in Section "B".

1988–90 2.8L (VIN S) – CAMARO AND FIREBIRD

CODE 23
MANIFOLD AIR TEMPERATURE (MAT) SENSOR CIRCUIT
(LOW TEMPERATURE INDICATED)
2.8L (VIN S) "F" CARLINE (PORT)

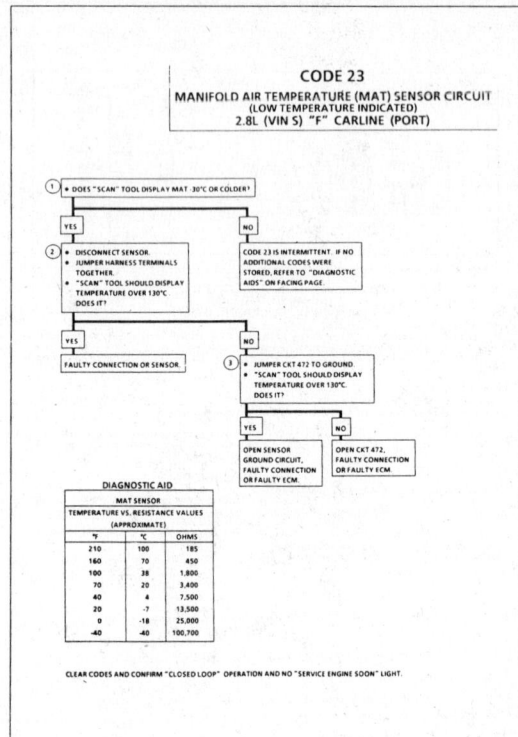

CLEAR CODES AND CONFIRM "CLOSED LOOP" OPERATION AND NO "SERVICE ENGINE SOON" LIGHT.

1988–90 2.8L (VIN S) – CAMARO AND FIREBIRD

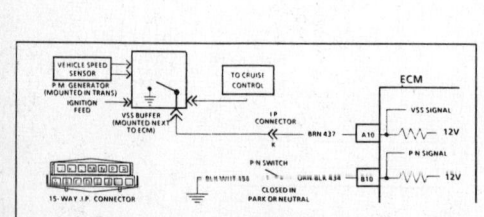

CODE 24
VEHICLE SPEED SENSOR (VSS) CIRCUIT
2.8L (VIN S) "F" CARLINE (PORT)

Circuit Description:

The ECM applies and monitors 12 volts on CKT 437. CKT 437 connects to the vehicle speed sensor buffer which alternately grounds CKT 437 when drive wheels are turning. This pulsing action takes place about 2000 times per mile and the ECM will calculate vehicle speed based on the time between "pulses".

A "SCAN" tool reading should closely match with speedometer reading with drive wheels turning.

Test Description: Numbers below refer to circled numbers on the diagnostic chart.
1. Code 24 will set if:
   - CKT 437 voltage is constant.
   - Engine speed between 1400 and 3600 rpm.
   - Less than 2% throttle opening.
   - Low load condition (low air flow).
   - Not in park or neutral.
   - All conditions must be met for 3 seconds.
   These conditions are met during a road load deceleration.
2. A voltage of less than 1 volt at the 15-way connector indicates that the CKT 437 wire may be shorted to ground. Disconnect CKT 437 at the VSS buffer. If voltage now reads above 10 volts, the VSS buffer is faulty.

If voltage remains less than 10 volt, then CKT 437 wire is grounded or open. If 437 is not grounded or open, check for a faulty ECM connector or ECM.

Diagnostic Aids:

If "Scan" displays vehicle speed, check park/neutral switch CHART C-1A on vehicle with auto trans. If switch is OK, check for intermittent connections. An open or short to ground in CKT 437 will result in a Code 24. If the customer also complained about a loss of mph on the I.P., check the P.M. generator circuit.

Refer to "Intermittents" in Section "B".

1988–90 2.8L (VIN S) – CAMARO AND FIREBIRD

CODE 24
VEHICLE SPEED SENSOR (VSS) CIRCUIT
2.8L (VIN S) "F" CARLINE (PORT)

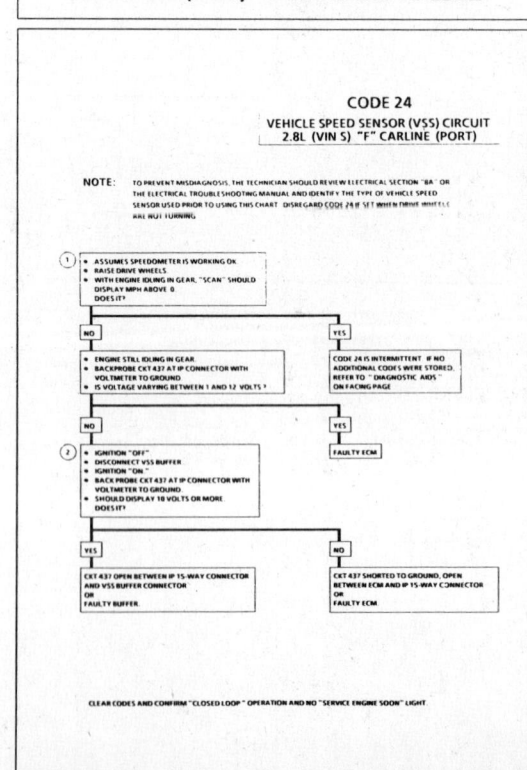

CLEAR CODES AND CONFIRM "CLOSED LOOP" OPERATION AND NO "SERVICE ENGINE SOON" LIGHT.

1988–90 2.8L (VIN S) — CAMARO AND FIREBIRD

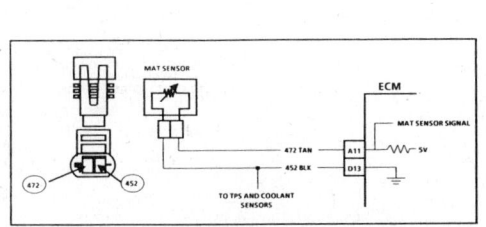

CODE 25
MANIFOLD AIR TEMPERATURE (MAT) SENSOR CIRCUIT
(HIGH TEMPERATURE INDICATED)
2.8L (VIN S) "F" CARLINE (PORT)

Circuit Description:
The manifold air temperature sensor uses a thermistor to control the signal voltage to the ECM. The ECM applies a voltage (about 5 volts) on CKT 472 to the sensor. When manifold air is cold, the sensor (thermistor) resistance is high, therefore the ECM will see a high signal voltage. As the air warms, the sensor resistance becomes less, and the voltage drops.

Test Description: Numbers below refer to circled numbers on the diagnostic chart.
1. Code 25 will set if:
   * Signal voltage indicates a manifold air temperature greater than 145°C (293° F) for 3 seconds.
   * Time since engine start is 8 minutes or longer.
   * A vehicle speed is present.

Diagnostic Aids:
A "SCAN" tool reads temperature of the air entering the engine and should read close to ambient air temperature, when engine is cold, and rises as underhood temperature increases.
A short to ground in CKT 472 will result in a Code 25.
Refer to "Intermittents" in Section "B".

1988–90 2.8L (VIN S) — CAMARO AND FIREBIRD

CODE 25
MANIFOLD AIR TEMPERATURE (MAT) SENSOR CIRCUIT
(HIGH TEMPERATURE INDICATED)
2.8L (VIN S) "F" CARLINE (PORT)

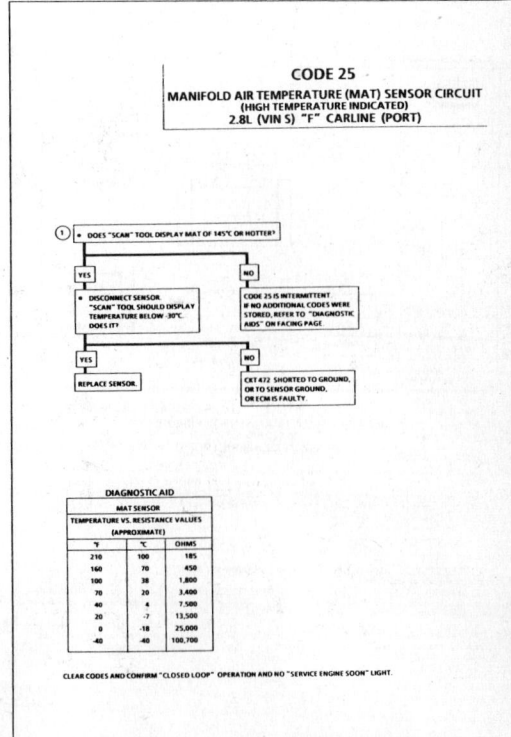

DIAGNOSTIC AID		
MAT SENSOR		
TEMPERATURE VS. RESISTANCE VALUES (APPROXIMATE)		
°F	°C	OHMS
210	100	185
160	70	450
100	38	1,800
70	20	3,400
40	4	7,500
20	-7	13,500
0	-18	25,000
-40	-40	100,700

CLEAR CODES AND CONFIRM "CLOSED LOOP" OPERATION AND NO "SERVICE ENGINE SOON" LIGHT.

1988–90 2.8L (VIN S) — CAMARO AND FIREBIRD

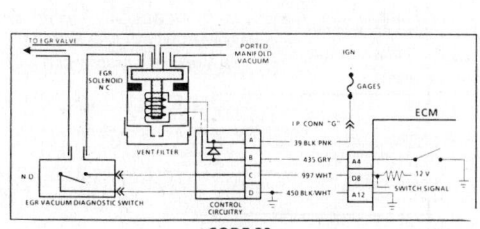

CODE 32
EXHAUST GAS RECIRCULATION (EGR) CIRCUIT
2.8L (VIN S) "F" CARLINE (PORT)

Circuit Description:
The EGR vacuum control uses an ECM controlled solenoid. The solenoid is normally closed and the vacuum source is a ported signal. The ECM will turn the EGR "ON" and "OFF" (Duty Cycle) by grounding CKT 435. The duty cycle is calculated by the ECM based on information from the coolant and mass airflow sensor and engine rpm. The duty cycle should be 0% (no EGR) when in park or neutral, TPS input below a specified value, or TPS indicating WOT.
With the ignition "ON", engine stopped, the EGR solenoid is de-energized unless the diagnostic terminal is grounded.
Code 32 means that the EGR vacuum diagnostic switch was closed during start-up, or that the switch was not detected closed under the following conditions:
* Coolant temperature greater than 80°C (176°F).
* EGR duty cycle commanded by the ECM is greater than 55%.
* TPS less than half throttle, but not at idle.
* All conditions above must be met for 5 seconds.
If the switch is detected closed during start-up, or, if the switch is detected open when the above conditions are met, the "Service Engine Soon" light will remain "ON" unless the switch changes state.

Test Description: Numbers below refer to circled numbers on the diagnostic chart.
1. If the first step caused Code 32 to set, then the ECM has recognized a closed vacuum switch on start up. This test will determine whether the EGR vacuum diagnostic switch is the cause or if the wiring or the ECM is the cause.
2. With the ignition "ON", the solenoid should not be energized and vacuum should not pass to the EGR valve.
3. To this point the EGR solenoid and valve are OK and the following check will check the diagnostic vacuum switch portion of the system.
4. The diagnostic switch should close at about 2" of vacuum. With vacuum applied, the switch should close and resistance go to near zero ohms and the vacuum should hold.

1988–90 2.8L (VIN S) — CAMARO AND FIREBIRD

CODE 32
EXHAUST GAS RECIRCULATION (EGR) CIRCUIT
2.8L (VIN S) "F" CARLINE (PORT)

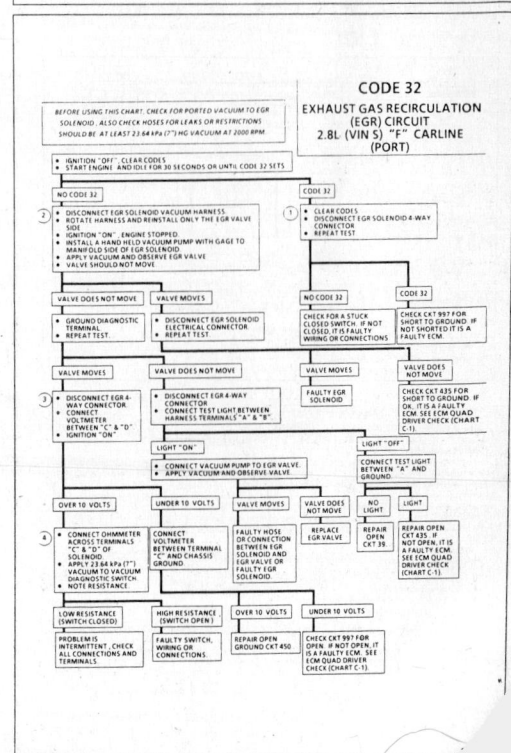

1988–90 2.8L (VIN S) – CAMARO AND FIREBIRD

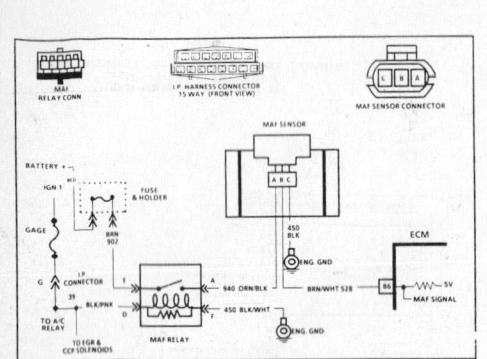

CODE 33
MASS AIR FLOW (MAF) SENSOR CIRCUIT
(GM/SEC HIGH)
2.8L (VIN S) "F" CARLINE (PORT)

Circuit Description:
The MAF sensor measures the flow of air entering the engine. The sensor produces a frequency output between 32 and 150 hertz (3gm/sec to 150gm/sec). A large quantity (high frequency) indicates acceleration, and a small quantity (low frequency) indicates deceleration or idle. This information is used by the ECM for fuel control and is converted by a "Scan" tool to read out the air flow in grams per second. A normal reading is about 4-7 grams per second at idle and increases with rpm.
The MAF sensor is powered up by the MAF sensor relay and the sensor should have power supplied to it anytime the ignition is "ON". The MAF sensor is located in the air duct.

Test Description: Numbers below refer to circled numbers on the diagnostic chart.
1. Code 33 will set if:
   * Ign "ON" and air flow exceeds 20gm/sec
     OR
   * Engine is running less than 1300 rpm.
   * TPS is 8% or less.
   * Air flow greater than 20 grams per second (high frequency)
   * All of the above are met for 2 seconds

Diagnostic Aids:
The "Scan" tool is not of much use in diagnosing this code because when the code sets gm/sec will be displaying the default value. However, the "Scan" may be useful in comparing the signal of a problem vehicle with that of a known good running one.
Refer to "Intermittents" in Section "B".

1988–90 2.8L (VIN S) – CAMARO AND FIREBIRD

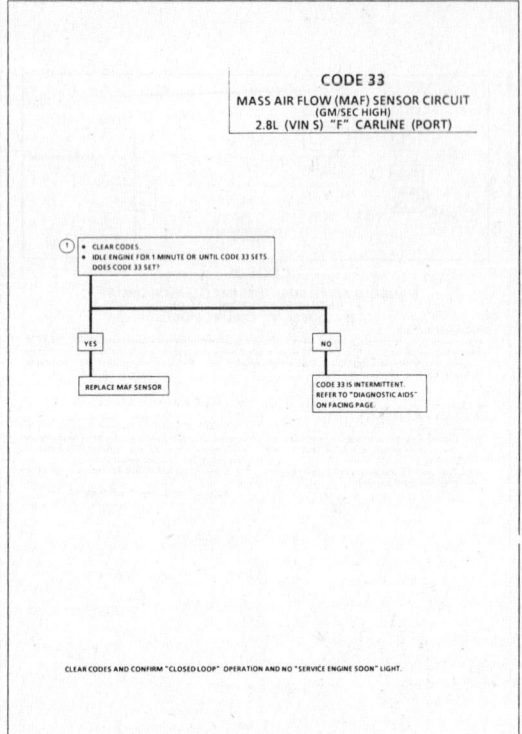

CODE 33
MASS AIR FLOW (MAF) SENSOR CIRCUIT
(GM/SEC HIGH)
2.8L (VIN S) "F" CARLINE (PORT)

CLEAR CODES AND CONFIRM "CLOSED LOOP" OPERATION AND NO "SERVICE ENGINE SOON" LIGHT.

1988–90 2.8L (VIN S) – CAMARO AND FIREBIRD

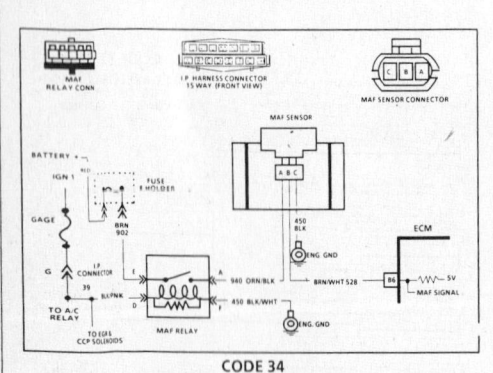

CODE 34
MASS AIR FLOW (MAF) SENSOR CIRCUIT
(GM/SEC LOW)
2.8L (VIN S) "F" CARLINE (PORT)

Circuit Description:
The MAF sensor measures the flow of air entering the engine. The sensor produces a frequency output between 32 and 150 hertz (3gm/sec to 150gm/sec). A large quantity (high frequency) indicates acceleration, and a small quantity (low frequency) indicates deceleration or idle. This information is used by the ECM for fuel control and is converted by a "SCAN" tool to read out the air flow in grams per second. A normal reading is about 4-7 grams per second at idle and increase with rpm.
The MAF sensor is powered up by the MAF sensor relay and the sensor should have power supplied to it anytime the ignition is "ON". The MAF sensor is located in the air duct.

Test Description: Numbers below refer to circled numbers on the diagnostic chart.
1. Code 34 will set if:
   * Engine running
   * MAF sensor disconnected, faulty relay, or MAF signal circuit shorted to ground.
     OR
   * less than 2 grams per second flow

   A bad air duct can set Code 34 . . .
   . . . see if ECM recognizes a . . . at this point indicates an . . .
   . . . wire from ECM . . .
   . . . connector

3. Checks for 12 volt supply to MAF sensor
4. Checks for open in 12 volt supply to relay

Diagnostic Aids:
The "Scan" tool is not of much use in diagnosing this code because when the code sets gm/sec will be displaying the default value. However, the "Scan" may be useful in comparing the signal of a problem vehicle with that of a known good running one.
Check for loose or damaged air duct.
Inspect sensor and relay connections as an open will result in a Code 34.
Refer to "Intermittents" in Section "B".

1988–90 2.8L (VIN S) – CAMARO AND FIREBIRD

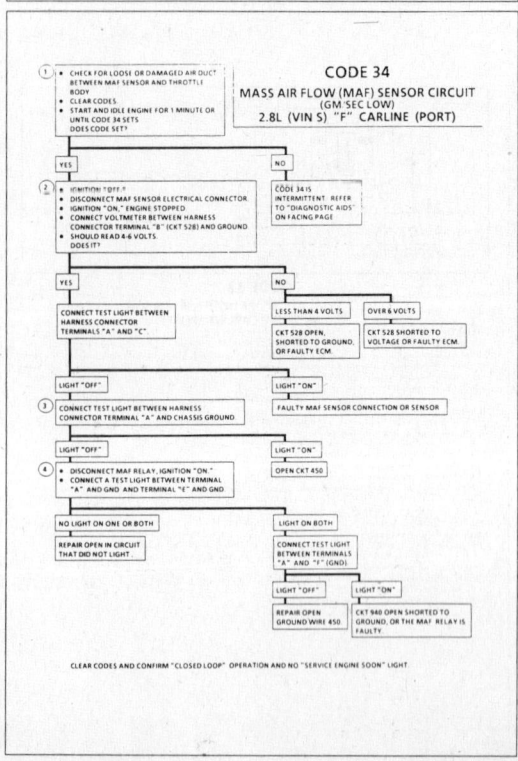

CODE 34
MASS AIR FLOW (MAF) SENSOR CIRCUIT
(GM/SEC LOW)
2.8L (VIN S) "F" CARLINE (PORT)

CLEAR CODES AND CONFIRM "CLOSED LOOP" OPERATION AND NO "SERVICE ENGINE SOON" LIGHT.

1988–90 2.8L (VIN S) – CAMARO AND FIREBIRD

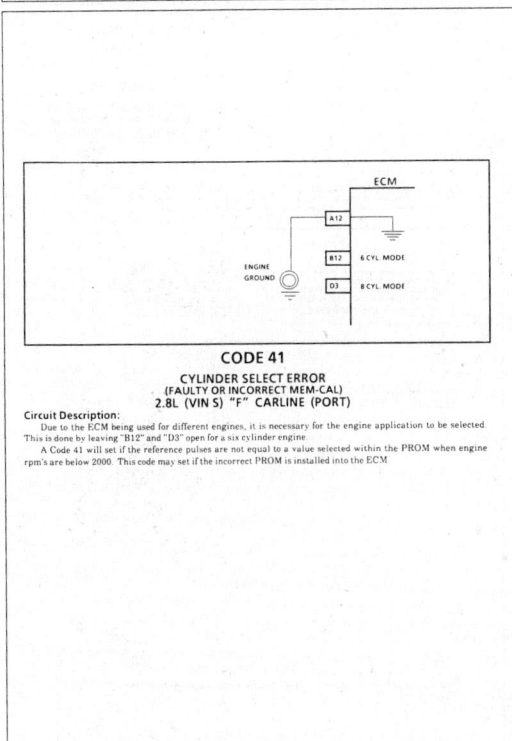

CODE 41
CYLINDER SELECT ERROR
(FAULTY OR INCORRECT MEM-CAL)
2.8L (VIN S) "F" CARLINE (PORT)

Circuit Description:

Due to the ECM being used for different engines, it is necessary for the engine application to be selected. This is done by leaving "B12" and "D3" open for a six cylinder engine.

A Code 41 will set if the reference pulses are not equal to a value selected within the PROM when engine rpm's are below 2000. This code may set if the incorrect PROM is installed into the ECM.

1988–90 2.8L (VIN S) – CAMARO AND FIREBIRD

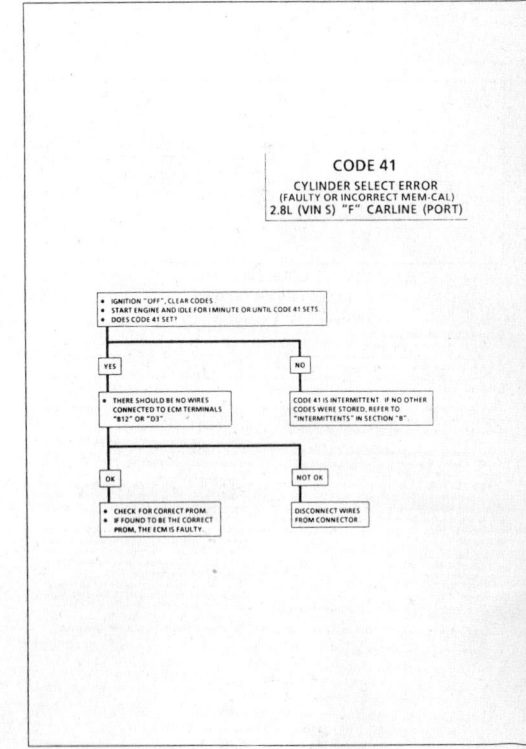

CODE 41
CYLINDER SELECT ERROR
(FAULTY OR INCORRECT MEM-CAL)
2.8L (VIN S) "F" CARLINE (PORT)

1988–90 2.8L (VIN S) – CAMARO AND FIREBIRD

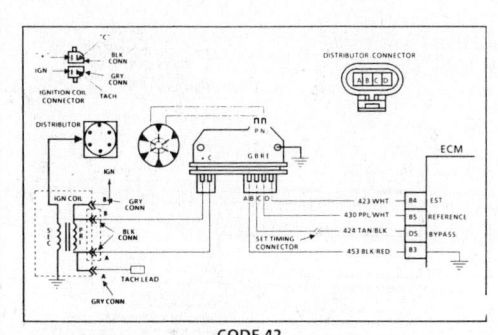

CODE 42
ELECTRONIC SPARK TIMING (EST) CIRCUIT
2.8L (VIN S) "F" CARLINE (PORT)

Circuit Description:

When the system is running on the ignition module, that is, no voltage on the bypass line, the ignition module grounds the EST signal. The ECM expects to see no voltage on the EST Line during this condition. If it sees a voltage, it sets Code 42 and will not go into the EST mode.

When the rpm for EST is reached (about 400 rpm), and bypass voltage applied, the EST should no longer be grounded in the ignition module so the EST voltage should be varying.

If the bypass line is open or grounded, the ignition module will not switch to EST mode so the EST voltage will be low and Code 42 will be set.

If the EST line is grounded, the ignition module will switch to EST, but because the line is grounded there will be no EST signal. A Code 42 will be set.

Test Description: Numbers below refer to circled numbers on the diagnostic chart.

1. Code 42 means the ECM has seen an open or short to ground in the EST or bypass circuits. This test confirms Code 42 and that the fault causing the code is present.
2. Checks for a normal EST ground path through the ignition module. An EST CKT 423 shorted to ground will also read less than 500 ohms, however, this will be checked later.
3. As the test light touches CKT 424, the module should switch causing the ohmmeter to "overrange" if the meter is in the 1000-2000 ohms position. Selecting the 10-20,000 ohms position will indicate above 5000 ohms. The important thing is that the module "switched"

4. The module did not switch and this step checks for
   - EST CKT 423 shorted to ground
   - Bypass CKT 424 open
   - Faulty ignition module connection or module.
5. Confirms that Code 42 is a faulty ECM and not an intermittent in CKTs 423 or 424

Diagnostic Aids:

The "Scan" tool does not have any ability to help diagnose a Code 42 problem.

A PROM not fully seated in the ECM can result in a Code 42

Refer to "Intermittents" in Section "B"

1988–90 2.8L (VIN S) – CAMARO AND FIREBIRD

CODE 42
ELECTRONIC SPARK TIMING (EST) CIRCUIT
2.8L (VIN S) "F" CARLINE (PORT)

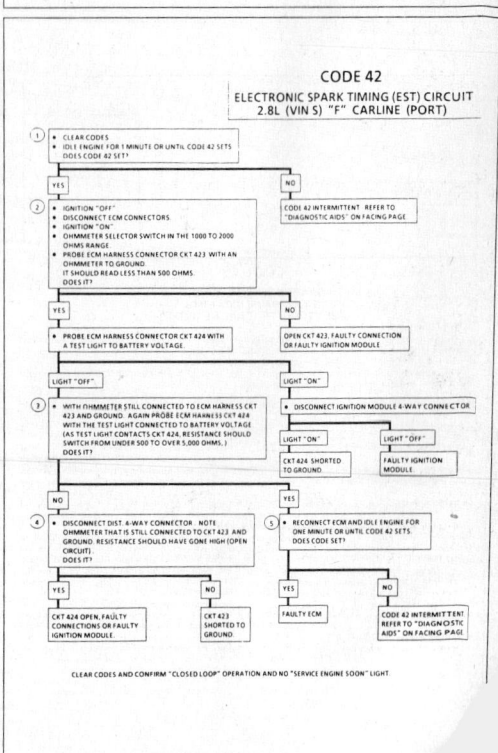

1988–90 2.8L (VIN S) – CAMARO AND FIREBIRD

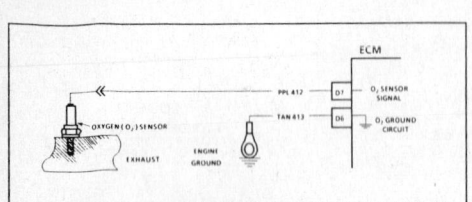

CODE 44
OXYGEN SENSOR CIRCUIT
(LEAN EXHAUST INDICATED)
2.8L (VIN S) "F" CARLINE (PORT)

Circuit Description:
The ECM supplies a voltage of about .45 volt between terminals "D6" and "D7". (If measured with a 10 megohm digital voltmeter, this may read as low as .32 volt.) The O_2 sensor varies the voltage within a range of about 1 volt if the exhaust is rich, down through about .10 volt if exhaust is lean.
The sensor is like an open circuit and produces no voltage when it is below about 360°C (600°F). An open sensor circuit or cold sensor causes "Open Loop" operation.

Test Description: Numbers below refer to circled numbers on the diagnostic chart.
1. Code 44 is set when the O_2 sensor signal voltage on CKT 412
   - Remains below .2 volt for 60 seconds or more
   - And the system is operating in "Closed Loop".

Diagnostic Aids:

Using the "Scan", observe the block learn values at different rpm and air flow conditions. If the conditions for Code 44 exists the block learn values will be around 150
- O_2 Sensor Wire Sensor pigtail may be mispositioned and contacting the exhaust manifold.
- Check for intermittent ground in wire between connector and sensor.
- MAF Sensor A mass air flow (MAF) sensor output that causes the ECM to sense a lower than normal air flow will cause the system to go lean. Disconnect the MAF sensor and if the lean condition is gone, check for a Code 34.

- Lean Injector(s) Perform injector balance test. CHART C-2A
- Fuel Contamination Water, even in small amounts, near the in-tank fuel pump inlet can be delivered to the injectors. The water causes a lean exhaust and can set a Code 44.
- Fuel Pressure System will be lean if pressure is too low. It may be necessary to monitor fuel pressure while driving the car at various road speeds and/or loads to confirm. See "Fuel System Diagnosis", CHART A-7.
- Exhaust Leaks If there is an exhaust leak, the engine can cause outside air to be pulled into the exhaust and past the sensor. Vacuum or crankcase leaks can cause a lean condition.
- Air System (manual trans only)
  Be sure air is not being directed to the exhaust ports while in "Closed Loop". If the block learn value goes down while squeezing air hose to exhaust ports, refer to CHART C-6.
- If the above are OK, it is a faulty oxygen sensor.

1988–90 2.8L (VIN S) – CAMARO AND FIREBIRD

CODE 44
OXYGEN SENSOR CIRCUIT
(LEAN EXHAUST INDICATED)
2.8L (VIN S) "F" CARLINE (PORT)

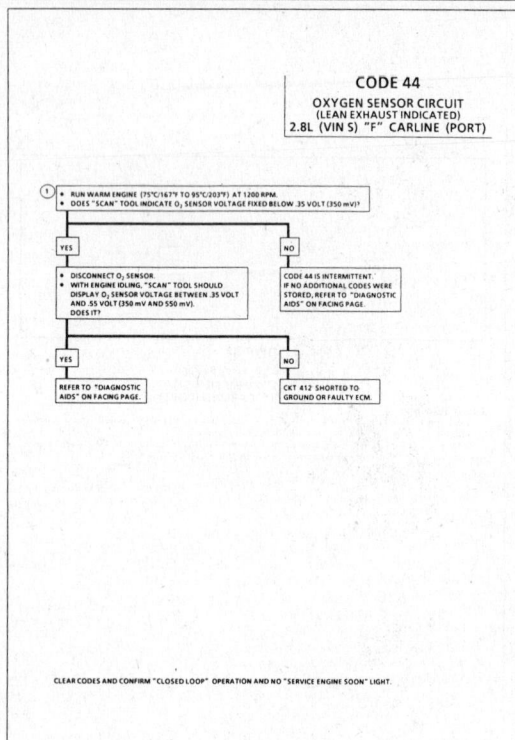

CLEAR CODES AND CONFIRM "CLOSED LOOP" OPERATION AND NO "SERVICE ENGINE SOON" LIGHT.

1988–90 2.8L (VIN S) – CAMARO AND FIREBIRD

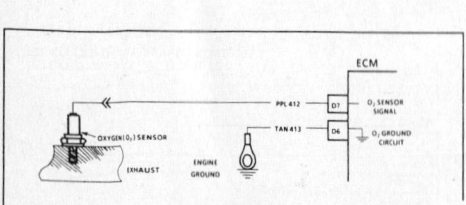

CODE 45
OXYGEN SENSOR CIRCUIT
(RICH EXHAUST INDICATED)
2.8L (VIN S) "F" CARLINE (PORT)

Circuit Description:
The ECM supplies a voltage of about .45 volt between terminals "D6" and "D7". (If measured with a 10 megohm digital voltmeter, this may read as low as .32 volt.) The O_2 sensor varies the voltage within a range of about 1 volt if the exhaust is rich, down through about .10 volt if exhaust is lean.
The sensor is like an open circuit and produces no voltage when it is below about 315°C (600°F). An open sensor circuit or cold sensor causes "Open Loop" operation.

Test Description: Numbers below refer to the circled numbers on the diagnostic chart.
1. Code 45 is set when the O_2 sensor signal voltage on CKT 412.
   - Remains above .7 volt for 30 seconds; and in "Closed Loop".
   - Engine time after start is 1 minute or more
   - Throttle less than 1/2 open but not at idle.

Diagnostic Aids:

Using the "Scan", observe the block learn values at different rpm and air flow conditions. If the conditions for Code 45 exists, the block learn values will be around 115.
- Fuel Pressure System will go rich if pressure is too high. The ECM can compensate for some increase [...] if it gets too high, a Code 45 may be set [...] System Diagnosis", CHART A-7.
  [...] Perform injector balance test CHART [...] CHART A-7. [...]

- HEI Shielding An open ground CKT 453 (ignition system reflow) may result in EMI, or induced electrical "noise". The ECM looks at this "noise" as reference pulses. The additional pulses result in a higher than actual engine speed signal. The ECM then delivers too much fuel, causing system to go rich. Engine tachometer will also show higher than actual engine speed, which can help in diagnosing this problem.
- Canister Purge Check for fuel saturation. If full of fuel, check canister control and hoses.
- MAF Sensor An output that causes the ECM to sense a higher than normal airflow can cause the system to go rich. Disconnecting the MAF sensor will allow the ECM to set a fixed value for the sensor. Substitute a different MAF sensor if the rich condition is gone while the sensor is disconnected. Check for a Code 34.
- Check for leaking fuel pressure regulator diaphragm by checking vacuum line to regulator for fuel.
- TPS An intermittent TPS output will cause the system to go rich, due to a false indication of the engine accelerating.
- EGR An EGR staying open (especially at idle) will cause the O_2 sensor to indicate a rich exhaust.

1988–90 2.8L (VIN S) – CAMARO AND FIREBIRD

CODE 45
OXYGEN SENSOR CIRCUIT
(RICH EXHAUST INDICATED)
2.8L (VIN S) "F" CARLINE (PORT)

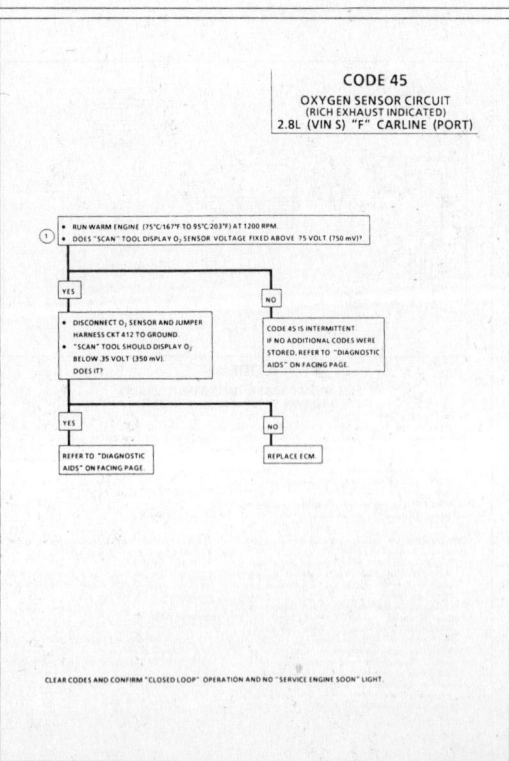

CLEAR CODES AND CONFIRM "CLOSED LOOP" OPERATION AND NO "SERVICE ENGINE SOON" LIGHT.

1988–90 2.8L (VIN S) – CAMARO AND FIREBIRD

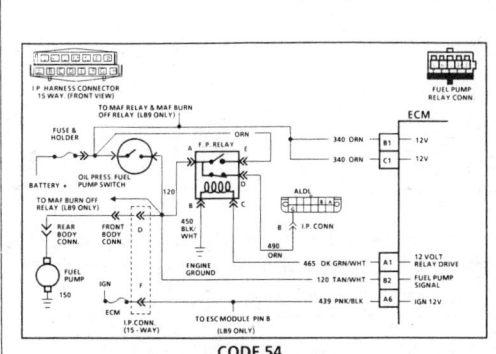

CODE 54
FUEL PUMP CIRCUIT
(LOW VOLTAGE)
2.8L (VIN S) "F" CARLINE (PORT)

Circuit Description:

The status of the fuel pump CKT 120 is monitored by the ECM at terminal "B2", and is used to compensate fuel delivery based on system voltage. This signal is also used to store a trouble code if the fuel pump relay is defective or fuel pump voltage is lost while the engine is running. There should be about 12 volts on CKT 120 for 2 seconds after the ignition is turned or any time references pulses are being received by the ECM.

Diagnostic Aids:

Code 54 will set if the voltage at terminal "B2" is less than 2 volts for 1.5 seconds since the last reference pulse was received. This will help in detecting a faulty relay, causing extended crank time and the code will help the diagnosis of an engine that "CRANKS BUT WILL NOT RUN".

If a fault is detected during start-up the "Service Engine Soon" light will stay "ON" until the ignition is cycled off. However, if the voltage is detected below 2 volts with the engine running the light will only remain on while the condition exists.

1988–90 2.8L (VIN S) – CAMARO AND FIREBIRD

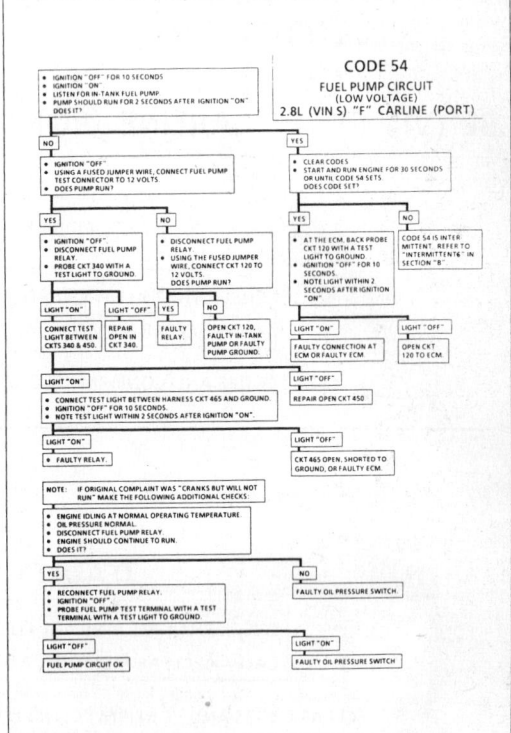

1988–90 2.8L (VIN S) – CAMARO AND FIREBIRD

SECTION B
SYMPTOMS

TABLE OF CONTENTS

Before Starting
Intermittents
Hard Start
Hesitation, Sag, Stumble
Surges and/or Chuggle
Lack of Power, Sluggish, or Spongy
Detonation/Spark Knock
Cuts Out, Misses
Backfire
Poor Fuel Economy
Dieseling, Run-On
Rough, Unstable, or Incorrect Idle, Stalling
Excessive Exhaust Emissions or Odors
Restricted Exhaust System Check (Chart B-1)

BEFORE STARTING

Before using this section you should have performed the DIAGNOSTIC CIRCUIT CHECK and found out that
1 The ECM and "Service Engine Soon" light are operating
2 There are no trouble codes stored, or there is a trouble code but no "Service Engine Soon" light. Verify the customer complaint, and locate the correct SYMPTOM below. Check the items indicated under that symptom.
If the ENGINE CRANKS BUT WILL NOT RUN, see CHART A-3.
Several of the symptom procedures below call for a careful visual check. This check should include

- ECM grounds for being clean and tight
- Vacuum hoses for splits, kinks, and proper connections, as shown on Emission Control Information label
- Air leaks at throttle body mounting and intake manifold
- Air leaks between MAF sensor and throttle body
- Ignition wires for cracking, hardness, proper routing, and carbon tracking
- Wiring for proper connections, pinches, and cuts. The importance of this step cannot be stressed too strongly - it can lead to correcting a problem without further checks and can save valuable time.

1988–90 2.8L (VIN S) – CAMARO AND FIREBIRD

INTERMITTENTS

Problem may or may not turn "ON" the "Service Engine Soon" light, or store a code

DO NOT use the trouble code charts in Section "A" for intermittent problems. The fault must be present to locate the problem. If the fault is intermittent, use of trouble code charts may result in replacement of good parts.

- Most intermittent problems are caused by faulty electrical connections or wiring. Perform careful check as described at start of Section "B". Check for:
  - Poor mating of the connector halves, or terminals not fully seated in the connector body (backed out)
  - Improperly formed or damaged terminals. All connector terminals in problem circuit should be carefully reformed to increase contact tension
  - Poor terminal to wire connection. This requires removing the terminal from the connector body to check
- If a visual check does not find the cause of the problem, the car can be driven with a voltmeter connected to a suspected circuit. A "Scan" tool can also be used for monitoring input signals to the ECM to help detect intermittent conditions. An abnormal voltage, or "Scan" reading, when the problem occurs, indicates the problem may be in that circuit. If the wiring and connectors check OK and a trouble code was stored for a circuit having a sensor, except for Codes 43, 44, and 45, substitute a known good sensor and recheck

An intermittent "Service Engine Soon" light with no stored code may be caused by
- Ignition coil shorted to ground and arcing at spark plug wires or plugs
- "Service Engine Soon" light wire to ECM shorted to ground. (CKT 419)
- Diagnostic "Test" terminal wire to ECM, shorted to ground. (CKT 451)
- ECM power grounds. See ECM wiring diagrams
- Loss of trouble memory. To check, disconnect TPS and idle engine until "Service Engine Soon" light comes "ON". Code 22 should be stored, and kept in memory when ignition is turned "OFF". If not, the ECM is faulty
- Check for an electrical system interference caused by a defective relay, ECM driven solenoid, or switch. They can cause a sharp electrical surge. Normally, the problem will occur when the faulty component is operated
- Check for improper installation of electrical options, such as lights, 2-way radios, etc.
- EST wires should be kept away from spark plug wires, distributor wires, distributor housing, coil, and generator. Wire from ECM to distributor (CKT 453) should be a good connection
- Check for open diode across A/C compressor clutch, and for other open diodes (see wiring diagrams)

HARD START

Definition: Engine cranks OK, but does not start for a long time. Does eventually run, or may start but immediately dies

- Perform careful check as described at start of Section "B".
- Make sure driver is using correct starting procedure
- CHECK:
  - TPS for sticking or binding or a high TPS voltage with the throttle closed (should read less than 700 volts)
  - High resistance in coolant sensor circuit or sensor itself. See Code 15 chart or with a "Scan" tool compare coolant temperature with ambient temperature on a cold engine
  - Fuel pressure CHART A-7
  - Water contaminated fuel
  - EGR operation. Be sure valve seats properly and is not staying open. See CHART C-7
  - Both injector fuses (visually inspect)
  - Ignition system - Check distributor for:
    - Proper output with ST-125
    - Worn shaft
    - Bare and shorted wires
    - Pickup coil resistance and connections
    - Loose ignition coil ground
    - Moisture in distributor cap
  - If problem exists in cold weather, check cold start valve. See CHART A-9

CODE 51
CODE 52
CODE 53
CODE 55
2.8L (VIN S) "F" CARLINE (PORT)

CODE 51

PROM ERROR
(FAULTY OR INCORRECT PROM)

CHECK THAT ALL PINS ARE FULLY INSERTED IN THE SOCKET. IF OK , REPLACE PROM , CLEAR MEMORY, AND RECHECK. IF CODE 51 REAPPEARS, REPLACE ECM.

CLEAR CODES AND CONFIRM "CLOSED LOOP" OPERATION AND NO "SERVICE ENGINE SOON" LIGHT.

CODE 52

CALPAK ERROR
(FAULTY OR INCORRECT CALPAK)

CHECK THAT ALL PINS ARE FULLY INSERTED IN THE SOCKET. IF OK , REPLACE CALPAK , CLEAR MEMORY, AND RECHECK. IF CODE 52 REAPPEARS, REPLACE ECM.

CLEAR CODES AND CONFIRM "CLOSED LOOP" OPERATION AND NO "SERVICE ENGINE SOON" LIGHT.

CODE 53

SYSTEM OVER VOLTAGE

THIS CODE INDICATES THERE IS A BASIC GENERATOR PROBLEM .
- CODE 53 WILL SET IF VOLTAGE AT ECM TERMINAL "B2" IS GREATER THAN 17.1 VOLTS FOR 2 SECONDS.
- CHECK AND REPAIR CHARGING SYSTEM. SEE SECTION "6D" .

CLEAR CODES AND CONFIRM "CLOSED LOOP" OPERATION AND NO "SERVICE ENGINE SOON" LIGHT.

CODE 55

ECM ERROR

BE SURE ECM GROUNDS ARE OK. IF OK
REPLACE ELECTRONIC CONTROL MODULE (ECM)

...ODES AND CONFIRM "CLOSED LOOP" OPERATION AND NO "SERVICE ENGINE SOON" LIGHT.

1988–90 2.8L (VIN S)–CAMARO AND FIREBIRD

- A faulty in-tank fuel pump check valve will allow the fuel in the lines to drain back to the tank after the engine is stopped. To check for this condition.
  Perform Fuel System Diagnosis, CHART A-7.
- Remove spark plugs. Check for wet plugs, cracks, wear, improper gap, burned electrodes, or heavy deposits. Repair or replace as necessary.
- If engine starts but then immediately stalls, open distributor by pass line. If engine then starts and runs OK, replace pickup coil.
- If engine starts and stalls, disconnect MAF sensor. If engine then runs and sensor connections are OK, replace the sensor.

HESITATION, SAG, STUMBLE

Definition: Momentary lack of response as the accelerator is pushed down. Can occur at all car speeds. Usually most severe when first trying to make the car move, as from a stop sign. May cause the engine to stall if severe enough.

- Perform careful visual check as described at start of Section "B".
- CHECK:
  - Fuel pressure. See CHART A-7. Also check for water contaminated fuel.
  - Air leaks at air duct between MAF sensor and throttle body.
  - Spark plugs for being fouled or faulty wiring.
  - PROM (2.8L) or MEM-CAL (5.0L & 5.7L) number. Also check service bulletins for latest MEM-CAL or PROM.
  - TPS for binding or sticking. Voltage should increase at a steady rate as throttle is moved toward WOT.
  - Ignition timing. See Emission Control Information label.
  - Generator output voltage. Repair if less than 9 or more than 16 volts.
  - HEI ground, CKT 453.
  - Canister purge system for proper operation. See CHART C-3.
  - EGR - See CHART C-7.
- Perform injector balance test CHART C-2A.

SURGES AND/OR CHUGGLE

Definition: Engine power variation under steady throttle or cruise. Feels like the car speeds up and slows down with no change in the accelerator pedal.

- Be sure driver understands transmission converter clutch and A/C compressor operation in Owner's Manual.
- Perform careful visual inspection as described at start of Section "B".
- CHECK:
  - TCC and 4th gear switch operation - See CHART C-8A.
  - Loose or leaking air duct between MAF sensor and throttle body.
  - Generator output voltage. Repair if less than 9 or more than 16 volts.
  - EGR - There should be no EGR at idle. See CHART C-7. Also check for plugged EGR solenoid filter.
  - Vacuum lines for kinks or leaks.
  - Ignition timing. See Emission Control Information label.
  - In-line fuel filter. Replace if dirty or plugged.
  - Fuel pressure while condition exists. See CHART A-7.
- Inspect oxygen sensor for silicone contamination from fuel, or use of improper RTV sealant. The sensor may have a white, powdery coating and result in a high but false signal voltage (rich exhaust indication). The ECM will then reduce the amount of fuel delivered to the engine, causing a severe driveability problem.
- Remove spark plugs. Check for cracks, wear, improper gap, burned electrodes, or heavy deposits. Also check condition of distributor cap, rotor, and spark plug wires.
- To help determine if the condition is caused by a rich or lean system, the car should be driven at the speed of the complaint. Monitoring block learn at the complaint speed will help identify the cause of the problem. If the system is runnig lean (block learn greater than 138), refer to "Diagnostic Aids" on facing page of Code 44. If the system is running rich (block learn less than 118), refer to "Diagnostic Aids" on facing page of Code 45.

1988–90 2.8L (VIN S)–CAMARO AND FIREBIRD

LACK OF POWER, SLUGGISH, OR SPONGY

Definition: Engine delivers less than expected power. Little or no increase in speed when accelerator pedal is pushed down part way.

- Perform careful visual check as described at start of Section "B".
- Compare customer's car to similar unit. Make sure the customer's car has an actual problem.
- Remove air cleaner and check air filter for dirt, or for being plugged. Replace as necessary.
- CHECK:
  - For loose or leaking air duct between MAF Sensor and throttle body.
  - Ignition timing. See Emission Control Information label.
  - Restricted fuel filter, contaminated fuel or improper fuel pressure. See CHART A-7.
  - ECM ground circuits - See ECM wiring diagrams.
  - EGR operation for being open or partly open all the time - CHART C-7.
  - Exhaust system for possible restriction. See CHART B-1.
    - Inspect exhaust system for damaged or collapsed pipes.
    - Inspect muffler for heat distress or possible internal failure.
  - Generator output voltage. Repair if less than 9 or more than 16 volts.
  - Engine valve timing and compression.
  - Engine for proper or worn camshaft.
  - Secondary voltage using a shop ocilliscope or a spark tester J-26792 (ST-125) or equivalent.

DETONATION /SPARK KNOCK

Definition: A mild to severe ping, usually worse under acceleration. The engine makes sharp metallic knocks that change with throttle opening. Sounds like popcorn popping.

- Check for obvious overheating problems:
  - Low coolant.
  - Loose water pump belt.
  - Restricted air flow to radiator, or restricted water flow thru radiator.
  - Inoperative electric cooling fan circuit. See CHART C-12.
- CHECK:
  - Ignition timing. See Vehicle Emission Control Information label.
  - EGR system for not opening - CHART C-7.
  - TCC operation - CHART C-8.
  - Fuel system pressure. See CHART A-7.
  - PROM or MEM-CAL - Be sure it's the correct one.
  - Valve oil seals for leaking.
- Check for incorrect basic engine parts such as cam, heads, pistons, etc.
- Check for poor fuel quality.
- Remove carbon with top engine cleaner. Follow instructions on can.
- Check ESC system (5.0L & 5.7L) See CHART C-5.
- To help determine if the condition is caused by a rich or lean system, the car should be driven at the speed of the complaint. Monitoring block learn at the complaint speed will help identify the cause of the problem. If the system is runnig lean (block learn greater than 138), refer to "Diagnostic Aids" on facing page of Code 44. If the system is running rich (block learn less than 118), refer to "Diagnostic Aids" on facing page of Code 45.

1988–90 2.8L (VIN S)–CAMARO AND FIREBIRD

CUTS OUT, MISSES

Definition: Steady pulsation or jerking that follows engine speed, usually more pronounced as engine load increases. The exhaust has a steady spitting sound at idle or low speed.

- Perform careful visual check as described at start of Section "B".
- Check for missing cylinder by:
  1. Disconnect IAC valve. Start engine. Remove one spark plug wire at a time using insulated pliers.
  2. If there is an rpm drop on all cylinders (equal to within 50 rpm), go to "ROUGH, UNSTABLE, OR INCORRECT IDLE, STALLING" symptom. Reconnect IAC valve.
  3. If there is no rpm drop on one or more cylinders, or excessive variation in drop, check for spark on the suspected cylinder(s) with J-26792 (ST-125) Spark Gap Tool or equivalent.

     If there is spark, remove spark plug(s) in these cylinders and check for:
     - Cracks
     - Wear
     - Improper gap
     - Burned electrodes
     - Heavy deposits
- Perform compression check on questionable cylinder(s) found above. If compression is low, repair as necessary.
- Disconnect all injector harness connectors. Connect J-34730-2 injector test light or equivalent 6 volt test light between the harness terms, of each injector connector and note light while cranking. If test light fails to blink at any connector, it is a faulty injector drive circuit harness, connector, or terminal.
- Perform the injector balance test. See CHART C-2A.
- CHECK:
  - Spark plug wires by connecting ohmmeter to ends of each wire in question. If meter reads over 30,000 ohms, replace wire(s).
  - Fuel System - Plugged fuel filter, water, low pressure. See CHART A-7.
  - Valve timing.
  - Secondary voltage using a shop ocilliscope or a spark tester J-26792 (ST-125) or equivalent.
- Visually inspect distributor cap and rotor for moisture, dust, cracks, burns, etc. Spray cap and plug wires with fine water mist to check for shorts.
- A miss condition can be caused by EMI (Electromagnetic Interference) on the reference circuit. EMI can usually be detected by monitoring engine rpm with a "Scan" tool. A sudden increase in rpm with little change in actual engine rpm change, indicates EMI is present.
  If the problem exists, check routing of secondary wires, check all distributor ground circuits.
- Remove rocker covers. Check for bent pushrods, worn rocker arms, broken valve springs, worn camshaft lobes. Repair as necessary.

BACKFIRE

Definition: Fuel ignites in intake manifold, or in exhaust system, making a loud popping noise.

- CHECK:
  - Loose wiring connector or air duct at MAF sensor.
  - Compression - Look for sticking or leaking valves.
  - EGR operation for being open all the time. See CHART C-7.
  - EGR gasket for faulty or loose fit.
  - Valve timing.
  - Output voltage of ignition coil using a shop ocilliscope or spark tester J-26792 (ST-125) or equivalent.
  - Spark plugs for crossfire also inspect (distributor cap, spark plug wires, and proper routing of plug wires).
  - Ignition system for intermittent condition.
  - Engine timing - see Emission Control Information label.
  - Perform fuel system diagnosis check, CHART A-7A.
  - Perform injector balance test, CHART C-2A.
  - Deceleration valve (2.8L manual/trans).
  - AIR system check valves.

1988–90 2.8L (VIN S)–CAMARO AND FIREBIRD

POOR FUEL ECONOMY

Definition: Fuel economy, as measured by an actual road test, is noticeably lower than expected. Also, economy is noticeably lower than it was on this car at one time, as previously shown by an actual road test.

- Perform careful visual check as described at start of Section "B".
- CHECK:
  - Coolant level.
  - Engine thermostat for faulty part (always open) or for wrong heat range.
  - Compression.
  - Ignition timing. See Emission Control Information label.
  - TCC for proper operation. A "Scan" should indicate an rpm drop when the TCC is commanded "ON". See CHART C-8.
  - Induction system and crankcase for air leaks.
  - Check for exhaust restriction. See CHART B-1.

DIESELING, RUN-ON

Definition: Engine continues to run after key is turned "OFF", but runs very roughly. If engine runs smoothly, check ignition switch and adjustment.

- Check injectors for leaking. See CHART A-7.

ROUGH, UNSTABLE, OR INCORRECT IDLE, STALLING

Definition: The engine runs unevenly at idle. If bad enough, the car may shake. Also, the idle may vary in rpm (called "hunting"). Either condition may be bad enough to cause stalling. Engine idles at incorrect speed.

- Perform careful visual check as described at start of Section "B".
- CHECK:
  - Throttle linkage for sticking or binding. Also check TPS adjustment.
  - Ignition timing. See Emission Control Information label.
  - ECM ground circuits.
  - IAC system. See CHART C-2C.
  - Generator output voltage. Repair if less than 9 or more than 16 volts.
  - P/N switch circuit. See CHART C-1A, or use "Scan" tool.
  - Injector balance. See CHART C-2A.
  - PCV valve for proper operation by placing finger over inlet hole in valve end several times. Valve should snap back. If not, replace valve.
  - Evaporative Emission Control System. CHART C-3.
  - A/C signal to ECM terminal "B8" (5.0L & 5.7L). "Scan" tool should indicate A/C is being requested when ever A/C is selected and the pressure cycling switch is closed.
  - A/C system operation (2.8L) - See CHART C-10.
  - Loose or damaged MAF sensor duct between sensor and throttle body.
  - Power Steering Pressure Switch (2.8L - See CHART C-1E.
  - Check AIR system. There should be no AIR to ports while in "Closed Loop". See CHART C-6.
  - EGR valve. There should be no EGR at idle.
- Run a cylinder compression check.
- Inspect oxygen sensor for silicon contamination from fuel, or use of improper RTV sealant. The sensor will have a white, powdery coating, and will result in a high but false signal voltage (rich exhaust indication). The ECM will then reduce the amount of fuel delivered to the engine, causing a severe driveability problem.
- Check for fuel in pressure regulator hose. If present replace regulator assembly.
- Check ignition system, wires, plugs, rotor, etc.
- Check for loose or damaged air duct between MAF sensor and throttle body.
- Disconnect MAF sensor and if condition is corrected replace sensor.
- Clean injectors.
- Monitoring block learn will help identify the cause of the problem. If the system is runnig lean (block learn greater than 138), refer to "Diagnostic Aids" on facing page of Code 44. If the system is running rich (block learn less than 118), refer to "Diagnostic Aids" on facing page of Code 45.

1988–90 2.8L (VIN S) – CAMARO AND FIREBIRD

EXCESSIVE EXHAUST EMISSIONS OR ODORS
Definition: Vehicle fails an emission test. Vehicle has excessive "rotten egg" smell.
Excessive odors do not necessarily indicate excessive emission.

- Perform "Diagnostic Circuit Check."
- IF TEST SHOWS EXCESSIVE CO AND HC, (or also has excessive odors)
  - Check items which cause car to run RICH
    - Make sure engine is at normal operating temperature.
- CHECK:
  - Fuel pressure. See CHART A-7
  - Incorrect timing. See vehicle emission control information label
  - Canister for fuel loading. See CHART C-3
  - Injector balance. See CHART C-2A
  - PCV valve for being plugged, stuck, or blocked PCV hose, or fuel in the crankcase
  - Spark plugs, plug wires, and ignition components.
  - Check for lead contamination of catalytic converter (look for removal of fuel filler neck restrictor).
  - Check for properly installed fuel cap.

- If the system is running rich, (block learn less than 118), refer to "Diagnostic Aids" on facing page of Code 45.
- IF TEST SHOWS EXCESSIVE NOx
  - Check items which cause car to run LEAN, or to run too hot
    - EGR valve for not opening. See CHART C-7
    - Vacuum leaks
    - Coolant system and coolant fan for proper operation. See CHART C-12
    - Remove carbon with top engine cleaner. Follow instructions on can.
    - Check ignition timing for excessive base advance. See emission control information label.
- If the system is running lean, (block learn greater than 138), refer to "Diagnostic Aids" on facing page of Code 44.

1988–90 2.8L (VIN S) – CAMARO AND FIREBIRD

CHART B-1
RESTRICTED EXHAUST SYSTEM CHECK
ALL ENGINES

Proper diagnosis for a restricted exhaust system is essential before any components are replaced. Either of the following procedures may be used for diagnosis, depending upon engine or tool used.

CHECK AT A.I.R. PIPE:
1. Remove the rubber hose at the exhaust manifold A.I.R. pipe check valve. Remove check valve.
2. Connect a fuel pump pressure gauge to a hose and nipple from a Propane Enrichment Device (J 26911) (see illustration).
3. Insert the nipple into the exhaust manifold A.I.R. pipe.

OR CHECK AT O₂ SENSOR:
1. Carefully remove O_2 sensor.
2. Install Borroughs exhaust backpressure tester (BT 8515 or BT 8603) or equivalent in place of O_2 sensor (see illustration).
3. After completing test described below, be sure to coat threads of O_2 sensor with anti-seize compound P/N 5613695 or equivalent prior to re-installation.

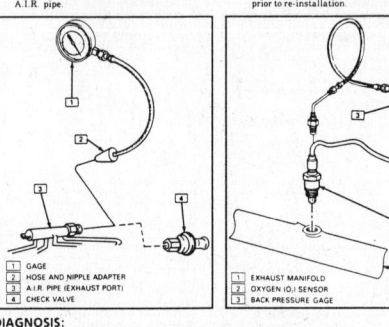

1	GAGE
2	HOSE AND NIPPLE ADAPTER
3	A.I.R. PIPE (EXHAUST PORT)
4	CHECK VALVE

1	EXHAUST MANIFOLD
2	OXYGEN (O₂) SENSOR
3	BACK PRESSURE GAGE

DIAGNOSIS:
1. With the engine idling at normal operating temperature, observe the exhaust system backpressure reading on the gage. Reading should not exceed 8.6 kPa (1.25 psi).
2. Increase engine speed to 2000 rpm and observe gage. Reading should not exceed 20.7 kPa (3 psi).
3. If the backpressure at either speed exceeds specification, a restricted exhaust system is indicated.
4. Inspect the entire exhaust system for a collapsed pipe, heat distress, or possible internal muffler failure.
5. If there are no obvious reasons for the excessive backpressure, the catalytic converter is suspected to be restricted and should be replaced using current recommended procedures.

1988–90 2.8L (VIN S) – CAMARO AND FIREBIRD

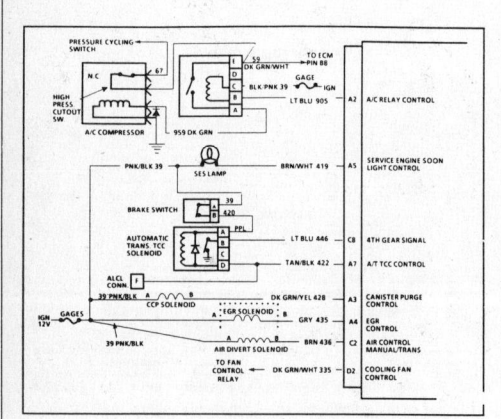

CHART C-1
ECM QDR CHECK
2.8L (VIN S) "F" CARLINE (PORT)

ECM Quad Driver (QDR) Check

The ECM uses an integrated circuit (IC) called a quad driver (QDR) in place of separate transistors to turn "ON" or "OFF" different circuits controlled by the ECM. Each QDR has four separate outputs that can independently turn "ON" or "OFF" four different circuits.

ECM service part number 1227302, used with this engine, does not have fault protection, therefore, a single faulty circuit many times causes all four QDR outputs to be inoperative or "ON" all the time. A failed QDR usually results in either a shorted or open ECM output. Because of the increased current flow, two QDR outputs are used to drive the TCC solenoid.

Refer to the ECM QDR check procedure on the facing page. This check will not test all ECM functions, but it will determine if a specific circuit has caused a specific QDR to fail in the ECM. A faulty QDR, therefore, is the largest cause of a failed QDR, therefore, the check procedure should be used whenever ECM replacement is indicated, especially if the removed ECM exhibits characteristics of a damaged QDR such as:

- SES light with no code stored.
- Engine will not start and/or ECM will not flash Code 12.
- Flickering, intermittent, or dim SES light.
- Output, such as TCC circuit, is inoperative or "ON" all times.
- Engine misfires, surges or stalls.
- "Scan" tool is erratic or inoperative.

1988–90 2.8L (VIN S) – CAMARO AND FIREBIRD

USE THIS CHECK PROCEDURE ONLY AFTER OTHER DIAGNOSTIC CHARTS IN THIS SERVICE MANUAL HAVE DETERMINED THAT THERE WAS AN ECM FAILURE

CHART C-1
ECM QDR CHECK
2.8L (VIN S) "F" CARLINE (PORT)

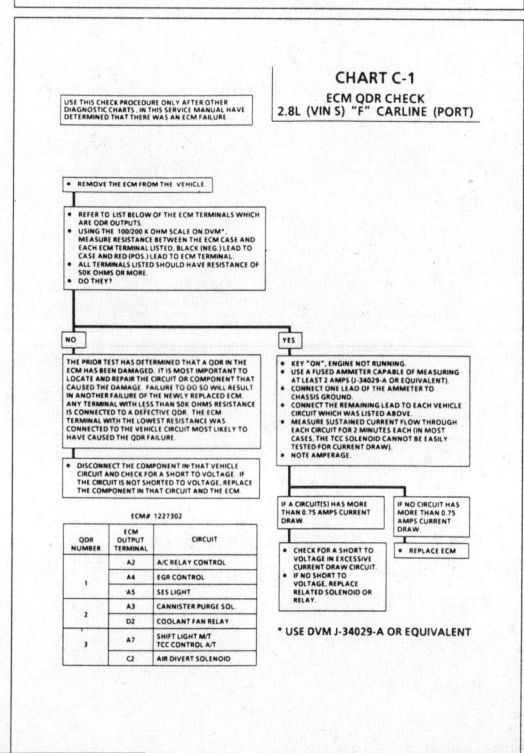

* REMOVE THE ECM FROM THE VEHICLE

* REFER TO LIST BELOW OF THE ECM TERMINALS WHICH ARE QDR OUTPUTS.
* USING THE 100/200 K OHM SCALE ON DVM*, MEASURE RESISTANCE BETWEEN THE ECM CASE AND EACH ECM TERMINAL LISTED. BLACK (NEG.) LEAD TO CASE AND RED (POS.) LEAD TO ECM TERMINAL.
* ALL TERMINALS LISTED SHOULD HAVE RESISTANCE OF 50K OHMS OR MORE.
DO THEY?

NO

THE PRIOR TEST HAS DETERMINED THAT A QDR IN THE ECM HAS BEEN DAMAGED. IT IS MOST IMPORTANT TO LOCATE AND REPAIR THE CIRCUIT OR COMPONENT THAT CAUSED THE DAMAGE. FAILURE TO DO SO WILL RESULT IN ANOTHER FAILURE OF THE NEWLY REPLACED ECM. ANY TERMINAL WITH LESS THAN 50K OHMS RESISTANCE IS CONNECTED TO A DEFECTIVE QDR. THE ECM TERMINAL WITH THE LOWEST RESISTANCE WAS CONNECTED TO THE VEHICLE CIRCUIT MOST LIKELY TO HAVE CAUSED THE QDR FAILURE.

* DISCONNECT THE COMPONENT IN THAT VEHICLE CIRCUIT AND CHECK FOR A SHORT TO VOLTAGE. IF THE CIRCUIT IS NOT SHORTED TO VOLTAGE, REPLACE THE COMPONENT IN THAT CIRCUIT AND THE ECM.

QDR NUMBER	ECM OUTPUT TERMINAL	CIRCUIT
1	A2	A/C RELAY CONTROL
	A4	EGR CONTROL
	A5	SES LIGHT
2	A3	CANNISTER PURGE SOL.
	D2	COOLANT FAN RELAY
3	A7	SHIFT LIGHT A/T TCC CONTROL A/T
	C2	AIR DIVERT SOLENOID

YES

* KEY "ON", ENGINE NOT RUNNING.
* USE A FUSED AMMETER CAPABLE OF MEASURING AT LEAST 2 AMPS (J-34029-A OR EQUIVALENT).
* CONNECT ONE LEAD OF THE AMMETER TO CHASSIS GROUND.
* CONNECT THE REMAINING LEAD TO EACH VEHICLE CIRCUIT WHICH WAS LISTED ABOVE.
* MEASURE SUSTAINED CURRENT FLOW THROUGH EACH CIRCUIT FOR 2 MINUTES EACH (IN MOST CASES, THE TCC SOLENOID CANNOT BE EASILY TESTED FOR CURRENT DRAW).
* NOTE AMPERAGE.

IF A CIRCUIT(S) HAS MORE THAN 0.75 AMPS CURRENT DRAW.	IF NO CIRCUIT HAS MORE THAN 0.75 AMPS CURRENT DRAW.
* CHECK FOR A SHORT TO VOLTAGE IN EXCESSIVE CURRENT DRAW CIRCUIT. IF NO SHORT TO VOLTAGE, REPLACE RELATED SOLENOID OR RELAY.	* REPLACE ECM

* USE DVM J-34029-A OR EQUIVALENT

1988–90 2.8L (VIN S) – CAMARO AND FIREBIRD

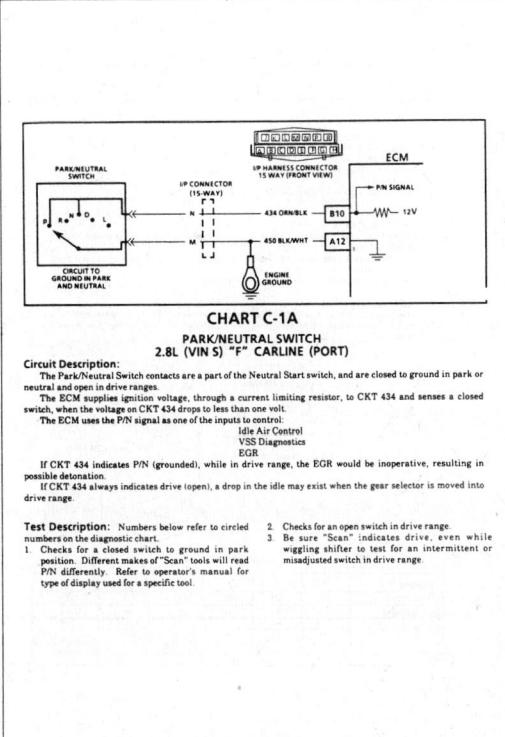

CHART C-1A
PARK/NEUTRAL SWITCH
2.8L (VIN S) "F" CARLINE (PORT)

Circuit Description:
The Park/Neutral Switch contacts are a part of the Neutral Start switch, and are closed to ground in park or neutral and open in drive ranges.
The ECM supplies ignition voltage, through a current limiting resistor, to CKT 434 and senses a closed switch, when the voltage on CKT 434 drops to less than one volt.
The ECM uses the P/N signal as one of the inputs to control:
Idle Air Control
VSS Diagnostics
EGR
If CKT 434 indicates P/N (grounded), while in drive range, the EGR would be inoperative, resulting in possible detonation.
If CKT 434 always indicates drive (open), a drop in the idle may exist when the gear selector is moved into drive range.

Test Description: Numbers below refer to circled numbers on the diagnostic chart.
1. Checks for a closed switch to ground in park position. Different makes of "Scan" tools will read P/N differently. Refer to operator's manual for type of display used for a specific tool.
2. Checks for an open switch in drive range.
3. Be sure "Scan" indicates drive, even while wiggling shifter to test for an intermittent or misadjusted switch in drive range.

1988–90 2.8L (VIN S) – CAMARO AND FIREBIRD

CHART C-1A
PARK/NEUTRAL SWITCH
2.8L (VIN S) "F" CARLINE (PORT)

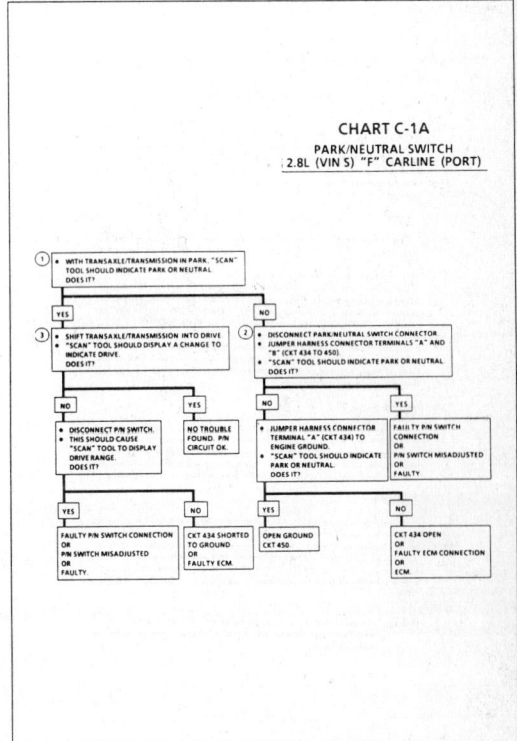

1988–90 2.8L (VIN S) – CAMARO AND FIREBIRD

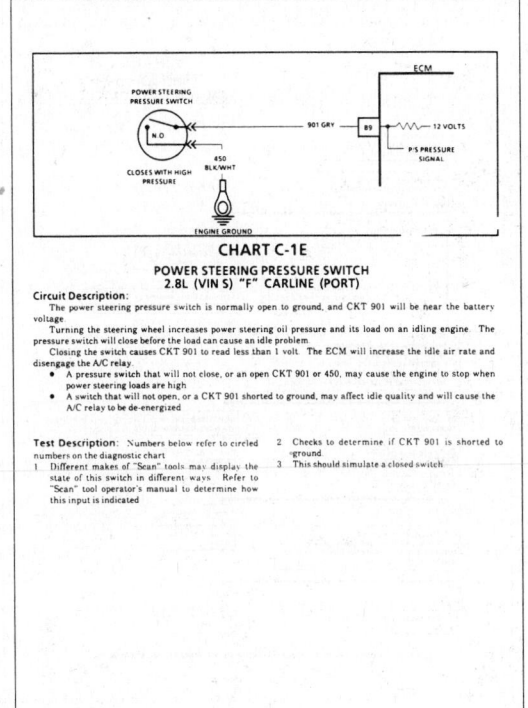

CHART C-1E
POWER STEERING PRESSURE SWITCH
2.8L (VIN S) "F" CARLINE (PORT)

Circuit Description:
The power steering pressure switch is normally open to ground, and CKT 901 will be near the battery voltage.
Turning the steering wheel increases power steering oil pressure and its load on an idling engine. The pressure switch will close before the load can cause an idle problem.
Closing the switch causes CKT 901 to read less than 1 volt. The ECM will increase the idle air rate and disengage the A/C relay.
• A pressure switch that will not close, or an open CKT 901 or 450, may cause the engine to stop when power steering loads are high
• A switch that will not open, or a CKT 901 shorted to ground, may affect idle quality and will cause the A/C relay to be de-energized

Test Description: Numbers below refer to circled numbers on the diagnostic chart.
1. Different makes of "Scan" tools may display the state of this switch in different ways. Refer to "Scan" tool operator's manual to determine how this input is indicated
2. Checks to determine if CKT 901 is shorted to ground.
3. This should simulate a closed switch.

1988–90 2.8L (VIN S) – CAMARO AND FIREBIRD

CHART C-1E
POWER STEERING PRESSURE SWITCH
2.8L (VIN S) "F" CARLINE (PORT)

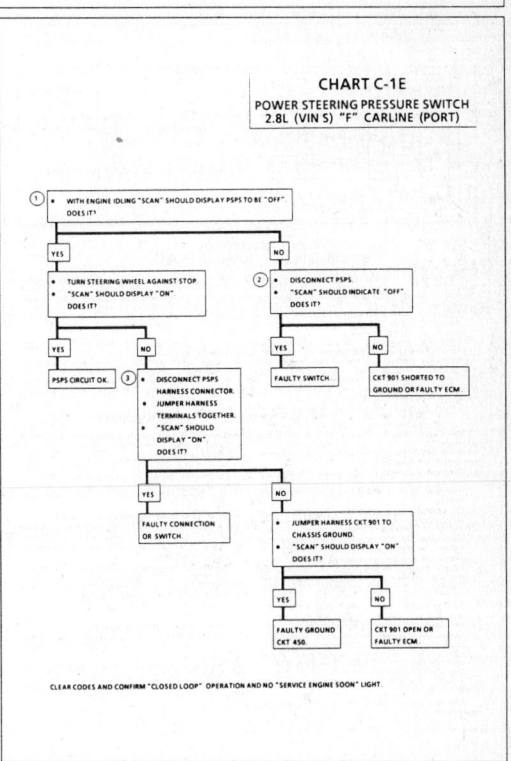

1988–90 2.8L (VIN S) – CAMARO AND FIREBIRD

CHART C-2A
INJECTOR BALANCE TEST

The injector balance tester is a tool used to turn the injector on for a precise amount of time, thus spraying a measured amount of fuel into the manifold. This causes a drop in fuel rail pressure that we can record and compare between each injector. All injectors should have the same amount of pressure drop (± 10 kPa). Any injector with a pressure drop that is 10 kPa (or more) greater or less than the average drop of the other injectors should be considered faulty and replaced.

STEP 1

Engine "cool down" period (10 minutes) is necessary to avoid irregular readings due to "Hot Soak" fuel boiling. With ignition "OFF" connect fuel gauge J-347301 or equivalent to fuel pressure tap. Wrap a shop towel around fitting while connecting gage to avoid fuel spillage.

Disconnect harness connectors at all injectors, and connect injector tester J-34730-3, or equivalent, to one injector. On Turbo equipped engines, use adaptor harness furnished with injector tester to energize injectors that are not accessible. Follow manufacturers instructions for use of adaptor harness. Ignition must be "OFF" at least 10 seconds to complete ECM shutdown cycle. Fuel pump should run about 2 seconds after ignition is turned "ON". At this point, insert clear tubing attached to vent valve into a suitable container and bleed air from gauge and hose to insure accurate gauge operation. Repeat this step until all air is bled from gauge.

STEP 2

Turn ignition "OFF" for 10 seconds and then "ON" again to get fuel pressure to its maximum. Record this initial pressure reading. Energize tester one time and note pressure drop at its lowest point. (Disregard any slight pressure increase after drop hits low point.) By subtracting this second pressure reading from the initial pressure, we have the actual amount of injector pressure drop.

STEP 3

Repeat step 2 on each injector and compare the amount of drop. Usually, good injectors will have virtually the same drop. Retest any injector that has a pressure difference of 10 kPa, either more or less than the average of the other injectors on the engine. Replace any injector that also fails the retest. If the pressure drop of all injectors is within 10 kPa of this average, the injectors appear to be flowing properly. Reconnect them and review "Symptoms," Section "B".

NOTE: The entire test should not be repeated more than once without running the engine to prevent flooding. (This includes any retest on faulty injectors).

1988–90 2.8L (VIN S) – CAMARO AND FIREBIRD

NOTE: If injectors are suspected of being dirty, they should be cleaned using an approved tool and procedure prior to performing this test. The fuel pressure test in Section "A", Chart A-7, should be completed prior to this test.

CHART C-2A
INJECTOR BALANCE TEST
2.8L (VIN S) "F" CARLINE (PORT)

Step 1. If engine is at operating temperature, allow a 10 minute "cool down" period then connect fuel pressure gauge and injector tester.
1. Ignition "OFF."
2. Connect fuel pressure gauge and injector tester.
3. Ignition "ON."
4. Bleed off air in gauge. Repeat until all air is bled from gauge.

Step 2. Run test:
1. Ignition "OFF" for 10 seconds.
2. Ignition "ON", Record gauge pressure. (Pressure must hold steady, if not see the Fuel System diagnosis, Chart A-7, in Section "A")
3. Turn injector on by depressing button on injector tester, and note pressure at the instant the gauge needle stops.

Step 3.
1. Repeat step 2 on all injectors and record pressure drop on each. Retest injectors that appear faulty (Any injectors that have a 10 kPa difference, either more or less, in pressure from the average). If no problem is found, review "Symptoms" Section "B".

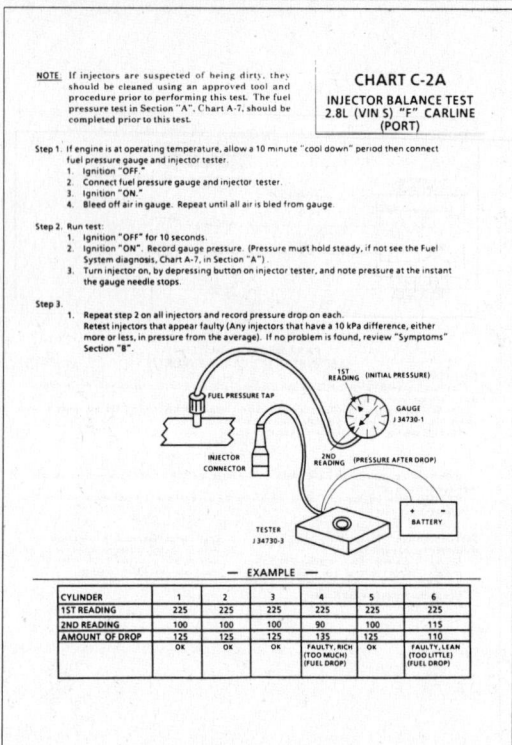

— EXAMPLE —

CYLINDER	1	2	3	4	5	6
1ST READING	225	225	225	225	225	225
2ND READING	100	100	100	90	100	115
AMOUNT OF DROP	125	125	125	135	125	110
	OK	OK	OK	FAULTY, RICH (TOO MUCH) (FUEL DROP)	OK	FAULTY, LEAN (TOO LITTLE) (FUEL DROP)

1988–90 2.8L (VIN S) – CAMARO AND FIREBIRD

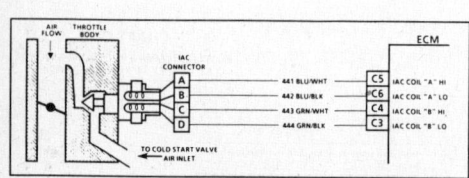

CHART C-2C
IDLE AIR CONTROL
2.8L (VIN S) "F" CARLINE (PORT)

Circuit Description:
The ECM will control engine idle speed by moving the IAC valve to control air flow around the throttle plate. It does this by sending voltage pulses to the proper motor winding for each IAC motor. This will cause the motor shaft and pintle to move in or out of the motor a given distance for each pulse received. The pulses are referred to as "counts".
• To increase idle speed. ECM will send enough counts to retract the IAC valve and allow more air to flow through the idle air passage and bypass the throttle plate until idle speed reaches the proper RPM. This will increase the ECM counts.
• To decrease idle speed. ECM will send enough counts to extend the IAC valve and reduce air flow through the idle passage around the throttle plate. This will reduce the ECM counts.
Each time the engine is started and then the ignition is turned "OFF" the ECM will reset the IAC valve. This is done by sending enough counts to seat the valve. The fully seated valve is the ECM reference zero. A given number of counts are then added to open the valve, and normal ECM control of IAC will begin from this point. The number of counts are then calculated by the ECM. This is how the ECM knows what the motor position is for a given idle speed.
The ECM uses the following information to control idle speed.
• Battery voltage   • Engine Speed   • Throttle Position Sensor
• Coolant Temperature   • A/C clutch signal
Don't apply battery voltage across the IAC motor terminals. It will permanently damage the IAC motor windings.

Test Description: Numbers below refer to circled numbers on the diagnostic chart.
1   Continue with test even if engine will not idle. If idle is too low. "Scan" will display 80 or more counts, or steps. If idle is high it will display "0" counts.
Occasionally an erratic or unstable idle may occur. Engine speed may vary 200 rpm or more up and down. Disconnect IAC. If the condition is unchanged, the IAC is not at fault. There is a system problem. Proceed to diagnostic aids below.
2   When the engine was stopped, the IAC valve retracted (more air) to a fixed "Park" position for increased air flow and idle speed during the next engine start. A "Scan" will display 140 or more counts.
3   Be sure to disconnect the IAC valve prior to this test. The test light will confirm the ECM signals by a steady or flashing light on all circuits.
4   There is a remote possibility that one of the CKTs is shorted to voltage which would have been indicated by a steady light. Disconnect ECM and turn the Ignition "ON" and probe terminals to check for this condition.

Diagnostic Aids:
Engine idle speed can be adversely affected by the following

• Park/Neutral Switch - If ECM thinks the car is always in neutral, then idle will not be controlled to the specified rpm when in drive range.
• Leaking injector(s) will cause fuel imbalance and poor idle quality due to excess fuel. See CHART A-7.
• Vacuum or crankcase leaks can affect idle.
• When the throttle shaft or throttle position sensor is binding or sticking in an open throttle position, the ECM does not know if the vehicle has stopped and does not control idle.
• Check AIR management system for intermittent air to ports while in "Closed Loop".
• In addition to electrical control of EGR, be sure to examine the EGR valve for proper seating.
• Faulty battery cables can result in voltage variations. The ECM will try to compensate, which results in erratic idle speeds.
• The ECM will compensate for A/C compressor clutch loads. Loss of this signal would be most apparent in neutral.
• Contaminated fuel can adversely affect idle.
• Perform injector balance test CHART C-2A.
If all OK, refer to "Rough, Unstable, Incorrect Idle or Stalling" "Symptoms" in Section "B".

1988–90 2.8L (VIN S) – CAMARO AND FIREBIRD

CHART C-2C
IDLE AIR CONTROL
2.8L (VIN S) "F" CARLINE (PORT)

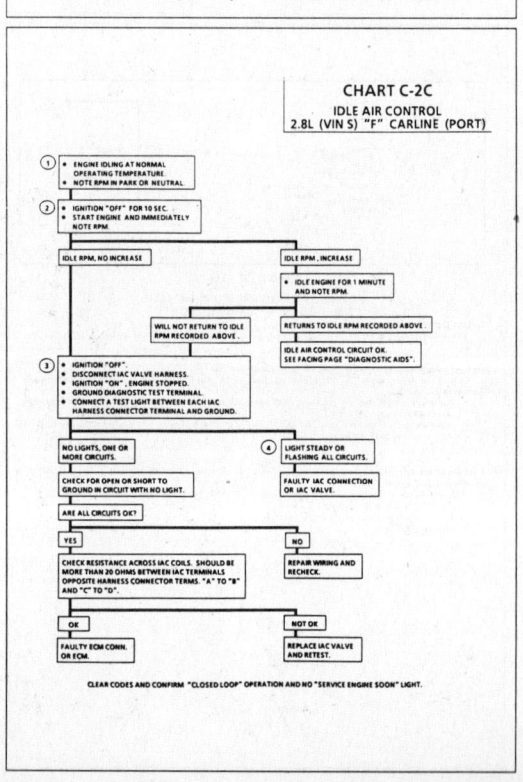

1988–90 2.8L (VIN S) – CAMARO AND FIREBIRD

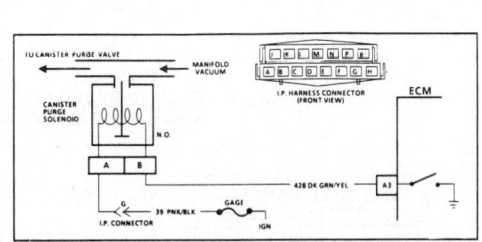

CHART C-3
CANISTER PURGE VALVE CHECK
2.8L (VIN S) "F" CARLINE (PORT)

Circuit Description:
Canister purge is controlled by a solenoid that allows manifold vacuum to purge the canister when de-energized. The ECM supplies a ground to energize the solenoid (purge "OFF")
If the diagnostic test terminal is ungrounded with the engine stopped or the following is met with the engine running, the purge solenoid is de-energized (purge "ON").
- Engine run time after start more than 1 minute.
- Coolant temperature above 75°C.
- Vehicle speed above 15 mph.
- Throttle off idle.

Test Description: Numbers below refer to circled numbers on the diagnostic chart.
1. Checks to see if the solenoid is opened or close. The solenoid is normally energized in this step, so it should be close.
2. Checks for a complete circuit. Normally there is ignition voltage on CKT 39 and the ECM provides a ground on CKT 428.

A shorted solenoid could cause an open circuit in the ECM.
3. Completes functional check by ungrounding test terminal. This should normally de-energize the solenoid and allow the vacuum to drop (purge "ON")

1988–90 2.8L (VIN S) – CAMARO AND FIREBIRD

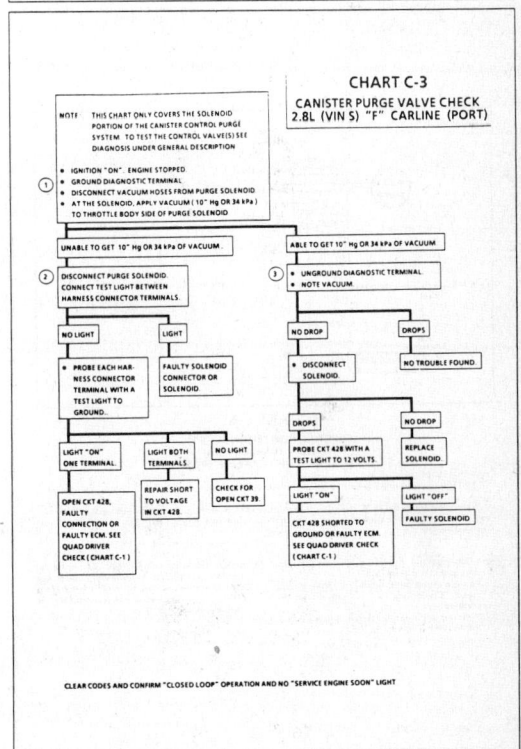

CHART C-3
CANISTER PURGE VALVE CHECK
2.8L (VIN S) "F" CARLINE (PORT)

1988–90 2.8L (VIN S) – CAMARO AND FIREBIRD

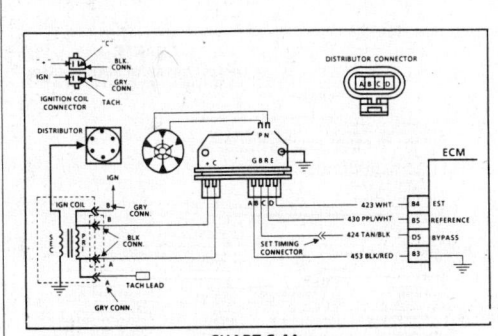

CHART C-4A
IGNITION SYSTEM CHECK
(REMOTE COIL / SEALED MODULE CONNECTOR DISTRIBUTOR)
2.8L (VIN S) "F" CARLINE (PORT)

Test Description: Numbers below refer to circled numbers on the diagnostic chart.
1. Two wires are checked, to ensure that an open is not present in a spark plug wire.
1A. If spark occurs with EST connector disconnected, pick-up coil output is too low for EST operation.
2. A spark indicates the problem must be the distributor cap or rotor.
3. Normally, there should be battery voltage at the "C" and "+" terminals. Low voltage would indicate an open or a high resistance circuit from the distributor to the coil or ignition switch. If "C" term. voltage was low, but "+" term voltage is 10 volts or more, circuit from "C" term. to Ign. coil or ignition coil primary winding is open.
4. Checks for a shorted module or grounded circuit from the ignition coil to the module. The distributor module should be turned "OFF", so normal voltage should be about 12 volts. If the module is turned "ON", the voltage would be low, but above 1 volt. This could cause the ignition coil to fail from excessive heat. With an open ignition coil primary winding, a small amount of voltage will leak through the module from the "Bat." to the tach terminal.
5. Applying a voltage (1.5 to 8V) to module terminal "P" should turn the module "ON" and the tach term. voltage should drop to about 7-9 volts. This test will determine whether the module or coil is faulty or if the pick-up coil is not generating the proper signal to turn the module "ON". This test can be performed by connecting a DC battery with a rating of 1.5 to 8 volts. The use of the test light is mainly to allow the "P" terminal to be probed more easily.
Some digital multi-meters can also be used to trigger the module by selecting ohms, usually the diode position. In this position the meter may have a voltage across it's terminals which can be used to trigger the module. The ohm's position can be checked by using a second meter or by checking the manufacture's specification of the tool being used.
6. This should turn "OFF" the module and cause a spark. If no spark occurs, the fault is most likely in the ignition coil because most module problems would have been found before this point in the procedure. A module tester could determine which is at fault.

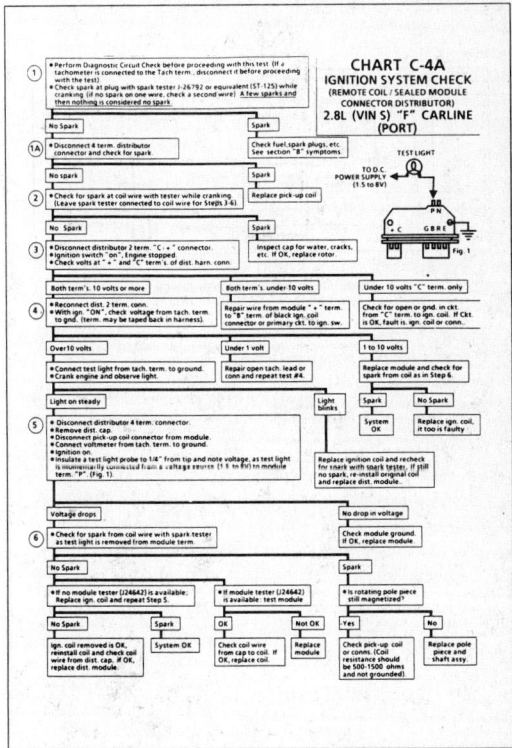

1988–90 2.8L (VIN S) – CAMARO AND FIREBIRD

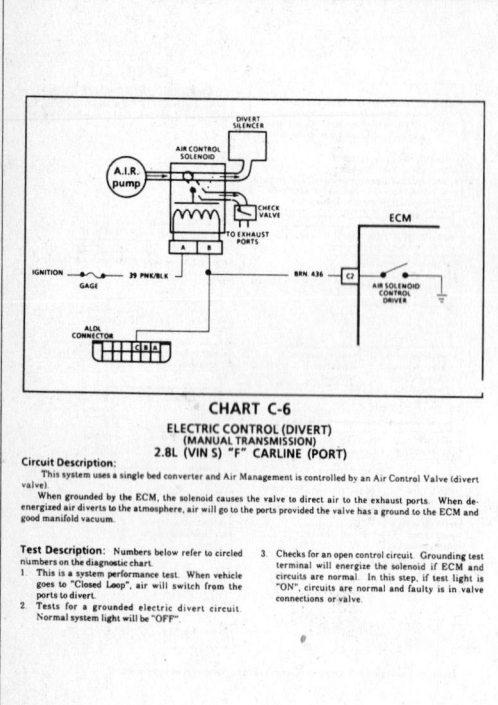

CHART C-6
ELECTRIC CONTROL (DIVERT)
(MANUAL TRANSMISSION)
2.8L (VIN S) "F" CARLINE (PORT)

Circuit Description:
This system uses a single bed converter and Air Management is controlled by an Air Control Valve (divert valve).
When grounded by the ECM, the solenoid causes the valve to direct air to the exhaust ports. When de-energized air diverts to the atmosphere, air will go to the ports provided the valve has a ground to the ECM and good manifold vacuum.

Test Description: Numbers below refer to circled numbers on the diagnostic chart.
1. This is a system performance test. When vehicle goes to "Closed Loop", air will switch from the ports to divert.
2. Tests for a grounded electric divert circuit. Normal system light will be "OFF".
3. Checks for an open control circuit. Grounding test terminal will energize the solenoid if ECM and circuits are normal. In this step, if test light is "ON", circuits are normal and faulty is in valve connections or valve.

1988–90 2.8L (VIN S) – CAMARO AND FIREBIRD

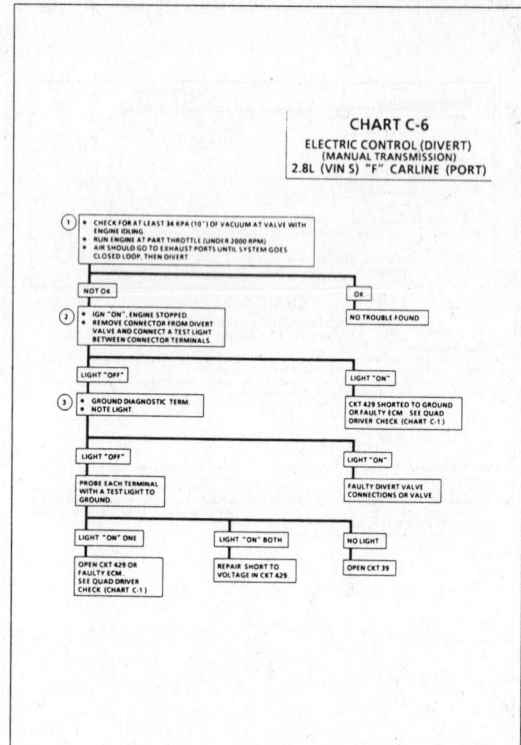

CHART C-6
ELECTRIC CONTROL (DIVERT)
(MANUAL TRANSMISSION)
2.8L (VIN S) "F" CARLINE (PORT)

1988–90 2.8L (VIN S) – CAMARO AND FIREBIRD

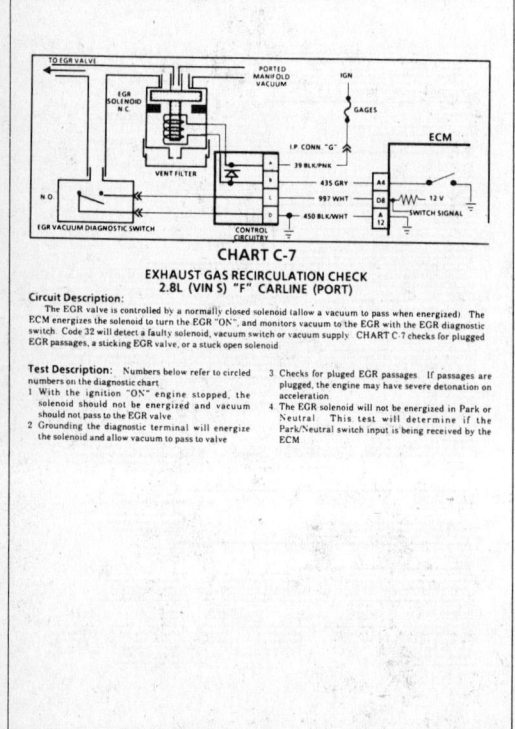

CHART C-7
EXHAUST GAS RECIRCULATION CHECK
2.8L (VIN S) "F" CARLINE (PORT)

Circuit Description:
The EGR valve is controlled by a normally closed solenoid (allow a vacuum to pass when energized). The ECM energizes the solenoid to turn the EGR "ON", and monitors vacuum to the EGR with the EGR diagnostic switch. Code 32 will detect a faulty solenoid, vacuum switch or vacuum supply. CHART C-7 checks for plugged EGR passages, a sticking EGR valve, or a stuck open solenoid.

Test Description: Numbers below refer to circled numbers on the diagnostic chart.
1. With the ignition "ON", engine stopped, the solenoid should not be energized and vacuum should not pass to the EGR valve
2. Grounding the diagnostic terminal will energize the solenoid and allow vacuum to pass to valve
3. Checks for plugged EGR passages. If passages are plugged, the engine may have severe detonation on acceleration.
4. The EGR solenoid will not be energized in Park or Neutral. This test will determine if the Park/Neutral switch input is being received by the ECM

1988–90 2.8L (VIN S) – CAMARO AND FIREBIRD

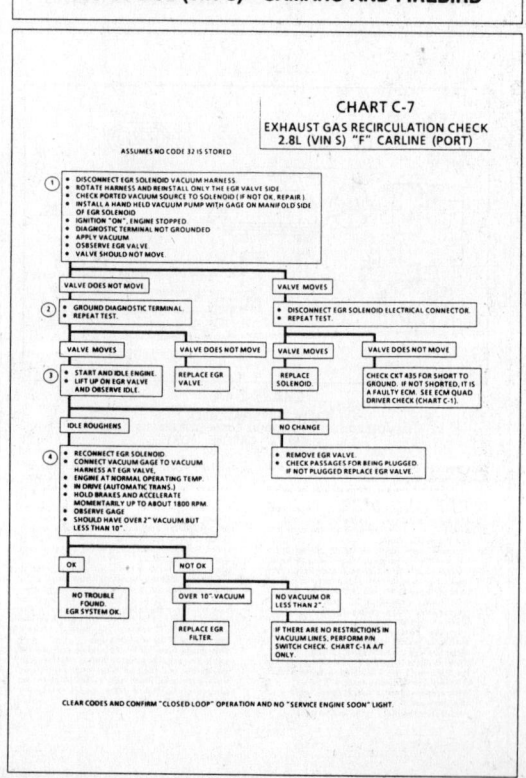

CHART C-7
EXHAUST GAS RECIRCULATION CHECK
2.8L (VIN S) "F" CARLINE (PORT)

1988–90 2.8L (VIN S) – CAMARO AND FIREBIRD

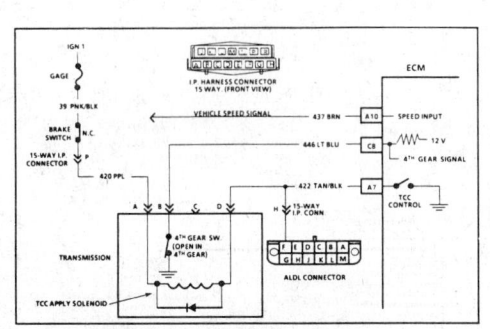

CHART C-8
(Page 1 of 2)
AUTOMATIC TRANSMISSION
CONVERTER CLUTCH (TCC)
2.8L (VIN S) "F" CARLINE (PORT)

Circuit Description:
The purpose of the automatic transmission torque converter clutch feature is to eliminate the power loss of the torque converter stage when the vehicle is in a cruise condition. This allows the convenience of the automatic transmission and the fuel economy of a manual transmission. The heart of the system is a solenoid located inside the automatic transmission which is controlled by the ECM.

When the solenoid coil is activated ("ON"), the torque converter clutch is applied which results in straight through mechanical coupling from the engine to transmission. When the transmission solenoid is deactivated, the torque converter clutch is released which allows the torque converter to operate in the conventional manner (fluidic coupling between engine and transmission).

The ECM turns "ON" the TCC when coolant temperature is above 65°C (149°F), TPS not changing, and vehicle speed above a specified value.

Test Description: Numbers below refer to circled numbers on the diagnostic chart.
1. When a test light is connected from ALDL terminal "F" to ground, a light "ON" indicates battery voltage is OK and the TCC solenoid is disengaged.
2. When the diagnostic terminal is grounded, the ECM should energise the TCC solenoid and the test light should go out.

Diagnostic Aids:
A "Scan" tool only indicates when the ECM has turned "ON" the TCC driver (grounded CKT 422), but this does not confirm that the TCC has engaged. To determine if TCC is functioning properly, engine rpm should decrease when the "Scan" indicates the TCC driver has turned "ON". To determine if the 4th gear switch is functioning properly, perform the checks in CHART C-8 (Page 2 of 2). The switches will not prevent TCC from functioning but will affect TCC lock and unlock points. If the 4th gear switch CKT is always open the TCC may engage as soon as sufficient oil pressure is reached.

1988–90 2.8L (VIN S) – CAMARO AND FIREBIRD

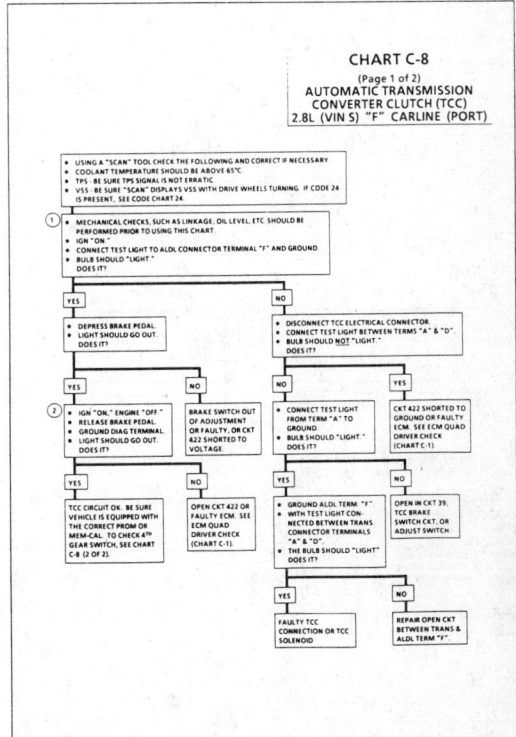

1988–90 2.8L (VIN S) – CAMARO AND FIREBIRD

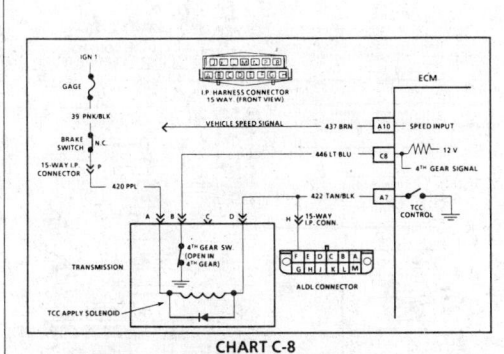

CHART C-8
(Page 2 of 2)
AUTOMATIC TRANSMISSION
CONVERTER CLUTCH (TCC)
2.8L (VIN S) "F" CARLINE (PORT)

Circuit Description:
A 4th gear switch (mounted in the trans.) opens when the trans. shifts into 4th gear, and this switch is used by the ECM to modify TCC lock and unlock points, when in a 4-3 downshift maneuver.

Test Description: Numbers below refer to the diagnostic chart.
1. Unless the switch or CKT 446 is open the "Scan" should display "NO," indicating the trans. is not in 4th gear. The 4th gear switch should only be open while in 4th gear.
2. This step determines if the ECM and wiring are OK. Grounding CKT 446 should cause the "Scan" to display "NO," indicating the trans. is not in 4th gear.
3. Checks the operation of the 4th gear switch. When the trans. shifts into 4th gear the switch should open and the "Scan" should display "YES".
4. Disconnecting the TCC connector simulates an open switch to determine if CKT 446 is shorted to ground or the problem is in the transmission.

Diagnostic Aids:
A road test may be necessary to verify the customer complaint. If the "Scan" indicates TCC is turning "ON" and "OFF" erratically, check the state of the 4th gear switch to be sure it is not changing states under a steady throttle position. If the switch is changing states, check connections and wire routing carefully. Also if the 4th gear switch is always open the TCC may engage as soon as sufficient oil pressure is reached.

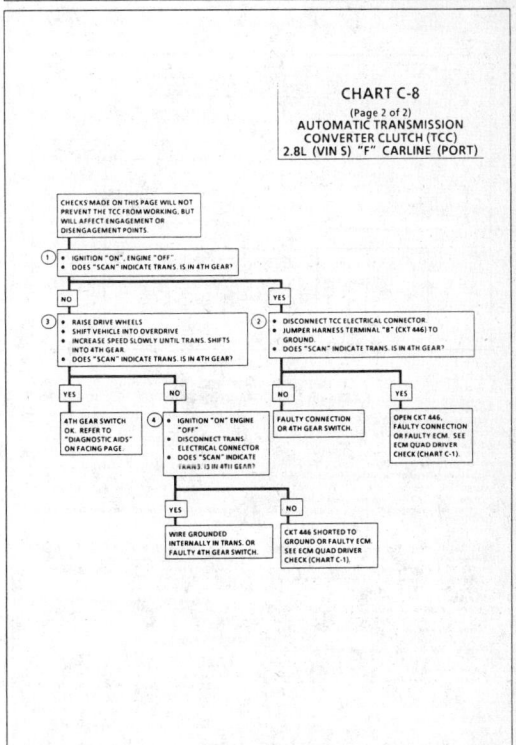

1988–90 2.8L (VIN S) – CAMARO AND FIREBIRD

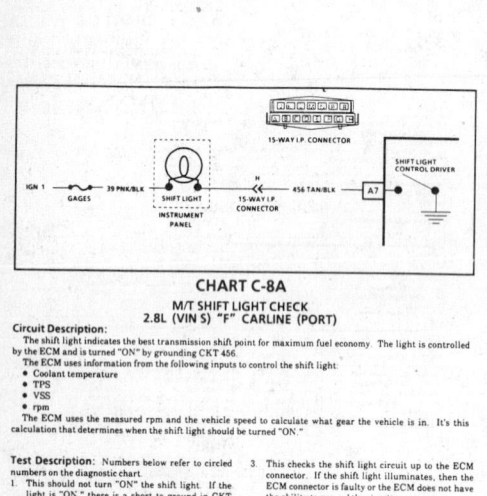

CHART C-8A
M/T SHIFT LIGHT CHECK
2.8L (VIN S) "F" CARLINE (PORT)

Circuit Description:
The shift light indicates the best transmission shift point for maximum fuel economy. The light is controlled by the ECM and is turned "ON" by grounding CKT 456.
The ECM uses information from the following inputs to control the shift light:
- Coolant temperature
- TPS
- VSS
- rpm

The ECM uses the measured rpm and the vehicle speed to calculate what gear the vehicle is in. It's this calculation that determines when the shift light should be turned "ON."

Test Description: Numbers below refer to circled numbers on the diagnostic chart.
1. This should not turn "ON" the shift light. If the light is "ON," there is a short to ground in CKT 456 wiring or a fault in the ECM.
2. When the diagnostic terminal is grounded, the ECM should ground CKT 456 and the shift light should come "ON."
3. This checks the shift light circuit up to the ECM connector. If the shift light illuminates, then the ECM connector is faulty or the ECM does not have the ability to ground the circuit.

1988–90 2.8L (VIN S) – CAMARO AND FIREBIRD

CHART C-8A
M/T SHIFT LIGHT CHECK
2.8L (VIN S) "F" CARLINE (PORT)

1. IGNITION "ON", ENGINE "OFF".
   OBSERVE SHIFT LIGHT.

   LIGHT "OFF" | **LIGHT "ON"**

2. GROUND ALDL DIAGNOSTIC TERMINAL AND OBSERVE SHIFT LIGHT.

   IGNITION "ON".
   DISCONNECT ECM CONNECTORS.
   TURN IGNITION "ON" AND OBSERVE SHIFT LIGHT.

   LIGHT "OFF" | **LIGHT "ON"** | **LIGHT "ON"** | **LIGHT "OFF"**

3. IGNITION "OFF".
   DISCONNECT ECM CONNECTORS.
   IGNITION "ON".
   JUMPER CKT 456 TO GROUND AND OBSERVE SHIFT LIGHT.

   CHECK FOR:
   - CODE 24, (NO VSS).
   - THERMOSTAT FAULTY OR HAS INCORRECT HEAT RANGE. IF OK, REVIEW SYMPTOMS IN SECTION "B".

   REPAIR SHORT TO GROUND IN CKT 456.

   REPLACE ECM.

   LIGHT "OFF" | **LIGHT "ON"**

   OPEN IGNITION CKT 39.
   OPEN CKT 456, OR FAULTY BULB.

   POOR CONNECTION AT ECM OR FAULTY ECM.

1988–90 2.8L (VIN S) – CAMARO AND FIREBIRD

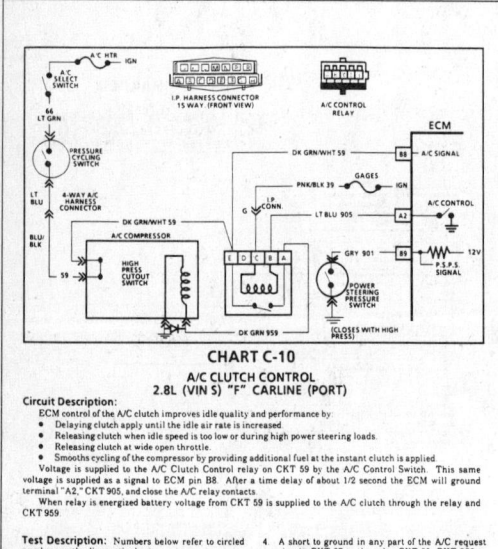

CHART C-10
A/C CLUTCH CONTROL
2.8L (VIN S) "F" CARLINE (PORT)

Circuit Description:
ECM control of the A/C clutch improves idle quality and performance by:
- Delaying clutch apply until the idle air rate is increased.
- Releasing clutch when idle speed is too low or during high power steering loads.
- Releasing clutch at wide open throttle.
- Smooths cycling of the compressor by providing additional fuel at the instant clutch is applied.

Voltage is supplied to the A/C Clutch Control relay on CKT 59 by the A/C Control Switch. This same voltage is supplied as a signal to ECM pin B8. After a time delay of about 1/2 second the ECM will ground terminal "A2," CKT 905, and close the A/C relay contacts.

When relay is energized battery voltage from CKT 59 is supplied to the A/C clutch through the relay and CKT 959.

Test Description: Numbers below refer to circled numbers on the diagnostic chart.
1. The ECM will only energize the A/C relay, when the engine is running. This test will determine if the relay, or CKT 905, is faulty.
2. In order for the clutch to properly be engaged, the pressure cycling switch must be closed to provide 12 volts to the relay, and the high pressure switch must be closed, so the A/C request (12 volts) will be present at the ECM.
3. Determines if the signal is reaching the ECM on CKT 59 from the A/C control panel. Signal should only be present when the A/C mode or defrost mode has been selected.

4. A short to ground in any part of the A/C request circuit, CKT 67 to the relay CKT 59, CKT 959 to the A/C clutch, or the A/C clutch, could be the cause of the blown fuse.
5. With the ignition "ON" and the diagnostic terminal grounded, the ECM should be grounding CKT 905 which should cause the test light to be "ON."

Diagnostic Aids:
If complaint was insufficient cooling, the problem may be caused by an inoperative cooling fan. The engine cooling fan should turn "ON," when A/C is "ON" and A/C head pressure exceeds about 233 psi. If not, see CHART C-12 for diagnosing the cooling fan.

1988–90 2.8L (VIN S) – CAMARO AND FIREBIRD

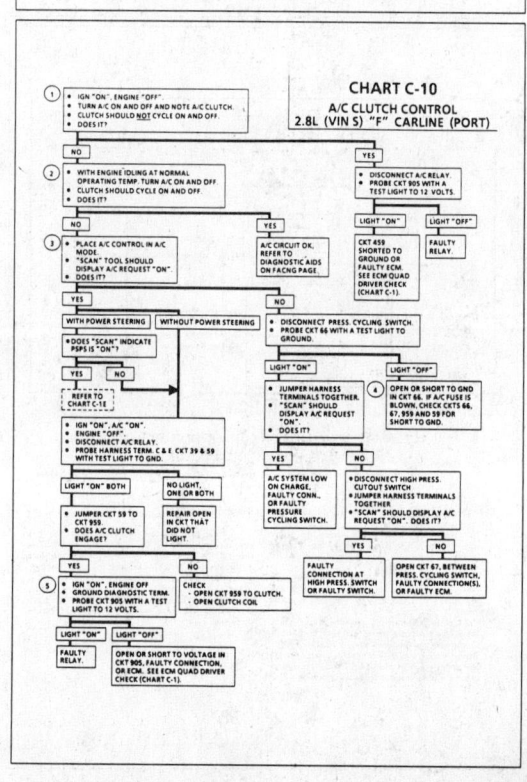

CHART C-10
A/C CLUTCH CONTROL
2.8L (VIN S) "F" CARLINE (PORT)

1988–90 2.8L (VIN S) – CAMARO AND FIREBIRD

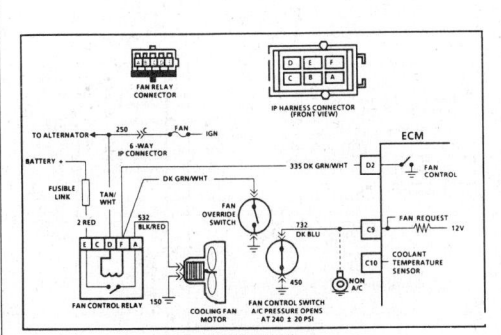

CHART C-12
(Page 1 of 2)
ELECTRIC COOLING FAN CONTROL CIRCUIT
2.8L (VIN S) "F" CARLINE (PORT)

Circuit Description:
The electric cooling fan is controlled by the ECM, based on inputs from the coolant temperature sensor, the A/C fan control switch, and vehicle speed. The ECM controls the fan by grounding CKT 335, which energizes the fan control relay. Battery voltage is then supplied to the fan motor.
The ECM grounds CKT 335, when coolant temp. is over about 107°C (225°F), or when A/C has been requested, and the fan control switch opens with high A/C pressure, about 240 psi (1655 kPa). Once the ECM turns the relay "ON", it will keep it "ON" for a minimum of 30 seconds, or until vehicle speed exceeds 70 mph.
Also, if Code 14 or 15 sets, or the ECM is in throttle body back up, the fan will run at all times.
On a vehicle not equipped with A/C, CKT 732 is jumpered to ground so that the fan does not run at all times.

Test Description: Numbers below refer to circled numbers on the diagnostic chart.
1. With the diagnostic terminal grounded, the cooling fan control driver will close, which should energize the fan control relay.
2. If the A/C fan control circuit or circuit is open, the fan would run whenever the engine is running.
3. With A/C clutch engaged, the A/C fan control switch should open, when A/C high pressure exceeds about 200 psi (1380 kPa). This signal should cause the ECM to energize the fan control relay.

Diagnostic Aids:
If the owner complained of an overheating problem, it must be determined if the complaint was due to an actual boilover, or the hot light, or temperature gage indicated over heating.
If the gage, or light, indicates overheating, but no boilover is detected, the gage circuit should be checked. The gage accuracy can, also, be checked by comparing the coolant sensor reading using a "Scan" tool and comparing its reading with the gage reading.
If the engine is actually overheating, and the gage indicates overheating, but the cooling fan is not coming "ON", the coolant sensor has probably shifted out of calibration and should be replaced.
If the engine is overheating, and the cooling fan is "ON", the cooling system should be checked.

1988–90 2.8L (VIN S) – CAMARO AND FIREBIRD

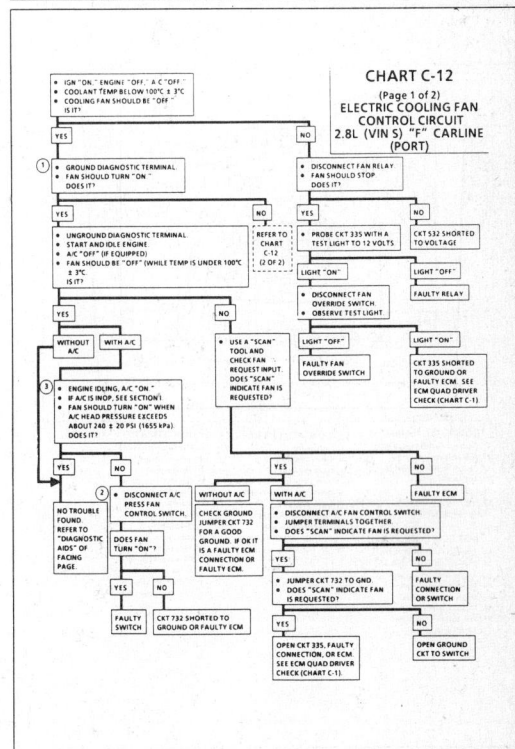

CHART C-12
(Page 1 of 2)
ELECTRIC COOLING FAN CONTROL CIRCUIT
2.8L (VIN S) "F" CARLINE (PORT)

1988–90 2.8L (VIN S) – CAMARO AND FIREBIRD

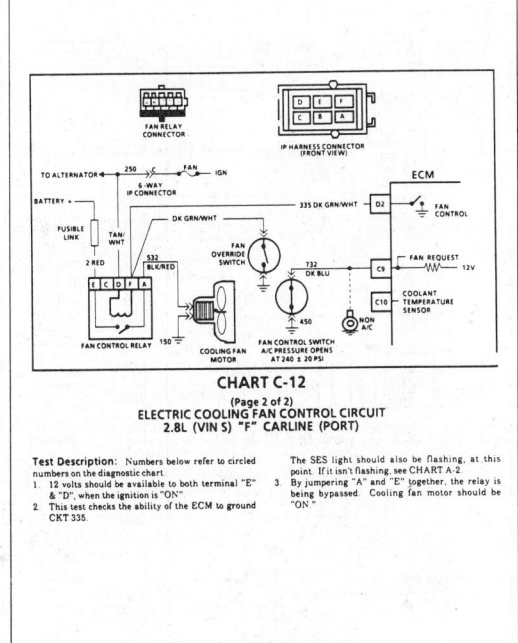

CHART C-12
(Page 2 of 2)
ELECTRIC COOLING FAN CONTROL CIRCUIT
2.8L (VIN S) "F" CARLINE (PORT)

Test Description: Numbers below refer to circled numbers on the diagnostic chart.
1. 12 volts should be available to both terminal "E" & "D", when the ignition is "ON".
2. This test checks the ability of the ECM to ground CKT 335.

The SES light should also be flashing, at this point. If it isn't flashing, see CHART A-2.
3. By jumpering "A" & "E" together, the relay is being bypassed. Cooling fan motor should be "ON".

1988–90 2.8L (VIN S) – CAMARO AND FIREBIRD

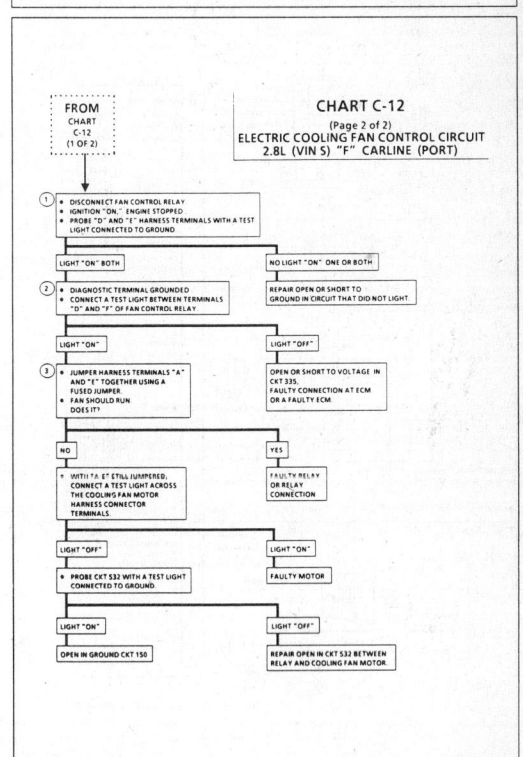

CHART C-12
(Page 2 of 2)
ELECTRIC COOLING FAN CONTROL CIRCUIT
2.8L (VIN S) "F" CARLINE (PORT)

**2.8L (VIN W) COMPONENT LOCATIONS
CORSICA AND BERETTA**

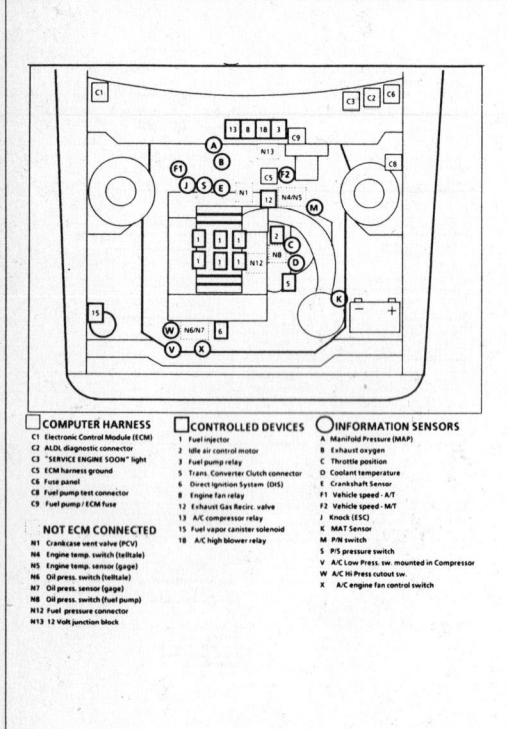

**2.8L (VIN W) ECM WIRING DIAGRAM
CORSICA AND BERETTA**

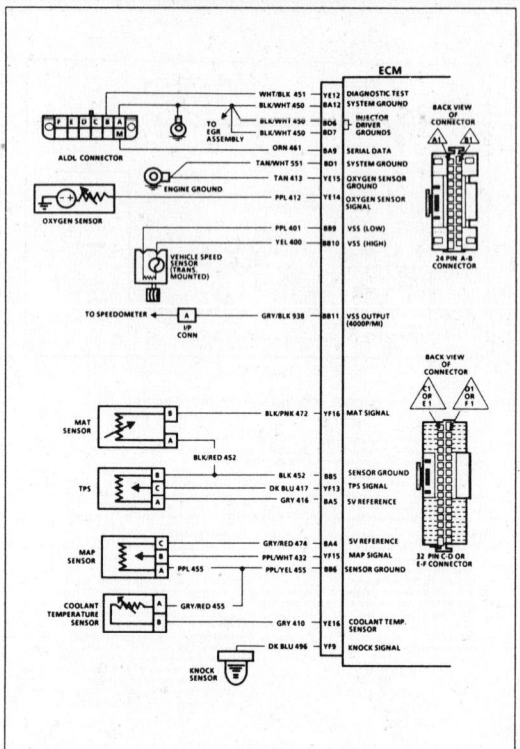

**2.8L (VIN W) ECM WIRING DIAGRAM
CORSICA AND BERETTA (CONT.)**

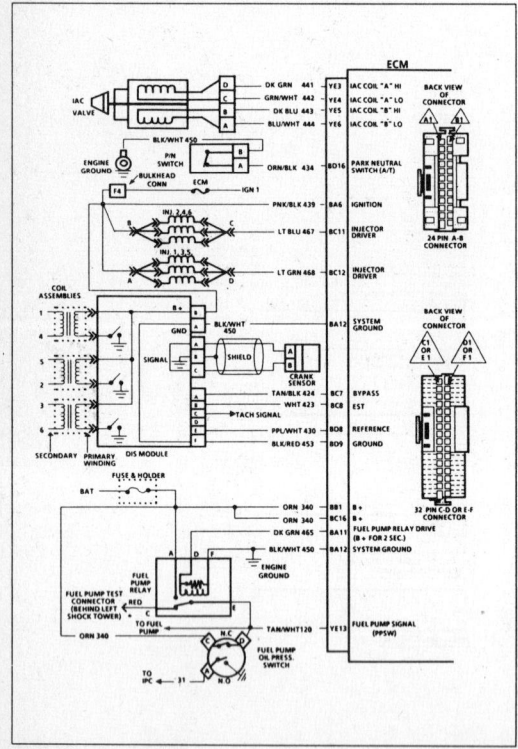

**2.8L (VIN W) ECM WIRING DIAGRAM
CORSICA AND BERETTA (CONT.)**

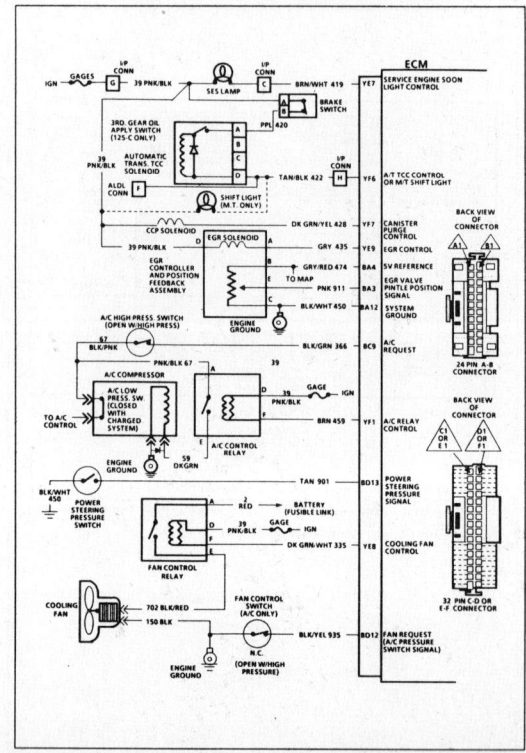

2.8L (VIN W) ECM CONNECTOR TERMINAL END VIEW—CORSICA AND BERETTA

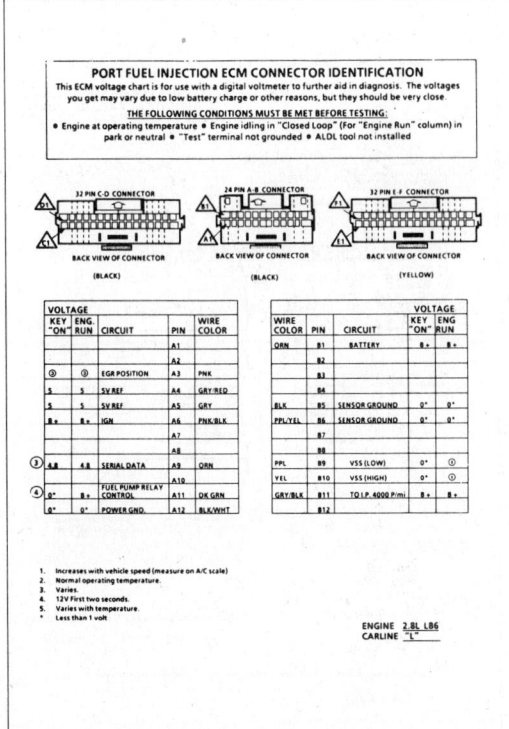

PORT FUEL INJECTION ECM CONNECTOR IDENTIFICATION

This ECM voltage chart is for use with a digital voltmeter to further aid in diagnosis. The voltages you get may vary due to low battery charge or other reasons, but they should be very close.

THE FOLLOWING CONDITIONS MUST BE MET BEFORE TESTING:
- Engine at operating temperature • Engine idling in "Closed Loop" (For "Engine Run" column) in park or neutral • "Test" terminal not grounded • ALDL tool not installed

32 PIN C-D CONNECTOR   24 PIN A-B CONNECTOR   32 PIN E-F CONNECTOR
BACK VIEW OF CONNECTOR   BACK VIEW OF CONNECTOR   BACK VIEW OF CONNECTOR
(BLACK)   (BLACK)   (YELLOW)

VOLTAGE

KEY "ON"	ENG. RUN	CIRCUIT	PIN	WIRE COLOR
			A1	
			A2	
②		EGR POSITION	A3	PNK
5		5V REF	A4	GRY/RED
5	5	5V REF	A5	GRY
B+	B+	IGN	A6	PNK/BLK
			A7	
			A8	
③ 4.8	4.8	SERIAL DATA	A9	ORN
			A10	
④ 0*	0*	FUEL PUMP RELAY CONTROL	A11	DK GRN
0*	0*	POWER GND.	A12	BLK/WHT

WIRE COLOR	PIN	CIRCUIT	KEY "ON"	ENG. RUN
ORN	B1	BATTERY	B+	B+
	B2			
	B3			
	B4			
BLK	B5	SENSOR GROUND	0*	0*
PPL/YEL	B6	SENSOR GROUND	0*	0*
	B7			
	B8			
PPL	B9	VSS (LOW)	0*	①
YEL	B10	VSS (HIGH)	0*	①
GRY/BLK	B11	TO I.P. 4000 P(mi)	B+	B+
	B12			

1. Increases with vehicle speed (measure on A/C scale)
2. Normal operating temperature.
3. Varies.
4. 12V First two seconds.
5. Varies with temperature.
* Less than 1 volt.

ENGINE 2.8L LB6
CARLINE "L"

2.8L (VIN W) ECM CONNECTOR TERMINAL END VIEW—CORSICA AND BERETTA (CONT.)

VOLTAGE

KEY "ON"	ENG. RUN	CIRCUIT	PIN	WIRE COLOR
			C1	
			C2	
			C3	
			C4	
			C5	
			C6	
0*	4.7	BYPASS	C7	TAN/BLK
0*	1.3	EST	C8	TAN
0*	0*	WITH A/C "ON" A/C REQUEST	C9	BLK/GRN
			C10	
B+	B+	INJECTOR 2,4,6	C11	LT BLU
B+	B+	INJECTOR 1,3,5	C12	LT GRN
			C13	
			C14	
			C15	
B+	B+	BATTERY	C16	ORN

WIRE COLOR	PIN	CIRCUIT	KEY "ON"	ENG. RUN
TAN/WHT	D1	POWER GROUND	0*	0*
	D2			
	D3			
	D4			
	D5			
BLK/WHT	D6	INJ DRIVE LOW	0*	0*
BLK/WHT	D7	INJ DRIVE LOW	0*	0*
PPL/WHT	D8	REFERENCE	0*	2.3
BLK/RED	D9	REFERENCE LOW	0*	0*
	D10			
	D11			
BLK/YEL	D12	A/C PRESS FAN SW	0*	0*
TAN	D13	PSPS	B+	B+
	D14			
	D15			
ORN/BLK	D16	P/N SWITCH	0*	0*

VOLTAGE

KEY "ON"	ENG. RUN	CIRCUIT	PIN	WIRE COLOR
			E1	
			E2	
NOT USEABLE		IAC "A" HI	E3	DK GRN
NOT USEABLE		IAC "A" LO	E4	GRN/WHT
NOT USEABLE		IAC "B" HI	E5	DK BLU
NOT USEABLE		IAC "B" LO	E6	BLU/WHT
0*	B+	"SERVICE ENGINE SOON" LIGHT	E7	BRN/WHT
B+	B+	FAN RELAY CONTROL	E8	DK GRN/WHT
B+	B+	EGR CONTROL	E9	GRY
			E10	
			E11	
5	5	DIAG. TERMINAL	E12	WHT/BLK
35-5.5	⑤	O₂ SIGNAL	E13	TAN/WHT
⑤	⑤	O₂ SIGNAL	E14	PPL
0*	0*	O₂ GROUND	E15	TAN
⑤	⑤	COOLANT TEMP.	E16	GRY

WIRE COLOR	PIN	CIRCUIT	KEY "ON"	ENG. RUN
BRN	F1	A/C RELAY CONTROL	B+	B+
	F2			
	F3			
	F4			
	F5			
TAN/BLK	F6	TCC CONTROL A/T / SHIFT LIGHT M/T	0* / B+	0* / B+
DK GRN/ YEL	F7	PURGE CONTROL	0*	25
DK BLU	F8	ESC SIGNAL	2.5	2.5
	F9			
	F10			
	F11			
DK BLU	F13	TPS SIGNAL	.65	.65
	F14			
PPL/WHT	F15	MAP SIGNAL	4.57	1.7
BLK/PNK	F16	MAT SIGNAL	3.1	3.2

1. Increases with vehicle speed (measure on A/C scale).
2. Normal operating temperature.
3. Varies.
4. 12 volts first two seconds.
5. Varies with temperature.
* Less than 1 volt.

1988–90 2.8L (VIN W)—CORSICA AND BERETTA

DIAGNOSTIC CIRCUIT CHECK

The Diagnostic Circuit Check must be the starting point for any driveability complaint diagnosis.

The Diagnostic Circuit Check is an organized approach to identifying a problem created by an Electronic Engine Control System malfunction because it directs the Service Technician to the next logical step in diagnosing the complaint.

If after completing the Diagnostic Circuit Check and finding the on-board diagnostics functioning properly and no trouble codes displayed, a comparison of "Typical Scan Values", for the appropriate engine, may be used for comparison. The "Typical Values" are an average of display values recorded from normally operating vehicles and are intended to represent what a normally functioning system would display.

A "SCAN" TOOL THAT DISPLAYS FAULTY DATA SHOULD NOT BE USED, AND THE PROBLEM SHOULD BE REPORTED TO THE MANUFACTURER. THE USE OF A FAULTY "SCAN" CAN RESULT IN MISDIAGNOSIS AND UNNECESSARY PARTS REPLACEMENT.

Only the parameters listed below are used in this manual for diagnosis. If a "Scan" reads other parameters, the values are not recommended by General Motors for use in diagnosis. For more description on the values and use of the "Scan" to diagnose ECM inputs, refer to the applicable diagnosis section. If all values are within the range illustrated, refer to "Symptoms" in Section "B".

"SCAN" DATA

Idle / Upper Radiator Hose Hot / Closed Throttle / Park or Neutral / Closed Loop / Acc. off

"SCAN" Position	Units Displayed	Typical Data Value
Desired RPM	RPM	ECM idle command (varies with temp.)
RPM	RPM	± 100 RPM from desired RPM (± 50 in drive)
Coolant Temp.	C°	85°- 105°
MAT Temp.	C°	10°- 80° (depends on underhood temp.)
MAP	Volts	1 - 2 (depends on Vac. & Baro pressure)
BARO	Volts	2.5 - 5.5 (depends on altitude & Baro pressure)
BPW (base pulse width)	M/Sec	1 - 4, and varying
O₂	Volts	.1-1.0, and varying
TPS	Volts	.65
Throttle Angle	0 - 100%	0
IAC	Counts (steps)	5 - 50
P/N Switch	P/N and RDL	Park/Neutral (P/N)
INT (Integrator)	Counts	Varies
BLM (Block Learn)	Counts	118 - 138
Open/Closed Loop	Open/Closed	Closed Loop (may go open with extended idle)
BLM Cell	Cell Number	0 or 1 (depends on Air Flow & RPM)
VSS	MPH	0
TCC	On/Off	Off/ (on with TCC commanded)
EGRDC	0 - 100%	0 at idle
EGR Position	Volts	3 - 2 Volts
Spark Advance	# of Degrees	Varies
Knock Retard	Degrees of Retard	0 °
Knock Signal	Yes/No	No
Battery	Volts	13.5 - 14.5
Fan	On/Off	Off (below 106°C)
P/S Switch	Normal/Hi Press.	Normal
A/C Request	Yes/No	No (yes, with A/C requested)
A/C Clutch	On/Off	Off (on, with A/C commanded on)
Fan Request	Yes/No	No (yes, with A/C high pressure)
Shift Light (M/T)	On/Off	Off
PPSW (Fuel Pump)	Volts	13.5 - 14.5

NOTE: If maximum retard is indicated, go to CHART C-5.

1988–90 2.8L (VIN W)—CORSICA AND BERETTA

DIAGNOSTIC CIRCUIT CHECK
2.8L (VIN W) "L" CARLINE (PORT)

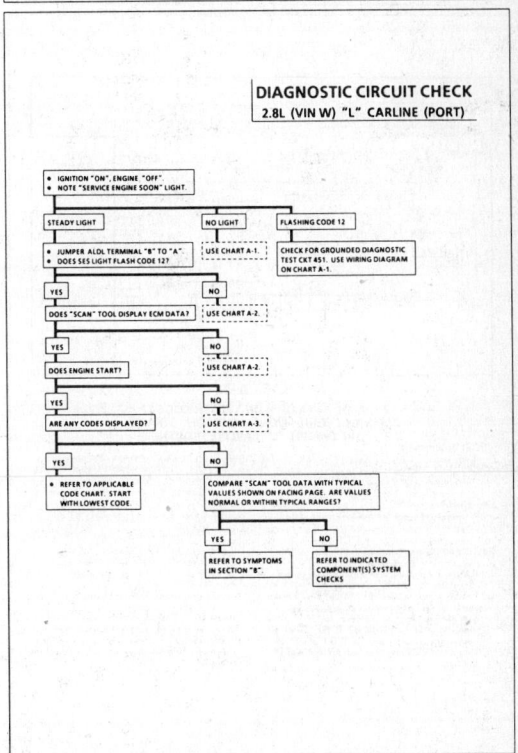

1988–90 2.8L (VIN W) – CORSICA AND BERETTA

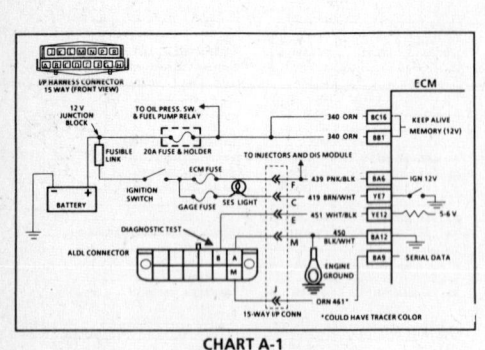

CHART A-1
NO "SERVICE ENGINE SOON" LIGHT
2.8L (VIN W) "L" CARLINE (PORT)

Circuit Description:

There should always be a steady "Service Engine Soon" light when the ignition is "ON" and engine stopped. Battery is supplied directly to the light bulb. The electronic control module (ECM) will control the light and turn it "ON" by providing a ground path through CKT 419 to the ECM.

Test Description: Numbers below refer to circled numbers on the diagnostic chart.
1. If the fuse in holder is blown refer to facing page of Code 54 for complete circuit.
2. Using a test light connected to 12 volts probe each of the system ground circuits to be sure a good ground is present. See ECM terminal end view in front of this section for ECM pin locations of ground circuits.

Diagnostic Aids:

Engine runs OK, check
- Faulty light bulb
- CKT 419 open.
- Gage fuse blown. This will result in no oil or generator lights, seat belt reminder, etc.
Engine cranks but will not run.
- Continuous battery - fuse or fusible link open.
- ECM ignition fuse open.
- Battery CKT 340 to ECM open.
- Ignition CKT 439 to ECM open.
- Poor connection to ECM

1988–90 2.8L (VIN W) – CORSICA AND BERETTA

CHART A-1
NO "SERVICE ENGINE SOON" LIGHT
2.8L (VIN W) "L" CARLINE (PORT)

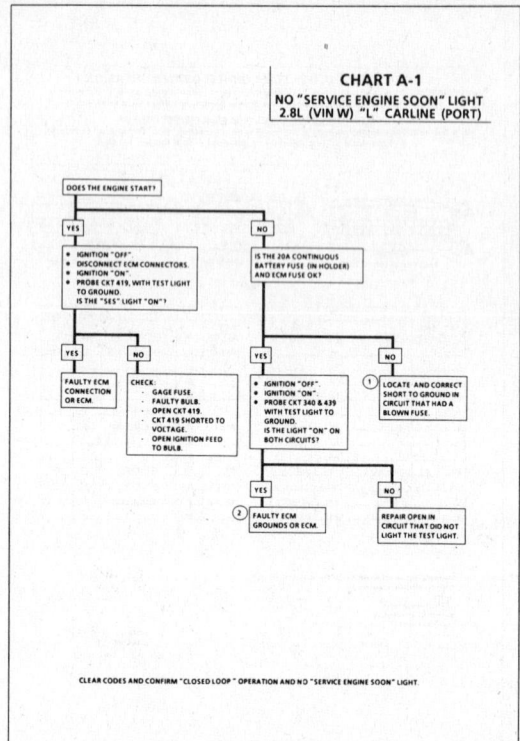

CLEAR CODES AND CONFIRM "CLOSED LOOP" OPERATION AND NO "SERVICE ENGINE SOON" LIGHT.

1988–90 2.8L (VIN W) – CORSICA AND BERETTA

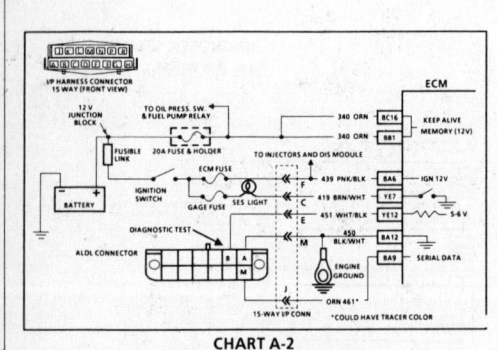

CHART A-2
NO ALDL DATA OR WON'T FLASH CODE 12
"SERVICE ENGINE SOON" LIGHT "ON" STEADY
2.8L (VIN W) "L" CARLINE (PORT)

Circuit Description:

There should always be a steady "Service Engine Soon" light when the ignition is "ON" and engine stopped. Battery ignition voltage is supplied to the light bulb. The electronic control module (ECM) will turn the light "ON" by grounding CKT 419 at the ECM.

With the diagnostic terminal grounded, the light should flash a Code 12, followed by any trouble code(s) stored in memory.

A steady light suggests a short to ground in the light control CKT 419, or an open in diagnostic CKT 451.

Test Description: Numbers below refer to circled numbers on the diagnostic chart.
1. If there is a problem with the ECM that causes a "Scan" tool to not read serial data, the ECM should not flash a Code 12. If Code 12 is flashing check for CKT 451 short to ground. If Code 12 does flash, be sure that the "Scan" tool is working properly on another vehicle. If the "Scan" is functioning properly and CKT 461 is OK, the Mem-Cal or ECM may be at fault for the No ALDL symptom.

2. If the light goes "OFF" when the ECM connector is disconnected, CKT 419 is not shorted to ground.
3. This step will check for an open diagnostic CKT 451.
4. At this point, the "Service Engine Soon" light wiring is OK. The problem is a faulty ECM or Mem-Cal. The ECM should be replaced using the original Mem-Cal. Replace the Mem-Cal only after trying an ECM, as a defective Mem-Cal is an unlikely cause of the problem.

1988–90 2.8L (VIN W) – CORSICA AND BERETTA

CHART A-2
NO ALDL DATA OR WON'T FLASH CODE 12
"SERVICE ENGINE SOON" LIGHT "ON" STEADY
2.8L (VIN W) "L" CARLINE (PORT)

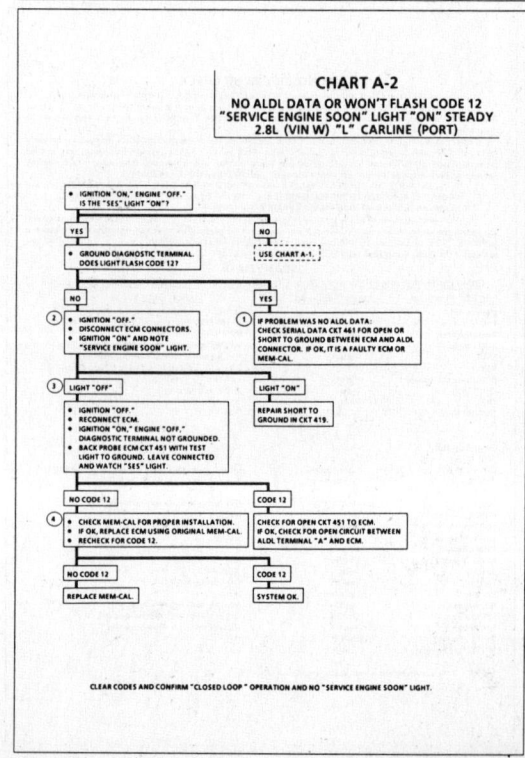

CLEAR CODES AND CONFIRM "CLOSED LOOP" OPERATION AND NO "SERVICE ENGINE SOON" LIGHT.

1988–90 2.8L (VIN W) – CORSICA AND BERETTA

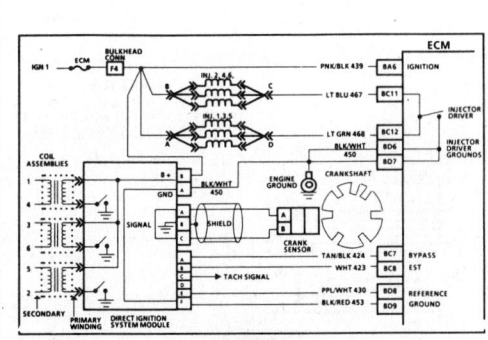

CHART A-3
(Page 1 of 3)
ENGINE CRANKS BUT WILL NOT RUN
2.8L (VIN W) "L" CARLINE (PORT)

Circuit Description:

This chart assumes that battery condition and engine cranking speed are OK, and there is adequate fuel in the tank.

Test Description: Numbers below refer to circled numbers on the diagnostic chart.

1. A "Service Engine Soon" light "ON" is a basic test to determine if there is a 12 volt supply and ignition 12 volts to ECM. No ALDL may be due to an ECM problem and CHART A-2 will diagnose the ECM. If TPS is over 2.5 volts the engine may be in the clear flood mode which will cause starting problems. The engine will not start without reference pulses and therefore the "Scan" should read rpm (reference) during crank.

2. For the first two seconds with ignition "ON" or whenever reference pulses are being received, PPSW should indicate fuel pump circuit voltage (8 to 12 volts).

3. Because the direct ignition system uses two plugs and wires to complete the circuit of each coil, the opposite spark should be left connected. If rpm was indicated during crank, the ignition module is receiving a crank signal, but no spark at this test

indicates the ignition module is not triggering the coils.

4. The test light should blink indicating the ignition is controlling the injectors OK. How bright the light blinks is not important. However, the test light should be a J-34730-3 or equivalent.

5. Use fuel pressure gage J-34730-1 or equivalent. Wrap a shop towel around the fuel pressure tap to absorb any small amount of fuel leakage that may occur when installing the gage.

6. This test will determine if the ignition module is not generating the reference pulse or if the wiring or ECM is at fault. By touching and removing a test light to 12 volts on CKT 430, a reference pulse should be generated. If rpm is indicated, the ECM and wiring are OK.

7. This test will determine if the ignition module is not triggering the problem coil or if the tested coil is at fault. This test could also be performed by using another known good coil.

1988–90 2.8L (VIN W) – CORSICA AND BERETTA

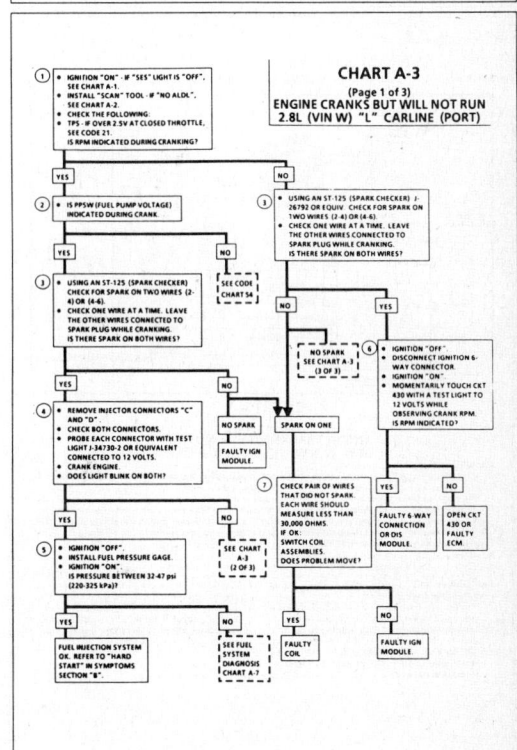

CHART A-3
(Page 1 of 3)
ENGINE CRANKS BUT WILL NOT RUN
2.8L (VIN W) "L" CARLINE (PORT)

1988–90 2.8L (VIN W) – CORSICA AND BERETTA

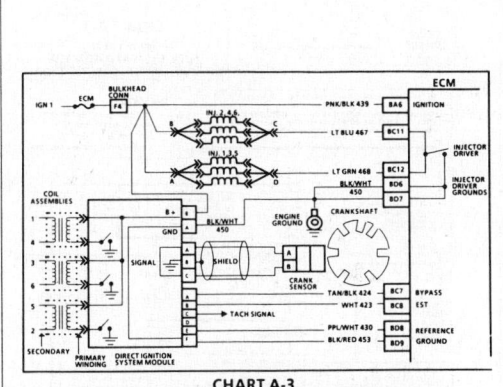

CHART A-3
(Page 2 of 3)
ENGINE CRANKS BUT WILL NOT RUN
2.8L (VIN W) "L" CARLINE (PORT)

Test Description: Numbers below refer to circled numbers on the diagnostic chart.

1. Checks for 12 volt supply to injectors. Because the injectors are wired in parallel, there should be a light "ON" on both terminals.

2. Checks continuity of CKTs 467 and 468.

3. All checks made to this point would indicate that the ECM is at fault. However, there is a possibility of CKT 467 or 468 being shorted to a voltage source either in the engine harness or in the injector harness.
   - To test for this condition:
     - Disconnect the injector 4-way connector
     - Ignition "ON."

   - Probe CKTs 467 and 468 on the ECM side of harness with a test light connected to ground. There should be no light. If light is "ON" repair short to voltage
   - If OK, check the resistance of the injector harness between terminals "A" & "C", "A" & "D", "B" & "D" and "B" & "C".
     - Should be more than 4 ohms.
     - If less than 4 ohms, check harness for wires shorted together and check each injector resistance. (Resistance should be 8 ohms or more.)
   - If all OK, replace ECM.

1988–90 2.8L (VIN W) – CORSICA AND BERETTA

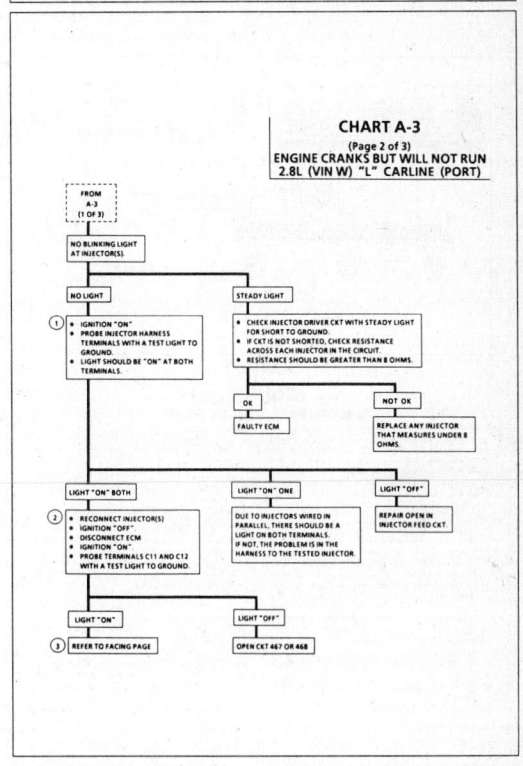

CHART A-3
(Page 2 of 3)
ENGINE CRANKS BUT WILL NOT RUN
2.8L (VIN W) "L" CARLINE (PORT)

1988–90 2.8L (VIN W) – CORSICA AND BERETTA

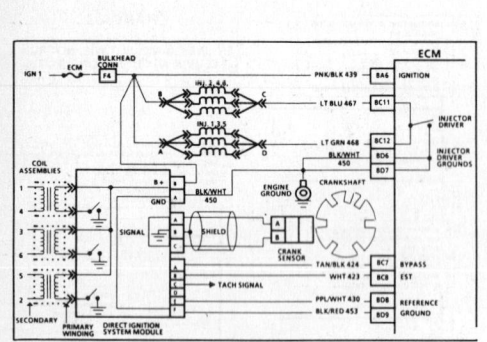

CHART A-3
(Page 3 of 3)
ENGINE CRANKS BUT WILL NOT RUN
2.8L (VIN W) "L" CARLINE (PORT)

Circuit Description:
If the "Scan" tool did not indicate a cranking rpm and there is no spark present at the plugs, the problem lies in the direct ignition system or the power and ground supplies to the module.

The magnetic crank sensor is used to determine engine crankshaft position much the same way as the pick-up coil did in distributor type systems. The sensor is mounted in the block near a seven slot wheel on the crank shaft. The rotation of the wheel creates a flux change in the sensor which produces a voltage signal. The ignition module then processes this signal and creates the reference pulses needed by the ECM and the signal triggers the correct coil at the correct time.

Test Description: Numbers below refer to circled numbers on the diagnostic chart.
1. This test will determine if the 12 volt supply and a good ground is available at the ignition module.
2. Tests for continuity of CKT 439 to the ignition module. If test light does not light but the "Service Engine Soon" light is "ON" with ignition repair open in CKT 439 between DIS ignition module and splice.
3. Checks for continuity of the crank sensor and connections.
4. Voltage will vary in this test depending on cranking speed of engine. The voltage will vary from about 500 mV, at very slow cranking speeds to about 100 mV at high speeds.

1988–90 2.8L (VIN W) – CORSICA AND BERETTA

CHART A-3
(Page 3 of 3)
ENGINE CRANKS BUT WILL NOT RUN
2.8L (VIN W) "L" CARLINE (PORT)

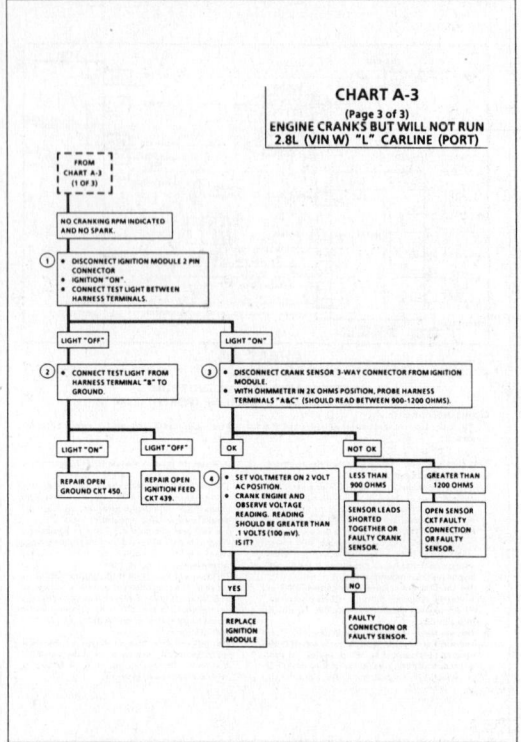

1988–90 2.8L (VIN W) – CORSICA AND BERETTA

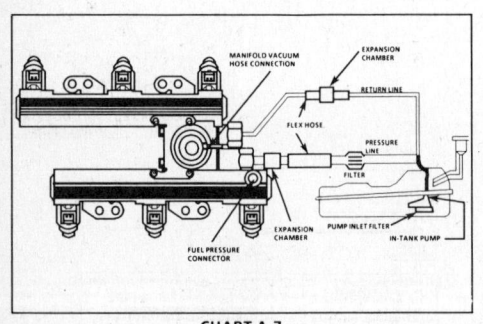

CHART A-7
(Page 1 of 2)
FUEL SYSTEM DIAGNOSIS
2.8L (VIN W) "L" CARLINE (PORT)

Circuit Description:
When the ignition switch is turned "ON," the electronic control module (ECM) will turn "ON" the in-tank fuel pump. It will remain "ON" as long as the engine is cranking or running, and the ECM is receiving reference pulses. If there are no reference pulses, the ECM will shut "OFF" the fuel pump within 2 seconds after ignition "ON" or engine stops.

The pump will deliver fuel to the fuel rail and injectors, then to the pressure regulator, where the system pressure is controlled to about 234 to 325 kPa (34 to 47 psi). Excess fuel is then returned to the fuel tank.

Test Description: Numbers below refer to circled numbers on the diagnostic chart.
1. Wrap a shop towel around the fuel pressure connector to absorb any small amount of fuel leakage that may occur when installing the gage. Ignition "ON" pump pressure should be 280-325 kPa (40.5-47 psi). This pressure is controlled by spring pressure within the regulator assembly.
2. When the engine is idling, the manifold pressure is low (high vacuum) and is applied to the fuel regulator diaphragm. This will offset the spring and result in a lower fuel pressure. This idle pressure will vary somewhat depending on barometric pressure, however, the pressure idling should be less indicating pressure regulator control.
3. Pressure that continues to fall is caused by one of the following:
   - In-tank fuel pump check valve not holding.

- Pump coupling hose or pulsator leaking
- Fuel pressure regulator valve leaking
- Injector(s) sticking open.
4. An injector sticking open can best be determined by checking for a fouled or saturated spark plug(s). If a leaking injector can not be determined by a fouled or saturated spark plug the following procedure should be used.
   - Remove plenum and fuel rail bolts. Follow the procedures in the Fuel Control Section of this manual but leave fuel lines connected.
   - Lift fuel rail out just enough to leave injector nozzles in the ports.

CAUTION: Be sure injector(s) are not allowed to spray on engine and that injector retaining clips are intact. This should be carefully followed to prevent fuel spray on engine which would cause a fire hazard.
   - Pressurize the fuel system and observe injector nozzles.

1988–90 2.8L (VIN W) – CORSICA AND BERETTA

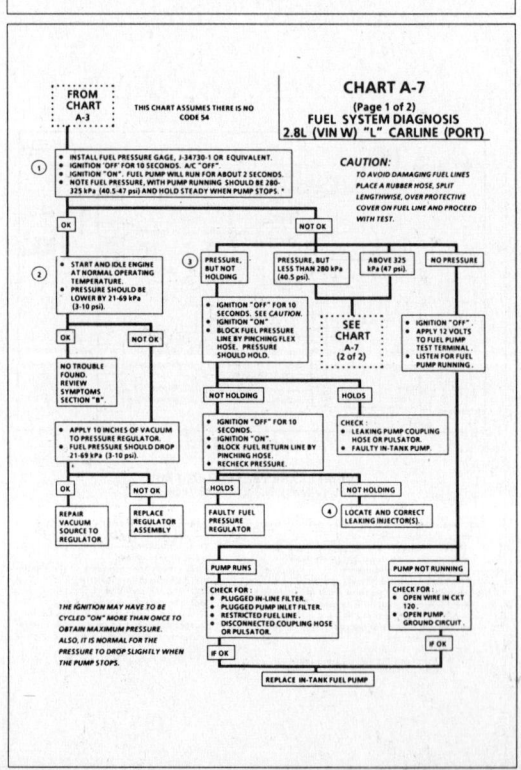

CHART A-7
(Page 1 of 2)
FUEL SYSTEM DIAGNOSIS
2.8L (VIN W) "L" CARLINE (PORT)

1988–90 2.8L (VIN W) – CORSICA AND BERETTA

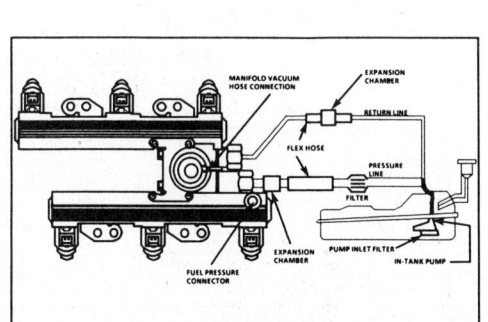

MANIFOLD VACUUM HOSE CONNECTION — EXPANSION CHAMBER — RETURN LINE — FLEX HOSE — PRESSURE LINE — FILTER — EXPANSION CHAMBER — FUEL PRESSURE CONNECTOR — PUMP INLET FILTER — IN-TANK PUMP

CHART A-7
(Page-2 of 2)
FUEL SYSTEM DIAGNOSIS
2.8L (VIN W) "L" CARLINE (PORT)

Test Description: Numbers below refer to circled numbers on the diagnostic chart.

1. Pressure but less than 280 kPa (40.5 psi) falls into two areas.
   - Regulated pressure but less than 280 kPa (40.5 psi). Amount of fuel to injectors OK, but pressure is too low. System will be lean running and may set Code 44. Also, hard starting cold and overall poor performance.
   - Restricted flow causing pressure drop. Normally, a vehicle with a fuel pressure of less than 165 kPa (24 psi) at idle will not be driveable. However, if the pressure drop occurs only while driving, the engine will normally surge then stop running as pressure begins to drop rapidly. This is most likely caused by a restricted fuel line or plugged filter.

2. Restricting the the fuel return line allows fuel prssure to build above regulated pressure. With battery applied to the pump "test" terminal, pressure should rise above 325 kPa (47 psi) as the fuel return hose is gradually pinched.
   NOTICE: Do not allow pressure to exceed 414 kPa (60 psi), as damage to the regulator may result.

3. This test determines if the high fuel pressure is due to a restricted fuel return line or a pressure regulator problem.

1988–90 2.8L (VIN W) – CORSICA AND BERETTA

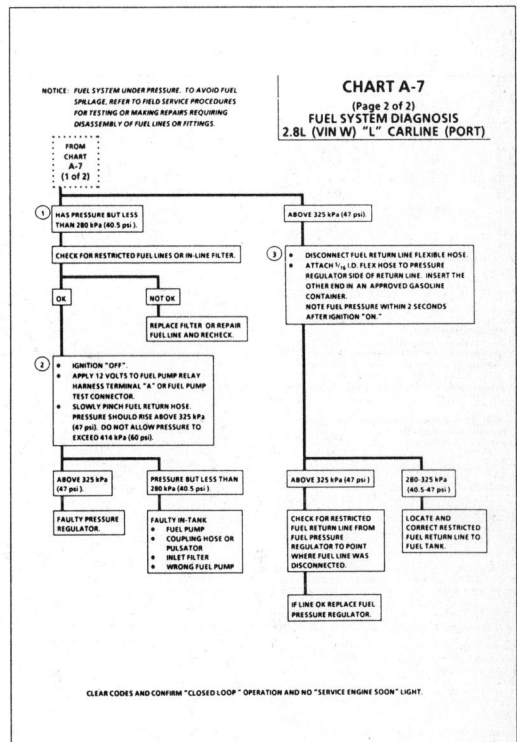

NOTICE: *FUEL SYSTEM UNDER PRESSURE. TO AVOID FUEL SPILLAGE, REFER TO FIELD SERVICE PROCEDURES FOR TESTING OR MAKING REPAIRS REQUIRING DISASSEMBLY OF FUEL LINES OR FITTINGS.*

CHART A-7
(Page 2 of 2)
FUEL SYSTEM DIAGNOSIS
2.8L (VIN W) "L" CARLINE (PORT)

FROM CHART A-7 (1 of 2)

① HAS PRESSURE BUT LESS THAN 280 kPa (40.5 psi).

CHECK FOR RESTRICTED FUEL LINES OR IN-LINE FILTER.

OK / NOT OK

REPLACE FILTER OR REPAIR FUEL LINE AND RECHECK.

② • IGNITION "OFF". • APPLY 12 VOLTS TO FUEL PUMP RELAY HARNESS TERMINAL "A" OR FUEL PUMP TEST CONNECTOR. • SLOWLY PINCH FUEL RETURN HOSE. PRESSURE SHOULD RISE ABOVE 325 kPa (47 psi). DO NOT ALLOW PRESSURE TO EXCEED 414 kPa (60 psi).

ABOVE 325 kPa (47 psi). / PRESSURE BUT LESS THAN 280 kPa (40.5 psi).

FAULTY PRESSURE REGULATOR.

FAULTY IN-TANK
- FUEL PUMP
- COUPLING HOSE OR PULSATOR
- INLET FILTER
- WRONG FUEL PUMP

③ • DISCONNECT FUEL RETURN LINE FLEXIBLE HOSE. • ATTACH 5/16 I.D. FLEX HOSE TO PRESSURE REGULATOR SIDE OF RETURN LINE. INSERT THE OTHER END IN AN APPROVED GASOLINE CONTAINER. • NOTE FUEL PRESSURE WITHIN 2 SECONDS AFTER IGNITION "ON."

ABOVE 325 kPa (47 psi) / 280-325 kPa (40.5-47 psi)

CHECK FOR RESTRICTED FUEL LINE FROM FUEL PRESSURE REGULATOR TO POINT WHERE FUEL LINE WAS DISCONNECTED.

LOCATE AND CORRECT RESTRICTED FUEL RETURN LINE TO FUEL TANK.

IF LINE OK REPLACE FUEL PRESSURE REGULATOR.

CLEAR CODES AND CONFIRM "CLOSED LOOP" OPERATION AND NO "SERVICE ENGINE SOON" LIGHT.

1988–90 2.8L (VIN W) – CORSICA AND BERETTA

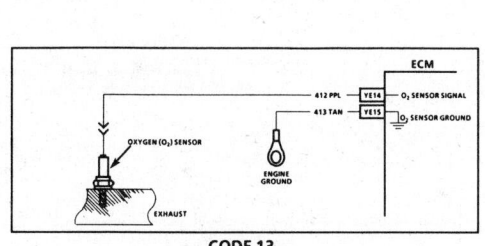

ECM — 412 PPL — YE14 — O₂ SENSOR SIGNAL — 413 TAN — YE15 — O₂ SENSOR GROUND — OXYGEN (O₂) SENSOR — ENGINE GROUND — EXHAUST

CODE 13
OXYGEN SENSOR CIRCUIT
(OPEN CIRCUIT)
2.8L (VIN W) "L" CARLINE (PORT)

Circuit Description:
The ECM supplies a voltage of about .45 volt between terminals "YE14" and "YE15". (If measured with a 10 megohm digital voltmeter, this may read as low as .32 volt.) The O_2 sensor varies the voltage within a range of about 1 volt if the exhaust is rich, down through about .10 volt if exhaust is lean.

The sensor is like an open circuit and produces no voltage when it is below 315°C (600°F). An open sensor circuit or cold sensor causes "Open Loop" operation.

Test Description: Numbers below refer to circled numbers on the diagnostic chart.

1. Code 13 will set
   - Engine at normal operating temperature
   - At least 2 minutes engine time after start
   - O_2 signal voltage steady between .35 and .55 volt
   - Throttle position sensor signal above 4%
   - All conditions must be met for about 60 seconds

   If the conditions for a Code 13 exist, the system will not go "Closed Loop."

2. This will determine if the sensor is at fault, or the wiring, or ECM is the cause of the Code 13.

3. In doing this test use only a high impedance digital volt ohm meter. This test checks the continuity of CKTs 412 and 413 because if CKT 413 is open the ECM voltage on CKT 412 will be over .6 volt (600 mV).

Diagnostic Aids:

Normal "Scan" voltage varies between 100 mV to 999 mV (.1 and 1.0 volt) while in "Closed Loop." Code 13 sets in one minute if voltage remains between .35 and .55 volt, but the system will go "Open Loop" in about 15 seconds. Refer to "Intermittents" in Section "B".

1988–90 2.8L (VIN W) – CORSICA AND BERETTA

CODE 13
OXYGEN SENSOR CIRCUIT
(OPEN CIRCUIT)
2.8L (VIN W) "L" CARLINE (PORT)

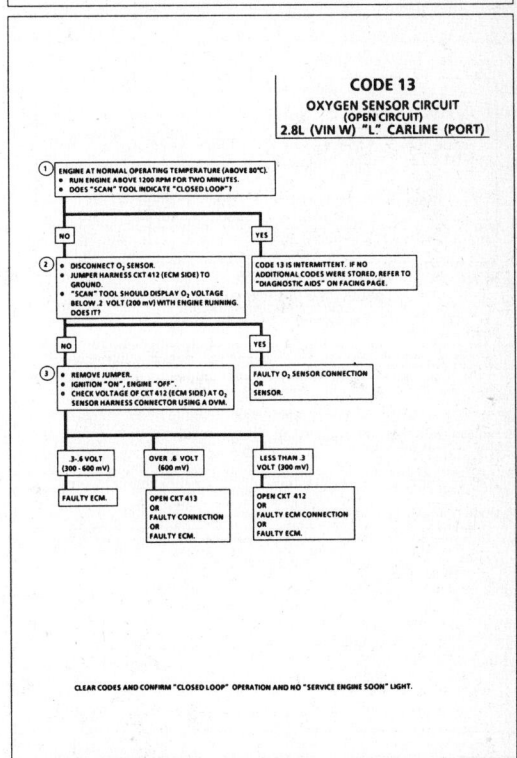

① • ENGINE AT NORMAL OPERATING TEMPERATURE (ABOVE 80°C). • RUN ENGINE ABOVE 1200 RPM FOR TWO MINUTES. • DOES "SCAN" TOOL INDICATE "CLOSED LOOP"?

NO / YES

② • DISCONNECT O₂ SENSOR. • JUMPER HARNESS CKT 412 (ECM SIDE) TO GROUND. • "SCAN" TOOL SHOULD DISPLAY O₂ VOLTAGE BELOW .2 VOLT (200 mV) WITH ENGINE RUNNING. DOES IT?

CODE 13 IS INTERMITTENT. IF NO ADDITIONAL CODES WERE STORED, REFER TO "DIAGNOSTIC AIDS" ON FACING PAGE.

NO / YES

③ • REMOVE JUMPER. • IGNITION "ON", ENGINE "OFF". • CHECK VOLTAGE OF CKT 412 (ECM SIDE) AT O₂ SENSOR HARNESS CONNECTOR USING A DVM.

FAULTY O₂ SENSOR CONNECTION OR SENSOR.

.3 - .6 VOLT (300 - 600 mV) / OVER .6 VOLT (600 mV) / LESS THAN .3 VOLT (300 mV)

FAULTY ECM.

OPEN CKT 413 OR FAULTY CONNECTION OR FAULTY ECM.

OPEN CKT 412 OR FAULTY ECM CONNECTION OR FAULTY ECM.

CLEAR CODES AND CONFIRM "CLOSED LOOP" OPERATION AND NO "SERVICE ENGINE SOON" LIGHT.

1988–90 2.8L (VIN W) – CORSICA AND BERETTA

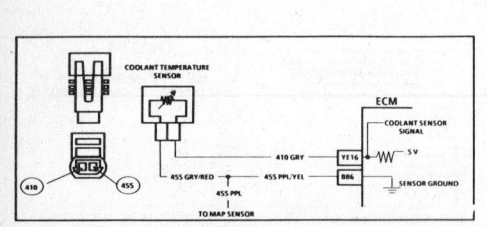

CODE 14
COOLANT TEMPERATURE SENSOR CIRCUIT
(HIGH TEMPERATURE INDICATED)
2.8L (VIN W) "L" CARLINE (PORT)

Circuit Description:

The coolant temperature sensor uses a thermistor to control the signal voltage to the ECM. The ECM applies a voltage on CKT 410 to the sensor. When the engine is cold, the sensor (thermistor) resistance is high, therefore, the ECM will see high signal voltage.

As the engine warms, the sensor resistance becomes less and the voltage drops. At normal engine operating temperature (85°C to 95°C), the voltage will measure about 1.5 to 2.0 volts.

Test Description: Numbers below refer to circled numbers on the diagnostic chart.
1. Code 14 will set if:
   - Signal voltage indicates a coolant temperature above 135°C (275°F) for 3 seconds.
2. This test will determine if CKT 410 is shorted to ground which will cause the conditions for Code 14.

Diagnostic Aids:

Check harness routing for a potential short to ground in CKT 410. "Scan" tool displays engine temperature in degrees centigrade. After engine is started, the temperature should rise steadily to about 90°C then stabilize when thermostat opens. Refer to "Intermittents" in Section "B".

1988–90 2.8L (VIN W) – CORSICA AND BERETTA

CODE 14
COOLANT TEMPERATURE SENSOR CIRCUIT
(HIGH TEMPERATURE INDICATED)
2.8L (VIN W) "L" CARLINE (PORT)

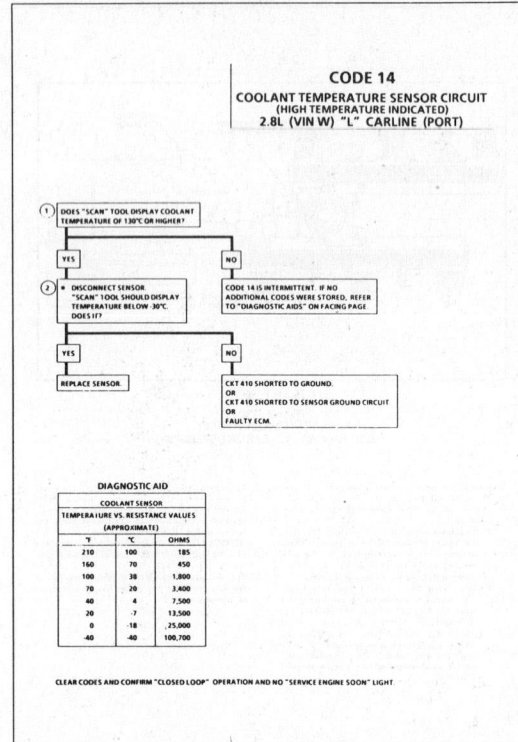

DIAGNOSTIC AID

COOLANT SENSOR		
TEMPERATURE VS. RESISTANCE VALUES (APPROXIMATE)		
°F	°C	OHMS
210	100	185
160	70	450
100	38	1,800
70	20	3,400
40	4	7,500
20	-7	13,500
0	-18	25,000
-40	-40	100,700

CLEAR CODES AND CONFIRM "CLOSED LOOP" OPERATION AND NO "SERVICE ENGINE SOON" LIGHT.

1988–90 2.8L (VIN W) – CORSICA AND BERETTA

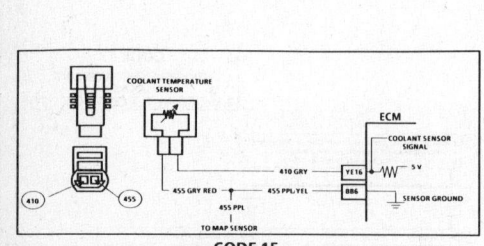

CODE 15
COOLANT TEMPERATURE SENSOR CIRCUIT
(LOW TEMPERATURE INDICATED)
2.8L (VIN W) "L" CARLINE (PORT)

Circuit Description:

The coolant temperature sensor uses a thermistor to control the signal voltage to the ECM. The ECM applies a voltage on CKT 410 to the sensor. When the engine is cold, the sensor (thermistor) resistance is high, therefore, the ECM will see high signal voltage.

As the engine warms, the sensor resistance becomes less and the voltage drops. At normal engine operating temperature (85°C to 95°C), the voltage will measure about 1.5 to 2.0 volts at the ECM.

Test Description: Numbers below refer to circled numbers on the diagnostic chart.
1. Code 15 will set if:
   - Signal voltage indicates a coolant temperature less than -44°C (-47°F) for 3 seconds.
2. This test simulates a Code 14. If the ECM recognizes the low signal voltage (high temperature) and the "Scan" reads 130°C, the ECM and wiring are OK.
3. This test will determine if CKT 410 is open. There should be 5 volts present at sensor connector if measured with a DVM.

Diagnostic Aids:

A "Scan" tool reads engine temperature in degrees centigrade. After engine is started, the temperature should rise steadily to about 90°C then stabilize when thermostat opens.

A faulty connection or an open in CKT 410 or 455 will result in a Code 15.

If Code 21 or 23 is also set, check CKT 455 for faulty wiring or connections. Check terminals at sensor for good contact. Refer to "Intermittents" in Section "B".

1988–90 2.8L (VIN W) – CORSICA AND BERETTA

CODE 15
COOLANT TEMPERATURE SENSOR CIRCUIT
(LOW TEMPERATURE INDICATED)
2.8L (VIN W) "L" CARLINE (PORT)

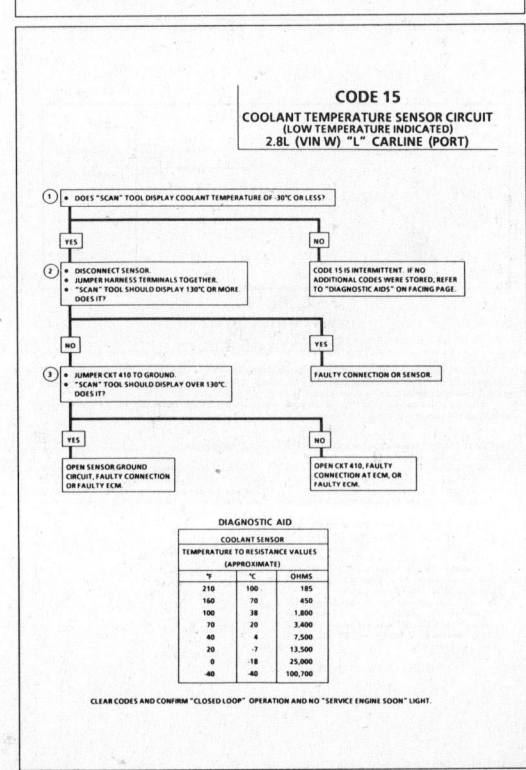

DIAGNOSTIC AID

COOLANT SENSOR		
TEMPERATURE TO RESISTANCE VALUES (APPROXIMATE)		
°F	°C	OHMS
210	100	185
160	70	450
100	38	1,800
70	20	3,400
40	4	7,500
20	-7	13,500
0	-18	25,000
-40	-40	100,700

CLEAR CODES AND CONFIRM "CLOSED LOOP" OPERATION AND NO "SERVICE ENGINE SOON" LIGHT.

1988–90 2.8L (VIN W) – CORSICA AND BERETTA

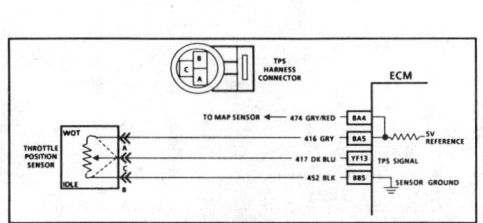

CODE 21
THROTTLE POSITION SENSOR (TPS) CIRCUIT
(SIGNAL VOLTAGE HIGH)
2.8L (VIN W) "L" CARLINE (PORT)

Circuit Description:
The throttle position sensor (TPS) provides a voltage signal that changes relative to the throttle blade. Signal voltage will vary from about .5 at idle to about 5 volts at wide open throttle, and is nonadjustable.
The TPS signal is one of the most important inputs used by the ECM for fuel control and for most of the ECM control outputs.

Test Description: Numbers below refer to circled numbers on the diagnostic chart.
1. Code 21 will set if
   - Engine is running
   - TPS signal voltage is greater than 4.3 volts
   - Air flow is less than 17 GM/sec
   - All conditions met for 10 seconds
   OR
   - TPS signal voltage over 4.5 volts with ignition "ON."
   - TPS check: The TPS has an auto zeroing feature. If the voltage reading is within the range of 0.45 to 0.85 volt, the ECM will use that value as closed throttle. If TPS is out of range, make sure cruise control and throttle cables are not being held open.

2. With the TPS sensor disconnected, the TPS voltage should go low if the ECM and wiring are OK.
3. Probing CKT 452 with a test light checks the 5 volt return circuit because a faulty 5 volt return will cause a Code 21.

Diagnostic Aids:
A "Scan" tool reads throttle position in volts. Voltage should increase at a steady rate as throttle is moved toward WOT.
Also some "Scan" tools will read throttle angle .0% = closed throttle 100% = WOT.
An open in CKT 452 will result in a Code 21 and may also set Codes 15 and 23. Refer to "Intermittents" in Section "B".

1988–90 2.8L (VIN W) – CORSICA AND BERETTA

CODE 21
THROTTLE POSITION SENSOR (TPS) CIRCUIT
(SIGNAL VOLTAGE HIGH)
2.8L (VIN W) "L" CARLINE (PORT)

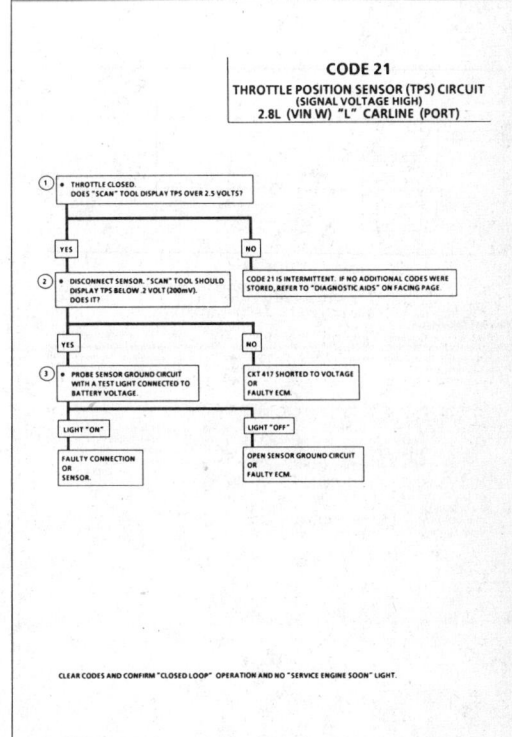

CLEAR CODES AND CONFIRM "CLOSED LOOP" OPERATION AND NO "SERVICE ENGINE SOON" LIGHT.

1988–90 2.8L (VIN W) – CORSICA AND BERETTA

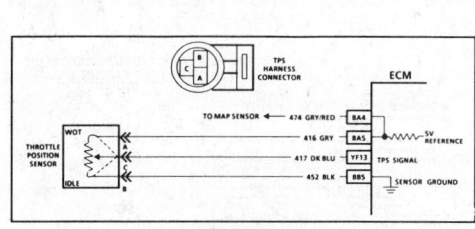

CODE 22
THROTTLE POSITION SENSOR (TPS) CIRCUIT
(SIGNAL VOLTAGE LOW)
2.8L (VIN W) "L" CARLINE (PORT)

Circuit Description:
The throttle position sensor (TPS) provides a voltage signal that changes relative to the throttle blade. Signal voltage will vary from about .5 at idle to about 5 volts at wide open throttle, and is nonadjustable.
The TPS signal is one of the most important inputs used by the ECM for fuel control and for most of the ECM control outputs.

Test Description: Numbers below refer to circled numbers on the diagnostic chart.
1. Code 22 will set if
   - Engine running
   - TPS signal voltage is less than about 2 volt for 3 seconds
2. Simulates Code 21 (high voltage) if the ECM recognizes the high signal voltage, the ECM and wiring are OK.
3. TPS check: The TPS has an auto zeroing feature. If the voltage reading is within the range of 0.45 to 0.85 volt, the ECM will use that value as closed throttle.
4. This simulates a high signal voltage to check for an open in CKT 417.

5. CKTs 416 and 474 share a common 5 volts buffered reference signal. If either circuit is shorted to ground, Code 22 will set. To determine if the MAP sensor is causing the 22 problem, disconnect it to see if the conditions for Code 22 still exist.

Diagnostic Aids:
A "Scan" tool reads throttle position in volts. Voltage should increase at a steady rate as throttle is moved toward WOT.
Also some "Scan" tools will read throttle angle 0% = closed throttle 100% = WOT.
An open or short to ground in CKT 416 or 417 will result in a Code 22. Also, a short to ground in CKT 474 will result in a Code 22. Refer to "Intermittents" in Section "B".

1988–90 2.8L (VIN W) – CORSICA AND BERETTA

CODE 22
THROTTLE POSITION SENSOR (TPS) CIRCUIT
(SIGNAL VOLTAGE LOW)
2.8L (VIN W) "L" CARLINE (PORT)

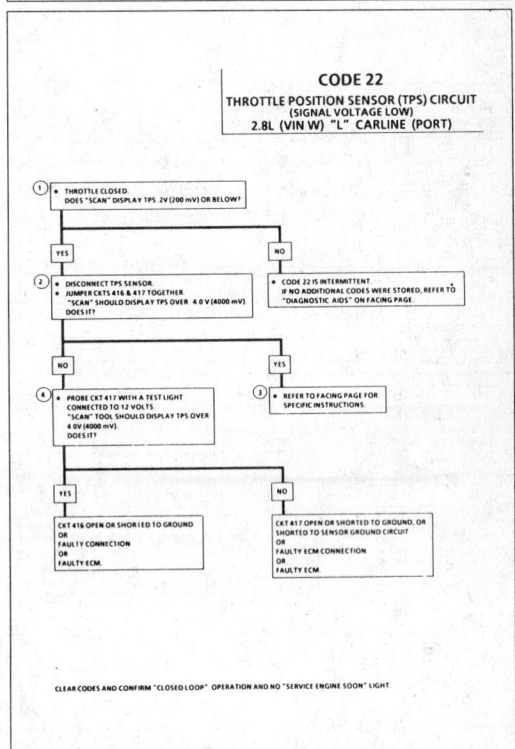

CLEAR CODES AND CONFIRM "CLOSED LOOP" OPERATION AND NO "SERVICE ENGINE SOON" LIGHT.

1988–90 2.8L (VIN W) – CORSICA AND BERETTA

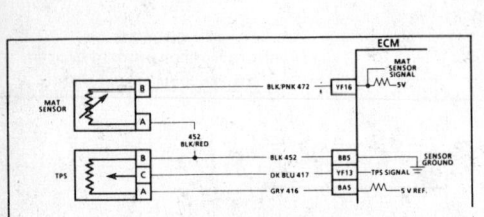

CODE 23
MANIFOLD AIR TEMPERATURE (MAT) SENSOR CIRCUIT
(LOW TEMPERATURE INDICATED)
2.8L (VIN W) "L" CARLINE (PORT)

Circuit Description:
The MAT sensor uses a thermistor to control the signal voltage to the ECM. The ECM applies a voltage (about 5 volts) on CKT 472 to the sensor. When the air is cold, the sensor (thermistor) resistance is high, therefore, the ECM will see a high signal voltage. If the air is warm the sensor resistance is low, the ECM will see a low voltage.

Test Description: Numbers below refer to circled numbers on the diagnostic chart.
1. Code 23 will set if
   - A signal voltage indicates a manifold air temperature below –35°C (–31°F) for 3 seconds
   - Time since engine start is 4 minutes or longer
   - No VSS
2. A Code 23 will set due to an open sensor, wire, or connection. This test will determine if the wiring and ECM are OK.
3. This will determine if the signal CKT 472 or the sensor ground CKT 452 is open.

Diagnostic Aids:
A "Scan" tool reads temperature of the air entering the engine and should read close to ambient air temperature when engine is cold, and rises as underhood temperature increases.
A faulty connection or an open in CKT 472 or 452 will result in a Code 23.
Refer to "Intermittents" in Section "B".

1988–90 2.8L (VIN W) – CORSICA AND BERETTA

CODE 23
MANIFOLD AIR TEMPERATURE (MAT) SENSOR CIRCUIT
(LOW TEMPERATURE INDICATED)
2.8L (VIN W) "L" CARLINE (PORT)

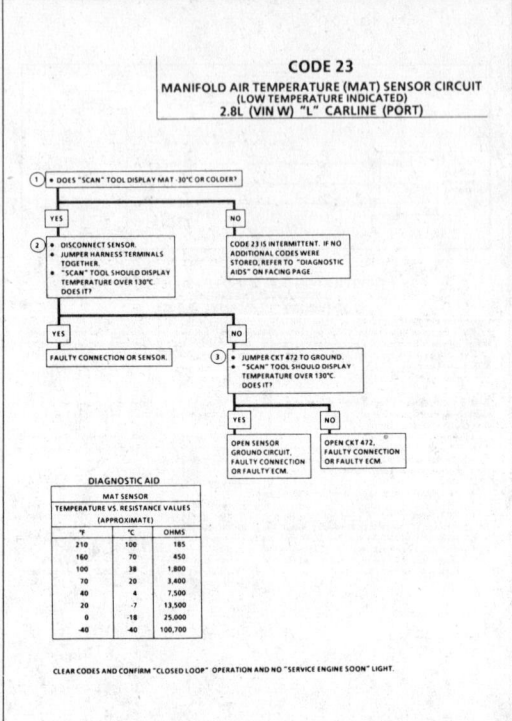

DIAGNOSTIC AID		
MAT SENSOR		
TEMPERATURE VS. RESISTANCE VALUES (APPROXIMATE)		
°F	°C	OHMS
210	100	185
160	70	450
100	38	1,800
70	20	3,400
40	4	7,500
20	–7	13,500
0	–18	25,000
–40	–40	100,700

CLEAR CODES AND CONFIRM "CLOSED LOOP" OPERATION AND NO "SERVICE ENGINE SOON" LIGHT.

1988–90 2.8L (VIN W) – CORSICA AND BERETTA

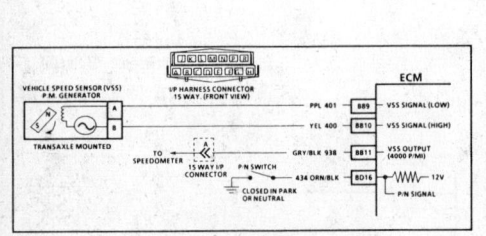

CODE 24
VEHICLE SPEED SENSOR (VSS) CIRCUIT
2.8L (VIN W) "L" CARLINE (PORT)

Circuit Description:
Vehicle speed information is provided to the ECM by the vehicle speed sensor which is a permanent magnet (PM) generator and it is mounted in the transaxle. The PM generator produces a pulsing voltage whenever vehicle speed is over about 3 mph. The voltage level and the number of pulses increases with vehicle speed. The ECM then converts the pulsing voltage to mph which is used for calculations, and the mph can be displayed with a "Scan" tool.
The function of VSS buffer used in past model years has been incorporated into the ECM. The ECM then supplies the necessary signal for the instrument panel (4000 pulses per mile) for operating the speedometer and the odometer.

Test Description: Numbers below refer to circled numbers on the diagnostic chart.
1. Code 24 will set if vehicle speed equals 0 mph when
   - Engine speed is between 1400 and 3600 rpm
   - TPS is less than 2%
   - Low load condition (low air flow)
   - Not in park or neutral
   - All conditions met for 5 seconds
   These conditions are met during a road load deceleration. Disregard Code 24 that sets when drive wheels are not turning.
   - The PM generator only produces a signal if drive wheels are turning greater than 3 mph.

Diagnostic Aids:
"Scan" should indicate a vehicle speed whenever the drive wheels are turning greater than 3 mph.
A problem in CKT 938 will not affect the VSS input or the readings on a "Scan."
Check CKTs 400 and 401 for proper connections to be sure they're clean and tight and the harness is routed correctly. Refer to "Intermittents" in Section "B".
(A/T) A faulty or misadjusted Park/Neutral switch can result in a false Code 24. Use a "Scan" and check for proper signal while in drive (125C). Refer to CHART C-1A for P/N switch diagnosis check.

1988–90 2.8L (VIN W) – CORSICA AND BERETTA

CODE 24
VEHICLE SPEED SENSOR (VSS) CIRCUIT
2.8L (VIN W) "L" CARLINE (PORT)

DISREGARD CODE 24 IF SET WHILE DRIVE WHEELS ARE NOT TURNING.

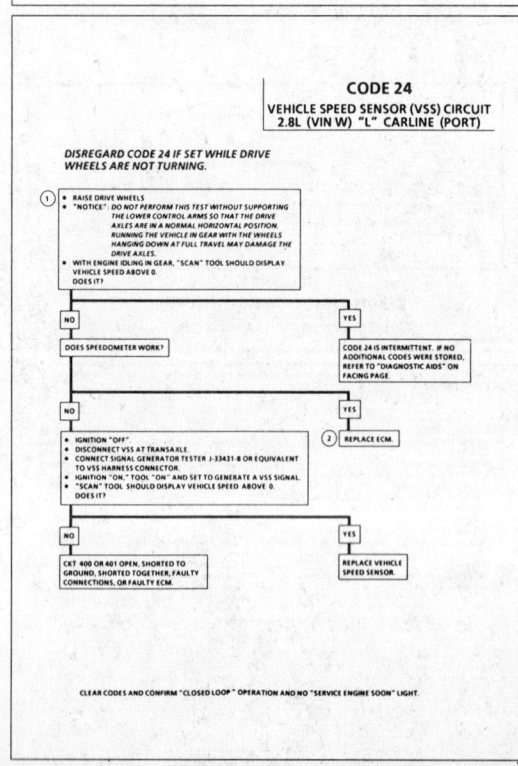

CLEAR CODES AND CONFIRM "CLOSED LOOP" OPERATION AND NO "SERVICE ENGINE SOON" LIGHT.

1988–90 2.8L (VIN W) – CORSICA AND BERETTA

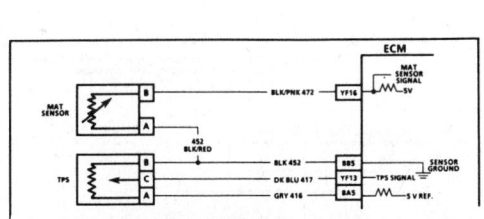

CODE 25
MANIFOLD AIR TEMPERATURE (MAT) SENSOR CIRCUIT
(HIGH TEMPERATURE INDICATED)
2.8L (VIN W) "L" CARLINE (PORT)

Circuit Description:
The manifold air temperature sensor uses a thermistor to control the signal voltage of the ECM. The ECM applies a voltage (about 5 volts) on CKT 472 to the sensor. When manifold air is cold the sensor (Thermistor) resistance is high, the ECM will see a high signal voltage. As the air warms, the sensor resistance becomes less and the voltage drops.

Test Description: Numbers below refer to circled numbers on the diagnostic chart.
1. Code 25 will be set if
- Signal voltage indicates a manifold air temperature greater than 145°C (293°F) for 3 seconds.
- Time since engine start is 4 minutes or longer.
- A vehicle speed is present.

Diagnostic Aids:
A "Scan" tool reads temperature of the air entering the engine and should read close to ambient air temperature when engine is cold, and rises as underhood temperature increases.
A short to ground in CKT 472 will result in a Code 25.
Refer to "Intermittents" in Section "B".

1988–90 2.8L (VIN W) – CORSICA AND BERETTA

CODE 25
MANIFOLD AIR TEMPERATURE (MAT) SENSOR CIRCUIT
(HIGH TEMPERATURE INDICATED)
2.8L (VIN W) "L" CARLINE (PORT)

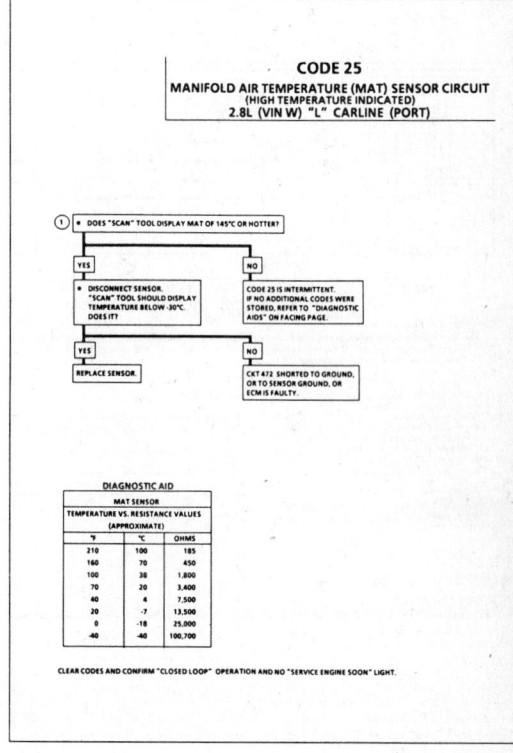

1988–90 2.8L (VIN W) – CORSICA AND BERETTA

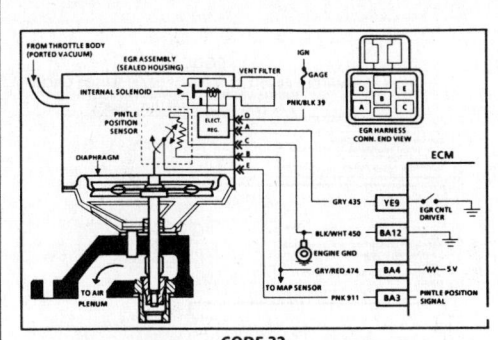

CODE 32
EXHAUST GAS RECIRCULATION (EGR) CIRCUIT
2.8L (VIN W) "L" CARLINE (PORT)

Circuit Description:
The Integrated Electronic EGR Valve functions similar to a port valve with a remote vacuum regulator. The internal solenoid is normally open, which causes the vacuum signal to be vented off to the atmosphere when EGR is not being commanded by the ECM. This EGR valve has a sealed cap and the solenoid valve opens and closes the vacuum signal, which controls the amount of vacuum vented to atmosphere. The electronic EGR valve contains a vacuum regulator, which converts the ECM signal, to provide different amounts of EGR flow by regulating the current to the solenoid. The ECM controls EGR flow with a pulse width modulated signal (turns "ON" and "OFF" many times a second) based on Airflow, TPS, and RPM.
This system, also, contains a pintle position sensor, which works similiar to a TPS sensor, and as EGR flow is increased, the sensor output also increases.
If the ECM does not see the 12 volt signal on circuit 435, when the solenoid is not being energized, a code 26 may be set. Also, if the ECM is not able to ground CKT 435, due to a shorted solenoid or CKT 435 shorted to voltage, a Code 26 may also be set.

Test Description: Step numbers refer to step numbers on diagnostic chart.
1. With the engine running and the transmission in gear, with brakes applied, increasing engine rpm will put a load on the engine. Engine vacuum will be applied to the EGR diaphragm and cause the EGR pintle to open increasing pintle position voltage.
2. This test will determine if the EGR filter is plugged, or if the EGR itself is faulty. Use care, when removing the filter, to avoid damaging the EGR assembly. See ON-CAR Service for procedure.
3. If the valve moves in this test, it's probably due to CKT 435 being shorted to ground.

4. Grounding the diagnostic terminal should energize the solenoid which closes "OFF" the vent and allows the vacuum to move the diaphragm.
5. The EGR assembly is designed to have some leak and, therefore, 7" of vacuum is all that should be able to be held on the assembly. However, if too much of a leak exists (less than 4"), the EGR assembly is leaking and must be replaced.

Diagnostic Aids:
The EGR position voltage can be used to determine that the pintle is moving. When no EGR is commanded (0% duty cycle), the position sensor should read between .5 volt and 1.5 volts, and increase with the commanded EGR duty cycle.

1988–90 2.8L (VIN W) – CORSICA AND BERETTA

CODE 32
EXHAUST GAS RECIRCULATION (EGR) CIRCUIT
2.8L (VIN W) "L" CARLINE (PORT)

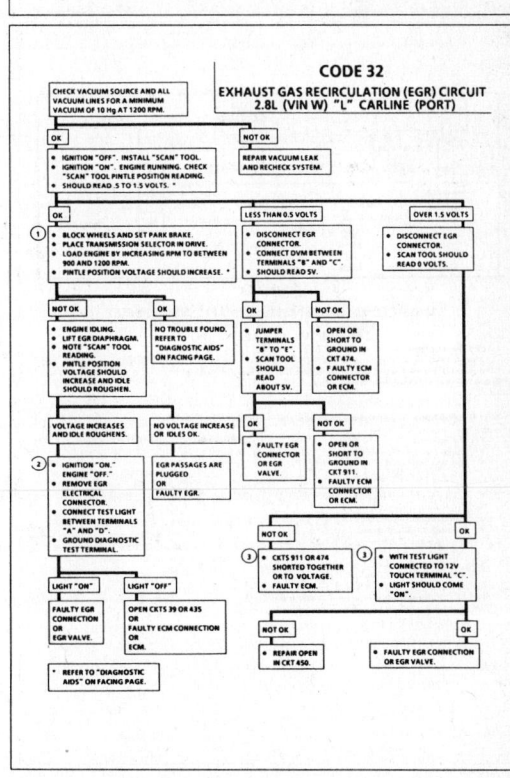

1988–90 2.8L (VIN W) – CORSICA AND BERETTA

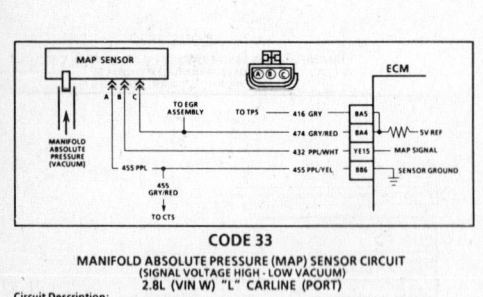

CODE 33
MANIFOLD ABSOLUTE PRESSURE (MAP) SENSOR CIRCUIT
(SIGNAL VOLTAGE HIGH - LOW VACUUM)
2.8L (VIN W) "L" CARLINE (PORT)

Circuit Description:
The manifold absolute pressure sensor (MAP) responds to changes in manifold pressure (vacuum). The ECM receives this information as a signal voltage that will vary from about 1 to 1.5 volts at idle (high vacuum) to 4–4.5 volts at wide open throttle (low vacuum).

Test Description: Numbers below refer to circled numbers on the diagnostic chart.
1. Code 33 will set when
   - Engine Running
   - Manifold pressure greater than 75.3 kPa (A/C "OFF") 81.2 kPa (A/C "ON")
   - Throttle angle less than 2%
   - Conditions met for 2 seconds
   Engine misfire or a low unstable idle may set Code 33
2. With the MAP sensor disconnected, the ECM should see a low voltage if the ECM and wiring are OK.

Diagnostic Aids:
If idle is rough or unstable refer to "Symptoms" in Section "B" for items which can cause an unstable idle.
An open in CKT 455 or the connection will result in a Code 33.
Ignition "ON" engine "OFF," voltage should be within the values shown in the table on the chart. Also, CHART C-1D can be used to test the MAP sensor.
Refer to "Intermittents" in Section "B".

1988–90 2.8L (VIN W) – CORSICA AND BERETTA

CODE 33
MANIFOLD ABSOLUTE PRESSURE (MAP) SENSOR CIRCUIT
(SIGNAL VOLTAGE HIGH - LOW VACUUM)
2.8L (VIN W) "L" CARLINE (PORT)

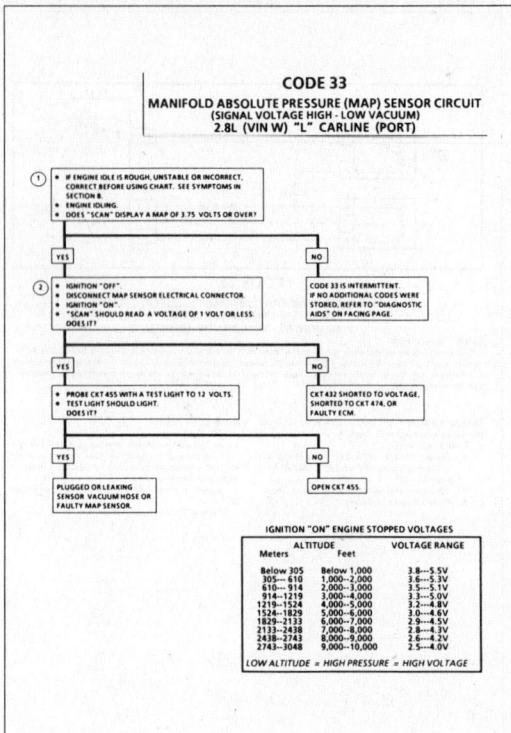

IGNITION "ON" ENGINE STOPPED VOLTAGES

ALTITUDE		VOLTAGE RANGE
Meters	Feet	
Below 305	Below 1,000	3.8—5.5V
305—610	1,000—2,000	3.6—5.3V
610—914	2,000—3,000	3.5—5.1V
914—1219	3,000—4,000	3.3—5.0V
1219—1524	4,000—5,000	3.2—4.8V
1524—1829	5,000—6,000	3.0—4.6V
1829—2133	6,000—7,000	2.9—4.5V
2133—2438	7,000—8,000	2.8—4.3V
2438—2743	8,000—9,000	2.6—4.2V
2743—3048	9,000—10,000	2.5—4.0V

LOW ALTITUDE = HIGH PRESSURE = HIGH VOLTAGE

1988–90 2.8L (VIN W) – CORSICA AND BERETTA

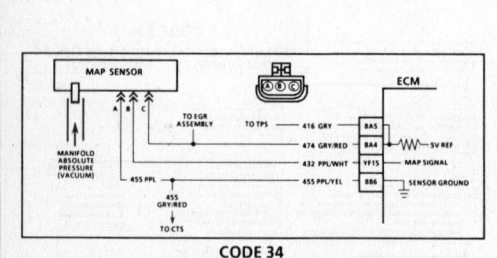

CODE 34
MANIFOLD ABSOLUTE PRESSURE (MAP) SENSOR CIRCUIT
(SIGNAL VOLTAGE LOW - HIGH VACUUM)
2.8L (VIN W) "L" CARLINE (PORT)

Circuit Description:
The manifold absolute pressure sensor (MAP) responds to changes in manifold pressure (vacuum). The ECM receives this information as a signal voltage that will vary from about 1 to 1.5 volts at idle (high vacuum) to 4–4.5 volts at wide open throttle (low vacuum).

Test Description: Numbers below refer to circled numbers on the diagnostic chart.
1. Code 34 will set if
   - Engine rpm less than 600
   - Manifold pressure reading less than 13 kPa.
   - Conditions met for 1 second
   or
   - Engine rpm greater than 600
   - Throttle angle over 20%
   - Manifold pressure less than 13 kPa
   - Conditions met for 1 second
2. This test is to see if the sensor is at fault for the low voltage or if there is a ECM or wiring problem.
3. This simulates a high signal voltage to check for an open in CKT 432. If the test light is bright during this test, CKT 432 is probably shorted to ground. If "Scan" reads over 4 volts at this test, CKT 474 can be checked by measuring the voltage at terminal "C" (should be 5 volts).

Diagnostic Aids:
An intermittent open in CKT 432 or 474 will result in a Code 34.
Ignition "ON" engine "OFF," voltages should be within the values shown in the table on the chart. Also CHART C-1D can be used to test MAP sensor.
Refer to "Intermittents" in Section "B".

1988–90 2.8L (VIN W) – CORSICA AND BERETTA

CODE 34
MANIFOLD ABSOLUTE PRESSURE (MAP) SENSOR CIRCUIT
(SIGNAL VOLTAGE LOW - HIGH VACUUM)
2.8L (VIN W) "L" CARLINE (PORT)

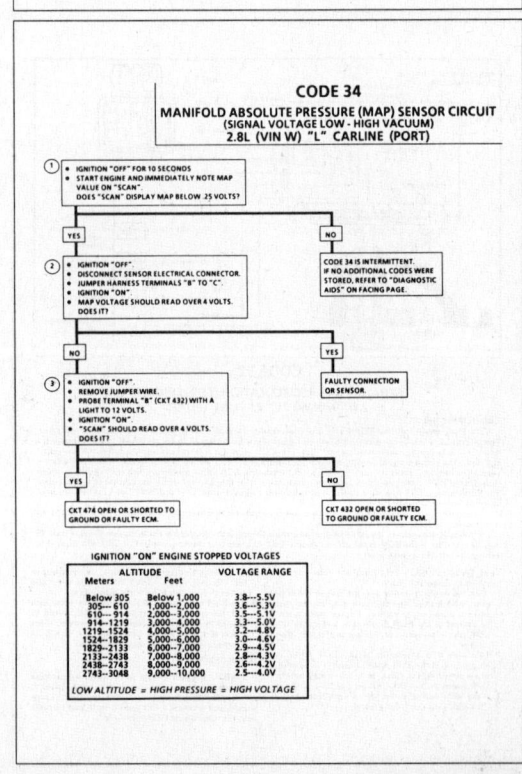

IGNITION "ON" ENGINE STOPPED VOLTAGES

ALTITUDE		VOLTAGE RANGE
Meters	Feet	
Below 305	Below 1,000	3.8—5.5V
305—610	1,000—2,000	3.6—5.3V
610—914	2,000—3,000	3.5—5.1V
914—1219	3,000—4,000	3.3—5.0V
1219—1524	4,000—5,000	3.2—4.8V
1524—1829	5,000—6,000	3.0—4.6V
1829—2133	6,000—7,000	2.9—4.5V
2133—2438	7,000—8,000	2.8—4.3V
2438—2743	8,000—9,000	2.6—4.2V
2743—3048	9,000—10,000	2.5—4.0V

LOW ALTITUDE = HIGH PRESSURE = HIGH VOLTAGE

1988–90 2.8L (VIN W) – CORSICA AND BERETTA

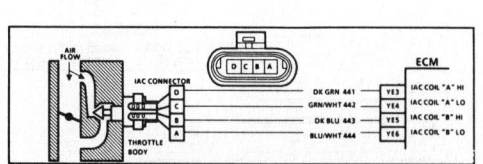

CODE 35
IDLE SPEED ERROR
2.8L (VIN W) "L" CARLINE (PORT)

Circuit Description:
Code 35 will set when the closed throttle engine speed is 300 rpm above or below the desired (commanded) idle speed for 50 seconds.

Test Description: Numbers below refer to circled numbers on the diagnostic chart.
1. Continue with test even if engine will not idle. If idle is too low, "Scan" will display 80 or more counts or steps. If idle is high, it will display "0" counts. Occasionally an erratic or unstable idle may occur. Engine speed may vary 200 rpm or more up and down. Disconnect IAC. If the condition is unchanged, the IAC is not at fault. There is a system problem. Proceed to diagnostic aids below.
2. When the engine was stopped, the IAC Valve retracted (more air) to a fixed "Park" position for increased air flow and idle speed during the next engine start. A "Scan" will display 80 or more counts. Observe idle immediately as on a warm engine, the idle speed should decrease rapidly.
3. Be sure to disconnect the IAC valve prior to this test. The test light will confirm the ECM signals by a steady or flashing light on all circuits.
4. There is a remote possibility that one of the circuits is shorted to voltage which would have been indicated by a steady light. Disconnect ECM and turn the ignition "ON" and probe terminals to check for this condition.

Diagnostic Aids:
A slow unstable idle may be caused by a system problem that cannot be controlled by the IAC. "Scan" counts will be above 80 counts if idle is too low, and "0" counts if it is too high.
If idle is too high, stop engine. Ignition "ON." Ground diagnostic terminal. Wait a few seconds for IAC to seat, then disconnect IAC. Start engine. If idle speed is above 600-700 rpm in drive with an A/T or 800-900 in neutral with a M/T, locate and correct vacuum leak. If rpm is below spec., check for foreign material around throttle plates and if OK.
- System too lean (High Air/fuel ratio)
  Idle speed may be too high or too low. Engine speed may vary up and down, disconnecting IAC does not help. This may set Code 44.
  "Scan" and/or Voltmeter will read an oxygen sensor output less than 300 mV (.3 volt). Check for low regulated fuel pressure or water in fuel. A lean exhaust with an oxygen sensor output fixed above 800 mV (.8 volt) will be a contaminated sensor, usually silicone. This may also set a Code 45 or 61.
- System too rich (Low Air/fuel ratio)
  Idle speed too low. "Scan" counts usually above 80. System obviously rich and may exhibit black smoke exhaust.
  "Scan" tool and/or Voltmeter will read an oxygen sensor signal fixed above 800 mV (.8 volt).
  Check
  - For fuel in pressure regulator hose
  - High fuel pressure
  - Injector leaking or sticking
- Throttle body. Remove IAC and inspect bore for foreign material or evidence of IAC valve dragging the bore.
- A/C Compressor or Relay failure. See CHART C-10 if the A/C control relay drive circuit is shorted to ground or if the relay is faulty, an idle problem may exist.
- If above are all OK, refer to "Rough, Unstable, Incorrect Idle or Stalling" in "Symptoms" in Section "B".

1988–90 2.8L (VIN W) – CORSICA AND BERETTA

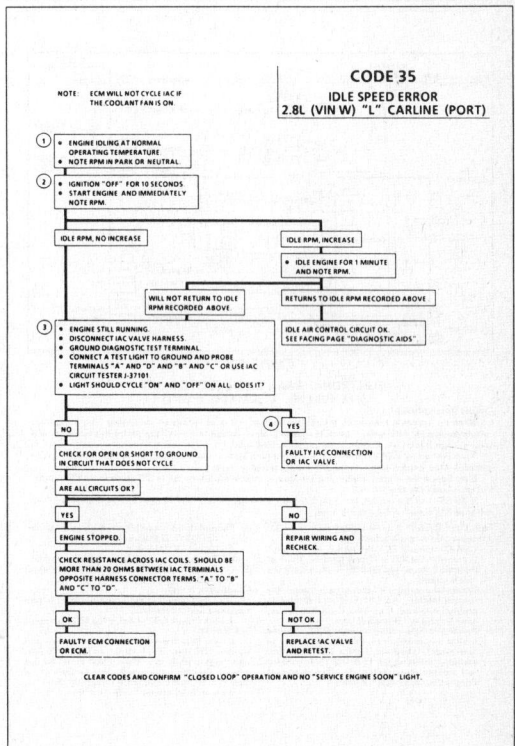

CODE 35
IDLE SPEED ERROR
2.8L (VIN W) "L" CARLINE (PORT)

1988–90 2.8L (VIN W) – CORSICA AND BERETTA

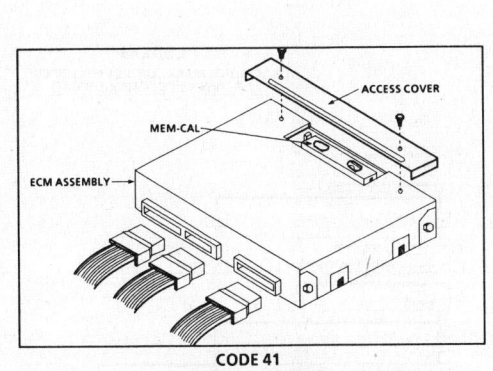

CODE 41
CYLINDER SELECT ERROR
(FAULTY OR INCORRECT MEM-CAL)
2.0L (VIN W) "L" CARLINE (PORT)

Test Description: Numbers below refer to circled numbers on the diagnostic chart.
1. The ECM used for this engine can also be used for other engines, and the difference is in the Mem-Cal. If a Code 41 sets, the incorrect Mem-Cal has been installed or it is faulty, and it must be replaced.

Diagnostic Aids:
Check Mem-Cal to be sure locking tabs are secure. Also check the pins on both the Mem-Cal and ECM to assure they are making proper contact. Check the Mem-Cal part number to assure it is the correct part. If the Mem-Cal is faulty, it must be replaced. It is also possible that the ECM is faulty. However, it should not be replaced until all of the above have been checked. For additional information, refer to "Intermittents" on page 6E3-B-2.

1988–90 2.8L (VIN W) – CORSICA AND BERETTA

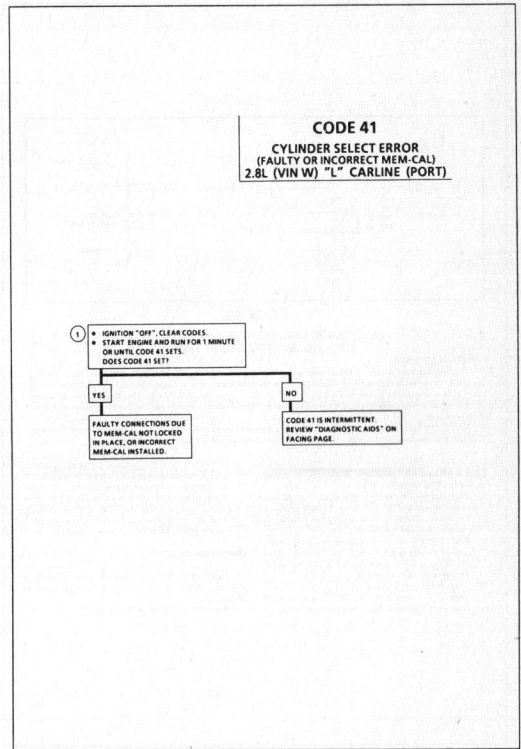

CODE 41
CYLINDER SELECT ERROR
(FAULTY OR INCORRECT MEM-CAL)
2.8L (VIN W) "L" CARLINE (PORT)

1988–90 2.8L (VIN W) – CORSICA AND BERETTA

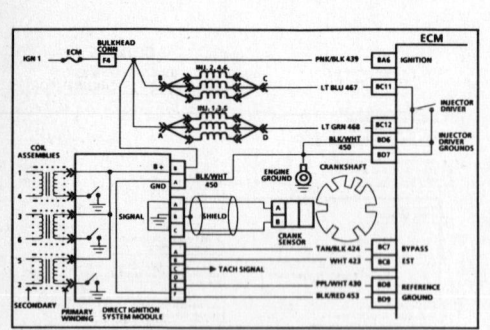

CODE 42
ELECTRONIC SPARK TIMING (EST) CIRCUIT
2.8L (VIN W) "L" CARLINE (PORT)

Circuit Description:

When the system is running on the ignition module, that is no voltage on the Bypass line, the ignition module grounds the EST signal. The ECM expects to see no voltage on the EST line during this condition. If it sees a voltage, it sets Code 42 and will not go into the EST mode.

When the rpm for EST is reached (about 400 rpm) and Bypass voltage applied, the EST should on longer be grounded in the ignition module so the EST voltage should be varying.

If the Bypass line is open or grounded, the ignition module will not switch to EST mode so the EST voltage will be low and Code 42 will be set.

If the EST line is grounded, the ignition module will switch to EST, but because the line is grounded there will be no EST signal. A Code 42 will be set.

Test Description: Numbers below refer to circled numbers on the diagnostic chart.
1. Code 42 means the ECM has seen an open or short to ground in the EST or Bypass circuits. This test confirms Code 42 and that the fault causing the code is present.
2. Checks for a normal EST ground path through the ignition module. An EST CKT 423 shorted to ground will also read less than 500 ohms, however, this will be checked later.
3. As the test light voltage touches CKT 424, the module should switch causing the ohmmeter to "overrange" if the meter is in the 1000-2000 ohms position. Selecting the 10-20,000 ohms position will indicate above 5000 ohms. The important thing is that the module "switched."

4. The module did not switch and this step checks for
   • EST CKT 423 shorted to ground
   • Bypass CKT 424 open
   • Faulty ignition module connection or module
5. Confirms that Code 42 is a faulty ECM and not an intermittent in CKT 423 or 424.

Diagnostic Aids:

The "Scan" tool does not have any ability to help diagnose a Code 42 problem.

A Mem-Cal not fully seated in the ECM can result in a Code 42.

If Code 42 is intermittent, there is a possibility of an open EST line. To check for an open EST line, crank engine while in a "Clear Flood Mode" for five seconds, then start engine and check for Code 42. If Code 42 is set, repair open in CKT 423 (EST line). Refer to "Intermittents" in Section "B".

1988–90 2.8L (VIN W) – CORSICA AND BERETTA

CODE 42
ELECTRONIC SPARK TIMING (EST) CIRCUIT
2.8L (VIN W) "L" CARLINE (PORT)

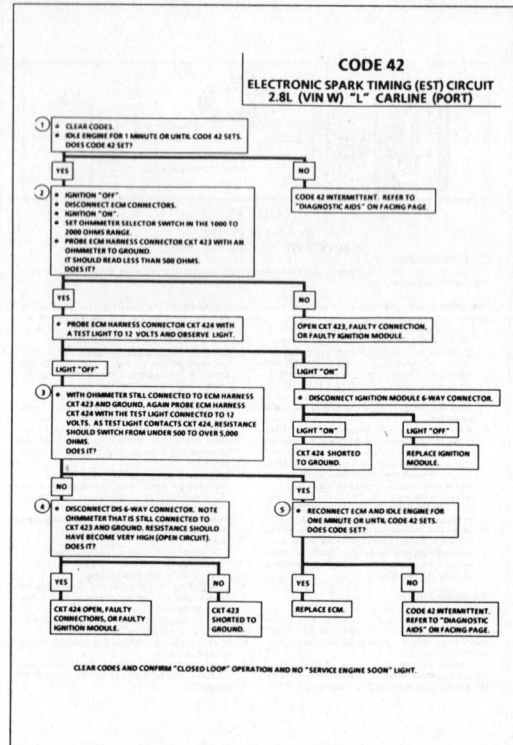

1988–90 2.8L (VIN W) – CORSICA AND BERETTA

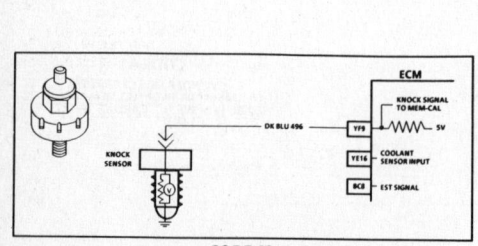

CODE 43
ELECTRONIC SPARK CONTROL (ESC) CIRCUIT
2.8L (VIN W) "L" CARLINE (PORT)

Circuit Description:

The knock sensor is used to detect engine detonation and the ECM will retard the electronic spark timing based on the signal being received. The circuitry within the knock sensor causes the ECM 5 volts to be pulled down so that under a no knock condition, CKT 496 would measure about 2.5 volts. The knock sensor produces an A/C signal which rides on the 2.5 volts DC voltage. The amplitude and signal frequency is dependent upon the knock level.

If CKT 496 becomes open or shorted to ground, the voltage will either go above 3.5 volts or below 1.5 volts. If either of these conditions are met for about ½ second, a Code 43 will be stored.

Test Description: Numbers below refer to circled numbers on the diagnostic chart.
1. This step determines if conditions for Code 43 still exist (voltage on CKT 496 above 3.5 volts or below 1.5 volts). The system is designed to retard the spark 6°, if either condition exists.
2. The ECM has a 5 volt pull-up resistor, which applies 5 volts to CKT 496. The five volt signal should be present at the knock sensor terminal during these test conditions.
3. This step determines if the knock sensor resistance is 3900 ohms ± 15% If the resistance is between 3300 and 4500 ohms, the sensor is OK.

4. If CKT 496 is not open or shorted to ground and the voltage reading is below 4 volts, the most likely cause is an open circuit in the ECM. It is possible that a faulty Mem-Cal could be drawing the 5 volt signal down and it should be replaced, if a replacement ECM did not correct the problem.

Diagnostic Aids:

Check CKT 496 for a potential open or short to ground. Also, check for proper installation of Mem-Cal.

Refer to "Intermittents" in Section "B".

1988–90 2.8L (VIN W) – CORSICA AND BERETTA

CODE 43
ELECTRONIC SPARK CONTROL (ESC) CIRCUIT
2.8L (VIN W) "L" CARLINE (PORT)

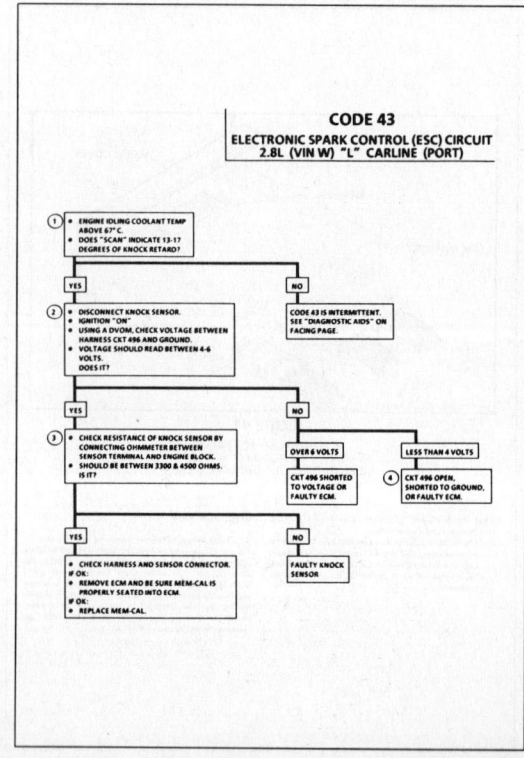

1988–90 2.8L (VIN W) – CORSICA AND BERETTA

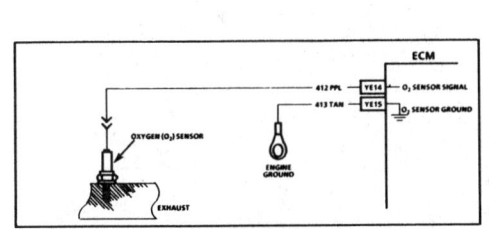

CODE 44
OXYGEN SENSOR CIRCUIT
(LEAN EXHAUST INDICATED)
2.8L (VIN W) "L" CARLINE (PORT)

Circuit Description:
The ECM supplies a voltage of about .45 volt between terminals "YE14" and "YE15". (If measured with a 10 megohm digital voltmeter, this may read as low as .32 volt.) The O_2 sensor varies the voltage within a range of about 1 volt if the exhaust is rich, down through about .10 volt if exhaust is lean.
The sensor is like an open circuit and produces no voltage when it is below about 315°C (600°F). An open sensor circuit or cold sensor causes "Open Loop" operation.

Test Description: Numbers below refer to circled numbers on the diagnostic chart.
1. Code 44 is set when the O_2 sensor signal voltage on CKT 412
   - Remains below .2 volt for 60 seconds or more
   - And the system is operating in "Closed Loop"

Diagnostic Aids:

Using the "Scan," observe the block learn values at different rpm and air flow conditions. The "Scan" also displays the block cells, so the block learn values can be checked in each of the cells to determine when the Code 44 may have been set. If the conditions for Code 44 exist, the block learn values will be around 150.
- O_2 Sensor Wire. Sensor pigtail may be mispositioned and contacting the exhaust manifold.

- Check for intermittent ground in wire between connector and sensor.
- Lean Injector(s). Perform injector balance test CHART C-2A.
- Fuel Contamination. Water, even in small amounts near the in-tank fuel pump inlet can be delivered to the injectors. The water causes a lean exhaust and can set a Code 44.
- Fuel Pressure. System will be lean if pressure is too low. It may be necessary to monitor fuel pressure while driving the car at various road speeds and/or loads to confirm. See Fuel System diagnosis CHART A-7.
- Exhaust Leaks. If there is an exhaust leak, the engine can cause outside air to be pulled into the exhaust and past the sensor. Vacuum or crankcase leaks can cause a lean condition.
- If the above are OK, it is a faulty oxygen sensor.

1988–90 2.8L (VIN W) – CORSICA AND BERETTA

CODE 44
OXYGEN SENSOR CIRCUIT
(LEAN EXHAUST INDICATED)
2.8L (VIN W) "L" CARLINE (PORT)

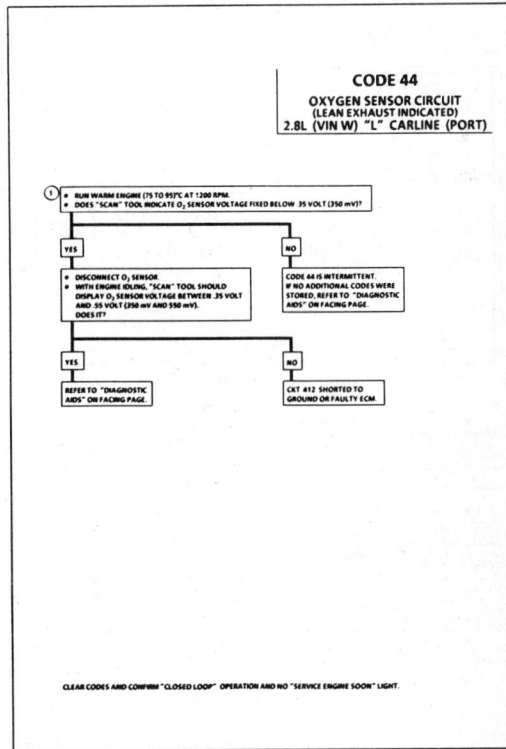

CLEAR CODES AND CONFIRM "CLOSED LOOP" OPERATION AND NO "SERVICE ENGINE SOON" LIGHT.

1988–90 2.8L (VIN W) – CORSICA AND BERETTA

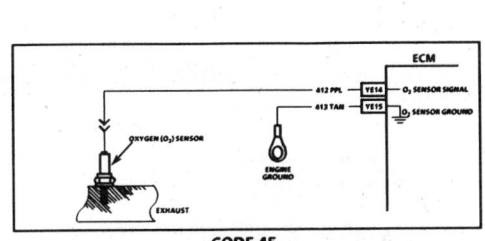

CODE 45
OXYGEN SENSOR CIRCUIT
(RICH EXHAUST INDICATED)
2.8L (VIN W) "L" CARLINE (PORT)

Circuit Description:
The ECM supplies a voltage of about .45 volt between terminals "YE14" and "YE15". (If measured with a 10 megohm digital voltmeter, this may read as low as .32 volt.) The O_2 sensor varies the voltage within a range of about 1 volt if the exhaust is rich, down through about .10 volt if exhaust is lean.
The sensor is like an open circuit and produces no voltage when it is below about 315°C (600°F). An open sensor circuit or cold sensor causes "Open Loop" operation.

Test Description: Numbers below refer to circled numbers on the diagnostic chart.
1. Code 45 is set when the O_2 sensor signal voltage or CKT 412
   - Remains above .7 volt for 30 seconds, and in "Closed Loop"
   - Engine time after start is 1 minute or more
   - Throttle angle between 3% and 45%

Diagnostic Aids:

Using the "Scan," observe the block learn values at different rpm and air flow conditions. The "Scan" also displays the block cells, so the block learn values can be checked in each of the cells to determine when the Code 45 may have been set. If the conditions for Code 45 exists, the block learn values will be around 115.
- Fuel Pressure. System will go rich if pressure is too high. The ECM can compensate for some increase. However, if it gets too high, a Code 45 may be set. See Fuel System diagnosis CHART A-7.
- Rich injector. Perform injector balance test CHART C-2A.

- Leaking injector, see CHART A-7.
- Check for fuel contaminated oil.
- Check for short to voltage on CKT 412.
- HEI Shielding. An open ground CKT 453 (ignition system reflow) may result in EMI, or induced electrical "noise." The ECM looks at this "noise" as reference pulses. The additional pulses result in a higher than actual engine speed signal. The ECM then delivers too much fuel, causing system to go rich. Engine tachometer will also show higher than actual engine speed, which can help in diagnosing this problem.
- Canister purge. Check for fuel saturation. If full of fuel, check canister control and hoses.

- Check for leaking fuel pressure regulator diaphragm by checking vacuum line to regulator for fuel.
- TPS. An intermittent TPS output will cause the system to go rich, due to a false indication of the engine accelerating.
- EGR. An EGR staying open (especially at idle) will cause the O_2 sensor to indicate a rich exhaust.

1988–90 2.8L (VIN W) – CORSICA AND BERETTA

CODE 45
OXYGEN SENSOR CIRCUIT
(RICH EXHAUST INDICATED)
2.8L (VIN W) "L" CARLINE (PORT)

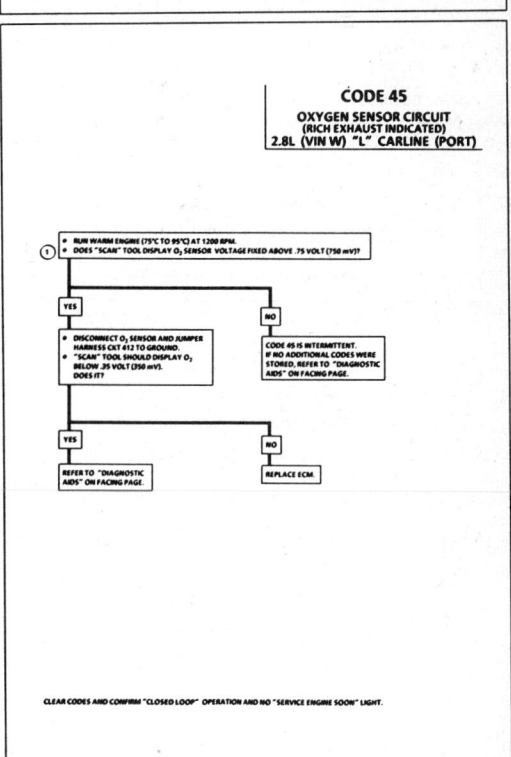

CLEAR CODES AND CONFIRM "CLOSED LOOP" OPERATION AND NO "SERVICE ENGINE SOON" LIGHT.

1988–90 2.8L (VIN W) – CORSICA AND BERETTA

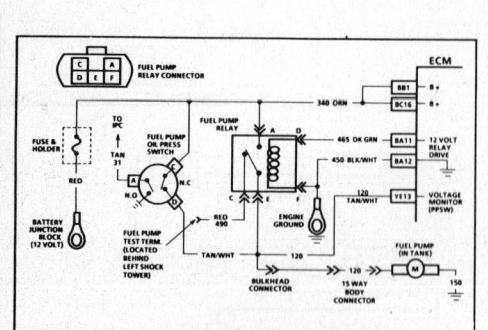

CODE 54
FUEL PUMP CIRCUIT
(LOW VOLTAGE)
2.8L (VIN W) "L" CARLINE (PORT)

Circuit Description:

The status of the fuel pump CKT 120 is monitored by the ECM at terminal "YE13" and is used to compensate fuel delivery based on system voltage. This signal is also used to store a trouble code if the fuel pump relay is defective or fuel pump voltage is lost while the engine is running. There should be about 12 volts on CKT 120 for 2 seconds after the ignition is turned, or any time references pulses are being received by the ECM.

Code 54 will set if the voltage at terminal "YE13" is less than 2 volts for 1.5 seconds since the last reference pulse was received. This code is designed to detect a faulty relay, causing extended crank time, and the code will help the diagnosis of an engine that "CRANKS BUT WILL NOT RUN."

If a fault is detected during start-up the "Service Engine Soon" light will stay "ON" until the ignition is cycled "OFF." However, if the voltage is detected below 2 volts with the engine running, the light will only remain "ON" while the condition exists.

1988–90 2.8L (VIN W) – CORSICA AND BERETTA

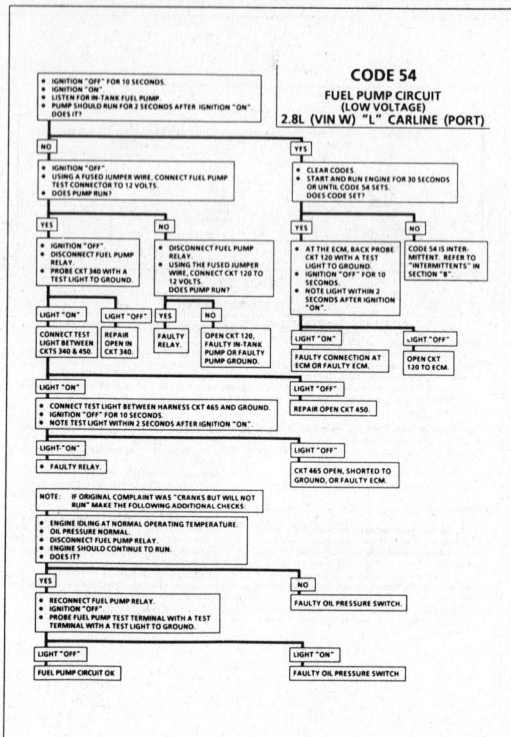

CODE 54
FUEL PUMP CIRCUIT
(LOW VOLTAGE)
2.8L (VIN W) "L" CARLINE (PORT)

1988–90 2.8L (VIN W) – CORSICA AND BERETTA

CODE 51
CODE 52
CODE 53

2.8L (VIN W) "L" CARLINE (PORT)

CODE 51
MEM-CAL ERROR
(FAULTY OR INCORRECT MEM-CAL)

CHECK THAT ALL PINS ARE FULLY INSERTED IN THE SOCKET AND THAT MEM-CAL IS PROPERLY LATCHED. IF OK, REPLACE MEM-CAL. CLEAR MEMORY, AND RECHECK. IF CODE 51 REAPPEARS, REPLACE ECM.

CLEAR CODES AND CONFIRM "CLOSED LOOP" OPERATION AND NO "SERVICE ENGINE SOON" LIGHT.

CODE 52
CALPAK ERROR
(FAULTY OR INCORRECT CALPAK)

CHECK THAT THE MEM-CAL IS FULLY SEATED AND LATCHED INTO THE MEM-CAL SOCKET. IF OK, REPLACE MEM-CAL, CLEAR MEMORY, AND RECHECK. IF CODE 52 REAPPEARS, REPLACE ECM.

CLEAR CODES AND CONFIRM "CLOSED LOOP" OPERATION AND NO "SERVICE ENGINE SOON" LIGHT.

CODE 53
SYSTEM OVER VOLTAGE

THIS CODE INDICATES THERE IS A BASIC GENERATOR PROBLEM.
- CODE 53 WILL SET, IF VOLTAGE AT ECM IGNITION INPUT PIN IS GREATER THAN 17.1 VOLTS FOR 2 SECONDS.
- CHECK AND REPAIR CHARGING SYSTEM. REFER TO SECTION "6D".

CLEAR CODES AND CONFIRM "CLOSED LOOP" OPERATION AND NO "SERVICE ENGINE SOON" LIGHT.

1988–90 2.8L (VIN W) – CORSICA AND BERETTA

CODE 61
DEGRADED OXYGEN SENSOR
2.8L (VIN W) "L" CARLINE (PORT)

IF A CODE 61 IS STORED IN MEMORY THE ECM HAS DETERMINED THE OXYGEN SENSOR IS CONTAMINATED OR DEGRADED, BECAUSE THE VOLTAGE CHANGE TIME IS SLOW OR SLUGGISH.

THE ECM PERFORMS THE OXYGEN SENSOR RESPONSE TIME TEST WHEN:

COOLANT TEMPERATURE IS GREATER THAN 85°C.

MAT TEMPERATURE IS GREATER THAN 10°C.

IN CLOSED LOOP.

IN DECEL FUEL CUT-OFF MODE.

IF A CODE 61 IS STORED THE OXYGEN SENSOR SHOULD BE REPLACED. A CONTAMINATED SENSOR CAN BE CAUSED BY FUEL ADDITIVES, SUCH AS SILICON, OR BY USE OF NON-GM APPROVED LUBRICANTS OR SEALANTS. SILICON CONTAMINATION IS USUALLY INDICATED BY A WHITE POWDERY SUBSTANCE ON THE SENSOR FINS.

CLEAR CODES AND CONFIRM "CLOSED LOOP" OPERATION AND NO "SERVICE ENGINE SOON" LIGHT.

1988–90 2.8L (VIN W) – CORSICA AND BERETTA

SECTION B
SYMPTOMS

TABLE OF CONTENTS

Before Starting
Intermittents
Hard Start
Hesitation, Sag, Stumble
Surges and/or Chuggle
Lack of Power, Sluggish, or Spongy
Detonation/Spark Knock
Cuts Out, Misses
Backfire
Poor Fuel Economy
Dieseling, Run-On
Rough, Unstable, or Incorrect Idle, Stalling
Excessive Exhaust Emissions or Odors
Restricted Exhaust System Check - Chart B-1

BEFORE STARTING

Before using this section you should have performed the DIAGNOSTIC CIRCUIT CHECK and found out that:
1. The ECM and "Service Engine Soon" light are operating.
2. There are no trouble codes.

Verify the customer complaint, and locate the correct SYMPTOM below. Check the items indicated under that symptom.

If the ENGINE CRANKS BUT WILL NOT RUN, see CHART A-3.

Several of the symptom procedures below call for a Careful Visual Check. This check should include:
- ECM grounds for being clean and tight.
- Vacuum hoses for splits, kinks, and proper connections, as shown on Emission Control Information label.
- Air leaks at throttle body mounting and intake manifold.
- Ignition wires for cracking, hardness, proper routing, and carbon tracking.
- Wiring for proper connections, pinches, and cuts. The importance of this step cannot be stressed too strongly - it can lead to correcting a problem without further checks and can save valuable time.

1988–90 2.8L (VIN W) – CORSICA AND BERETTA

INTERMITTENTS

Problem may or may not turn "ON" the "Service Engine Soon" light, or store a code.

DO NOT use the Trouble Code Charts in Section "A" for intermittent problems. The fault must be present to locate the problem. If a fault is intermittent, use of trouble code charts may result in replacement of good parts.
- Most intermittent problems are caused by faulty electrical connections or wiring. Perform careful check as described at start of Section "B". Check for
  - Poor mating of the connector halves, or terminals not fully seated in the connector body (backed out).
  - Improperly formed or damaged terminals. All connector terminals in problem circuit should be carefully reformed to increase contact tension.
  - Poor terminal to wire connection. This requires removing the terminal from the connector body to check.
- If a visual check does not find the cause of the problem, the car can be driven with a voltmeter connected to a suspected circuit. A "Scan" tool can, also, be used for monitoring input signals to help detect intermittent conditions. An abnormal voltage, or "Scan" reading, when the problem occurs, indicates the problem may be in that circuit. If the wiring and connectors check OK and a trouble code was stored for a circuit having a sensor, except for Codes 43, 44, and 45, substitute a known good sensor and recheck.

An intermittent "Service Engine Soon" light with no stored code may be caused by
- Ignition coil shorted to ground and arcing at spark plug wires or plugs.
- "Service Engine Soon" light wire to ECM shorted to ground. (CKT 419)
- Diagnostic "Test" terminal wire to ECM, shorted to ground. (CKT 451)
- ECM power grounds. See ECM wiring diagrams.
- Loss of trouble code memory. To check, disconnect TPS and idle engine until "Service Engine Soon" light comes "ON." Code 22 should be stored, and kept in memory when ignition is turned "OFF." If not, the ECM is faulty.
- Check for an electrical system interference caused by a defective relay, ECM driven solenoid, or switch. They can cause a sharp electrical surge. Normally the problem will occur when the faulty component is operated.
- Check for improper installation of electrical options, such as lights, 2-way radios, etc.
- EST wires should be kept away from spark plug wires, coils and generator. Wire from ECM to ignition system (CKT 453) should be a good connection.
- Check for open diode across A/C compressor clutch, and for other open diodes (see wiring diagrams)

HARD START

Definition: Engine cranks OK, but does not start for a long time. Does eventually run, or may start but immediately dies.
- Perform careful check as described at start of Section "B".
- Make sure driver is using correct starting procedure.
- CHECK:
  - TPS for sticking or binding or a high TPS voltage with the throttle closed.
  - High resistance in coolant sensor circuit or sensor itself. See Code 15 chart or with a "Scan" tool compare coolant temperature with ambient temperature on a cold engine.
  - Fuel pressure CHART A-7.
  - Water contaminated fuel.
  - EGR operation. Be sure valve seats properly and is not staying open. See CHART C-7.
- Ignition system - Check for
  Bare and shorted wires and proper output with spark tester J-26792 (ST-125) or equivalent.
  - IAC operation - See Code 35 chart.
- A faulty in-tank fuel pump check valve will allow the fuel in the lines to drain back to the tank after the engine is stopped. To check for this condition:
  Perform Fuel System Diagnosis, CHART A-7.
- Remove spark plugs. Check for wet plugs, cracks, wear, improper gap, burned electrodes, or heavy deposits. Repair or replace as necessary.

1988–90 2.8L (VIN W) – CORSICA AND BERETTA

HESITATION, SAG, STUMBLE

Definition: Momentary lack of response as the accelerator is pushed down. Can occur at all car speeds. Usually most severe when first trying to make the car move, as from a stop sign. May cause the engine to stall if severe enough.
- Perform careful visual check as described at start of Section "B".
- CHECK:
  - Fuel pressure. See CHART A-7. Also Check for water contaminated fuel.
  - Spark plugs for being fouled or faulty wiring.
  - Mem-Cal number and Service Bulletins for latest Mem-Cal.
- TPS for binding or sticking. Voltage should increase at a steady rate as throttle is moved toward WOT.
- Generator output voltage. Repair, if less than 9 or more than 16 volts.
- Ignition system ground, CKT 453.
- Canister purge system for proper operation. See CHART C-3.
- EGR - See CHART C-7.
- Check for Codes 33 and 34. (MAP sensor circuit.)
- Perform injector balance test, CHART C-2A.

SURGES AND/OR CHUGGLE

Definition: Engine power variation under steady throttle or cruise. Feels like the car speeds up and slows down with no change in the accelerator pedal.
- Be sure driver understands Torque Converter Clutch and A/C compressor operation in Owner's Manual.
- Perform careful visual inspection as described at start of Section "B".
- To help determine if the condition is caused by a rich or lean system, the car should be driven at the speed of the complaint. Monitoring block learn at the complaint speed will help identify the cause of the problem. If the system is running lean (block learn greater than 138), refer to diagnostic aids on facing page of Code 44. If the system is running rich (block learn less than 118), refer to diagnostic aids on facing page of Code 45.
- CHECK:
  - Generator output voltage. Repair if less than 9 or more than 16 volts.
  - EGR: There should be no EGR at idle. See CHART C-7.
  - EGR filter for being plugged. See CHART C-7.
  - Vacuum lines for kinks or leaks.
  - In-line fuel filter. Replace if dirty or plugged.
  - Fuel pressure while condition exists. See CHART A-7.
- Remove spark plugs. Check for cracks, wear, improper gap, burned electrodes, or heavy deposits. Also check condition of spark plug wires and check for proper output voltage using spark tester J-26792 (ST-125) or equivalent.

1988–90 2.8L (VIN W) – CORSICA AND BERETTA

LACK OF POWER, SLUGGISH, OR SPONGY

Definition: Engine delivers less than expected power. Little or no increase in speed when accelerator pedal is pushed down part way.
- Perform careful visual check as described at start of Section "B".
- Compare customer's car has a similar unit. Make sure the customer's car has an actual problem.
- Remove air filter element for dirt, or for being plugged. Replace as necessary.
- CHECK:
  - Restricted fuel filter, contaminated fuel or improper fuel pressure. See CHART A-7.
  - ECM power grounds, see wiring diagrams.
  - EGR operation for being open or partly open all the time - See CHART C-7.
- Exhaust system for possible restriction: See CHART A-7.
  - Inspect exhaust system for damaged or collapsed pipes.
  - Inspect muffler for heat distress or possible internal failure.
- Generator output voltage. Repair if less than 9 or more than 16 volts.
- Engine valve timing and compression.
- Engine for proper or worn camshaft.
- Secondary voltage using a shop ocilliscope or a spark tester J-26792 (ST-125) or equivalent. Check A/C operation. A/C clutch should cut out at WOT. See A/C CHART C-10.

DETONATION/SPARK KNOCK

Definition: A mild to severe ping, usually worse under acceleration. The engine makes sharp metallic knocks that change with throttle opening. Sounds like popcorn popping.
- Check for obvious overheating problems:
  - Low coolant.
  - Loose belt.
  - Restricted air flow to radiator, or restricted water flow through radiator.
  - Inoperative electric cooling fan circuit. See CHART C-12.
- To help determine if the conditions is caused by a rich or lean system, the car should be driven at the speed of the complaint. Monitoring block learn, at the complaint speed, will help identify the cause of the problem. If the system is running lean (block learn greater than 138), refer to diagnostic aids on facing page of Code 44. If the system is running rich (block learn less than 118), refer to facing page of Code 45.
- CHECK:
  - EGR system for not opening or plugged EGR passages - See CHART C-7.
  - ESC system for no retard - See CHART C-5.
  - Park/Neutral switch. Be sure "Scan" indicates drive with gear selector in drive or overdrive. - See CHART C-1A.
  - TCC operation, TCC applying too soon - see CHART C-8.
  - Fuel system pressure - See CHART A-7.
  - Engine for excessive carbon build-up. Remove carbon with top engine cleaner. Follow instructions on can.
  - Mem-Cal for correct part.
  - Combustion chamber for excessive oil.
  - Engine for incorrect basic parts such as cam, heads, pistons, etc.
  - Fuel for poor quality, proper octane rating.

1988–90 2.8L (VIN W) – CORSICA AND BERETTA

CUTS OUT, MISSES

Definition: Steady pulsation or jerking that follows engine speed, usually more pronounced as engine load increases. The exhaust has a steady spitting sound at idle or under a load.

- Perform careful visual (physical) check as described at start of Section "B".
- If ignition system is suspected of causing a miss at idle or cutting out under load, refer to appropriate ignition "Misfire" chart.
- If the previous checks did not find the problem:
  - Visually inspect ignition system for moisture, dust, cracks, burns, etc. Spray plug wires with fine water mist to check for shorts.
- If above checks did not correct problem, check the following:
  - To determine if one injector or connector is faulty, disconnect injector 4-way connector. On the injector side of the harness, connect ohmmeter between the B+ circuit and the injector drive side of the harness for each bank of injectors. Because each bank of injectors is wired in parallel, an ohmmeter should measure about 4 ohms. If the reading is 6 ohms, or greater, an open circuit is likely the cause. If 2 ohms or less, a shorted circuit or injector is likely. Reconnect injector 4-way connector.

- If either ohm reading is out of range:
  - Disconnect all injector harness connectors. Connect J-34730-2 Injector Test Light or equivalent 6 volt test light between the harness terminals of each injector connector and note light while cranking. If test light fails to blink at any connector, it is a faulty injector drive circuit harness, connector, or terminal.
- CHECK:
  - Injector Balance. See CHART C-2A.
  - Fuel System - Plugged fuel filter, water, low pressure. See CHART A-7.
  - Valve Timing
  - Remove rocker covers. Check for bent pushrods, worn rocker arms, broken valve springs, worn camshaft lobes. Repair as necessary.
- Perform compression check.

BACKFIRE

Definition: Fuel ignites in intake manifold, or in exhaust system, making a loud popping noise.

- CHECK:
  - Compression - Look for sticking or leaking valves.
  - EGR operation for being open all the time. See CHART C-7.
  - EGR gasket for faulty or loose fit.
- Output voltage of ignition coils, using a shop oscilliscope or spark tester J-26792 (ST-125), or equivalent.
- Valve timing
- Spark plugs, spark plug wires, and proper routing of plug wires.

1988–90 2.8L (VIN W) – CORSICA AND BERETTA

POOR FUEL ECONOMY

Definition: Fuel economy, as measured by an actual road test, is noticeably lower than expected. Also, economy is noticeably lower than it was on this car at one time, as previously shown by an actual road test

- Check owner's driving habits
  - Is A/C "ON" full time? (Defroster mode "ON"?)
  - Are tires at correct pressure?
  - Are excessively heavy loads being carried?
  - Is acceleration too much, too often?
  - Suggest owner fill fuel tank and recheck fuel economy.
  - Suggest driver read "Important Facts on Fuel Economy" in Owner's Manual.
- Check for proper calibration of speedometer.
- Visually (physically) Check.
  - Vacuum hoses for splits, kinks, and proper connections as shown on Vehicle Emission Control Information label.
  - Ignition wires for cracking, hardness, and proper connections.
  - Air cleaner element (filter) for dirt or being plugged.
- Remove spark plugs. Check for cracks, wear, improper gap, burned electrodes, or heavy deposits. Repair or replace as necessary.

- CHECK:
  - Engine thermostat for faulty part (always open) or for wrong heat range. Using a "Scan" tool, monitor engine temperature. A "Scan" displays engine temp. in degrees centigrade. After engine is started, the temperature should rise steadily to about 90°C, then stabilize, when thermostat opens
  - Fuel Pressure. See CHART A-7.
  - Compression.
  - TCC for proper operation. See CHART C-8A. A "Scan" should indicate an rpm drop, when the TCC is commanded "ON".
  - Exhaust system restriction. See CHART B-1.
  - Using a "Scan" tool, crank engine for five seconds while in clear flood mode. Start engine and check for a Code 42. An open EST line will cause poor fuel economy. Refer to Code 42 diagnostic chart for repair procedure.

DIESELING, RUN-ON

Definition: Engine continues to run after key is turned "OFF" but runs very roughly.

- Check injectors for leaking. See CHART A-7.
- If engine runs smoothly, check ignition switch and adjustment.

1988–90 2.8L (VIN W) – CORSICA AND BERETTA

ROUGH, UNSTABLE, OR INCORRECT IDLE, STALLING

Definition: The engine runs unevenly at idle. If bad enough, the car may shake. Also, the idle may vary in rpm (called "hunting"). Either condition may be bad enough to cause stalling. Engine idles at incorrect speed.

- Perform careful visual check as described at start of Section "B".
- CHECK:
  - Motor mounts for damage, grounding out on frame, or mispositioned, etc.
  - Throttle linkage for sticking or binding.
  - TPS for sticking or binding, be sure output is stable at idle and adjustment specification is correct.
  - IAC system. See Code 35 chart.
  - Generator output voltage. Repair if less than 9 or more than 16 volts.
  - P/N switch circuit. See CHART C-1A, or use "Scan" tool, and be sure tool indicates vehicle is in drive while gear selector in drive (125C).
  - Injector balance. See CHART C-2A.
  - PCV valve for proper operation by placing finger over inlet hole in valve and several times. Valve should be snap back. If not, replace valve.
  - Evaporative Emission Control System. CHART C-3.

- Engine will run rough or stall if loss of battery power to the ECM (Electronic Control Module) has occurred. This ECM has an idle learn feature. (See Section "C2" for IDLE Learn Procedure.)
- Power Steering Pressure switch input. The state of the switch should only change when wheels are turned up against the stops. See CHART C-1E.
- ECM ground circuits.
- EGR valve: There should be no EGR at idle.
- Monitoring block learn values may help identify the cause of the problem. If the system is running lean (block learn greater than 138) refer to Diagnostic Aids on facing page of Code 44. If the system is running rich (block learn values less than 118) refer to Diagnostic Aids on facing page of Code 45.
- Run a cylinder compression check.

- Check for fuel in pressure regulator hose. If present, replace regulator assembly.
- Check ignition system; wires and plugs.
- If problem exists with A/C "ON", check A/C system operation CHART C-10.

EXCESSIVE EXHAUST EMISSIONS OR ODORS

Definition: Vehicle fails an emission test. Vehicle has excessive "rotten egg" smell. Excessive odors do not necessarily indicate excessive emissions.

- Perform "Diagnostic Circuit Check."
- IF TEST SHOWS EXCESSIVE CO AND HC, (or also has excessive odors):
  - Check items which cause car to run RICH.
    - Make sure engine is at normal operating temperature.
  - CHECK:
    - Fuel pressure. See CHART A-7.
    - Canister for fuel loading. See CHART C-3.
    - Injector balance. See CHART C-2A.
    - PCV valve for being plugged, stuck, or blocked PCV hose, or fuel in the crankcase.
    - Spark plugs, plug wires, and ignition components.
    - Check for lead contamination of catalytic converter (look for removal of fuel filler neck restrictor).
    - Check for properly installed fuel cap.

- If the system is running rich, (block learn less than 118), refer to "Diagnostic Aids" on facing page of Code 45.
- IF TEST SHOWS EXCESSIVE NOx:
  - Check items which cause car to run LEAN, or run too hot.
    - EGR valve for not opening. See CHART C-7.
    - Vacuum leaks.
    - Coolant system and coolant fan for proper operation. See CHART C-12.
    - Remove carbon with top engine cleaner. Follow instructions on can.
- If the system is running lean, (block learn greater than 138), refer to "Diagnostic Aids" on facing page of Code 44.

1988–90 2.8L (VIN W) – CORSICA AND BERETTA

CHART B-1
RESTRICTED EXHAUST SYSTEM CHECK
ALL ENGINES

Proper diagnosis for a restricted exhaust system is essential before any components are replaced. Either of the following procedures may be used for diagnosis, depending upon engine or tool used:

CHECK AT A. I. R. PIPE:
1. Remove the rubber hose at the exhaust manifold A.I.R. pipe check valve. Remove check valve.
2. Connect a fuel pump pressure gauge to a hose and nipple from a Propane Enrichment Device (J26911) (see illustration)
3. Insert the nipple into the exhaust manifold A.I.R. pipe.

OR CHECK AT O₂ SENSOR:
1. Carefully remove O₂ sensor.
2. Install Borroughs exhaust backpressure tester (BT 8515 or BT 8603) or equivalent in place of O₂ sensor (see illustration)
3. After completing test described below, be sure to coat threads of O₂ sensor with anti-seize compound P/N 5613695 or equivalent prior to re-installation.

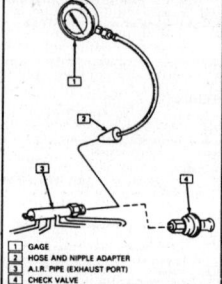

1	GAGE
2	HOSE AND NIPPLE ADAPTER
3	A.I.R. PIPE (EXHAUST PORT)
4	CHECK VALVE

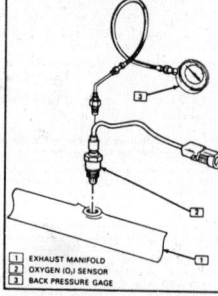

1	EXHAUST MANIFOLD
2	OXYGEN (O₂) SENSOR
3	BACK PRESSURE GAGE

DIAGNOSIS:
1. With the engine idling at normal operating temperature, observe the exhaust system backpressure reading on the gauge. Reading should not exceed 8.6 kPa (1.25 psi).
2. Increase engine speed to 2000 rpm and observe gauge. Reading should not exceed 20.7 kPa (3 psi).
3. If the backpressure at either speed exceeds specification, a restricted exhaust system is indicated.
4. Inspect the entire exhaust system for a collapsed pipe, heat distress, or possible internal muffler failure.
5. If there are no obvious reasons for the excessive backpressure, the catalytic converter is suspected to be restricted and should be replaced using current recommended procedures.

1988–90 2.8L (VIN W) – CORSICA AND BERETTA

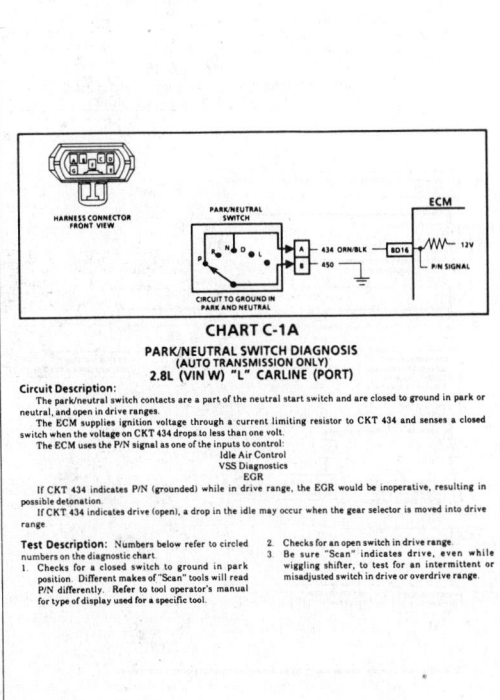

HARNESS CONNECTOR FRONT VIEW

CHART C-1A
PARK/NEUTRAL SWITCH DIAGNOSIS
(AUTO TRANSMISSION ONLY)
2.8L (VIN W) "L" CARLINE (PORT)

Circuit Description:
The park/neutral switch contacts are a part of the neutral start switch and are closed to ground in park or neutral, and open in drive ranges.
The ECM supplies ignition voltage through a current limiting resistor to CKT 434 and senses a closed switch when the voltage on CKT 434 drops to less than one volt.
The ECM uses the P/N signal as one of the inputs to control:
Idle Air Control
VSS Diagnostics
EGR
If CKT 434 indicates P/N (grounded) while in drive range, the EGR would be inoperative, resulting in possible detonation.
If CKT 434 indicates drive (open), a drop in the idle may occur when the gear selector is moved into drive range

Test Description: Numbers below refer to circled numbers on the diagnostic chart.
1. Checks for a closed switch to ground in park position. Different makes of "Scan" tools will read P/N differently. Refer to tool operator's manual for type of display used for a specific tool.

2. Checks for an open switch in drive range.
3. Be sure "Scan" indicates drive, even while wiggling shifter, to test for an intermittent or misadjusted switch in drive or overdrive range.

1988–90 2.8L (VIN W) – CORSICA AND BERETTA

CHART C-1A
PARK/NEUTRAL SWITCH DIAGNOSIS
(AUTO TRANSMISSION ONLY)
2.8L (VIN W) "L" CARLINE (PORT)

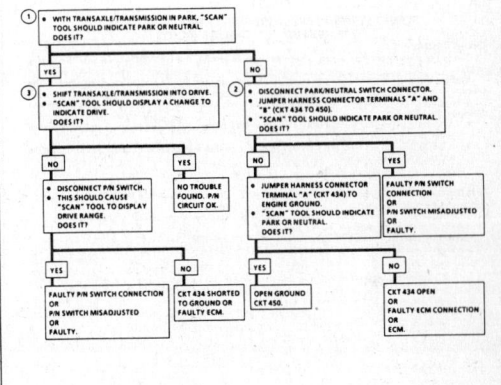

1988–90 2.8L (VIN W) – CORSICA AND BERETTA

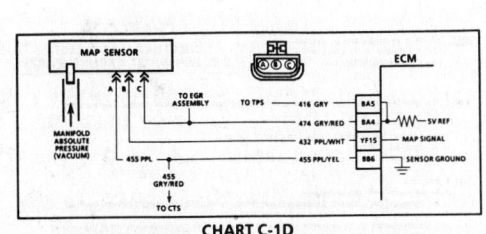

CHART C-1D
MANIFOLD ABSOLUTE PRESSURE (MAP) OUTPUT CHECK
2.8L (VIN W) "L" CARLINE (PORT)

Circuit Description:
The manifold absolute pressure sensor (MAP) measures manifold pressure (vacuum) and sends that signal to the ECM. The MAP sensor is mainly used for fuel calculation when the ECM is running in the throttle body backup mode. The MAP sensor is also used to determine the barometric pressure and to help calculate fuel delivery.

Test Description: Numbers below refer to circled numbers on the diagnostic chart.
1. Checks MAP sensor output voltage to the ECM. This voltage without engine running, represents a barometer reading to the ECM.
2. Applying 34 kPa (10 inches Hg) vacuum to the MAP sensor should cause the voltage to be 1.2 volts less than the voltage at Step 1. Upon applying vacuum to the sensor, the change in voltage should be instantaneous. A slow voltage change indicates a faulty sensor.

The engine must be running in this step or the "Scanner" will not indicate a change in voltage. It is normal for the "Service Engine Soon" light to come "ON" and for the system to set a Code 33 during this step. Make sure the code is cleared when this test is completed.
3. Check vacuum hose to sensor for leaking or restriction. Be sure no other vacuum devices are connected to the MAP hose.

1988–90 2.8L (VIN W) – CORSICA AND BERETTA

CHART C-1D
MANIFOLD ABSOLUTE PRESSURE (MAP) OUTPUT CHECK
2.8L (VIN W) "L" CARLINE (PORT)

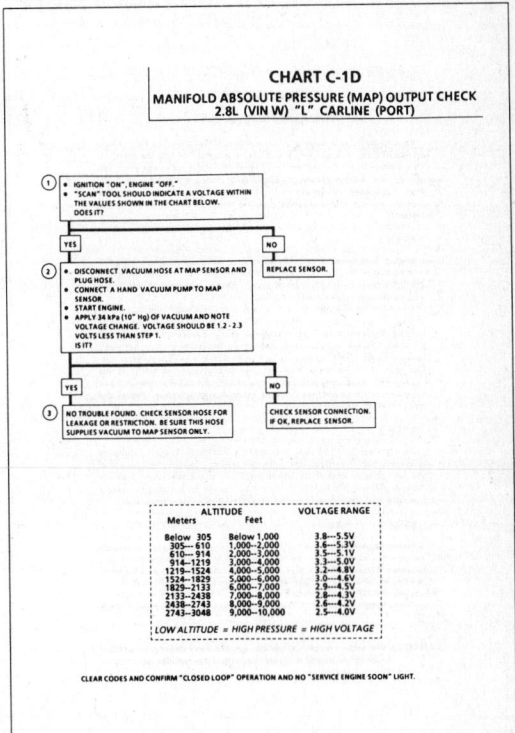

	ALTITUDE		VOLTAGE RANGE
	Meters	Feet	
	Below 305	Below 1,000	3.8–-5.5V
	305— 610	1,000–2,000	3.6–-5.3V
	610— 914	2,000–3,000	3.5–-5.1V
	914—1219	3,000–4,000	3.3–-5.0V
	1219—1524	4,000–5,000	3.2–-4.8V
	1524—1829	5,000–6,000	3.0–-4.6V
	1829—2133	6,000–7,000	2.9–-4.5V
	2133—2438	7,000–8,000	2.8–-4.3V
	2438—2743	8,000–9,000	2.6–-4.2V
	2743—3048	9,000–10,000	2.5–-4.0V

LOW ALTITUDE = HIGH PRESSURE = HIGH VOLTAGE

CLEAR CODES AND CONFIRM "CLOSED LOOP" OPERATION AND NO "SERVICE ENGINE SOON" LIGHT.

1988–90 2.8L (VIN W) – CORSICA AND BERETTA

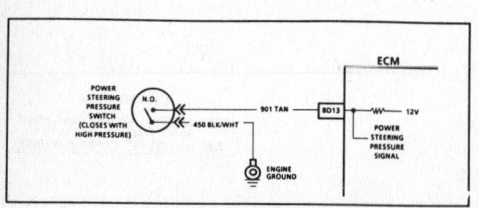

CHART C-1E
POWER STEERING PRESSURE SWITCH (PSPS) DIAGNOSIS
2.8L (VIN W) "L" CARLINE (PORT)

Circuit Description:
The power steering pressure switch is normally open to ground, and CKT 901 will be near the battery voltage.
Turning the steering wheel increases power steering oil pressure and its load on an idling engine. The pressure switch will close before the load can cause an idle problem.
Closing the switch causes CKT 901 to read less than 1 volt. The ECM will increase the idle air rate and disengage the A/C relay.

- A pressure switch that will not close, or an open CKT 901 or 450, may cause the engine to stop when power steering loads are high.
- A switch that will not open, or a CKT 901 shorted to ground, may affect idle quality and will cause the A/C relay to be de-energized.

Test Description: Numbers below refer to circled numbers on the diagnostic chart.

1. Different makes of "Scan" tools may display the state of this switch in different ways. Refer to "Scan" tool operator's manual to determine how this input is indicated.
2. Checks to determine if CKT 901 is shorted to ground.
3. This should simulate a closed switch.

1988–90 2.8L (VIN W) – CORSICA AND BERETTA

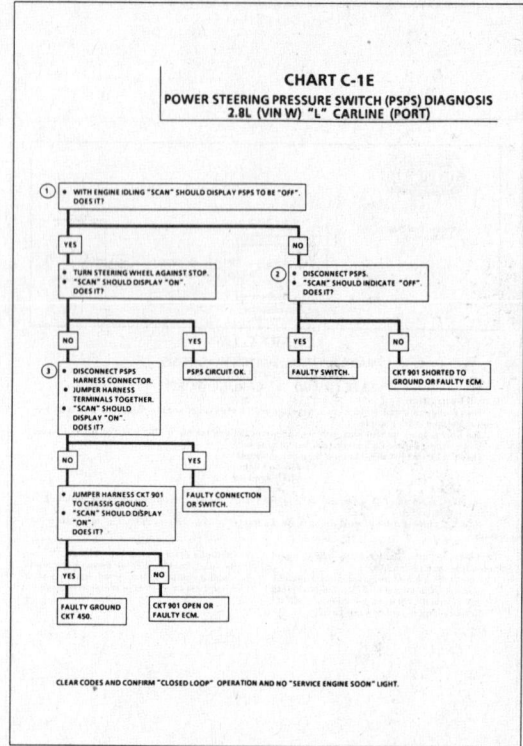

CHART C-1E
POWER STEERING PRESSURE SWITCH (PSPS) DIAGNOSIS
2.8L (VIN W) "L" CARLINE (PORT)

1988–90 2.8L (VIN W) – CORSICA AND BERETTA

CHART C-2A
INJECTOR BALANCE TEST

The injector balance tester is a tool used to turn the injector on for a precise amount of time, thus spraying a measured amount of fuel into the manifold. This causes a drop in fuel rail pressure that we can record and compare between each injector. All injectors should have the same amount of pressure drop (± 10 kPa). Any injector with a pressure drop that is 10 kPa (or more) greater or less than the average drop of the other injectors should be considered faulty and replaced.

STEP 1

Engine "cool down" period (10 minutes) is necessary to avoid irregular readings due to "Hot Soak" fuel boiling. With ignition "OFF" connect fuel gauge J 347301 or equivalent to fuel pressure tap. Wrap a shop towel around fitting while connecting gage to avoid fuel spillage.
Disconnect harness connectors at all injectors, and connect injector tester J-34730-3, or equivalent, to one injector. On Turbo equipped engines, use adaptor harness furnished with injector tester to energize injectors that are not accessible. Follow manufacturers instructions for use of adaptor harness. Ignition must be "OFF" at least 10 seconds to complete ECM shutdown cycle. Fuel pump should run about 2 seconds after ignition is turned "ON". At this point, insert clear tubing attached to vent valve into a suitable container and bleed air from gauge and hose to insure accurate gauge operation. Repeat this step until all air is bled from gauge.

STEP 2

Turn ignition "OFF" for 10 seconds and then "ON" again to get fuel pressure to its maximum. Record this initial pressure reading. Energize tester one time and note pressure drop at its lowest point (Disregard any slight pressure increase after drop hits low point.) By subtracting this second pressure reading from the initial pressure, we have the actual amount of injector pressure drop.

STEP 3

Repeat step 2 on each injector and compare the amount of drop. Usually, good injectors will have virtually the same drop. Retest any injector that has a pressure difference of 10 kPa, either more or less than the average of the other injectors on the engine. Replace any injector that also fails the retest. If the pressure drop of all injectors is within 10 kPa of this average, the injectors appear to be flowing properly. Reconnect them and review Symptoms, Section "B".

NOTE: The entire test should not be repeated more than once without running the engine to prevent flooding. (This includes any retest on faulty injectors).

1988–90 2.8L (VIN W) – CORSICA AND BERETTA

NOTE: The fuel pressure test in Section "A", CHART A-7, should be completed prior to this test.

CHART C-2A
INJECTOR BALANCE TEST
2.8L (VIN W) "L" CARLINE (PORT)

Step 1. If engine is at operating temperature, allow a 10 minute "cool down" period then connect fuel pressure gauge and injector tester.
1. Ignition "OFF".
2. Connect fuel pressure gauge and injector tester.
3. Ignition "ON".
4. Bleed off air in gauge. Repeat until all air is bled from gauge.

Step 2. Run test:
1. Ignition "OFF" for 10 seconds.
2. Ignition "ON". Record gauge pressure. (Pressure must hold steady, if not see the Fuel System diagnosis, Chart A-7, in Section "A")
3. Turn injector on, by depressing button on injector tester, and note pressure at the instant the gauge needle stops.

Step 3.
1. Repeat step 2 on all injectors and record pressure drop on each.
   Retest injectors that appear faulty (Any injectors that have a 10 kPa difference, either more or less, in pressure from the average). If no problem is found, review "Symptoms" Section "B".

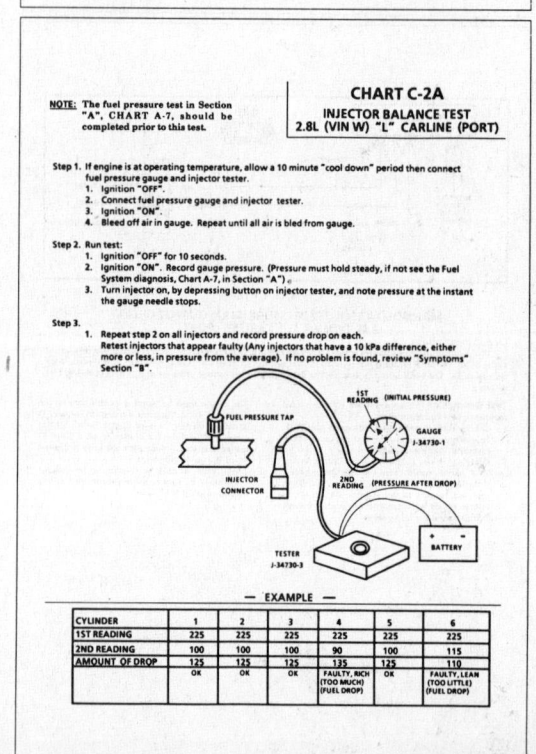

— EXAMPLE —

CYLINDER	1	2	3	4	5	6
1ST READING	225	225	225	225	225	225
2ND READING	100	100	100	90	100	115
AMOUNT OF DROP	125	125	125	135	125	110
	OK	OK	OK	FAULTY, RICH (TOO MUCH) (FUEL DROP)	OK	FAULTY, LEAN (TOO LITTLE) (FUEL DROP)

1988–90 2.8L (VIN W) – CORSICA AND BERETTA

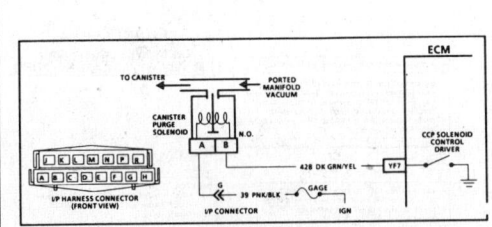

CHART C-3
CANISTER PURGE VALVE CHECK
2.8L (VIN W) "L" CARLINE (PORT)

Circuit Description:

Canister purge is controlled by a solenoid that allows manifold vacuum to purge the canister when de-energized. The ECM supplies a ground to energize the solenoid (purge "OFF.") The purge solenoid control by the ECM is pulse width modulated (turned "ON" and "OFF" several times a second). The duty cycle (pulse width) is determined by the amount of air flow, and the engine vacuum as determined by the MAP sensor input. The duty cycle is calculated by the ECM and the output commanded when the following conditions have been met.

- Engine run time after start more than 3 minutes
- Coolant temperature above 80°C
- Vehicle speed above 5 mph
- Throttle off idle (about 3%)

Also, if the diagnostic "test" terminal is grounded with the engine stopped, the purge solenoid is de-energized (purge "ON.")

Test Description: Numbers below refer to circled numbers on the diagnostic chart.

1. Checks to see if the solenoid is opened or closed. The solenoid is normally energized in this step; so it should be closed.
2. Checks for a complete circuit. Normally there is ignition voltage on CKT 39 and the ECM provides a ground on CKT 428.
3. Completes functional check by grounding the "test" terminal. This should normally de-energize the solenoid opening the valve which should allow the vacuum to drop (purge "ON.")

1988–90 2.8L (VIN W) – CORSICA AND BERETTA

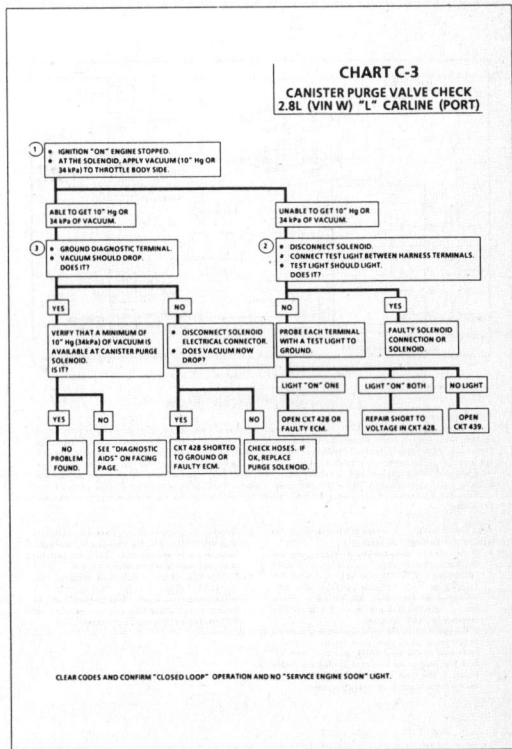

CHART C-3
CANISTER PURGE VALVE CHECK
2.8L (VIN W) "L" CARLINE (PORT)

CLEAR CODES AND CONFIRM "CLOSED LOOP" OPERATION AND NO "SERVICE ENGINE SOON" LIGHT.

1988–90 2.8L (VIN W) – CORSICA AND BERETTA

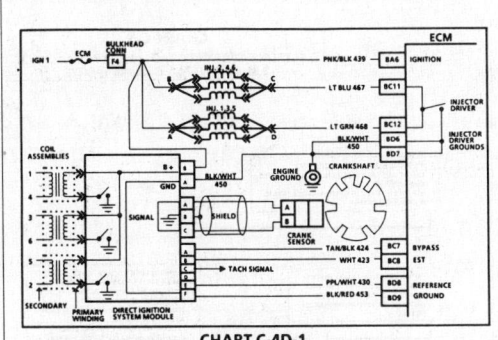

CHART C-4D-1
DIS MISFIRE AT IDLE
2.8L (VIN W) "L" CARLINE (PORT)

Circuit Description:

The direct ignition system (DIS) uses a waste spark method of distribution. For example, in this type of system the ignition module triggers the #1/4 coil pair resulting in both #1 and #4 spark plugs firing at the same time. #1 cylinder is on the compression stroke at the same time the #4 is on the exhaust stroke, resulting in a lower energy requirement to fire #4 spark plug. This leaves the remainder of the high voltage to be used to fire #1 spark plug. On this application, the crank sensor is mounted to the engine block and protrudes through the block to within approximately .050" of the crankshaft reluctor. Since the reluctor is a machined portion of the crankshaft and the crank sensor is mounted in a fixed position on the block, timing adjustments are not possible or necessary.

Test Description: Numbers below refer to circled numbers on the diagnostic chart.

1. If the "Misfire" complaint exists under load only, the diagnostic chart on page 2 must be used. Engine rpm should drop approximately equally on all plug leads.
2. A spark test such as a ST-125 must be used because it is essential to verify adequate available secondary voltage at the spark plug. (25,000 volts).
3. If the spark jumps the test gap after grounding the opposite plug wire, it indicates excessive resistance in the plug which was Bypassed. A faulty or poor connection at that plug could also result in the miss condition. Also, check for carbon deposits inside the spark plug boot.
4. If carbon tracking is evident, replace coil and be sure plug wires relating to that coil are clean and tight. Excessive wire resistance or faulty connections could have caused the coil to be damaged.
5. If no spark condition follows the suspected coil, that coil is faulty, otherwise, the ignition module is the cause of no spark. This test could also be performed by substituting a known good coil for the one causing the no spark condition.

1988–90 2.8L (VIN W) – CORSICA AND BERETTA

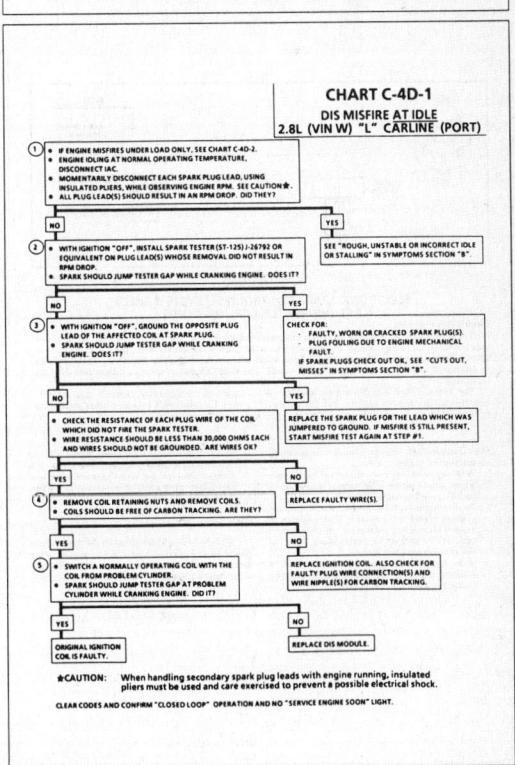

CHART C-4D-1
DIS MISFIRE AT IDLE
2.8L (VIN W) "L" CARLINE (PORT)

★CAUTION: When handling secondary spark plug leads with engine running, insulated pliers must be used and care exercised to prevent a possible electrical shock.

CLEAR CODES AND CONFIRM "CLOSED LOOP" OPERATION AND NO "SERVICE ENGINE SOON" LIGHT.

1988–90 2.8L (VIN W) – CORSICA AND BERETTA

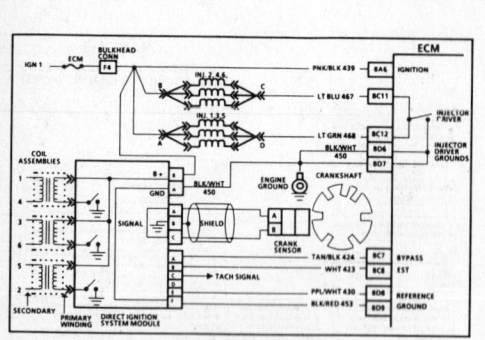

CHART C-4D-2
DIS MISFIRE UNDER LOAD
2.8L (VIN W) "L" CARLINE (PORT)

Circuit Description:

The direct ignition system (DIS) uses a waste spark method of distribution. For example, in this type of system, the ignition module triggers the #1/4 coil pair resulting in both #1 and #4 spark plugs firing at the same time. #1 cylinder is on the compression stroke at the same time #4 is on the exhaust stroke, resulting in a lower energy requirement to fire #4 spark plug. This leaves the remainder of the high voltage to be used to fire #1 spark plug. On this application, the crank sensor is mounted to the engine block and protrudes through the block to within approximately .050" of the crankshaft reluctor. Since the reluctor is a machined portion of the crankshaft and the crank sensor is mounted in a fixed position on the block, timing adjustments are not possible or necessary.

Test Description: Numbers below refer to circled numbers on the diagnostic chart.

1. If the "Misfire" complaint exists at idle only, the diagnostic chart on page 1 must be used. A spark tester such as a ST-125 must be used because it is essential to verify adequate available secondary voltage at the spark plug. (25,000 volts). Spark should jump the test gap on all 4 leads. This simulates a "load" condition.
2. If the spark jumps the tester gap after grounding the opposite plug wire, it indicates excessive resistance in the plug which was Bypassed. A faulty or poor connection at that plug could also result in the miss condition. Also, check for carbon deposits inside the spark plug boot.

3. If carbon tracing is evident replace coil and be sure plug wires relating to that coil are clean and tight. Excessive wire resistance or faulty connections could have caused the coil to be damaged.
4. If the no spark condition follows the suspected coil, that coil is faulty, otherwise, the ignition module is the cause of no spark. This test could also be performed by substituting a known good coil for the one causing the no spark condition.

1988–90 2.8L (VIN W) – CORSICA AND BERETTA

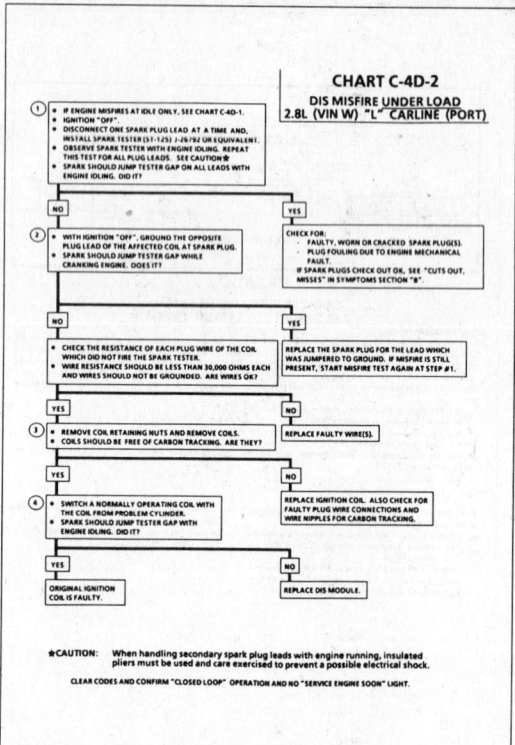

★CAUTION: When handling secondary spark plug leads with engine running, insulated pliers must be used and care exercised to prevent a possible electrical shock.

CLEAR CODES AND CONFIRM "CLOSED LOOP" OPERATION AND NO "SERVICE ENGINE SOON" LIGHT.

1988–90 2.8L (VIN W) – CORSICA AND BERETTA

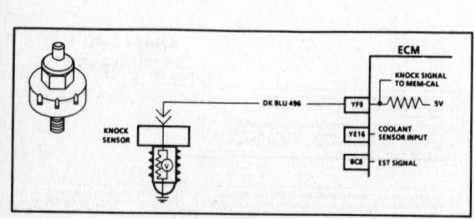

CHART C-5
ELECTRONIC SPARK CONTROL (ESC) SYSTEM CHECK
2.8L (VIN W) "L" CARLINE (PORT)

Circuit Description:

The knock sensor is used to detect engine detonation and the ECM will retard the electronic spark timing based on the signal being received. The circuitry within the knock sensor causes the ECM's 5 volts to be pulled down so that under a no knock condition, CKT 496 would measure about 2.5 volts. The knock sensor produces an A/C signal which rides on the 2.5 volts DC voltage. The amplitude and frequency are dependent upon the knock level.

The Mem-Cal used with this engine, contains the functions which were part of remotely mounted ESC modules used on other GM vehicles. The ESC portion of the Mem-Cal, then sends a signal to other parts of the ECM which adjusts the spark timing to retard the spark and reduce the detonation.

Test Description: Numbers below refer to circled numbers on the diagnostic chart.

1. With engine idling, there should not be a knock signal present at the ECM because detonation is not likely under a no load condition.
2. Tapping on the engine lift bracket should simulate a knock signal to determine if the sensor is capable of detecting detonation. If no knock is detected, try tapping on engine block closer to sensor before replacing sensor.
3. If the engine has an internal problem which is creating a knock, the knock sensor may be responding to the internal failure.

4. This test determines if the knock sensor is faulty or if the ESC portion of the Mem-Cal is faulty. If it is determined that the Mem-Cal is faulty, be sure that it is properly installed and latched into place. If not properly installed, repair and retest.

Diagnostic Aids:

While observing knock signal on the "Scan," there should be an indication that knock is present when detonation can be heard. Detonation is most likely to occur under high engine load conditions.

1988–90 2.8L (VIN W) – CORSICA AND BERETTA

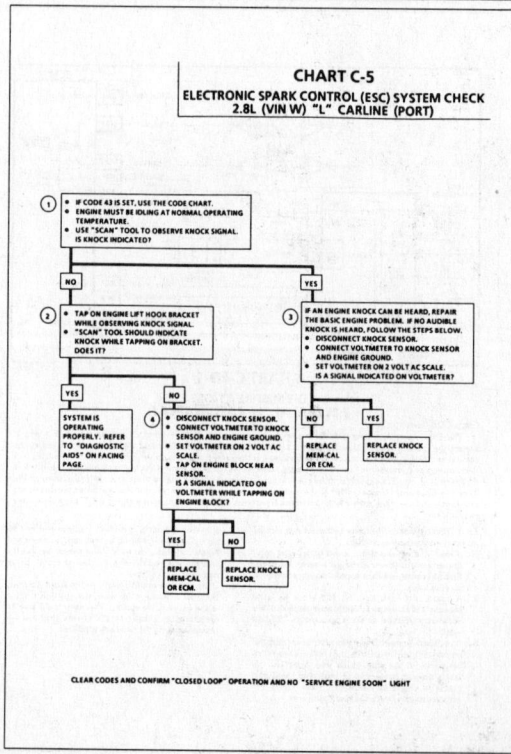

CHART C-5
ELECTRONIC SPARK CONTROL (ESC) SYSTEM CHECK
2.8L (VIN W) "L" CARLINE (PORT)

CLEAR CODES AND CONFIRM "CLOSED LOOP" OPERATION AND NO "SERVICE ENGINE SOON" LIGHT

1988–90 2.8L (VIN W) – CORSICA AND BERETTA

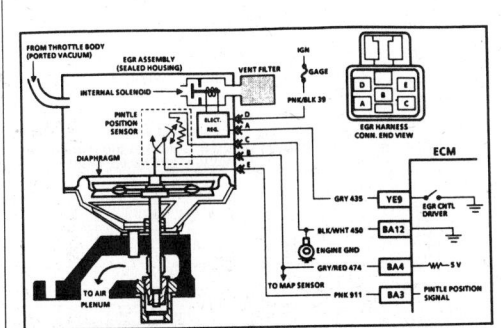

CHART C-7
EGR VALVE CHECK
2.8L (VIN W) "L" CARLINE (PORT)

Circuit Description:
The integrated electronic EGR valve functions similar to a port EGR valve with a remote vacuum regulator. The internal solenoid is normally open, which causes the vacuum signal to be vented off to the atmosphere when EGR is not being commanded by the ECM. This EGR valve has a sealed cap and the internal solenoid valve opens and closes the vacuum signal, which controls the amount of vacuum vented to atmosphere, and this controls the amount of vacuum applied to the diaphragm. The electronic EGR valve contains a voltage regulator, which converts the ECM signal, to provide different amounts of EGR flow by regulating the current to the solenoid. The ECM controls EGR flow with a pulse width modulated signal (turns "ON" and "OFF" many times a second) based on airflow, TPS, and rpm.
This system also contains a pintle position sensor, which works similar to a TPS sensor, and as EGR flow is increased, the sensor output also increases.

Test Description: Numbers below refer to circled numbers on the diagnostic chart.
1. Whenever the internal solenoid is de-energized, the solenoid valve should be open, which should not allow the vacuum to move the EGR diaphragm. However, if the filter is plugged, the vacuum applied with the hand held vacuum pump will cause the diaphragm to move because the vacuum will not be vented to the atmosphere.
2. This test will determine if the EGR filter is plugged or if the EGR itself is faulty. Use care, when removing the filter, to avoid damaging the EGR assembly. See On-Car Service for procedure.
3. If the valve moves in this test, it's probably due to CKT 435 being shorted to ground.
4. Grounding the diagnostic terminal should energize the solenoid which closes off the vent and allows the vacuum to move the diaphragm.
5. The EGR assembly is designed to have some leak and, therefore, 7" of vacuum is all that should be able to be held on the assembly. However, if too much of a leak exists (less than 3"), the EGR assembly is leaking and must be replaced.

Diagnostic Aids:
The EGR position voltage can be used to determine that the pintle is moving. When no EGR is commanded (0% duty cycle), the position sensor should read between .5 volt and 1.5 volts, and increase with the commanded EGR duty cycle.

1988–90 2.8L (VIN W) – CORSICA AND BERETTA

CHART C-7
EGR VALVE CHECK
2.8L (VIN W) "L" CARLINE (PORT)

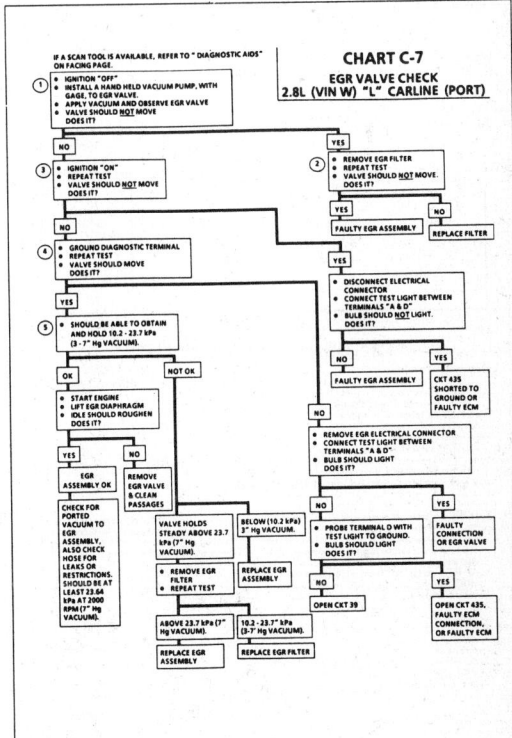

1988–90 2.8L (VIN W) – CORSICA AND BERETTA

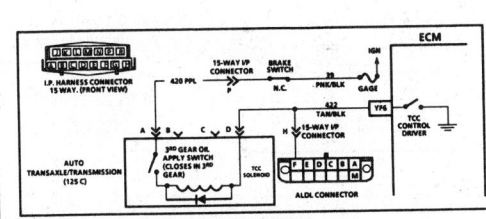

CHART C-8A
125C TORQUE CONVERTER CLUTCH (TCC)
2.8L (VIN W) "L" CARLINE (PORT)

Circuit Description:
The purpose of the torque converter clutch feature is to eliminate the power loss of the transmission converter stage when the vehicle is in a cruise condition. This allows the convenience of the automatic transmission and the fuel economy of a manual transmission.
Fused battery ignition is supplied to the TCC solenoid through the brake switch, and transmission third gear apply switch. The ECM will engage TCC by grounding CKT 422 to energize the solenoid.
TCC will engage when:
- Engine warmed up
- Vehicle speed above a calibrated value. (about 32 mph 51 km/h)
- Throttle position sensor output not changing, indicating a steady road speed
- Transmission third gear switch closed
- Brake switch closed

Test Description: Numbers below refer to circled numbers on the diagnostic chart.
1. Light "OFF" confirms transmission third gear apply switch is open.
2. At 25 mph the transmission third gear apply switch should close. Test light will come "ON" and confirm battery supply and closed brake switch.
3. Grounding the diagnostic terminal with ignition "ON," engine "OFF," should energize the TCC solenoid by grounding CKT 422. This test checks the ability of the ECM to supply a ground to the TCC solenoid. The test light connected from 12 volts to ALDL terminal "F" will turn "ON" as CKT 422 is grounded.

Diagnostic Aids:
A "Scan" tool only indicates when the ECM has turned on the TCC driver and this does not confirm that the TCC has engaged. To determine if TCC is functioning properly, engine rpm should decrease when the "Scan" indicates the TCC driver has turned "ON."

1988–90 2.8L (VIN W) – CORSICA AND BERETTA

CHART C-8A
125C TORQUE CONVERTER CLUTCH (TCC)
2.8L (VIN W) "L" CARLINE (PORT)

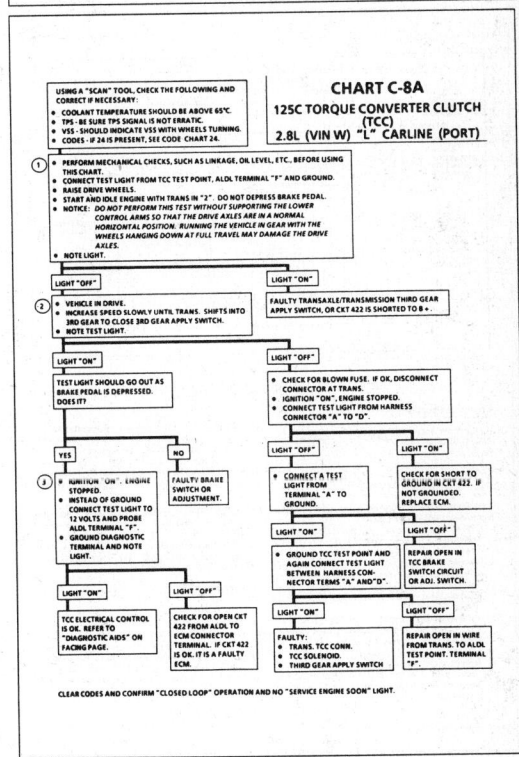

1988–90 2.8L (VIN W) – CORSICA AND BERETTA

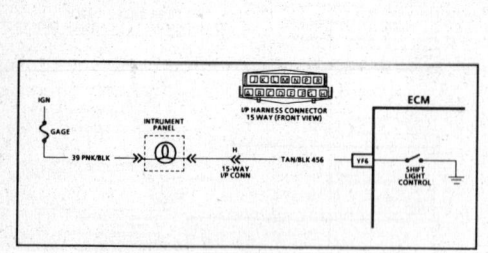

CHART C-8C
MANUAL TRANSMISSION (M/T) SHIFT LIGHT CHECK
2.8L (VIN W) "L" CARLINE (PORT)

Circuit Description:
The shift light indicates the best transmission shift point for maximum fuel economy. The light is controlled by the ECM and is turned "ON" by grounding CKT 456.
The ECM uses information from the following inputs to control the shift light:
- Coolant temperature must be above 16°C (61°F)
- TPS above 4%
- VSS
- RPM above about 1900
- Air Flow - The ECM uses rpm, airflow VSS to calculate what gear the vehicle is in.
It's this calculation that determines when the shift light should be turned "ON." The shift light will only stay "ON" 5 seconds after the conditions were met to turn it on.

Test Description: Numbers below refer to circled numbers on the diagnostic chart.
1. This should not turn "ON" the shift light. If the light is "ON," there is a short to ground in CKT 456 wiring or a fault in the ECM.
2. When the diagnostic terminal is grounded, the ECM should ground CKT 456 and the shift light should come "ON."
3. This checks the shift light circuit up to the ECM connector. If the shift light illuminates, then the ECM connector is faulty or the ECM does not have the ability to ground the circuit.

1988–90 2.8L (VIN W) – CORSICA AND BERETTA

CHART C-8C
MANUAL TRANSMISSION (M/T) SHIFT LIGHT CHECK
2.8L (VIN W) "L" CARLINE (PORT)

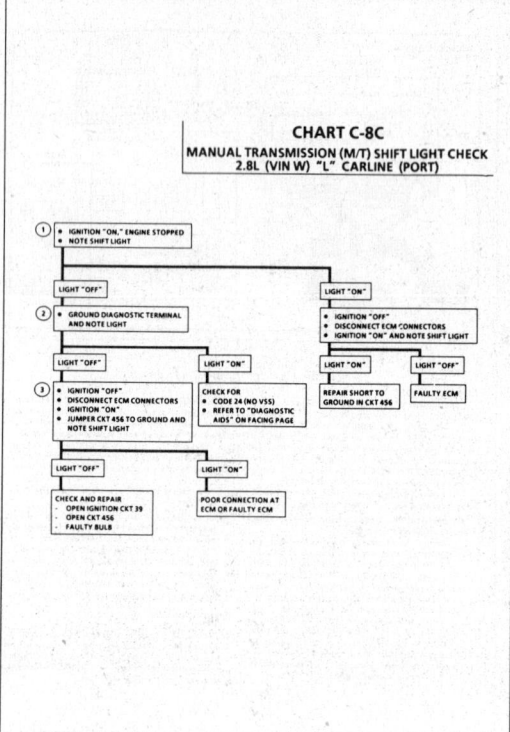

1988–90 2.8L (VIN W) – CORSICA AND BERETTA

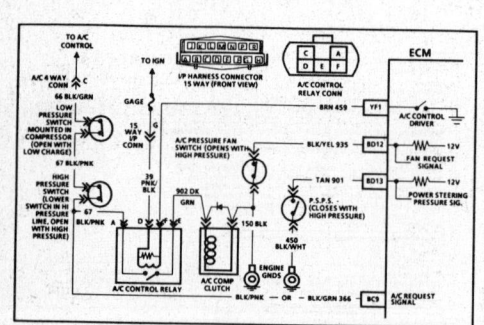

CHART C-10
A/C CLUTCH CONTROL CIRCUIT DIAGNOSIS
2.8L (VIN W) "L" CARLINE (PORT)

Circuit Description:
The A/C clutch control relay is ECM controlled to delay A/C clutch engagement about .4 second after A/C is turned "ON." This allows the IAC to adjust engine rpm before the A/C clutch engages. The ECM also causes the relay to disengage the A/C clutch during WOT when high power steering pressure is present, or if engine is overheating. The A/C clutch control relay is energized when the ECM provides a ground path for CKT 459. The low pressure switch will open if A/C pressure is less than 40 psi (276 kPa). The high pressure switch will open if A/C pressure exceeds about 440 psi (3034 kPa). The high pressure fan switch opens when A/C pressure exceeds about 200 psi (1380 kPa).

Test Description: Numbers below refer to circled numbers on the diagnostic chart.
1. The ECM will only energize the A/C relay when the engine is running. This test will determine if the relay or CKT 459 is faulty.
2. In order for the clutch to properly be engaged, the low pressure switch must be closed to provide 12 volts to the relay and the high pressure switch must be closed so the A/C request (12 volts) will be present at the ECM.
3. Determines if the signal is reaching the ECM on CKT 366 from the A/C control panel. Signal should only be present when the A/C mode or defrost mode has been selected.
4. A short to ground in any part of the A/C request circuit, CKT 67 to the relay, CKT 902 to the A/C clutch, or the A/C clutch, could be the cause of the blown fuse.
5. If the ECM is seeing a high power steering pressure signal the A/C clutch will be disengaged by the ECM.
6. With the engine idling and A/C "ON," the ECM should be grounding CKT 459, which should cause the test light to be "ON."

Diagnostic Aids:
If complaint was insufficient cooling, the problem may be caused by an inoperative cooling fan or A/C pressure fan switch. The engine cooling fan should turn "ON" when A/C pressure exceeds a value to open the switch which causes the ECM to energize the cooling fan relay. See CHART C-12 for cooling fan diagnosis. If fan operates correctly, see A/C diagnosis in Section "1".

1988–90 2.8L (VIN W) – CORSICA AND BERETTA

CHART C-10
A/C CLUTCH CONTROL CIRCUIT DIAGNOSIS
2.8L (VIN W) "L" CARLINE (PORT)

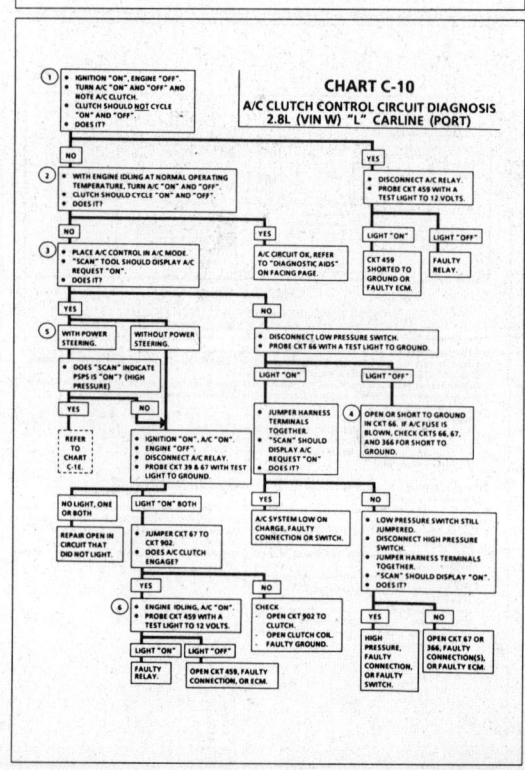

FUEL INJECTION SYSTEMS
MULTI-PORT INJECTION (MPI)/TUNED PORT INJECTION (TPI)

5 SECTION

1988–90 2.8L (VIN W) – CORSICA AND BERETTA

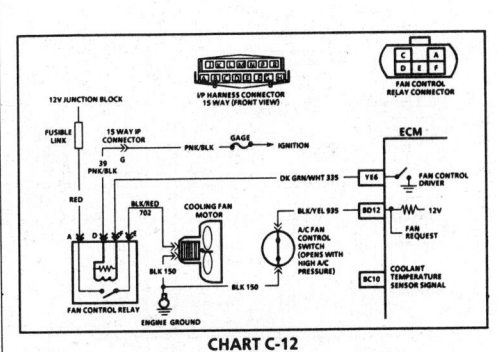

CHART C-12
(Page 1 of 2)
FAN CONTROL CIRCUIT
2.8L (VIN W) "L" CARLINE (PORT)

Circuit Description:
The electric cooling fan is controlled by the ECM, based on inputs from the coolant temperature sensor, the A/C fan control switch and vehicle speed. The ECM controls the fan by grounding CKT 335 which energizes the fan control relay. Battery voltage is then supplied to the fan motor.

The ECM grounds CKT 335 when coolant temperature is over about 106°C (223°F), or when A/C has been requested and the fan control switch opens with high A/C pressure about 200 psi (1380 kPa). Once the ECM turns the relay "ON," it will keep it "ON" for a minimum of 30 seconds, or until vehicle speed exceeds 70 mph.

Also, if Code 14 or 15 sets or the ECM is in throttle body back up, the fan will run at all times.

Test Description: Numbers below refer to circled numbers on the diagnostic chart.
1. With the diagnostic terminal grounded, the cooling fan control driver will close, which should energize the fan control relay.
2. If the A/C fan control switch or circuit is open, the fan would run whenever A/C is requested.
3. With A/C clutch engaged, the A/C fan control switch should open when A/C high pressure exceeds about 200 psi (1380 kPa). This signal should cause the ECM to energize the fan control relay.

Diagnostic Aids:
If the owner complained of an overheating problem, it must be determined if the complaint was due to an actual boil over, or the hot light, or temperature gage indicated over heating.

If the gage or light indicates overheating, but no boil over is detected, the gage circuit should be checked. The gage accuracy can also be checked by comparing the coolant sensor reading using a "Scan" tool and comparing its reading with the gage reading.

If the engine is actually overheating, and the gage indicates overheating, but the cooling fan is not coming "ON," the coolant sensor has probably shifted out of calibration and should be replaced.

If the engine is overheating and the cooling fan is "ON," the cooling system should be checked.

1988–90 2.8L (VIN W) – CORSICA AND BERETTA

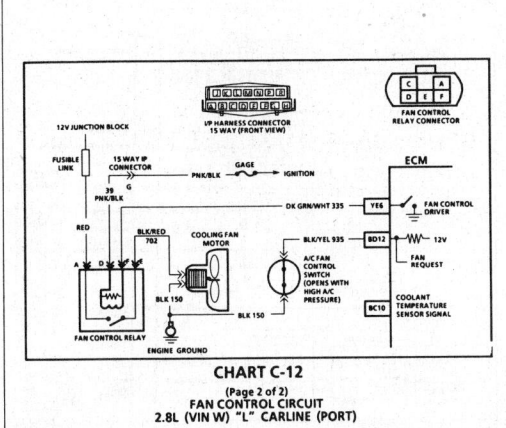

CHART C-12
(Page 2 of 2)
FAN CONTROL CIRCUIT
2.8L (VIN W) "L" CARLINE (PORT)

Test Description: Numbers below refer to circled numbers on the diagnostic chart.
1. 12 volts should be available to both terminals "A" & "D" when the ignition is "ON."
2. This test checks the ability of the ECM to ground CKT 335. The "Service Engine Soon" light should also be flashing at this point. If it isn't flashing, see CHART A-2.
3. If the fan does not turn "ON" at this point, CKT 702 or CKT 150 is open or the cooling fan motor is faulty.

1988–90 2.8L (VIN W) – CORSICA AND BERETTA

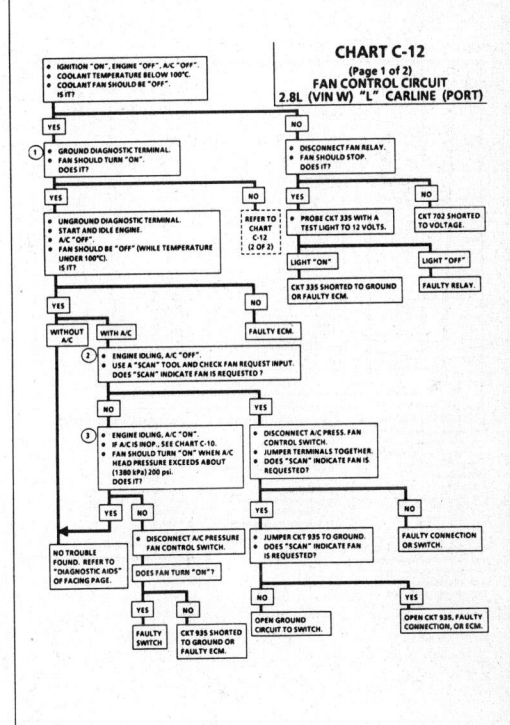

1988–90 2.8L (VIN W) – CORSICA AND BERETTA

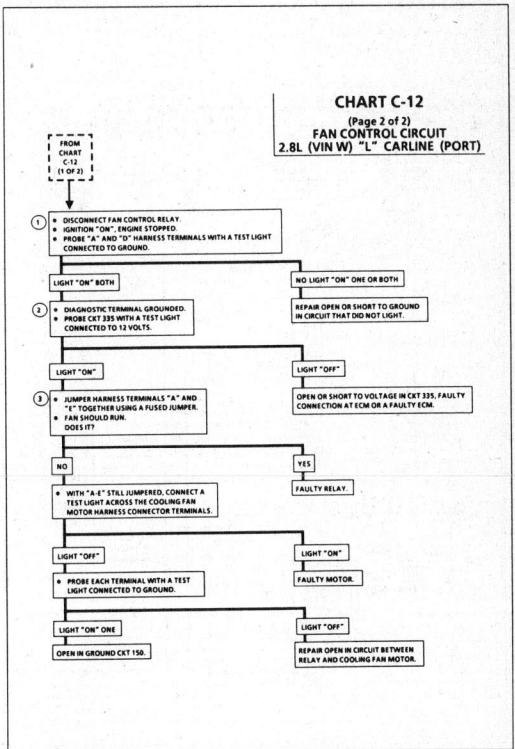

2.8L (VIN W) COMPONENT LOCATIONS CELEBRITY AND 6000

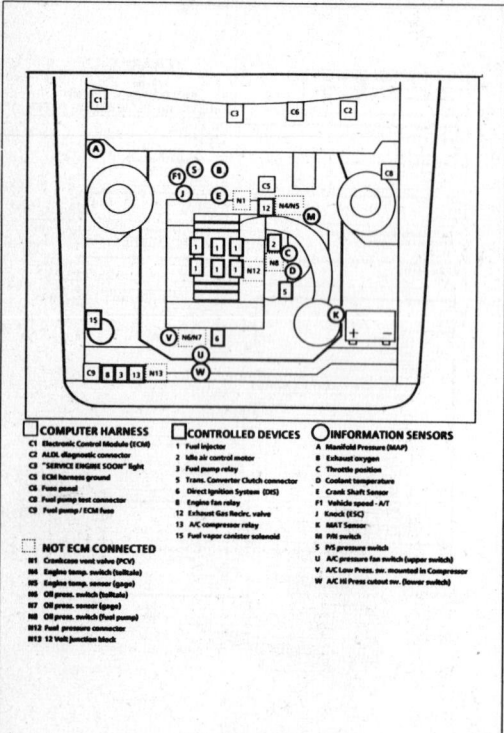

☐ COMPUTER HARNESS
C1 Electronic Control Module (ECM)
C3 ALDL diagnostic connector
C3 "SERVICE ENGINE SOON" light
C5 ECM harness ground
C6 Fuse panel
C8 Fuel pump test connector
C9 Fuel pump / ECM fuse

☐ NOT ECM CONNECTED
N1 Crankcase vent valve (PCV)
N4 Engine temp. switch (telltale)
N5 Engine temp. sensor (gage)
N6 Oil press. switch (telltale)
N7 Oil press. sensor (gage)
N8 Oil press. switch (fuel pump)
N12 Fuel pressure connector
N13 12 Volt junction block

☐ CONTROLLED DEVICES
1 Fuel injector
2 Idle air control motor
3 Fuel pump relay
4 Trans. Converter Clutch connector
6 Direct Ignition System (DIS)
8 Engine fan relay
12 Exhaust Gas Recirc. valve
13 A/C compressor relay
14 Fuel vapor canister solenoid

○ INFORMATION SENSORS
A Manifold Pressure (MAP)
B Exhaust oxygen
C Throttle position
D Coolant temperature
E Crank Shaft Sensor
F1 Vehicle speed - A/T
J Knock (ESC)
K MAT Sensor
N PN switch
M PS pressure switch
U A/C pressure fan switch (upper switch)
V A/C Low Press. sw. mounted in Compressor
W A/C HI Press cutout sw. (lower switch)

2.8L (VIN W) ECM WIRING DIAGRAM CELEBRITY AND 6000

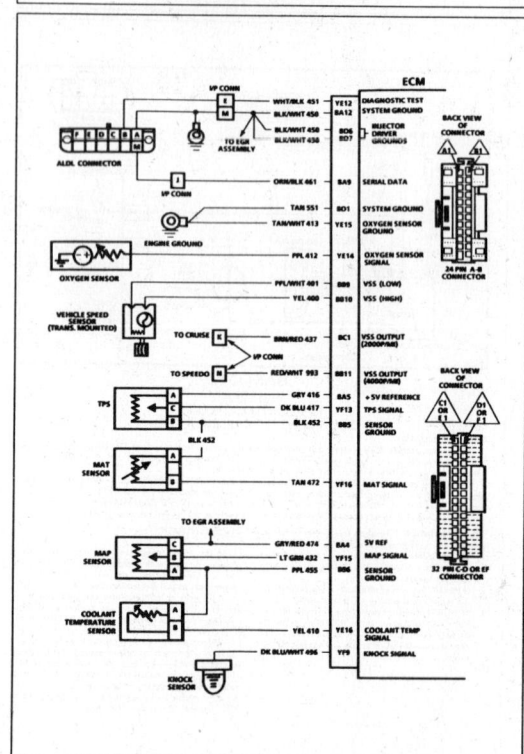

2.8L (VIN W) ECM WIRING DIAGRAM CELEBRITY AND 6000 (CONT.)

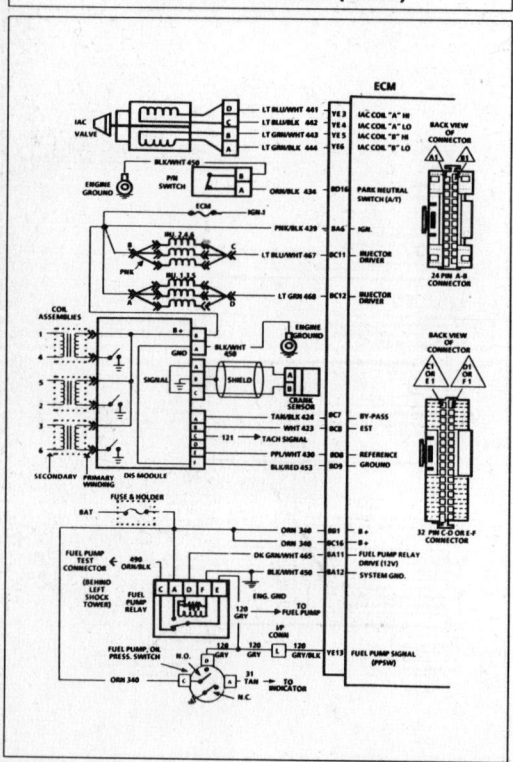

2.8L (VIN W) ECM WIRING DIAGRAM CELEBRITY AND 6000 (CONT.)

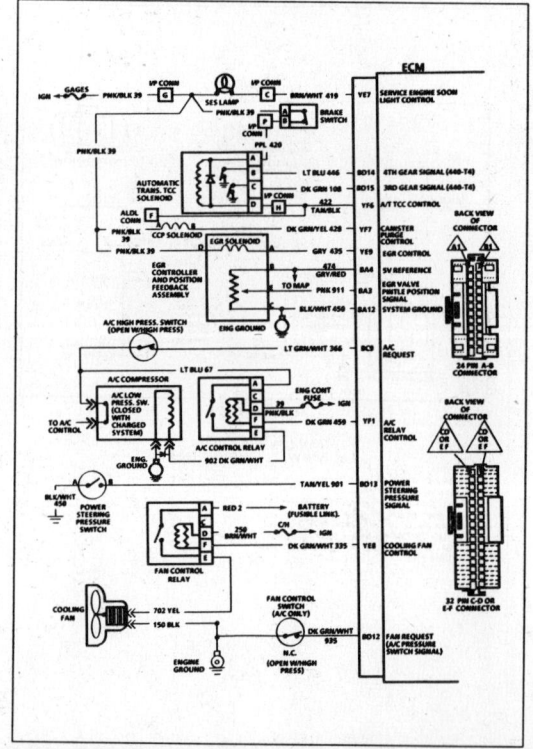

2.8L (VIN W) ECM CONNECTOR TERMINAL END VIEW—CELEBRITY AND 6000

PORT FUEL INJECTION ECM CONNECTOR IDENTIFICATION

This ECM voltage chart is for use with a digital voltmeter to further aid in diagnosis. The voltages you get may vary due to low battery charge or other reasons, but they should be very close.

THE FOLLOWING CONDITIONS MUST BE MET BEFORE TESTING:
- Engine at operating temperature • Engine idling in closed loop (For "Engine Run" column) in park or neutral • Test terminal not grounded • ALDL tool not installed

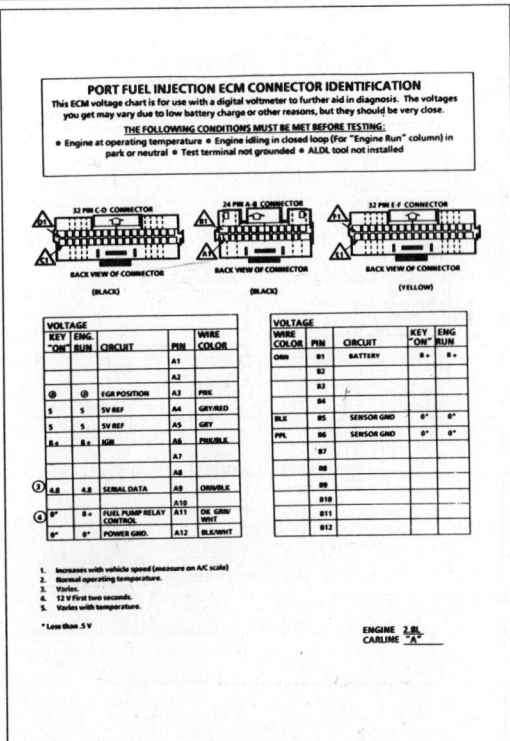

1. Increases with vehicle speed (measure on A/C scale)
2. Normal operating temperature.
3. Varies.
4. 12 V first two seconds.
5. Varies with temperature.

* Less than .5 V

ENGINE 2.8L
CARLINE "A"

2.8L (VIN W) ECM CONNECTOR TERMINAL END VIEW—CELEBRITY AND 6000 (CONT.)

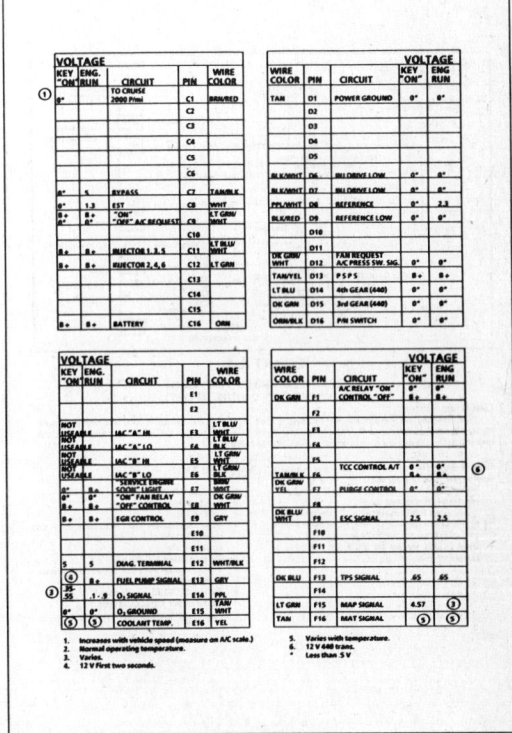

1. Increases with vehicle speed (measure on A/C scale).
2. Normal operating temperature.
3. Varies.
4. 12 V first two seconds.
5. Varies with temperature.
6. 12 V 440 trans.
* Less than .5 V

1988–90 2.8L (VIN W)—CELEBRITY AND 6000

DIAGNOSTIC CIRCUIT CHECK

The Diagnostic Circuit Check is an organized approach to identifying a problem created by an Electronic Engine Control System malfunction. It must be the starting point for any driveability complaint diagnosis, because it directs the Service Technician to the next logical step in diagnosing the complaint.

The "Scan Data" listed in the table may be used for comparison, after completing the Diagnostic Circuit Check and finding the on-board diagnostics functioning properly and no trouble codes displayed. The "Typical Values" are an average of display values recorded from normally operating vehicles and are intended to represent what a normally functioning vehicle would typically display.

A "SCAN" TOOL THAT DISPLAYS FAULTY DATA SHOULD NOT BE USED, AND THE PROBLEM SHOULD BE REPORTED TO THE MANUFACTURER. THE USE OF A FAULTY "SCAN" CAN RESULT IN MISDIAGNOSIS AND UNNECESSARY PARTS REPLACEMENT.

Only the parameters listed below are used in this manual for diagnosing. If a "Scan" reads other parameters, the values are not recommended by General Motors for use in diagnosing. For more description on the values and use of the "Scan" to diagnose ECM inputs, refer to the applicable diagnosis section. If all values are within the range illustrated, refer to symptoms in Section "B."

"SCAN" DATA
Idle / Upper Radiator Hose Hot / Closed Throttle / Park or Neutral / Closed Loop / Acc. off

"SCAN" Position	Units Displayed	Typical Data Value
Desired RPM	RPM	ECM idle command (varies with temp.)
RPM	RPM	± 100 RPM from desired RPM (± 50 in drive)
Coolant Temp.	C°	85°- 105°
MAT Temp.	C°	10°- 80° (depends on underhood temp.)
MAP	Volts	1 - 2 (depends on Vac. & Baro pressure)
BARO	Volts	2.5 - 5.5 (depends on altitude & Baro pressure)
BPW (base pulse width)	M/Sec	1 - 4 and varying
O₂	Volts	.1 - 1.0 V and varying
TPS	Volts	.65
Throttle Angle	0 - 100%	0
IAC	Counts (steps)	5 - 50
P/N Switch	P/N and RDL	Park/Neutral (P/N)
INT (Integrator)	Counts	Varies
BLM (Block Learn)	Counts	118 - 138
Open/Closed Loop	Open/Closed	Closed Loop (may go open with extended idle)
BLM Cell	Cell Number	0 or 1 (depends on Air Flow & RPM)
VSS	MPH	0
TCC	On/Off	Off/ (on with TCC commanded)
EGRDC	0 - 100%	0 at idle
EGR Position	Volts	3 - 2 Volts
Spark Advance	# of Degrees	Varies
Knock Retard	Degrees of Retard	0
Knock Signal	Yes/No	No
Battery	Volts	13.5 - 14.5
Fan	On/Off	Off (below 106°C)
P/S Switch	Normal/Hi Press.	Normal
3rd Gear (440-T4)	Yes/No	No (yes, when in 3rd or 4th gear)
4th Gear (440-T4)	Yes/No	No (yes, when in 4th gear)
A/C Request	Yes/No	No (yes, with A/C requested)
A/C Clutch	On/Off	Off (on, when A/C commanded on)
Fan Request	Yes/No	No (yes, with A/C high pressure)
PPSW (Fuel Pump)	Volts	13.5 - 14.5

1988–90 2.8L (VIN W)—CELEBRITY AND 6000

DIAGNOSTIC CIRCUIT CHECK
2.8L (VIN W) "A" CARLINE (PORT)

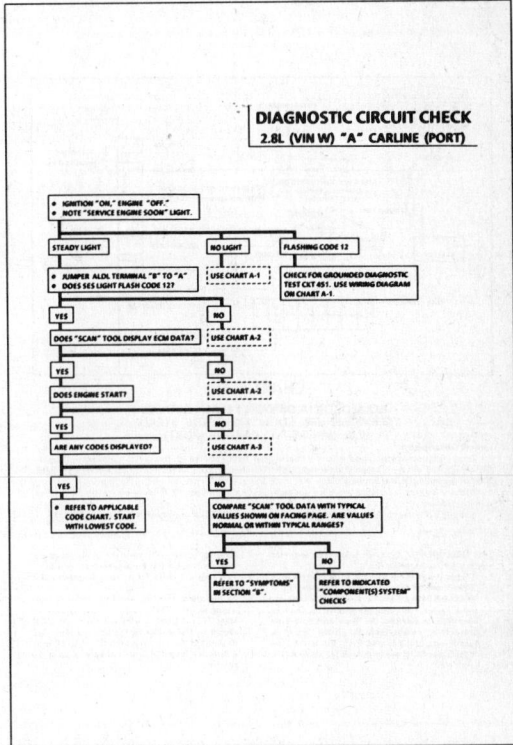

1988–90 2.8L (VIN W) – CELEBRITY AND 6000

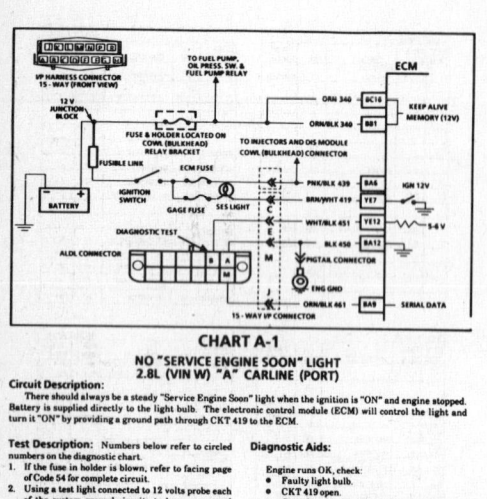

CHART A-1
NO "SERVICE ENGINE SOON" LIGHT
2.8L (VIN W) "A" CARLINE (PORT)

Circuit Description:
There should always be a steady "Service Engine Soon" light when the ignition is "ON" and engine stopped. Battery is supplied directly to the light bulb. The electronic control module (ECM) will control the light and turn it "ON" by providing a ground path through CKT 419 to the ECM.

Test Description: Numbers below refer to circled numbers on the diagnostic chart.
1. If the fuse in holder is blown, refer to facing page of Code 54 for complete circuit.
2. Using a test light connected to 12 volts probe each of the system ground circuits to be sure a good ground is present. See ECM terminal end view in front of this section for ECM pin locations of ground circuits.

Diagnostic Aids:
Engine runs OK, check:
• Faulty light bulb.
• CKT 419 open.
• Gage fuse blown. This will result in no stop lights, oil or generator lights, seat belt reminder, etc.
Engine cranks but will not run:
• Continuous battery - fuse or fusible link open.
• ECM ignition fuse open.
• Battery CKT 340 to ECM open.
• Ignition CKT 439 to ECM open.
• Poor connection to ECM.

1988–90 2.8L (VIN W) – CELEBRITY AND 6000

CHART A-1
NO "SERVICE ENGINE SOON" LIGHT
2.8L (VIN W) "A" CARLINE (PORT)

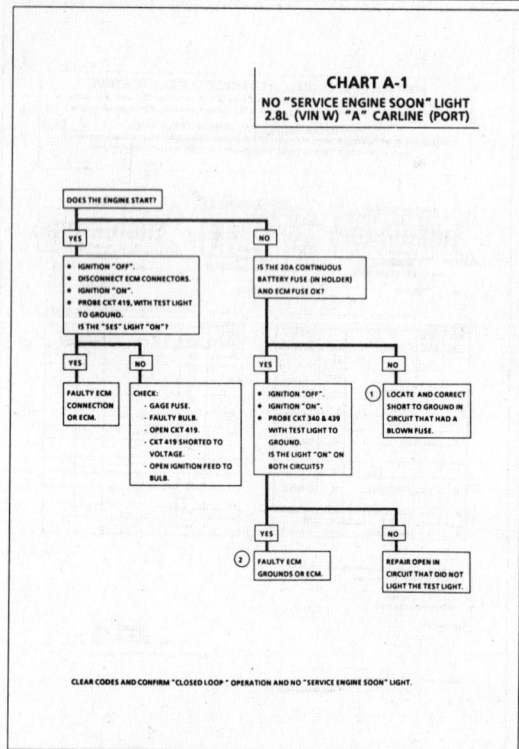

CLEAR CODES AND CONFIRM "CLOSED LOOP " OPERATION AND NO "SERVICE ENGINE SOON" LIGHT.

1988–90 2.8L (VIN W) – CELEBRITY AND 6000

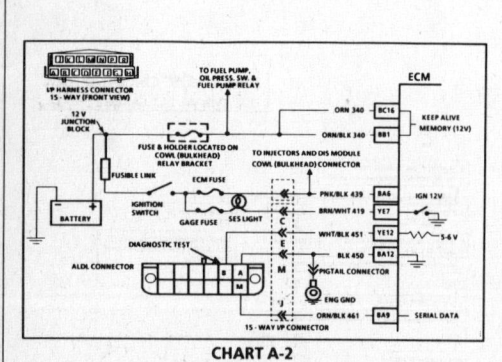

CHART A-2
NO ALDL DATA OR WON'T FLASH CODE 12
"SERVICE ENGINE SOON" LIGHT "ON" STEADY
2.8L (VIN W) "A" CARLINE (PORT)

Circuit Description:
There should always be a steady "Service Engine Soon" light when the ignition is "ON" and engine stopped. Battery ignition voltage is supplied to the light bulb. The electronic control module (ECM) will turn the light "ON" by grounding CKT 419 at the ECM.
With the diagnostic terminal grounded, the light should flash a Code 12, followed by any trouble code(s) stored in memory.
A steady light suggests a short to ground in the light control CKT 419, or an open in diagnostic CKT 451.

Test Description: Numbers below refer to circled numbers on the diagnostic chart.
1. If there is a problem with the ECM that causes a "Scan" tool to not read serial data, the ECM should not flash a Code 12. If Code 12 is flashing, check for CKT 451 short to ground. If Code 12 does flash, be sure that the "Scan" tool is working properly on another vehicle. If the "Scan" is functioning properly and CKT 461 is OK, the Mem-Cal or ECM may be at fault for the no ALDL symptom.
2. If the light goes "OFF" when the ECM connector is disconnected, CKT 419 is not shorted to ground.
3. This step will check for an open diagnostic CKT 451.
4. At this point the "Service Engine Soon" light wiring is OK. The problem is a faulty ECM or Mem-Cal. If Code 12 does not flash, the ECM should be replaced using the original Mem-Cal. Replace the Mem-Cal only after trying an ECM, as a defective Mem-Cal is an unlikely cause of the problem.

1988–90 2.8L (VIN W) – CELEBRITY AND 6000

CHART A-2
NO ALDL DATA OR WON'T FLASH CODE 12
"SERVICE ENGINE SOON" LIGHT "ON" STEADY
2.8L (VIN W) "A" CARLINE (PORT)

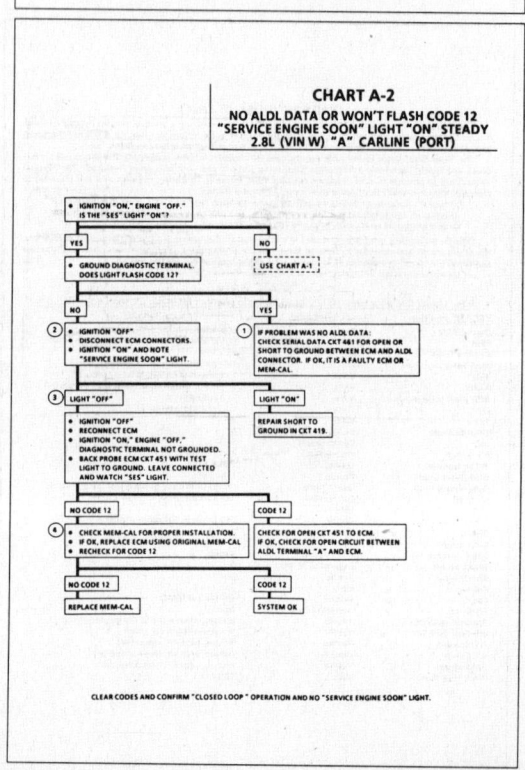

CLEAR CODES AND CONFIRM "CLOSED LOOP " OPERATION AND NO "SERVICE ENGINE SOON" LIGHT.

1988–90 2.8L (VIN W) – CELEBRITY AND 6000

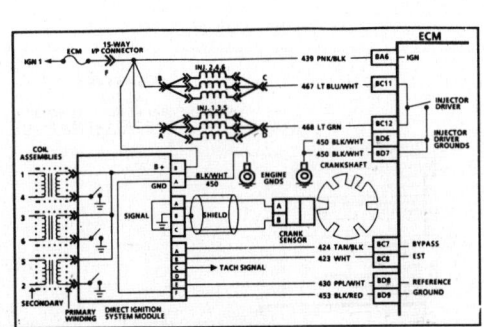

CHART A-3
(Page 1 of 3)
ENGINE CRANKS BUT WILL NOT RUN
2.8L (VIN W) "A" CARLINE (PORT)

Circuit Description:
This chart assumes that battery condition and engine cranking speed are OK, and there is adequate fuel in the tank.

Test Description: Numbers below refer to circled numbers on the diagnostic chart.

1. A "Service Engine Soon" light "ON" is a basic test to determine if there is a 12 volt supply and ignition 12 volts to ECM. No ALDL may be due to an ECM problem and CHART A-2 will diagnose the ECM. If TPS is over 2.5 volts the engine may be in the clear flood mode which will cause starting problems. The engine will not start without reference pulses and therefore the "Scan" should read rpm (reference) during crank.

2. For the first two seconds with ignition "ON", or whenever reference pulses are being received, PPSW should indicate fuel pump circuit voltage. (8 to 12 volts).

3. Because the direct ignition system uses two plugs and wires to complete the circuit of each coil, the opposite spark should be left connected. If rpm was indicated during crank, the ignition module is receiving a crank signal, but no spark at this test indicates the ignition module is not triggering the coils.

4. The test light should blink, indicating the ECM is controlling the injectors OK. How bright the light blinks is not important. However, the test light should be a J 34730-3 or equivalent.

5. Use fuel pressure gage J 34730-1 or equivalent. Wrap a shop towel around the fuel pressure tap to absorb any small amount of fuel leakage that may occur when installing the gage.

6. This test will determine if the ignition module is not generating the reference pulse or if the wiring or ECM are at fault. By touching and removing a test light to 12 volts on CKT 430, a reference pulse should be generated. If rpm is indicated the ECM and wiring are OK.

7. This test will determine if the ignition module is not triggering the problem coil or if the tested coil is at fault. This test could also be performed by using another known good coil.

1988–90 2.8L (VIN W) – CELEBRITY AND 6000

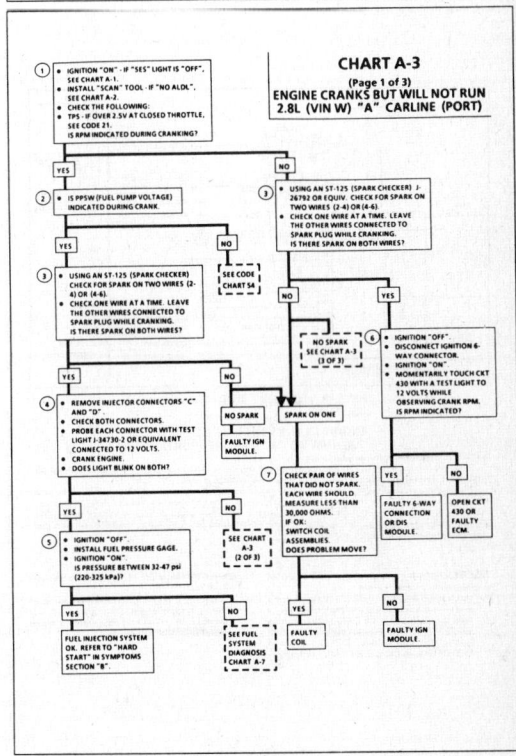

CHART A-3
(Page 1 of 3)
ENGINE CRANKS BUT WILL NOT RUN
2.8L (VIN W) "A" CARLINE (PORT)

1988–90 2.8L (VIN W) – CELEBRITY AND 6000

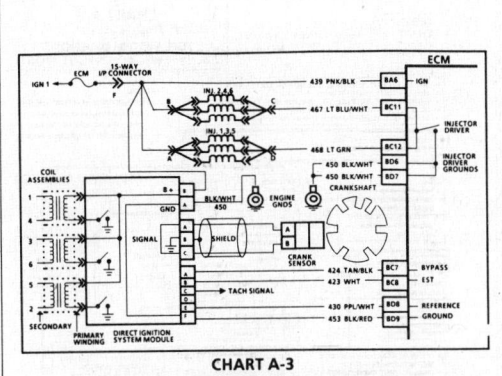

CHART A-3
(Page 2 of 3)
ENGINE CRANKS BUT WILL NOT RUN
2.8L (VIN W) "A" CARLINE (PORT)

Test Description: Numbers below refer to circled numbers on the diagnostic chart.

1. Checks for 12 volt supply to injectors. Due to the injectors wired in parallel there should be a light "ON" on both terminals.

2. Checks continuity of CKTs 467 and 468.

3. All checks made to this point would indicate that the ECM is at fault. However, there is a possibility of CKT 467 or 468 being shorted to a voltage source either in the engine harness or in the injector harness.
   - To test for this condition:
     - Disconnect the injector 4-way connector.
     - Ignition "ON."

- Probe CKTs 467 and 468 on the ECM side of harness with a test light connected to ground. There should be no light. If light is "ON" repair short to voltage.
- If OK, check the resistance of the injector harness between terminals "A" & "C", "A" & "D", "B" & "D", and "B" & "C".
- Should be more than 4 ohms.
- If less than 4 ohms, check harness for wires shorted together and check each injector resistance. (Resistance should be 8 ohms or more.)
- If all OK, replace ECM

1988–90 2.8L (VIN W) – CELEBRITY AND 6000

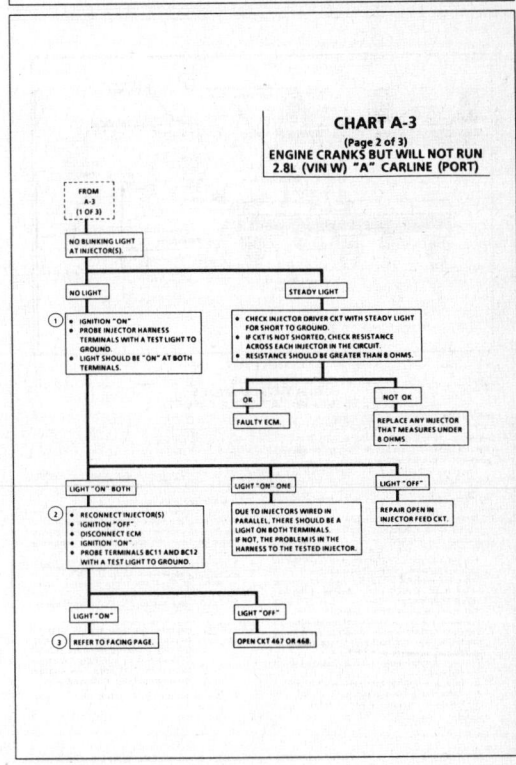

CHART A-3
(Page 2 of 3)
ENGINE CRANKS BUT WILL NOT RUN
2.8L (VIN W) "A" CARLINE (PORT)

SECTION 5

FUEL INJECTION SYSTEMS
MULTI-PORT INJECTION (MPI)/TUNED PORT INJECTION (TPI)

1988–90 2.8L (VIN W) – CELEBRITY AND 6000

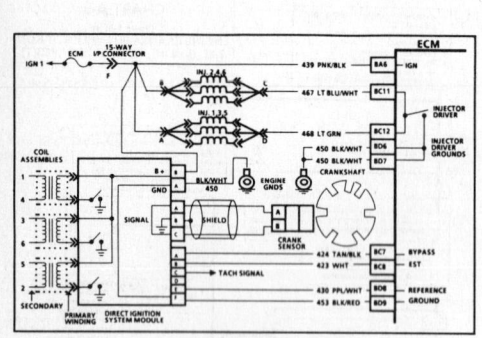

CHART A-3
(Page 3 of 3)
ENGINE CRANKS BUT WILL NOT RUN
2.8L (VIN W) "A" CARLINE (PORT)

Circuit Description:
If the "Scan" tool did not indicate a cranking rpm and there is no spark present at the plugs, the problem lies in the direct ignition system or the power and ground supplies to the module.

The magnetic crank sensor is used to determine engine crankshaft position much the same way as the pick-up coil did in distributor type systems. The sensor is mounted in the block near a seven slot wheel on the crank shaft. The rotation of the wheel creates a flux change in the sensor which produces a voltage signal. The ignition module then processes this signal and creates the reference pulses needed by the ECM and the signal triggers the correct coil at the correct time.

Test Description: Numbers below refer to circled numbers on the diagnostic chart.
1. This test will determine if the 12 volt supply and a good ground is available at the ignition module.
2. Tests for continuity of CKT 439 to the ignition module. If test light does not light but the SES light is "ON" with ignition "ON," repair open in CKT 439 between DIS ignition module and splice.
3. Checks for continuity of the crank sensor and connections.
4. Voltage will vary in this test depending on cranking speed of engine. The voltage will vary from about 500 mV at very slow cranking speeds to about 100 mV at high speeds.

1988–90 2.8L (VIN W) – CELEBRITY AND 6000

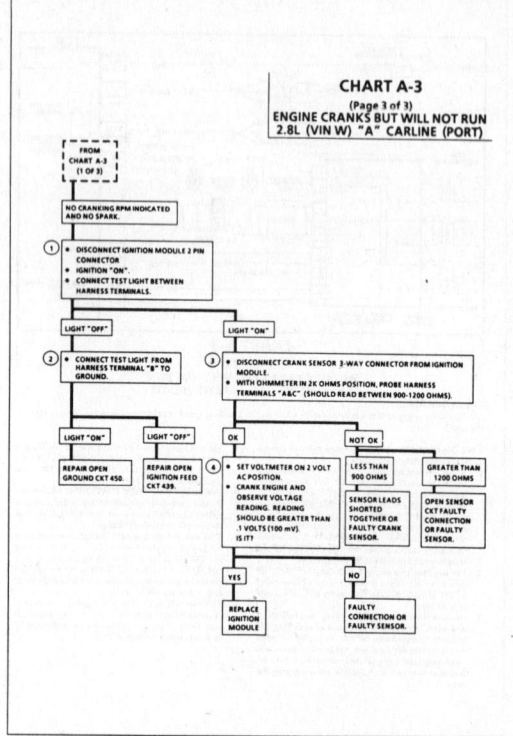

CHART A-3
(Page 3 of 3)
ENGINE CRANKS BUT WILL NOT RUN
2.8L (VIN W) "A" CARLINE (PORT)

1988–90 2.8L (VIN W) – CELEBRITY AND 6000

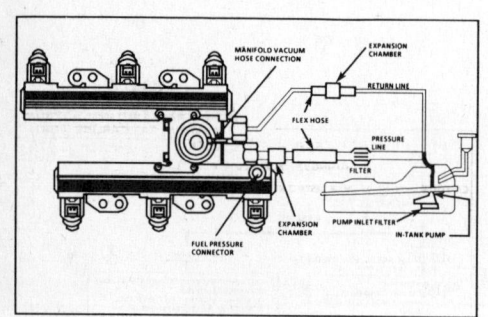

CHART A-7
(Page 1 of 3)
FUEL SYSTEM DIAGNOSIS
2.8L (VIN W) "A" CARLINE (PORT)

Circuit Description:
When the ignition switch is turned "ON," the electronic control module (ECM) will turn "ON" the in-tank fuel pump. It will remain "ON" as long as the engine is cranking or running, and the ECM is receiving reference pulses. If there are no reference pulses, the ECM will shut "OFF" the fuel pump within 2 seconds after ignition "ON" or engine stops.

The pump will deliver fuel to the fuel rail and injectors, then to the pressure regulator, where the system pressure is controlled to about 234 to 325 kPa (34 to 47 psi). Excess fuel is then returned to the fuel tank.

Test Description: Numbers below refer to circled numbers on the diagnostic chart.
1. Wrap a shop towel around the fuel pressure connector to absorb any small amount of fuel leakage that may occur when installing the gage. Ignition "ON" pump pressure should be 280-325 kPa (40.5-47 psi). This pressure is controlled by spring pressure within the regulator assembly.
2. When the engine is idling, the manifold pressure is low (high vacuum) and is applied to the fuel regulator diaphragm. This will offset the spring and result in a lower fuel pressure. This idle pressure will vary somewhat depending on barometric pressure, however, the pressure idling should be less, indicating pressure regulator control.
3. Pressure that continues to fall is caused by one of the following:
   • In-tank fuel pump check valve not holding
   • Pump coupling hose or pulsator leaking
   • Fuel pressure regulator valve leaking

4. Injector(s) sticking open.
   An injector sticking open can best be determined by checking for a fouled or saturated spark plug(s). If a leaking injector can not be determined by a fouled or saturated spark plug the following procedure should be used.
   • Remove plenum and remove fuel rail bolts. Follow the procedures in the fuel control section of this manual, but leave fuel lines connected
   • Lift fuel rail out just enough to clear injector nozzles in the ports.

CAUTION: Be sure injector(s) are not allowed to spray on engine and that injector retaining clips are intact. This should be carefully followed to prevent fuel spray on engine which would cause a fire hazard.

   • Pressurize the fuel system and observe injector nozzles.

1988–90 2.8L (VIN W) – CELEBRITY AND 6000

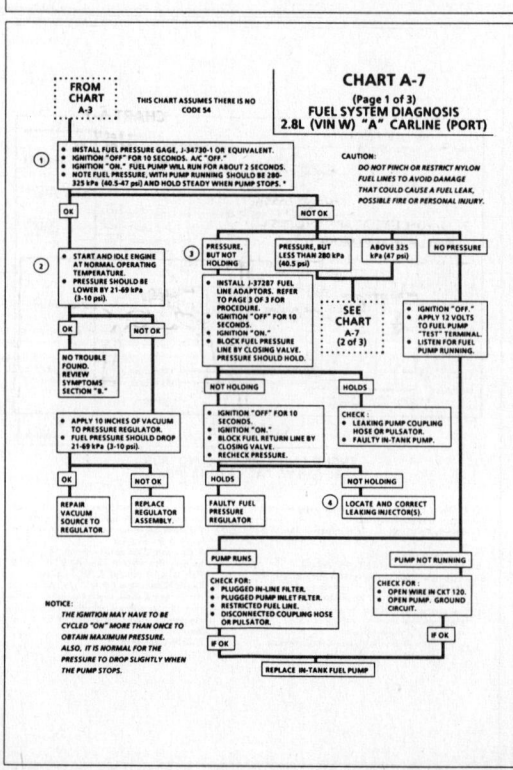

CHART A-7
(Page 1 of 3)
FUEL SYSTEM DIAGNOSIS
2.8L (VIN W) "A" CARLINE (PORT)

1988–90 2.8L (VIN W) – CELEBRITY AND 6000

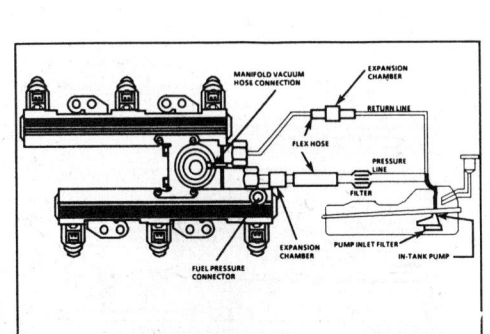

CHART A-7
(Page 2 of 3)
FUEL SYSTEM DIAGNOSIS
2.8L (VIN W) "A" CARLINE (PORT)

Test Description: Numbers below refer to circled numbers on the diagnostic chart.

1. Pressure but less than 280 kPa (40.5 psi) falls into two areas:
   - Regulated pressure but less than 280 kPa (40.5 psi). Amount of fuel to injectors OK, but pressure is too low. System will be lean running and may set Code 44. Also, hard starting cold and overall poor performance.
   - Restricted flow causing pressure drop - Normally, a vehicle with a fuel pressure of less than 165 kPa (24 psi) at idle will not be driveable. However, if the pressure drop occurs only while driving, the engine will normally surge then stop running as pressure begins to drop rapidly. This is most likely caused by a restricted fuel line or plugged filter.

2. Restricting the fuel return line allows fuel pressure to build above regulated pressure. With battery voltage applied to the pump test terminal, pressure should rise above 325 kPa (47 psi) as the valve in the return line is partially closed.

   NOTICE: Do Not allow pressure to exceed 414 kPa (60 psi), as damage to the regulator may result.

3. This test determines if the high fuel pressure is due to a restricted fuel return line or a pressure regulator problem.

1988–90 2.8L (VIN W) – CELEBRITY AND 6000

NOTICE: *FUEL SYSTEM UNDER PRESSURE. TO AVOID FUEL SPILLAGE, REFER TO FIELD SERVICE PROCEDURES FOR TESTING OR REPAIRS REQUIRING DISASSEMBLY OF FUEL LINES OR FITTINGS.*

CHART A-7
(Page 2 of 3)
FUEL SYSTEM DIAGNOSIS
2.8L (VIN W) "A" CARLINE (PORT)

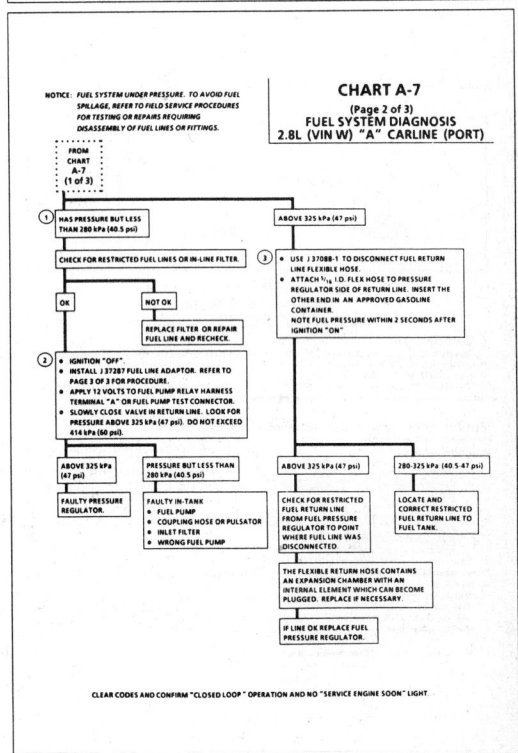

CLEAR CODES AND CONFIRM "CLOSED LOOP" OPERATION AND NO "SERVICE ENGINE SOON" LIGHT.

1988–90 2.8L (VIN W) – CELEBRITY AND 6000

CHART A-7
(Page 3 of 3)
FUEL SYSTEM DIAGNOSIS
2.8L (VIN W) "A" CARLINE (PORT)

FUEL PRESSURE CHECK TOOLS

37287 FUEL LINE ADAPTER TOOLS
37088 FUEL LINE REMOVE TOOL

J 37287 — FUEL RETURN LINE — 3/8" FUEL LINE — 95 5294-6E

J 37287 — FUEL FEED LINE — 5/16" FUEL LINE — 95 5287-6E

J 37088 – 1 — 95 5263-6E

1988–90 2.8L (VIN W) – CELEBRITY AND 6000

CHART A-7
(Page 3 of 3)
FUEL SYSTEM DIAGNOSIS
2.8L (VIN W) "A" CARLINE (PORT)

FUEL PRESSURE CHECK
(USING J 37287 FUEL LINE ADAPTOR)

CAUTION: DO NOT PINCH OR RESTRICT NYLON FUEL LINES TO AVOID SEVERING, WHICH COULD CAUSE A FUEL LEAK.

REMOVE OR DISCONNECT

1. REMOVE FUEL CAP.
2. INSTALL GAGE BLEED HOSE INTO AN APPROVED CONTAINER AND OPEN GAGE VALVE TO BLEED SYSTEM PRESSURE.
3. REMOVE FUEL LINES. USE J 37088-1 TOOL. USE A SHOP CLOTH AND APPROVED CONTAINER TO COLLECT FUEL FROM LINES.
4. INSTALL J 37287 FUEL LINE ADAPTORS TO ENGINE AND BODY SIDE FUEL LINE. (LEAVE VALVE OPEN ON ADAPTOR LINES.)
5. INSTALL FUEL CAP.
6. INSTALL GAGE BLEED HOSE INTO AN APPROVED CONTAINER AND OPEN GAGE VALVE TO BLEED AIR FROM SYSTEM.
7. THE IGNITION MAY HAVE TO BE CYCLED "ON" MORE THAN ONCE TO REMOVE ALL AIR FROM FUEL LINES AND OBTAIN MAXIMUM PRESSURE. IT IS NORMAL FOR THE PRESSURE TO DROP SLIGHTLY WHEN PUMP STOPS.

RECONNECT OR INSTALL

1. REMOVE FUEL CAP.
2. INSTALL GAGE BLEED HOSE INTO AN APPROVED CONTAINER AND OPEN GAGE VALVE TO BLEED SYSTEM PRESSURE.
3. REMOVE FUEL LINE ADAPTORS (J 37287) WITH J 37088-1 TOOL. USE A SHOP CLOTH AND APPROVED CONTAINER TO COLLECT FUEL FROM LINES.
4. INSTALL FUEL LINES TO ENGINE.
5. INSTALL FUEL CAP.
6. INSTALL GAGE BLEED HOSE INTO AN APPROVED CONTAINER AND OPEN GAGE VALVE TO BLEED AIR FROM SYSTEM.
7. THE IGNITION MAY HAVE TO BE CYCLED "ON" MORE THAN ONCE TO REMOVE ALL AIR FROM FUEL LINES AND OBTAIN MAXIMUM PRESSURE. IT IS NORMAL FOR THE PRESSURE TO DROP SLIGHTLY WHEN PUMP STOPS.
8. CHECK FOR LEAKS IN FUEL LINE.

1988–90 2.8L (VIN W) – CELEBRITY AND 6000

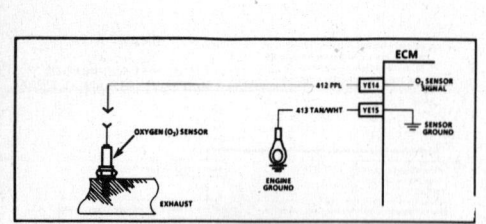

CODE 13
OXYGEN SENSOR CIRCUIT
(OPEN CIRCUIT)
2.8L (VIN W) "A" CARLINE (PORT)

Circuit Description:

The ECM supplies a voltage of about .45 volt between terminals "YE14" and "YE15". (If measured with a 10 megohm digital voltmeter, this may read as low as .32 volt.) The O₂ sensor varies the voltage within a range of about 1 volt if the exhaust is rich, down through about .10 volt if exhaust is lean.

The sensor is like an open circuit and produces no voltage when it is below 315°C (600°F). An open sensor circuit or cold sensor causes "Open Loop" operation.

Test Description: Numbers below refer to circled numbers on the diagnostic chart.
1. Code 13 will be set:
   - Engine at normal operating temperature.
   - At least 2 minutes engine time after start.
   - O₂ signal voltage steady between .35 and .55 volt.
   - Throttle position sensor signal above 4%.
   - All conditions must be met for about 60 seconds.
   If the conditions for a Code 13 exist, the system will not go "Closed Loop."
2. This will determine if the sensor is at fault or the wiring or ECM is the cause of the Code 13.
3. In doing this test use only a high impedance digital volt ohmmeter. This test checks the continuity of CKTs 412 and 413 because if CKT 413 is open the ECM voltage on CKT 412 will be over .6 volt (600 mV).

Diagnostic Aids:

Normal "Scan" voltage varies between 100 mV to 999 mV (.1 and 1.0 volt) while in "Closed Loop." Code 13 sets in one minute if voltage remains between .35 and .55 volt, but the system will go "Open Loop" in about 15 seconds. Refer to "Intermittents" in Section "B".

1988–90 2.8L (VIN W) – CELEBRITY AND 6000

CODE 13
OXYGEN SENSOR CIRCUIT
(OPEN CIRCUIT)
2.8L (VIN W) "A" CARLINE (PORT)

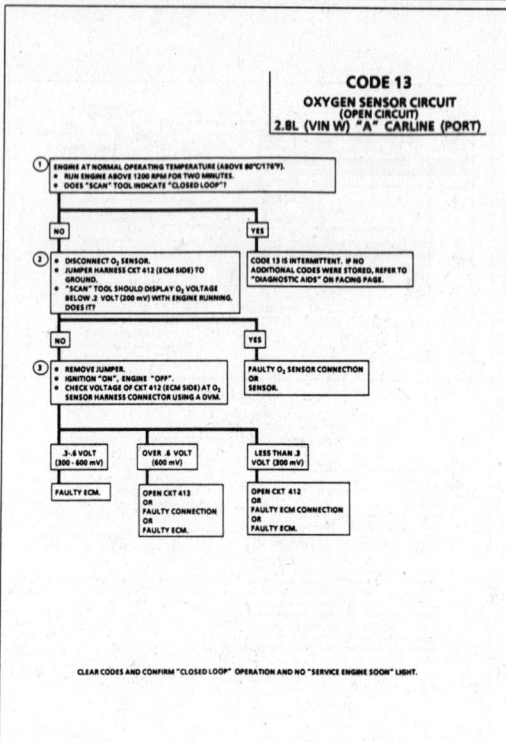

1988–90 2.8L (VIN W) – CELEBRITY AND 6000

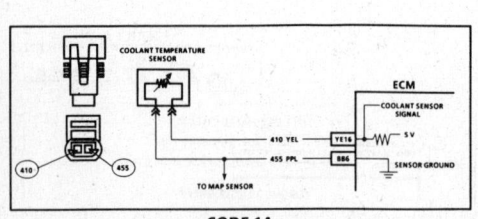

CODE 14
COOLANT TEMPERATURE SENSOR CIRCUIT
(HIGH TEMPERATURE INDICATED)
2.8L (VIN W) "A" CARLINE (PORT)

Circuit Description:

The coolant temperature sensor uses a thermistor to control the signal voltage to the ECM. The ECM applies a voltage on CKT 410 to the sensor. When the engine is cold, the sensor (thermistor) resistance is high, therefore the ECM will see high signal voltage.

As the engine warms, the sensor resistance becomes less, and the voltage drops. At normal engine operating temperature (85°C to 95°C), the voltage will measure about 1.5 to 2.0 volts.

Test Description: Numbers below refer to circled numbers on the diagnostic chart.
1. Code 14 will set if:
   - Signal voltage indicates a coolant temperature above 135°C (275°F) for 3 seconds.
2. This test will determine if CKT 410 is shorted to ground which will cause the conditions for Code 14.

Diagnostic Aids:

Check harness routing for a potential short to ground in CKT 410. "Scan" tool displays engine temperature in degrees centigrade. After engine is started, the temperature should rise steadily to about 90°C, then stabilize when thermostat opens. Refer to "Intermittents" in Section "B"

1988–90 2.8L (VIN W) – CELEBRITY AND 6000

CODE 14
COOLANT TEMPERATURE SENSOR CIRCUIT
(HIGH TEMPERATURE INDICATED)
2.8L (VIN W) "A" CARLINE (PORT)

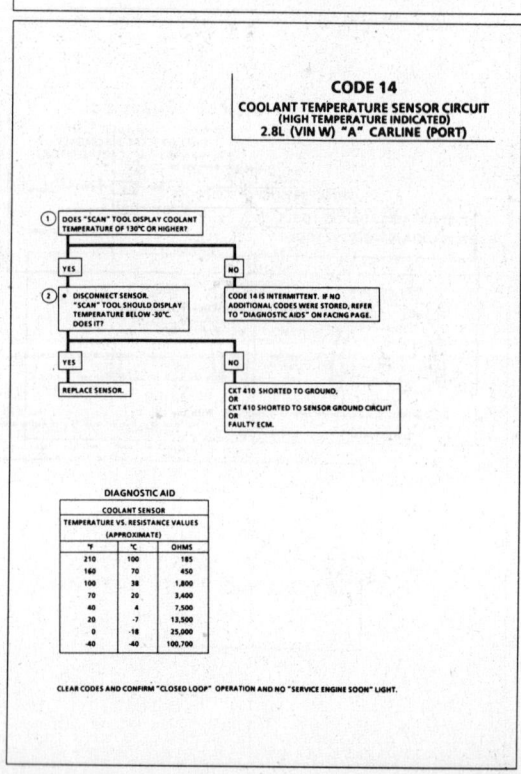

DIAGNOSTIC AID

COOLANT SENSOR
TEMPERATURE VS. RESISTANCE VALUES
(APPROXIMATE)

°F	°C	OHMS
210	100	185
160	70	450
100	38	1,800
70	20	3,400
40	4	7,500
20	-7	13,500
0	-18	25,000
-40	-40	100,700

CLEAR CODES AND CONFIRM "CLOSED LOOP" OPERATION AND NO "SERVICE ENGINE SOON" LIGHT.

1988–90 2.8L (VIN W) – CELEBRITY AND 6000

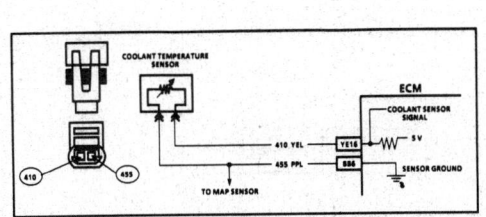

CODE 15
COOLANT TEMPERATURE SENSOR CIRCUIT
(LOW TEMPERATURE INDICATED)
2.8L (VIN W) "A" CARLINE (PORT)

Circuit Description:
The coolant temperature sensor uses a thermistor to control the signal voltage to the ECM. The ECM applies a voltage on CKT 410 to the sensor. When the engine is cold the sensor (thermistor) resistance is high, therefore, the ECM will see high signal voltage.
As the engine warms, the sensor resistance becomes less and the voltage drops. At normal engine operating temperature (85°C to 95°C), the voltage will measure about 1.5 to 2.0 volts at the ECM.

Test Description: Numbers below refer to circled numbers on the diagnostic chart.
1. Code 15 will set if:
   - Signal voltage indicates a coolant temperature less than -44°C (-47°F) for 3 seconds.
2. This test simulates a Code 14. If the ECM recognises the low signal voltage, (high temperature) and the "Scan" reads 130°C, the ECM and wiring are OK.
3. This test will determine if CKT 410 is open. There should be 5 volts present at sensor connector, if measured with a DVOM.

Diagnostic Aids:
A "Scan" tool reads engine temperature in degrees centigrade. After engine is started the temperature should rise steadily to about 90°C then stabilize when thermostat opens.
A faulty connection, or an open in CKT 410 or 455 will result in a Code 15.
If Code 23 or 33 is also set, check CKT 455 for faulty wiring or connections. Check terminals at sensor for good contact. Refer to "Intermittents" in Section "B".

1988–90 2.8L (VIN W) – CELEBRITY AND 6000

CODE 15
COOLANT TEMPERATURE SENSOR CIRCUIT
(LOW TEMPERATURE INDICATED)
2.8L (VIN W) "A" CARLINE (PORT)

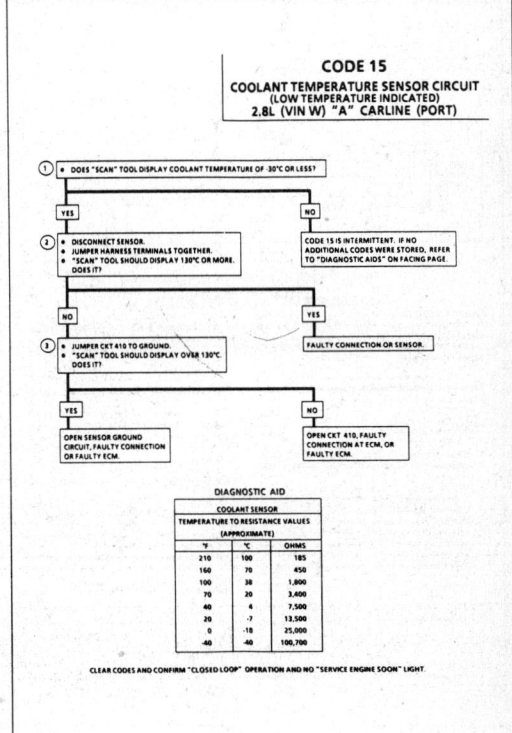

DIAGNOSTIC AID		
COOLANT SENSOR		
TEMPERATURE TO RESISTANCE VALUES (APPROXIMATE)		
°F	°C	OHMS
210	100	185
160	70	450
100	38	1,800
70	20	3,400
40	4	7,500
20	-7	13,500
0	-18	25,000
-40	-40	100,700

CLEAR CODES AND CONFIRM "CLOSED LOOP" OPERATION AND NO "SERVICE ENGINE SOON" LIGHT.

1988–90 2.8L (VIN W) – CELEBRITY AND 6000

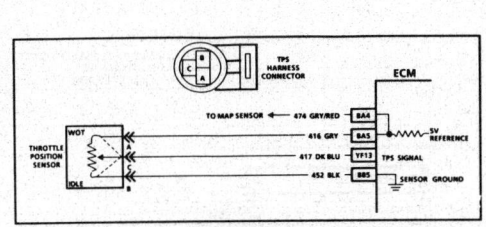

CODE 21
THROTTLE POSITION SENSOR (TPS) CIRCUIT
(SIGNAL VOLTAGE HIGH)
2.8L (VIN W) "A" CARLINE (PORT)

Circuit Description:
The throttle position sensor (TPS) provides a voltage signal that changes relative to the throttle blade. Signal voltage will vary from about .5 at idle to about 5 volts at wide open throttle.
The TPS signal is one of the most important inputs used by the ECM for fuel control and for most of the ECM control outputs.

Test Description: Numbers below refer to circled numbers on the diagnostic chart.
1. Code 21 will set if:
   - Engine is running.
   - TPS signal voltage is greater than 4.3 volts.
   - Air flow is less than 17 GM/sec.
   - All conditions met for 10 seconds.
   or
   - TPS signal voltage over 4.5 volts with ignition "ON."
   - TPS check. The TPS has an auto zeroing feature. If the voltage reading is within the range of 0.45 to 0.85 volt, the ECM will use that value as closed throttle. If the voltage reading is out of the auto zero range on an existing or replacement TPS, make sure the cruise control and throttle cables are not being held open.

2. With the TPS sensor disconnected, the TPS voltage should go low if the ECM and wiring are OK.
3. Probing CKT 452 with a test light checks the 5 volts return circuit, because a faulty 5 volts return will cause a Code 21.

Diagnostic Aids:
A "Scan" tool reads throttle position in volts. Voltage should increase at a steady rate as throttle is moved toward WOT.
Also, some "Scan" tools will read throttle angle 0% = closed throttle, 100% = WOT.
An open in CKT 452 will result in a Code 21. Refer to "Intermittents" in Section "B".

1988–90 2.8L (VIN W) – CELEBRITY AND 6000

CODE 21
THROTTLE POSITION SENSOR (TPS) CIRCUIT
(SIGNAL VOLTAGE HIGH)
2.8L (VIN W) "A" CARLINE (PORT)

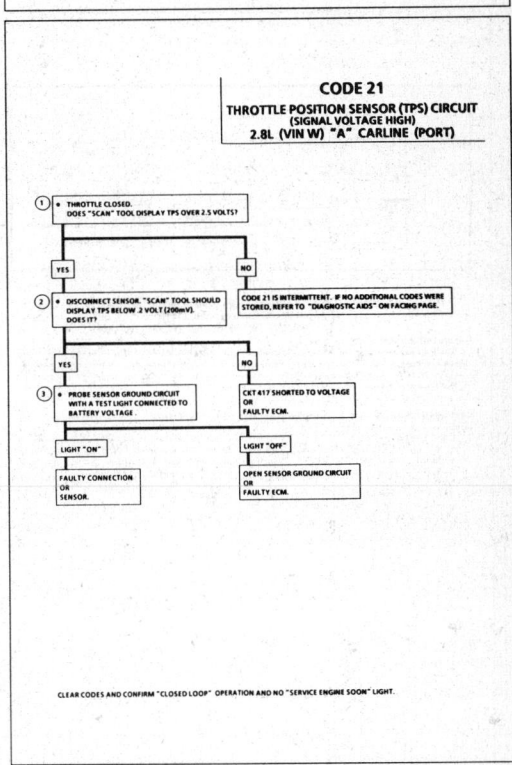

CLEAR CODES AND CONFIRM "CLOSED LOOP" OPERATION AND NO "SERVICE ENGINE SOON" LIGHT.

1988–90 2.8L (VIN W) – CELEBRITY AND 6000

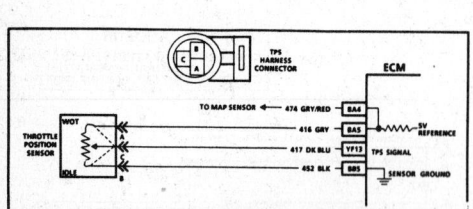

CODE 22
THROTTLE POSITION SENSOR (TPS) CIRCUIT
(SIGNAL VOLTAGE LOW)
2.8L (VIN W) "A" CARLINE (PORT)

Circuit Description:
The throttle position sensor (TPS) provides a voltage signal that changes relative to the throttle blade. Signal voltage will vary from about .5 at idle to about 5 volts at wide open throttle.
The TPS signal is one of the most important inputs used by the ECM for fuel control and for most of the ECM control outputs.

Test Description: Numbers below refer to circled numbers on the diagnostic chart.
1. Code 22 will set if:
   - Engine running
   - TPS signal voltage is less than about .2 volt for 3 seconds.
2. Simulates Code 21: (high voltage) if the ECM recognizes the high signal voltage, the ECM and wiring are OK.
3. TPS check: The TPS has an auto zeroing feature. If the voltage reading is within the range of 0.45 to 0.86 volt, the ECM will use that value as closed throttle. If the voltage reading is out of the auto zero range on an existing or replacement TPS, make sure the cruise control and throttle cables are not being held open.
4. This simulates a high signal voltage to check for an open in CKT 417.

5. CKTs 416 and 474 share a common sensor ground buffered reference signal. If either of these circuits is shorted to ground, Code 22 will set. To determine if the MAP sensor is causing the 22 problem, disconnect it to see if Code 22 resets. Be sure TPS is connected and clear codes before testing.

Diagnostic Aids:

A "Scan" tool reads throttle position in volts. Voltage should increase at a steady rate as throttle is moved toward WOT.
Also, some "Scan" tools will read: throttle angle 0% = closed throttle, 100% = WOT.
An open or short to ground in CKT 416 or 417 will result in a Code 22. Also, a short to ground in CKT 474 will result in a Code 22.
Refer to "Intermittents" in Section "B".

1988–90 2.8L (VIN W) – CELEBRITY AND 6000

CODE 22
THROTTLE POSITION SENSOR (TPS) CIRCUIT
(SIGNAL VOLTAGE LOW)
2.8L (VIN W) "A" CARLINE (PORT)

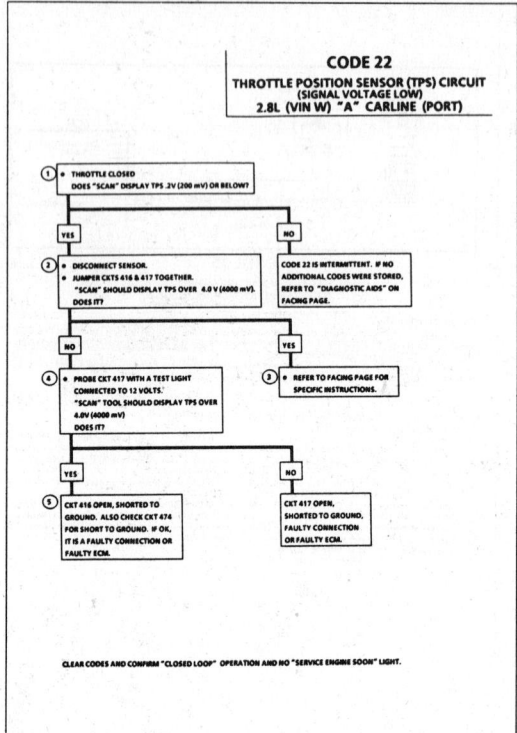

CLEAR CODES AND CONFIRM "CLOSED LOOP" OPERATION AND NO "SERVICE ENGINE SOON" LIGHT.

1988–90 2.8L (VIN W) – CELEBRITY AND 6000

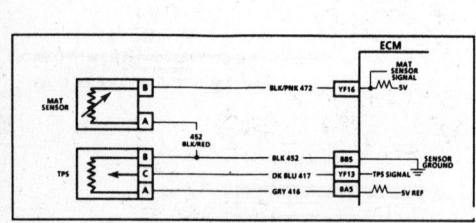

CODE 23
MANIFOLD AIR TEMPERATURE (MAT) SENSOR CIRCUIT
(LOW TEMPERATURE INDICATED)
2.8L (VIN W) "A" CARLINE (PORT)

Circuit Description:
The MAT sensor uses a thermistor to control the signal voltage to the ECM. The ECM applies a voltage (about 5 volts) on CKT 472 to the sensor. When the air is cold, the sensor (thermistor) resistance is high, therefore the ECM will see a high signal voltage. If the air is warm, the sensor resistance is low, therefore, the ECM will see a low voltage.

Test Description: Numbers below refer to circled numbers on the diagnostic chart.
1. Code 23 will set if:
   - A signal voltage indicates a manifold air temperature below −35°C (−31°F) for 3 seconds.
   - Time since engine start is 4 minutes or longer.
   - No VSS.
2. A Code 23 will set, due to an open sensor, wire, or connection. This test will determine if the wiring and ECM are OK.
3. This will determine if the signal CKT 472 or the sensor ground return CKT 452 is open.

Diagnostic Aids:

A "Scan" tool reads temperature of the air entering the engine and should read close to ambient air temperature when engine is cold, and rises as underhood temperature increases.
A faulty connection, or an open in CKT 472 or 452 will result in a Code 23.
Refer to "Intermittents" in Section "B".

1988–90 2.8L (VIN W) – CELEBRITY AND 6000

CODE 23
MANIFOLD AIR TEMPERATURE (MAT) SENSOR CIRCUIT
(LOW TEMPERATURE INDICATED)
2.8L (VIN W) "A" CARLINE (PORT)

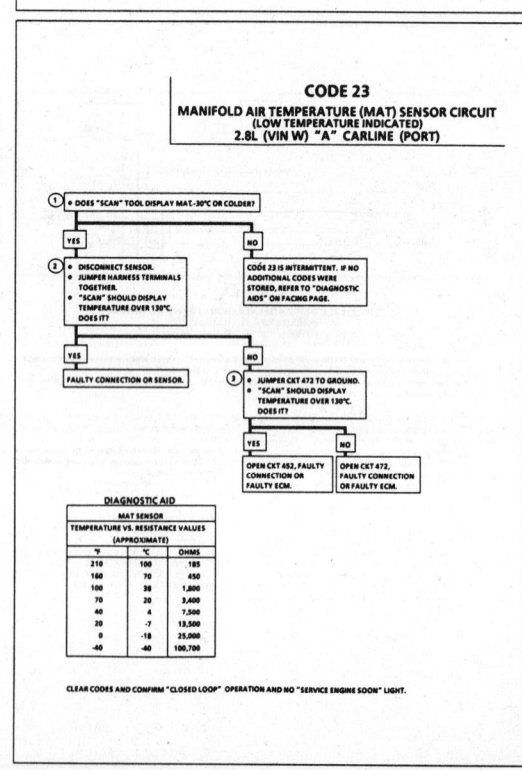

DIAGNOSTIC AID

MAT SENSOR		
TEMPERATURE VS. RESISTANCE VALUES (APPROXIMATE)		
°F	°C	OHMS
210	100	185
160	70	450
100	38	1,800
70	20	3,400
40	4	7,500
20	−7	13,500
0	−18	25,000
−40	−40	100,700

CLEAR CODES AND CONFIRM "CLOSED LOOP" OPERATION AND NO "SERVICE ENGINE SOON" LIGHT.

1988–90 2.8L (VIN W) – CELEBRITY AND 6000

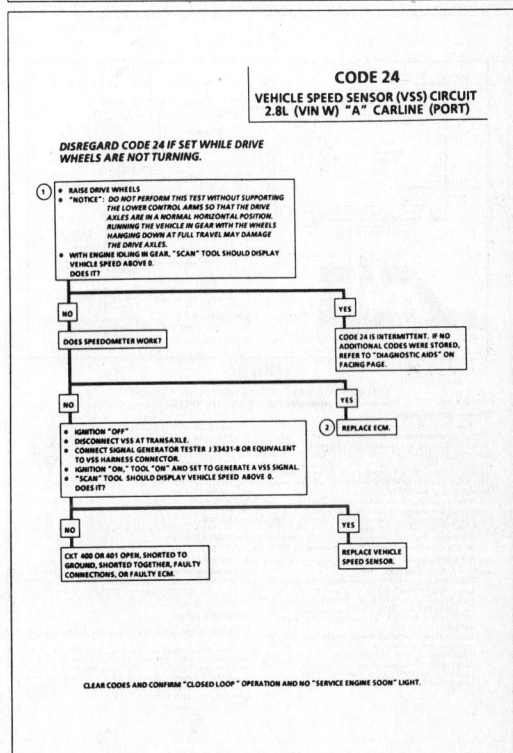

CODE 24
VEHICLE SPEED SENSOR (VSS) CIRCUIT
2.8L (VIN W) "A" CARLINE (PORT)

Circuit Description:
Vehicle speed information is provided to the ECM by the vehicle speed sensor, which is a permanent magnet (PM) generator and it is mounted in the transaxle. The PM generator produces a pulsing voltage, whenever vehicle speed is over about 3 mph. The A/C voltage level and the number of pulses increases with vehicle speed. The ECM then converts the pulsing voltage to mph which is used for calculations, and the mph can be displayed with a "Scan" tool.

The function of VSS buffer used in past model years has been incorporated into the ECM. The ECM then supplies the necessary signal for the instrument panel (4000 pulses per mile) for operating the speedometer and the odometer. If the vehicle is equipped with cruise control, the ECM also provides a signal (2000 pulses per mile) to the cruise control module.

Test Description: Numbers below refer to circled numbers on the diagnostic chart.
1. Code 24 will set if vehicle speed equals 0 mph when:
   - Engine speed is between 1400 and 3600 rpm
   - TPS is less than 2%
   - Low load condition (low air flow)
   - Not in park or neutral
   - All conditions met for 5 seconds
   These conditions are met during a road load deceleration. Disregard Code 24 that sets when drive wheels are not turning.
   - The PM generator only produces a signal if drive wheels are turning greater than 3 mph.
2. CKTs 400, 401 and 993 are OK if the speedometer works properly. Code 24 is being caused by a faulty ECM, faulty Mem-Cal or an incorrect Mem-Cal.

Diagnostic Aids:
"Scan" should indicate a vehicle speed whenever the drive wheels are turning greater than 3 mph

A problem in CKT 437 or 993 will not affect the VSS input or the readings on a "Scan."

Check CKTs 400 and 401 for proper connections to be sure they're clean and tight and the harness is routed correctly. Refer to "Intermittents" in Section "B".

(A/T) A faulty or misadjusted park/neutral switch can result in a false Code 24. Use a "Scan" and check for proper signal while in drive (125C) or overdrive (440-T4). Refer to CHART C-1A for P/N switch diagnosis check.

1988–90 2.8L (VIN W) – CELEBRITY AND 6000

CODE 24
VEHICLE SPEED SENSOR (VSS) CIRCUIT
2.8L (VIN W) "A" CARLINE (PORT)

DISREGARD CODE 24 IF SET WHILE DRIVE WHEELS ARE NOT TURNING.

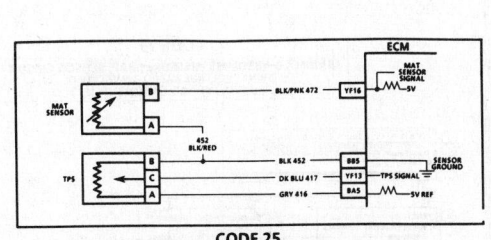

CLEAR CODES AND CONFIRM "CLOSED LOOP" OPERATION AND NO "SERVICE ENGINE SOON" LIGHT.

1988–90 2.8L (VIN W) – CELEBRITY AND 6000

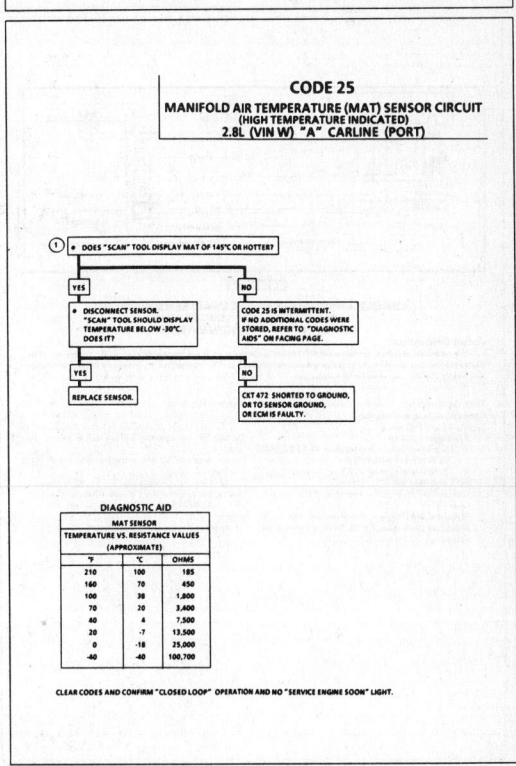

CODE 25
MANIFOLD AIR TEMPERATURE (MAT) SENSOR CIRCUIT
(HIGH TEMPERATURE INDICATED)
2.8L (VIN W) "A" CARLINE (PORT)

Circuit Description:
The manifold air temperature sensor uses a thermistor to control the signal voltage to the ECM. The ECM applies a voltage (about 5 volts) on CKT 472 to the sensor. When manifold air is cold, the sensor (thermistor) resistance is high, therefore, the ECM will see a high signal voltage. As the air warms, the sensor resistance becomes less, and the voltage drops.

Test Description: Numbers below refer to circled numbers on the diagnostic chart.
1. Code 25 will set if:
   - Signal voltage indicates a manifold air temperature greater than 145°C (293°F) for 3 seconds
   - Time since engine start is 4 minutes or longer.
   - A vehicle speed is present.

Diagnostic Aids:
A "Scan" tool reads temperature of the air entering the engine and should read close to ambient air temperature when engine is cold and rises as underhood temperature increases.

A short to ground in CKT 472 will result in a Code 25.

Refer to "Intermittents" in Section "B".

1988–90 2.8L (VIN W) – CELEBRITY AND 6000

CODE 25
MANIFOLD AIR TEMPERATURE (MAT) SENSOR CIRCUIT
(HIGH TEMPERATURE INDICATED)
2.8L (VIN W) "A" CARLINE (PORT)

DIAGNOSTIC AID

MAT SENSOR		
TEMPERATURE VS. RESISTANCE VALUES (APPROXIMATE)		
°F	°C	OHMS
210	100	185
160	70	450
100	38	1,800
70	20	3,400
40	4	7,500
20	-7	13,500
0	-18	25,000
-40	-40	100,700

CLEAR CODES AND CONFIRM "CLOSED LOOP" OPERATION AND NO "SERVICE ENGINE SOON" LIGHT.

1988–90 2.8L (VIN W) – CELEBRITY AND 6000

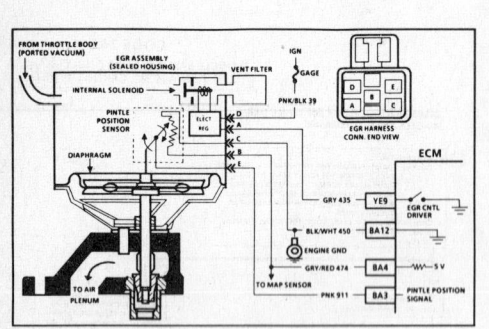

CODE 32
EXHAUST GAS RECIRCULATION (EGR) CIRCUIT
2.8L (VIN W) "A" CARLINE (PORT)

Circuit Description:
The integrated electronic EGR valve functions similar to a port EGR valve with a remote vacuum regulator. The internal solenoid is normally open, which causes the vacuum signal to be vented "OFF" to the atmosphere when EGR is not being commanded by the ECM. This EGR valve has a sealed cap and the solenoid valve opens and closes the vacuum signal which controls the amount of vacuum vented to atmosphere, and this controls the amount of vacuum applied to the diaphragm. The electronic EGR valve contains a voltage regulator which converts the ECM signal to provide different amounts of EGR flow by regulating the current to the solenoid. The ECM controls EGR flow with a pulse width modulated signal (turns "ON" and "OFF" many times a second) based on air flow, TPS, and rpm.

This system also contains a pintle position sensor which works similar to a sensor, and as EGR flow is increased, the sensor output also increases.

Code 32 means that there has been an EGR system faulty detected.

Code 32 will set under two conditions:
• Coolant temperature above a specified amount, EGR should be "ON" or;
• EGR pintle position does not match duty cycle.

Test Description: Numbers below refer to circled numbers on the diagnostic chart.
1. With the engine running and the transmission in gear, with brakes applied, increasing engine rpm will put a load on the engine. Engine vacuum will be applied to the EGR diaphragm and cause the EGR pintle to open increasing pintle position voltage.
2. Grounding the diagnostic terminal should energize the internal solenoid which closes off the vent and allows the vacuum to move the diaphragm. This test determines that the ECM is capable of controlling the solenoid. When EGR is commanded on by the ECM, the test light should be "ON."

3. If CKT 450 is open, the pintle signal will go high (showing a 5 volts signal). This will set a Code 32. If CKT 911 becomes shorted to 12 volts or to CKT 474, the signal voltage will go high causing a Code 32.

Diagnostic Aids:
Some "Scan" tools will read pintle position in volts.
The EGR position voltage can be used to determine that the pintle is moving. When no EGR is commanded 0-% duty cycle), the position sensor should read between .5 volt and 1.5 volts and increase with the commanded EGR duty cycle. If system operates correctly, refer to "Intermittents" in Section "B"

1988–90 2.8L (VIN W) – CELEBRITY AND 6000

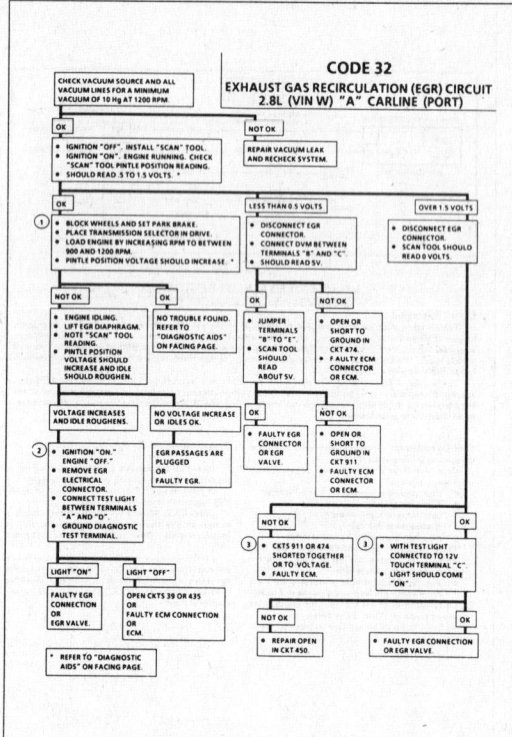

CODE 32
EXHAUST GAS RECIRCULATION (EGR) CIRCUIT
2.8L (VIN W) "A" CARLINE (PORT)

1988–90 2.8L (VIN W) – CELEBRITY AND 6000

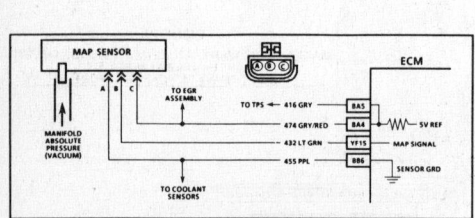

CODE 33
MANIFOLD ABSOLUTE PRESSURE (MAP) SENSOR CIRCUIT
(SIGNAL VOLTAGE HIGH - LOW VACUUM)
2.8L (VIN W) "A" CARLINE (PORT)

Circuit Description:
The manifold absolute pressure (MAP) sensor responds to changes in manifold pressure (vacuum). The ECM receives this information as a signal voltage that will vary from about 1–1.5 volts at idle (high vacuum) to 4–4.5 volts at wide open throttle (low vacuum).

Test Description: Numbers below refer to numbers on the diagnostic chart.
1. Code 33 will set when:
   • Engine running
   • Manifold pressure greater than 75.3 kPa (A/C "OFF") 81.2 kPa (A/C "ON")
   • Throttle angle less than 2%
   • Conditions met for 2 seconds
   Engine misfire or a low unstable idle may set Code 33.
2. With the MAP sensor disconnected, the ECM should see a low voltage if the ECM and wiring are OK.

Diagnostic Aids:
If idle is rough or unstable, refer to "Symptoms" in Section "B" for items which can cause an unstable idle.
An open in CKT 455 or the connection will result in a Code 33.
Ignition "ON" engine "OFF," voltages should be within the values shown in the table on the chart. Also, CHART C-1D can be used to test the MAP sensor.
Refer to "Intermittents" in Section "B".

1988–90 2.8L (VIN W) – CELEBRITY AND 6000

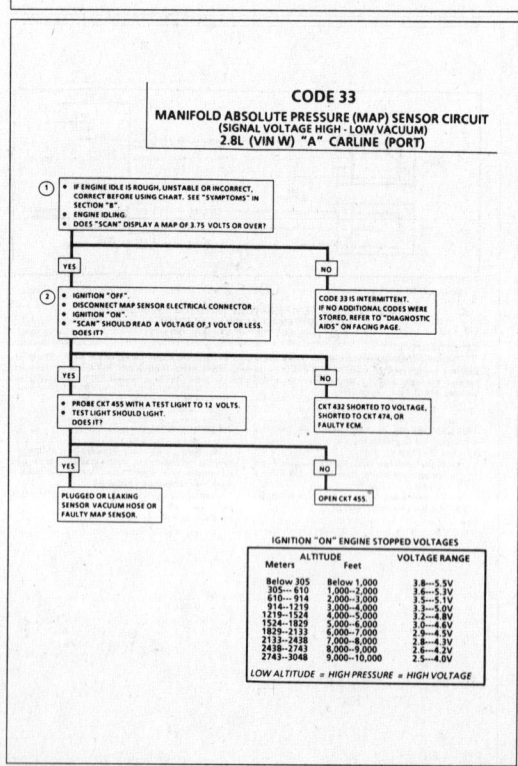

CODE 33
MANIFOLD ABSOLUTE PRESSURE (MAP) SENSOR CIRCUIT
(SIGNAL VOLTAGE HIGH - LOW VACUUM)
2.8L (VIN W) "A" CARLINE (PORT)

ALTITUDE		VOLTAGE RANGE
Meters	Feet	
Below 305	Below 1,000	3.8—5.5V
305—610	1,000–2,000	3.6—5.3V
610—914	2,000–3,000	3.5—5.1V
914—1219	3,000–4,000	3.3—5.0V
1219—1524	4,000–5,000	3.2—4.8V
1524—1829	5,000–6,000	3.0—4.6V
1829—2133	6,000–7,000	2.9—4.5V
2133—2438	7,000–8,000	2.8—4.3V
2438—2743	8,000–9,000	2.6—4.2V
2743—3048	9,000–10,000	2.5—4.0V

LOW ALTITUDE = HIGH PRESSURE = HIGH VOLTAGE

1988–90 2.8L (VIN W) – CELEBRITY AND 6000

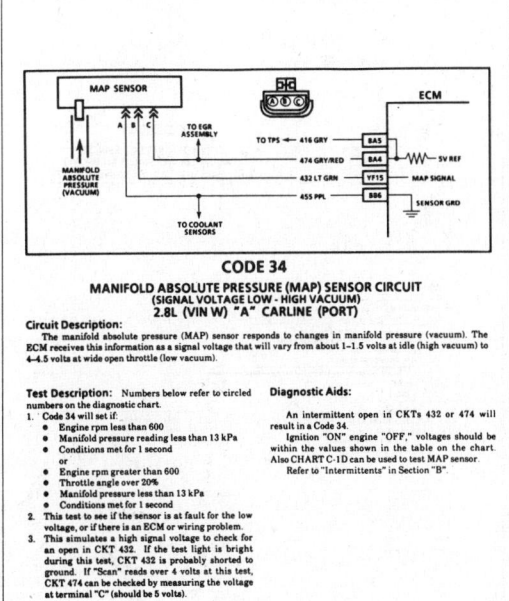

CODE 34
MANIFOLD ABSOLUTE PRESSURE (MAP) SENSOR CIRCUIT
(SIGNAL VOLTAGE LOW - HIGH VACUUM)
2.8L (VIN W) "A" CARLINE (PORT)

Circuit Description:
The manifold absolute pressure (MAP) sensor responds to changes in manifold pressure (vacuum). The ECM receives this information as a signal voltage that will vary from about 1–1.5 volts at idle (high vacuum) to 4–4.5 volts at wide open throttle (low vacuum).

Test Description: Numbers below refer to circled numbers on the diagnostic chart.
1. Code 34 will set if:
   - Engine rpm less than 600
   - Manifold pressure reading less than 13 kPa
   - Conditions met for 1 second
   or
   - Engine rpm greater than 600
   - Throttle angle over 20%
   - Manifold pressure less than 13 kPa
   - Conditions met for 1 second
2. This test see if the sensor is at fault for the low voltage, or if there is an ECM or wiring problem.
3. This simulates a high signal voltage to check for an open in CKT 432. If the test light is bright during this test, CKT 432 is probably shorted to ground. If "Scan" reads over 4 volts at this test, CKT 474 can be checked by measuring the voltage at terminal "C" (should be 5 volts).

Diagnostic Aids:
An intermittent open in CKTs 432 or 474 will result in a Code 34.
Ignition "ON" engine "OFF," voltages should be within the values shown in the table on the chart. Also CHART C-1D can be used to test MAP sensor.
Refer to "Intermittents" in Section "B".

1988–90 2.8L (VIN W) – CELEBRITY AND 6000

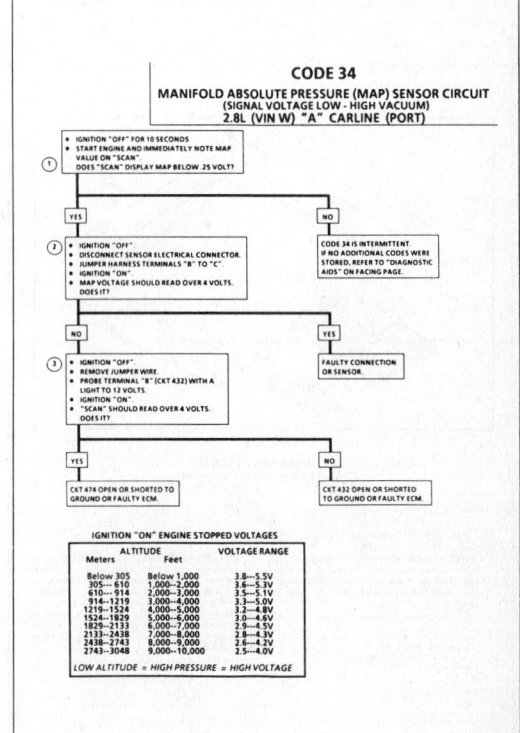

CODE 34
MANIFOLD ABSOLUTE PRESSURE (MAP) SENSOR CIRCUIT
(SIGNAL VOLTAGE LOW - HIGH VACUUM)
2.8L (VIN W) "A" CARLINE (PORT)

	IGNITION "ON" ENGINE STOPPED VOLTAGES	
Meters	ALTITUDE **Feet**	**VOLTAGE RANGE**
Below 305	Below 1,000	3.8—5.5V
305— 610	1,000—2,000	3.6—5.3V
610— 914	2,000—3,000	3.5—5.1V
914—1219	3,000—4,000	3.3—5.0V
1219—1524	4,000—5,000	3.2—4.8V
1524—1829	5,000—6,000	3.0—4.5V
1829—2133	6,000—7,000	2.9—4.5V
2133—2438	7,000—8,000	2.8—4.3V
2438—2743	8,000—9,000	2.6—4.2V
2743—3048	9,000—10,000	2.5—4.0V

LOW ALTITUDE = HIGH PRESSURE = HIGH VOLTAGE

1988–90 2.8L (VIN W) – CELEBRITY AND 6000

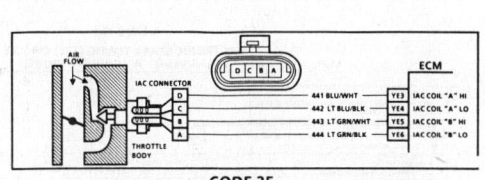

CODE 35
IDLE SPEED ERROR
2.8L (VIN W) "A" CARLINE (PORT)

Circuit Description:
Code 35 will set when the closed throttle engine speed is 300 rpm above or below the desired (commanded) idle speed for 50 seconds.

Test Description: Numbers below refer to circled numbers on the diagnostic chart.
1. Continue with test even if engine will not idle. If idle is too low, "Scan" will display 80 or more counts, or steps. If idle is high it will display "0" counts.
   Occasionally an erratic or unstable idle may occur. Engine speed may vary 200 rpm or more up and down. Disconnect IAC. If the condition is unchanged, the IAC is not at fault. There is a system problem. Proceed to diagnostic aids below.
2. When the engine was stopped, the IAC was retracted (more air) to a fixed "park" position for increased air flow and idle speed during the next engine start. A "Scan" will display 80 or more counts. Observe idle speed as on a warm engine, the idle speed should decrease rapidly.
3. Be sure to disconnect the IAC valve prior to this test. The test light will confirm the ECM signals by a steady or flashing light on all circuits.
4. There is a remote possibility that one of the circuits is shorted to voltage which would have been indicated by a steady light. Disconnect ECM and turn the ignition "ON" and probe terminals to check for this condition.

Diagnostic Aids:
A slow unstable idle may be caused by a system problem that cannot be controlled by the IAC. "Scan" counts will be above 80 counts if idle is too low and "0" counts if it is too high.
If idle is too high, stop engine. Ignition "ON." Ground diagnostic terminal. Wait a few seconds for IAC to seat then disconnect IAC. Start engine. If idle speed is above 600–700 rpm in drive with an A/T, locate and correct vacuum leak. If rpm is below spec., check for foreign material around throttle plates. Refer to Section "C2"
- **System too lean (high air/fuel ratio)**
  Idle speed may be too high or too low. Engine speed may vary up and down, disconnecting IAC does not help. This may set Code 44.
  "Scan" and/or voltmeter will read an oxygen sensor output less than 300 mV (.3 volt). Check for low regulated fuel pressure or water in fuel. A lean exhaust with an oxygen sensor output fixed above 800 mV (.8 volt) will be a contaminated sensor, usually silicone. This may also set a Code 45 or 61.
- **System too rich (low air/fuel ratio)**
  Idle speed too low. "Scan" counts usually above 80. System obviously rich and may exhibit black smoke exhaust.
  "Scan" tool and/or voltmeter will read an oxygen sensor signal fixed above 800 mV (.8 volt). Check.
    For fuel in pressure regulator hose.
    High fuel pressure.
    Injector leaking or sticking.
- **Throttle body** Remove IAC and inspect bore for foreign material or evidence of IAC valve dragging the bore.
- **A/C compressor or relay failure** See CHART C-10 if the A/C control relay drive circuit is shorted to ground or if the relay is faulty, an idle problem may exist.
  If above are all OK, refer to "Rough, Unstable, Incorrect Idle, or Stalling" in "Symptoms" in Section "B".

1988–90 2.8L (VIN W) – CELEBRITY AND 6000

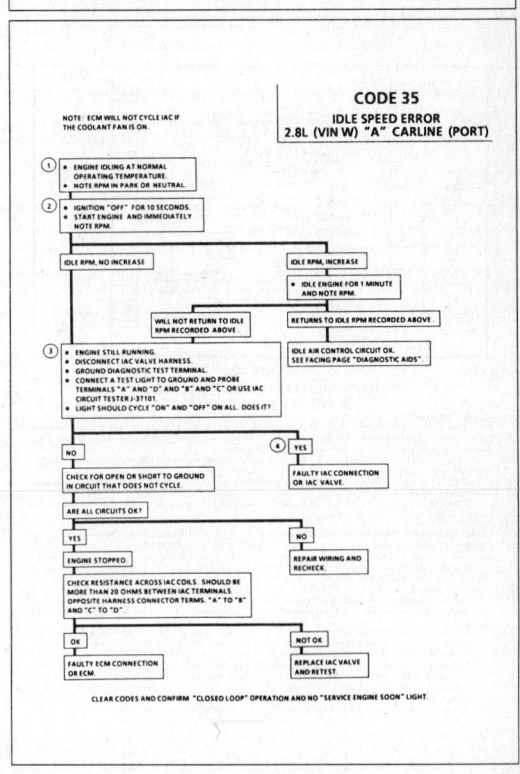

CODE 35
IDLE SPEED ERROR
2.8L (VIN W) "A" CARLINE (PORT)

1988-90 2.8L (VIN W) – CELEBRITY AND 6000

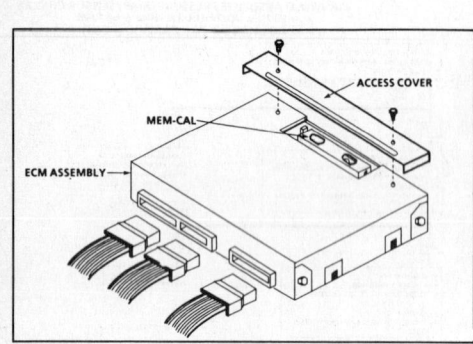

CODE 41
CYLINDER SELECT ERROR
(FAULTY OR INCORRECT MEM-CAL)
2.8L (VIN W) "A" CARLINE (PORT)

Test Description: Numbers below refer to circled numbers on the diagnostic chart.

1. The ECM used for this engine can also be used for other engines, and the difference is in the Mem-Cal. If a Code 41 sets, the incorrect Mem-Cal has been installed, or it is faulty and must be replaced.

Diagnostic Aids:

Check Mem-Cal to be sure locking tabs are secure. Also check the pins on both the Mem-Cal and ECM to assure they are making proper contact. Check the Mem-Cal part number to assure it is the correct part. If the Mem-Cal is faulty it must be replaced. It is also possible that the ECM is faulty, however, it should not be replaced until all of the above have been checked. For additional information, refer to "Intermittents" on page "6E3-B-2".

1988-90 2.8L (VIN W) – CELEBRITY AND 6000

CODE 41
CYLINDER SELECT ERROR
(FAULTY OR INCORRECT MEM-CAL)
2.8L (VIN W) "A" CARLINE (PORT)

① • IGNITION "OFF", CLEAR CODES.
   • START ENGINE AND RUN FOR 1 MINUTE OR UNTIL CODE 41 SETS.
     DOES CODE 41 SET?

YES → FAULTY CONNECTIONS DUE TO MEM-CAL NOT LOCKED IN PLACE, OR INCORRECT MEM-CAL INSTALLED.

NO → CODE 41 IS INTERMITTENT. REVIEW "DIAGNOSTIC AIDS" ON FACING PAGE.

1988-90 2.8L (VIN W) – CELEBRITY AND 6000

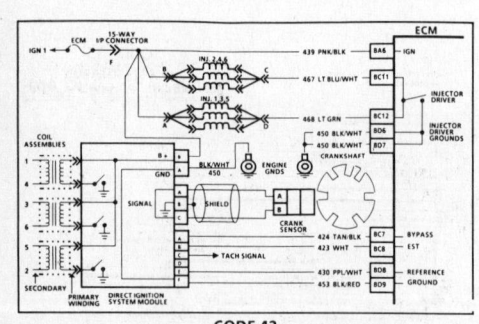

CODE 42
ELECTRONIC SPARK TIMING (EST) CIRCUIT
2.8L (VIN W) "A" CARLINE (PORT)

Circuit Description:

When the system is running on the ignition module, that is, no voltage on the bypass line, the ignition module grounds the EST signal. The ECM expects to see no voltage on the EST line during this condition. If it sees a voltage, it sets Code 42 and will not go into the EST mode.

When the rpm for EST is reached (about 400 rpm), and bypass voltage applied, the EST should no longer be grounded in the ignition module so the EST voltage should be varying.

If the bypass line is open or grounded, the ignition module will not switch to EST mode, so the EST voltage will be low and Code 42 will be set.

If the EST line is grounded, the ignition module will switch to EST, but because the line is grounded, there will be no EST signal. A Code 42 will be set.

Test Description: Numbers below refer to circled numbers on the diagnostic chart.

1. Code 42 means the ECM has seen an open or short to ground in the EST or bypass circuits. This test confirms Code 42 and that the fault causing the code is present.
2. Checks for a normal EST ground path through the ignition module. An EST CKT 423 shorted to ground will also read less than 500 ohms, however, this will be checked later.
3. As the test light voltage touches CKT 424, the module should switch causing the ohmmeter to "overrange" if the meter is in the 1000-2000 ohms position. Selecting the 10-20,000 ohms position will indicate above 5000 ohms. The important thing is that the module "switched."
4. The module did not switch and this step checks for

• EST CKT 423 shorted to ground
• Bypass CKT 424 open
• Faulty ignition module connection or module
5. Confirms that Code 42 is a faulty ECM and not an intermittent in CKTs 423 or 424.

Diagnostic Aids:

The "Scan" tool does not have any ability to help diagnose a Code 42 problem.

A Mem-Cal not fully seated in the ECM can result in a Code 42. If Code 42 is intermittent, there is a possibility of an open EST line. To check for an open EST line, crank engine while in a "clear flood mode" for five seconds, then start engine and check for Code 42. If Code 42 is set, repair open in CKT 423 (EST line)

Refer to "Intermittents" in Section "H"

1988-90 2.8L (VIN W) – CELEBRITY AND 6000

CODE 42
ELECTRONIC SPARK TIMING (EST) CIRCUIT
2.8L (VIN W) "A" CARLINE (PORT)

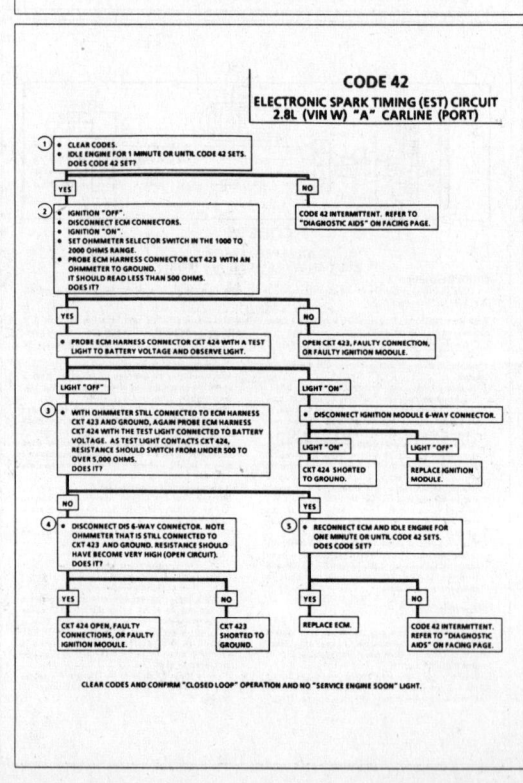

CLEAR CODES AND CONFIRM "CLOSED LOOP" OPERATION AND NO "SERVICE ENGINE SOON" LIGHT.

1988–90 2.8L (VIN W) – CELEBRITY AND 6000

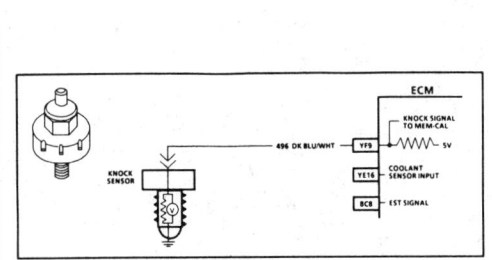

CODE 43
ELECTRONIC SPARK CONTROL (ESC) CIRCUIT
2.8L (VIN W) "A" CARLINE (PORT)

Circuit Description:
The knock sensor is used to detect engine detonation and the ECM will retard the electronic spark timing based on the signal being received. The circuitry within the knock sensor causes the ECM 5 volts to be pulled down so that, under a no knock condition, CKT 496 would measure about 2.5 volts. The knock sensor produces an A/C signal which rides on the 2.5 volts DC voltage. The amplitude and signal frequency is dependent upon the knock level.

If CKT 496 becomes open or shorted to ground, the voltage will either go above 3.5 volts or below 1.5 volts. If either of these conditions are met for about 1/2 second, a Code 43 will be stored.

Test Description: Numbers below refer to circled numbers on the diagnostic chart.
1. This step determines if conditions for Code 43 still exist (voltage on CKT 496 above 3.5 volts or below 1.5 volts). The system is designed to retard the spark 15° if either condition exists.
2. The ECM has a 5 volt pullup resistor, which applies 5 volts to CKT 496. The 5 volt signal should be present at the knock sensor terminal during these test conditions.
3. This step determines if the knock sensor resistance is 3300 to 4500 ohms the sensor is OK.

4. If CKT 496 is not open or shorted to ground and the voltage reading is below 4 volts, the most likely cause is an open circuit in the ECM. It is possible that a faulty Mem-Cal could be drawing the 5 volt signal down, and it should be replaced, if a replacement ECM did not correct the problem.

Diagnostic Aids:
Check CKT 496 for a potential open or short to ground. Also, check for proper installation of Mem-Cal.

Refer to "Intermittents" in Section "B".

1988–90 2.8L (VIN W) – CELEBRITY AND 6000

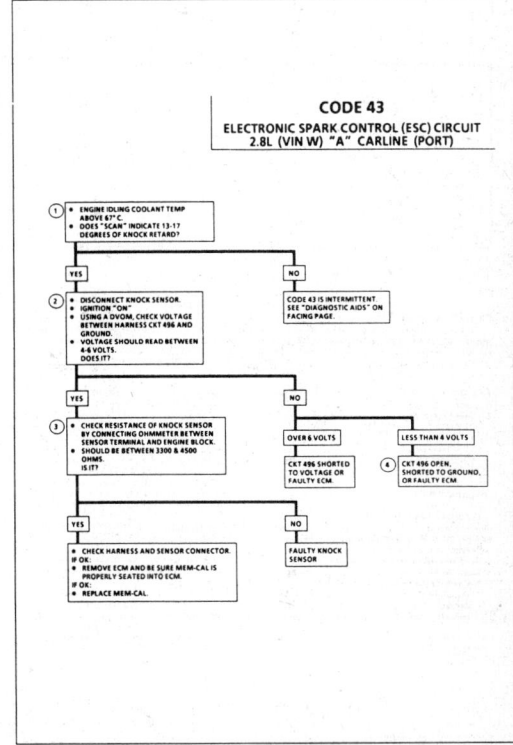

1988–90 2.8L (VIN W) – CELEBRITY AND 6000

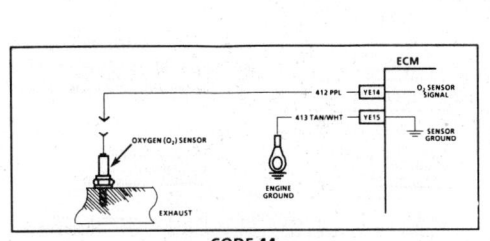

CODE 44
OXYGEN SENSOR CIRCUIT
(LEAN EXHAUST INDICATED)
2.8L (VIN W) "A" CARLINE (PORT)

Circuit Description:
The ECM supplies a voltage of about .45 volt between terminals "E14" and "E15". (If measured with a 10 megohm digital voltmeter, this may read as low as .32 volt). The O₂ sensor varies the voltage within a range of about 1 volt if the exhaust is rich, down through about .10 volt if the exhaust is lean.

The sensor is like an open circuit and produces no voltage when it is below about 315°C (600°F). An open sensor circuit or cold sensor causes "Open Loop" operation.

Test Description: Numbers below refer to circled numbers on the diagnostic chart.
1. Code 44 is set when the O₂ sensor signal voltage on CKT 412:
   - Remains below .2 volt for 60 seconds or more.
   - And the system is operating "Closed Loop."

Diagnostic Aids:
Using the "Scan," observe the block learn values at different rpm and air flow conditions. The "Scan" also displays the block cells, so the block learn values can be checked in each of the cells to determine when the Code 44 may have been set. If the conditions for Code 44 exist, the block learn values will be around 150.
- O₂ Sensor Wire. Sensor pigtail may be mispositioned and contacting the exhaust manifold.
- Check for intermittent ground in wire between connector and sensor.

- Lean Injector(s). Perform injector balance test CHART C-2A.
- Fuel Contamination. Water, even in small amounts, near the in-tank fuel pump inlet can be delivered to the injectors. The water causes a lean exhaust and can set a Code 44.
- Fuel Pressure. System will be lean if pressure is too low. It may be necessary to monitor fuel pressure while driving the car at various road speeds and/or loads to confirm. See "Fuel System Diagnosis" CHART A-7.
- Exhaust Leaks. If there is an exhaust leak, the engine can cause outside air to be pulled into the exhaust and past the sensor. Vacuum or crankcase leaks can cause a lean condition.
- Air System (manual trans "A" series only). Be sure air is not being directed to the exhaust ports while in "Closed Loop." If the block learn value goes down while squeezing air hose to exhaust ports, refer to CHART C-6.
- If the above are OK, it is a faulty oxygen sensor.

1988–90 2.8L (VIN W) – CELEBRITY AND 6000

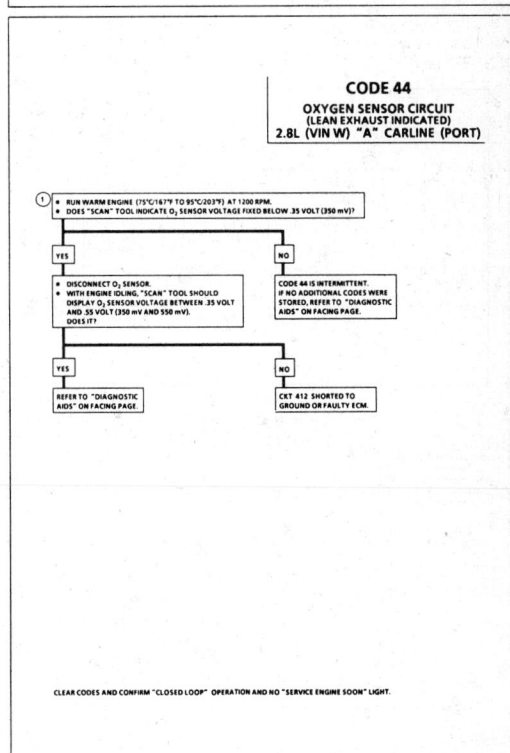

1988–90 2.8L (VIN W) – CELEBRITY AND 6000

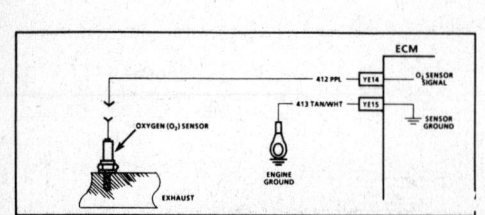

CODE 45
OXYGEN SENSOR CIRCUIT
(RICH EXHAUST INDICATED)
2.8L (VIN W) "A" CARLINE (PORT)

Circuit Description:

The ECM supplies a voltage of about .45 volt between terminals "E14" and "E15". (If measured with a 10 megohm digital voltmeter, this may read as low as .32 volt.) The O_2 sensor varies the voltage within a range of about 1 volt if the exhaust is rich, down through about .10 volt if exhaust is lean.

The sensor is like an open circuit and produces no voltage when it is below about 315°C (600°F). An open sensor circuit or cold sensor causes "Open Loop" operation.

Test Description: Numbers below refer to circled numbers on the diagnostic chart.
1. Code 45 is set when the O_2 sensor signal voltage or CKT 412.
   - Remains above .7 volt for 30 seconds, and in "Closed Loop."
   - Engine time after start is 1 minute or more.
   - Throttle angle between 3% and 45%.

Diagnostic Aids:

Using the "Scan," observe the block learn values at different rpm and air flow conditions. The "Scan" also displays the block cells, so the block learn values can be checked in each of the cells to determine when the Code 45 may have been set. If the conditions for Code 45 exist, the block learn values will be around 115.
- Check for short to voltage on CKT 412.
- Fuel Pressure. System will go rich if pressure is too high. The ECM can compensate for some increase. However, if it gets too high, a Code 45 may be set. See "Fuel System Diagnosis" CHART A-7.
- Rich Injector. Perform injector balance test Chart C-2A.

- Leaking Injector. See CHART A-7.
- Check for fuel contaminated oil.
- HEI Shielding. An open ground CKT 453 (ignition system reflow) may result in EMI, or induced electrical "noise." The ECM looks at this "noise" as reference pulses. The additional pulses result in a higher than actual engine speed signal. The ECM then delivers too much fuel, causing system to go rich. Engine tachometer will also show higher than actual engine speed, which can help in diagnosing this problem.
- Canister Purge. Check for fuel saturation. If full of fuel, check canister control and hoses.
- Check for leaking fuel pressure regulator diaphragm by checking vacuum line to regulator for fuel.
- TPS. An intermittent TPS output will cause the system to go rich, due to a false indication of the engine accelerating.
- EGR. An EGR staying open (especially at idle) will cause the O_2 sensor to indicate a rich exhaust, and this could result in a Code 45.

1988–90 2.8L (VIN W) – CELEBRITY AND 6000

CODE 45
OXYGEN SENSOR CIRCUIT
(RICH EXHAUST INDICATED)
2.8L (VIN W) "A" CARLINE (PORT)

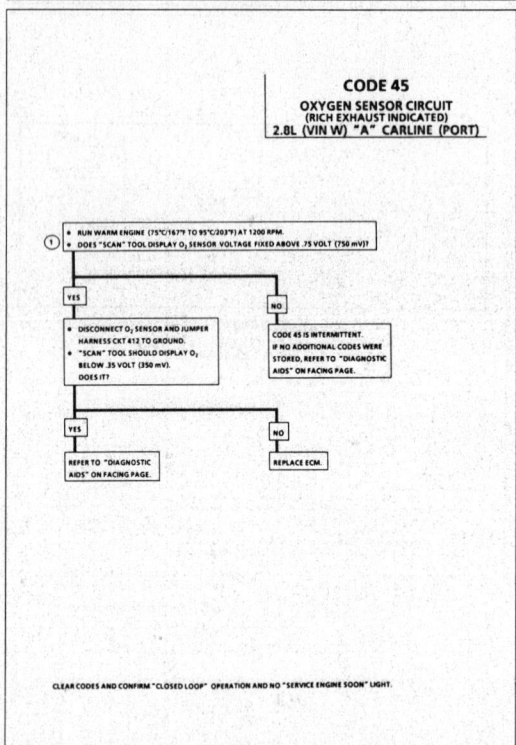

1988–90 2.8L (VIN W) – CELEBRITY AND 6000

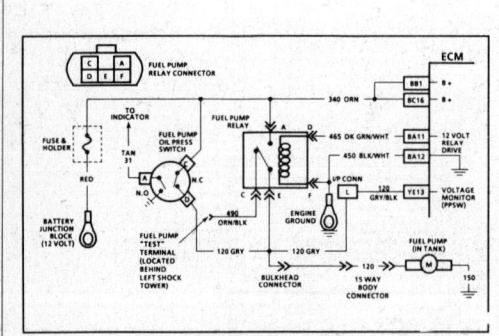

CODE 54
FUEL PUMP CIRCUIT
(LOW VOLTAGE)
2.8L (VIN W) "A" CARLINE (PORT)

Circuit Description:

The status of the fuel pump CKT 120 is monitored by the ECM at terminal "YE13" and is used to compensate fuel delivery based on system voltage. This signal is also used to store a trouble code if the fuel pump relay is defective or fuel pump voltage is lost while the engine is running. There should be about 12 volts on CKT 120 for 2 seconds after the ignition is turned or any time references pulses are being received by the ECM.

Code 54 will set if the voltage at terminal "YE13" is less than 2 volts for 1.5 seconds since the last reference pulse was received. This code is designed to detect a faulty relay, causing extended crank time, and the code will help the diagnosis of an engine that "Cranks But Will Not Run."

If a fault is detected during start-up, the "Service Engine Soon" light will stay "ON" until the ignition is cycled "OFF." However, if the voltage is detected below 2 volts with the engine running, the light will only remain "ON" while the condition exists.

1988–90 2.8L (VIN W) – CELEBRITY AND 6000

CODE 54
FUEL PUMP CIRCUIT
(LOW VOLTAGE)
2.8L (VIN W) "A" CARLINE (PORT)

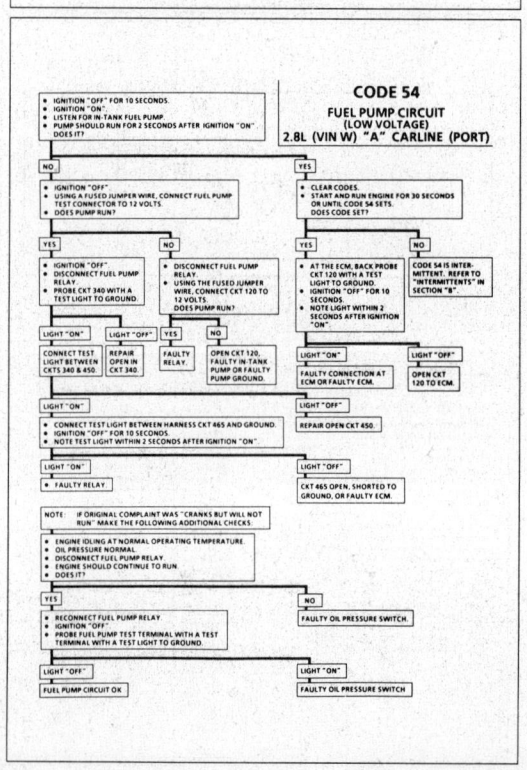

1988–90 2.8L (VIN W) – CELEBRITY AND 6000

> **CODE 51**
> **CODE 52**
> **CODE 53**
> 2.8L (VIN W) "A" CARLINE (PORT)
>
> #### CODE 51
> **MEM-CAL ERROR**
> **(FAULTY OR INCORRECT MEM-CAL)**
>
> CHECK THAT ALL PINS ARE FULLY INSERTED IN THE SOCKET AND THAT MEM-CAL IS PROPERLY LATCHED. IF OK, REPLACE MEM-CAL, CLEAR MEMORY, AND RECHECK. IF CODE 51 REAPPEARS, REPLACE ECM.
>
> CLEAR CODES AND CONFIRM "CLOSED LOOP" OPERATION AND NO "SERVICE ENGINE SOON" LIGHT.
>
> #### CODE 52
> **(FAULTY OR INCORRECT CALPAK)**
>
> CHECK THAT THE MEM-CAL IS FULLY SEATED AND LATCHED INTO THE MEM-CAL SOCKET. IF OK, REPLACE MEM-CAL, CLEAR MEMORY, AND RECHECK. IF CODE 52 REAPPEARS, REPLACE ECM.
>
> CLEAR CODES AND CONFIRM "CLOSED LOOP" OPERATION AND NO "SERVICE ENGINE SOON" LIGHT.
>
> #### CODE 53
> **SYSTEM OVER VOLTAGE**
>
> THIS CODE INDICATES THERE IS A BASIC GENERATOR PROBLEM.
> - CODE 53 WILL SET, IF VOLTAGE AT ECM IGNITION INPUT PIN IS GREATER THAN 17.1 VOLTS FOR 2 SECONDS.
> - CHECK AND REPAIR CHARGING SYSTEM. REFER TO SECTION "6D".
>
> CLEAR CODES AND CONFIRM "CLOSED LOOP" OPERATION AND NO "SERVICE ENGINE SOON" LIGHT.

1988–90 2.8L (VIN W) – CELEBRITY AND 6000

> **CODE 61**
> DEGRADED OXYGEN SENSOR
> 2.8L (VIN W) "A" CARLINE (PORT)
>
> IF A CODE 61 IS STORED IN MEMORY THE ECM HAS DETERMINED THE OXYGEN SENSOR IS CONTAMINATED OR DEGRADED, BECAUSE THE VOLTAGE CHANGE TIME IS SLOW OR SLUGGISH.
>
> THE ECM PERFORMS THE OXYGEN SENSOR RESPONSE TIME TEST WHEN:
>
> COOLANT TEMPERATURE IS GREATER THAN 85°C.
>
> MAT TEMPERATURE IS GREATER THAN 10°C.
>
> IN CLOSED LOOP.
>
> IN DECEL FUEL CUT-OFF MODE.
>
> IF A CODE 61 IS STORED THE OXYGEN SENSOR SHOULD BE REPLACED. A CONTAMINATED SENSOR CAN BE CAUSED BY FUEL ADDITIVES, SUCH AS SILICON, OR BY USE OF NON-GM APPROVED LUBRICANTS OR SEALANTS. SILICON CONTAMINATION IS USUALLY INDICATED BY A WHITE POWDERY SUBSTANCE ON THE SENSOR FINS.
>
> CLEAR CODES AND CONFIRM "CLOSED LOOP" OPERATION AND NO "SERVICE ENGINE SOON" LIGHT.

1988–90 2.8L (VIN W) – CELEBRITY AND 6000

SECTION B
SYMPTOMS

TABLE OF CONTENTS

Before Starting
Intermittents
Hard Start
Hesitation, Sag, Stumble
Surges and/or Chuggle
Lack of Power, Sluggish, or Spongy
Detonation/Spark Knock
Cuts Out, Misses
Backfire
Poor Fuel Economy
Dieseling, Run-On
Rough, Unstable, or Incorrect Idle, Stalling
Excessive Exhaust Emissions or Odors
CHART B-1 - Restricted Exhaust System Check

BEFORE STARTING

Before using this section you should have performed the DIAGNOSTIC CIRCUIT CHECK and found out that:
1. The ECM and "Service Engine Soon" light are operating.
2. There are no trouble codes.

Verify the customer complaint, and locate the correct SYMPTOM below. Check the items indicated under that symptom.

If the ENGINE CRANKS BUT WILL NOT RUN, see CHART A-3.

Several of the symptom procedures below call for a Careful Visual Check. This check should include:
- ECM grounds for being clean and tight.

- Vacuum hoses for splits, kinks, and proper connections, as shown on Emission Control Information label.
- Air leaks at throttle body mounting and intake manifold.
- Ignition wires for cracking, hardness, proper routing, and carbon tracking.
- Wiring for proper connections, pinches, and cuts.

The importance of this step cannot be stressed too strongly - it can lead to correcting a problem without further checks and can save valuable time.

1988–90 2.8L (VIN W) – CELEBRITY AND 6000

INTERMITTENTS

Problem may or may not turn "ON" the "Service Engine Soon" light or store a code.

DO NOT use the trouble code charts in Section "A" for intermittent problems. The fault must be present to locate the problem. If a fault is intermittent, use of trouble code charts may result in replacement of good parts.
- Most intermittent problems are caused by faulty electrical connections or wiring. Perform careful check as described at start of Section "B". Check for:
  - Poor mating of the connector halves, or terminals not fully seated in the connector body (backed out).
  - Improperly formed or damaged terminals. All connector terminals in problem circuit should be carefully reformed to increase contact tension.
  - Poor terminal to wire connection. This requires removing the terminal from the connector body to check.
- If a visual check does not find the cause of the problem, the car can be driven with a voltmeter connected to a suspected circuit. A "Scan" tool can, also, be used for monitoring input signals to help detect intermittent conditions. An abnormal voltage, or "Scan" reading, when the problem occurs, indicates the problem may be in that circuit. If the voltage and connectors check OK and a trouble code was stored for a circuit having a sensor, except for Codes 43, 44, and 45, substitute a known good sensor and recheck.

An intermittent "Service Engine Soon" light with no stored code may be caused by:
- Ignition coil shorted to ground and arcing at spark plug wires or plugs.
- "Service Engine Soon" light wire to ECM shorted to ground. (CKT 419)
- Diagnostic "Test" terminal wire to ECM, shorted to ground. (CKT 451)
- ECM power grounds. See ECM wiring diagrams.
- Loss of trouble code memory. To check, disconnect TPS and idle engine until "Service Engine Soon" light comes "ON" Code 22 should be stored, and kept in memory when ignition is turned "OFF". If not, the ECM is faulty.
- Check for an electrical system interference caused by a defective relay, ECM driven solenoid, or switch. They can cause a sharp electrical surge. Normally, the problem will occur when the faulty component is operated.
- Check for improper installation of electrical options, such as lights, 2-way radios, etc.
- EST wires should be kept away from spark plug wires, coils and generator. Wire from ECM to ignition system (CKT 453) should be a good connection.
- Check for open diode across A/C compressor clutch, and for other open diodes (see wiring diagrams).

HARD START

Definition: Engine cranks OK, but does not start for a long time. Does eventually run, or may start but immediately dies.

- Perform careful check as described at start of Section "B".
- Make sure driver is using correct starting procedure.
- CHECK:
  - TPS for sticking or binding or a high TPS voltage with the throttle closed (should read less than .85 volt).
  - High resistance in coolant sensor circuit or sensor itself. See Code 15 chart or with a "Scan" tool compare coolant temperature with ambient temperature on a cold engine.
  - Fuel pressure CHART A-7.
  - Water contaminated fuel.
  - EGR operation. Be sure valve seats properly and is not staying open. See CHART C-7.

- Ignition system - Check for:
  - Bare and shorted wires and proper output with spark tester J 26792 (ST-125) or equivalent.
  - IAC operation - See Code 35 chart.
- A faulty in-tank fuel pump check valve will allow the fuel in the lines to drain back to the tank after the engine is stopped. To check for this condition:
  Perform fuel system diagnosis, CHART A-7.
- Remove spark plugs. Check for wet plugs, cracks, wear, improper gap, burned electrodes, or heavy deposits. Repair or replace as necessary.

1988–90 2.8L (VIN W) – CELEBRITY AND 6000

HESITATION, SAG, STUMBLE

Definition: Momentary lack of response as the accelerator is pushed down. Can occur at all car speeds. Usually most severe when first trying to make the car move, as from a stop sign. May cause the engine to stall if severe enough.

- Perform careful visual check as described at start of Section "B".
- CHECK:
  - Fuel pressure. See CHART A-7. Also check for water contaminated fuel.
  - Spark plugs for being fouled or faulty wiring.
  - Mem-Cal number and Service Bulletins for latest Mem-Cal.
- TPS for binding or sticking. Voltage should increase at a steady rate as throttle is moved toward WOT.
- Generator output voltage. Repair, if less than 9 or more than 16 volts.
- Ignition system ground, CKT 453.
- Canister purge system for proper operation. See CHART C-3.
- EGR - See CHART C-7.
- MAP Sensor - See CHART C-1D, or Codes 33 & 34.
- Perform injector balance test, CHART C-2A.

SURGES AND/OR CHUGGLE

Definition: Engine power variation under steady throttle or cruise. Feels like the car speeds up and slows down with no change in the accelerator pedal.

- Be sure driver understands Transmission/Transaxle Converter Clutch and A/C compressor operation in Owner's Manual.
- Perform careful visual inspection as described at start of Section "B".
- To help determine if the condition is caused by a rich or lean system, the car should be driven at the speed of the complaint. Monitoring block learn at the complaint speed will help identify the cause of the problem. If the system is running lean (block learn greater than 138), refer to "Diagnostic Aids" on facing page of Code 44. If the system is running rich (block learn less than 118), refer to "Diagnostic Aids" on facing page of Code 45.
- CHECK:
  - Generator output voltage. Repair if less than 9 or more than 16 volts.
  - EGR - There should be no EGR at idle. See CHART C-7.
  - EGR filter for being plugged, see CHART C-7.
  - Vacuum lines for kinks or leaks.
  - In-line fuel filter. Replace if dirty or plugged.
  - Fuel pressure while condition exists. See CHART A-7.
- Remove spark plugs. Check for cracks, wear, improper gap, burned electrodes, or heavy deposits. Also check condition of spark plug wires and check for proper output voltage using spark tester J 26792 (ST-125) or equivalent.

1988–90 2.8L (VIN W) – CELEBRITY AND 6000

LACK OF POWER, SLUGGISH, OR SPONGY

Definition: Engine delivers less than expected power. Little or no increase in speed when accelerator pedal is pushed down part way.

- Perform careful visual check as described at start of Section "B".
- Compare customer's car to similar unit. Make sure the customer's car has an actual problem.
- Remove air filter and check for dirt, or for being plugged. Replace as necessary.
- CHECK:
  - Restricted fuel filter, contaminated fuel or improper fuel pressure. See CHART A-7.
  - ECM power grounds, see wiring diagrams.
  - EGR operation for being open or partly open all the time. See CHART C-7.
- Exhaust system for possible restriction: See CHART B-1.
  - Inspect exhaust system for damaged or collapsed pipes.
  - Inspect muffler for heat distress or possible internal failure.
- Generator output voltage. Repair if less than 9 or more than 16 volts.
- Engine valve timing and compression.
- Engine for proper or worn camshaft.
- Secondary voltage using a shop oscilloscope or a spark tester J 26792 (ST-125) or equivalent.
- Check A/C operation. A/C clutch should cut out at WOT. See A/C CHART C-10.

DETONATION /SPARK KNOCK

Definition: A mild to severe ping, usually worse under acceleration. The engine makes sharp metallic knocks that change with throttle opening. Sounds like popcorn popping.

- Check for obvious overheating problems:
  - Low coolant.
  - Loose belt.
  - Restricted air flow to radiator, or restricted water flow through radiator.
  - Inoperative electric cooling fan circuit. See CHART C-12.
- To help determine if the conditions is caused by a rich or lean system, the car should be driven at the speed of the complaint. Monitoring block learn, at the complaint speed, will help identify the cause of the complaint. If the system is running lean (block learn greater than 138), refer to "Diagnostic Aids" on facing page of Code 44. If the system is running rich (block learn less than 118), refer to facing page of Code 45.
- CHECK:
  - EGR system for not opening or plugged EGR passages. See CHART C-7.
  - ESC system for no retard. See CHART C-5.
  - Park/Neutral switch. Be sure "Scan" indicates drive with gear selector in drive or overdrive. See CHART C-1A.
  - TCC operation, TCC applying too soon. See CHART C-8.
  - Fuel system pressure - See CHART A-7.
  - Engine for excessive carbon build-up. Remove carbon with top engine cleaner. Follow instructions on can.
  - Mem-Cal for correct part.
  - Combustion chamber for excessive oil.
  - Engine for incorrect basic parts such as cam, heads, pistons, etc.
  - Fuel for poor quality, proper octane rating.

1988–90 2.8L (VIN W) – CELEBRITY AND 6000

CUTS OUT, MISSES

Definition: Steady pulsation or jerking that follows engine speed, usually more pronounced as engine load increases. The exhaust has a steady spitting sound at idle or under a load.

- Perform careful visual (physical) check as described at start of Section "B".
- If ignition system is suspected of causing a miss at idle or cutting out under load, refer to appropriate ignition "Misfire" chart.
- If the previous checks did not find the problem:
  - Visually inspect ignition system for moisture, dust, cracks, burns, etc. Spray plug wires with fine water mist to check for shorts.
- If above checks did not correct problem, check the following:
  - To determine if one injector or connector is faulty, disconnect injector 4-way connector. On the injector side of the harness, connect ohmmeter between the B+ circuit and the injector drive side of the harness for each bank of injectors. Because each bank of injectors is wired in parallel, an ohmmeter should measure about 4 ohms. If the reading is 6 ohms, or greater, an open circuit is likely the cause. If 2 ohms, or less, a shorted circuit or injector is likely. Reconnect injector 4-way connector.
- If either ohm reading is out of range:
  - Disconnect all injector harness connectors. Connect J 34730-2 Injector test light, or equivalent 6 volt test light, between the harness terminals of each injector connector and note light while cranking. If test light fails to blink at any connector, it is a faulty injector drive circuit harness, connector, or terminal.
- CHECK:
  - Injector balance. See CHART C-2A.
  - Fuel System - Plugged fuel filter, water, low pressure. See CHART A-7.
  - Valve Timing.
- Remove rocker covers. Check for bent pushrods, worn rocker arms, broken valve springs, worn camshaft lobes. Repair as necessary.
- Perform compression check.

BACKFIRE

Definition: Fuel ignites in intake manifold, or in exhaust system, making a loud popping noise.

- CHECK:
  - Compression - Look for sticking or leaking valves.
  - EGR operation for being open all the time. See CHART C-7.
  - EGR gasket for faulty or loose fit.
  - Output voltage of ignition coils, using a shop oscilloscope or spark tester J 26792 (ST-125), or equivalent.
  - Valve timing.
  - Spark plugs, spark plug wires, and proper routing of plug wires.

1988–90 2.8L (VIN W) – CELEBRITY AND 6000

POOR FUEL ECONOMY

Definition: Fuel economy, as measured by an actual road test, is noticeably lower than expected. Also, economy is noticeably lower than it was on this car at one time, as previously shown by an actual road test.

- Check owner's driving habits.
  - Is A/C "ON" full time (defroster mode "ON")?
  - Are tires at correct pressure?
  - Are excessively heavy loads being carried?
  - Is acceleration too much, too often?
  - Suggest owner fill fuel tank and recheck fuel economy.
  - Suggest driver read "Important Facts on Fuel Economy" in Owner's Manual.
- Check for proper calibration of speedometer.
- Visually (physically) check:
  - Vacuum hoses for splits, kinks, and proper connections as shown on Vehicle Emission Control Information label.
  - Ignition wires for cracking, hardness, and proper connections.
  - Air cleaner element (filter) for dirt or being plugged.
- Remove spark plugs. Check for cracks, wear, improper gap, burned electrodes, or heavy deposits. Repair or replace as necessary.
- CHECK:
  - Engine thermostat for faulty part (always open) or for wrong heat range. Using a "Scan" tool, monitor engine temperature. A "Scan" displays engine temperature in degrees centigrade. After engine is started, the temperature should rise steadily to about 90°C, then stabilize, when thermostat opens.
  - Fuel Pressure. See CHART A-7.
  - Compression.
  - TCC for proper operation. See CHART C-8.
  - Exhaust system restriction. See CHART B-1. Using a "Scan" tool, crank engine for five seconds while in clear flood mode. Start engine and check for a Code 42. An open EST line will cause poor fuel economy. Refer to Code 42 diagnostic chart for repair procedure.

DIESELING, RUN-ON

Definition: Engine continues to run after key is turned "OFF," but runs very roughly.

- Check injectors for leaking. See CHART A-7.
- If engine runs smoothly, check ignition switch and adjustment.

1988–90 2.8L (VIN W) – CELEBRITY AND 6000

ROUGH, UNSTABLE, OR INCORRECT IDLE, STALLING

Definition: The engine runs unevenly at idle. If bad enough, the car may shake. Also, the idle may vary in rpm (called "hunting"). Either condition may be bad enough to cause stalling. Engine idles at incorrect speed. Engine will run rough or stall if loss of battery power to the ECM (Electronic Control Module) has occurred. This ECM has an idle learn feature. (See Section "C2" for Idle Learn Procedure.)

- Perform careful visual check as described at start of Section "B".
- **CHECK:**
  - For damaged, mispositioned or grounded motor mounts.
  - Throttle linkage for sticking or binding.
  - TPS for sticking or binding, be sure output is stable at idle and adjustment specification is correct.
  - IAC system. See Code 35 chart.
  - Generator output voltage. Repair if less than 9 or more than 16 volts.
  - P/N switch circuit. See CHART C-1A, or use "Scan" tool, and be sure tool indicates vehicle is in drive with gear selector in drive (125C), or overdrive (440-T4).
  - Injector balance. See CHART C-2A.
  - PCV valve for proper operation by placing finger over inlet hole in valve end several times. Valve should snap back. If not, replace valve.

- Evaporative Emission Control System. CHART C-3.
- Power Steering Pressure switch input. The state of the switch should only change when wheels are turned up against the stops. See CHART C-1E.
- ECM ground circuits.
- EGR valve: There should be no EGR at idle.
- Monitoring block learn values may help identify the cause of the problem. If the system is running lean (block learn greater than 138) refer to "Diagnostic Aids" on facing page of Code 44. If the system is running rich (block learn values less than 118) refer to "Diagnostic Aids" on facing page of Code 45.
- Run a cylinder compression check.

- Check for fuel pressure regulator hose. If present, replace regulator assembly.
- Check ignition system; wires and plugs
- If problem exists with A/C "ON", check A/C system operation CHART C-10.

EXCESSIVE EXHAUST EMISSIONS OR ODORS

Definition: Vehicle fails an emission test. Vehicle has excessive "rotten egg" smell. Excessive odors do not necessarily indicate excessive emissions.

- Perform "Diagnostic Circuit Check."
- IF TEST SHOWS EXCESSIVE CO AND HC, (or also has excessive odors):
- Check items which cause car to run RICH.
- Make sure engine is at normal operating temperature.
- **CHECK:**
  - Fuel pressure. See CHART A-7.
  - Canister for fuel loading. See CHART C-3.
  - Injector balance. See CHART C-2A.
  - PCV valve for being plugged, stuck, or blocked PCV hose, or fuel in the crankcase.
  - Spark plugs, plug wires, and ignition components.
  - Check for lead contamination of catalytic converter (look for removal of fuel filler neck restrictor).
  - Check for properly installed fuel cap.

- If the system is running rich, (block learn less than 118), refer to "Diagnostic Aids" on facing page of Code 45.
- IF TEST SHOWS EXCESSIVE NOx:
- Check items which cause car to run LEAN, or to run too hot:
  - EGR valve for not opening. See CHART C-7.
  - Vacuum leaks.
  - Coolant system and coolant fan for proper operation. See CHART C-12.
  - Remove carbon with top engine cleaner. Follow instructions on can.
- If the system is running lean, (block learn greater than 138), refer to "Diagnostic Aids" on facing page of Code 44.

1988–90 2.8L (VIN W) – CELEBRITY AND 6000

CHART B-1
RESTRICTED EXHAUST SYSTEM CHECK
ALL ENGINES

Proper diagnosis for a restricted exhaust system is essential before any components are replaced. Either of the following procedures may be used for diagnosis, depending upon engine or tool used:

CHECK AT A.I.R. PIPE:
1. Remove the rubber hose at the exhaust manifold A.I.R. pipe check valve. Remove check valve
2. Connect a fuel pump pressure gauge to a hose and nipple from a Propane Enrichment Device (J 26911) (see illustration).
3. Insert the nipple into the exhaust manifold A.I.R. pipe.

OR CHECK AT O₂ SENSOR:
1. Carefully remove O₂ sensor
2. Install Borroughs exhaust backpressure tester (BT 8515 or BT 8603) or equivalent in place of O₂ sensor (see illustration).
3. After completing test described below, be sure to coat threads of O₂ sensor with anti-seize compound P/N 5613695 or equivalent prior to re-installation.

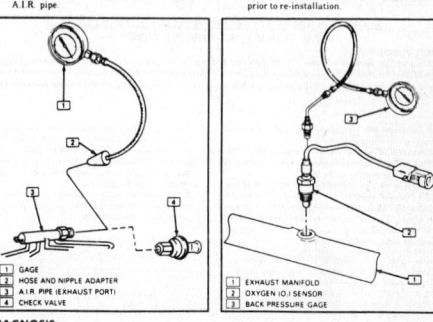

1	GAGE
2	HOSE AND NIPPLE ADAPTER
3	A.I.R. PIPE (EXHAUST PORT)
4	CHECK VALVE

1	EXHAUST MANIFOLD
2	OXYGEN (O₂) SENSOR
3	BACK PRESSURE GAGE

DIAGNOSIS:
1. With the engine idling at normal operating temperature, observe the exhaust system backpressure reading on the gage. Reading should not exceed 8.6 kPa (1.25 psi).
2. Increase engine speed to 2000 rpm and observe gage. Reading should not exceed 20.7 kPa (3 psi).
3. If the backpressure at either speed exceeds specification, a restricted exhaust system is indicated.
4. Inspect the entire exhaust system for a collapsed pipe, heat distress, or possible internal muffler failure.
5. If there are no obvious reasons for the excessive backpressure, the catalytic converter is suspected to be restricted and should be replaced using current recommended procedures.

1988–90 2.8L (VIN W) – CELEBRITY AND 6000

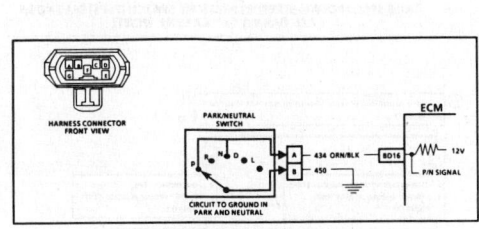

HARNESS CONNECTOR FRONT VIEW
PARK/NEUTRAL SWITCH
ECM
434 ORN/BLK
12v
P/N SIGNAL
CIRCUIT TO GROUND IN PARK AND NEUTRAL

CHART C-1A
PARK/NEUTRAL SWITCH DIAGNOSIS
2.8L (VIN W) "A" CARLINE (PORT)

Circuit Description:
The park/neutral switch contacts are a part of the neutral start switch and are closed to ground in park or neutral, and open in drive ranges.
The ECM supplies ignition voltage through a current limiting resistor to CKT 434 and senses a closed switch when the voltage on CKT 434 drops to less than one volt.
The ECM uses the P/N signal as one of the inputs to control:
- Idle air control
- VSS diagnostics
- EGR
If CKT 434 indicates P/N (grounded) while in drive range, the EGR would be inoperative, resulting in possible detonation.
If CKT 434 indicates drive (open) a dip in the idle may exist when the gear selector is moved into drive range.

Test Description: Numbers below refer to circled numbers on the diagnostic chart.
1. Checks for a closed switch to ground in park position. Different makes of "Scan" tools will read P/N differently. Refer to tool operator's manual for type of display used for a specific tool.
2. Checks for an open switch in drive range.
3. Be sure "Scan" indicates drive, even while wiggling shifter, to test for an intermittent or misadjusted switch in drive or overdrive range.

1988–90 2.8L (VIN W) – CELEBRITY AND 6000

CHART C-1A
PARK/NEUTRAL SWITCH DIAGNOSIS
2.8L (VIN W) "A" CARLINE (PORT)

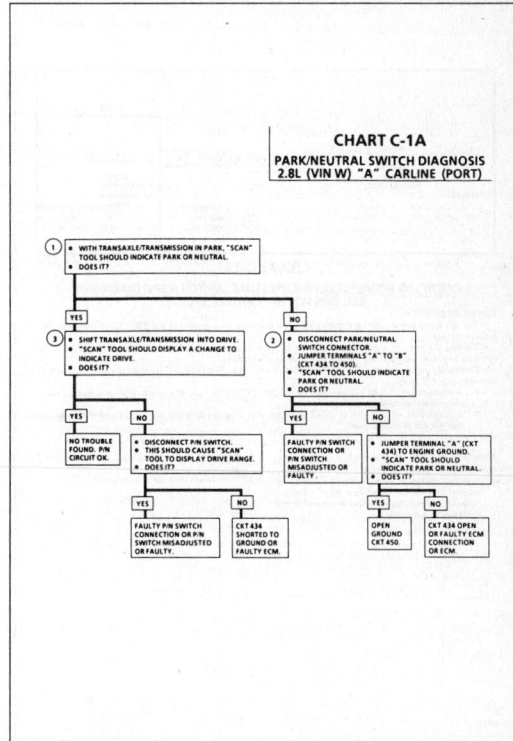

1988–90 2.8L (VIN W) – CELEBRITY AND 6000

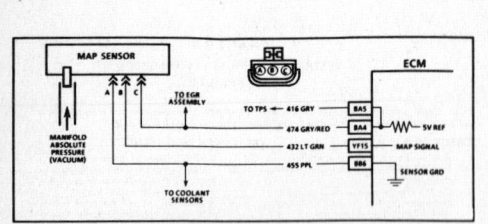

CHART C-1D
MANIFOLD ABSOLUTE PRESSURE (MAP) OUTPUT CHECK
2.8L (VIN W) "A" CARLINE (PORT)

Circuit Description:
The manifold absolute pressure (MAP) sensor measures manifold pressure (vacuum) and sends that signal to the ECM. The MAP sensor is mainly used for fuel calculation when the ECM is running in the throttle body backup mode. The MAP sensor is also used to determine the barometric pressure and to help calculate fuel delivery.

Test Description: Numbers below refer to circled numbers on the diagnostic chart.
1. Checks MAP sensor output voltage to the ECM. This voltage, without engine running, represents a barometer reading to the ECM.
2. Applying 34 kPa (10 inches Hg) vacuum to the MAP sensor should cause the voltage to be 1.2 volts less than the voltage at step 1. Upon applying vacuum to the sensor, the change in voltage should be instantaneous. A slow voltage change indicates a faulty sensor.

The engine must be running in this step or the "scanner" will not indicate a change in voltage. It is normal for the "Service Engine Soon" light to come "ON" and for the system to set a Code 63 during this step. Make sure the code is cleared when this test is completed.
3. Check vacuum hose to sensor for leaking or restriction. Be sure no other vacuum devices are connected to the MAP hose.

1988–90 2.8L (VIN W) – CELEBRITY AND 6000

CHART C-1D
MANIFOLD ABSOLUTE PRESSURE (MAP) OUTPUT CHECK
2.8L (VIN W) "A" CARLINE (PORT)

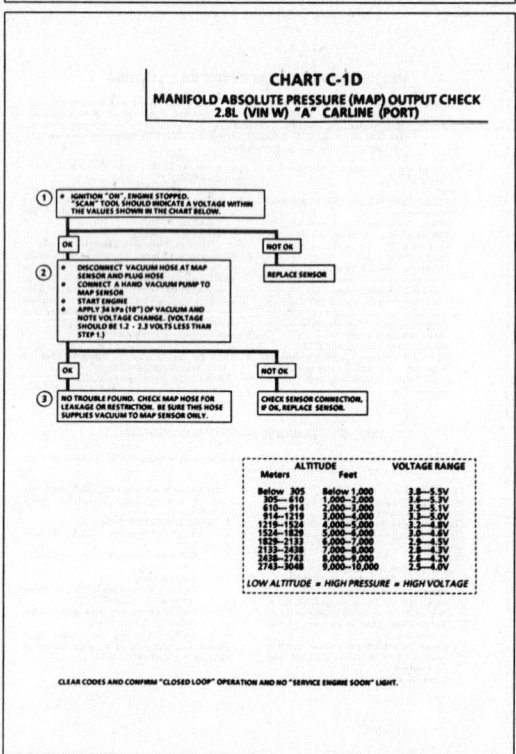

ALTITUDE		VOLTAGE RANGE
Meters	Feet	
Below 305	Below 1,000	3.8–5.5V
305–610	1,000–2,000	3.6–5.3V
610–914	2,000–3,000	3.5–5.1V
914–1219	3,000–4,000	3.3–5.0V
1219–1524	4,000–5,000	3.2–4.8V
1524–1829	5,000–6,000	3.0–4.6V
1829–2133	6,000–7,000	2.9–4.5V
2133–2438	7,000–8,000	2.8–4.3V
2438–2743	8,000–9,000	2.6–4.2V
2743–3048	9,000–10,000	2.5–4.0V

LOW ALTITUDE = HIGH PRESSURE = HIGH VOLTAGE

CLEAR CODES AND CONFIRM "CLOSED LOOP" OPERATION AND NO "SERVICE ENGINE SOON" LIGHT.

1988–90 2.8L (VIN W) – CELEBRITY AND 6000

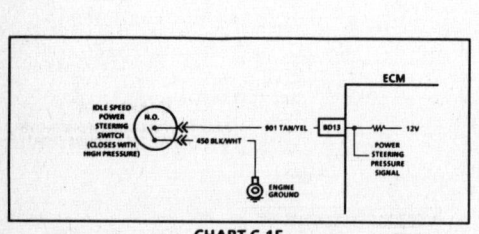

CHART C-1E
IDLE SPEED POWER STEERING PRESSURE SWITCH (PSPS) DIAGNOSIS
2.8L (VIN W) "A" CARLINE (PORT)

Circuit Description:
The idle speed power steering pressure switch is normally open to ground, and CKT 901 will be near the battery voltage.
Turning the steering wheel increases power steering oil pressure and its load on an idling engine. The pressure switch will close before the load can cause an idle problem.
Closing the switch causes CKT 901 to read less than 1 volt. The ECM will increase the idle air rate and disengage the A/C relay.
- A pressure switch that will not close, or an open CKT 901 or 450, may cause the engine to stop when power steering loads are high.
- A switch that will not open or a CKT 901 shorted to ground may affect idle quality and will cause the A/C relay to be de-energized.

Test Description: Numbers below refer to circled numbers on the diagnostic chart.
1. Different makes of "Scan" tools may display the state of this switch in different ways. Refer to "Scan" tool operator's manual to determine how this input is indicated.
2. Checks to determine if CKT 901 is shorted to ground.
3. This should simulate a closed switch.

1988–90 2.8L (VIN W) – CELEBRITY AND 6000

CHART C-1E
IDLE SPEED POWER STEERING PRESSURE SWITCH (PSPS) DIAGNOSIS
2.8L (VIN W) "A" CARLINE (PORT)

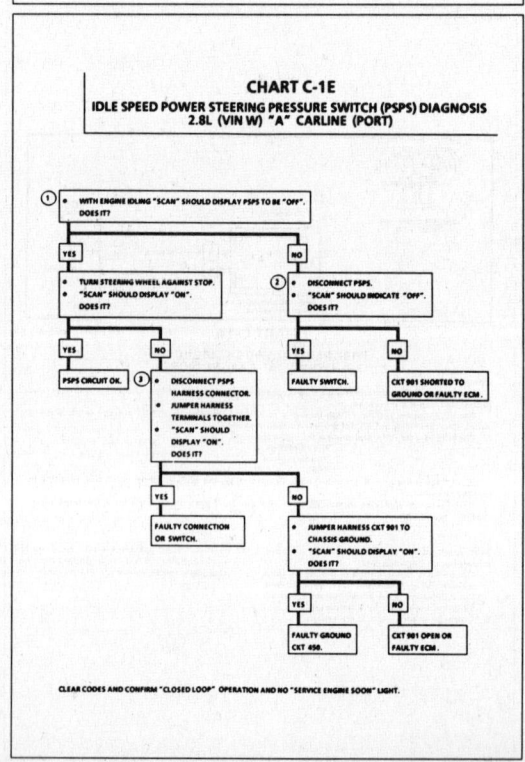

CLEAR CODES AND CONFIRM "CLOSED LOOP" OPERATION AND NO "SERVICE ENGINE SOON" LIGHT.

1988–90 2.8L (VIN W) – CELEBRITY AND 6000

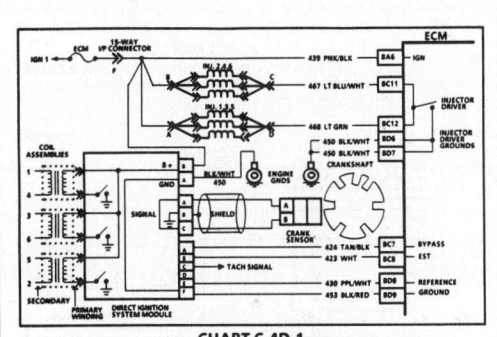

CHART C-3
CANISTER PURGE VALVE DIAGNOSIS
2.8L (VIN W) "A" CARLINE (PORT)

Circuit Description:
Canister purge is controlled by a solenoid that allows manifold vacuum to purge the canister when de-energized. The ECM supplies a ground to energize the solenoid (purge "OFF"). The purge solenoid controlled by the ECM is pulse width modulated (turned "ON" and "OFF" several times a second). The duty cycle (pulse width) is determined by the amount of air flow, and the engine vacuum as determined by the MAP sensor input. The duty cycle is calculated by the ECM and the output commanded when the following conditions have been met:

- Engine run time after start more than 3 minutes
- Coolant temperature above 80°C
- Vehicle speed above 15 mph
- Throttle "OFF" idle

Also, if the diagnostic test terminal is grounded, with the engine stopped, the purge solenoid is de-energized (purge "ON").

Test Description: Numbers below refer to circled numbers on the diagnostic chart.
1. Checks to see if the solenoid is opened or closed. The solenoid is normally energized in this step; so it should be closed.
2. Checks for a complete circuit. Normally there is ignition voltage on CKT 39 and the ECM provides a ground on CKT 428.
3. Completes functional check by grounding test terminal. This should normally de-energize the solenoid opening the valve which should allow the vacuum to drop (purge "ON").

1988–90 2.8L (VIN W) – CELEBRITY AND 6000

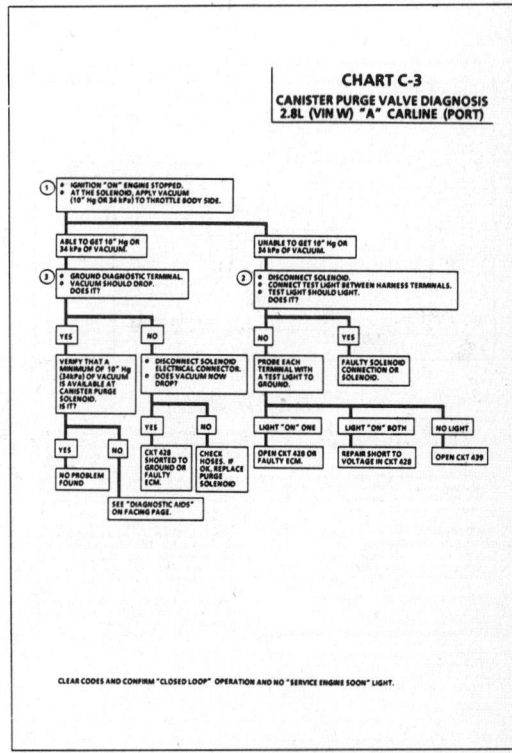

CHART C-3
CANISTER PURGE VALVE DIAGNOSIS
2.8L (VIN W) "A" CARLINE (PORT)

CLEAR CODES AND CONFIRM "CLOSED LOOP" OPERATION AND NO "SERVICE ENGINE SOON" LIGHT.

1988–90 2.8L (VIN W) – CELEBRITY AND 6000

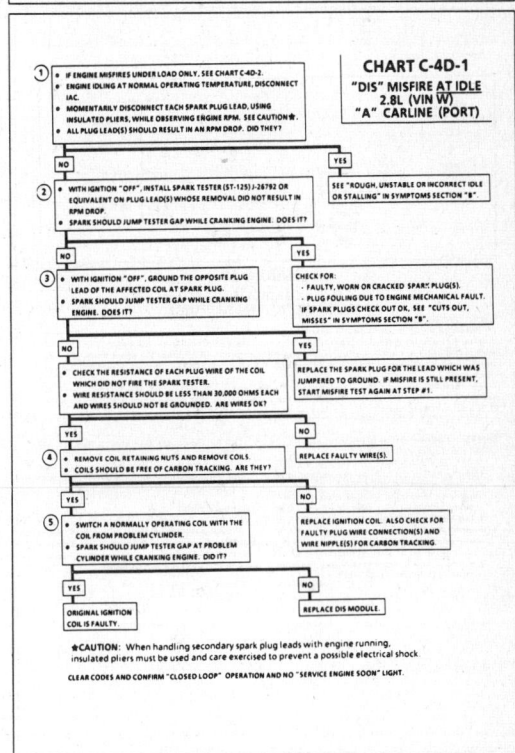

CHART C-4D-1
"DIS" MISFIRE AT IDLE
2.8L (VIN W) "A" CARLINE (PORT)

Circuit Description:
The direct ignition system (DIS) uses a waste spark method of distribution. In this type of system, the ignition module triggers the #1/4 coil pair resulting in both #1 and #4 spark plugs firing at the same time. #1 cylinder is on the compression stroke at the same time #4 is on the exhaust stroke, resulting in a lower energy requirement to fire #4 spark plug. This leaves the remainder of the high voltage to be used to fire #1 spark plug. On this application, the crank sensor is mounted to the engine block and protrudes through the block to within approximately 050" of the crankshaft reluctor. Since the reluctor is a machined portion of the crankshaft and the crank sensor is mounted in a fixed position on the block, timing adjustments are not possible or necessary.

Test Description: Numbers below refer to circled numbers on the diagnostic chart.
1. If the "misfire" complaint exists under load only, the diagnostic chart on page 2 must be used. Engine rpm should drop approximately equally on all plug leads.
2. A spark test such as a ST-125 must be used because it is essential to verify adequate available secondary voltage at the spark plug. (25,000 volts).
3. If the spark jumps the test gap after grounding the opposite plug wire, it indicates excessive resistance in the plug which was bypassed. A faulty or poor connection at that plug could also result in the miss condition. Also check for carbon deposits inside the spark plug boot.
4. If carbon tracking is evident, replace coil and be sure plug wires relating to that coil are clean and tight. Excessive wire resistance or faulty connections could have caused the coil to be damaged.
5. If the no spark condition follows the suspected coil, that coil is faulty, otherwise, the ignition module is the cause of no spark. This test could also be performed by substituting a known good coil for the one causing the no spark condition.

CHART C-4D-1
"DIS" MISFIRE AT IDLE
2.8L (VIN W) "A" CARLINE (PORT)

★CAUTION: When handling secondary spark plug leads with engine running, insulated pliers must be used and care exercised to prevent a possible electrical shock.

CLEAR CODES AND CONFIRM "CLOSED LOOP" OPERATION AND NO "SERVICE ENGINE SOON" LIGHT.

1988–90 2.8L (VIN W) – CELEBRITY AND 6000

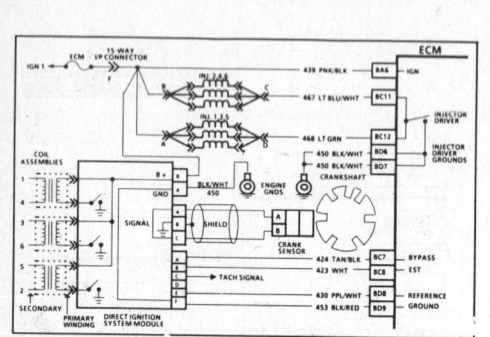

CHART C-4D-2
"DIS" MISFIRE UNDER LOAD
2.8L (VIN W) "A" CARLINE (PORT)

Circuit Description:
The direct ignition system (DIS) uses a waste spark method of distribution. In this type of system, the ignition module triggers the #1/4 coil pair resulting in both #1 and #4 spark plugs firing at the same time. #1 cylinder is on the compression stroke at the same time #4 is on the exhaust stroke, resulting in a lower energy requirement to fire #4 spark plug. This leaves the remainder of the high voltage to be used to fire #1 spark plug. On this application, the crank sensor is mounted on the engine block and protrudes through the block to within approximately .050" of the crankshaft reluctor. Since the reluctor is a machined portion of the crankshaft and the crank sensor is mounted in a fixed position on the block, timing adjustments are not possible or necessary.

Test Description: Numbers below refer to circled numbers on the diagnostic chart.
1. If the "misfire" complaint exists at idle only, the diagnostic chart on page 1 must be used. A spark tester such as a ST-125 must be used because it is essential to verify adequate available secondary voltage at the spark plug (25,000 volts). Spark should jump the test gap on all 4 leads. This simulates a "load" condition.
2. If the spark jumps the tester gap after grounding the opposite plug wire, it indicates excessive resistance in the plug which was bypassed. A faulty or poor connection at that plug could also result in the miss condition. Also check for carbon deposits inside the spark plug boot.
3. If carbon tracing is evident replace coil and be sure plug wires relating to that coil are clean and tight. Excessive wire resistance or faulty connections could have caused the coil to be damaged.
4. If the no spark condition follows the suspected coil, that coil is faulty. Otherwise, the ignition module is the cause of no spark. This test could also be performed by substituting a known good coil for the one causing the no spark condition.

1988–90 2.8L (VIN W) – CELEBRITY AND 6000

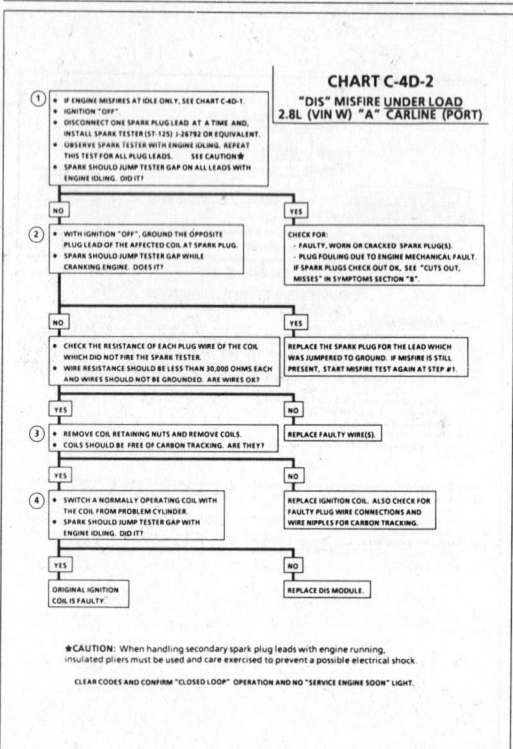

CHART C-4D-2
"DIS" MISFIRE UNDER LOAD
2.8L (VIN W) "A" CARLINE (PORT)

★CAUTION: When handling secondary spark plug leads with engine running, insulated pliers must be used and care exercised to prevent a possible electrical shock.

CLEAR CODES AND CONFIRM "CLOSED LOOP" OPERATION AND NO "SERVICE ENGINE SOON" LIGHT.

1988–90 2.8L (VIN W) – CELEBRITY AND 6000

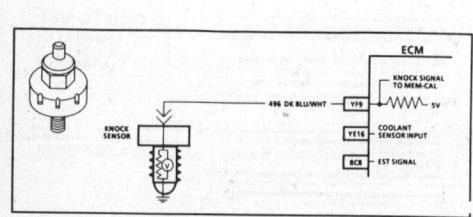

CHART C-5
ELECTRONIC SPARK CONTROL (ESC) SYSTEM CHECK
2.8L (VIN W) "A" CARLINE (PORT)

Circuit Description:
The knock sensor is used to detect engine detonation and the ECM will retard the electronic spark timing based on the signal being received. The circuitry within the knock sensor causes the ECM's 5 volts to be pulled down so that under a no knock condition, CKT 496 would measure about 2.5 volts. The knock sensor produces an A/C signal which rides on the 2.5 volt DC voltage. The amplitude and frequency are dependent upon the knock level.
The Mem-Cal used with this engine, contains the functions which were part of remotely mounted ESC modules used on other GM vehicles. The ESC portion of the Mem-Cal then sends a signal to other parts of the ECM which adjusts the spark timing to retard the spark and reduce the detonation.

Test Description: Numbers below refer to circled numbers on the diagnostic chart.
1. With engine idling, there should not be a knock signal present at the ECM because detonation is not likely under a no load condition.
2. Tapping on the engine lift hook bracket should simulate a knock signal to determine if the sensor is capable of detecting detonation. If no knock is detected, try tapping on engine block closer to sensor before replacing sensor.
3. If the engine has an internal problem which is creating a knock, the knock sensor may be responding to the internal failure.
4. This test determines if the knock sensor is faulty or if the ESC portion of the Mem-Cal is faulty. If it is determined that the Mem-Cal is faulty, be sure that is is properly installed and latched into place. If not properly installed, repair and retest.

Diagnostic Aids:
While observing knock signal on the "Scan," there should be an indication that knock is present when detonation can be heard. Detonation is most likely to occur under high engine load conditions.

1988–90 2.8L (VIN W) – CELEBRITY AND 6000

CHART C-5
ELECTRONIC SPARK CONTROL (ESC) SYSTEM CHECK
2.8L (VIN W) "A" CARLINE (PORT)

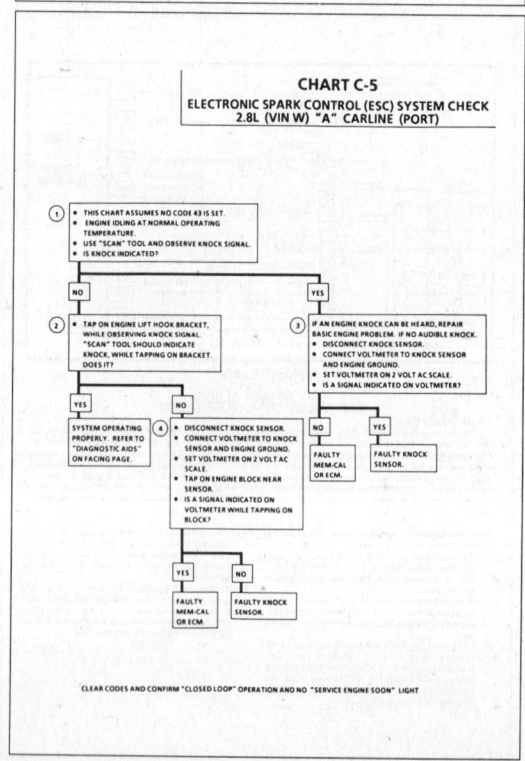

CLEAR CODES AND CONFIRM "CLOSED LOOP" OPERATION AND NO "SERVICE ENGINE SOON" LIGHT.

1988–90 2.8L (VIN W) – CELEBRITY AND 6000

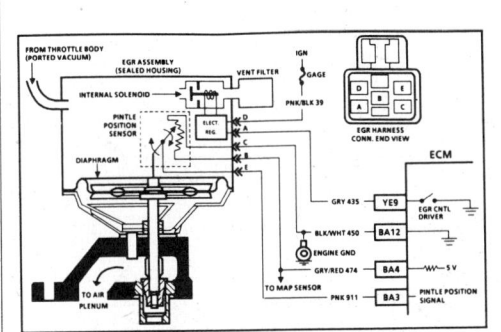

CHART C-7
EGR VALVE CHECK
2.8L (VIN W) "A" CARLINE (PORT)

Circuit Description:

The integrated electronic EGR valve functions similar to a port valve with a remote vacuum regulator. The internal solenoid is normally open, which causes the vacuum signal to be vented "OFF" to the atmosphere when EGR is not being commanded by the ECM. This EGR valve has a sealed cap and the solenoid valve opens and closes the vacuum signal, which controls the amount of vacuum vented to atmosphere, and this controls the amount of vacuum applied to the diaphragm. The electronic EGR valve contains a voltage regulator, which converts the ECM signal, to provide different amounts of EGR flow by regulating the current to the solenoid. The ECM controls EGR flow with a pulse width modulated signal (turns "ON" and "OFF" many times a second) based on airflow, TPS, and rpm.

This system also contains a pintle position sensor, which works similar to a TPS sensor, and as EGR flow is increased, the sensor output also increases.

Test Description: Numbers below refer to circled numbers on the diagnostic chart.

1. Whenever the solenoid is de-energized, the solenoid valve should be closed, which should not allow the vacuum to move the EGR diaphragm. However, if the filter is plugged, the vacuum applied with the hand held vacuum pump will cause the diaphragm to move because the vacuum will not be vented to the atmosphere.
2. This test will determine if the EGR filter is plugged, or if the EGR itself is faulty. Use care when removing the filter to avoid damaging the EGR assembly. See On-Car Service for procedure.
3. If the valve moves in this test, it's probably due to CKT 435 being shorted to ground.
4. Grounding the diagnostic terminal should energize the solenoid which closes off the vent and allows the vacuum to move the diaphragm.
5. The EGR assembly is designed to have some leakage and, therefore, 7" of vacuum is all that should be able to be held on the assembly. However, if too much of a leak exists (less than 3"), the EGR assembly is leaking and must be replaced.

Diagnostic Aids:

Some "Scan" tools will read EGR pintle position in volts. The EGR position voltage can be used to determine that the pintle is moving. When no EGR is commanded (0% duty cycle), the position sensor should read between .5 volt and 1.5 volts, and increase with the commanded EGR duty cycle.

1988–90 2.8L (VIN W) – CELEBRITY AND 6000

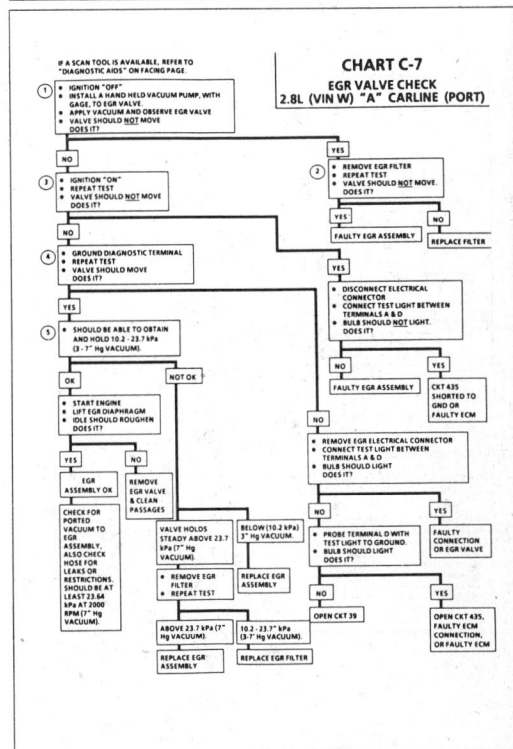

CHART C-7
EGR VALVE CHECK
2.8L (VIN W) "A" CARLINE (PORT)

1988–90 2.8L (VIN W) – CELEBRITY AND 6000

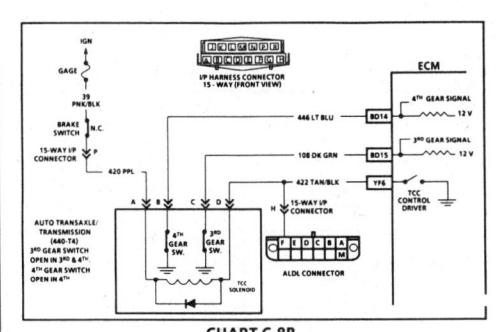

CHART C-8B
(Page 1 of 2)
440-T4 TORQUE CONVERTER CLUTCH (TCC)
(ELECTRICAL DIAGNOSIS)
2.8L (VIN W) "A" CARLINE (PORT)

Circuit Description:

The purpose of the automatic transmission torque converter clutch feature is to eliminate the power loss of the torque converter stage when the vehicle is in a cruise condition. This allows the convenience of an automatic transmission and the fuel economy of a manual transmission. The heart of the system is a solenoid located inside the automatic transmission which is controlled by the ECM.

When the solenoid coil is activated ("ON"), the torque converter clutch is applied which results in straight through mechanical coupling from the engine to transmission. When the transmission solenoid is deactivated, the torque converter clutch is released which allows the torque converter to operate in the conventional manner (fluidic coupling between engine and transmission).

The TCC will engage on a warm engine under given road load in 3rd and 4th gears.

TCC will engage when:
- Engine warmed up
- Vehicle speed above a calibrated value (about 28 mph 45 km/h)
- Throttle position sensor output not changing, indicating a steady road speed
- Brake switch closed

Test Description: Numbers below refer to circled numbers on the diagnostic chart.

1. This test checks the continuity of the TCC circuit from the fuse to the ALDL connector.
2. When the brake pedal is released, the light should come back "ON" and then go "OFF" when the diagnostic terminal is grounded. This tests CKT 422 and the TCC driver in the ECM.

Diagnostic Aids:

The "Scan" tool only indicates when the ECM has turned "ON" the TCC driver, and this does not confirm that the TCC has engaged. To determine if TCC is functioning properly, engine rpm should decrease when the "Scan" indicates the TCC driver has turned "ON."

1988–90 2.8L (VIN W) – CELEBRITY AND 6000

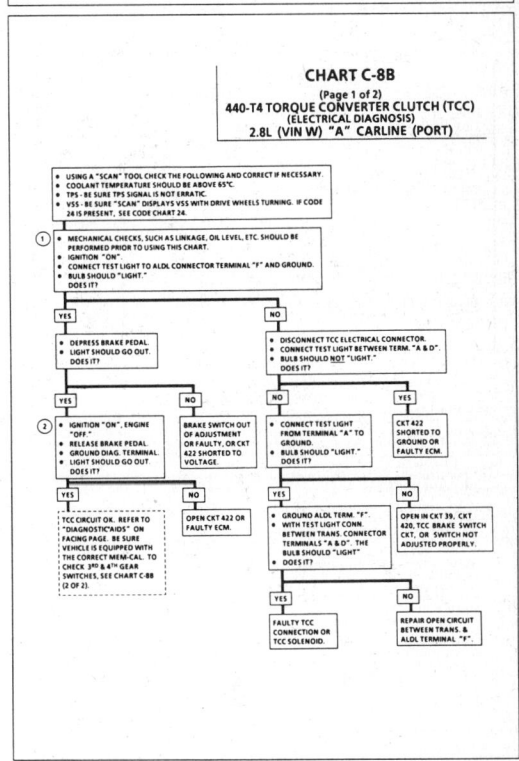

CHART C-8B
(Page 1 of 2)
440-T4 TORQUE CONVERTER CLUTCH (TCC)
(ELECTRICAL DIAGNOSIS)
2.8L (VIN W) "A" CARLINE (PORT)

1988–90 2.8L (VIN W) – CELEBRITY AND 6000

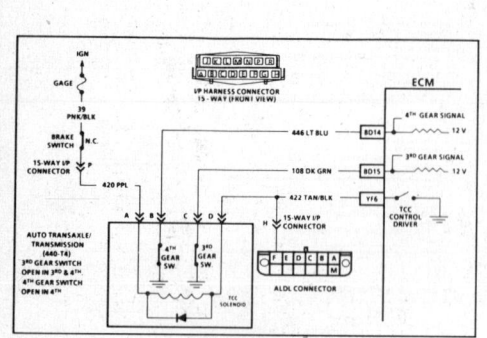

CHART C-8B
(Page 2 of 2)
440-T4 TORQUE CONVERTER CLUTCH (TCC)
(ELECTRICAL DIAGNOSIS)
2.8L (VIN W) "A" CARLINE (PORT)

Circuit Description:
The 3rd gear switch in this vehicle is open in 3rd and 4th gear. The ECM uses this signal to disengage the TCC when going into a downshift.
The fourth gear switch is open in fourth gear.

Test Description: Numbers below refer to circled numbers on the diagnostic chart.
1. Some "Scan" tools display the state of these switches in different ways. Be familiar with the type of tool being used. Since both switches should be in the closed state during this test, the tool should read the same for either the 3rd or 4th gear switch.
2. Determines whether the switch or signal circuit is open. The circuit can be checked for an open by measuring the voltage (with a voltmeter) at the TCC connector. Should be about 12 volts.

3. Because the switch(s) should be grounded in this step, disconnecting the TCC connector should cause the "Scan" switch state to change.
4. The switch state should change when the vehicle shifts into 3rd gear.

Diagnostic Aids:
If vehicle is road tested because of a TCC related problem, be sure the switch states do not change while in 4th gear because the TCC will disengage. If switches change state, carefully check wire routing and connections.

1988–90 2.8L (VIN W) – CELEBRITY AND 6000

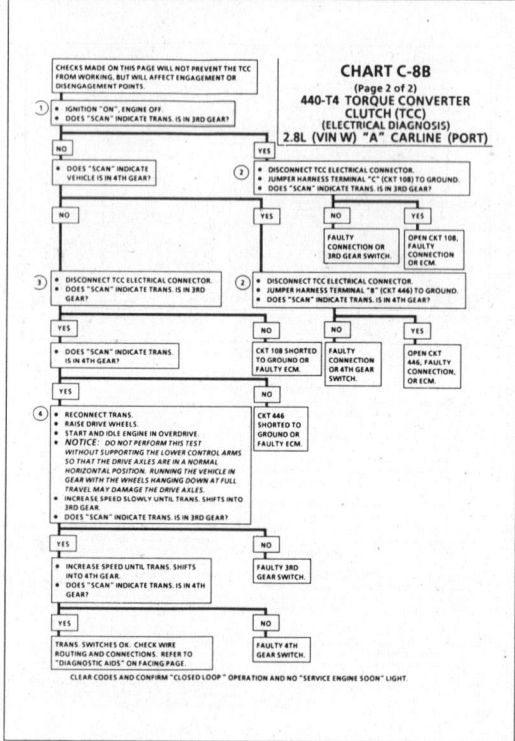

CHART C-8B
(Page 2 of 2)
440-T4 TORQUE CONVERTER CLUTCH (TCC)
(ELECTRICAL DIAGNOSIS)
2.8L (VIN W) "A" CARLINE (PORT)

1988–90 2.8L (VIN W) – CELEBRITY AND 6000

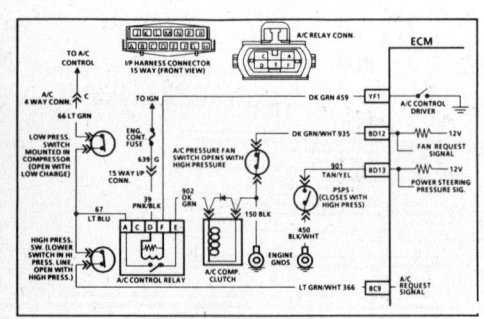

CHART C-10
A/C CLUTCH CONTROL CIRCUIT DIAGNOSIS
2.8L (VIN W) "A" CARLINE (PORT)

Circuit Description:
The A/C clutch control relay is ECM controlled to delay A/C clutch engagement about .4 second after A/C is turned "ON." This allows the IAC to adjust engine rpm, before the A/C clutch engages. The ECM, also, causes the relay to disengage the A/C clutch during WOT, when high power steering pressure is present, or if engine is overheating. The A/C clutch control relay is energized, when the ECM provides a ground path for CKT 459. The low pressure switch will open if A/C pressure is less than about 50 psi (276 kPa). The high pressure switch will open, if A/C pressure exceeds about 440 psi (3034 kPa). The A/C pressure fan switch opens when A/C pressure exceeds about 200 psi (1380 kPa).

Test Description: Numbers below refer to circled numbers on the diagnostic chart.
1. The ECM will only energize the A/C relay, when the engine is running. This test will determine if the relay or CKT 459 is faulty.
2. In order for the clutch to properly be engaged, the low pressure switch must be closed to provide 12 volts to the relay, and the high pressure switch must be closed, so the A/C request (12 volts) will be present at the ECM.
3. Determines if the signal is reaching the ECM on CKT 366 from the A/C control panel. Signal should only be present when the A/C mode or defrost mode has been selected.
4. A short to ground in any part of the A/C request circuit, CKT 67 to the relay, CKT 902 to the A/C clutch, or the A/C clutch, could be the cause of the blown fuse.

5. If the ECM is seeing a high power steering pressure signal, the A/C clutch will be disengaged by the ECM.
6. With the engine idling and A/C "ON," the ECM should be grounding CKT 459, which should cause the test light to be "ON."

Diagnostic Aids:
If complaint was insufficient cooling, the problem may be caused by a inoperative cooling fan, or A/C pressure fan switch. The cooling fan should turn "ON," when A/C pressure exceeds a value to open the switch, which causes the ECM to energize the cooling fan relay. If fan operates correctly, see "A/C Diagnosis" in Section "1." See CHART C-12 for cooling fan diagnosis.

1988–90 2.8L (VIN W) – CELEBRITY AND 6000

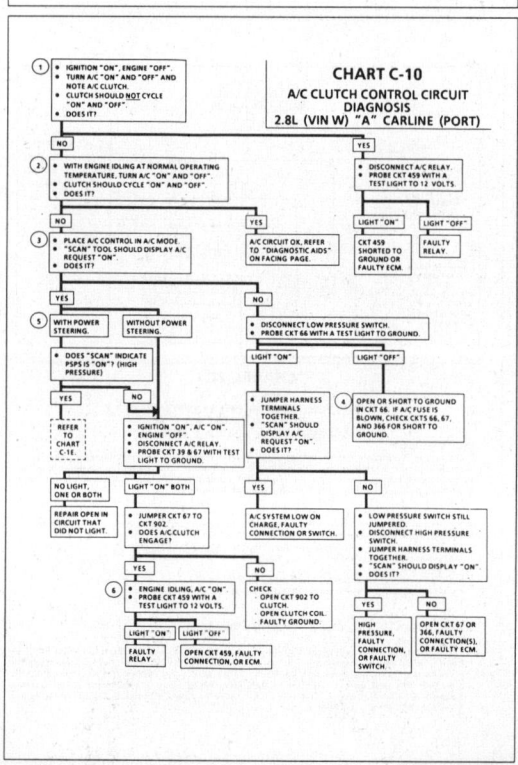

CHART C-10
A/C CLUTCH CONTROL CIRCUIT DIAGNOSIS
2.8L (VIN W) "A" CARLINE (PORT)

1988–90 2.8L (VIN W) – CELEBRITY AND 6000

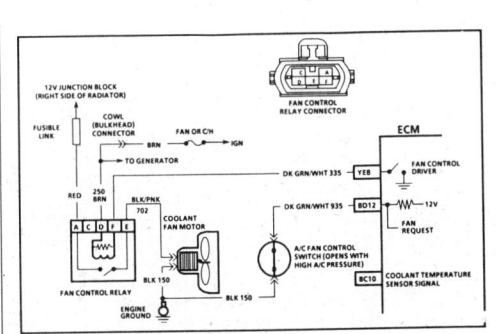

CHART C-12
(Page 1 of 2)
COOLANT FAN CONTROL CIRCUIT DIAGNOSIS
2.8L (VIN W) "A" CARLINE (PORT)

Circuit Description:
The electric cooling fan is controlled by the ECM based on inputs from the coolant temperature sensor, the A/C fan control switch, and vehicle speed. The ECM controls the fan by grounding CKT 335, which energizes the fan control relay. Battery voltage is then supplied to the fan motor.

The ECM grounds CKT 335, when coolant temperature is over about 106°C (223°F), or when A/C has been requested, and the fan control switch opens with high A/C pressure, about 200 psi (1380 kPa). Once the ECM turns the relay "ON," it will keep it "ON" for a minimum of 30 seconds, or until vehicle speed exceeds 70 mph.

Also, if Code 14 or 15 sets, or the ECM is in throttle body back up, the fan will run at all times.

Test Description: Numbers below refer to circled numbers on the diagnostic chart.
1. With the diagnostic terminal grounded, the cooling fan control driver will close, which should energize the fan control relay.
2. If the A/C fan control switch or circuit is open, the fan would run whenever A/C is requested.
3. With A/C clutch engaged, the A/C fan control switch should open, when A/C high pressure exceeds about 200 psi (1380 kPa). This signal should cause the ECM to energize the fan control relay.

Diagnostic Aids:
If the owner complained of an overheating problem, it must be determined if the complaint was due to an actual boil over, or the hot light, or temperature gage indicated over heating.

If the gage, or light, indicates overheating, but no boil over is detected, the gage circuit should be checked. The gage accuracy can also be checked by comparing the coolant sensor reading using a "Scan" tool and comparing its reading with the gage reading.

If the engine is actually overheating, and the gage indicates overheating, but the cooling fan is not coming "ON," the coolant sensor has probably shifted out of calibration and should be replaced.

If the engine is overheating, and the cooling fan is "ON," the cooling system should be checked.

1988–90 2.8L (VIN W) – CELEBRITY AND 6000

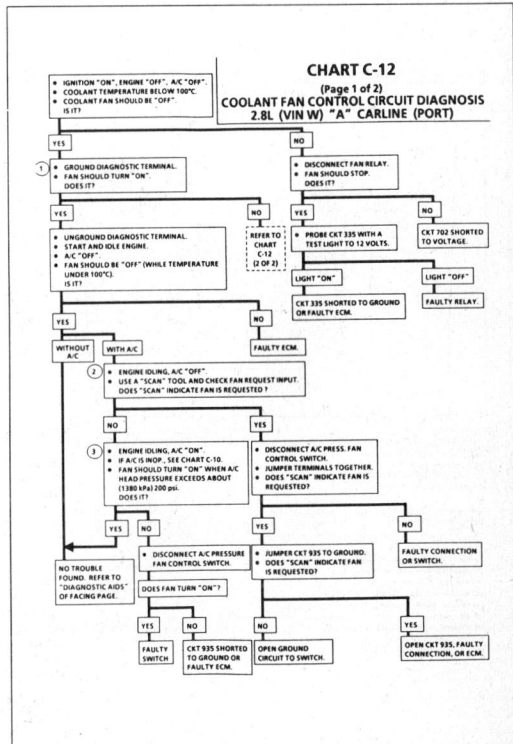

1988–90 2.8L (VIN W) – CELEBRITY AND 6000

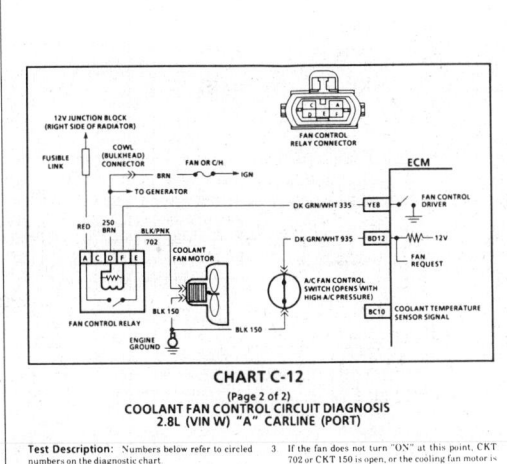

CHART C-12
(Page 2 of 2)
COOLANT FAN CONTROL CIRCUIT DIAGNOSIS
2.8L (VIN W) "A" CARLINE (PORT)

Test Description: Numbers below refer to circled numbers on the diagnostic chart.
1. 12 volts should be available to both terminals "E" & "C," when the ignition is "ON."
2. This test checks the ability of the ECM to ground CKT 335. The "Service Engine Soon" light should also be flashing at this point. If it isn't flashing, see CHART A-2.

3. If the fan does not turn "ON" at this point, CKT 702 or CKT 150 is open, or the cooling fan motor is faulty.

1988–90 2.8L (VIN W) – CELEBRITY AND 6000

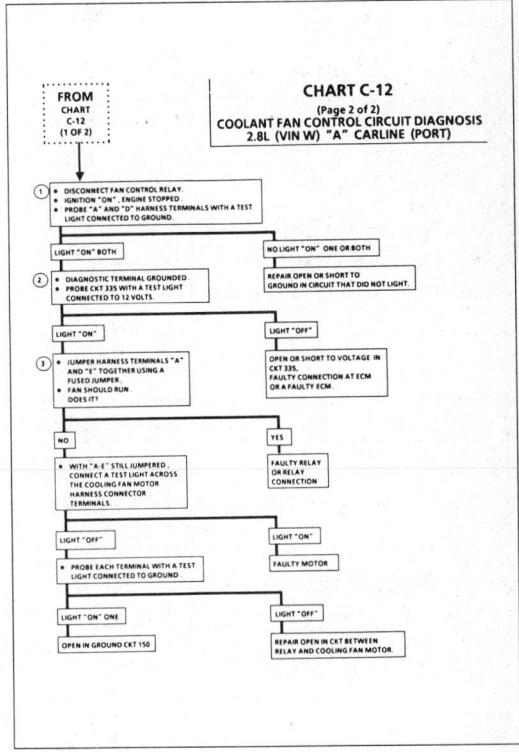

2.8L (VIN W) COMPONENT LOCATIONS CENTURY AND CUTLASS CIERA

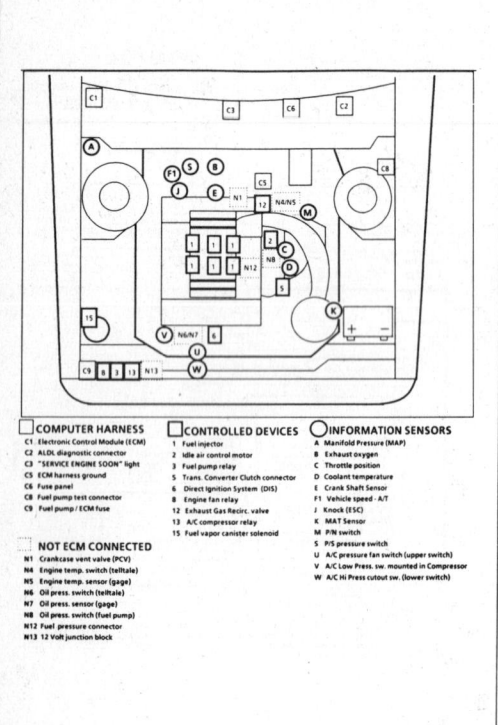

2.8L (VIN W) ECM WIRING DIAGRAM CENTURY AND CUTLASS CIERA

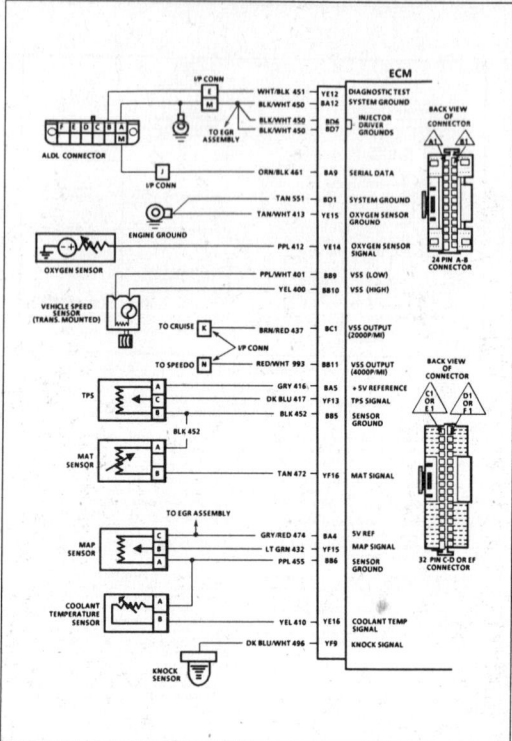

2.8L (VIN W) ECM WIRING DIAGRAM CENTURY AND CUTLASS CIERA (CONT.)

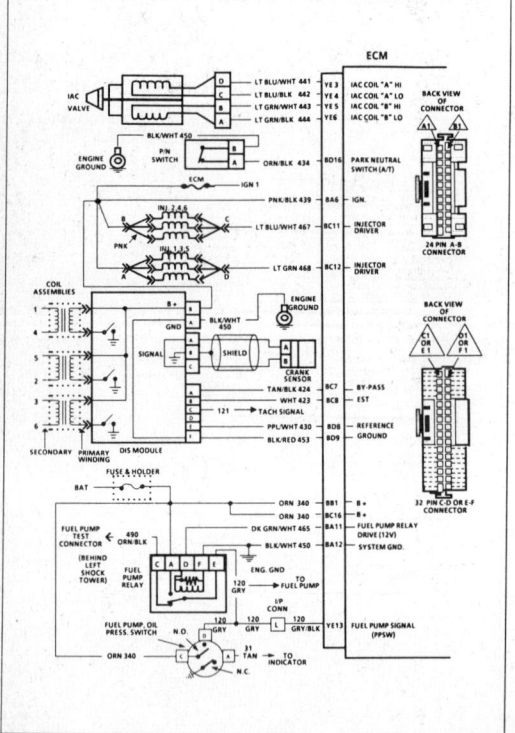

2.8L (VIN W) ECM WIRING DIAGRAM CENTURY AND CUTLASS CIERA (CONT.)

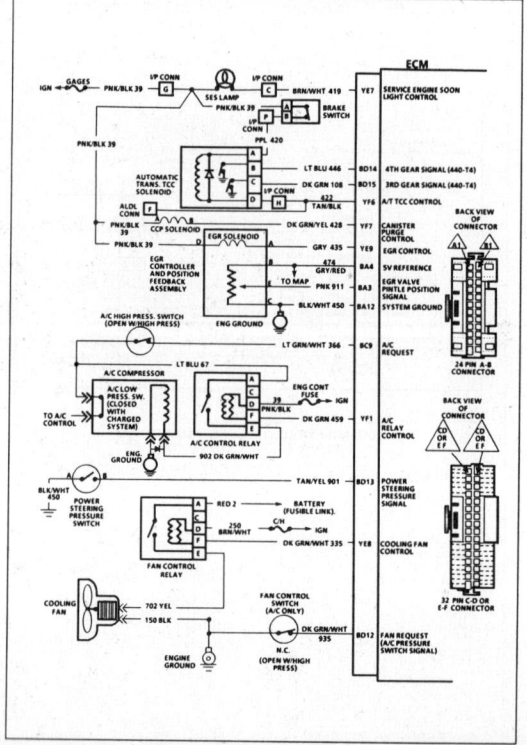

2.8L (VIN W) ECM CONNECTOR TERMINAL END VIEW—CENTURY AND CUTLASS CIERA

PORT FUEL INJECTION ECM CONNECTOR IDENTIFICATION

This ECM voltage chart is for use with a digital voltmeter to further aid in diagnosis. The voltages you get may vary due to low battery charge or other reasons, but they should be very close.

THE FOLLOWING CONDITIONS MUST BE MET BEFORE TESTING:

- Engine at operating temperature • Engine idling in closed loop (For "Engine Run" column) in park or neutral • Test terminal not grounded • ALDL tool not installed

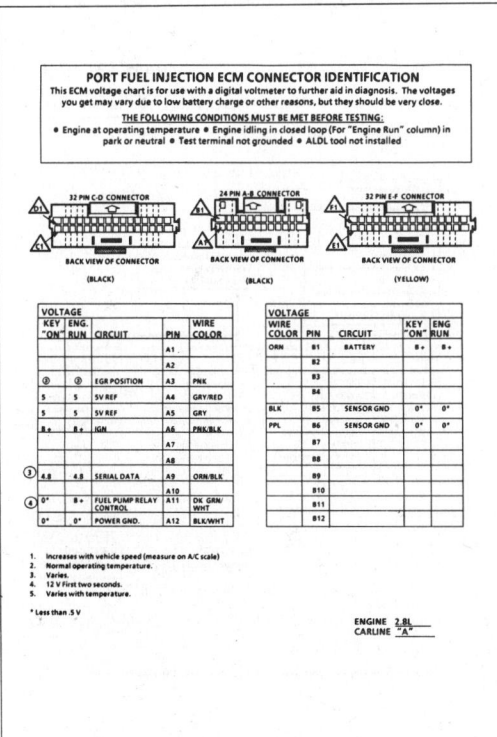

32 PIN C-D CONNECTOR — BACK VIEW OF CONNECTOR (BLACK)

24 PIN A-B CONNECTOR — BACK VIEW OF CONNECTOR (BLACK)

32 PIN E-F CONNECTOR — BACK VIEW OF CONNECTOR (YELLOW)

VOLTAGE

KEY "ON"	ENG. RUN	CIRCUIT	PIN	WIRE COLOR
			A1	
			A2	
②	②	EGR POSITION	A3	PNK
5	5	5V REF	A4	GRY/RED
5	5	5V REF	A5	GRY
B+	B+	IGN	A6	PNK/BLK
			A7	
			A8	
③ 4.8	4.8	SERIAL DATA	A9	ORN/BLK
			A10	
④ 0*	0*	FUEL PUMP RELAY CONTROL	A11	DK GRN/WHT
0*	0*	POWER GND.	A12	BLK/WHT

WIRE COLOR	PIN	CIRCUIT	KEY "ON"	ENG. RUN
ORN	B1	BATTERY	B+	B+
	B2			
	B3			
	B4			
BLK	B5	SENSOR GND	0*	0*
PPL	B6	SENSOR GND	0*	0*
	B7			
	B8			
	B9			
	B10			
	B11			
	B12			

1. Increases with vehicle speed (measure on A/C scale)
2. Normal operating temperature.
3. Varies.
4. 12 V first two seconds.
5. Varies with temperature.
* Less than .5 V

ENGINE 2.8L
CARLINE "A"

2.8L (VIN W) ECM CONNECTOR TERMINAL END VIEW—CENTURY AND CUTLASS CIERA (CONT.)

	VOLTAGE			
KEY "ON"	ENG. RUN	CIRCUIT	PIN	WIRE COLOR
① 0*		TO CRUISE 2000 Pmi	C1	BRN/RED
			C2	
			C3	
			C4	
			C5	
			C6	
0*	5	BYPASS	C7	TAN/BLK
0*	1.3	EST	C8	WHT
B+	B+	"ON" "OFF" A/C REQUEST	C9	LT GRN/WHT
			C10	
B+	B+	INJECTOR 1,3,5	C11	LT GRN
B+	B+	INJECTOR 2,4,6	C12	LT GRN
			C13	
			C14	
			C15	
B+	B+	BATTERY	C16	ORN

WIRE COLOR	PIN	CIRCUIT	KEY "ON"	ENG. RUN
TAN	D1	POWER GROUND	0*	0*
	D2			
	D3			
	D4			
	D5			
BLK/WHT	D6	INJ DRIVE LOW	0*	0*
BLK/WHT	D7	INJ DRIVE LOW	0*	0*
PPL/WHT	D8	REFERENCE	0*	2.3
BLK/RED	D9	REFERENCE LOW	0*	0*
	D10			
DK GRN/WHT	D11	FAN REQUEST	0*	0*
	D12	A/C PRESS SW. SIG.	0*	0*
TAN/YEL	D13	P S P S	B+	B+
LT BLU	D14	4th GEAR (440)	0*	0*
DK GRN	D15	3rd GEAR (440)	0*	0*
ORN/BLK	D16	P/N SWITCH	0*	0*

	VOLTAGE			
KEY "ON"	ENG. RUN	CIRCUIT	PIN	WIRE COLOR
			E1	
			E2	
NOT USEABLE		IAC "A" HI	E3	LT BLU/WHT
NOT USEABLE		IAC "A" LO	E4	LT BLU/BLK
NOT USEABLE		IAC "B" HI	E5	LT GRN/WHT
NOT USEABLE		IAC "B" LO	E6	LT GRN/BLK
0*		"SERVICE ENGINE SOON" LIGHT	E7	BRN/WHT
B+	B+	"ON" FAN RELAY	E7	DK GRN/WHT
B+	B+	"OFF" CONTROL	E8	DK GRN/WHT
B+	B+	EGR CONTROL	E9	GRY
			E10	
			E11	
5	5	DIAG. TERMINAL	E12	WHT/BLK
④ B+		FUEL PUMP SIGNAL	E13	GRY
③ .1-.9		O₂ SIGNAL	E14	PPL
35-55		O₂ GROUND	E15	TAN/WHT
⑤	⑤	COOLANT TEMP.	E16	YEL

WIRE COLOR	PIN	CIRCUIT	KEY "ON"	ENG. RUN
DK GRN	F1	A/C RELAY "ON" CONTROL "OFF"	0* B+	0* B+
	F2			
	F3			
	F4			
TAN/BLK	F5	TCC CONTROL A/T	0*	0*
DK GRN/YEL	F7	PURGE CONTROL	0*	0*
DK BLU/WHT	F9	ESC SIGNAL	2.5	2.5
	F10			
	F11			
	F12			
DK BLU	F13	TPS SIGNAL	.65	.65
	F14			
LT GRN	F15	MAP SIGNAL	4.57	⑤
TAN	F16	MAT SIGNAL	⑤	⑤

1. Increases with vehicle speed (measure on A/C scale).
2. Normal operating temperature.
3. Varies.
4. 12 V first two seconds.
5. Varies with temperature.
6. 12 V 440 trans.
* Less than .5 V

1988–90 2.8L (VIN W) CENTURY AND CUTLASS CIERA

DIAGNOSTIC CIRCUIT CHECK

The Diagnostic Circuit Check is an organized approach to identifying a problem created by an Electronic Engine Control System malfunction. **It must be the starting point for any driveability complaint diagnosis,** because it directs the Service Technician to the next logical step in diagnosing the complaint.

The "Scan Data" listed in the table may be used for comparison, after completing the Diagnostic Circuit Check and finding the on-board diagnostics functioning properly and no trouble codes displayed. The "Typical Values" are an average of display values recorded from normally operating vehicles and are intended to represent what a normally functioning system would typically display.

A "SCAN" TOOL THAT DISPLAYS FAULTY DATA SHOULD NOT BE USED, AND THE PROBLEM SHOULD BE REPORTED TO THE MANUFACTURER. THE USE OF A FAULTY "SCAN" CAN RESULT IN MISDIAGNOSIS AND UNNECESSARY PARTS REPLACEMENT.

Only the parameters listed below are used in this manual for diagnosing. If a "Scan" reads other parameters, the values are not recommended by General Motors for use in diagnosing. For more description on the values and use of the "Scan" to diagnosis ECM inputs, refer to the applicable diagnosis section "B."

"SCAN" DATA
Idle / Upper Radiator Hose Hot / Closed Throttle / Park or Neutral / Closed Loop / Acc. off

"SCAN" Position	Units Displayed	Typical Data Value
Desired RPM	RPM	ECM idle command (varies with temp.)
RPM	RPM	± 100 RPM from desired RPM (± 50 in drive)
Coolant Temp.	C°	85° - 105°
MAT Temp.	C°	10° - 80° (depends on underhood temp.)
MAP	Volts	1 - 2 (depends on Vac. & Baro pressure)
BARO	Volts	2.5 - 5.5 (depends on altitude & Baro pressure)
BPW (base pulse width)	M/Sec	1 - 4 and varying
O₂	Volts	1 - 1.0 V and varying
TPS	Volts	.65
Throttle Angle	0 - 100%	0
IAC	Counts (steps)	5 - 50
P/N Switch	P/N and RDL	Park/Neutral (P/N)
INT (Integrator)	Counts	Varies
BLM (Block Learn)	Counts	118 - 138
Open/Closed Loop	Open/Closed	Closed Loop (may go open with extended idle)
BLM Cell	Cell Number	0 or 1 (depends on Air Flow & RPM)
VSS	MPH	0
TCC	On/Off	Off/ (on with TCC commanded)
EGRDC	0 - 100%	0 at idle
EGR Position	Volts	3 - 2 Volts
Spark Advance	# of Degrees	Varies
Knock Retard	Degrees of Retard	0
Knock Signal	Yes/No	No
Battery	Volts	13.5 - 14.5
Fan	On/Off	Off (below 106°C)
P/S Switch	Normal/Hi Press.	Normal
3rd Gear (440-T4)	Yes/No	No (yes, when in 3rd or 4th gear)
4th Gear (440-T4)	Yes/No	No (yes, when in 4th gear)
A/C Request	Yes/No	No (yes, with A/C requested)
A/C Clutch	On/Off	Off (on, with A/C commanded)
Fan Request	Yes/No	No (yes, with A/C high pressure)
PPSW (Fuel Pump)	Volts	13.5 - 14.5

1988–90 2.8L (VIN W) CENTURY AND CUTLASS CIERA

DIAGNOSTIC CIRCUIT CHECK
2.8L (VIN W) "A" CARLINE (PORT)

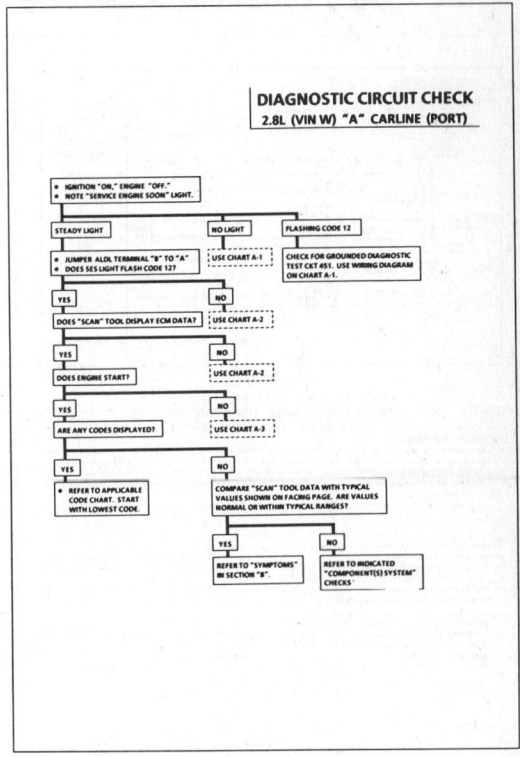

1988–90 2.8L (VIN W) CENTURY AND CUTLASS CIERA

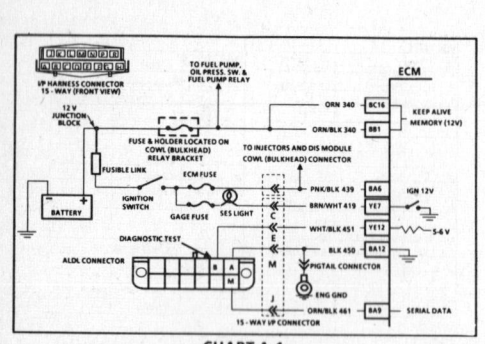

CHART A-1
NO "SERVICE ENGINE SOON" LIGHT
2.8L (VIN W) "A" CARLINE (PORT)

Circuit Description:
There should always be a steady "Service Engine Soon" light when the ignition is "ON" and engine stopped. Battery is supplied directly to the light bulb. The electronic control module (ECM) will control the light and turn it "ON" by providing a ground path through CKT 419 to the ECM.

Test Description: Numbers below refer to circled numbers on the diagnostic chart.
1. If the fuse in holder is blown, refer to facing page of Code 54 for complete circuit.
2. Using a test light connected to 12 volts probe each of the system ground circuits to be sure a good ground is present. See ECM terminal end view in front of this section for ECM pin locations of ground circuits.

Diagnostic Aids:
Engine runs OK, check.
- Faulty light bulb.
- CKT 419 open.
- Gage fuse blown. This will result in no stop lights, oil or generator lights, seat belt reminder, etc.
Engine cranks but will not run:
- Continuous battery - fuse or fusible link open.
- ECM ignition fuse open.
- Battery CKT 340 to ECM open.
- Ignition CKT 439 to ECM open.
- Poor connection to ECM.

1988–90 2.8L (VIN W) CENTURY AND CUTLASS CIERA

CHART A-1
NO "SERVICE ENGINE SOON" LIGHT
2.8L (VIN W) "A" CARLINE (PORT)

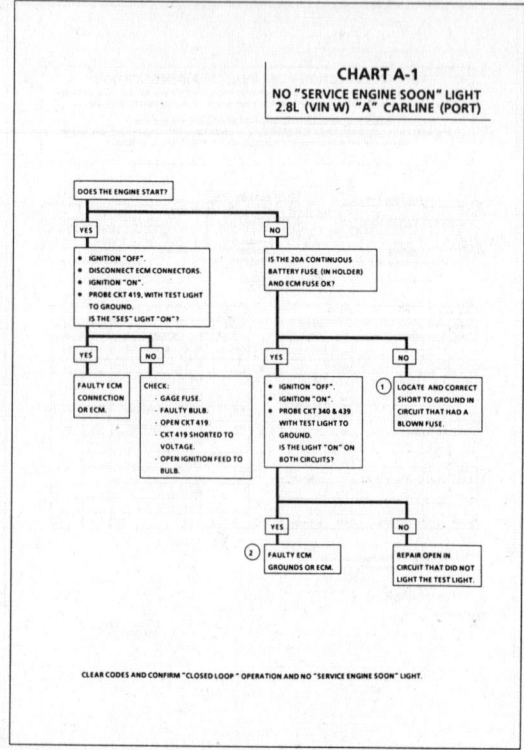

CLEAR CODES AND CONFIRM "CLOSED LOOP" OPERATION AND NO "SERVICE ENGINE SOON" LIGHT.

1988–90 2.8L (VIN W) CENTURY AND CUTLASS CIERA

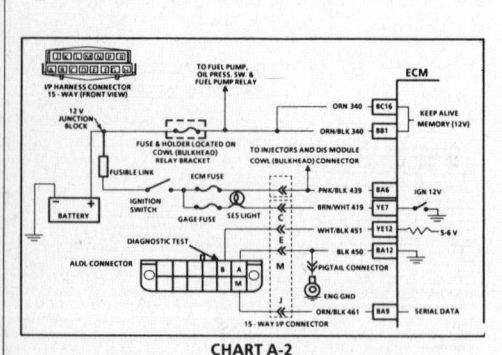

CHART A-2
NO ALDL DATA OR WON'T FLASH CODE 12
"SERVICE ENGINE SOON" LIGHT "ON" STEADY
2.8L (VIN W) "A" CARLINE (PORT)

Circuit Description:
There should always be a steady "Service Engine Soon" light when the ignition is "ON" and engine stopped. Battery ignition voltage is supplied to the light bulb. The electronic control module (ECM) will turn the light "ON" by grounding CKT 419 at the ECM.
With the diagnostic terminal grounded, the light should flash a Code 12, followed by any trouble code(s) stored in memory.
A steady light suggests a short to ground in the light control CKT 419, or an open in diagnostic CKT 451.

Test Description: Numbers below refer to circled numbers on the diagnostic chart.
1. If there is a problem with the ECM that causes a "Scan" tool to not read serial data, the ECM should not flash a Code 12. If Code 12 is flashing, check for CKT 451 short to ground. If Code 12 does flash, be sure that the "Scan" tool is working properly on another vehicle. If the "Scan" is functioning properly and CKT 461 is OK, the Mem-Cal or ECM may be at fault for the no ALDL symptom.

2. If the light goes "OFF" when the ECM connector is disconnected, CKT 419 is not shorted to ground.
3. This step will check for an open diagnostic CKT 451.
4. At this point the "Service Engine Soon" light wiring is OK. The problem is a faulty ECM or Mem-Cal. If Code 12 does not flash, the ECM should be replaced using the original Mem-Cal. Replace Mem-Cal only after trying an ECM, as a defective Mem-Cal is an unlikely cause of the problem.

1988–90 2.8L (VIN W) CENTURY AND CUTLASS CIERA

CHART A-2
NO ALDL DATA OR WON'T FLASH CODE 12
"SERVICE ENGINE SOON" LIGHT "ON" STEADY
2.8L (VIN W) "A" CARLINE (PORT)

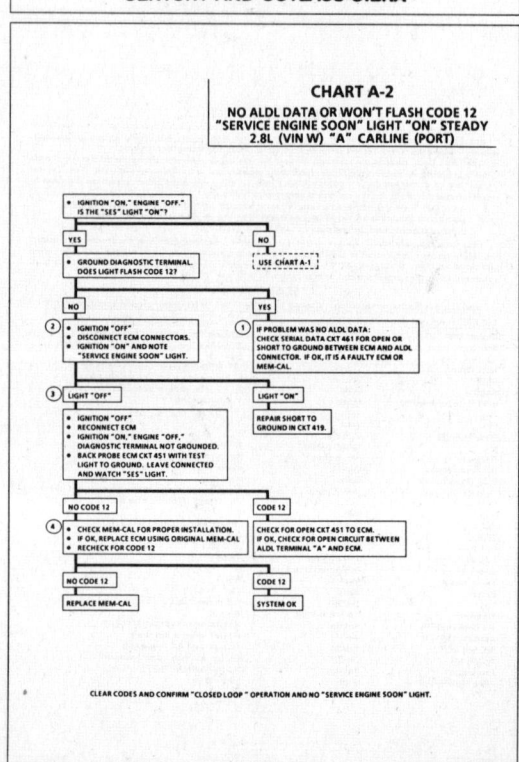

CLEAR CODES AND CONFIRM "CLOSED LOOP" OPERATION AND NO "SERVICE ENGINE SOON" LIGHT.

1988–90 2.8L (VIN W)
CENTURY AND CUTLASS CIERA

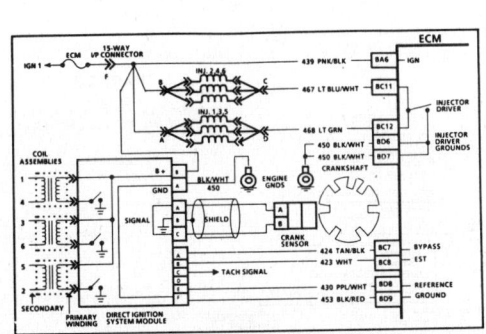

CHART A-3
(Page 1 of 3)
ENGINE CRANKS BUT WILL NOT RUN
2.8L (VIN W) "A" CARLINE (PORT)

Circuit Description:
This chart assumes that battery condition and engine cranking speed are OK, and there is adequate fuel in the tank.

Test Description: Numbers below refer to circled numbers on the diagnostic chart.

1. A "Service Engine Soon" light "ON" is a basic test to determine if there is a 12 volt supply and ignition 12 volts to ECM. No ALDL may be due to an ECM problem and CHART A-2 will diagnose the ECM. If TPS is over 2.5 volts the engine may be in the clear flood mode which will cause starting problems. The engine will not start without reference pulses and therefore the "Scan" should read rpm (reference) during crank.

2. For the first two seconds with ignition "ON," or whenever reference pulses are being received, PPSW should indicate fuel pump circuit voltage (8 to 12 volts).

3. Because the direct ignition system uses two plugs and wires to complete the circuit of each coil, the opposite spark should be left connected. If rpm was indicated during crank, the ignition module is receiving a crank signal, but no spark at this test indicates the ignition module is not triggering the coils.

4. The test light should blink, indicating the ECM is controlling the injectors OK. How bright the light blinks is not important. However, the test light should be a J 34730-3 or equivalent.

5. Use fuel pressure gage J 34730-1 or equivalent. Wrap a shop towel around the fuel pressure tap to absorb any small amount of fuel leakage that may occur when installing the gage.

6. This test will determine if the ignition module is not generating the reference pulse or if the wiring or ECM are at fault. By touching and removing a test light to 12 volts on CKT 430, a reference pulse should be generated. If rpm is indicated the ECM and wiring are OK.

7. This test will determine if the ignition module is not triggering the problem coil or if the tested coil is at fault. This test could also be performed by using another known good coil.

1988–90 2.8L (VIN W)
CENTURY AND CUTLASS CIERA

CHART A-3
(Page 1 of 3)
ENGINE CRANKS BUT WILL NOT RUN
2.8L (VIN W) "A" CARLINE (PORT)

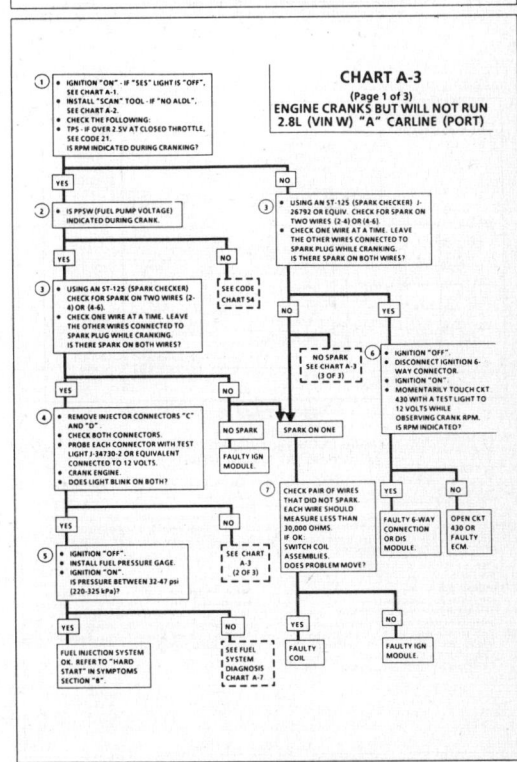

1988–90 2.8L (VIN W)
CENTURY AND CUTLASS CIERA

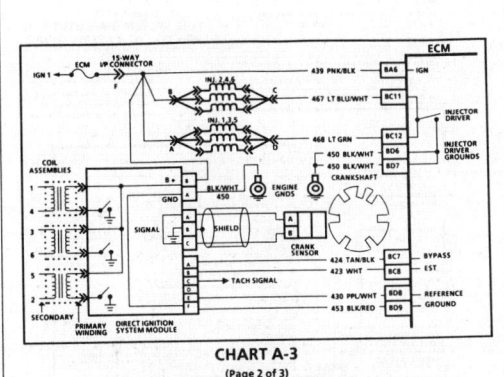

CHART A-3
(Page 2 of 3)
ENGINE CRANKS BUT WILL NOT RUN
2.8L (VIN W) "A" CARLINE (PORT)

Test Description: Numbers below refer to circled numbers on the diagnostic chart.

1. Checks for 12 volt supply to injectors. Due to the injectors wired in parallel there should be a light "ON" on both terminals.

2. Checks continuity of CKTs 467 and 468.

3. All checks made to this point would indicate that the ECM is at fault. However, there is a possibility of CKT 467 or 468 being shorted to a voltage source either in the engine harness or in the injector harness.
   - To test for this condition:
     - Disconnect the injector 4-way connector.
     - Ignition "ON."

- Probe CKTs 467 and 468 on the ECM side of harness with a test light connected to ground. There should be no light. If light is "ON" repair short to voltage.
- If OK, check the resistance of the injector harness between terminals "A" & "C", "A" & "D", "B" & "D", and "B" & "C."
- Should be more than 4 ohms.
- If less than 4 ohms, check harness for wires shorted together and check each injector resistance. (Resistance should be 8 ohms or more.)
- If all OK, replace ECM.

1988–90 2.8L (VIN W)
CENTURY AND CUTLASS CIERA

CHART A-3
(Page 2 of 3)
ENGINE CRANKS BUT WILL NOT RUN
2.8L (VIN W) "A" CARLINE (PORT)

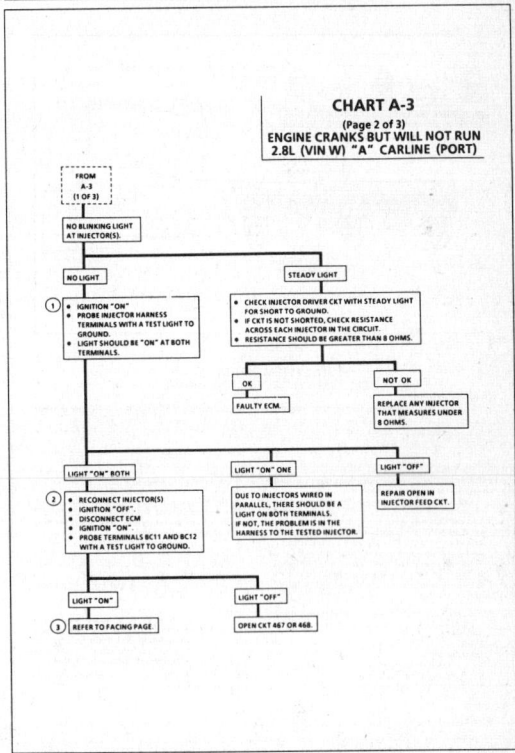

1988–90 2.8L (VIN W)
CENTURY AND CUTLASS CIERA

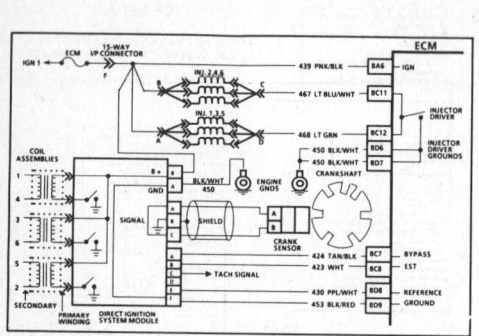

CHART A-3
(Page 3 of 3)
ENGINE CRANKS BUT WILL NOT RUN
2.8L (VIN W) "A" CARLINE (PORT)

Circuit Description:
If the "Scan" tool did not indicate a cranking rpm and there is no spark present at the plugs, the problem lies in the direct ignition system or the power and ground supplies to the module.

The magnetic crank sensor is used to determine engine crankshaft position much the same way as the pick-up coil did in distributor type systems. The sensor is mounted in the block near a seven slot wheel on the crank shaft. The rotation of the wheel creates a flux change in the sensor which produces a voltage signal. The ignition module then processes this signal and creates the reference pulses needed by the ECM and the signal triggers the correct coil at the correct time.

Test Description: Numbers below refer to circled numbers on the diagnostic chart.
1. This test will determine if the 12 volt supply and a good ground is available at the ignition module.
2. Tests for continuity of CKT 439 to the ignition module. If test light does not light but the SES light is "ON" with ignition "ON," repair open in CKT 439 between DIS ignition module and splice.

3. Checks for continuity of the crank sensor and connections.
4. Voltage will vary in this test depending on cranking speed of engine. The voltage will vary from about 500 mV at very slow cranking speeds to about 100 mV at high speeds.

1988–90 2.8L (VIN W)
CENTURY AND CUTLASS CIERA

CHART A-3
(Page 3 of 3)
ENGINE CRANKS BUT WILL NOT RUN
2.8L (VIN W) "A" CARLINE (PORT)

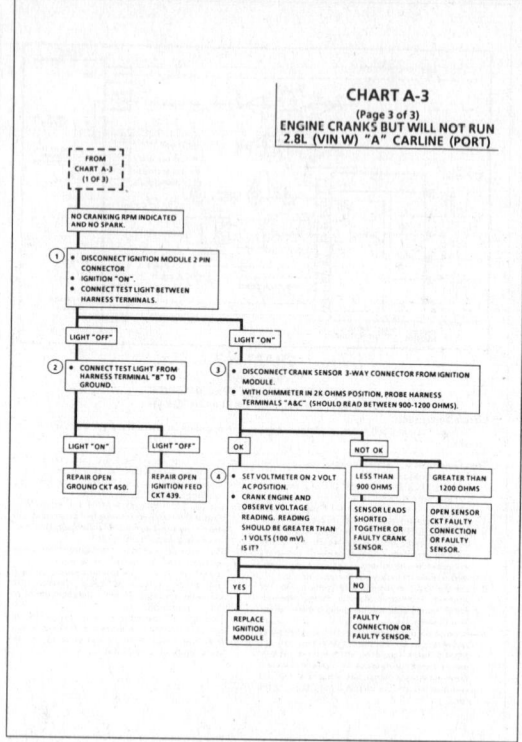

1988–90 2.8L (VIN W)
CENTURY AND CUTLASS CIERA

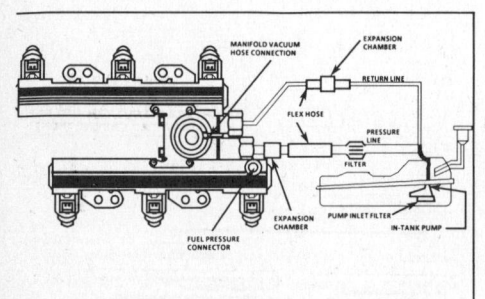

CHART A-7
(Page 1 of 3)
FUEL SYSTEM DIAGNOSIS
2.8L (VIN W) "A" CARLINE (PORT)

Circuit Description:
When the ignition switch is turned "ON," the electronic control module (ECM) will turn "ON" the in-tank fuel pump. It will remain "ON" as long as the engine is cranking or running, and the ECM is receiving reference pulses. If there are no reference pulses, the ECM will shut "OFF" the fuel pump within 2 seconds after ignition "ON" or engine stops.

The pump will deliver fuel to the fuel rail and injectors, then to the pressure regulator, where the system pressure is controlled to about 234 to 325 kPa (34 to 47 psi). Excess fuel is then returned to the fuel tank.

Test Description: Numbers below refer to circled numbers on the diagnostic chart.
Wrap a shop towel around the fuel pressure connector to absorb any small amount of fuel leakage that may occur when installing the gage. Ignition "ON," the pump pressure should be 280-325 kPa (40.5-47 psi). This pressure is controlled by spring pressure within the regulator assembly.
When the engine is idling, the manifold pressure is low (high vacuum) and is applied to the fuel regulator diaphragm. This will offset the spring and result in a lower fuel pressure. This idle pressure will vary somewhat depending on barometric pressure, however, the pressure idling should be less, indicating pressure regulator control.

Pressure that continues to fall is caused by one of the following:
- In-tank fuel pump check valve not holding.
- Pump coupling hose or pulsator leaking.
- Fuel pressure regulator valve leaking.

- Injector(s) sticking open.
4. An injector sticking open can best be determined by checking for a fouled or saturated spark plug(s). If a leaking injector can not be determined by a fouled or saturated spark plug the following procedure should be used.
- Remove plenum and remove fuel rail bolts. Follow the procedures in the fuel control section of this manual, but leave fuel lines connected.
- Lift fuel rail out just enough to leave injector nozzles in the ports.

CAUTION: Be sure injector(s) are not allowed to spray on engine and that injector retaining clips are intact. This should be carefully followed to prevent fuel spray on engine which would cause a fire hazard.

- Pressurize the fuel system and observe injector nozzles.

1988–90 2.8L (VIN W)
CENTURY AND CUTLASS CIERA

CHART A-7
(Page 1 of 3)
FUEL SYSTEM DIAGNOSIS
2.8L (VIN W) "A" CARLINE (PORT)

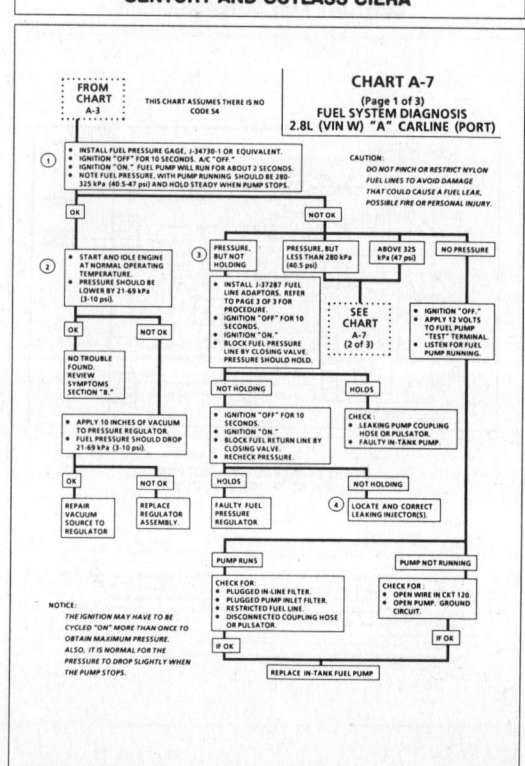

1988–90 2.8L (VIN W)
CENTURY AND CUTLASS CIERA

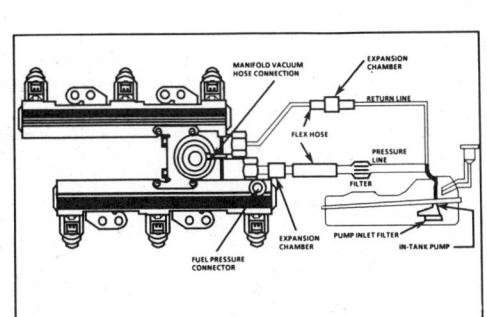

MANIFOLD VACUUM HOSE CONNECTION — EXPANSION CHAMBER — RETURN LINE — FLEX HOSE — PRESSURE LINE — FILTER — EXPANSION CHAMBER — FUEL PRESSURE CONNECTOR — PUMP INLET FILTER — IN-TANK PUMP

CHART A-7
(Page 2 of 3)
FUEL SYSTEM DIAGNOSIS
2.8L (VIN W) "A" CARLINE (PORT)

Test Description: Numbers below refer to circled numbers on the diagnostic chart.

1. Pressure but less than 280 kPa (40.5 psi) falls into two areas:
   - Regulated pressure but less than 280 kPa (40.5 psi). Amount of fuel to injectors OK, but pressure is too low. System will be lean running and may set Code 44. Also, hard starting cold and overall poor performance.
   - Restricted flow causing pressure drop - Normally, a vehicle with a fuel pressure of less than 165 kPa (24 psi) at idle will not be driveable. However, if the pressure drop occurs only while driving, the engine will normally surge then stop running as pressure begins to drop rapidly. This is most likely caused by a restricted fuel line or plugged filter.

2. Restricting the fuel return line allows fuel pressure to build above regulated pressure. With battery voltage applied to the pump test terminal, pressure should rise above 325 kPa (47 psi) as the valve in the return line is partially closed.

   NOTICE: Do Not allow pressure to exceed 414 kPa (60 psi), as damage to the regulator may result.

3. This test determines if the high fuel pressure is due to a restricted fuel return line or a pressure regulator problem.

1988–90 2.8L (VIN W)
CENTURY AND CUTLASS CIERA

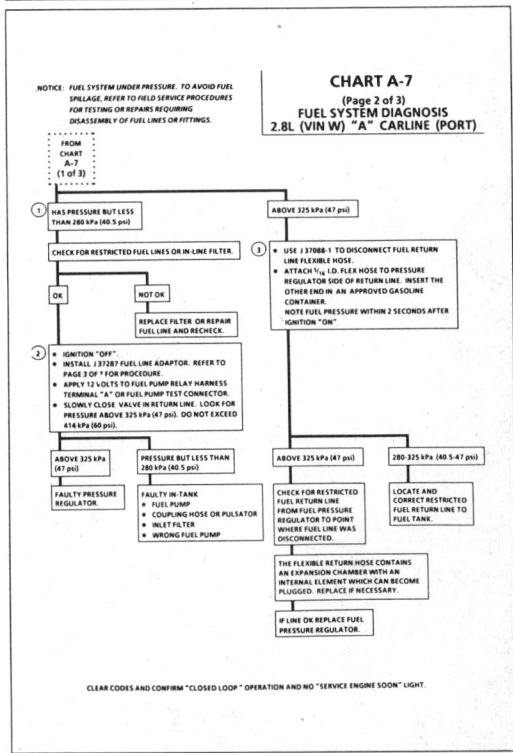

NOTICE: FUEL SYSTEM UNDER PRESSURE. TO AVOID FUEL SPILLAGE, REFER TO FIELD SERVICE PROCEDURES FOR TESTING OR REPAIRS REQUIRING DISASSEMBLY OF FUEL LINES OR FITTINGS.

CHART A-7
(Page 2 of 3)
FUEL SYSTEM DIAGNOSIS
2.8L (VIN W) "A" CARLINE (PORT)

CLEAR CODES AND CONFIRM "CLOSED LOOP" OPERATION AND NO "SERVICE ENGINE SOON" LIGHT.

1988–90 2.8L (VIN W)
CENTURY AND CUTLASS CIERA

CHART A-7
(Page 3 of 3)
FUEL SYSTEM DIAGNOSIS
2.8L (VIN W) "A" CARLINE (PORT)

FUEL PRESSURE CHECK TOOLS

37287 FUEL LINE ADAPTER TOOLS
37088 FUEL LINE REMOVE TOOL

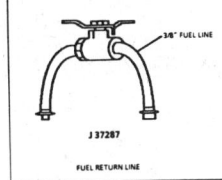

3/8" FUEL LINE

J 37287

FUEL RETURN LINE

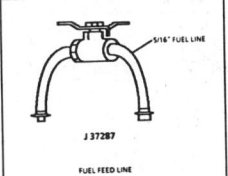

5/16" FUEL LINE

J 37287

FUEL FEED LINE

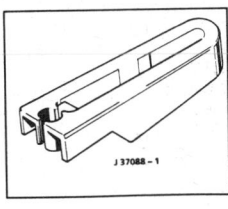

J 37088-1

1988–90 2.8L (VIN W)
CENTURY AND CUTLASS CIERA

CHART A-7
(Page 3 of 3)
FUEL SYSTEM DIAGNOSIS
2.8L (VIN W) "A" CARLINE (PORT)

FUEL PRESSURE CHECK
(USING J 37287 FUEL LINE ADAPTOR)

CAUTION: DO NOT PINCH OR RESTRICT NYLON FUEL LINES TO AVOID SEVERING, WHICH COULD CAUSE A FUEL LEAK.

REMOVE OR DISCONNECT

1. REMOVE FUEL CAP.
2. INSTALL GAGE BLEED HOSE INTO AN APPROVED CONTAINER AND OPEN GAGE VALVE TO BLEED SYSTEM PRESSURE.
3. REMOVE FUEL LINES. USE J 37088-1 TOOL. USE A SHOP CLOTH AND APPROVED CONTAINER TO COLLECT FUEL FROM LINES.
4. INSTALL J 37287 FUEL LINE ADAPTORS TO ENGINE AND BODY SIDE FUEL LINE. (LEAVE VALVE OPEN ON ADAPTOR LINES.)
5. INSTALL FUEL CAP.
6. INSTALL GAGE BLEED HOSE INTO AN APPROVED CONTAINER AND OPEN GAGE VALVE TO BLEED AIR FROM SYSTEM.
7. THE IGNITION MAY HAVE TO BE CYCLED "ON" MORE THAN ONCE TO REMOVE ALL AIR FROM FUEL LINES AND OBTAIN MAXIMUM PRESSURE. IT IS NORMAL FOR THE PRESSURE TO DROP SLIGHTLY WHEN PUMP STOPS.

RECONNECT OR INSTALL

1. REMOVE FUEL CAP.
2. INSTALL GAGE BLEED HOSE INTO AN APPROVED CONTAINER AND OPEN GAGE VALVE TO BLEED SYSTEM PRESSURE.
3. REMOVE FUEL LINE ADAPTORS (J 37287) WITH J 37088-1 TOOL. USE A SHOP CLOTH AND APPROVED CONTAINER TO COLLECT FUEL FROM LINES.
4. INSTALL FUEL LINES TO ENGINE.
5. INSTALL FUEL CAP.
6. INSTALL GAGE BLEED HOSE INTO AN APPROVED CONTAINER AND OPEN GAGE VALVE TO BLEED AIR FROM SYSTEM.
7. THE IGNITION MAY HAVE TO BE CYCLED "ON" MORE THAN ONCE TO REMOVE ALL AIR FROM FUEL L:NES AND OBTAIN MAXIMUM PRESSURE. IT IS NORMAL FOR THE PRESSURE TO DROP SLIGHTLY WHEN PUMP STOPS.
8. CHECK FOR LEAKS IN FUEL LINE.

1988–90 2.8L (VIN W)
CENTURY AND CUTLASS CIERA

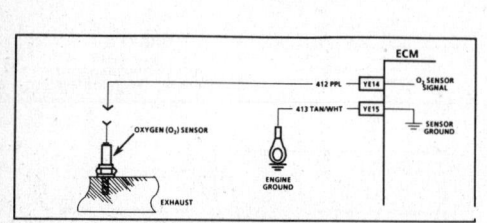

CODE 13
OXYGEN SENSOR CIRCUIT
(OPEN CIRCUIT)
2.8L (VIN W) "A" CARLINE (PORT)

Circuit Description:

The ECM supplies a voltage of about .45 volt between terminals "YE14" and "YE15". (If measured with a 10 megohm digital voltmeter, this may read as low as .32 volt.) The O₂ sensor varies the voltage within a range of about 1 volt if the exhaust is rich, down through about .10 volt if exhaust is lean.

The sensor is like an open circuit and produces no voltage when it is below 315°C (600°F). An open sensor circuit or cold sensor causes "Open Loop" operation.

Test Description: Numbers below refer to circled numbers on the diagnostic chart.
1. Code 13 will set:
   • Engine at normal operating temperature.
   • At least 2 minutes time after start.
   • O₂ signal voltage steady between .35 and .55 volt.
   • Throttle position sensor signal above 4%.
   • All conditions must be met for about 60 seconds.
   If the conditions for a Code 13 exist, the system will not go "Closed Loop."
2. This will determine if the sensor is at fault or the wiring or ECM is the cause of the Code 13.
3. In doing this test use only a high impedence digital volt ohmmeter. This test checks the continuity of CKTs 412 and 413 because if CKT 413 is open the ECM voltage on CKT 412 will be over .6 volt (600 mV).

Diagnostic Aids:

Normal "Scan" voltage varies between 100 mV to 999 mV (.1 and 1.0 volt) while in "Closed Loop." Code 13 sets in one minute if voltage remains between .35 and .55 volt, but the system will go "Open Loop" in about 15 seconds. Refer to "Intermittents" in Section "B".

1988–90 2.8L (VIN W)
CENTURY AND CUTLASS CIERA

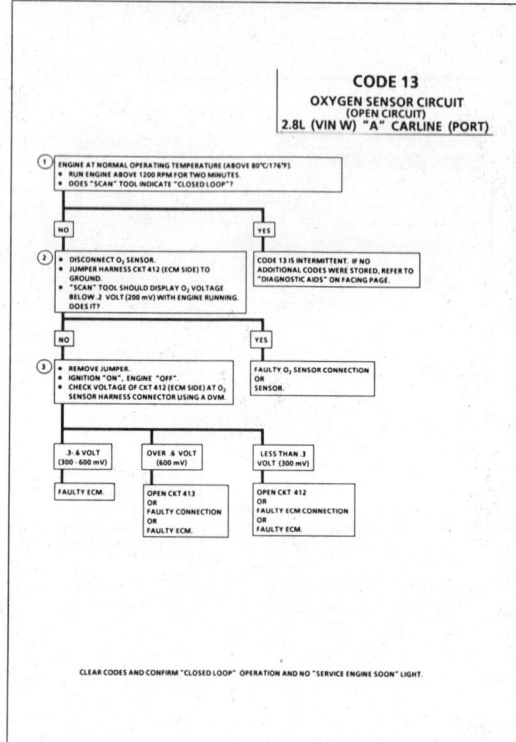

1988–90 2.8L (VIN W)
CENTURY AND CUTLASS CIERA

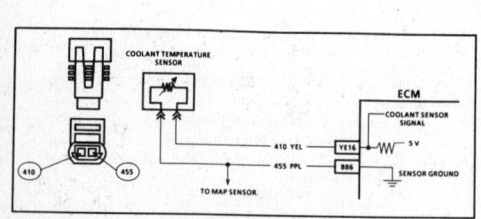

CODE 14
COOLANT TEMPERATURE SENSOR CIRCUIT
(HIGH TEMPERATURE INDICATED)
2.8L (VIN W) "A" CARLINE (PORT)

Circuit Description:

The coolant temperature sensor uses a thermistor to control the signal voltage to the ECM. The ECM applies a voltage on CKT 410 to the sensor. When the engine is cold, the sensor (thermistor) resistance is high, therefore the ECM will see high signal voltage.

As the engine warms, the sensor resistance becomes less, and the voltage drops. At normal engine operating temperature (85°C to 95°C), the voltage will measure about 1.5 to 2.0 volts.

Test Description: Numbers below refer to circled numbers on the diagnostic chart.
1. Code 14 will set if:
   • Signal voltage indicates a coolant temperature above 135°C (275°F) for 3 seconds.
2. This test will determine if CKT 410 is shorted to ground which will cause the conditions for Code 14.

Diagnostic Aids:

Check harness routing for a potential short to ground in CKT 410. "Scan" tool displays engine temperature in degrees centigrade. After engine is started, the temperature should rise steadily to about 90°C, then stabilize when thermostat opens. Refer to "Intermittents" in Section "B".

1988–90 2.8L (VIN W)
CENTURY AND CUTLASS CIERA

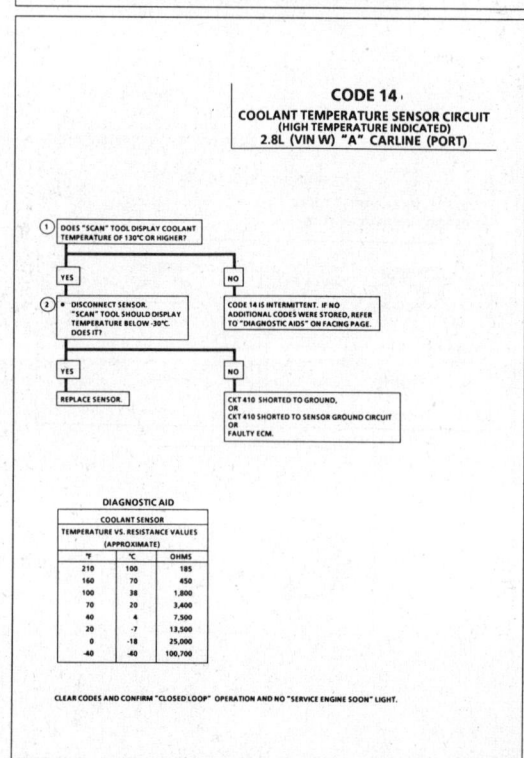

1988–90 2.8L (VIN W)
CENTURY AND CUTLASS CIERA

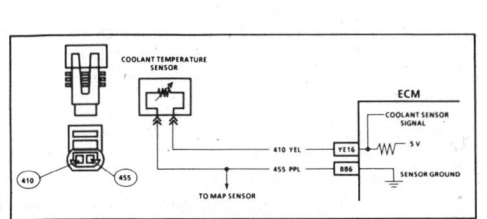

CODE 15
COOLANT TEMPERATURE SENSOR CIRCUIT
(LOW TEMPERATURE INDICATED)
2.8L (VIN W) "A" CARLINE (PORT)

Circuit Description:

The coolant temperature sensor uses a thermistor to control the signal voltage to the ECM. The ECM applies a voltage on CKT 410 to the sensor. When the engine is cold the sensor (thermistor) resistance is high, therefore, the ECM will see high signal voltage.

As the engine warms, the sensor resistance becomes less and the voltage drops. At normal engine operating temperature (85°C to 95°C), the voltage will measure about 1.5 to 2.0 volts at the ECM.

Test Description: Numbers below refer to circled numbers on the diagnostic chart.

1. Code 15 will set if:
   - Signal voltage indicates a coolant temperature less than -44°C (-47°F) for 3 seconds.
2. This test simulates a Code 14. If the ECM recognizes the low signal voltage, (high temperature) and the "Scan" reads 130°C, the ECM and wiring are OK.
3. This test will determine if CKT 410 is open. There should be 5 volts present at sensor connector, if measured with a DVOM.

Diagnostic Aids:

A "Scan" tool reads engine temperature in degrees centigrade. After engine is started the temperature should rise steadily to about 90°C then stabilize when thermostat opens.

A faulty connection, or an open in CKT 410 or 455 will result in a Code 15.

If Code 23 or 33 is also set, check CKT 455 for faulty wiring or connections. Check terminals at sensor for good contact. Refer to "Intermittents" in Section "B".

1988–90 2.8L (VIN W)
CENTURY AND CUTLASS CIERA

CODE 15
COOLANT TEMPERATURE SENSOR CIRCUIT
(LOW TEMPERATURE INDICATED)
2.8L (VIN W) "A" CARLINE (PORT)

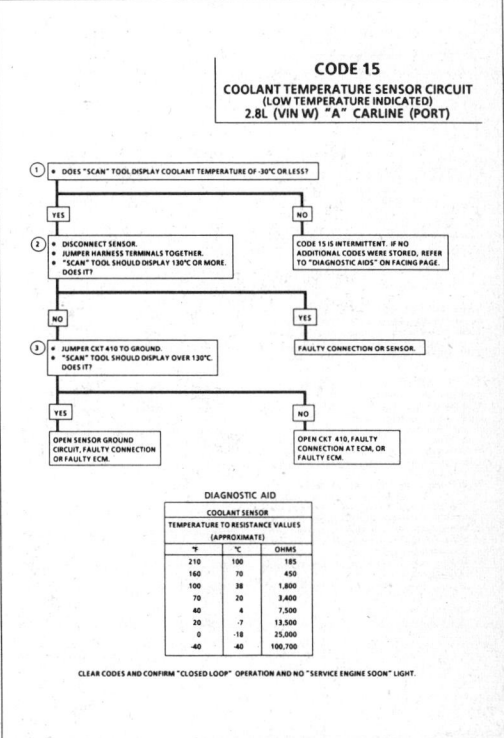

DIAGNOSTIC AID		
COOLANT SENSOR		
TEMPERATURE TO RESISTANCE VALUES (APPROXIMATE)		
°F	°C	OHMS
210	100	185
160	70	450
100	38	1,800
70	20	3,400
40	4	7,500
20	-7	13,500
0	-18	25,000
-40	-40	100,700

CLEAR CODES AND CONFIRM "CLOSED LOOP" OPERATION AND NO "SERVICE ENGINE SOON" LIGHT.

1988–90 2.8L (VIN W)
CENTURY AND CUTLASS CIERA

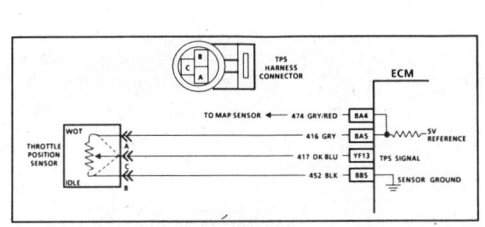

CODE 21
THROTTLE POSITION SENSOR (TPS) CIRCUIT
(SIGNAL VOLTAGE HIGH)
2.8L (VIN W) "A" CARLINE (PORT)

Circuit Description:

The throttle position sensor (TPS) provides a voltage signal that changes relative to the throttle blade. Signal voltage will vary from about .5 at idle to about 5 volts at wide open throttle.

The TPS signal is one of the most important inputs used by the ECM for fuel control and for most of the ECM control outputs.

Test Description: Numbers below refer to circled numbers on the diagnostic chart.

1. Code 21 will set if:
   - Engine is running.
   - TPS signal voltage is greater than 4.3 volts.
   - Air flow is less than 17 GM/sec.
   - All conditions met for 10 seconds.

   or
   - TPS signal voltage over 4.5 volts with ignition "ON."
   - TPS check: The TPS has an auto zeroing feature. If the voltage reading is within the range of 0.45 to 0.85 volt, the ECM will use that value as closed throttle. If the voltage reading is out of the auto zero range on an existing or replacement TPS, make sure the cruise control and throttle cables are not being held open.

2. With the TPS sensor disconnected, the TPS voltage should go low if the ECM and wiring are OK.
3. Probing CKT 452 with a test light checks the 5 volts return circuit, because a faulty 5 volts return will cause a Code 21.

Diagnostic Aids:

A "Scan" tool reads throttle position in volts. Voltage should increase at a steady rate as throttle is moved toward WOT.

Also, some "Scan" tools will read throttle angle 0% = closed throttle, 100% = WOT.

An open in CKT 452 will result in a Code 21. Refer to "Intermittents" in Section "B".

1988–90 2.8L (VIN W)
CENTURY AND CUTLASS CIERA

CODE 21
THROTTLE POSITION SENSOR (TPS) CIRCUIT
(SIGNAL VOLTAGE HIGH)
2.8L (VIN W) "A" CARLINE (PORT)

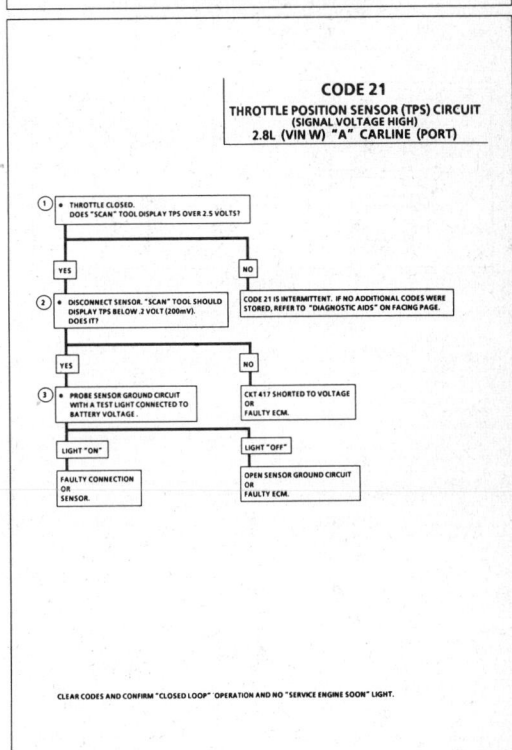

CLEAR CODES AND CONFIRM "CLOSED LOOP" OPERATION AND NO "SERVICE ENGINE SOON" LIGHT.

1988–90 2.8L (VIN W)
CENTURY AND CUTLASS CIERA

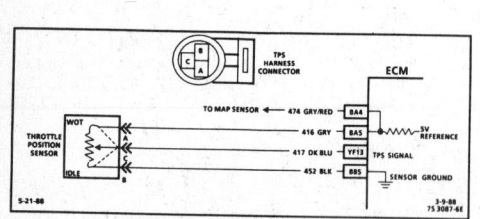

CODE 22
THROTTLE POSITION SENSOR (TPS) CIRCUIT
(SIGNAL VOLTAGE LOW)
2.8L (VIN W) "A" CARLINE (PORT)

Circuit Description:
The throttle position sensor (TPS) provides a voltage signal that changes relative to the throttle blade. Signal voltage will vary from about .5 at idle to about 5 volts at wide open throttle.
The TPS signal is one of the most important inputs used by the ECM for fuel control and for most of the ECM control outputs.

Test Description: Numbers below refer to circled numbers on the diagnostic chart.
1. Code 22 will set if:
   - Engine running
   - TPS signal voltage is less than about .2 volt for 3 seconds.
2. Simulates Code 21: (high voltage) if the ECM recognizes the high signal voltage, the ECM and wiring are OK.
3. TPS check: The TPS has an auto zeroing feature. If the voltage reading is within the range of 0.45 to 0.85 volt, the ECM will use that value as closed throttle. If the voltage reading is out of the auto zero range on an existing or replacement TPS, make sure the cruise control and throttle cables are not being held open.
4. This simulates a high signal voltage to check for an open in CKT 417.
5. CKTs 416 and 474 share a common sensor ground buffered reference signal. If either of these circuits are shorted to ground, Code 22 will set. To determine if the MAP sensor is causing the 22 problem, disconnect it to see if Code 22 resets. Be sure TPS is connected and clear codes before testing.

Diagnostic Aids:
A "Scan" tool reads throttle position in volts. Voltage should increase at a steady rate as throttle is moved toward WOT.
Also, some "Scan" tools will read: throttle angle 0% = closed throttle, 100% = WOT.
An open or short to ground in CKT 416 or 417 will result in a Code 22. Also, a short to ground in CKT 474 will result in a Code 22.
Refer to "Intermittents" in Section "B".

1988–90 2.8L (VIN W)
CENTURY AND CUTLASS CIERA

CODE 22
THROTTLE POSITION SENSOR (TPS) CIRCUIT
(SIGNAL VOLTAGE LOW)
2.8L (VIN W) "A" CARLINE (PORT)

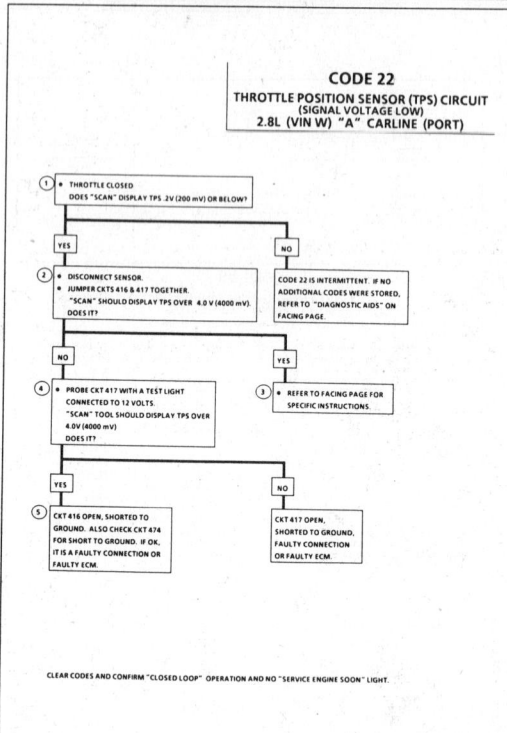

CLEAR CODES AND CONFIRM "CLOSED LOOP" OPERATION AND NO "SERVICE ENGINE SOON" LIGHT.

1988–90 2.8L (VIN W)
CENTURY AND CUTLASS CIERA

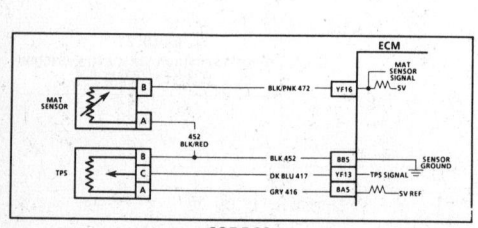

CODE 23
MANIFOLD AIR TEMPERATURE (MAT) SENSOR CIRCUIT
(LOW TEMPERATURE INDICATED)
2.8L (VIN W) "A" CARLINE (PORT)

Circuit Description:
The MAT sensor uses a thermistor to control the signal voltage to the ECM. The ECM applies a voltage (about 5 volts) on CKT 472 to the sensor. When the air is cold, the sensor (thermistor) resistance is high, therefore the ECM will see a high signal voltage. If the air is warm, the sensor resistance is low, therefore, the ECM will see a low voltage.

Test Description: Numbers below refer to circled numbers on the diagnostic chart.
1. Code 23 will set if:
   - A signal voltage indicates a manifold air temperature below –35°C (–31°F) for 3 seconds.
   - Time since engine start is 4 minutes or longer.
   - No VSS.
2. A Code 23 will set, due to an open sensor, wire, or connection. This test will determine if the wiring and ECM are OK.
3. This will determine if the signal CKT 472 or the sensor ground return CKT 452 is open.

Diagnostic Aids:
A "Scan" tool reads temperature of the air entering the engine and should read close to ambient air temperature when engine is cold, and rises as underhood temperature increases.
A faulty connection, or an open in CKT 472 or 452 will result in a Code 23.
Refer to "Intermittents" in Section "B".

1988–90 2.8L (VIN W)
CENTURY AND CUTLASS CIERA

CODE 23
MANIFOLD AIR TEMPERATURE (MAT) SENSOR CIRCUIT
(LOW TEMPERATURE INDICATED)
2.8L (VIN W) "A" CARLINE (PORT)

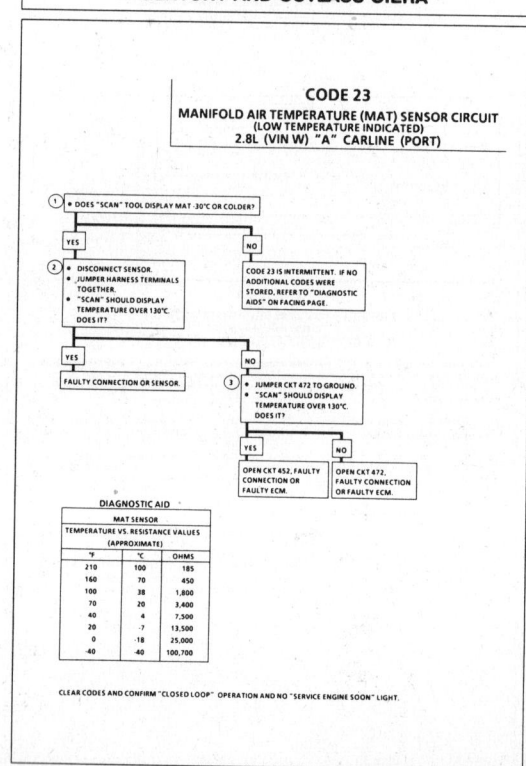

DIAGNOSTIC AID
MAT SENSOR
TEMPERATURE VS. RESISTANCE VALUES
(APPROXIMATE)

°F	°C	OHMS
210	100	185
160	70	450
100	38	1,800
70	20	3,400
40	4	7,500
20	–7	13,500
0	–18	25,000
–40	–40	100,700

CLEAR CODES AND CONFIRM "CLOSED LOOP" OPERATION AND NO "SERVICE ENGINE SOON" LIGHT.

1988–90 2.8L (VIN W)
CENTURY AND CUTLASS CIERA

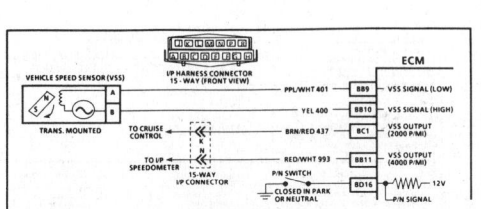

CODE 24
VEHICLE SPEED SENSOR (VSS) CIRCUIT
2.8L (VIN W) "A" CARLINE (PORT)

Circuit Description:
Vehicle speed information is provided to the ECM by the vehicle speed sensor, which is a permanent magnet (PM) generator and it is mounted in the transaxle. The PM generator produces a pulsing voltage, whenever vehicle speed is over about 3 mph. The A/C voltage level and the number of pulses increases with vehicle speed. The ECM then converts the pulsing voltage to mph which is used for calculations, and the mph can be displayed with a "Scan" tool.
The function of VSS buffer used in past model years has been incorporated into the ECM. The ECM then supplies the necessary signal for the instrument panel (4000 pulses per mile) for operating the speedometer and the odometer. If the vehicle is equipped with cruise control, the ECM also provides a signal (2000 pulses per mile) to the cruise control module.

Test Description: Numbers below refer to circled numbers on the diagnostic chart.
1. Code 24 will set if vehicle speed equals 0 mph when:
   - Engine speed is between 1400 and 3600 rpm
   - TPS is less than 2%
   - Low load condition (low air flow)
   - Not in park or neutral
   - All conditions met for 5 seconds
   These conditions are met during a road load deceleration. Disregard Code 24 that sets when drive wheels are not turning.
   - The PM generator only produces a signal if drive wheels are turning greater than 3 mph.
2. CKTs 400, 401 and 993 are OK if the speedometer works properly. Code 24 is being caused by a faulty ECM, faulty Mem-Cal or an incorrect Mem-Cal.

Diagnostic Aids:
"Scan" should indicate a vehicle speed whenever the drive wheels are turning greater than 3 mph.
A problem in CKT 437 or 993 will not affect the VSS input or the readings on a "Scan."
Check CKTs 400 and 401 for proper connections to be sure they're clean and tight and the harness is routed correctly. Refer to "Intermittents" in Section "B."
(A/T) A faulty or misadjusted park/neutral switch can result in a false Code 24. Use a "Scan" and check for proper signal while in drive (125C) or overdrive (440-T4). Refer to CHART C-1A for P/N switch diagnosis check.

1988–90 2.8L (VIN W)
CENTURY AND CUTLASS CIERA

CODE 24
VEHICLE SPEED SENSOR (VSS) CIRCUIT
2.8L (VIN W) "A" CARLINE (PORT)

DISREGARD CODE 24 IF SET WHILE DRIVE WHEELS ARE NOT TURNING.

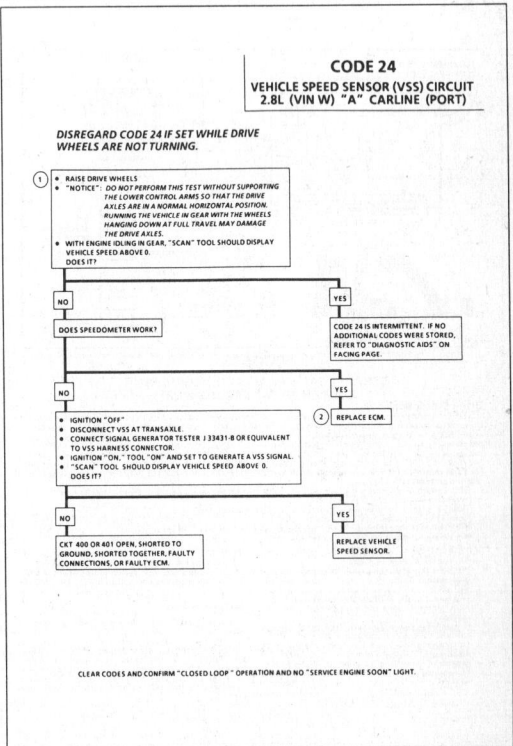

CLEAR CODES AND CONFIRM "CLOSED LOOP" OPERATION AND NO "SERVICE ENGINE SOON" LIGHT.

1988–90 2.8L (VIN W)
CENTURY AND CUTLASS CIERA

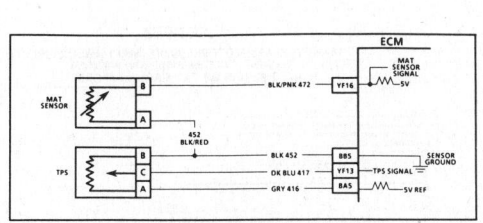

CODE 25
MANIFOLD AIR TEMPERATURE (MAT) SENSOR CIRCUIT
(HIGH TEMPERATURE INDICATED)
2.8L (VIN W) "A" CARLINE (PORT)

Circuit Description:
The manifold air temperature sensor uses a thermistor to control the signal voltage to the ECM. The ECM applies a voltage (about 5 volts) on CKT 472 to the sensor. When manifold air is cold, the sensor (thermistor) resistance is high, therefore, the ECM will see a high signal voltage. As the air warms, the sensor resistance becomes less, and the voltage drops.

Test Description: Numbers below refer to circled numbers on the diagnostic chart.
1. Code 25 will set if:
   - Signal voltage indicates a manifold air temperature greater than 145°C (293°F) for 3 seconds
   - Time since engine start is 4 minutes or longer
   - A vehicle speed is present.

Diagnostic Aids:
A "Scan" tool reads temperature of the air entering the engine and should read close to ambient air temperature when engine is cold and rises as underhood temperature increases.
A short to ground on CKT 472 will result in a Code 25.
Refer to "Intermittents" in Section "B".

1988–90 2.8L (VIN W)
CENTURY AND CUTLASS CIERA

CODE 25
MANIFOLD AIR TEMPERATURE (MAT) SENSOR CIRCUIT
(HIGH TEMPERATURE INDICATED)
2.8L (VIN W) "A" CARLINE (PORT)

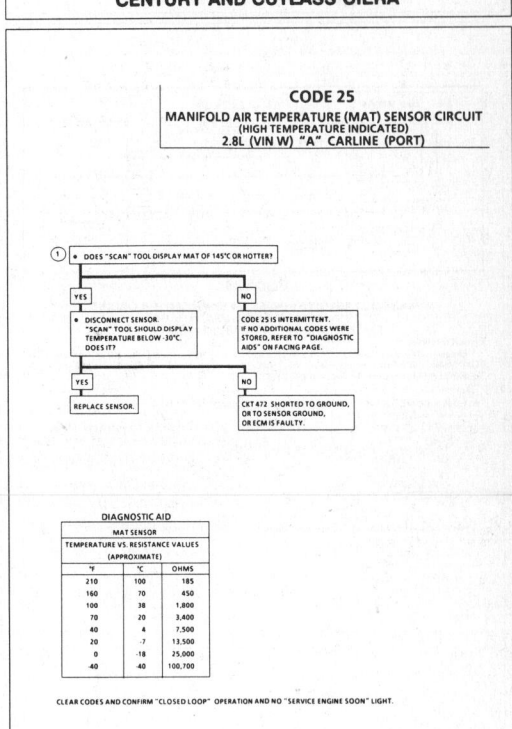

DIAGNOSTIC AID		
MAT SENSOR		
TEMPERATURE VS. RESISTANCE VALUES (APPROXIMATE)		
°F	°C	OHMS
210	100	185
160	70	450
100	38	1,800
70	20	3,400
40	4	7,500
20	-7	13,500
0	-18	25,000
-40	-40	100,700

CLEAR CODES AND CONFIRM "CLOSED LOOP" OPERATION AND NO "SERVICE ENGINE SOON" LIGHT.

1988–90 2.8L (VIN W) CENTURY AND CUTLASS CIERA

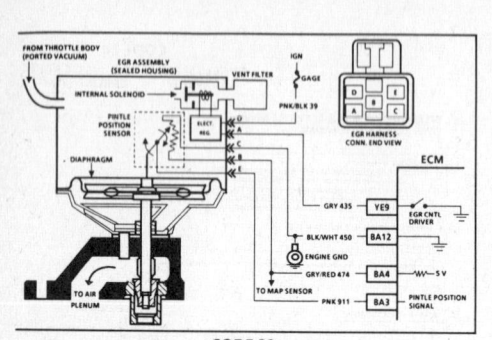

CODE 32
EXHAUST GAS RECIRCULATION (EGR) CIRCUIT
2.8L (VIN W) "A" CARLINE (PORT)

Circuit Description:
The integrated electronic EGR valve functions similar to a port EGR valve with a remote vacuum regulator. The internal solenoid is normally open, which causes the vacuum signal to be vented "OFF" to the atmosphere when EGR is not being commanded by the ECM. This EGR valve has a sealed cap and the solenoid valve opens and closes the vacuum signal which controls the amount of vacuum vented to atmosphere, and this controls the amount of vacuum applied to the diaphragm. The electronic EGR valve contains a voltage regulator which converts the ECM signal to provide different amounts of EGR flow by regulating the current to the solenoid. The ECM controls EGR flow with a pulse width modulated signal (turns "ON" and "OFF" many times a second) based on air flow, TPS, and rpm.

This system also contains a pintle position sensor which works similar to a sensor, and as EGR flow is increased, the sensor output also increases.

Code 32 means that there has been an EGR system faulty detected.

Code 32 will set under two conditions:
- Coolant temperature above a specified amount, EGR should be "ON" or;
- EGR pintle position does not match duty cycle.

Test Description: Numbers below refer to circled numbers on the diagnostic chart.
1. With the engine running and the transmission in gear, with brakes applied, increase engine rpm will put a load on the engine. Engine vacuum will be applied to the EGR diaphragm and cause the EGR pintle to open increasing pintle position voltage.
2. Grounding the diagnostic terminal should energize the internal solenoid which closes off the vent and allows the vacuum to move the diaphragm. This test determines that the ECM is capable of controlling the solenoid. When EGR is commanded on by the ECM, the test light should be "ON".

3. If CKT 450 is open, the pintle signal will go high (showing a 5 volts signal). This will set a Code 32. If CKT 911 becomes shorted to 12 volts or to CKT 474, the signal voltage will go high causing a Code 32.

Diagnostic Aids:
Some "Scan" tools will read pintle position in volts.

The EGR position voltage can be used to determine that the pintle is moving. When no EGR is commanded 0% duty cycle, the position sensor should read between .5 volt and 1.5 volts and increase with the commanded EGR duty cycle. If system operates correctly, refer to "Intermittents" in Section "B".

1988–90 2.8L (VIN W) CENTURY AND CUTLASS CIERA

CODE 32
EXHAUST GAS RECIRCULATION (EGR) CIRCUIT
2.8L (VIN W) "A" CARLINE (PORT)

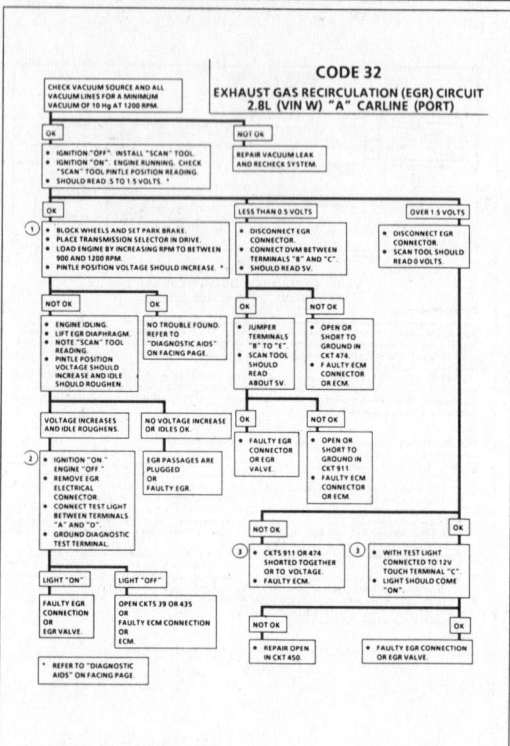

1988–90 2.8L (VIN W) CENTURY AND CUTLASS CIERA

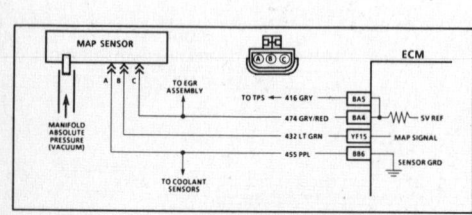

CODE 33
MANIFOLD ABSOLUTE PRESSURE (MAP) SENSOR CIRCUIT
(SIGNAL VOLTAGE HIGH - LOW VACUUM)
2.8L (VIN W) "A" CARLINE (PORT)

Circuit Description:
The manifold absolute pressure (MAP) sensor responds to changes in manifold pressure (vacuum). The ECM receives this information as a signal voltage that will vary from about 1–1.5 volts at idle (high vacuum) to 4–4.5 volts at wide open throttle (low vacuum).

Test Description: Numbers below refer to circled numbers on the diagnostic chart.
1. Code 33 will set when:
   - Engine running
   - Manifold pressure greater than 75.3 kPa (A/C "OFF") 81.2 kPa (A/C "ON")
   - Throttle angle less than 2%
   - Conditions met for 2 seconds
   Engine misfire or a low unstable idle may set Code 33.
2. With the MAP sensor disconnected, the ECM should see a low voltage if the ECM and wiring are OK.

Diagnostic Aids:
If idle is rough or unstable, refer to "Symptoms" in Section "B" for items which can cause an unstable idle.

An open in CKT 455 or the connection will result in a Code 33.

Ignition "ON" engine "OFF," voltages should be within the values shown in the table on the chart. Also, CHART C-1D can be used to test the MAP sensor.

Refer to "Intermittents" in Section "B".

1988–90 2.8L (VIN W) CENTURY AND CUTLASS CIERA

CODE 33
MANIFOLD ABSOLUTE PRESSURE (MAP) SENSOR CIRCUIT
(SIGNAL VOLTAGE HIGH - LOW VACUUM)
2.8L (VIN W) "A" CARLINE (PORT)

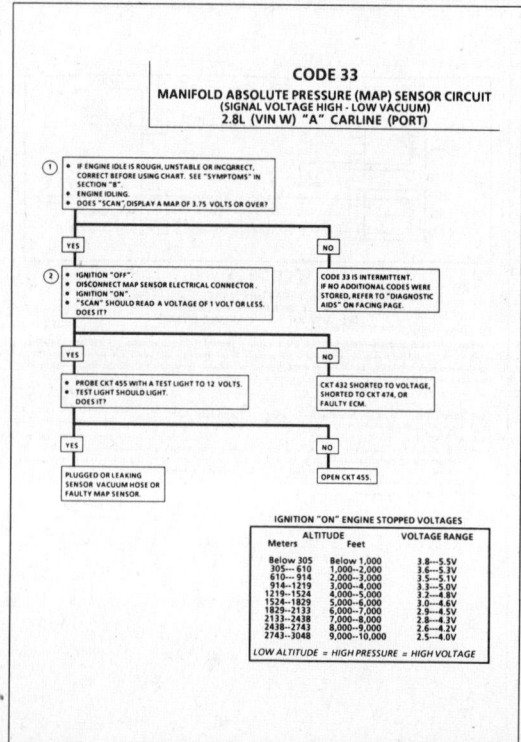

ALTITUDE		VOLTAGE RANGE
Meters	Feet	
Below 305	Below 1,000	3.8—5.5V
305—610	1,000—2,000	3.6—5.3V
610—914	2,000—3,000	3.5—5.1V
914—1219	3,000—4,000	3.3—5.0V
1219—1524	4,000—5,000	3.2—4.8V
1524—1829	5,000—6,000	3.0—4.6V
1829—2133	6,000—7,000	2.9—4.5V
2133—2438	7,000—8,000	2.8—4.3V
2438—2743	8,000—9,000	2.6—4.2V
2743—3048	9,000—10,000	2.5—4.0V

LOW ALTITUDE = HIGH PRESSURE = HIGH VOLTAGE

1988–90 2.8L (VIN W)
CENTURY AND CUTLASS CIERA

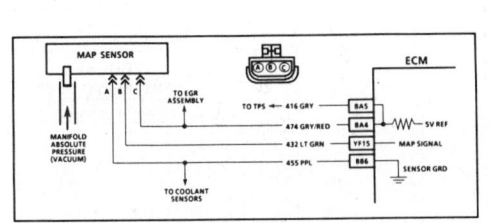

CODE 34
MANIFOLD ABSOLUTE PRESSURE (MAP) SENSOR CIRCUIT
(SIGNAL VOLTAGE LOW - HIGH VACUUM)
2.8L (VIN W) "A" CARLINE (PORT)

Circuit Description:
The manifold absolute pressure (MAP) sensor responds to changes in manifold pressure (vacuum). The ECM receives this information as a signal voltage that will vary from about 1–1.5 volts at idle (high vacuum) to 4–4.5 volts at wide open throttle (low vacuum).

Test Description: Numbers below refer to circled numbers on the diagnostic chart.
1. Code 34 will set if:
   - Engine rpm less than 600
   - Manifold pressure reading less than 13 kPa
   - Conditions met for 1 second
   or
   - Engine rpm greater than 600
   - Throttle angle over 20%
   - Manifold pressure less than 13 kPa
   - Conditions met for 1 second
2. This test is to see if the sensor is at fault for the low voltage, or if there is an ECM or wiring problem.
3. This simulates a high signal voltage to check for an open in CKT 432. If the test light is bright during this test, CKT 432 is probably shorted to ground. If "Scan" reads over 4 volts at this test, CKT 474 can be checked by measuring the voltage at terminal "C" (should be 5 volts).

Diagnostic Aids:
An intermittent open in CKTs 432 or 474 will result in a Code 34.
Ignition "ON" engine "OFF," voltages should be within the values shown in the table on the chart. Also CHART C-1D can be used to test MAP sensor. Refer to "Intermittents" in Section "B".

1988–90 2.8L (VIN W)
CENTURY AND CUTLASS CIERA

CODE 34
MANIFOLD ABSOLUTE PRESSURE (MAP) SENSOR CIRCUIT
(SIGNAL VOLTAGE LOW - HIGH VACUUM)
2.8L (VIN W) "A" CARLINE (PORT)

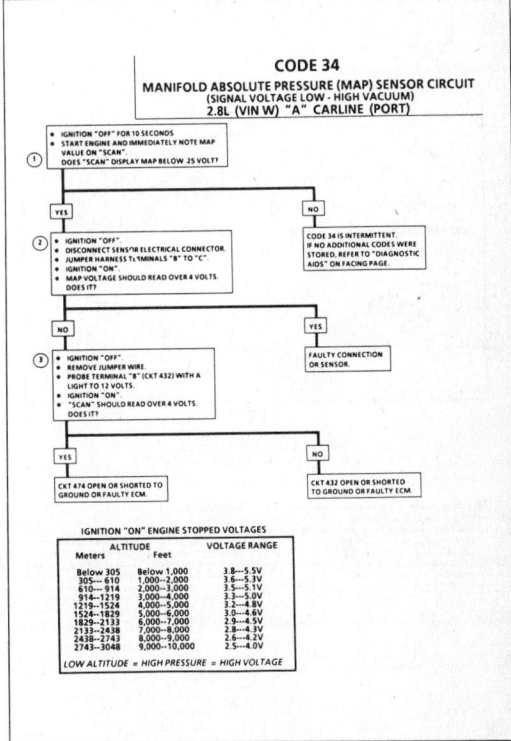

IGNITION "ON" ENGINE STOPPED VOLTAGES

Altitude		VOLTAGE RANGE
Meters	Feet	
Below 305	Below 1,000	3.8—5.5V
305— 610	1,000—2,000	3.6—5.3V
610— 914	2,000—3,000	3.5—5.1V
914—1219	3,000—4,000	3.3—5.0V
1219—1524	4,000—5,000	3.2—4.8V
1524—1829	5,000—6,000	3.0—4.6V
1829—2133	6,000—7,000	2.9—4.5V
2133—2438	7,000—8,000	2.8—4.3V
2438—2743	8,000—9,000	2.6—4.2V
2743—3048	9,000—10,000	2.5—4.0V

LOW ALTITUDE = HIGH PRESSURE = HIGH VOLTAGE

1988–90 2.8L (VIN W)
CENTURY AND CUTLASS CIERA

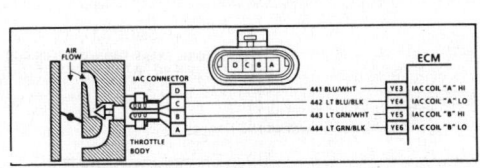

CODE 35
IDLE SPEED ERROR
2.8L (VIN W) "A" CARLINE (PORT)

Circuit Description:
Code 35 will be set when the closed throttle engine speed is 300 rpm above or below the desired (commanded) idle speed for 50 seconds. Review the general description of the IAC operation in Section "C2".

Test Description: Numbers below refer to circled numbers on the diagnostic chart.
1. Continue with test even if engine will not idle. If idle is too low, "Scan" will display 80 or more counts, or steps. If idle is high it will display "0" counts.
   Occasionally an erratic or unstable idle may occur. Engine speed may vary 200 rpm or more up and down. Disconnect IAC. If the condition is unchanged, the IAC is not at fault. There is a system problem. Proceed to diagnostic aids below.
2. When the engine was stopped, the IAC valve retracted (more air) to a fixed "park" position for increased air flow and idle speed during the next engine start. A "Scan" will display 80 or more counts. Observe idle immediately as on a warm engine, the idle speed should decrease rapidly
3. Be sure to disconnect the IAC valve prior to this test. The test light will confirm the ECM signals by a steady or flashing light on all circuits.
4. There is a remote possibility that one of the circuits is shorted to voltage which would have been indicated by a steady light. Disconnect ECM and turn the ignition "ON" and probe terminals to check for this condition.

Diagnostic Aids:
A slow unstable idle may be caused by a system problem that cannot be controlled by the IAC. "Scan" counts will be above 80 counts if idle is too low and "0" counts if it is too high.
If idle is too high, stop engine. Ignition "ON." Ground diagnostic terminal. Wait a few seconds for IAC to seat then disconnect IAC. Start engine. If idle

speed is above 600–700 rpm in drive with an A/T, locate and correct vacuum leak. If rpm is below spec., check for foreign material around throttle plates. Refer to Section "C2".

- **System too lean (high air/fuel ratio)**
  Idle speed may be too high or too low. Engine speed may vary up and down, disconnecting IAC does not help. This may set Code 44.
  "Scan" and/or voltmeter will read an oxygen sensor output less than 300 mV (.3 volt). Check for low regulated fuel pressure or water in fuel. A lean exhaust with an oxygen sensor output fixed above 800 mV (.8 volt) will be a contaminated sensor, usually silicone. This may also set a Code 45 or 61.
- **System too rich (low air/fuel ratio)**
  Idle speed too low. "Scan" counts usually above 80. System obviously rich and may exhibit black smoke exhaust.
  "Scan" tool and/or voltmeter will read an oxygen sensor signal fixed above 800 mV (.8 volt). Check:
  - For fuel pressure regulator hose.
  - High fuel pressure.
  - Injector leaking or sticking.
- **Throttle body.** Remove IAC and inspect bore for foreign material or evidence of IAC valve dragging the bore.
- **A/C compressor or relay failure.** See CHART C-10 if the A/C control relay drive circuit is shorted to ground or if the relay is faulty, an idle problem may exist.
- If above are all OK, refer to "Rough, Unstable, Incorrect Idle, or Stalling" in "Symptoms" in Section "B".

1988–90 2.8L (VIN W)
CENTURY AND CUTLASS CIERA

CODE 35
IDLE SPEED ERROR
2.8L (VIN W) "A" CARLINE (PORT)

NOTE: ECM WILL NOT CYCLE IAC IF THE COOLANT FAN IS ON.

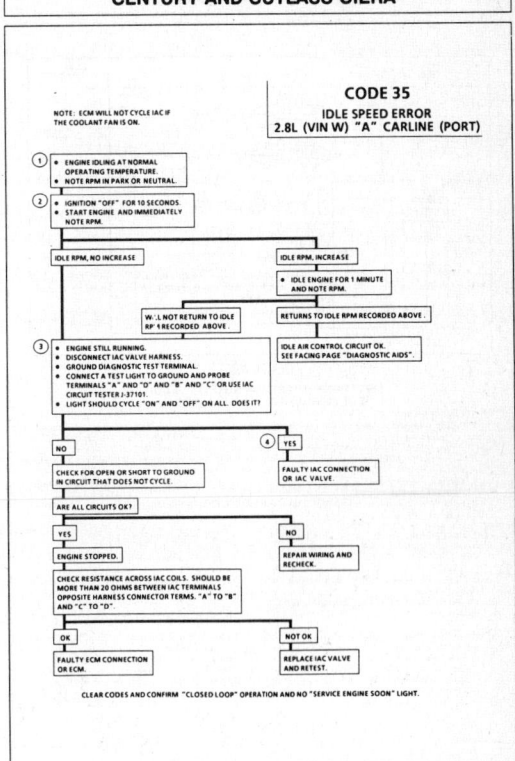

1988–90 2.8L (VIN W)
CENTURY AND CUTLASS CIERA

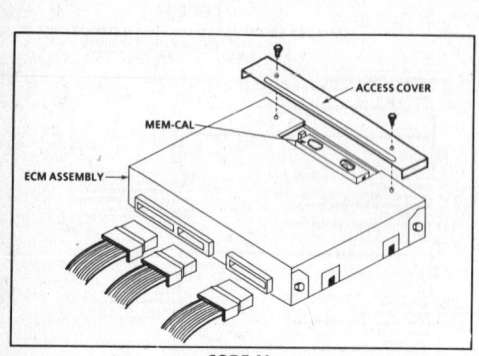

ACCESS COVER

MEM-CAL

ECM ASSEMBLY

CODE 41
CYLINDER SELECT ERROR
(FAULTY OR INCORRECT MEM-CAL)
2.8L (VIN W) "A" CARLINE (PORT)

Test Description: Numbers below refer to circled numbers on the diagnostic chart.
1. The ECM used for this engine can also be used for other engines, and the difference is in the Mem-Cal. If a Code 41 sets, the incorrect Mem-Cal has been installed, or it is faulty and must be replaced.

Diagnostic Aids:

Check Mem-Cal to be sure locking tabs are secure. Also check the pins on both the Mem-Cal and ECM to assure they are making proper contact. Check the Mem-Cal part number to assure it is the correct part. If the Mem-Cal is faulty it must be replaced. It is also possible that the ECM is faulty, however, it should not be replaced until all of the above have been checked. For additional information, refer to "Intermittents" on page "6E3-B-2".

1988–90 2.8L (VIN W)
CENTURY AND CUTLASS CIERA

CODE 41
CYLINDER SELECT ERROR
(FAULTY OR INCORRECT MEM-CAL)
2.8L (VIN W) "A" CARLINE (PORT)

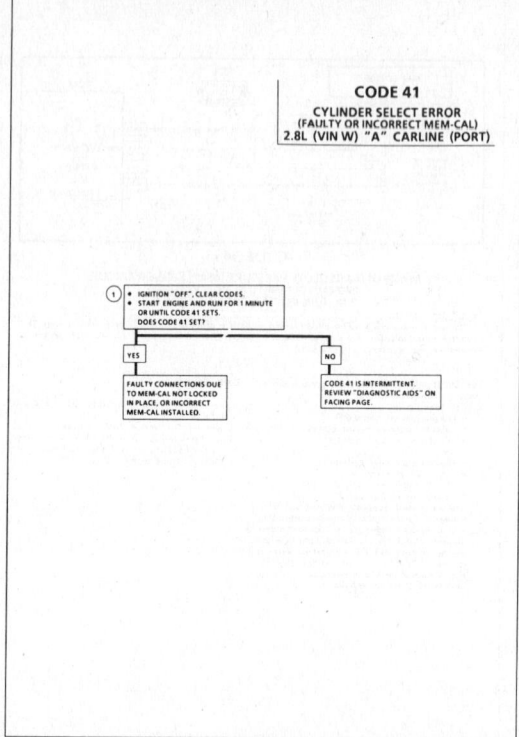

1988–90 2.8L (VIN W)
CENTURY AND CUTLASS CIERA

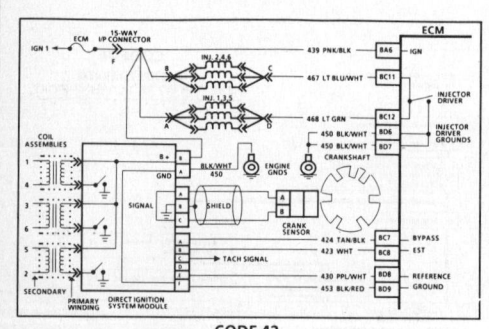

CODE 42
ELECTRONIC SPARK TIMING (EST) CIRCUIT
2.8L (VIN W) "A" CARLINE (PORT)

Circuit Description:
When the system is running on the ignition module, that is, no voltage on the bypass line, the ignition module grounds the EST signal. The ECM expects to see no voltage on the EST line during this condition. If it sees a voltage, it sets Code 42 and will not go into the EST mode.
When the rpm for EST is reached (about 400 rpm), and bypass voltage applied, the EST should no longer be grounded in the ignition module so the EST voltage should be varying.
If the bypass line is open or grounded, the ignition module will not switch to EST mode, so the EST voltage will be low and Code 42 will be set.
If the EST line is grounded, the ignition module will switch to EST, but because the line is grounded, there will be no EST signal. A Code 42 will be set.

Test Description: Numbers below refer to circled numbers on the diagnostic chart.
1. Code 42 means the ECM has seen an open or short to ground in the EST or bypass circuits. This test confirms Code 42 and that the fault causing the code is present.
2. Checks for a normal EST ground path through the ignition module. An EST CKT 423 shorted to ground will also read less than 500 ohms, however, this will be checked luter.
3. As the test light voltage touches CKT 424, the module should switch causing the ohmmeter to "overrange" if the meter is in the 1000 to 20,000 ohms position. Selecting the 10-20,000 ohms position will indicate above 5000 ohms. The important thing is that the module "switched."
4. The module did not switch and this step checks for

- EST CKT 423 shorted to ground
- Bypass CKT 424 open
- Faulty ignition module connection or module
5. Confirms that Code 42 is a faulty ECM and not an intermittent in CKTs 423 or 424.

Diagnostic Aids:

The "Scan" tool does not have any ability to help diagnose a Code 42 problem.
A Mem-Cal not fully seated in the ECM can result in a Code 42. If Code 42 is intermittent, there is a possibility of an open EST line. To check for an open EST line, crank engine while in a "clear flood mode" for five seconds, then start engine and check for Code 42. If Code 42 is set, repair open in CKT 423 (EST line)
Refer to "Intermittents" in Section "H"

1988–90 2.8L (VIN W)
CENTURY AND CUTLASS CIERA

CODE 42
ELECTRONIC SPARK TIMING (EST) CIRCUIT
2.8L (VIN W) "A" CARLINE (PORT)

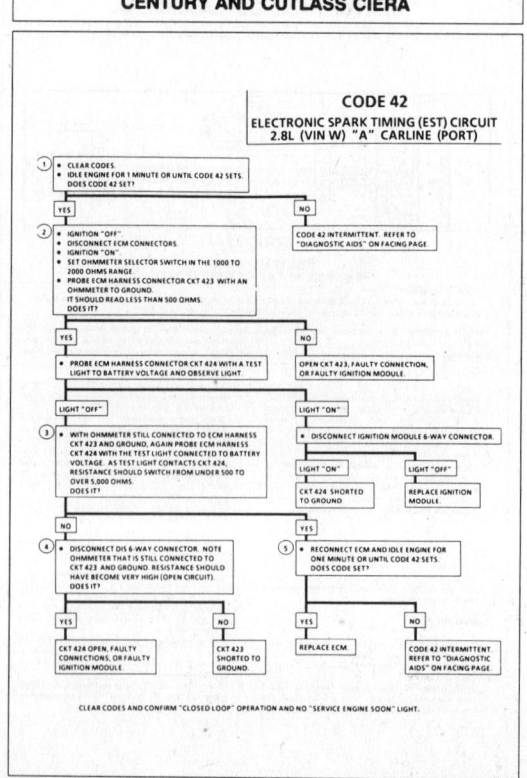

1988–90 2.8L (VIN W)
CENTURY AND CUTLASS CIERA

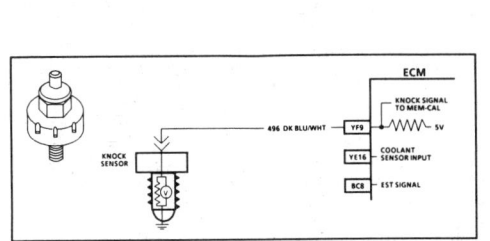

CODE 43
ELECTRONIC SPARK CONTROL (ESC) CIRCUIT
2.8L (VIN W) "A" CARLINE (PORT)

Circuit Description:

The knock sensor is used to detect engine detonation and the ECM will retard the electronic spark timing based on the signal being received. The circuitry within the knock sensor causes the ECM 5 volts to be pulled down so that, under a no knock condition, CKT 496 would measure about 2.5 volts. The knock sensor produces an A/C signal which rides on the 2.5 volts DC voltage. The amplitude and signal frequency is dependent upon the knock level.

If CKT 496 becomes open or shorted to ground, the voltage will either go above 3.5 volts or below 1.5 volts. If either of these conditions are met for about 1/2 second, a Code 43 will be stored.

Test Description: Numbers below refer to circled numbers on the diagnostic chart.

1. This step determines if conditions for Code 43 still exist (voltage on CKT 496 above 3.5 volts or below 1.5 volts). The system is designed to retard the spark 15° if either condition exists.
2. The ECM has a 5 volt pullup resistor, which applies 5 volts to CKT 496. The 5 volt signal should be present at the knock sensor terminal during these test conditions.
3. This step determines if the knock sensor resistance is 3300 to 4500 ohms the sensor is OK.

4. If CKT 496 is not open or shorted to ground and the voltage reading is below 4 volts, the most likely cause is an open circuit in the ECM. It is possible that a faulty Mem-Cal could be drawing the 5 volt signal down, and it should be replaced, if a replacement ECM did not correct the problem.

Diagnostic Aids:

Check CKT 496 for a potential open or short to ground. Also, check for proper installation of Mem-Cal.

Refer to "Intermittents" in Section "B".

1988–90 2.8L (VIN W)
CENTURY AND CUTLASS CIERA

CODE 43
ELECTRONIC SPARK CONTROL (ESC) CIRCUIT
2.8L (VIN W) "A" CARLINE (PORT)

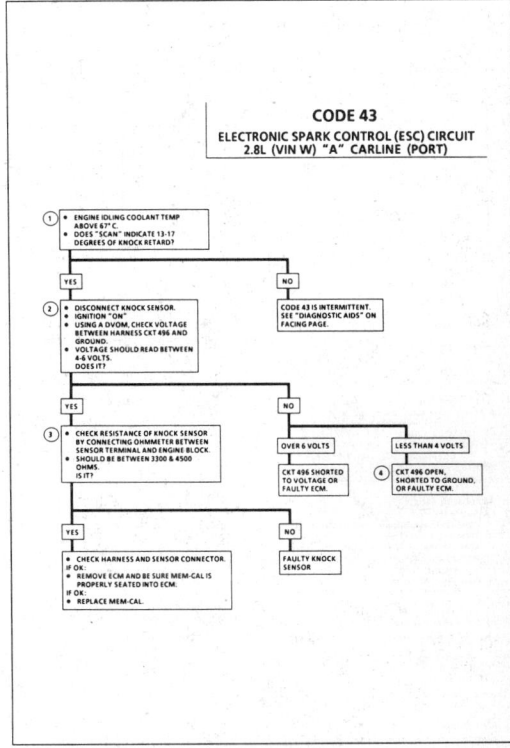

1988–90 2.8L (VIN W)
CENTURY AND CUTLASS CIERA

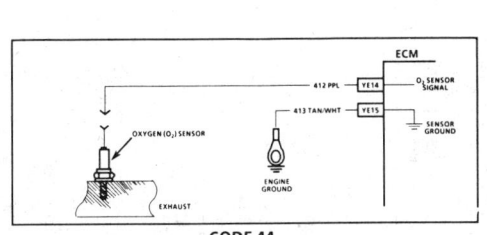

CODE 44
OXYGEN SENSOR CIRCUIT
(LEAN EXHAUST INDICATED)
2.8L (VIN W) "A" CARLINE (PORT)

Circuit Description:

The ECM supplies a voltage of about .45 volt between terminals "E14" and "E15". (If measured with a 10 megohm digital voltmeter, this may read as low as .32 volt). The O₂ sensor varies the voltage within a range of about 1 volt if the exhaust is rich, down through about .10 volt if exhaust is lean. The sensor is like an open circuit and produces no voltage when it is below 315°C (600°F). An open sensor circuit or cold sensor causes "Open Loop" operation.

Test Description: Numbers below refer to circled numbers on the diagnostic chart.

1. Code 44 is set when the O₂ sensor signal voltage on CKT 412
   - Remains below .2 volt for 60 seconds or more.
   - And the system is operating "Closed Loop."

Diagnostic Aids:

Using the "Scan," observe the block learn values at different rpm and air flow conditions. The "Scan" also displays the block cells, so the block learn values can be checked in each of the cells to determine when the Code 44 may have been set. If the conditions for Code 44 exist, the block learn values will be around 150.

- O₂ Sensor Wire Sensor pigtail may be mispositioned and contacting the exhaust manifold
- Check for intermittent ground in wire between connector and sensor.

- Lean Injector(s) Perform injector balance test CHART C-2A
- Fuel Contamination Water, even in small amounts, near the in-tank fuel pump inlet can be delivered to the injectors. The water causes a lean exhaust and can set a Code 44.
- Fuel Pressure System will be lean if pressure is too low. It may be necessary to monitor fuel pressure while driving the car at various road speeds and/or loads to confirm. See "Fuel System Diagnosis" CHART A-7.
- Exhaust Leaks If there is an exhaust leak, the engine can cause outside air to be pulled into the exhaust and past the sensor. Vacuum or crankcase leaks can cause a lean condition.
- Air System (manual trans "A" series only) Be sure air is not being directed to the exhaust ports while in "Closed Loop." If the block learn value goes down while squeezing air hose to exhaust ports, refer to CHART C-6
- If the above are OK, it is a faulty oxygen sensor.

1988–90 2.8L (VIN W)
CENTURY AND CUTLASS CIERA

CODE 44
OXYGEN SENSOR CIRCUIT
(LEAN EXHAUST INDICATED)
2.8L (VIN W) "A" CARLINE (PORT)

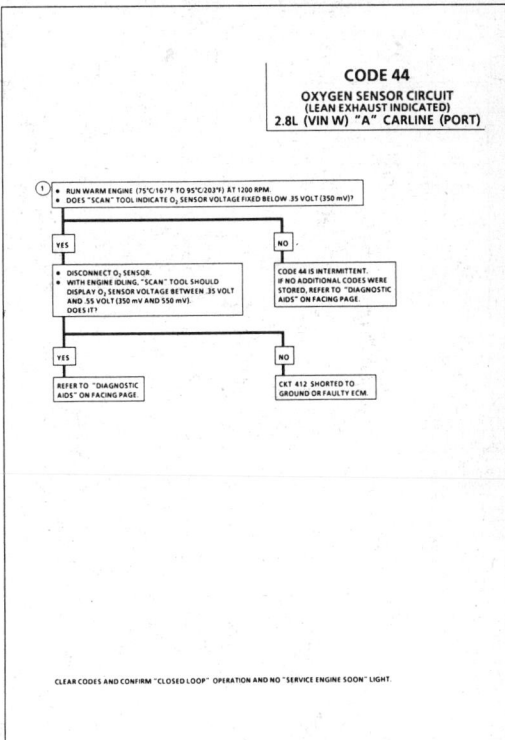

1988–90 2.8L (VIN W)
CENTURY AND CUTLASS CIERA

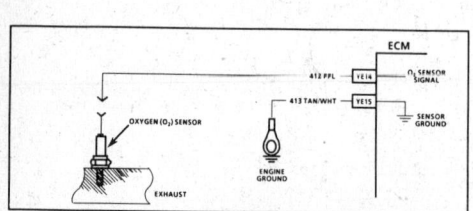

CODE 45
OXYGEN SENSOR CIRCUIT
(RICH EXHAUST INDICATED)
2.8L (VIN W) "A" CARLINE (PORT)

Circuit Description:

The ECM supplies a voltage of about .45 volt between terminals "E14" and "E15". (If measured with a 10 megohm digital voltmeter, this may read as low as .32 volt). The O_2 sensor varies the voltage within a range of about 1 volt if the exhaust is rich, down through about .10 volt if exhaust is lean.

The sensor is like an open circuit and produces no voltage when it is below about 315°C (600°F). An open sensor circuit or cold sensor causes "Open Loop" operation.

Test Description: Numbers below refer to circled numbers on the diagnostic chart.

1. Code 45 is set when the O_2 sensor signal voltage or CKT 412:
   - Remains above .7 volt for 30 seconds, and in "Closed Loop."
   - Engine time after start is 1 minute or more.
   - Throttle angle between 3% and 45%.

Diagnostic Aids:

Using the "Scan," observe the block learn values at different rpm and air flow conditions. The "Scan" also displays the block cells, so the block learn values can be checked in each of the cells to determine when the Code 45 may have been set. If the conditions for Code 45 exist, the block learn values will be around 115.

- Check for short to voltage on CKT 412.
- Fuel Pressure. System will go rich if pressure is too high. The ECM can compensate for some increase. However, if it gets too high, a Code 45 may be set. See "Fuel System Diagnosis" CHART A-7.
- Rich Injector. Perform injector balance test Chart C-2A.

- Leaking Injector. See CHART A-7.
- Check for fuel contaminated oil.
- HEI Shielding. An open ground CKT 453 (ignition system reflow) may result in EMI, or induced electrical "noise." The ECM looks at this "noise" as reference pulses. The additional pulses result in a higher than actual engine speed signal. The ECM then delivers too much fuel, causing system to go rich. Engine tachometer will also show higher than actual engine speed, which can help in diagnosing this problem.
- Canister Purge. Check for fuel saturation. If full of fuel, check canister control and hoses.
- Check for leaking fuel pressure regulator diaphragm by checking vacuum line to regulator for fuel.
- TPS. An intermittent TPS output will cause the system to go rich, due to a false signal the engine accelerating.
- EGR. An EGR staying open (especially at idle) will cause the O_2 sensor to indicate a rich exhaust, and this could result in a Code 45.

1988–90 2.8L (VIN W)
CENTURY AND CUTLASS CIERA

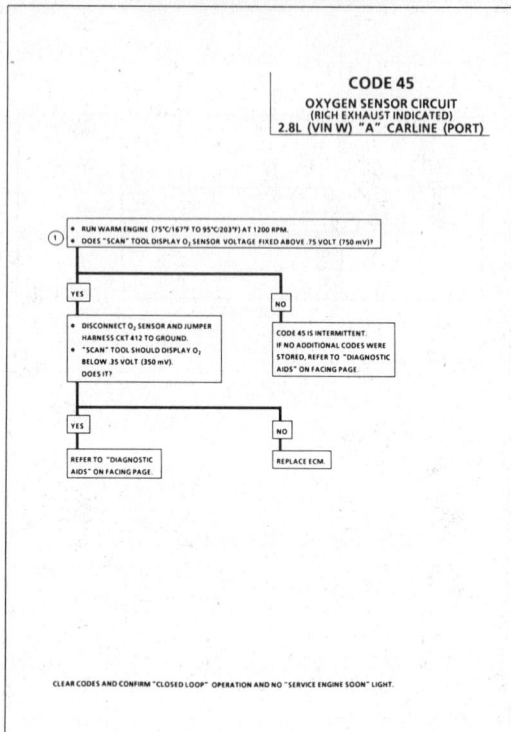

1988–90 2.8L (VIN W)
CENTURY AND CUTLASS CIERA

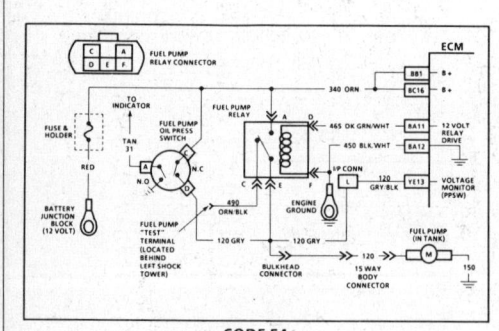

CODE 54
FUEL PUMP CIRCUIT
(LOW VOLTAGE)
2.8L (VIN W) "A" CARLINE (PORT)

Circuit Description:

The status of the fuel pump CKT 120 is monitored by the ECM at terminal "YE13" and is used to compensate fuel delivery based on system voltage. This signal is also used to store a trouble code if the fuel pump relay is defective or fuel pump voltage is lost while the engine is running. There should be about 12 volts on CKT 120 for 2 seconds after the ignition is turned on any time references pulses are being received by the ECM.

Code 54 will set if the voltage at terminal "YE13" is less than 2 volts for 1.5 seconds since the last reference pulse was received. This code is designed to detect a faulty relay, causing extended crank time, and the code will help the diagnosis of an engine that "Cranks But Will Not Run."

If a fault is detected during start-up, the "Service Engine Soon" light will stay "ON" until the ignition is cycled "OFF." However, if the voltage is detected below 2 volts with the engine running, the light will only remain "ON" while the condition exists.

1988–90 2.8L (VIN W)
CENTURY AND CUTLASS CIERA

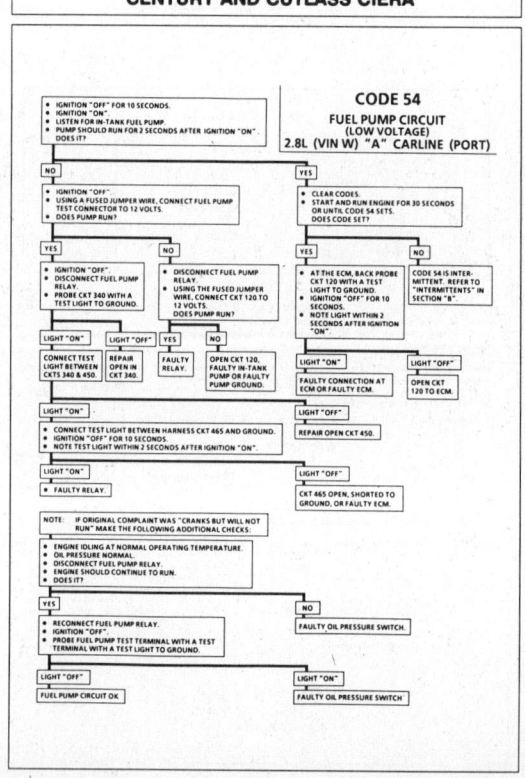

1988–90 2.8L (VIN W)
CENTURY AND CUTLASS CIERA

CODE 51
CODE 52
CODE 53
2.8L (VIN W) "A" CARLINE (PORT)

CODE 51
MEM-CAL ERROR
(FAULTY OR INCORRECT MEM-CAL)

CHECK THAT ALL PINS ARE FULLY INSERTED IN THE SOCKET AND THAT MEM-CAL IS PROPERLY LATCHED. IF OK, REPLACE MEMCAL, CLEAR MEMORY, AND RECHECK. IF CODE 51 REAPPEARS, REPLACE ECM.

CLEAR CODES AND CONFIRM "CLOSED LOOP" OPERATION AND NO "SERVICE ENGINE SOON" LIGHT.

CODE 52
CALPAK ERROR
(FAULTY OR INCORRECT CALPAK)

CHECK THAT THE MEM-CAL IS FULLY SEATED AND LATCHED INTO THE MEM-CAL SOCKET. IF OK, REPLACE MEM-CAL, CLEAR MEMORY, AND RECHECK. IF CODE 52 REAPPEARS, REPLACE ECM.

CLEAR CODES AND CONFIRM "CLOSED LOOP" OPERATION AND NO "SERVICE ENGINE SOON" LIGHT.

CODE 53
SYSTEM OVER VOLTAGE

THIS CODE INDICATES THERE IS A BASIC GENERATOR PROBLEM.
- CODE 53 WILL SET, IF VOLTAGE AT ECM IGNITION INPUT PIN IS GREATER THAN 17.1 VOLTS FOR 2 SECONDS.
- CHECK AND REPAIR CHARGING SYSTEM. REFER TO SECTION "6D".

CLEAR CODES AND CONFIRM "CLOSED LOOP" OPERATION AND NO "SERVICE ENGINE SOON" LIGHT.

1988–90 2.8L (VIN W)
CENTURY AND CUTLASS CIERA

CODE 61
DEGRADED OXYGEN SENSOR
2.8L (VIN W) "A" CARLINE (PORT)

IF A CODE 61 IS STORED IN MEMORY THE ECM HAS DETERMINED THE OXYGEN SENSOR IS CONTAMINATED OR DEGRADED, BECAUSE THE VOLTAGE CHANGE TIME IS SLOW OR SLUGGISH.

THE ECM PERFORMS THE OXYGEN SENSOR RESPONSE TIME TEST WHEN:

COOLANT TEMPERATURE IS GREATER THAN 85°C.

MAT TEMPERATURE IS GREATER THAN 10°C.

IN CLOSED LOOP.

IN DECEL FUEL CUT-OFF MODE.

IF A CODE 61 IS STORED THE OXYGEN SENSOR SHOULD BE REPLACED. A CONTAMINATED SENSOR CAN BE CAUSED BY FUEL ADDITIVES, SUCH AS SILICON, OR BY USE OF NON-GM APPROVED LUBRICANTS OR SEALANTS. SILICON CONTAMINATION IS USUALLY INDICATED BY A WHITE POWDERY SUBSTANCE ON THE SENSOR FINS.

CLEAR CODES AND CONFIRM "CLOSED LOOP" OPERATION AND NO "SERVICE ENGINE SOON" LIGHT.

1988–90 2.8L (VIN W)
CENTURY AND CUTLASS CIERA

SECTION B
SYMPTOMS

TABLE OF CONTENTS

Before Starting
Intermittents
Hard Start
Hesitation, Sag, Stumble
Surges and/or Chuggle
Lack of Power, Sluggish, or Spongy
Detonation/Spark Knock
Cuts Out, Misses
Backfire
Poor Fuel Economy
Dieseling, Run-On
Rough, Unstable, or Incorrect Idle, Stalling
Excessive Exhaust Emissions or Odors
CHART B-1 - Restricted Exhaust System Check

BEFORE STARTING

Before using this section you should have performed the DIAGNOSTIC CIRCUIT CHECK and found out that:
1. The ECM and "Service Engine Soon" light are operating.
2. There are no trouble codes.
   Verify the customer complaint, and locate the correct SYMPTOM below. Check the items indicated under that symptom.
   If the ENGINE CRANKS BUT WILL NOT RUN, see CHART A-3.
   Several of the symptom procedures below call for a Careful Visual Check. This check should include:
- ECM grounds for being clean and tight.
- Vacuum hoses for splits, kinks, and proper connections, as shown on Emission Control Information label.
- Air leaks at throttle body mounting and intake manifold.
- Ignition wires for cracking, hardness, proper routing, and carbon tracking.
- Wiring for proper connections, pinches, and cuts. The importance of this step cannot be stressed too strongly - it can lead to correcting a problem without further checks and can save valuable time.

1988–90 2.8L (VIN W)
CENTURY AND CUTLASS CIERA

INTERMITTENTS
Problem may or may not turn "ON" the "Service Engine Soon" light or store a code.

DO NOT use the trouble code charts in Section "A" for intermittent problems. The fault must be present to locate the problem. If a fault is intermittent, use of trouble code charts may result in replacement of good parts.
- Most intermittent problems are caused by faulty electrical connections or wiring. Perform careful check as described at start of Section "B". Check for:
  - Poor mating of the connector halves, or terminals not fully seated in the connector body (backed out).
  - Improperly formed or damaged terminals. All connector terminals in problem circuit should be carefully reformed to increase contact tension.
  - Poor terminal to wire connection. This requires removing the terminal from the connector body to check.
- If a visual check does not find the cause of the problem, the car can be driven with a voltmeter connected to a suspected circuit. A "Scan" tool can, also, be used for monitoring input signals to help detect intermittent conditions. An abnormal voltage, or "Scan" reading, when the problem occurs, indicates the problem may be in that circuit. If the wiring and connectors check OK and a trouble code was stored for a circuit having a sensor, except for Codes 43, 44, and 45, substitute a known good sensor and recheck.

An intermittent "Service Engine Soon" light with no stored code may be caused by:
- Ignition coil shorted to ground and arcing at spark plug wires or plugs.
- "Service Engine Soon" light wire to ECM shorted to ground. (CKT 419)
- Diagnostic "Test" terminal wire to ECM, shorted to ground. (CKT 451)
- ECM power grounds. See ECM wiring diagrams.
- Loss of trouble code memory. To check, disconnect TPS and idle engine until "Service Engine Soon" light comes "ON." Code 22 should be stored, and kept in memory when ignition is turned "OFF." If not, the ECM is faulty.
- Check for an electrical system interference caused by a defective relay, ECM driven solenoid, or switch. They can cause a sharp electrical surge. Normally, the problem will occur when the faulty component is operated.
- Check for improper installation of electrical options, such as lights, 2-way radios, etc.
- EST wires should be kept away from spark plug wires, coils and generator. Wire from ECM to ignition system (CKT 453) should be a good connection.
- Check for open diode across A/C compressor clutch, and for other open diodes (see wiring diagrams).

HARD START
Definition: Engine cranks OK, but does not start for a long time. Does eventually run, or may start but immediately dies.

- Perform careful check as described at start of Section "B".
- Make sure driver is using correct starting procedure.
- CHECK:
  - TPS for sticking or binding or a high TPS voltage with the throttle closed (should read less than .85 volt).
  - High resistance in coolant sensor circuit or sensor itself. See Code 15 chart or with a "Scan" tool compare coolant temperature with ambient temperature on a cold engine.
  - Fuel pressure CHART A-7.
  - Water contaminated fuel.
  - EGR operation. Be sure valve seats properly and is not staying open. See CHART C-7.
- Ignition system - Check for:
  Bare and shorted wires and proper output with spark tester J 26792 (ST-125) or equivalent.
  IAC operation - See Code 35 chart.
- A faulty in-tank fuel pump check valve will allow the fuel in the lines to drain back to the tank after the engine is stopped. To check for this condition:
  Perform fuel system diagnosis, CHART A-7.
  Remove spark plugs. Check for wet plugs, cracks, wear, improper gap, burned electrodes, or heavy deposits. Repair or replace as necessary.

1988–90 2.8L (VIN W)
CENTURY AND CUTLASS CIERA

HESITATION, SAG, STUMBLE

Definition: Momentary lack of response as the accelerator is pushed down. Can occur at all car speeds. Usually most severe when first trying to make the car move, as from a stop sign. May cause the engine to stall if severe enough.

- Perform careful visual check as described at start of Section "B".
- CHECK:
  - Fuel pressure. See CHART A-7. Also check for water contaminated fuel.
  - Spark plugs for being fouled or faulty wiring.
  - Mem-Cal number and Service Bulletins for latest Mem-Cal.

- TPS for binding or sticking. Voltage should increase at a steady rate as throttle is moved toward WOT.
- Generator output voltage. Repair, if less than 9 or more than 16 volts.
- Ignition system ground, CKT 453.
- Canister purge system for proper operation. See CHART C-3.
- EGR - See CHART C-7.
- MAP Sensor - See CHART C-1D, or Codes 33 & 34.
- Perform injector balance test, CHART C-2A.

SURGES AND/OR CHUGGLE

Definition: Engine power variation under steady throttle or cruise. Feels like the car speeds up and slows down with no change in the accelerator pedal.

- Be sure driver understands Transmission/Transaxle Converter Clutch and A/C compressor operation in Owner's Manual.
- Perform careful visual inspection as described at start of Section "B".
- To help determine if the condition is caused by a rich or lean system, the car should be driven at the speed of the complaint. Monitoring block learn at the complaint speed will help identify the cause of the problem. If the system is running lean (block learn greater than 138), refer to "Diagnostic Aids" on facing page of Code 44. If the system is running rich (block learn less than 118), refer to "Diagnostic Aids" on facing page of Code 45.

- CHECK:
  - Generator output voltage. Repair if less than 9 or more than 16 volts.
  - EGR - There should be no EGR at idle. See CHART C-7.
  - EGR filter for being plugged, see CHART C-7.
  - Vacuum lines for kinks or leaks.
  - In-line fuel filter. Replace if dirty or plugged.
  - Fuel pressure while condition exists. See CHART A-7.
- Remove spark plugs. Check for cracks, wear, improper gap, burned electrodes, or heavy deposits. Also check condition of spark plug wires and check for proper output voltage using spark tester J 26792 (ST-125) or equivalent.

1988–90 2.8L (VIN W)
CENTURY AND CUTLASS CIERA

LACK OF POWER, SLUGGISH, OR SPONGY

Definition: Engine delivers less than expected power. Little or no increase in speed when accelerator pedal is pushed down part way.

- Perform careful visual check as described at start of Section "B".
- Compare customer's car to similar unit. Make sure the customer's car has an actual problem.
- Remove air filter and check for dirt, or for being plugged. Replace as necessary.
- CHECK:
  - Restricted fuel filter, contaminated fuel or improper fuel pressure. See CHART A-7.
  - ECM power grounds, see wiring diagrams.
  - EGR operation for being open or partly open all the time. See CHART C-7.

- Exhaust system for possible restriction. See CHART B-1.
  - Inspect exhaust system for damaged or collapsed pipes.
  - Inspect muffler for heat distress or possible internal failure.
- Generator output voltage. Repair if less than 9 or more than 16 volts.
- Engine valve timing and compression.
- Engine for proper or worn camshaft.

- Secondary voltage using a shop ocilliscope or a spark tester J 26792 (ST-125) or equivalent.
- Check A/C operation. A/C clutch should cut out at WOT. See A/C CHART C-10.

DETONATION /SPARK KNOCK

Definition: A mild to severe ping, usually worse under acceleration. The engine makes sharp metallic knocks that change with throttle opening. Sounds like popcorn popping.

- Check for obvious overheating problems.
  - Low coolant.
  - Loose belt.
  - Restricted air flow to radiator, or restricted water flow through radiator.
  - Inoperative electric cooling fan circuit. See CHART C-12.
- To help determine if the condition is caused by a rich or lean system, the car should be driven at the speed of the complaint. Monitoring block learn, at the complaint speed, will help identify the cause of the problem. If the system is running lean (block learn greater than 138), refer to "Diagnostic Aids" on facing page of Code 44. If the system is running rich (block learn less than 118), refer to facing page of Code 45.

- CHECK:
  - EGR system for not opening or plugged EGR passages. See CHART C-7.
  - ESC system for no retard. See CHART C-5.
  - Park/Neutral switch. Be sure "Scan" indicates drive with gear selector in drive or overdrive. See CHART C-1A.
  - TCC operation, TCC applying too soon. See CHART C-8.
  - Fuel system pressure - See CHART A-7.
  - Engine for excessive carbon build-up. Remove carbon with top engine cleaner. Follow instructions on can.
  - Mem-Cal for correct part.
  - Combustion chamber for excessive oil.
  - Engine for incorrect basic parts such as cam, heads, pistons, etc.
  - Fuel for poor quality, proper octane rating.

1988–90 2.8L (VIN W)
CENTURY AND CUTLASS CIERA

CUTS OUT, MISSES

Definition: Steady pulsation or jerking that follows engine speed, usually more pronounced as engine load increases. The exhaust has a steady spitting sound at idle or under a load.

- Perform careful visual (physical) check as described at start of Section "B".
- If ignition system is suspected of causing a miss at idle or cutting out under load, refer to appropriate ignition "Misfire" chart in Section "C4".
- If the previous checks did not find the problem:
  - Visually inspect ignition system for moisture, dust, cracks, burns, etc. Spray plug wires with fine water mist to check for shorts.
- If above checks did not correct problem, check the following:
  - To determine if one injector or connector is faulty, disconnect injector 4-way connector. On the injector side of the harness, connect ohmmeter between the B+ circuit and the injector drive circuit for each bank of injectors. Because each bank of injectors is wired in parallel, an ohmmeter should measure about 4 ohms. If the reading is 6 ohms or greater, an open circuit is likely the cause. If 2 ohms or less, a shorted circuit or injector is likely. Reconnect injector 4-way connector.

- If either ohm reading is out of range:
  - Disconnect all injector harness connectors. Connect J 34730-2 Injector test light, or equivalent 6 volt test light, between the harness terminals of each injector connector and note light while cranking. If test light fails to blink at any connector, it is a faulty injector drive circuit harness, connector, or terminal.
- CHECK:
  - Injector balance. See CHART C-2A.
  - Fuel System - Plugged fuel filter, water, low pressure. See CHART A-7.
  - Valve Timing
- Remove rocker covers. Check for bent pushrods, worn rocker arms, broken valve springs, worn camshaft lobes. Repair as necessary.

- Perform compression check.

BACKFIRE

Definition: Fuel ignites in intake manifold, or in exhaust system, making a loud popping noise.

- CHECK:
  - Compression. Look for sticking or leaking valves.
  - EGR operation for being open all the time. See CHART C-7.
  - EGR gasket for faulty or loose fit.

- Output voltage of ignition coils, using a shop ocilliscope or spark tester J 26792 (ST-125), or equivalent.
- Valve timing.
- Spark plugs, spark plug wires, and proper routing of plug wires.

1988–90 2.8L (VIN W)
CENTURY AND CUTLASS CIERA

POOR FUEL ECONOMY

Definition: Fuel economy, as measured by an actual road test, is noticeably lower than expected. Also, economy is noticeably lower than it was on this car at one time, as previously shown by an actual road test.

- Check owner's driving habits.
  - Is A/C "ON" full time (defroster mode "ON")?
  - Are tires at correct pressure?
  - Is acceleration too much, too often?
  - Are excessively heavy loads being carried?
  - Suggest owner fill fuel tank and recheck fuel economy.
  - Suggest driver read "Important Facts on Fuel Economy" in Owner's Manual.
- Check for proper calibration of speedometer.
- Visually (physically) check:
  - Vacuum hoses for splits, kinks, and proper connections as shown on Vehicle Emission Control Information label.
  - Ignition wires for cracking, hardness, and proper connections.
  - Air cleaner element (filter) for dirt or being plugged.

- Remove spark plugs. Check for cracks, wear, improper gap, burned electrodes, or heavy deposits. Repair or replace as necessary.
- CHECK:
  - Engine thermostat for faulty part (always open) or for wrong heat range. Using a "Scan" tool, monitor engine temperature. A "Scan" displays engine temperature in degrees centigrade. After engine is started, the temperature should rise steadily to about 90°C, then stabilize, when thermostat opens.
  - Fuel Pressure. See CHART A-7.
  - Compression.
  - TCC for proper operation. See CHART C-8.
  - Exhaust system restriction. See CHART B-1.
  - Using a "Scan" tool, crank engine for five seconds while in clear flood mode. Start engine and check for a Code 42. An open EST line will cause poor fuel economy. Refer to Code 42 diagnostic chart for repair procedure.

DIESELING, RUN-ON

Definition: Engine continues to run after key is turned "OFF," but runs very roughly.

- Check injectors for leaking. See CHART A-7.
- If engine runs smoothly, check ignition switch and adjustment.

1988–90 2.8L (VIN W)
CENTURY AND CUTLASS CIERA

ROUGH, UNSTABLE, OR INCORRECT IDLE, STALLING

Definition: The engine runs unevenly at idle. If bad enough, the car may shake. Also, the idle may vary in rpm (called "hunting"). Either condition may be bad enough to cause stalling. Engine idles at incorrect speed. Engine will run rough or stall if loss of battery power to the ECM (Electronic Control Module) has occurred. This ECM has an idle learn feature. (See Section "C2" for Idle Learn Procedure.)

- Perform careful visual check as described at start of Section "B".
- CHECK:
  - For damaged, mispositioned or grounded motor mounts.
  - Throttle linkage for sticking or binding.
  - TPS for sticking or binding, be sure output is stable at idle and adjustment specification is correct.
  - IAC system. See Code 35 chart.
  - Generator output voltage. Repair if less than 9 or more than 16 volts.
  - P/N switch circuit. See CHART C-1A, or use "Scan" tool, and be sure tool indicates vehicle is in drive with gear selector in drive (125C), or overdrive (440-T4).
  - Injector balance. See CHART C-2A.
  - PCV valve for proper operation by placing finger over inlet hole in valve end several times. Valve should snap back. If not, replace valve.

- Evaporative Emission Control System. CHART C-3.
- Power Steering Pressure switch input. The state of the switch should only change when wheels are turned up against the stops. See CHART C-1E.
- ECM ground circuits.
- EGR valve. There should be no EGR at idle.
- Monitoring block learn values may help identify the cause of the problem. If the system is running lean (block learn greater than 138) refer to "Diagnostic Aids" on facing page of Code 44. If the system is running rich (block learn values less than 118) refer to "Diagnostic Aids" on facing page of Code 45.
- Run a cylinder compression check.

- Check for fuel in pressure regulator hose. If present, replace regulator assembly.
- Check ignition system; wires and plugs.
- If problem exists with A/C "ON," check A/C system operation CHART C-10.

EXCESSIVE EXHAUST EMISSIONS OR ODORS

Definition: Vehicle fails an emission test. Vehicle has excessive "rotten egg" smell. Excessive odors do not necessarily indicate excessive emissions.

- Perform "Diagnostic Circuit Check."
- IF TEST SHOWS EXCESSIVE CO AND HC, (or also has excessive odors):
- Check items which cause car to run RICH.
- Make sure engine is at normal operating temperature.
- CHECK:
  - Fuel pressure. See CHART A-7.
  - Canister for fuel loading. See CHART C-3.
  - Injector balance. See CHART C-2A.
  - PCV valve for being plugged, stuck, or blocked PCV hose, or fuel in the crankcase.
  - Spark plugs, plug wires, and ignition components.
  - Check for lead contamination of catalytic converter (look for removal of fuel filler neck restrictor).
  - Check for properly installed fuel cap.

- If the system is running rich, (block learn less than 118), refer to "Diagnostic Aids" on facing page of Code 45.
- IF TEST SHOWS EXCESSIVE NOx:
- Check items which cause car to run LEAN, or to run too hot.
  - EGR valve for not opening. See CHART C-7.
  - Vacuum leaks.
  - Coolant system and coolant fan for proper operation. See CHART C-12.
  - Remove carbon with top engine cleaner. Follow instructions on can.
- If the system is running lean (block learn greater than 138), refer to "Diagnostic Aids" on facing page of Code 44.

1988–90 2.8L (VIN W)
CENTURY AND CUTLASS CIERA

CHART B-1
RESTRICTED EXHAUST SYSTEM CHECK
ALL ENGINES

Proper diagnosis for a restricted exhaust system is essential before any components are replaced. Either of the following procedures may be used for diagnosis, depending upon engine or tool used.

CHECK AT A.I.R. PIPE:
1. Remove the rubber hose at the exhaust manifold A.I.R. pipe check valve. Remove check valve.
2. Connect a fuel pump pressure gauge to a hose and nipple from a Propane Enrichment Device (J 26911) (see illustration).
3. Insert the nipple into the exhaust manifold A.I.R. pipe.

OR   CHECK AT O₂ SENSOR:
1. Carefully remove O₂ sensor.
2. Install Borroughs exhaust backpressure tester (BT 8515 or BT 8603) or equivalent in place of O₂ sensor (see illustration).
3. After completing test described below, be sure to coat threads of O₂ sensor with anti-seize compound P/N 5613695 or equivalent prior to re-installation.

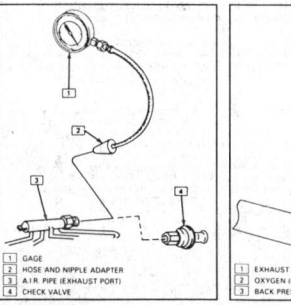

1	GAGE
2	HOSE AND NIPPLE ADAPTER
3	A.I.R. PIPE (EXHAUST PORT)
4	CHECK VALVE

1	EXHAUST MANIFOLD
2	OXYGEN (O₂) SENSOR
3	BACK PRESSURE GAGE

DIAGNOSIS:
1. With the engine idling at normal operating temperature, observe the exhaust system backpressure reading on the gage. Reading should not exceed 8.6 kPa (1.25 psi).
2. Increase engine speed to 2000 rpm and observe gage. Reading should not exceed 20.7 kPa (3 psi).
3. If the backpressure at either speed exceeds specification, a restricted exhaust system is indicated.
4. Inspect the entire exhaust system for a collapsed pipe, heat distress, or possible internal muffler failure.
5. If there are no obvious reasons for the excessive backpressure, the catalytic converter is suspected to be restricted and should be replaced using current recommended procedures.

1988–90 2.8L (VIN W)
CENTURY AND CUTLASS CIERA

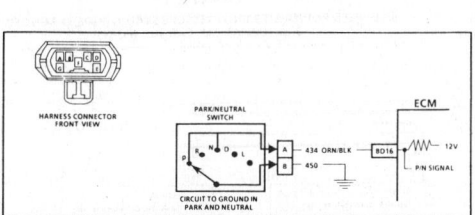

CHART C-1A
PARK/NEUTRAL SWITCH DIAGNOSIS
2.8L (VIN W) "A" CARLINE (PORT)

Circuit Description:
The park/neutral switch contacts are a part of the neutral start switch and are closed to ground in park or neutral, and open in drive ranges.

The ECM supplies ignition voltage through a current limiting resistor to CKT 434 and senses a closed switch when the voltage on CKT 434 drops to less than one volt.

The ECM uses the P/N signal as one of the inputs to control
- Idle air control
- VSS diagnostics
- EGR

If CKT 434 indicates P/N (grounded) while in drive range, the EGR would be inoperative, resulting in possible detonation.

If CKT 434 indicates drive (open) a dip in the idle may exist when the gear selector is moved into drive range.

Test Description: Numbers below refer to circled numbers on the diagnostic chart.
1. Checks for a closed switch to ground in park position. Different makes of "Scan" tools will read P/N differently. Refer to tool operator's manual for type of display used for a specific tool.
2. Checks for an open switch in drive range.
3. Be sure "Scan" indicates drive, even while wiggling shifter, to test for an intermittent or misadjusted switch in drive or overdrive range.

1988–90 2.8L (VIN W)
CENTURY AND CUTLASS CIERA

CHART C-1A
PARK/NEUTRAL SWITCH DIAGNOSIS
2.8L (VIN W) "A" CARLINE (PORT)

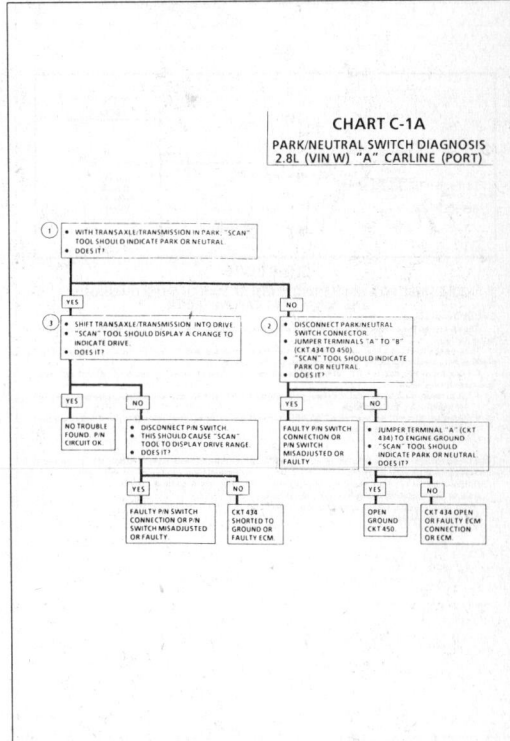

1988–90 2.8L (VIN W)
CENTURY AND CUTLASS CIERA

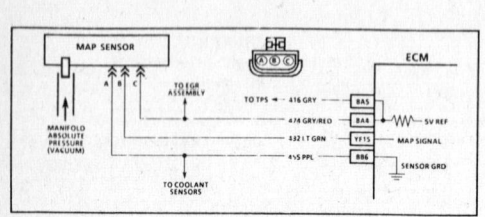

CHART C-1D
MANIFOLD ABSOLUTE PRESSURE (MAP) OUTPUT CHECK
2.8L (VIN W) "A" CARLINE (PORT)

Circuit Description:
The manifold absolute pressure (MAP) sensor measures manifold pressure (vacuum) and sends that signal to the ECM. The MAP sensor is mainly used for fuel calculation when the ECM is running in the throttle body backup mode. The MAP sensor is also used to determine the barometric pressure and to help calculate fuel delivery.

Test Description: Numbers below refer to circled numbers on the diagnostic chart.
1. Checks MAP sensor output voltage to the ECM. This voltage, without engine running, represents a barometer reading to the ECM.
2. Applying 34 kPa (10 inches Hg) vacuum to the MAP sensor should cause the voltage to be 1.2 volts less than the voltage at step 1. Upon applying vacuum to the sensor, the change in voltage should be instantaneous. A slow voltage change indicates a faulty sensor.

The engine must be running in this step or the "scanner" will not indicate a change in voltage. It is normal for the "Service Engine Soon" light to come "ON" and for the system to set a Code 63 during this step. Make sure the code is cleared when this test is completed.
3. Check vacuum hose to sensor for leaking or restriction. Be sure no other vacuum devices are connected to the MAP hose.

1988–90 2.8L (VIN W)
CENTURY AND CUTLASS CIERA

CHART C-1D
MANIFOLD ABSOLUTE PRESSURE (MAP) OUTPUT CHECK
2.8L (VIN W) "A" CARLINE (PORT)

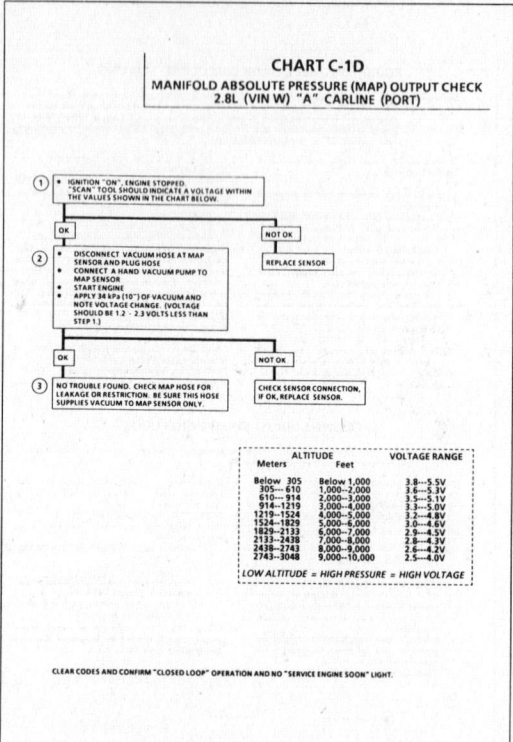

	ALTITUDE		VOLTAGE RANGE
Meters		Feet	
Below 305		Below 1,000	3.8–-5.5V
305— 610		1,000–2,000	3.6–-5.3V
610— 914		2,000–3,000	3.5—5.1V
914—1219		3,000–4,000	3.3—5.0V
1219—1524		4,000–5,000	3.2—4.8V
1524—1829		5,000–6,000	3.0—4.6V
1829—2133		6,000–7,000	2.9—4.5V
2133—2438		7,000–8,000	2.8—4.3V
2438—2743		8,000–9,000	2.6—4.2V
2743—3048		9,000–10,000	2.5—4.0V

LOW ALTITUDE = HIGH PRESSURE = HIGH VOLTAGE

CLEAR CODES AND CONFIRM "CLOSED LOOP" OPERATION AND NO "SERVICE ENGINE SOON" LIGHT.

1988–90 2.8L (VIN W)
CENTURY AND CUTLASS CIERA

CHART C-1E
IDLE SPEED POWER STEERING PRESSURE SWITCH (PSPS) DIAGNOSIS
2.8L (VIN W) "A" CARLINE (PORT)

Circuit Description:
The idle speed power steering pressure switch is normally open to ground, and CKT 901 will be near the battery voltage.
Turning the steering wheel increases power steering oil pressure and its load on an idling engine. The pressure switch will close before the load can cause an idle problem.
Closing the switch causes CKT 901 to read less than 1 volt. The ECM will increase the idle air rate and disengage the A/C relay.
- A pressure switch that will not close, or an open CKT 901 or 450, may cause the engine to stop when power steering loads are high.
- A switch that will not open or a CKT 901 shorted to ground may affect idle quality and will cause the A/C relay to be de-energized.

Test Description: Numbers below refer to circled numbers on the diagnostic chart.
1. Different makes of "Scan" tools may display the state of this switch in different ways. Refer to "Scan" tool operator's manual to determine how this input is indicated.
2. Checks to determine if CKT 901 is shorted to ground.
3. This should simulate a closed switch.

1988–90 2.8L (VIN W)
CENTURY AND CUTLASS CIERA

CHART C-1E
IDLE SPEED POWER STEERING PRESSURE SWITCH (PSPS) DIAGNOSIS
2.8L (VIN W) "A" CARLINE (PORT)

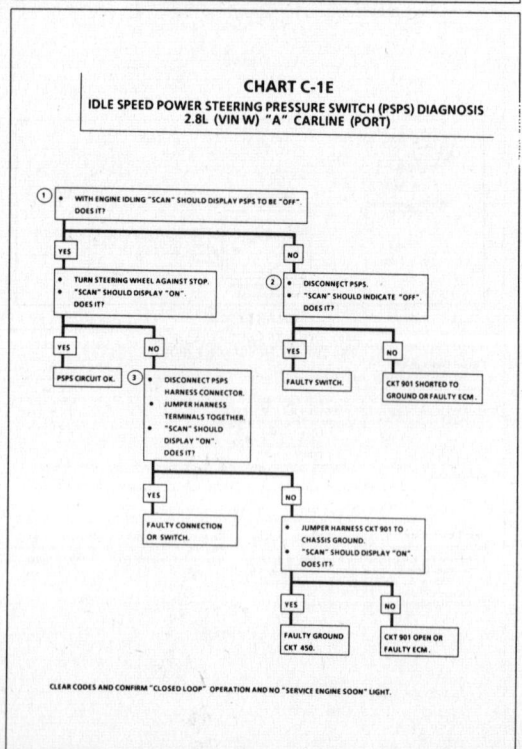

CLEAR CODES AND CONFIRM "CLOSED LOOP" OPERATION AND NO "SERVICE ENGINE SOON" LIGHT.

1988–90 2.8L (VIN W)
CENTURY AND CUTLASS CIERA

CHART C-2A

INJECTOR BALANCE TEST

The injector balance tester is a tool used to turn the injector on for a precise amount of time, thus spraying a measured amount of fuel into the manifold. This causes a drop in fuel rail pressure that we can record and compare between each injector. All injectors should have the same amount of pressure drop (± 10 kpa). Any injector with a pressure drop that is 10 kpa (or more) greater or less than the average drop of the other injectors should be considered faulty and replaced.

STEP 1

Engine "cool down" period (10 minutes) is necessary to avoid irregular readings due to "Hot Soak" fuel boiling. With ignition "OFF" connect fuel gauge J347301 or equivalent to fuel pressure tap. Wrap a shop towel around fitting while connecting gage to avoid fuel spillage.

Disconnect harness connectors at all injectors, and connect injector tester J-34730-3, or equivalent, to one injector. On Turbo equipped engines, use adaptor harness furnished with injector tester to energize injectors that are not accessible. Follow manufacturers instructions for use of adaptor harness. Ignition must be "OFF" at least 10 seconds to complete ECM shutdown cycle. Fuel pump should run about 2 seconds after ignition is turned "ON". At this point, insert clear tubing attached to vent valve into a suitable container and bleed air from gauge and hose to insure accurate gauge operation. Repeat this step until all air is bled from gauge

STEP 2

Turn ignition "OFF" for 10 seconds and then "ON" again to get fuel pressure to its maximum. Record this initial pressure reading. Energize tester one time and note pressure drop at its lowest point (Disregard any slight pressure increase after drop hits low point.) By subtracting this second pressure reading from the initial pressure, we have the actual amount of injector pressure drop.

STEP 3

Repeat step 2 on each injector and compare the amount of drop. Usually, good injectors will have virtually the same drop. Retest any injector that has a pressure difference of 10kPa, either more or less than the average of the other injectors on the engine. Replace any injector that also fails the retest. If the pressure drop of all injectors is within 10kPa of this average, the injectors appear to be flowing properly. Reconnect them and review Symptoms, Section "B".

NOTE: *The entire test should not be repeated more than once without running the engine to prevent flooding. (This includes any retest on faulty injectors).*

1988–90 2.8L (VIN W)
CENTURY AND CUTLASS CIERA

<u>NOTE:</u> The fuel pressure test in Section "A", CHART A-7, should be completed prior to this test.

CHART C-2A
INJECTOR BALANCE TEST
2.8L (VIN W) "A" CARLINE (PORT)

Step 1. If engine is at operating temperature, allow a 10 minute "cool down" period then connect fuel pressure gauge and injector tester.
1. Ignition "OFF".
2. Connect fuel pressure gauge and injector tester.
3. Ignition "ON".
4. Bleed off air in gauge. Repeat until all air is bled from gauge.

Step 2. Run test:
1. Ignition "OFF" for 10 seconds.
2. Ignition "ON". Record gauge pressure. (Pressure must hold steady, if not see the Fuel System diagnosis, Chart A-7, in Section "A").
3. Turn injector on, by depressing button on injector tester, and note pressure at the instant the gauge needle stops.

Step 3.
1. Repeat step 2 on all injectors and record pressure drop on each. Retest injectors that appear faulty (Any injectors that have a 10 kPa difference, either more or less, in pressure from the average. If no problem is found, review "Symptoms" Section "B".

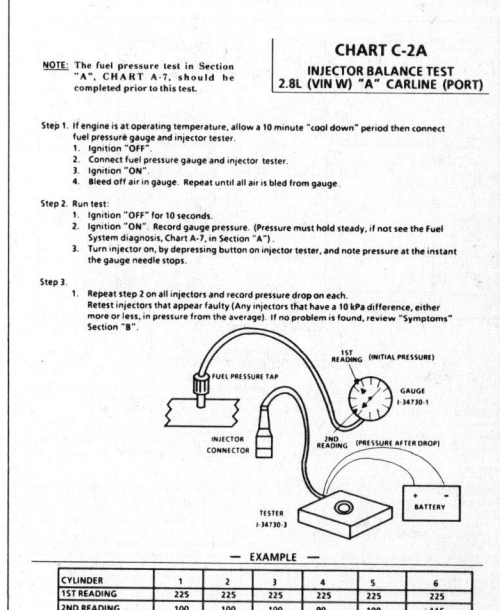

— EXAMPLE —

CYLINDER	1	2	3	4	5	6
1ST READING	225	225	225	225	225	225
2ND READING	100	100	100	90	100	115
AMOUNT OF DROP	125	125	125	135	125	110
	OK	OK	OK	FAULTY, RICH (TOO MUCH) (FUEL DROP)	OK	FAULTY, LEAN (TOO LITTLE) (FUEL DROP)

1988–90 2.8L (VIN W)
CENTURY AND CUTLASS CIERA

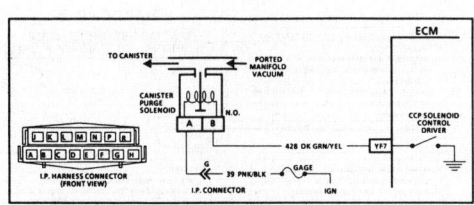

CHART C-3
CANISTER PURGE VALVE DIAGNOSIS
2.8L (VIN W) "A" CARLINE (PORT)

Circuit Description:
Canister purge is controlled by a solenoid that allows manifold vacuum to purge the canister when de-energized. The ECM supplies a ground to energize the solenoid (purge "OFF"). The purge solenoid controlled by the ECM is pulse width modulated (turned "ON" and "OFF" several times a second). The duty cycle (pulse width) is determined by the amount of air flow, and the engine vacuum as determined by the MAP sensor input. The duty cycle is calculated by the ECM and the output commanded when the following conditions have been met:
- Engine run time after start more than 3 minutes
- Coolant temperature above 80°C
- Vehicle speed above 15 mph
- Throttle "OFF" idle

Also, if the diagnostic test terminal is grounded, with the engine stopped, the purge solenoid is de-energized (purge "ON").

Test Description: Numbers below refer to circled numbers on the diagnostic chart.
1. Checks to see if the solenoid is opened or closed. The solenoid is normally energized in this step; so it should be closed.
2. Checks for a complete circuit. Normally there is ignition voltage on CKT 39 and the ECM provides a ground on CKT 428.
3. Completes functional check by grounding test terminal. This should normally de-energize the solenoid opening the valve which should allow the vacuum to drop (purge "ON").

1988–90 2.8L (VIN W)
CENTURY AND CUTLASS CIERA

CHART C-3
CANISTER PURGE VALVE DIAGNOSIS
2.8L (VIN W) "A" CARLINE (PORT)

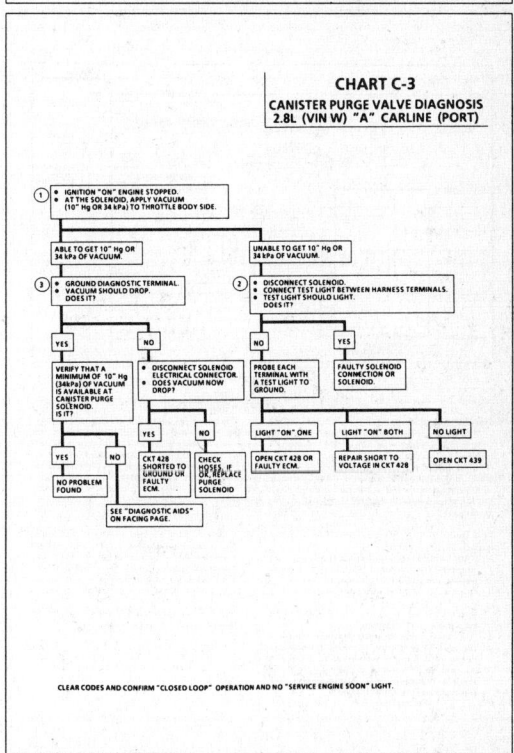

CLEAR CODES AND CONFIRM "CLOSED LOOP" OPERATION AND NO "SERVICE ENGINE SOON" LIGHT.

1988–90 2.8L (VIN W) CENTURY AND CUTLASS CIERA

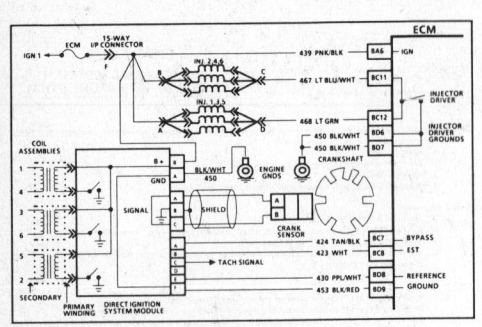

CHART C-4D-1
"DIS" MISFIRE AT IDLE
2.8L (VIN W) "A" CARLINE (PORT)

Circuit Description:
The direct ignition system (DIS) uses a waste spark method of distribution. In this type of system, the ignition module triggers the #1/4 coil pair resulting in both #1 and #4 spark plugs firing at the same time. #1 cylinder is on the compression stroke at the same time #4 is on the exhaust stroke, resulting in a lower energy requirement to fire #4 spark plug. This leaves the remainder of the high voltage to be used to fire #1 spark plug. On this application, the crank sensor is mounted to the engine block and protrudes through the block to within approximately .050" of the crankshaft reluctor. Since the reluctor is a machined portion of the crankshaft and the crank sensor is mounted in a fixed position on the block, timing adjustments are not possible or necessary.

Test Description: Numbers below refer to circled numbers on the diagnostic chart.
1. If the "misfire" complaint exists under load only, the diagnostic chart on page 2 must be used. Engine rpm should drop approximately equally on all plug leads.
2. A spark test such as a ST-125 must be used because it is essential to verify adequate available secondary voltage at the spark plug. (25,000 volts).
3. If the spark jumps the test gap after grounding the opposite plug wire, it indicates excessive resistance in the plug which was bypassed. A faulty or poor connection at that plug could also result in the miss condition. Also check for carbon deposits inside the spark plug boot.
4. If carbon tracking is evident, replace coil and be sure plug wires relating to that coil are clean and tight. Excessive wire resistance or faulty connections could have caused the coil to be damaged.
5. If the no spark condition follows the suspected coil, that coil is faulty, otherwise, the ignition module is the cause of no spark. This test could also be performed by substituting a known good coil for the one causing the no spark condition.

1988–90 2.8L (VIN W) CENTURY AND CUTLASS CIERA

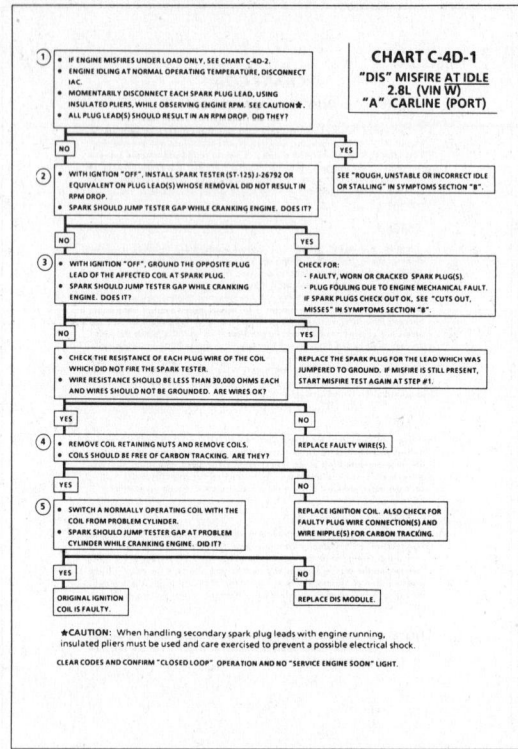

CHART C-4D-1
"DIS" MISFIRE AT IDLE
2.8L (VIN W) "A" CARLINE (PORT)

1988–90 2.8L (VIN W) CENTURY AND CUTLASS CIERA

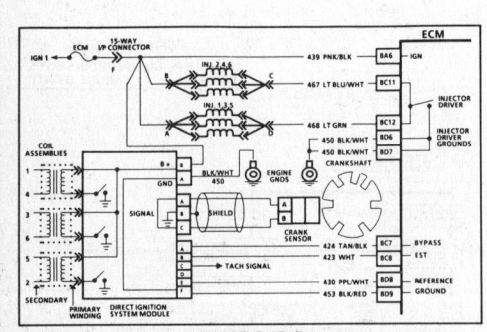

CHART C-4D-2
"DIS" MISFIRE UNDER LOAD
2.8L (VIN W) "A" CARLINE (PORT)

Circuit Description:
The direct ignition system (DIS) uses a waste spark method of distribution. In this type of system, the ignition module triggers the #1/4 coil pair resulting in both #1 and #4 spark plugs firing at the same time. #1 cylinder is on the compression stroke at the same time #4 is on the exhaust stroke, resulting in a lower energy requirement to fire #4 spark plug. This leaves the remainder of the high voltage to be used to fire #1 spark plug. On this application, the crank sensor is mounted to the engine block and protrudes through the block to within approximately .050" of the crankshaft reluctor. Since the reluctor is a machined portion of the crankshaft and the crank sensor is mounted in a fixed position on the block, timing adjustments are not possible or necessary.

Test Description: Numbers below refer to circled numbers on the diagnostic chart.
1. If the "misfire" complaint exists at idle only, the diagnostic chart on page 1 must be used. A spark tester such as a ST-125 must be used because it is essential to verify adequate available secondary voltage at the spark plug. (25,000 volts). Spark should jump the test gap on all 4 leads. This simulates a "load" condition.
2. If the spark jumps the tester gap after grounding the opposite plug wire, it indicates excessive resistance in the plug which was bypassed. A faulty or poor connection at that plug could also result in the miss condition. Also check for carbon deposits inside the spark plug boot.
3. If carbon tracing is evident replace coil and be sure plug wires relating to that coil are clean and tight. Excessive wire resistance or faulty connections could have caused the coil to be damaged.
4. If the no spark condition follows the suspected coil, that coil is faulty. Otherwise, the ignition module is the cause of no spark. This test could also be performed by substituting a known good coil for the one causing the no spark condition.

1988–90 2.8L (VIN W) CENTURY AND CUTLASS CIERA

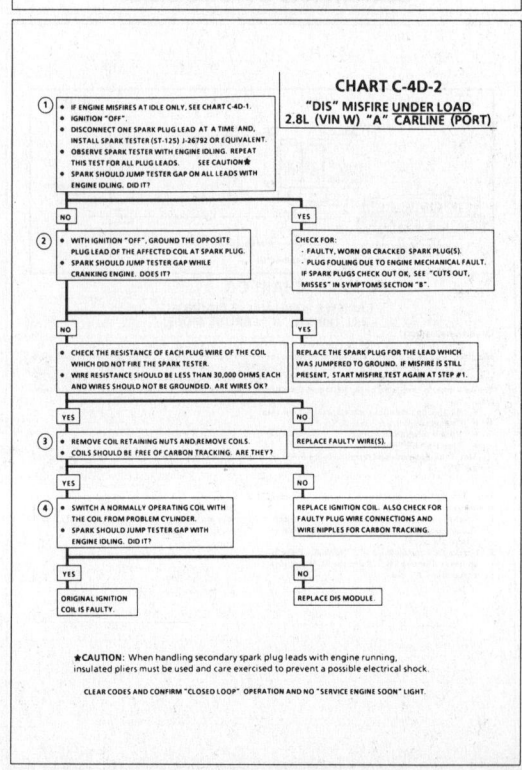

CHART C-4D-2
"DIS" MISFIRE UNDER LOAD
2.8L (VIN W) "A" CARLINE (PORT)

1988–90 2.8L (VIN W)
CENTURY AND CUTLASS CIERA

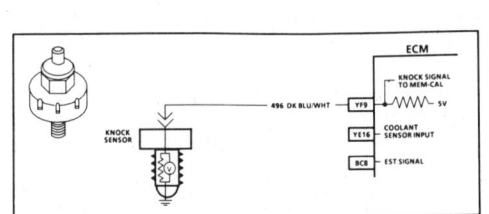

```
                                    ECM
                              KNOCK SIGNAL
                              TO MEM-CAL
         496 DK BLU/WHT  YF9  ─── 5V
  KNOCK                         COOLANT
  SENSOR                  YE16  SENSOR INPUT

                          BC8   EST SIGNAL
```

CHART C-5
ELECTRONIC SPARK CONTROL (ESC) SYSTEM CHECK
2.8L (VIN W) "A" CARLINE (PORT)

Circuit Description:

The knock sensor is used to detect engine detonation and the ECM will retard the electronic spark timing based on the signal being received. The circuitry within the knock sensor causes the ECM's 5 volts to be pulled down so that under a no knock condition, CKT 496 would measure about 2.5 volts. The knock sensor produces an A/C signal which rides on the 2.5 volt DC voltage. The amplitude and frequency are dependent upon the knock level.

The Mem-Cal used with this engine, contains the functions which were part of remotely mounted ESC modules used on other GM vehicles. The ESC portion of the Mem-Cal then sends a signal to other parts of the ECM which adjusts the spark timing to retard the spark and reduce the detonation.

Test Description: Numbers below refer to circled numbers on the diagnostic chart.

1. With engine idling, there should not be a knock signal present at the ECM because detonation is not likely under a no load condition.
2. Tapping on the engine lift hood bracket should simulate a knock signal to determine if the sensor is capable of detecting detonation. If no knock is detected, try tapping on engine block closer to sensor before replacing sensor.
3. If the engine has an internal problem which is creating a knock, the knock sensor may be responding to the internal failure.

4. This test determines if the knock sensor is faulty or if the ESC portion of the Mem-Cal is faulty. If it is determined that the Mem-Cal is faulty, be sure that is is properly installed and latched into place. If not properly installed, repair and retest.

Diagnostic Aids:

While observing knock signal on the "Scan," there should be an indication that knock is present when detonation can be heard. Detonation is most likely to occur under high engine load conditions.

1988–90 2.8L (VIN W)
CENTURY AND CUTLASS CIERA

CHART C-5
ELECTRONIC SPARK CONTROL (ESC) SYSTEM CHECK
2.8L (VIN W) "A" CARLINE (PORT)

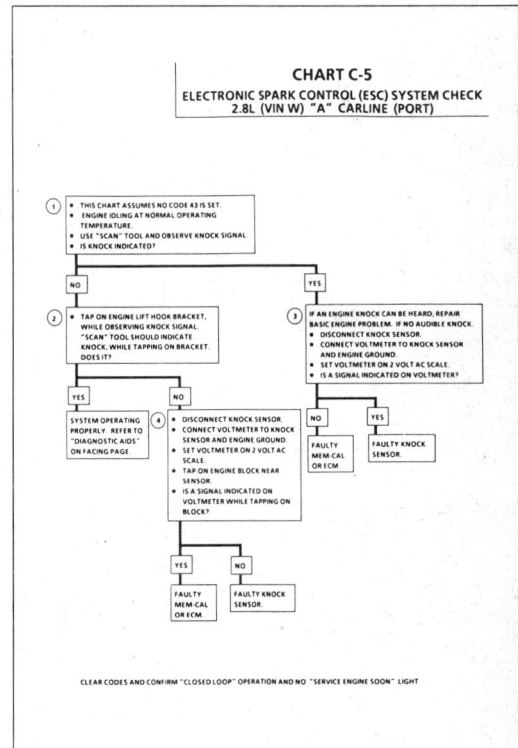

1988–90 2.8L (VIN W)
CENTURY AND CUTLASS CIERA

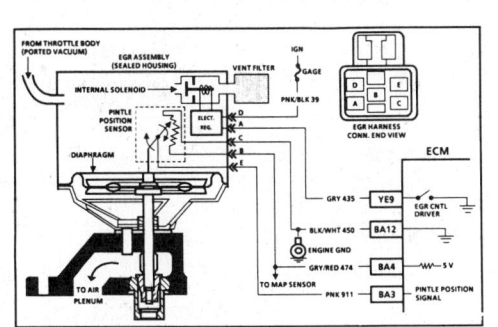

```
                                            ECM
                          GRY 435   YE9   EGR CNTL
                                           DRIVER
                          BLK/WHT 450 BA12
                          GRY/RED 474  BA4  ─── 5 V
                                           ENGINE GND
                          PNK 911   BA3   PINTLE POSITION
                                           SIGNAL
```

CHART C-7
EGR VALVE CHECK
2.8L (VIN W) "A" CARLINE (PORT)

Circuit Description:

The integrated electronic EGR valve functions similiar to a port valve with a remote vacuum regulator. The internal solenoid is normally open, which causes the vacuum signal to be vented "OFF" to the atmosphere when EGR is not being commanded by the ECM. This EGR valve has a sealed cap and the solenoid valve opens and closes the vacuum signal, which controls the amount of vacuum vented to atmosphere, and this controls the amount of vacuum applied to the diaphragm. The electronic EGR valve contains a voltage regulator, which converts the ECM signal, to provide different amounts of EGR flow by regulating the current to the solenoid. The ECM controls EGR flow with a pulse width modulated signal (turns "ON" and "OFF" many times a second) based on airflow, TPS, and rpm.

This system also contains a pintle position sensor, which works similiar to a TPS sensor, and as EGR flow is increased, the sensor output also increases.

Test Description: Numbers below refer to circled numbers on the diagnostic chart.

1. Whenever the solenoid is denergized, the solenoid valve should be closed, which should not allow the vacuum to move the EGR diaphragm. However, if the filter is plugged, the vacuum applied with the hand held vacuum pump will cause the diaphragm to move because the vacuum will not be vented to the atmosphere.
2. This test will determine if the EGR filter is plugged, or if the EGR itself is faulty. Use care when removing the filter to avoid damaging the EGR assembly. See On-Car Service for procedure.
3. If the valve moves in this test, it's probably due to CKT 435 being shorted to ground.

4. Grounding the diagnostic terminal should energize the solenoid which closes off the vent and allows the vacuum to move the diaphragm.
5. The EGR assembly is designed to have some leakage and, therefore, 7" of vacuum is all that should be able to be held on the assembly. However, if too much of a leak exists (less than 3"), the EGR assembly is leaking and must be replaced.

Diagnostic Aids:

Some "Scan" tools will read EGR pintle position in volts. The EGR position voltage can be used to determine that the pintle is moving. When no EGR is commanded (0% duty cycle), the position sensor should read between .5 volt and 1.5 volts, and increase with the commanded EGR duty cycle.

1988–90 2.8L (VIN W)
CENTURY AND CUTLASS CIERA

CHART C-7
EGR VALVE CHECK
2.8L (VIN W) "A" CARLINE (PORT)

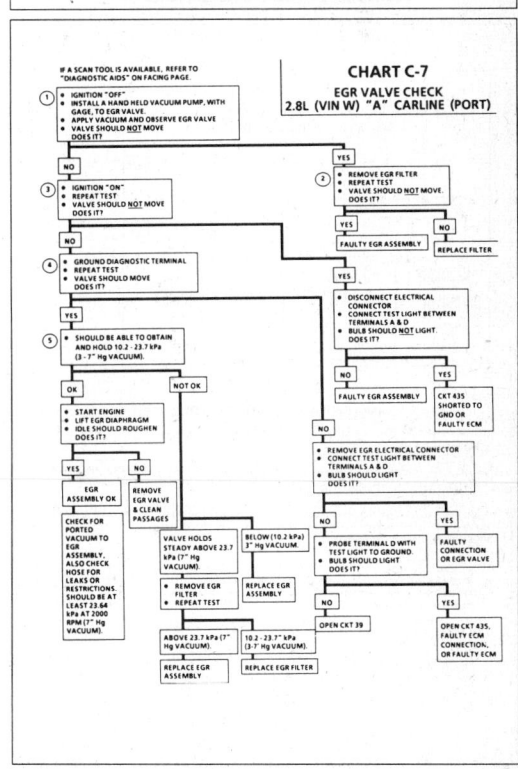

1988–90 2.8L (VIN W) CENTURY AND CUTLASS CIERA

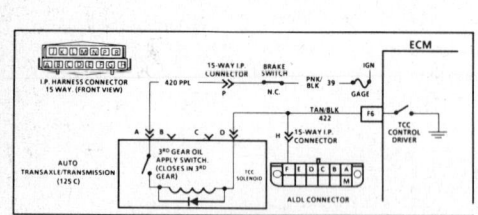

CHART C-8A
125C TORQUE CONVERTER CLUTCH (TCC)
(ELECTRICAL DIAGNOSIS)
2.8L (VIN W) "A" CARLINE (PORT)

Circuit Description:
The purpose of the transmission converter clutch feature is to eliminate the power loss of the transmission converter stage when the vehicle is in a cruise condition. This allows the convenience of the automatic transmission and the fuel economy of a manual transmission.

Fused battery ignition is supplied to the TCC solenoid through the brake switch, and transmission third gear apply switch. The ECM will engage the TCC by grounding CKT 422 to energize the solenoid.

TCC will engage when:
• Engine warmed up
• Vehicle speed above a calibrated value. (about 32 mph 51 km/h)
• Throttle position sensor output not changing, indicating a steady road speed
• Transmission third gear switch closed
• Brake switch closed

Test Description: Numbers below refer to circled numbers on the diagnostic chart.
1. Light "OFF" confirms transmission third gear apply switch is open.
2. At 25 mph the transmission third gear apply switch should close. Test light will come "ON" and confirm battery supply and closed brake switch.
3. Grounding the diagnostic terminal with ignition "ON," engine "OFF," should energize the TCC solenoid by grounding CKT 422. This test checks the ability of the ECM to supply a ground to the TCC solenoid. The test light connected from 12 volts to ALDL terminal "F" will turn "ON" as CKT 422 is grounded.

Diagnostic Aids:
A "Scan" tool only indicates when the ECM has turned "ON" the TCC driver, and this does not confirm that the TCC has engaged. To determine if TCC is functioning properly, engine rpm should decrease when the "Scan" indicates the TCC driver has turned "ON."

1988–90 2.8L (VIN W) CENTURY AND CUTLASS CIERA

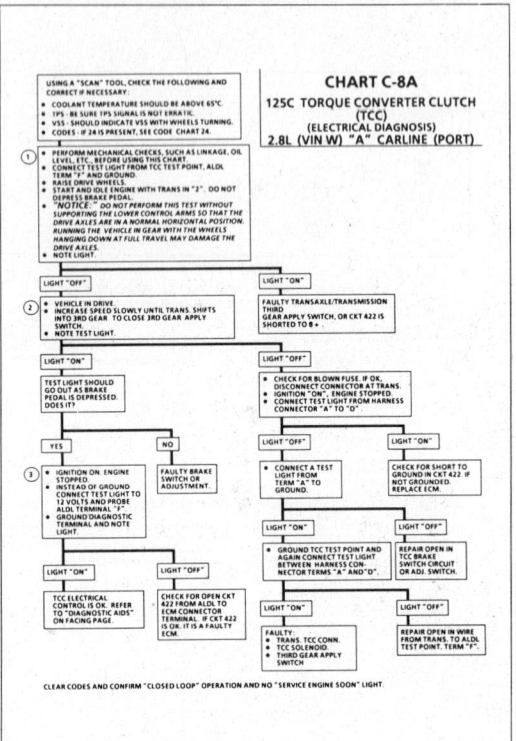

1988–90 2.8L (VIN W) CENTURY AND CUTLASS CIERA

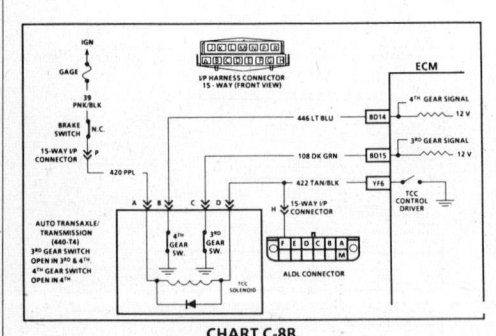

CHART C-8B
(Page 1 of 2)
440-T4 TORQUE CONVERTER CLUTCH (TCC)
(ELECTRICAL DIAGNOSIS)
2.8L (VIN W) "A" CARLINE (PORT)

Circuit Description:
The purpose of the automatic transmission torque converter clutch feature is to eliminate the power loss of the torque converter stage when the vehicle is in a cruise condition. This allows the convenience of the automatic transmission and the fuel economy of a manual transmission. The heart of the system is a solenoid located inside the automatic transmission which is controlled by the ECM.

When the solenoid coil is activated ("ON"), the torque converter clutch is applied which results in straight through mechanical coupling from the engine to transmission. When the transmission solenoid is deactivated, the torque converter clutch is released which allows the torque converter to operate in the conventional manner (fluidic coupling between engine and transmission).

The TCC will engage on a warm engine under given road load in 3rd and 4th gears.
TCC will engage when:
• Engine warmed up
• Vehicle speed above a calibrated value (about 28 mph 45 km/h)
• Throttle position sensor output not changing, indicating a steady road speed
• Brake switch closed

Test Description: Numbers below refer to circled numbers on the diagnostic chart.
1. This test checks the continuity of the TCC circuit from the fuse to the ALDL connector.
2. When the brake pedal is released, the light should come back "ON" and then go "OFF" when the diagnostic terminal is grounded. This tests CKT 422 and the TCC driver in the ECM.

Diagnostic Aids:
The "Scan" tool only indicates when the ECM has turned "ON" the TCC driver, and this does not confirm that the TCC has engaged. To determine if TCC is functioning properly, engine rpm should decrease when the "Scan" indicates the TCC driver has turned "ON."

1988–90 2.8L (VIN W) CENTURY AND CUTLASS CIERA

CHART C-8B
(Page 1 of 2)
440-T4 TORQUE CONVERTER CLUTCH (TCC)
(ELECTRICAL DIAGNOSIS)
2.8L (VIN W) "A" CARLINE (PORT)

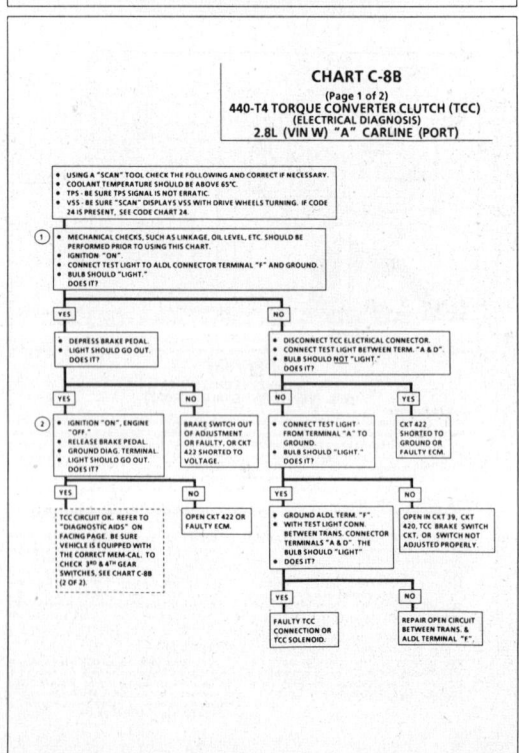

1988–90 2.8L (VIN W)
CENTURY AND CUTLASS CIERA

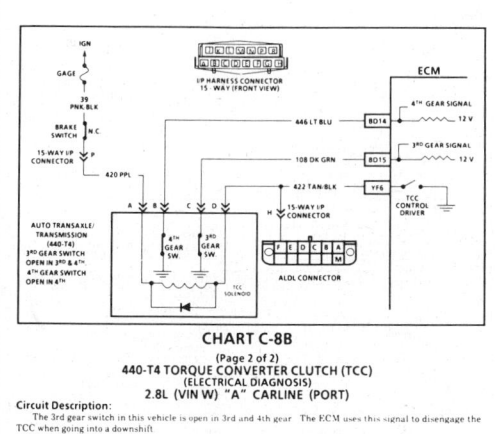

CHART C-8B
(Page 2 of 2)
440-T4 TORQUE CONVERTER CLUTCH (TCC)
(ELECTRICAL DIAGNOSIS)
2.8L (VIN W) "A" CARLINE (PORT)

Circuit Description:
The 3rd gear switch in this vehicle is open in 3rd and 4th gear. The ECM uses this signal to disengage the TCC when going into a downshift.
The fourth gear switch is open in fourth gear.

Test Description: Numbers below refer to circled numbers on the diagnostic chart.
1. Some "Scan" tools display the state of these switches in different ways. Be familiar with the type of tool being used. Since both switches should be in the closed state during this test, the tool should read the same for either the 3rd or 4th gear switch.
2. Determines whether the switch or signal circuit is open. The circuit can be checked for an open by measuring the voltage (with a voltmeter) at the TCC connector. Should be about 12 volts.
3. Because the switch(es) should be grounded in this step, disconnecting the TCC connector should cause the "Scan" switch state to change.
4. The switch state should change when the vehicle shifts into 3rd gear.

Diagnostic Aids:
If vehicle is road tested because of a TCC related problem, be sure the switch states do not change while in 4th gear because the TCC will disengage. If switches change state, carefully check wire routing and connections.

1988–90 2.8L (VIN W)
CENTURY AND CUTLASS CIERA

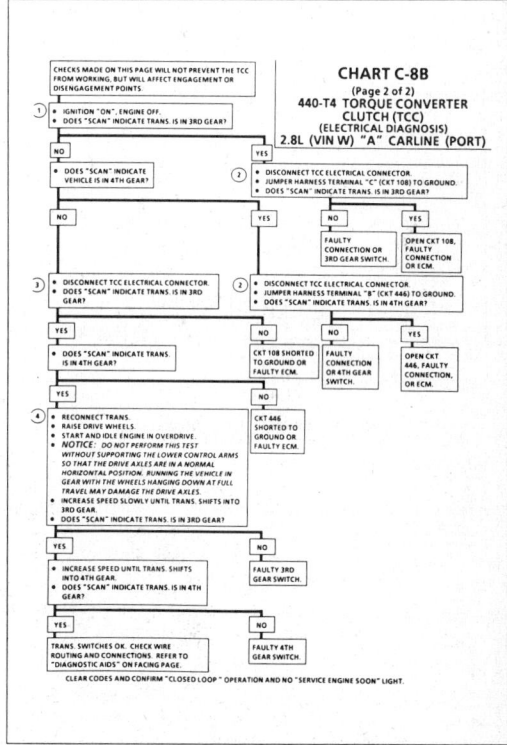

CHART C-8B
(Page 2 of 2)
440-T4 TORQUE CONVERTER CLUTCH (TCC)
(ELECTRICAL DIAGNOSIS)
2.8L (VIN W) "A" CARLINE (PORT)

1988–90 2.8L (VIN W)
CENTURY AND CUTLASS CIERA

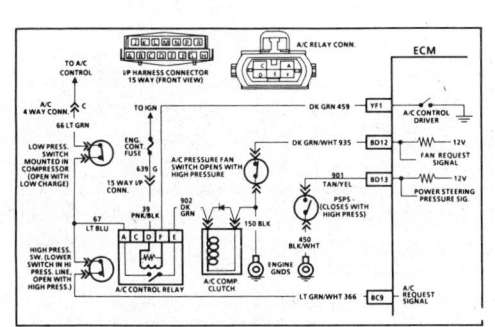

CHART C-10
A/C CLUTCH CONTROL CIRCUIT DIAGNOSIS
2.8L (VIN W) "A" CARLINE (PORT)

Circuit Description:
The A/C clutch control relay is ECM controlled to delay A/C clutch engagement about .4 second after A/C is turned "ON." This allows the IAC to adjust engine rpm, before the A/C clutch engages. The ECM, also, causes the relay to disengage the A/C clutch during WOT, when high power steering pressure is present, or if engine is overheating. The A/C clutch control relay is energized, when the ECM provides a ground path for CKT 459. The low pressure switch will open if A/C pressure is less than 40 psi (276 kPa). The high pressure switch will open, if A/C pressure exceeds about 440 psi (3034 kPa). The A/C pressure fan switch opens when A/C pressure exceeds about 200 psi (1380 kPa).

Test Description: Numbers below refer to circled numbers on the diagnostic chart.
1. The ECM will only energize the A/C relay, when the engine is running. This test will determine if the relay or CKT 459 is faulty.
2. In order for the clutch to properly be engaged, the low pressure switch must be closed to provide 12 volts to the relay, and the high pressure switch must be closed, so the A/C request (12 volts) will be present at the ECM.
3. Determines if the signal is reaching the ECM on CKT 366 from the A/C control panel. Signal should only be present when the A/C mode or defrost mode has been selected.
4. A short to ground in any part of the A/C request circuit, CKT 67 to the relay, CKT 902 to the A/C clutch, or the A/C clutch, could be the cause of the blown fuse.
5. If the ECM is seeing a high power steering pressure signal, the A/C clutch will be disengaged by the ECM.
6. With the engine idling and A/C "ON," the ECM should be grounding CKT 459, which should cause the test light to be "ON."

Diagnostic Aids:
If complaint was insufficient cooling, the problem may be caused by a inoperative cooling fan, or A/C pressure fan switch. The engine cooling fan should turn "ON," if A/C pressure exceeds a value to open the switch, which causes the ECM to energize the cooling fan relay. See CHART C-12, for cooling fan diagnosis.

1988–90 2.8L (VIN W)
CENTURY AND CUTLASS CIERA

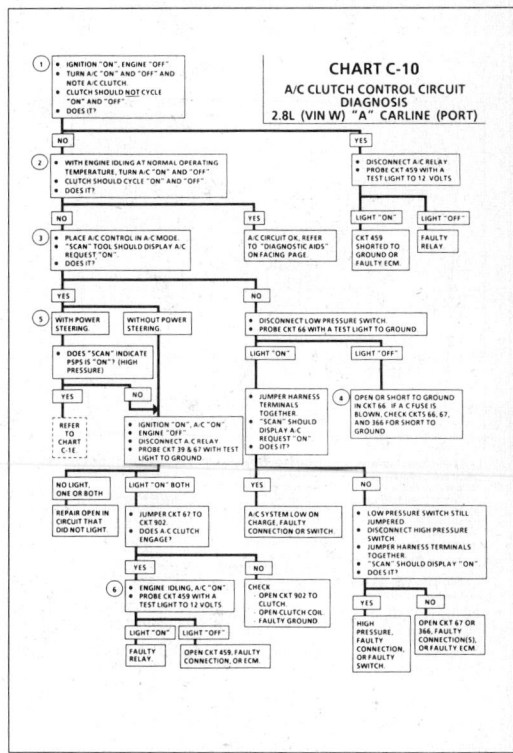

CHART C-10
A/C CLUTCH CONTROL CIRCUIT DIAGNOSIS
2.8L (VIN W) "A" CARLINE (PORT)

1988–90 2.8L (VIN W)
CENTURY AND CUTLASS CIERA

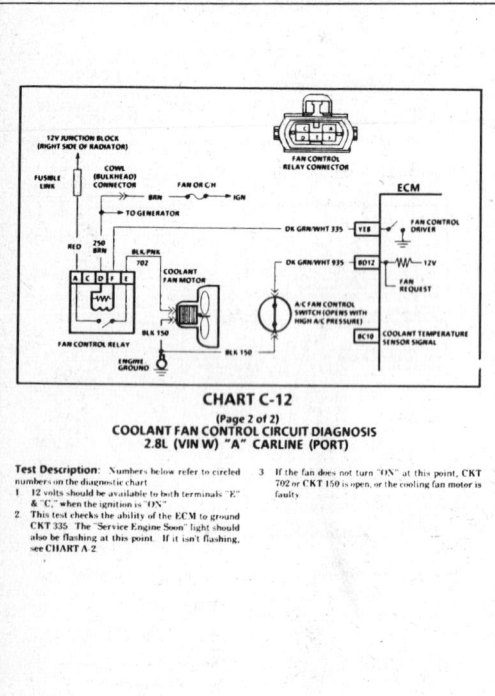

CHART C-12
(Page 2 of 2)
COOLANT FAN CONTROL CIRCUIT DIAGNOSIS
2.8L (VIN W) "A" CARLINE (PORT)

Test Description: Numbers below refer to circled numbers on the diagnostic chart.

1. 12 volts should be available to both terminals "E" & "C" when the ignition is "ON".
2. This test checks the ability of the ECM to ground CKT 335. The "Service Engine Soon" light should also be flashing at this point. If it isn't flashing, see CHART A-2.
3. If the fan does not turn "ON" at this point, CKT 702 or CKT 150 is open, or the cooling fan motor is faulty.

1988–90 2.8L (VIN W)
CENTURY AND CUTLASS CIERA

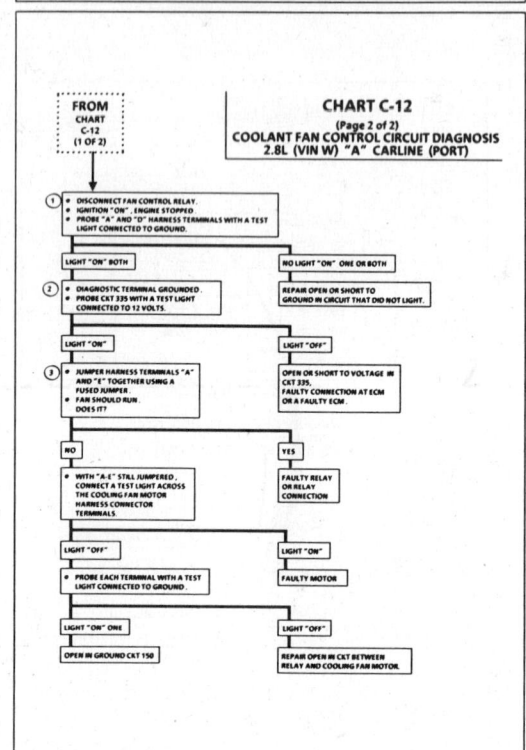

2.8L (VIN W) COMPONENT LOCATIONS—CAVALIER

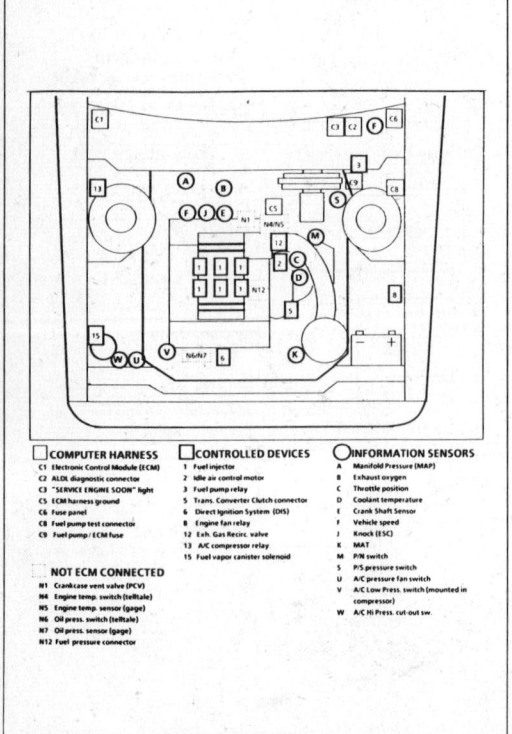

☐ **COMPUTER HARNESS**
C1 Electronic Control Module (ECM)
C2 ALDL diagnostic connector
C3 "SERVICE ENGINE SOON" light
C5 ECM harness ground
C6 Fuse panel
C8 Fuel pump test connector
C9 Fuel pump / ECM fuse

☐ **CONTROLLED DEVICES**
1 Fuel injector
2 Idle air control motor
3 Fuel pump relay
5 Trans. Converter Clutch connector
6 Direct Ignition System (DIS)
8 Engine fan relay
12 Exh. Gas Recirc. valve
13 A/C compressor relay
15 Fuel vapor canister solenoid

○ **INFORMATION SENSORS**
A Manifold Pressure (MAP)
B Exhaust oxygen
C Throttle position
D Coolant temperature
E Crank Shaft Sensor
F Vehicle speed
J Knock (ESC)
K MAT
M P/N switch
S P/S pressure switch
U A/C pressure switch
V A/C Low Press. switch (mounted in compressor)
W A/C Hi Press. cut-out sw.

NOT ECM CONNECTED
N1 Crankcase vent valve (PCV)
N4 Engine temp. switch (telltale)
N5 Engine temp. sensor (gage)
N6 Oil press. switch (telltale)
N7 Oil press. sensor (gage)
N12 Fuel pressure connector

2.8L (VIN W) ECM WIRING DIAGRAM—CAVALIER

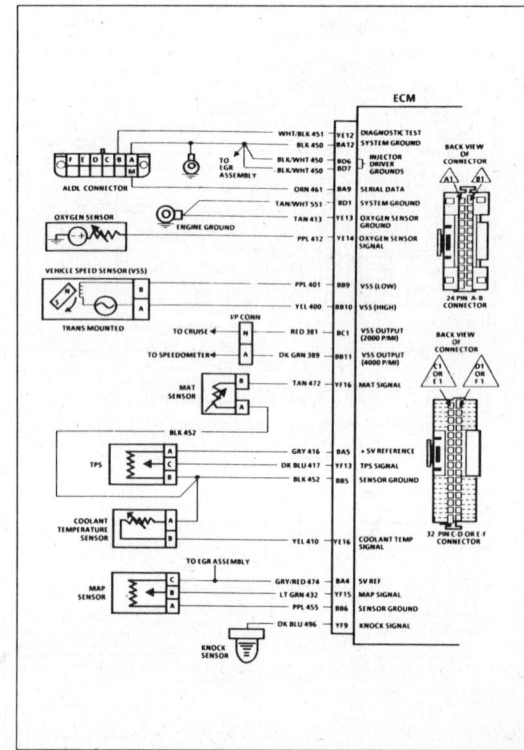

2.8L (VIN W) ECM WIRING DIAGRAM CAVALIER (CONT.)

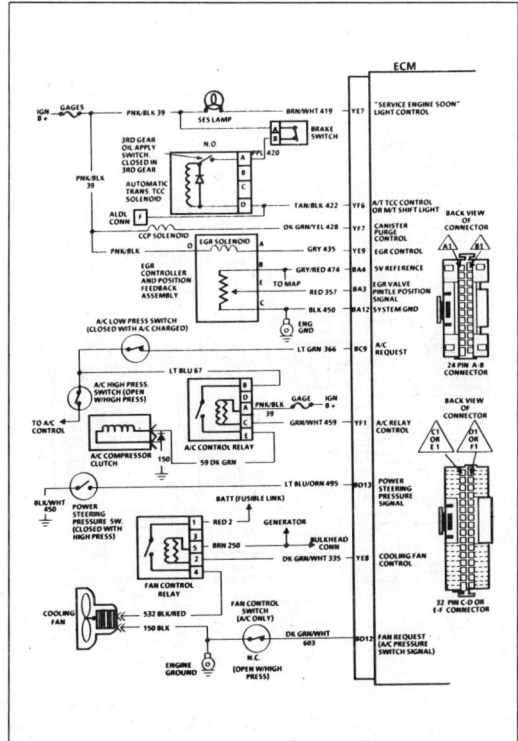

2.8L (VIN W) ECM WIRING DIAGRAM CAVALIER (CONT.)

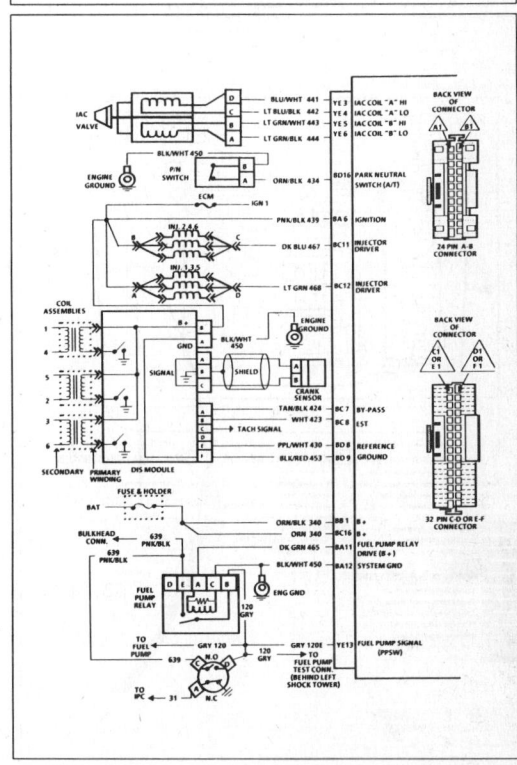

2.8L (VIN W) ECM CONNECTOR TERMINAL END VIEW — CAVALIER

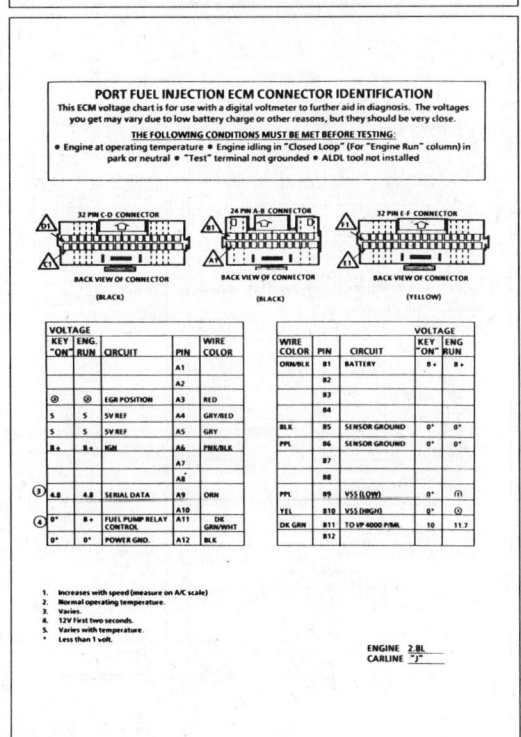

PORT FUEL INJECTION ECM CONNECTOR IDENTIFICATION

This ECM voltage chart is for use with a digital voltmeter to further aid in diagnosis. The voltages you get may vary due to low battery charge or other reasons, but they should be very close.

THE FOLLOWING CONDITIONS MUST BE MET BEFORE TESTING:
- Engine at operating temperature • Engine idling in "Closed Loop" (For "Engine Run" column) in park or neutral • "Test" terminal not grounded • ALDL tool not installed

Voltage KEY "ON"	Voltage ENG. RUN	CIRCUIT	PIN	WIRE COLOR
			A1	
			A2	
⊘	⊘	EGR POSITION	A3	RED
5	5	5V REF	A4	GRY/RED
5	5	5V REF	A5	GRY
B+	B+	IGN	A6	PNK/BLK
			A7	
			A8	
③ 4.8	③ 4.8	SERIAL DATA	A9	ORN
			A10	
④ 0*	B+	FUEL PUMP RELAY CONTROL	A11	DK GRN/WHT
0*	0*	POWER GND.	A12	BLK

WIRE COLOR	PIN	CIRCUIT	Voltage KEY "ON"	Voltage ENG. RUN
ORN/BLK	B1	BATTERY	B+	B+
	B2			
	B3			
	B4			
BLK	B5	SENSOR GROUND	0*	0*
PPL	B6	SENSOR GROUND	0*	0*
	B7			
	B8			
PPL	B9	VSS (LOW)	0*	①
YEL	B10	VSS (HIGH)	0*	①
DK GRN	B11	TO VP 4000 P/MI	10	11.7
	B12			

1. Increases with speed (measure on A/C scale)
2. Normal operating temperature.
3. Varies.
4. 12V First two seconds.
5. Varies with temperature.
• Less than 1 volt.

ENGINE 2.8L
CARLINE "J"

2.8L (VIN W) ECM CONNECTOR TERMINAL END VIEW — CAVALIER (CONT.)

Voltage KEY "ON"	Voltage ENG. RUN	CIRCUIT	PIN	WIRE COLOR
① 0*		TO CRUISE 2000 P/MI	C1	RED
			C2	
			C3	
B+	B+	IF USED	C4	PPL
			C5	
			C6	
0*	5	BYPASS	C7	TAN/BLK
0*	1.3	EST	C8	WHT
0*	0*	A/C REQUEST	C9	LT GRN
			C10	
			C11	LT BLU
B+	B+	INJECTOR 1,3,5	C11	LT BLU
B+	B+	INJECTOR 2,4,6	C12	LT GRN
			C13	
			C14	
			C15	
B+	B+	BATTERY	C16	ORN

WIRE COLOR	PIN	CIRCUIT	Voltage KEY "ON"	Voltage ENG. RUN
TAN/WHT	D1	POWER GND.	0*	0*
	D2			
	D3			
	D4			
	D5			
BLK/WHT	D6	INJ. DRIVE LOW	0*	0*
BLK/WHT	D7	INJ. DRIVE LOW	0*	0*
PPL/WHT	D8	REFERENCE	0*	2.3
BLK/RED	D9	REF. LOW	0*	0*
	D10			
	D11			
DK GRN/WHT LT BLU/ORN	D12 D13	A/C PRESS FAN SW. (WITH HI PS PRESS) PSPS	0* B+	0* B+
	D14			
	D15			
ORN/BLK	D16	P/N SWITCH	0*	0*

Voltage KEY "ON"	Voltage ENG. RUN	CIRCUIT	PIN	WIRE COLOR
			E1	
			E2	
NOT USEABLE		IAC "A" HI	E3	LT BLU/ WHT
NOT USEABLE		IAC "A" LO	E4	LT BLU/ WHT
NOT USEABLE		IAC "B" HI	E5	LT GRN/ WHT
NOT USEABLE		IAC "B" LO	E6	LT GRN/ WHT
0*	B+	SERVICE ENGINE SOON LIGHT	E7	BRN/ WHT
0*	B+	"ON" FAN CONTROL	E8	DK GRN/ WHT
B+	B+	"OFF" RELAY	E8	WHT
B+	B+	EGR CONTROL	E9	GRY
			E10	
			E11	
5	5	DIAG. TERM.	E12	WHT/BLK
④	④	FUEL PUMP SIGNAL	E13	GRY
35- 35	35- 35	O₂ SIGNAL	E14	PPL
0*	0*	O₂ GND.	E15	TAN
⑤	⑤	COOLANT TEMP.	E16	YEL

WIRE COLOR	PIN	CIRCUIT	Voltage KEY "ON"	Voltage ENG. RUN
GRN/WHT	F1	A/C RELAY "ON" CONTROL "OFF"	0* B+	0* B+
	F2			
	F3			
	F4			
	F5			
TAN/BLK DK GRN YEL	F6	TCC CONTROL A/T SHIFT LIGHT M/T	0* B+	0* B+
DK GRN/ YEL	F7	PURGE CONTROL	0*	0*
	F8			
DK BLU	F9	ISC SIGNAL	2.5	2.5
	F10			
	F11			
	F12			
DK BLU	F13	TPS SIGNAL	65	65
	F14			
LT GRN	F15	MAP SIGNAL	4.57	③
TAN	F16	MAT SIGNAL	⑤	⑤

1. Increases with speed (measure on A/C scale)
2. Normal operating temperature.
3. Varies.

4. 12V First two seconds.
5. Varies with temperature.
• Less than 1 volt.

2.8L (VIN W) COMPONENT LOCATIONS – CIMARRON

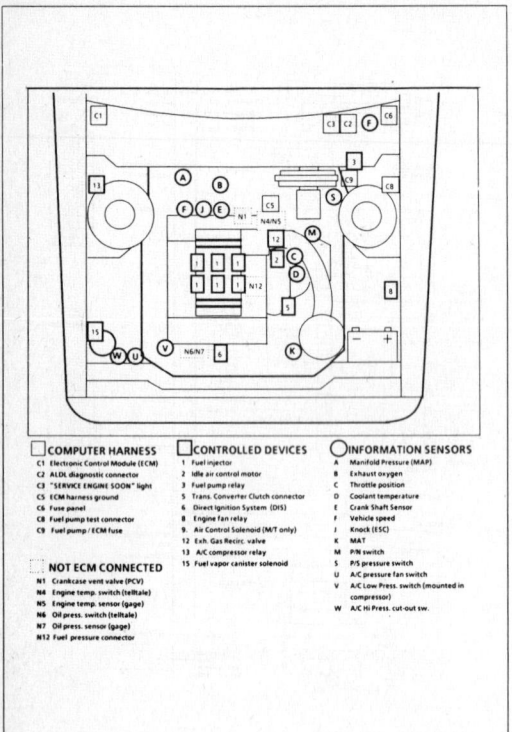

2.8L (VIN W) ECM WIRING DIAGRAM – CIMARRON

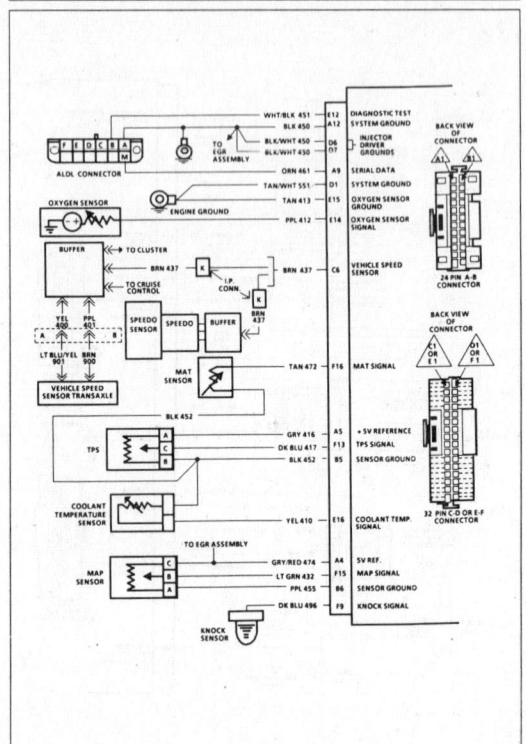

2.8L (VIN W) ECM WIRING DIAGRAM CIMARRON (CONT.)

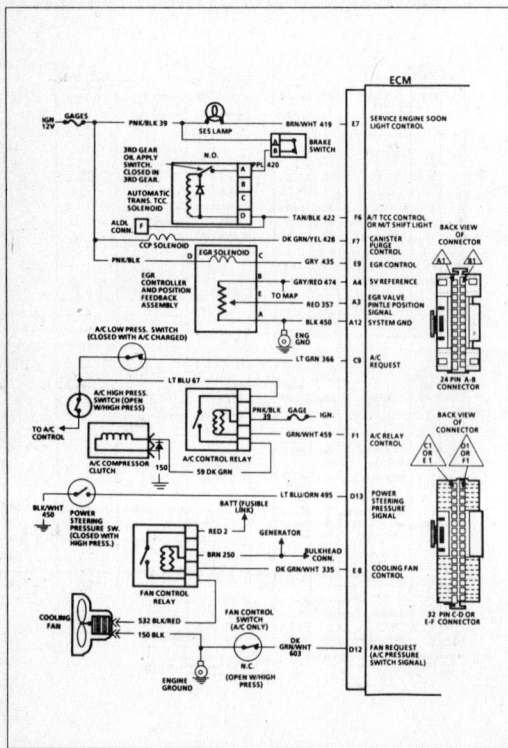

2.8L (VIN W) ECM WIRING DIAGRAM CIMARRON (CONT.)

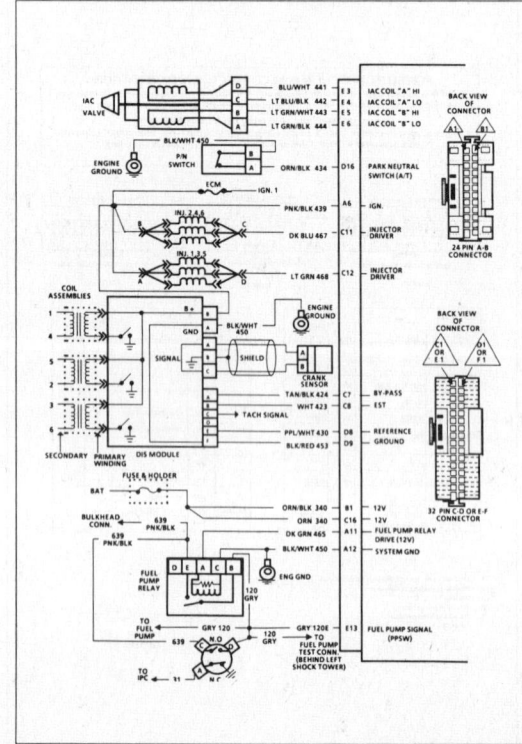

2.8L (VIN W) ECM CONNECTOR TERMINAL END VIEW – CIMARRON

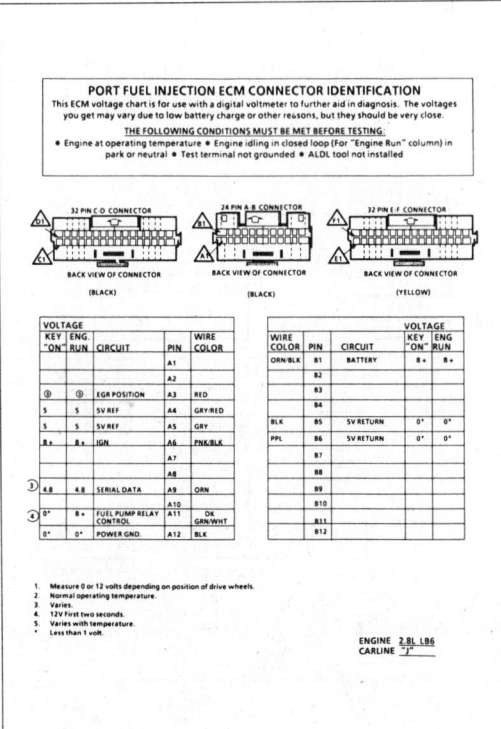

PORT FUEL INJECTION ECM CONNECTOR IDENTIFICATION

This ECM voltage chart is for use with a digital voltmeter as further aid in diagnosis. The voltages you get may vary due to low battery charge or other reasons, but they should be very close.

THE FOLLOWING CONDITIONS MUST BE MET BEFORE TESTING:
• Engine at operating temperature • Engine idling in closed loop (For "Engine Run" column) in park or neutral • Test terminal not grounded • ALDL tool not installed

32 PIN C-D CONNECTOR   24 PIN A-B CONNECTOR   32 PIN E-F CONNECTOR
BACK VIEW OF CONNECTOR (BLACK)   BACK VIEW OF CONNECTOR (BLACK)   BACK VIEW OF CONNECTOR (YELLOW)

VOLTAGE KEY "ON"	ENG. "RUN"	CIRCUIT	PIN	WIRE COLOR
			A1	
			A2	
⑤	⑤	EGR POSITION	A3	RED
5	5	5V REF	A4	GRY/RED
5	5	5V REF	A5	GRY
B+	B+	IGN	A6	PNK/BLK
			A7	
			A8	
4.8 ③	4.8	SERIAL DATA	A9	ORN
			A10	
0* ④	B+	FUEL PUMP RELAY CONTROL	A11	DK GRN/WHT
0*	0*	POWER GND.	A12	BLK

WIRE COLOR	PIN	CIRCUIT	VOLTAGE KEY "ON"	ENG. "RUN"
ORN/BLK	B1	BATTERY	B+	B+
	B2			
	B3			
	B4			
BLK	B5	5V RETURN	0*	0*
PPL	B6	5V RETURN	0*	0*
	B7			
	B8			
	B9			
	B10			
	B11			
	B12			

1. Measure 0 or 12 volts depending on position of drive wheels.
2. Normal operating temperature.
3. Varies.
4. 12V First two seconds.
5. Varies with temperature.
* Less than 1 volt.

ENGINE 2.8L LB6
CARLINE "J"

2.8L (VIN W) ECM CONNECTOR TERMINAL END VIEW – CIMARRON (CONT.)

VOLTAGE KEY "ON"	ENG RUN	CIRCUIT	PIN	WIRE COLOR
			C1	
			C2	
			C3	
B+	B+	IF USED	C4	PPL
			C5	
①	①	VSS SIGNAL	C6	BRN
0*		BYPASS	C7	TAN/BLK
0*	1.3	EST	C8	WHT
0*	0*	A/C REQUEST	C9	LT GRN
			C10	
B+	B+	INJECTOR 1,3,5	C11	BLU
B+	B+	INJECTOR 2,4,6	C12	LT GRN
			C13	
			C14	
			C15	
B+		BATTERY	C16	ORN

WIRE COLOR	PIN	CIRCUIT	VOLTAGE KEY "ON"	ENG RUN
TAN/WHT	D1	POWER GND.	0*	0*
	D2			
	D3			
	D4			
	D5			
BLK/WHT	D6	INJ. DRIVE LOW	0*	0*
BLK/WHT	D7	INJ. DRIVE LOW	0*	0*
PPL/WHT	D8	REFERENCE	0*	2.3
BLK/RED	D9	REF. LOW	0*	0*
	D10			
	D11			
DK GRN/WHT	D12	A/C PRESS FAN SW.	0*	0*
LT BLU/ORN	D13	(WITH HI PS PRESS) P S P S	B+	B+
	D14			
	D15			
ORN/BLK	D16	P/N SWITCH	0*	0*

VOLTAGE KEY "ON"	ENG RUN	CIRCUIT	PIN	WIRE COLOR
			E1	
			E2	
NOT USEABLE		IAC "A" HI	E3	LT BLU/WHT
NOT USEABLE		IAC "A" LO	E4	LT BLU/BLK
NOT USEABLE		IAC "B" HI	E5	LT GRN/WHT
NOT USEABLE		IAC "B" LO	E6	LT GRN/BLK
0*	B+	SERVICE ENGINE SOON LIGHT	E7	BRN/WHT
0*	B+	"ON" FAN CONTROL "OFF" RELAY	E8	DK GRN/WHT
B+		EGR CONTROL	E9	GRY
			E10	
			E11	
5	5	DIAG. TERM.	E12	WHT/BLK
④	④	FUEL PUMP SIGNAL	E13	GRY
35-55	35-55	O₂ SIGNAL	E14	PPL
0*	0*	O₂ GND.	E15	TAN
⑤	⑤	COOLANT TEMP.	E16	YEL

WIRE COLOR	PIN	CIRCUIT	VOLTAGE KEY "ON"	ENG RUN
GRN/WHT	F1	A/C RELAY "ON" CONTROL "OFF"	0*	B+
	F2			
	F3			
	F4			
	F5			
TAN/BLK	F6	TCC CONTROL A/T SHIFT LIGHT M/T	0*	0*
DK GRN/YEL		PURGE CONTROL	0*	0*
	F8			
DK BLU	F9	ESC SIGNAL	2.5	2.5 ③
	F10			
	F11			
	F12			
DK BLU	F13	TPS SIGNAL	.55	8.1
	F14			
LT GRN	F15	MAP SIGNAL	4.57	③
TAN	F16	MAT SIGNAL	⑤	⑤

1. Measures 0 or 12 volts depending on position of drive wheels.
2. Normal operating temperature.
3. Varies.
4. 12V First two seconds.
5. Varies with temperature.
* Less than 1 volt.

1988-90 2.8L (VIN W) – CAVALIER
1988 2.8L (VIN W) – CIMARRON

DIAGNOSTIC CIRCUIT CHECK

The Diagnostic Circuit Check must be the starting point for any driveability complaint diagnosis.

The Diagnostic Circuit Check is an organized approach to identifying a problem created by an Electronic Engine Control System malfunction because it directs the Service Technician to the next logical step in diagnosing the complaint.

If after completing the Diagnostic Circuit Check and finding the on-board diagnostics functioning properly and no trouble codes displayed, a comparison of "Typical Scan Values", for the appropriate engine, may be used for comparison. The "Typical Values" are an average of display values recorded from normally operating vehicles and are intended to represent what a normally functioning system would display.

A "SCAN" TOOL THAT DISPLAYS FAULTY DATA SHOULD NOT BE USED, AND THE PROBLEM SHOULD BE REPORTED TO THE MANUFACTURER. THE USE OF A FAULTY "SCAN" CAN RESULT IN MISDIAGNOSIS AND UNNECESSARY PARTS REPLACEMENT.

Only the parameters listed below are used in this manual for diagnosis. If a "Scan" reads other parameters, the values are not recommended by General Motors for use in diagnosis. For more description on the values and use of the "Scan" to diagnose ECM inputs, refer to the applicable diagnosis section. If all values are within the range illustrated, refer to symptoms in Section "B".

"SCAN" DATA
Idle / Upper Radiator Hose Hot / Closed Throttle / Park or Neutral / Closed Loop / Acc. off

"SCAN" Position	Units Displayed	Typical Data Value
Desired RPM	RPM	ECM idle command (varies with temp.)
RPM	RPM	± 100 RPM from desired RPM (± 50 in drive)
Coolant Temp.	C°	85° - 105°
MAT Temp.	C°	10° - 80° (depends on underhood temp.)
MAP	Volts	1 - 2 (depends on Vac. & Baro pressure)
BARO	Volts	2.5 - 5.5 (depends on altitude & Baro pressure)
BPW (base pulse width)	M/Sec	1 - 4, and varying
O₂	Volts	1-1.0, and varying
TPS	Volts	.45 - .65
Throttle Angle	0 - 100%	0
IAC	Counts (steps)	5 - 50
P/N Switch	P/N and RDL	Park/Neutral (P/N)
INT (Integrator)	Counts	Varies
BLM (Block Learn)	Counts	118 - 138
Open/Closed Loop	Open/Closed	Closed Loop (may go open with extended idle)
BLM Cell	Cell Number	0 or 1 (depends on Air Flow & RPM)
VSS	MPH	0
TCC	On/Off	Off (on with TCC commanded)
EGRDC	0 - 100%	0 at idle
EGR Position	Volts	.3 - 2 Volts
Spark Advance	# of Degrees	Varies
Knock Retard	Degrees of Retard	0 °
Knock Signal	Yes/No	No
Battery	Volts	13.5 - 14.5
Fan	On/Off	Off (below 106°C)
P/S Switch	Normal/Hi Press.	Normal
A/C Request	Yes/No	No (yes, with A/C requested)
A/C Clutch	On/Off	Off (on, with A/C commanded on)
Fan Request	Yes/No	No (yes, with A/C high pressure)
Shift Light (M/T)	On/Off	Off
PPSW (Fuel Pump)	Volts	13.5 - 14.5

NOTE: If maximum retard is indicated, go to CHART C-5.

1988-90 2.8L (VIN W) – CAVALIER
1988 2.8L (VIN W) – CIMARRON

DIAGNOSTIC CIRCUIT CHECK
2.8L (VIN W) "J" SERIES (PORT)

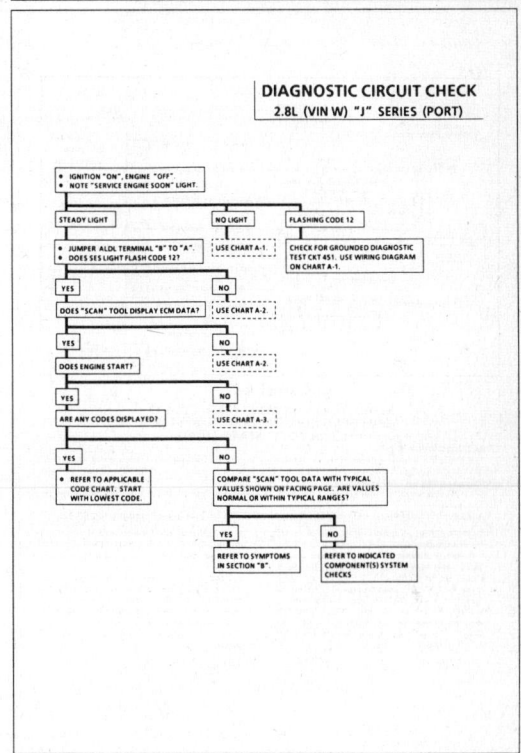

1988–90 2.8L (VIN W) – CAVALIER
1988 2.8L (VIN W) – CIMARRON

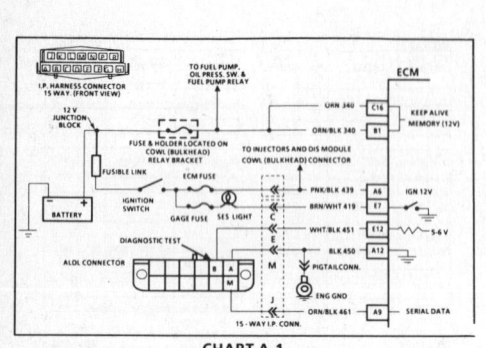

CHART A-1
NO "SERVICE ENGINE SOON" LIGHT
2.8L (VIN W) "J" SERIES (PORT)

Circuit Description:

There should always be a steady "Service Engine Soon" light when the ignition is "ON" and engine stopped. Battery is supplied directly to the light bulb. The Electronic Control Module (ECM) will control the light and turn it "ON", by providing a ground path through CKT 419 to the ECM.

Test Description: Numbers below refer to circled numbers on the diagnostic chart.

1. If the fuse in holder is blown, refer to facing page of Code 54 for complete circuit.
2. Using a test light connected to 12 volts, probe each of the system ground circuits to be sure a good ground is present. See ECM terminal end view in front of this section for ECM pin locations of ground circuits.

Diagnostic Aids:

Engine runs OK, check
- Faulty light bulb
- CKT 419 open
- Gage fuse blown. This will result in no oil or generator lights, seat belt reminder, etc.

Engine cranks but will not run.
- Continuous battery - fuse or fusible link open
- ECM ignition fuse open
- Battery CKT 340 to ECM open
- Ignition CKT 439 to ECM open
- Poor connection to ECM

1988–90 2.8L (VIN W) – CAVALIER
1988 2.8L (VIN W) – CIMARRON

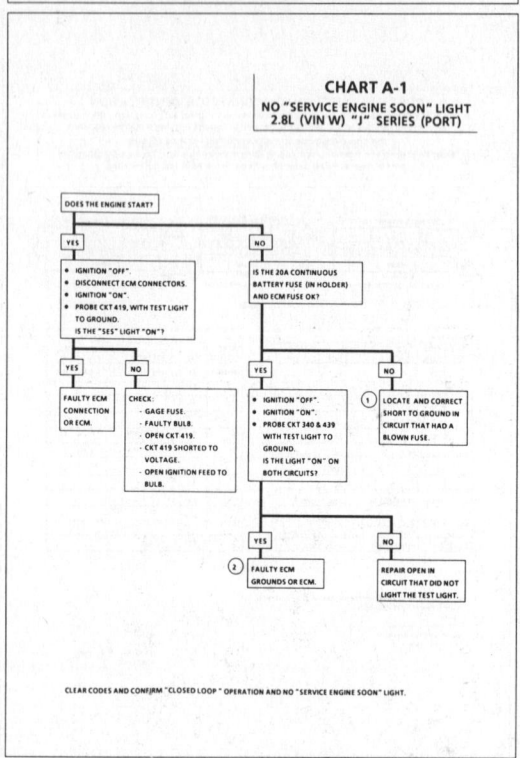

1988–90 2.8L (VIN W) – CAVALIER
1988 2.8L (VIN W) – CIMARRON

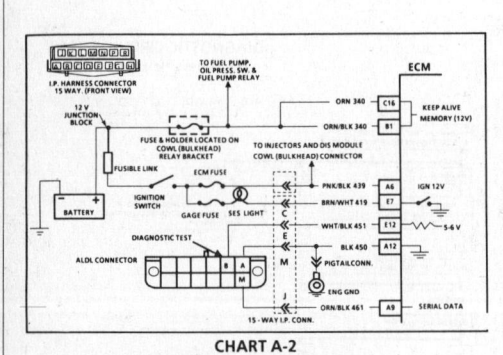

CHART A-2
NO ALDL DATA OR WON'T FLASH CODE 12
"SERVICE ENGINE SOON" LIGHT "ON" STEADY
2.8L (VIN W) "J" SERIES (PORT)

Circuit Description:

There should always be a steady "Service Engine Soon" light when the ignition is "ON" and engine stopped. Battery ignition voltage is supplied to the light bulb. The Electronic Control Module (ECM) will turn the light "ON" by grounding CKT 419 at the ECM.

With the diagnostic terminal grounded, the light should flash a Code 12, followed by any trouble code(s) stored in memory.

A steady light suggests a short to ground in the light control CKT 419, or an open in diagnostic CKT 451.

Test Description: Numbers below refer to circled numbers on the diagnostic chart.

1. If there is a problem with the ECM that causes a "Scan" tool to not display Serial data, the ECM should not flash a Code 12. If Code 12 is flashing check for CKT 451 short to ground. If Code 12 does flash be sure that the "Scan" tool is working properly on another vehicle. If the "Scan" is functioning properly and CKT 461 is OK the Mem-Cal or ECM may be at fault for the NO ALDL symptom.

2. If the light goes "OFF" when the ECM connector is disconnected, CKT 419 is not shorted to ground.
3. This step will check for an open diagnostic CKT 451.
4. At this point the "Service Engine Soon" light wiring is OK. The problem is a faulty ECM or Mem-Cal. If Code 12 does not flash, the ECM should be replaced using the original Mem-Cal. Replace the Mem-Cal only after trying an ECM, as a defective Mem-Cal is an unlikely cause of the problem.

1988–90 2.8L (VIN W) – CAVALIER
1988 2.8L (VIN W) – CIMARRON

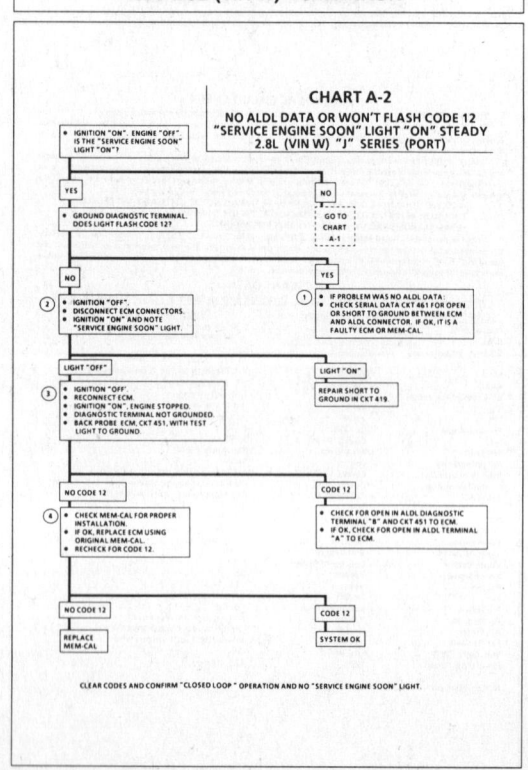

1988–90 2.8L (VIN W) – CAVALIER
1988 2.8L (VIN W) – CIMARRON

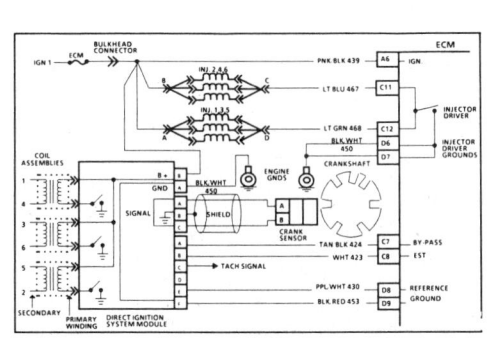

CHART A-3
(Page 1 of 3)
ENGINE CRANKS BUT WILL NOT RUN
2.8L (VIN W) "J" SERIES (PORT)

Circuit Description:

This chart assumes that battery condition and engine cranking speed are OK, and there is adequate fuel in the tank.

Test Description: Numbers below refer to circled numbers on the diagnostic chart.

1. A "Service Engine Soon" light "ON" is a basic test to determine if there is a 12 volt supply and ignition 12 volts to ECM. No ALDL may be due to an ECM problem and CHART A-2 will diagnose the ECM. If TPS is over 2.5 volts the engine may be in the clear flood mode which will cause starting problems. The engine will not start without reference pulses and therefore the "Scan" should read rpm (reference) during crank.

2. For the first two seconds with Ign "ON" or whenever reference pulses are being received, PPSW should indicate fuel pump circuit voltage (8 to 12 volts).

3. Because the Direct Ignition System uses two plugs and wires to complete the CKT of each coil, the opposite spark should be left connected. If rpm was indicated during crank, the ignition module is receiving a crank signal, but no spark at this test indicates the ignition module is not triggering the coils.

4. The test light should blink, indicating the ECM is controlling the injectors OK. How bright the light blinks is not important. However, the test light should be J-34730-3 or equivalent.

5. Use fuel pressure gage J-34730-1 or equivalent. Wrap a shop towel around the fuel pressure tap to absorb any small amount of fuel leakage that may occur when installing the gage.

6. This test will determine if the ignition module is not generating the reference pulse or if the wiring or ECM are at fault. By touching and removing a test light to 12 volts on CKT 430, a reference pulse should be generated. If rpm is indicated the ECM and wiring are OK.

7. This test will determine if the ignition module is not triggering the problem coil or if the tested coil is at fault. This test could also be performed by using another known good coil.

1988–90 2.8L (VIN W) – CAVALIER
1988 2.8L (VIN W) – CIMARRON

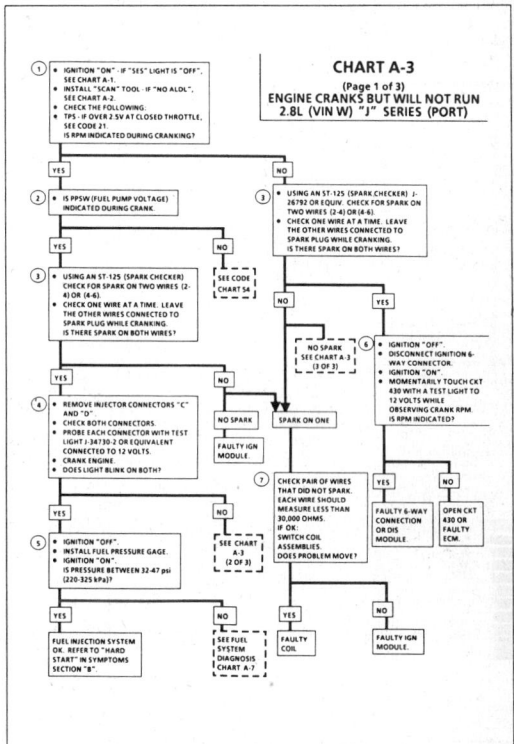

CHART A-3
(Page 1 of 3)
ENGINE CRANKS BUT WILL NOT RUN
2.8L (VIN W) "J" SERIES (PORT)

1988–90 2.8L (VIN W) – CAVALIER
1988 2.8L (VIN W) – CIMARRON

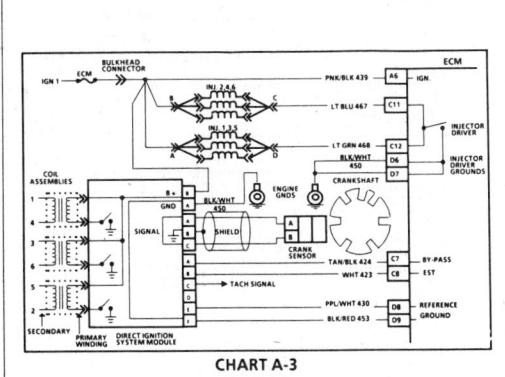

CHART A-3
(Page 2 of 3)
ENGINE CRANKS BUT WILL NOT RUN
2.8L (VIN W) "J" SERIES (PORT)

Test Description: Numbers below refer to circled numbers on the diagnostic chart.

1. Checks for 12 volt supply to injectors. Because the injectors are wired in parallel there should be a light "ON" on both terminals.

2. Checks continuity of CKT 467 and 468.

3. All checks made to this point would indicate that the ECM is at fault. However, there is a possibility of CKT 467 or 468 being shorted to a voltage source either in the engine harness or in the injector harness.
   - To test for this condition:
     - Disconnect the injector 4-way connector
     - Ignition "ON"

- Probe CKTs 467 and 468 on the ECM side of harness with a test light connected to ground. There should be no light. If light is "ON" repair short to voltage.
- If OK, check resistance of the injector harness between terminals "A & C", "A & D", "B & D" and "B & C"
  - Should be more than 4 ohms
  - If less than 4 ohms, check harness for wires shorted together and check each injector resistance. (Resistance should be 8 ohms or more.)
  - If all OK, replace ECM

1988–90 2.8L (VIN W) – CAVALIER
1988 2.8L (VIN W) – CIMARRON

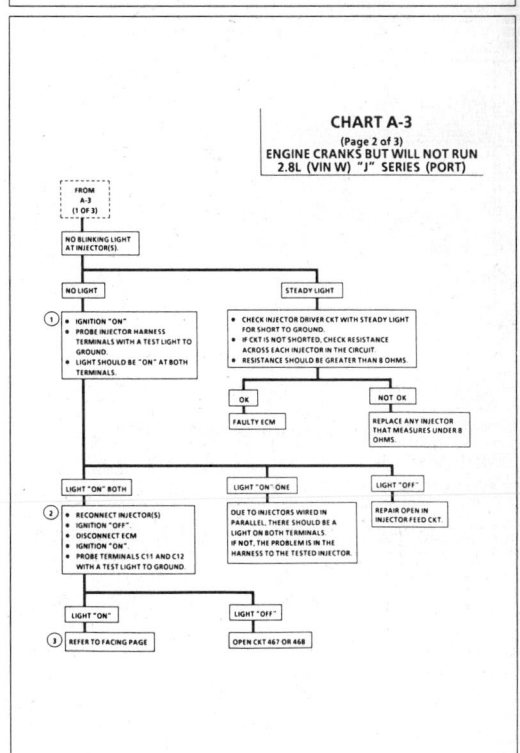

CHART A-3
(Page 2 of 3)
ENGINE CRANKS BUT WILL NOT RUN
2.8L (VIN W) "J" SERIES (PORT)

1988–90 2.8L (VIN W) – CAVALIER
1988 2.8L (VIN W) – CIMARRON

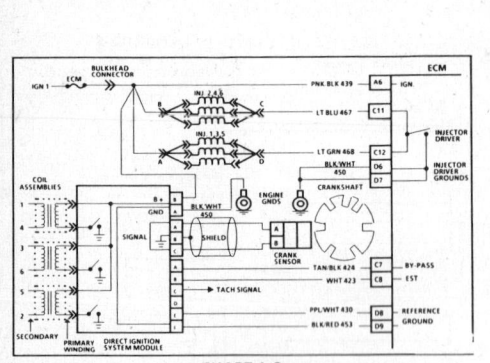

CHART A-3
(Page 3 of 3)
ENGINE CRANKS BUT WILL NOT RUN
2.8L (VIN W) "J" SERIES (PORT)

Test Description:
If the "Scan" tool did not indicate a cranking rpm and there is no spark present at the plugs, the problem lies in the Direct Ignition System or the power and ground supplies to the module.

The magnetic crank sensor is used to determine engine crankshaft position much the same way as the pick-up coil did in distributor type systems. The sensor is mounted in the block near a seven slot wheel on the crank shaft. The rotation of the wheel creates a flux change in the sensor which produces a voltage signal. The ignition module then processes this signal and creates the reference pulses needed by the ECM and the signal triggers the correct coil at the correct time.

Test Description: Numbers below refer to circled numbers on the diagnostic chart.
1 This test will determine if the 12 volt supply and a good ground is available at the ignition module
2 Tests for continuity of CKT 439 to the ignition module. If test light does not light but the SES light is "ON" with ignition "ON," repair open in CKT 439 between DIS ignition module and splice.
3 Checks for continuity of the crank sensor and connections.
4 Voltage will vary in this test depending on cranking speed of engine.

1988–90 2.8L (VIN W) – CAVALIER
1988 2.8L (VIN W) – CIMARRON

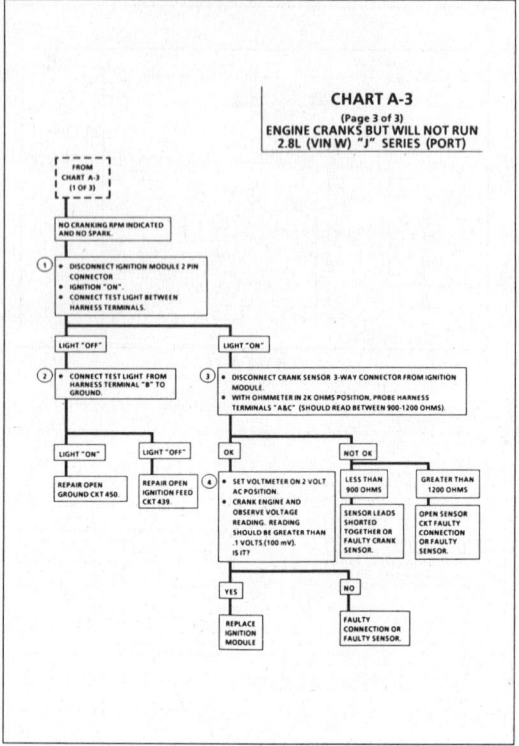

CHART A-3
(Page 3 of 3)
ENGINE CRANKS BUT WILL NOT RUN
2.8L (VIN W) "J" SERIES (PORT)

1988–90 2.8L (VIN W) – CAVALIER
1988 2.8L(VIN W) – CIMARRON

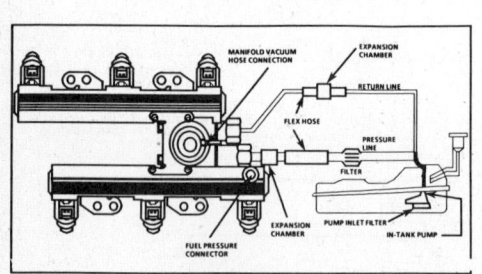

CHART A-7
(Page 1 of 2)
FUEL SYSTEM DIAGNOSIS
2.8L (VIN W) "J" SERIES (PORT)

Circuit Description:
When the ignition switch is turned "ON", the Electronic Control Module (ECM) will turn "ON" the in-tank fuel pump. It will remain "ON" as long as the engine is cranking or running, and the ECM is receiving reference pulses. If there are no reference pulses, the ECM will shut "OFF" the fuel pump within 2 seconds after ignition "ON" or engine stops.

The pump will deliver fuel to the fuel rail and injectors, then to the pressure regulator, where the system pressure is controlled to about 234 to 325 kPa (34 to 47 psi).

Test Description: Numbers below refer to circled numbers on the diagnostic chart.
1. Wrap a shop towel around the fuel pressure connector to absorb any small amount of fuel leakage that may occur when installing the gage. Ignition "ON" pump pressure should be 280-325 kPa (40.5-47 psi). This pressure is controlled by spring pressure within the regulator assembly.
2. When the engine is idling, the manifold pressure is low (high vacuum). This vacuum is applied to the fuel regulator diaphragm. This will offset the spring and result in a lower fuel pressure. This idle pressure will vary somewhat depending on barometric pressure, however, the pressure idling should be less indicating pressure regulator control.
3. Pressure that continues to fall is caused by one of the following
  • In-tank fuel pump check valve not holding.
  • Pump coupling hose or pulsator leaking.
  • Fuel pressure regulator valve leaking.
  • Injector(s) sticking open.

4. An injector sticking open can best be determined by checking for a fouled or saturated spark plug(s). If a leaking injector can not be determined by a fouled or saturated spark plug the following procedure should be used.
  • Remove Plenum, and remove fuel rail bolts. Follow the procedures in the Fuel Control Section of this manual, but leave fuel lines connected.
  • Lift fuel rail out just enough to leave injector nozzles in the ports.

CAUTION: Be sure injector(s) are not allowed to spray on engine and that injector retaining clips are intact. This should be carefully followed to prevent fuel spray on engine which would cause a fire hazard.

  • Pressurize the fuel system and observe injector nozzles.

1988–90 2.8L (VIN W) – CAVALIER
1988 2.8L (VIN W) – CIMARRON

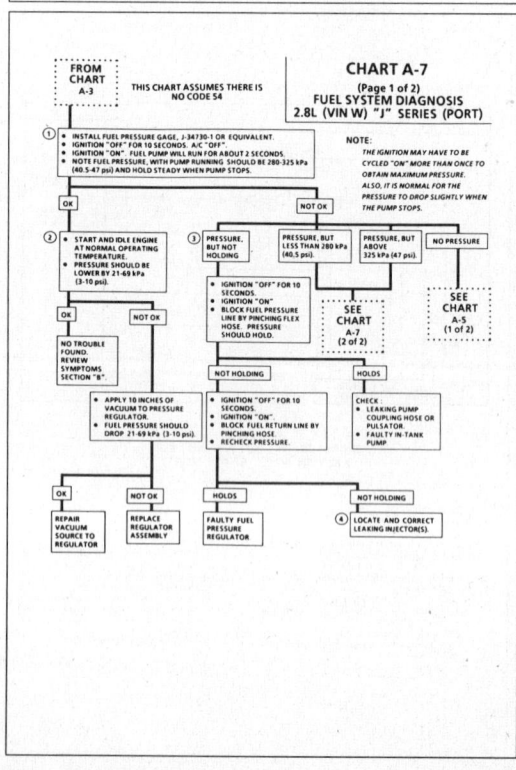

CHART A-7
(Page 1 of 2)
FUEL SYSTEM DIAGNOSIS
2.8L (VIN W) "J" SERIES (PORT)

1988–90 2.8L (VIN W) – CAVALIER
1988 2.8L (VIN W) – CIMARRON

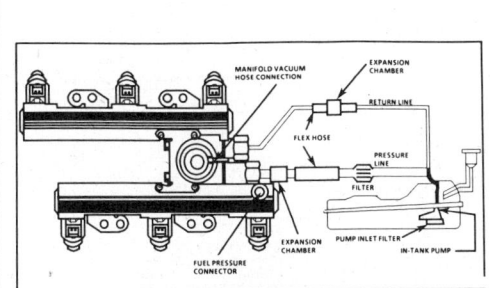

CHART A-7
(Page 2 of 2)
FUEL SYSTEM DIAGNOSIS
2.8L (VIN W) "J" SERIES (PORT)

Test Description: Numbers below refer to circled numbers on the diagnostic chart.

1. Pressure but less than 280 kPa (40.5 psi) falls into two areas:
   - Regulated pressure but less than 280 kPa (40.5 psi) Amount of fuel to injectors OK but pressure is too low. System will be lean running and may set Code 44. Also, hard starting cold and overall poor performance.
   - Restricted flow causing pressure drop - Normally, a vehicle with a fuel pressure of less than 165 kPa (24 psi) at idle will not be driveable. However, if the pressure drop occurs only, while driving, the engine will normally surge then stop running as pressure begins to drop rapidly. This is most likely caused by a restricted fuel line or plugged filter.

2. Restricting the the fuel return line allows the fuel pump to develop its maximum pressure (dead head pressure) When battery voltage is applied to the pump test terminal, pressure should be above 414 kPa (60 psi).

3. This test determines if the high fuel pressure is due to a restricted fuel return line or a pressure regulator problem.

1988–90 2.8L (VIN W) – CAVALIER
1988 2.8L (VIN W) – CIMARRON

NOTICE: FUEL SYSTEM UNDER PRESSURE. TO AVOID FUEL SPILLAGE, REFER TO FIELD SERVICE PROCEDURES FOR TESTING OR MAKING REPAIRS REQUIRING DISASSEMBLY OF FUEL LINES OR FITTINGS.

CHART A-7
(Page 2 of 2)
FUEL SYSTEM DIAGNOSIS
2.8L (VIN W) "J" SERIES (PORT)

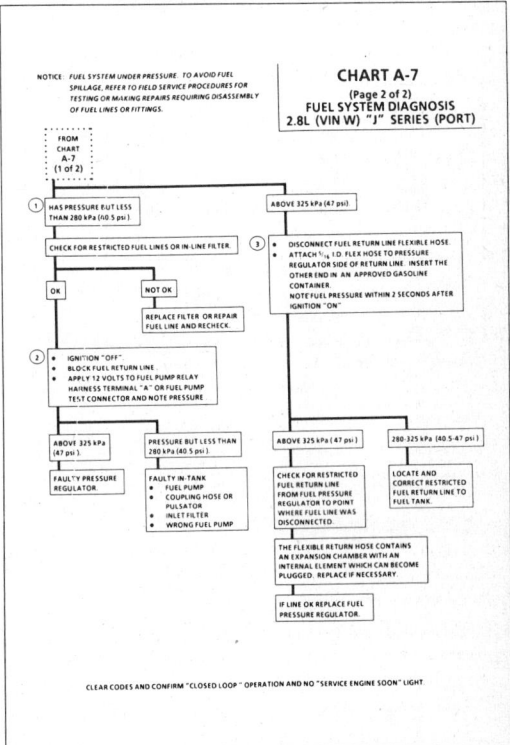

CLEAR CODES AND CONFIRM "CLOSED LOOP" OPERATION AND NO "SERVICE ENGINE SOON" LIGHT.

1988–90 2.8L (VIN W) – CAVALIER
1988 2.8L (VIN W) – CIMARRON

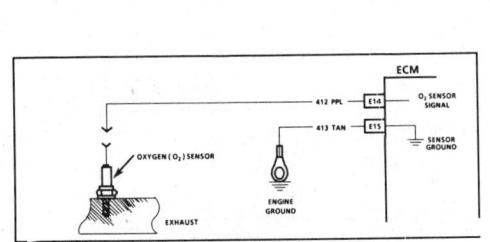

CODE 13
OXYGEN SENSOR CIRCUIT
(OPEN CIRCUIT)
2.8L (VIN W) "J" SERIES (PORT)

Circuit Description:
The ECM supplies a voltage of about .45 volt between terminals "E14" and "E15". (If measured with a 10 megohm digital voltmeter, this may read as low as .32 volts.) The O₂ sensor varies the voltage within a range of about 1 volt if the exhaust is rich, down through about .10 volt if exhaust is lean.

The sensor is like an open circuit and produces no voltage when it is below 315°C (600°F). An open sensor circuit or cold sensor causes "Open Loop" operation.

Test Description: Numbers below refer to circled numbers on the diagnostic chart.
1. Code 13 WILL SET:
   - Engine at normal operating temperature
   - At least 2 minutes engine time after start
   - O₂ signal voltage steady between .35 and .55 volts
   - Throttle position sensor signal above 4%
   - All conditions must be met for about 60 seconds.
   If the conditions for a code 13 exist the system will not go "Closed Loop".
2. This will determine if the sensor is at fault or the wiring or ECM is the cause of the Code 13.
3. In doing this test use only a high impedance digital volt ohm meter. This test checks the continuity of CKTs 412 and 413 because if CKT 413 is open the ECM voltage on CKT 412 will be over .6 volts (600 mv).

Diagnostic Aids:
Normal "Scan" voltage varies between 100 mv to 999 mv (.1 and 1.0 volt) while in "Closed Loop". Code 13 sets in one minute if voltage remains between .35 and .55 volts, but the system will go "Open Loop" in about 15 seconds.
Refer to "Intermittents" in Section "B".

1988–90 2.8L (VIN W) – CAVALIER
1988 2.8L (VIN W) – CIMARRON

CODE 13
OXYGEN SENSOR CIRCUIT
(OPEN CIRCUIT)
2.8L (VIN W) "J" SERIES (PORT)

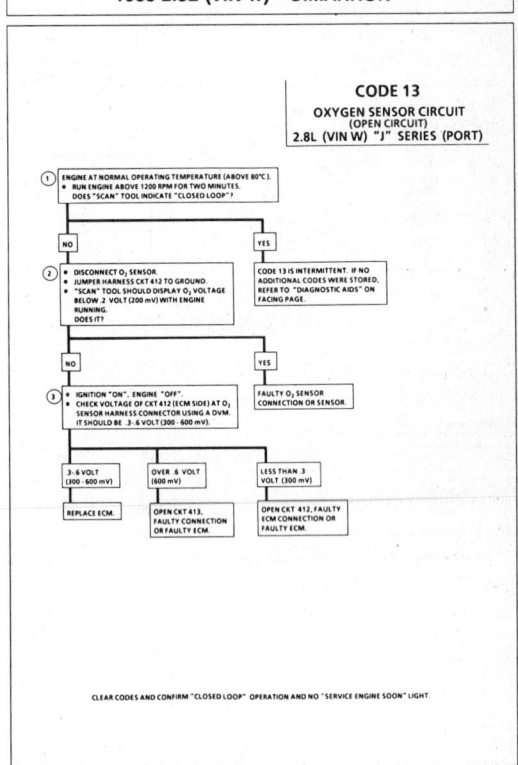

CLEAR CODES AND CONFIRM "CLOSED LOOP" OPERATION AND NO "SERVICE ENGINE SOON" LIGHT.

1988–90 2.8L (VIN W) – CAVALIER
1988 2.8L (VIN W) – CIMARRON

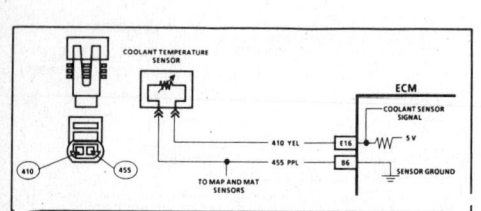

CODE 14
COOLANT TEMPERATURE SENSOR CIRCUIT
(HIGH TEMPERATURE INDICATED)
2.8L (VIN W) "J" SERIES (PORT)

Circuit Description:

The coolant temperature sensor uses a thermistor to control the signal voltage to the ECM. The ECM applies a voltage on CKT 410 to the sensor. When the engine is cold, the sensor (thermistor) resistance is high, therefore, the ECM will see high signal voltage.

As the engine warms, the sensor resistance becomes less, and the voltage drops. At normal engine operating temperature (85°C to 95°C) the voltage will measure about 1.5 to 2.0 volts.

Test Description: Numbers below refer to circled numbers on the diagnostic chart.

1. Code 14 will set if
   - Signal voltage indicates a coolant temperature above 135°C (275°F) for 3 seconds.
2. This test will determine if CKT 410 is shorted to ground which will cause the conditions for Code 14.

Diagnostic Aids:

Check harness routing for a potential short to ground in CKT 410. "Scan" tool displays engine temperature in degrees centigrade. After engine is started, the temperature should rise steadily to about 90°C then stabilize when thermostat opens. Refer to "Intermittents" in Section "B".

1988–90 2.8L (VIN W) – CAVALIER
1988 2.8L (VIN W) – CIMARRON

CODE 14
COOLANT TEMPERATURE SENSOR CIRCUIT
(HIGH TEMPERATURE INDICATED)
2.8L (VIN W) "J" SERIES (PORT)

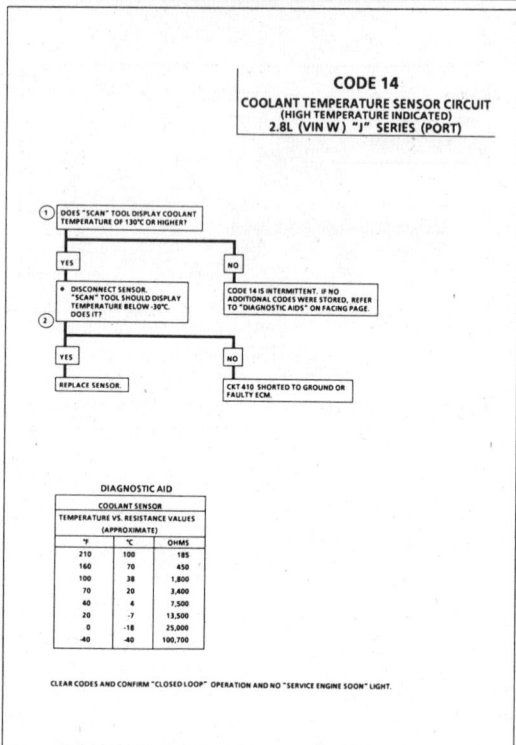

DIAGNOSTIC AID		
COOLANT SENSOR		
TEMPERATURE VS. RESISTANCE VALUES		
(APPROXIMATE)		
°F	°C	OHMS
210	100	185
160	70	450
100	38	1,800
70	20	3,400
40	4	7,500
20	-7	13,500
0	-18	25,000
-40	-40	100,700

CLEAR CODES AND CONFIRM "CLOSED LOOP" OPERATION AND NO "SERVICE ENGINE SOON" LIGHT.

1988–90 2.8L (VIN W) – CAVALIER
1988 2.8L (VIN W) – CIMARRON

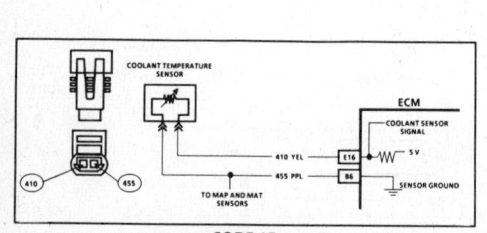

CODE 15
COOLANT TEMPERATURE SENSOR CIRCUIT
(LOW TEMPERATURE INDICATED)
2.8L (VIN W) "J" SERIES (PORT)

Circuit Description:

The Coolant Temperature Sensor uses a thermistor to control the signal voltage to the ECM. The ECM applies a voltage on CKT 410 to the sensor. When the engine is cold the sensor (thermistor) resistance is high, therefore the ECM will see high signal voltage.

As the engine warms, the sensor resistance becomes less, and the voltage drops. At normal engine operating temperature (85°C to 95°C) the voltage will measure about 1.5 to 2.0 volts at the ECM.

Test Description: Numbers below refer to circled numbers on the diagnostic chart.

1. Code 15 will set if:
   - Signal voltage indicates an open circuit (a coolant temperature less than -39°C) for 3 seconds at normal ambient temperature.
2. This test simulates a Code 14. If the ECM recognizes the low signal voltage, (high temp.) and the "Scan" reads 130°C, the ECM and wiring are OK.
3. This test will determine if CKT 410 is open. There should be 5 volts present at sensor connector if measured with a DVOM.

Diagnostic Aids:

A "Scan" tool reads engine temperature in degrees centigrade. After engine is started the temperature should rise steadily to about 90°C then stabilize when thermostat opens.

A faulty connection, or an open in CKT 410 or 452 will result in a Code 15.

If Code 21 or 23 is also set, check CKT 452 for faulty wiring or connections. Check terminals at sensor for good contact.

Refer to "Intermittents" in Section "B".

1988–90 2.8L (VIN W) – CAVALIER
1988 2.8L (VIN W) – CIMARRON

CODE 15
COOLANT TEMPERATURE SENSOR CIRCUIT
(LOW TEMPERATURE INDICATED)
2.8L (VIN W) "J" SERIES (PORT)

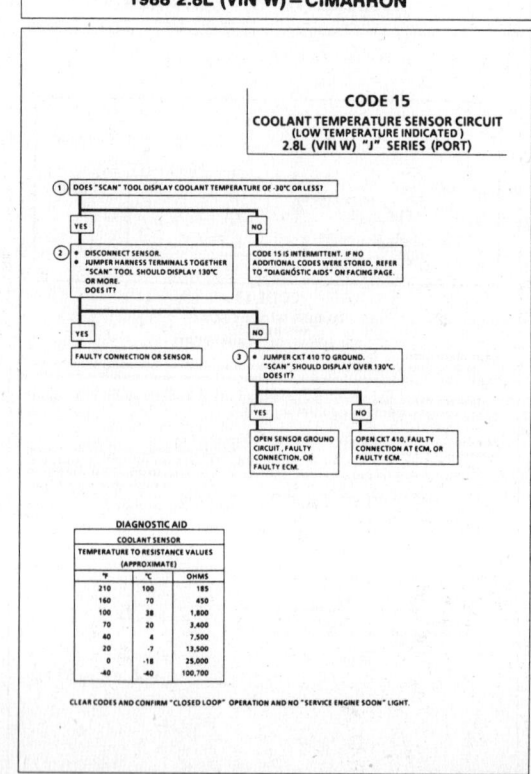

DIAGNOSTIC AID		
COOLANT SENSOR		
TEMPERATURE TO RESISTANCE VALUES		
(APPROXIMATE)		
°F	°C	OHMS
210	100	185
160	70	450
100	38	1,800
70	20	3,400
40	4	7,500
20	-7	13,500
0	-18	25,000
-40	-40	100,700

CLEAR CODES AND CONFIRM "CLOSED LOOP" OPERATION AND NO "SERVICE ENGINE SOON" LIGHT.

1988–90 2.8L (VIN W) – CAVALIER
1988 2.8L (VIN W) – CIMARRON

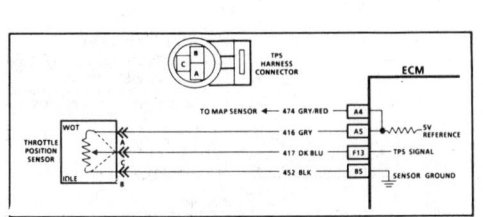

CODE 21
THROTTLE POSITION SENSOR (TPS) CIRCUIT
(SIGNAL VOLTAGE HIGH)
2.8L (VIN W) "J" SERIES (PORT)

Circuit Description:

The throttle position sensor (TPS) provides a voltage signal that changes relative to the throttle blade. Signal voltage will vary from about .5 at idle to about 4.5 volts at wide open throttle.

The TPS signal is one of the most important inputs used by the ECM for fuel control and for most of the ECM control outputs.

Test Description: Numbers below refer to circled numbers on the diagnostic chart.
1. Code 21 will set if
   - Engine is running
   - TPS signal voltage is greater than 4.3 volts
   - Air flow is less than 17 GM/sec
   - All conditions met for 5 seconds
   OR
   - TPS signal voltage over 4.5 volts with ignition "ON"

TPS: The TPS has an auto zeroing feature. If the voltage reading is within the range of 0.35 to 0.70 volts, the ECM will use that value as closed throttle. If the voltage reading is out of the auto zero range on an existing or replacement TPS, the TPS should be adjusted.

2. With the TPS sensor disconnected, the TPS voltage should go low, if the ECM and wiring is OK

3. Probing CKT 452, with a test light, checks the 5 volt return circuit. Faulty sensor ground circuit will cause a Code 21

Diagnostic Aids:

A "Scan" tool reads throttle position in volts. Voltage should increase at a steady rate as throttle is moved toward WOT.

Also, some "Scan" tools will read throttle angle. 0% = closed throttle, 100% = WOT.

An open in CKT 452 will result in a Code 21. Refer to "Intermittents" in Section "B".

1988–90 2.8L (VIN W) – CAVALIER
1988 2.8L (VIN W) – CIMARRON

CODE 21
THROTTLE POSITION SENSOR (TPS) CIRCUIT
(SIGNAL VOLTAGE HIGH)
2.8L (VIN W) "J" SERIES (PORT)

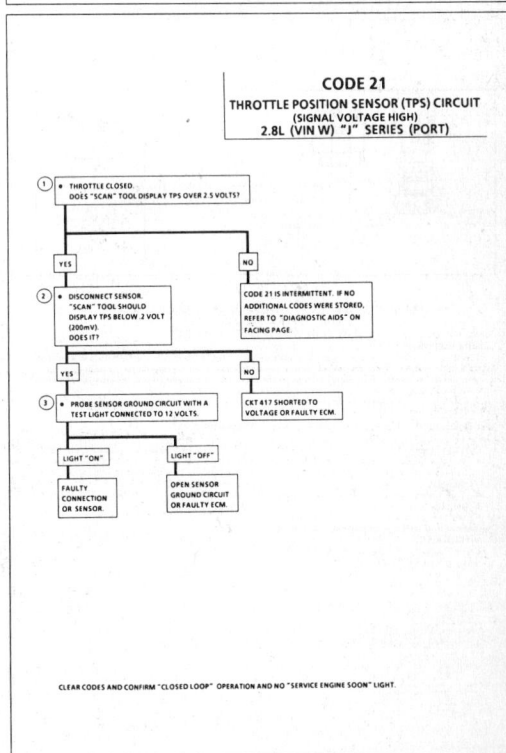

CLEAR CODES AND CONFIRM "CLOSED LOOP" OPERATION AND NO "SERVICE ENGINE SOON" LIGHT.

1988–90 2.8L (VIN W) – CAVALIER
1988 2.8L (VIN W) – CIMARRON

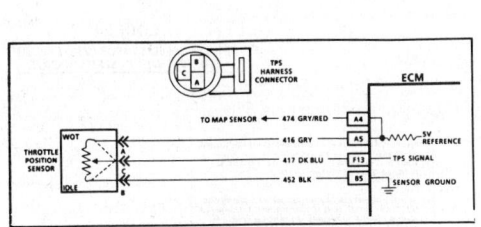

CODE 22
THROTTLE POSITION SENSOR (TPS) CIRCUIT
(SIGNAL VOLTAGE LOW)
2.8L (VIN W) "J" SERIES (PORT)

Circuit Description:

The Throttle Position Sensor (TPS) provides a voltage signal that changes relative to the throttle blade. Signal voltage will vary from about .5 at idle to about 5 volts at wide open throttle.

The TPS signal is one of the most important inputs used by the ECM for fuel control and for most of the ECM control outputs.

Test Description: Numbers below refer to circled numbers on the diagnostic chart.
1. Code 22, will set if
   - Engine running
   - TPS signal voltage is less than about .2 volt for 3 seconds
2. Simulates Code 21: (high voltage) If the ECM recognizes the high signal voltage, the ECM and wiring are OK
3. TPS check: The TPS has an auto zeroing feature. If the voltage reading is within the range of 0.35 to 0.70 volts, the ECM will use that value as closed throttle. If the voltage reading is out of the auto zero range on an existing or replacement TPS, the TPS should be adjusted.
4. This simulates a high signal voltage to check for an open in CKT 417.

5. CKTs 416 and 474 share a common 5 volts buffered reference signal. If either of these circuits is shorted to ground, Code 22 will set. To determine if the MAP sensor is causing the 22 problem, disconnect it to see if Code 22 resets. Be sure TPS is connected and clear codes before testing

Diagnostic Aids:

A "Scan" tool reads throttle position in volts. Voltage should increase at a steady rate as throttle is moved toward WOT.

Also, some "Scan" tools will read throttle angle. 0% = closed throttle, 100% = WOT.

An open, or short to ground, in CKTs 416 or 417 will result in a Code 22. Also, a short to ground, in CKT 474, will result in a Code 22. Refer to "Intermittents" in Section "B".

1988–90 2.8L (VIN W) – CAVALIER
1988 2.8L (VIN W) – CIMARRON

CODE 22
THROTTLE POSITION SENSOR (TPS) CIRCUIT
(SIGNAL VOLTAGE LOW)
2.8L (VIN W) "J" SERIES (PORT)

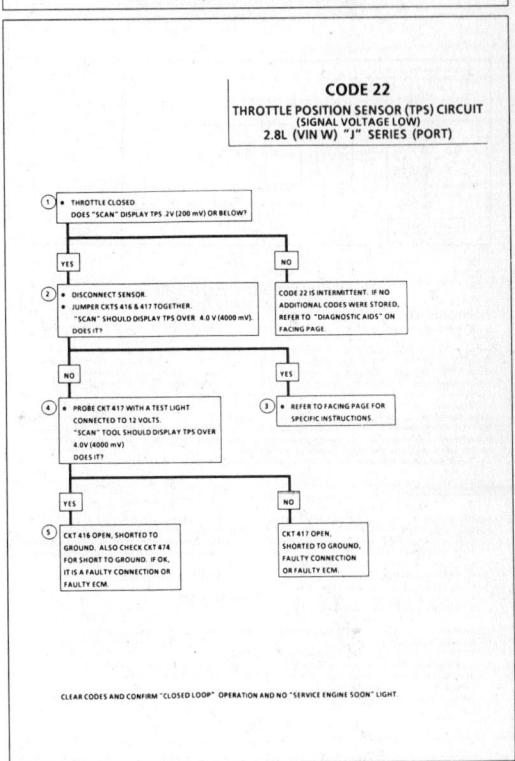

CLEAR CODES AND CONFIRM "CLOSED LOOP" OPERATION AND NO "SERVICE ENGINE SOON" LIGHT.

1988–90 2.8L (VIN W) – CAVALIER
1988 2.8L (VIN W) – CIMARRON

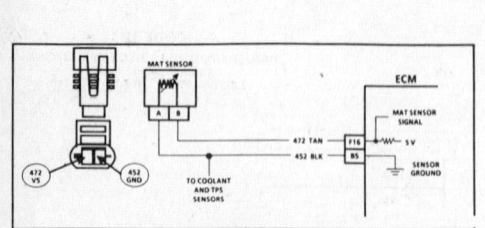

CODE 23
MANIFOLD AIR TEMPERATURE (MAT) SENSOR CIRCUIT
(LOW TEMPERATURE INDICATED)
2.8L (VIN W) "J" SERIES (PORT)

Circuit Description:
The MAT sensor uses a thermistor to control the signal voltage to the ECM. The ECM applies a voltage (about 5 volts) on CKT 472 to the sensor. When the air is cold the sensor (thermistor) resistance is high, therefore the ECM will see a high signal voltage. If the air is warm the sensor resistance is low therefore the ECM will see a low voltage.
The MAT sensor is located in the air cleaner.

Test Description: Numbers below refer to circled numbers on the diagnostic chart.
1. Code 23 will set if
   - A signal voltage indicates a manifold air temperature below −35°C (−31°F) for 3 seconds.
   - Time since engine start is 4 minutes or longer
   - No VSS
2. A Code 23 will set, due to an open sensor, wire, or connection. This test will determine if the wiring and ECM are OK.
3. This will determine if the signal CKT 472 or the sensor ground, CKT 452 is open.

Diagnostic Aids:
A "Scan" tool reads temperature of the air entering the engine and should read close to ambient air temperature when engine is cold, and rises as underhood temperature increases.
A faulty connection, or an open in CKT 472 or 452 will result in a Code 23.
Refer to "Intermittents" in Section "B".

1988–90 2.8L (VIN W) – CAVALIER
1988 2.8L (VIN W) – CIMARRON

CODE 23
**MANIFOLD AIR TEMPERATURE (MAT)
SENSOR CIRCUIT**
(LOW TEMPERATURE INDICATED)
2.8L (VIN W) "J" SERIES (PORT)

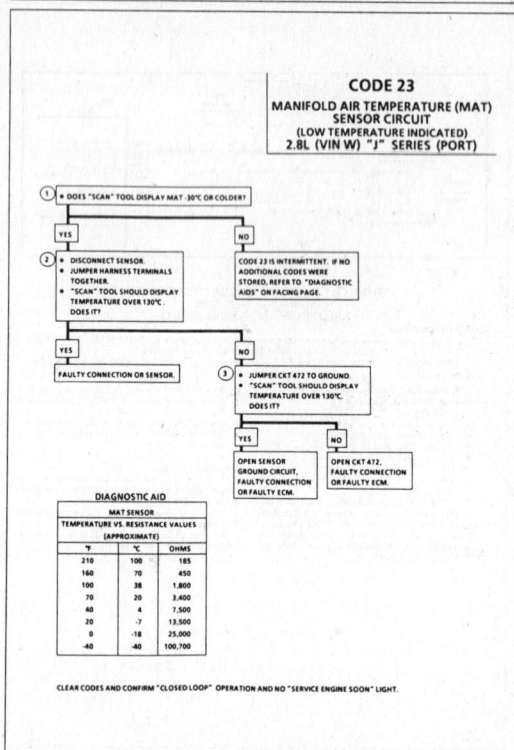

DIAGNOSTIC AID		
MAT SENSOR		
TEMPERATURE VS. RESISTANCE VALUES (APPROXIMATE)		
°F	°C	OHMS
210	100	185
160	70	450
100	38	1,800
70	20	3,400
40	4	7,500
20	−7	13,500
0	−18	25,000
−40	−40	100,700

CLEAR CODES AND CONFIRM "CLOSED LOOP" OPERATION AND NO "SERVICE ENGINE SOON" LIGHT.

1988–90 2.8L (VIN W) – CAVALIER
1988 2.8L (VIN W) – CIMARRON

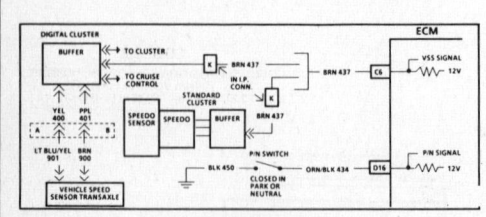

CODE 24
VEHICLE SPEED SENSOR (VSS) CIRCUIT
2.8L (VIN W) "J" SERIES (PORT)

Circuit Description:
This vehicle will have one of two types of vehicle speed sensors depending on the type of instrument cluster used. When equipped with digital cluster, vehicle speed information is sensed by a permanent magnet (PM) generator located in the transaxle. The PM generator produces a pulsing output whenever vehicle speed is over about 3 miles per hour. The pulsing output of the PM generator is applied to the input of the vehicle speed sensor buffer. The output of the buffer is a pulse at a rate of about 2000 times per mile. This pulse is fed to the vehicle speed input of the ECM and toggles the 12 volt vehicle speed input of the ECM low between pulses. The ECM then calculates vehicle speed based on the time between these pulses.
When equipped with non-digital cluster, vehicle speed information is sensed by an optical scanner in the speedometer head, driven by the speedometer cable. The square wave output of the sensor is then buffered and fed to the vehicle speed input of the ECM. The ECM then calculates vehicle speed based on the time between pulses in the same manner as with the optional cluster described above. With key on and buffer disconnected, 12 volts will be present at the vehicle speed input of the ECM.
"Scan" reading should closely match with speedometer reading with drive wheels turning.

Test Description: Numbers below refer to circled numbers on the diagnostic chart.
1. Code 24 will set if indicated vehicle speed equals 3 mph or less when
   - Engine speed is between 2200 and 4400 rpm.
   - TPS is less than 2%
   - Low load condition (MAP less than 30 kPa)
   - Not in park or neutral
   - All conditions met for 5 seconds
   These conditions are met during a road load deceleration. Disregard Code 24 that sets when drive wheels are not turning.
2. 8–12 volts, at the I.P. connector, indicates CKT 437 is open between the I.P. connector and the VSS, or there is a faulty vehicle speed sensor, or buffer. A voltage of less than 1 volt, at the I.P. connector, indicates that CKT 437 wire is shorted to ground.

If, after disconnecting CKT 437 at the vehicle speed sensor, the voltage reads above 10 volts, the vehicle speed sensor is faulty. If voltage remains less than 8 volts, then CKT 437 wire is grounded. If 437 is not grounded, there is a faulty connection at the ECM, or a faulty ECM.

Diagnostic Aids:
"Scan" should indicate a vehicle speed whenever the drive wheels are turning above 3 mph.
A faulty or misadjusted park/neutral switch can result in a false Code 24. Use a "Scan" and check for proper signal while in drive. Refer to CHART C-1A for P/N switch diagnosis check.

If all OK, refer to "Intermittents" in Section "B".

1988–90 2.8L (VIN W) – CAVALIER
1988 2.8L (VIN W) – CIMARRON

CODE 24
VEHICLE SPEED SENSOR (VSS) CIRCUIT
2.8L (VIN W) "J" SERIES (PORT)

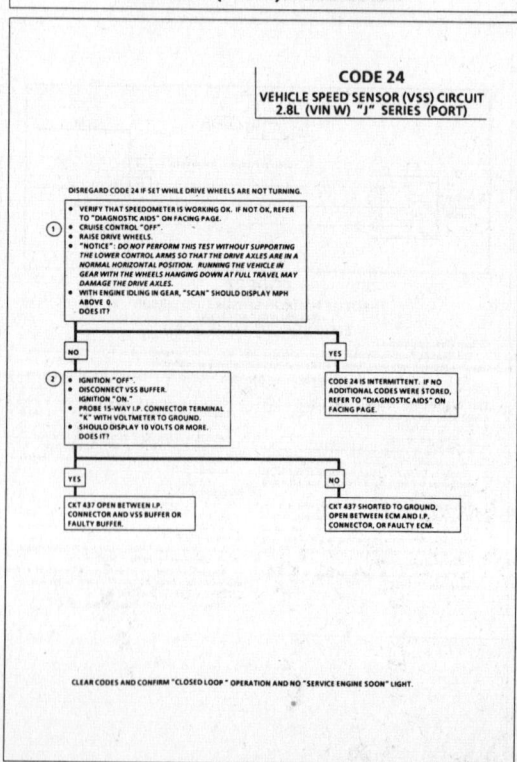

CLEAR CODES AND CONFIRM "CLOSED LOOP" OPERATION AND NO "SERVICE ENGINE SOON" LIGHT.

1988–90 2.8L (VIN W) – CAVALIER
1988 2.8L (VIN W) – CIMARRON

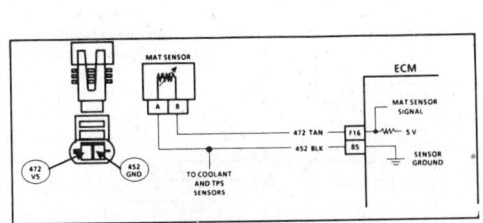

CODE 25
MANIFOLD AIR TEMPERATURE (MAT) SENSOR CIRCUIT
(HIGH TEMPERATURE INDICATED)
2.8L (VIN W) "J" SERIES (PORT)

Circuit Description:

The manifold air temperature sensor uses a thermistor to control the signal voltage to the ECM. The ECM applies a voltage (about 5 volts) on CKT 472 to the sensor. When manifold air is cold, the sensor (thermistor) resistance is high, therefore, the ECM will see a high signal voltage. As the air warms, the sensor resistance becomes less, and the voltage drops.

The MAT sensor is located in the air cleaner.

Test Description: Numbers below refer to circled numbers on the diagnostic chart.
1. Code 25 will set if:
   - Signal voltage indicates a manifold air temperature greater than 145°C (293° F) for 3 seconds
   - Time since engine start is 4 minutes or longer
   - A vehicle speed is present

Diagnostic Aids:

A "Scan" tool reads temperature of the air entering the engine and should read close to ambient air temperature, when engine is cold, and rises as underhood temperature increases.

A faulty connection, or an open in CKT 472 or 452 will result in a Code 23.

Refer to "Intermittents" in Section "B".

1988–90 2.8L (VIN W) – CAVALIER
1988 2.8L (VIN W) – CIMARRON

CODE 25
MANIFOLD AIR TEMPERATURE (MAT) SENSOR CIRCUIT
(HIGH TEMPERATURE INDICATED)
2.8L (VIN W) "J" SERIES (PORT)

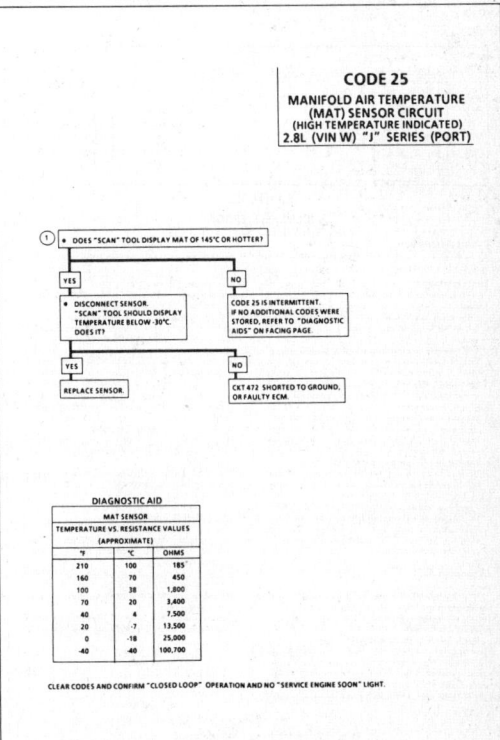

DIAGNOSTIC AID

MAT SENSOR
TEMPERATURE VS. RESISTANCE VALUES
(APPROXIMATE)

°F	°C	OHMS
210	100	185
160	70	450
100	38	1,800
70	20	3,400
40	4	7,500
20	-7	13,500
0	-18	25,000
-40	-40	100,700

CLEAR CODES AND CONFIRM "CLOSED LOOP" OPERATION AND NO "SERVICE ENGINE SOON" LIGHT.

1988–90 2.8L (VIN W) – CAVALIER
1988 2.8L (VIN W) – CIMARRON

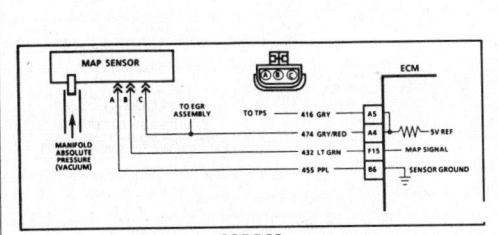

CODE 33
MANIFOLD ABSOLUTE PRESSURE (MAP) SENSOR CIRCUIT
(SIGNAL VOLTAGE HIGH - LOW VACUUM)
2.8L (VIN W) "J" SERIES (PORT)

Circuit Description:

The manifold absolute pressure (MAP) Sensor responds to changes in manifold pressure (vacuum). The ECM receives this information as a signal voltage that will vary from about 1 to 1.5 volts at idle (high vacuum) to 4- 4.5 volts at wide open throttle (low vacuum).

Test Description: Numbers below refer to circled numbers on the diagnostic chart.
1. Code 33 will set when:
   - Engine Running
   - Manifold pressure greater than 74 kPa (A/C "OFF"), 83.4 kPa (A/C "ON")
   - Throttle angle less than 2%
   - Conditions met for 4.8 seconds
   Engine misfire, or a low unstable idle, may set Code 33.
2. With the MAP sensor disconnected, the ECM should see a low voltage, if the ECM and wiring is OK.

Diagnostic Aids:

If idle is rough or unstable, refer to symptoms in Section "B" for items which can cause an unstable idle.

An open in CKT 455 or the connection, will result in a Code 33.

Ignition "ON", engine "OFF", voltages should be within the values shown in the table on the chart. Also, CHART C-1D can be used to test the MAP sensor.

Refer to "Intermittents" in Section "B".

1988–90 2.8L (VIN W) – CAVALIER
1988 2.8L (VIN W) – CIMARRON

CODE 33
MANIFOLD ABSOLUTE PRESSURE (MAP) SENSOR CIRCUIT
(SIGNAL VOLTAGE HIGH - LOW VACUUM)
2.8L (VIN W) "J" SERIES (PORT)

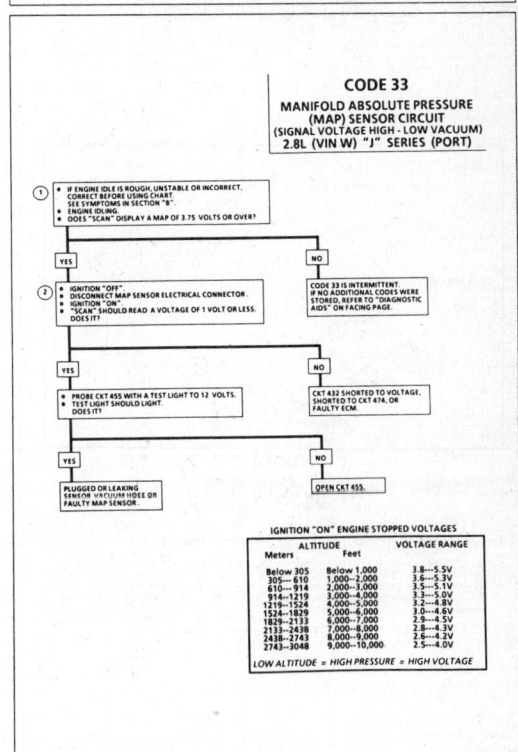

IGNITION "ON" ENGINE STOPPED VOLTAGES

ALTITUDE		VOLTAGE RANGE
Meters	Feet	
Below 305	Below 1,000	3.8—5.5V
305— 610	1,000—2,000	3.6—5.3V
610— 914	2,000—3,000	3.5—5.1V
914—1219	3,000—4,000	3.3—5.0V
1219—1524	4,000—5,000	3.2—4.8V
1524—1829	5,000—6,000	3.0—4.6V
1829—2133	6,000—7,000	2.9—4.5V
2133—2438	7,000—8,000	2.8—4.3V
2438—2743	8,000—9,000	2.6—4.2V
2743—3048	9,000—10,000	2.5—4.0V

LOW ALTITUDE = HIGH PRESSURE = HIGH VOLTAGE

1988–90 2.8L (VIN W) – CAVALIER
1988 2.8L (VIN W) – CIMARRON

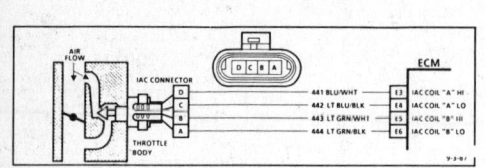

CODE 35
IDLE SPEED ERROR
2.8L (VIN W) "J" SERIES (PORT)

Circuit Description:

Code 35 will be set when the closed throttle engine speed is 300 rpm above or below the desired (commanded) idle speed for 15 seconds.

Test Description: Numbers below refer to circled numbers on the diagnostic chart.

1. Continue with test even if engine will not idle. If idle is too low, "Scan" will display 80 or more counts, or steps. If idle is high it will display "0" counts.
   Occasionally an erratic or unstable idle may occur. Engine speed may vary 200 rpm or more up and down. Disconnect IAC. If the condition is unchanged, the IAC is not at fault. There is a system problem. Proceed to diagnostic aids below.
2. When the engine was stopped, the IAC Valve retracted (more air) to a fixed "Park" position for increased air flow and idle speed during the next engine start. A "Scan" will display 80 or more counts. Observe idle immediately as on a warm engine, the idle speed should decrease rapidly.
3. Be sure to disconnect the IAC valve prior to this test. The test light will confirm the ECM signals by cycling each circuit "ON" and "OFF".
4. There is a remote possibility that one of the CKT's is shorted to voltage which would have been indicated by a steady light. Disconnect ECM and turn the ignition "ON" and probe terminals to check for this condition.

Diagnostic Aids:

A slow unstable idle may be caused by a system problem that cannot be controlled by the IAC. "Scan" counts will be above 80 counts if idle is too low and "0" counts if it is too high.
   If idle is too high, stop engine. Ignition "ON". Ground diagnostic terminal. Wait a few seconds for IAC to seat then disconnect IAC. Start engine. If idle speed is above 550 rpm in drive with an A/T or 675 in neutral with a M/T, locate and correct vacuum leak. If rpm is below spec , check for foreign material around throttle plates and clean as needed. If OK, reset minimum idle speed. Refer to Section "C2".

• System too lean (High Air/fuel ratio)
   Idle speed may be too high or too low. Engine speed may vary up and down, disconnecting IAC does not help. This may set Code 44.
   "Scan" and/or voltmeter will read an oxygen sensor output less than 300 mV (.3 volts). Check for low regulated fuel pressure or water in fuel. A lean exhaust with an oxygen sensor output fixed above 800 mv (.8 volts) will be a contaminated sensor, usually silicone. This may also set a Code 45 or 61.
• System too rich (Low Air/fuel ratio)
   Idle speed too low. "Scan" counts usually above 80. System obviously rich and may exhibit black smoke exhaust.
   "Scan" tool and/or Voltmeter will read an oxygen sensor signal fixed above 800 mv (.8 volts).
   Check
      - For fuel in pressure regulator hose
      - High fuel pressure
      - Injector leaking or sticking
• Throttle body. Remove IAC and inspect bore for foreign material or evidence of IAC valve dragging the bore.
• A/C Compressor or Relay failure. See CHART C-10. if the A/C control relay drive circuit is shorted to gnd. or if the relay is faulty an idle problem may exist.
• If above are all OK, refer to "Rough, Unstable, Incorrect Idle or Stalling" in Symptoms in Section "B"

1988–90 2.8L (VIN W) – CAVALIER
1988 2.8L (VIN W) – CIMARRON

NOTE: ECM WILL NOT CYCLE IAC IF THE COOLANT FAN IS ON.

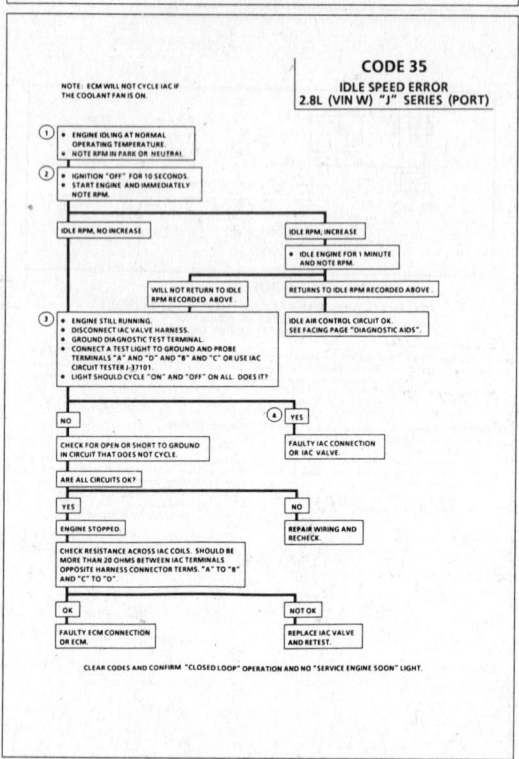

CODE 35
IDLE SPEED ERROR
2.8L (VIN W) "J" SERIES (PORT)

1988–90 2.8L (VIN W) – CAVALIER
1988 2.8L (VIN W) – CIMARRON

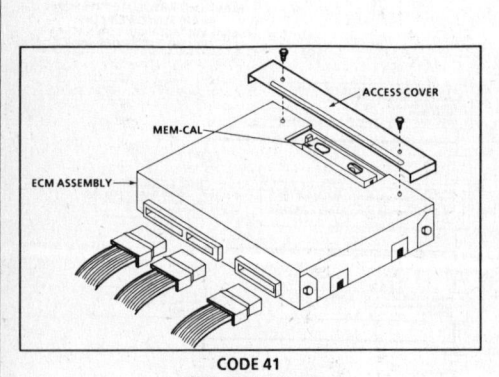

CODE 41
CYLINDER SELECT ERROR
(FAULTY OR INCORRECT MEM-CAL)
2.8L (VIN W) "J" SERIES (PORT)

Test Description: Numbers below refer to circled numbers on the diagnostic chart.

1. The ECM used for this engine can also be used for other engines, and the difference is in the Mem-Cal. If a Code 41 sets, the incorrect Mem-Cal has been installed or it is faulty and it must be replaced.

Diagnostic Aids:

Check Mem-Cal to be sure locking tabs are secure. Also check the pins on both the Mem-Cal and ECM to assure they are making proper contact. Check the Mem-Cal part number to assure it is the correct part. If the Mem-Cal is faulty it must be replaced. It is also possible that the ECM is faulty, however it should not be replaced until all of the above have been checked.

1988–90 2.8L (VIN W) – CAVALIER
1988 2.8L (VIN W) – CIMARRON

CODE 41
CYLINDER SELECT ERROR
(FAULTY OR INCORRECT MEM-CAL)
2.8L (VIN W) "J" SERIES (PORT)

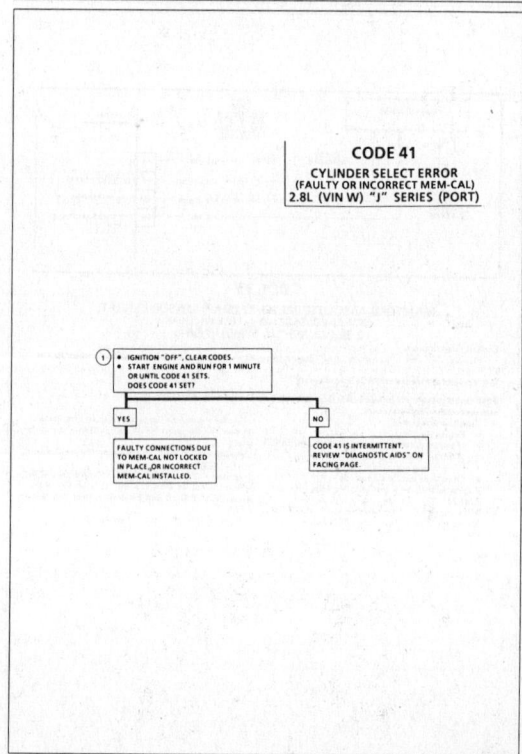

1988–90 2.8L (VIN W) – CAVALIER
1988 2.8L (VIN W) – CIMARRON

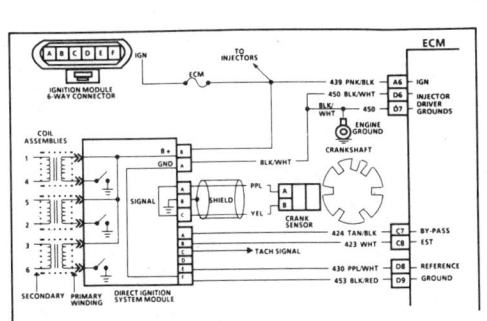

CODE 42
ELECTRONIC SPARK TIMING (EST) CIRCUIT
2.8L (VIN W) "J" SERIES (PORT)

Circuit Description:

When the system is running on the ignition module, that is, no voltage on the by-pass line, the ignition module grounds the EST signal. The ECM expects to see no voltage on the EST line during this condition. If it sees a voltage, it sets Code 42, and will not go into the EST mode.

When the rpm for EST is reached (about 400 rpm), and by-pass voltage applied, the EST should no longer be grounded in the ignition module, so the EST voltage should be varying.

If the by-pass line is open, or grounded, the ignition module will not switch to EST mode, so the EST voltage will be low and Code 42 will be set.

If the EST line is grounded, the ignition module will switch to EST, but because the line is grounded, there will be no EST signal. A Code 42 will be set.

Test Description: Numbers below refer to circled numbers on the diagnostic chart.
1. Code 42 means the ECM has seen an open, or short to ground, in the EST or by-pass circuits. This test confirms Code 42, and that the fault causing the code is present.
2. Checks for a normal EST ground path through the ignition module. An EST CKT 423, shorted to ground, will also read less than 500 ohms; however, this will be checked later.
3. As the test light touches CKT 424, the module should switch causing the ohmmeter to "overrange", if the meter is in the 1000-2000 ohms position. Selecting the 10-20,000 ohms position will indicate above 5000 ohms. The important thing is that the module "switched".

4. The module did not switch and this step checks for
   - EST CKT 423 shorted to ground
   - Bypass CKT 424 open
   - Faulty ignition module connection or module
5. Confirms that Code 42 is a faulty ECM and not an intermittent in CKTs 423 or 424.

Diagnostic Aids:

The "Scan" tool does not have any ability to help diagnose a Code 42 problem.

A Mem-Cal not fully seated in the ECM can result in a Code 42.

Refer to "Intermittents" in Section "B".

1988–90 2.8L (VIN W) – CAVALIER
1988 2.8L (VIN W) – CIMARRON

CODE 42
ELECTRONIC SPARK TIMING (EST) CIRCUIT
2.8L (VIN W) "J" SERIES (PORT)

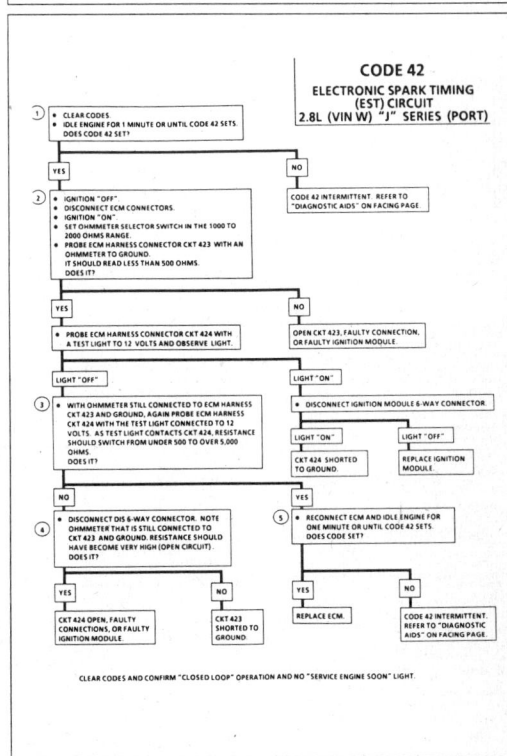

CLEAR CODES AND CONFIRM "CLOSED LOOP" OPERATION AND NO "SERVICE ENGINE SOON" LIGHT.

1988–90 2.8L (VIN W) – CAVALIER
1988 2.8L (VIN W) – CIMARRON

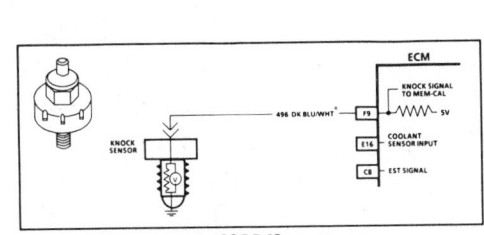

CODE 43
ELECTRONIC SPARK CONTROL (ESC) CIRCUIT
2.8L (VIN W) "J" SERIES (PORT)

Circuit Description:

The Knock Sensor is used to detect engine detonation, and the ECM will retard the electronic spark timing based on the signal being received. The circuitry, within the knock sensor, causes the ECM 5 volts to be pulled down so that under a no knock condition, CKT 496 would measure about 2.5 volts. The knock sensor produces an A/C signal, which rides on the 2.5 volts DC voltage. The amplitude and signal frequency is dependent upon the knock level.

If CKT 496 becomes open or shorted to ground, the voltage will either go above 3.5 volts, or below 1.5 volts. If either of these conditions are met for about ½second, a Code 43 will be stored.

Test Description: Numbers below refer to circled numbers on the diagnostic chart.
1. This step determines if conditions for Code 43 still exist (voltage on CKT 496 above 3.5 volts or below 1.5 volts). The system is designed to retard the spark 6°, if either condition exists
2. The ECM has a 5 volt pull-up resistor, which applies 5 volts to CKT 496. The five volt signal should be present at the knock sensor terminal during these test conditions.
3. This step determines if the knock sensor resistance is 3900 ohms ± 15%. If the resistance is between 3300 and 4500 ohms, the sensor is OK.

4. If CKT 496 is not open or shorted to ground and the voltage reading is below 4 volts, the most likely cause is an open circuit in the ECM. It is possible that a faulty Mem-Cal could be drawing the 5 volt signal down and it should be replaced, if a replacement ECM did not correct the problem.

Diagnostic Aids:

Check CKT 496 for a potential open or short to ground. Also, check for proper installation of Mem-Cal.

Refer to "Intermittents" in Section "B".

1988–90 2.8L (VIN W) – CAVALIER
1988 2.8L (VIN W) – CIMARRON

CODE 43
ELECTRONIC SPARK CONTROL (ESC) CIRCUIT
2.8L (VIN W) "J" SERIES (PORT)

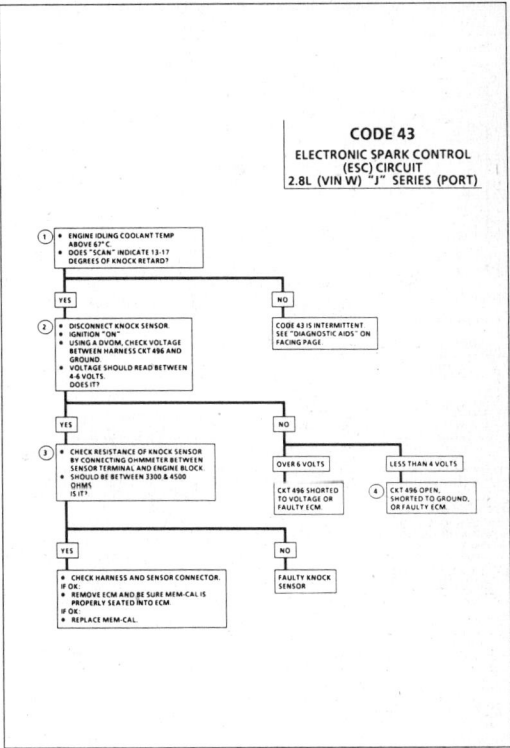

1988–90 2.8L (VIN W) – CAVALIER
1988 2.8L (VIN W) – CIMARRON

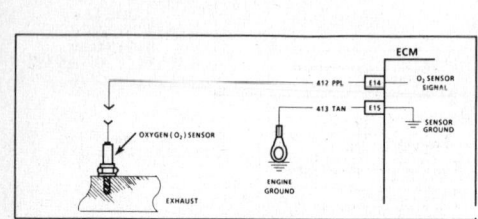

CODE 44
OXYGEN SENSOR CIRCUIT
(LEAN EXHAUST INDICATED)
2.8L (VIN W) "J" SERIES (PORT)

Circuit Description:

The ECM supplies a voltage of about .45 volt between terminals "E14" and "E15". (If measured with a 10 megohm digital voltmeter, this may read as low as .32 volts.) The O₂ sensor varies the voltage within a range of about 1 volt if the exhaust is rich, down through about .10 volt if exhaust is lean.

The sensor is like an open circuit and produces no voltage, when it is below about 315°C (600°F). An open sensor circuit or cold sensor causes "Open Loop" operation.

Test Description: Numbers below refer to circled numbers on the diagnostic chart.
1. Code 44 is set when the O₂ sensor signal voltage on CKT 412
   - Remains below .2 volt for 60 seconds or more.
   - And the system is operating in "Closed Loop".

Diagnostic Aids:

Using the "Scan", observe the block learn values at different rpm and air flow conditions. The "Scan" also displays the block cells, so the block learn values can be checked in each of the cells to determine when the Code 44 may have been set. If the conditions for Code 44 exists, the block learn values will be around 150.
- **O₂ Sensor Wire.** Sensor pigtail may be mispositioned and contacting the exhaust manifold.
- Check for intermittent ground in wire between connector and sensor.

- **MAP Sensor.** A shifted "Low" MAP sensor could cause the fuel system to go lean.
- **Lean Injector(s).** Perform injector balance test CHART C-2A.
- **Fuel Contamination.** Water, even in small amounts, near the in-tank fuel pump inlet can be delivered to the injectors. The water causes a lean exhaust and can set a Code 44.
- **Fuel Pressure.** System will be lean if pressure is too low. It may be necessary to monitor fuel pressure while driving the car at various road speeds and/or loads to confirm. See Fuel System diagnosis CHART A-7.
- **Exhaust Leaks.** If there is an exhaust leak, the engine can cause outside air to be pulled into the exhaust and past the sensor. Vacuum or crankcase leaks can cause a lean condition.
- If the above are OK, it is a faulty oxygen sensor.

1988–90 2.8L (VIN W) – CAVALIER
1988 2.8L (VIN W) – CIMARRON

CODE 44
OXYGEN SENSOR CIRCUIT
(LEAN EXHAUST INDICATED)
2.8L (VIN W) "J" SERIES (PORT)

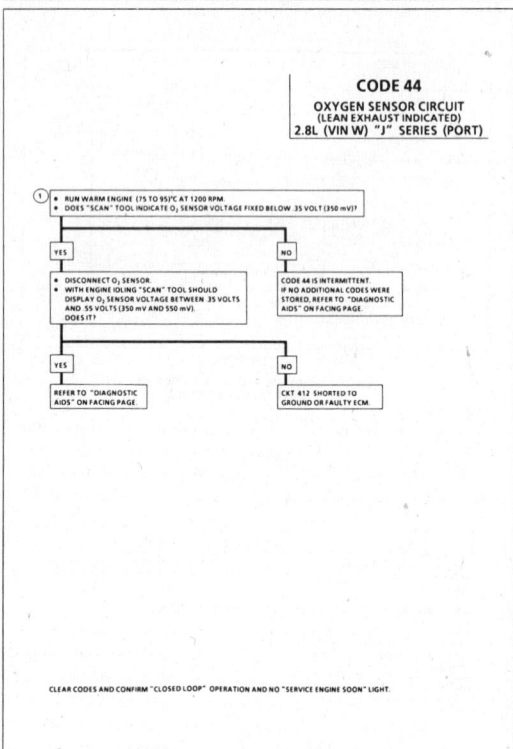

① • RUN WARM ENGINE (75 TO 95°C) AT 1200 RPM.
• DOES "SCAN" TOOL INDICATE O₂ SENSOR VOLTAGE FIXED BELOW .35 VOLT (350 mV)?

YES → • DISCONNECT O₂ SENSOR.
• WITH ENGINE IDLING "SCAN" TOOL SHOULD DISPLAY O₂ SENSOR VOLTAGE BETWEEN .35 VOLTS AND .55 VOLTS (350 mV AND 550 mV). DOES IT?

NO → CODE 44 IS INTERMITTENT. IF NO ADDITIONAL CODES WERE STORED, REFER TO "DIAGNOSTIC AIDS" ON FACING PAGE.

YES → REFER TO "DIAGNOSTIC AIDS" ON FACING PAGE.

NO → CKT 412 SHORTED TO GROUND OR FAULTY ECM.

CLEAR CODES AND CONFIRM "CLOSED LOOP" OPERATION AND NO "SERVICE ENGINE SOON" LIGHT.

1988–90 2.8L (VIN W) – CAVALIER
1988 2.8L (VIN W) – CIMARRON

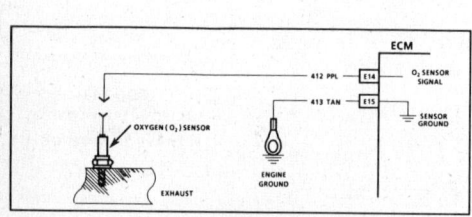

CODE 45
OXYGEN SENSOR CIRCUIT
(RICH EXHAUST INDICATED)
2.8L (VIN W) "J" SERIES (PORT)

Circuit Description:

The ECM supplies a voltage of about .45 volt between terminals "E14" and "E15". (If measured with a 10 megohm digital voltmeter, this may read as low as .32 volts.) The O₂ sensor varies the voltage within a range of about 1 volt, if the exhaust is rich, down through about .10 volt, if exhaust is lean.

The sensor produces no voltage when it is below about 315°C (600°F). An open sensor circuit or cold sensor causes "Open Loop" operation.

Test Description: Numbers below refer to circled numbers on the diagnostic chart.
1. Code 45 is set when the O₂ sensor signal voltage or CKT 412
   - Remains above .7 volt for 50 seconds; and in "Closed Loop".
   - Engine time after start is 1 minute or more.
   - Throttle angle between 3% and 45%.

Diagnostic Aids:

Using the "Scan", observe the block learn values at different rpm and air flow conditions. The "Scan" also displays the block cells, so the block learn values can be checked in each of the cells to determine when the Code 45 may have been set. If the conditions for Code 45 exists, the block learn values will be around 115.
- **Fuel Pressure.** System will go rich, if pressure is too high. The ECM can compensate for some increase. However, if it gets too high, a Code 45 will be set. See Fuel System diagnosis CHART A-7.
- **Rich injector.** Perform injector balance test CHART C-2A.
- **Leaking injector.** See CHART A-7.
- Check for fuel contaminated oil.

Check For
• Short to voltage on CKT 412
- **HEI Shielding.** An open ground CKT 453 (ignition system) may result in EMI, or induced electrical "noise". The ECM looks at this "noise" as reference pulses. The additional pulses result in a higher than actual engine speed signal. The ECM then delivers too much fuel, causing system to go rich. Engine tachometer will also show higher than actual engine speed, which can help in diagnosing this problem.
- **Canister purge.** Check for fuel saturation. If full of fuel, check canister control and hoses.

- **MAP sensor.** A shifted "High" MAP sensor could cause the fuel system to go rich.
- Check for leaking fuel pressure regulator diaphragm by checking vacuum line to regulator for fuel.
- **TPS.** An intermittent TPS output will cause the system to go rich, due to a false indication of the engine accelerating.
- **EGR.** An EGR staying open (especially at idle) will cause the O₂ sensor to indicate a rich exhaust.

1988–90 2.8L (VIN W) – CAVALIER
1988 2.8L (VIN W) – CIMARRON

CODE 45
OXYGEN SENSOR CIRCUIT
(RICH EXHAUST INDICATED)
2.8L (VIN W) "J" SERIES (PORT)

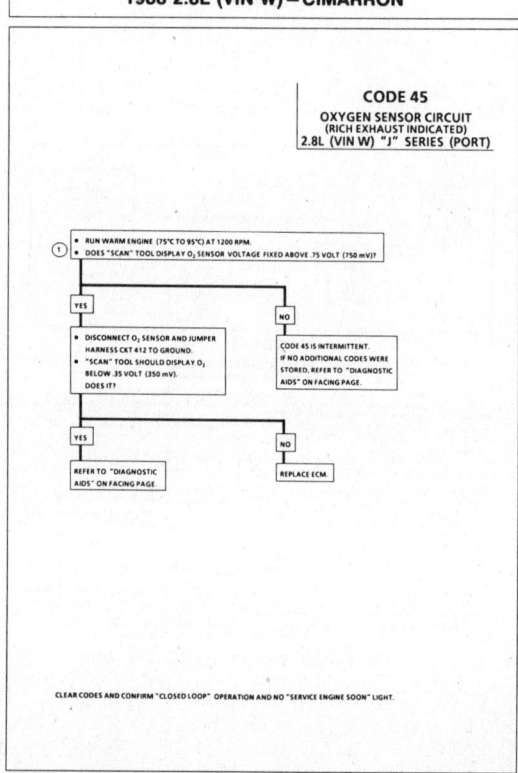

① • RUN WARM ENGINE (75°C TO 95°C) AT 1200 RPM.
• DOES "SCAN" TOOL DISPLAY O₂ SENSOR VOLTAGE FIXED ABOVE .75 VOLT (750 mV)?

YES → • DISCONNECT O₂ SENSOR AND JUMPER HARNESS CKT 412 TO GROUND. "SCAN" TOOL SHOULD DISPLAY O₂ BELOW .35 VOLT (350 mV). DOES IT?

NO → CODE 45 IS INTERMITTENT. IF NO ADDITIONAL CODES WERE STORED, REFER TO "DIAGNOSTIC AIDS" ON FACING PAGE.

YES → REFER TO "DIAGNOSTIC AIDS" ON FACING PAGE.

NO → REPLACE ECM.

CLEAR CODES AND CONFIRM "CLOSED LOOP" OPERATION AND NO "SERVICE ENGINE SOON" LIGHT.

1988–90 2.8L (VIN W) – CAVALIER
1988 2.8L (VIN W) – CIMARRON

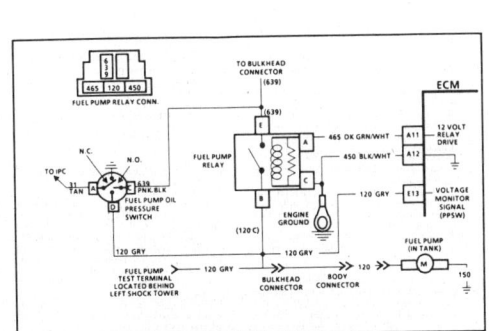

CODE 54
FUEL PUMP CIRCUIT
(LOW VOLTAGE)
2.8L (VIN W) "J" SERIES (PORT)

Circuit Description:

The status of the fuel pump CKT 120 is monitored by the ECM at terminal "E13" and is used to compensate fuel delivery based on system voltage. This signal is also used to store a trouble code, if the fuel pump relay is defective or fuel pump voltage is lost, while the engine is running. There should be about 12 volts on CKT 120 for 2 seconds, after the ignition is turned "ON", or any time references pulses are being received by the ECM.

Diagnostic Aids:

Code 54 will set, if the voltage at terminal "E13" is less than 4 volts for .5 seconds since the last reference pulse was received. This code is designed to detect a faulty relay, causing extended crank time, and the code will help the diagnosis of an engine that "CRANKS BUT WILL NOT RUN".

If a fault is detected during start-up, the "Service Engine Soon" light will stay "ON" until the ignition is cycled "OFF". However, if the voltage is detected below 2 volts with the engine running, the light will only remain "ON" while the condition exists.

Check sources of intermittent low system voltage such as; loose battery connections, poor generator belt or belt tension or loose generator connections. If OK, refer to "Intermittents" in Section "B".

1988–90 2.8L (VIN W) – CAVALIER
1988 2.8L (VIN W) – CIMARRON

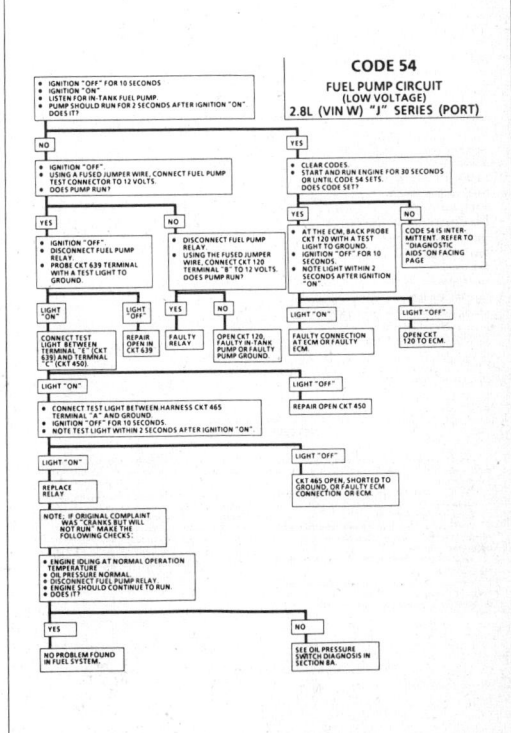

CODE 54
FUEL PUMP CIRCUIT
(LOW VOLTAGE)
2.8L (VIN W) "J" SERIES (PORT)

1988–90 2.8L (VIN W) – CAVALIER
1988 2.8L (VIN W) – CIMARRON

CODE 51
CODE 52
CODE 53
2.8L (VIN W) "J" SERIES (PORT)

CODE 51
MEM-CAL ERROR
(FAULTY OR INCORRECT MEM-CAL)

CHECK THAT ALL PINS ARE FULLY INSERTED IN THE SOCKET AND THAT MEM-CAL IS PROPERLY LATCHED.
IF OK, REPLACE MEM-CAL, CLEAR MEMORY, AND RECHECK. IF CODE 51 REAPPEARS, REPLACE ECM.

CLEAR CODES AND CONFIRM "CLOSED LOOP" OPERATION AND NO "SERVICE ENGINE SOON" LIGHT.

CODE 52
CALPAK ERROR
(FAULTY OR INCORRECT CALPAK)

CHECK THAT THE MEM-CAL IS FULLY SEATED AND LATCHED INTO THE MEM-CAL
SOCKET. IF OK, REPLACE MEM-CAL, CLEAR MEMORY, AND RECHECK.
IF CODE 52 REAPPEARS, REPLACE ECM.

CLEAR CODES AND CONFIRM "CLOSED LOOP" OPERATION AND NO "SERVICE ENGINE SOON" LIGHT.

CODE 53
SYSTEM OVER VOLTAGE

THIS CODE INDICATES THERE IS A BASIC GENERATOR PROBLEM.
• CODE 53 WILL SET, IF VOLTAGE AT ECM IGNITION INPUT PIN IS GREATER THAN
  17.1 VOLTS FOR 2 SECONDS.
• CHECK AND REPAIR CHARGING SYSTEM. REFER TO SECTION "6D".

CLEAR CODES AND CONFIRM "CLOSED LOOP" OPERATION AND NO "SERVICE ENGINE SOON" LIGHT.

1988–90 2.8L (VIN W) – CAVALIER
1988 2.8L (VIN W) – CIMARRON

CODE 61
DEGRADED OXYGEN SENSOR
2.8L (VIN W) "J" SERIES (PORT)

IF A CODE 61 IS STORED IN MEMORY THE ECM HAS DETERMINED THE OXYGEN SENSOR IS CONTAMINATED OR DEGRADED, BECAUSE THE VOLTAGE CHANGE TIME IS SLOW OR SLUGGISH.

THE ECM PERFORMS THE OXYGEN SENSOR RESPONSE TIME TEST WHEN:

COOLANT TEMPERATURE IS GREATER THAN 85°C.

MAT TEMPERATURE IS GREATER THAN 10°C.

IN CLOSED LOOP.

IN DECEL FUEL CUT-OFF MODE.

IF A CODE 61 IS STORED THE OXYGEN SENSOR SHOULD BE REPLACED. A CONTAMINATED SENSOR CAN BE CAUSED BY FUEL ADDITIVES, SUCH AS SILICON, OR BY USE OF NON-GM APPROVED LUBRICANTS OR SEALANTS. SILICON CONTAMINATION IS USUALLY INDICATED BY A WHITE POWDERY SUBSTANCE ON THE SENSOR FINS.

CLEAR CODES AND CONFIRM "CLOSED LOOP" OPERATION AND NO "SERVICE ENGINE SOON" LIGHT.

1988–90 2.8L (VIN W) – CAVALIER
1988 2.8L (VIN W) – CIMARRON

SECTION B
SYMPTOMS

TABLE OF CONTENTS

Before Starting
Intermittents
Hard Start
Hesitation, Sag, Stumble
Surges and/or Chuggle
Lack of Power, Sluggish, or Spongy
Detonation/Spark Knock
Cuts Out, Misses
Backfire
Poor Fuel Economy
Dieseling, Run-On
Rough, Unstable, or Incorrect Idle, Stalling
Excessive Exhaust Emissions or Odors
Restricted Exhaust System Check (all engines)

BEFORE STARTING

Before using this section you should have performed the DIAGNOSTIC CIRCUIT CHECK and found out that:

1. The ECM and "Service Engine Soon" light are operating.
2. There are no trouble codes.

Verify the customer complaint, and locate the correct SYMPTOM below. Check the items indicated under that symptom.

If the ENGINE CRANKS BUT WILL NOT RUN, see CHART A-3.

Several of the symptom procedures below call for a careful visual check. This check should include:

- ECM grounds for being clean and tight.
- Vacuum hoses for splits, kinks, and proper connections, as shown on Emission Control Information label.
- Air leaks at throttle body mounting and intake manifold.
- Ignition wires for cracking, hardness, proper routing, and carbon tracking.
- Wiring for proper connections, pinches, and cuts.

The importance of this step cannot be stressed too strongly - it can lead to correcting a problem without further checks and can save valuable time.

1988–90 2.8L (VIN W) – CAVALIER
1988 2.8L (VIN W) – CIMARRON

INTERMITTENTS

Problem may or may not turn "ON" the "Service Engine Soon" light, or store a code.

DO NOT use the trouble code charts for intermittent problems. The fault must be present to locate the problem. If a fault is intermittent, use of trouble code charts may result in replacement of good parts.

- Most intermittent problems are caused by faulty electrical connections or wiring. Perform careful check as described at start of Section "B".
  Check for:
  - Poor mating of the connector halves, or terminals not fully seated in the connector body (backed out).
  - Improperly formed or damaged terminals. All connector terminals in problem circuit should be carefully reformed to increase contact tension.
  - Poor terminal to wire connection. This requires removing the terminal from the connector body to check.
- If a visual check does not find the cause of the problem, the car can be driven with a voltmeter connected to a suspected circuit. A "Scan" tool can, also, be used for monitoring input signals to help detect intermittent conditions. An abnormal voltage, or "Scan" reading, when the problem occurs, indicates the problem may be in that circuit. If the wiring and connectors check OK and a trouble code was stored for a circuit having a sensor, except for Codes 43, 44, and 45, substitute a known good sensor and recheck.

An intermittent "Service Engine Soon" light with no stored code may be caused by;

- Ignition coil shorted to ground and arcing at spark plug wires or plugs.
- "Service Engine Soon" light wire to ECM shorted to ground. (CKT 419).
- Diagnostic "Test" terminal wire to ECM, shorted to ground. (CKT 451)
- ECM power grounds. See ECM wiring diagrams.

- <u>Loss of trouble code memory.</u> To check, disconnect TPS and idle engine until "Service Engine Soon" light comes "ON". Code 22 should be stored, and kept in memory when ignition is turned "OFF". If not, the ECM is faulty.
- Check for an electrical system interference caused by a defective relay, ECM driven solenoid, or switch. They can cause a sharp electrical surge. Normally, the problem will occur when the faulty component is operated.
- Check for improper installation of electrical options, such as lights, 2-way radios, etc.
- EST wires should be kept away from spark plug wires, coils and generator. Wire from ECM to ignition system (CKT 453) should be a good connection.
- Check for open diode across A/C compressor clutch, and for other open diodes (see wiring diagrams).

HARD START

Definition: Engine cranks OK, but does not start for a long time. Does eventually run, or may start but immediately dies.

- Perform careful check as described at start of Section "B".
- Make sure driver is using correct starting procedure.
- CHECK:
  - TPS for sticking or binding or a high TPS voltage with the throttle closed (should read less than .700 volts).
  - High resistance in coolant sensor circuit or sensor itself. See Code 15 chart or with a "Scan" tool compare coolant temperature with ambient temperature on a cold engine.
  - Fuel pressure CHART A-7.
  - Water contaminated fuel.
  - EGR operation. Be sure valve seats properly and is not staying open. See CHART C-7.
- Ignition system - Check for:
  - Bare and shorted wires and proper output with spark tester J-26792 (ST-125) or equivalent.
  - IAC operation - See Code 35 chart.
- A faulty in-tank fuel pump check valve will allow the fuel in the lines to drain back to the tank after the engine is stopped. To check for this condition: Perform fuel system diagnosis, CHART A-7.
- Remove spark plugs. Check for wet plugs, cracks, wear, improper gap, burned electrodes, or heavy deposits. Repair or replace as necessary.

1988–90 2.8L (VIN W) – CAVALIER
1988 2.8L (VIN W) – CIMARRON

HESITATION, SAG, STUMBLE

Definition: Momentary lack of response as the accelerator is pushed down. Can occur at all car speeds. Usually most severe when first trying to make the car move, as from a stop sign. May cause the engine to stall if severe enough.

- Perform careful visual check as described at start of Section "B".
- CHECK:
  - Fuel pressure. See CHART A-7. Also Check for water contaminated fuel.
  - Spark plugs for being fouled or faulty wiring.
  - Mem-Cal number and Service Bulletins for latest Mem-Cal.
  - Electronic Spark Control - See CHART C-5.
  - TPS for binding or sticking. Voltage should increase at a steady rate as throttle is moved toward WOT.
  - Generator output voltage. Repair, if less than 9 or more than 16 volts.
  - Ignition system ground, CKT 453.
  - Canister purge system for proper operation. See CHART C-3.
  - EGR - See CHART C-7.
  - MAP sensor - See CHART C-1D.
- Perform injector balance test, CHART C-2A.

SURGES AND/OR CHUGGLE

Definition: Engine power variation under steady throttle or cruise. Feels like the car speeds up and slows down with no change in the accelerator pedal.

- Be sure driver understands transmission/transaxle converter clutch and A/C compressor operation in Owner's Manual.
- Perform careful visual inspection as described at start of Section "B".
- To help determine if the condition is caused by a rich or lean system, the car should be driven at the speed of the complaint. Monitoring block learn at the complaint speed will help identify the cause of the problem. If the system is running lean (block learn greater than 138), refer to "Diagnostic Aid" on facing page of Code 44. If the system is running rich (block learn less than 118), refer to "Diagnostic Aids" on facing page of Code 45.
- CHECK:
  - Generator output voltage. Repair if less than 9 or more than 16 volts.
  - EGR - There should be no EGR at idle. See CHART C-7.
  - EGR filter for being plugged, see CHART C-7
  - Vacuum lines for kinks or leaks.
  - In-line fuel filter. Replace if dirty or plugged
  - Fuel pressure while condition exists. See CHART A-7
- Remove spark plugs. Check for cracks, wear, improper gap, burned electrodes, or heavy deposits. Also check condition of spark plug wires and check for proper output voltage using spark tester J-26792 (ST-125) or equivalent.

LACK OF POWER, SLUGGISH, OR SPONGY

Definition: Engine delivers less than expected power. Little or no increase in speed when accelerator pedal is pushed down part way.

- Perform careful visual check as described at start of Section "B".
- Compare customer's car to similar unit. Make sure the customer's car has an actual problem.
- Remove air filter and check for dirt, or for being plugged. Replace as necessary.
- CHECK:
  - Restricted fuel filter, contaminated fuel or improper fuel pressure - See CHART A-7
  - ECM power grounds, see wiring diagrams.
  - EGR operation for being open or partly open all the time - See CHART C-7
  - ECS for incorrect retard - See CHART C-5
  - Exhaust system for possible restriction: See CHART B-1.
    - Inspect exhaust system for damaged or collapsed pipes.
    - Inspect muffler for heat distress or possible internal failure.
  - Generator output voltage. Repair if less than 9 or more than 16 volts.
  - Engine valve timing and compression.
  - Engine for proper or worn camshaft. See Section "6A".
  - Secondary voltage using a shop ocilliscope or a spark tester J-26792 (ST-125) or equivalent.
  - Check A/C operation. A/C clutch should cut out at WOT. See A/C CHART C-10.

1988–90 2.8L (VIN W) – CAVALIER
1988 2.8L (VIN W) – CIMARRON

DETONATION /SPARK KNOCK

Definition: A mild to severe ping, usually worse under acceleration. The engine makes sharp metallic knocks that change with throttle opening. Sounds like popcorn popping.

- Check for obvious overheating problems:
  - Low coolant.
  - Loose belt.
  - Restricted air flow to radiator, or restricted water flow through radiator.
  - Inoperative electric cooling fan circuit. See CHART C-12.
- To help determine if the conditions is caused by a rich or lean system, the car should be driven at the speed of the complaint. Monitoring block learn, at the complaint speed, will help identify the cause of the problem. If the system is running lean (block learn greater than 138), refer to "Diagnostic Aids" on facing page of Code 44. If the system is running rich (block learn less than 118), refer to facing page of Code 45.
- CHECK:
  - EGR system for not opening or plugged EGR passages - See CHART C-7.
  - ESC system for no retard - See CHART C-5.
  - Park/Neutral switch. Be sure "Scan" indicates drive with gear selector in drive or overdrive. See CHART C-1A.
  - TCC operation, TCC applying too soon - See CHART C-8.
  - Fuel system pressure - See CHART A-7.
  - Engine for excessive carbon build-up. Remove carbon with top engine cleaner. Follow instructions on can.
  - Mem-Cal for correct part.
  - Combustion chamber for excessive oil.
  - Engine for incorrect basic parts such as cam, heads, pistons, etc.
  - Fuel for poor quality, proper octane rating.

CUTS OUT, MISSES

Definition: Steady pulsation or jerking that follows engine speed, usually more pronounced as engine load increases. The exhaust has a steady spitting sound at idle or under a load.

- Perform careful visual (physical) check as described at start of Section "B".
- If ignition system is suspected of causing a miss at idle or cutting out under load, refer to appropriate ignition "Misfire" chart
- If the previous checks did not find the problem:
  - Visually inspect ignition system for moisture, dust, cracks, burns, etc. Spray plug wires with fine water mist to check for shorts.
- If above checks did not correct problem, check the following:
  - To determine if one injector or connector is faulty, disconnect injector 4-way connector. On the injector side of the harness, connect ohmmeter between the B+ circuit and the injector drive side of the harness for each bank of injectors. Because each bank of injectors is wired in parallel, an ohmmeter should measure about 4 ohms. If the reading is 6 ohms, or greater, an open circuit is likely the cause. If 2 ohms, or less, a shorted circuit or injector is likely. Reconnect injector 4-way connector.
- If either ohm reading is out of range:
  - Disconnect all injector harness connectors. Connect J-34730-2 injector test light or equivalent 6 volt test light between the harness terminals of each injector connector and note light while cranking. If test light fails to blink at any connector, it is a faulty injector drive circuit harness, connector, or terminal.
- CHECK:
  - Injector Balance - See CHART C-2A.
  - Fuel System - Plugged fuel filter, water, low pressure - See CHART A-7.
  - Valve Timing
- Remove rocker covers. Check for bent pushrods, worn rocker arms, broken valve springs, worn camshaft lobes. Repair as necessary.
- Perform compression check.

CHART B-1
RESTRICTED EXHAUST SYSTEM CHECK
ALL ENGINES

Proper diagnosis for a restricted exhaust system is essential before any components are replaced. Either of the following procedures may be used for diagnosis, depending upon engine or tool used:

CHECK AT A. I. R. PIPE:	OR	CHECK AT O₂ SENSOR:

CHECK AT A. I. R. PIPE:

1. Remove the rubber hose at the exhaust manifold A.I.R. pipe check valve. Remove check valve.
2. Connect a fuel pump pressure gauge to a hose and nipple from a Propane Enrichment Device (J26911) (see illustration).
3. Insert the nipple into the exhaust manifold A.I.R. pipe.

OR

CHECK AT O₂ SENSOR:

1. Carefully remove O_2 sensor.
2. Install Borroughs exhaust backpressure tester (BT 8515 or BT 8603) or equivalent in place of O_2 sensor (see illustration).
3. After completing test described below, be sure to coat threads of O_2 sensor with anti-seize compound P/N 5613695 or equivalent prior to re-installation.

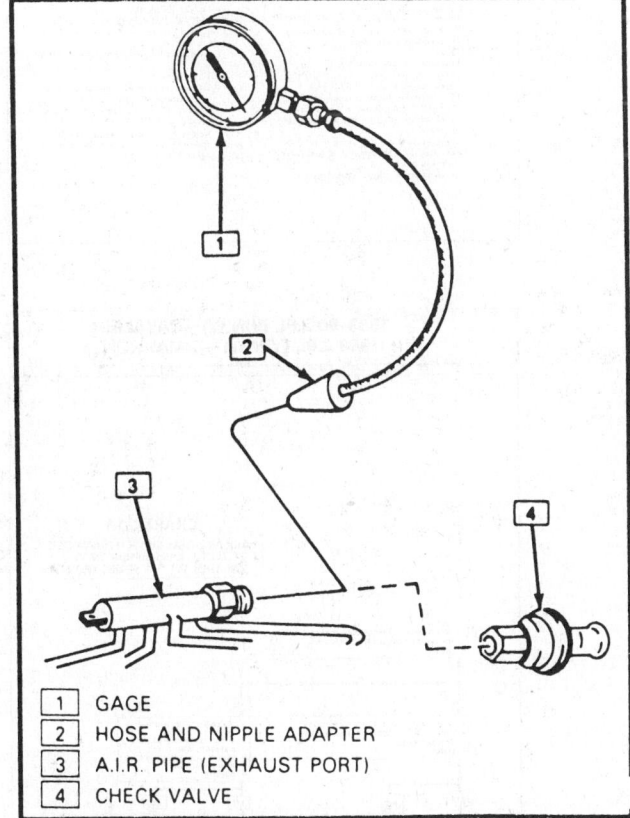

1	GAGE
2	HOSE AND NIPPLE ADAPTER
3	A.I.R. PIPE (EXHAUST PORT)
4	CHECK VALVE

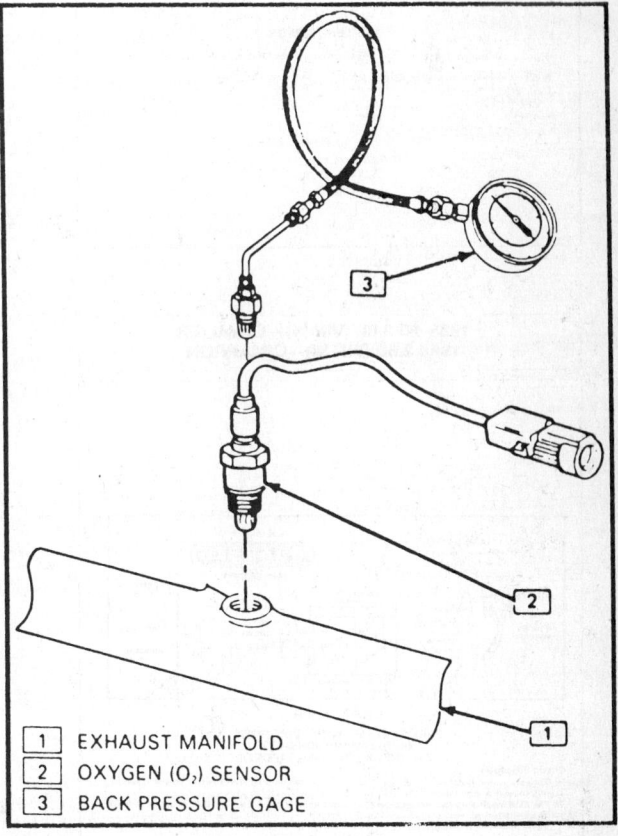

1	EXHAUST MANIFOLD
2	OXYGEN (O₂) SENSOR
3	BACK PRESSURE GAGE

DIAGNOSIS:

1. With the engine idling at normal operating temperature, observe the exhaust system backpressure reading on the gauge. Reading should not exceed 8.6 kPa (1.25 psi).
2. Increase engine speed to 2000 rpm and observe gauge. Reading should not exceed 20.7 kPa (3 psi).
3. If the backpressure at either speed exceeds specification, a restricted exhaust system is indicated.
4. Inspect the entire exhaust system for a collapsed pipe, heat distress, or possible internal muffler failure.
5. If there are no obvious reasons for the excessive backpressure, the catalytic converter is suspected to be restricted and should be replaced using current recommended procedures.

1988-90 2.8L (VIN W) — CAVALIER
1988 2.8L (VIN W) — CIMARRON

BACKFIRE

Definition: Fuel ignites in intake manifold, or in exhaust system, making a loud popping noise.

- **CHECK:**
  - Compression - Look for sticking or leaking valves.
  - EGR operation for being open all the time. See CHART C-7.
  - EGR gasket for faulty or loose fit.
- Output voltage of ignition coils, using a shop oscilloscope or spark tester J-26792 (ST-125), or equivalent.
- Valve timing.
- Spark plugs, spark plug wires, and proper routing of plug wires.

POOR FUEL ECONOMY

Definition: Fuel economy, as measured by an actual road test, is noticeably lower than expected. Also, economy is noticeably lower than it was on this car at one time, as previously shown by an actual road test.

- Check owner's driving habits.
  - Is A/C "ON" full time (defroster mode "ON")?
  - Are tires at correct pressure?
  - Are excessively heavy loads being carried?
  - Is acceleration too much, too often?
  - Suggest owner fill fuel tank and recheck fuel economy.
  - Suggest driver read "Important Facts on Fuel Economy" in Owner's Manual.
- Check for proper calibration of speedometer.
- Visually (physically) Check.
  - Vacuum hoses for splits, kinks, and proper connections as shown on Vehicle Emission Control Information label.
  - Ignition wires for cracking, hardness, and proper connections.
  - Air cleaner element (filter) for dirt or being plugged.
- Remove spark plugs. Check for cracks, wear, improper gap, burned electrodes, or heavy deposits. Repair or replace as necessary.
- **CHECK:**
  - Engine thermostat for faulty part (always open) or for wrong heat range. Using a "Scan" tool, monitor engine temperature. A "Scan" displays engine temp. in degrees centigrade. After engine is started, the temperature should rise steadily to about 90°C, then stabilize, when thermostat opens.
  - Fuel Pressure. See CHART A-7.
  - Compression.
  - TCC for proper operation. See CHART C-8A. A "Scan" should indicate an rpm drop, when the TCC is commanded "ON".
  - Exhaust system restriction. See CHART B-1
  - ESC for incorrect retard. See CHART C-5

DIESELING, RUN-ON

Definition: Engine continues to run after key is turned "OFF", but runs very roughly.

- Check injectors for leaking. See CHART A-7.
- If engine runs smoothly, check ignition switch and adjustment.

1988-90 2.8L (VIN W) — CAVALIER
1988 2.8L (VIN W) — CIMARRON

ROUGH, UNSTABLE, OR INCORRECT IDLE, STALLING

Definition: The engine runs unevenly at idle. If bad enough, the car may shake. Also, the idle may vary in rpm (called "hunting"). Either condition may be bad enough to cause stalling. Engine idles at incorrect speed.

- Perform careful visual check as described at start of Section "B"
- Clean Injectors.
- **CHECK:**
  - Throttle linkage for sticking or binding.
  - TPS for sticking or binding, be sure output is stable at idle and adjustment specification is correct.
  - IAC system. See Code 35 chart.
  - Generator output voltage. Repair if less than 9 or more than 16 volts.
  - P/N switch circuit. See CHART C-1A, or use "Scan" tool, and be sure tool indicates vehicle is in drive with gear selector in drive (125C).
  - Injector balance. See CHART C-2A.
  - PCV valve for proper operation by placing finger over inlet hole in valve end several times. Valve should snap back. If not, replace valve.
  - Evaporative Emission Control System CHART C-3.
- Power Steering Pressure switch input. The state of the switch should only change when wheels are turned up against the stops. See CHART C-1E.
- Minimum Idle Speed. Incorrect minimum idle speed may be caused by foreign material accumulation in the throttle bore, on the throttle valve or on the throttle shaft
- ECM ground circuits
- Monitoring block learn values may help identify the cause of the problem. If the system is running lean (block learn greater than 138) refer to "Diagnostic Aids" on facing page of Code 44. If the system is running rich (block learn values less than 118) refer to "Diagnostic Aids" on facing page of Code 45.
- Run a cylinder compression check
- Check for fuel in pressure regulator hose. If present, replace regulator assembly.
- Check ignition system; wires and plugs
- If problem exists with A/C "ON," check A/C system operation CHART C-10.

EXCESSIVE EXHAUST EMISSIONS OR ODORS

Definition: Vehicle fails an emission test. Vehicle has excessive "rotten egg" smell. Excessive odors do not necessarily indicate excessive emissions.

- Perform "Diagnostic Circuit Check."
- IF TEST SHOWS EXCESSIVE CO AND HC, (or also has excessive odors):
  - Check items which cause car to run RICH.
    - Make sure engine is at normal operating temperature.
- **CHECK:**
  - Fuel pressure. See CHART A-7.
  - Canister for fuel loading. See CHART C-3.
  - Injector balance. See CHART C-2A.
  - PCV valve for being plugged, stuck, or blocked PCV hose, or fuel in the crankcase.
  - Spark plugs, plug wires, and ignition components.
  - Check for lead contamination of catalytic converter (look for removal of fuel filler neck restrictor).
  - Check for properly installed fuel cap.
- If the system is running rich, (block learn less than 118), refer to "Diagnostic Aids" on facing page of Code 45.
- IF TEST SHOWS EXCESSIVE NOx
  - Check items which cause car to run LEAN, or to run too hot.
    - EGR valve for not opening. See CHART C-7.
    - Vacuum leaks.
    - Coolant system and coolant fan for proper operation. See CHART C-12.
    - Remove carbon with top engine cleaner. Follow instructions on can.
- If the system is running lean, (block learn greater than 138) refer to "Diagnostic Aids" on facing page of Code 44

1988-90 2.8L (VIN W) — CAVALIER
1988 2.8L (VIN W) — CIMARRON

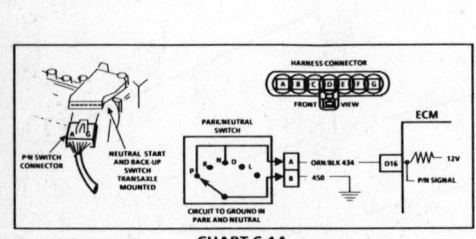

CHART C-1A

PARK/NEUTRAL SWITCH DIAGNOSIS
(AUTO TRANSMISSION ONLY)
2.8L (VIN W) "J" SERIES (PORT)

Circuit Description:

The park/neutral switch contacts are a part of the neutral start switch and are closed to ground in park or neutral, and open in drive ranges.

The ECM supplies ignition voltage through a current limiting resistor to CKT 434 and senses a closed switch when the voltage on CKT 434 drops to less than one volt.

The ECM uses the P/N signal as one of the inputs to control:
  Idle Air Control
  VSS Diagnostics
  EGR

If CKT 434 indicates P/N (grounded), while in drive range, the EGR would be inoperative, resulting in possible detonation.

If CKT 434 indicates drive (open) a drop in the idle may occur when the gear selector is moved into drive range.

Test Description: Numbers below refer to circled numbers on the diagnostic chart.

1. Checks for a closed switch to ground in park position. Different makes of "Scan" tools will read P/N differently. Refer to tool operator's manual for type of display used for a specific tool.
2. Checks for an open switch in drive range.
3. Be sure "Scan" indicates drive, even while wiggling shifter, to test for an intermittent or misadjusted switch in drive range.

1988-90 2.8L (VIN W) — CAVALIER
1988 2.8L (VIN W) — CIMARRON

CHART C-1A

PARK/NEUTRAL SWITCH DIAGNOSIS
(AUTO TRANSMISSION ONLY)
2.8L (VIN W) "J" SERIES (PORT)

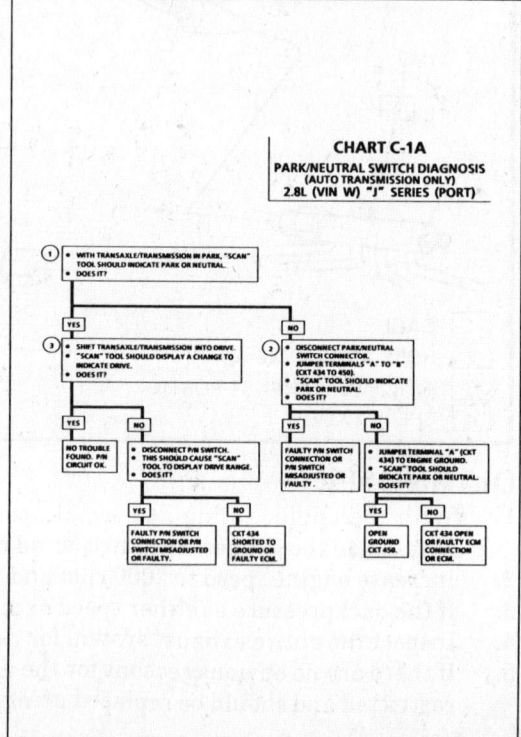

1988–90 2.8L (VIN W)–CAVALIER
1988 2.8L (VIN W)–CIMARRON

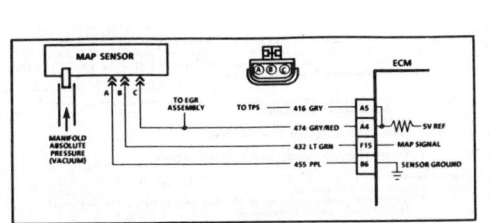

CHART C-1D
MAP OUTPUT CHECK
2.8L (VIN W) "J" SERIES (PORT)

Circuit Description:
The manifold absolute pressure sensor (MAP) measures manifold pressure (vacuum) and sends that signal to the ECM. The MAP sensor is mainly used for fuel calculation, when the ECM is running in the throttle body backup mode. The MAP sensor is also used to determine the barometric pressure and to help calculate fuel delivery.

Test Description: Numbers below refer to circled numbers on the diagnostic chart.
1. Checks MAP sensor output voltage to the ECM. This voltage, without engine running, represents a barometer reading to the ECM.
2. Applying 34 kPa (10 inches Hg) vacuum to the MAP sensor should cause the voltage to be 1.2 volts less than the voltage at Step 1. Upon applying vacuum to the sensor, the change in voltage should be instantaneous.

A slow voltage change indicates a faulty sensor. The engine must be running in this step or the "Scanner" will not indicate a change in voltage. It is normal for the "Service Engine Soon" light to come "ON" and for the system to set a Code 33 during this step. Make sure the code is cleared when this test is completed.
3. Check vacuum hose to sensor for leaking or restriction. Be sure no other vacuum devices are connected to the MAP hose.

1988–90 2.8L (VIN W)–CAVALIER
1988 2.8L (VIN W)–CIMARRON

CHART C-1D
MAP OUTPUT CHECK
2.8L (VIN W) "J" SERIES (PORT)

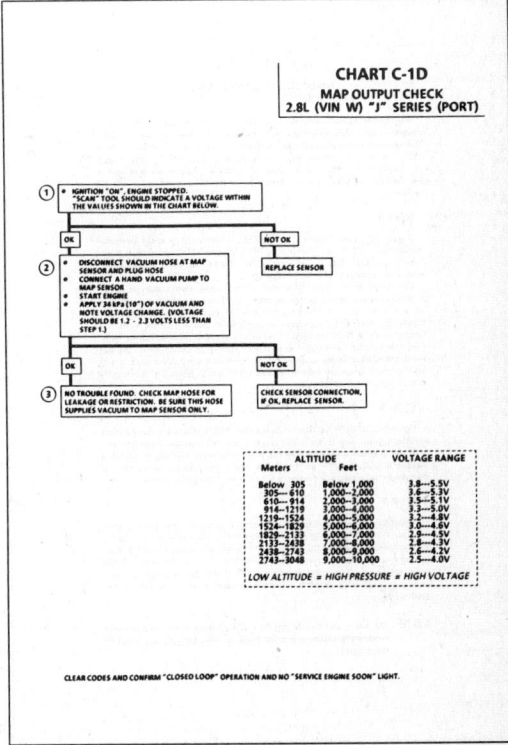

CLEAR CODES AND CONFIRM "CLOSED LOOP" OPERATION AND NO "SERVICE ENGINE SOON" LIGHT.

1988–90 2.8L (VIN W)–CAVALIER
1988 2.8L (VIN W)–CIMARRON

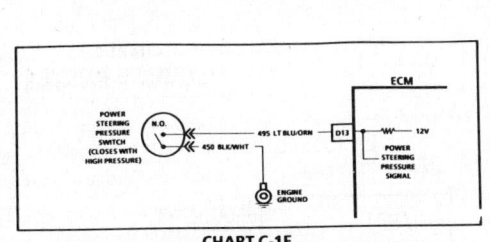

CHART C-1E
POWER STEERING PRESSURE SWITCH (PSPS) DIAGNOSIS
2.8L (VIN W) "J" SERIES (PORT)

Circuit Description:
The power steering pressure switch is normally open to ground, and CKT 495 will be near the battery voltage.
Turning the steering wheel increases power steering oil pressure and its load on an idling engine. The pressure switch will close before the load can cause an idle problem.
Closing the switch causes CKT 495 to read less than 1 volt. The ECM will increase the idle air rate and disengage the A/C relay.
- A pressure switch that will not close or an open CKT 495 or 450, may cause the engine to stop when power steering loads are high.
- A switch that will not open, or a CKT 495 shorted to ground, may affect idle quality and will cause the A/C relay to be de-energized.

Test Description: Numbers below refer to circled numbers on the diagnostic chart.
1. Different makes of "Scan" tools may display the state of this switch in different ways. Refer to "Scan" tool operator's manual to determine how this input is indicated.
2. Checks to determine if CKT 495 is shorted to ground.
3. This should simulate a closed switch.

1988–90 2.8L (VIN W)–CAVALIER
1988 2.8L (VIN W)–CIMARRON

CHART C-1E
POWER STEERING PRESSURE SWITCH (PSPS) DIAGNOSIS
2.8L (VIN W) "J" SERIES (PORT)

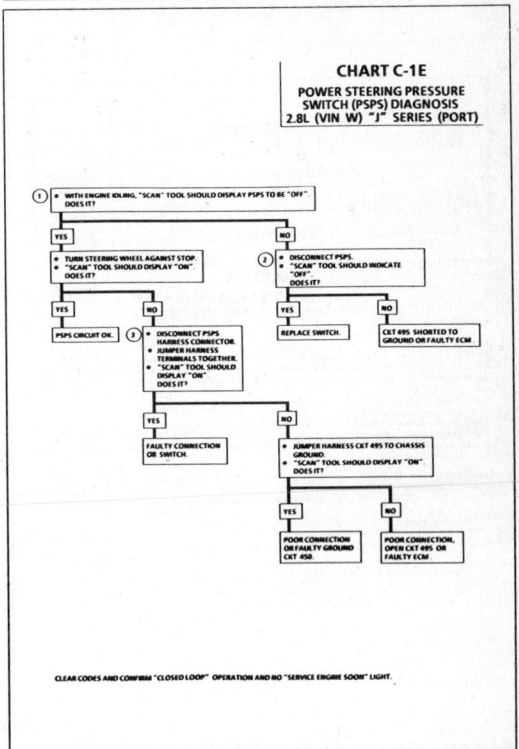

CLEAR CODES AND CONFIRM "CLOSED LOOP" OPERATION AND NO "SERVICE ENGINE SOON" LIGHT.

1988–90 2.8L (VIN W)–CAVALIER
1988 2.8L (VIN W)–CIMARRON

CHART C-2A
INJECTOR BALANCE TEST
2.8L (VIN W) "J" SERIES (PORT)

The injector balance tester is a tool used to turn the injector on for a precise amount of time, thus spraying a measured amount of fuel into the manifold. This causes a drop in fuel rail pressure that we can record and compare between each injector. All injectors should have the same amount of pressure drop (± 10 kpas). Any injector with a pressure drop that is 10 kpa (or more) greater or less than the average drop of the other injectors should be considered faulty and replaced.

STEP 1

Engine "cool down" period (10 minutes) is necessary to avoid irregular readings due to "Hot Soak" fuel boiling. With ignition "OFF" connect fuel gauge J-34730-1 or equivalent to fuel pressure tap. Wrap a shop towel around fitting while connecting gage to avoid fuel spillage.

Disconnect harness connectors at all injectors, and connect injector tester J-34730-3, or equivalent, to one injector. On Turbo equipped engines, use adaptor harness furnished with injector tester to energize injectors that are not accessible. Follow manufacturers instructions for use of adaptor harness. Ignition must be "OFF" at least 10 seconds to complete ECM shutdown cycle. Fuel pump should run about 2 seconds after ignition is turned "ON". At this point, insert clear tubing attached to vent valve into a suitable container and bleed air from gauge and hose to insure accurate gauge operation. Repeat this step until all air is bled from gauge.

STEP 2

Turn ignition "OFF" for 10 seconds and then "ON" again to get fuel pressure to its maximum. Record this initial pressure reading. Energize tester one time and note pressure drop at its lowest point (Disregard any slight pressure increase after drop hits low point.). By subtracting this second pressure reading from the initial pressure, we have the actual amount of injector pressure drop.

STEP 3

Repeat step 2 on each injector and compare the amount of drop. Usually, good injectors will have virtually the same drop. Retest any injector that has a pressure difference of 10 kPa, either more or less than the average of the other injectors on the engine. Replace any injector that also fails the retest. If the pressure drop of all injectors is within 10kPa of this average, the injectors appear to be flowing properly. Reconnect them and review Symptoms, Section "B".

NOTE: *The entire test should not be repeated more than once without running the engine to prevent flooding. (This includes any retest on faulty injectors).*

1988–90 2.8L (VIN W)–CAVALIER
1988 2.8L (VIN W)–CIMARRON

NOTE: If injectors are suspected of being dirty, they should be cleaned using an approved tool and procedure prior to performing this test. The fuel pressure test in Section A, Chart A-7, should be completed prior to this test.

CHART C-2A
INJECTOR BALANCE TEST
2.8L (VIN W) "J" SERIES (PORT)

Step 1. If engine is at operating temperature, allow a 10 minute "cool down" period then connect fuel pressure gauge and injector tester.
1. Ignition "OFF".
2. Connect fuel pressure gauge and injector tester.
3. Ignition "ON".
4. Bleed off air in gauge. Repeat until all air is bled from gauge.

Step 2. Run test:
1. Ignition "OFF" for 10 seconds.
2. Ignition "ON". Record gauge pressure. (Pressure must hold steady, if not see the Fuel System diagnosis, Chart A-7, in Section A)
3. Turn injector on, by depressing button on injector tester, and note pressure at the instant the gauge needle stops.

Step 3.
1. Repeat step 2 on all injectors and record pressure drop on each. Retest injectors that appear faulty (Any injectors that have a 10 kPa difference, either more or less, in pressure from the average. If no problem is found, review Symptoms Section B.

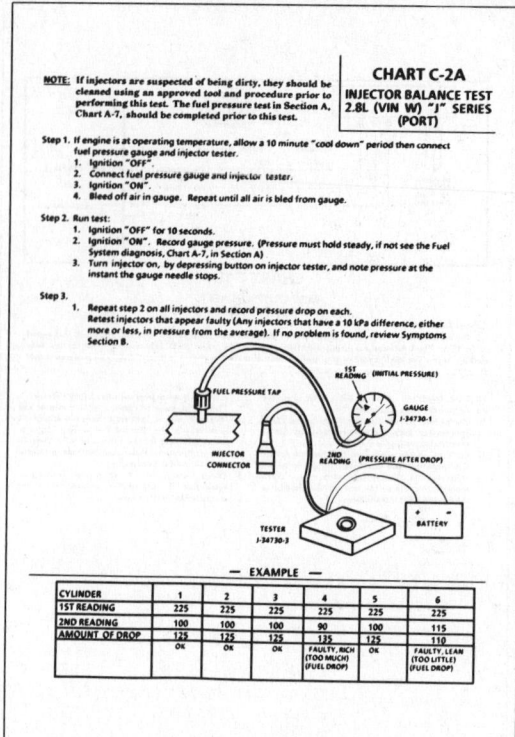

— EXAMPLE —

CYLINDER	1	2	3	4	5	6
1ST READING	225	225	225	225	225	225
2ND READING	100	100	100	90	100	115
AMOUNT OF DROP	125	125	125	135	125	110
	OK	OK	OK	FAULTY, RICH (TOO MUCH) (FUEL DROP)	OK	FAULTY, LEAN (TOO LITTLE) (FUEL DROP)

1988–90 2.8L (VIN W)–CAVALIER
1988 2.8L (VIN W)–CIMARRON

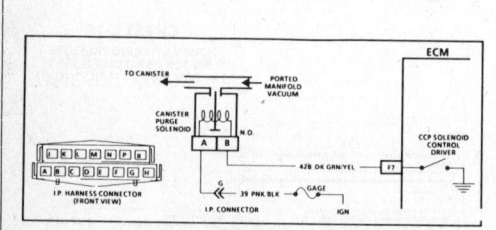

CHART C-3
CANISTER PURGE VALVE CHECK
2.8L (VIN W) "J" SERIES (PORT)

Circuit Description:
Canister purge is controlled by a solenoid that allows manifold vacuum to purge the canister when de-energized. The ECM supplies a ground to energize the solenoid (purge "OFF"). The purge solenoid control by the ECM is pulse width modulated (turned "ON" and "OFF" several times a second). The duty cycle (pulse width) is determined by the amount of air flow, and the engine vacuum as determined by the MAP sensor input. The duty cycle is calculated by the ECM and the output commanded when the following conditions have been met.
* Engine run time after start more than 3 minutes.
* Coolant temperature above 80°C.
* Vehicle speed above 5 mph.
* Throttle off idle (about 3%).
Also, if the diagnostic test terminal is grounded with the engine stopped, the purge solenoid is de-energized (purge "ON")

Test Description: Numbers below refer to circled numbers on the diagnostic chart.
1. Checks to see if the solenoid is opened or closed. The solenoid is normally energized in this step, so it should be closed.
2. Checks for a complete circuit. Normally there is ignition voltage on CKT 39 and the ECM provides a ground on CKT 428.
3. Completes functional check by grounding test terminal. This should normally de-energize the solenoid opening the valve which should allow the vacuum to drop (purge "ON")

1988–90 2.8L (VIN W)–CAVALIER
1988 2.8L (VIN W)–CIMARRON

CHART C-3
CANISTER PURGE VALVE CHECK
2.8L (VIN W) "J" SERIES (PORT)

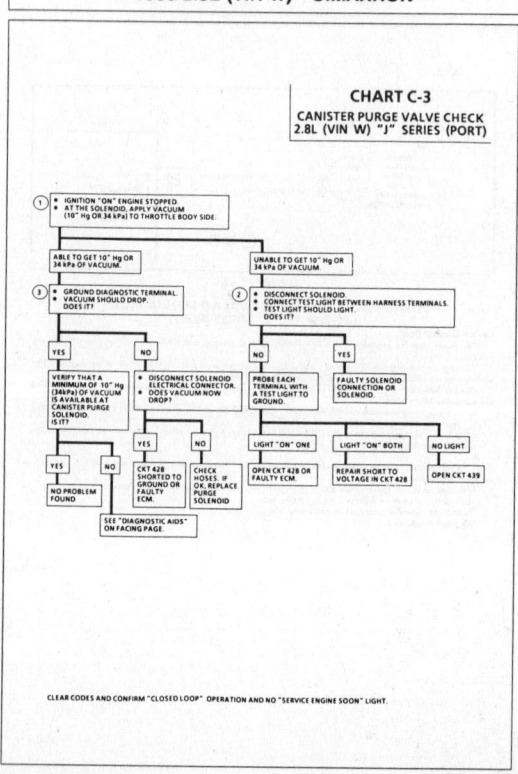

CLEAR CODES AND CONFIRM "CLOSED LOOP" OPERATION AND NO "SERVICE ENGINE SOON" LIGHT.

1988–90 2.8L (VIN W) – CAVALIER
1988 2.8L (VIN W) – CIMARRON

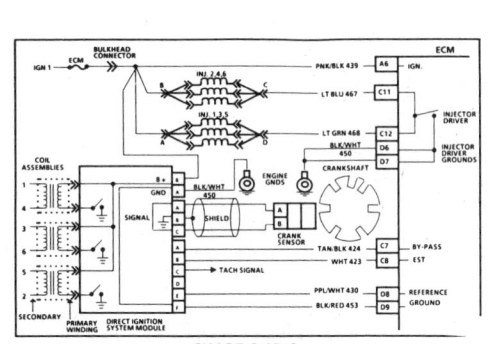

CHART C-4D-2
"DIS" MISFIRE UNDER LOAD
2.8L (VIN W) "J" SERIES (PORT)

Circuit Description:

The direct ignition system (DIS) uses a waste spark method of distribution. In this type of system, the ignition module triggers the #1/4 coil pair resulting in both #1 and #4 spark plugs firing at the same time. #1 cylinder is on the compression stroke at the same time #4 is on the exhaust stroke, resulting in a lower energy requirement to fire #4 spark plug. This leaves the remainder of the high voltage to be used to fire #1 spark plug. On this application, the crank sensor is mounted to the engine block and protrudes through the block to within approximately .050˝ of the crankshaft reluctor. Since the reluctor is a machined portion of the crankshaft and the crank sensor is mounted in a fixed position on the block, timing adjustments are not possible or necessary.

Test Description: Numbers below refer to circled numbers on the diagnostic chart.
1. If the "Misfire" complaint exists at idle only, diagnostic chart C-4D-1 must be used. A spark tester such as ST-125 must be used because it is essential to verify adequate available secondary voltage at the spark plug. (25,000 volts) Spark should jump the test gap on all 4 leads. This simulates a "load" condition.
2. If the spark jumps the tester gap after grounding the opposite plug wire, it indicates excessive resistance in the plug which was bypassed.

A faulty or poor connection at that plug could also result in the miss condition. Also check for carbon deposits inside the spark plug boot.
3. If carbon tracing is evident replace coil and be sure plug wires relating to that coil are clean and tight. Excessive wire resistance or faulty connections could have caused the coil to be damaged.
4. If the no spark condition follows the suspected coil, that coil is faulty. Otherwise, the ignition module is the cause of no spark. This test could also be performed by substituting a known good coil for the one causing the no spark condition.

1988–90 2.8L (VIN W) – CAVALIER
1988 2.8L (VIN W) – CIMARRON

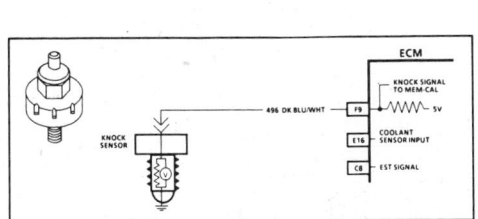

CHART C-5
ELECTRONIC SPARK CONTROL (ESC) SYSTEM CHECK
2.8L (VIN W) "J" SERIES (PORT)

Circuit Description:

The knock sensor is used to detect engine detonation and the ECM will retard the electronic spark timing based on the signal being received. The circuitry, within the knock sensor, causes the ECM's 5 V to be pulled down so that under a no knock condition, CKT 496 would measure about 2.5 V. The knock sensor produces an A/C signal, which rides on the 2.5 V DC voltage. The amplitude and frequency are dependent upon the knock level.

The Mem-Cal used with this engine, contains the functions which were part of remotely mounted ESC modules used on other GM vehicles. The ESC portion of the Mem-Cal, then sends a signal to other parts of the ECM which adjusts the spark timing to retard the spark and reduce the detonation.

Test Description: Numbers below refer to circled numbers on the diagnostic chart.
1. With engine idling, there should not be a knock signal present at the ECM, because detonation is not likely under a no load condition.
2. Tapping on the engine lift bracket should simulate a knock signal to determine if the sensor is capable of detecting detonation. If no knock is detected, try tapping on engine block closer to sensor before replacing sensor.
3. If the engine has an internal problem which is creating a knock, the knock sensor may be responding to the internal failure.
4. This test determines if the knock sensor is faulty or if the ESC portion of the Mem-Cal is faulty. If it is determined that the Mem-Cal is faulty, be sure that it is properly installed and latched into place. If not properly installed, repair and replace.

Diagnostic Aids:

While observing knock signal on the "Scan," there should be an indication that knock is present, when detonation can be heard. Detonation is most likely to occur under high engine load conditions.

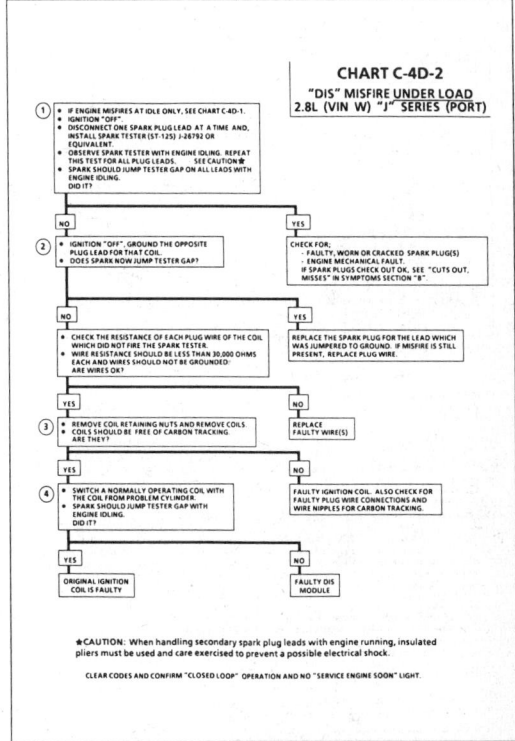

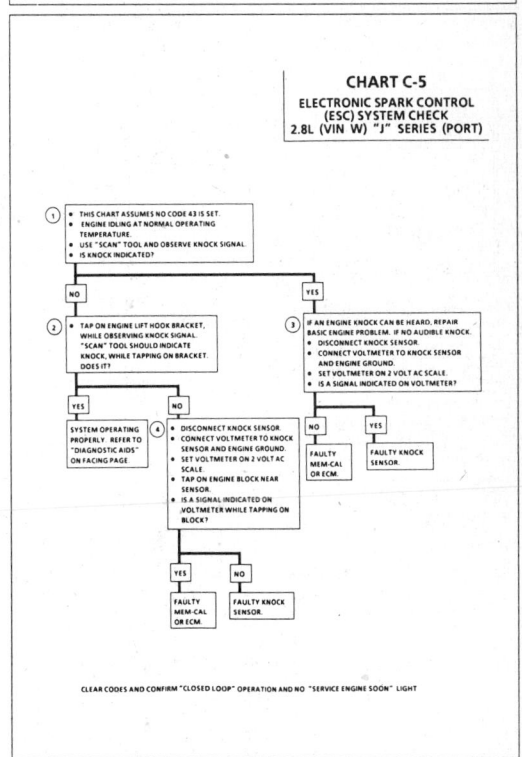

1988–90 2.8L (VIN W) – CAVALIER
1988 2.8L (VIN W) – CIMARRON

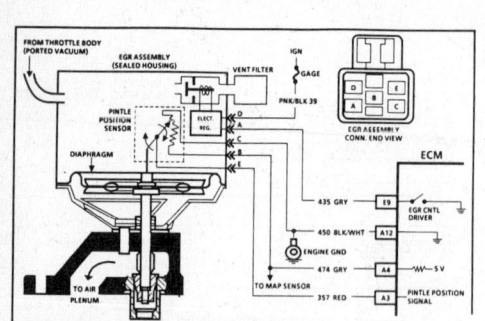

CHART C-7
EGR VALVE CHECK
2.8L (VIN W) "J" SERIES (PORT)

Circuit Description:
The integrated electronic EGR valve functions similiar to a port valve with a remote vacuum regulator. The internal solenoid is normally open, which causes the vacuum signal to be vented off to the atmosphere when EGR is not being commanded by the ECM. This EGR valve has a sealed cap and the solenoid valve opens and closes the vacuum signal, which controls the amount of vacuum vented to atmosphere, and this controls the amount of vacuum applied to the diaphragm. The electronic EGR valve contains a voltage regulator, which converts the ECM signal, to provide different amounts of EGR flow by regulating the current to the solenoid. The ECM controls EGR flow with a pulse width modulated signal (turns "ON" and "OFF" many times a second) based on MAP, TPS, and rpm.
This system also contains a pintle position sensor, which works similiar to a TPS sensor, and as EGR flow is increased, the sensor output also increases.

Test Description: Numbers below refer to circled numbers on the diagnostic chart.
1. Whenever the solenoid is denergized, the solenoid valve should be closed, which should not allow the vacuum to move the EGR diaphragm. However, if the filter is plugged, the vacuum applied with the hand held vacuum pump will cause the diaphragm to move because the vacuum will not be vented to the atmosphere.
2. This test will determine if the EGR filter is plugged or if the EGR itself is faulty. Use care when removing the filter to avoid damaging the EGR assembly. See On-Car Service for details.
3. If the valve moves in this test, it's probably due to CKT 435 being shorted to ground.

4. Grounding the diagnostic terminal should energize the solenoid which closes off the vent and allows the vacuum to move the diaphragm.
5. The EGR assembly is designed to have some leakage and therefore, 7" of vacuum is all that should be able to be held on the assembly. However, if too much of a leak exists (less than 3"), the EGR assembly is leaking and must be replaced.

Diagnostic Aids:

Some "Scan" tools will read EGR pintle position in volts. The EGR position voltage can be used to determine that the pintle is moving. When no EGR is commanded (0% duty cycle), the position sensor

1988–90 2.8L (VIN W) – CAVALIER
1988 2.8L (VIN W) – CIMARRON

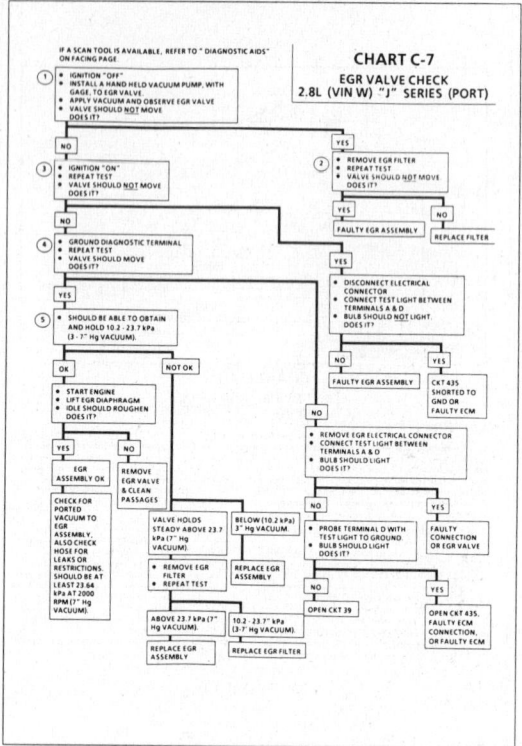

CHART C-7
EGR VALVE CHECK
2.8L (VIN W) "J" SERIES (PORT)

1988–90 2.8L (VIN W) – CAVALIER
1988 2.8L (VIN W) – CIMARRON

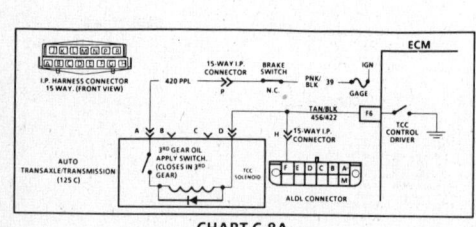

CHART C-8A
TORQUE CONVERTER CLUTCH (TCC)
2.8L (VIN W) "J" SERIES (PORT)

Circuit Description:
The purpose of the torque converter clutch feature is to eliminate the power loss of the transmission converter stage when the vehicle is in a cruise condition. This allows the convenience of the automatic transmission and the fuel economy of a manual transmission.
Fused battery ignition is supplied to the TCC solenoid through the brake switch, and transmission third gear apply switch. The ECM will engage TCC by grounding CKT 422 to energize the solenoid.
TCC will engage when:
- Engine warmed up
- Vehicle speed above a calibrated value (about 32 mph 51 km/h)
- Throttle position sensor output not changing, indicating a steady road speed
- Transmission third gear switch closed
- Brake switch closed

Test Description: Numbers below refer to circled numbers on the diagnostic chart.
1. Light "OFF" confirms transmission third gear apply switch is open.
2. At 25 mph, the transmission third gear apply switch should close. Test light will come "ON" and confirm battery supply and closed brake switch.
3. Grounding the diagnostic terminal with ignition "ON," engine "OFF," should energize the TCC solenoid by grounding CKT 422. This test checks the ability of the ECM to supply a ground to the TCC solenoid. The test light connected from 12 V to ALDL terminal "F" will turn "ON" as CKT 422 is grounded.

Diagnostic Aids:

A "Scan" tool only indicates when the ECM has turned on the TCC driver, and this does not confirm that the TCC is functioning properly. To determine if TCC is functioning properly, engine rpm should decrease when the "Scan" indicates the TCC driver has turned "ON." Check for correct Mem-Cal. If all electrical checks are OK and TCC is inoperative, the fault is internal to the transmission. See Section "7" for transmission repairs.

1988–90 2.8L (VIN W) – CAVALIER
1988 2.8L (VIN W) – CIMARRON

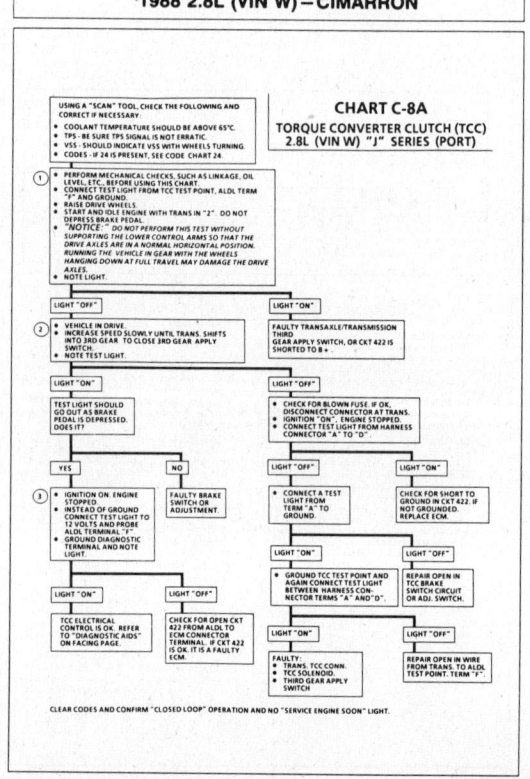

CHART C-8A
TORQUE CONVERTER CLUTCH (TCC)
2.8L (VIN W) "J" SERIES (PORT)

1988–90 2.8L (VIN W) – CAVALIER
1988 2.8L (VIN W) – CIMARRON

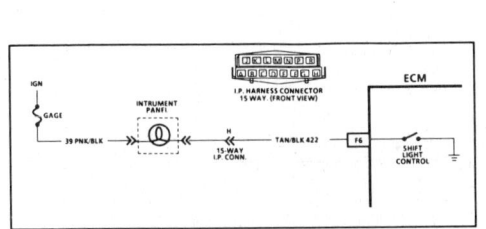

CHART C-8C
MANUAL TRANSMISSION (M/T) SHIFT LIGHT CHECK
2.8L (VIN W) "J" SERIES (PORT)

Circuit Description:

The shift light indicates the best transmission shift point for maximum fuel economy. The light is controlled by the ECM and is turned "ON" by grounding CKT 456.

The ECM uses information from the following inputs to control the shift light.
- Coolant temperature must be above 16°C (61°F)
- TPS above 4%
- VSS
- RPM above about 1900
- Air Flow - The ECM uses rpm, MAP and VSS to calculate what gear the vehicle is in.

It's this calculation that determines when the shift light should be turned "ON." The shift light will only stay "ON" 5 seconds after the conditions were met to turn it "ON."

Test Description: Numbers below refer to circled numbers on the diagnostic chart.
1. This should not turn "ON" the shift light. If the light is "ON," there is a short to ground in CKT 422 wiring or a fault in the ECM.
2. When the diagnostic terminal is grounded, the ECM should ground CKT 422 and the shift light should come "ON."
3. This checks the shift light circuit up to the ECM connector. If the shift light illuminates, then the ECM connector is faulty or the ECM does not have the ability to ground the circuit, or has incorrect or faulty PROM.

Diagnositc Aids:

A loss of vehicle speed input to the ECM should set a Code 24.

1988–90 2.8L (VIN W) – CAVALIER
1988 2.8L (VIN W) – CIMARRON

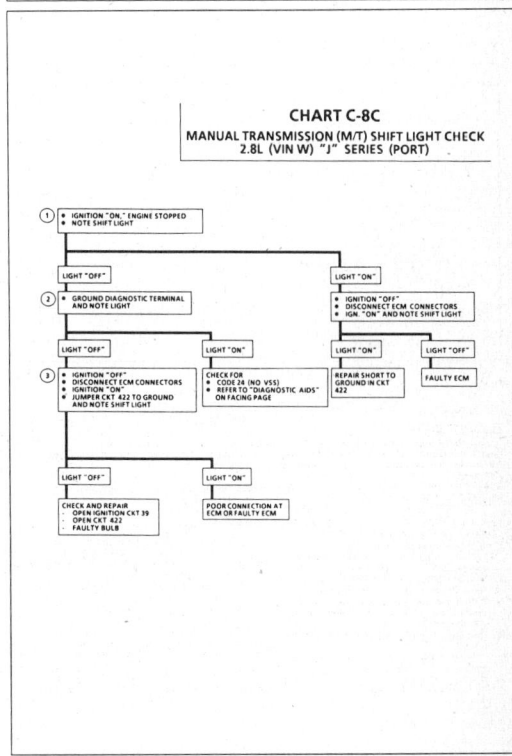

CHART C-8C
MANUAL TRANSMISSION (M/T) SHIFT LIGHT CHECK
2.8L (VIN W) "J" SERIES (PORT)

1988–90 2.8L (VIN W) – CAVALIER
1988 2.8L (VIN W) – CIMARRON

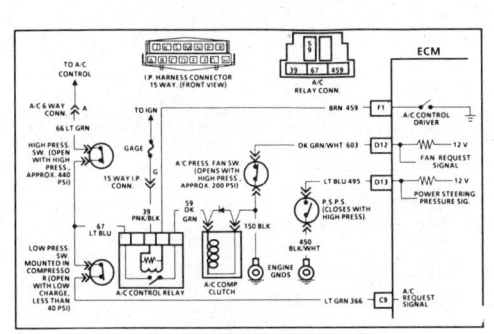

CHART C-10
A/C CLUTCH CONTROL CIRCUIT DIAGNOSIS
2.8L (VIN W) "J" SERIES (PORT)

Circuit Description:

The A/C clutch control relay is ECM controlled to delay A/C clutch engagement about .4 second after A/C is turned "ON." This allows the IAC to adjust engine rpm before the A/C clutch engages. The ECM also causes the relay to disengage the A/C clutch during WOT, when high power steering pressure is present, or if engine is overheating. The A/C clutch control relay is energized when the ECM provides a ground path for CKT 459. The low pressure switch will open if A/C pressure is less than 40 psi (276 kPa). The high pressure switch will open if A/C pressure exceeds about 440 psi (3034 kPa). The A/C pressure fan switch opens when A/C pressure exceeds about 200 psi (1380 kPa). The A/C clutch should engage only if the engine is running.

Test Description: Numbers below refer to circled numbers on the diagnostic chart.
1. The ECM will only energize the A/C relay when the engine is running. This test will determine if the relay or CKT 459 is faulty.
2. In order for the clutch to properly be engaged, the low pressure switch must be closed to provide 12 V to the relay, and the high pressure switch must be closed, so the A/C request (12 V) will be present at the ECM.
3. Determines if the signal is reaching the ECM on CKT 366 from the A/C control panel. Signal should only be present when the A/C mode or defrost mode has been selected.
4. A short to ground in any part of the A/C request circuit, CKT 67 to the relay, CKT 59 to the A/C clutch, or the A/C clutch, could be the cause of blown fuse.
5. If the ECM is seeing a high power steering pressure signal, the A/C clutch will be disengaged by the ECM.
6. With the diagnostic test terminal grounded, the ECM should be grounding CKT 459, which should cause the test light to be "ON."

Diagnostic Aids:

If complaint was insufficient cooling, the problem may be caused by a inoperative cooling fan, or A/C pressure fan switch. The engine cooling fan should turn "ON" when A/C pressure exceeds a value to open the switch, which causes the ECM to energize the cooling fan relay. See CHART C-12 for cooling fan diagnosis. If fan operates correctly, see A/C diagnosis in Section "1". Also, it could be a high pressure switch opening too soon or a low pressure switch opening intermittently, especially with a low charge.

1988–90 2.8L (VIN W) – CAVALIER
1988 2.8L (VIN W) – CIMARRON

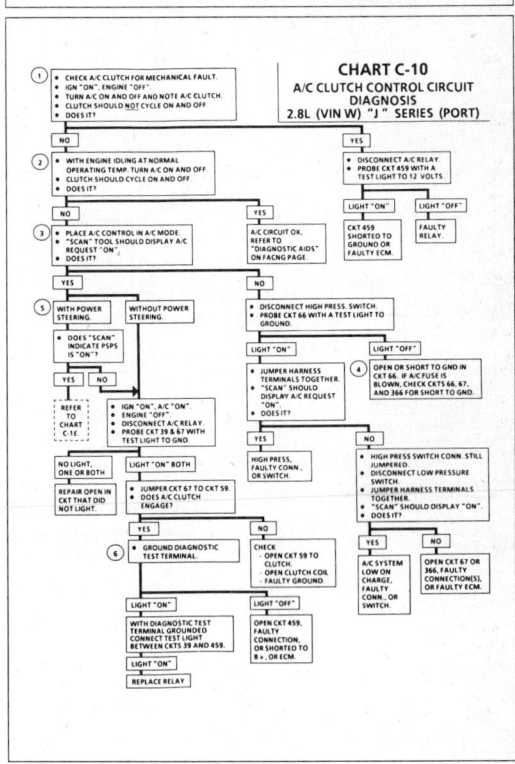

CHART C-10
A/C CLUTCH CONTROL CIRCUIT DIAGNOSIS
2.8L (VIN W) "J" SERIES (PORT)

1988–90 2.8L (VIN W) – CAVALIER
1988 2.8L (VIN W) – CIMARRON

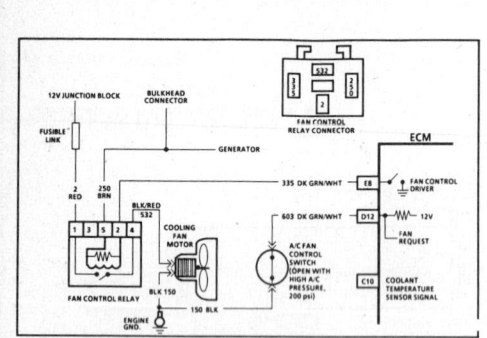

CHART C-12
(Page 1 of 2)
COOLANT FAN CONTROL CIRCUIT DIAGNOSIS
2.8L (VIN W) "J" SERIES (PORT)

Circuit Description:

The electric cooling fan is controlled by the ECM based on inputs from the coolant temperature sensor, the A/C fan control switch, and vehicle speed. The ECM controls the fan by grounding CKT 335, which energizes the fan control relay. Battery voltage is then supplied to the fan motor.

The ECM grounds CKT 335 when coolant temperature is over about 106°C (223°F), or when A/C has been requested and the fan control switch indicates high A/C pressure, 200 psi (1380 kPa). Once the ECM turns the relay "ON," it will keep it "ON" for a minimum of 30 seconds or until vehicle speed exceeds 70 mph.

Also, if Code 14 or 15 sets or the ECM is in back up, the fan will run at all times.

Test Description: Numbers below refer to circled numbers on the diagnostic chart.

1. With the diagnostic terminal grounded, the cooling fan control driver will close, which should energize the fan control relay.
2. If the A/C fan control switch or circuit is open, the fan would run whenever A/C is requested.
3. With A/C clutch engaged, the A/C fan control switch should open when A/C high pressure exceeds about 200 psi (1380 kPa). This signal should cause the ECM to energize the fan control relay.

Diagnostic Aids:

If the owner complained of an overheating problem, it must be determined if the complaint was due to an actual boilover or the hot light (temperature gage) indicated over heating.

If the gage (light) indicates overheating but no boilover is detected, the gage circuit should be checked. The gage accuracy can also be checked by comparing the coolant sensor reading using a "Scan" tool and comparing its reading with the gage reading.

If the engine is actually overheating and the gage indicates overheating but the cooling fan is not coming "ON," the coolant sensor has probably shifted out of calibration and should be replaced.

If the engine is overheating and the cooling fan is "ON," the cooling system should be checked.

1988–90 2.8L (VIN W) – CAVALIER
1988 2.8L (VIN W) – CIMARRON

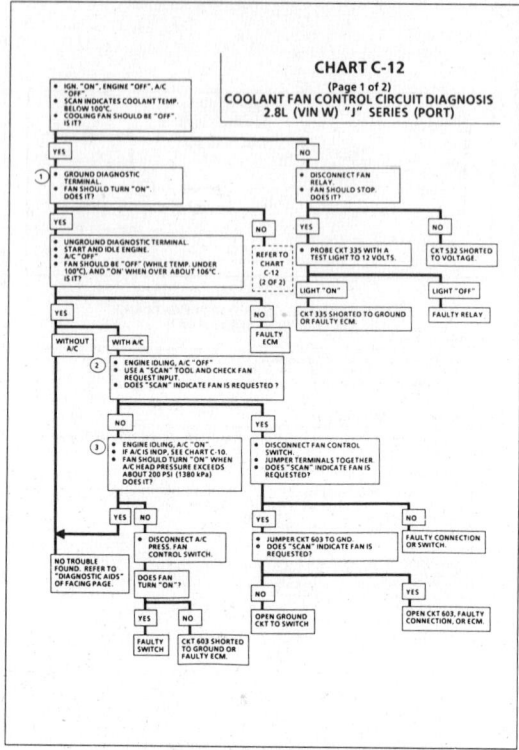

1988–90 2.8L (VIN W) – CAVALIER
1988 2.8L (VIN W) – CIMARRON

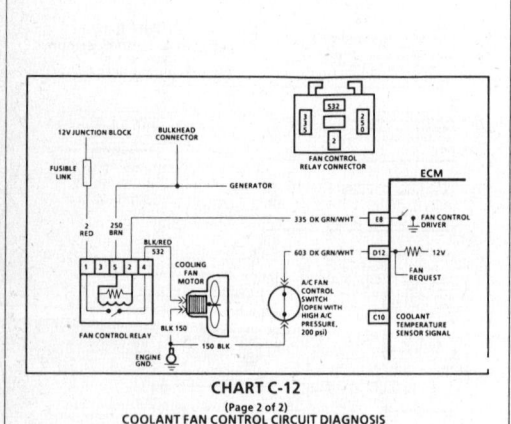

CHART C-12
(Page 2 of 2)
COOLANT FAN CONTROL CIRCUIT DIAGNOSIS
2.8L (VIN W) "J" SERIES (PORT)

Test Description: Numbers below refer to circled numbers on the diagnostic chart.

1. 12 V should be available to both terminal "E" & "C" when the ignition is "ON."
2. This test checks the ability of the ECM to ground CKT 335

The "Service Engine Soon" light should also be flashing at this point. If it isn't flashing, see CHART A-2.

3. If the fan does not turn "ON" at this point, CKT 702 or CKT 150 is open or the cooling fan motor is faulty.

1988–90 2.8L (VIN W) – CAVALIER
1988 2.8L (VIN W) – CIMARRON

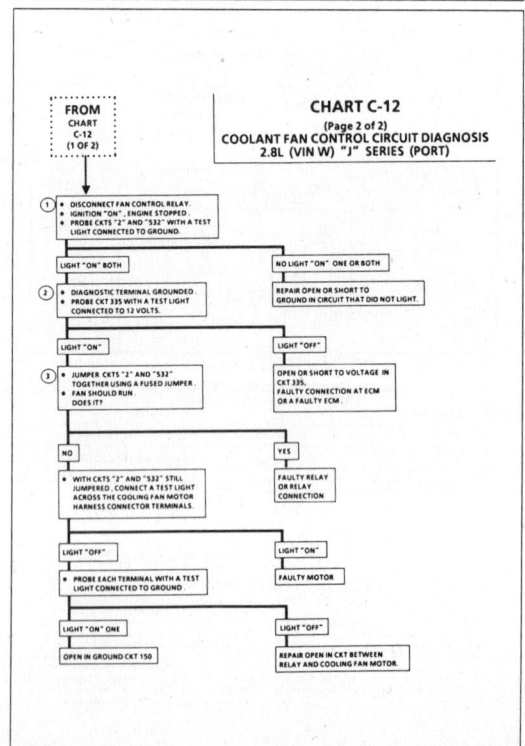

2.8L (VIN W) COMPONENT LOCATIONS REGAL AND CUTLASS

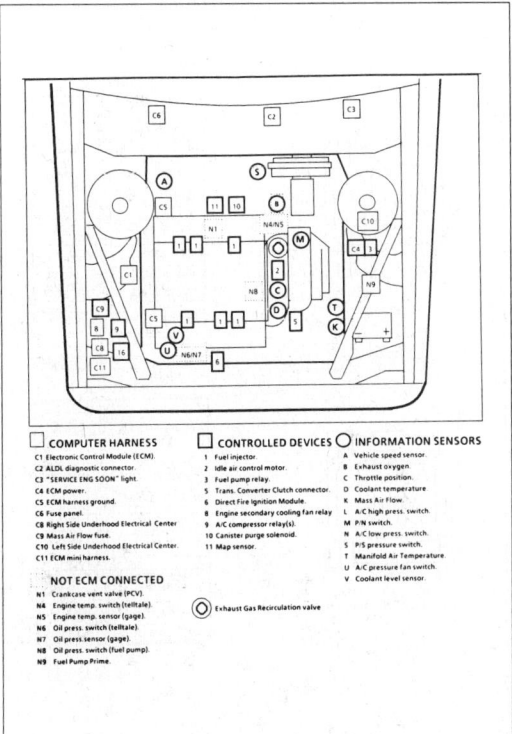

2.8L (VIN W) ECM WIRING DIAGRAM REGAL AND CUTLASS

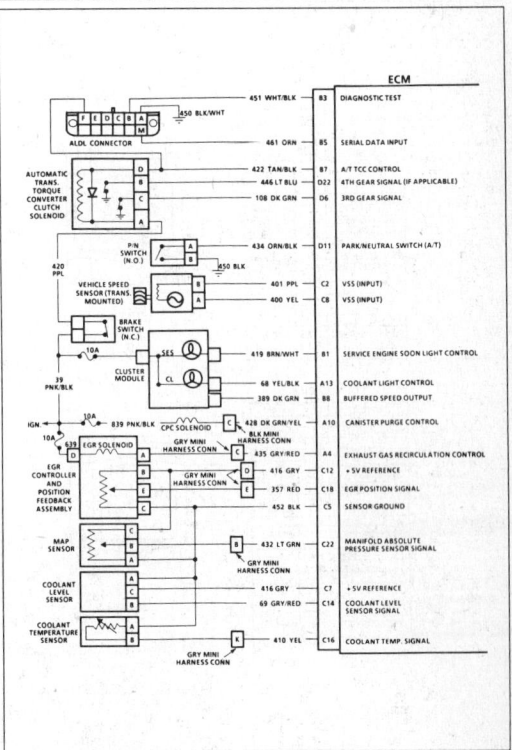

2.8L (VIN W) ECM WIRING DIAGRAM REGAL AND CUTLASS (CONT.)

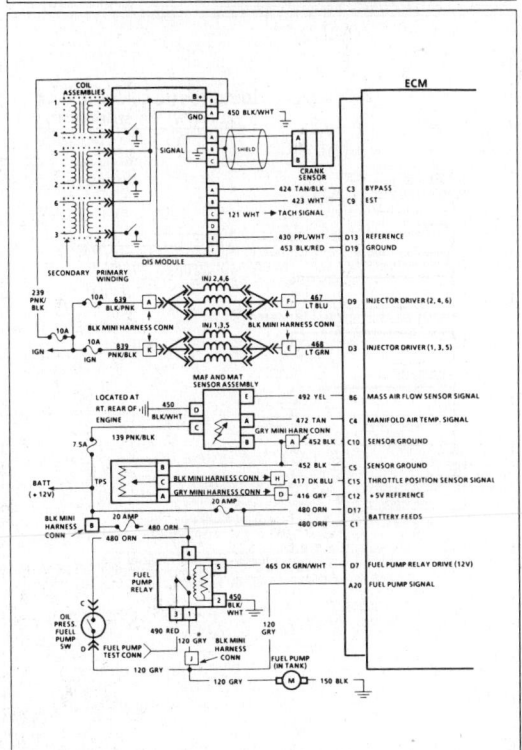

2.8L (VIN W) ECM WIRING DIAGRAM REGAL AND CUTLASS (CONT.)

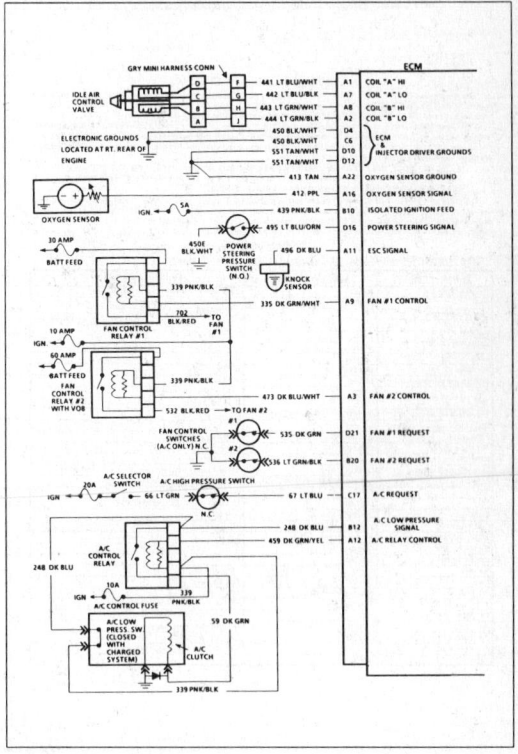

2.8L (VIN W) ECM CONNECTOR TERMINAL END VIEW – REGAL AND CUTLASS

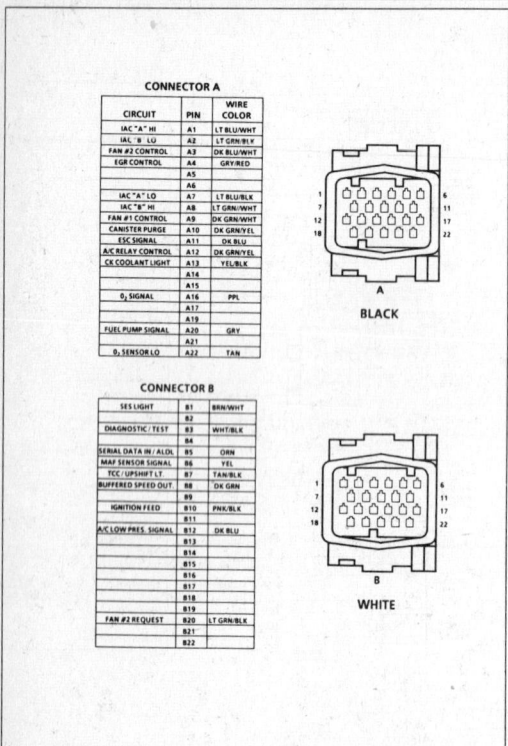

CONNECTOR A

CIRCUIT	PIN	WIRE COLOR
IAC "A" HI	A1	LT BLU/WHT
IAL "B" LO	A2	LT GRN/BLK
FAN #2 CONTROL	A3	DK BLU/WHT
EGR CONTROL	A4	GRY/RED
	A5	
	A6	
IAC "A" LO	A7	LT BLU/BLK
IAC "B" HI	A8	LT GRN/WHT
FAN #1 CONTROL	A9	DK GRN/WHT
CANISTER PURGE	A10	DK GRN/YEL
ESC SIGNAL	A11	DK BLU
A/C RELAY CONTROL	A12	DK GRN/YEL
CK COOLANT LIGHT	A13	YEL/BLK
	A14	
	A15	
O₂ SIGNAL	A16	PPL
	A17	
	A19	
FUEL PUMP SIGNAL	A20	GRY
	A21	
O₂ SENSOR LO	A22	TAN

BLACK

CONNECTOR B

CIRCUIT	PIN	WIRE COLOR
SES LIGHT	B1	BRN/WHT
	B2	
DIAGNOSTIC / TEST	B3	WHT/BLK
	B4	
SERIAL DATA IN / ALDL	B5	ORN
MAF SENSOR SIGNAL	B6	YEL
TCC / UPSHIFT LT.	B7	TAN/BLK
BUFFERED SPEED OUT.	B8	DK GRN
	B9	
IGNITION FEED	B10	PNK/BLK
	B11	
A/C LOW PRES. SIGNAL	B12	DK BLU
	B13	
	B14	
	B15	
	B16	
	B17	
	B18	
	B19	
FAN #2 REQUEST	B20	LT GRN/BLK
	B21	
	B22	

WHITE

2.8L (VIN W) ECM CONNECTOR TERMINAL END VIEW – REGAL AND CUTLASS (CONT.)

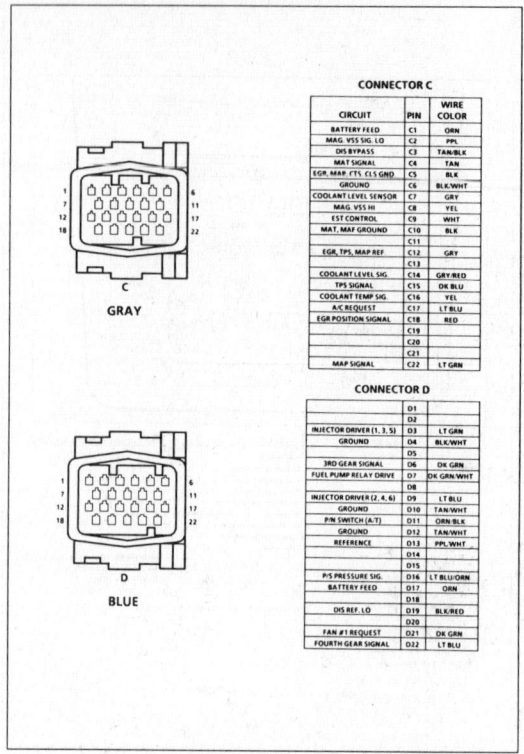

CONNECTOR C

CIRCUIT	PIN	WIRE COLOR
BATTERY FEED	C1	ORN
MAG. VSS SIG. LO	C2	PPL
DIS BYPASS	C3	TAN/BLK
MAT SIGNAL	C4	TAN
EGR, MAP, CTS, CLS GND	C5	BLK
GROUND	C6	BLK/WHT
COOLANT LEVEL SENSOR	C7	GRY
MAG. VSS HI	C8	YEL
EST CONTROL	C9	WHT
MAT, MAF GROUND	C10	BLK
	C11	
EGR, TPS, MAP REF	C12	GRY
	C13	
COOLANT LEVEL SIG.	C14	GRY/RED
TPS SIGNAL	C15	DK BLU
COOLANT TEMP SIG.	C16	YEL
A/C REQUEST	C17	LT BLU
EGR POSITION SIGNAL	C18	RED
	C19	
	C20	
	C21	
MAP SIGNAL	C22	LT GRN

GRAY

CONNECTOR D

CIRCUIT	PIN	WIRE COLOR
	D1	
	D2	
INJECTOR DRIVER (1, 3, 5)	D3	LT GRN
GROUND	D4	BLK/WHT
	D5	
3RD GEAR SIGNAL	D6	DK GRN
FUEL PUMP RELAY DRIVE	D7	DK GRN/WHT
	D8	
INJECTOR DRIVER (2, 4, 6)	D9	LT BLU
	D10	TAN/WHT
P/N SWITCH (A/T)	D11	ORN/BLK
GROUND	D12	TAN/WHT
REFERENCE	D13	PPL/WHT
	D14	
	D15	
P/S PRESSURE SIG.	D16	LT BLU/ORN
BATTERY FEED	D17	ORN
	D18	
DIS REF. LO	D19	BLK/RED
	D20	
FAN #1 REQUEST	D21	DK GRN
FOURTH GEAR SIGNAL	D22	LT BLU

BLUE

1988–90 2.8L (VIN W)
REGAL AND CUTLASS SUPREME

DIAGNOSTIC CIRCUIT CHECK

The diagnostic circuit check is an organized approach to identifying a problem created by an electronic engine control system malfunction. It must be the starting point for any driveability complaint diagnosis, because it directs the service technician to the next logical step in diagnosing the complaint.

The "Scan Data" listed in the table may be used for comparison, after completing the diagnostic circuit check and finding the on-board diagnostics functioning properly and no trouble codes displayed. The "Typical Values" are an average of display values recorded from normally operating vehicles and are intended to represent what a normally functioning system would typically display.

A "SCAN" TOOL THAT DISPLAYS FAULTY DATA SHOULD NOT BE USED, AND THE PROBLEM SHOULD BE REPORTED TO THE MANUFACTURER. THE USE OF A FAULTY "SCAN" CAN RESULT IN MISDIAGNOSIS AND UNNECESSARY PARTS REPLACEMENT.

Only the parameters listed below are used in this manual for diagnosing. If a "Scan" reads other parameters, the values are not recommended by General Motors for use in diagnosing. For more description on the values and use of the "Scan" to diagnose ECM inputs, refer to the applicable diagnosis section. If all values are within the range illustrated, refer to symptoms in Section "B".

"SCAN" DATA
Idle / Upper Radiator Hose Hot / Closed Throttle / Park or Neutral / Closed Loop / Acc. off

"SCAN" Position	Units Displayed	Typical Data Value
Desired RPM	RPM	ECM idle command (varies with temp.)
RPM	RPM	± 100 RPM from desired RPM (± 50 in drive)
Coolant Temp.	C°	85° - 105°
MAT Temp.	C°	10° - 80° (depends on underhood temp)
MAP	Volts	1 - 2 (depends on Vac. & Baro pressure)
BARO	Volts	2.5 - 5.5 (depends on altitude & Baro pressure)
MAF	Gm/Sec	4 - 7
Air Flow	Gm/Sec	4 - 7
BPW (base pulse width)	M/Sec	1 - 4 and varying
O₂	Volts	1-1000 and varying
TPS	Volts	0.35 - 0.7
Throttle Angle	0 - 100%	0
IAC	Counts (steps)	10 - 50
P/N Switch	P/N and RDL	Park/Neutral (P/N)
INT (Integrator)	Counts	Varies
BLM (Block Learn)	Counts	118 - 138
Open/Closed Loop	Open/Closed	Closed Loop (may go open with extended idle)
BLM Cell	Cell Number	0 or 1 (depends on Air Flow & RPM)
VSS	MPH	0
TCC	On/Off	Off (on with TCC commanded)
EGRDC	0 - 100%	0 at idle
EGR Position	Volts	3 - 2 Volts
Spark Advance	# of Degrees	Varies
Knock Retard	Degrees of Retard	0
Knock Signal	Yes/No	No
Battery	Volts	13.5 - 14.5
Fan	On/Off	Off (below 106°C)
P/S Switch	Normal/Hi Press.	Normal
3rd Gear (440-T4)	Yes/No	No (yes, when in 3rd or 4th gear)
4th Gear (440-T4)	Yes/No	No (yes, when in 4th gear)
A/C Request	Yes/No	No (yes, when A/C requested)
A/C Clutch	On/Off	Off (on, with A/C commanded on)
Fan Request	Yes/No	No (yes, with A/C high pressure)
PPSW	Volts	13.5 - 14.5

1988–90 2.8L (VIN W)
REGAL AND CUTLASS SUPREME

DIAGNOSTIC CIRCUIT CHECK
2.8L (VIN W) "W" SERIES (PORT)

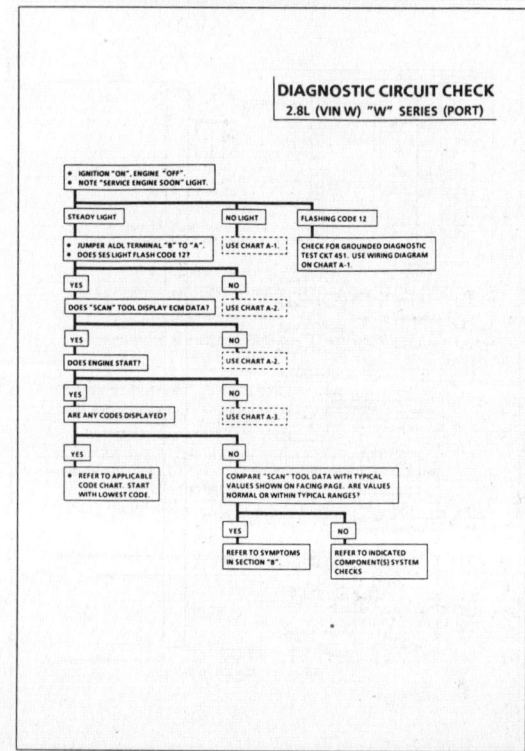

1988–90 2.8L (VIN W)
REGAL AND CUTLASS SUPREME

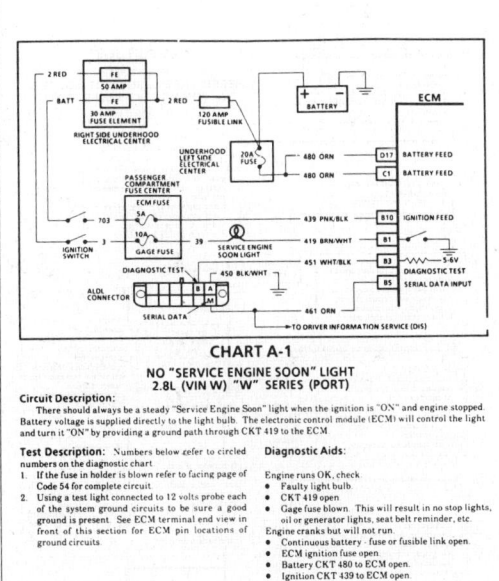

CHART A-1
NO "SERVICE ENGINE SOON" LIGHT
2.8L (VIN W) "W" SERIES (PORT)

Circuit Description:

There should always be a steady "Service Engine Soon" light when the ignition is "ON" and engine stopped. Battery voltage is supplied directly to the light bulb. The electronic control module (ECM) will control the light and turn it "ON" by providing a ground path through CKT 419 to the ECM.

Test Description: Numbers below refer to circled numbers on the diagnostic chart.

1. If the fuse in holder is blown refer to facing page of Code 54 for complete circuit.
2. Using a test light connected to 12 volts probe each of the system ground circuits to be sure a good ground is present. See ECM terminal end view in front of this section for ECM pin locations of ground circuits.

Diagnostic Aids:

Engine runs OK, check.
- Faulty light bulb.
- CKT 419 open.
- Gage fuse blown. This will result in no stop lights, oil or generator lights, seat belt reminder, etc.

Engine cranks but will not run.
- Continuous battery - fuse or fusible link open.
- ECM ignition fuse open.
- Battery CKT 480 to ECM open.
- Ignition CKT 439 to ECM open.
- Poor connection to ECM.

1988–90 2.8L (VIN W)
REGAL AND CUTLASS SUPREME

CHART A-1
NO "SERVICE ENGINE SOON" LIGHT
2.8L (VIN W) "W" SERIES (PORT)

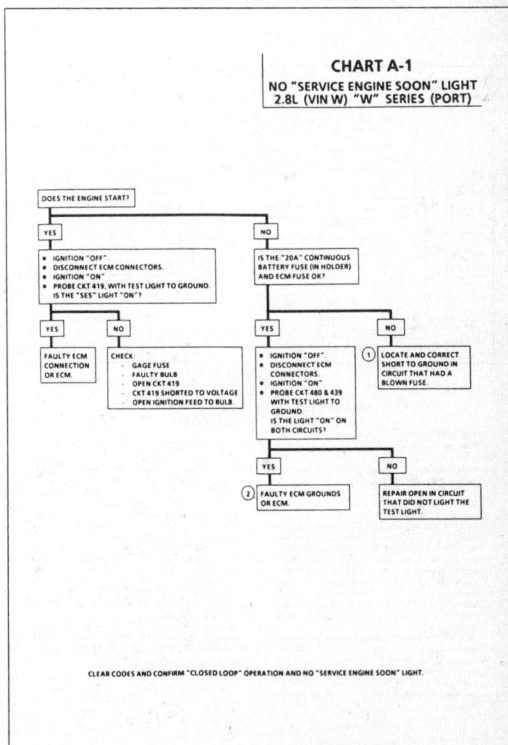

1988–90 2.8L (VIN W)
REGAL AND CUTLASS SUPREME

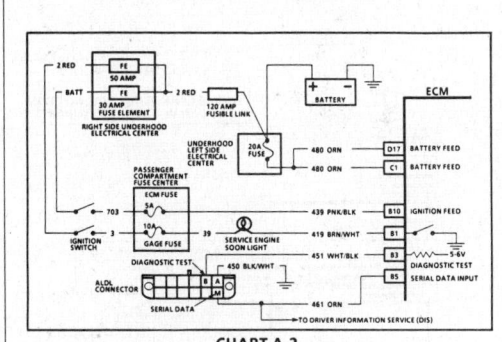

CHART A-2
NO ALDL DATA OR WON'T FLASH CODE 12
"SERVICE ENGINE SOON" LIGHT "ON" STEADY
2.8L (VIN W) "W" SERIES (PORT)

Circuit Description:

There should always be a steady "Service Engine Soon" light when the ignition is "ON" and engine stopped. Battery ignition voltage is supplied to the light bulb. The electronic control module (ECM) will turn the light "ON" by grounding CKT 419 at the ECM.

With the diagnostic terminal grounded, the light should flash a Code 12, followed by any trouble code(s) stored in memory.

A steady light suggests a short to ground in the light control CKT 419, or an open in diagnostic CKT 451.

Test Description: Numbers below refer to circled numbers on the diagnostic chart.

1. If there is a problem with the ECM that causes a "Scan" tool not to read serial data, the light should not flash a Code 12. If Code 12 is flashing check for CKT 451 short to ground. If Code 12 does flash be sure that the "Scan" tool is working properly on another vehicle. If the "Scan" is functioning properly and CKT 461 is OK the Mem-Cal or ECM may be at fault for the no ALDL symptom.

2. If the light goes "OFF" when the ECM connector is disconnected, CKT 419 is not shorted to ground.
3. This step will check for an open diagnostic CKT 451.
4. At this point the "Service Engine Soon" light wiring is OK. The problem is a faulty ECM or Mem-Cal. If Code 12 does not flash, the ECM should be replaced using the original Mem-Cal. Replace CKT 461 only after trying an ECM, as a defective Mem-Cal is an unlikely cause of the problem.

1988–90 2.8L (VIN W)
REGAL AND CUTLASS SUPREME

CHART A-2
NO ALDL DATA OR WON'T FLASH CODE 12
"SERVICE ENGINE SOON" LIGHT "ON" STEADY
2.8L (VIN W) "W" SERIES (PORT)

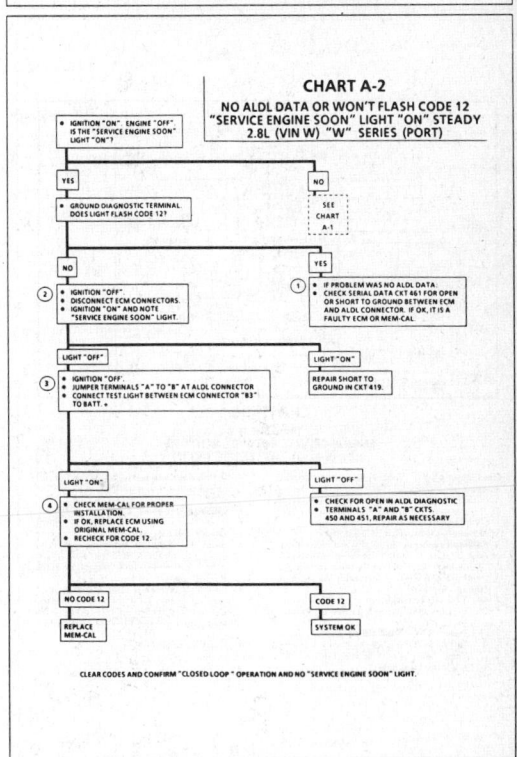

1988–90 2.8L (VIN W) REGAL AND CUTLASS SUPREME

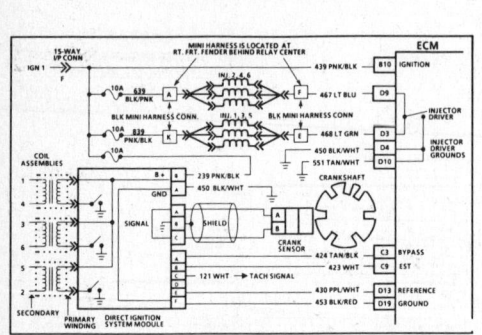

CHART A-3
(Page 1 of 3)
ENGINE CRANKS BUT WILL NOT RUN
2.8L (VIN W) "W" SERIES (PORT)

Circuit Description:
This chart assumes that battery condition and engine cranking speed are OK, and there is adequate fuel in the tank.

Test Description: Numbers below refer to circled numbers on the diagnostic chart.
1. A "Service Engine Soon" light "ON" is a basic test to determine if there is a 12 volt supply and ignition 12 volts to ECM. No ALDL may be due to an ECM problem and CHART A-2 will diagnose the ECM. If TPS is over 2.5 volts the engine may be in the clear flood mode which will cause starting problems. The engine will not start without reference pulses and therefore the "Scan" should read rpm (reference) during crank.
2. For the first two seconds with ignition "ON" or whenever reference pulses are being received, PPSW should indicate fuel pump circuit voltage (8 to 12 volts).
3. Because the direct ignition system uses two plugs and wires to complete the circuit of each coil, the opposite spark should be left connected. If rpm was indicated during crank, the ignition module is receiving a crank signal, but no spark at this test indicates the ignition module is not triggering the coils.

4. The test light should blink, indicating the ECM is in control of the injectors. How bright the light blinks is not important. However, the test light should be a J-34730-3 or equivalent.
5. Use fuel pressure gage J-34730-1 or equivalent. Wrap a shop towel around the fuel pressure tap to absorb any small amount of fuel leakage that may occur when installing the gage.
6. This test will determine if the ignition module is not generating the reference pulse or if the wiring or ECM are at fault. By touching and removing a test light to 12 volts on CKT 430, a reference pulse should be generated. If rpm is indicated, the ECM and wiring are OK.
7. This test will determine if the ignition module is not triggering the problem coil or if the tested coil is at fault. This test could also be performed by using another known good coil.

1988–90 2.8L (VIN W) REGAL AND CUTLASS SUPREME

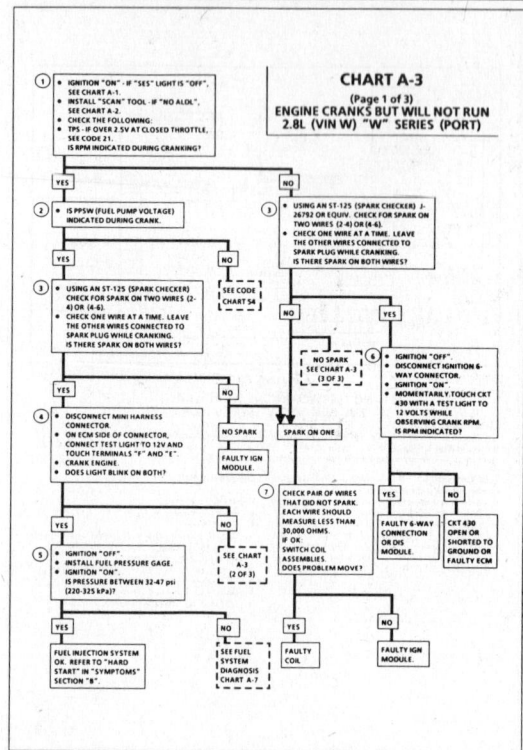

CHART A-3
(Page 1 of 3)
ENGINE CRANKS BUT WILL NOT RUN
2.8L (VIN W) "W" SERIES (PORT)

1988–90 2.8L (VIN W) REGAL AND CUTLASS SUPREME

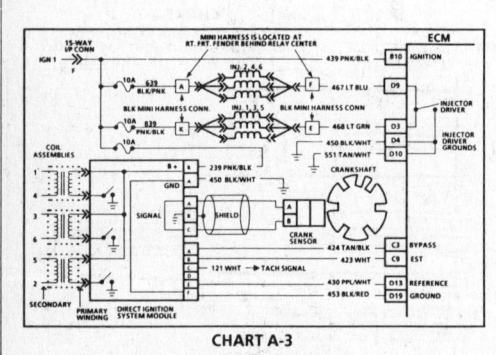

CHART A-3
(Page 2 of 3)
ENGINE CRANKS BUT WILL NOT RUN
2.8L (VIN W) "W" SERIES (PORT)

Test Description: Numbers below refer to circled numbers on the diagnostic chart.
1. Checks for 12 volt supply to injectors. Due to the injectors wired in parallel there should be a light "ON" on both terminals.
2. Checks continuity of CKT 467 and 468.
3. All checks made to this point would indicate that the ECM is at fault. However, there is a possibility of CKT 467 or 468 being shorted to a voltage source either in the engine harness or in the mini harness.
   - To test for this condition:
     • Disconnect gray mini harness connector.
     • Ignition "ON"
     • Probe CKTs 467 and 468 on the ECM side of harness with a test light connected to ground. There should be no light. If light is "ON", repair short to voltage.

   • If OK, check the resistance of the injector harness between terminals "A" & "F" and "K" & "E".
   • Should be 4 to 5 ohms.
   • If less than 4 ohms, check harness for wires shorted together and check each injector resistance. There should be no continuity between terminals "K" & "E" or "A" & "F". If there is continuity, check harness for wires shorted together.
   • Resistance of each injector should be 12.2 ohms ± .2 ohms (at ambient temperature). Injector resistance will increase with temperature and could read as high as 14.4 ohms at 93°C.
   • If all OK, replace ECM.

1988–90 2.8L (VIN W) REGAL AND CUTLASS SUPREME

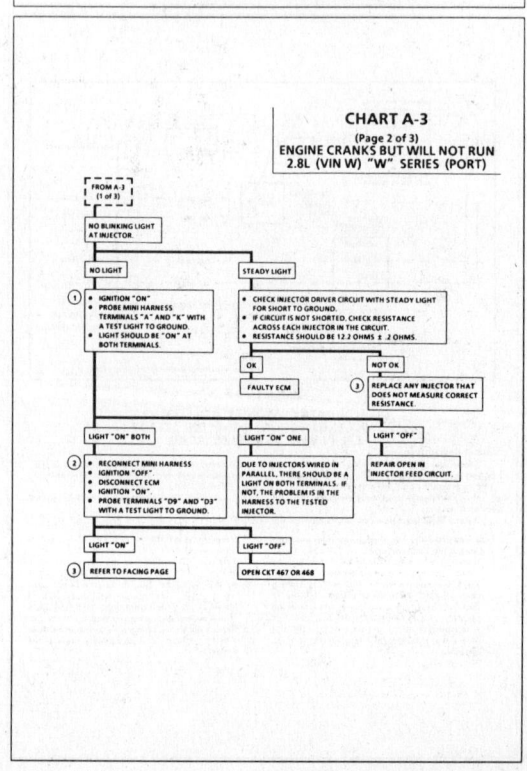

CHART A-3
(Page 2 of 3)
ENGINE CRANKS BUT WILL NOT RUN
2.8L (VIN W) "W" SERIES (PORT)

1988–90 2.8L (VIN W) REGAL AND CUTLASS SUPREME

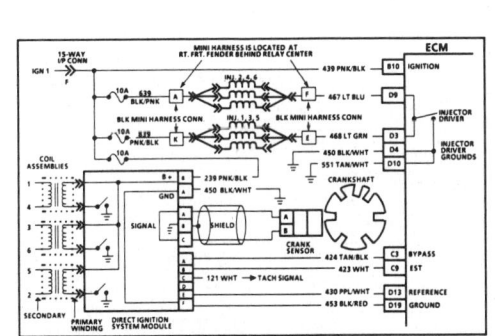

CHART A-3
(Page 3 of 3)
ENGINE CRANKS BUT WILL NOT RUN
2.8L (VIN W) "W" SERIES (PORT)

Circuit Description:
If the "Scan" tool did not indicate a cranking rpm and there is no spark present at the plugs, the problem lies in the direct ignition system or the power and ground supplies to the module.

The magnetic crank sensor is used to determine engine crankshaft position much the same way as the pick-up coil did in distributor type systems. The sensor is mounted in the block near a seven slot wheel on the crank shaft. The rotation of the wheel creates a flux change in the sensor which produces a voltage signal. The ignition module then processes this signal and creates the reference pulses needed by the ECM and the signal triggers the correct coil at the correct time.

Test Description: Numbers below refer to circled numbers on the diagnostic chart.
1. This test will determine if the 12 volt supply and a good ground is available at the ignition module.
2. Tests for continuity of CKT 439 to the ignition module. If test light does not light but the "SES" light is "ON" with ignition "ON", repair open in CKT 439 between DIS ignition module and splice.
3. Checks for continuity of the crank sensor and connections.
4. Voltage will vary in this test depending on cranking speed of engine.

1988–90 2.8L (VIN W) REGAL AND CUTLASS SUPREME

CHART A-3
(Page 3 of 3)
ENGINE CRANKS BUT WILL NOT RUN
2.8L (VIN W) "W" SERIES (PORT)

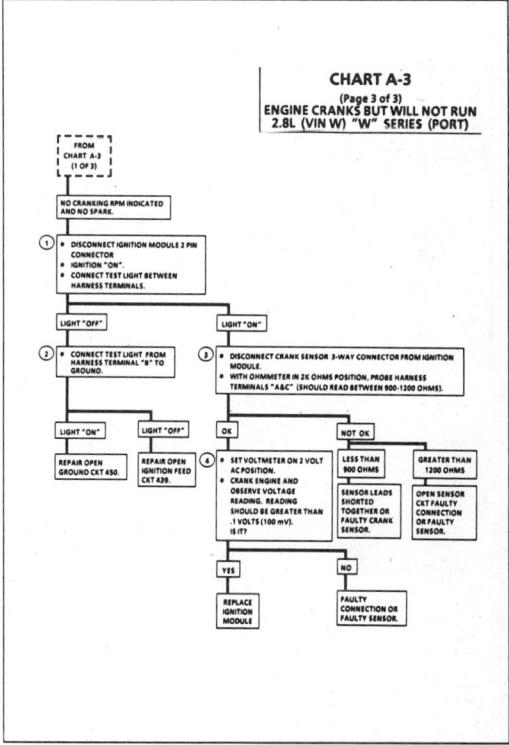

1988–90 2.8L (VIN W) REGAL AND CUTLASS SUPREME

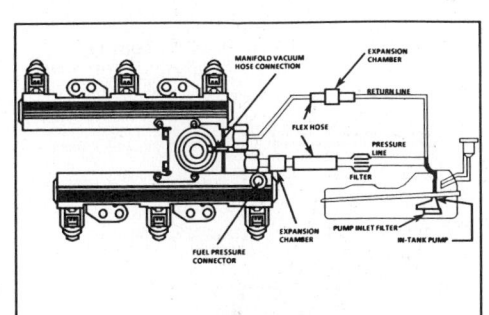

CHART A-7
(Page 1 of 2)
FUEL SYSTEM DIAGNOSIS
2.8L (VIN W) "W" SERIES (PORT)

Circuit Description:
When the ignition switch is turned "ON", the electronic control module (ECM) will turn "ON" the in-tank fuel pump. It will remain "ON" as long as the engine is cranking or running, and the ECM is receiving reference pulses. If there are no reference pulses, the ECM will shut "OFF" the fuel pump within 2 seconds after ignition "ON" or engine stops.

The pump will deliver fuel to the fuel rail and injectors, then to the pressure regulator, where the system pressure is controlled to about 225 to 325 kPa (33 to 47 psi). Excess fuel is then returned to the fuel tank.

Test Description: Numbers below refer to circled numbers on the diagnostic chart.
1. Wrap a shop towel around the fuel pressure connector to absorb any small amount of fuel leakage that may occur when installing the gage. Ignition "ON", pump pressure should be 280-320 kPa (40-46 psi). This pressure is controlled by spring pressure within the regulator assembly.
2. When the engine is idling, the manifold pressure is low (high vacuum) and is applied to the fuel regulator diaphragm. This will offset the spring and result in a lower fuel pressure. This idle pressure will vary somewhat depending on barometric pressure, however, the pressure idling should be less indicating pressure regulator control.
3. Pressure that continues to fall is caused by one of the following:
   • In-tank fuel pump check valve not holding.
   • Pump coupling hose or pulsator leaking.

• Fuel pressure regulator valve leaking.
• Injector(s) sticking open.
4. An injector sticking open can best be determined by checking for a fouled or saturated spark plug(s). If a leaking injector can not be determined by a fouled or saturated spark plug the following procedure should be used.
   • Remove plenum, and remove fuel rail bolts. Follow the procedures in the fuel control section of this manual, but leave fuel lines connected.
   • Lift fuel rail out just enough to leave injector nozzles in the ports.
CAUTION: *BE SURE INJECTOR(S) ARE NOT ALLOWED TO SPRAY ON ENGINE AND THAT INJECTOR RETAINING CLIPS ARE INTACT. THIS SHOULD BE CAREFULLY FOLLOWED TO PREVENT FUEL SPRAY ON ENGINE WHICH WOULD CAUSE A FIRE HAZARD.*
   • Pressurize the fuel system and observe for injector(s) leaking.

CHART A-7
(Page 1 of 2)
FUEL SYSTEM DIAGNOSIS
2.8L (VIN W) "W" SERIES (PORT)

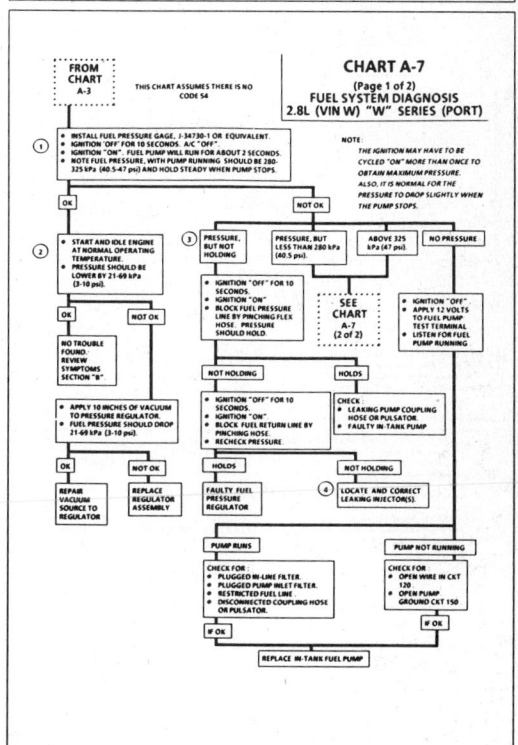

1988–90 2.8L (VIN W)
REGAL AND CUTLASS SUPREME

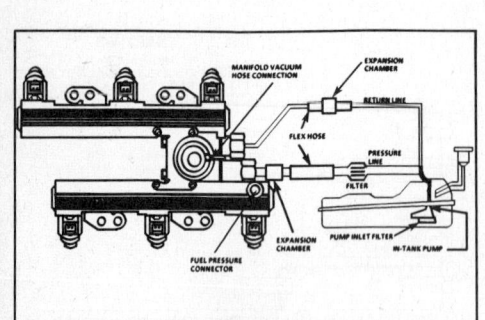

CHART A-7
(Page 2 of 2)
FUEL SYSTEM DIAGNOSIS
2.8L (VIN W) "W" SERIES (PORT)

Test Description: Numbers below refer to circled numbers on the diagnostic chart.

1. Pressure present, but less than 280 kPa (40.5 psi) falls into two areas:
   - Regulated pressure but less than 280 kPa (40.5 psi). Amount of fuel to injectors OK but pressure is too low. System will be lean running and may set Code 44. Also, hard starting cold and overall poor performance.
   - Restricted flow causing pressure drop – Normally, a vehicle with a fuel pressure of less than 165 kPa (24 psi) at idle will not be driveable.

However, if the pressure drop occurs only while driving, the engine will normally surge then stop running as pressure begins to drop rapidly. This is most likely caused by a restricted fuel line or plugged filter.

2. Restricting the fuel return line allows the fuel pump to develop its maximum pressure (dead head pressure). When battery voltage is applied to the pump test terminal, pressure should be above 414 kPa (60 psi).

3. This test determines if the high fuel pressure is due to a restricted fuel return line or a pressure regulator problem.

1988–90 2.8L (VIN W)
REGAL AND CUTLASS SUPREME

CHART A-7
(Page 2 of 2)
FUEL SYSTEM DIAGNOSIS
2.8L (VIN W) "W" SERIES (PORT)

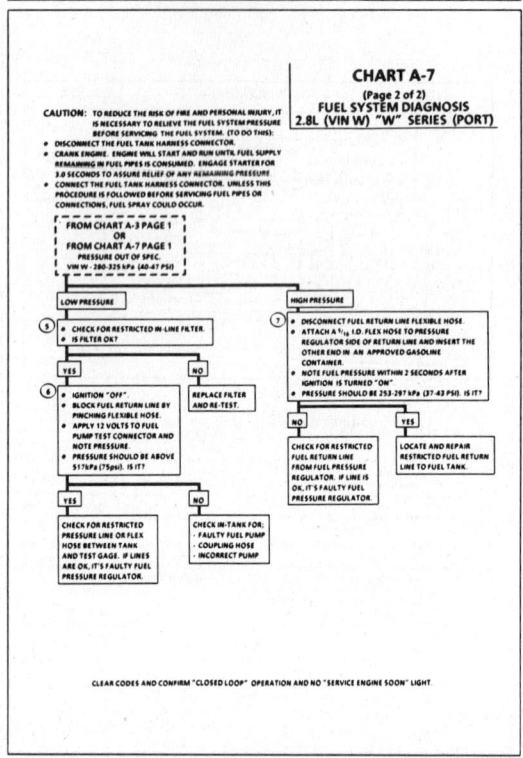

CLEAR CODES AND CONFIRM "CLOSED LOOP" OPERATION AND NO "SERVICE ENGINE SOON" LIGHT.

1988–90 2.8L (VIN W)
REGAL AND CUTLASS SUPREME

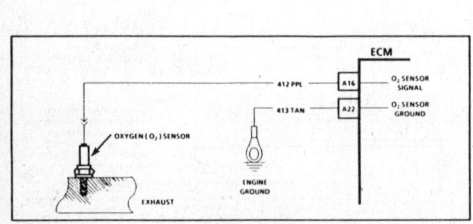

CODE 13
OXYGEN SENSOR CIRCUIT
(OPEN CIRCUIT)
2.8L (VIN W) "W" SERIES (PORT)

Circuit Description:

The ECM supplies a voltage of about .45 volt between terminals "A16" and "A22". (If measured with a 10 megohm digital voltmeter, this may read as low as .32 volts.) The O₂ sensor varies the voltage within a range of about 1 volt if the exhaust is rich, down through about .10 volt if exhaust is lean.

The sensor is like an open circuit and produces no voltage when it is below 315°C (600°F). An open sensor circuit or cold sensor causes "Open Loop" operation.

Test Description: Numbers below refer to circled numbers on the diagnostic chart.

1. Code 13 will set under the following conditions:
   - Engine running at least 2 minutes after start.
   - Coolant temperature at least 50°C.
   - No Code 21 or 22.
   - O₂ signal voltage steady between .35 and .55 volts.
   - Throttle position sensor signal above 6% for more time than TPS was below 6%. (About .3 volts above closed throttle voltage.)
   - All conditions must be met and held for at least 60 seconds.

If the conditions for a Code 13 exist the system will not go "Closed Loop".

2. This will determine if the sensor is at fault or the wiring or ECM is the cause of the Code 13.

3. In doing this test use only a high impedance digital volt ohmmeter. This test checks the continuity of CKT's 412 and 413 because if CKT 413 is open the ECM voltage on CKT 412 will be over 6 volts (600 mV).

Diagnostic Aids:

Normal "Scan" voltage varies between 100 mV to 999 mV (.1 and 1.0 volts) while in "Closed Loop". Code 13 sets in one minute if voltage remains between .35 and .55 volts, but the system will go "Open Loop" in about 15 seconds. Refer to "Intermittents" in Section "B".

1988–90 2.8L (VIN W)
REGAL AND CUTLASS SUPREME

CODE 13
OXYGEN SENSOR CIRCUIT
(OPEN CIRCUIT)
2.8L (VIN W) "W" SERIES (PORT)

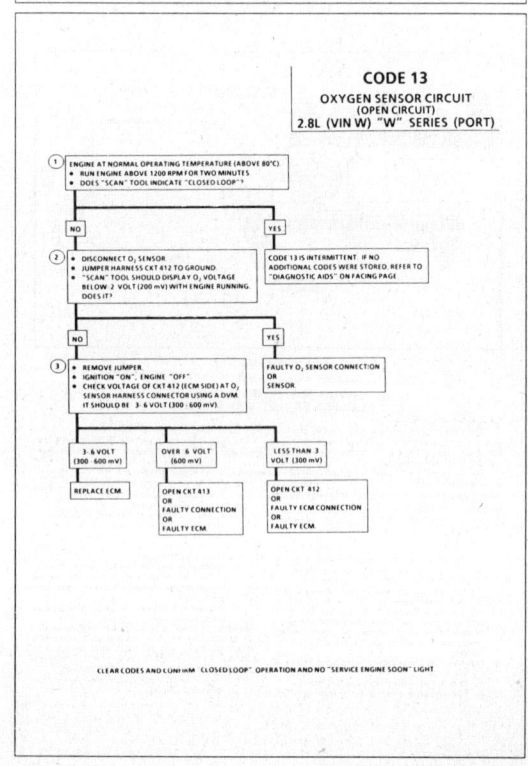

CLEAR CODES AND CONFIRM "CLOSED LOOP" OPERATION AND NO "SERVICE ENGINE SOON" LIGHT.

1988–90 2.8L (VIN W)
REGAL AND CUTLASS SUPREME

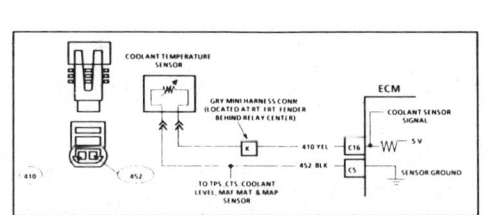

CODE 14
COOLANT TEMPERATURE SENSOR CIRCUIT
(HIGH TEMPERATURE INDICATED)
2.8L (VIN W) "W" SERIES (PORT)

Circuit Description:

The coolant temperature sensor uses a thermistor to control the signal voltage to the ECM. The ECM supplies a voltage on CKT 410 to the sensor. When the engine is cold the sensor (thermistor) resistance is high, therefore the ECM will see high signal voltage.

As the engine warms, the sensor resistance becomes less, and the voltage drops. At normal engine operating temperature the voltage will measure about 1.5 to 2.0 volts at the ECM terminal "C16".

Coolant temperature is one of the inputs used to control:
- Fuel delivery
- Engine spark timing (EST)
- Idle (IAC)
- Converter clutch (TCC)
- Canister purge (CCP)
- EGR
- Cooling fan

Test Description: Numbers below refer to circled numbers on the diagnostic chart.
1. Code 14 will set if:
   - Signal voltage indicates a coolant temperature above 130°C (285°F)
   - Engine running longer than 128 seconds.
2. This test will determine if CKT 410 is shorted to ground which will cause the conditions for Code 14.

Diagnostic Aids:

Check harness routing for a potential short to ground on CKT 410. CKT is routed from the ECM to a mini harness, and then to the coolant temperature sensor.

"Scan" tool displays engine temperature in degrees centigrade. After engine is started, the temperature should rise steadily to about 90°C then stabilize when thermostat opens. Refer to "Intermittents" in Section "B".

Verify that engine is not overheating and has not been subjected to conditions which could create an overheating condition (i.e. overload, trailer towing, hilly terrain, heavy stop and go traffic, etc.). The "Temperature To Resistance Value" scale at the right may be used to test the coolant sensor at various temperature levels to evaluate the possibility of a "shifted" (mis-scaled) sensor. A "shifted" sensor could result in poor driveability complaints.

1988–90 2.8L (VIN W)
REGAL AND CUTLASS SUPREME

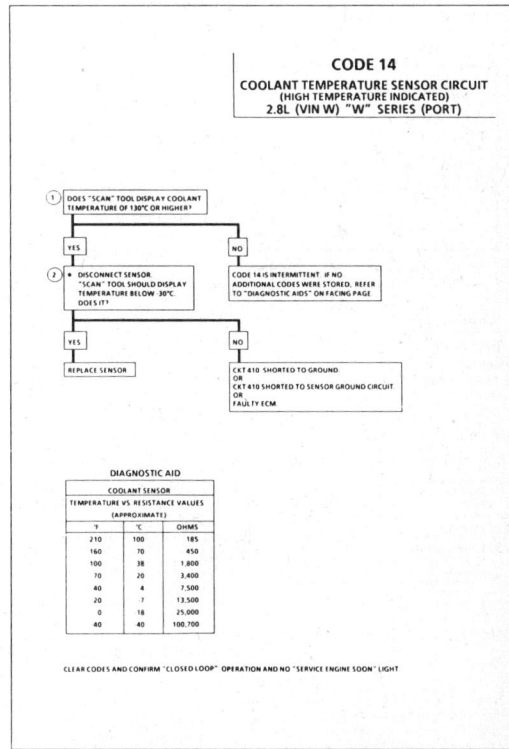

CODE 14
COOLANT TEMPERATURE SENSOR CIRCUIT
(HIGH TEMPERATURE INDICATED)
2.8L (VIN W) "W" SERIES (PORT)

DIAGNOSTIC AID		
COOLANT SENSOR		
TEMPERATURE VS RESISTANCE VALUES		
(APPROXIMATE)		
°F	°C	OHMS
210	100	185
160	70	450
100	38	1,800
70	20	3,400
40	4	7,500
20	-7	13,500
0	-18	25,000
-40	-40	100,700

CLEAR CODES AND CONFIRM "CLOSED LOOP" OPERATION AND NO "SERVICE ENGINE SOON" LIGHT

1988–90 2.8L (VIN W)
REGAL AND CUTLASS SUPREME

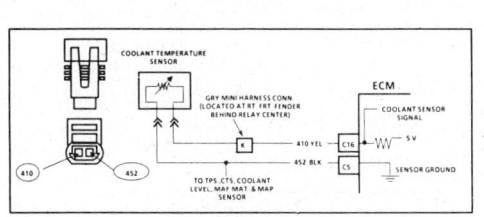

CODE 15
COOLANT TEMPERATURE SENSOR CIRCUIT
(LOW TEMPERATURE INDICATED)
2.8L (VIN W) "W" SERIES (PORT)

Circuit Description:

The coolant temperature sensor uses a thermistor to control the signal voltage to the ECM. The ECM applies a voltage on CKT 410 to the sensor. When the engine is cold the sensor (thermistor) resistance is high, therefore the ECM will see high signal voltage.

As the engine warms, the sensor resistance becomes less, and the voltage drops. At normal engine operating temperature the voltage will measure about 1.5 to 2.0 volts at the ECM terminal "C16".

Coolant temperature is one of the inputs used to control:
- Fuel delivery
- Engine spark timing (EST)
- Idle (IAC)
- Converter clutch (TCC)
- Canister purge (CCP)
- EGR
- Cooling fan

Test Description: Numbers below refer to circled numbers on the diagnostic chart.
1. Code 15 will set if:
   - Signal voltage indicates a coolant temperature less than -44°C (-47°F) for 20 seconds.
2. This test simulates a Code 14. If the ECM recognizes the low signal voltage, (high temperature) and the "Scan" reads 130°C, the ECM and wiring are OK.
3. This test will determine if CKT 410 is open. There should be 5 volts present at sensor connector if measured with a DVM.

Diagnostic Aids:

A "Scan" tool reads engine temperature in degrees centigrade.

After engine is started the temperature should rise steadily to about 95°C then stabilize when thermostat opens. CKT 410 is routed from the ECM to a mini harness, and then to the coolant temperature sensor.

A faulty connection, or an open in CKT 410 or 452 will result in a Code 15.

If Code 23 is also set, check CKT 452 for faulty wiring or connections. Check terminals at sensor for good contact.

Codes 15 and 21 are stored at the same time could be the result of an open CKT 452 which would also turn the temperature warning indicator "ON". The "Temperature to Resistance Value" scale at the right may be used to test the coolant sensor at various temperature levels to evaluate the possibility of a "shifted" (mis-scaled) sensor. A "shifted" sensor could result in poor driveability complaints.

Refer to "Intermittents" in Section "B".

1988–90 2.8L (VIN W)
REGAL AND CUTLASS SUPREME

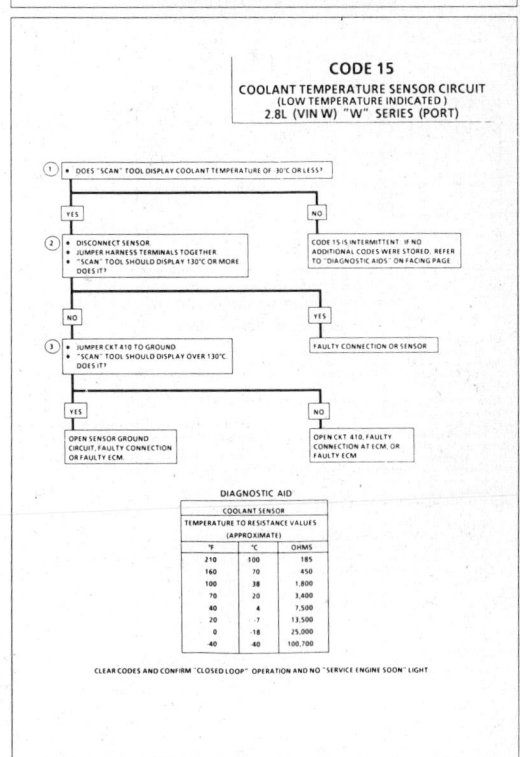

CODE 15
COOLANT TEMPERATURE SENSOR CIRCUIT
(LOW TEMPERATURE INDICATED)
2.8L (VIN W) "W" SERIES (PORT)

DIAGNOSTIC AID		
COOLANT SENSOR		
TEMPERATURE TO RESISTANCE VALUES		
(APPROXIMATE)		
°F	°C	OHMS
210	100	185
160	70	450
100	38	1,800
70	20	3,400
40	4	7,500
20	-7	13,500
0	-18	25,000
-40	-40	100,700

CLEAR CODES AND CONFIRM "CLOSED LOOP" OPERATION AND NO "SERVICE ENGINE SOON" LIGHT

1988–90 2.8L (VIN W)
REGAL AND CUTLASS SUPREME

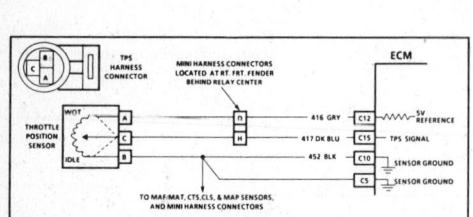

CODE 21
THROTTLE POSITION SENSOR (TPS) CIRCUIT
(SIGNAL VOLTAGE HIGH)
2.8L (VIN W) "W" SERIES (PORT)

Circuit Description:
The throttle position sensor (TPS) provides a voltage signal that changes relative to the throttle blade. Signal voltage will vary from about .5 at idle to about 5 volts at wide open throttle.
The TPS signal is one of the most important inputs used by the ECM for fuel control and for most of the ECM control outputs.

Test Description: Numbers below refer to circled numbers on the diagnostic chart.
1. Code 21 will set if:
   • Engine is running
   • No Code 33 or 34
   • MAF less than 12 gm/sec
   • TPS voltage greater than 4.9 volts.
   • Above conditions exist for over 5 seconds.
   OR
   • TPS voltage greater than 4.9 volts.
   With throttle closed, the TPS should read less than .70 volts. If it doesn't, check TPS adjustment.
2. With the TPS sensor disconnected, the TPS voltage should go low if the ECM and wiring is OK.
3. Probing CKT 452 with a test light checks the 5 volts return circuit, because a faulty 5 volts return will cause a Code 21.

Diagnostic Aids:
A "Scan" tool reads throttle position in volts. Voltage should increase at a steady rate as throttle is moved toward WOT.
Also some "Scan" tools will read throttle angle 0% = closed throttle 100% = WOT.
An open in CKT 452 will result in a Code 21.
Codes 15 and 21 are stored at the same time could be the result of an open CKT 452 which would also turn the temperature warning indicator "ON". "Scan" TPS while depressing accelerator pedal with engine stopped and ignition "ON". Display should vary from below 2500 mV (2.5 volts) when throttle was closed, to over 4500 mV (4.5 volts) when throttle is held at wide open throttle position.
Refer to "Intermittents" in Section "B".

1988–90 2.8L (VIN W)
REGAL AND CUTLASS SUPREME

CODE 21
THROTTLE POSITION SENSOR (TPS) CIRCUIT
(SIGNAL VOLTAGE HIGH)
2.8L (VIN W) "W" SERIES (PORT)

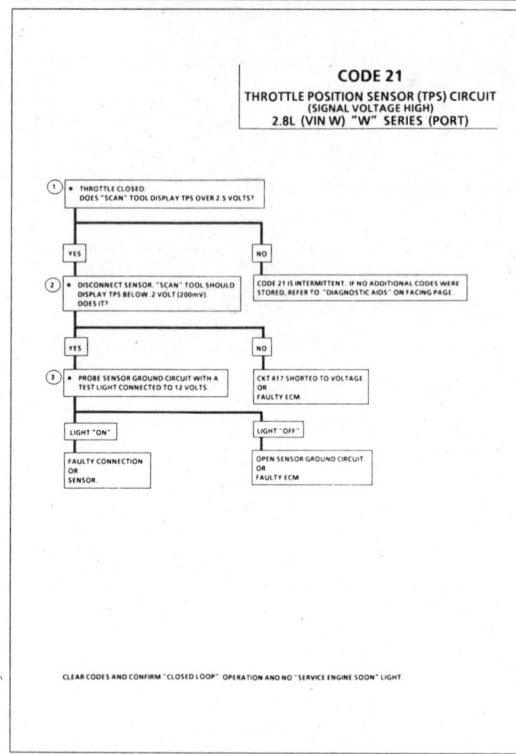

1988–90 2.8L (VIN W)
REGAL AND CUTLASS SUPREME

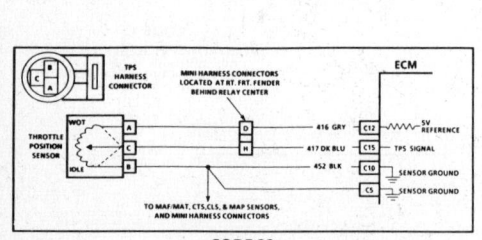

CODE 22
THROTTLE POSITION SENSOR (TPS) CIRCUIT
(SIGNAL VOLTAGE LOW)
2.8L (VIN W) "W" SERIES (PORT)

Circuit Description:
The throttle position sensor (TPS) provides a voltage signal that changes relative to the throttle blade. Signal voltage will vary from about .5 at idle to about 5 volts at wide open throttle.
The TPS signal is one of the most important inputs used by the ECM for fuel control and for most of the ECM control outputs.

Test Description: Numbers below refer to circled numbers on the diagnostic chart.
1. Code 22 will set if:
   • Engine running
   • TPS signal voltage is less than about .25 volt for 3 seconds.
2. Simulates Code 21. (High Voltage) If the ECM recognizes the high signal voltage the ECM and wiring are OK.
3. TPS check. The TPS has an auto zeroing feature. If the voltage reading is within the range of 0.35 to 0.70 volts, the ECM will use that value as closed throttle. If the voltage reading is out of the auto zero range on an existing or replacement TPS, the TPS should be adjusted.
4. This simulates a high signal voltage to check for an open in CKT 417.
5. CKTs 416 and 432 share a common 5 volts buffered reference signal. If either of these circuits is shorted to ground, Code 22 will set. To determine if the MAP sensor is causing the 22 problem disconnect it to see if Code 22 resets. Be sure TPS is connected and clear codes before testing.

Diagnostic Aids:
A "Scan" tool reads throttle position in volts. Voltage should increase at a steady rate as throttle is moved toward WOT.
Also some "Scan" tools will read throttle angle 0% = closed throttle 100% = WOT.
An open or short to ground in CKTs 416 or 417 will result in a Code 22.
CKTs 416 and 417 are routed through a mini harness. CKT 416 is connected to terminal "D" at the gray connector, CKT 417 is connected to terminal "H" at the black connector.
"Scan" TPS while depressing accelerator pedal with engine stopped and ignition "ON". Display should vary from below 500 mV (.5V) when throttle was closed, to over 4500 mV (4.5V) when throttle is held at wide open throttle position.
Also some "Scan" tools will read throttle angle.
0% = closed throttle
100% = open throttle
If Code 22 is set, check CKT 416 for faulty wiring or connections.
Refer to "Intermittents" in Section "B".

1988–90 2.8L (VIN W)
REGAL AND CUTLASS SUPREME

CODE 22
THROTTLE POSITION SENSOR (TPS) CIRCUIT
(SIGNAL VOLTAGE LOW)
2.8L (VIN W) "W" SERIES (PORT)

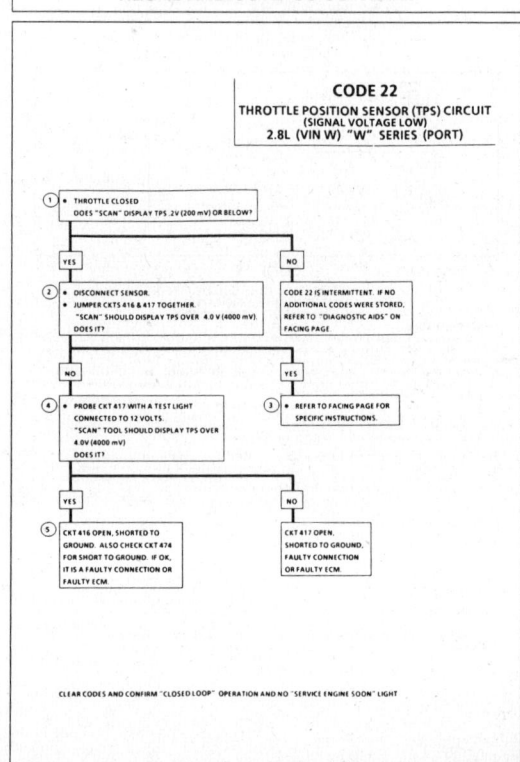

1988–90 2.8L (VIN W) REGAL AND CUTLASS SUPREME

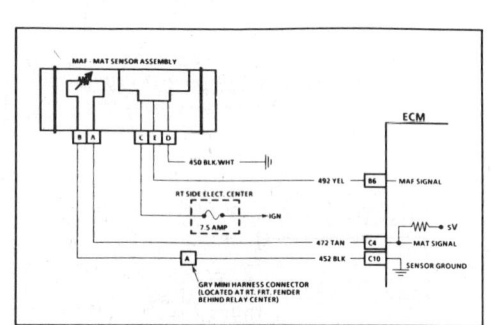

CODE 23
MANIFOLD AIR TEMPERATURE (MAT) SENSOR CIRCUIT
(LOW TEMPERATURE INDICATED)
2.8L (VIN W) "W" SERIES (PORT)

Circuit Description:

The MAT sensor uses a thermistor to control the signal voltage to the ECM. The ECM applies a voltage (about 5 volts) on CKT 472 to the sensor. When the air is cold the sensor (thermistor) resistance is high, therefore the ECM will see a high signal voltage. If the air is warm the sensor resistance is low therefore the ECM will see a low voltage.

The MAT sensor is part of the MAF sensor assembly so the ECM can accurately compensate the air flow reading based on temperature.

Test Description: Numbers below refer to circled numbers on the diagnostic chart.

1. Code 23 will set if:
   • A signal voltage indicates a manifold air temperature below -35°C (-31°F).
   • Time since engine start is 4 minutes or longer.
   • No VSS
2. A Code 23 will set, due to an open sensor, wire, or connection. This test will determine if the wiring and ECM are OK.
3. This will determine if the signal CKT 472 or the 5 volts return CKT 452 is open.

Diagnostic Aids:

A "Scan" tool reads temperature of the air entering the engine and should read close to ambient air temperature when engine is cold, and rises as underhood temperature increases.

A faulty connection, or an open in CKT 472 or 452 will result in a Code 23.

Codes 23 and 34 stored at the same time, could be the result of an open CKT 452 which would also turn the temperature warning indicator "ON". CKT 452 is routed through a mini harness. A faulty connection could result in intermittent failures. The "Temperature to Resistance Values" scale at the right may be used to test the MAT sensor at various temperature levels to evaluate the possibility of a "shifted" (mis-scaled) sensor. A "slewed" sensor could result in poor driveability complaints.

Refer to "Intermittents" in Section "B"

1988–90 2.8L (VIN W) REGAL AND CUTLASS SUPREME

CODE 23
MANIFOLD AIR TEMPERATURE (MAT) SENSOR CIRCUIT
(LOW TEMPERATURE INDICATED)
2.8L (VIN W) "W" SERIES (PORT)

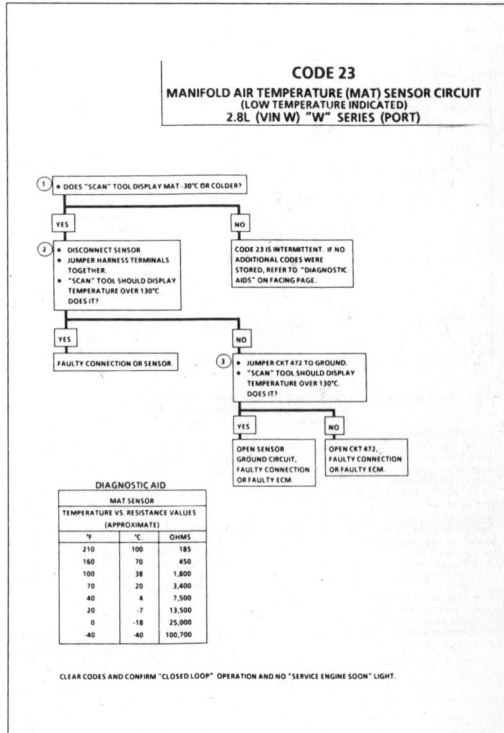

DIAGNOSTIC AID		
MAT SENSOR		
TEMPERATURE VS. RESISTANCE VALUES (APPROXIMATE)		
°F	°C	OHMS
210	100	185
160	70	450
100	38	1,800
70	20	3,400
40	4	7,500
20	-7	13,500
0	-18	25,000
-40	-40	100,700

CLEAR CODES AND CONFIRM "CLOSED LOOP" OPERATION AND NO "SERVICE ENGINE SOON" LIGHT.

1988–90 2.8L (VIN W) REGAL AND CUTLASS SUPREME

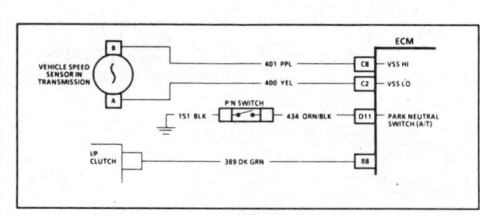

CODE 24
VEHICLE SPEED SENSOR (VSS) CIRCUIT
2.8L (VIN W) "W" SERIES (PORT)

Circuit Description:

Vehicle speed information is provided to the ECM by the vehicle speed sensor which uses a permanent magnet (PM) generator and it is mounted in the transaxle. The PM generator produces a pulsing voltage whenever vehicle speed is over about 3 mph. The AC voltage level and the number of pulses increases with vehicle speed. The ECM then converts the pulsing voltage to mph which is used for calculations, and the mph can be displayed with a "Scan" tool. Output of the generator can also be seen by using a digital voltmeter on the AC scale while rotating the generator.

The function of VSS buffer used in past model years has been incorporated into the ECM. The ECM then supplies the necessary signal for the instrument panel (4000 pulses per mile) for operating the speedometer and the odometer. If the vehicle is equipped with cruise control the ECM also provides a signal (2000 pulses per mile) to the cruise control module.

Test Description: Numbers below refer to circled numbers on the diagnostic chart.

Code 24 will set if vehicle speed equals 0 mph when:
   • Engine speed is between 1400 and 3600 rpm
   • TPS is less than 2%.
   • Low load condition (LV8 below 35).
   • Not in park or neutral.
   • No Code 21, 22, 33 or 34
   • All conditions met for 10 seconds

These conditions are met during a road load deceleration. Disregard Code 24 that sets when drive wheels are not turning.
   • The PM generator only produces a signal if drive wheels are turning greater than 3 mph.

Diagnostic Aids:

"Scan" should indicate a vehicle speed whenever the drive wheels are turning greater than 3 mph.

A problem in CKT 389 will not affect the VSS input or the readings on a "Scan".

Check CKT 400 and 401 for proper connections to be sure there are clean and tight and the harness is routed correctly. Refer to "Intermittents" in Section "B".

(A/T) A faulty or misadjusted park/neutral switch can result in a false Code 24. Use a "Scan" and check for proper signal while in overdrive (440-T4). Refer to CHART C-1A for P/N switch diagnosis check.

1988–90 2.8L (VIN W) REGAL AND CUTLASS SUPREME

CODE 24
VEHICLE SPEED SENSOR (VSS) CIRCUIT
2.8L (VIN W) "W" SERIES (PORT)

DISREGARD CODE 24 IF SET WHILE DRIVE WHEELS ARE NOT TURNING.

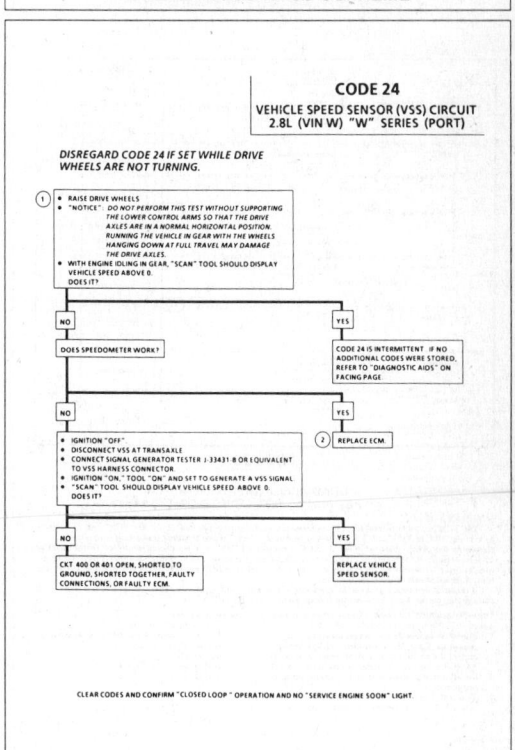

CLEAR CODES AND CONFIRM "CLOSED LOOP" OPERATION AND NO "SERVICE ENGINE SOON" LIGHT.

1988–90 2.8L (VIN W)
REGAL AND CUTLASS SUPREME

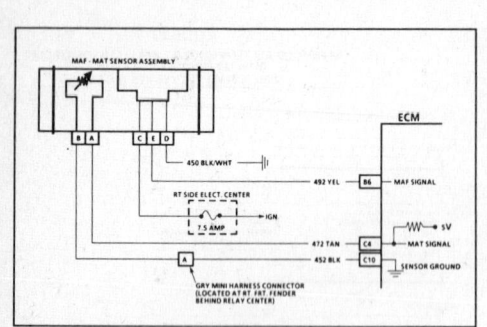

CODE 25

MANIFOLD AIR TEMPERATURE (MAT) SENSOR CIRCUIT
(HIGH TEMPERATURE INDICATED)
2.8L (VIN W) "W" SERIES (PORT)

Circuit Description:

The MAT sensor uses a thermistor to control the signal voltage to the ECM. The ECM applies a voltage (about 5 volts) on CKT 472 to the sensor. When the air is cold the sensor (thermistor) resistance is high, therefore the ECM will see a high signal voltage. If the air is warm the sensor resistance is low therefore the ECM will see a low voltage.

The MAT sensor is part of the MAF sensor assembly so the ECM can accurately compensate the air flow reading based on temperature.

Test Description: Numbers below refer to circled numbers on the diagnostic chart.

Code 25 will set if:
- Signal voltage indicates a manifold air temperature greater than 135°C (293°F) for 3 seconds.
- Time since engine start is 8 minutes or longer.
- A vehicle speed over 3 mph is present.

Due to the conditions necessary to set a Code 25 the "Service Engine Soon" light will remain "ON" while the signal is low and vehicle speed is present.

Diagnostic Aids:

A "Scan" tool reads temperature of the air entering the engine and should read close to ambient air temperature when engine is cold, and rises as underhood temperature increases.

A short to ground in CKT 472 will result in a Code 25.

The "Temperature to Resistance Values" scale at the right may be used to test the MAT sensor at various temperature levels to evaluate the possibility of a "shifted" (mis-scaled) sensor. A "slewed" sensor could result in poor driveability complaints.

Refer to "Intermittents" in Section "B".

1988–90 2.8L (VIN W)
REGAL AND CUTLASS SUPREME

CODE 25
MANIFOLD AIR TEMPERATURE (MAT) SENSOR CIRCUIT
(HIGH TEMPERATURE INDICATED)
2.8L (VIN W) "W" SERIES (PORT)

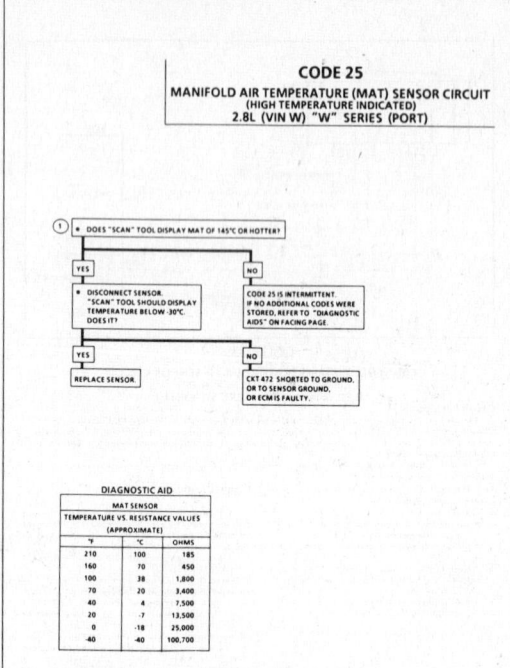

DIAGNOSTIC AID		
MAT SENSOR		
TEMPERATURE VS. RESISTANCE VALUES (APPROXIMATE)		
°F	°C	OHMS
210	100	185
160	70	450
100	38	1,800
70	20	3,400
40	4	7,500
20	-7	13,500
0	-18	25,000
-40	-40	100,700

CLEAR CODES AND CONFIRM "CLOSED LOOP" OPERATION AND NO "SERVICE ENGINE SOON" LIGHT.

1988–90 2.8L (VIN W)
REGAL AND CUTLASS SUPREME

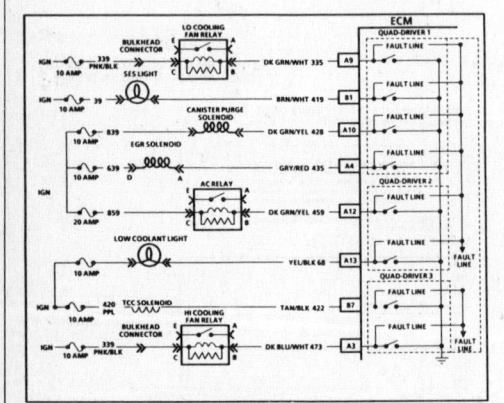

CODE 26
(Page 1 of 3)
QUAD-DRIVER (QDM) ERROR
2.8L (VIN W) "W" SERIES (PORT)

Circuit Description:

The ECM is used to control several components such as those illustrated above. The ECM controls a device by turning "ON" or "OFF" a device called a quad-driver. The ECM also monitors the state of each of the control circuits by way of the fault line. When the ECM is not turning "ON" the circuit being controlled, there should be 12 volts present on the fault line. If 12 volts is not present due to an open or shorted control circuit, or an open component coil, a failure will be logged in memory. Also when the ECM is commanding the component "ON" the control circuit should be low (near 0 volts).

If the fault line does not sense the low state, a failure will be logged. This could be caused by a short to 12 volts on the control circuit by a shorted component or by a faulty ECM driver.

Test Description: Numbers below refer to circled numbers on the diagnostic chart.

1. The ECM does not know which controlled circuit caused the Code 26 so this chart will go through each of the circuits to determine which is at fault. An EGR circuit problem should also set a Code 32 but after that problem is fixed this chart should be performed.
2. This test checks the "Service Engine Soon" light driver and the "Service Engine Soon" light circuit.

QDM symptoms:
- Cooling fan(s) inoperative.
- Poor driveability due to 100% canister purge.
- Coolant light "ON" all the time, "OFF" during bulb check.
- EGR inoperative - Code 32.
- TCC inoperative.

1988–90 2.8L (VIN W)
REGAL AND CUTLASS SUPREME

CODE 26
(Page 1 of 3)
QUAD-DRIVER (QDM) ERROR
2.8L (VIN W) "W" SERIES (PORT)

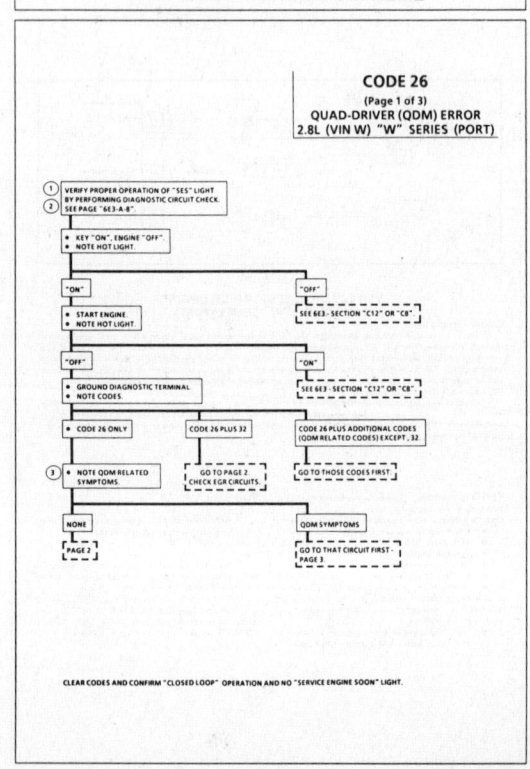

CLEAR CODES AND CONFIRM "CLOSED LOOP" OPERATION AND NO "SERVICE ENGINE SOON" LIGHT.

1988–90 2.8L (VIN W) REGAL AND CUTLASS SUPREME

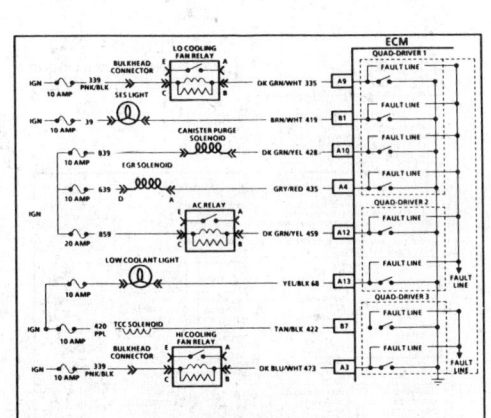

CODE 26
(Page 2 of 3)
QUAD-DRIVER (QDM) ERROR
2.8L (VIN W) "W" SERIES (PORT)

Circuit Description:
The ECM is used to control several components such as those illustrated above. The ECM controls a device by turning "ON" or "OFF" a device called a quad-driver. The ECM also monitors the state of each of the control circuits by way of the fault line. When the ECM is not turning "ON" the circuit being controlled, there should be 12 volts present on the fault line. If 12 volts is not present due to an open or shorted control circuit, or an open component coil, a failure will be logged in memory. Also when the ECM is commanding the component "ON", the control circuit should be low (near 0 volts).
If the fault line does not sense the low state, a failure will be logged. This could be caused by a short to 12 volts on the control circuit by a shorted component or by a faulty ECM driver.

Test Description: Numbers below refer to circled numbers on the diagnostic chart.
4. This test will determine which circuit is out of specifications. All circuits *EXCEPT* "B1", the "SES" light, "A18", and "hot light" should be B + when key is "ON", engine not running. The diagnostic test terminal is not grounded.

1988–90 2.8L (VIN W) REGAL AND CUTLASS SUPREME

CODE 26
(Page 2 of 3)
QUAD-DRIVER (QDM) ERROR
2.8L (VIN W) "W" SERIES (PORT)

CIRCUIT NOT ISOLATED BY PRIOR STEPS
(AMP TEST)

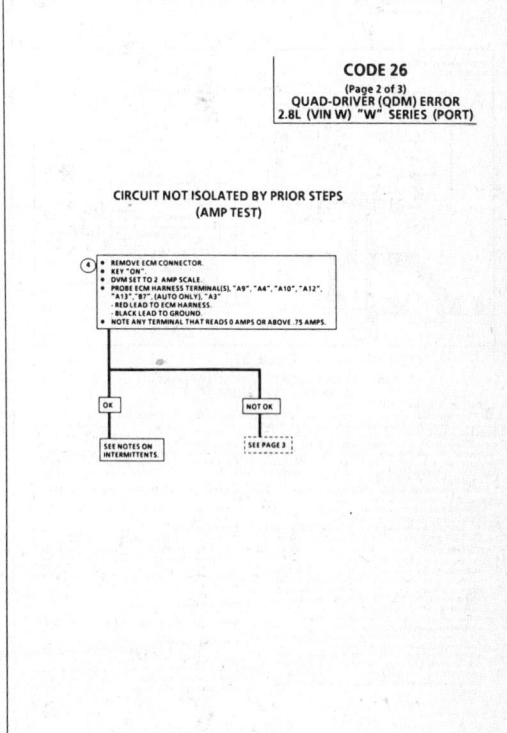

1988–90 2.8L (VIN W) REGAL AND CUTLASS SUPREME

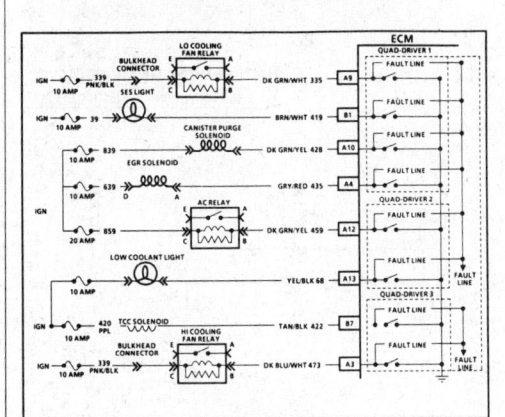

CODE 26
(Page 3 of 3)
QUAD-DRIVER (QDM) ERROR
2.8L (VIN W) "W" SERIES (PORT)

Circuit Description:
The ECM is used to control several components such as those illustrated above. The ECM controls a device by turning "ON" or "OFF" a device called a quad-driver. The ECM also monitors the state of each of the control circuits by way of the fault line. When the ECM is commanding the component "ON", the control circuit should be low (near 0 volts).
If the fault line does not sense the low state when a device is commanded "ON", a failure will be logged. This could be caused by a short to 12 volts on the control circuit by a shorted component or by a faulty ECM driver.

Test Description: Numbers below refer to circled numbers on the diagnostic chart.
5. This test will determine if the problem is the circuit or the component. As the factory installed ECM is protected with an internal fuse, it is highly unlikely that the ECM needs to be replaced.

1988–90 2.8L (VIN W) REGAL AND CUTLASS SUPREME

CODE 26
(Page 3 of 3)
QUAD-DRIVER (QDM) ERROR
2.8L (VIN W) "W" SERIES (PORT)

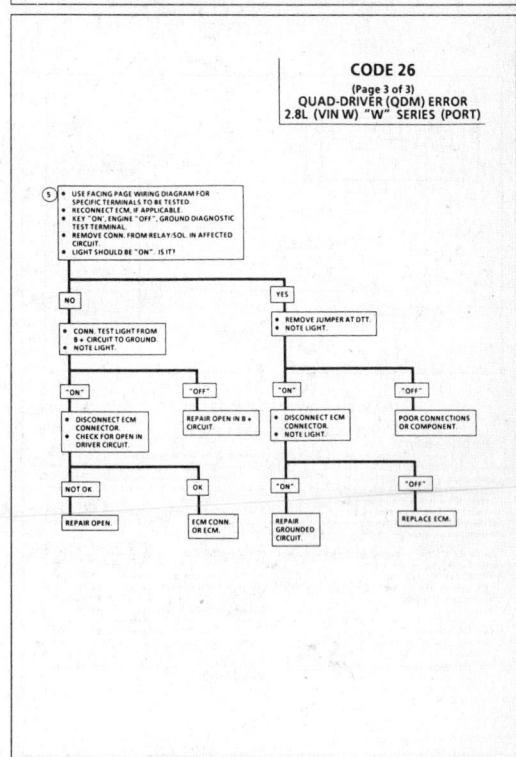

FUEL INJECTION SYSTEMS
MULTI-PORT INJECTION (MPI)/TUNED PORT INJECTION (TPI)

1988–90 2.8L (VIN W)
REGAL AND CUTLASS SUPREME

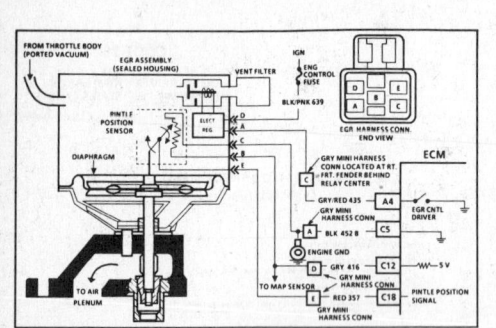

CODE 32
EXHAUST GAS RECIRCULATION (EGR) CIRCUIT
2.8L (VIN W) "W" SERIES (PORT)

Circuit Description:

The integrated electronic EGR valve functions similar to a port valve with a remote vacuum regulator. The internal solenoid is normally open, which causes the vacuum signal to be vented "OFF" to the atmosphere when EGR is not being commanded by the ECM. This EGR valve has a sealed cap and the solenoid valve opens and closes the vacuum signal which controls the amount of vacuum vented to atmosphere, and this controls the amount of vacuum applied to the diaphragm. The electronic EGR valve contains a voltage regulator which converts the ECM signal to provide different amounts of EGR flow by regulating the current to the solenoid. The ECM controls EGR flow with a pulse width modulated signal (turns "ON" and "OFF" many times a second) based on air flow, TPS, and rpm.

This system also contains a pintle position sensor which works similar to a TPS sensor, and as EGR flow is increased, the sensor output also increases.

Code 32 means that there has been an EGR system fault detected.

Code 32 will set under two conditions:
• Coolant temperature above a specified amount, EGR should be "ON" or.
• EGR pintle position does not match duty cycle.

Test Description: Numbers below refer to circled numbers on the diagnostic chart.

1. Whenever the solenoid is de-energized, the solenoid valve should be closed which should not allow the vacuum to move the EGR diaphragm. However, if the filter is plugged, the vacuum applied with the hand held vacuum pump is about to cause the diaphragm to move because the vacuum will not be vented to the atmosphere.

2. Grounding the diagnostic terminal should energize the solenoid which closes off the vent and allows the vacuum to move the diaphragm. This test determines that the ECM is capable of controlling the solenoid. When EGR is commanded on by the ECM, the test light should be "ON."

3. If CKT 452 is open, the pintle signal will go high (showing a 5 volts signal). This will set a Code 32. If CKT 357 becomes shorted to 12 volts or to CKT 416, the signal voltage will go high causing a Code 32.

Diagnostic Aids:

Some "Scan" tools will read pintle position in volts.

The EGR position voltage can be used to determine that the pintle is moving. When no EGR is commanded (0% duty cycle), the position sensor should read between .5 volt and 1.5 volts and increase with the commanded EGR duty cycle. If system operates correctly, refer to "Intermittents" in Section "B."

1988–90 2.8L (VIN W)
REGAL AND CUTLASS SUPREME

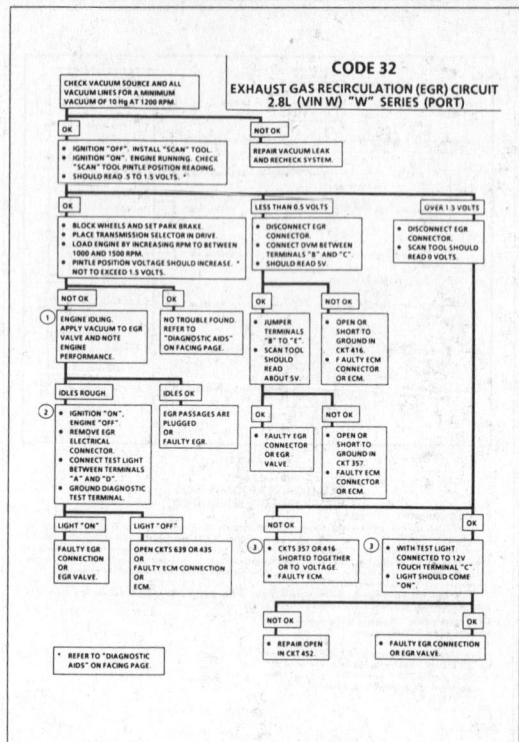

CODE 32
EXHAUST GAS RECIRCULATION (EGR) CIRCUIT
2.8L (VIN W) "W" SERIES (PORT)

1988–90 2.8L (VIN W)
REGAL AND CUTLASS SUPREME

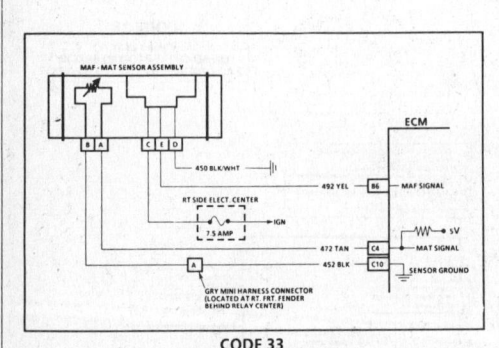

CODE 33
MASS AIR FLOW (MAF) SENSOR CIRCUIT
(GM/SEC HIGH)
2.8L (VIN W) "W" SERIES (PORT)

Circuit Description:

The MAF sensor measures the flow of air entering the engine. The sensor produces a frequency output between 32 and 150 hertz (3gm/sec to 150gm/sec). A large quantity (high frequency) indicates acceleration, and a small quantity (low frequency) indicates deceleration or idle. This information is used by the ECM for fuel control and is converted by a "Scan" tool to read out the air flow in grams per second. A normal reading is about 4-7 grams per second at idle and increases with rpm.

The MAF sensor is powered up by a 7.5 amp fuse which is located underhood in the right side relay center and the sensor should have power supplied to it anytime the ignition is "ON." The MAF and MAT sensor are combined into one assembly located in the air duct.

Test Description: Numbers below refer to circled numbers on the diagnostic chart.

Code 33 will set if:
• Ignition "ON," engine "OFF," and air flow exceeds 20 gm/sec.
  OR
• Engine is running at less than 1300 rpm.
• TPS is 8% or less.
• Air flow greater than 20 gm/sec. (high frequency).
• The above three conditions are met for 2 seconds.

Diagnostic Aids:

The "Scan" tool is not of much use in diagnosing this code because when the code sets, the value displayed will be the default value. However, it may be useful in comparing the signal of a problem vehicle with that of a known good running one.

Refer to "Intermittents" in Section "B."

Inspect wire routing of high voltage wires such as spark plug wires. Such wires routed too closely to MAF wiring harness could possibly cause an intermittent Code 33.

1988–90 2.8L (VIN W)
REGAL AND CUTLASS SUPREME

CODE 33
MASS AIR FLOW (MAF) SENSOR CIRCUIT
(GM/SEC HIGH)
2.8L (VIN W) "W" SERIES (PORT)

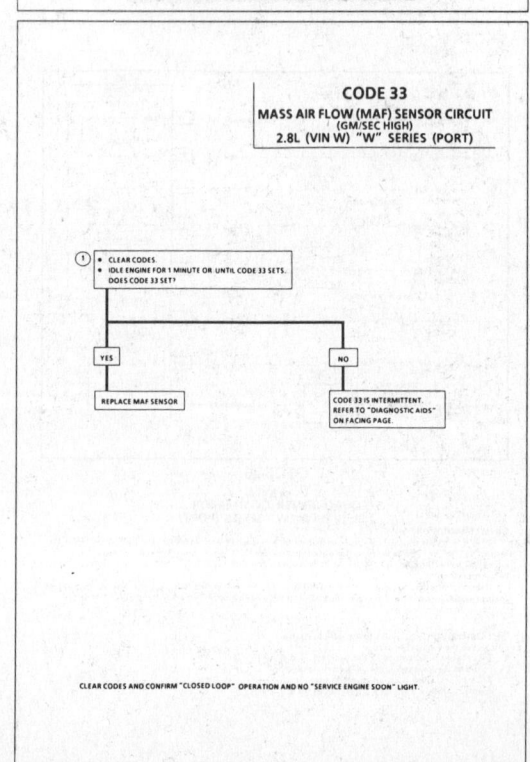

CLEAR CODES AND CONFIRM "CLOSED LOOP" OPERATION AND NO "SERVICE ENGINE SOON" LIGHT.

1988–90 2.8L (VIN W)
REGAL AND CUTLASS SUPREME

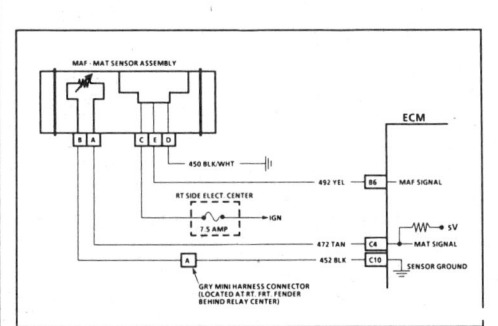

CODE 34
MASS AIR FLOW (MAF) SENSOR CIRCUIT
(GM/SEC LOW)
2.8L (VIN W) "W" SERIES (PORT)

Circuit Description:

The MAF sensor measures the flow of air entering the engine. The sensor produces a frequency output between 32 and 150 hertz (3gm/sec to 150gm/sec). A large quantity (high frequency) indicates acceleration, and a small quantity (low frequency) indicates deceleration or idle. This information is used by the ECM for fuel control and is converted by a "Scan" tool to read out the air flow in grams per second. A normal reading is about 4-7 grams per second at idle and increase with rpm.

The MAF sensor is powered up by a 7.5 amp fuse which is located underhood in the right side relay center and the sensor should have power supplied to it anytime the ignition is "ON." The MAF and MAT sensor are combined into one assembly located in the air duct.

Test Description: Numbers below refer to circled numbers on the diagnostic chart.

1. Code 34 will set if:
   - Engine running
   - Check for vacuum leaks
   - MAF sensor disconnected, or MAF signal circuit shorted to ground
      OR
   - Air flow less than 2 gm/sec (low frequency) with engine running.

   A loose or damaged air duct can set Code 34.

   This test checks to see if ECM recognizes a problem. A light "OFF" at this point indicates an intermittent problem.

2. Checks to see if 5 volt reference signal from ECM is at MAF sensor harness connector.

3. Checks continuity of electrical circuit at MAF sensor.

4. Checks for open in 12 volt supply.

Diagnostic Aids:

The "Scan" tool is not of much use in diagnosing this code because when the code sets, the value displayed will be the default value. It may be useful in comparing the signal of a problem vehicle with that of a known good running one.

Check for loose or damaged air duct.
Check for any vacuum leaks.
Inspect sensor connections as an open will result in a Code 34.

Refer to "Intermittents" in Section "B".

1988–90 2.8L (VIN W)
REGAL AND CUTLASS SUPREME

CODE 34
MASS AIR FLOW (MAF) SENSOR CIRCUIT
(GM/SEC LOW)
2.8L (VIN W) "W" SERIES (PORT)

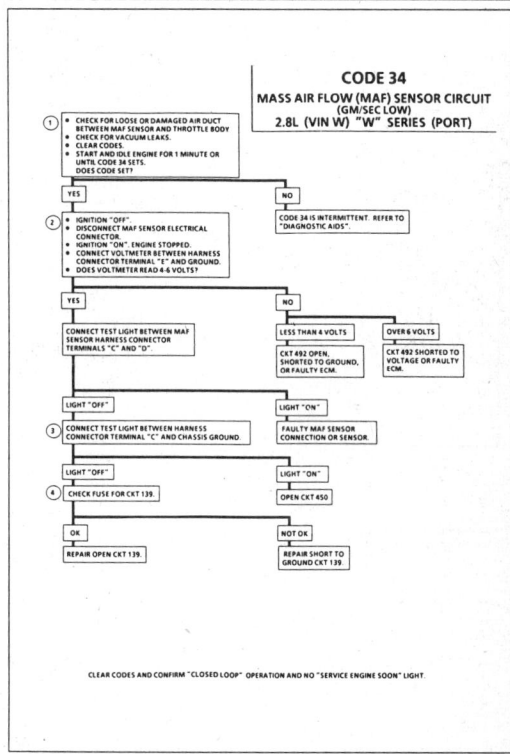

CLEAR CODES AND CONFIRM "CLOSED LOOP" OPERATION AND NO "SERVICE ENGINE SOON" LIGHT.

1988–90 2.8L (VIN W)
REGAL AND CUTLASS SUPREME

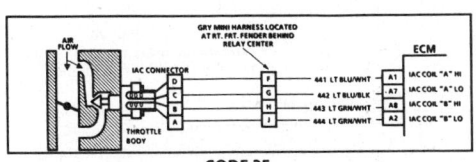

CODE 35
IDLE SPEED CONTROL (ISC) CIRCUIT
2.8L (VIN W) "W" SERIES (PORT)

Circuit Description:

Code 35 will set when the closed throttle engine speed is 100 rpm above or below the desired (commanded) idle speed for 45 seconds.

Test Description: Numbers below refer to circled numbers on the diagnostic chart.

1. Continue with test even if engine will not idle. If idle is too low, "Scan" will display 80 or more counts, or steps. If idle is high it will display "0" counts.

   Occasionally an erratic or unstable idle may occur. Engine speed may vary 200 rpm or more up and down. Disconnect IAC. If the condition is unchanged, the IAC is not at fault. There is a system problem. Proceed to diagnostic aids below.

2. When the engine was stopped, the IAC valve retracted (more air) to a fixed "park" position for increased air flow and idle speed during the next engine start. A "Scan" will display 80 or more counts.

3. Be sure to disconnect the IAC valve prior to this test. The test light will confirm the ECM signals by a steady or flashing light on all circuits.

4. There is a remote possibility that some of the circuits is shorted to voltage which would have been indicated by a steady light. Disconnect ECM and turn the ignition "ON" and probe terminals to check for this condition.

5. IAC wiring is routed from the ECM through a mini harness. Faulty connection or shorted wires could result in poor IAC operation.

Diagnostic Aids:

A slow unstable idle may be caused by a system problem that cannot be overcome by the IAC. "Scan" counts will be above 80 counts if idle is too low and "0" counts if it is too high.

If idle is too high, stop engine. Ignition "ON." Ground diagnostic terminal. Wait a few seconds for IAC to seat then disconnect IAC. Start engine. If idle speed is above rpm listed above, locate and correct vacuum leak.

- **System too lean (high air/fuel ratio)** Idle speed may be too high or too low. Engine speed may vary up and down, disconnecting IAC does not help. This may set Code 44. "Scan" and/or voltmeter will read an oxygen sensor output less than 300 mV(.3volt). Check for low regulated fuel pressure or water in fuel. A lean exhaust with an oxygen sensor output fixed above 800 mV (.8volt) will be a contaminated sensor, usually silicone. This may also set a Code 45 or 61.
- **System too rich (low air/fuel ratio)** Idle speed too low. "Scan" counts usually above 80. System obviously rich and may exhibit black smoke exhaust. "Scan" tool and/or voltmeter will read an oxygen sensor signal fixed above 800 mV (.8volt). Check:
   - For fuel in pressure regulator hose
   - High fuel pressure
   - Injector leaking or sticking.
- **MAF sensor** If idle is rough or unstable, disconnect MAF sensor. If idle improves, check for Code 34.
- **Throttle body** Remove IAC and inspect bore for foreign material or evidence of IAC valve dragging the bore.
- **A/C compressor or relay failure** See CHART C-10 if the A/C control relay drive circuit is shorted to ground or, if the relay is faulty, an idle problem may exist.
- Refer to "Rough, Unstable, Incorrect Idle or Stalling" in "Symptoms" in Section "B".

1988–90 2.8L (VIN W)
REGAL AND CUTLASS SUPREME

CODE 35
IDLE SPEED CONTROL (ISC) CIRCUIT
2.8L (VIN W) "W" SERIES (PORT)

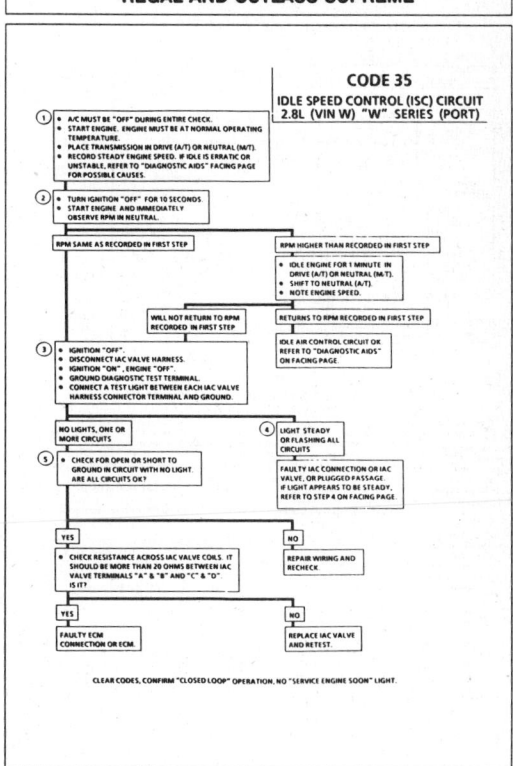

CLEAR CODES, CONFIRM "CLOSED LOOP" OPERATION, NO "SERVICE ENGINE SOON" LIGHT.

1988–90 2.8L (VIN W)
REGAL AND CUTLASS SUPREME

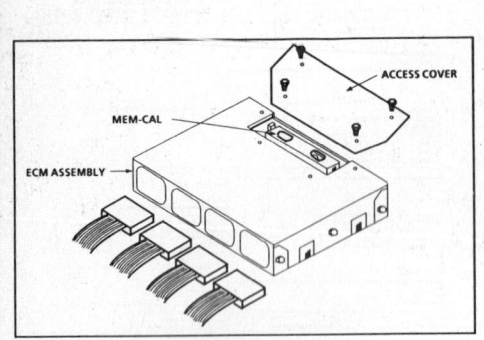

ACCESS COVER

MEM-CAL

ECM ASSEMBLY

CODE 41
CYLINDER SELECT ERROR
(FAULTY OR INCORRECT MEM-CAL)
2.8L (VIN W) "W" SERIES (PORT)

Test Description: Numbers below refer to circled numbers on the diagnostic chart.
1. The ECM used for this engine can also be used for other engines and the difference is in the Mem-Cal. If a Code 41 sets, the incorrect Mem-Cal has been installed or it is faulty and it must be replaced.

Diagnostic Aids:

Check Mem-Cal to be sure locking tabs are secure.

1988–90 2.8L (VIN W)
REGAL AND CUTLASS SUPREME

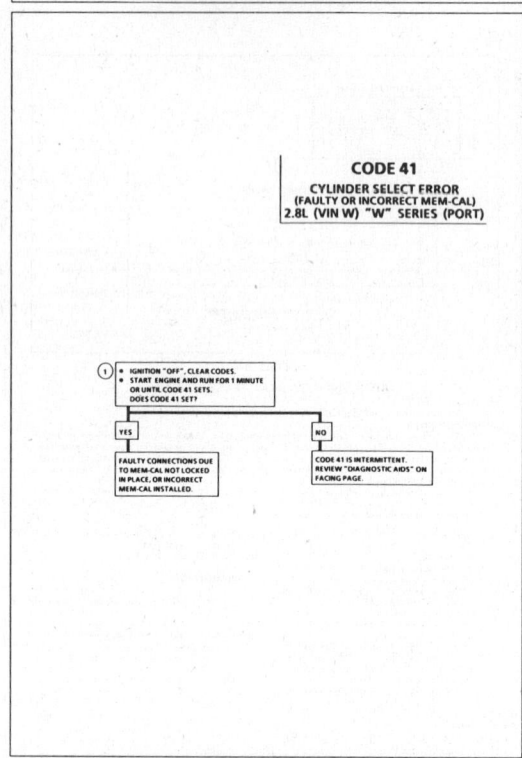

CODE 41
CYLINDER SELECT ERROR
(FAULTY OR INCORRECT MEM-CAL)
2.8L (VIN W) "W" SERIES (PORT)

1
• IGNITION "OFF", CLEAR CODES.
• START ENGINE AND RUN FOR 1 MINUTE OR UNTIL CODE 41 SETS.
DOES CODE 41 SET?

YES

FAULTY CONNECTIONS DUE TO MEM-CAL NOT LOCKED IN PLACE, OR INCORRECT MEM-CAL INSTALLED.

NO

CODE 41 IS INTERMITTENT. REVIEW "DIAGNOSTIC AIDS" ON FACING PAGE.

1988–90 2.8L (VIN W)
REGAL AND CUTLASS SUPREME

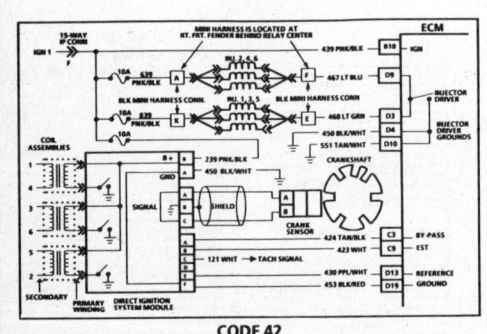

CODE 42
ELECTRONIC SPARK TIMING (EST) CIRCUIT
2.8L (VIN W) "W" SERIES (PORT)

Circuit Description:
When the system is running on the ignition module, that is, no voltage on the bypass line, the ignition module grounds the EST signal. The ECM expects to see no voltage on the EST line during this condition. If it sees a voltage, it sets Code 42 and will not go into the EST mode.
When the rpm for EST is reached (about 400 rpm), and bypass voltage applied, the EST should no longer be grounded in the ignition module so the EST voltage should be varying.
If the bypass line is open or grounded, the ignition module will not switch to EST mode so the EST voltage will be low and Code 42 will be set.
If the EST line is grounded, the ignition module will switch to EST, but because the line is grounded there will be no EST signal. A Code 42 will be set.

Test Description: Numbers below refer to circled numbers on the diagnostic chart.
1. Code 42 means the ECM has seen an open or short to ground in the EST or bypass circuits. This test confirms Code 42 and that the fault causing the code is present.
2. Checks for a normal EST ground path through the ignition module. An EST CKT 423 shorted to ground will also read less than 500 ohms; however, this will be checked later.
3. As the test light voltage touches CKT 424, the module should switch causing the ohmmeter to "overrange" if the meter is in the 1000-2000 ohms position. Selecting the 10-20,000 ohms position will indicate above 5000 ohms. The important thing is that the module "switched".

4. The module did not switch and this step checks for:
• EST CKT 423 shorted to ground.
• Bypass CKT 424 open.
• Faulty ignition module connection or module.
5. Confirms that Code 42 is a faulty ECM and not an intermittent in CKTs 423 or 424.

Diagnostic Aids:

The "Scan" tool does not have any ability to help diagnose a Code 42 problem.
A Mem-Cal not fully seated in the ECM can result in a Code 42.
Refer to "Intermittents" in Section "B".

1988–90 2.8L (VIN W)
REGAL AND CUTLASS SUPREME

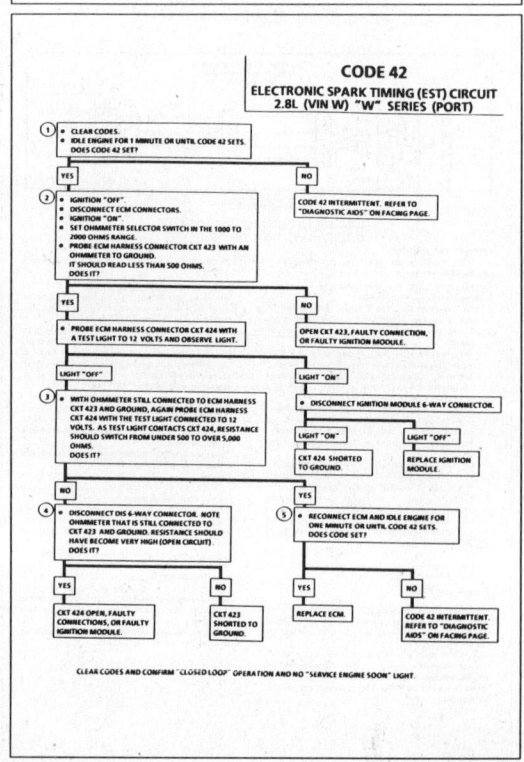

CODE 42
ELECTRONIC SPARK TIMING (EST) CIRCUIT
2.8L (VIN W) "W" SERIES (PORT)

1
• CLEAR CODES.
• IDLE ENGINE FOR 1 MINUTE OR UNTIL CODE 42 SETS.
DOES CODE 42 SET?

YES / NO

CODE 42 INTERMITTENT. REFER TO "DIAGNOSTIC AIDS" ON FACING PAGE.

2
• IGNITION "OFF".
• DISCONNECT ECM CONNECTORS.
• IGNITION "ON".
• SET OHMMETER SELECTOR SWITCH IN THE 1000 TO 2000 OHMS RANGE.
• PROBE ECM HARNESS CONNECTOR CKT 423 WITH AN OHMMETER TO GROUND.
IT SHOULD READ LESS THAN 500 OHMS.
DOES IT?

YES

• PROBE ECM HARNESS CONNECTOR CKT 424 WITH A TEST LIGHT TO 12 VOLTS AND OBSERVE LIGHT.

NO

OPEN CKT 423, FAULTY CONNECTION, OR FAULTY IGNITION MODULE.

LIGHT "OFF"

3
• WITH OHMMETER STILL CONNECTED TO ECM HARNESS CKT 423 AND GROUND, AGAIN PROBE ECM HARNESS CKT 424 WITH THE TEST LIGHT CONNECTED TO 12 VOLTS. AS TEST LIGHT CONTACTS CKT 424, RESISTANCE SHOULD SWITCH FROM UNDER 500 TO OVER 5,000 OHMS.
DOES IT?

LIGHT "ON"

• DISCONNECT IGNITION MODULE 6-WAY CONNECTOR.

LIGHT "ON"

CKT 424 SHORTED TO GROUND.

LIGHT "OFF"

REPLACE IGNITION MODULE.

NO

4
• DISCONNECT DIS 6-WAY CONNECTOR. NOTE OHMMETER THAT IS STILL CONNECTED TO CKT 423 AND GROUND. RESISTANCE SHOULD HAVE BECOME VERY HIGH (OPEN CIRCUIT).
DOES IT?

YES

CKT 424 OPEN, FAULTY CONNECTIONS, OR FAULTY IGNITION MODULE.

NO

CKT 423 SHORTED TO GROUND.

5
• RECONNECT ECM AND IDLE ENGINE FOR ONE MINUTE OR UNTIL CODE 42 SETS.
DOES CODE SET?

YES

REPLACE ECM.

NO

CODE 42 INTERMITTENT. REFER TO "DIAGNOSTIC AIDS" ON FACING PAGE.

CLEAR CODES AND CONFIRM "CLOSED LOOP" OPERATION AND NO "SERVICE ENGINE SOON" LIGHT.

1988–90 2.8L (VIN W)
REGAL AND CUTLASS SUPREME

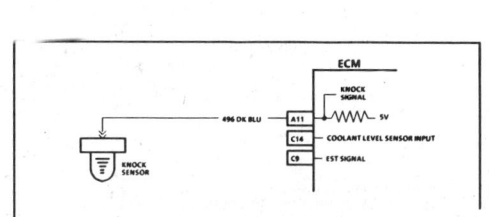

CODE 43
ELECTRONIC SPARK CONTROL (ESC) CIRCUIT
2.8L (VIN W) "W" SERIES (PORT)

Circuit Description:

The knock sensor is used to detect engine detonation and the ECM will retard the electronic spark timing based on the signal being received. The circuitry within the knock sensor causes the ECM 5 volts to be pulled down so that under a no knock condition, CKT 496 would measure about 2.5 volts. The knock sensor produces an AC signal which rides on the 2.5volts DC voltage. The amplitude and signal frequency is dependent upon the knock level.

There are two tests run on this circuit to determine if it is operating correctly. If either of the tests fail, a Code 43 will be set.

43A If CKT 496 becomes open or shorted to ground the voltage will either go above 3.5 volts or below 1.5 volts. If either of these conditions are met for about ½ second a Code 43 will be stored.

43B This system also performs a functional test to determine if the knock sensor is responding to engine detonation. To perform this test, the ECM will advance the spark under certain load conditions and look for a knock signal response. If knock is detected before conditions are met to run the test, the test is bypassed, but if the test is run and no knock detected the Code 43 will be set.

The test is performed when.
Coolant temperature is over 90°C.
MAT temperature is over 0°C.
High engine load based on air flow and rpm between 3400 and 4400.

Test Description: Numbers below refer to circled numbers on the diagnostic chart.

1. If the conditions for the A test, as described above, are being met the "Scan" tool will always indicate "yes" when the knock signal position is selected. If an audible knock is heard from the engine, repair the internal engine problem, as normally no knock should be detected at idle.
2. If tapping on the engine lift hook does not produce a knock signal, try tapping engine closer to sensor before proceeding.
3. The ECM has a 5 volt pull-up resistor which should be present at the knock sensor terminal.
4. This test determines if the knock sensor is faulty or if the ESC portion of the Mem-Cal is faulty.

Diagnostic Aids:

Check CKT 496 for a potential open or short to ground.
Also check for proper installation of Mem-Cal.
Refer to "Intermittents" in Section "B".
If the customer's complaint is the "Service Engine Soon" light comes "ON" in acceleration, the B portion of the code is failing. There is a possibility that the direct ignition system was in bypass mode when the 43 test was run. An intermittent open in the EST circuit will put the DIS module in bypass which will not allow the spark to be advanced so the 43B test would fail. If ECM also had a 42 stored, then the EST circuit is likely the cause of the Code 43.

1988–90 2.8L (VIN W)
REGAL AND CUTLASS SUPREME

CODE 43
ELECTRONIC SPARK CONTROL (ESC) CIRCUIT
2.8L (VIN W) "W" SERIES (PORT)

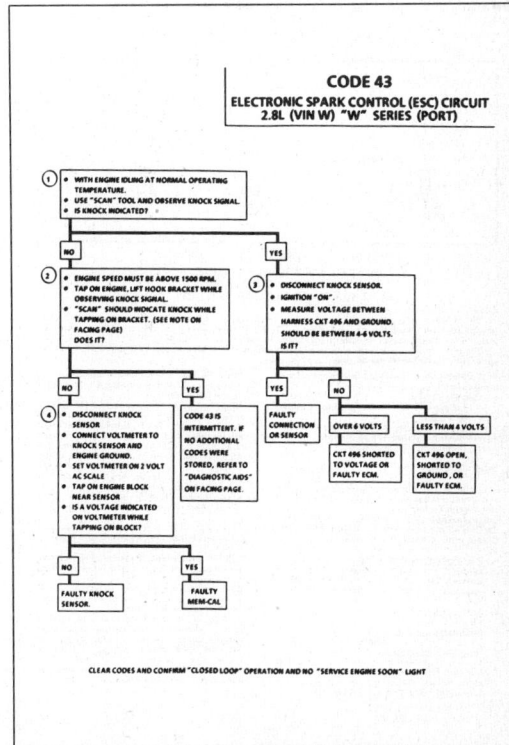

CLEAR CODES AND CONFIRM "CLOSED LOOP" OPERATION AND NO "SERVICE ENGINE SOON" LIGHT

1988–90 2.8L (VIN W)
REGAL AND CUTLASS SUPREME

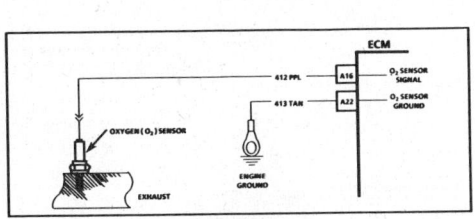

CODE 44
OXYGEN SENSOR CIRCUIT
(LEAN EXHAUST INDICATED)
2.8L (VIN W) "W" SERIES (PORT)

Circuit Description:

The ECM supplies a voltage of about .45 volt between terminals "A16" and "A22". (If measured with a 10 megohm digital voltmeter, this may read as low as .32 volt). The O_2 sensor varies the voltage within a range of about 1 volt if the exhaust is rich, down through about .10 volt if exhaust is lean.

The sensor is like an open circuit and produces no voltage when it is below about 315°C (600°F). An open sensor circuit or cold sensor causes "Open Loop" operation.

Test Description: Numbers below refer to circled numbers on the diagnostic chart.

Code 44 will be set if.
• Voltage on CKT 412 remains below .2 volt for 60 seconds or more.
• The system is operating in "Closed Loop".
• No Code 33 or 34.

Diagnostic Aids:

Using the "Scan", observe the block learn values at different rpm and air flow conditions. The "Scan" also displays the block cells, so the block learn values can be checked in each of the cells to determine when the code 44 may have been met. If the conditions for Code 44 exists the block learn values will be around 150.

• O_2 Sensor Wire Sensor pigtail may be mispositioned and contacting the exhaust manifold.
• Check for intermittent ground in wire between connector and sensor.

• Lean Injector(s) Perform injector balance test CHART C-2A.
• Fuel Contamination Water, even in small amounts, near the in-tank fuel pump inlet can be delivered to the injectors. The water causes a lean exhaust and can set a Code 44.
• Fuel Pressure System will be lean if pressure is too low. It may be necessary to monitor fuel pressure while driving the car at various road speeds and/or loads to confirm. See "Fuel System Diagnosis" CHART A-7.
• Exhaust Leaks If there is an exhaust leak, the engine can cause outside air to be pulled into the exhaust and past the sensor. Vacuum or crankcase leaks can cause a lean condition.
If the above are OK, it is a faulty oxygen sensor.
• MAF Sensor A mass air flow (MAF) sensor output that causes the ECM to sense a lower than normal air flow will cause the system to go lean. Disconnect the MAF sensor and if the lean condition is gone, check for Code 34.

1988–90 2.8L (VIN W)
REGAL AND CUTLASS SUPREME

CODE 44
OXYGEN SENSOR CIRCUIT
(LEAN EXHAUST INDICATED)
2.8L (VIN W) "W" SERIES (PORT)

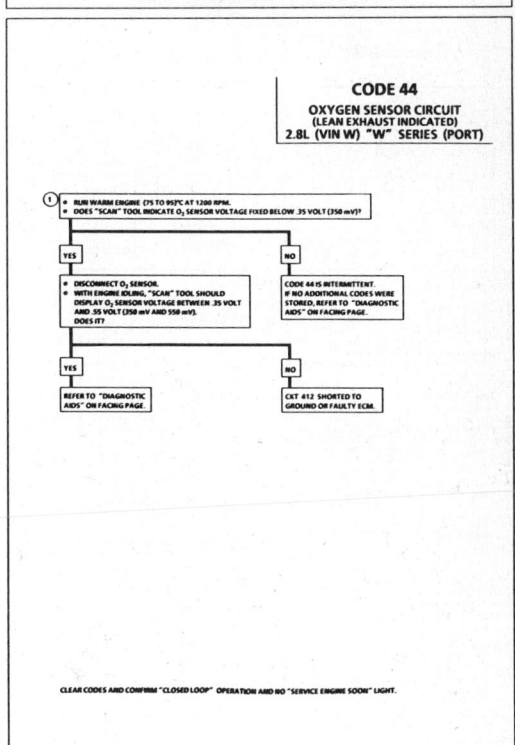

CLEAR CODES AND CONFIRM "CLOSED LOOP" OPERATION AND NO "SERVICE ENGINE SOON" LIGHT.

1988–90 2.8L (VIN W) REGAL AND CUTLASS SUPREME

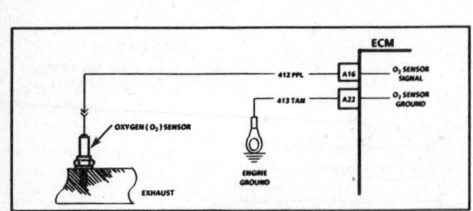

CODE 45
OXYGEN SENSOR CIRCUIT
(RICH EXHAUST INDICATED)
2.8L (VIN W) "W" SERIES (PORT)

Circuit Description:

The ECM supplies a voltage of about .45 volt between terminals "A16" and "A22". (If measured with a 10 megohm digital voltmeter, this may read as low as .32 volt.) The O₂ sensor varies the voltage within a range of about 1 volt if the exhaust is rich, down through about .10 volt if exhaust is lean.

The sensor is like an open circuit and produces no voltage when it is below about 315°C (600°F). An open sensor circuit or cold sensor causes "Open Loop" operation.

Test Description: Numbers below refer to circled numbers on the diagnostic chart.

Code 45 will set if:
- Voltage on CKT 412 remains above .7 volt for 30 seconds.
- Engine time after start is 1 minute or more.
- Throttle angle between 3% and 45%.
- Operation is in "Closed Loop".

Diagnostic Aids:

Using the "Scan", observe the block learn values at different rpm and air flow conditions. The "Scan" also displays the block cells, so the block learn values can be checked in each of the cells to determine when the Code 45 may have been set. If the conditions for Code 45 exists, the block learn values will be around 115.

- **Fuel Pressure**. System will go rich if pressure is too high. The ECM can compensate for some increase. However, if it gets too high, a Code 45 may be set. See fuel system diagnosis CHART A-7.
- **Rich Injector**. Perform injector balance test CHART C-2A.
- **Leaking Injector** See CHART A-7.
- Check for fuel contaminated oil.
- **O₂ Sensor Contamination** Inspect oxygen sensor for silicone contamination from fuel, or use of improper RTV sealant. The sensor may have a white, powdery coating and result in a high but false signal voltage (rich exhaust indication).

The ECM will then reduce the amount of fuel delivered to the engine, causing a severe surge driveability problem.

- **HEI Shielding** An open ground CKT 453 (ignition system reflow) may result in EMI, or induced electrical "noise". The ECM looks at this "noise" as reference pulses. The additional pulses result in a higher than actual engine speed signal. The ECM then delivers too much fuel, causing system to go rich. Engine tachometer will also show higher than actual engine speed, which can help in diagnosing this problem.
- **Canister Purge** Check for fuel saturation. If full of fuel, check canister control and hoses.
- Check for leaking fuel pressure regulator diaphragm by checking vacuum line to regulator for fuel.
- **TPS** An intermittent TPS output will cause the system to go rich, due to a false indication of the engine accelerating.
- **EGR** An EGR staying open (especially at idle) will cause the O₂ sensor to indicate a rich exhaust, and this could result in a Code 45.
- **MAF Sensor** An output that causes the ECM to sense a higher than normal airflow can cause the system to go rich. Disconnecting the MAF sensor will allow the ECM to set a fixed value for the sensor. Substitute a different MAF sensor if the the rich condition is gone while the sensor is disconnected, check for a Code 34.

1988–90 2.8L (VIN W) REGAL AND CUTLASS SUPREME

CODE 45
OXYGEN SENSOR CIRCUIT
(RICH EXHAUST INDICATED)
2.8L (VIN W) "W" SERIES (PORT)

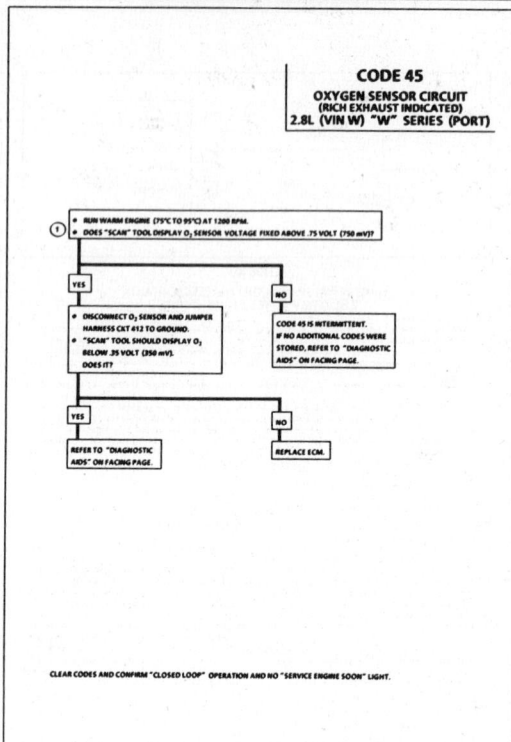

CLEAR CODES AND CONFIRM "CLOSED LOOP" OPERATION AND NO "SERVICE ENGINE SOON" LIGHT.

1988–90 2.8L (VIN W) REGAL AND CUTLASS SUPREME

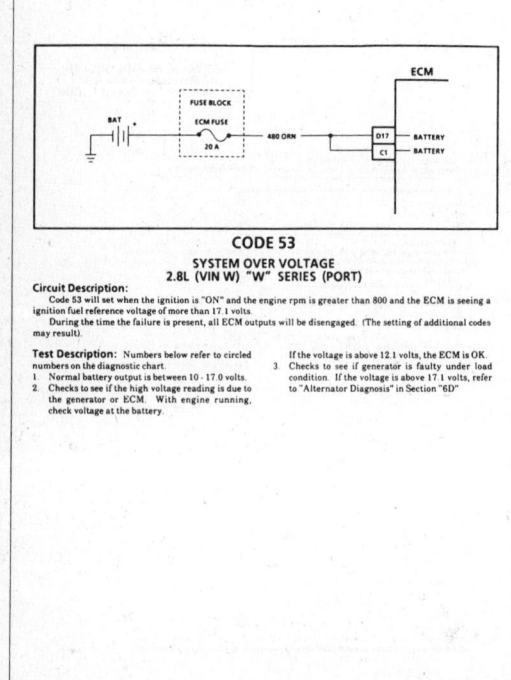

CODE 53
SYSTEM OVER VOLTAGE
2.8L (VIN W) "W" SERIES (PORT)

Circuit Description:

Code 53 will set when the ignition is "ON" and the engine rpm is greater than 800 and the ECM is seeing a ignition fuel reference voltage of more than 17.1 volts.

During the time the failure is present, all ECM outputs will be disengaged. (The setting of additional codes may result).

Test Description: Numbers below refer to circled numbers on the diagnostic chart.

1. Normal battery output is between 10 - 17.0 volts.
2. Checks to see if the high voltage reading is due to the generator or ECM. With engine running, check voltage at the battery.

If the voltage is above 12.1 volts, the ECM is OK.
3. Checks to see if generator is faulty under load condition. If the voltage is above 17.1 volts, refer to "Alternator Diagnosis" in Section "6D".

1988–90 2.8L (VIN W) REGAL AND CUTLASS SUPREME

CODE 53
SYSTEM OVER VOLTAGE
2.8L (VIN W) "W" SERIES (PORT)

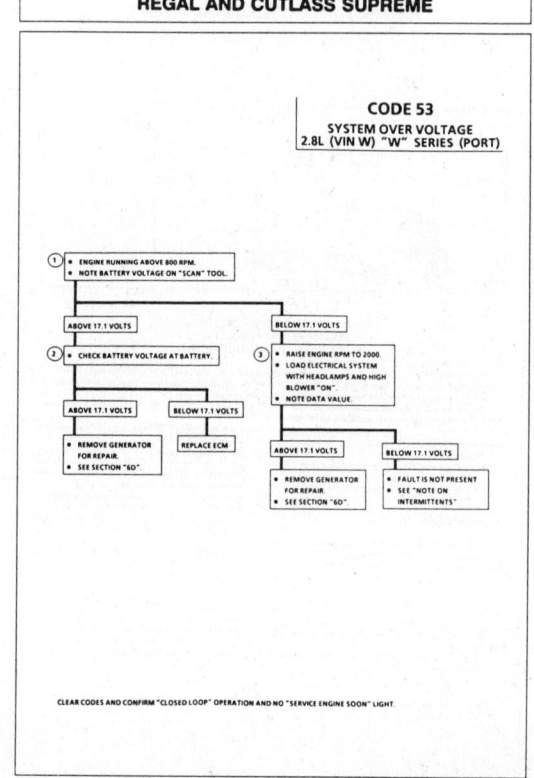

CLEAR CODES AND CONFIRM "CLOSED LOOP" OPERATION AND NO "SERVICE ENGINE SOON" LIGHT.

1988-90 2.8L (VIN W) REGAL AND CUTLASS SUPREME

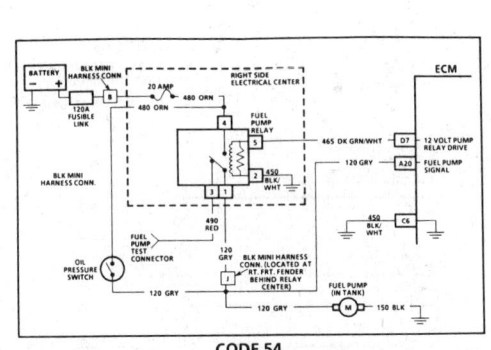

CODE 54
FUEL PUMP CIRCUIT
(LOW VOLTAGE)
2.8L (VIN W) "W" SERIES (PORT)

Circuit Description:

The status of the fuel pump CKT 120 is monitored by the ECM at terminal "A20" and is used to compensate fuel delivery based on system voltage. This signal is also used to store a trouble code if the fuel pump relay is defective or fuel pump voltage is lost while the engine is running. There should be about 12 volts on CKT 120 for 2 seconds after the ignition is turned, or any time references pulses are being received by the ECM.

Code 54 will be set, if the voltage at terminal "A20" is less than 2 volts for 1.5 seconds since the last reference pulse was received. This code is designed to detect a faulty relay, causing extended crank time, and the code will help the diagnosis of an engine that "CRANKS BUT WILL NOT RUN".

If a fault is detected during start-up the "Service Engine Soon" light will stay "ON" until the ignition is cycled "OFF". However, if the voltage is detected below 2 volts with the engine running, the light will only remain "ON" while the condition exists.

CKTs 120 and 480 are routed from the ECM through a mini harness connector to the fuel pump relay. A faulty connection at terminals "B" and "J" at the black connector could cause a Code 54.

1988-90 2.8L (VIN W) REGAL AND CUTLASS SUPREME

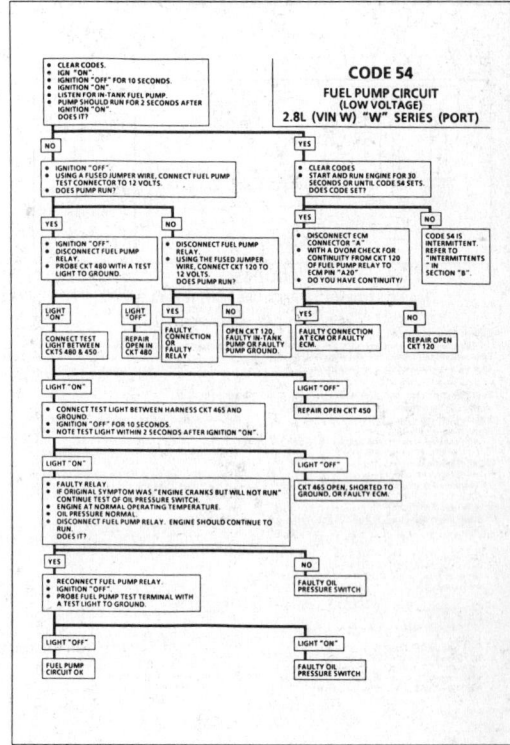

CODE 54
FUEL PUMP CIRCUIT
(LOW VOLTAGE)
2.8L (VIN W) "W" SERIES (PORT)

1988-90 2.8L (VIN W) REGAL AND CUTLASS SUPREME

CODE 51
MEM-CAL ERROR
(FAULTY OR INCORRECT MEM-CAL)
2.8L (VIN W) "W" SERIES (PORT)

CHECK THAT ALL PINS ARE FULLY INSERTED IN THE SOCKET AND THAT MEM-CAL IS PROPERLY LATCHED. IF OK, REPLACE, MEM-CAL, CLEAR MEMORY, AND RECHECK. IF CODE 51 REAPPEARS, REPLACE ECM.

CODE 55
ECM ERROR
2.8L (VIN W) "W" SERIES (PORT)

BE SURE ECM GROUNDS ARE OK AND THAT MEM-CAL IS PROPERLY LATCHED. IF OK REPLACE ELECTRONIC CONTROL MODULE (ECM)

CLEAR CODES AND CONFIRM "CLOSED LOOP" OPERATION AND NO "SERVICE ENGINE SOON" LIGHT.

1988-90 2.8L (VIN W) REGAL AND CUTLASS SUPREME

CODE 61
DEGRADED OXYGEN SENSOR
2.8L (VIN W) "W" SERIES (PORT)

IF A CODE 61 IS STORED IN MEMORY THE ECM HAS DETERMINED THE OXYGEN SENSOR IS CONTAMINATED OR DEGRADED, BECAUSE THE VOLTAGE CHANGE TIME IS SLOW OR SLUGGISH.

THE ECM PERFORMS THE OXYGEN SENSOR RESPONSE TIME TEST WHEN:

COOLANT TEMPERATURE IS GREATER THAN 85°C.

MAT TEMPERATURE IS GREATER THAN 10°C.

IN CLOSED LOOP.

IN DECEL FUEL CUT-OFF MODE.

IF A CODE 61 IS STORED THE OXYGEN SENSOR SHOULD BE REPLACED. A CONTAMINATED SENSOR CAN BE CAUSED BY FUEL ADDITIVES, SUCH AS SILICON, OR BY USE OF NON-GM APPROVED LUBRICANTS OR SEALANTS. SILICON CONTAMINATION IS USUALLY INDICATED BY A WHITE POWDERY SUBSTANCE ON THE SENSOR FINS.

CLEAR CODES AND CONFIRM "CLOSED LOOP" OPERATION AND NO "SERVICE ENGINE SOON" LIGHT.

1988–90 2.8L (VIN W)
REGAL AND CUTLASS SUPREME

CODE 63
MANIFOLD ABSOLUTE PRESSURE (MAP) SENSOR CIRCUIT
(SIGNAL VOLTAGE HIGH - LOW VACUUM)
2.8L (VIN W) "W" SERIES (PORT)

Circuit Description:
The manifold absolute pressure sensor (MAP) responds to changes in manifold pressure (vacuum). The ECM recieves this information as a signal voltage that will vary from about 1/1.5 volts at idle (high vacuum) to 4/4.5 volts at wide open throttle (low vacuum).

Test Description: Numbers below refer to circled numbers on the diagnostic chart.
1. Code 63 will set when:
   - Engine running
   - Manifold pressure greater than 75.3 kPa (A/C "OFF") 81.2 kPa (A/C "ON")
   - Throttle angle less than 2%
   - Conditions met for 2 seconds
   Engine misfire or a low unstable idle may set Code 63.
2. With the MAP sensor disconnected, the ECM should see a low voltage if the ECM and wiring are OK.

Diagnostic Aids:
If idle is rough or unstable, refer to "Symptoms" in Section "B" for items which can cause an unstable idle.
An open in CKT 452 or the connection will result in a Code 63.
Ignition "ON" engine "OFF," voltages should be within the values shown in the table on the chart. Also CHART C-1D can be used to test the MAP sensor. Refer to "Intermittents" in Section "B".

1988–90 2.8L (VIN W)
REGAL AND CUTLASS SUPREME

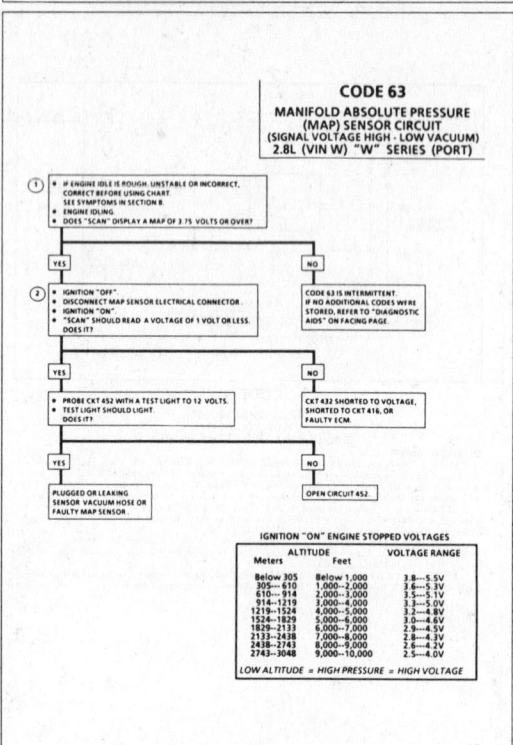

1988–90 2.8L (VIN W)
REGAL AND CUTLASS SUPREME

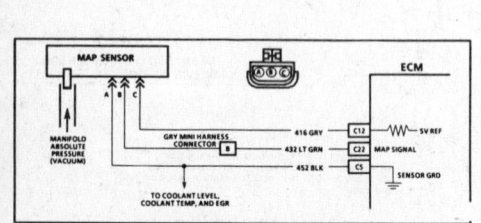

CODE 64
MANIFOLD ABSOLUTE PRESSURE (MAP) SENSOR CIRCUIT
(SIGNAL VOLTAGE LOW - HIGH VACUUM)
2.8L (VIN W) "W" SERIES (PORT)

Circuit Description:
The manifold absolute pressure sensor (MAP) responds to changes in manifold pressure (vacuum). The ECM recieves this information as a signal voltage that will vary from about 1/1.5 volts at idle (high vacuum) to 4/4.5 volts at wide open throttle (low vacuum).

Test Description: Numbers below refer to circled numbers on the diagnostic chart.
1. Code 64 will set if:
   - Engine rpm less than 600
   - Manifold pressure reading less than 13 kPa
   - Conditions met for 1 second
   or
   - Engine rpm greater than 600
   - Throttle angle over 20%
   - Manifold pressure less than 13 kPa
   - Conditions met for 1 second
2. This test to see if the sensor is at fault for the low voltage, or if there is an ECM or wiring problem.

3. This simulates a high signal voltage to check for an open in CKT 432. If the test light is bright during this test, CKT 432 is probably shorted to ground. If "Scan" reads over 4 volts at this test, CKT 416 can be checked by measuring the voltage at terminal "C" (should be 5 volts.)

Diagnostic Aids:
An intermittent open in CKTs 432 or 416 will result in a Code 64.
Ignition "ON" engine "OFF," voltages should be within the values shown in the table on the chart. Also CHART C-1D can be used to test MAP sensor. Refer to "Intermittents" in Section "B".

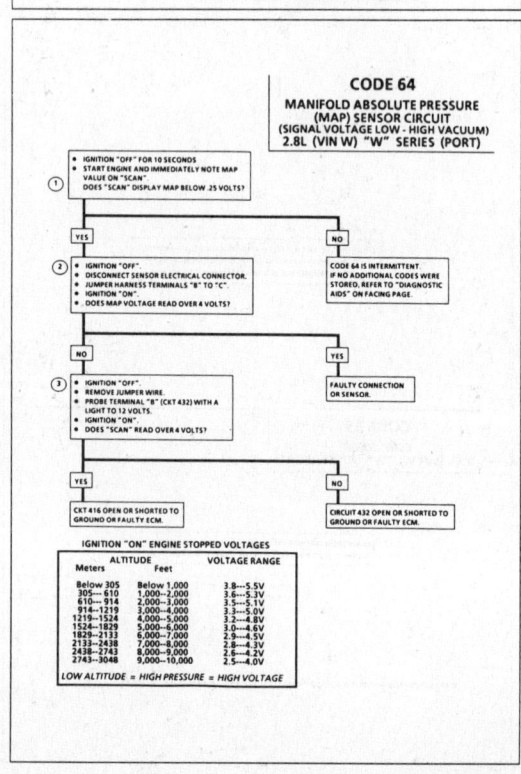

1988–90 2.8L (VIN W)
REGAL AND CUTLASS SUPREME

SECTION B
SYMPTOMS

TABLE OF CONTENTS

Before Starting
Intermittents
Hard Start
Hesitation, Sag, Stumble
Surges and/or Chuggle
Lack of Power, Sluggish, or Spongy
Detonation/Spark Knock
Cuts Out, Misses
Backfire
Poor Fuel Economy
Dieseling, Run-On
Rough, Unstable, or Incorrect Idle, Stalling
Excessive Exhaust Emissions Or Odors
Chart B-1 Restricted Exhaust System Check (all engines)

BEFORE STARTING

Before using this section you should have performed the DIAGNOSTIC CIRCUIT CHECK and found out that:
1. The ECM and "Service Engine Soon" light are operating
2. There are no trouble codes.

Verify the customer complaint, and locate the correct SYMPTOM below. Check the items indicated under that symptom.

If the ENGINE CRANKS BUT WILL NOT RUN, see CHART A-3.

Several of the symptom procedures below call for a careful visual check. This check should include:

- ECM grounds for being clean and tight
- Vacuum hoses for splits, kinks, and proper connections, as shown on Emission Control Information label.
- Air leaks at throttle body mounting and intake manifold.
- Air leaks between MAF sensor and throttle body
- Ignition wires for cracking, hardness, proper routing, and carbon tracking
- Wiring for proper connections, pinches, and cuts.

The importance of this step cannot be stressed too strongly - it can lead to correcting a problem without further checks and can save valuable time.

1988–90 2.8L (VIN W)
REGAL AND CUTLASS SUPREME

INTERMITTENTS

Problem may or may not turn "ON" the "Service Engine Soon" light, or store a code.

DO NOT use the trouble code charts in Section "A" for intermittent problems. The fault must be present to locate the problem. If a fault is intermittent, use of trouble code charts may result in replacement of good parts.

- Most intermittent problems are caused by faulty electrical connections or wiring. Perform careful check as described at start of Section "B." Check for:
  - Poor mating of the connector halves, or terminals not fully seated in the connector body (backed out).
  - Improperly formed or damaged terminals. All connector terminals in problem circuit should be carefully reformed to increase contact tension.
  - Poor terminal to wire connection. This requires removing the terminal from the connector body to check.
- If a visual check does not find the cause of the problem, the car can be driven with a voltmeter connected to a suspected circuit. A "Scan" tool can, also, be used for monitoring input signals to help detect intermittent conditions. An abnormal voltage, or "Scan" reading, when the problem occurs, indicates the problem may be in that circuit. If the wiring and connectors check OK and a trouble code was stored for a circuit having a sensor, except for Codes 43, 44, and 45, substitute a known good sensor and recheck.

An intermittent "Service Engine Soon" light with no stored code may be caused by:
- Ignition coil shorted to ground and arcing at spark plug wires or plugs.
- "Service Engine Soon" light wire to ECM shorted to ground. (CKT 419)
- Diagnostic "Test" terminal wire to ECM, shorted to ground. (CKT 451)
- ECM power grounds. See ECM wiring diagrams.
- Loss of trouble code memory. To check, disconnect TPS and idle engine until "Service Engine Soon" light comes "ON." Code 22 should be stored and kept in memory when ignition is turned "OFF." If not, the ECM is faulty.
- Check for an electrical system interference caused by a defective relay, ECM driven solenoid, or switch. They can cause a sharp electrical surge. Normally, the problem will occur when the faulty component is operated.
- Check for improper installation of electrical options, such as lights, 2-way radios, etc.
- EST wires should be kept away from spark plug wires, coils and generator. Wire from ECM to ignition system (CKT 453) should be a good connection.
- Check for open diode across A/C compressor clutch, and for other open diodes (see wiring diagrams).

HARD START

Definition: Engine cranks OK, but does not start for a long time. Does eventually run, or may start but immediately dies.

- Perform careful check as described at start of Section "B."
- Make sure driver is using correct starting procedure.
- CHECK:
  - TPS for sticking or binding or a high TPS voltage with the throttle closed (should read less than .700 volts).
  - High resistance in coolant sensor circuit or sensor itself. See Code 15 CHART OR with a "Scan" tool compare coolant temperature with ambient temperature on a cold engine.
  - Fuel pressure CHART A-7.
  - Water contaminated fuel.
  - EGR operation. Be sure valve seats properly and is not staying open. See CHART C-7.
- Ignition system - Check for:
  - Bare and shorted wires and proper output with spark tester J-26792 or equivalent (ST-125).
  - IAC operation - See Code 35 chart.
- A faulty in-tank fuel pump check valve will allow the fuel in the lines to drain back to the tank after the engine is stopped. To check for this condition:
  - Perform fuel system diagnosis, CHART A-7.
- Remove spark plugs. Check for wet plugs, cracks, wear, improper gap, burned electrodes, or heavy deposits. Repair or replace as necessary.
- If engine starts and stalls, disconnect MAF sensor. If engine then runs, and sensor connections are OK, replace the sensor.

1988–90 2.8L (VIN W)
REGAL AND CUTLASS SUPREME

HESITATION, SAG, STUMBLE

Definition: Momentary lack of response as the accelerator is pushed down. Can occur at all car speeds. Usually most severe when first trying to make the car move, as from a stop sign. May cause the engine to stall if severe enough.

- Perform careful visual check as described at start of Section "B."
- CHECK:
  - Fuel pressure. See CHART A-7. Also check for water contaminated fuel.
  - Air leaks at air duct between MAF sensor and throttle body.
  - Spark plugs for being fouled or faulty wiring.
  - Mem-Cal number and Service Bulletins for latest Mem-Cal.
  - TPS for binding or sticking. Voltage should increase at a steady rate as throttle is moved toward W.O.T.
  - Generator output voltage. Repair if less than 9 or more than 16 volts.
  - Ignition system ground, CKT 453.
  - Canister purge system for proper operation. See CHART C-3.
  - EGR - See CHART C-7.
  - MAP sensor - See CHART C-1D
- Perform injector balance test, CHART C-2A.

SURGES AND/OR CHUGGLE

Definition: Engine power variation under steady throttle or cruise. Feels like the car speeds up and slows down with no change in the accelerator pedal.

- Be sure driver understands transmission/transaxle converter clutch and A/C compressor operation in Owner's Manual.
- Perform careful visual inspection as described at start of Section "B."
- To help determine if the condition is caused by a rich or lean system, the car should be driven at the speed of the complaint. Monitoring block learn at the complaint speed will help identify the cause of the problem. If the system is running lean (block learn greater than 138), refer to "Diagnostic Aids" on facing page of Code 44. If the system is running rich (block learn less than 118), refer to "Diagnostic Aids" on facing page of Code 45.
- CHECK:
  - Loose or leaking air duct between MAF sensor and throttle body.
  - Generator output voltage. Repair if less than 9 or more than 16 volts.
  - EGR (2.8L). There should be no EGR at idle. See CHART C-7.
  - EGR filter for being plugged (2.8L). See CHART C-7.
  - Vacuum lines for kinks or leaks.
  - In-line fuel filter. Replace if dirty or plugged.
  - Fuel pressure while condition exists. See CHART A-7.
- Remove spark plugs. Check for cracks, wear, improper gap, burned electrodes, or heavy deposits. Also check condition of spark plug wires and check for proper output voltage using spark tester (ST-125) J-26792 or equivalent.

1988–90 2.8L (VIN W)
REGAL AND CUTLASS SUPREME

LACK OF POWER, SLUGGISH, OR SPONGY

Definition: Engine delivers less than expected power. Little or no increase in speed when accelerator pedal is pushed down part way.

- Perform careful visual check as described at start of Section "B."
- Compare customer's car to similar unit. Make sure the customer's car has an actual problem.
- Remove air filter and check air filter for dirt, or for being plugged. Replace as necessary.
- CHECK:
  - For loose or leaking air duct between MAF Sensor and throttle body.
  - Restricted fuel filter, contaminated fuel or improper fuel pressure. See CHART A-7.
  - ECM power grounds, see wiring diagrams.
  - EGR operation for being open or partly open all the time - See CHART C-7.
  - Exhaust system for possible restriction: See CHART B-1.
    - Inspect exhaust system for damaged or collapsed pipes.
    - Inspect muffler for heat distress or possible internal failure.
  - Generator output voltage. Repair if less than 9 or more than 16 volts.
  - Engine valve timing and compression.
  - Engine for proper or worn camshaft.
  - Secondary voltage using a shop ocilliscope or a spark tester J-26792 (ST-125) or equivalent.
  - Check A/C operation. A/C clutch should cut out at WOT. See A/C CHART C-10.

DETONATION /SPARK KNOCK

Definition: A mild to severe ping, usually worse under acceleration. The engine makes sharp metallic knocks that change with throttle opening. Sounds like popcorn popping.

- Check for obvious overheating problems:
  - Low coolant.
  - Loose belt.
  - Restricted air flow to radiator, or restricted water flow through radiator.
  - Inoperative electric cooling fan circuit. See CHART C-12.
- To help determine if the conditions is caused by a rich or lean system, the car should be driven at the speed of the complaint. Monitoring block learn, at the complaint speed, will help identify the cause of the problem. If the system is running lean (block learn greater than 138), refer to diagnostic aids on facing page of Code 44. If the system is running rich (block learn less than 118), refer to facing page of Code 45.
- CHECK:
  - EGR system for not opening or plugged EGR passages - See CHART C-7.
  - ESC system for no retard - See CHART C-5.
  - Park/Neutral switch. Be sure "Scan" indicates drive with gear selector in drive or overdrive. See CHART C-1A.
  - TCC operation, TCC applying too soon - see CHART C-8.
  - Fuel system pressure - See CHART A-7.
  - Remove carbon with top engine cleaner. Follow instructions on can.
    - Check for correct Mem-Cal.
    - Check for excessive oil in combustion chamber
    - Check for incorrect basic engine parts such as cam, heads, pistons, etc.
    - Check for poor fuel quality, proper octane rating

1988–90 2.8L (VIN W)
REGAL AND CUTLASS SUPREME

CUTS OUT, MISSES

Definition: Steady pulsation or jerking that follows engine speed, usually more pronounced as engine load increases. Not normally felt above 1500 rpm or 30 mph (48 km/h). The exhaust has a steady spitting sound at idle or low speed.

- Perform careful visual check as described at start of Section "B".
- Check for missing cylinder by:
  1. Disconnect IAC motor. Start engine. Remove one spark plug wire at a time using insulated pliers.
  2. If there is an rpm drop on all cylinders (equal to within 50 rpm), go to ROUGH, UNSTABLE, OR INCORRECT IDLE, STALLING symptom. Reconnect IAC motor.
  3. If there is no rpm drop on one or more cylinders, or excessive variation in drop, check for spark on the suspected cylinder(s) with J-26792 (ST-125) spark gap tool or equivalent. If no spark, see Section "6D" for Intermittent Operation or Miss. If there is spark, remove spark plug(s) in these cylinders and check for:
     - Cracks
     - Wear
     - Improper Gap
     - Burned Electrodes
     - Heavy Deposits
- Perform compression check on questionable cylinder(s) found above. If compression is low, repair as necessary.

- Disconnect all injector harness connectors. Connect J-34730-2 injector test light or equivalent 6 volt test light between the harness terms. of each injector connector and note light while cranking. If test light fails to blink at any connector, it is a faulty injector drive circuit harness, connector, terminal, or ECM.
- Perform the injector balance test. See CHART C-2A.
- Check spark plug wires by connecting ohmmeter to ends of each wire in question. If meter reads over 30,000 ohms, replace wire(s).
- Spray plug wires with fine water mist to check for shorts.
- Check for restricted fuel filter. Also check fuel tank for water.
- Check for low fuel pressure. See CHART A-7.
- Remove rocker covers. Check for bent pushrods, worn rocker arms, broken valve springs, worn camshaft lobes. Repair as necessary. See Section "6A."
- Check for proper valve timing.
- Check secondary voltage using a shop oscilliscope or a spark tester J-26792 (ST-125) or equivalent.

BACKFIRE

Definition: Fuel ignites in intake manifold, or in exhaust system, making a loud popping noise.

- CHECK:
  - Loose wiring connector or air duct at MAF Sensor.
  - Compression - Look for sticking or leaking valves.
  - EGR operation for being open all the time. See CHART C-7.
  - EGR gasket for faulty or loose fit.
  - Output voltage of ignition coils, using a shop oscilliscope or spark tester J-26792 (ST-125), or equivalent.

  - Valve timing.
  - Spark plugs, spark plug wires, and proper routing of plug wires.
  - AIR check valve (manual trans. "A" Series only).
  - Intermittent condition in ignition system.

1988–90 2.8L (VIN W)
REGAL AND CUTLASS SUPREME

POOR FUEL ECONOMY

Definition: Fuel economy, as measured by an actual road test, is noticeably lower than expected. Also, economy is noticeably lower than it was on this car at one time, as previously shown by an actual road test.

- CHECK:
  - Engine thermostat for faulty part (always open or for wrong heat range. Remove check valve.) Using a "Scan" tool, monitor engine temperature. A "Scan" displays engine temp. in degrees centigrade. After engine is started, the temperature should rise steadily to about 90°C, then stabilize, when thermostat opens.
  - Fuel Pressure. See CHART A-5.
- Check owner's driving habits.
  - Is A/C "ON" full time (Defroster mode "ON")?
  - Are tires at correct pressure?
  - Are excessively heavy loads being carried?
  - Is acceleration too much, too often?
  - Suggest driver read "Important Facts on Fuel Economy" in Owner's Manual.
- Check air cleaner element (filter) for dirt or being plugged.
- Check for proper calibration of speedometer.

- Visually (physically) check:
  - Vacuum hoses for splits, kinks, and proper connections as shown on Vehicle Emission Control Information label.
  - Ignition wires for cracking, hardness, and proper connections.
- Remove spark plugs. Check for cracks, wear, improper gap, burned electrodes, or heavy deposits. Repair or replace as necessary.
- Check compression.
- Check TCC for proper operation. See CHART C-8. A "Scan" should indicate an rpm drop, when the TCC is commanded "ON".
- Suggest owner fill fuel tank and recheck fuel economy.
- Check for exhaust system restriction. See CHART B-1.

DIESELING, RUN-ON

Definition: Engine continues to run after key is turned "OFF," but runs very roughly. If engine runs smoothly, check ignition switch and adjustment.

- Check injectors for leaking. See CHART A-7.

1988–90 2.8L (VIN W)
REGAL AND CUTLASS SUPREME

ROUGH, UNSTABLE, OR INCORRECT IDLE, STALLING

Definition: The engine runs unevenly at idle. If bad enough, the car may shake. Also, the idle may vary in rpm (called "hunting"). Either condition may be bad enough to cause stalling. Engine idles at incorrect speed.

- Perform careful visual check as described at start of Section "B".
- CHECK:
  - Throttle linkage for sticking or binding.
  - TPS for sticking or binding, and be sure output is stable at idle.
  - IAC system. See Code 35 Chart.
  - Generator output voltage. Repair if less than 9 or more than 16 volts.
  - P/N switch circuit. See CHART C-1A, or use "Scan" tool, and be sure tool indicates vehicle is in drive with gear selector in drive (125C), or overdrive (440-T4).
  - Injector balance. See CHART C-2A.
  - PCV valve for proper operation by placing finger over inlet hole in valve and several times. Valve should snap back. If not, replace valve.
  - Evaporative Emission Control System. CHART C-3.
  - Power steering pressure switch input. The state of the switch should only change when wheels are turned up against the stops. See CHART C-1E.
  - Minimum Idle Speed.
  - ECM ground circuits.
  - EGR valve: There should be no EGR at idle.

- Monitoring block learn values may help identify the cause of the problem. If the system is running lean (block learn greater than 138) refer to "Diagnostic Aids" on facing page of Code 44. If the system is running rich (block learn values less than 118) refer to "Diagnostic Aids" on facing page of Code 45.
- Run a cylinder compression check. See Section "6".
- Check for fuel in pressure regulator hose. If present, replace regulator assembly.
- Check ignition system; wires and plugs.
- Check for loose or damaged MAF duct between sensor and throttle body.
- Disconnect MAF sensor and if condition is corrected, replace sensor. "Scan" tool should read about 4-8 grams per second at idle.
- If problem exists with A/C "ON," check A/C system operation CHART C-10.
- M/T "A" series - check AIR system for intermittent air to exhaust ports, while in "Closed Loop." See CHART C-6.
- Inspect oxygen sensor for silicon contamination from fuel, or use of improper RTV sealant. The sensor will have a white, powdery coating, and will result in a high but false signal voltage (rich exhaust indication). The ECM will then reduce the amount of fuel delivered to the engine causing a severe driveability problem.

EXCESSIVE EXHAUST EMISSIONS OR ODORS

Definition: Vehicle fails an emission test. Vehicle has excessive "rotten egg" smell. Excessive odors do not necessarily indicate excessive emissions.

- Perform "Diagnostic Circuit Check."
- IF TEST SHOWS EXCESSIVE CO AND HC, (or also has excessive odors):
  - Check items which cause car to run RICH.
    - Make sure engine is at normal operating temperature.
  - CHECK:
    - Fuel pressure. See CHART A-7.
    - Canister for fuel loading. See CHART C-3.
    - Injector balance. See CHART C-2A.
    - PCV valve for being plugged, stuck, or blocked PCV hose, or fuel in the crankcase. Spark plugs, plug wires, and ignition components.
    - Check for lead contamination of catalytic converter (look for removal of fuel filler neck restrictor).
    - Check for properly installed fuel cap.

- If the system is running rich, (block learn less than 118), refer to "Diagnostic Aids" on facing page of Code 45.
- IF TEST SHOWS EXCESSIVE NOx:
  - Check items which cause car to run LEAN, or to run too hot.
    - EGR valve for not opening. See CHART C-7.
    - Vacuum leaks.
    - Coolant system and coolant fan for proper operation. See CHART C-12.
    - Remove carbon with top engine cleaner. Follow instructions on can.
- If the system is running lean (block learn greater than 138), refer to "Diagnostic Aids" on facing page of Code 44.

1988–90 2.8L (VIN W)
REGAL AND CUTLASS SUPREME

CHART B-1
RESTRICTED EXHAUST SYSTEM CHECK
ALL ENGINES

Proper diagnosis for a restricted exhaust system is essential before any components are replaced. Either of the following procedures may be used for diagnosis, depending upon engine or tool used:

CHECK AT A.I.R. PIPE:

1. Remove the rubber hose at the exhaust manifold A.I.R. pipe check valve. Remove check valve.
2. Connect a fuel pump pressure gauge to a hose and nipple from a Propane Enrichment Device (J26911) (see illustration).
3. Insert the nipple into the exhaust manifold A.I.R. pipe.

OR CHECK AT O₂ SENSOR:

1. Carefully remove O₂ sensor.
2. Install Borroughs exhaust backpressure tester (BT 8515 or BT 8603) or equivalent in place of O₂ sensor (see illustration).
3. After completing test described below, be sure to coat threads of O₂ sensor with anti-seize compound P/N 5613695 or equivalent prior to re-installation.

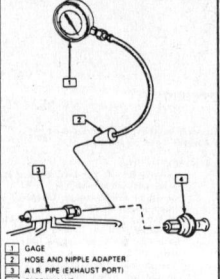

1	GAGE
2	HOSE AND NIPPLE ADAPTER
3	A.I.R. PIPE (EXHAUST PORT)
4	CHECK VALVE

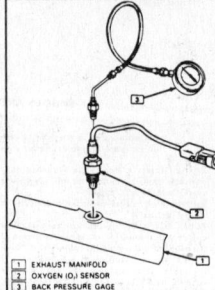

1	EXHAUST MANIFOLD
2	OXYGEN (O₂) SENSOR
3	BACK PRESSURE GAGE

DIAGNOSIS:

1. With the engine idling at normal operating temperature, observe the exhaust system backpressure reading on the gauge. Reading should not exceed 8.6 kPa (1.25 psi).
2. Increase engine speed to 2000 rpm and observe gauge. Reading should not exceed 20.7 kPa (3 psi).
3. If the backpressure at either speed exceeds specification, a restricted exhaust system is indicated.
4. Inspect the entire exhaust system for a collapsed pipe, heat distress, or possible internal muffler failure.
5. If there are no obvious reasons for the excessive backpressure, the catalytic converter is suspected to be restricted and should be replaced using current recommended procedures.

1988–90 2.8L (VIN W)
REGAL AND CUTLASS SUPREME

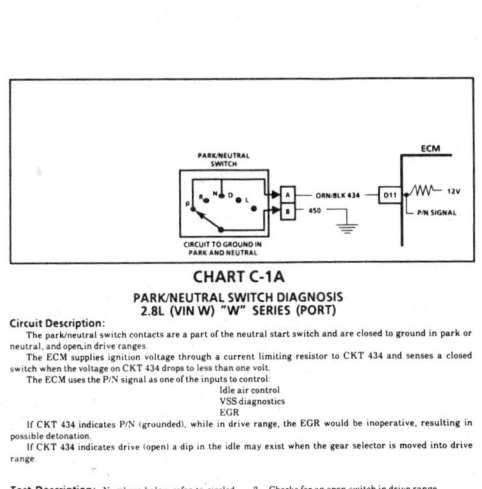

CHART C-1A
PARK/NEUTRAL SWITCH DIAGNOSIS
2.8L (VIN W) "W" SERIES (PORT)

Circuit Description:
The park/neutral switch contacts are a part of the neutral start switch and are closed to ground in park or neutral, and open in drive ranges.

The ECM supplies ignition voltage through a current limiting resistor to CKT 434 and senses a closed switch when the voltage on CKT 434 drops to less than one volt.

The ECM uses the P/N signal as one of the inputs to control:
- Idle air control
- VSS diagnostics
- EGR

If CKT 434 indicates P/N (grounded), while in drive range, the EGR would be inoperative, resulting in possible detonation.

If CKT 434 indicates drive (open) a dip in the idle may exist when the gear selector is moved into drive range.

Test Description: Numbers below refer to circled numbers on the diagnostic chart.
1. Checks for a closed switch to ground in park position. Different makes of "Scan" tools will read P/N differently. Refer to tool operator's manual for type of display used for a specific tool.
2. Checks for an open switch in drive range.
3. Be sure "Scan" indicates drive, even while wiggling shifter, to test for an intermittent or misadjusted switch in drive or overdrive range.

1988–90 2.8L (VIN W)
REGAL AND CUTLASS SUPREME

CHART C-1A
PARK/NEUTRAL SWITCH DIAGNOSIS
2.8L (VIN W) "W" SERIES (PORT)

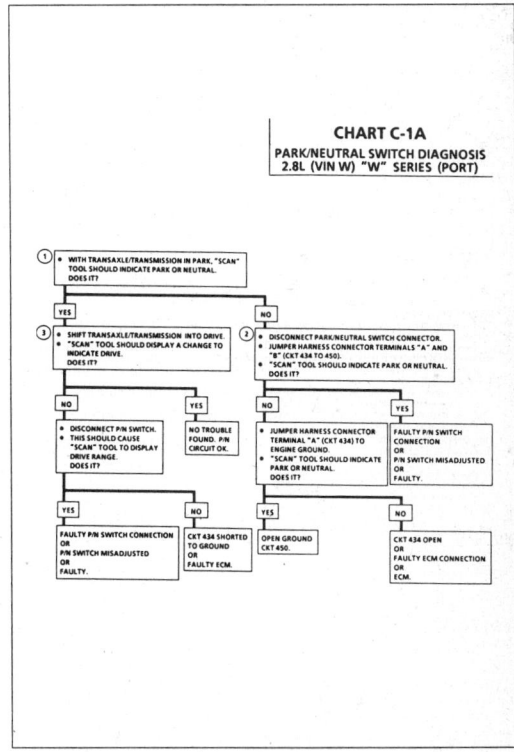

1988–90 2.8L (VIN W)
REGAL AND CUTLASS SUPREME

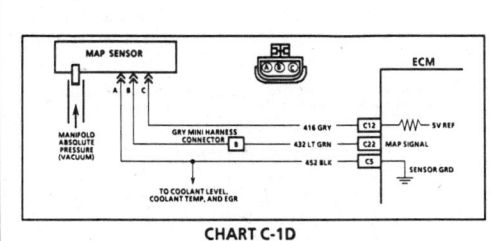

CHART C-1D
MANIFOLD ABSOLUTE PRESSURE (MAP) OUTPUT CHECK
2.8L (VIN W) "W" SERIES (PORT)

Circuit Description:
The manifold absolute pressure (MAP) sensor measures manifold pressure (vacuum) and sends that signal to the ECM. The MAP sensor is mainly used for fuel calculation when the ECM is running in the throttle body backup mode. The MAP sensor is also used to determine the barometric pressure and to help calculate fuel delivery.

Test Description: Numbers below refer to circled numbers on the diagnostic chart.
1. Checks MAP sensor output voltage to the ECM. This voltage, without engine running, represents a barometer reading to the ECM.
2. Applying 34 kPa (10 inches Hg) vacuum to the MAP sensor should cause the voltage to be 1.2 volts less than the voltage at Step 1. Upon applying vacuum to the sensor, the change in voltage should be instantaneous. A slow voltage change indicates a faulty sensor.

The engine must be running in this step or the "scanner" will not indicate a change in voltage. It is normal for the "Service Engine Soon" light to come "ON" and for the system to set a Code 63 during this step. Make sure the code is cleared when this test is completed.

3. Check vacuum hose to sensor for leaking or restriction. Be sure no other vacuum devices are connected to the MAP hose.

1988–90 2.8L (VIN W)
REGAL AND CUTLASS SUPREME

CHART C-1D
MANIFOLD ABSOLUTE PRESSURE (MAP) OUTPUT CHECK
2.8L (VIN W) "W" SERIES (PORT)

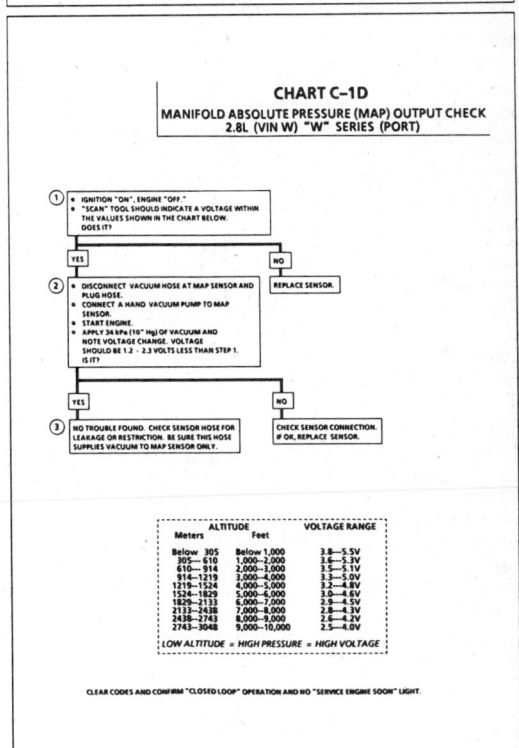

ALTITUDE		VOLTAGE RANGE
Meters	Feet	
Below 305	Below 1,000	3.8–5.5V
305– 610	1,000–2,000	3.6–5.3V
610– 914	2,000–3,000	3.5–5.1V
914–1219	3,000–4,000	3.3–5.0V
1219–1524	4,000–5,000	3.2–4.8V
1524–1829	5,000–6,000	3.0–4.6V
1829–2133	6,000–7,000	2.9–4.5V
2133–2438	7,000–8,000	2.8–4.3V
2438–2743	8,000–9,000	2.6–4.2V
2743–3048	9,000–10,000	2.5–4.0V

LOW ALTITUDE = HIGH PRESSURE = HIGH VOLTAGE

CLEAR CODES AND CONFIRM "CLOSED LOOP" OPERATION AND NO "SERVICE ENGINE SOON" LIGHT.

1988–90 2.8L (VIN W)
REGAL AND CUTLASS SUPREME

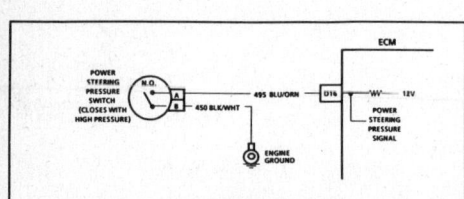

CHART C-1E
POWER STEERING PRESSURE SWITCH (PSPS) DIAGNOSIS
2.8L (VIN W) "W" SERIES (PORT)

Circuit Description:

The power steering pressure switch is normally open to ground, and CKT 495 will be near the battery voltage.

Turning the steering wheel increases power steering oil pressure and its load on an idling engine. The pressure switch will close before the load can cause an idle problem.

Closing the switch causes CKT 495 to read less than 1 volt. The ECM will increase the idle air rate and disengage the A/C relay.
- A pressure switch that will not close, or an open CKT 495 or 450, may cause the engine to stop when power steering loads are high.
- A switch that will not open, or a CKT 495 shorted to ground, may affect idle quality and will cause the A/C relay to be de-energized.

Test Description: Numbers below refer to circled numbers on the diagnostic chart.
1. Different makes of "Scan" tools may display the state of this switch in different ways. Refer to "Scan" tool operator's manual to determine how this input is indicated.
2. Checks to determine if CKT 495 is shorted to ground.
3. This should simulate a closed switch.

1988–90 2.8L (VIN W)
REGAL AND CUTLASS SUPREME

CHART C-1E
POWER STEERING PRESSURE SWITCH (PSPS) DIAGNOSIS
2.8L (VIN W) "W" SERIES (PORT)

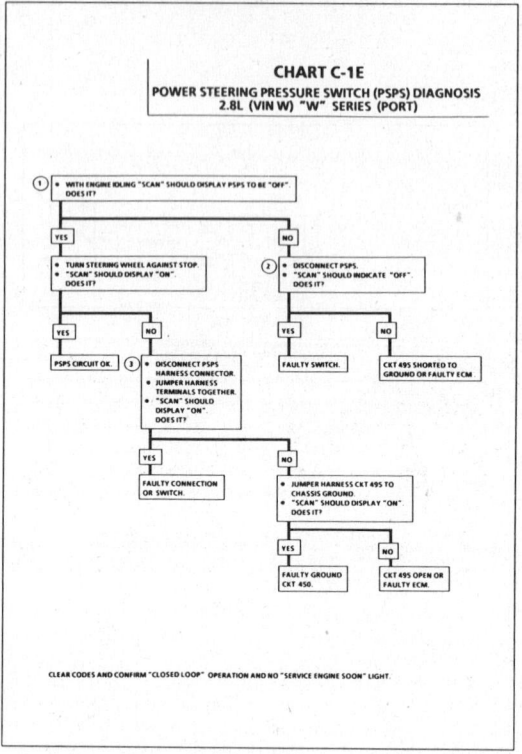

CLEAR CODES AND CONFIRM "CLOSED LOOP" OPERATION AND NO "SERVICE ENGINE SOON" LIGHT.

1988–90 2.8L (VIN W)
REGAL AND CUTLASS SUPREME

CHART C-2A
INJECTOR BALANCE TEST

The injector balance tester is a tool used to turn the injector on for a precise amount of time, thus spraying a measured amount of fuel into the manifold. This causes a drop in fuel rail pressure that we can record and compare between each injector. All injectors should have the same amount of pressure drop (± 10 kPa). Any injector with a pressure drop that is 10 kPa (or more) greater or less than the average drop of the other injectors should be considered faulty and replaced.

STEP 1

Engine "cool down" period (10 minutes) is necessary to avoid irregular readings due to "Hot Soak" fuel boiling. With ignition "OFF" connect fuel gauge J347301 or equivalent to fuel pressure tap. Wrap a shop towel around fitting while connecting gage to avoid fuel spillage.

Disconnect harness connectors at all injectors, and connect injector tester J-34730-3, or equivalent, to one injector. On Turbo equipped engines, use adaptor harness furnished with injector tester to energize injectors that are not accessible. Follow manufacturers instructions for use of adaptor harness. Ignition must be "OFF" at least 10 seconds to complete ECM shutdown cycle. Fuel pump should run about 2 seconds after ignition is turned "ON". At this point, insert clear tubing attached to vent valve into a suitable container and bleed air from gauge and hose to insure accurate gauge operation. Repeat this step until all air is bled from gauge.

STEP 2

Turn ignition "OFF" for 10 seconds and then "ON" again to get fuel pressure to its maximum. Record this initial pressure reading. Energize tester one time and note pressure drop at its lowest point (Disregard any slight pressure increase after drop hits low point). By subtracting this second pressure reading from the initial pressure, we have the actual amount of injector pressure drop.

STEP 3

Repeat step 2 on each injector and compare the amount of drop. Usually, good injectors will have virtually the same drop. Retest any injector that has a pressure difference of 10 kPa, either more or less than the average of the other injectors on the engine. Replace any injector that also fails the retest. If the pressure drop of all injectors is within 10 kPa of this average, the injectors appear to be flowing properly. Reconnect them and review Symptoms, Section "B".

NOTE: The entire test should not be repeated more than once without running the engine to prevent flooding. (This includes any retest on faulty injectors.)

1988–90 2.8L (VIN W)
REGAL AND CUTLASS SUPREME

NOTE: The fuel pressure test in Section "A", CHART A-7, should be completed prior to this test.

CHART C-2A
INJECTOR BALANCE TEST
2.8L "W" SERIES (PORT)

Step 1. If engine is at operating temperature, allow a 10 minute "cool down" period then connect fuel pressure gauge and injector tester.
1. Ignition "OFF".
2. Connect fuel pressure gauge and injector tester.
3. Ignition "ON".
4. Bleed off air in gauge. Repeat until all air is bled from gauge.

Step 2. Run test:
1. Ignition "OFF" for 10 seconds.
2. Ignition "ON". Record gauge pressure. (Pressure must hold steady, if not see the Fuel System diagnosis, Chart A-7, in Section "A")
3. Turn injector on, by depressing button on injector tester, and note pressure at the instant the gauge needle stops.

Step 3.
1. Repeat step 2 on all injectors and record pressure drop on each.
   Retest injectors that appear faulty (Any injectors that have a 10 kPa difference, either more or less, in pressure from the average). If no problem is found, review "Symptoms" Section "B".

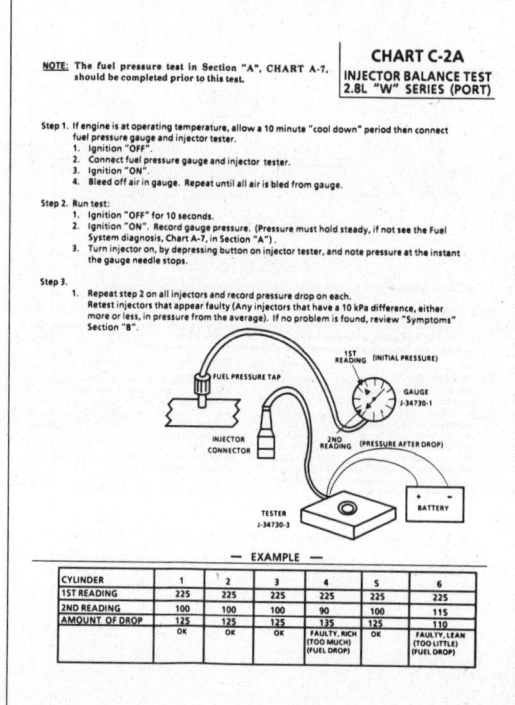

— EXAMPLE —

CYLINDER	1	2	3	4	5	6
1ST READING	225	225	225	225	225	225
2ND READING	100	100	100	90	100	115
AMOUNT OF DROP	125	125	125	135	125	110
	OK	OK	OK	FAULTY, RICH (TOO MUCH) (FUEL DROP)	OK	FAULTY, LEAN (TOO LITTLE) (FUEL DROP)

1988–90 2.8L (VIN W)
REGAL AND CUTLASS SUPREME

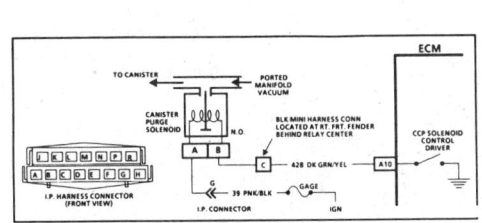

CHART C-3
CANISTER PURGE VALVE CHECK
2.8L (VIN W) "W" SERIES (PORT)

Circuit Description:
Canister purge is controlled by a solenoid that allows manifold vacuum to purge the canister when de-energized. The ECM supplies a ground to energize the solenoid (purge "OFF"). The purge solenoid control by the ECM is pulse width modulated (turned "ON" and "OFF" several times a second). The duty cycle (pulse width) is determined by the amount of air flow, and the engine vacuum as determined by the MAP sensor input. The duty cycle is calculated by the ECM and the output commanded when the following conditions have been met:

- Engine run time after start more than 3 minutes.
- Coolant temperature above 80°C.
- Vehicle speed above 15 mph.
- Throttle off idle (about 3%).

Also, if the diagnostic test terminal is grounded with the engine stopped, the purge solenoid is de-energized (purge "ON").

Test Description: Numbers below refer to the circled numbers on the diagnostic chart.
1. Checks to see if the solenoid is opened or closed. The solenoid is normally energized in this step, so it should be closed.
2. Checks for a complete circuit. Normally there is ignition voltage on CKT 39 and the ECM provides a ground on CKT 428.

3. Completes functional check by grounding test terminal. This should normally de-energize the solenoid opening the valve which should allow the vacuum to drop (purge "ON").

1988–90 2.8L (VIN W)
REGAL AND CUTLASS SUPREME

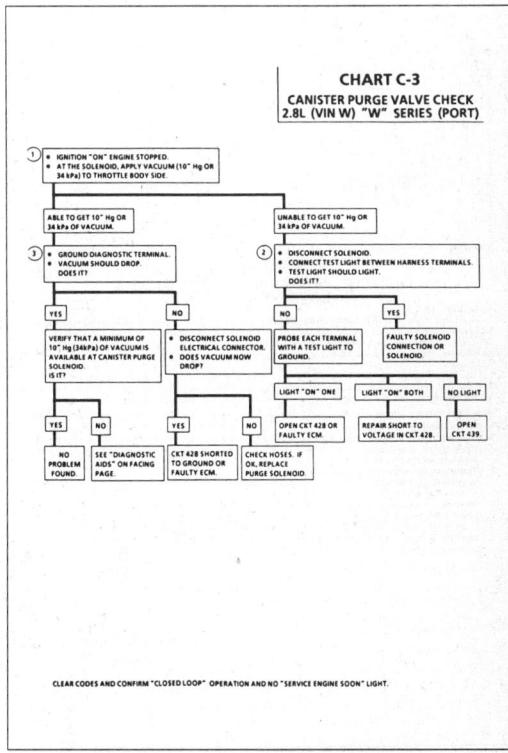

CHART C-3
CANISTER PURGE VALVE CHECK
2.8L (VIN W) "W" SERIES (PORT)

CLEAR CODES AND CONFIRM "CLOSED LOOP" OPERATION AND NO "SERVICE ENGINE SOON" LIGHT.

1988–90 2.8L (VIN W)
REGAL AND CUTLASS SUPREME

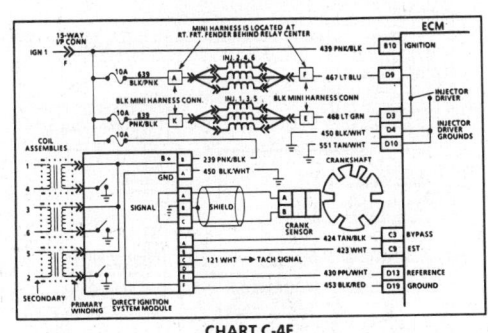

CHART C-4F
"DIS" MISFIRE
2.8L (VIN W) "W" SERIES (PORT)

Circuit Description:
The "Direct Ignition System" (DIS) uses a waste spark method of distribution. In this type of system, the ignition module triggers the #1/4 coil pair resulting in both #1 and #4 spark plugs firing at the same time. #1 cylinder is on the compression stroke at the same time #4 is on the exhaust stroke, resulting in a lower energy requirement to fire #4 spark plug. This leaves the remainder of the high voltage to be used to fire #1 spark plug. On this application, the crank sensor is mounted to the engine block and protrudes through the block to within approximately .050" of the crankshaft reluctor. Since the reluctor is a machined portion of the crankshaft and the crank sensor is mounted in a fixed position on the block, timing adjustments are not possible or necessary.

Test Description: Numbers below refer to circled numbers on the diagnostic chart.
1. Checks for voltage output of ignition system. The spark tester must be used, as this tool requires 25,000 volts to trigger. This checks for a potential weak coil.
2. If the spark tester fires on all wires, the ignition system, with the exception of the spark plugs, may be considered in good working order. If the spark plugs show no evidence of wear, damage or fouling, an engine mechanical fault should be suspected. Refer to Section "B", "Cuts out, Misses".
3. If the spark jumps the tester gap after grounding the opposite plug wire, it indicates excessive resistance in the plug which was bypassed.

A faulty or poor connection at that plug could also result in the miss condition. Also check for carbon deposits inside the spark plug boot.
4. If carbon tracking is evident replace coil and be sure plug wires relating to that coil are clean and tight. Excessive wire resistance or faulty connections could have caused the coil to be damaged.
5. If the no spark condition follows the suspected coil, that coil is faulty. Otherwise, the ignition module is the cause of no spark. This test could also be performed by substituting a known good coil for the one causing the no spark condition.

1988–90 2.8L (VIN W)
REGAL AND CUTLASS SUPREME

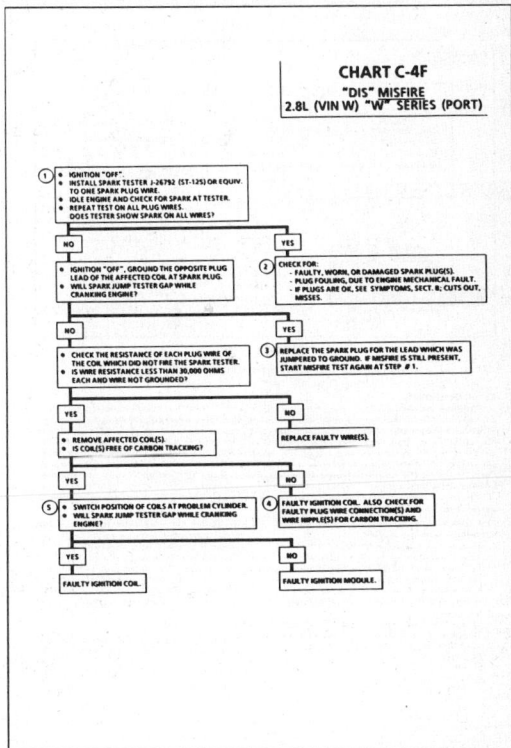

CHART C-4F
"DIS" MISFIRE
2.8L (VIN W) "W" SERIES (PORT)

1988–90 2.8L (VIN W)
REGAL AND CUTLASS SUPREME

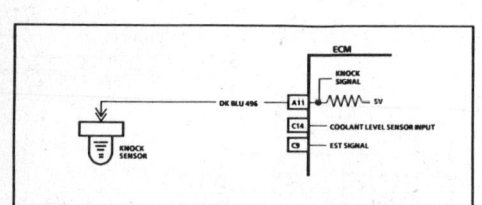

CHART C-5
ELECTRONIC SPARK CONTROL (ESC) SYSTEM CHECK
2.8L (VIN W) "W" SERIES (PORT)

Circuit Description:

The knock sensor is used to detect engine detonation and the ECM will retard the electronic spark timing based on the signal being received. The circuitry within the knock sensor causes the ECM's 5 volts to be pulled down so that under a no knock condition, CKT 496 would measure about 2.5 volts. The knock sensor produces an AC signal which rides on the 2.5 volt DC voltage. The amplitude and frequency are dependent upon the knock level.

The Mem-Cal used with this engine, contains the functions which were part of remotely mounted ESC modules used on other GM vehicles. The ESC portion of the Mem-Cal, then sends a signal to other parts of the ECM which adjusts the spark timing to retard the spark and reduce the detonation.

Test Description: Numbers below refer to circled numbers on the diagnostic chart.

1. With engine idling, there should not be a knock signal present at the ECM, because detonation is not likely under a no load condition.
2. Tapping on the engine lift hood bracket should simulate a knock signal to determine if the sensor is capable of detecting detonation. If no knock is detected, try tapping on engine block closer to sensor before replacing sensor.
3. If the engine has an internal problem which is creating a knock, the knock sensor may be responding to the internal failure.

4. This test determines if the knock sensor is faulty or if the ESC portion of the Mem-Cal is faulty. If it is determined that the Mem-Cal is faulty, be sure that is is properly installed and latched into place. If not properly installed, repair and retest.

Diagnostic Aids:

While observing knock signal on the "Scan," there should be an indication that knock is present when detonation can be heard. Detonation is most likely to occur under high engine load conditions.

1988–90 2.8L (VIN W)
REGAL AND CUTLASS SUPREME

CHART C-5
ELECTRONIC SPARK CONTROL (ESC) SYSTEM CHECK
2.8L (VIN W) "W" SERIES (PORT)

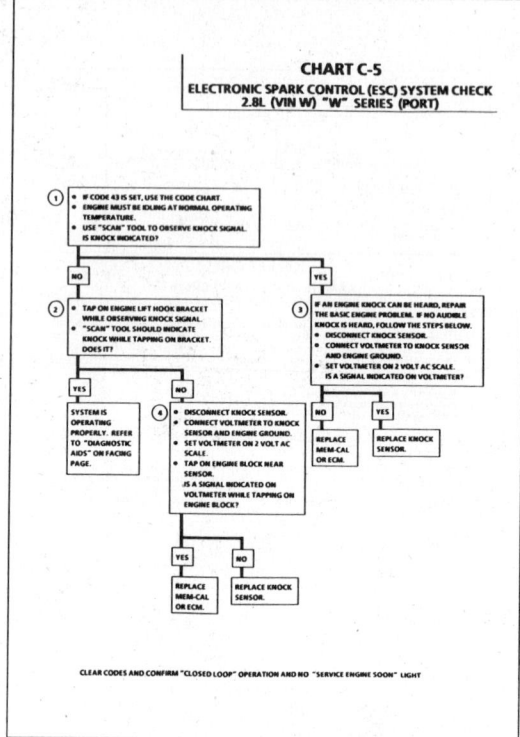

1988–90 2.8L (VIN W)
REGAL AND CUTLASS SUPREME

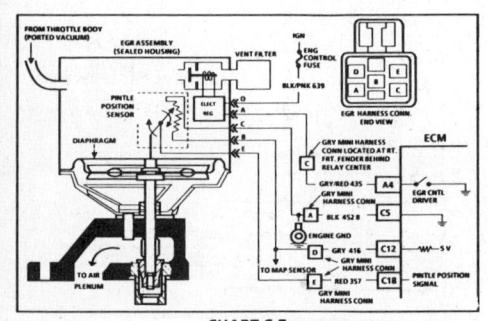

CHART C-7
EXHAUST GAS RECIRCULATION (EGR) VALVE CHECK
2.8L (VIN W) "W" SERIES (PORT)

Circuit Description:

The integrated electronic EGR valve functions similar to a port valve with a remote vacuum regulator. The internal solenoid is normally open, which causes the vacuum signal to be vented "OFF" to the atmosphere when EGR is not being commanded by the ECM. This EGR valve has a sealed cap and the solenoid valve opens and closes the vacuum signal which controls the amount of vacuum vented to atmosphere, and this controls the amount of vacuum applied to the diaphragm. The electronic EGR valve contains a voltage regulator which converts the ECM signal to provide different amounts of EGR flow by regulating the current to the solenoid. The ECM controls EGR flow with a pulse width modualted signal (turns "ON" and "OFF" many times a second) based on airflow, TPS, and rpm.

This system also contains a pintle position sensor which works similar to a TPS sensor, and as EGR flow is increased, the sensor output also increases.

Test Description: Numbers below refer to circled numbers on the diagnostic chart.

1. Whenever the solenoid is denergized, the solenoid valve should be closed which should not allow the vacuum to move the EGR diaphragm. However, if the filter is plugged, the vacuum applied with the hand held vacuum pump will cause the diaphragm to move because the vacuum will not be vented to the atmosphere.
2. This test will determine if the EGR filter is plugged or if the EGR itself is faulty. Use care when removing the filter to avoid damaging the EGR assembly. See "On-Car Service" for procedure.
3. If the valve moves when it's probably due to CKT 435 being shorted to ground.

4. Grounding the diagnostic terminal should energize the solenoid which closes off the vent and allows the vacuum to move the diaphragm.
5. The EGR assembly is designed to have some leak and therefore, 7" of vacuum is all that should be able to be held on the assembly. However, if toomuch of a leak exists (less than 3") the EGR assembly is leaking and must be replaced.

Diagnostic Aids:

The EGR position voltage can be used to determine that the pintle is moving. When no EGR is commanded (0% duty cycle), the position sensor should read between .5 volts and 1.5 volts, and increase with the commanded EGR duty cycle.

1988–90 2.8L (VIN W)
REGAL AND CUTLASS SUPREME

CHART C-7
EXHAUST GAS RECIRCULATION (EGR) VALVE CHECK
2.8L (VIN W) "W" SERIES (PORT)

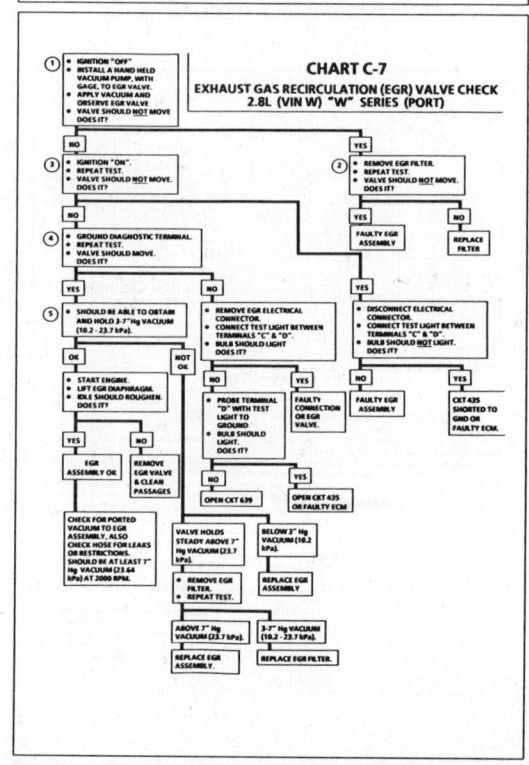

1988–90 2.8L (VIN W)
REGAL AND CUTLASS SUPREME

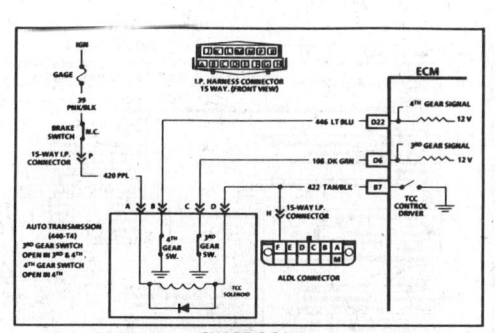

CHART C-8A
(Page 1 of 2)
440-T4 TORQUE CONVERTER CLUTCH (TCC)
(ELECTRICAL DIAGNOSIS)
2.8L (VIN W) "W" SERIES (PORT)

Circuit Description:
The purpose of the automatic transmission torque converter clutch feature is to eliminate the power loss of the torque converter stage when the vehicle is in a cruise condition. This allows the convenience of the automatic transmission and the fuel economy of a manual transmission. The heart of the system is a solenoid located inside the automatic transmission which is controlled by the ECM.

When the solenoid coil is activated ("ON"), the torque converter clutch is applied which results in straight through mechanical coupling from the engine to transmission. When the transmission solenoid is deactivated, the torque converter clutch is released which allows the torque converter to operate in the conventional manner (fluid coupling between engine and transmission).

The TCC will engage on a warm engine under given road load in 3rd and 4th gears.
TCC will engage when:
- Engine warmed up
- Vehicle speed above a calibrated value (about 28 mph 45 km/h)
- Throttle position sensor output not changing, indicating a steady road speed.
- Brake switch closed.

Test Description: Numbers below refer to circled numbers on the diagnostic chart.
1. This test checks the continuity of the TCC circuit from the fuse to the ALDL connector.
2. When the brake pedal is released, the light should come back "ON" and then go "OFF" when the diagnostic terminal is grounded. This tests CKT 422 and the TCC driver in the ECM.

Diagnostic Aids:
The "Scan" tool only indicates when the ECM has turned "ON" the TCC driver, and this does not confirm that the TCC has engaged. To determine if TCC is functioning properly, engine rpm should decrease when the "Scan" indicates the TCC driver has turned "ON".

1988–90 2.8L (VIN W)
REGAL AND CUTLASS SUPREME

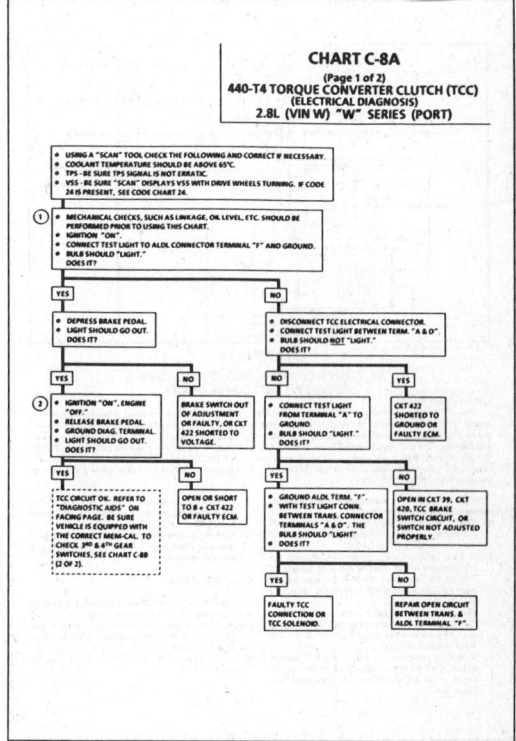

CHART C-8A
(Page 1 of 2)
440-T4 TORQUE CONVERTER CLUTCH (TCC)
(ELECTRICAL DIAGNOSIS)
2.8L (VIN W) "W" SERIES (PORT)

1988–90 2.8L (VIN W)
REGAL AND CUTLASS SUPREME

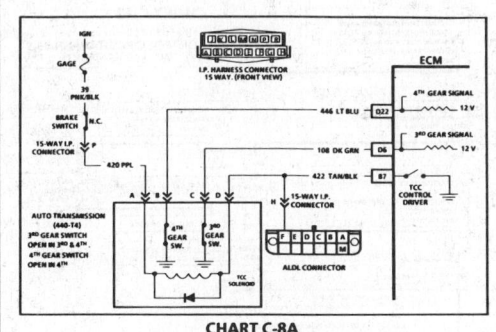

CHART C-8A
(Page 2 of 2)
440-T4 TORQUE CONVERTER CLUTCH (TCC)
(ELECTRICAL DIAGNOSIS)
2.8L (VIN W) "W" SERIES (PORT)

Circuit Description:
The 3rd gear switch in this vehicle is open in 3rd and 4th gear. The ECM uses this signal to disengage the TCC when going into a downshift.
The fourth gear switch is open in fourth gear.

Test Description: Numbers below refer to circled numbers on the diagnostic chart.
1. Some "Scan" tools display the state of these switches in different ways. Be familiar with the type of tool being used. Since both switches should be in the closed state during this test, the tool should read the same for either the 3rd or 4th gear switch.
2. Determines whether the switch or signal circuit is open. The circuit can be checked for an open by measuring the voltage (with a voltmeter) at the TCC connector. Should be about 12 volts.

3. Because the switch(es) should be grounded in this step, disconnecting the TCC connector should cause the "Scan" switch state to change.
4. The switch state should change when the vehicle shifts into 3rd gear.

Diagnostic Aids:
If vehicle is road tested because of a TCC related problem, be sure the switch states do not change while in 4th gear because the TCC will disengage. If switches change state, carefully check wire routing and connections.

1988–90 2.8L (VIN W)
REGAL AND CUTLASS SUPREME

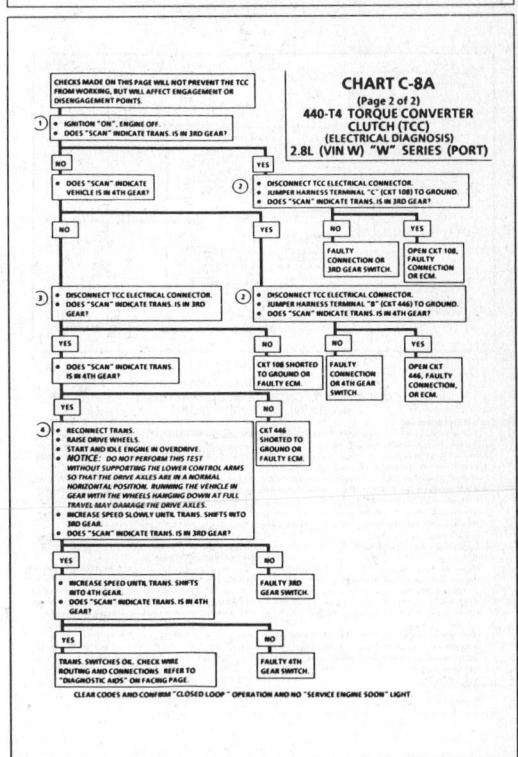

CHART C-8A
(Page 2 of 2)
440-T4 TORQUE CONVERTER CLUTCH (TCC)
(ELECTRICAL DIAGNOSIS)
2.8L (VIN W) "W" SERIES (PORT)

1988–90 2.8L (VIN W) REGAL AND CUTLASS SUPREME

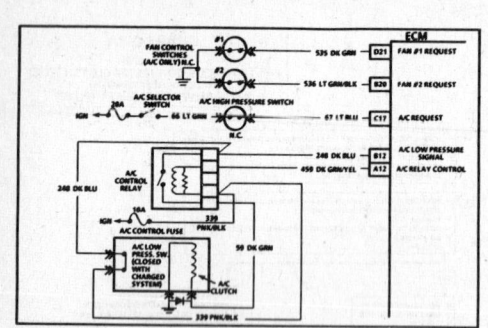

CHART C-10
A/C CLUTCH CONTROL CIRCUIT DIAGNOSIS
2.8L (VIN W) "W" SERIES (PORT)

Circuit Description:

The A/C clutch control relay is ECM controlled to delay A/C clutch engagement about 4 second after A/C is turned "ON". This allows the IAC to adjust engine rpm, before the A/C clutch engages. The ECM, also, causes the relay to disengage the A/C clutch during WOT, when high power steering pressure is present, or if engine is overheating. When the A/C clutch control relay is energized, when the ECM provides a ground path for CKT 459. The low pressure switch will open, if A/C pressure is less than 40 psi (276 kPa). The high pressure switch will open, if A/C pressure exceeds 440 psi (3034 kPa). The A/C pressure fan switch opens, when A/C pressure exceeds about 200 psi (1380 kPa).

Test Description: Numbers below refer to circled numbers on the diagnostic chart.
1. The ECM will only energize the A/C relay, when the engine is running. This test will determine if the relay, or CKT 459, is faulty.
2. In order for the clutch to properly be engaged, the low pressure switch must be closed to provide 12 volts to the relay, and the high pressure switch must be closed, so the A/C request (12 volts) will be present at the ECM.
3. Determines if the signal is reaching the ECM on CKT 66 from the A/C control panel. Signal should only be present when the A/C mode or defrost mode has been selected.
4. A short to ground in any part of the A/C request circuit, CKT 67 to the relay, CKT 59 to the A/C clutch, or the A/C clutch, could be the cause of the blown fuse.

5. If the ECM is seeing a high power steering pressure signal, the A/C clutch will be disengaged by the ECM.
6. With the engine idling and A/C "ON," the ECM should be grounding CKT 459, which should cause the test light to be "ON."

Diagnostic Aids:

If complaint was insufficient cooling, the problem may be caused by a inoperative cooling fan, or A/C pressure fan switch. The engine cooling fan should turn "ON," when A/C pressure exceeds a value to open the switch, which causes the ECM to energize the cooling fan relay. See CHART C-12, for cooling fan diagnosis.

1988–90 2.8L (VIN W) REGAL AND CUTLASS SUPREME

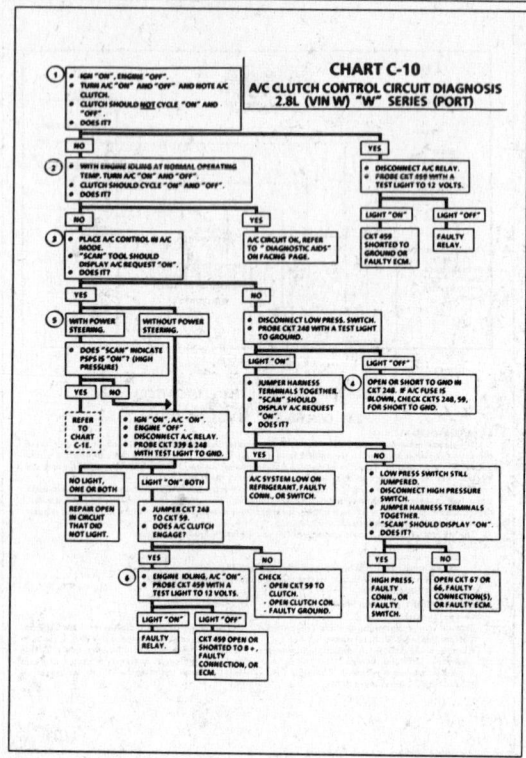

CHART C-10
A/C CLUTCH CONTROL CIRCUIT DIAGNOSIS
2.8L (VIN W) "W" SERIES (PORT)

1988–90 2.8L (VIN W) REGAL AND CUTLASS SUPREME

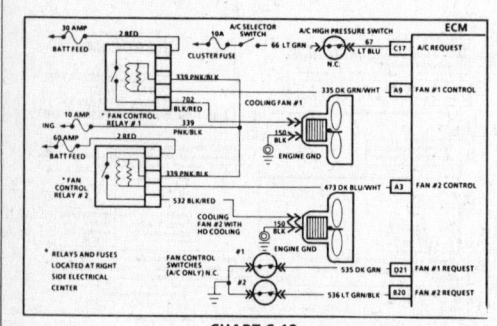

CHART C-12
(Page 1 of 2)
COOLANT FAN CONTROL CIRCUIT DIAGNOSIS
2.8L (VIN W) "W" SERIES (PORT)

Circuit Description:

The primary and secondary electric cooling fan(s) are controlled by the ECM, based on inputs from the coolant temperature sensor, the A/C control switches, vehicle speed and state of the A/C intermediate pressure switches. The ECM controls the fan(s) by grounding CKT 335 and/or CKT 473, which energizes the fan control relay. Battery voltage is then supplied to the fan motor.

The ECM grounds CKT 335 and/or CKT 473, when coolant temperature is over about 106°C (223°F), or when A/C has been requested, and the fan control switch(es) open with high A/C pressure, about 200 psi (1380 kPa). Once the ECM turns the fan "ON", it will keep it "ON" for a minimum of 30 seconds, or until vehicle speed exceeds 70 mph (40 mph for secondary fan).

Also, if Code 14 or 15 sets, the ECM is in throttle body back up, the primary fan will run at all times.

Test Description: Numbers below refer to circled numbers on the diagnostic chart.
1. With the diagnostic terminal grounded, the cooling fan control driver(s) will close, which should energize the fan control relay(s).
2. If the A/C fan control switch or circuit is open, the fan would run whenever A/C is requested.
3. With A/C clutch engaged, the A/C fan control switches should open, when A/C intermediate pressure exceeds about 200 psi (1380 kPa). This signal should cause the ECM to energize the fan control relay(s).

Diagnostic Aids:

If the owner complained of an overheating problem, it must be determined if the complaint was due to an actual boilover, or the hot light, or temp gage indicated over heating.

If the gage, or light, indicates overheating, but no boilover is detected, the gage circuit should be checked. The gage accuracy can, also, be checked by comparing the coolant sensor reading using a "Scan" tool and comparing its reading with the gage reading.

If the engine is actually overheating, and the gage indicates overheating, but the cooling fan is not coming "ON", the coolant sensor has probably shifted out of calibration and should be replaced.

If the engine is overheating, and the cooling fan is "ON", the cooling system should be checked.

1988–90 2.8L (VIN W) REGAL AND CUTLASS SUPREME

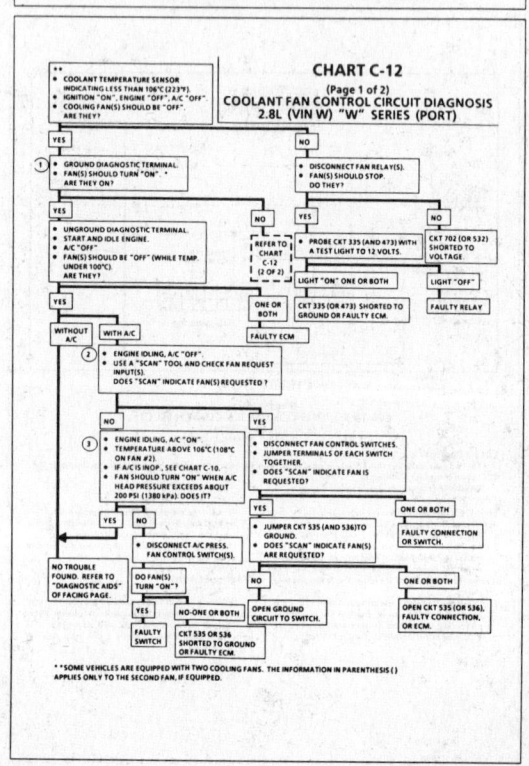

CHART C-12
(Page 1 of 2)
COOLANT FAN CONTROL CIRCUIT DIAGNOSIS
2.8L (VIN W) "W" SERIES (PORT)

1988–90 2.8L (VIN W)
REGAL AND CUTLASS SUPREME

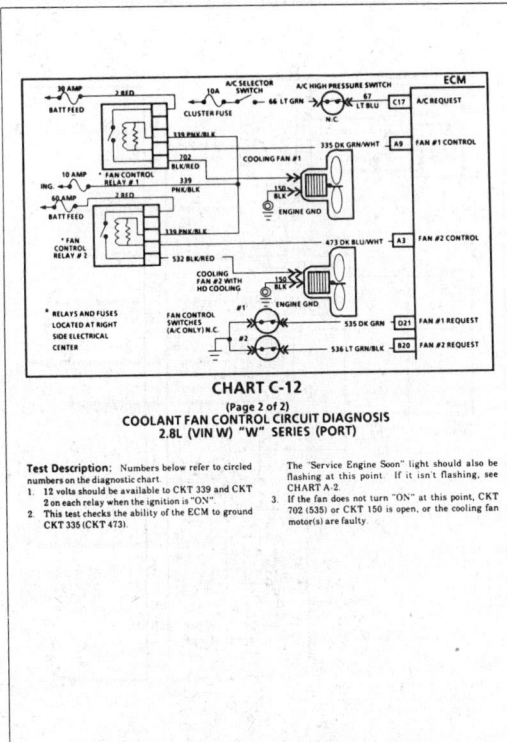

CHART C-12
(Page 2 of 2)
COOLANT FAN CONTROL CIRCUIT DIAGNOSIS
2.8L (VIN W) "W" SERIES (PORT)

Test Description: Numbers below refer to circled numbers on the diagnostic chart.
1. 12 volts should be available to CKT 339 and CKT 2 on each relay when the ignition is "ON".
2. This test checks the ability of the ECM to ground CKT 335 (CKT 473).

The "Service Engine Soon" light should also be flashing at this point. If it isn't flashing, see CHART A-2.
3. If the fan does not turn "ON" at this point, CKT 702 (535) or CKT 150 is open, or the cooling fan motor(s) are faulty.

1988–90 2.8L (VIN W)
REGAL AND CUTLASS SUPREME

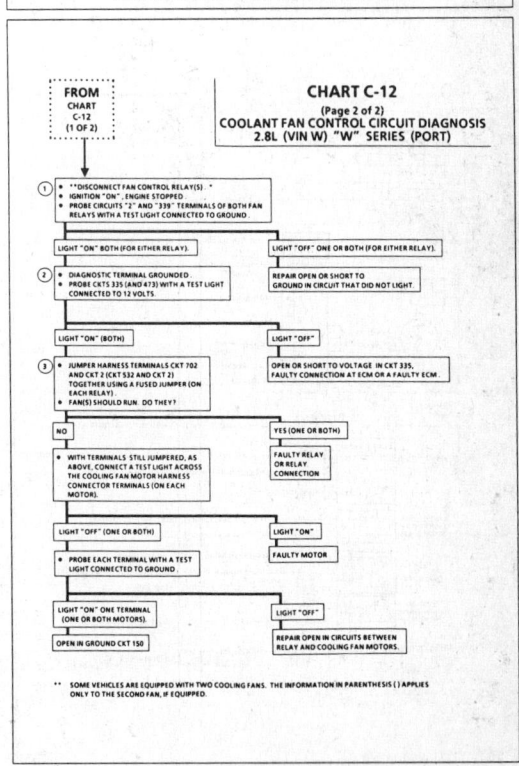

3.0L (VIN L) COMPONENT LOCATIONS
CALAIS AND SKYLARK

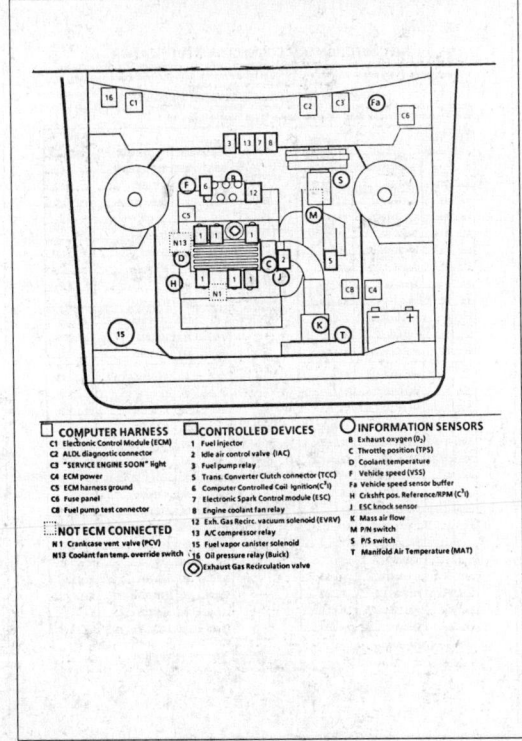

□ **COMPUTER HARNESS**
C1 Electronic Control Module (ECM)
C2 ALDL diagnostic connector
C3 "SERVICE ENGINE SOON" light
C4 ECM power
C5 ECM harness ground
C6 Fuse panel
C8 Fuel pump test connector

□ **NOT ECM CONNECTED**
N1 Crankcase vent valve (PCV)
N13 Coolant fan temp. override switch

□ **CONTROLLED DEVICES**
1 Fuel injector
2 Idle air control valve (IAC)
3 Fuel pump relay
4 Trans. Converter Clutch connector (TCC)
5 Computer Controlled Coil Ignition(C^{3}I)
6 Electronic Spark Control module (ESC)
7 Engine coolant fan relay
12 Exh. Gas Recirc. vacuum solenoid (EVRV)
14 A/C compressor relay
15 Fuel vapor canister solenoid
16 Oil pressure relay (Buick)
◎ Exhaust Gas Recirculation valve

○ **INFORMATION SENSORS**
B Exhaust oxygen (O₂)
C Throttle position (TPS)
D Coolant temperature
E Vehicle speed (VSS)
Fa Vehicle speed sensor buffer
H Crkshft pos. Reference/RPM (C³I)
J ESC knock sensor
K Mass air flow
M P/N switch
S P/S switch
T Manifold Air Temperature (MAT)

3.0L (VIN L) ENGINE COMPONENT LOCATIONS
CALAIS AND SKYLARK

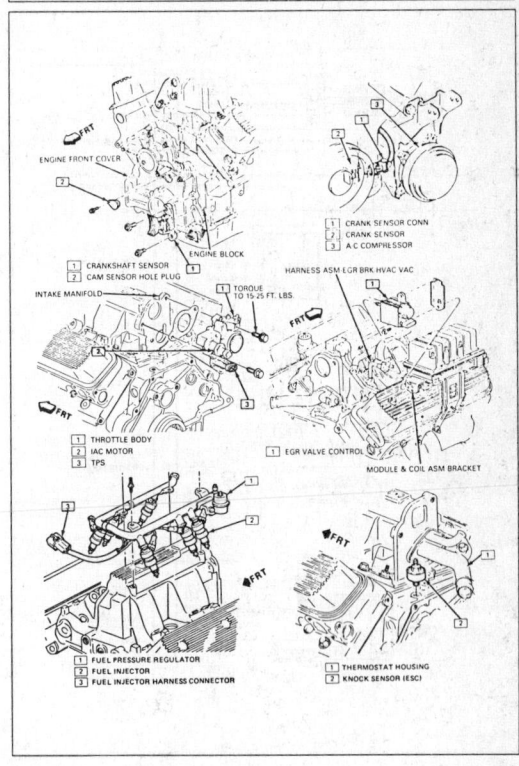

3.0L (VIN L) ECM WIRING DIAGRAM CALAIS AND SKYLARK

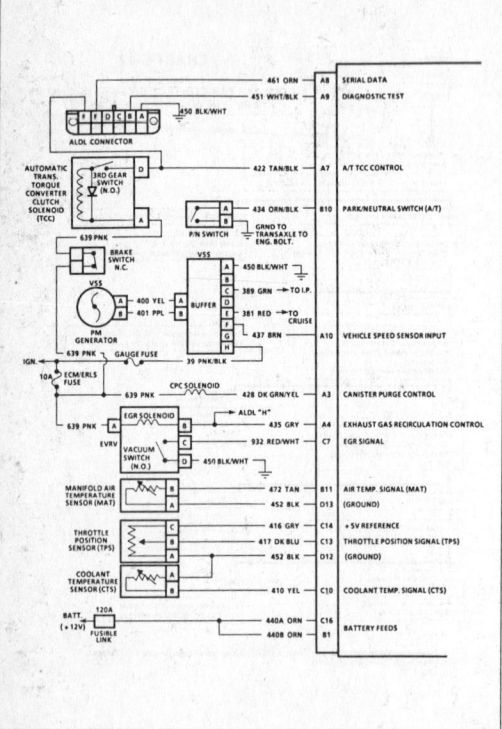

3.0L (VIN L) ECM WIRING DIAGRAM CALAIS AND SKYLARK (CONT.)

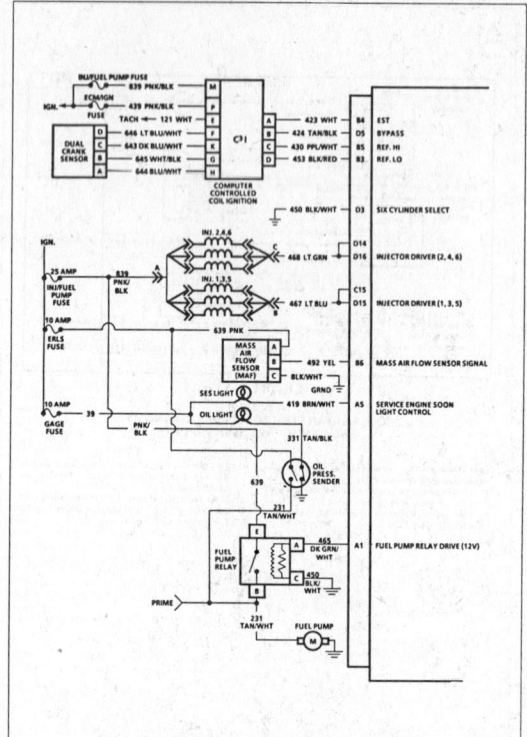

3.0L (VIN L) ECM WIRING DIAGRAM CALAIS AND SKYLARK (CONT.)

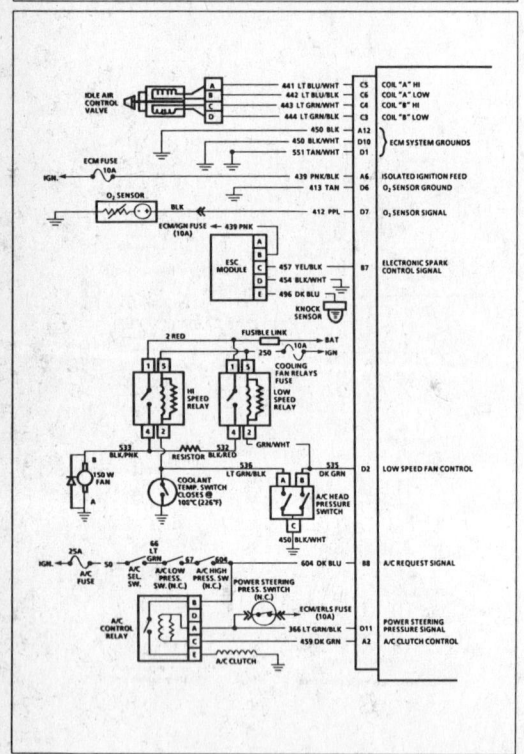

3.0L (VIN L) ECM CONNECTOR TERMINAL END VIEW—CALAIS AND SKYLARK

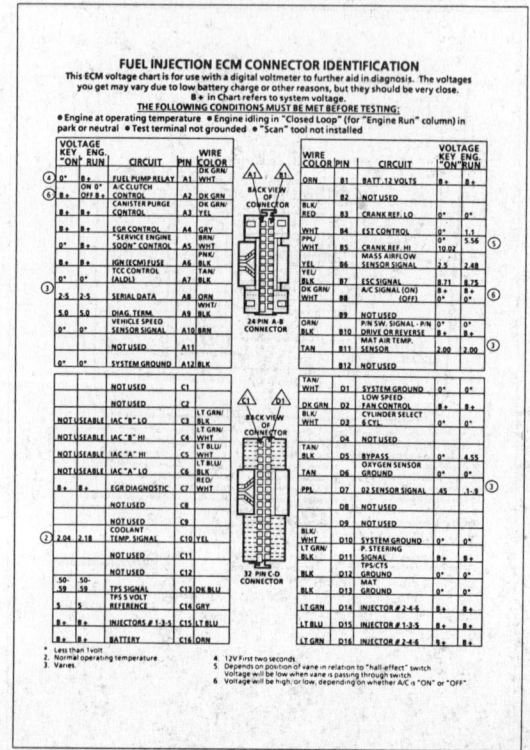

1988–90 3.0L (VIN L)
SKYLARK AND CUTLASS CALAIS

DIAGNOSTIC CIRCUIT CHECK

The Diagnostic Circuit Check must be the starting point for any driveability complaint diagnosis.

The diagnostic circuit check is an organized approach to identifying a problem created by an electronic engine control system malfunction because it directs the service technician to the next logical step in diagnosing the complaint.

If after completing the diagnostic circuit check and finding the on-board diagnostics functioning properly and no trouble codes displayed, a comparison of "Typical Values" for the appropriate engine may be used for comparison. The "Typical Values" are an average of display values recorded from normally operating vehicles and are intended to represent what a normally functioning system would display.

A "SCAN" TOOL THAT DISPLAYS FAULTY DATA SHOULD NOT BE USED, AND THE PROBLEM SHOULD BE REPORTED TO THE MANUFACTURER. THE USE OF A FAULTY "SCAN" CAN RESULT IN MISDIAGNOSIS AND UNNECESSARY PARTS REPLACEMENT.

Only the parameters listed below are used in this manual for diagnosis. If a "Scan" reads other parameters, the values are not recommended by General Motors for use in diagnosis. For more description on the values and use of the "Scan" to diagnose ECM inputs, refer to the applicable diagnosis section in Section "B". If all values are within the range illustrated, refer to symptoms in Section "B".

TYPICAL "SCAN" DATA VALUES
3.0L (VIN L)

Idle / Upper Radiator Hose Hot / Closed Throttle / Park or Neutral / Closed Loop / A/C off

"SCAN" Position	Units Displayed	Typical Data Value
Engine Speed	RPM	Varies
Coolant Temp.	C°	85° - 105°
MAT (Mani Air Temp.)	C°	Varies with Air Temperature
Air Flow	Gm/Sec	4 - 7
Oxygen Sensor	Millivolts	.1 - .9
Throt. Position	Volts	.50 - .59
Idle Air Control	Counts (steps)	10 - 50
Park/Neutral	P/N and R/DL	P-N
Fuel Integ.	Counts	118 - 138
Block Learn	Counts	118 - 138
Closed Loop Flag.	Open/Closed	Closed Loop
Vehicle Speed	MPH	0 (Zero)
Torque Conv. Cl.	On/Off	Off
EGR Diag. Switch	On/Off	Off
Spark Advance	Degrees	Approximately 30°
Knock Signal	No/Yes	No
Power Steering	Normal	Normal
3rd Gear	On/Off	Off
A/C Request	Yes/No	No
PROM I.D.	Numbers	Internal I.D. Only
LV8	0	60 - 70
Exhaust Recirculation	%	0%

NOTE: IF ALL VALUES ARE WITHIN THE RANGE ILLUSTRATED, REFER TO SYMPTOMS IN SECTION "B".

1988–90 3.0L (VIN L)
SKYLARK AND CUTLASS CALAIS

DIAGNOSTIC CIRCUIT CHECK
3.0L (VIN L) "N" SERIES (PORT)

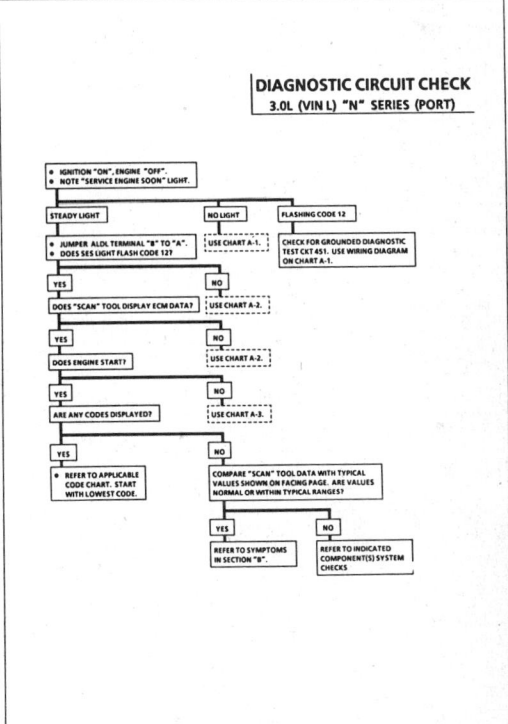

1988–90 3.0L (VIN L)
SKYLARK AND CUTLASS CALAIS

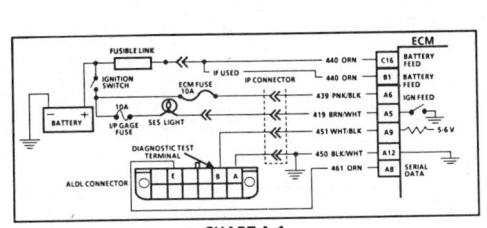

CHART A-1
NO "SERVICE ENGINE SOON" LIGHT
3.0L (VIN L) "N" SERIES (PORT)

Circuit Description:
There should always be a steady "Service Engine Soon" light when the ignition is "ON" and engine stopped. Battery is supplied directly to the light bulb. The electronic control module (ECM) will control the light and turn it "ON" by providing a ground path through CKT 419 to the ECM.

Test Description: Numbers below refer to circled numbers on the diagnostic chart.
1. "SES" light should be "ON" as the test light provides the ground.
2. Using a test light connected to 12 volts, probe each of the system ground circuits to be sure a good ground is present. See ECM terminal end view in front of this Section for ECM pin locations of ground circuits.

Diagnostic Aids:

Engine runs ok, check:
- Faulty light bulb
- CKT 419 open

Engine cranks, but will not run.
- Fuse or fusible link open
- ECM ignition fuse open
- Ignition CKT 439 or 440 to ECM open
- Poor connection to ECM

1988–90 3.0L (VIN L)
SKYLARK AND CUTLASS CALAIS

CHART A-1
NO "SERVICE ENGINE SOON" LIGHT
3.0L (VIN L) "N" SERIES (PORT)

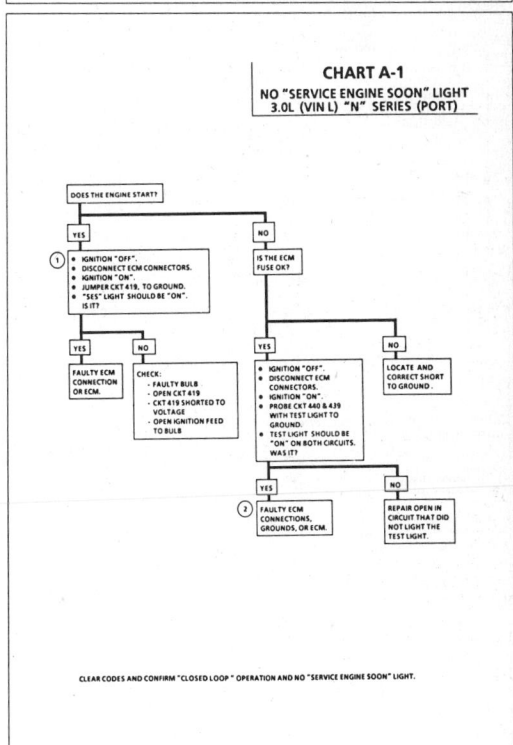

CLEAR CODES AND CONFIRM "CLOSED LOOP" OPERATION AND NO "SERVICE ENGINE SOON" LIGHT.

1988–90 3.0L (VIN L) SKYLARK AND CUTLASS CALAIS

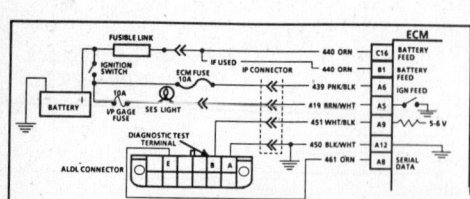

CHART A-2
WON'T FLASH CODE 12
"SERVICE ENGINE SOON" LIGHT "ON" STEADY
3.0L (VIN L) "N" SERIES (PORT)

Circuit Description:
There should always be a steady "Service Engine Soon" light when the ignition is "ON" and engine stopped. Battery ignition voltage is supplied to the light bulb. The electronic control module (ECM) will turn the light "ON" by grounding CKT 419 at the ECM.

With the diagnostic terminal grounded, the light should flash a Code 12, followed by any trouble code(s) stored in memory.

A steady light suggests a short to ground in the light control CKT 419 or an open in diagnostic CKT 451.

Test Description: Numbers below refer to circled numbers on the diagnostic chart.
1. If the light goes "OFF" when the ECM connector is disconnected, CKT 419 is not shorted to ground.
2. If there is a problem with the ECM that causes a "Scan" tool to not read serial data, the ECM should not flash a Code 12. If Code 12 is flashing, check for CKT 451 short to ground. If Code 12 does flash be sure that the "Scan" tool is working properly on another vehicle. If the "Scan" is

functioning properly and CKT 461 is OK, the PROM or ECM may be at fault for the NO ALDL symptom.
3. This step will check for an open diagnostic CKT 451.
4. At this point the "Service Engine Soon" light wiring is OK. The problem is a faulty ECM or PROM. Replace the PROM only after trying an ECM, as a defective PROM is an unlikely cause of the problem

---

1988–90 3.0L (VIN L) SKYLARK AND CUTLASS CALAIS

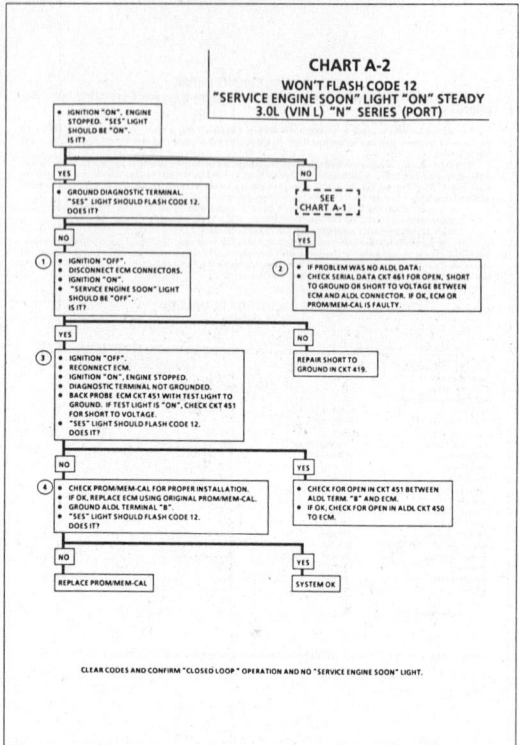

CHART A-2
WON'T FLASH CODE 12
"SERVICE ENGINE SOON" LIGHT "ON" STEADY
3.0L (VIN L) "N" SERIES (PORT)

CLEAR CODES AND CONFIRM "CLOSED LOOP" OPERATION AND NO "SERVICE ENGINE SOON" LIGHT.

1988–90 3.0L (VIN L) SKYLARK AND CUTLASS CALAIS

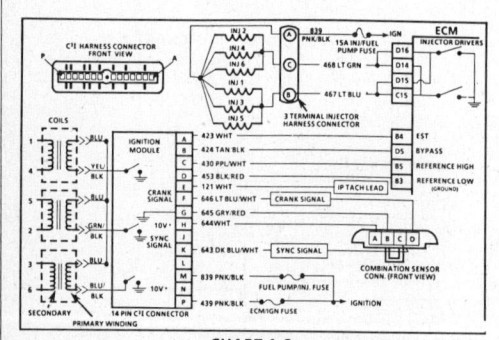

CHART A-3
(Page 1 of 3)
ENGINE CRANKS BUT WON'T RUN
3.0L (VIN L) "N" SERIES (PORT)

Circuit Description:
The C³I uses a waste spark method of spark distribution. In this type of ignition system the ignition module triggers the #1/4 coil pair resulting in both #1 and #4 spark plugs firing at the same time. #1 cylinder is on the compression stroke at the same time #4 is on the exhaust stroke, resulting in a lower energy requirement to fire #4 spark plug. This leaves the remainding high voltage to fire #1 spark plug.

This simultaneous fuel injection type of fuel delivery system utilizes two injector driver circuits in parallel, to activate the six individual fuel injectors. The ECM activates all six injectors simultaneously (all at one time).

Test Description: Numbers below refer to circled numbers on the diagnostic chart.
1. This step verifies "SES" light operation, TPS and coolant sensor signals are normal. A blinking injector test light verifies that the ECM is monitoring the C³I reference signal and attempting to activate the injectors.
2. Both the "Sync-Pulse" and crank signals have been verified as is evidenced by the blinking injector test light. A fuel pressure test at this point will seperate the diagnostic path into either a fuel related fault or ignition system malfunction.

3. The 3 terminal injector harness connector must be disconnected to avoid flooding of the engine and fouling of the spark plugs. By testing for spark on plug leads 1, 3 and 5, each ignition coil's ability to produce at least 25,000 volts is verified.
4. By testing the problem coil's control circuit with a test light, a determination can be made as to the problem coil being faulty or the module's internal driver for that coil being the source of the complaint.
5. An injector with a resistance value of less than 10 ohms (shorted) must be replaced.
6. Tests for battery voltage on CKT 839. If voltage was present, the "light off" test result was caused by no activation pulse reaching the injector connector from the ECM.

1988–90 3.0L (VIN L) SKYLARK AND CUTLASS CALAIS

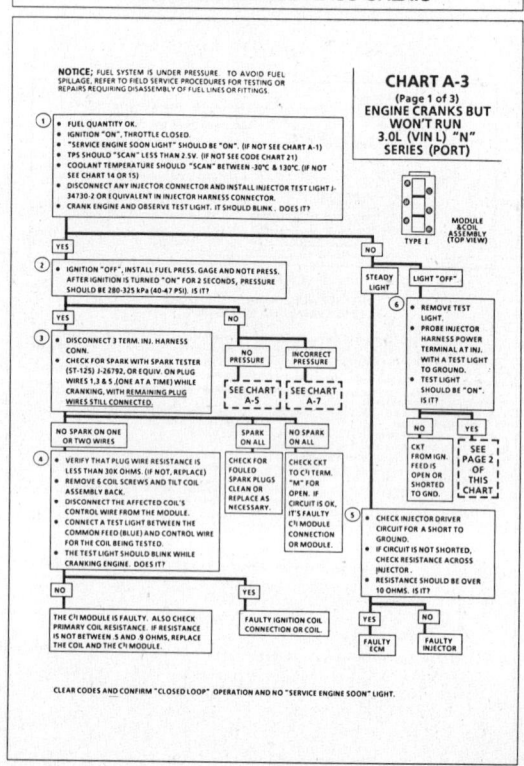

NOTICE: FUEL SYSTEM IS UNDER PRESSURE. TO AVOID FUEL SPILLAGE, REFER TO FIELD SERVICE PROCEDURES FOR TESTING OR REPAIRS REQUIRING DISASSEMBLY OF FUEL LINES OR FITTINGS.

CHART A-3
(Page 1 of 3)
ENGINE CRANKS BUT WON'T RUN
3.0L (VIN L) "N" SERIES (PORT)

CLEAR CODES AND CONFIRM "CLOSED LOOP" OPERATION AND NO "SERVICE ENGINE SOON" LIGHT.

1988–90 3.0L (VIN L) SKYLARK AND CUTLASS CALAIS

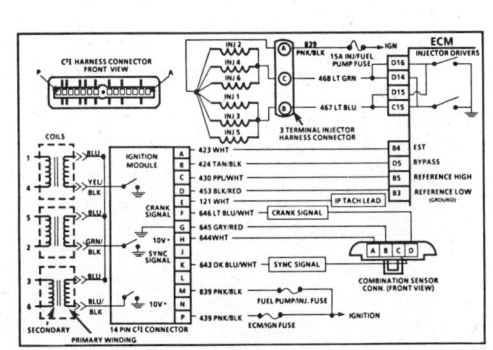

CHART A-3
(Page 2 of 3)
ENGINE CRANKS BUT WON'T RUN
3.0L (VIN L) "N" SERIES (PORT)

Circuit Description:
For synchronization of spark plug firing, a "Sync-Pulse" is created by the combination sensor "hall effect" switch. The sensor sends the "Sync-Pulse" to the ignition module when cylinders #1 and #4 are 25° after top dead center. This signal is used to start the correct sequence of coil firing with the #3/6 ignition coil.

The crank signal portion of the combination sensor sends a signal to the ignition module for coil activation and then to the ECM for reference rpm and crankshaft position. There are three windows in a disc (interruptor) which is mounted to the harmonic balancer. As these windows pass through the slot in the sensor, the next coil is triggered.

Test Description: Numbers below refer to circled numbers on the diagnostic chart.
7. Verifies ignition feed voltage at terminal "P" of the C³I module. Less than battery voltage would be an indication of a CKT 439 fault.
8. The test light to 12 volts simulates a reference signal to the ECM which will result in an injector test light blink. This will validate CKT 430, the ECM and the injector driver circuit are all OK.
9. Jumping the combination sensor harness terminals "B" and "C" together simulates a "Sync-Pulse" signal to the C³I module.

Then, by repeatedly jumping the combination sensor harness terminals "B" and "D" together a crank signal is simulated which should result in the injector test light blinking.
10. Verifies a proper "Sync-Pulse" signal circuit voltage of 6 to 9 volts and a good ground from the C³I module to terminal "B" of the sensor connector.
11. Determines if reason for incorrect voltage reading was due to a fault in CKT 643, an open in CKT 645, or a faulty C³I module. If the C³I module was faulty, also verify that CKT 453 to ECM terminal "B3" is not open.

1988–90 3.0L (VIN L) SKYLARK AND CUTLASS CALAIS

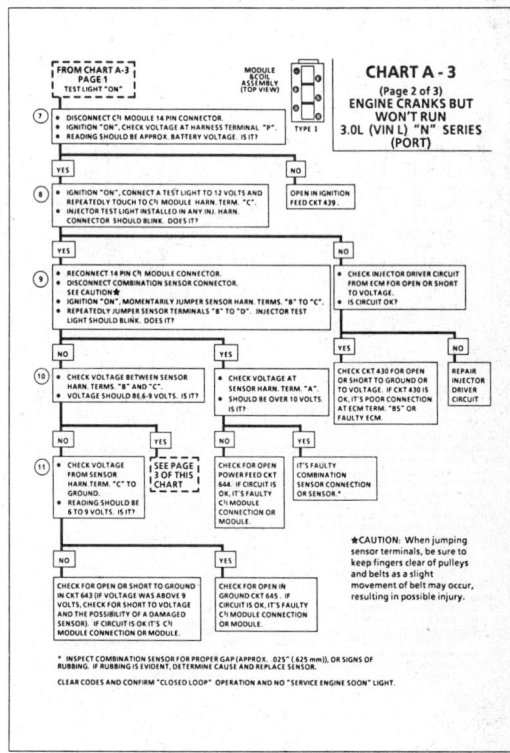

1988–90 3.0L (VIN L) SKYLARK AND CUTLASS CALAIS

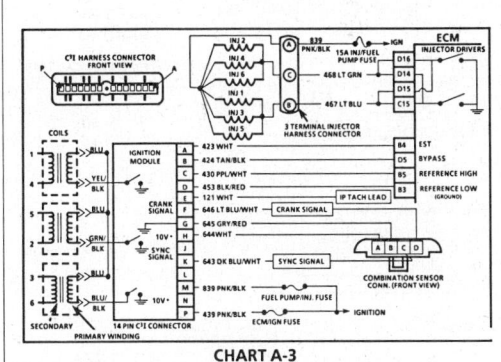

CHART A-3
(Page 3 of 3)
ENGINE CRANKS BUT WON'T RUN
3.0L (VIN L) "N" SERIES (PORT)

Circuit Description:
For synchronization of spark plug firing, a "Sync-Pulse" is created by the combination sensor "hall effect" switch. The sensor sends the "Sync-Pulse" to the ignition module when cylinders #1 and #4 are 25° after top dead center. This signal is used to start the correct sequence of coil firing with the #3/6 ignition coil.

The crank signal portion of the combination sensor sends a signal to the ignition module for coil activation and then to the ECM for reference rpm and crankshaft position. There are three windows in a disc (interruptor) which is mounted to the harmonic balancer. As these windows pass through the slot in the sensor, the next coil is triggered.

Test Description: Numbers below refer to circled numbers on the diagnostic chart.
12. Verifies a proper crank signal circuit voltage of 7 to 9 volts and a good ground from the C³I module to terminal "B" of the sensor connector.

13. Determines if reason for incorrect voltage reading was due to a fault in CKT 646, an open in CKT 645, or a faulty C³I module.

1988–90 3.0L (VIN L) SKYLARK AND CUTLASS CALAIS

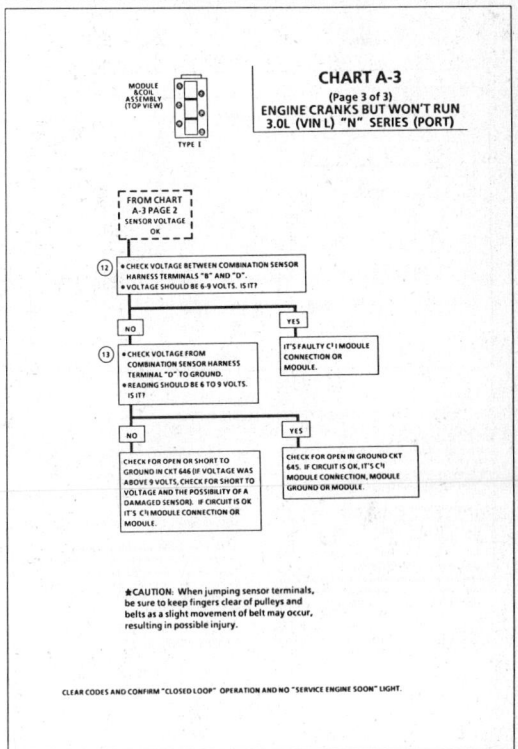

1988–90 3.0L (VIN L)
SKYLARK AND CUTLASS CALAIS

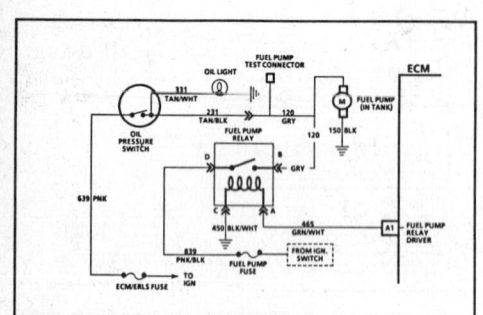

CHART A-5
(Page 1 of 2)
FUEL SYSTEM ELECTRICAL TEST
3.0L (VIN L) "N" SERIES (PORT)

Circuit Description:

When the ignition switch is turned "ON", the electronic control module (ECM) will energize the fuel pump relay which completes the circuit to the in-tank fuel pump. It will remain "ON" as long as the engine is cranking or running, and the ECM is receiving "C3I" reference pulses. If there are no reference pulses, the ECM will de-energize the fuel pump relay within 2 seconds after key "ON", or the engine is stopped.

The fuel pump will deliver fuel to the fuel rail and injectors, then to the pressure regulator, where the system pressure is controlled. Excess fuel pressure is bypassed back to the fuel tank. When the engine is stopped, the pump can be turned "ON" by applying battery voltage to the test terminal located in the engine compartment.

Improper fuel system pressure may contribute to one or all of the following symptoms:
- Cranks but won't run
- Code 44 or 45
- Cuts out, may feel like ignition problem
- Hesitation, loss of power or poor fuel economy

Test Description: Numbers below refer to circled numbers on the diagnostic chart.
1. If the fuse is blown, a short to ground in CKTs 120, 839, or the fuel pump itself is the cause.
2. Determines if the fuel pump circuit is being controlled by the ECM. The ECM should energize the fuel pump relay. The engine is not cranking or running so the ECM should de-energize the relay within 2 seconds after ignition is turned "ON".

3. Turns "ON" the fuel pump if CKT 120 wiring is OK. If the pump runs, it is a basic fuel delivery problem.
4. This test will determine if a short to ground on CKT 120 caused the fuse to blow. To prevent a mis-diagnosis, be sure the fuel pump is disconnected before the test.
5. Checks for a short to ground in the fuel pump relay harness CKT 839.

1988–90 3.0L (VIN L)
SKYLARK AND CUTLASS CALAIS

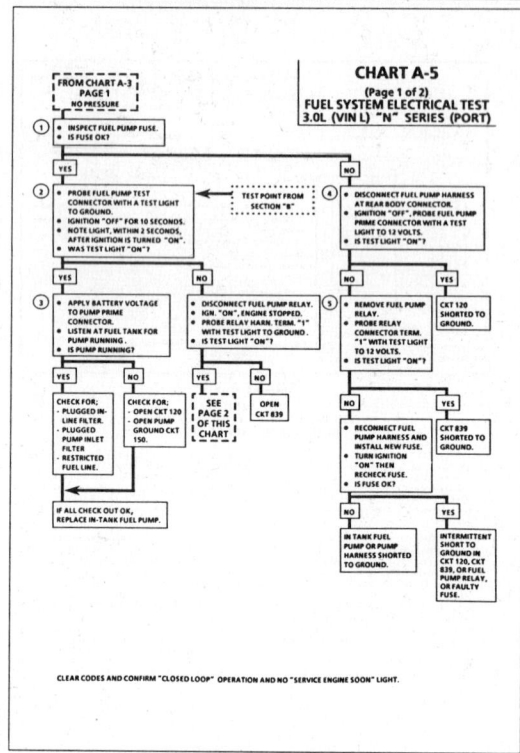

CLEAR CODES AND CONFIRM "CLOSED LOOP" OPERATION AND NO "SERVICE ENGINE SOON" LIGHT.

1988–90 3.0L (VIN L)
SKYLARK AND CUTLASS CALAIS

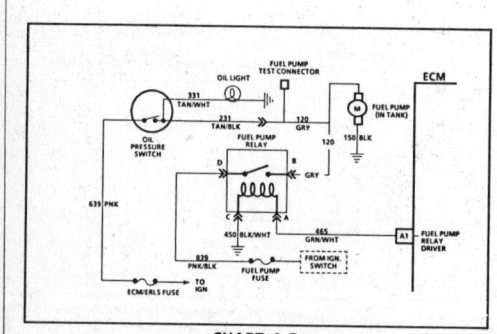

CHART A-5
(Page 2 of 2)
FUEL SYSTEM ELECTRICAL TEST
3.0L (VIN L) "N" SERIES (PORT)

Circuit Description:

When the ignition switch is turned "ON", the electronic control module (ECM) will energize the fuel pump relay which completes the circuit to the in-tank fuel pump. It will remain "ON" as long as the engine is cranking or running, and the ECM is receiving "C3I" reference pulses. If there are no reference pulses, the ECM will de-energize the fuel pump relay within 2 seconds after key "ON", or the engine is stopped.

Test Description: Numbers below refer to circled numbers on the diagnostic chart.
6. Checks for an open in the fuel pump relay ground, CKT 450.
7. Determines if the ECM is in control of the fuel pump relay through CKT 465.
8. The fuel pump control circuit includes an engine oil pressure switch with a separate set of normally open contacts. The switch closes at about 4 (lbs) 28 kPa of oil pressure and provides a second battery feed path to the fuel pump. If the relay fails, the pump will continue to run using the battery feed supplied by the closed oil pressure switch. This

step checks the oil pressure switch to be sure it provides battery feed to the fuel pump should the pump relay fail.

A failed pump relay will result in extended engine crank time, because of the time required to build enough oil pressure to close the oil pressure switch and turn "ON" the fuel pump. There may be instances when the relay has failed, but the engine will not crank fast enough to build enough oil pressure to close the switch. This, or a faulty oil pressure switch, can result in "Engine Cranks But Will Not Run".

1988–90 3.0L (VIN L)
SKYLARK AND CUTLASS CALAIS

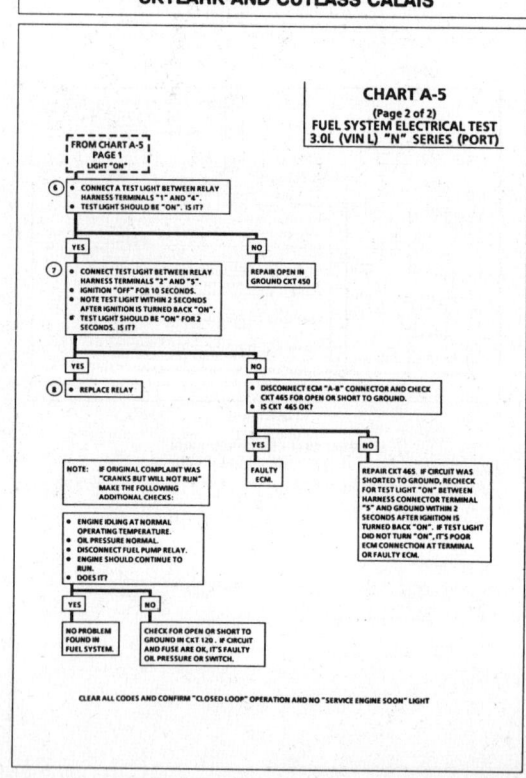

CLEAR ALL CODES AND CONFIRM "CLOSED LOOP" OPERATION AND NO "SERVICE ENGINE SOON" LIGHT

1988–90 3.0L (VIN L) SKYLARK AND CUTLASS CALAIS

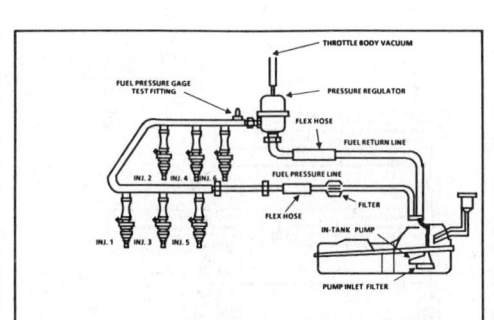

CHART A-7
(Page 1 of 2)
FUEL SYSTEM PRESSURE TEST
3.0L (VIN L) "N" SERIES (PORT)

Circuit Description:
The fuel pump will deliver fuel to the fuel rail and injectors, then to the pressure regulator, where the system pressure is controlled. Excess fuel pressure is bypassed back to the fuel tank. When the engine is stopped, the pump can be turned "ON" by applying battery voltage to the test terminal located in the engine compartment.
Improper fuel system pressure may contribute to one or all of the following symptoms:
- Cranks but won't run
- Code 44 or 45
- Cuts out, may feel like ignition problem
- Hesitation, Loss of power or poor fuel economy

Test Description: Numbers below refer to circled numbers on the diagnostic chart.
1. Use pressure gage J-34730-1. Wrap a shop towel around the fuel pressure tap to absorb any small amount of fuel leakage that may occur when installing the gage.
   With ignition "ON", pump pressure is controlled by spring pressure and throttle body vacuum within the pressure regulator assembly. Pressure should not leak down after the fuel pump is shut "OFF".
2. When the engine is idling, the throttle body vacuum is high and is applied to the fuel regulator diaphragm.

This will offset the spring and result in a lower fuel pressure.
3. The application of 12-14 inches of vacuum to the pressure regulator should result in a fuel pressure drop.
4. Pressure that leaks down may be caused by one of the following:
   - In-tank fuel pump check valve not holding
   - Pump coupling hose leaking
   - Fuel pressure regulator valve leaking
   - Injector sticking open

1988–90 3.0L (VIN L) SKYLARK AND CUTLASS CALAIS

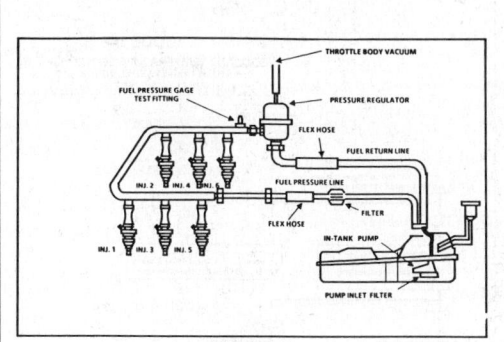

CHART A-7
(Page 2 of 2)
FUEL SYSTEM PRESSURE TEST
3.0L (VIN L) "N" SERIES (PORT)

Circuit Description:
The fuel pump will deliver fuel to the fuel rail and injectors, then to the pressure regulator, where the system pressure is controlled. Excess fuel pressure is bypassed back to the fuel tank. When the engine is stopped, the pump can be turned "ON" by applying battery voltage to the test terminal located in the engine compartment.
Improper fuel system pressure may contribute to one or all of the following symptoms:
- Cranks but won't run
- Code 44 or 45
- Cuts out, may feel like ignition problem
- Hesitation, Loss of power or poor fuel economy

Test Description: Numbers below refer to circled numbers on the diagnostic chart.
5. Pressure, but less than specification, falls into two areas:
- Regulated pressure, but less than specification: The amount of fuel to injectors is OK but pressure is too low. The system will be lean running and may set Code E044. Also, possible hard starting cold and overall poor performance.
- Restricted flow causing pressure drop; Normally, a vehicle with a fuel pressure of less than 165 kPa (24 psi) at idle will not be driveable.

However, if the pressure drop occurs only while driving, the engine will normally surge then stop as pressure begins to drop rapidly.
6. Restricting the fuel return line allows the fuel pump to develop its maximum pressure (dead head pressure). When battery voltage is applied to the pump test terminal, pressure should be above 517 kPa (75 psi).
7. This test determines if the high fuel pressure is due to a restricted fuel return line or a pressure regulator problem.

1988–90 3.0L (VIN L) SKYLARK AND CUTLASS CALAIS

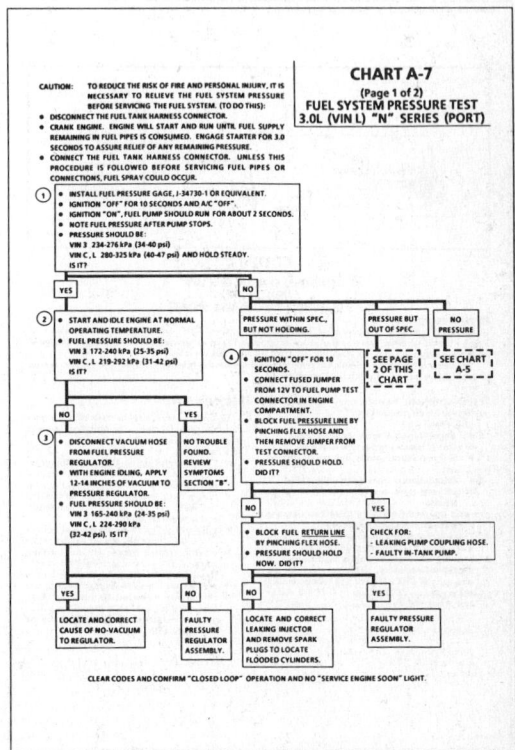

CHART A-7
(Page 1 of 2)
FUEL SYSTEM PRESSURE TEST
3.0L (VIN L) "N" SERIES (PORT)

CAUTION: TO REDUCE THE RISK OF FIRE AND PERSONAL INJURY, IT IS NECESSARY TO RELIEVE THE FUEL SYSTEM PRESSURE BEFORE SERVICING THE FUEL SYSTEM. (TO DO THIS):
- DISCONNECT THE FUEL TANK HARNESS CONNECTOR.
- CRANK ENGINE. ENGINE WILL START AND RUN UNTIL FUEL SUPPLY REMAINING IN FUEL PIPES IS CONSUMED. ENGAGE STARTER FOR 3.0 SECONDS TO ASSURE RELIEF OF ANY REMAINING PRESSURE.
- CONNECT THE FUEL TANK HARNESS CONNECTOR. UNLESS THIS PROCEDURE IS FOLLOWED BEFORE SERVICING FUEL PIPES OR CONNECTIONS, FUEL SPRAY COULD OCCUR.

CLEAR CODES AND CONFIRM "CLOSED LOOP" OPERATION AND NO "SERVICE ENGINE SOON" LIGHT.

1988–90 3.0L (VIN L) SKYLARK AND CUTLASS CALAIS

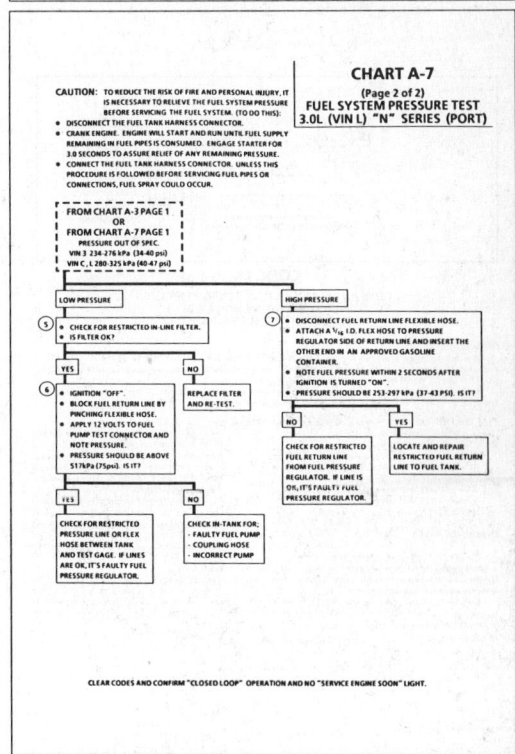

CHART A-7
(Page 2 of 2)
FUEL SYSTEM PRESSURE TEST
3.0L (VIN L) "N" SERIES (PORT)

CAUTION: TO REDUCE THE RISK OF FIRE AND PERSONAL INJURY, IT IS NECESSARY TO RELIEVE THE FUEL SYSTEM PRESSURE BEFORE SERVICING THE FUEL SYSTEM. (TO DO THIS):
- DISCONNECT THE FUEL TANK HARNESS CONNECTOR.
- CRANK ENGINE. ENGINE WILL START AND RUN UNTIL FUEL SUPPLY REMAINING IN FUEL PIPES IS CONSUMED. ENGAGE STARTER FOR 3.0 SECONDS TO ASSURE RELIEF OF ANY REMAINING PRESSURE.
- CONNECT THE FUEL TANK HARNESS CONNECTOR. UNLESS THIS PROCEDURE IS FOLLOWED BEFORE SERVICING FUEL PIPES OR CONNECTIONS, FUEL SPRAY COULD OCCUR.

CLEAR CODES AND CONFIRM "CLOSED LOOP" OPERATION AND NO "SERVICE ENGINE SOON" LIGHT.

1988–90 3.0L (VIN L)
SKYLARK AND CUTLASS CALAIS

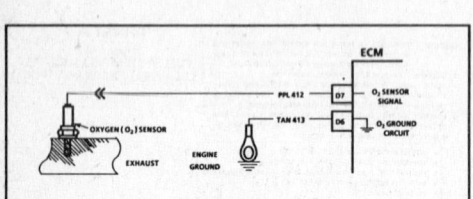

CODE 13
OXYGEN SENSOR CIRCUIT
(OPEN CIRCUIT)
3.0L (VIN L) "N" SERIES (PORT)

Circuit Description:

The ECM supplies a voltage of about .45 volt between terminals "D6" and "D7". (If measured with a 10 megohm digital voltmeter, this may read as low as .32 volt.) The O₂ sensor varies the voltage within a range of about 1 volt if the exhaust is rich, down through about .10 volt if exhaust is lean.

The sensor is like an open circuit and produces no voltage when it is below 360°C (600°F). An open sensor circuit or cold sensor causes "Open Loop" operation.

Test Description: Numbers below refer to circled numbers on the diagnostic chart.
1. Code 13 will set if:
   - Engine at normal operating temperature
   - Up to 2 minutes engine time after start
   - O₂ signal voltage steady between .35 and .55 volt
   - Throttle position sensor signal above idle
   - All conditions must be met for about 60 seconds.
   If the conditions for a Code 13 exist, the system will not go "Closed Loop".
2. This will determine if the sensor is at fault, or the wiring or ECM is the cause of the Code 13.
3. In doing this test use only a high impedence digital volt ohmeter. This test checks the continuity of CKTs 412 and 413 because if CKT 413 is open the ECM voltage on CKT 412 will be over .6 volt (600 mV).

Diagnostic Aids:

An intermittent may be caused by a poor connection, rubbed through wire insulation, or a wire broken inside the insulation.
Check For:
- Poor Connection or Damaged Harness Inspect ECM harness connectors for backed out terminals "D7" or "D6," improper mating, broken locks, improperly formed or damaged terminals, poor terminal to wire connection, and damaged harness.
- Intermittent Test If connections and harness check OK, "Scan" O₂ sensor voltage while moving related connectors and wiring harness, with warm engine running at part throttle in "Closed Loop." If the failure is induced, the "O₂ sensor voltage" reading will change from its normal fluctuating voltage (above 600 mV and below 300 mV) to a fixed value around 450 mV. This will help to isolate the location of the malfunction.

1988–90 3.0L (VIN L)
SKYLARK AND CUTLASS CALAIS

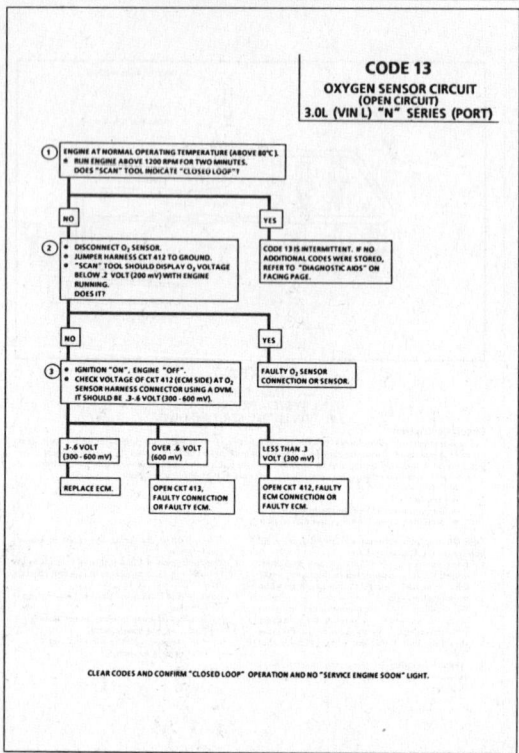

1988–90 3.0L (VIN L)
SKYLARK AND CUTLASS CALAIS

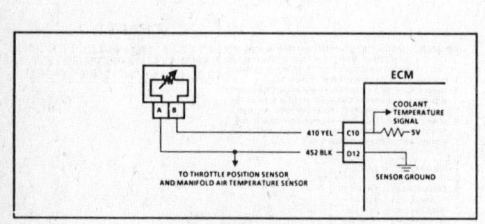

CODE 14
COOLANT TEMPERATURE SENSOR CIRCUIT
(HIGH TEMPERATURE INDICATED)
3.0L (VIN L) "N" SERIES (PORT)

Circuit Description:

The coolant temperature sensor uses a thermistor to control the signal voltage to the ECM. The ECM applies a voltage on CKT 410 to the sensor. When the engine is cold the sensor (thermistor) resistance is high, therefore, the ECM will see high signal voltage.

As the engine warms, the sensor resistance becomes less and the voltage drops. At normal engine operating temperature (85°C to 95°C) the voltage will measure about 1.5 to 2.0 volts.

Test Description: Numbers below refer to numbers on the diagnostic chart.
1. Code 14 will set if:
   - Signal voltage indicates a coolant temperature above 135°C (275°F) for calibrated time.
2. This test will determine if CKT 410 is shorted to ground which will cause the conditions for Code 14.

Diagnostic Aids:

"Scan" tool displays engine temperature in degrees centigrade. After engine is started, the temperature should rise steadily to about 90°C then stabilize when thermostat opens.

An intermittent may be caused by a poor connection, rubbed through wire insulation, or a wire broken inside the insulation.

Check For:
- Poor Connection or Damaged Harness Inspect ECM harness connectors for backed out terminals "C10" or "D12," improper mating, broken locks, improperly formed or damaged terminals, poor terminal to wire connection, and damaged harness.
- Intermittent Test If connections and harness check OK, "Scan" coolant temperature while moving related connectors and wiring harness. If the failure is induced, the "coolant temperature" display will abruptly change. This may help to isolate the location of the malfunction.
- Shifted Sensor The "Temperature To Resistance Value" scale may be used to test the coolant sensor at various temperature levels to evaluate the possibility of a "shifted" (mis-scaled) sensor which may result in driveability complaints.

1988–90 3.0L (VIN L)
SKYLARK AND CUTLASS CALAIS

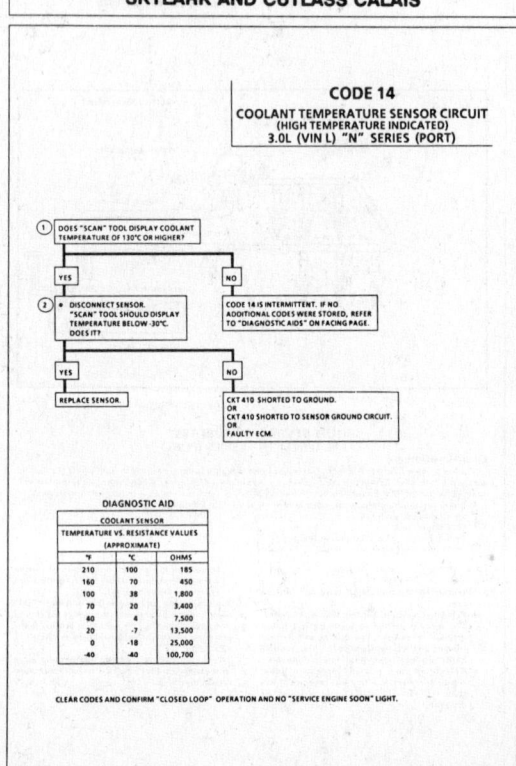

1988–90 3.0L (VIN L) SKYLARK AND CUTLASS CALAIS

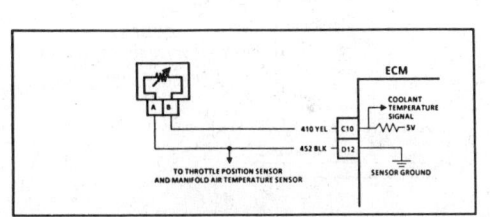

CODE 15
COOLANT TEMPERATURE SENSOR CIRCUIT
(LOW TEMPERATURE INDICATED)
3.0L (VIN L) "N" SERIES (PORT)

Circuit Description:
The coolant temperature sensor uses a thermistor to control the signal voltage to the ECM. The ECM applies a voltage on CKT 410 to the sensor. When the engine is cold the sensor (thermistor) resistance is high, therefore, the ECM will see high signal voltage.

As the engine warms, the sensor resistance becomes less and the voltage drops. At normal engine operating temperature (85°C to 95°C) the voltage will measure about 1.5 to 2.0 volts at the ECM.

Test Description: Numbers below refer to circled numbers on the diagnostic chart.
1. Code 15 will set if:
   - Signal voltage indicates a coolant temperature less than -44°C (-47°F) for at least 3 seconds.
2. This test simulates a Code 14. If the ECM recognizes the low signal voltage, (high temperature) and the "Scan" reads 130°C, the ECM and wiring are OK.
3. This test will determine if CKT 410 is open. There should be 5 volts present at sensor connector if measured with a DVM.

Diagnostic Aids:
A "Scan" tool reads engine temperature in degrees centigrade. After engine is started the temperature should rise steadily to about 90°C then stabilize when thermostat opens.

An intermittent may be caused by a poor connection, rubbed through wire insulation, or a wire broken inside the insulation.

Check For:
- **Poor Connection or Damaged Harness** Inspect ECM harness connectors for backed out terminals "C10" or "D12," improper mating, broken locks, improperly formed or damaged terminals, poor terminal to wire connection, and damaged harness.
- **Intermittent Test** If connections and harness check OK, "Scan" coolant temperature while moving related connectors and wiring harness. If the failure is induced, the "coolant temperature" display will abruptly change. This will help to isolate the location of the malfunction.
- **Shifted Sensor** The "Temperature To Resistance Value" scale may be used to test the coolant sensor at various temperature levels to evaluate the possibility of a "shifted" (mis-scaled) sensor which may result in driveability complaints.

A faulty connection, or an open in CKT 410 or 452 will result in a Code 15.
If Code 23 or 63 is also set, check CKT 452 for faulty wiring or connections. Check terminals at sensor for good contact. Refer to "Intermittents" in Section "B".

1988–90 3.0L (VIN L) SKYLARK AND CUTLASS CALAIS

CODE 15
COOLANT TEMPERATURE SENSOR CIRCUIT
(LOW TEMPERATURE INDICATED)
3.0L (VIN L) "N" SERIES (PORT)

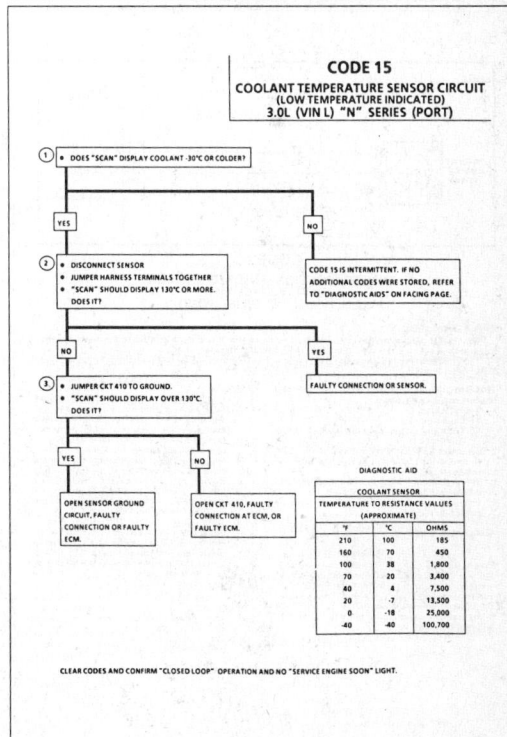

CLEAR CODES AND CONFIRM "CLOSED LOOP" OPERATION AND NO "SERVICE ENGINE SOON" LIGHT.

1988–90 3.0L (VIN L) SKYLARK AND CUTLASS CALAIS

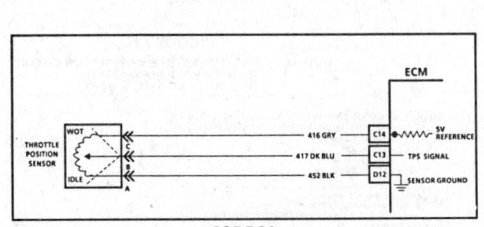

CODE 21
THROTTLE POSITION SENSOR (TPS) CIRCUIT
(SIGNAL VOLTAGE HIGH)
3.0L (VIN L) "N" SERIES (PORT)

Circuit Description:
The throttle position sensor (TPS) provides a voltage signal that changes relative to the throttle blade. Signal voltage will vary from about .5 at idle to about 5 volts at wide open throttle.
The TPS signal is one of the most important inputs used by the ECM for fuel control and for most of the ECM control outputs.

Test Description: Numbers below refer to circled numbers on the diagnostic chart.
1. Code 21 will set if:
   - Engine is running
   - TPS signal voltage is greater than 2.5 volts
   - Code 33 or 34 not present at first start up
   - All conditions met for 5 seconds
   With throttle closed the TPS should read less than .70 volt. If it doesn't, check adjustment.
2. With the TPS sensor disconnected, the TPS voltage should go low if the ECM and wiring are OK.
3. Probing CKT 452 with a test light checks the 5 volt return circuit because a faulty 5 volt return will cause a Code 21.

Diagnostic Aids:
A "Scan" tool reads throttle position in volts. Should read .50 - .59 volt with throttle closed and ignition "ON" or at idle. Voltage should increase at a steady rate as throttle is moved toward WOT.
Also some "Scan" tools will read throttle angle .0% = closed throttle 100% = WOT.

Check For:
- **Poor Connection or Damaged Harness** Inspect ECM harness connectors for backed out terminals "C13," "C14," or "D12," improper mating, broken locks, improperly formed or damaged terminals, poor terminal to wire connection, and damaged harness.
- **Intermittent Test** If connections and harness check OK, monitor TPS voltage display while moving related connectors and wiring harness. If the failure is induced, display will abruptly change. This may help to isolate the location of the malfunction.
- **TPS Scaling** Observe TPS voltage display while depressing accelerator pedal with engine stopped and ignition "ON". Display should vary from closed throttle TPS voltage to over 4500 mV when throttle is held at wide open throttle position.

1988–90 3.0L (VIN L) SKYLARK AND CUTLASS CALAIS

CODE 21
THROTTLE POSITION SENSOR (TPS) CIRCUIT
(SIGNAL VOLTAGE HIGH)
3.0L (VIN L) "N" SERIES (PORT)

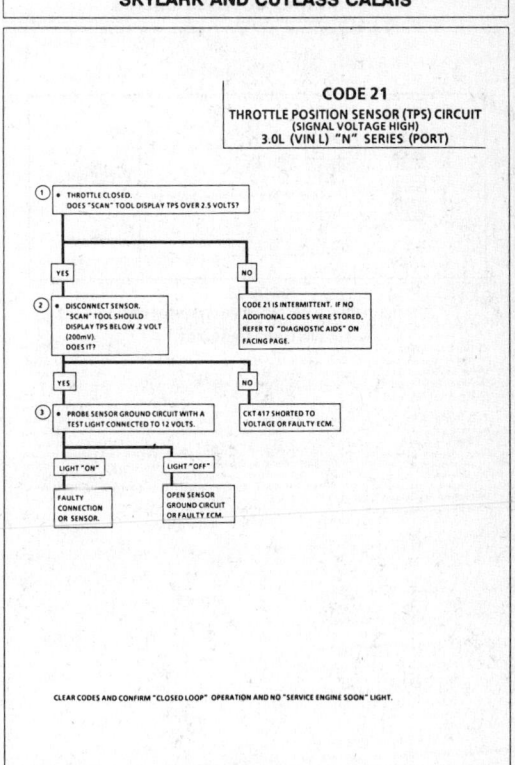

CLEAR CODES AND CONFIRM "CLOSED LOOP" OPERATION AND NO "SERVICE ENGINE SOON" LIGHT.

1988–90 3.0L (VIN L)
SKYLARK AND CUTLASS CALAIS

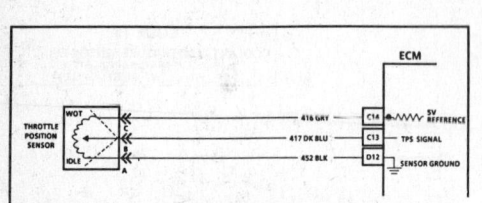

CODE 22
THROTTLE POSITION SENSOR (TPS) CIRCUIT
(SIGNAL VOLTAGE LOW)
3.0L (VIN L) "N" SERIES (PORT)

Circuit Description:
The throttle position sensor (TPS) provides a voltage signal that changes relative to the throttle blade. Signal voltage will vary from about .5 at idle to about 5 volts at wide open throttle.
The TPS signal is one of the most important inputs used by the ECM for fuel control and for most of the ECM control outputs.

Test Description: Numbers below refer to circled numbers on the diagnostic chart.
1. Code 22 will set if:
   - Engine running
   - TPS signal voltage is less than about .2 volt for 3 seconds
2. Simulates Code 21: (high voltage) If the ECM recognizes the high signal voltage, the ECM and wiring are OK.
3. TPS adjustment: With throttle closed, the TPS voltage reading should be .50 - .59 volt.
4. This simulates a high signal voltage to check for an open in CKT 417.

Diagnostic Aids:
A "Scan" tool reads throttle position in volts. Should read .50 - .59 volt with throttle closed and ignition "ON" or at idle. Voltage should increase at a steady rate as throttle is moved toward WOT.
Also some "Scan" tools will read throttle angle 0% = closed throttle 100% = WOT.
An open or short to ground in CKTs 416 or 417 will result in a Code 22.

Check For:
- Poor Connection or Damaged Harness Inspect ECM harness connectors for backed out terminals "C13," "C14," or "D12," improper mating, broken locks, improperly formed or damaged terminals, poor terminal to wire connection, and damaged harness.
- Intermittent Test If connections and harness check OK, monitor TPS voltage display while moving related connectors and wiring harness. If the failure is induced, display will abruptly change. This may help to isolate the location of the malfunction.
- TPS Scaling Observe TPS voltage display while depressing accelerator pedal with engine stopped and ignition "ON". Display should vary from closed throttle TPS voltage to over 4500 mV when throttle is held at wide open throttle position.

1988–90 3.0L (VIN L)
SKYLARK AND CUTLASS CALAIS

CODE 22
THROTTLE POSITION SENSOR (TPS) CIRCUIT
(SIGNAL VOLTAGE LOW)
3.0L (VIN L) "N" SERIES (PORT)

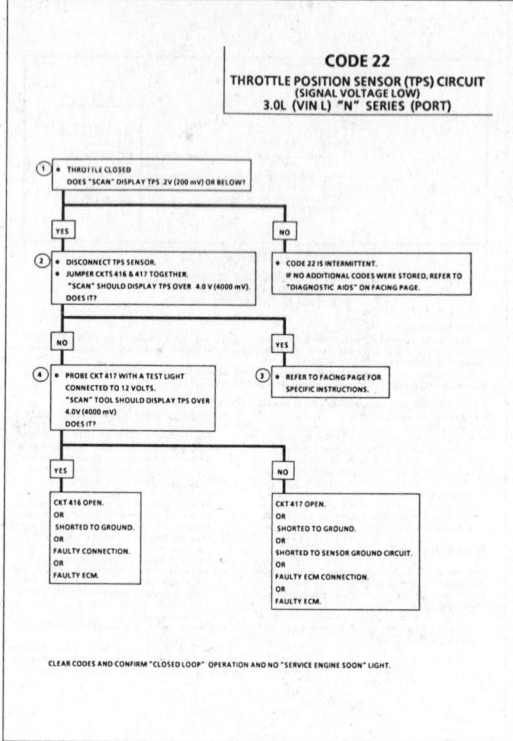

CLEAR CODES AND CONFIRM "CLOSED LOOP" OPERATION AND NO "SERVICE ENGINE SOON" LIGHT.

1988–90 3.0L (VIN L)
SKYLARK AND CUTLASS CALAIS

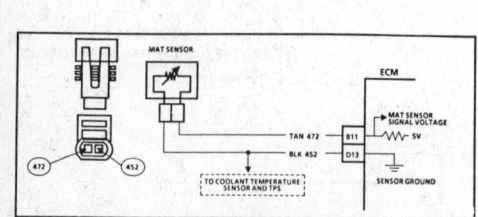

CODE 23
MANIFOLD AIR TEMPERATURE (MAT) SENSOR CIRCUIT
(LOW TEMPERATURE INDICATED)
3.0L (VIN L) "N" SERIES (PORT)

Circuit Description:
The MAT sensor uses a thermistor to control the signal voltage to the ECM. The ECM applies a voltage (about 5 volts) on CKT 472 to the sensor. When the air is cold, the sensor (thermistor) resistance is high, therefore, the ECM will see a high signal voltage. If the air is warm, the sensor resistance is low; therefore, the ECM will see a low voltage.

Test Description: Numbers below refer to numbers on the diagnostic chart.
Code 23 will set if:
   - A signal voltage indicates a manifold air temperature below −40°C (-40°F) for 4 seconds.

Due to the conditions necessary to set a Code 23, the "Service Engine Soon" light will only stay "ON" while the fault is present.

1. A "Scan" tool may not be used to diagnose this fault, due to the ECM transmitting "default" (substitute) values while the fault is present. A Code 23 will set, due to an open sensor, wire, or connection. This test will determine if the wiring and ECM are OK.
2. If the resistance is greater than 25,000 ohms, replace the sensor.

1988–90 3.0L (VIN L)
SKYLARK AND CUTLASS CALAIS

CODE 23
MANIFOLD AIR TEMPERATURE (MAT) SENSOR CIRCUIT
(LOW TEMPERATURE INDICATED)
3.0L (VIN L) "N" SERIES (PORT)

NOTE: A "SCAN" TOOL MAY NOT BE USED TO DIAGNOSE THIS FAULT, DUE TO THE ECM TRANSMITTING "DEFAULT" (SUBSTITUTE) VALUES WHEN THE FAULT IS PRESENT.

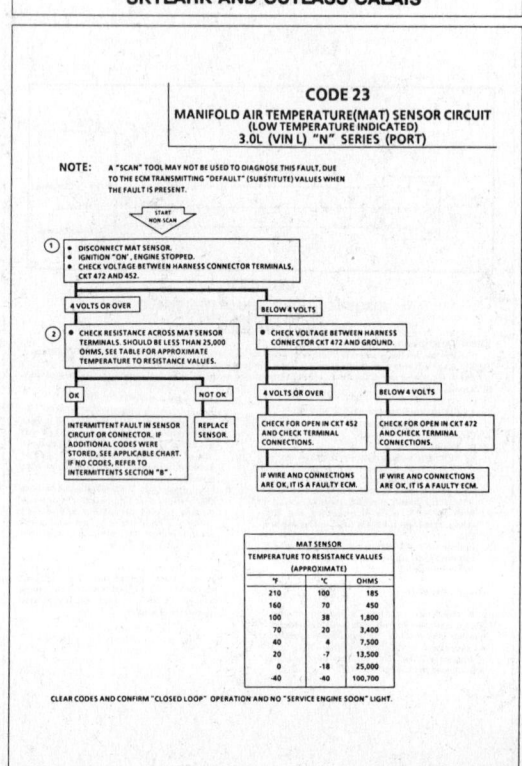

MAT SENSOR		
TEMPERATURE TO RESISTANCE VALUES		
(APPROXIMATE)		
°F	°C	OHMS
210	100	185
160	70	450
100	38	1,800
70	20	3,400
40	4	7,500
20	-7	13,500
0	-18	25,000
-40	-40	100,700

CLEAR CODES AND CONFIRM "CLOSED LOOP" OPERATION AND NO "SERVICE ENGINE SOON" LIGHT.

1988–90 3.0L (VIN L)
SKYLARK AND CUTLASS CALAIS

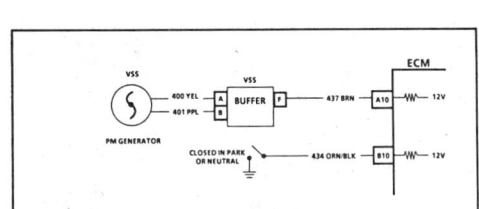

CODE 24
VEHICLE SPEED SENSOR (VSS) CIRCUIT
3.0L (VIN L) "N" SERIES (PORT)

Circuit Description:
The vehicle speed sensor consists of a PM generator, buffer, speedometer and ECM. The PM generator is a permanent magnet assembly attached to the transaxle. As the vehicle moves, the generator creates a "sine wave" electrical pulse, which is routed to the buffer. In the buffer, the signal is changed from a "sine wave" to a "square wave" and amplified. The square wave is an "ON/OFF" signal. The length of time between pulses determines vehicle speed. The ECM sends a 12 volt signal out on CKT 437. The frequency by which the signal is pulsed low is used by the ECM to determine vehicle speed.

Test Description: Numbers below refer to circled numbers on the diagnostic chart.
1. Code 24 will set if vehicle speed equals 0 mph when:
   - Engine speed is between 1400 and 3600 rpm
   - TPS voltage shows closed throttle
   - Low load condition (low air flow)
   - Not in park or neutral
   - All conditions met for 5 seconds
2. This step checks to see if the fault is in CKT 437, including the ECM
   The ECM is the source of 12 volts via CKT 437 to the buffer in a normal working system.

NOTE: Disregard a Code 24 that sets when the drive wheels are not turning.

Diagnostic Aids:
An intermittent may be caused by a poor connection, rubbed through wire insulation, or a wire broken inside the insulation.
Check for:
- Poor Connection or Damaged Harness Inspect ECM harness connector terminal "A10" for improper mating, broken locks, improperly formed or damaged terminals, poor terminal to wire connection, and damaged harness.
- Intermittent Test If connections and harness check OK, raise front wheels, block other wheels, idle engine in low gear, and "Scan" vehicle speed while moving related connectors and wiring harness. If the failure is induced, the "Vehicle Speed" display will abruptly change. This may help to isolate the location of the malfunction.

1988–90 3.0L (VIN L)
SKYLARK AND CUTLASS CALAIS

CODE 24
VEHICLE SPEED SENSOR (VSS) CIRCUIT
3.0L (VIN L) "N" SERIES (PORT)

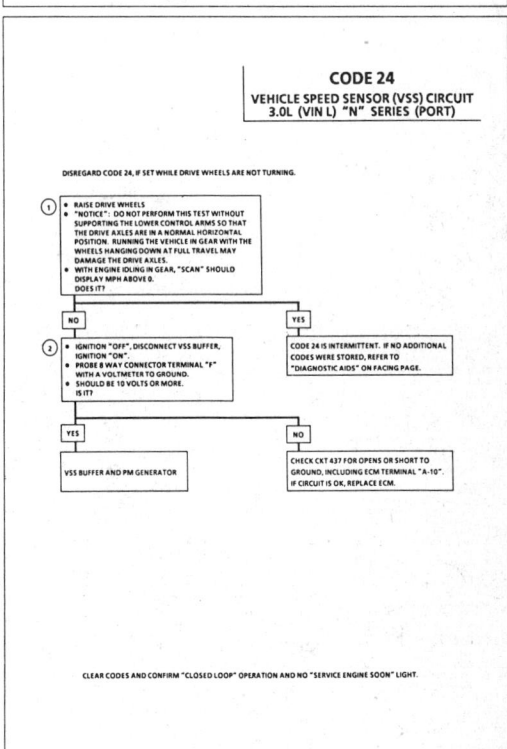

1988–90 3.0L (VIN L)
SKYLARK AND CUTLASS CALAIS

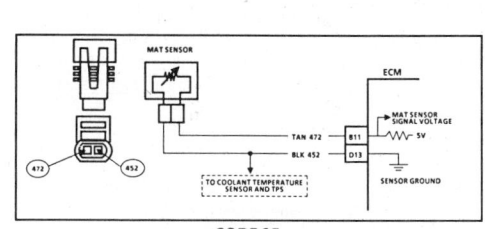

CODE 25
MANIFOLD AIR TEMPERATURE (MAT) SENSOR CIRCUIT
(HIGH TEMPERATURE INDICATED)
3.0L (VIN L) "N" SERIES (PORT)

Circuit Description:
The MAT sensor uses a thermistor to control the signal voltage to the ECM. The ECM applies a voltage (4–6) on CKT 472 to the sensor. When manifold air is cold, the sensor (thermistor) resistance is high, therefore, the ECM will see a high signal voltage. If the air warms, the sensor resistance becomes less, and the voltage drops.

Test Description: Numbers below refer to circled numbers on the diagnostic chart.
Code 25 will set if:
- Signal voltage indicates a manifold air temperature greater than 135°C (275°F)
- A vehicle speed is present
- Both of the above requirements are met for at least 30 seconds
Due to the conditions necessary to set a Code 25, the "Service Engine Soon" light will only stay "ON" while the fault is present.

1. A "Scan" tool may not be used to diagnose this fault due to the ECM transmitting "default" (substitute) values while the fault is present. If voltage is above 4 volts, the ECM and wiring are OK.
2. If the resistance is less than 100 ohms, replace the sensor.

1988–90 3.0L (VIN L)
SKYLARK AND CUTLASS CALAIS

CODE 25
MANIFOLD AIR TEMPERATURE (MAT) SENSOR CIRCUIT
(HIGH TEMPERATURE INDICATED)
3.0L (VIN L) "N" SERIES (PORT)

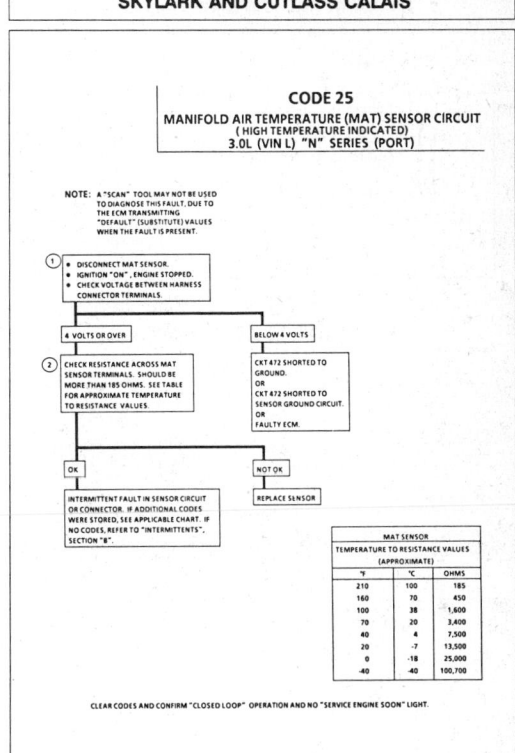

MAT SENSOR		
TEMPERATURE TO RESISTANCE VALUES		
(APPROXIMATE)		
°F	°C	OHMS
210	100	185
160	70	450
100	38	1,600
70	20	3,400
40	4	7,500
20	-7	13,500
0	-18	25,000
-40	-40	100,700

CLEAR CODES AND CONFIRM "CLOSED LOOP" OPERATION AND NO "SERVICE ENGINE SOON" LIGHT.

1988–90 3.0L (VIN L) SKYLARK AND CUTLASS CALAIS

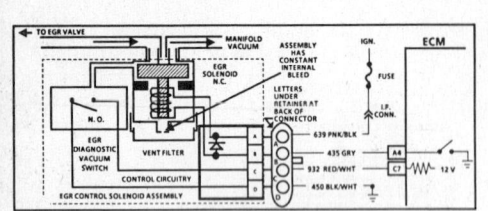

CODE 32
EXHAUST GAS RECIRCULATION (EGR) CIRCUIT
3.0L (VIN L) "N" SERIES (PORT)

Circuit Description:

The EGR valve is opened by engine vacuum. In order to control and monitor EGR application an electronic vacuum regulator valve is used (EVRV). The EVRV is composed of two devices: 1. EGR solenoid, normally closed (vacuum blocked). 2. EGR vacuum switch, normally open (no current flow).

EGR vacuum control is accomplished by the ECM grounding CKT 435. This energizes the EGR solenoid. This is done thousands of times a second. By varying the length of "ON" time, as compared to "OFF" time, pulse width modulation (PWM). The ECM controls the vacuum source to the EGR valve.

EGR is monitored by the ECM through the EGR vacuum switch. The EGR vacuum switch, a normally open electrical switch, has an orifice built in which restricts the vacuum signal to the EGR vacuum switch, when sufficent vacuum reaches the EGR vacuum switch to close the electrical switch. There should also be sufficient vacuum to open the EGR valve.

Code 32 will set, if the vacuum switch closes at idle, or if it does not close under load (less than WOT).
- Engine running
- Code 33 or 34 not present
- Coolant temperature above 42.5°C (108°F)
- LV8 reading less than 144 cts
- No vacuum to EGR (switch open)
- Conditions exist over 5 seconds

Test Description: Numbers below refer to circled numbers on the diagnostic chart.
1. "Scan" displays the condition of the EGR diagnostic switch. In park or neutral, the display should read "NO" (open switch).
2. Under moderate engine load, the display will switch from "NO" to "YES".
3. Checks the integrity of the 12 volts feed and ground circuits. If these circuits check OK, the fault is elsewhere in the EVRV/EGR control circuit.
4. A test light connected between terminals "A" and "B" will verify the EVRV/EGR wiring and check for proper ECM operation.
5. If "YES" was displayed at idle, disconnect the EVRV harness. If display remains "YES," the fault is either a short to ground in CKT 932 or the ECM.
6. If the EGR display switches from "YES" to "NO" when the EVRV is disconnected, the fault is either

in the EVRV/EGR solenoid, CKT 435 or the ECM. Probing at terminal "B" will further isolate the fault. If the test light is "ON," disconnect ECM A-B connector before checking CKT 435 for a short to ground since the short could be inside the ECM.

Diagnostic Aids:

An intermittent may be caused by a poor connection, rubbed through wire insulation, or a wire broken inside the insulation. Check For:
- Poor Connection or Damaged Harness Inspect ECM harness connectors for backed out terminal "C7," improper mating, broken locks, improperly formed or damaged terminals, poor terminal to wire connection, and damaged harness.
- Intermittent Test If connections and harness check OK, "Scan" EVRV switch, while moving related connectors and wiring harness. If the failure is induced, the display will change. This may help to isolate the location of the malfunction.

1988–90 3.0L (VIN L) SKYLARK AND CUTLASS CALAIS

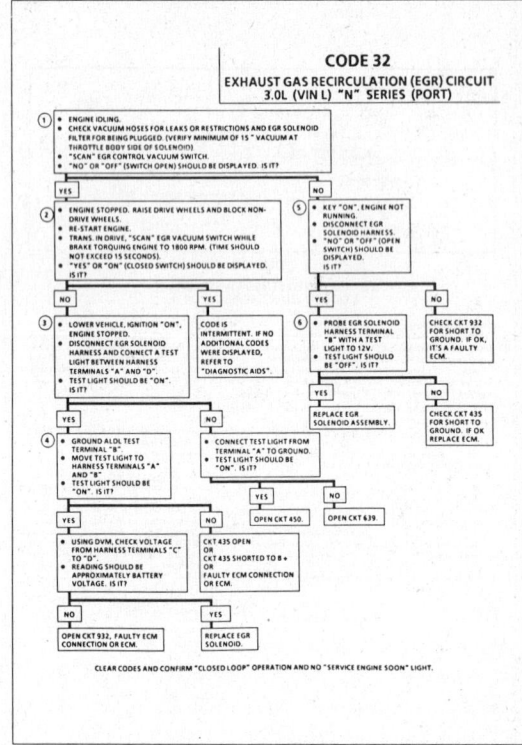

CODE 32
EXHAUST GAS RECIRCULATION (EGR) CIRCUIT
3.0L (VIN L) "N" SERIES (PORT)

CLEAR CODES AND CONFIRM "CLOSED LOOP" OPERATION AND NO "SERVICE ENGINE SOON" LIGHT.

1988–90 3.0L (VIN L) SKYLARK AND CUTLASS CALAIS

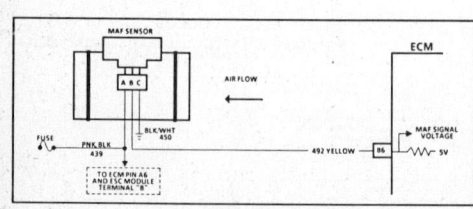

CODE 33
MASS AIR FLOW (MAF) SENSOR CIRCUIT
(GM/SEC HIGH)
3.0L (VIN L) "N" SERIES (PORT)

Circuit Description:

The MAF sensor measures the flow of air entering the engine. The sensor produces a frequency output between 32 and 150 hertz (3gm/sec to 150 gm/sec). A large quantity (high frequency) indicates acceleration, and a small quantity (low frequency) indicates deceleration or idle. This information is used by the ECM for fuel control and is converted by a "Scan" tool to read out the air flow in grams per second. A normal reading is about 4-7 grams per second at idle and increases with rpm.

Test Description: Numbers below refer to circled numbers on the diagnostic chart.
1. Code 33 will set if:
   - Ignition "ON" and air flow exceeds 20 gms/sec or
   - Engine is running less than 1300 rpm
   - TPS is 10% or less
   - Air flow greater than 20 grams per second (high frequency)
   - All of the above are met for 2 seconds or more

Diagnostic Aids:

The "Scan" tool is not of much use in diagnosing this code, because when the code sets, gm/sec will be displaying the default value. However, it may be useful in comparing the signal of a problem vehicle with that of a known good running one.

Check For:
- Poor Connection or Damaged Harness Inspect ECM harness connectors for backed out terminal "B6," improper mating, broken locks, improperly formed or damaged terminals, poor terminal to wire conection, and damaged harness.
- Intermittent Test If connections and harness check OK, "Scan" MAF while moving related connectors and wiring harness. If the failure is induced, the MAF display will abruptly change. This will help to isolate the location of the malfunction.
- Mis-routed Harness Inspect MAF sensor harness to insure that it is not too close to high voltage wires, such as spark plug leads.

1988–90 3.0L (VIN L) SKYLARK AND CUTLASS CALAIS

CODE 33
MASS AIR FLOW (MAF) SENSOR CIRCUIT
(GM/SEC HIGH)
3.0L (VIN L) "N" SERIES (PORT)

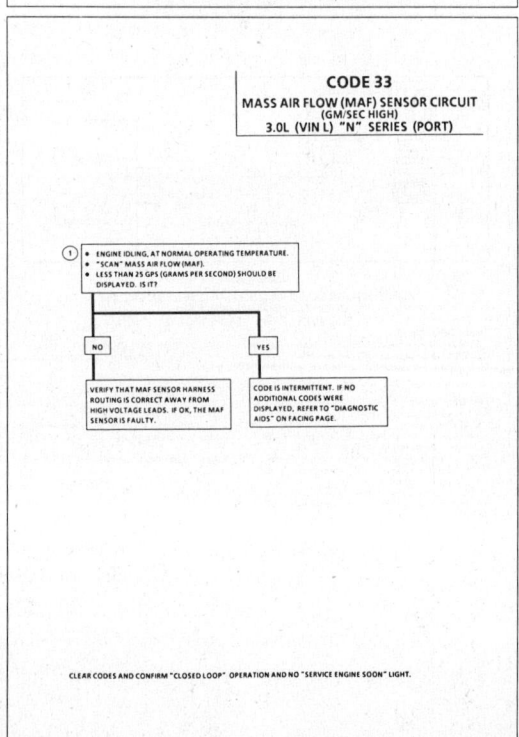

CLEAR CODES AND CONFIRM "CLOSED LOOP" OPERATION AND NO "SERVICE ENGINE SOON" LIGHT.

1988–90 3.0L (VIN L)
SKYLARK AND CUTLASS CALAIS

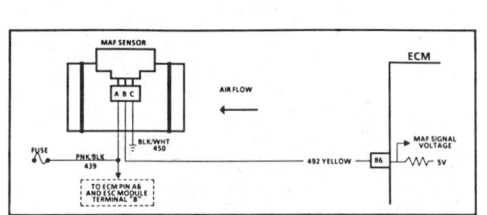

CODE 34
MASS AIR FLOW (MAF) SENSOR CIRCUIT
(GM/SEC LOW)
3.0L (VIN L) "N" SERIES (PORT)

Circuit Description:

The mass air flow (MAF) sensor measures the flow of air which passes through it in a given time. The ECM uses this information to monitor the operating condition of the engine in calculating fuel delivery. A large quantity of air movement indicates acceleration, while a small quantity indicates deceleration or idle.

The MAF sensor produces a frequency signal which cannot be easily measured. The sensor can be diagnosed using the procedures on this chart.

Code 34 will set, when either of the following sets of conditions exists:
- Engine running
- No MAF sensor signal for 250 mS

or
- Engine running over 1400 rpm
- TPS over 2.5 volts
- Air flow less than 10 grams per second (low frequency)
- Above conditions for over 10 seconds

Test Description: Numbers below refer to circled numbers on the diagnostic chart.

1. This step checks for a loose or damaged air duct, which could set Code 34, and also checks to see if ECM recognizes a problem. A light "OFF" at this point indicates an intermittent fault.
2. A voltage reading at sensor harness connector terminal "B" of less than 4 or over 6 volts indicates a fault in CKT 492 or poor connection.
3. Verifies that both ignition voltage and a good ground circuit are available.

Diagnostic Aids:

An intermittent may be caused by a poor connection, mis-routed harness, rubbed through wire insulation, or a wire broken inside the insulation.

Check For:
- **Poor Connection at ECM pin "B-6"** Inspect harness connectors for backed out terminals, improper mating, broken locks, improperly formed or damaged terminals, and poor terminal to wire connection.
- **Mis-routed Harness** Inspect MAF sensor harness to insure that it is not too close to high voltage wires, such as spark plug leads.
- **Damaged Harness** Inspect harness for damage. If harness appears OK, "Scan" while moving related connectors and wiring harness. A change in display would indicate the intermittent fault location.

1988–90 3.0L (VIN L)
SKYLARK AND CUTLASS CALAIS

CODE 34
MASS AIR FLOW (MAF) SENSOR CIRCUIT
(GM/SEC LOW)
3.0L (VIN L) "N" SERIES (PORT)

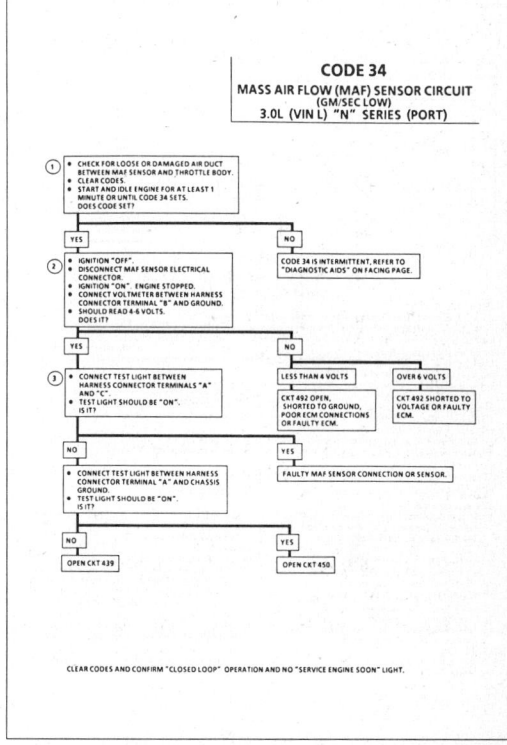

CLEAR CODES AND CONFIRM "CLOSED LOOP" OPERATION AND NO "SERVICE ENGINE SOON" LIGHT.

1988–90 3.0L (VIN L)
SKYLARK AND CUTLASS CALAIS

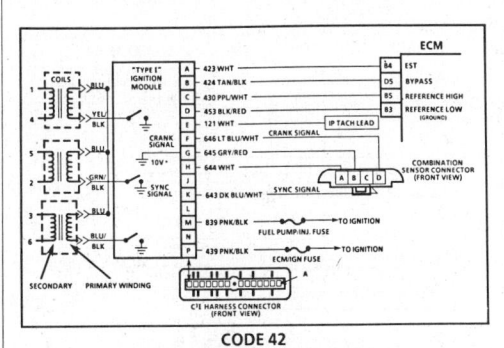

CODE 42
ELECTRONIC SPARK TIMING (EST) CIRCUIT
3.0L (VIN L) "N" SERIES (PORT)

Circuit Description:

The C³I module sends a reference signal to the ECM when the engine is cranking. While the engine is under 400 rpm, the C³I module controls the ignition timing. When the engine speed exceeds 400 rpm, the ECM sends a 5 volt signal on the bypass CKT 424 to switch the timing to ECM control through the EST CKT 423. An open or ground in the EST circuit will stall the engine and set a Code 42. The engine can be re-started but will run on module timing.

To set a Code 42 the following conditions must be met.
- Engine speed greater than 900 rpm with no EST pulse for 200 mS (open or grounded CKT 423), or
- ECM commanding bypass mode (open or grounded CKT 424)

Test Description: Numbers below refer to circled numbers on the diagnostic chart.

1. Checks to see if ECM recognizes a problem. If it does not set Code 42, at this point, it is an intermittent problem and could be due to a loose connection.
2. With the ECM disconnected, the ohmmeter should be reading less than 200 ohms, which is the normal resistance of the EST circuit through the C³I module. A higher resistance would indicate a fault in CKT 423, a poor C³I module connection, or a faulty C³I module.
3. If test light was "ON" when connected from 12 volts to ECM harness terminal "D5", either CKT 423 is shorted to ground or the C³I module is faulty.
4. Checks to see if the C³I module switches when the bypass circuit is energized by 12 volts, through the test light. If the C³I module actually switches, the ohmmeter reading should shift to over 8,000 ohms.

5. Disconnecting the ignition module should make the ohmmeter read as if it were monitoring an open circuit (infinite reading). Otherwise, CKT 423 is shorted to ground.

Diagnostic Aids:

An intermittent may be caused by a poor connection, rubbed through wire insulation, or a wire broken inside the insulation. Check For:
- **Poor Connection or Damaged Harness** Inspect ECM harness connectors for backed out terminals "B4" or "D5", improper mating, broken locks, improperly formed or damaged terminals, poor terminal to wire connection, and damaged harness.
- **Intermittent Test** If connections and harness check OK, observe a digital voltmeter connected from the affected terminal to ground while moving related connectors and wiring harness. If the failure is induced, the voltage reading will abruptly change.

1988–90 3.0L (VIN L)
SKYLARK AND CUTLASS CALAIS

CODE 42
ELECTRONIC SPARK TIMING (EST) CIRCUIT
3.0L (VIN L) "N" SERIES (PORT)

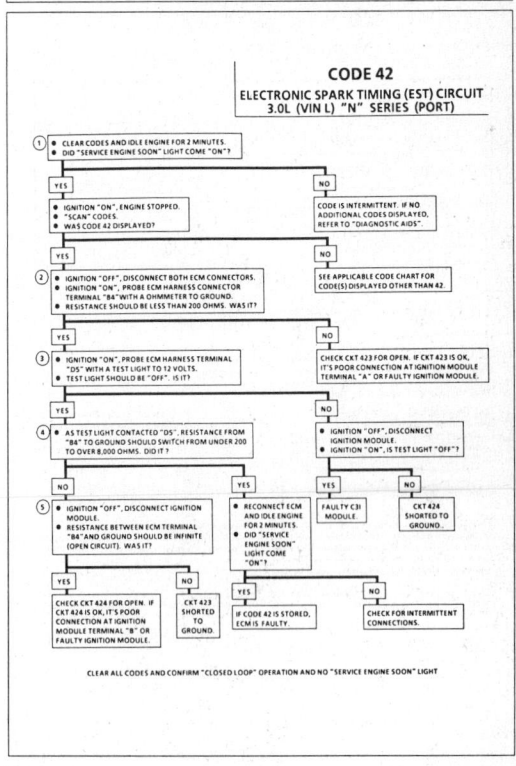

CLEAR ALL CODES AND CONFIRM "CLOSED LOOP" OPERATION AND NO "SERVICE ENGINE SOON" LIGHT

1988–90 3.0L (VIN L)
SKYLARK AND CUTLASS CALAIS

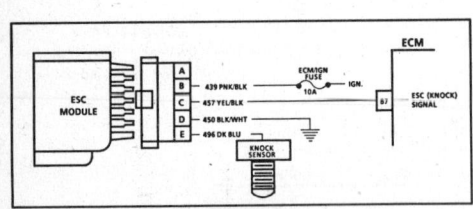

CODE 43
ELECTRONIC SPARK CONTROL (ESC) CIRCUIT
3.0L (VIN L) "N" SERIES (PORT)

Circuit Description:

The ESC system is comprised of a knock sensor and an ESC module.

As long as the ESC module is sending a voltage signal (8 to 10 volts) to the ECM (no detonation detected by the ESC sensor) the calculated spark advance (EST) remains unaffected by the ESC input.

When the sensor detects detonation, the module turns "OFF" the circuit to the ECM, and the voltage at ECM terminal "B7" drops to 0 volts. The ECM then retards EST as much as 20° in one (1) degree increments to reduce detonation. This happens fast and frequently enough that if looking at this signal with a DVM, you will not see 0 volts, but an average voltage somewhat less than what is normal with no detonation.

A loss of the knock sensor signal or a loss of ground at ESC module would cause the signal at the ECM to remain high. This condition would result in the ECM controlling EST as if no detonation were occuring. The EST would not be retarded, and detonation could become severe enough under heavy engine load conditions to result in pre-ignition and potential engine damage.

Loss of the ESC signal to the ECM would cause the ECM to constantly retard the EST to its max retard of 20° from the spark table. This could result in sluggish performance and cause a Code 43 to set.

Code 43 will set when:
- Engine Running
- ESC input signal has been low more than 2.2 seconds

Test Description: Numbers below refer to circled numbers on the diagnostic chart.
1. If the ECM data (knock signal) display is fluctuating widely, the ECM is monitoring a low voltage signal on CKT (457) at ECM terminal "B7".
2. Probing ESC harness terminal "C" with a test light connected to 12 volts should result in the "OLD PA3" (knock signal) display holding a steady reading over 8 volts having been applied to ECM terminal "B7" through CKT 457.
3. If over 6 volts is measured at ECM terminal "B7", CKT 457 is OK, the fault is due to a poor connection at the ECM, or the ECM is faulty.

Diagnostic Aids:

An intermittent may be caused by a poor connection, rubbed through wire insulation, or a wire broken inside the insulation.

Check For:
- **Poor Connection or Damaged Harness** Inspect ECM harness connectors for backed out terminal "B7," improper mating, broken locks, improperly formed or damaged terminals, poor terminal to wire connection, and damaged harness.
- **Intermittent Test** If connections and harness check OK, "Scan" knock signal (OLD PA3) while moving related connectors and wiring harness. If the failure is induced, the display will abruptly change. This may help to isolate the location of the malfunction.

1988–90 3.0L (VIN L)
SKYLARK AND CUTLASS CALAIS

CODE 43
ELECTRONIC SPARK CONTROL (ESC) CIRCUIT
3.0L (VIN L) "N" SERIES (PORT)

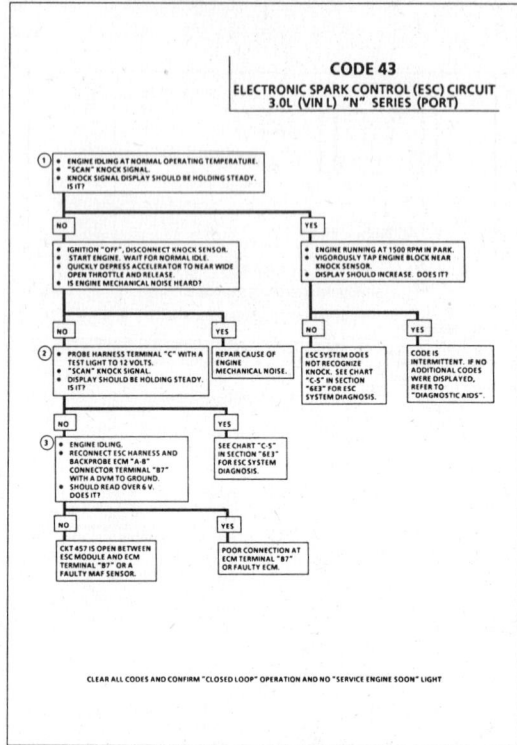

CLEAR ALL CODES AND CONFIRM "CLOSED LOOP" OPERATION AND NO "SERVICE ENGINE SOON" LIGHT

1988–90 3.0L (VIN L)
SKYLARK AND CUTLASS CALAIS

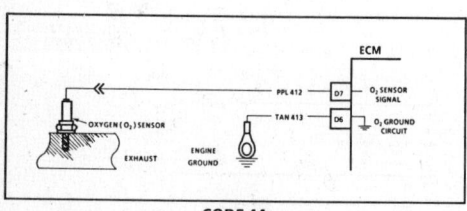

CODE 44
OXYGEN SENSOR CIRCUIT
(LEAN EXHAUST INDICATED)
3.0L (VIN L) "N" SERIES (PORT)

Circuit Description:

The ECM supplies a voltage of about .45 volt between terminals "D6" and "D7". (If measured with a 10 megohm digital voltmeter, this may read as low as .32 volt). The O_2 sensor varies the voltage within a range of about 1 volt if the exhaust is rich, down through about .10 volts if exhaust is lean.

The sensor is like an open circuit and produces no voltage when it is below about 360°C (600°F). An open sensor circuit or cold sensor causes "Open Loop" operation.

Test Description: Numbers below refer to circled numbers on the diagnostic chart.
1. Code 44 is set when the O_2 sensor signal voltage on CKT 412
   - Remains below .2 volt for 60 seconds or more
   - And the system is operating in "Closed Loop"

Diagnostic Aids:

Using the "Scan," observe the block learn values at different rpm and air flow conditions. The "Scan" also displays the block cells, so the block learn values can be checked in each of the cells to determine when the Code 44 may have been set. If the conditions for Code 44 exists, the block learn values will be around 150.
- **O_2 Sensor Wire** Sensor pigtail may be mispositioned and contacting the exhaust manifold.
- Check for intermittent ground in wire between connector and sensor.

- **MAF Sensor** A mass air flow (MAF) sensor output that causes the ECM to sense a lower than normal air flow will cause the system to go lean. Disconnect the MAF sensor, and if the lean condition is gone, replace the MAF sensor.
- **Lean Injector(s)** Perform injector balance test CHART C-2A.
- **Fuel Contamination** Water, even in small amounts, near the in-tank fuel pump inlet can be delivered to the injectors. The water causes a lean exhaust and can set a Code 44.
- **Fuel Pressure** System will be lean if pressure is too low. It may be necessary to monitor fuel pressure while driving the car at various road speeds and/or loads to confirm. See fuel system diagnosis CHART A-7.
- **Exhaust Leaks** If there is an exhaust leak, the engine can cause outside air to be pulled into the exhaust and past the sensor. Vacuum or crankcase leaks can cause a lean condition.
- If the above are OK, it is a faulty oxygen sensor.

1988–90 3.0L (VIN L)
SKYLARK AND CUTLASS CALAIS

CODE 44
OXYGEN SENSOR CIRCUIT
(LEAN EXHAUST INDICATED)
3.0L (VIN L) "N" SERIES (PORT)

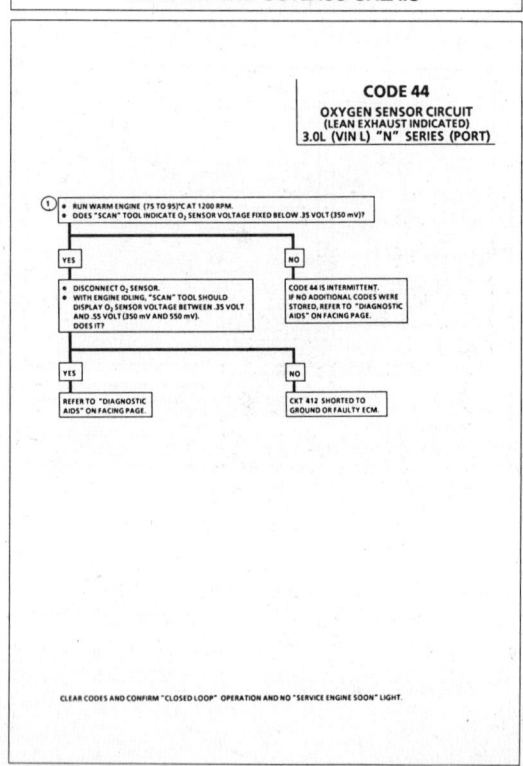

CLEAR CODES AND CONFIRM "CLOSED LOOP" OPERATION AND NO "SERVICE ENGINE SOON" LIGHT.

1988–90 3.0L (VIN L)
SKYLARK AND CUTLASS CALAIS

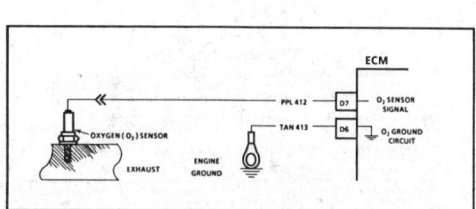

CODE 45
OXYGEN SENSOR CIRCUIT
(RICH EXHAUST INDICATED)
3.0L (VIN L) "N" SERIES (PORT)

Circuit Description:
The ECM supplies a voltage of about .45 volt between terminals "D6" and "D7". (If measured with a 10 megohm digital voltmeter, this may read as low as .32 volt). The O_2 sensor varies the voltage within a range of about 1 volt if the exhaust is rich, down through about .10 volt if exhaust is lean.
The sensor is like an open circuit and produces no voltage when it is below about 360°C (600°F). An open sensor circuit or cold sensor causes "Open Loop" operation.

Test Description: Numbers below refer to circled numbers on the diagnostic chart.
1. Code 45 is set when the O_2 sensor signal voltage or CKT 412
   • Remains above .7 volt for 30 seconds and in "Closed Loop"
   • Engine time after start is 1 minute or more
   • Throttle angle between 3% and 45%

Diagnostic Aids:

Using the "Scan," observe the block learn values at different rpm and air flow conditions. The "Scan" also displays the block cells, so the block learn values can be checked in each of the cells to determine when the Code 45 may have been set. If the conditions for Code 45 exists, the block learn values will be around 115.
• **Fuel Pressure System** will go rich if pressure is too high. The ECM can compensate for some increase. However, if it gets too high, a Code 45 may be set. See fuel system diagnosis CHART A-7.
• **Rich Injector** Perform injector balance test CHART C-2A.
• **Leaking Injector** See CHART A-7.
• Check for fuel contaminated oil.

• **HEI Shielding** An open ground CKT 453 (ignition system reflow) may result in EMI, or induced electrical "noise". The ECM looks at this "noise" as reference pulses. The additional pulses result in a higher than actual engine speed signal. The ECM then delivers too much fuel causing system to go rich. Engine tachometer will also show higher than actual engine speed, which can help in diagnosing this problem.
• **Canister Purge** Check for fuel saturation. If full of fuel, check canister control and hoses.
• **MAF Sensor** An output that causes the ECM to sense a higher than normal airflow can cause the system to go rich. Disconnecting the MAF sensor will allow the ECM to set a fixed value for the sensor. Substitute a different MAF sensor if the rich condition is gone while the sensor is isconnected or replace MAF sensor.
• Check for leaking fuel pressure regulator diaphragm by checking vacuum line to regulator for fuel.
• **TPS** An intermittent TPS output will cause the system to go rich due to a false indication of the engine accelerating.
• **EGR** An EGR staying open (especially at idle) will cause the O_2 sensor to indicate a rich exhaust, and this could result in a Code 45.

1988–90 3.0L (VIN L)
SKYLARK AND CUTLASS CALAIS

CODE 45
OXYGEN SENSOR CIRCUIT
(RICH EXHAUST INDICATED)
3.0L (VIN L) "N" SERIES (PORT)

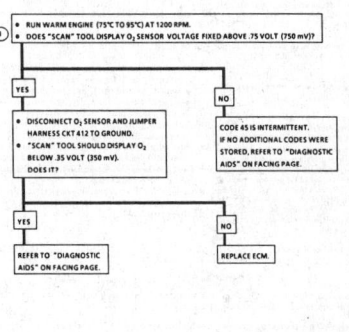

• RUN WARM ENGINE (75°C TO 95°C) AT 1200 RPM.
• DOES "SCAN" TOOL DISPLAY O_2 SENSOR VOLTAGE FIXED ABOVE .75 VOLT (750 mV)?

YES
• DISCONNECT O_2 SENSOR AND JUMPER HARNESS CKT 412 TO GROUND.
• "SCAN" TOOL SHOULD DISPLAY O_2 BELOW .35 VOLT (350 mV).
  DOES IT?

YES
REFER TO "DIAGNOSTIC AIDS" ON FACING PAGE.

NO
REPLACE ECM.

NO
CODE 45 IS INTERMITTENT. IF NO ADDITIONAL CODES WERE STORED, REFER TO "DIAGNOSTIC AIDS" ON FACING PAGE.

CLEAR CODES AND CONFIRM "CLOSED LOOP" OPERATION AND NO "SERVICE ENGINE SOON" LIGHT.

1988–90 3.0L (VIN L)
SKYLARK AND CUTLASS CALAIS

CODE 51
PROM ERROR
(FAULTY OR INCORRECT CALPAK)

CHECK THAT ALL PINS ARE FULLY INSERTED IN THE SOCKET. IF OK, REPLACE PROM, CLEAR MEMORY AND RECHECK. IF CODE 51 REAPPEARS, REPLACE ECM.

CLEAR ALL CODES AND CONFIRM "CLOSED LOOP" OPERATION AND NO "SERVICE ENGINE SOON" LIGHT

CODE 52
CALPAK ERROR
(FAULTY OR INCORRECT CALPAK)

CHECK THAT ALL PINS ARE FULLY INSERTED IN THE SOCKET. IF OK, REPLACE CALPAK, CLEAR MEMORY AND RECHECK. IF CODE 52 REAPPEARS, REPLACE ECM.

CLEAR ALL CODES AND CONFIRM "CLOSED LOOP" OPERATION AND NO "SERVICE ENGINE SOON" LIGHT

CODE 55
ECM ERROR

REPLACE ELECTRONIC CONTROL MODULE (ECM).

CLEAR ALL CODES AND CONFIRM "CLOSED LOOP" OPERATION AND NO "SERVICE ENGINE SOON" LIGHT

1988–90 3.0L (VIN L)
SKYLARK AND CUTLASS CALAIS

SECTION B
SYMPTOMS

TABLE OF CONTENTS

Before Starting
Intermittents
Hard Start
Hesitation, Sag, Stumble
Surges and/or Chuggle
Lack of Power, Sluggish, or Spongy
Detonation/Spark Knock
Cuts Out, Misses
Backfire
Poor Fuel Economy
Dieseling, Run On
Intermittent Stalling on Deceleration or when Vehicle is Stopped
Rough, Unstable, or Incorrect Idle, Stalling
Excessive Exhaust Emissions or Odors
Chart B-1 - Restricted Exhaust System Check

BEFORE STARTING

Before using this section you should have performed the diagnostic circuit check and found out that:
1. The ECM and "Service Engine Soon" light are operating.
2. There are no trouble codes stored, or there is a trouble code but no "Service Engine Soon" light.
3. The fuel control system is operating OK.
   Verify the customer complaint, and locate the correct symptom below. Check the items indicated under that symptom.
   If the "ENGINE CRANKS BUT WILL NOT RUN", see CHART A-3.

Several of the symptom procedures below call for a careful visual check. This check should include:
• ECM grounds for being clean and tight.
• Vacuum hoses for splits, kinks, and proper connections, as shown on emission control information label.
• Air leaks at throttle body mounting and intake manifold.
• Air leaks between MAF sensor and throttle body.
• Ignition wires for cracking, hardness, proper routing, and carbon tracking.
• Wiring for proper connections, pinches, and cuts ECM power grounds and battery feed circuits.
The importance of this step cannot be stressed too strongly - it can lead to correcting a problem without further checks and can save valuable time.

1988–90 3.0L (VIN L) SKYLARK AND CUTLASS CALAIS

INTERMITTENTS

Problem may or may not turn "ON" the "Service Engine Soon" light, or store a code.

Do not use the trouble code charts for intermittent problems. The fault must be present to locate the problem. If a fault is intermittent, use of trouble code charts may result in replacement of good parts.

- Most intermittent problems are caused by faulty electrical connections or wiring. Perform careful check as described at start of Section "B". Check for:
  - Poor mating of the connector halves, or terminals not fully seated in the connector body (backed out).
  - Improperly formed or damaged terminals. All connector terminals in problem circuit should be carefully reformed to increase contact tension.
  - Poor terminal to wire connection. This requires removing the terminal from the connector body to check.
- If a visual (physical) check does not find the cause of the problem, the car can be driven with a voltmeter connected to a suspected circuit. An abnormal voltage reading, when the problem occurs, indicates the problem may be in that circuit. If the wiring and connectors check OK and a trouble code was stored for a circuit having a sensor, except for Codes 43, 44 and 45, substitute a known good sensor and recheck.
  A "Scan" tool can, also, be used by monitoring the suspected problem circuit and moving related wiring and connectors.

An intermittent "Service Engine Soon" light, with no stored code, may be caused by:

- Ignition coil shorted to ground and arcing at spark plug wires or plugs.
- "Service Engine Soon" light wire to ECM shorted to ground (CKT 419).
- Diagnostic "Test" terminal wire to ECM, shorted to ground (CKT 451).
- ECM grounds. See ECM wiring diagrams.
- Loss of trouble code memory. To check, disconnect TPS and idle engine until "Service Engine Soon" light comes "ON". Code 22 should be stored, and kept in memory when ignition is turned "OFF". If not, the ECM is faulty.
- Check for an electrical system interference caused by a defective relay, ECM driven solenoid, or switch. They can cause a sharp electrical surge. Normally, the problem will occur when the faulty component is operated.
- Check for improper installation of electrical options, such as lights, 2-way radios, etc.
- EST wires should be kept away from spark plug wires, distributor wires, distributor housing, coil, and generator. Wire from CKT 453 to distributor should be a good ground.
- Check for open diode across A/C compressor clutch, and for other open diodes (see wiring diagrams).

HARD START

Definition: Engine cranks OK, but does not start for a long time. Does eventually run, or may start but immediately dies.

- Perform careful check as described at start of Section "B".
- Make sure driver is using correct starting procedure.
- **CHECK:**
  - TPS for sticking, binding, or a high TPS voltage with the throttle closed (should read less than .700 volts).
  - High resistance in coolant sensor circuit or faulty sensor. With a "Scan" tool, compare coolant temperature with MAT temperature on a cold engine.
  - Fuel pressure CHART A-7.
  - Water contaminated fuel.
  - Fuel pump relay - CHART A-5.
- Ignition system - Check distributor for:
  - Proper output with spark checker J-26792, ST-125 or equivalent.
  - Worn shaft.
  - Bare and shorted wires.
  - Pickup coil resistance and connections.
  - Loose ignition coil ground.
  - Moisture in distributor cap.
- A faulty in-tank fuel pump check valve will allow the fuel in the lines to drain back to the tank after the engine is stopped. This occurs only in very hot weather on re-starts. To check for this condition: See CHART A-7.
- Remove spark plugs. Check for wet plugs, cracks, wear, improper gap, burned electrodes, or heavy deposits. Repair or replace as necessary.

1988–90 3.0L (VIN L) SKYLARK AND CUTLASS CALAIS

HESITATION, SAG, STUMBLE

Definition: Momentary lack of response as the accelerator is pushed down. Can occur at all car speeds. Usually most severe when first trying to make the car move, as from a stop sign. May cause the engine to stall if severe enough.

- Perform careful visual check as described at start of Section "B".
- **CHECK:**
  - Fuel pressure. See CHART A-7. Also, for water contaminated fuel.
  - Spark plugs for being fouled or faulty wiring.
  - Mem-cal or PROM number. Also service bulletins for latest PROM or mem-cal.
  - TPS for binding or sticking. Voltage should increase at a steady rate as throttle is moved toward WOT.
  - Map sensor - CHART C-1D.
  - Perform injector balance test CHART C-2A.
  - Generator output voltage at rpm greater than idle. Repair, if less than 9, or more than 16 volts.
  - Ignition ground, CKT 453.
  - Canister purge system for proper operation.
  - Engine thermostat - functioning correctly & proper heat range.

SURGES AND/OR CHUGGLE

Definition: Engine power variation under steady throttle or cruise. Feels like the car speeds up and slows down with no change in the accelerator pedal.

- Be sure driver understands transmission converter clutch and A/C compressor operation in owner's manual.
- Perform careful visual inspection as described at start of Section "B".
- **CHECK:**
  - Generator output voltage at rpm greater than idle. Repair, if less than 9, or more than 16 volts.
    "Scan" reading of VSS. Does it match vehicle speedometer?
  - Vacuum lines for kinks or leaks.
  - In-line fuel filter. Replace if dirty or plugged.
  - Fuel pressure while condition exists. See CHART A-7.
  - Oxygen sensor for silicone contamination, from fuel, or use of improper RTV sealant. The sensor may have a white powdery coating which results in a high but false signal voltage (rich exhaust indication). The ECM will then reduce the amount of fuel delivered to the engine, causing a severe driveability problem.
  - Remove spark plugs. Check for cracks, wear, improper gap, burned electrodes, or heavy deposits.

LACK OF POWER, SLUGGISH, OR SPONGY

Definition: Engine delivers less than expected power. Little or no increase in speed when accelerator pedal is pushed down part way.

- Compare customer's car to similar unit. Make sure the customer's car has an actual problem.
- Remove air cleaner and check air filter for dirt, or for being plugged. Replace as necessary.
- **CHECK:**
  - Restricted fuel filter, contaminated fuel or improper fuel pressure. See CHART A-7.
  - ECM grounds. See wiring diagrams.
  - Exhaust system for possible restriction: Inspect exhaust system for damaged, or collapsed pipes. Inspect muffler for heat distress or possible internal failure.
  - Generator output voltage at rpm greater than idle. Repair, if less than 9 or more than 16 volts.
  - Secondary voltage, using a shop ocilliscope, or a spark tester J-26792 (ST-125), or equivalent.
  - Transmission torque convertor operation.
  - Engine valve timing and compression.
  - Engine for proper or worn camshaft.

1988–90 3.0L (VIN L) SKYLARK AND CUTLASS CALAIS

DETONATION /SPARK KNOCK

Definition: A mild to severe ping, usually worse under acceleration. The engine makes sharp metallic knocks that change with throttle opening. Sounds like popcorn popping.

- Check for obvious overheating problems:
  - Low coolant.
  - Loose water pump belt.
  - Restricted air/water flow thru radiator.
  - Faulty or incorrect thermostat.
  - Inoperative electric cooling fan circuit. See CHART C-12.
  - Coolant sensor which has shifted in value.
  - Correct coolant solution - should be a 50/50 mix of GM #1052753 anti-freeze coolant (or equiv.) and water.
- **CHECK:**
  - PROM - Be sure it's the correct one.
  - Check for poor fuel quality, proper octane rating.
  - Ignition timing. See vehicle emission control information label.
  - TCC operation - CHART C-8.
  - Fuel system pressure. See CHART A-7.
  - Remove carbon with top engine cleaner. Follow instructions on can.
  - Valve oil seals for leaking.
  - Check for incorrect basic engine parts such as cam, heads, pistons, etc.

CUTS OUT OR MISSES

Definition: Steady pulsation or jerking that follows engine speed, usually more pronounced as engine load increases. The exhaust has a steady spitting sound at idle or low speed.

- Perform careful visual check as described at start of section "B".
- Misses at idle:
  1. With engine idling, disconnect IAC motor. Remove one spark plug wire at a time using insulated pliers.
  2. If there is an rpm drop on all cylinders, go to ROUGH, UNSTABLE, OR INCORRECT IDLE, OR STALLING symptom. Reconnect IAC motor.
  3. If there is no rpm drop on one or more cylinders, or excessive variation in drop, check for spark on the suspected cylinder(s) with J-26792 (ST-125) spark tester or equivalent.

If spark exists:
  Inspect plugs for:
  - Cracks, wear, improper gap.
  - Burned electrodes, or heavy deposits.
  - Perform compression check on questionable cylinder.

If no spark:
  ON DISTRIBUTOR TYPE SYSTEM
    Check wire resistance (should not exceed 30,000 ohms), also, check rotor and distributor cap.
  ON NON-DISTRIBUTOR SYSTEMS
    Ground the opposite plug wire of the affected coil pair. If tester now sparks, replace the spark plug of the wire which was jumpered to ground. If tester still does not spark, refer to appropriate ignition chart

- Cuts Out Under Load:
  Test for spark on each plug wire with a J-26792 spark tester or equivalent.

If there is spark on all cylinders, remove plugs and check for:
  - Cracks, wear, improper gap.
  - Burned electrodes, or heavy deposits.
- If above checks did not correct problem, check the following:
  - Disconnect all injector harness connectors. Connect J-34730-2 injector test light or equivalent 6 volt test light between the harness terms, of each injector connector and note light while cranking. If test light fails to blink at any connector, it is a faulty injector drive circuit harness, connector, or terminal.
  - Perform the injector balance test. See CHART C-2A.
  - Visually inspect distributor cap and rotor for moisture, dust, cracks, burns, etc. Spray cap and plug wires with fine water mist to check for shorts.
  - Fuel system - Plugged fuel filter, water, low pressure. See CHART A-7.
  - Valve timing.
  - Remove rocker covers. Check for bent pushrods, worn rocker arms, broken valve springs, worn camshaft lobes. Repair as necessary. See Section "6A".
  - Perform compression check.

1988–90 3.0L (VIN L) SKYLARK AND CUTLASS CALAIS

BACKFIRE

Definition: Fuel ignites in intake manifold, or in exhaust system, making a loud popping noise.

- **CHECK:**
  - Loose wiring connector or air duct at MAF sensor.
  - Compression - Look for sticking or leaking valves.
  - Valve timing.
  - Output voltage of ignition coil using a shop ocilliscope or spark tester J-26792 (ST-125) or equivalent.
  - Spark plugs for crossfire also inspect (distributor cap, spark plug wires, and proper routing of plug wires).
  - Ignition system for intermittent condition.
  - Engine timing - see emission control information label.

POOR FUEL ECONOMY

Definition: Fuel economy, as measured by an actual road test, is noticeably lower than expected. Also, economy is noticeably lower than it was on this car at one time, as previously shown by an actual road test.

- Perform careful visual check as described at start of Section "B".
- **CHECK:**
  - Engine thermostat for faulty part (always open) or for wrong heat range.
  - Ignition timing. See emission control information label.
  - TCC for proper operation. See CHART C-8.

DIESELING, RUN-ON

Definition: Engine continues to run after key is turned "OFF", but runs very roughly. If engine runs smoothly, check ignition switch and adjustment.

- Check injectors for leaking. See CHART A-7.

INTERMITTENT STALLING ON DECELERATION OR WHEN VEHICLE IS STOPPED

Definition: If this condition is experienced, it may be caused by residue from the PCV system accumulating inside the throttle body and limiting air flow past the throttle plate in the idle position.

1. Following the procedure check the minimum air rate.
2. If below 500 rpm in drive, turn the engine "OFF" and remove the air intake duct from the throttle body. Clean the throttle body bore in the area behind the throttle plate using a shop towel with "GM Top Engine Cleaner" (AC-Delco part #1052626) or "AC-Delco Carburetor Tune-Up Conditioner" (part #X66-P) or equivalent product that doesn't contain methyl ethyl ketone. It should be noted that what appears to be just a small amount of accumulation can be enough to limit air flow.
3. Recheck minimum air rate and adjust, if necessary, to 500 rpm.
4. Adjust throttle position sensor to .50 - .59 volts.
5. Inspect for crankshaft sensor-to-crankshaft balancer interrupter vane clearance. If there are any signs of contact, the sensor should be replaced
6. With the engine at operating temperature, reset the IAC motor by starting the engine, running at least 10-15 seconds, then shutting it "OFF" for 30 seconds. This should be done at least twice.

NOTE: If the vehicle will intermittently not stay running while driving at highway speed and not with closed throttle, the cause would more likely be a component failure (i.e. MAF sensor, C3I system, etc.).

1988–90 3.0L (VIN L)
SKYLARK AND CUTLASS CALAIS

ROUGH, UNSTABLE, OR INCORRECT IDLE, STALLING

Definition: The engine runs unevenly at idle. If bad enough, the car may shake. Also, the idle may vary in rpm (called "hunting"). Either condition may be bad enough to cause stalling. Engine idles at incorrect speed.

- Perform careful visual check as described at start of Section "B".
- CHECK:
  - Throttle linkage for sticking or binding.
  - Ignition timing. See emission control information label.
  - Generator output voltage at above idle rpm. Repair if less than 9 or more than 16 volts.
  - P/N switch circuit. See CHART C-1A.
  - Injector balance. See CHART C-2A.
  - PCV valve for proper operation by placing finger over inlet hole in valve end several times. Valve should snap back. If not, replace valve.
  - Evaporative emission control system. CHART C-3.

- Power steering pressure switch input. See CHART C-1E or use "Scan" tool.
- Run a cylinder compression check.

- Inspect oxygen sensor for silicone contamination from fuel or use of improper RTV sealant. The sensor will have a white powdery coating, and will result in a high, but false, signal voltage (rich exhaust indication). The ECM will reduce the amount of fuel delivered to the engine, causing a severe driveability problem.
- Check for fuel in pressure regulator hose. If present, replace regulator assembly.
- Check ignition system, wires, plugs, rotor, etc.

EXCESSIVE EXHAUST EMISSIONS OR ODORS
Definition: Vehicle fails an emission test. Vehicle has excessive "rotten egg" smell.
Excessive odors do not necessarily indicate excessive emissions.

- Perform "Diagnostic Circuit Check."
- IF TEST SHOWS EXCESSIVE CO AND HC, (or also has excessive odors):
  - Check items which cause car to run RICH.
    - Make sure engine is at normal operating temperature.
- CHECK:
  - Fuel pressure. See CHART A-7.
  - Canister for fuel loading. See CHART C-3.
  - Injector balance. See CHART C-2A.
  - PCV valve for being plugged, stuck, or blocked PCV hose, or fuel in the crankcase.
  - Spark plugs, plug wires, and ignition components.
  - Check for lead contamination of catalytic converter (look for removal of fuel filler neck restrictor).
  - Check for properly installed fuel cap.

- If the system is running rich, (block learn less than 118), refer to "Diagnostic Aids" on facing page of Code 45.
- IF TEST SHOWS EXCESSIVE NOx:
  - Check items which cause car to run LEAN, or to run too hot.
    - EGR valve for not opening. See CHART C-7.
    - Vacuum leaks.
    - Coolant system and coolant fan for proper operation. See CHART C-12.
    - Remove carbon with top engine cleaner. Follow instructions on can.
- If the system is running lean, (block learn greater than 138), refer to "Diagnostic Aids" on facing page of Code 44.

1988–90 3.0L (VIN L)
SKYLARK AND CUTLASS CALAIS

CHART B-1
RESTRICTED EXHAUST SYSTEM CHECK
ALL ENGINES

Proper diagnosis for a restricted exhaust system is essential before any components are replaced. Either of the following procedures may be used for diagnosis, depending upon engine or tool used:

CHECK AT A.I.R. PIPE:

1. Remove the rubber hose at the exhaust manifold A.I.R. pipe check valve. Remove check valve.
2. Connect a fuel pump pressure gauge to a hose and nipple from a Propane Enrichment Device (J26911) (see illustration).
3. Insert the nipple into the exhaust manifold A.I.R. pipe.

OR **CHECK AT O₂ SENSOR:**

1. Carefully remove O₂ sensor.
2. Install Borroughs Exhaust Backpressure Tester (BT 8515 or BT 8603) or equivalent in place of O₂ sensor (see illustration).
3. After completing test described below, be sure to coat threads of O₂ sensor with anti-seize compound P/N 5613695 or equivalent prior to re-installation.

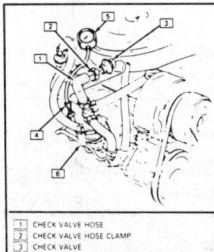

1	CHECK VALVE HOSE
2	CHECK VALVE HOSE CLAMP
3	CHECK VALVE
4	HOSE & NIPPLE FROM PROPANE ENRICHMENT DEVICE J 26911
5	FUEL PUMP PRESSURE GAGE
6	RIGHT A.I.R. MANIFOLD

1	EXHAUST MANIFOLD
2	O₂ SENSOR
3	BACK PRESSURE TESTER

DIAGNOSIS:

1. With the engine at normal operating temperature and running at 2500 rpm, observe the exhaust system backpressure reading on the gauge.
2. If the backpressure exceeds 1 1/4 psi (8.62 kPa), a restricted exhaust system is indicated.
3. Inspect the entire exhaust system for a collapsed pipe, heat distress, or possible internal muffler failure.
4. If there are no obvious reasons for the excessive backpressure, a restricted catalytic converter should be suspected, and replaced using current recommended procedures.

1988–90 3.0L (VIN L)
SKYLARK AND CUTLASS CALAIS

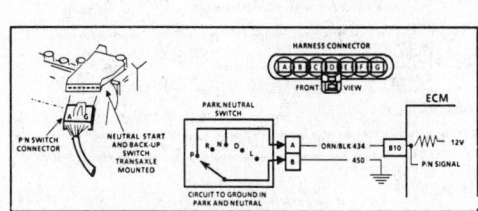

CHART C-1A
PARK/NEUTRAL SWITCH DIAGNOSIS
(AUTO TRANSMISSION ONLY)
3.0L (VIN L) "N" SERIES (PORT)

Circuit Description:
The park/neutral switch contacts are a part of the neutral start switch and are closed to ground in park or neutral, and open in drive ranges.

The ECM supplies ignition voltage through a current limiting resistor to CKT 434 and senses a closed switch when the voltage on CKT 434 drops to less than one volt.

The ECM uses the P/N signal as one of the inputs to control:
Idle Air Control
VSS Diagnostics

Test Description: Numbers below refer to circled numbers on the diagnostic chart.

1. Checks for a closed switch to ground in park position. If using an ohmmeter instead of a test light the resistance will be low, indicating continuity to ground.
2. Checks for an open switch in drive range. If using an ohmmeter instead of a test light to 12 volts, the resistance will be high or infinity, indicating an open switch.

3. Checks to this point indicate the P/N switch and wiring are OK, however, the ECM signal voltage on CKT 434 may be missing. To check, reconnect ECM. Either back probe ECM connector CKT 434 with selector in drive or disconnect P/N switch and probe harness connector CKT 434 with a voltmeter to ground.

1988–90 3.0L (VIN L)
SKYLARK AND CUTLASS CALAIS

CHART C-1A
PARK/NEUTRAL SWITCH DIAGNOSIS
(AUTO TRANSAXLE ONLY)
3.0L (VIN L) "N" SERIES (PORT)

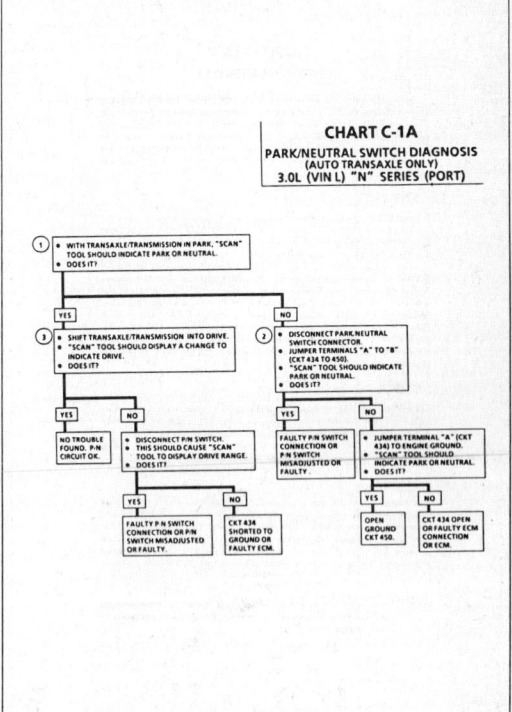

1988–90 3.0L (VIN L)
SKYLARK AND CUTLASS CALAIS

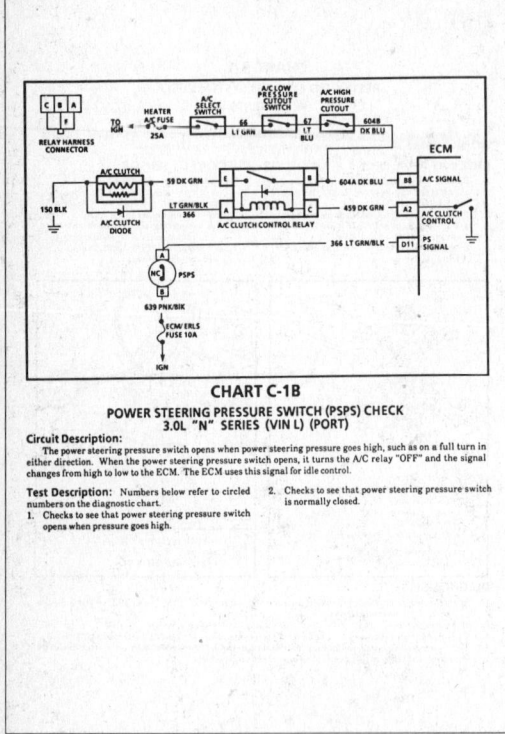

CHART C-1B
POWER STEERING PRESSURE SWITCH (PSPS) CHECK
3.0L "N" SERIES (VIN L) (PORT)

Circuit Description:
The power steering pressure switch opens when power steering pressure goes high, such as on a full turn in either direction. When the power steering pressure switch opens, it turns the A/C relay "OFF" and the signal changes from high to low to the ECM. The ECM uses this signal for idle control.

Test Description: Numbers below refer to circled numbers on the diagnostic chart.
1. Checks to see that power steering pressure switch opens when pressure goes high.

2. Checks to see that power steering pressure switch is normally closed.

1988–90 3.0L (VIN L)
SKYLARK AND CUTLASS CALAIS

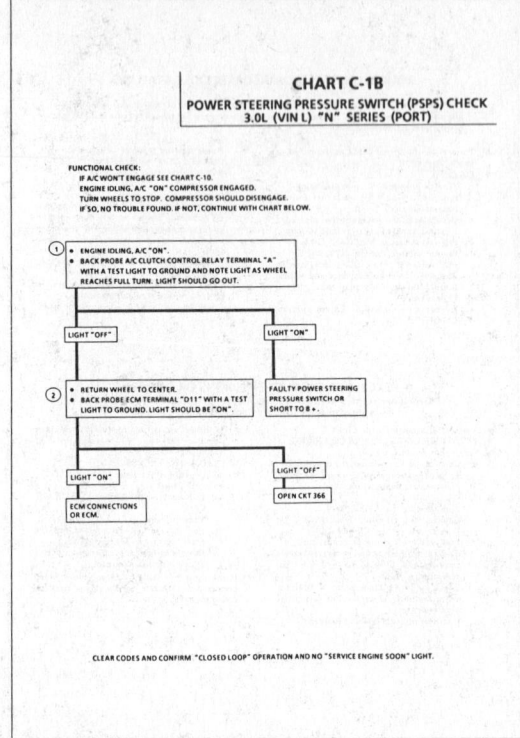

CHART C-1B
POWER STEERING PRESSURE SWITCH (PSPS) CHECK
3.0L (VIN L) "N" SERIES (PORT)

1988–90 3.0L (VIN L)
SKYLARK AND CUTLASS CALAIS

CHART C-2A
INJECTOR BALANCE TEST

The injector balance tester is a tool used to turn the injector on for a precise amount of time, thus spraying a measured amount of fuel into the manifold. This causes a drop in fuel rail pressure that we can record and compare between each injector. All injectors should have the same amount of pressure drop (± 10 kpa). Any injector with a pressure drop that is 10 kpa (or more) greater or less than the average drop of the other injectors should be considered faulty and replaced.

STEP 1

Engine "cool down" period (10 minutes) is necessary to avoid irregular readings due to "Hot Soak" fuel boiling. With ignition "OFF" connect fuel gauge J-347301 or equivalent to fuel pressure tap. Wrap a shop towel around fitting while connecting gage to avoid fuel spillage.

Disconnect harness connectors at all injectors, and connect injector tester J-34730-3, or equivalent, to one injector. On Turbo equipped engines, use adaptor harness furnished with injector tester to energize injectors that are not accessible. Follow manufacturers instructions for use of adaptor harness. Ignition must be "OFF" at least 10 seconds to complete ECM shutdown cycle. Fuel pump should run about 2 seconds after ignition is turned "ON". At this point, insert clear tubing attached to vent valve into a suitable container and bleed air from gauge and hose to insure accurate gauge operation. Repeat this step until all air is bled from gauge.

STEP 2

Turn ignition "OFF" for 10 seconds and then "ON" again to get fuel pressure to its maximum. Record this initial pressure reading. Energize tester one time and note pressure drop at its lowest point (Disregard any slight pressure increase after drop hits low point). By subtracting this second pressure reading from the initial pressure, we have the actual amount of injector pressure drop.

STEP 3

Repeat step 2 on each injector and compare the amount of drop. Usually, good injectors will have virtually the same drop. Retest any injector that has a pressure difference of 10kPa, either more or less than the average of the other injectors on the engine. Replace any injector that also fails the retest. If the pressure drop of all injectors is within 10kPa of this average, the injectors appear to be flowing properly. Reconnect them and review "Symptoms" Section "B".

NOTE: The entire test should not be repeated more than once without running the engine to prevent flooding. (This includes any retest on faulty injectors).

1988–90 3.0L (VIN L)
SKYLARK AND CUTLASS CALAIS

NOTE: If injectors are suspected of being dirty, they should be cleaned using an approved tool and procedure prior to performing this test. The fuel pressure test in Section "A", Chart A-7, should be completed prior to this test.

CHART C-2A
INJECTOR BALANCE TEST
3.0L (VIN L) "N" SERIES (PORT)

Step 1. If engine is at operating temperature, allow a 10 minute "cool down" period then connect fuel pressure gauge and injector tester.
1. Ignition "OFF".
2. Connect fuel pressure gauge and injector tester.
3. Ignition "ON".
4. Bleed off air in gauge. Repeat until all air is bled from gauge.

Step 2. Run test:
1. Ignition "OFF" for 10 seconds.
2. Ignition "ON". Record gauge pressure. (Pressure must hold steady, if not see the Fuel System diagnosis, Chart A-7, in Section "A").
3. Turn injector on, by depressing button on injector tester, and note pressure at the instant the gauge needle stops.

Step 3.
1. Repeat step 2 on all injectors and record pressure drop on each. Retest injectors that appear faulty (Any injectors that have a 10 kPa difference, either more or less, in pressure from the average). If no problem is found, review "Symptoms" Section "B".

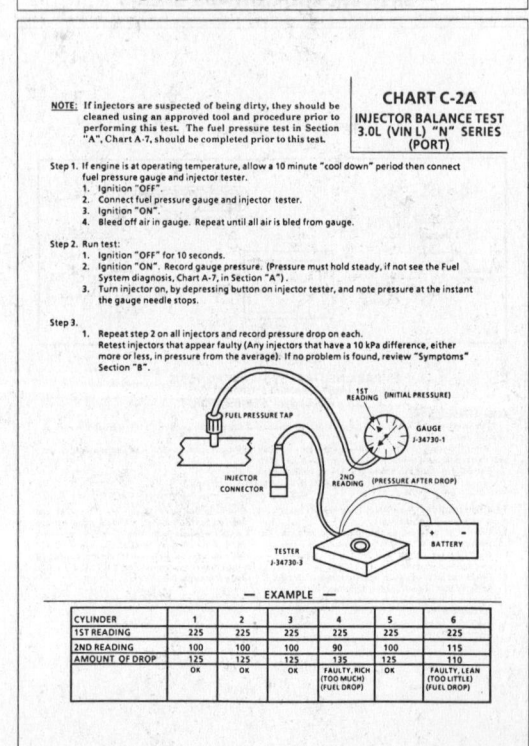

— EXAMPLE —

CYLINDER	1	2	3	4	5	6
1ST READING	225	225	225	225	225	225
2ND READING	100	100	100	90	100	115
AMOUNT OF DROP	125	125	125	135	125	110
	OK	OK	OK	FAULTY, RICH (TOO MUCH) (FUEL DROP)	OK	FAULTY, LEAN (TOO LITTLE) (FUEL DROP)

1988–90 3.0L (VIN L) SKYLARK AND CUTLASS CALAIS

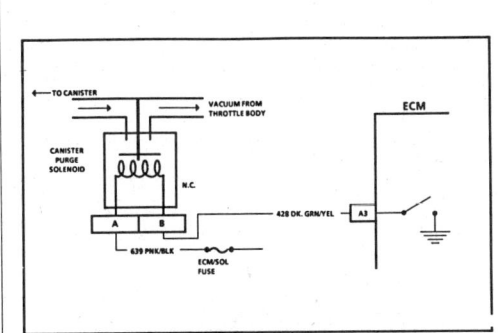

CHART C-3
CANISTER PURGE VALVE CHECK
3.0L "N" SERIES (VIN L) (PORT)

Circuit Description:
Canister purge is controlled by a solenoid that allows manifold vacuum to purge the canister when energized. The ECM supplies a ground to energize the solenoid (purge "ON").
If the diagnostic test terminal is grounded with the engine stopped or the following is met with the engine running, the purge solenoid is energized (purge "ON").
- Engine run time after start more than 1 minute.
- Coolant temperature above 80°C (176°F).
- Vehicle speed above 5 mph (8 km/h).
- Throttle off idle. TPS signal above .75 volt.

Test Description: Numbers below refer to circled numbers on the diagnostic chart.
1. Checks to see if the solenoid is opened or closed. The solenoid is normally de-energized in this step, so it should be closed.
2. Completes functional check by grounding test terminal. This should normally energize the solenoid and allow the vacuum to drop (purge "ON").
3. Checks for open or shorted solenoid circuit.
4. Checks to see if ECM control circuit or solenoid is at fault.
   Solenoid coil resistance must measure more than 20 ohms.
5. Checks to see if the short to voltage damaged original ECM.

1988–90 3.0L (VIN L) SKYLARK AND CUTLASS CALAIS

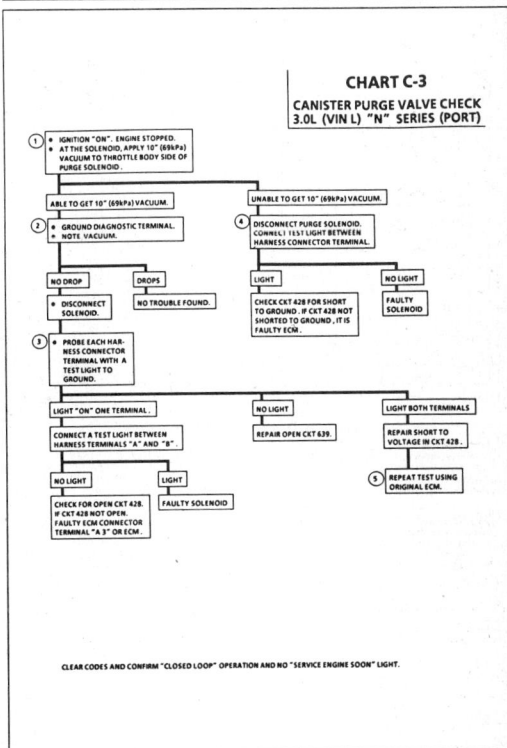

1988–90 3.0L (VIN L) SKYLARK AND CUTLASS CALAIS

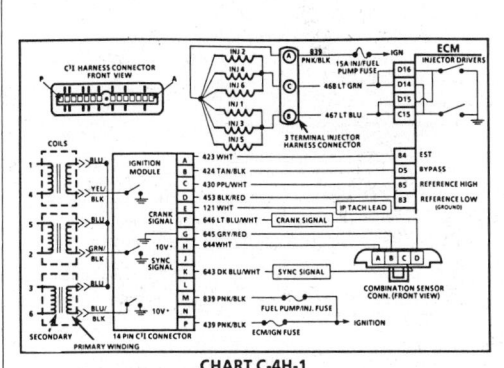

CHART C-4H-1
C3I MISFIRE AT IDLE
3.0L (VIN L) "N" SERIES (PORT)

Circuit Description:
The C3I uses a waste spark method of spark distribution. In this type of ignition system the ignition module triggers the #1/4 coil pair resulting in both #1 and #4 spark plugs firing at the same time. #1 cylinder is on the compression stroke at the same time #4 is on the exhaust stroke, resulting in a lower energy requirement to fire #4 spark plug. This leaves the remaining high voltage to fire #1 spark plug.

Test Description: Numbers below refer to circled numbers on the diagnostic chart.
1. If the "misfire" complaint exists under load only, the diagnostic chart C-4H-2 must be used. Engine rpm should drop approximately equally on all plug leads.
2. A spark tester such as a ST-125 must be used because it is essential to verify adequate available secondary voltage at the spark plug. (25,000 volts). Secondary voltage of at least (25,000 volts) must be present to jump the gap of a ST-125.
3. If ignition coils are carbon tracked, the coil tower spark plug wire nipples may be damaged.
4. By checking the secondary resistance, a coil with an open secondary may be located.
5. By switching a normally operating coil into the position of the malfunctioning one, a determination can be made as to fault being the coil or C3I module.

1988–90 3.0L (VIN L) SKYLARK AND CUTLASS CALAIS

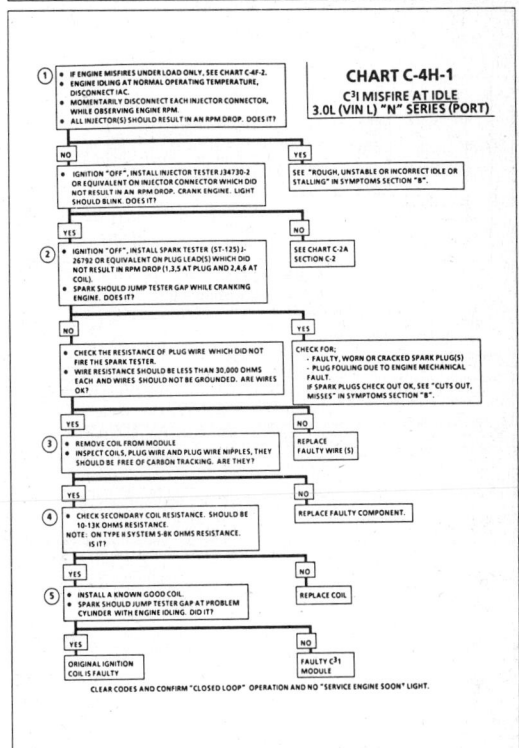

1988–90 3.0L (VIN L) SKYLARK AND CUTLASS CALAIS

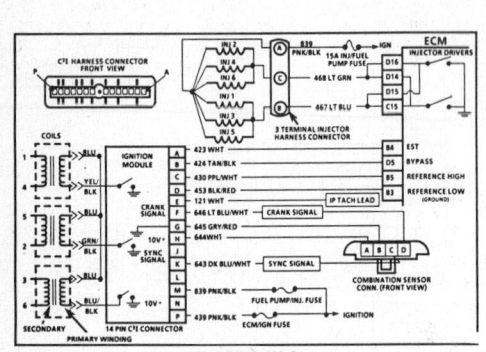

CHART C-4H-2
C³I MISFIRE UNDER LOAD
3.0L (VIN L) "N" SERIES (PORT)

Circuit Description:

The C³I uses a waste spark method of spark distribution. In this type of ignition system the ignition module triggers the #1/4 coil pair resulting in both #1 and #4 spark plugs firing at the same time. #1 cylinder is on the compression stroke at the same time #4 is on the exhaust stroke, resulting in a lower energy requirement to fire #4 spark plug. This leaves the remaining high voltage to fire #1 spark plug.

Test Description: Numbers below refer to circled numbers on the diagnostic chart.

1. If the "misfire" complaint exists at idle only, the diagnostic chart C-4H-1 must be used.
2. A spark tester such as a ST-125 must be used because it is essential to verify adequate available secondary voltage at the spark plug. (25,000 volts). Spark should jump tester gap on all 6 leads. This simulates a "load" condition.
3. If ignition coils are carbon tracked, the coil tower spark plug wire nipples may be damaged.
4. By switching a normally operating coil into the position of the malfunctioning one, a determination can be made as to fault being the coil or C³I module.

1988–90 3.0L (VIN L) SKYLARK AND CUTLASS CALAIS

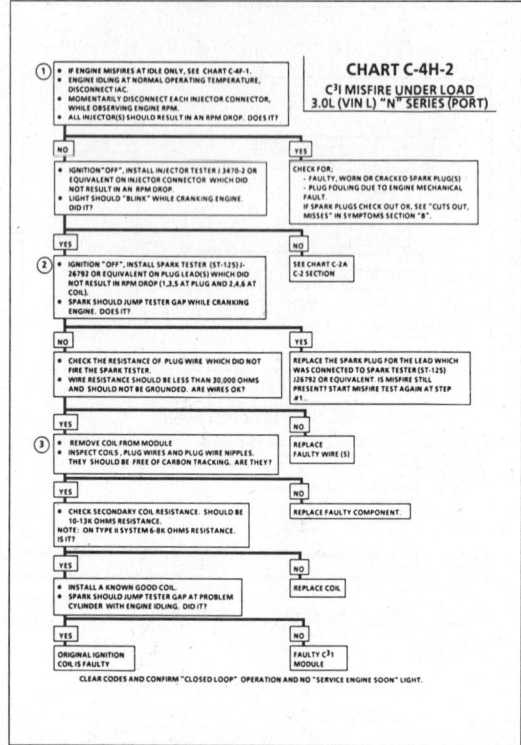

CHART C-4H-2
C³I MISFIRE UNDER LOAD
3.0L (VIN L) "N" SERIES (PORT)

1988–90 3.0L (VIN L) SKYLARK AND CUTLASS CALAIS

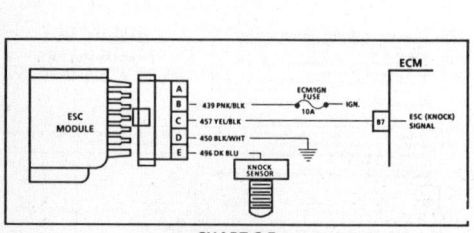

CHART C-5
ELECTRONIC SPARK CONTROL (ESC)
3.0L (VIN L) "N" SERIES (PORT)

Circuit Description:

The ESC system is comprised of a knock sensor and an ESC module.

As long as the ESC module is sending a voltage signal (8 to 10 volts) to the ECM (no detonation detected by the ESC sensor) the ECM provides normal spark advance.

When the sensor detects detonation, the module turns "OFF" the circuit to the ECM and the voltage at ECM terminal "B7" drops to 0 volts. The ECM then retards EST as much as 20° to reduce detonation. This happens fast and frequently enough that if looking at this signal with a DVM, you won't see 0 volts, but an average voltage somewhat less than what is normal with no detonation.

A loss of the knock sensor signal or a loss of ground at ESC module would cause the signal at the ECM to remain high. This condition would result in the ECM controlling EST as if no detonation were occurring. The EST would not be retarded, and detonation could become severe enough under heavy engine load conditions to result in pre-ignition and potential engine damage.

Loss of the ESC signal to the ECM would cause the ECM to constantly retard EST. This could result in sluggish performance and cause a Code 43 to set.

Test Description: Numbers below refer to circled numbers on the diagnostic chart.

1. Tests ESC system's ability to detect detonation and retard the ignition timing.
2. By disconnecting the ESC module, the ECM monitors a low voltage at terminal "B7" and should retard the ignition timing.
3. After approximately 4 seconds, the "Service Engine Soon" light will come "ON" and Code 43 will be stored.
4. Checks for proper voltage output (measured on A.C. scale) of knock sensor. Low or no voltage would indicate an open circuit to terminal "E" or faulty sensor.
5. Checks to see if constant retard is due to a faulty knock sensor or module, or if a false voltage signal is being transmitted on the wire from the knock sensor by induction from an adjacent wire, such as a spark plug wire, ignition wire, etc. Reroute wires as necessary.

1988–90 3.0L (VIN L) SKYLARK AND CUTLASS CALAIS

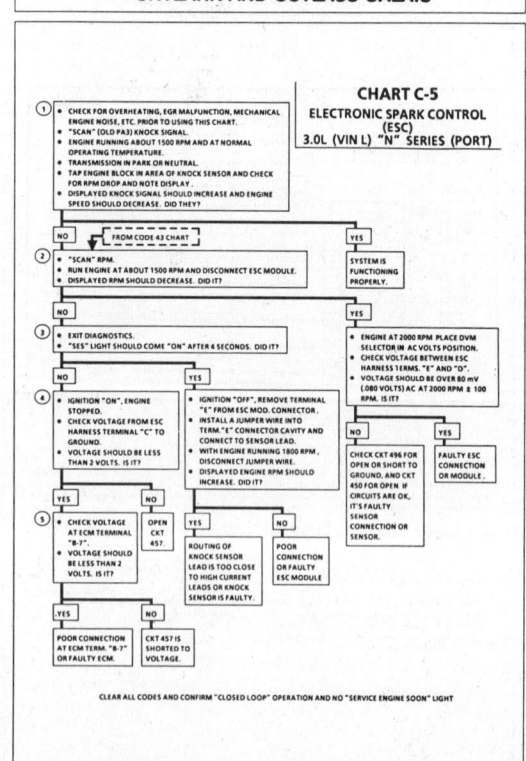

CHART C-5
ELECTRONIC SPARK CONTROL (ESC)
3.0L (VIN L) "N" SERIES (PORT)

1988–90 3.0L (VIN L) SKYLARK AND CUTLASS CALAIS

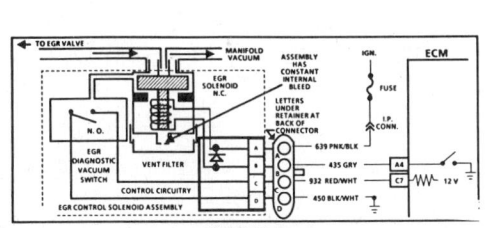

CHART C-7
EXHAUST GAS RECIRCULATION (EGR) CHECK
3.0L (VIN L) "N" SERIES (PORT)

Circuit Description:
The EGR valve is opened by manifold vacuum to let exhaust gas flow into the intake manifold. The exhaust gas then moves with the air/fuel mixture into the combustion chamber. If too much exhaust gas enters combustion will not occur, for this reason, very little exhaust gas is allowed to pass through the valve, especially at idle. The EGR valve is usually open under the following conditions:
- Warm engine operation.
- Above idle speed.

The amount of exhaust gas recirculated is controlled by variations in vacuum and the EGR vacuum control solenoid.

Test Description: Numbers below refer to circled numbers on the diagnostic chart.
1. Checks for a sticking EGR valve. If sticking, remove and examine valve to determine whether it can be cleaned, or must be replaced. A sticking EGR valve will most likely cause a rough idle condition.
2. Checks for plugged EGR passages. If passages are plugged, the engine may have severe detonation on acceleration.

1988–90 3.0L (VIN L) SKYLARK AND CUTLASS CALAIS

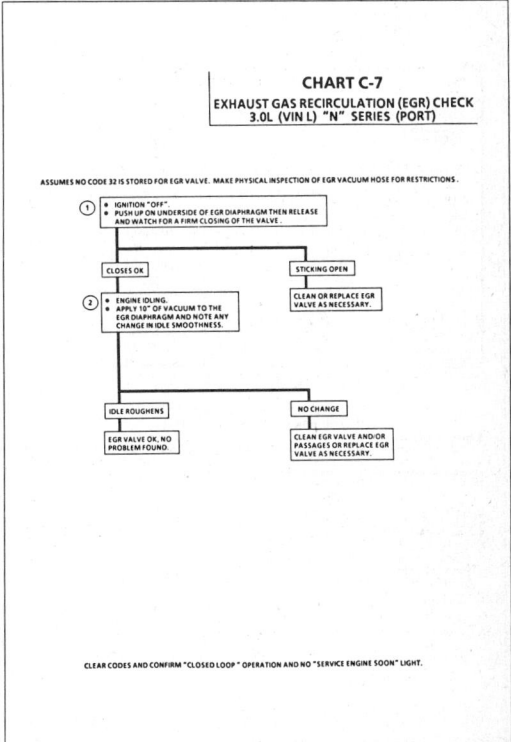

1988–90 3.0L (VIN L) SKYLARK AND CUTLASS CALAIS

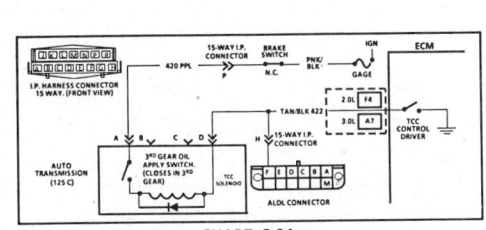

CHART C-8A
TORQUE CONVERTER CLUTCH (TCC)
3.0L (VIN L) "N" SERIES (PORT)

Circuit Description:
The purpose of the torque converter clutch is to eliminate the power loss of the torque converter when the vehicle is in a cruise condition. This allows the convenience of the automatic and the fuel economy of a manual. Fused battery ignition is supplied to the TCC solenoid through the brake switch and third gear apply switch. The ECM will engage TCC by grounding CKT 422 to energize the solenoid.

TCC will engage when:
- Vehicle speed above 45 mph (73 km/h).
- Engine at normal operating temperature (above 70°C, 156°F).
- Throttle position sensor output not changing, indicating a steady road speed.
- Third gear switch closed.
- Brake switch closed.

Test Description: Numbers below refer to circled numbers on the diagnostic chart.
1. Light "OFF" confirms third gear apply switch is open.
2. At 48 km/h (30 mph), the third gear switch should close. Test light will come "ON" and confirm battery supply and close brake switch.
3. Grounding the diagnostic terminal with engine "OFF" should energize the TCC solenoid. This test checks the capability of the ECM to control the solenoid.

Check TCC solenoid resistance as follows:
A. Disconnect TCC wiring harness at transaxle.
B. Connect ohmmeter between transaxle connector opposite harness connector terminals "A" and "D".
C. Raise drive wheels.
D. Run engine in drive about 48 km/h (30 mph) to close third gear apply switch.
E. Replace the TCC solenoid and ECM if resistance measures less than 20 ohms when switch is closed.

Diagnostic Aids:
An engine coolant thermostat that is stuck open or opens at too low a temperature, may result in an inoperative TCC.

1988–90 3.0L (VIN L) SKYLARK AND CUTLASS CALAIS

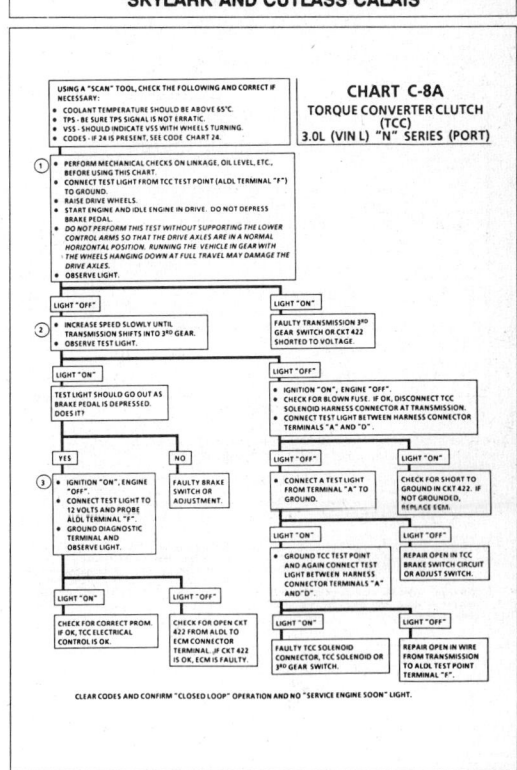

1988–90 3.0L (VIN L)
SKYLARK AND CUTLASS CALAIS

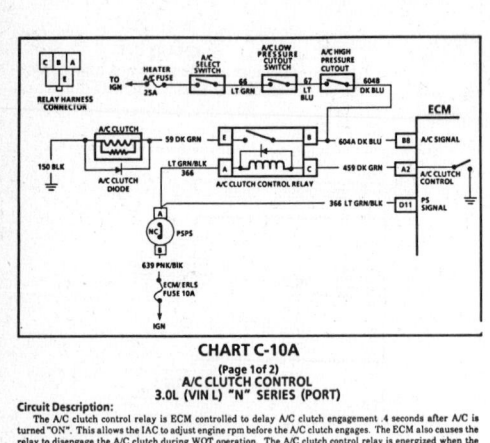

CHART C-10A
(Page 1of 2)
A/C CLUTCH CONTROL
3.0L (VIN L) "N" SERIES (PORT)

Circuit Description:
The A/C clutch control relay is ECM controlled to delay A/C clutch engagement .4 seconds after A/C is turned "ON". This allows the IAC to adjust engine rpm before the A/C clutch engages. The ECM also causes the relay to disengage the A/C clutch during WOT operation. The A/C clutch control relay is energized when the ECM provides a ground path for CKT 459.

Test Description: Numbers below refer to circled numbers on the diagnostic chart.
1. Checks to see if ECM is controlling A/C clutch control relay.
2. Checks operation of A/C cycling switch.
3. Checks for open circuit on either side of relay coil.

1988–90 3.0L (VIN L)
SKYLARK AND CUTLASS CALAIS

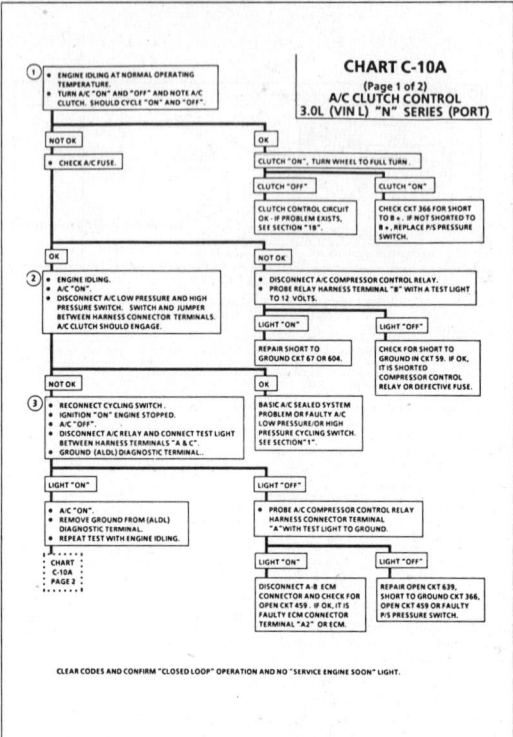

CHART C-10A
(Page 1 of 2)
A/C CLUTCH CONTROL
3.0L (VIN L) "N" SERIES (PORT)

CLEAR CODES AND CONFIRM "CLOSED LOOP" OPERATION AND NO "SERVICE ENGINE SOON" LIGHT.

1988–90 3.0L (VIN L)
SKYLARK AND CUTLASS CALAIS

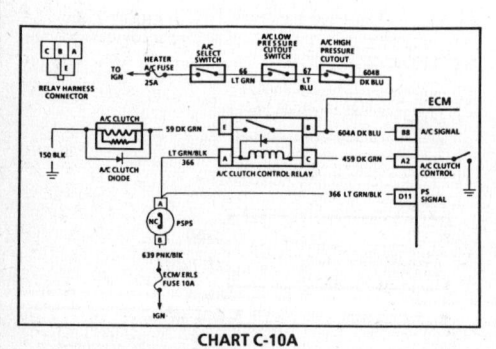

CHART C-10A
(Page 2 of 2)
A/C CLUTCH CONTROL
3.0L (VIN L) "N" SERIES (PORT)

Circuit Description:
The A/C clutch control relay is ECM controlled to delay A/C clutch engagement .4 seconds after A/C is turned "ON". This allows the IAC to adjust engine rpm before the A/C clutch engages. The ECM also causes the relay to disengage the A/C clutch during WOT operation. The A/C clutch control relay is energized when the ECM provides a ground path for CKT 366.

Test Description: Numbers below refer to circled numbers on the diagnostic chart.
1. Checks for battery voltage to relay through CKT 67.
2. Substitutes for relay to determine if problem is in relay or in CKT 59, A/C clutch coil, high press, switch, or ground.
3. Checks for open in CKT 67 between cycling switch and A/C fuse, or open CKT 67 to relay.
4. Checks to see that A/C "ON" signal is getting to ECM through CKT 67. A test light "OFF" at this time indicates CKT 67 is open between the cycling switch and the ECM.

1988–90 3.0L (VIN L)
SKYLARK AND CUTLASS CALAIS

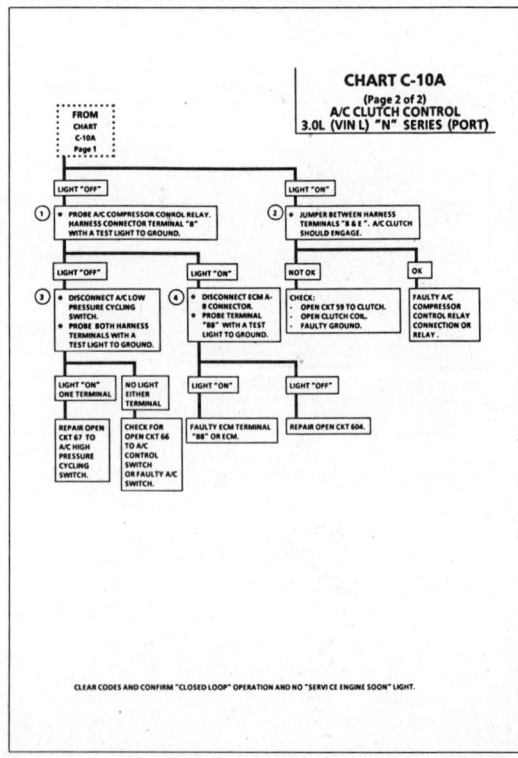

CHART C-10A
(Page 2 of 2)
A/C CLUTCH CONTROL
3.0L (VIN L) "N" SERIES (PORT)

CLEAR CODES AND CONFIRM "CLOSED LOOP" OPERATION AND NO "SERVICE ENGINE SOON" LIGHT.

1988–90 3.0L (VIN L)
SKYLARK AND CUTLASS CALAIS

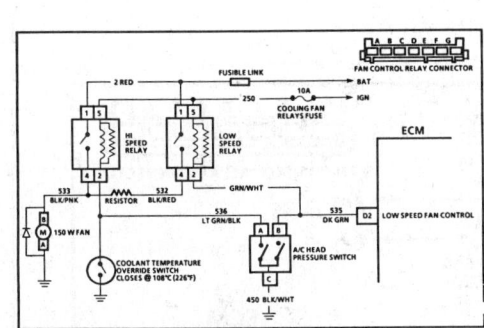

CHART C-12A
COOLANT FAN CHECK
3.0L (VIN L) "N" SERIES (PORT)

Circuit Description:
The coolant fan is energized through a low speed, high speed relay. Power for the fan(s) comes through the fusible link to terminal "1" on all relays. The relays are energized when current flows to ground through the activation of the A/C pressure, cool temp. switches, and/or the ECM.
Low Speed Relay - The low speed relay is energized by the ECM or the A/C pressure switch. The ECM energizes the relay through terminal "D2" when the coolant temperature reaches 98°C (208°F) and vehicle speed is below 45 mph. The low speed relay is also energized through the A/C pressure switch through terminal "B" when refrigerant pressure reaches 150 psi (1034 kPa).
High Speed Relay - The high speed relay is energized by the A/C high pressure and coolant override switches. If the A/C refrigerant pressure reaches 275 psi (1896 kPa) or the coolant temperature reaches 108°C (226°F) the high speed fan relay is energized. The ECM has no control of the high speed fan relay.

Test Description: Numbers below refer to circled numbers on the diagnostic chart.
1. Grounding the (ALDL) diagnostic test terminal should cause the ECM to ground CKT 535 and the fan should run in low speed.
2. Grounding the coolant temperature override switch harness terminal will check CKT 536 and the high speed relay.
3. Checks CKT 533 between the high speed relay terminal "4" and the motor. If the fan does not operate, CKT 533 is open.
4. This step checks to see if the coolant temperature override switch is grounding and is grounded when the light comes "ON". The switch should close at 108°C (226°F).
5. If the vehicle is equipped with A/C, the following steps will check the A/C high pressure switch and related wiring from the switch to the fan control relays. If poor A/C performance is noted, the A/C pressure switch should be checked by a qualified A/C repair person. The low speed fan should come "ON" if A/C system pressure exceeds 150 psi (1036 kPa).

1988–90 3.0L (VIN L)
SKYLARK AND CUTLASS CALAIS

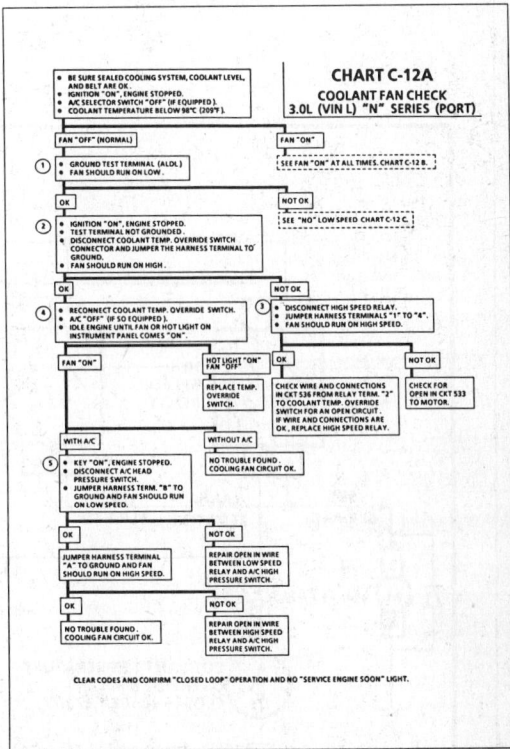

CHART C-12A
COOLANT FAN CHECK
3.0L (VIN L) "N" SERIES (PORT)

1988–90 3.0L (VIN L)
SKYLARK AND CUTLASS CALAIS

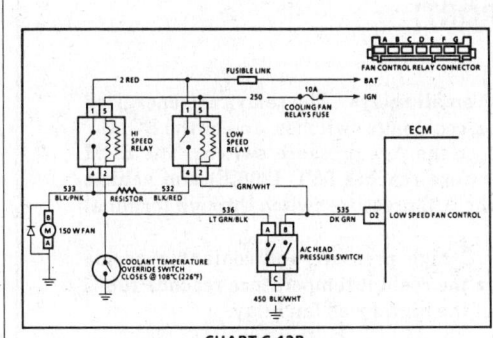

CHART C-12B
FAN "ON" AT ALL TIMES
3.0L (VIN L) "N" SERIES (PORT)

Circuit Description:
The coolant fan is energized through a low speed, high speed relay. Power for the fan(s) comes through the fusible link to terminal "1" on all relays. The relays are energized when current flows to ground through the activation of the A/C pressure, cool temp. switches, and/or the ECM.
Low Speed Relay - The low speed relay is energized by the ECM or the A/C pressure switch. The ECM energizes the relay through terminal "D2" when the coolant temperature reaches 98°C (208°F) and vehicle speed is below 45 mph. The low speed relay is also energized through the A/C pressure switch through terminal "B" when refrigerant pressure reaches 150 psi (1034 kPa).
High Speed Relay - The high speed relay is energized by the A/C high pressure and coolant override switches. If the A/C refrigerant pressure reaches 275 psi (1896 kPa) or the coolant temperature reaches 108°C (226°F) the high speed fan relay is energized. The ECM has no control of the high speed fan relay.

Test Description: Numbers below refer to circled numbers on the diagnostic chart.
1. Check to see if CKT 535 is shorted to ground which would keep the relay grounded at all times.
2. Check to see if CKT 536 is shorted to ground. A light indicates the wire is shorted to ground and the following steps will isolate the short.
3. If the test light is "OFF" after disconnecting the ECM is shorted internally. Before replacing the ECM, be sure and check the resistance value of low speed side of the fan control relay. Replace if resistance is less than 20 ohms.

1988–90 3.0L (VIN L)
SKYLARK AND CUTLASS CALAIS

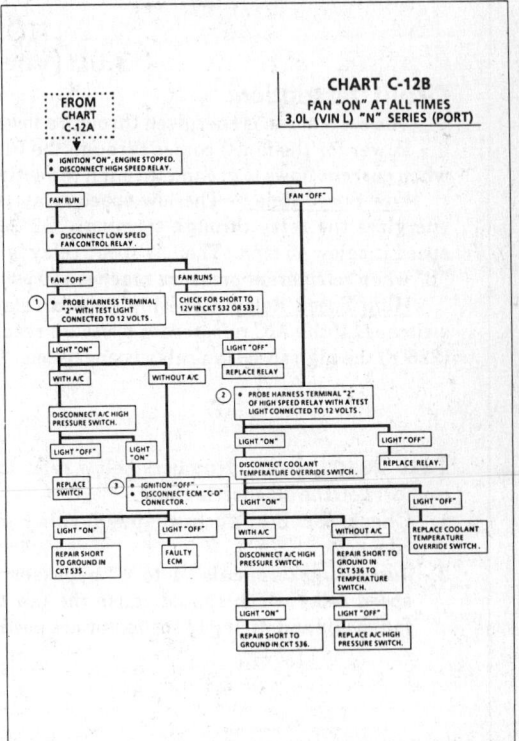

CHART C-12B
FAN "ON" AT ALL TIMES
3.0L (VIN L) "N" SERIES (PORT)

1988–90 3.0L (VIN L)
SKYLARK AND CUTLASS CALAIS

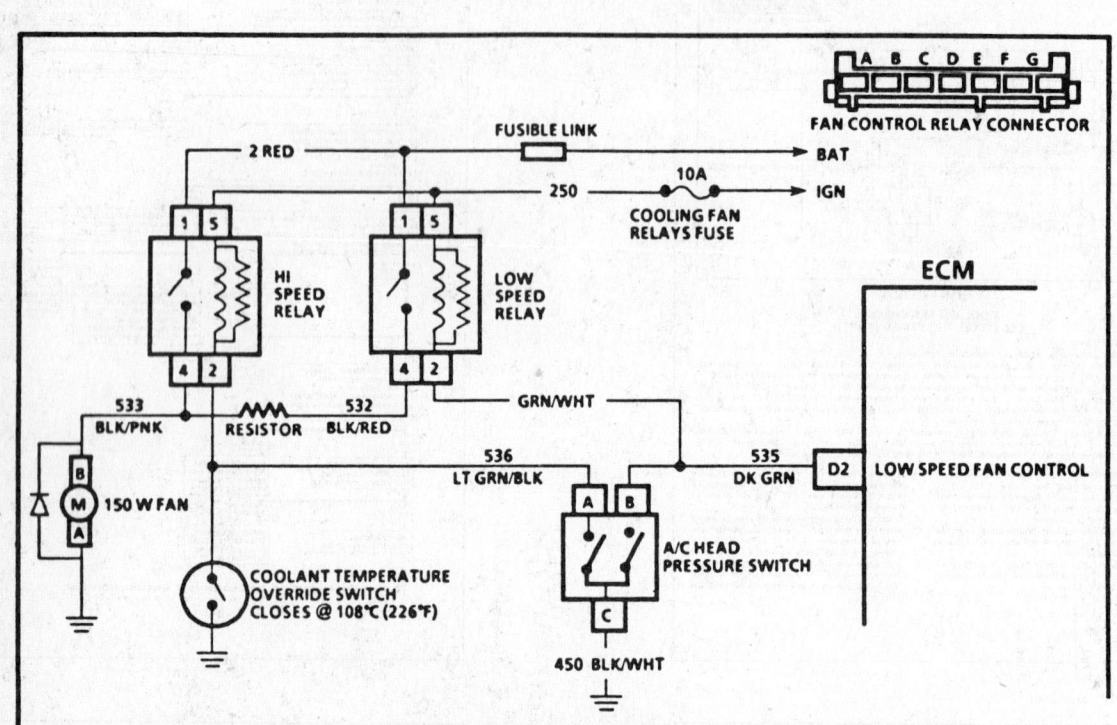

CHART C-12C

NO LOW SPEED FAN
3.0L (VIN L) "N" SERIES (PORT)

Circuit Description:

The coolant fan is energized through a low speed, high speed relay.

Power for the fan(s) comes through the fusible link to terminal "1" on all relays. The relays are energized when current flows to ground through the activation of the A/C pressure, cool temp. switches, and/or the ECM.

Low Speed Relay - The low speed relay is energized by the ECM or the A/C pressure switch. The ECM energizes the relay through terminal "D2" when the coolant temperature reaches 98°C (208°F) and vehicle speed is below 45 mph. The low speed relay is also energized through the A/C pressure switch through terminal "B" when refrigerant pressure reaches 150 psi (1034 kPa).

High Speed Relay - The high speed relay is energized by the A/C high pressure and coolant override switches. If the A/C refrigerant pressure reaches 275 psi (1896 kPa) or the coolant temperature reaches 108°C (226°F) the high speed fan relay is energized. The ECM has no control of the high speed fan relay.

Test Description: Numbers below refer to circled numbers on the diagnostic chart.

1. Check for B+ at low speed relay harness connector.
2. Jumpering terminals "1 to 4" bypasses the low speed relay which should cause the fan to run if fan motor and wiring to the motor are good.
3. Grounding the (ALDL) test terminal should cause the ECM to ground CKT 535. At this point, the test light should light if the ECM is good and CKT 535 isn't open.
4. This checks for B+ and ground to the fan motor. A test light "ON" at this point indicates a faulty fan motor connection or motor.

1988–90 3.0L (VIN L)
SKYLARK AND CUTLASS CALAIS

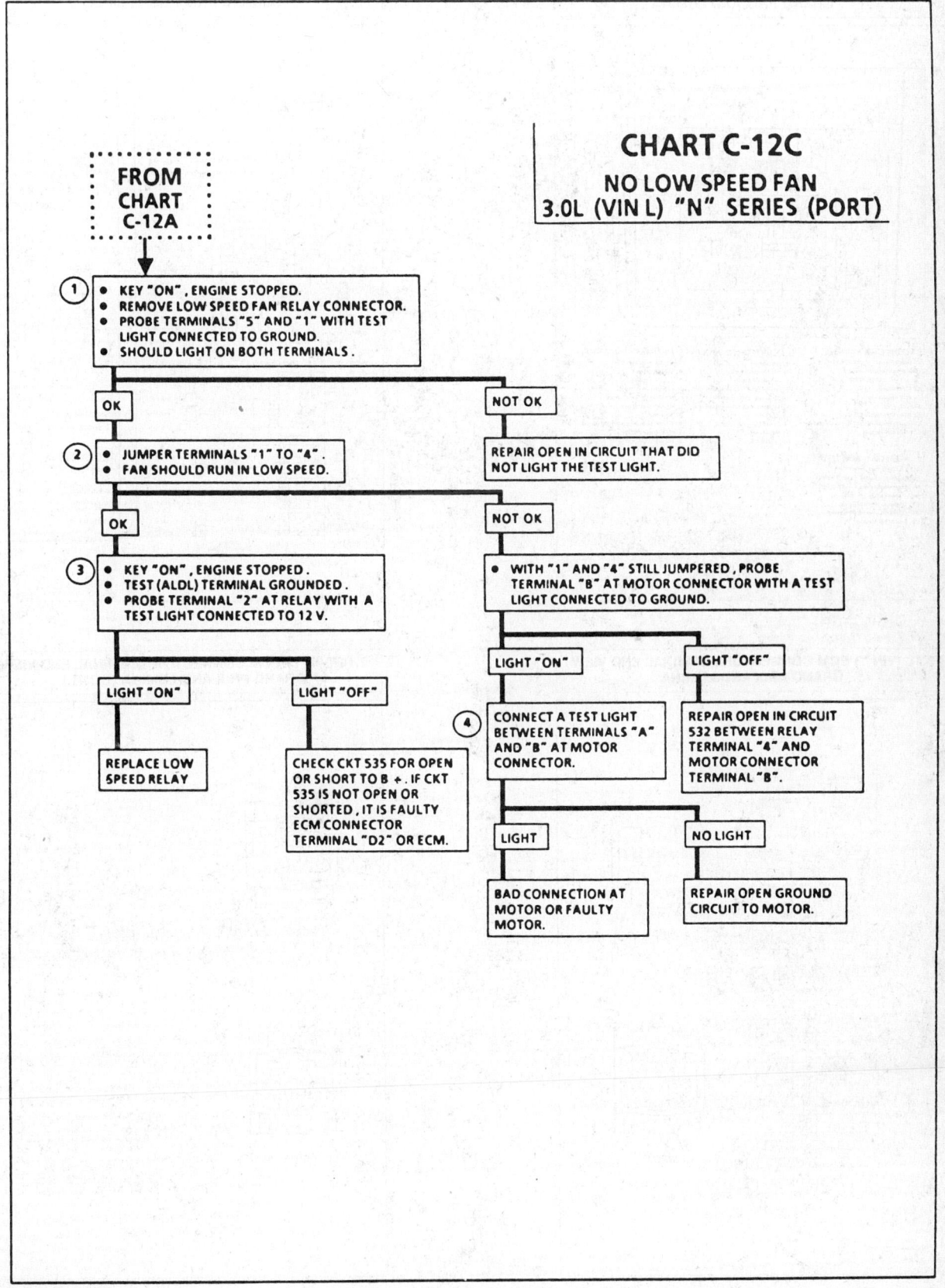

FROM CHART C-12A

CHART C-12C
NO LOW SPEED FAN
3.0L (VIN L) "N" SERIES (PORT)

1
- KEY "ON", ENGINE STOPPED.
- REMOVE LOW SPEED FAN RELAY CONNECTOR.
- PROBE TERMINALS "5" AND "1" WITH TEST LIGHT CONNECTED TO GROUND.
- SHOULD LIGHT ON BOTH TERMINALS.

OK | **NOT OK**

REPAIR OPEN IN CIRCUIT THAT DID NOT LIGHT THE TEST LIGHT.

2
- JUMPER TERMINALS "1" TO "4".
- FAN SHOULD RUN IN LOW SPEED.

OK | **NOT OK**

- WITH "1" AND "4" STILL JUMPERED, PROBE TERMINAL "B" AT MOTOR CONNECTOR WITH A TEST LIGHT CONNECTED TO GROUND.

3
- KEY "ON", ENGINE STOPPED.
- TEST (ALDL) TERMINAL GROUNDED.
- PROBE TERMINAL "2" AT RELAY WITH A TEST LIGHT CONNECTED TO 12 V.

LIGHT "ON" | **LIGHT "OFF"**

REPLACE LOW SPEED RELAY

CHECK CKT 535 FOR OPEN OR SHORT TO B +. IF CKT 535 IS NOT OPEN OR SHORTED, IT IS FAULTY ECM CONNECTOR TERMINAL "D2" OR ECM.

LIGHT "ON" | **LIGHT "OFF"**

4
CONNECT A TEST LIGHT BETWEEN TERMINALS "A" AND "B" AT MOTOR CONNECTOR.

REPAIR OPEN IN CIRCUIT 532 BETWEEN RELAY TERMINAL "4" AND MOTOR CONNECTOR TERMINAL "B".

LIGHT | **NO LIGHT**

BAD CONNECTION AT MOTOR OR FAULTY MOTOR.

REPAIR OPEN GROUND CIRCUIT TO MOTOR.

3.1L (VIN T) COMPONENT LOCATIONS GRAND PRIX AND LUMINA

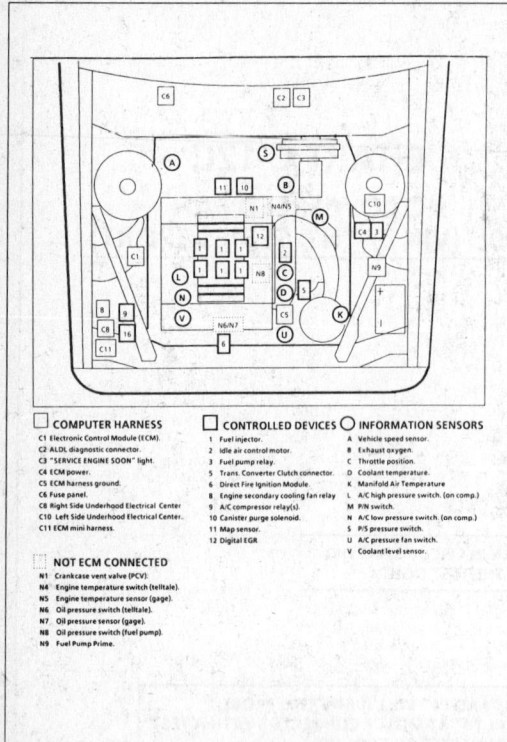

COMPUTER HARNESS
C1 Electronic Control Module (ECM).
C2 ALDL diagnostic connector.
C3 "SERVICE ENGINE SOON" light.
C4 ECM power.
C5 ECM harness ground.
C6 Fuse panel.
C8 Right Side Underhood Electrical Center.
C10 Left Side Underhood Electrical Center.
C11 ECM mini harness.

NOT ECM CONNECTED
N1 Crankcase vent valve (PCV).
N4 Engine temperature switch (telltale).
N5 Engine temperature sensor (gage).
N6 Oil pressure switch (telltale).
N7 Oil pressure sensor (gage).
N8 Oil pressure switch (fuel pump).
N9 Fuel Pump Prime.

CONTROLLED DEVICES
1 Fuel injector.
2 Idle air control motor.
3 Fuel pump relay.
4 Trans. Converter Clutch connector.
5 Direct Fire Ignition Module.
6 Engine secondary cooling fan relay.
7 A/C compressor relay(s).
8 Canister purge solenoid.
11 Map sensor.
12 Digital EGR

INFORMATION SENSORS
A Vehicle speed sensor.
B Exhaust oxygen.
C Throttle position.
D Coolant temperature.
H Manifold Air Temperature.
J A/C high pressure switch. (on comp.)
M P/N switch.
N A/C low pressure switch. (on comp.)
S P/S pressure switch.
U A/C pressure fan switch.
V Coolant level sensor.

3.1L (VIN T) ECM WIRING DIAGRAM GRAND PRIX AND LUMINA

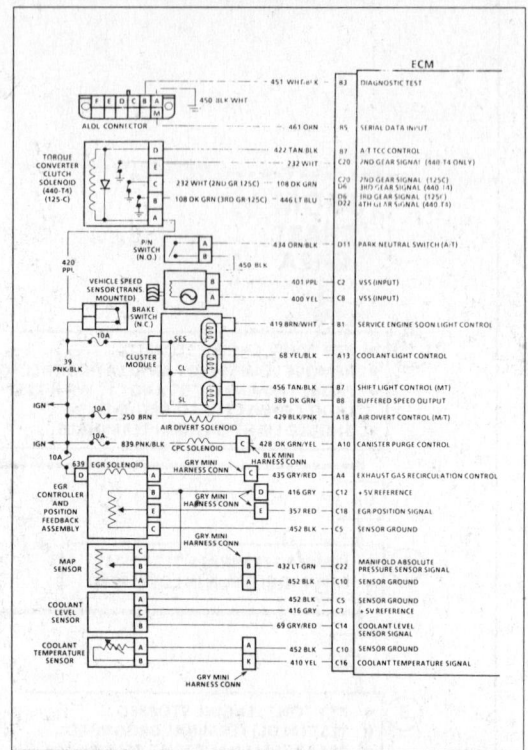

3.1L (VIN T) ECM CONNECTOR TERMINAL END VIEW GRAND PRIX AND LUMINA

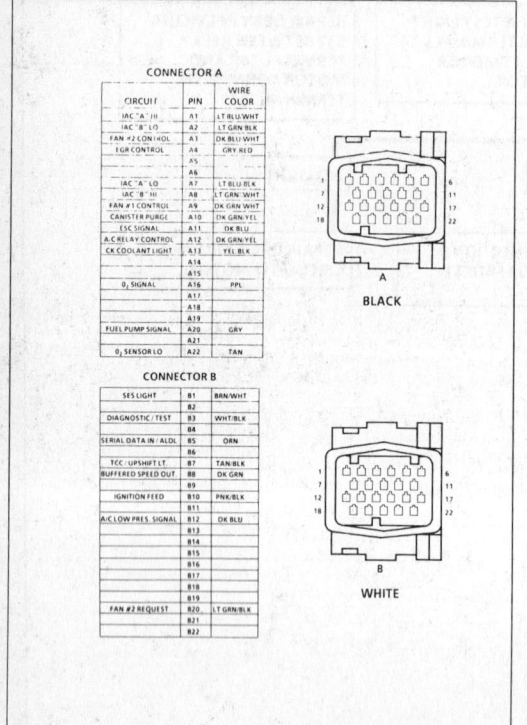

CONNECTOR A

CIRCUIT	PIN	WIRE COLOR
IAC "A" HI	A1	LT BLU/WHT
IAC "B" LO	A2	LT GRN/BLK
FAN #2 CONTROL	A3	DK BLU/WHT
EGR CONTROL	A4	GRY/RED
	A5	
	A6	
IAC "A" LO	A7	LT BLU/BLK
IAC "B" HI	A8	LT GRN/WHT
FAN #1 CONTROL	A9	DK GRN/WHT
CANISTER PURGE	A10	DK GRN/YEL
ESC SIGNAL	A11	DK BLU
A/C RELAY CONTROL	A12	DK GRN/YEL
CK COOLANT LIGHT	A13	YEL/BLK
	A14	
	A15	
O₂ SIGNAL	A16	PPL
	A17	
	A18	
	A19	
FUEL PUMP SIGNAL	A20	GRY
	A21	
O₂ SENSOR LO	A22	TAN

A — BLACK

CONNECTOR B

CIRCUIT	PIN	WIRE COLOR
SES LIGHT	B1	BRN/WHT
	B2	
DIAGNOSTIC / TEST	B3	WHT/BLK
	B4	
SERIAL DATA IN / ALDL	B5	ORN
	B6	
TCC / UPSHIFT LT.	B7	TAN/BLK
BUFFERED SPEED OUT.	B8	DK GRN
	B9	
IGNITION FEED	B10	PNK/BLK
	B11	
A/C LOW PRES. SIGNAL	B12	DK BLU
	B13	
	B14	
	B15	
	B16	
	B17	
	B18	
	B19	
FAN #2 REQUEST	B20	LT GRN/BLK
	B21	
	B22	

B — WHITE

3.1L (VIN T) ECM CONNECTOR TERMINAL END VIEW GRAND PRIX AND LUMINA (CONT.)

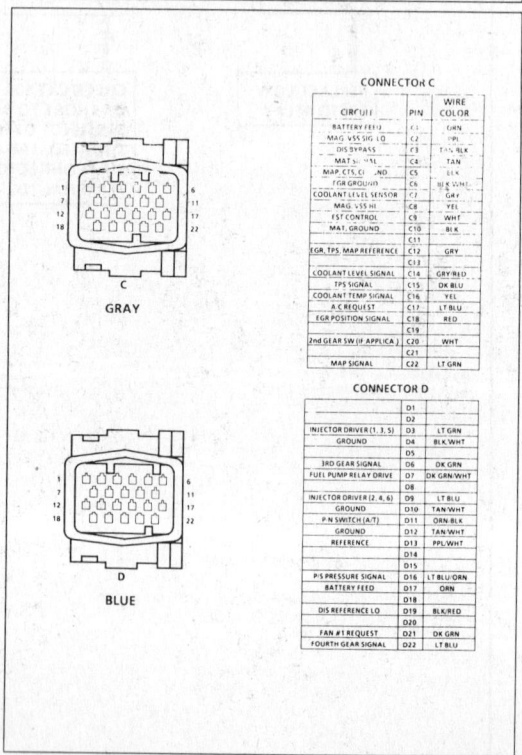

CONNECTOR C

CIRCUIT	PIN	WIRE COLOR
BATTERY FEED	C1	ORN
MAG VSS SIG LO	C2	PPL
DIS BYPASS	C3	TAN/BLK
MAT SIGNAL	C4	TAN
MAP, CTS, CL GND	C5	BLK
EGR GROUND	C6	BLK/WHT
COOLANT LEVEL SENSOR	C7	GRY
MAG VSS HI	C8	YEL
EST CONTROL	C9	WHT
MAT GROUND	C10	BLK
	C11	
EGR, TPS, MAP REFERENCE	C12	GRY
	C13	
COOLANT LEVEL SIGNAL	C14	GRY/RED
TPS SIGNAL	C15	DK BLU
COOLANT TEMP SIGNAL	C16	YEL
A/C REQUEST	C17	LT BLU
EGR POSITION SIGNAL	C18	RED
	C19	
2nd GEAR SW (IF APPLICA)	C20	WHT
	C21	
MAP SIGNAL	C22	LT GRN

C — GRAY

CONNECTOR D

CIRCUIT	PIN	WIRE COLOR
	D1	
	D2	
INJECTOR DRIVER (1, 3, 5)	D3	LT GRN
GROUND	D4	BLK/WHT
	D5	
3RD GEAR SIGNAL	D6	DK GRN
FUEL PUMP RELAY DRIVE	D7	DK GRN/WHT
	D8	
INJECTOR DRIVER (2, 4, 6)	D9	LT BLU
GROUND	D10	TAN/WHT
P-N SWITCH (A/T)	D11	ORN/BLK
GROUND	D12	TAN/WHT
REFERENCE	D13	PPL/WHT
	D14	
	D15	
P/S PRESSURE SIGNAL	D16	LT BLU/ORN
BATTERY FEED	D17	ORN
	D18	
DIS REFERENCE LO	D19	BLK/RED
	D20	
FAN #1 REQUEST	D21	DK GRN
FOURTH GEAR SIGNAL	D22	LT BLU

D — BLUE

3.1L (VIN T) ECM WIRING DIAGRAM GRAND PRIX AND LUMINA (CONT.)

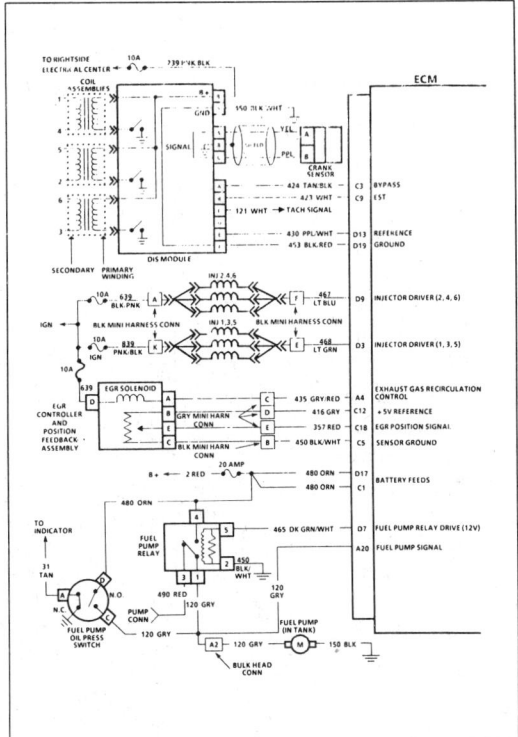

3.1L (VIN T) ECM WIRING DIAGRAM GRAND PRIX AND LUMINA (CONT.)

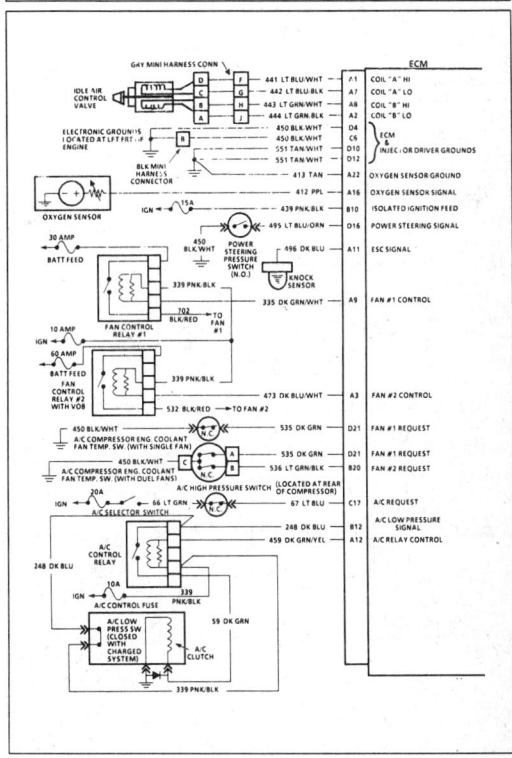

3.1L (VIN T) COMPONENT LOCATIONS CUTLASS AND REGAL

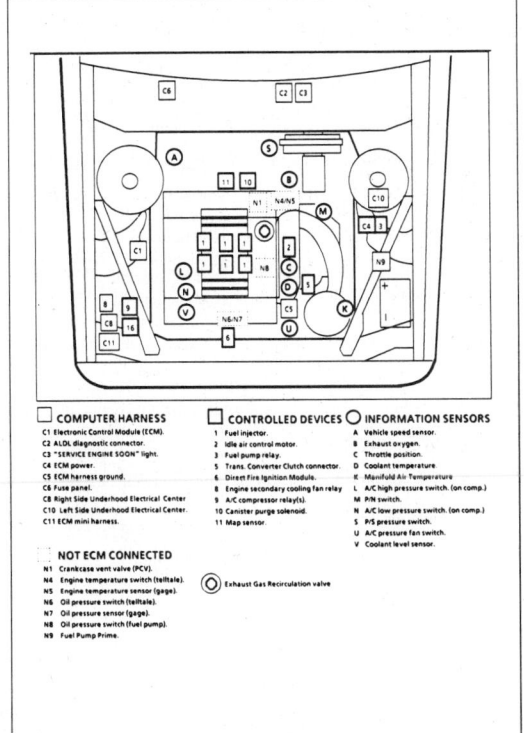

☐ **COMPUTER HARNESS**

C1 Electronic Control Module (ECM).
C2 ALDL diagnostic connector.
C3 "SERVICE ENGINE SOON" light.
C4 ECM power.
C5 ECM harness ground.
C6 Fuse panel.
C8 Right Side Underhood Electrical Center.
C10 Left Side Underhood Electrical Center.
C11 ECM mini harness.

NOT ECM CONNECTED

N1 Crankcase vent valve (PCV).
N4 Engine temperature switch (telltale).
N5 Engine temperature sensor (gage).
N6 Oil pressure switch (telltale).
N7 Oil pressure sensor (gage).
N8 Oil pressure switch (fuel pump).
N9 Fuel Pump Prime.

☐ **CONTROLLED DEVICES**

1. Fuel injector.
2. Idle air control motor.
3. Fuel pump relay.
4. Trans. Converter Clutch connector.
5. Direct Fire Ignition Module.
6. Engine secondary cooling fan relay.
7. A/C compressor relay(s).
10 Canister purge solenoid.
11 Map sensor.

○ **INFORMATION SENSORS**

A Vehicle speed sensor.
B Exhaust oxygen.
C Throttle position.
D Coolant temperature.
E Manifold Air Temperature.
L A/C high pressure switch. (on comp.)
M P/N switch.
N A/C low pressure switch. (on comp.)
S P/S pressure switch.
U A/C pressure fan switch.
V Coolant level sensor.

○ Exhaust Gas Recirculation valve.

3.1L (VIN T) ECM WIRING DIAGRAM CUTLASS AND REGAL

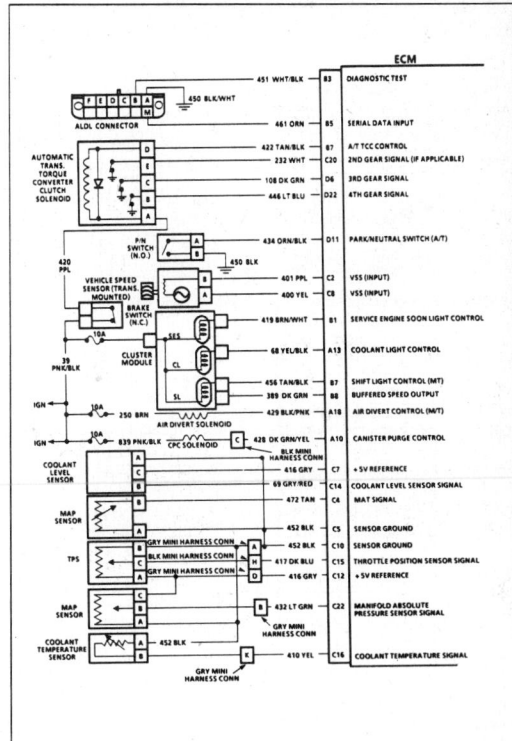

3.1L (VIN T) ECM WIRING DIAGRAM CUTLASS AND REGAL (CONT.)

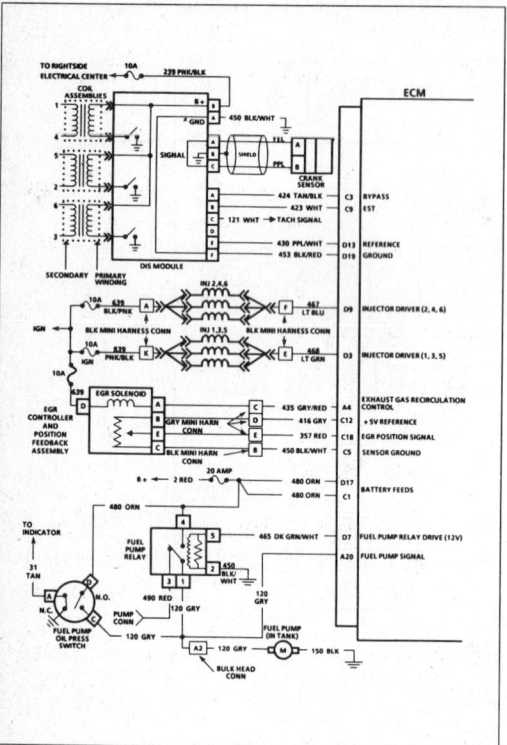

3.1L (VIN T) ECM WIRING DIAGRAM CUTLASS AND REGAL (CONT.)

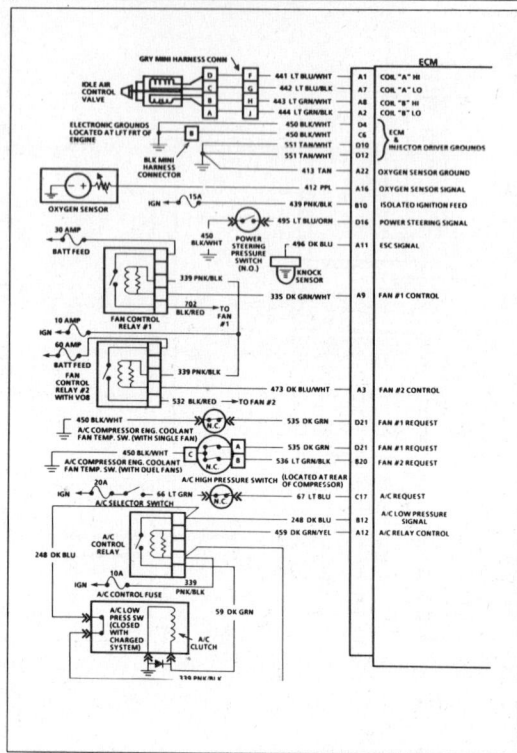

3.1L (VIN T) ECM CONNECTOR TERMINAL END VIEW—CUTLASS AND REGAL

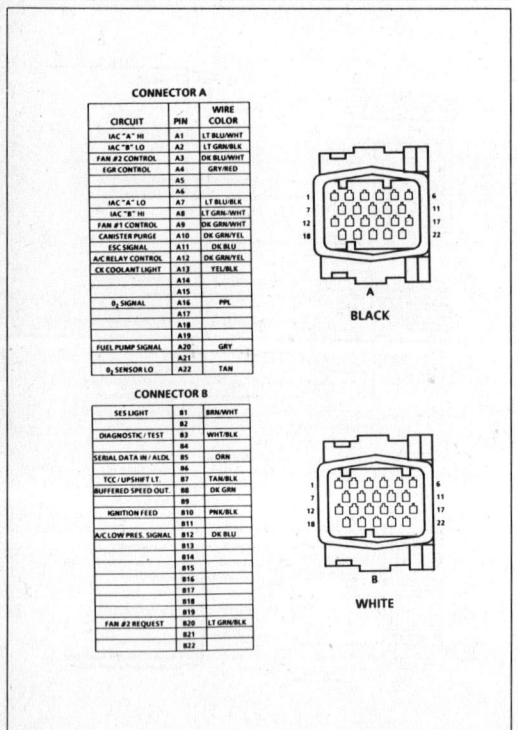

3.1L (VIN T) ECM CONNECTOR TERMINAL END VIEW—CUTLASS AND REGAL (CONT.)

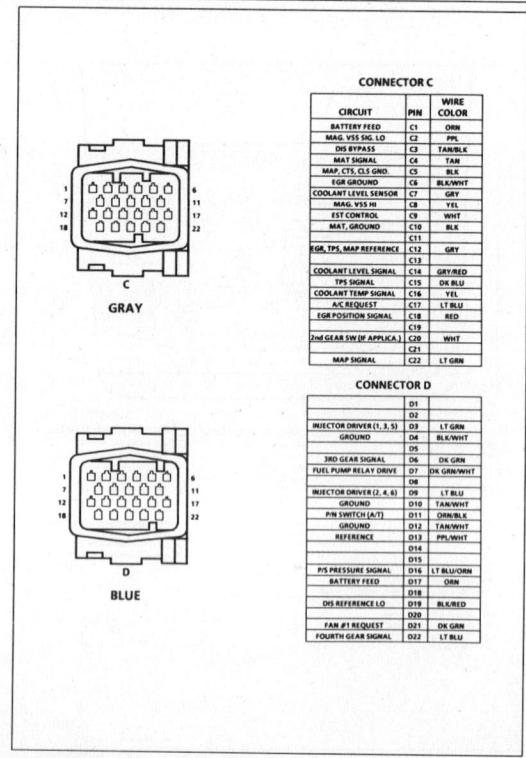

1989–90 3.1L (VIN T) – ALL W-BODY VEHICLES

DIAGNOSTIC CIRCUIT CHECK

The Diagnostic Circuit Check is an organized approach to identifying a problem created by an Electronic Engine Control System malfunction. It must be the starting point for any driveability complaint diagnosis, because it directs the Service Technician to the next logical step in diagnosing the complaint.

The "Scan Data" listed in the table may be used for comparison, after completing the Diagnostic Circuit Check and finding the on-board diagnostics functioning properly and no trouble codes displayed. The "Typical Values" are an average of display values recorded from normally operating vehicles and are intended to represent what a normally functioning system would typically display.

A "SCAN" TOOL THAT DISPLAYS FAULTY DATA SHOULD NOT BE USED, AND THE PROBLEM SHOULD BE REPORTED TO THE MANUFACTURER. THE USE OF A FAULTY "SCAN" CAN RESULT IN MISDIAGNOSIS AND UNNECESSARY PARTS REPLACEMENT.

Only the parameters listed below are used in this manual for diagnosing. If a "Scan" reads other parameters, the values are not recommended by General Motors for use in diagnosing. For more description on the values and use of the "Scan" to diagnosis ECM inputs, refer to the applicable diagnosis section. If all values are within the range illustrated, refer to "Symptoms" in Section "B"

"SCAN" DATA
Idle / Upper Radiator Hose Hot / Closed Throttle / Park or Neutral / Closed Loop / Acc. off

"SCAN" Position	Units Displayed	Typical Data Value
Desired RPM	RPM	ECM idle command (varies with temp.)
RPM	RPM	± 100 RPM from desired RPM (± 50 in drive)
Coolant Temp.	C°	85° - 105°
MAT Temp.	C°	10° - 80° (depends on underhood temp.)
MAP	Volts	1 - 2 (depends on Vac. & Baro pressure)
BARO	Volts	2.5 - 5.5 (depends on altitude & Baro pressure)
BPW (base pulse width)	M/Sec	1 - 4 and varying
O₂	Volts	1-1000 and varying
TPS	Volts	0.65
Throttle Angle	0 - 100%	0
IAC	Counts (steps)	10 - 50
P/N Switch	P/N and RDL	Park/Neutral (P/N)
INT (Integrator)	Counts	Varies
BLM (Block Learn)	Counts	118 - 13B
Open/Closed Loop	Open/Closed	Closed Loop (may go open with extended idle)
BLM Cell	Cell Number	0 or 1 (depends on Air Flow & RPM)
VSS	MPH	0
TCC	On/Off	Off/ (on with TCC commanded)
EGRDC	0 - 100%	0 at idle
EGR Position	Volts	3 - 2 Volts
Spark Advance	# of Degrees	Varies
Knock Retard	Degrees of Retard	0
Knock Signal	Yes/No	No
Battery	Volts	13.5 - 14.5
Fan	On/Off	Off (below 106°C)
P/S Switch	Normal/Hi Press.	Normal
2nd Gear (440-T4-THM)	Yes/No	No (yes when in 2nd, 3rd or 4th gear)
3rd Gear (440-T4)	Yes/No	No (yes, when in 3rd or 4th gear)
4th Gear (440-T4)	Yes/No	No (yes, when in 4th gear)
A/C Request	Yes/No	No (yes, with A/C requested)
A/C Clutch	On/Off	Off (on, with A/C commanded on)
Fan Request	Yes/No	No (yes, with A/C high pressure)
Shift Light (M/T) On/Off	On/Off	Off
PPSW	Volts	13.5 - 14.5
Purge Duty Cycle	0-100%	0%

1989–90 3.1L (VIN T) – ALL W-BODY VEHICLES

DIAGNOSTIC CIRCUIT CHECK
3.1L (VIN T) "W" CARLINE (PORT)

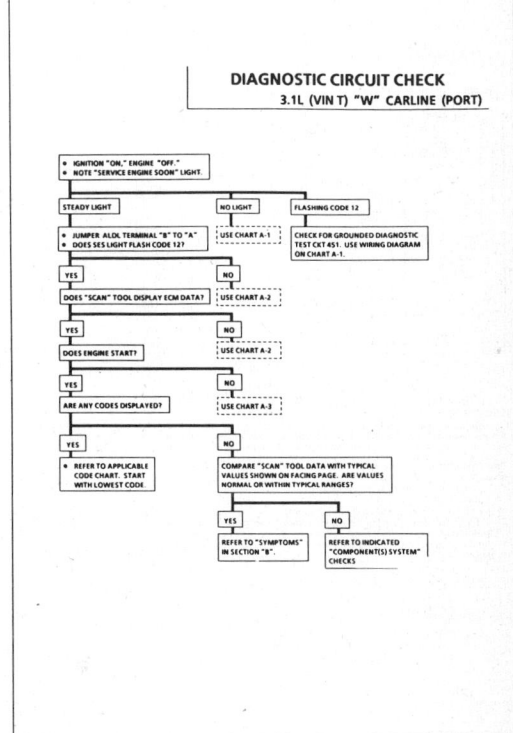

1989–90 3.1L (VIN T) – ALL W-BODY VEHICLES

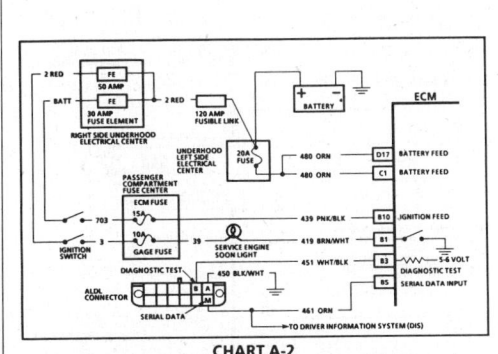

CHART A-2
NO ALDL DATA OR WON'T FLASH CODE 12
"SERVICE ENGINE SOON" LIGHT "ON" STEADY
3.1L (VIN T) "W" CARLINE (PORT)

Circuit Description:
There should always be a steady "Service Engine Soon" light when the ignition is "ON" and engine stopped. Battery ignition voltage is supplied to the light bulb. The electronic control module (ECM) will turn the light "ON" by grounding CKT 419 at the ECM.

With the diagnostic terminal grounded, the light should flash a Code 12, followed by any trouble code(s) stored in memory.

A steady light suggests a short to ground in the light control CKT 419, or an open in diagnostic CKT 451.

Test Description: Numbers below refer to circled numbers on the diagnostic chart.

1. If there is a problem with the ECM that causes a "Scan" tool not to read serial data, the ECM should not flash a Code 12. If Code 12 is flashing check for CKT 451 short to ground. If Code 12 does flash be sure that the "Scan" tool is working properly on another vehicle. If the "Scan" is functioning properly and CKT 461 is OK the Mem-Cal or ECM may be at fault for the no ALDL symptom.

2. If the light goes "OFF" when the ECM connector is disconnected, CKT 419 is not shorted to ground.

3. This step will check for an open diagnostic CKT 451.

4. At this point the "Service Engine Soon" light wiring is OK. The problem is a faulty ECM or Mem-Cal. If Code 12 does not flash, the ECM should be replaced using the original Mem-Cal. Replace the Mem-Cal only after trying an ECM, as a defective Mem-Cal is an unlikely cause of the problem.

1989–90 3.1L (VIN T) – ALL W-BODY VEHICLES

CHART A-2
NO ALDL DATA OR WON'T FLASH CODE 12
"SERVICE ENGINE SOON" LIGHT "ON" STEADY
3.1L (VIN T) "W" CARLINE (PORT)

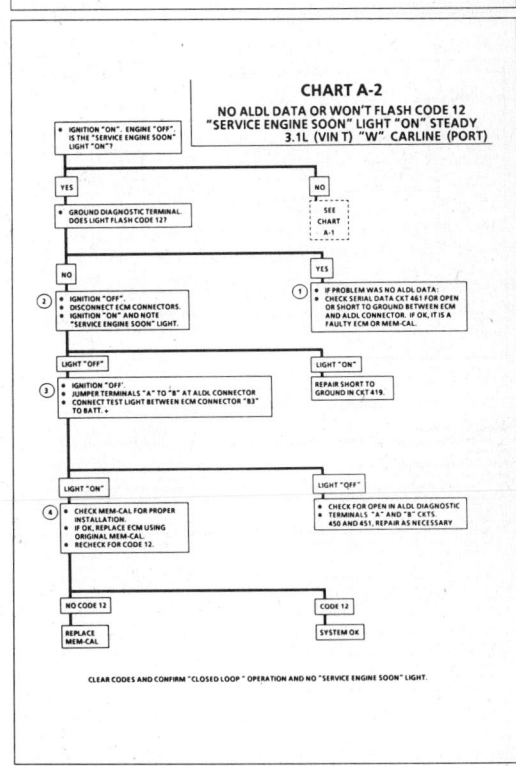

1989–90 3.1L (VIN T) – ALL W-BODY VEHICLES

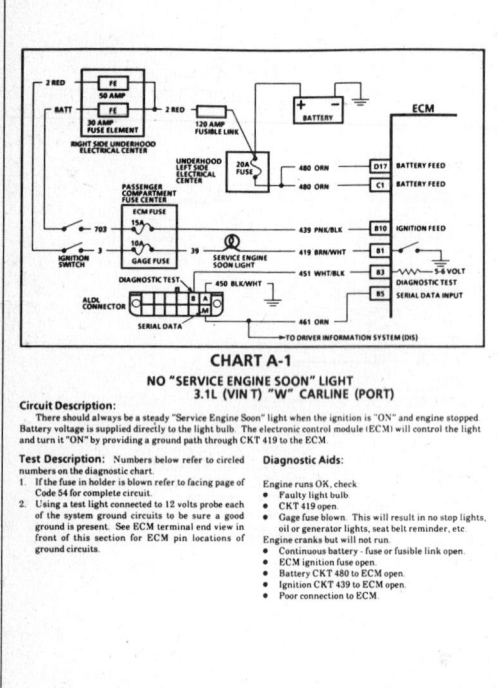

CHART A-1
NO "SERVICE ENGINE SOON" LIGHT
3.1L (VIN T) "W" CARLINE (PORT)

Circuit Description:
There should always be a steady "Service Engine Soon" light when the ignition is "ON" and engine stopped. Battery voltage is supplied directly to the light bulb. The electronic control module (ECM) will control the light and turn it "ON" by providing a ground path through CKT 419 to the ECM.

Test Description: Numbers below refer to circled numbers on the diagnostic chart.
1. If the fuse in holder is blown refer to facing page of Code 54 for complete circuit.
2. Using a test light connected to 12 volts probe each of the system ground circuits to be sure a good ground is present. See ECM terminal end view in front of this section for ECM pin locations of ground circuits.

Diagnostic Aids:

Engine runs OK, check
- Faulty light bulb.
- CKT 419 open.
- Gage fuse blown. This will result in no stop lights, oil or generator lights, seat belt reminder, etc.

Engine cranks but will not run.
- Continuous battery - fuse or fusible link open.
- ECM ignition fuse open.
- Battery CKT 480 to ECM open.
- Ignition CKT 439 to ECM open.
- Poor connection to ECM.

1989–90 3.1L (VIN T) – ALL W-BODY VEHICLES

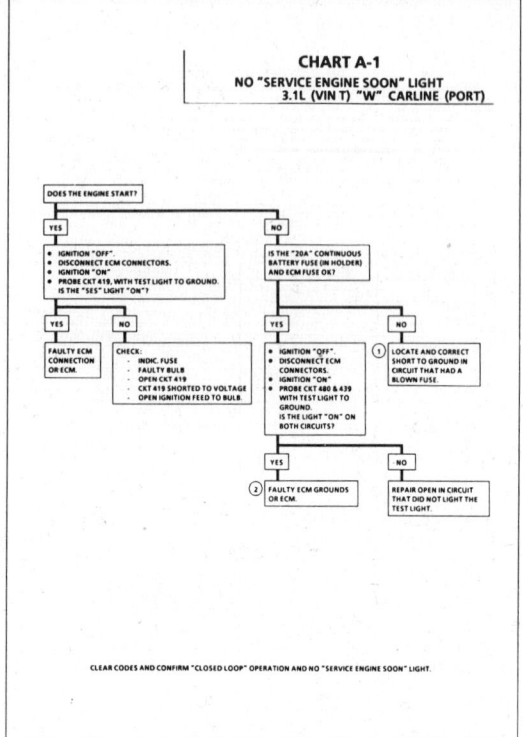

CHART A-1
NO "SERVICE ENGINE SOON" LIGHT
3.1L (VIN T) "W" CARLINE (PORT)

1989–90 3.1L (VIN T) – ALL W-BODY VEHICLES

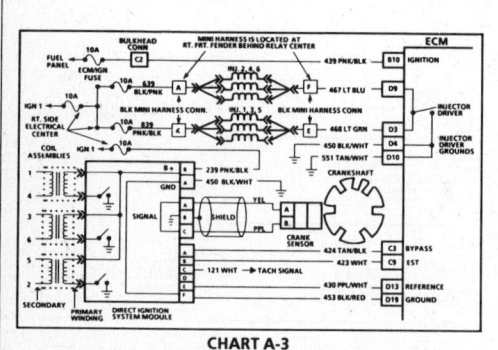

CHART A-3
(Page 1 of 3)
ENGINE CRANKS BUT WILL NOT RUN
3.1L (VIN T) "W" CARLINE (PORT)

Circuit Description:
This chart assumes that battery condition and engine cranking speed are OK, and there is adequate fuel in the tank.

Test Description: Numbers below refer to circled numbers on the diagnostic chart.
1. A "Service Engine Soon" light "ON" is a basic test to determine if there is a 12 volt supply and ignition 12 volts to ECM. No ALDL will be due to an ECM problem and CHART A-2 will diagnose the ECM. If TPS is over 2.5 volts the engine may be in the clear flood mode which will cause starting problems. The engine will not start without reference pulses and therefore the "Scan" should read rpm (reference) during crank.
2. For the first two seconds with ignition "ON" or whenever reference pulses are being received, PPSW should indicate fuel pump circuit voltage (8 to 12 volts).
3. Because the direct ignition system uses two plugs and wires to complete the circuit of each coil, the opposite spark should be left connected. If rpm was indicated during crank, but no spark at this test indicates the ignition module is not triggering the coils

4. The test light should blink, indicating the ECM is in control of the injectors. How bright the light blinks is not important. However, the test light should be a J-34730-3 or equivalent.
5. Use fuel pressure gage J-34730-1 or equivalent. Wrap a shop towel around the fuel pressure tap to absorb any small amount of fuel leakage that may occur when installing the gage.
6. This test will determine if the ignition module is not generating the reference pulse or if the wiring or ECM are at fault. By touching and removing a test light to 12 volts on CKT 430, a reference pulse should be generated. If rpm is indicated, the ECM and wiring are OK.
7. This test will determine if the ignition module is not triggering the problem coil or if the tested coil is at fault. This test could also be performed by using another known good coil.

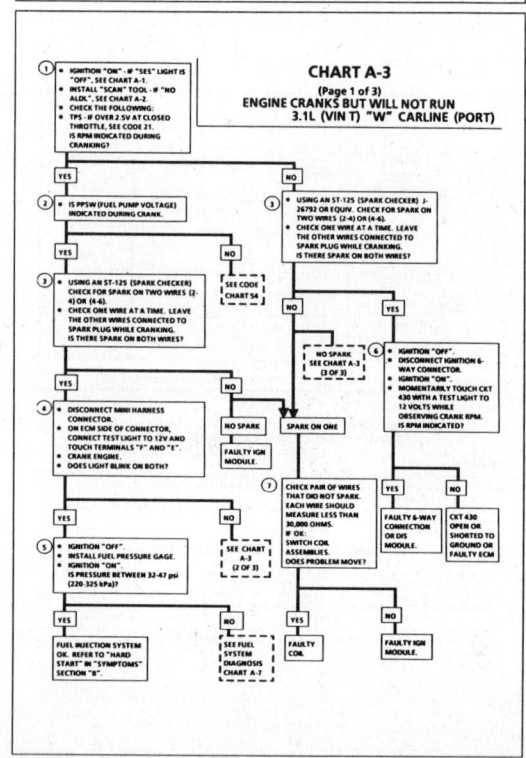

1989–90 3.1L (VIN T) – ALL W-BODY VEHICLES

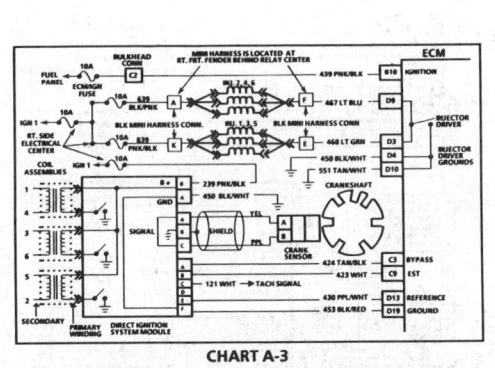

CHART A-3
(Page 2 of 3)
ENGINE CRANKS BUT WILL NOT RUN
3.1L (VIN T) "W" CARLINE (PORT)

Test Description: Numbers below refer to circled numbers on the diagnostic chart.

1. Checks for 12 volt supply to injectors. Due to the injectors wired in parallel there should be a light "ON" on both terminals.
2. Checks continuity of CKT 467 and 468.
3. All checks made to this point would indicate that the ECM is at fault. However, there is a possibility of CKT 467 or 468 being shorted to a voltage source either in the engine harness or in the mini harness.
   - To test for this condition:
   - Disconnect gray mini harness connector.
   - Ignition "ON".
   - Probe CKTs 467 and 468 on the ECM side of harness with a test light connected to ground. There should be no light. If light is "ON," repair short to voltage.

- If OK, check the resistance of the injector harness between terminals "A" & "F" and "K" & "E".
- Should be 4 to 5 ohms.
- If less than 4 ohms, check harness for wires shorted together and check each injector resistance. There should be no continuity between terminals "K" & "E" or "A" & "F". If there is continuity, check harness for wires shorted together.
- Resistance of each injector should be 12.2 ohms ± 2 ohms (at ambient temperature). Injector resistance will increase with temperature and could read as high as 14.4 ohms at 93°C.
- If all OK, replace ECM.

1989–90 3.1L (VIN T) – ALL W-BODY VEHICLES

CHART A-3
(Page 2 of 3)
ENGINE CRANKS BUT WILL NOT RUN
3.1L (VIN T) "W" CARLINE (PORT)

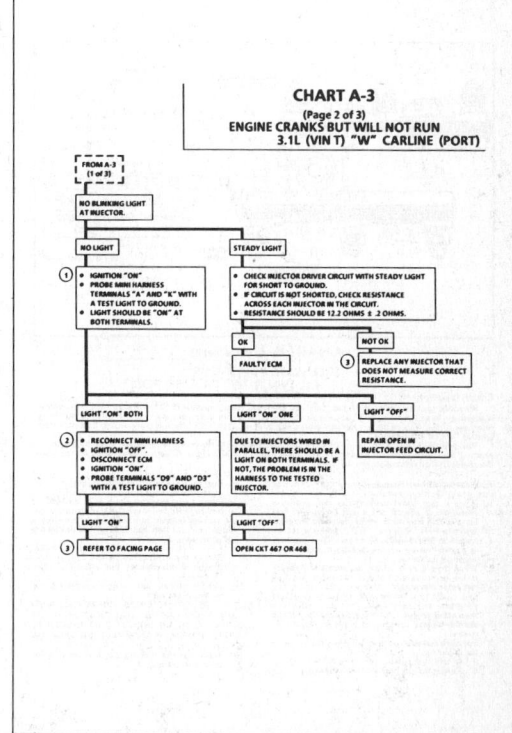

1989–90 3.1L (VIN T) – ALL W-BODY VEHICLES

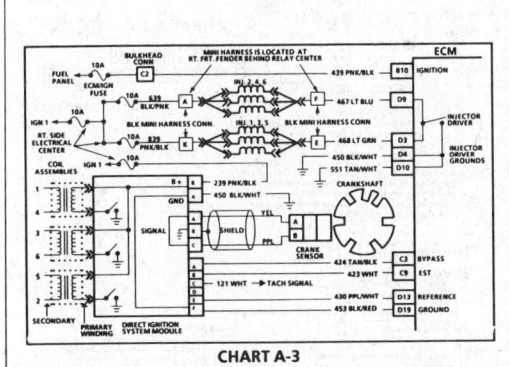

CHART A-3
(Page 3 of 3)
ENGINE CRANKS BUT WILL NOT RUN
3.1L (VIN T) "W" CARLINE (PORT)

Circuit Description:
If the "Scan" tool did not indicate a cranking rpm and there is no spark present at the plugs, the problem lies in the direct ignition system or the power and ground supplies to the module.

The magnetic crank sensor is used to determine engine crankshaft position much the same way as the pick up coil did in distributor type systems. The sensor is mounted in the block near a seven slot wheel on the crank shaft. The rotation of the wheel creates a flux change in the sensor which produces a voltage signal. The ignition module then processes this signal and creates the reference pulses needed by the ECM and the signal triggers the correct coil at the correct time.

Test Description: Numbers below refer to circled numbers on the diagnostic chart.

1. This test will determine if the 12 volt supply and a good ground is available at the ignition module.
2. Tests for continuity of CKT 439 to the ignition module. If test light does not light but the "SES" light is "ON" with ignition "ON," repair open in CKT 439 between DIS ignition module and splice.
3. Checks for continuity of the crank sensor and connections.
4. Voltage will vary in this test depending on cranking speed of engine.

1989–90 3.1L (VIN T) – ALL W-BODY VEHICLES

CHART A-3
(Page 3 of 3)
ENGINE CRANKS BUT WILL NOT RUN
3.1L (VIN T) "W" CARLINE (PORT)

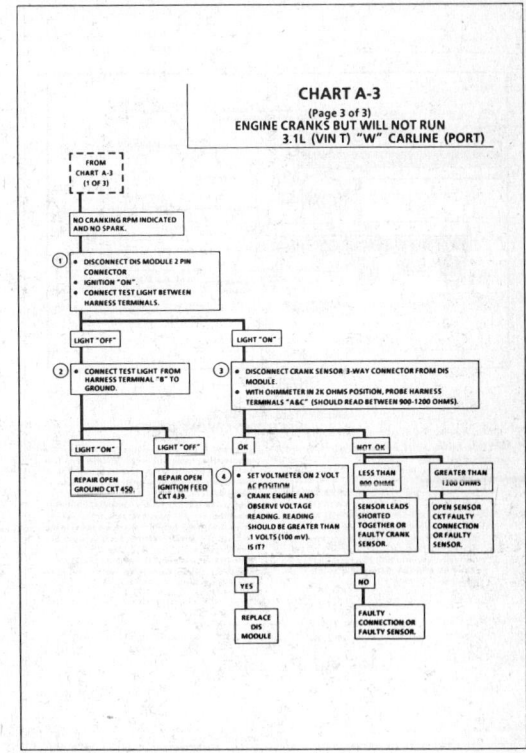

1989–90 3.1L (VIN T) – ALL W-BODY VEHICLES

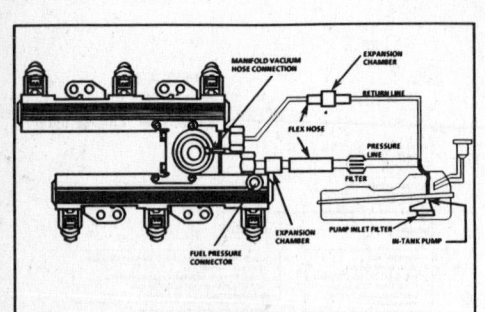

CHART A-7 (Page 1 of 2)
FUEL SYSTEM DIAGNOSIS
3.1L (VIN T) "W" CARLINE (PORT)

Circuit Description:
When the ignition switch is turned "ON," the electronic control module (ECM) will turn "ON" the in-tank fuel pump. It will remain "ON" as long as the engine is cranking or running, and the ECM is receiving reference pulses. If there are no reference pulses, the ECM will shut "OFF" the fuel pump within 2 seconds after ignition "ON" or engine stops.

The pump will deliver fuel to the fuel rail and injectors, then to the pressure regulator, when the engine is running the system pressure is controlled to about 225 to 325 kPa (34 to 47 psi), depending on engine load. Excess fuel is then returned to the fuel tank.

Test Description: Numbers below refer to circled numbers on the diagnostic chart.

1. Wrap a shop towel around the fuel pressure connector to absorb any small amount of fuel leakage that may occur when installing the gage. Ignition "ON," Engine "OFF," pump pressure should be 280-320 kPa (40-47 psi). This pressure is controlled by spring pressure within the regulator assembly.
2. When the engine is idling, the manifold pressure is low (high vacuum) and is applied to the fuel regulator diaphragm. This will offset the spring and result in a lower fuel pressure. This idle pressure will vary somewhat depending on barometric pressure, however, the pressure idling should be less indicating pressure regulator control.
3. Pressure that continues to fall is caused by one of the following:
   - In-tank fuel pump check valve not holding.
   - Pump coupling hose or pulsator leaking.

- Fuel pressure regulator valve leaking.
- Injector(s) sticking open.
4. An injector sticking open can best be determined by checking for a fouled or saturated spark plug(s). If a leaking injector can not be determined by a fouled or saturated spark plug the following procedure should be used.
   - Remove plenum, and remove fuel rail bolts. Follow the procedures in the fuel control section of this manual, but leave fuel lines connected.
   - Lift fuel rail out just enough to leave injector nozzles in the ports.

CAUTION: BE SURE INJECTOR(S) ARE NOT ALLOWED TO SPRAY ON ENGINE AND THAT INJECTOR RETAINING CLIPS ARE INTACT. THIS SHOULD BE CAREFULLY FOLLOWED TO PREVENT FUEL SPRAY ON ENGINE WHICH WOULD CAUSE A FIRE HAZARD.
   - Pressurize the fuel system and observe for injector(s) leaking.

1989–90 3.1L (VIN T) – ALL W-BODY VEHICLES

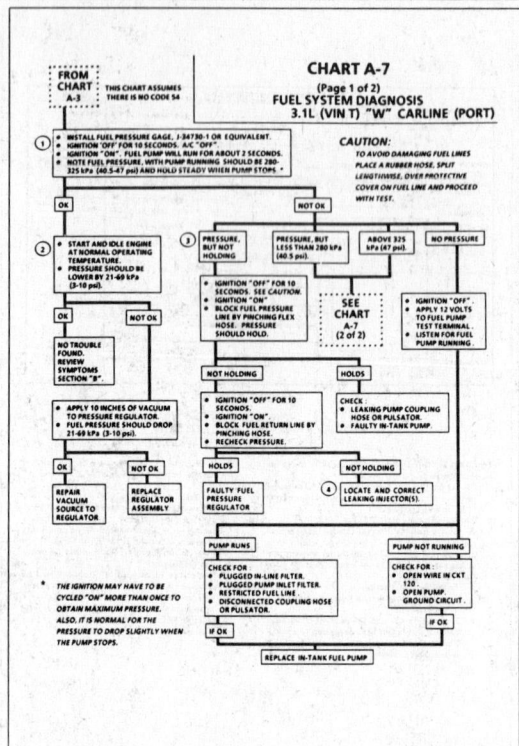

CHART A-7
(Page 1 of 2)
FUEL SYSTEM DIAGNOSIS
3.1L (VIN T) "W" CARLINE (PORT)

FROM CHART A-3 — THIS CHART ASSUMES THERE IS NO CODE 54

CAUTION: TO AVOID DAMAGING FUEL LINES PLACE A RUBBER HOSE, SPLIT LENGTHWISE, OVER PROTECTIVE COVER ON FUEL LINE AND PROCEED WITH TEST.

1989–90 3.1L (VIN T) – ALL W-BODY VEHICLES

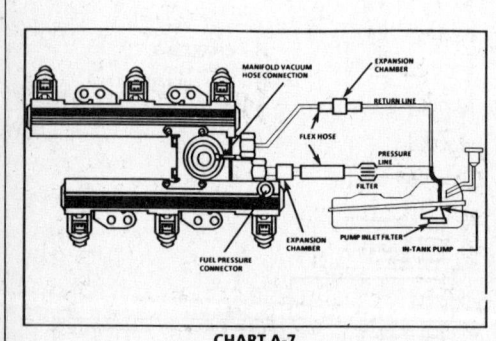

CHART A-7
(Page 2 of 2)
FUEL SYSTEM DIAGNOSIS
3.1L (VIN T) "W" CARLINE (PORT)

Test Description: Numbers below refer to circled numbers on the diagnostic chart.

5. Pressure present, but less than 280 kPa (40 psi) falls into two areas.
   - Regulated pressure but less than 280 kPa (40 psi). Amount of fuel to injectors OK but pressure is too low. System will be lean running and may set Code 44. Also, hard starting cold and overall poor performance.
   - Restricted flow causing pressure drop. Normally, a vehicle with a fuel pressure of less than 165 kPa (24 psi) at idle will not be driveable.

However, if the pressure drop occurs only while driving, the engine will normally surge then stop running as pressure begins to drop rapidly. This is most likely caused by a restricted fuel line or plugged filter.
6. Restricting the fuel return line allows the fuel pump to develop its maximum pressure (dead head pressure). When battery voltage is applied to the pump test terminal, pressure should be above 517 kPa (75 psi).
7. This test determines if the high fuel pressure is due to a restricted fuel return line or a pressure regulator problem.

1989–90 3.1L (VIN T) – ALL W-BODY VEHICLES

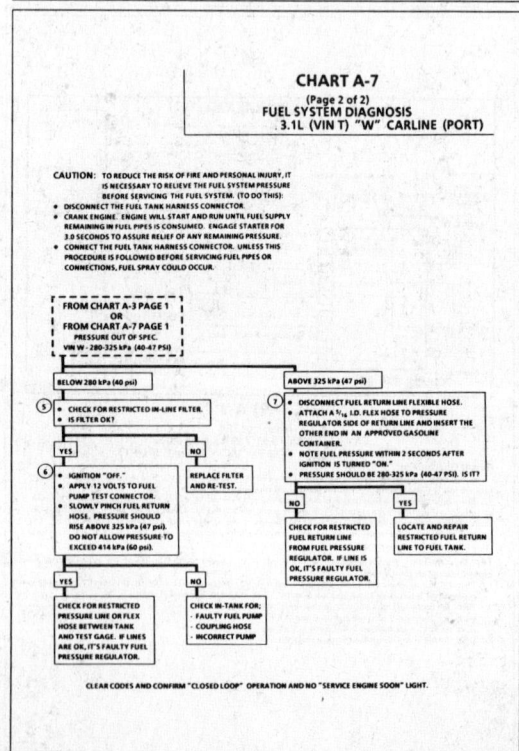

CHART A-7
(Page 2 of 2)
FUEL SYSTEM DIAGNOSIS
3.1L (VIN T) "W" CARLINE (PORT)

CAUTION: TO REDUCE THE RISK OF FIRE AND PERSONAL INJURY, IT IS NECESSARY TO RELIEVE THE FUEL SYSTEM PRESSURE BEFORE SERVICING THE FUEL SYSTEM. (TO DO THIS):
- DISCONNECT THE FUEL TANK HARNESS CONNECTOR.
- CRANK ENGINE. ENGINE WILL START AND RUN UNTIL FUEL SUPPLY REMAINING IN FUEL PIPES IS CONSUMED. ENGAGE STARTER FOR 3.0 SECONDS TO ASSURE RELIEF OF ANY REMAINING PRESSURE.
- CONNECT THE FUEL TANK HARNESS CONNECTOR. UNLESS THIS PROCEDURE IS FOLLOWED BEFORE SERVICING FUEL PIPES OR CONNECTIONS, FUEL SPRAY COULD OCCUR.

1989–90 3.1L (VIN T) – ALL W-BODY VEHICLES

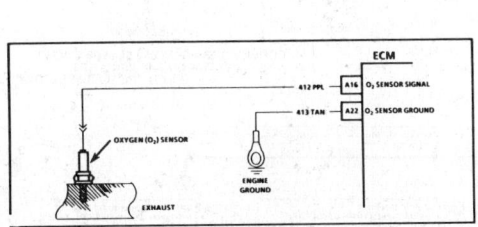

CODE 13
OXYGEN SENSOR CIRCUIT
(OPEN CIRCUIT)
3.1L (VIN T) "W" CARLINE (PORT)

Circuit Description:

The ECM supplies a voltage of about .45 volt between terminals "A16" and "A22". (If measured with a 10 megohm digital voltmeter, this may read as low as .32 volts.) The O_2 sensor varies the voltage within a range of about 1 volt if the exhaust is rich, down through about .10 volt if exhaust is lean.

The sensor is like an open circuit and produces no voltage when it is below 315°C (600°F). An open sensor circuit or cold sensor causes "Open Loop" operation.

Test Description: Numbers below refer to circled numbers on the diagnostic chart.
1. Code 13 will set under the following conditions:
   - Engine running at least 2 minutes after start.
   - Coolant temperature at least 50°C.
   - No Code 21 or 22.
   - O_2 signal voltage steady between 35 and 55 volts.
   - Throttle position sensor signal above 6% for more time than TPS was below 6%. (About 3 volts above closed throttle voltage)
   - All conditions must be met and held for at least 60 seconds
   If the conditions for a Code 13 exist the system will not go "Closed Loop"

2. This will determine if the sensor is at fault or the wiring or ECM is the cause of the Code 13.
3. In doing this test use only a high impedance digital volt ohmmeter. This test checks the continuity of CKTs 412 and 413 because if CKT 413 is open the ECM voltage on CKT 412 will be over 6 volts (600 mV).

Diagnostic Aids:

Normal "Scan" voltage varies between 100 mV to 999 mV (.1 and 1.0 volt) while in "Closed Loop." Code 13 sets in one minute if voltage remains between 35 and .55 volts, but the system will go "Open Loop" in about 15 seconds. Refer to "Intermittents" in Section "B".

1989–90 3.1L (VIN T) – ALL W-BODY VEHICLES

CODE 13
OXYGEN SENSOR CIRCUIT
(OPEN CIRCUIT)
3.1L (VIN T) "W" CARLINE (PORT)

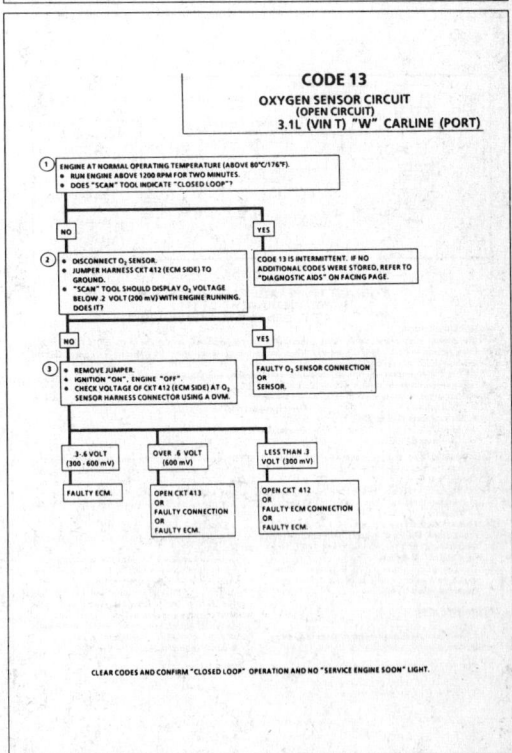

CLEAR CODES AND CONFIRM "CLOSED LOOP" OPERATION AND NO "SERVICE ENGINE SOON" LIGHT.

1989–90 3.1L (VIN T) – ALL W-BODY VEHICLES

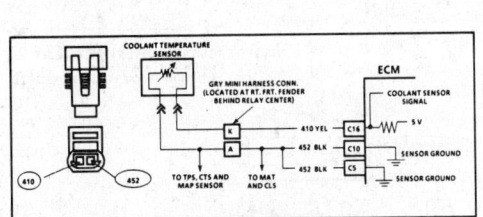

CODE 14
COOLANT TEMPERATURE SENSOR CIRCUIT
(HIGH TEMPERATURE INDICATED)
3.1L (VIN T) "W" CARLINE (PORT)

Circuit Description:

The coolant temperature sensor uses a thermistor to control the signal voltage to the ECM. The ECM applies a voltage on CKT 410 to the sensor. When the engine is cold the sensor (thermistor) resistance is high, therefore the ECM will see high signal voltage.

As the engine warms, the sensor resistance becomes less, and the voltage drops. At normal engine operating temperature the voltage will measure about 1.5 to 2.0 volts at the ECM terminal "C16".

Coolant temperature is one of the inputs used to control:
- Fuel delivery
- Engine spark timing (EST)
- Idle (IAC)
- Converter clutch (TCC)
- Canister purge (CCP)
- EGR
- Cooling fan

Test Description: Numbers refer to circled numbers on the diagnostic chart.
1. Code 14 will set if:
   - Signal voltage indicates a coolant temperature above 140°C (285°F).
   - Engine running longer than 128 seconds
2. This test will determine if CKT 410 is shorted to ground which will cause the conditions for Code 14.

Diagnostic Aids:

Check harness routing for a potential short to ground in CKT 410. CKT is routed from the ECM to a mini harness, and then to the coolant temperature sensor.

"Scan" tool displays engine temperature in degrees centigrade. After engine is started, the temperature should rise steadily to about 90°C then stabilize when thermostat opens. Refer to "Intermittents" in Section "B".

Verify that engine is not overheating and has not been subjected to conditions which could create an overheating condition (i.e. overload, trailer towing, hilly terrain, heavy stop and go traffic, etc.). The "Temperature To Resistance Value" scale at the right may be used to test the coolant sensor at various temperature levels to evaluate the possibility of a "shifted" (mis-scaled) sensor. A "shifted" sensor could result in poor driveability complaints.

1989–90 3.1L (VIN T) – ALL W-BODY VEHICLES

CODE 14
COOLANT TEMPERATURE SENSOR CIRCUIT
(HIGH TEMPERATURE INDICATED)
3.1L (VIN T) "W" CARLINE (PORT)

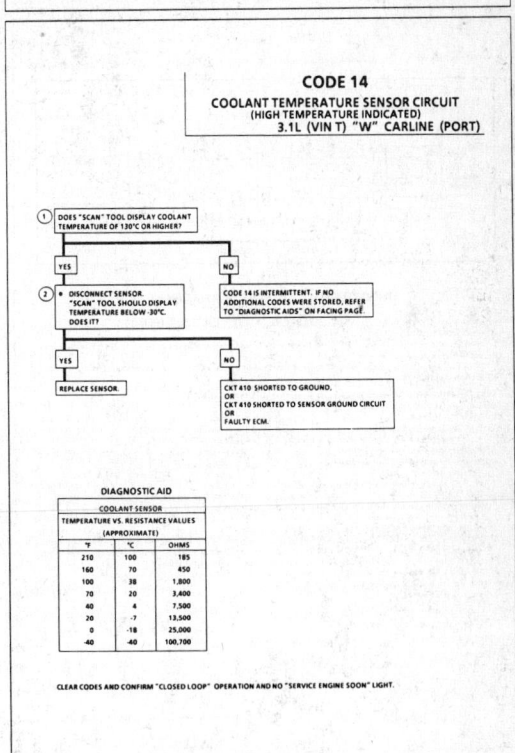

DIAGNOSTIC AID

COOLANT SENSOR
TEMPERATURE VS. RESISTANCE VALUES
(APPROXIMATE)

°F	°C	OHMS
210	100	185
160	70	450
100	38	1,800
70	20	3,400
40	4	7,500
20	-7	13,500
0	-18	25,000
-40	-40	100,700

CLEAR CODES AND CONFIRM "CLOSED LOOP" OPERATION AND NO "SERVICE ENGINE SOON" LIGHT.

1989–90 3.1L (VIN T) – ALL W-BODY VEHICLES

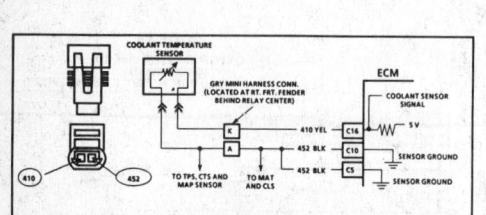

CODE 15
COOLANT TEMPERATURE SENSOR CIRCUIT
(LOW TEMPERATURE INDICATED)
3.1L (VIN T) "W" CARLINE (PORT)

Circuit Description:

The coolant temperature sensor uses a thermistor to control the signal voltage to the ECM. The ECM applies a voltage on CKT 410 to the sensor. When the engine is cold the sensor (thermistor) resistance is high, therefore the ECM will see high signal voltage.

As the engine warms, the sensor resistance becomes less, and the voltage drops. At normal engine operating temperature the voltage will measure about 1.5 to 2.0 volts at the ECM terminal "C16".

Coolant temperature is one of the inputs used to control:

- Fuel delivery
- Engine spark timing (EST)
- Idle (IAC)
- Converter clutch (TCC)
- Canister purge (CCP)
- EGR
- Cooling fan

Test Description: Numbers below refer to circled numbers on the diagnostic chart.

1. Code 15 will set if:
   - Signal voltage indicates a coolant temperature less than -44°C (-47°F) for 20 seconds.
2. This test simulates a Code 14. If the ECM recognizes the low signal voltage, (high temperature) and the "Scan" reads 130°C, the ECM and wiring are OK.
3. This test will determine if CKT 410 is open. There should be 5 volts present at sensor connector if measured with a DVM.

Diagnostic Aids:

A "Scan" tool reads engine temperature in degrees centigrade.

After engine is started the temperature should rise steadily to about 95°C then stabilize when thermostat opens. CKT 410 is routed from the ECM to a mini harness, and then to the coolant temperature sensor.

A faulty connection, or an open in CKT 410 or 452 will result in a Code 15.

If Code 23 is also set, check CKT 452 for faulty wiring or connections. Check terminals at sensor for good contact.

Codes 15 and 21 are stored at the same time could be the result of an open CKT 452 which would also turn the temperature warning indicator "ON." The "Temperature to Resistance Value" scale at the right may be used to test the coolant sensor at various temperature levels to evaluate the possibility of a "shifted" (mis-scaled) sensor. A "shifted" sensor could result in poor driveability complaints.

Refer to "Intermittents" in Section "B".

1989–90 3.1L (VIN T) – ALL W-BODY VEHICLES

CODE 15
COOLANT TEMPERATURE SENSOR CIRCUIT
(LOW TEMPERATURE INDICATED)
3.1L (VIN T) "W" CARLINE (PORT)

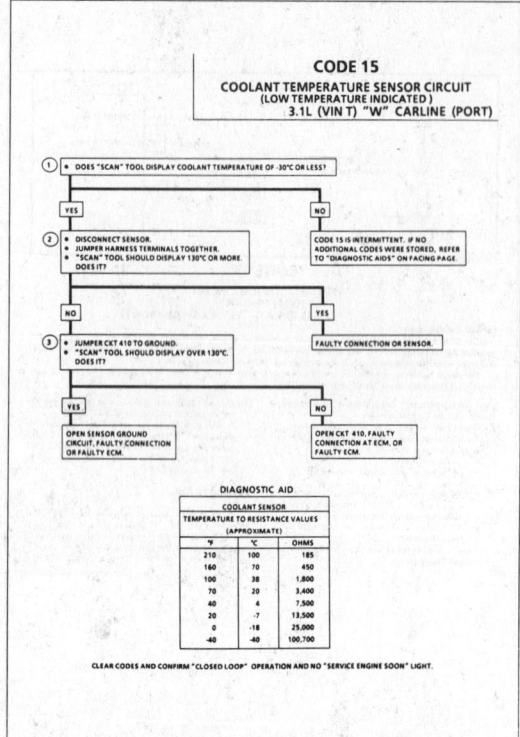

COOLANT SENSOR		
TEMPERATURE TO RESISTANCE VALUES		
(APPROXIMATE)		
°F	°C	OHMS
210	100	185
160	70	450
100	38	1,800
70	20	3,400
40	4	7,500
20	-7	13,500
0	-18	25,000
-40	-40	100,700

CLEAR CODES AND CONFIRM "CLOSED LOOP" OPERATION AND NO "SERVICE ENGINE SOON" LIGHT.

1989–90 3.1L (VIN T) – ALL W-BODY VEHICLES

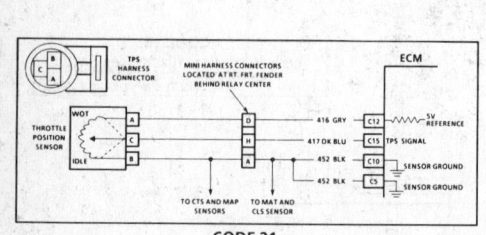

CODE 21
THROTTLE POSITION SENSOR (TPS) CIRCUIT
(SIGNAL VOLTAGE HIGH)
3.1L (VIN T) "W" CARLINE (PORT)

Circuit Description:

The throttle position sensor (TPS) provides a voltage signal that changes relative to the throttle blade. Signal voltage will vary from about .5 at idle to about 5 volts at wide open throttle.

The TPS signal is one of the most important inputs used by the ECM for fuel control and for most of the ECM control outputs.

Test Description: Numbers below refer to circled numbers on the diagnostic chart.

1. Code 21 will set if:
   - Engine is running
   - TPS signal voltage is greater than 4.3 volts
   - Air flow is less than 17 GM/sec.
   - All conditions met for 10 seconds.
   OR
   - TPS signal voltage over 4.5 volts with ignition "ON."
   With throttle closed, the TPS should read less than .85 volt. If it doesn't, make sure cruise control and throttle cables are not being held open.
2. With the TPS sensor disconnected, the TPS voltage should go low, if the ECM and wiring is OK.
3. Probing CKT 452 with a test light, checks the 5 volt return circuit. Faulty sensor ground circuit will cause a Code 21.

Diagnostic Aids:

A "Scan" tool reads throttle position in volts. Voltage should increase at a steady rate as throttle is moved toward WOT.

Also some "Scan" tools will read throttle angle 0% = closed throttle, 100% = WOT.

An open in CKT 452 will result in a Code 21.

Codes 15 and 21 are stored at the same time could be the result of an open CKT 452 which would also turn the temperature warning indicator "ON." "Scan" TPS while depressing accelerator pedal with engine stopped and ignition "ON." Display should vary from below 2500 mV (2.5 volts) when throttle was closed, to over 4500 mV (4.5 volts) when throttle is held at wide open throttle position.

Refer to "Intermittents" in Section "B".

1989–90 3.1L (VIN T) – ALL W-BODY VEHICLES

CODE 21
THROTTLE POSITION SENSOR (TPS) CIRCUIT
(SIGNAL VOLTAGE HIGH)
3.1L (VIN T) "W" CARLINE (PORT)

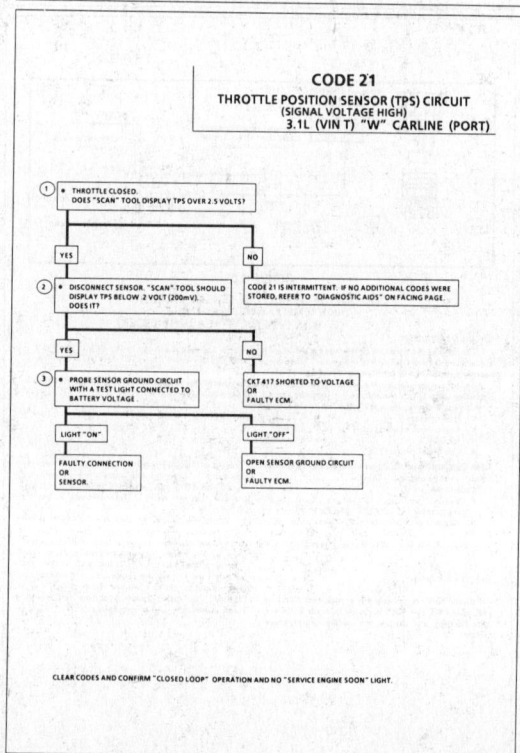

CLEAR CODES AND CONFIRM "CLOSED LOOP" OPERATION AND NO "SERVICE ENGINE SOON" LIGHT.

1989–90 3.1L (VIN T) – ALL W-BODY VEHICLES

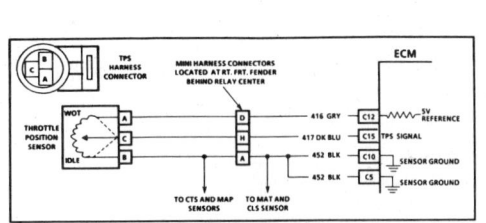

CODE 22
THROTTLE POSITION SENSOR (TPS) CIRCUIT
(SIGNAL VOLTAGE LOW)
3.1L (VIN T) "W" CARLINE (PORT)

Circuit Description:

The throttle position sensor (TPS) provides a voltage signal that changes relative to the throttle blade. Signal voltage will vary from about .5 at idle to about 5 volts at wide open throttle.

The TPS signal is one of the most important inputs used by the ECM for fuel control and for most of the ECM control outputs.

Test Description: Numbers below refer to circled numbers on the diagnostic chart.
1. Code 22 will set if:
   - Engine running
   - TPS signal voltage is less than about .25 volt for 3 seconds.
2. Simulates Code 21: (High Voltage) If the ECM recognizes the high signal voltage the ECM and wiring are OK.
3. TPS check: The TPS has an auto zeroing feature. If the voltage reading is within the range of 0.45 to 0.85 volt, the ECM will use that value as closed throttle. If the voltage reading is out of the auto zero range on an existing or replacement TPS; check for cruise control and throttle cables for being held open.
4. This simulates a high signal voltage to check for an open in CKT 417.
5. CKTs 416 and 432 share a common 5volts buffered reference signal. If either of these circuits is shorted to ground, Code 22 will set. To determine if the MAP sensor is causing the 22 problem disconnect it to see if Code 22 resets. Be sure TPS is connected and clear codes before testing.

Diagnostic Aids:

A "Scan" tool reads throttle position in volts. Voltage should increase at a steady rate as throttle is moved toward WOT.

Also some "Scan" tools will read throttle angle 0% = closed throttle 100% = WOT.

An open or short to ground in CKTs 416 or 417 will result in a Code 22.

CKTs 416 and 417 are routed through a mini harness. CKT 416 is connected to terminal "D" at the gray connector, CKT 417 is connected to terminal "H" at the black connector.

"Scan" TPS while depressing accelerator pedal with engine stopped and ignition "ON." Display should vary from below 500 mV (.5V) when throttle was closed, to over 4500 mV (4.5V) when throttle is held at wide open throttle position.

Also some "Scan" tools will read throttle angle. 0% = closed throttle. 100% = open throttle.

If Code 22 is set, check CKT 416 for faulty wiring or connections.

Refer to "Intermittents" in Section "B".

1989–90 3.1L (VIN T) – ALL W-BODY VEHICLES

CODE 22
THROTTLE POSITION SENSOR (TPS) CIRCUIT
(SIGNAL VOLTAGE LOW)
3.1L (VIN T) "W" CARLINE (PORT)

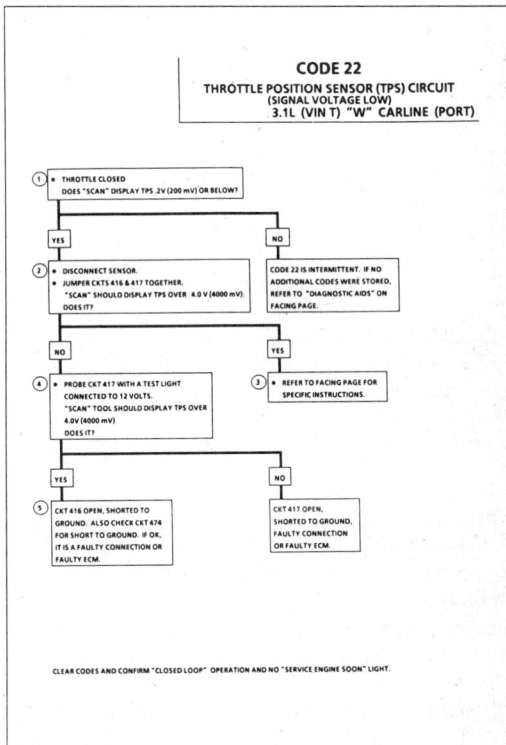

CLEAR CODES AND CONFIRM "CLOSED LOOP" OPERATION AND NO "SERVICE ENGINE SOON" LIGHT.

1989–90 3.1L (VIN T) – ALL W-BODY VEHICLES

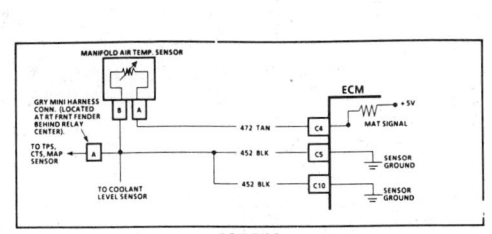

CODE 23
MANIFOLD AIR TEMPERATURE (MAT) SENSOR CIRCUIT
(LOW TEMPERATURE INDICATED)
3.1L (VIN T) "W" CARLINE (PORT)

Circuit Description:

The MAT sensor uses a thermistor to control the signal voltage to the ECM. The ECM applies a voltage (about 5 volts) on CKT 472 to the sensor. When the air is cold the sensor (thermistor) resistance is high, therefore the ECM will see a high signal voltage. If the air is warm the sensor resistance is low therefore the ECM will see a low voltage.

The MAT sensor is located in the air cleaner.

Test Description: Numbers below refer to circled numbers on the diagnostic chart.
1. Code 23 will set if:
   - A signal voltage indicates a manifold air temperature below -35°C (-31°F).
   - Time since engine start is 4 minutes or longer.
   - No VSS.
2. A Code 23 will set, due to an open sensor, wire, or connection. This test will determine if the wiring and ECM are OK.
3. This will determine if the signal CKT 472 or the 5volts return CKT 452 is open.

Diagnostic Aids:

A "Scan" tool reads temperature of the air entering the engine and should read close to ambient air temperature when engine is cold, and rises as underhood temperature increases.

A faulty connection, or an open in CKT 472 or 452 will result in a Code 23.

Codes 23 and 34 stored at the same time, could be the result of an open CKT 452 which would also turn the temperature warning indicator "ON." CKT 452 is routed through a mini harness. A faulty connection could result in intermittent failures. The "Temperature vs. Resistance Values" scale at the right may be used to test the MAT sensor at various temperature levels to evaluate the possibility of a "shifted" (mis-scaled) sensor. A "slewed" sensor could result in poor driveability complaints.

Refer to "Intermittents" in Section "B"

1989–90 3.1L (VIN T) – ALL W-BODY VEHICLES

CODE 23
MANIFOLD AIR TEMPERATURE (MAT) SENSOR CIRCUIT
(LOW TEMPERATURE INDICATED)
3.1L (VIN T) "W" CARLINE (PORT)

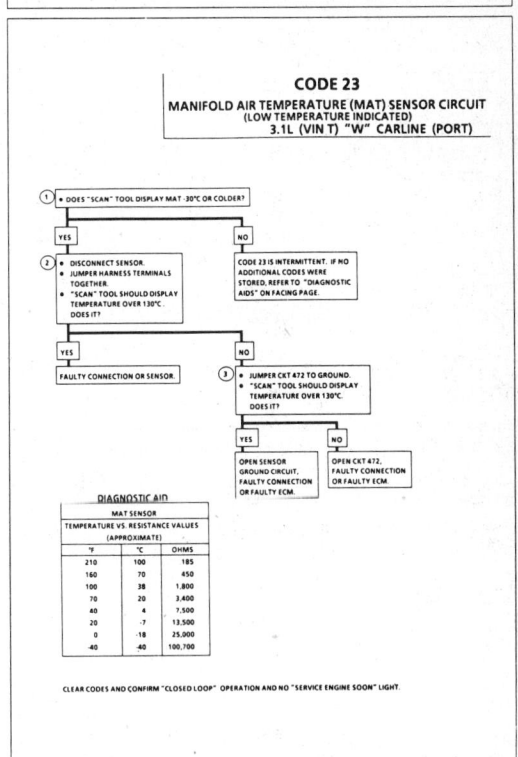

DIAGNOSTIC AID

MAT SENSOR

TEMPERATURE VS. RESISTANCE VALUES (APPROXIMATE)		
°F	°C	OHMS
210	100	185
160	70	450
100	38	1,800
70	20	3,400
40	4	7,500
20	-7	13,500
0	-18	25,000
-40	-40	100,700

CLEAR CODES AND CONFIRM "CLOSED LOOP" OPERATION AND NO "SERVICE ENGINE SOON" LIGHT.

1989–90 3.1L (VIN T) – ALL W-BODY VEHICLES

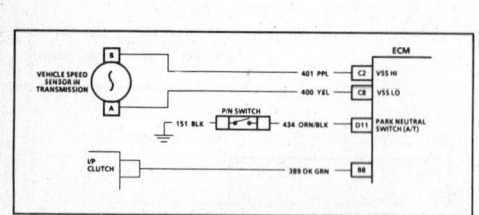

CODE 24
VEHICLE SPEED SENSOR (VSS) CIRCUIT
3.1L (VIN T) "W" CARLINE (PORT)

Circuit Description:

Vehicle speed information is provided to the ECM by the vehicle speed sensor which uses a permanent magnet (PM) generator and it is mounted in the transaxle. The PM generator produces a pulsing voltage whenever vehicle speed is over about 3 mph. The AC voltage level and the number of pulses increases with vehicle speed. The ECM then converts the pulsing voltage to mph which is used for calculations, and the mph can be displayed with a "Scan" tool. Output of the generator can also be seen by using a digital voltmeter on the AC scale while rotating the generator.

The function of VSS buffer used in past model years has been incorporated into the ECM. The ECM then supplies the necessary signal for the instrument panel (4000 pulses per mile) for operating the speedometer and the odometer. If the vehicle is equipped with cruise control the ECM also provides a signal (2000 pulses per mile) to the cruise control module.

Test Description: Numbers below refer to circled numbers on the diagnostic chart.

Code 24 will set if vehicle speed equals 0 mph when:
- Engine speed is between 1400 and 3600 rpm.
- TPS is less than 2%.
- Low load condition (LV8 below 35).
- Not in park or neutral.
- No Code 21, 22, 33 or 34.
- All conditions met for 10 seconds.

These conditions are met during a road load deceleration. Disregard Code 24 that sets when drive wheels are not turning.
- The PM generator only produces a signal if drive wheels are turning greater than 3 mph.

Diagnostic Aids:

"Scan" should indicate a vehicle speed whenever the drive wheels are turning greater than 3 mph.

A problem in CKT 400 will not affect the VSS input or the readings on a "Scan."

Check CKT 400 and 401 for proper connections to be sure there clean and tight and the harness is routed correctly. Refer to "Intermittents" in Section "B"

(A/T) A faulty or misadjusted park/neutral switch can result in a false Code 24. Use a "Scan" and check for proper signal while in overdrive (440-T4). Refer to CHART C-1A for P/N switch diagnosis check.

1989–90 3.1L (VIN T) – ALL W-BODY VEHICLES

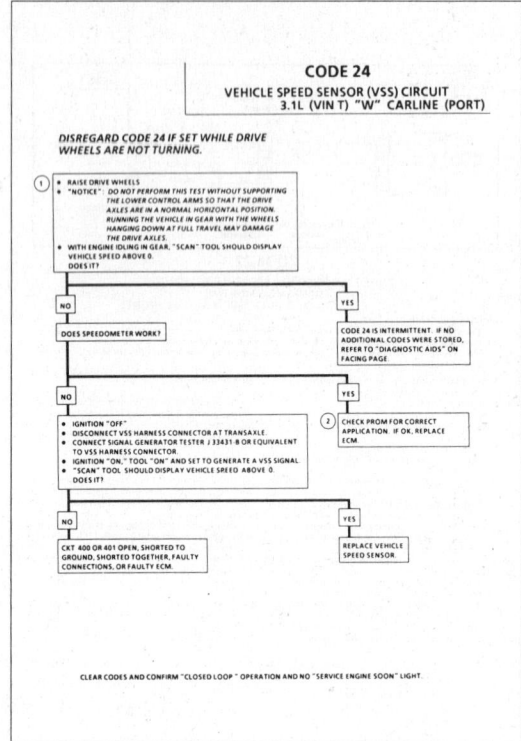

CODE 24
VEHICLE SPEED SENSOR (VSS) CIRCUIT
3.1L (VIN T) "W" CARLINE (PORT)

DISREGARD CODE 24 IF SET WHILE DRIVE WHEELS ARE NOT TURNING.

CLEAR CODES AND CONFIRM "CLOSED LOOP" OPERATION AND NO "SERVICE ENGINE SOON" LIGHT.

1989–90 3.1L (VIN T) – ALL W-BODY VEHICLES

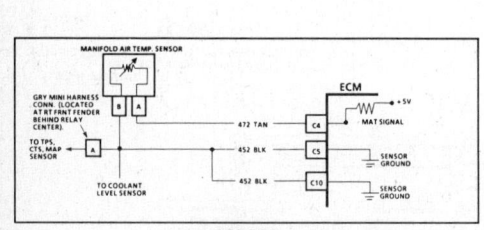

CODE 25
MANIFOLD AIR TEMPERATURE (MAT) SENSOR CIRCUIT
(HIGH TEMPERATURE INDICATED)
3.1L (VIN T) "W" CARLINE (PORT)

Circuit Description:

The MAT sensor uses a thermistor to control the signal voltage to the ECM. The ECM applies a voltage (about 5 volts) on CKT 472 to the sensor. When the air is cold the sensor (thermistor) resistance is high, therefore the ECM will see a high signal voltage. If the air is warm the sensor resistance is low therefore the ECM will see a low voltage.

The MAT sensor is located in the air cleaner.

Test Description: Numbers below refer to circled numbers on the diagnostic chart.
1. Code 25 will set if:
- A signal voltage indicates a manifold air temperature greater than 135°C (293°F) for 3 seconds.
- Time since engine start is 8 minutes or longer.
- A vehicle speed over 3 mph is present.

Due to the conditions necessary to set a Code 25 the "Service Engine Soon" light will remain "ON" while the signal is low and vehicle speed is present.

Diagnostic Aids:

A "Scan" tool reads temperature of the air entering the engine and should read close to ambient air temperature when engine is cold, and rises as underhood temperature increases.

A short to ground in CKT 472 will result in a Code 25.

The "Temperature to Resistance Values" scale at the right may be used to test the MAT sensor at various temperature levels to evaluate the possibility of a "shifted" (mis-scaled) sensor. A "slewed" sensor could result in poor driveability complaints.

Refer to "Intermittents" in Section "B"

1989–90 3.1L (VIN T) – ALL W-BODY VEHICLES

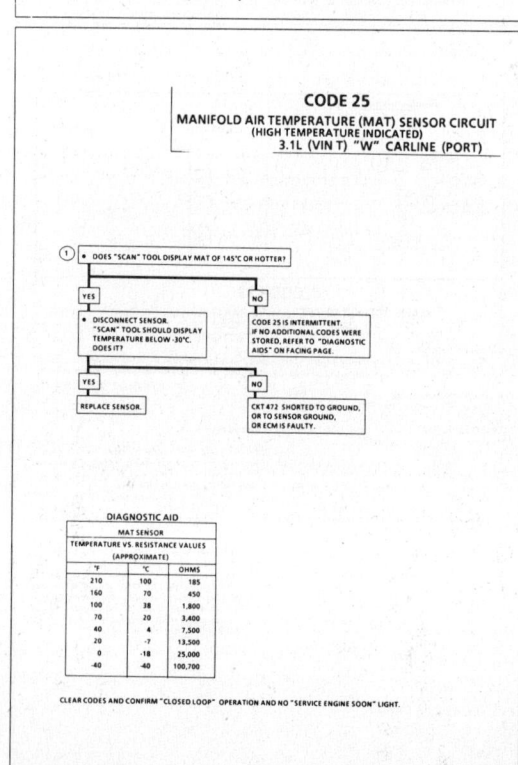

CODE 25
MANIFOLD AIR TEMPERATURE (MAT) SENSOR CIRCUIT
(HIGH TEMPERATURE INDICATED)
3.1L (VIN T) "W" CARLINE (PORT)

DIAGNOSTIC AID

MAT SENSOR
TEMPERATURE VS. RESISTANCE VALUES
(APPROXIMATE)

°F	°C	OHMS
210	100	185
160	70	450
100	38	1,800
70	20	3,400
40	4	7,500
20	-7	13,500
0	-18	25,000
-40	-40	100,700

CLEAR CODES AND CONFIRM "CLOSED LOOP" OPERATION AND NO "SERVICE ENGINE SOON" LIGHT.

1989–90 3.1L (VIN T) – ALL W-BODY VEHICLES

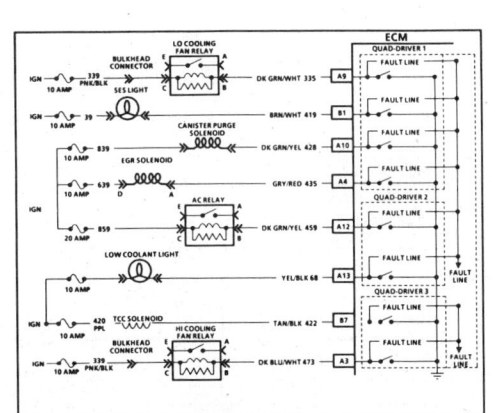

CODE 26 (Page 1 of 3)
QUAD-DRIVER (QDM) ERROR
3.1L (VIN T) "W" CARLINE (PORT)

Circuit Description:
The ECM is used to control several components such as those illustrated above. The ECM controls a device by turning "ON" or "OFF" a device called a quad-driver. The ECM also monitors the state of each of the control circuits by way of the fault line. When the ECM is not turning "ON" the circuit being controlled, there should be 12 volts present on the fault line. If 12 volts is not present due to an open or shorted control circuit, or an open component coil, a failure will be logged in memory. Also when the ECM is commanding the component "ON" the control circuit should be low (near 0 volts).

If the fault line does not sense the low state, a failure will be logged. This could be caused by a short to 12 volts on the control circuit by a shorted component or by a faulty ECM driver.

Test Description: Numbers below refer to circled numbers on the diagnostic chart.
1. The ECM does not know which controlled circuit caused the Code 26 so this chart will go through each of the circuits to determine which is at fault. An EGR circuit problem should also set a Code 32 but after that problem is fixed this chart should be performed.
2. This test checks the "Service Engine Soon" light driver and the "Service Engine Soon" light circuit.

QDM symptoms:
- Cooling fan(s) inoperative
- Poor driveability due to 100% canister purge.
- Coolant light "ON" all the time, "OFF" during bulb check.
- EGR inoperative - Code 32.
- TCC inoperative

1989–90 3.1L (VIN T) – ALL W-BODY VEHICLES

CODE 26
(Page 1 of 3)
QUAD-DRIVER (QDM) ERROR
3.1L (VIN T) "W" CARLINE (PORT)

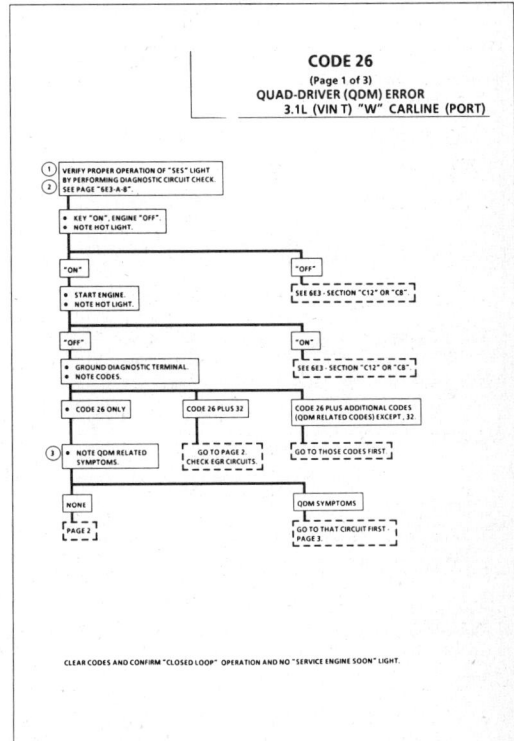

CLEAR CODES AND CONFIRM "CLOSED LOOP" OPERATION AND NO "SERVICE ENGINE SOON" LIGHT.

1989–90 3.1L (VIN T) – ALL W-BODY VEHICLES

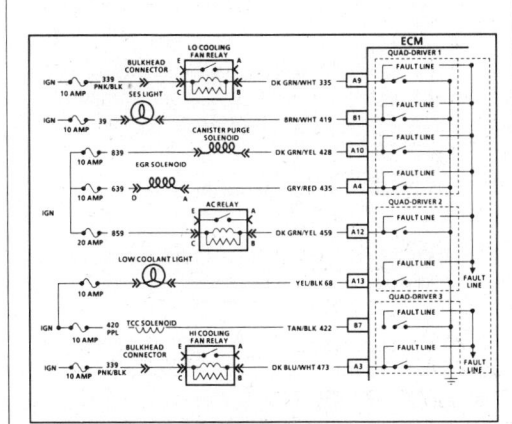

CODE 26
(Page 2 of 3)
QUAD-DRIVER (QDM) ERROR
3.1L (VIN T) "W" CARLINE (PORT)

Circuit Description:
The ECM is used to control several components such as those illustrated above. The ECM controls a device by turning "ON" or "OFF" a device called a quad-driver. The ECM also monitors the state of each of the control circuits by way of the fault line. When the ECM is not turning "ON" the circuit being controlled, there should be 12 volts present on the fault line. If 12 volts is not present due to an open or shorted control circuit, or an open component coil, a failure will be logged in memory. Also when the ECM is commanding the component "ON," the control circuit should be low (near 0 volts).

If the fault line does not sense the low state, a failure will be logged. This could be caused by a short to 12 volts on the control circuit by a shorted component or by a faulty ECM driver.

Test Description: Numbers below refer to circled numbers on the diagnostic chart.
4. This test will determine which circuit is out of specifications. All circuits _EXCEPT_, "B1", the "SES" light, "A18", and "hot light" should be B+ when key is "ON", engine not running. The Diagnostic "Test" Terminal is not grounded.

1989–90 3.1L (VIN T) – ALL W-BODY VEHICLES

CODE 26
(Page 2 of 3)
QUAD-DRIVER (QDM) ERROR
3.1L (VIN T) "W" CARLINE (PORT)

CIRCUIT NOT ISOLATED BY PRIOR STEPS
(AMP TEST)

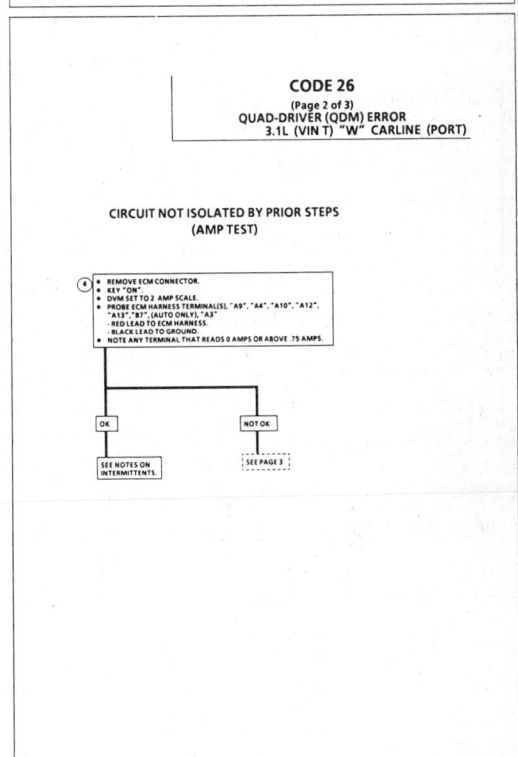

1989–90 3.1L (VIN T) – ALL W-BODY VEHICLES

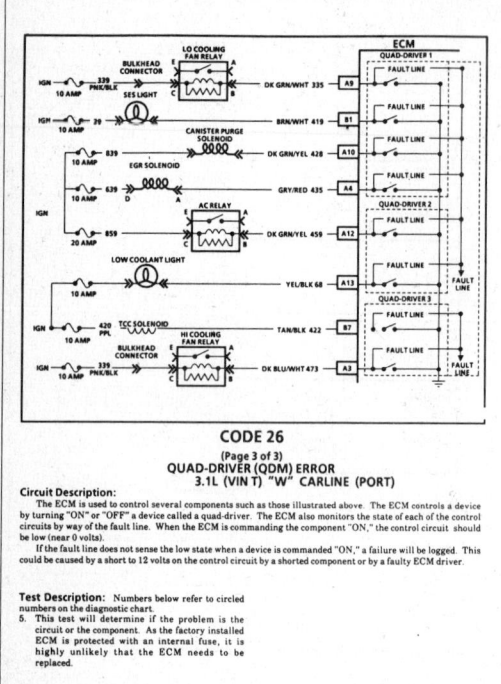

CODE 26
(Page 3 of 3)
QUAD-DRIVER (QDM) ERROR
3.1L (VIN T) "W" CARLINE (PORT)

Circuit Description:
The ECM is used to control several components such as those illustrated above. The ECM controls a device by turning "ON" or "OFF" a device called a quad-driver. The ECM also monitors the state of each of the control circuits by way of the fault line. When the ECM is commanding the component "ON," the control circuit should be low (near 0 volts).
If the fault line does not sense the low state when a device is commanded "ON," a failure will be logged. This could be caused by a short to 12 volts on the control circuit by a shorted component or by a faulty ECM driver.

Test Description: Numbers below refer to circled numbers on the diagnostic chart.
5. This test will determine if the problem is the circuit or the component. As the factory installed ECM is protected with an internal fuse, it is highly unlikely that the ECM needs to be replaced.

1989–90 3.1L (VIN T) – ALL W-BODY VEHICLES

CODE 26
(Page 3 of 3)
QUAD-DRIVER (QDM) ERROR
3.1L (VIN T) "W" CARLINE (PORT)

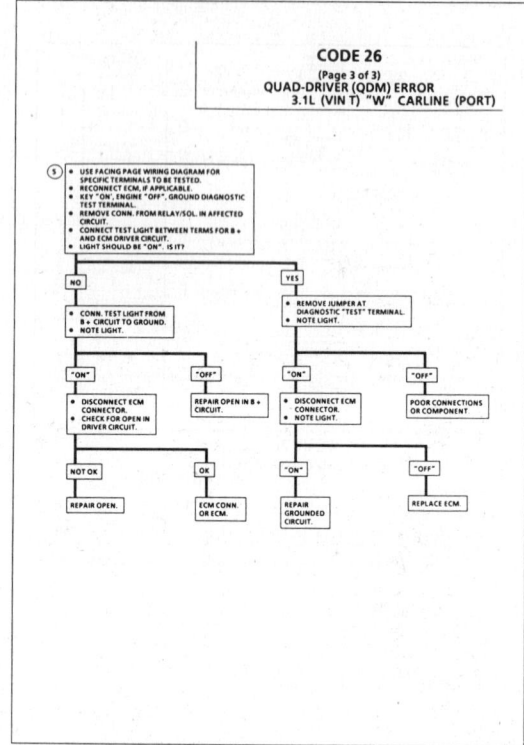

1989–90 3.1L (VIN T) – ALL W-BODY VEHICLES

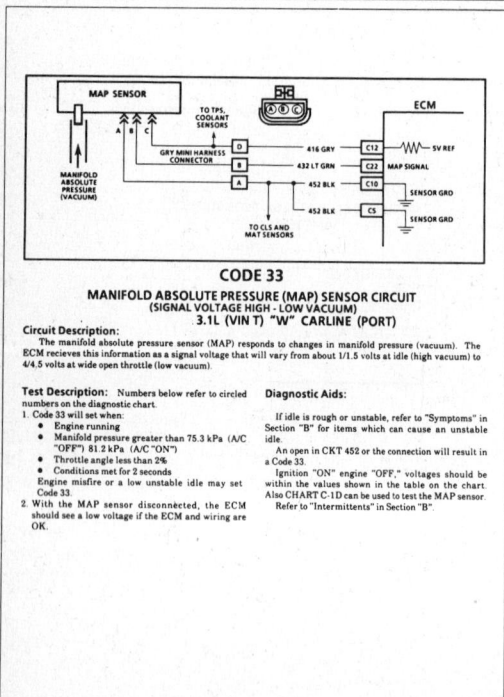

CODE 33
MANIFOLD ABSOLUTE PRESSURE (MAP) SENSOR CIRCUIT
(SIGNAL VOLTAGE HIGH - LOW VACUUM)
3.1L (VIN T) "W" CARLINE (PORT)

Circuit Description:
The manifold absolute pressure sensor (MAP) responds to changes in manifold pressure (vacuum). The ECM recieves this information as a signal voltage that will vary from about 1/1.5 volts at idle (high vacuum) to 4/4.5 volts at wide open throttle (low vacuum).

Test Description: Numbers below refer to circled numbers on the diagnostic chart.
1. Code 33 will set when:
   - Engine running
   - Manifold pressure greater than 75.3 kPa (A/C "OFF") 81.2 kPa (A/C "ON")
   - Throttle angle less than 2%
   - Conditions met for 2 seconds
   Engine misfire or a low unstable idle may set Code 33.
2. With the MAP sensor disconnected, the ECM should see a low voltage if the ECM and wiring are OK.

Diagnostic Aids:
If idle is rough or unstable, refer to "Symptoms" in Section "B" for items which can cause an unstable idle.
An open in CKT 452 or the connection will result in a Code 33.
Ignition "ON" engine "OFF," voltages should be within the values shown in the table on the chart. Also CHART C-1D can be used to test the MAP sensor. Refer to "Intermittents" in Section "B".

1989–90 3.1L (VIN T) – ALL W-BODY VEHICLES

CODE 33
MANIFOLD ABSOLUTE PRESSURE (MAP) SENSOR CIRCUIT
(SIGNAL VOLTAGE HIGH - LOW VACUUM)
3.1L (VIN T) "W" CARLINE (PORT)

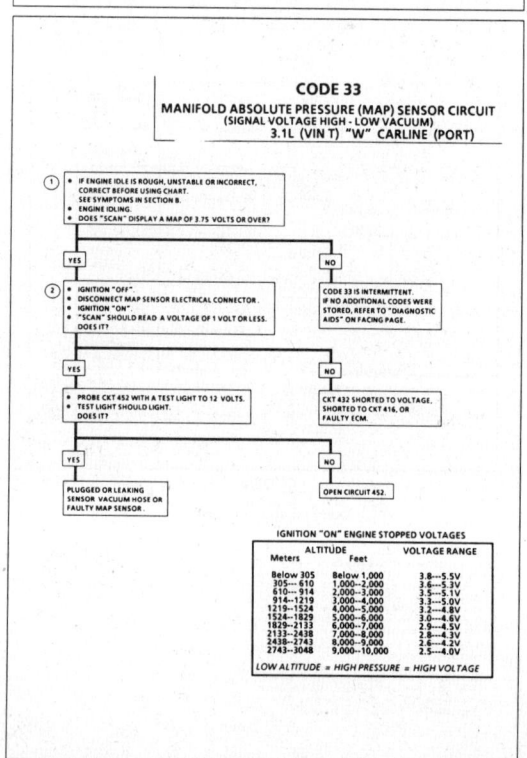

1989–90 3.1L (VIN T) – ALL W-BODY VEHICLES

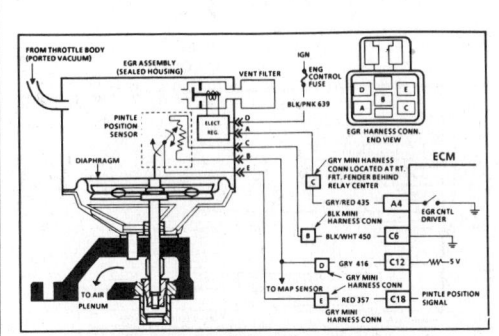

CODE 32
EXHAUST GAS RECIRCULATION (EGR) CIRCUIT
3.1L (VIN T) "W" CARLINE (PORT)

Circuit Description:

The integrated electronic EGR valve functions similar to a port valve with a remote vacuum regulator. The internal solenoid is normally open, which causes the vacuum signal to be vented "OFF" to the atmosphere when EGR is not being commanded by the ECM. This EGR valve has a sealed cap and the solenoid valve opens and closes the vacuum signal which controls the amount of vacuum vented to atmosphere, and this controls the amount of vacuum applied to the diaphragm. The electronic EGR valve contains a voltage regulator which converts the ECM signal to provide different amounts of EGR flow by regulating the current to the solenoid. The ECM controls EGR flow with a pulse width modulated signal (turns "ON" and "OFF" many times a second) based on air flow, TPS, and rpm.

This system also contains a pintle position sensor which works similar to a TPS sensor, and as EGR flow is increased, the sensor output also increases.

Code 32 means that there has been an EGR system fault detected.

Code 32 will set under two conditions:
- Coolant temperature above a specified amount, EGR should be "ON", or
- EGR pintle position does not match duty cycle.

Test Description: Numbers below refer to circled numbers on the diagnostic chart.

1. With the engine running and the transmission in gear, with brakes applied, increasing engine rpm will put a load on the engine. Engine vacuum will be applied to the EGR diaphragm and cause the EGR pintle to open increasing pintle position voltage.

2. Grounding the diagnostic terminal should energize the solenoid which closes off the vent and allows the vacuum to move the diaphragm. This test determines that the ECM is capable of controlling the solenoid. When EGR is commanded on by the ECM, the test light should be "ON."

3. If CKT 450 is open, the pintle signal will go high (showing a 5 volts signal). This will set a Code 32. If CKT 357 becomes shorted to 12 volts or to CKT 416, the signal voltage will go high causing a Code 32.

Diagnostic Aids:

Some "Scan" tools will read pintle position in volts.

The EGR position voltage can be used to determine that the pintle is moving. When no EGR is commanded (0% duty cycle), the position sensor should read between .5 volts and 1.5 volts and increase with the commanded EGR duty cycle. If system operates correctly, refer to "Intermittents" in Section "B".

1989–90 3.1L (VIN T) – ALL W-BODY VEHICLES

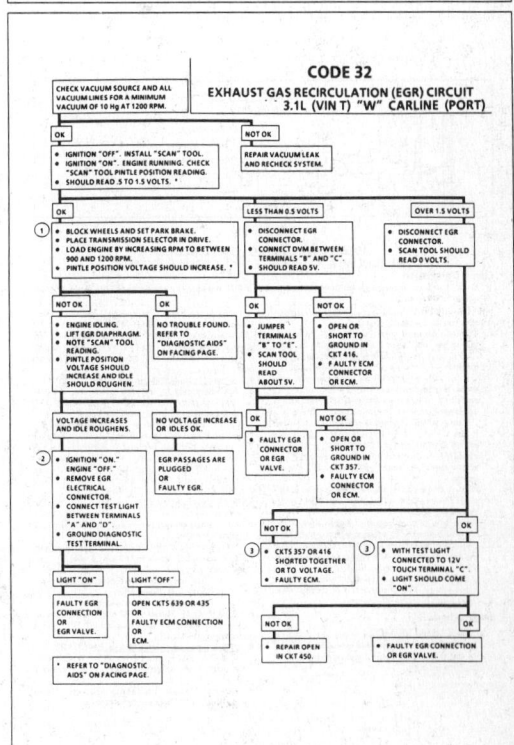

CODE 32
EXHAUST GAS RECIRCULATION (EGR) CIRCUIT
3.1L (VIN T) "W" CARLINE (PORT)

1989–90 3.1L (VIN T) – ALL W-BODY VEHICLES

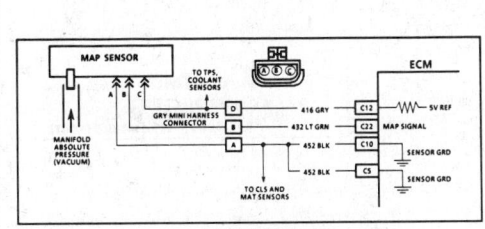

CODE 34
MANIFOLD ABSOLUTE PRESSURE (MAP) SENSOR CIRCUIT
(SIGNAL VOLTAGE LOW - HIGH VACUUM)
3.1L (VIN T) "W" CARLINE (PORT)

Circuit Description:

The manifold absolute pressure sensor (MAP) responds to changes in manifold pressure (vacuum). The ECM recieves this information as a signal voltage that will vary from about 1/1.5 volts at idle (high vacuum) to 4/4.5 volts at wide open throttle (low vacuum).

Test Description: Numbers below refer to circled numbers on the diagnostic chart.

1. Code 34 will set if:
- Engine rpm less than 600
- Manifold pressure reading less than 13 kPa
- Conditions met for 1 second
or
- Engine rpm greater than 600
- Throttle angle over 20%
- Manifold pressure less than 13 kPa
- Conditions met for 1 second

2. This test to see if the sensor is at fault for the low voltage, or if there is an ECM or wiring problem.

3. This simulates a high signal voltage to check for an open in CKT 432. If the test light is bright during this test, CKT 432 is probably shorted to ground. If "Scan" reads over 4 volts at this test, CKT 416 can be checked by measuring the voltage at terminal "C" (should be 5 volts.)

Diagnostic Aids:

An intermittent open in CKTs 432 or 416 will result in a Code 34.

Ignition "ON," engine "OFF," voltages should be within the values shown in the table on the chart. Also CHART C-1D can be used to test MAP sensor. Refer to "Intermittents" in Section "B".

1989–90 3.1L (VIN T) – ALL W-BODY VEHICLES

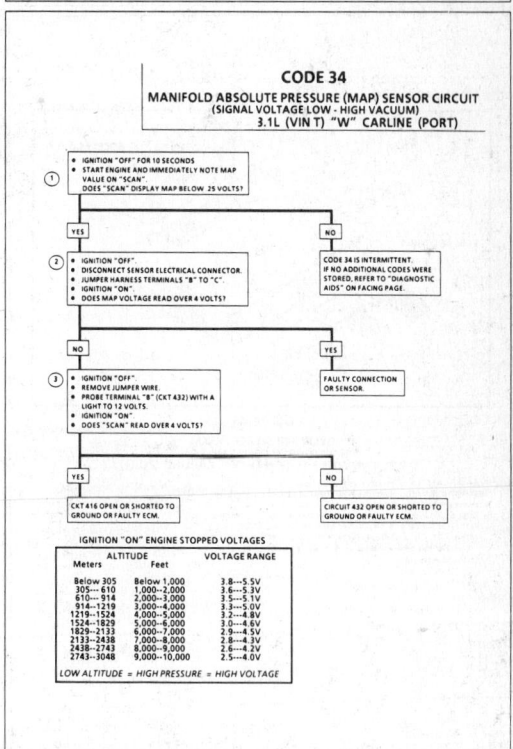

CODE 34
MANIFOLD ABSOLUTE PRESSURE (MAP) SENSOR CIRCUIT
(SIGNAL VOLTAGE LOW - HIGH VACUUM)
3.1L (VIN T) "W" CARLINE (PORT)

IGNITION "ON" ENGINE STOPPED VOLTAGES

ALTITUDE Meters	Feet	VOLTAGE RANGE
Below 305	Below 1,000	3.8—5.5V
305— 610	1,000–2,000	3.6—5.3V
610— 914	2,000–3,000	3.5—5.1V
914—1219	3,000–4,000	3.3—5.0V
1219—1524	4,000–5,000	3.2—4.8V
1524—1829	5,000–6,000	3.0—4.6V
1829—2133	6,000–7,000	2.9—4.5V
2133—2438	7,000–8,000	2.8—4.3V
2438—2743	8,000–9,000	2.6—4.2V
2743—3048	9,000–10,000	2.5—4.0V

LOW ALTITUDE = HIGH PRESSURE = HIGH VOLTAGE

1989–90 3.1L (VIN T) – ALL W-BODY VEHICLES

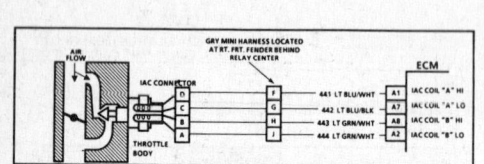

CODE 35
IDLE SPEED CONTROL (ISC) CIRCUIT
2.8L (VIN W) & 3.1L (VIN T) "W" CARLINE (PORT)

Circuit Description:
Code 35 will set when the closed throttle engine speed is 300 rpm above or below the desired (commanded) idle speed for 50 seconds.

Test Description: Numbers below refer to circled numbers on the diagnostic chart.
1. Continue with test even if engine will not idle. If idle is too low, "Scan" will display 80 or more counts, or steps. If idle is high it will display "0" counts.
   Occasionally an erratic or unstable idle may occur. Engine speed may vary 200 rpm or more up and down. Disconnect IAC. If the condition is unchanged, the IAC is not at fault. There is a system problem. Proceed to diagnostic aids below.
2. When the engine was stopped, the IAC valve retracted (more air) to a fixed "park" position for increased air flow and idle speed during the next engine start. A "Scan" will display 80 or more counts.
3. Be sure to disconnect the IAC valve prior to this test. The test light will confirm the ECM signals by a steady or flashing light on all circuits.
4. There is a remote possibility that one of the circuits is shorted to voltage which would have been indicated by a steady light. Disconnect ECM and turn the ignition "ON" and probe terminals to check for this condition.
5. IAC wiring is routed from the ECM through a mini harness. Faulty connection or shorted wires could result in poor IAC operation.

Diagnostic Aids:
A slow unstable idle may be caused by a system problem that cannot be overcome by the IAC. "Scan" counts will be above 80 counts if idle is too low and "0" counts if it is too high.

If idle is too high, stop engine. Ignition "ON." Ground diagnostic terminal. Wait a few seconds for IAC to seat then disconnect IAC. Start engine. If idle speed is above 600-700 rpm in drive for an A/T or 800-900 rpm in neutral for M/T, locate and correct vacuum leak.
- **System too lean (high air/fuel ratio)**
  Idle speed may be too high or too low. Engine speed may vary up and down, disconnecting IAC does not help. This may set Code 44.
  "Scan" and/or voltmeter will read an oxygen sensor output less than 300 mV(.3volt). Check for low regulated fuel pressure or water in fuel. A lean exhaust with an oxygen sensor output fixed above 800 mV (.8 volt) will be a contaminated sensor, usually silicone. This may also set a Code 45 or 61.
- **System too rich (low air/fuel ratio)**
  Idle speed too low. "Scan" counts usually above 80. System obviously rich and may exhibit black smoke exhaust.
  "Scan" tool and/or voltmeter will read an oxygen sensor signal fixed above 800 mV (.8 volt).
  Check:
  - For fuel in pressure regulator hose
  - High fuel pressure
  - Injector leaking or sticking.
- **Throttle body** Remove IAC and inspect bore for foreign material or evidence of IAC valve dragging the bore.
- **A/C compressor or relay failure** See CHART C-10 if the A/C control relay drive circuit is shorted to ground or, if the relay is faulty, an idle problem may exist.
- Refer to "Rough, Unstable, Incorrect Idle or Stalling" in "Symptoms" in Section "B".

1989–90 3.1L (VIN T) – ALL W-BODY VEHICLES

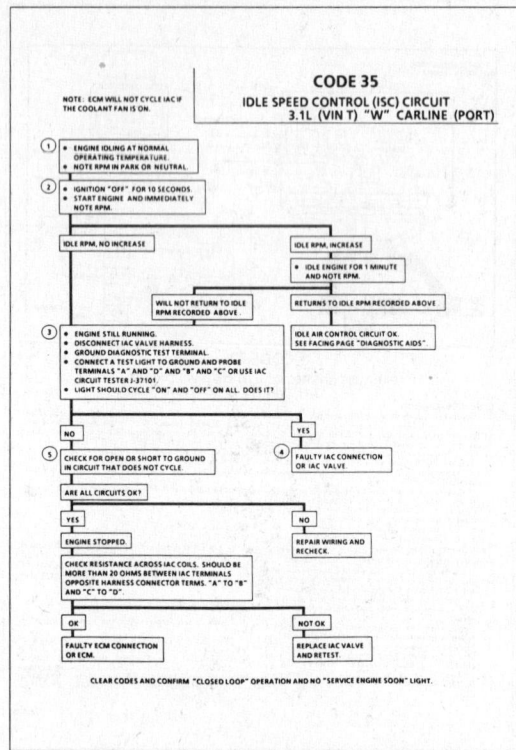

CODE 35
IDLE SPEED CONTROL (ISC) CIRCUIT
3.1L (VIN T) "W" CARLINE (PORT)

CLEAR CODES AND CONFIRM "CLOSED LOOP" OPERATION AND NO "SERVICE ENGINE SOON" LIGHT.

1989–90 3.1L (VIN T) – ALL W-BODY VEHICLES

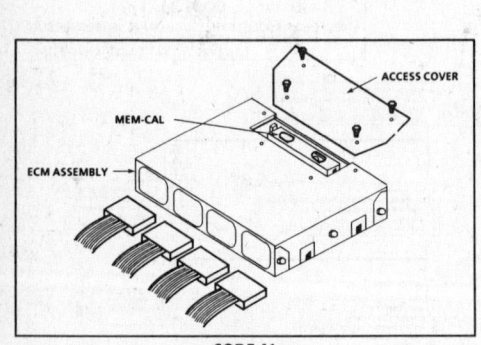

CODE 41
CYLINDER SELECT ERROR
(FAULTY OR INCORRECT MEM-CAL)
2.8L (VIN W) & 3.1L (VIN T) "W" CARLINE (PORT)

Test Description: Numbers below refer to circled numbers on the diagnostic chart.
1. The ECM used for this engine can also be used for other engines and the difference is in the Mem-Cal. If a Code 41 sets, the incorrect Mem-Cal has been installed or it is faulty and it must be replaced.

Diagnostic Aids:

Check Mem-Cal to be sure locking tabs are secure.

1989–90 3.1L (VIN T) – ALL W-BODY VEHICLES

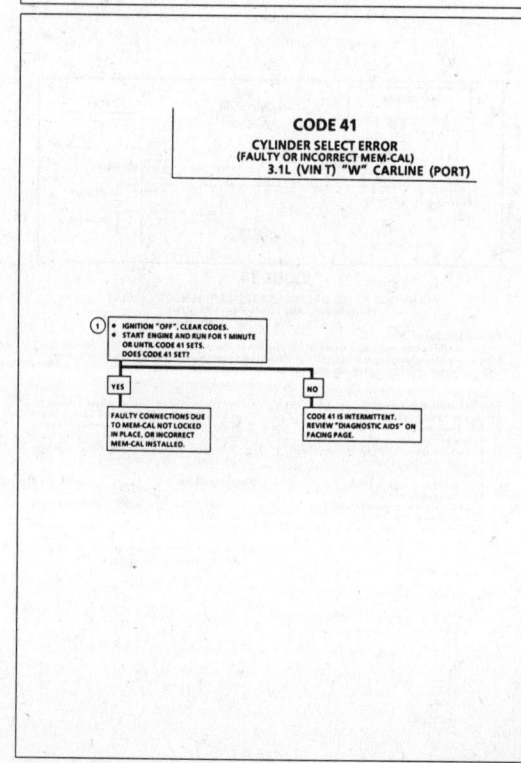

CODE 41
CYLINDER SELECT ERROR
(FAULTY OR INCORRECT MEM-CAL)
3.1L (VIN T) "W" CARLINE (PORT)

1989–90 3.1L (VIN T) – ALL W-BODY VEHICLES

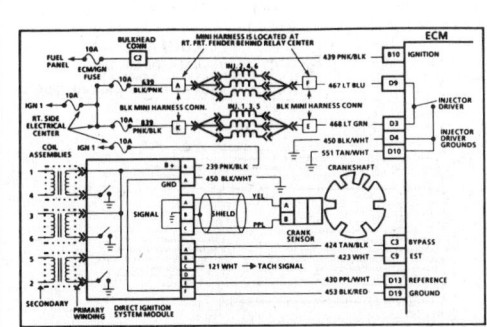

CODE 42
ELECTRONIC SPARK TIMING (EST) CIRCUIT
3.1L (VIN T) "W" CARLINE (PORT)

Circuit Description:

When the system is running on the ignition module, that is, no voltage on the bypass line, the ignition module grounds the EST signal. The ECM expects to see no voltage on the EST line during this condition. If it sees a voltage, it sets Code 42 and will not go into the EST mode.

When the rpm for EST is reached (about 400 rpm), and bypass voltage applied, the EST should no longer be grounded in the ignition module so the EST voltage should be varying.

If the bypass line is open or grounded, the ignition module will not switch to EST mode so the EST voltage will be low and Code 42 will be set.

If the EST line is grounded, the ignition module will switch to EST, but because the line is grounded there will be no EST signal. A Code 42 will be set.

Test Description: Numbers below refer to circled numbers on the diagnostic chart.
1. Code 42 means the ECM has seen an open or short to ground in the EST or bypass circuits. This test confirms Code 42 and that the fault causing the code is present.
2. Checks for a normal EST ground path through the ignition module. An EST CKT 423 shorted to ground will also read less than 500 ohms; however, this will be checked later.
3. As the test light voltage touches CKT 424, the module should switch causing the ohmmeter to "overrange" if the meter is in the 1000-2000 ohms position. Selecting the 10-20,000 ohms position will indicate above 5000 ohms. The important thing is that the module "switched."
4. The module did not switch and this step checks for:

- EST CKT 423 shorted to ground.
- Bypass CKT 424 open.
- Faulty ignition module connection or module.
5. Confirms that Code 42 is a faulty ECM and not an intermittent in CKTs 423 or 424.

Diagnostic Aids:

The "Scan" tool does not have any ability to help diagnose a Code 42 problem.

A Mem-Cal not fully seated in the ECM can result in a Code 42.

If Code 42 is intermittent, there is a possibility of an open EST line. To check for an open EST line, crank engine while in a "Clear Flood Mode" for five seconds, then start engine and check for Code 42. If Code 42 is set, repair open in CKT 423 (EST line). Refer to "Intermittents" in Section "B".

1989–90 3.1L (VIN T) – ALL W-BODY VEHICLES

CODE 42
ELECTRONIC SPARK TIMING (EST) CIRCUIT
3.1L (VIN T) "W" CARLINE (PORT)

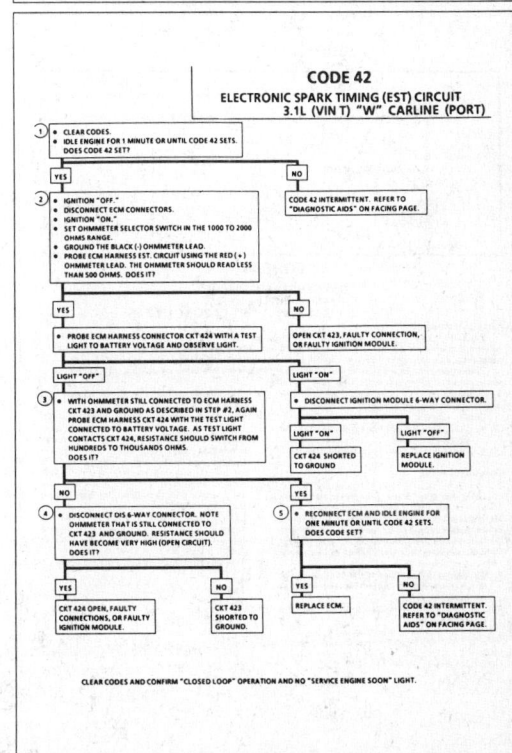

CLEAR CODES AND CONFIRM "CLOSED LOOP" OPERATION AND NO "SERVICE ENGINE SOON" LIGHT.

1989–90 3.1L (VIN T) – ALL W-BODY VEHICLES

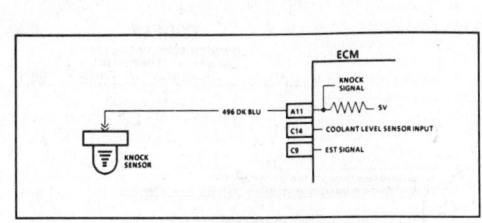

CODE 43
ELECTRONIC SPARK CONTROL (ESC) CIRCUIT
3.1L (VIN T) "W" CARLINE (PORT)

Circuit Description:

The knock sensor is used to detect engine detonation and the ECM will retard the electronic spark timing based on the signal being received. The circuitry within the knock sensor causes the ECM 5 volts to be pulled down so that under a no knock condition, CKT 496 would measure about 2.5 volts. The knock sensor produces an AC signal which rides on the 2.5 volts DC voltage. The amplitude and signal frequency is dependent upon the knock level.

There are two tests run on this circuit to determine if it is operating correctly. If either of the tests fail, a Code 43 will be set.

43A. If CKT 496 becomes open or shorted to ground the voltage will either go above 3.5 volts or below 1.5 volts. If either of these conditions are met for about ¼ second a Code 43 will be stored.

43B. This system also performs a functional test to determine if the knock sensor is responding to engine detonation. To perform this test, the ECM will advance the spark under certain load conditions and look for a knock signal response. If knock is detected before conditions are met to run the test, the test is bypassed. If the test is run and no knock detected the Code 43 will be set.

The test is performed when:
- Coolant temperature is over 90°C.
- MAT temperature is over 0°C.
- High engine load based on air flow and rpm between 3400 and 4400.

Test Description: Numbers below refer to circled numbers on the diagnostic chart.
1. If the conditions for the A test, as described above, are being met the "Scan" tool will always indicate "yes" when the knock sensor position is selected. If an audible knock is heard from the engine, repair the internal engine problem, as normally no knock should be detected at idle.
2. If tapping on the engine lift hook does not produce a knock signal, try tapping engine closer to sensor before proceeding.
3. The ECM has a 5 volt pull-up resistor which should be present at the knock sensor terminal.
4. This test determines if the knock sensor is faulty or if the ESC portion of the Mem-Cal is faulty.

Diagnostic Aids:

Check CKT 496 for a potential open or short to ground.

Also check for proper installation of Mem-Cal.

Refer to "Intermittents" in Section "B".

If the customer's complaint is the "Service Engine Soon" light comes "ON" in acceleration, the B portion of the code is failing. There is a possibility that the direct ignition system was in bypass mode when the 43 test was run. An intermittent open in the EST circuit will put the DIS module in bypass which will not allow the spark to be advanced so the 43B test would fail. If ECM also had a 42 stored, then the EST circuit is likely the cause of the Code 43.

1989–90 3.1L (VIN T) – ALL W-BODY VEHICLES

CODE 43
ELECTRONIC SPARK CONTROL (ESC) CIRCUIT
3.1L (VIN T) "W" CARLINE (PORT)

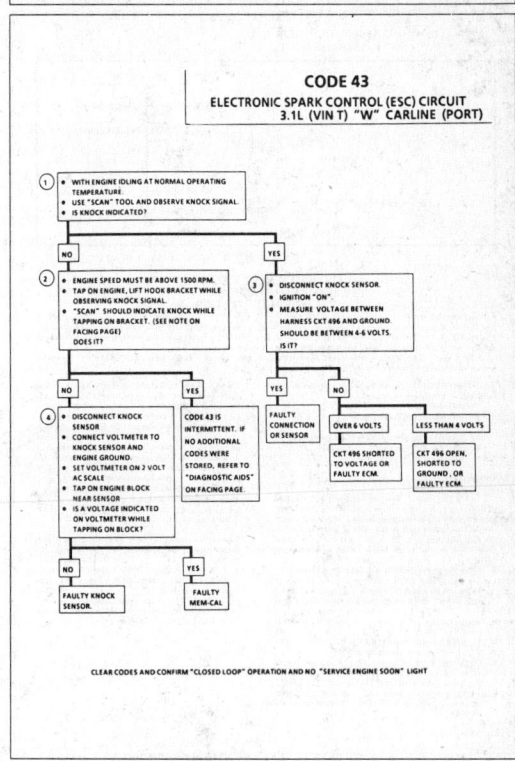

CLEAR CODES AND CONFIRM "CLOSED LOOP" OPERATION AND NO "SERVICE ENGINE SOON" LIGHT

1989–90 3.1L (VIN T) – ALL W-BODY VEHICLES

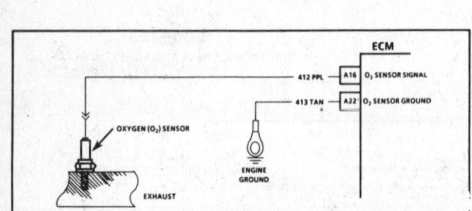

ECM
412 PPL — A16 — O₂ SENSOR SIGNAL
413 TAN — A22 — O₂ SENSOR GROUND

OXYGEN (O₂) SENSOR
EXHAUST
ENGINE GROUND

CODE 44
OXYGEN SENSOR CIRCUIT
(LEAN EXHAUST INDICATED)
3.1L (VIN T) "W" CARLINE (PORT)

Circuit Description:
The ECM supplies a voltage of about .45 volt between terminals "A16" and "A22". (If measured with a 10 megohm digital voltmeter, this may read as low as .32 volt.) The O₂ sensor varies the voltage within a range of about 1 volt if the exhaust is rich, down through about .10 volt if exhaust is lean.
The sensor is like an open circuit and produces no voltage when it is below about 315°C (600°F). An open sensor circuit or cold sensor causes "Open Loop" operation.

Test Description: Numbers below refer to circled numbers on the diagnostic chart.
Code 44 will set if:
- Voltage on CKT 412 remains below .2 volt for 60 seconds or more.
- The system is operating in "Closed Loop."

Diagnostic Aids:
Using the "Scan," observe the block learn values at different rpm and air flow conditions. The "Scan" also displays the block cells, so the block learn values can be checked in each of the cells to determine when the Code 44 may have been set. If the conditions for Code 44 exists the block learn values will be around 150.
- O₂ Sensor Wire Sensor pigtail may be mispositioned and contacting the exhaust manifold.
- Check for intermittent ground in wire between connector and sensor.

- Lean Injector(s) Perform injector balance test CHART C-2A.
- Fuel Contamination Water, even in small amounts, near the in-tank fuel pump inlet can be delivered to the injectors. The water causes a lean exhaust and can set a Code 44.
- Fuel Pressure System will be lean if pressure is too low. It may be necessary to monitor fuel pressure while driving the car at various road speeds and/or loads to confirm. See "Fuel System Diagnosis" CHART A-7.
- Exhaust Leaks If there is an exhaust leak, the engine can cause outside air to be pulled into the exhaust and past the sensor. Vacuum or crankcase leaks can cause a lean condition.
- If the above are OK, it is a faulty oxygen sensor.

1989–90 3.1L (VIN T) – ALL W-BODY VEHICLES

CODE 44
OXYGEN SENSOR CIRCUIT
(LEAN EXHAUST INDICATED)
3.1L (VIN T) "W" CARLINE (PORT)

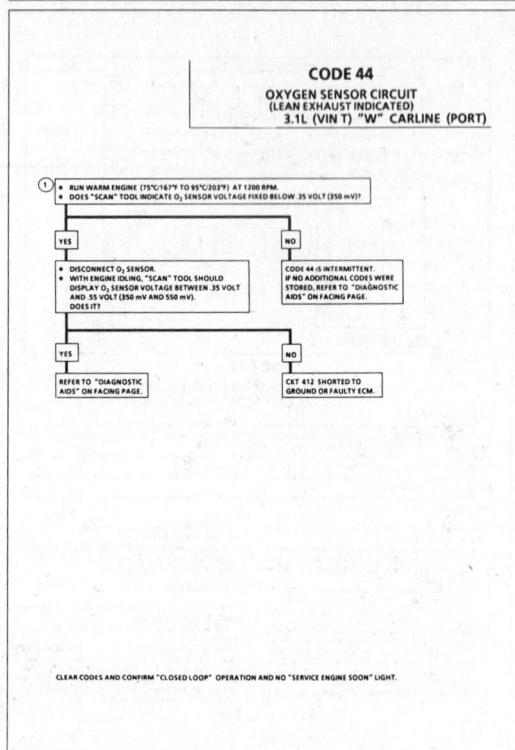

CLEAR CODES AND CONFIRM "CLOSED LOOP" OPERATION AND NO "SERVICE ENGINE SOON" LIGHT.

1989–90 3.1L (VIN T) – ALL W-BODY VEHICLES

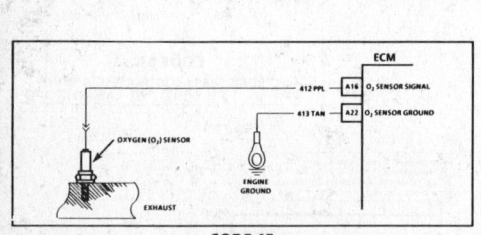

ECM
412 PPL — A16 — O₂ SENSOR SIGNAL
413 TAN — A22 — O₂ SENSOR GROUND

OXYGEN (O₂) SENSOR
EXHAUST
ENGINE GROUND

CODE 45
OXYGEN SENSOR CIRCUIT
(RICH EXHAUST INDICATED)
3.1L (VIN T) "W" CARLINE (PORT)

Circuit Description:
The ECM supplies a voltage of about .45 volt between terminals "A16" and "A22". (If measured with a 10 megohm digital voltmeter, this may read as low as .32 volt.) The O₂ sensor varies the voltage within a range of about 1 volt if the exhaust is rich, down through .10 volt if exhaust is lean.
The sensor is like an open circuit and produces no voltage when it is below about 315°C (600°F). An open sensor circuit or cold sensor causes "Open Loop" operation.

Test Description: Numbers below refer to circled numbers on the diagnostic chart.
Code 45 will set if:
- Voltage on CKT 412 remains above .7 volt for 30 seconds.
- Engine time after start is 1 minute or more.
- Throttle angle between 3% and 45%.
- Operation is in "Closed Loop."

Diagnostic Aids:
Using the "Scan," observe the block learn values at different rpm and air flow conditions. The "Scan" also displays the block cells, so the block learn values can be checked in each of the cells to determine when the Code 45 exists, the block learn values will be around 115.
- Fuel Pressure System will go rich if pressure is too high. The ECM can compensate for some increase. However, if it gets too high, a Code 45 can be set. See fuel system diagnosis CHART A-7.
- Rich Injector Perform injector balance test CHART C 2A.
- Leaking Injector See CHART A 7.
- Check for fuel contaminated oil.
- O₂ Sensor Contamination Inspect oxygen sensor for silicone contamination from fuel, or use of improper RTV sealant. The sensor may have a white, powdery coating and result in a high but false signal voltage (rich exhaust indication).

The ECM will then reduce the amount of fuel delivered to the engine, causing a severe surge driveability problem
- HEI Shielding An open ground CKT 453 (ignition system reflow) may result in EMI, or induced electrical "noise." The ECM looks at this "noise" as reference pulses. The additional pulses result in a higher than actual engine speed signal. The ECM then delivers too much fuel, causing system to go rich. Engine tachometer will also show higher than actual engine speed, which can help in diagnosing this problem.
- Canister Purge Check for fuel saturation. If full of fuel, check canister control and hoses.
- Check for leaking fuel pressure regulator diaphragm by checking vacuum line to regulator for fuel
- TPS An intermittent TPS output will cause the system to go rich, due to a false indication of the engine accelerating.
- EGR An EGR staying open (especially at idle) will cause the O₂ sensor to indicate a rich exhaust, and this could result in a Code 45.

1989–90 3.1L (VIN T) – ALL W-BODY VEHICLES

CODE 45
OXYGEN SENSOR CIRCUIT
(RICH EXHAUST INDICATED)
3.1L (VIN T) "W" CARLINE (PORT)

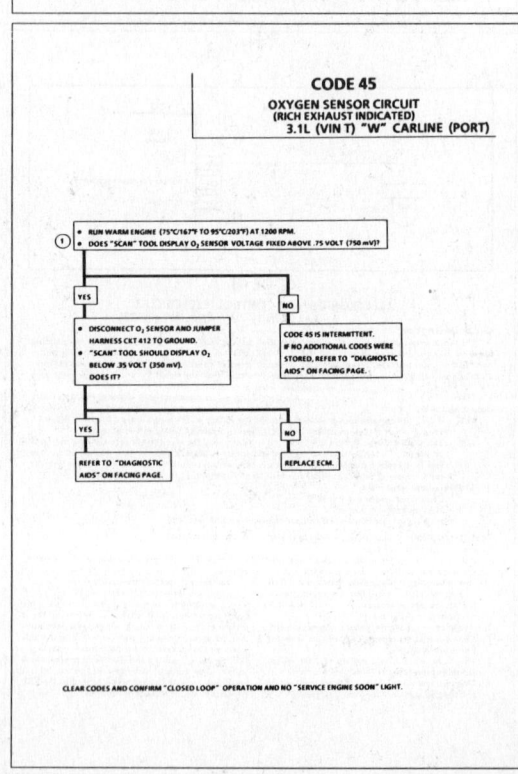

CLEAR CODES AND CONFIRM "CLOSED LOOP" OPERATION AND NO "SERVICE ENGINE SOON" LIGHT.

1989–90 3.1L (VIN T) – ALL W-BODY VEHICLES

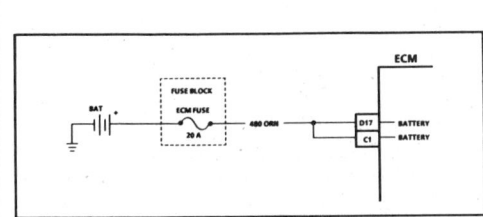

CODE 53
SYSTEM OVER VOLTAGE
3.1L (VIN T) "W" CARLINE (PORT)

Circuit Description:
Code 53 will set when the ignition is "ON" and the engine rpm is greater than 800 and the ECM is seeing a ignition fuel reference voltage of more than 17.1 volts.
During the time the failure is present, all ECM outputs will be disengaged. (The setting of additional codes may result).

Test Description: Numbers below refer to circled numbers on the diagnostic chart.
1. Normal battery output is between 10 - 17.0 volts.
2. Checks to see if the high voltage reading is due to the generator or ECM. With engine running, check voltage at the battery.

If the voltage is above 12.1 volts, the ECM is OK.
3. Checks to see if generator is faulty under load condition. If the voltage is above 17.1 volts, refer to "Alternator Diagnosis" in Section "6D".

1989–90 3.1L (VIN T) – ALL W-BODY VEHICLES

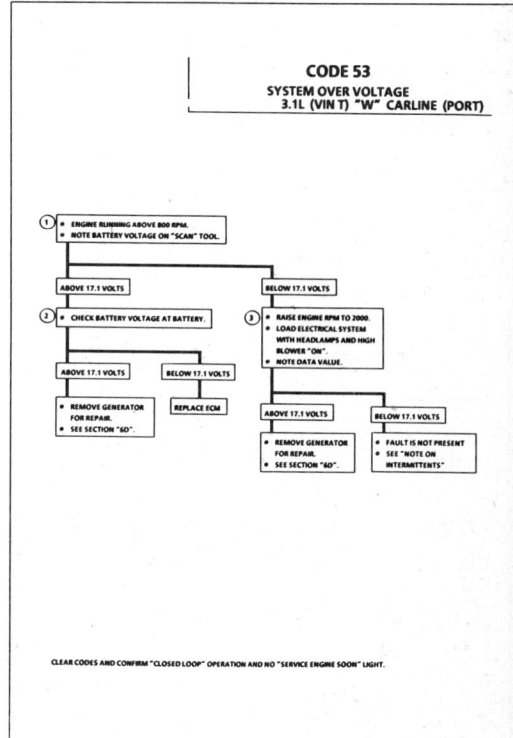

CODE 53
SYSTEM OVER VOLTAGE
3.1L (VIN T) "W" CARLINE (PORT)

1989–90 3.1L (VIN T) – ALL W-BODY VEHICLES

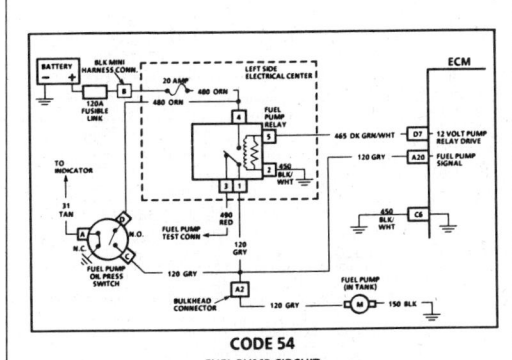

CODE 54
FUEL PUMP CIRCUIT
(LOW VOLTAGE)
3.1L (VIN T) "W" CARLINE (PORT)

Circuit Description:
The status of the fuel pump CKT 120 is monitored by the ECM at terminal "A20" and is used to compensate fuel delivery based on system voltage. This signal is also used to store a trouble code if the fuel pump relay is defective or fuel pump voltage is lost while the engine is running. There should be about 12 volts on CKT 120 for 2 seconds after the ignition is turned, or any time references pulses are being received by the ECM.
Code 54 will set, if the voltage at terminal "A20" is less than 2 volts for 1.5 seconds since the last reference pulse was received. This code is designed to detect a faulty relay, causing extended crank time, and the code will help the diagnosis of an engine that "CRANKS BUT WILL NOT RUN."
If a fault is detected during start-up the "Service Engine Soon" light will stay "ON" until the ignition is cycled "OFF." However, if the voltage is detected below 2 volts with the engine running, the light will only remain "ON" while the condition exists.
CKT 480 is routed from the ECM through a mini harness connector to the fuel pump relay. A faulty connection at terminal "B" at the black connector could cause a Code 54.

1989–90 3.1L (VIN T) – ALL W-BODY VEHICLES

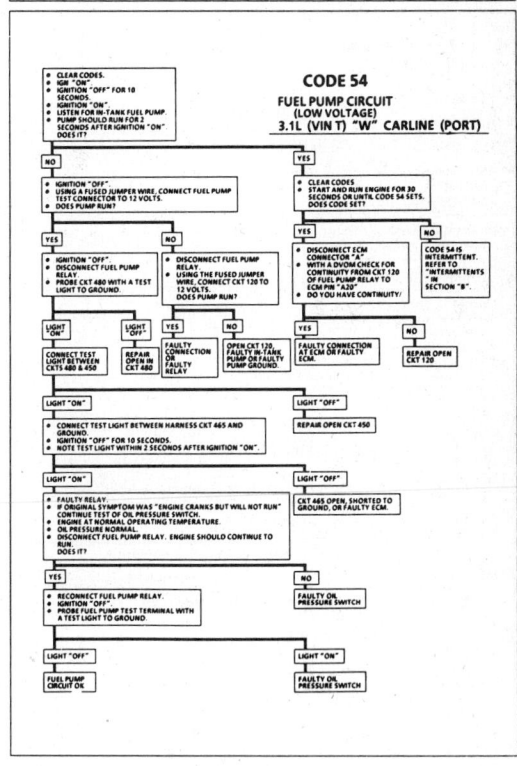

CODE 54
FUEL PUMP CIRCUIT
(LOW VOLTAGE)
3.1L (VIN T) "W" CARLINE (PORT)

1989–90 3.1L (VIN T) – ALL W-BODY VEHICLES

CODE 51
MEM-CAL ERROR
(FAULTY OR INCORRECT MEM-CAL)
3.1L (VIN-T) "W" CARLINE (PORT)

CHECK THAT ALL PINS ARE FULLY INSERTED IN THE SOCKET AND THAT MEM-CAL IS PROPERLY LATCHED. IF OK, REPLACE MEM-CAL, CLEAR MEMORY, AND RECHECK. IF CODE 51 REAPPEARS, REPLACE ECM.

CODE 55
ECM ERROR
3.1L (VIN T) "W" CARLINE (PORT)

BE SURE ECM GROUNDS ARE OK AND THAT MEM-CAL IS PROPERLY LATCHED. IF OK REPLACE ELECTRONIC CONTROL MODULE (ECM)

CLEAR CODES AND CONFIRM "CLOSED LOOP" OPERATION AND NO "SERVICE ENGINE SOON" LIGHT.

1989–90 3.1L (VIN T) – ALL W-BODY VEHICLES

CODE 61
DEGRADED OXYGEN SENSOR
3.1L (VIN T) "W" CARLINE (PORT)

IF A CODE 61 IS STORED IN MEMORY THE ECM HAS DETERMINED THE OXYGEN SENSOR IS CONTAMINATED OR DEGRADED, BECAUSE THE VOLTAGE CHANGE TIME IS SLOW OR SLUGGISH.

THE ECM PERFORMS THE OXYGEN SENSOR RESPONSE TIME TEST WHEN:

COOLANT TEMPERATURE IS GREATER THAN 85°C.

MAT TEMPERATURE IS GREATER THAN 10°C.

IN CLOSED LOOP.

IN DECEL FUEL CUT-OFF MODE.

IF A CODE 61 IS STORED THE OXYGEN SENSOR SHOULD BE REPLACED. A CONTAMINATED SENSOR CAN BE CAUSED BY FUEL ADDITIVES, SUCH AS SILICON, OR BY USE OF NON-GM APPROVED LUBRICANTS OR SEALANTS. SILICON CONTAMINATION IS USUALLY INDICATED BY A WHITE POWDERY SUBSTANCE ON THE SENSOR FINS.

CLEAR CODES AND CONFIRM "CLOSED LOOP" OPERATION AND NO "SERVICE ENGINE SOON" LIGHT.

1989–90 3.1L (VIN T) – ALL W-BODY VEHICLES

SECTION B
SYMPTOMS

TABLE OF CONTENTS

Before Starting
Intermittents
Hard Start
Hesitation, Sag, Stumble
Surges and/or Chuggle
Lack of Power, Sluggish, or Spongy
Detonation/Spark Knock
Cuts Out, Misses
Backfire
Poor Fuel Economy
Dieseling, Run-On
Rough, Unstable, or Incorrect Idle, Stalling
Excessive Exhaust Emissions Or Odors
Chart B-1 Restricted Exhaust System Check (all engines)

BEFORE STARTING

Before using this section you should have performed the DIAGNOSTIC CIRCUIT CHECK and found out that:
1. The ECM and "Service Engine Soon" light are operating.
2. There are no trouble codes.
Verify the customer complaint, and locate the correct SYMPTOM below. Check the items indicated under that symptom.
If the ENGINE CRANKS BUT WILL NOT RUN, see CHART A-3.
Several of the symptom procedures below call for a careful visual check. This check should include:

- ECM grounds for being clean and tight
- Vacuum hoses for splits, kinks, and proper connections, as shown on Emission Control Information label.
- Air leaks at throttle body mounting and intake manifold.
- Ignition wires for cracking, hardness, proper routing, and carbon tracking.
- Wiring for proper connections, pinches, and cuts.
The importance of this step cannot be stressed too strongly - it can lead to correcting a problem without further checks and can save valuable time.

1989–90 3.1L (VIN T) – ALL W-BODY VEHICLES

INTERMITTENTS
Problem may or may not turn "ON" the "Service Engine Soon" light, or store a code.

DO NOT use the trouble code charts in Section "A" for intermittent problems. The fault must be present to locate the problem. If a fault is intermittent, use of trouble code charts may result in replacement of good parts.
- Most intermittent problems are caused by faulty electrical connections or wiring. Perform careful check as described at start of Section "B". Check for:
  - Poor mating of the connector halves, or terminals not fully seated in the connector body (backed out).
  - Improperly formed or damaged terminals. All connector terminals in problem circuit should be carefully reformed to increase contact tension.
  - Poor terminal to wire connection. This requires removing the terminal from the connector body to check.
- If a visual check does not find the cause of the problem, the car can be driven with a voltmeter connected to a suspected circuit. A "Scan" tool can, also, be used for monitoring input signals to help detect intermittent conditions. An abnormal voltage, or "Scan" reading, when the problem occurs, indicates the problem may be in that circuit. If the wiring and connectors check OK and a trouble code was stored for a circuit having a sensor, except for Codes 43, 44, and 45, substitute a known good sensor and recheck.

An intermittent "Service Engine Soon" light with no stored code may be caused by:
- Ignition coil shorted to ground and arcing at spark plug wires or plugs.
- "Service Engine Soon" light wire to ECM shorted to ground. (CKT 419)
- Diagnostic "Test" terminal wire to ECM, shorted to ground. (CKT 451)
- ECM power grounds. See ECM wiring diagrams.
- Loss of trouble code memory. To check, disconnect TPS and idle engine until "Service Engine Soon" light comes "ON." Code 22 should be stored and kept in memory when ignition is turned "OFF." If not, the ECM is faulty.
- Check for an electrical system interference caused by a defective relay, ECM driven solenoid, or switch. They can cause a sharp electrical surge. Normally, the problem will occur when the faulty component is operated.
- Check for improper installation of electrical options, such as lights, 2-way radios, etc.
- EST wires should be kept away from spark plug wires, coils and generator. Wire from ECM to ignition system (CKT 453) should be a good connection.
- Check for open diode across A/C compressor clutch, and for other open diodes (see wiring diagrams).

HARD START
Definition: Engine cranks OK, but does not start for a long time. Does eventually run, or may start but immediately dies.

- Perform careful check as described at start of Section "B".
- Make sure driver is using correct starting procedure.
- CHECK:
  - TPS for sticking or binding or a high TPS voltage with the throttle closed (should read less than .85 volt).
  - High resistance in coolant sensor circuit or sensor itself. See Code 15 CHART OR with a "Scan" tool compare coolant temperature with ambient temperature on a cold engine.
  - Fuel pressure CHART A-7.
  - Water contaminated fuel.
  - EGR operation. Be sure valve seats properly and is not staying open. See CHART C-7.

- Ignition system - Check for:
  - Bare and shorted wires and proper output with spark tester J-26792 or equivalent (ST-125).
  - IAC operation - See Code 35 chart.
- A faulty in-tank fuel pump check valve will allow the fuel in the lines to drain back to the tank after the engine is stopped. To check for this condition:
    Perform fuel system diagnosis, CHART A-7.
- Remove spark plugs. Check for wet plugs, cracks, wear, improper gap, burned electrodes, or heavy deposits. Repair or replace as necessary.

1989–90 3.1L (VIN T) – ALL W-BODY VEHICLES

HESITATION, SAG, STUMBLE

Definition: Momentary lack of response as the accelerator is pushed down. Can occur at all car speeds. Usually most severe when first trying to make the car move, as from a stop sign. May cause the engine to stall if severe enough.

- Perform careful visual check as described at start of Section "B".
- CHECK:
  - Fuel pressure. See CHART A-7. Also check for water contaminated fuel.
  - Spark plugs for being fouled or faulty wiring.
  - Mem-Cal number and Service Bulletins for latest Mem-Cal.
- TPS for binding or sticking. Voltage should increase at a steady rate as throttle is moved toward WOT.
- Generator output voltage. Repair, if less than 9 or more than 16 volts.
- Ignition system ground, CKT 453.
- Canister purge system for proper operation. See CHART C-3.
- EGR - See CHART C-7.
- MAP sensor - See CHART C-1D.
- Perform injector balance test, CHART C-2A.

SURGES AND/OR CHUGGLE

Definition: Engine power variation under steady throttle or cruise. Feels like the car speeds up and slows down with no change in the accelerator pedal.

- Be sure driver understands transmission/transaxle converter clutch and A/C compressor operation in Owner's Manual.
- Perform careful visual inspection as described at start of Section "B".
- To help determine if the condition is caused by a rich or lean system, the car should be driven at the speed of the complaint. Monitoring block learn at the complaint speed will help identify the cause of the problem. If the system is running lean (block learn greater than 138), refer to "Diagnostic Aids" on facing page of Code 44. If the system is running rich (block learn less than 118), refer to "Diagnostic Aids" on facing page of Code 45.
- CHECK:
  - Generator output voltage. Repair if less than 9 or more than 16 volts.
  - EGR. There should be no EGR at idle. See CHART C-7.
  - EGR filter for being plugged. See CHART C-7.
  - Vacuum lines for kinks or leaks.
  - In-line fuel filter. Replace if dirty or plugged.
  - Fuel pressure while condition exists. See CHART A-7.
- Remove spark plugs. Check for cracks, wear, improper gap, burned electrodes, or heavy deposits. Also check condition of spark plug wires and check for proper output voltage using spark tester (ST-125) J-26792 or equivalent.

1989–90 3.1L (VIN T) – ALL W-BODY VEHICLES

LACK OF POWER, SLUGGISH, OR SPONGY

Definition: Engine delivers less than expected power. Little or no increase in speed when accelerator pedal is pushed down part way.

- Perform careful visual check as described at start of Section "B".
- Compare customer's car to similar unit. Make sure the customer's car has an actual problem.
- Remove air filter and check air filter for dirt, or for being plugged. Replace as necessary.
- CHECK:
  - Restricted fuel filter, contaminated fuel or improper fuel pressure. See CHART A-7.
  - ECM power grounds, see wiring diagrams.
  - EGR operation for being open or partly open all the time - See CHART C-7.
- Exhaust system for possible restriction: See CHART B-1.
  - Inspect exhaust system for damaged or collapsed pipes.
  - Inspect muffler for heat distress or possible internal failure.
- Generator output voltage. Repair if less than 9 or more than 16 volts.
- Engine valve timing and compression.
- Engine for proper or worn camshaft.
- Secondary voltage using a shop ocilliscope or a spark tester J-26792 (ST-125) or equivalent.
- Check A/C operation. A/C clutch should cut out at WOT. See A/C CHART C-10.

DETONATION /SPARK KNOCK

Definition: A mild to severe ping, usually worse under acceleration. The engine makes sharp metallic knocks that change with throttle opening. Sounds like popcorn popping.

- Check for obvious overheating problems:
  - Low coolant.
  - Loose belt.
  - Restricted air flow to radiator, or restricted water flow through radiator.
  - Inoperative electric cooling fan circuit. See CHART C-12.
- To help determine if the conditions is caused by a rich or lean system, the car should be driven at the speed of the complaint. Monitoring block learn, at the complaint speed, will help identify the cause of the problem. If the system is running lean (block learn greater than 138), refer to "Diagnostic Aids" on facing page of Code 44. If the system is running rich (block learn less than 118), refer to facing page of Code 45.
- CHECK:
  - EGR system for not opening or plugged EGR passages - See CHART C-7.
  - ESC system for no retard - See CHART C-5.
  - Park/neutral switch. Be sure "Scan" indicates drive with gear selector in drive or overdrive. See CHART C-1A.
  - TCC operation, TCC applying too soon - See CHART C-8.
- Fuel system pressure - See CHART A-7.
- Remove carbon with top engine cleaner. Follow instructions on can.
  - Check for correct Mem-Cal.
  - Check for excessive oil in combustion chamber.
  - Check for incorrect basic engine parts such as cam, heads, pistons, etc.
  - Check for poor fuel quality, proper octane rating.

1989–90 3.1L (VIN T) – ALL W-BODY VEHICLES

CUTS OUT, MISSES

Definition: Steady pulsation or jerking that follows engine speed, usually more pronounced as engine load increases. Not normally felt above 1500 rpm or 30 mph (48 km/h). The exhaust has a steady spitting sound at idle or low speed.

- Perform careful visual check as described at start of Section "B".
- Check for missing cylinder by:
  1. Disconnect IAC motor. Start engine. Remove one spark plug wire at a time using insulated pliers.
  2. If there is an rpm drop on all cylinders (equal to within 50 rpm), go to ROUGH, UNSTABLE, OR INCORRECT IDLE, STALLING symptom. Reconnect IAC motor.
  3. If there is no rpm drop on one or more cylinders, or excessive variation in drop, check for spark on the suspected cylinder(s) with J-26792 (ST-125) spark gap tool or equivalent.
     If there is spark, remove spark plug(s) in these cylinders and check for:
     - Cracks
     - Wear
     - Improper gap
     - Burned electrodes
     - Heavy deposits
- Perform compression check on questionable cylinder(s) found above. If compression is low, repair as necessary.
- Disconnect all injector harness connectors. Connect J-34730-2 injector test light or equivalent 6 volt test light between the harness terms. of each injector connector and note light while cranking. If test light fails to blink at any connector, it is a faulty injector drive circuit harness, connector, terminal, or ECM.
- Perform the injector balance test. See CHART C-2A.
- Check spark plug wires by connecting ohmmeter to ends of each wire in question. If meter reads over 30,000 ohms, replace wire(s).
- Spray plug wires with fine water mist to check for shorts.
- Check for restricted fuel filter. Also check fuel tank for water.
- Check for low fuel pressure. See CHART A-7.
- Remove rocker covers. Check for bent pushrods, worn rocker arms, broken valve springs, worn camshaft lobes. Repair as necessary.
- Check for proper valve timing.
- Check secondary voltage using a shop oscilliscope or a spark tester J-26792 (ST-125) or equivalent.

BACKFIRE

Definition: Fuel ignites in intake manifold, or in exhaust system, making a loud popping noise.

- CHECK:
  - Compression - Look for sticking or leaking valves.
  - EGR operation for being open all the time. See CHART C-7.
  - EGR gasket for faulty or loose fit.
  - Output voltage of ignition coils, using a shop ocilliscope or spark tester J-26792 (ST-125), or equivalent.
  - Valve timing.
  - Spark plugs, spark plug wires, and proper routing of plug wires.
  - AIR check valve (manual trans. only).
  - Intermittent condition in ignition system.

1989–90 3.1L (VIN T) – ALL W-BODY VEHICLES

POOR FUEL ECONOMY

Definition: Fuel economy, as measured by an actual road test, is noticeably lower than expected. Also, economy is noticeably lower than it was on this car at one time, as previously shown by an actual road test.

- CHECK:
  - Engine thermostat for faulty part (always open) or for wrong heat range. Using a "Scan" tool, monitor engine temperature. A "Scan" tool displays engine temp. in degrees centigrade. After engine is started, the temperature should rise steadily to about 90°C, then stabilize, when thermostat opens.
  - Fuel Pressure. See CHART A-7.
- Check owner's driving habits.
  - Is A/C "ON" all time (Defroster mode "ON")?
  - Are tires at correct pressure?
  - Are excessively heavy loads being carried?
  - Is acceleration too much, too often?
  - Suggest driver read "Important Facts on Fuel Economy" in Owner's Manual.
- Check air cleaner element (filter) for dirt or being plugged.
- Check for proper calibration of speedometer.
- Visually (physically) check:
  - Vacuum hoses for splits, kinks, and proper connections as shown on Vehicle Emission Control Information label.
  - Ignition wires for cracking, hardness, and proper connections.
- Remove spark plugs. Check for cracks, wear, improper gap, burned electrodes, or heavy deposits. Repair or replace as necessary.
- Check compression.
- Check TCC for proper operation. See CHART C-8. A "Scan" should indicate an rpm drop, when the TCC is commanded "ON."
- Using a "Scan" tool, crank engine for five seconds while in a clear flood mode. Start engine and check for a Code 42. An open EST line will cause poor fuel economy. Refer to Code 42 diagnostic chart for repair procedure.
- Check for exhaust system restriction. See CHART B-1.

DIESELING, RUN-ON

Definition: Engine continues to run after key is turned "OFF," but runs very roughly. If engine runs smoothly, check ignition switch and adjustment.

- Check injectors for leaking. See CHART A-7.

1989–90 3.1L (VIN T) – ALL W-BODY VEHICLES

ROUGH, UNSTABLE, OR INCORRECT IDLE, STALLING

Definition: The engine runs unevenly at idle. If bad enough, the car may shake. Also, the idle may vary in rpm (called "hunting"). Either condition may be bad enough to cause stalling. Engine idles at incorrect speed. Engine will run rough or stall if loss of battery power to the ECM (electronic control module) has occurred. This ECM has an idle learn feature. (See Section C2 for Idle Learn Procedure).

- Perform careful visual check as described at start of Section "B".
- CHECK:
  - For damaged, mispositioned, or grounded motor mounts.
  - Throttle linkage for sticking or binding.
  - TPS for sticking or binding, and be sure output is stable at idle.
  - IAC system. See Code 35 Chart.
  - Generator output voltage. Repair if less than 9 or more than 16 volts.
  - P/N switch circuit. See CHART C-1A, or use "Scan" tool, and be sure tool indicates vehicle is in drive with gear selector in drive (125C), or overdrive (440-T4).
  - Injector balance. See CHART C-2A.
  - PCV valve for proper operation by placing finger over inlet hole in valve and several times. Valve should snap back. If not, replace valve.
  - Evaporative Emission Control System. CHART C-3.
  - Power steering pressure switch input. The state of the switch should only change when wheels are turned up against the stops. See CHART C-1E.

- ECM ground circuits.
- EGR valve: There should be no EGR at idle.
- Monitoring block learn values may help identify the cause of the problem. If the system is running lean (block learn greater than 138) refer to "Diagnostic Aids" on facing page of Code 44. If the system is running rich (block learn values less than 118) refer to "Diagnostic Aids" on facing page of Code 45.
- Run a cylinder compression check. See Section "6".
- Check for fuel in pressure regulator hose. If present, replace regulator assembly.
- Check ignition system, wires and plugs.
- If problem exists with A/C "ON," check A/C system operation CHART C-10.
- M/T series - check AIR system for intermittent air to exhaust ports, while in "Closed Loop." See CHART C-6.
- Inspect oxygen sensor for silicon contamination from fuel, or use of improper RTV sealant. The sensor will have a white, powdery coating, and will result in a high but false signal voltage (rich exhaust indication). The ECM will then reduce the amount of fuel delivered to the engine causing a severe driveability problem.

EXCESSIVE EXHAUST EMISSIONS OR ODORS

Definition: Vehicle fails an emission test. Vehicle has excessive "rotten egg" smell. Excessive odors do not necessarily indicate excessive emissions.

- Perform "Diagnostic Circuit Check."
- IF TEST SHOWS EXCESSIVE CO AND HC, (or also has excessive odors):
  - Check items which cause car to run RICH.
    - Make sure engine is at normal operating temperature.
  - CHECK:
    - Fuel pressure. See CHART A-7.
    - Canister for fuel loading. See CHART C-3.
    - Injector balance. See CHART C-2A.
    - PCV valve for being plugged, stuck, or blocked PCV hose, or fuel in the crankcase.
    - Spark plugs, plug wires, and ignition components.
    - Check for lead contamination of catalytic converter (look for removal of fuel filler neck restrictor).
    - Check for properly installed fuel cap.

- If the system is running rich, (block learn less than 118), refer to "Diagnostic Aids" on facing page of Code 45.
- IF TEST SHOWS EXCESSIVE NOx:
  - Check items which cause car to run LEAN, or to run too hot.
    - EGR valve for not opening. See CHART C-7.
    - Vacuum leaks.
    - Coolant system and coolant fan for proper operation. See CHART C-12.
    - Remove carbon with top engine cleaner. Follow instructions on can.
  - If the system is running lean, (block learn greater than 138), refer to "Diagnostic Aids" on facing page of Code 44.

1989–90 3.1L (VIN T) – ALL W-BODY VEHICLES

CHART B-1
RESTRICTED EXHAUST SYSTEM CHECK
ALL ENGINES

Proper diagnosis for a restricted exhaust system is essential before any components are replaced. Either of the following procedures may be used for diagnosis, depending upon engine or tool used.

CHECK AT A. I. R. PIPE:
1. Remove the rubber hose at the exhaust manifold A.I.R. pipe check valve. Remove check valve.
2. Connect a fuel pump pressure gauge to a hose and nipple from a Propane Enrichment Device (J 26911) (see illustration).
3. Insert the nipple into the exhaust manifold A.I.R. pipe.

OR CHECK AT O₂ SENSOR:
1. Carefully remove O₂ sensor.
2. Install Borroughs exhaust backpressure tester (BT 8515 or BT 8603) or equivalent in place of O₂ sensor (see illustration).
3. After completing test described below, be sure to coat threads of O₂ sensor with anti-seize compound P/N 5613695 or equivalent prior to re-installation.

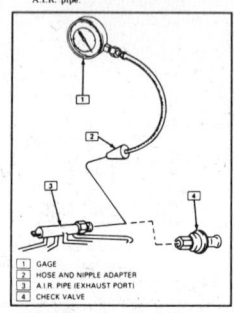

1	GAGE
2	HOSE AND NIPPLE ADAPTER
3	A.I.R. PIPE (EXHAUST PORT)
4	CHECK VALVE

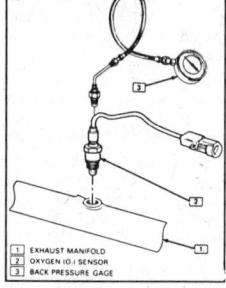

1	EXHAUST MANIFOLD
2	OXYGEN (O₂) SENSOR
3	BACK PRESSURE GAGE

DIAGNOSIS:
1. With the engine idling at normal operating temperature, observe the exhaust system backpressure reading on the gage. Reading should not exceed 8.6 kPa (1.25 psi).
2. Increase engine speed to 2000 rpm and observe gage. Reading should not exceed 20.7 kPa (3 psi).
3. If the backpressure at either speed exceeds specification, a restricted exhaust system is indicated.
4. Inspect the entire exhaust system for a collapsed pipe, heat distress, or possible internal muffler failure.
5. If there are no obvious reasons for the excessive backpressure, the catalytic converter is suspected to be restricted and should be replaced using current recommended procedures.

1989–90 3.1L (VIN T) – ALL W-BODY VEHICLES

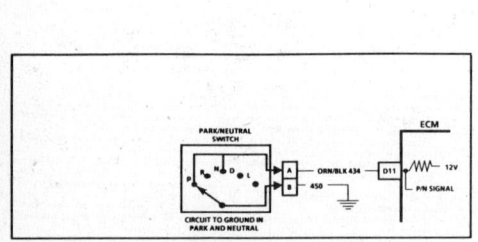

CHART C-1A
PARK/NEUTRAL SWITCH DIAGNOSIS
3.1L (VIN T) "W" CARLINE (PORT)

Circuit Description:
The park/neutral switch contacts are a part of the neutral start switch and are closed to ground in park or neutral, and open in drive ranges.

The ECM supplies ignition voltage through a current limiting resistor to CKT 434 and senses a closed switch when the voltage on CKT 434 drops to less than one volt.

The ECM uses the P/N signal as one of the inputs to control:
Idle air control
VSS diagnostics
EGR

If CKT 434 indicates P/N (grounded), while in drive range, the EGR would be inoperative, resulting in possible detonation.

If CKT 434 indicates drive (open) a dip in the idle may exist when the gear selector is moved into drive range.

Test Description: Numbers below refer to circled numbers on the diagnostic chart.
1. Checks for a closed switch to ground in park position. Different makes of "Scan" tools will read P/N differently. Refer to tool operator's manual for type of display used for a specific tool.

2. Checks for an open switch in drive range.
3. Be sure "Scan" indicates drive, even while wiggling shifter, to test for an intermittent or misadjusted switch in drive or overdrive range.

1989–90 3.1L (VIN T) – ALL W-BODY VEHICLES

CHART C-1A
PARK/NEUTRAL SWITCH DIAGNOSIS
3.1L (VIN T) "W" CARLINE (PORT)

1. WITH TRANSAXLE/TRANSMISSION IN PARK, "SCAN" TOOL SHOULD INDICATE PARK OR NEUTRAL. DOES IT?
   - YES →
   - NO →

3. SHIFT TRANSAXLE/TRANSMISSION INTO DRIVE. "SCAN" TOOL SHOULD DISPLAY A CHANGE TO INDICATE DRIVE. DOES IT?
   - NO →
     - DISCONNECT P/N SWITCH. THIS SHOULD CAUSE "SCAN" TOOL TO DISPLAY DRIVE RANGE. DOES IT?
       - YES → FAULTY P/N SWITCH CONNECTION OR P/N SWITCH MISADJUSTED OR FAULTY.
       - NO → CKT 434 SHORTED TO GROUND OR FAULTY ECM.
   - YES → NO TROUBLE FOUND. P/N CIRCUIT OK.

2. DISCONNECT PARK/NEUTRAL SWITCH CONNECTOR. JUMPER HARNESS CONNECTOR TERMINALS "A" AND "B" (CKT 434 TO 450). "SCAN" TOOL SHOULD INDICATE PARK OR NEUTRAL. DOES IT?
   - NO →
     - JUMPER HARNESS CONNECTOR TERMINAL "A" (CKT 434) TO ENGINE GROUND. "SCAN" TOOL SHOULD INDICATE PARK OR NEUTRAL. DOES IT?
       - YES → OPEN GROUND CKT 450.
       - NO → CKT 434 OPEN OR FAULTY ECM CONNECTION OR ECM.
   - YES → FAULTY P/N SWITCH CONNECTION OR P/N SWITCH MISADJUSTED OR FAULTY.

1989–90 3.1L (VIN T) – ALL W-BODY VEHICLES

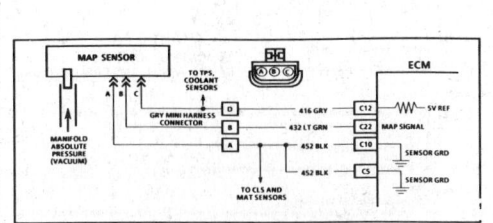

CHART C-1D
MANIFOLD ABSOLUTE PRESSURE (MAP) OUTPUT CHECK
3.1L (VIN T) "W" CARLINE (PORT)

Circuit Description:
The manifold absolute pressure (MAP) sensor measures manifold pressure (vacuum) and sends that signal to the ECM. The MAP sensor is mainly used for fuel calculation when the ECM is running in the throttle body backup mode. The MAP sensor is also used to determine the barometric pressure and to help calculate fuel delivery.

Test Description: Numbers below refer to circled numbers on the diagnostic chart.
1. Checks MAP sensor output voltage to the ECM. This voltage, without engine running, represents a barometer reading to the ECM.
2. Applying 34 kPa (10 inches Hg) vacuum to the MAP sensor should cause the voltage to be 1.2 volts less than the voltage at Step 1. Upon applying vacuum to the sensor, the change in voltage should be instantaneous. A slow voltage change indicates a faulty sensor.

The engine must be running in this step or the "scanner" will not indicate a change in voltage. It is normal for the "Service Engine Soon" light to come "ON" and for the system to set a Code 33 during this step. Make sure the code is cleared when this test is completed.
3. Check vacuum hose to sensor for leaking or restriction. Be sure no other vacuum devices are connected to the MAP hose.

1989–90 3.1L (VIN T) – ALL W-BODY VEHICLES

CHART C-1D
MANIFOLD ABSOLUTE PRESSURE (MAP) OUTPUT CHECK
3.1L (VIN T) "W" CARLINE (PORT)

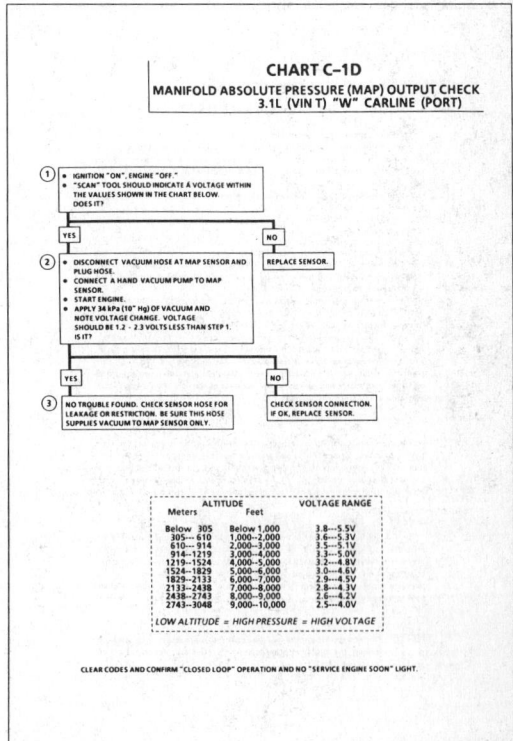

	ALTITUDE	VOLTAGE RANGE
Meters	Feet	
Below 305	Below 1,000	3.8—5.5V
305— 610	1,000—2,000	3.6—5.3V
610— 914	2,000—3,000	3.5—5.1V
914—1219	3,000—4,000	3.3—5.0V
1219—1524	4,000—5,000	3.2—4.8V
1524—1829	5,000—6,000	3.0—4.6V
1829—2133	6,000—7,000	2.9—4.5V
2133—2438	7,000—8,000	2.8—4.3V
2438—2743	8,000—9,000	2.6—4.2V
2743—3048	9,000—10,000	2.5—4.0V

LOW ALTITUDE = HIGH PRESSURE = HIGH VOLTAGE

CLEAR CODES AND CONFIRM "CLOSED LOOP" OPERATION AND NO "SERVICE ENGINE SOON" LIGHT.

1989–90 3.1L (VIN T) – ALL W-BODY VEHICLES

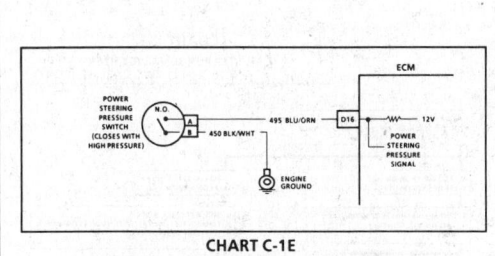

CHART C-1E
POWER STEERING PRESSURE SWITCH (PSPS) DIAGNOSIS
3.1L (VIN T) "W" CARLINE (PORT)

Circuit Description:
The power steering pressure switch is normally open to ground, and CKT 495 will be near the battery voltage.
Turning the steering wheel increases power steering oil pressure and its load on an idling engine. The pressure switch will close before the load can cause an idle problem.
Closing the switch causes CKT 495 to read less than 1 volt. The ECM will increase the idle air rate and disengage the A/C relay.
- A pressure switch that will not close, or an open CKT 495 or 450, may cause the engine to stop when power steering loads are high.
- A switch that will not open, or a CKT 495 shorted to ground, may affect idle quality and will cause the A/C relay to be de-energized.

Test Description: Numbers below refer to circled numbers on the diagnostic chart.
1. Different makes of "Scan" tools may display the state of this switch in different ways. Refer to "Scan" tool operator's manual to determine how this input is indicated.
2. Checks to determine if CKT 495 is shorted to ground.
3. This should simulate a closed switch.

1989–90 3.1L (VIN T) – ALL W-BODY VEHICLES

CHART C-1E
POWER STEERING PRESSURE SWITCH (PSPS) DIAGNOSIS
3.1L (VIN T) "W" CARLINE (PORT)

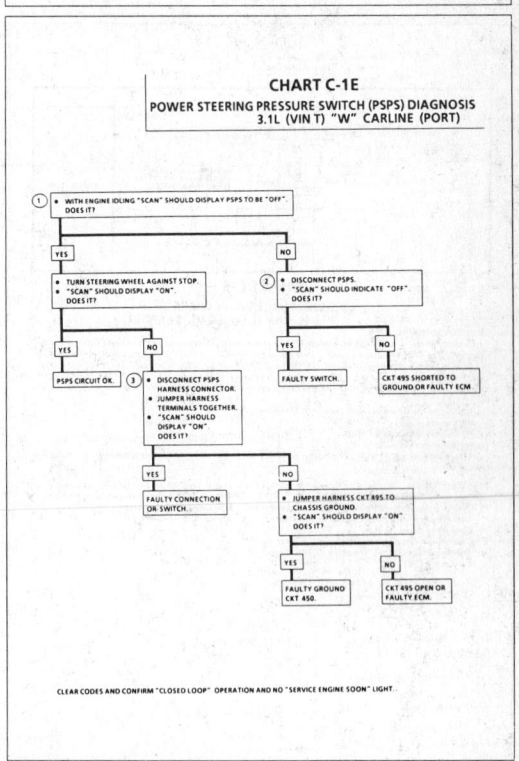

CLEAR CODES AND CONFIRM "CLOSED LOOP" OPERATION AND NO "SERVICE ENGINE SOON" LIGHT.

1989–90 3.1L (VIN T) – ALL W-BODY VEHICLES

CHART C-2A
INJECTOR BALANCE TEST

The injector balance tester is a tool used to turn the injector on for a precise amount of time, thus spraying a measured amount of fuel into the manifold. This causes a drop in fuel rail pressure that we can record and compare between each injector. All injectors should have the same amount of pressure drop (± 10 kPa). Any injector with a pressure drop that is 10 kPa (or more) greater or less than the average drop of the other injectors should be considered faulty and replaced.

STEP 1

Engine "cool down" period (10 minutes) is necessary to avoid irregular readings due to "Hot Soak" fuel boiling. With ignition "OFF" connect fuel gauge J347301 or equivalent to fuel pressure tap. Wrap a shop towel around fitting while connecting gage to avoid fuel spillage.

Disconnect harness connectors at all injectors, and connect injector tester J-34730-3, or equivalent, to one injector. On Turbo equipped engines, use adaptor harness furnished with injector tester to energize injectors that are not accessible. Follow manufacturers instructions for use of adaptor harness. Ignition must be "OFF" at least 10 seconds to complete ECM shutdown cycle. Fuel pump should run about 2 seconds after ignition is turned "ON". At this point, insert clear tubing attached to vent valve into a suitable container and bleed air from gauge and hose to insure accurate gauge operation. Repeat this step until all air is bled from gauge.

STEP 2

Turn ignition "OFF" for 10 seconds and then "ON" again to get fuel pressure to its maximum. Record this initial pressure reading. Energize tester one time and note pressure drop at its lowest point (Disregard any slight pressure increase after drop hits low point.) By subtracting this second pressure reading from the initial pressure, we have the actual amount of injector pressure drop.

STEP 3

Repeat step 2 on each injector and compare the amount of drop. Usually, good injectors will have virtually the same drop. Retest any injector that has a pressure difference of 10 kPa, either more or less than the average of the other injectors on the engine. Replace any injector that also fails the retest. If the pressure drop of all injectors is within 10 kPa of this average, the injectors appear to be flowing properly. Reconnect them and review Symptoms, Section "B".

NOTE: *The entire test should not be repeated more than once without running the engine to prevent flooding. (This includes any retest on faulty injectors).*

1989–90 3.1L (VIN T) – ALL W-BODY VEHICLES

NOTE: The fuel pressure test in Section "A", CHART A-7, should be completed prior to this test.

CHART C-2A
INJECTOR BALANCE TEST
3.1L (VIN T) "W" CARLINE (PORT)

Step 1. If engine is at operating temperature, allow a 10 minute "cool down" period then connect fuel pressure gauge and injector tester.
1. Ignition "OFF".
2. Connect fuel pressure gauge and injector tester.
3. Ignition "ON".
4. Bleed off air in gauge. Repeat until all air is bled from gauge.

Step 2. Run test:
1. Ignition "OFF" for 10 seconds.
2. Ignition "ON". Record gauge pressure. (Pressure must hold steady, if not see the Fuel System diagnosis, Chart A-7, in Section "A").
3. Turn injector on, by depressing button on injector tester, and note pressure at the instant the gauge needle stops.

Step 3.
1. Repeat step 2 on all injectors and record pressure drop on each.
Retest injectors that appear faulty (Any injectors that have a 10 kPa difference, either more or less, in pressure from the average). If no problem is found, review "Symptoms" Section "B".

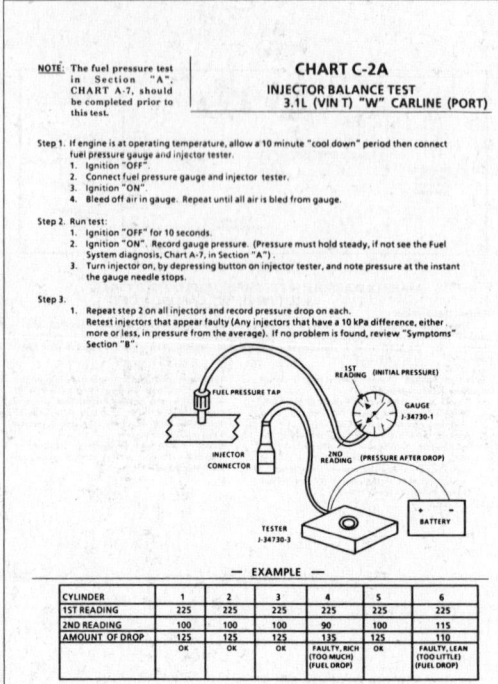

— EXAMPLE —

CYLINDER	1	2	3	4	5	6
1ST READING	225	225	225	225	225	225
2ND READING	100	100	100	90	100	115
AMOUNT OF DROP	125	125	125	135	125	110
	OK	OK	OK	FAULTY, RICH (TOO MUCH) (FUEL DROP)	OK	FAULTY, LEAN (TOO LITTLE) (FUEL DROP)

1989–90 3.1L (VIN T) – ALL W-BODY VEHICLES

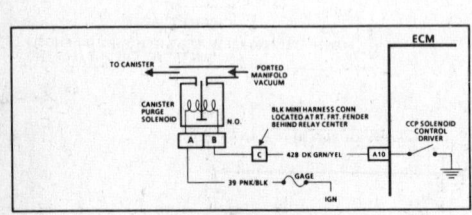

CHART C-3
CANISTER PURGE VALVE CHECK
3.1L (VIN T) "W" CARLINE (PORT)

Circuit Description:
Canister purge is controlled by a solenoid that allows manifold vacuum to purge the canister when de-energized. The ECM supplies a ground to energize the solenoid (purge "OFF"). The purge solenoid control by the ECM is pulse width modulated (turned "ON" and "OFF" several times a second). The duty cycle (pulse width) is determined by the amount of air flow, and the engine vacuum is determined by the MAP sensor input. The duty cycle is calculated by the ECM and the output commanded when the following conditions have been met:
- Engine run time after start more than 3 minutes.
- Coolant temperature above 80°C.
- Vehicle speed above 15 mph.
- Throttle off idle (about 3%).

Also, if the diagnostic test terminal is grounded with the engine stopped, the purge solenoid is de-energized (purge "ON").

Test Description: Numbers below refer to circled numbers on the diagnostic chart.
1. Checks to see if the solenoid is opened or closed. The solenoid is normally energized in this step; so it should be closed.
2. Checks for a complete circuit. Normally there is ignition voltage on CKT 39 and the ECM provides a ground on CKT 428.
3. Completes functional check by grounding test terminal. This should normally de-energize the solenoid opening the valve which should allow the vacuum to drop (purge "ON").

1989–90 3.1L (VIN T) – ALL W-BODY VEHICLES

CHART C-3
CANISTER PURGE VALVE CHECK
3.1L (VIN T) "W" CARLINE (PORT)

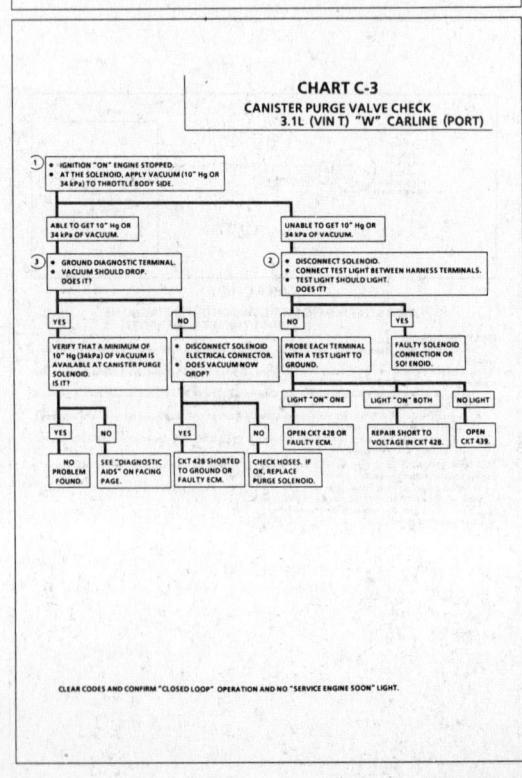

CLEAR CODES AND CONFIRM "CLOSED LOOP" OPERATION AND NO "SERVICE ENGINE SOON" LIGHT.

1989–90 3.1L (VIN T) – ALL W-BODY VEHICLES

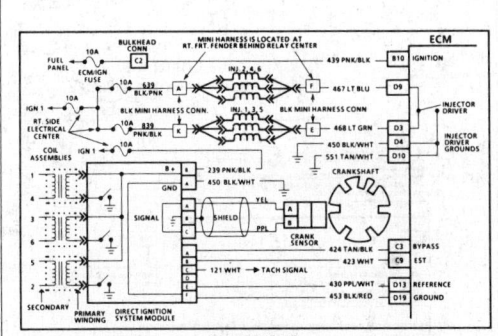

CHART C-4F
"DIS" MISFIRE
3.1L (VIN T) "W" CARLINE (PORT)

Circuit Description:

The "direct ignition system" (DIS) uses a waste spark method of distribution. In this type of system, the ignition module triggers the #1/4 coil pair resulting in both #1 and #4 spark plugs firing at the same time. #1 cylinder is on the compression stroke at the same time #4 is on the exhaust stroke, resulting in a lower energy requirement to fire #4 spark plug. This leaves the remainder of the high voltage to be used to fire #1 spark plug. On this application, the crank sensor is mounted to the engine block and protrudes through the block to within approximately .050" of the crankshaft reluctor. Since the reluctor is a machined portion of the crankshaft and the crank sensor is mounted in a fixed position on the block, timing adjustments are not possible or necessary.

Test Description: Numbers below refer to circled numbers on the diagnostic chart.

1. Checks for voltage output of ignition system. The spark tester must be used, as this tool requires 25,000 volts to trigger. This checks for a potential weak coil.
2. If the spark tester fires on all wires, the ignition system, with the exception of the spark plugs, may be considered in good working order. If the spark plugs show no evidence of wear, damage or fouling, an engine mechanical fault should be suspected. Refer to Section "B", "Cuts out, Misses"
3. If the spark jumps the tester gap after grounding the opposite plug wire, it indicates excessive resistance in the plug which was bypassed.

A faulty or poor connection at that plug could also result in the miss condition. Also check for carbon deposits inside the spark plug boot.

4. If carbon tracking is evident replace coil and be sure plug wires relating to that coil are clean and tight. Excessive wire resistance or faulty connections could have caused the coil to be damaged.

5. If the no spark condition follows the suspected coil, that coil is faulty. Otherwise, the ignition module is the cause of no spark. This test could also be performed by substituting a known good coil for the one causing the no spark condition.

1989–90 3.1L (VIN T) – ALL W-BODY VEHICLES

CHART C-4F
"DIS" MISFIRE
3.1L (VIN T) "W" CARLINE (PORT)

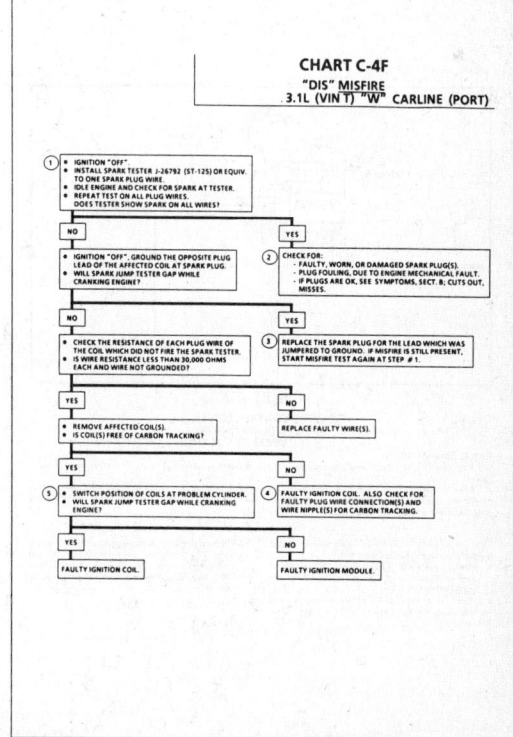

1989–90 3.1L (VIN T) – ALL W-BODY VEHICLES

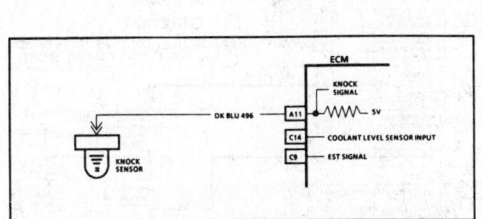

CHART C-5
ELECTRONIC SPARK CONTROL (ESC) SYSTEM CHECK
3.1L (VIN T) "W" CARLINE (PORT)

Circuit Description:

The knock sensor is used to detect engine detonation and the ECM will retard the electronic spark timing based on the signal being received. The circuitry within the knock sensor causes the ECM's 5 volts to be pulled down so that under a no knock condition, CKT 496 would measure about 2.5 volts. The knock sensor produces an AC signal which rides on the 2.5 volt DC voltage. The amplitude and frequency are dependent upon the knock level.

The Mem-Cal used with this engine, contains the functions which were part of remotely mounted ESC modules on other GM vehicles. The ESC portion of the Mem-Cal, then sends a signal to other parts of the ECM which adjusts the spark timing to retard the spark and reduce the detonation.

Test Description: Numbers below refer to circled numbers on the diagnostic chart.

1. With engine idling, there should not be a knock signal present at the ECM, because detonation is not likely under a no load condition.
2. Tapping on the engine lift hood bracket should simulate a knock signal to determine if the sensor is capable of detecting detonation. If no knock is detected, try tapping on engine block closer to sensor before replacing sensor.
3. If the engine has an internal problem which is creating a knock, the knock sensor may be responding to the internal failure.

4. This test determines if the knock sensor is faulty or if the ESC portion of the Mem-Cal is faulty. If it is determined that the Mem-Cal is faulty, be sure that is is properly installed and latched into place. If not properly installed, repair and retest.

Diagnostic Aids:

While observing knock signal on the "Scan," there should be an indication that knock is present when detonation can be heard. Detonation is most likely to occur under high engine load conditions.

1989–90 3.1L (VIN T) – ALL W-BODY VEHICLES

CHART C-5
ELECTRONIC SPARK CONTROL (ESC) SYSTEM CHECK
3.1L (VIN T) "W" CARLINE (PORT)

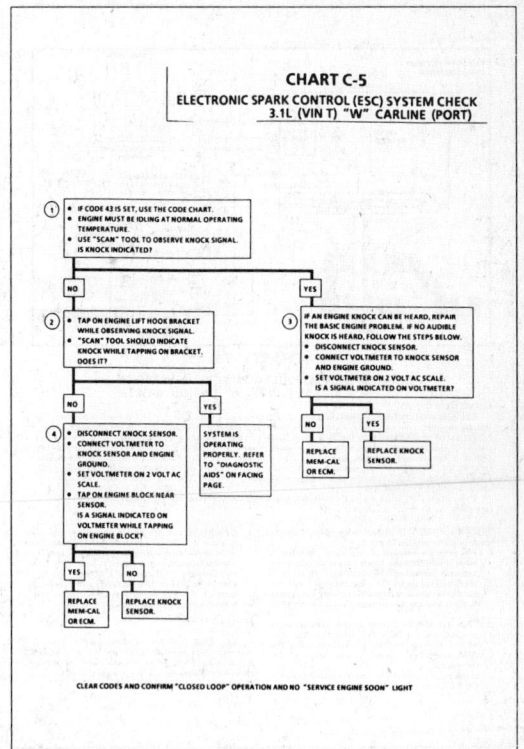

1989-90 3.1L (VIN T) – ALL W-BODY VEHICLES

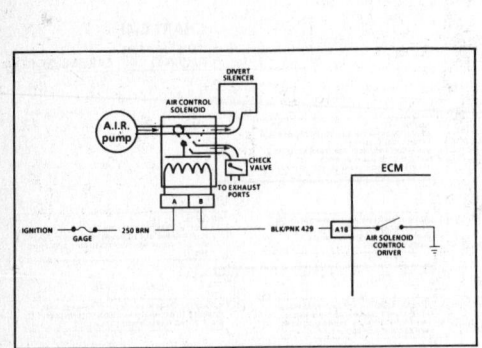

CHART C-6
ELECTRIC CONTROL (DIVERT)
(MANUAL TRANSMISSION)
3.1L (VIN T) "W" CARLINE (PORT)

Circuit Description:

This system uses a single bed converter and air management is controlled by an air control valve (divert valve).

When grounded by the ECM, the solenoid causes the valve to direct air to the exhaust ports. When de-energized air diverts to the atmosphere. Air will go to the ports provided the valve has a ground to the ECM and good manifold vacuum.

Test Description: Numbers below refer to circled numbers on the diagnostic chart.
1. This is a system performance test. When vehicle goes to "Closed Loop", air will switch from the ports to divert.
2. Tests for a grounded electric divert circuit. Normal system light will be "OFF".
3. Checks for an open control circuit. Grounding test terminal will energize the solenoid if ECM and circuits are normal. In this step, if test light is "ON", circuits are normal and faulty is in valve connections or valve.

1989-90 3.1L (VIN T) – ALL W-BODY VEHICLES

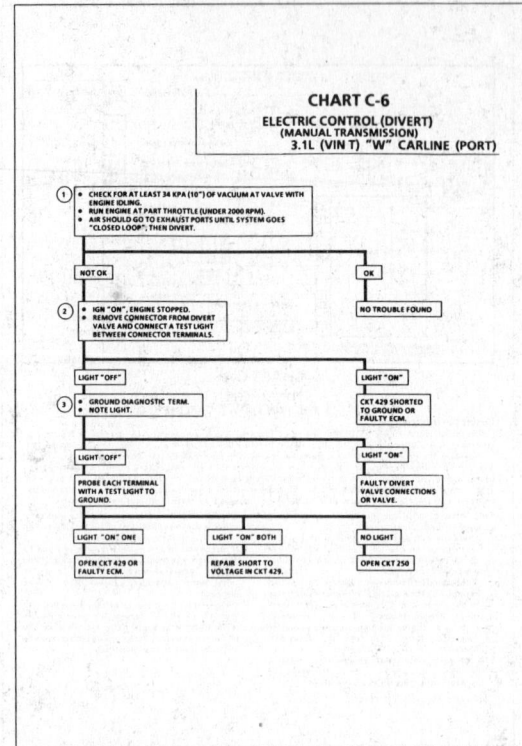

1989-90 3.1L (VIN T) – ALL W-BODY VEHICLES

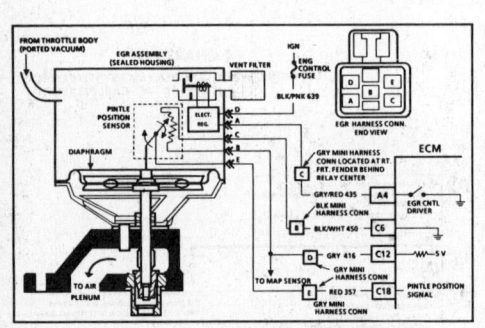

CHART C-7
EXHAUST GAS RECIRCULATION (EGR) VALVE CHECK
3.1L (VIN T) "W" CARLINE (PORT)

Circuit Description:

The integrated electronic EGR valve functions similar to a port valve with a remote vacuum regulator. The internal solenoid is normally open, which causes the vacuum signal to be vented "OFF" to the atmosphere when EGR is not being commanded by the ECM. This EGR valve has a sealed cap and the solenoid valve opens and closes the vacuum signal which controls the amount of vacuum vented to atmosphere, and this controls the amount of vacuum applied to the diaphragm. The electronic EGR valve contains a voltage regulator which converts the ECM signal to provide different amounts of EGR flow by regulating the current to the solenoid. The ECM controls EGR flow with a pulse width modulated signal (turns "ON" and "OFF" many times a second) based on airflow, TPS, and rpm.

This system also contains a pintle position sensor which works similar to a TPS sensor, and as EGR flow is increased, the sensor output also increases.

Test Description: Numbers below refer to circled numbers on the diagnostic chart.
1. Whenever the solenoid is denergized, the solenoid valve should be open which does not allow the vacuum to move the EGR diaphragm. However, if the filter is plugged, the vacuum applied with the hand held vacuum pump will cause the diaphragm to move because the vacuum will not be vented to the atmosphere.
2. This test will determine if the EGR filter is plugged or if the EGR itself is faulty. Use care when removing the filter to avoid damaging the EGR assembly. See "On-Car Service" for procedure.
3. If the valve moves in this test, it's probably due to CKT 435 being shorted to ground.
4. Grounding the diagnostic terminal should energize the solenoid which closes "OFF" the vent and allows the vacuum to move the diaphragm.
5. The EGR assembly is designed to have some leak or restrictions, 7" of vacuum is all that should be able to be held on the assembly. However, if too much of a leak exists (less than 3") the EGR assembly is leaking and must be replaced.

Diagnostic Aids:

The EGR position voltage can be used to determine that the pintle is moving. When no EGR is commanded (0% duty cycle), the position sensor should read between .5 volts and 1.5 volts, and increase with the commanded EGR duty cycle.

1989-90 3.1L (VIN T) – ALL W-BODY VEHICLES

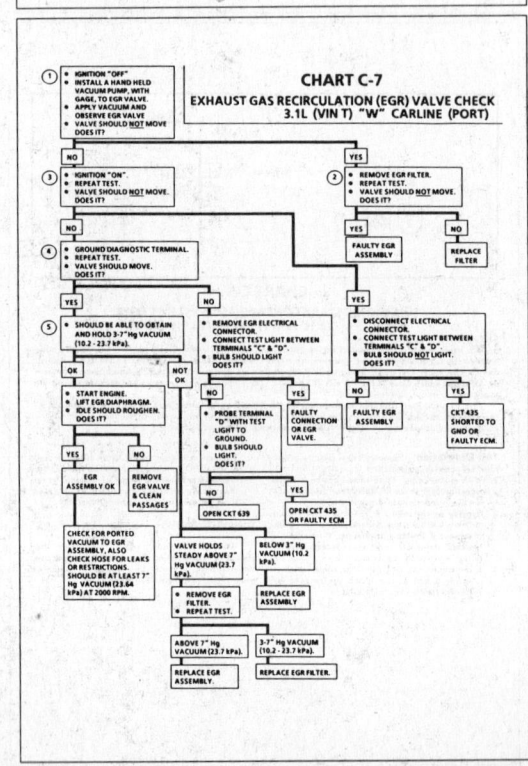

1989–90 3.1L (VIN T) – ALL W-BODY VEHICLES

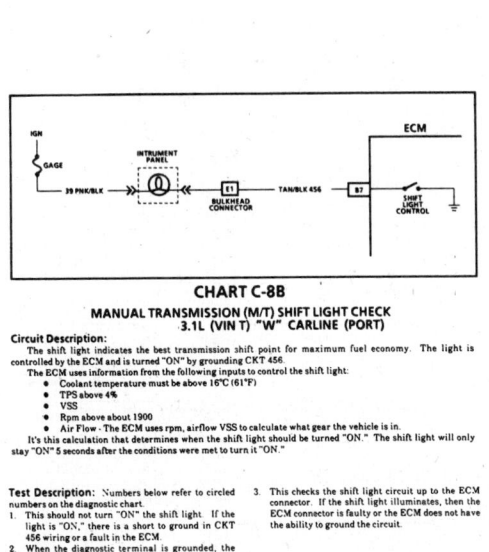

CHART C-8B
MANUAL TRANSMISSION (M/T) SHIFT LIGHT CHECK
3.1L (VIN T) "W" CARLINE (PORT)

Circuit Description:
The shift light indicates the best transmission shift point for maximum fuel economy. The light is controlled by the ECM and is turned "ON" by grounding CKT 456.
The ECM uses information from the following inputs to control the shift light:
- Coolant temperature must be above 16°C (61°F)
- TPS above 4%
- VSS
- Rpm above about 1900
- Air Flow - The ECM uses rpm, airflow VSS to calculate what gear the vehicle is in.

It's this calculation that determines when the shift light should be turned "ON." The shift light will only stay "ON" 5 seconds after the conditions were met to turn it "ON."

Test Description: Numbers below refer to circled numbers on the diagnostic chart.
1. This should not turn "ON" the shift light. If the light is "ON," there is a short to ground in CKT 456 wiring or a fault in the ECM.
2. When the diagnostic terminal is grounded, the ECM should ground CKT 456 and the shift light should come "ON."

3. This checks the shift light circuit up to the ECM connector. If the shift light illuminates, then the ECM connector is faulty or the ECM does not have the ability to ground the circuit.

1989–90 3.1L (VIN T) – ALL W-BODY VEHICLES

CHART C-8B
MANUAL TRANSMISSION (M/T) SHIFT LIGHT CHECK
3.1L (VIN T) "W" CARLINE (PORT)

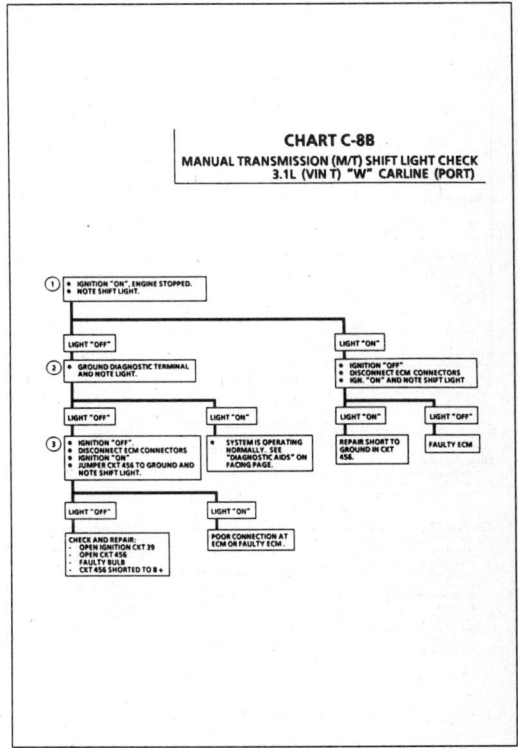

1989–90 3.1L (VIN T) – ALL W-BODY VEHICLES

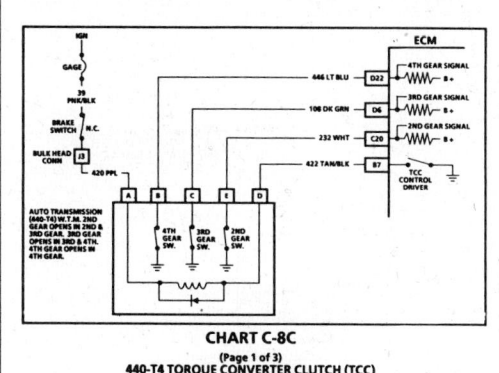

CHART C-8C
(Page 1 of 3)
440-T4 TORQUE CONVERTER CLUTCH (TCC)
WITH TORQUE MANAGEMENT
(ELECTRICAL DIAGNOSIS)
3.1L (VIN T) "W" CARLINE (PORT)

Circuit Description:
The purpose of the automatic transmission torque converter clutch feature is to eliminate the power loss of the torque converter stage when the vehicle is in a cruise condition. This allows the convenience of the automatic transmission and the fuel economy of a manual transmission. The heart of the system is a solenoid located inside the automatic transmission which is controlled by the ECM.
When the solenoid coil is activated ("ON"), the torque converter clutch is applied which results in straight through mechanical coupling from the engine to transmission. When the transmission solenoid is deactivated, the torque converter clutch is released which allows the torque converter to operate in the conventional manner (fluidic coupling between engine and transmission).
The TCC will engage on a warm engine under given road load in 2nd, 3rd and 4th gears.
TCC will engage when:
- Engine warmed up
- Vehicle speed above a calibrated value (about 28 mph 45 km/h)
- Throttle position sensor output not changing, indicating a steady road speed.
- Brake switch closed

Test Description: Numbers below refer to circled numbers on the diagnostic chart.
1. This test checks the continuity of the TCC circuit from the fuse to the ALDL connector.
2. When the brake pedal is released, the light should come back "ON" and then go "OFF" when the diagnostic terminal is grounded. This tests CKT 422 and the TCC driver in the ECM.

Diagnostic Aids:
The "Scan" tool only indicates when the ECM has turned "ON" the TCC driver, and this does not confirm that the TCC has engaged. To determine if TCC is functioning properly, engine rpm should decrease when the "Scan" indicates the TCC driver has turned "ON."

1989–90 3.1L (VIN T) – ALL W-BODY VEHICLES

CHART C-8C
(Page 1 of 3)
440-T4 TORQUE CONVERTER CLUTCH (TCC)
WITH TORQUE MANAGEMENT
(ELECTRICAL DIAGNOSIS)
3.1L (VIN T) "W" CARLINE (PORT)

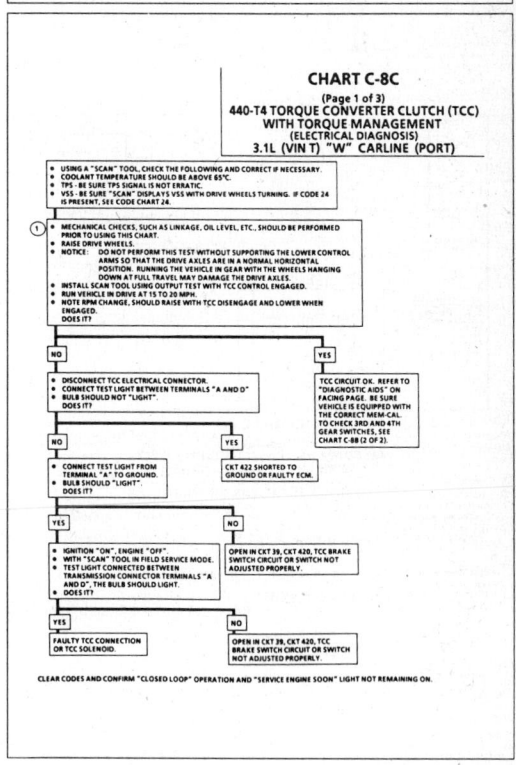

1989–90 3.1L (VIN T) – ALL W-BODY VEHICLES

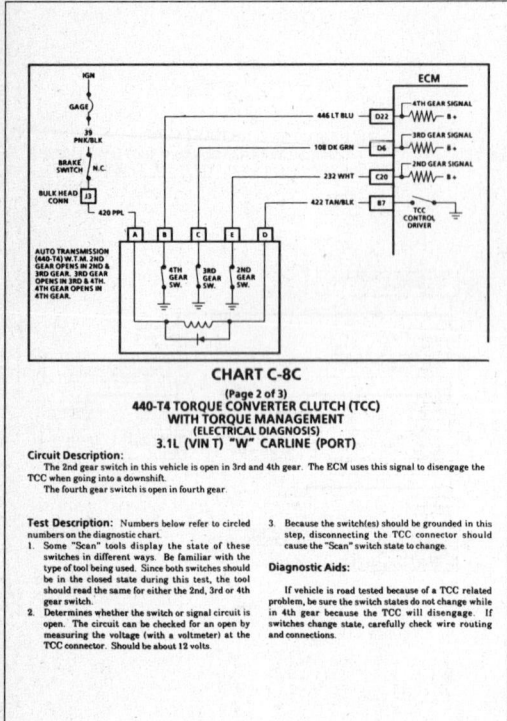

CHART C-8C
(Page 2 of 3)
440-T4 TORQUE CONVERTER CLUTCH (TCC)
WITH TORQUE MANAGEMENT
(ELECTRICAL DIAGNOSIS)
3.1L (VIN T) "W" CARLINE (PORT)

Circuit Description:
The 2nd gear switch in this vehicle is open in 3rd and 4th gear. The ECM uses this signal to disengage the TCC when going into a downshift.
The fourth gear switch is open in fourth gear.

Test Description: Numbers below refer to circled numbers on the diagnostic chart.
1. Some "Scan" tools display the state of these switches in different ways. Be familiar with the type of tool being used. Since both switches should be in the closed state during this test, the tool should read the same for either the 2nd, 3rd or 4th gear switch.
2. Determines whether the switch or signal circuit is open. The circuit can be checked for an open by measuring the voltage (with a voltmeter) at the TCC connector. Should be about 12 volts.

3. Because the switch(es) should be grounded in this step, disconnecting the TCC connector should cause the "Scan" switch state to change.

Diagnostic Aids:
If vehicle is road tested because of a TCC related problem, be sure the switch states do not change while in 4th gear because the TCC will disengage. If switches change state, carefully check wire routing and connections.

1989–90 3.1L (VIN T) – ALL W-BODY VEHICLES

CHART C-8C
(Page 2 of 3)
440-T4 TORQUE CONVERTER CLUTCH (TCC)
WITH TORQUE MANAGEMENT
(ELECTRICAL DIAGNOSIS)
3.1L (VIN T) "W" CARLINE (PORT)

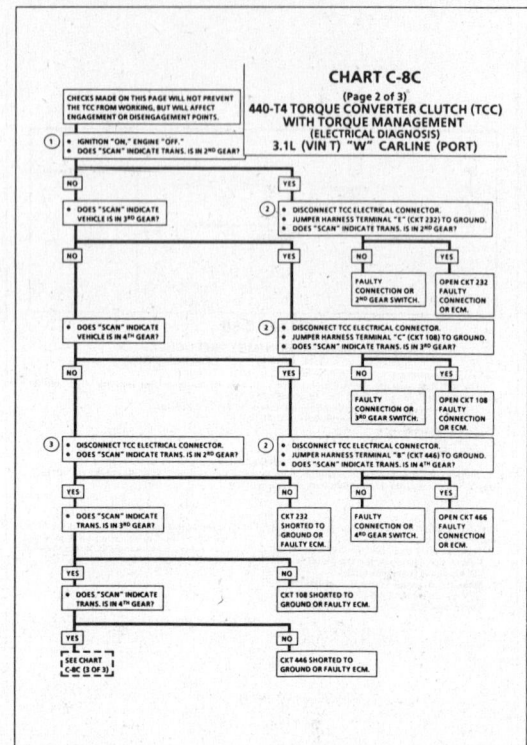

1989–90 3.1L (VIN T) – ALL W-BODY VEHICLES

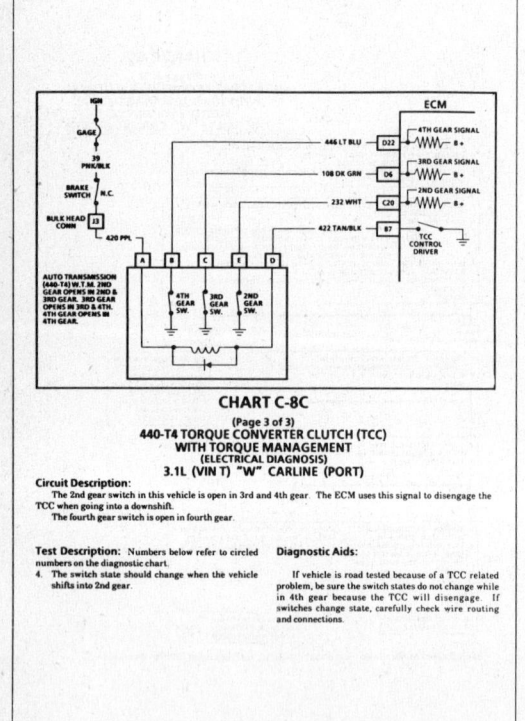

CHART C-8C
(Page 3 of 3)
440-T4 TORQUE CONVERTER CLUTCH (TCC)
WITH TORQUE MANAGEMENT
(ELECTRICAL DIAGNOSIS)
3.1L (VIN T) "W" CARLINE (PORT)

Circuit Description:
The 2nd gear switch in this vehicle is open in 3rd and 4th gear. The ECM uses this signal to disengage the TCC when going into a downshift.
The fourth gear switch is open in fourth gear.

Test Description: Numbers below refer to circled numbers on the diagnostic chart.
4. The switch state should change when the vehicle shifts into 2nd gear.

Diagnostic Aids:
If vehicle is road tested because of a TCC related problem, be sure the switch states do not change while in 4th gear because the TCC will disengage. If switches change state, carefully check wire routing and connections.

1989–90 3.1L (VIN T) – ALL W-BODY VEHICLES

CHART C-8C
(Page 3 of 3)
440-T4 TORQUE CONVERTER CLUTCH (TCC)
WITH TORQUE MANAGEMENT
(ELECTRICAL DIAGNOSIS)
3.1L (VIN T) "W" CARLINE (PORT)

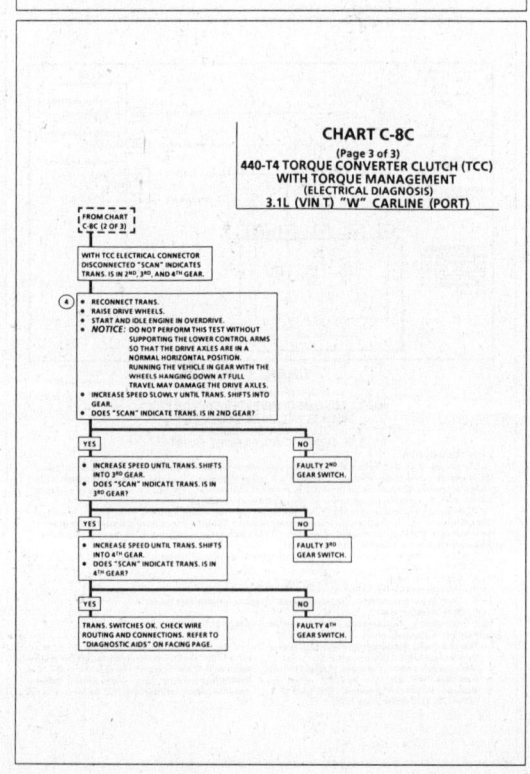

1989–90 3.1L (VIN T) – ALL W-BODY VEHICLES

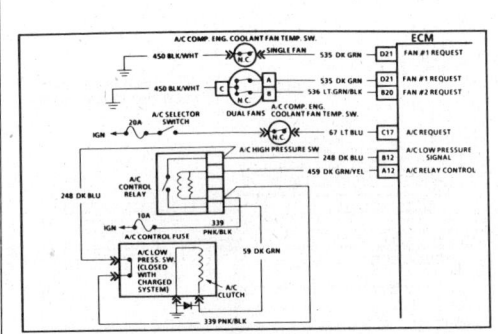

CHART C-10
A/C CLUTCH CONTROL CIRCUIT DIAGNOSIS
3.1L (VIN T) "W" CARLINE (PORT)

Circuit Description:
The A/C clutch control relay is ECM controlled to delay A/C clutch engagement about .4 second after A/C is turned "ON." This allows the IAC to adjust engine rpm, before the A/C clutch engages. The ECM, also, causes the relay to disengage the A/C clutch during WOT, when high power steering pressure is present, or if engine is overheating. The A/C clutch control relay is energized, when the ECM provides a ground path for CKT 459. The low pressure switch will open, if A/C pressure is less than 40 psi (276 kPa). The high pressure switch will open, if A/C pressure exceeds about 440 psi (3034 kPa). The A/C pressure fan switch opens, when A/C pressure exceeds about 200 psi (1380 kPa).

Test Description: Numbers below refer to circled numbers on the diagnostic chart.
1. The ECM will only energize the A/C relay, when the engine is running. This test will determine if the relay, or CKT 459, is faulty.
2. In order for the clutch to properly be engaged, the low pressure switch must be closed to provide 12 volts to the relay, and the high pressure switch must be closed, so the A/C request (12 volts) will be present at the ECM.
3. Determines if the signal is reaching the ECM on CKT 66 from the A/C control panel. Signal should only be present when the A/C mode or defrost mode has been selected.
4. A short to ground in any part of the A/C request circuit, CKT 67 to the relay, CKT 59 to the A/C clutch, or blown fuse.

5. If the ECM is seeing a high power steering pressure signal, the A/C clutch will be disengaged by the ECM
6. With the engine idling and A/C "ON," the ECM should be grounding CKT 459, which should cause the test light to be "ON."

Diagnostic Aids:
If complaint was insufficient cooling, the problem may be caused by a inoperative cooling fan, or A/C pressure fan switch. The engine cooling fan should turn "ON," when A/C pressure exceeds a value to open the switch, which causes the ECM to energize the cooling fan relay. See CHART C-12 for cooling fan diagnosis

1989–90 3.1L (VIN T) – ALL W-BODY VEHICLES

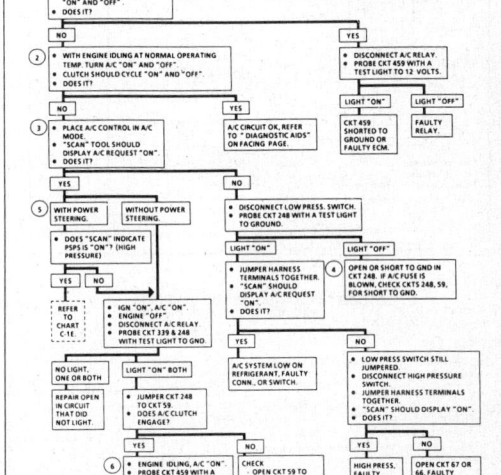

CHART C-10
A/C CLUTCH CONTROL CIRCUIT DIAGNOSIS
3.1L (VIN T) "W" CARLINE (PORT)

1989–90 3.1L (VIN T) – ALL W-BODY VEHICLES

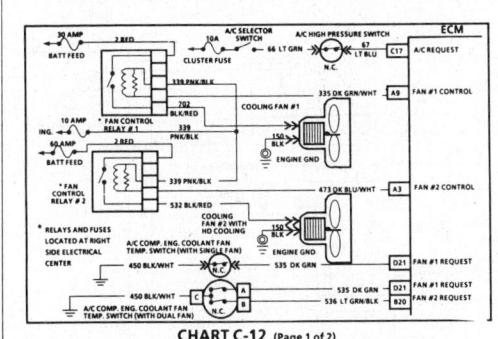

CHART C-12 (Page 1 of 2)
COOLANT FAN CONTROL CIRCUIT DIAGNOSIS
3.1L (VIN T) "W" CARLINE (PORT)

Circuit Description:
The primary and secondary electric cooling fan(s) are controlled by the ECM, based on inputs from the coolant temperature sensor, the A/C control switches, vehicle speed and state of the A/C intermediate pressure switches. The ECM controls the fan(s) by grounding CKT 335 and/or CKT 473, which energizes the fan control relay. Battery voltage is then supplied to the fan motor.
The ECM grounds CKT 335 and/or CKT 473, when coolant temperature is over about 106°C (223°F), or when A/C has been requested, and the fan control switche(s) open with high A/C pressure, about 200 psi (1380 kPa). Once the ECM turns the relay "ON," it will keep it "ON" for a minimum of 30 seconds, or until vehicle speed exceeds 70 mph (40 mph for secondary fan).
Also, if Code 14 or 15 sets, or the ECM is in throttle body back up, the primary fan will run at all times.

Test Description: Numbers below refer to circled numbers on the diagnostic chart.
1. With the diagnostic terminal grounded, the cooling fan control driver(s) will close, which should energize the fan control relay(s).
2. If the A/C fan control switch or circuit is open, the fan would run whenever A/C is requested.
3. With A/C clutch engaged, the A/C fan control switches should open, when A/C intermediate pressure exceeds about 200 psi (1380 kPa). This signal should cause the ECM to energize the fan control relay(s).

Diagnostic Aids:
If the owner complained of an overheating

problem, it must be determined if the complaint was due to an actual boilover, or the hot light, or temp. gage indicated over heating.
If the gage, or light, indicates overheating, but no boilover is detected, the gage circuit should be checked. The gage accuracy can, also, be checked by comparing the coolant sensor reading using a "Scan" tool and comparing its reading with the gage reading.
If the engine is actually overheating, and the gage indicates overheating, but the cooling fan is not coming "ON," the coolant sensor has probably shifted out of calibration and should be replaced.
If the engine is overheating, and the cooling fan is "ON," the cooling system should be checked.

1989–90 3.1L (VIN T) – ALL W-BODY VEHICLES

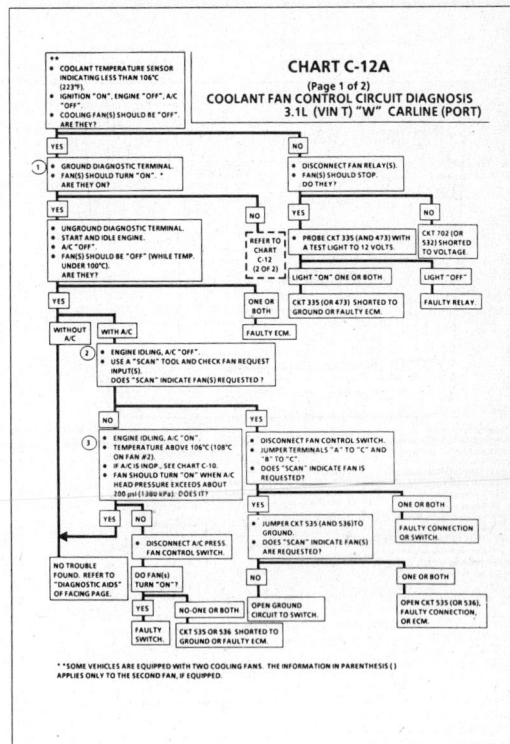

CHART C-12A
(Page 1 of 2)
COOLANT FAN CONTROL CIRCUIT DIAGNOSIS
3.1L (VIN T) "W" CARLINE (PORT)

1989–90 3.1L (VIN T) – ALL W-BODY VEHICLES

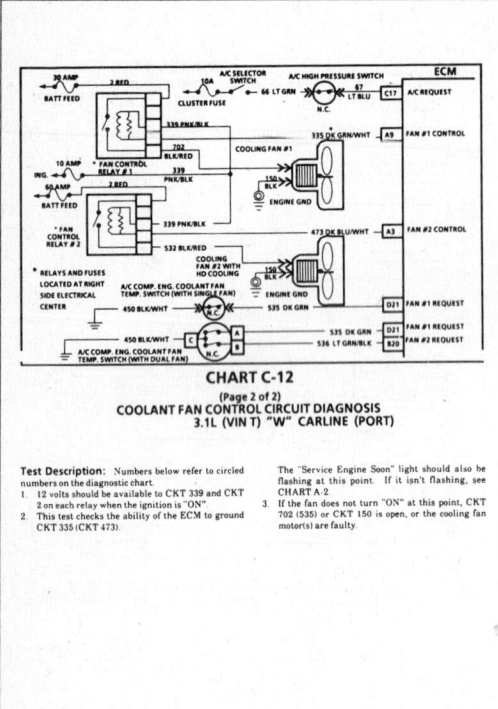

CHART C-12
(Page 2 of 2)
COOLANT FAN CONTROL CIRCUIT DIAGNOSIS
3.1L (VIN T) "W" CARLINE (PORT)

Test Description: Numbers below refer to circled numbers on the diagnostic chart.

1. 12 volts should be available to CKT 339 and CKT 2 on each relay when the ignition is "ON"
2. This test checks the ability of the ECM to ground CKT 335 (CKT 473).

The "Service Engine Soon" light should also be flashing at this point. If it isn't flashing, see CHART A-2.

3. If the fan does not turn "ON" at this point, CKT 702 (535) or CKT 150 is open, or the cooling fan motor(s) are faulty.

1989–90 3.1L (VIN T) – ALL W-BODY VEHICLES

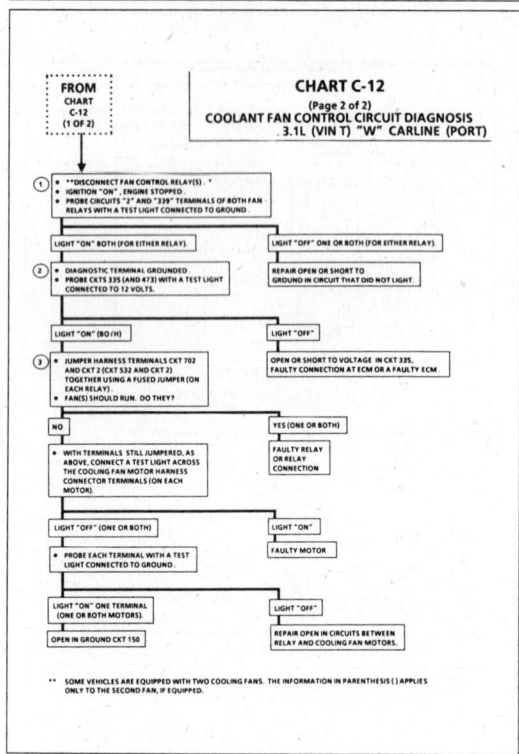

3.3L (VIN N) COMPONENT LOCATIONS CIERA AND CUTLASS CRUISER

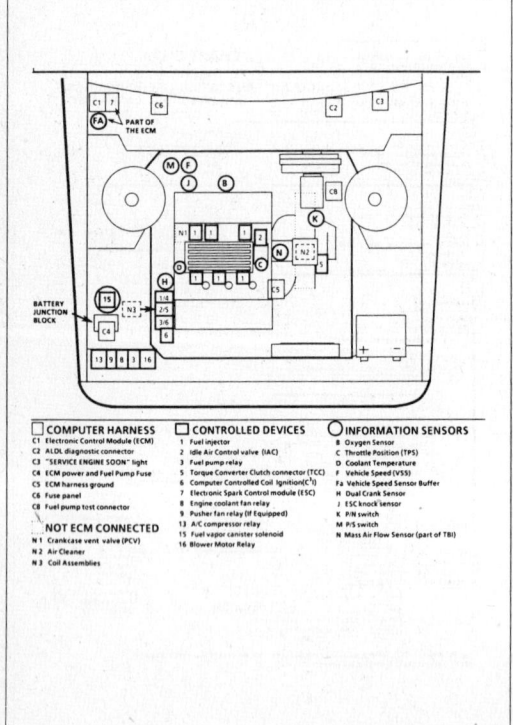

☐ **COMPUTER HARNESS**
C1 Electronic Control Module (ECM)
C2 ALDL diagnostic connector
C3 "SERVICE ENGINE SOON" light
C4 ECM power and Fuel Pump Fuse
C5 ECM harness ground
C6 Fuse panel
C8 Fuel pump test connector

NOT ECM CONNECTED
N1 Crankcase vent valve (PCV)
N2 Air Cleaner
N3 Coil Assemblies

☐ **CONTROLLED DEVICES**
1 Fuel injector
2 Idle Air Control valve (IAC)
3 Fuel pump relay
5 Torque Converter Clutch connector (TCC)
6 Computer Controlled Coil Ignition(C³I)
7 Electronic Spark Control module (ESC)
8 Engine coolant fan relay
9 Pusher fan relay (if equipped)
13 A/C compressor relay
14 P/N switch
15 Fuel vapor canister solenoid
16 Blower motor relay

○ **INFORMATION SENSORS**
B Oxygen Sensor
C Throttle Position (TPS)
D Coolant Temperature
F Vehicle Speed (VSS)
Fa Vehicle Speed Sensor Buffer
H Dual Crank Sensor
J ESC knock sensor
K P/N switch
M P/S switch
N Mass Air Flow Sensor (part of TBI)

3.3L (VIN N) ECM WIRING DIAGRAM CIERA AND CUTLASS CRUISER

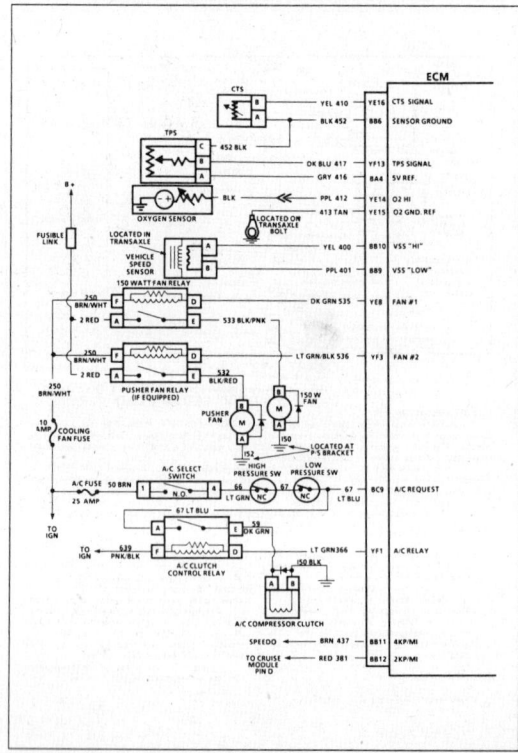

3.3L (VIN N) ECM WIRING DIAGRAM CIERA AND CUTLASS CRUISER (CONT).

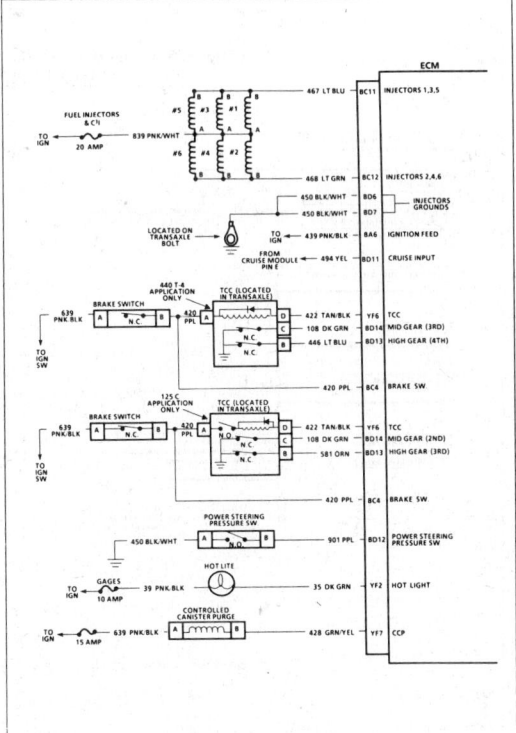

3.3L (VIN N) ECM WIRING DIAGRAM CIERA AND CUTLASS CRUISER (CONT).

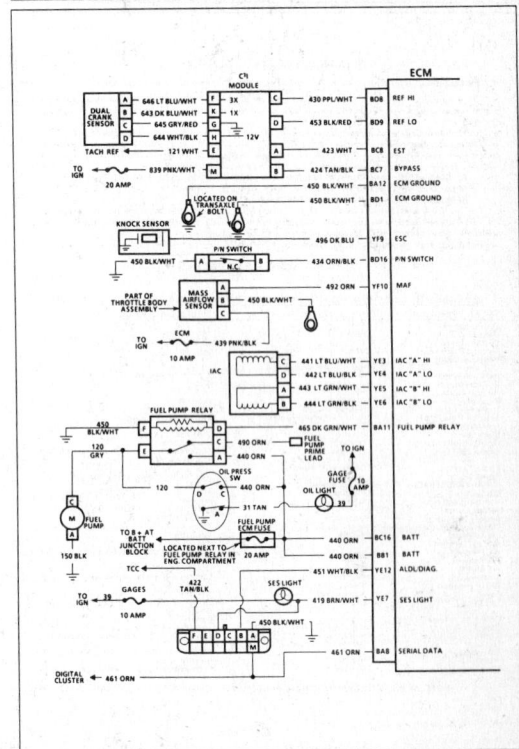

3.3L (VIN N) ECM CONNECTOR TERMINAL END VIEW – CIERA AND CUTLASS CRUISER

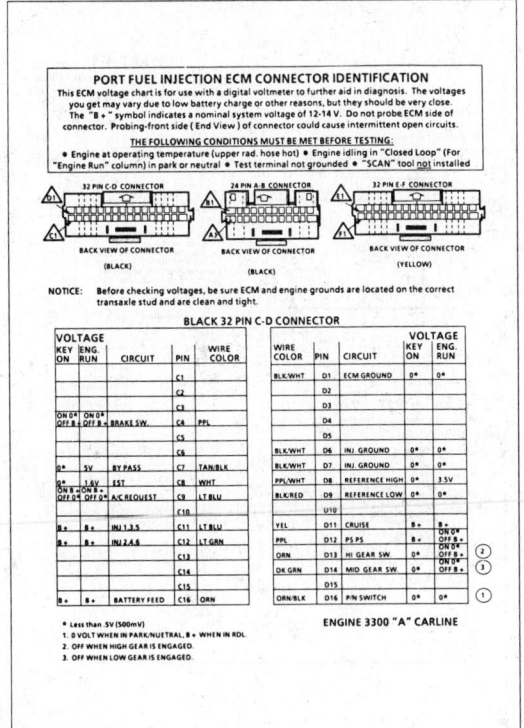

3.3L (VIN N) ECM CONNECTOR TERMINAL END VIEW – CIERA AND CUTLASS CRUISER (CONT.)

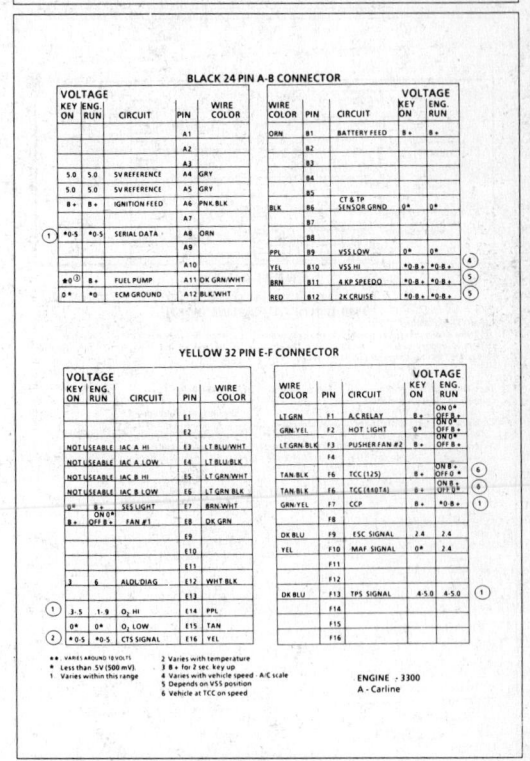

1989–90 3.3L (VIN N)
CUTLASS CIERA AND CUTLASS CRUISER

DIAGNOSTIC CIRCUIT CHECK

The Diagnostic Circuit Check must be the starting point for any driveability complaint diagnosis.

The diagnostic circuit check is an organized approach to identifying a problem created by an electronic engine control system malfunction because it directs the service technician to the next logical step in diagnosing the complaint.

If after completing the diagnostic circuit check and finding the on-board diagnostics functioning properly and no trouble codes displayed, a comparison of "Typical Scan Values," for the appropriate engine, may be used for comparison. The "Typical Values" are an average of display values recorded from normally operating vehicles and are intended to represent what a normally functioning system would display.

A "SCAN" TOOL THAT DISPLAYS FAULTY DATA SHOULD NOT BE USED, AND THE PROBLEM SHOULD BE REPORTED TO THE MANUFACTURER. THE USE OF A FAULTY "SCAN" CAN RESULT IN MISDIAGNOSIS AND UNNECESSARY PARTS REPLACEMENT.

Only the parameters listed below are used in this manual for diagnosis. If a "Scan" reads other parameters, the values are not recommended by General Motors for use in diagnosis. For more description on the values and use of the "Scan" to diagnose ECM inputs, refer to the applicable diagnosis section   If all values are within the range illustrated, refer to "Symptoms" in Section "B".

TYPICAL "Scan" DATA VALUES

Idle / Upper Radiator Hose Hot / Closed Throttle / Park or Neutral / Closed Loop / Acc. off

"SCAN" Position	Units Displayed	Typical Data Value
Engine Speed	RPM	650 - 750
Desired Idle	RPM	650 - 750
Coolant Temp.	C°/F°	85°C - 105°C (185°F - 221°F)
Throttle Position	Volts	.33 - .46 v
LV8 (Eng. Load)	Number	60 - 50
O_2 Sensor	mV	100-900 mV
Inj. Pulse Width	m Sec	4.0 V - 5.0 V (Varies)
O_2 Cross counts	# number	0-20
Air Fuel Ratio	Ratio	14.7
Rich/Lean	Lean/Rich	Lean - Rich
Spark Advance	Degrees	20° varies
Mass Air Flow (MAF)	Gram Per Second (Gm/Sec)	4 - 7 varies
Fuel Integrator	Counts	118 - 138
Block Learn	Counts	118 - 138
Open/Closed Loop	Open/Closed	Closed Loop
Block Learn Cell	Cell Number	Cell number varies with engine RPM and Mass Air Flow
Knock Retard	Degrees of Retard	0°
Knock Signal	Yes/No	No (Yes, if detonation is present)
Idle Air Control (IAC)	Counts (Steps)	10 - 30
P/N Switch	P/N and RDL	P/N
Vehicle Speed (VSS)	MPH	0(mph)
Torque Converter Clutch (TCC)	On/Off	Off
Battery Voltage	Volts	13.8 volts (varies)
Purge Duty Cycle	%	15%
IAC Learned	Yes/No	Yes
Air Fuel Learned	Yes/No	Yes
A/C Pressure	Volts/kPa	.2 - 4.5V/139-399 kPa (varies with Pressure)
A/C Request	Off/On	Off
A/C Clutch	Off/On	Off
Brake Switch	Yes/No	No (Yes with pedal depressed)
Cruise Engaged	Yes/No	No.
Power Steering (PSPS)	Normal/High	Normal
Fan 1/Fan 2	Off/On	Off (On if temperature is above 100°C) (212°F)
QDM1	High/Low	Low
QDM2	High/Low	Low (High if A/C head pressure's high)
QDM3	High/Low	Low (High with brake pedal depressed)
2nd Gear	Off/On	Off
3rd/4th Gear	Off/On	Off
Prom ID	Number	Internal ID only
Time From Start	Min/Sec	Start when engine is running and varies

NOTE: IF ALL VALUES ARE WITHIN THE RANGE ILLUSTRATED, REFER TO "SYMPTOMS" IN SECTION "B".

1989–90 3.3L (VIN N)
CUTLASS CIERA AND CUTLASS CRUISER

DIAGNOSTIC CIRCUIT CHECK
3300 (VIN N) "A" CARLINE (PORT)

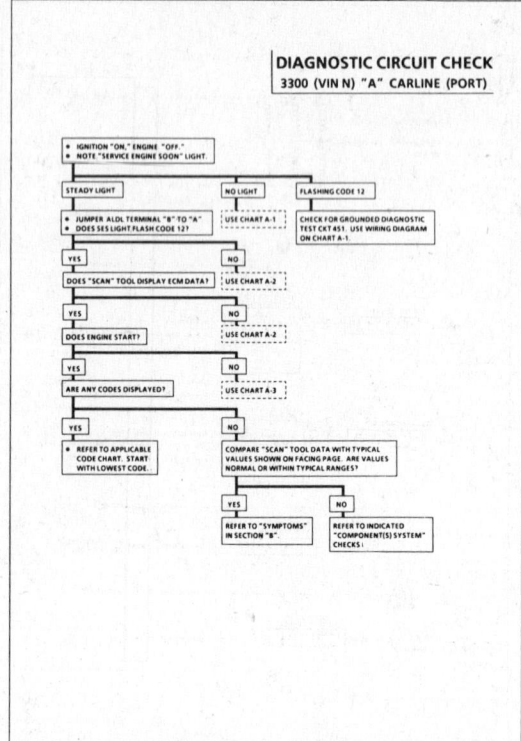

1989–90 3.3L (VIN N)
CUTLASS CIERA AND CUTLASS CRUISER

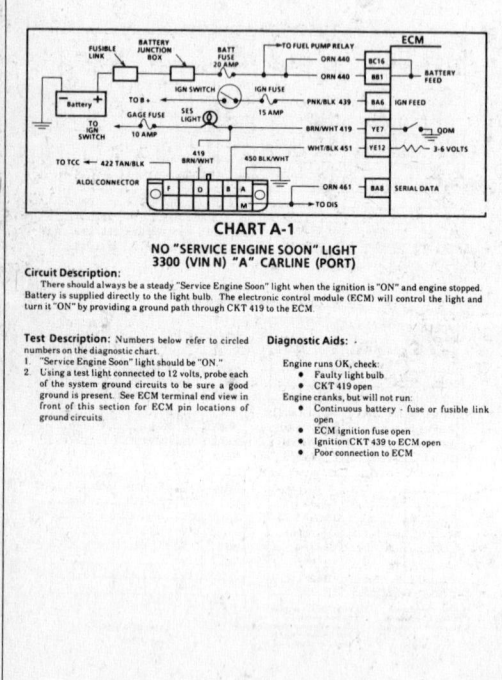

CHART A-1
NO "SERVICE ENGINE SOON" LIGHT
3300 (VIN N) "A" CARLINE (PORT)

Circuit Description:
There should always be a steady "Service Engine Soon" light when the ignition is "ON" and engine stopped. Battery is supplied directly to the light bulb. The electronic control module (ECM) will control the light and turn it "ON" by providing a ground path through CKT 419 to the ECM.

Test Description: Numbers below refer to circled numbers on the diagnostic chart.
1. "Service Engine Soon" light should be "ON."
2. Using a test light connected to 12 volts, probe each of the system ground circuits to be sure a good ground is present. See ECM terminal end view in front of this section for ECM pin locations of ground circuits.

Diagnostic Aids:
Engine runs OK, check:
- Faulty light bulb.
- CKT 419 open.
Engine cranks, but will not run:
- Continuous battery - fuse or fusible link open
- ECM ignition fuse open
- Ignition CKT 439 to ECM open
- Poor connection to ECM

1989–90 3.3L (VIN N)
CUTLASS CIERA AND CUTLASS CRUISER

CHART A-1
NO "SERVICE ENGINE SOON" LIGHT
3300 (VIN N) "A" CARLINE (PORT)

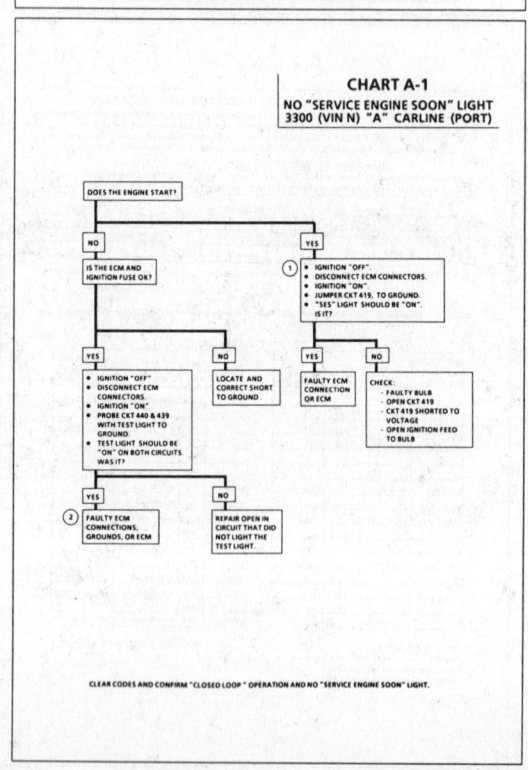

1989–90 3.3L (VIN N) CUTLASS CIERA AND CUTLASS CRUISER

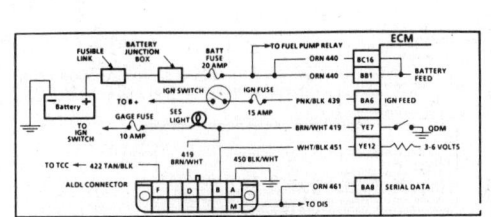

CHART A-2
WON'T FLASH CODE 12 - NO SERIAL DATA
"SERVICE ENGINE SOON" LIGHT "ON" STEADY
3300 (VIN N) "A" CARLINE (PORT)

Circuit Description:
There should always be a steady "Service Engine Soon" light when the ignition is "ON" and engine not running. Battery is supplied directly to the light bulb. The electronic control module (ECM) will turn the light "ON" by grounding CKT 419 at the ECM.

With the diagnostic terminal grounded, the light should flash a Code 12, followed by any trouble code(s) stored in memory.

A steady light suggests a short to ground in the light control CKT 419 or an open in diagnostic CKT 451.

Test Description: Numbers below refer to circled numbers on the diagnostic chart.
1. If the light goes "OFF" when the ECM connector is disconnected, then CKT 419 is not shorted to ground.
2. If there is a problem with the ECM that causes a "Scan" tool to not read serial data, then the ECM should not flash a Code 12. If Code 12 does flash, be sure that the "Scan" tool is working properly on another vehicle.

If the "Scan" is functioning properly and CKT 461 is OK, the Mem-Cal or ECM may be at fault for the No ALDL symptom.
3. This step will check for an open diagnostic CKT 451.
4. At this point, the "Service Engine Soon" light wiring is OK. The problem is a faulty ECM or Mem-Cal. If Code 12 does not flash, the ECM should be replaced using the original Mem-Cal. Replace the Mem-Cal only after trying an ECM, as a defective Mem-Cal is an unlikely cause of the problem.

1989–90 3.3L (VIN N) CUTLASS CIERA AND CUTLASS CRUISER

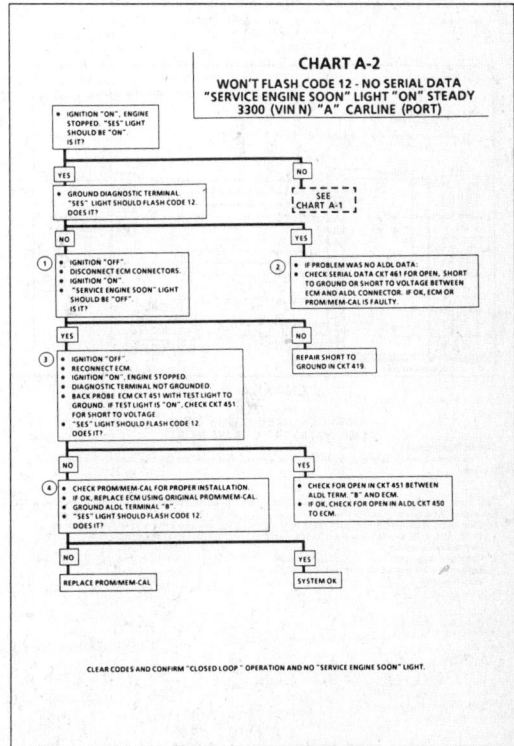

1989–90 3.3L (VIN N) CUTLASS CIERA AND CUTLASS CRUISER

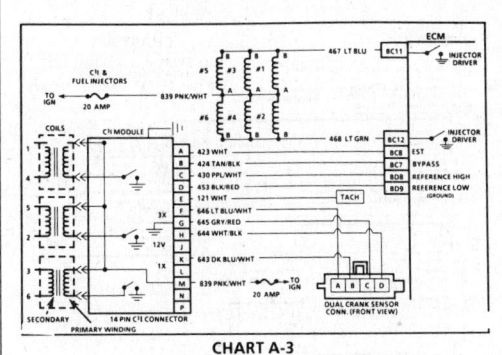

CHART A-3
(Page 1 of 2)
ENGINE CRANKS BUT WON'T RUN
3300 (VIN N) "A" CARLINE (PORT)

Circuit Description:
This chart assumes that battery condition and engine cranking speed are OK, and there is adequate fuel in the tank.

Test Description: Numbers below refer to circled numbers on the diagnostic chart.
1. A "Service Engine Soon" light "ON" is a basic test to determine if there is a 12 volt supply and ignition 12 volts to ECM. The injector test light should blink, indicating the ECM is in control of the injectors. How bright the light blinks is not important. However, the test light should be a J-34730-2A or equivalent. The engine will not start without reference pulses and therefore the "Scan" should read rpm (reference) during crank.
2. Use fuel pressure gage J-34730-1 or equivalent. Wrap a shop towel around the fuel pressure tap to absorb any small amount of fuel leakage that may occur when installing the gage.
This test will determine if the ignition module is not generating the reference pulse or if the wiring or ECM are at fault. By touching and removing a test light to 12 volts on CKT 430, a reference pulse should be generated. If rpm is indicated, the ECM and wiring are OK.
3. Because the direct ignition system uses two plugs and wires to complete the circuit of each coil, the opposite spark should be left connected. If rpm was indicated during crank, the ignition module is receiving a crank signal, but no spark at this test indicates the ignition module is not triggering the coils.
4. This test will determine if the ignition module is not triggering the problem coil or if the tested coil is at fault. This test could also be performed by using another known good coil.
5. If test light is "OFF", the 20 amp fuse could be blown or CKT 839 open or shorted to ground.

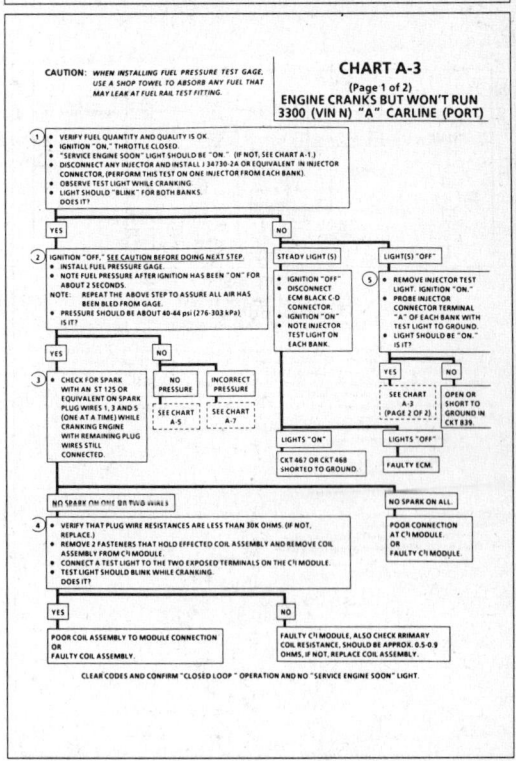

1989–90 3.3L (VIN N)
CUTLASS CIERA AND CUTLASS CRUISER

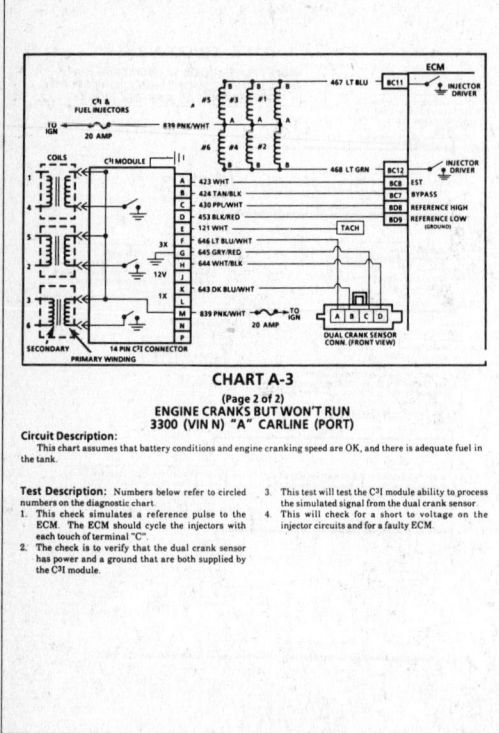

CHART A-3
(Page 2 of 2)
ENGINE CRANKS BUT WON'T RUN
3300 (VIN N) "A" CARLINE (PORT)

Circuit Description:
This chart assumes that battery conditions and engine cranking speed are OK, and there is adequate fuel in the tank.

Test Description: Numbers below refer to circled numbers on the diagnostic chart.
1. This check simulates a reference pulse to the ECM. The ECM should cycle the injectors with each touch of terminal "C".
2. The check is to verify that the dual crank sensor has power and a ground that are both supplied by the C³I module.

3. This test will test the C³I module ability to process the simulated signal from the dual crank sensor.
4. This will check for a short to voltage on the injector circuits and for a faulty ECM.

1989–90 3.3L (VIN N)
CUTLASS CIERA AND CUTLASS CRUISER

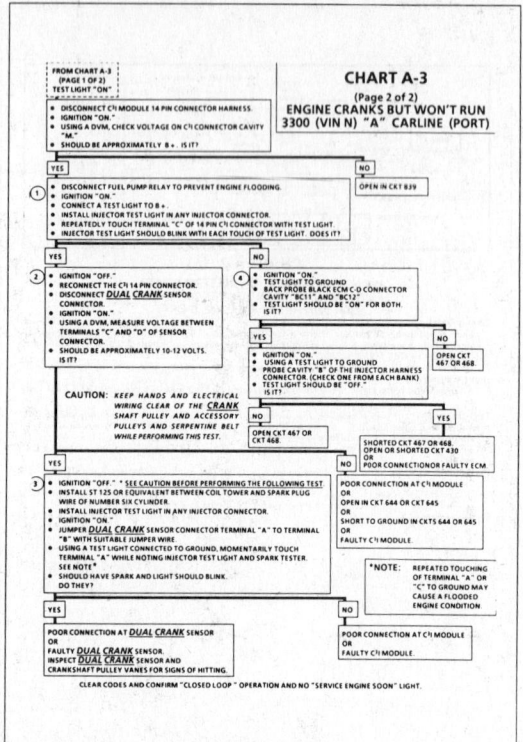

1989–90 3.3L (VIN N)
CUTLASS CIERA AND CUTLASS CRUISER

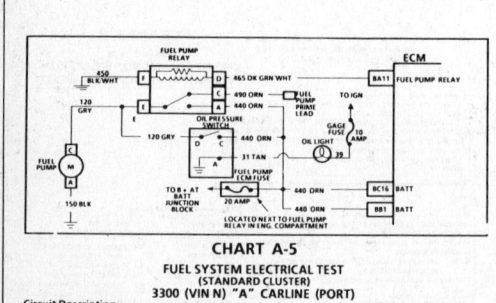

CHART A-5
FUEL SYSTEM ELECTRICAL TEST
(STANDARD CLUSTER)
3300 (VIN N) "A" CARLINE (PORT)

Circuit Description:
When the ignition switch is turned "ON," the electronic control module (ECM) will energize the fuel pump relay which completes the circuit to the in-tank fuel pump. It will remain "ON" as long as the engine is cranking or running and the ECM is receiving C³I reference pulses. If there are no reference pulses, the ECM will de-energize the fuel pump relay within 2 seconds after key "ON" or the engine is stopped.

The fuel pump will deliver fuel to the fuel rail and injectors, then to the pressure regulator where the system pressure is controlled. Excess fuel pressure is bypassed back to the fuel tank. When the engine is stopped, the pump can be turned "ON" by applying battery voltage to the test terminal located in the engine compartment.

Improper fuel system pressure may contribute to one or all of the following symptoms:
- Cranks but won't run.
- Code 44 or 45
- Cut-out, may feel like ignition problem.
- Hesitation, loss of power or poor fuel economy.

Test Description: Numbers below refer to circled numbers on the diagnostic chart.
1. Check for B+ and ground at the fuel pump relay.
2. Check that ECM is supplying voltage to fuel pump relay.

3. This checks the fuel pump circuit to the tank for opens or shorts.

1989–90 3.3L (VIN N)
CUTLASS CIERA AND CUTLASS CRUISER

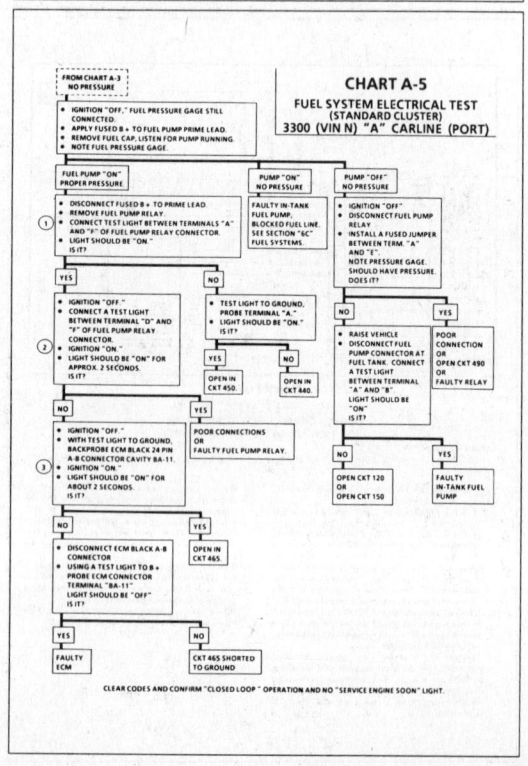

1989–90 3.3L (VIN N) CUTLASS CIERA AND CUTLASS CRUISER

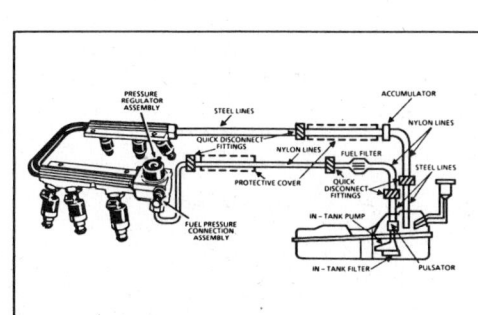

CHART A-7
(Page 1 of 3)
FUEL SYSTEM PRESSURE TEST
3300 (VIN N) "A" CARLINE (PORT)

Circuit Description:
The fuel pump will deliver fuel to the fuel rail and injectors, then to the pressure regulator, where the system pressure is controlled to 256 to 296 kPa (37 to 43 psi). Excess fuel pressure is bypassed back to the fuel tank. The fuel pump test terminal is located in the engine compartment. When the engine is stopped, the pump can be turned "ON" by applying battery voltage to the test terminal.

Improper fuel system pressure may contribute to one or all of the following symptoms:
- Cranks but won't run.
- Code E044 or E045.
- Cuts out, may feel like ignition problem.
- Hesitation, loss of power, or poor fuel economy.

Test Description: Numbers below refer to circled numbers on the diagnostic chart.
1. Use pressure gage J-34730-1. Wrap a shop towel around the fuel pressure tap to absorb any small amount of fuel leakage that may occur when installing the gage.
   With ignition "ON" pump pressure should be 256-296 kPa (34-43 psi). This pressure is controlled by spring pressure and throttle body vacuum within the pressure regulator assembly. Pressure should not leak down after the fuel pump is shut "OFF."

2. When the engine is idling, the throttle body vacuum is high and is applied to the fuel regulator diaphragm. This will offset the spring and result in a lower fuel pressure, 222-234 kPa (29-34 psi).
3. The application of 12-14 inches of vacuum to the pressure regulator should result in a fuel pressure drop of at least 21-41 kPa (3-6 psi).
4. Pressure that leaks down may be caused by one of the following:
   - In-tank fuel pump check valve not holding.
   - Pump coupling hose leaking.
   - Fuel pressure regulator valve leaking.
   - Injector sticking open.

1989–90 3.3L (VIN N) CUTLASS CIERA AND CUTLASS CRUISER

CAUTION: TO REDUCE THE RISK OF FIRE AND PERSONAL INJURY, IT IS NECESSARY TO RELIEVE THE FUEL SYSTEM PRESSURE BEFORE SERVICING THE FUEL SYSTEM (TO DO THIS):
- DISCONNECT FUEL TANK HARNESS CONNECTOR AT FUEL TANK.
- CRANK ENGINE. ENGINE WILL START AND RUN UNTIL FUEL SUPPLY REMAINING IN PIPES AND FUEL RAIL IS CONSUMED. ENGAGE STARTER FOR 3.0 SECONDS TO ASSURE TO RELIEVE ANY REMAINING PRESSURE.
- RECONNECT THE FUEL TANK HARNESS CONNECTOR. UNLESS THIS PROCEDURE IS FOLLOWED BEFORE SERVICING FUEL PIPES OR CONNECTIONS, FUEL SPRAY COULD OCCUR.

CHART A-7
(Page 1 of 3)
FUEL SYSTEM PRESSURE TEST
3300 (VIN N) "A" CARLINE (PORT)

CAUTION:
DO NOT PINCH OR RESTRICT NYLON FUEL LINES TO AVOID SEVERING AS THEY COULD CAUSE A FUEL LEAK.

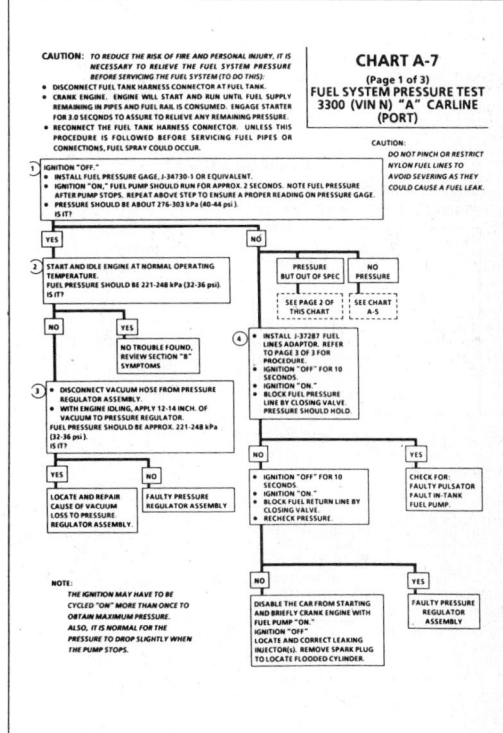

NOTE:
THE IGNITION MAY HAVE TO BE CYCLED "ON" MORE THAN ONCE TO OBTAIN MAXIMUM PRESSURE.
ALSO, IT IS NORMAL FOR THE PRESSURE TO DROP SLIGHTLY WHEN THE PUMP STOPS.

1989–90 3.3L (VIN N) CUTLASS CIERA AND CUTLASS CRUISER

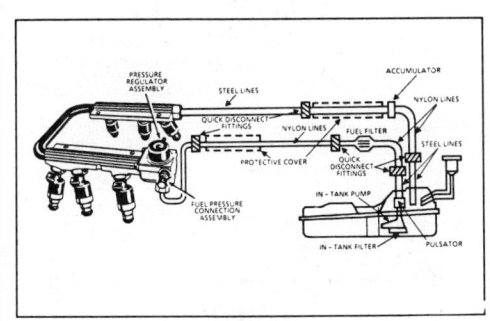

CHART A-7
(Page 2 of 3)
FUEL SYSTEM PRESSURE TEST
3300 (VIN N) "A" CARLINE (PORT)

Circuit Description:
The fuel pump will deliver fuel to the fuel rail and injectors, then to the pressure regulator, where the system pressure is controlled. Excess fuel pressure is bypassed back to the fuel tank. When the engine is stopped, the pump can be turned "ON" by applying battery voltage to the test terminal located in the engine compartment.

Improper fuel system pressure may contribute to one or all of the following symptoms:
- Cranks but won't run.
- Code 44 or 45.
- Cuts out, may feel like ignition problem.
- Hesitation, loss of power, or poor fuel economy.

Test Description: Numbers below refer to circled numbers on the diagnostic chart.
5. Pressure but less than specifications falls into two areas:
   - Regulated pressure but less than specifications; the amount of fuel to injectors is OK, but pressure is too low. The system will be lean running and may set Code 44. Also, possible hard starting cold and overall poor performance.
   - Restricted flow causing pressure drop: Normally, a vehicle with a fuel pressure of less than 165 kPa (24 psi) at idle will not be driveable.

However, if the pressure drop occurs only while driving, the engine will normally surge then stop as pressure begins to drop rapidly.
6. Do not allow fuel pressure to build above regulated pressure. With battery voltage applied to the pump fuel test terminal, pressure should rise above 325 kPa (47 psi), as the valve in the return line is partially closed.
   NOTICE: Do Not allow pressure to exceed 414 kPa (60 psi), as damage to the regulator may result.
7. This test determines if the high fuel pressure is due to a restricted fuel return line or a pressure regulator problem.

1989–90 3.3L (VIN N) CUTLASS CIERA AND CUTLASS CRUISER

CAUTION: TO REDUCE THE RISK OF FIRE AND PERSONAL INJURY, IT IS NECESSARY TO RELIEVE THE FUEL SYSTEM PRESSURE BEFORE SERVICING THE FUEL SYSTEM. (TO DO THIS):
- DISCONNECT THE FUEL TANK HARNESS CONNECTOR.
- CRANK ENGINE. ENGINE WILL START AND RUN UNTIL FUEL SUPPLY REMAINING IN FUEL PIPES IS CONSUMED. ENGAGE STARTER FOR 3.0 SECONDS TO ASSURE RELIEF OF ANY REMAINING PRESSURE.
- CONNECT THE FUEL TANK HARNESS CONNECTOR. UNLESS THIS PROCEDURE IS FOLLOWED BEFORE SERVICING FUEL PIPES OR CONNECTIONS, FUEL SPRAY COULD OCCUR.

CHART A-7
(Page 2 of 3)
FUEL SYSTEM PRESSURE TEST
3300 (VIN N) "A" CARLINE (PORT)

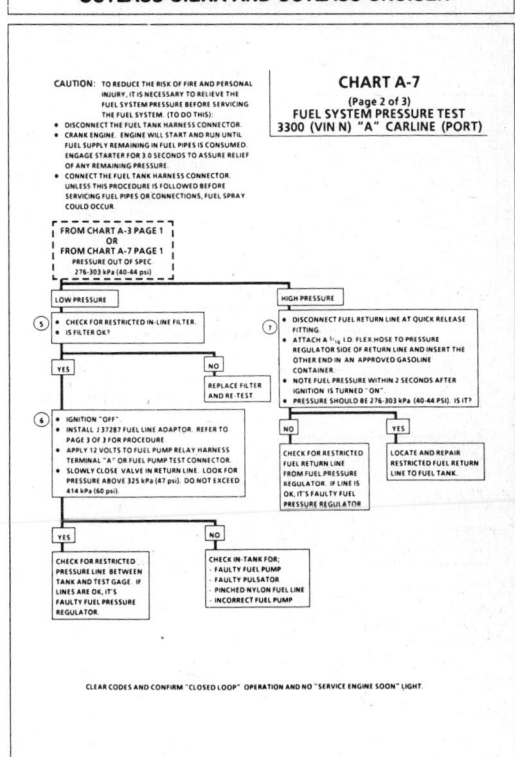

CLEAR CODES AND CONFIRM "CLOSED LOOP" OPERATION AND NO "SERVICE ENGINE SOON" LIGHT.

1989–90 3.3L (VIN N)
CUTLASS CIERA AND CUTLASS CRUISER

CHART A-7
(Page 3 of 3)
FUEL SYSTEM PRESSURE TEST
3300 (VIN N) "A" CARLINE (PORT)

FUEL PRESSURE CHECK TOOLS

37287 FUEL LINE ADAPTER TOOLS
37088 FUEL LINE REMOVE TOOL

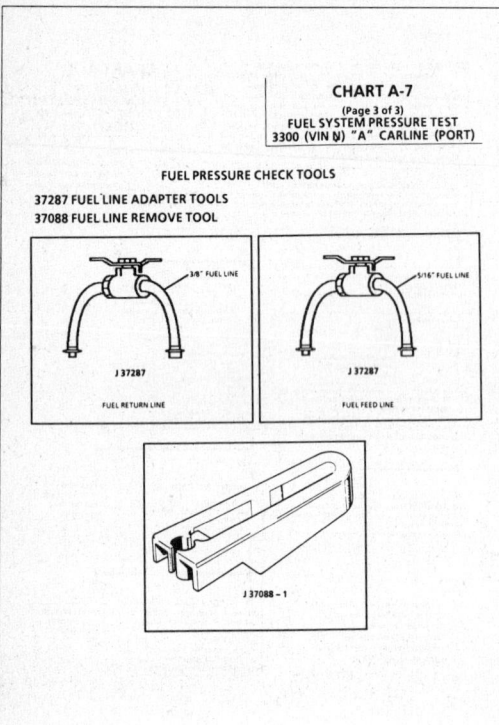

J 37287 — FUEL RETURN LINE · 3/8" FUEL LINE
J 37287 — FUEL FEED LINE · 5/16" FUEL LINE

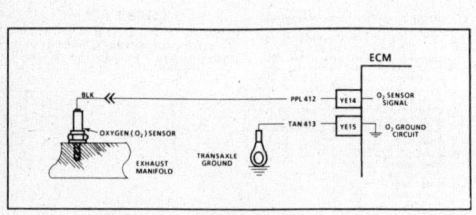

J 37088 - 1

1989–90 3.3L (VIN N)
CUTLASS CIERA AND CUTLASS CRUISER

CHART A-7
(Page 3 of 3)
FUEL SYSTEM PRESSURE TEST
3300 (VIN N) "A" CARLINE (PORT)

FUEL PRESSURE CHECK
(USING J 37287 FUEL LINE ADAPTOR)

CAUTION: DO NOT PINCH OR RESTRICT NYLON FUEL LINES TO AVOID SEVERING, WHICH COULD CAUSE A FUEL LEAK.

REMOVE OR DISCONNECT

1. REMOVE FUEL CAP
2. INSTALL GAGE BLEED HOSE INTO AN APPROVED CONTAINER AND OPEN GAGE VALVE TO BLEED SYSTEM PRESSURE.
3. REMOVE FUEL LINES. USE J 37088-1 TOOL. USE A SHOP CLOTH AND APPROVED CONTAINER TO COLLECT FUEL FROM LINES
4. INSTALL J 37287 FUEL LINE ADAPTORS TO ENGINE AND BODY SIDE FUEL LINE. (LEAVE VALVE OPEN ON ADAPTOR LINES)
5. INSTALL FUEL CAP
6. INSTALL GAGE BLEED HOSE INTO AN APPROVED CONTAINER AND OPEN GAGE VALVE TO BLEED AIR FROM SYSTEM
7. THE IGNITION MAY HAVE TO BE CYCLED "ON" MORE THAN ONCE TO REMOVE ALL AIR FROM FUEL LINES AND OBTAIN MAXIMUM PRESSURE. IT IS NORMAL FOR THE PRESSURE TO DROP SLIGHTLY WHEN PUMP STOPS.

RECONNECT OR INSTALL

1. REMOVE FUEL CAP
2. INSTALL GAGE BLEED HOSE INTO AN APPROVED CONTAINER AND OPEN GAGE VALVE TO BLEED SYSTEM PRESSURE.
3. REMOVE FUEL LINE ADAPTORS (J 37287) WITH J 37088-1 TOOL. USE A SHOP CLOTH AND APPROVED CONTAINER TO COLLECT FUEL FROM LINES.
4. INSTALL FUEL LINES TO ENGINE
5. INSTALL FUEL CAP
6. INSTALL GAGE BLEED HOSE INTO AN APPROVED CONTAINER AND OPEN GAGE VALVE TO BLEED AIR FROM SYSTEM.
7. THE IGNITION MAY HAVE TO BE CYCLED "ON" MORE THAN ONCE TO REMOVE ALL AIR FROM FUEL LINES AND OBTAIN MAXIMUM PRESSURE. IT IS NORMAL FOR THE PRESSURE TO DROP SLIGHTLY WHEN PUMP STOPS.
8. CHECK FOR LEAKS IN FUEL LINE

1989–90 3.3L (VIN N)
CUTLASS CIERA AND CUTLASS CRUISER

CODE 13
OXYGEN SENSOR CIRCUIT
(OPEN CIRCUIT)
3300 (VIN N) "A" CARLINE (PORT)

Circuit Description:
The electronic control module (ECM) supplies a voltage of about .45 volt between terminals "YE14" and "YE15". (If measured with a 10 megohm digital voltmeter, this may read as low as .32 volt.) The O_2 sensor varies the voltage within a range of about 1 volt if the exhaust is rich, down through about .10 volt if exhaust is lean.

The sensor is like an open circuit and produces no voltage when it is below 360°C (600°F). An open oxygen sensor circuit or cold oxygen sensor causes "Open Loop" operation.

Code 13 will set:
- Engine at normal operating temperature above 50°C (122°F).
- No Codes 21 or 22.
- Engine running at least 40 seconds.
- O_2 signal voltage is steady between .35 and .55 volt.
- Throttle position sensor signal above 4% (about .58 volt above closed throttle voltage).
- All conditions must be met for about 30 seconds.

Test Description: Numbers below refer to circled numbers on the diagnostic chart.
1. If the conditions for a Code 13 exists, the system will not go to "Closed Loop."
2. This will determine if the sensor or the wiring is the cause of the Code 13.
3. In doing this test use only a high impedence digital volt ohmmeter. This test checks the continuity of CKTs 412 and 413. If CKT 413 is open the ECM voltage on CKT 412 will be over .6 volt (600 mV).

Diagnostic Aids:
Normal "Scan" voltage varies between 100 mV to 999mV (.1 and 1.0 volt) while in "Closed Loop." Code 13 sets in 30 seconds if voltage remains between .35 and .55 volt, but the system will go "Open Loop" in about 15 seconds.

Refer to "Intermittents" in Section "B".

1989–90 3.3L (VIN N)
CUTLASS CIERA AND CUTLASS CRUISER

CODE 13
OXYGEN SENSOR CIRCUIT
(OPEN CIRCUIT)
3300 (VIN N) "A" CARLINE (PORT)

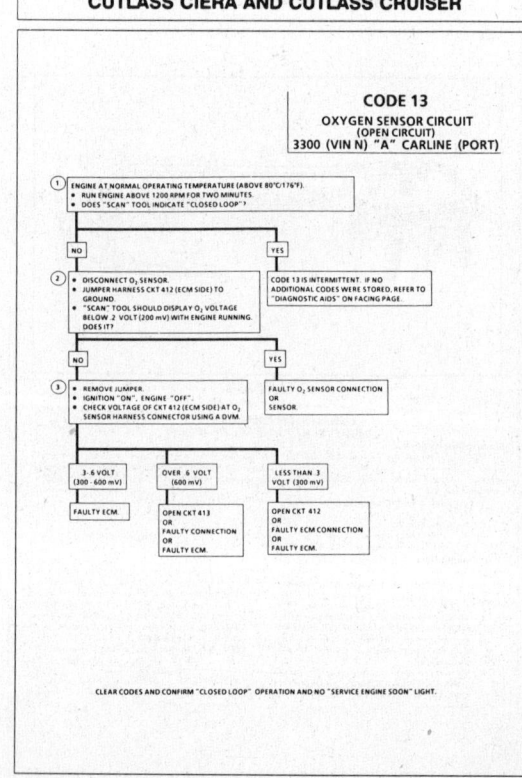

CLEAR CODES AND CONFIRM "CLOSED LOOP" OPERATION AND NO "SERVICE ENGINE SOON" LIGHT.

1989–90 3.3L (VIN N) CUTLASS CIERA AND CUTLASS CRUISER

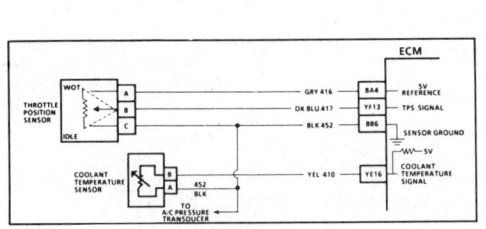

CODE 14
COOLANT TEMPERATURE SENSOR CIRCUIT
(HIGH TEMPERATURE INDICATED)
3300 (VIN N) "A" CARLINE (PORT)

Circuit Description:

The coolant temperature sensor uses a thermistor to control the signal voltage to the ECM. The ECM applies a voltage on CKT 410 to the sensor. When the engine is cold the sensor (thermistor) resistance is high, therefore, the ECM will see high signal voltage.

As the engine warms, the sensor resistance becomes less, and the voltage drops. At normal engine operating temperature (85°C to 95°C), the voltage will measure about 1.5 to 2.0 volts.

Code 14 will set if:
- Engine run time 10 seconds or more.
- Signal voltage indicates a coolant temperature above 140°F (284°F) for .4 seconds.

Test Description: Numbers below refer to circled numbers on the diagnostic chart.

1. This will determine if the coolant sensor is indicating a high temperature to the ECM.
2. This test will determine if CKT 410 is shorted to ground which will cause the conditions for Code 14.

If Code 14 is set, the ECM will use a default coolant temperature value of 49°C (120°F) for fuel control. A "Scan" tool will display actual value.

Diagnostic Aids:

"Scan" tool displays engine temperature in degrees centigrade. After engine is started, the temperature should rise steadily to about 90°C then stabilize when thermostat opens.

An intermittent may be caused by a poor connection, rubbed through wire insulation, or a wire broken inside the insulation.

Check For:
- **Poor Connection or Damaged Harness** Inspect ECM harness connectors for backed out terminal "YE16", improper mating, broken locks, improperly formed or damaged terminals, poor terminal to wire connection, and damaged harness.
- **Intermittent Test** If connections and harness check OK, "Scan" coolant temperature while moving related connectors and wiring harness. If the failure is induced, the "coolant temperature" display will change. This may help to isolate the location of the malfunction.
- **Shifted Sensor** The "Temperature To Resistance Value" scale may be used to test the coolant sensor at various temperature levels to evaluate the possibility of a "shifted" (mis-scaled) sensor, which may result in driveability complaints.

1989–90 3.3L (VIN N) CUTLASS CIERA AND CUTLASS CRUISER

CODE 14
COOLANT TEMPERATURE SENSOR CIRCUIT
(HIGH TEMPERATURE INDICATED)
3300 (VIN N) "A" CARLINE (PORT)

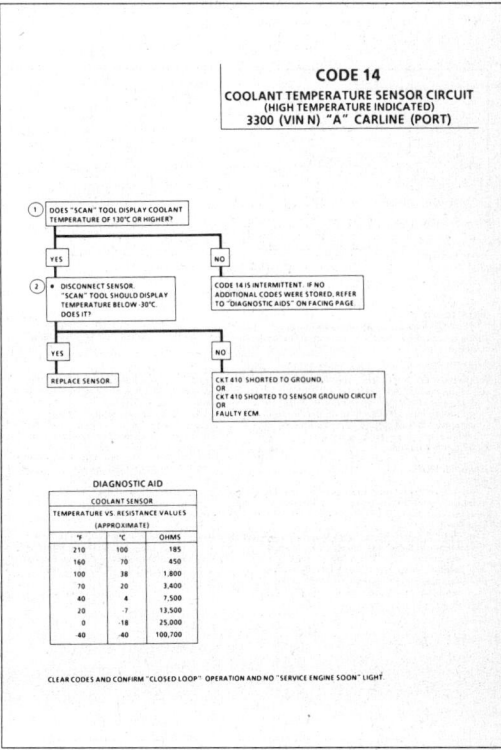

COOLANT SENSOR		
TEMPERATURE VS. RESISTANCE VALUES (APPROXIMATE)		
°F	°C	OHMS
210	100	185
160	70	450
100	38	1,800
70	20	3,400
40	4	7,500
20	-7	13,500
0	-18	25,000
-40	-40	100,700

CLEAR CODES AND CONFIRM "CLOSED LOOP" OPERATION AND NO "SERVICE ENGINE SOON" LIGHT.

1989–90 3.3L (VIN N) CUTLASS CIERA AND CUTLASS CRUISER

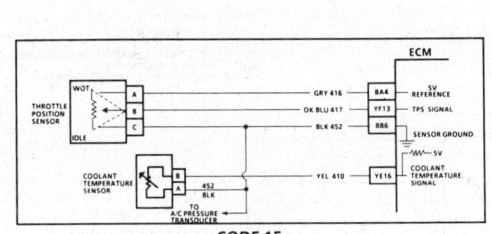

CODE 15
COOLANT TEMPERATURE SENSOR CIRCUIT
(LOW TEMPERATURE INDICATED)
3300 (VIN N) "A" CARLINE (PORT)

Circuit Description:

The coolant temperature sensor uses a thermistor to control the signal voltage to the ECM. The ECM applies a voltage on CKT 410 to the sensor. When the engine is cold, the sensor (thermistor) resistance is high, therefore, the ECM will see high signal voltage.

As the engine warms, the sensor resistance becomes less, and the voltage drops. At normal engine operating temperature (85°C to 95°C) the voltage will measure about 1.5 to 2.0 volts at the ECM.

Code 15 will set if:
- Engine is running.
- Signal voltage indicates a coolant temperature less than -38°C (-38°F) for at least .4 seconds.

Test Description: Numbers below refer to circled numbers on the diagnostic chart.

1. This will determine if the coolant sensor is indicating a low temperature to the ECM.
2. This test simulates a Code 14. If the ECM recognizes the low signal voltage (high temperature) and the "Scan" reads 140°C, the ECM and wiring are OK.
3. This test will determine if CKT 410 is open. There should be 5 volts present at sensor connector if measured with a DVM.

Note: If Code 15 is set, the ECM will use 49°C (120°F) for fuel control. "Scan" tool will read the actual value.

Diagnostic Aids:

A "Scan" tool reads engine temperature in degrees centigrade. After engine is started the temperature should rise steadily to about 90°C then stabilize when thermostat opens.

An intermittent may be caused by a poor connection, rubbed through wire insulation or a wire broken inside the insulation.

Check For:
- **Poor Connection or Damaged Harness** Inspect ECM harness connectors for backed out terminal "YE16", improper mating, broken locks, improperly formed or damaged terminals, poor terminal to wire connection and damaged harness.
- **Intermittent Test** If connections and harness check OK, "Scan" coolant temperature while moving related connectors and wiring harness. If the failure is induced, the display will change. This may help to isolate the location of the malfunction.
- **Shifted Sensor** The "Temperature To Resistance Value" scale may be used to test the coolant sensor at various temperature levels to evaluate the possibility of a "shifted" (mis-scaled) sensor which may result in driveability complaints.

A faulty connection, or an open in CKTs 410 or 452 will result in a Code 15.

1989–90 3.3L (VIN N) CUTLASS CIERA AND CUTLASS CRUISER

CODE 15
COOLANT TEMPERATURE SENSOR CIRCUIT
(LOW TEMPERATURE INDICATED)
3300 (VIN N) "A" CARLINE (PORT)

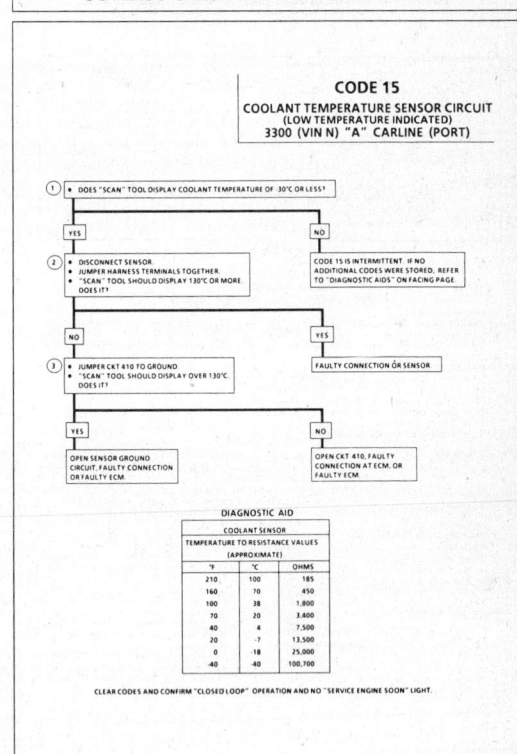

COOLANT SENSOR		
TEMPERATURE TO RESISTANCE VALUES (APPROXIMATE)		
°F	°C	OHMS
210	100	185
160	70	450
100	38	1,800
70	20	3,400
40	4	7,500
20	-7	13,500
0	-18	25,000
-40	-40	100,700

CLEAR CODES AND CONFIRM "CLOSED LOOP" OPERATION AND NO "SERVICE ENGINE SOON" LIGHT.

1989–90 3.3L (VIN N)
CUTLASS CIERA AND CUTLASS CRUISER

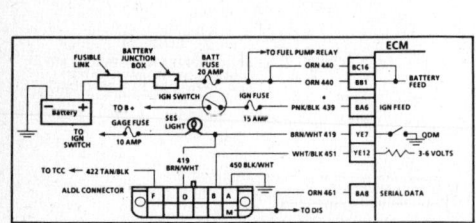

CODE 16
SYSTEM VOLTAGE HIGH
3300 (VIN N) "A" CARLINE (PORT)

Circuit Description:
The ECM monitors battery or system voltage on CKT 440 to terminals "BC16" and "BB1". If the ECM detects voltage above 16 volts for more than 10 seconds, it will turn the SES light "ON" and set Code 16 in memory.

Test Description: Numbers below refer to circled numbers on the diagnostic chart.
1. Test generator output as outlined in "6D3" to determine proper operation of the voltage regulator. Run engine at moderate speed and measure voltage across the battery. If over 16 volts, repair generator

Diagnostic Aids:

An intermittent may be caused by a poor connection, rubbed through insulation, a wire broken inside the insulation or poor ECM grounds.
Check For:
- **Poor Connection or Damaged Harness.** Inspect ECM harness connectors for backed out terminal "BC16" or "BB1", improper mating, broken locks, improperly formed or damaged terminals, poor terminal to wire connection and damaged harness.

- **Intermittent Test.** If connections and harness checks OK, monitor battery voltage display while moving related connectors. If the failure is induced, the battery voltage will abruptly change. This may help to isolate the location of the malfunction. An engine stall while manipulating the harness indicates that the ECM has lost voltage at terminal "BC16" or "BB1". Check for loose connectors in CKT 440.

Note: Charging battery with a battery charger and starting the engine may set a Code 16.

Important: If ECM and sensor grounds are located on the same stud at the transaxle as the battery ground you could have improper data to ECM or false codes.

1989–90 3.3L (VIN N)
CUTLASS CIERA AND CUTLASS CRUISER

CODE 16
SYSTEM VOLTAGE HIGH
3300 (VIN N) "A" CARLINE (PORT)

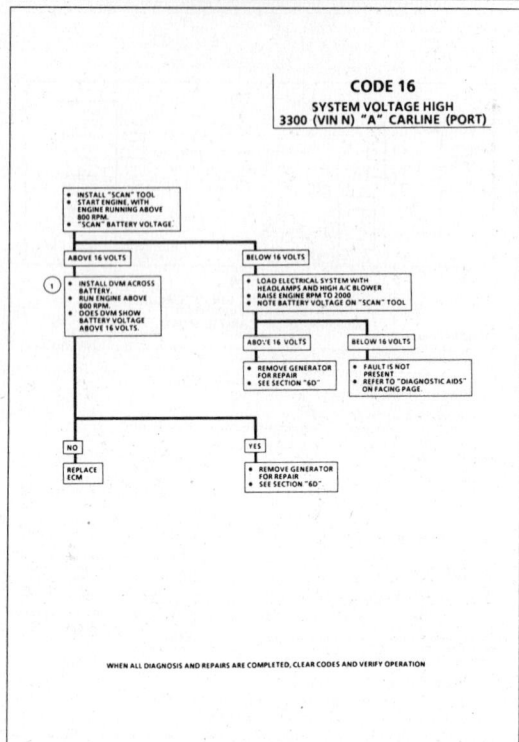

WHEN ALL DIAGNOSIS AND REPAIRS ARE COMPLETED, CLEAR CODES AND VERIFY OPERATION

1989–90 3.3L (VIN N)
CUTLASS CIERA AND CUTLASS CRUISER

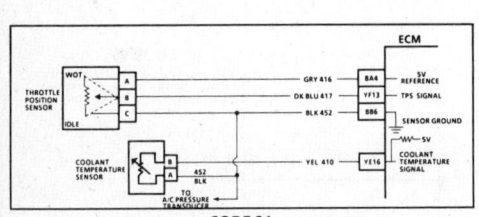

CODE 21
THROTTLE POSITION SENSOR (TPS) CIRCUIT
(SIGNAL VOLTAGE HIGH)
3300 (VIN N) "A" CARLINE (PORT)

Circuit Description:
The throttle position sensor (TPS) provides a voltage signal that changes relative to throttle blade angle. Signal voltage will vary from about 4 at idle to about 5 volts at wide open throttle.
The TPS signal is one of the most important inputs used by the ECM for fuel control and for most of the ECM control outputs.
Code 21 will set if:
- TPS signal voltage is greater than 4.9 volts at any time or
- if
- Engine is running and air flow is less than 15 gm/sec while
- TPS signal voltage is greater than 1.95 volts (30%)
- Code 34 not present
- All conditions met for 5 seconds

Test Description: Numbers below refer to circled numbers on the diagnostic chart.
1. With closed throttle, ignition "ON," or at idle, voltage at "YF13" should be .38-.42 volt.
2. When the TPS sensor is disconnected, the TPS voltage will go low and a Code 22 will set. Therefore, the ECM and wiring are OK.
3. Probing CKT 452 with a test light checks the sensor ground CKT. A faulty sensor ground circuit will cause a Code 21.
Note: If a Code 21 is set, the ECM will use a defaulted value for TPS of about .5 volt.

Diagnostic Aids:

A "Scan" tool reads throttle position in volts. With closed throttle, ignition "ON" or at idle, voltage should be .38-.42 volt.

Also some "Scan" tools will read throttle angle 0% = closed throttle 100% = WOT.

An open in CKT 452 will result in a Code 21. Refer to "Intermittents" in Section "B".
Check For:
- **Poor Connection or Damaged Harness.** Inspect ECM harness connectors for backed out terminal "YF13", improper mating, broken locks, improperly formed or damaged terminals, poor terminal to wire connection, and damaged harness.
- **Intermittent Test.** If connections and harness check OK, monitor TPS voltage while moving related connectors and wiring harness. If the failure is induced, the display will change. This may help to isolate the location of the malfunction.
- **TPS Scaling.** Observe TPS voltage display while depressing accelerator pedal with engine stopped and ignition "ON." Display should vary from closed throttle TPS voltage when throttle was closed, to over (4.5 volts) 4500 mV when throttle is held at wide open throttle position.

1989–90 3.3L (VIN N)
CUTLASS CIERA AND CUTLASS CRUISER

CODE 21
THROTTLE POSITION SENSOR (TPS) CIRCUIT
(SIGNAL VOLTAGE HIGH)
3300 (VIN N) "A" CARLINE (PORT)

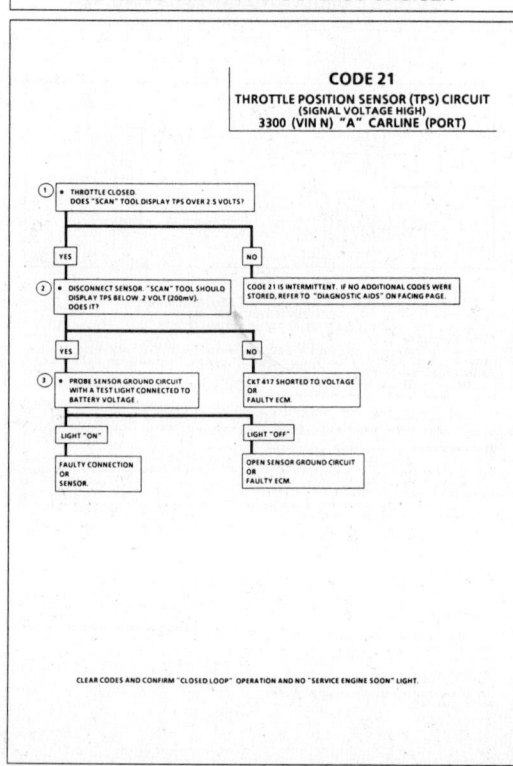

CLEAR CODES AND CONFIRM "CLOSED LOOP" OPERATION AND NO "SERVICE ENGINE SOON" LIGHT.

1989–90 3.3L (VIN N)
CUTLASS CIERA AND CUTLASS CRUISER

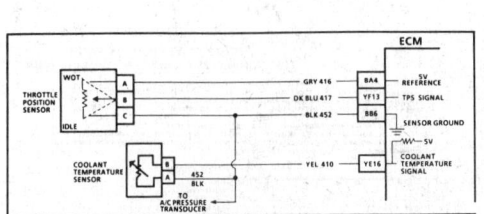

CODE 22
THROTTLE POSITION SENSOR (TPS) CIRCUIT
(SIGNAL VOLTAGE LOW)
3300 (VIN N) "A" CARLINE (PORT)

Circuit Description:

The throttle position sensor (TPS) provides a voltage signal that changes relative to throttle blade angle. Signal voltage will vary from about .4 at idle to about 5 volts at wide open throttle.

The TPS signal is one of the most important inputs used by the ECM for fuel control and for most of the ECM control outputs.

Code 22 will set if:
- The ignition key is "ON"
- TPS signal voltage is less than .24 volts for 4 seconds

Test Description: Numbers below refer to circled numbers on the diagnostic chart.
1. This, check what signal the TPS is indicating.
2. Simulates Code 21. (high voltage). If ECM recognizes the high signal voltage the ECM and wiring are OK.
3. With closed throttle, ignition "ON" or at idle, voltage at "YF13" should be .38-.42 volt. If not, check adjustment.
4. Simulates a high signal voltage. Checks CKT 417 for an open.

Diagnostic Aids:

A "Scan" tool reads throttle position in volts. Voltage should increase at a steady rate as throttle is moved toward WOT.

Also some "Scan" tools will read throttle angle 0% = closed throttle 100% = WOT.

An open or short to ground in CKTs 416 or 417 will result in a Code 22.

Check For
- Poor Connection or Damaged Harness. Inspect ECM harness connectors for backed out terminal "YF13", improper mating, broken locks, improperly formed or damaged terminals, poor terminal to wire connection, and damaged harness.
- Intermittent Test. If connections and harness check OK, monitor TPS voltage display while moving related connectors and wiring harness. If the failure is induced, the display will change. This may help to isolate the location of the malfunction.
- TPS Scaling. Observe TPS voltage display while depressing accelerator pedal with engine stopped and ignition "ON." Display should vary from closed throttle TPS when throttle was closed, to over 4.2 volts (4500 mV) when throttle is held at wide open throttle position.

1989–90 3.3L (VIN N)
CUTLASS CIERA AND CUTLASS CRUISER

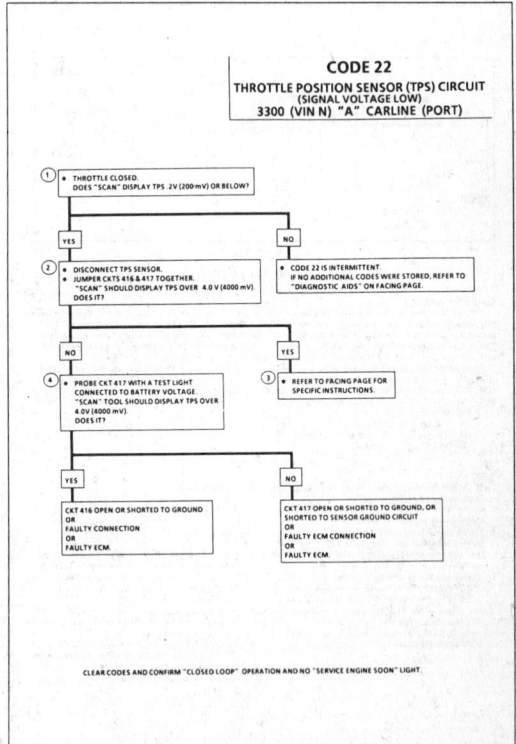

1989–90 3.3L (VIN N)
CUTLASS CIERA AND CUTLASS CRUISER

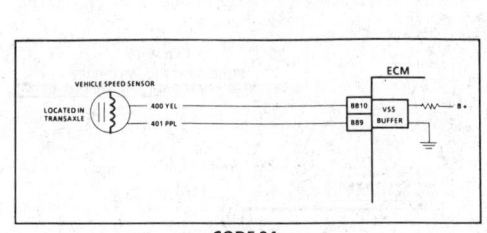

CODE 24
VEHICLE SPEED SENSOR (VSS) CIRCUIT
3300 (VIN N) "A" CARLINE (PORT)

Circuit Description:

Vehicle speed information is provided to the ECM by the vehicle speed sensor, a permanent magnet (PM) generator mounted in the transmission. The PM generator produces a pulsing voltage whenever vehicle speed is over about 3 mph. The A/C voltage level and the number of pulses increases with vehicle speed. The ECM converts the pulsing voltage to mph, and the mph can be displayed with a "Scan" tool.

The function of the VSS buffer, used in past model years, has been incorporated into the ECM. The ECM supplies the necessary signal for the instrument panel (4004 pulses per mile) for operating the speedometer and the odometer.

Code 24 will set if vehicle speed signal equals less than 3 mph when:
- Engine is running
- No Code 31
- When P/N switch is open (indicating drive range)
- When high gear switch is open
- All conditions met for 40 seconds

Test Description: Numbers below refer to circled numbers on the diagnostic chart.
1. The PM generator only produces a signal if drive wheels are turning greater then 3 mph.
2. Before replacing the ECM, check the Mem Cal for correct application.

Diagnostic Aids:

"Scan" should indicate a vehicle speed whenever the drive wheels are turning greater than 3 mph.

Check CKT 400 and 401 for proper connections to be sure they are clean and tight and the harness is routed correctly. Refer to "Intermittents" in Section "B".

1989–90 3.3L (VIN N)
CUTLASS CIERA AND CUTLASS CRUISER

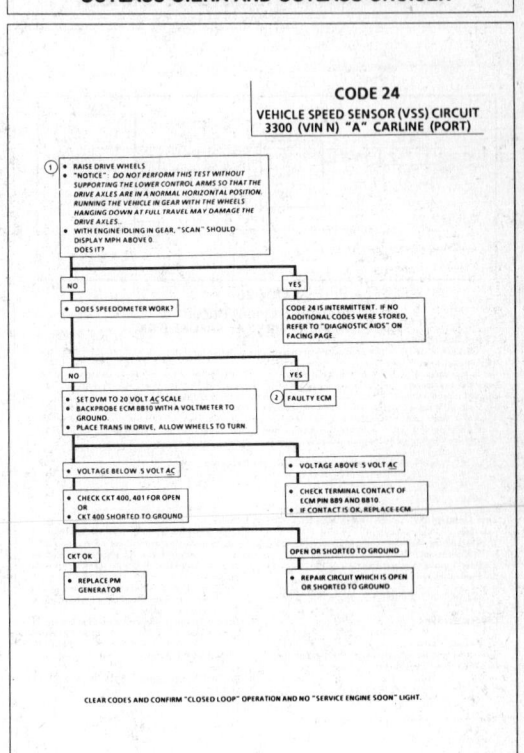

1989–90 3.3L (VIN N)
CUTLASS CIERA AND CUTLASS CRUISER

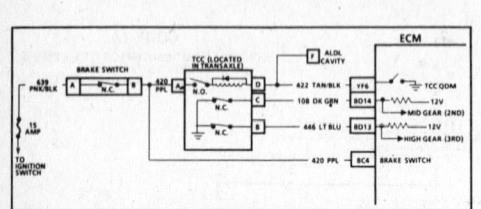

CODE 26
QUAD-DRIVER (QDM) CIRCUIT
3300 (VIN N) 125-C ONLY "A" CARLINE (PORT)

Circuit Description:

The ECM is used to control several components such as the torque converter clutch. The ECM controls the TCC through a device called a quad-driver module (QDM). When the ECM is commanding a component "ON", the voltage potential of the output circuit will be "low" (near 0 volt). When the ECM is commanding the output circuit to a component "OFF", the voltage potential of the circuit will be "high" (near battery voltage). The primary function of the QDM is to supply a ground for the device being controlled.

Each QDM has a fault line which is monitored by the ECM. The fault line signal is available on the data stream for "Scan" tool test equipment. The ECM will compare the voltage at the QDM based on accepted values of the fault line. If the QDM fault detection circuit senses a voltage other than the accepted value, the fault line will go from "low" signal on the data stream to a "high" signal and a Code 26 will set if applicable.

A Code 26 will be set if:
- A fault is detected by the QDM that controls the TCC circuit.
- The transmission is in mid or high gear.
- The brake is not applied.

Test Description: Numbers below refer to circled numbers on the diagnostic chart.
1. Light "OFF" confirms transmission second gear apply switch is open.
2. At 25 mph, the transmission second gear apply switch should close. Test light will come "ON" and confirm battery supply and closed brake switch.
3. Grounding the diagnostic terminal with ignition "ON", engine "OFF", should energize the TCC solenoid by grounding CKT 422. This test checks the ability of the ECM to supply a ground to the TCC solenoid. The test light connected from 12 volts to ALDL terminal "F" will turn "ON" as CKT 422 is grounded.

Diagnostic Aids:

An intermittent may be caused by a poor connection, mis-routed harness, rubbed through wire insulation, or a wire broken inside the insulation. Check For:
- **Poor Connection** at ECM pins. Inspect harness connectors for backed out terminals, improper mating, broken locks, improperly formed or damaged terminals, and poor terminal to wire connection.
- **Mis-routed Harness** Inspect wiring harness to insure that it is not too close to high voltage wires, such as spark plug leads.
- **Damaged Harness** Inspect harness for damage. If harness appears OK, "Scan" while moving related connectors and wiring harness. A change in display would indicate the intermittent fault location.

1989–90 3.3L (VIN N)
CUTLASS CIERA AND CUTLASS CRUISER

CODE 26
QUAD-DRIVER (QDM) CIRCUIT
3300 (VIN N) 125-C ONLY "A" CARLINE (PORT)

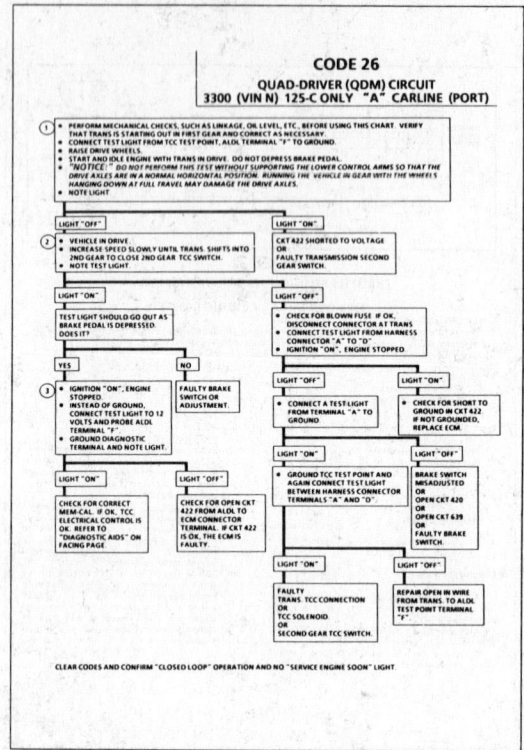

1989–90 3.3L (VIN N)
CUTLASS CIERA AND CUTLASS CRUISER

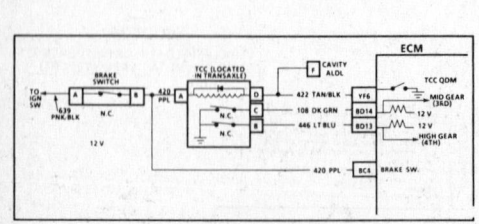

CODE 26
QUAD-DRIVER (QDM) CIRCUIT
3300 (VIN N) 440-T4 ONLY "A" CARLINE (PORT)

Circuit Description:

The ECM is used to control several components such as the torque converter clutch. The ECM controls the TCC through a device called a quad-driver module (QDM). When the ECM is commanding a component "ON", the voltage potential of the output circuit will be "low" (near 0 volt). When the ECM is commanding the output circuit to a component "OFF", the voltage potential of the circuit will be "high" (near battery voltage). The primary function of the QDM is to supply a ground for the device being controlled.

Each QDM has a fault line which is monitored by the ECM. The fault line signal is available on the data stream for "Scan" tool test equipment. The ECM will compare the voltage at the QDM based on accepted values of the fault line. If the QDM fault detection circuit senses a voltage other than the accepted value, the fault line will go from "low" signal on the data stream to a "high" signal and a Code 26 will set if applicable.

A Code 26 will be set if:
- A fault is detected by the QDM that controls the TCC circuit.
- The transmission is in mid or high gear.
- The brake is not applied.

Test Description: Numbers below refer to circled numbers on the diagnostic chart.
1. Checks fuse, brake switch and B+ circuit to the TCC solenoid.
2. Checks availability of B+ on CKT 420
3. Electrical circuits have checked out.

Diagnostic Aids:

An intermittent may be caused by a poor connection, mis-routed harness, rubbed through wire insulation, or a wire broken inside the insulation.

Check For:
- **Poor Connection** at ECM pins. Inspect harness connectors for backed out terminals, improper mating, broken locks, improperly formed or damaged terminals, and poor terminal to wire connection.
- **Mis-routed Harness** Inspect wiring harness to insure that it is not too close to high voltage wires, such as spark plug leads.
- **Damaged Harness** Inspect harness for damage. If harness appears OK, "Scan" while moving related connectors and wiring harness. A change in display would indicate the intermittent fault location.
- **System OK Electrically** Check for a hydraulic problem.

1989–90 3.3L (VIN N)
CUTLASS CIERA AND CUTLASS CRUISER

CODE 26
QUAD-DRIVER (QDM) CIRCUIT
3300 (VIN N) 440-T4 ONLY "A" CARLINE (PORT)

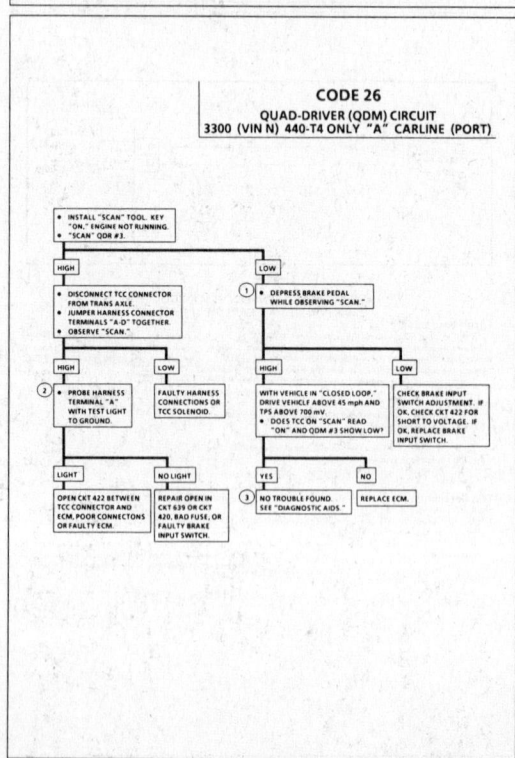

1989–90 3.3L (VIN N) CUTLASS CIERA AND CUTLASS CRUISER

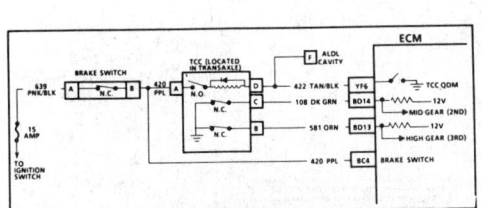

CODE 27, 28
GEAR SWITCH CIRCUITS
3300 (VIN N) 125-C ONLY "A" CARLINE (PORT)

Circuit Description:
The gear switches are located inside the transaxle. They are pressure operated switches, normally closed. The ECM supplies 12 volts through each selected circuit to the switch. As road speed increases, hydraulic pressure applies the specific gear clutches and the gear switch opens. At this time, the ECM monitors a high, 12 volt potential, and interprets this to indicate that gear is applied. The ECM uses the gear signals to control fuel delivery (and TCC).

Code 27 will set if:
- CKT 108 indicates ground or closed switch for 10 seconds when vehicle is in 3rd gear operation.
- CKT 108 indicates an open (drive) when the engine is first started.

Code 28 will set if:
- CKT 581 indicates ground or closed switch for 10 seconds when vehicle is in 3rd gear operation.
- CKT 581 indicates an open (drive) when the engine is first started.

Test Description: Numbers below refer to circled numbers on the diagnostic chart.
1. Must use a DVM. A test light will not light due to the very low current being supplied by the ECM.
2. Checks to see if circuit is grounded through the switch.
3. Checks for a good, properly operating switch and checks circuit within transaxle for an improper ground.

Diagnostic Aids:
An intermittent may be caused by a poor connection, mis-routed harness, rubbed through wire insulation, or a wire broken inside the insulation.

Check For:
- Poor Connection at ECM pins. Inspect harness connectors for backed out terminals, improper mating, broken locks, improperly formed or damaged terminals, and poor terminal to wire connection.
- Mis-routed Harness. Inspect wiring harness to insure that it is not too close to high voltage wires, such as spark plug leads.
- Damaged Harness. Inspect harness for damage. If harness appears OK, "Scan" while moving related connectors and wiring harness. A change in display would indicate the intermittent fault location.

1989–90 3.3L (VIN N) CUTLASS CIERA AND CUTLASS CRUISER

CODE 27, 28
GEAR SWITCH CIRCUITS
3300 (VIN N) 125-C ONLY "A" CARLINE (PORT)

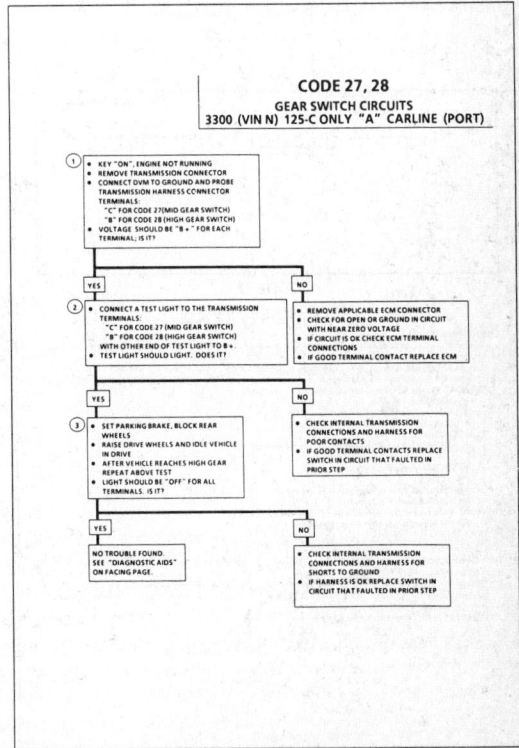

1989–90 3.3L (VIN N) CUTLASS CIERA AND CUTLASS CRUISER

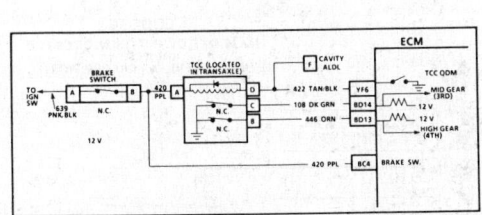

CODE 28, 29
GEAR SWITCH CIRCUITS
3300 (VIN N) 440-T4 ONLY "A" CARLINE (PORT)

Circuit Description:
The gear switches are located inside the transaxle. They are pressure operated switches, normally closed. The ECM supplies 12 volts through each selected circuit to the switch. As road speed increases, hydraulic pressure applies the specific gear clutches and the gear switch opens. At this time, the ECM monitors a high, 12 volt potential, and interprets this to indicate that gear is applied. The ECM uses the gear signals to control fuel delivery (and TCC).

Code 28 will set if:
- CKT 108 indicates ground or closed switch for 10 seconds when vehicle is in 4th gear operation.
- CKT 108 indicates an open (drive) when the engine is first started.

Code 29 will set if:
- CKT 446 indicates ground or closed switch for 10 seconds when vehicle is in 4th gear operation.
- CKT 446 indicates an open (drive) when the engine is first started.

Test Description: Numbers below refer to circled numbers on the diagnostic chart.
1. Must use a DVM. A test light will not light due to the very low current being supplied by the ECM.
2. Checks to see if circuit is grounded through the switch.
3. Checks for a good, properly operating switch and checks circuit within transaxle for an improper ground.

Diagnostic Aids:
An intermittent may be caused by a poor connection, mis-routed harness, rubbed through wire insulation, or a wire broken inside the insulation.

Check For:
- Poor Connection at ECM pins. Inspect harness connectors for backed out terminals, improper mating, broken locks, improperly formed or damaged terminals, and poor terminal to wire connection.
- Mis-routed Harness. Inspect wiring harness to insure that it is not too close to high voltage wires, such as spark plug leads.
- Damaged Harness. Inspect harness for damage. If harness appears OK, "Scan" while moving related connectors and wiring harness. A change in display would indicate the intermittent fault location.

1989–90 3.3L (VIN N) CUTLASS CIERA AND CUTLASS CRUISER

CODE 28, 29
GEAR SWITCH CIRCUITS
3300 (VIN N) 440-T4 ONLY "A" CARLINE (PORT)

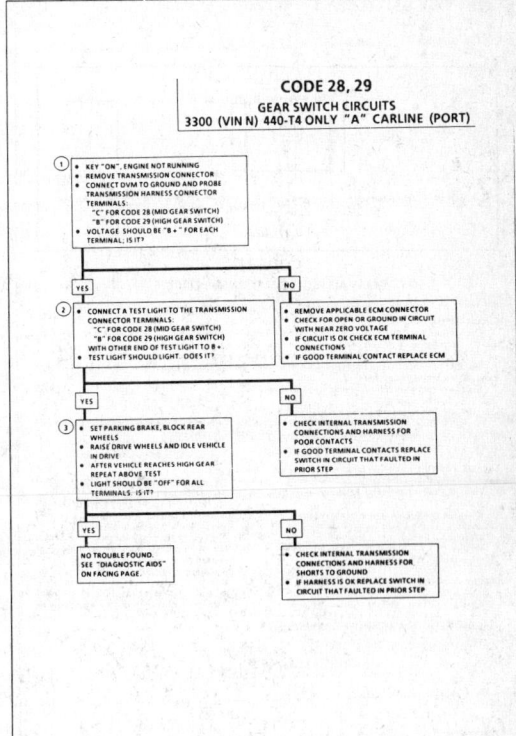

1989–90 3.3L (VIN N)
CUTLASS CIERA AND CUTLASS CRUISER

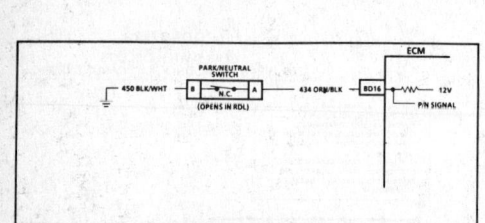

CODE 31
PARK/NEUTRAL SWITCH CIRCUIT
3300 (VIN N) "A" CARLINE (PORT)

Circuit Description:

The park/neutral switch contacts are a part of the neutral start switch and are closed to ground in park or neutral and open in drive ranges.

The ECM supplies ignition voltage through a current limiting resistor to CKT 434 and senses a closed switch when the voltage on CKT 434 drops to less than one volt.

The ECM uses the P/N signal as one of the inputs to control:
- Idle Speed (IAC)
- Vehicle Speed Sensor Diagnostics (VSS)
- Spark Advance

Code 31 will set if:
- CKT 434 indicates an open for 4 consecutive starts

or if:
- CKT 434 indicates a ground
- No Code 38
- Transmission is in high gear
- TCC is locked (4-speed only)
- VSS greater than 45 mph and TPS less than 15% (.94 volt) - 3 speed only
- Above conditions are met for 12 seconds

Test Description: Numbers below refer to circled numbers on the diagnostic chart.
1. Checks for a closed switch to ground in park position. Different makes of "Scan" tools will read P/N differently. Refer to "Tool Operator's" manual for type of display used for a specific tool.
2. Checks for an open switch in drive range.
3. Be sure "Scan" indicates drive, even while wiggling shifter, to test for an intermittent or misadjusted switch in drive or overdrive range.

1989–90 3.3L (VIN N)
CUTLASS CIERA AND CUTLASS CRUISER

CODE 31
PARK/NEUTRAL SWITCH CIRCUIT
3300 (VIN N) "A" CARLINE (PORT)

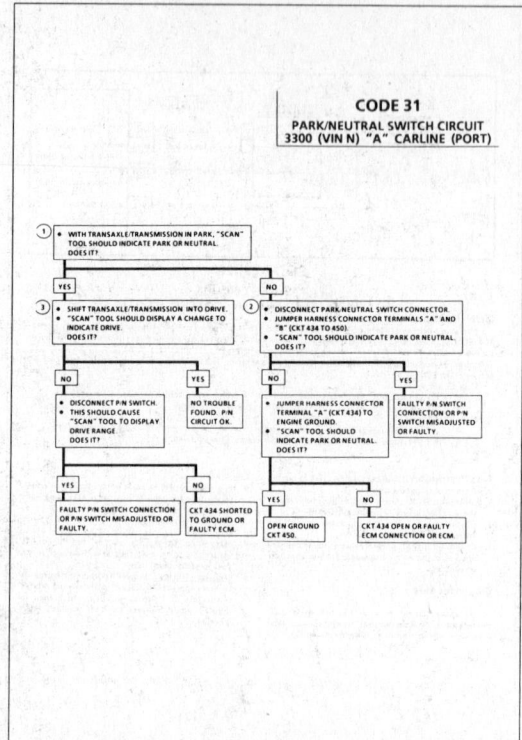

1989–90 3.3L (VIN N)
CUTLASS CIERA AND CUTLASS CRUISER

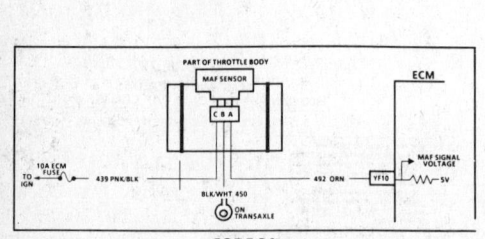

CODE 34
MASS AIR FLOW (MAF) SENSOR CIRCUIT
(GM/SEC LOW)
3300 (VIN N) "A" CARLINE (PORT)

Circuit Description:

The mass air flow (MAF) sensor measures the flow of air which passes through it in a given time. The ECM uses this information to monitor the operating condition of the engine for fuel delivery calculations. A large quantity of air movement indicates acceleration, while a small quantity indicates deceleration or idle. The MAF sensor produces a frequency signal which cannot be easily measured. The sensor can be diagnosed using the procedures on this chart.

Code 34 will set when of the following conditions exists:
- Engine running
- If MAF sensor signal frequency is less than 960 Hz

Note: If the MAF sensor signal frequency is too low (Code 34 is set), a substitute value for airflow is calculated based on engine rpm, TPS, and IAC motor position.

Test Description: Numbers below refer to circled numbers on the diagnostic chart.
1. This step checks to see if ECM recognizes a problem.
2. A voltage reading at sensor harness connector terminal "A" of less than 4 or over 6 volts indicates a fault in CKT 492 or poor connection.
3. Verifies that both ignition voltage and a good ground circuit are available.

Diagnostic Aids:

An intermittent may be caused by a poor connection, mis-routed harness, rubbed through wire insulation, or a wire broken inside the insulation.

Check For:
- **Poor connection** at ECM pin "YF10". Inspect harness connectors for backed out terminals, improper mating, broken locks, improperly formed or damaged terminals, and poor terminal to wire connection.
- **Mis-routed Harness** Inspect MAF sensor harness to insure that it is not too close to high voltage wires, such as spark plug leads.
- **Damaged Harness** Inspect harness for damage. If harness appears OK, "Scan" while moving related connectors and wiring harness. A change in display would indicate the intermittent fault location.

1989–90 3.3L (VIN N)
CUTLASS CIERA AND CUTLASS CRUISER

CODE 34
MASS AIR FLOW (MAF) SENSOR CIRCUIT
(GM/SEC LOW)
3300 (VIN N) "A" CARLINE (PORT)

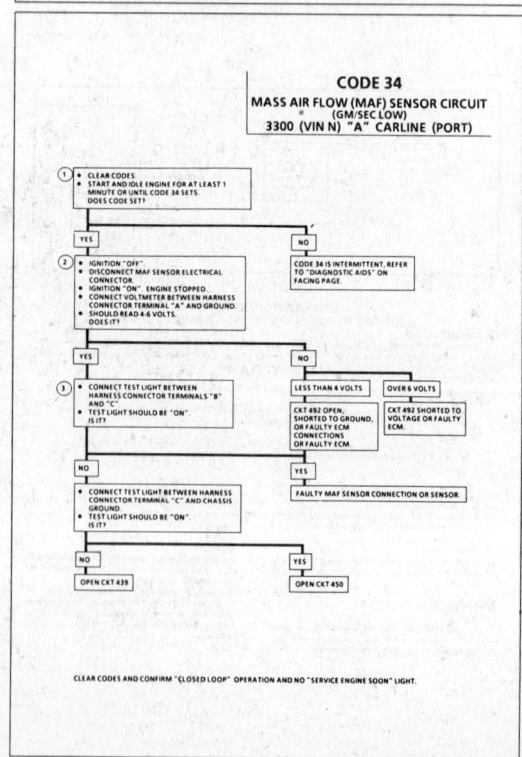

CLEAR CODES AND CONFIRM "CLOSED LOOP" OPERATION AND NO "SERVICE ENGINE SOON" LIGHT.

1989–90 3.3L (VIN N) CUTLASS CIERA AND CUTLASS CRUISER

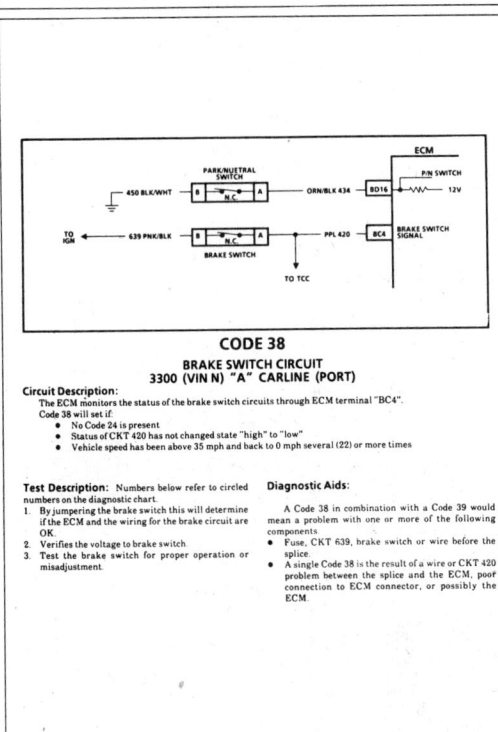

CODE 38
BRAKE SWITCH CIRCUIT
3300 (VIN N) "A" CARLINE (PORT)

Circuit Description:
The ECM monitors the status of the brake switch circuits through ECM terminal "BC4".
Code 38 will set if:
- No Code 24 is present
- Status of CKT 420 has not changed state "high" to "low"
- Vehicle speed has been above 35 mph and back to 0 mph several (22) or more times

Test Description: Numbers below refer to circled numbers on the diagnostic chart.
1. By jumpering the brake switch this will determine if the ECM and the wiring for the brake circuit are OK.
2. Verifies the voltage to brake switch.
3. Test the brake switch for proper operation or misadjustment.

Diagnostic Aids:
A Code 38 in combination with a Code 39 would mean a problem with one or more of the following components.
- Fuse, CKT 639, brake switch or wire before the splice.
- A single Code 38 is the result of a wire or CKT 420 problem between the splice and the ECM, poor connection to ECM connector, or possibly the ECM.

1989–90 3.3L (VIN N) CUTLASS CIERA AND CUTLASS CRUISER

CODE 38
BRAKE SWITCH CIRCUIT
3300 (VIN N) "A" CARLINE (PORT)

NOTICE: IF CODES 26, 38, & 39, ARE PRESENT, SEE CODE 26 CHART.

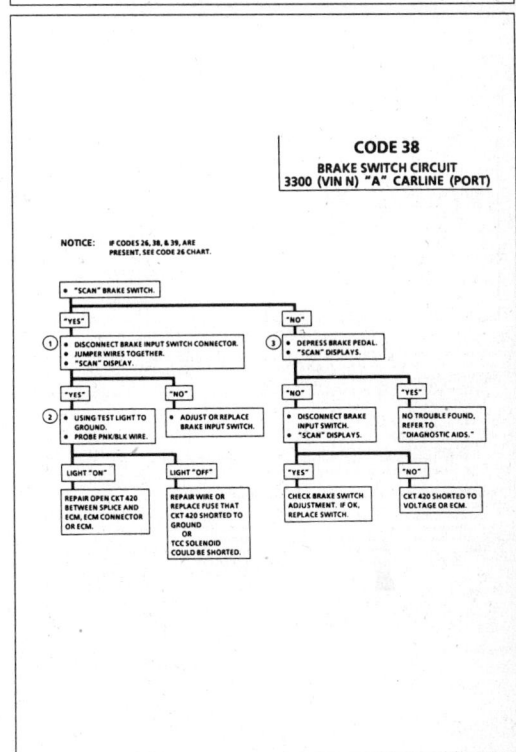

1989–90 3.3L (VIN N) CUTLASS CIERA AND CUTLASS CRUISER

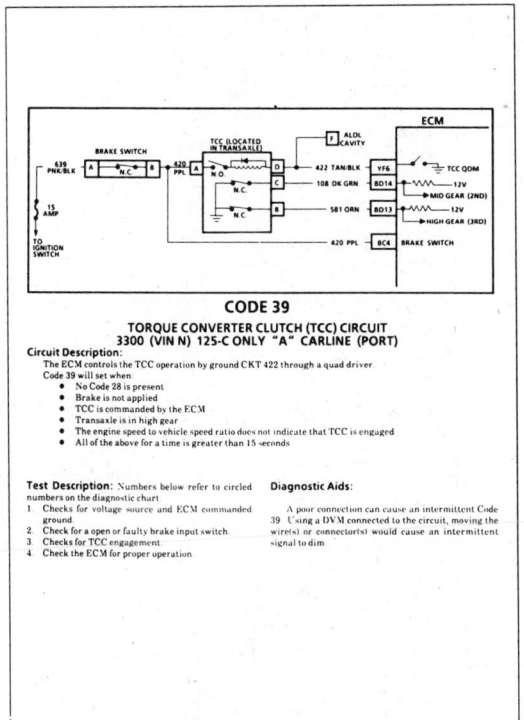

CODE 39
TORQUE CONVERTER CLUTCH (TCC) CIRCUIT
3300 (VIN N) 125-C ONLY "A" CARLINE (PORT)

Circuit Description:
The ECM controls the TCC operation by ground CKT 422 through a quad driver.
Code 39 will be set when:
- No Code 28 is present
- Brake is not applied
- TCC is commanded by the ECM
- Transaxle is in high gear
- The engine speed to vehicle speed ratio does not indicate that TCC is engaged
- All of the above for a time is greater than 15 seconds

Test Description: Numbers below refer to circled numbers on the diagnostic chart.
1. Checks for voltage source and ECM commanded ground.
2. Check for an open or faulty brake input switch.
3. Checks for TCC engagement.
4. Check the ECM for proper operation.

Diagnostic Aids:
A poor connection can cause an intermittent Code 39. Using a DVM connected to the circuit, moving the wire(s) or connector(s) would cause an intermittent signal to dim.

1989–90 3.3L (VIN N) CUTLASS CIERA AND CUTLASS CRUISER

CODE 39
TORQUE CONVERTER CLUTCH (TCC) CIRCUIT
3300 (VIN N) 125-C ONLY "A" CARLINE (PORT)

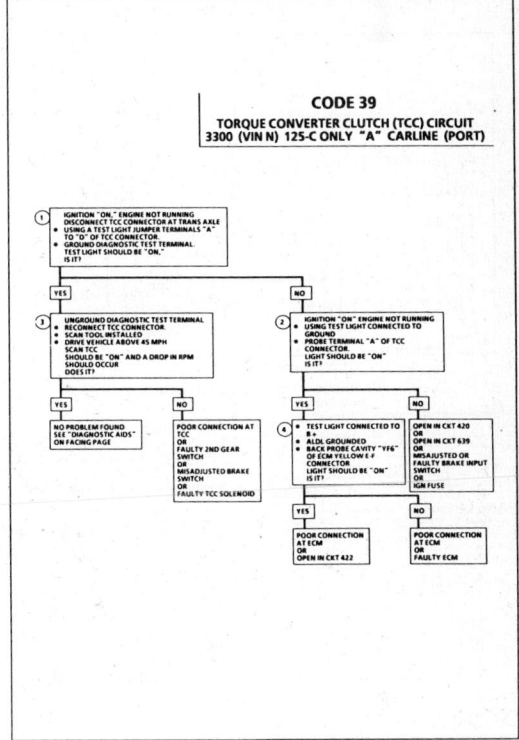

1989–90 3.3L (VIN N)
CUTLASS CIERA AND CUTLASS CRUISER

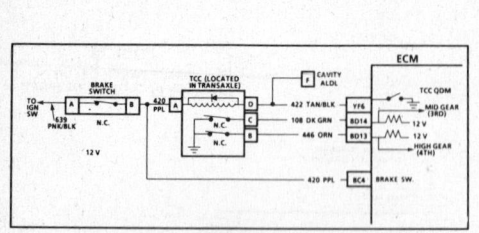

CODE 39
TORQUE CONVERTER CLUTCH (TCC) CIRCUIT
3300 (VIN N) 440-T4 ONLY "A" CARLINE (PORT)

Circuit Description:
The ECM controls TCC operation by grounding CKT 422 through the quad-driver.
Code 39 will set when
- No Code 28 & 29 is present
- Brake switch is closed, "OFF"
- TCC is commanded by the ECM
- Vehicle is in high gear (4th)
- The engine speed to vehicle speed ratio does not indicate that TCC is engaged
- All of the above for a time greater than 15 seconds

Test Description: Numbers below refer to the circled numbers on the diagnostic chart.
1. Checks fuse, brake switch and B+ circuit to the TCC solenoid.
2. Checks availability of B+ on CKT 420.
3. Electrical circuits have checked out.

Diagnostic Aids:

A Code 39 in combination with a Code 38 would mean a problem with one or more of the following components. Fuse, CKT 639, brake switch or wire before the splice. A single Code 39 is the result of a wire or CKT 420 problem between the splice and the TCC solenoid, CKT 422 problem between TCC solenoid and ECM, poor connection to ECM connector, or possibly the ECM.

1989–90 3.3L (VIN N)
CUTLASS CIERA AND CUTLASS CRUISER

CODE 39
TORQUE CONVERTER CLUTCH (TCC) CIRCUIT
3300 (VIN N) 440-T4 ONLY "A" CARLINE (PORT)

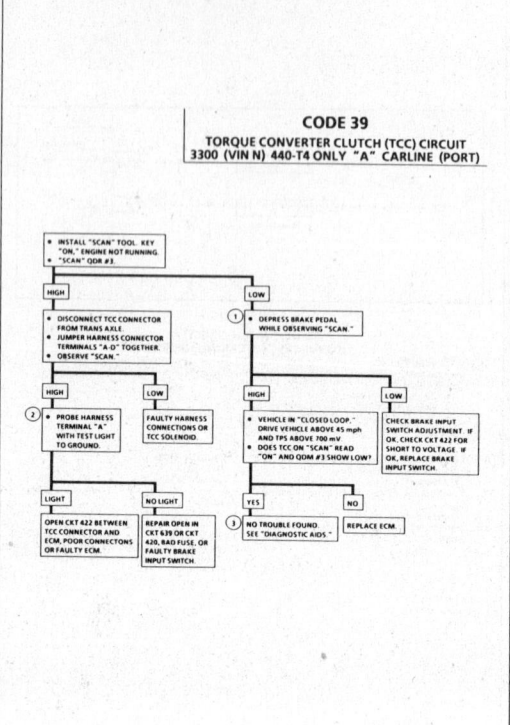

1989–90 3.3L (VIN N)
CUTLASS CIERA AND CUTLASS CRUISER

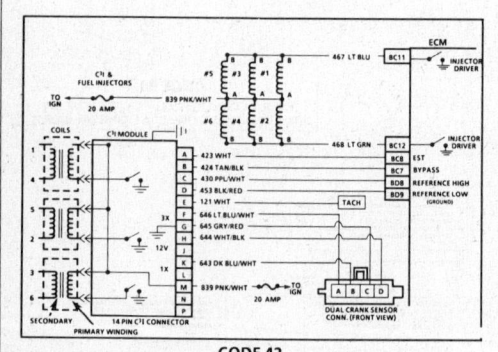

CODE 42
ELECTRONIC SPARK TIMING (EST) CIRCUIT
3300 (VIN N) "A" CARLINE (PORT)

Circuit Description:
The C3I module sends a reference signal to the ECM when the engine is cranking. Under 400 rpm, the C3I module controls ignition timing. When the engine speed exceeds 400 rpm, the ECM sends a 5 volt signal on the bypass CKT 424 to switch timing to ECM control. If an open or ground occurs while engine is running in EST mode, the engine will stall and set a Code 42. If CKT 424 is open or grounded while engine is cranking, the engine will start and remain in backup.
To set a Code 42 the following conditions must be met.
- Engine speed greater than 600 rpm with no EST pulse for 200 mS (open or grounded CKT 423), or
- ECM commanding bypass mode (open or grounded CKT 424)

Test Description: Numbers below refer to circled numbers on the diagnostic chart.
1. Checks to see if ECM recognizes a problem. If it does not set Code 42, it is an intermittent problem and could be due to a loose connection.
2. With the ECM disconnected, the ohmmeter should be reading less than 200 ohms, which is the normal resistance of the EST circuit through the C3I module. A higher resistance would indicate a fault in CKT 423, a poor C3I module connection, or a faulty C3I module.
3. If test light was "ON" when connected from 12 volts to ECM harness terminal "BC7", and CKT 424 is shorted to ground or the C3I module is faulty.
4. Checks to see if C3I module switches when the bypass circuit is energized by 12 volts through the test light. If the C3I module actually switches, the ohmmeter reading should shift to over 6,000 ohms.

5. Disconnecting the ignition module should make the ohmmeter read as if it were monitoring an open circuit (infinite reading). Otherwise, CKT 423 is shorted to ground.

Diagnostic Aids:
An intermittent may be caused by a poor connection, rubbed through wire insulation, or a wire broken inside the insulation. Check For:
- **Poor Connection or Damaged Harness** Inspect ECM harness connectors for backed out terminals "BC7" or "BC8", improper mating, broken locks, improperly formed or damaged terminals, poor terminal to wire connection, and damaged harness.
- **Intermittent Test** If connections and harness check OK, a digital voltmeter connected from affected terminal to ground while moving related connectors and wiring harness. If the failure is induced, the voltage reading will change.

1989–90 3.3L (VIN N)
CUTLASS CIERA AND CUTLASS CRUISER

CODE 42
ELECTRONIC SPARK TIMING (EST) CIRCUIT
3300 (VIN N) "A" CARLINE (PORT)

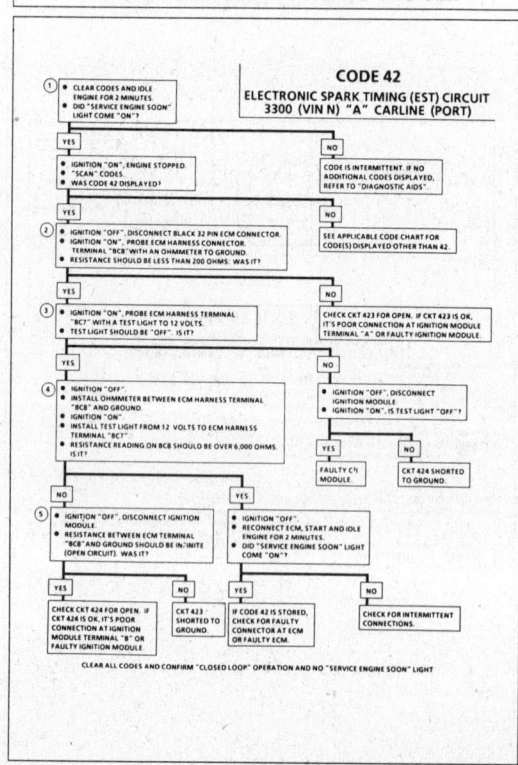

1989–90 3.3L (VIN N)
CUTLASS CIERA AND CUTLASS CRUISER

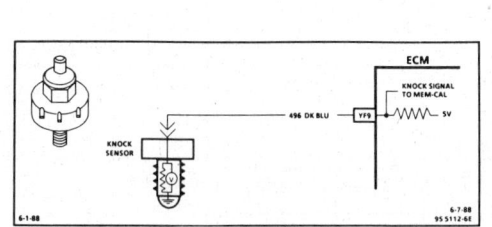

CODE 43
ELECTRONIC SPARK CONTROL (ESC) CIRCUIT
3300 (VIN N) "A" CARLINE (PORT)

Circuit Description:

The knock sensor is used to detect engine detonation and the ECM will retard the electronic spark timing based on the signal being received. The circuitry within the knock sensor causes the ECM's supplied 5 volt signal to be pulled down, so that under a no knock condition CKT 496 would measure about 2.5 volts. The knock sensor produces an A/C signal which rides on the 2.5 volts DC voltage. The amplitude and signal frequency is dependent upon the knock level.

If CKT 496 becomes open or shorted to ground, the voltage will either go above 3.5 volts or below 1.5 volts. If either of these conditions are met for 20 seconds, a Code 43 will be stored.

Code 43 will set if:
- Voltage on CKT 496 goes above 3.5 volts or below 1.5 volts
- Condition present for 20 seconds

Test Description: Numbers below refer to circled numbers on the diagnostic chart.
1. If a Code 43 is detected, the ECM will retard spark timing 10 degrees.
   If an audible knock is heard from the engine, repair the internal engine problem, normally no knock should be detected at idle.
2. If tapping on the engine lift hook does not produce a knock signal, try tapping engine closer to sensor before proceeding.

3. The ECM has a 5 volt pull-up resistor which should be present at the knock sensor terminal.
4. This test determines if the knock sensor is faulty or if the ESC portion of the Mem-Cal is faulty.

Diagnostic Aids:

Check CKT 496 for a potential open or short to ground. Also check for proper installation of Mem-Cal.

Refer to "Intermittents" in Section "B".

1989–90 3.3L (VIN N)
CUTLASS CIERA AND CUTLASS CRUISER

CODE 43
ELECTRONIC SPARK CONTROL (ESC) CIRCUIT
3300 (VIN N) "A" CARLINE (PORT)

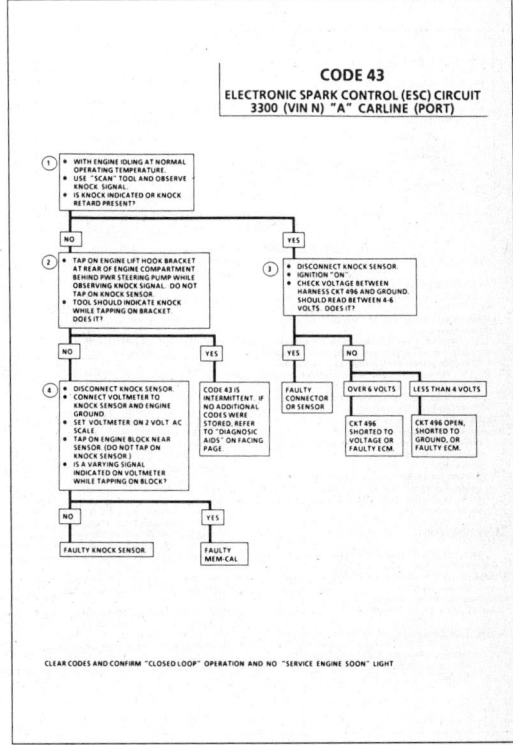

1989–90 3.3L (VIN N)
CUTLASS CIERA AND CUTLASS CRUISER

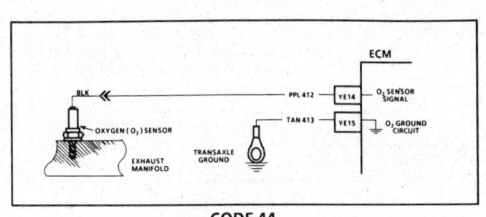

CODE 44
OXYGEN SENSOR CIRCUIT
(LEAN EXHAUST INDICATED)
3300 (VIN N) "A" CARLINE (PORT)

Circuit Description:

The ECM supplies a voltage of about .45 volt (450 mV) between terminals "YE14" and "YE15". (If measured with a 10 megohm digital voltmeter, this may read as low as .32 volt.) The O_2 sensor varies the voltage within a range of about 1 volt, (1000 mV) if the exhaust is rich, down through about .10 volt (100 mV) if exhaust is lean.

The sensor is like an open circuit and produces no voltage when it is below about 360°C (600°F). An open sensor circuit or cold sensor causes "Open Loop" operation.

Code 44 is set when the O_2 sensor signal voltage on CKT 412
- Remains below .25 volt for up to 4.5 minutes
- The system is operating in "Closed Loop"

Test Description: Numbers below refer to circled numbers on the diagnostic chart.
1. Running the engine at 1200 rpm keeps the O_2 sensor hot, so an accurate display voltage is maintained.
   Opening the O_2 sensor wire should result in a voltage display of between 350 and 550 mV. If the display is still fixed below 350 mV, the fault is a short to ground in CKT 412 or the ECM is faulty.

Diagnostic Aids:

Using the "Scan", observe the block learn values at different rpm and air flow conditions. The "Scan" also displays the block cells, so the block learn values can be checked in each of the cells to determine when the Code 44 may have been set. If the conditions for Code 44 exists, the block learn values will be around 150.
- O_2 Sensor Wire Sensor pigtail may be mispositioned and contacting the exhaust manifold.
- Check for intermittent ground in wire between connector and sensor.

- MAF Sensor A mass air flow (MAF) sensor output that causes the ECM to sense a lower than normal air flow will cause the system to go lean. Disconnect the MAF sensor and if the lean condition is gone, replace the MAF sensor
- Lean Injector(s) Perform injector balance test CHART C-2A
- Fuel Contamination Water, even in small amounts, near the in-tank fuel pump inlet can be delivered to the injectors. The water causes a lean exhaust and can set a Code 44
- Fuel Pressure System will be lean if pressure is too low. It may be necessary to monitor fuel pressure while driving the car at various road speeds and/or loads to confirm. See "Fuel System Diagnosis", CHART A-7.
- Exhaust Leaks If there is an exhaust leak, the engine can cause outside air to be pulled into the exhaust and past the sensor
- If the above are OK, it is a faulty oxygen sensor
- Vacuum or crankcase leaks can cause a lean condition.

1989–90 3.3L (VIN N)
CUTLASS CIERA AND CUTLASS CRUISER

CODE 44
OXYGEN SENSOR CIRCUIT
(LEAN EXHAUST INDICATED)
3300 (VIN N) "A" CARLINE (PORT)

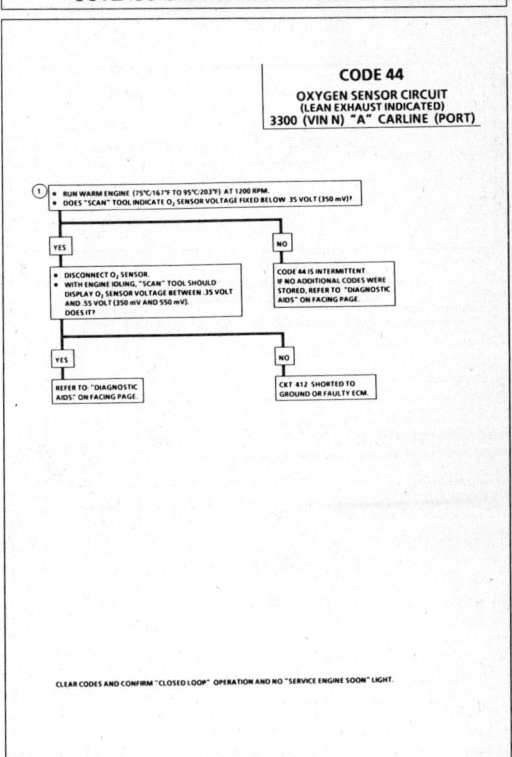

1989–90 3.3L (VIN N)
CUTLASS CIERA AND CUTLASS CRUISER

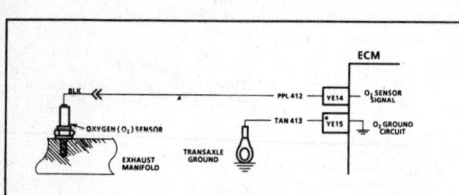

CODE 45
OXYGEN SENSOR CIRCUIT
(RICH EXHAUST INDICATED)
3300 (VIN N) "A" CARLINE (PORT)

Circuit Description:
The ECM supplies a voltage of about .45 volt (450 mV) between terminals "YE14" and "YE15". (If measured with a 10 megohm digital voltmeter, this may read as low as .32 volt.) The O_2 sensor varies the voltage within a range of about 1 volt (1000 mV) if the exhaust is rich, down through about .10 volt (100 mV) if exhaust is lean.

The sensor is like an open circuit and produces no voltage when it is below about 360°C (600°F). An open sensor circuit or cold sensor causes "Open Loop" operation.

Code 45 is set when the O_2 sensor signal voltage or CKT 412:
- remains above .75 volt for 2 minutes and in "Closed Loop"
- Throttle angle between .52 and 1.1 volts
- Engine time after start is 1 minute or more
- No Code 21 or Code 22

Test Description: Numbers below refer to circled numbers on the diagnostic chart.
1. Running the engine at 1200 rpm keeps the O_2 sensor hot, so an accurate display voltage is maintained.
   Opening the O_2 sensor wire should result in a voltage display of between 350 and 550 mV. If the display is still fixed below .350 mV, the fault is a short to ground in CKT 412 or the ECM is faulty.

Diagnostic Aids:

Using the "Scan," observe the block learn values at different rpm and air flow conditions. The "Scan" also displays the block cells, so the block learn values can be checked in each of the cells to determine when the Code 45 may have been set. If the conditions for Code 45 exists, the block learn values will be around 115.
- **Fuel Pressure** System will go rich if pressure is too high. The ECM can compensate for some increase. However, if it gets too high, a Code 45 may be set. See "Fuel System Diagnosis," CHART A-7.

- **Rich Injector** Perform injector balance test CHART C-2A.
- **Leaking Injector** See CHART A-7.
- Check for fuel contaminated oil.
- **Canister Purge** Check for fuel saturation. If full of fuel, check canister control and hoses.

- **MAF Sensor** An output that causes the ECM to sense a higher than normal airflow can cause the system to go rich. Disconnecting the MAF sensor will allow the ECM to set a fixed value for the sensor. Substitute a different MAF sensor if the rich condition is gone while the sensor is disconnected.
- Check for leaking fuel pressure regulator diaphragm by checking vacuum line to regulator for fuel.
- **TPS** An intermittent TPS output will cause the system to go rich due to a false indication of the engine accelerating.
- **EGR** An EGR staying open (especially at idle) will cause the O_2 sensor to indicate a rich exhaust and this could result in a Code 45.

1989–90 3.3L (VIN N)
CUTLASS CIERA AND CUTLASS CRUISER

CODE 45
OXYGEN SENSOR CIRCUIT
(RICH EXHAUST INDICATED)
3300 (VIN N) "A" CARLINE (PORT)

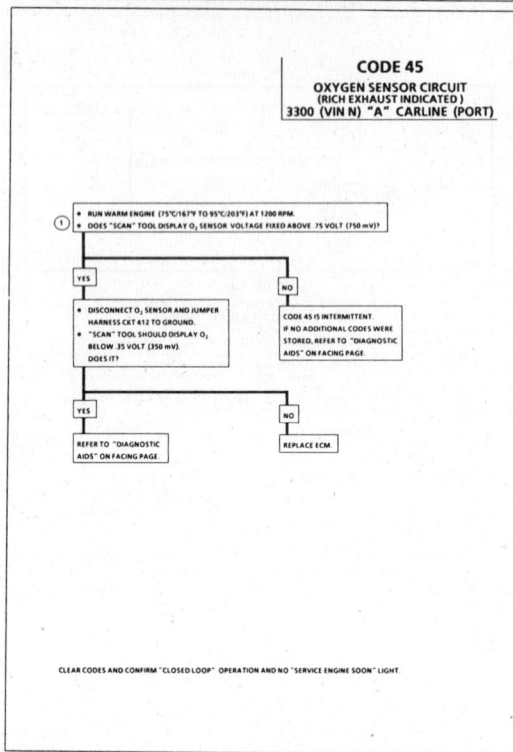

CLEAR CODES AND CONFIRM "CLOSED LOOP" OPERATION AND NO "SERVICE ENGINE SOON" LIGHT.

1989–90 3.3L (VIN N)
CUTLASS CIERA AND CUTLASS CRUISER

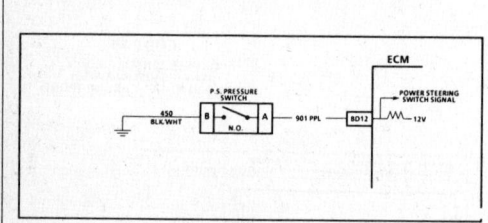

CODE 46
POWER STEERING PRESSURE SWITCH (PSPS) CIRCUIT
3300 (VIN N) "A" CARLINE (PORT)

Circuit Description:
The power steering pressure switch is normally open, and CKT 901 will be near battery voltage. Turning the steering wheel increases power steering oil pressure and its load on an idling engine. The pressure switch will close before the load can cause an idle problem.

Closing the switch causes CKT 901 to read less than 1 volt. The electronic control module (ECM) will increase the idle air rate and disengage the A/C clutch.
- A pressure switch that will not close, or an open CKT 901 or 450, may cause the engine to stop when power steering loads are high.
- A switch that will not open, or a CKT 901 shorted to ground, may affect idle quality and will cause the A/C relay to be de-energized.

To set a Code 46, the following conditions must be met:
- Closed power steering switch. "LO" (low voltage potential)
- Vehicle speed is greater than 40 mph
- Both conditions existing for a time greater than 25 seconds

Test Description: Numbers refer to circled numbers on the diagnostic chart.
1. Different makes of "Scan" tools may display the state of this switch in different ways. Refer to "Scan" tool operator's manual to determine how this input is indicated.

2. Checks to determine if CKT 901 is shorted to ground.
3. This should simulate a closed switch.

1989–90 3.3L (VIN N)
CUTLASS CIERA AND CUTLASS CRUISER

CODE 46
POWER STEERING PRESSURE SWITCH (PSPS) CIRCUIT
3300 (VIN N) "A" CARLINE (PORT)

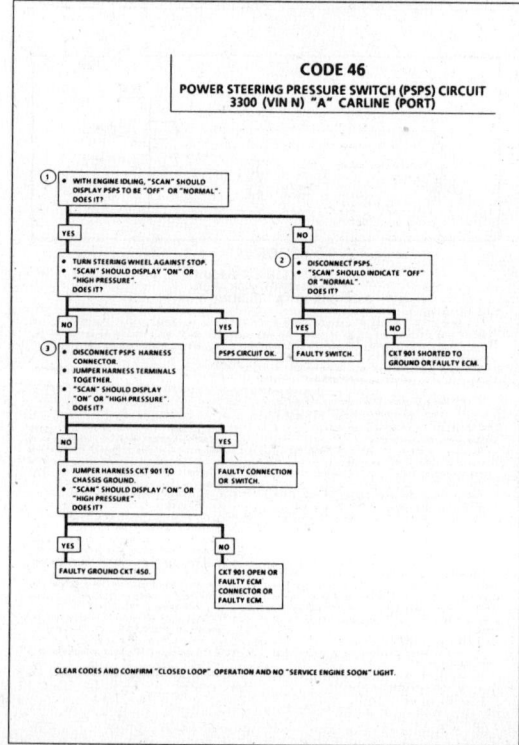

CLEAR CODES AND CONFIRM "CLOSED LOOP" OPERATION AND NO "SERVICE ENGINE SOON" LIGHT.

1989–90 3.3L (VIN N)
CUTLASS CIERA AND CUTLASS CRUISER

CODE 48
MISFIRE DIAGNOSIS
3300 (VIN N) "A" CARLINE (PORT)

If multiple codes are set, go to the lowest code first.
Repairing for a Code 13, 44, or 45 may correct Code 48.

Test Description:
Code 48 will set if the following:
- TPS is between .58 and .93 volts
- Rpm is between 1500 and 2500
- Mph is between 50 and 60
- O_2 cross counts greater than 32-125C and 26-440 T_4
- All of the above for 30 seconds

O_2 Sensor Test:
Code 48 could be set if the O_2 sensor is degraded and cannot travel over the full rich to lean voltage range. This narrowed range could allow O_2 cross counts to be above the value necessary to set the code.

> - WITH "SCAN" TOOL INSTALLED, VERIFY ENGINE IS AT NORMAL OPERATING TEMPERATURE AND IN "CLOSED LOOP".

> - ENGINE IDLING IN PARK.
> - SELECT O_2 SENSOR POSITION ON "SCAN".
> - RAPIDLY FLASH THE THROTTLE FROM IDLE TO NEAR WIDE OPEN THROTTLE AND BACK WHILE OBSERVING O_2 VOLTAGE.
> - REPEAT IF NECESSARY TO CONFIRM VOLTAGE RANGE, AND "CLOSED LOOP".

VOLTAGE EXCEEDS 250-750 mV RANGE.	VOLTAGE REMAINS WITHIN 250-750 mV RANGE.
O_2 SENSOR OK, SEE "DIAGNOSTIC AIDS".	REPLACE O_2 SENSOR.

Diagnostic Aids:

1. **Ignition system checks:**
   Remove each spark plug and inspect (fouled, cracked, worn)
   Fouled -- check ignition wires (high resistance, damage, poor connections, grounds)
      check coil and module operation (see .... CHART C-4)
      check basic engine problem (see 3 below)
   Cracked or worn -- replace as necessary

2. **Fuel system checks:**
   Restricted fuel system (injectors, fuel pump, lines, and filter)
   Injectors -- perform injector balance test (..................... CHART C-2)
      verify each injector circuit with tool J-34730-2 or equivalent
   Fuel Pump -- verify proper fuel pressure and fuel quality
   Lines and Filter -- verify no restrictions in lines or filter

3. **Basic Engine Checks:**
   Unless spark plug(s) inspection identifies a specific cylinder(s), road test vehicle under test conditions to reverify Code 48 prior to engine disassembly.
   Basic engine (valves, compression, camshaft, lifters)
   Compression -- check rings, pistons, valves
   Valves -- check for burned, weak springs, broken parts, worn or loose guide
   Camshaft -- check for worn or broken
   Lifters -- check for worn, broken
      For additional items see Section "B" under "Rough Unstable Idle", "Hard Start", or "Hesitation, Sag, or Stumble".

1989–90 3.3L (VIN N)
CUTLASS CIERA AND CUTLASS CRUISER

CODE 51
MEM-CAL ERROR
(FAULTY OR INCORRECT MEM-CAL)
3300 (VIN N) "A" CARLINE (PORT)

> CHECK THAT ALL PINS ARE FULLY INSERTED IN THE SOCKET. IF OK, REPLACE MEM-CAL, CLEAR MEMORY AND RECHECK. IF CODE 51 REAPPEARS, REPLACE ECM.

NOTICE: To prevent possible Electrostatic Discharge damage to the ECM or MEM-CAL, Do Not touch the component leads, and Do Not remove integrated circuit from carrier.

1989–90 3.3L (VIN N)
CUTLASS CIERA AND CUTLASS CRUISER

SECTION B
SYMPTOMS
TABLE OF CONTENTS

Before Starting
Intermittents
Hard Start
Hesitation, Sag, Stumble
Surges and/or Chuggle
Lack of Power, Sluggish, or Spongy
Detonation/Spark Knock
Cuts Out, Misses
Backfire
Poor Fuel Economy
Dieseling, Run-On
Rough, Unstable, or Incorrect Idle, Stalling
Excessive Exhaust Emissions Or Odors
Chart B-1 - Restricted Exhaust System Check

BEFORE STARTING

Before using this section you should have performed the DIAGNOSTIC CIRCUIT CHECK and found out that:
1. The ECM and "Service Engine Soon" light are operating.
2. There are no trouble codes stored, or there is a trouble code but no "Service Engine Soon" light.
3. The fuel control system is operating OK (by performing field service mode check).
   Verify the customer complaint, and locate the correct SYMPTOM. Check the items indicated under that symptom.
   If the ENGINE CRANKS BUT WILL NOT RUN, see CHART A-3.

Several of the symptom procedures below call for a careful visual/physical check. This check should include:
- ECM B+ wires for being clean and tight at starter and/or junction block.
- ECM grounds for being clean and tight.
- Vacuum hoses for splits, kinks, and proper connections, as shown on emission control information label.
- Air leaks at throttle body mounting and intake manifold.
- Ignition wires for cracking, hardness, proper routing, and carbon tracking.
- Wiring for proper connections, pinches, and cuts.
   The importance of this step cannot be stressed too strongly - it can lead to correcting a problem without further checks and can save valuable time.

1989–90 3.3L (VIN N)
CUTLASS CIERA AND CUTLASS CRUISER

INTERMITTENTS

Problem may or may not turn "ON" the "Service Engine Soon" light or store a code.

DO NOT use the trouble charts in Section "A" for intermittent problems. The fault must be present to locate the problem. If a fault is intermittent, use of trouble code charts may result in replacement of good parts.
- Most intermittent problems are caused by faulty electrical connections or wiring. Perform a careful check as described at start of this section. Check for:
  - Poor mating of the connector halves, or terminals not fully seated in the connector body (backed out).
  - Improperly formed or damaged terminals. All connector terminals in problem circuit should be carefully reformed to increase contact tension.
  - Poor terminal to wire connection. This requires removing the terminal from the connector body to check.
- If a visual/physical check does not find the cause of the problem, the car may be driven with a voltmeter connected to a suspected circuit. An abnormal voltage reading, when the problem occurs, indicates the problem may be in that circuit. If the wiring and connectors check OK and a trouble code was stored for a circuit having a sensor, except for Codes 43, 44, and 45, substitute a known good sensor and recheck.

An intermittent "Service Engine Soon" light with no stored code may be caused by:
- Check that the ECM and sensor ground wires are not located on the same stud at the transaxle as the battery ground.
- Ignition coil shorted to ground, arcing at spark plug wires or plugs.
- "Service Engine Soon" light wire to ECM shorted to ground (CKT 419).
- Diagnostic "Test" terminal wire to ECM shorted to ground (CKT 451).
- ECM power grounds. See ECM wiring diagrams.
- Loss of trouble code memory. To check, disconnect TPS and idle engine until "Service Engine Soon" light comes "ON." Code 22 should be stored, and kept in memory when ignition is turned "OFF". If not, the ECM is faulty.
- Check for an electrical system interference caused by a defective relay, ECM driven solenoid, or switch. They can cause a sharp electrical surge. Normally, the problem will occur when the faulty component is operated.
- Check for improper installation of electrical options, such as lights, 2-way radios, etc.
- EST wires should be kept away from spark plug wires, coil and generator.
- Check for open diode across A/C compressor clutch, and for other open diodes (see wiring diagrams).

HARD START

Definition: Engine cranks OK, but does not start for a long time. Does eventually run, or may start but immediately dies.

- Perform careful check as described at start of Section "B".
- Make sure driver is using correct starting procedure.
- CHECK:
  - TPS for sticking or binding or a high TPS voltage with the throttle closed.
  - High resistance in coolant sensor circuit or sensor itself. See CODE 15 CHART or with a "Scan" tool compare coolant temperature with ambient temperature on a cold engine.
  - Fuel pressure for being too low.
  - Water contaminated fuel.
  - Fuel pump relay - See CHART A-5.
  - PCV valve for hose off or damaged valve.

Ignition system - Check for:
   Proper output with ST-125
   Bare and shorted wires.
   Loose C3I module ground, mounting screws.
   Dual crank sensor for damage from crankshaft pulley vanes rubbing.
- A faulty in-tank fuel pump check valve will allow the fuel in the lines to drain back to the tank after the engine is stopped. To check for this condition:
   Perform fuel system diagnosis, CHART A-7.
- Remove spark plugs. Check for wet plugs, cracks, wear, improper gap, burned electrodes, or heavy deposits. Repair or replace as necessary.

1989–90 3.3L (VIN N)
CUTLASS CIERA AND CUTLASS CRUISER

HESITATION, SAG, STUMBLE

Definition: Momentary lack of response as the accelerator is pushed down. Can occur at all car speeds. Usually most severe when first trying to make the car move, as from a stop sign. May cause the engine to stall if severe enough.

- Perform careful visual check as described at start of Section "B".
- **CHECK:**
  - Fuel pressure. See CHART A-7. Also Check for water contaminated fuel.
  - Spark plugs for being fouled or faulty wiring.
  - PROM number. Also check service bulletins for latest PROM.
  - TPS for binding or sticking. Voltage should increase at a steady rate as throttle is moved toward WOT.
- Generator output voltage. Repair if less than 9 or more than 16 volts.
- CH ground, CKT 453.
- Canister purge system for proper operation. See CHART C-3.
- Engine Thermostat - functioning correctly and proper heat range.
- Perform injector balance test CHART C-2A.

SURGES AND/OR CHUGGLE

Definition: Engine power variation under steady throttle or cruise. Feels like the car speeds up and slows down with no change in the accelerator pedal.

- Be sure driver understands transmission converter clutch and A/C compressor operation in owner's manual.
- Perform careful visual inspection as described at start of Section "B".
- **CHECK:**
  - Generator output voltage. Repair if less than 9 or more than 16 volts.
  - If a "Scan" tool is available which plugs in to the ALDL connector, make sure reading of VSS matches vehicle speedometer. See introduction explaining "Scan" tool positions.
  - Check for PCV valve for being blown out of intake or blown off at the PCV valve. Also inspect PCV valve for kinks or leaks.
- In-line fuel filter. Replace if dirty or plugged.
- Fuel pressure while condition exists. See CHART A-7.
- Inspect oxygen sensor for silicone contamination, from fuel or use of improper RTV sealant. The sensor may have a white powdery coating. This will result in a high but false signal voltage (rich exhaust indication). The ECM will then reduce the amount of fuel delivered to the engine, causing a severe driveability problem.
- Remove spark plugs. Check for cracks, wear, improper gap, burned electrodes, or heavy deposits.
- Check spark plug leads.

LACK OF POWER, SLUGGISH, OR SPONGY

Definition: Engine delivers less than expected power. Little or no increase in speed when accelerator pedal is pushed down part way.

- Perform careful visual check as described at start of Section "B".
- Compare customer's car to similar unit. Make sure the customer's car has an actual problem.
- Remove air cleaner and check air filter for dirt, or for being plugged. Replace as necessary.
- **CHECK:**
  - Restricted fuel filter, contaminated fuel or improper fuel pressure. See CHART A-7.
  - ECM power grounds - See wiring diagrams.
  - ESC system for false retard due to mechanical noise.
- Exhaust system for possible restriction.
  - Inspect exhaust system for damaged or collapsed pipes.
  - Inspect muffler for heat distress or possible internal failure.
- Generator output voltage. Repair if less than 9 or more than 16 volts.
- Engine valve timing and compression.
- Engine for proper or worn camshaft.
- Secondary voltage using a shop ocilliscope or a spark tester J-26792 (ST-125) or equivalent.

1989–90 3.3L (VIN N)
CUTLASS CIERA AND CUTLASS CRUISER

DETONATION /SPARK KNOCK

Definition: A mild to severe ping, usually worse under acceleration. The engine makes sharp metallic knocks that change with throttle opening. Sounds like popcorn popping.

- Check for obvious overheating problems.
  - Low/weak coolant mixture.
  - Inoperative thermostat.
  - Restricted air/water flow through radiator.
  - Inoperative electric coolant fan circuit. See CHART C-12.
- **CHECK:**
  - ESC system for no retard - see CHART C-5.
  - TCC operation - see CHART C-8.
- Fuel system pressure. See CHART A-7.
- Remove carbon with top engine cleaner. Follow instructions on can.
  - Check for leaking valve oil seals
  - Check for poor fuel quality, proper octane rating
  - Check for correct PROM
- Check for incorrect basic engine parts such as cam, heads, pistons, etc.

CUTS OUT, MISSES

Definition: Steady pulsation or jerking that follows engine speed, usually more pronounced as engine load increases. The exhaust has a steady spitting sound at idle or low speed.

- Perform careful visual (physical) check as described at start of Section "B".
- If engine "Misses At Idle", see CHART C-4II-1.
- If engine "Misses Under Load", see Chart C-4II-2.
- If above checks did not discover cause of problem, check the following
  - Disconnect all injector harness connectors. Connect J-34730-2 injector test light or equivalent 12 volt test light between the harness terminals of each injector connector and note while cranking. If test light fails to blink at any connector, it is a faulty injector drive circuit harness, connector, or terminal.
  - Fuel system - Plugged fuel filter, water, low pressure. See CHART A-7.
- Perform the injector balance test. See CHART C-2A
- On CH systems, a misfire may be caused by a misaligned crank sensor or bent vane on rotating interrupter. Inspect for proper clearance at each vane, using tool J-36179 or equivalent
  - Sensor should be replaced if it shows evidence of having been rubbed by the interrupter
- Perform compression check
- Remove rocker covers. Check for bent pushrods, worn rocker arms, broken valve springs, worn camshaft lobes. Repair as necessary
- Valve timing

1989–90 3.3L (VIN N)
CUTLASS CIERA AND CUTLASS CRUISER

BACKFIRE

Definition: Fuel ignites in intake manifold, or in exhaust system, making a loud popping noise

- **CHECK:**
  - Compression - Look for sticking or leaking valves.
  - Valve timing.
  - Output voltage of ignition coil using a shop ocilliscope or spark tester J-26792 (ST-125) or equivalent.
- Spark plugs for crossfire. Also inspect spark plug wires, and proper routing of plug wires.
- Ignition system for intermittent condition.
- Engine timing - see emission control information label.

POOR FUEL ECONOMY

Definition: Fuel economy, as measured by an actual road test, is noticeably lower than expected. Also, economy is noticeably lower than it was on this car at one time, as previously shown by an actual road test.

- **CHECK:**
  - Engine thermostat for faulty part (always open) or for wrong heat range.
  - Fuel pressure. See CHART A-5.
- Check owner's driving habits.
  - Is A/C "ON" full time defroster mode "ON"?
  - Are tires at correct pressure?
  - Are excessively heavy loads being carried?
  - Is acceleration too much, too often?
  - Suggest driver read "Important Facts on Fuel Economy" in owner's manual.
- Perform "Diagnostic Circuit Check."
- Check air cleaner element (filter) for dirt or being plugged.
- Check for proper calibration of speedometer.
- Visually (physically) Check
  - Vacuum hoses for splits, kinks and proper connections as shown on Vehicle Emissions Control Information label.
  - Ignition wires for cracking, hardness, and proper connections.
- Remove spark plugs. Check for cracks, wear, in proper gap, burned electrodes or heavy deposits. Repair or replace as necessary
- Check TCC for proper operation. See CHART C-8, use "Scan" tool if available.
- Check for dragging brakes.
- Suggest owner fill fuel tank and recheck fuel economy.
- Check for exhaust system restriction.

DIESELING, RUN-ON

Definition: Engine continues to run after key is turned "OFF", but runs very roughly. If engine runs smoothly, check ignition switch and adjustment.

- Check injectors for leaking. See CHART A-7.

1989–90 3.3L (VIN N)
CUTLASS CIERA AND CUTLASS CRUISER

ROUGH, UNSTABLE, OR INCORRECT IDLE, STALLING

Definition: The engine runs unevenly at idle. If bad enough, the car may shake. Also, the idle may vary in rpm (called "hunting"). Either condition may be bad enough to cause stalling. Engine idles at incorrect speed.

- Perform careful visual check as described at start of Section "B".
- Clean injectors.
- **CHECK:**
  - Throttle linkage for sticking or binding.
  - TPS for sticking or binding, be sure output is stable at idle and adjustment specification is correct.
  - IAC system.
  - Generator output voltage. Repair if less than 9 or more than 16 volts.
  - P/N switch circuit. Code 31, 3300 (VIN N), or use "Scan" tool, and be sure tool indicates vehicle is in drive with gear selector in drive.
  - Injector balance. See CHART C-2A.
  - PCV valve for proper operation by placing finger over inlet hole in valve and several times. Valve should snap back. If not, replace valve. Check valve for damage or hose not fully seated at intake.
- Evaporative emission control system. CHART C-3.
- ECM ground circuits.
- Monitoring block learn values may help identify the cause of the problem. If the system is running lean (block learn greater than 138) refer to "Diagnostic Aids" on facing page of Code 44. If the system is running rich (block learn values less than 118) refer to "Diagnostic Aids" on facing page of Code 45.
- Run a cylinder compression check.
- Check for fuel in pressure regulator hose. If present, replace regulator assembly.
- Check ignition system; wires and plugs.
- Check for loose or damaged gaskets MAF between sensor and throttle body intake.
- Disconnect MAF sensor and if condition is corrected, replace sensor. "Scan" tool should read about 4-8 grams per second at idle.
- If problem exists, only with A/C "ON", check A/C system operation CHART C-10.
- For damaged, grounded out, or mispositioned motor mounts.

EXCESSIVE EXHAUST EMISSIONS OR ODORS

Definition: Vehicle fails an emission test. Vehicle has excessive "rotten egg" smell. Excessive odors do not necessarily indicate excessive emissions.

- Perform "Diagnostic Circuit Check."
- IF TEST SHOWS EXCESSIVE CO AND HC, (or also has excessive odors).
  - Check items which cause car to run RICH.
    - Make sure engine is at normal operating temperature.
- **CHECK:**
  - Fuel pressure. See CHART A-7.
  - Canister for fuel loading. See CHART C-3.
  - Injector balance. See CHART C-2A.
  - PCV valve for being plugged, stuck, or blocked PCV hose or fuel in the crankcase.
  - Spark plugs, plug wires, and ignition components.
  - Check for lead contamination of catalytic converter (look for removal of fuel filler neck restrictor).
- Check for properly installed fuel cap.
- If the system is running rich, (block learn less than 118), refer to "Diagnostic Aids" on facing page of Code 45.
- IF TEST SHOWS EXCESSIVE NOx:
  - Check items which cause car to run LEAN, or to run too hot.
    - Vacuum leaks.
    - Coolant system and coolant fan for proper operation. See CHART C-12.
  - Remove carbon with top engine cleaner. Follow instructions on can.
- If the system is running lean (block learn greater than 138), refer to "Diagnostic Aids" on facing page of Code 44.

CHART B-1
RESTRICTED EXHAUST SYSTEM CHECK

Proper diagnosis for a Restricted Exhaust System is essential before any components are replaced. The following procedure may be used for diagnosis:

CHECK AT O_2 SENSOR:

1. Carefully remove O_2 sensor.
2. Install exhaust backpressure tester (BT 8515 or BT 8603) or equivalent in place of O_2 sensor (see illustration).
3. After completing test described below, be sure to coat threads of O_2 sensor with anti-seize compound P/N 5613695 or equivalent prior to re-installation.

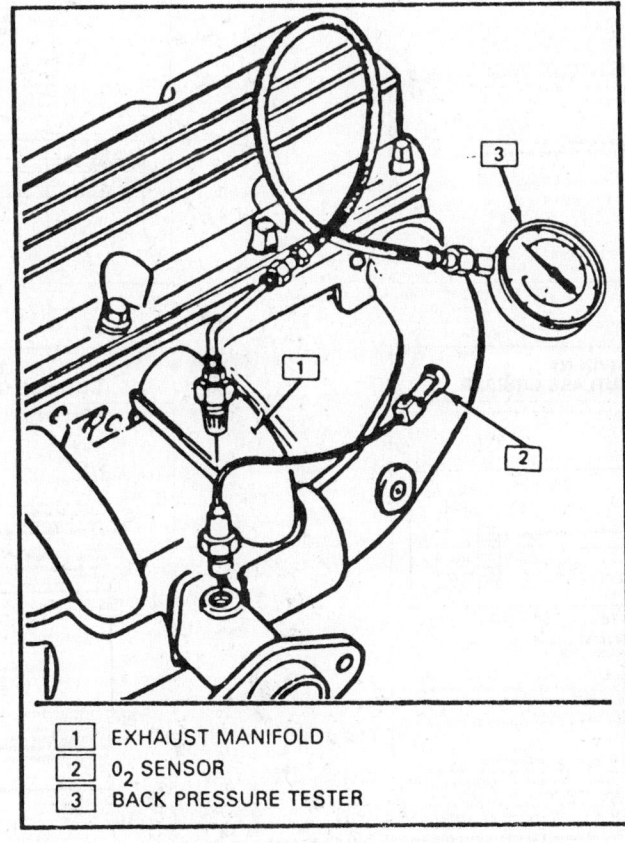

1	EXHAUST MANIFOLD
2	O_2 SENSOR
3	BACK PRESSURE TESTER

DIAGNOSIS:

1. With the engine at normal operating temperature and running at 2500 rpm, observe the exhaust system backpressure reading on the gauge.
2. If the backpressure exceeds 1 1/4 psi (8.62 kPa), a restricted exhaust system is indicated.
3. Inspect the entire exhaust system for a collapsed pipe, heat distress, or possible internal muffler failure.
4. If there are no obvious reasons for the excessive backpressure, a restricted catalytic converter should be suspected, and replaced using current recommended procedures.

1989–90 3.3L (VIN N)
CUTLASS CIERA AND CUTLASS CRUISER

CHART C-2A
INJECTOR BALANCE TEST

The injector balance tester is a tool used to turn the injector on for a precise amount of time, thus spraying a measured amount of fuel into the manifold. This causes a drop in fuel rail pressure that we can record and compare between each injector. All injectors should have the same amount of pressure drop (± 10 kpa). Any injector with a pressure drop that is 10 kpa (or more) greater than the average drop of the other injectors should be considered faulty and replaced. Any injector with a pressure drop of less than 10 kpa from the average drop of the other injectors should be cleaned and retested.

STEP 1

Engine "cool down" period (10 minutes) is necessary to avoid irregular readings due to "Hot Soak" fuel boiling.
Disconnect harness connectors at all injectors, and connect injector tester J-34730-3, or equivalent, to one injector. With ignition "OFF" connect fuel gauge J347301 or equivalent to fuel pressure tap. Wrap a shop towel around fitting while connecting gage to avoid fuel spillage. At this point, insert clear tubing attached to vent valve into a suitable container and bleed air from gauge and hose to insure accurate gauge operation, by applying B + to fuel pump test terminal. Repeat this step until all air is bled from gauge

STEP 2

Apply B + to fuel pump test terminal until fuel pressure reaches its maximum, about 3 seconds. Record this initial pressure reading. Energize tester one time and note pressure at its lowest point (Disregard any slight pressure increase after drop hits low point.) By subtracting this second pressure reading from the initial pressure, we have the actual amount of injector pressure drop. See note below.

STEP 3

Repeat step 2 on each injector and compare the amount of drop. Usually, good injectors will have virtually the same drop. Clean and retest any injector that has a pressure difference of 10kPa, either more or less than the average of the other injectors on the engine. Replace any injector that also fails the retest. If the pressure drop of all injectors is within 10kPa of this average, the injectors appear to be flowing properly. Reconnect them and review "Symptoms," Section "B".

NOTE: *The entire test should not be repeated more than once without running the engine to prevent flooding. (This includes any retest on faulty injectors.)*

1989–90 3.3L (VIN N)
CUTLASS CIERA AND CUTLASS CRUISER

NOTE: If injectors are suspected of being dirty, they should be cleaned using an approved tool and procedure prior to performing this test. The fuel pressure test in Section A, Chart A-7, should be completed prior to this test.

CHART C-2A
INJECTOR BALANCE TEST
3300 (VIN N) "A" CARLINE (PORT)

Step 1. If engine is at operation temperature start at A, otherwise you can start at B
  A. Engine "cool down" period (10 minutes) is necessary to avoid Irregular Readings due to "Hot Soak" Fuel Boiling.
  B. Disconnect harness connectors at all injectors and connect injector tester J-34730-3, or equivalent, to one injector.
  C. With ignition "OFF" connect fuel gauge J-34730-1 or equivalent to Fuel Pressure Tap. NOTE: Wrap a shop towel around fitting while connecting gauge to avoid fuel spray or spillage.
  D. Apply B + to fuel pump test terminal and bleed off to fuel pressure test gauge.

Step 2. Run test:
  A. Apply B + to Fuel Pump Test Terminal. Allow about 3 seconds for fuel pressure to reach its maximum.
  B. Record gauge pressure. (Pressure must hold steady, if not see the Fuel System diagnosis, Chart A-7, in Section A.)
  C. Turn injector on, by depressing button on injector tester, and note pressure at the instant the gauge needle stops.

Step 3.
  A. Repeat step 2 on all injectors and record pressure drop on each. Clean and retest injectors that appear faulty (Any injectors that have a 10 kPa difference, either more or less, in pressure from the average). If no problem is found, review Symptoms Section B.

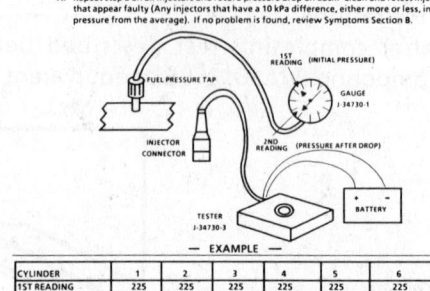

— EXAMPLE —

CYLINDER	1	2	3	4	5	6
1ST READING	225	225	225	225	225	225
2ND READING	100	100	100	90	100	115
AMOUNT OF DROP	125	125	125	135	125	110
	OK	OK	OK	FAULTY, RICH (TOO MUCH) (FUEL DROP)	OK	FAULTY, LEAN (TOO LITTLE) (FUEL DROP)

1989–90 3.3L (VIN N)
CUTLASS CIERA AND CUTLASS CRUISER

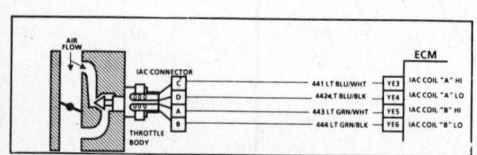

			ECM
	C	441 LT BLU/WHT	YE3 IAC COIL "A" HI
IAC CONNECTOR	D	442 LT BLU/BLK	YE4 IAC COIL "A" LO
	A	443 LT GRN/WHT	YE5 IAC COIL "B" HI
	B	444 LT GRN/BLK	YE6 IAC COIL "B" LO

CHART C-2B
IDLE AIR CONTROL (IAC) VALVE CHECK
3300 (VIN N) "A" CARLINE (PORT)

Circuit Description:

The ECM controls idle rpm with the IAC valve. To increase idle rpm, the ECM retracts the IAC pintle, allowing more air to bypass the throttle plate. To decrease rpm, it extends the IAC pintle valve, reducing air flow through the IAC valve port in the throttle body. A "Scan" tool will read the ECM commands to the IAC valve in counts. The higher the counts, the more air allowed (higher idle). The lower the counts, the less air allowed (lower idle). The ECM learns a new IAC position every time the ignition is cycled. This is to control engine speed at different idle conditions.

Test Description: Numbers below refer to circled numbers on the diagnostic chart.
1. Continue with test, even if engine will not idle. If idle is too low, "Scan" will display 80 or more counts, or steps. If idle is high, it will display "0" counts. Occasionally, an erratic or unstable idle may occur. Engine speed may vary 200 rpm or more up and down. Disconnect IAC. If the condition is unchanged, the IAC is not at fault.
2. When the engine was stopped, the IAC pintle retracted (more air) to a fixed "park" position for increased air flow and idle speed during the next engine start. A "Scan" will display 100 or more counts.
3. Be sure to disconnect the IAC valve prior to this test. The test light will confirm the ECM signals by a steady or flashing light on all circuits.
4. There is a remote possibility that one of the circuits is shorted to voltage, which would have been indicated by a steady light. Disconnect ECM and turn the ignition "ON" and probe terminals to check for this condition.

Diagnostic Aids:

A slow, unstable idle may be caused by a system problem that cannot be overcome by the IAC. "Scan" counts will be above 60 counts, if idle speed is too low. If idle speed is too high, IAC counts will be "0".

- **System lean (High Air/Fuel Ratio)**
  Idle speed may be too high or too low. Engine speed may vary up and down, disconnecting IAC does not help. May set Code 44.
  "Scan" and/or voltmeter will read an oxygen sensor output less than 300 mV (.3 volt). Check for low regulated fuel pressure or water in fuel. A lean exhaust, with an oxygen sensor output fixed above 800 mV (.8 volt) indicates rich exhaust (reading to ECM), will be a contaminated sensor, usually silicone. This may also set Code 45.
- **System rich (Low Air/Fuel Ratio)**
  Idle speed too low. "Scan" counts usually above 80. System obviously rich and may exhibit black smoke exhaust.
  "Scan" tool and/or voltmeter will read an oxygen sensor signal fixed above 800 mV (.8 volt).
  Check:
  - High fuel pressure
  - Injector leaking or sticking
- Throttle Body - Remove IAC and inspect bore for foreign material or evidence of IAC pintle dragging the bore.
- Refer to "Rough, Unstable, Incorrect Idle or Stalling" in "Symptoms" in Section "B".

1989–90 3.3L (VIN N)
CUTLASS CIERA AND CUTLASS CRUISER

CHART C-2B
IDLE AIR CONTROL (IAC) VALVE CHECK
3300 (VIN N) "A" CARLINE (PORT)

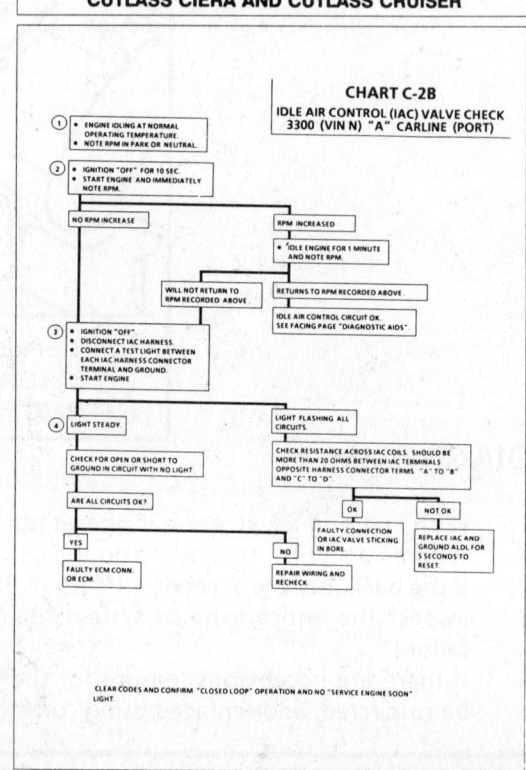

1989–90 3.3L (VIN N)
CUTLASS CIERA AND CUTLASS CRUISER

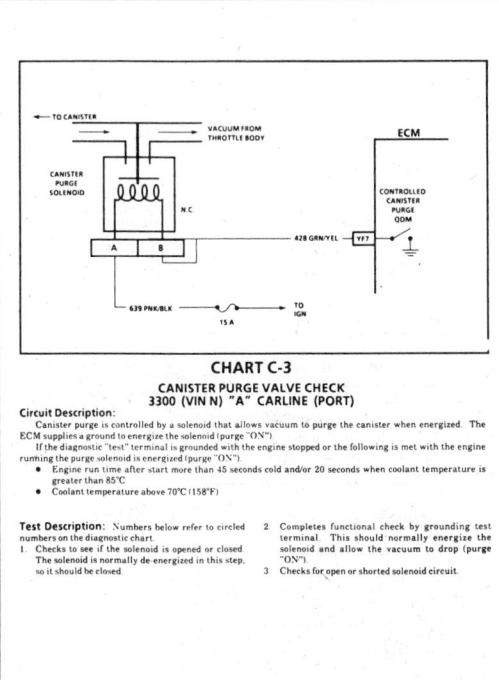

CHART C-3
CANISTER PURGE VALVE CHECK
3300 (VIN N) "A" CARLINE (PORT)

Circuit Description:
Canister purge is controlled by a solenoid that allows vacuum to purge the canister when energized. The ECM supplies a ground to energize the solenoid (purge "ON").
If the diagnostic "test" terminal is grounded with the engine stopped or the following is met with the engine running the purge solenoid is energized (purge "ON").
- Engine run time after start more than 45 seconds cold and/or 20 seconds when coolant temperature is greater than 85°C.
- Coolant temperature above 70°C (158°F).

Test Description: Numbers below refer to circled numbers on the diagnostic chart.
1. Checks to see if the solenoid is opened or closed. The solenoid is normally de-energized in this step, so it should be closed.
2. Completes functional check by grounding test terminal. This should normally energize the solenoid and allow the vacuum to drop (purge "ON").
3. Checks for open or shorted solenoid circuit.

1989–90 3.3L (VIN N)
CUTLASS CIERA AND CUTLASS CRUISER

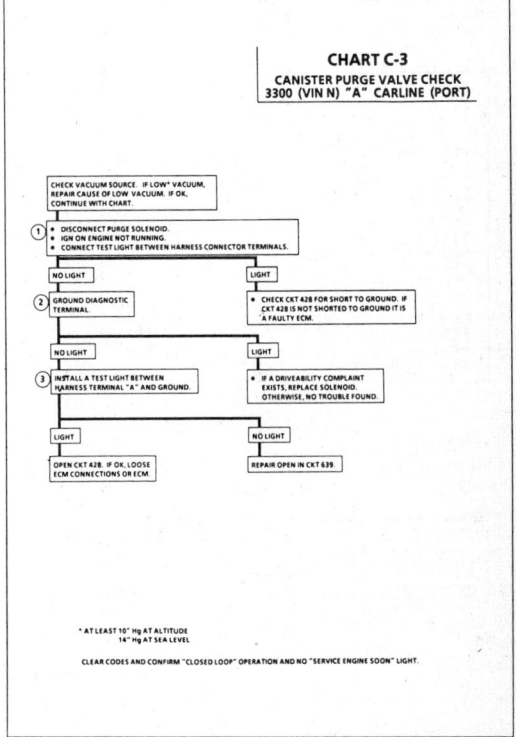

1989–90 3.3L (VIN N)
CUTLASS CIERA AND CUTLASS CRUISER

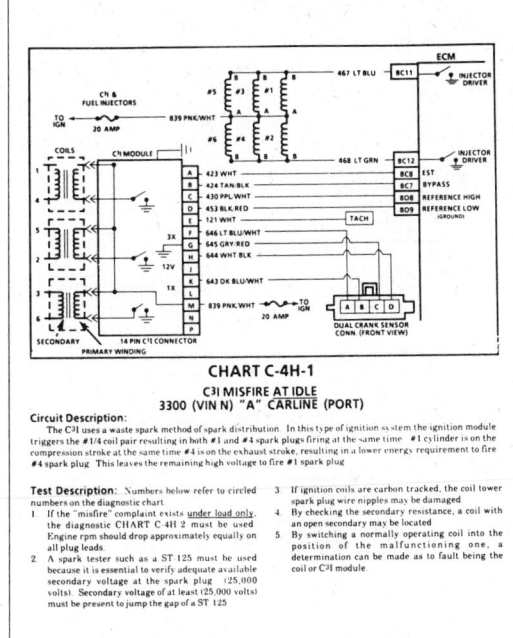

CHART C-4H-1
C³I MISFIRE AT IDLE
3300 (VIN N) "A" CARLINE (PORT)

Circuit Description:
The C³I uses a waste spark method of spark distribution. In this type of ignition system the ignition module triggers the #1/4 coil pair resulting in both #1 and #4 spark plugs firing at the same time. #1 cylinder is on the compression stroke at the same time #4 is on the exhaust stroke, resulting in a lower energy requirement to fire the #4 spark plug. This leaves the remaining high voltage to fire the #1 spark plug.

Test Description: Numbers below refer to circled numbers on the diagnostic chart.
1. If the "misfire" complaint exists under load only, the diagnostic CHART C-4H-2 must be used. Engine rpm should drop approximately equally on all plug leads.
2. A spark tester such as a ST-125 must be used because it is essential to verify adequate available secondary voltage at the spark plug. (25,000 volts). Secondary voltage of at least (25,000 volts) must be present to jump the gap of a ST-125.
3. If ignition coils are carbon tracked, the coil tower spark plug wire nipples may be damaged.
4. By checking the secondary resistance, a coil with an open secondary may be located.
5. By switching a normally operating coil into the position of the malfunctioning one, a determination can be made as to fault being the coil or C³I module.

1989–90 3.3L (VIN N)
CUTLASS CIERA AND CUTLASS CRUISER

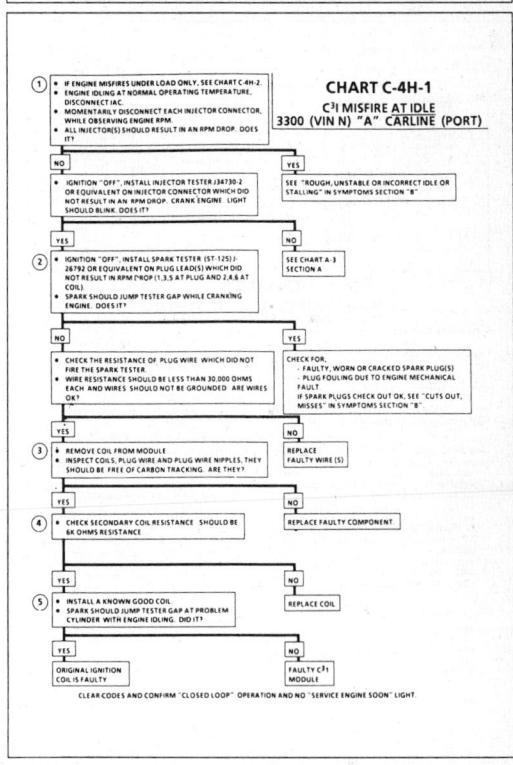

1989–90 3.3L (VIN N)
CUTLASS CIERA AND CUTLASS CRUISER

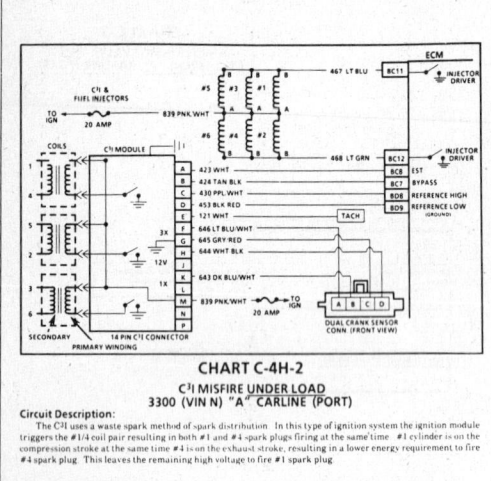

CHART C-4H-2
C³I MISFIRE UNDER LOAD
3300 (VIN N) "A" CARLINE (PORT)

Circuit Description:
The C³I uses a waste spark method of spark distribution. In this type of ignition system the ignition module triggers the #1/4 coil pair resulting in both #1 and #4 spark plugs firing at the same time. #1 cylinder is on the compression stroke at the same time #4 is on the exhaust stroke, resulting in a lower energy requirement to fire #4 spark plug. This leaves the remaining high voltage to fire #1 spark plug.

Test Description: Numbers below refer to circled numbers on the diagnostic chart.
1. If the "misfire" complaint exists at idle only, the diagnostic CHART C-4H-1 must be used.
2. A spark tester such as a ST-125 must be used because it is essential to verify adequate available secondary voltage at the spark plug (25,000 volts). Spark should jump the tester gap on all 6 leads. This simulates a "load" condition.

3. If ignition coils are carbon tracked, the coil tower spark plug wire nipples may be damaged.
4. By switching a normally operating coil into the position of the malfunctioning one, a determination can be made as to fault being the coil or C³I module.

1989–90 3.3L (VIN N)
CUTLASS CIERA AND CUTLASS CRUISER

CHART C-4H-2
C³I MISFIRE UNDER LOAD
3300 (VIN N) "A" CARLINE (PORT)

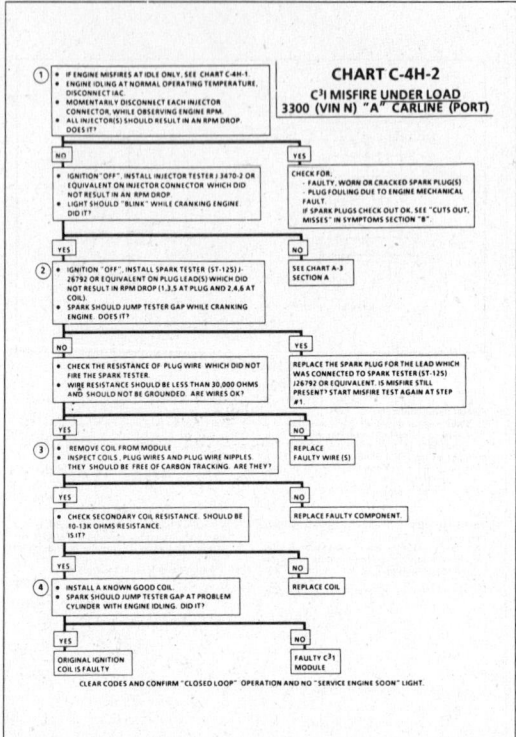

1989–90 3.3L (VIN N)
CUTLASS CIERA AND CUTLASS CRUISER

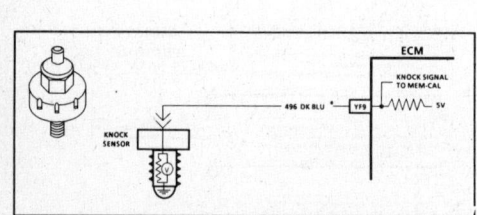

CHART C-5
ELECTRONIC SPARK CONTROL (ESC) SYSTEM CHECK
3300 (VIN N) "A" CARLINE (PORT)

Circuit Description:
The knock sensor is used to detect engine detonation and the ECM will retard the electronic spark timing based on the signal being received. The circuitry within the knock sensor causes the ECM's supplied 5 volt signal to be pulled down so that under a no knock condition, CKT 496 would measure about 2.5 volts. The knock sensor produces an A/C signal which rides on the 2.5 volts DC voltage. The amplitude and frequency are dependent upon the knock level.
The Mem-Cal used with this engine contains the functions which were part of the remotely mounted ESC modules used on other GM vehicles. The ESC portion of the Mem-Cal then sends a signal to other parts of the ECM which adjusts the spark timing to retard the spark and reduce the detonation.

Test Description: Numbers below refer to circled numbers on the diagnostic chart.
1. With engine idling, there should not be a knock signal present at the ECM because detonation is not likely under a no load condition.
2. Tapping on the engine lift hook should simulate a knock signal to determine if the sensor is capable of detecting detonation. If no knock is detected, try tapping on engine block closer to sensor before replacing sensor.
3. If the engine has an internal problem which is creating a knock, the knock sensor may be responding to the internal failure.

4. This test determines if the knock sensor is faulty or if the ESC portion of the Mem-Cal is faulty. If it is determined that the Mem-Cal is faulty, be sure that it is properly installed and latched into place. If not properly installed, repair and retest.

Diagnostic Aids:
While observing knock signal on the "Scan," there should be an indication that knock is present when detonation can be heard. Detonation is most likely to occur under high engine load conditions.

1989–90 3.3L (VIN N)
CUTLASS CIERA AND CUTLASS CRUISER

CHART C-5
ELECTRONIC SPARK CONTROL (ESC) SYSTEM CHECK
3300 (VIN N) "A" CARLINE (PORT)

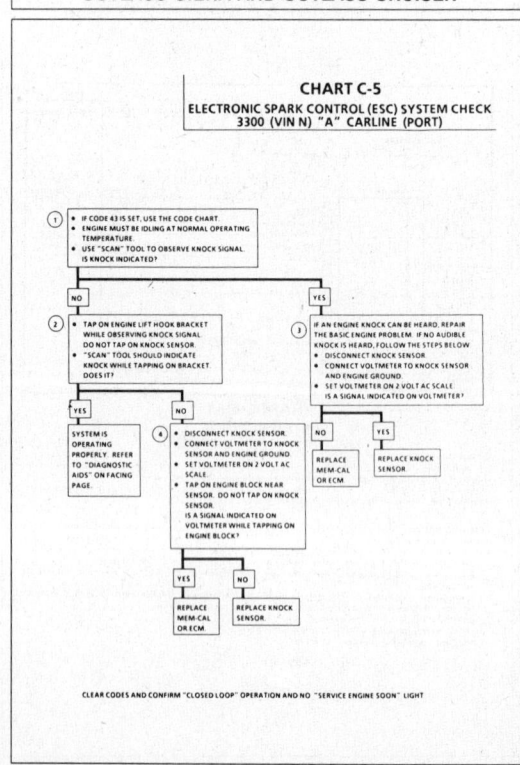

1989–90 3.3L (VIN N)
CUTLASS CIERA AND CUTLASS CRUISER

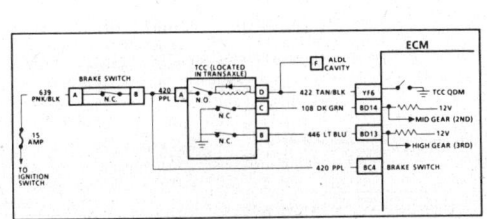

CHART C-8A
TORQUE CONVERTER CLUTCH (TCC) 125C
3300 (VIN N) "A" CARLINE (PORT)

Circuit Description:
The purpose of the transmission converter clutch feature is to eliminate the power loss of the transmission converter stage when the vehicle is in a cruise condition. This allows the convenience of the automatic transmission and the fuel economy of a manual transaxle.

Fused battery ignition is supplied to the TCC solenoid through the brake switch, and transmission third gear apply switch. The ECM will engage TCC by grounding CKT 422 to energize the solenoid.

TCC will engage when
- Engine warmed up
- Vehicle speed above a calibrated value (about 32 mph 51 km/h)
- Throttle position sensor output not changing, indicating a steady road speed
- Transmission third gear switch closed
- Brake switch closed

Test Description: Numbers below refer to circled numbers on the diagnostic chart.
1. Light "OFF" confirms transmission third gear apply switch is open.
2. At 25 mph the transmission 2nd gear apply switch should close. Test light will come "ON" and confirm battery supply and closed brake switch.
3. Grounding the diagnostic terminal with ignition "ON," engine "OFF," should energize the TCC solenoid by grounding CKT 422. This test checks the ability of the ECM to supply a ground to the TCC solenoid. The test light connected from 12 volts to ALDL terminal "F" should turn "ON" as CKT 422 is grounded.

Diagnostic Aids:
The "Scan" tool only indicates when the ECM has turned "ON" the TCC driver and this does not confirm that the TCC has engaged. To determine if TCC is functioning properly, engine rpm should decrease when the "Scan" indicates the TCC driver has turned "ON."

1989–90 3.3L (VIN N)
CUTLASS CIERA AND CUTLASS CRUISER

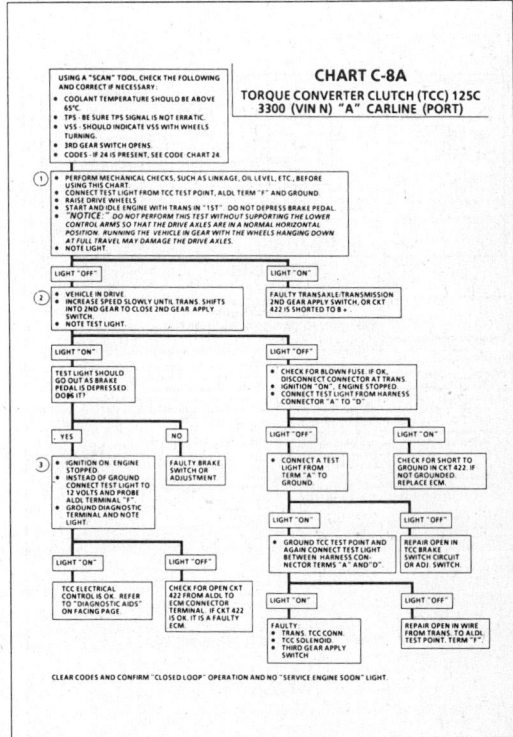

1989–90 3.3L (VIN N)
CUTLASS CIERA AND CUTLASS CRUISER

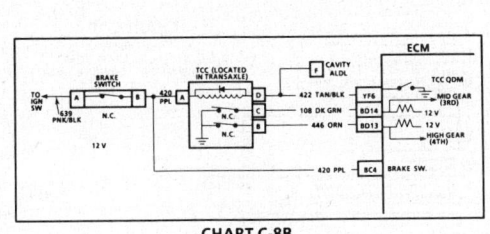

CHART C-8B
(Page 1 of 2)
TORQUE CONVERTER CLUTCH (TCC) 440-T4
(ELECTRICAL DIAGNOSIS)
3300 (VIN N) "A" CARLINE (PORT)

Circuit Description:
Each gear switch opens when the appropriate clutch is applied. All gear switches are open in fourth gear.

Test Description: Numbers below refer to circled numbers on the diagnostic chart.
1. Some "Scan" tools display the state of these switches in different ways. Be familiar with the type of tool being used. All switches should be in the closed state during this test, the tool should read the same for 3rd or 4th gear switches.
2. Determines whether the switch or signal circuit is open. The circuit can be checked for an open by measuring the voltage (with a voltmeter) at the TCC connector. Should be about 12 volts.

Diagnostic Aids:
If vehicle is road tested because of a TCC related problem, be sure the switch states do not change while in 4th gear because the TCC will disengage. If switches change state, carefully check wire routing and connections.

1989–90 3.3L (VIN N)
CUTLASS CIERA AND CUTLASS CRUISER

CHART C-8B
(Page 1 of 2)
TORQUE CONVERTER CLUTCH (TCC) 440-T4
(ELECTRICAL DIAGNOSIS)
3300 (VIN N) "A" CARLINE (PORT)

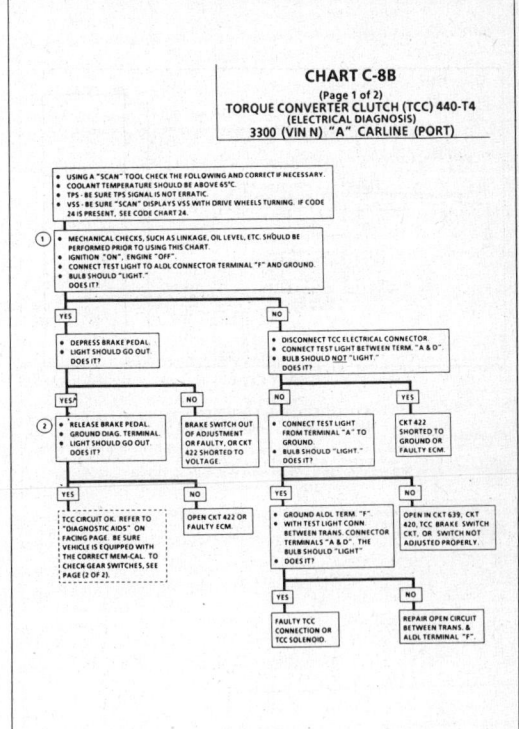

1989–90 3.3L (VIN N) CUTLASS CIERA AND CUTLASS CRUISER

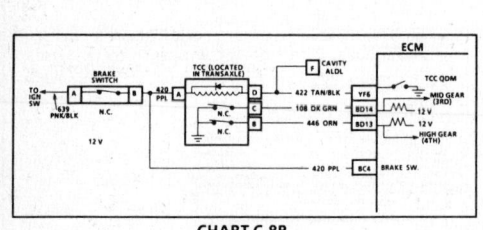

CHART C-8B
(Page 2 of 2)
TORQUE CONVERTER CLUTCH (TCC) 440-T4
(ELECTRICAL DIAGNOSIS)
3300 (VIN N) "A" CARLINE (PORT)

Circuit Description:
Each gear switch opens when the appropriate clutch is applied. All gear switches are open in fourth gear.

Test Description: Numbers below refer to circled numbers on the diagnostic chart.
1. Some "Scan" tools display the state of these switches in different ways. Be familiar with the type of tool being used. All switches should be in the closed state during this test, the tool should read the same for 2nd, 3rd or 4th gear switches.
2. Determines whether the switch or signal circuit is open. The circuit can be checked for an open by measuring the voltage (with a voltmeter) at the TCC connector. Should be about 12 volts.

3. Because the switch(s) should be grounded in this step, disconnecting the TCC connector should cause the "Scan" switch state to change.
4. The switch state should change when the vehicle shifts into 3rd gear.

Diagnostic Aids:
If vehicle is road tested because of a TCC related problem, be sure the switch states do not change while in 4th gear because the TCC will disengage. If switches change state, carefully check wire routing and connections.

1989–90 3.3L (VIN N) CUTLASS CIERA AND CUTLASS CRUISER

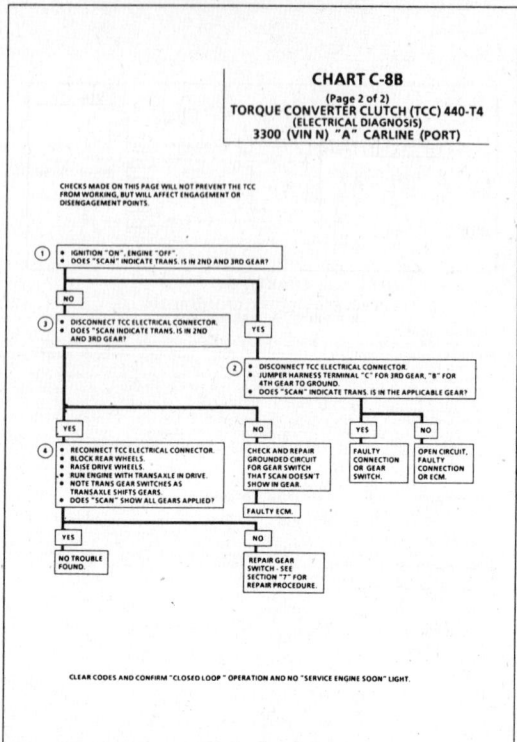

CHART C-8B
(Page 2 of 2)
TORQUE CONVERTER CLUTCH (TCC)
(ELECTRICAL DIAGNOSIS)
3300 (VIN N) "A" CARLINE (PORT)

CHECKS MADE ON THIS PAGE WILL NOT PREVENT THE TCC FROM WORKING, BUT WILL AFFECT ENGAGEMENT OR DISENGAGEMENT POINTS.

CLEAR CODES AND CONFIRM "CLOSED LOOP" OPERATION AND NO "SERVICE ENGINE SOON" LIGHT.

1989–90 3.3L (VIN N) CUTLASS CIERA AND CUTLASS CRUISER

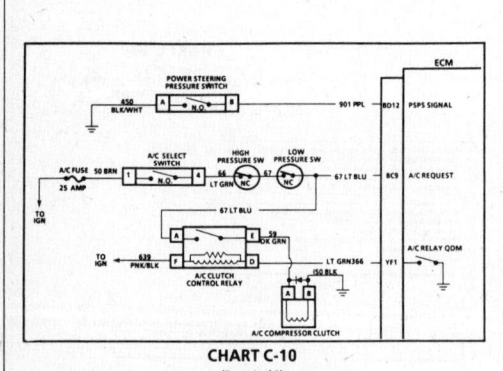

CHART C-10
(Page 1 of 2)
A/C CLUTCH CIRCUIT DIAGNOSIS
3300 (VIN N) "A" CARLINE (PORT)

Circuit Description:
The A/C relay is ECM controlled to delay A/C clutch engagement .3 seconds after A/C is turned "ON." This allows the IAC to adjust engine rpm before the A/C clutch engages. The ECM also causes the relay to disengage the A/C clutch during WOT operation. The A/C relay is energized when the ECM provides a ground path for CKT 366.

Test Description: Numbers refer to circled numbers on the diagnostic chart.
1. Checks operation of A/C cycling switch.

2. Checks to see if ECM is controlling A/C clutch control relay.
3. Checks for open circuit on either side of relay coil.

1989–90 3.3L (VIN N) CUTLASS CIERA AND CUTLASS CRUISER

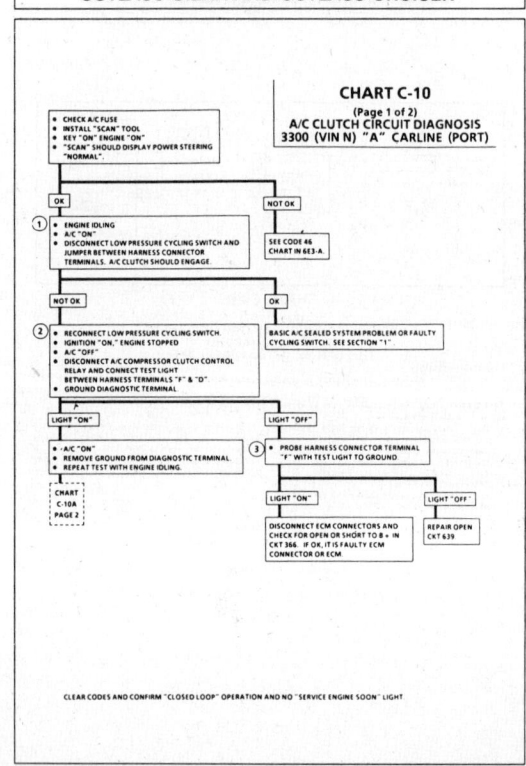

CHART C-10
(Page 1 of 2)
A/C CLUTCH CIRCUIT DIAGNOSIS
3300 (VIN N) "A" CARLINE (PORT)

CLEAR CODES AND CONFIRM "CLOSED LOOP" OPERATION AND NO "SERVICE ENGINE SOON" LIGHT.

1989-90 3.3L (VIN N)
CUTLASS CIERA AND CUTLASS CRUISER

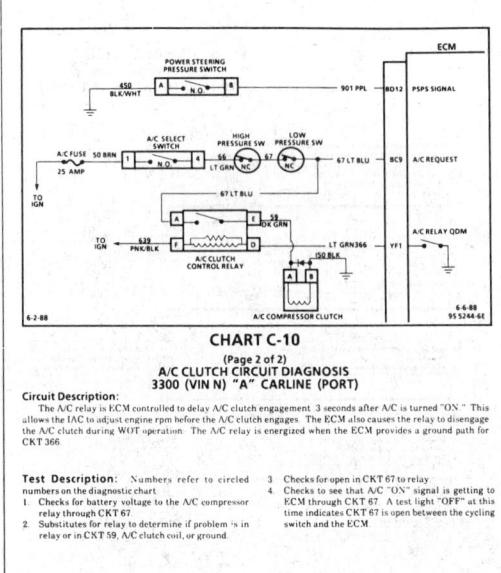

CHART C-10
(Page 2 of 2)
A/C CLUTCH CIRCUIT DIAGNOSIS
3300 (VIN N) "A" CARLINE (PORT)

Circuit Description:
The A/C relay is ECM controlled to delay A/C clutch engagement 3 seconds after A/C is turned "ON." This allows the IAC to adjust engine rpm before the A/C clutch engages. The ECM also causes the relay to disengage the A/C clutch during WOT operation. The A/C relay is energized when the ECM provides a ground path for CKT 366.

Test Description: Numbers refer to circled numbers on the diagnostic chart.
1. Checks for battery voltage to the A/C compressor relay through CKT 67.
2. Substitutes for relay to determine if problem is in relay or in CKT 59, A/C clutch coil, or ground.
3. Checks for open in CKT 67 to relay.
4. Checks to see that A/C "ON" signal is getting to ECM through CKT 67. A test light "OFF" at this time indicates CKT 67 is open between the cycling switch and the ECM.

1989-90 3.3L (VIN N)
CUTLASS CIERA AND CUTLASS CRUISER

CHART C-10
(Page 2 of 2)
A/C CLUTCH CIRCUIT DIAGNOSIS
3300 (VIN N) "A" CARLINE (PORT)

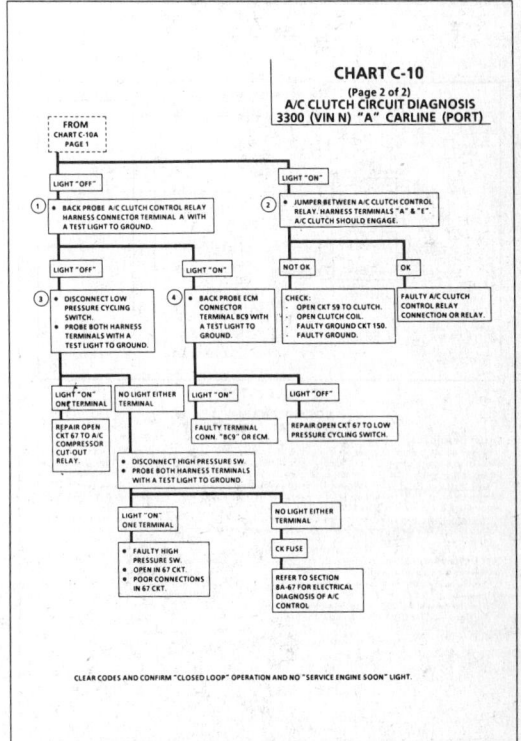

1989-90 3.3L (VIN N)
CUTLASS CIERA AND CUTLASS CRUISER

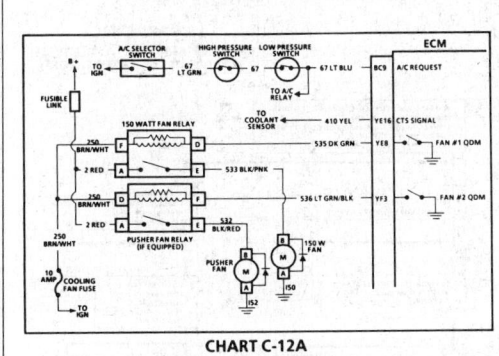

CHART C-12A
COOLANT FAN FUNCTIONAL CHECK
3300 (VIN N) "A" CARLINE (PORT)

Circuit Description:
Power for the coolant fan motor comes through the fusible link to terminal "A" on both relays. The relays are energized when current flows to ground through the quad-driver module inside the ECM. The ECM controls the fan(s) after ignition "OFF" if coolant temperature is above value. The fan(s) could run up to ten minutes.
150 Watt Fan Relay - The 150 watt fan relay is energized by the ECM. The ECM energizes the relay through terminal "YE8" when the coolant temperature reaches 98°C (208°F). The fan relay is also energized when refrigerant pressure reaches 150 psi (1034 kPa).
Pusher Coolant Fan - The optional pusher fan relay is energized by the ECM when coolant temperature reaches 108°C (226°F) or refrigerant pressure reaches 245 psi (1896 kPa).

Test Description: Numbers below refer to circled numbers on the diagnostic chart.
1. Grounding the diagnostic "test" terminal bypasses all the timers and turns on the QDM so the fan(s) should be "ON."
2. Fan should be "ON" when A/C is requested at idle.
3. If ECM connector is OK, then the QDM inside the ECM is faulty and the ECM has to be replaced.

1989-90 3.3L (VIN N)
CUTLASS CIERA AND CUTLASS CRUISER

CHART C-12A
COOLANT FAN FUNCTIONAL CHECK
3300 (VIN N) "A" CARLINE (PORT)

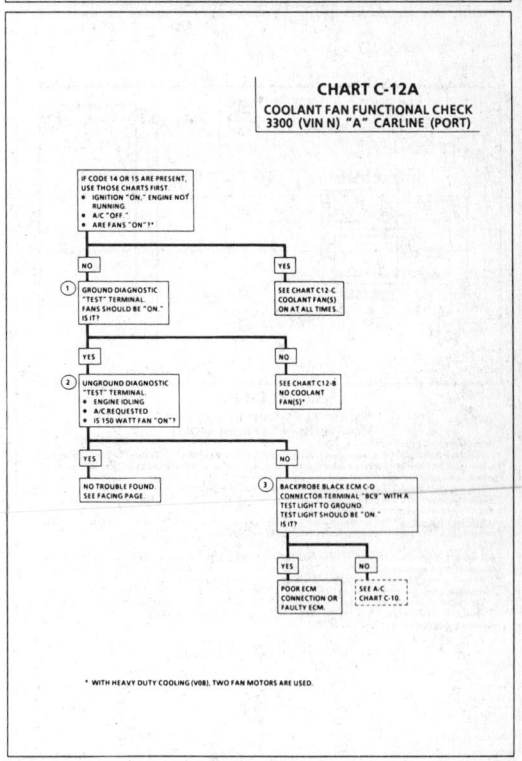

1989–90 3.3L (VIN N)
CUTLASS CIERA AND CUTLASS CRUISER

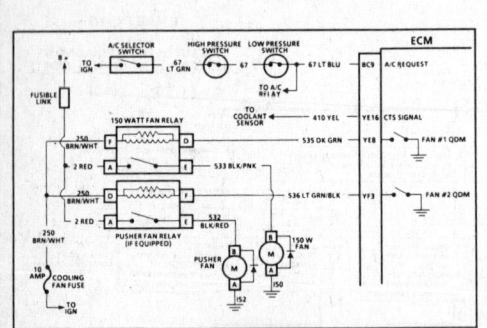

CHART C-12B
NO COOLANT FAN(S)
3300 (VIN N) "A" CARLINE (PORT)

Circuit Description:
Power for the coolant fan motor comes through the fusible link to terminal "A" on both relays. The relays are energized when current flows to ground through the quad-driver module inside the ECM. The ECM controls the fan(s) after ignition "OFF" if coolant temperature is above value. The fan(s) could run up to ten minutes.
150 Watt Fan Relay - The 150 watt fan relay is energized by the ECM. The ECM energizes the relay through terminal "YE8" when the coolant temperature reaches 98°C (208°F). The fan relay is also energized when refrigerant pressure reaches 150 psi (1034 kPa).
Pusher Coolant Fan - The optional pusher fan relay is energized by the ECM when coolant temperature reaches 108°C (226°F) or refrigerant pressure reaches 245 psi (1896 kPa).

Test Description: Numbers below refer to circled numbers on the diagnostic chart.
1. Test light should be "ON" because the "F" and "D" terminals are fed directly from the ignition.
2. Test light should be "ON", as grounding the ALDL will turn "ON" the QDM allowing current to flow.
3. Jumpering terminals "A" and "E" feeds the fan motor directly and it should run.
4. Test light should be "ON" as terminal(s) "A" are connected directly to B+.

1989–90 3.3L (VIN N)
CUTLASS CIERA AND CUTLASS CRUISER

CHART C-12B
NO COOLANT FAN(S)
3300 (VIN N) "A" CARLINE (PORT)

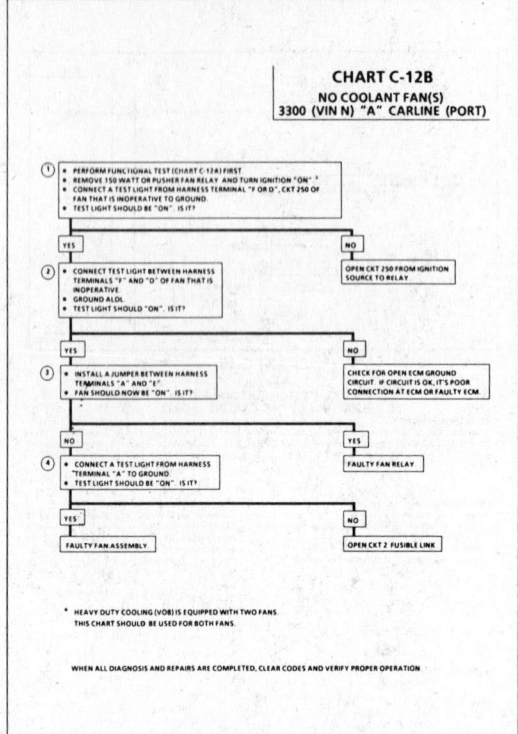

* HEAVY DUTY COOLING (V08) IS EQUIPPED WITH TWO FANS. THIS CHART SHOULD BE USED FOR BOTH FANS.

WHEN ALL DIAGNOSIS AND REPAIRS ARE COMPLETED, CLEAR CODES AND VERIFY PROPER OPERATION

1989–90 3.3L (VIN N)
CUTLASS CIERA AND CUTLASS CRUISER

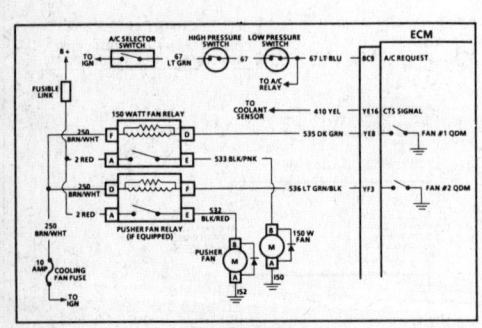

CHART C-12C
COOLANT FAN(S) "ON" AT ALL TIMES
3300 (VIN N) "A" CARLINE (PORT)

Circuit Description:
Power for the coolant fan motor comes through the fusible link to terminal "A" on both relays. The relays are energized when current flows to ground through the quad-driver module inside the ECM. The ECM controls the fan(s) after ignition "OFF" if coolant temperature is above value. The fan(s) could run up to ten minutes.
150 Watt Fan Relay - The 150 watt fan relay is energized by the ECM. The ECM energizes the relay through terminal "YE8" when the coolant temperature reaches 98°C (208°F). The fan relay is also energized when refrigerant pressure reaches 150 psi (1034 kPa).
Pusher Coolant Fan - The optional pusher fan relay is energized by the ECM when coolant temperature reaches 108°C (226°F) or refrigerant pressure reaches 245 psi (1896 kPa).

Test Description: Numbers below refer to circled numbers on the diagnostic chart.
1. Removing the relay interrupts current flow to the fan motor so the fan(s) should not be "ON".
2. Test light should be "OFF", as the QDM should be "OFF", when A/C is not requested and coolant temperature is below 98°C (208°F).
3. Test light should be "OFF", as the A/C system should not have high pressure, due to A/C not being requested.
4. If the test light goes "OFF" when black C-D connector is disconnected, the QDM is faulty and ECM has to be replaced.

1989–90 3.3L (VIN N)
CUTLASS CIERA AND CUTLASS CRUISER

CHART C-12C
COOLANT FAN(S) "ON" AT ALL TIMES
3300 (VIN N) "A" CARLINE (PORT)

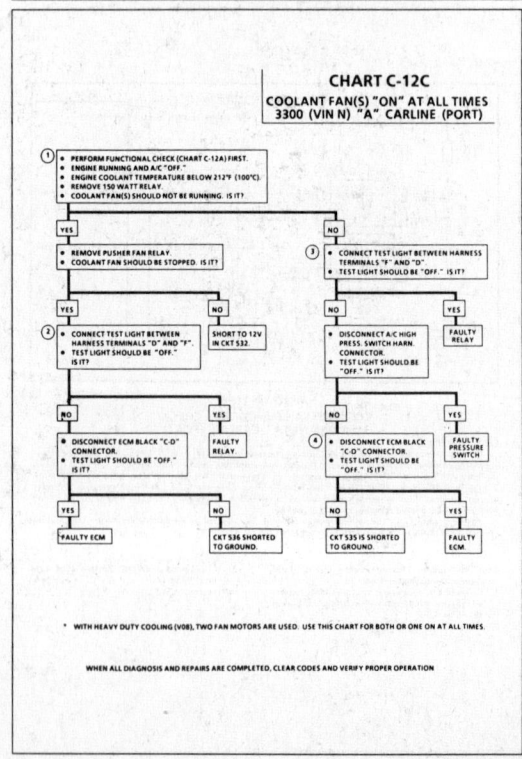

* WITH HEAVY DUTY COOLING (V08), TWO FAN MOTORS ARE USED. USE THIS CHART FOR BOTH OR ONE ON AT ALL TIMES.

WHEN ALL DIAGNOSIS AND REPAIRS ARE COMPLETED, CLEAR CODES AND VERIFY PROPER OPERATION

3.3L (VIN N) COMPONENT LOCATIONS CALAIS AND SKYLARK

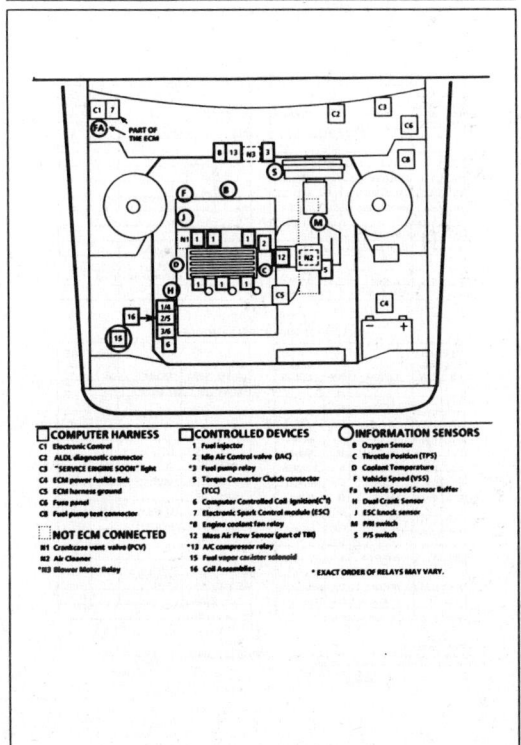

3.3L (VIN N) ECM WIRING DIAGRAM CALAIS AND SKYLARK

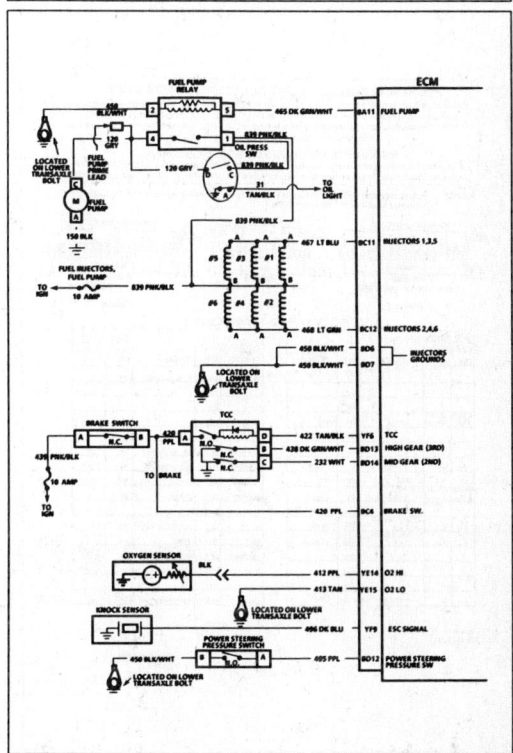

3.3L (VIN N) ECM WIRING DIAGRAM CALAIS AND SKYLARK (CONT.)

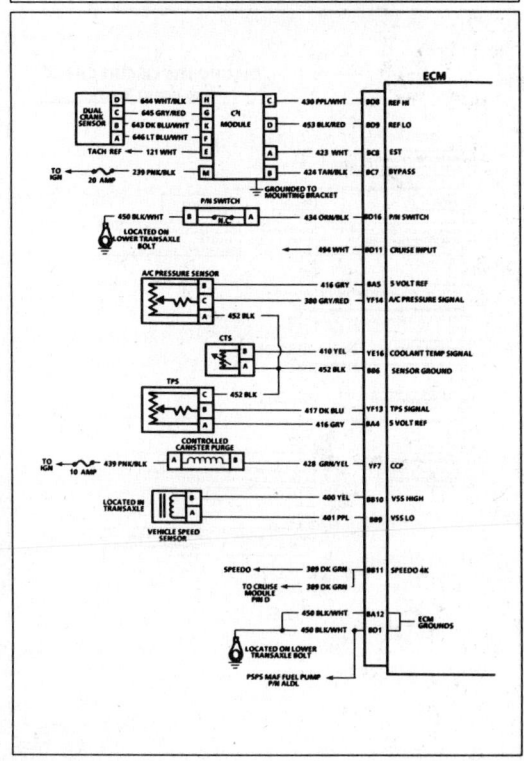

3.3L (VIN N) ECM WIRING DIAGRAM CALAIS AND SKYLARK (CONT.)

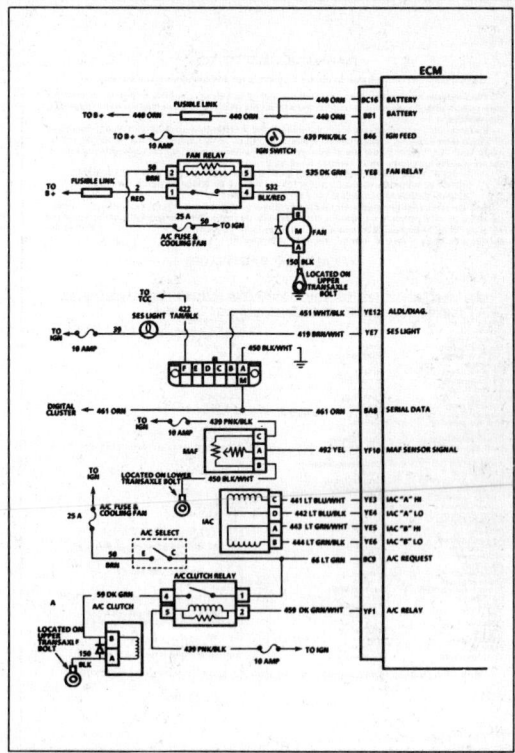

3.3L (VIN N) ECM CONNECTOR TERMINAL END VIEW—CALAIS AND SKYLARK

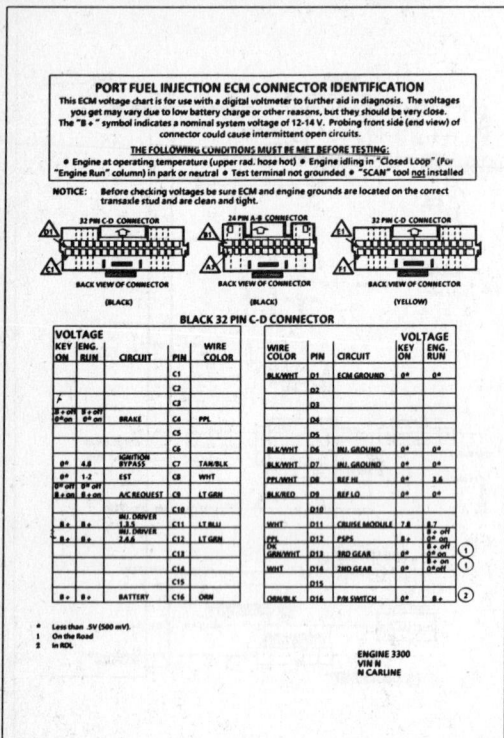

PORT FUEL INJECTION ECM CONNECTOR IDENTIFICATION
This ECM voltage chart is for use with a digital voltmeter to further aid in diagnosis. The voltages you get may vary due to low battery charge or other reasons, but they should be very close. The "B+" symbol indicates a nominal system voltage of 12-14 V. Probing front side (end view) of connector could cause intermittent open circuits.

THE FOLLOWING CONDITIONS MUST BE MET BEFORE TESTING:
• Engine at operating temperature (upper rad. hose hot) • Engine idling in "Closed Loop" (For "Engine Run" column) in park or neutral • Test terminal not grounded • "SCAN" tool not installed

NOTICE: Before checking voltages be sure ECM and engine grounds are located on the correct transaxle stud and are clean and tight.

3.3L (VIN N) ECM CONNECTOR TERMINAL END VIEW—CALAIS AND SKYLARK (CONT.)

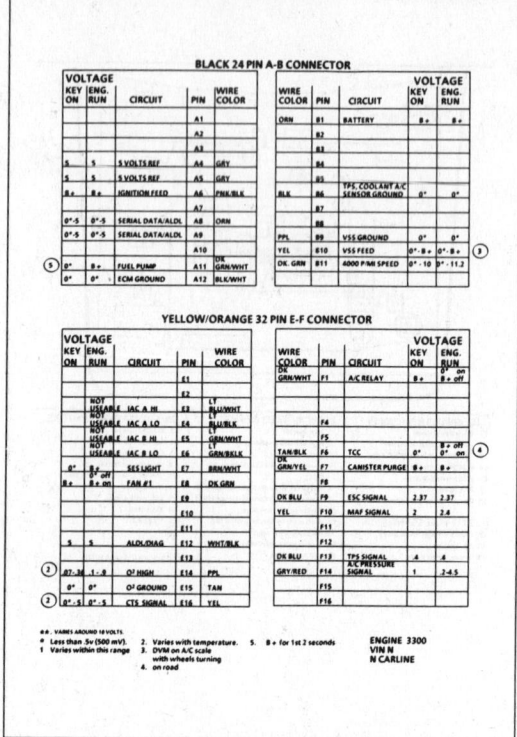

1989–90 3.3L (VIN N) CUTLASS CALAIS AND SKYLARK

DIAGNOSTIC CIRCUIT CHECK

The Diagnostic Circuit Check must be the starting point for any driveability complaint diagnosis.

The diagnostic circuit check is an organized approach to identifying a problem created by an electronic engine control system malfunction because it directs the service technician to the next logical step in diagnosing the complaint.

If after completing the diagnostic circuit check and finding the on-board diagnostics functioning properly and no trouble codes displayed, a comparison of "Typical Scan Values," for the appropriate engine, may be used for comparison. The "Typical Values" are an average of display values recorded from normally operating vehicles and are intended to represent what a normally functioning system would display.

A "SCAN" TOOL THAT DISPLAYS FAULTY DATA SHOULD NOT BE USED, AND THE PROBLEM SHOULD BE REPORTED TO THE MANUFACTURER. THE USE OF A FAULTY "SCAN" CAN RESULT IN MISDIAGNOSIS AND UNNECESSARY PARTS REPLACEMENT.

Only the parameters listed below are used in this manual for diagnosis. If a "Scan" reads other parameters, the values are not recommended by General Motors for use in diagnosis. For more description on the values and use of the "Scan" to diagnose ECM inputs, refer to the applicable diagnosis section. If all values are within the range illustrated, refer to "Symptoms" in Section "B".

TYPICAL "SCAN" DATA VALUES

Idle / Upper Radiator Hose Hot / Closed Throttle / Park or Neutral / Closed Loop / Acc. off

"SCAN" Position	Units Displayed	Typical Data Value
Engine Speed	RPM	900 - 1000 P/N
Desired Idle	RPM	675 - 750 In Drive
Coolant Temp.	C° F°	85°C - 105°C (185°F - 221°F)
Throttle Position	Volts	.33 - .46 V
LV8 (Eng. Load)	Number	60 - 90
O2 Sensor	mV	100-900 mV
Inj. Pulse Width	m Sec	4.0-5.0 mSec (varies)
O2 Cross counts	# number	0-20
Air/Fuel Ratio	Ratio	14.7
Rich/Lean	Lean/Rich	Lean - Rich
Spark Advance	Degrees	20° varies
Mass Air Flow (MAF)	Gram Per Second (Gm/Sec)	4 - 7 varies
Fuel Integrator	Counts	118 - 138
Block Learn	Counts	118 - 138
Open/Closed Loop	Open/Closed	Closed Loop
Block Learn Cell	Cell Number	Cell number varies with engine RPM and Mass Air Flow
Knock Retard	Degrees of Retard	0°
Knock Signal	Yes/No	No (Yes, if detonation is present)
Idle Air Control (IAC)	Counts (Steps)	10 - 30
P/N Switch	P/N and RDL	P/N
Vehicle Speed (VSS)	MPH	0(mph)
Torque Converter Clutch (TCC)	On/Off	Off
Battery Voltage	Volts	13.8 volts (varies)
Purge Duty Cycle	%	15%
IAC Learned	Yes/No	Yes
Air Fuel Learned	Yes/No	Yes
A/C Pressure	Volt/PSI	2 - 4.5V/139-399 PSI (varies with Pressure)
A/C Request	Off/On	Off/On
A/C Clutch	Yes/No	No (Yes when A/C on)
Brake Switch	Yes/No	No (Yes with pedal depressed)
Cruise Engaged	Yes/No	No
Power Steering (PSPS)	Normal/High	Normal
Fan 1/Fan 2	Off/On	Off (On if temperature is above 100°C (212°F)
QDM1	High/Low	Low
QDM2	High/Low	Low (High if A/C head pressure's high)
QDM3	High/Low	Low (High with brake pedal depressed)
2nd Gear	Off/On	Off
3rd/4th Gear	Off/On	Off
Prom ID	Number	internal ID only
Time From Start	Min/Sec	Start when engine is running and varies

NOTE: IF ALL VALUES ARE WITHIN THE RANGE ILLUSTRATED, REFER TO "SYMPTOMS" IN SECTION "B".

1989–90 3.3L (VIN N) CUTLASS CALAIS AND SKYLARK

DIAGNOSTIC CIRCUIT CHECK
3300 (VIN N) (PORT)

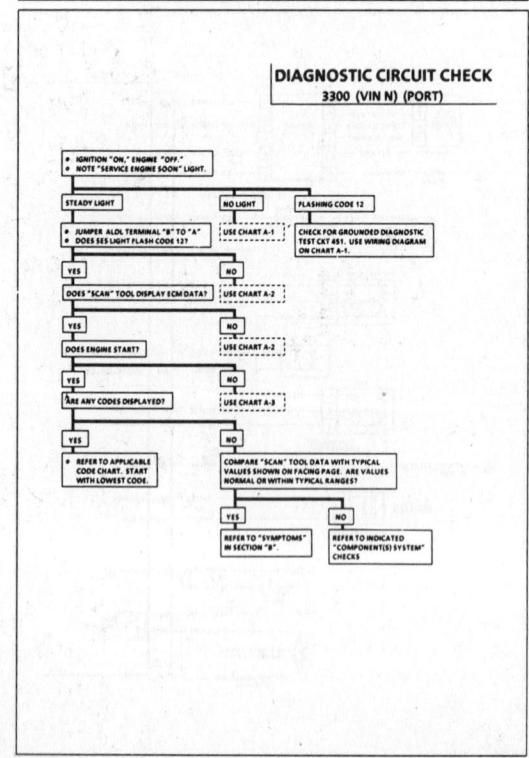

1989–90 3.3L (VIN N)
CUTLASS CALAIS AND SKYLARK

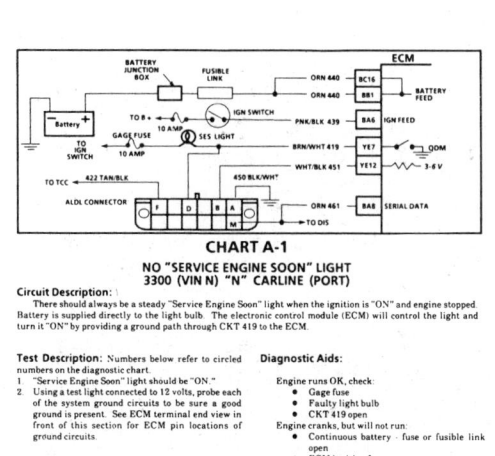

CHART A-1
NO "SERVICE ENGINE SOON" LIGHT
3300 (VIN N) "N" CARLINE (PORT)

Circuit Description:
There should always be a steady "Service Engine Soon" light when the ignition is "ON" and engine stopped. Battery is supplied directly to the light bulb. The electronic control module (ECM) will control the light and turn it "ON" by providing a ground path through CKT 419 to the ECM.

Test Description: Numbers below refer to circled numbers on the diagnostic chart.
1. "Service Engine Soon" light should be "ON."
2. Using a test light connected to 12 volts, probe each of the system ground circuits to be sure a good ground is present. See ECM terminal end view in front of this section for ECM pin locations of ground circuits.

Diagnostic Aids:
Engine runs OK, check:
- Gage fuse
- Faulty light bulb
- CKT 419 open

Engine cranks, but will not run:
- Continuous battery - fuse or fusible link open
- ECM ignition fuse open
- Ignition CKT 439 to ECM open
- Poor connection to ECM

1989–90 3.3L (VIN N)
CUTLASS CALAIS AND SKYLARK

CHART A-1
NO "SERVICE ENGINE SOON" LIGHT
3300 (VIN N) "N" CARLINE (PORT)

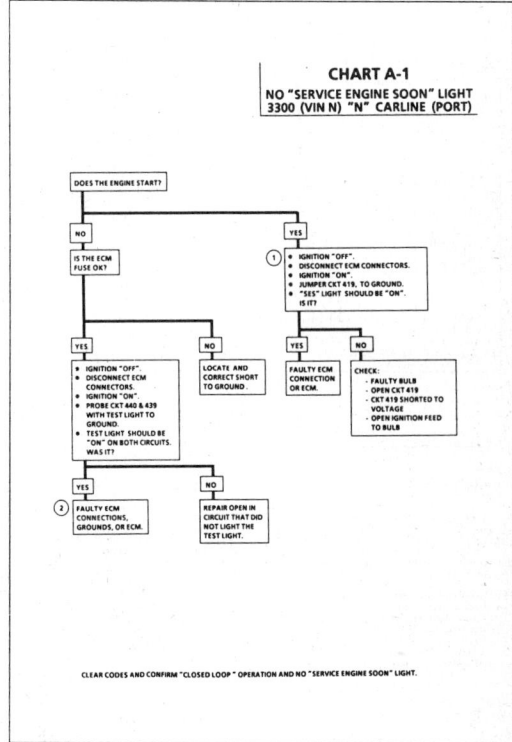

1989–90 3.3L (VIN N)
CUTLASS CALAIS AND SKYLARK

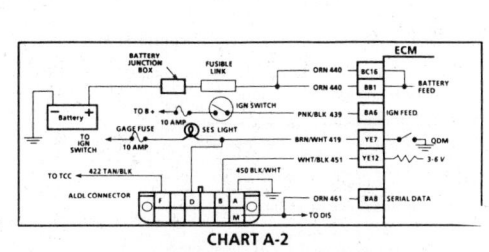

CHART A-2
WON'T FLASH CODE 12 - NO SERIAL DATA
"SERVICE ENGINE SOON" LIGHT "ON" STEADY
3300 (VIN N) "N" CARLINE (PORT)

Circuit Description:
There should be a steady "Service Engine Soon" light when the ignition is "ON" and engine not running. Ignition is supplied directly to the light bulb. The electronic control module (ECM) will turn the light "ON" by grounding CKT 419 at the ECM.

With the diagnostic terminal grounded, the light should flash a Code 12, followed by any trouble code(s) stored in memory.

A steady light suggests a short to ground in the light control CKT 419 or an open in diagnostic CKT 451 or an open in the 450 circuit.

Test Description: Numbers below refer to circled numbers on the diagnostic chart.
1. If the light goes "OFF" when the ECM connector is disconnected, then CKT 419 is not shorted to ground.
2. If there is a problem with the ECM that causes a "Scan" tool to not read serial data, then the ECM should not flash a Code 12. If Code 12 does flash, be sure that the "Scan" tool is working properly on another vehicle.

If the "Scan" is functioning properly and CKT 461 is OK, the Mem-Cal or ECM may be at fault for the NO ALDL symptom.
3. This step will check for an open diagnostic CKT 451
4. At this point, the "Service Engine Soon" light wiring is OK. The problem is a faulty ECM or Mem Cal. If Code 12 does not flash, the ECM should be replaced using the original Mem-Cal. Replace the Mem-Cal only after trying an ECM, as a defective Mem-Cal is an unlikely cause of the problem.

1989–90 3.3L (VIN N)
CUTLASS CALAIS AND SKYLARK

CHART A-2
WON'T FLASH CODE 12 - NO SERIAL DATA
"SERVICE ENGINE SOON" LIGHT "ON" STEADY
3300 (VIN N) "N" CARLINE (PORT)

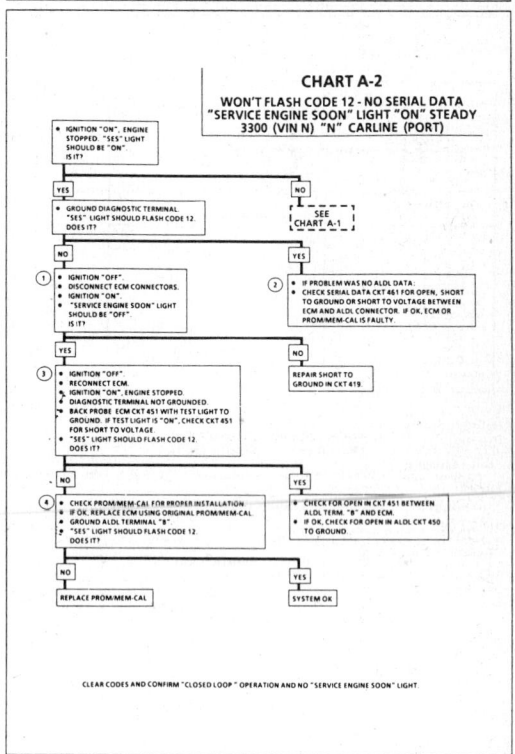

1989–90 3.3L (VIN N)
CUTLASS CALAIS AND SKYLARK

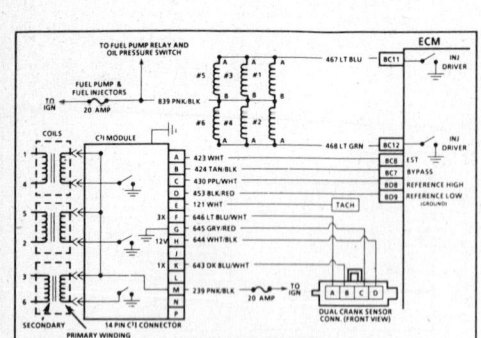

CHART A-3
(Page 1 of 2)
ENGINE CRANKS BUT WON'T RUN
3300 (VIN N) "N" CARLINE (PORT)

Circuit Description:
This chart assumes that battery condition and engine cranking speed are OK and there is adequate fuel in the tank.

Test Description: Numbers below refer to circled numbers on the diagnostic chart.
1. A "Service Engine Soon" light "ON" is a basic test to determine if there is a 12 volt supply and ignition 12 volts to ECM. The injector test light should blink, indicating the ECM is in control of the injectors. How bright the light blinks is not important. However, the test light should be a J 34730-2A or equivalent. The engine will not start without reference pulses and therefore the "Scan" should read rpm (reference) during crank.
2. Use fuel pressure gage J 34730-1 or equivalent. Wrap a shop towel around the fuel pressure tap to absorb any small amount of fuel leakage that may occur when installing the gage.
This test will determine if the ignition module is not generating the reference pulse or if the wiring or ECM are at fault.

By touching and removing test light to 12 volts on CKT 430, a reference pulse should be generated. If rpm is indicated, the ECM and wiring are OK.
3. Because the direct ignition system uses two plugs and wires to complete the circuit of each coil, the opposite spark should be left connected. If rpm was indicated during crank, the ignition module is receiving a crank signal, but no spark at this test indicates the ignition module is not triggering the coils.
4. This test will determine if the ignition module is not triggering the problem coil or if the tested coil is at fault. This test could also be performed by using another known good coil.
5. If test light is "OFF," the 20 amp fuse could be blown or CKT 839 open or shorted to ground.

1989–90 3.3L (VIN N)
CUTLASS CALAIS AND SKYLARK

CAUTION: WHEN INSTALLING FUEL PRESSURE TEST GAGE, USE A SHOP TOWEL TO ABSORB ANY FUEL THAT MAY LEAK AT FUEL RAIL TEST FITTING.

CHART A-3
(Page 1 of 2)
ENGINE CRANKS BUT WILL NOT RUN
3300 (VIN N) "N" CARLINE (PORT)

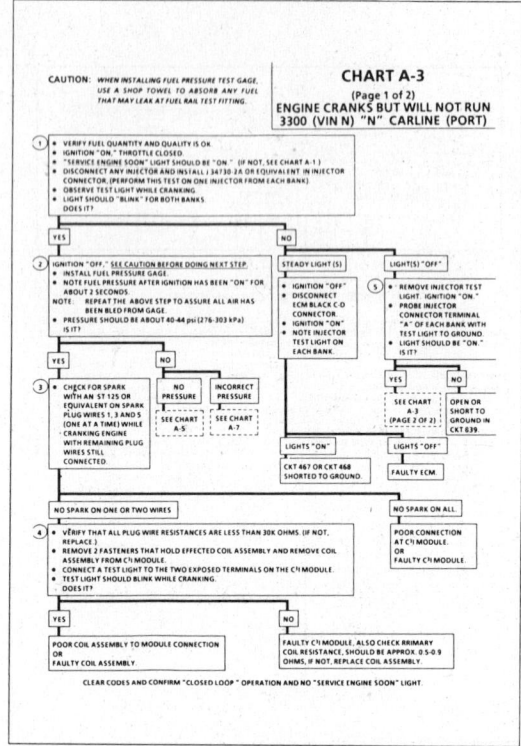

1989–90 3.3L (VIN N)
CUTLASS CALAIS AND SKYLARK

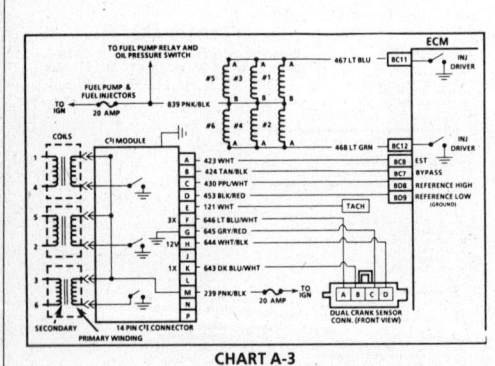

CHART A-3
(Page 2 of 2)
ENGINE CRANKS BUT WON'T RUN
3300 (VIN N) "N" CARLINE (PORT)

Circuit Description:
This chart assumes that battery conditions and engine cranking speed are OK, and there is adequate fuel in the tank.

Test Description: Numbers below refer to circled numbers on the diagnostic chart.
1. This check simulates a reference pulse to the ECM. The ECM should cycle the injectors with each touch of terminal "C".
2. The check is to verify that the dual crank sensor has power and a ground that are both supplied by the C³I module.

3. This test will test the C³I module ability to process the simulated signal from the dual crank sensor.
4. This will check for a short to voltage on the injector circuits and for a faulty ECM.

Diagnostic Aids:

Intermittents
- Poor connection at the C³I module, dual crank sensor, or the ECM could cause a starting problem.

1989–90 3.3L (VIN N)
CUTLASS CALAIS AND SKYLARK

FROM CHART A-3
(PAGE 1 OF 2)
TEST LIGHT "ON"

CHART A-3
(Page 2 of 2)
ENGINE CRANKS BUT WON'T RUN
3300 (VIN N) "N" CARLINE (PORT)

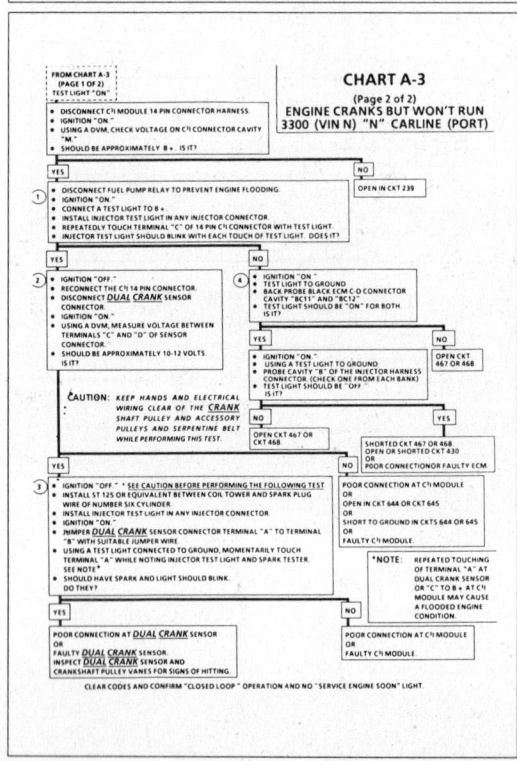

1989-90 3.3L (VIN N)
CUTLASS CALAIS AND SKYLARK

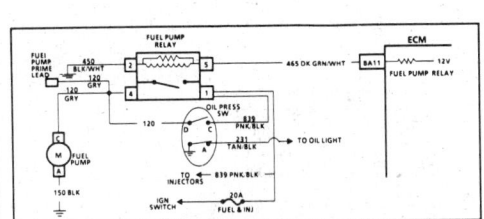

CHART A-5
(Page 1 of 2)
FUEL SYSTEM ELECTRICAL TEST
3300 (VIN N) "N" CARLINE (PORT)

Circuit Description:

When the ignition switch is turned "ON," the electronic control module (ECM) will energize the fuel pump relay which completes the circuit to the in-tank fuel pump. It will remain "ON" as long as the engine is cranking or running and the ECM is receiving C3I reference pulses. If there are no reference pulses, the ECM will de-energize the fuel pump relay within 2 seconds after key "ON" or the engine is stopped.

The fuel pump will deliver fuel to the fuel rail and injectors, then to the pressure regulator where the system pressure is controlled. Excess fuel pressure is bypassed back to the fuel tank. When the engine is stopped, the pump can be turned "ON" by applying battery voltage to the "test" terminal located in the engine compartment.

Improper fuel system pressure may contribute to one or all of the following symptoms:

- Cranks but won't run
- Code 44 or 45
- Cuts out, may feel like ignition problem
- Hesitation, loss of power or poor fuel economy

Test Description: Numbers below refer to circled numbers on the diagnostic chart.

1. If the fuse is blown, a short to ground in CKTs 120, 839, or the fuel pump itself is the cause.
2. This step determines if the fuel pump circuit is being controlled by the ECM. The ECM should energize the fuel pump relay and turn the fuel pump "ON." If the engine is not cranking or running, the ECM should de-energize the relay and/or fuel pump within 2 seconds after the ignition is turned "ON."

3. Applying B+ to the pump prime connector turns "ON" the fuel pump. This validates CKT 120 wiring. If the pump runs, it is a basic fuel delivery problem.
4. This test will determine if a short to ground on CKT 120 caused the fuse to blow. To prevent a mis-diagnosis, be sure the fuel pump is disconnected before the test.
5. Checks for a short to ground in the fuel pump relay harness CKT 839.

1989-90 3.3L (VIN N)
CUTLASS CALAIS AND SKYLARK

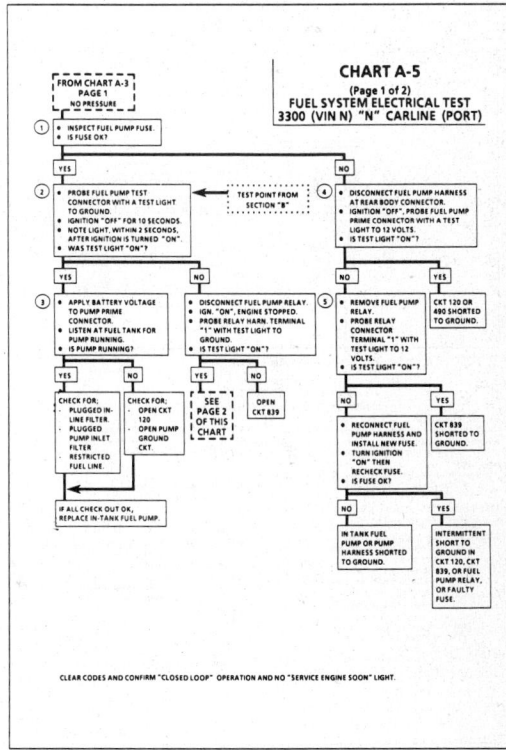

CLEAR CODES AND CONFIRM "CLOSED LOOP" OPERATION AND NO "SERVICE ENGINE SOON" LIGHT.

1989-90 3.3L (VIN N)
CUTLASS CALAIS AND SKYLARK

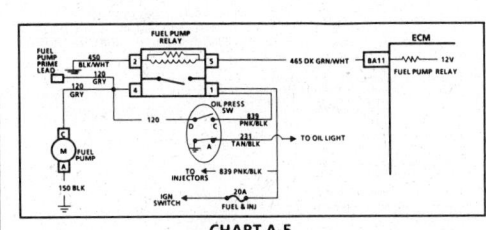

CHART A-5
(Page 2 of 2)
FUEL SYSTEM ELECTRICAL TEST
3300 (VIN N) "N" CARLINE (PORT)

Circuit Description:

When the ignition switch is turned "ON," the electronic control module (ECM) will energize the fuel pump relay which completes the circuit to the in-tank fuel pump. It will remain "ON" as long as the engine is cranking or running and the ECM is receiving C3I reference pulses. If there are no reference pulses, the ECM will de-energize the fuel pump relay within 2 seconds after key "ON" or the engine is stopped.

Test Description: Numbers below refer to circled numbers on the diagnostic chart.

6. Checks for open in the fuel pump relay ground CKT 450.
7. Determines if the ECM is in control of the fuel pump relay through CKT 465 (terminal "A").
8. The fuel pump control circuit includes an engine oil pressure switch with a separate set of normally open contacts. The switch closes at about (4 lbs) 28 kPa of oil pressure and provides a second ignition feed path to the fuel pump. If the relay fails, the pump will run using the ignition feed supplied by the closed oil pressure switch.

This step checks the oil pressure switch to be sure it provides ignition feed to the fuel pump should the pump relay fail. A failed pump relay will result in extended engine crank time because of the time required to build enough oil pressure to close the oil pressure switch and turn "ON" the fuel pump. There may be instances when the relay has failed but the engine will not crank fast enough to build enough oil pressure to close the switch. This or a faulty oil pressure switch can result in "Engine Cranks But Won't Run."

1989-90 3.3L (VIN N)
CUTLASS CALAIS AND SKYLARK

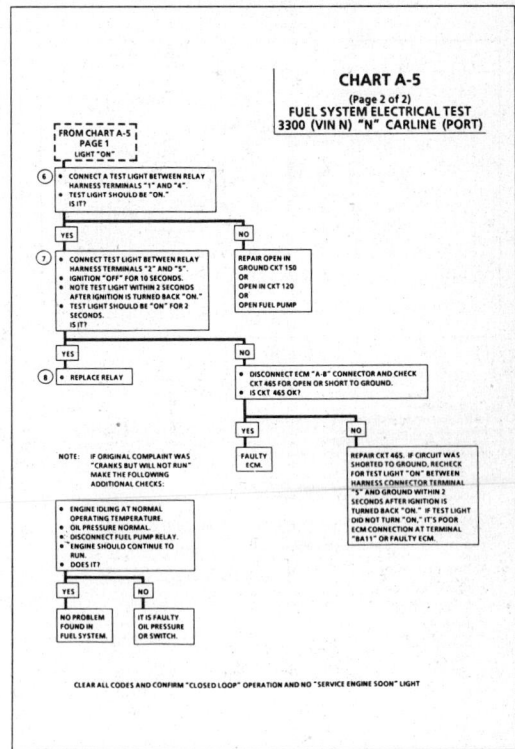

CLEAR ALL CODES AND CONFIRM "CLOSED LOOP" OPERATION AND NO "SERVICE ENGINE SOON" LIGHT

1989–90 3.3L (VIN N)
CUTLASS CALAIS AND SKYLARK

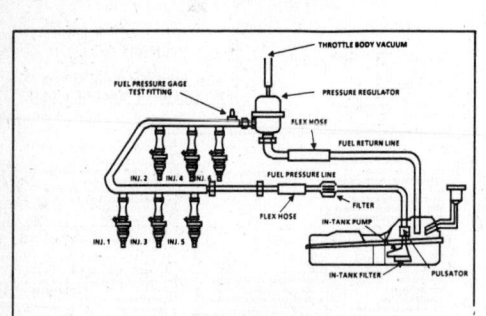

CHART A-7
(Page 1 of 2)
FUEL SYSTEM PRESSURE TEST
3300 (VIN N) "N" CARLINE (PORT)

Circuit Description:
The fuel pump will deliver fuel to the fuel rail and injectors, then to the pressure regulator, where the system pressure is controlled. Excess fuel pressure is bypassed back to the fuel tank. When the engine is stopped, the pump can be turned "ON" by applying battery voltage to the test terminal located in the engine compartment.

Improper fuel system pressure may contribute to one or all of the following symptoms:
- Cranks but won't run
- Code 44 or 45
- Cuts out, may feel like ignition problem
- Hesitation, loss of power or poor fuel economy

Test Description: Numbers below refer to circled numbers on the diagnostic chart.

CAUTION: To reduce the risk of fire and personal injury, wrap a shop towel around the fuel pressure connection to absorb any fuel leakage that may occur when installing the fuel pressure gage. Place towel in approved container.

1. Install pressure gage J 34730-1 to fuel pressure connection.
   - Connect the fuel tank harness connector
   - Start engine. With ignition "ON", pump pressure is controlled by spring pressure and throttle body vacuum within the pressure regulator assembly.

- Ignition "OFF" for 10 seconds. Pressure should not leak down after the fuel pump is shut "OFF".
2. When the engine is idling, the throttle body vacuum is high and is applied to the fuel regulator diaphragm. This will offset the spring and result in a lower fuel pressure.
3. The application of 12-14 inches of vacuum to the pressure regulator should result in a fuel pressure less than step 1.
4. Pressure that leaks down may be caused by one of the following:
   - In-tank fuel pump check valve not holding
   - Pump coupling hose leaking
   - Fuel pressure regulator valve leaking
   - Injector sticking open

1989–90 3.3L (VIN N)
CUTLASS CALAIS AND SKYLARK

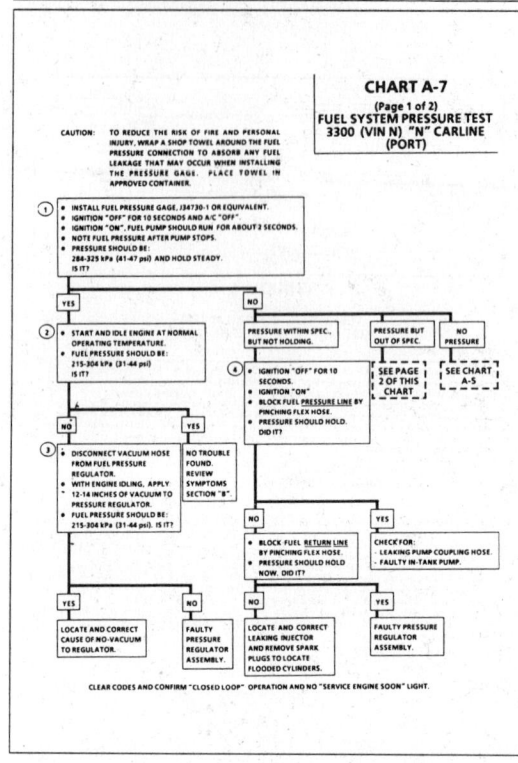

CHART A-7
(Page 1 of 2)
FUEL SYSTEM PRESSURE TEST
3300 (VIN N) "N" CARLINE (PORT)

CLEAR CODES AND CONFIRM "CLOSED LOOP" OPERATION AND NO "SERVICE ENGINE SOON" LIGHT.

1989–90 3.3L (VIN N)
CUTLASS CALAIS AND SKYLARK

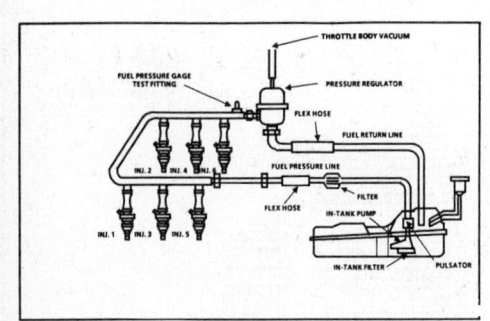

CHART A-7
(Page 2 of 2)
FUEL SYSTEM PRESSURE TEST
3300 (VIN N) "N" CARLINE (PORT)

Circuit Description:
The fuel pump will deliver fuel to the fuel rail and injectors, then to the pressure regulator, where the system pressure is controlled. Excess fuel pressure is bypassed back to the fuel tank. When the engine is stopped, the pump can be turned "ON" by applying battery voltage to the test terminal located in the engine compartment.

Improper fuel system pressure may contribute to one or all of the following symptoms:
- Cranks but won't run
- Code 44 or 45
- Cuts out, may feel like ignition problem
- Hesitation, loss of power, or poor fuel economy

Test Description: Numbers below refer to circled numbers on the diagnostic chart.
5. Pressure but less than specifications falls into two areas:
   - Regulated pressure but less than specifications; the amount of fuel to injectors is OK, but pressure is too low. The system will be lean running and may set Code 44. Also, possible hard starting cold and overall poor performance.
   - Restricted flow causing pressure drop; Normally, a vehicle with a fuel pressure of less than 165 kPa (24 psi) at idle will not be driveable.

However, if the pressure drop occurs only while driving, the engine will normally surge then stop as pressure begins to drop rapidly.
6. Restricting the fuel return line allows the fuel pressure to build above regulated pressure. With battery voltage applied to the pump test terminal, pressure should rise above 325 kPa (47 psi) as the return line is slowly pinched off.

NOTICE: Do Not allow pressure to exceed 414 kPa (60 psi), as damage to the regulator may result.

7. This test determines if the high fuel pressure is due to a restricted fuel return line or a pressure regulator problem.

1989–90 3.3L (VIN N)
CUTLASS CALAIS AND SKYLARK

CHART A-7
(Page 2 of 2)
FUEL SYSTEM PRESSURE TEST
3300 (VIN N) "N" CARLINE (PORT)

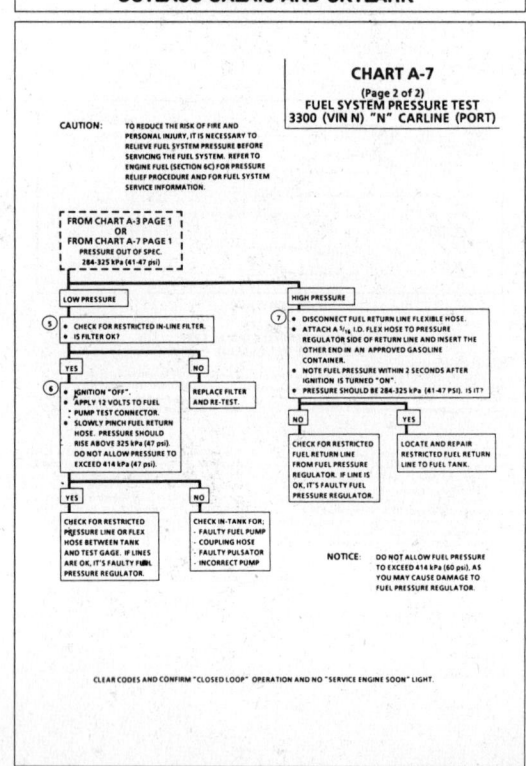

CLEAR CODES AND CONFIRM "CLOSED LOOP" OPERATION AND NO "SERVICE ENGINE SOON" LIGHT.

1989–90 3.3L (VIN N)
CUTLASS CALAIS AND SKYLARK

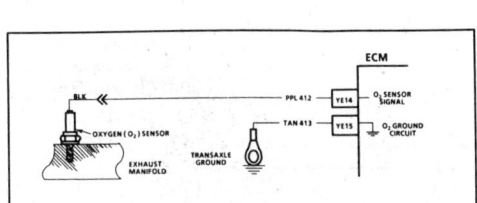

CODE 13
OXYGEN SENSOR CIRCUIT
(OPEN CIRCUIT)
3300 (VIN N) "N" CARLINE (PORT)

Circuit Description:

The electronic control module (ECM) supplies a voltage of about .45 volt between terminals "YE14" and "YE15". (If measured with a 10 megohm digital voltmeter, this may read as low as .32 volt.) The O_2 sensor varies the voltage within a range of about 1 volt if the exhaust is rich, down through about .10 volt if exhaust is lean.

The sensor is like an open circuit and produces no voltage when it is below 360°C (600°F). An open oxygen sensor circuit or cold oxygen sensor causes "Open Loop" operation.

Code 13 will set:
- Engine at normal operating temperature above 50°C (122°F)
- No Codes 21 or 22
- At least 40 seconds engine time after start
- O_2 signal voltage is steady between .30 and .55 volt
- Throttle position sensor signal above 4% (about .58 volt above closed throttle voltage)
- All conditions must be met for about 30 seconds

Test Description: Numbers below refer to circled numbers on the diagnostic chart.
1. If the conditions for a Code 13 exists, the system will not go to "Closed Loop."
2. This will determine if the sensor or the wiring is the cause of the Code 13.
3. In doing this test use only a high impedance digital volt ohmmeter. This test checks the continuity of CKTs 412 and 413. If CKT 413 is open the ECM voltage on CKT 412 will be over .6 volt (600 mV).

Diagnostic Aids:

Normal "Scan" voltage varies between 100mV to 999mV (.1 and 1.0 volt) while in "Closed Loop." Code 13 sets in 30 seconds if voltage remains between .30 and .55 volt, but the system will go "Open Loop" in about 15 seconds.

Refer to "Intermittents" in Section "B".

1989–90 3.3L (VIN N)
CUTLASS CALAIS AND SKYLARK

CODE 13
OXYGEN SENSOR CIRCUIT
(OPEN CIRCUIT)
3300 (VIN N) "N" CARLINE (PORT)

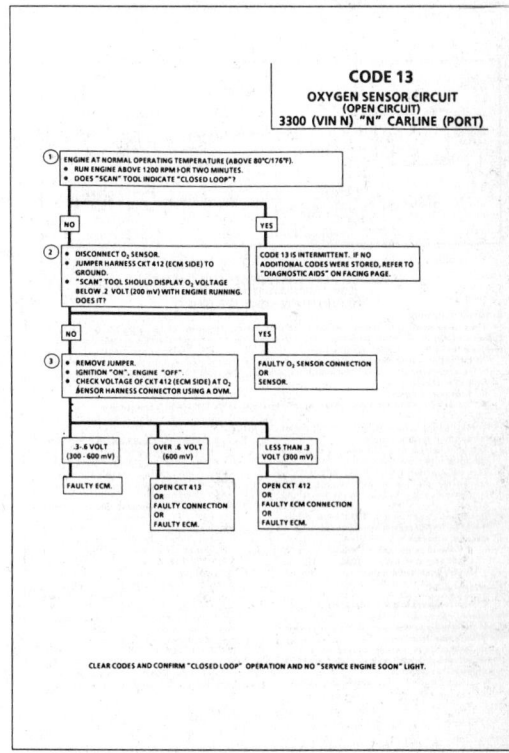

CLEAR CODES AND CONFIRM "CLOSED LOOP" OPERATION AND NO "SERVICE ENGINE SOON" LIGHT.

1989–90 3.3L (VIN N)
CUTLASS CALAIS AND SKYLARK

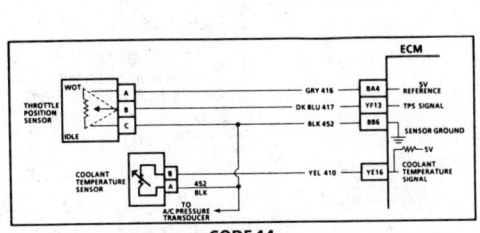

CODE 14
COOLANT TEMPERATURE SENSOR CIRCUIT
(HIGH TEMPERATURE INDICATED)
3300 (VIN N) "N" CARLINE (PORT)

Circuit Description:

The coolant temperature sensor uses a thermistor to control the signal voltage to the ECM. The ECM applies a voltage on CKT 410 to the sensor. When the engine is cold the sensor (thermistor) resistance is high, therefore, the ECM will see high signal voltage.

As the engine warms, the sensor resistance becomes less, and the voltage drops. At normal engine operating temperature (85°C to 95°C), the voltage will measure about 1.5 to 2.0 volts.

Code 14 will set if:
- Engine run time is 10 seconds or more.
- Signal voltage indicates a coolant temperature above 140°C (284°F) for .4 seconds.

Test Description: Numbers below refer to circled numbers on the diagnostic chart.
1. This will determine if the coolant sensor is indicating a high temperature to the ECM.
2. This test will determine if CKT 410 is shorted to ground which will cause the conditions for Code 14.
If Code 14 is set, the ECM will use a default coolant temperature value of 49°C (120°F) for fuel control. "Scan" tool will read actual value - not default value.

Diagnostic Aids:

"Scan" tool displays engine temperature in degrees centigrade. After engine is started, the temperature should rise steadily to about 90°C then stabilize when thermostat opens.

An intermittent may be caused by a poor connection, rubbed through wire insulation, or a wire broken inside the insulation.

Check For:
- **Poor Connection or Damaged Harness** Inspect ECM harness connectors for backed out terminal "YE16", improper mating, broken locks, improperly formed or damaged terminals, poor terminal to wire connection, and damaged harness.
- **Intermittent Test** If connections and harness check OK, "Scan" coolant temperature while moving related connectors and wiring harness. If the failure is induced, the "coolant temperature" display will change. This may help to isolate the location of the malfunction.
- **Shifted Sensor** The "Temperature To Resistance Value" scale may be used to test the coolant sensor at various temperature levels to evaluate the possibility of a "shifted" (mis-scaled) sensor, which may result in driveability complaints.

1989–90 3.3L (VIN N)
CUTLASS CALAIS AND SKYLARK

CODE 14
COOLANT TEMPERATURE SENSOR CIRCUIT
(HIGH TEMPERATURE INDICATED)
3300 (VIN N) "N" CARLINE (PORT)

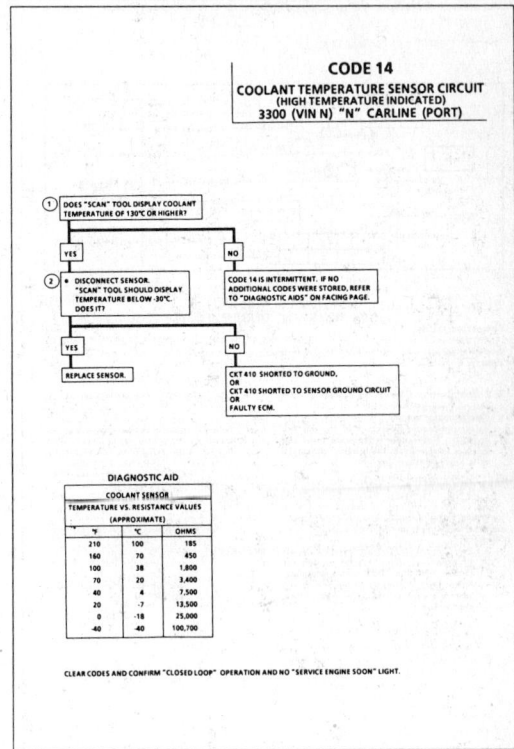

DIAGNOSTIC AID

COOLANT SENSOR
TEMPERATURE VS. RESISTANCE VALUES
(APPROXIMATE)

°F	°C	OHMS
210	100	185
160	70	450
100	38	1,800
70	20	3,400
40	4	7,500
20	-7	13,500
0	-18	25,000
-40	-40	100,700

CLEAR CODES AND CONFIRM "CLOSED LOOP" OPERATION AND NO "SERVICE ENGINE SOON" LIGHT.

1989–90 3.3L (VIN N)
CUTLASS CALAIS AND SKYLARK

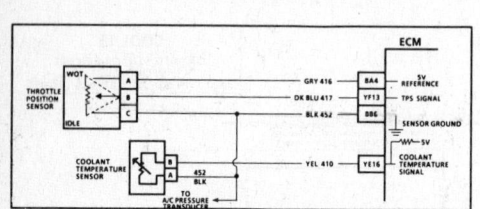

CODE 15
(LOW TEMPERATURE INDICATED)
COOLANT TEMPERATURE SENSOR CIRCUIT
3300 (VIN N) "N" CARLINE (PORT)

Circuit Description:
The coolant temperature sensor uses a thermistor to control the signal voltage to the ECM. The ECM applies a voltage on CKT 410 to the sensor. When the engine is cold, the sensor (thermistor) resistance is high, therefore, the ECM will see high signal voltage.

As the engine warms, the sensor resistance becomes less, and the voltage drops. At normal engine operating temperature (85°C to 95°C) the voltage will measure about 1.5 to 2.0 volts at the ECM.

Code 15 will be set if:
- Engine is running
- Signal voltage indicates a coolant temperature less than -38°C (-38°F) for at least 4 seconds

Test Description: Numbers below refer to circled numbers on the diagnostic chart.
1. This will determine if the coolant sensor is indicating a low temperature to the ECM.
2. This test simulates a Code 14. If the ECM recognizes the low signal voltage (high temperature) and the "Scan" reads 140°C, the ECM and wiring are OK.
3. This test will determine if CKT 410 is open. There should be 5 volts present at sensor connector if measured with a DVM.

Note: If Code 15 is set, the ECM will use 32°C (120°F) for fuel control. "Scan" tool will display actual sensor value.

Diagnostic Aids:
A "Scan" tool reads engine temperature in degrees centigrade. After engine is started the temperature should rise steadily to about 90°C then stabilize when thermostat opens.

An intermittent may be caused by a poor connection, rubbed through wire insulation or a wire broken inside the insulation.

Check For:
- **Poor Connection or Damaged Harness** Inspect ECM harness connectors for backed out terminal "YE16", improper mating, broken locks, improperly formed or damaged terminals, poor terminal to wire connection and damaged harness.
- **Intermittent Test** If connections and harness check OK, "Scan" coolant temperature while moving related connectors and wiring harness. If the failure is induced, the display will change. This may help to isolate the location of the malfunction.
- **Shifted Sensor** The "Temperature To Resistance Value" scale may be used to test the coolant sensor at various temperature levels to evaluate the possibility of a "shifted" (mis-scaled) sensor which may result in driveability complaints.

A faulty connection, or an open in CKTs 410 or 452 will result in a Code 15.

If Code 23 is also set, check CKT 452 for faulty wiring or connections. Check terminals at sensor for good contact. Refer to "Intermittents" in Section "B".

1989–90 3.3L (VIN N)
CUTLASS CALAIS AND SKYLARK

CODE 15
COOLANT TEMPERATURE SENSOR CIRCUIT
(LOW TEMPERATURE INDICATED)
3300 (VIN N) "N" CARLINE (PORT)

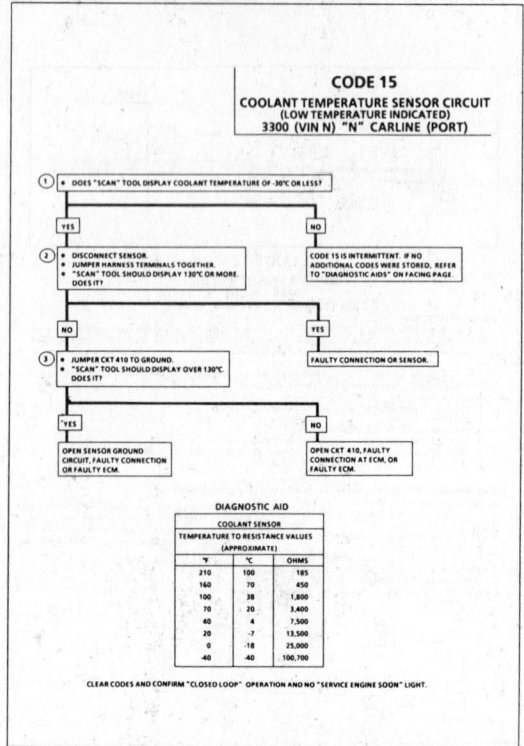

DIAGNOSTIC AID

COOLANT SENSOR		
TEMPERATURE TO RESISTANCE VALUES		
(APPROXIMATE)		
°F	°C	OHMS
210	100	185
160	70	450
100	38	1,800
70	20	3,400
40	4	7,500
20	-7	13,500
0	-18	25,000
-40	-40	100,700

CLEAR CODES AND CONFIRM "CLOSED LOOP" OPERATION AND NO "SERVICE ENGINE SOON" LIGHT.

1989–90 3.3L (VIN N)
CUTLASS CALAIS AND SKYLARK

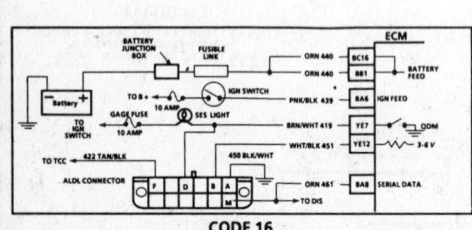

CODE 16
SYSTEM VOLTAGE HIGH
3300 (VIN N) "N" CARLINE (PORT)

Circuit Description:
The ECM monitors battery or system voltage on CKT 440 to terminals "BC16" and "BB1". If the ECM detects voltage above 16 volts for more than 10 seconds, it will turn the SES light "ON" and set Code 16 in memory.

Test Description: Numbers below refer to circled numbers on the diagnostic chart.
1. Test generator output to determine proper operation of the voltage regulator. Run engine at moderate speed and measure voltage across the battery. If over 16 volts, repair generator.

Diagnostic Aids:
An intermittent may be caused by a poor connection, rubbed through insulation, a wire broken inside the insulation or poor ECM grounds.

Check For:
- **Poor Connection or Damaged Harness** Inspect ECM harness connectors for backed out terminal "BC16" or "BB1", improper mating, broken locks, improperly formed or damaged terminals, poor terminal to wire connection and damaged harness.

- **Intermittent Test** If connections and harness checks OK, monitor battery voltage display while moving related connectors. If the failure is induced, the battery voltage will abruptly change. This may help to isolate the location of the malfunction. An engine stall while manipulating the harness indicates that the ECM has lost voltage at terminal "BC16" or "BB1". Check for loose connectors in CKT 440.

NOTE: Charging battery with a battery charger and starting the engine may set a Code 16.

IMPORTANT: If ECM grounds and battery grounds are located on the same stud at transaxle, this could cause intermittent or improper sensor data to ECM.

1989–90 3.3L (VIN N)
CUTLASS CALAIS AND SKYLARK

CODE 16
SYSTEM VOLTAGE HIGH
3300 (VIN N) "N" CARLINE (PORT)

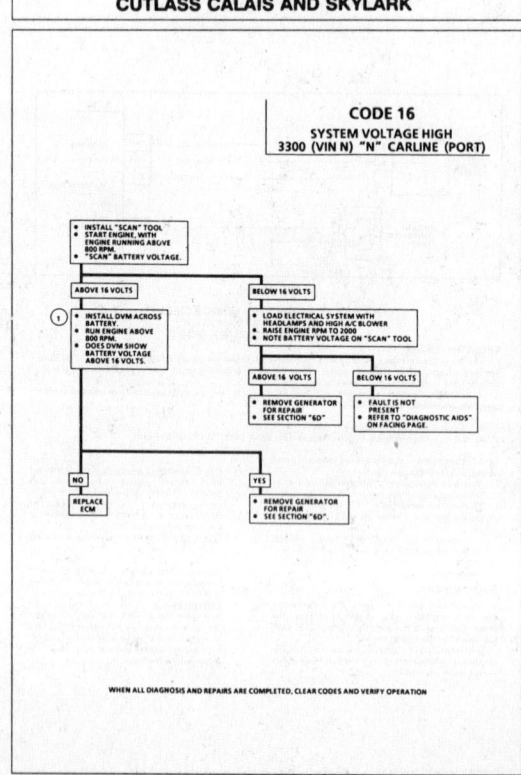

WHEN ALL DIAGNOSIS AND REPAIRS ARE COMPLETED, CLEAR CODES AND VERIFY OPERATION

1989–90 3.3L (VIN N) CUTLASS CALAIS AND SKYLARK

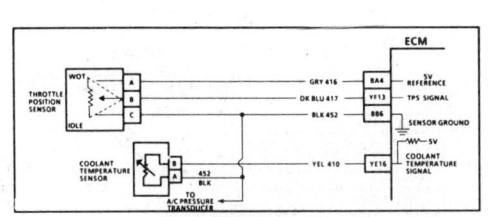

CODE 21
THROTTLE POSITION SENSOR (TPS) CIRCUIT
(SIGNAL VOLTAGE HIGH)
3300 (VIN N) "N" CARLINE (PORT)

Circuit Description:

The throttle position sensor (TPS) provides a voltage signal that changes relative to throttle blade angle. Signal voltage will vary from about 4 at idle to about 5 volts at wide open throttle.

The TPS signal is one of the most important inputs used by the ECM for fuel control and for most of the ECM control outputs.

Code 21 will set if:
- TPS voltage is greater than 4.9 volts at any time or
- Engine is running and air flow is less than 15 gm/sec
- TPS signal voltage is greater than 1.95 volts (30%)
- Code 34 not present
- All conditions met for 5 seconds.

Test Description: Numbers below refer to circled numbers on the diagnostic chart.
1. With closed throttle, ignition "ON," or at idle, voltage at YD13 should be .33-.46 volt.
2. When the TPS sensor is disconnected, the TPS voltage will go low and a Code 22 will set. Therefore, the ECM and wiring are OK.
3. Probing CKT 452 with a test light checks the sensor ground CKT. A faulty sensor ground circuit will cause a Code 21.

NOTE: If a Code 21 is set, the ECM will use a defaulted value for TPS of about 5 volt.

Diagnostic Aids:

A "Scan" tool reads throttle position in volts. With closed throttle, ignition "ON" or at idle, voltage should be .33-.46 volt.

Also some "Scan" tools will read throttle angle 0% = closed throttle 100% = WOT

An open in CKT 452 will result in a Code 21. Refer to "Intermittents" in Section "B".

Check For:
- **Poor Connection or Damaged Harness** Inspect ECM harness connectors for backed out terminal "YF13", improper mating, broken locks, improperly formed or damaged terminals, poor terminal to wire connection, and damaged harness.
- **Intermittent Test** If connections and harness check OK, monitor TPS voltage while moving related connectors and wiring harness. If the failure is induced, the display will change. This may help to isolate the location of the malfunction.
- **TPS Scaling** Observe TPS voltage display while depressing accelerator pedal with engine stopped and ignition "ON." Display should vary from closed throttle TPS voltage when throttle was closed, to over (4.5 volts) 4500 mV when throttle is held at wide open throttle position.

1989–90 3.3L (VIN N) CUTLASS CALAIS AND SKYLARK

CODE 21
THROTTLE POSITION SENSOR (TPS) CIRCUIT
(SIGNAL VOLTAGE HIGH)
3300 (VIN N) "N" CARLINE (PORT)

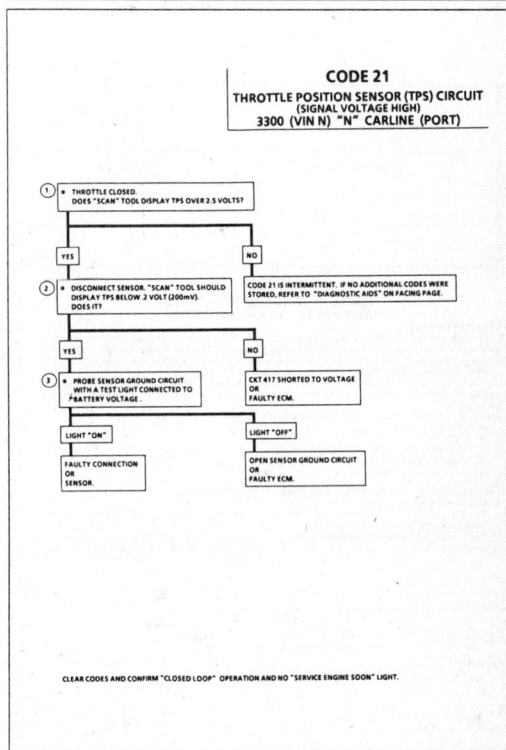

CLEAR CODES AND CONFIRM "CLOSED LOOP" OPERATION AND NO "SERVICE ENGINE SOON" LIGHT.

1989–90 3.3L (VIN N) CUTLASS CALAIS AND SKYLARK

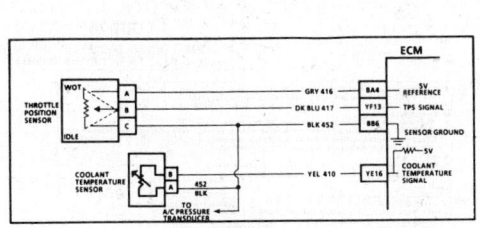

CODE 22
THROTTLE POSITION SENSOR (TPS) CIRCUIT
(SIGNAL VOLTAGE LOW)
3300 (VIN N) "N" CARLINE (PORT)

Circuit Description:

The throttle position sensor (TPS) provides a voltage signal that changes relative to throttle blade angle. Signal voltage will vary from about .4 at idle to about 5 volts at wide open throttle

The TPS signal is one of the most important inputs used by the ECM for fuel control and for most of the ECM control outputs.

Code 22 will set if:
- The ignition key is "ON"
- TPS signal voltage is less than .24 volt for 4 seconds.

Test Description: Numbers below refer to circled numbers on the diagnostic chart.
1. With closed throttle, ignition "ON" or at idle voltage at "YD13" should be .33-.46 volt.
2. Simulates Code 21: (high voltage) If ECM recognizes the high signal voltage the ECM and wiring are OK.
3. With closed throttle, ignition "ON" or at idle, voltage at "YF13" should be .33-.46 volt. If not, check adjustment.
4. Simulates a high signal voltage. Checks CKT 417 for an open.

Diagnostic Aids:

A "Scan" tool reads throttle position in volts. Voltage should increase at a steady rate as throttle is moved toward WOT.

Also some "Scan" tools will read throttle angle 0% = closed throttle 100% = WOT.

An open or short to ground in CKTs 416 or 417 will result in a Code 22.

Check For:
- **Poor Connection or Damaged Harness** Inspect ECM harness connectors for backed out terminal "YF13", improper mating, broken locks, improperly formed or damaged terminals, poor terminal to wire connection, and damaged harness.
- **Intermittent Test** If connections and harness check OK, monitor TPS voltage display while moving related connectors and wiring harness. If the failure is induced, the display will change. This may help to isolate the location of the malfunction.
- **TPS Scaling** Observe TPS voltage display while depressing accelerator pedal with engine stopped and ignition "ON." Display should vary from closed throttle TPS voltage when throttle was closed, to approximately 4.5 volts (4500 mV) when throttle is held at wide open throttle position.

1989–90 3.3L (VIN N) CUTLASS CALAIS AND SKYLARK

CODE 22
THROTTLE POSITION SENSOR (TPS) CIRCUIT
(SIGNAL VOLTAGE LOW)
3300 (VIN N) "N" CARLINE (PORT)

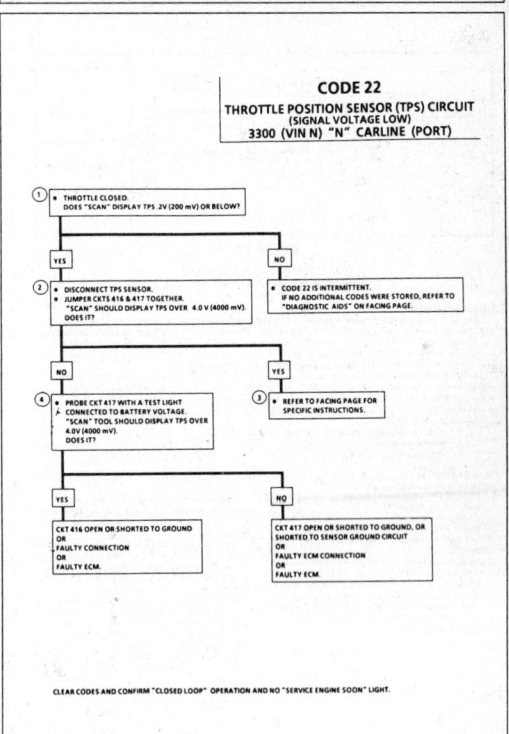

CLEAR CODES AND CONFIRM "CLOSED LOOP" OPERATION AND NO "SERVICE ENGINE SOON" LIGHT.

1989–90 3.3L (VIN N)
CUTLASS CALAIS AND SKYLARK

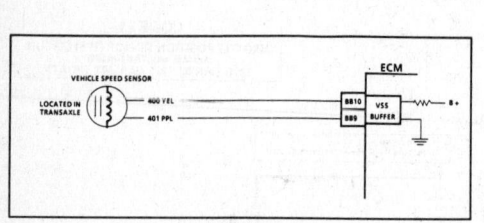

CODE 24
VEHICLE SPEED SENSOR (VSS) CIRCUIT
3300 (VIN N) "N" CARLINE (PORT)

Circuit Description:

Vehicle speed information is provided to the ECM by the vehicle speed sensor, a permanent magnet (PM) generator mounted in the transmission. The PM generator produces a pulsing voltage whenever vehicle speed is over about 3 mph. The A/C voltage level and the number of pulses increases with vehicle speed. The ECM converts the pulsing voltage to mph, and the mph can be displayed with a "Scan" tool.

The function of the VSS buffer, used in past model years, has been incorporated into the ECM. The ECM supplies the necessary signal for the instrument panel (4004 pulses per mile) for operating the speedometer and the odometer.

Code 24 will set if vehicle speed signal equals less than 3 mph when:
* Engine is running
* No Code 31
* When P/N switch is open (indicating drive range).
* When high gear switch is open
* All conditions met for 40 seconds

Test Description: Numbers below refer to circled numbers on the diagnostic chart.
1. The PM generator only produces a signal if drive wheels are turning greater than 3 mph.
2. Before replacing the ECM, check the Mem-Cal for correct application.

Diagnostic Aids:

"Scan" should indicate a vehicle speed whenever the drive wheels are turning greater than 3 mph.

Check CKT 400 and 401 for proper connections to be sure they are clean and tight and the harness is routed correctly. Refer to "Intermittents" in Section "B".

1989–90 3.3L (VIN N)
CUTLASS CALAIS AND SKYLARK

CODE 24
VEHICLE SPEED SENSOR (VSS) CIRCUIT
3300 (VIN N) "N" CARLINE (PORT)

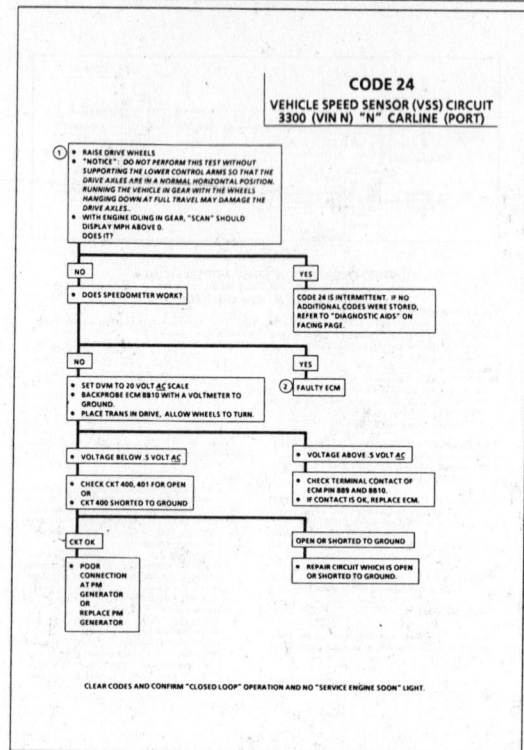

1989–90 3.3L (VIN N)
CUTLASS CALAIS AND SKYLARK

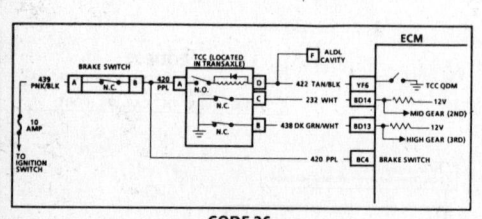

CODE 26
QUAD-DRIVER (QDM) CIRCUIT
3300 (VIN N) "N" CARLINE (PORT)

Circuit Description:

The ECM is used to control several components such as the torque converter clutch. The ECM controls the TCC through a device called a quad-driver module (QDM). When the ECM is commanding a component "ON," the voltage potential of the output circuit will be "low" (near 0 volt). When the ECM is commanding the output circuit to a component "OFF," the voltage potential of the circuit will be "high" (near battery voltage). The primary function of the QDM is to supply a ground for the device being controlled.

Each QDM has a fault line which is monitored by the ECM. The fault line signal is available on the data stream for "Scan" tool test equipment. The ECM will compare the voltage at the QDM based on accepted values of the fault line. If the QDM fault detection circuit senses a voltage other than the accepted value, the fault line will go from "low" signal on the data stream to a "high" signal and a Code 26 will set if applicable.

A Code 26 will be set if:
* A fault is detected by the QDM that controls the TCC circuit.
* The transmission is in mid or high gear.
* The brake is not applied.

Test Description: Numbers below refer to circled numbers on the diagnostic chart.
1. Light "OFF" confirms transmission second gear apply switch is open.
2. At 25 mph, the transmission second gear apply switch should close. Test light will come "ON" and confirm battery supply and closed brake switch.
3. Grounding the diagnostic terminal with ignition "ON," engine "OFF," should energize the TCC solenoid by grounding CKT 422. This test checks the ability of the ECM to supply a ground to the TCC solenoid. The test light connected from 12 volts to ALDL terminal "F" will turn "ON" as CKT 422 is grounded.

Diagnostic Aids:

An intermittent may be caused by a poor connection, mis-routed harness, rubbed through wire insulation, or a wire broken inside the insulation.
Check For:
* Poor Connection at ECM pins. Inspect harness connectors for backed out terminals, improper mating, broken locks, improperly formed or damaged terminals, and poor terminal to wire connection.
* Mis-routed Harness Inspect wiring harness to insure that it is not too close to high voltage wires, such as spark plug leads.
* Damaged Harness Inspect harness for damage. If harness appears OK, "Scan" while moving related connectors and wiring harness. A change in display would indicate the intermittent fault location.
* System OK Electrically Check for a hydraulic problem.

1989–90 3.3L (VIN N)
CUTLASS CALAIS AND SKYLARK

CODE 26
QUAD-DRIVER (QDM) CIRCUIT
3300 (VIN N) "N" CARLINE (PORT)

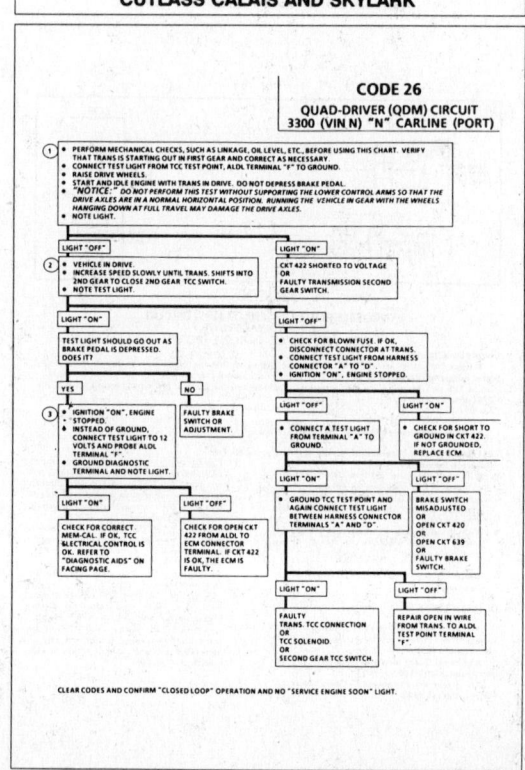

1989–90 3.3L (VIN N)
CUTLASS CALAIS AND SKYLARK

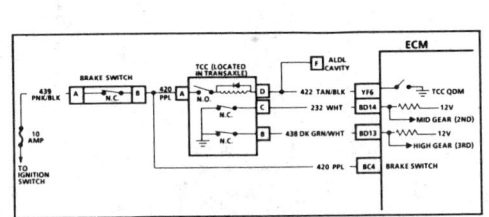

CODES 27, 28
GEAR SWITCHES CIRCUITS
3300 (VIN N) "N" CARLINE (PORT)

Circuit Description:

The gear switches are located inside the transaxle. They are pressure operated switches, normally closed. The ECM supplies 12 volts through each selected circuit to the switch. In any condition other than the specific gear application, the signal line monitors low voltage, low potential. As road speed increases, hydraulic pressure applies the specific gear clutches and the gear switch opens. At this time, the ECM monitors a high, 12 volt potential, and interprets this to indicate that gear is applied. The ECM uses the gear signals to control fuel delivery (and TCC).

Code 27 will set if:
- CKT 232 indicates ground or closed switch for 10 seconds when vehicle is in 2nd gear operation.
- CKT 232 indicates an open (drive) when the engine is first started.

Code 28 will set if:
- CKT 438 indicates ground or closed switch for 10 seconds when vehicle is in 3rd gear operation.
- CKT 438 indicates an open (drive) when the engine is first started.

Test Description: Numbers below refer to circled numbers on the diagnostic chart.
1. Must use a DVM. A test light will not light due to the very low current being supplied by the ECM.
2. Checks to see if CKT is grounded through the switch.
3. Checks for a good, properly operating switch and checks CKT within transaxle for an improper ground.

Diagnostic Aids:

An intermittent may be caused by a poor connection, mis-routed harness, rubbed through wire insulation, or a wire broken inside the insulation.

Check For:
- **Poor Connection** at ECM pins. Inspect harness connectors for backed out terminals, improper mating, broken locks, improperly formed or damaged terminals, and poor terminal to wire connection.
- **Mis-routed Harness** Inspect wiring harness to insure that it is not too close to high voltage wires, such as spark plug leads.
- **Damaged Harness** Inspect harness for damage. If harness appears OK, "Scan" while moving related connectors and wiring harness. A change in display would indicate the intermittent fault location.

1989–90 3.3L (VIN N)
CUTLASS CALAIS AND SKYLARK

CODES 27, 28
GEAR SWITCHES CIRCUITS
3300 (VIN N) "N" CARLINE (PORT)

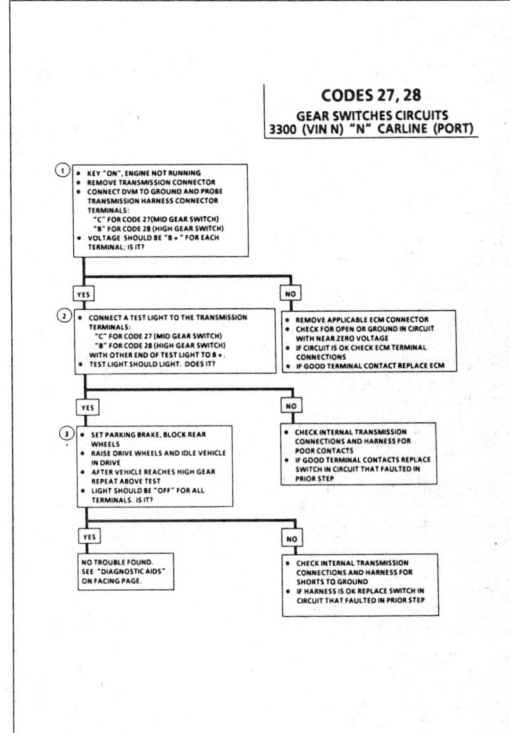

1989–90 3.3L (VIN N)
CUTLASS CALAIS AND SKYLARK

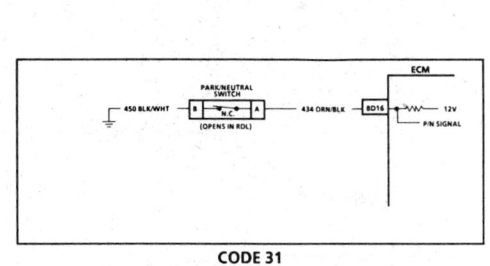

CODE 31
PARK/NEUTRAL SWITCH CIRCUIT
3300 (VIN N) "N" CARLINE (PORT)

Circuit Description:

The park/neutral switch contacts are a part of the neutral start switch and are closed to ground in park or neutral and open in drive ranges.

The ECM supplies ignition voltage through a current limiting resistor to CKT 434 and senses a closed switch when the voltage on CKT 434 drops to less than one volt.

The ECM uses the P/N signal as one of the inputs to control:
- Idle Speed (IAC)
- Vehicle Speed Sensor Diagnostics (VSS)
- Spark Advance

Code 31 will set if:
- CKT 434 indicates an open for 4 consecutive starts

Or if:
- CKT 434 indicates a ground
- No Code 38
- Transmission is in high gear
- VSS greater than 45 mph and TPS less than 15% (.94 volt)
- Above conditions are met for 12 seconds

Test Description: Numbers below refer to circled numbers on the diagnostic chart.
1. Checks for a closed switch to ground in park position. Different makes of "Scan" tools will read P/N differently. Refer to "Tool Operator's" manual for type of display used for a specific tool.
2. Checks for an open switch in drive range
3. Be sure "Scan" indicates drive, even while wiggling shifter, to test for an intermittent or misadjusted switch in drive or overdrive range.

1989–90 3.3L (VIN N)
CUTLASS CALAIS AND SKYLARK

CODE 31
PARK/NEUTRAL SWITCH CIRCUIT
3300 (VIN N) "N" CARLINE (PORT)

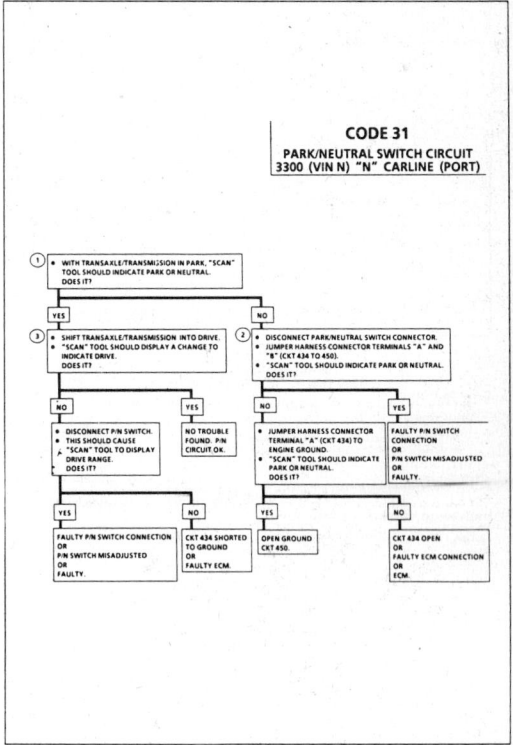

1989–90 3.3L (VIN N)
CUTLASS CALAIS AND SKYLARK

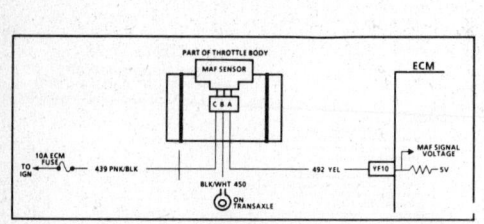

CODE 34

MASS AIR FLOW (MAF) SENSOR CIRCUIT
(GM/SEC LOW)
3300 (VIN N) "N" CARLINE (PORT)

Circuit Description:

The mass air flow (MAF) sensor measures the flow of air which passes through it in a given time. The ECM uses this information to monitor the operating condition of the engine for fuel delivery calculations. A large quantity of air movement indicates acceleration, while a small quantity indicates deceleration or idle.

The MAF sensor produces a frequency signal which cannot be easily measured. The sensor can be diagnosed using the procedures on this chart.

Code 34 will set when of the following conditions exists:
- Engine running
- If MAF sensor signal frequency is less than 960 Hz

NOTE: If the MAF sensor signal frequency is too low (Code 34 is set), a substitute value for airflow is calculated based on engine rpm, TPS, and IAC motor position.

Test Description: Numbers below refer to circled numbers on the diagnostic chart.
1. This step checks to see if ECM recognizes a problem.
2. A voltage reading at sensor harness connector terminal "A" of less than 4 or over 6 volts indicates a fault in CKT 492 or poor connection.
3. Verifies that both ignition voltage and a good ground circuit are available.

Diagnostic Aids:

An intermittent may be caused by a poor connection, mis-routed harness, rubbed through wire insulation, or a wire broken inside the insulation.

Check For:
- <u>Poor connection</u> at ECM pin "YD10". Inspect harness connectors for backed out terminals, improper mating, broken locks, improperly formed or damaged terminals, and poor terminal to wire connection.
- <u>Mis-routed Harness</u> Inspect MAF sensor harness to insure that it is not too close to high voltage wires, such as spark plug leads.
- <u>Damaged Harness</u> Inspect harness for damage. If harness appears OK, "Scan" while moving related connectors and wiring harness. A change in display would indicate the intermittent fault location.

1989–90 3.3L (VIN N)
CUTLASS CALAIS AND SKYLARK

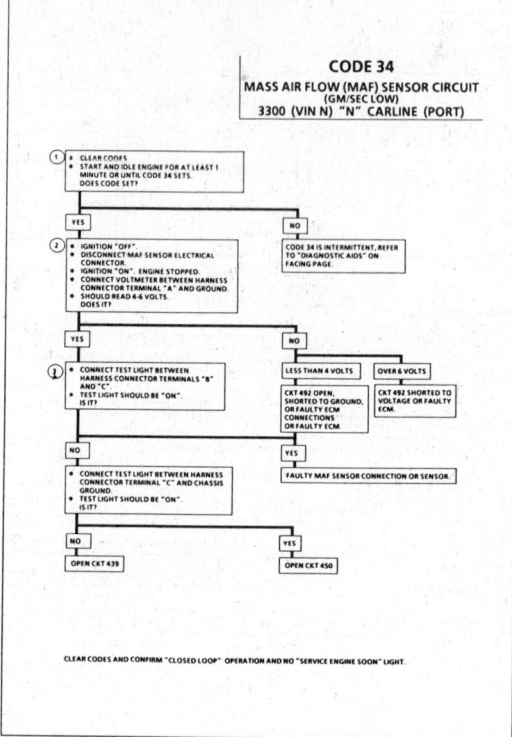

1989–90 3.3L (VIN N)
CUTLASS CALAIS AND SKYLARK

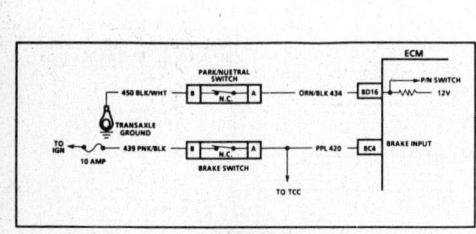

CODE 38

BRAKE SWITCH CIRCUIT
3300 (VIN N) "N" CARLINE (PORT)

Circuit Description:

The ECM monitors the status of the brake switch circuits through ECM terminal "BC4".

Code 38 will be set if:
- No Code 24 is present
- Vehicle speed has been above 35 mph and back to 0 mph several (22) times
- Status of CKT 420 has not changed state "high" to "low"

Diagnostic Aids:

A Code 38 in combination with a Code 39 would mean a problem with one or more of the following components.

Fuse, CKT 439, brake switch or wire before the splice.

A single Code 38 is the result of a wire or CKT 420 problem between the splice and the ECM, poor connection to ECM connector, or possibly the ECM.

1989–90 3.3L (VIN N)
CUTLASS CALAIS AND SKYLARK

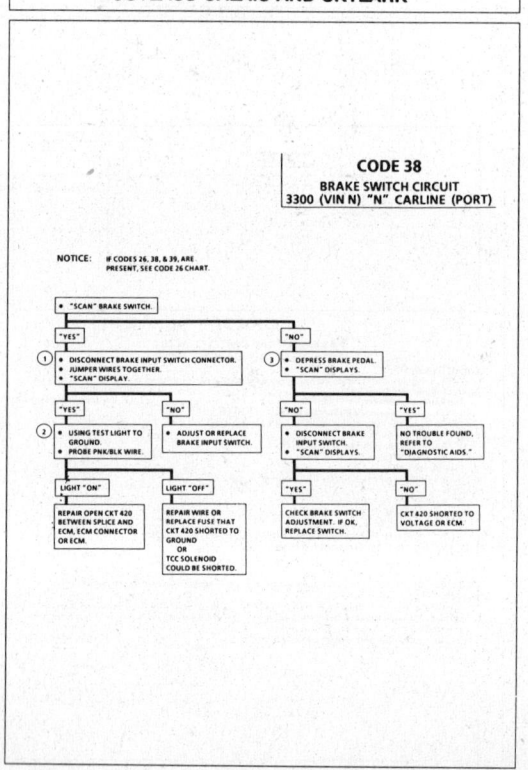

1989–90 3.3L (VIN N)
CUTLASS CALAIS AND SKYLARK

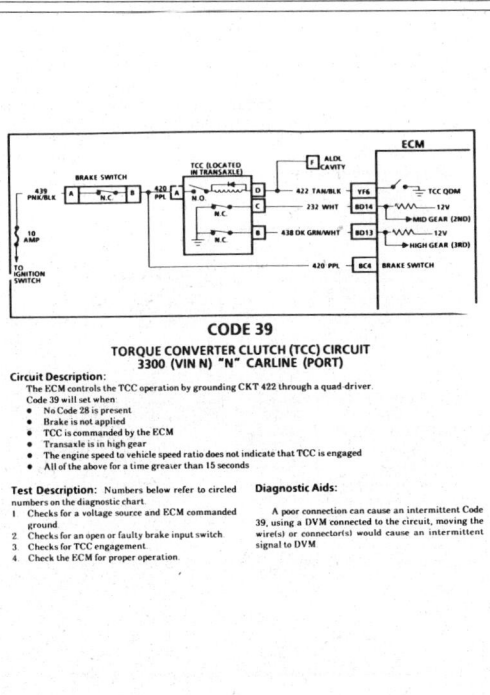

CODE 39
TORQUE CONVERTER CLUTCH (TCC) CIRCUIT
3300 (VIN N) "N" CARLINE (PORT)

Circuit Description:
The ECM controls the TCC operation by grounding CKT 422 through a quad-driver. Code 39 will set when:
- No Code 28 is present
- Brake is not applied
- TCC is commanded by the ECM
- Transaxle is in high gear
- The engine speed to vehicle speed ratio does not indicate that TCC is engaged
- All of the above for a time greater than 15 seconds

Test Description: Numbers below refer to circled numbers on the diagnostic chart.
1. Checks for a voltage source and ECM commanded ground.
2. Checks for an open or faulty brake input switch.
3. Checks for TCC engagement.
4. Check the ECM for proper operation.

Diagnostic Aids:
A poor connection can cause an intermittent Code 39, using a DVM connected to the circuit, moving the wire(s) or connector(s) would cause an intermittent signal to DVM.

1989–90 3.3L (VIN N)
CUTLASS CALAIS AND SKYLARK

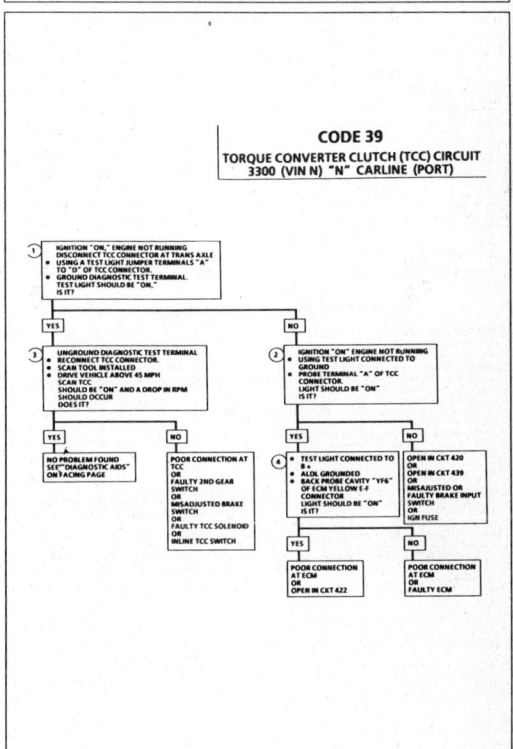

CODE 39
TORQUE CONVERTER CLUTCH (TCC) CIRCUIT
3300 (VIN N) "N" CARLINE (PORT)

1989–90 3.3L (VIN N)
CUTLASS CALAIS AND SKYLARK

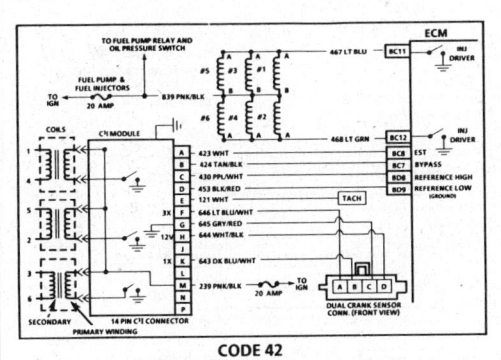

CODE 42
ELECTRONIC SPARK TIMING (EST) CIRCUIT
3300 (VIN N) "N" CARLINE (PORT)

Circuit Description:
The C^{3}I module sends a reference signal to the ECM when the engine is cranking. Under 400 rpm, the C^{3}I module controls ignition timing. When the engine speed exceeds 400 rpm, the ECM sends a 5 volt signal on the bypass CKT 424 to switch timing to ECM control. If an open or ground occurs while engine is running in EST mode, the engine will stall and set a Code 42. If circuit 424 is open or grounded while engine is cranking, the engine will start and remain in backup.
To set a Code 42 the following conditions must be met:
- Engine speed greater than 600 rpm with no EST pulse for 200 mS (open or grounded CKT 423), or
- ECM commanding bypass mode (open or grounded CKT 424)

Test Description: Numbers below refer to circled numbers on the diagnostic chart.
1. Checks to see if ECM recognizes a problem. If it does not set Code 42, it is an intermittent problem and could be due to a loose connection.
2. With the ECM disconnected, the ohmmeter should be reading less than 200 ohms, which is the normal resistance of the EST circuit through the C^{3}I module. A higher resistance would indicate a fault in CKT 423, a poor C^{3}I module connection, or a faulty C^{3}I module.
3. If test light was "ON" when connected from 12 volts to ECM harness terminal "BC7" either CKT 424 is shorted to ground or the C^{3}I module is faulty.
4. Checks to see if C^{3}I module switches when the bypass circuit is energized by 12 volts through the test light. If the C^{3}I module actually switches, the ohmmeter reading should shift to over 6,000 ohms.

5. Disconnecting the ignition module should make the ohmmeter read as if it were monitoring an open circuit (infinite reading). Otherwise, CKT 423 is shorted to ground.

Diagnostic Aids:
An intermittent may be caused by a poor connection, rubbed through wire insulation, or a wire broken inside the insulation. Check For:
- **Poor Connection or Damaged Harness** Inspect ECM harness connectors for backed out terminals "BC7" or "BC8", improper mating, broken locks, improperly formed or damaged terminals, poor terminal to wire connection, and damaged harness.
- **Intermittent Test** If connections and harness check OK, a digital voltmeter connected to affected terminal to ground while moving related connectors and wiring harness. If the failure is induced, the voltage reading will change.

1989–90 3.3L (VIN N)
CUTLASS CALAIS AND SKYLARK

CODE 42
ELECTRONIC SPARK TIMING (EST) CIRCUIT
3300 (VIN N) "N" CARLINE (PORT)

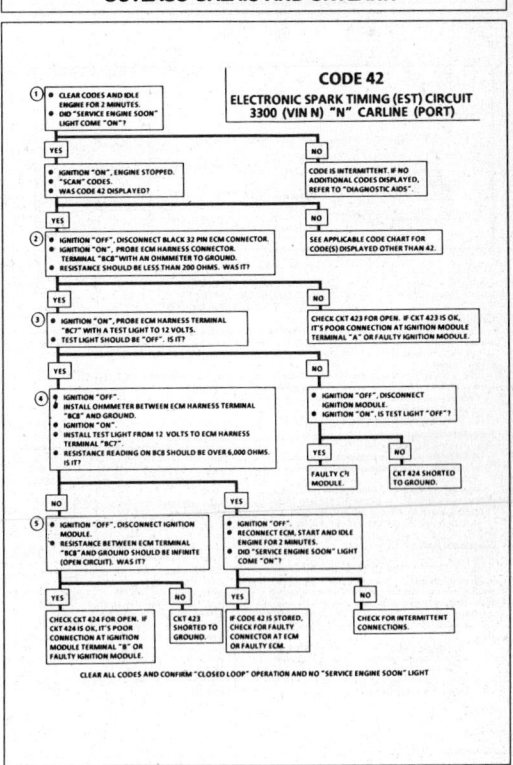

1989–90 3.3L (VIN N) CUTLASS CALAIS AND SKYLARK

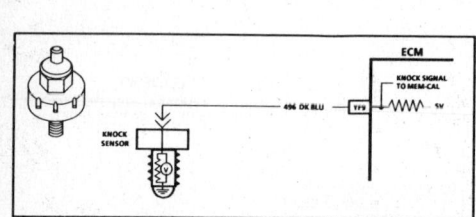

CODE 43
ELECTRONIC SPARK CONTROL (ESC) CIRCUIT
3300 (VIN N) "N" CARLINE (PORT)

Circuit Description:
The knock sensor is used to detect engine detonation and the ECM will retard the electronic spark timing based on the signal being received. The circuitry within the knock sensor causes the ECM's supplied 5 volt signal to be pulled down, so that under a no knock condition CKT 496 would measure about 2.5 volts. The knock sensor produces an A/C signal which rides on the 2.5 volts DC voltage. The amplitude and signal frequency is dependent upon the knock level.

If CKT 496 becomes open or shorted to ground, the voltage will either go above 3.5 volts or below 1.5 volts. If either of these conditions are met for 20 seconds, a Code 43 will be stored.

Code 43 will set if:
- Voltage on CKT 496 goes above 3.5 volts or below 1.5 volts.
- Condition present for 20 seconds

Test Description: Numbers below refer to circled numbers on the diagnostic chart.
1. If a Code 43 is detected, the ECM will retard spark timing by 10 degrees.
   If an audible knock is heard from the engine, repair the internal engine problem, normally no knock should be detected at idle.
2. If tapping on the engine lift hook does not produce a knock signal, try tapping engine closer to sensor before proceeding.

3. The ECM has a 5 volt pull-up resistor which should be present at the knock sensor terminal.
4. This test determines if the knock sensor is faulty or if the ESC portion of the Mem-Cal is faulty.

Diagnostic Aids:

Check CKT 496 for a potential open or short to ground. Also check for proper installation of Mem-Cal.

Refer to "Intermittents" in Section "B".

1989–90 3.3L (VIN N) CUTLASS CALAIS AND SKYLARK

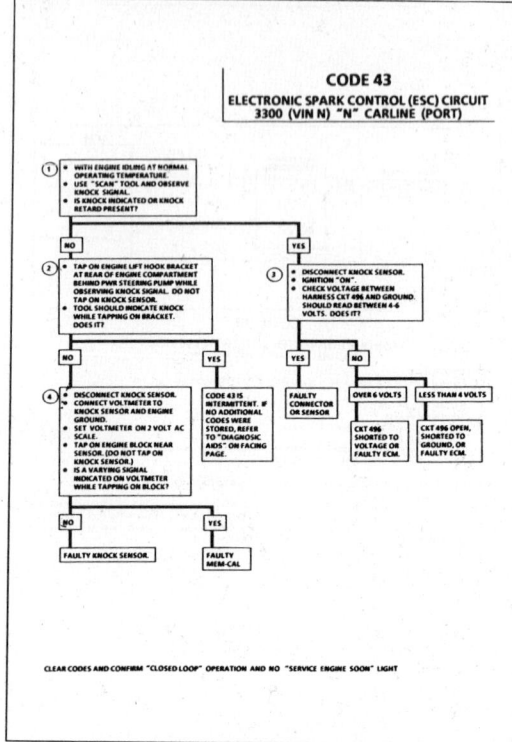

CODE 43
ELECTRONIC SPARK CONTROL (ESC) CIRCUIT
3300 (VIN N) "N" CARLINE (PORT)

CLEAR CODES AND CONFIRM "CLOSED LOOP" OPERATION AND NO "SERVICE ENGINE SOON" LIGHT

1989–90 3.3L (VIN N) CUTLASS CALAIS AND SKYLARK

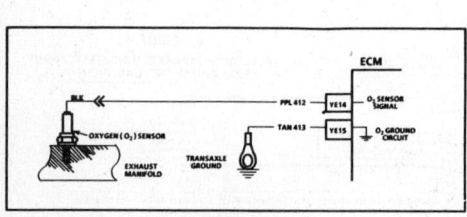

CODE 44
OXYGEN SENSOR CIRCUIT
(LEAN EXHAUST INDICATED)
3300 (VIN N) "N" CARLINE (PORT)

Circuit Description:
The ECM supplies a voltage of about .45 volt (450 mV) between terminals "YE14" and "YE15". (If measured with a 10 megohm digital voltmeter, this may read as low as .32 volt.) The O_2 sensor varies the voltage within a range of about 1 volt, (1000 mV) if the exhaust is rich, down through about .10 volt (100 mV) if exhaust is lean.

The sensor is like an open circuit and produces no voltage when it is below about 360°C (600°F). An open sensor circuit or cold sensor causes "Open Loop" operation.

Code 44 is set when the O_2 sensor signal voltage on CKT 412
- Remains below .25 volts for up to 4.5 minutes
- The system is operating in "Closed Loop"

Test Description: Numbers below refer to circled numbers on the diagnostic chart.
1. Running the engine at 1000 rpm keeps the O_2 sensor hot, so an accurate display voltage is maintained.
   Opening the O_2 sensor wire should result in a voltage display of between 350 and 550 mV. If the display is still fixed below 350 mV, the fault is a short to ground in CKT 412 or the ECM is faulty.

Diagnostic Aids:

Using the "Scan," observe the block learn values at different rpm and air flow conditions. The "Scan" also displays the block learn values, so the block learn values can be checked in each of the cells to determine when the Code 44 may have been set. If the conditions for Code 44 exists, the block learn values will be around 150.
- O_2 Sensor Wire Sensor pigtail may be mispositioned and contacting the exhaust manifold.
- Check for intermittent ground in wire between connector and sensor.

- MAF Sensor A mass air flow (MAF) sensor output that causes the ECM to sense a lower than normal air flow will cause the system to go lean. Disconnect the MAF sensor and if the lean condition is gone, replace the MAF sensor.
- Lean Injector(s) Perform injector balance test CHART C-2A.
- Fuel Contamination Water, even in small amounts, near the in-tank fuel pump inlet can be delivered to the injectors. The water causes a lean exhaust and can set a Code 44.
- Fuel Pressure System will be lean if pressure is too low. It may be necessary to monitor fuel pressure while driving the car at various road speeds and/or loads to confirm. See "Fuel System Diagnosis," CHART A-7.
- Exhaust Leaks If there is an exhaust leak, the engine can cause outside air to be pulled into the exhaust and past the sensor.
- If the above are OK, it is a faulty oxygen sensor.
- Vacuum or crankcase leaks can cause a lean condition.

1989–90 3.3L (VIN N) CUTLASS CALAIS AND SKYLARK

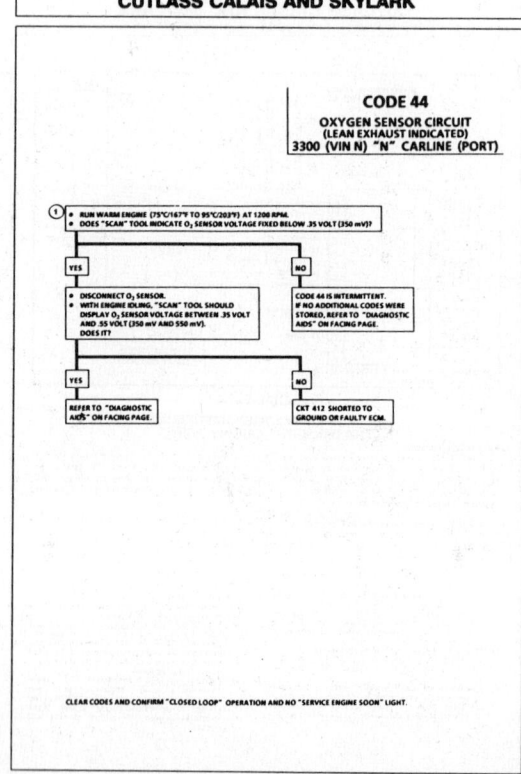

CODE 44
OXYGEN SENSOR CIRCUIT
(LEAN EXHAUST INDICATED)
3300 (VIN N) "N" CARLINE (PORT)

CLEAR CODES AND CONFIRM "CLOSED LOOP" OPERATION AND NO "SERVICE ENGINE SOON" LIGHT.

1989–90 3.3L (VIN N) CUTLASS CALAIS AND SKYLARK

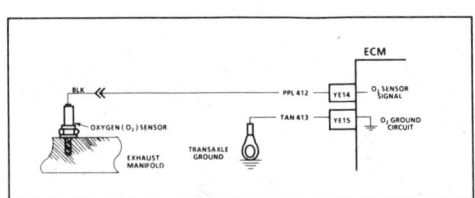

CODE 45
OXYGEN SENSOR CIRCUIT
(RICH EXHAUST INDICATED)
3300 (VIN N) "N" CARLINE (PORT)

Circuit Description:

The ECM supplies a voltage of about .45 volt (450 mV) between terminals "YE14" and "YE15" (If measured with a 10 megohm digital voltmeter, this may read as low as .32 volt.) The O_2 sensor varies the voltage within a range of about 1 volt (1000 mV) if the exhaust is rich, down through about .10 volt (100 mV) if exhaust is lean.

The sensor is like an open circuit and produces no voltage when it is below about 360°C (600°F). An open sensor circuit or cold sensor causes "Open Loop" operation.

Code 45 is set when the O_2 sensor signal voltage or CKT 412

- Remains above .75 volt for 2 minutes and in "Closed Loop"
- Throttle angle between .52 and 1.1 volts
- Engine time after start is 1 minute or more
- No Code 21 or Code 22

Test Description: Numbers below refer to circled numbers on the diagnostic chart.

1. Running the engine at 1200 rpm keeps the O_2 sensor hot, so an accurate display voltage is maintained.
   Opening the O_2 sensor wire should result in a voltage display of between 350 and 550 mV. If the display is still fixed below 350 mV, the fault is a short to ground in CKT 412 or the ECM is faulty.

Diagnostic Aids:

Using the "Scan," observe the block learn values at different rpm and air flow conditions. The "Scan" also displays the block cells, so the block learn values can be checked in each of the cells to determine when the Code 45 may have been set. If the conditions for Code 45 exists, the block learn values will be around 115.

- *Fuel Pressure* System will go rich if pressure is too high. The ECM can compensate for some increase. However, if it gets too high, a Code 45 may be set. See "Fuel System Diagnosis," CHART A-7.

- *Rich Injector* Perform injector balance test CHART C-2A.
- *Leaking Injector* See CHART A-7.
  Check for fuel contaminated oil.
- *Canister Purge* Check for fuel saturation. If full of fuel, check canister control and hoses.
- *MAF Sensor* An output that causes the ECM to sense a higher than normal airflow can cause the system to go rich. Disconnecting the MAF sensor will allow the ECM to set a fixed value for the sensor. Substitute a different MAF sensor if the rich condition is gone while the sensor is disconnected.
- Check for leaking fuel pressure regulator diaphragm by checking vacuum line to regulator for fuel.
- *TPS* An intermittent TPS output will cause the system to go rich due to a false indication of the engine accelerating.

1989–90 3.3L (VIN N) CUTLASS CALAIS AND SKYLARK

CODE 45
OXYGEN SENSOR CIRCUIT
(RICH EXHAUST INDICATED)
3300 (VIN N) "N" CARLINE (PORT)

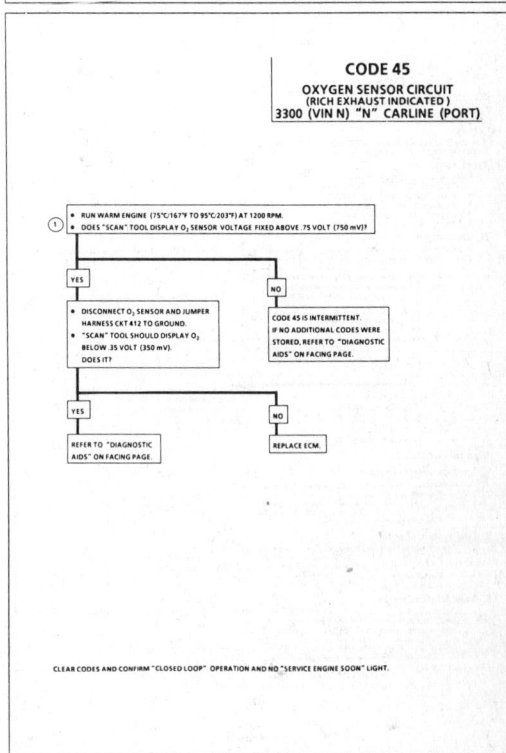

1989–90 3.3L (VIN N) CUTLASS CALAIS AND SKYLARK

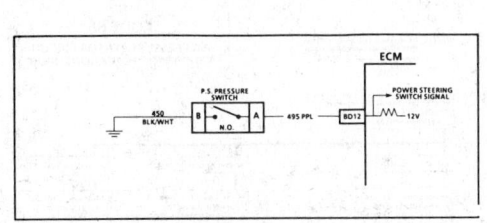

CODE 46
POWER STEERING PRESSURE SWITCH (PSPS) CIRCUIT
3300 (VIN N) "N" CARLINE (PORT)

Circuit Description:

The power steering pressure switch is incorporated as an ECM discrete input signal representing the parasitic load placed on the engine during high power steering demand periods, such as parking. This load may cause the engine to stall under high power steering pump pressure conditions, therefore, the ECM compensates by automatically increasing the engine idle speed, via IAC whenever the switch closes and a low voltage is monitored at ECM terminal "BD12". This low voltage indicates a high pressure within the power steering system resulting from high demand. The ECM turns the current "OFF" to the A/C compressor relay so the A/C clutch is disengaged to further reduce engine load.

To set a Code 46; the following conditions must be met.
- Closed power steering switch. "LO" (low voltage potential)
- Vehicle speed is greater than 40 mph
- Both conditions existing for a time greater than 25 seconds

Test Description: Numbers below refer to circled numbers on the diagnostic chart.

1. Tests the power steering pressure switch for proper operation by using the "self-diagnostics" system "LO" should be displayed when high power steering pressure loads are created.
2. This step determines whether the fault is in the switch or the circuit. A jumpered switch connector should normally indicate "HI" if the circuit is complete. If so, the power steering pressure switch or connection is faulty.
3. If the display switches to "HI" when CKT 495 is grounded, the fault is an open in CKT 450.

Diagnostic Aids:

An intermittent may be caused by a poor connection, rubbed through wire insulation or a wire broken inside the insulation.

Check For

- *Poor Connection or Damaged Harness* Inspect ECM harness connectors for backed out terminal "BD12", improper mating, broken locks, improperly formed or damaged terminals, poor terminal to wire connection, and damaged harness.
- *Intermittent Test* If connections and harness check OK, monitor power steering pressure switch display while moving related connectors and wiring harness. If the failure is induced, the power steering pressure switch display will abruptly change. This may help to isolate the location of the malfunction.

1989–90 3.3L (VIN N) CUTLASS CALAIS AND SKYLARK

CODE 46
POWER STEERING PRESSURE SWITCH (PSPS) CIRCUIT
3300 (VIN N) "N" CARLINE (PORT)

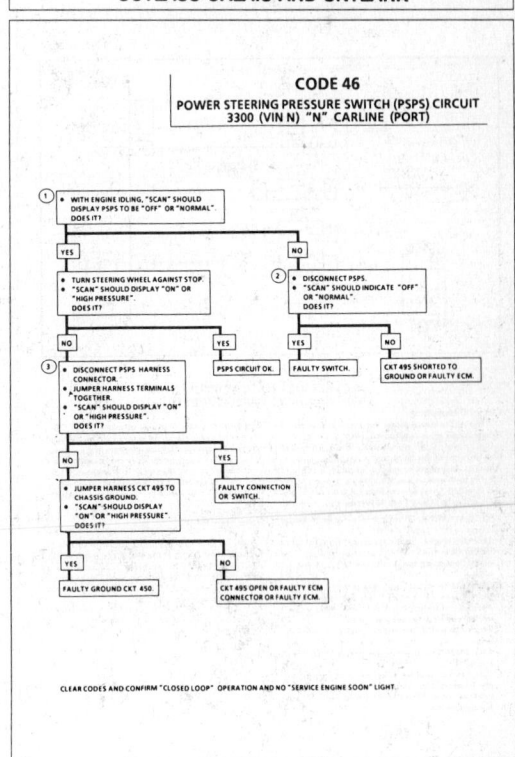

1989–90 3.3L (VIN N)
CUTLASS CALAIS AND SKYLARK

CODE 48
MISFIRE DIAGNOSIS
3300 (VIN N) "N" CARLINE (PORT)

If multiple codes are set, go to the lowest code first.
Repairing for a Code 13, 44, or 45 may correct Code 48.

Test Description:
Code 48 will set if the following:
- TPS is between .58 and 1.02 volts.
- Rpm is between 1500 and 2500.
- Mph is between 50 and 60.
- O_2 cross counts greater than 30.
- All of the above for 30 seconds.

O_2 Sensor Test:
Code 48 could be set if the O_2 sensor is degraded and cannot travel over the full rich to lean voltage range. This narrowed range could allow O_2 cross counts to be above the value necessary to set the code.

> - WITH "SCAN" TOOL INSTALLED, VERIFY ENGINE IS AT NORMAL OPERATING TEMPERATURE AND IN "CLOSED LOOP".

> - ENGINE IDLING IN PARK.
> - SELECT O_2 SENSOR POSITION ON "SCAN".
> - RAPIDLY FLASH THE THROTTLE FROM IDLE TO NEAR WIDE OPEN THROTTLE AND BACK WHILE OBSERVING O_2 VOLTAGE.
> - REPEAT IF NECESSARY TO CONFIRM VOLTAGE RANGE, AND "CLOSED LOOP".

> VOLTAGE EXCEEDS 250-750 mV RANGE. → O_2 SENSOR OK, SEE "DIAGNOSTIC AIDS"

> VOLTAGE REMAINS WITHIN 250-750 mV RANGE. → REPLACE O_2 SENSOR.

Diagnostic Aids:

1. **Ignition system checks:**
   Remove each spark plug and inspect (fouled, cracked, worn)
   Fouled -- check ignition wires (hi resistance, damage, poor connections, grounds)
   check coil and module operation (see -- CHART C-4)
   check basic engine problem (see 3 below)
   Cracked or worn -- replace as necessary

2. **Fuel system checks:**
   Restricted fuel system (injectors, fuel pump, lines, and filter)
   Injectors -- perform injector balance test (see : CHART C-2)
   verify each injector circuit with tool J-34730-2 or equivalent
   Fuel Pump --verify proper fuel pressure and fuel quality
   Lines and Filter --verify no restrictions in lines or filter

3. **Basic Engine Checks:**
   Unless spark plug(s) inspection identifies a specific cylinder(s), road test vehicle under test conditions to reverify Code 48 prior to engine disassembly.
   Basic engine (valves, compression, camshaft, lifters)
   Compression -- check rings, pistons, valves
   Valves -- check for burned, weak springs, broken parts, worn or loose guide
   Camshaft -- check for worn or broken
   Lifters -- check for worn, broken
   For additional items see Section "B" under "Rough Unstable Idle", "Hard Start", or "Hesitation, Sag, or Stumble"

1989–90 3.3L (VIN N)
CUTLASS CALAIS AND SKYLARK

CODE 51
MEM-CAL ERROR
(FAULTY OR INCORRECT MEM-CAL)
3300 (VIN N) "N" CARLINE (PORT)

> CHECK THAT ALL PINS ARE FULLY INSERTED IN THE SOCKET. IF OK, REPLACE MEM-CAL, CLEAR MEMORY AND RECHECK. IF CODE 51 REAPPEARS, REPLACE ECM.

NOTICE: To prevent possible Electrostatic Discharge damage to the ECM or MEM-CAL, Do Not touch the component leads. Do Not remove integrated circuit from carrier.

1989–90 3.3L (VIN N)
CUTLASS CALAIS AND SKYLARK

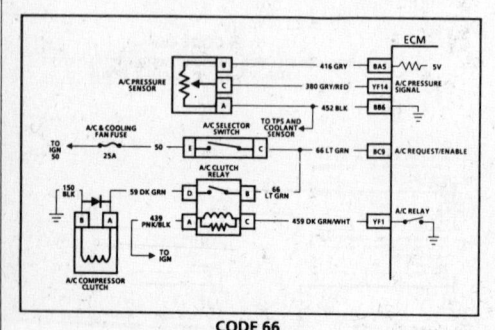

CODE 66
A/C PRESSURE SENSOR CIRCUIT
3300 (VIN N) "N" CARLINE (PORT)

Circuit Description:
The A/C pressure sensor responds to changes in A/C refrigerant system high side pressure. This input indicates how much load the A/C compressor is putting on the engine and is one of the factors used by the ECM to determine IAC valve position for idle speed control. The circuit consists of a 5 volts reference and a ground, both provided by the ECM, and a signal line to the ECM. The signal is a voltage which is proportional to the pressure. The sensor's range of operation is 165 to 399 psi. At 165 psi or less, the signal will be about .5 volt, varying up to about 4.5 volts at 399 psi or above. Code 66 sets if the voltage falls outside of calibrated minimum and less than .1 volt or more than 4.9 volts for 25 seconds, while the A/C is requested on. The A/C compressor is disabled by the ECM if Code 66 is present.

Test Description: Numbers below refer to circled numbers on the diagnostic chart.
1. This step checks the voltage signal being received by the ECM from the A/C pressure sensor. The normal operating range is between .5 volt and 4.5 volts.
2. Checks to see if the high voltage signal is from a shorted sensor or a short to voltage in the circuit. Normally, disconnecting the sensor would make a normal circuit go to near zero volts.
3. Checks to see if low voltage signal is from the sensor or the circuit. Jumpering the sensor signal CKT 380 to 5 volts, checks the circuit, connections, and ECM.
4. This step checks to see if the low voltage signal was due to an open in the sensor circuit or the 5 volts reference circuit since the prior step eliminated the pressure switch.

Diagnostic Aids:
Code 66 sets when signal voltage falls outside the normal possible range of the sensor and is not due to a refrigerant system problem. If problem is intermittent, check for opens or shorts in harness or poor connections. If OK, replace A/C pressure sensor. If Code 66 re-sets, replace ECM.

1989–90 3.3L (VIN N)
CUTLASS CALAIS AND SKYLARK

CODE 66
A/C PRESSURE SENSOR CIRCUIT
3300 (VIN N) "N" CARLINE (PORT)

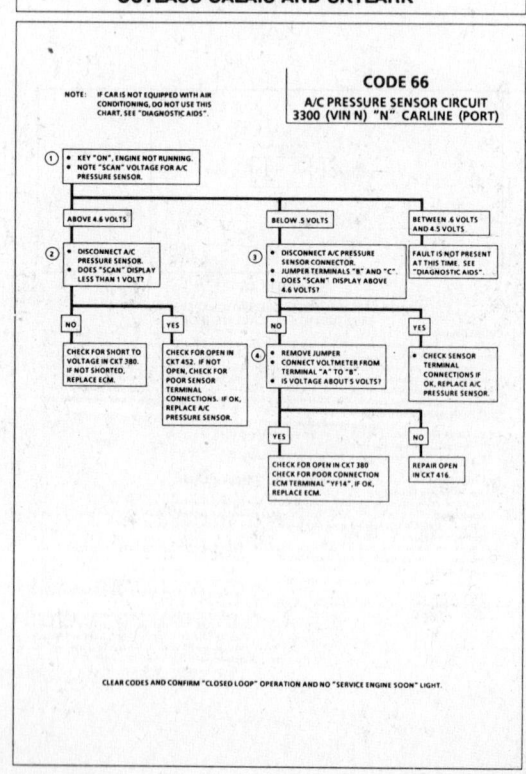

1989–90 3.3L (VIN N)
CUTLASS CALAIS AND SKYLARK

SECTION B
SYMPTOMS
TABLE OF CONTENTS

Before Starting
Intermittents
Hard Start
Hesitation, Sag, Stumble
Surges and/or Chuggle
Lack of Power, Sluggish, or Spongy
Detonation/Spark Knock
Cuts Out, Misses
Backfire
Poor Fuel Economy
Dieseling, Run On
Rough, Unstable, or incorrect Idle, Stalling
Excessive Exhaust Emissions Or Odors
Chart B-1 Restricted Exhaust System Check

BEFORE STARTING

Before using this section you should have performed the "DIAGNOSTIC CIRCUIT CHECK" and found out that
1. The ECM and "Service Engine Soon" light are operating.
2. There are no trouble codes stored, or there is a trouble code but no "Service Engine Soon" light.
3. The fuel control system is operating OK (by performing field service mode check).
Verify the customer complaint, and locate the correct SYMPTOM. Check the items indicated under that symptom.
If the ENGINE CRANKS BUT WILL NOT RUN, see CHART A-3.

Several of the symptom procedures below call for a careful visual/physical check. This check should include
• ECM and + wires for being clean and tight at starter and/or junction block.
• ECM grounds for being clean and tight.
• Vacuum hoses for splits, kinks, and proper connections, as shown on emission control information label.
• Air leaks at throttle body mounting and intake manifold.
• Air leaks between MAF sensor and throttle body.
• Ignition wires for cracking, hardness, proper routing, and carbon tracking.
• Wiring for proper connections, pinches, and cuts.
The importance of this step cannot be stressed too strongly - it can lead to correcting a problem without further checks and can save valuable time.

1989–90 3.3L (VIN N)
CUTLASS CALAIS AND SKYLARK

INTERMITTENTS

Problem may or may not turn "ON" the "Service Engine Soon" light or store a code.

DO NOT use the trouble code charts in Section "A" for intermittent problems. The fault must be present to locate the problem. If a fault is intermittent, use of trouble code charts may result in replacement of good parts.
• Most intermittent problems are caused by faulty electrical connections or wiring. Perform a careful check as described at start of this section. Check for:
  • Poor mating of the connector halves, or terminals not fully seated in the connector body (backed out).
  • Improperly formed or damaged terminals. All connector terminals in problem circuit should be carefully reformed to increase contact tension.
  • Poor terminal to wire connection. This requires removing the terminal from the connector body to check.
• If a visual/physical check does not find the cause of the problem, the car may be driven with a voltmeter connected to a suspected circuit. An abnormal voltage reading, when the problem occurs, indicates the problem may be in that circuit. If the wiring and connectors check OK and a trouble code was stored for a circuit having a sensor, except for Codes 43, 44, and 45, substitute a known good sensor and recheck.

An intermittent "Service Engine Soon" light with no stored code may be caused by
• Check that the ECM and sensor ground wires are not located on the same stud at the transaxle as the battery ground.
• Ignition coil shorted to ground, arcing at spark plug wires or plugs.
• "Service Engine Soon" light wire to ECM shorted to ground (CKT 419).
• Diagnostic "Test" terminal wire to ECM, shorted to ground (CKT 451).
• ECM power grounds. See ECM wiring diagrams.
• Loss of trouble code memory. To check, disconnect TPS and idle engine until "Service Engine Soon" light comes "ON." Code 22 should be stored, and kept in memory when ignition is turned "OFF." If not, the ECM is faulty.
• Check for an electrical system interference caused by a defective relay, ECM driven solenoid, or switch. They can cause a sharp electrical surge. Normally, the problem will occur when the faulty component is operated.
• Check for improper installation of electrical options, such as lights, 2 way radios, etc.
• EST wires should be kept away from spark plug wires, coil and generator.
• Check for open diode across A/C compressor clutch, and for other open diodes (see wiring diagrams).

HARD START

Definition: Engine cranks OK, but does not start for a long time. Does eventually run, or may start but immediately dies.

• Perform careful check as described at start of Section "B".
• Make sure driver is using correct starting procedure.
• CHECK:
  — TPS for sticking or binding or a high TPS voltage with the throttle closed.
  — High resistance in coolant sensor circuit or sensor itself. See CODE 15 CHART or with a "Scan" tool compare coolant temperature with ambient temperature on a cold engine.
  — Fuel pressure CHART A-7.
  — Water contaminated fuel.
  — Fuel pump relay - See CHART A-5.
  — PCV valve for hose off or damaged valve.

Ignition system - Check for
  Proper output with ST 125.
  Bare and shorted wires.
  Loose C³I module ground, mounting screws.
  Dual crank sensor for damage from crankshaft pulley vanes rubbing.
• A faulty in-tank fuel pump check valve will allow the fuel in the lines to drain back to the tank after the engine is stopped. To check for this condition:
  Perform fuel system diagnosis, CHART A-7.
• Remove spark plugs. Check for wet plugs, cracks, wear, improper gap, burned electrodes, or heavy deposits. Repair or replace as necessary.

1989–90 3.3L (VIN N)
CUTLASS CALAIS AND SKYLARK

HESITATION, SAG, STUMBLE

Definition: Momentary lack of response as the accelerator is pushed down. Can occur at all car speeds. Usually most severe when first trying to make the car move, as from a stop sign. May cause the engine to stall if severe enough.

• Perform careful visual check as described at start of Section "B".
• CHECK:
  — Fuel pressure. See CHART A-7. Also Check for water contaminated fuel.
  — Spark plugs for being fouled or faulty wiring.
  — PROM number. Also check service bulletins for latest PROM.
  — TPS for binding or sticking. Voltage should increase at a steady rate as throttle is moved toward WOT.
  — Generator output voltage. Repair if less than 9 or more than 16 volts.
  — C³I ground, CKT 453.
  — Canister purge system for proper operation. See CHART C-3.
  — Engine Thermostat - functioning correctly and proper heat range.
• Perform injector balance test CHART C-2A.

SURGES AND/OR CHUGGLE

Definition: Engine power variation under steady throttle or cruise. Feels like the car speeds up and slows down with no change in the accelerator pedal.

• Be sure driver understands transmission converter clutch and A/C compressor operation in owner's manual.
• Perform careful visual inspection as described at start of Section "B".
• CHECK:
  — Generator output voltage. Repair if less than 9 or more than 16 volts.
  — If a "Scan" tool is available which plugs in to the ALDL connector, make sure reading of VSS matches vehicle speedometer. See introduction explaining "Scan" tool positions.
  Check for PCV valve for being blown out of intake or blown off at the PCV valve. Also inspect PCV valve for damage.
  Vacuum lines for kinks or leaks.
  — In line fuel filter. Replace if dirty or plugged.
  — Fuel pressure while condition exists. See CHART A-7.
• Inspect oxygen sensor for silicone contamination, from fuel or use of improper RTV sealant. The sensor may have a white powdery coating. This will result in a high but false signal voltage (rich exhaust indication). The ECM will then reduce the amount of fuel delivered to the engine, causing a severe driveability problem.
• Remove spark plugs. Check for cracks, wear, improper gap, burned electrodes, or heavy deposits.
• Check spark plug leads.

LACK OF POWER, SLUGGISH, OR SPONGY

Definition: Engine delivers less than expected power. Little or no increase in speed when accelerator pedal is pushed down part way.

• Perform careful visual check as described at start of Section "B".
• Compare customer's car to similar unit. Make sure the customer's car has an actual problem.
• Remove air cleaner and check air filter for dirt, or for being plugged. Replace as necessary.
• CHECK:
  — Restricted fuel filter, contaminated fuel or improper fuel pressure. See CHART A-7.
  — ECM power grounds. See wiring diagrams.
  — ESC system for false retard due to mechanical noise.

Exhaust system for possible restriction.
  — Inspect exhaust system for damaged or collapsed pipes.
  — Inspect muffler for heat distress or possible internal failure.
• Generator output voltage. Repair if less than 9 or more than 16 volts.
• Engine valve timing and compression.
• Engine for proper or worn camshaft.

Secondary voltage using a shop oscilloscope or a spark tester J-26792 (ST-125) or equivalent.

1989–90 3.3L (VIN N)
CUTLASS CALAIS AND SKYLARK

DETONATION /SPARK KNOCK

Definition: A mild to severe ping, usually worse under acceleration. The engine makes sharp metallic knocks that change with throttle opening. Sounds like popcorn popping.

• Check for obvious overheating problems.
  — Low/lean coolant mixture.
  — Inoperative thermostat.
  — Restricted air/water flow through radiator.
  — Inoperative electric coolant fan circuit. See CHART C-12.
• CHECK:
  — ESC system for no retard - see CHART C-5.
  — TCC operation - see CHART C-8.

• Fuel system pressure. See CHART A-7.
• Remove carbon with top engine cleaner. Follow instructions on can.
  • Check for leaking valve oil seals.
  • Check for poor fuel quality, proper octane rating.
  • Check for correct PROM.
  • Check for incorrect basic engine parts such as cam, heads, pistons, etc.

CUTS OUT, MISSES

Definition: Steady pulsation or jerking that follows engine speed, usually more pronounced as engine load increases. The exhaust has a steady spitting sound at idle or low speed.

• Perform careful visual (physical) check as described at start of Section "B".
• If engine "Misses At Idle" see CHART C-4II.1
• If engine "Misses Under Load," see Chart C-4II.
• If above checks did not discover cause of problem, check the following:
  Disconnect all injector harness connectors. Connect J-34730-2 injector test light or equivalent 12 volt test light between the harness terminals of each injector connector and note light while cranking. See introduction "Scan" tool positions. If test light fails to blink at any connector, it is a faulty injector drive circuit harness, connector, or terminal.
  Fuel system - Plugged fuel filter, water, low pressure. See CHART A-7.

Perform the injector balance test. See CHART C-2A.
On C³I systems, a misfire may be caused by a misaligned crank sensor or bent vane on rotating interrupter. Inspect for proper clearance at each vane, using tool J-36179 or equivalent.
  Sensor should be replaced if it shows evidence of having been rubbed by the interrupter.
Perform compression check.
Remove rocker covers. Check for bent pushrods, worn rocker arms, broken valve springs, worn camshaft lobes. Repair as necessary.
Valve timing.

1989–90 3.3L (VIN N)
CUTLASS CALAIS AND SKYLARK

BACKFIRE

Definition: Fuel ignites in intake manifold, or in exhaust system, making a loud popping noise.

- **CHECK:**
  - Compression - Look for sticking or leaking valves.
  - Valve timing.
  - Output voltage of ignition coil using a shop oscilliscope or spark tester J-26792 (ST-125) or equivalent.
- Spark plugs for crossfire. Also inspect spark plug wires, and proper routing of plug wires.
- Ignition system for intermittent condition.
- Engine timing - see emission control information label.

POOR FUEL ECONOMY

Definition: Fuel economy, as measured by an actual road test, is noticeably lower than expected. Also, economy is noticeably lower than it was on this car at one time, as previously shown by an actual road test.

- **CHECK:**
  - Engine thermostat for faulty part (always open) or for wrong heat range.
  - Fuel pressure. See CHART A-5.
  - Check owner's driving habits.
    - Is A/C "ON" full time (defroster mode "ON")?
    - Are tires at correct pressure?
    - Are excessively heavy loads being carried?
    - Is acceleration too much, too often?
    - Suggest driver read "Important Facts on Fuel Economy" in owner's manual.
  - Perform "Diagnostic Circuit Check".
  - Check air cleaner element (filter) for dirt or being plugged.
  - Check for proper calibration of speedometer.
- Visually (physically) Check:
  - Vacuum hoses for splits, kinks and proper connections as shown on Vehicle Emissions Control Information label.
  - Ignition wires for cracking, hardness, and proper connections.
- Remove spark plugs. Check for cracks, wear, improper gap, burned electrodes or heavy deposits. Repair or replace as necessary.
- Check compression.
- Check TCC for proper operation. See CHART C-8, use "Scan" tool if available.
- Check for dragging brakes.
- Suggest owner fill fuel tank and recheck fuel economy.
- Check for exhaust system restriction.

DIESELING, RUN-ON

Definition: Engine continues to run after key is turned "OFF", but runs very roughly. If engine runs smoothly, check ignition switch and adjustment.

- Check injectors for leaking. See CHART A-7.

1989–90 3.3L (VIN N)
CUTLASS CALAIS AND SKYLARK

ROUGH, UNSTABLE, OR INCORRECT IDLE, STALLING

Definition: The engine runs unevenly at idle. If bad enough, the car may shake. Also, the idle may vary in rpm (called "hunting"). Either condition may be bad enough to cause stalling. Engine idles at incorrect speed.

- Perform careful visual check as described at start of Section "B".
- Clean injectors.
- **CHECK:**
  - Throttle linkage for sticking or binding.
  - TPS for sticking or binding, be sure output is stable at idle and adjustment specification is correct.
  - IAC system.
  - Generator output voltage. Repair if less than 9 or more than 16 volts.
  - P/N switch circuit. Code 31, 3300 (VIN N), or use "Scan" tool, but be sure tool indicates vehicle is in drive with gear selector in drive.
  - Injector balance. See CHART C-2A.
  - PCV valve for proper operation by placing finger over inlet hole in valve end several times. Valve should snap back. If not, replace valve. Check valve for damage or hose not fully seated at intake.
- Evaporative emission control system. CHART C-3.
- ECM ground circuits.
- Monitoring block learn values may help identify the cause of the problem. If the system is running lean (block learn greater than 138) refer to "Diagnostic Aids" on facing page of Code 44. If the system is running rich (block learn values less than 118) refer to "Diagnostic Aids" on facing page of Code 45.
- Run a cylinder compression check.
- Check for fuel in pressure regulator hose. If present, replace regulator assembly.
- Check ignition system, wires and plugs.
- Check for loose or damaged gaskets MAF between sensor and throttle body intake.
- Disconnect MAF sensor and if condition is corrected, replace sensor. "Scan" tool should read about 4-8 grams per second at idle.
- If problem exists, only with A/C "ON," check A/C system operation CHART C 10.
- For damaged, grounded out, or mispositioned motor mounts.

EXCESSIVE EXHAUST EMISSIONS OR ODORS

Definition: Vehicle fails an emission test. Vehicle has excessive "rotten egg" smell. Excessive odors do not necessarily indicate excessive emissions.

- Perform "Diagnostic Circuit Check."
- IF TEST SHOWS EXCESSIVE CO AND HC, (or also has excessive odors)
  - Check items which cause car to run RICH
    - Make sure engine is at normal operating temperature.
- CHECK
  - Fuel pressure. See CHART A 7.
  - Canister for fuel loading. See CHART C 3.
  - Injector balance. See CHART C-2A.
  - PCV valve for being plugged, stuck, or blocked PCV hose to fuel in the crankcase.
  - Spark plugs, plug wires, and ignition components
  - Check for lead contamination of catalytic converter (look for removal of fuel filler neck restrictor).
- Check for properly installed fuel cap.
- If the system is running rich, (block learn less than 118), refer to "Diagnostic Aids" on facing page of Code 45.
- IF TEST SHOWS EXCESSIVE NOx
  - Check items which cause car to run LEAN, or to run too hot
    - Vacuum leaks.
    - Coolant system and coolant fan for proper operation. See CHART C 12.
    - Remove carbon with top engine cleaner. Follow instructions on can.
  - If the system is running lean (block learn greater than 138), refer to "Diagnostic Aids" on facing page of Code 44.

CHART B-1
RESTRICTED EXHAUST SYSTEM CHECK

Proper diagnosis for a Restricted Exhaust System is essential before any components are replaced. The following procedure may be used for diagnosis:

CHECK AT O₂ SENSOR:

1. Carefully remove O₂ sensor.
2. Install exhaust backpressure tester (BT 8515 or BT 8603) or equivalent in place of O₂ sensor (see illustration).
3. After completing test described below, be sure to coat threads of O₂ sensor with anti-seize compound P/N 5613695 or equivalent prior to re-installation.

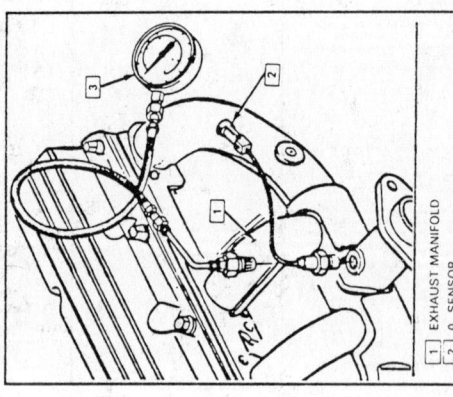

1	EXHAUST MANIFOLD
2	O₂ SENSOR
3	BACK PRESSURE TESTER

DIAGNOSIS:

1. With the engine at normal operating temperature and running at 2500 rpm, observe the exhaust system backpressure reading on the gauge.
2. If the backpressure exceeds 1 1/4 psi (8.62 kPa), a restricted exhaust system is indicated.
3. Inspect the entire exhaust system for a collapsed pipe, heat distress, or possible internal muffler failure.
4. If there are no obvious reasons for the excessive backpressure, a restricted catalytic converter should be suspected, and replaced using current recommended procedures.

1989–90 3.3L (VIN N)
CUTLASS CALAIS AND SKYLARK

CHART C-2A
INJECTOR BALANCE TEST

The injector balance tester is a tool used to turn the injector on for a precise amount of time, thus spraying a measured amount of fuel into the manifold. This causes a drop in fuel rail pressure that we can record and compare between each injector. All injectors should have the same amount of pressure drop (± 10 kPa). Any injector with a pressure drop that is 10 kPa (or more) greater than the average drop of the other injectors should be considered faulty and replaced. Any injector with a pressure drop of less than 10 kPa from the average drop of the other injectors should be cleaned and retested.

STEP 1

Engine "cool down" period (10 minutes) is necessary to avoid irregular readings due to "Hot Soak" fuel boiling.

Disconnect harness connectors at all injectors, and connect injector tester J 34730-3, or equivalent, to one injector. With ignition "OFF" connect fuel gauge J 347301 or equivalent to fuel pressure tap. Wrap a shop towel around fitting while connecting gage to avoid fuel spillage. At this point, insert clear tubing attached to vent valve into a suitable container and bleed air from gauge and hose to insure accurate gauge operation, by applying B+ to fuel pump test terminal. Repeat this step until all air is bled from gauge.

STEP 2

Apply B+ to fuel pump test terminal until fuel pressure reaches its maximum, about 3 seconds. Record this initial pressure reading. Energize tester one time and note pressure at its lowest point (Disregard any slight pressure increase after drop hits low point.) By subtracting this second pressure reading from the initial pressure, we have the actual amount of injector pressure drop. See note below.

STEP 3

Repeat step 2 on each injector and compare the amount of drop. Usually, good injectors will have virtually the same drop. Clean and retest any injector that has a pressure difference of 10 kPa, either more or less than the average of the other injectors on the engine. Replace any injector that also fails the retest. If the pressure drop of all injectors is within 10kPa of this average, the injectors appear to be flowing properly. Reconnect them and review "Symptoms," Section "B"

NOTE: *The entire test should not be repeated more than once without running the engine to prevent flooding. (This includes any retest on faulty injectors).*

1989–90 3.3L (VIN N)
CUTLASS CALAIS AND SKYLARK

NOTE: If injectors are suspected of being dirty, they should be cleaned using an approved tool and procedure prior to performing this test. The fuel pressure test in Chart A-7, should be completed prior to this test.

> ### CHART C-2A
> INJECTOR BALANCE TEST
> 3300 (VIN N) (PORT)

Step 1. If engine is at operation temperature start at A, otherwise you can start at B
  A. Engine "cool down" period (10 minutes) is necessary to avoid irregular Readings due to "Hot Soak" Fuel Boiling.
  B. Disconnect harness connectors at all injectors and connect injector tester J-34730-3, or equivalent, to one injector.
  C. With ignition "OFF" connect fuel gauge J-347301 or equivalent to Fuel Pressure Tap. NOTE: Wrap a shop towel around fitting while connecting gauge to avoid fuel spray or spillage.
  D. Apply B+ to fuel pump test terminal and bleed off to fuel pressure test gauge.

Step 2. Run test:
  A. Apply B+ to Fuel Pump Test Terminal. Allow about 3 seconds for fuel pressure to reach its maximum.
  B. Record gauge pressure. (Pressure must hold steady, if you don't see the Fuel System diagnosis, Chart A-7, in Section A).
  C. Turn injector on, by depressing button on injector tester, and note pressure at the instant the gauge needle stops.

Step 3.
  A. Repeat step 2 on all injectors and record pressure drop on each. Clean and retest injectors that appear faulty (Any injectors that have a 10 kPa difference, either more or less, in pressure from the average). If no problem is found, review Symptoms Section B.

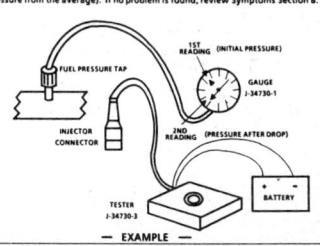

— EXAMPLE —

CYLINDER	1	2	3	4	5	6
1ST READING	225	225	225	225	225	225
2ND READING	100	100	100	90	100	115
AMOUNT OF DROP	125	125	125	135	125	110
	OK	OK	OK	FAULTY, RICH (TOO MUCH) (FUEL DROP)	OK	FAULTY, LEAN (TOO LITTLE) (FUEL DROP)

1989–90 3.3L (VIN N)
CUTLASS CALAIS AND SKYLARK

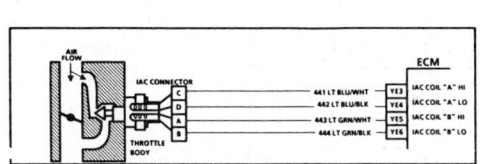

CHART C-2B
IDLE AIR CONTROL (IAC) VALVE CHECK
3300 (VIN N) (PORT)

Circuit Description:

The ECM controls idle rpm with the IAC valve. To increase idle rpm, the ECM retracts the IAC pintle, allowing more air to bypass the throttle plate. To decrease rpm, it extends the IAC pintle valve, reducing air flow through the IAC valve port in the throttle body. A "Scan" tool will read the ECM commands to the IAC valve in counts. The higher the counts, the more air allowed (higher idle). The lower the counts, the less air allowed (lower idle). The ECM learns a new IAC position every time the ignition is cycled and in "Closed Loop" with the transmission in drive. This is to control engine speed at different idle conditions.

Test Description: Numbers below refer to circled numbers on the diagnostic chart.
1. Continue with test, even if engine will not idle. If idle is too low, "Scan" will display 80 or more counts, or steps. If idle is high, it will display "0" counts. Occasionally, an erratic or unstable idle may occur. Engine speed may vary 200 rpm or more up and down. Disconnect IAC. If the condition is unchanged, the IAC is not at fault.
2. When the engine was stopped, the IAC pintle retracted (more air) to a fixed "Park" position for increased air flow and idle speed during the next engine start. A "Scan" will display 100 or more counts.
3. Be sure to disconnect the IAC valve prior to this test. The test light will confirm the ECM signals by a steady or flashing light on all circuits.
4. There is a remote possibility that one of the circuits is shorted to voltage, which would have been indicated by a steady light. Disconnect ECM and turn the ignition "ON" and probe terminals to check for this condition.

Diagnostic Aids:

A slow, unstable idle may be caused by a system problem that cannot be overcome by the IAC. "Scan" counts will be above 60 counts, if idle speed is too low. If idle speed is too high, IAC counts will be "0".

- **System lean (High Air/Fuel Ratio)**
  Idle speed may be too high or too low. Engine speed may vary up and down, disconnecting IAC does not help. May set Code 44.
  "Scan" and/or voltmeter will read an oxygen sensor output less than 300 mV (.3 volt). Check for low regulated fuel pressure or water in fuel. A lean exhaust, with an oxygen sensor output fixed above 800 mV (.8 volt indicating rich exhaust to ECM), will be a contaminated sensor, usually silicone. This may also set Code 45.
- **System rich (Low Air/Fuel Ratio)**
  Idle speed too low. "Scan" counts usually above 80. System obviously rich and may exhibit black smoke exhaust.
  "Scan" tool and/or voltmeter will read an oxygen sensor signal fixed above 800 mV (.8 volt).
  Check:
  - High fuel pressure
  - Injector leaking or sticking
  - Throttle Body - Remove IAC and inspect bore for foreign material or evidence of IAC pintle dragging the bore.
- Refer to "Rough, Unstable, Incorrect Idle or Stalling" in "Symptoms" in Section "B"

1989–90 3.3L (VIN N)
CUTLASS CALAIS AND SKYLARK

> ### CHART C-2B
> IDLE AIR CONTROL (IAC) VALVE CHECK
> 3300 (VIN N) (PORT)

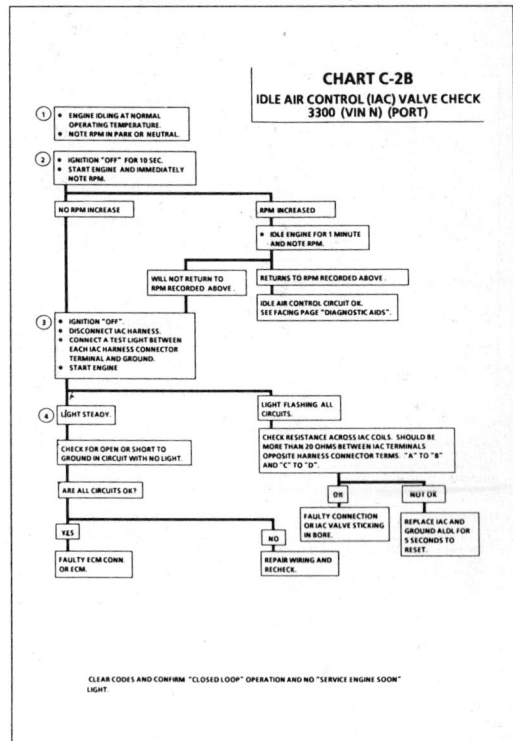

1989–90 3.3L (VIN N)
CUTLASS CALAIS AND SKYLARK

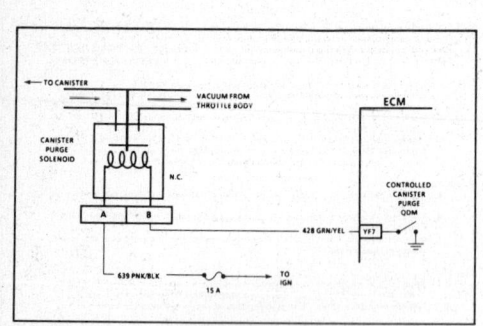

CHART C-3
CANISTER PURGE VALVE CHECK
3300 (VIN N) (PORT)

Circuit Description:
Canister purge is controlled by a solenoid that allows vacuum to purge the canister when energized. The ECM supplies a ground to energize the solenoid (purge "ON").
The purge solenoid is energized (purge "ON"), if the diagnostic "test" terminal is grounded with the engine stopped or the following is met with the engine running:
- Engine run time after start more than 45 seconds cold and/or 25 seconds when coolant temperature is greater than 85°C.
- Coolant temperature above 70°C (158°F).

Test Description: Numbers below refer to circled numbers on the diagnostic chart.
1. Checks to see if the solenoid is opened or closed. The solenoid is normally de-energized in this step, so it should be closed.

2. Completes functional check by grounding "test" terminal. This should normally energize the solenoid and allow the vacuum to drop (purge "ON").
3. Checks for open or shorted solenoid circuit.

1989–90 3.3L (VIN N)
CUTLASS CALAIS AND SKYLARK

CHART C-3
CANISTER PURGE VALVE CHECK
3300 (VIN N) (PORT)

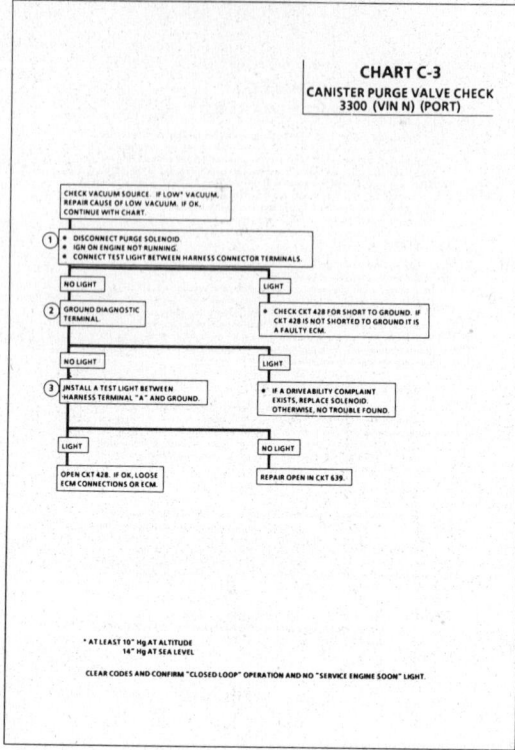

* AT LEAST 10" Hg AT ALTITUDE
14" Hg AT SEA LEVEL

CLEAR CODES AND CONFIRM "CLOSED LOOP" OPERATION AND NO "SERVICE ENGINE SOON" LIGHT.

1989–90 3.3L (VIN N)
CUTLASS CALAIS AND SKYLARK

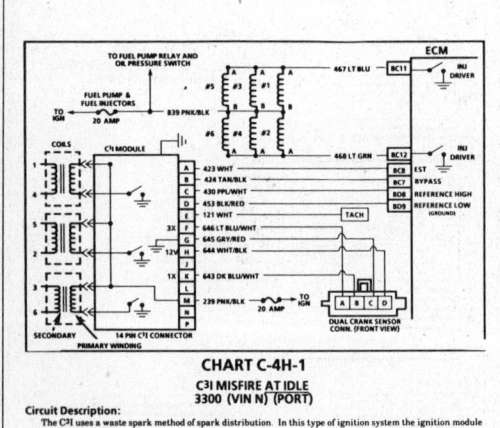

CHART C-4H-1
C³I MISFIRE AT IDLE
3300 (VIN N) (PORT)

Circuit Description:
The C³I uses a waste spark method of spark distribution. In this type of ignition system the ignition module triggers the #1/4 coils coil pair resulting in both #1 and #4 spark plugs firing at the same time. #1 cylinder is on the compression stroke at the same time #4 is on the exhaust stroke, resulting in a lower energy requirement to fire #4 spark plug. This leaves the remaining high voltage to fire #1 spark plug.

Test Description: Numbers below refer to circled numbers on the diagnostic chart.
1. If the "misfire" complaint exists under load only, the diagnostic CHART C-4H-2 must be used. Engine rpm should drop approximately equally on all plug leads.
2. A spark tester such as a ST-125 must be used because it is essential to verify adequate available secondary voltage at the spark plug. (25,000 volts). Secondary voltage of at least (25,000 volts) must be present to jump the gap of a ST-125.

3. If ignition coils are carbon tracked, the coil tower spark plug wire nipples may be damaged.
4. By checking the secondary resistance, a coil with an open secondary may be located.
5. By switching a normally operating coil into the position of the malfunctioning one, a determination can be made as to fault being the coil or C³I module.

1989–90 3.3L (VIN N)
CUTLASS CALAIS AND SKYLARK

CHART C-4H-1
C³I MISFIRE AT IDLE
3300 (VIN N) (PORT)

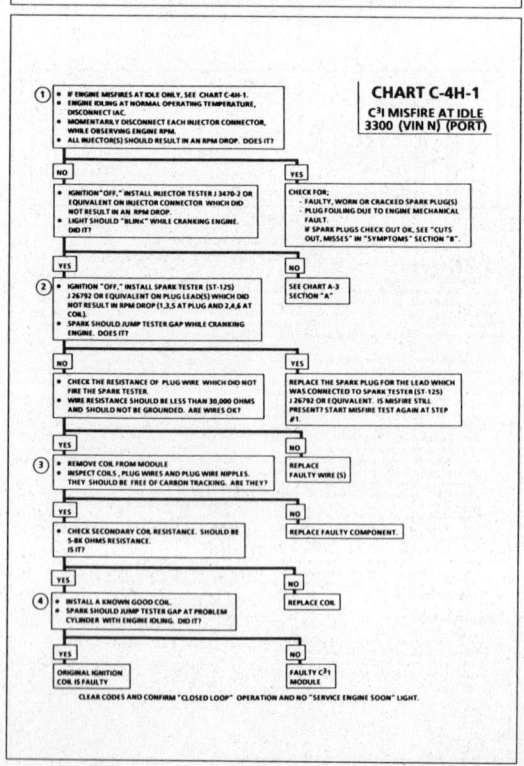

CLEAR CODES AND CONFIRM "CLOSED LOOP" OPERATION AND NO "SERVICE ENGINE SOON" LIGHT.

1989–90 3.3L (ViN N)
CUTLASS CALAIS AND SKYLARK

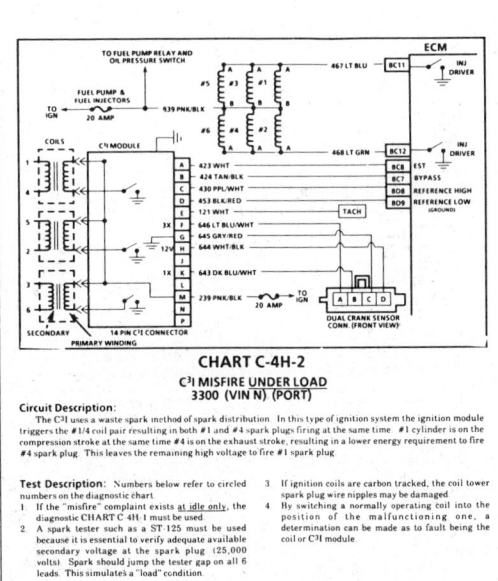

CHART C-4H-2
C³I MISFIRE UNDER LOAD
3300 (VIN N) (PORT)

Circuit Description:

The C³I uses a waste spark method of spark distribution. In this type of ignition system the ignition module triggers the #1/4 coil pair resulting in both #1 and #4 spark plugs firing at the same time. #1 cylinder is on the compression stroke at the same time #4 is on the exhaust stroke, resulting in a lower energy requirement to fire #4 spark plug. This leaves the remaining high voltage to fire #1 spark plug.

Test Description: Numbers below refer to circled numbers on the diagnostic chart.

1. If the "misfire" complaint exists at idle only, the diagnostic CHART C-4H-1 must be used.
2. A spark tester such as a ST-125 must be used because it is essential to verify adequate available secondary voltage at the spark plug. (25,000 volts). Spark should jump the tester gap on all 6 leads. This simulates a "load" condition.

3. If ignition coils are carbon tracked, the coil tower spark plug wire nipples may be damaged.
4. By switching a normally operating coil into the position of the malfunctioning one, a determination can be made as to fault being the coil or C³I module.

1989–90 3.3L (ViN N)
CUTLASS CALAIS AND SKYLARK

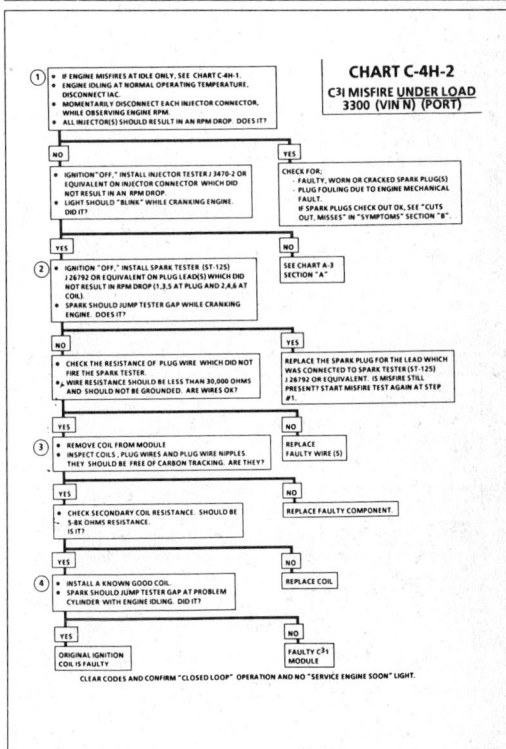

CHART C-4H-2
C³I MISFIRE UNDER LOAD
3300 (VIN N) (PORT)

1989–90 3.3L (ViN N)
CUTLASS CALAIS AND SKYLARK

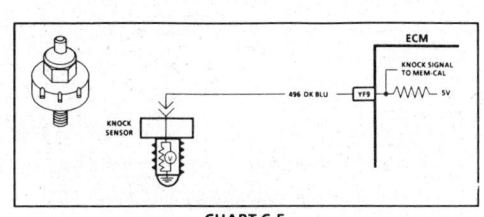

CHART C-5
ELECTRONIC SPARK CONTROL (ESC) SYSTEM CHECK
3300 (VIN N) (PORT)

Circuit Description:

The knock sensor is used to detect engine detonation and the ECM will retard the electronic spark timing based on the signal being received. The circuitry within the knock sensor causes the ECM's supplied 5 volt signal to be pulled down so that under a no knock condition, CKT 496 would measure about 2.5 volts. The knock sensor produces an A/C signal which rides on the 2.5 volts DC voltage. The amplitude and frequency are dependent upon the knock level.

The Mem-Cal used with this engine contains the functions which were part of the remotely mounted ESC modules used on other GM vehicles. The ESC portion of the Mem-Cal then sends a signal to other parts of the ECM which adjusts the spark timing to retard the spark and reduce the detonation.

Test Description: Numbers below refer to circled numbers on the diagnostic chart.

1. With engine idling, there should not be a knock signal present at the ECM because detonation is not likely under a no load condition.
2. Tapping on the engine lift hook should simulate a knock signal to determine if the sensor is capable of detecting detonation. If no knock is detected, try tapping on engine block closer to sensor before replacing sensor.
3. If the engine has an internal problem which is creating a knock, the knock sensor may be responding to the internal failure.

4. This test determines if the knock sensor is faulty or if the ESC portion of the Mem-Cal is faulty. If it is determined that the Mem-Cal is faulty, be sure that it is properly installed and latched into place. If not properly installed, repair and retest.

Diagnostic Aids:

While observing knock signal on the "Scan", there should be an indication that knock is present when detonation can be heard. Detonation is most likely to occur under high engine load conditions.

1989–90 3.3L (ViN N)
CUTLASS CALAIS AND SKYLARK

CHART C-5
ELECTRONIC SPARK CONTROL (ESC) SYSTEM CHECK
3300 (VIN N) (PORT)

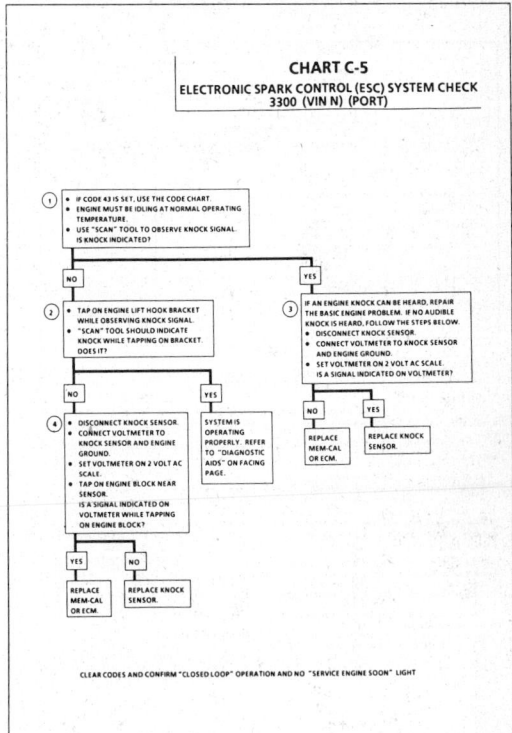

1989–90 3.3L (VIN N)
CUTLASS CALAIS AND SKYLARK

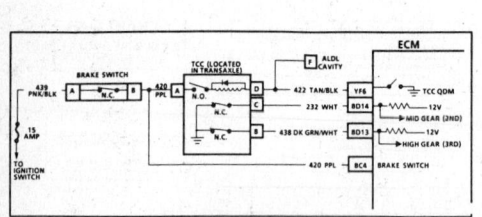

CHART C-8A
TRANSMISSION CONVERTER CLUTCH (TCC)
3300 (VIN N) (PORT)

Circuit Description:
The purpose of the transmission converter clutch feature is to eliminate the power loss of the transmission converter stage when the vehicle is in a cruise condition. This allows the convenience of the automatic transmission and the fuel economy of a manual transaxle.
Fused battery ignition is supplied to the TCC solenoid through the brake switch, and transmission 2nd gear apply switch. The ECM will engage TCC by grounding CKT 422 to energize the solenoid.

TCC will engage when:
- Engine warmed up.
- Vehicle speed above a calibrated value (about 32 mph 51 km/h).
- Throttle position sensor output not changing, indicating a steady road speed.
- Transmission third gear switch closed.
- Brake switch closed.
- TCC in-line apply switch closed.

Test Description: Numbers below refer to circled numbers on the diagnostic chart.
1. Light "OFF" confirms transmission third gear apply switch is open.
2. At 25 mph the transmission TCC in-line apply switch should close. Test light will come "ON" and confirm battery supply and closed brake switch.
3. Grounding the diagnostic terminal with ignition "ON," engine "OFF," checks the ability of the ECM to supply a ground to the TCC solenoid. The test light connected from 12 volts to ALDL terminal "F" should turn "ON" as CKT 422 is grounded.

Diagnostic Aids:

The "Scan" tool only indicates when the ECM has turned "ON" the TCC driver and this does not confirm that the TCC has engaged. To determine if TCC is functioning properly, engine rpm should decrease when the "Scan" indicates the TCC driver has turned "ON."

1989–90 3.3L (VIN N)
CUTLASS CALAIS AND SKYLARK

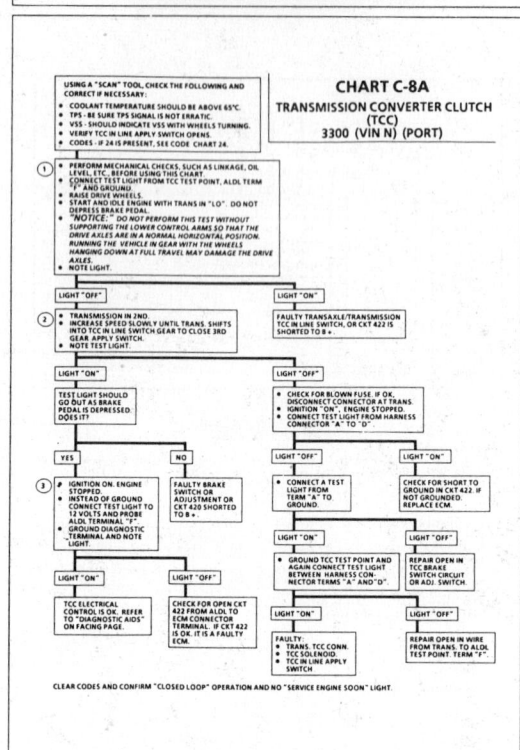

CHART C-8A
TRANSMISSION CONVERTER CLUTCH (TCC)
3300 (VIN N) (PORT)

1989–90 3.3L (VIN N)
CUTLASS CALAIS AND SKYLARK

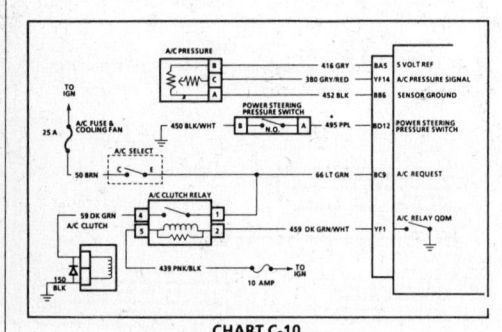

CHART C-10
A/C CLUTCH CONTROL CIRCUIT DIAGNOSIS
3300 (VIN N) (PORT)

Circuit Description:
The A/C clutch control relay is energized when the ECM provides a ground path through CKT 459 and A/C is requested. A/C clutch is delayed about .3 seconds after A/C is requested. This will allow the IAC to adjust engine rpm for the additional load.
The ECM will temporarily disengage the A/C clutch relay for a calibrated time for one or more of the following:
- Hot engine restart.
- Wide open throttle (TPS over 90%).
- Power steering pressure high (open power steering pressure switch).
- Engine rpm greater than about 6000 rpm.
- During IAC reset.
The A/C clutch relay will remain disengaged when a Code 66 is present or there is no A/C request signal due to an open A/C select switch or low pressure signal.

Test Description: Numbers below refer to circled numbers on the diagnostic chart.
1. The ECM will only energize the A/C relay when the engine is running. This test will determine if the relay or CKT 459 or ECM is faulty.
2. Determines if the signal is reaching the ECM through CKTs 50 and 66 from the A/C control panel. Signal should only be present when the A/C mode or defrost mode has been selected.
3. A short to ground in any part of the A/C request (CKTs 50 and 66). CKT 66 to the relay, CKT 59 to the A/C clutch, or the A/C clutch, could be the cause of a blown fuse.

4. If the ECM is receiving a high power steering pressure signal, the A/C clutch will be disengaged by the ECM.
5. With the engine idling and A/C "ON," the ECM should be grounding CKT 459, which should cause the test light to be "ON."

Diagnostic Aids:

If complaint is insufficient cooling, the problem may be caused by an inoperative cooling fan. See CHART C-12 for cooling fan diagnosis.

1989–90 3.3L (VIN N)
CUTLASS CALAIS AND SKYLARK

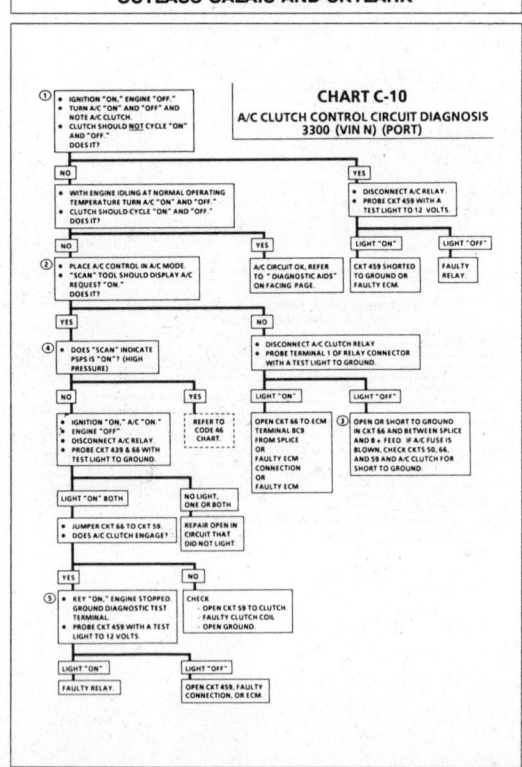

CHART C-10
A/C CLUTCH CONTROL CIRCUIT DIAGNOSIS
3300 (VIN N) (PORT)

1989–90 3.3L (VIN N)
CUTLASS CALAIS AND SKYLARK

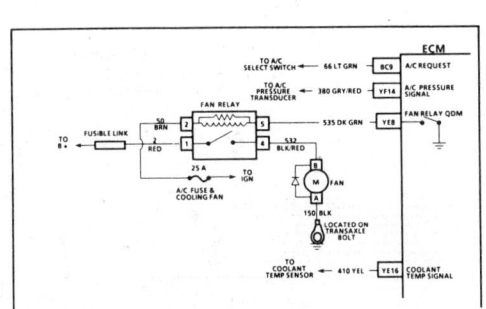

CHART C-12
FUNCTIONAL CHECK
3300 (VIN N) (PORT)

Circuit Description:

Battery voltage to operate the cooling fan motor is supplied to relay terminal "1". Ignition voltage to energize the relay is supplied to relay terminal "2". When the ECM grounds CKT 535, the relay is energized and the cooling fan is turned "ON." When the engine is running, the ECM will energize the cooling fan relay if a coolant temperature sensor code (14 or 15) has been set, or under the following conditions:

- A/C pressure is at a calibrated value
- Coolant temperature is greater than 100°C (212°F).

Test Description: Numbers below refer to circled numbers on the diagnostic chart.

1. Fan should be on as grounding the ALDL turns the QDM on providing a ground path for current to flow through relay coil.
2. The fan should be running if the A/C compressor is running as the ECM monitors B+ at terminal BC9 when A/C is requested.

Diagnostic Aids:

If the vehicle has an overheating problem, it must be determined if the complaint was due to an actual boilover, the coolant temperature warning light, or the temperature gage indicated over heating.

If the gage or light indicates overheating but no boilover is detected, the gage circuit should be checked. The gage accuracy can be checked by comparing the coolant sensor reading using a "Scan" tool with the gage reading.

If the engine is actually overheating and the gage indicates overheating, but the cooling fan is not turning "ON," the coolant sensor has probably shifted out of calibration and should be replaced.

1989–90 3.3L (VIN N)
CUTLASS CALAIS AND SKYLARK

CHART C-12
FUNCTIONAL CHECK
3300 (VIN N) (PORT)

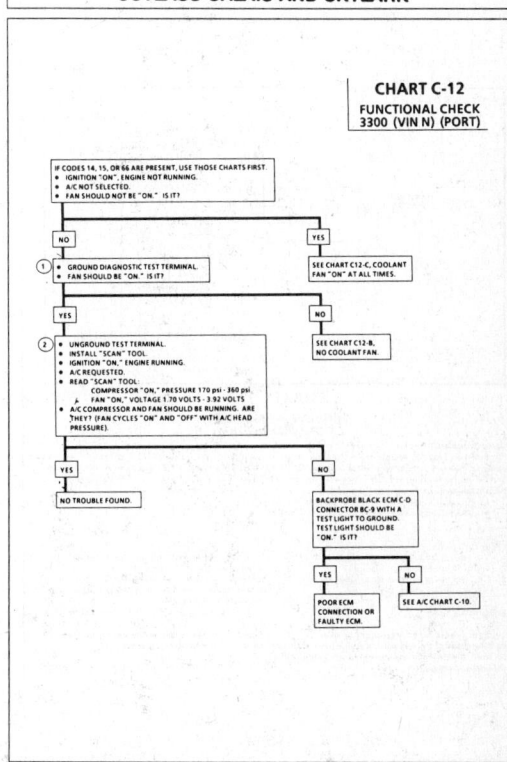

1989–90 3.3L (VIN N)
CUTLASS CALAIS AND SKYLARK

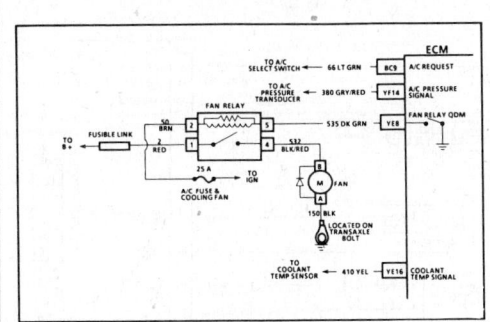

CHART C-12A
NO COOLANT FAN
3300 (VIN N) (PORT)

Circuit Description:

Battery voltage to operate the cooling fan motor is supplied to relay terminal "1". Ignition voltage to energize the relay is supplied to relay terminal "2". When the ECM grounds CKT 535, the relay is energized and the cooling fan is turned "ON." When the engine is running, the ECM will energize the cooling fan relay if a coolant temperature sensor code (14 or 15) has been set, or under the following conditions:

- A/C is "ON" and vehicle speed is less than 30 mph (48 km/h).
- Coolant temperature is greater than 100°C (212°F).

Test Description: Numbers below refer to circled numbers on the diagnostic chart.

1. Test light should be "ON" as the ignition switch is turned "ON."
2. Test light should be "ON" as grounding the ALDL turns the QDM "ON" providing current flow to ground.
3. Test light should be "ON" as terminal is direct from B+.

Diagnostic Aids

If the vehicle has an overheating problem, it must be determined if the complaint was due to an actual boilover, the coolant temperature warning light, or the temperature gage indicated overheating.

If the gage or light indicates overheating but no boilover is detected, the gage circuit should be checked. The gage accuracy can be checked by comparing the coolant sensor reading using a "Scan" tool with the gage reading.

If the engine is actually overheating and the gage indicates overheating, but the cooling fan is not turning "ON," the coolant sensor has probably shifted out of calibration and should be replaced.

1989–90 3.3L (VIN N)
CUTLASS CALAIS AND SKYLARK

CHART C-12A
NO COOLANT FAN
3300 (VIN N) (PORT)

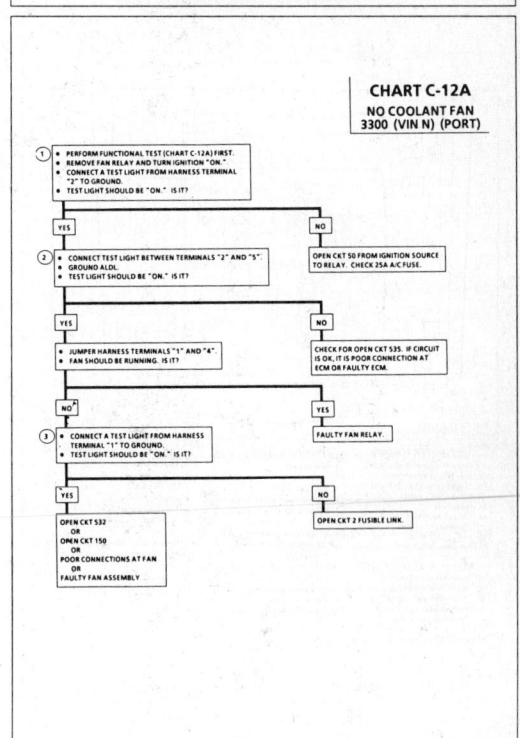

1989-90 3.3L (VIN N) CUTLASS CALAIS AND SKYLARK

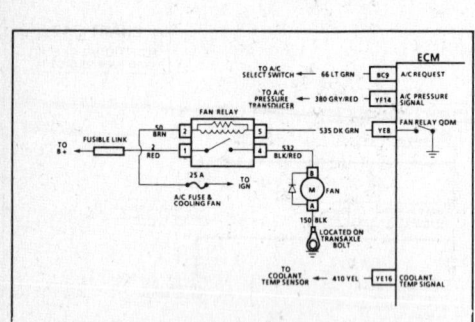

CHART C-12B
FAN "ON" AT ALL TIMES
3300 (VIN N) (PORT)

Circuit Description:

Battery voltage to operate the cooling fan motor is supplied to relay terminal "1". Ignition voltage to energize the relay is supplied to relay terminal "2". When the ECM grounds CKT 535, the relay is energized and the cooling fan is turned "ON." When the engine is running, the ECM will energize the cooling fan relay if a coolant temperature sensor code (14 or 15) has been set, or under the following conditions:

- A/C is "ON" and vehicle speed is less than 30 mph (48 km/h).
- Coolant temperature is greater than 100°C (212°F).

1. The coolant fan runs continuously if codes 14, 15 and/or 66 are present.
2. CKT 535 has to be grounded or QDM inside ECM is faulty if test light is "ON."

Diagnostic Aids:

If the vehicle has an overheating problem, it must be determined if the complaint was due to an actual boilover, the coolant temperature warning light, or the temperature gage indicated overheating.

If the gage or light indicates overheating but no boilover is detected, the gage circuit should be checked. The gage accuracy can be checked by comparing the coolant sensor reading using a "Scan" tool with the gage reading.

If the engine is actually overheating and the gage indicates overheating, but the cooling fan is not turning "ON", the coolant sensor has probably shifted out of calibration and should be replaced.

1989-90 3.3L (VIN N) CUTLASS CALAIS AND SKYLARK

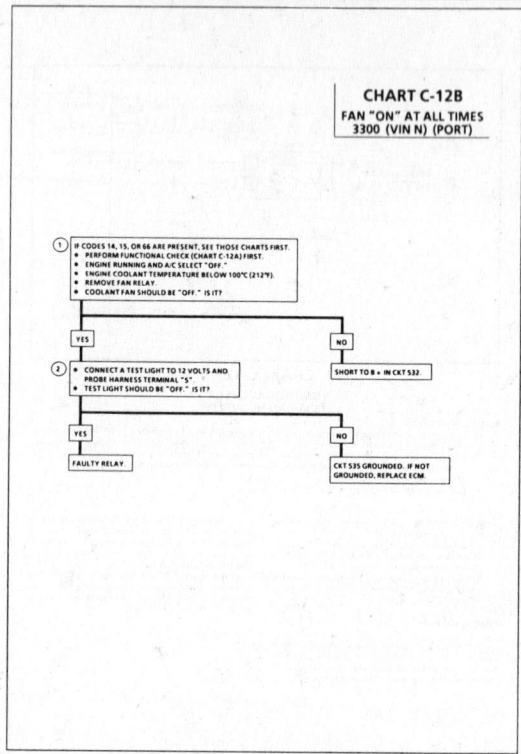

CHART C-12B
FAN "ON" AT ALL TIMES
3300 (VIN N) (PORT)

5.0L (VIN F) AND 5.7L (VIN 8) COMPONENT LOCATIONS—CAMARO AND FIREBIRD

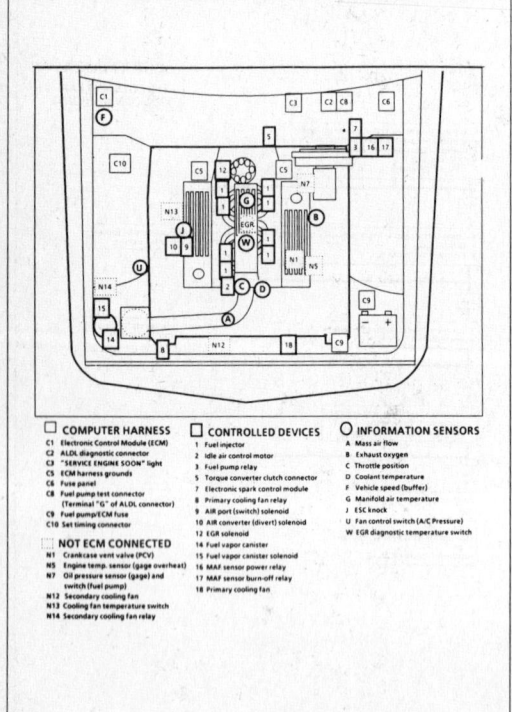

☐ **COMPUTER HARNESS**
C1 Electronic Control Module (ECM)
C3 ALDL diagnostic connector
C3 "SERVICE ENGINE SOON" light
C5 ECM harness grounds
C6 Fuse panel
C8 Fuel pump test connector
  (Terminal "G" of ALDL connector)
C9 Fuel pump/ECM fuse
C10 Set timing connector

☐ **NOT ECM CONNECTED**
N1 Crankcase vent valve (PCV)
N5 Engine temp. sensor (gage overheat)
N7 Oil pressure sensor (gage) and
  switch (fuel pump)
N12 Secondary cooling fan
N13 Cooling fan temperature switch
N14 Secondary cooling fan relay

◻ **CONTROLLED DEVICES**
1 Fuel injector
2 Idle air control motor
3 Fuel pump relay
4 Torque converter clutch connector
7 Electronic spark control module
8 Primary cooling fan relay
9 AIR port (switch) solenoid
10 AIR converter (divert) solenoid
12 EGR solenoid
14 Fuel vapor canister
15 Fuel vapor canister solenoid
16 MAF sensor power relay
17 MAF sensor burn-off relay
18 Primary cooling fan

○ **INFORMATION SENSORS**
A Mass air flow
B Exhaust oxygen
C Throttle position
D Coolant temperature
V Vehicle speed (buffer)
G Manifold air temperature
J ESC knock
U Fan control switch (A/C Pressure)
W EGR diagnostic temperature switch

5.0L (VIN F) AND 5.7L (VIN 8) ECM WIRING DIAGRAM CAMARO AND FIREBIRD

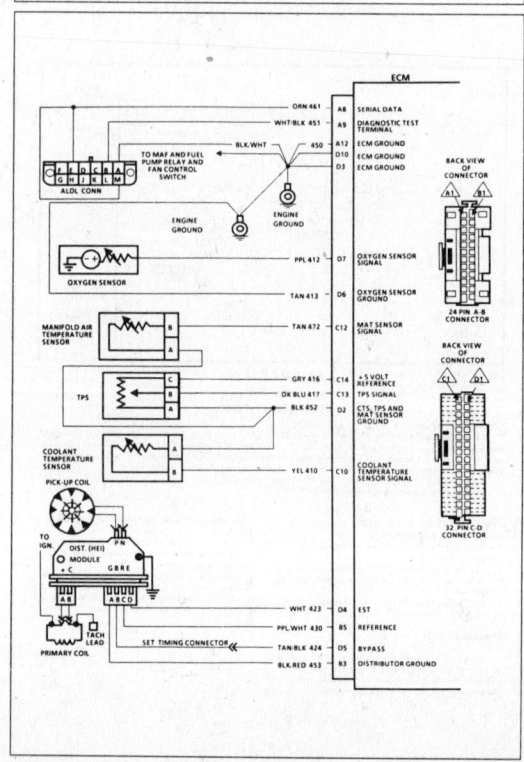

5.0L (VIN F) AND 5.7L (VIN 8) ECM WIRING DIAGRAM CAMARO AND FIREBIRD (CONT.)

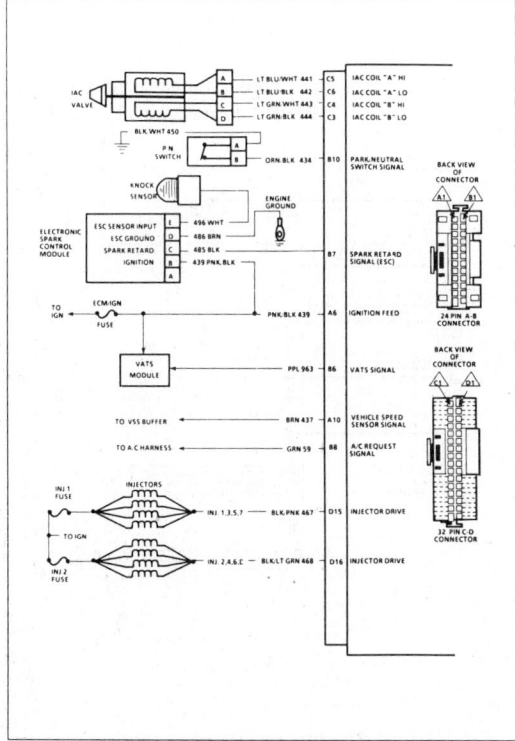

5.0L (VIN F) AND 5.7L (VIN 8) ECM WIRING DIAGRAM CAMARO AND FIREBIRD (CONT.)

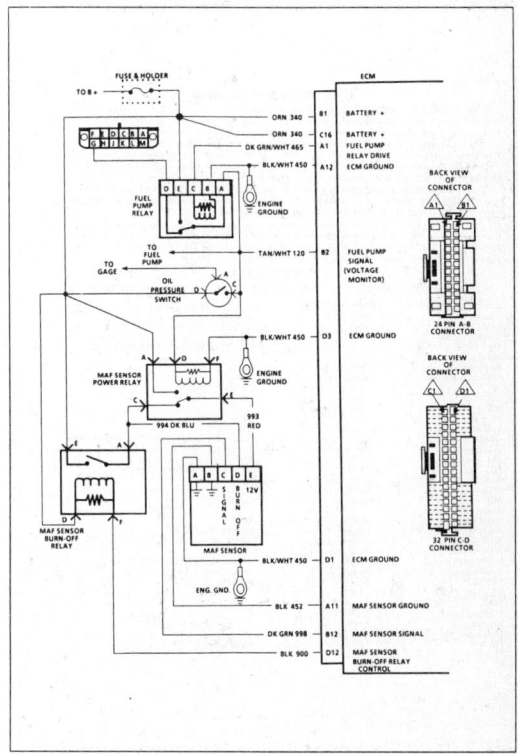

5.0L (VIN F) AND 5.7L (VIN 8) ECM WIRING DIAGRAM CAMARO AND FIREBIRD (CONT.)

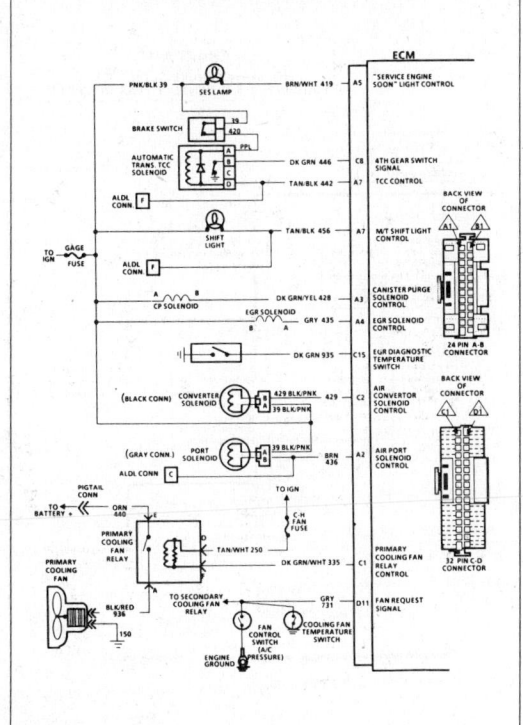

5.0L (VIN F) AND 5.7L (VIN 8) ECM CONNECTOR TERMINAL END VIEW—CAMARO AND FIREBIRD

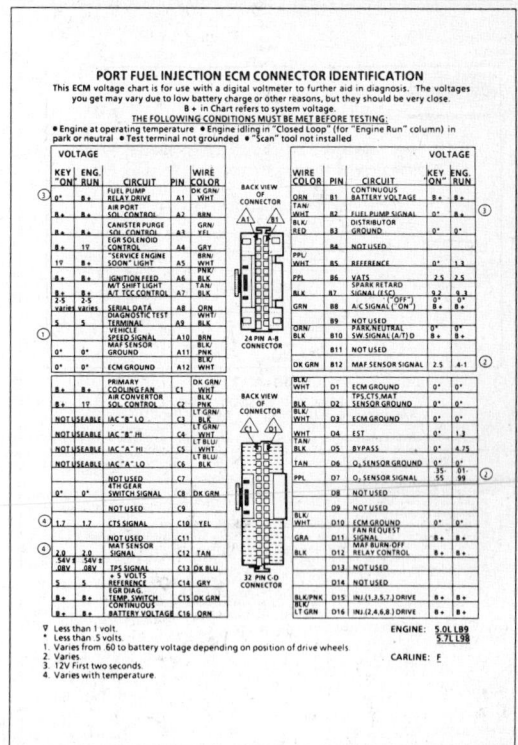

1988–90 5.0L (VIN F) AND 5.7L (VIN 8) CAMARO AND FIREBIRD

DIAGNOSTIC CIRCUIT CHECK

The Diagnostic Circuit Check is an organized approach to identifying a problem created by an Electronic Engine Control System malfunction. It must be the starting point for any driveability complaint diagnosis, because it directs the Service Technician to the next logical step in diagnosing the complaint.

The "Scan Data" listed in the table may be used for comparison, after completing the Diagnostic Circuit Check and finding the on-board diagnostics functioning properly and no trouble codes displayed. The "Typical Values" are an average of display values recorded from normally operating vehicles and are intended to represent what a normally functioning system would typically display.

A "SCAN" TOOL THAT DISPLAYS FAULTY DATA SHOULD NOT BE USED, AND THE PROBLEM SHOULD BE REPORTED TO THE MANUFACTURER. THE USE OF A FAULTY "SCAN" CAN RESULT IN MISDIAGNOSIS AND UNNECESSARY PARTS REPLACEMENT.

Only the parameters listed below are used in this manual for diagnosing. If a "Scan" reads other parameters, the values are not recommended by General Motors for use in diagnosing. For more description on the values and use of the "Scan" to diagnose ECM inputs, refer to the applicable symptoms in Section B. If all values are within the range illustrated, refer to symptoms in Section B.

"SCAN" DATA

Idle / Upper Radiator Hose Hot / Closed Throttle / Park or Neutral / Closed Loop / Acc. off

"SCAN" Position	Units Displayed	Typical Data Value
Desired RPM	RPM	ECM idle command (varies with temp.)
RPM	RPM	± 100 RPM from desired RPM (± 50 in drive)
Coolant Temp.	C°	85° - 105°
MAT Temp.	C°	10° - 90° (depends on underhood temp.)
MAF	Gm/Sec	4 - 7
Air Flow	Gm/Sec	4 - 7
BPW (base pulse width)	M/Sec	1 - 4 and varying
O₂	Volts	1-1000 and varying
TPS	Volts	.46 - .62
IAC	Counts (steps)	5 - 50
INT (Integrator)	Counts	Varies
BLM (Block Learn)	Counts	118 - 138
Open/Closed Loop	Open/Closed	Closed Loop (may go open with extended idle)
BLM Cell	Cell Number	0 or 1 (depends on Air Flow & RPM)
VSS	MPH	0
TCC	On/Off	Off (on with TCC commanded)
Battery	Volts	13.5 - 14.5
PPSW	Volts	13.5 - 14.5
LV8	Counts	30 - 60
Knock Retard	Degrees of Retard	0
Spark Advance	# of Degrees	Varies
P/N Switch	P/N and RDL	Park/Neutral (P/N)
A.I.R. Control	Normal/Divert	Normal
A.I.R. Switch	Port/Converter	Converter
A/C Request	Yes/No	No (yes, with A/C requested)
Fan Request	Yes/No	No (yes, with A/C high pressure)
EGRDC	0 - 100%	0 at idle
EGR Diagnostic	On/Off	off
Fan	On/Off	Off (below 108°C)
CCP duty cycle	0 - 100%	0
Knock Signal	Yes/No	No (yes, when knock is detected)
Shift Light (M/T)	On/Off	Off
4th Gear	Yes/No	No (yes, when in 4th gear)

1988–90 5.0L (VIN F) AND 5.7L (VIN 8) CAMARO AND FIREBIRD

DIAGNOSTIC CIRCUIT CHECK
5.0L (VIN F) & 5.7L (VIN 8) "F" CARLINE (PORT)

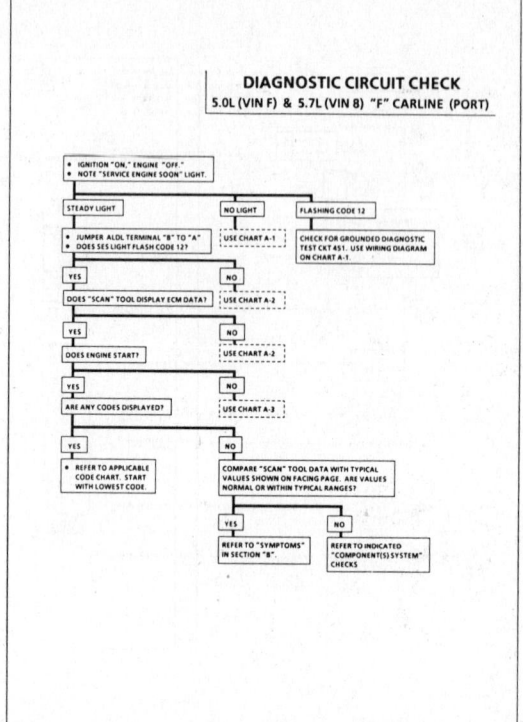

1988–90 5.0L (VIN F) AND 5.7L (VIN 8) CAMARO AND FIREBIRD

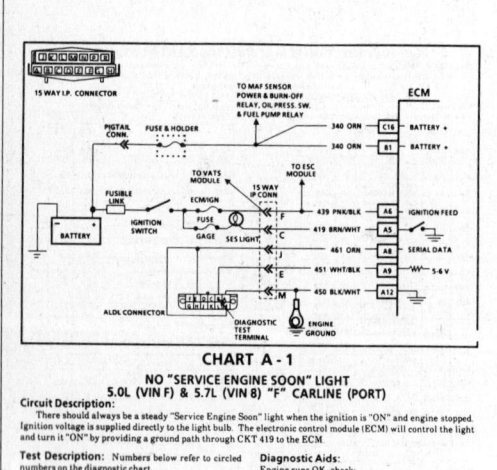

CHART A - 1
NO "SERVICE ENGINE SOON" LIGHT
5.0L (VIN F) & 5.7L (VIN 8) "F" CARLINE (PORT)

Circuit Description:
There should always be a steady "Service Engine Soon" light when the ignition is "ON" and engine stopped. Ignition voltage is supplied directly to the light bulb. The electronic control module (ECM) will control the light and turn it "ON" by providing a ground path through CKT 419 to the ECM.

Test Description: Numbers below refer to circled numbers on the diagnostic chart.
1. If the fuse in holder is blown, refer to facing page of Code 54 for complete circuit.
2. Using a test light connected to 12 volts probe each of the system ground circuits to be sure a good ground is present. Refer to the ECM terminal end view in front of this section for ECM pin locations of ground circuits.

Diagnostic Aids:
Engine runs OK, check:
• Faulty light bulb.
• CKT 419 open.
• Gage fuse blown. This will result in no oil or generator lights, seat belt reminder, etc.
Engine cranks but will not run, check:
• Continuous battery - fuse or fusible link open.
• ECM/Ignition fuse open.
• Battery CKT 340 to ECM open.
• Ignition CKT 439 to ECM open.
• Poor connection to ECM.
• Faulty ECM ground circuit(s).

1988–90 5.0L (VIN F) AND 5.7L (VIN 8) CAMARO AND FIREBIRD

CHART A-1
NO "SERVICE ENGINE SOON" LIGHT
5.0L (VIN F) & 5.7L (VIN 8) "F" CARLINE (PORT)

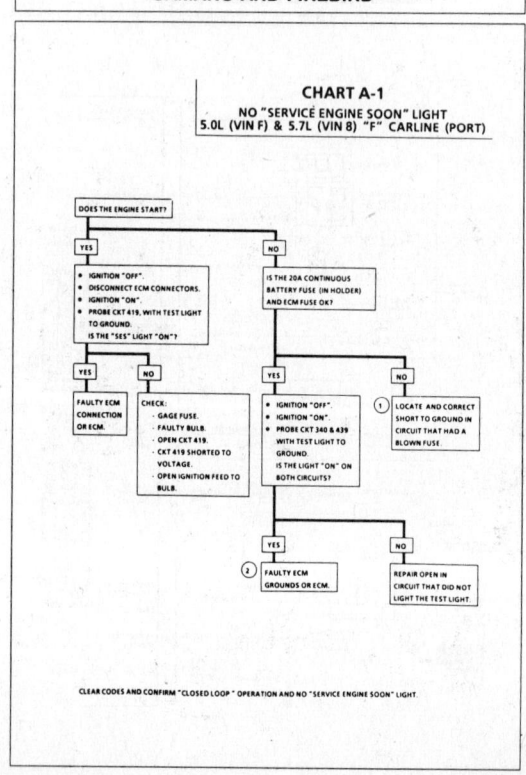

1988–90 5.0L (VIN F) AND 5.7L (VIN 8) CAMARO AND FIREBIRD

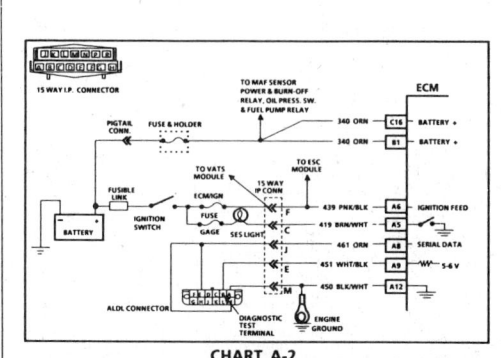

CHART A-2
NO ALDL DATA OR WILL NOT FLASH CODE 12
"SERVICE ENGINE SOON" LIGHT "ON" STEADY
5.0L (VIN F) & 5.7L (VIN 8) "F" CARLINE (PORT)

Circuit Description:
There should always be a steady "Service Engine Soon" light when the ignition is "ON" and engine stopped. Ignition voltage is supplied to the light bulb. The electronic control module (ECM) will turn the light "ON" by grounding CKT 419 in the ECM.

With the diagnostic "test" terminal grounded, the light should flash a Code 12, followed by any trouble code(s) stored in memory.

A steady light suggests a short to ground in the light control CKT 419, or an open in diagnostic CKT 451.

Test Description: Numbers below refer to circled numbers on the diagnostic chart.
1. If there is a problem with the ECM that causes a "Scan" tool to not read Serial data, the ECM should not flash a Code 12. If Code 12 is flashing check for CKT 451 short to ground. If Code 12 does flash be sure that the "Scan" tool is working properly on another vehicle. If the "Scan" is functioning properly and CKT 461 is OK, the Mem-Cal or ECM may be at fault for the NO ALDL symptom.
2. If the light goes "OFF" when the ECM connector is disconnected, CKT 419 is not shorted to ground.
3. This step will check for an open diagnostic CKT 451
4. At this point the "Service Engine Soon" light wiring is OK. The problem is a faulty ECM or Mem-Cal. If Code 12 does not flash, the ECM should be replaced using the original Mem-Cal. Replace the Mem-Cal only after trying an ECM, as a defective Mem-Cal is an unlikely cause of the problem.

1988–90 5.0L (VIN F) AND 5.7L (VIN 8) CAMARO AND FIREBIRD

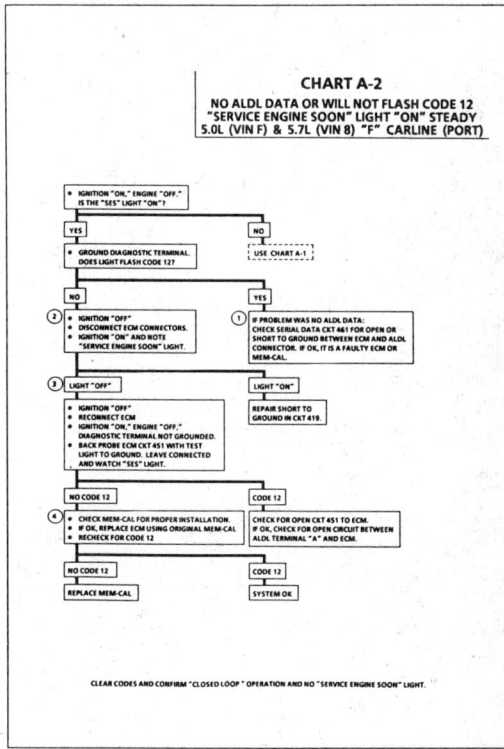

1988–90 5.0L (VIN F) AND 5.7L (VIN 8) CAMARO AND FIREBIRD

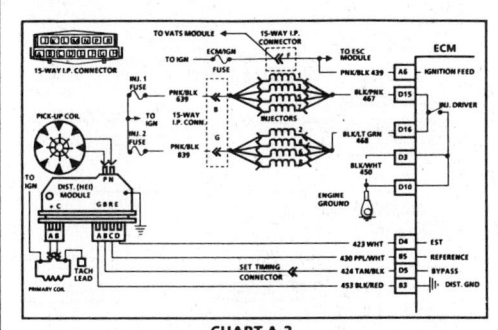

CHART A-3
(Page 1 of 2)
ENGINE CRANKS BUT WILL NOT RUN
5.0L (VIN F) & 5.7L (VIN 8) "F" CARLINE (PORT)

Circuit Description:
This chart assumes that battery condition and engine cranking speed are OK, and there is adequate fuel in the tank.

Test Description: Numbers below refer to circled numbers on the diagnostic chart.
1. A "Service Engine Soon" light "ON" is a basic test to determine if there is a 12 volt supply and ignition 12 volts to ECM. No ALDL may be due to an ECM problem and CHART A-2 will diagnose the ECM. If TPS is over 2.5 volts the engine may be in the clear flood mode which will cause no starting problems. The engine will not start without reference pulses and therefore the "Scan" should read engine rpm (reference) during crank.
2. No spark may be caused by one of several components related to the Ignition System. CHART C-4 will address all problems related to the causes of a no spark condition.
3. The test light should blink, indicating the ECM is controlling the injectors OK. How bright the light blinks is not important. However, the test light should be a J-34730-3 or equivalent.
4. Use fuel pressure gage J-34730-1 or equivalent. Wrap a shop towel around the fuel pressure tap to absorb any small amount of fuel leakage that may occur when installing the gage.

Diagnostic Aids:
- An EGR valve sticking open can cause a low air/fuel ratio during cranking. Unless engine enters "Clear Flood" at the first indication of a flooding condition, it can result in a no start.
- Check for fouled plugs.
- Water or foreign material in fuel line can cause a no start in freezing weather.
- A defective MAF Sensor may cause a no start or a stall after start. To determine if the sensor is causing the problem, disconnect it. The ECM will then use a default value for the sensor, and if the condition is corrected and the connections are OK, replace the sensor.
- Also check that injectors on both sides of engine will cause a test light to "blink". If not OK, check injector fuses.
  If above are all OK, refer to "Symptoms" in Section "B", Hard Start.

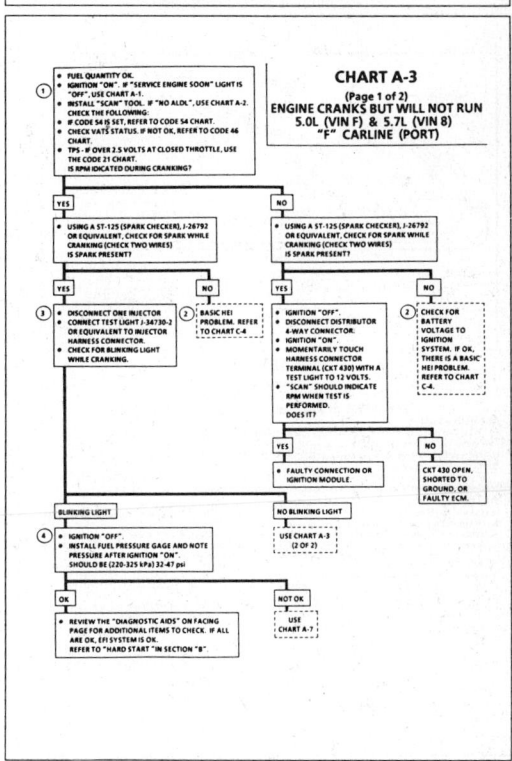

CHART A-3
(Page 1 of 2)
ENGINE CRANKS BUT WILL NOT RUN
5.0L (VIN F) & 5.7L (VIN 8) "F" CARLINE (PORT)

**1988–90 5.0L (VIN F) AND 5.7L (VIN 8)
CAMARO AND FIREBIRD**

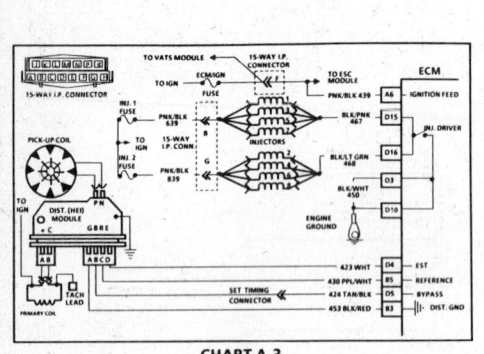

CHART A-3

(Page 2 of 2)
ENGINE CRANKS BUT WILL NOT RUN
5.0L (VIN F) & 5.7L (VIN 8) "F" CARLINE (PORT)

Test Description: Numbers below refer to circled numbers on the diagnostic chart.

1. Checks for 12 volt supply to injectors. Due to the Injectors wired in parallel there should be a light "ON" on both terminals.
2. Checks continuity of CKT 467 and CKT 468.
3. All checks made to this point would indicate that the ECM is at fault. However, there is a possibility of CKT 467 or CKT 468 being shorted to a voltage source either in the engine harness or in the injector harness.

To test for this condition:
- Disconnect all injectors
- Ignition "ON."
- Probe CKT 467 and CKT 468 on the ECM side of injector harness with a test light connected to ground. (Test one injector harness on each side of engine.) There should be no light. If light is "ON" repair short to voltage.
- If OK, check the resistance of the injectors.
- Should be 10 ohms or more.
- Check injector harness connector. Be sure terminals are not backed out of connector and contacting each other.
- If all OK, replace ECM.

**1988–90 5.0L (VIN F) AND 5.7L (VIN 8)
CAMARO AND FIREBIRD**

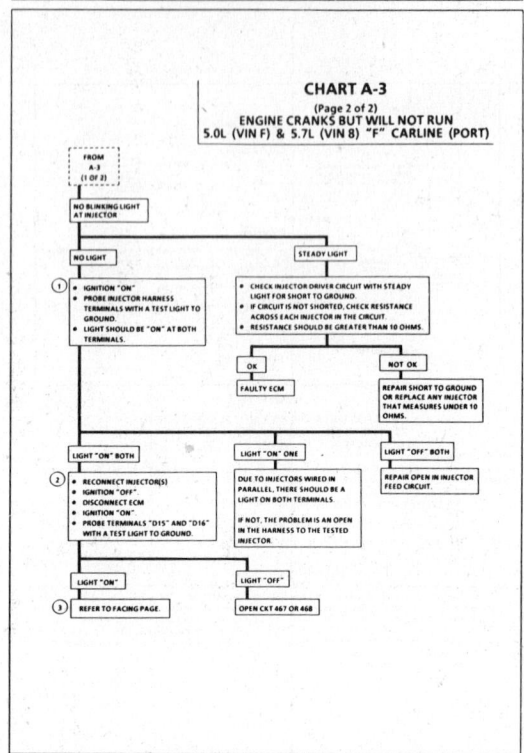

CHART A-3

(Page 2 of 2)
ENGINE CRANKS BUT WILL NOT RUN
5.0L (VIN F) & 5.7L (VIN 8) "F" CARLINE (PORT)

**1988–90 5.0L (VIN F) AND 5.7L (VIN 8)
CAMARO AND FIREBIRD**

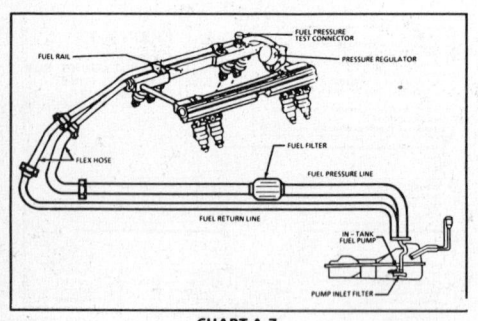

CHART A-7

(Page 1 of 2)
FUEL SYSTEM DIAGNOSIS
5.0L (VIN F) & 5.7L (VIN 8) "F" CARLINE (PORT)

Circuit Description:

When the ignition switch is turned "ON," the electronic control module (ECM) will turn "ON" the in-tank fuel pump. It will remain "ON" as long as the engine is cranking or running, and the ECM is receiving reference pulses. If there are no reference pulses, the ECM will shut "OFF" the fuel pump within 2 seconds after ignition "ON" or engine stops.

The pump will deliver fuel to the fuel rail and injectors, then to the pressure regulator, where the system pressure is controlled at 234 to 325 kPa (34 to 47 psi). Excess fuel is then returned to the fuel tank.

Test Description: Numbers below refer to circled numbers on the diagnostic chart.

1. Wrap a shop towel around the fuel pressure connector to absorb any small amount of fuel leakage that may occur when installing the gage. Ignition "ON," pump pressure should be 280-325 KPa (40.5-47 psi). This pressure is controlled by spring pressure within the regulator assembly.
2. When the engine is idling, the manifold pressure is low (high vacuum) and is applied to the fuel regulator diaphragm. This will offset the spring and result in a lower fuel pressure. This idle pressure will vary somewhat depending on barometric pressure; however, the pressure idling should be less, indicating pressure regulator control.
3. Pressure that continues to fall is caused by one of the following:
   - In-tank fuel pump check valve not holding.
   - Pump coupling hose or pulsator leaking.
   - Fuel pressure regulator valve leaking.

- Injector(s) sticking open.
4. An injector sticking open can best be determined by checking for a fouled or saturated spark plug(s). If a leaking injector cannot be determined by a fouled or saturated spark plug, the following procedure should be used:
   - Remove Plenum, and remove fuel rail bolts. Follow the procedures in the Fuel Control Section of this manual, but leave fuel lines connected.
   - Lift fuel rail out just enough to observe injector nozzles in the ports.

CAUTION: *BE SURE INJECTOR(S) ARE NOT ALLOWED TO SPRAY ON ENGINE AND THAT INJECTOR RETAINING CLIPS ARE INTACT. THIS SHOULD BE CAREFULLY FOLLOWED TO PREVENT FUEL SPRAY ON ENGINE WHICH WOULD CAUSE A FIRE HAZARD.*

- Pressurize the fuel system and observe injector nozzles.

**1988–90 5.0L (VIN F) AND 5.7L (VIN 8)
CAMARO AND FIREBIRD**

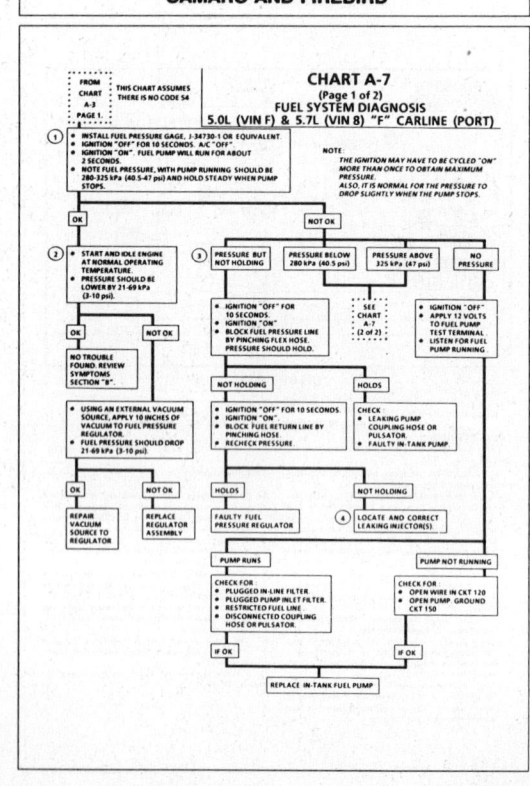

CHART A-7

(Page 1 of 2)
FUEL SYSTEM DIAGNOSIS
5.0L (VIN F) & 5.7L (VIN 8) "F" CARLINE (PORT)

1988–90 5.0L (VIN F) AND 5.7L (VIN 8) CAMARO AND FIREBIRD

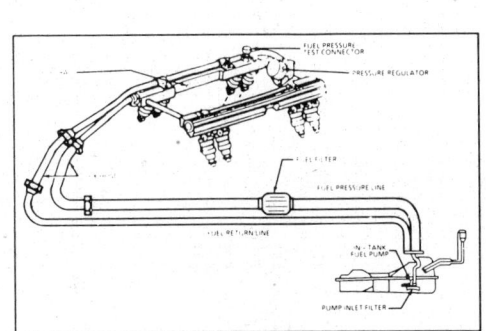

CHART A-7
(Page 2 of 2)
FUEL SYSTEM DIAGNOSIS
5.0L (VIN F) & 5.7L (VIN 8) "F" CARLINE (PORT)

Test Description: Numbers below refer to circled numbers on the diagnostic chart.

1. Fuel pressure less than 280 kPa (40.5 psi) falls into two areas.
   - Regulated pressure less than 280 kPa (40.5 psi). System will be lean and may set Code 44. Also, hard starting cold and overall poor performance.
   - Restricted flow causing pressure drop. Normally, a vehicle with a fuel pressure of less than 165 kPa (24 psi) at idle will not be driveable. However, if the pressure drop occurs only while driving, the engine will normally surge then stop running as pressure begins to drop rapidly. This is most likely caused by a restricted fuel line or plugged filter.

2. Restricting the fuel return line allows the fuel pressure to build above regulated pressure. With battery voltage applied to the pump test terminal, pressure should rise to 414 kPa (60 psi) as the fuel return hose is gradually pinched.

   NOTICE: Do Not allow fuel pressure to exceed 414 kPa (60 psi), damage to the pressure regulator may result.

3. This test determines if the high fuel pressure is due to a restricted fuel return line or a pressure regulator problem.

1988–90 5.0L (VIN F) AND 5.7L (VIN 8) CAMARO AND FIREBIRD

NOTICE: FUEL SYSTEM UNDER PRESSURE. TO AVOID FUEL SPILLAGE, REFER TO FIELD SERVICE PROCEDURES FOR TESTING OR MAKING REPAIRS REQUIRING DISASSEMBLY OF FUEL LINES OR FITTINGS.

CHART A-7
(Page 2 of 2)
FUEL SYSTEM DIAGNOSIS
5.0L (VIN F) & 5.7L (VIN 8) "F" CARLINE (PORT)

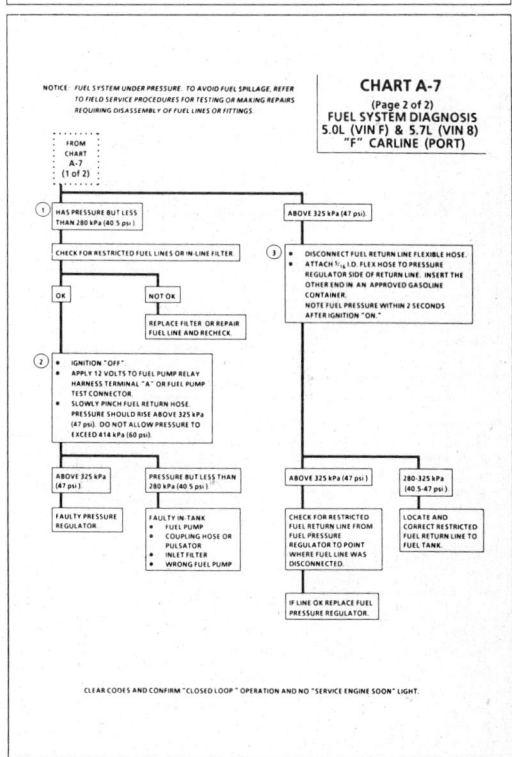

CLEAR CODES AND CONFIRM "CLOSED LOOP" OPERATION AND NO "SERVICE ENGINE SOON" LIGHT.

1988–90 5.0L (VIN F) AND 5.7L (VIN 8) CAMARO AND FIREBIRD

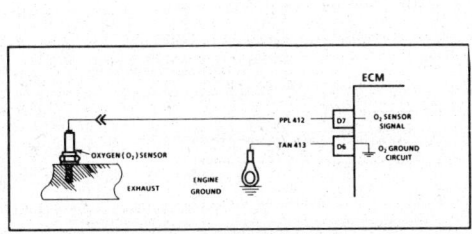

CODE 13
OXYGEN SENSOR CIRCUIT
(OPEN CIRCUIT)
5.0L (VIN F) & 5.7L (VIN 8) "F" CARLINE (PORT)

Circuit Description:
The ECM supplies a voltage of about .45 volt between terminals "D7" and "D6". (If measured with a 10 megohm digital voltmeter, this may read as low as .32 volt.) The O₂ sensor varies the voltage within a range of about 1 volt if the exhaust is rich, down through about .10 volt if exhaust is lean.

The sensor is like an open circuit and produces no voltage when it is below 360° C (600°F). An open sensor circuit or cold sensor causes "Open Loop" operation.

Test Description: Numbers below refer to circled numbers on the diagnostic chart.

1. Code 13 WILL SET:
   - Engine at normal operating temperature (above 80°C/176°F).
   - At least 2 minutes engine time after start.
   - O₂ signal voltage steady between .35 and .55 volt.
   - Throttle position sensor signal above 5% (about .3 volt above closed throttle voltage).
   - All conditions must be met for about 60 seconds.
     If the conditions for a Code 13 exist, the system will not go "Closed Loop."

2. This will determine if the sensor is at fault or the wiring or ECM is the cause of the Code 13.

3. For this test use only a high impedance digital volt ohmmeter. This test checks the continuity of CKT 412 and CKT 413. If CKT 413 is open, the ECM voltage on CKT 412 will be over .6 volt (600 mV).

Diagnostic Aids:
Normal "Scan" voltage varies between 100mV to 999mV (.1 and 1.0 volt) while in "Closed Loop." Code 13 sets in one minute if voltage remains between .35 and .55 volts, but the system will go "Open Loop" in about 15 seconds.

Refer to "Intermittents" in Section "B."

1988–90 5.0L (VIN F) AND 5.7L (VIN 8) CAMARO AND FIREBIRD

CODE 13
OXYGEN SENSOR CIRCUIT
(OPEN CIRCUIT)
5.0L (VIN F) & 5.7L (VIN 8) "F" CARLINE (PORT)

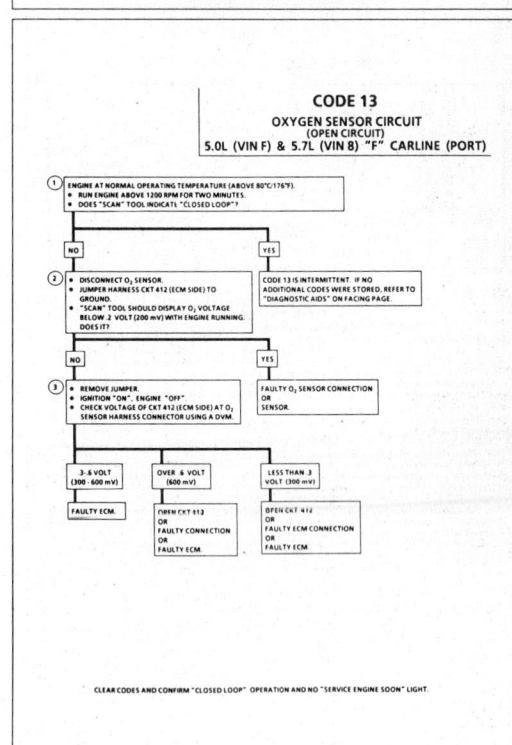

CLEAR CODES AND CONFIRM "CLOSED LOOP" OPERATION AND NO "SERVICE ENGINE SOON" LIGHT.

1988–90 5.0L (VIN F) AND 5.7L (VIN 8) CAMARO AND FIREBIRD

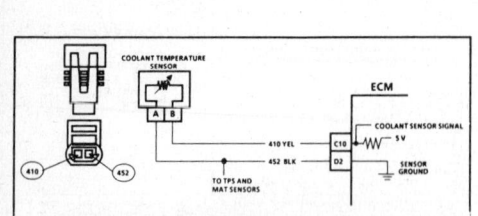

CODE 14
COOLANT TEMPERATURE SENSOR CIRCUIT
(HIGH TEMPERATURE INDICATED)
5.0L (VIN F) & 5.7L (VIN 8) "F" CARLINE (PORT)

Circuit Description:
The coolant temperature sensor uses a thermistor to control the signal voltage to the ECM. The ECM applies a voltage on CKT 410 to the sensor. When the engine coolant is cold, the sensor (thermistor) resistance is high, therefore, the ECM will see high signal voltage.
As the engine coolant warms, the sensor resistance becomes less, and the voltage drops. At normal engine operating temperature (85°C to 95°C or 185°F to 203°F), the voltage will measure about 1.5 to 2.0 volts.

Test Description: Numbers below refer to circled numbers on the diagnostic chart.
1. Code 14 will set if:
   - signal voltage indicates a coolant temperature above 130°C (266°F) for 3 seconds
2. This test will determine if CKT 410 is shorted to ground which will cause the conditions for Code 14.

Diagnostic Aids:
Check harness routing for a potential short to ground in CKT 410.
"Scan" tool displays engine temperature in degrees centigrade. After engine is started, the temperature should rise steadily to about 90°C (194°F), then stabilize when thermostat opens.
Refer to "Intermittents" in Section "B".

1988–90 5.0L (VIN F) AND 5.7L (VIN 8) CAMARO AND FIREBIRD

CODE 14
COOLANT TEMPERATURE SENSOR CIRCUIT
(HIGH TEMPERATURE INDICATED)
5.0L (VIN F) & 5.7L (VIN 8) "F" CARLINE (PORT)

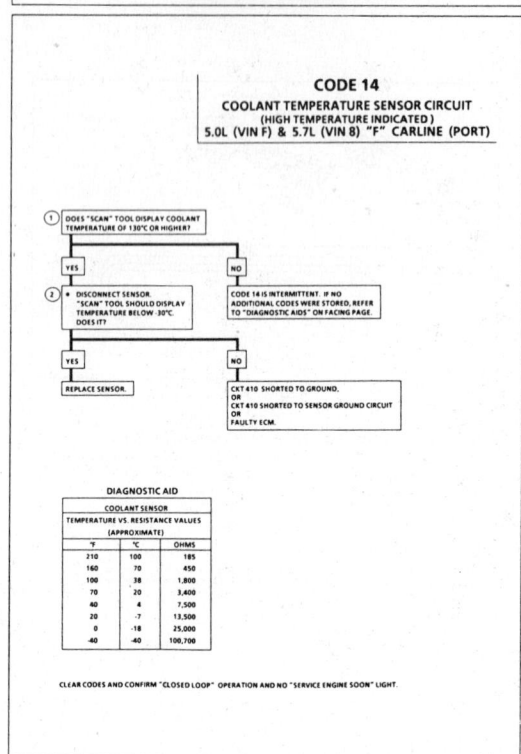

DIAGNOSTIC AID		
COOLANT SENSOR		
TEMPERATURE VS. RESISTANCE VALUES (APPROXIMATE)		
°F	°C	OHMS
210	100	185
160	70	450
100	38	1,800
70	20	3,400
40	4	7,500
20	-7	13,500
0	-18	25,000
-40	-40	100,700

CLEAR CODES AND CONFIRM "CLOSED LOOP" OPERATION AND NO "SERVICE ENGINE SOON" LIGHT.

1988–90 5.0L (VIN F) AND 5.7L (VIN 8) CAMARO AND FIREBIRD

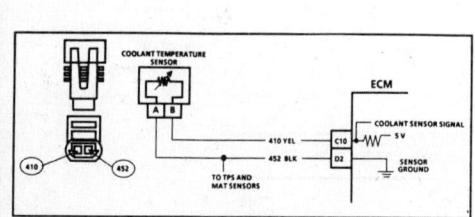

CODE 15
COOLANT TEMPERATURE SENSOR CIRCUIT
(LOW TEMPERATURE INDICATED)
5.0L (VIN F) & 5.7L (VIN 8) "F" CARLINE (PORT)

Circuit Description:
The coolant temperature sensor uses a thermistor to control the signal voltage to the ECM. The ECM applies a voltage on CKT 410 to the sensor. When the engine coolant is cold, the sensor (thermistor) resistance is high, therefore, the ECM will see high signal voltage.
As the engine coolant warms, the sensor resistance becomes less, and the voltage drops. At normal engine operating temperature (85°C to 95°C or 185°F to 203°F), the voltage will measure about 1.5 to 2.0 volts at the ECM.

Test Description: Numbers below refer to circled numbers on the diagnostic chart.
1. Code 15 will set if:
   - Engine coolant temperature less than -39° C (-38° F) for 3 seconds.
2. This test simulates a Code 14. If the ECM recognizes the low signal voltage, (high temperature) and the "Scan" reads 130°C (266°F) or above, the ECM and wiring are OK.
3. This test will determine if CKT 410 is open. There should be 5 volts present at sensor connector if measured with a DVM.

Diagnostic Aids:
A "Scan" tool reads engine coolant temperature in degrees centigrade. After engine is started the temperature should rise steadily to about 90°C (190°F) then stabilize when thermostat opens.
A faulty connection, or an open in CKT 410 or 452 will result in a Code 15.
If Code 21 or 23 is also set, check CKT 452 for faulty wiring or connections. Check terminals at sensor for good contact.
Refer to "Intermittents" in Section "B".

1988–90 5.0L (VIN F) AND 5.7L (VIN 8) CAMARO AND FIREBIRD

CODE 15
COOLANT TEMPERATURE SENSOR CIRCUIT
(LOW TEMPERATURE INDICATED)
5.0L (VIN F) & 5.7L (VIN 8) "F" CARLINE (PORT)

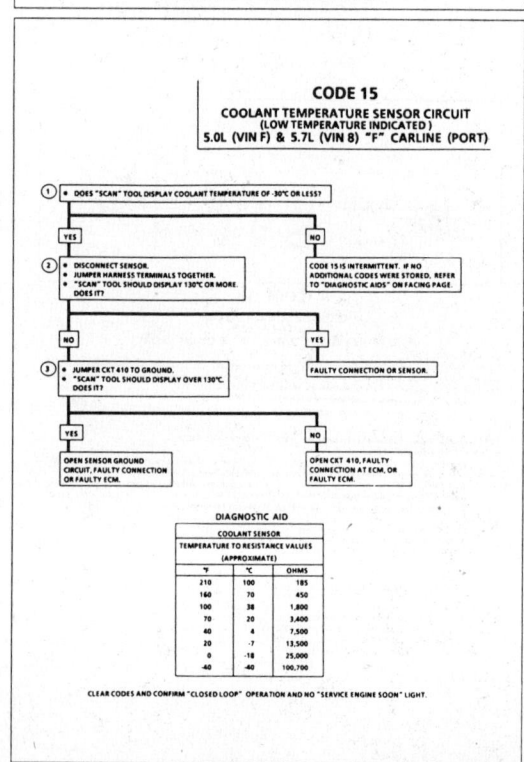

DIAGNOSTIC AID		
COOLANT SENSOR		
TEMPERATURE TO RESISTANCE VALUES (APPROXIMATE)		
°F	°C	OHMS
210	100	185
160	70	450
100	38	1,800
70	20	3,400
40	4	7,500
20	-7	13,500
0	-18	25,000
-40	-40	100,700

CLEAR CODES AND CONFIRM "CLOSED LOOP" OPERATION AND NO "SERVICE ENGINE SOON" LIGHT.

1988–90 5.0L (VIN F) AND 5.7L (VIN 8) CAMARO AND FIREBIRD

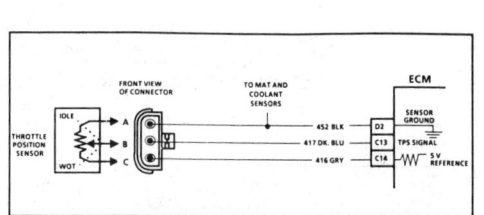

CODE 21
THROTTLE POSITION SENSOR (TPS) CIRCUIT
(SIGNAL VOLTAGE HIGH)
5.0L (VIN F) & 5.7L (VIN 8) "F" CARLINE (PORT)

Circuit Description:

The throttle position sensor (TPS) provides a voltage signal that changes relative to the throttle blade. Signal voltage will vary from about .5 at idle to about 5 volts at wide open throttle.

The TPS signal is one of the most important inputs used by the ECM for fuel control and for most of the ECM controlled outputs.

Test Description: Numbers below refer to circled numbers on the diagnostic chart.

1. Code 21 will set if:
   - TPS signal voltage is greater than 2.5 volts
   - Engine is running
   - Mass air flow is less than 12 GM/sec
   - All conditions met for 3 seconds
     OR
   - TPS signal voltage over 4.8 volts with ignition "ON"

   With throttle closed, the TPS should read less than .62 volt. If it doesn't check adjustment.
2. With the TPS sensor disconnected, the TPS voltage should go low if the ECM and wiring are OK.
3. Probing CKT 452 with a test light checks the 5V return CKT, because a faulty 5 volts return will cause a Code 21.

Diagnostic Aids:

A "Scan" tool reads throttle position in volts. With ignition "ON" or at idle, voltage should read .54V ± .08V with throttle closed and increase at a steady rate as throttle is moved toward WOT.

An open in CKT 452 will result in a Code 21.
Refer to "Intermittents" in Section "B".

1988–90 5.0L (VIN F) AND 5.7L (VIN 8) CAMARO AND FIREBIRD

CODE 21
THROTTLE POSITION SENSOR (TPS) CIRCUIT
(SIGNAL VOLTAGE HIGH)
5.0L (VIN F) & 5.7L (VIN 8) "F" CARLINE (PORT)

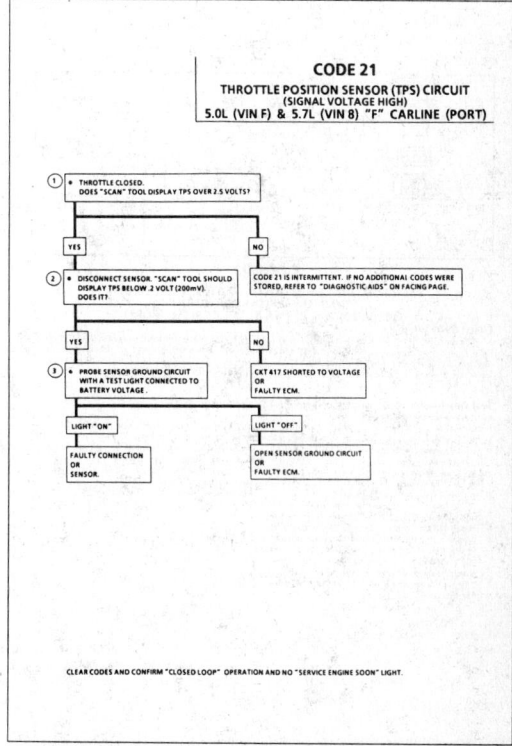

CLEAR CODES AND CONFIRM "CLOSED LOOP" OPERATION AND NO "SERVICE ENGINE SOON" LIGHT.

1988–90 5.0L (VIN F) AND 5.7L (VIN 8) CAMARO AND FIREBIRD

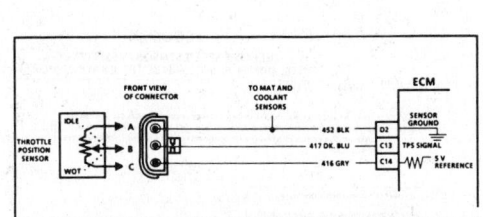

CODE 22
THROTTLE POSITION SENSOR (TPS) CIRCUIT
(SIGNAL VOLTAGE LOW)
5.0L (VIN F) & 5.7L (VIN 8) "F" CARLINE (PORT)

Circuit Description:

The throttle position sensor (TPS) provides a voltage signal that changes relative to the throttle blade. Signal voltage will vary from about .5 at idle to about 5 volts at wide open throttle.

The TPS signal is one of the most important inputs used by the ECM for fuel control and for most of the ECM controlled outputs.

Test Description: Numbers below refer to circled numbers on the diagnostic chart.

1. Code 22 will set if:
   - Engine running
   - TPS signal voltage is less than about .2 volt for 3 seconds.
2. Simulates Code 21 (high voltage) If the ECM recognizes the high signal voltage the ECM and wiring are OK.
3. TPS adjustment: With throttle closed, the TPS voltage reading should be .54V ± .08V.
4. This simulates a high signal voltage to check for an open in CKT 417.

Diagnostic Aids:

A "Scan" tool reads throttle position in volts. With ignition "ON" or at idle, voltage should read .54V ± .08V with throttle closed and increase at a steady rate as throttle is moved toward WOT.

An open or short to ground in CKTs 416 or 417 will result in a Code 22.
Refer to "Intermittents" in Section "B".

1988–90 5.0L (VIN F) AND 5.7L (VIN 8) CAMARO AND FIREBIRD

CODE 22
THROTTLE POSITION SENSOR (TPS) CIRCUIT
(SIGNAL VOLTAGE LOW)
5.0L (VIN F) & 5.7L (VIN 8) "F" CARLINE (PORT)

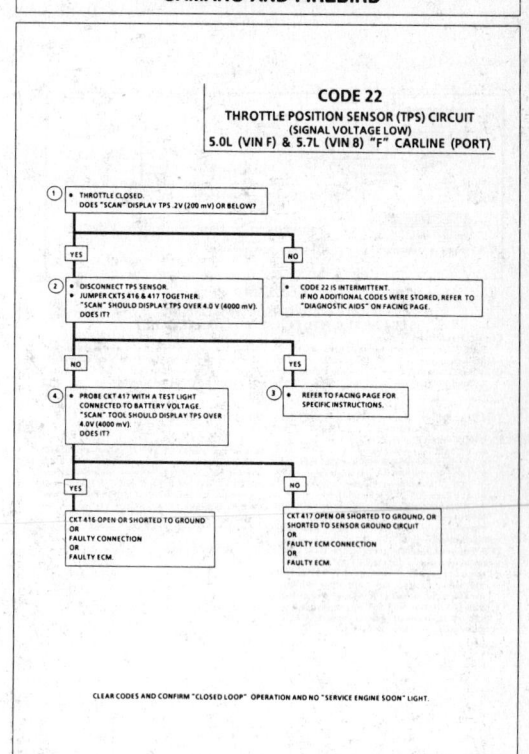

CLEAR CODES AND CONFIRM "CLOSED LOOP" OPERATION AND NO "SERVICE ENGINE SOON" LIGHT.

1988–90 5.0L (VIN F) AND 5.7L (VIN 8) CAMARO AND FIREBIRD

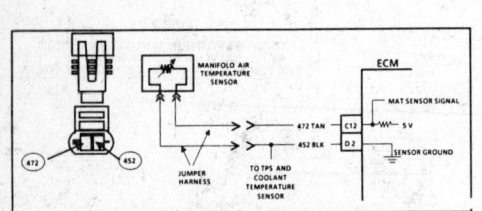

CODE 23
MANIFOLD AIR TEMPERATURE (MAT) SENSOR CIRCUIT
(LOW TEMPERATURE INDICATED)
5.0L (VIN F) & 5.7L (VIN 8) "F" CARLINE (PORT)

Circuit Description:
The manifold air temperature (MAT) sensor uses a thermistor to control the signal voltage to the ECM. The ECM applies a voltage (about 5 volts) on CKT 472 to the sensor. When the manifold air is cold, the sensor (thermistor) resistance is high, therefore, the ECM will see a high signal voltage. If the manifold air is warm, the sensor (thermistor) resistance is low, therefore, the ECM will see a low voltage.

Test Description: Numbers below refer to circled numbers on the diagnostic chart.
1. Code 23 will set if:
   - A signal voltage indicates a manifold air temperature below −30°C (−22°F) for 12 seconds.
   - Time since engine start is 1 minute or longer.
   - No VSS (vehicle not moving)
2. A Code 23 will set, due to an open sensor, wire or connection. This test will determine if the wiring and ECM are OK. The MAT sensor is difficult to reach and this test can be performed by disconnecting the MAT jumper harness connector. If the "Scan" indicates a temperature of over 130°C (266°F) the jumper harness to the sensor should be checked before replacing the sensor.
3. This will determine if the signal CKT 472 or the 5V return CKT 452 is open.

Diagnostic Aids:
A "Scan" tool reads temperature of the air entering the engine and should read close to ambient air temperature when engine is cold, and rises as underhood temperature increases.
Carefully check harness and connections for possible open CKT 472 or CKT 452.
Refer to "Intermittents" in Section "B".

1988–90 5.0L (VIN F) AND 5.7L (VIN 8) CAMARO AND FIREBIRD

CODE 23
MANIFOLD AIR TEMPERATURE (MAT) SENSOR CIRCUIT
(LOW TEMPERATURE INDICATED)
5.0L (VIN F) & 5.7L (VIN 8) "F" CARLINE (PORT)

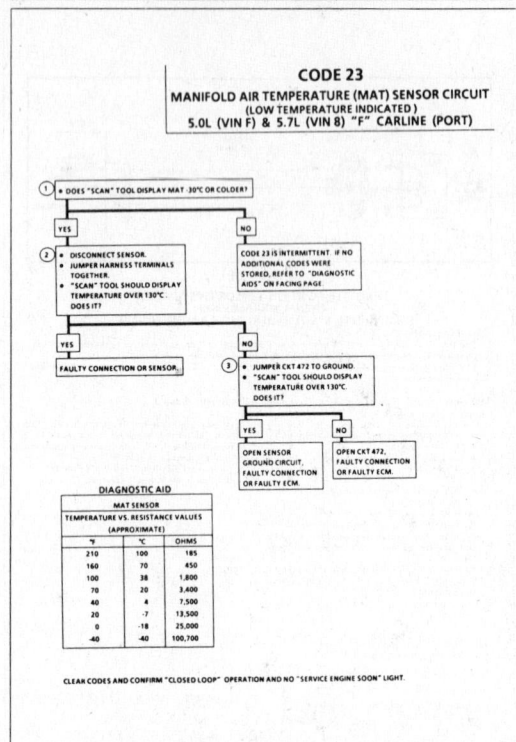

MAT SENSOR		
TEMPERATURE VS. RESISTANCE VALUES (APPROXIMATE)		
°F	°C	OHMS
210	100	185
160	70	450
100	38	1,800
70	20	3,400
40	4	7,500
20	-7	13,500
0	-18	25,000
-40	-40	100,700

CLEAR CODES AND CONFIRM "CLOSED LOOP" OPERATION AND NO "SERVICE ENGINE SOON" LIGHT.

1988–90 5.0L (VIN F) AND 5.7L (VIN 8) CAMARO AND FIREBIRD

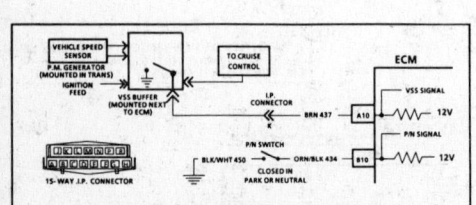

CODE 24
VEHICLE SPEED SENSOR (VSS) CIRCUIT
5.0L (VIN F) & 5.7L (VIN 8) "F" CARLINE (PORT)

Circuit Description:
The ECM applies and monitors 12 volts on CKT 437. CKT 437 connects to the vehicle speed sensor buffer which alternately grounds CKT 437 when drive wheels are turning. This pulsing action takes place about 2000 times per mile and the ECM will calculate vehicle speed based on the time between "pulses."
A "Scan" tool reading should closely match with speedometer reading with drive wheels turning.

Test Description: Numbers below refer to circled numbers on the diagnostic chart.
1. Code 24 will set if:
   - CKT 437 voltage is constant.
   - Engine speed between 1400 and 3600 rpm.
   - Less than 2% throttle opening, about .10V (100mV) above closed throttle.
   - Low load condition (low air flow).
   - Not in park or neutral.
   - All conditions must be met for 4 seconds.
   These conditions are met during a road load deceleration.
2. A voltage of less than 1 volt at the 15-way I/P connector indicates that the CKT 437 wire may be shorted to ground. Disconnect CKT 437 at the VSS buffer. If voltage now reads above 10 volts, the VSS buffer is faulty. If voltage remains less than 10 volt, then CKT 437 wire is grounded or open. If 437 is not grounded or open, check for a faulty ECM connector or ECM.

Diagnostic Aids:
If "Scan" displays vehicle speed, check park/neutral switch CHART C-1A on vehicle with auto transmission. If switch is OK, check for intermittent connections. An open or short to ground in CKT 437 will result in a Code 24. If the customer also complained about a loss of mph on the I.P., check the P.M. generator circuit. Refer to Section "8A" for complete wiring diagram.
Refer to "Intermittents" in Section "B".

1988–90 5.0L (VIN F) AND 5.7L (VIN 8) CAMARO AND FIREBIRD

CODE 24
VEHICLE SPEED SENSOR (VSS) CIRCUIT
5.0L (VIN F) & 5.7L (VIN 8) "F" CARLINE (PORT)

NOTE: TO PREVENT MISDIAGNOSIS, THE TECHNICIAN SHOULD REVIEW ELECTRICAL SECTION "8A" OR THE ELECTRICAL TROUBLESHOOTING MANUAL AND IDENTIFY THE TYPE OF VEHICLE SPEED SENSOR USED PRIOR TO USING THIS CHART. DISREGARD CODE 24 IF SET WHEN DRIVE WHEELS ARE NOT TURNING.

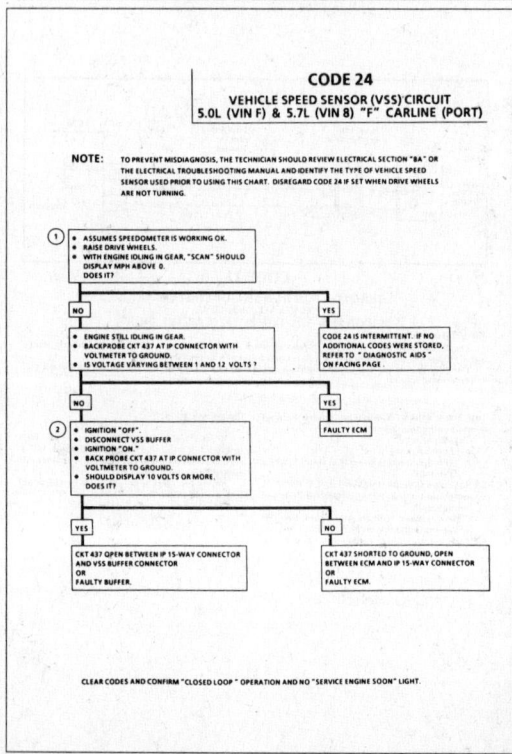

CLEAR CODES AND CONFIRM "CLOSED LOOP" OPERATION AND NO "SERVICE ENGINE SOON" LIGHT.

1988–90 5.0L (VIN F) AND 5.7L (VIN 8) CAMARO AND FIREBIRD

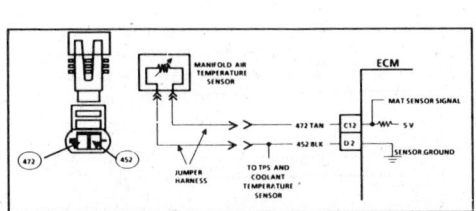

CODE 25
MANIFOLD AIR TEMPERATURE (MAT) SENSOR CIRCUIT
(HIGH TEMPERATURE INDICATED)
5.0L (VIN F) & 5.7L (VIN 8) "F" CARLINE (PORT)

Circuit Description:
The manifold air temperature (MAT) sensor uses a thermistor to control the signal voltage to the ECM. The ECM applies a voltage (about 5 volts) on CKT 472 to the sensor. When manifold air is cold, the sensor (thermistor) resistance is high, therefore, the ECM will see a high signal voltage. If the sensor (thermistor) resistance is low, therefore, the ECM will see a low signal voltage.

Test Description: Numbers below refer to circled numbers on the diagnostic chart.
1 Code 25 will set if:
- Signal voltage indicates a manifold air temperature greater than 150°C (302°F) for 2 seconds.
- Time since engine start is 2 minutes or longer
- A vehicle speed is present, greater than 5 MPH

Diagnostic Aids:
A "Scan" tool reads temperature of the air entering the engine and should read close to ambient air temperature when engine is cold, and rise as underhood temperature increases.
Check harness routing for possible short to ground in CKT 472.
Refer to "Intermittents" in Section "B".

1988–90 5.0L (VIN F) AND 5.7L (VIN 8) CAMARO AND FIREBIRD

CODE 25
MANIFOLD AIR TEMPERATURE (MAT) SENSOR CIRCUIT
(HIGH TEMPERATURE INDICATED)
5.0L (VIN F) & 5.7L (VIN 8) "F" CARLINE (PORT)

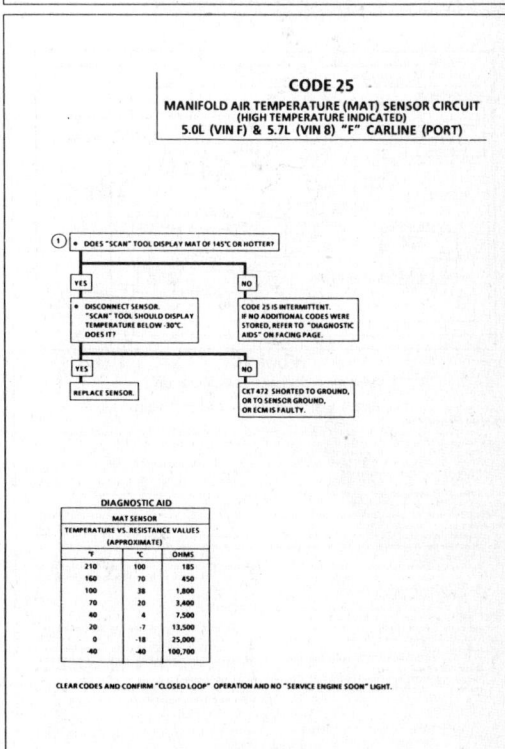

DIAGNOSTIC AID

MAT SENSOR		
TEMPERATURE VS. RESISTANCE VALUES (APPROXIMATE)		
°F	°C	OHMS
210	100	185
160	70	450
100	38	1,800
70	20	3,400
40	4	7,500
20	-7	13,500
0	-18	25,000
-40	-40	100,700

CLEAR CODES AND CONFIRM "CLOSED LOOP" OPERATION AND NO "SERVICE ENGINE SOON" LIGHT.

1988–90 5.0L (VIN F) AND 5.7L (VIN 8) CAMARO AND FIREBIRD

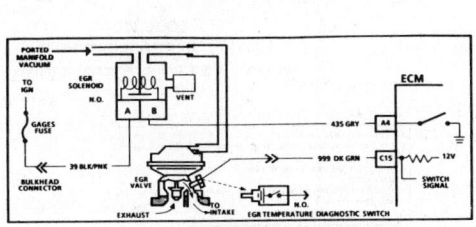

CODE 32
EXHAUST GAS RECIRCULATION (EGR) CIRCUIT
5.0L (VIN F) & 5.7L (VIN 8) "F" CARLINE (PORT)

Circuit Description:
The EGR valve vacuum is controlled by an ECM controlled solenoid. The ECM will turn the EGR "ON" and "OFF" (Duty Cycle) by grounding CKT 435. The duty cycle is calculated by the ECM, based on information from the coolant and mass air flow sensor and engine rpms. There should be (NO EGR) when in park or neutral, TPS input below a specified value or TPS indicating wide open throttle (WOT).
With the ignition "ON", engine stopped, the EGR solenoid is de-energized and, by grounding the diagnostic "test" terminal, the solenoid is energized.

Test Description: Numbers below refer to circled numbers on the diagnostic chart.
Code 32 means that the EGR diagnostic switch was closed during start-up or that the switch was not detected closed under the following conditions:
- Coolant temperature greater than 80°C (176°F).
- EGR duty cycle commanded by the ECM is greater than 48%.
- TPS less than wide open throttle (WOT), but not at idle.
- Codes 21,22,33,34 not present
- All conditions above must be met for about 4 minutes.

If the switch is detected closed during start up, or if the switch is detected open when the above conditions are met, the "Service Engine Soon" light will remain "ON" unless the switch changes state.
1 This test will determine if the ECM set the code due to CKT 999 being grounded on start up. If the "Scan" does not indicate the switch is closed but the customer complained of a "Service Engine Soon" light after start-up, then this circuit should be checked carefully for an intermittent grounded condition.

2 If the "Scan" indicates the switch is no longer closed after disconnecting it, be sure the switch is not closed due to heat. (EGR being "ON" prior to test).
3 This test will check for a possible open in CKT 999. The ECM supplies 9-12 volts to CKT 999 and the "Scan" should indicate switch being closed when CKT 999 is grounded.
4 By grounding the diagnostic "test" terminal, the EGR solenoid should close, and allow vacuum to be applied and the vacuum should hold.
5 This test will determine if the electrical control part of the system is at fault or if the connector or solenoid are at fault.
6 By plugging the EGR valve side and ungrounding the diagnostic "test" terminal, the solenoid valve should open and allow vacuum to bleed off through the vent.
7 With the engine not running and vacuum is applied to the valve, the valve should move to the fully open position.
8 This engine uses a negative back pressure valve and the valve should close when the engine is cranked over.

1988–90 5.0L (VIN F) AND 5.7L (VIN 8) CAMARO AND FIREBIRD

CODE 32
EXHAUST GAS RECIRCULATION (EGR) CIRCUIT
5.0L (VIN F) & 5.7L (VIN 8) "F" CARLINE (PORT)

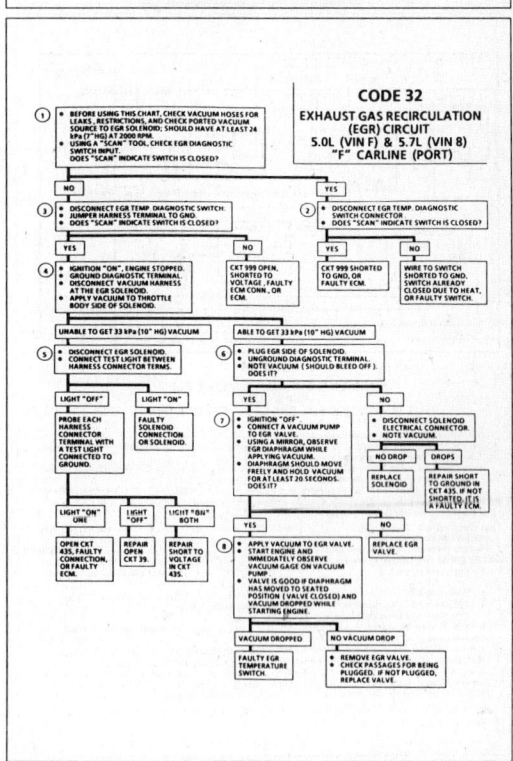

1988–90 5.0L (VIN F) AND 5.7L (VIN 8) CAMARO AND FIREBIRD

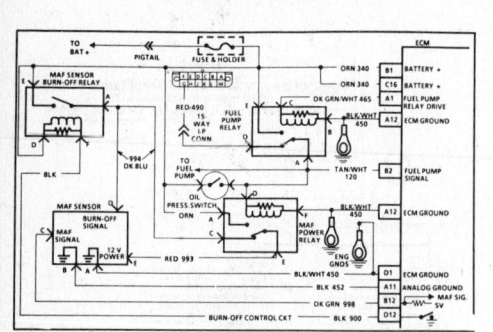

CODE 33
MASS AIR FLOW (MAF) SENSOR CIRCUIT
(GM/SEC HIGH)
5.0L (VIN F) & 5.7L (VIN 8) "F" CARLINE (PORT)

Circuit Description:

The mass air flow (MAF) sensor measures the amount of air which passes through it. The ECM uses this information to determine the operating condition of the engine to control fuel delivery. For a detailed description of the MAF sensor operation refer to Section "C".

The oil pressure switch or the ECM, through control of the fuel pump relay, will provide 12 volts for the MAF power relay which provides the 12 volts needed by the MAF sensor.

The ECM provides a current limiting 5 volts on the signal line (CKT 998). The MAF sensor then changes the signal by dropping the voltage, so that with low air flow the ECM sees a low voltage and a high air flow will cause the ECM to see near the 5 volts supply.

Test Description: Numbers below refer to circled numbers on the diagnostic chart.

Code 33 indicates: ECM has seen flow in excess of 45 grams per second (above about 2.2 volts) for one second when:
- Engine is first started
  OR
- TPS is less than ⅓ throttle.
- RPM is less than 2000.

Due to the 5 volt pull-up resistor in the ECM if CKT 998 becomes open, the ECM will see a high voltage signal and set a Code 33.

1. This test will determine if the conditions to set the code still exist.
2. With the ALDL terminal "G" jumpered to 12 volts, there should be 12 volts at the sensor. If no voltage is present, make sure that the fuel pump is running. If not, repair fuel pump circuit.

3. If a burn-off signal is present at the MAF sensor with the engine running, a Code 33 will set. Be sure no voltage is present on CKT 994 for the first 2 seconds after the ignition is turned "ON," or after the 2-second period.
4. The ECM sources a voltage (about 5 volts) to the MAF sensor on CKT 998. This test checks for that voltage.

Diagnostic Aids:

Intermittent: By jumpering the fuel pump test terminal (G term. of ALDL) to 12 volts, the MAF sensor will stay powered up and the signal line should see a low voltage, less than 250 mV or low grams per second on a "Scan" tool. By wiggling the related wiring the intermittent may be detected. Also, an erratic signal with the engine running may indicate faulty wiring or components.

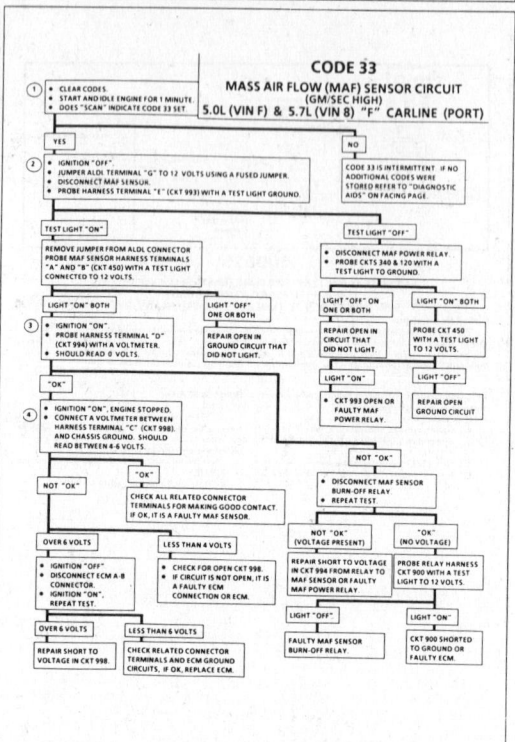

1988–90 5.0L (VIN F) AND 5.7L (VIN 8) CAMARO AND FIREBIRD

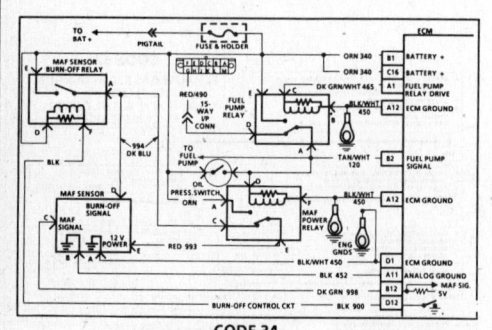

CODE 34
MASS AIR FLOW (MAF) SENSOR CIRCUIT
(GM/SEC LOW)
5.0L (VIN F) & 5.7L (VIN 8) "F" CARLINE (PORT)

Circuit Description:

The mass air flow (MAF) sensor measures the amount of air which passes through it. The ECM uses this information to determine the operating condition of the engine to control fuel delivery. For a detailed description of the MAF sensor operation refer to Section "C".

The oil pressure switch or the ECM, through control of the fuel pump relay, will provide 12 volts for the MAF power relay which provides the 12 volts needed by the MAF sensor.

The ECM provides a current limiting 5 volts on the signal line (CKT 998). The MAF sensor then changes the signal by dropping the voltage so that with low air flow the ECM sees a low voltage and a high air flow will cause the ECM to see near the 5 volts supply.

Test Description: Numbers below refer to circled numbers on the diagnostic chart.

Code 34 indicates: ECM has seen low air flow less than 2.5 gm/sec. (low voltage) for one second when:
- Engine is first started
  OR
- Rpm above 600
- TPS above 6%. To obtain 6%, the engine has to be running at about 2300 rpm in neutral.)

1. A Code 34 may be caused by an engine that exhibits a low, rough, unstable or incorrect idle problem. If this condition exists, disconnect the MAF sensor. If the unstable idle still exists, refer to Symptoms in Section "B". (Rough, unstable, incorrect idle, or stalling.) If the idle improved with the sensor disconnected, replace it.

2. This test will determine if the conditions still exist to set a code or if the problem is intermittent.
3. With the MAF sensor disconnected, the ECM should see a high signal voltage and set a Code 33. If a Code 34 resets then the wiring or the ECM is at fault.

Diagnostic Aids:

A low, rough or unstable idle could result in a Code 34. Be sure air ducts are tight and not cracked, thoroughly inspect the induction for vacuum leaks. Check CKT 998 for short to ground.

Code 34 could also result from a dirty or misadjusted throttle body. Refer to Section "C2" for minimum idle speed check.

1988–90 5.0L (VIN F) AND 5.7L (VIN 8) CAMARO AND FIREBIRD

CODE 34
MASS AIR FLOW (MAF) SENSOR CIRCUIT
(GM/SEC LOW)
5.0L (VIN F) & 5.7L (VIN 8) "F" CARLINE (PORT)

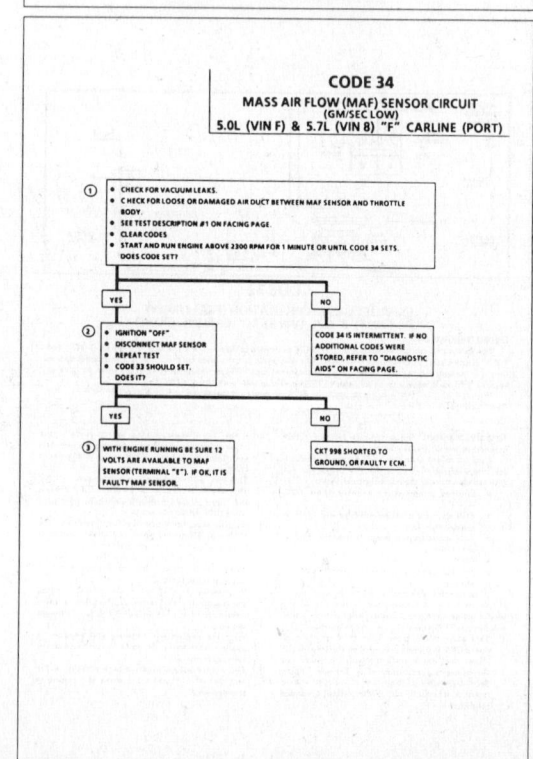

1988–90 5.0L (VIN F) AND 5.7L (VIN 8) CAMARO AND FIREBIRD

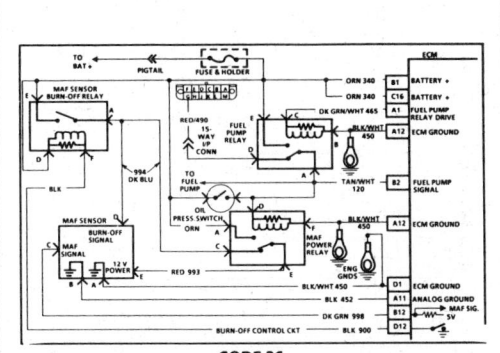

CODE 36
MASS AIR FLOW (MAF) BURN-OFF CIRCUIT
5.0L (VIN F) & 5.7L (VIN 8) "F" CARLINE (PORT)

Circuit Description:

The mass air flow (MAF) sensor measures the amount of air which passes through it. The ECM uses this information to determine the operating condition of the engine to control fuel delivery.

Due to contaminates in the atmosphere, a residue may build up on the MAF sensor sensing wire. To maintain an accurate reading from the sensor, a "burn-off" cycle will occur when the ignition is turned "OFF" after the engine had been running a specified amount of time and engine warmed up. The burn-off function is enabled when the ECM grounds CKT 900 which energizes the MAF sensor burn-off relay. With the MAF sensor burn off relay energized, voltage will be supplied to the MAF sensor terminal "D". Voltage will also be supplied through the normally closed set of contacts in the MAF power relay which will supply 12 volts to terminal "E" of the MAF sensor.

Test Description: Numbers below refer to circled numbers on the diagnostic chart.

1. This test will determine if the burn-off function is operative or if the Code was set due to an intermittent condition.
2. Check for continuous 12 volt supply to burn-off relay
3. Grounding CKT 900 should energize the relay and close the contacts. CKT 900 should be grounded by using a jumper wire at ECM connector "D12". If the test light is dim, check for corroded or faulty connections. If OK, replace relay.
4. With the burn-off relay energized there should be 12 volts supplied to the MAF sensor on terminal "D" & "E" (CKTs 993 and 994). If the test light is dim, check for corroded or faulty connections. If OK, replace relay.

Diagnostic Aids:

The Code 36 could have been set due to a poor connection at any of the relays or the MAF sensor. Be sure that these connections and terminals are OK. A faulty MAF sensor should not be considered as the cause if Code 36 is set.

Refer to "Intermittents" in Section "B".

1988–90 5.0L (VIN F) AND 5.7L (VIN 8) CAMARO AND FIREBIRD

CODE 36
MASS AIR FLOW (MAF) BURN-OFF CIRCUIT
5.0L (VIN F) & 5.7L (VIN 8) "F" CARLINE (PORT)

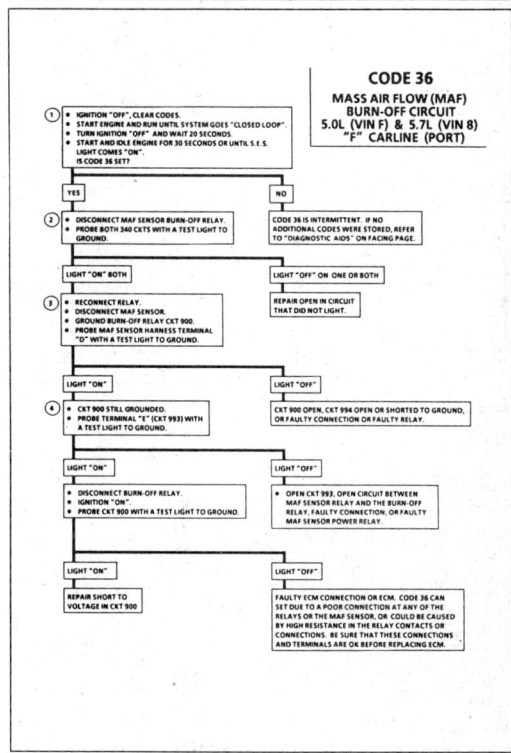

1988–90 5.0L (VIN F) AND 5.7L (VIN 8) CAMARO AND FIREBIRD

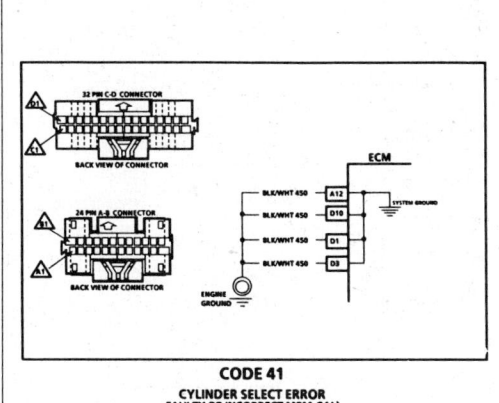

CODE 41
CYLINDER SELECT ERROR
(FAULTY OR INCORRECT MEM-CAL)
5.0L (VIN F) & 5.7L (VIN 8) "F" CARLINE (PORT)

Test Description: Numbers below refer to circled numbers on the diagnostic chart.

1. The ECM used for this engine can also be used for other engines, and the difference is in the Mem-Cal. If a Code 41 sets, the incorrect Mem-Cal has been installed, may not be installed properly, or it is faulty and it must be replaced.

Diagnostic Aids:

Check Mem-Cal to be sure locking tabs are secure. Also check the pins on both the Mem-Cal and ECM to assure they are making proper contact. Check the Mem-Cal part number to assure it is the correct part. If the Mem-Cal is faulty, it must be replaced. It is also possible that the ECM is faulty, however, it should not be replaced until all of the above have been checked. For additional information, refer to "Intermittents" in Section "B".

1988–90 5.0L (VIN F) AND 5.7L (VIN 8) CAMARO AND FIREBIRD

CODE 41
CYLINDER SELECT ERROR
(FAULTY OR INCORRECT MEM-CAL)
5.0L (VIN F) & 5.7L (VIN 8) "F" CARLINE (PORT)

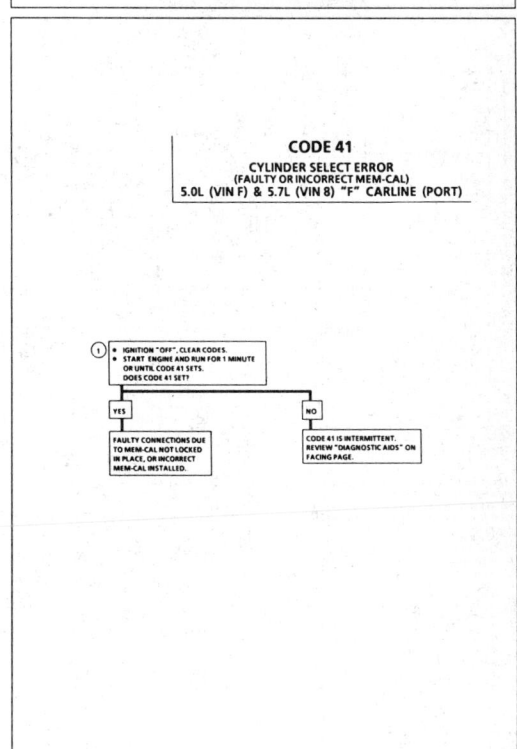

1988–90 5.0L (VIN F) AND 5.7L (VIN 8) CAMARO AND FIREBIRD

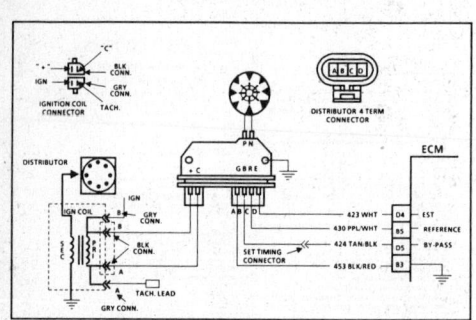

CODE 42
ELECTRONIC SPARK TIMING (EST) CIRCUIT
5.0L (VIN F) & 5.7L (VIN 8) "F" CARLINE (PORT)

Circuit Description:

When the system is running on the ignition module, that is, no voltage on the bypass line, the ignition module grounds the EST signal. The ECM expects to see no voltage on the EST line during this condition. If it sees a voltage, it sets Code 42 and will not go into the EST mode.

When the rpm for EST is reached (about 400 rpm), and bypass voltage applied, the EST should no longer be grounded in the ignition module, so the EST voltage should be varying.

If the bypass line is open or grounded, the ignition module will not switch to EST mode so the EST voltage will be low and Code 42 will be set.

If the EST line is grounded, the ignition module will switch to EST but, because the line is grounded there will be no EST signal. A Code 42 will be set.

Test Description: Numbers below refer to circled numbers on the diagnostic chart.

1. Code 42 means the ECM has seen an open or short to ground in the EST or bypass circuits. This test confirms Code 42 and that the fault causing the code is present.
2. Checks for a normal EST ground path through the ignition module. An EST CKT 423 shorted to ground will also read less than 500 ohms; however, this will be checked later.
3. As the test light voltage touches CKT 424 the module should switch. The ohmmeter may "overrange" if the meter is in the 1000-2000 ohms position. The important thing is that the module "switched."
4. The module did not switch and this step checks for:
   - EST CKT 423 shorted to ground.
   - Bypass CKT 424 open.
   - Faulty ignition module connection or module.
5. Confirms that Code 42 is a faulty ECM and not an intermittent in CKTs 423 or 424.

Diagnostic Aids:

The "Scan" tool does not have any ability to help diagnose a Code 42 problem.

A Mem-Cal not fully seated in the ECM can result in a Code 42.

Refer to "Intermittents" in Section "B".

1988–90 5.0L (VIN F) AND 5.7L (VIN 8) CAMARO AND FIREBIRD

CODE 42
ELECTRONIC SPARK TIMING (EST) CIRCUIT
5.0L (VIN F) & 5.7L (VIN 8) "F" CARLINE (PORT)

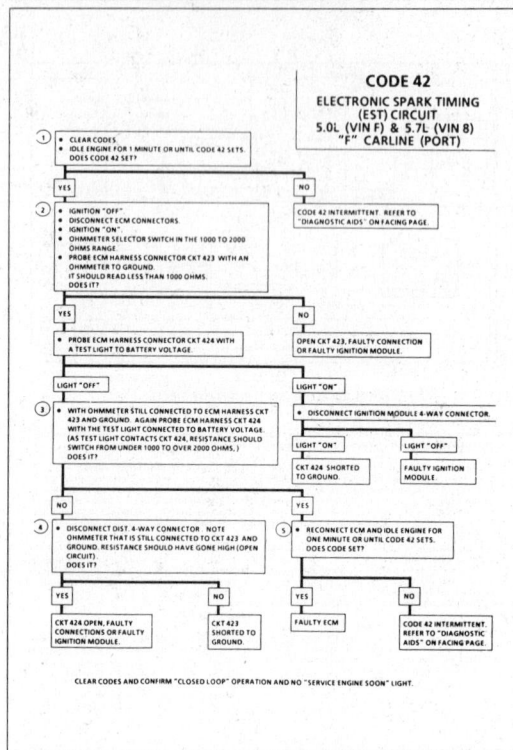

1988–90 5.0L (VIN F) AND 5.7L (VIN 8) CAMARO AND FIREBIRD

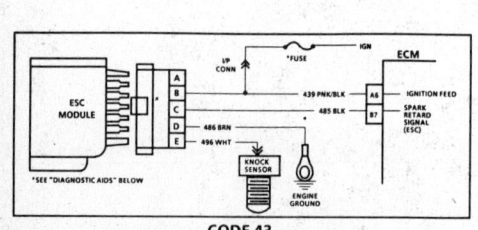

CODE 43
ELECTRONIC SPARK CONTROL (ESC) CIRCUIT
5.0L (VIN F) & 5.7L (VIN 8) "F" CARLINE (PORT)

Circuit Description:

Electronic spark control is accomplished with a module that sends a voltage signal to the ECM. As the knock sensor detects engine knock, the voltage from the ESC module to the ECM drops, and this signals the ECM to retard timing. The ECM will retard the timing when knock is detected and rpm is above about 900 rpm.

Code 43 means the ECM has read low voltage on CKT 485 for longer than 5 seconds, with the engine running, or the system has failed the functional check.

This system performs a functional check once per start up to check the ESC system. To perform this test the ECM will advance the spark when coolant is above 95°C (194°F) at a high load condition (near WOT). The ECM then checks the signal on CKT 485 to see if a knock is detected. The functional check is performed once per start up, if knock is detected when coolant is below 95°C (194°F) the test has passed and the functional check will not be run. If the functional check fails, the "Service Engine Soon" light will remain "ON" until ignition is turned "OFF," or until a knock signal is detected.

Test Description: Numbers below refer to circled numbers on the diagnostic chart.

1. If the conditions for a Code 43 are present, the "Scan" will always display "yes." There should not be a knock at idle unless there is an internal engine problem, or a system problem exists.
2. This test will determine if the system is functioning at this time. Usually a knock signal can be generated by tapping on the right exhaust manifold. If no knock signal is generated try tapping on block close to the area of the sensor.
3. Because Code 43 sets when the signal voltage on CKT 485 remains low, this test should cause the signal on CKT 485 to go high. The 12 volts signal should be seen by the ECM as "no knock" if the ECM and wiring are OK.
4. This test will determine if the knock signal is being detected on CKT 496 or if the ESC module is at fault.
5. If CKT 496 is routed to close to secondary ignition wires, the ESC module may see the interference as a knock signal.
6. This checks the ground circuit to the module. An open ground will cause the voltage on CKT 485 to be about 12 volts, which would cause the Code 43 functional test to fail.
7. Contacting CKT 496 with a test light to 12 volts should generate a knock signal. This will determine if the ESC module is operating correctly.

Diagnostic Aids:

* = ECM/IGN fuse

Code 43 can be caused by a faulty connection at the knock sensor or at the ECM. Also check CKT 485 for possible open or short to ground.

Refer to "Intermittents" in Section "B".

CODE 43
ELECTRONIC SPARK CONTROL (ESC) CIRCUIT
5.0L (VIN F) & 5.7L (VIN 8) "F" CARLINE (PORT)

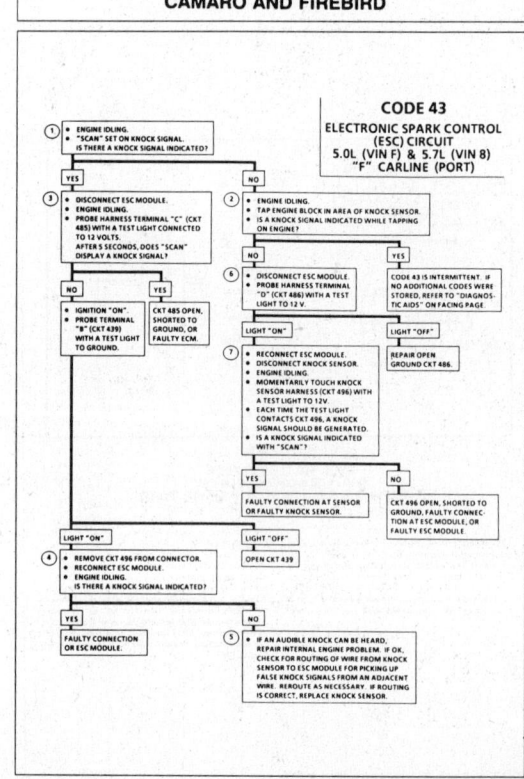

1988–90 5.0L (VIN F) AND 5.7L (VIN 8) CAMARO AND FIREBIRD

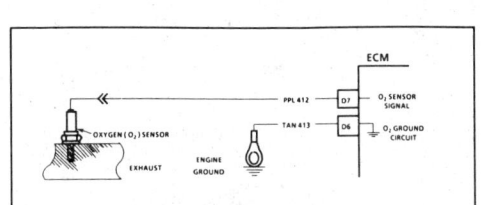

CODE 44
OXYGEN SENSOR CIRCUIT
(LEAN EXHAUST INDICATED)
5.0L (VIN F) & 5.7L (VIN 8) "F" CARLINE (PORT)

Circuit Description:
The ECM supplies a voltage of about .45 volt between terminals "D6" and "D7." (If measured with a 10 megohm digital voltmeter, this may read as low as .32 volt.) The O₂ sensor varies the voltage within a range of about 1 volt if the exhaust is rich, down through about .10 volt if exhaust is lean.
The sensor is like an open circuit and produces no voltage when it is below about 360°C (600°F). An open sensor circuit or cold sensor causes "Open Loop" operation.

Test Description: Numbers below refer to circled numbers on the diagnostic chart.
1. Code 44 is set when the O₂ sensor signal voltage on CKT 412.
   - Remains below .2 volt for 50 seconds
   - And the system is operating in "Closed Loop."

Diagnostic Aids:

Using the "Scan," observe the block learn values at different rpm and air flow conditions. The "Scan" also displays the block cells, so the block learn values can be checked in each of the cells to determine when the Code 44 may have been set. If the conditions for Code 44 exist, the block learn values will be around 150.
- **O₂ Sensor wire.** Sensor pigtail may be mispositioned and contacting the exhaust manifold.
  - Check for intermittent ground in wire between connector and sensor.
- **MAF Sensor.** A mass air flow (MAF) sensor output that causes the ECM to sense a lower than normal air flow will cause the system to go lean. Disconnect the MAF sensor and, if the lean condition is gone, replace the MAF sensor.

- **Lean Injector(s).** Perform injector balance test CHART C-2A.
- **Fuel Contamination.** Water, even in small amounts, near the in-tank fuel pump inlet can be delivered to the injectors. The water causes a lean exhaust and can set a Code 44.
- **Fuel Pressure.** System will be lean if pressure is too low. It may be necessary to monitor fuel pressure while driving the car at various road speeds and/or loads to confirm. See "Fuel System Diagnosis" CHART A-7.
- **Exhaust Leaks.** If there is an exhaust leak, outside air can be pulled into the exhaust and past the sensor. Vacuum or crankcase leaks can cause a lean condition.
- **AIR System.** Be sure air is not being directed to the exhaust ports while in "Closed Loop." If the block learn value goes down while squeezing air hose to left side exhaust ports, refer to CHART C-6.
- If the above are OK, it is a faulty oxygen sensor.

1988–90 5.0L (VIN F) AND 5.7L (VIN 8) CAMARO AND FIREBIRD

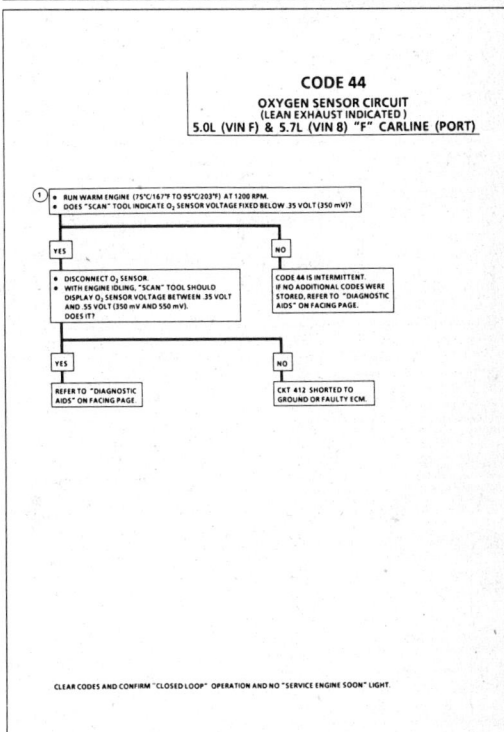

CODE 44
OXYGEN SENSOR CIRCUIT
(LEAN EXHAUST INDICATED)
5.0L (VIN F) & 5.7L (VIN 8) "F" CARLINE (PORT)

CLEAR CODES AND CONFIRM "CLOSED LOOP" OPERATION AND NO "SERVICE ENGINE SOON" LIGHT.

1988–90 5.0L (VIN F) AND 5.7L (VIN 8) CAMARO AND FIREBIRD

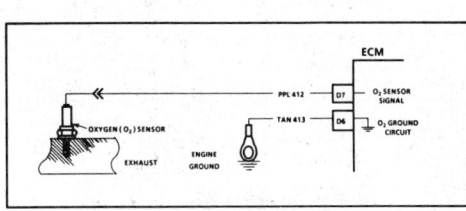

CODE 45
OXYGEN SENSOR CIRCUIT
(RICH EXHAUST INDICATED)
5.0L (VIN F) & 5.7L (VIN 8) "F" CARLINE (PORT)

Circuit Description:
The ECM supplies a voltage of about .45 volt between terminals "D6" and "D7." (If measured with a 10 megohm digital voltmeter, this may read as low as .32 volts.) The O₂ sensor varies the voltage within a range of about 1 volt if the exhaust is rich, down through about .10 volt if exhaust is lean.
The sensor is like an open circuit and produces no voltage when it is below about 360°C (600°F). An open sensor circuit or cold sensor causes "Open Loop" operation.

Test Description: Numbers below refer to circled numbers on the diagnostic chart.
1. Code 45 is set when the O₂ sensor signal voltage or CKT 412.
   - Remains above .7 volt for 50 seconds; and in "Closed Loop."
   - Engine time after start is 1 minute or more.
   - Throttle angle greater than 2% (about .2 volts above idle voltage)

Diagnostic Aids:

Using the "Scan," observe the block learn values at different rpm and air flow conditions. The "Scan" also displays the block cells, so the block learn values can be checked in each of the cells to determine when the Code 45 may have been set. If the conditions for Code 45 exists, the block learn values will be around 115.
- **Fuel Pressure.** System will go rich if pressure is too high. The ECM can compensate for some increase. However, if it gets too high, a Code 45 may be set.
Use the Fuel System diagnosis CHART A-7.
- **Rich injector.** Perform injector balance test CHART C-2A.
- **Leaking injector.** See CHART A-7.

- Check for fuel contaminated oil.
- **HEI Shielding.** An open ground CKT 453 (ignition system reflow) may result in EMI, or induced electrical "noise." The ECM looks at this "noise" as reference pulses. The additional pulses result in a higher than actual engine speed signal. The ECM then delivers too much fuel, causing system to go rich. Engine tachometer will also show higher than actual engine speed, which can help in diagnosing this problem.
- **Canister purge.** Check for fuel saturation. If full of fuel, check canister control and hoses.
- **MAF sensor.** An output that causes the ECM to sense a higher than normal airflow can cause the system to go rich. Disconnecting the MAF sensor will allow the ECM to set a fixed value for the sensor. Substitute a different MAF sensor if the rich condition is gone while the sensor is disconnected.
- Check for leaking fuel pressure regulator diaphram by checking vacuum line to regulator for fuel.
- **TPS.** An intermittent TPS output will cause the system to go rich, due to a false indication of the engine accelerating.

1988–90 5.0L (VIN F) AND 5.7L (VIN 8) CAMARO AND FIREBIRD

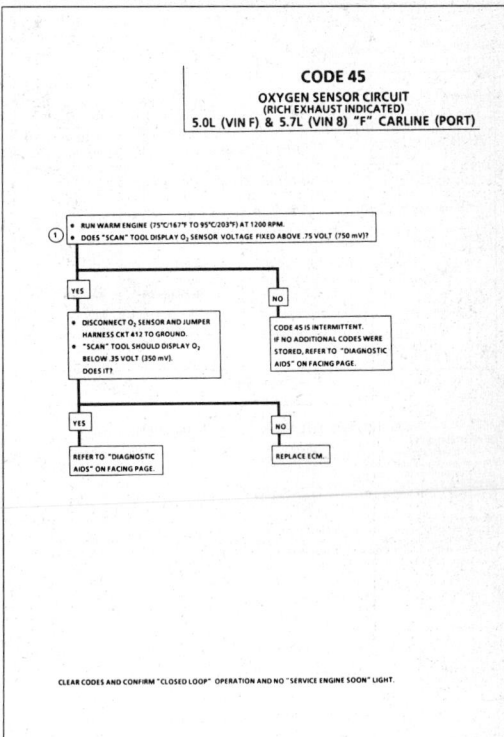

CODE 45
OXYGEN SENSOR CIRCUIT
(RICH EXHAUST INDICATED)
5.0L (VIN F) & 5.7L (VIN 8) "F" CARLINE (PORT)

CLEAR CODES AND CONFIRM "CLOSED LOOP" OPERATION AND NO "SERVICE ENGINE SOON" LIGHT.

1988–90 5.0L (VIN F) AND 5.7L (VIN 8) CAMARO AND FIREBIRD

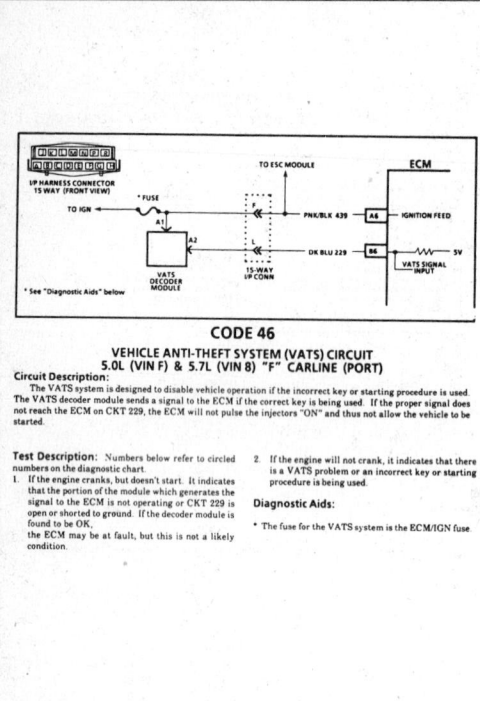

CODE 46
VEHICLE ANTI-THEFT SYSTEM (VATS) CIRCUIT
5.0L (VIN F) & 5.7L (VIN 8) "F" CARLINE (PORT)

Circuit Description:
The VATS system is designed to disable vehicle operation if the incorrect key or starting procedure is used. The VATS decoder module sends a signal to the ECM if the correct key is being used. If the proper signal does not reach the ECM on CKT 229, the ECM will not pulse the injectors "ON" and thus not allow the vehicle to be started.

Test Description: Numbers below refer to circled numbers on the diagnostic chart.
1. If the engine cranks, but doesn't start. It indicates that the portion of the module which generates the signal to the ECM is not operating or CKT 229 is open or shorted to ground. If the decoder module is found to be OK, the ECM may be at fault, but this is not a likely condition.

2. If the engine will not crank, it indicates that there is a VATS problem or an incorrect key or starting procedure is being used.

Diagnostic Aids:
- The fuse for the VATS system is the ECM/IGN fuse.

1988–90 5.0L (VIN F) AND 5.7L (VIN 8) CAMARO AND FIREBIRD

CODE 46
VEHICLE ANTI-THEFT SYSTEM (VATS) CIRCUIT
5.0L (VIN F) & 5.7L (VIN 8) "F" CARLINE (PORT)

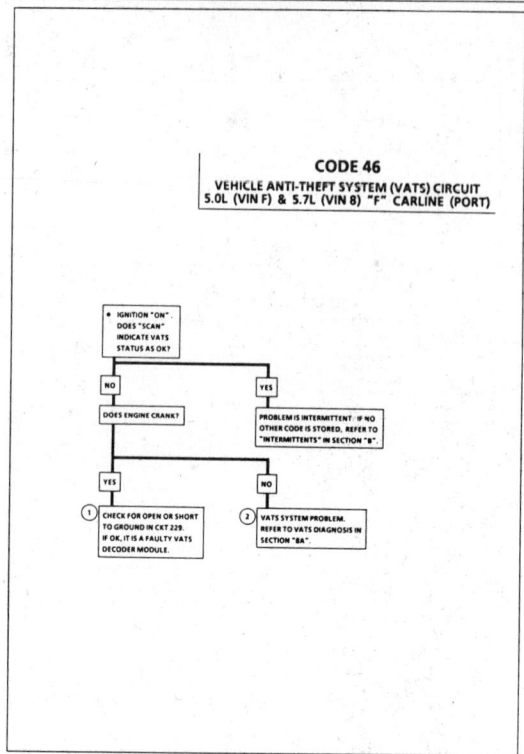

1988–90 5.0L (VIN F) AND 5.7L (VIN 8) CAMARO AND FIREBIRD

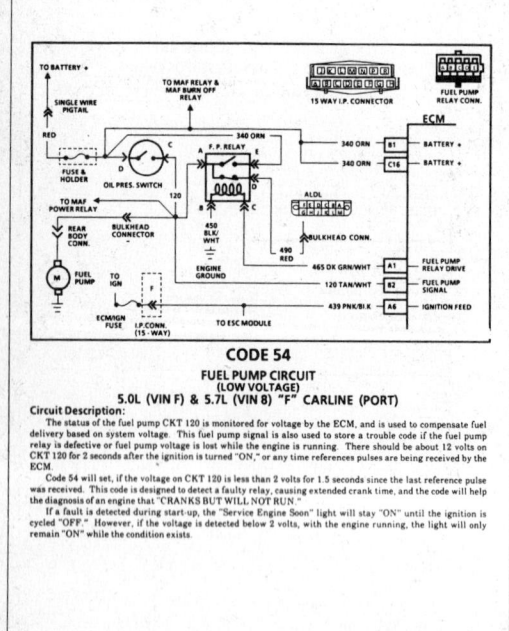

CODE 54
FUEL PUMP CIRCUIT
(LOW VOLTAGE)
5.0L (VIN F) & 5.7L (VIN 8) "F" CARLINE (PORT)

Circuit Description:
The status of the fuel pump CKT 120 is monitored for voltage by the ECM, and is used to compensate fuel delivery based on system voltage. This fuel pump signal is also used to store a trouble code if the fuel pump relay is defective or fuel pump voltage is lost while the engine is running. There should be about 12 volts on CKT 120 for 2 seconds after the ignition is turned "ON," or any time references pulses are being received by the ECM.
Code 54 will set, if the voltage on CKT 120 is less than 2 volts for 1.5 seconds since the last reference pulse was received. This code is designed to detect a faulty relay, causing extended crank time, and the code will help the diagnosis of an engine that "CRANKS BUT WILL NOT RUN."
If a fault is detected during start-up, the "Service Engine Soon" light will stay "ON" until the ignition is cycled "OFF." However, if the voltage is detected below 2 volts, with the engine running, the light will only remain "ON" while the condition exists.

1988–90 5.0L (VIN F) AND 5.7L (VIN 8) CAMARO AND FIREBIRD

CODE 54
FUEL PUMP CIRCUIT
(LOW VOLTAGE)
5.0L (VIN F) & 5.7L (VIN 8) "F" CARLINE (PORT)

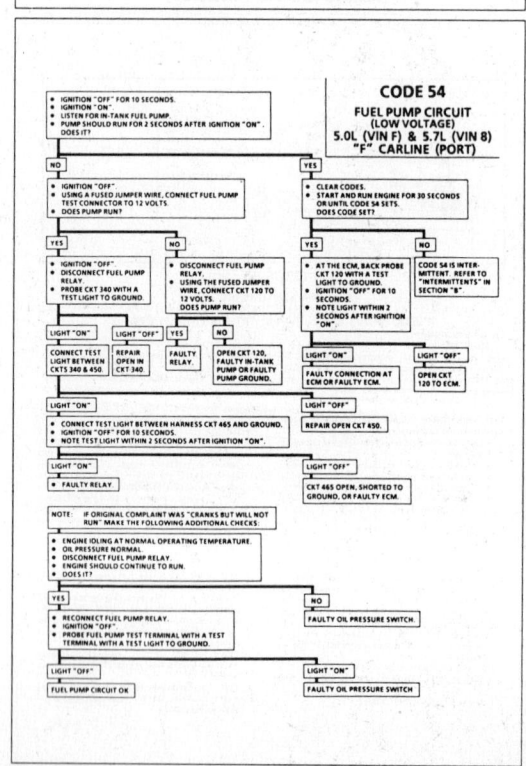

1988–90 5.0L (VIN F) AND 5.7L (VIN 8) CAMARO AND FIREBIRD

CODE 51
CODE 52
CODE 53

5.0L (VIN F) & 5.7L (VIN 8)
"F" CARLINE (PORT)

CODE 51

MEM-CAL ERROR
(FAULTY OR INCORRECT MEM-CAL)

CHECK THAT ALL PINS ARE FULLY INSERTED IN THE SOCKET AND THAT MEM-CAL IS PROPERLY LATCHED.
IF OK, REPLACE MEM-CAL, CLEAR MEMORY, AND RECHECK. IF CODE 51 REAPPEARS, REPLACE ECM.

CLEAR CODES AND CONFIRM "CLOSED LOOP" OPERATION AND NO "SERVICE ENGINE SOON" LIGHT.

CODE 52

CALPAK ERROR
(FAULTY OR INCORRECT CALPAK)

CHECK THAT THE MEM-CAL IS FULLY SEATED AND LATCHED INTO THE MEM-CAL
SOCKET. IF OK, REPLACE MEM-CAL, CLEAR MEMORY, AND RECHECK.
IF CODE 52 REAPPEARS, REPLACE ECM.

CLEAR CODES AND CONFIRM "CLOSED LOOP" OPERATION AND NO "SERVICE ENGINE SOON" LIGHT.

CODE 53

SYSTEM OVER VOLTAGE

THIS CODE INDICATES THERE IS A BASIC GENERATOR PROBLEM.
• CODE 53 WILL SET, IF VOLTAGE AT ECM IGNITION INPUT PIN IS GREATER THAN
17.1 VOLTS FOR 2 SECONDS.
• CHECK AND REPAIR CHARGING SYSTEM. REFER TO SECTION "6D".

CLEAR CODES AND CONFIRM "CLOSED LOOP" OPERATION AND NO "SERVICE ENGINE SOON" LIGHT.

1988–90 5.0L (VIN F) AND 5.7L (VIN 8) CAMARO AND FIREBIRD

SECTION B
SYMPTOMS
TABLE OF CONTENTS

Before Starting
Intermittents
Hard Start
Surges and/or Chuggle
Lack of Power, Sluggish, or Spongy
Detonation/Spark Knock
Hesitation, Sag, Stumble
Cuts Out, Misses
Poor Fuel Economy
Rough, Unstable or Incorrect Idle, Stalling
Excessive Exhaust Emissions or Odors
Dieseling, Run-On
Backfire
Restricted Exhaust System Check (CHART B-1) (All Engines)

BEFORE STARTING

Before using this section you should have performed the "Diagnostic Circuit Check" and found out that:
1. The ECM and "Service Engine Soon" light are operating.
2. There are no trouble codes stored, or there is a trouble code but no "Service Engine Soon" light. Verify the customer complaint, and locate the correct SYMPTOM below. Check the items indicated under that symptom.

If the ENGINE CRANKS BUT WILL NOT RUN, see CHART A-3.

Several of the symptom procedures below call for a careful visual physical check. This check should include:

• ECM grounds for being clean and tight.
• Vacuum hoses for splits, kinks, and proper connections, as shown on Vehicle Emission Control Information label.
• Air leaks at throttle body mounting and intake manifold.
• Air leaks between MAF sensor and throttle body.
• Ignition wires for cracking, hardness, proper routing, and carbon tracking.
• Wiring for proper connections, pinches, and cuts. The importance of this step cannot be stressed too strongly - it can lead to correcting a problem without further checks and can save valuable time.

1988–90 5.0L (VIN F) AND 5.7L (VIN 8) CAMARO AND FIREBIRD

INTERMITTENTS

Definition: Problem may or may not turn "ON" the "Service Engine Soon" light, or may store a code.

DO NOT use the Trouble Code Charts in Section "A" for intermittent problems. The fault must be present to locate the problem. If a fault is intermittent, use of Trouble Code Charts may result in replacement of good parts.

• Most intermittent problems are caused by faulty electrical connections or wiring. Perform careful check as described at start of Section "B". Check for:
  • Poor mating of the connector halves, or terminals not fully seated in the connector body (backed out).
  • Improperly formed or damaged terminals. All connector terminals in problem circuit should be carefully reformed or replaced to insure proper contact tension.
  • Poor terminal to wire connection. This requires removing the terminal from the connector body to check.

• If a visual physical check does not find the cause of the problem, the car can be driven with a voltmeter connected to a suspected circuit. A "Scan" tool can also be used for monitoring input signals to the ECM to help detect intermittent conditions. An abnormal voltage, or "Scan" reading when the problem occurs, indicates the problem may be in that circuit. If the wiring and connectors check OK and a Trouble Code was stored for a circuit having a sensor, except for Codes 43, 44, and 45, substitute a known good sensor and recheck.

• Loss of trouble code memory. To check, disconnect TPS and idle engine until "Service Engine Soon" light comes "ON". Code 22 should be stored, and kept in memory when ignition is turned "OFF". If not, the ECM is faulty. An intermittent "Service Engine Soon" light with no stored code may be caused by:
  • Ignition coil shorted to ground and arcing at spark plug wires or plugs.
  • "Service Engine Soon" light wire to ECM shorted to ground. (CKT 419).
  • Diagnostic "test" terminal wire to ECM, shorted to ground. (CKT 451)
  • ECM power grounds. See ECM wiring diagrams.

• Check for an electrical system interference caused by a defective relay, ECM driven solenoid, or switch. They can cause a sharp electrical surge. Normally, the problem will occur when the faulty component is operated.

• Check for improper installation of electrical options, such as lights, 2-way radios, etc.

• EST wires should be routed away from spark plug wires, distributor wires, distributor housing, coil, and generator. Wire from distributor (CKT 453) should be a good ground.

• Check for open diode across A/C compressor clutch, and for other open diodes (see wiring diagrams).

HARD START

Definition: Engine cranks OK, but does not start for a long time. Does eventually run, or may start but immediately dies.

• Perform careful check as described at start of Section "B".
• Make sure driver is using correct starting procedure.
• CHECK:
  - TPS for sticking or binding or a high TPS voltage with the throttle closed.
  - High resistance in coolant sensor circuit or coolant temperature sensor itself. See Code 15 chart or with a "Scan" tool compare coolant temperature with ambient temperature on a cold engine.
  - Fuel system pressure CHART A-7. For contaminated fuel.
  - EGR operation. Be sure valve seats properly and is not staying open. See CHART C-7.
  - Both injector fuses (visually inspect).
  - Ignition system - Check distributor for Proper Output with ST-125.
    Worn distributor shaft.
    Bare and shorted wires.
  - Pickup coil resistance and connections.
    Loose ignition coil ground.
    Moisture in distributor cap.

1988–90 5.0L (VIN F) AND 5.7L (VIN 8) CAMARO AND FIREBIRD

• A faulty in-tank fuel pump check valve will allow the fuel in the lines to drain back to the tank after the engine is stopped. To check for this condition:
  Perform Fuel System Diagnosis, see CHART A-7.
• Remove spark plugs. Check for wet plugs, cracks, wear, improper gap, burned electrodes, or heavy deposits. Repair or replace as necessary.

• If engine starts but then immediately stalls open distributor, set timing connector. If engine then starts and runs OK, replace pickup coil.
• If engine starts and stalls, disconnect MAF sensor. If engine then runs and MAF sensor connections are OK, replace the sensor.
• Basic engine problem.

SURGES AND/OR CHUGGLE

Definition: Engine power variation under steady throttle or cruise. Feels like the car speeds up and slows down with no change in the accelerator pedal.

• Be sure driver understands Torque Converter Clutch and A/C compressor operation in owner's manual.
• Perform careful visual physical check as described at start of Section "B".
• CHECK:
  - Loose or leaking air duct between MAF sensor and throttle body.
  - Generator output voltage. Repair if less than 9 or more than 16 volts.
  - EGR - There should be no EGR at idle, see CHART C-7.
  - Vacuum lines for kinks or leaks.
  - In-line fuel filter. Replace if dirty or plugged.
  - Fuel pressure while condition exists. See CHART A-7.
  - Ignition timing. See Vehicle Emission Control Information label.
  - Inspect oxygen sensor for silicon contamination from fuel, or use of improper RTV sealant.

The sensor may have a white, powdery coating and result in a high but false signal voltage (rich exhaust indication). The ECM will then reduce the amount of fuel delivered to the engine, causing a severe driveability problem.
• Remove spark plugs. Check for cracks, wear, improper gap, burned electrodes, or heavy deposits. Also check condition of distributor cap, rotor, and spark plug wires.
• To help determine if the condition is caused by a rich or lean system, the car should be driven at the speed of the complaint. Monitoring block learn at the complaint speed will help identify the cause of the problem. If the system is lean (block learn greater than 138), refer to "Diagnostic Aids" on facing page of Code 44. If the system is running rich (block learn less than 118), refer to "Diagnostic Aids" on facing page of Code 45.

LACK OF POWER, SLUGGISH, OR SPONGY

Definition: Engine delivers less than expected power. Little or no increase in speed when accelerator pedal is pushed down part way.

• Perform careful visual check as described at start of Section "B".
• Compare customer's car to similar unit. Make sure the customer's car has an actual problem.
• Remove air cleaner and check air filter for dirt, or for being plugged. Replace as necessary.
• CHECK:
  - For loose or leaking air duct between MAF sensor and throttle body.
  - Restricted fuel filter, contaminated fuel or improper fuel pressure. See CHART A-7.
  - ECM Ground circuits - See ECM wiring diagrams.
  - EGR valve for being open, or partly open all the time - CHART C-7.
  - Ignition timing - See Vehicle Emission Control Information label.
  - Exhaust system for possible restriction, see CHART B-1.

  - Inspect exhaust system for damaged or collapsed pipes.
  - Inspect muffler for heat distress or possible internal failure.
  - For possible plugged catalytic converter by comparing exhaust system backpressure on each side at engine. Check backpressure by removing A.I.R. check valves near exhaust manifolds. See CHART B-1 for procedure.
  - Generator output voltage. Repair if less than 9 or more than 16 volts.
  - Engine valve timing and compression.
  - Engine for correct or worn camshaft.

  - Secondary voltage using a shop oscilliscope or a spark tester J 26792 (ST-125) or equivalent.
  - Check for excessive retarded ignition timing. See CHART C-5.

1988–90 5.0L (VIN F) AND 5.7L (VIN 8) CAMARO AND FIREBIRD

DETONATION/SPARK KNOCK

Definition: A mild to severe ping, usually worse under acceleration. The engine makes sharp metallic knocks that change with throttle opening. Sounds like popcorn popping.

- Check for obvious overheating problems:
  - Low engine coolant.
  - Loose water pump belt.
  - Restricted air flow to radiator, or restricted water flow thru radiator.
  - Inoperative electric cooling fan circuit, see CHART C-12.
- CHECK:
  - Ignition timing. See Vehicle Emission Control Information Label.
  - EGR system for not opening - CHART C-7.
  - TCC operation, see CHART C-8.
  - Fuel system pressure, see CHART A-7.
  - Mem-Cal - Be sure it's the correct one.

  - Valve oil seals for leaking.

- Check for incorrect basic engine parts such as cam, heads, pistons, etc.
- Check for poor fuel quality, proper octane rating
- Remove carbon with top engine cleaner. Follow instructions on can.
- Check ESC system, see CHART C-5.
- To help determine if the condition is caused by a rich or lean system, the car should be driven at the speed of the complaint. Monitoring block learn at the complaint speed will help identify the cause of the problem. If the system is running lean (block learn greater than 138), refer to "Diagnostic Aids" on facing page of Code 44. If the system is running rich (block learn less than 118), refer to "Diagnostic Aids" on facing page of Code 45.

HESITATION, SAG, STUMBLE

Definition: Momentary lack of response as the accelerator is pushed down. Can occur at all car speeds. Usually most severe when first trying to make the car move, as from a stop sign. May cause the engine to stall if severe enough.

- Perform careful visual physical check as described at start of Section "B".
- CHECK:
  - Fuel system pressure. See CHART A-7. Also, check for water or contaminated fuel.
  - Air leaks at air duct between MAF sensor and throttle body.
  - Spark plugs for being fouled or faulty wiring.
  - Mem-Cal number. Also check service bulletins for latest Mem-Cal.
  - TPS for binding or sticking. Voltage should increase at a steady rate as throttle is moved toward WOT.

- Ignition timing. See Vehicle Emission Control Information label.
- Generator output voltage. Repair if less than 9 or more than 16 volts.
- Distributor (HEI) ground, CKT 453.
- Canister purge system for proper operation. See CHART C-3.
- EGR - See CHART C-7.
- Perform injector balance test CHART C-2A.

1988–90 5.0L (VIN F) AND 5.7L (VIN 8) CAMARO AND FIREBIRD

CUTS OUT, MISSES

Definition: Steady pulsation or jerking that follows engine speed, usually more pronounced as engine load increases. The exhaust has a steady spitting sound at idle or low speed.

- Perform careful visual check as described at start of Section "B".
- Check for missing cylinder by:
  1. Start engine, allow engine to stabilize then disconnect IAC motor. Remove one spark plug wire at a time, using insulated pliers.
  CAUTION: Do not perform this test for more than 2 minutes, as this may cause damage to the catalytic converter.
  2. If there is an rpm drop on all cylinders (equal to within 50 rpm), go to "ROUGH, UNSTABLE, OR INCORRECT IDLE, STALLING" symptom. Reconnect IAC valve.
  3. If there is no rpm drop on one or more cylinders, or excessive variation in drop, check for spark on the suspected cylinder(s) with J 26792 (ST-125) Spark or equivalent.

     If there is spark, remove spark plug(s) in these cylinders and check for:
     - Cracks
     - Wear
     - Improper Gap
     - Burned Electrodes
     - Heavy Deposits
- Perform compression check on questionable cylinder(s) found above. If compression is low, repair as necessary.

- Disconnect all injector harness connectors. Connect J 34730-2 Injector Test Light or equivalent 6 volts test light between the harness terms, of each injector connector and note light while cranking. If test light fails to blink at any connector, it is a faulty injector drive circuit harness, connector, or terminal.
- Perform the Injector Balance Test. See CHART C-2A.
- CHECK:
  - Spark plug wires by connecting ohmmeter to ends of each wire in question. If meter reads over 30,000 ohms, replace wire(s).
  - Fuel System - Plugged fuel filter, water, low pressure. See CHART A-7.
  - Secondary voltage using a shop oscilliscope or a spark tester J-26792 (ST-125) or equivalent.
  - Visually inspect distributor cap and rotor for moisture, dust, cracks, burns, etc. Spray cap and plug wires with fine water mist to check for shorts.
  - A miss condition can be caused by EMI (Electromagnetic Interference) on the reference circuit. EMI can usually be detected by monitoring engine rpm with a "Scan" tool. A sudden increase in rpm with little change in actual engine rpm change, indicates EMI is present.
  - If the problem exists, check routing of secondary wires, check all distributor ground circuits.
  - Remove rocker covers. Check for bent pushrods, worn rocker arms, broken valve springs, worn camshaft lobes. Repair as necessary.

  - Valve timing

POOR FUEL ECONOMY

Definition: Fuel economy, as measured by an actual road test, is noticeably lower than expected. Also, economy is noticeably lower than it was on this car at one time, as previously shown by an actual road test.

- Perform careful visual check as described at start of Section "B".
- CHECK:
  - Engine coolant level.
  - Engine thermostat for faulty part (always open) or for wrong heat range.

- Ignition timing. See Vehicle Emission Control Information label.
- TCC for proper operation. A "Scan" should indicate an rpm drop when the TCC is commanded "ON," see CHART C-8.
- Induction system and crankcase for air leaks.
- Check for exhaust restriction, see CHART B-1
- Compression

1988–90 5.0L (VIN F) AND 5.7L (VIN 8) CAMARO AND FIREBIRD

ROUGH, UNSTABLE OR INCORRECT IDLE, STALLING

Definition: The engine runs unevenly at idle. If bad enough, the car may shake. Also, the idle may vary in rpm (called "hunting"). Either condition may be bad enough to cause stalling. Engine idles at incorrect speed.

- Perform careful visual check at start of Section "B".
- CHECK:
  - Vacuum leaks.
  - Throttle linkage for sticking or binding.
  - ECM ground circuits.
  - IAC system. See CHART C-2C.
  - Generator output voltage. Repair if less than 9 or more than 16 volts.
  - P/N switch circuit. See CHART C-1A, or use "Scan" tool.
  - Injector balance. See CHART C-2A.
  - PCV valve for proper operation by placing finger over inlet hole in valve and several times. Valve should snap back. If not, replace valve.
  - Evaporative emission control system. CHART C-3.
  - A/C signal to ECM terminal "B8". "Scan" tool should indicate A/C is being requested whenever A/C is selected and the pressure cycling switch is closed.
  - Check A.I.R. system. There should be no A.I.R to ports while in "Closed Loop." See CHART C-6.

- EGR valve: There should be no EGR at idle.
- Run a cylinder compression check

- Inspect oxygen sensor for silicon contamination from fuel, or use of improper RTV sealant. The sensor will have a white, powdery coating, and will result in a high but false signal voltage (rich exhaust indication). The ECM will then reduce the amount of fuel delivered to the engine, causing a severe driveability problem.
- Check for fuel in pressure regulator hose. If present replace regulator assembly.
- Check ignition system; wires, plugs, rotor, etc.
- Check for loose or damaged air duct between MAF sensor and throttle body.
- Disconnect MAF sensor and if condition is corrected replace sensor.
- Clean injectors.
- Monitoring block learn will help identify the cause of the problem. If the system is running lean (block learn greater than 138), refer to "Diagnostic Aids" on facing page of Code 44. If the system is running rich (block learn less than 118), refer to "Diagnostic Aids" on facing page of Code 45.

EXCESSIVE EXHAUST EMISSIONS OR ODORS

Definition: Vehicle fails an emission test. Vehicle has excessive "rotten egg" smell. Excessive odors do not necessarily indicate excessive emissions.

- Perform "Diagnostic Circuit Check."
- IF TEST SHOWS EXCESSIVE CO AND HC, (or also has excessive odors!)
  - Check items which cause engine to run RICH.
  - Make sure engine is at normal operating temperature.
- CHECK:
  - Fuel pressure. See CHART A-7.
  - Incorrect timing. See Vehicle Emission Control Information Label.
  - Fuel vapor canister for fuel loading. See CHART C-3.
  - Injector balance. See CHART C-2A.
  - PCV valve for being plugged, stuck or blocked PCV hose or fuel in the crankcase.
  - Spark plugs, plug wires, and ignition components.
  - Check for lead contamination of catalytic converter (look for removal of fuel filler neck restrictor).
  - Check for properly installed fuel cap.

- If the system is running rich, (block learn less than 118), refer to "Diagnostic Aids" on facing page of Code 45.
- IF TEST SHOWS EXCESSIVE NOx:
  - Check items which cause car to run LEAN, or run too Hot.
    - EGR valve for not opening. See CHART C-7.
    - Vacuum leaks.
    - Coolant system and coolant fan for proper operation. See CHART C-12.
    - Remove carbon with top engine cleaner, follow instructions on can.
    - Check ignition timing for excessive base advance. See Vehicle Emission Control Information Label.
  - If the system is running lean, (block learn greater than 138), refer to "Diagnostic Aids" on facing page of Code 44.

1988–90 5.0L (VIN F) AND 5.7L (VIN 8) CAMARO AND FIREBIRD

DIESELING, RUN-ON

Definition: Engine continues to run after key is turned "OFF," but runs very roughly. If engine runs smoothly, check ignition switch and adjustment.

- Check injectors for leaking. See CHART A-7.

BACKFIRE

Definition: Fuel ignites in intake manifold, or in exhaust system making a loud popping noise.

- CHECK:
  - Loose wiring connector or air duct at MAF sensor.
  - EGR operation for being open all the time. See CHART C-7.
  - EGR gasket for failure or loose fit.
  - Spark plugs for crossfire also inspect (distributor cap, spark plug wires, and proper routing of plug wires).
  - Ignition system for intermittent condition.

  - Ignition timing, see Vehicle Emission Control Information Label.

- Perform fuel system diagnosis check, see CHART A-7A.
- Perform injector balance test, see CHART C-2A.
- A.I.R. system check valves
- Compression - Look for sticking or leaking valves
- Valve timing
- Output voltage of ignition coil using a shop oscilliscope or spark tester J 26792 (ST-125) or equivalent.

CHART B-1
RESTRICTED EXHAUST SYSTEM CHECK
ALL ENGINES

Proper diagnosis for a restricted exhaust system is essential before any components are replaced. Either of the following procedures may be used for diagnosis, depending upon engine or tool used.

CHECK AT A.I.R. PIPE:

1. Remove the rubber hose at the exhaust manifold A.I.R. pipe check valve. Remove check valve.
2. Connect a fuel pump pressure gauge to a hose and nipple from a Propane Enrichment Device (J 26911) (see illustration).
3. Insert the nipple into the exhaust manifold A.I.R. pipe.

OR

CHECK AT O₂ SENSOR:

1. Carefully remove O₂ sensor.
2. Install Borroughs Exhaust Backpressure Tester (BT 8515 or BT 8603) or equivalent in place of O₂ sensor (see illustration).
3. After completing test described below, be sure to coat threads of O₂ sensor with anti-seize compound P/N 5613695 or equivalent prior to re-installation.

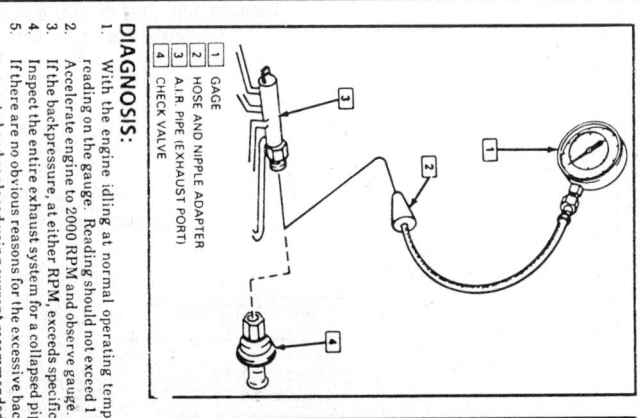

1	GAGE
2	HOSE AND NIPPLE ADAPTER
3	A.I.R. PIPE (EXHAUST PORT)
4	CHECK VALVE

1	EXHAUST MANIFOLD
2	OXYGEN (O₂) SENSOR
3	BACK PRESSURE GAGE

DIAGNOSIS:

1. With the engine idling at normal operating temperature, observe the exhaust system backpressure reading on the gauge. Reading should not exceed 1¼ psi (8.6 kPa).
2. Accelerate engine to 2000 RPM and observe gauge. Reading should not exceed 3 psi (20.7 kPa).
3. If the backpressure, at either RPM, exceeds specification, a restricted exhaust system is indicated.
4. Inspect the entire exhaust system for a collapsed pipe, heat distress, or possible internal muffler failure.
5. If there are no obvious reasons for the excessive backpressure, a restricted catalytic converter should be suspected and replaced using current recommended procedures.

1988–90 5.0L (VIN F) AND 5.7L (VIN 8) CAMARO AND FIREBIRD

CHART C-2A
INJECTOR BALANCE TEST

The injector balance tester is a tool used to turn the injector on for a precise amount of time, thus spraying a measured amount of fuel into the manifold. This causes a drop in fuel pressure that we can record and compare between each injector. All injectors should have the same amount of pressure drop (± 10 kpa). Any injector with a pressure drop that is 10 kpa (or more) greater or less than the average drop of the other injectors should be considered faulty and replaced.

STEP 1

Engine "cool down" period (10 minutes) is necessary to avoid irregular readings due to "Hot Soak" fuel boiling. With ignition "OFF" connect fuel gauge J 347301 or equivalent to fuel pressure tap. Wrap a shop towel around fitting while connecting gage to avoid fuel spillage.

Disconnect harness connectors at all injectors, and connect injector tester J 34730-3, or equivalent, to one injector. On Turbo equipped engines, use adaptor harness furnished with injector tester to energize injectors that are not accessible. Follow manufacturers instructions for use of adaptor harness. Ignition must be "OFF" at least 10 seconds to complete ECM shutdown cycle. Fuel pump should run about 2 seconds after ignition is turned "ON". At this point, insert clear tubing attached to vent valve in a suitable container and bleed air from gauge and hose to insure accurate gauge operation. Repeat this step until all air is bled from gauge.

STEP 2

Turn ignition "OFF" for 10 seconds and then "ON" again to get fuel pressure to its maximum. Record this initial pressure reading. Energize tester one time and note pressure drop at its lowest point (Disregard any slight pressure increase after drop hits low point). By subtracting this second pressure reading from the initial pressure, we have the actual amount of injector pressure drop.

STEP 3

Repeat step 2 on each injector and compare the amount of drop. Usually, good injectors will have virtually the same drop. Retest any injector that has a pressure difference of 10kPa, either more or less than the average of the other injectors on the engine. Replace any injector that also fails the retest. If the pressure drop of all injectors is within 10kPa of this average, the injectors appear to be flowing properly. Reconnect them and review "Symptoms", Section "B".

NOTE: The entire test should not be repeated more than once without running the engine to prevent flooding. (This includes any retest on faulty injectors).

1988–90 5.0L (VIN F) AND 5.7L (VIN 8) CAMARO AND FIREBIRD

NOTE: If injectors are suspected of being dirty, they should be cleaned using an approved tool and procedure prior to performing this test. The fuel pressure test in Section "A", Chart A-7, should be completed prior to this test.

CHART C-2A
INJECTOR BALANCE TEST
5.0L (VIN F) & 5.7L (VIN 8) "F" CARLINE (PORT)

Step 1. If engine is at operating temperature, allow a 10 minute "cool down" period then connect fuel pressure gauge and injector tester.
1. Ignition "OFF."
2. Connect fuel pressure gauge and injector tester.
3. Ignition "ON."
4. Bleed off air in gauge. Repeat until all air is bled from gauge.

Step 2. Run test:
1. Ignition "OFF" for 10 seconds.
2. Ignition "ON". Record gauge pressure. (Pressure must hold steady, if not see the Fuel System diagnosis, Chart A-7, in Section "A")
3. Turn injector on by, depressing button on injector tester, and note pressure at the instant the gauge needle stops.

Step 3.
1. Repeat step 2 on all injectors and record pressure drop on each.

Retest injectors that appear faulty (Any injectors that have a 10 kPa difference, either more or less, in pressure from the average). If no problem is found, review "Symptoms" Section "B".

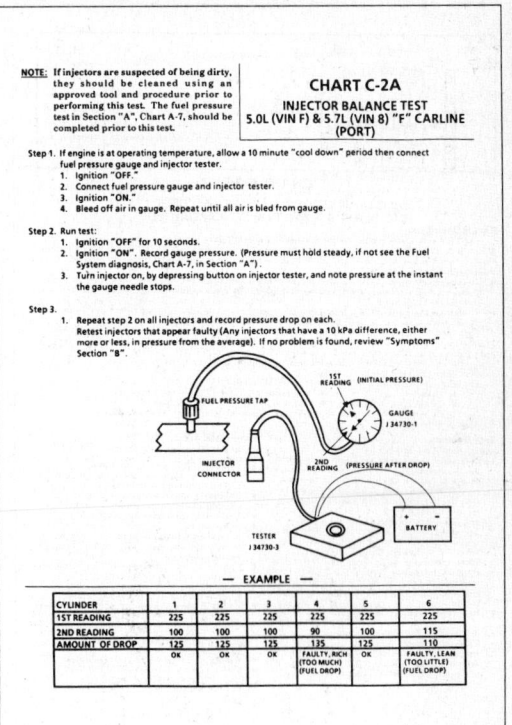

— EXAMPLE —

CYLINDER	1	2	3	4	5	6
1ST READING	225	225	225	225	225	225
2ND READING	100	100	100	90	100	115
AMOUNT OF DROP	125	125	125	135	125	110
	OK	OK	OK	FAULTY, RICH (TOO MUCH) (FUEL DROP)	OK	FAULTY, LEAN (TOO LITTLE) (FUEL DROP)

1988–90 5.0L (VIN F) AND 5.7L (VIN 8) CAMARO AND FIREBIRD

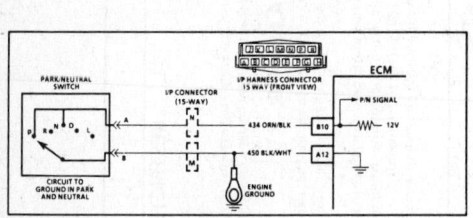

CHART C-1
PARK/NEUTRAL SWITCH DIAGNOSIS
(AUTO TRANSMISSION ONLY)
5.0L (VIN F) & 5.7L (VIN 8) "F" CARLINE (PORT)

Circuit Description:
The park/neutral switch contacts are a part of the neutral start switch, and are closed to ground in park or neutral and open in drive ranges.
The ECM supplies ignition voltage through a current limiting resistor to CKT 434 and senses a closed switch when the voltage on CKT 434 drops to less than one volt.
The ECM uses the P/N signal as one of the inputs to control:
- Idle air control
- VSS diagnostics
- EGR

If CKT 434 indicates P/N (grounded), while in drive range, the EGR would be inoperative, resulting in possible detonation.
If CKT 434 always indicates drive (open), a drop in the idle speed may exist when the gear selector is moved into drive range.

Test Description: Numbers below refer to circled numbers on the diagnostic chart.
1. Checks for a closed switch to ground in park position. Different makes of "Scan" tools will read P/N differently. Refer to "Operators Manual" for type of display used for a specific tool.
2. Checks for an open switch in drive range.
3. Be sure "Scan" indicates drive, even while wiggling shifter to test for an intermittent or misadjusted switch in drive range.

1988–90 5.0L (VIN F) AND 5.7L (VIN 8) CAMARO AND FIREBIRD

CHART C-1
PARK/NEUTRAL SWITCH DIAGNOSIS
(AUTO TRANSMISSION ONLY)
5.0L (VIN F) & 5.7L (VIN 8) "F" CARLINE (PORT)

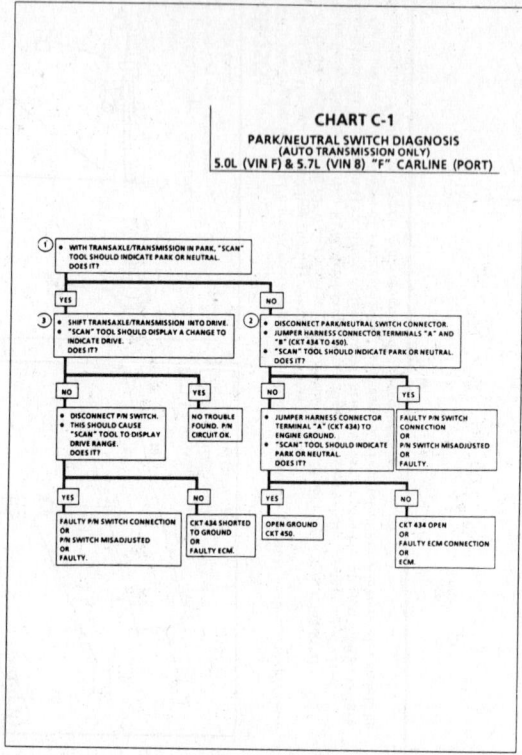

1988–90 5.0L (VIN F) AND 5.7L (VIN 8) CAMARO AND FIREBIRD

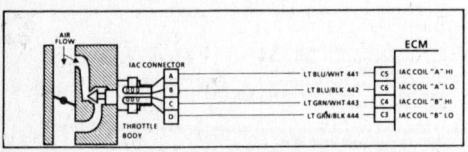

CHART C-2C
IDLE AIR CONTROL (IAC) SYSTEM CHECK
5.0L (VIN F) & 5.7L (VIN 8) "F" CARLINE (PORT)

Circuit Description:
The ECM will control engine idle speed by moving the IAC valve to control air flow around the throttle plates. It does this by sending voltage pulses to the proper motor winding for each IAC motor. This will cause the motor shaft and valve to move "IN" or "OUT" of the motor a given distance for each pulse received. ECM pulses are referred to as "counts".
- To increase idle speed - ECM will send enough counts to retract the IAC valve and allow more air to flow through the idle air passage and bypass the throttle plates until idle speed reaches the proper rpm. This will increase the ECM counts.
- To decrease idle speed - ECM will send enough counts to extend the IAC valve and reduce air flow through the idle passage around the throttle plates. This will reduce the ECM counts.

Each time the engine is started and then the ignition is turned "OFF", the ECM will reset the IAC valve. This is done by sending enough counts to seat the valve. The fully seated valve is the ECM reference zero. A given number of counts are then issued to open the valve, and normal ECM control of IAC will begin from this point. The number of counts are then calculated by the ECM. This is how the ECM knows what the motor position is for a given idle speed.
The ECM uses the following information to control idle speed:
- Battery voltage
- Engine speed
- Throttle position sensor
- P/N switch
- Coolant temperature
- A/C clutch signal

Don't apply battery voltage across the IAC motor terminals. It will permanently damage the IAC motor windings.

Test Description: Numbers below refer to circled numbers on the diagnostic chart.
1. Continue with test, even if engine will not idle. If idle is too low, "Scan" will display 80 or more counts, or steps. If idle is high, it will display "0" counts.
Occasionally an erratic or unstable idle may occur. Engine speed may vary 200 rpm or more up and down. Disconnect IAC. If the condition is unchanged, the IAC is not at fault. There is a system problem. Proceed to "Diagnostic Aids" below.
2. When the engine was stopped, the IAC valve retracted (more air) to a fixed "Park" position for increased air flow and idle speed during the next engine start. A "Scan" will display 140 or more counts.
3. Be sure to disconnect the IAC valve prior to this test. The test light will confirm the ECM signals by a steady or flashing light on all circuits.
4. There is a remote possibility that one of the circuits is shorted to voltage which would have been indicated by a steady light. Disconnect ECM and turn the ignition "ON" and probe terminals to check for this condition.

Diagnostic Aids:
Engine idle speed can be adversely affected by the following:
- Park/Neutral switch - If ECM thinks the car is always in neutral, then idle will not be controlled to the specified rpm when in drive range.
- Leaking injector(s) will cause fuel imbalance and poor idle quality due to excess fuel. See CHT A-7.
- Vacuum or crankcase leaks can affect idle.
- When the throttle shaft or throttle position sensor is binding or sticking in an open throttle position, the ECM does not know if the vehicle has stopped and does not control idle.
- Check A.I.R. management system for intermittent air to ports while in "Closed Loop".
- In addition to electrical control of EGR, be sure to examine the EGR valve for proper seating.
- Faulty battery cables can result in voltage variations. The ECM will try to compensate, which results in erratic idle speeds.
- The ECM will compensate for A/C compressor clutch loads. Loss of the A/C request signal would be most apparent in neutral.
- Contaminated fuel can adversely affect idle. Perform injector balance test CHART C 2A.

If all OK, refer to "Rough, Unstable, Incorrect Idle or Stalling" symptoms in Section "B".

1988–90 5.0L (VIN F) AND 5.7L (VIN 8) CAMARO AND FIREBIRD

CHART C-2C
IDLE AIR CONTROL (IAC) SYSTEM CHECK
5.0L (VIN F) & 5.7L (VIN 8) "F" CARLINE (PORT)

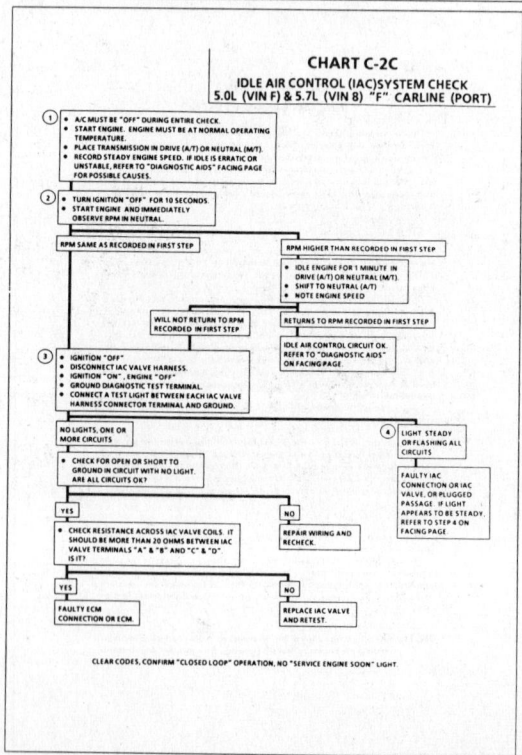

1988–90 5.0L (VIN F) AND 5.7L (VIN 8) CAMARO AND FIREBIRD

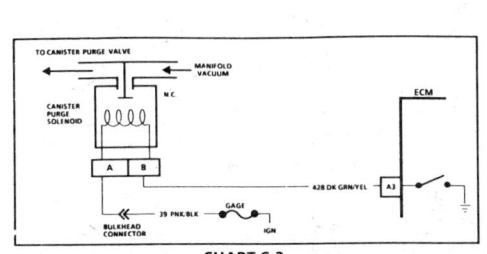

CHART C-3
CANISTER PURGE SOLENOID CHECK
5.0L (VIN F) & 5.7L (VIN 8) "F" CARLINE (PORT)

Circuit Description:
Canister purge is controlled by a solenoid that allows ported manifold vacuum to purge the fuel vapor canister when de-energized. The ECM supplies a ground to energize the solenoid (purge "ON").

If the diagnostic "test" terminal is grounded, with the engine stopped, or the following conditions are met with the engine running, the canister purge solenoid will be energized (purge "ON").
- Engine run time after start more than 1 minute.
- Coolant temperature above 75°C (167°F).
- Vehicle speed above 15 mph.
- Throttle position is above idle.

Test Description: Numbers below refer to circled numbers on the diagnostic chart.
1. The external vacuum source must be applied to the canister purge solenoid at the canister

2. Grounding the diagnostic "test" terminal will energize the canister purge solenoid and allow vacuum to pass.
3. Some canister purge solenoids may have a large enough bleed built into them to appear to be operating incorrectly

1988–90 5.0L (VIN F) AND 5.7L (VIN 8) CAMARO AND FIREBIRD

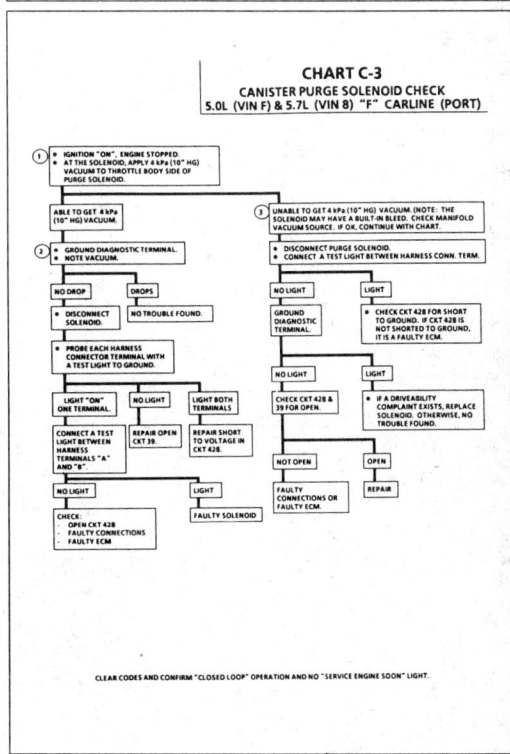

1988–90 5.0L (VIN F) AND 5.7L (VIN 8) CAMARO AND FIREBIRD

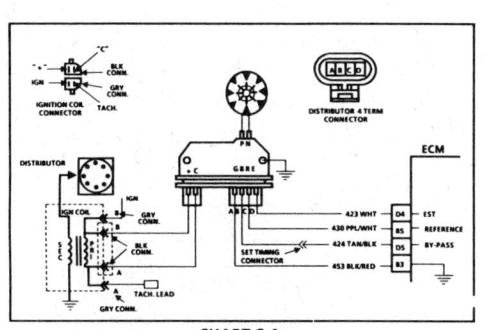

CHART C-4
IGNITION SYSTEM CHECK
(REMOTE COIL/SEALED MODULE CONNECTOR DISTRIBUTOR)
5.0L (VIN F) & 5.7L (VIN 8) "F" CARLINE (PORT)

Test Description: Numbers below refer to circled numbers on the diagnostic chart.
1. Two wires are checked, to ensure that an open is not present in a spark plug wire.
1A. If spark occurs with EST connector disconnected, pick-up coil output is too low for EST operation.
2. A spark indicates the problem must be the distributor cap or rotor.
3. Normally, there should be battery voltage at the "C" and "+" terminals. Low voltage would indicate an open or a high resistance circuit from the distributor to the coil or ignition switch. If "C" terminal voltage was low, but "+" terminal voltage is 10 volts or more, circuit from "C" terminal to ignition coil or ignition coil primary winding is open.
4. Checks for a shorted module or grounded circuit from the ignition coil to the module. The distributor should be turned "OFF", so normal voltage should be about 12 volts.
   If the module is turned "ON", the voltage would be low, but above 1 volt. This could cause the ignition coil to fail from excessive heat.
   With an open ignition coil primary winding, a small amount of voltage will leak through the module from the Batt + to the tach terminal.

5. Applying a voltage (1.5 to 8 volts) to module terminal "P" should turn the module ON and the tach terminal voltage should drop to about 7-9 volts. This test will determine whether the module or coil is faulty or if the pick-up coil is not generating the proper signal to turn the module "ON". This test can be performed by using a DC battery with a rating of 1.5 to 8 volts. The use of the test light is mainly to allow the "P" terminal to be probed more easily. Some digital multi-meters can also be used to trigger the module by selecting ohms, usually the diode position. In this position the meter may have a voltage across it's terminals which can be used to trigger the module. The voltage in the ohm's position can be checked by using a second meter or by checking the manufacturer's specification of the tool being used.
6. This should turn "OFF" the module and cause a spark. If no spark occurs, the fault is most likely in the ignition coil, because most module problems would have been found before this point in the procedure. A module tester could determine which is at fault.

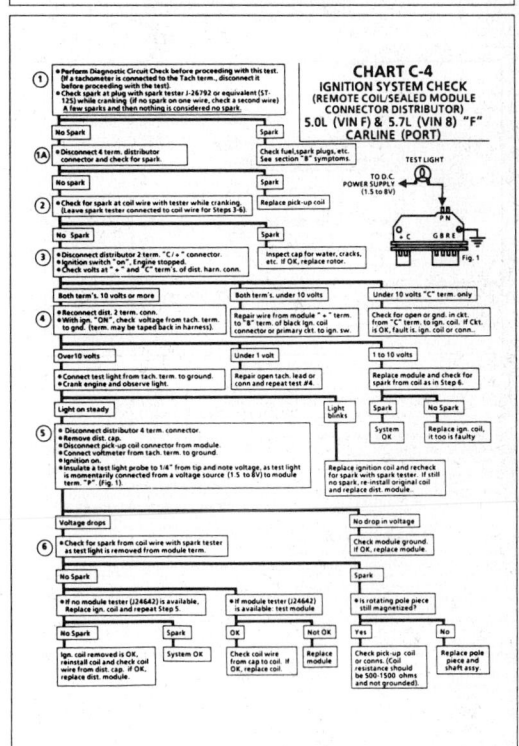

1988–90 5.0L (VIN F) AND 5.7L (VIN 8) CAMARO AND FIREBIRD

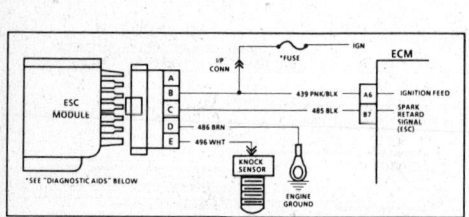

CHART C-5
ELECTRONIC SPARK CONTROL
5.0L (VIN F) & 5.7L (VIN 8) "F" CARLINE (PORT)

Circuit Description:
Electronic Spark Control (ESC) is accomplished with a module that sends a voltage signal to the ECM. As the knock sensor detects engine knock, the voltage from the ESC module to the ECM is shut "OFF" and this signals the ECM to retard timing, if engine rpm is over about 900.

Test Description: Numbers below refer to circled numbers on the diagnostic chart.

1. If a Code 43 is not set, but a knock signal is indicated while running at 1500 rpm, listen for an internal engine noise. Under a no load condition, there should not be any detonation, and if knock is indicated, an internal engine problem may exist.
2. Usually a knock signal can be generated by tapping on the exhaust manifold. This test can also be performed at idle. Test number 1 was run at 1500 rpm to determine if a constant knock signal was present, which would affect engine performance.
3. This tests whether the knock signal is due to the sensor, a basic engine problem, or the ESC module.
4. If the module ground circuit is faulty, the ESC module will not function correctly. The test light should light indicating the ground circuit is OK.

5. Contacting CKT 496, with a test light to 12 volts, should generate a knock signal to determine whether the knock sensor is faulty, or the ESC module can't recognize a knock signal.

Diagnostic Aids:

* **ECM/IGN FUSE**
"Scan" tools have two positions to diagnose the ESC system. The knock signal can be monitored to see if the knock sensor is detecting a knock condition and if the ESC module is functioning, knock signal should display "YES", whenever detonation is present. The knock retard position on the "Scan" displays the amount of spark retard the ECM is commanding. The ECM can retard the timing up to 20 degrees.
If the ESC system checks OK, but detonation is the complaint, refer to "Detonation/Spark Knock" in Section "B".

1988–90 5.0L (VIN F) AND 5.7L (VIN 8) CAMARO AND FIREBIRD

CHART C-5
ELECTRONIC SPARK CONTROL
5.0L (VIN F) & 5.7L (VIN 8) "F" CARLINE (PORT)

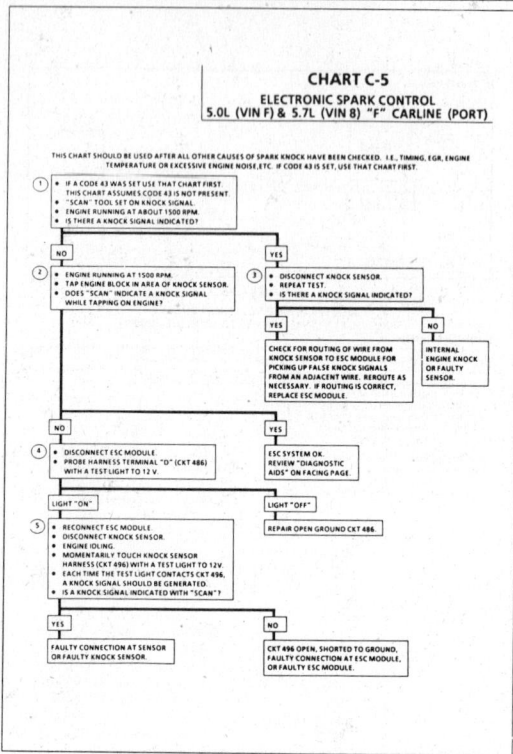

1988–90 5.0L (VIN F) AND 5.7L (VIN 8) CAMARO AND FIREBIRD

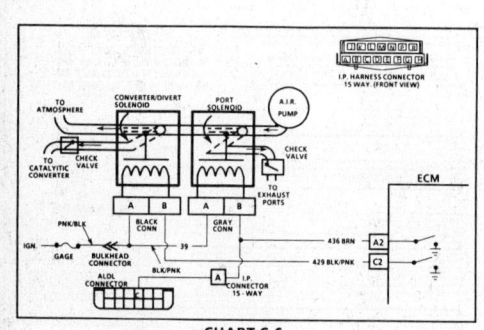

CHART C-6
AIR MANAGEMENT CHECK - CONTROL PEDES VALVE
(PRESSURE OPERATED ELECTRICAL DIVERT/ELECTRIC SWITCHING)
5.0L (VIN F) & 5.7L (VIN 8) "F" CARLINE (PORT)

Circuit Description:
Air management is controlled by a pressure operated divert valve and a converter valve, each with an ECM controlled solenoid. When the solenoid is grounded by the ECM, AIR pressure will activate the valve and allow pump air to be directed as follows:

* **Cold Mode** - The port solenoid is energized which in turn opens the port valve and allows flow to the exhaust ports.
* **Warm Mode** - The port solenoid is de-energized and the converter solenoid energized which closes the port valve and keeps the converter valve seated, thus forcing flow past the converter valve and to the converter.
* **Divert Mode** - Both solenoids are de-energized, which opens the converter valve, allowing air to take the path of least resistance, i.e., out the divert/relief tube to atmosphere

Test Description: Numbers below refer to circled numbers on the diagnostic chart.

1. This is a system functional check. Air is directed to ports during "Open Loop" and all engines start in "Open Loop" even on a warm engine. Since the air to the ports may be very short, be prepared to observe port air prior to engine start up. This can be done by squeezing a hose.
2. This should normally be a Code 22. When any code is set, the ECM opens the ground to the converter solenoid and allows air to divert. This checks for ECM response to a fault. A ground in the control valve circuit to the ECM would prevent divert action.

3. This checks for a grounded circuit to the ECM. Test light "OFF" is normal and would indicate the circuit is not grounded.
4. Checks for an open in the solenoid control circuits. Grounding the "test" terminal should ground both solenoid circuits. Normally, the test light should be "ON", which indicates the problem is not in the ECM or wiring but at the solenoid connections or valve itself.
5. This checks for a grounded solenoid circuit. Test light "OFF" would indicate the circuit is normal and fault is in the valve.

1988–90 5.0L (VIN F) AND 5.7L (VIN 8) CAMARO AND FIREBIRD

CHART C-6
AIR MANAGEMENT CHECK-CONTROL PEDES VALVE
(PRESSURE OPERATED ELECTRIC DIVERT/ELECTRIC SWITCHING)
5.0L (VIN F) & 5.7L (VIN 8) "F" CARLINE (PORT)

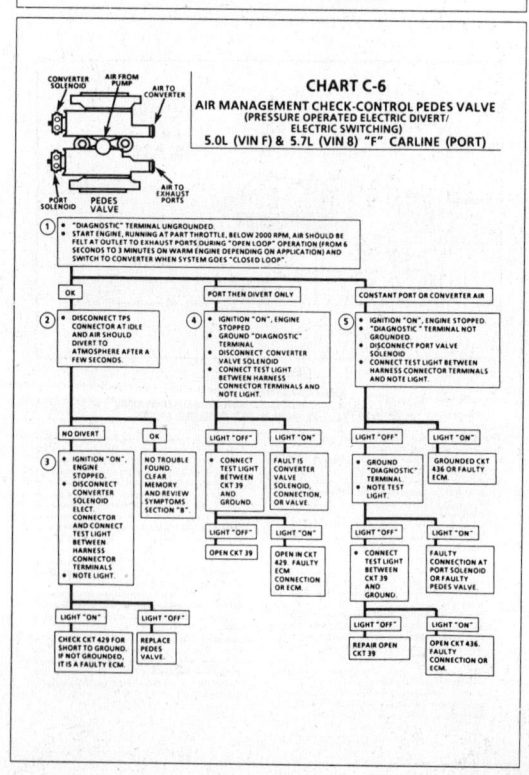

1988–90 5.0L (VIN F) AND 5.7L (VIN 8) CAMARO AND FIREBIRD

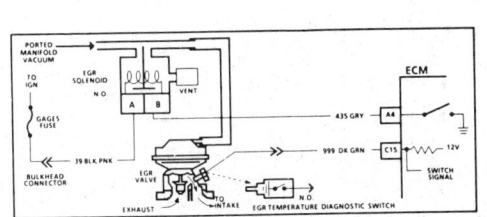

CHART C-7
EXHAUST GAS RECIRCULATION (EGR) CHECK
5.0L (VIN F) & 5.7L (VIN 8) "F" CARLINE (PORT)

Circuit Description:

The Exhaust Gas Recirculation (EGR) valve is controlled by, a normally open, Pulse Width Modulated (PWM) solenoid. The ECM turns the solenoid "OFF" to allow vacuum to pass to the EGR and turns the solenoid "ON" to prohibit EGR operation. When EGR is commanded, the solenoid is turned "ON" and "OFF" many times a second (duty cycle).

The duty cycle is calculated by the ECM based on information from the coolant, MAT, TPS, and MAF sensors. Also, engine rpm's and the P/N switch input affect EGR. There should be no EGR when in park or neutral, TPS below a calibrated value or TPS indicating WOT.

With the ignition "ON" and engine stopped, the EGR solenoid is de-energized. The solenoid, however, should be energized if the diagnostic terminal is grounded with the ignition "ON" and engine not running.

Test Description: Numbers below refer to circled numbers on the diagnostic chart.

1. This will test the solenoid and the manifold vacuum from the EGR valve. The vacuum may bleed off slowly but this should not be considered a fault.

2. As soon as back pressure is available at the EGR valve, the bleed portion in the valve should open and cause the valve to go to its heated position.

3. The EGR will be inoperative if the P/N switch is misadjusted or faulty. Use "Scan" tool and check P/N switch. Refer to CHART C-1.

1988–90 5.0L (VIN F) AND 5.7L (VIN 8) CAMARO AND FIREBIRD

CHART C-7
EXHAUST GAS RECIRCULATION (EGR) CHECK
5.0L (VIN F) & 5.7L (VIN 8) "F" CARLINE (PORT)

BEFORE USING THIS CHART, CHECK FOR PORTED VACUUM TO EGR SOLENOID, ALSO CHECK HOSES FOR LEAKS OR RESTRICTIONS. SHOULD BE AT LEAST 7" HG VACUUM AT 2000 RPM. THIS CHART ASSUMES THERE IS NO CODE 32.

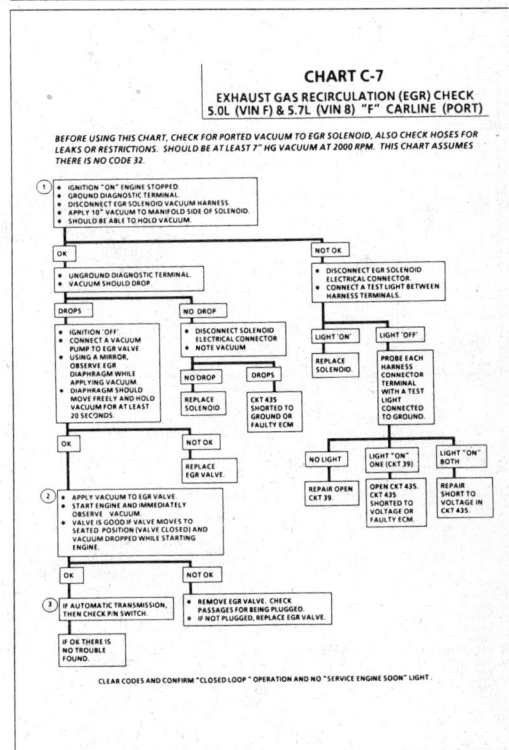

1988–90 5.0L (VIN F) AND 5.7L (VIN 8) CAMARO AND FIREBIRD

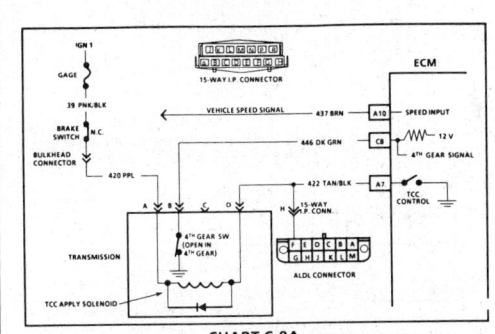

CHART C-8A
(Page 1 of 2)
TORQUE CONVERTER CLUTCH (TCC) SYSTEM
5.0L (VIN F) & 5.7L (VIN 8) "F" CARLINE (PORT)

Circuit Description:

The purpose of the automatic transmission torque converter clutch feature is to eliminate the power loss of the torque converter stage when the vehicle is in a cruise condition. This allows the convenience of the automatic transmission and the fuel economy of a manual transmission. The heart of the system is a solenoid located inside the automatic transmission which is controlled by the ECM.

When the solenoid coil is activated ("ON"), the torque converter clutch is applied which results in 100% mechanical coupling from the engine to transmission. When the transmission solenoid is deactivated, the torque converter clutch is released, which allows the torque converter to operate in the conventional manner (fluid coupling between engine and transmission).

The ECM turns "ON" the TCC when coolant temperature is above 65°C (149°F), TPS not changing, and vehicle speed above a specified value.

Test Description: Numbers below refer to circled numbers on the diagnostic chart.

1. When a test light is connected from ALDL terminal "F" to ground, a test light "ON" indicates battery voltage is OK and the TCC solenoid is disengaged.

2. When the diagnostic terminal is grounded, the ECM should energize the TCC solenoid and the test light should go out.

Diagnostic Aids:

A "Scan" tool only indicates when the ECM has turned "ON" the TCC driver (grounded CKT 422) but this does not confirm that the TCC has engaged. To determine if TCC is functioning properly, engine rpm should decrease when the "Scan" indicates the TCC driver has turned "ON". To determine if the 4th gear switch is functioning properly, perform the checks in CHART C-8A (Page 2 of 2). The switches will not prevent TCC from functioning but will affect TCC lock and unlock points. If the 4th gear switch circuit is always open, the TCC may engage as soon as sufficient oil pressure is reached.

1988–90 5.0L (VIN F) AND 5.7L (VIN 8) CAMARO AND FIREBIRD

CHART C-8A
(Page 1 of 2)
TORQUE CONVERTER CLUTCH (TCC) SYSTEM
5.0L (VIN F) & 5.7L (VIN 8) "F" CARLINE (PORT)

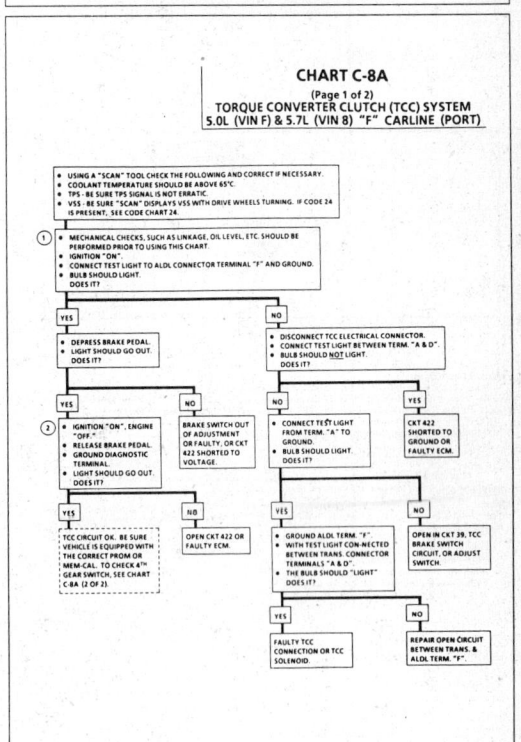

1988–90 5.0L (VIN F) AND 5.7L (VIN 8) CAMARO AND FIREBIRD

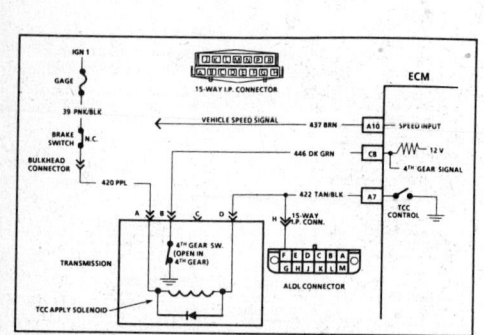

CHART C-8A
(Page 2 of 2)
TORQUE CONVERTER CLUTCH (TCC) ELECTRICAL DIAGNOSIS
5.0L (VIN F) & 5.7L (VIN 8) "F" CARLINE (PORT)

Circuit Description:
A 4th gear switch (mounted in the trans.) opens when the transmission shifts into 4th gear, and this switch is used by the ECM to modify TCC lock and unlock points, when in a 4-3 downshift maneuver.

Test Description: Numbers below refer to circled numbers on the diagnostic chart.
1. Unless the switch or CKT 446 is open, the "Scan" should display "NO", indicating the transmission is not in 4th gear. The 4th gear switch should only be open while in 4th gear.
2. This step determines if the ECM and wiring are OK. Grounding CKT 446 should cause the "Scan" to display "NO", indicating the trans. is not in 4th gear.
3. Checks the operation of the 4th gear switch. When the transmission shifts into 4th gear the switch should open and the "Scan" should display "YES".
4. Disconnecting the TCC connector simulates an open switch to determine if CKT 446 is shorted to ground or the problem is in the transmission.

Diagnostic Aids:
A road test may be necessary to verify the customer complaint. If the "Scan" indicates TCC is turning "ON" and "OFF" erratically, check the state of the 4th gear switch to be sure it is not changing states under a steady throttle position. If the switch is changing states, check connections and wire routing carefully. Also, if the 4th gear switch is always open the TCC may engage as soon as sufficient oil pressure is reached.

1988–90 5.0L (VIN F) AND 5.7L (VIN 8) CAMARO AND FIREBIRD

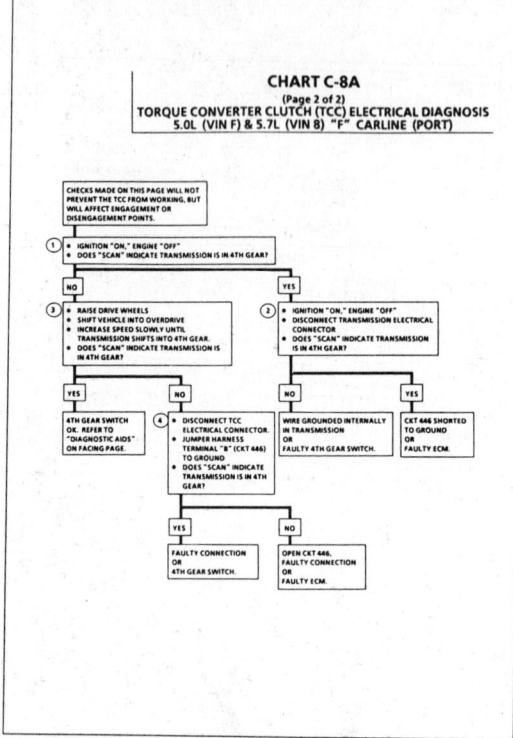

CHART C-8A
(Page 2 of 2)
TORQUE CONVERTER CLUTCH (TCC) ELECTRICAL DIAGNOSIS
5.0L (VIN F) & 5.7L (VIN 8) "F" CARLINE (PORT)

1988–90 5.0L (VIN F) AND 5.7L (VIN 8) CAMARO AND FIREBIRD

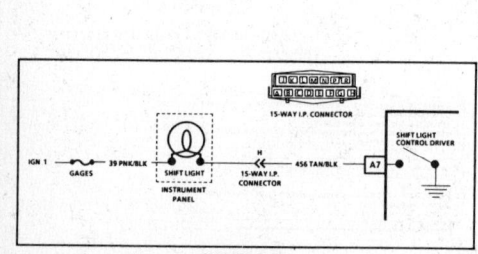

CHART C-8B
MANUAL TRANSMISSION SHIFT LIGHT DIAGNOSIS
5.0L (VIN F) & 5.7L (VIN 8) "F" CARLINE (PORT)

Circuit Description:
The shift light lamp display is in the instrument panel. The purpose of the shift light is to provide a display, to the driver, which indicates the best transmission shift point for maximum fuel economy based on engine speed and load. The light is controlled by the ECM and is turned "ON" by grounding CKT 456.
The ECM uses information from the following inputs to control the shift light:
• Coolant temperature
• TPS
• VSS
• rpm
The ECM uses the measured rpm and the vehicle speed to calculate what gear the vehicle is in. It's this calculation that determines when the shift light should be turned "ON".

Test Description: Numbers below refer to circled numbers on the diagnostic chart.
1. This should not turn "ON" the shift light. If the light is "ON", there is a short to ground in CKT 456 wiring or a fault in the ECM.
2. When the diagnostic terminal is grounded, the ECM should ground CKT 456 and the shift light should come "ON".
3. This checks the shift light circuit up to the ECM connector. If the shift light illuminates, then the ECM connector is faulty or the ECM does not have the ability to ground the circuit.

1988–90 5.0L (VIN F) AND 5.7L (VIN 8) CAMARO AND FIREBIRD

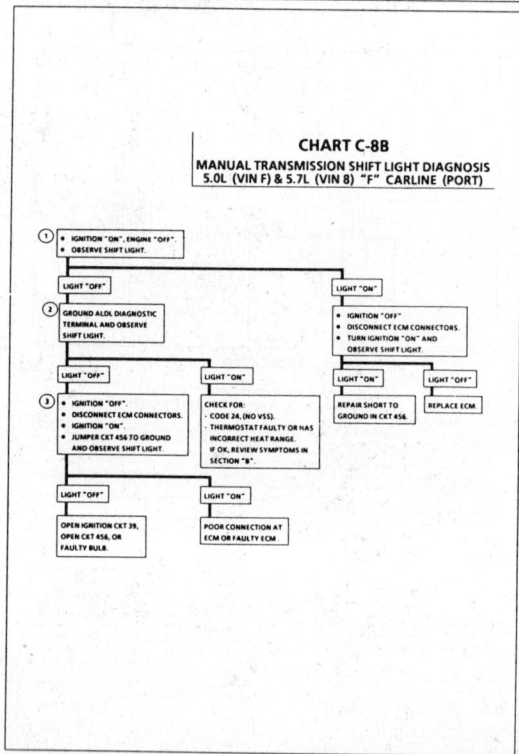

CHART C-8B
MANUAL TRANSMISSION SHIFT LIGHT DIAGNOSIS
5.0L (VIN F) & 5.7L (VIN 8) "F" CARLINE (PORT)

1988–90 5.0L (VIN F) AND 5.7L (VIN 8) CAMARO AND FIREBIRD

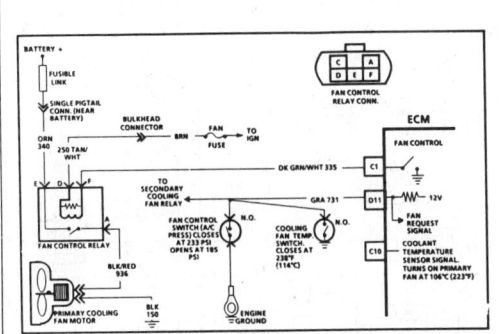

CHART C-12
(Page 1 of 2)
COOLING FAN CONTROL CIRCUIT
5.0L (VIN F) & 5.7L (VIN 8) "F" CARLINE (PORT)

Circuit Description:
- The primary cooling fan is totally controlled by the ECM based on inputs from the coolant sensor and fan control switch. The fan should run, if coolant temperature is greater than 106°C (223°F).
- Battery voltage is supplied to the fan relay on terminal "E" and ignition voltage to terminal "D".
- Grounding CKT 335 (relay terminal "F") will energize the relay and supply battery voltage to the fan motor. Once the fan relay is energized by the ECM, it will remain "ON" for a minimum of 15 seconds.
- The ECM will remove the ground to CKT 335 if vehicle speed is over 40 mph unless the engine is overheating.
- The fan control switch, mounted in the A/C high pressure line, will close when head pressure exceeds 1600 kPa (233 psi) and this input causes the ECM to ground CKT 335.
- If a Code 14 or 15 sets, or if the ECM is operating in the fuel back-up mode, the ECM will turn "ON" the cooling fan.

Diagnostic Aids:

If the owner complained of an overheating problem, it must be determined if the complaint was due to an actual boil over or the hot light or temperature gage indicated overheating.

If the gage or light indicates overheating, but no boilover is detected, the gage circuit should be checked. The gage accuracy can also be checked by comparing the coolant sensor reading using a "Scan" tool and comparing its reading with the gage reading.

If the engine is actually overheating and the gage indicates overheating, but the cooling fan is not coming "ON", the coolant sensor has probably shifted out of calibration and should be replaced.

If the engine is overheating, and the cooling fan is "ON", the cooling system should be checked.

1988–90 5.0L (VIN F) AND 5.7L (VIN 8) CAMARO AND FIREBIRD

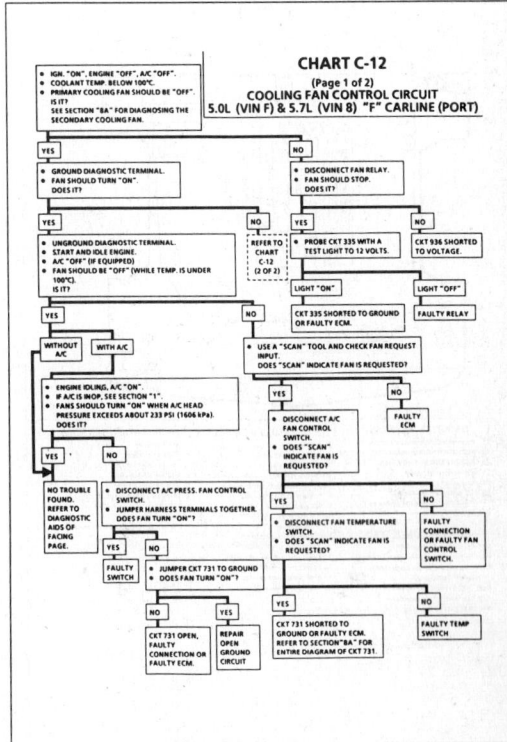

CHART C-12
(Page 1 of 2)
COOLING FAN CONTROL CIRCUIT
5.0L (VIN F) & 5.7L (VIN 8) "F" CARLINE (PORT)

1988–90 5.0L (VIN F) AND 5.7L (VIN 8) CAMARO AND FIREBIRD

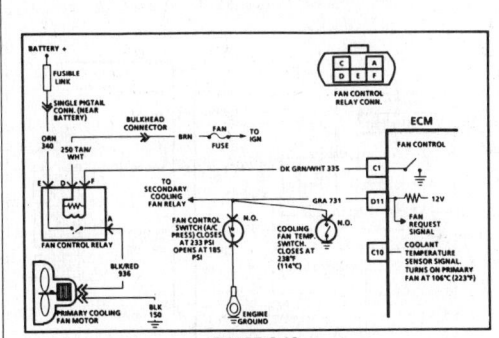

CHART C-12
(Page 2 of 2)
COOLING FAN CONTROL CIRCUIT
5.0L (VIN F) & 5.7L (VIN 8) "F" CARLINE (PORT)

Circuit Description:
- The primary cooling fan is totally controlled by the ECM based on inputs from the coolant sensor and fan control switch. The fan should run, if coolant temperature is greater than 106°C (223°F).
- Battery voltage is supplied to the fan relay on terminal "E" and ignition voltage to terminal "D".
- Grounding CKT 335 (relay terminal "F") will energize the relay and supply battery voltage to the fan motor. Once the fan relay is energized by the ECM, it will remain "ON" for a minimum of 15 seconds.
- The ECM will remove the ground to CKT 335 if vehicle speed is over 40 mph unless the engine is overheating.
- The fan control switch, mounted in the A/C high pressure line, will close when head pressure exceeds 1600 kPa (233 psi) and this input causes the ECM to ground CKT 335.
- If a Code 14 or 15 sets, or if the ECM is operating in the fuel back-up mode, the ECM will turn "ON" the cooling fan.

Diagnostic Aids:

If the owner complained of an overheating problem, it must be determined if the complaint was due to an actual boil over or the hot light or temperature gage indicated overheating.

If the gage or light indicates overheating, but no boilover is detected, the gage circuit should be checked. The gage accuracy can also be checked by comparing the coolant sensor reading using a "Scan" tool and comparing its reading with the gage reading.

If the engine is actually overheating and the gage indicates overheating, but the cooling fan is not coming "ON", the coolant sensor has probably shifted out of calibration and should be replaced.

If the engine is overheating, and the cooling fan is "ON", the cooling system should be checked.

1988–90 5.0L (VIN F) AND 5.7L (VIN 8) CAMARO AND FIREBIRD

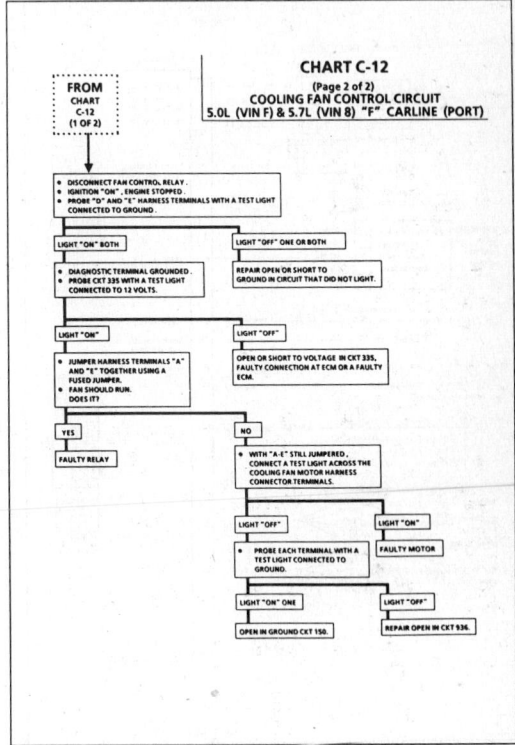

CHART C-12
(Page 2 of 2)
COOLING FAN CONTROL CIRCUIT
5.0L (VIN F) & 5.7L (VIN 8) "F" CARLINE (PORT)

5.7L (VIN 8) COMPONENT LOCATIONS—CORVETTE

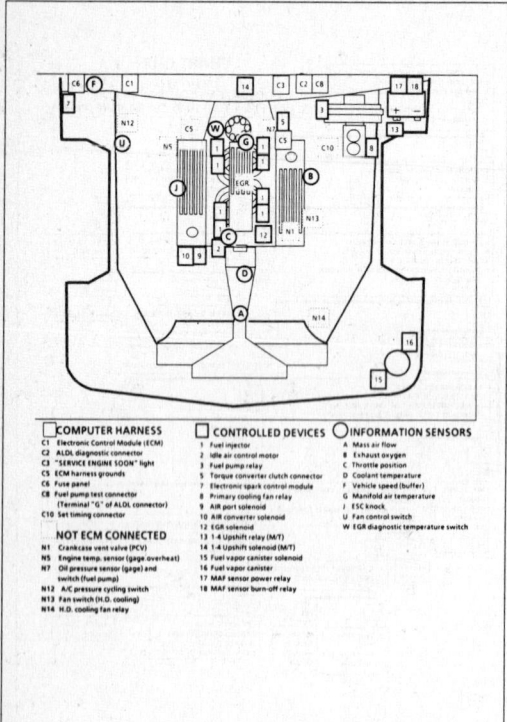

□ COMPUTER HARNESS
- C1 Electronic Control Module (ECM)
- C2 ALDL diagnostic connector
- C3 "SERVICE ENGINE SOON" light
- C5 ECM harness grounds
- C6 Fuse panel
- C8 Fuel pump test connector (Terminal "G" of ALDL connector)
- C10 Set timing connector

NOT ECM CONNECTED
- N1 Crankcase vent valve (PCV)
- N5 Engine temp. sensor (gage overheat)
- N7 Oil pressure sensor (gage) and switch (fuel pump)
- N12 A/C pressure cycling switch
- N13 Fan switch (H.D. cooling)
- N14 H.D. cooling fan relay

□ CONTROLLED DEVICES
1. Fuel injector
2. Idle air control motor
3. Fuel pump relay
4. Torque converter clutch connector
5. Electronic spark control module
6. Primary cooling fan relay
8. Air port solenoid
9. Air converter solenoid
12. EGR solenoid
13. 1-4 Upshift relay (M/T)
14. 1-4 Upshift solenoid (M/T)
15. Fuel vapor canister solenoid
16. Fuel vapor canister
17. MAF sensor power relay
18. MAF sensor burn-off relay

○ INFORMATION SENSORS
- A Mass air flow
- B Exhaust oxygen
- C Throttle position
- D Coolant temperature
- F Vehicle speed (buffer)
- H Manifold air temperature
- J ESC knock
- U Fan control switch
- W EGR diagnostic temperature switch

5.7L (VIN 8) ECM WIRING DIAGRAM—CORVETTE

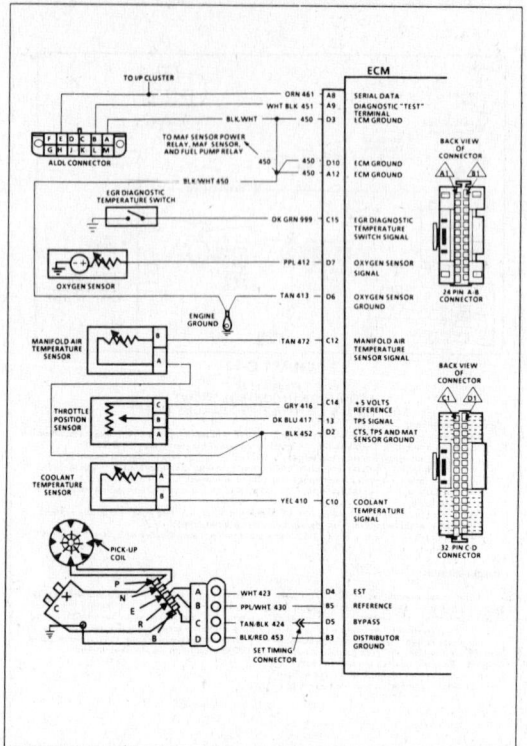

5.7L (VIN 8) ECM WIRING DIAGRAM CORVETTE (CONT.)

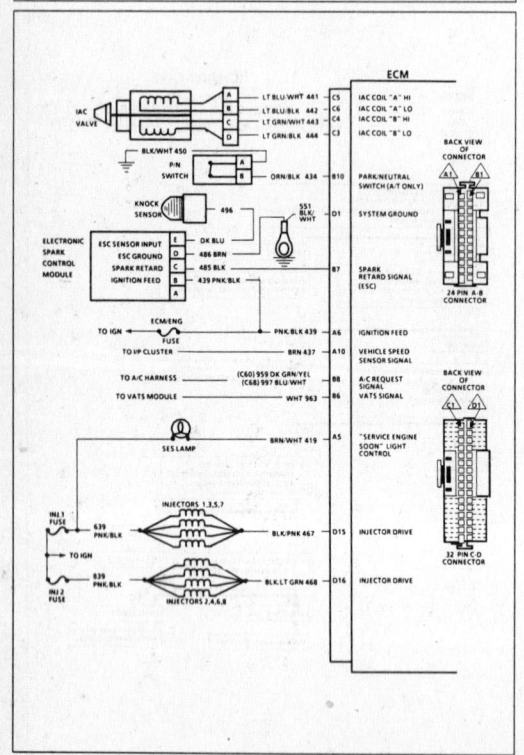

5.7L (VIN 8) ECM WIRING DIAGRAM CORVETTE (CONT.)

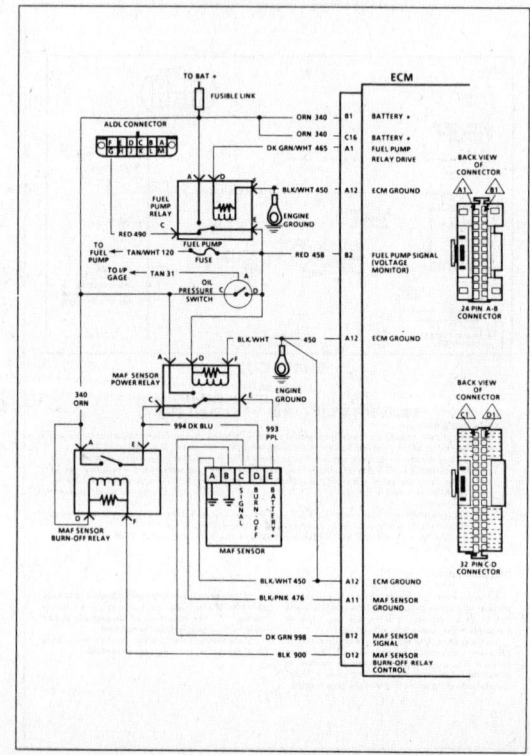

5.7L (VIN 8) ECM WIRING DIAGRAM CORVETTE (CONT.)

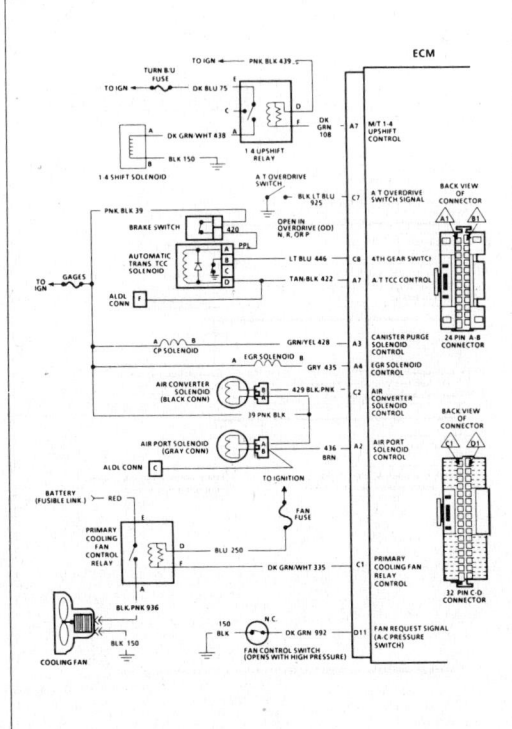

5.7L (VIN 8) ECM CONNECTOR TERMINAL END VIEW — CORVETTE

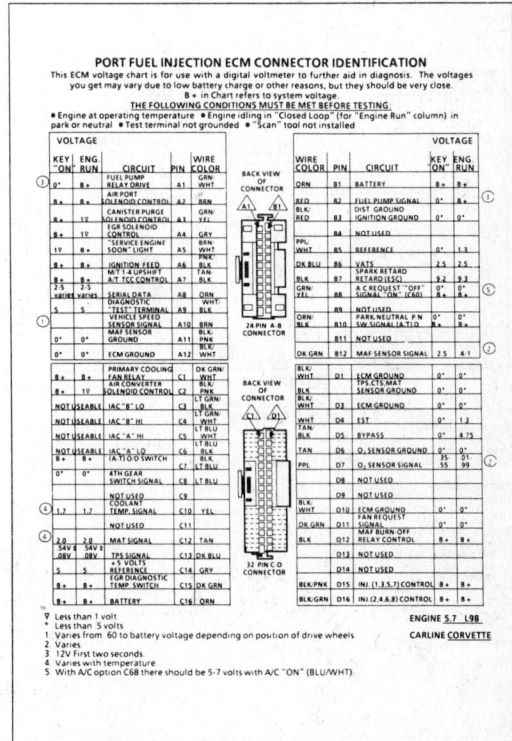

PORT FUEL INJECTION ECM CONNECTOR IDENTIFICATION

This ECM voltage chart is for use with a digital voltmeter to further aid in diagnosis. The voltages you get may vary due to low battery charge or other reasons, but they should be very close.
B + in Chart refers to system voltage.

THE FOLLOWING CONDITIONS MUST BE MET BEFORE TESTING:
• Engine at operating temperature • Engine idling in "Closed Loop" (for "Engine Run" column) in park or neutral • Test terminal not grounded • "Scan" tool not installed

▽ Less than 1 volt
* Less than 5 volts
1 Varies from 60 to battery voltage depending on position of drive wheels
2 Varies
3 12V first two seconds
4 Varies with temperature
5 With A/C option C68 there should be 5-7 volts with A/C "ON" (BLU/WHT).

ENGINE 5.7 L98
CARLINE CORVETTE

1988–90 5.7L (VIN 8) — CORVETTE

DIAGNOSTIC CIRCUIT CHECK

The Diagnostic Circuit Check is an organized approach to identifying a problem created by an Electronic Engine Control System malfunction. It must be the starting point for any driveability complaint diagnosis, because it directs the Service Technician to the next logical step in diagnosing the complaint.

The "Scan Data" listed in the table may be used for comparison, after completing the Diagnostic Circuit Check and finding the on board diagnostics functioning properly and no trouble codes displayed. The "Typical Values" are an average of display values recorded from normally operating vehicles and are intended to represent what a normally functioning system would typically display.

A "SCAN" TOOL THAT DISPLAYS FAULTY DATA SHOULD NOT BE USED, AND THE PROBLEM SHOULD BE REPORTED TO THE MANUFACTURER. THE USE OF A FAULTY "SCAN" CAN RESULT IN MISDIAGNOSIS AND UNNECESSARY PARTS REPLACEMENT.

Only the parameters listed below are used in this manual for diagnosing. If a "Scan" reads other parameters, the values are not recommended by General Motors for use in diagnosing. For more description on the values and use of the "Scan" to diagnosis ECM inputs, refer to the applicable diagnosis section. If all values are within the range illustrated, refer to symptoms in Section B.

"SCAN" DATA

Idle / Upper Radiator Hose Hot / Closed Throttle / Park or Neutral / Closed Loop / Acc. off

"SCAN" Position	Units Displayed	Typical Data Value
Desired RPM	RPM	ECM idle command (varies with temp.)
RPM	RPM	± 100 RPM from desired RPM (± 50 in drive)
Coolant Temp	C°	85° - 105°
MAT Temp	C°	10° - 90° (depends on underhood temp.)
MAF	Gm/Sec	4 - 7
Air Flow	Gm/Sec	4 - 7
BPW (base pulse width)	M/Sec	1 - 4 and varying
O₂	Volts	1 - 1000 and varying
TPS	Volts	.46 - .62
IAC	Counts (steps)	1 - 40
INT (Integrator)	Counts	Varies
BLM (Block Learn)	Counts	118 - 138
Open/Closed Loop	Open/Closed	Closed Loop (may go open with extended idle)
BLM Cell	Cell Number	0 or 1 (depends on Air Flow & RPM)
VSS	MPH	0
TCC	On/Off	Off (on with TCC commanded)
Battery	Volts	13.5 - 14.5
PPSW (fuel pump voltage)	Volts	13.5 - 14.5
LV8	Counts	30 - 60
Knock Retard	Degrees of Retard	0
Spark Advance	# of Degrees	Varies
P/N Switch	P/N and R/DL	Park/Neutral (P/N)
Overdrive Switch (A/T)	Yes/No	Yes (no, when in 1, 2, or D range)
A.I.R Control	Normal/Divert	Normal
A.I.R Switch	Port/Converter	Converter
A/C Request	Yes/No	No (yes, with A/C requested)
Fan Request	Yes/No	No (yes, with A/C high pressure)
EGRDC	0 - 100%	0 at idle
EGR Diagnostic	On/Off	off
Fan	On/Off	Off (below 108°)
CCP duty cycle	0 - 100%	0
Knock Signal	Yes/No	No (yes, when knock is detected)
Shift Light (M/T)	On/Off	Off
4th Gear (A/T)	Yes/No	No (yes, when in 4th gear)
Vats Status	Ok/Not Ok	Ok
Skip Shift Enabled	Yes/No	No
Skip Shift Active	Yes/No	No

1988–90 5.7L (VIN 8) — CORVETTE

DIAGNOSTIC CIRCUIT CHECK
5.7L (VIN 8) "Y" CARLINE (PORT)

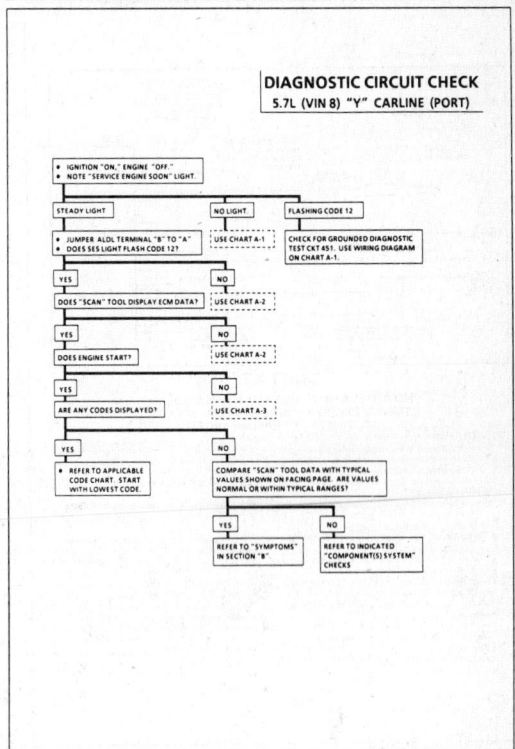

1988-90 5.7L (VIN 8) – CORVETTE

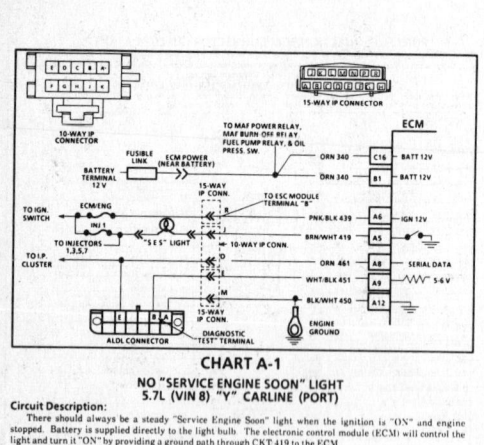

CHART A-1
NO "SERVICE ENGINE SOON" LIGHT
5.7L (VIN 8) "Y" CARLINE (PORT)

Circuit Description:
There should always be a steady "Service Engine Soon" light when the ignition is "ON" and engine stopped. Battery is supplied directly to the light bulb. The electronic control module (ECM) will control the light and turn it "ON" by providing a ground path through CKT 419 to the ECM.

Test Description: Numbers below refer to circled numbers on the diagnostic chart.
1. This test will determine if the ECM has the ability to ground CKT 419
2. If both the continuous battery supply voltages are lost at terminals "B1" and "C16" or the ignition feed to terminal "A6" is not present, the "Service Engine Soon" light will not come "ON" with the ignition "ON."

Diagnostic Aids:
Engine runs OK, check:
- Faulty light bulb.
- CKT 419 open.
Engine cranks but will not run.
- Continuous battery - fusible link open.
- ECM fuse open.
- Battery CKT 340 to ECM open.
- Ignition CKT 439 to ECM open.
- Poor connection to ECM.
- Faulty ECM ground circuit(s).

1988-90 5.7L (VIN 8) – CORVETTE

CHART A-1
NO "SERVICE ENGINE SOON" LIGHT
5.7L (VIN 8) "Y" CARLINE (PORT)

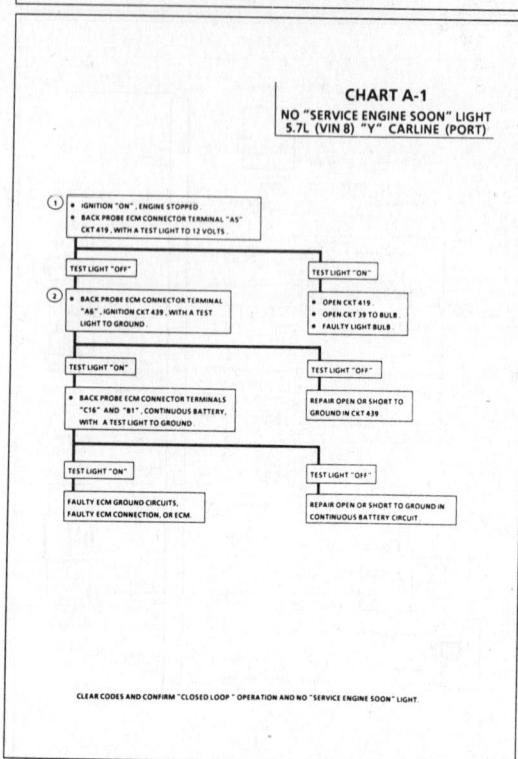

CLEAR CODES AND CONFIRM "CLOSED LOOP" OPERATION AND NO "SERVICE ENGINE SOON" LIGHT.

1988-90 5.7L (VIN 8) – CORVETTE

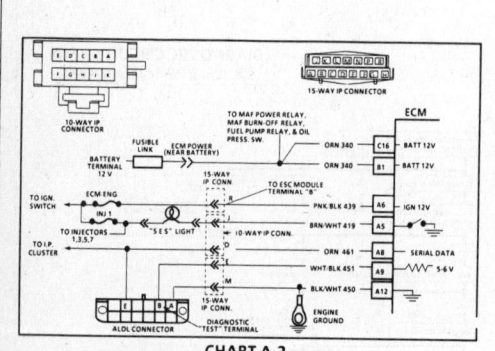

CHART A-2
NO ALDL DATA OR WILL NOT FLASH CODE 12
"SERVICE ENGINE SOON" LIGHT "ON" STEADY
5.7L (VIN 8) "Y" CARLINE (PORT)

Circuit Description:
There should always be a steady "Service Engine Soon" light when the ignition is "ON" and engine stopped. Battery ignition voltage is supplied to the light bulb. The electronic control module (ECM) will turn the light "ON" by grounding CKT 419 in the ECM.
With the diagnostic "test" terminal grounded, the light should flash a Code 12, followed by any trouble code(s) stored in memory.
A steady light suggests a short to ground in the light control CKT 419, or an open in diagnostic CKT 451.

Test Description: Numbers below refer to circled numbers on the diagnostic chart.
1. If there is a problem with the ECM that causes a "Scan" tool to not read serial data, the ECM should not flash a Code 12. If Code 12 is flashing check CKT 451 for an open to ground. If Code 12 does not flash, be sure that the "Scan" tool is working properly on another vehicle. If the "Scan" is functioning properly and CKT 461 is OK, the Mem-Cal or ECM may be at fault for the NO ALDL symptom. Also, be sure CKT 461 is OK going to the instrument panel cluster.
2. If the light goes "OFF," when the ECM connector is disconnected, CKT 419 is not shorted to ground.
3. This step will check for an open diagnostic CKT 451
4. At this point, the "Service Engine Soon" light wiring is OK. The problem is a faulty ECM or Mem-Cal. If Code 12 does not flash, the ECM should be replaced using the original Mem-Cal. Replace the Mem-Cal only after trying an ECM, as a defective Mem-Cal is an unlikely cause of the problem.

1988-90 5.7L (VIN 8) – CORVETTE

CHART A-2
NO ALDL DATA OR WILL NOT FLASH CODE 12
"SERVICE ENGINE SOON" LIGHT "ON" STEADY
5.7L (VIN 8) "Y" CARLINE (PORT)

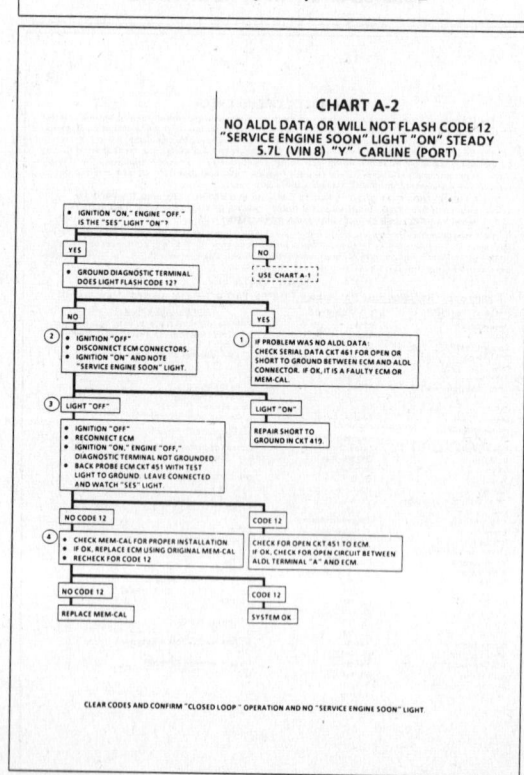

CLEAR CODES AND CONFIRM "CLOSED LOOP" OPERATION AND NO "SERVICE ENGINE SOON" LIGHT.

1988–90 5.7L (VIN 8) – CORVETTE

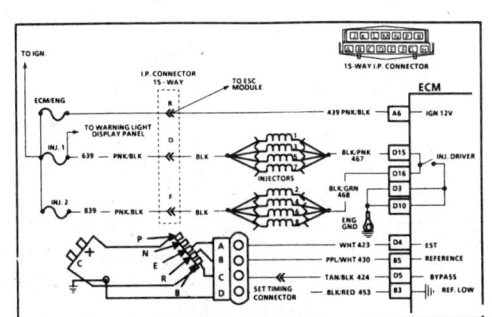

CHART A-3
(Page 1 of 2)
ENGINE CRANKS BUT WILL NOT RUN
5.7L (VIN 8) "Y" CARLINE (PORT)

Circuit Description:
This chart assumes that battery condition and engine cranking speed are OK, and there is adequate fuel in the tank.

Test Description: Numbers below refer to circled numbers on the diagnostic chart.

1. A "Service Engine Soon" light "ON" is a basic test to determine if there is a 12 volt supply and ignition 12 volts to ECM. No ALDL may be due to an ECM problem and CHART A-2 will diagnose the ECM. If TPS is over 2.5 volts, the engine may be in the clear flood mode, which will cause starting problems. The engine will not start without reference pulses and, therefore, the "Scan" should read engine rpm (reference) during crank.

2. No spark may be caused by one of several components related to the ignition system. CHART C-4 will address all problems related to the causes of a no spark condition.

3. The test light should blink, indicating the ECM is controlling the injectors OK. How bright the light blinks is not important. However, the test light should be a J-34730-3 or equivalent.

4. Use fuel pressure gage J-34730-1 or equivalent. Wrap a shop towel around the fuel pressure tap to absorb any small amount of fuel leakage that may occur when installing the gage

Diagnostic Aids:

- An EGR valve sticking open can cause a low air/fuel ratio during cranking. Unless engine enters clear flood at the first indication of a flooding condition, it can result in a no start.
- Check for fouled plugs
- Water or foreign material in fuel line can cause a no start in cold weather
- A defective MAF sensor may cause a no start or a stall after start. To determine if the sensor is causing the problem, disconnect it. The ECM will then use a default value for the sensor, and if the condition is corrected and the connections are OK, replace the sensor.
- Also check that injectors on both sides of engine will cause a test light to "blink." If not OK, check injector fuses.
If above are all OK, refer to "Symptoms" in Section "B", "Hard Start."

1988–90 5.7L (VIN 8) – CORVETTE

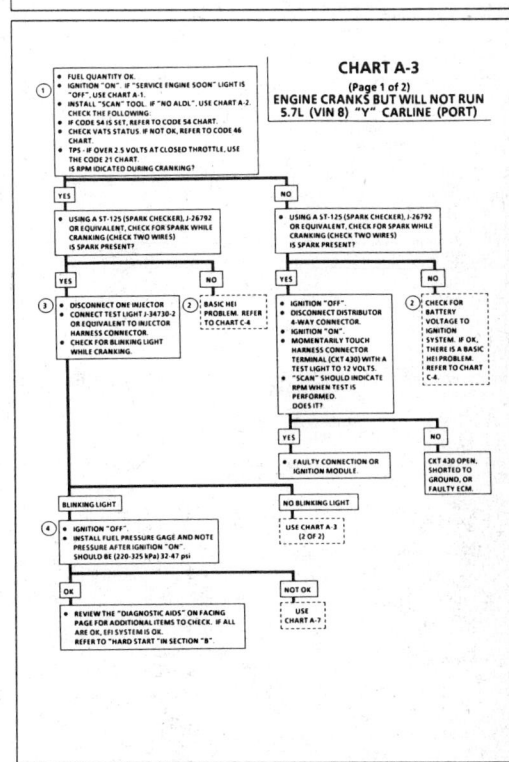

CHART A-3
(Page 1 of 2)
ENGINE CRANKS BUT WILL NOT RUN
5.7L (VIN 8) "Y" CARLINE (PORT)

1988–90 5.7L (VIN 8) – CORVETTE

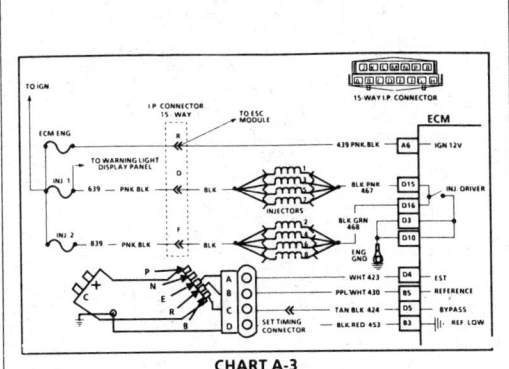

CHART A-3
(Page 2 of 2)
ENGINE CRANKS BUT WILL NOT RUN
5.7L (VIN 8) "Y" CARLINE (PORT)

Test Description: Numbers below refer to circled numbers on the diagnostic chart.

1. Checks for 12 volt supply to injectors. Due to the injectors wired in parallel there should be a light "ON" on both terminals
2. Checks continuity of CKT 467 and CKT 468
3. All checks made to this point would indicate that the ECM is at fault. However, there is a possibility of CKT 467 or CKT 468 being shorted to a voltage source either in the engine harness or in the injector harness.
   - Test for this condition.
     - Disconnect all injectors
     - Ignition "ON"

- Probe CKT 467 and CKT 468 on the ECM side of the injector harness with a test light connected to ground. (Test one injector harness on each side of engine) There should be no light. If light is "ON" repair short to voltage
- If OK, check the resistance of the injectors.
- Should be 10 ohms or more.
- Check injector harness connector. Be sure terminals are not backed out of connector and contacting each other
- If all OK, replace ECM

1988–90 5.7L (VIN 8) – CORVETTE

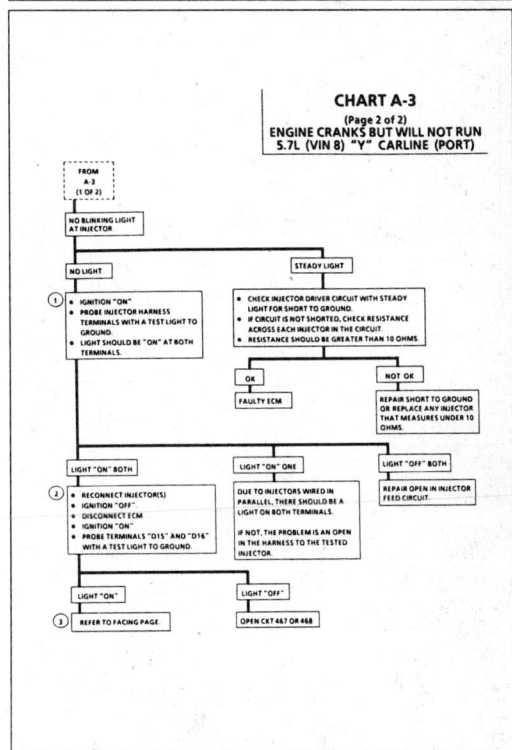

CHART A-3
(Page 2 of 2)
ENGINE CRANKS BUT WILL NOT RUN
5.7L (VIN 8) "Y" CARLINE (PORT)

SECTION 5

1988–90 5.7L (VIN 8) – CORVETTE

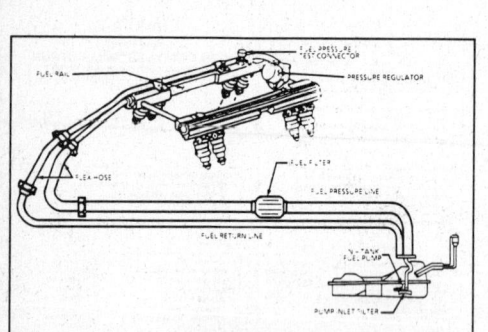

CHART A-7
(Page 2 of 2)
FUEL SYSTEM DIAGNOSIS
5.7L (VIN 8) "Y" CARLINE (PORT)

Test Description: Numbers below refer to circled numbers on the diagnostic chart.

1. Pressure but less than 280 kPa (40.5 psi) falls into two areas.
   - Regulated pressure but less than 280 kPa (40.5 psi). System will be lean running and may set Code 44. Also, hard starting cold and overall poor performance.
   - Restricted flow causing pressure drop. Normally, a vehicle with a fuel pressure of less than 165 kPa (24 psi) at idle will not be driveable. However, if the pressure drop occurs only while driving, the engine will normally surge then stop running as pressure begins to drop rapidly. This is most likely caused by a restricted fuel line or plugged filter.

2. Restricting the fuel return line allows the fuel pressure to build above regulated pressure. With battery voltage applied to the pump test terminal, pressure should rise to 414 kPa (60 psi) as the fuel return hose is gradually pinched.

   NOTICE: Do not allow fuel pressure to exceed 414 kPa (60 psi), damage to the pressure regulator may result.

3. This test determines if the high fuel pressure is due to a restricted fuel return line or a pressure regulator problem.

1988–90 5.7L (VIN 8) – CORVETTE

NOTICE: *FUEL SYSTEM UNDER PRESSURE. TO AVOID FUEL SPILLAGE, REFER TO FIELD SERVICE PROCEDURES FOR TESTING OR MAKING REPAIRS REQUIRING DISASSEMBLY OF FUEL LINES OR FITTINGS.*

CHART A-7
(Page 2 of 2)
FUEL SYSTEM DIAGNOSIS
5.7L (VIN 8) "Y" CARLINE (PORT)

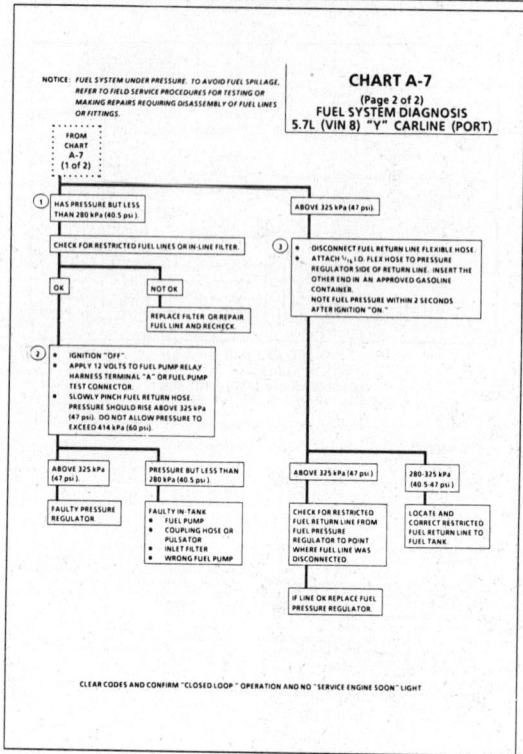

CLEAR CODES AND CONFIRM "CLOSED LOOP" OPERATION AND NO "SERVICE ENGINE SOON" LIGHT

1988–90 5.7L (VIN 8) – CORVETTE

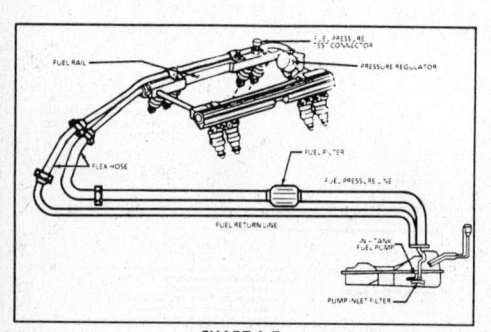

CHART A-7
(Page 1 of 2)
FUEL SYSTEM DIAGNOSIS
5.7L (VIN 8) "Y" CARLINE (PORT)

Circuit Description:

When the ignition switch is turned "ON," the electronic control module (ECM) will turn "ON" the in-tank fuel pump. It will remain "ON" as long as the engine is cranking or running, and the ECM is receiving reference pulses. If there are no reference pulses, the ECM will shut "OFF" the fuel pump within 2 seconds after ignition "ON" or engine stops.

The pump will deliver fuel to the fuel rail and injectors, then to the pressure regulator, where the system pressure is controlled to about 234 to 325 kPa (34 to 47 psi). Excess fuel is then returned to the fuel tank.

Test Description: Numbers below refer to circled numbers on the diagnostic chart.

1. Wrap a shop towel around the fuel pressure connector to absorb any small amount of fuel leakage that may occur when installing the gage. Ignition "ON" pump pressure should be 280 kPa (40.5-47 psi). This pressure is controlled by spring pressure within the regulator assembly.

2. When the engine is idling, the manifold pressure is low (high vacuum) and is applied to the fuel regulator diaphragm. This will offset the spring and result in a lower fuel pressure. This idle pressure will vary somewhat depending on barometric pressure, however, the pressure idling should be less indicating pressure regulator control.

3. Pressure that continues to fall is caused by one of the following:
   - In-tank pump check valve not holding.
   - Pump coupling hose or pulsator leaking.

- Fuel pressure regulator valve leaking.
- Injector(s) sticking open.

4. An injector sticking open can best be determined by checking for a fouled or saturated spark plug(s). If a leaking injector can not be determined by a fouled or saturated spark plug, the following procedure should be used.
   - Remove plenum, and remove fuel rail bolts. Follow the procedures in the fuel control section of this manual, but leave fuel lines connected.
   - Lift fuel rail out just enough to leave injector nozzles in the ports.

   CAUTION: Be sure injector(s) are not allowed to spray on engine and that injector retaining clips are intact. This should be carefully followed to prevent fuel spray on engine which would cause a fire hazard.

   - Pressurize the fuel system and observe injector nozzles.

1988–90 5.7L (VIN 8) – CORVETTE

CHART A-7
(Page 1 of 2)
FUEL SYSTEM DIAGNOSIS
5.7L (VIN 8) "Y" CARLINE (PORT)

THIS CHART ASSUMES THERE IS NO CODE 54

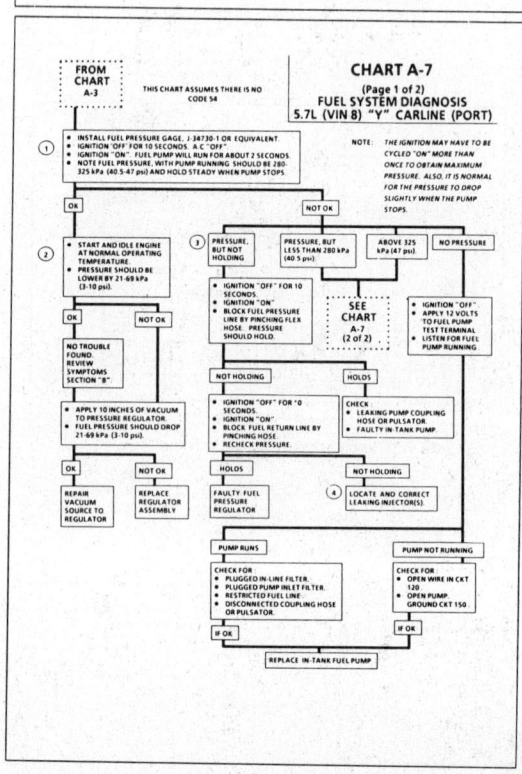

1988–90 5.7L (VIN 8) – CORVETTE

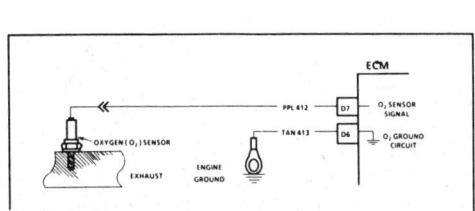

CODE 13
OXYGEN SENSOR CIRCUIT
(OPEN CIRCUIT)
5.7L (VIN 8) "Y" CARLINE (PORT)

Circuit Description:

The ECM supplies a voltage of about .45 volt between terminals "D7" and "D6". (If measured with a 10 megohm digital voltmeter, this may read as low as .32 volt). The O₂ sensor varies the voltage within a range of about 1 volt if the exhaust is rich, down through about .10 volt if exhaust is lean.

The sensor is like an open circuit and produces no voltage when it is below 360°C (600°F). An open sensor circuit or cold sensor causes "Open Loop" operation.

Test Description: Numbers below refer to circled numbers on the diagnostic chart.

1. Code 13 will set.
   - Engine at normal operating temperature (above 80°C/176°F)
   - At least 2 minutes engine time after start.
   - O₂ signal voltage steady between .35 and .55 volt.
   - Throttle position sensor signal above 5% (about .3 volt above closed throttle voltage).
   - All conditions must be met for about 60 seconds.
     If the conditions for a Code 13 exist, the system will not go "Closed Loop."
2. This will determine if the sensor is at fault or the wiring or ECM is the cause of the Code 13.
3. For this test use only a high impedence digital volt ohmmeter. This test checks the continuity of CKT 412 and CKT 413. If CKT 413 is open, the ECM voltage on CKT 412 will be over .6 volt (600 mV).

Diagnostic Aids:

Normal "Scan" voltage varies between 100 mV to 999 mV (.1 and 1.0 volt), while in "Closed Loop." Code 13 sets in one minute, if voltage remains between .35 and .55 volt, but the system will go "Open Loop" in about 15 seconds.

Refer to "Intermittents" in Section "B".

1988–90 5.7L (VIN 8) – CORVETTE

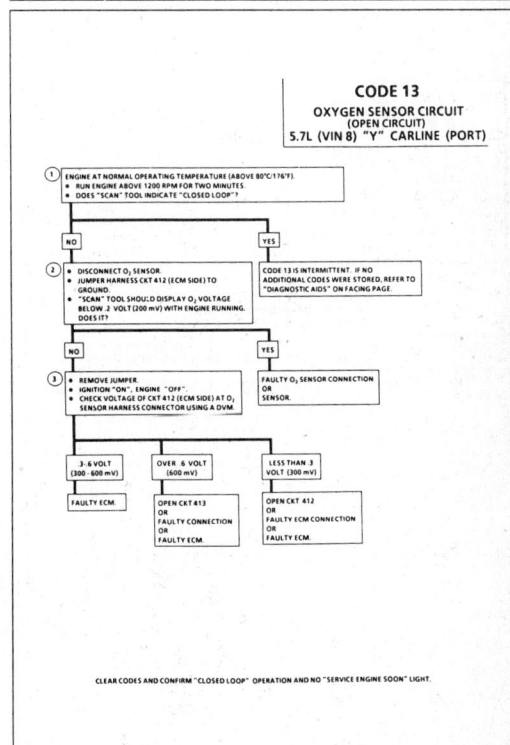

CODE 13
OXYGEN SENSOR CIRCUIT
(OPEN CIRCUIT)
5.7L (VIN 8) "Y" CARLINE (PORT)

1988–90 5.7L (VIN 8) – CORVETTE

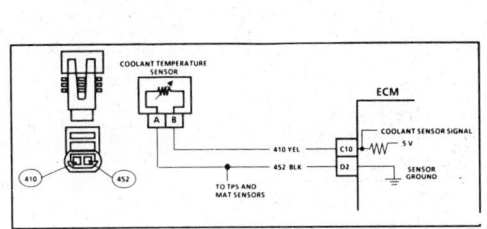

CODE 14
COOLANT TEMPERATURE SENSOR CIRCUIT
(HIGH TEMPERATURE INDICATED)
5.7L (VIN 8) "Y" CARLINE (PORT)

Circuit Description:

The coolant temperature sensor uses a thermistor to control the signal voltage to the ECM. The ECM applies a voltage on CKT 410 to the sensor. When the engine coolant is cold, the sensor (thermistor) resistance is high, therefore, the ECM will see high signal voltage.

As the engine coolant warms, the sensor resistance becomes less, and the voltage drops. At normal engine operating temperature (85°C to 95°C or 185°F to 203°F) the voltage will measure about 1.5 to 2.0 volts.

Test Description: Numbers below refer to circled numbers on the diagnostic chart.

1. Code 14 will set if.
   - Signal voltage indicates a coolant temperature above 130°C (266°F) for 3 seconds
2. This test will determine if CKT 410 is shorted to ground, which will cause the condition for Code 14.

Diagnostic Aids:

Check harness routing for a potential short to ground in CKT 410.

"Scan" tool displays engine temperature in degrees centigrade. After engine is started, the temperature should rise steadily to about 90°C (194°F), then stabilize when thermostat opens.

See "Intermittents" in Section "B".

1988–90 5.7L (VIN 8) – CORVETTE

CODE 14
COOLANT TEMPERATURE SENSOR CIRCUIT
(HIGH TEMPERATURE INDICATED)
5.7L (VIN 8) "Y" CARLINE (PORT)

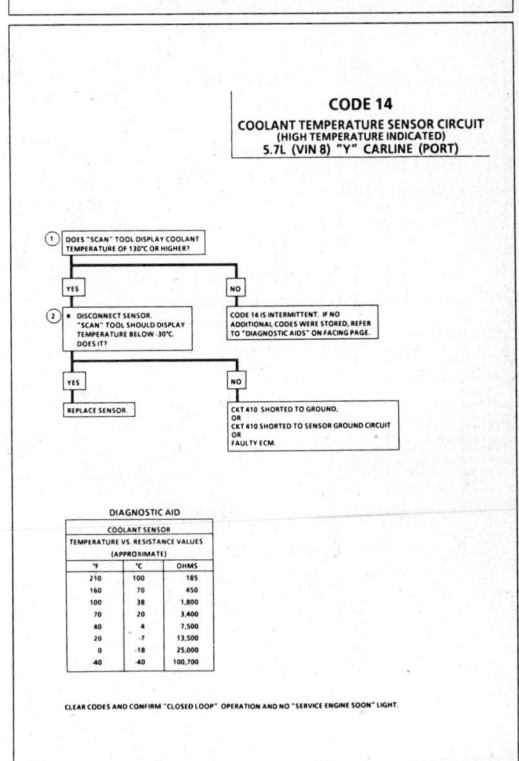

DIAGNOSTIC AID

COOLANT SENSOR TEMPERATURE VS. RESISTANCE VALUES (APPROXIMATE)		
°F	°C	OHMS
210	100	185
160	70	450
100	38	1,800
70	20	3,400
40	4	7,500
20	-7	13,500
0	-18	25,000
-40	-40	100,700

CLEAR CODES AND CONFIRM "CLOSED LOOP" OPERATION AND NO "SERVICE ENGINE SOON" LIGHT.

1988–90 5.7L (VIN 8) – CORVETTE

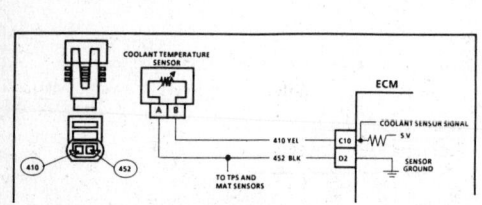

CODE 15
COOLANT TEMPERATURE SENSOR CIRCUIT
(LOW TEMPERATURE INDICATED)
5.7L (VIN 8) "Y" CARLINE (PORT)

Circuit Description:

The coolant temperature sensor uses a thermistor to control the signal voltage to the ECM. The ECM applies a voltage on CKT 410 to the sensor. When the engine coolant is cold, the sensor (thermistor) resistance is high, therefore the ECM will see high signal voltage.

As the engine coolant warms, the sensor resistance becomes less, and the voltage drops. At normal engine operating temperature (85°C to 95°C or 185°F to 203°F), the voltage will measure about 1.5 to 2.0 volts at the ECM.

Test Description: Numbers below refer to circled numbers on the diagnostic chart.
1. Code 15 will set if:
   - Engine coolant temperature less than -39°C (-38°F) for 3 seconds.
2. This test simulates a Code 14. If the ECM recognizes the low signal voltage, (high temperature) and the "Scan" reads 130°C (266°F) or above, the ECM and wiring are OK.
3. This test will determine if CKT 410 is open. There should be 5 volts present at sensor connector, if measured with a DVM.

Diagnostic Aids:

A "Scan" tool reads engine coolant temperature in degrees centigrade. After engine is started, the temperature should rise steadily to about 90°C (190°F), then stabilize when thermostat opens.

A faulty connection, or an open in CKT 410 or 452, will result in a Code 15.

If Code 21 or 23 is also set, check CKT 452 for faulty wiring or connections. Check terminals at sensor for good contact.

Refer to "Intermittents" in Section "B".

1988–90 5.7L (VIN 8) – CORVETTE

CODE 15
COOLANT TEMPERATURE SENSOR CIRCUIT
(LOW TEMPERATURE INDICATED)
5.7L (VIN 8) "Y" CARLINE (PORT)

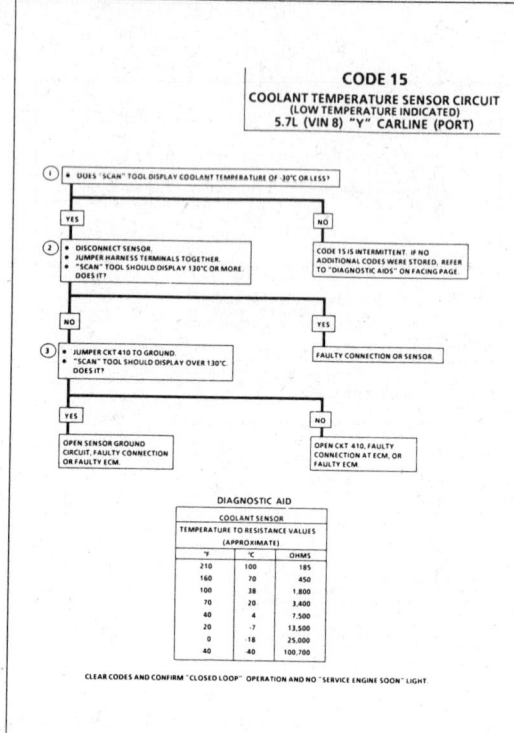

DIAGNOSTIC AID		
COOLANT SENSOR		
TEMPERATURE TO RESISTANCE VALUES		
(APPROXIMATE)		
°F	°C	OHMS
210	100	185
160	70	450
100	38	1,800
70	20	3,400
40	4	7,500
20	-7	13,500
0	-18	25,000
-40	-40	100,700

CLEAR CODES AND CONFIRM "CLOSED LOOP" OPERATION AND NO "SERVICE ENGINE SOON" LIGHT.

1988–90 5.7L (VIN 8) – CORVETTE

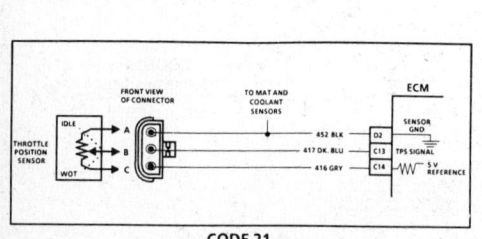

CODE 21
THROTTLE POSITION SENSOR (TPS) CIRCUIT
(SIGNAL VOLTAGE HIGH)
5.7L (VIN 8) "Y" CARLINE (PORT)

Circuit Description:

The Throttle Position Sensor (TPS) provides a voltage signal that changes, relative to the throttle blade. Signal voltage will vary from about .5 at idle to about 5 volts at wide open throttle.

The TPS signal is one of the most important inputs used by the ECM for fuel control and for most of the ECM controled outputs.

Test Description: Numbers below refer to circled numbers on the diagnostic chart.
1. Code 21 will set if:
   - TPS signal voltage is greater than 2.5 volts
   - Engine is running.
   - Mass air flow is less than 15 gm/sec
   - All conditions met for 3 seconds.
     OR
   - TPS signal voltage over 4.5 volts with ignition "ON."
   With throttle closed the TPS should read less than .62 volt. If it doesn't, check adjustment.
2. With the TPS sensor disconnected, the TPS voltage should go low, if the ECM and wiring are OK.
3. Probing CKT 452 with a test light checks the 5 volt return circuit, because a faulty 5 volts return will cause a Code 21.

Diagnostic Aids:

A "Scan" tool reads throttle position in volts. With ignition "ON" or at idle, voltage should read .54V ± .08V with throttle closed and increase at a steady rate as throttle is moved toward WOT.

An open in CKT 452 will result in a Code 21.

Refer to "Intermittents" in Section "B".

1988–90 5.7L (VIN 8) – CORVETTE

CODE 21
THROTTLE POSITION SENSOR (TPS) CIRCUIT
(SIGNAL VOLTAGE HIGH)
5.7L (VIN 8) "Y" CARLINE (PORT)

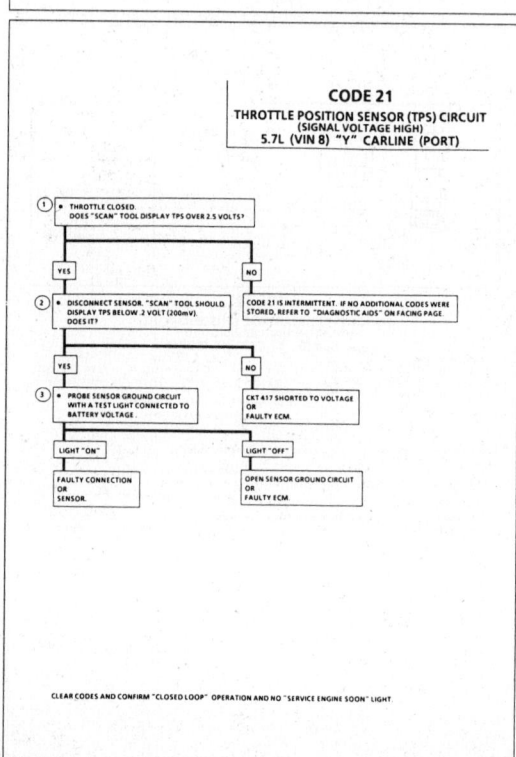

CLEAR CODES AND CONFIRM "CLOSED LOOP" OPERATION AND NO "SERVICE ENGINE SOON" LIGHT.

1988–90 5.7L (VIN 8) – CORVETTE

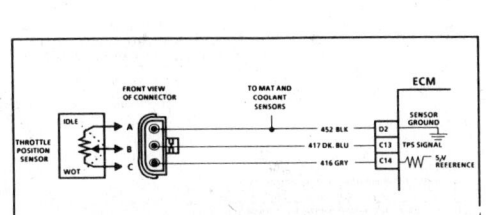

CODE 22
THROTTLE POSITION SENSOR (TPS) CIRCUIT
(SIGNAL VOLTAGE LOW)
5.7L (VIN 8) "Y" CARLINE (PORT)

Circuit Description:
The Throttle Position Sensor (TPS) provides a voltage signal that changes, relative to the throttle blade. Signal voltage will vary from about .5 at idle to about 5 volts at wide open throttle.
The TPS signal is one of the most important inputs used by the ECM for fuel control and for most of the ECM controlled outputs.

Test Description: Numbers below refer to circled numbers on the diagnostic chart.
1. Code 22, will set if:
   - Engine running.
   - TPS signal voltage is less than about .2 volt for 3 seconds.
2. Simulates Code 21 (high voltage) If the ECM recognizes the high signal voltage, the ECM and wiring are OK.
3. Adjust TPS. With throttle closed, the TPS voltage reading should be .54V ± .08V. If TPS cannot be adjusted to specification, the TPS should be replaced.
4. This simulates a high signal voltage to check for an open in CKT 417.

Diagnostic Aids:
A "Scan" tool reads throttle position in volts. With ignition "ON" or at idle, voltage should read .54V ± .08V with throttle closed and increase at a steady rate as throttle is moved toward WOT.
An open or short to ground in CKTs 416 or 417 will result in a Code 22.
Refer to "Intermittents" in Section "B".

1988–90 5.7L (VIN 8) – CORVETTE

CODE 22
THROTTLE POSITION SENSOR (TPS) CIRCUIT
(SIGNAL VOLTAGE LOW)
5.7L (VIN 8) "Y" CARLINE (PORT)

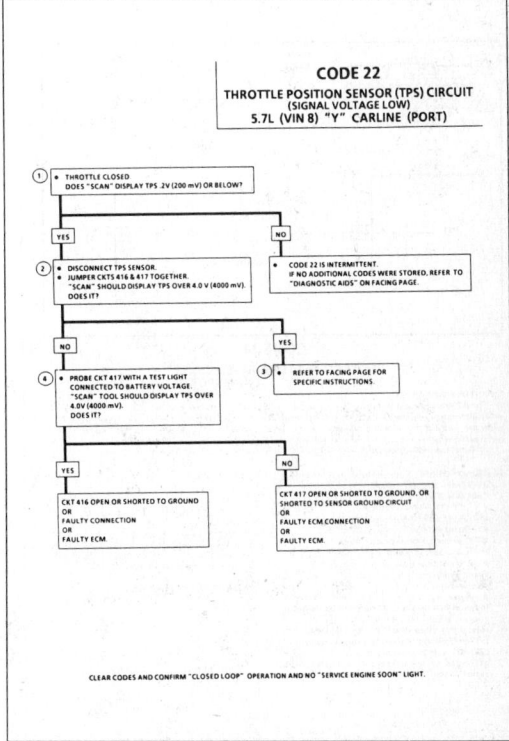

CLEAR CODES AND CONFIRM "CLOSED LOOP" OPERATION AND NO "SERVICE ENGINE SOON" LIGHT.

1988–90 5.7L (VIN 8) – CORVETTE

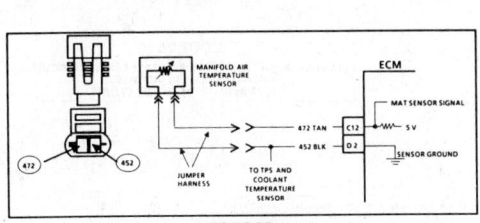

CODE 23
MANIFOLD AIR TEMPERATURE (MAT) SENSOR CIRCUIT
(LOW TEMPERATURE INDICATED)
5.7L (VIN 8) "Y" CARLINE (PORT)

Circuit Description:
The Manifold Air Temperature (MAT) sensor uses a thermistor to control the signal voltage to the ECM. The ECM applies a voltage (about 5 volts) on CKT 472 to the sensor. When the manifold air is cold, the sensor (thermistor) resistance is high, therefore, the ECM will see a high signal voltage. If the manifold air is warm, the sensor (thermistor) resistance is low, therefore the ECM will see a low signal voltage.

Test Description: Numbers below refer to circled numbers on the diagnostic chart.
1. Code 23 will set if:
   - A signal voltage indicates a manifold air temperature below −30°C (−22°F) for 12 seconds.
   - Time since engine start is 4 minutes or longer.
   - No VSS (vehicle not moving)
2. A Code 23 will set, due to an open sensor, wire or connection. This test will determine if the wiring and ECM are OK. The MAT sensor is difficult to reach and this test can be performed by disconnecting the MAT jumper harness connector. If the "Scan" indicates a temperature of over 130°C (266°F), the jumper harness to the sensor should be checked before replacing the sensor.
3. This will determine if the signal CKT 472 or the sensor ground (CKT 452) is open.

Diagnostic Aids:
A "Scan" tool reads temperature of the air entering the engine and should read close to ambient air temperature when engine is cold, and rises as underhood temperature increases.
Carefully check harness and connections for possible open CKT 472 or CKT 452.
Refer to "Intermittents" in Section "B".

1988–90 5.7L (VIN 8) – CORVETTE

CODE 23
MANIFOLD AIR TEMPERATURE (MAT) SENSOR CIRCUIT
(LOW TEMPERATURE INDICATED)
5.7L (VIN 8) "Y" CARLINE (PORT)

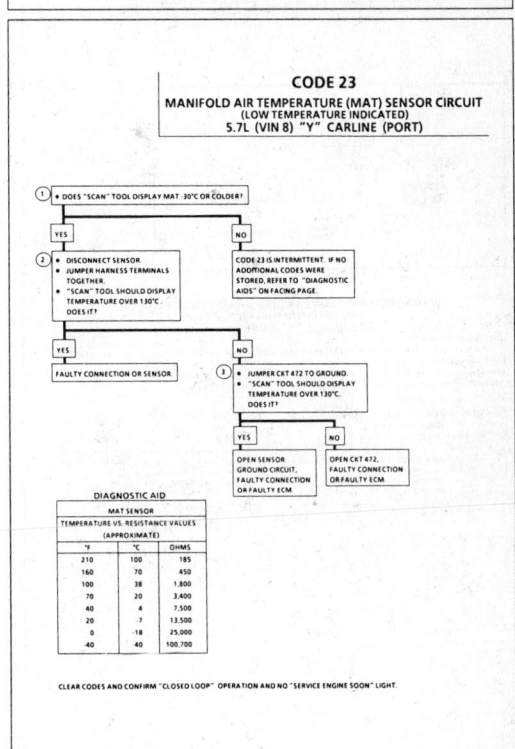

DIAGNOSTIC AID		
MAT SENSOR		
TEMPERATURE VS. RESISTANCE VALUES (APPROXIMATE)		
°F	°C	OHMS
210	100	185
160	70	450
100	38	1,800
70	20	3,400
40	4	7,500
20	-7	13,500
0	-18	25,000
-40	-40	100,700

CLEAR CODES AND CONFIRM "CLOSED LOOP" OPERATION AND NO "SERVICE ENGINE SOON" LIGHT.

1988–90 5.7L (VIN 8) – CORVETTE

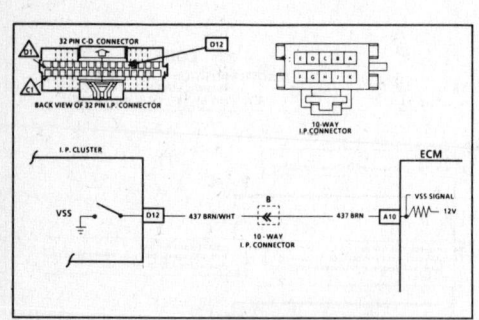

CODE 24
VEHICLE SPEED SENSOR (VSS) CIRCUIT
5.7L (VIN 8) "Y" CARLINE (PORT)

Circuit Description:
The ECM applies and monitors 12 volts on CKT 437. CKT 437 connects to the IP vehicle speed sensor, which alternately grounds CKT 437 when drive wheels are turning. This pulsing action takes place about 2000 times per mile and the ECM will calculate vehicle speed based on the time between "pulses." The IP cluster receives the vehicle speed signal from a permanent magnetic generator mounted in the transmission.

A "Scan" tool reading should closely match with speedometer reading with drive wheels turning.

Test Description: Numbers below refer to circled numbers on the diagnostic chart.
1. Code 24 will set if:
   - CKT 437 voltage is constant.
   - Engine speed between 900 and 4400 rpm.
   - Closed throttle.
   - Low load condition (low air flow).
   - Not in park or neutral.
   - All conditions must be met for 2 seconds. These conditions are met during a road load deceleration.
2. A voltage of less than 1 volt at the IP connector indicates that the CKT 437 wire may be shorted to ground. Disconnect CKT 437 at the IP cluster. If voltage now reads above 10 volts, the IP cluster is faulty. If voltage remains less than 10 volts, then CKT 437 wire is grounded or open. If 437 is not grounded or open, there is a faulty ECM connection or ECM.

Diagnostic Aids:
If "Scan" displays vehicle speed, check park/neutral switch CHART C-1A on vehicle with auto transmission. If switch is OK, check for intermittent connections. An open or short to ground in CKT 437 will result in a Code 24. If the customer also complained about a loss of mph on the IP, check the permanent magnetic generator circuit.

Refer to "Intermittents" in Section "B".

1988–90 5.7L (VIN 8) – CORVETTE

CODE 24
VEHICLE SPEED SENSOR (VSS) CIRCUIT
5.7L (VIN 8) "Y" CARLINE (PORT)

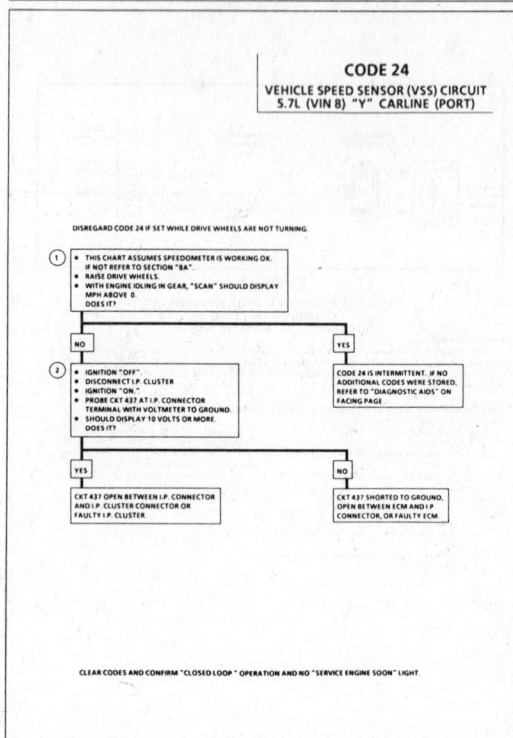

1988–90 5.7L (VIN 8) – CORVETTE

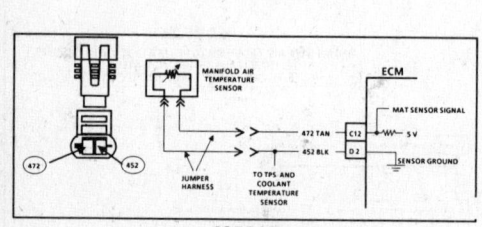

CODE 25
MANIFOLD AIR TEMPERATURE (MAT) SENSOR CIRCUIT
(HIGH TEMPERATURE INDICATED)
5.7L (VIN 8) "Y" CARLINE (PORT)

Circuit Description:
The Manifold Air Temperature (MAT) sensor uses a thermistor to control the signal-voltage to the ECM. The ECM applies a voltage (about 5 volts) on CKT 472 to the sensor. When manifold air is cold, the sensor (thermistor) resistance is high, therefore, the ECM will see a high signal voltage. If the manifold air is warm, the sensor (thermistor) resistance is low, therefore, the ECM will see a low signal voltage.

Test Description: Numbers below refer to circled numbers on the diagnostic chart.
1. Code 25 will set if:
   - Signal voltage indicates a manifold air temperature greater than 150°C (302°F) for 12 seconds.
   - Time since engine start is 4 minutes or longer.
   - A vehicle speed is present, greater than 5 mph.

Diagnostic Aids:
A "Scan" tool reads temperature of the air entering the engine and should read close to ambient air temperature, when engine is cold, and rise as underhood temperature increases.

Check harness routing for possible short to ground in CKT 472.

Refer to "Intermittents" in Section "B".

CODE 25
MANIFOLD AIR TEMPERATURE (MAT) SENSOR CIRCUIT
(HIGH TEMPERATURE INDICATED)
5.7L (VIN 8) "Y" CARLINE (PORT)

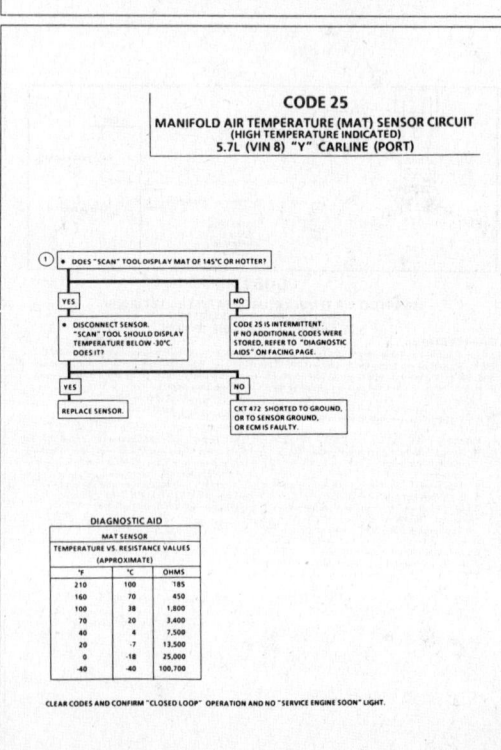

1988–90 5.7L (VIN 8) – CORVETTE

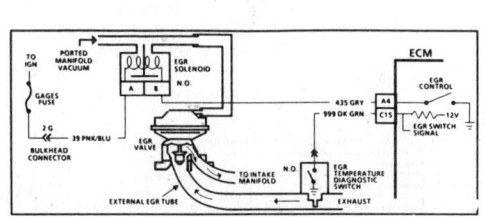

CODE 32
EXHAUST GAS RECIRCULATION (EGR) CIRCUIT
5.7L (VIN 8) "Y" CARLINE (PORT)

Circuit Description:

The EGR valve vacuum is controlled by an ECM controlled solenoid. The ECM will turn the EGR "ON" and "OFF" (Duty Cycle) by grounding CKT 435. The duty cycle is calculated by the ECM based on information from the coolant and mass air flow sensor and engine rpms. There should be (NO EGR) when in park or neutral, TPS input below a specified value, or TPS indicating wide open throttle (WOT).

With the ignition "ON," engine stopped, the EGR solenoid is de-energized and, by grounding the diagnostic "test" terminal, the solenoid is energized.

Test Description: Numbers below refer to circled numbers on the diagnostic chart.

Code 32 means that the EGR diagnostic switch was closed during start up or that the switch was not detected closed under the following conditions:

- Coolant temperature greater than 80°C (176°F).
- EGR duty cycle commanded by the ECM is greater than 75%.
- TPS less than ⅓ throttle, but not at idle.
- Codes 21, 22, 33, 34 not present.
- All conditions above must be met for about 4 minutes.

If the switch is detected closed during start-up, or if the switch is detected open when the above conditions are met, the "Service Engine Soon" light will remain "ON," unless the switch changes state.

1. This test will determine if the ECM set the code due to CKT 999 being grounded on start up. If the "Scan" does not indicate the switch is closed, but the customer complained of a "Service Engine Soon" light after start up, then this circuit should be checked carefully for an intermittent grounded condition.

2. If the "Scan" indicates the switch is no longer closed after disconnecting it, be sure the switch is not closed due to heat. (EGR being "ON" prior to test)

3. This test will check for a possible open in CKT 999. The ECM supplies 9-12 volts to CKT 999 and the "Scan" should indicate switch being closed when CKT 999 is grounded.

4. By grounding the diagnostic terminal, the EGR solenoid should close, and allow vacuum to be applied and the vacuum should hold.

5. This test will determine if the electrical control part of the system is at fault, or if the connector and solenoid are at fault.

6. By grounding the diagnostic "test" terminal, the solenoid valve should open and allow vacuum to bleed off through the vent.

7. With the engine not running and vacuum is applied to the valve, the valve should move to the fully open position.

8. Due to this engine using a negative back pressure valve, the valve should close when the engine is cranked over.

1988–90 5.7L (VIN 8) – CORVETTE

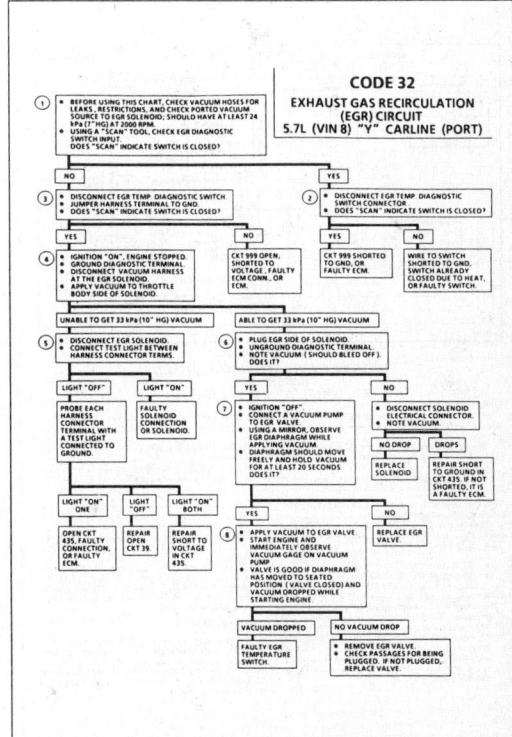

CODE 32
EXHAUST GAS RECIRCULATION
(EGR) CIRCUIT
5.7L (VIN 8) "Y" CARLINE (PORT)

1988–90 5.7L (VIN 8) – CORVETTE

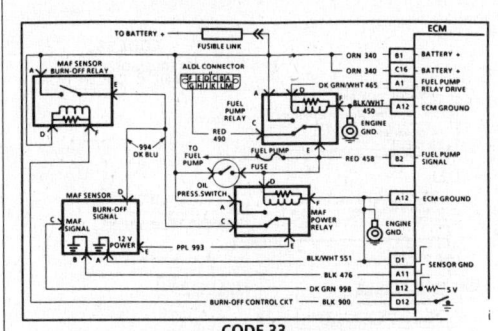

CODE 33
MASS AIR FLOW (MAF) SENSOR CIRCUIT
(GM/SEC HIGH)
5.7L (VIN 8) "Y" CARLINE (PORT)

Circuit Description:

The Mass Air Flow (MAF) sensor measures the amount of air which passes through it. The ECM uses this information to determine the operating condition of the engine, to control fuel delivery.

The oil pressure switch or the ECM, through control of the fuel pump relay, will provide 12 volts for the MAF power relay which provides the 12 volts needed by the MAF sensor.

The ECM provides a current limiting 5 volts on the signal line (CKT 998). The MAF sensor then changes the signal by dropping the voltage so that with low air flow the ECM sees a low voltage and a high air flow will cause the ECM to see near the 5 volts shown.

Test Description: Numbers below refer to circled numbers on the diagnostic chart.

Code 33 indicates: ECM has seen flow in excess of 45 grams per second (above about 2.2 volts) for one second when:
- Engine is first started.
OR
- TPS is less than ⅓ throttle.
- RPM is less than 2000.

Due to the 5 volt pull-up resistor in the ECM, if CKT 998 becomes open, the ECM will see a high voltage and set a Code 33.

1. This test will determine if the conditions to set the code still exist.

2. With the ALDL terminal "G" jumpered to 12 volts, there should be 12 volts at the sensor. If no voltage is present, make sure that the fuel pump is running. If not, repair fuel pump circuit.

3. If a burn off signal is present at the MAF sensor with the engine running, a Code 33 will set.

Be sure no voltage is present on CKT 994 during the first 25 seconds that the fuel pump runs or after the 2 second period.

4. The ECM sources a voltage (about 5 volts) to the MAF sensor on CKT 998. This test checks for that voltage.

Diagnostic Aids:

Intermittent: By jumpering the fuel pump test terminal "G" of ALDL connector to 12 volts, the MAF sensor will stay powered up and the signal line should see a low voltage, less than 250 mV or displayed as low grams per second on a "Scan" tool. By wiggling the related wiring the intermittent may be detected. Also, an erratic signal while the engine running may indicate faulty wiring or components. A burn-off signal to the MAF sensor, while running, may result in a Code 33. Be sure relay connections are clean and check CKT 994 for a potential short to voltage.

1988–90 5.7L (VIN 8) – CORVETTE

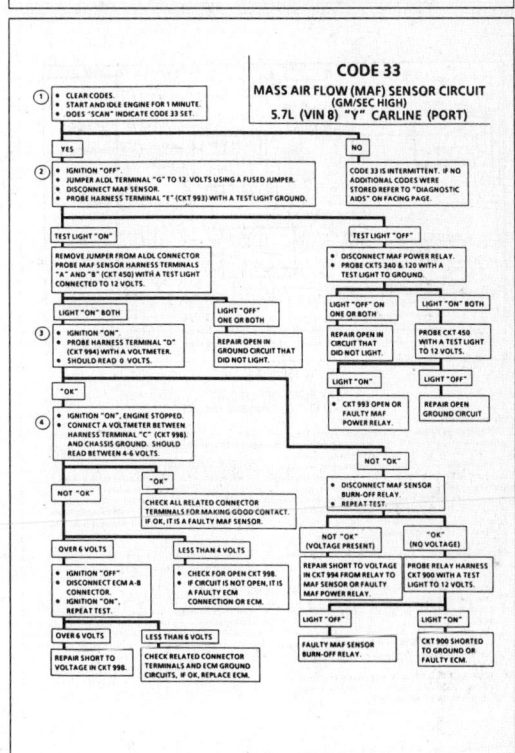

CODE 33
MASS AIR FLOW (MAF) SENSOR CIRCUIT
(GM/SEC HIGH)
5.7L (VIN 8) "Y" CARLINE (PORT)

1988–90 5.7L (VIN 8) – CORVETTE

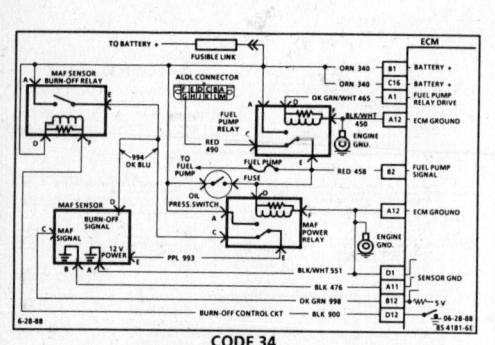

CODE 34
MASS AIR FLOW (MAF) SENSOR CIRCUIT
(GM/SEC LOW)
5.7L (VIN 8) "Y" CARLINE (PORT)

Circuit Description:

The Mass Air Flow (MAF) sensor measures the amount of air which passes through it. The ECM uses this information to determine the operating condition of the engine to control fuel delivery.

The oil pressure switch or the ECM, through control of the fuel pump relay, will provide 12 volts for the MAF power relay which provides the 12 volts needed by the MAF sensor.

The ECM provides a current limiting 5 volts on the signal line (CKT 998). The MAF sensor then changes the signal by dropping the voltage so that with low air flow the ECM sees a low voltage and a high air flow will cause the ECM to see near the 5 volts supply.

Test Description: Numbers below refer to circled numbers on the diagnostic chart.

Code 34 indicates: ECM has seen low air flow less than 2.5 gm/sec (low voltage) for one second when:
- Engine is first started
  OR
- RPM above 600
- TPS above 6%. To obtain 6%, the engine has to be running at about 2300 rpm in neutral.

1. A Code 34 may be caused by an engine that exhibits a low, rough, unstable or incorrect idle problem. If this condition exists, disconnect the MAF sensor. If the unstable idle still exists, see "Symptoms" in Section "B" (Rough, Unstable, Incorrect Idle, or Stalling). If the idle improved with the sensor disconnected, replace it. Code 34 could also result from a dirty or misadjusted throttle body.

2. This test will determine if the conditions still exist to set a code or if the problem is intermittent. With the MAF sensor disconnected, the ECM should see a high signal voltage and set a Code 33. If a Code 34 resets, then the wiring or the ECM is at fault.

3. With engine running and MAF sensor disconnected probe harness terminal "E" (CKT 993) with a test light to ground. Light should be bright which indicates proper voltage (12 volts) available to MAF sensor.

Diagnostic Aids:

A low, rough, or unstable idle could result in a Code 34. Air ducts are light and not cracked, throughly inspect the induction system for vacuum leaks. Check CKT 998 for a potential short to ground. Code 34 could also result from a dirty or misadjusted throttle body.

1988–90 5.7L (VIN 8) – CORVETTE

CODE 34
MASS AIR FLOW (MAF) SENSOR CIRCUIT
(GM/SEC LOW)
5.7L (VIN 8) "Y" CARLINE (PORT)

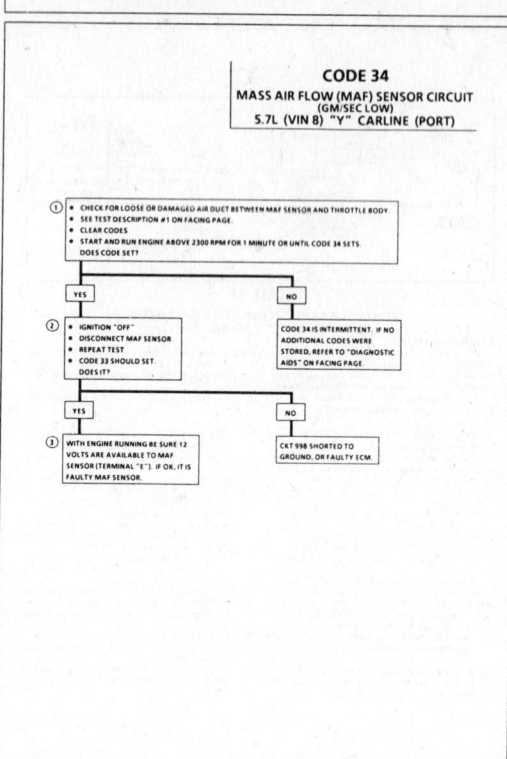

1988–90 5.7L (VIN 8) – CORVETTE

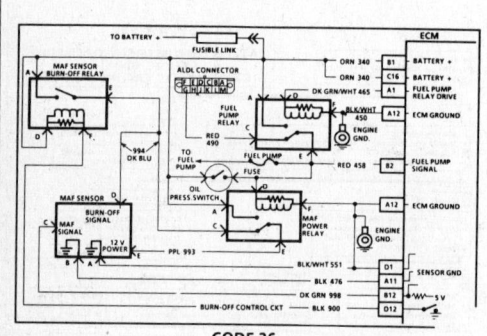

CODE 36
MAF BURN-OFF CIRCUIT
5.7L (VIN 8) "Y" CARLINE (PORT)

Circuit Description:

The Mass Air Flow (MAF) sensor measures the amount of air which passes through it. The ECM uses this information to determine the operating condition of the engine to control fuel delivery.

Due to contaminates in the atmosphere, a residue may build up on the MAF sensor sensing wire. To maintain an accurate reading from the sensor, a burn-off cycle will occur, when the ignition is turned "OFF," after the engine had been running a specified amount of time and engine warmed up. The burn-off function is enabled when the ECM grounds CKT 900, which energizes the MAF sensor burn-off relay. With the MAF sensor burn-off relay energized, voltage will be supplied to the MAF sensor terminal "D". Voltage will also be supplied through the normally closed set of contacts in the MAF power relay, which will supply 12 volts to terminal "E" of the MAF sensor.

Test Description: Numbers below refer to circled numbers on the diagnostic chart.

1. This test will determine if the burn-off function is operative or if the code was set due to an intermittent condition.

2. Check for continuous 12 volt supply to burn-off relay.

3. Grounding CKT 900 should energize the relay and close the contacts. If the test light is dim, check for corroded or faulty connections. If OK replace relay.

4. With the burn-off relay energized there should be 12 volts supplied to the MAF sensor on terminal "D" & "E" (CKTs 993 and 994).

If the test light is dim, check for corroded or faulty connections. If OK replace relay.

Diagnostic Aids:

The Code 36 could have been set due to a poor connection at any of the relays or the MAF sensor. Be sure that these connections and terminals are OK. A faulty MAF sensor should not be considered as the cause if Code 36 is set.

Refer to "Intermittents" in Section "B".

1988–90 5.7L (VIN 8) – CORVETTE

CODE 36
MAF BURN-OFF CIRCUIT
5.7L (VIN 8) "Y" CARLINE (PORT)

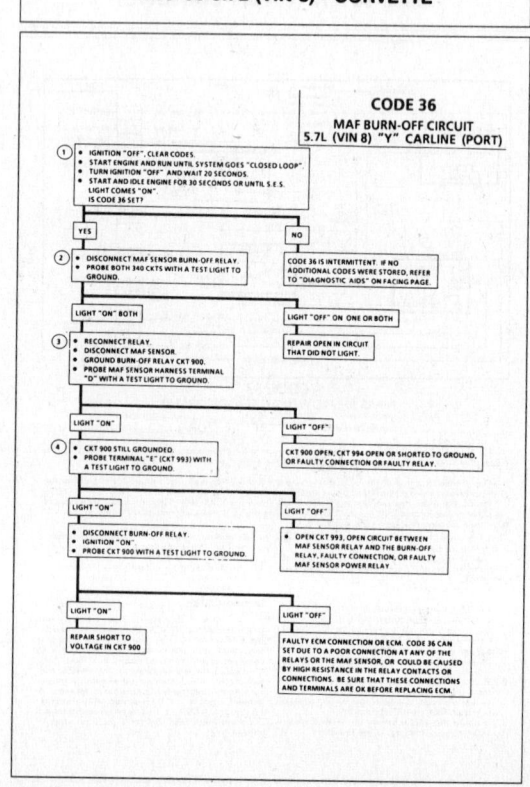

1988–90 5.7L (VIN 8) – CORVETTE

CODE 41
CYLINDER SELECT ERROR
(FAULTY OR INCORRECT MEM-CAL)
5.7L (VIN 8) "Y" CARLINE (PORT)

Test Description: Numbers below refer to circled numbers on the diagnostic chart.
1. The ECM used for this engine can also be used for other engines, and the difference is in the Mem-Cal. If a Code 41 sets, the incorrect Mem-Cal has been installed, or may not be installed properly, or is faulty and it must be replaced.

Diagnostic Aids:

Check Mem-Cal to be sure locking tabs are secure. Also check the pins on both the Mem-Cal and ECM to assure they are making proper contact. Check the Mem-Cal part number to assure it is the correct part. If the Mem-Cal is faulty is must be replaced. It is also possible that the ECM is faulty, however it should not be replaced until all of the above have been checked. For additional information, refer to "Intermittents" in Section "B".

1988–90 5.7L (VIN 8) – CORVETTE

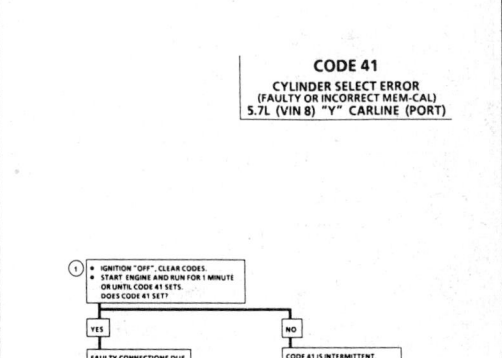

CODE 41
CYLINDER SELECT ERROR
(FAULTY OR INCORRECT MEM-CAL)
5.7L (VIN 8) "Y" CARLINE (PORT)

① • IGNITION "OFF", CLEAR CODES.
• START ENGINE AND RUN FOR 1 MINUTE OR UNTIL CODE 41 SETS.
DOES CODE 41 SET?

YES → FAULTY CONNECTIONS DUE TO MEM-CAL NOT LOCKED IN PLACE, OR INCORRECT MEM-CAL INSTALLED.

NO → CODE 41 IS INTERMITTENT. REVIEW "DIAGNOSTIC AIDS" ON FACING PAGE.

1988–90 5.7L (VIN 8) – CORVETTE

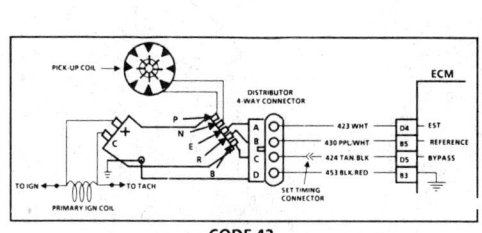

CODE 42
ELECTRONIC SPARK TIMING (EST) CIRCUIT
5.7L (VIN 8) "Y" CARLINE (PORT)

Circuit Description:
When the system is running on the ignition module, that is, no voltage on the bypass line, the ignition module grounds the EST signal. The ECM expects to see no voltage on the EST line during this condition. If it sees a voltage, it sets Code 42 and will not go into the EST mode.

When the rpm for EST is reached (about 400 rpm), and bypass voltage applied, the EST should no longer be grounded in the ignition module so the EST voltage should be varying.

If the bypass line is open or grounded, the ignition module will not switch to EST mode, so the EST voltage will be low and Code 42 will be set.

If the EST line is grounded, the ignition module will switch to EST, but because the line is grounded there will be no EST signal. A Code 42 will be set.

Test Description: Numbers below refer to circled numbers on the diagnostic chart.
1. Code 42 means the ECM has seen an open or short to ground in the EST or bypass circuits. This is a confirm Code 42 and that the fault causing the code is present.
2. Checks for a normal EST ground path through the ignition module. An EST CKT 423 shorted to ground will also read less than 1000 ohms, however, this will be checked later.
3. As the test light voltage reaches CKT 424, the module should switch. The ohmmeter may "overrange" if the meter is on the 1000-2000 ohms position. The important thing is that the module "switched."

4. The module did not switch and this step checks for
• EST CKT 423 shorted to ground.
• Bypass CKT 424 open.
• Faulty ignition module connection or module
5. Confirms that Code 42 is a faulty ECM and not an intermittent in CKTs 423 or 424.

Diagnostic Aids:

The "Scan" tool does not have any ability to help diagnose a Code 42 problem.

A Mem-Cal not fully seated in the ECM can result in a Code 42.

Refer to "Intermittents" in Section "B".

1988–90 5.7L (VIN 8) – CORVETTE

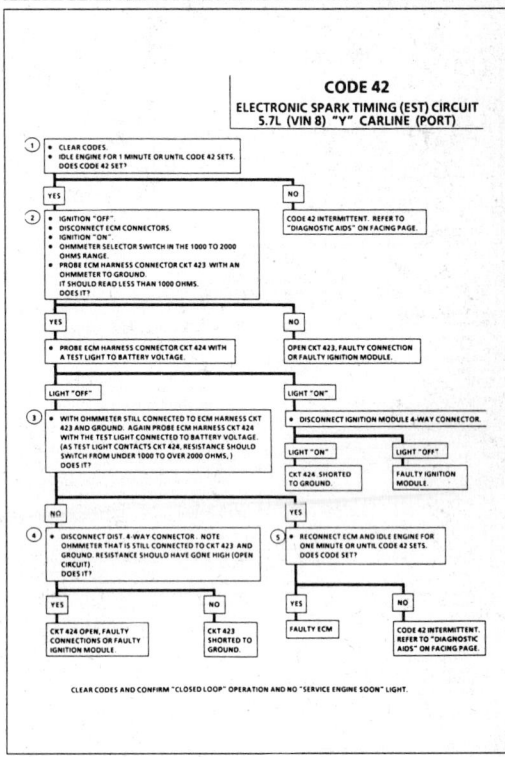

CODE 42
ELECTRONIC SPARK TIMING (EST) CIRCUIT
5.7L (VIN 8) "Y" CARLINE (PORT)

CLEAR CODES AND CONFIRM "CLOSED LOOP" OPERATION AND NO "SERVICE ENGINE SOON" LIGHT.

1988–90 5.7L (VIN 8) – CORVETTE

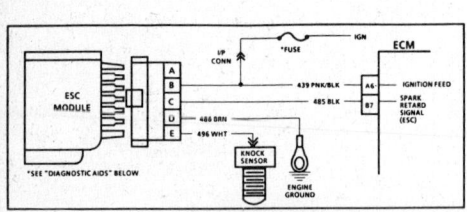

CODE 43
ELECTRONIC SPARK CONTROL (ESC) CIRCUIT
5.7L (VIN 8) "Y" CARLINE (PORT)

Circuit Description:

Electronic spark control is accomplished with a module that sends a voltage signal to the ECM. As the knock sensor detects engine knock, the voltage from the ESC module to the ECM drops, and this signals the ECM to retard timing. The ECM will retard the timing, when knock is detected and rpm is above about 900 rpm.

Code 43 means the ECM has read low voltage or CKT 485 for longer than 5 seconds with the engine running or the system has failed the functional check.

This system performs a functional check once per start up to check the ESC system. To perform this test, the ECM will advance the spark when coolant is above 95°C (194°F) and at a high load condition (near WOT). The ECM then checks the signal on CKT 485 to see if a knock is detected. The functional check is performed once per start up. If knock is detected when coolant is below 95°C (194°F), the test has passed and the functional check will not be run. If the functional check fails, the "Service Engine Soon" light will remain "ON" until ignition is turned "OFF," or until a knock signal is detected.

Test Description: Numbers below refer to circled numbers on the diagnostic chart.
1. If the conditions for a Code 43 are present, the "Scan" will always display "YES." There should not be a knock at idle unless an internal engine problem, or a system problem, exists.
2. This test will determine if. the system is functioning at this time. Usually, a knock signal can be generated by tapping on the right exhaust manifold. If no knock signal is generated, try tapping on block close to the area of the sensor.
3. Because Code 43 sets when the signal voltage on CKT 485 remains low, the test should cause the signal on CKT 485 to go too high. The 12 volts signal should be seen by the ECM as "no knock," if the ECM and wiring are OK.
4. This test will determine if the knock signal is being detected on CKT 496, or if the ESC module is at fault.

5. If CKT 496 is routed to close to secondary ignition wires, the ESC module may see the interference as a knock signal.
6. This checks the ground circuit to the module. An open ground will cause the voltage on CKT 485 to be about 12 volts, which would cause the Code 43 functional test to fail.
7. Contacting CKT 496 with a test light to 12 volts should generate a knock signal. This will determine if the ESC module is operating correctly.

Diagnostic Aids:

* ECM/ENG fuse
  Code 43 can be caused by a faulty connection at the knock sensor at the ESC module or at the ECM. Also, check CKT 485 for possible open or short to ground.
  Refer to "Intermittents" in Section "B".

1988–90 5.7L (VIN 8) – CORVETTE

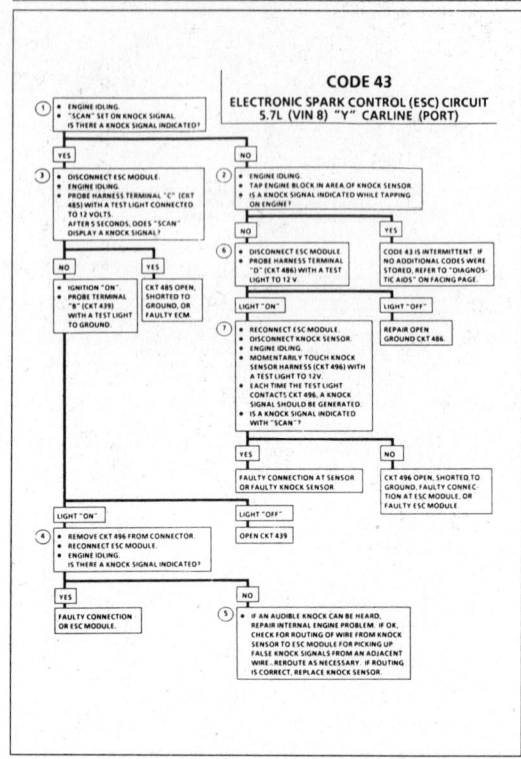

CODE 43
ELECTRONIC SPARK CONTROL (ESC) CIRCUIT
5.7L (VIN 8) "Y" CARLINE (PORT)

1988–90 5.7L (VIN 8) – CORVETTE

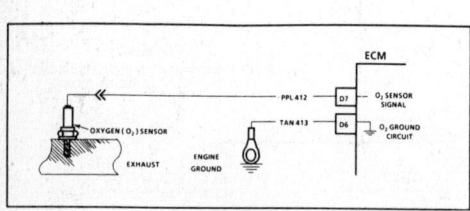

CODE 44
OXYGEN SENSOR CIRCUIT
(LEAN EXHAUST INDICATED)
5.7L (VIN 8) "Y" CARLINE (PORT)

Circuit Description:

The ECM supplies a voltage of about .45 volt between terminals "D6" and "D7". (If measured with a 10 megohm digital voltmeter, this may read as low as .32 volt.) The O_2 sensor varies the voltage within a range of about 1 volt, if the exhaust is rich, down through about .10 volt, if exhaust is lean.

The sensor is like an open circuit and produces no voltage when it is below about 360°C (600°F). An open sensor circuit or cold sensor causes "Open Loop" operation.

Test Description: Numbers below refer to circled numbers on the diagnostic chart.
1. Code 44 is set when the O_2 sensor signal voltage on CKT 412:
   * Remains below .2 volt for 50 seconds.
   * And the system is operating in "Closed Loop."

Diagnostic Aids:

Using the "Scan," observe the block learn values at different rpm and air flow conditions. The "Scan," also, displays the block cells, so the block learn values can be checked in each of the cells to determine when the Code 44 may have been set. If the conditions for Code 44 exist, the block learn value will be around 150

* O_2 Sensor Wire: Sensor pigtail may be mispositioned and contacting the exhaust manifold.
* Check for intermittent ground in wire between connector and sensor.
* MAF Sensor: A mass air flow (MAF) sensor output that causes the ECM to sense a lower than normal air flow will cause the system to go lean. Disconnect the MAF sensor and, if the lean condition is gone, replace the MAF sensor.

* Lean Injector(s): Perform injector balance test CHART C-2A.
* Fuel Contamination: Water, even in small amounts, near the in tank fuel pump inlet can be delivered to the injectors. The water causes a lean exhaust and can set a Code 44.
* Fuel Pressure: System will be lean if pressure is too low. It may be necessary to monitor fuel pressure while driving the car at various road speeds and/or loads to confirm. See Fuel System diagnosis CHART A-7.
* Exhaust Leaks: If there is an exhaust leak, the engine can cause outside air to be pulled into the exhaust and past the sensor. Vacuum or crankcase leaks can cause a lean condition.
* AIR System: Be sure air is not being directed to the exhaust ports while in "Closed Loop." If the block learn value goes down, while squeezing air hose to left side exhaust ports, refer to CHART C-6
* If the above is OK, it is a faulty oxygen sensor.

1988–90 5.7L (VIN 8) – CORVETTE

CODE 44
OXYGEN SENSOR CIRCUIT
(LEAN EXHAUST INDICATED)
5.7L (VIN 8) "Y" CARLINE (PORT)

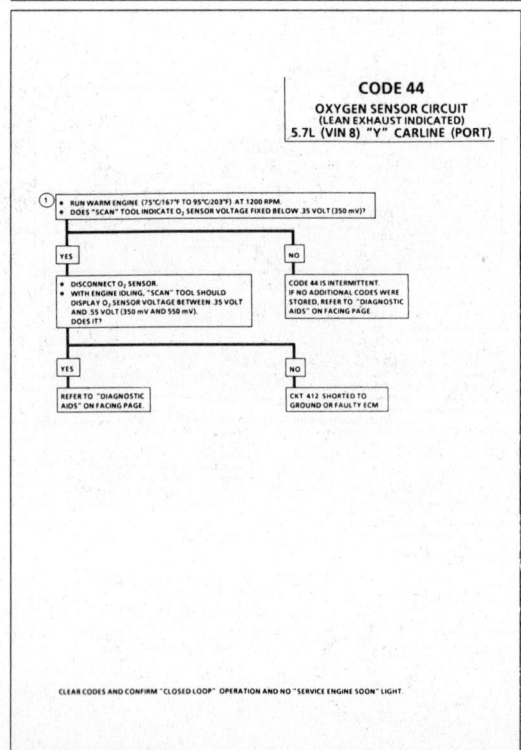

CLEAR CODES AND CONFIRM "CLOSED LOOP" OPERATION AND NO "SERVICE ENGINE SOON" LIGHT.

1988–90 5.7L (VIN 8) – CORVETTE

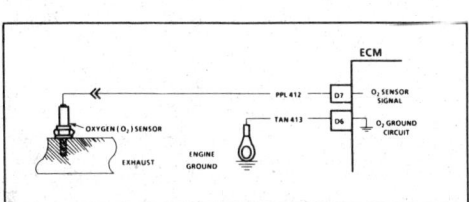

CODE 45
OXYGEN SENSOR CIRCUIT
(RICH EXHAUST INDICATED)
5.7L (VIN 8) "Y" CARLINE (PORT)

Circuit Description:
The ECM supplies a voltage of about .45 volt between terminals "D6" and "D7". (If measured with a 10 megohm digital voltmeter, this may read as low as .32 volt). The O₂ sensor varies the voltage within a range of about 1 volt, if the exhaust is rich, down through about .10 volt, if exhaust is lean.
The sensor is like an open circuit and produces no voltage when it is below about 360°C (600°F). An open sensor circuit, or cold sensor, causes "Open Loop" operation.

Test Description: Numbers below refer to circled numbers on the diagnostic chart.
1 Code 45 is set when the O₂ sensor signal voltage or CKT 412:
   - Remains above .7 volt for 50 seconds, and in "Closed Loop."
   - Engine time after start is 1 minute or more
   - Throttle angle greater than 2% (about 2 volts above idle voltage).

Diagnostic Aids:

Using the "Scan," observe the block learn values at different rpm and air flow conditions. The "Scan" also displays the block learn values, so the block learn values can be checked in each of the cells to determine when the Code 45 may have been set. If the conditions for Code 45 exist, the block learn values will be around 115.
   - Fuel Pressure: System will go rich if pressure is too high. The ECM can compensate for some increase. However, if it gets too high, a Code 45 may be set. See Fuel System Diagnosis, CHART A-7.
   - Rich Injector: Perform injector balance test, CHART C-2A.
   - Leaking Injector: See CHART A-7.

   - Check for fuel contaminated oil.
   - HEI Shielding: An open ground CKT 453 (ignition system reflow) may result in EMI, or induced electrical "noise." The ECM looks at this "noise" as reference pulses. The additional pulses result in a higher than actual engine speed signal. The ECM then delivers too much fuel, causing system to go rich. Engine tachometer will also show higher than actual engine speed, which can help in diagnosing this problem.
   - Canister Purge: Check for fuel saturation. If full of fuel, check canister control and hoses.
   - MAF Sensor: An output that causes the ECM to sense a higher than normal airflow can cause the system to go rich. Disconnecting the MAF sensor will allow the ECM to set a fixed value for the sensor. Substitute a different MAF sensor, if the the rich condition is gone while the sensor is disconnected.
   - Check for leaking fuel pressure regulator diaphram by checking vacuum line to regulator for fuel.
   - TPS: An intermittent TPS output will cause the system to go rich, due to a false indication of the engine accelerating.

1988–90 5.7L (VIN 8) – CORVETTE

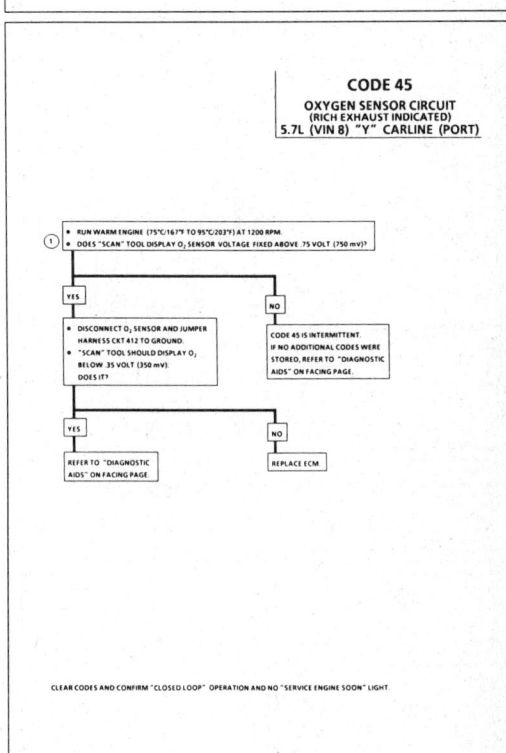

CLEAR CODES AND CONFIRM "CLOSED LOOP" OPERATION AND NO "SERVICE ENGINE SOON" LIGHT.

1988–90 5.7L (VIN 8) – CORVETTE

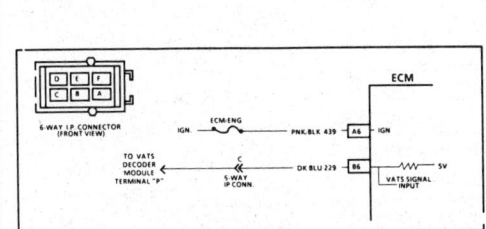

CODE 46
VEHICLE ANTI-THEFT SYSTEM (VATS) CIRCUIT
5.7L (VIN 8) "Y" CARLINE (PORT)

Circuit Description:
The VATS system is designed to disable vehicle operation if the incorrect key or starting procedure is used. The VATS decoder module sends a signal to the ECM if the correct key is being used. If the proper signal does not reach the ECM on CKT 229, the ECM will not pulse the injectors "ON" and thus not allow the vehicle to be started.

Test Description: Numbers below refer to circled numbers on the diagnostic chart.
1 If the engine cranks but doesn't start, it indicates that the portion of the module which generates the signal to the ECM is not operating or the 229 CKT is open or shorted to ground. If the decoder module is found to be OK, the ECM may be at fault, but this is not a likely condition.

2 If the engine will not crank, it indicates that there is a VATS problem or an incorrect key or starting procedure is being used.

1988–90 5.7L (VIN 8) – CORVETTE

CODE 46
VEHICLE ANTI-THEFT SYSTEM (VATS) CIRCUIT
5.7L (VIN 8) "Y" CARLINE (PORT)

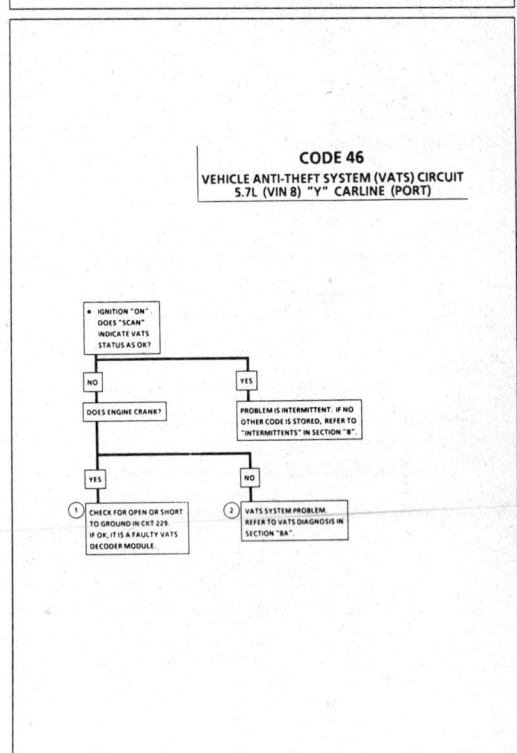

1988–90 5.7L (VIN 8) – CORVETTE

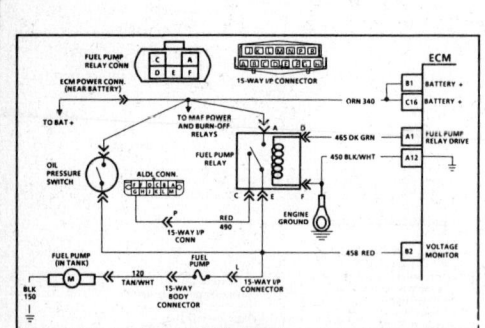

CODE 54
FUEL PUMP CIRCUIT
(LOW VOLTAGE)
5.7L (VIN 8) "Y" CARLINE (PORT)

Circuit Description:

The status of the fuel pump CKT 120 is monitored for voltage by the ECM, and is used to compensate fuel delivery based on system voltage. This fuel pump signal is also used to store a trouble code, if the fuel pump relay is defective or fuel pump voltage is lost while the engine is running. There should be about 12 volts on CKT 120 for 2 seconds after the ignition is turned "ON" or any time reference pulses are being received by the ECM.

Code 54 will set, if the voltage on CKT 120 is less than 2 volts for 1.5 seconds since the last reference pulse was received. This code is designed to detect a faulty relay, causing extended crank time, and the code will help the diagnosis of an engine that "CRANKS BUT WILL NOT RUN."

If a fault is detected during start-up the "Service Engine Soon" light will stay "ON" until the ignition is cycled "OFF." However, if the voltage is detected below 2 volts, with the engine running, the light will only remain "ON" while the condition exists.

The fuel pump test connector is located in terminal "G" of the ALDL connector.

1988–90 5.7L (VIN 8) – CORVETTE

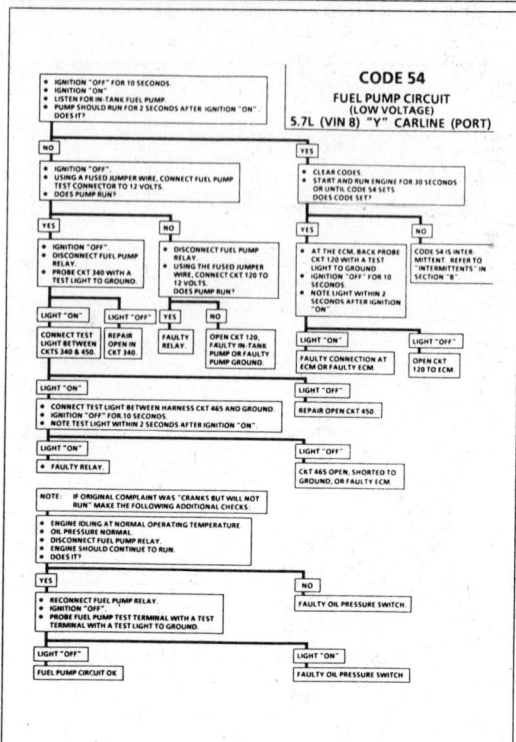

CODE 54
FUEL PUMP CIRCUIT
(LOW VOLTAGE)
5.7L (VIN 8) "Y" CARLINE (PORT)

1988–90 5.7L (VIN 8) – CORVETTE

CODE 51
CODE 52
CODE 53
5.7L (VIN 8) "Y" CARLINE (PORT)

CODE 51
MEM-CAL ERROR
(FAULTY OR INCORRECT MEM-CAL)

CHECK THAT ALL PINS ARE FULLY INSERTED IN THE SOCKET AND THAT MEM-CAL IS PROPERLY LATCHED. IF OK, REPLACE MEM-CAL, CLEAR MEMORY, AND RECHECK. IF CODE 51 REAPPEARS, REPLACE ECM.

CLEAR CODES AND CONFIRM "CLOSED LOOP" OPERATION AND NO "SERVICE ENGINE SOON" LIGHT.

CODE 52
CALPAK ERROR
(FAULTY OR INCORRECT CALPAK)

CHECK THAT THE MEM-CAL IS FULLY SEATED AND LATCHED INTO THE MEM-CAL SOCKET. IF OK, REPLACE MEM-CAL, CLEAR MEMORY, AND RECHECK. IF CODE 52 REAPPEARS, REPLACE ECM.

CLEAR CODES AND CONFIRM "CLOSED LOOP" OPERATION AND NO "SERVICE ENGINE SOON" LIGHT.

CODE 53
SYSTEM OVER VOLTAGE

THIS CODE INDICATES THERE IS A BASIC GENERATOR PROBLEM.
• CODE 53 WILL SET, IF VOLTAGE AT ECM IGNITION INPUT PIN IS GREATER THAN 17.1 VOLTS FOR 2 SECONDS.
• CHECK AND REPAIR CHARGING SYSTEM.

CLEAR CODES AND CONFIRM "CLOSED LOOP" OPERATION AND NO "SERVICE ENGINE SOON" LIGHT.

1988–90 5.7L (VIN 8) – CORVETTE

SECTION B
SYMPTOMS
TABLE OF CONTENTS

Before Starting

Intermittents

Hard Start

Surges and/or Chuggle

Lack of Power, Sluggish, or Spongy

Detonation/Spark Knock

Hesitation, Sag, Stumble

Cuts Out, Misses

Poor Fuel Economy

Rough, Unstable or Incorrect Idle, Stalling

Excessive Exhaust Emissions or Odors

Dieseling, Run-On

Backfire

Restricted Exhaust System Check (CHART B-1) (All Engines)

BEFORE STARTING

Before using this section you should have performed the "Diagnostic Circuit Check" and found out that:
1. The ECM and "Service Engine Soon" light are operating.
2. There are no trouble codes stored, or there is a trouble code but no "Service Engine Soon" light.

Verify the customer complaint, and locate the correct SYMPTOM below. Check the items indicated under that symptom.

If the ENGINE CRANKS BUT WILL NOT RUN, see CHART A-3.

Several of the symptom procedures below call for a careful visual physical check. This check should include:

• ECM grounds for being clean and tight
• Vacuum hoses for splits, kinks, and proper connections, as shown on Vehicle Emission Control Information label.
• Air leaks at throttle body mounting and intake manifold
• Air leaks between MAF sensor and throttle body
• Ignition wires for cracking, hardness, proper routing, and carbon tracking
• Wiring for proper connections, pinches, and cuts

The importance of this step cannot be stressed too strongly - it can lead to correcting a problem without further checks and can save valuable time.

1988-90 5.7L (VIN 8) – CORVETTE

INTERMITTENTS

Definition: Problem may or may not turn "ON" the "Service Engine Soon" light, or store a code.

DO NOT use the Trouble Code Charts in Section "A" for intermittent problems. The fault must be present to locate the problem. If a fault is intermittent, use of Trouble Code Charts may result in replacement of good parts.

- Most intermittent problems are caused by faulty electrical connections or wiring. Perform careful check as described at start of Section "B". Check for:
  - Poor mating of the connector halves, or terminals not fully seated in the connector body (backed out).
  - Improperly formed or damaged terminals. All connector terminals in problem circuit should be carefully reformed or replaced to insure proper contact tension.
  - Poor terminal to wire connection. This requires removing the terminal from the connector body to check.
- If a visual physical check does not find the cause of the problem, the car can be driven with a voltmeter connected to a suspected circuit. A "Scan" tool can also be used for monitoring input signals to the ECM to help detect intermittent conditions. An abnormal voltage, or "Scan" reading, when the problem occurs, indicates the problem may be in that circuit. If the wiring and connectors check OK and a Trouble Code was stored for a circuit having a sensor, except for Codes 43, 44, and 45, substitute a known good sensor and recheck

- Loss of trouble code memory. To check, disconnect TPS and idle engine until "Service Engine Soon" light comes "ON". Code 22 should be stored, and kept in memory when ignition is turned "OFF". If not, the ECM is faulty. An intermittent "Service Engine Soon" light with no stored code may be caused by:
  - Ignition coil shorted to ground and arcing at spark plug wires or plugs.
  - "Service Engine Soon" light wire to ECM shorted to ground. (CKT 419)
  - Diagnostic "test" terminal wire to ECM, shorted to ground. (CKT 451)
  - ECM power grounds. See ECM wiring diagrams.
- Check for an electrical system interference caused by a defective relay, ECM driven solenoid, or switch. They can cause a sharp electrical surge. Normally, the problem will occur when the faulty component is operated.
- EST wires should be routed away from spark plug wires, distributor wires, distributor housing, coil, and generator. Wire from ECM to distributor (CKT 453) should be a good connection.
- Check for open diode across A/C compressor clutch, and for other open diodes (see wiring diagrams)

HARD START

Definition: Engine cranks OK, but does not start for a long time. Does eventually run, or may start but immediately dies.

- Perform careful check as described at start of Section "B".
- Make sure driver is using correct starting procedure.
- **CHECK:**
  - TPS for sticking or binding or a high TPS voltage with the throttle closed.
  - High resistance in coolant sensor circuit or coolant temperature sensor itself. See Code 15 chart or with a "Scan" tool compare coolant temperature with ambient temperature on a cold engine.

- Fuel system pressure CHART A-7.
- For contaminated fuel.
- EGR operation. Be sure valve seats properly and is not staying open. See CHART C-7.
- Both injector fuses (visually inspect).
- Ignition system - Check distributor for:
  - Proper Output with ST-125.
  - Worn distributor shaft.
  - Bare and shorted wires.
  - Pickup coil resistance and connections.
  - Loose ignition coil ground.
  - Moisture in distributor cap.

1988-90 5.7L (VIN 8) – CORVETTE

- A faulty in-tank fuel pump check valve will allow the fuel in the lines to drain back to the tank after the engine is stopped. To check for this condition, Perform Fuel System Diagnosis, see CHART A-7.
- Remove spark plugs. Check for wet plugs, cracks, wear, improper gap, burned electrodes, or heavy deposits. Repair or replace as necessary

- If engine starts but then immediately stalls open distributor, set timing connector. If engine then starts and runs OK, replace pickup coil.
- If engine starts and stalls, disconnect MAF sensor. If engine then runs and MAF sensor connections are OK, replace the sensor.
- Basic engine problem.

SURGES AND/OR CHUGGLE

Definition: Engine power variation under steady throttle or cruise. Feels like the car speeds up and slows down with no change in the accelerator pedal

- Be sure driver understands Torque Converter Clutch and A/C compressor operation in owner's manual.
- Perform careful visual physical check as described at start of Section "B".
- **CHECK:**
  - Loose or leaking air duct between MAF sensor and throttle body.
  - Generator output voltage. Repair if less than 9 or more than 16 volts.
  - EGR - There should be no EGR at idle, see CHART C-7.
  - Vacuum lines for kinks or leaks.
  - In-line fuel filter. Replace if dirty or plugged.
  - Fuel pressure while condition exists. See CHART A-7.
  - Ignition timing. See Vehicle Emission Control Information label.
- Inspect oxygen sensor for silicon contamination from fuel, or use of improper RTV sealant.

The sensor may have a white, powdery coating and result in a high but false signal voltage (rich exhaust indication). The ECM will then reduce the amount of fuel delivered to the engine, causing a severe driveability problem.

- Remove spark plugs. Check for cracks, wear, improper gap, burned electrodes, or heavy deposits. Also check condition of distributor cap, rotor, and spark plug wires.
- To help determine if the condition is caused by a rich or lean system, the car should be driven at the speed of the complaint. Monitoring block learn at the complaint speed will help identify the cause of the problem. If the system is lean (block learn greater than 138), refer to "Diagnostic Aids" on facing page of Code 44. If the system is running rich (block learn less than 118), refer to "Diagnostic Aids" on facing page of Code 45.

LACK OF POWER, SLUGGISH, OR SPONGY

Definition: Engine delivers less than expected power. Little or no increase in speed when accelerator pedal is pushed down part way

- Perform careful visual check as described at start of Section "B".
- Compare customer's car to similar unit. Make sure the customer's car has an actual problem.
- Remove air cleaner and check air filter for dirt, or for being plugged. Replace as necessary.
- **CHECK:**
  - For loose or leaking air duct between MAF sensor and throttle body.
  - Restricted fuel filter, contaminated fuel or improper fuel pressure. See CHART A-7.
  - ECM Ground circuits - See ECM wiring diagrams.
  - EGR operation for being open, or partly open all the time - CHART C-7.
  - Ignition timing. See Vehicle Emission Control Information label.
  - Exhaust system for possible restriction, see CHART B-1.

  - Inspect exhaust system for damaged or collapsed pipes
  - Inspect muffler for heat distress or possible internal failure
  - For possible plugged catalytic converter by comparing exhaust system backpressure on each bank. Then reduce the backpressure by removing A.I.R. check valves near exhaust manifolds. See CHART B-1 for procedure.
  - Generator output voltage. Repair if less than 9 or more than 16 volts.
  - Engine valve timing and compression.
  - Engine for correct or worn camshaft.

  - Secondary voltage using a shop oscilloscope or a spark tester J-26792 (ST-125) or equivalent. Check for excessive retarded ignition timing. See CHART C-5.

1988-90 5.7L (VIN 8) – CORVETTE

DETONATION/SPARK KNOCK

Definition: A mild to severe ping, usually worse under acceleration. The engine makes sharp metallic knocks that change with throttle opening. Sounds like popcorn popping

- Check for obvious overheating problems
  - Low engine coolant
  - Loose water pump belt
  - Restricted air flow to radiator, or restricted water flow thru radiator
  - Inoperative electric cooling fan circuit, see CHART C-12.
- **CHECK:**
  - Ignition timing. See Vehicle Emission Control Information Label.
  - EGR system for not opening. CHART C-7.
  - TCC operation, see CHART C-8.
  - Fuel system pressure, see CHART A-7.
  - Mem Cal. Be sure it's the correct one.
  - Valve oil seals for leaking

- Check for incorrect basic engine parts such as cam, heads, pistons, etc.
- Check for poor fuel quality, proper octane rating
- Remove carbon with top engine cleaner. Follow instructions on can
- Check ESC system, see CHART C-5.
- To help determine if the condition is caused by a rich or lean system, the car should be driven at the speed of the complaint. Monitoring block learn at the complaint speed will help identify the cause of the problem. If the system is running lean (block learn greater than 138), refer to "Diagnostic Aids" on facing page of Code 44. If the system is running rich (block learn less than 118), refer to "Diagnostic Aids" on facing page of Code 45

HESITATION, SAG, STUMBLE

Definition: Momentary lack of response as the accelerator is pushed down. Can occur at all car speeds. Usually most severe when first trying to make the car move, as from a stop sign. May cause the engine to stall if severe enough.

- Perform careful visual check as described at start of Section "B".
- **CHECK:**
  - Fuel system pressure. See CHART A-7. Also, check for water or contaminated fuel
  - Air leaks at air duct between MAF sensor and throttle body
  - Spark plugs for being fouled or faulty wiring
  - Mem Cal number. Also check service bulletins for latest Mem Cal
  - TPS for binding or sticking. Voltage should increase at a steady rate as throttle is moved toward WOT

- Ignition timing. See Vehicle Emission Control Information label
- Generator output voltage. Repair if less than 9 or more than 16 volts.
- Distributor (HEI) ground, CKT 453
- Canister purge system for proper operation. See CHART C-4
- EGR. See CHART C-7.
- Perform injector balance test CHART C-2A.

1988-90 5.7L (VIN 8) – CORVETTE

CUTS OUT, MISSES

Definition: Steady pulsation or jerking that follows engine speed, usually more pronounced as engine load increases. The exhaust has a steady spitting sound at idle or low speed.

- Perform careful visual check as described at start of Section "B"
- Check for missing cylinder
  1. Start engine, allow engine to stabilize then disconnect IAC motor. Remove one spark plug wire at a time, using insulated pliers.
  CAUTION: Don't do this for more than 2 minutes, as this may cause damage to the catalytic converter.
  2. If there is an rpm drop on all cylinders (equal to within 50 rpm), go to "ROUGH, UNSTABLE, OR INCORRECT IDLE, STALLING" symptom. Reconnect IAC valve.
  3. If there is no rpm drop on one or more cylinders, or excessive variation in drop, check for spark on the suspected cylinder(s) with J 26792 (ST-125) Spark or equivalent.
     (If there is spark, remove spark plug(s) in these cylinders and check for:)
     - Cracks
     - Wear
     - Improper Gap
     - Burned Electrodes
     - Heavy Deposits
  - Perform compression check on questionable cylinder(s) found above. If compression is low, repair as necessary

- Disconnect all injector harness connectors. Connect J 34730-2 Injector Test Light or equivalent 6 volts test light between the harness terms, of each injector connector and note light on cranking. If test light fails to blink at any connector, it is a faulty injector drive circuit harness, connector, or terminal.
- Perform the Injector Balance Test. See CHART C-2A.
- **CHECK:**
  - Spark plug wires by connecting ohmmeter to ends of each wire in question. If meter reads over 30,000 ohms, replace wire(s).
  - Fuel System - Plugged fuel filter, water, low pressure.
  - Secondary voltage using a shop oscilloscope or a spark tester J-26792 (ST-125) or equivalent.
  - Visually inspect distributor cap and rotor for moisture, dust, cracks, burns, etc. Spray cap and plug wires with fine water mist to check for shorts.
- A miss condition can be caused by EMI (Electromagnetic Interference) on the reference circuit. EMI can usually be detected by monitoring engine rpm with a "Scan" tool. A sudden increase in rpm with little change in actual engine rpm change, indicates EMI is present.
  If the problem exists, check routing of secondary wires, check all distributor ground circuits
- Remove rocker covers. Check for bent pushrods, worn rocker arms, broken valve springs, worn camshaft lobes. Repair as necessary.
- Valve timing

POOR FUEL ECONOMY

Definition: Fuel economy, as measured by an actual road test, is noticeably lower than expected. Also, economy is noticeably lower than it was on this car at one time, as previously shown by an actual road test.

- Perform careful visual check as described at start of Section "B".
- **CHECK:**
  - Engine coolant level.
  - Engine thermostat for faulty part (always open) or for wrong heat range.

- Ignition timing. See Vehicle Emission Control Information label
- TCC for proper operation. A "Scan" should indicate an rpm drop when the TCC is commanded "ON", see CHART C-8.
- Induction system and crankcase for air leaks.
- Check for exhaust restriction, see CHART B-1.
- Compression

1988–90 5.7L (VIN 8) — CORVETTE

ROUGH, UNSTABLE OR INCORRECT IDLE, STALLING

Definition: The engine runs unevenly at idle. If bad enough, the car may shake. Also, the idle may vary in rpm (called "hunting"). Either condition may be bad enough to cause stalling. Engine idles at incorrect speed.

- Perform careful visual check as described at start of Section "B".
- **CHECK:**
  - Vacuum leaks.
  - Throttle linkage for sticking or binding.
  - ECM ground circuits.
  - IAC system. See CHART C-2C.
  - Generator output voltage. Repair if less than 9 or more than 16 volts.
  - P/N switch circuit. See CHART C-1A, or use "Scan" tool.
  - Injector balance. See CHART C-2A.
  - PCV valve for proper operation by placing finger over inlet hole in valve end several times. Valve should snap back. If not, replace valve.
  - Evaporative emission control system. CHART C-3.
  - A/C signal to ECM terminal "B8". "Scan" tool should indicate A/C is being requested whenever A/C is selected and the pressure cycling switch is closed.
  - Check A.I.R. system. There should be no A.I.R. to ports while in "Closed Loop." See CHART C-6.

  - EGR valve. There should be no EGR at idle.
  - Run a cylinder compression check.
- Inspect oxygen sensor for silicon contamination (from fuel, or use of improper RTV sealant. The sensor will have a white, powdery coating, and will result in a high but false signal voltage (rich exhaust indication). The ECM will then reduce the amount of fuel delivered to the engine, causing a severe driveability problem.
- Check for fuel in pressure regulator hose. If present replace regulator assembly.
- Check ignition system, wires, plugs, rotor, etc.
- Check for loose or damaged air duct between MAF sensor and throttle body.
- Disconnect MAF sensor and if condition is corrected replace sensor.
- Clean injectors.
- Monitoring block learn will help identify the cause of the problem. If the system is running lean (block learn greater than 138), refer to "Diagnostic Aids" on facing page of Code 44. If the system is running rich (block learn less than 118), refer to "Diagnostic Aids" on facing page of Code 45.

EXCESSIVE EXHAUST EMISSIONS OR ODORS

Definition: Vehicle fails an emission test. Vehicle has excessive "rotten egg" smell. Excessive odors do not necessarily indicate excessive emissions.

- Perform "Diagnostic Circuit Check."
- **IF TEST SHOWS EXCESSIVE CO AND HC,** (or also has excessive odors)
  - Check items which cause engine to run RICH.
  - Make sure engine is at normal operating temperature.
- **CHECK:**
  - Fuel pressure. See CHART A-7.
  - Incorrect timing. See Vehicle Emission Control Information Label.
  - Fuel vapor canister for fuel loading. See CHART C-3.
  - Injector balance. See CHART C-2A.
  - PCV valve for being plugged, stuck or blocked PCV hose or fuel in the crankcase.
  - Spark plugs, plug wires, and ignition components.
  - Check for lead contamination of catalytic converter (look for removal of fuel filler neck restrictor).
  - Check for properly installed fuel cap.

- If the system is running rich, (block learn less than 118), refer to "Diagnostic Aids" on facing page of Code 45.
- **IF TEST SHOWS EXCESSIVE NOx,**
  - Check items which cause car to run LEAN, or run too Hot.
    - EGR valve for not opening. See CHART C-7.
    - Vacuum leaks.
    - Coolant system and coolant fan for proper operation. See CHART C-12.
    - Remove carbon with top engine cleaner, follow instructions on can.
    - Check ignition timing for excessive base advance. See Vehicle Emission Control Information Label.
- If the system is running lean, (block learn greater than 138) on "Diagnostic Aids" on facing page of Code 44.

1988–90 5.7L (VIN 8) — CORVETTE

DIESELING, RUN-ON

Definition: Engine continues to run after key is turned "OFF," but runs very roughly. If engine runs smoothly, check ignition switch and adjustment.

- Check injectors for leaking. See CHART A-7.

BACKFIRE

Definition: Fuel ignites in intake manifold, or in exhaust system making a loud popping noise.

- **CHECK:**
  - Loose wiring connector or air duct at MAF sensor.
  - EGR operation for being open all the time. See CHART C-7.
  - EGR gasket for faulty or loose fit.
  - Spark plugs for crossfire also inspect distributor cap, spark plug wires, and proper routing of plug wires).
  - Ignition system for intermittent condition.

  - Ignition timing, see Vehicle Emission Control Information Label.

  - Perform fuel system diagnosis check, see CHART A-7A.
  - Perform injector balance test, see CHART C-2A.
  - A.I.R. system check valves.
  - Compression : Look for sticking or leaking valves.
  - Valve timing.
  - Output voltage of ignition coil using a shop oscilliscope or spark tester J 26792 (ST-125) or equivalent.

CHART B-1
RESTRICTED EXHAUST SYSTEM CHECK
ALL ENGINES

Proper diagnosis for a restricted exhaust system is essential before any components are replaced. Either of the following procedures may be used for diagnosis, depending upon engine or tool used:

CHECK AT A.I.R. PIPE:

1. Remove the rubber hose at the exhaust manifold A.I.R. pipe check valve. Remove check valve.
2. Connect a fuel pump pressure gauge to a hose and nipple from a Propane Enrichment Device (J 26911) (see illustration).
3. Insert the nipple into the exhaust manifold A.I.R. pipe.

OR CHECK AT O₂ SENSOR:

1. Carefully remove O₂ sensor.
2. Install Borroughs exhaust backpressure tester (BT 8515 or BT 8603) or equivalent in place of O₂ sensor (see illustration).
3. After completing test described below, be sure to coat threads of O₂ sensor with anti-seize compound P/N 5613695 or equivalent prior to re-installation.

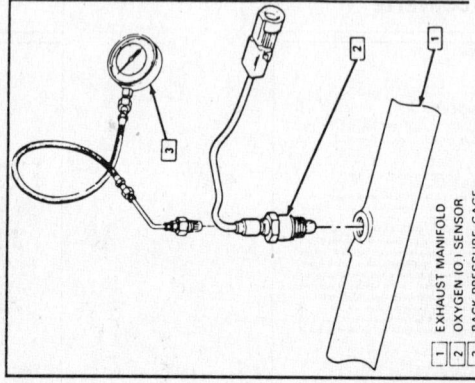

1	EXHAUST MANIFOLD
2	OXYGEN (O₂) SENSOR
3	BACK PRESSURE GAGE

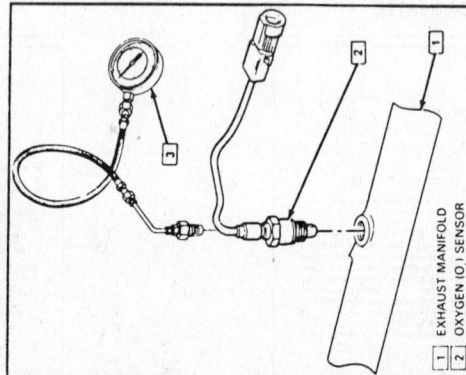

1	GAGE
2	HOSE AND NIPPLE ADAPTER
3	A.I.R. PIPE (EXHAUST PORT)
4	CHECK VALVE

DIAGNOSIS:

1. With the engine idling at normal operating temperature, observe the exhaust system backpressure reading on the gage. Reading should not exceed 8.6 kPa (1.25 psi).
2. Increase engine speed to 2000 rpm and observe gage. Reading should not exceed 20.7 kPa (3 psi).
3. If the backpressure at either speed exceeds specification, a restricted exhaust system is indicated.
4. Inspect the entire exhaust system for a collapsed pipe, heat distress, or possible internal muffler failure.
5. If there are no obvious reasons for the excessive backpressure, the catalytic converter is suspected to be restricted and should be replaced using current recommended procedures.

1988–90 5.7L (VIN 8) – CORVETTE

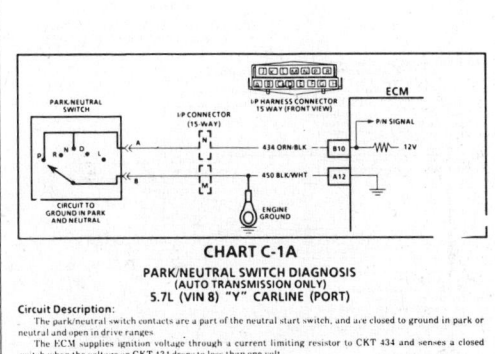

CHART C-1A
PARK/NEUTRAL SWITCH DIAGNOSIS
(AUTO TRANSMISSION ONLY)
5.7L (VIN 8) "Y" CARLINE (PORT)

Circuit Description:
The park/neutral switch contacts are a part of the neutral start switch, and are closed to ground in park or neutral and open in drive ranges.

The ECM supplies ignition voltage through a current limiting resistor to CKT 434 and senses a closed switch when the voltage on CKT 434 drops to less than one volt.

The ECM uses the P/N signal as one of the inputs to control:
- Idle air control
- VSS diagnostics
- EGR

If CKT 434 indicates P/N (grounded), while in drive range, the EGR would be inoperative, resulting in possible detonation.

If CKT 434 always indicates drive (open), a drop in the idle may exist when the gear selector is moved into drive range.

Test Description: Numbers below refer to circled numbers on the diagnostic chart.
1. Checks for a closed switch to ground in park position. Different makes of "Scan" tools will read P/N differently. Refer to "Operators Manual" for type of display used for a specific tool.
2. Checks for an open switch in drive range.
3. Be sure "Scan" indicates drive, even while wiggling shifter to test for an intermittent or misadjusted switch in drive range.

1988–90 5.7L (VIN 8) – CORVETTE

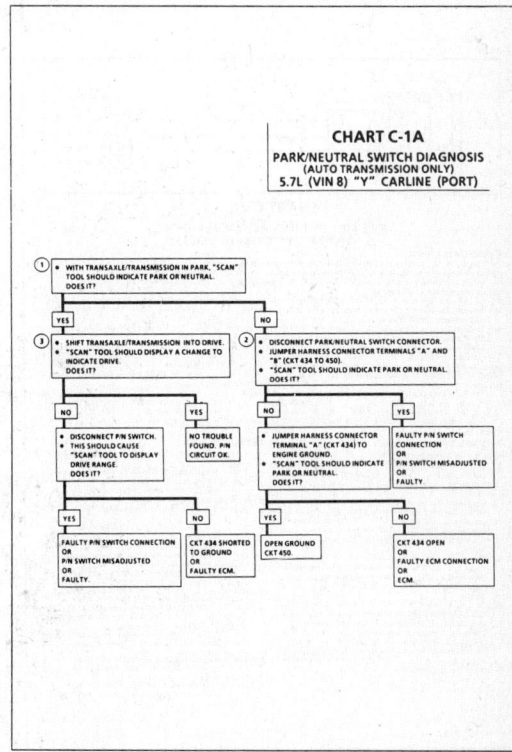

1988–90 5.7L (VIN 8) – CORVETTE

CHART C-2A
INJECTOR BALANCE TEST

The injector balance tester is a tool used to turn the injector on for a precise amount of time, thus spraying a measured amount of fuel into the manifold. This causes a drop in fuel rail pressure that we can record and compare between each injector. All injectors should have the same amount of pressure drop (± 10 kpa). Any injector with a pressure drop that is 10 kpa (or more) greater or less than the average drop of the other injectors should be considered faulty and replaced.

STEP 1
Engine "cool down" period (10 minutes) is necessary to avoid irregular readings due to "Hot Soak" fuel boiling. With ignition "OFF" connect fuel gauge J 347301 or equivalent to fuel pressure tap. Wrap a shop towel around fitting while connecting gage to avoid fuel spillage.

Disconnect harness connectors at all injectors, and connect injector tester J 34730-3, or equivalent, to one injector. On Turbo equipped engines, use adaptor harness furnished with injector tester to energize injectors that are not accessible. Follow manufacturers instructions for use of adaptor harness. Ignition must be "OFF" at least 10 seconds to complete ECM shutdown cycle. Fuel pump should run about 2 seconds after ignition is turned "ON". At this point, insert clear tubing attached to vent valve into a suitable container and bleed air from gauge and hose to insure accurate gauge operation. Repeat this step until all air is bled from gauge.

STEP 2
Turn ignition "OFF" for 10 seconds and then "ON" again to get fuel pressure to its maximum. Record this initial pressure reading. Energize tester one time and note pressure drop at its lowest point (Disregard any slight pressure increase after drop hits low point). By subtracting this second pressure reading from the initial pressure, we have the actual amount of injector pressure drop.

STEP 3
Repeat step 2 on each injector and compare the amount of drop. Usually, good injectors will have virtually the same drop. Retest any injector that has a pressure difference of 10kPa, either more or less than the average of the other injectors on the engine. Replace any injector that also fails the retest. If the pressure drop of all injectors is within 10kPa of this average, the injectors appear to be flowing properly. Reconnect them and review "Symptoms", Section "B".

NOTE: The entire test should <u>not</u> be repeated more than once without running the engine to prevent flooding. (This includes any retest on faulty injectors).

1988–90 5.7L (VIN 8) – CORVETTE

NOTE: If injectors are suspected of being dirty, they should be cleaned using an approved tool and procedure prior to performing this test. The fuel pressure test in Section "A", Chart A-7, should be completed prior to this test.

CHART C-2A
INJECTOR BALANCE TEST
5.7L (VIN 8) "Y" CARLINE (PORT)

Step 1. If engine is at operating temperature, allow a 10 minute "cool down" period then connect fuel pressure gauge and injector tester.
1. Ignition "OFF."
2. Connect fuel pressure gauge and injector tester.
3. Ignition "ON."
4. Bleed off air in gauge. Repeat until all air is bled from gauge.

Step 2. Run test:
1. Ignition "OFF" for 10 seconds.
2. Ignition "ON." Record gauge pressure. (Pressure must hold steady, if not see the Fuel System diagnosis, Chart A-7).
3. Turn injector on, by depressing button on injector tester, and note pressure at the instant the gauge needle stops.

Step 3.
1. Repeat step 2 on all injectors and record pressure drop on each. Retest injectors that appear faulty (Any injectors that have a 10 kPa difference, either more or less, in pressure from the average). If no problem is found, review "Symptoms" Section "B".

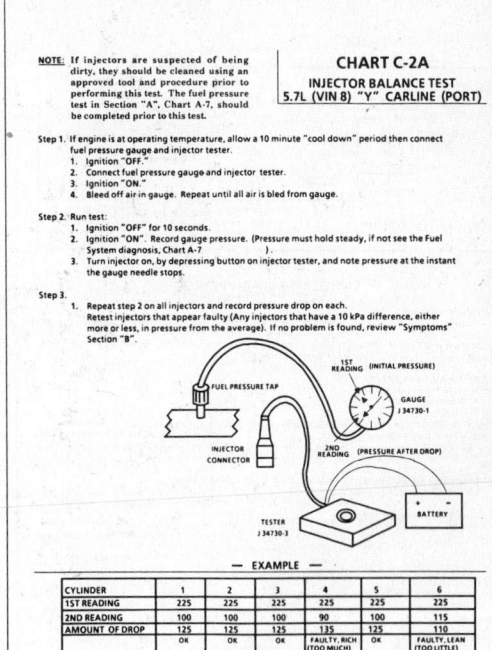

— EXAMPLE —

CYLINDER	1	2	3	4	5	6
1ST READING	225	225	225	225	225	225
2ND READING	100	100	100	90	100	115
AMOUNT OF DROP	125	125	125	135	125	110
	OK	OK	OK	FAULTY, RICH (TOO MUCH) (FUEL DROP)	OK	FAULTY, LEAN (TOO LITTLE) (FUEL DROP)

1988–90 5.7L (VIN 8) – CORVETTE

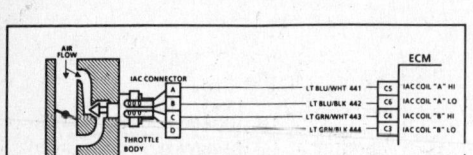

CHART C-2C
IDLE AIR CONTROL (IAC) VALVE CHECK
5.7L (VIN 8) "Y" CARLINE (PORT)

Circuit Description:
The ECM will control engine idle speed by moving the IAC valve to control air flow around the throttle plates. It does this by sending voltage pulses to the proper motor winding for each IAC motor. This will cause the motor shaft and valve to move "IN" or "OUT" of the motor a given distance for each pulse received. ECM pulses are referred to as "counts".
• To increase idle speed - ECM will send enough counts to retract the IAC valve and allow more air to flow through the idle air passage and bypass the throttle plates until idle speed reaches the proper rpm. This will increase the ECM counts.
• To decrease idle speed - ECM will send enough counts to extend the IAC valve and reduce air flow through the idle passage around the throttle plates. This will reduce the ECM counts.
Each time the engine is started and then the ignition is turned "OFF", the ECM will reset the IAC valve. This is done by sending enough counts to seat the valve. The fully seated valve is the ECM reference zero. A given number of counts are then issued to open the valve, and normal ECM control of IAC will begin from this point. The number of counts are then calculated by the ECM. This is how the ECM knows what the motor position is for a given idle speed.
The ECM uses the following information to control idle speed:
• Battery voltage • Engine speed • Coolant temperature
• Throttle position sensor • P/N switch • A/C clutch signal
Don't apply battery voltage across the IAC motor terminals. It will permanently damage the IAC motor windings.

Test Description: Numbers below refer to circled numbers on the diagnostic chart.
1. Continue with test, even if engine will not idle. If idle is too low, "Scan" will display 80 or more counts, or steps. If idle is high, it will display "0" counts.
Occasionally an erratic or unstable idle may occur. Engine speed may vary 200 rpm or more up and down. Disconnect IAC. If the condition is unchanged, the IAC is not at fault. There is a system problem. Proceed to "Diagnostic Aids" below.
2. When the engine was stopped, the IAC valve retracted (more air) to a fixed "park" position for increased air flow and idle during the next engine start. A "Scan" will display 140 or more counts.
3. Be sure to disconnect IAC valve prior to this test. The test light will confirm the ECM signals by a steady or flashing light on all circuits.
4. There is a remote possibility that one of the circuits is shorted to voltage which would've been indicated by a steady light. Disconnect ECM and turn the ignition "ON" and probe terminals to check for this condition.

Diagnostic Aids:
Engine idle speed can be adversely affected by the following:
• Park/Neutral switch - If ECM thinks the car is always in neutral, then idle will not be controlled to the specified rpm when in drive range
• Leaking injector(s) will cause fuel imbalance and poor idle quality due to excess fuel. See CHART A-7.
• Vacuum or crankcase leaks can affect idle.
• When the throttle shaft or throttle position sensor is binding or sticking in an open throttle position, the ECM does not know if the vehicle has stopped and does not control idle.
• Check A.I.R. management system for intermittent air to ports while in "Closed Loop"
• In addition to electrical control of EGR, be sure to examine the EGR valve for proper seating.
• Faulty battery cables can result in voltage variations. The ECM will try to compensate, which results in erratic idle speeds.
• The ECM will compensate for A/C compressor clutch loads. Loss of the A/C request signal would be most apparent in neutral.
• Contaminated fuel can adversely affect idle.
• Perform injector balance test CHART C-2A.
If all OK, refer to "Rough, Unstable, Incorrect Idle or Stalling" symptoms in Section "B"

1988–90 5.7L (VIN 8) – CORVETTE

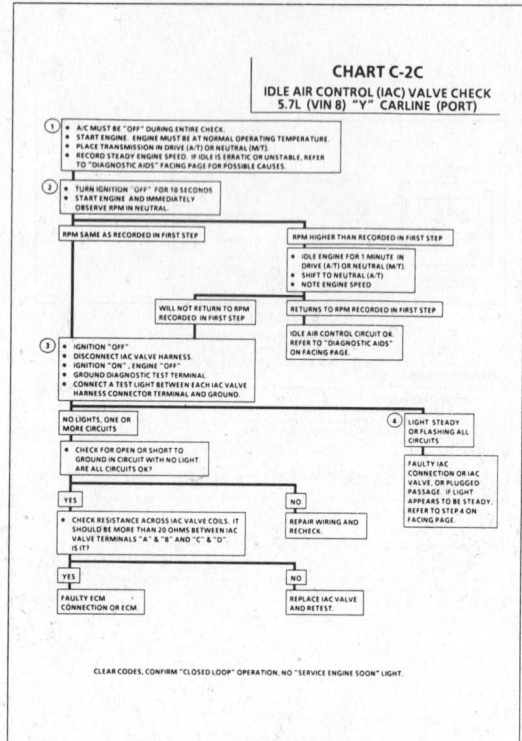

CHART C-2C
IDLE AIR CONTROL (IAC) VALVE CHECK
5.7L (VIN 8) "Y" CARLINE (PORT)

1988–90 5.7L (VIN 8) – CORVETTE

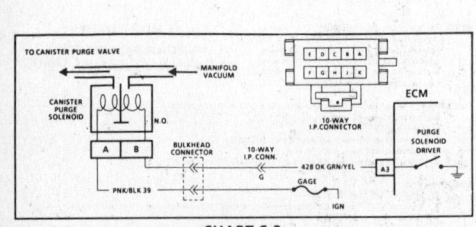

CHART C-3
CANISTER PURGE VALVE CHECK
5.7L (VIN 8) "Y" CARLINE (PORT)

Circuit Description:
Canister purge is controlled by a solenoid that allows manifold vacuum to purge the canister when de-energized. The ECM supplies a ground to energize the solenoid (purge "OFF").
If the diagnostic "test" terminal is ungrounded with the engine stopped, or the following is met with the engine running, the purge solenoid is de-energized (purge "ON").
• Engine run time after start more than 1 minute.
• Coolant temperature above 75°C.
• Vehicle speed above 2 mph.
• Throttle off idle.

Test Description: Numbers below refer to circled numbers on the diagnostic chart.
1. Checks to see if the solenoid is opened or closed. The solenoid is normally energized in this step, so it should be closed.
2. Checks for a complete circuit. Normally, there is ignition voltage on CKT 39 and the ECM provides a ground on CKT 428. A shorted solenoid could cause an open circuit in the ECM.
3. Completes functional check by ungrounding test terminal. This should normally de-energize the solenoid and allow the vacuum to drop (purge "ON").

Diagnostic Aids:
Normal operation of the canister purge solenoid is described as follows:
With the ignition "ON", engine "OFF", diagnostic "test" terminal ungrounded, the canister purge solenoid will be energized.

1988–90 5.7L (VIN 8) – CORVETTE

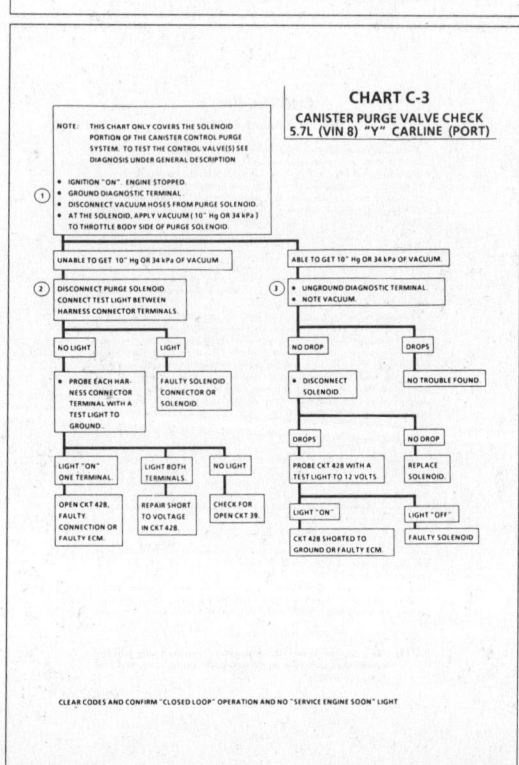

CHART C-3
CANISTER PURGE VALVE CHECK
5.7L (VIN 8) "Y" CARLINE (PORT)

1988–90 5.7L (VIN 8) – CORVETTE

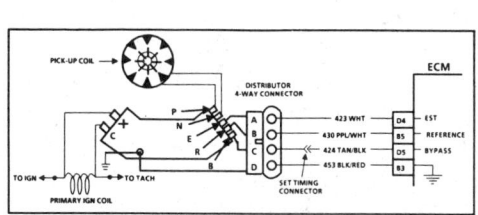

CHART C-4A
IGNITION SYSTEM CHECK
(INTEGRAL COIL)
5.7L (VIN 8) "Y" CARLINE (PORT)

Test Description: Numbers below refer to circled numbers on the diagnostic chart.

1. Checks for proper output from the ignition system. The spark tester requires a minimum of 25,000 volts to fire. This check can be used in case of an ignition miss, because the system may provide enough voltage to run the engine but not enough to fire a spark plug under heavy load.

1A. If spark occurs with EST connector disconnected, pick-up coil output is too low for EST operation.

2. Normal reading during cranking is about 8-10 volts.

3. Checks for a shorted module or grounded circuit from the ignition coil to the module. The distributor module should be turned "OFF," so normal voltage should be about 12 volts. If the module is turned "ON," the voltage would be low, but above 1 volt. This could cause the ignition coil to fail from excessive heat. With an open ignition coil primary winding, a small mount of voltage will leak through the module from the battery to the tach. terminal.

4. Checks the voltage output with the pick-up coil triggering the module. A spark indicates that the ignition system has sufficient output, however, intermittent no-starts or poor performance could be the result of incorrect polarity between the ignition coil and the pick-up coil.
   The color of the pick-up coil connector has to be yellow, if one of the ignition coil leads is yellow. If the ignition has a white lead, any pick-up coil connector color, except yellow, is OK.

5. Checks for an open module or circuit to it. 12 volts applied to the module "P" terminal should turn the module "ON" and the voltage should drop to about 7-9 volts.

6. This should turn "OFF" the module and cause a spark. If no spark occurs, the fault is most likely in the ignition coil, because most module problems would have been found before this point in the procedure. A module tester could determine which is at fault.

1988–90 5.7L (VIN 8) – CORVETTE

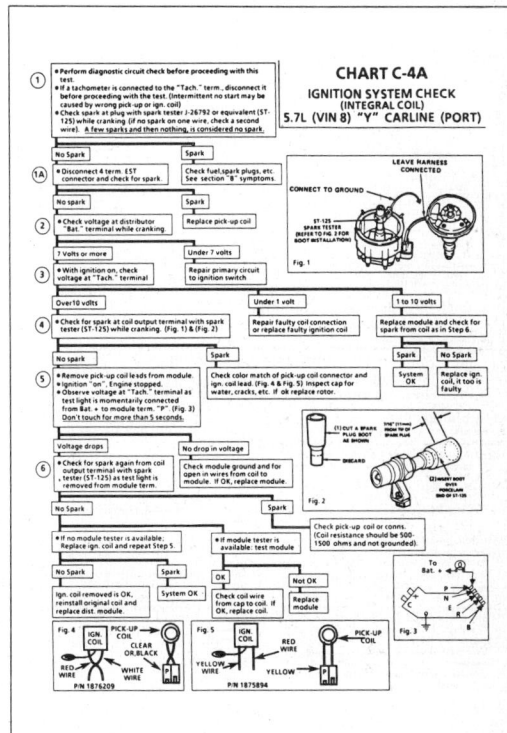

CHART C-4A
IGNITION SYSTEM CHECK
(INTEGRAL COIL)
5.7L (VIN 8) "Y" CARLINE (PORT)

1988–90 5.7L (VIN 8) – CORVETTE

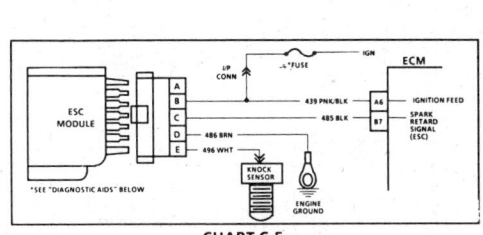

CHART C-5
ELECTRONIC SPARK CONTROL (ESC) SYSTEM CHECK
(ENGINE KNOCK, POOR PERFORMANCE, OR POOR FUEL ECONOMY)
5.7L (VIN 8) "Y" CARLINE (PORT)

Circuit Description:

Electronic Spark Control (ESC) is accomplished with a module that sends a voltage signal to the ECM. As the knock sensor detects engine knock, the voltage from the ESC module to the ECM is shut "OFF" and this signals the ECM to retard timing, if engine rpm is over about 850.

Test Description: Numbers below refer to circled numbers on the diagnostic chart.

1. If a Code 43 is not set, but a knock signal is indicated while running at 1500 rpm, listen for an internal engine noise. Under a no load condition, there should not be any detonation, and if knock is indicated, an internal engine problem may exist.

2. Usually a knock signal can be generated by tapping on the exhaust manifold. This test can also be performed at idle. Test number 1 was run at 1500 rpm to determine if a constant knock signal was present, which would affect engine performance.

3. This tests whether the knock signal is due to the sensor, a basic engine problem or the ESC module.

4. If the module ground circuit is faulty, the ESC module will not function correctly. The test light should light, indicating the ground circuit is OK.

5. Contacting CKT 496, with a test light to 12 volts, should generate a knock signal to determine whether the knock sensor is faulty, or the ESC module can't recognize a knock signal.

Diagnostic Aids:

*ECM/IGN fuse

"Scan" tools have two positions to diagnose the ESC system. The knock signal can be monitored to see if the knock sensor is detecting a knock condition and, if the ESC module is functioning, a knock signal should display "YES" whenever detonation is present. The knock retard position on the "Scan" displays the amount of spark retard the ECM is commanding. The ECM can retard the timing up to 20 degrees.

If the ESC system checks OK, but detonation is the complaint, refer to "Detonation/Spark Knock" in Section "B".

1988–90 5.7L (VIN 8) – CORVETTE

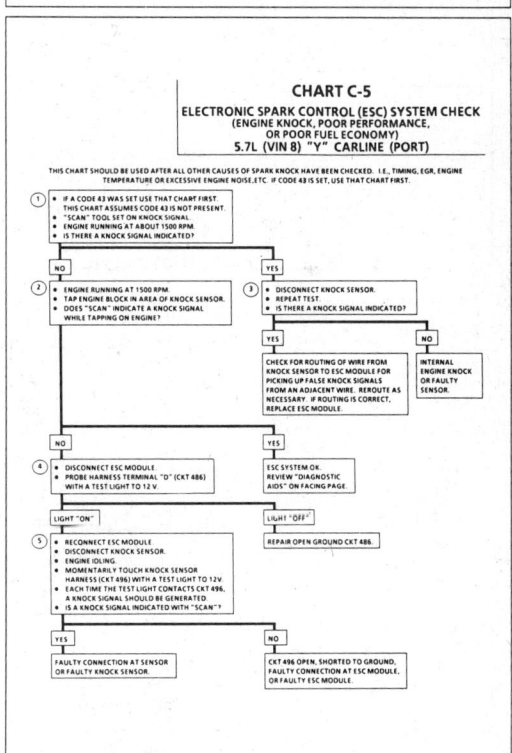

CHART C-5
ELECTRONIC SPARK CONTROL (ESC) SYSTEM CHECK
(ENGINE KNOCK, POOR PERFORMANCE, OR POOR FUEL ECONOMY)
5.7L (VIN 8) "Y" CARLINE (PORT)

THIS CHART SHOULD BE USED AFTER ALL OTHER CAUSES OF SPARK KNOCK HAVE BEEN CHECKED. I.E., TIMING, EGR, ENGINE TEMPERATURE OR EXCESSIVE ENGINE NOISE,ETC. IF CODE 43 IS SET, USE THAT CHART FIRST.

1988–90 5.7L (VIN 8) – CORVETTE

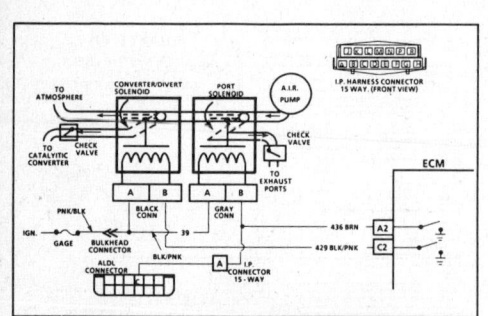

CHART C-6
AIR MANAGEMENT CHECK - CONTROL (PEDES) VALVE
(PRESSURE OPERATED ELECTRIC DIVERT/ELECTRIC SWITCHING)
5.7L (VIN 8) "Y" CARLINE (PORT)

Circuit Description:

Air management is controlled by a pressure operated port valve and a converter valve, each with an ECM controlled solenoid. When the solenoid is grounded by the ECM, A.I.R. pump pressure will activate the valve and allow pump air to be directed as follows:

- **Cold Mode** - The port (switch) solenoid is energized which in turn opens the port valve and allows flow to the exhaust ports.
- **Warm Mode** - The port solenoid is de-energized which closes the port valve and keeps the converter valve seated, thus forcing flow past the converter valve and to the catalytic converter.
- **Divert Mode** - Both solenoids are de-energized which opens the converter valve, allowing air to take the path of least resistance, i.e., out the divert/relief tube to atmosphere.

Test Description: Numbers below refer to circled numbers on the diagnostic chart.

1. This is a system functional check. Air is directed to ports during "Open Loop" and all engines start in "Open Loop" even on a warm engine. Since the air to the ports may be very short, prepare to observe port air prior to engine start up. This can be done by squeezing a hose.
2. This should normally set a Code 22. When any code is set, the ECM opens the ground to the converter solenoid and allows air to divert. This checks for ECM response to a fault. A ground in the control valve circuit to the ECM would prevent divert action.
3. This checks for a grounded circuit to the ECM Test light "OFF" is normal and would indicate the circuit is not grounded.
4. Checks for an open in the solenoid control circuits. Grounding the diagnostic "test" terminal should ground both solenoid circuits. Normally, the test light should be "ON" which indicates the problem is not in the ECM or wiring but at the solenoid connections or valve itself.
5. Checks for a grounded port solenoid circuit. Test light "OFF" would indicate the circuit is normal and fault is in the valve.

1988–90 5.7L (VIN 8) – CORVETTE

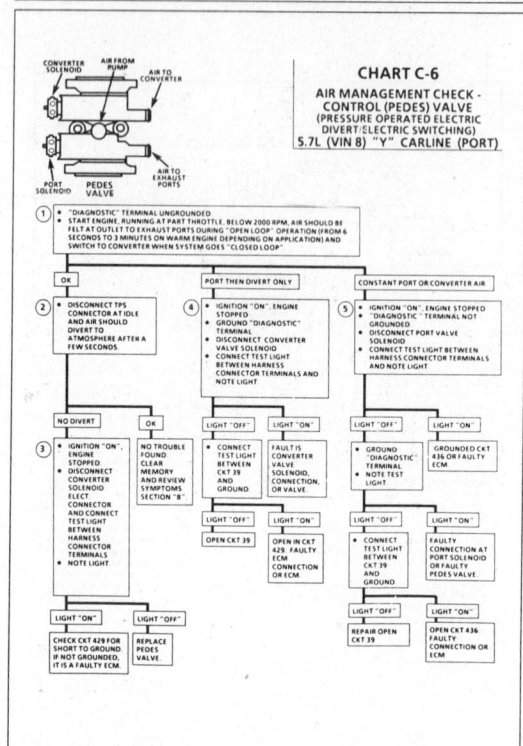

CHART C-6
AIR MANAGEMENT CHECK - CONTROL (PEDES) VALVE
(PRESSURE OPERATED ELECTRIC DIVERT/ELECTRIC SWITCHING)
5.7L (VIN 8) "Y" CARLINE (PORT)

1988–90 5.7L (VIN 8) – CORVETTE

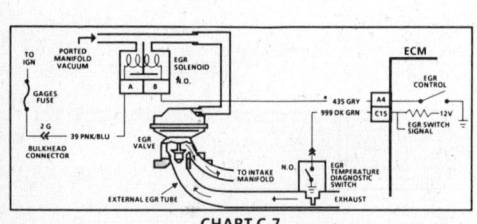

CHART C-7
EXHAUST GAS RECIRCULATION (EGR) CHECK
5.7L (VIN 8) "Y" CARLINE (PORT)

Circuit Description:

The Exhaust Gas Recirculation (EGR) valve is controlled by a normally open Pulse Width Modulated (PWM) solenoid. The ECM turns the solenoid "OFF", to allow vacuum to pass to the EGR, and turns the solenoid "ON" to prohibit EGR operation. When EGR is commanded, the solenoid is turned "ON" and "OFF" many times a second (duty cycle).

The duty cycle is calculated by the ECM based on information from the coolant, MAT, TPS, and MAF sensors. Also, engine rpms and the P/N switch input affect EGR. There should be no EGR when in park or neutral, TPS below a calibrated value, or TPS indicating WOT.

With the ignition "ON" and engine stopped, the EGR solenoid is de-energized. The solenoid, however, should be energized, if the diagnostic terminal is grounded with the ignition "ON" and engine not running.

Test Description: Numbers below refer to circled numbers on the diagnostic chart.

1. This will test the solenoid value to determine if it is capable of closing off the manifold vacuum from the EGR valve. The vacuum may bleed off slowly, but this should not be considered a fault.
2. As soon as back pressure is available at the EGR valve, the bleed portion in the valve should open and cause the valve to go to its seated position.
3. The EGR will be inoperative if the P/N switch is misadjusted or faulty. The EGR is disabled when in park or neutral. Use "Scan" tool and check P/N switch. Refer to CHART C-1A.

1988–90 5.7L (VIN 8) – CORVETTE

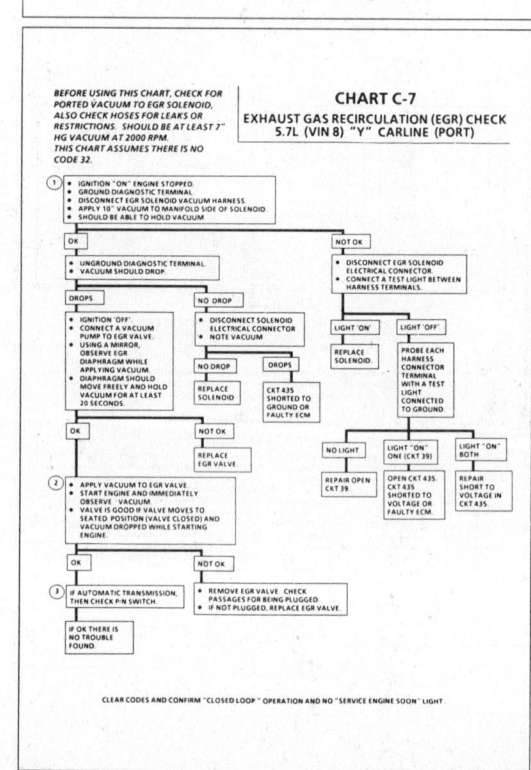

CHART C-7
EXHAUST GAS RECIRCULATION (EGR) CHECK
5.7L (VIN 8) "Y" CARLINE (PORT)

1988–90 5.7L (VIN 8) – CORVETTE

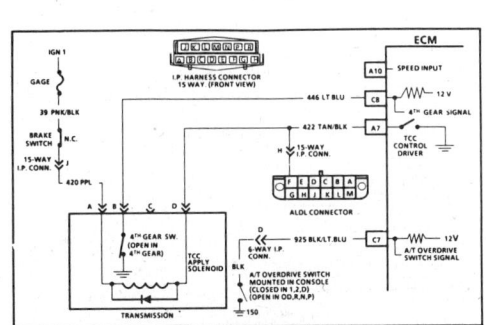

CHART C-8A
(Page 1 of 2)
TORQUE CONVERTER CLUTCH (TCC)
(ELECTRICAL DIAGNOSIS)
5.7L (VIN 8) "Y" CARLINE (PORT)

Circuit Description:
The purpose of the automatic transmission Torque Converter Clutch (TCC) feature is to eliminate the power loss of the torque converter stage, when the vehicle is in a cruise condition. This allows the convenience of the automatic transmission and the fuel economy of a manual transmission. The heart of the system is a solenoid, located inside the automatic transmission, which is controlled by the ECM.

When the solenoid coil is activated ("ON"), the torque converter clutch is applied, which results in straight through mechanical coupling from the engine to transmission. When the transmission solenoid is deactivated, the torque converter clutch is released which allows the torque converter to operate in the conventional manner (fluid coupling between engine and transmission).

The ECM turns "ON" the TCC when coolant temperature is above 65°C (149°F), TPS not changing, and vehicle speed above a specified value. For TCC lock-up speed and function of the gear switches, refer to circuit description on the facing page of CHART C-8A (Page 2 of 2).

Test Description: Numbers below refer to circled numbers on the diagnostic chart.
1. When a test light is connected from ALDL terminal "F" to ground, a test light "ON" indicates battery voltage is OK and the TCC solenoid is disengaged.
2. When the diagnostic terminal is grounded, the ECM should energize the TCC solenoid and the test light should go out.

Diagnostic Aids:
A "Scan" tool only indicates when the ECM has turned "ON" the TCC driver (grounded CKT 422) but this does not confirm that the TCC has engaged. To determine if the TCC is functioning properly, engine rpm should decrease when the "Scan" indicates the TCC driver has turned "ON." To determine if the gear switches are functioning properly, perform the checks in CHART C-8A (Page 2 of 2). The switches will not prevent the TCC from functioning, but will affect TCC lock and unlock points.

1988–90 5.7L (VIN 8) – CORVETTE

CHART C-8A
(Page 1 of 2)
TORQUE CONVERTER CLUTCH (TCC)
(ELECTRICAL DIAGNOSIS)
5.7L (VIN 8) "Y" CARLINE (PORT)

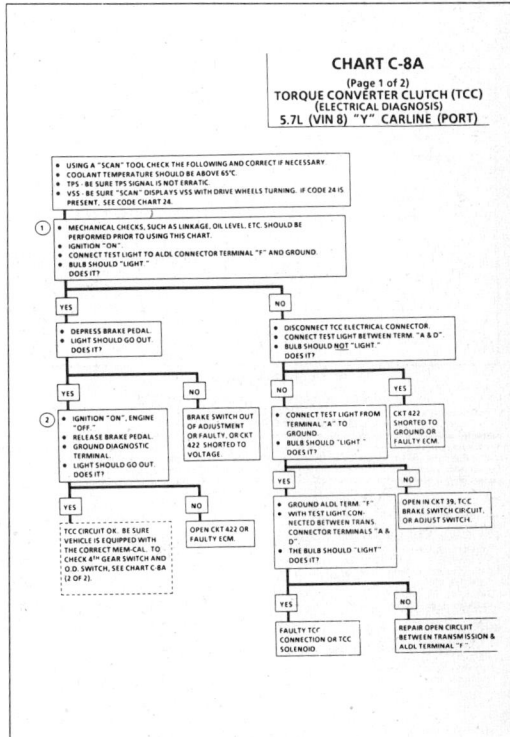

1988–90 5.7L (VIN 8) – CORVETTE

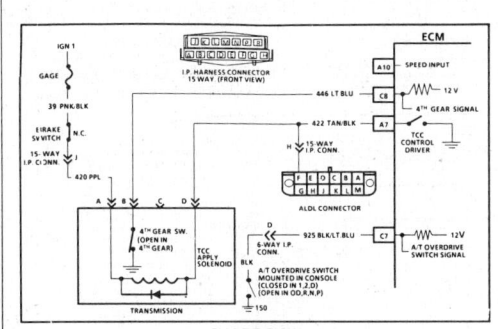

CHART C-8A
(Page 2 of 2)
TORQUE CONVERTER CLUTCH (TCC)
(ELECTRICAL DIAGNOSIS)
5.7L (VIN 8) "Y" CARLINE (PORT)

Circuit Description:
The overdrive switch (mounted in the console) is used to modify TCC engagement speeds. The TCC will engage, on a warm engine, under a given road load, in 2nd, 3rd, and 4th gear at about 30 mph, if the gear selector is in overdrive (OD). If 2nd or drive (D) range is selected, the overdrive switch should be closed and the ECM will not engage TCC until vehicle speed reaches about 40 mph. This switch will help eliminate the TCC from engaging and disengaging while in city traffic.

A 4th gear switch (mounted in the trans.) opens when the transmission shifts into 4th gear, and this switch is used by the ECM to modify TCC lock and unlock points, when in a 4-3 downshift maneuver.

Test Description: Numbers below refer to circled numbers on the diagnostic chart.
1. With the gear shifter in overdrive, the "Scan" should display "YES". In this position, TCC will engage at about 30 mph.
2. CKT 925 can be checked for continuity by measuring for the 12 volt signal from the ECM with a DVOM. If the 12 volts is present, the switch is misadjusted or faulty.
3. This step should cause the switch to close and the "Scan" should display "NO".
4. Unless the switch or CKT 446 is open, the "Scan" should display "NO", indicating the transmission is not in 4th gear. The 4th gear switch should only be open while in 4th gear.
5. This step determines if the ECM and wiring are OK.

Grounding CKT 446 should cause the "Scan" to display "NO", indicating the transmission is not in 4th gear.
6. Checks the operation of the 4th gear switch. When the transmission shifts into 4th gear, the switch should open and the "Scan" should display "YES."
7. Disconnecting the TCC connector simulates an open switch, to determine if CKT 446 is shorted to ground or the problem is in the transmission.

Diagnostic Aids:
A road test may be necessary to verify the customer complaint. If the "Scan" indicates TCC is turning "ON" and "OFF" erratically, check the state of the 4th gear switch and the overdrive switch to be sure they're not changing states under a steady throttle position. If the switches are changing states, check connections and wire routing carefully.

1988–90 5.7L (VIN 8) – CORVETTE

CHART C-8A
(Page 2 of 2)
TORQUE CONVERTER CLUTCH (TCC)
(ELECTRICAL DIAGNOSIS)
5.7L (VIN 8) "Y" CARLINE (PORT)

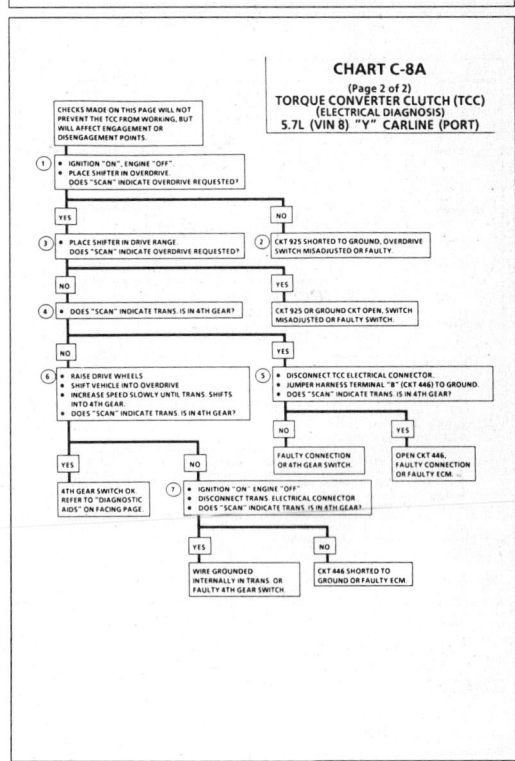

1988–90 5.7L (VIN 8) – CORVETTE

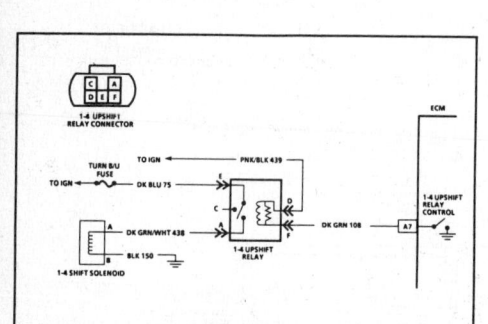

CHART C-8B
MANUAL TRANSMISSION 1-4 UPSHIFT SYSTEM
(ELECTRICAL DIAGNOSIS)
5.7L (VIN 8) "Y" CARLINE (PORT)

Circuit Description:
The 1-4 upshift relay is energized when the ECM grounds CKT 108, which provides battery voltage to the 1-4 upshift solenoid and activates it. When the 1-4 upshift solenoid (located in the transmission) is activated, it mechanically locks out the transmission linkage from being shifted from 1st gear to 2nd or 3rd gears. Under certain low load conditions the vehicle can only be shifted from 1st to 4th gear to improve fuel economy.

Test Description: Numbers below refer to circled numbers on the diagnostic chart.
1. A mechanical transmission problem is indicated if the transmission can only be put in 1st or 4th gear in this step.
2. Grounding the diagnostic "test" terminal with the ignition "ON" and engine not running, will energize the 1-4 upshift relay and solenoid indicating normal system operation.

3. If the fuse for CKT 75 is blown, check circuit 438 for a short to ground.

NOTE: Do Not keep the diagnostic "test" terminal grounded for more than 1 minute, as damage to the upshift solenoid may occur.

1988–90 5.7L (VIN 8) – CORVETTE

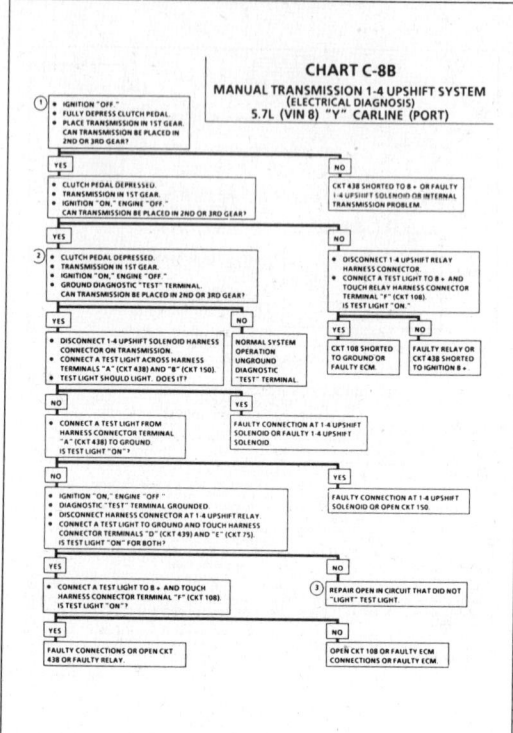

CHART C-8B
MANUAL TRANSMISSION 1-4 UPSHIFT SYSTEM
(ELECTRICAL DIAGNOSIS)
5.7L (VIN 8) "Y" CARLINE (PORT)

1988–90 5.7L (VIN 8) – CORVETTE

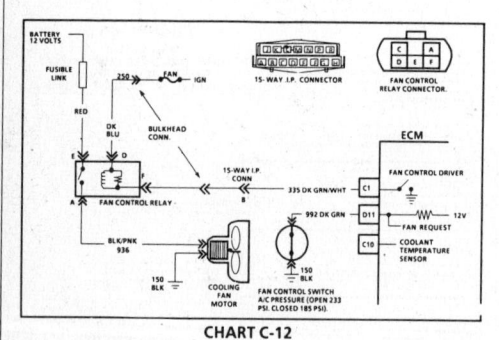

CHART C-12
(Page 1 of 2)
ELECTRIC COOLING FAN CONTROL CIRCUIT DIAGNOSIS
5.7L (VIN 8) "Y" CARLINE (PORT)

Circuit Description:
- The cooling fan is totally controlled by the ECM based on inputs from the coolant temperature sensor and fan control switch. The fan should run if coolant temperature is greater than 108°C (226°F).
- Battery voltage is supplied to the fan relay on terminal "E" and ignition voltage to terminal "D".
- Grounding CKT 335 (relay terminal "F") will energize the relay and supply battery voltage to the fan motor. Once the fan relay is energized by the ECM, it will remain "ON" for a minimum of 30 seconds.
- The ECM will remove the ground to CKT 335 if vehicle speed is over 40 mph unless the engine is overheating.
- The fan control switch, mounted in the A/C high pressure line, will open when head pressure exceeds 1600 kPa (233 psi) and this input causes the ECM to ground CKT 335.
- If a Code 14 or 15 sets or the ECM is operating in the fuel back-up mode, the ECM will turn "ON" the cooling fan.

Diagnostic Aids:

If the owner complained of an overheating problem, it must be determined if the complaint was due to an actual boil over or the hot light or temperature gage indicated overheating.

If the gate or light indicates overheating, but no boil over is detected, the gage circuit should be checked. The gage accuracy can also be checked by comparing the coolant sensor reading using a "Scan" tool and comparing its reading with the gage reading.

If the engine is actually overheating, but the cooling fan is not coming "ON," the coolant sensor has probably shifted out of calibration and should be replaced.

If the engine is overheating and the cooling fan is "ON," the cooling system should be checked.

1988–90 5.7L (VIN 8) – CORVETTE

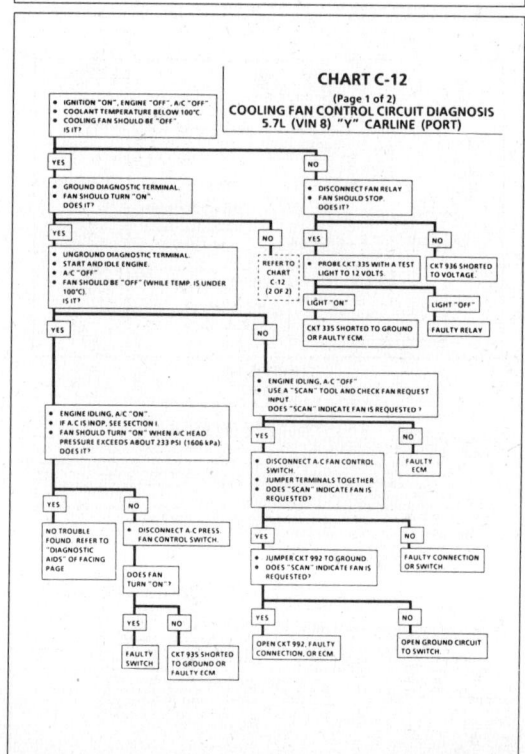

CHART C-12
(Page 1 of 2)
COOLING FAN CONTROL CIRCUIT DIAGNOSIS
5.7L (VIN 8) "Y" CARLINE (PORT)

1988–90 5.7L (VIN 8) – CORVETTE

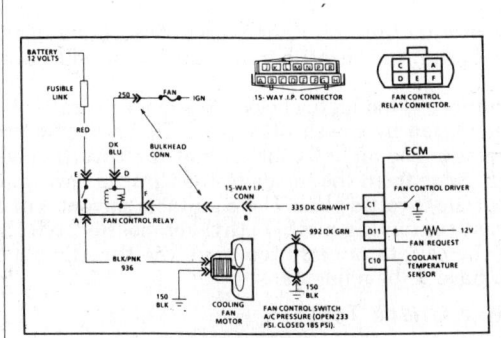

CHART C-12
(Page 2 of 2)
ELECTRIC COOLING FAN CONTROL CIRCUIT DIAGNOSIS
5.7L (VIN 8) "Y" CARLINE (PORT)

Circuit Description:
- The cooling fan is totally controlled by the ECM based on inputs from the coolant temperature sensor and fan control switch. The fan should run if coolant temperature is greater than 108°C (226°F).
- Battery voltage is supplied to the fan relay on terminal "E" and ignition voltage to terminal "D".
- Grounding CKT 335 (relay terminal "F") will energize the relay and supply battery voltage to the fan motor. Once the fan relay is energized by the ECM, it will remain "ON" for a minimum of 30 seconds.
- The ECM will remove the ground to CKT 335 if vehicle speed is over 40 mph unless the engine is overheating.
- The fan control switch, mounted in the A/C high pressure line, will open when head pressure exceeds 1600 kPa (233 psi) and this input causes the ECM to ground CKT 335.
- If a Code 14 or 15 sets or the ECM is operating in the fuel back-up mode, the ECM will turn "ON" the cooling fan.

Diagnostic Aids:

If the owner complained of an overheating problem, it must be determined if the complaint was due to an actual boil over or the hot light or temperature gage indicated overheating.

If the gate or light indicates overheating, but no boilover is detected, the gage circuit should be checked. The gage accuracy can also be checked by comparing the coolant sensor reading using a "Scan" tool and comparing its reading with the gage reading.

If the engine is actually overheating and the gage indicates overheating, but the cooling fan is not coming "ON", the coolant sensor has probably shifted out of calibration and should be replaced.

If the engine is overheating and the cooling fan is "ON", the cooling system should be checked.

1988–90 5.7L (VIN 8) – CORVETTE

CHART C-12
(Page 2 of 2)
COOLING FAN CONTROL CIRCUIT DIAGNOSIS
5.7L (VIN 8) "Y" CARLINE (PORT)

FROM CHART C-12 (1 OF 2)

- DISCONNECT FAN CONTROL RELAY
- IGNITION "ON", ENGINE STOPPED
- PROBE "D" AND "E" HARNESS TERMINALS WITH A TEST LIGHT CONNECTED TO GROUND

| LIGHT "ON" BOTH | LIGHT "OFF" ONE OR BOTH |

LIGHT "ON" BOTH:
- DIAGNOSTIC TERMINAL GROUNDED
- PROBE CKT 335 WITH A TEST LIGHT CONNECTED TO 12 VOLTS

LIGHT "OFF" ONE OR BOTH:
REPAIR OPEN OR SHORT TO GROUND IN CIRCUIT THAT DID NOT LIGHT

| LIGHT "ON" | LIGHT "OFF" |

LIGHT "ON":
- JUMPER HARNESS TERMINALS "A" AND "E" TOGETHER USING A FUSED JUMPER
- FAN SHOULD RUN. DOES IT?

LIGHT "OFF":
OPEN OR SHORT TO VOLTAGE IN CKT 335. FAULTY CONNECTION AT ECM OR A FAULTY ECM

| YES | NO |

YES: FAULTY RELAY

NO:
- WITH "A-E" STILL JUMPERED, CONNECT A TEST LIGHT ACROSS THE COOLING FAN MOTOR HARNESS CONNECTOR TERMINALS

| LIGHT "OFF" | LIGHT "ON" |

LIGHT "OFF":
- PROBE EACH TERMINAL WITH A TEST LIGHT CONNECTED TO GROUND

LIGHT "ON": FAULTY MOTOR

| LIGHT "ON" ONE | LIGHT "OFF" |

LIGHT "ON" ONE: OPEN IN GROUND CKT 150

LIGHT "OFF": REPAIR OPEN IN CKT 936

COMPONENT TESTING

Cold Start Injector Valve Testing

1. Remove the screws holding the injector in the intake manifold. DO NOT disconnect the fuel lines or electrical connector.
2. Place the cold start injector in a container to catch fuel. Wrap a clean rag around the mouth of the container.
3. Operate the starter and note the injection time. injector should spray fuel for 1–12 seconds if the coolant temperature is lower than approximately 95°F (35°F). Above this temperature, no drip or spray should be noted.
4. If the cold start injector sprays continuously or drips, replace it.
5. If the cold start injector fails to function below 95°F (35°C), replace it.

NOTE: Perform this test as quickly as possible. Avoid energizing the injector for any length of time.

6. Disconnect the cold start injector and hook up a test light across its connector. Ground the No. 1 coil terminal and run the starter. The light should glow for several seconds and then go out. If not, replace the thermo-time switch. Measure the resistance of the coils start injector using an ohmmeter. Correct resistance is 3–5 ohms. Check continuity across the cold start injector terminals.

NOTE: No starts or poor cold starting can be caused by a malfunctioning cold start injector. Cranking a cold engine with the coil wire grounded should produce a cone-shaped spray from the cold start injector. Cranking a warm engine should produce no fuel; if the injector dribbles fuel, replace it.

Throttle Plate Stop Screw (Minimum Air Rate) Adjustment Check

2.3L, 3.1L AND 1989–90 2.8L ENGINES

The throttle plate stop screw or minimum air rate adjustment for the 2.3L engine should be considered the minimum idle speed as on other fuel injected engine. Low internal friction and provision for slight production variations and various operation altitudes resulted in a calibrated minimum air rate which is too low allow most engines to idle. The adjustment is preset at the factory and no further adjustment should be necessary.

If there is a complaint of high idle speed, vacuum leaks should be considered the most likely cause. Because the Electronic Control Module (ECM) learns idle air control, it is even less likely that a stalling complaint would be due to incorrect minimum air rate.

If it is determined that the minimum air rate is suspect, be sure the IAC valve is not lost (not actually at the location indicated by current IAC counts). The IAC valve could be lost if the ECM power has been interrupted with the ignition in the **ON** position or the IAC valve has been disconnected with the engine running since the last reset. The minimum air rate may be checked using the following procedure.

1. Block the drive wheels and apply the parking brake.
2. Connect a suitable scan tool the ALCL.
3. Start the engine and allow it to reach normal operating temperature and go into the **CLOSED LOOP** mode.
4. On vehicles equipped with automatic transaxles, shift to **D** and then back to the **N** position.
5. With the A/C and all electrical accessories **OFF** and the transaxle in the **N** position, scan the power steering pressure input. It should be **OFF** or normal. If it is not repair the fault in the power steering pump circuit and allow the engine idle to stabilize.
6. Scan the IAC counts, if the counts are between 5–45 counts (10–15 counts on the 3.1L engine), the throttle plate stop screw adjustment is acceptable. It is important to allow the en-

gine idle speed to stabilize to assure correct counts are determined.
7. If the counts are below specifications, check the intake manifold for vacuum leaks at the hoses, throttle body and intake manifold or damaged throttle lever and correct as necessary.
8. If no vacuum leaks or other causes of excessive air into the intake are found, the throttle plate stop screw will have to be adjusted.

If the counts are too high, check for damaged throttle lever or airflow restriction by the throttle plate. If there is no problem evident, remove the air inlet air duct at the throttle body and clean the residue from the inside of the throttle bore and from the edges of the throttle plate. Use a suitable solvent, but do not use a solvent containing methyl ethyl ketone. Recheck the IAC counts, if the counts are still too high, the throttle plate stop screw will have to be adjusted.

TPS Output Check Test

WITH SCAN TOOL

The following check should be performed only when the throttle body or TPS has been replaced.
1. Use a suitable scan tool to read the TPS voltage.
2. With the ignition switch **ON** and the engine **OFF**, the TPS voltage should be less, than 1.25 volts.
3. If the voltage reading is higher than specified, replace the throttle position sensor.

WITHOUT SCAN TOOL

This check should only be performed, when the throttle body or the TPS has been replaced or after the minimum idle speed has been adjusted.
1. Disconnect the TPS harness from the TPS.
2. Using suitable jumper wires, connect a digital voltmeter J–29125–A or equivalent to the correct TPS terminals, terminals **A** and **B** on all models.
3. With the ignition **ON** and the engine running, The TPS voltage should be 0.450–1.25 volts at base idle to approximately 4.5 volts at wide open throttle.
4. If the reading on the TPS is out of specification, adjust the TPS and check the minimum idle speed before replacing the TPS.
5. If the voltage reading is correct, remove the voltmeter and jumper wires and reconnect the TPS connector to the sensor. Re-install the air cleaner.

Crankshaft Sensor Inspection

3.0L ENGINE

1. Disconnect the negative battery cable.

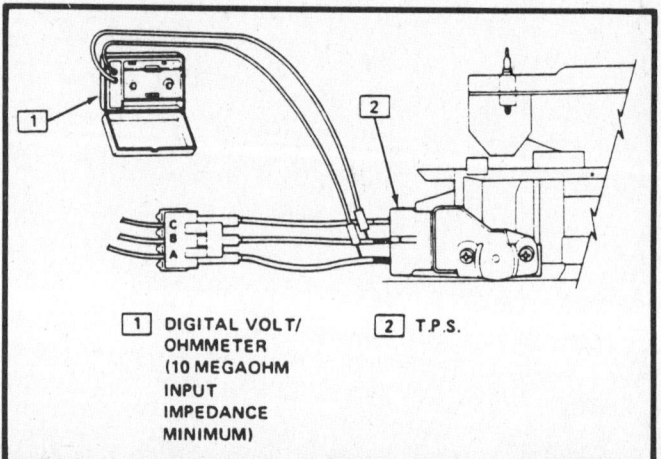

| 1 | DIGITAL VOLT/ OHMMETER (10 MEGAOHM INPUT IMPEDANCE MINIMUM) | 2 | T.P.S. |

Throttle position sensor adjustment

2. Rotate the harmonic balancer using a 28mm socket and pull on the handle until the interrupter ring fills the sensor slots and edge of the interrupter window is aligned with the edge of the deflector on the pedestal.

3. Insert feeler gauge adjustment tool J–36179 or equivalent into the gap between the sensor and the interrupter on each side of the interrupter ring.

4. If the gauge will not slide past the sensor on either side of the interrupter ring, the sensor is out of adjustment or the interrupter ring is bent.

5. The clearance should be checked again, at 3 positions around the interrupter ring approximately 120 degrees apart.

6. If found out of adjustment, the sensor should be removed and inspected for potential damage.

Fuel System Pressure Test

When the ignition switch is turned **ON**, the in-tank fuel pump is energized for as long as the engine is cranking or running and the control unit is receiVINg signals from the HEI distributor or DIS. If there are no reference pulses, the control unit will shut off the fuel pump within 2 seconds. The pump will deliver fuel to the fuel rail and injectors, then the pressure regulator where the system pressure is controlled to maintain 26–46 psi.

1. Connect pressure gauge J–34730–1, or equivalent, to fuel pressure test point on the fuel rail. Wrap a rag around the pressure tap to absorb any leakage that may occur when installing the gauge.

2. Turn the ignition **ON** and check that pump pressure is 24–40 psi. This pressure is controlled by spring pressure within the regulator assembly.

3. Start the engine and allow it to idle. The fuel pressure should drop to 28–32 psi due to the lower manifold pressure.

NOTE: The idle pressure will vary somewhat depending on barometric pressure. Check for a drop in pressure indicating regulator control, rather than specific values.

4. On turbocharged models, use a low pressure air pump to apply air pressure to the regulator to simulate turbocharger boost pressure. Boost pressure should increase fuel pressure 1 lb. for every lb. of boost. Again, look for changes rather than specific pressures. The maximum fuel pressure should not exceed 46 psi.

5. If the fuel pressure drops, check the operation of the check valve, the pump coupling connection, fuel pressure regulator valve and the injectors. A restricted fuel line or filter may also cause a pressure drip. To check the fuel pump output, restrict the fuel return line and run 12 volts to the pump. The fuel pressure should rise to approximately 75 psi with the return line restricted.

—————————— CAUTION ——————————
Before attempting to remove or service any fuel system component, it is necessary to relieve the fuel system pressure.

Relieving Fuel System Pressure

1. Remove the fuel pump fuse from the fuse block.

2. Start the engine. It should run and then stall when the fuel in the lines is exhausted. When the engine stops, crank the starter for about 3 seconds to make sure all pressure in the fuel lines is released.

3. Replace the fuel pump fuse.

4. On some models a pressure relief valve is located on the fuel rail. To relive the fuel pressure using the relief valve use the following procedure.

a. Disconnect the negative battery cable to avoid possible fuel discharge if an accidental attempt is made to start the engine. Loosen the fuel filler cap to relieve fuel tank pressure.

b. Connect fuel gauge J–34730–1, or equivalent, to fuel pressure relief valve on the fuel rail. Wrap a rag around the pressure tap to absorb any leakage that may occur when installing the gauge.

c. Install bleed hose into a suitable container and open the the valve to bleed off the fuel pressure in the fuel system.

Component Replacement

Due to the varied application of components, a general component replacement procedure section is outlined. The removal steps can be altered as required by the technician.

CLEANING AND INSPECTION

All MPI component parts, with the exception of those noted below, should be cleaned in a cold immersion cleaner such as Carbon X (X–55) or equivalent. The throttle position sensor, idle air control valve, pressure regulator diaphragm assembly, fuel injectors or other components containing rubber should not be placed in a solvent or cleaner bath. A chemical reaction will cause these parts to swell, harden or distort. Do not soak the throttle body with the above parts attached. If the throttle body assembly requires cleaning, soak time in the cleaner should be kept to a minimum. Some models have hidden throttle shaft dust seals that could lose their effectiveness by extended soaking.

1. Clean all parts thoroughly and blow dry with compressed air. Be sure that all fuel and air passages are free of dirt and burrs.

2. Inspect the mating casting surfaces for damage that could affect gasket sealing.

THREAD LOCKING COMPOUND

Service repair kits are supplied with a small vial of thread locking compound with directions for use. If material is not available, use Loctite® 262 or GM part No. 10522624 or equivalent. In precoating the screws, do not use a higher strength locking compound than recommended, since to do so could make removing the screw extremely difficult, or result in damaging the screw head.

FUEL RAIL ASSEMBLY

When servicing the fuel rail assembly, be careful to prevent dirt and other contaminants from entering the fuel passages. Fitting should be capped and holes plugged during servicing. At any time the fuel system is opened for service, the O-ring seals and retainers used with related components should be replaced.

Before removing the fuel rail, the fuel rail assembly may be cleaned with a spray type cleaner, GM–30A or equivalent, following package instructions. Do not immerse fuel rails in liquid cleaning solvent. Be sure to always use new O-rings and seals when reinstalling the fuel rail assemblies.

There is an 8 digit number stamped on the under side of the fuel rail assembly on 4 cylinder engines and on the left hand fuel rail on dual rail assemblies (fueling even cylinders No. 2, 4, 6). Refer to this number if servicing or part replacement is required.

Removal and Installation
2.0L AND 2.3L ENGINES

1. Disconnect the negative battery cable. Relieve fuel system pressure.

2. Remove the crankcase ventilation oil/air separator.

3. Disconnect the fuel feed line and return line from the fuel rail assembly, be sure to use a backup wrench on the inlet fitting to prevent turning.

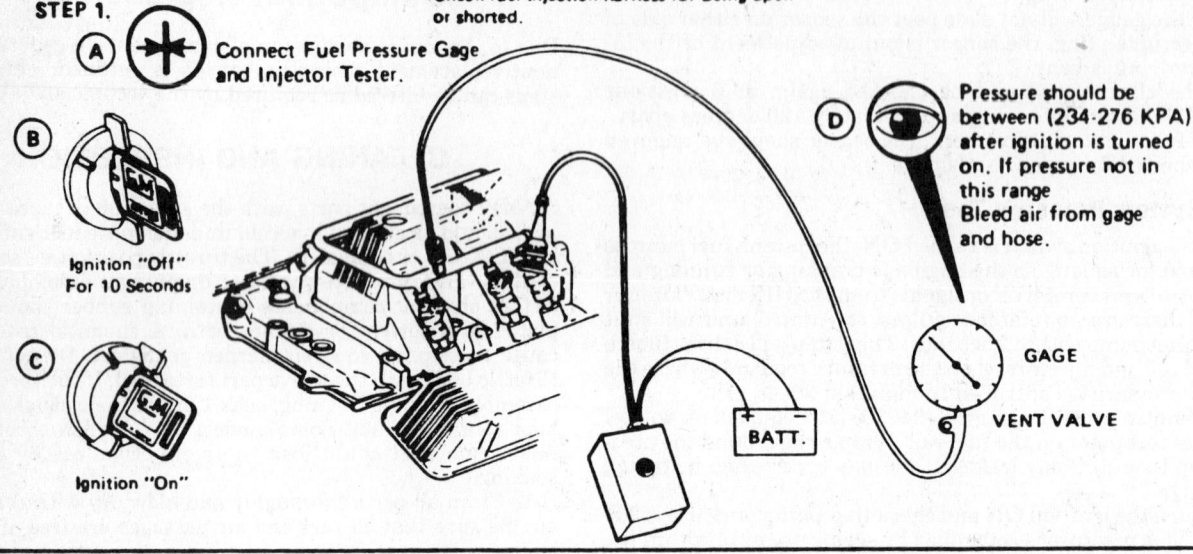

PORT FUEL INJECTION—INJECTOR BALANCE TEST
Before performing this test, the items listed below must be done.

- Check spark plugs and wires.
- Check compression.
- Check fuel injection harness for being open or shorted.

STEP 1.

Ⓐ Connect Fuel Pressure Gage and Injector Tester.

Ⓑ Ignition "Off" For 10 Seconds

Ⓒ Ignition "On"

Ⓓ Pressure should be between (234-276 KPA) after ignition is turned on. If pressure not in this range Bleed air from gage and hose.

GAGE
VENT VALVE
BATT.

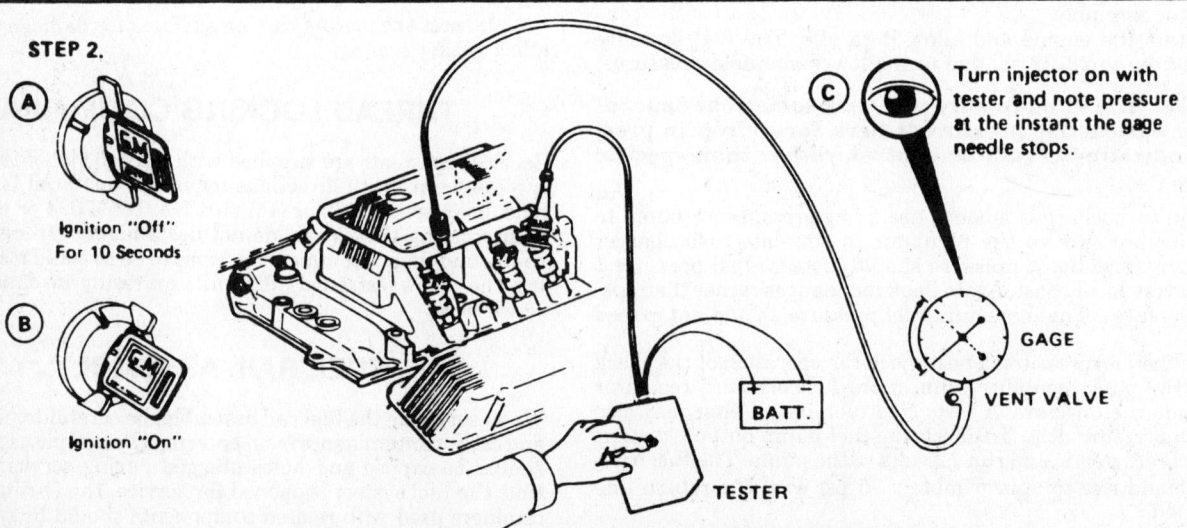

STEP 2.

Ⓐ Ignition "Off" For 10 Seconds

Ⓑ Ignition "On"

Ⓒ Turn injector on with tester and note pressure at the instant the gage needle stops.

GAGE
VENT VALVE
BATT.
TESTER

STEP 3.

Repeat test as in step 2 on all injectors and record pressure drop on each.

Retest injectors that appear faulty. Replace any injectors that have a 10 KPA difference either (more or less) in pressure.

— EXAMPLE —

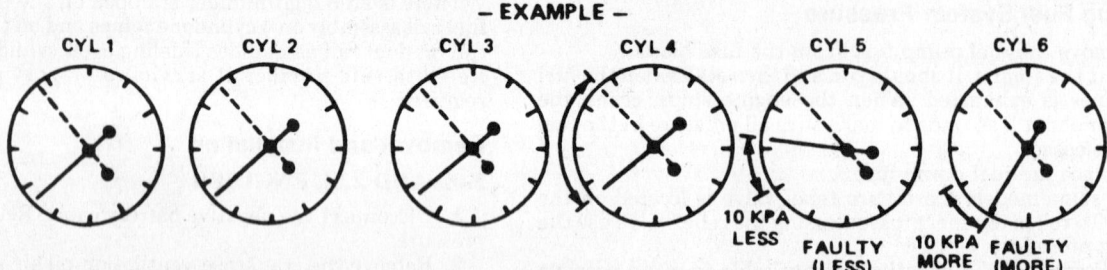

CYL 1   CYL 2   CYL 3   CYL 4   CYL 5   CYL 6

10 KPA LESS   FAULTY (LESS)   10 KPA MORE   FAULTY (MORE)

Fuel injection balance test

4. Remove the vacuum line at the pressure regulator. Remove the 2 fuel rail assembly retaining bolts.

5. Push in the wire connector clip, while pulling the connector away from the injector.

6. Remove the fuel rail assembly and cover all openings with masking tape to prevent dirt entry.

NOTE: If any injectors become separated from the fuel rail and remain in the intake manifold, both O-ring seals and injector retaining clip must be replaced. Use care in removing the fuel rail assembly, to prevent damage to the injector electrical connector terminals and the injector spray tips. When removed, support the the fuel rail to avoid damaging its components. The fuel injector is serviced as a complete unit only. Since it is an electrical component, it should not be immersed in any type of cleaner.

7. Installation is the reverse order of the removal procedure. Be sure to lubricate all the O-rings and seals with clean engine oil. Carefully push the injectors into the cylinder head intake ports until the bolt holes on the fuel rail and manifold are aligned. Torque the fuel rail retaining bolts to 19 ft. lbs. (26 Nm) and the fuel feed line nut to 22 ft. lbs. (30 Nm). Energize the fuel pump and check for leaks.

2.8L, 3.0L, 3.1L AND 3.3L ENGINES

1. Disconnect the negative battery cable. Relieve fuel system pressure.

2. Remove the intake manifold plenum (if necessary). Remove the fuel line bracket bolt.

3. Disconnect the fuel feed line and return line from the fuel rail assembly, be sure to use a backup wrench on the inlet fitting to prevent turning.

4. Remove the vacuum line at the pressure regulator. Remove the fuel rail assembly retaining bolts.

5. Push in the wire connector clip, while pulling the connector away from the injector.

6. Remove the fuel rail assembly and cover all openings with masking tape to prevent dirt entry.

NOTE: If any injectors become separated from the fuel rail and remain in the intake manifold, both O-ring seals and injector retaining clip must be replaced. Use care in removing the fuel rail assembly, to prevent damage to the injector electrical connector terminals and the injector spray tips. When removed, support the the fuel rail to avoid damaging its components. The fuel injector is serviced as a complete unit only. Since it is an electrical component, it should not be immersed in any type of cleaner.

7. Installation is the reverse order of the removal procedure. Be sure to lubricate all the O-rings and seals with clean engine oil. Carefully tilt the fuel rail assembly and push the injectors into the cylinder head intake ports until the bolt holes on the fuel rail and manifold are aligned. Torque the fuel rail retaining bolts to 88 inch. lbs. (10 Nm) and the fuel feed line nut to 17 ft. lbs. (23 Nm). Energize the fuel pump and check for leaks.

5.0L AND 5.7L ENGINES

1. Disconnect the negative battery cable. Relieve fuel system pressure.

2. Remove the intake manifold plenum. Remove the cold start valve fuel line and plenum runners. Remove the cold start valve.

3. Disconnect the fuel feed line and return line from the fuel rail assembly, be sure to use a backup wrench on the inlet fitting to prevent turning.

4. Disconnect the injector wire harness connectors.

5. Remove the vacuum line from the fuel pressure regulator. Remove the fuel rail assembly retaining bolts.

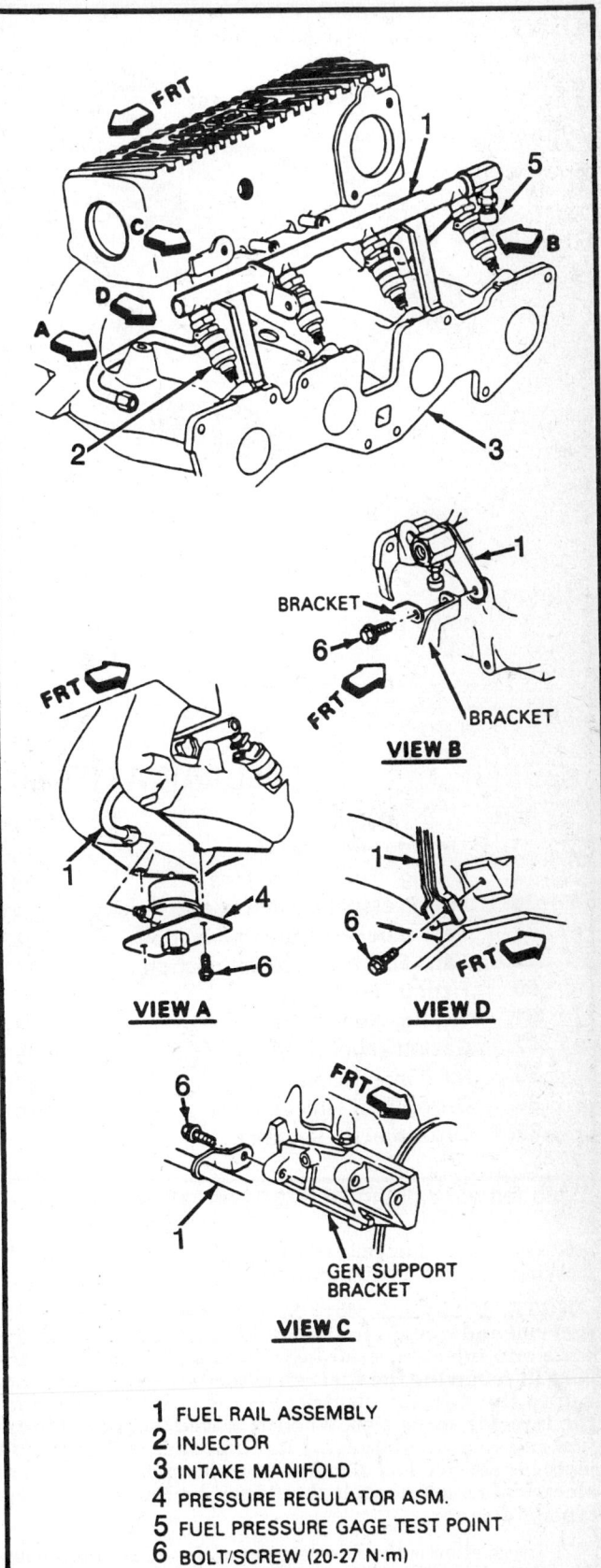

1 FUEL RAIL ASSEMBLY
2 INJECTOR
3 INTAKE MANIFOLD
4 PRESSURE REGULATOR ASM.
5 FUEL PRESSURE GAGE TEST POINT
6 BOLT/SCREW (20-27 N·m)

Fuel rail assembly removal and installation — 2.0L engine

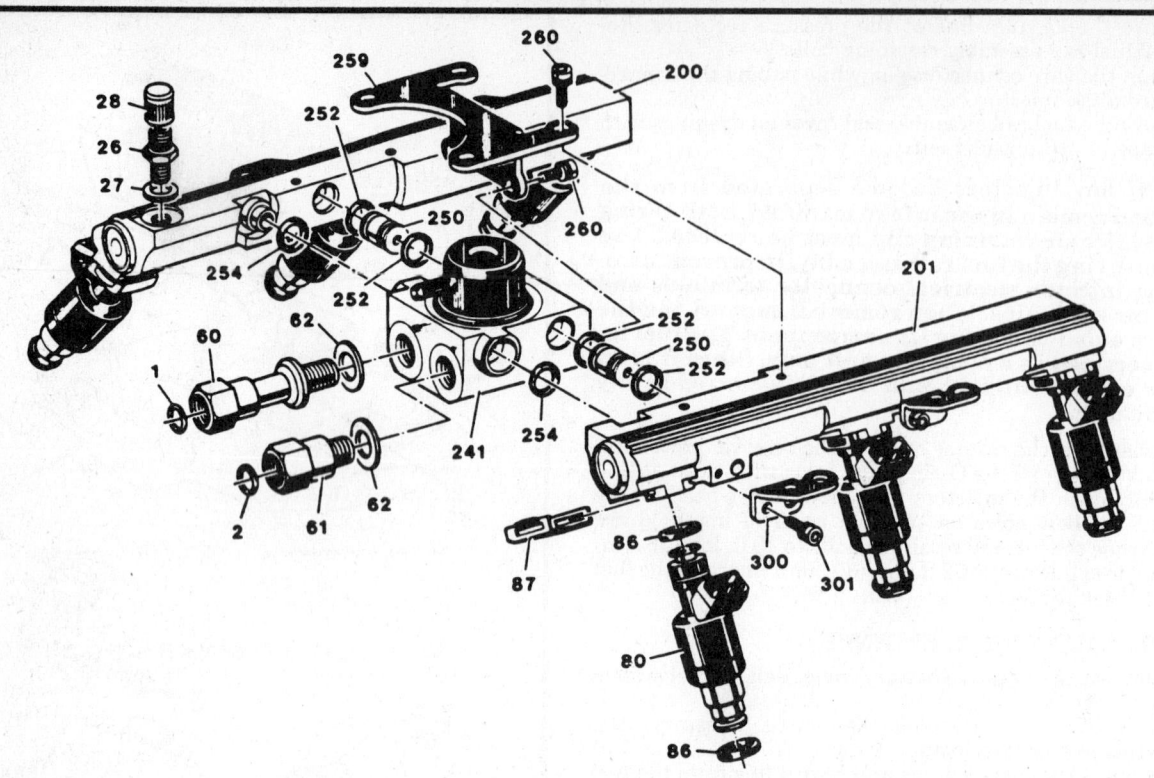

FUEL RAIL ASSEMBLY PARTS IDENTIFICATION

Part	Name	Part	Name
1.	O-ring - Fuel Inlet Line	200.	Fuel Rail and Plug Assembly - Left Hand
2.	O-ring - Fuel Return Line	201.	Fuel Rail and Plug Assembly - Right Hand
26.	Fuel Pressure Connection Assembly	241.	Pressure Regulator Assembly
27.	Seal - Fuel Pressure Connection	250.	Connector - Base to Rail
28.	Cap - Fuel Pressure Connection	252.	O-ring Seal - Connector
60.	Fitting - Fuel Inlet	254.	O-ring Seal - Fuel Return
61.	Fitting - Fuel Outlet	259.	Bracket - Pressure Regulator Mounting
62.	Gasket - Fuel Fitting	260.	Screw Assembly - Pressure Regulator (Attaching)
80.	MPFI Injector Assembly	300.	Bracket - Rail Mounting
86.	O-ring Seal - Injector	301.	Screw Assembly - Bracket Attaching
87.	Clip - Injector Retainer		

Exploded view of model R260 fuel rail assembly

6. Remove the fuel rail assembly and cover all openings with masking tape to prevent dirt entry.

NOTE: If any injectors become separated from the fuel rail and remain in the intake manifold, both O-ring seals and injector retaining clip must be replaced. Use care in removing the fuel rail assembly, to prevent damage to the injector electrical connector terminals and the injector spray tips. When removed, support the the fuel rail to avoid damaging its components. The fuel injector is serviced as a complete unit only. Since it is an electrical component, it should not be immersed in any type of cleaner.

7. Installation is the reverse order of the removal procedure. Be sure to lubricate all the O-rings and seals with clean engine oil. Carefully tilt the fuel rail assembly and push the injectors into the cylinder head intake ports until the bolt holes on the fuel rail and manifold are aligned. Torque the fuel rail retaining bolts to 15–22 ft. lbs. (front lower fuel rail lines to engine block bolt to 20–30 ft. lbs.) and the fuel feed line nut to 17 ft. lbs. (23 Nm). Energize the fuel pump and check for leaks.

FUEL INJECTORS

Removal and Installation

Use care in removing the fuel injectors to prevent damage to the electrical connector pins on the injector and the nozzle. The fuel injector is serviced as a complete assembly only and should not be immersed in any kind of cleaner. Support the fuel rail to avoid damaging other components while removing the injector. Be sure to note that different injectors are calibrated for differ-

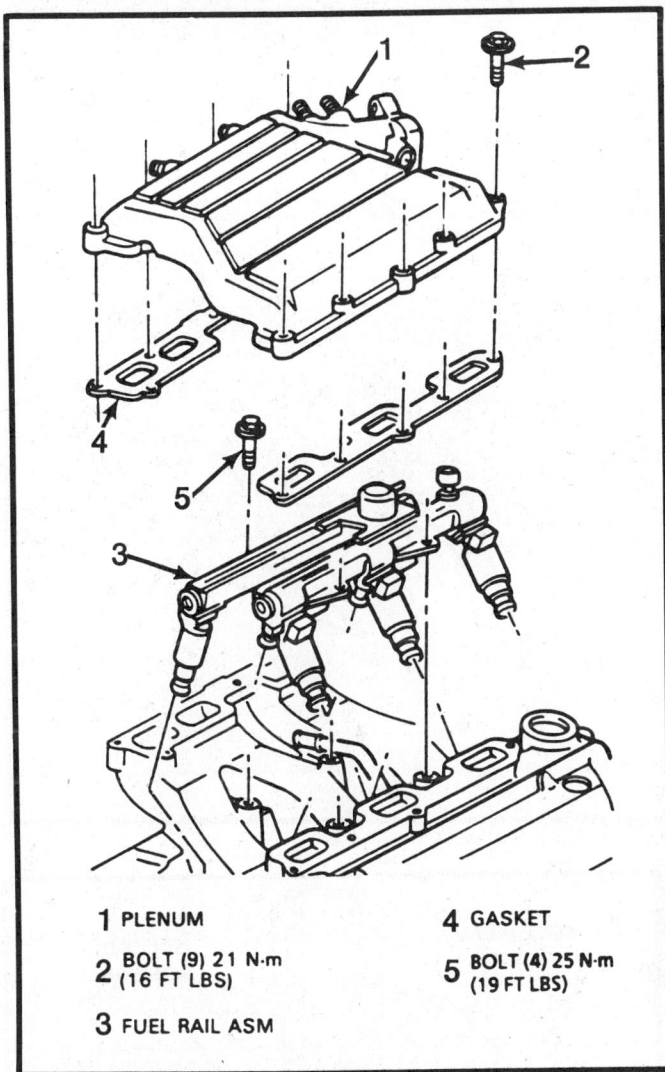

1 PLENUM	**4** GASKET
2 BOLT (9) 21 N·m (16 FT LBS)	**5** BOLT (4) 25 N·m (19 FT LBS)
3 FUEL RAIL ASM	

Removing the plenum and fuel rail

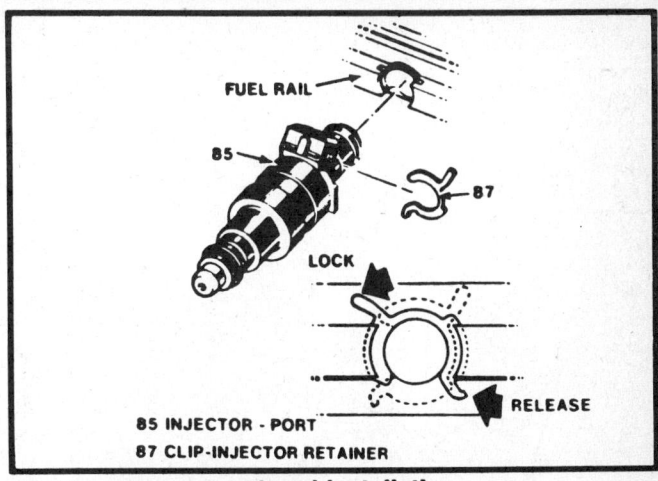

85 INJECTOR - PORT
87 CLIP-INJECTOR RETAINER

Fuel injector removal and installation

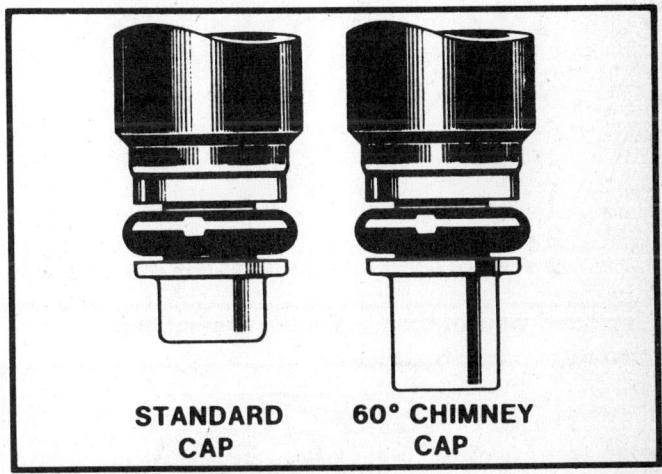

STANDARD CAP   60° CHIMNEY CAP

Typical Multec (top feed) injector

ent flow rates. When ordering new fuel injectors, be sure to order the identical part number that is inscribed on the bottom of the old injector.

NOTE: The most widely used injector for the MPI units are Bosch injectors, but starting in 1987 there will also be a new injector being used. This new injector will be a Multec MPI injector (a Rochester product). It is classified as a top feed design because the fuel enters the top of the injector and then flows through the entire length of the injector. It is designed to operate with the system fuel pressures ranging from 36–51 psi (250–350 kPa), and uses a high impedance (12.2 ohms) solenoid coil. It is used in the multi-port fuel injection system on some Chevrolet produced 2.8L V6 engines.

1. Relieve fuel system pressure.
2. Remove the intake manifold plenum.
3. Remove the injector electrical connections. Remove the fuel rail assembly.
4. Remove the injector retaining clip (if used). Separate the injector from the fuel rail.
5. Installation is the reverse of removal. Replace the O-rings when installing injectors into intake manifold.

NOTE: As a running change during the 1987 model year, the model F8A fuel rails, used on the 5.0L and 5.7L V8 MPI engines, will be produced with Bosch injectors having aluminum CHIMNEY CAPS with a 60 degrees taper opening. The chimney cap replaces the plastic protective caps earlier designs, and shrouds the pintle to reduce chance of plugging. Both engines use a new intake manifold to accept the 60 degrees chimney cap injector, so this injector will not fit in the earlier engines.

FUEL PRESSURE REGULATOR

Removal and Installation

ALL MODELS

1. Disconnect the negative battery cable and relieve fuel system pressure.
2. Remove the intake manifold plenum (if necessary). Remove the fuel line bracket bolt.
3. Disconnect the fuel feed line and return line from the fuel rail assembly, be sure to use a backup wrench on the inlet fitting to prevent turning.
4. Remove the vacuum line at the pressure regulator. Remove the fuel rail assembly retaining bolts.
5. Push in the wire connector clip, while pulling the connector away from the injector.
6. Remove the fuel rail assembly and cover all openings with masking tape to prevent dirt entry.

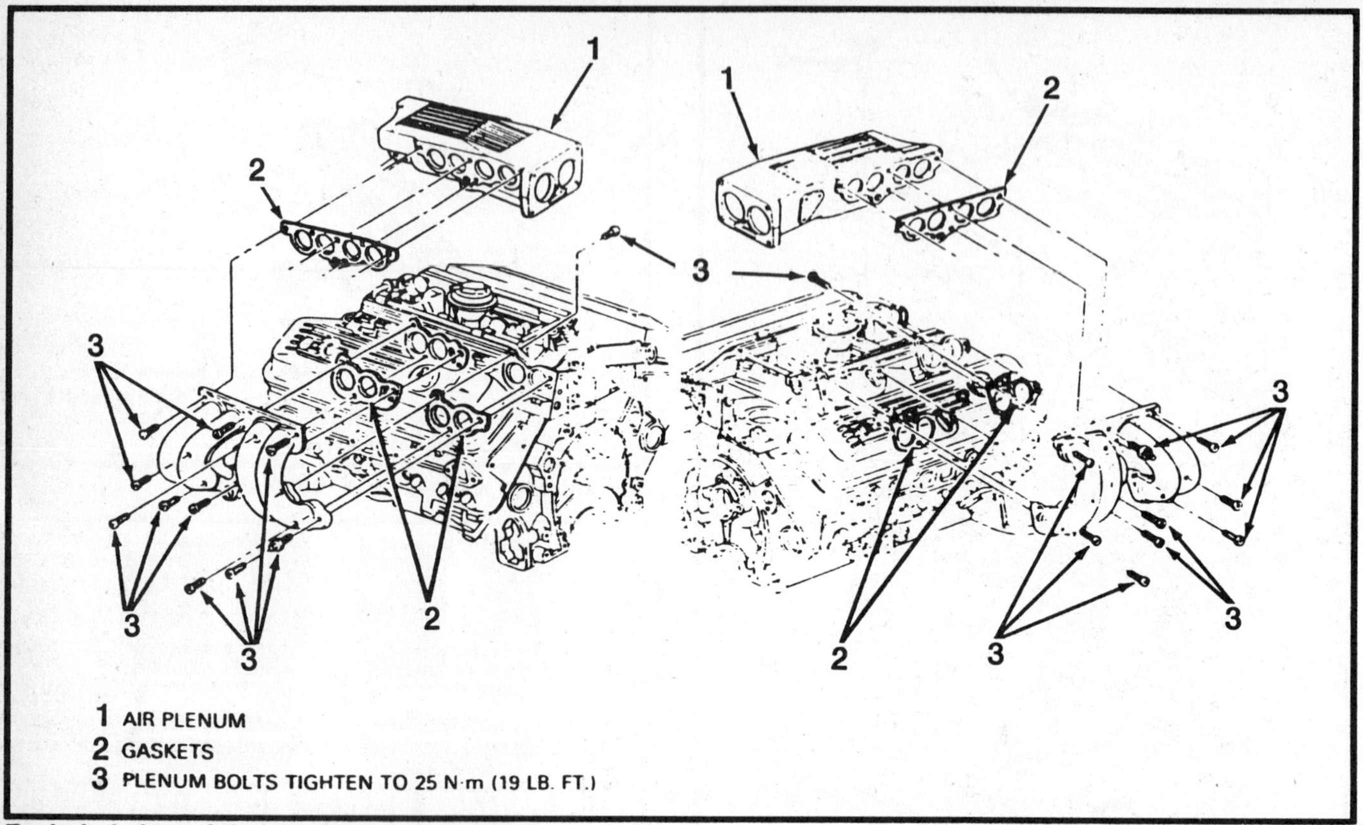

1 AIR PLENUM
2 GASKETS
3 PLENUM BOLTS TIGHTEN TO 25 N·m (19 LB. FT.)

Exploded view of the air plenum assemblies

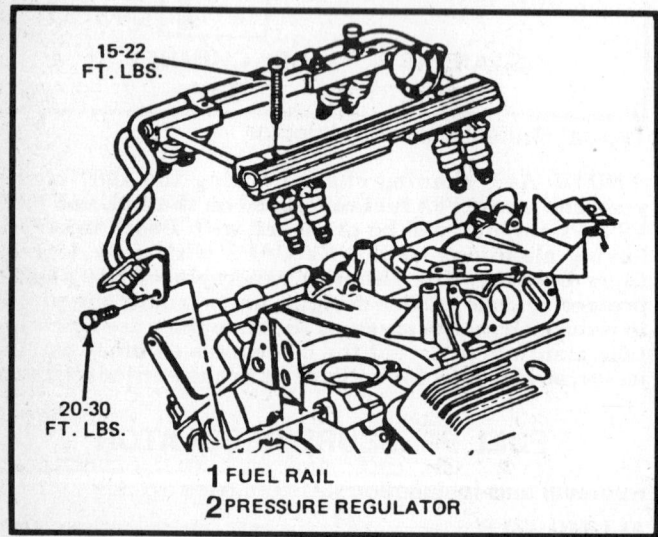

1 FUEL RAIL
2 PRESSURE REGULATOR

15-22 FT. LBS.

20-30 FT. LBS.

Fuel rail assembly removal and installation—5.0L and 5.7L engines

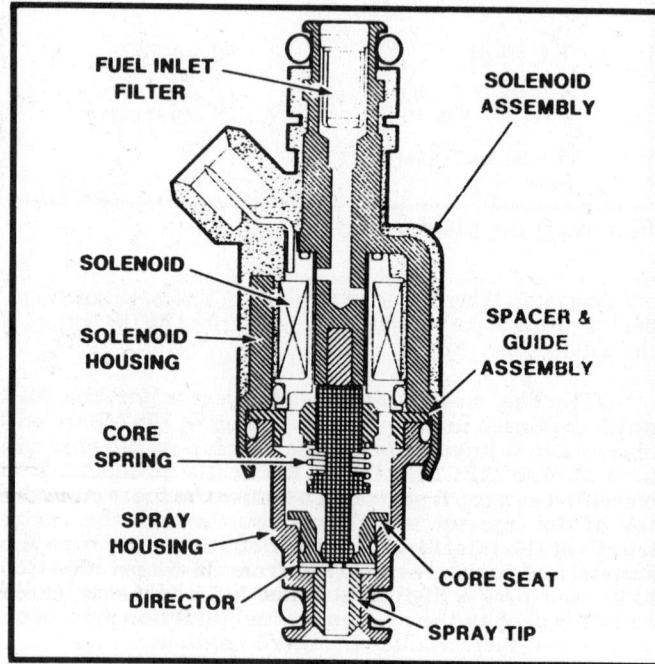

FUEL INLET FILTER

SOLENOID ASSEMBLY

SOLENOID

SOLENOID HOUSING

CORE SPRING

SPRAY HOUSING

DIRECTOR

SPACER & GUIDE ASSEMBLY

CORE SEAT

SPRAY TIP

Standard and Chimney-Bosch injector caps

THROTTLE BODY ASSEMBLY
Removal and Installation
2.0L AND 2.3L ENGINES

1. Disconnect the negative battery cable. Drain the top half of the engine coolant into a suitable drain pan.

7. Remove the fuel inlet fitting and fuel outlet fitting, along with the gaskets from the fuel pressure regulator.

8. Remove the fuel pressure regulator bracket attaching screws and then remove the bracket.

9. Remove the left hand fuel rail assembly and the right hand fuel rail assembly, from the fuel pressure regulator assembly.

10. Remove the base rail connectors from the pressure regulator or rails.

11. Installation is the reverse order of the removal procedure. Be sure to use new O-rings and gaskets.

2. Remove the air inlet duct. Disconnect the idle air control valve and throttle position sensor connectors.

3. Remove and mark all necessary vacuum lines. Remove and plug the 2 coolant hoses.

4. Remove the throttle, T.V. and cruise control cables. Remove the power steering pump brace.

5. Remove the throttle body retaining bolts and then remove the throttle body assembly. Discard the flange gasket.

6. Installation is the reverse order of the removal procedure. Torque the retaining bolts to 16–22 ft. lbs.

2.8L, 3.0L, 3.1L AND 3.3L ENGINES

1. Disconnect the negative battery cable. Drain the top half of the engine coolant into a suitable drain pan.

2. Remove the air inlet duct assembly. Disconnect the idle air control valve, throttle position sensor connectors and mass airflow sensor.

3. Remove and mark all necessary vacuum lines. Remove and plug the 2 coolant hoses.

4. Remove the 10mm screw holding the fuel lines to the throttle cable bracket. Remove the throttle, T.V. and cruise control cables. Remove the power steering pump brace.

5. Remove the throttle body retaining bolts and then remove the throttle body assembly. Discard the flange gasket.

6. Installation is the reverse order of the removal procedure. Torque the retaining bolts to 11–18 ft. lbs.

5.0 AND 5.7L ENGINES

1. Disconnect the negative battery cable. Drain the top half of the engine coolant into a suitable drain pan.

2. Remove the air inlet duct assembly. Disconnect the idle air control valve and throttle position sensor connectors.

3. Remove and mark all necessary vacuum lines. Remove and plug the 2 coolant hoses.

4. Remove the throttle, T.V. and cruise control cables. Remove the power steering pump brace.

5. Remove the throttle body retaining bolts and then remove the throttle body assembly. Discard the flange gasket.

6. Installation is the reverse order of the removal procedure. Torque the retaining bolts to 18 ft. lbs.

COLD START TUBE AND VALVE ASSEMBLY

Removal and Installation

5.0L AND 5.7L ENGINES

1. Disconnect the negative battery cable. Relieve fuel system pressure.

2. Remove the air intake plenum assembly. Disconnect the brake booster line.

3. Provide a clean container to catch any fuel, or wrap some clean rags around the electrical connections.

—————————— CAUTION ——————————

Be careful not to let any dirt enter the fuel system and take precautions to avoid the risk of fire.

4. Disconnect the electrical connector from the cold start injector and clean off any dirt or grease from the injector. Disconnect the PCV line.

5. Remove the fuel line from the injector and be careful because the injector body is plastic. After removing the fuel hose from the injector, inspect the hose from cracks and/or leaks.

6. Remove the 2 fasteners holding the cold start injector in the base plate, remove the injector and discard the old O-ring or gasket.

7. Installation is the reverse order of the removal procedure. Torque the retaining bolts to 20 ft. lbs. After installation make sure the system is tight and free from leaks. Replace the rubber sealing ring, or gasket and hose clamp is necessary.

NOTE: The fuel injections systems are very susceptible to dirt in the system. Be sure that all components are clean and free from dirt and grease before reinstalling them.

IDLE AIR CONTROL VALVE

Removal and Installation

1. Disconnect the negative battery cable. Removal electrical connector from idle air control valve.

2. Remove the idle air control valve using a suitable (1¼) wrench.

3. Installation is the reverse of removal. Before installing the idle air control valve, measure the distance that the valve is extended. Measurement should be made from the motor housing to the end of the cone. The distance should not exceed 1⅛ in. (28mm), or damage to the valve may occur when installed. Use a new gasket and turn the ignition **ON** then **OFF** again to allow the ECM to reset the idle air control valve. Torque the IAC valve to 13 ft. lbs. (18 Nm).

NOTE: Identify replacement IAC valve as being either. Type 1 (with collar at electric terminal end) or Type 2 (without collar). If measuring distance is greater than specified above, proceed as follows:

a. Type 1: Press on valve firmly to retract it, using a slight side to side motion to help it retract easier.

b. Type 2: Compress retaining spring from valve while turning valve in with a clockwise motion. Return spring to original position with straight portion of spring end aligned with flat surface of valve.

NOTE: The 2.0L and the 3.3L MPI engines do not have a minimum idle speed adjustment procedure. On the 2.0L engines, the ECM will reset the IAC valve when the vehicle is driven above 35 mph. On the 3.3L engine, the IAC valve is reset by grounding the ALCL (or ALDL) for 5–10 seconds.

Minimum Idle Speed Adjustment

The throttle stop screw that is used to adjust the idle speed of the vehicle, is adjusted to specifications at the factory. The throttle stop screw is then covered with a steel plug to prevent the unnecessary readjustment in the field. If it is necessary to gain access to the throttle stop screw, the following procedure will allow access to the throttle stop screw without removing the throttle body unit from the manifold.

3.0L, 5.0L, 5.7L AND 1988 2.8L ENGINES

1. Apply the parking brake and block the drive wheels. Remove the plug from the idle stop screw by piercing it first with a suitable tool, then applying leverage to the tool to lift the plug out.

2. Leave the Idle Air Control (IAC) valve connected and ground the diagnostic terminal (ALDL connector).

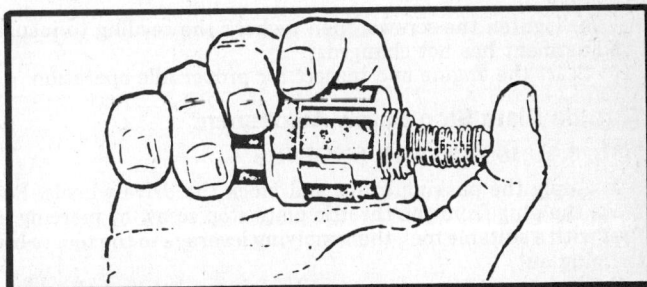

Adjusting the idle air control valve—Type 1 with a collar at the electrical terminal end

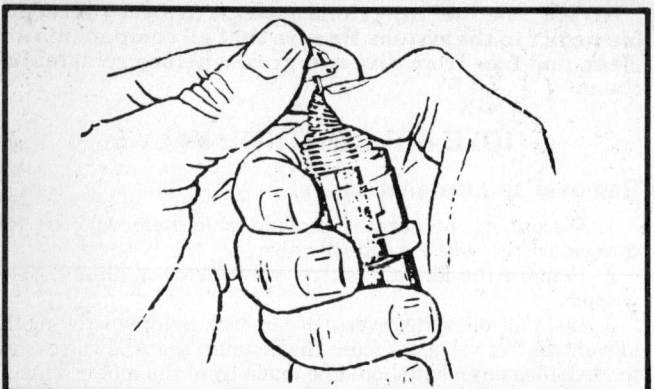

Adjusting the idle air control valve—Type 2 without a collar at the electrical terminal end

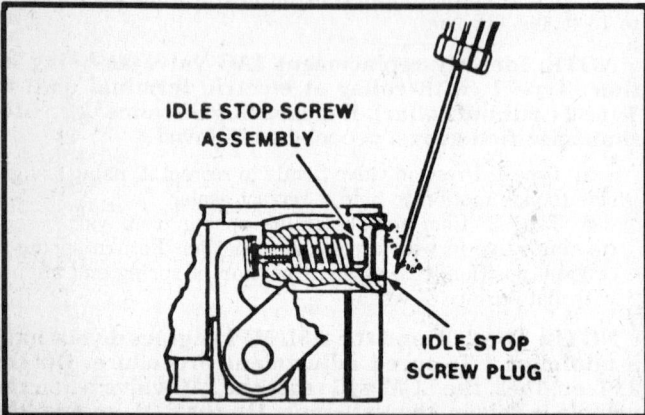

Idle stop screw plug removal

3. Turn the ignition switch to the **ON** position, do not start the engine. Wait for at least 30 seconds (this allows the IAC valve pintle to extend and seat in the throttle body).

4. With the ignition switch still in the **ON** position, disconnect IAC electrical connector.

5. Remove the ground from the diagnostic terminal. Disconnect the distributor set-timing connector (5.0L and 5.7L only). Start the engine and allow the engine to reach normal operating temperature.

6. Apply the parking brake and block the drive wheels.

7. With the engine in the drive position adjust the idle stop screw to obtain the correct specifications.

8. Turn the ignition **OFF** and reconnect the connector at the IAC motor.

9. Adjust the TPS if necessary as follows:

a. With the ignition **ON**, use a suitable scan tool or 3 jumper wires and adjust the TPS to obtain the specified voltage.

b. Tighten the screws, then recheck the reading to insure adjustment has not changed.

10. Start the engine and inspect for proper idle operation.

Throttle Plate Stop Screw Adjustment

2.3L, 3.1L 1989-90 2.8L ENGINES

1. Apply the parking brake and block the drive wheels. Remove the plug from the throttle plate stop screw by piercing it first with a suitable tool, then applying leverage to the tool to lift the plug out.

2. Back the stop screw out until an air gap between the screw and the throttle lever can be seen by looking downward from above the throttle body.

3. Turn the stop screw out until it just contacts the throttle lever. Turn the stop screw 1½ turns further.

4. Recheck the adjustment as previously outlined in this section.

5. If the IAC valve counts are still out of the acceptable range after the idle has been allowed to stabilize, the adjustment may be tailored as follows:

a. For more throttle plate opening—lower the IAC counts.

b. For less throttle plate opening—raise the IAC counts.

6. If it is necessary to change the adjustment more than a ½ a turn either way, other causes of incorrect idle speed should be investigated.

7. Adjust the TPS as follows:

a. Install a suitable scan tool to the ALCL and place it on TPS.

b. Turn the ignition to the **ON** position and adjust the TPS to obtain the specified voltage.

c. Tighten the screws, then recheck the reading to insure adjustment has not changed.

8. Start the engine and inspect for proper idle operation.

THROTTLE POSITION SENSOR

The throttle position sensor on some models are not adjustable. If the sensor is found out of specifications and the sensor is at fault it cannot be adjusted and should be replaced.

Removal and Installation

1. Disconnect the electrical connector from the sensor.

2. Remove the attaching screws, lock washers and retainers.

3. Remove the throttle position sensor. If necessary, remove the screw holding the actuator to the end of the throttle shaft.

4. With the throttle valve in the normal closed idle position, install the throttle position sensor on the throttle body assembly, making sure the sensor pickup lever is located above the tang on the throttle actuator lever.

5. Install the retainers, screws and lock washers using a thread locking compound. DO NOT tighten the screws until the throttle position switch is adjusted.

TPS Output Check Test

WITH SCAN TOOL

The following check should be performed only when the throttle body or TPS has been replaced.

1. Use a suitable scan tool to read the TPS voltage.

2. With the ignition switch **ON** and the engine **OFF**, the TPS voltage should be less, than 1.25 volts.

3. If the voltage reading is higher than specified, replace the throttle position sensor.

WITHOUT SCAN TOOL

This check should only be performed, when the throttle body or the TPS has been replaced or after the minimum idle speed has been adjusted.

1. Disconnect the TPS harness from the TPS.

2. Using suitable jumper wires, connect a digital voltmeter J–29125–A or equivalent to the correct TPS terminals, terminals **A** and **B** on all models.

3. With the ignition **ON** and the engine running, The TPS voltage should be 0.450–1.25 volts at base idle to approximately 4.5 volts at wide open throttle.

4. If the reading on the TPS is out of specification, adjust the TPS as necessary and check the minimum idle speed before replacing the TPS.

5. If the TPS is out of specifications, loosen the 2 TPS attaching screws and rotate throttle position sensor to obtain a correct voltage reading.

6. Tighten the mounting screws, then recheck the reading to insure that the adjustment has not changed.

7. Turn ignition **OFF**, remove jumper wires, then reconnect harness to throttle position switch.

OXYGEN SENSOR

Removal and Installation

The oxygen sensor uses a permanently attached pigtail and connector. This pigtail should not be removed from the oxygen sensor. Damage or removal of the pigtail connector could effect the proper operation of the oxygen sensor. Use caution when handling the oxygen sensor. The in-line electrical connector and louvered end must be kept free of grease, dirt or other contaminants. Also avoid using cleaner solvent of any type. Do not drop or roughly handle the oxygen sensor. It is not recommended to clean or wire brush an oxygen sensor when in question, it is recommended that a new oxygen sensor be used.

NOTE: The oxygen sensor may be difficult to remove when the engine temperature is below 120°F (48°C). Excessive force may damage the threads in the exhaust manifold of exhaust pipe.

1. Start the engine and let it warm up to 120°F (48°C), stop the engine and disconnect the negative battery cable.
2. Disconnect the electrical connector from the oxygen sensor.
3. Using a special oxygen sensor socket, remove the oxygen sensor from the exhaust manifold or exhaust pipe.

NOTE: A special anti-sieze compound is used on the oxygen sensor threads. The compound consists of a liquid graphite and glass beads. The graphite will burn away, but the glass beads will remain, making the sensor easier to remove. New or service sensors will already have the compound applied to the threads. If a oxygen sensor is removed from the engine and if for any reason it is to be reinstalled, the threads must have this anti-sieze compound applied before installation.

4. Coat the threads of the oxygen sensor with anti-sieze compound 5613695 or equivalent, if necessary.
5. Install the sensor and torque it to 30 ft. lbs (41 Nm).
6. Reconnect the electrical connector to the sensor and the negative battery cable.

CRANKSHAFT SENSOR OR COMBINATION SENSOR

On the 3.0L engine, the crankshaft sensor and camshaft sensor functions are combined into one sensor called the combination sensor which is mounted at the harmonic balancer.

Removal and Installation

2.3L, 2.8L, 3.1L ENGINE

1. Disconnect the negative battery cable.
2. Remove the sensor harness connector at the sensor.
3. Remove the sensor to the block retaining bolt and remove the sensor from the block.
4. Installation is the reverse order of the removal procedure and be sure to use a new O-ring on the sensor. Torque the sensor retaining bolt to 71–88 inch. lbs. (8–10 Nm).

3.0L ENGINE

1. Disconnect the negative battery cable.
2. Remove the right side lower engine compartment filler panel and the right lower wheel house to engine compartment bolt.
3. Disconnect the combination sensor harness connector.
4. Rotate the harmonic balancer, using a 28mm socket and pull handle, until any window in the interrupter is aligned with the combination sensor.

5. Loosen the pinch bolt on the sensor pedestal until the sensor is free to slide into the pedestal.
6. Remove the pedestal to engine mounting bolt.
7. While manipulating the sensor within the pedestal, carefully remove the sensor and pedestal as a unit.
8. To install the new sensor, loosen the pinch bolt on the new sensor pedestal until the sensor is free to slide in the pedestal.
9. Verify that the window in the interrupter is still properly positioned and install the pedestal and sensor as a unit while making sure that the interrupter ring is aligned with the proper slot.
10. Install the pedestal to engine mounting bolt and torque it to 22 ft. lbs. (30 Nm).
11. Replace the lower wheel house to engine compartment bolt and reinstall the right lower filler panel. Adjust the sensor as follows:

    a. Disconnect the negative battery cable.

    b. Loosen the pinch bolt on the sensor pedestal and insert feeler gauge adjustment tool J–36179 or equivalent into the gap between the sensor and the interrupter on each side of the interrupter ring.

    c. Be sure that the interrupter is sandwiched between the blades of the adjustment tool and both blades are properly inserted into the sensor slot.

    d. Torque the sensor retaining pinch bolt to 30 inch. lbs. (3.4 Nm) while maintaining light pressure on the sensor against the gauge and interrupter ring. This clearance should be checked again, at 3 positions around the interrupter ring approximately 120 degrees apart.

    e. If the interrupter ring contacts the sensor at any point during harmonic balancer rotation, the interrupter ring has excessive runout and must be replaced.

3.3L ENGINE

1. Disconnect the negative battery cable.
2. Disconnect the serpentine belt from crankshaft pulley.

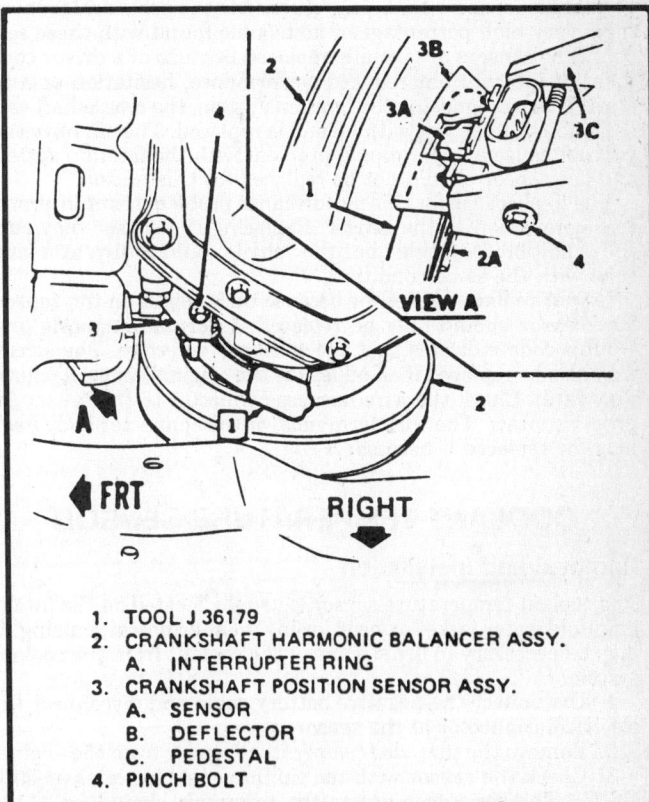

1. TOOL J 36179
2. CRANKSHAFT HARMONIC BALANCER ASSY.
   A. INTERRUPTER RING
3. CRANKSHAFT POSITION SENSOR ASSY.
   A. SENSOR
   B. DEFLECTOR
   C. PEDESTAL
4. PINCH BOLT

Combination sensor adjustment – 3.0L engine

3. Raise and support the vehicle safely.

4. Remove the right front tire and wheel assembly.

5. Remove the right inner fender access cover.

6. Using a 28mm socket and remove the crankshaft harmonic balancer retaining bolt.

7. Remove the harmonic balancer and disconnect the electrical connector. Remove the sensor and pedestal from the block face.

8. Remove the sensor from the pedestal.

9. To install and adjust us the following procedure:

a. Loosely install the crankshaft sensor on the pedestal. Position the sensor with the pedestal attached on special crankshaft sensor adjustment tool J–37089.

b. Position the special tool on the crankshaft. Install the bolts to hold the pedestal to block and torque them to 14–28 ft. lbs. (20–40 Nm).

c. Torque the pedestal pinch bolt to 30–35 inch. lbs. (3–4 Nm). Remove special tool J–37089.

d. Place special tool J–37089 onto the harmonic balancer and turn it. If any vane of the harmonic balancer touches the tool, replace the balancer assembly.

e. Install the balancer onto the crankshaft and torque the bolt to 200–239 ft. lbs. (270–325 Nm).

f. Install the inner fender shield. Install the tire and wheel assembly. Lower the vehicle and install the serpentine belt and negative battery cable.

CHANGES AND CORRECTIONS

Improper Crankshaft Sensor Replacements Due to Oil On the Sensor

1988–89 CUTLASS CIERA (2.8L) CUTLASS SUPREME (2.8L/3.1L) AND 1988 FIRENZA (2.0L) ENGINES WITH DIRECT IGNITION SYSTEM

Crankshaft sensors are being returned to the parts dealer showing a very high percentage of no trouble found with these sensors. The sensors are usually replaced because of a driver comment of intermittent reduced performance, hesitation or a no start. When diagnosing the ignition system, the crankshaft sensor is found to be wet with oil and is replaced. The oil, however, will not cause an operational problem with the ignition system and the sensor should not be replaced for this reason.

The likely cause for this performance problem is ant intermittent connection to the sensor. Replacing the sensor may cure the condition for a time, but the vehicle could return at a later time with the same condition.

Do not replace the sensor because oil is found on the sensor. The sensor should only be replaced if normal diagnosis or a trouble code indicates that the sensor is defective. The sensor may also be replaced if an oil leak is the original concern (this is very rare). Check the wire harness connector to the sensor for proper contact. The wire terminals, or the connector body itself may, be replaced if necessary.

COOLANT TEMPERATURE SENSOR

Removal and Installation

The coolant temperature sensor is usually located on the intake manifold water jacket or near (or in) the thermostat housing. It may be necessary to drain some of the coolant from the coolant system.

1. Disconnect the negative battery cable and disconnect the electrical connector at the sensor.

2. Remove the threaded temperature sensor from the engine.

3. Check the sensor with the tip immersed in water at 50°F (15°C). The resistance across the terminals should be 4114–4743 ohms. If not within specifications, replace the sensor.

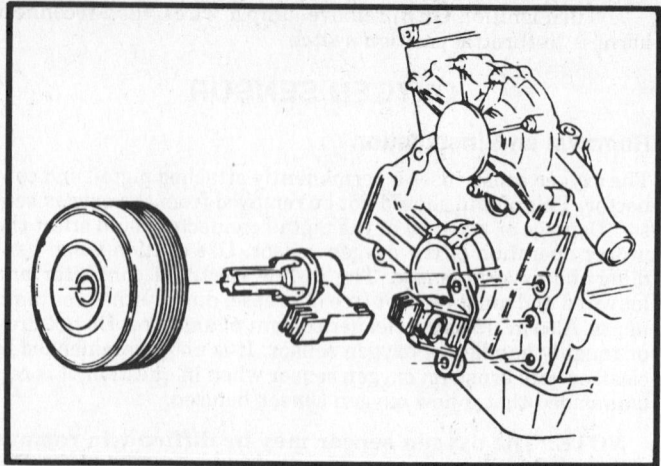

Position special tool J–37089 on the crankshaft

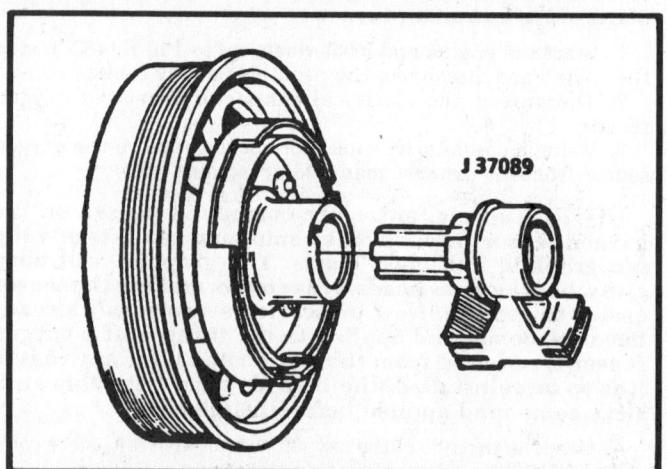

Checking the harmonic balancer vanes

4. Apply some pipe tape to the threaded sensor and install the senor. Torque the sensor to 22 ft. lbs. (30 Nm).

5. Reconnect the sensor connector and the negative battery cable.

TORQUE CONVERTER CLUTCH (TCC) SOLENOID

Removal and Installation

1. Remove the negative battery cable. Raise and support the vehicle safely.

2. Drain the transmission fluid into a suitable drain pan. Remove the transmission pan.

3. Remove the TCC solenoid retaining screws and then remove the electrical connector, solenoid and check ball.

4. Clean and inspect all parts. Replace defective parts as necessary.

5. Install the check ball, TCC solenoid and electrical connector. Install the solenoid retaining screws and torque them to 10 ft. lbs. (14 Nm).

6. Install the transmission pan with a new gasket and torque the pan retaining bolts to 10 ft. lbs. (14 Nm).

7. Lower the vehicle and refill the transmission with the proper amount of the recommended automatic transmission fluid.

VEHICLE SPEED SENSOR (VSS)

Removal and Installation
INSTRUMENT PANEL MOUNTED

1. Disconnect the negative battery cable.
2. Remove the instrument cluster from the dash board.
3. Remove the 2 screws on the back side of the instrument cluster, 1 in the optic head and 1 in the buffer amplifier.
4. Disconnect the electrical connections from the foil to the buffer amplifier.
5. Remove the vehicle speed sensor.
6. Installation is the reverse order of the removal procedures.

NOTE: If the vehicle is equipped with a digital cluster, just pull the cluster out far enough from the dash so that the optic head can be disconnected. Then remove the left side sound insulator and reach up to the instrument panel harness and disconnect the buffer amplifier connection.

TRANSMISSION MOUNTED

1. Disconnect the negative battery cable.
2. Remove the vehicle speed sensor electrical lead from the transmission.
3. Remove the VSS sensor bolt and screw retainer.
4. Remove the vehicle speed sensor assembly and O-ring.
5. Installation is the reverse order of the removal procedure. Use a new O-ring and coat with transmission fluid.

MANIFOLD ABSOLUTE PRESSURE (MAP) SENSOR

Removal and Installation

1. Disconnect the negative battery cable and remove the vacuum hose from the MAP sensor.
2. Disconnect the electrical connector to the MAP sensor.
3. Remove the MAP sensor retaining screws and remove the MAP sensor.
4. Installation is the reverse order of the removal procedure.

POWER STEERING PRESSURE SWITCH

Removal and Installation

1. Disconnect the negative battery cable.
2. Disconnect the electrical connector to the switch and place a drip pan under the switch.
3. Unscrew switch and remove the switch from the power steering line.
4. Installation is the reverse order of the removal procedure. Be sure to replace the power steering fluid lost during removal of the switch.

MASS AIR FLOW (MAF) SENSOR

Removal and Installation

1. Disconnect the negative battery cable.
2. Disconnect the electrical connector to the MAF sensor.
3. Remove the air cleaner and duct assembly. Remove the sensor retaining clamps and remove the sensor.
4. Installation is the reverse order of the removal procedure.

MASS AIR TEMPERATURE (MAT) SENSOR

Removal and Installation

1. Disconnect the negative battery cable.
2. Disconnect the electrical connector to the MAT sensor.
3. Unscrew the MAT sensor from the intake manifold.
4. Installation is the reverse order of the removal procedure.

ELECTRONIC CONTROL MODULE (ECM)

To prevent possible electrostatic discharge damage to the ECM, do not touch the connector pins or soldered components on the circuit board. Service of the ECM should normally consists of either replacement of the ECM or a Mem-Cal (PROM) change.

If the diagnostic procedures call for the ECM to be replaced, the engine calibrator (Mem-Cal or PROM) and the ECM should be checked first to see if they are the correct parts. If they are, remove the Mem-Cal or PROM from the faulty ECM and install it in the new service ECM. The service ECM will not contain a new Mem-Cal or PROM.

The trouble Code 51 indicates that the Mem-Cal or PROM is installed improperly or has malfunctioned. When Code 51 is obtained, check the ECM installation for bent pins or pins not fully seated in the socket. If it is installed correctly and the Code 51 still appears, replace the Mem-Cal or PROM.

Removal and Installation

NOTE: When replacing the production ECM with a service ECM (controller), it is important to transfer the broadcast code and production ECM number to the service ECM label. Do not record the number on the ECM cover. This will allow positive identification of the ECM parts throughout the service life of the vehicle. To prevent internal ECM damage, the ignition must be OFF when disconnecting and reconnecting the power the ECM (for example , the battery cable, ECM pigtail, ECM fuse, jumper cables, etc.).

1. Disconnect the negative battery cable.
2. Remove the glove box assembly, if necessary or remove the right hand side kick panel or what ever is necessary to gain access to the ECM.
3. Disconnect the connectors from the ECM.
4. Remove the ECM mounting hardware and remove the ECM from the passenger compartment.
5. Installation is the reverse order of the removal procedure. Be sure to remove the new ECM from its packaging and check the service number to make sure it is the same at the defective ECM.

MEM-CAL

Removal and Installation

1. Remove the ECM.
2. Remove the Mem-Cal access panel.
3. Using 2 fingers, push both retaining clips back away from the Mem-Cal. At the same time, grasp it at both ends and lift it out of the socket. Do not remove the cover of the Mem-Cal. Use of an unapproved Mem-Cal removal procedures may cause damage to the Mem-Cal or the socket.
4. Install the new Mem-Cal by pressing only the ends of the Mem-Cal. Small notches in the Mem-Cal must be aligned with the small notches in the Mem-Cal socket.
5. Press on the ends of the Mem-Cal until the retaining clips snap into the ends of the Mem-Cal. Do not press on the middle of the Mem-Cal, only on the ends.
6. Install the Mem-Cal access cover and reinstall the ECM.

Functional Check

1. Turn the ignition switch to the **ON** position.
2. Enter the diagnostic mode.
3. Allow a Code 12 to flash 4 times to verify that no other codes are present. This indicates that the Mem-Cal is installed properly and the ECM is working correctly.
4. If trouble Codes 41, 42, 43, 51 or 55 occurs or if the "CHECK ENGINE" light in on constantly with no codes, the Mem-Cal is not fully seated or is defective.
5. If not fully seated press down firmly on the Mem-Cal carrier If found to be defective, replace it with a new one.

NOTE: To prevent possible electrostatic discharge damage to the Mem-Cal, do not touch the component leads and do not remove the integrated circuit from the carrier.

PROM

Removal and Installation

1. Remove the ECM.
2. Remove the PROM access panel.
3. Using the rocker type PROM removal tool, engage 1 end of the PROM carrier with the hook end of the tool. Press on the vertical bar end of the tool and rock the engaged end of the PROM carrier up as far as possible.
4. Engage the opposite end of the PROM carrier in the same manner and rock this end up as far as possible.
5. Repeat this process until the PROM carrier and PROM are free of the PROM socket. The PROM carrier with PROM in it should lift off of the PROM socket easily.
6. The PROM carrier should only be removed by using the special PROM removal tool. Other methods could cause damage to the PROM or PROM socket.
7. Before installing a new PROM, be sure the new PROM part number is the same as the old one or the as the updated number per a service bulletin.
8. Install the PROM with the small notch of the carrier aligned with the small notch in the socket. Press on the PROM carrier until the PROM is firmly seated, do not press on the PROM itself only on the PROM carrier.
9. Install the PROM access cover and reinstall the ECM.

Functional Check

1. Turn the ignition switch to the ON position.
2. Enter the diagnostic mode.
3. Allow a Code 12 to flash four times to verify that no other codes are present. This indicates that the PROM is installed properly.

4. If trouble Code 51 occurs or if the "CHECK ENGINE" light is on constantly with no codes, the PROM is not fully seated, installed backwards, has bent pins or is defective.
5. If not fully seated press down firmly on the PROM carrier.
6. If installed backwards, replace the PROM.

NOTE: Any time the PROM is installed backwards and the ignition switch is turned ON, the PROM is destroyed.

7. If the pins are bent, remove the PROM straighten the pins and reinstall the PROM. If the bent pins break or crack during straightening, discard the PROM and replace with a new PROM.

NOTE: To prevent possible electrostatic discharge damage to the PROM or Cal-Pak, do not touch the component leads and do not remove the integrated circuit from the carrier.

CAL-PAK

Removal and Installation

1. Remove the ECM.
2. Remove the Cal-Pak access panel.
3. Using the special Cal-Pak removal tool, grasp the Cal-Pak carrier at the narrow end and gently rock back and fourth while applying a firm upward force.
4. Inspect the reference end of the Cal-Pak carrier and set it aside. Do not remove the Cal-Pak from the carrier to confirm the Cal-Pak correctness.
5. Before installing a new Cal-Pak, be sure the new Cal-Pak part number is the same as the old one or the as the updated number per a service bulletin.
6. Install the Cal-Pak with the small notch of the carrier aligned with the small notch in the socket. Press on the Cal-Pak carrier until the Cal-Pak is firmly seated, do not press on the Cal-Pak itself only on the carrier.
7. Install the Cal-Pak access cover and reinstall the ECM.

1988-90 MULTI-PORT INJECTION SPECIFICATIONS

Model	Engine VIN Code	Engine	Minimum Idle rpm	TPS Voltage	Fuel Pressure (psi)
1988 BUICK					
Century	W	2.8L	550±50 ①	0.55±1V	40–47
Regal	W	2.8L	550±50	0.55±1V	40–47
Skylark	D	2.3L	②	0.54±0.05V	40–47
Skylark	L	3.0L	550±50	0.50–0.59	40–47
1989-90 BUICK					
Century	W	2.8L	550±50 ①	0.55±1V	40–47
Century	N	3.3L	③	0.40±0.02V	40–47
Regal	W	2.8L	550±50	0.55±1V	40–47
Regal	T	3.1L	④	1.25 ⑤	40–47
Skylark	D	2.3L	②	0.54±0.05V	40–47
Skylark	N	3.3L	③	0.40±0.02V	40–47
1989-90 CADILLAC					
Cimarron	W	2.8L	550±50	0.55±1V	40–47
1989-90 CHEVROLET					
Beretta	W	2.8L	550±50 ①	0.55±1V	40–47
Camaro (89–90)	S	2.8L	④	0.55±1V	40–47

1988-90 MULTI-PORT INJECTION SPECIFICATIONS

Model	Engine VIN Code	Engine	Minimum Idle rpm	TPS Voltage	Fuel Pressure (psi)
Camaro	F	5.0L	425 ± 25	0.54 ± 0.08V	40–47
Camaro	8	5.7L	425 ± 25	0.54 ± 0.08V	40–47
Cavalier	W	2.8L	④	1.25 ⑤	40–47
Celebrity	W	2.8L	④	1.25 ⑤	40–47
Corsica	W	2.8L	550 ± 50 ①	0.55 ± 1V	40–47
Corvette	8	5.7L	425 ± 25	0.54 ± 0.08V	40–47
1988-90 OLDSMOBILE					
Calais	D	2.3L	②	0.54 ± 0.05V	40–47
Calais	A	2.3L	②	0.54 ± 0.05V	40–47
Calais	L	3.0L	550 ± 50	0.50 ± 0.59V	40–47
Calais	N	3.3L	③	0.40 ± 0.02V	40–47
Ciera	W	2.8L	550 ± 50	0.55 ± 1V	40–47
Ciera	N	3.3L	③	0.40 ± 0.02V	40–47
Cutlass Supreme	W	2.8L	550 ± 50	0.55 ± 1V	40–47
Cutlass Supreme	T	3.1L	④	1.25 ⑤	40–47
1988-90 PONTIAC					
Firebird	8	5.7L	425 ± 25	0.54 ± 0.08V	40–47
Grand Am	M	2.0L Turbo	⑥	1.25 ⑤	30–39
Grand Am	D	2.3L	②	0.54 ± 0.05V	40–47
Grand Am	A	2.3L	②	0.54 ± 0.05V	40–47
Grand Prix	W	2.8L	550 ± 50	0.55 ± 1V	40–47
Grand Prix	T	3.1L	④	1.25 ⑤	40–47
Sunbird	M	2.0L Turbo	⑥	1.25 ⑤	30–39
6000	W	2.8L	550 ± 50	0.55 ± 1V	40–47
6000	T	3.1L	④	1.25 ⑤	40–47

NOTE: The underhood specifications sticker often reflects tune-up changes made during the production run. The sticker specifications must always be used first if they disagree with those in this chart.

① 650 ± 50 rpm on manual transaxles.
② Select the IAC valve display on a suitable SCAN tool and read the IAC counts. If the counts are between 5–45 counts, the minimum throttle valve position adjustment is acceptable.
③ The IAC valve is reset by grounding the ALCL (or ALDL) or 5–10 seconds.
④ With the engine stabilized, select the IAC valve display on a suitable SCAN tool and read the IAC counts. If the counts are between 10–15 counts, the minimum throttle valve position adjustment is acceptable.
⑤ With the ignition ON and the engine OFF. The TBI unit will be at closed throttle, the TPS voltage should read less than 1.25 volts. If the reading is higher than specified, replace the TPS.
⑥ The ECM will reset the IAC valve when the vehicle is driven above 35 mph.

THROTTLE BODY INJECTION SYSTEMS

APPLICATION CHART
1988–90 Throttle Body Injection Vehicles

Year	Models	Series VIN	Engine	Engine VIN Code
	BUICK			
1988–90	Century	A	2.5L	R
	Skyhawk	J	2.0L	K
	Skyhawk	J	2.0L	1
	Skylark	N	2.5L	U
	CHEVROLET			
1988–90	Beretta	L	2.0L	1
	Camaro	F	5.0L	E
	Caprice	B	4.3L	Z
	Cavalier	J	2.0L	1
	Celebrity	A	2.5L	R
	Corsica	L	2.0L	1
	Lumina	W	2.5L	R
	Monte Carlo	G	4.3L	Z
	Astro Van	M	2.5L	E
	Light Truck	C/K	2.5L	E
	S-10 & S-15	S/T	2.5L	E
	Van	G	2.5L	E
	Astro Van	M	2.8L	R
	Light Truck	C/K	2.8L	R

APPLICATION CHART
1988–90 Throttle Body Injection Vehicles

Year	Models	Series VIN	Engine	Engine VIN Code
1988–90	S-10 & S-15	S/T	4.3L	Z
	Van	G	4.3L	Z
	Light Truck	C/K	5.0L	H
	Van	G	5.0L	H
	Light Truck	C/K	5.7L	K
	Van	G	5.7L	K
	Light Truck	C/K	7.4L	N
	OLDSMOBILE			
1988–90	Calais	N	2.5L	U
	Cutlass Ciera	A	2.5L	R
	Cutlass Cruiser	A	2.5L	R
	Firenza	J	2.0L	1
	Firenza	J	2.0L	K
	PONTIAC			
1988–90	Fiero	P	2.5L	R
	Firebird	F	5.0L	E
	Grand Am	N	2.5L	U
	Sunbird	J	2.0L	K
	Transport	U	3.1L	D

NOTE: This manual has the ability to save time in diagnosis and prevent the replacement of good parts. The key to using this manual successfully for diagnosis lies in the technician's ability to understand the system he is trying to diagnose as well as an understanding of the manual's layout and limitations. The technician should review this manual to become familiar with the way this manual should be used.

General Information

The electronic throttle body fuel injection system is a fuel metering system with the amount of fuel delivered by the throttle body injector(s) (TBI) determined by an electronic signal supplied by the Electronic Control Module (ECM). The ECM monitors various engine and vehicle conditions to calculate the fuel delivery time (pulse width) of the injector(s). The fuel pulse may be modified by the ECM to account for special operating conditions, such as cranking, cold starting, altitude, acceleration, and deceleration.

The ECM controls the exhaust emissions by modifying fuel delivery to achieve, as near as possible, and air/fuel ratio of 14.7:1. The injector on-time is determined by various inputs to the ECM. By increasing the injector pulse, more fuel is delivered, enriching the air/fuel ratio. Decreasing the injector pulse, leans the air/fuel ratio. Pulses are sent to the injector in 2 different modes: synchronized and nonsynchronized.

SYNCHRONIZED MODE

In synchronized mode operation, the injector is pulsed once for each distributor reference pulse. In dual injector throttle body systems, the injectors are pulse alternately.

NONSYNCHRONIZED MODE

In nonsynchronized mode operation, the injector is pulsed once every 12.5 milliseconds or 6.25 milliseconds depending on calibration. This pulse time is totally independent of distributor reference pulses.

Nonsynchronized mode results only under the following conditions:

1. The fuel pulse width is too small to be delivered accurately by the injector (approximately 1.5 milliseconds)
2. During the delivery of prime pulses (prime pulses charge the intake manifold with fuel during or just prior to engine starting)
3. During acceleration enrichment
4. During deceleration leanout

The basic TBI unit is made up of 2 major casting assemblies: (1) a throttle body with a valve to control airflow and (2) a fuel body assembly with an integral pressure regulator and fuel injector to supply the required fuel. An electronically operated device to control the idle speed and a device to provide information regarding throttle valve position are included as part of the TBI unit.

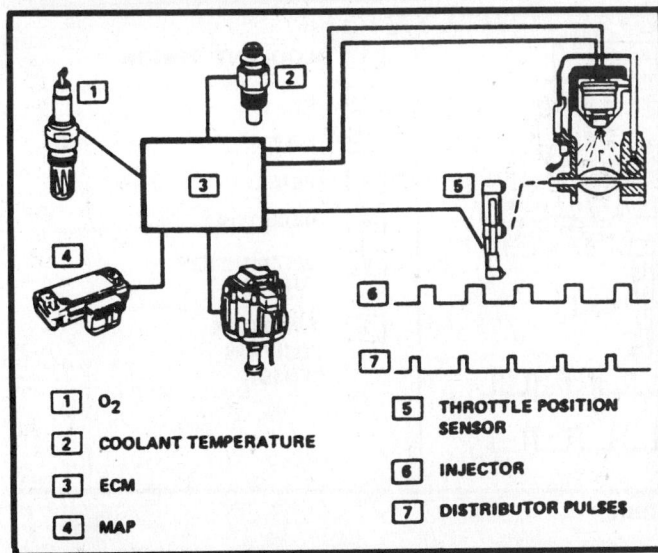

Typical synchronized mode for single throttle body system

1 O₂	5 THROTTLE POSITION SENSOR
2 COOLANT TEMPERATURE	6 INJECTOR
3 ECM	7 DISTRIBUTOR PULSES
4 MAP	

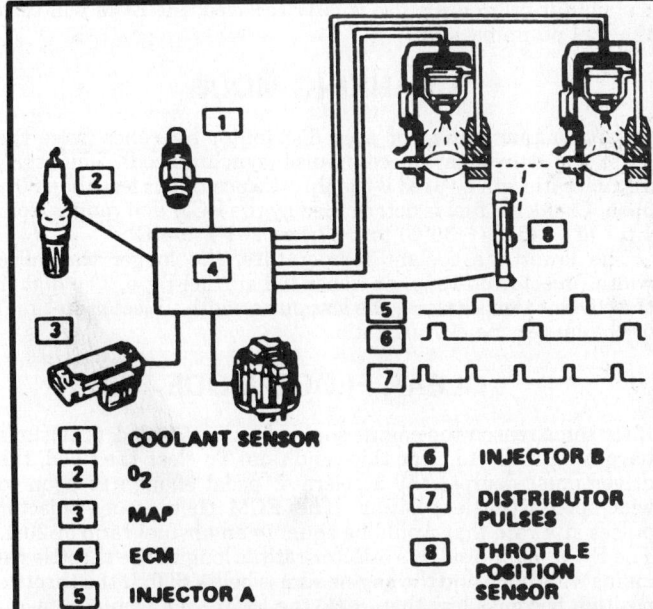

Typical synchronized mode for dual throttle body system

1 COOLANT SENSOR	6 INJECTOR B
2 O₂	7 DISTRIBUTOR PULSES
3 MAP	8 THROTTLE POSITION SENSOR
4 ECM	
5 INJECTOR A	

The fuel injector(s) is a solenoid-operated device controlled by the ECM. The incoming fuel is directed to the lower end of the injector assembly which has a fine screen filter surrounding the injector inlet. The ECM actuates the solenoid, which lifts a normally closed ball valve off a seat. The fuel under pressure is injected in a conical spray pattern at the walls of the throttle body bore above the throttle valve. The excess fuel passes through a pressure regulator before being returned to the vehicle's fuel tank.

The pressure regulator is a diaphragm-operated relief valve with injector pressure on one side and air cleaner pressure on the other. The function of the regulator is to maintain a constant pressure drop across the injector throughout the operating load and speed range of the engine.

The throttle body portion of the TBI may contain ports located at, above, or below the throttle valve. These ports generate

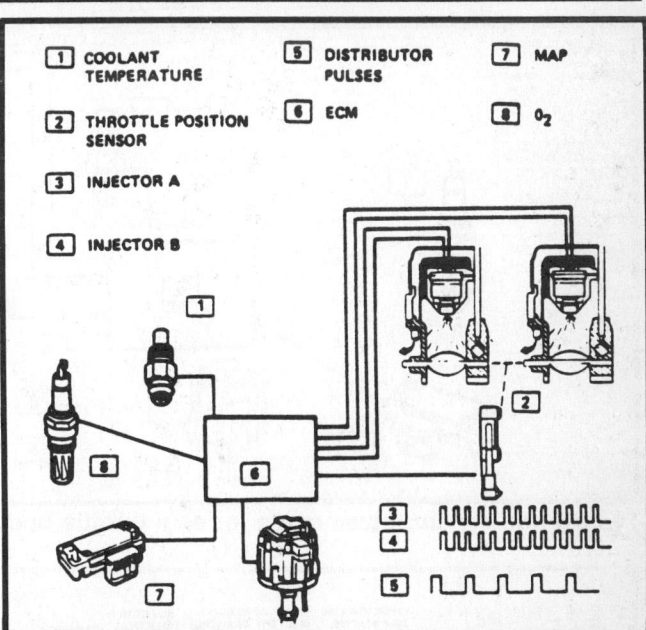

Typical non-synchronized mode for single throttle body system

the vacuum signals for the EGR valve, MAP sensor, and the canister purge system.

The Throttle Position Sensor (TPS) is a variable resistor used to convert the degree of throttle plate opening to an electrical signal to the ECM. The ECM uses this signal as a reference point of throttle valve position. In addition, an Idle Air Control (IAC) assembly, mounted in the throttle body is used to control idle speeds. A cone-shaped valve in the IAC assembly is located in an air passage in the throttle body that leads from the point beneath the air cleaner to below the throttle valve. The ECM monitors idle speeds and, depending on engine load, moves the IAC cone in the air passage to increase or decrease air bypassing the throttle valve to the intake manifold for control of idle speeds.

Components and Operation

The Throttle Body Injection (TBI) system provides a means of fuel distribution for controlling exhaust emissions within legislated limits by precisely controlling the air/fuel mixture and under all operating conditions for, as near as possible, complete combustion.

This is accomplished by using an Electronic Control Module (ECM) (a small on-board microcomputer) that receives electrical inputs from various sensors about engine operating conditions. An oxygen sensor in the main exhaust stream functions to provide feedback information to the ECM as to the oxygen content, lean or rich, in the exhaust. The ECM uses this information from the oxygen sensor, and other sensors, to modify fuel delivery to achieve, as near as possible, an ideal air/fuel ratio of 14.7:1. This air/fuel ratio allows the 3-way catalytic converter to be more efficient in the conversion process of reducing exhaust emissions while at the same time providing acceptable levels of driveability and fuel economy.

The ECM program electronically signals the fuel injector in the TBI assembly to provide the correct quantity of fuel for a wide range of operating conditions. Several sensors are used to determine existing operating conditions and the ECM then signals the injector to provide the precise amount of fuel required.

The ECM used on EFI vehicles has a learning capability. If the battery is disconnected to clear diagnostic codes, or for re-

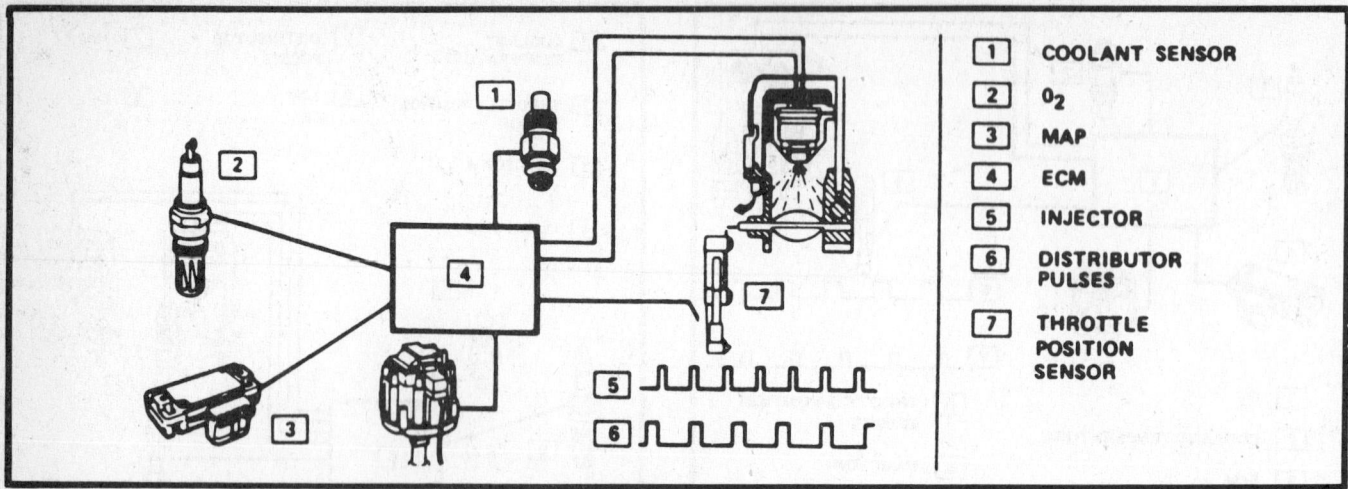

Typical non-synchronized mode for dual throttle body system

1	COOLANT SENSOR
2	O₂
3	MAP
4	ECM
5	INJECTOR
6	DISTRIBUTOR PULSES
7	THROTTLE POSITION SENSOR

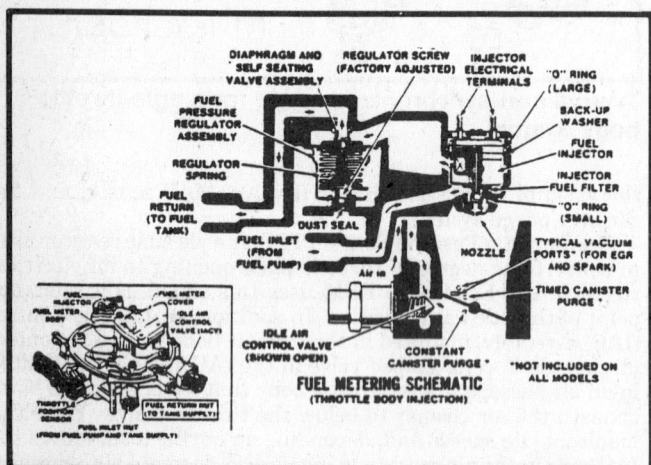

Exploded view of a typical throttle body injector assembly

pair, the learning process has to begin all over again. A change may be noted in vehicle performance. To teach the vehicle, make sure the vehicle is at operating temperature and drive at part throttle, under moderate acceleration and idle conditions, until performance returns.

With the EFI system, the TBI assembly is centrally located on the intake manifold where air and fuel are distributed through a single bore in the throttle body, similar to a carbureted engine. Air for combustion is controlled by a single throttle valve which is connected to the accelerator pedal linkage by a throttle shaft and lever assembly. A special plate is located directly beneath the throttle valve to aid in mixture distribution.

Fuel for combustion is supplied by 1 or 2 fuel injector(s), mounted on the TBI assembly, whose metering tip is located directly above the throttle valve. The injector is pulsed or timed open or closed by an electronic output signal received from the ECM. The ECM receives inputs concerning engine operating conditions from the various sensors (coolant temperature sensor, oxygen sensor, etc.). The ECM, using this information, performs high speed calculations of engine fuel requirements and pulses or times the injector, open or closed, thereby controlling fuel and air mixtures to achieve, as near as possible, ideal air/fuel mixture ratios.

When the ignition key is turned **ON**, the ECM will initialize (start program running) and energize the fuel pump relay. The fuel pump pressurizes the system to approximately 10 psi. If the

ECM does not receive a distributor reference pulse (telling the ECM the engine is turning) within 2 seconds, the ECM will then de-energize the fuel pump relay, turning off the fuel pump. If a distributor reference pulse is later received, the ECM will turn the fuel pump back on.

CRANKING MODE

During engine crank, for each distributor reference pulse the ECM will deliver an injector pulse (synchronized). The crank air/fuel ratio will be used if the throttle position is less than 80% open. Crank air fuel is determined by the ECM and ranges from 1.5:1 at −33°F (−36°C) to 14.7:1 at 201°F (94°C).

The lower the coolant temperature, the longer the pulse width (injector on-time) or richer the air/fuel ratio. The higher the coolant temperature, the less pulse width (injector on-time) or the leaner the air/fuel ratio.

CLEAR FLOOD MODE

If for some reason the engine should become flooded, provisions have been made to clear this condition. To clear the flood, the driver must depress the accelerator pedal enough to open to wide-open throttle position. The ECM then issues injector pulses at a rate that would be equal to an air/fuel ratio of 20:1. The ECM maintains this injector rate as long as the throttle remains wide open and the engine rpm is below 600. If the throttle position becomes less than 80%, the ECM then would immediately start issuing crank pulses to the injector calculated by the ECM based on the coolant temperature.

RUN MODE

There are 2 different run modes. When the engine rpm is above 400, the system goes into open loop operation. In open loop operation, the ECM will ignore the signal from the oxygen (O₂) sensor and calculate the injector on-time based upon inputs from the coolant and manifold absolute pressure sensors.

During open loop operation, the ECM analyzes the following items to determine when the system is ready to go to the closed loop mode:

1. The oxygen sensor varying voltage output. (This is dependent on temperature).
2. The coolant sensor must be above specified temperature.
3. A specific amount of time must elapse after starting the engine. These values are stored in the PROM.

When these conditions have been met, the system goes into closed loop operation In closed loop operation, the ECM will

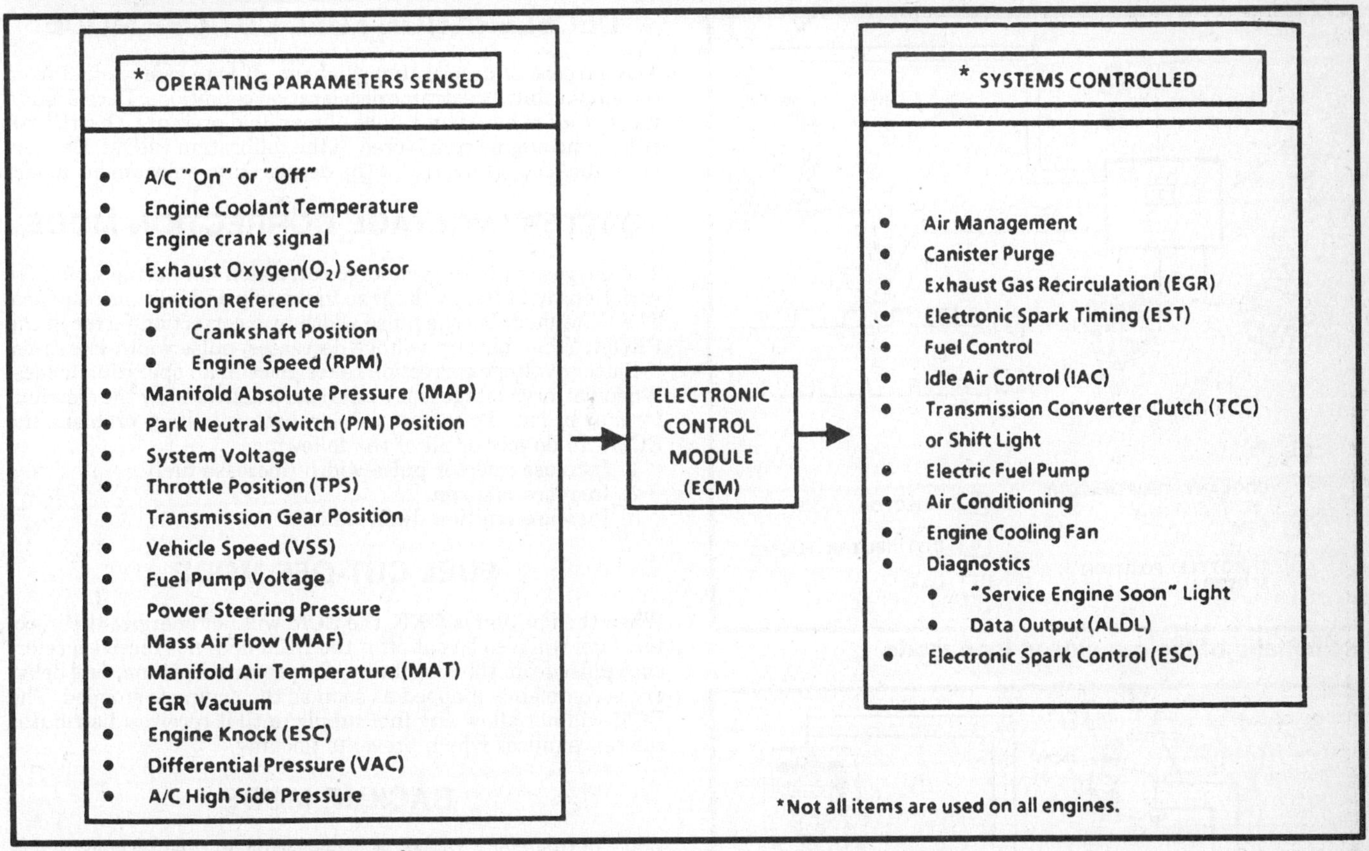

* OPERATING PARAMETERS SENSED	* SYSTEMS CONTROLLED

- A/C "On" or "Off"
- Engine Coolant Temperature
- Engine crank signal
- Exhaust Oxygen(O_2) Sensor
- Ignition Reference
  - Crankshaft Position
  - Engine Speed (RPM)
- Manifold Absolute Pressure (MAP)
- Park Neutral Switch (P/N) Position
- System Voltage
- Throttle Position (TPS)
- Transmission Gear Position
- Vehicle Speed (VSS)
- Fuel Pump Voltage
- Power Steering Pressure
- Mass Air Flow (MAF)
- Manifold Air Temperature (MAT)
- EGR Vacuum
- Engine Knock (ESC)
- Differential Pressure (VAC)
- A/C High Side Pressure

ELECTRONIC CONTROL MODULE (ECM)

- Air Management
- Canister Purge
- Exhaust Gas Recirculation (EGR)
- Electronic Spark Timing (EST)
- Fuel Control
- Idle Air Control (IAC)
- Transmission Converter Clutch (TCC) or Shift Light
- Electric Fuel Pump
- Air Conditioning
- Engine Cooling Fan
- Diagnostics
  - "Service Engine Soon" Light
  - Data Output (ALDL)
- Electronic Spark Control (ESC)

*Not all items are used on all engines.

Engine control systems parameters

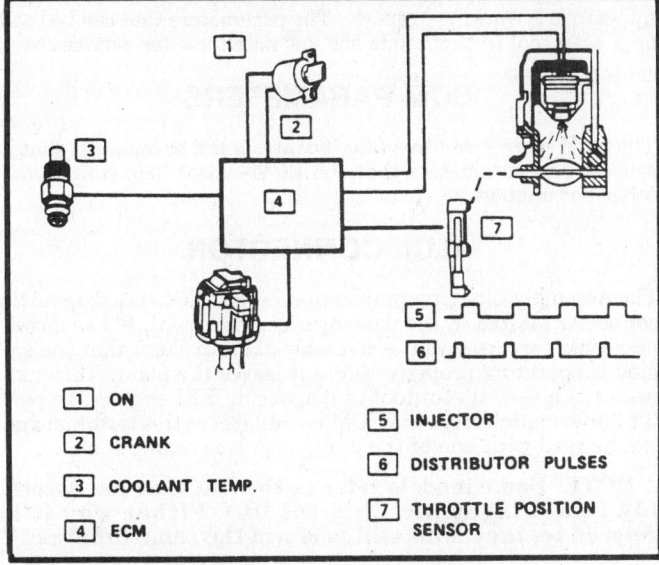

1	ON
2	CRANK
3	COOLANT TEMP.
4	ECM
5	INJECTOR
6	DISTRIBUTOR PULSES
7	THROTTLE POSITION SENSOR

Cranking air/fuel ratio

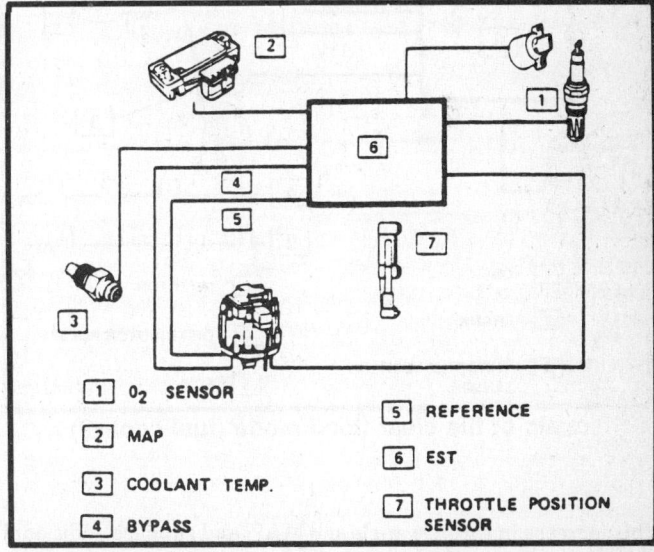

1	O_2 SENSOR
2	MAP
3	COOLANT TEMP.
4	BYPASS
5	REFERENCE
6	EST
7	THROTTLE POSITION SENSOR

Schematic of the run open loop mode

modify the pulse width (injector on-time) based upon the signal from the oxygen sensor. The ECM will decrease the on-time if the air/fuel ratio is too rich, and will increase the on-time if the air/fuel ratio is too lean.

The pulse width, thus the amount of enrichment, is determined by manifold pressure change, throttle angle change, and coolant temperature. The higher the manifold pressure and the wider the throttle opening, the wider the pulse width. The acceleration enrichment pulses are delivered nonsynchronized.

Any reduction in throttle angle will cancel the enrichment pulses. This way, quick movements of the accelerator will not over-enrich the mixture.

ACCELERATION ENRICHMENT MODE

When the engine is required to accelerate, the opening of the throttle valve(s) causes a rapid increase in Manifold Absolute Pressure (MAP). This rapid increase in the manifold pressure causes fuel to condense on the manifold walls. The ECM senses

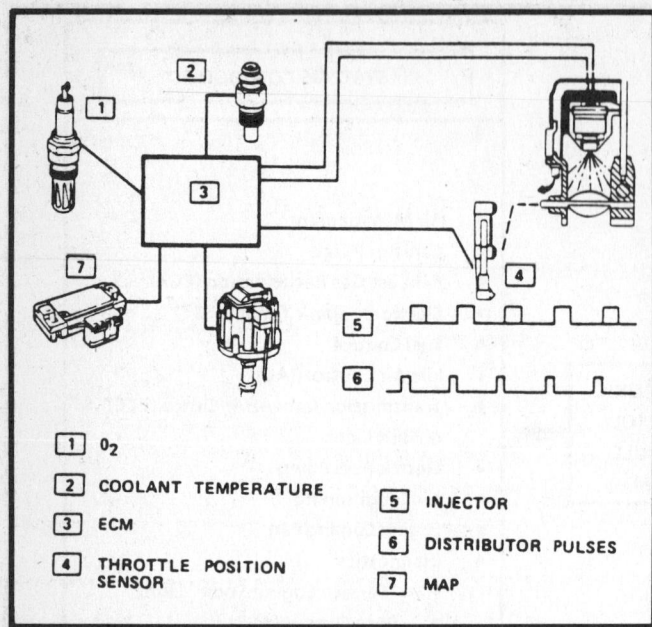

1 O₂

2 COOLANT TEMPERATURE

3 ECM

4 THROTTLE POSITION SENSOR

5 INJECTOR

6 DISTRIBUTOR PULSES

7 MAP

Schematic of the run closed loop mode

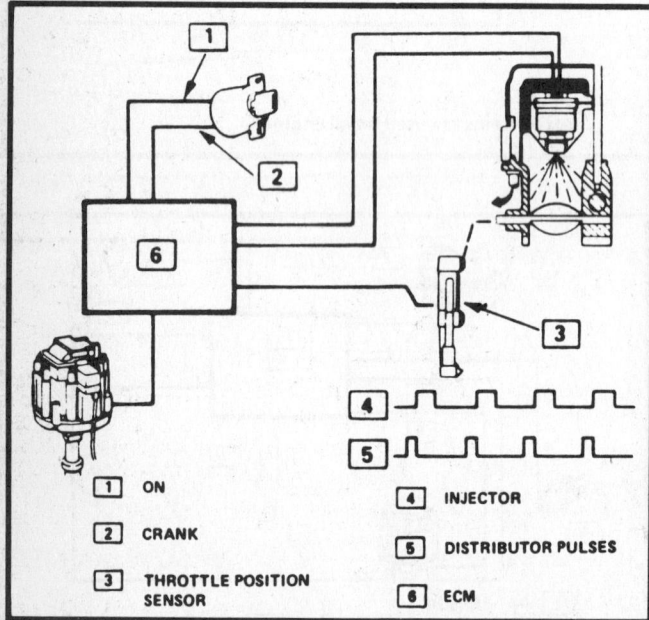

1 ON

2 CRANK

3 THROTTLE POSITION SENSOR

4 INJECTOR

5 DISTRIBUTOR PULSES

6 ECM

Schematic of the clear flood mode (fuel control)

this increase in throttle angle and MAP, and supplies additional fuel for a short period of time. This prevents the engine from stumbling due to too lean a mixture.

DECELERATION LEANOUT MODE

Upon deceleration, a leaner fuel mixture is required to reduce emission of hydrocarbons (HC) and carbon monoxide (CO). To adjust the injection on-time, the ECM uses the decrease in manifold pressure and the decrease in throttle position to calculate a decrease in pulse width. To maintain an idle fuel ratio of 14.7:1, fuel output is momentarily reduced. This is done because of the fuel remaining in the intake manifold during deceleration.

DECELERATION FUEL CUT-OFF MODE

The purpose of deceleration fuel cut-off is to remove fuel from the engine during extreme deceleration conditions. Deceleration fuel cut-off is based on values of manifold pressure, throttle position, and engine rpm stored in the calibration PROM. Deceleration fuel cut-off overrides the deceleration enleanment mode.

BATTERY VOLTAGE CORRECTION MODE

The purpose of battery voltage correction is to compensate for variations in battery voltage to fuel pump and injector response. The ECM modifies the pulse width by a correction factor in the PROM. When battery voltage decreases, pulse width increases.

Battery voltage correction takes place in all operating modes. When battery voltage is low, the spark delivered by the distributor may be low. To correct this low battery voltage problem, the ECM can do any or all of the following:
1. Increase injector pulse width (increase fuel)
2. Increase idle rpm
3. Increase ignition dwell time

FUEL CUT-OFF MODE

When the ignition is **OFF**, the ECM will not energize the injector. Fuel will also be cut off if the ECM does not receive a reference pulse from the distributor. To prevent dieseling, fuel delivery is completely stopped as soon as the engine is stopped. The ECM will not allow any fuel supply until it receives distributor reference pulses which prevents flooding.

BACKUP MODE

When in this mode, the ECM is operating on the fuel backup logic calibrated by the CalPak. The CalPak is used to control the fuel delivery if the ECM fails. This mode verifies that the backup feature is working properly. The parameters that can be read on a scan tool in this mode are not much use for service.

ECM PARAMETERS

The ECM has parameters that it controls and parameters that it senses. The parameters that the ECM senses help control the other parameters.

ALCL CONNECTOR

The Assembly Line Communication Link (ALCL) is a diagnostic connector located in the passenger compartment. It has terminals which are used in the assembly plant to check that the engine is operating properly before it leaves the plant. This connector is a very useful tool in diagnosing EFI engines. Important information from the ECM is available at this terminal and can be read with one of the many popular scanner tools.

NOTE: Some models refer to the ALCL as the Assembly Line Diagnostic Link (ALDL). Either way it is refered to, they both still perform the same function.

Electronic Fuel Injection Subsystems

Electronic Fuel Injection (EFI) is the name given to the entire fuel injection system. Various subsystems are combined to form the overall system. These subsystems are:
1. Fuel supply system
2. Throttle Body Injector (TBI) assembly
3. Idle Air Control (IAC)
4. Electronic Control Module (ECM)
5. Data sensors

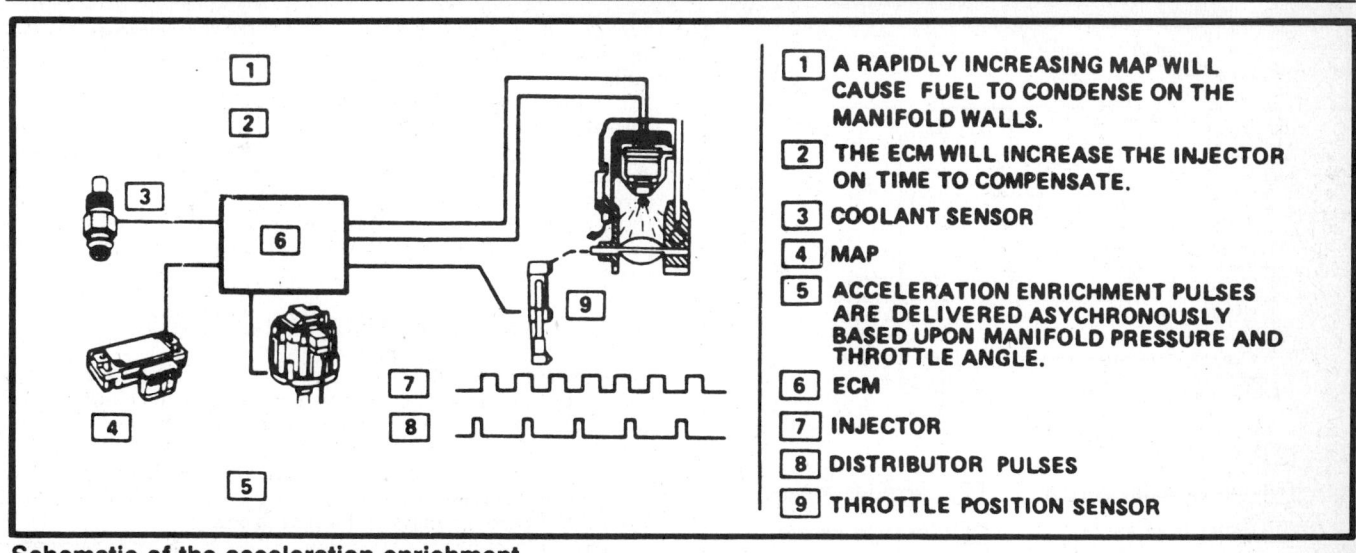

1. A RAPIDLY INCREASING MAP WILL CAUSE FUEL TO CONDENSE ON THE MANIFOLD WALLS.
2. THE ECM WILL INCREASE THE INJECTOR ON TIME TO COMPENSATE.
3. COOLANT SENSOR
4. MAP
5. ACCELERATION ENRICHMENT PULSES ARE DELIVERED ASYCHRONOUSLY BASED UPON MANIFOLD PRESSURE AND THROTTLE ANGLE.
6. ECM
7. INJECTOR
8. DISTRIBUTOR PULSES
9. THROTTLE POSITION SENSOR

Schematic of the acceleration enrichment

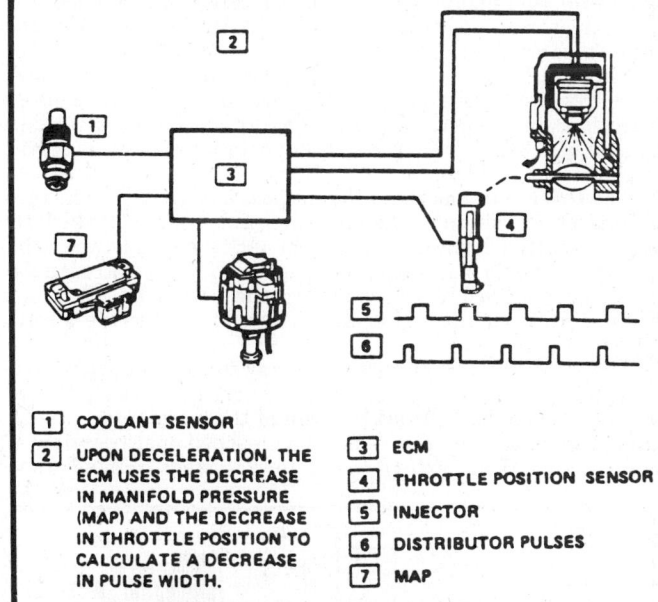

1. COOLANT SENSOR
2. UPON DECELERATION, THE ECM USES THE DECREASE IN MANIFOLD PRESSURE (MAP) AND THE DECREASE IN THROTTLE POSITION TO CALCULATE A DECREASE IN PULSE WIDTH.
3. ECM
4. THROTTLE POSITION SENSOR
5. INJECTOR
6. DISTRIBUTOR PULSES
7. MAP

Schematic of the deceleration leanout

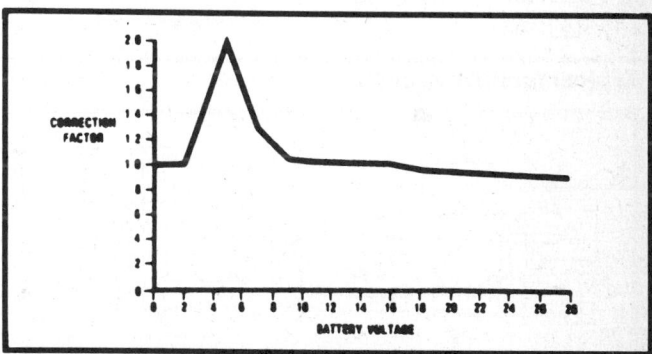

Battery voltage correction graph

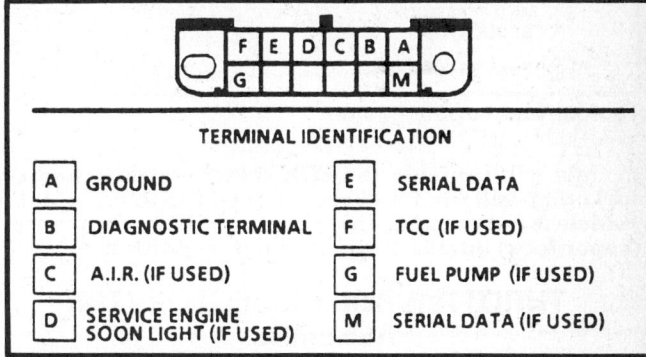

TERMINAL IDENTIFICATION

A	GROUND	E	SERIAL DATA
B	DIAGNOSTIC TERMINAL	F	TCC (IF USED)
C	A.I.R. (IF USED)	G	FUEL PUMP (IF USED)
D	SERVICE ENGINE SOON LIGHT (IF USED)	M	SERIAL DATA (IF USED)

Typical ALCL connector terminal identification

6. Electronic Spark Timing (EST)
7. Emission controls

Each subsystem is described in the following paragraphs.

FUEL SUPPLY SYSTEM

Fuel, supplied by an electric fuel pump mounted in the fuel tank, passes through an in-line fuel filter to the TBI assembly. To control fuel pump operation, a fuel pump relay is used.

When the ignition switch is turned to the **ON** position, the fuel pump relay activates the electric fuel pump for 1.5–2.0 seconds to prime the injector. If the ECM does not receive reference pulses from the distributor, the ECM signals the relay to turn the fuel pump off. The relay will once again activate the fuel pump when the ECM receives distributor reference pulses.

The oil pressure sender is the backup for the fuel pump relay. The sender has 2 circuits, 1 for the instrument cluster light or gauge, the other to activate the fuel pump if the relay fails. If the fuse relay has failed, the sender activates the fuel pump when oil pressure reaches 4 psi. Thus a failed fuel pump relay would cause a longer crank, especially in cold weather. If the fuel pump fails, a no start condition exists.

On the model 300, the fuel feed and return lines are attached with nuts and O-rings. They should be replaced if the lines are disconnected. The Model 500 fuel feed line uses a special banjo bolt for attachment to the fuel meter. The shank of the bolt has a hole drilled through it for fuel passage. It attaches to the fuel meter body fitting with 2 special O-ring washers. This system design is not interchangeable with any other type of fitting.

NOTE: On the 7.4L engine or a G van with the 5.7L engine and all other 5.7L engines over 8500 GVW, a fuel

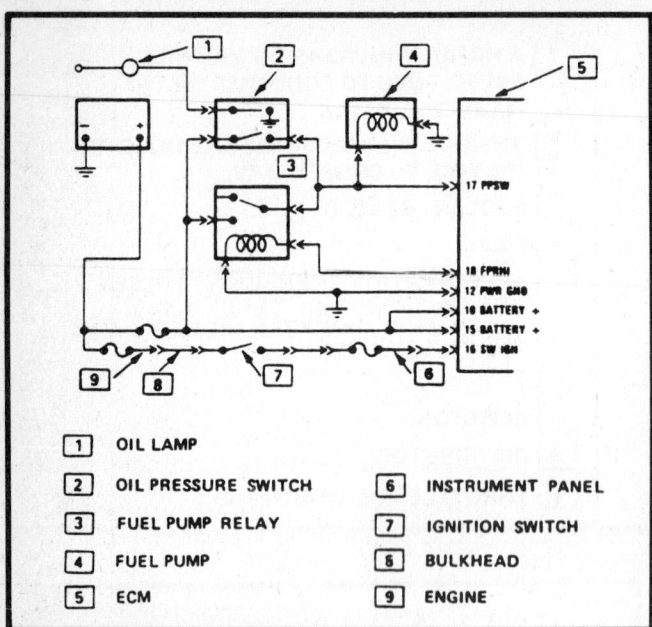

Typical fuel pump circuit

1	OIL LAMP		
2	OIL PRESSURE SWITCH	6	INSTRUMENT PANEL
3	FUEL PUMP RELAY	7	IGNITION SWITCH
4	FUEL PUMP	8	BULKHEAD
5	ECM	9	ENGINE

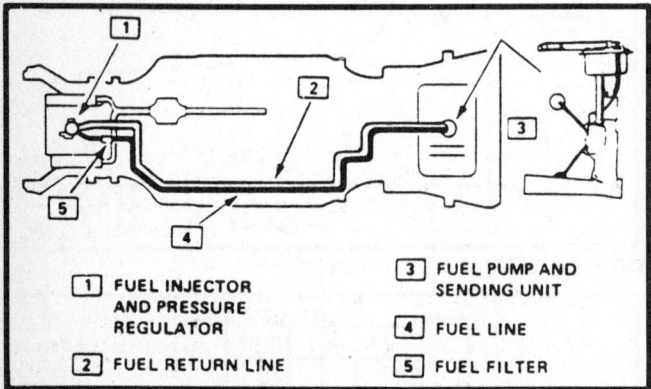

Typical fuel supply system

1	FUEL INJECTOR AND PRESSURE REGULATOR	3	FUEL PUMP AND SENDING UNIT
2	FUEL RETURN LINE	4	FUEL LINE
		5	FUEL FILTER

module will override the ECM 2 second timer and the fuel pump will run for 20 seconds and then shut off if the vehicle is not started. This circuit corrects a hot restart (vapor lock) during a high ambient condition.

THROTTLE BODY INJECTOR (TBI) ASSEMBLY

The basic TBI unit is made up of 2 major casting assemblies: (1) a throttle body with a valve to control airflow and (2) a fuel body assembly with an integral pressure regulator and fuel injector to supply the required fuel. A device to control idle speed (IAC) and a device to provide information about throttle valve position (TPS) are included as part of the TBI unit.

The throttle body portion of the TBI unit may contain ports located at, above, or below the throttle valve. These ports generate the vacuum signals for the EGR valve, MAP sensor, and the canister purge system.

The fuel injector is a solenoid-operated device controlled by the ECM. The incoming fuel is directed to the lower end of the injector assembly which has a fine screen filter surrounding the injector inlet. The ECM turns on the solenoid, which lifts a normally closed ball valve off a seat. The fuel, under pressure, is injected in a conical spray pattern at the walls of the throttle body

bore above the throttle valve. The excess fuel passes through a pressure regulator before being returned to the vehicle fuel tank.

The pressure regulator is a diaphragm-operated relief valve with the injector pressure on one side, and the air cleaner pressure on the other. The function of the regulator is to maintain constant pressure (approximately 11 psi) to the injector throughout the operating loads and speed ranges of the engine. If the regulator pressure is too low, below 9 psi, it can cause poor performance. Too high a pressure could cause detonation and a strong fuel odor.

IDLE AIR CONTROL (IAC)

The purpose of the Idle Air Control (IAC) system is to control engine idle speeds while preventing stalls due to changes in engine load. The IAC assembly, mounted on the throttle body, controls bypass air around the throttle plate. By extending or retracting a conical valve, a controlled amount of air can move around the throttle plate. If rpm is too low, more air is diverted around the throttle plate to increase rpm.

During idle, the proper position of the IAC valve is calculated by the ECM based on battery voltage, coolant temperature, engine load, and engine rpm. If the rpm drops below a specified rate, the throttle plate is closed. The ECM will then calculate a new valve position.

Three different designs are used for the IAC conical valve. The first design used is single 35 taper while the second design used is a dual taper. The third design is a blunt valve. Care should be taken to insure use of the correct design when service replacement is required.

The IAC motor has 255 different positions or steps. The zero, or reference position, is the fully extended position at which the pintle is seated in the air bypass seat and no air is allowed to bypass the throttle plate. When the motor is fully retracted, maximum air is allowed to bypass the throttle plate. When the motor is fully retracted, maximum air is allowed to bypass the throttle plate.

The ECM always monitors how many steps it has extended or retracted the pintle from the zero or reference position; thus, it always calculates the exact position of the motor. Once the engine has started and the vehicle has reached approximately 40 mph, the ECM will extend the motor 255 steps from whatever

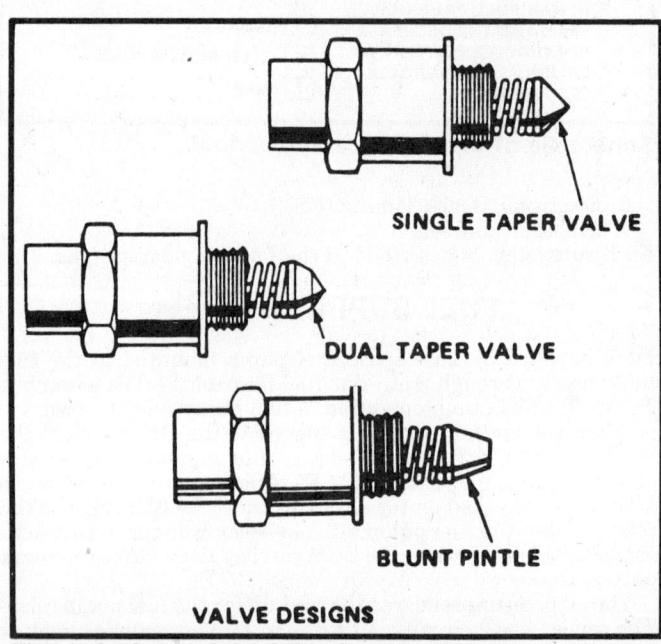

The 3 designs of the idle air control valve

position it is in. This will bottom out the pintle against the seat. The ECM will call this position 0 and thus keep its zero reference updated.

The IAC only affects the engine's idle characteristics. If it is stuck fully open, idle speed is too high (too much air enters the throttle bore) If it is stuck closed, idle speed is too low (not enough air entering). If it is stuck somewhere in the middle, idle may be rough, and the engine won't respond to load changes.

IDLE SPEED CONTROL

Incorrect diagnosis and/or misunderstanding of the idle speed control systems used on EFI engines may lead to unnecessary replacement of the IAC valve. Engine idle speed is controlled by the ECM which changes the idle speed by moving the IAC valve. The ECM adjusts idle speed in response to fluctuations in engine load (A/C, power steering, electrical loads, etc.) to maintain acceptable idle quality and proper exhaust emission performance.

The following is provided to assist the technician to better understand the system and correctly respond to the following customer concerns:
1. Rough idle/low idle speed
2. High idle speed/warm-up idle speed; no kickdown

ROUGH IDLE/LOW IDLE SPEED

The ECM will respond to increases in engine load, which would cause a drop in idle speed, by moving the IAC valve to maintain proper idle speed. After the induced load is removed the ECM will return the idle speed to the proper level.

During A/C compressor operation (**MAX, BI-LEVEL, NORM** or **DEFROST** mode) the ECM will increase idle speed in response to an **A/C-ON** signal, thereby compensating for any drop in idle speed due to compressor load. On some models, the ECM will also increase the idle speed in response to high power steering loads.

During periods of especially heavy loads (**A/C-ON** plus parking maneuvers) significant effects on idle quality may be experienced. These effects are more pronounced on 4-cylinder engines. Abnormally low idle, rough idle and idle shake may occur if the ECM does not receive the proper signals from the monitored systems.

HIGH IDLE SPEED/WARM-UP IDLE SPEED (NO KICKDOWN)

Engine idle speeds as high as 2100 rpm may be experienced during cold starts to quickly raise the catalytic converter to operating temperature for proper exhaust emissions performance. The idle speed attained after a cold start is ECM controlled and will not drop for 45 seconds regardless of driver attempts to kickdown.

It is important to recognize the EFI engines have no accelerator pump or choke. Idle speed during warm-up is entirely ECM controlled and cannot be changed by accelerator kickdown or pumping.

Diagnosis

Abnormally low idle speeds are usually caused by an ECM system-controlled or monitored irregularity, while the most common cause for abnormally high idle speed is an induction (intake air) leak. The idle air control valve may occasionally lose its memory function, and it has an ECM programmed method of relearning the correct idle position. This reset, when required, will occur the next time the car exceeds 35 mph. At this time the ECM seats the pintle of the IAC valve in the throttle body to determine a reference point. Then it backs out a fixed distance to maintain proper idle speed.

ELECTRONIC CONTROL MODULE (ECM)

The throttle body injection system is controlled by an on-board computer, the ECM, usually located in the passenger compartment. The ECM monitors engine operations and environmental conditions (ambient temperature, barometric pressure, etc.) needed to calculate the fuel delivery time (pulse width) of the fuel injector(s). The fuel pulse may be modified by the ECM to account for special operating conditions, such as cranking, cold starting, altitude, acceleration and deceleration.

The ECM controls the exhaust emissions by modifying fuel delivery to achieve, as nearly as possible an air/fuel ratio of 14.7:1. The injector on-time is determined by the various sensor inputs to the ECM. By increasing the injector pulse, more fuel is delivered, enriching the air/fuel ratio. Pulses are sent to the injector in 2 different modes, synchonized and non-synchronized.

In synchronized mode operation, the injector is pulsed once for each distributor reference pulse. In dual throttle body systems, the injectors are pulse alternately. In nonsynchronized mode operation, the injector is pulsed once every 12.5 milliseconds or 6.25 milliseconds depending on calibration. This pulse time is totally independent of distributor reference.

The ECM constantly monitors the input information, processes this information from various sensors, and generates output commands to the various systems that affect vehicle performance.

The ability of the ECM to recognize and adjust for vehicle variations (engine transmission, vehicle weight, axle ratio, etc.) is provided by a removable calibration unit (PROM) that is programmed to tailor the ECM for the particular vehicle. There is a specific ECM/PROM combination for each specific vehicle, and the combinations are not interchangeable with those of other vehicles.

The ECM also performs the diagnostic function of the system. It can recognize operational problems, alert the driver through the "CHECK ENGINE" light, and store a code or codes, which identify the problem areas to aid the technician in making repairs.

NOTE: Instead of an edgeboard connector, newer ECM's have a header connector which attaches solidly to the ECM case. Like the edgeboard connectors, the header connectors have a different pinout identification for different engine designs.

Data Sensors

A variety of sensors provide information to the ECM regarding engine operating characteristics. These sensors and their functions are described below. Be sure to take note that not every sensor described is used with every GM engine application.

ENGINE COOLANT TEMPERATURE

The coolant sensor is a thermister (a resistor which changes value based on temperature) mounted on the engine coolant stream. As the temperature of the engine coolant changes, the resistance of the coolant sensor changes. Low coolant temperature produces a high resistance (100,000 ohms at −40°C/ −40°F), while high temperature causes low resistance (70 ohms at 130°C/266°F).

The ECM supplies a 5 volt signal to the coolant sensor and measures the voltage that returns. By measuring the voltage change, the ECM determines the engine coolant temperature. The voltage will be high when the engine is cold and low when the engine is hot. This information is used to control fuel management, IAC, spark timing, EGR, canister purge and other engine operating conditions.

A failure in the coolant sensor circuit should either set a Code 14 or 15. These codes indicate a failure in the coolant temperature sensor circuit.

OXYGEN SENSOR

The exhaust oxygen sensor is mounted in the exhaust system where it can monitor the oxygen content of the exhaust gas stream. The oxygen content in the exhaust reacts with the oxygen sensor to produce a voltage output. This voltage ranges from approximately 100 millivolts (high oxygen — lean mixture) to 900 millivolts (low oxygen — rich mixture).

By monitoring the voltage output of the oxygen sensor, the ECM will determine what fuel mixture command to give to the injector (lean mixture — low voltage — rich command, rich mixture — high voltage — lean command).

Remember that oxygen sensor indicates to the ECM what is happening in the exhaust. It does not cause things to happen. It is a type of gauge: high oxygen content = lean mixture; low oxygen content = rich mixture. The ECM adjust fuel to keep the system working.

The oxygen sensor, if open should set a Code 13. A constant low voltage in the sensor circuit should set a Code 44 while a constant high voltage in the circuit should set a Code 45. Codes 44 and 45 could also be set as a result of fuel system problems.

MANIFOLD ABSOLUTE PRESSURE SENSOR

The Manifold Absolute Pressure (MAP) sensor measures the changes in the intake manifold pressure which result from engine load and speed changes. The pressure measured by the MAP sensor is the difference between barometric pressure (outside air) and manifold pressure (vacuum). A closed throttle engine coastdown would produce a relatively low MAP value (approximately 20–35 kPa), while wide-open throttle would produce a high value (100 kPa). This high value is produced when the pressure inside the manifold is the same as outside the manifold, and 100% of outside air (or 100 kPa) is being measured. This MAP output is the opposite of what you would measure on a vacuum gauge. The use of this sensor also allows the ECM to adjust automatically for different altitude.

The ECM sends a 5 volt reference signal to the MAP sensor. As the MAP changes, the electrical resistance of the sensor also changes. By monitoring the sensor output voltage the ECM can determine the manifold pressure. A higher pressure, lower vacuum (high voltage) requires more fuel, while a lower pressure, higher vacuum (low voltage) requires less fuel. The ECM uses the MAP sensor to control fuel delivery and ignition timing. A failure in the MAP sensor circuit should set a Code 33 or Code 34.

MANIFOLD AIR TEMPERATURE SENSOR

The Manifold Air Temperature (MAT) sensor is a thermistor mounted in the intake manifold. A thermistor is a resistor which changes resistance based on temperature. Low manifold air temperature produces a high resistance (100,000 ohms at −40°F/−40°C), while high temperature cause low resistance (70 ohms at 266°F/130°C).

The ECM supplies a 5 volt signal to the MAT sensor through a resistor in the ECM and monitors the voltage. The voltage will be high when the manifold air is cold and low when the air is hot. By monitoring the voltage, the ECM calculates the air temperature and uses this data to help determine the fuel delivery and spark advance. A failure in the MAT circuit should set either a Code 23 or Code 25.

VEHICLE SPEED SENSOR (VSS)

NOTE: A vehicle equipped with a speed sensor, should not be driven without a the speed sensor connected, as idle quality may be affected.

The Vehicle Speed Sensor (VSS) is mounted behind the speedometer in the instrument cluster or on the transmission/speed-

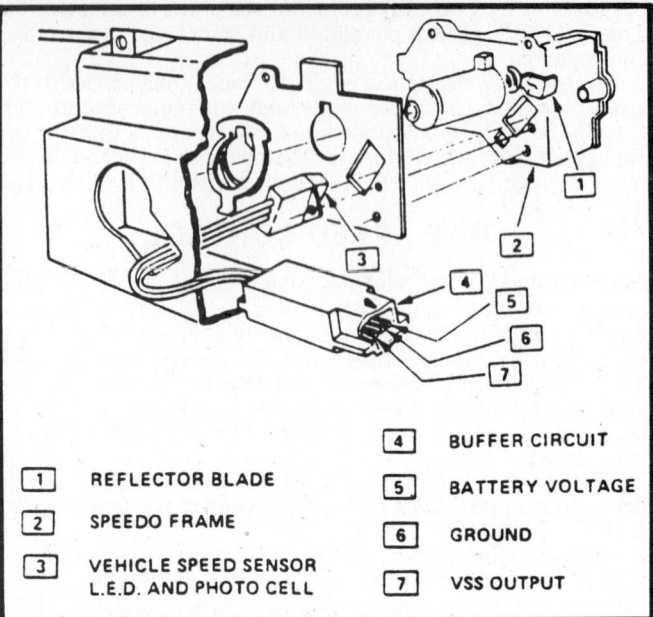

1	REFLECTOR BLADE	4	BUFFER CIRCUIT
2	SPEEDO FRAME	5	BATTERY VOLTAGE
3	VEHICLE SPEED SENSOR L.E.D. AND PHOTO CELL	6	GROUND
		7	VSS OUTPUT

Typical vehicle speed sensor

ometer drive gear. It provides electrical pulses to the ECM from the speedometer head. The pulses indicate the road speed. The ECM uses this information to operate the IAC, canister purge, and TCC.

Some vehicles equipped with digital instrument clusters use a Permanent Magnet (PM) generator to provide the VSS signal. The PM generator is located in the transmission and replaces the speedometer cable. The signal from the PM generator drives a stepper motor which drives the odometer. A failure in the VSS circuit should set a Code 24.

THROTTLE POSITION SENSOR (TPS)

The Throttle Position Sensor (TPS) is connected to the throttle shaft and is controlled by the throttle mechanism. A 5 volt reference signal is sent to the TPS from the ECM. As the throttle valve angle is changed (accelerator pedal moved), the resistance of the TPS also changes. At a closed throttle position, the resistance of the TPS is high, so the output voltage to the ECM will be low (approximately 0.5 volts). As the throttle plate opens, the resistance decreases so that, at wide open throttle, the output voltage should be approximately 5 volts. At closed throttle position, the voltage at the TPS should be less than 1.25 volts.

By monitoring the output voltage from the TPS, the ECM can determine fuel delivery based on throttle valve angle (driver demand). The TPS can either be misadjusted, shorted, open or loose. Misadjustment might result in poor idle or poor wide-open throttle performance. An open TPS signals the ECM that the throttle is always closed, resulting in poor performance. This usually sets a Code 22. A shorted TPS gives the ECM a constant wide-open throttle signal and should set a Code 21. A loose TPS indicates to the ECM that the throttle is moving. This causes intermittent bursts of fuel from the injector and an unstable idle. On some vehicles, the TPS is adjustable and therefore can be adjusted to correct any complications caused by to high or to low of a voltage signal.

CRANKSHAFT AND CAMSHAFT SENSOR

These sensors are mounted on the engine block, near the engine crankshaft, and also near the camshaft on some engines. The sensors are used to send a signal through the Direct Ignition

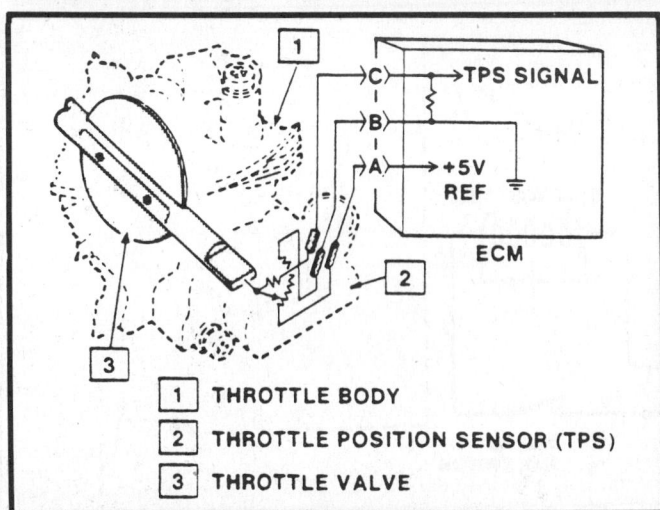

1 THROTTLE BODY
2 THROTTLE POSITION SENSOR (TPS)
3 THROTTLE VALVE

Typical throttle position sensor schematic

System (DIS) module to the ECM. The ECM uses this reference signal to calculate engine speed and crankshaft position.

In a typical 4 cylinder engine application, a sensor is mounted with the ignition module and 2 ignition coils to comprise the direct ignition assembly. When mounted on the engine block, the sensor tip is very close to a metal disk wheel with slots which is mounted on the crankshaft.

The sensor tip contains a small magnet and a small coil of wire. As the metal disk wheel with the slots rotates past the sensor tip, the magnetic field of the permanent magnet is changed and a voltage is induced into the coil. This voltage signal is sent to the ignition module. The ignition module is able to determine engine speed from the frequency of the voltage curve, which changes with engine speed.

A 6 cylinder engine may use a different type of sensor called a hall effect switch. With the direct ignition connected to the vehicle electrical system, the system voltage is applied to the hall effect switch located near the tip of the sensor. A small permanent magnet creates a magnetic field in the hall effect switch circuit. As the disc wheel with the slots rotates past the sensor tip, the magnetic field in the hall effect switch changes and a change in the voltage occurs at the hall effect switch output terminal.

Since this terminal is connect to the ignition module, the module senses this change in voltage and correlates the frequency of the voltage curve to determine the engine speed. The ignition module then uses this voltage input to help determine when to close and open the ignition coil primary circuit and fire the spark plug.

PARK/NEUTRAL SWITCH

NOTE: Vehicle should not be driven with the park/ neutral switch disconnected as idle quality may be affected in PARK or NEUTRAL and a Code 24 (VSS) may be set.

This switch indicates to the ECM when the transmission is in **P** or **N**. The information is used by the ECM for control on the torque converter clutch, EGR, and the idle air control valve operation.

AIR CONDITIONER REQUEST SIGNAL

This signal indicates to the ECM that an air conditioning mode is selected at the switch and that the A/C low pressure switch is closed. The ECM controls the A/C and adjusts the idle speed in response to this signal.

TORQUE CONVERTER CLUTCH SOLENOID

The purpose of the torque converter clutch system is designed to eliminate power loss by the converter (slippage) to increase fuel economy. By locking the converter clutch, a more effective coupling to the flywheel is achieved. The converter clutch is operated by the ECM controlled torque converter clutch solenoid.

NOTE: There is a downshift solenoid use with some THM 400 transmissions. The ECM controls the downshift solenoid for a wide open throttle downshift. On vehicles equipped with a manual transmission and a shift light, the ECM will control the shift light to indicate the best shift point for maximum fuel economy.

POWER STEERING PRESSURE SWITCH

The power steering pressure switch is used so that the power steering oil pressure pump load will not effect the engine idle. Turning the steering wheel increase the power steering oil pressure and pump load on the engine. The power steering pressure switch will close before the load can cause an idle problem.

OIL PRESSURE SWITCH

The oil pressure switch is usually mounted on the back of the engine, just below the intake manifold. Some vehicles use the oil pressure switch as a parallel power supply, with the fuel pump relay and will provide voltage to the fuel pump, after approximately 4 psi (28 kPa) of oil pressure is reached. This switch will also help prevent engine seizure by shutting off the power to the fuel pump and causing the engine to stop when the oil pressure is lower than 4 psi.

ELECTRONIC SPARK TIMING (EST)

Electronic spark timing (EST) is used on all engines. The EST distributor contains no vacuum or centrifugal advance and uses a 7-terminal distributor module. It also has 4 wires going to a 4-terminal connector in addition to the connectors normally found on HEI distributors. A reference pulse, indicating both engine rpm and crankshaft position, is sent to the ECM. The ECM determines the proper spark advance for the engine operating conditions and sends an EST pulse to the distributor.

The EST system is designed to optimize spark timing for better control of exhaust emissions and for fuel economy improvements. The ECM monitors information from various engine sensors, computes the desired spark timing and changes the timing accordingly. A backup spark advance system is incorporated in the module in case of EST failure.

ELECTRONIC SPARK CONTROL (ESC)

When engines are equipped with ESC in conjunction with EST, ESC is used to reduce spark advance under conditions of detonation. A knock sensor signals a separate ESC controller to retard the timing when it senses engine knock. The ESC controller signals the ECM which reduces spark advance until no more signals are received from the knock sensor.

DIRECT IGNITION SYSTEM (DIS)

Components of the Direct Ignition System (DIS) are a coil pack, ignition module, crankshaft reluctor ring, magnetic sensor and the ECM. The coil pack consists of 2 separate, interchangeable, ignition coils. These coils operate in the same manner as previous coils. Two coils are needed because each coil fires for 2 cylinders. The ignition module is located under the coil pack and is

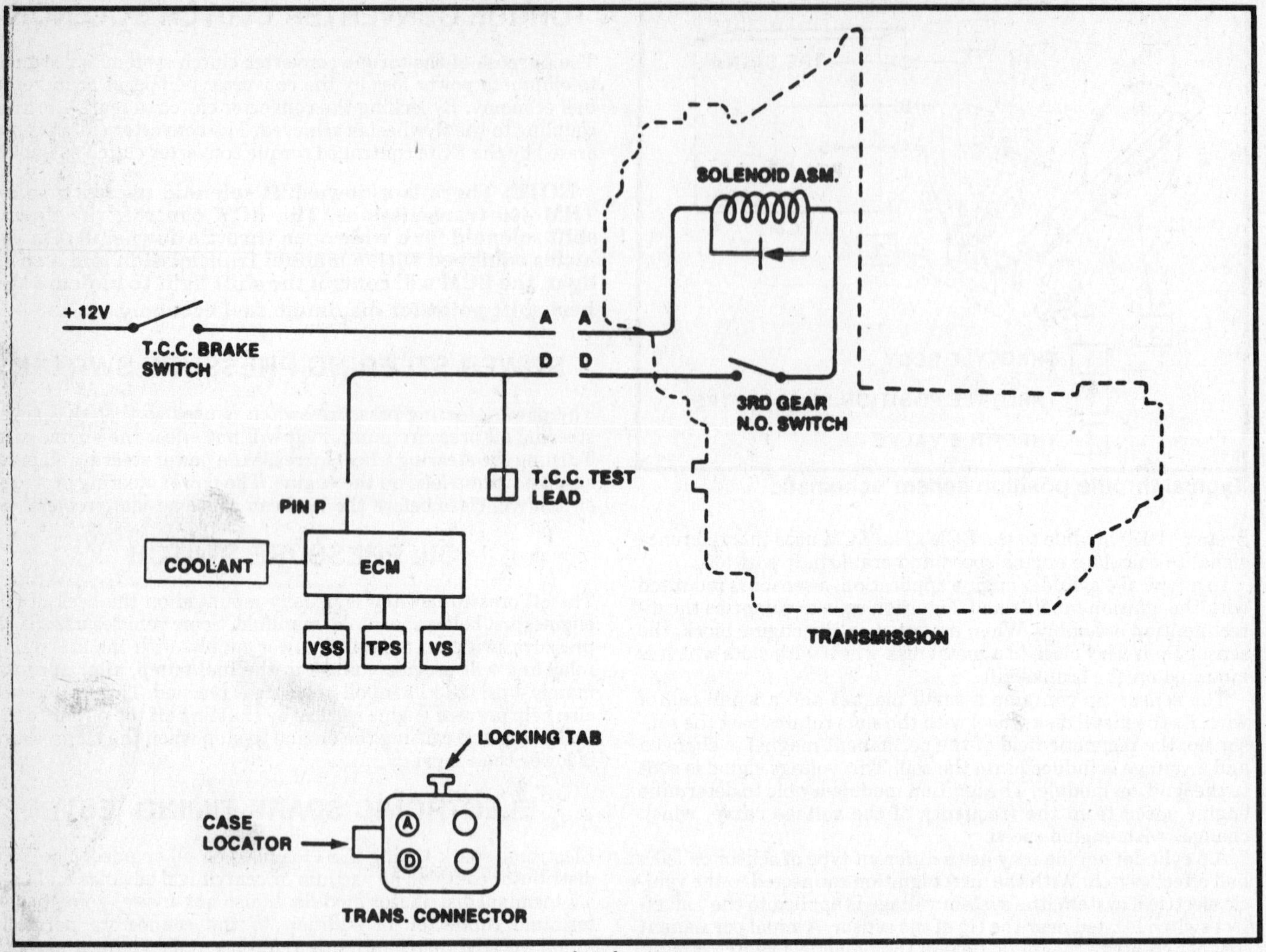

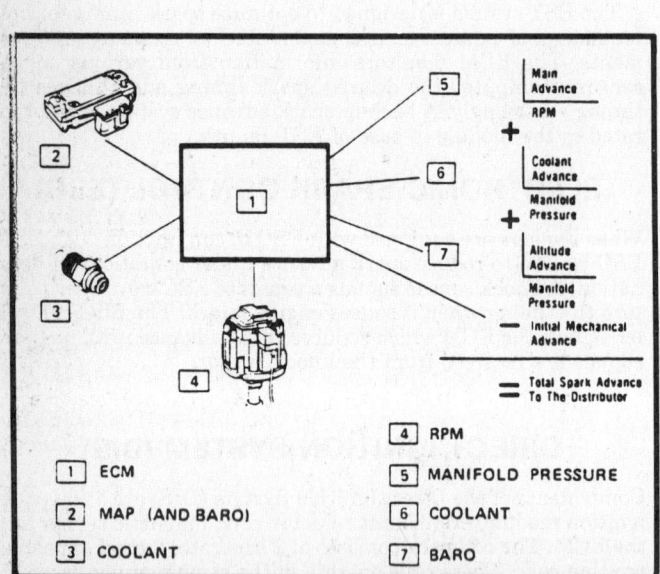

SOLENOID ASM.

+12V

T.C.C. BRAKE
SWITCH

A A

D D

3RD GEAR
N.O. SWITCH

T.C.C. TEST
LEAD

PIN P

COOLANT   ECM

VSS TPS VS

TRANSMISSION

LOCKING TAB

CASE
LOCATOR

Ⓐ
Ⓓ

TRANS. CONNECTOR

Typical wiring schematic of the torque converter clutch system

5 Main Advance
RPM

+

6 Coolant Advance
Manifold Pressure

+

7 Altitude Advance
Manifold Pressure

— Initial Mechanical Advance

— Total Spark Advance To The Distributor

1 ECM

4 RPM

2 MAP (AND BARO)

5 MANIFOLD PRESSURE

3 COOLANT

6 COOLANT

7 BARO

Schematic of the electronic spark timing

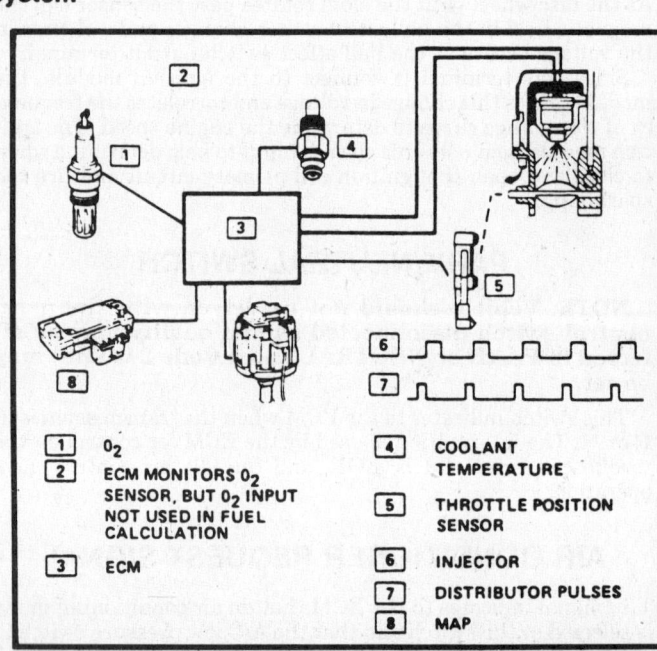

1 O_2

2 ECM MONITORS O_2 SENSOR, BUT O_2 INPUT NOT USED IN FUEL CALCULATION

3 ECM

4 COOLANT TEMPERATURE

5 THROTTLE POSITION SENSOR

6 INJECTOR

7 DISTRIBUTOR PULSES

8 MAP

Schematic of the electronic spark timing inputs

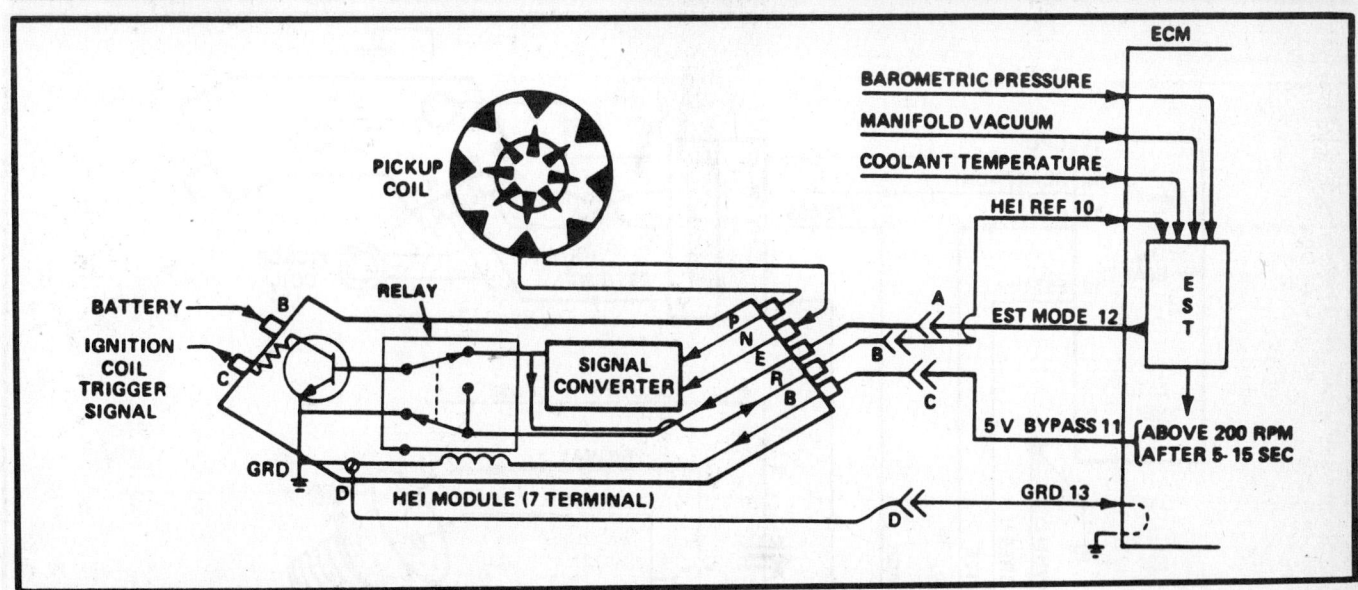

Typical wiring schematic of the electronic spark timing in the cranking mode

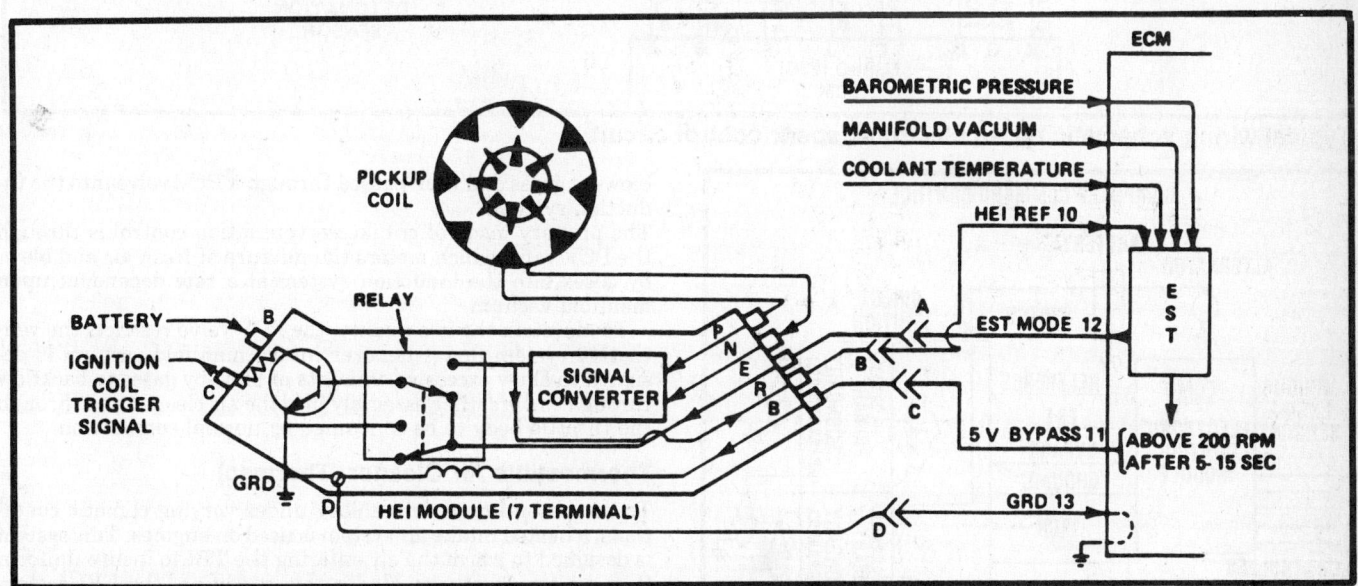

Typical wiring schematic of the electronic spark timing in the running mode

connected to the ECM by a 6 pin connector. The ignition module controls the primary circuits to the coils, turning them on and off and controls spark timing below 400 rpm and if the ECM bypass circuit becomes open or grounded.

The magnetic pickup sensor inserts through the engine block, just above the pan rail in proximity to the crankshaft reluctor ring. Notches in the crankshaft reluctor ring trigger the magnetic pickup sensor to provide timing information to the ECM. The magnetic pickup sensor provides a cam signal to identify correct firing sequence and crank signals to trigger each coil at the proper time.

This system uses EST and control wires from the ECM, as with the distributor systems. The ECM controls the timing using crankshaft position, engine rpm, engine temperature and manifold absolute pressure sensing.

EMISSION CONTROL

Various components are used to control exhaust emissions from a vehicle. These components are controlled by the ECM based on different engine operating conditions. These components are described in the following paragraphs. Not all components are used on all engines.

Exhaust Gas Recirculation (EGR)

EGR is a oxides of nitrogen (NOx) control which recycles exhaust gases through the combustion cycle by admitting exhaust gases into the intake manifold. The amount of exhaust gas admitted is adjusted by a vacuum controlled valve in response to engine operating conditions. If the valve is open, the recirculated exhaust gas is released into the intake manifold to be drawn into the combustion chamber.

The integral exhaust pressure modulated EGR valve uses a transducer responsive to exhaust pressure to modulate the vacuum signal to the EGR valve. The vacuum signal is provided by an EGR vacuum port in the throttle body valve. Under conditions when exhaust pressure is lower than the control pressure, the EGR signal is reduced by an air bleed within the transducer.

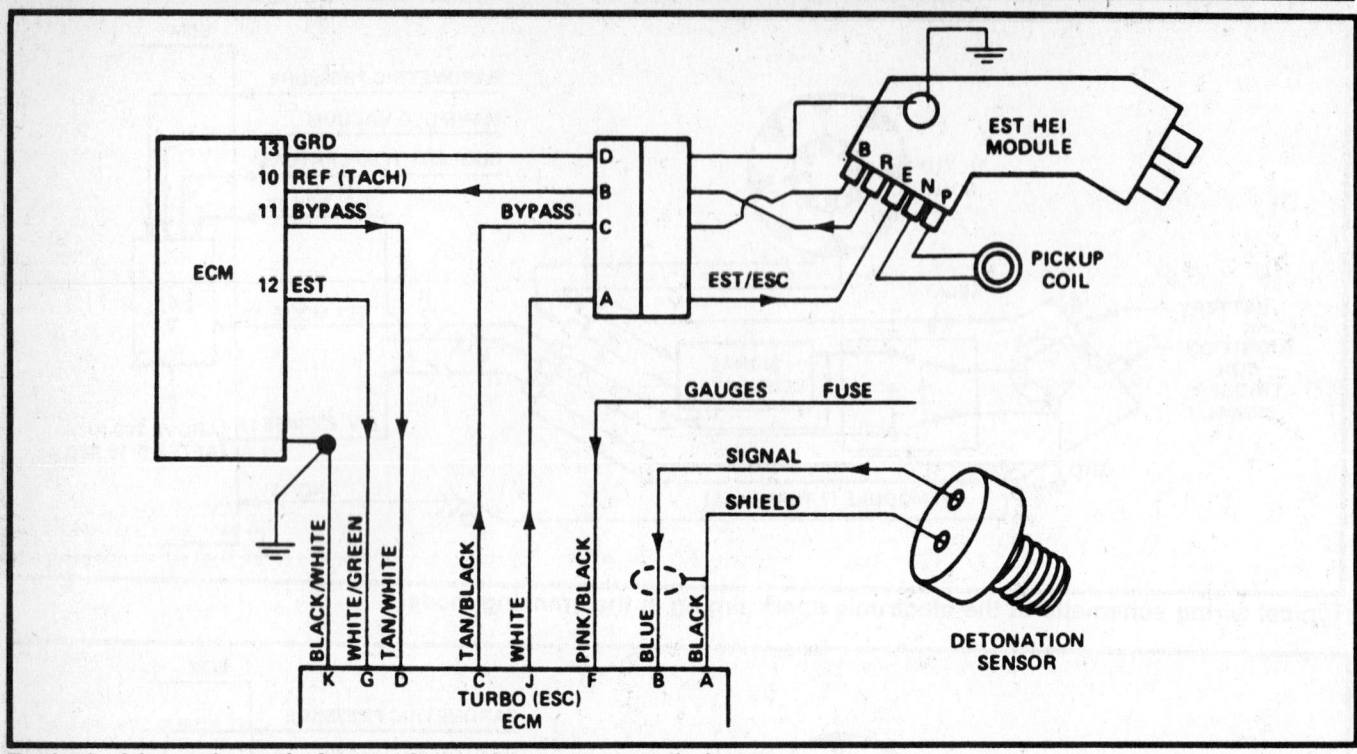

Typical wiring schematic for an electronic spark control circuit

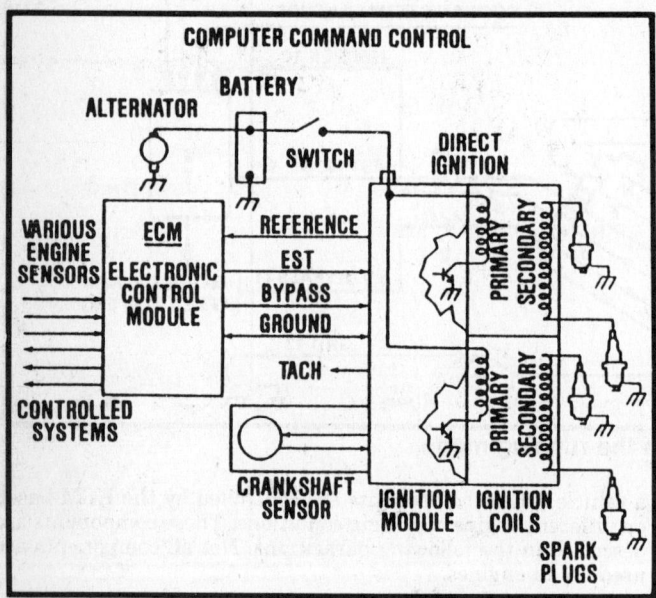

Typical wiring schematic for the direct ignition system circuit

Under conditions when exhaust pressure is higher than the control pressure, the air bleed is closed and the EGR valve responds to an unmodified vacuum signal. Physical arrangement of the valve components will vary depending on whether the control pressure is positive or negative.

Positive Crankcase Ventilation (PCV) System

A closed Positive Crankcase Ventilation (PCV) system is used to provide more complete scavenging of crankcase vapors. Fresh air from the air cleaner is supplied to the crankcase, mixed with blow-by gases and then passed through a PCV valve into the induction system.

The primary mode of crankcase ventilation control is through the PCV valve which meters the mixture of fresh air and blow-by gases into the induction system at a rate dependent upon manifold vacuum.

To maintain the idle quality, the PCV valve restricts the ventilation system flow whenever intake manifold vacuum is designed to allow excessive amounts of blow-by gases to backflow through the breather assembly into the air cleaner and through the throttle body to be consumed by normal combustion.

Thermostatic Air Cleaner (Thermac)

To assure optimum driveability under varying climatic conditions, a heated intake air system is used on engines. This system is designed to warm the air entering the TBI to insure uniform inlet air temperatures. Under this condition, the EFI system can be calibrated to efficiently reduce exhaust emission and to eliminate throttle blade icing. The Thermac system used on vehicles equipped with EFI operates identical to other Thermac systems.

EVAPORATIVE EMISSION CONTROL (EEC) SYSTEMS

The basic evaporative emission control system used on all vehicles uses the carbon canister storage method. This method transfers fuel vapor to an activated carbon storage device for retention when the vehicle is not operating. A ported vacuum signal is used for purging vapors stored in the canister.

Controlled Canister Purge

The ECM controls a solenoid valve which controls vacuum to the purge valve in the charcoal canister. In open loop, before a specified time has expired and below a specified rpm, the solenoid valve is energized and blocks vacuum to the purge valve. When the system is in closed loop, after a specified time and above a specified rpm, the solenoid valve is de-energized and vacuum can be applied to the purge valve. This releases the collected vapors into the intake manifold. On systems not using an ECM controlled solenoid, a Thermo Vacuum Valve (TVV) is used to control purge. See the appropriate vehicle sections for checking procedures.

Air Management Control

The air management system aids in the reduction of exhaust emissions by supplying air to either the catalytic converter, engine exhaust manifold, or to the air cleaner. The ECM controls the air management system by energizing or deenergizing an air switching valve. Operation of the air switching valve is dependent upon such engine operating characteristics as coolant temperature, engine load, and acceleration (or deceleration), all of which are sensed by the ECM.

Pulsair Reactor System

The Pulsair Injection Reactor (PAIR) system utilizes exhaust pressure pulsations to draw air into the exhaust system. Fresh air from the clean side of the air cleaner supplies filtered air to avoid dirt build-up on the check valve seat. The air cleaner also serves as a muffler for noise reduction.
The internal mechanism of the Pulsair valve reacts to 3 distinct conditions.

The firing of the engine creates a pulsating flow of exhaust gases which are of positive (+) or negative (−) pressure. This pressure or vacuum is transmitted through external tubes to the Pulsair valve.

1. If the pressure is positive, the disc is forced to the closed position and no exhaust gas is allowed to flow past the valve and into the air supply line.

2. If there is a negative pressure (vacuum) in the exhaust system at the valve, the disc will open, allowing fresh air to mix with the exhaust gases.

3. Due to the inertia of the system, the disc ceases to follow the pressure pulsations at high engine rpm. At this point, the disc remains closed, preventing any further fresh air flow.

Catalytic Converter

Of all emission control devices available, the catalytic converter is the most effective in reducing tailpipe emissions. The major tailpipe pollutants are hydrocabons (HC), carbon monoxide (CO), and oxides of nitrogen (NOx).

SERVICE PRECAUTIONS

When working around any part of the fuel system, take precautionary steps to prevent fire and/or explosion:
- Disconnect negative terminal from battery (except when testing with battery voltage is required).
- When ever possible, use a flashlight instead of a drop light.
- Keep all open flame and smoking material out of the area.
- Use a shop cloth or similar to catch fuel when opening a fuel system.
- Relieve fuel system pressure before servicing.
- Use eye protection.
- Always keep a dry chemical (class B) fire extinguisher near the area.

NOTE: Due to the amount of fuel pressure in the fuel lines, before doing any work to the fuel system, the fuel system should be de-pressurized. To de-pressurize the fuel system, remove the fuel tank cap to relieve tank va-

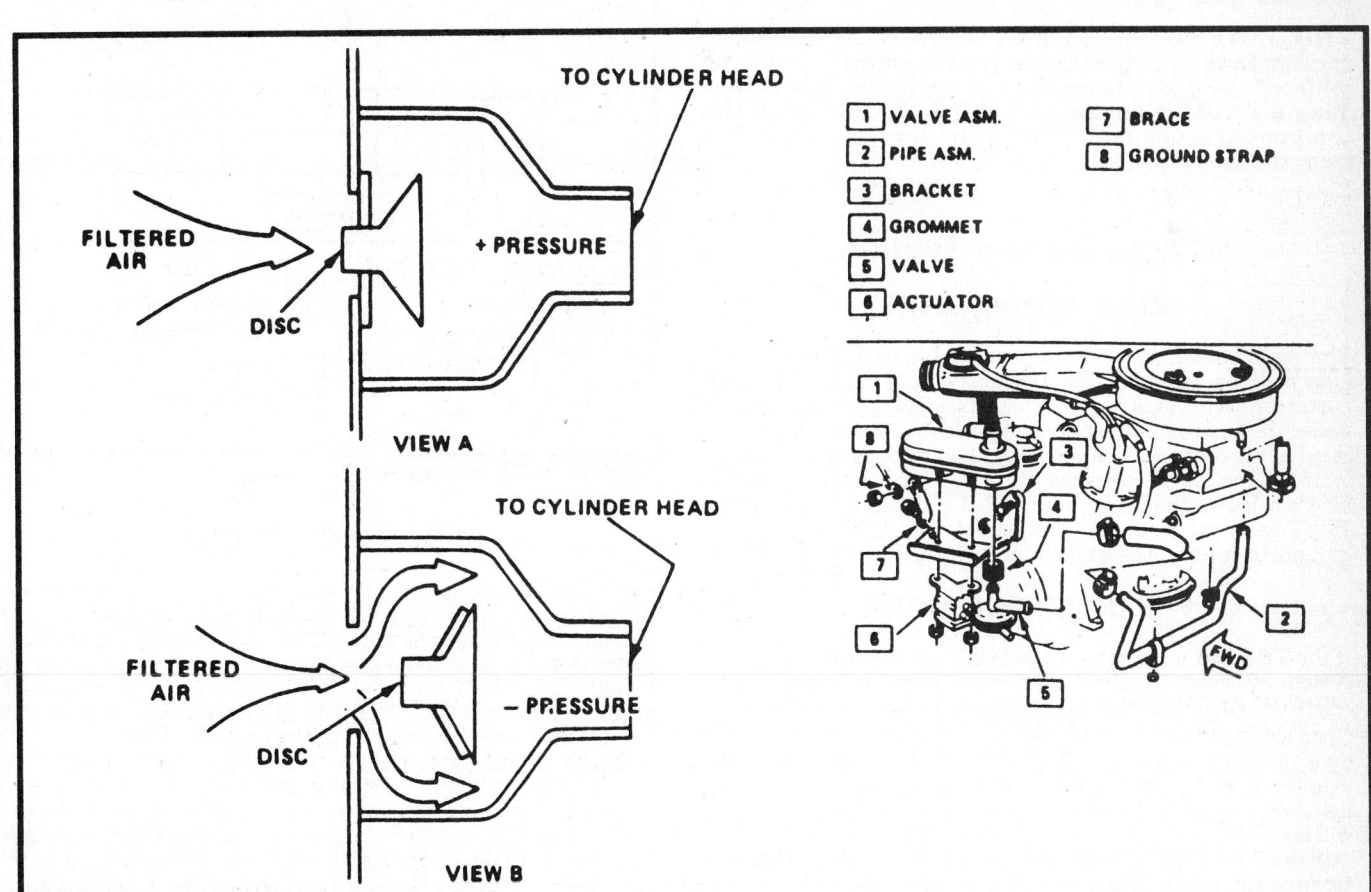

Typical pulsair injector reactor valve

por pressure. Remove the fuel pump fuse and disconnect the fuel pump electrical connections at the fuel pump (if necessary) and start the vehicle. Let the vehicle run until it burns up the remaining fuel in the fuel lines. This way there will be no pressure left in the fuel system and the repair work can be performed. Some engines (like the 2.0L engine) may have a bleed valve located in the fuel pressure regulator or on the fuel line. This bleed valve may be used to bleed the fuel pressure from the system instead of using the other method. The TBI 220 model has an internal constant bleed feature and relieves the fuel pump system pressure when the engine is turned OFF. Therefore, no further pressure relief procedure is required for that model.

Electrocstatic Discharge Damage

Electronic components used in the control system are often design to carry very low voltage and are very susceptible to damage caused by electrostatic discharge. It is possible for less than 100 volts of static electricity to cause damage to some electronic components. By comparison it takes as much as 4000 volts for a person to even feel the zap of a static discharge.

There are several ways for a person to become statically charged. The most common methods of charging are by friction and induction. An example of charging by friction is a person sliding across a car seat, in which a charge as much as 25000 volts can build up. Charging by induction occurs when a person with well insulated shoes stands near a highly charged object and momentarily touches ground. Charges of the same polarity are drained off, leaving the person highly charged with the opposite polarity. Static charges of either type can cause damage, therefore, it is important to use care when handling and testing electronic components.

NOTE: To prevent possible electrostatic discharge damage to the ECM, do not touch the connector pins or soldered components on the circuit board. When handling a PROM, Mem-Cal or Cal-Pak, do not touch the component leads and remove the integrated circuit from the carrier.

Diagnosis and Testing

ALCL CONNECTOR

The Assembly Line Communication Link (ALCL) (or also known as the Assembly Line Diagnostic Link (ALDL) is a diagnostic connector located in the passenger compartment usually under the instrument panel (except Pontiac Fiero which is located in the console). The assembly plant were the vehicles originate use the connector to check the engine for proper operation before it leaves the plant. Terminal **B** is usually the diagnostic **TEST** terminal (lead) and it can be connected to terminal **A**, or ground, to enter the Diagnostic mode or the Field Service Mode.

FIELD SERVICE MODE

If the **TEST** terminal is grounded with the engine running, the system will enter the the Field Service mode. In this mode, the "CHECK ENGINE" light will show whether the system is in Open loop or Closed loop. In Open loop the "CHECK ENGINE" light flashes 2 times and one half times per second. In Closed loop the light flashes once per second. Also in closed loop, the light will stay OUT most of the time if the system is too lean. It will stay ON most of the time if the system is too rich. In either case the Field Service mode check, which is part of the Diagnostic circuit check, will lead the technician into choosing the correct diagnostic chart to refer to.

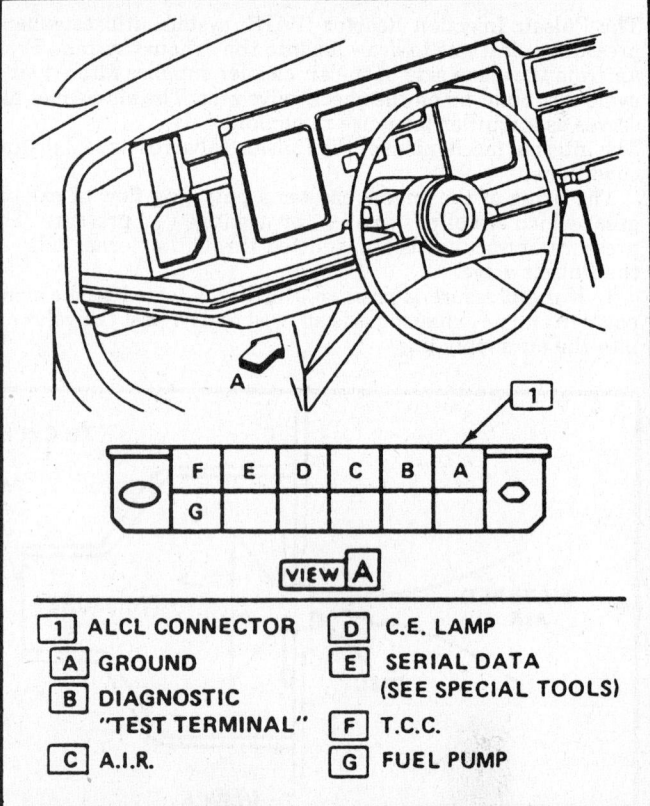

TERMINAL IDENTIFICATION

A	GROUND		F	T.C.C. (IF USED)
B	DIAGNOSTIC TERMINAL		G	FUEL PUMP (IF USED)
C	A.I.R. (IF USED)		H	BRAKE SENSE SPEED INPUT
D	SERVICE ENGINE SOON LAMP - IF USED		M	SERIAL DATA (L4) (SEE SPECIAL TOOLS)
E	SERIAL DATA (SEE SPECIAL TOOLS)			

Typical ALCL connector—trucks

VIEW **A**

1	ALCL CONNECTOR		D	C.E. LAMP
A	GROUND		E	SERIAL DATA (SEE SPECIAL TOOLS)
B	DIAGNOSTIC "TEST TERMINAL"		F	T.C.C.
C	A.I.R.		G	FUEL PUMP

Typical ALCL connector

DIAGNOSTIC MODE

A built-in, self-diagnostic system catches the problems most likely to occur in the Computer Command Control system. The diagnostic system turns on a "CHECK ENGINE" light in the instrument panel when a problem is detected. By grounding a trouble code **TEST** terminal under the dash, (ignition **ON** engine not running) the "CHECK ENGINE" light will flash a trouble code or codes indicating the problem areas.

As a bulb and system check, the "CHECK ENGINE" light will come on with the ignition switch **ON** and the engine not running. If the **TEST** terminal is then grounded, the light will flash a Code 12, which indicates the self-diagnostic system is working. A Code 12 consists of 1 flash, followed by a short pause, then 2 flashes in quick succession. After a longer pause, the code will repeat 2 more times.

When the engine is started, the "CHECK ENGINE" light will turn off. If the "CHECK ENGINE" light remains on, the self-diagnostic system has detected a problem. If the **TEST** terminal is then grounded with the ignition **ON**, engine not running, each trouble code will flash and repeat 3 times. If more than 1 problem has been detected, each trouble code will flash 3 times. Trouble codes will flash in numeric order (lowest number first). The trouble code series will repeat as long as the **TEST** terminal is grounded.

A trouble code indicates a problem in a given circuit (Code 14, for example, indicates a problem in the coolant sensor circuit; this includes the coolant sensor, connector harness, and ECM). The procedure for pinpointing the problem can be found in diagnosis. Similar charts are provided for each code.

Also in this mode all ECM controlled relays and solenoids except the fuel pump relay. This allows checking the circuits which may be difficult to energize without driving the vehicle and being under particular operating conditions. The IAC valve will move to its fully extended position on most models, block the idle air passage. This is useful in checking the minimum idle speed.

ECM LEARNING ABILITY

The ECM has a learning capability. If the battery is disconnected the learning process has to begin all over again. A change may be noted in the vehicle's performance. To teach the ECM, insure the vehicle is at operating temperature and drive at part throttle, with moderate acceleration and idle conditions, until performance returns.

READING THE DATA STREAM

This information is able to be read by putting the ECM into 1 of 3 different modes. These modes are entered by inserting a specific amount of resistance between the the ALCL (ALDL) connector terminals **A** and **B**. The modes and resistances needed to enter these modes are as follows:

0 OHMS DIAGNOSTIC MODES

When 0 resistance is between terminals **A** and **B** of the ALCL connector, the diagnostic mode is entered. There are 2 positions to this mode. One with the engine **OFF**, but the ignition **ON**; the other is when the engine is running.

If the diagnostic mode is entered with the engine in the **OFF** position, trouble codes will flash and the idle air control motor will pulsate in and out. Also, the relays and solenoids are energized with the exception of the fuel pump and injector.

In the event the ALCL connector terminal **B** is grounded with the engine running, the ECM goes into the field service mode. In this mode, the "CHECK ENGINE" light flashes closed or open loop and indicates the rich/lean status of the engine. The ECM runs the engine at a fixed ignition timing advanced above the base.

3.9 KILO-OHMS BACKUP MODE

The backup mode is entered by applying 3.9 kilo-ohms resistance between terminals **A** and **B** of the ALCL connector with the ignition switch in the **ON** position. The ALCL scanner tool can now read 5 of the 20 parameters on the data stream. These parameters are as mode status, oxygen sensor voltage, rpm, block learn and idle air control. There are 2 ways to enter the backup mode. Using a scan tool is one way; putting a 3.9 kilo-ohms resistor across terminals **A** and **B** of the ALCL is another.

10K OHMS SPECIAL MODE

This special mode is entered by applying a 10K ohms resistor across terminals **A** and **B**. When this happens the ECM does the following:
1. Allows all of the serial data to be read

2. Bypasses all timers
3. Add a calibrated spark advance
4. Enables the canister purge solenoid on some engines
5. Idles at 1000 rpm fixed idle air control and fixed base pulse width on the injector
6. Forces the idle air control to reset at part throttle (approximately 2000 rpm)
7. Disables the park/neutral restrict functions

20K OHMS OPEN OR ROAD TEST MODE

The system is in this mode during normal operation.

INTEGRATOR AND BLOCK LEARN

The integrator and block learn functions of the ECM are responsible for making minor adjustments to the air/fuel ratio on the fuel injected GM vehicles. These small adjustments are necessary to compensate for pinpoint air leaks and normal wear.

The integrator and block learn are 2 separate ECM memory functions which control fuel delivery. The integrator makes a temporary change and the block learn makes a more permanent change. Both of these functions apply only while the engine is in **CLOSED LOOP**. They represent the on-time of the injector. Also, integrator and block learn controls fuel delivery on the fuel injected engines as does the MC solenoid dwell on the CCC carbureted engines.

INTEGRATOR

Integrator is the term applied to a means of temporary change in fuel delivery. Integrator is displayed through the ALCL data line and monitored with a scanner as a number between 0 and 255 with an average of 128. The integrator monitors the oxygen sensor output voltage and adds and subtracts fuel depending on the lean or rich condition of the oxygen sensor. When the integrator is displaying 128, it indicates a neutral condition. This means that the oxygen sensor is seeing results of the 14.7:1 air/fuel mixture burned in the cylinders.

NOTE: An air leak in the system (a lean condition) would cause the oxygen sensor voltage to decrease while the integrator would increase (add more fuel) to temporarily correct for the lean condition. If this happened the injector pulse width would increase.

BLOCK LEARN

Although the integrator can correct fuel delivery over a wide range, it is only for a temporary correction. Therefore, another control called block learn was added. Although it cannot make as many corrections as the integrator, it does so for a longer period of time. It gets its name from the fact that the operating range of the engine for any given combinations of rpm and load is divided into 16 cell or blocks.

The computer has a given fuel delivery stored in each block. As the operating range gets into a given block the fuel delivery will be based on what value is stored in the memory in that block. Again, just like the integrator, the number represents the on-time of the injector. Also, just like the integrator, the number 128 represents no correction to the value that is stored in the cell or block. When the integrator increases or decreases, block learn which is also watching the integrator will make corrections in the same direction. As the block learn makes corrections, the integrator correction will be reduced until finally the integrator will return to 128 if the block learn has corrected the fuel delivery.

BLOCK LEARN MEMORY

Block learn operates on 1 of 2 types of memories depending on application, non-volatile and volatile. The non-volatile memories retain the value in the block learn cells even when the ignition switch is turned **OFF**. When the engine is restarted, the fuel delivery for a given block will be based on information stored in memory.

The volatile memories lose the numbers stored in the block learn cells when the ignition is turned to the **OFF** position. Upon restarting, the block learn starts at 128 in every block and corrects from that point as necessary.

INTEGRATOR/BLOCK LEARN LIMITS

Both the integrator and block learn have limits which will vary from engine to engine. If the mixture is off enough so that the block learn reaches the limit of its control and still cannot correct the condition, the integrator would also go to its limit of control in the same direction and the engine would then begin to run poorly. If the integrators and block learn are close to or at their limits of control, the engine hardware should be checked to determine the cause of the limits being reached, vacuum leaks, sticking injectors, etc.

If the integrator is lied to, for example, if the oxygen sensor lead was grounded (lean signal) the integrator and block learn would add fuel to the engine to cause it to run rich. However, with the oxygen sensor lead grounded, the ECM would continue seeing a lean condition eventually setting a Code 44 and the fuel control system would change to open loop operations.

CLOSED LOOP FUEL CONTROL

The purpose of closed loop fuel control is to precisely maintain an air/fuel mixture 14.7:1. When the air/fuel mixture is maintained at 14.7:1, the catalytic converter is able to operate at maximum efficiency which results in lower emission levels.

Since the ECM controls the air/fuel mixture, it needs to check its output and correct the fuel mixture for deviations from the ideal ratio. The oxygen sensor feeds this output information back to the ECM.

READING CODES

The codes stores in the ECM's memory can be read either through a hand held diagnostic scanner plugged into the ALCL connector or by counting the number of flashes on the "CHECK ENGINE" light when the diagnostic terminal of the ALCL connector is grounded. The ALCL connector terminal **B** (diagnostic terminal) is usually the second terminal from the right of the top row in the ALCL connector. The terminal is most easily grounded by connecting to terminal **A** (internal ECM ground), the terminal to the right of the terminal **B** on the top row in the ALCL connector.

Once terminals **A** and **B** are grounded, the ignition switch must be turned **ON** with the engine not running. At this pointthe "CHECK ENGINE" light should flash a Code 12 (3) times consecutively. This would be the following flash sequence; flash, pause, flash-flash, long pause, flash, pause, flash-flash, long pause, flash, pause, flash-flash.

Code 12 indicates that the ECM's diagnostic system is operating. If Code 12 is not indicated, a problem is present within the diagnostic system itself and should by addresses by using the appropriate diagnostic chart.

Following the output of Code 12, the "CHECK ENGINE" light will indicate a diagnostic trouble code 3 times, if a code is present or it will simply continue to output Code 12. If more than 1 diagnostic code has been stored in the ECM's memory, the codes will be output from the lowest to the highest, with each code being displayed 3 times.

CLEARING THE TROUBLE CODES

When the ECM finds a problem with the system, the "CHECK ENGINE" light will come on and a trouble code will be recorded in the ECM memory. If the problem is intermittent, the "CHECK ENGINE" light will go out after 10 seconds, when the fault goes away. However the trouble code will stay in the ECM memory until the battery voltage to the ECM is removed. Removing the battery voltage for 10 seconds will clear all trouble codes. Do this by disconnecting the ECM harness from the positive battery terminal pigtail for 10 seconds with the key in the **OFF** position, or by removing the ECM fuse for 10 seconds with the key **OFF**.

NOTE: To prevent ECM damage, the key must be OFF when disconnecting and reconnecting ECM power.

TROUBLE CODE TEST LEAD

The trouble code **TEST** lead terminal is mounted in a multi-terminal connector located under the dash. Grounding this terminal signals the ECM to flash any trouble codes stored in the memory. This is easily done by jumping to the adjacent ground terminal.

If the **TEST** terminal is grounded with the ignition **ON** and the engine stopped, the system will enter the diagnostic mode. In the diagnostic mode the ECM will:
1. Flash a Code 12 (indicating system is operating)
2. Energize all ECM controlled relays

If the **TEST** terminal is grounded with the engine running, the system will enter the field service mode. In this mode, the "CHECK ENGINE" light will indicate whether the system is in open or closed loop. Open loop is indicated by the **CHECK ENGINE** light flashing approximately twice per second. In closed loop, the light flashes approximately once per second.

ENGINE PERFORMANCE DIAGNOSIS

Engine performance diagnosis procedures are guides that will lead to the most probable causes of engine performance complaints. They consider the components of the fuel, ignition, and mechanical systems that could cause a particular complaint, and then outline repairs in a logical sequence.

It is important to determine if the "CHECK ENGINE" light is on or has come on for a short interval while driving. If the "CHECK ENGINE" light has come on, the Computer Command Control System should be checked for stored **TROUBLE CODES** which may indicate the cause for the performance complaint.

All of the symptoms can be caused by worn out or defective parts such as spark plugs, ignition wiring, etc. If time and/or mileage indicate that parts should be replaced, it is recommended that it be done.

NOTE: Before checking any system controlled by the Electronic Fuel Injection (EFI) system, the Diagnostic Circuit Check must be performed or misdiagnosis may occur. If the complaint involves the "CHECK ENGINE" light, go directly to the Diagnostic Circuit Check.

Basic Troubleshooting

NOTE: The following explains how to activate the trouble code signal light in the instrument cluster and gives an explanation of what each code means. This is not a full system troubleshooting and isolation procedure.

Before suspecting the system or any of its components as faulty, check the ignition system including distributor, timing, spark plugs and wires. Check the engine compression, air cleaner, and emission control components not controlled by the ECM. Also check the intake manifold, vacuum hoses and hose connectors for leaks.

The following symptoms could indicate a possible problem with the system:
1. Detonation
2. Stalls or rough idle-cold
3. Stalls or rough idle-hot

4. Missing
5. Hesitation
6. Surges
7. Poor gasoline mileage
8. Sluggish or spongy performance
9. Hard starting-cold
10. Objectionable exhaust odors (that rotten egg smell)
11. Cuts out
12. Improper idle speed

As a bulb and system check, the "CHECK ENGINE" light will come on when the ignition switch is turned to the **ON** position but the engine is not started. The "CHECK ENGINE" light will also produce the trouble code or codes by a series of flashes which translate as follows. When the diagnostic test terminal under the dash is grounded, with the ignition in the **ON** position and the engine not running, the "CHECK ENGINE" light will flash once, pause, then flash twice in rapid succession. This is a Code 12, which indicates that the diagnostic system is working. After a long pause, the Code 12 will repeat itself 2 more times. The cycle will then repeat itself until the engine is started or the ignition is turned off.

When the engine is started, the "CHECK ENGINE" light will remain on for a few seconds, then turn off. If the **CHECK ENGINE** light remains on, the self-diagnostic system has detected a problem. If the test terminal is then grounded, the trouble code will flash 3 times. If more than 1 problem is found, each trouble code will flash 3 times. Trouble codes will flash in numerical order (lowest code number to highest). The trouble codes series will repeat as long as the test terminal is grounded.

A trouble code indicates a problem with a given circuit. For example, trouble Code 14 indicates a problem in the cooling sensor circuit. This includes the coolant sensor, its electrical harness, and the ECM. Since the self-diagnostic system cannot diagnose every possible fault in the system, the absence of a trouble code does not mean the system is trouble-free. To determine problems within the system which do not activate a trouble code, a system performance check must be made.

In the case of an intermittent fault in the system, the "CHECK ENGINE" light will go out when the fault goes away, but the trouble code will remain in the memory of the ECM. Therefore, it a trouble code can be obtained even though the "CHECK ENGINE" light is not on, the trouble code must be evaluated. It must be determined if the fault is intermittent or if the engine must be at certain operating conditions (under load, etc.) before the "CHECK ENGINE" light will come on. Some trouble codes will not be recorded in the ECM until the engine has been operated at part throttle for about 5–18 minutes. On the CCC System, a trouble code will be stored until terminal **R** of the ECM has been disconnected from the battery for 10 seconds.

An easy way to erase the computer memory on the CCC System is to disconnect the battery terminals from the battery. If this method is used, don't forget to reset clocks and electronic pre-programmable radios. Another method is to remove the fuse marked ECM in the fuse panel. Not all models have such a fuse.

SYSTEM DIAGNOSTIC CIRCUIT CHECK

Begin the Diagnostic Circuit Check by making sure that the diagnostic system itself is working. Turn the ignition to **ON** with the engine stopped. If the "CHECK ENGINE" or "SERVICE ENGINE SOON" light comes on, ground the diagnostic code terminal (test lead) under the dash. If the "CHECK ENGINE" or "SERVICE ENGINE SOON" light flashes a Code 12, the self-diagnostic system is working and can detect a faulty circuit. If there is no Code 12, see the appropriate chart in this section. If any additional codes flash, record them for later use.

If a Code 51 flashes, use chart 51 to diagnose that condition before proceeding with the Diagnostic Circuit Check. A Code 51

means that the "CHECK ENGINE" or "SERVICE ENGINE SOON" light flashes 5 times, pauses, then flashes once. After a longer pause, Code 51 will flash again twice in this same way. To find out what diagnostic step to follow, look up the chart for Code 51 in this section. If there is not a Code 51, follow the NO CODE 51 branch of the chart.

Clear the ECM memory by disconnecting the voltage lead either at the fuse panel or the ECM letter connector for 10 seconds. This clears any codes remaining from previous repairs, or codes for troubles not present at this time. Remember, even though a code is stored, if the trouble is not present the diagnostic charts cannot be used. The charts are designed only to locate present faults.

NOTE: An easy way to erase the computer memory on the CCC System is to disconnect the battery terminals from the battery. If this method is used, don't forget to reset clocks and electronic pre-programmable radios. Not all models have an ECM fuse.

Next, remove the **TEST** terminal ground, set the parking brake and put the transmission in **P**. Run the warm engine for 2 minutes, making sure you run it at the specified curb idle for the 2 minutes. Then, if the "CHECK ENGINE" or "SERVICE ENGINE SOON" light comes on while the engine is idling, ground the test lead again and not the flashing trouble code.

If the "CHECK ENGINE" or "SERVICE ENGINE SOON" light does not come on, check the codes which were recorded earlier. If there were no additional codes, road test the car for the problem being diagnosed to make sure it still exists.

The purpose of the Diagnostic Circuit check is to make sure the "CHECK ENGINE" or "SERVICE SOON SOON" light works, that the ECM is operating and can recognize a fault and to determine if any trouble codes are stored in the ECM memory.

If trouble codes are stored, it also checks to see if they indicate an intermittent problem. This is the starting point of any diagnosis. If there are no codes stored, move on to the System Performance Check.

The codes obtained from the "CHECK ENGINE" or "SERVICE ENGINE SOON" light display method indicate which diagnostic charts provide in the section are to be used. For example, Code 23 can be diagnosed by following the step-by-step procedures on chart 23.

NOTE: If more than 1 code is stored in the ECM, the lowest code No. must be diagnosed first. Then proceed to the next highest code. The only exception is when a 50 series flashes. Fifty series code take precedence over all other trouble codes and must be dealt with first, since they point to a fault in the PROM unit or the ECM.

If the diagnostic procedures call for the ECM to be replaced, the calibration unit (PROM) should be checked first to see if it is functioning correctly. If it is correct, the PROM should be removed from the defective ECM and installed in the new service ECM. The service ECM will not contain a PROM. Trouble Code 51 indicates the PROM is installed improperly or has malfunctioned. When Code 51 is obtained, the PROM installation should be checked for bent pins or pins not fully seated in the socket. If the PROM is installed correctly and Code 51 still shows, the PROM should be replaced.

NOTE: To prevent internal ECM damage, the ignition switch must be in the OFF position when reconnecting power to the ECM (for example, battery positive cable, ECM pigtail, ECM fuse, jumper cables, etc.).

INTERMITTENT CHECK ENGINE LIGHT

An intermittent open in the ground circuit would cause loss of power through the ECM and intermittent "CHECK ENGINE"

TROUBLE CODE IDENTIFICATION

The "SERVICE ENGINE SOON" light will only be "ON" if the malfunction exists under the conditions listed below. It takes up to five seconds minimum for the light to come on when a problem occurs. If the malfunction clears, the light will go out and a trouble code will be set in the ECM. Code 12 does not store in memory. If the light comes "on" intermittently, but no code is stored, go to section B - Symptoms. Any codes stored will be erased if no problem reoccurs within 50 engine starts. A specific engine may not use all available codes.

The trouble codes indicate problems as follows:

TROUBLE CODE 12 No distributor reference signal to the ECM. This code is not stored in memory and will only flash while the fault is present. Normal code with ignition "on," engine not running.

TROUBLE CODE 13 Oxygen Sensor Circuit - The engine must run up to four minutes at part throttle, under road load, before this code will set.

TROUBLE CODE 14 Shorted coolant sensor circuit - The engine must run two minutes before this code will set.

TROUBLE CODE 15 Open coolant sensor circuit - The engine must run five minutes before this code will set.

TROUBLE CODE 21 Throttle Position Sensor (TPS) circuit voltage high (open circuit or misadjusted TPS). The engine must run 10 seconds, at specified curb idle speed, before this code will set.

TROUBLE CODE 22 Throttle Position Sensor (TPS) circuit voltage low (grounded circuit or misadjusted TPS). Engine must run 20 seconds at specified curb idle speed, to set code.

TROUBLE CODE 23 M/C solenoid circuit open or grounded.

TROUBLE CODE 24 Vehicle speed sensor (VSS) circuit - The vehicle must operate up to two minutes, at road speed, before this code will set.

TROUBLE CODE 32 Barometric pressure sensor (BARO) circuit low.

TROUBLE CODE 34 Vacuum sensor or Manifold Absolute Pressure (MAP) circuit - The engine must run up to two minutes, at specified curb idle, before this code will set.

TROUBLE CODE 35 Idle speed control (ISC) switch circuit shorted. (Up to 70% TPS for over 5 seconds.)

TROUBLE CODE 41 No distributor reference signal to the ECM at specified engine vacuum. This code will store in memory.

TROUBLE CODE 42 Electronic spark timing (EST) bypass circuit or EST circuit grounded or open.

TROUBLE CODE 43 Electronic Spark Control (ESC) retard signal for too long a time; causes retard in EST signal.

TROUBLE CODE 44 Lean exhaust indication - The engine must run two minutes, in closed loop and at part throttle, before this code will set.

TROUBLE CODE 45 Rich exhaust indication - The engine must run two minutes, in closed loop and at part throttle, before this code will set.

TROUBLE CODE 51 Faulty or improperly installed calibration unit (PROM). It takes up to 30 seconds before this code will set.

TROUBLE CODE 53 Exhaust Gas Recirculation (EGR) valve vacuum sensor has seen improper EGR control vacuum.

TROUBLE CODE 54 M/C solenoid voltage high at ECM as a result of a shorted M/C solenoid circuit and/or faulty ECM.

Typical ECM diagnosis code numbers—Identification chart—passenger vehicles

CODE IDENTIFICATION

The "Service Engine Soon" light will only be "ON" if the malfunction exists under the conditions listed below. If the malfunction clears, the light will go out and the code will be stored in the ECM. Any codes stored will be erased if no problem reoccurs within 50 engine starts.

CODE AND CIRCUIT	PROBABLE CAUSE	CODE AND CIRCUIT	PROBABLE CAUSE
Code 13 - O_2 Sensor Open Oxygen Sensor Circuit	Indicates that the oxygen sensor circuit or sensor was open for one minute while off idle.	Code 33 - MAP Sensor Low Vacuum	MAP sensor output to high for 5 seconds or an open signal circuit.
Code 14 - Coolant Sensor High Temperature Indication	Sets if the sensor or signal line becomes grounded for 3 seconds.	Code 34 - MAP Sensor High Vacuum	Low or no output from sensor with engine running.
Code 15 - Coolant Sensor Low Temperature Indication	Sets if the sensor, connections, or wires open for 3 seconds.	Code 35 - IAC	IAC error
		Code 42 - EST	ECM has seen an open or grounded EST or Bypass circuit.
Code 21 - TPS Signal Voltage High	TPS voltage greater than 2.5 volts for 3 seconds with less than 1200 RPM.	Code 43 - ESC	Signal to the ECM has remained low for too long or the system has failed a functional check.
Code 22 - TPS Signal Voltage Low	A shorted to ground or open signal circuit will set code in 3 seconds.	Code 44 Lean Exhaust Indication	Sets if oxygen sensor voltage remains below .2 volts for about 20 seconds.
Code 23 - MAT Low Temperature Indication	Sets if the sensor, connections, or wires open for 3 seconds.	Code 45 Rich Exhaust Indication	Sets if oxygen sensor voltage remains above .7 volts for about 1 minute.
Code 24 - VSS No Vehicle Speed Indication	No vehicle speed present during a road load decel.		
Code 25 - MAT High Temperature Indication	Sets if the sensor or signal line becomes grounded for 3 seconds.	Code 51	Faulty MEM-CAL, PROM, or ECM.
		Code 52	Fuel CALPAK missing or faulty.
		Code 53	System overvoltage. Indicates a basic generator problem.
Code 32 - EGR	Vacuum switch shorted to ground on start up OR Switch not closed after the ECM has commanded EGR for a specified period of time. OR EGR solenoid circuit open for a specified period of time.	Code 54 - Fuel Pump Low voltage	Sets when the fuel pump voltage is less than 2 volts when reference pulses are being received.
		Code 55	Faulty ECM

Typical ECM diagnosis code numbers—identification chart—light trucks and vans

light operation. When the ECM loses ground, distributor ignition is lost. An intermittent open in the ground circuit would be described as an engine miss.

Therefore, an intermittent "CHECK ENGINE" light, no code stored and a driveability comment described as similar to a miss will require checking the grounding circuit and the Code 12 circuit as it originates at the ignition coil.

UNDERVOLTAGE TO THE ECM

A circuit breaker will shut off the ECM if the power supply falls below 9 volts. The undervoltage condition will cause the "CHECK ENGINE" light to come on as long as the condition exist.

Therefore, an intermittent "CHECK ENGINE" light, no code stored and a driveability comment described as similar to a miss will require checking the grounding circuit, Code 12 circuit and the ignition feed circuit to terminal C of the ECM. This does nor eliminate the necessity of checking the normal vehicle electrical system for possible cause such as a loose battery cable.

OVERVOLTAGE TO THE ECM

The ECM will also shut off by the circuit breaker when the power supply rises above 16 volts. The overvoltage condition will also cause the "CHECK ENGINE" to come on as long as this condition exist.

A momentary voltage surge in a vehicle's electrical system is a common occurrence. These voltage surges have never presented any problems because the entire electrical system acted as a shock absorber until the surge dissipated. Voltage surges or spikes in the vehicle's electrical system have been known, on occasion, to exceed 100 volts.

The system is a low voltage (between 9 and 16 volts) system and will not tolerate these surges. The ECM will be shut off by any surge in excess of 16 volts and will come back on, only after the surge has dissipated sufficiently to bring the voltage under 16 volts.

A surge will usually occur when an accessory requiring a high voltage supply is turned off or down. The voltage regulator in the vehicle's charging system cannot react to the changes in the voltage demands quickly enough and surge occurs. The driver should be questioned to determine which accessory circuit was turned off that caused the "CHECK ENGINE" light to come on.

Therefore, intermittent "CHECK ENGINE" light operation, with no trouble code stored, will require installation of a diode in the appropriate accessory circuit.

ELECTRONIC FUEL INJECTION – ALCL SCAN TESTER INFORMATION

An ALCL display unit (ALCL tester, scanner, monitor, etc), allows a technician to read the engine control system information from the ALCL connector under the instrument panel. It can provide information faster than a digital voltmeter or ohmmeter can. The scan tool does not diagnose the exact location of the problem. The tool supplies information about the ECM, the information that it is receiving and the commands that it is sending plus special information such as integrator and block learn. To use an ALCL display tool you should understand throughly how an engine control system operates.

An ALCL scanner or monitor puts a fuel injection system into a special test mode. This mode commands an idle speed of 1000 rpm. The idle quality cannot be evaluated with a tester plugged in. Also the test mode commands a fixed spark with no advance. On vehicles with Electronic Spark Control (ESC), there will be a fixed spark, but it will be advanced. On vehicles with ESC, there might be a serious spark knock, this spark knock could be bad

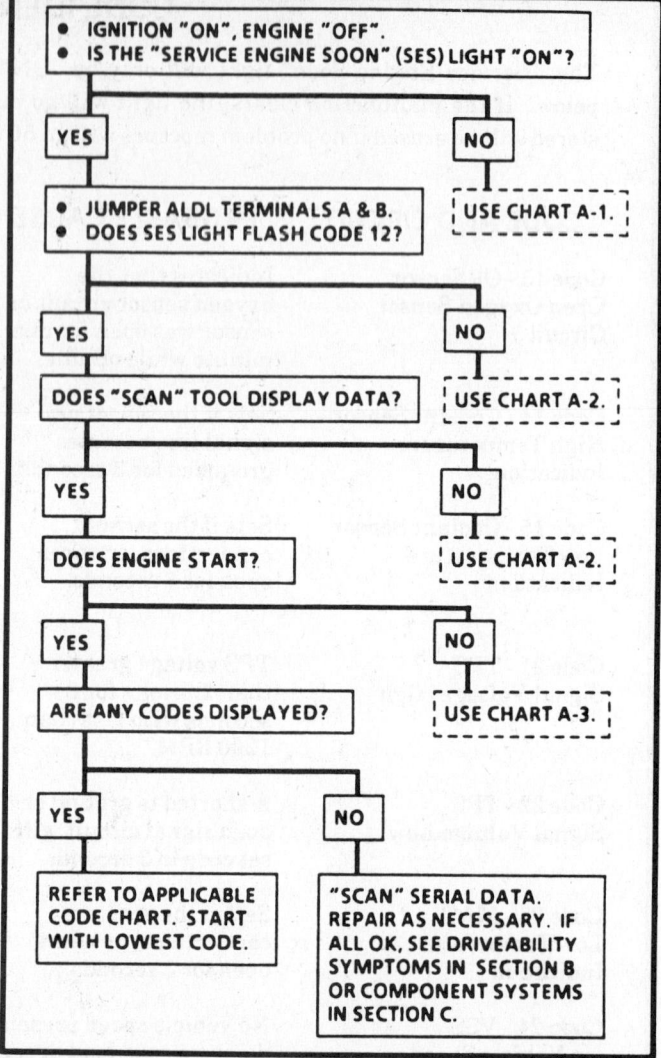

Typical diagnosis circuit check trouble tree chart

enough so as not being able to road test the vehicle in the ALCL test mode. Be sure to check the tool manufacturer for instructions on special test modes which should overcome these limitations.

When a tester is used with a fuel injected engine, it bypasses the timer that keeps the system in Open loop for a certain period of time. When all Closed loop conditions are met, the engine will go into Closed loop as soon as the vehicle is started. This means that the air management system will not function properly and air may go directly to the converter as soon as the engine is started.

These tools cannot diagnose everything. They do not tell the technician where a problem is located in a circuit. The diagnostic charts to pinpoint the problems must still be used. These tester's do not let a technician know if a solenoid or relay has been turned on. They only tell the technician the ECM command. To find out if a solenoid has been turned on, check it with a suitable test light or digital voltmeter, or see if vacuum through the solenoid changes.

SCAN TOOLS USED WITH INTERMITTENTS

In some scan tool applications, the data update rate makes the tool less effective than a voltmeter, such as when trying to detect an intermittent problem which lasts for a very short time. However, the scan tool does allow one to manipulate the wiring harness or components under the hood with the engine not running while observing the scan tool's readout.

The scan tool can be plugged in and observed while driving the vehicle under the condition when the "CHECK ENGINE" light turns on momentarily or when the engine driveability is momentarily poor. If the problem seems to be related to certain parameters that can be checked on the scan tool, they should be checked while driving the vehicle. If there does not seem to be any correlation between the problem and any specific circuit, the scan tool can be checked on each position. Watching for a period of time to see if there is any change in the reading that indicates intermittent operation.

The scan tool is also an easy way to compare the operating parameters of a poorly operating engine with those of a known good one. For example, a sensor may shift in value but not set a trouble code. Comparing the sensor's reading with those of a known good vehicle may uncover the problem.

The scan tool has the ability to save time in diagnosis and prevent the replacement of good parts. The key to using the scan tool successfully for diagnosis lies in the technician's ability to understand the system he is trying to diagnose as well as an understanding of the scan tool's operation and limitations. The technician should read the tool manufacturer's operating manual to become familiar with the tool's operation.

WIRE HARNESS AND CONNECTOR SERVICE

A wire harness should be replaced with the proper part number harnesses. When the signal wires are spliced into a harness, use wire with high temperature insulation only. With low current

and voltage levels found in the system, it is important that the best possible bond be made at all wire splices by soldering the splices carefully and correctly.

Use care when probing a connector or replacing connector terminals. It is possible to short between the opposite terminals. If this happens, certain components can be damaged. Always use a jumper wire between connectors for circuit checking. NEVER probe through connector seals, wire insulation, secondary ignition wires, boots, nipples or covers. Even microscopic damage or holes may result in eventual water intrusion, corrosion and or component or circuit failure.

When diagnosing, open circuits are often difficult to locate by sight because oxidation or terminal misalignment are hidden by the sealed connectors. Merely wiggling a connector on a sensor or in the wiring harness may locate the open circuit condition. This should always be considered when an open circuit or failed sensor is indicated. Intermittent problems may also be caused by oxidized or loss connections.

METRI-PACK SERIES 150 TERMINALS

Some ECM harness connectors contain terminals called metri-pack. These may be used at the coolant sensors as well as at the ignition modules. Metri-pack terminal are also called pull to seat terminals, because to install a terminal on a wire, the wire must first be inserted through the seal and connector. The terminal is then crimped on the wire and the terminal pulled back into the connector to seat it in place. To remove one of these metri-pack terminals, proceed as follows:

1. Slide the seal back on the wire.
2. Insert connector tool BT–8518/J–35689 or equivalent, to release the terminal locking tang.
3. Push the wire and terminal out through the connector.
4. If the terminal is being reused, reshape the locking tangs.

TWISTED/SHIELDED CABLE	TWISTED LEADS
DRAIN WIRE / OUTER JACKET / MYLAR	
1. REMOVE OUTER JACKET. 2. UNWRAP ALUMINUM/MYLAR TAPE. DO NOT REMOVE MYLAR.	1. LOCATE DAMAGED WIRE. 2. REMOVE INSULATION AS REQUIRED.
	SPLICE & SOLDER
3. UNTWIST CONDUCTORS. STRIP INSULATION AS NECESSARY.	3. SPLICE TWO WIRES TOGETHER USING SPLICE CLIPS AND ROSIN CORE SOLDER.
DRAIN WIRE	
4. SPLICE WIRES USING SPLICE CLIPS AND ROSIN CORE SOLDER. WRAP EACH SPLICE TO INSULATE. 5. WRAP WITH MYLAR AND DRAIN (UNINSULATED) WIRE.	4. COVER SPLICE WITH TAPE TO INSULATE FROM OTHER WIRES. 5. RETWIST AS BEFORE AND TAPE WITH ELECTRICAL TAPE AND HOLD IN PLACE.
6. TAPE OVER WHOLE BUNDLE TO SECURE AS BEFORE.	

Typical wiring harness repair

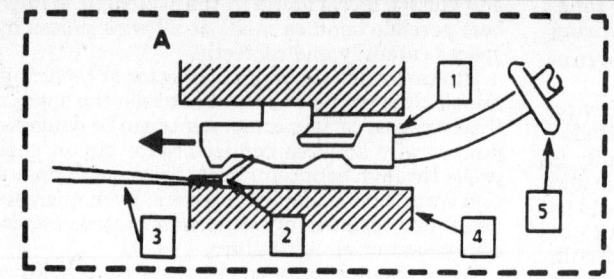

1. **METRI-PACK SERIES 150 FEMALE TERMINAL**	3. **TOOL J35689 OR BT-8446**
2. **LOCKING TANG**	4. **CONNECTOR BODY**
	5. **SEAL**

Removing the Metri-pack series 150 terminal

WEATHER-PACK CONNECTORS

Special connector tools BT–8234–A/J–28742 or equivalents are needed to service the weather-pack connectors. This tool is used to remove the pin and sleeve terminals. If terminal removal is attempted with an ordinary pick, there is a good chance that the terminal will be bent or deformed an unlike standard blade type terminals, these terminals cannot be straightened once they are bent.

Make certain that the connectors are properly seated and all of the sealing rings in place when connecting the leads. The hinge-type flap provides a secondary locking feature for the connector. It improves the connector reliability by retaining the terminals if the small terminal lock tangs are not positioned properly.

Weather-pack connections cannot be replaced with standard connections. Instructions are provided with the weather-pack connector and terminal packages.

COMPACT THREE CONNECTORS

The compact 3 connector, which looks similar to a weather-pack connector, is not sealed and is used where resistance to the environment is not required. This type connector most likely is used at the AIR control solenoid. Use the standard method when repairing a terminal. Do not use the weather-pack terminal tools on this connector.

MICRO-PACK CONNECTORS

The micro-pack connector terminal replacement requires the use of special connector tool BT–8234–A/J–33095 or equivalent.

TOOLS NEEDED TO SERVICE THE SYSTEM

The system requires a scan tool, tachometer, test light, ohmmeter, digital voltmeter with a 10 megohms impedance, vacuum gauge and jumper wires for diagnosis. A test light or voltmeter must be used when specified in the procedures.

NOTE: Some vehicles will use more sensors than others. Also, a complete general diagnostic section is outlined. The steps and procedures can be altered as required (if necessary) by the technician according to the specific model being diagnosed and the sensors it is equipped with. The wiring diagrams and schematics may not coincide with every GM vehicle in use. If this situation should arise, use the wiring diagram or schematic as a general guide. On some models, the electronic

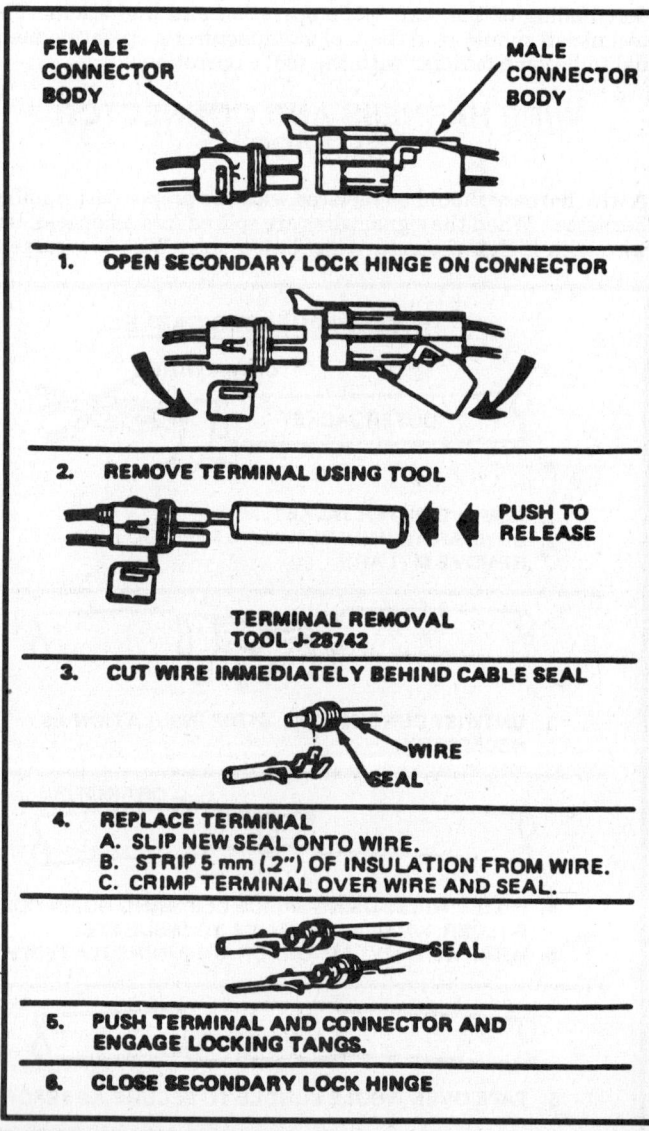

Typical weather-pack terminal repair

module may have a Learning Ability which allows them to make corrections for minor variations in the fuel system to improve driveability. If the battery power is disconnected for any reason, the volatile memory resets and the learning process begins again. A change may be noted in the performance of the vehicle. To teach the vehicle, ensure that the engine is at normal operating temperature. Then, the vehicle should be driven at part throttle, with moderate acceleration and idle conditions until normal performance returns.

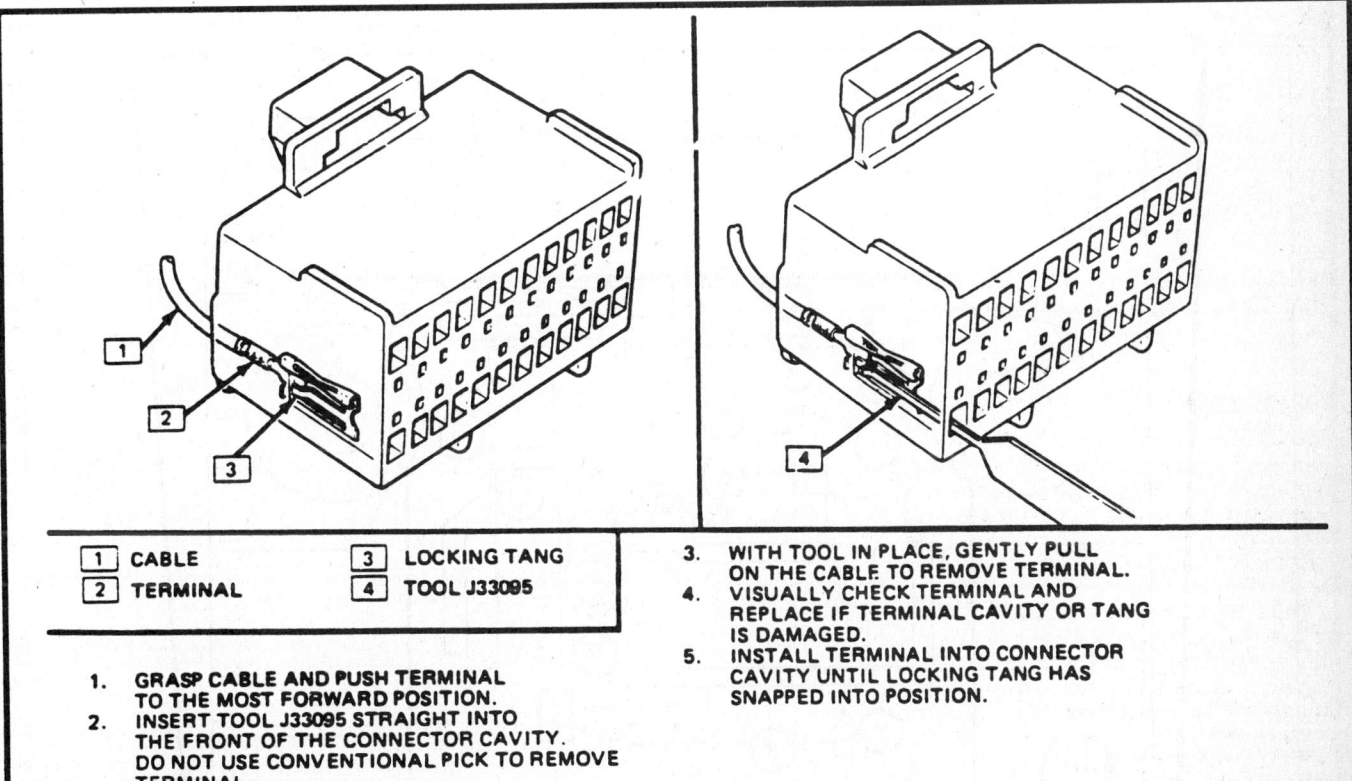

1	CABLE	3	LOCKING TANG
2	TERMINAL	4	TOOL J33095

3. WITH TOOL IN PLACE, GENTLY PULL ON THE CABLE TO REMOVE TERMINAL.
4. VISUALLY CHECK TERMINAL AND REPLACE IF TERMINAL CAVITY OR TANG IS DAMAGED.
5. INSTALL TERMINAL INTO CONNECTOR CAVITY UNTIL LOCKING TANG HAS SNAPPED INTO POSITION.

1. GRASP CABLE AND PUSH TERMINAL TO THE MOST FORWARD POSITION.
2. INSERT TOOL J33095 STRAIGHT INTO THE FRONT OF THE CONNECTOR CAVITY. DO NOT USE CONVENTIONAL PICK TO REMOVE TERMINAL.

Typical micro-pack terminal replacement

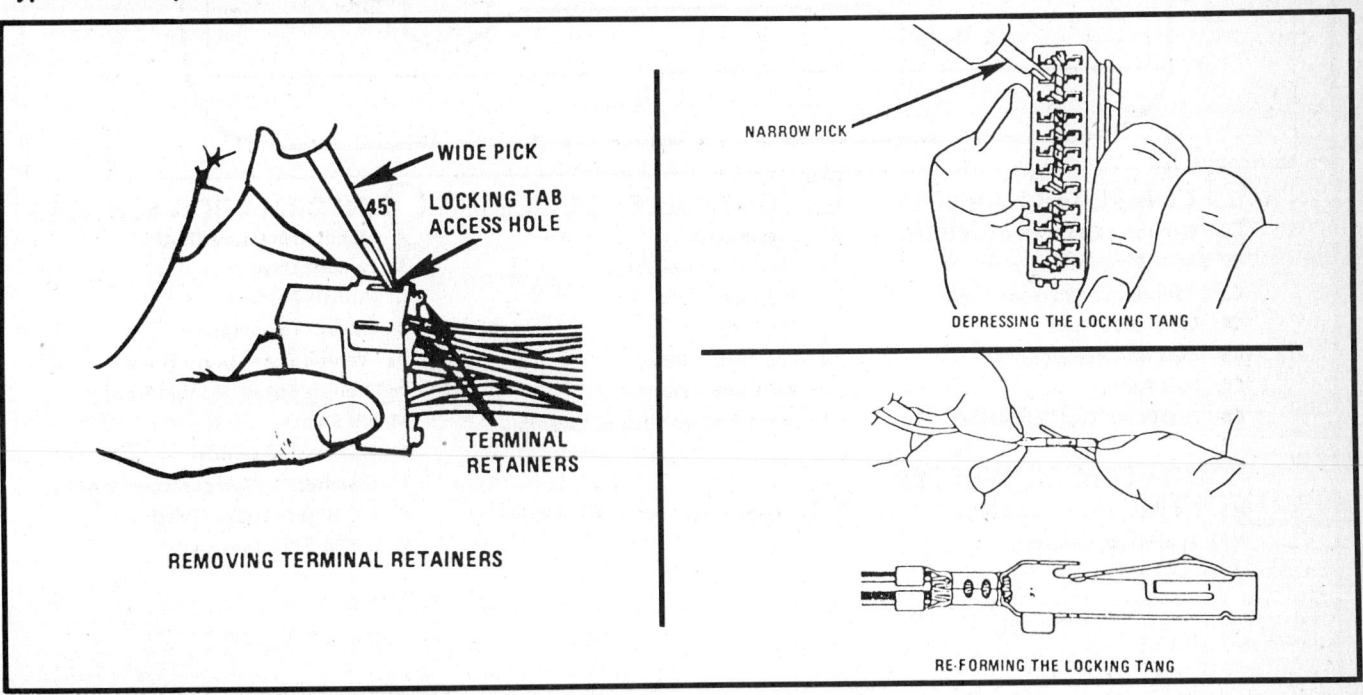

A High density connector servicing

2.0L (VIN 1, L-CARLINE) COMPONENT LOCATIONS
CORSICA AND BERETTA

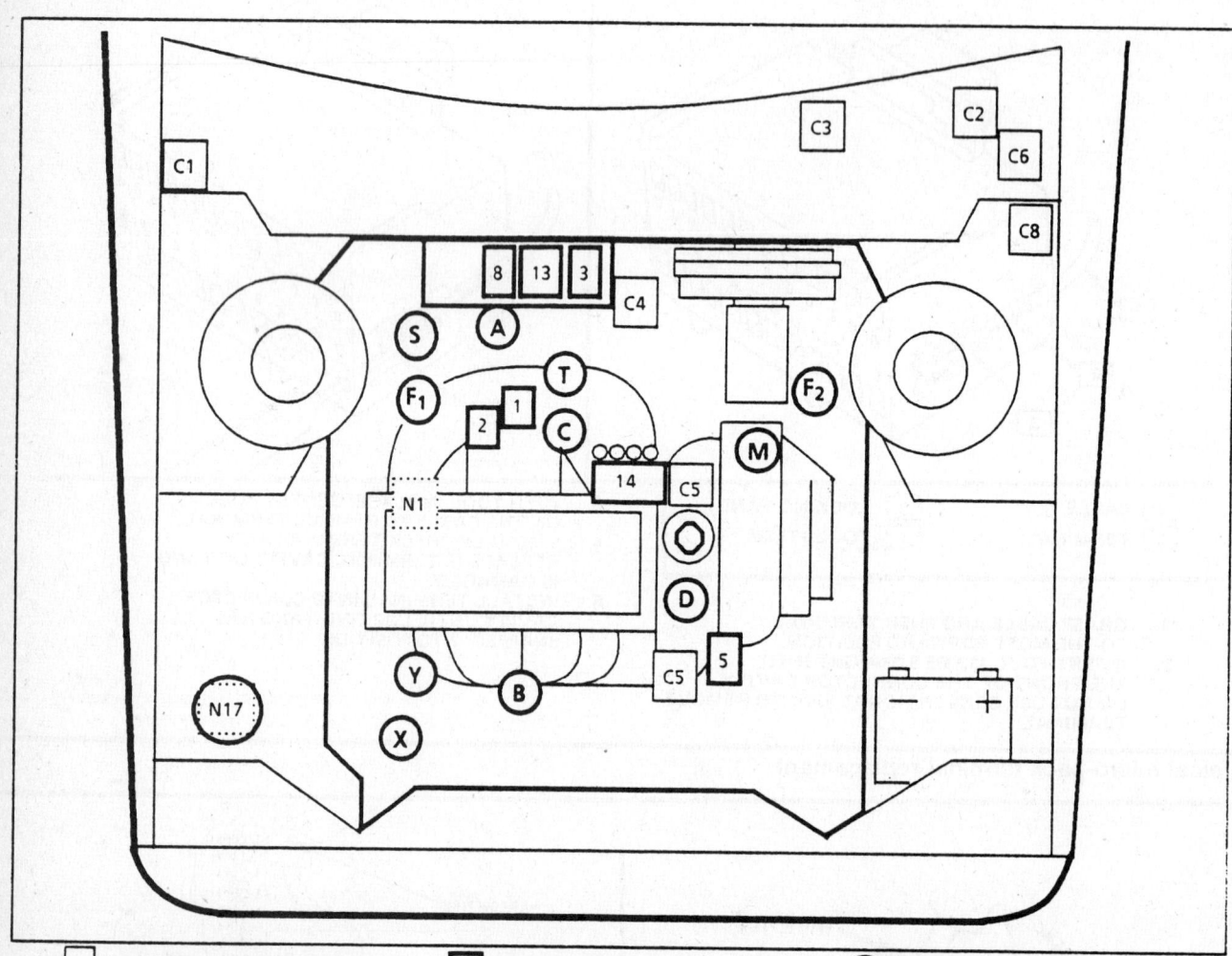

COMPUTER HARNESS

C1 Electronic Control Module (ECM)
C2 ALDL Diagnostic Connector
C3 "Service Engine Soon" light
C4 ECM Power Fuse
C5 ECM Harness Grounds
C6 Fuse Panel
C8 Fuel Pump Test Connector

NOT ECM CONNECTED

N1 Crankcase Vent Valve (PCV)
N17 Fuel Vapor Canister

CONTROLLED DEVICES

1 Fuel Injector
2 Idle Air Control Valve
3 Fuel Pump Relay
5 TCC Solenoid
8 Cooling Fan Relay
13 A/C Compressor Relay
14 Direct Ignition System Assembly

Exhaust Gas Recirculation Valve

INFORMATION SENSORS

A Manifold Pressure (MAP)
B Exhaust Oxygen
C Throttle Position
D Coolant Temperature
F₁ Vehicle Speed (Auto Trans.)
F₂ Vehicle Speed (Manual Trans.)
M P/N Switch
S P/S Pressure Switch
T Manifold Air Temperature (MAT)
X A/C High Pressure Switch
Y A/C Low Pressure Switch

2.0L (VIN 1, L-CARLINE) ECM WIRING DIAGRAM CORSICA AND BERETTA

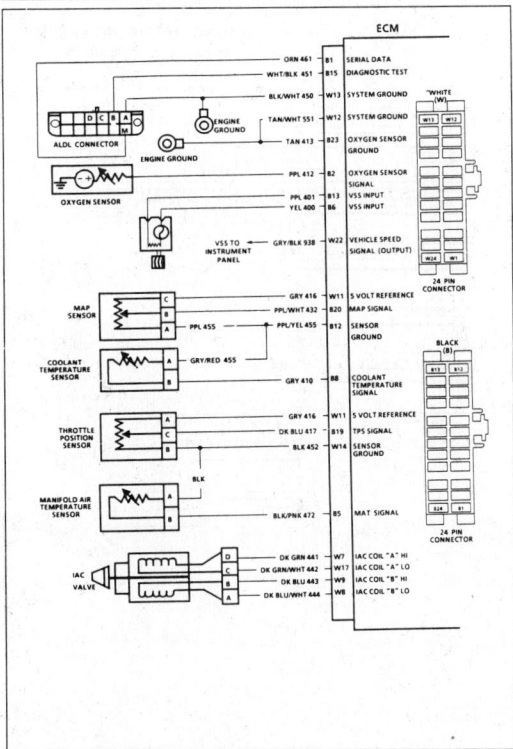

2.0L (VIN 1, L-CARLINE) ECM WIRING DIAGRAM CORSICA AND BERETTA (CONT.)

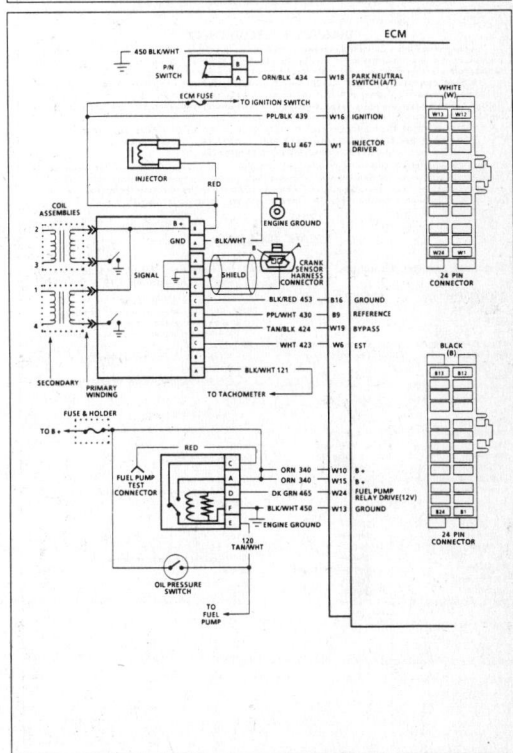

2.0L (VIN 1, L-CARLINE) ECM WIRING DIAGRAM CORSICA AND BERETTA (CONT.)

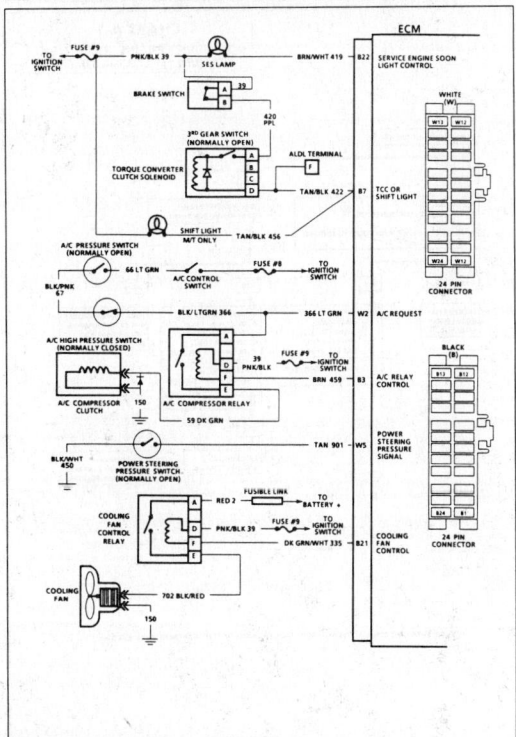

2.0L (VIN 1, L-CARLINE) ECM CONNECTOR TERMINAL END VIEW—CORSICA AND BERETTA

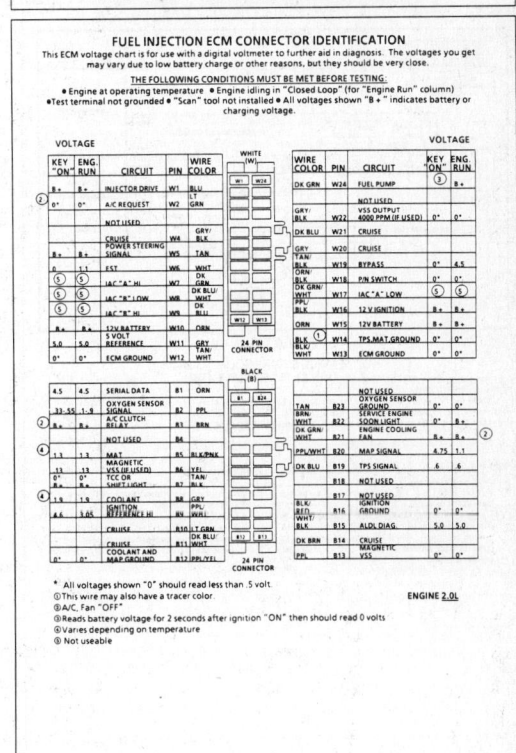

1988–90 2.0L (VIN 1) – CORSICA AND BERETTA

DIAGNOSTIC CIRCUIT CHECK

The Diagnostic Circuit Check is an organized approach to identifying a problem created by an electronic engine control system malfunction. It must be the starting point for any driveability complaint diagnosis because it directs the service technician to the next logical step in diagnosing the complaint.

The "Scan" data listed in the table may be used for comparison after completing the diagnostic circuit check and finding the on-board diagnostics functioning properly with no trouble codes displayed. The "Typical Data Values" are an average of display values recorded from normally operating vehicles and are intended to represent what a normally functioning system would typically display

A "SCAN" TOOL THAT DISPLAYS FAULTY DATA SHOULD NOT BE USED, AND THE PROBLEM SHOULD BE REPORTED TO THE MANUFACTURER. THE USE OF A FAULTY "SCAN" TOOL CAN RESULT IN MISDIAGNOSIS AND UNNECESSARY PARTS REPLACEMENT.

Only the parameters listed below are used in this manual for diagnosis. If a "Scan" tool reads other parameters, the values are not recommended by General Motors for use in diagnosis. For more description on the values and use of the "Scan" tool to diagnosis ECM inputs, refer to the applicable component diagnosis section in Section "C". If all values are within the range illustrated, refer to symptoms in Section "B".

"SCAN" TOOL DATA

Test Under Following Conditions: Idle, Upper Radiator Hose Hot, Closed Throttle, Park or Neutral, "Closed Loop", All Accessories "OFF".

"SCAN" Position	Units Displayed	Typical Data Value
Desired RPM	RPM	ECM idle command (varies with temperature)
RPM	RPM	± 50 RPM from desired rpm in drive (AUTO)
		± 100 RPM from desired rpm in neutral (MANUAL)
Coolant Temperature	Degrees Celsius	85 - 105
MAT Temperature	Degrees Celsius	10 - 90 (varies with underhood temperature and sensor location)
MAP	Volts	1 - 2 (varies with manifold and barometric pressures)
E (base pulse width)	Milliseconds	.8 - 3.0
C, C,	Volts	1 - 1 (varies continuously)
TPS	Volts	.4 - 1.25
Throttle Angle	0 - 100%	0
IAC	Counts (steps)	1 - 50
P/N Switch	P-N and R-D-L	Park/Neutral (P/N)
INT (Integrator)	Counts	110 - 145
BLM (Block Learn Memory)	Counts	118 - 138
Open/Closed Loop	Open/Closed	"Closed Loop" (may enter "Open Loop" with extended idle)
SS	MPH	0
C	ON/OFF	"OFF"
Spark Advance	Degrees	Varies
Battery	Volts	13.5 - 14.5
Fan	ON/OFF	"OFF" (coolant temperature below 102°C)
P/S Switch	Normal/Hi Pressure	Normal
A/C Request	Yes/No	No
A/C Clutch	ON/OFF	"OFF"
Shift Light (M/T)	ON/OFF	"OFF"

1988–90 2.0L (VIN 1) – CORSICA AND BERETTA

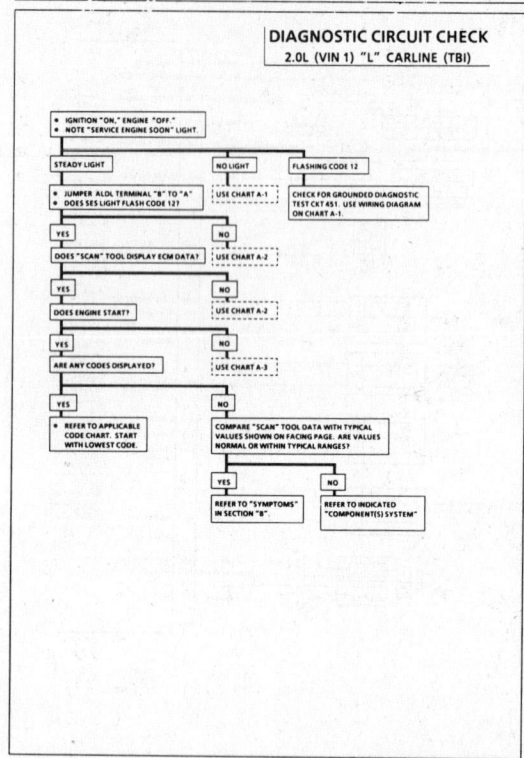

DIAGNOSTIC CIRCUIT CHECK
2.0L (VIN 1) "L" Carline (TBI)

1988–90 2.0L (VIN 1) – CORSICA AND BERETTA

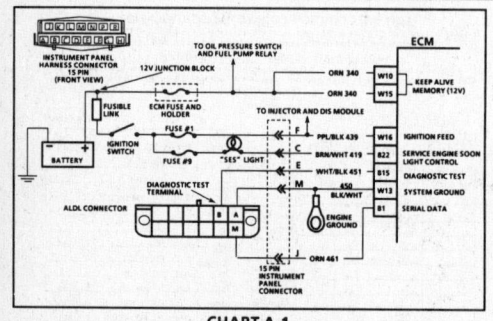

CHART A-1
NO "SERVICE ENGINE SOON" LIGHT
2.0L (VIN 1) "L" CARLINE (TBI)

Circuit Description:
There should always be a steady "Service Engine Soon" light, when the ignition is "ON" and engine "OFF." Battery voltage is supplied directly to the light bulb. The electronic control module (ECM) will control the light and turn it "ON" by providing a ground path through CKT 419 to the ECM.

Test Description: Numbers below refer to circled numbers on the diagnostic chart.
1. Battery feed CKT 340 is protected by a fusible link, at the battery.
2. Using a test light connected to B+, probe each of the system ground circuits to be sure a good ground is present. See ECM terminal end view in front of this section for ECM pin locations of ground circuits.

Diagnostic Aids:
If engine runs correctly, check for the following:
- Faulty light bulb.
- CKT 419 open.
- Gages fuse blown. This will result in no oil or generator lights, seat belt reminder, etc.
If engine cranks but will not run, use CHART A-3.

1988–90 2.0L (VIN 1) – CORSICA AND BERETTA

CHART A-1
NO "SERVICE ENGINE SOON" LIGHT
2.0L (VIN 1) "L" Carline (TBI)

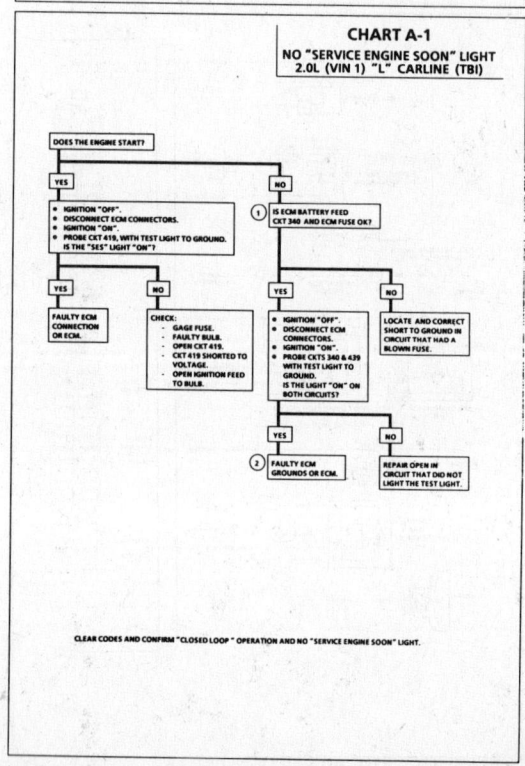

1988–90 2.0L (VIN 1) – CORSICA AND BERETTA

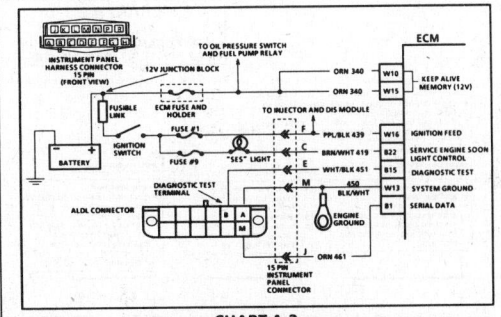

CHART A-2
NO ALDL DATA OR WON'T FLASH CODE 12
NO "SERVICE ENGINE SOON" LIGHT
2.0L (VIN 1) "L" CARLINE (TBI)

Circuit Description:
There should always be a steady "Service Engine Soon" light when the ignition is "ON" and the engine is "OFF." Battery voltage is supplied directly to the light bulb. The electronic control module (ECM) will control the light and turn it "ON" by providing a ground path through CKT 419 to the ECM.

With the diagnostic terminal grounded, the light should flash a Code 12, followed by any trouble code(s) stored in memory. A steady light suggests a short to ground in the light control CKT 419, or an open in diagnostic CKT 451.

Test Description: Numbers below refer to circled numbers on the diagnostic chart.
1. If there is a problem with the ECM that causes a "Scan" tool to not read data from the ECM, then the ECM should not flash a Code 12. If Code 12 does flash, be sure that the "Scan" tool is working properly on another vehicle. If the "Scan" is functioning properly and CKT 461 is OK, the PROM or ECM may be at fault for the "NO ALDL" symptom.
2. If the light turns "OFF" when the ECM connector is disconnected, then CKT 419 is not shorted to ground.
3. This step will check for an open diagnostic CKT 451.
4. At this point, the "Service Engine Soon" light wiring is OK. The problem is a faulty ECM or PROM. If Code 12 does not flash, the ECM should be replaced using the original PROM. Replace the PROM only after trying an ECM, as a defective PROM is an unlikely cause of the problem.

1988–90 2.0L (VIN 1) – CORSICA AND BERETTA

CHART A-2
NO ALDL DATA OR WON'T FLASH CODE 12
"SERVICE ENGINE SOON" LIGHT "ON" STEADY
2.0L (VIN 1) "L" CARLINE (TBI)

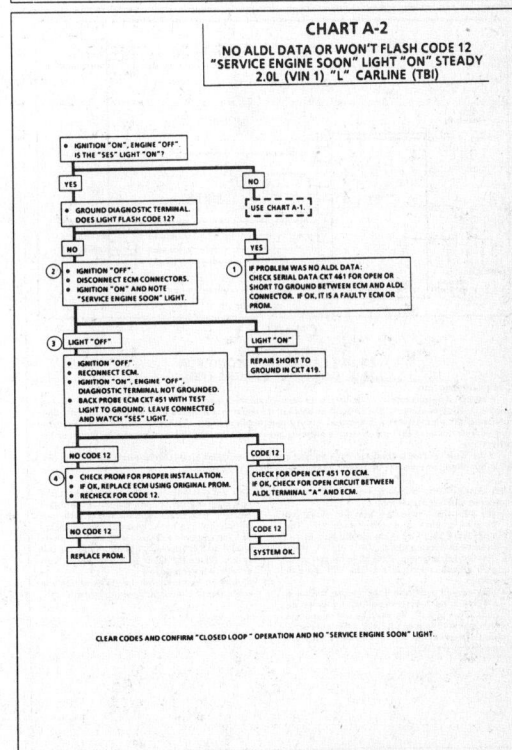

1988–90 2.0L (VIN 1) – CORSICA AND BERETTA

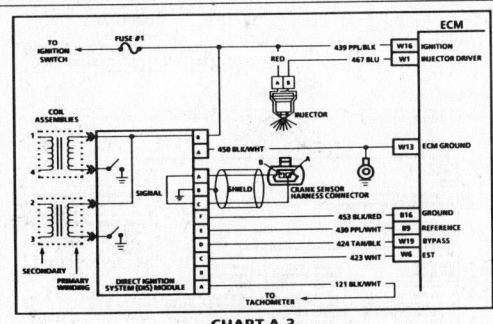

CHART A-3
(Page 1 of 3)
ENGINE CRANKS BUT WON'T RUN
2.0L (VIN 1) "L" CARLINE (TBI)

Circuit Description:
Before using this chart, battery condition, engine cranking speed, and fuel quantity should be checked and verified as being OK.

Test Description: Numbers below refer to circled numbers on the diagnostic chart.
1. A "Service Engine Soon" light "ON" is a basic test to determine if there is battery and ignition voltage at the ECM. No ALDL data may be due to an ECM problem, and CHART A-2 will diagnose the ECM. If TPS is over 2.5 volts, the engine may be in the clear flood mode, which will cause starting problems. The engine will not start without crank sensor reference pulses. The "Scan" tool should display rpm during cranking if pulses are received at the ECM.
2. Because the direct ignition system uses two plugs and wires to complete the circuit of each coil, the opposite spark plug wire should be left connected. If rpm was indicated during crank, the ignition module is receiving a crank signal, but "No Spark" at this test indicates the ignition module is not triggering the coil.
3. While cranking the engine, there should be no fuel spray with the injector electrical connector disconnected. Replace the injector if it sprays fuel or drips.

4. The test light should flash, indicating the ECM is controlling the injector. How bright the light flashes is not important. However, the test light should be a BT 8329 or equivalent.
5. Fuel spray from the injector indicates that fuel is available. However, the engine could be severely flooded due to too much fuel. No fuel spray from injector indicates a faulty fuel system or no ECM control of injector.

Diagnostic Aids:
- Water or foreign material can cause a no start condition during freezing weather. The engine may start after approximately 5 minutes in a heated shop. The problem may not re-occur until an overnight park in freezing temperatures.
- An EGR valve sticking open can cause a low air/fuel ratio during cranking. Unless engine enters "Clear Flood" at the first indication of a flooding condition, it can result in a no start.
- **Fuel Pressure.** Low fuel pressure can result in a very lean air/fuel ratio. See CHART A-7.

1988–90 2.0L (VIN 1) – CORSICA AND BERETTA

CHART A-3
(Page 1 of 3)
ENGINE CRANKS BUT WON'T RUN
2.0L (VIN 1) "L" CARLINE (TBI)

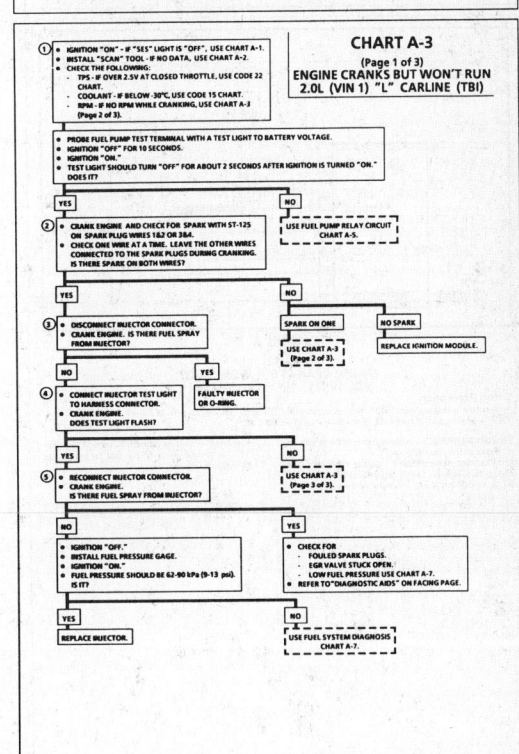

1988–90 2.0L (VIN 1) – CORSICA AND BERETTA

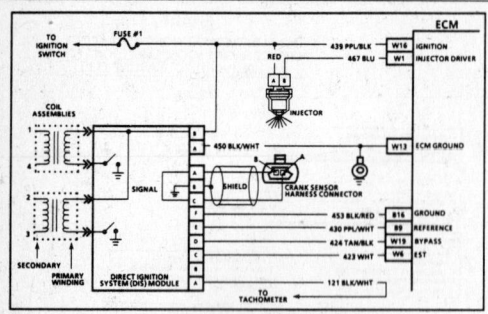

CHART A-3
(Page 2 of 3)
ENGINE CRANKS BUT WON'T RUN
2.0L (VIN 1) "L" CARLINE (TBI)

Circuit Description:

A magnetic crank sensor is used to determine engine crankshaft position, much the same way as the pick-up coil did in HEI type systems. The sensor is mounted in the block, near a slotted wheel on the crankshaft. The rotation of the wheel creates a flux change in the sensor, which produces a voltage signal. The DIS ignition module processes this signal and creates the reference pulses needed by the ECM to trigger the correct coil at the correct time.

If the "Scan" tool did not indicate cranking rpm, and there iano spark present at the plugs, the problem lies in the direct ignition system or the power and ground supplies to the module.

Test Description: Numbers below refer to circled numbers on the diagnostic chart.

1. The direct ignition system uses two plugs and wires to complete the circuit of each coil. The other spark plug wire in the circuit must be left connected to create a spark.
2. This test will determine if the 12 volt supply and a good ground is available at the DIS ignition module.
3. This test will determine if the ignition module is not generating the reference pulse, or if the wiring or ECM are at fault. By touching and removing a test light to 12 volts on CKT 430, a reference pulse should be generated. If rpm is indicated, the ECM and wiring are OK.
4. This test will determine if the ignition module is not triggering the problem coil, or if the tested coul is at fault. This test could also be performed by substituting a known good coil. The secondary coil winding can be checked with a DVM. There should be 5,000 to 10,000 ohms across the coil towers. There should not be any continuity from either coil tower to ground.
5. Checks for continuity of the crank sensor and connections. Also checks sensor magnetism.

1988–90 2.0L (VIN 1) – CORSICA AND BERETTA

CHART A-3
(Page 2 of 3)
ENGINE CRANKS BUT WON'T RUN
2.0L (VIN 1) "L" CARLINE (TBI)

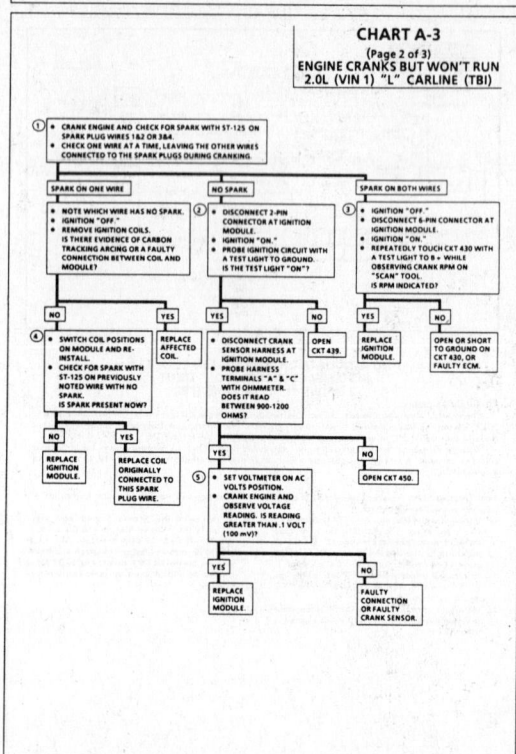

1988–90 2.0L (VIN 1) – CORSICA AND BERETTA

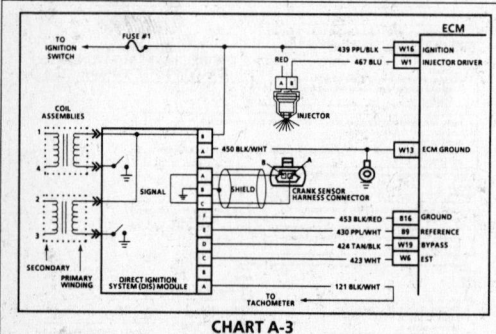

CHART A-3
(Page 3 of 3)
ENGINE CRANKS BUT WON'T RUN
2.0L (VIN 1) "L" CARLINE (TBI)

Circuit Description:

Ignition voltage is supplied to the fuel injector on CKT 439. The injector will be pulsed (turned "ON" and "OFF"), when the ECM opens and grounds injector drive CKT 467.

Test Description: Numbers below refer to circled numbers on the diagnostic chart.

1. This test determines if injector connector has ignition voltage, and on only one terminal.
2. A faulty ECM may result in damage to the injector.
3. A test light connected from ECM harness terminal "W1" to ground should light due to continuity through the injector.

1988–90 2.0L (VIN 1) – CORSICA AND BERETTA

CHART A-3
(Page 3 of 3)
ENGINE CRANKS BUT WON'T RUN
2.0L (VIN 1) "L" CARLINE (TBI)

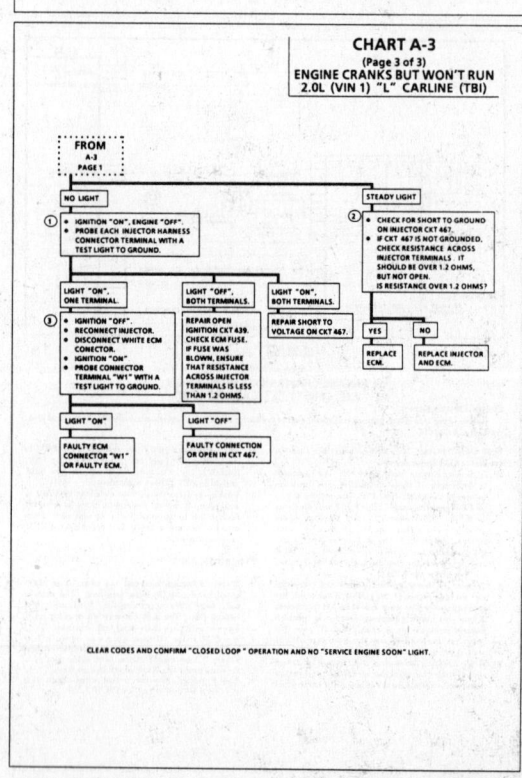

CLEAR CODES AND CONFIRM "CLOSED LOOP" OPERATION AND NO "SERVICE ENGINE SOON" LIGHT.

1988–90 2.0L (VIN 1) – CORSICA AND BERETTA

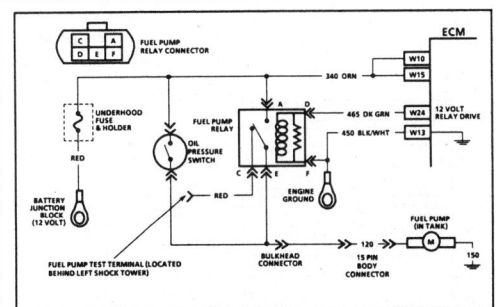

CHART A-5
FUEL PUMP RELAY CIRCUIT
2.0L (VIN 1) "L" CARLINE (TBI)

Circuit Description:

When the ignition switch is turned "ON," the engine control module (ECM) will activate the fuel pump relay with a 12 volt signal and run the in-tank fuel pump. The fuel pump will operate as long as the engine is cranking or running and the ECM is receiving ignition reference pulses. If there are no ignition reference pulses, the ECM will no longer supply the fuel pump relay signal within 2 seconds after key "ON."

Should the fuel pump relay or the 12 volt relay drive from the ECM fail, the fuel pump will receive electrical current through the oil pressure switch back-up circuit.

The fuel pump test terminal is located in the left side of the engine compartment. When the engine is stopped, the pump can be turned "ON" by applying battery voltage to the test terminal.

Diagnostic Aids:

An inoperative fuel pump relay can result in long cranking times. The extended crank period is caused by the time necessary for oil pressure to reach the pressure required to close the oil pressure switch and turn "ON" the fuel pump.

A "Scan" tool set in the PPSW position may be used to check the status of the fuel pump relay circuit. Normally, PPSW will read battery voltage for 2 seconds after the ignition is turned "ON." A reading of zero volts for the first two seconds after ignition "ON" would indicate a problem with the fuel pump relay or the relay drive circuit.

1988–90 2.0L (VIN 1) – CORSICA AND BERETTA

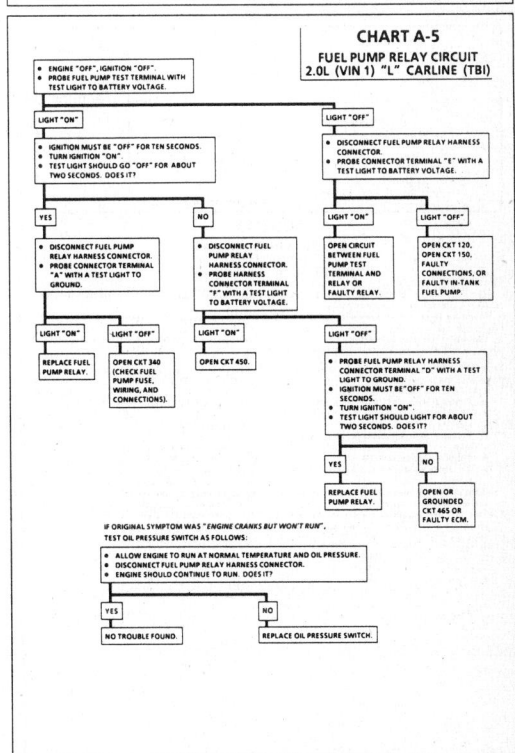

CHART A-5
FUEL PUMP RELAY CIRCUIT
2.0L (VIN 1) "L" CARLINE (TBI)

1988–90 2.0L (VIN 1) – CORSICA AND BERETTA

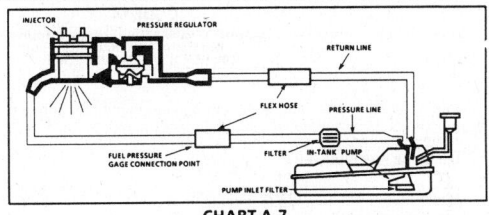

CHART A-7
FUEL SYSTEM DIAGNOSIS
2.0L (VIN 1) "L" CARLINE (TBI)

Circuit Description:

The fuel pump test terminal delivers fuel to the TBI unit where the system pressure is controlled to 62 to 90 kPa (9 to 13 psi). Excess fuel is returned to the fuel tank.

When the ignition switch is turned "ON," the electronic control module (ECM) will activate the fuel pump relay with a 12 volt signal and run the in-tank fuel pump. The fuel pump will operate as long as the engine is cranking or running and the ECM is receiving ignition reference pulses. If there are no ignition reference pulses, the ECM will no longer supply the fuel pump relay signal within 2 seconds after key "ON."

Should the fuel pump relay or the 12 volt relay drive from the ECM fail, the fuel pump will receive electrical current through the oil pressure switch back-up circuit.

Test Description: Numbers below refer to circled numbers on the diagnostic chart.

1. If fuse in jumper wire blows, check CKT 120, between relay and fuel pump, for a short to ground.

2. Fuel flow at less than 62 kPa (9 psi) can cause the following
   System will run lean and may set Code 44. Also, hard starting and poor overall performance will result.
   Engine surging and possible stalling during driving with satisfactory idle quality. This would be caused by a restriction in the fuel flow. As the fuel flow increases, the pressure drop across the restriction becomes great enough to starve the engine of fuel. At idle, the pressure may still be adequate. Normally, an engine with a fuel pressure less than 62 kPa (9 psi) at idle will not be driveable.

3. Gradually restricting the fuel return line allows the fuel pump to develop its maximum pressure (dead head pressure). When battery voltage is applied to the pump test terminal, pressure should be between 90 and 124 kPa (13 to 18 psi).

4. This test determines if the high fuel pressure is due to a restricted fuel return line or a throttle body pressure regulator problem.

Diagnostic Aids:

Improper fuel system pressure can result in one of the following symptoms:
- Cranks, but won't run.
- Code 44.
- Code 45.
- Cuts out, may feel like ignition problem.
- Poor fuel economy, loss of power.
- Hesitation.

1988–90 2.0L (VIN 1) – CORSICA AND BERETTA

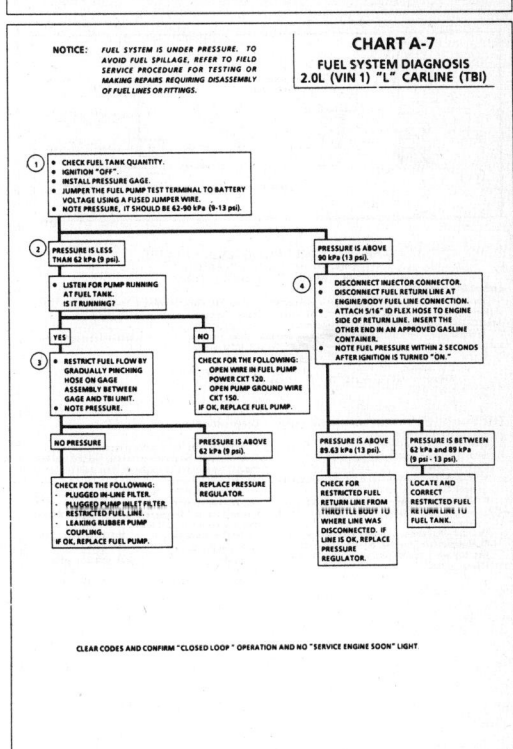

CHART A-7
FUEL SYSTEM DIAGNOSIS
2.0L (VIN 1) "L" CARLINE (TBI)

1988–90 2.0L (VIN 1) – CORSICA AND BERETTA

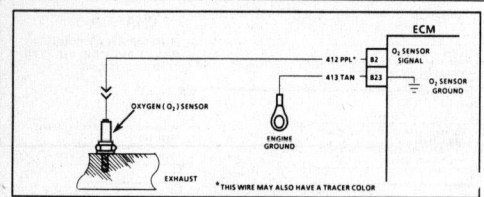

CODE 13
OXYGEN SENSOR CIRCUIT
(OPEN CIRCUIT)
2.0L (VIN 1) "L" CARLINE (TBI)

Circuit Description:

The ECM supplies a voltage of about .45 volt between terminals "B2" and "B23". (If measured with a 10 megohm digital voltmeter, this may read as low as .32 volt).

When the O_2 sensor reaches operating temperature, it varies this voltage from about .1 volt (exhaust is lean) to about .9 volt (exhaust is rich).

The sensor is like an open circuit and produces no voltage when it is below 360°C (600°F). An open sensor circuit, or cold sensor, causes "Open Loop" operation.

Test Description: Numbers below refer to circled numbers on the diagnostic chart.
1. Code 13 will be set under the following conditions:
   - Engine at normal operating temperature
   - At least 1 minutes have elapsed since engine start-up
   - O_2 signal voltage is steady between .35 and .55 volt
   - Throttle angle is above 7%
   - All above conditions are met for about 20 seconds.
   If the conditions for a Code 13 exist, the system will not operate in "Closed Loop."
2. This test determines if the O_2 sensor is the problem or if the ECM and wiring are at fault.
3. In doing this test, use only a 10 megohm digital voltmeter. This test checks the continuity of CKTs 412 and 413. If CKT 413 is open, the ECM voltage on CKT 412 will be over 6 volt (600 mV).

Diagnostic Aids:

Normal "Scan" tool O_2 sensor voltage varies between 100mV to 999 mV (.1 and 1.0 volt). While in "Closed Loop." Code 13 sets in one minute if sensor signal voltage remains between .35 and .55 volt, but the system will go to "Open Loop" in about 15 seconds.

Verify a clean, tight ground connection for CKT 413. Open CKT(s) 412 or 413 will result in a Code 13. If Code 13 is intermittent, refer to Section "B".

1988–90 2.0L (VIN 1) – CORSICA AND BERETTA

CODE 13
OXYGEN SENSOR CIRCUIT
(OPEN CIRCUIT)
2.0L (VIN 1) "L" CARLINE (TBI)

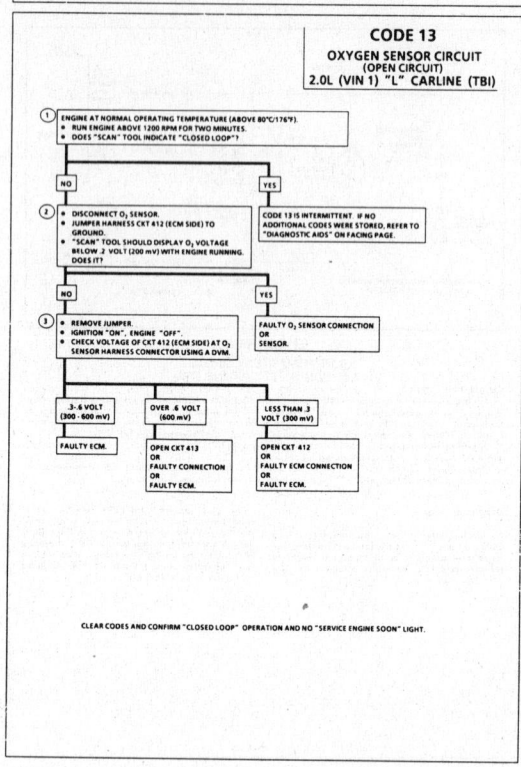

1988–90 2.0L (VIN 1) – CORSICA AND BERETTA

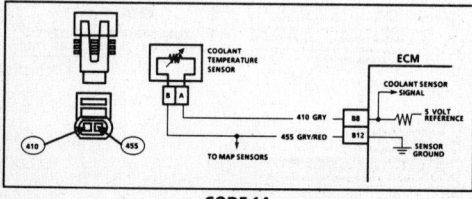

CODE 14
COOLANT TEMPERATURE SENSOR CIRCUIT
(HIGH TEMPERATURE INDICATED)
2.0L (VIN 1) "L" CARLINE (TBI)

Circuit Description:

The coolant temperature sensor uses a thermistor to control the signal voltage to the ECM. The ECM applies a voltage on CKT 410 to the sensor. When the engine is cold, the sensor (thermistor) resistance is high. The ECM will then sense a high signal voltage.

As the engine warms up, the sensor resistance decreases and the voltage drops. At normal engine operating temperature, the voltage will measure about 1.5 to 2.0 volts at ECM terminal "B8".

Coolant temperature is one of the inputs used to control the following:
- Fuel delivery
- Electronic spark timing (EST)
- Cooling fan
- Torque converter clutch (TCC)
- Idle air control (IAC)

Test Description: Numbers below refer to circled numbers on the diagnostic chart.
1. Checks to see if code was set as result of hard failure or intermittent condition.
   Code 14 will be set if:
   - Engine has been running for more than 10 seconds
   - Signal voltage indicates a coolant temperature above 135°C (275°F) for 3 seconds
2. This test simulates conditions for a Code 15. If the ECM recognizes the open circuit (high voltage), and displays a low temperature, the ECM and wiring are OK.

Diagnostic Aids:

A "Scan" tool reads engine temperature in degrees Celsius. After the engine is started, the temperature should rise steadily to about 90°, then stabilize when the thermostat opens.

If the engine has been allowed to cool to an ambient temperature (overnight), coolant and MAT temperature may be checked with a "Scan" tool and should read close to each other.

When a Code 14 is set, the ECM will turn "ON" the engine cooling fan.

A Code 14 will result if CKT 410 is shorted to ground.

If Code 14 is intermittent refer to Section "B".

1988–90 2.0L (VIN 1) – CORSICA AND BERETTA

CODE 14
COOLANT TEMPERATURE SENSOR CIRCUIT
(HIGH TEMPERATURE INDICATED)
2.0L (VIN 1) "L" CARLINE (TBI)

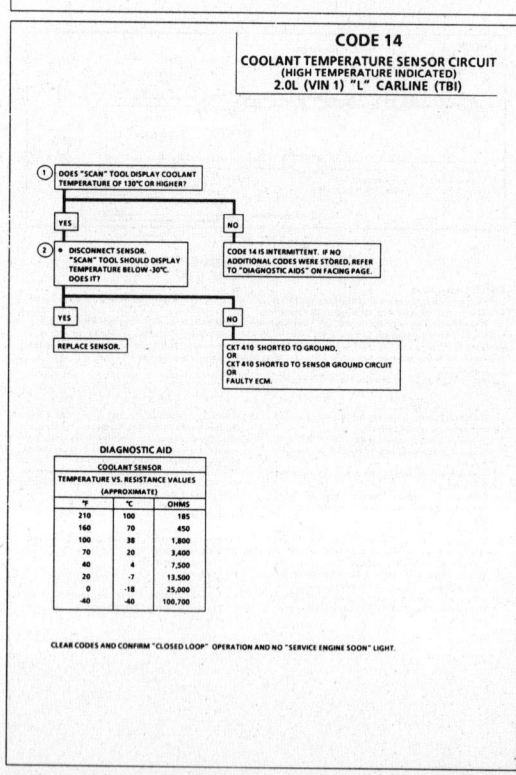

DIAGNOSTIC AID
COOLANT SENSOR
TEMPERATURE VS. RESISTANCE VALUES
(APPROXIMATE)

°F	°C	OHMS
210	100	185
160	70	450
100	38	1,800
70	20	3,400
40	4	7,500
20	-7	13,500
0	-18	25,000
-40	-40	100,700

CLEAR CODES AND CONFIRM "CLOSED LOOP" OPERATION AND NO "SERVICE ENGINE SOON" LIGHT.

1988–90 2.0L (VIN 1) – CORSICA AND BERETTA

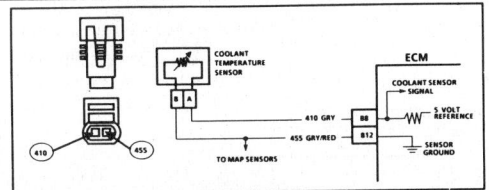

CODE 15
COOLANT TEMPERATURE SENSOR CIRCUIT
(LOW TEMPERATURE INDICATED)
2.0L (VIN 1) "L" CARLINE (TBI)

Circuit Description:
The coolant temperature sensor uses a thermistor to control the signal voltage to the ECM. The ECM applies a voltage on CKT 410 to the sensor. When the engine is cold, the sensor (thermistor) resistance is high. The ECM will then sense a high signal voltage.
As the engine warms up, the sensor resistance decreases and the voltage drops. At normal engine operating temperature, the voltage will measure about 1.5 to 2.0 volts at ECM terminal "B8".
Coolant temperature is one of the inputs used to control the following:
- Fuel delivery
- Torque converter clutch (TCC)
- Electronic spark timing (EST)
- Idle air control (IAC)
- Cooling fan

Test Description: Numbers below refer to circled numbers on the diagnostic chart.
1. Check to see if code was set as result of hard failure or intermittent condition.
   Code 15 will set if:
   - Engine has been running for more than 50 seconds
   - Signal voltage indicates a coolant temperature below -30°C (-22°F)
2. This test simulates conditions for a Code 14. If the ECM recognizes the grounded circuit (low voltage), and displays a high temperature, the ECM and wiring are OK.
3. This test will determine if there is a wiring problem or a faulty ECM. If CKT 452 is open, there may also abe a Code 33 stored.

Diagnostic Aids:
A "Scan" tool reads engine temperature in degrees Celsius. After the engine is started, the temperature should rise steadily to about 90°, then stabilize, when the thermostat opens.
If the engine has been allowed to cool to an ambient temperature (overnight), coolant and MAT temperature may be checked with a "Scan" tool and should read close to each other.
When a Code 15 is set, the ECM will turn "ON" the engine cooling fan.
A Code 15 will result if CKTs 410 or 455 are open. If Code 15 is intermittent, refer to Section "B".

1988–90 2.0L (VIN 1) – CORSICA AND BERETTA

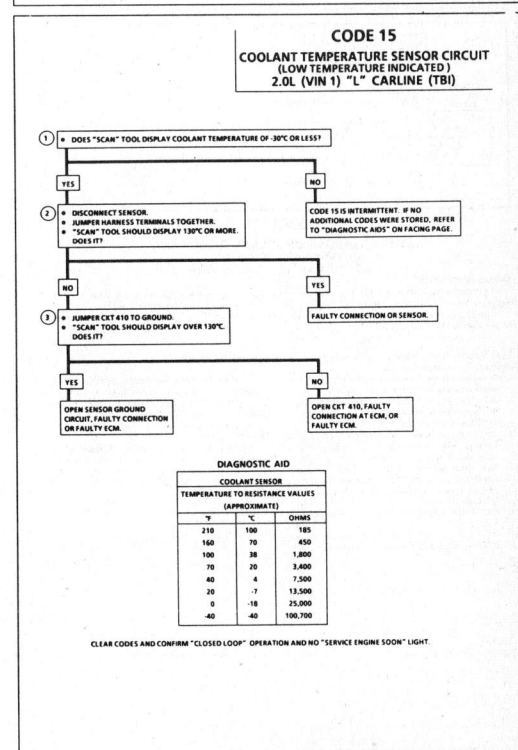

CODE 15
COOLANT TEMPERATURE SENSOR CIRCUIT
(LOW TEMPERATURE INDICATED)
2.0L (VIN 1) "L" CARLINE (TBI)

DIAGNOSTIC AID

COOLANT SENSOR		
TEMPERATURE TO RESISTANCE VALUES		
(APPROXIMATE)		
°F	°C	OHMS
210	100	185
160	70	450
100	38	1,800
70	20	3,400
40	4	7,500
20	-7	13,500
0	-18	25,000
-40	-40	100,700

CLEAR CODES AND CONFIRM "CLOSED LOOP" OPERATION AND NO "SERVICE ENGINE SOON" LIGHT.

1988–90 2.0L (VIN 1) – CORSICA AND BERETTA

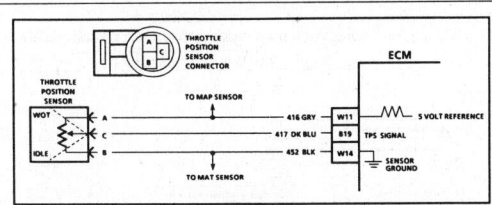

CODE 21
THROTTLE POSITION SENSOR (TPS) CIRCUIT
(SIGNAL VOLTAGE HIGH)
2.0L (VIN 1) "L" CARLINE (TBI)

Circuit Description:
The throttle position sensor (TPS) provides a voltage signal that changes relative to the throttle valve. Signal voltage will vary from less than 1.25 volts at idle to about 5 volts at wide open throttle (WOT).
The TPS signal is one of the most important inputs used by the ECM for fuel control and for many of the ECM controlled outputs.

Test Description: Numbers below refer to circled numbers on the diagnostic chart.
1. This step checks to see if Code 21 is the result of a hard failure or an intermittent condition.
   A Code 21 is set under the following conditions:
   - TPS reading above 2.5 volts.
   - MAP reading below 55 kPa.
   - All of the above conditions present for 5 seconds.
2. This step simulates conditions for a Code 22. If the ECM recognizes the change of state, the ECM and CKTs 416 and 417 are OK.
3. This step isolates a faulty sensor, ECM, or an open CKT 452. If CKT 452 is open, there may also be a Code 23 stored.

Diagnostic Aids:
A "Scan" tool displays throttle position in volts. Closed throttle voltage should be less than 1.25 volts. TPS voltage should increase at a steady rate as throttle is moved to WOT.
A Code 21 will result if CKT 452 is open or CKT 417 is shorted to voltage. If Code 21 is intermittent, refer to Section "B".

1988–90 2.0L (VIN 1) – CORSICA AND BERETTA

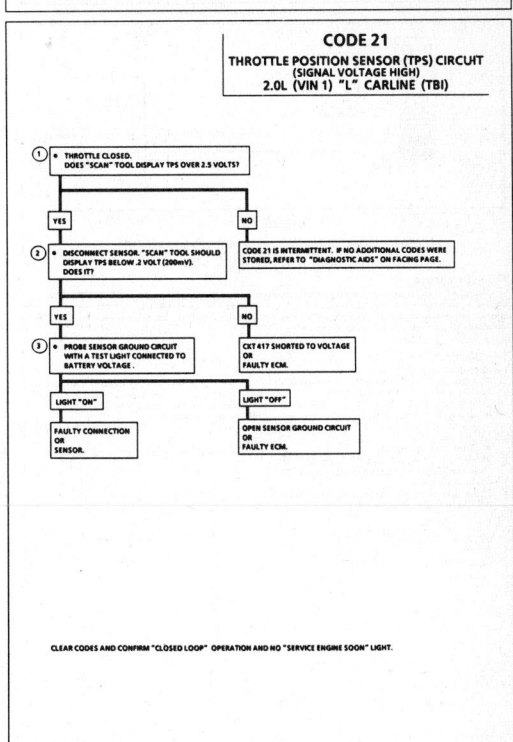

CODE 21
THROTTLE POSITION SENSOR (TPS) CIRCUIT
(SIGNAL VOLTAGE HIGH)
2.0L (VIN 1) "L" CARLINE (TBI)

CLEAR CODES AND CONFIRM "CLOSED LOOP" OPERATION AND NO "SERVICE ENGINE SOON" LIGHT.

1988–90 2.0L (VIN 1) – CORSICA AND BERETTA

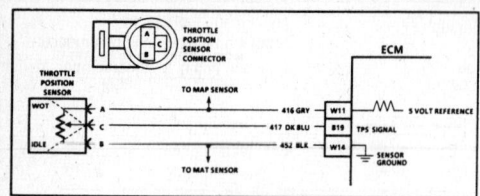

CODE 22
THROTTLE POSITION SENSOR (TPS) CIRCUIT
(SIGNAL VOLTAGE LOW)
2.0L (VIN 1) "L" CARLINE (TBI)

Circuit Description:
The throttle position sensor (TPS) provides a voltage signal that changes relative to the throttle valve. Signal voltage will vary from less than 1.25 volts at idle to about 5 volts at wide open throttle (WOT).
The TPS signal is one of the most important inputs used by the ECM for fuel control and for many of the ECM controlled outputs.

Test Description: Numbers below refer to circled numbers on the diagnostic chart.
1. This step checks to see if Code 22 is the result of a hard failure or an intermittent condition.
   A Code 22 will set under the following conditions:
   - The engine is running.
   - TPS voltage is below .2 volt (200 mV).
2. This step simulates conditions for a Code 21. If a Code 21 is set or the "Scan" tool displays over 4 volts, the ECM and wiring are OK.
3. The "Scan" tool may not display 12 volts. What is important is that the ECM recognizes the voltage as over 4 volts, indicating that CKT 417 and the ECM are OK.
4. If CKT 416 is shorted to ground, there may also be a stored Code 34.
   If CKT 416 is NOT shorted to ground and a Code 34 is stored, check CKT 432 for a short to ground.

Diagnostic Aids:
A "Scan" tool displays throttle position in volts. Closed throttle voltage should be less than 1.25 volts. TPS voltage should increase at a steady rate as throttle is moved to WOT.
An open or grounded 416 or 417 will result in a Code 22.
If Code 22 is intermittent, refer to Section "B".

1988–90 2.0L (VIN 1) – CORSICA AND BERETTA

CODE 22
THROTTLE POSITION SENSOR (TPS) CIRCUIT
(SIGNAL VOLTAGE LOW)
2.0L (VIN 1) "L" CARLINE (TBI)

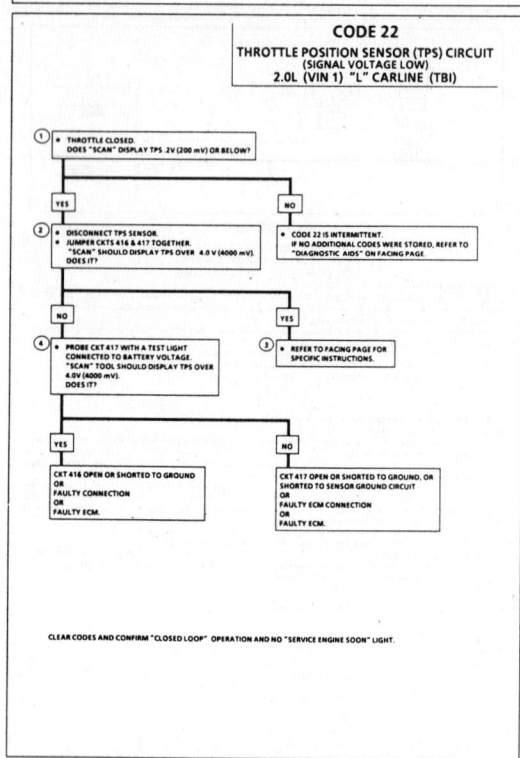

CLEAR CODES AND CONFIRM "CLOSED LOOP" OPERATION AND NO "SERVICE ENGINE SOON" LIGHT.

1988–90 2.0L (VIN 1) – CORSICA AND BERETTA

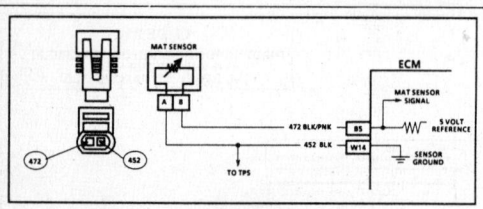

CODE 23
MANIFOLD AIR TEMPERATURE (MAT) SENSOR CIRCUIT
(LOW TEMPERATURE INDICATED)
2.0L (VIN 1) "L" CARLINE (TBI)

Circuit Description:
The manifold air temperature sensor uses a thermistor to control the signal voltage to the ECM. The ECM applies a reference voltage (4-6 volts) on CKT 472 to the sensor. When manifold air is cold, the sensor (thermistor) resistance is high. The ECM will then sense a high signal voltage. As the air warms, the sensor resistance becomes less and the voltage drops.

Test Description: Numbers below refer to circled numbers on the diagnostic chart.
1. This step checks to see if Code 23 is the result of a hard failure or an intermittent condition.
   Code 23 will set under the following conditions:
   - Engine is running for longer than 8.5 minutes.
   - Signal voltage indicates a MAT temperature less than -30°C.
   - There is no VSS signal.
2. This test simulates conditions for a Code 25. If the "Scan" tool displays a high temperature, the ECM and wiring are OK.
3. This step checks continuity of CKTs 472 and 452. If CKT 452 is open, there may also be a Code 21.

Diagnostic Aids:
If the engine has been allowed to cool to an ambient temperature (overnight), coolant and MAT temperatures may be checked with a "Scan" tool and should read close to each other.
A Code 23 will result if CKTs 472 or 452 become open.
If Code 23 is intermittent, refer to Section "B".

1988–90 2.0L (VIN 1) – CORSICA AND BERETTA

CODE 23
MANIFOLD AIR TEMPERATURE (MAT) SENSOR CIRCUIT
(LOW TEMPERATURE INDICATED)
2.0L (VIN 1) "L" CARLINE (TBI)

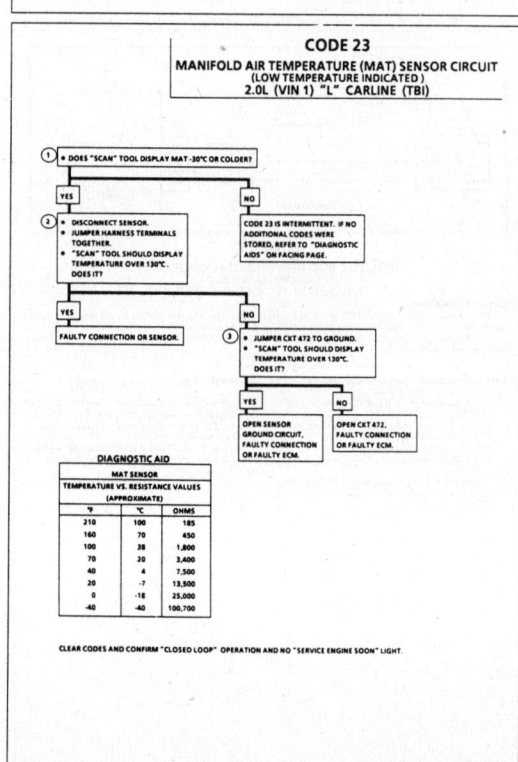

DIAGNOSTIC AID		
MAT SENSOR		
TEMPERATURE VS. RESISTANCE VALUES (APPROXIMATE)		
°F	°C	OHMS
210	100	185
160	70	450
100	38	1,800
70	20	3,400
40	4	7,500
20	-7	13,500
0	-18	25,000
-40	-40	100,700

CLEAR CODES AND CONFIRM "CLOSED LOOP" OPERATION AND NO "SERVICE ENGINE SOON" LIGHT.

1988–90 2.0L (VIN 1) – CORSICA AND BERETTA

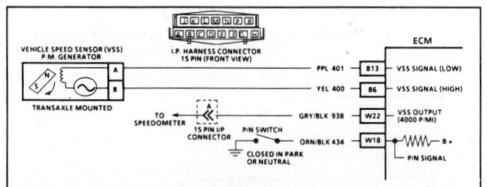

CODE 24
VEHICLE SPEED SENSOR (VSS) CIRCUIT
2.0L (VIN 1) "L" CARLINE (TBI)

Circuit Description:

Vehicle speed information is provided to the ECM by the vehicle speed sensor, which is a permanent magnet (PM) generator, and it is mounted in the transaxle. The PM generator produces a pulsing voltage, whenever vehicle speed is over about 3 mph. The AC voltage level and the number of pulses increases with vehicle speed. The ECM, then, converts the pulsing voltage to mph, which is used for calculations, and the mph can be displayed with a "Scan" tool.

The function of VSS buffer used in past model years has been incorporated into the ECM. The ECM then supplies the necessary signal for the instrument panel (4000 pulses per mile) for operating the speedometer and the odometer. If the vehicle is equipped with cruise control, the ECM also provides a signal (2000 pulses per mile) to the cruise control module.

Test Description:
1. Code 24 will set if vehicle speed equals 0 mph when:
   - Engine speed is between 1400 and 3600 rpm
   - TPS is less than 2%
   - Low load condition (low MAP voltage, high manifold vacuum)
   - Transmission not in park or neutral
   - All above conditions are met for 5 seconds

   These conditions are met during a road load deceleration.
   Disregard a Code 24 that sets when the drive wheels are not turning. This can be caused by a faulty park/neutral switch circuit.
   The PM generator only produces a signal if the drive wheels are turning greater than 3 mph.
2. Before replacing ECM, make sure that the correct PROM is installed for the application.

Diagnostic Aids:

"Scan" tool should indicate a vehicle speed whenever the drive wheels are turning greater than 3 mph.

A problem in CKT 938 will not affect the VSS input or the readings on a "Scan" tool.

Check CKTs 400 and 401 for proper connections to be sure they are clean and tight and the harness is routed correctly. Refer to intermittents in Section "B".

(A/T) - A faulty or misadjusted park/neutral switch can result in a false Code 24. Use a "Scan" tool and check for the proper signal while in a drive range. Refer to CHART C-1A for the P/N switch check.

1988–90 2.0L (VIN 1) – CORSICA AND BERETTA

CODE 24
VEHICLE SPEED SENSOR (VSS) CIRCUIT
2.0L (VIN 1) "L" CARLINE (TBI)

DISREGARD CODE 24 IF SET WHILE DRIVE WHEELS ARE NOT TURNING.

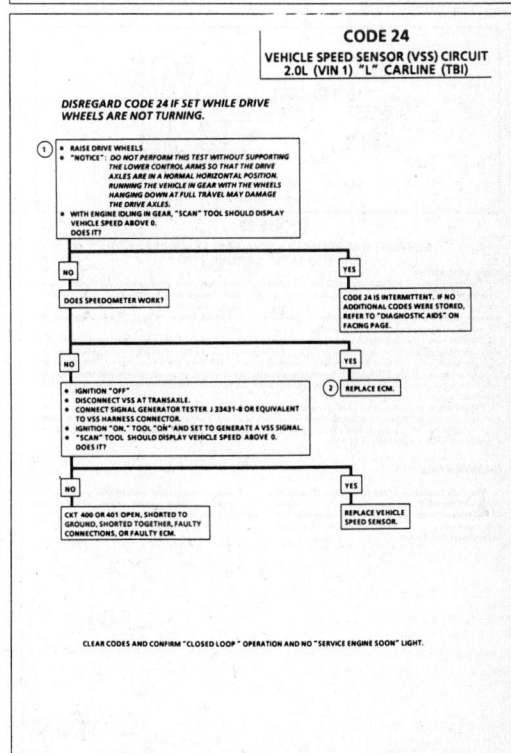

CLEAR CODES AND CONFIRM "CLOSED LOOP" OPERATION AND NO "SERVICE ENGINE SOON" LIGHT.

1988–90 2.0L (VIN 1) – CORSICA AND BERETTA

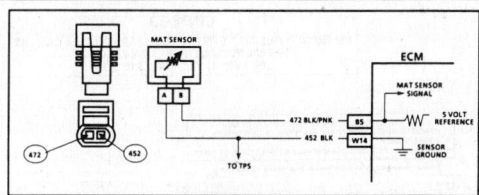

CODE 25
MANIFOLD AIR TEMPERATURE (MAT) SENSOR CIRCUIT
(HIGH TEMPERATURE INDICATED)
2.0L (VIN 1) "L" CARLINE (TBI)

Circuit Description:

The manifold air temperature sensor uses a thermistor to control the signal voltage to the ECM. The ECM applies a reference voltage (4-6 volts) on CKT 472 to the sensor. When manifold air is cold, the sensor (thermistor) resistance is high. The ECM will then sense a high signal voltage. As the air warms, the sensor resistance becomes less and the voltage drops.

Test Description: Numbers below refer to circled numbers on the diagnostic chart.
1. This check determines if the Code 25 is the result of a hard failure or an intermittent condition.
   A Code 25 will set under the following conditions:
   - Engine has been running longer than 8.5 minutes.
   - A MAT temperature greater than 135°C is detected for a time longer than 2 seconds.
   - VSS signal present.

Diagnostic Aids:

If the engine has been allowed to cool to an ambient temperature (overnight), coolant and MAT temperatures may be checked with a "Scan" tool and should read close to each other.

A Code 25 will result if CKT 472 is shorted to ground.

If Code 25 is intermittent, refer to Section "B".

1988–90 2.0L (VIN 1) – CORSICA AND BERETTA

CODE 25
MANIFOLD AIR TEMPERATURE (MAT) SENSOR CIRCUIT
(HIGH TEMPERATURE INDICATED)
2.0L (VIN 1) "L" CARLINE (TBI)

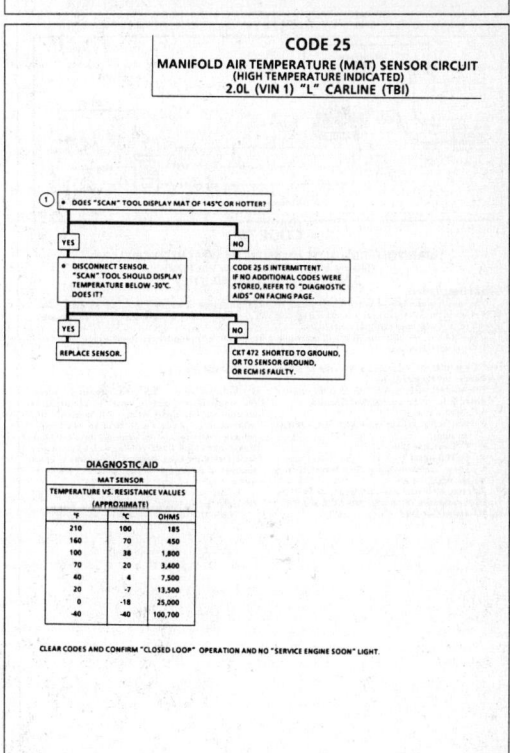

DIAGNOSTIC AID
MAT SENSOR
TEMPERATURE VS. RESISTANCE VALUES
(APPROXIMATE)

°F	°C	OHMS
210	100	185
160	70	450
100	38	1,800
70	20	3,400
40	4	7,500
20	-7	13,500
0	-18	25,000
-40	-40	100,700

CLEAR CODES AND CONFIRM "CLOSED LOOP" OPERATION AND NO "SERVICE ENGINE SOON" LIGHT.

1988–90 2.0L (VIN 1) – CORSICA AND BERETTA

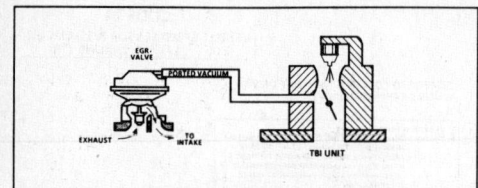

CODE 32
EXHAUST GAS RECIRCULATION (EGR) SYSTEM FAILURE
2.0L (VIN 1) "L" CARLINE (TBI)

Code Description:

A properly operating EGR system will directly affect the air/fuel mixture requirements of the engine. Since the exhaust gas introduced into the air/fuel mixture cannot be used in combustion (contains very little oxygen), less fuel is required to maintain a correct air/fuel ratio. If the EGR system were to fail in a closed position, the exhaust gas would be replaced with air, and the air/fuel mixture would be leaner. The ECM would compensate for the lean condition by adding fuel, resulting in higher block learn values.

The fuel control on this engine is conducted within 16 block learn cells. Since EGR is not used at idle, the closed throttle cell would not be affected by EGR system operation. The other block learn cells are affected by EGR operation, and, when the EGR system is operating properly, the block learn values in all cells should be close to the same. If the EGR system becomes inoperative, the block learn values in the open throttle cells would change to compensate for the resulting lean mixtures, but the block learn value in the closed throttle cell would not change.

The difference in block learn values between the idle (closed throttle) cell and cell 10 is used to monitor EGR system performance. When the difference between the two block learn values is greater than 12 and the block learn value in cell 10 is greater than 140, Code 32 is set. The system operates in block learn cell 10 during a cruise condition at approximately 55 mph.

Diagnostic Aids:

The Code 32 chart is a functional check of the EGR system. If the EGR system works properly but a Code 32 has been set, check other items that could result in high block learn values in block learn cell 10, but not in the closed throttle cell.

Check for restricted or blocked EGR passages.
Perform a MAP output check. Follow the procedure in CHART C-10.

1988–90 2.0L (VIN 1) – CORSICA AND BERETTA

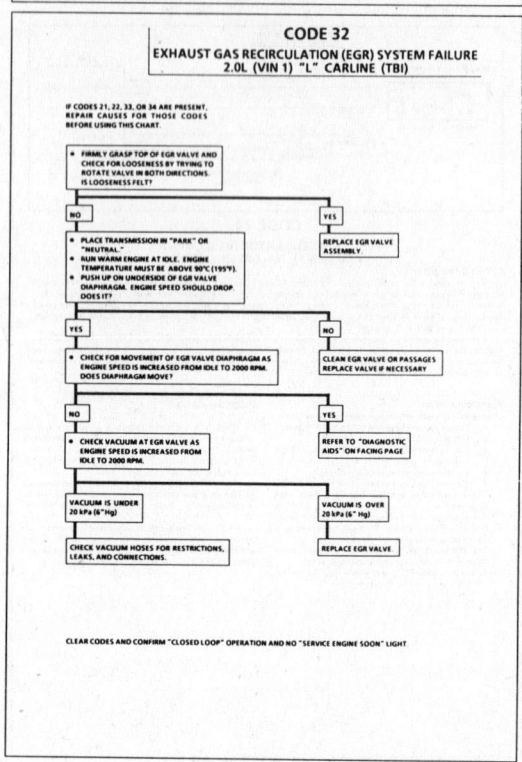

CODE 32
EXHAUST GAS RECIRCULATION (EGR) SYSTEM FAILURE
2.0L (VIN 1) "L" CARLINE (TBI)

1988–90 2.0L (VIN 1) – CORSICA AND BERETTA

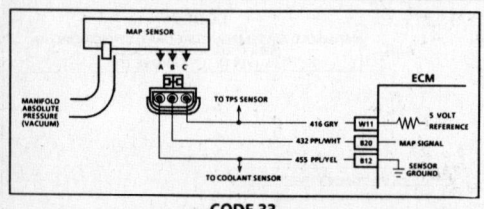

CODE 33
MANIFOLD ABSOLUTE PRESSURE (MAP) SENSOR CIRCUIT
(SIGNAL VOLTAGE HIGH - LOW VACUUM)
2.0L (VIN 1) "L" CARLINE (TBI)

Circuit Description:

The manifold absolute pressure (MAP) sensor responds to changes in manifold pressure (vacuum). The ECM receives this information as a signal voltage that will vary from about 1 to 1.5 volts at closed throttle idle, to 4-4.5 volts at wide open throttle (low vacuum).

If the MAP sensor fails, the ECM will substitute a fixed MAP value and use the throttle position sensor (TPS) to control fuel delivery.

Test Description: Numbers below refer to circled numbers on the diagnostic chart.
1. This step will determine if Code 33 is the result of a hard failure or an intermittent condition.
   A Code 33 will set if:
   - MAP signal voltage is too high (low manifold vacuum)
   - TPS less than 12%
   - No VSS signal
   - These conditions for a time longer than 5 seconds.
2. This step simulates conditions for a Code 34. If the ECM recognizes the change, the ECM and CKTs 416 and 432 are OK.

Diagnostic Aids:

With the ignition "ON" and the engine stopped, the manifold pressure is equal to atmospheric pressure and the signal voltage will be high. This information is used by the ECM as an indication of vehicle altitude and is referred to as BARO. Comparison of this BARO reading with a known good vehicle with the same sensor is a good way to check accuracy of a "suspect" sensor. Reading should be within .4 volt.

A Code 33 will result if CKT 455 is open, or if CKT 432 is shorted to voltage or to CKT 416.

If Code 33 is intermittent, refer to Section "B".

1988–90 2.0L (VIN 1) – CORSICA AND BERETTA

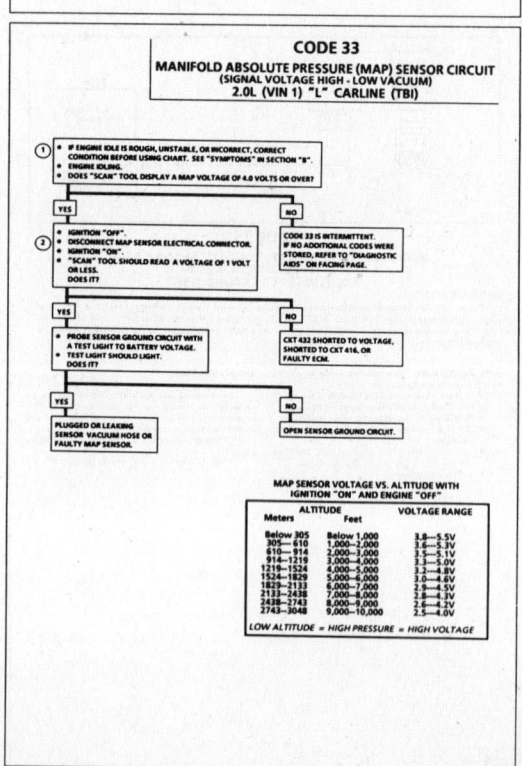

CODE 33
MANIFOLD ABSOLUTE PRESSURE (MAP) SENSOR CIRCUIT
(SIGNAL VOLTAGE HIGH - LOW VACUUM)
2.0L (VIN 1) "L" CARLINE (TBI)

1988–90 2.0L (VIN 1) — CORSICA AND BERETTA

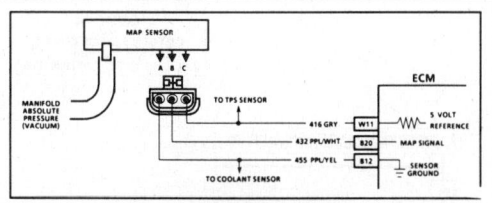

CODE 34
MANIFOLD ABSOLUTE PRESSURE (MAP) SENSOR CIRCUIT
(SIGNAL VOLTAGE LOW - HIGH VACUUM)
2.0L (VIN 1) "L" CARLINE (TBI)

Circuit Description:
The manifold absolute pressure (MAP) sensor responds to changes in manifold pressure (vacuum). The ECM receives this information as a signal voltage that will vary from about 1 to 1.5 volts at closed throttle (idle), to 4-4.5 volts at wide open throttle (low vacuum).
If the MAP sensor fails, the ECM will substitute a fixed MAP value and use the throttle position sensor (TPS) to control fuel delivery.

Test Description: Numbers below refer to circled numbers on the diagnostic chart.
1. This step determines if Code 34 is the result of a hard failure or an intermittent condition.
A Code 34 will set under the following conditions:
   • MAP signal voltage is too low
   • Engine speed is over 1200 rpm
2. Jumpering harness terminals "B" to "C", 5 volt to signal, will determine if the sensor is at fault, or if there is a problem with the ECM or wiring.
3. The "Scan" tool may not display 12 volts. What is important is that the ECM recognizes the voltage as more than 4 volts, indicating that the ECM and CKT 432 are OK.

Diagnostic Aids:
With the ignition "ON" and the engine stopped, the manifold pressure is equal to atmospheric pressure and the signal voltage will be high. This information is used by the ECM as an indication of vehicle altitude and is referred to as BARO. Comparison of this BARO reading with a known good vehicle with the same sensor is a good way to check accuracy of a "suspect" sensor. Reading should be within .4 volt.
A Code 34 will result if CKTs 416 or 432 are open or shorted to ground.
If CKT 416 is NOT shorted to ground and there is also a Code 22 stored, check CKT 432 for a short to ground.
If Code 34 is intermittent, refer to Section "B".

1988–90 2.0L (VIN 1) — CORSICA AND BERETTA

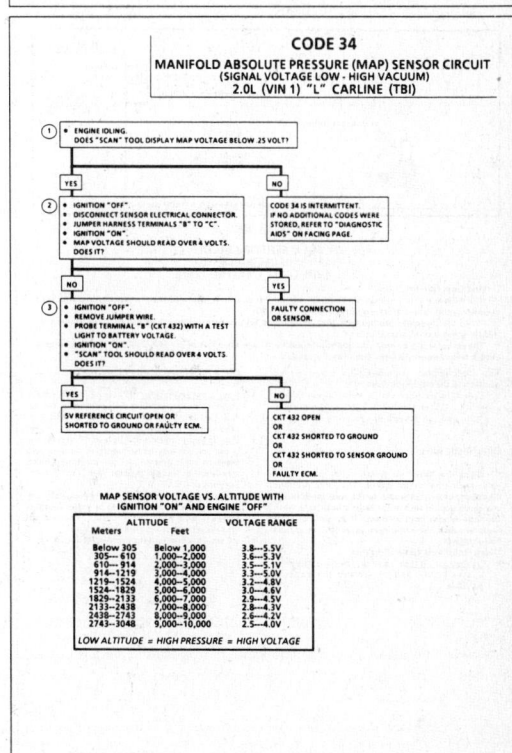

CODE 34
MANIFOLD ABSOLUTE PRESSURE (MAP) SENSOR CIRCUIT
(SIGNAL VOLTAGE LOW - HIGH VACUUM)
2.0L (VIN 1) "L" CARLINE (TBI)

MAP SENSOR VOLTAGE VS. ALTITUDE WITH IGNITION "ON" AND ENGINE "OFF"

ALTITUDE		VOLTAGE RANGE
Meters	Feet	
Below 305	Below 1,000	3.8–5.5V
305— 610	1,000–2,000	3.6–5.3V
610— 914	2,000–3,000	3.5–5.1V
914—1219	3,000–4,000	3.3–5.0V
1219—1524	4,000–5,000	3.2–4.8V
1524—1829	5,000–6,000	3.0–4.6V
1829—2133	6,000–7,000	2.9–4.5V
2133—2438	7,000–8,000	2.8–4.3V
2438—2743	8,000–9,000	2.6–4.2V
2743—3048	9,000–10,000	2.5–4.0V

LOW ALTITUDE = HIGH PRESSURE = HIGH VOLTAGE

1988–90 2.0L (VIN 1) — CORSICA AND BERETTA

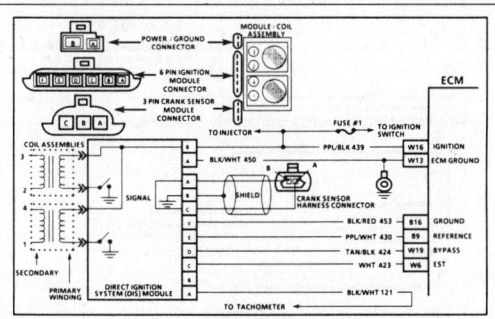

CODE 42
ELECTRONIC SPARK TIMING (EST) CIRCUIT
2.0L (VIN 1) "L" CARLINE (TBI)

Circuit Description:
The DIS module sends a reference signal to the ECM when the engine is cranking. While the engine speed is under 400 rpm, the DIS module controls the ignition timing. When the system is running on the ignition module (no voltage on the bypass line), the ignition module grounds the EST signal. The ECM expects to sense no voltage on the EST line during this condition. If it senses a voltage, it sets Code 42 and will not enter the EST mode.
When the engine speed exceeds 400 rpm, the ECM applies 5 volts to the bypass line to switch the timing to ECM control (EST). If the bypass line is open or grounded, once the rpm for EST control is reached, the ignition module will not switch to EST mode. This results in low EST voltage and the setting of Code 42. If the EST line is grounded, the ignition module will switch to EST, but because the line is grounded, there will be no EST signal. A Code 42 will be set.

Test Description: Numbers below refer to circled numbers on the diagnostic chart.
1. Code 42 means the ECM has sensed an open or short to ground in the EST or bypass circuits. This test confirms Code 42 and that the fault causing the code is present.
2. Checks for a normal EST ground path through the ignition module. An EST CKT 423, shorted to ground, will also read less than 500 ohms, but this will be checked later.
3. As the test light voltage contacts CKT 424, the module should switch, causing the ohmmeter to "overrange" if the meter is in the 1000-2000 ohms position. Selecting the 10 - 20,000 ohms position will indicate a reading above 5000 ohms.

The important thing is that the module "switched."
4. The module did not switch and this step checks for:
   • EST CKT 423 shorted to ground
   • Bypass CKT 424 open
   • Faulty ignition module connection or module
5. Confirms that Code 42 is a faulty ECM and not an intermittent in CKTs 423 or 424.

Diagnostic Aids:
The "Scan" tool does not have any ability to help diagnose a Code 42 problem.
If Code 42 is intermittent, refer to Section "B".

1988–90 2.0L (VIN 1) — CORSICA AND BERETTA

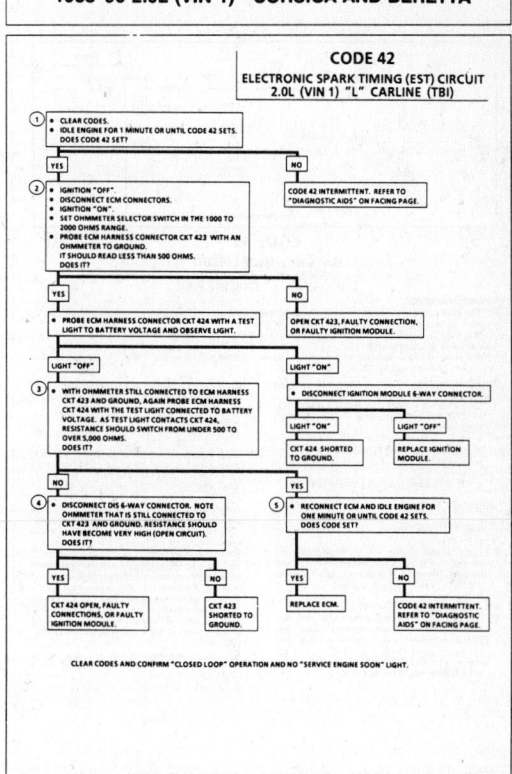

CODE 42
ELECTRONIC SPARK TIMING (EST) CIRCUIT
2.0L (VIN 1) "L" CARLINE (TBI)

1988–90 2.0L (VIN 1) – CORSICA AND BERETTA

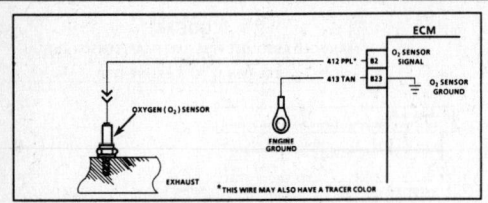

CODE 44
OXYGEN SENSOR CIRCUIT
(LEAN EXHAUST INDICATED)
2.0L (VIN 1) "L" CARLINE (TBI)

Circuit Description:

The ECM supplies a voltage of about .45 volt between terminals "B2" and "B23". (If measured with a 10 megohm digital voltmeter, this may read as low as .32 volt).

When the O₂ sensor reaches operating temperature, it varies this voltage from about .1 volt (exhaust is lean) to about .9 volt (exhaust is rich).

The sensor is like an open circuit and produces no voltage when it is below 360° C (600°F). An open sensor circuit, or cold sensor, causes "Open Loop" operation.

Test Description: Numbers below refer to circled numbers on the diagnostic chart.
1. Code 44 is set when the O₂ sensor signal voltage on CKT 412 remains below .3 volt for 50 seconds or more and the system is operating in "Closed Loop."

Diagnostic Aids:

Using the "Scan" tool, observe the block learn value at different engine speeds. The "Scan" tool also displays the block learn values can be checked in each of the cells to determine when the Code 44 may have been set. If the conditions for Code 44 exists, the block learn values will be around 150 or higher.

Check the following possible causes:
- O₂ Sensor Wire. Sensor pigtail may be mispositioned and contacting the exhaust manifold.

Check for ground in wire between connector and sensor.
- Fuel Contamination. Water, even in small amounts, near the in-tank fuel pump inlet can be delivered to the injector. The water causes a lean exhaust and can set a Code 44.
- Fuel Pressure. System will be lean if fuel pressure is too low. It may be necessary to monitor fuel pressure while driving the car at various road speeds and/or loads to confirm. See "Fuel System Diagnosis," CHART A-7.
- Exhaust Leaks. If there is an exhaust leak, the engine can cause outside air to be pulled into the exhaust and past the sensor. Vacuum or crankcase leaks can cause a lean condition.
- If Code 44 is intermittent, refer to Section "B".

1988–90 2.0L (VIN 1) – CORSICA AND BERETTA

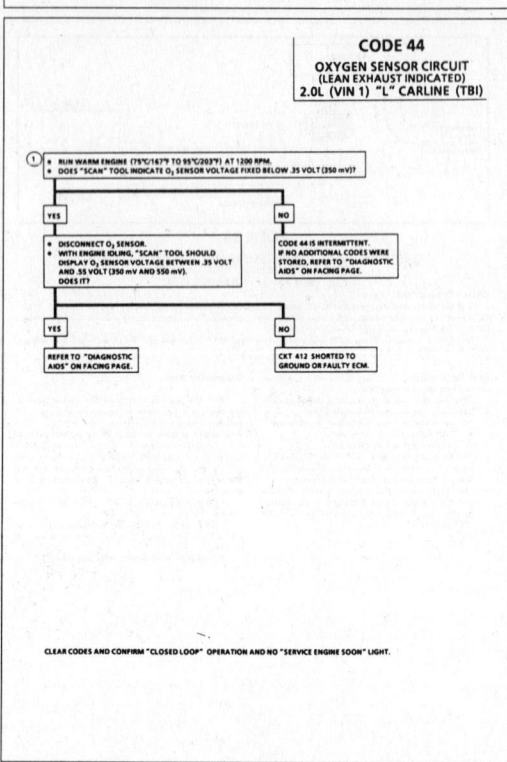

CODE 44
OXYGEN SENSOR CIRCUIT
(LEAN EXHAUST INDICATED)
2.0L (VIN 1) "L" CARLINE (TBI)

① • RUN WARM ENGINE (75°C/167°F TO 95°C/203°F) AT 1200 RPM.
• DOES "SCAN" TOOL INDICATE O₂ SENSOR VOLTAGE FIXED BELOW .35 VOLT (350 mV)?

YES | **NO**

- DISCONNECT O₂ SENSOR.
- WITH ENGINE IDLING, "SCAN" TOOL SHOULD DISPLAY O₂ SENSOR VOLTAGE BETWEEN .35 VOLT AND .55 VOLT (350 mV AND 550 mV).
DOES IT?

CODE 44 IS INTERMITTENT. IF NO ADDITIONAL CODES WERE STORED, REFER TO "DIAGNOSTIC AIDS" ON FACING PAGE.

YES | **NO**

REFER TO "DIAGNOSTIC AIDS" ON FACING PAGE.

CKT 412 SHORTED TO GROUND OR FAULTY ECM.

CLEAR CODES AND CONFIRM "CLOSED LOOP" OPERATION AND NO "SERVICE ENGINE SOON" LIGHT.

1988–90 2.0L (VIN 1) – CORSICA AND BERETTA

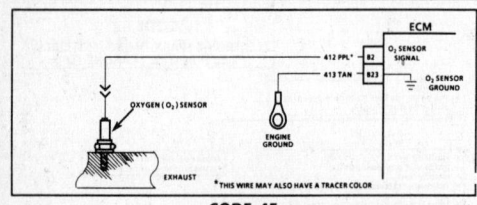

CODE 45
OXYGEN SENSOR CIRCUIT
(RICH EXHAUST INDICATED)
2.0L (VIN 1) "L" CARLINE (TBI)

Circuit Description:

The ECM supplies a voltage of about .45 volt between terminals "B2" and "B23". (If measured with a 10 megohm digital voltmeter, this may read as low as .32 volt).

When the O₂ sensor reaches operating temperature, it varies this voltage from about .1 volt (exhaust is lean) to about .9 volt (exhaust is rich).

The sensor is like an open circuit and produces no voltage when it is below 360° C (600°F). An open sensor circuit, or cold sensor, causes "Open Loop" operation.

Test Description: Numbers below refer to circled numbers on the diagnostic chart.
1. Code 45 is set when the O₂ sensor signal voltage on CKT 412 remains above .7 volt under the following conditions:
   - 30 seconds or more.
   - System is operating in "Closed Loop."
   - Engine run time after start is 1 minute or more.
   - Throttle angle is between 2% and 20%

Diagnostic Aids:

Code 45, or rich exhaust, is most likely caused by one of the following:
- Fuel Pressure. System will go rich, if pressure is too high. The ECM can compensate for some increase. However, if it gets too high, a Code 45 will be set. See "Fuel System Diagnosis," CHART A-7.
- Leaking Injector. See CHART A-7.
- HEI Shielding. An open ground CKT 453 may result in EMI, or induced electrical "noise." The ECM looks at this "noise" as reference pulses. The additional pulses result in a higher than actual engine speed signal. The ECM then delivers too much fuel causing the system to go rich.

The engine tachometer will also show higher than actual engine speed, which can help in diagnosing this problem.
- Canister Purge. Check for fuel saturation. If full of fuel, check canister control and hoses. See "Canister Purge," Section "C3".
- MAP Sensor. An output that causes the ECM to sense a higher than normal manifold pressure (low vacuum) can cause the system to go rich. Disconnecting the MAP sensor will allow the ECM to set a fixed value for the MAP sensor. Substitute a different MAP sensor if the rich condition is gone, while the sensor is disconnected.
- TPS. An intermittent TPS output will cause the system to operate richly due to a false indication of the engine accelerating.
- O₂ Sensor Contamination. Inspect oxygen sensor for silicone contamination from fuel, or use of improper RTV sealant. The sensor may have a white, powdery coating and result in a high but false signal voltage (rich exhaust indication). The ECM will then reduce the amount of fuel delivered to the engine causing a severe surge driveability problem.
- EGR Valve. EGR sticking open at idle is usually accompanied by a rough idle and/or stall condition.

If Code 45 is intermittent, refer to Section "B".

1988–90 2.0L (VIN 1) – CORSICA AND BERETTA

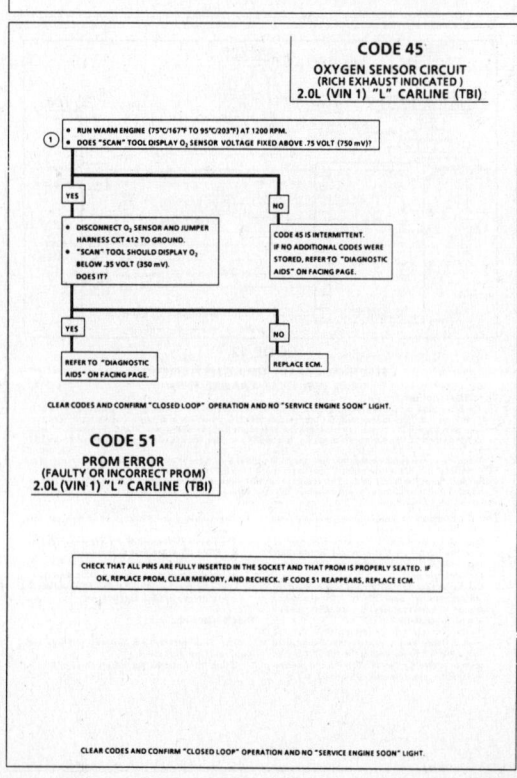

CODE 45
OXYGEN SENSOR CIRCUIT
(RICH EXHAUST INDICATED)
2.0L (VIN 1) "L" CARLINE (TBI)

① • RUN WARM ENGINE (75°C/167°F TO 95°C/203°F) AT 1200 RPM.
• DOES "SCAN" TOOL DISPLAY O₂ SENSOR VOLTAGE FIXED ABOVE .75 VOLT (750 mV)?

YES | **NO**

- DISCONNECT O₂ SENSOR AND JUMPER HARNESS CKT 412 TO GROUND.
- "SCAN" TOOL SHOULD DISPLAY O₂ BELOW .35 VOLT (350 mV).
DOES IT?

CODE 45 IS INTERMITTENT. IF NO ADDITIONAL CODES WERE STORED, REFER TO "DIAGNOSTIC AIDS" ON FACING PAGE.

YES | **NO**

REFER TO "DIAGNOSTIC AIDS" ON FACING PAGE.

REPLACE ECM.

CLEAR CODES AND CONFIRM "CLOSED LOOP" OPERATION AND NO "SERVICE ENGINE SOON" LIGHT.

CODE 51
PROM ERROR
(FAULTY OR INCORRECT PROM)
2.0L (VIN 1) "L" CARLINE (TBI)

CHECK THAT ALL PINS ARE FULLY INSERTED IN THE SOCKET AND THAT PROM IS PROPERLY SEATED. IF OK, REPLACE PROM, CLEAR MEMORY, AND RECHECK. IF CODE 51 REAPPEARS, REPLACE ECM.

CLEAR CODES AND CONFIRM "CLOSED LOOP" OPERATION AND NO "SERVICE ENGINE SOON" LIGHT.

1988-90 2.0L (VIN 1) – CORSICA AND BERETTA

SECTION B
SYMPTOMS
TABLE OF CONTENTS

Performing Symptom Diagnosis
Intermittents
Hard Start
Rough, Unstable, or Incorrect Idle, Stalling
Poor Gas Mileage
Detonation/Spark Knock
Lack of Power, Sluggish, or Spongy
Surges and/or Chuggle
Cuts Out, Misses
Hesitation, Sag, Stumble
Excessive Exhaust Emissions or Odors
Dieseling, Run-On
Backfire

PERFORMING SYMPTOM DIAGNOSIS

The DIAGNOSTIC CIRCUIT CHECK should be performed before using this section. The purpose of this section is to locate the source of a driveability or emissions problem when other diagnostic procedures cannot be used. This may be because of difficulties in locating a suspected sub-system or component.

Many driveability related problems can be eliminated by following the procedures found in Service Bulletins. These bulletins supersede this manual. Be sure to check all bulletins related to the complaint or suspected system.

If the engine cranks but will not run, use CHART A-3.

The sequence of the checks listed in this section is not intended to be followed as an a step-by-step procedure. The checks are listed such that the less difficult and time consuming operations are performed before more difficult ones.

Most of the symptom procedures call for a careful visual and physical check. *The importance of this step cannot be stressed too strongly.* It can lead to correcting a problem without further checks, and can save valuable time. This procedure includes checking the following.

- Vacuum hoses for splits, kinks, and proper connections, as shown on the underhood Emission Control Information label.
- Throttle body and intake manifold for leaks.
- Ignition wires for cracking, hardness, proper routing, and carbon tracking.
- Wiring for proper connections, pinches, and cuts.

1988-90 2.0L (VIN 1) – CORSICA AND BERETTA

INTERMITTENTS

Definition: Problem may or may not activate the "Service Engine Soon" light or store a trouble code.

DO NOT use the trouble code charts in Section "A" for intermittent problems. The fault must be present to locate the problem. If a fault is intermittent, the use of trouble code charts may result in the replacement of good parts.
- Most intermittent problems are caused by faulty electrical connections or wiring. Perform careful checks of suspected circuits for
  Poor mating of the connector halves and terminals not fully seated in the connector body (backed out)
  Improperly formed or damaged terminals
  All connector terminals in problem circuit should be carefully reformed to increase contact tension
  Poor terminal to wire connection. This requires removing the terminal from the connector body to check
- If a visual and physical check does not locate the cause of the problem, the car can be driven with a voltmeter connected to a suspected circuit or a "Scan" tool may be used An abnormal voltage reading while the problem occurs indicates that the problem may be in that circuit.

- Check for loss of trouble code memory. To check, disconnect the TPS and allow the engine to idle until the "Service Engine Soon" light turns "ON." Code 22 should be stored and kept in memory when the ignition is turned "OFF" for at least 10 seconds. If not, the ECM is faulty.
- An intermittent SES light and no trouble codes may be caused by
  Electrical system interference caused by a defective relay, ECM driven solenoid, or switch. They can cause a sharp electrical surge. Normally, the problem will occur when the faulty component is operated
  Improper installation of electrical options, such as lights, 2-way radios, etc.
  EST wires which should be routed away from spark plug wires, ignition system components, and generator. Ground wire from ECM to ignition system which may be faulty.
  Ignition secondary wire shorted to ground.
  "Service Engine Soon" light and diagnostic "test" terminal circuits intermittently shorted to ground.
  Faulty ECM grounds.

HARD START

Definition: Engine cranks well but does not start for a long time. Engine does eventually start, but may or may not continue to run.

Perform careful visual and physical check as described at the beginning of Section "B"
Perform "Diagnostic Circuit Check."
- CHECK
  - Fuel for poor quality, "stale" fuel, and water contamination
  - Ignition wires for shorts or faulty insulation
  - Ignition coil connections
  - Fuel pump relay. Connect test light between pump test terminal and battery voltage. Light should be "OFF" for 2 seconds following ignition "ON." If not, use CHART A-5
  - Secondary ignition voltage output with ST-125 tester
  - Spark plugs. Look for wetness, cracks, improper gap, burned electrodes, and heavy deposits. Visually inspect ignition system for moisture, dust, cracks, burns, etc
  - For faulty ECM and ignition grounds

- PROM for correct application.
  Spray plug wires with fine water mist to check for shorts.
- For possibility of misfiring, crossfiring, or cutting out under load or at idle. If possible, refer to the "Misfire" Chart
- For improper crank sensor resistance or faulty connections
- EGR operation. Use CHART C-7
- Idle Air Control system. Use Code 35 chart.
- Fuel system for restricted filter or improper pressure. Use CHART A-7.
- Injector and TBI assembly for leakage. Pressurize system by energizing fuel pump through the underhood fuel pump test connector
- Coolant sensor for a shift in calibration. Use Code 14 or Code 15 chart.

1988-90 2.0L (VIN 1) – CORSICA AND BERETTA

- TPS for sticking or binding. TPS voltage should read less than 1.25 V on a "Scan" tool.
- In-tank fuel pump check valve. A faulty valve would allow the fuel in the lines to drain back to the tank after the engine is stopped. To check for this condition, conduct the following test.
  1. Ignition "OFF."
  2. Disconnect fuel line at the filter.
  3. Remove the fuel tank cap.
  4. Connect a radiator test pump to the line and apply 103 kPa (15 psi) pressure. If the pressure will hold for 60 seconds, the check valve is OK.
- For the possibility of an exhaust restriction or improper valve timing by performing the following test.
  1. With engine at normal operating temperature, connect a vacuum gauge to any convenient vacuum port on intake manifold.

2. Run engine at 1000 rpm and record vacuum reading.
3. Increase engine speed slowly to 2500 rpm. Note vacuum reading at steady 2500 rpm.
4. If vacuum at 2500 rpm decreases more than 3" Hg from reading at 1000 rpm, the exhaust system should be inspected for restrictions.
5. Disconnect exhaust pipe from engine and repeat Steps 3 & 4. If vacuum still drops more than 3" Hg with exhaust disconnected, check valve timing.

- Engine valve timing and compression.

ROUGH, UNSTABLE, OR INCORRECT IDLE, STALLING

Definition: The engine runs unevenly at idle. If severe, the car may shake. Also, the idle speed may vary (called "hunting"). Either condition may be severe enough to cause stalling. Engine idles at incorrect speed.

Perform careful visual and physical check as described at the beginning of Section "B".
Perform "Diagnostic Circuit Check."
- CHECK
  - MAP sensor. Use CHART C1-D.
  - Throttle for sticking shaft or binding linkage. This will cause a high TPS voltage (open throttle indication) and the ECM will not control idle. TPS voltage should be less than 1.25 volts with throttle closed.
  - Battery cables and ground straps for poor contact. Erratic voltage will cause the IAC valve to change its position, resulting in poor idle quality.
  - Ignition wires for shorts or faulty insulation.
  - Ignition system for moisture, dust, cracks, burns, etc. Spray plug wires with fine water mist to check for shorts.
  - For possibility of misfiring, crossfiring, or cutting out under load or at idle. If present, refer to the "Misfire" Chart
  - Secondary ignition voltage output with ST-1 tester.
  - Ignition coil connections.
  - ECM and ignition system for faulty grounds.
  - Proper operation of EST

- Spark plugs. Look for wetness, cracks, improper gap, burned electrodes, and heavy deposits.
- Fuel system for restricted filter or improper pressure. Use CHART A-7.
- Injector and TBI assembly for leakage. Pressurize system by energizing fuel pump through the underhood fuel pump test connector.
- EGR operation. Use CHART C-7.
- For vacuum leaks at intake manifold gasket.
- Idle Air Control system. Use Code 35 chart.
- Electrical system voltage IAC valve will not move if voltage is below 9 volts or greater than 17.8 volts. Also check battery cables and ground straps for poor contact. Erratic voltage will cause the IAC valve to change its position, resulting in poor idle quality
- PCV valve for proper operation by placing finger over inlet hole in valve and several times. Valve should snap back. If not, replace valve. Ensure that valve is correct part. Also check PCV hose.

1988–90 2.0L (VIN 1)—CORSICA AND BERETTA

- Canister purge system for proper operation. Use CHART C-3.
- PROM for correct application.
- Throttle shaft or TPS for sticking or binding. TPS voltage should read less than 1.25 V on a "Scan" tool with the throttle closed.
- MAP sensor output. Use CHART C1-D and/or check sensor by comparing it to the output on a similar vehicle if possible.
- Oxygen sensor for silicone contamination from contaminated fuel or use of improper RTV sealant. The sensor will have a white, powdery coating and will cause a high but false signal voltage (rich exhaust indication). The ECM will reduce the amount of fuel delivered to the engine, causing a severe driveability problem.
- Coolant sensor for a shift in calibration. Use Code 14 or Code 15 chart.
- A/C refrigerant pressure for high pressure. Check for overcharging or faulty pressure switch.
- P/N switch circuit on vehicle with automatic transmission. Use CHART C-1A.
- Generator output voltage. Repair if less than 9 V or more than 16 V.
- Power steering. Use CHART C-1E. The ECM should compensate for power steering loads. Loss of this signal would be most noticeable when steering loads are high such as during parking.
- Engine valve timing and compression.
- For worn or incorrect basic engine parts such as cam, heads, pistons, etc. Also check for bent pushrods, worn rocker arms, and broken or weak valve springs.

For the possibility of an exhaust restriction or improper valve timing, perform the following test.

1. With engine at normal operating temperature, connect a vacuum gauge to any convenient vacuum port on intake manifold.
2. Run engine at 1000 rpm and record vacuum reading.
3. Increase engine speed slowly to 2500 rpm. Note vacuum reading at steady 2500 rpm.
4. If vacuum at 2500 rpm decreases more than 3" Hg from reading at 1000 rpm, the exhaust system should be inspected for restrictions.
5. Disconnect exhaust pipe from engine and repeat Steps 3 & 4. If vacuum still drops more than 3" Hg with exhaust disconnected, check valve timing.

- For overheating and possible causes. Look for the following.
  - Low or incorrect coolant solution. It should be a 50/50 mix of GM #1052753 anti-freeze coolant (or equivalent) and water.
  - Loose water pump belt
  - Restricted air flow to radiator, or restricted water flow through radiator
  - Faulty or incorrect thermostat
  - Inoperative electric cooling fan circuit
- If the system is running RICH, (block learn less than 118), refer to "Diagnostic Aids" on facing page of Code 45.
- If the system is running LEAN, (block learn greater than 138), refer to "Diagnostic Aids" on facing page of Code 44.

POOR GAS MILEAGE

Definition: Gas mileage, as measured by an actual road test, is noticeably lower than expected. Gas mileage is noticeably lower than it was during a previous actual road test.

Perform careful visual and physical check as described at the beginning of Section "B".

Perform "Diagnostic Circuit Check.

- CHECK
  - Proper operation of EST
  - For possibility of misfiring, crossfiring, or cutting out under load or at idle. If present, refer to the "Misfire" Chart
  - Spark plugs. Look for wetness, cracks, improper gap, burned electrodes, and heavy deposits.
  - Spark plugs for correct heat range
  - Fuel for poor quality, "stale" fuel, and water contamination.
  - Fuel system for restricted filter or improper pressure. Use CHART A-7.
  - Injector and TBI assembly for leakage. Pressurize system by energizing fuel pump through the underhood fuel pump test connector
  - EGR operation. Use CHART C-7.

1988–90 2.0L (VIN 1)—CORSICA AND BERETTA

- For vacuum leaks at intake manifold gasket
- Air cleaner element (filter) for dirt or plugging
- Idle Air Control system. Use Code 35 chart.
- Canister purge system for proper operation. Use CHART C-3.
- PROM for correct application
- Throttle shaft or TPS for sticking or binding. TPS voltage should read less than 1.25 V on a "Scan" tool with the throttle closed.
- MAP sensor output. Use CHART C1-D and/or check sensor by comparing it to the output on a similar vehicle if possible.
- Oxygen sensor for silicone contamination from contaminated fuel or use of improper RTV sealant. The sensor will have a white, powdery coating and will cause a high but false signal voltage (rich exhaust indication). The ECM will reduce the amount of fuel delivered to the engine, causing a severe driveability problem
- Coolant sensor for a shift in calibration. Use Code 14 or Code 15 chart.
- Vehicle speed sensor (VSS) input with a "Scan" tool to make sure reading of VSS matches that of vehicle speedometer.
- A/C relay operation. A/C should cut out at wide open throttle. Use CHART C-10.
- A/C refrigerant pressure for high pressure. Check for overcharging or faulty pressure switch.
- Generator output voltage. Repair if less than 9 V or more than 16 V.
- Cooling fan operation. Use CHART C-12.
- Power steering. Use CHART C-1E. The ECM should compensate for power steering loads. Loss of this signal would be most noticeable when steering loads are high such as during parking.
- Transmission torque converter operation.
- Transmission for proper shift points.
- Transmission torque converter clutch operation. Use CHART C-8.
- Engine valve timing and compression
- For worn or incorrect basic engine parts such as cam, heads, pistons, etc. Also check for bent pushrods, worn rocker arms, and broken or weak valve springs
- For the possibility of an exhaust restriction or improper valve timing by performing the following test
  1. With engine at normal operating temperature, connect a vacuum gauge to any convenient vacuum port on intake manifold
  2. Run engine at 1000 rpm and record vacuum reading
  3. Increase engine speed slowly to 2500 rpm. Note vacuum reading at steady 2500 rpm.
  4. If vacuum at 2500 rpm decreases more than 3" Hg from reading at 1000 rpm, the exhaust system should be inspected for restrictions.
  5. Disconnect exhaust pipe from engine and repeat Steps 3 & 4. If vacuum still drops more than 3" Hg with exhaust disconnected, check valve timing
- Thermostat for incorrect heat range or being inoperative
- Check driver's driving habits and vehicle conditions which affect gas mileage
  - Suggest driver read "Important Facts on Fuel Economy" in Owner's Manual
  - Is A/C "ON" full time (Defroster mode "ON")?
  - Are tires at correct pressure?
  - Are excessively heavy loads being carried?
  - Is acceleration often heavy?
  - Are the wheels aligned correctly?
  - Is the speedometer calibrated correctly?
  - Are the vehicle brakes dragging?
  - Is the brake switch applying excessive force on the brake pedal?
- If the system is running RICH, (block learn less than 118), refer to "Diagnostic Aids" on facing page of Code 45

1988–90 2.0L (VIN 1)—CORSICA AND BERETTA

DETONATION/SPARK KNOCK

Definition: A mild to severe ping, usually worse under acceleration. The engine makes sharp metallic knocks that change with throttle opening.

Perform careful visual and physical check as described at the beginning of Section "B".

Perform "Diagnostic Circuit Check."

- CHECK
  - Ignition wires for shorts or faulty insulation
  - For possibility of misfiring, crossfiring, or cutting out under load or at idle. If present, refer to the "Misfire" Chart
  - Spark plugs for correct heat range
  - Fuel for poor quality, "stale" fuel, and water contamination
  - Fuel system for restricted filter or improper pressure. Use CHART A-7.
  - For excessive oil entering combustion chamber. Oil will reduce the effective octane of fuel.
  - EGR operation. Use CHART C-7.
  - For vacuum leaks at intake manifold gasket
  - PCV valve for proper operation by placing finger over inlet hole in valve end several times. Valve should snap back. If not, replace valve. Ensure that valve is correct part. Also check PCV hose.
  - MAP sensor output. Use CHART C1-D and/or check sensor by comparing it to the output on a Similar vehicle, if possible.
  - Coolant sensor for a shift in calibration
  - Oxygen sensor for silicone contamination from contaminated fuel or use of improper RTV sealant. The sensor will have a white, powdery coating and will cause a high but false signal voltage (rich exhaust indication). The ECM will reduce the amount of fuel delivered to the engine, causing a severe driveability problem
  - Vehicle speed sensor (VSS) input with a "Scan" tool to make sure reading of VSS matches that of vehicle speedometer.
  - Transmission torque converter operation.
  - Transmission for proper shift points.
  - Transmission torque converter clutch operation. Use CHART C-8.
  - Vehicle brakes for dragging
  - PROM for correct application.
  - For overheating and possible causes. Look for the following.
    - Low or incorrect coolant solution. It should be a 50/50 mix of GM #1052753 anti-freeze coolant (or equivalent) and water.
    - Loose water pump belt
    - Restricted air flow to radiator or restricted water flow through radiator
    - Faulty or incorrect thermostat
    - Inoperative electric cooling fan circuit
  - Engine valve timing and compression.
  - For worn or incorrect basic engine parts such as cam, heads, pistons, etc. Also check for bent pushrods, worn rocker arms, and broken or weak valve springs.
  - For the possibility of an exhaust restriction or improper valve timing by performing the following test.
    1. With engine at normal operating temperature, connect a vacuum gauge to any convenient vacuum port on intake manifold.
    2. Run engine at 1000 rpm and record vacuum reading.
    3. Increase engine speed slowly to 2500 rpm. Note vacuum reading at steady 2500 rpm.
    4. If vacuum at 2500 rpm decreases more than 3" Hg from reading at 1000 rpm, the exhaust system should be inspected for restrictions.
    5. Disconnect exhaust pipe from engine and repeat Steps 3 & 4. If vacuum still drops more than 3" Hg with exhaust disconnected, check valve timing.
- Remove internal engine carbon with top engine cleaner.
- If the system is running LEAN, (block learn greater than 138), refer to "Diagnostic Aids" on facing page of Code 44.

1988–90 2.0L (VIN 1)—CORSICA AND BERETTA

LACK OF POWER, SLUGGISH, OR SPONGY

Definition: Engine delivers less than expected power. There is little or no increase in speed when the accelerator pedal is depressed partially.

Perform careful visual and physical check as described at the beginning of Section "B".

Perform "Diagnostic Circuit Check."

- CHECK
  - Ignition wires for shorts or faulty insulation
  - Ignition system for moisture, dust, cracks, burns, etc. Spray plug wires with fine water mist to check for shorts.
  - For possibility of misfiring, crossfiring, or cutting out under load or at idle. If present, refer to the "Misfire" Chart
  - Secondary ignition voltage output with ST-125 tester.
  - Ignition coil connections
  - ECM and ignition system for faulty grounds
  - Proper operation of EST.
  - Spark plugs. Look for wetness, cracks, improper gap, burned electrodes, and heavy deposits.
  - Spark plugs for correct heat range
  - Fuel for poor quality, "stale" fuel, and water contamination
  - Fuel system for restricted filter or improper pressure. Use CHART A-7.
  - EGR operation. Use CHART C-7.
  - For vacuum leaks at intake manifold gasket
  - Air cleaner element (filter) for dirt or plugging
  - PROM for correct application.
  - Throttle shaft or TPS for sticking or binding. TPS voltage should read less than 1.25 V on a "Scan" tool with the throttle closed.
  - MAP sensor output. Use CHART C1-D and/or check sensor by comparing it to the output on a similar vehicle if possible.
  - Oxygen sensor for silicone contamination from contaminated fuel or use of improper RTV sealant. The sensor will have a white, powdery coating and will cause a high but false signal voltage (rich exhaust indication). The ECM will reduce the amount of fuel delivered to the engine, causing a severe driveability problem.
  - Coolant sensor for a shift in calibration. Use Code 14 or Code 15 chart.
  - Vehicle speed sensor (VSS) input with a "Scan" tool to make sure reading of VSS matches that of vehicle speedometer.
  - Engine for improper or worn camshaft.
  - A/C relay operation. A/C should cut out at wide open throttle. Use CHART C-10.
  - A/C refrigerant pressure for high pressure. Check for overcharging or faulty pressure switch.
  - Generator output voltage. Repair if less than 9 V or more than 16 V.
  - Cooling fan operation. Use CHART C-12.
  - Power steering. Use CHART C-1E. The ECM should compensate for power steering loads. Loss of this signal would be most noticeable when steering loads are high such as during parking.
  - Transmission torque converter operation
  - Transmission for proper shift points.
  - Transmission torque converter clutch operation. Use CHART C-8.
  - Vehicle brakes for dragging
  - Engine valve timing and compression.
  - For worn or incorrect basic engine parts such as cam, heads, pistons, etc. Also check for bent pushrods, worn rocker arms, and broken or weak valve springs.
  - For the possibility of an exhaust restriction or improper valve timing by performing the following test.
    1. With engine at normal operating temperature, connect a vacuum gauge to any convenient vacuum port on intake manifold.
    2. Run engine at 1000 rpm and record vacuum reading.
    3. Increase engine speed slowly to 2500 rpm. Note vacuum reading at steady 2500 rpm.
    4. If vacuum at 2500 rpm decreases more than 3" Hg from reading at 1000 rpm, the exhaust system should be inspected for restrictions.

1988–90 2.0L (VIN 1)–CORSICA AND BERETTA

5. Disconnect exhaust pipe from engine and repeat Steps 3 & 4. If vacuum still drops more than 3" Hg with exhaust disconnected, check valve timing.

For overheating and possible causes. Look for the following:
- Low or incorrect coolant solution. It should be a 50/50 mix of GM #1052753 anti-freeze coolant (or equivalent) and water.
- Loose water pump belt
- Restricted air flow to radiator, or restricted water flow through radiator
- Faulty or incorrect thermostat
- Inoperative electric cooling fan circuit. See CHART C-12

- • If the system is running RICH (block learn less than 118), refer to "Diagnostic Aids" on facing page of Code 45.
- • If the system is running LEAN (block learn greater than 138), refer to "Diagnostic Aids" on facing page of Code 44.

SURGES AND/OR CHUGGLE

Definition: Engine power variation under steady throttle or cruise. Feels like the car speeds up and slows down with no change in the accelerator pedal.

Perform careful visual and physical check as described at the beginning of Section "B".

Perform "Diagnostic Circuit Check."

- • **CHECK**
  - Ignition wires for shorts or faulty insulation
  - Ignition system for moisture, dust, cracks, burns, etc. Spray plug wires with fine water mist to check for shorts.
  - For possibility of misfiring, crossfiring, or cutting out under load or at idle. If present, refer to the "Misfire" Chart
  - Secondary ignition voltage output with ST-125 tester
  - Ignition coil connections
  - ECM and ignition system for faulty grounds.
  - Proper operation of EST
  - Spark plugs. Look for wetness, cracks, improper gap, burned electrodes, and heavy deposits.
  - Spark plugs for correct heat range.
  - Fuel for poor quality, "stale" fuel, and water contamination.
  - Fuel system for restricted filter or improper pressure. Use CHART A-7
  - Injector and TBI assembly for leakage. Pressurize system by energizing fuel pump through the underhood fuel pump test connector.
  - EGR operation. Use CHART C-7
  - For vacuum leaks at intake manifold gasket.
  - Idle Air Control system. Use Code 35 chart
  - Electrical system voltage. IAC valve will not move if voltage is below 9 V or greater than 17.8 V. Also check battery cables and ground straps for poor contact. Erratic voltage will cause the IAC valve to change its position, resulting in poor idle quality.
  - PCV valve for proper operation by placing finger over inlet hole in valve end several times. Valve should snap back. If not, replace valve. Ensure that valve is correct part. Also check PCV hose
  - Canister purge system for proper operation. Use CHART C-3.
  - PROM for correct application
  - Throttle shaft or TPS for sticking or binding. TPS voltage should read less than 1.25 volts on a "Scan" tool with the throttle closed.
  - MAP sensor output. Use CHART C1-D and/or check sensor by comparing it to the output on a similar vehicle, if possible.
  - Oxygen sensor for silicone contamination from contaminated fuel or use of improper RTV sealant. The sensor will have a white, powdery coating and will cause a high but false signal voltage (rich exhaust indication). The ECM will reduce the amount of fuel delivered to the engine, causing a severe driveability problem.
  - Coolant sensor for a shift in calibration. Use Code 14 or Code 15 chart.
  - Vehicle speed sensor (VSS) input with a "Scan" tool to make sure reading of VSS matches that of vehicle speedometer.
  - A/C relay operation. A/C should cut out at wide open throttle. Use CHART C-10.
  - P/N switch circuit on vehicle with automatic transmission. Use CHART C-1A
  - Transmission torque converter clutch operation. Use CHART C-8.

1988–90 2.0L (VIN 1)–CORSICA AND BERETTA

For the possibility of an exhaust restriction or improper valve timing by performing the following test:

1. With engine at normal operating temperature, connect a vacuum gauge to any convenient vacuum port on intake manifold.
2. Run engine at 1000 rpm and record vacuum reading.
3. Increase engine speed slowly to 2500 rpm. Note vacuum reading at steady 2500 rpm.
4. If vacuum at 2500 rpm decreases more than 3" Hg from reading at 1000 rpm, the exhaust system should be inspected for restrictions.
5. Disconnect exhaust pipe from engine and repeat Steps 3 & 4. If vacuum still drops more than 3" Hg with exhaust disconnected, check valve timing.

Engine valve timing and compression.

- For worn or incorrect basic engine parts such as cam, heads, pistons, etc. Also check for bent pushrods, worn rocker arms, and broken or weak valve springs.

- • If the system is running RICH, (block learn less than 118), refer to "Diagnostic Aids" on facing page of Code 45.
- • If the system is running LEAN, (block learn greater than 138), refer to "Diagnostic Aids" on facing page of Code 44.

CUTS OUT, MISSES

Definition: Steady pulsation or jerking that follows engine speed, usually more pronounced as engine load increases. The exhaust has a steady spitting sound at idle or low speed.

Perform careful visual and physical check as described at the beginning of Section "B".

Perform "Diagnostic Circuit Check."

- • **CHECK**
  - Ignition wires for shorts or faulty insulation
  - Ignition system for moisture, dust, cracks, burns, etc. Spray plug wires with fine water mist to check for shorts.
  - For possibility of misfiring, crossfiring, or cutting out under load or at idle. If present, refer to the "Misfire" Chart
  - Secondary ignition voltage output with ST-125 tester
  - Ignition coil connections
  - ECM and ignition system for faulty grounds
  - Proper operation of EST.
  - Spark plugs. Look for wetness, cracks, improper gap, burned electrodes, and heavy deposits.
  - Spark plugs for correct heat range
  - For improper crank sensor resistance or faulty connections
  - Fuel for poor quality, "stale" fuel, and water contamination
  - Fuel system for restricted filter or improper pressure. Use CHART A-7.
  - Throttle shaft or TPS for sticking or binding. TPS voltage should read less than 1.25 V on a "Scan" tool with the throttle closed.

HESITATION, SAG, STUMBLE

Definition: Momentary lack of response as the accelerator is pushed down. Can occur at all vehicle speeds. Usually most severe when first trying to make the car move, as from a stop sign. May cause the engine to stall if severe enough.

Perform careful visual and physical check as described at the beginning of Section "B".

Perform "Diagnostic Circuit Check."

- • **CHECK**
  - Ignition wires for shorts or faulty insulation
  - Ignition system for moisture, dust, cracks, burns, etc. Spray plug wires with fine water mist to check for shorts.
  - For possibility of misfiring, crossfiring, or cutting out under load or at idle. If present, refer to the "Misfire" Chart
  - Secondary ignition voltage output with ST-125 tester
  - Ignition coil connections
  - ECM and ignition system for faulty grounds
  - Proper operation of EST. See Section "C4".
  - Spark plugs. Look for wetness, cracks, improper gap, burned electrodes, and heavy deposits.
  - Spark plugs for correct heat range
  - Fuel for poor quality, "stale" fuel, and water contamination
  - Fuel system for restricted filter or improper pressure. Use CHART A-7.

1988–90 2.0L (VIN 1)–CORSICA AND BERETTA

- EGR operation. Use CHART C-7
- For vacuum leaks at intake manifold gasket
- Air cleaner element (filter) for dirt or plugging
- Idle Air Control system
- Check electrical system voltage. IAC valve will not move if voltage is below 8.7 volts. Also check battery cables and ground straps for poor contact. Erratic voltage will cause the IAC valve to change its position, resulting in poor idle quality.
- PCV valve for proper operation by placing finger over inlet hole in valve end several times. Valve should snap back. If not, replace valve. Ensure that valve is correct part. Also check PCV hose.
- Canister purge system for proper operation. Use CHART C-3.
- PROM for correct application
- Throttle shaft or TPS for sticking or binding. TPS voltage should read less than 1.25 volts on a "Scan" tool with the throttle closed.
- MAP sensor output. Use CHART C1-D and/or check sensor by comparing it to the output on a similar vehicle, if possible.
- Oxygen sensor for silicone contamination from contaminated fuel or use of improper RTV sealant. The sensor will have a white, powdery coating and will cause a high but false signal voltage (rich exhaust indication). The ECM will reduce the amount of fuel delivered to the engine, causing a severe driveability problem.
- Coolant sensor for a shift in calibration. Use Code 14 or Code 15 chart.
- A/C relay operation. A/C should cut out at wide open throttle. Use CHART C-10.
- A/C refrigerant pressure for high pressure. Check for overcharging or faulty pressure switch.
- P/N switch circuit on vehicle with automatic transmission. Use CHART C-1A.
- Generator output voltage. Repair if less than 9 volts or more than 16 volts.
- Transmission torque converter operation.
- Transmission for proper shift points.
- Transmission torque converter clutch operation. Use CHART C-8.
- Vehicle brakes for dragging
- Engine valve timing and compression.

For the possibility of an exhaust restriction or improper valve timing by performing the following test:

1. With engine at normal operating temperature, connect a vacuum gauge to any convenient vacuum port on intake manifold.
2. Run engine at 1000 rpm and record vacuum reading.
3. Increase engine speed slowly to 2500 rpm. Note vacuum reading at steady 200 rpm.
4. If vacuum at 2500 rpm decreases more than 3" Hg from reading at 1000 rpm, the exhaust system should be inspected for restrictions.
5. Disconnect exhaust pipe from engine and repeat steps 3 & 4. If vacuum still drops more than 3" Hg with exhaust disconnected, check valve timing.

For worn or incorrect basic engine parts such as cam, heads, pistons, etc. Also check for bent pushrods, worn rocker arms, and broken or weak valve springs.

For overheating and possible causes. Look for the following:
- Low or incorrect coolant solution. It should be a 50/50 mix of GM #1052753 anti freeze coolant (or equivalent) and water.
- Loose water pump belt
- Restricted air flow to radiator, or restricted water flow through radiator
- Faulty or incorrect thermostat
- Inoperative electric cooling fan circuit. See CHART C-12.

- • If the system is running RICH (block learn less than 118), refer to "Diagnostic Aids" on facing page of Code 45.
- • If the system is running LEAN (block learn greater than 145), refer to "Diagnostic Aids" on facing page of Code 44.

1988–90 2.0L (VIN 1)–CORSICA AND BERETTA

EXCESSIVE EXHAUST EMISSIONS OR ODORS

Definition: Vehicle fails an emission test or vehicle has excessive "rotten egg" smell. (Excessive odors do not necessarily indicate excessive emissions).

Perform careful visual and physical check as described at the beginning of Section "B".

Perform "Diagnostic Circuit Check."

- • **CHECK**
  - EGR valve not opening. Use CHART C-7.
  - Vacuum leaks
  - Faulty coolant system and/or coolant fan operation. Use CHART C-12.
  - Remove carbon with top engine cleaner. Follow instructions on can.
- • If the system is running RICH (block learn less than 118), refer to "Diagnostic Aids" on facing page of Code 45.
- • If the system is running LEAN (block learn greater than 138), refer to "Diagnostic Aids" on facing page of Code 44.
- • If emission test indicates excessive NO_x, check for items which cause car to run lean or too hot.
- • If emission test indicates excessive HC and CO or exhaust has excessive odors, check for items which cause car to run RICH.
  - Incorrect fuel pressure. Use CHART A-7.
  - Fuel loading of evaporative vapor canister. Use CHART C-3.
  - PCV valve plugging, sticking, or blocked PCV hose. Check for fuel in crankcase.
  - Catalytic converter lead contamination (Look for removal of fuel filler neck restrictor.)
  - Improper fuel cap installation
  - Faulty spark plugs, plug wires, or ignition components.

DIESELING, RUN-ON

Definition: Engine continues to run after key is turned "OFF," but runs very roughly. (If engine runs smoothly, check ignition switch).

Perform careful visual and physical check as described at the beginning of Section "B".

Perform "Diagnostic Circuit Check."

- • **CHECK**
  - Injector and TBI assembly for leakage. Pressurize system by energizing fuel pump through the fuel pump test connector.

BACKFIRE

Definition: Fuel ignites in intake manifold or in exhaust system, making a loud popping sound.

Perform careful visual and physical check as described at the beginning of Section "B".

Perform "Diagnostic Circuit Check."

- • **CHECK**
  - EGR operation for valve being open all the time. Use CHART C-7.
  - Intake manifold gasket for leaks
  - For possibility of misfiring, crossfiring, or cutting out under load or at idle. If present, refer to the "Misfire" Chart
  - Spark plugs. Look for wetness, cracks, improper gap, burned electrodes, and heavy deposits.
  - Ignition coil connections
  - Ignition system for moisture, dust, cracks, burns, etc. Spray plug wires with fine water mist to check for shorts.
  - ECM and ignition system for faulty grounds
  - Secondary ignition voltage output with ST-125 tester
  - For vacuum leaks at intake manifold gasket
  - Engine valve timing and compression.
  - For worn or incorrect basic engine parts such as cam, heads, pistons, etc. Also check for bent pushrods, worn rocker arms, and broken or weak valve springs.

CHART B-1
RESTRICTED EXHAUST SYSTEM CHECK
ALL ENGINES

Proper diagnosis for a restricted exhaust system is essential before any components are replaced. Either of the following procedures may be used for diagnosis, depending upon engine or tool used:

CHECK AT A. I. R. PIPE:

1. Remove the rubber hose at the exhaust manifold A.I.R. pipe check valve. Remove check valve.
2. Connect a fuel pump pressure gauge to a hose and nipple from a Propane Enrichment Device (J26911) (see illustration).
3. Insert the nipple into the exhaust manifold A.I.R. pipe.

OR CHECK AT O₂ SENSOR:

1. Carefully remove O₂ sensor.
2. Install Borroughs Exhaust Backpressure Tester (BT 8515 or BT 8603) or equivalent in place of O₂ sensor (see illustration).
3. After completing test described below, be sure to coat threads of O₂ sensor with anti-seize compound P/N 5613695 or equivalent prior to re-installation.

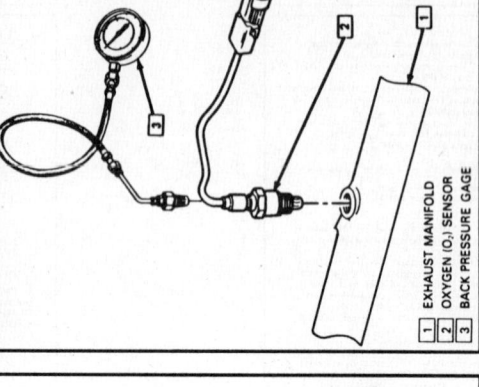

1	EXHAUST MANIFOLD
2	OXYGEN (O₂) SENSOR
3	BACK PRESSURE GAGE

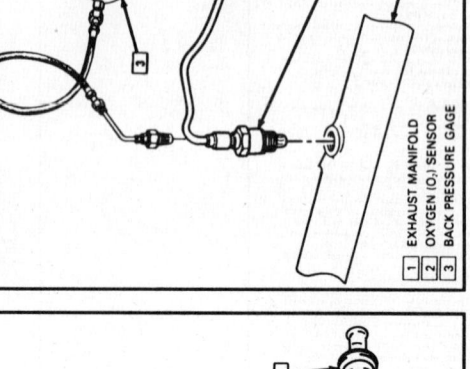

1	GAGE
2	HOSE AND NIPPLE ADAPTER
3	A.I.R. PIPE (EXHAUST PORT)
4	CHECK VALVE

DIAGNOSIS:

1. With the engine idling at normal operating temperature, observe the exhaust system backpressure reading on the gauge. Reading should not exceed 1¼ psi (8.6 kPa).
2. Accelerate engine to 2000 RPM and observe gauge. Reading should not exceed 3 psi (20.7 kPa).
3. If the backpressure, at either RPM, exceeds specification, a restricted exhaust system is indicated.
4. Inspect the entire exhaust system for a collapsed pipe, heat distress, or possible internal muffler failure.
5. If there are no obvious reasons for the excessive backpressure, a restricted catalytic converter should be suspected and replaced using current recommended procedures.

1988–90 2.0L (VIN 1) – CORSICA AND BERETTA

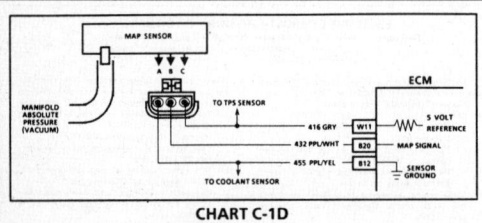

CHART C-1D
MAP OUTPUT CHECK
2.0L (VIN 1) "L" CARLINE (TBI)

Circuit Description:

The manifold absolute pressure sensor (MAP) measures manifold pressure (vacuum) and sends that signal to the ECM. The MAP sensor is mainly used to calculate engine load, which is fundamental input for spark and fuel calculations. The MAP sensor is also used to determine the barometric pressure.

Test Description: Numbers below refer to circled numbers on the diagnostic chart.

1. Checks MAP sensor output voltage to the ECM. This voltage, without engine running, represents a barometer reading to the ECM.
2. Applying 34 kPa (10" Hg) vacuum to the MAP sensor should cause the voltage to be 1.2 to 2.3 volts less than the voltage at Step 1. Upon applying vacuum to the sensor, the change in voltage should be instantaneous. A slow voltage change indicates a faulty sensor.

3. Check vacuum hose to sensor for leaking or restriction. Be sure no other vacuum devices are connected to the MAP hose

---

1988–90 2.0L (VIN 1) – CORSICA AND BERETTA

CHART C-1D
MAP OUTPUT CHECK
2.0L (VIN 1) "L" CARLINE (TBI)

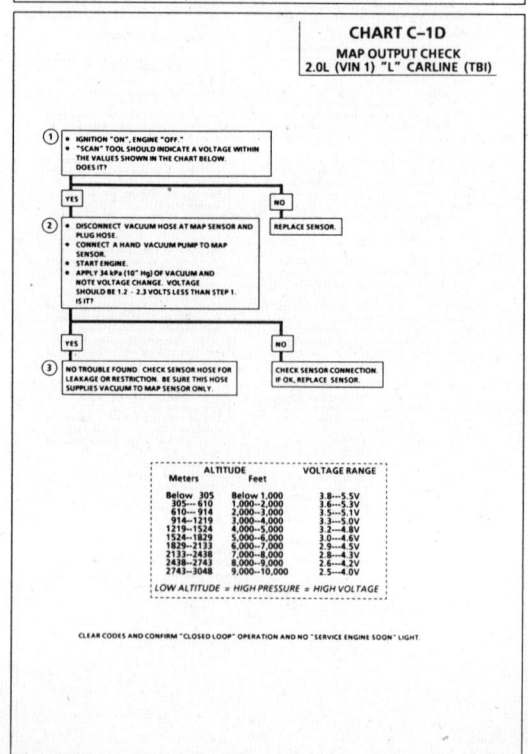

Meters	ALTITUDE Feet	VOLTAGE RANGE
Below 305	Below 1,000	3.8—5.5V
305—610	1,000—2,000	3.6—5.3V
610—914	2,000—3,000	3.5—5.1V
914—1219	3,000—4,000	3.3—5.0V
1219—1524	4,000—5,000	3.2—4.8V
1524—1829	5,000—6,000	3.0—4.6V
1829—2133	6,000—7,000	2.9—4.5V
2133—2438	7,000—8,000	2.8—4.3V
2438—2743	8,000—9,000	2.6—4.2V
2743—3048	9,000—10,000	2.5—4.0V

LOW ALTITUDE = HIGH PRESSURE = HIGH VOLTAGE

CLEAR CODES AND CONFIRM "CLOSED LOOP" OPERATION AND NO "SERVICE ENGINE SOON" LIGHT.

1988–90 2.0L (VIN 1) – CORSICA AND BERETTA

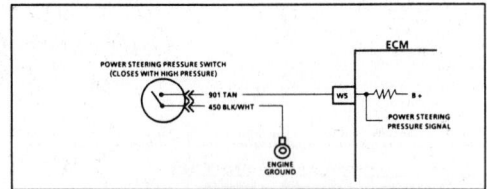

CHART C-1E
POWER STEERING PRESSURE SWITCH (PSPS) DIAGNOSIS
2.0L (VIN 1) "L" CARLINE (TBI)

Circuit Description:
The power steering pressure switch is normally open to ground, with CKT 901 supplying battery voltage to the switch.

Turning the steering wheel increases power steering oil pressure and its load on an idling engine. The pressure switch will close before the load can cause an idle problem.

Closing the switch causes CKT 901 to read less than 1 volt and the ECM will increase the idle air rate and de-energize the A/C relay

- A pressure switch that will not close, or an open CKT 901 or 450, may cause the engine to stall when power steering loads are high.
- A switch that will not open, or a CKT 901 shorted to ground, may affect idle quality, and will cause the A/C relay to be de-energized

Test Description: Numbers below refer to circled numbers on the diagnostic chart.
1. Different makes of "Scan" tools may display the state of this switch in different ways. Refer to "Scan" tool upgrading to determine how this input is indicated.

2. Checks to determine if CKT 901 is shorted to ground.
3. This should simulate a closed switch.

1988–90 2.0L (VIN 1) – CORSICA AND BERETTA

CHART C-1E
POWER STEERING PRESSURE SWITCH (PSPS) DIAGNOSIS
2.0L (VIN 1) "L" CARLINE (TBI)

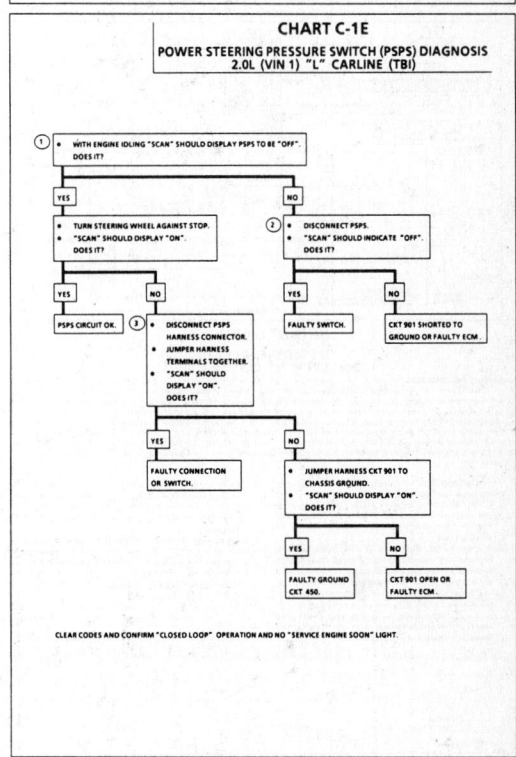

1988–90 2.0L (VIN 1) – CORSICA AND BERETTA

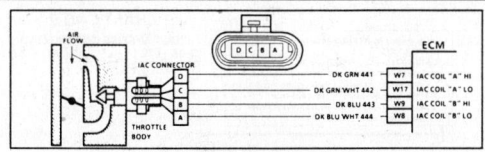

CHART C-2C
IDLE AIR CONTROL (IAC) SYSTEM CHECK
2.0L (VIN 1) "L" CARLINE (TBI)

Circuit Description:
The ECM controls idle rpm with the IAC valve. To increase idle rpm, the ECM moves the IAC valve out allowing more air to pass by the throttle plate. To decrease rpm, it moves the IAC valve in, reducing air flow by the throttle plate. A "Scan" tool will read the ECM commands to the IAC valve in counts. The higher the counts, the more air allowed (higher idle). The lower the counts, the less air allowed (lower idle).

Test Description: Numbers below refer to numbers on the diagnostic chart.
1. Continue with test, even if engine will not idle. If idle is too low, "Scan" will display 80 or more counts, or steps. If idle is high, it will display "0" counts. Occasionally an erratic or unstable idle may occur. Engine speed may vary 200 rpm or more up and down. Disconnect IAC. If the condition is unchanged, the IAC is not at fault
2. When the engine was stopped, the IAC valve retracted (more air) to a fixed "Park" position for increased air flow and idle speed during the next engine start. A "Scan" will display 100 or more counts
3. Be sure to disconnect the IAC valve prior to this test. The test light will confirm the ECM signals by a steady or flashing light on all circuits.
4. There is a remote possibility that one of the circuits is shorted to voltage which would have been indicated by a steady light. Disconnect ECM and turn the ignition "ON" and probe terminals to check for this condition.

Diagnostic Aids:
An incorrect idle may be caused by a system problem that cannot be controlled by the ECM.
A "Scan" tool may be used to monitor desired idle, actual engine speed, and IAC counts to help isolate a system problem.

For example, a vacuum leak may be indicated if the desired idle is 900 rpm, IAC counts are at 0, but the actual engine speed is 1500 rpm.
- **System too lean** (High Air/Fuel Ratio)
  Idle speed may be too high or too low. Engine speed may vary up and down, disconnecting IAC does not help. May set Code 44
  "Scan" and/or voltmeter will read an oxygen sensor output less than 300 mV (.3 volt). Check for low regulated fuel pressure or water in fuel. A lean exhaust, with an oxygen sensor output fixed above 800 mV (.8 volt) will be a contaminated sensor, usually silicone. This may also set a Code 45
- **System too rich** (Low Air/Fuel Ratio)
  Idle speed too low. "Scan" counts usually above 80. System obviously rich and may exhibit black smoke exhaust.
  "Scan" tool and/or voltmeter will read an oxygen sensor signal fixed above 800 mV (.8 volt)
  Check
    High fuel pressure
    Injector leaking or sticking
- **Throttle Body**
  Remove IAC and inspect bore for foreign material or evidence of IAC valve dragging the bore
- Refer to "Rough, Unstable, Incorrect Idle or Stalling" in "Symptoms" in Section "B"

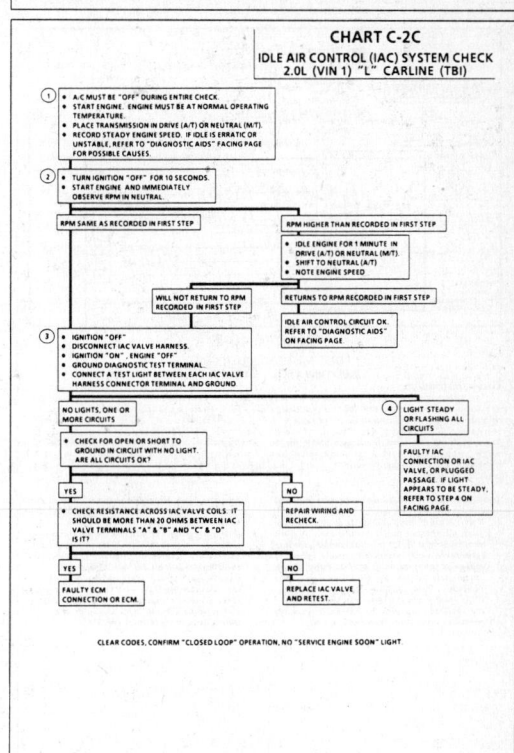

1988–90 2.0L (VIN 1) – CORSICA AND BERETTA

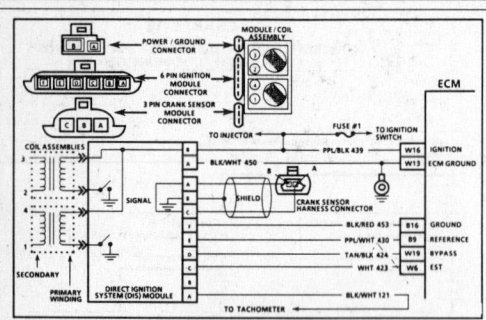

CHART C-4D-1
"DIS" MISFIRE AT IDLE
2.0L (VIN 1) "L" CARLINE (TBI)

Circuit Description:
The "direct ignition system" (DIS) uses a waste spark method of distribution. In this type of system, the ignition module triggers the #1/4 coil pair resulting in both #1 and #4 spark plugs firing at the same time. #1 cylinder is on the compression stroke at the same time #4 is on the exhaust stroke, resulting in a lower energy requirement to fire #4 spark plug. This leaves the remainder of the high voltage to be used to fire #1 spark plug. The crank sensor is remotely mounted beside the module/coil assembly and protrudes through the block to within approximately .050" of the crankshaft reluctor. Since the reluctor is a machined portion of the crankshaft and the crankshaft sensor is mounted in a fixed position on the block, timing adjustments are not possible or necessary.

Test Description: Numbers below refer to circled numbers on the diagnostic chart.
1. If the "Misfire" complaint exists under load only, the diagnostic chart on page 2 must be used. Engine rpm should drop approximately equally on all plug leads.
2. A spark tester, such as a ST-125, must be used because it is essential to verify adequate available secondary voltage at the spark plug (25,000 volts).
3. If the spark jumps the tester gap after grounding the opposite plug wire, it indicates excessive resistance in the plug which was bypassed. A faulty or poor connection at that plug could also result in the miss condition. Also, check for carbon deposits inside the spark plug boot.

4. If carbon tracking is evident, replace coil and be sure plug wires relating to that coil are clean and tight. Excessive wire resistance or faulty connections could have caused the coil to be damaged.
5. If the no spark condition follows the suspected coil, that coil is faulty. Otherwise, the ignition module is the cause of no spark. This test could also be performed by substituting a known good coil for the one causing the no spark condition.

1988–90 2.0L (VIN 1) – CORSICA AND BERETTA

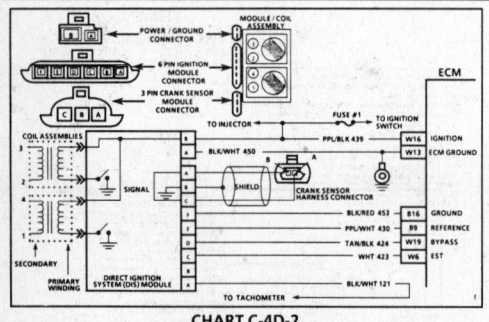

CHART C-4D-2
"DIS" MISFIRE UNDER LOAD
2.0L (VIN 1) "L" CARLINE (TBI)

Circuit Description:
The "direct ignition system" (DIS) uses a waste spark method of distribution. In this type of system, the ignition module triggers the #1/4 coil pair resulting in both #1 and #4 spark plugs firing at the same time. #1 cylinder is on the compression stroke at the same time #4 is on the exhaust stroke, resulting in a lower energy requirement to fire #4 spark plug. This leaves the remainder of the high voltage to be used to fire #1 spark plug. The crank sensor is remotely mounted beside the module/coil assembly and protrudes through the block to within approximately .050" of the crankshaft reluctor. Since the reluctor is a machined portion of the crankshaft, and the crankshaft sensor is mounted in a fixed position on the block, timing adjustments are not possible or necessary.

Test Description: Numbers below refer to circled numbers on the diagnostic chart.
1. If the "Misfire" complaint exists at idle only, the diagnostic chart on page 1 must be used. A spark tester such as a ST-125 must be used because it is essential to verify adequate available secondary voltage at the spark plug, (25,000 volts). Spark should jump the test gap on all 4 leads. This simulates a "load" condition.
2. If spark jumps the tester gap after grounding the opposite plug wire, it indicates excessive resistance in the plug which was bypassed.

A faulty or poor connection at that plug could also result in the miss condition. Also, check for carbon deposits inside the spark plug boot.
3. If carbon tracing is evident replace coil and be sure plug wires relating to that coil are clean and tight. Excessive wire resistance or faulty connections could have caused the coil to be damaged.
4. If the no spark condition follows the suspected coil, that coil is faulty. Otherwise, the ignition module is the cause of no spark. This test could also be performed by substituting a known good coil for the one causing the no spark condition.

1988–90 2.0L (VIN 1) – CORSICA AND BERETTA

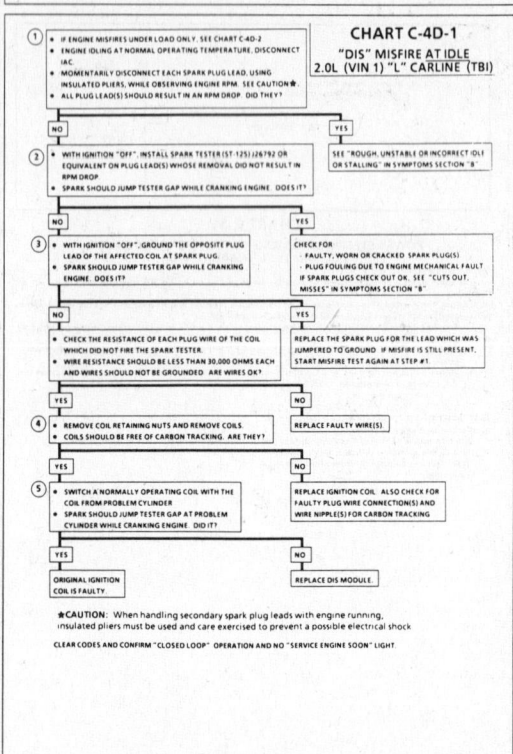

CHART C-4D-1
"DIS" MISFIRE AT IDLE
2.0L (VIN 1) "L" CARLINE (TBI)

1. • IF ENGINE MISFIRES UNDER LOAD ONLY, SEE CHART C-4D-2
   • ENGINE IDLING AT NORMAL OPERATING TEMPERATURE, DISCONNECT IAC.
   • MOMENTARILY DISCONNECT EACH SPARK PLUG LEAD, USING INSULATED PLIERS, WHILE OBSERVING ENGINE RPM. SEE CAUTION★.
   • ALL PLUG LEAD(S) SHOULD RESULT IN AN RPM DROP. DID THEY?

   NO → | YES → SEE "ROUGH, UNSTABLE OR INCORRECT IDLE OR STALLING" IN SYMPTOMS SECTION "B"

2. • WITH IGNITION "OFF", INSTALL SPARK TESTER (ST-125) J26792 OR EQUIVALENT ON PLUG LEAD(S) WHOSE REMOVAL DID NOT RESULT IN RPM DROP.
   • SPARK SHOULD JUMP TESTER GAP WHILE CRANKING ENGINE. DOES IT?

   NO → | YES →

3. • WITH IGNITION "OFF", GROUND THE OPPOSITE PLUG LEAD OF THE AFFECTED COIL AT SPARK PLUG.
   • SPARK SHOULD JUMP TESTER GAP WHILE CRANKING ENGINE. DOES IT?

   CHECK FOR:
   • FAULTY, WORN OR CRACKED SPARK PLUG(S)
   • PLUG FOULING DUE TO ENGINE MECHANICAL FAULT.
   IF SPARK PLUGS CHECK OUT OK, SEE "CUTS OUT, MISSES" IN SYMPTOMS SECTION "B"

   NO → | YES →

   • CHECK THE RESISTANCE OF EACH PLUG WIRE OF THE COIL WHICH DID NOT FIRE THE SPARK TESTER.
   • WIRE RESISTANCE SHOULD BE LESS THAN 30,000 OHMS EACH AND WIRES SHOULD NOT BE GROUNDED. ARE WIRES OK?

   REPLACE THE SPARK PLUG FOR THE LEAD WHICH WAS JUMPERED TO GROUND. IF MISFIRE IS STILL PRESENT, START MISFIRE TEST AGAIN AT STEP #1

   YES → | NO →

4. • REMOVE COIL RETAINING NUTS AND REMOVE COILS.
   • COILS SHOULD BE FREE OF CARBON TRACKING. ARE THEY?

   REPLACE FAULTY WIRE(S).

   YES → | NO →

5. • SWITCH A NORMALLY OPERATING COIL WITH THE COIL FROM PROBLEM CYLINDER.
   • SPARK SHOULD JUMP TESTER GAP AT PROBLEM CYLINDER WHILE CRANKING ENGINE. DID IT?

   REPLACE IGNITION COIL. ALSO CHECK FOR FAULTY PLUG WIRE CONNECTION(S) AND WIRE NIPPLE(S) FOR CARBON TRACKING

   YES → | NO →

   ORIGINAL IGNITION COIL IS FAULTY.

   REPLACE DIS MODULE.

★CAUTION: When handling secondary spark plug leads with engine running, insulated pliers must be used and care exercised to prevent a possible electrical shock

CLEAR CODES AND CONFIRM "CLOSED LOOP" OPERATION AND NO "SERVICE ENGINE SOON" LIGHT.

1988–90 2.0L (VIN 1) – CORSICA AND BERETTA

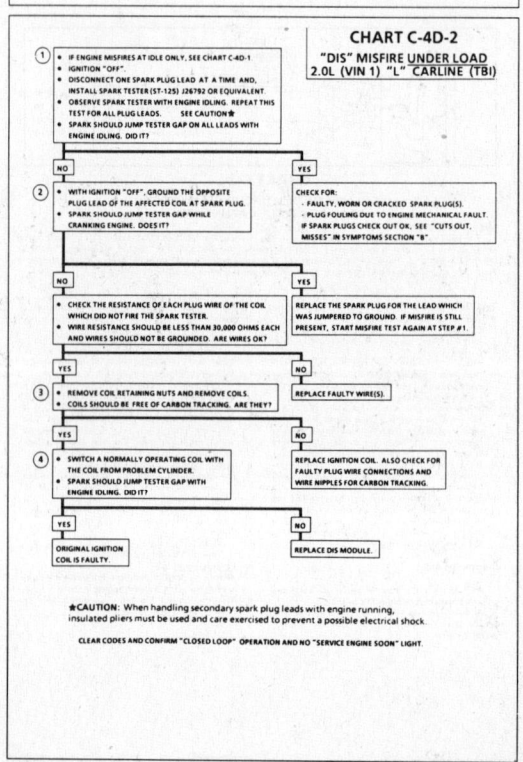

CHART C-4D-2
"DIS" MISFIRE UNDER LOAD
2.0L (VIN 1) "L" CARLINE (TBI)

1. • IF ENGINE MISFIRES AT IDLE ONLY, SEE CHART C-4D-1
   • WITH IGNITION "OFF",
   • DISCONNECT ONE SPARK PLUG LEAD AT A TIME AND, INSTALL SPARK TESTER (ST-125) J26792 OR EQUIVALENT.
   • OBSERVE SPARK TESTER WITH ENGINE IDLING. REPEAT THIS TEST FOR ALL PLUG LEADS. SEE CAUTION★
   • SPARK SHOULD JUMP TESTER GAP ON ALL LEADS WITH ENGINE IDLING. DID IT?

   NO → | YES →

2. • WITH IGNITION "OFF", GROUND THE OPPOSITE PLUG LEAD OF THE AFFECTED COIL AT SPARK PLUG.
   • SPARK SHOULD JUMP TESTER GAP WHILE CRANKING ENGINE. DOES IT?

   CHECK FOR:
   • FAULTY, WORN OR CRACKED SPARK PLUG(S).
   • PLUG FOULING DUE TO ENGINE MECHANICAL FAULT.
   IF SPARK PLUGS CHECK OUT OK, SEE "CUTS OUT, MISSES" IN SYMPTOMS SECTION "B"

   NO → | YES →

   • CHECK THE RESISTANCE OF EACH PLUG WIRE OF THE COIL WHICH DID NOT FIRE THE SPARK TESTER.
   • WIRE RESISTANCE SHOULD BE LESS THAN 30,000 OHMS EACH AND WIRES SHOULD NOT BE GROUNDED. ARE WIRES OK?

   REPLACE THE SPARK PLUG FOR THE LEAD WHICH WAS JUMPERED TO GROUND. IF MISFIRE IS STILL PRESENT, START MISFIRE TEST AGAIN AT STEP #1.

   YES → | NO →

3. • REMOVE COIL RETAINING NUTS AND REMOVE COILS.
   • COILS SHOULD BE FREE OF CARBON TRACKING. ARE THEY?

   REPLACE FAULTY WIRE(S).

   YES → | NO →

4. • SWITCH A NORMALLY OPERATING COIL WITH THE COIL FROM PROBLEM CYLINDER.
   • SPARK SHOULD JUMP TESTER GAP WITH ENGINE IDLING. DID IT?

   REPLACE IGNITION COIL. ALSO CHECK FOR FAULTY PLUG WIRE CONNECTIONS AND WIRE NIPPLES FOR CARBON TRACKING.

   YES → | NO →

   ORIGINAL IGNITION COIL IS FAULTY.

   REPLACE DIS MODULE.

★CAUTION: When handling secondary spark plug leads with engine running, insulated pliers must be used and care exercised to prevent a possible electrical shock.

CLEAR CODES AND CONFIRM "CLOSED LOOP" OPERATION AND NO "SERVICE ENGINE SOON" LIGHT.

1988–90 2.0L (VIN 1) – CORSICA AND BERETTA

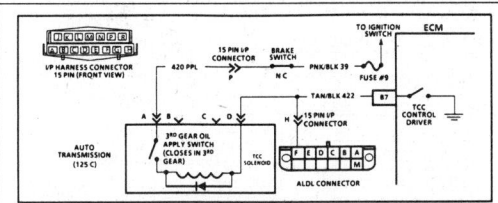

CHART C-8A
TORQUE CONVERTER CLUTCH (TCC)
(ELECTRICAL DIAGNOSIS)
2.0L (VIN 1) "L" CARLINE (TBI)

Circuit Description:
The purpose of the automatic transmission torque converter clutch is to eliminate the power loss of the torque converter when the vehicle is in a cruise condition. This allows the convenience of the automatic transmission and the fuel economy of a manual transmission.

Fused battery ignition is supplied to the TCC solenoid through the brake switch and transmission third gear apply switch. The ECM will engage TCC by grounding CKT 422 to energize the solenoid.

TCC will engage when:
- Vehicle speed above 30 mph (48 km/h).
- Engine at normal operating temperature (above 70°C, 156°F).
- Throttle position sensor output not changing, indicating a steady road speed.
- Transmission third gear switch closed.
- Brake switch closed.

Test Description: Numbers below refer to circled numbers on the diagnostic chart.
1. Light "OFF" confirms transmission third gear apply switch is open.
2. At 48 km/h (30 mph), the transmission third gear switch should close. Test light will come "ON" and confirm battery supply and close brake switch.
3. Grounding the diagnostic terminal with engine "OFF" should energize the TCC solenoid. This test checks the capability of the ECM to control the solenoid.

Check TCC solenoid resistance as follows:
1. Disconnect TCC at transmission.
2. Connect ohmmeter between transmission connector opposite harness connector terminal "A" and "D".

3. Raise drive wheels.
4. Run engine in drive about 48 km/h (30 mph) to close third gear apply switch.
5. Replace the TCC solenoid and ECM if resistance measures less than 20 ohms when switch is closed.

Diagnostic Aids:
An engine coolant thermostat that is stuck open, or opens at too low a temperature may result in an inoperative TCC.

1988–90 2.0L (VIN 1) – CORSICA AND BERETTA

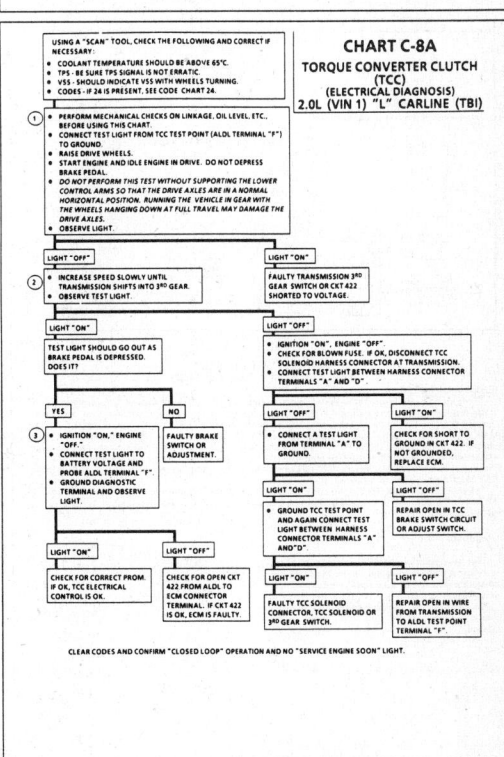

CHART C-8A
TORQUE CONVERTER CLUTCH (TCC)
(ELECTRICAL DIAGNOSIS)
2.0L (VIN 1) "L" CARLINE (TBI)

1988–90 2.0L (VIN 1) – CORSICA AND BERETTA

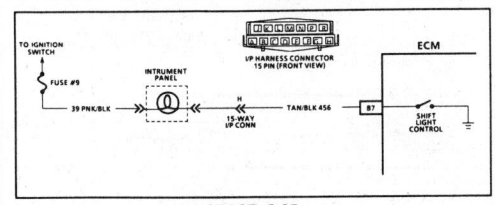

CHART C-8B
MANUAL TRANSMISSION (M/T) SHIFT LIGHT CHECK
2.0L (VIN 1) "L" CARLINE (TBI)

Circuit Description:
The shift light indicates the best transmission shift point for maximum fuel economy. The light is controlled by the ECM and is turned "ON" by grounding CKT 456.

The ECM uses information from the following inputs to control the shift light.
- Coolant temperature
- TPS
- VSS
- rpm

The ECM uses the measured rpm and the vehicle speed to calculate what gear the vehicle is in. It's this calculation that determines when the shift light should be turned "ON."

Test Description: Numbers below refer to circled numbers on the diagnostic chart.
1. This should not turn "ON" the shift light. If the light is "ON," there is a short to ground in CKT 456 wiring or a fault in the ECM.
2. When the diagnostic terminal is grounded, the ECM should ground CKT 456 and the shift light should come "ON."

3. This checks the shift light circuit up to the ECM connector. If the shift light illuminates, then the ECM connector is faulty or the ECM does not have the ability to ground the circuit.

1988–90 2.0L (VIN 1) – CORSICA AND BERETTA

CHART C-8B
MANUAL TRANSMISSION (M/T) SHIFT LIGHT CHECK
2.0L (VIN 1) "L" CARLINE (TBI)

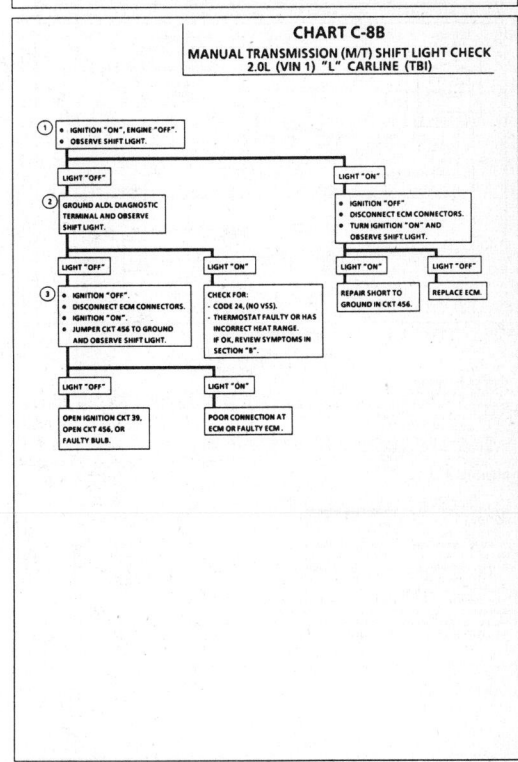

1988–90 2.0L (VIN 1) – CORSICA AND BERETTA

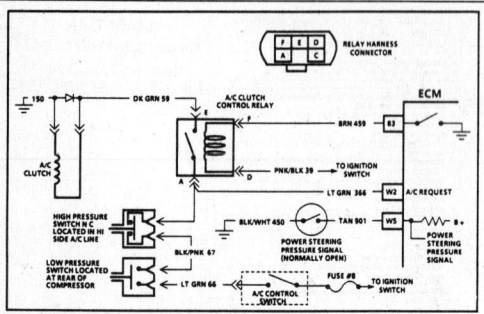

CHART C-10

A/C CLUTCH CONTROL
2.0L (VIN 1) "L" CARLINE (TBI)

Circuit Description:

When an A/C mode is selected on the A/C control switch, ignition voltage is supplied to the compressor low pressure switch. If there is sufficient A/C refrigerant charge, the low pressure switch will be closed and complete the circuit to the closed high pressure cut-off switch and to CKTs 67 and 366. The voltage on CKT 366 to the ECM is shown by the "Scan" tool as A/C request "ON" (voltage present), "OFF" (no voltage). When a request for A/C is seen by the ECM, the ECM will ground CKT 459 of the A/C clutch control relay, the relay contact will close, and current will flow from CKT 366 to CKT 59 and engage the A/C compressor clutch. A "Scan" tool will show the grounding or CKT 459 as A/C clutch "ON." Also, when voltage is seen by the ECM on CKT 366, the cooling fan will be turned "ON."

When power steering hydraulic pressure increases, the power steering pressure switch will close, grounding CKT 901. The ECM will then open CKT 459 which de-energizes the A/C clutch control relay, disengaging the A/C compressor clutch:

Diagnostic Aids:

The low pressure switch will be closed at 40-47 psi and allow A/C clutch operation. Below 37 psi, the low pressure switch will be open and the A/C clutch will not operate.

At about 430 psi, the high pressure switch will open to disengage the A/C clutch and prevent system damage.

1988–90 2.0L (VIN 1) – CORSICA AND BERETTA

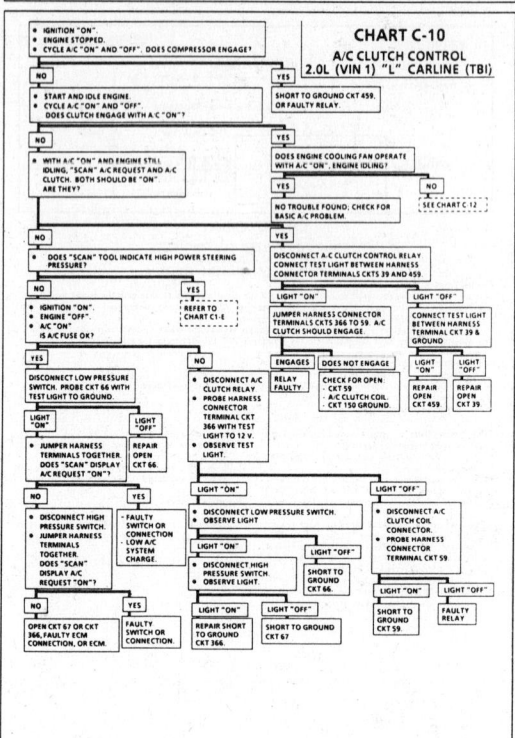

CHART C-10
A/C CLUTCH CONTROL
2.0L (VIN 1) "L" CARLINE (TBI)

1988–90 2.0L (VIN 1) – CORSICA AND BERETTA

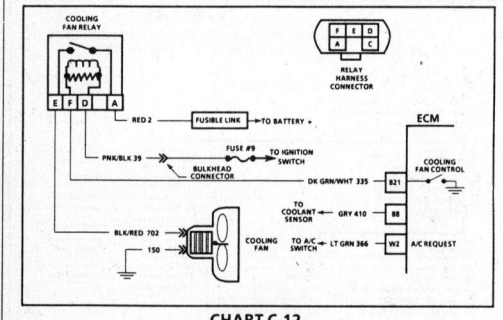

CHART C-12

ENGINE COOLING FAN
2.0L (VIN 1) "L" CARLINE (TBI)

Circuit Description:

Battery voltage to operate the cooling fan motor is supplied to relay by CKT 2. Ignition voltage to energize the relay is supplied to relay by CKT 39. When the ECM grounds CKT 335, the relay is energized and the cooling fan is turned "ON." When the engine is running, the ECM will turn the cooling fan "ON" if:

- A/C is "ON."
- Coolant temperature greater than 108°C (230°F).
- Code 14 or 15, coolant sensor failure.

Diagnostic Aids:

If the owner complained of an overheating problem, it must be determined if the complaint was due to an actual boil over, or the hot light, or temperature gage indicated over heating.

If the gage or light indicates overheating, but no boil over is detected, the gage circuit should be checked. The gage accuracy can also be checked by comparing the coolant sensor reading using a "Scan" tool and comparing its reading with the gage reading.

If the engine is actually overheating and the gage indicates overheating, but the cooling fan is not coming "ON," the coolant sensor has probably shifted out of calibration and should be replaced.

1988–90 2.0L (VIN 1) – CORSICA AND BERETTA

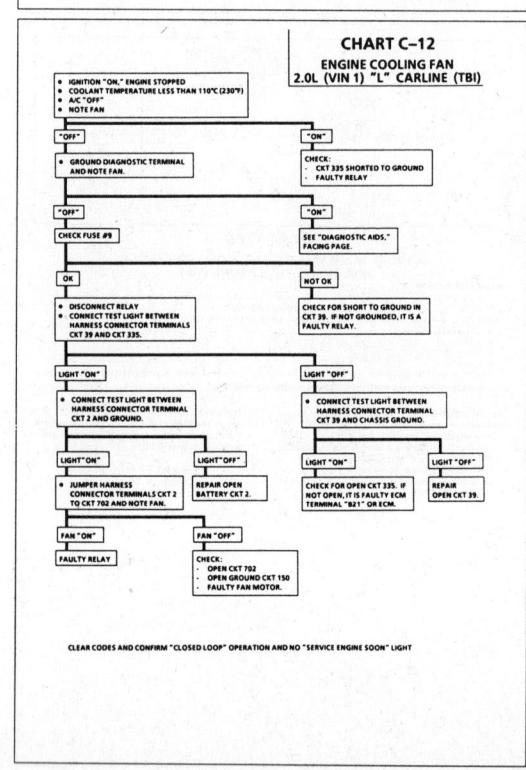

CHART C-12
ENGINE COOLING FAN
2.0L (VIN 1) "L" CARLINE (TBI)

CLEAR CODES AND CONFIRM "CLOSED LOOP" OPERATION AND NO "SERVICE ENGINE SOON" LIGHT

2.0L (VIN 1) COMPONENT LOCATIONS — CAVALIER

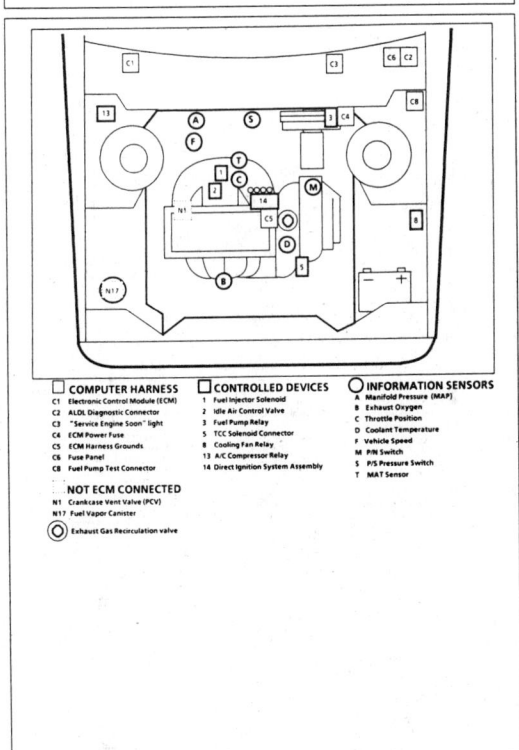

2.0L (VIN 1) ECM WIRING DIAGRAM — CAVALIER

2.0L (VIN 1) ECM WIRING DIAGRAM CAVALIER (CONT.)

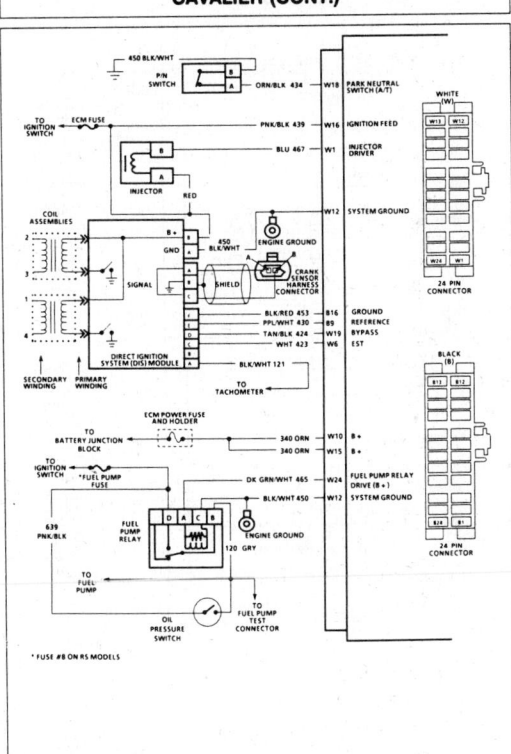

2.0L (VIN 1) ECM WIRING DIAGRAM CAVALIER (CONT.)

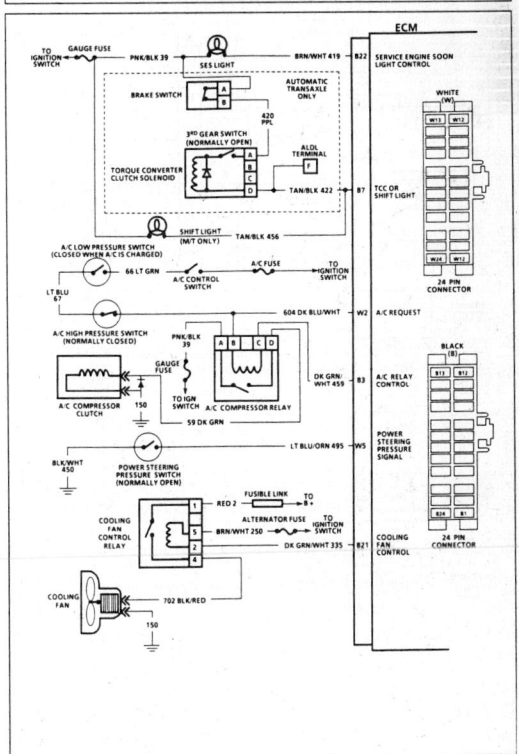

2.0L (VIN 1) ECM CONNECTOR TERMINAL END VIEW — CAVALIER

FUEL INJECTION ECM CONNECTOR IDENTIFICATION

This ECM voltage chart is for use with a digital voltmeter to further aid in diagnosis. The voltages you get may vary due to low battery charge or other reasons, but they should be very close.

THE FOLLOWING CONDITIONS MUST BE MET BEFORE TESTING:
- Engine at operating temperature • Engine idling in "Closed Loop" (for "Engine Run" column)
- "Test" terminal not grounded • Scan tool not installed • All voltages shown "B +" indicates battery or charging voltage.

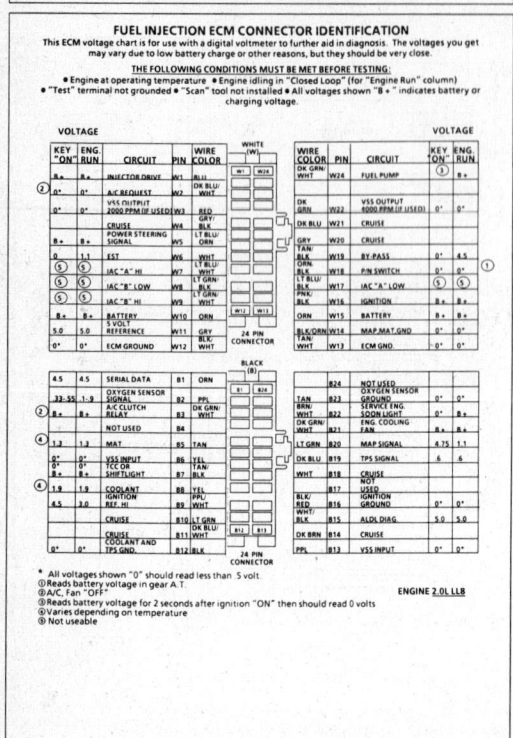

* All voltages shown "0" should read less than .5 volt.
① Reads battery voltage in gear A.T.
② A/C, Fan "OFF"
③ Reads battery voltage for 2 seconds after ignition "ON" then should read 0 volts
④ Varies depending on temperature
⑤ Not useable

ENGINE 2.0L LL8

2.0L (VIN 1) COMPONENT LOCATIONS — FIRENZA

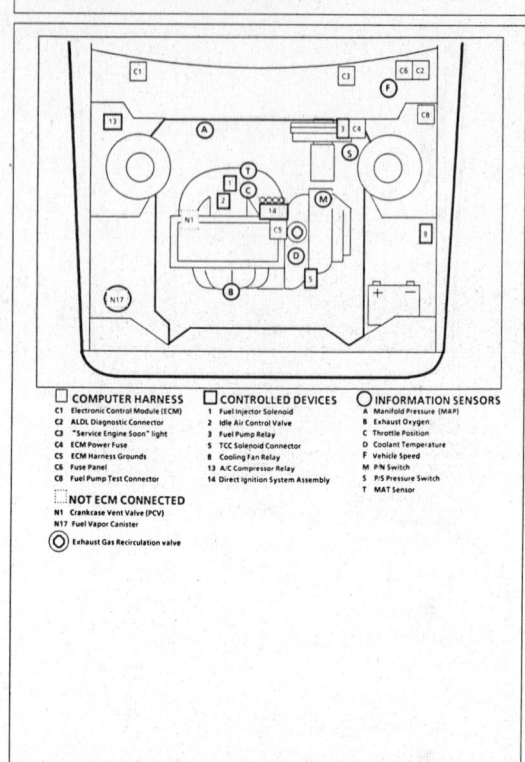

☐ **COMPUTER HARNESS**
C1 Electronic Control Module (ECM)
C2 ALDL Diagnostic Connector
C3 "Service Engine Soon" light
C4 ECM Power Fuse
C5 ECM Harness Grounds
C6 Fuse Panel
C8 Fuel Pump Test Connector

☐ **CONTROLLED DEVICES**
1 Fuel Injector Solenoid
2 Idle Air Control Valve
3 Fuel Pump Relay
5 TCC Solenoid Connector
8 Cooling Fan Relay
13 A/C Compressor Relay
14 Direct Ignition System Assembly

○ **INFORMATION SENSORS**
A Manifold Pressure (MAP)
B Exhaust Oxygen
C Throttle Position
D Coolant Temperature
F Vehicle Speed
M P/N Switch
S P/S Pressure Switch
T MAT Sensor

☐ **NOT ECM CONNECTED**
N1 Crankcase Vent Valve (PCV)
N17 Fuel Vapor Canister

○ Exhaust Gas Recirculation valve

2.0L (VIN 1) ECM WIRING DIAGRAM — FIRENZA

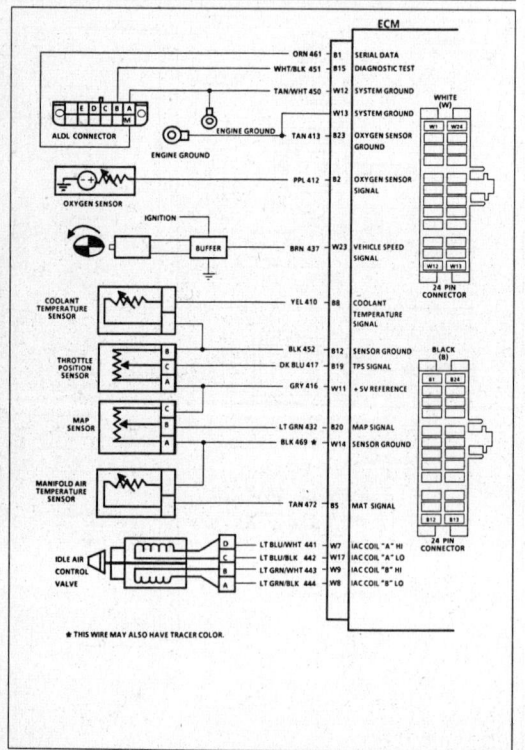

★ THIS WIRE MAY ALSO HAVE TRACER COLOR.

2.0L (VIN 1) ECM WIRING DIAGRAM FIRENZA (CONT.)

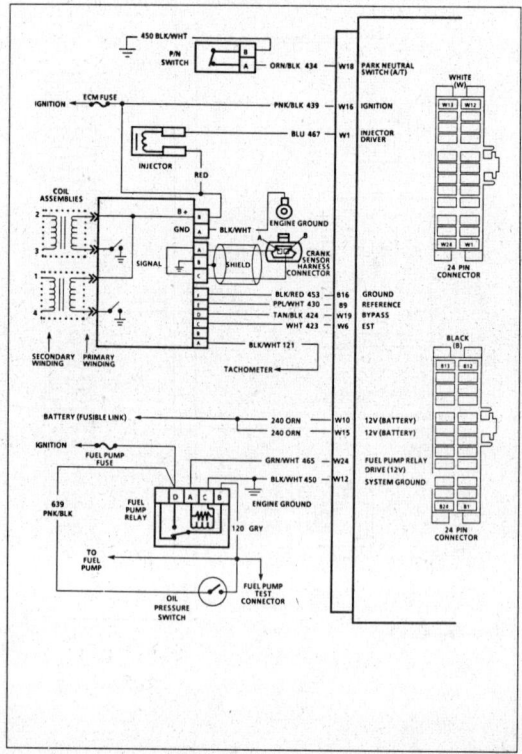

2.0L (VIN 1) ECM WIRING DIAGRAM FIRENZA (CONT.)

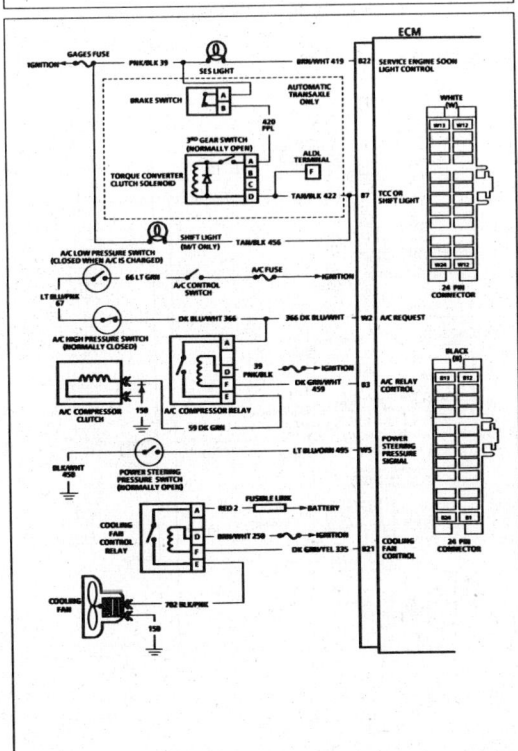

2.0L (VIN 1) ECM CONNECTOR TERMINAL END VIEW—FIRENZA

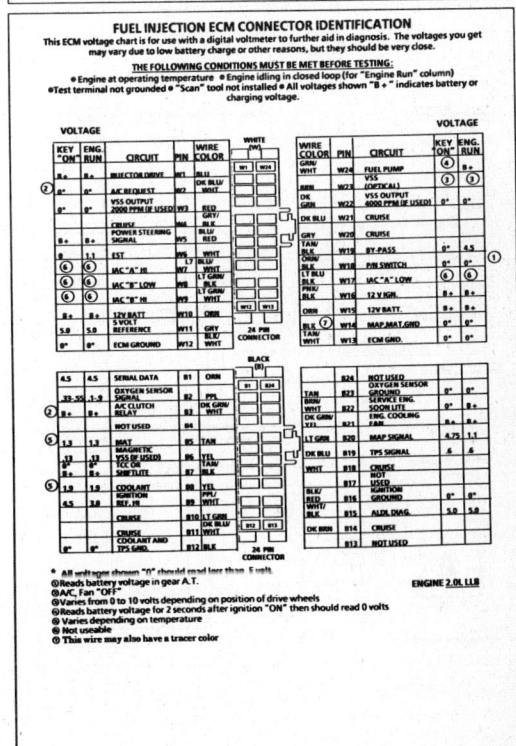

2.0L (VIN 1) COMPONENT LOCATIONS—SKYHAWK

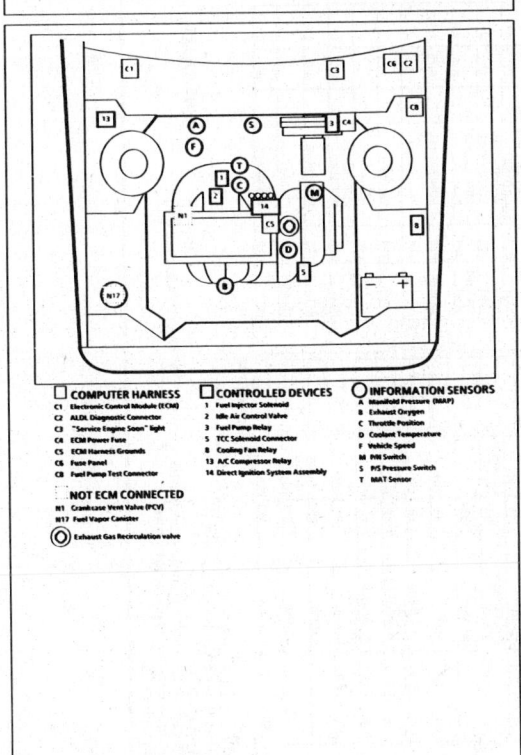

2.0L (VIN 1) ECM WIRING DIAGRAM—SKYHAWK

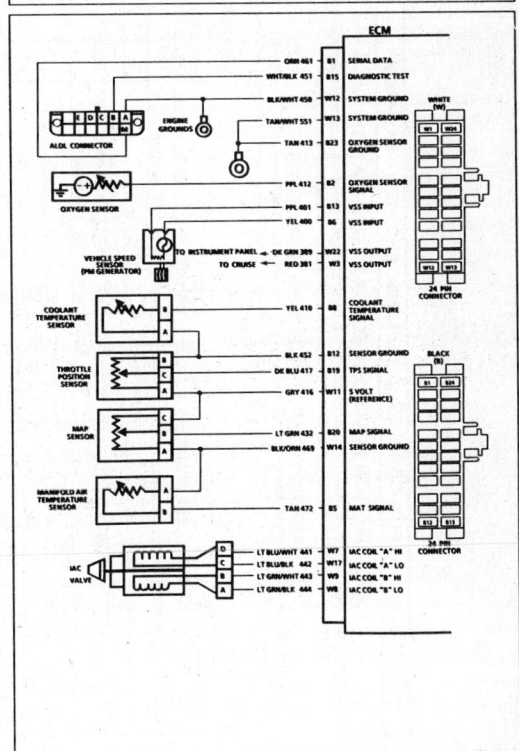

2.0L (VIN 1) ECM WIRING DIAGRAM SKYHAWK (CONT.)

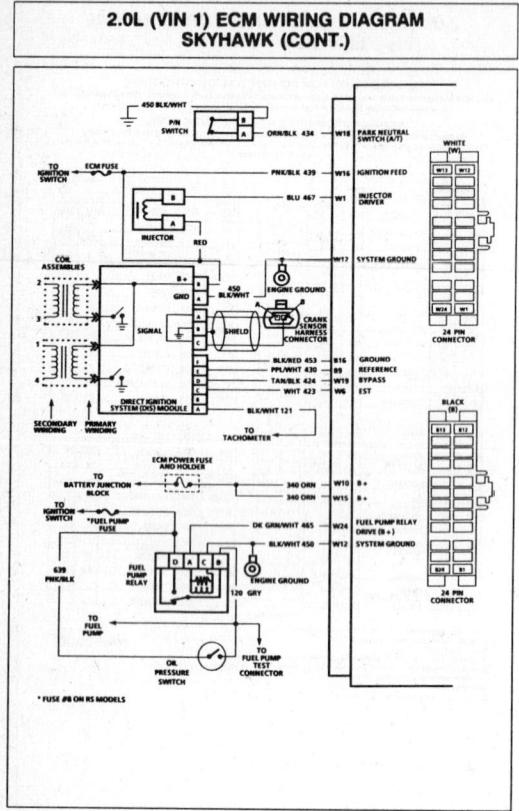

2.0L (VIN 1) ECM WIRING DIAGRAM SKYHAWK (CONT.)

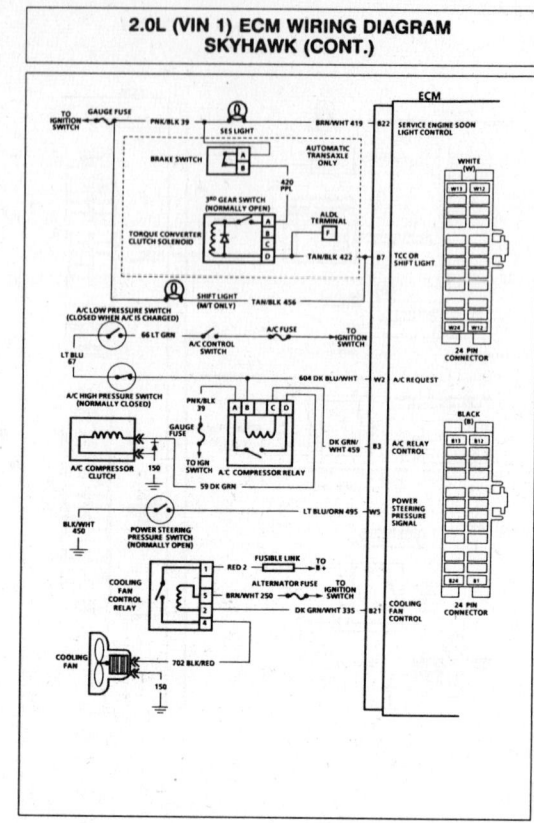

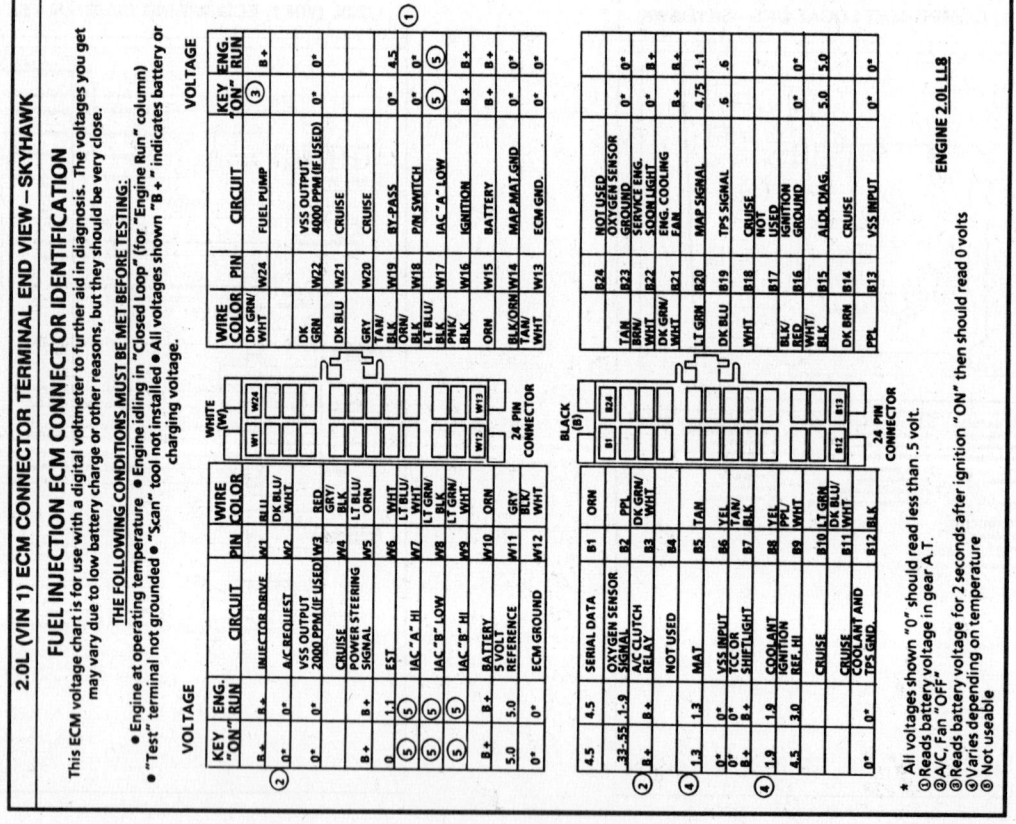

1988–90 2.0L (VIN 1)
ALL MODELS EXCEPT CORSICA AND BERETTA

DIAGNOSTIC CIRCUIT CHECK

The Diagnostic Circuit Check is an organized approach to identifying a problem created by an electronic engine control system malfunction. It must be the starting point for any driveability complaint diagnosis because it directs the service technician to the next logical step in diagnosing the complaint.

The "Scan" data listed in the table may be used for comparison after completing the diagnostic circuit check and finding the on-board diagnostics functioning properly with no trouble codes displayed. The "Typical Data Values" are an average of display values recorded from normally operating vehicles and are intended to represent what a normally functioning system would typically display.

A "SCAN" TOOL THAT DISPLAYS FAULTY DATA SHOULD NOT BE USED, AND THE PROBLEM SHOULD BE REPORTED TO THE MANUFACTURER. THE USE OF A FAULTY "SCAN" TOOL CAN RESULT IN MISDIAGNOSIS AND UNNECESSARY PARTS REPLACEMENT.

Only the parameters listed below are used in this manual for diagnosis. If a "Scan" tool reads other parameters, the values are not recommended by General Motors for use in diagnosis. For more description on the values and use of the "Scan" tool to diagnosis ECM inputs, refer to the applicable component diagnosis section in Section "C". If all values are within the range illustrated, refer to symptoms in Section "B".

"SCAN" TOOL DATA

Test Under Following Conditions: Idle, Upper Radiator Hose Hot, Closed Throttle, Park or Neutral, "Closed Loop", All Accessories "OFF".

"SCAN" Position	Units Displayed	Typical Data Value
Desired RPM	RPM	ECM idle command (varies with temperature)
RPM	RPM	± 50 RPM from desired rpm in drive (AUTO)
		± 100 RPM from desired rpm in neutral (MANUAL)
Coolant Temperature	Degrees Celsius	85 - 105
MAT Temperature	Degrees Celsius	10 - 90 (varies with underhood temperature and sensor location)
MAP	Volts	1 - 2 (varies with manifold and barometric pressures)
BPW (base pulse width)	Milliseconds	.8 - 3.0
O2	Volts	.1 - 1 (varies continuously)
TPS	Volts	.4 - 1.25
Throttle Angle	0 - 100%	0
IAC	Counts (steps)	1 - 50
P/N Switch	P-N and R-D-L	Park/Neutral (P/N)
INT (Integrator)	Counts	110 - 145
BLM (Block Learn Memory)	Counts	118 - 138
Open/Closed Loop	Open/Closed	"Closed Loop" (may enter "Open Loop" with extended idle)
VSS	MPH	0
TCC	ON/OFF	"OFF"
Spark Advance	Degrees	Varies
Battery	Volts	13.5 - 14.5
Fan	ON/OFF	"OFF" (coolant temperature below 102°C)
P/S Switch	Normal/Hi Pressure	Normal
A/C Request	Yes/No	No
A/C Clutch	ON/OFF	"OFF"
Shift Light (M/T)	ON/OFF	"OFF"

1988–90 2.0L (VIN 1)
ALL MODELS EXCEPT CORSICA AND BERETTA

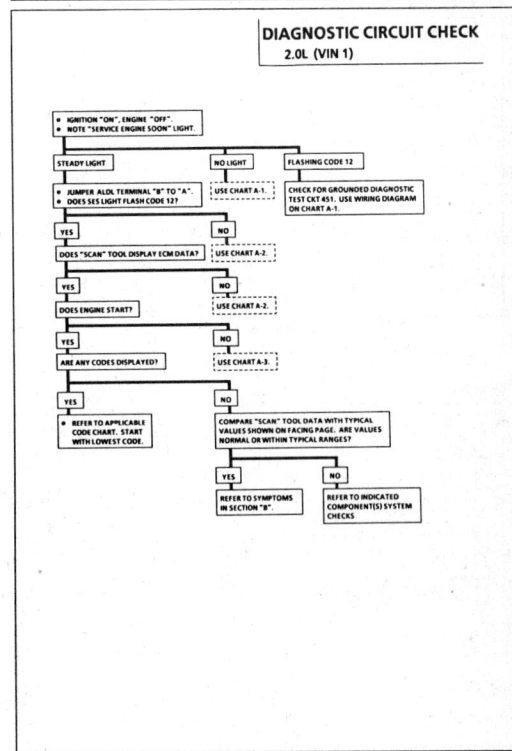

DIAGNOSTIC CIRCUIT CHECK
2.0L (VIN 1)

1988–90 2.0L (VIN 1)
ALL MODELS EXCEPT CORSICA AND BERETTA

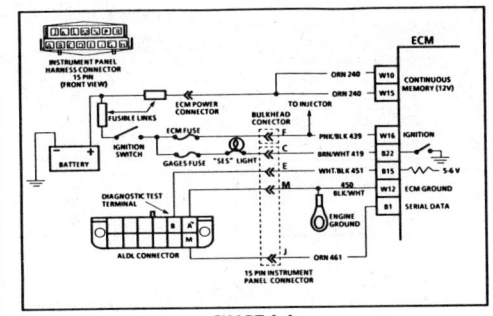

CHART A-1
NO "SERVICE ENGINE SOON" LIGHT
2.0L (VIN 1)

Circuit Description:
There should always be a steady "Service Engine Soon" light when the ignition is "ON" and engine stopped. Battery voltage is supplied directly to the light bulb. The electronic control module (ECM) will control the light and turn it "ON" by providing a ground path through CKT 419 to the ECM.

Test Description: Numbers below refer to circled numbers on the diagnostic chart.
1. This check determines if there is a fault in the "SES" light circuit or in the ECM.
2. If CKTs 240 and 439 have voltage, the ECM grounds or the ECM is faulty.
3. Using a test light connected to 12 volts, probe each of the system ground circuits to be sure a good ground is present. See ECM terminal end view in front of this section for the ECM terminal locations of ground circuits.

Diagnostic Aids:
If engine runs OK, check the following:
- Faulty light bulb.
- CKT 419 open.
- Gages fuse blown. This will result in no oil or generator lights, seat belt reminder, etc.
If engine cranks, but will not run, use CHART A-3.

1988–90 2.0L (VIN 1)
ALL MODELS EXCEPT CORSICA AND BERETTA

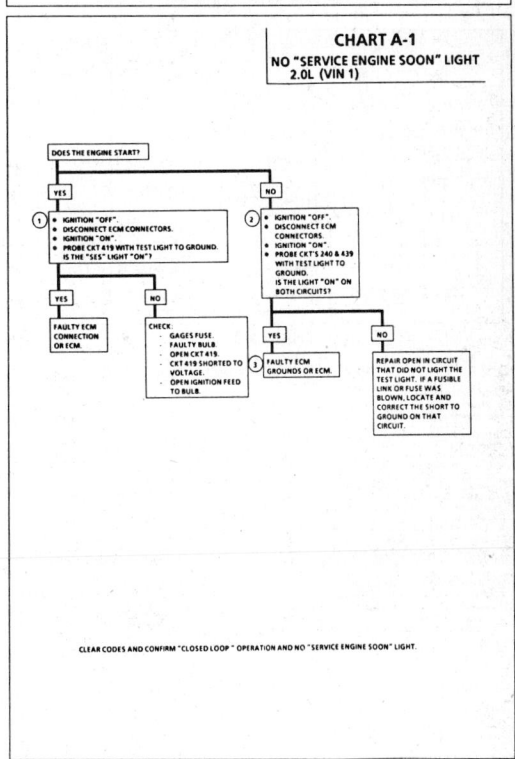

CHART A-1
NO "SERVICE ENGINE SOON" LIGHT
2.0L (VIN 1)

1988–90 2.0L (VIN 1)
ALL MODELS EXCEPT CORSICA AND BERETTA

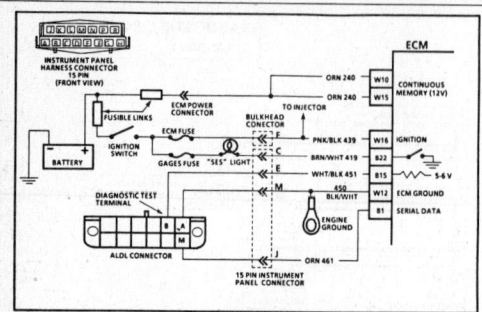

CHART A-2
"NO ALDL DATA OR WON'T FLASH CODE 12
"SERVICE ENGINE SOON" LIGHT "ON" STEADY
2.0L (VIN 1)

Circuit Description:
There should always be a steady "Service Engine Soon" light when the ignition is "ON" and the engine is "OFF". Battery voltage is supplied directly to the light bulb. The electronic control module (ECM) controls the light and turns it "ON" by providing a ground path through CKT 419 to the ECM.
With the diagnostic terminal grounded, the light should flash a Code 12, followed by any trouble code(s) stored in memory. A steady light suggests a short to ground in the light control CKT 419, or an open in diagnostic CKT 451.

Test Description: Numbers below refer to circled numbers on the diagnostic chart.
1. If there is a problem with the ECM that causes a "Scan" tool to not read data from the ECM, then the ECM should not flash a Code 12. If Code 12 does flash, be sure that the "Scan" tool is working properly on another vehicle. If the "Scan" tool is functioning properly and CKT 461 is OK, the PROM or ECM may be at fault for the "NO ALDL" symptom.
2. If the light turns "OFF" when the ECM connector is disconnected, then CKT 419 is not shorted to ground.

3. This step will check for an open diagnostic CKT 451.
4. At this point, the "Service Engine Soon" light wiring is OK. The problem is a faulty ECM or PROM. If Code 12 does not flash, the ECM should be replaced using the original PROM. Replace the PROM only after trying an ECM, as a defective PROM is an unlikely cause of the problem.

1988–90 2.0L (VIN 1)
ALL MODELS EXCEPT CORSICA AND BERETTA

CHART A-2
NO ALDL DATA OR WON'T FLASH CODE 12
"SERVICE ENGINE SOON" LIGHT "ON" STEADY
2.0L (VIN 1)

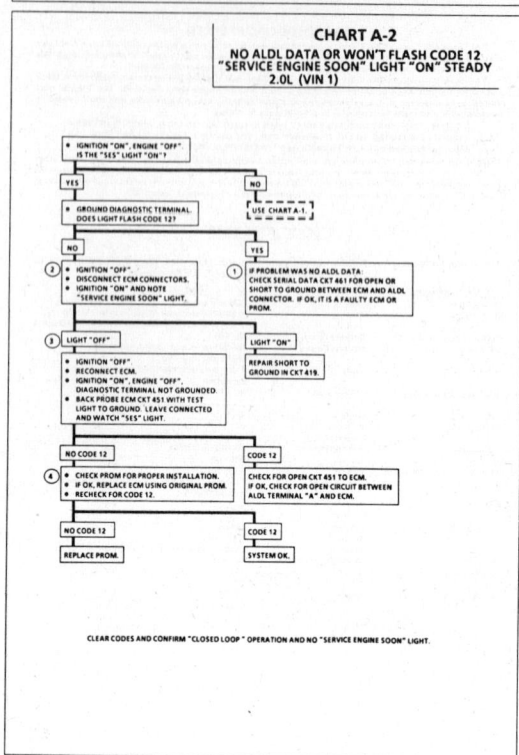

CLEAR CODES AND CONFIRM "CLOSED LOOP" OPERATION AND NO "SERVICE ENGINE SOON" LIGHT.

1988–90 2.0L (VIN 1)
ALL MODELS EXCEPT CORSICA AND BERETTA

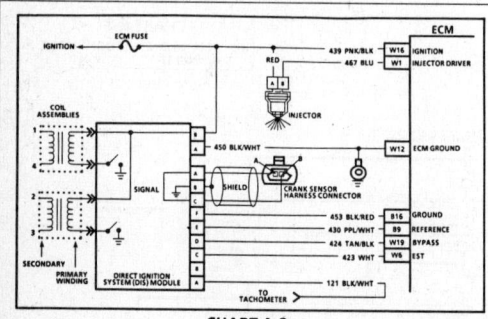

CHART A-3
(Page 1 of 3)
ENGINE CRANKS BUT WON'T RUN
2.0L (VIN 1)

Circuit Description:
Before using this chart, battery condition, engine cranking speed, and fuel quantity should be checked and verified as being OK.

Test Description: Numbers below refer to circled numbers on the diagnostic chart.
1. A "Service Engine Soon" light "ON" is a basic test to determine if there is a 12 volt supply and ignition 12 volts to ECM. No ALDL may be due to an ECM problem, and CHART A-2 will diagnose the ECM. If TPS voltage is over 2.5 volts, the engine may be in the clear flood mode, which will cause starting problems.
2. Because the direct ignition system uses two plugs and wires to complete the circuit of each coil, the opposite spark plug wire should be left connected.
3. While cranking engine, there should be no fuel spray with injector disconnected. Replace injector if it sprays fuel or drips.
4. Fuel spray from the injector indicates that fuel is available. However, the engine could be severely flooded due to too much fuel. No fuel spray from injector indicates a faulty fuel system or no ECM control of injector.

5. The test light should flash, indicating the ECM is controlling the injector. How bright the light flashes is not important. However, the test light should be a BT 8329 or equivalent.
6. This step checks for ECM control of the fuel pump relay circuit and checks the relay.

Diagnostic Aids:
- Water or foreign material can cause a no start during freezing weather. The engine may start after 5 or 6 minutes in a heated shop. The problem may not re-occur until an overnight park in freezing temperatures.
- An EGR sticking open can cause a low air/fuel ratio during cranking. Unless engine enters "Clear Flood" at the first indication of a flooding condition, it can result in a no start.
- Fuel Pressure: Low fuel pressure can result in a very lean air/fuel ratio. See CHART A-7.

1988–90 2.0L (VIN 1)
ALL MODELS EXCEPT CORSICA AND BERETTA

CHART A-3
(Page 1 of 3)
ENGINE CRANKS BUT WON'T RUN
2.0L (VIN 1)

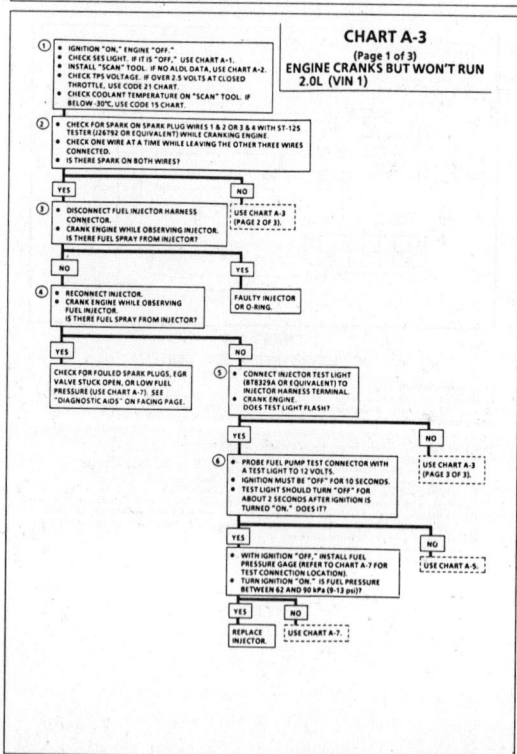

1988–90 2.0L (VIN 1)
ALL MODELS EXCEPT CORSICA AND BERETTA

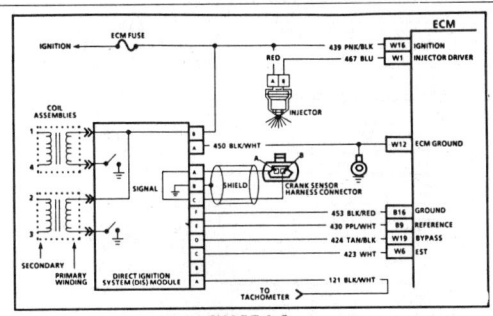

CHART A-3
(Page 2 of 3)
ENGINE CRANKS BUT WON'T RUN
2.0L (VIN 1)

Circuit Description:

A magnetic crank sensor is used to determine engine position, much the same way as the pick-up coil did in HEI type systems. The sensor is mounted in the block, near a slotted wheel on the crank shaft. The rotation of the wheel creates a flux change in the sensor, which produces a voltage signal. The DIS ignition module then processes this signal and creates the reference pulses needed by the ECM to trigger the correct coil at the correct time.

Test Description: Numbers below refer to circled numbers on the diagnostic chart.

1. will determine if the 12 volt supply and a good ground is available at the DIS ignition module.
2. This test will determine if the ignition module is not triggering the problem coil or if the tested coil is at fault. This test could also be performed by using another known good coil.
3. This test is for checking the crank sensor resistance.
4. This checks the output from the crank sensor due to the changing flux in the sensor. The voltmeter must be on the AC scale.

1988–90 2.0L (VIN 1)
ALL MODELS EXCEPT CORSICA AND BERETTA

CHART A-3
(Page 2 of 3)
ENGINE CRANKS BUT WON'T RUN
2.0L (VIN 1)

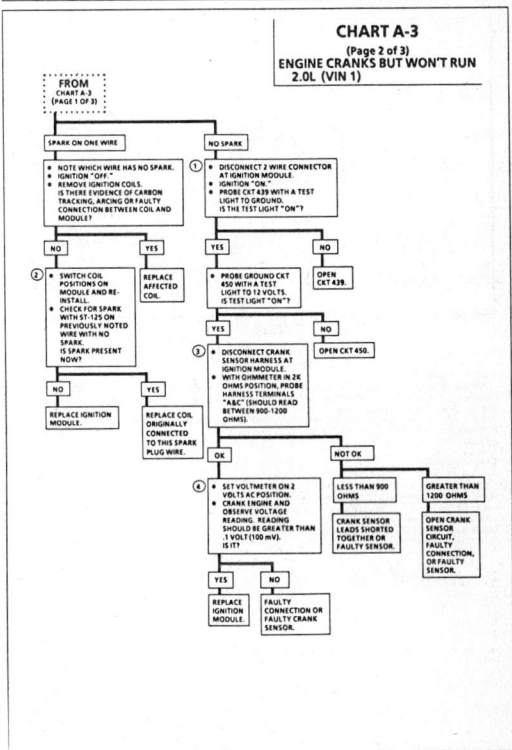

1988–90 2.0L (VIN 1)
ALL MODELS EXCEPT CORSICA AND BERETTA

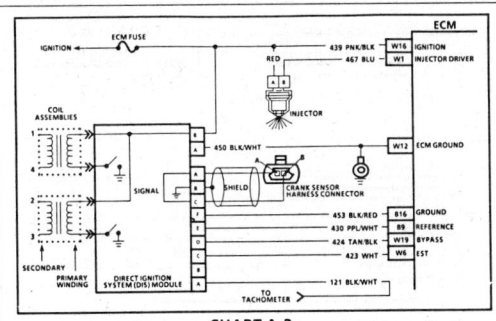

CHART A-3
(Page 3 of 3)
ENGINE CRANKS BUT WON'T RUN
2.0L (VIN 1)

Circuit Description:

Ignition voltage is supplied to the fuel injector on CKT 439. The injector will be pulsed (turned "ON" and "OFF"), when the ECM opens and grounds injector drive CKT 467.

Test Description: Numbers below refer to circled numbers on the diagnostic chart.

1. No light indicates no ECM control of the injector or a wiring problem.
2. A steady light indicates that injector driver CKT 467 is grounded.
3. This test is to check the injector control and voltage supply circuits.
4. This test checks for continuity through the injector control circuit wiring
5. This test will determine if the ignition module is not generating reference pulses or if the wiring and ECM are at fault. By touching and removing a test light to 12 volts on CKT 430, a reference pulse is created. If the injector test light flashes, the wiring and ECM are functioning correctly.

1988–90 2.0L (VIN 1)
ALL MODELS EXCEPT CORSICA AND BERETTA

CHART A-3
(Page 3 of 3)
ENGINE CRANKS BUT WON'T RUN
2.0L (VIN 1)

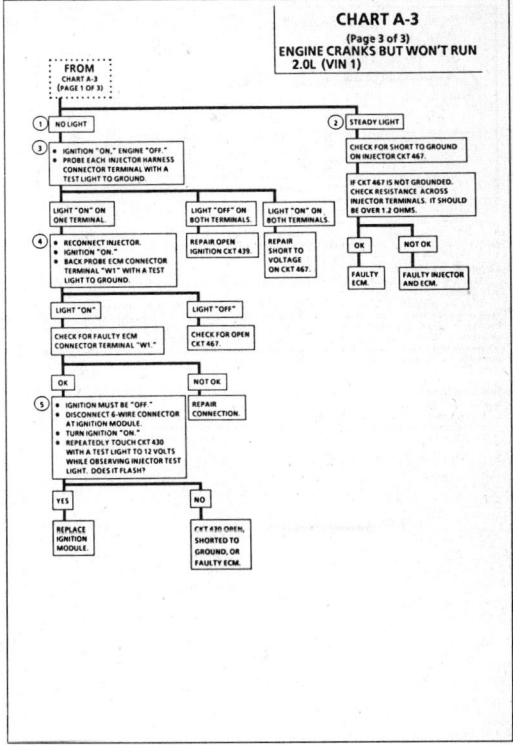

1988–90 2.0L (VIN 1)
ALL MODELS EXCEPT CORSICA AND BERETTA

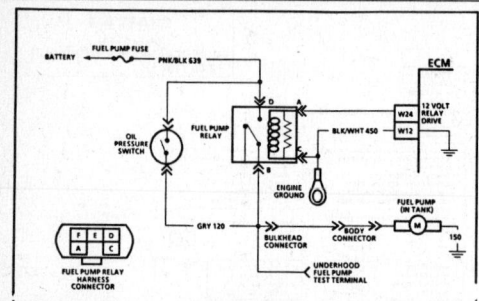

CHART A-5
FUEL PUMP RELAY CIRCUIT
2.0L (VIN 1)

Circuit Description:

When the ignition switch is turned "ON," the electronic control module (ECM) will activate the fuel pump relay and run the in-tank fuel pump. The fuel pump will operate as long as the engine is cranking or running and the ECM is receiving ignition reference pulses.

If there are no reference pulses, the ECM will shut "OFF" the fuel pump within 2 seconds after key "ON."

Should the fuel pump relay or the 12V relay drive from the ECM fail, the fuel pump will be run through an oil pressure switch back-up circuit.

Diagnostic Aids:

An inoperative fuel pump relay can result in long cranking times, particularly if the engine is cold or engine oil pressure is low. The extended crank period is caused by the time necessary for oil pressure to build enough to close the oil pressure switch and turn "ON" the fuel pump.

1988–90 2.0L (VIN 1)
ALL MODELS EXCEPT CORSICA AND BERETTA

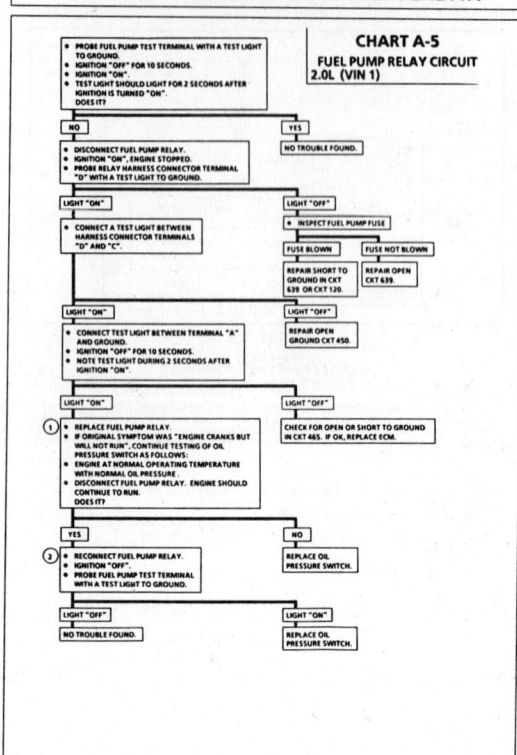

CHART A-5
FUEL PUMP RELAY CIRCUIT
2.0L (VIN 1)

1988–90 2.0L (VIN 1)
ALL MODELS EXCEPT CORSICA AND BERETTA

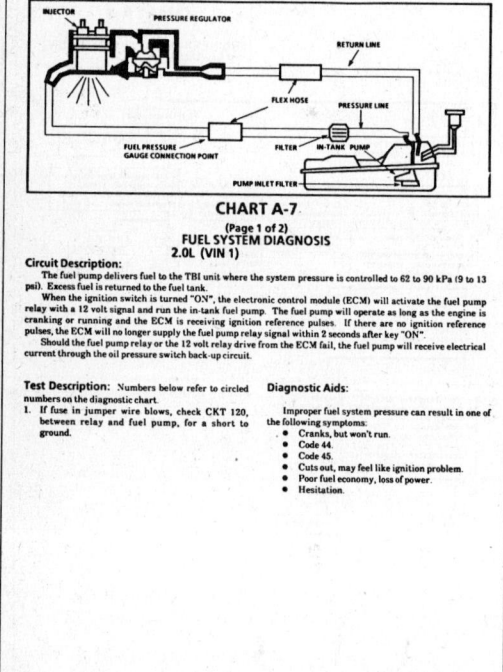

CHART A-7
(Page 1 of 2)
FUEL SYSTEM DIAGNOSIS
2.0L (VIN 1)

Circuit Description:

The fuel pump delivers fuel to the TBI unit where the system pressure is controlled to 62 to 90 kPa (9 to 13 psi). Excess fuel is returned to the fuel tank.

When the ignition switch is turned "ON," the electronic control module (ECM) will activate the fuel pump relay with a 12 volt signal and run the in-tank fuel pump. The fuel pump will operate as long as the engine is cranking or running and the ECM is receiving ignition reference pulses. If there are no ignition reference pulses, the ECM will no longer supply the fuel pump relay signal within 2 seconds after key "ON".

Should the fuel pump relay or the 12 volt relay drive from the ECM fail, the fuel pump will receive electrical current through the oil pressure switch back-up circuit.

Test Description: Numbers below refer to circled numbers on the diagnostic chart.

1. If fuse in jumper wire blows, check CKT 120, between relay and fuel pump, for a short to ground.

Diagnostic Aids:

Improper fuel system pressure can result in one of the following symptoms:
- Cranks, but won't run.
- Code 44.
- Code 45.
- Cuts out, may feel like ignition problem.
- Poor fuel economy, loss of power.
- Hesitation.

1988–90 2.0L (VIN 1)
ALL MODELS EXCEPT CORSICA AND BERETTA

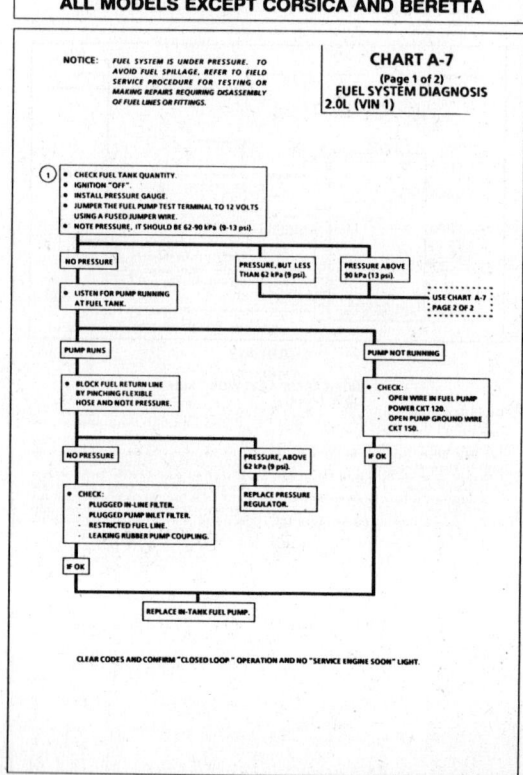

CHART A-7
(Page 1 of 2)
FUEL SYSTEM DIAGNOSIS
2.0L (VIN 1)

1988–90 2.0L (VIN 1)
ALL MODELS EXCEPT CORSICA AND BERETTA

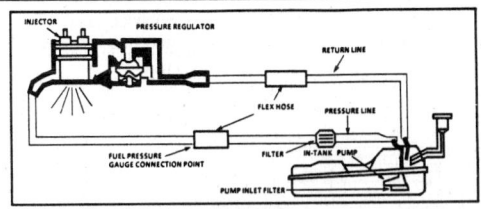

CHART A-7
(Page 2 of 2)
FUEL SYSTEM DIAGNOSIS
2.0L (VIN 1)

Test Description: Numbers below refer to circled numbers on the diagnostic chart.

1. Fuel flow at less than 62 kPa (9 psi) can cause the following:
   - Amount of fuel to injector is adequate, but pressure is too low. System will be running lean and may set Code 44. Also, hard starting and poor overall performance will result.
   - Engine surging and possible stalling during driving with satisfactory idle quality. This would be caused by a restriction in the fuel flow. As the fuel flow increases, the pressure drop across the restriction becomes great enough to starve the engine of fuel. At idle, the pressure may still be adequate. Normally, an engine with a fuel pressure less than 62 kPa (9 psi) at idle will not be driveable.

2. Restricting the fuel return line allows the fuel pump to develop its maximum pressure (dead head pressure). When battery voltage is applied to the pump test terminal, pressure should be from 90 to 124 kPa (13 to 18 psi).

3. This test determines if the high fuel pressure is due to a restricted fuel return line or a throttle body pressure regulator problem.

1988–90 2.0L (VIN 1)
ALL MODELS EXCEPT CORSICA AND BERETTA

CHART A-7
(Page 2 of 2)
FUEL SYSTEM DIAGNOSIS
2.0L (VIN 1)

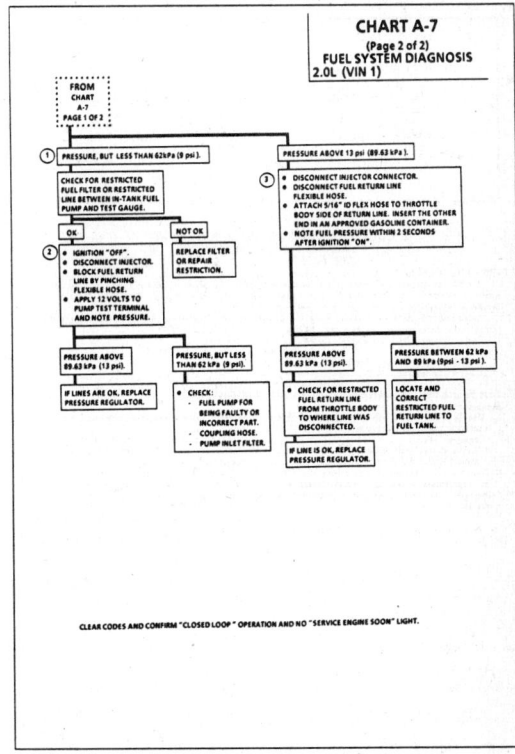

CLEAR CODES AND CONFIRM "CLOSED LOOP" OPERATION AND NO "SERVICE ENGINE SOON" LIGHT.

1988–90 2.0L (VIN 1)
ALL MODELS EXCEPT CORSICA AND BERETTA

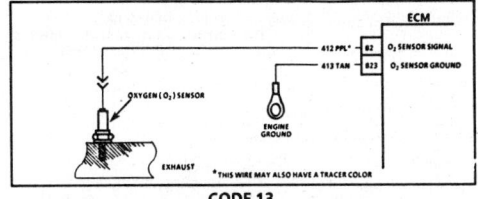

CODE 13
OXYGEN SENSOR CIRCUIT
(OPEN CIRCUIT)
2.0L (VIN 1)

Circuit Description:
The ECM supplies a voltage of about .45 volt between terminals "B2" and "B23". (If measured with a 10 megohm digital voltmeter, this may read as low as .32 volt).

When the O_2 sensor reaches operating temperature, it varies this voltage from about .1 volt (exhaust is lean) to about .9 volt (exhaust is rich).

The sensor is like an open circuit and produces no voltage when it is below 360°C (600°F). An open sensor circuit, or cold sensor, causes "Open Loop" operation.

Test Description: Numbers below refer to circled numbers on the diagnostic chart.

1. Code 13 will be set under the following conditions:
   - Engine at normal operating temperature.
   - At least 1 minute has elapsed since engine start-up.
   - O_2 signal voltage is steady between .35 and .55 volt.
   - Throttle angle is above 7%.
   - All above conditions are met for about 60 seconds.
   If the conditions for a Code 13 exist, the system will not operate in "Closed Loop".

2. This test determines if the O_2 sensor is the problem or if the ECM and wiring are at fault.

3. In doing this test, use only a 10 megohm digital voltmeter. This test checks the continuity of CKTs 412 and 413. If CKT 413 is open, the ECM voltage on CKT 412 will be over 6 volt (600 mV).

Diagnostic Aids:
Normal "Scan" tool O_2 sensor voltage varies between 100mV to 999 mV (.1 and 1.0 volt) while in "Closed Loop". Code 13 sets in one minute if sensor signal voltage remains between .35 and .55 volt, but the system will go to "Open Loop" in about 15 seconds.

Verify a clean, tight ground connection for CKT 413. Open CKT(s) 412 or 413 will result in a Code 13. If Code 13 is intermittent, refer to Section "B".

1988–90 2.0L (VIN 1)
ALL MODELS EXCEPT CORSICA AND BERETTA

CODE 13
OXYGEN SENSOR CIRCUIT
(OPEN CIRCUIT)
2.0L (VIN 1)

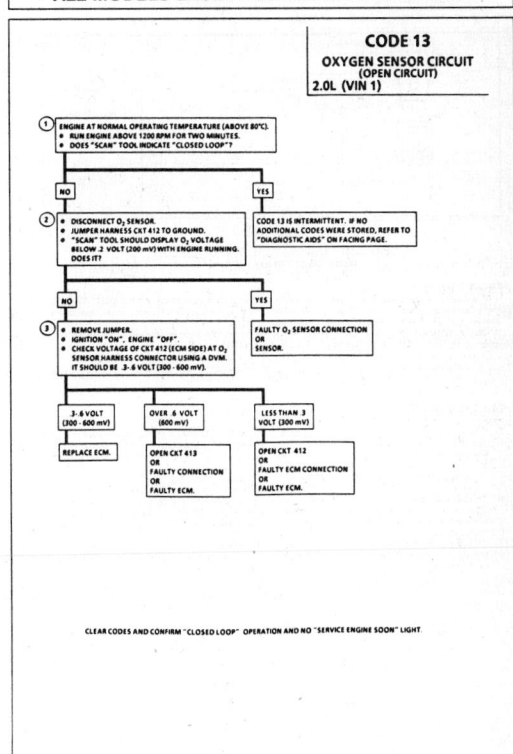

CLEAR CODES AND CONFIRM "CLOSED LOOP" OPERATION AND NO "SERVICE ENGINE SOON" LIGHT.

1988–90 2.0L (VIN 1)
ALL MODELS EXCEPT CORSICA AND BERETTA

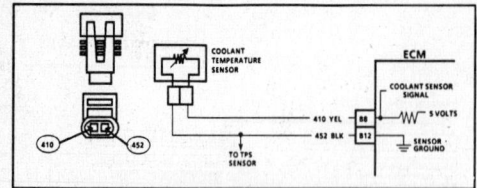

CODE 14
COOLANT TEMPERATURE SENSOR CIRCUIT
(HIGH TEMPERATURE INDICATED)
2.0L (VIN 1)

Circuit Description:

The coolant temperature sensor uses a thermistor to control the signal voltage to the ECM. The ECM applies a reference voltage on CKT 410 to the sensor. When the engine is cold, the sensor (thermistor) resistance is high. The ECM will then sense a high signal voltage.

As the engine warms up, the sensor resistance decreases and the voltage drops. At normal engine operating temperature, the voltage will measure about 1.5 to 2.0 volts at ECM terminal "B8".

Coolant temperature is one of the inputs used to control the following:
- Fuel Delivery.
- Electronic Spark Timing (EST).
- Cooling Fan.
- Torque Converter Clutch (TCC).
- Idle Air Control (IAC).

Test Description: Numbers below refer to circled numbers on the diagnostic chart.
1. Checks to see if code was set as result of hard failure or intermittent condition.
   Code 14 will set if:
   - Signal voltage indicates a coolant temperature above 135°C (275°F) for 3 seconds.
2. This test simulates conditions for a Code 15. If the ECM recognizes the open circuit (high voltage) and displays a low temperature, the ECM and wiring are OK.

Diagnostic Aids:

A "Scan" tool reads engine temperature in degrees Celsius.

After the engine is started, the temperature should rise steadily to about 90°C, then stabilize when the thermostat opens.

If the engine has been allowed to cool to an ambient temperature (overnight), coolant temperature and manifold air temperature (MAT) may be checked with a "Scan" tool and should read close to each other.

When a Code 14 is set, the ECM will turn "ON" the engine cooling fan.

A Code 14 will result if CKT 410 is shorted to ground.

If Code 14 is intermittent, refer to Section "B".

1988–90 2.0L (VIN 1)
ALL MODELS EXCEPT CORSICA AND BERETTA

CODE 14
COOLANT TEMPERATURE SENSOR CIRCUIT
(HIGH TEMPERATURE INDICATED)
2.0L (VIN 1)

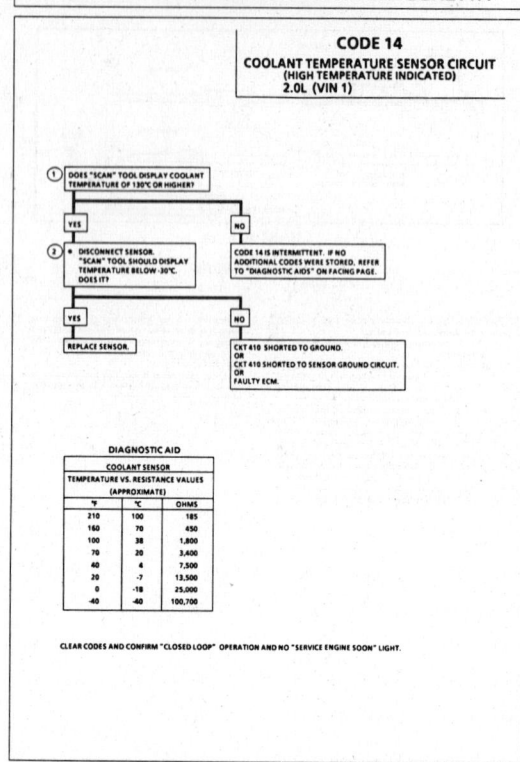

DIAGNOSTIC AID

COOLANT SENSOR		
TEMPERATURE VS. RESISTANCE VALUES		
(APPROXIMATE)		
°F	°C	OHMS
210	100	185
160	70	450
100	38	1,800
70	20	3,400
40	4	7,500
20	-7	13,500
0	-18	25,000
-40	-40	100,700

CLEAR CODES AND CONFIRM "CLOSED LOOP" OPERATION AND NO "SERVICE ENGINE SOON" LIGHT.

1988–90 2.0L (VIN 1)
ALL MODELS EXCEPT CORSICA AND BERETTA

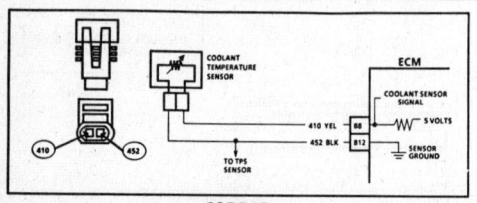

CODE 15
COOLANT TEMPERATURE SENSOR CIRCUIT
(LOW TEMPERATURE INDICATED)
2.0L (VIN 1)

Circuit Description:

The coolant temperature sensor uses a thermistor to control the signal voltage to the ECM. The ECM applies a reference voltage on CKT 410 to the sensor. When the engine is cold, the sensor (thermistor) resistance is high. The ECM will then sense a high signal voltage.

As the engine warms up, the sensor resistance decreases and the voltage drops. At normal engine operating temperature, the voltage will measure about 1.5 to 2.0 volts at ECM terminal "B8".

Coolant temperature is one of the inputs used to control the following:
- Fuel Delivery.
- Electronic Spark Timing (EST).
- Cooling Fan.
- Torque Converter Clutch (TCC).
- Idle Air Control (IAC).

Test Description: Numbers below refer to circled numbers on the diagnostic chart.
1. Checks to see if code was set as result of hard failure or intermittent condition.
   Code 15 will set if:
   - Signal voltage indicates a coolant temperature below -37°C (-35°F) for 60 seconds.
2. This test simulates conditions for a Code 14. If the ECM recognizes the grounded circuit (low voltage) and displays a high temperature, the ECM and wiring are OK.
3. This test will determine if there is a wiring problem or a faulty ECM. If CKT 452 is open, there may also be a Code 21 stored.

Diagnostic Aids:

A "Scan" tool reads engine temperature in degrees Celsius. After the engine is started, the temperature should rise steadily to about 90°C, then stabilize when the thermostat opens.

If the engine has been allowed to cool to an ambient temperature (overnight), coolant temperature and manifold air temperature (MAT) may be checked with a "Scan" tool and should read close to each other.

When a Code 15 is set, the ECM will turn "ON" the engine cooling fan.

A Code 15 will result if CKTs 410 or 452 are open.

If Code 15 is intermittent, refer to Section "B".

1988–90 2.0L (VIN 1)
ALL MODELS EXCEPT CORSICA AND BERETTA

CODE 15
COOLANT TEMPERATURE SENSOR CIRCUIT
(LOW TEMPERATURE INDICATED)
2.0L (VIN 1)

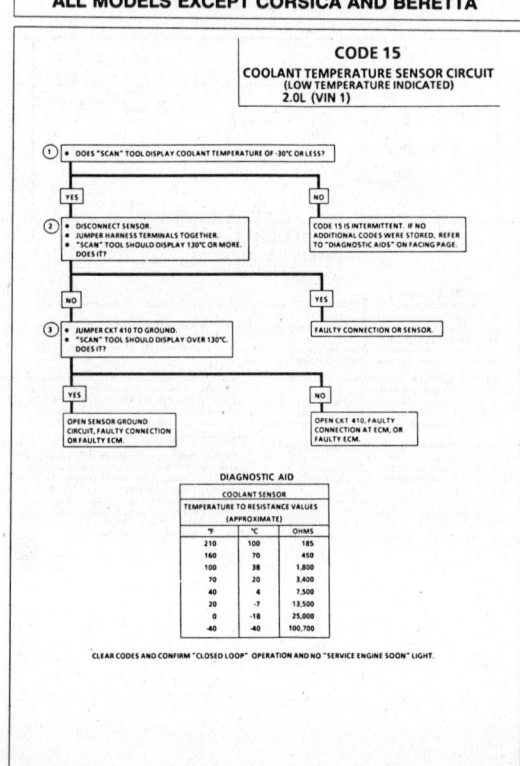

DIAGNOSTIC AID

COOLANT SENSOR		
TEMPERATURE TO RESISTANCE VALUES		
(APPROXIMATE)		
°F	°C	OHMS
210	100	185
160	70	450
100	38	1,800
70	20	3,400
40	4	7,500
20	-7	13,500
0	-18	25,000
-40	-40	100,700

CLEAR CODES AND CONFIRM "CLOSED LOOP" OPERATION AND NO "SERVICE ENGINE SOON" LIGHT.

1988–90 2.0L (VIN 1)
ALL MODELS EXCEPT CORSICA AND BERETTA

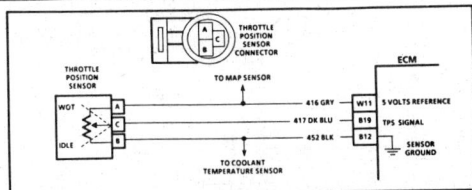

CODE 21
THROTTLE POSITION SENSOR (TPS) CIRCUIT
(SIGNAL VOLTAGE HIGH)
2.0L (VIN 1)

Circuit Description:
The throttle position sensor (TPS) provides a voltage signal that changes relative to the throttle valve. Signal voltage will vary from less than 1.25 volts at idle to about 5 volts at wide open throttle (WOT).
The TPS signal is one of the most important inputs used by the ECM for fuel control and for many of the ECM controlled outputs.

Test Description: Numbers below refer to circled numbers on the diagnostic chart.
1. This step checks to see if Code 21 is the result of a hard failure or an intermittent condition.
   A Code 21 will set under the following conditions:
   - TPS reading above 2.5 volts.
   - Engine speed less than 1800 rpm.
   - MAP reading below 55 kPa.
   - All of the above conditions present for at least 2 seconds.
2. This step simulates conditions for a Code 22. If the ECM recognizes the change of state, the ECM and CKTs 416 and 417 are OK.
3. This step isolates a faulty sensor, ECM, or an open CKT 452. If CKT 452 is open, there may also be a Code 15 stored.

Diagnostic Aids:
A "Scan" tool displays throttle position in volts. Closed throttle voltage should be less than 1.25 volts. TPS voltage should increase at a steady rate as throttle is moved to WOT.
A Code 21 will result if CKT 452 is open or CKT 417 is shorted to voltage. If Code 21 is intermittent, refer to Section "B".

1988–90 2.0L (VIN 1)
ALL MODELS EXCEPT CORSICA AND BERETTA

CODE 21
THROTTLE POSITION SENSOR (TPS) CIRCUIT
(SIGNAL VOLTAGE HIGH)
2.0L (VIN 1)

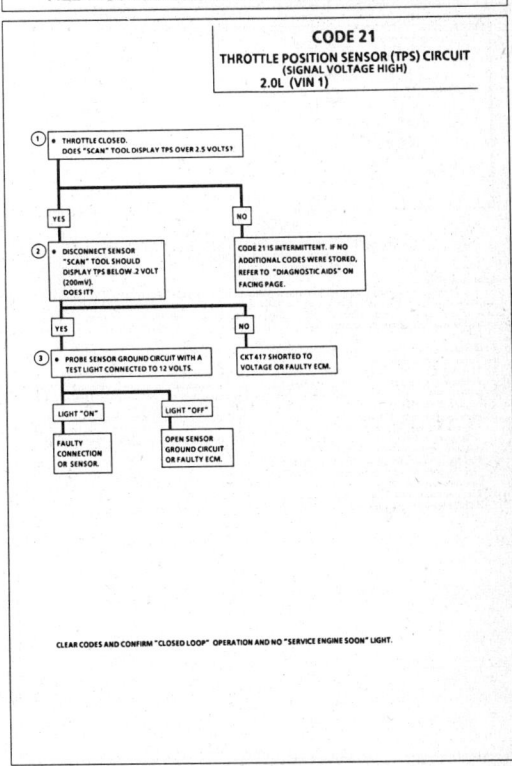

CLEAR CODES AND CONFIRM "CLOSED LOOP" OPERATION AND NO "SERVICE ENGINE SOON" LIGHT.

1988–90 2.0L (VIN 1)
ALL MODELS EXCEPT CORSICA AND BERETTA

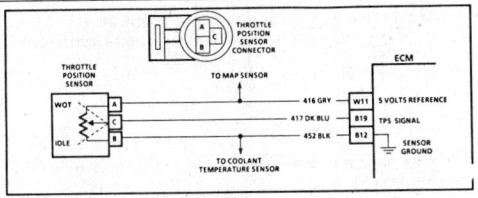

CODE 22
THROTTLE POSITION SENSOR (TPS) CIRCUIT
(SIGNAL VOLTAGE LOW)
2.0L (VIN 1)

Circuit Description:
The throttle position sensor (TPS) provides a voltage signal that changes relative to the throttle valve. Signal voltage will vary from less than 1.25 volts at idle to about 4.5 volts at wide open throttle (WOT).
The TPS signal is one of the most important inputs used by the ECM for fuel control and for many of the ECM controlled outputs.

Test Description: Numbers below refer to circled numbers on the diagnostic chart.
1. This step checks to see if Code 22 is the result of a hard failure or an intermittent condition. A Code 22 will set when the engine is running and TPS voltage is below .2 volts (200 mV).
2. This step simulates conditions for a Code 21. If a Code 21 is set, or the "Scan" tool displays over 4 volts, the ECM and wiring are OK.
3. The "Scan" tool may not display 12 volts. The important thing is that the ECM recognizes the voltage as over 4 volts, indicating that CKT 417 and the ECM are OK.
4. If CKT 416 is shorted to ground, there may also be a stored Code 34.

Diagnostic Aids:
A "Scan" tool displays throttle position in volts. Closed throttle voltage should be less than 1.25 volts. TPS voltage should increase at a steady rate as throttle is moved to WOT
An open or grounded CKT 416 or 417 will result in a Code 22.
If Code 22 is intermittent, refer to Section "B".

1988–90 2.0L (VIN 1)
ALL MODELS EXCEPT CORSICA AND BERETTA

CODE 22
THROTTLE POSITION SENSOR (TPS) CIRCUIT
(SIGNAL VOLTAGE LOW)
2.0L (VIN 1)

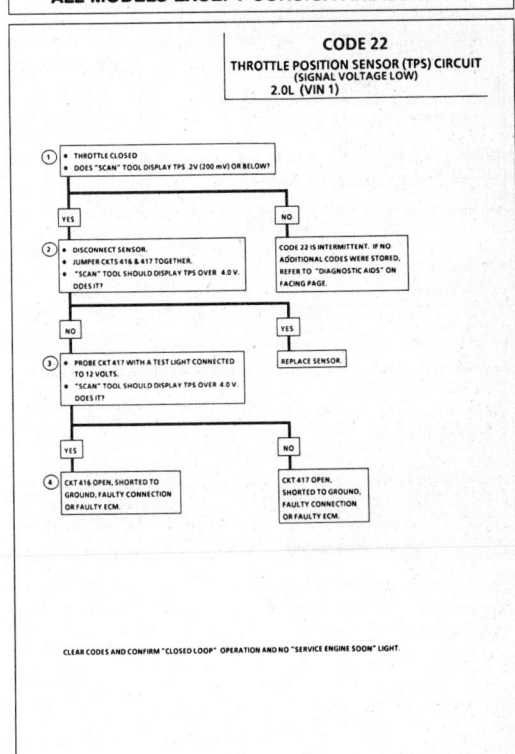

CLEAR CODES AND CONFIRM "CLOSED LOOP" OPERATION AND NO "SERVICE ENGINE SOON" LIGHT.

1988–90 2.0L (VIN 1)
ALL MODELS EXCEPT CORSICA AND BERETTA

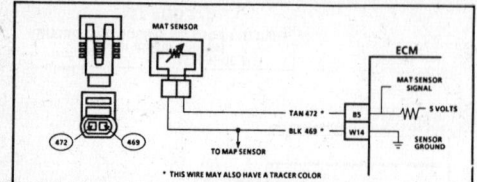

CODE 23
MANIFOLD AIR TEMPERATURE (MAT) SENSOR CIRCUIT
(LOW TEMPERATURE INDICATED)
2.0L (VIN 1)

Circuit Description:
The manifold air temperature sensor uses a thermistor to control the signal voltage to the ECM. The ECM applies a reference voltage (4-6 volts) on CKT 472 to the sensor. When manifold air is cold, the sensor (thermistor) resistance is high. The ECM will then sense a high signal voltage. As the air warms, the sensor resistance becomes less and the voltage drops.

Test Description: Numbers below refer to circled numbers on the diagnostic chart.
1. This step determines if Code 23 is the result of a hard failure or an intermittent condition. Code 23 will set when signal voltage indicates a MAT reading of less than -35°C (-31°F) with the engine running for longer than two minutes. Code 32 may also set with the engine not running when the MAT reading is less than -35°C (-31°F) while the coolant temperature is greater than -26°C (-15°F).
2. This test simulates conditions for a Code 25. If the "Scan" tool displays a high temperature, the ECM and wiring are OK.
3. This step checks continuity of CKTs 472 and 469. If CKT 469 is open there may also be a Code 33.

Diagnostic Aids:
If the engine has been allowed to cool to an ambient temperature (overnight), coolant and MAT temperatures may be checked with a "Scan" tool and should read close to each other.
A Code 23 will result if CKT 472 or 469 becomes open.
If Code 23 is intermittent, refer to Section "B".

1988–90 2.0L (VIN 1)
ALL MODELS EXCEPT CORSICA AND BERETTA

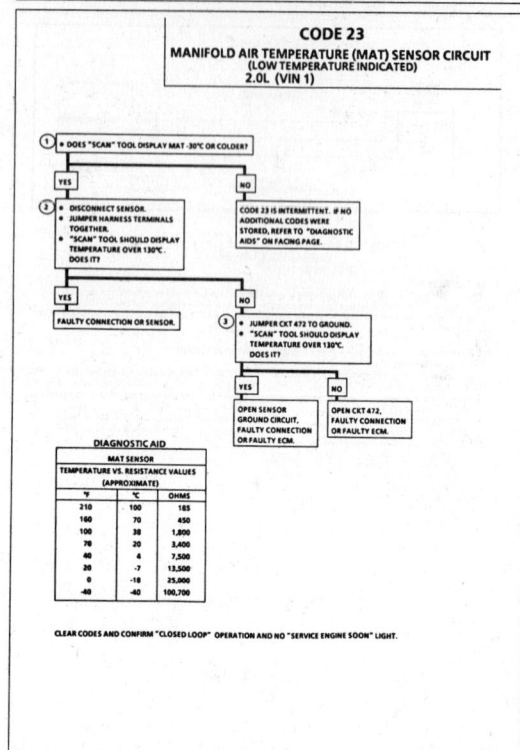

CODE 23
MANIFOLD AIR TEMPERATURE (MAT) SENSOR CIRCUIT
(LOW TEMPERATURE INDICATED)
2.0L (VIN 1)

DIAGNOSTIC AID

MAT SENSOR
TEMPERATURE VS. RESISTANCE VALUES
(APPROXIMATE)

°F	°C	OHMS
210	100	185
160	70	450
100	38	1,800
70	20	3,400
40	4	7,500
20	-7	13,500
0	-18	25,000
-40	-40	100,700

CLEAR CODES AND CONFIRM "CLOSED LOOP" OPERATION AND NO "SERVICE ENGINE SOON" LIGHT.

1988–90 2.0L (VIN 1)
ALL MODELS EXCEPT CORSICA AND BERETTA

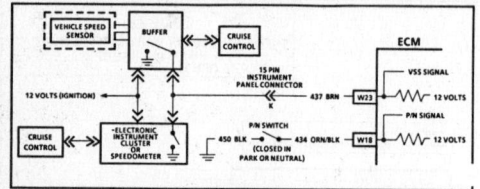

CODE 24
VEHICLE SPEED SENSOR (VSS) CIRCUIT
2.0L (VIN 1)

Circuit Description:
The ECM applies and monitors 12 volts on CKT 437. CKT 437 connects to the vehicle speed sensor which alternately grounds CKT 437 when drive wheels are turning. This pulsing action takes place about 2000 times per mile and the ECM will calculate vehicle speed based on the time between "pulses".
"Scan" tool reading should closely match the speedometer reading with the drive wheels turning.
Disregard a Code 24 set when drive wheels are not turning.

Test Description: Numbers below refer to circled numbers on the diagnostic chart.
1. Code 24 will set if vehicle speed equals 0 mph when:
   • Engine speed is between 1100 and 4400 rpm.
   • Throttle position is less than 2%.
   • Manifold vacuum is high.
   • Transaxle is not in park or neutral.
   • Above conditions met for 5 seconds.
   These conditions are met during a road load deceleration. Disregard Code 24 that sets when drive wheels are not turning.
2. 8-12 volts at the instrument panel connector indicates that CKT 437 is open between the connector and the vehicle speed sensor, or that the sensor is faulty. A voltage of less than 1 volt at the instrument panel connector indicates that CKT 437 is shorted to ground.

If the voltage reads above 10 volts after disconnecting CKT 437 at the vehicle speed sensor, the vehicle speed sensor is faulty. If voltage remains less than 8 volts, then CKT 437 is grounded. If 437 is not grounded, there is a faulty connection at the ECM, or the ECM is faulty.

Diagnostic Aids:
"Scan" tool should indicate a vehicle speed whenever the drive wheels are turning greater than 3 mph.
A faulty or maladjusted park/neutral switch can result in a false Code 24. Use a "Scan" tool and check for the proper signal while in drive. Refer to CHART C-1A for P/N switch check.
If Code 24 is intermittent, refer to Section "B".

1988–90 2.0L (VIN 1)
ALL MODELS EXCEPT CORSICA AND BERETTA

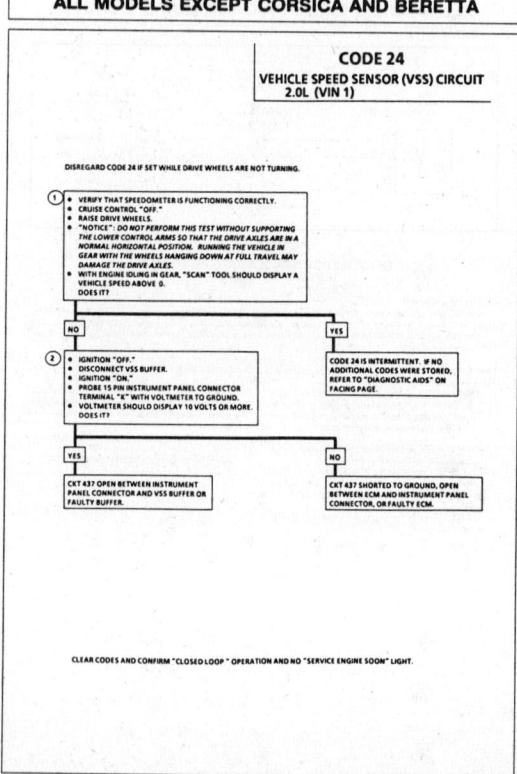

CODE 24
VEHICLE SPEED SENSOR (VSS) CIRCUIT
2.0L (VIN 1)

DISREGARD CODE 24 IF SET WHILE DRIVE WHEELS ARE NOT TURNING.

CLEAR CODES AND CONFIRM "CLOSED LOOP" OPERATION AND NO "SERVICE ENGINE SOON" LIGHT.

1988–90 2.0L (VIN 1)
ALL MODELS EXCEPT CORSICA AND BERETTA

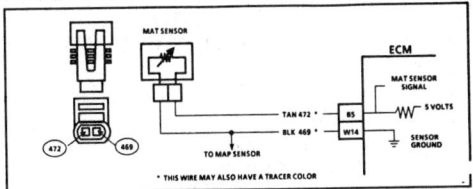

CODE 25
MANIFOLD AIR TEMPERATURE (MAT) SENSOR CIRCUIT
(HIGH TEMPERATURE INDICATED)
2.0L (VIN 1)

Circuit Description:

The manifold air temperature sensor uses a thermistor to control the signal voltage to the ECM. The ECM applies a reference voltage (4-6 volts) on CKT 472 to the sensor. When manifold air is cold, the sensor (thermistor) resistance is high. Therefore, the ECM will see a high signal voltage. As the air warms, the sensor resistance becomes less and the voltage drops.

Test Description: Numbers below refer to circled numbers on the diagnostic chart.

1. This check determines if the Code 25 is the result of a hard failure or an intermittent condition. A Code 25 will set if a MAT temperature greater than 135°C is detected for a time longer than 2 seconds and the ignition is "ON".

Diagnostic Aids:

If the engine has been allowed to cool to an ambient temperature (overnight), coolant and MAT temperatures may be checked with a "Scan" tool and should read close to each other.

A Code 25 will result if CKT 472 is shorted to ground.

If Code 25 is intermittent, refer to Section "B".

1988–90 2.0L (VIN 1)
ALL MODELS EXCEPT CORSICA AND BERETTA

CODE 25
MANIFOLD AIR TEMPERATURE (MAT) SENSOR CIRCUIT
(HIGH TEMPERATURE INDICATED)
2.0L (VIN 1)

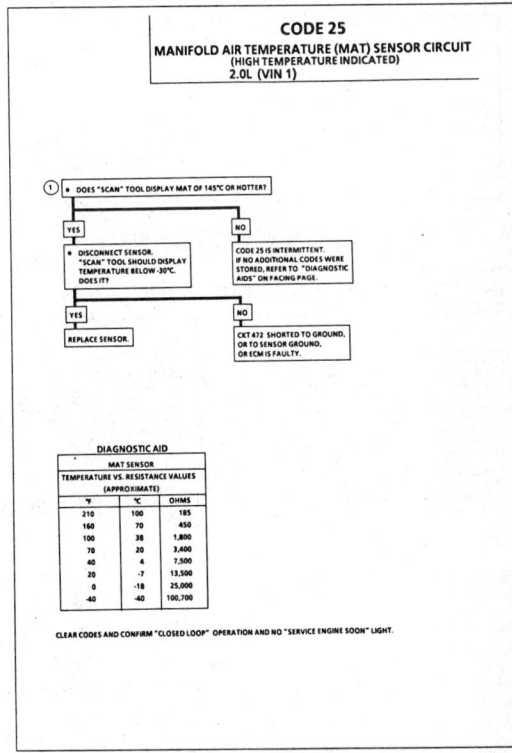

MAT SENSOR		
TEMPERATURE VS. RESISTANCE VALUES (APPROXIMATE)		
°F	°C	OHMS
210	100	185
160	70	450
100	38	1,800
70	20	3,400
40	4	7,500
20	-7	13,500
0	-18	25,000
-40	-40	100,700

CLEAR CODES AND CONFIRM "CLOSED LOOP" OPERATION AND NO "SERVICE ENGINE SOON" LIGHT.

1988–90 2.0L (VIN 1)
ALL MODELS EXCEPT CORSICA AND BERETTA

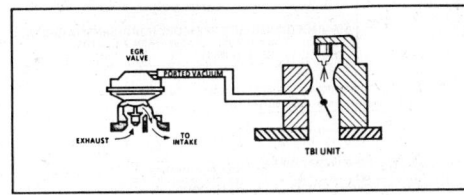

CODE 32
EXHAUST GAS RECIRCULATION (EGR) SYSTEM FAILURE
2.0L (VIN 1)

Code Description:

A properly operating EGR system will directly affect the air/fuel mixture requirements of the engine. Since the exhaust gas introduced into the air/fuel mixture cannot be used in combustion (contains very little oxygen), less fuel is required to maintain a correct air/fuel ratio. If the EGR system were to fail in a closed position, the exhaust gas would be replaced with air, and the air/fuel mixture would be leaner. The ECM would compensate for the lean condition by adding fuel, resulting in higher block learn values.

The fuel control on this engine is conducted within 16 block learn cells. Since EGR is not used at idle, the closed throttle cell would not be affected by EGR system operation. The other block learn cells are affected by EGR operation, and, when the EGR system is operating properly, the block learn values in all cells should be close to the same. If the EGR system becomes inoperative, the block learn values in the open throttle cells would change to compensate for the resulting lean mixtures, but the block learn value in the closed throttle cell would not change.

The difference in block learn values between the idle (closed throttle) cell and cell 10 is used to monitor EGR system performance. When the difference between the two block learn values is greater than 12 and the block learn value in cell 10 is greater than 140, Code 32 is set. The system operates in block learn cell 10 during a cruise condition at approximately 55 mph.

Diagnostic Aids:

The Code 32 chart is a functional check of the EGR system. If the EGR system works properly but a Code 32 has been set, check other items that could result in high block learn values in block learn cell 10, but not in the closed throttle cell.

Check for restricted or blocked EGR passages. Perform a MAP output check. Follow the procedure in CHART C-10.

1988–90 2.0L (VIN 1)
ALL MODELS EXCEPT CORSICA AND BERETTA

CODE 32
EXHAUST GAS RECIRCULATION (EGR) SYSTEM FAILURE
2.0L (VIN 1)

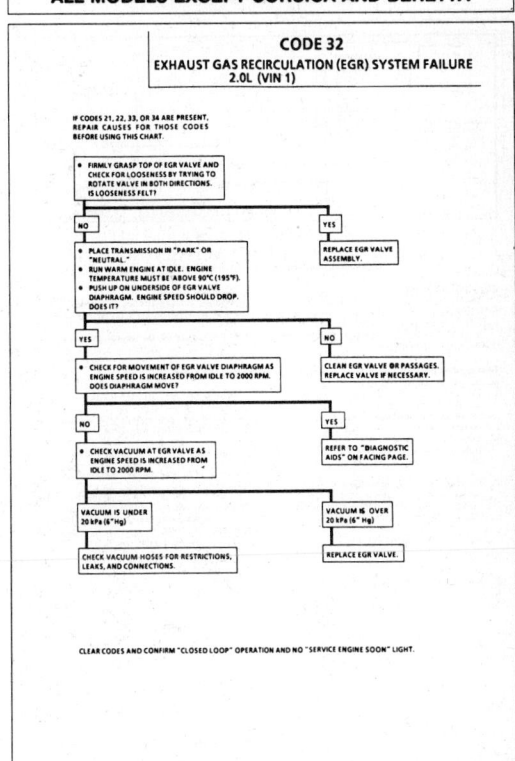

CLEAR CODES AND CONFIRM "CLOSED LOOP" OPERATION AND NO "SERVICE ENGINE SOON" LIGHT.

1988–90 2.0L (VIN 1)
ALL MODELS EXCEPT CORSICA AND BERETTA

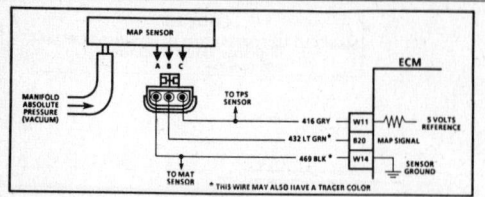

CODE 33
MANIFOLD ABSOLUTE PRESSURE (MAP) SENSOR CIRCUIT
(SIGNAL VOLTAGE HIGH - LOW VACUUM)
2.0L (VIN 1)

Circuit Description:

The manifold absolute pressure sensor (MAP) responds to changes in manifold pressure (vacuum). The ECM receives this information as a signal voltage that will vary from about 1 - 1.5 volt at closed throttle (idle), to 4 - 4.5 volts at wide open throttle (low vacuum).

If the MAP sensor fails, the ECM will substitute a fixed MAP value and use the throttle position sensor (TPS) to control fuel delivery.

Test Description: Numbers below refer to circled numbers on the diagnostic chart.
1. This step will determine if Code 33 is the result of a hard failure or an intermittent condition.
   A Code 33 will set under the following conditions:
   - The engine is running.
   - MAP signal indicates greater than 80 kPa (low vacuum).
   - Throttle position is less than 12%
   - These conditions exist for a time longer than 5 seconds
2. This step simulates conditions for a Code 34. If the ECM recognizes the change, the ECM and CKTs 416 and 432 are OK. If CKT 469 is open, there may also be a stored Code 23.

Diagnostic Aids:

With the ignition "ON" and the engine stopped, the manifold pressure is equal to atmospheric pressure and the signal voltage will be high. This information is used by the ECM as an indication of vehicle altitude and is referred to as BARO. Comparison of this BARO reading with a known good vehicle with the same sensor is a good way to check accuracy of a "suspect" sensor. Readings should be within .4 volt.

A Code 33 will result if CKT 469 is open or if CKT 432 is shorted to voltage or to CKT 416.

If Code 33 is intermittent, refer to Section "B".

1988–90 2.0L (VIN 1)
ALL MODELS EXCEPT CORSICA AND BERETTA

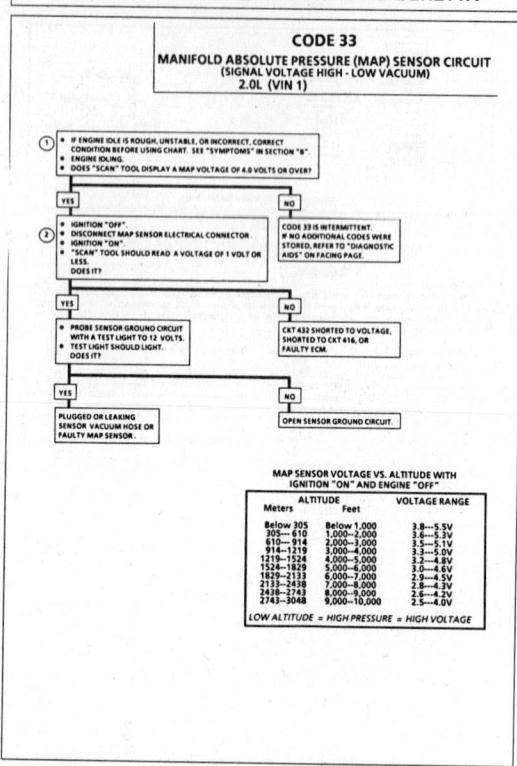

1988–90 2.0L (VIN 1)
ALL MODELS EXCEPT CORSICA AND BERETTA

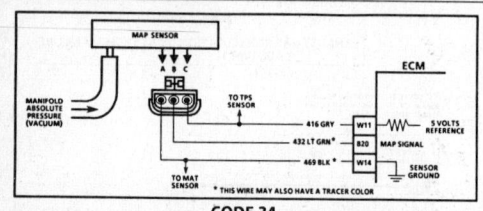

CODE 34
MANIFOLD ABSOLUTE PRESSURE (MAP) SENSOR CIRCUIT
(SIGNAL VOLTAGE LOW - HIGH VACUUM)
2.0L (VIN 1)

Circuit Description:

The manifold absolute pressure sensor (MAP) responds to changes in manifold pressure (vacuum). The ECM receives this information as a signal voltage that will vary from about 1 - 1.5 volt at closed throttle (idle) to 4 - 4.5 volts at wide open throttle (low vacuum).

If the MAP sensor fails, the ECM will substitute a fixed MAP value and use the throttle position sensor (TPS) to control fuel delivery.

Test Description: Numbers below refer to circled numbers on the diagnostic chart.
1. This step determines if Code 34 is the result of a hard failure or an intermittent condition. A Code 34 will set when MAP signal voltage is too low and the ignition is "ON".
2. Jumpering harness terminals "B" to "C" (5 volts to signal circuit) will determine if the sensor is at fault, or if there is a problem with the ECM or wiring.
3. The "Scan" tool may not display 12 volts. The important thing is that the ECM recognizes the voltage as more than 4 volts, indicating that the ECM and CKT 432 are OK.

Diagnostic Aids:

With the ignition "ON" and the engine stopped, the manifold pressure is equal to atmospheric pressure and the signal voltage will be high. This information is used by the ECM as an indication of vehicle altitude and is referred to as BARO. Comparison of this BARO reading with a known good vehicle with the same sensor is a good way to check accuracy of a "suspect" sensor. Readings should be within .4 volt.

A Code 34 will result if CKTs 416 or 432 are open or shorted to ground.

If CKT 416 is open or shorted to ground, there may also be a stored Code 22.

If Code 34 is intermittent, refer to Section "B".

1988–90 2.0L (VIN 1)
ALL MODELS EXCEPT CORSICA AND BERETTA

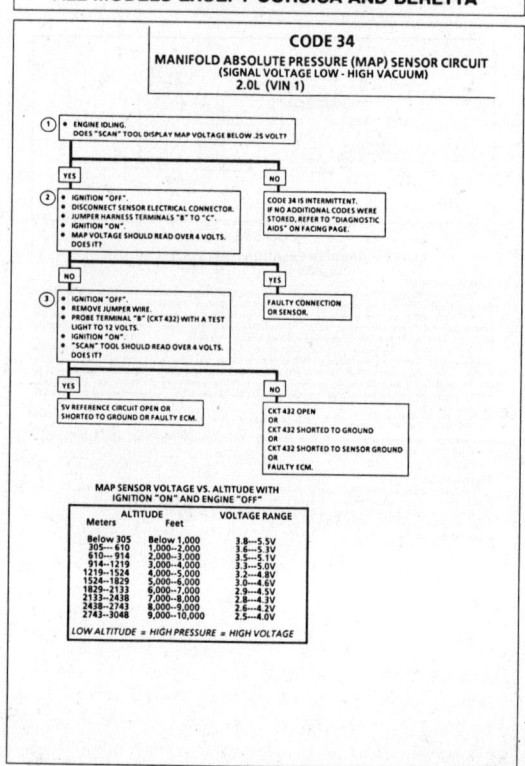

1988–90 2.0L (VIN 1)
ALL MODELS EXCEPT CORSICA AND BERETTA

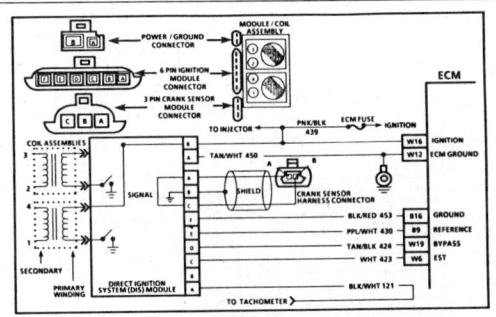

CODE 42
ELECTRONIC SPARK TIMING (EST) CIRCUIT
2.0L (VIN 1)

Circuit Description:
The DIS module sends a reference signal to the ECM when the engine is cranking. While the engine speed is under 400 rpm, the DIS module controls the ignition timing. When the system is running on the ignition module (no voltage on the bypass line), the ignition module grounds the EST signal. The ECM expects to sense no voltage on the EST line during this condition. If it senses a voltage, it sets Code 42 and will not enter the EST mode.

When the engine speed exceeds 400 rpm, the ECM applies 5 volts to the bypass line to switch the timing to ECM control (EST). If the bypass line is open, once the rpm for EST control is reached, the ignition module will not switch to EST mode. This results in low EST voltage and the setting of Code 42. If the EST line is grounded, the ignition module will switch to EST, but because the line is grounded, there will be no EST signal. A Code 42 will be set.

Test Description: Numbers below refer to circled numbers on the diagnostic chart.
1. Code 42 means the ECM has sensed an open or short to ground in the EST or bypass circuits. This test confirms Code 42 and that the fault causing the code is present.
2. Checks for a normal EST ground path through the ignition module. An EST CKT 423, shorted to ground, will also read less than 500 ohms, but this will be checked later.
3. As the test light voltage contacts CKT 424, the module should switch, causing the ohmmeter to "overrange" if the meter is in the 1000-2000 ohms position. Selecting the 10 - 20,000 ohms position will indicate a reading above 5000 ohms. The important thing is that the module "switched".

4. The module did not switch and this step checks for:
   • EST CKT 423 shorted to ground.
   • Bypass CKT 424 open.
   • Faulty ignition module connection or module.
5. Confirms that Code 42 is a faulty ECM and not an intermittent in CKT(s) 423 or 424.

Diagnostic Aids:
The "Scan" tool does not have any ability to help diagnose a Code 42 problem.

An open or ground in the EST circuit will result in the engine continuing to run, but in a back-up ignition timing mode (module timing) at a calculated timing value and the "Service Engine Soon" light. If the EST fault is still present the next time the engine is restarted, a Code 42 will be set and the engine will operate in module timing.

If Code 42 is intermittent, refer to Section "B".

1988–90 2.0L (VIN 1)
ALL MODELS EXCEPT CORSICA AND BERETTA

CODE 42
ELECTRONIC SPARK TIMING (EST) CIRCUIT
2.0L (VIN 1)

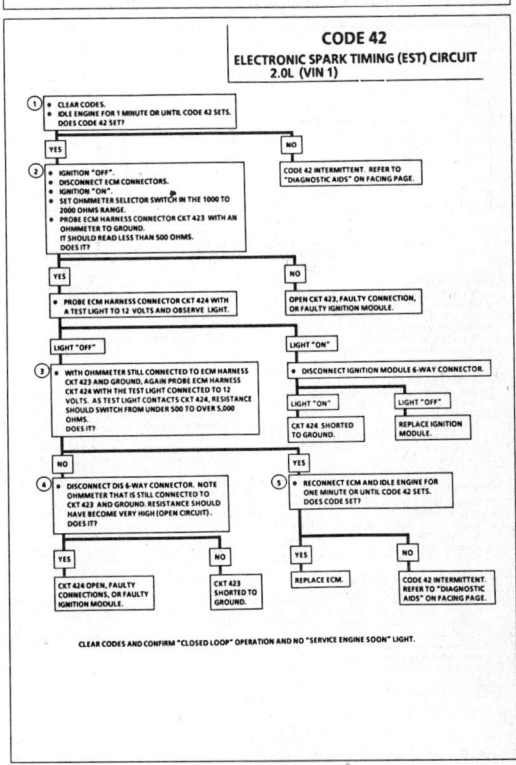

CLEAR CODES AND CONFIRM "CLOSED LOOP" OPERATION AND NO "SERVICE ENGINE SOON" LIGHT.

1988–90 2.0L (VIN 1)
ALL MODELS EXCEPT CORSICA AND BERETTA

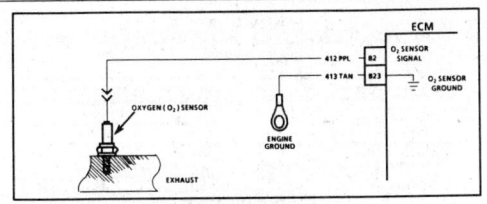

CODE 44
OXYGEN SENSOR CIRCUIT
(LEAN EXHAUST INDICATED)
2.0L (VIN 1)

Circuit Description:
The ECM supplies a voltage of about .45 volt between terminals "B2" and "B23". (If measured with a 10 megohm digital voltmeter, this may read as low as .32 volt.)

When the O₂ sensor reaches operating temperature, it varies this voltage from about .1 volt (exhaust is lean) to about .9 volt (exhaust is rich).

The sensor is like an open circuit and produces no voltage when it is below 360°C (600°F). An open sensor circuit, or cold sensor, causes "Open Loop" operation.

Test Description: Numbers below refer to circled numbers on the diagnostic chart.
1. Code 44 is set when the O₂ sensor signal voltage on CKT 412 remains below .3 volt for 50 seconds or more and the system is operating in "Closed Loop."

Diagnostic Aids:
Use the "Scan" tool to observe the block learn values at different engine speeds and loads. When the operating conditions for Code 44 exist, the block learn value will be around 150 or higher.

Also check the following:
• O₂ Sensor Wire. Sensor pigtail may be mispositioned and contacting the exhaust manifold. Check for ground in wire between connector and sensor.

• Fuel Contamination. Water, even in small amounts, near the in-tank fuel pump inlet can be delivered to the injector. The water causes a lean exhaust and can set a Code 44.
• Fuel Pressure. System will be lean if pressure is too low. It may be necessary to monitor fuel pressure while driving the car at various road speeds and/or loads to confirm. See "Fuel System" diagnosis CHART A-7.
• Exhaust Leaks. If there is an exhaust leak, the engine can cause outside air to be pulled into the exhaust and past the sensor. Vacuum or crankcase leaks can also cause a lean condition.
• If Code 44 intermittent, refer to Section "B".

1988–90 2.0L (VIN 1)
ALL MODELS EXCEPT CORSICA AND BERETTA

CODE 44
OXYGEN SENSOR CIRCUIT
(LEAN EXHAUST INDICATED)
2.0L (VIN 1)

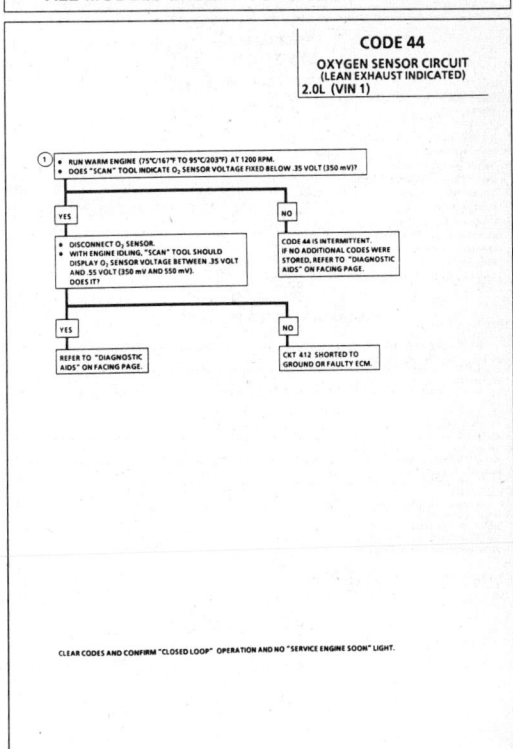

CLEAR CODES AND CONFIRM "CLOSED LOOP" OPERATION AND NO "SERVICE ENGINE SOON" LIGHT.

1988–90 2.0L (VIN 1)
ALL MODELS EXCEPT CORSICA AND BERETTA

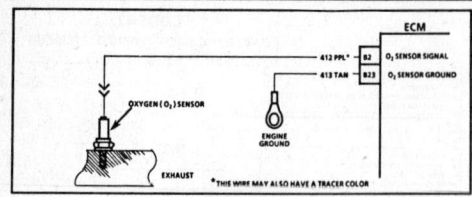

ECM
412 PPL — B2 — O₂ SENSOR SIGNAL
413 TAN — B23 — O₂ SENSOR GROUND

OXYGEN (O₂) SENSOR
ENGINE GROUND
EXHAUST
*THIS WIRE MAY ALSO HAVE A TRACER COLOR

CODE 45
OXYGEN SENSOR CIRCUIT
(RICH EXHAUST INDICATED)
2.0L (VIN 1)

Circuit Description:

The ECM supplies a voltage of about .45 volt between terminals "B2" and "B23". (If measured with a 10 megohm digital voltmeter, this may read as low as .32 volt).

When the O₂ sensor reaches operating temperature, it varies this voltage from about .1 volt (exhaust is lean) to about .9 volt (exhaust is rich).

The sensor is like an open circuit and produces no voltage when it is below 360° C (600°F). An open sensor circuit, or cold sensor, causes "Open Loop" operation.

Test Description: Numbers below refer to circled numbers on the diagnostic chart.

1. Code 45 is set when the O₂ sensor signal voltage on CKT 412 remains above .75 volt under the following conditions:
   - 30 seconds or more.
   - System is operating in "Closed Loop".
   - Engine run time after start is 30 seconds or more.
   - Throttle angle is not between 2% and 20%.

Diagnostic Aids:

Code 45, or rich exhaust, is most likely caused by one of the following:

- **Fuel Pressure.** System will go rich, if pressure is too high. The ECM can compensate for some increase. However, if it gets too high, a Code 45 will be set. See "Fuel System Diagnosis", CHART A-7
- **Leaking Injector.** See CHART A-7.
- An **open ground CKT 453** may result in induced electrical "noise". The ECM interprets this "noise" as reference pulses. The additional pulses result in a higher than actual engine speed signal. The ECM then delivers too much fuel causing the system to go rich. The engine tachometer will also show higher than actual engine speed, which can help in diagnosing this problem.

- **Canister Purge.** Check for fuel saturation. If full of fuel, check canister control and hoses.
- **MAP Sensor.** An output that causes the ECM to sense a higher than normal manifold pressure (low vacuum) can cause the system to go rich. Disconnecting the MAP sensor will allow the ECM to set a fixed value for the MAP sensor. Substitute a different MAP sensor if the rich condition is gone, while the sensor is disconnected.
- **TPS.** An intermittent TPS output will cause the system to operate richly due to a false indication of the engine accelerating.
- **O₂ Sensor Contamination.** Inspect oxygen sensor for silicone contamination from fuel, or use of improper RTV sealant. The sensor may have a white, powdery coating and result in a high but false signal voltage (rich exhaust indication). The ECM will then reduce the amount of fuel delivered to the engine causing a severe surge driveability problem.
- **EGR Valve.** EGR sticking open at idle is usually accompanied by a rough idle and/or stall condition.

If Code 45 is intermittent, refer to Section "B".

1988–90 2.0L (VIN 1)
ALL MODELS EXCEPT CORSICA AND BERETTA

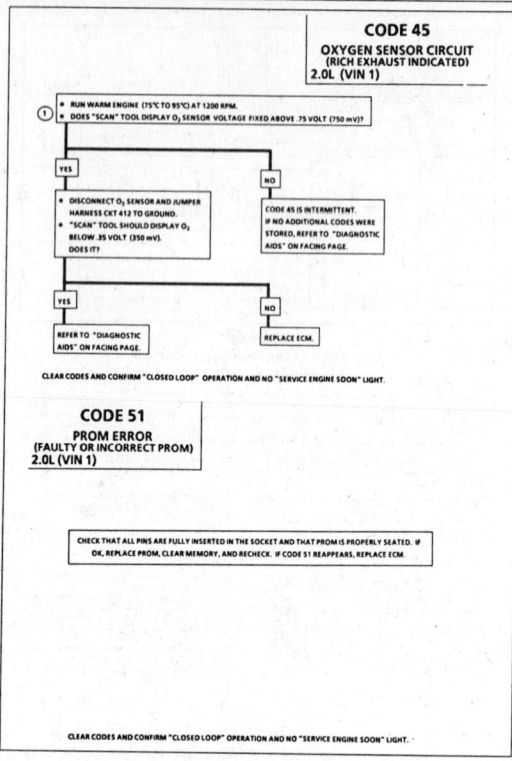

CODE 45
OXYGEN SENSOR CIRCUIT
(RICH EXHAUST INDICATED)
2.0L (VIN 1)

①
- RUN WARM ENGINE (75°C TO 95°C) AT 1200 RPM.
- DOES "SCAN" TOOL DISPLAY O₂ SENSOR VOLTAGE FIXED ABOVE .75 VOLT (750 mV)?

YES → NO

YES
- DISCONNECT O₂ SENSOR AND JUMPER HARNESS CKT 412 TO GROUND.
- "SCAN" TOOL SHOULD DISPLAY O₂ BELOW .35 VOLT (350 mV). DOES IT?

NO
CODE 45 IS INTERMITTENT. IF NO ADDITIONAL CODES WERE STORED, REFER TO "DIAGNOSTIC AIDS" ON FACING PAGE.

YES → REFER TO "DIAGNOSTIC AIDS" ON FACING PAGE.

NO → REPLACE ECM.

CLEAR CODES AND CONFIRM "CLOSED LOOP" OPERATION AND NO "SERVICE ENGINE SOON" LIGHT.

CODE 51
PROM ERROR
(FAULTY OR INCORRECT PROM)
2.0L (VIN 1)

CHECK THAT ALL PINS ARE FULLY INSERTED IN THE SOCKET AND THAT PROM IS PROPERLY SEATED. IF OK, REPLACE PROM, CLEAR MEMORY, AND RECHECK. IF CODE 51 REAPPEARS, REPLACE ECM.

CLEAR CODES AND CONFIRM "CLOSED LOOP" OPERATION AND NO "SERVICE ENGINE SOON" LIGHT.

1988–90 2.0L (VIN 1)
ALL MODELS EXCEPT CORSICA AND BERETTA

SECTION B
SYMPTOMS
TABLE OF CONTENTS

Performing Symptom Diagnosis
Intermittents
Hard Start
Rough, Unstable, or Incorrect Idle, Stalling
Poor Gas Mileage
Detonation/Spark Knock
Lack of Power, Sluggish, or Spongy
Surges and/or Chuggle
Cuts Out, Misses
Hesitation, Sag, Stumble
Excessive Exhaust Emissions or Odors
Dieseling, Run-On
Backfire
Restricted Exhaust System Check - CHART B-1

PERFORMING SYMPTOM DIAGNOSIS

The DIAGNOSTIC CIRCUIT CHECK should be performed before using this section. The purpose of this section is to locate the source of a driveability or emissions problem when other diagnostic procedures cannot be used. This may be because of difficulties in locating a suspected sub-system or component.

Many driveability related problems can be eliminated by following the procedures found in Service Bulletins. These bulletins provide the manual. Be sure to check all bulletins related to the complaint or suspected system.

If the engine cranks but will not run, use CHART A-3.

The sequence of the checks listed in this section is not intended to be followed as on a step-by-step procedure. The checks are listed such that the less difficult and time consuming operations are performed before more difficult ones.

Most of the symptom procedures call for a careful visual and physical check. *The importance of this step cannot be stressed too strongly. It can lead to correcting a problem without further checks, and can save valuable time.* This procedure includes checking the following:
- Vacuum hoses for splits, kinks, and proper connections, as shown on the underhood Emission Control Information label.
- Throttle body and intake manifold for leaks
- Ignition wires for cracking, hardness, proper routing, and carbon tracking
- Wiring for proper connections, pinches, and cuts

1988–90 2.0L (VIN 1)
ALL MODELS EXCEPT CORSICA AND BERETTA

INTERMITTENTS

Definition: Problem may or may not activate the "Service Engine Soon" light or store a trouble code.

DO NOT use the trouble code charts in Section "A" for intermittent problems. The fault must be present to locate the problem. If a fault is intermittent, the use of trouble code charts may result in the replacement of good parts.
- Most intermittent problems are caused by faulty electrical connections or wiring. Perform careful checks of suspected circuits for
  - Poor mating of the connector halves and terminals not fully seated in the connector body (backed out)
  - Improperly formed or damaged terminals. All connector terminals in problem circuit should be carefully reformed to increase contact tension
  - Poor terminal to wire connection. This requires removing the terminal from the connector body to check
- If a visual and physical check does not locate the cause of the problem, the car can be driven with a voltmeter connected to a suspected circuit or a "Scan" tool may be used. An abnormal voltage reading while the problem occurs indicates that the problem may be in that circuit.

- Check for loss of trouble code memory. To check, disconnect the TPS and allow the engine to idle until the "Service Engine Soon" light turns "ON." Code 22 should be stored and kept in memory when the ignition is turned "OFF" for at least 10 seconds. If not, the ECM is faulty.
- An intermittent SES light and no trouble codes may be caused by
  - Electrical system interference caused by a defective relay, ECM driven solenoid, or switch. They can cause a sharp electrical surge. Normally, the problem will occur when the faulty component is operated.
  - Improper installation of electrical options, such as lights, 2-way radios, etc.
  - EST wires which should be routed away from spark plug wires, ignition system components, and generator. Ground wire from ECM to ignition system which may be faulty.
  - Ignition secondary wire shorted to ground.
  - "Service Engine Soon" light and diagnostic test terminal circuits intermittently shorted to ground.
  - Faulty ECM grounds.

HARD START

Definition: Engine cranks well but does not start for a long time. Engine does eventually start, but may or may not continue to run.

Perform careful visual and physical check as described at the beginning of Section "B".
Perform "Diagnostic Circuit Check."
- CHECK
  - Fuel for poor quality, "stale" fuel, and water contamination
  - Ignition wires for shorts or faulty insulation
  - Ignition coil connections
  - Fuel pump relay. Connect test light between pump test terminal and ground. Light should be on for 2 seconds following ignition "ON." If not, use CHART A-5.
  - Secondary ignition voltage output with ST-125 tester
  - Spark plugs. Look for wetness, cracks, improper gap, burned electrodes, and heavy deposits. Visually inspect ignition system for moisture, dust, cracks, burns, etc.
  - For faulty ECM and ignition grounds

- PROM for correct application.
  - Spray plug wires with fine water mist to check for shorts.
- For possibility of misfiring, crossfiring, or cutting out under load or at idle. If possible, refer to the "Misfire" Chart.
- For improper crank sensor resistance or faulty connections
- EGR operation. Use CHART C-7.
- Idle Air Control system. Use CHART C2-C.
- Fuel system for restricted filter or improper pressure. Use CHART A-7.
- Injector and TBI assembly for leakage. Pressurize system by energizing fuel pump through the underhood fuel pump test connector.
- Coolant sensor for a shift in calibration. Use Code 14 or Code 15 chart.

1988–90 2.0L (VIN 1)
ALL MODELS EXCEPT CORSICA AND BERETTA

- TPS for sticking or binding. TPS voltage should read less than 1.25 volts on a "Scan" tool.
- In-tank fuel pump check valve. A faulty valve would allow the fuel in the lines to drain back to the tank after the engine is stopped. To check for this condition, conduct the following test.

1. Ignition "OFF."
2. Disconnect fuel line at the filter.
3. Remove the tank filler cap.
4. Connect a radiator test pump to the line and apply 103 kPa (15 psi) pressure. If the pressure will hold for 60 seconds, the check valve is OK.

- For the possibility of an exhaust restriction or improper valve timing by performing the following test.

1. With engine at normal operating temperature, connect a vacuum gauge to any convenient vacuum port on intake manifold.
2. Run engine at 1000 rpm and record vacuum reading.
3. Increase engine speed slowly to 2500 rpm. Note vacuum reading at steady 2500 rpm.
4. If vacuum at 2500 rpm decreases more than 3" Hg from reading at 1000 rpm, the exhaust system should be inspected for restrictions. Use CHART B-1.
5. Disconnect exhaust pipe from engine and repeat Steps 3 & 4. If vacuum still drops more than 3" Hg with exhaust disconnected, check valve timing.

Engine valve timing and compression.

ROUGH, UNSTABLE, OR INCORRECT IDLE, STALLING

Definition: The engine runs unevenly at idle. If severe, the car may shake. Also, the idle speed may vary (called "hunting"). Either condition may be severe enough to cause stalling. Engine idles at incorrect speed.

Perform careful visual and physical check as described at the beginning of Section "B".

Perform "Diagnostic Circuit Check."

- CHECK
  - MAP sensor. Use CHART C1-D
  - Throttle for sticking shaft or binding linkage. This will cause a high TPS voltage (open throttle indication) and the ECM will not control idle. TPS voltage should be less than 1.25 volts with throttle closed.
  - Battery cables and ground straps for poor contact. Erratic voltage will cause the IAC valve to change its position, resulting in poor idle quality.
  - Ignition wires for shorts or faulty insulation.
  - Ignition system for moisture, dust, cracks, burns, etc. Spray plug wires with fine water mist to check for shorts.
  - For possibility of misfiring, crossfiring, or cutting out under load or at idle. If present, refer to the "Misfire" chart
  - Secondary ignition voltage output with ST-1 tester.
  - Ignition coil connections.
  - ECM and ignition system for faulty grounds.
  - Proper operation of EST.
  - Spark plugs. Look for wetness, cracks, improper gap, burned electrodes, and heavy deposits.
  - Fuel system for restricted filter or improper pressure. Use CHART A-7.
  - Injector and TBI assembly for leakage. Pressurize system by energizing fuel pump through the underhood fuel pump test connector.
  - EGR operation. Use CHART C-7.
  - For vacuum leaks at intake manifold gasket.
  - Idle Air Control system. Use CHART C2-C.
  - Electrical system voltage. IAC valve will not move if voltage is below 8.7 volts.
  - PCV valve for proper operation by placing finger over inlet hole in valve end several times. Valve should snap back. If not, replace valve. Ensure that valve is correct part. Also check PCV hose.

1988–90 2.0L (VIN 1)
ALL MODELS EXCEPT CORSICA AND BERETTA

- Canister purge system for proper operation. Use CHART C-3.
- PROM for correct application
- Throttle shaft or TPS for sticking or binding. TPS voltage should read less than 1.25 volts on a "Scan" tool with the throttle closed.
- MAP sensor output. Use CHART C1-D and/or check sensor by comparing it to the output ona similar vehicle if possible.
- Oxygen sensor for silicone contamination from contaminated fuel or use of improper RTV sealant. The sensor will have a white, powdery coating and will cause a high but false signal voltage (rich exhaust indication). The ECM will reduce the amount of fuel delivered to the engine, causing a severe driveability problem.
- Coolant sensor for a shift in calibration. Use Code 14 or Code 15 chart.
- A/C refrigerant pressure for high pressure. Check for overcharging or faulty pressure switch.
- P/N switch circuit on vehicle with automatic transmission. Use CHART C-1A.
- Generator output voltage. Repair if less than 9 volts or more than 16 volts.
- Power steering. Use CHART C-1E. The ECM should compensate for power steering loads. Loss of this signal would be most noticeable when steering loads are high such as during parking.
- Engine valve timing and compression.
- For worn or incorrect basic engine parts such as cam, heads, pistons, etc. Also check for bent pushrods, worn rocker arms, and broken or weak valve springs.
- For the possibility of an exhaust restriction or improper valve timing, perform the following test.
  1. With engine at normal operating temperature, connect a vacuum gauge to any convenient vacuum port on intake manifold.
  2. Run engine at 1000 rpm and record vacuum reading.
  3. Increase engine speed slowly to 2500 rpm. Note vacuum reading at steady 2500 rpm.
  4. If vacuum at 2500 rpm decreases more than 3" Hg from reading at 1000 rpm, the exhaust system should be inspected for restrictions. Use CHART B-1.
  5. Disconnect exhaust pipe from engine and repeat Steps 3 & 4. If vacuum still drops more than 3" Hg with exhaust disconnected, check valve timing.

For overheating and possible causes. Look for the following.

- Low or incorrect coolant solution. It should be a 50/50 mix of GM #1052753 anti-freeze coolant (or equivalent) and water.
- Loose water pump belt
- Restricted air flow to radiator, or restricted water flow through radiator
- Faulty or incorrect thermostat
- Inoperative electric cooling fan circuit

- If the system is running RICH, (block learn less than 118), refer to "Diagnostic Aids" on facing page of Code 45
- If the system is running LEAN, (block learn greater than 145), refer to "Diagnostic Aids" on facing page of Code 44.

POOR GAS MILEAGE

Definition: Gas mileage, as measured by an actual road test, is noticeably lower than expected. Gas mileage is noticeably lower than it was during a previous actual road test.

Perform careful visual and physical check as described at the beginning of Section "B".

Perform "Diagnostic Circuit Check.

- CHECK
  - Proper operation of EST.
  - For possibility of misfiring, crossfiring, or cutting out under load or at idle. If present, refer to the "Misfire" chart
  - Spark plugs. Look for wetness, cracks, improper gap, burned electrodes, and heavy deposits.
  - Spark plugs for correct heat range
  - Fuel for poor quality, "stale" fuel, and water contamination.
  - Fuel system for restricted filter or improper pressure. Use CHART A-7.
  - Injector and TBI assembly for leakage. Pressurize system by energizing fuel pump through the underhood fuel pump test connector.
  - EGR operation. Use CHART C-7.

1988–90 2.0L (VIN 1)
ALL MODELS EXCEPT CORSICA AND BERETTA

- For vacuum leaks at intake manifold gasket
- Air cleaner element (filter) for dirt or plugging
- Idle Air Control system. Use CHART C2-C.
- Canister purge system for proper operation. Use CHART C-3
- PROM for correct application
- Throttle shaft or TPS for sticking or binding. TPS voltage should read less than 1.25 volts on a "Scan" tool with the throttle closed.
- MAP sensor output. Use CHART C1-D and/or check sensor by comparing it to the output on a similar vehicle if possible
- Oxygen sensor for silicone contamination from contaminated fuel or use of improper RTV sealant. The sensor will have a white, powdery coating and will cause a high but false signal voltage (rich exhaust indication). The ECM will reduce the amount of fuel delivered to the engine, causing a severe driveability problem.
- Coolant sensor for a shift in calibration. Use Code 14 or Code 15 chart.
- Vehicle speed sensor (VSS) input with a "Scan" tool to make sure reading of VSS matches that of vehicle speedometer. See "Special Information" in Section "6E."
- A/C relay operation. A/C should cut out at wide open throttle. Use CHART C-10.
- A/C refrigerant pressure for high pressure. Check for overcharging or faulty pressure switch
- Generator output voltage. Repair if less than 9 volts or more than 16 volts.
- Cooling fan operation. Use CHART C-12.
- Power steering. Use CHART C-1E. The ECM should compensate for power steering loads. Loss of this signal would be most noticeable when steering loads are high such as during parking.
- Transmission torque converter operation.
- Transmission for proper shift points.
- Transmission torque converter clutch operation. Use CHART C-8.
- Engine valve timing and compression.
- For worn or incorrect basic engine parts such as cam, heads, pistons, etc. Also check for bent pushrods, worn rocker arms, and broken or weak valve springs.
- For the possibility of an exhaust restriction or improper valve timing by performing the following test.
  1. With engine at normal operating temperature, connect a vacuum gauge to any convenient vacuum port on intake manifold.
  2. Run engine at 1000 rpm and record vacuum reading.
  3. Increase engine speed slowly to 2500 rpm. Note vacuum reading at steady 2500 rpm.
  4. If vacuum at 2500 rpm decreases more than 3" Hg from reading at 1000 rpm, the exhaust system should be inspected for restrictions. Use CHART B-1.
  5. Disconnect exhaust pipe from engine and repeat steps 3 & 4. If vacuum still drops more than 3" Hg with exhaust disconnected, check valve timing.
- Thermostat for incorrect heat range or being inoperative

- Check driver's driving habits and vehicle conditions which affect gas mileage.
  - Suggest driver read "Important Facts on Fuel Economy" in Owner's Manual.
  - Is A/C "ON" full time (Defroster mode "ON")?
  - Are tires at correct pressure?
  - Are excessively heavy loads being carried?
  - Is acceleration often heavy?
  - Are the wheels aligned correctly?
  - Is the speedometer calibrated correctly?
  - Are the vehicle brakes dragging?
  - Is the brake switch applying excessive force on the brake pedal?
- If the system is running RICH, (block learn less than 118), refer to "Diagnostic Aids" on facing page of Code 45.

1988–90 2.0L (VIN 1)
ALL MODELS EXCEPT CORSICA AND BERETTA

DETONATION/SPARK KNOCK

Definition: A mild to severe ping, usually worse under acceleration. The engine makes sharp metallic knocks that change with throttle opening.

Perform careful visual and physical check as described at the beginning of Section "B".

Perform "Diagnostic Circuit Check."

- CHECK
  - Ignition wires for shorts or faulty insulation
  - For possibility of misfiring, crossfiring, or cutting out under load or at idle. If present, refer to the "Misfire" Chart
  - Spark plugs for correct heat range
  - Fuel for poor quality, "stale" fuel, and water contamination
  - Fuel system for restricted filter or improper pressure. Use CHART A-7.
  - For excessive oil entering combustion chamber. Oil will reduce the effective octane of fuel.
  - EGR operation. Use CHART C-7.
  - For vacuum leaks at intake manifold gasket
  - PCV valve for proper operation by placing finger over inlet hole in valve end several times. Valve should snap back. If not, replace valve. Ensure that valve is correct part. Also check PCV hose.
  - MAP sensor output. Use CHART C1-D and/or check sensor by comparing it to the output on a Similar vehicle, if possible.
  - Coolant sensor for a shift in calibration
  - Oxygen sensor for silicone contamination from contaminated fuel or use of improper RTV sealant. The sensor will have a white, powdery coating and will cause a high but false signal voltage (rich exhaust indication). The ECM will reduce the amount of fuel delivered to the engine, causing a severe driveability problem
  - Vehicle speed sensor (VSS) input with a "Scan" tool to make sure reading of VSS matches that of vehicle speedometer.
  - Transmission torque converter operation.
  - Transmission for proper shift points.
  - Transmission torque converter clutch operation. Use CHART C-8.
  - Vehicle brakes for dragging
  - PROM for correct application
  - For overheating and possible causes. Look for the following.
    - Low or incorrect coolant solution. It should be a 50/50 mix of GM #1052753 anti-freeze coolant (or equivalent) and water.
    - Loose water pump belt
    - Restricted air flow to radiator or restricted water flow through radiator
    - Faulty or incorrect thermostat
    - Inoperative electric cooling fan circuit
  - Engine valve timing and compression.
  - For worn or incorrect basic engine parts such as cam, heads, pistons, etc. Also check for bent pushrods, worn rocker arms, and broken or weak valve springs.
  - For the possibility of an exhaust restriction or improper valve timing by performing the following test.
    1. With engine at normal operating temperature, connect a vacuum gauge to any convenient vacuum port on intake manifold.
    2. Run engine at 1000 rpm and record vacuum reading.
    3. Increase engine speed slowly to 2500 rpm. Note vacuum reading at steady 2500 rpm.
    4. If vacuum at 2500 rpm decreases more than 3" Hg from reading at 1000 rpm, the exhaust system should be inspected for restrictions. Use CHART B-1.
    5. Disconnect exhaust pipe from engine and repeat Steps 3 & 4. If vacuum still drops more than 3" Hg with exhaust disconnected, check valve timing.
- Remove internal engine carbon with top engine cleaner.
- If the system is running LEAN, (block learn greater than 145), refer to "Diagnostic Aids" on facing page of Code 44.

1988–90 2.0L (VIN 1) ALL MODELS EXCEPT CORSICA AND BERETTA

LACK OF POWER, SLUGGISH, OR SPONGY

Definition: Engine delivers less than expected power. There is little or no increase in speed when the accelerator pedal is depressed partially.

Perform careful visual and physical check as described at the beginning of Section "B".

Perform "Diagnostic Circuit Check."

- **CHECK**
  - Ignition wires for shorts or faulty insulation
  - Ignition system for moisture, dust, cracks, burns, etc. Spray plug wires with fine water mist to check for shorts.
  - For possibility of misfiring, crossfiring, or cutting out under load or at idle. If present, refer to the "Misfire" chart
  - Secondary ignition voltage output with ST-125 tester.
  - Ignition coil connections
  - ECM and ignition system for faulty grounds
  - Proper operation of EST.
  - Spark plugs. Look for wetness, cracks, improper gap, burned electrodes, and heavy deposits.
  - Spark plugs for correct heat range
  - Fuel for poor quality, "stale" fuel, and water contamination
  - Fuel system for restricted filter or improper pressure. Use CHART A-7.
  - EGR operation. Use CHART C-7.
  - For vacuum leaks at intake manifold gasket
  - Air cleaner element (filter) for dirt or plugging
  - PROM for correct application
  - Throttle shaft or TPS for sticking or binding. TPS voltage should read less than 1.25 volts on a "Scan" tool with the throttle closed.
  - MAP sensor output. Use CHART C1-D and/or check sensor by comparing it to the output on a similar vehicle if possible.
  - Oxygen sensor for silicone contamination from contaminated fuel or use of improper RTV sealant. The sensor will have a white, powdery coating and will cause a high but false signal voltage (rich exhaust indication). The ECM will reduce the amount of fuel delivered to the engine, causing a severe driveability problem.
  - Coolant sensor for a shift in calibration. Use Code 14 or Code 15 chart.
  - Vehicle speed sensor (VSS) input with a "Scan" tool to make sure reading of VSS matches that of vehicle speedometer.
  - Engine for improper or worn camshaft.
  - A/C relay operation. A/C should cut out at wide open throttle. Use CHART C-10.
  - A/C refrigerant pressure for high pressure. Check for overcharging or faulty pressure switch.
  - Generator output voltage. Repair if less than 9 volts or more than 16 volts.
  - Cooling fan operation. Use CHART C-12.
  - Power steering. Use CHART C-1E. The ECM should compensate for power steering loads. Loss of this signal would be most noticeable when steering loads are high such as during parking.
  - Transmission torque converter operation.
  - Transmission for proper shift points.
  - Transmission torque converter clutch operation. Use CHART C-8.
  - Vehicle brakes for dragging
  - Engine valve timing and compression.
  - For worn or incorrect basic engine parts such as cam, heads, pistons, etc. Also check for bent pushrods, worn rocker arms, and broken or weak valve springs.
  - For the possibility of an exhaust restriction or improper valve timing by performing the following test.
    1. With engine at normal operating temperature, connect a vacuum gauge to any convenient vacuum port on intake manifold.
    2. Run engine at 1000 rpm and record vacuum reading.
    3. Increase engine speed slowly to 2500 rpm. Note vacuum reading at steady 2500 rpm.
    4. If vacuum at 2500 rpm decreases more than 3" Hg from reading at 1000 rpm, the exhaust system should be inspected for restrictions. Use CHART B-1.

1988–90 2.0L (VIN 1) ALL MODELS EXCEPT CORSICA AND BERETTA

5. Disconnect exhaust pipe from engine and repeat Steps 3 & 4. If vacuum still drops more than 3" Hg with exhaust disconnected, check valve timing.

- For overheating and possible causes. Look for the following:
  - Low or incorrect coolant solution. It should be a 50/50 mix of GM #1052753 anti-freeze coolant (or equivalent) and water.
  - Loose water pump belt.
  - Restricted air flow to radiator, or restricted water flow through radiator.
  - Faulty or incorrect thermostat.
  - Inoperative electric cooling fan circuit. See CHART C-12.
- If the system is running RICH (block learn less than 118), refer to "Diagnostic Aids" on facing page of Code 45.
- If the system is running LEAN (block learn greater than 145), refer to "Diagnostic Aids" on facing page of Code 44.

SURGES AND/OR CHUGGLE

Definition: Engine power variation under steady throttle or cruise. Feels like the car speeds up and slows down with no change in the accelerator pedal.

Perform careful visual and physical check as described at the beginning of Section "B".

Perform "Diagnostic Circuit Check."

- **CHECK**
  - Ignition wires for shorts or faulty insulation.
  - Ignition system for moisture, dust, cracks, burns, etc. Spray plug wires with fine water mist to check for shorts.
  - For possibility of misfiring, crossfiring, or cutting out under load or at idle. If present, refer to the "Misfire" Chart
  - Secondary ignition voltage output with ST-125 tester.
  - Ignition coil connections.
  - ECM and ignition system for faulty grounds.
  - Proper operation of EST.
  - Spark plugs. Look for wetness, cracks, improper gap, burned electrodes, and heavy deposits.
  - Spark plugs for correct heat range.
  - Fuel for poor quality, "stale" fuel, and water contamination.
  - Fuel system for restricted filter or improper pressure. Use CHART A-7.
  - Injector and TBI assembly for leakage. Pressurize system by energizing fuel pump through the underhood fuel pump test connector.
  - EGR operation. Use CHART C-7.
  - For vacuum leaks at intake manifold gasket.
  - Idle Air Control system. Use CHART C2-C.
  - Electrical system voltage. IAC valve will not move if voltage is below 8.7 volts. Also check battery cables and ground straps for poor contact. Erratic voltage will cause the IAC valve to change its position, resulting in poor idle quality.
  - PCV valve for proper operation by placing finger over inlet hole in valve end several times. Valve should snap back. If not, replace valve. Ensure that valve is correct part. Also check PCV hose.
  - Canister purge system for proper operation. Use CHART C-3.
  - PROM for correct application.
  - Throttle shaft or TPS for sticking or binding. TPS voltage should read less than 1.25 volts on a "Scan" tool with the throttle closed.
  - MAP sensor output. Use CHART C1-D and/or check sensor by comparing it to the output on a similar vehicle, if possible.
  - Oxygen sensor for silicone contamination from contaminated fuel or use of improper RTV sealant. The sensor will have a white, powdery coating and will cause a high but false signal voltage (rich exhaust indication). The ECM will reduce the amount of fuel delivered to the engine, causing a severe driveability problem.
  - Coolant sensor for a shift in calibration. Use Code 14 or Code 15 chart.
  - Vehicle speed sensor (VSS) input with a "Scan" tool to make sure reading of VSS matches that of vehicle speedometer.
  - A/C relay operation. A/C should cut out at wide open throttle. Use CHART C-10.
  - P/N switch circuit on vehicle with automatic transmission. Use CHART C-1A.
  - Transmission torque converter clutch operation. Use CHART C-8.

1988–90 2.0L (VIN 1) ALL MODELS EXCEPT CORSICA AND BERETTA

- For the possibility of an exhaust restriction or improper valve timing by performing the following test:
  1. With engine at normal operating temperature, connect a vacuum gauge to any convenient vacuum port on intake manifold.
  2. Run engine at 1000 rpm and record vacuum reading.
  3. Increase engine speed slowly to 2500 rpm. Note vacuum reading at steady 2500 rpm.
  4. If vacuum at 2500 rpm decreases more than 3" Hg from reading at 1000 rpm, the exhaust system should be inspected for restrictions. Use CHART B-1.
  5. Disconnect exhaust pipe from engine and repeat Steps 3 & 4. If vacuum still drops more than 3" Hg with exhaust disconnected, check valve timing.
- Engine valve timing and compression.
- For worn or incorrect basic engine parts such as cam, heads, pistons, etc. Also check for bent pushrods, worn rocker arms, and broken or weak valve springs.
- If the system is running RICH, (block learn less than 118), refer to "Diagnostic Aids" on facing page of Code 45.
- If the system is running LEAN, (block learn greater than 145), refer to "Diagnostic Aids" on facing page of Code 44.

CUTS OUT, MISSES

Definition: Steady pulsation or jerking that follows engine speed, usually more pronounced as engine load increases. The exhaust has a steady spitting sound at idle or low speed.

Perform careful visual and physical check as described at the beginning of Section "B".

Perform "Diagnostic Circuit Check."

- **CHECK**
  - Ignition wires for shorts or faulty insulation
  - Ignition system for moisture, dust, cracks, burns, etc. Spray plug wires with fine water mist to check for shorts.
  - For possibility of misfiring, crossfiring, or cutting out under load or at idle. If present, refer to the "Misfire" chart
  - Secondary ignition voltage output with ST-125 tester
  - Ignition coil connections
  - ECM and ignition system for faulty grounds
  - Proper operation of EST.
  - Spark plugs. Look for wetness, cracks, improper gap, burned electrodes, and heavy deposits.
  - Spark plugs for correct heat range
  - For improper crank sensor resistance or faulty connections
  - Fuel for poor quality, "stale" fuel, and water contamination
  - Fuel system for restricted filter or improper pressure. Use CHART A-7.
  - Throttle shaft or TPS for sticking or binding. TPS voltage should read less than 1.25 volts on a "Scan" tool with the throttle closed.

HESITATION, SAG, STUMBLE

Definition: Momentary lack of response as the accelerator is pushed down. Can occur at all vehicle speeds. Usually most severe when first trying to make the car move, as from a stop sign. May cause the engine to stall if severe enough.

Perform careful visual and physical check as described at the beginning of Section "B".

Perform "Diagnostic Circuit Check."

- **CHECK**
  - Ignition wires for shorts or faulty insulation
  - Ignition system for moisture, dust, cracks, burns, etc. Spray plug wires with fine water mist to check for shorts.
  - For possibility of misfiring, crossfiring, or cutting out under load or at idle. If present, refer to the "Misfire" chart
  - Secondary ignition voltage output with ST-125 tester
  - Ignition coil connections
  - ECM and ignition system for faulty grounds
  - Proper operation of EST.
  - Spark plugs. Look for wetness, cracks, improper gap, burned electrodes, and heavy deposits.
  - Spark plugs for correct heat range
  - Fuel for poor quality, "stale" fuel, and water contamination
  - Fuel system for restricted filter or improper pressure. Use CHART A-7.

1988–90 2.0L (VIN 1) ALL MODELS EXCEPT CORSICA AND BERETTA

- EGR operation. Use CHART C-7.
- For vacuum leaks at intake manifold gasket
- Air cleaner element (filter) for dirt or plugging
- Idle Air Control system. Use CHART C2-C.
- Check electrical system voltage. IAC valve will not move if voltage is below 8.7 volts. Also check battery cables and ground straps for poor contact. Erratic voltage will cause the IAC valve to change its position, resulting in poor idle quality.
- PCV valve for proper operation by placing finger over inlet hole in valve end several times. Valve should snap back. If not, replace valve. Ensure that valve is correct part. Also check PCV hose.
- Canister purge system for proper operation. Use CHART C-3.
- PROM for correct application
- Throttle shaft or TPS for sticking or binding TPS voltage should read less than 1.25 volts on a "Scan" tool with the throttle closed.
- MAP sensor output. Use CHART C1-D and/or check sensor by comparing it to the output on a similar vehicle, if possible.
- Oxygen sensor for silicone contamination from contaminated fuel or use of improper RTV sealant. The sensor will have a white, powdery coating and will cause a high but false signal voltage (rich exhaust indication). The ECM will reduce the amount of fuel delivered to the engine, causing a severe driveability problem.
- Coolant sensor for a shift in calibration. Use Code 14 or Code 15 chart.
- A/C relay operation. A/C should cut out at wide open throttle. Use CHART C-10.
- A/C refrigerant pressure for high pressure. Check for overcharging or faulty pressure switch.
- P/N switch circuit on vehicle with automatic transmission. Use CHART C-1A.
- Generator output voltage. Repair if less than 9 volts or more than 16 volts.
- Transmission torque converter operation.
- Transmission for proper shift points.
- Transmission torque converter clutch operation. Use CHART C-8.
- Vehicle brakes for dragging
- Engine valve timing and compression.
- For the possibility of an exhaust restriction or improper valve timing by performing the following test.
  1. With engine at normal operating temperature, connect a vacuum gauge to any convenient vacuum port on intake manifold.
  2. Run engine at 1000 rpm and record vacuum reading.
  3. Increase engine speed slowly to 2500 rpm. Note vacuum reading at steady 200 rpm.
  4. If vacuum at 2500 rpm decreases more than 3" Hg from reading at 1000 rpm, the exhaust system should be inspected for restrictions. Use CHART B-1.
  5. Disconnect exhaust pipe from engine and repeat steps 3 & 4. If vacuum still drops more than 3" Hg with exhaust disconnected, check valve timing.
- For worn or incorrect basic engine parts such as cam, heads, pistons, etc. Also check for bent pushrods, worn rocker arms, and broken or weak valve springs.
- For overheating and possible causes. Look for the following.
  - Low or incorrect coolant solution. It should be a 50/50 mix of GM #1052753 anti-freeze coolant (or equivalent) and water.
  - Loose water pump belt
  - Restricted air flow to radiator, or restricted water flow through radiator
  - Faulty or incorrect thermostat
  - Inoperative electric cooling fan circuit. See CHART C-12.
- If the system is running RICH (block learn less than 118), refer to "Diagnostic Aids" on facing page of Code 45.
- If the system is running LEAN (block learn greater than 145), refer to "Diagnostic Aids" on facing page of Code 44.

1988–90 2.0L (VIN 1)
ALL MODELS EXCEPT CORSICA AND BERETTA

EXCESSIVE EXHAUST EMISSIONS OR ODORS

Definition: Vehicle fails an emission test or vehicle has excessive "rotten egg" smell. (Excessive odors do not necessarily indicate excessive emissions).

Perform careful visual and physical check as described at the beginning of Section "B". Perform "Diagnostic Circuit Check."

- **CHECK**
  - EGR valve not opening. Use CHART C-7.
  - Vacuum leaks.
  - Faulty coolant system and/or coolant fan operation. Use CHART C-12.
  - Remove carbon with top engine cleaner. Follow instructions on can.
- If the system is running RICH (block learn less than 118), refer to "Diagnostic Aids" on facing page of Code 45.
- If the system is running LEAN (block learn greater than 145), refer to "Diagnostic Aids" on facing page of Code 44.

- If emission test indicates excessive NO$_x$, check for items which cause car to run lean or too hot.
- If emission test indicates excessive HC and CO or exhaust has excessive odors, check for items which cause car to run RICH.
  - Incorrect fuel pressure. Use CHART A-7.
  - Fuel loading of evaporative vapor canister. Use CHART C-3.
  - PCV valve plugging, sticking, or blocked PCV hose. Check for fuel in crankcase.
  - Catalytic converter lead contamination (Look for removal of fuel filler neck restrictor.)
  - Improper fuel cap installation.
  - Faulty spark plugs, plug wires, or ignition components.

DIESELING, RUN-ON

Definition: Engine continues to run after key is turned "OFF", but runs very roughly. (If engine runs smoothly, check ignition switch).

Perform careful visual and physical check as described at the beginning of Section "B". Perform "Diagnostic Circuit Check."

- **CHECK**
  - Injector and TBI assembly for leakage. Pressurize system by energizing fuel pump through the fuel pump test connector.

BACKFIRE

Definition: Fuel ignites in intake manifold or in exhaust system, making a loud popping sound.

Perform careful visual and physical check as described at the beginning of Section "B". Perform "Diagnostic Circuit Check."

- **CHECK**
  - EGR operation for valve being open all the time. Use CHART C-7.
  - Intake manifold gasket for leaks.
  - For possibility of misfiring, crossfiring, or cutting out under load or at idle. If present, refer to "Misfire" chart.
  - Spark plugs. Look for wetness, cracks, improper gap, burned electrodes, and heavy deposits.
  - Ignition coil connections.

- Ignition system for moisture, dust, cracks, burns, etc. Spray plug wires with fine water mist to check for shorts.
- ECM and ignition system for faulty grounds.
- Secondary ignition voltage output with ST-125 tester.
- For vacuum leaks at intake manifold gasket.
- Engine valve timing and compression.

- For worn or incorrect basic engine parts such as cam, heads, pistons, etc. Also check for bent pushrods, worn rocker arms, and broken or weak valve springs.

1988–90 2.0L (VIN 1)
ALL MODELS EXCEPT CORSICA AND BERETTA

CHART B-1
RESTRICTED EXHAUST SYSTEM CHECK
ALL ENGINES

Proper diagnosis for a restricted exhaust system is essential before any components are replaced. Either of the following procedures may be used for diagnosis, depending upon engine or tool used.

CHECK AT A. I. R. PIPE: OR **CHECK AT O$_2$ SENSOR:**

1. Remove the rubber hose at the exhaust manifold A.I.R. pipe check valve. Remove check valve.
2. Connect a fuel pump pressure gauge to a hose and nipple from a propane enrichment device (J26911) (see illustration).
3. Insert the nipple into the exhaust manifold A.I.R. pipe.

1. Carefully remove O$_2$ sensor.
2. Install Borroughs exhaust backpressure tester (BT 8515 or BT 8603) or equivalent in place of O$_2$ sensor (see illustration).
3. After completing test described below, be sure to coat threads of O$_2$ sensor with anti-seize compound P/N 5613695 or equivalent prior to re-installation.

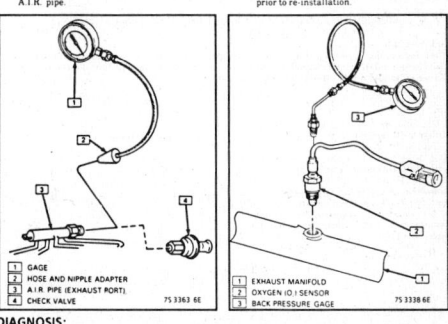

	GAGE
1	GAGE
2	HOSE AND NIPPLE ADAPTER
3	A.I.R. PIPE (EXHAUST PORT)
4	CHECK VALVE

7S 3363 6E

1	EXHAUST MANIFOLD
2	OXYGEN (O) SENSOR
3	BACK PRESSURE GAGE

7S 3338 6E

DIAGNOSIS:

1. With the engine idling at normal operating temperature, observe the exhaust system backpressure reading on the gauge. Reading should not exceed 8.6 kPa (1.25 psi).
2. Increase engine speed to 2000 rpm and observe gauge. Reading should not exceed 20.7 kPa (3 psi).
3. If the backpressure at either speed exceeds specification, a restricted exhaust system is indicated.
4. Inspect the entire exhaust system for a collapsed pipe, heat distress, or possible internal muffler failure.
5. If there are no obvious reasons for the excessive backpressure, the catalytic converter is suspected to be restricted and should be replaced using current recommended procedures.

1988–90 2.0L (VIN 1)
ALL MODELS EXCEPT CORSICA AND BERETTA

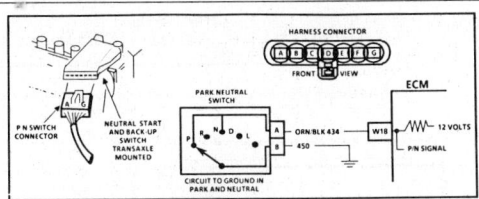

CHART C-1A
PARK/NEUTRAL SWITCH DIAGNOSIS
(AUTO TRANSAXLE)
2.0L (VIN 1)

Circuit Description:

The park/neutral switch contacts are a part of the neutral start switch and are closed to ground in park or neutral, and open in drive ranges.

The ECM supplies ignition voltage through a current limiting resistor to CKT 434 and senses a closed switch when the voltage on CKT 434 drops to less than one volt.

The ECM uses the P/N signal as one of the inputs to control idle air control and vehicle speed sensor diagnostics.

Test Description: Numbers below refer to circled numbers on the diagnostic chart.

1. Checks for a closed switch to ground in park position. Different "Scan" tools may display P/N differently. Refer to tool operator's manual for the display used.
2. Checks for an open switch in drive range.

3. Be sure "Scan" tool indicates drive even while wiggling shifter to test for an intermittent or maladjusted switch in drive range.

1988–90 2.0L (VIN 1)
ALL MODELS EXCEPT CORSICA AND BERETTA

CHART C-1A
PARK/NEUTRAL SWITCH DIAGNOSIS
(AUTO TRANSAXLE)
2.0L (VIN 1)

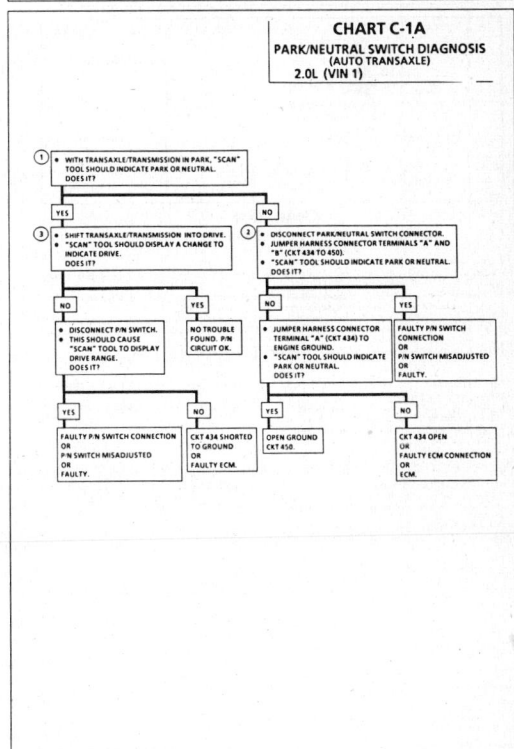

1988–90 2.0L (VIN 1)
ALL MODELS EXCEPT CORSICA AND BERETTA

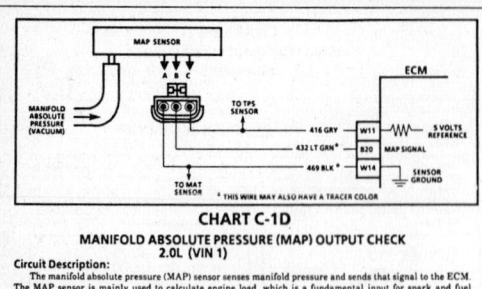

CHART C-1D
MANIFOLD ABSOLUTE PRESSURE (MAP) OUTPUT CHECK
2.0L (VIN 1)

Circuit Description:
The manifold absolute pressure (MAP) sensor senses manifold pressure and sends that signal to the ECM. The MAP sensor is mainly used to calculate engine load, which is a fundamental input for spark and fuel calculations. The MAP sensor is also used to determine barometric pressure.

Test Description: Numbers below refer to circled numbers on the diagnostic chart.
1. Checks MAP sensor output voltage to the ECM. This voltage, without engine running, represents a barometer reading to the ECM.
2. Applying 34 kPa (10" Hg) vacuum to the MAP sensor should cause the voltage to be 1.2 volts less than the voltage at step 1. Upon applying vacuum to the sensor, the change in voltage should be instantaneous. A slow voltage change indicates a faulty sensor.

3. Check vacuum hose to sensor for leaking or restriction. Be sure no other vacuum devices are connected to the MAP hose.

1988–90 2.0L (VIN 1)
ALL MODELS EXCEPT CORSICA AND BERETTA

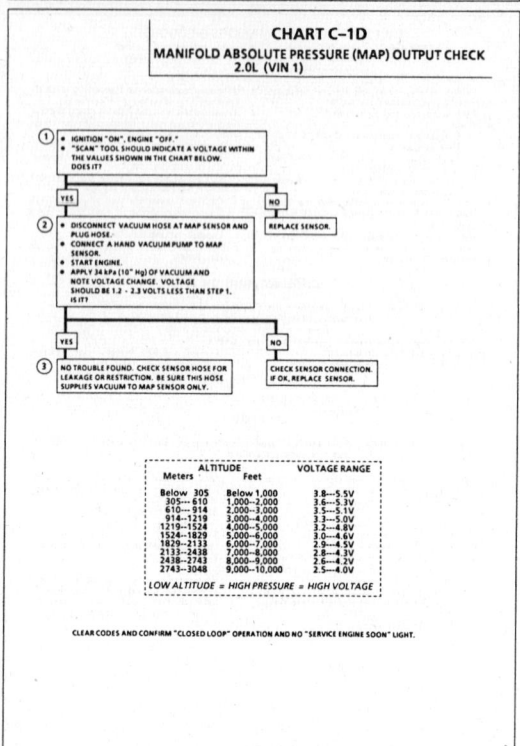

CHART C-1D
MANIFOLD ABSOLUTE PRESSURE (MAP) OUTPUT CHECK
2.0L (VIN 1)

ALTITUDE		VOLTAGE RANGE
Meters	Feet	
Below 305	Below 1,000	3.8---5.5V
305--- 610	1,000---2,000	3.6---5.3V
610--- 914	2,000---3,000	3.5---5.1V
914---1219	3,000---4,000	3.3---5.0V
1219---1524	4,000---5,000	3.2---4.8V
1524---1829	5,000---6,000	3.0---4.6V
1829---2133	6,000---7,000	2.9---4.5V
2133---2438	7,000---8,000	2.8---4.3V
2438---2743	8,000---9,000	2.6---4.2V
2743---3048	9,000---10,000	2.5---4.0V

LOW ALTITUDE = HIGH PRESSURE = HIGH VOLTAGE

CLEAR CODES AND CONFIRM "CLOSED LOOP" OPERATION AND NO "SERVICE ENGINE SOON" LIGHT.

1988–90 2.0L (VIN 1)
ALL MODELS EXCEPT CORSICA AND BERETTA

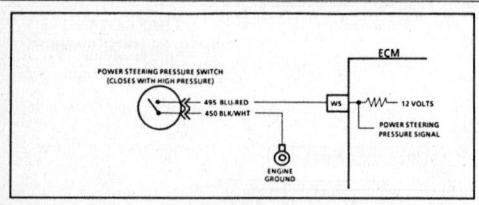

CHART C-1E
POWER STEERING PRESSURE SWITCH (PSPS) DIAGNOSIS
2.0L (VIN 1)

Circuit Description:
The power steering pressure switch is normally open to ground, and CKT 495 will be at approximately battery voltage.
Turning the steering wheel increases power steering oil pressure and its load on an idling engine. The pressure switch will close before the load becomes great enough to cause an idle problem.
Closing the switch causes CKT 495 to read less than 1 volt and the ECM will increase the idle air rate.
- A pressure switch that will not close, or an open CKT 495 or 450, may cause the engine to stall at idle when power steering loads are high.
- A switch that will not open, or a CKT 495 shorted to ground, may affect idle quality.

Test Description: Numbers below refer to circled numbers on the diagnostic chart.
1. Different "Scan" tools may display the state of this switch in different ways. Refer to the "Scan" tool operator's manual to determine how this input is indicated.

2. Checks to determine if CKT 495 is shorted to ground.
3. This step should simulate a closed switch.

1988–90 2.0L (VIN 1)
ALL MODELS EXCEPT CORSICA AND BERETTA

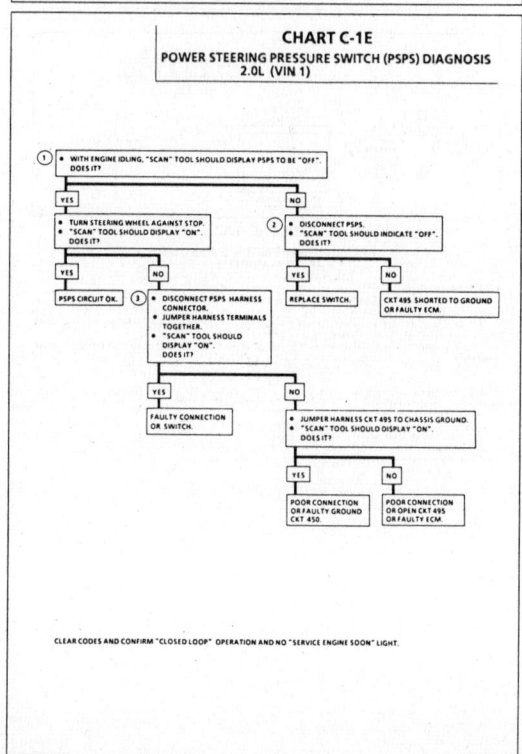

CHART C-1E
POWER STEERING PRESSURE SWITCH (PSPS) DIAGNOSIS
2.0L (VIN 1)

CLEAR CODES AND CONFIRM "CLOSED LOOP" OPERATION AND NO "SERVICE ENGINE SOON" LIGHT.

1988–90 2.0L (VIN 1)
ALL MODELS EXCEPT CORSICA AND BERETTA

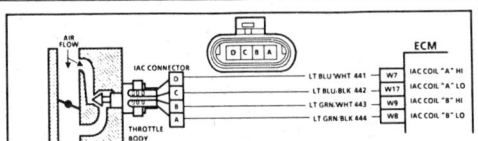

CHART C-2C
IDLE AIR CONTROL (IAC) SYSTEM CHECK
2.0L (VIN 1)

Test Description: Numbers below refer to circled numbers on the diagnostic chart.

1. Continue with test even if engine will not idle. If idle is too low, "Scan" tool will display 90 or more counts (steps). If idle is high it will display "0" counts. Occasionally an erratic or unstable idle may occur. Engine speed may vary 200 rpm or more. Disconnect IAC valve. If the condition is unchanged, the IAC valve is not at fault.
2. When the engine was stopped, the IAC valve retracted (more air) to a fixed "Park" position for increased air flow and idle speed during the next engine start. A "Scan" tool will display 100 or more counts.
3. Be sure to disconnect the IAC valve prior to this test. The test light will confirm the ECM signals by a steady or flashing light on all circuits.
4. There is a remote possibility that one of the circuits is shorted to voltage, which would have been indicated by a steady light. Disconnect ECM and turn the ignition "ON" and probe terminals to check for this condition.

Diagnostic Aids:

A slow unstable idle may be caused by a system problem that cannot be overcome by the IAC valve. "Scan" tool counts will be above 60 counts if too low, and "0" counts if too high.

If idle is too high, stop engine. Turn ignition "ON." Ground diagnostic terminal. Wait 45 seconds for IAC valve pintle to seat, then disconnect IAC valve.

Start engine. If idle speed is above 800 rpm locate and correct vacuum leak.

- **System too lean (High Air/Fuel Ratio).** Idle speed may be too high or too low. Engine speed may vary up and down, and disconnecting IAC valve does not help. May set Code 44.
  "Scan" tool and/or voltmeter will read an oxygen sensor output less than 300 mV (.3 volt). Check for low regulated fuel pressure or water in fuel. A lean exhaust with an oxygen sensor output fixed above 800 mV (.8 volt) will be a contaminated sensor, usually by silicone. This may also set a Code 45.
- **System too rich (Low Air/Fuel Ratio).** Idle speed too low. "Scan" counts are usually above 80. System obviously rich and may exhibit black smoke in exhaust.
  "Scan" tool or voltmeter will read an oxygen sensor signal fixed above 800 mV (.8 volt).
  Check for
  - High fuel pressure
  - Injector leaking or sticking
- **Throttle Body.** Remove IAC valve and inspect bore for foreign material or evidence of IAC valve dragging the bore.
- **PCV.** A faulty or incorrect PCV may result in an incorrect idle speed.
- **IAC Harness Connections.** IAC valve harness connectors should be carefully inspected for proper contact.
- Refer to Section "B", "Rough, Unstable, Incorrect Idle or Stalling."

1988–90 2.0L (VIN 1)
ALL MODELS EXCEPT CORSICA AND BERETTA

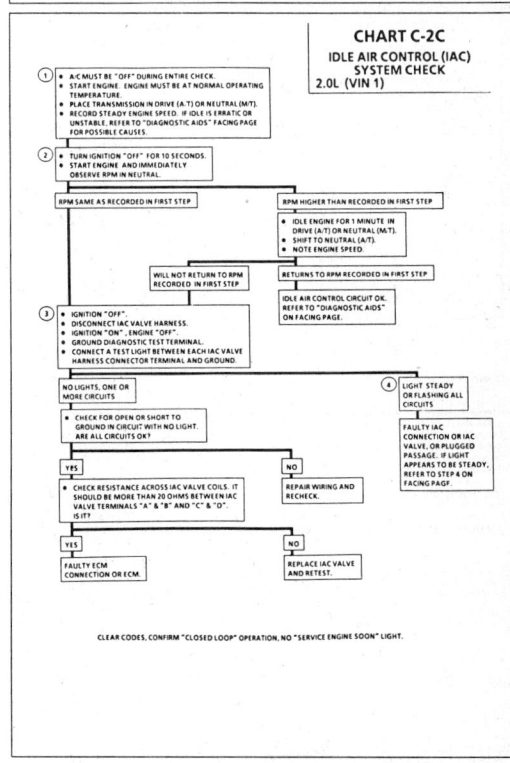

CHART C-2C
IDLE AIR CONTROL (IAC) SYSTEM CHECK
2.0L (VIN 1)

1988–90 2.0L (VIN 1)
ALL MODELS EXCEPT CORSICA AND BERETTA

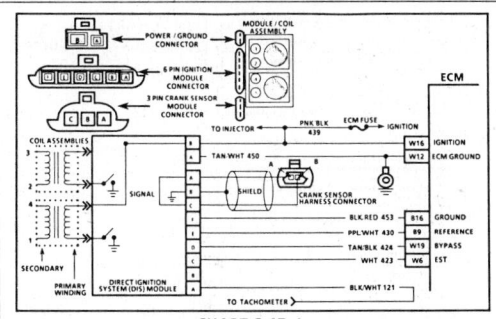

CHART C-4D-1
"DIS" MISFIRE AT IDLE
2.0L (VIN 1)

Circuit Description:

The "direct ignition system" (DIS) uses a waste spark method of distribution. In this type of system, the ignition module triggers the #1/#4 coil pair resulting in both #1 and #4 spark plugs firing at the same time. #1 cylinder is on the compression stroke at the same time #4 is on the exhaust stroke, resulting in a lower energy requirement to fire #4 spark plug. This leaves the remainder of the high voltage to be used to fire #1 spark plug. The crank sensor is remotely mounted beside the module/coil assembly and protrudes through the block to within approximately .050" of the crankshaft reluctor. Since the reluctor is a machined portion of the crankshaft and the crankshaft sensor is mounted in a fixed position on the block, timing adjustments are not possible or necessary.

Test Description: Numbers below refer to circled numbers on the diagnostic chart.

1. If the "Misfire" complaint exists under load only, the diagnostic chart on page 2 must be used. Engine rpm should drop approximately equally on all plug leads.
2. A spark tester, such as a ST-125, must be used because it is essential to verify adequate available secondary voltage at the spark plug (25,000 volts).
3. If the spark jumps the tester gap after grounding the opposite plug wire, it indicates excessive resistance in the plug which was bypassed. A faulty or poor connection at that plug could also result in the miss condition. Also, check for carbon deposits inside the spark plug boot.
4. If carbon tracking is evident, replace coil and be sure plug wires relating to that coil are clean and tight. Excessive wire resistance or faulty connections could have caused the coil to be damaged.
5. If the no spark condition follows that coil, that coil is faulty. Otherwise, the ignition module is the cause of no spark. This test could also be performed by substituting a known good coil for the one causing the no spark condition.

1988–90 2.0L (VIN 1)
ALL MODELS EXCEPT CORSICA AND BERETTA

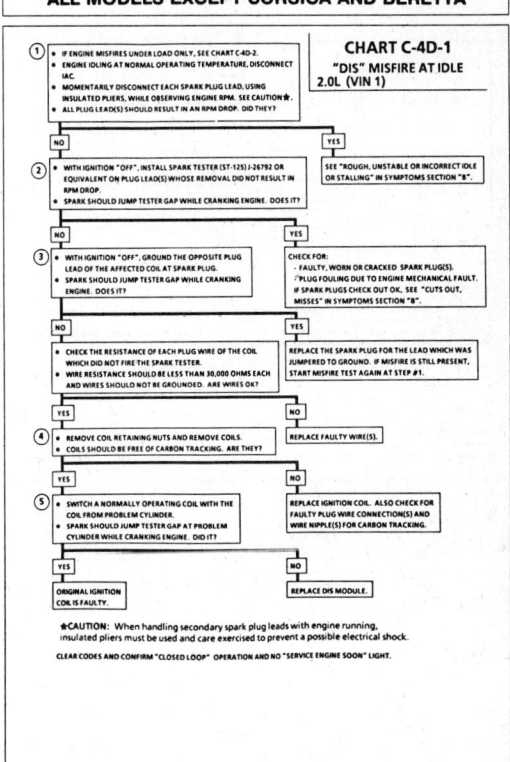

CHART C-4D-1
"DIS" MISFIRE AT IDLE
2.0L (VIN 1)

1988–90 2.0L (VIN 1)
ALL MODELS EXCEPT CORSICA AND BERETTA

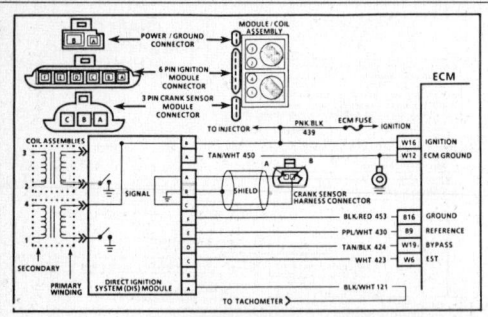

CHART C-4D-2
"DIS" MISFIRE UNDER LOAD
2.0L (VIN 1)

Circuit Description:

The "direct ignition system" (DIS) uses a waste spark method of distribution. In this type of system, the ignition module triggers the #1/4 coil pair resulting in both #1 and #4 spark plugs firing at the same time. #1 cylinder is on the compression stroke at the same time #4 is on the exhaust stroke, resulting in a lower energy requirement to fire #4 spark plug. This leaves the remainder of the high voltage to be used to fire #1 spark plug. The crank sensor is remotely mounted beside the module/coil assembly and protrudes through the block to within approximately .050" of the crankshaft reluctor. Since the reluctor is a machined portion of the crankshaft, and the crankshaft sensor is mounted in a fixed position on the block, timing adjustments are not possible or necessary.

Test Description: Numbers below refer to circled numbers on the diagnostic chart.

1. If the "Misfire" complaint exists at idle only, the diagnostic chart on page 1 must be used. A spark tester such as a ST-125 must be used because it is essential to verify adequate available secondary voltage at the spark plug. (25,000 volts). Spark should jump the test gap on all 4 leads. This simulates a "load" condition.
2. If the spark jumps the tester gap after grounding the opposite plug wire, it indicates excessive resistance in the plug which was bypassed.

3. A faulty or poor connection at that plug could also result in the miss condition. Also, check for carbon deposits inside the spark plug boot.
3. If carbon tracing is evident replace coil and be sure plug wires relating to that coil are clean and tight. Excessive wire resistance or faulty connections could have caused the coil to be damaged.
4. If the no spark condition follows the suspected coil, that coil is faulty. Otherwise, the ignition module is the cause of no spark. This test could also be performed by substituting a known good coil for the one causing the no spark condition.

1988–90 2.0L (VIN 1)
ALL MODELS EXCEPT CORSICA AND BERETTA

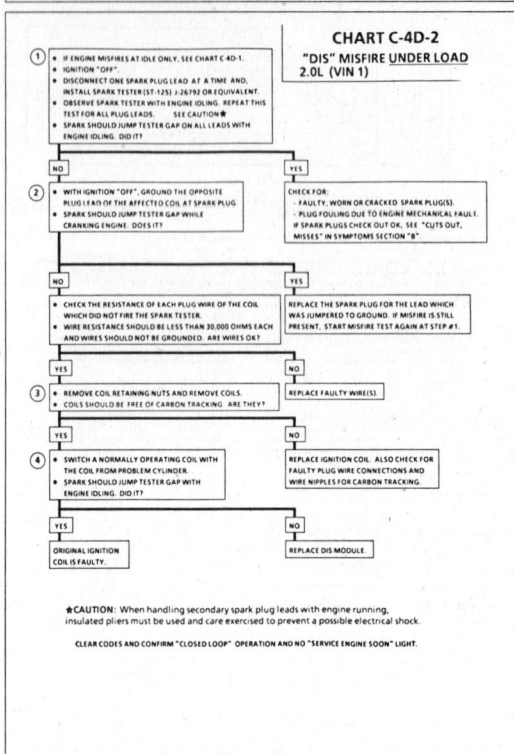

CHART C-4D-2
"DIS" MISFIRE UNDER LOAD
2.0L (VIN 1)

★ CAUTION: When handling secondary spark plug leads with engine running, insulated pliers must be used and care exercised to prevent a possible electrical shock.

CLEAR CODES AND CONFIRM "CLOSED LOOP" OPERATION AND NO "SERVICE ENGINE SOON" LIGHT.

1988–90 2.0L (VIN 1)
ALL MODELS EXCEPT CORSICA AND BERETTA

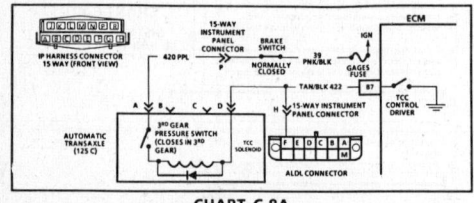

CHART C-8A
TORQUE CONVERTER CLUTCH (TCC)
2.0L (VIN 1)

Circuit Description:

The purpose of the automatic transaxle torque converter clutch is to eliminate the power loss of the torque converter when the vehicle is in a cruise condition. This allows the convenience of the automatic transaxle and the fuel economy of a manual transaxle.

Voltage is supplied to the TCC solenoid through the brake switch and transmission third gear apply switch. The ECM will engage TCC by grounding CKT 422 to energize the solenoid.

The TCC will engage under the following conditions:
- Vehicle speed exceeds 23 mph (37 kph) with A/C "OFF" (35 mph with A/C "ON")
- Engine coolant temperature exceeds 30°C (86°F)
- Throttle position sensor output is not changing faster than a calibrated rate (steady throttle)
- Transaxle third gear switch is closed
- Brake switch is closed

Test Description: Numbers below refer to circled numbers on the diagnostic chart.

1. Light "OFF" confirms that transaxle third gear apply switch is open.
2. At 48 km/h (30 mph), the transaxle third gear switch should close. Test light will light and confirm battery supply and closed brake switch.
3. Grounding the diagnostic terminal with engine "OFF" should energize the TCC solenoid. This test checks the capability of the ECM to control the solenoid.

Diagnostic Aids:

An engine coolant thermostat that is stuck open or opens at too low a temperature may result in an inoperative TCC.

1988–90 2.0L (VIN 1)
ALL MODELS EXCEPT CORSICA AND BERETTA

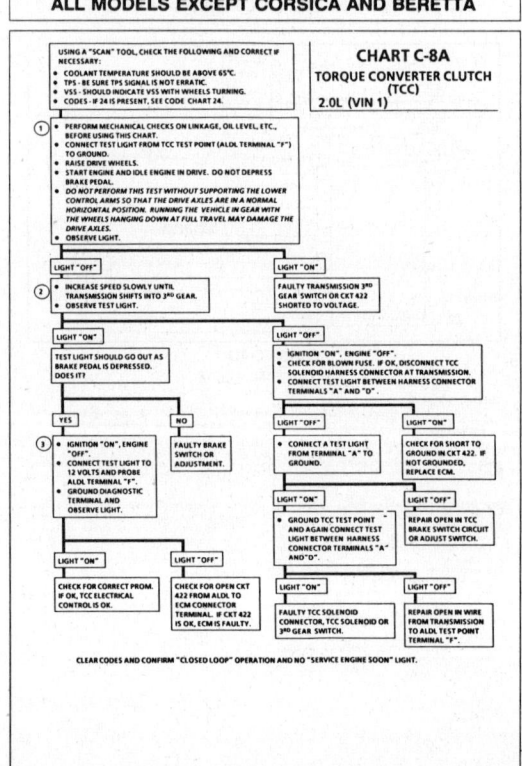

CHART C-8A
TORQUE CONVERTER CLUTCH (TCC)
2.0L (VIN 1)

CLEAR CODES AND CONFIRM "CLOSED LOOP" OPERATION AND NO "SERVICE ENGINE SOON" LIGHT.

1988–90 2.0L (VIN 1)
ALL MODELS EXCEPT CORSICA AND BERETTA

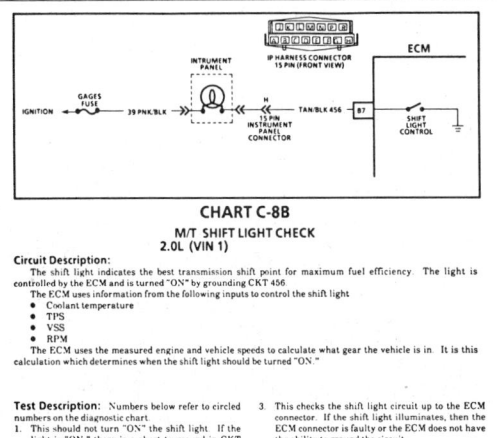

CHART C-8B
M/T SHIFT LIGHT CHECK
2.0L (VIN 1)

Circuit Description:
The shift light indicates the best transmission shift point for maximum fuel efficiency. The light is controlled by the ECM and is turned "ON" by grounding CKT 456.

The ECM uses information from the following inputs to control the shift light

- Coolant temperature
- TPS
- VSS
- RPM

The ECM uses the measured engine and vehicle speeds to calculate what gear the vehicle is in. It is this calculation which determines when the shift light should be turned "ON."

Test Description: Numbers below refer to circled numbers on the diagnostic chart.

1. This should not turn "ON" the shift light. If the light is "ON," there is a short to ground in CKT 456 wiring or a fault in the ECM.
2. When the diagnostic terminal is grounded, the ECM should ground CKT 456, and the shift light should be "ON."
3. This checks the shift light circuit up to the ECM connector. If the shift light illuminates, then the ECM connector is faulty or the ECM does not have the ability to ground the circuit.

1988–90 2.0L (VIN 1)
ALL MODELS EXCEPT CORSICA AND BERETTA

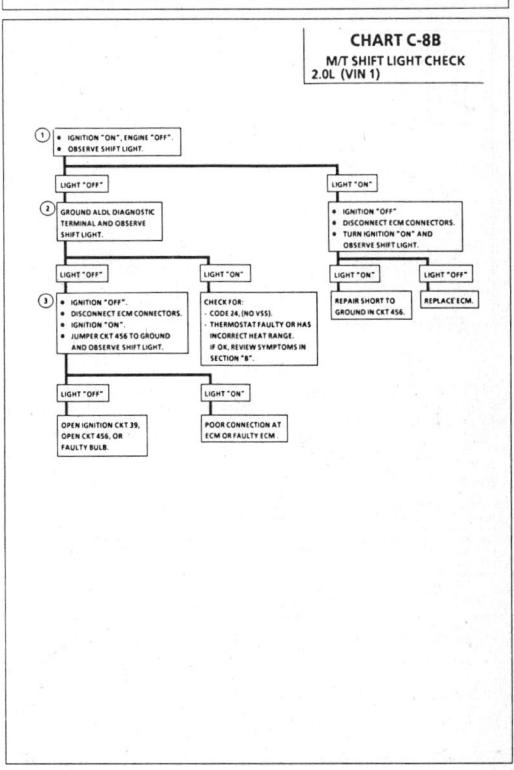

CHART C-8B
M/T SHIFT LIGHT CHECK
2.0L (VIN 1)

1988–90 2.0L (VIN 1)
ALL MODELS EXCEPT CORSICA AND BERETTA

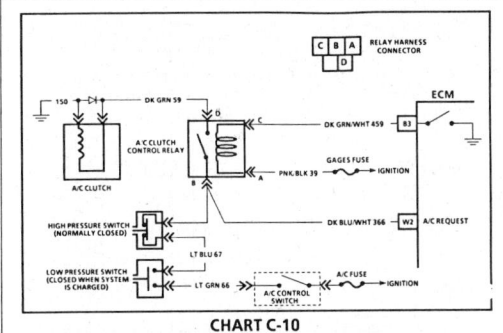

CHART C-10
A/C CLUTCH CONTROL
2.0L (VIN 1)

Circuit Description:
When an A/C mode is selected on the A/C control switch, ignition voltage is supplied to the compressor low pressure switch. If there is sufficient A/C refrigerant charge, the low pressure switch will be closed and complete the circuit to the closed high pressure cut off switch and to CKTs 67 and 366. The voltage on CKT 366 to the ECM is shown by the "Scan" tool as A/C request "ON" (voltage present), "OFF" (no voltage). When a request for A/C is seen by the ECM, the ECM will ground CKT 459 of the A/C clutch control relay, the relay contact will close, and current will flow from CKT 366 to CKT 59 and engage the A/C compressor clutch. A "Scan" tool will show the grounding of CKT 459 as A/C clutch "ON."

Diagnostic Aids:
Both pressure switches are located on the high side of the A/C system. The low pressure switch will be closed at 40-47 psi and allow A/C clutch operation. Below 37 psi, the low pressure switch will be open and the A/C clutch will not operate.

At about 430 psi, the high pressure switch will open to disengage the A/C clutch and prevent system damage.

1988–90 2.0L (VIN 1)
ALL MODELS EXCEPT CORSICA AND BERETTA

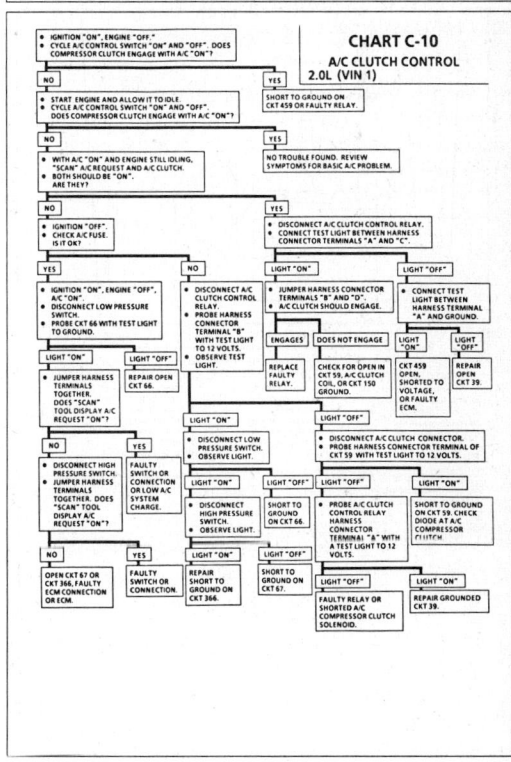

CHART C-10
A/C CLUTCH CONTROL
2.0L (VIN 1)

1988–90 2.0L (VIN 1)
ALL MODELS EXCEPT CORSICA AND BERETTA

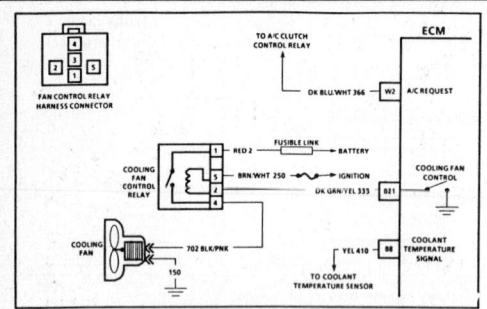

CHART C-12
ENGINE COOLING FAN
2.0L (VIN 1)

Circuit Description:

Battery voltage to operate the cooling fan motor is supplied to relay terminal "1". Ignition voltage to energize the relay is supplied to relay terminal "5". When the ECM grounds CKT 335, the relay is energized and the cooling fan is turned "ON." When the engine is running, the ECM will turn the cooling fan "ON" if the conditions for a coolant temperature sensor trouble code (Code 14 or 15) are present or under the following conditions:

- A/C is "ON"
- Coolant temperature is greater than 108°C (230°F)

Diagnostic Aids:

If the vehicle has an overheating problem, it must be determined if the complaint was due to an actual boil over, the coolant temperature warning light, or the temperature gage indicated over heating.

If the gage or light indicates overheating, but no boilover is detected, the gage circuit should be checked.

The gage accuracy can be checked by comparing the coolant sensor reading on a "Scan" tool with the gage reading.

If the engine is actually overheating and the gage indicates overheating but the cooling fan is not turning "ON", the coolant sensor has probably shifted out of calibration and should be replaced.

1988–90 2.0L (VIN 1)
ALL MODELS EXCEPT CORSICA AND BERETTA

CHART C-12
ENGINE COOLING FAN
2.0L (VIN 1)

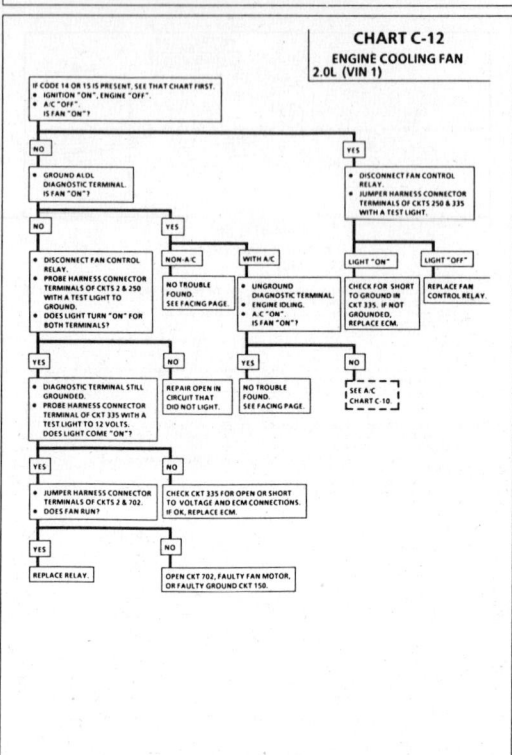

2.0L (VIN K, OHC) COMPONENT LOCATIONS

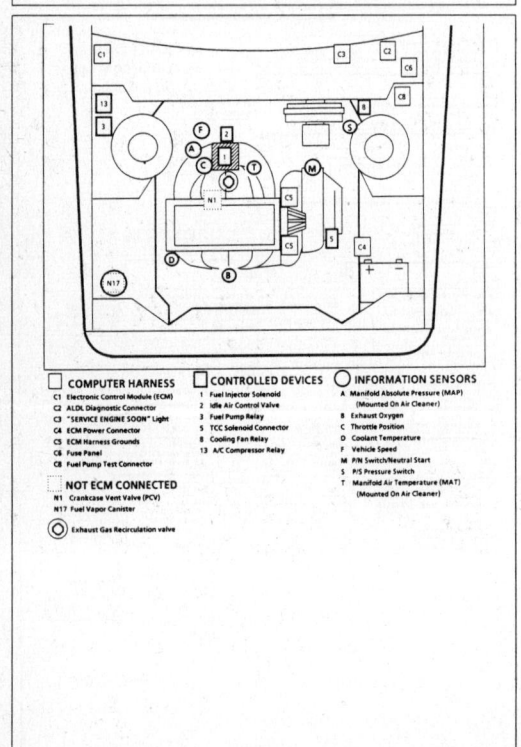

COMPUTER HARNESS
C1 Electronic Control Module (ECM)
C2 ALDL Diagnostic Connector
C3 "SERVICE ENGINE SOON" Light
C4 ECM Power Connector
C5 ECM Harness Grounds
C6 Fuse Panel
Fuel Pump Test Connector

NOT ECM CONNECTED
N1 Crankcase Vent Valve (PCV)
N17 Fuel Vapor Canister

Exhaust Gas Recirculation valve

CONTROLLED DEVICES
1 Fuel Injector Solenoid
2 Idle Air Control Valve
3 Fuel Pump Relay
5 TCC Solenoid Connector
6 Cooling Fan Relay
13 A/C Compressor Relay

INFORMATION SENSORS
A Manifold Absolute Pressure (MAP) (Mounted On Air Cleaner)
B Exhaust Oxygen
C Throttle Position
D Coolant Temperature
F Vehicle Speed
M P/N Switch/Neutral Start
S P/S Pressure Switch
T Manifold Air Temperature (MAT) (Mounted On Air Cleaner)

2.0L (VIN K, OHC) ECM WIRING DIAGRAM

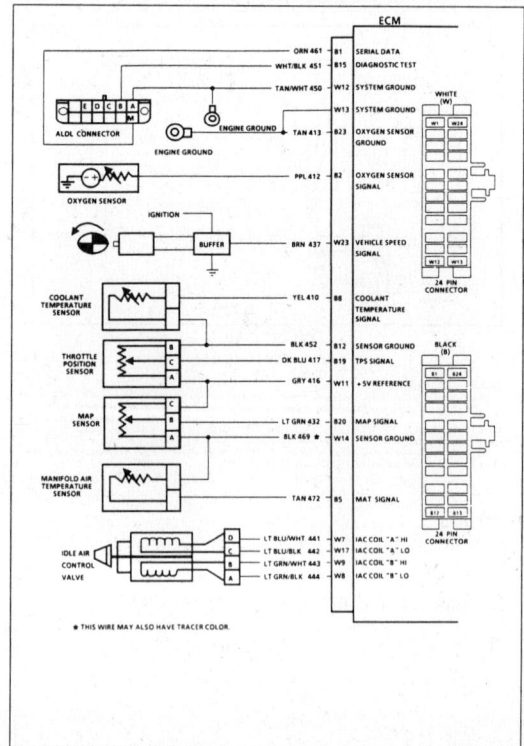

2.0L (VIN K, OHC) ECM WIRING DIAGRAM (CONT.)

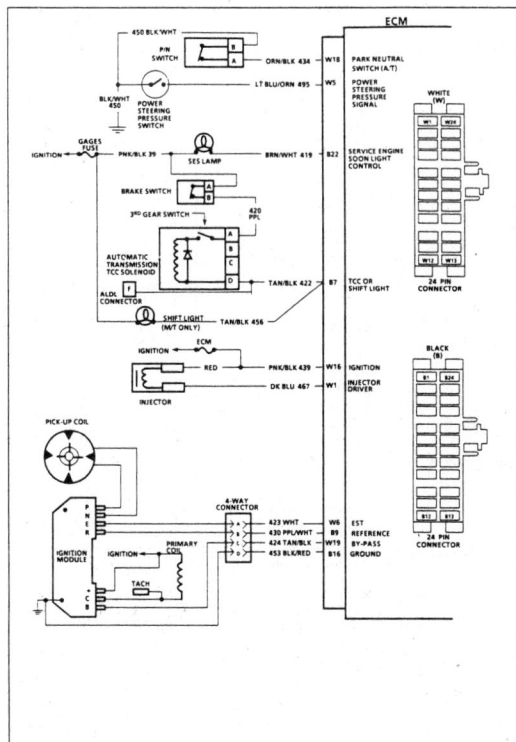

2.0L (VIN K, OHC) ECM WIRING DIAGRAM (CONT.)

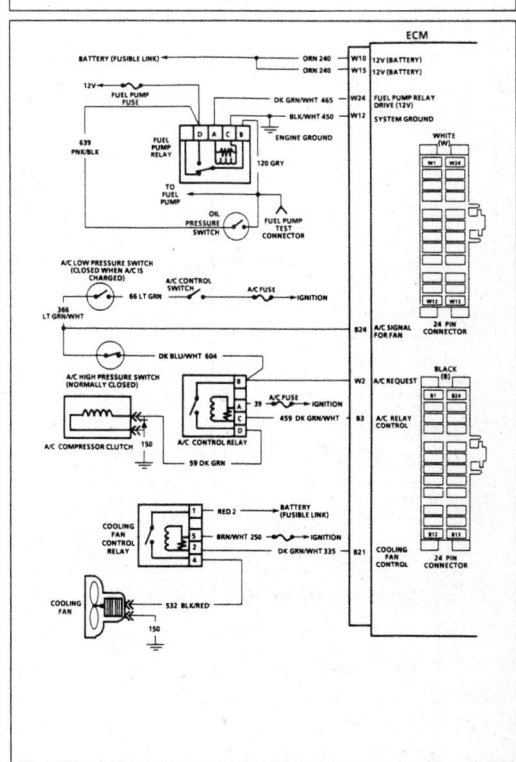

2.0L (VIN K, OHC) ECM CONNECTOR TERMINAL END VIEW

FUEL INJECTION ECM CONNECTOR IDENTIFICATION

This ECM voltage chart is for use with a digital voltmeter to further aid in diagnosis. The voltages you get may vary due to low battery charge or other reasons, but they should be very close.

THE FOLLOWING CONDITIONS MUST BE MET BEFORE TESTING:
- Engine at operating temperature • Engine idling in closed loop (for "Engine Run" column)
- Test terminal not grounded • "Scan" tool not installed • All voltages shown "B +" indicates battery or charging voltage.

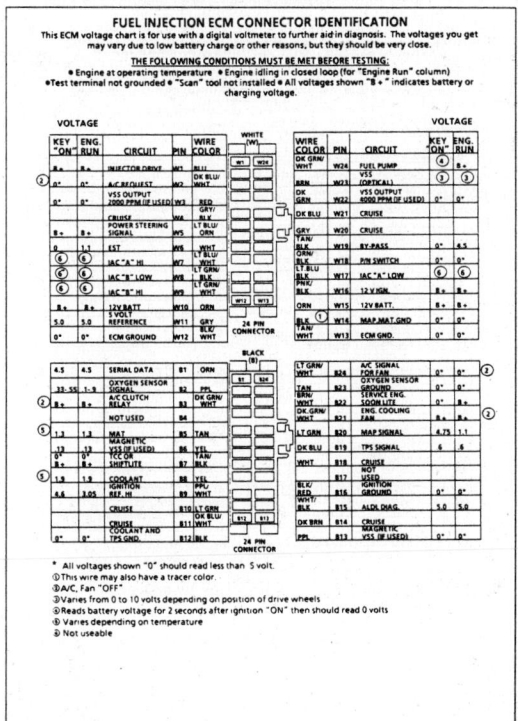

* All voltages shown "0" should read less than .5 volt.
① This wire may also have a tracer color.
② A/C, Fan "OFF"
③ Varies from 0 to 10 volts depending on position of drive wheels
④ Reads battery voltage for 2 seconds after ignition "ON" then should read 0 volts
⑤ Varies depending on temperature
⑥ Not useable

2.0L (VIN K, OHC) COMPONENT LOCATIONS SUNBIRD

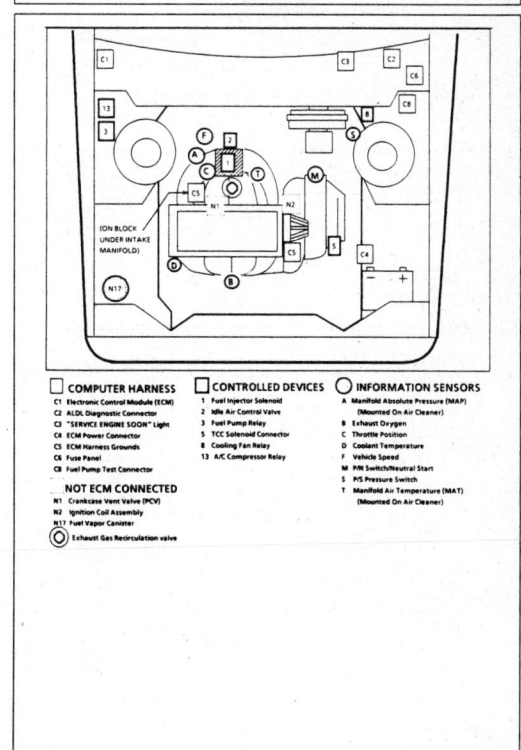

☐ **COMPUTER HARNESS**
C1 Electronic Control Module (ECM)
C2 ALDL Diagnostic Connector
C3 "SERVICE ENGINE SOON" Light
C4 ECM Power Connector
C5 ECM Harness Grounds
C6 Fuse Panel
C8 Fuel Pump Test Connector

NOT ECM CONNECTED
N1 Crankcase Vent Valve (PCV)
N2 Ignition Coil Assembly
N17 Fuel Vapor Canister
○ Exhaust Gas Recirculation Valve

☐ **CONTROLLED DEVICES**
1 Fuel Injector Solenoid
2 Idle Air Control Valve
3 Fuel Pump Relay
5 TCC Solenoid Connector
8 Cooling Fan Relay
13 A/C Compressor Relay

○ **INFORMATION SENSORS**
A Manifold Absolute Pressure (MAP)
  (Mounted On Air Cleaner)
B Exhaust Oxygen
C Throttle Position
D Coolant Temperature
F Vehicle Speed
M P/N Switch/Neutral Start
S P/S Pressure Switch
T Manifold Air Temperature (MAT)
  (Mounted On Air Cleaner)

2.0L (VIN K, OHC) ECM WIRING DIAGRAM SUNBIRD

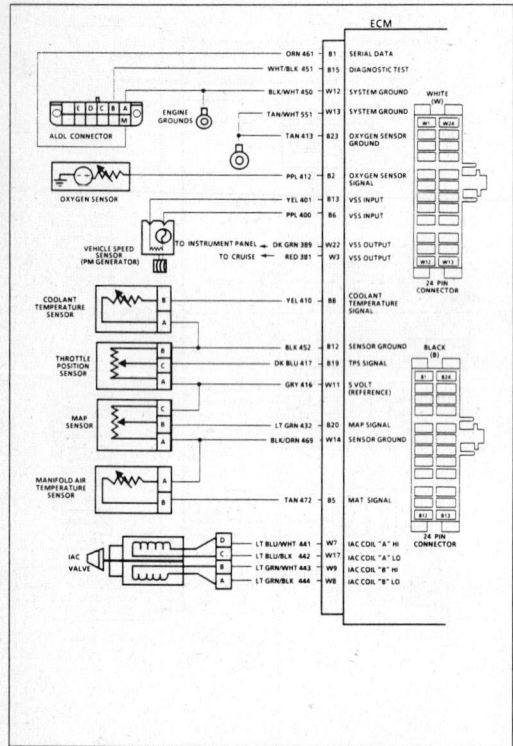

2.0L (VIN K, OHC) ECM WIRING DIAGRAM SUNBIRD (CONT.)

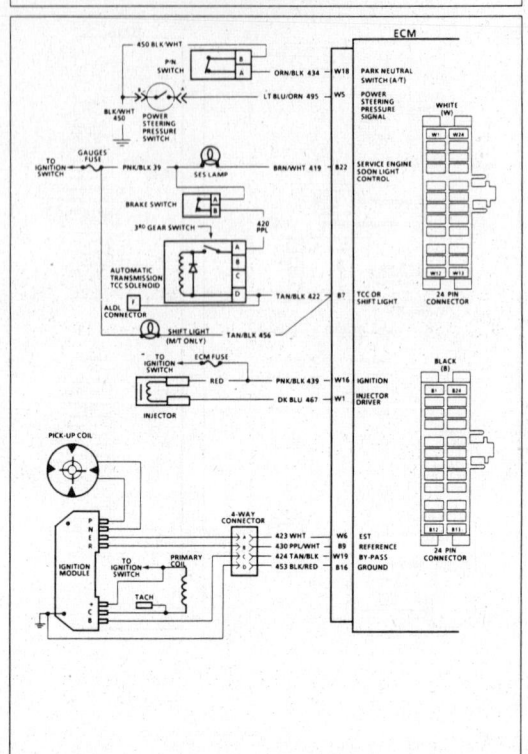

2.0L (VIN K, OHC) ECM WIRING DIAGRAM SUNBIRD (CONT.)

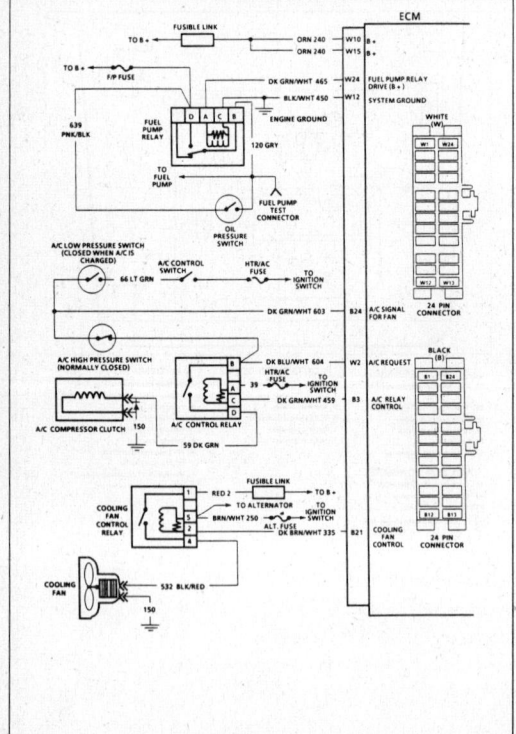

2.0L (VIN K, OHC) ECM CONNECTOR TERMINAL END VIEW—SUNBIRD

FUEL INJECTION ECM CONNECTOR IDENTIFICATION
This ECM voltage chart is for use with a digital voltmeter to further aid in diagnosis. The voltages you get may vary due to low battery charge or other reasons, but they should be very close.

THE FOLLOWING CONDITIONS MUST BE MET BEFORE TESTING:
• Engine at operating temperature • Engine idling in "Closed Loop" (for "Engine Run" column)
• "Test" terminal not grounded • "Scan" tool not installed • All voltages shown "B+" indicates battery or charging voltage.

VOLTAGE KEY "ON"	ENG. RUN	CIRCUIT	PIN	WIRE COLOR
B+	B+	INJECTOR DRIVE	W1	BLU
0*	0*	A/C REQUEST	W2	DK BLU
0*	0*	VSS OUTPUT 2000 PPM	W3	RED
		CRUISE	W4	GRY
B+	B+	POWER STEERING SIGNAL	W5	LT BLU/ORN
0	1.1	EST	W6	LT BLU
④	④	IAC "A" HI	W7	LT GRN
④	④	IAC "B" LOW	W8	LT GRN/WHT
④	④	IAC "B" HI	W9	LT GRN/WHT
B+	B+	12V BATT	W10	ORN
5.0	5.0	5 VOLT REFERENCE	W11	GRY
0*	0*	ECM GROUND	W12	BLK

VOLTAGE KEY "ON"	ENG. RUN	CIRCUIT	PIN	WIRE COLOR
3.9	3.9	SERIAL DATA	B1	ORN
33.55	1.9	OXYGEN SENSOR SIGNAL	B2	PPL
B+	B+	A/C CLUTCH RELAY	B3	DK GRN/WHT
		NOT USED	B4	
1.3	1.3	MAT	B5	TAN
.13	.13	MAGNETIC VSS (IF USED)	B6	PPL
0*	0*	TCC OR SHIFTLITE	B7	TAN/BLK
B+	B+	COOLANT	B8	YEL
1.9	1.9	IGNITION REF HI	B9	WHT
		CRUISE	B10	LT GRN
		CRUISE	B11	DK BLU/WHT
0*	0*	COOLANT AND TPS GND	B12	BLK

WIRE COLOR	PIN	CIRCUIT	KEY "ON"	ENG. RUN
DK GRN/WHT	W24	FUEL PUMP	②	B+
	W23	NOT USED		
DK GRN	W22	VSS OUTPUT 4000 PPM (IF USED)	0*	0*
DK BLU	W21	CRUISE		
GRY/TAN	W20	CRUISE		
BLK	W19	BY-PASS	0*	4.7
ORN	W18	P/N SWITCH	0*	0*
LT BLU	W17	IAC "A" LOW	④	④
PNK/BLK	W16	12V IGN	B+	B+
ORN	W15	12V BATT	B+	B+
BLK/ORN	W14	MAP, MAT, GND	0*	0*
BLK/WHT	W13	ECM GND	0*	0*

WIRE COLOR	PIN	CIRCUIT	KEY "ON"	ENG. RUN
DK GRN/WHT	B24	A/C SIGNAL	0*	0*
TAN	B23	OXYGEN SENSOR GROUND	0*	0*
BRN	B22	SERVICE ENG SOON LITE	0*	B+
DK GRN/WHT	B21	ENG COOLING FAN	0*	0*
LT GRN	B20	MAP SIGNAL	4.8	1.1
DK BLU	B19	TPS SIGNAL	.6	.6
	B18	NOT USED		
	B17	NOT USED		
BLK/RED	B16	IGNITION GROUND	0*	0*
WHT	B15	ALDL DIAG	5.0	5.0
YEL	B14	CRUISE MAGNETIC VSS (IF USED)	0*	0*

* All voltages shown "0" should read less than .5 volt
① A/C, Fan "OFF"
② Reads battery voltage for 2 seconds after ignition "ON" then should read 0 volts
③ Varies depending on temperature
④ Not usable

2.0L (VIN K, OHC) COMPONENT LOCATIONS FIRENZA AND SKYHAWK

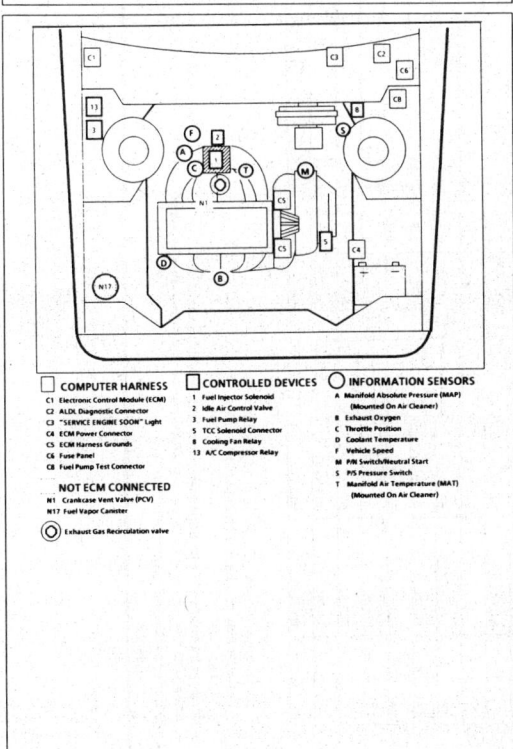

☐ **COMPUTER HARNESS**
C1 Electronic Control Module (ECM)
C2 ALDL Diagnostic Connector
C3 "SERVICE ENGINE SOON" Light
C4 ECM Power Connector
C5 ECM Harness Grounds
C6 Fuse Panel
C8 Fuel Pump Test Connector

NOT ECM CONNECTED
N1 Crankcase Vent Valve (PCV)
N17 Fuel Vapor Canister

◎ Exhaust Gas Recirculation valve

☐ **CONTROLLED DEVICES**
1 Fuel Injector Solenoid
2 Idle Air Control Valve
3 Fuel Pump Relay
5 TCC Solenoid Connector
6 Cooling Fan Relay
13 A/C Compressor Relay

○ **INFORMATION SENSORS**
A Manifold Absolute Pressure (MAP) (Mounted On Air Cleaner)
B Exhaust Oxygen
C Throttle Position
D Coolant Temperature
F Vehicle Speed
M P/N Switch/Neutral Start
S P/S Pressure Switch
T Manifold Air Temperature (MAT) (Mounted On Air Cleaner)

2.0L (VIN K, OHC) ECM WIRING DIAGRAM FIRENZA AND SKYHAWK

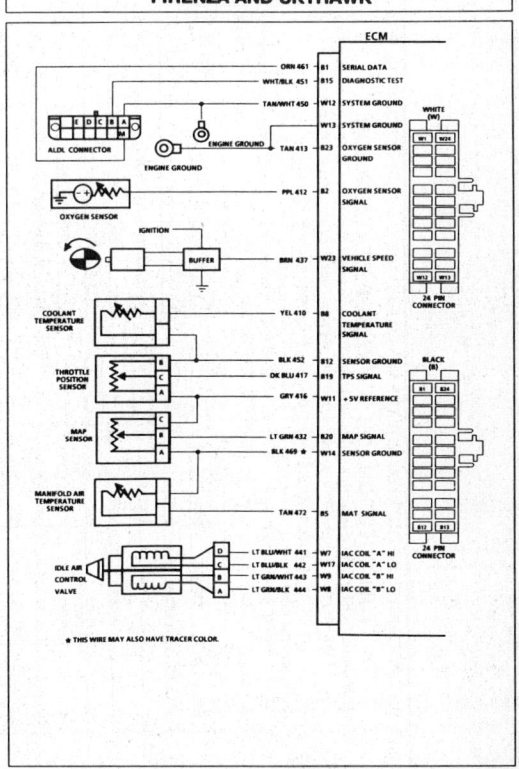

2.0L (VIN K, OHC) ECM WIRING DIAGRAM FIRENZA AND SKYHAWK (CONT.)

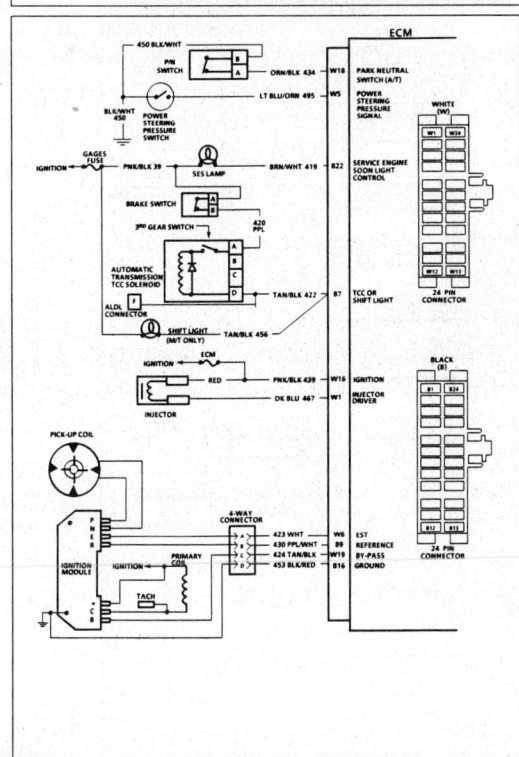

2.0L (VIN K, OHC) ECM WIRING DIAGRAM FIRENZA AND SKYHAWK (CONT.)

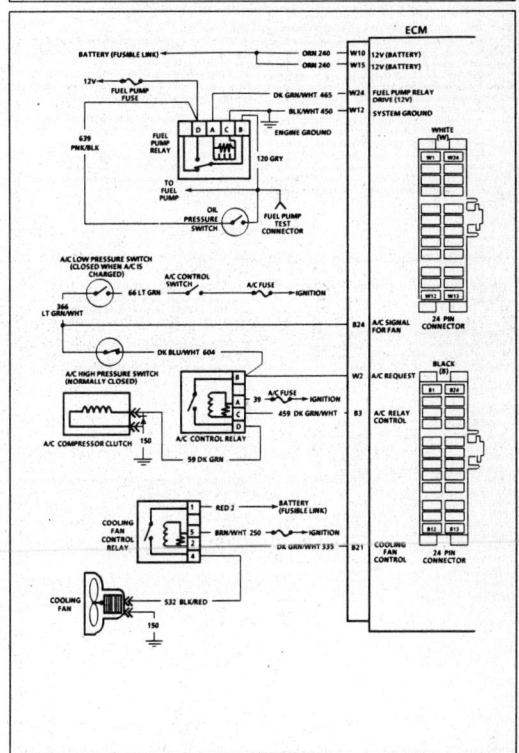

2.0L (VIN K, OHC) ECM CONNECTOR TERMINAL END VIEW – FIRENZA AND SKYHAWK
FUEL INJECTION ECM CONNECTOR IDENTIFICATION

This ECM voltage chart is for use with a digital voltmeter to further aid in diagnosis. The voltages you get may vary due to low battery charge or other reasons, but they should be very close.

THE FOLLOWING CONDITIONS MUST BE MET BEFORE TESTING:
● Engine at operating temperature ● Engine idling in closed loop (for "Engine Run" column) ● Test terminal not grounded ● "Scan" tool not installed ● All voltages shown "B +" indicates battery or charging voltage.

WHITE (W) — 24 PIN CONNECTOR

KEY "ON"	ENG "RUN"	WIRE COLOR	PIN	CIRCUIT
B+	B+	BLU	W1	INJECTOR DRIVE (2)
0*	0*	DK BLU/WHT	W2	A/C REQUEST
0*	0*	RED/GRY	W3	VSS OUTPUT 2000 RPM (IF USED)
		GRY/BLK	W4	CRUISE
B+	B+	LT BLU/ORN	W5	POWER STEERING SIGNAL
1.1	(6)	WHT	W6	EST
(6)	(6)	LT BLU/WHT	W7	IAC "A" HI
0*	(6)	LT GRN/BLK	W8	IAC "B" LOW
(6)	(6)	LT GRN/WHT	W9	IAC "B" HI
B+	B+	ORN	W10	12V BATT.
5.0	5.0	GRY/BLK	W11	5 VOLT REFERENCE
0*	0*	BLK/WHT	W12	ECM GROUND

WIRE COLOR	PIN	CIRCUIT	KEY "ON"	ENG "RUN"
DK GRN/WHT	W24	FUEL PUMP	(4)	B+
BRN	W23	VSS (OPTICAL)	(3)	(3)
DK GRN	W22	VSS OUTPUT 4000 RPM (IF USED)	0*	0*
DK BLU	W21	CRUISE		
GRY	W20	CRUISE		
TAN/BLK	W19	BY-PASS	0*	4.5
ORN/BLK	W18	P/N SWITCH	(6)	(6)
LT BLU/BLK	W17	IAC "A" LOW	0*	0*
PNK/BLK	W16	12 V IGN.	B+	B+
ORN	W15	12V BATT.	B+	B+
TAN/WHT	W14	MAP,MAT,GND	0*	0*
BLK/WHT	W13	ECM GND.	0*	0*

BLACK (B) — 24 PIN CONNECTOR

KEY "ON"	ENG "RUN"	WIRE COLOR	PIN	CIRCUIT
4.5	4.5	ORN	B1	SERIAL DATA
.33-.55	1-.9	PPL	B2	OXYGEN SENSOR SIGNAL (2)
B+	B+	DK GRN/WHT	B3	A/C CLUTCH RELAY
—	—		B4	NOT USED
1.3	1.3	TAN	B5	MAT
.13	.13	YEL/BLK	B6	MAGNETIC VSS (IF USED)
0*	0*	YEL/BLK	B7	TCC OR SHIFTLITE
B+	B+	YEL	B8	COOLANT REF. HI
1.9	1.9	IPU/WHT	B9	COOLANT
4.6	3.05	LT GRN/DK BLU WHT	B10	CRUISE
		WHT	B11	COOLANT AND TPS GND. (5)
0*	0*	BLK	B12	CRUISE

WIRE COLOR	PIN	CIRCUIT	KEY "ON"	ENG "RUN"
LT GRN/WHT	B24	A/C SIGNAL FOR FAN	0*	0* (2)
TAN/BRN/WHT	B23	OXYGEN SENSOR SIGNAL	0*	0*
DK GRN/WHT	B22	SERVICE ENG. SOON LIGHT	B+	B+
LT GRN	B21	MAP SIGNAL	4.75	1.1
DK BLU	B20	TPS SIGNAL	.6	.6
WHT	B19	CRUISE		
	B18	NOT USED		
BLK/RED	B17	USED IGNITION GROUND		
WHT/BLK	B16	IGNITION GROUND	0*	0*
BLK	B15	ALDL DIAG.	5.0	5.0
DK BRN	B14	CRUISE		
PPL	B13	MAGNETIC VSS (IF USED)	0*	0* (2)

* All voltages shown "0" should read less than .5 volt. This wire may also have a tracer color.
(1) A/C, Fan "OFF"
(2) Varies from 0 to 10 volts depending on position of drive wheels
(3) Reads battery voltage for 2 seconds after ignition "ON" then should read 0 volts
(4) Varies depending on temperature
(5) Not useable

1988–90 2.0L (VIN K, OHC) – ALL MODELS

DIAGNOSTIC CIRCUIT CHECK

The Diagnostic Circuit Check is an organized approach to identifying a problem created by an electronic engine control system malfunction. It must be the starting point for any driveability complaint diagnosis because it directs the service technician to the next logical step in diagnosing the problem.

The "Scan" data listed in the table may be used for comparison after completing the diagnostic circuit check and finding the on-board diagnostics functioning properly with no trouble codes displayed. The "Typical Data Values" are an average of display values recorded from normally operating vehicles and are intended to represent what a normally functioning system would typically display.

A "SCAN" TOOL THAT DISPLAYS FAULTY DATA SHOULD NOT BE USED, AND THE PROBLEM SHOULD BE REPORTED TO THE MANUFACTURER. THE USE OF A FAULTY "SCAN" TOOL CAN RESULT IN MISDIAGNOSIS AND UNNECESSARY PARTS REPLACEMENT.

Only the parameters listed below are used in this manual for diagnosis. If a "Scan" tool reads other parameters, the values are not recommended by General Motors for use in diagnosis. For more description on the values and use of the "Scan" tool to diagnosis ECM inputs, refer to the applicable component diagnosis section. If all values are within the range illustrated, refer to symptoms in Section "B".

"SCAN" TOOL DATA

Test Under Following Conditions: Idle, Upper Radiator Hose Hot, Closed Throttle, Park or Neutral, "Closed Loop", All Accessories "Off".

"SCAN" Position	Units Displayed	Typical Data Value
Desired RPM	RPM	ECM idle command (varies with temperature)
RPM	RPM	± 50 RPM from desired rpm in drive (AUTO) ± 100 RPM from desired rpm in neutral (MANUAL)
Coolant Temperature	Degrees Celsius	85 - 105
MAT Temperature	Degrees Celsius	10 - 90 (varies with underhood temperature and sensor location)
MAP	Volts	1 - 2 (varies with manifold and barometric pressures)
BPW (base pulse width)	Milliseconds	.8 - 3.0
O2	Volts	.1 - 1 (varies continuously)
TPS	Volts	.4 - 1.25
Throttle Angle	0 - 100%	0
IAC	Counts (steps)	1 - 50
P/N Switch	P-N and R-D-L	Park/Neutral (P/N)
INT (Integrator)	Counts	110 - 145
BLM (Block Learn Memory)	Counts	118 - 138
Open/Closed Loop	Open/Closed	"Closed Loop" (may enter "Open Loop" with extended idle)
VSS	MPH	0
TCC	ON/OFF	"OFF"
Spark Advance	Degrees	Varies
Battery	Volts	13.5 - 14.5
Fan	ON/OFF	"OFF" (coolant temperature below 102°C)
P/S Switch	Normal/Hi Pressure	Normal
A/C Request	Yes/No	No
A/C Clutch	ON/OFF	"OFF"
Shift Light (M/T)	ON/OFF	"OFF"

1988–90 2.0L (VIN K, OHC) – ALL MODELS

DIAGNOSTIC CIRCUIT CHECK
2.0L OHC (VIN K)

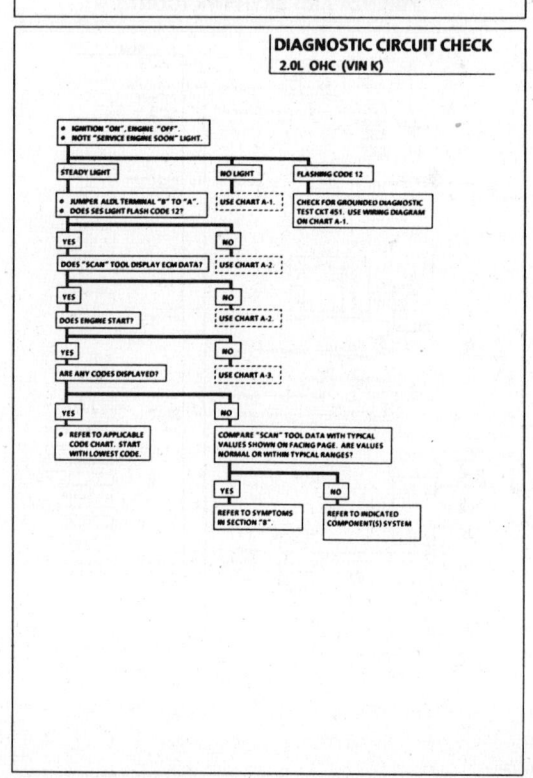

1988–90 2.0L (VIN K, OHC) – ALL MODELS

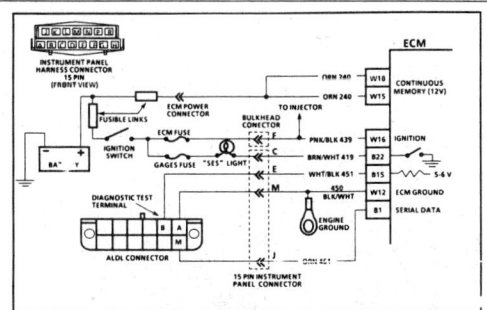

CHART A-1
NO "SERVICE ENGINE SOON" LIGHT
2.0L OHC (VIN K)

Circuit Description:
There should always be a steady "Service Engine Soon" light, when the ignition is "ON" and engine stopped. Battery is supplied directly to the light bulb. The electronic control module (ECM) will control the light and turn it "ON" by providing a ground path throught CKT 419 to the ECM

Test Description: Numbers below refer to circled numbers on the diagnostic chart
1. Battery feed CKT 240 is protected by a fusible link at the battery.
2. Using a test light connected to 12 volts, probe each of the system ground circuits to be sure a good ground is present. See ECM terminal end view in front of this section for ECM pin locations of ground circuits.

Diagnostic Aids:

Engine runs OK, check
- Faulty light bulb
- CKT 419 open
- Gage fuse blown. This will result in no oil or generator lights, seat belt reminder, etc

Engine cranks, but will not run.
- Continuous battery - fuse or fusible link open
- ECM ignition fuse open
- Battery CKT 240 to ECM open
- Ignition CKT 439 to ECM open
- Poor connection to ECM

1988–90 2.0L (VIN K, OHC) – ALL MODELS

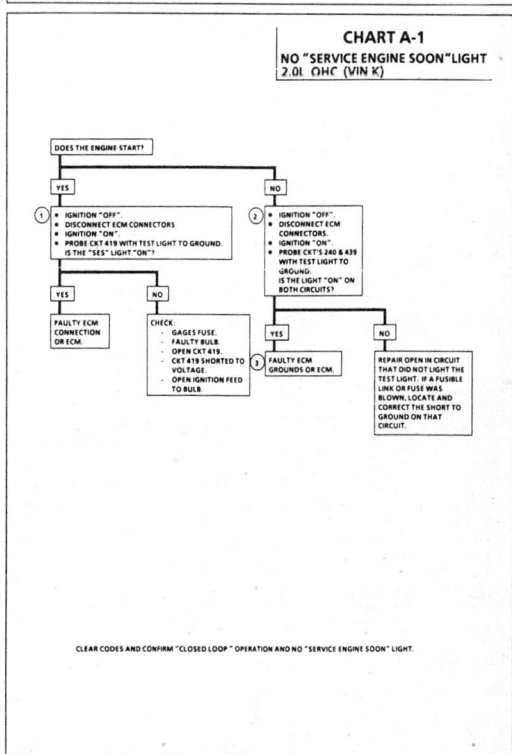

CHART A-1
NO "SERVICE ENGINE SOON" LIGHT
2.0L OHC (VIN K)

1988–90 2.0L (VIN K, OHC) – ALL MODELS

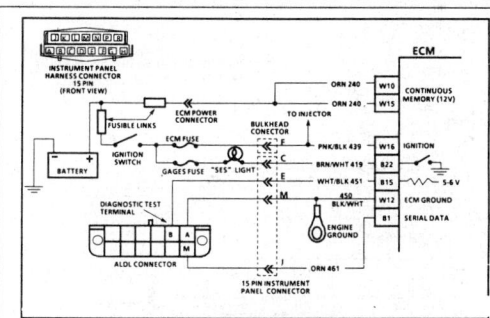

CHART A-2
NO ALDL DATA OR WON'T FLASH CODE 12
"SERVICE ENGINE SOON" LIGHT "ON" STEADY
2.0L OHC (VIN K)

Circuit Description:
There should always be a steady "Service Engine Soon" light, when the ignition is "ON" and engine stopped. Battery is supplied directly to the light bulb. The electronic control module (ECM) will turn the light "ON" by grounding CKT 419 at the ECM
With the diagnostic terminal grounded, the light should flash a Code 12, followed by any trouble code(s) stored in memory.
A steady light suggests a short to ground in the light control CKT 419, or an open in diagnostic CKT 451.

Test Description: Numbers below refer to circled numbers on the diagnostic chart
1. If there is a problem with the ECM that causes a "Scan" tool to not read Serial data, then the ECM should not flash a Code 12. If Code 12 does flash, be sure that the "Scan" tool is working properly on another vehicle. If the "Scan" is functioning properly and CKT 461 is OK, the PROM, or ECM, may be at fault for the NO ALDL symptom.
2. If the light goes "OFF" when the ECM connector is disconnected, then CKT 419 is not shorted to ground
3. This step will check for an open diagnostic CKT 451
4. At this point, the "Service Engine Soon" light wiring is OK. The problem is a faulty ECM or PROM. If Code 12 does not flash, the ECM should be replaced using the original PROM. Replace the PROM only after trying an ECM, as a defective PROM is an unlikely cause of the problem.

1988–90 2.0L (VIN K, OHC) – ALL MODELS

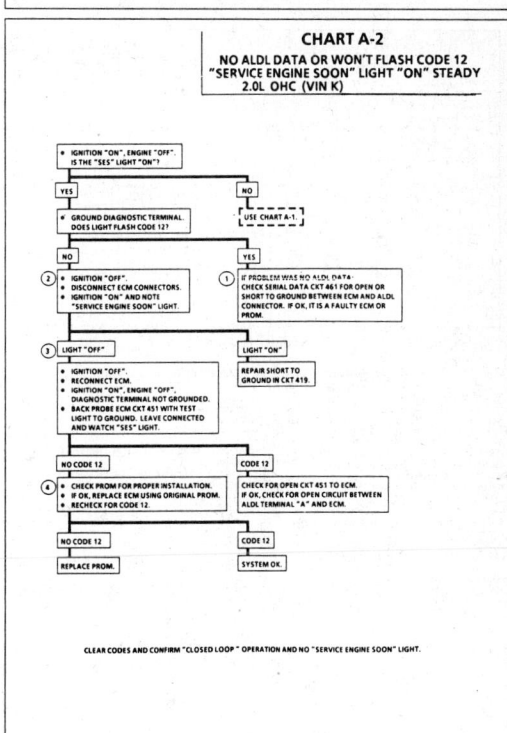

CHART A-2
NO ALDL DATA OR WON'T FLASH CODE 12
"SERVICE ENGINE SOON" LIGHT "ON" STEADY
2.0L OHC (VIN K)

1988–90 2.0L (VIN K, OHC) – ALL MODELS

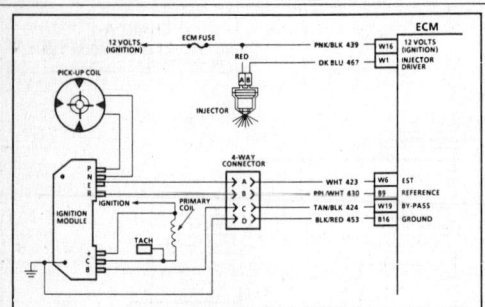

CHART A-3
(Page 1 of 2)
ENGINE CRANKS BUT WON'T RUN
2.0L OHC (VIN K)

Circuit Description:
Before using this chart, battery condition, engine cranking speed, and fuel quantity should be checked and verified as being OK.

Test Description: Numbers below refer to circled numbers on the diagnostic chart.

1. A "Service Engine Soon" light "ON" is a basic test to determine if there is a 12 volts supply and ignition 12 volts to ECM. No ALDL may be due to an ECM problem, and CHART A-2 will diagnose the ECM. If TPS is over 2.5 volts, the engine may be in the clear flood mode, which will cause starting problems. The engine will not start without reference pulses and, therefore, the "Scan" should read rpm (reference) during cranking.

2. If rpm was indicated during crank, the ignition module is receiving a crank signal, but "No Spark" at this test indicates the ignition module is not triggering the coil.

3. While cranking engine, there should be no fuel spray with injector disconnected. Replace the injector, if it sprays fuel or drips like a leaking water faucet.

4. The test light should blink, indicating the ECM is controlling the injector OK. How bright the light blinks is not important. However, the test light should be a BT 8329 or equivalent.

5. Fuel spray from the injector indicates that fuel is available. However, the engine could be severely flooded due to too much fuel. No fuel spray from injector indicates a faulty fuel system or no ECM control of injector.

6. This test will determine if the ignition module is not generating the reference pulse, or if the wiring or ECM are at fault. By touching and removing a test light to 12 volts on CKT 430, a reference pulse should be generated. If rpm is indicated, the ECM and wiring are OK.

Diagnostic Aids:
- Water or foreign material can cause a no start during freezing weather. The engine may start after 5 or 6 minutes in a heated shop. The problem may not re-occur until an overnight park in freezing temperatures.
- An EGR sticking open can cause a low air/fuel ratio during cranking. Unless engine enters "Clear Flood" at the first indication of a flooding condition, it can result in a no start.
- Fuel pressure: Low fuel pressure can result in a very lean air/fuel ratio. See CHART A-7.

1988–90 2.0L (VIN K, OHC) – ALL MODELS

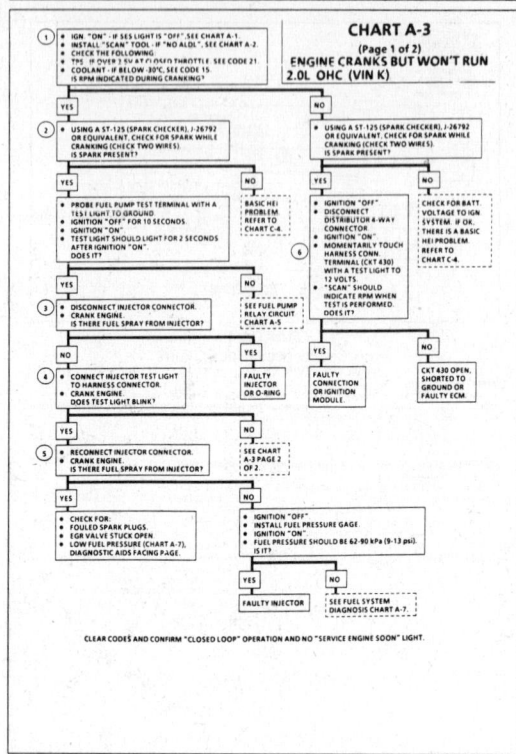

1988–90 2.0L (VIN K, OHC) – ALL MODELS

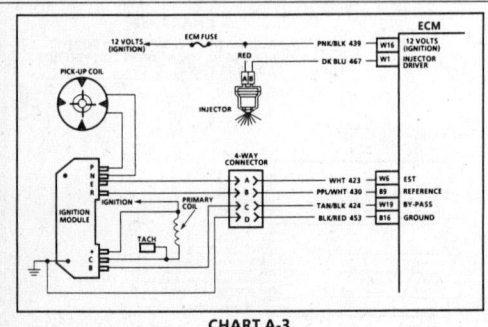

CHART A-3
(Page 2 of 2)
ENGINE CRANKS BUT WON'T RUN
2.0L OHC (VIN K)

Circuit Description:
Ignition voltage is supplied to the fuel injector on CKT 439. The injector will be pulsed (turned "ON" and "OFF") when the ECM opens and grounds injector drive CKT 467.

Test Description: Numbers below refer to circled numbers on the diagnostic chart.

1. This check determines if injector connector has ignition voltage and on only one terminal.

2. A faulty ECM may result in damage to the injector.

3. A test light connected from ECM harness terminal "W1" to ground should light due to continuity through the injector.

1988–90 2.0L (VIN K, OHC) – ALL MODELS

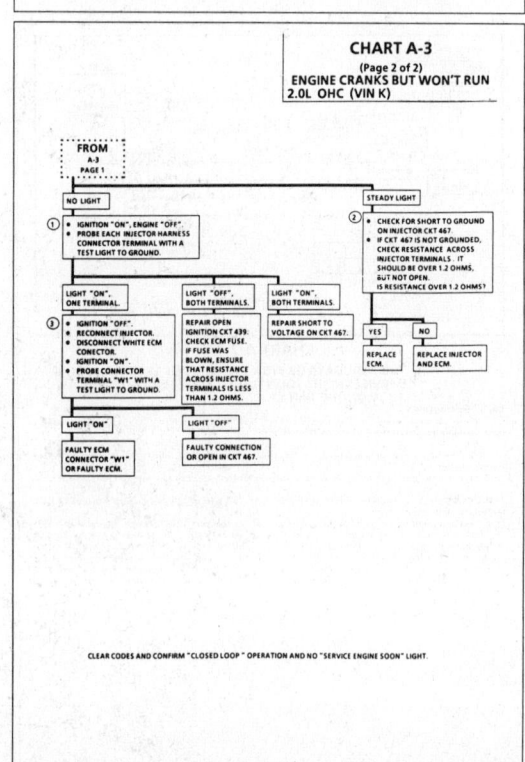

1988–90 2.0L (VIN K, OHC) – ALL MODELS

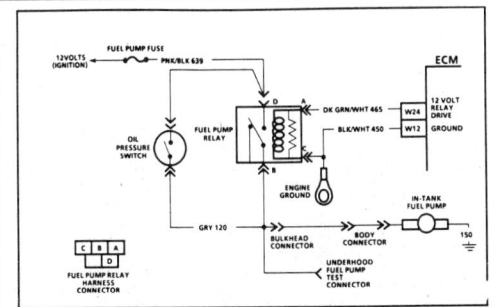

CHART A-5
FUEL PUMP RELAY CIRCUIT
2.0L OHC (VIN K)

Circuit Description:

When the ignition switch is turned "ON," the electronic control module (ECM) will activate the fuel pump relay and run the in-tank fuel pump. The fuel pump will operate as long as the engine is cranking or running and the ECM is receiving ignition reference pulses.

If there are no reference pulses, the ECM will shut "OFF" the fuel pump within 2 seconds after key "ON."

Should the fuel pump relay, or the 12 volts relay drive from the ECM fail, the fuel pump will be run through an oil pressure switch back-up circuit.

Diagnostic Aids:

An inoperative fuel pump relay can result in long cranking times, particularly if the engine is cold or engine oil pressure is low. The extended crank period is caused by the time necessary for oil pressure to build enough to close the oil pressure switch and turn "ON" the fuel pump. The fuel pump part of the oil pressure switch closes at about 28 kPa (4 psi).

1988–90 2.0L (VIN K, OHC) – ALL MODELS

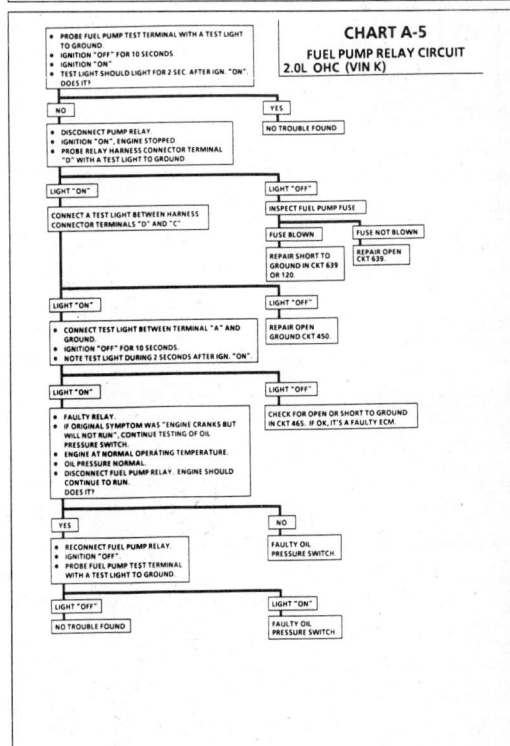

1988–90 2.0L (VIN K, OHC) – ALL MODELS

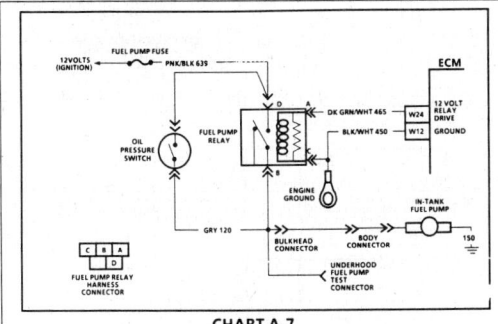

CHART A-7
(Page 1 of 2)
FUEL SYSTEM DIAGNOSIS
2.0L OHC (VIN K)

Circuit Description:

When the ignition switch is turned "ON," the electronic control module (ECM) will turn "ON" the in-tank fuel pump. It will remain "ON" as long as the engine is cranking or running, and the ECM is receiving ignition reference pulses.

If there are no reference pulses, the ECM will shut "OFF" the fuel pump within 2 seconds after "ON."

The pump will deliver fuel to the TBI unit, where the system pressure is controlled to 62 to 90 kPa (9 to 13 psi). Excess fuel is then returned to the fuel tank.

The fuel pump test terminal is located in the left side of the engine compartment. When the engine is stopped, the pump can be turned "ON" by applying battery voltage to the test terminal.

Test Description: Numbers below refer to circled numbers on the diagnostic chart.
1. If the fuse in the jumper wire blows, check CKT 120 for a short to ground.

Diagnostic Aids:

Improper fuel system pressure can result in one of the following symptoms:
- Cranks, but won't run
- Code 44
- Code 45
- Cuts out, may feel like ignition problem
- Poor fuel economy, loss of power
- Hesitation

1988–90 2.0L (VIN K, OHC) – ALL MODELS

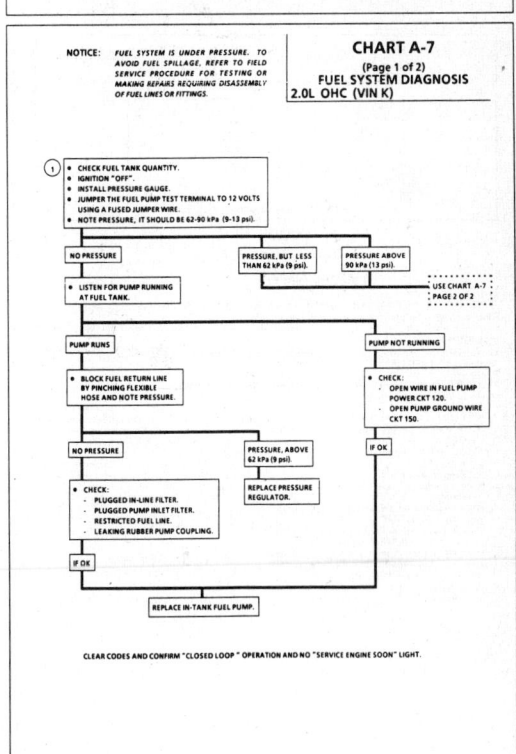

1988–90 2.0L (VIN K, OHC) – ALL MODELS

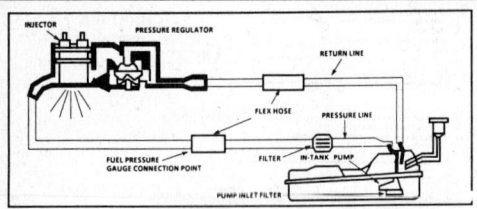

CHART A-7
(Page 2 of 2)
FUEL SYSTEM DIAGNOSIS
2.0L OHC (VIN K)

Test Description: Numbers below refer to circled numbers on the diagnostic chart.

1. Pressure, but less than 62 kPa (9 psi), falls into two areas:
   - Amount of fuel to injector OK, but pressure is too low. System will be lean running and may set Code 44. Also, hard starting cold and poor overall performance.
   - Restricted flow causing pressure drop. Normally, a vehicle with a fuel pressure of less than 62 kPa (9 psi) at idle will not be driveable. However, if the pressure drop occurs only while driving, the engine will normally surge then stop as pressure begins to drop rapidly to zero.

2. Restricting the fuel return line allows the fuel pump to develop its maximum pressure (dead head pressure). When battery voltage is applied to the pump test terminal, pressure should be from 90 to 124 kPa (13 to 18 psi).

3. This test determines if the high fuel pressure is due to a restricted fuel return line or a throttle body pressure regulator problem.

1988–90 2.0L (VIN K, OHC) – ALL MODELS

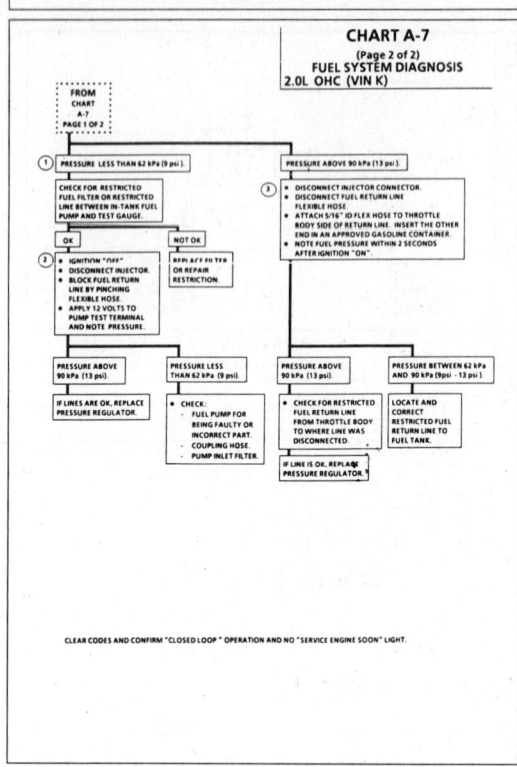

1988–90 2.0L (VIN K, OHC) – ALL MODELS

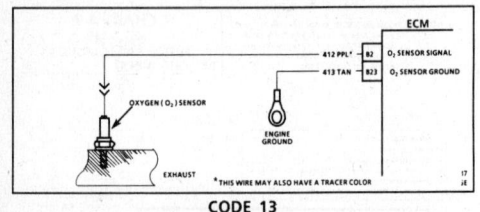

CODE 13
OXYGEN SENSOR CIRCUIT
(OPEN CIRCUIT)
2.0L OHC (VIN K)

Circuit Description:
The ECM supplies a voltage of about .45 volt between terminals "B2" and "B23". (If measured with 10 megohm digital voltmeter, this may read as low as .32 volt).

The O₂ sensor varies the voltage within a range of about 1 volt, if the exhaust is rich, down through about .10 volt, if exhaust is lean.

The sensor is like an open circuit and produces no voltage, when it is below 360°C (600°F). An open sensor circuit, or cold sensor, causes "Open Loop" operation.

Test Description: Numbers below refer to circled numbers on the diagnostic chart.

1. Code 13 WILL SET:
   - Engine at normal operating temperature.
   - At least 40 seconds engine run time after start.
   - O₂ signal voltage steady between .35 and .55 volt.
   - Throttle angle above 7%.
   - All conditions must be met for about 3 seconds.
   If the conditions for a Code 13 exist, the system will not go "Closed Loop".

2. This test determines if the O₂ sensor is the problem, or, if the ECM and wiring are at fault.

3. In doing this test, use only a high impedance digital volt ohmmeter. This test checks the continuity of CKTs 412 and 413. If CKT 413 is open, the ECM voltage on CKT 412 will be over .6 volt (600 mV).

Diagnostic Aids:

Normal "Scan" voltage varies between 100 mV to 999 mV (.1 and 1.0 volt), while in "Closed Loop". Code 13 sets in one minute, if voltage remains between .35 and .55 volt, but the system will go "Open Loop" in about 15 seconds.

Verify a clean, tight ground connection for CKT 413. Open CKT(s) 412 or 413 will result in a Code 13. If Code 13 is intermittent, refer to Section "B".

1988–90 2.0L (VIN K, OHC) – ALL MODELS

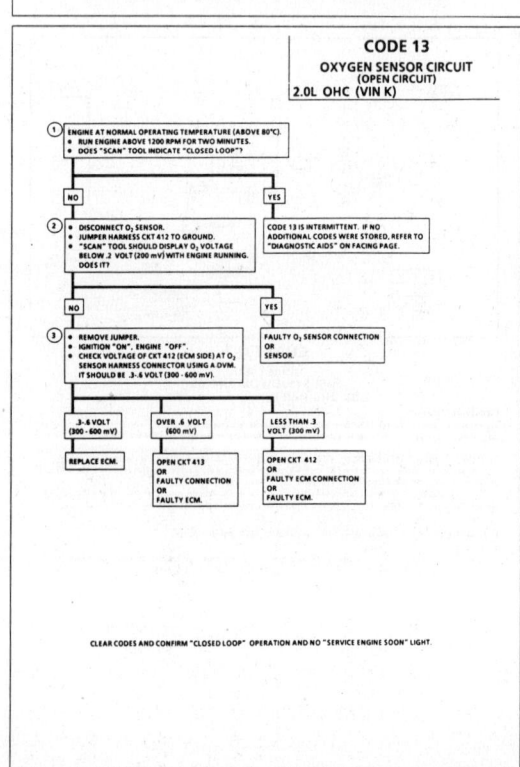

1988–90 2.0L (VIN K, OHC) – ALL MODELS

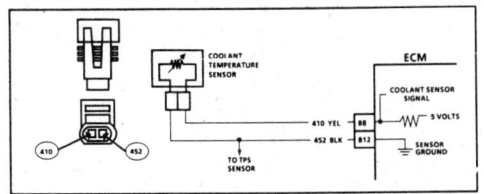

CODE 14
COOLANT TEMPERATURE SENSOR CIRCUIT
(HIGH TEMPERATURE INDICATED)
2.0L OHC (VIN K)

Circuit Description:

The coolant temperature sensor uses a thermistor to control the signal voltage to the ECM. The ECM applies a voltage on CKT 410 to the sensor. When the engine is cold, the sensor (thermistor) resistance is high, therefore, the ECM will see high signal voltage.

As the engine warms, the sensor resistance becomes less and the voltage drops. At normal engine operating temperature, the voltage will measure about 1.5 to 2.0 volts at the ECM terminal "B8".

Coolant temperature is one of the inputs used to control:
- Fuel Delivery
- Engine Spark Timing (EST)
- Cooling Fan
- Convertor Clutch (TCC)
- Idle (IAC)

Test Description: Numbers below refer to circled numbers on the diagnostic chart.
1. Checks to see if code was set as result of hard failure or intermittent condition.
   Code 14 will set if:
   - Signal voltage indicates a coolant temperature above 135°C (275°F) for 2 seconds.
2. This test simulates conditions for a Code 15. If the ECM recognizes the open circuit (high voltage), and displays a low temperature, the ECM and wiring are OK.

Diagnostic Aids:

A "Scan" tool reads engine temperature in degrees centigrade.

After the engine is started, the temperature should rise steadily to about 90°C, then stabilize when the thermostat opens.

If the engine has been allowed to cool to an ambient temperature (overnight), coolant and MAT temperature may be checked with a "Scan" tool and should read close to each other.

When a Code 14 is set, the ECM will turn "ON" the engine cooling fan.

A Code 14 will result if CKT 410 is shorted to ground.

If Code 14 is intermittent, refer to Section "B".

1988–90 2.0L (VIN K, OHC) – ALL MODELS

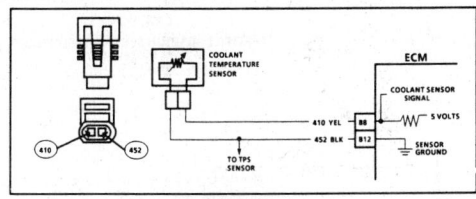

CODE 15
COOLANT TEMPERATURE SENSOR CIRCUIT
(LOW TEMPERATURE INDICATED)
2.0L OHC (VIN K)

Circuit Description:

The coolant temperature sensor uses a thermistor to control the signal voltage to the ECM. The ECM applies a voltage on CKT 410 to the sensor. When the engine is cold, the sensor (thermistor) resistance is high, therefore, the ECM will see high signal voltage.

As the engine warms, the sensor resistance becomes less and the voltage drops. At normal engine operating temperature, the voltage will measure about 1.5 to 2.0 volts at the ECM terminal "B8".

Coolant temperature is one of the inputs used to control:
- Fuel Delivery
- Engine Spark Timing (EST)
- Cooling Fan
- Convertor Clutch (TCC)
- Idle (IAC)

Test Description: Numbers below refer to circled numbers on the diagnostic chart.
1. Checks to see if code was set as result of hard failure or intermittent condition.
   Code 15 will set if:
   - The engine has been running for 2 minutes.
   - Signal voltage indicates a coolant temperature below -30°C (-22°F).
2. This test simulates conditions for a Code 14. If the ECM recognizes the grounded circuit (low voltage), and displays a high temperature, the ECM and wiring are OK.
3. This test will determine if there is a wiring problem or a faulty ECM. If CKT 452 is open, there may also be a Code 21 stored.

Diagnostic Aids:

A "Scan" tool reads engine temperature in degrees centigrade. After the engine is started, the temperature should rise steadily to about 90°C, then stabilize when the thermostat opens.

If the engine has been allowed to cool to an ambient temperature (overnight), coolant and MAT temperatures may be checked with a "Scan" tool and should read close to each other.

When a Code 15 is set, the ECM will turn "ON" the engine cooling fan.

A Code 15 will result if CKT(s) 410 or 452 are open.

If Code 15 is intermittent, refer to Section "B".

1988–90 2.0L (VIN K, OHC) – ALL MODELS

CODE 14
COOLANT TEMPERATURE SENSOR CIRCUIT
(HIGH TEMPERATURE INDICATED)
2.0L OHC (VIN K)

1. DOES "SCAN" TOOL DISPLAY COOLANT TEMPERATURE OF 130°C OR HIGHER?
   - YES
   - NO

2. DISCONNECT SENSOR. "SCAN" TOOL SHOULD DISPLAY TEMPERATURE BELOW -30°C. DOES IT?
   - YES → REPLACE SENSOR.
   - NO → CKT 410 SHORTED TO GROUND. OR CKT 410 SHORTED TO SENSOR GROUND CIRCUIT. OR FAULTY ECM.

   (NO branch from step 1) → CODE 14 IS INTERMITTENT. IF NO ADDITIONAL CODES WERE STORED, REFER TO "DIAGNOSTIC AIDS" ON FACING PAGE.

DIAGNOSTIC AID

COOLANT SENSOR		
TEMPERATURE VS. RESISTANCE VALUES (APPROXIMATE)		
°F	°C	OHMS
210	100	185
160	70	450
100	38	1,800
70	20	3,400
40	4	7,500
20	-7	13,500
0	-18	25,000
-40	-40	100,700

CLEAR CODES AND CONFIRM "CLOSED LOOP" OPERATION AND NO "SERVICE ENGINE SOON" LIGHT.

1988–90 2.0L (VIN K, OHC) – ALL MODELS

CODE 15
COOLANT TEMPERATURE SENSOR CIRCUIT
(LOW TEMPERATURE INDICATED)
2.0L OHC (VIN K)

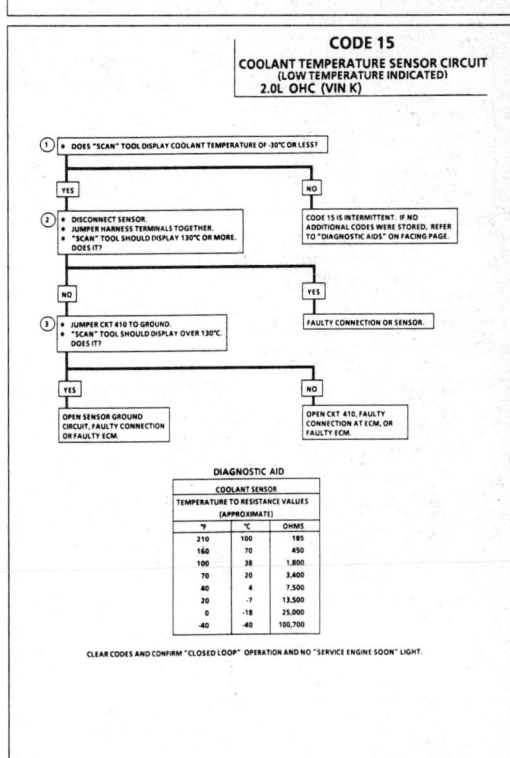

1. DOES "SCAN" TOOL DISPLAY COOLANT TEMPERATURE OF -30°C OR LESS?
   - YES
   - NO → CODE 15 IS INTERMITTENT. IF NO ADDITIONAL CODES WERE STORED, REFER TO "DIAGNOSTIC AIDS" ON FACING PAGE.

2. DISCONNECT SENSOR.
   - JUMPER HARNESS TERMINALS TOGETHER.
   - "SCAN" TOOL SHOULD DISPLAY 130°C OR MORE. DOES IT?
   - NO
   - YES → FAULTY CONNECTION OR SENSOR.

3. JUMPER CKT 410 TO GROUND.
   - "SCAN" TOOL SHOULD DISPLAY OVER 130°C. DOES IT?
   - YES → OPEN SENSOR GROUND CIRCUIT, FAULTY CONNECTION OR FAULTY ECM.
   - NO → OPEN CKT 410, FAULTY CONNECTION AT ECM, OR FAULTY ECM.

DIAGNOSTIC AID

COOLANT SENSOR		
TEMPERATURE TO RESISTANCE VALUES (APPROXIMATE)		
°F	°C	OHMS
210	100	185
160	70	450
100	38	1,800
70	20	3,400
40	4	7,500
20	-7	13,500
0	-18	25,000
-40	-40	100,700

CLEAR CODES AND CONFIRM "CLOSED LOOP" OPERATION AND NO "SERVICE ENGINE SOON" LIGHT.

1988–90 2.0L (VIN K, OHC) – ALL MODELS

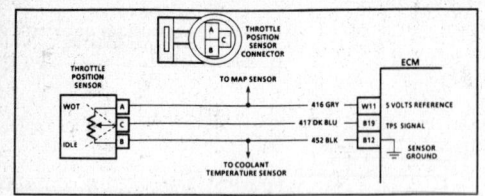

CODE 21
THROTTLE POSITION SENSOR (TPS) CIRCUIT
(SIGNAL VOLTAGE HIGH)
2.0L OHC (VIN K)

Circuit Description:
The throttle position sensor (TPS) provides a voltage signal that changes relative to the throttle valve. Signal voltage will vary from less than 1.25 volts at idle to about 5 volts at wide open throttle (WOT).
The TPS signal is one of the most important inputs used by the ECM for fuel control and for many of the ECM controlled outputs.

Test Description: Numbers below refer to circled numbers on the diagnostic chart.
1. This step checks to see if Code 21 is the result of a hard failure or an intermittent condition.
   A Code 21 will set if
   - TPS reading above 2.5 volts
   - Engine speed less than 1800 rpm
   - MAP reading below 65 kPa
   - All of the above conditions present for 2 seconds
2. This step simulates conditions for a Code 22. If the ECM recognizes the change of state, the ECM and CKTs 416 and 417 are OK.
3. This step isolates a faulty sensor, ECM, or an open CKT 452. If CKT 452 is open, there may also be a Code 15 stored.

Diagnostic Aids:
A "Scan" tool displays throttle position in volts. Closed throttle voltage should be less than 1.25 volts. TPS voltage should increase at a steady rate as throttle is moved to WOT.
A Code 21 will result if CKT 452 is open or CKT 417 is shorted to voltage. If Code 21 is intermittent, refer to Section "B".

1988–90 2.0L (VIN K, OHC) – ALL MODELS

CODE 21
THROTTLE POSITION SENSOR (TPS) CIRCUIT
(SIGNAL VOLTAGE HIGH)
2.0L OHC (VIN K)

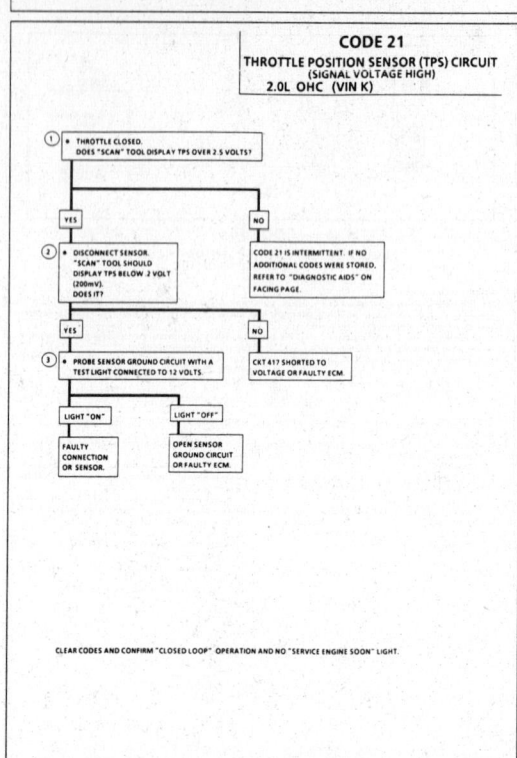

CLEAR CODES AND CONFIRM "CLOSED LOOP" OPERATION AND NO "SERVICE ENGINE SOON" LIGHT.

1988–90 2.0L (VIN K, OHC) – ALL MODELS

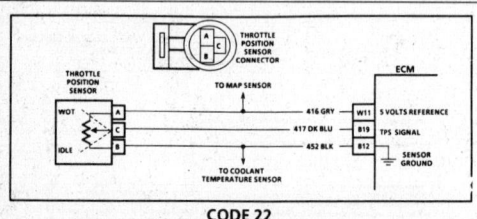

CODE 22
THROTTLE POSITION SENSOR (TPS) CIRCUIT
(SIGNAL VOLTAGE LOW)
2.0L OHC (VIN K)

Circuit Description:
The throttle position sensor (TPS) provides a voltage signal that changes, relative to the throttle valve. Signal voltage will vary from less than 1.25 volts at idle to about 4.5 volts at wide open throttle (WOT).
The TPS signal is one of the most important inputs used by the ECM for fuel control and for many of the ECM controlled outputs.

Test Description: Numbers below refer to circled numbers on the diagnostic chart.
1. This step checks to see if Code 22 is the result of a hard failure or an intermittent condition.
   A Code 22 will set if
   - The engine is running
   - TPS voltage is below .2 volt (200 mV)
2. This step simulates conditions for a Code 21. If a Code 21 is set, or the "Scan" tool displays over 4 volts, the ECM and wiring are OK.
3. The "Scan" tool may not display 12 volts. The important thing is that the ECM recognizes the voltage as over 4 volts, indicating that CKT 417 and the ECM are OK.
4. If CKT 416 is open or shorted to ground, there may also be a stored Code 34.

Diagnostic Aids:
A "Scan" tool displays throttle position in volts. Closed throttle voltage should be less than 1.25 volts. TPS voltage should increase at a steady rate as throttle is moved to WOT.
An open or grounded 416 or 417 will result in a Code 22.
If Code 22 is intermittent, refer to Section "B".

1988–90 2.0L (VIN K, OHC) – ALL MODELS

CODE 22
THROTTLE POSITION SENSOR (TPS) CIRCUIT
(SIGNAL VOLTAGE LOW)
2.0L OHC (VIN K)

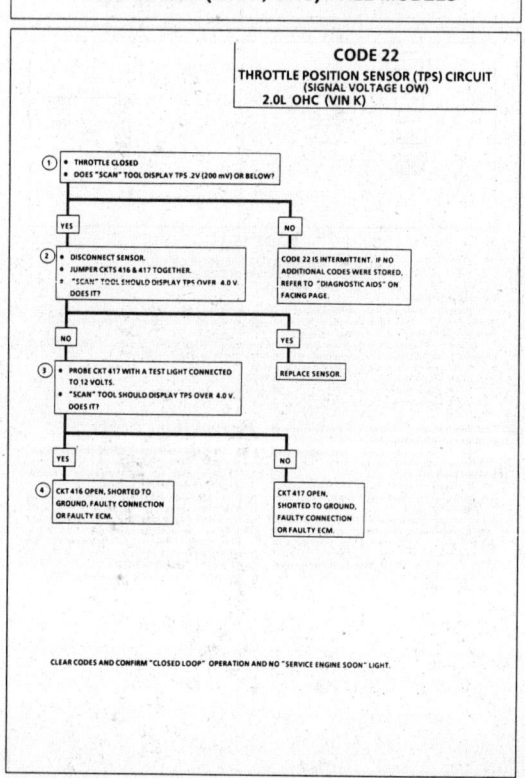

CLEAR CODES AND CONFIRM "CLOSED LOOP" OPERATION AND NO "SERVICE ENGINE SOON" LIGHT.

1988–90 2.0L (VIN K, OHC) – ALL MODELS

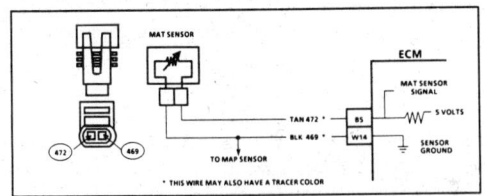

* THIS WIRE MAY ALSO HAVE A TRACER COLOR

CODE 23
MANIFOLD AIR TEMPERATURE (MAT) SENSOR CIRCUIT
(LOW TEMPERATURE INDICATED)
2.0L OHC (VIN K)

Circuit Description:
The manifold air temperature sensor uses a thermistor to control the signal voltage to the ECM. The ECM applies a voltage (4-6 volts) on CKT 472 to the sensor. When manifold air is cold, the sensor (thermistor) resistance is high, therefore, the ECM will see a high signal voltage. As the air warms, the sensor resistance becomes less and the voltage drops.

Test Description: Numbers below refer to circled numbers on the diagnostic chart.
1. This step checks to see if Code 23 is the result of a hard failure or, an intermittent condition.
   A Code 23 will set if:
   * Engine has been running for longer than 2 minutes.
   * Signal voltage indicates a MAT temperature less than -30°C.
2. This test simulates conditions for a Code 25. If the "Scan" tool displays a high temperature, the ECM and wiring are OK.
3. This step checks continuity of CKTs 472 and 469. If CKT 469 is open there may also be a Code 33.

Diagnostic Aids:
If the engine has been allowed to cool to an ambient temperature (overnight), coolant and MAT temperatures may be checked with a "Scan" tool and should read close to each other.
A Code 23 will result if CKT(s) 472 or 469 become open.
If Code 23 is intermittent, refer to Section "B".

1988–90 2.0L (VIN K, OHC) – ALL MODELS

CODE 23
MANIFOLD AIR TEMPERATURE (MAT) SENSOR CIRCUIT
(LOW TEMPERATURE INDICATED)
2.0L OHC (VIN K)

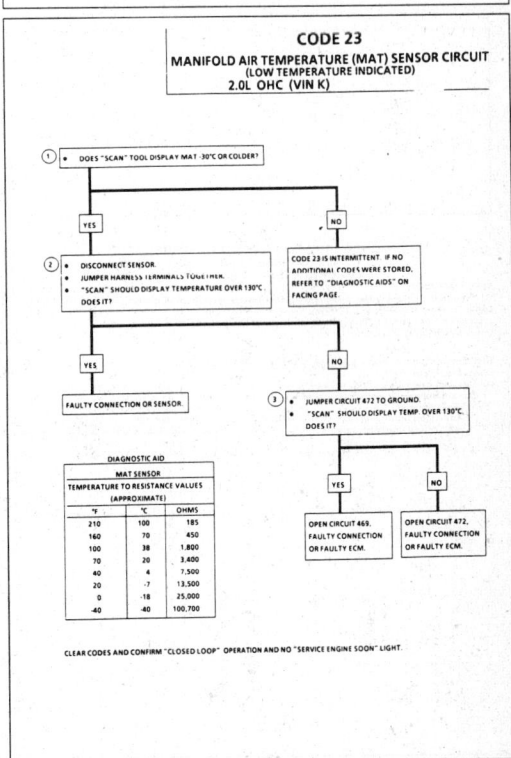

DIAGNOSTIC AID		
MAT SENSOR		
TEMPERATURE TO RESISTANCE VALUES (APPROXIMATE)		
°F	°C	OHMS
210	100	185
160	70	450
100	38	1,800
70	20	3,400
40	4	7,500
20	-7	13,500
0	-18	25,000
-40	-40	100,700

CLEAR CODES AND CONFIRM "CLOSED LOOP" OPERATION AND NO "SERVICE ENGINE SOON" LIGHT.

1988–90 2.0L (VIN K, OHC) – ALL MODELS

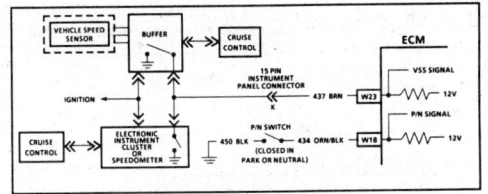

CODE 24
VEHICLE SPEED SENSOR (VSS) CIRCUIT
2.0L OHC (VIN K)

Circuit Description:
The ECM applies and monitors 12 volts on CKT 437. CKT 437 connects to the vehicle speed sensor which alternately grounds CKT 437 when drive wheels are turning. This pulsing action takes place about 2000 times per mile and the ECM will calculate vehicle speed based on the time between "pulses".
"Scan" tool reading should closely match the speedometer reading with the drive wheels turning.
Disregard a Code 24 set when drive wheels are not turning.

Test Description: Numbers below refer to circled numbers on the diagnostic chart.
1. Code 24 will set if vehicle speed equals 0 mph when:
   * Engine speed is between 1400 and 4400 rpm.
   * Throttle position is less than 2%.
   * Manifold vacuum is high.
   * Transaxle is not in park or neutral.
   * Above conditions met for 5 seconds.
   These conditions are met during a road load deceleration. Disregard Code 24 that sets when drive wheels are not turning.
2. 8-12 volts at the instrument panel connector indicates that CKT 437 is open between the connector and the vehicle speed sensor, or that the sensor is faulty. A voltage of less than 1 volt at the instrument panel connector indicates that CKT 437 is shorted to ground.

If the voltage reads above 10 volts after disconnecting CKT 437 at the vehicle speed sensor, the vehicle speed sensor is faulty. If voltage remains less than 8 volts, then CKT 437 is grounded. If 437 is not grounded, there is a faulty connection at the ECM, or the ECM is faulty.

Diagnostic Aids:
"Scan" tool should indicate a vehicle speed whenever the drive wheels are turning greater than 3 mph.
A faulty or maladjusted park/neutral switch can result in a false Code 24. Use a "Scan" tool and check for the proper signal while in drive. Refer to CHART C-1A for P/N switch check.
If Code 24 is intermittent, refer to Section "B".

1988–90 2.0L (VIN K, OHC) – ALL MODELS

CODE 24
VEHICLE SPEED SENSOR (VSS) CIRCUIT
2.0L OHC (VIN K)

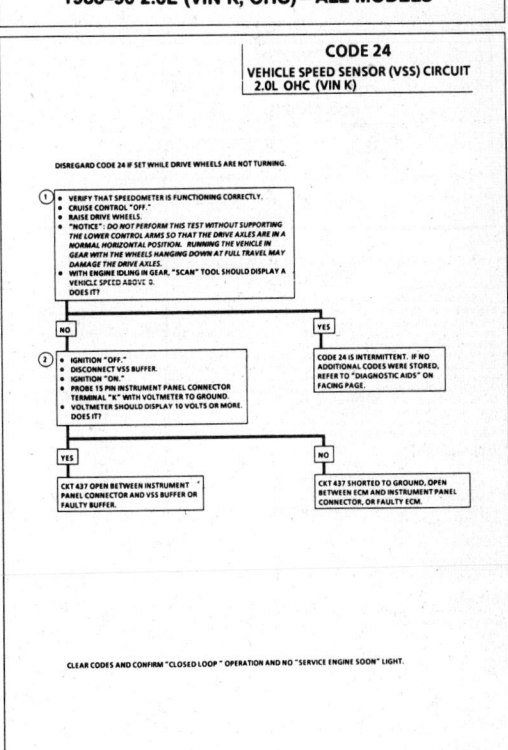

CLEAR CODES AND CONFIRM "CLOSED LOOP" OPERATION AND NO "SERVICE ENGINE SOON" LIGHT.

1988–90 2.0L (VIN K, OHC) – ALL MODELS

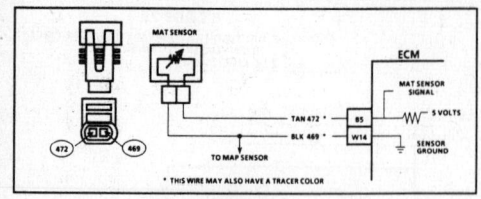

CODE 25
MANIFOLD AIR TEMPERATURE (MAT) SENSOR CIRCUIT
(HIGH TEMPERATURE INDICATED)
2.0L OHC (VIN K)

Circuit Description:

The manifold air temperature sensor uses a thermistor to control the signal voltage to the ECM. The ECM applies a voltage (4-6 volts) on CKT 472 to the sensor. When manifold air is cold, the sensor (thermistor) resistance is high, therefore, the ECM will see a high signal voltage. As the air warms, the sensor resistance becomes less and the voltage drops.

Test Description: Numbers below refer to circled numbers on the diagnostic chart.
1. This check determines if the Code 25 is the result of a hard failure or an intermittent condition.
   A Code 25 will set if
   - A MAT temperature greater than 135°C is detected for a time longer than 2 seconds.

Diagnostic Aids:

If the engine has been allowed to cool to an ambient temperature (overnight), coolant and MAT temperatures may be checked with a "Scan" tool and should read close to each other.

A Code 25 will result if CKT 472 is shorted to ground.

If Code 25 is intermittent, refer to Section "B".

1988–90 2.0L (VIN K, OHC) – ALL MODELS

CODE 25
MANIFOLD AIR TEMPERATURE (MAT) SENSOR CIRCUIT
(HIGH TEMPERATURE INDICATED)
2.0L OHC (VIN K)

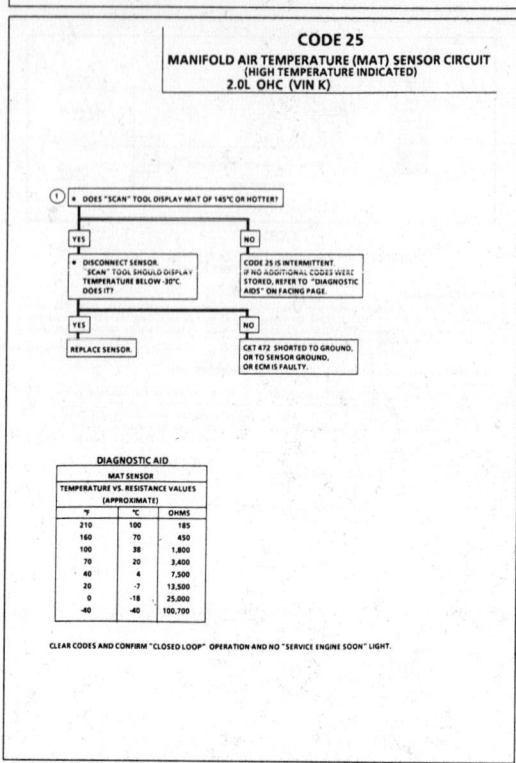

MAT SENSOR		
TEMPERATURE VS. RESISTANCE VALUES (APPROXIMATE)		
°F	°C	OHMS
210	100	185
160	70	450
100	38	1,800
70	20	3,400
40	4	7,500
20	-7	13,500
0	-18	25,000
-40	-40	100,700

CLEAR CODES AND CONFIRM "CLOSED LOOP" OPERATION AND NO "SERVICE ENGINE SOON" LIGHT.

1988–90 2.0L (VIN K, OHC) – ALL MODELS

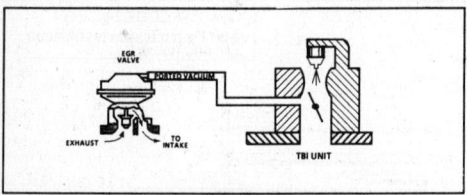

CODE 32
EXHAUST GAS RECIRCULATION (EGR) SYSTEM FAILURE
2.0L OHC (VIN K)

Code Description:

A properly operating EGR system will directly affect the air/fuel requirements of the engine. Since the exhaust gas introduced into the air/fuel mixture is an inert gas (contains very little or no oxygen), less fuel is required to maintain a correct air/fuel ratio. If the EGR system were to become inoperative, the inert exhaust gas would be replaced with air and the air/fuel mixture would be leaner. The ECM would compensate for the lean condition by adding fuel, resulting in higher block learn values.

The engine control system operates within two block learn cells, a closed throttle cell, and an open throttle cell. Since EGR is not used at idle, the closed throttle cell would not be affected by EGR system operation. The open throttle cell is affected by EGR operation and, when the EGR system is operating properly, the block learn values in both cells should be close to being the same. If the EGR system was inoperative, the block learn value in the open throttle cell would change (become higher) to compensate for the resulting lean system, but the block learn value in the closed throttle cell would not change.

This change or difference in block learn values is used to monitor EGR system performance. When the change becomes too great, a Code 32 is set.

Diagnostic Aids:

The Code 32 chart is a functional check of the EGR system. If the EGR system works properly, but a Code 32 has been set, check other items that could result in high block learn values in the open throttle cell, but not in the closed throttle cell.

CHECK:

EGR Passages
   Restricted or blocked

MAP Sensor
   A MAP sensor may shift in calibration enough to affect fuel delivery. Use CHART C-1D, MAP output check.

1988–90 2.0L (VIN K, OHC) – ALL MODELS

CODE 32
EXHAUST GAS RECIRCULATION (EGR) SYSTEM FAILURE
2.0L OHC (VIN K)

IF CODES 21, 22, 33, OR 34 ARE PRESENT, REPAIR CAUSES FOR THOSE CODES BEFORE USING THIS CHART.

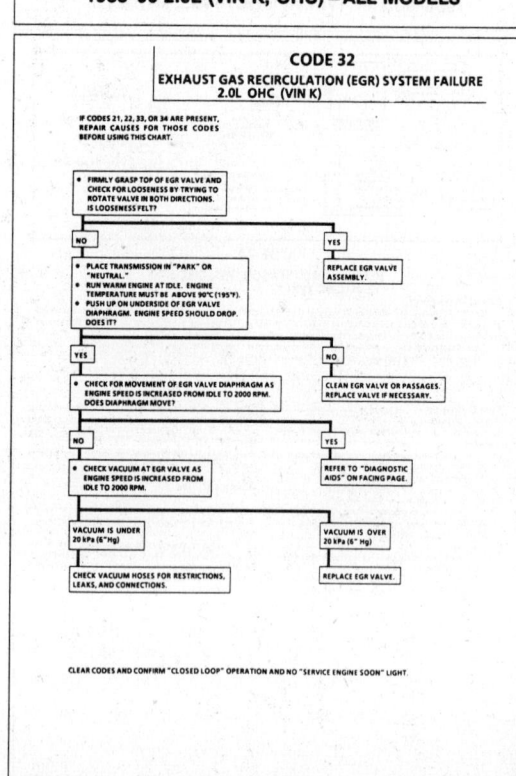

CLEAR CODES AND CONFIRM "CLOSED LOOP" OPERATION AND NO "SERVICE ENGINE SOON" LIGHT.

1988–90 2.0L (VIN K, OHC) – ALL MODELS

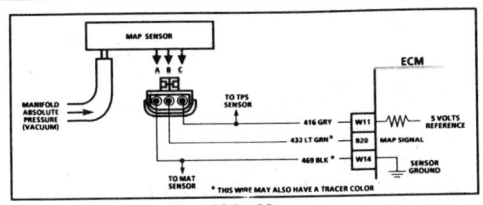

CODE 33
MANIFOLD ABSOLUTE PRESSURE (MAP) SENSOR CIRCUIT
(SIGNAL VOLTAGE HIGH - LOW VACUUM)
2.0L OHC (VIN K)

Circuit Description:

The manifold absolute pressure (MAP) sensor responds to changes in manifold pressure (vacuum). The ECM receives this information as a signal voltage that will vary from about 1 to 1.5 volts, at closed throttle idle, to 4 - 4.5 volts at wide open throttle (low vacuum).

If the MAP sensor fails, the ECM will substitute a fixed MAP value and use the throttle position sensor (TPS) to control fuel delivery.

Test Description: Numbers below refer to circled numbers on the diagnostic chart.

1. This step will determine if Code 33 is the result of a hard failure or an intermittent condition.

   A Code 33 will set if:
   - MAP signal indicates greater than 84 kPa (low vacuum)
   - TPS less than 2%
   - These conditions for a time longer than 5 seconds

2. This step simulates conditions for a Code 34. If the ECM recognizes the change, the ECM and CKTs 416 and 432, are OK. If CKT 469 is open, there may, also, be a stored Code 23.

Diagnostic Aids:

With the ignition "ON" and the engine stopped, the manifold pressure is equal to atmospheric pressure and the signal voltage will be high. This information is used by the ECM as an indication of vehicle altitude and is referred to as BARO. Comparison of this BARO reading with a known good vehicle with the same sensor is a good way to check accuracy of a "suspect" sensor. Readings should be the same ± .4 volts.

A Code 33 will result if CKT 469 is open, or if CKT 432 is shorted to voltage or to CKT 416.

If Code 33 is intermittent, refer to Section "B".

1988–90 2.0L (VIN K, OHC) – ALL MODELS

CODE 33
MANIFOLD ABSOLUTE PRESSURE (MAP) SENSOR CIRCUIT
(SIGNAL VOLTAGE HIGH - LOW VACUUM)
2.0L OHC (VIN K)

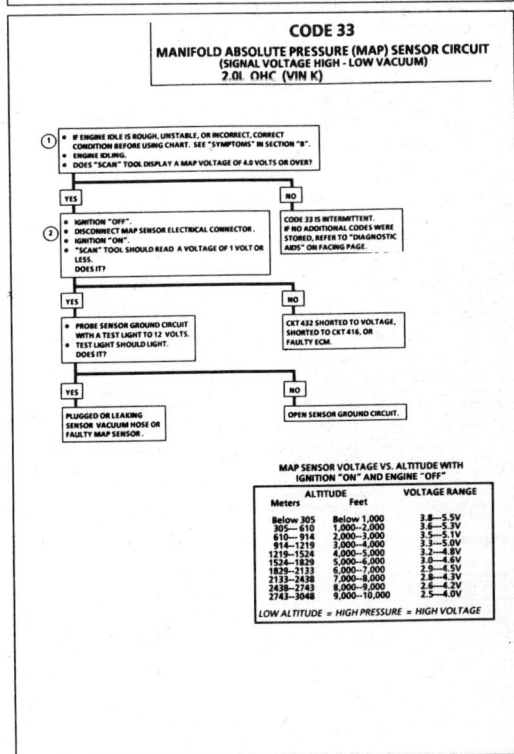

1988–90 2.0L (VIN K, OHC) – ALL MODELS

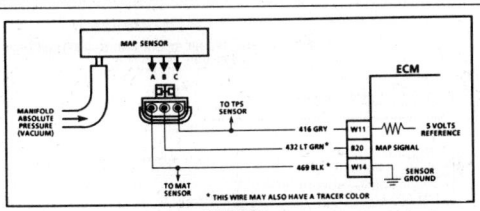

CODE 34
MANIFOLD ABSOLUTE PRESSURE (MAP) SENSOR CIRCUIT
(SIGNAL VOLTAGE LOW - HIGH VACUUM)
2.0L OHC (VIN K)

Circuit Description:

The manifold absolute pressure (MAP) sensor responds to changes in manifold pressure (vacuum). The ECM receives this information as a signal voltage that will vary from about 1 to 1.5 volts at closed throttle idle, to 4 - 4.5 volts at wide open throttle.

If the MAP sensor fails, the ECM will substitute a fixed MAP value and use the throttle position sensor (TPS) to control fuel delivery.

Test Description: Numbers below refer to circled numbers on the diagnostic chart.

1. This step determines if Code 34 is the result of a hard failure or an intermittent condition.

   A Code 34 will set when:
   - Engine rpm is greater than 1200 rpm
   - TPS is greater the 20%
   - MAP signal voltage is too low

2. Jumpering harness terminals "B" to "C," 5 volts to signal, will determine if the sensor is at fault, or if there is a problem with the ECM or wiring.

3. The "Scan" tool may not display 12 volts. The important thing is that the ECM recognizes the voltage as more than 4 volts, indicating that the ECM and CKT 432 are OK.

Diagnostic Aids:

With the ignition "ON" and the engine stopped, the manifold pressure is equal to atmospheric pressure and the signal voltage will be high. This information is used by the ECM as an indication of vehicle altitude and is referred to as BARO. Comparison of this BARO reading with a known good vehicle with the same sensor is a good way to check accuracy of a "suspect" sensor. Readings should be the same ± .4 volts.

A Code 34 will result if CKT(s) 416 or 432 are open or shorted to ground.

If Code 34 is intermittent, refer to Section "B".

1988–90 2.0L (VIN K, OHC) – ALL MODELS

CODE 34
MANIFOLD ABSOLUTE PRESSURE (MAP) SENSOR CIRCUIT
(SIGNAL VOLTAGE LOW - HIGH VACUUM)
2.0L OHC (VIN K)

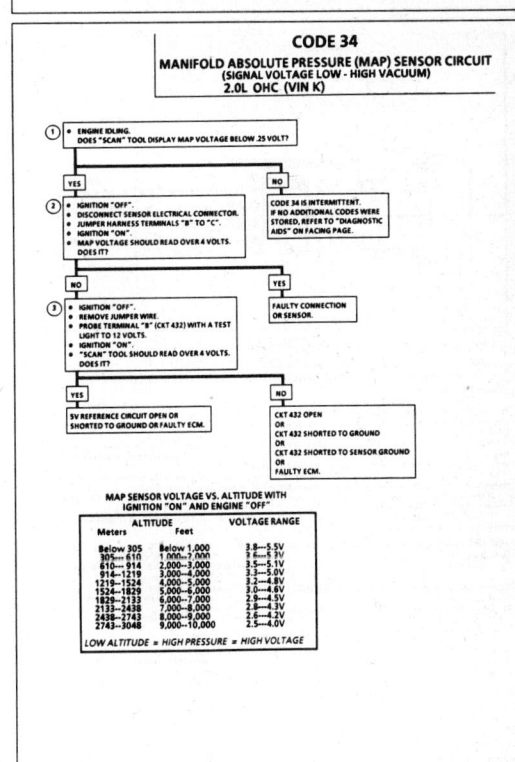

1988–90 2.0L (VIN K, OHC) – ALL MODELS

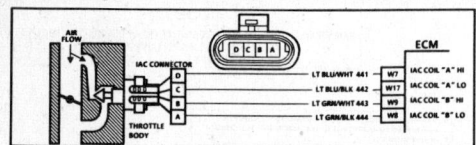

CODE 35
IDLE SPEED ERROR
2.0L OHC (VIN K)

Circuit Description:
Code 35 will be set, when the closed throttle engine speed is 300 rpm above or below the desired idle speed for 20 seconds. Review general description in Section "C".

Test Description: Numbers below refer to circled numbers on the diagnostic chart.
1. Continue with test, even if engine will not idle. If idle is too low, "Scan" will display 80 or more counts, or steps. If idle is high, it will display "0" counts. Occasionally, an erratic or unstable idle may occur. Engine speed may vary 200 rpm or more up and down. Disconnect IAC. If the condition is unchanged, the IAC is not at fault.
2. When the engine was stopped, the IAC Valve retracted (more air) to a fixed "Park" position for increased air flow and idle speed during the next engine start. A "Scan" will display 100 or more counts.
3. Be sure to disconnect the IAC valve prior to this test. The test light will confirm the ECM signals by a steady or flashing light on all circuits.
4. There is a remote possibility that one of the CKTs is shorted to voltage, which would have been indicated by a steady light. Disconnect ECM and turn the ignition "ON" and probe terminals to check for this condition.

Diagnostic Aids:

A slow unstable idle may be caused by a system problem that cannot be overcome by the IAC. "Scan" counts will be above 60 counts, if too low, and "0" counts, if too high.
If idle is too high, stop engine. Ignition "ON". Ground diagnostic terminal. Wait 45 seconds for IAC

to seat, then, disconnect IAC. Start engine. If idle speed is above 800 rpm, locate and correct vacuum leak.
- **System too lean (High Air/Fuel Ratio)**
Idle speed may be too high or too low. Engine speed may vary up and down, disconnecting IAC does not help. May set Code 44. "Scan" and/or voltmeter will read an oxygen sensor output less than 300 mV (.3 volt). Check for low regulated fuel pressure or water in fuel. A lean exhaust, with an oxygen sensor output fixed above 800 mV (.8 volt), will be a contaminated sensor, usually silicone. This may also set a Code 45.
- **System too rich (Low Air/Fuel Ratio)**
Idle speed too low. "Scan" counts usually above 80. System obviously rich and may exhibit black smoke exhaust.
"Scan" tool and/or voltmeter will read an oxygen sensor signal fixed above 800 mV (.8 volt).
Check:
 - High fuel pressure
 - Injector leaking or sticking
- **Throttle Body** - Remove IAC and inspect bore for foreign material or evidence of IAC valve dragging the bore.
- **PCV Valve**
A faulty or incorrect PCV may result in an incorrect idle speed.
- **IAC Harness Connections**
Carefully check IAC harness connections for proper contact.
- Refer to Rough, Unstable, Incorrect Idle or Stalling, in Symptoms in Section "B".

1988–90 2.0L (VIN K, OHC) – ALL MODELS

CODE 35
IDLE SPEED ERROR
2.0L OHC (VIN K)

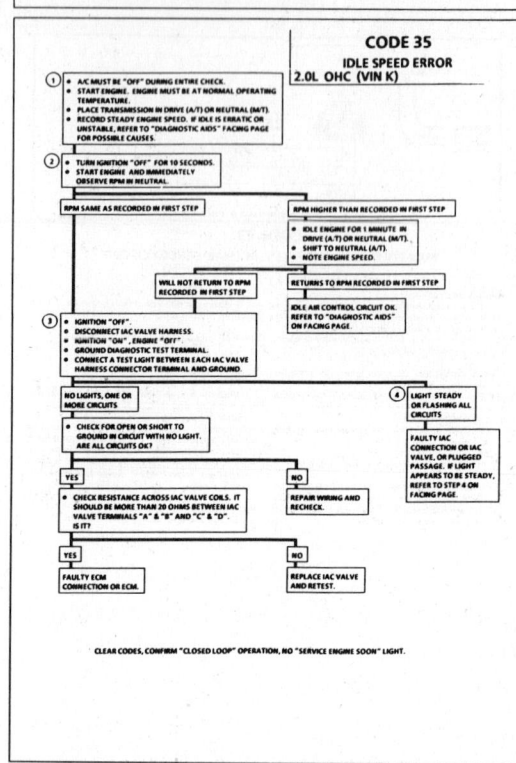

1988–90 2.0L (VIN K, OHC) – ALL MODELS

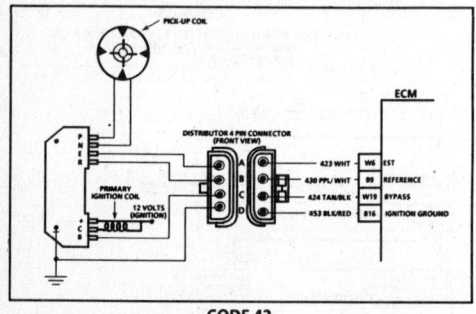

CODE 42
ELECTRONIC SPARK TIMING (EST) CIRCUIT
2.0L OHC (VIN K)

Circuit Description:
The ignition module sends a reference signal (CKT 430) to the ECM, when the engine is cranking. While the engine speed is under 400 rpm, the ignition module will control ignition timing. When the engine speed exceeds 400 rpm, the ECM applies 5 volts to the bypass line (CKT 424) to switch the timing to ECM control (EST CKT 423).
When the system is running on the ignition module, that is, no voltage on the bypass line, the ignition module grounds the EST signal. The ECM expects to see no voltage on the EST line during this condition. If it sees a voltage, it sets Code 42 and will not go into the EST mode.
When the rpm for EST is reached (about 400 rpm), voltage will be applied to the bypass line. The EST should no longer be grounded in the ignition module, so the EST voltage should be varying.
If the bypass line is open or grounded, the ignition module will not switch to EST mode, so the EST voltage will be low and Code 42 will be set.
If the EST line is grounded, the ignition module will switch to EST but, because the line is grounded, there will be no EST signal. A Code 42 will be set.

Test Description: Numbers below refer to circled numbers on the diagnostic chart.
1. Code 42 means the ECM has seen an open or short to ground in the EST or bypass circuits. This test confirms Code 42 and that the fault causing the code is present.
2. Checks for a normal EST ground path through the ignition module. An EST CKT 423, shorted to ground, will also read less than 500 ohms, however, this will be checked later.
3. As the test light voltage touches CKT 424, the module should switch, causing the ohmmeter to "overrange" if the meter is in the 10,000-2000 ohms position. Selecting the 10-20,000 ohms position

will indicate above 5000 ohms. The important thing is that the module "switched".
4. The module did not switch and this step checks for:
 - EST CKT 423 shorted to ground
 - Bypass CKT 424 open
 - Faulty ignition module connection or module
5. Confirms that CKT(s) is a faulty ECM and not an intermittent in CKT(s) 423 or 424.

Diagnostic Aids:

The "Scan" tool does not have any ability to help diagnose a Code 42 problem.
If Code 42 is intermittent, refer to Section "B".

1988–90 2.0L (VIN K, OHC) – ALL MODELS

CODE 42
ELECTRONIC SPARK TIMING (EST) CIRCUIT
2.0L OHC (VIN K)

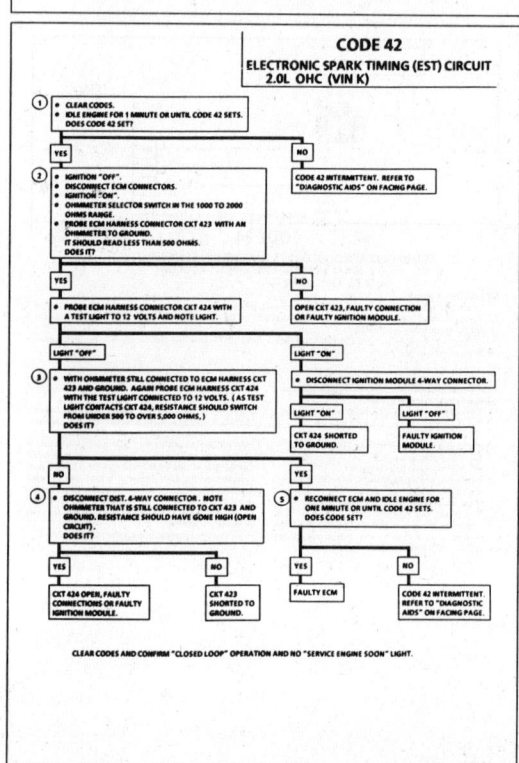

1988–90 2.0L (VIN K, OHC) – ALL MODELS

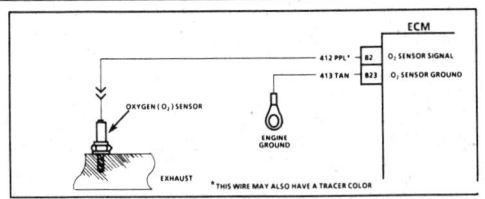

CODE 44
OXYGEN SENSOR CIRCUIT
(LEAN EXHAUST INDICATED)
2.0L OHC (VIN K)

Circuit Description:

The ECM supplies a voltage of about .45 volt between terminals "B2" and "B23". (If measured with a 10 megohm digital voltmeter, this may read as low as .32 volt.) The O₂ sensor varies the voltage within a range of about 1 volt, if the exhaust is rich, down through about .10 volt, if exhaust is lean.

The sensor is like an open circuit and produces no voltage, when it is below about 360°C (600°F). An open sensor circuit, or cold sensor, causes "Open Loop" operation.

Test Description: Numbers below refer to circled numbers on the diagnostic chart.
1. Code 44 is set, when the O₂ sensor signal voltage on CKT 412:
   - Remains below .2 volt for 50 seconds or more.
   - And the system is operating in "Closed Loop".

Diagnostic Aids:

Using the "Scan," observe the block learn value at different rpms. The "Scan" also displays the block cells, so the block learn values can be checked in each of the cells, to determine when the Code 44 may have been set. If the conditions for Code 44 exists, the block learn values will be around 150.
- O₂ Sensor Wire - Sensor pigtail may be mispositioned and contacting the exhaust manifold.
- Check for ground in wire between connector and sensor.

- **Fuel Contamination** - Water, even in small amounts, near the in-tank fuel pump inlet can be delivered to the injector. The water causes a lean exhaust and can set a Code 44.
- **Fuel Pressure** - System will be lean if pressure is too low. It may be necessary to monitor fuel pressure, while driving the car at various road speeds and/or loads to confirm. See "Fuel System Diagnosis," CHART A-7.
- **Exhaust Leaks** - If there is an exhaust leak, the engine can cause outside air to be pulled into the exhaust and past the sensor. Vacuum or crankcase leaks can cause a lean condition.
- If Code 44 intermittent, refer to Section "B".

1988–90 2.0L (VIN K, OHC) – ALL MODELS

CODE 44
OXYGEN SENSOR CIRCUIT
(LEAN EXHAUST INDICATED)
2.0L OHC (VIN K)

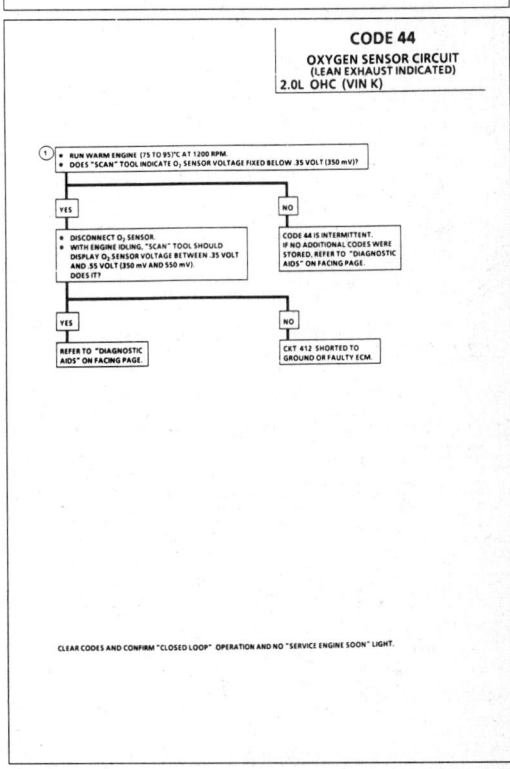

1988–90 2.0L (VIN K, OHC) – ALL MODELS

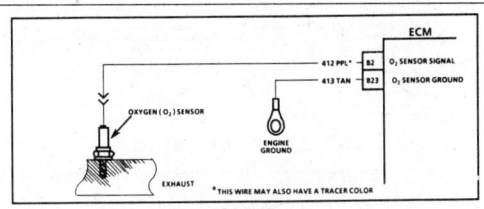

CODE 45
OXYGEN SENSOR CIRCUIT
(RICH EXHAUST INDICATED)
2.0L OHC (VIN K)

Circuit Description:

The ECM supplies a voltage of about .45 volt between terminals "B2" and "B23". (If measured with a 10 megohm digital voltmeter, this may read as low as .32 volt.) The O₂ sensor varies the voltage within a range of about 1 volt, if the exhaust is rich, down through about .10 volt, if exhaust is lean.

The sensor is like an open circuit and produces no voltage, when it is below about 360°C (600°F). An open sensor circuit, or cold sensor, causes "Open Loop" operation.

Test Description: Numbers below refer to circled numbers on the diagnostic chart.
1. Code 45 is set, when the O₂ sensor signal voltage on CKT 412:
   - Remains above .75 volt for 50 seconds or more, and in "Closed Loop".

Diagnostic Aids:

The Code 45, or rich exhaust, is most likely caused by one of the flowing:
- **Fuel Pressure** - System will go rich if pressure is too high. The ECM can compensate for some increase. However, if it gets too high, a Code 45 will be set. See "Fuel System Diagnosis" CHART A-7.
- **HEI Shielding** - An open ground CKT 453 may result in EMI, or induced electrical "noise." The ECM looks at this "noise" as reference pulses. The additional pulses result in a higher than actual engine speed signal. The ECM then delivers too much fuel, causing system to go rich. Engine tachometer will also show higher than actual engine speed, which can help in diagnosing this problem.

- **Canister Purge** - Check for fuel saturation. If full of fuel, check canister control and hoses.
- **MAP Sensor** - An output that causes the ECM to sense a higher than normal manifold pressure (low vacuum) can cause the system to go rich. Disconnecting the MAP sensor will allow the ECM to set a fixed value for the MAP sensor. Substitute a different MAP sensor if the rich condition is gone, while the sensor is disconnected.
- **TPS** - An intermittent TPS output will cause the system to go rich, due to a false indication of the engine accelerating.
- **O₂ Sensor Contamination** - Inspect oxygen sensor for silicone contamination from fuel, or use of improper RTV sealant. The sensor may have a white, powdery coating and result in a high, but false signal (rich exhaust indication). The ECM will then reduce the amount of fuel delivered to the engine, causing a severe surge driveability problem.
- **EGR** - Valve sticking open at idle, usually accompanied by a rough idle, stall complaint. If Code 45 is intermittent, refer to Section "B".

1988–90 2.0L (VIN K, OHC) – ALL MODELS

CODE 45
OXYGEN SENSOR CIRCUIT
(RICH EXHAUST INDICATED)
2.0L OHC (VIN K)

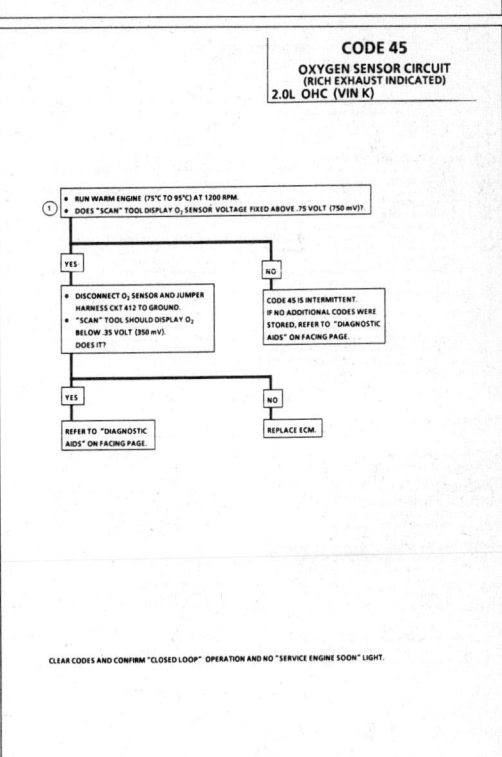

1988–90 2.0L (VIN K, OHC) – ALL MODELS

CODE 51
CODE 53
2.0L OHC (VIN K)

CODE 51
PROM ERROR
(FAULTY OR INCORRECT PROM)

CHECK THAT ALL PINS ARE FULLY INSERTED IN THE SOCKET AND THAT PROM IS PROPERLY SEATED. IF OK, REPLACE PROM, CLEAR MEMORY, AND RECHECK. IF CODE 51 REAPPEARS, REPLACE ECM.

CLEAR CODES AND CONFIRM "CLOSED LOOP" OPERATION AND NO "SERVICE ENGINE SOON" LIGHT.

CODE 53
SYSTEM OVERVOLTAGE

THIS CODE INDICATES THAT THERE IS A BASIC GENERATOR PROBLEM.
• CODE 53 WILL SET IF BATTERY VOLTAGE AT THE ECM IS GREATER THAN 16.9 VOLTS FOR AT LEAST 50 SECONDS.
• CHECK AND REPAIR CHARGING SYSTEM.

CLEAR CODES AND CONFIRM "CLOSED LOOP" OPERATION AND NO "SERVICE ENGINE SOON" LIGHT.

1988–90 2.0L (VIN K, OHC) – ALL MODELS

SECTION B
SYMPTOMS
TABLE OF CONTENTS

Performing Symptom Diagnosis
Intermittents
Hard Start
Rough, Unstable, or Incorrect Idle, Stalling
Poor Gas Mileage
Detonation/Spark Knock
Lack of Power, Sluggish, or Spongy
Surges and/or Chuggle
Cuts Out, Misses
Hesitation, Sag, Stumble
Excessive Exhaust Emissions or Odors
Dieseling, Run-On
Backfire
Restricted Exhaust System Check - Chart B-1

PERFORMING SYMPTOM DIAGNOSIS

The DIAGNOSTIC CIRCUIT CHECK should be performed before using this section. The purpose of this section is to locate the source of a driveability or emissions problem when other diagnostic procedures cannot be used. This may be because of difficulties in locating a suspected sub-system or component.

Many driveability related problems can be eliminated by following the procedures found in Service Bulletins. These bulletins supersede this manual. Be sure to check all bulletins related to the complaint or suspected system.

If the engine cranks but will not run, use CHART A-3.

The sequence of the checks listed in this section is not intended to be followed as a step-by-step procedure. The checks are listed such that the less difficult and time consuming operations are performed before more difficult ones.

Most of the symptom procedures call for a careful visual and physical check. *The importance of this step cannot be stressed too strongly.* It can lead to correcting a problem without further checks, and *can save valuable time.* This procedure includes checking the following:

- Vacuum hoses for splits, kinks, and proper connections, as shown on the underhood Emission Control Information label.
- Throttle body and intake manifold for leaks
- Ignition wires for cracking, hardness, proper routing, and carbon tracking
- Wiring for proper connections, pinches, and cuts

1988–90 2.0L (VIN K, OHC) – ALL MODELS

INTERMITTENTS

Definition: Problem may or may not activate the "Service Engine Soon" light or store a trouble code.

DO NOT use the trouble code charts in Section "A" for intermittent problems. The fault must be present to locate the problem. If a fault is intermittent, the use of trouble code charts may result in the replacement of good parts.

• Most intermittent problems are caused by faulty electrical connections or wiring. Perform careful checks of suspected circuits for
   Poor mating of the connector halves and terminals not fully seated in the connector body (backed out)
   Improperly formed or damaged terminals. All connector terminals in problem circuit should be carefully reformed to increase contact tension
   Poor terminal to wire connection. This requires removing the terminal from the connector body to check

• If a visual and physical check does not locate the cause of the problem, the car can be driven with a voltmeter connected to a suspected circuit or a "Scan" tool may be used. An abnormal voltage reading while the problem occurs indicates that the problem may be in that circuit.

• Check for loss of trouble code memory. To check, disconnect the TPS and allow the engine to idle until the "Service Engine Soon" light turns "ON." Code 22 should be stored and kept in memory when the ignition is turned "OFF" for at least 10 seconds. If not, the ECM is faulty.

• An intermittent SES light and no trouble codes may be caused by
   Electrical system interference caused by a defective relay, ECM driven solenoid, or switch. They can cause a sharp electrical surge. Normally, the problem will occur when the faulty component is operated.
   Improper installation of electrical options, such as lights, 2-way radios, etc.
   EST wires which should be routed away from spark plug wires, ignition system components, and generator. Ground wire from ECM to ignition system which may be faulty.
   Ignition secondary wire shorted to ground.
   "Service Engine Soon" light and diagnostic test terminal circuits intermittently shorted to ground.
   Faulty ECM grounds.

HARD START

Definition: Engine cranks well but does not start for a long time. Engine does eventually start, but may or may not continue to run.

Perform careful visual and physical check as described at the beginning of Section "B". Perform "Diagnostic Circuit Check."

• CHECK
For possibility of misfiring, crossfiring, or cutting under load or at idle. Locate misfiring cylinder(s) by performing the following test.
1. Start engine. Disconnect idle air control valve. Remove one spark plug wire from a spark plug and ground it against the engine.
2. Note drop in engine speed.
3. Repeat for all four cylinders.
4. Stop engine and reconnect idle air control valve.
If the engine speed dropped equally (within 50 rpm) on all cylinders, refer to "Rough, Unstable, or Incorrect Idle, Stalling" symptom.

If there was no drop or excessive variation in engine speed on one or more cylinders, check for spark on the respective cylinder(s) with J26792 (ST-125) spark tester or equivalent.
   If spark is present, remove the spark plugs from the cylinder(s) and check for the following:
   Cracks
   Wear
   Improper gap
   Burned electrode
   Heavy deposits
For worn distributor shaft
For moisture in distributor cap
Ignition pickup coil resistance and connections
Ignition timing. See underhood emission control information label.

1988–90 2.0L (VIN K, OHC) – ALL MODELS

- Fuel for poor quality, "stale" fuel, and water contamination
- Ignition wires for shorts or faulty insulation
- Ignition coil connections
- Fuel pump relay. Connect test light between pump test terminal and ground. Light should be on for 2 seconds following ignition "ON." If not, use CHART A-5.
- Secondary ignition voltage output with ST-125 tester.
- Spark plugs. Look for wetness, cracks, improper gap, burned electrodes, and heavy deposits. Visually inspect ignition system for moisture, dust, cracks, burns, etc.
- For faulty ECM and ignition grounds.
- PROM for correct application. (Consult Service Bulletins.) Spray plug wires with fine water mist to check for shorts.
- EGR operation. Use CHART C-7.
- Idle Air Control system. Use Code 35 chart.
- Fuel system for restricted filter or improper pressure. Use CHART A-7.
- Injector and TBI assembly for leakage. Pressurize system by energizing fuel pump through the underhood fuel pump test connector.
- Coolant sensor for a shift in calibration. Use Code 14 or Code 15 chart.
- TPS for sticking or binding. TPS voltage should read less than 1.25 volts on a "Scan" tool.

- In-tank fuel pump check valve. A faulty valve would allow the fuel in the lines to drain back to the tank after the engine is stopped. To check for this condition, conduct the following test.
1. Ignition "OFF."
2. Disconnect fuel line at the filter.
3. Remove the tank filler cap.
4. Connect a radiator test pump to the line and apply 103 kPa (15 psi) pressure. If the pressure will hold for 60 seconds, the check valve is OK.

- For the possibility of an exhaust restriction or improper valve timing by performing the following test.
1. With engine at normal operating temperature, connect a vacuum gauge to any convenient vacuum port on intake manifold.
2. Run engine at 1000 rpm and record vacuum reading.
3. Increase engine speed slowly to 2500 rpm. Note vacuum reading at steady 2500 rpm.
4. If vacuum at 2500 rpm decreases more than 3" Hg from reading at 1000 rpm, the exhaust system should be inspected for restrictions. Use CHART B-1.
5. Disconnect exhaust pipe from engine and repeat Steps 3 & 4. If vacuum still drops more than 3" Hg with exhaust disconnected, check valve timing.

Engine valve timing and compression.

ROUGH, UNSTABLE, OR INCORRECT IDLE, STALLING

Definition: The engine runs unevenly at idle. If severe, the car may shake. Also, the idle speed may vary (called "hunting"). Either condition may be severe enough to cause stalling. Engine idles at incorrect speed.

Perform careful visual and physical check as described at the beginning of Section "B". Perform "Diagnostic Circuit Check."

• CHECK
For possibility of misfiring, crossfiring, or cutting under load or at idle. Locate misfiring cylinder(s) by performing the following test.
1. Start engine. Disconnect idle air control valve. Remove one spark plug wire from a spark plug and ground it against the engine.
2. Note drop in engine speed.
3. Repeat for all four cylinders.

4. Stop engine and reconnect idle air control valve.
If the engine speed dropped equally (within 50 rpm) on all cylinders, proceed through the causes listed. If there was no drop or excessive variation in engine speed on one or more cylinders, check for spark on the respective cylinder(s) with J26792 (ST-125) spark tester or equivalent.
   If spark is present, remove the spark plugs from the cylinder(s) and check for the following.
   Cracks
   Wear

1988–90 2.0L (VIN K, OHC) – ALL MODELS

Improper gap
Burned electrode
Heavy deposits

- For worn distributor shaft
- For moisture in distributor cap
- Ignition pickup coil resistance and connections
- Ignition timing. See underhood emission control information label.
- MAP sensor. Use CHART C1-D.
- Throttle for sticking shaft or binding linkage. This will cause a high TPS voltage (open throttle indication) and the ECM will not control idle. TPS voltage should be less than 1.25 volts with throttle closed.
- Battery cables and ground straps for poor contact. Erratic voltage will cause the IAC valve to change its position, resulting in poor idle quality.
- Ignition wires for shorts or faulty insulation
- Ignition system for moisture, dust, cracks, burns, etc. Spray plug wires with fine water mist to check for shorts.
- Secondary ignition voltage output with ST-1 tester
- Ignition coil connections
- ECM and ignition system for faulty grounds
- Proper operation of EST. See Section "C4."
- Spark plugs. Look for wetness, cracks, improper gap, burned electrodes, and heavy deposits.
- Fuel system for restricted filter or improper pressure. Use CHART A-7.
- Injector and TBI assembly for leakage. Pressurize system by energizing fuel pump through the underhood fuel pump test connector.
- EGR operation. Use CHART C-7.
- For vacuum leaks at intake manifold gasket
- Idle Air Control system. Use Code 35 chart.
- Electrical system voltage. IAC valve will not move if voltage is below 9 volts or greater than 17.8 volts. Also check battery cables and ground straps for poor contact. Erratic voltage will cause the IAC valve to change its position, resulting in poor idle quality.
- PCV valve for proper operation by placing finger over inlet hole in valve end several times. Valve should snap back. If not, replace valve. Ensure that valve is correct part. Also, check PCV hose.
- Canister purge system for proper operation. Use CHART C-3.
- PROM for correct application
- Throttle shaft or TPS for sticking or binding. TPS voltage should read less than 1.25 volts on a "Scan" tool with the throttle closed.
- MAP sensor output. Use CHART C1-D and/or check sensor by comparing it to the output on a similar vehicle if possible.
- Oxygen sensor for silicone contamination from contaminated fuel or use of improper RTV sealant. The sensor will have a white, powdery coating and will cause a high but false signal voltage (rich exhaust indication). The ECM will reduce the amount of fuel delivered to the engine, causing a severe driveability problem.
- Coolant sensor for a shift in calibration. Use Code 14 or Code 15 chart.
- A/C refrigerant pressure for high pressure. Check for overcharging or faulty pressure switch.
- P/N switch circuit on vehicle with automatic transmission. Use CHART C-1A.
- Generator output voltage. Repair if less than 9 volts or more than 16 volts.
- Power steering. Use CHART C-1E. The ECM should compensate for power steering loads. Loss of this signal would be most noticeable when steering loads are high such as during parking.
- Engine valve timing and compression.
- For worn or incorrect basic engine parts such as cam, heads, pistons, etc. Also check for bent pushrods, worn rocker arms, and broken or weak valve springs.
- For the possibility of an exhaust restriction or improper valve timing, perform the following test.
  1. With engine at normal operating temperature, connect a vacuum gauge to any convenient vacuum port on intake manifold.
  2. Run engine at 1000 rpm and record vacuum reading.
  3. Increase engine speed slowly to 2500 rpm. Note vacuum reading at steady 2500 rpm.
  4. If vacuum at 2500 rpm decreases more than 3" Hg from reading at 1000 rpm, the exhaust system should be inspected for restrictions. Use CHART B-1.

1988–90 2.0L (VIN K, OHC) – ALL MODELS

5. Disconnect exhaust pipe from engine and repeat Steps 3 & 4. If vacuum still drops more than 3" Hg with exhaust disconnected, check valve timing.

For overheating and possible causes. Look for the following.
- Low or incorrect coolant solution. It should be a 50/50 mix of GM #1052753 anti-freeze coolant (or equivalent) and water.
- Loose water pump belt
- Restricted air flow to radiator, or restricted water flow through radiator
- Faulty or incorrect thermostat
- Inoperative electric cooling fan circuit

- If the system is running RICH (block learn less than 118), refer to "Diagnostic Aids" on facing page of Code 45.
- If the system is running LEAN (block learn greater than 138), refer to "Diagnostic Aids" on facing page of Code 44.

POOR GAS MILEAGE

Definition: Gas mileage, as measured by an actual road test, is noticeably lower than expected. Gas mileage is noticeably lower than it was during a previous actual road test.

Perform careful visual and physical check as described at the beginning of Section "B".

Perform "Diagnostic Circuit Check.

- **CHECK**
  - For possibility of misfiring, crossfiring, or cutting under load or at idle. Locate misfiring cylinder(s) by performing the following test.
    1. Start engine. Disconnect idle air control valve. Remove one spark plug wire from a spark plug and ground it against the engine.
    2. Note drop in engine speed.
    3. Repeat for all four cylinders.
    4. Stop engine and reconnect idle air control valve.

    If the engine speed dropped equally (within 50 rpm) on all cylinders, refer to "Rough, Unstable, or Incorrect Idle, Stalling" symptom. If there was no drop or excessive variation in engine speed on one or more cylinders, check for spark on the respective cylinder(s) with J26792 (ST-125) spark tester or equivalent. If there is no spark, see Section "6D." If spark is present, remove the spark plugs from the cylinder(s) and check for the following.
    Cracks
    Wear
    Improper gap
    Burned electrode
    Heavy deposits
  - For worn distributor shaft
  - Ignition timing. See underhood emission control information label.
  - Proper operation of EST.
  - Spark plugs. Look for wetness, cracks, improper gap, burned electrodes, and heavy deposits.
  - Spark plugs for correct heat range
  - Fuel for poor quality, "stale" fuel, and water contamination
  - Fuel system for restricted filter or improper pressure. Use CHART A-7.
  - Injector and TBI assembly for leakage. Pressurize system by energizing fuel pump through the underhood fuel pump test connector.
  - EGR operation. Use CHART C-7.
  - For vacuum leaks at intake manifold gasket
  - Air cleaner element (filter) for dirt or plugging
  - Idle Air Control system. Use Code 35 chart.
  - Canister purge system for proper operation. Use CHART C-3.
  - PROM for correct application
  - Throttle shaft or TPS for sticking or binding. TPS voltage should read less than 1.25 volts on a "Scan" tool with the throttle closed.
  - MAP sensor output. Use CHART C1-D and/or check sensor by comparing it to the output on a similar vehicle if possible.
  - Oxygen sensor for silicone contamination from contaminated fuel or use of improper RTV sealant. The sensor will have a white, powdery coating and will cause a high but false signal voltage (rich exhaust indication). The ECM will reduce the amount of fuel delivered to the engine, causing a severe driveability problem.
  - Coolant sensor for a shift in calibration. Use Code 14 or Code 15 chart.
  - Vehicle speed sensor (VSS) input with a "Scan" tool to make sure reading of VSS matches that of vehicle speedometer.

1988–90 2.0L (VIN K, OHC) – ALL MODELS

- A/C relay operation. A/C should cut out at wide open throttle. Use CHART C-10.
- A/C refrigerant pressure for high pressure. Check for overcharging or faulty pressure switch.
- Generator output voltage. Repair if less than 9 volts or more than 16 volts.
- Cooling fan operation. Use CHART C-12.
- Power steering. Use CHART C-1E. The ECM should compensate for power steering loads. Loss of this signal would be most noticeable when steering loads are high such as during parking.
- Transmission torque converter operation
- Transmission for proper shift points
- Transmission torque converter clutch operation. Use CHART C-8.
- Engine valve timing and compression.
- For worn or incorrect basic engine parts such as cam, heads, pistons, etc. Also check for bent pushrods, worn rocker arms, and broken or weak valve springs.
- For the possibility of an exhaust restriction or improper valve timing by performing the following test.
  1. With engine at normal operating temperature, connect a vacuum gauge to any convenient vacuum port on intake manifold.
  2. Run engine at 1000 rpm and record vacuum reading.
  3. Increase engine speed slowly to 2500 rpm. Note vacuum reading at steady 2500 rpm.
  4. If vacuum at 2500 rpm decreases more than 3" Hg from reading at 1000 rpm, the exhaust system should be inspected for restrictions. Use CHART B-1.
  5. Disconnect exhaust pipe from engine and repeat Steps 3 & 4. If vacuum still drops more than 3" Hg with exhaust disconnected, check valve timing.
- Thermostat for incorrect heat range or being inoperative
- Check driver's driving habits and vehicle conditions which affect gas mileage.
  - Suggest driver read "Important Facts on Fuel Economy" in Owner's Manual.
  - Is A/C "ON" full time (Defroster mode "ON")?
  - Are tires at correct pressure?
  - Are excessively heavy loads being carried?
  - Is acceleration often heavy?
  - Are the wheels aligned correctly?
  - Is the speedometer calibrated correctly?
  - Are the vehicle brakes dragging?
  - Is the brake switch applying excessive force on the brake pedal?
- If the system is running RICH, (block learn less than 118), refer to "Diagnostic Aids" on facing page of Code 45.

DETONATION/SPARK KNOCK

Definition. A mild to severe ping, usually worse under acceleration. The engine makes sharp metallic knocks that change with throttle opening.

Perform careful visual and physical check as described at the beginning of Section "B".

Perform "Diagnostic Circuit Check."

- **CHECK**

  For possibility of misfiring, crossfiring, or cutting under load or at idle. Locate misfiring cylinder(s) by performing the following test.
  1. Start engine. Disconnect idle air control valve. Remove one spark plug wire from a spark plug and ground it against the engine.
  2. Note drop in engine speed.
  3. Repeat for all four cylinders.
  4. Stop engine and reconnect idle air control valve.

  If the engine speed dropped equally (within 50 rpm) on all cylinders, refer to "Rough, Unstable, or Incorrect Idle, Stalling" symptom. If there was no drop or excessive variation in engine speed on one or more cylinders, check for spark on the respective cylinder(s) with J26792 (ST-125) spark tester or equivalent.
  If spark is present, remove the spark plugs from the cylinder(s) and check for the following.
  Cracks
  Wear

1988–90 2.0L (VIN K, OHC) – ALL MODELS

Improper gap
Burned electrode
Heavy deposits

- For worn distributor shaft
- Ignition pickup coil resistance and connections
- Ignition timing. See underhood emission control information label.
- Ignition wires for shorts or faulty insulation
- Spark plugs for correct heat range
- Fuel for poor quality, "stale" fuel, and water contamination
- Fuel system for restricted filter or improper pressure. Use CHART A-7.
- For excessive oil entering combustion chamber. Oil will reduce the effective octane of fuel.
- EGR operation. Use CHART C-7.
- For vacuum leaks at intake manifold gasket
- PCV valve for proper operation by placing finger over inlet hole in valve end several times. Valve should snap back. If not, replace valve. Ensure that valve is correct part. Also check PCV hose.
- MAP sensor output. Use CHART C1-D and/or check sensor by comparing it to the output on a similar vehicle, if possible.
- Coolant sensor for a shift in calibration
- Oxygen sensor for silicone contamination from contaminated fuel or use of improper RTV sealant. The sensor will have a white, powdery coating and will cause a high but false signal voltage (rich exhaust indication). The ECM will reduce the amount of fuel delivered to the engine, causing a severe driveability problem
- Vehicle speed sensor (VSS) input with a "Scan" tool to make sure reading of VSS matches that of vehicle speedometer.
- Transmission torque converter operation.
- Transmission for proper shift points.
- Transmission torque converter clutch operation. Use CHART C-8.
- Vehicle brakes for dragging
- PROM for correct application
- For overheating and possible causes. Look for the following.
  - Low or incorrect coolant solution. It should be a 50/50 mix of GM #1052753 anti-freeze coolant (or equivalent) and water.
  - Loose water pump belt
  - Restricted air flow to radiator or restricted water flow through radiator
  - Faulty or incorrect thermostat
  - Inoperative electric cooling fan circuit
- Engine valve timing and compression.
- For worn or incorrect basic engine parts such as cam, heads, pistons, etc. Also check for bent pushrods, worn rocker arms, and broken or weak valve springs.
- For the possibility of an exhaust restriction or improper valve timing by performing the following test.
  1. With engine at normal operating temperature, connect a vacuum gauge to any convenient vacuum port on intake manifold.
  2. Run engine at 1000 rpm and record vacuum reading.
  3. Increase engine speed slowly to 2500 rpm. Note vacuum reading at steady 2500 rpm.
  4. If vacuum at 2500 rpm decreases more than 3" Hg from reading at 1000 rpm, the exhaust system should be inspected for restrictions. Use CHART B-1.
  5. Disconnect exhaust pipe from engine and repeat Steps 3 & 4. If vacuum still drops more than 3" Hg with exhaust disconnected, check valve timing.
- Remove internal engine carbon with top engine cleaner.
- If the system is running LEAN (block learn greater than 138), refer to "Diagnostic Aids" on facing page of Code 44.

1988–90 2.0L (VIN K, OHC) – ALL MODELS

LACK OF POWER, SLUGGISH, OR SPONGY

Definition: Engine delivers less than expected power. There is little or no increase in speed when the accelerator pedal is depressed partially.

Perform careful visual and physical check as described at the beginning of Section "B".

Perform "Diagnostic Circuit Check."

- **CHECK**
  For possibility of misfiring, crossfiring, or cutting under load or at idle. Locate misfiring cylinder(s) by performing the following test.
  1. Start engine. Disconnect idle air control valve. Remove one spark plug wire from a spark plug and ground it against the engine.
  2. Note drop in engine speed.
  3. Repeat for all four cylinders.
  4. Stop engine and reconnect idle air control valve.

  If the engine speed dropped equally (within 50 rpm) on all cylinders, refer to "Rough, Unstable, or Incorrect Idle, Stalling" symptom. If there was no drop or excessive variation in engine speed on one or more cylinders, check for spark on the respective cylinder(s) with J26792 (ST-125) spark tester or equivalent.

  If spark is present, remove the spark plugs from the cylinder(s) and check for the following.
  Cracks
  Wear
  Improper gap
  Burned electrode
  Heavy deposits
  - For worn distributor shaft
  - Ignition pickup coil resistance and connections
  - Ignition timing. See underhood emission control information label.
  - Ignition wires for shorts or faulty insulation
  - Ignition system for moisture, dust, cracks, burns, etc. Spray plug wires with fine water mist to check for shorts.
  - Secondary ignition voltage output with ST-125 tester.
  - Ignition coil connections
  - ECM and ignition system for faulty grounds
  - Proper operation of EST.
  - Spark plugs. Look for wetness, cracks, improper gap, burned electrodes, and heavy deposits.
  - Spark plugs for correct heat range
  - Fuel for poor quality, "stale" fuel, and water contamination
  - Fuel system for restricted filter or improper pressure. Use CHART A-7.
  - EGR operation. Use CHART C-7.
  - For vacuum leaks at intake manifold gasket
  - Air cleaner element (filter) for dirt or plugging
  - PROM for correct application
  - Throttle shaft or TPS for sticking or binding. TPS voltage should read less than 1.25 volts on a "Scan" tool with the throttle closed.
  - MAP sensor output. Use CHART C1-D and/or check sensor by comparing it to the output on a similar vehicle if possible.
  - Oxygen sensor for silicone contamination from contaminated fuel or use of improper RTV sealant. The sensor will have a white, powdery coating and will cause a high but false signal voltage (rich exhaust indication). The ECM will reduce the amount of fuel delivered to the engine, causing a severe driveability problem.
  - Coolant sensor for a shift in calibration. Use Code 14 or Code 15 chart.
  - Vehicle speed sensor (VSS) input with a "Scan" tool to make sure reading of VSS matches that of vehicle speedometer.
  - Engine for improper or worn camshaft.
  - A/C relay operation. A/C should cut out at wide open throttle. Use CHART C-10.
  - A/C refrigerant pressure for high pressure. Check for overcharging or faulty pressure switch.
  - Generator output voltage. Repair if less than 9 volts or more than 16 volts.
  - Cooling fan operation. Use CHART C-12.
  - Power steering. Use CHART C-1E. The ECM should compensate for power steering loads. Loss of this signal would be most noticeable when steering loads are high such as during parking.
  - Transmission torque converter operation.
  - Transmission for proper shift points.

1988–90 2.0L (VIN K, OHC) – ALL MODELS

- Transmission torque converter clutch operation. Use CHART C-8.
- Vehicle brakes for dragging
- Engine valve timing and compression
- For worn or incorrect basic engine parts such as cam, heads, pistons, etc. Also check for bent pushrods, worn rocker arms, and broken or weak valve springs. See Section "6A."
- For the possibility of an exhaust restriction or improper valve timing by performing the following test.
  1. With engine at normal operating temperature, connect a vacuum gauge to any convenient vacuum port on intake manifold.
  2. Run engine at 1000 rpm and record vacuum reading.
  3. Increase engine speed slowly to 2500 rpm. Note vacuum reading at steady 2500 rpm.
  4. If vacuum at 2500 rpm decreases more than 3" Hg from reading at 1000 rpm, the exhaust system should be inspected for restrictions. Use CHART B-1.
  5. Disconnect exhaust pipe from engine and repeat Steps 3 & 4. If vacuum still drops more than 3" Hg with exhaust disconnected, check valve timing.
- For overheating and possible causes. Look for the following
  - Low or incorrect coolant solution. It should be a 50/50 mix of GM #1052753 anti-freeze coolant (or equivalent) and water.
  - Loose water pump belt
  - Restricted air flow to radiator, or restricted water flow through radiator
  - Faulty or incorrect thermostat.
  - Inoperative electric cooling fan circuit. See CHART C-12.
- If the system is running RICH (block learn less than 118), refer to "Diagnostic Aids" on facing page of Code 45.
- If the system is running LEAN (block learn greater than 138), refer to "Diagnostic Aids" on facing page of Code 44.

SURGES AND/OR CHUGGLE

Definition: Engine power variation under steady throttle or cruise. Feels like the car speeds up and slows down with no change in the accelerator pedal.

Perform careful visual and physical check as described at the beginning of Section "B".

Perform "Diagnostic Circuit Check."

CHECK

- For possibility of misfiring, crossfiring, or cutting under load or at idle. Locate misfiring cylinder(s) by performing the following test.
  1. Start engine. Disconnect idle air control valve. Remove one spark plug wire from a spark plug and ground it against the engine.
  2. Note drop in engine speed.
  3. Repeat for all four cylinders.
  4. Stop engine and reconnect idle air control valve.

  If the engine speed dropped equally (within 50 rpm) on all cylinders, refer to "Rough, Unstable, or Incorrect Idle, Stalling" symptom. If there was no drop or excessive variation in engine speed on one or more cylinders, check for spark on the respective cylinder(s) with J26792 (ST-125) spark tester or equivalent.

  If spark is present, remove the spark plugs from the cylinder(s) and check for the following.
  Cracks
  Wear
  Improper gap
  Burned electrode
  Heavy deposits
- For worn distributor shaft
- Ignition pickup coil resistance and connections
- Ignition timing. See underhood emission control information label.
- Ignition wires for shorts or faulty insulation.
- Ignition system for moisture, dust, cracks, burns, etc. Spray plug wires with fine water mist to check for shorts.
- Secondary ignition voltage output with ST-125 tester.
- Ignition coil connections.

1988–90 2.0L (VIN K, OHC) – ALL MODELS

- ECM and ignition system for faulty grounds.
- Proper operation of EST.
- Spark plugs. Look for wetness, cracks, improper gap, burned electrodes, and heavy deposits.
- Spark plugs for correct heat range.
- Fuel for poor quality, "stale" fuel, and water contamination.
- Fuel system for restricted filter or improper pressure. Use CHART A-7.
- Injector and TBI assembly for leakage. Pressurize system by energizing fuel pump through the underhood fuel pump test connector.
- EGR operation. Use CHART C-7.
- For vacuum leaks at intake manifold gasket.
- Idle Air Control system. Use Code 35 chart.
- Electrical system voltage. IAC valve will not move if voltage is below 9 volts or greater than 17.8 volts. Also check batterv cables and ground straps for poor contact. Erratic voltage will cause the IAC valve to change its position, resulting in poor idle quality.
- PCV valve for proper operation by placing finger over inlet hole in valve end several times. Valve should snap back. If not, replace valve. Ensure that valve is correct part. Also check PCV hose.
- Canister purge system for proper operation. Use CHART C-3.
- PROM for correct application.
- Throttle shaft or TPS for sticking or binding. TPS voltage should read less than 1.25 volts on a "Scan" tool with the throttle closed.
- MAP sensor output. Use CHART C1-D and/or check sensor by comparing it to the output on a similar vehicle, if possible.
- Oxygen sensor for silicone contamination from contaminated fuel or use of improper RTV sealant. The sensor will have a white, powdery coating and will cause a high but false signal voltage (rich exhaust indication). The ECM will reduce the amount of fuel delivered to the engine, causing a severe driveability problem.
- Coolant sensor for a shift in calibration. Use Code 14 or Code 15 chart.
- Vehicle speed sensor (VSS) input with a "Scan" tool to make sure reading of VSS matches that of vehicle speedometer.
- A/C relay operation. A/C should cut out at wide open throttle. Use CHART C-10.
- P/N switch circuit on vehicle with automatic transmission. Use CHART C-1A.
- Transmission torque converter clutch operation. Use CHART C-8.
- For the possibility of an exhaust restriction or improper valve timing by performing the following test:
  1. With engine at normal operating temperature, connect a vacuum gauge to any convenient vacuum port on intake manifold.
  2. Run engine at 1000 rpm and record vacuum reading.
  3. Increase engine speed slowly to 2500 rpm. Note vacuum reading at steady 2500 rpm.
  4. If vacuum at 2500 rpm decreases more than 3" Hg from reading at 1000 rpm, the exhaust system should be inspected for restrictions. Use CHART B-1.
  5. Disconnect exhaust pipe from engine and repeat Steps 3 & 4. If vacuum still drops more than 3" Hg with exhaust disconnected, check valve timing.
- Engine valve timing and compression.
- For worn or incorrect basic engine parts such as cam, heads, pistons, etc. Also check for bent pushrods, worn rocker arms, and broken or weak valve springs.

- If the system is running RICH (block learn less than 118), refer to "Diagnostic Aids" on facing page of Code 45.
- If the system is running LEAN (block learn greater than 138), refer to "Diagnostic Aids" on facing page of Code 44.

1988–90 2.0L (VIN K, OHC) – ALL MODELS

CUTS OUT, MISSES

Definition: Steady pulsation or jerking that follows engine speed, usually more pronounced as engine load increases. The exhaust has a steady spitting sound at idle or low speed.

Perform careful visual and physical check as described at the beginning of Section "B".

Perform "Diagnostic Circuit Check."

- **CHECK**
  - For possibility of misfiring, crossfiring, or cutting under load or at idle. Locate misfiring cylinder(s) by performing the following test.
    1. Start engine. Disconnect idle air control valve. Remove one spark plug wire from a spark plug and ground it against the engine.
    2. Note drop in engine speed.
    3. Repeat for all four cylinders.
    4. Stop engine and reconnect idle air control valve.

    If the engine speed dropped equally (within 50 rpm) on all cylinders, refer to "Rough, Unstable, or Incorrect Idle, Stalling" symptom. If there was no drop or excessive variation in engine speed on one or more cylinders, check for spark on the respective cylinder(s) with J26792 (ST-125) spark tester or equivalent. If there is no spark, see Section "6D." If spark is present, remove the spark plugs from the cylinder(s) and check for the following.
    Cracks
    Wear
    Improper gap
    Burned electrode
    Heavy deposits
  - For worn distributor shaft
  - Ignition pickup coil resistance and connections
  - Ignition timing. See underhood emission control information label.
  - Ignition wires for shorts or faulty insulation
  - Ignition system for moisture, dust, cracks, burns, etc. Spray plug wires with fine water mist to check for shorts.
  - Secondary ignition voltage output with ST-125 tester
  - Ignition coil connections
  - ECM and ignition system for faulty grounds
  - Proper operation of EST. See Section "C4".
  - Spark plugs. Look for wetness, cracks, improper gap, burned electrodes, and heavy deposits.
  - Spark plugs for correct heat range
  - Fuel for poor quality, "stale" fuel, and water contamination
  - Fuel system for restricted filter or improper pressure. Use CHART A-7.
  - Throttle shaft or TPS for sticking or binding. TPS voltage should read less than 1.25 volts on a "Scan" tool with the throttle closed.

HESITATION, SAG, STUMBLE

Definition: Momentary lack of response as the accelerator is pushed down. Can occur at all vehicle speeds. Usually most severe when first trying to make the car move, as from a stop sign. May cause the engine to stall if severe enough.

Perform careful visual and physical check as described at the beginning of Section "B".

Perform "Diagnostic Circuit Check."

- **CHECK**
  - For possibility of misfiring, crossfiring, or cutting under load or at idle. Locate misfiring cylinder(s) by performing the following test.
    1. Start engine. Disconnect idle air control valve. Remove one spark plug wire from a spark plug and ground it against the engine.
    2. Note drop in engine speed.
    3. Repeat for all four cylinders.
    4. Stop engine and reconnect idle air control valve.

    If the engine speed dropped equally (within 50 rpm) on all cylinders, refer to "Rough, Unstable, or Incorrect Idle, Stalling" symptom. If there was no drop or excessive variation in engine speed on one or more cylinders, check for spark on the respective cylinder(s) with J26792 (ST-125) spark tester or equivalent.

    If spark is present, remove the spark plugs from the cylinder(s) and check for the following.
    Cracks
    Wear

1988–90 2.0L (VIN K, OHC) – ALL MODELS

Improper gap
Burned electrode
Heavy deposits
For worn distributor shaft
Ignition pickup coil resistance and
connections
Ignition timing. See underhood emission
control information label.
Ignition wires for shorts or faulty insulation
Ignition system for moisture, dust, cracks,
burns, etc. Spray plug wires with fine water
mist to check for shorts.
Secondary ignition voltage output with ST
125 tester
Ignition coil connections
ECM and ignition system for faulty grounds
Proper operation of EST
Spark plugs. Look for wetness, cracks,
improper gap, burned electrodes, and heavy
deposits.
Spark plugs for correct heat range
Fuel for poor quality, "stale" fuel, and water
contamination
Fuel system for restricted filter or improper
pressure. Use CHART A-7.
EGR operation. Use CHART C-7.
For vacuum leaks at intake manifold gasket
Air cleaner element (filter) for dirt or
plugging
Idle Air Control system. Use Code 35 chart.
Check electrical system. IAC valve
will not move if voltage is below 9 volts or
greater than 17.8 volts. Also check battery
cables and ground straps for poor contact.
Erratic voltage will cause the IAC valve to
change its position, resulting in poor idle
quality
PCV valve for proper operation by placing
finger over inlet hole in valve end several
times. Valve should snap back. If not,
replace valve. Ensure that valve is correct
part. Also check PCV hose.
Canister purge system for proper operation.
Use CHART C-3.
PROM for correct application

Throttle shaft or TPS for sticking or binding
TPS voltage should read less than 1.25 volts
on a "Scan" tool with the throttle closed.
MAP sensor output. Use CHART C1-D
and/or check sensor by comparing it to the
output on a similar vehicle, if possible.

Oxygen sensor for silicone contamination
from contaminated fuel or use of improper
RTV sealant. The sensor will have a white,
powdery coating and will cause a high but
false signal voltage (rich exhaust indication).
The ECM will reduce the amount of fuel
delivered to the engine, causing a severe
driveability problem
Coolant sensor for a shift in calibration. Use
Code 14 or Code 15 chart.
A/C relay operation. Use CHART C-10.
A/C refrigerant pressure for high pressure
Check for overcharging or faulty pressure
switch.
P/N switch circuit on vehicle with automatic
transmission. Use CHART C-1A.
Generator output voltage. Repair if less than
9 volts or more than 16 volts.
Transmission torque converter operation

Transmission for proper shift points.

Transmission torque converter clutch
operation. Use CHART C-8
Vehicle brakes for dragging
Engine valve timing and compression. See
Section "6"
For the possibility of an exhaust restriction
or improper valve timing by performing the
following test.
1. With engine at normal operating
temperature, connect a vacuum gauge to
any convenient vacuum port on intake
manifold
2. Run engine at 1000 rpm and record
vacuum reading.
3. Increase engine speed slowly to 2500
rpm. Note vacuum reading at steady 200
rpm.
4. If vacuum at 2500 rpm decreases more
than 3" Hg from reading at 1000 rpm, the
exhaust system should be inspected for
restrictions. Use CHART B-1.
5. Disconnect exhaust pipe from engine and
repeat Steps 3 & 4. If vacuum still drops
more than 3" Hg with exhaust
disconnected, check valve timing.
For worn or incorrect basic engine parts such
as cam, heads, pistons, etc. Also check for
bent pushrods, worn rocker arms, and broken
or weak valve springs.

1988–90 2.0L (VIN K, OHC) – ALL MODELS

- For overheating and possible causes.
  Look for the following.
- Low or incorrect coolant solution. It
  should be a 50/50 mix of GM #1052753
  anti-freeze coolant (or equivalent) and
  water.
- Loose water pump belt
- Restricted air flow to radiator, or
  restricted water flow through radiator

- Faulty or incorrect thermostat
- Inoperative electric cooling fan circuit.
  See CHART C-12.
- If the system is running RICH (block learn
  less than 118), refer to "Diagnostic Aids" on
  facing page of Code 45.
- If the system is running LEAN (block learn
  greater than 138), refer to "Diagnostic Aids" on
  facing page of Code 44.

EXCESSIVE EXHAUST EMISSIONS OR ODORS

Definition: Vehicle fails an emission test or vehicle has excessive "rotten
egg" smell. (Excessive odors do not necessarily indicate
excessive emissions).

Perform careful visual and physical check as
described at the beginning of Section "B"
Perform "Diagnostic Circuit Check."
- **CHECK**
  - EGR valve not opening. Use CHART C-7.
  - Vacuum leaks
  - Faulty coolant system and/or coolant fan
    operation. Use CHART C-12.
  - Remove carbon with top engine cleaner.
    Follow instructions on can.
- If the system is running RICH (block learn
  less than 118), refer to "Diagnostic Aids" on
  facing page of Code 45.
- If the system is running LEAN (block learn
  greater than 138), refer to "Diagnostic Aids" on
  facing page of Code 44.

- If emission test indicates excessive NO_x, check for
  items which cause car to run lean or too hot.
- If emission test indicates excessive HC and CO or
  exhaust has excessive odors, check for items
  which cause car to run RICH.
  - Incorrect fuel pressure. Use CHART A-7.
  - Fuel loading of evaporative vapor canister.
    Use CHART C-3.
  - PCV valve plugging, sticking, or blocked PCV
    hose. Check for fuel in crankcase
  - Catalytic converter lead contamination
    (Look for removal of fuel filler neck
    restrictor.)
  - Improper fuel cap installation
  - Faulty spark plugs, plug wires, or ignition
    components.

DIESELING, RUN-ON

Definition: Engine continues to run after key is turned "OFF", but runs
very roughly. (If engine runs smoothly, check ignition switch).

Perform careful visual and physical check as
described at the beginning of Section "B"
Perform "Diagnostic Circuit Check."
- **CHECK**
  Injector and TBI assembly for leakage.
  Pressurize system by energizing fuel pump
  through the fuel pump test connector.

1988–90 2.0L (VIN K, OHC) – ALL MODELS

BACKFIRE

Definition: Fuel ignites in intake manifold or in exhaust system,
making a loud popping sound.

Perform careful visual and physical check as
described at the beginning of Section "B"
Perform "Diagnostic Circuit Check."
- **CHECK**
  For possibility of misfiring, crossfiring, or
  cutting under load or at idle. Locate misfiring
  cylinder(s) by performing the following test.
  1. Start engine. Disconnect idle air control
     valve. Remove one spark plug wire from
     a spark plug and ground it against the
     engine.
  2. Note drop in engine speed.
  3. Repeat for all four cylinders.
  4. Stop engine and reconnect idle air control
     valve.
  If the engine speed dropped equally (within
  50 rpm) on all cylinders, refer to "Rough,
  Unstable, or Incorrect Idle, Stalling"
  symptom. If there was no drop or excessive
  variation in engine speed on one or more
  cylinders, check for spark on the respective
  cylinders with J26792 (ST-125) spark tester
  or equivalent. If there is no spark, see Section
  "6D." If spark is present, remove the spark
  plugs from the cylinder(s) and check for the
  following.
  Cracks
  Wear
  Improper gap
  Burned electrode
  Heavy deposits

- For worn distributor shaft
- Ignition timing. See underhood emission
  control information label.
- EGR valve being open all the
  time. Use CHART C-7.
- Intake manifold gasket for leaks
- Spark plugs. Look for wetness, cracks,
  improper gap, burned electrodes, and heavy
  deposits.
- Ignition coil connections
- Ignition system for moisture, dust, cracks,
  burns, etc. Spray plug wires with fine water
  mist to check for shorts.
- ECM and ignition system for faulty grounds
- Secondary ignition voltage output with
  ST-125 tester
- For vacuum leaks at intake manifold gasket
- Engine valve timing and compression.

- For worn or incorrect basic engine parts such
  as cam, heads, pistons, etc. Also check for
  bent pushrods, worn rocker arms, and
  broken or weak valve springs.

1988–90 2.0L (VIN K, OHC) – ALL MODELS

CHART B-1
RESTRICTED EXHAUST SYSTEM CHECK
ALL ENGINES

Proper diagnosis for a restricted exhaust system is essential before any components are replaced. Either of
the following procedures may be used for diagnosis, depending upon engine or tool used.

CHECK AT A. I. R. PIPE: **OR**   **CHECK AT O_2 SENSOR:**

1. Remove the rubber hose at the exhaust
   manifold A.I.R. pipe check valve. Remove
   check valve.
2. Connect a fuel pump pressure gauge to a hose
   and nipple from a propane enrichment device
   (J26911) (see illustration).
3. Insert the nipple into the exhaust manifold
   A.I.R. pipe.

1. Carefully remove O_2 sensor.
2. Install Borroughs exhaust backpressure
   tester (BT 8515 or BT 8603) or equivalent in
   place of O_2 sensor (see illustration).
3. After completing test described below, be
   sure to coat threads of O_2 sensor with anti-
   seize compound P/N 5613695 or equivalent
   prior to re-installation.

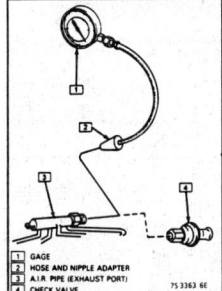

1	GAGE
2	HOSE AND NIPPLE ADAPTER
3	A.I.R. PIPE (EXHAUST PORT)
4	CHECK VALVE

75 3363 6E

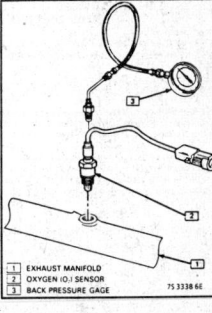

1	EXHAUST MANIFOLD
2	OXYGEN (O.) SENSOR
3	BACK PRESSURE GAGE

75 3338 6E

DIAGNOSIS:
1. With the engine idling at normal operating temperature, observe the exhaust system backpressure
   reading on the gauge. Reading should not exceed 8.6 kPa (1.25 psi).
2. Increase engine speed to 2000 rpm and observe gauge. Reading should not exceed 20.7 kPa (3 psi).
3. If the backpressure at either speed exceeds specification, a restricted exhaust system is indicated.
4. Inspect the entire exhaust system for a collapsed pipe, heat distress, or possible internal muffler failure.
5. If there are no obvious reasons for the excessive backpressure, the catalytic converter is suspected to be
   restricted and should be replaced using current recommended procedures.

1988–90 2.0L (VIN K, OHC) – ALL MODELS

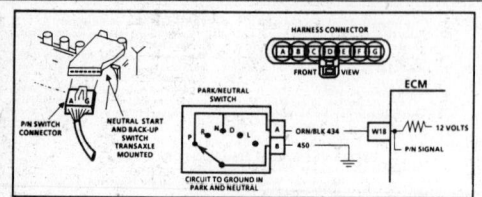

HARNESS CONNECTOR
FRONT VIEW
ECM

P/N SWITCH CONNECTOR
PARK/NEUTRAL SWITCH
NEUTRAL START AND BACK-UP SWITCH TRANSAXLE MOUNTED
CIRCUIT TO GROUND IN PARK AND NEUTRAL
ORN/BLK 434
W18
12 VOLTS
450
P/N SIGNAL

CHART C-1A
PARK/NEUTRAL SWITCH DIAGNOSIS
2.0L OHC (VIN K)

Circuit Description:
The park/neutral switch contacts are a part of the neutral start switch and are closed to ground in park or neutral, and open in drive ranges.
The ECM supplies ignition voltage through a current limiting resistor to CKT 434 and senses a closed switch when the voltage on CKT 434 drops to less than one volt.
The ECM uses the P/N signal as one of the inputs to control idle air control and VSS diagnostics.

Test Description: Numbers below refer to circled numbers on the diagnostic chart.
1. Checks for a closed switch to ground in park position. Different "Scan" tools may display P/N differently. Refer to tool operator's manual for the display used.
2. Checks for an open switch in drive range.

3. Be sure "Scan" tool indicates drive even while wiggling shifter to test for an intermittent or maladjusted switch in drive range.

1988–90 2.0L (VIN K, OHC) – ALL MODELS

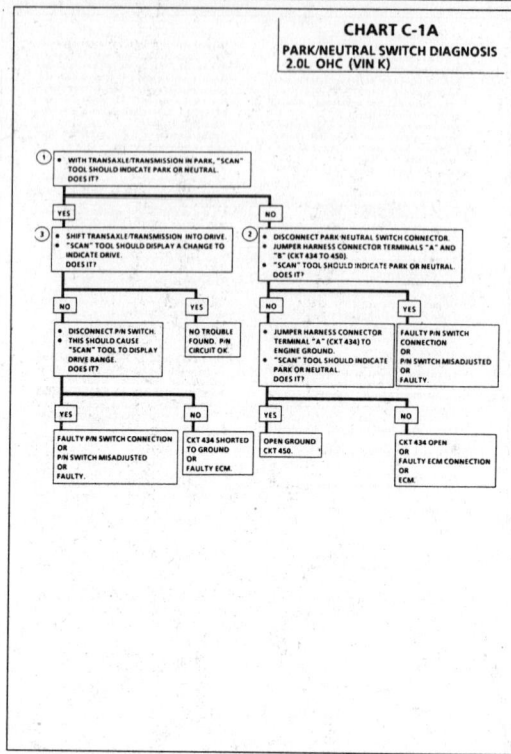

CHART C-1A
PARK/NEUTRAL SWITCH DIAGNOSIS
2.0L OHC (VIN K)

1988–90 2.0L (VIN K, OHC) – ALL MODELS

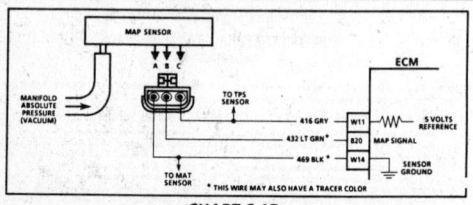

MAP SENSOR
A B C
TO TPS SENSOR
ECM
MANIFOLD ABSOLUTE PRESSURE (VACUUM)
416 GRY W11 5 VOLTS REFERENCE
432 LT GRN* 820 MAP SIGNAL
469 BLK* W14 SENSOR GROUND
TO MAT SENSOR
* THIS WIRE MAY ALSO HAVE A TRACER COLOR

CHART C-1D
MANIFOLD ABSOLUTE PRESSURE (MAP) OUTPUT CHECK
2.0L OHC (VIN K)

Circuit Description:
The manifold absolute pressure (MAP) sensor senses manifold pressure and sends that signal to the ECM. The MAP sensor is mainly used to calculate engine load, which is a fundamental input for spark and fuel calculations. The MAP sensor is also used to determine barometric pressure.

Test Description: Numbers below refer to circled numbers on the diagnostic chart.
1. Checks MAP sensor output voltage to the ECM. This voltage, without engine running, represents a barometer reading to the ECM. Comparison of this BARO reading with a known good vehicle with the same sensor is a good way to check the accuracy of a "suspect" sensor. Readings should be within .4 volt.

2. Applying 34 kPa (10" Hg) vacuum to the MAP sensor should cause the voltage to be 1.2 volts less than the voltage at step 1. Upon applying vacuum to the sensor, the change in voltage should be instantaneous. A slow voltage change indicates a faulty sensor.
3. Check vacuum hose to sensor for leaking or restriction. Be sure no other vacuum devices are connected to the MAP hose.

1988–90 2.0L (VIN K, OHC) – ALL MODELS

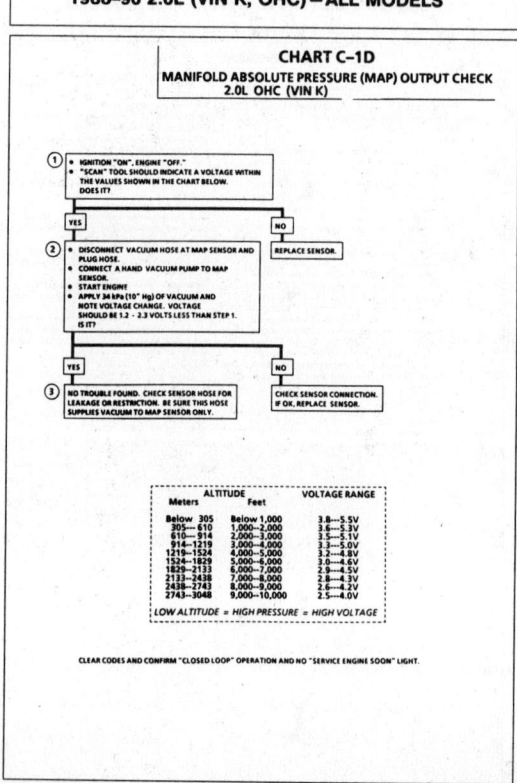

CHART C-1D
MANIFOLD ABSOLUTE PRESSURE (MAP) OUTPUT CHECK
2.0L OHC (VIN K)

ALTITUDE		VOLTAGE RANGE
Meters	Feet	
Below 305	Below 1,000	3.8–5.5V
305– 610	1,000–2,000	3.6–5.3V
610– 914	2,000–3,000	3.5–5.1V
914–1219	3,000–4,000	3.3–5.0V
1219–1524	4,000–5,000	3.2–4.8V
1524–1829	5,000–6,000	3.0–4.6V
1829–2133	6,000–7,000	2.9–4.5V
2133–2438	7,000–8,000	2.8–4.3V
2438–2743	8,000–9,000	2.6–4.3V
2743–3048	9,000–10,000	2.5–4.0V

LOW ALTITUDE = HIGH PRESSURE = HIGH VOLTAGE

CLEAR CODES AND CONFIRM "CLOSED LOOP" OPERATION AND NO "SERVICE ENGINE SOON" LIGHT.

1988–90 2.0L (VIN K, OHC) – ALL MODELS

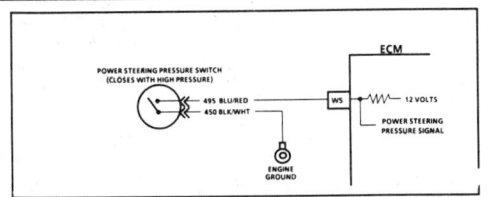

CHART C-1E
POWER STEERING PRESSURE SWITCH (PSPS) DIAGNOSIS
2.0L OHC (VIN K)

Circuit Description:
The power steering pressure switch is normally open to ground, and CKT 495 will be at approximately battery voltage.
Turning the steering wheel increases power steering oil pressure and its load on an idling engine. The pressure switch will close before the load becomes great enough to cause an idle problem.
Closing the switch causes CKT 495 to read less than 1 volt and the ECM will increase the idle air rate.
- A pressure switch that will not close, or an open CKT 495 or 450, may cause the engine to stall at idle when power steering loads are high.
- A switch that will not open, or a CKT 495 shorted to ground, may affect idle quality.

Test Description: Numbers below refer to circled numbers on the diagnostic chart.
1. Different "Scan" tools may display the state of this switch in different ways. Refer to the "Scan" tool operator's manual to determine how this input is indicated.

.2 Checks to determine if CKT 495 is shorted to ground.
3. This step should simulate a closed switch.

1988–90 2.0L (VIN K, OHC) – ALL MODELS

CHART C-1E
POWER STEERING PRESSURE SWITCH (PSPS) DIAGNOSIS
2.0L OHC (VIN K)

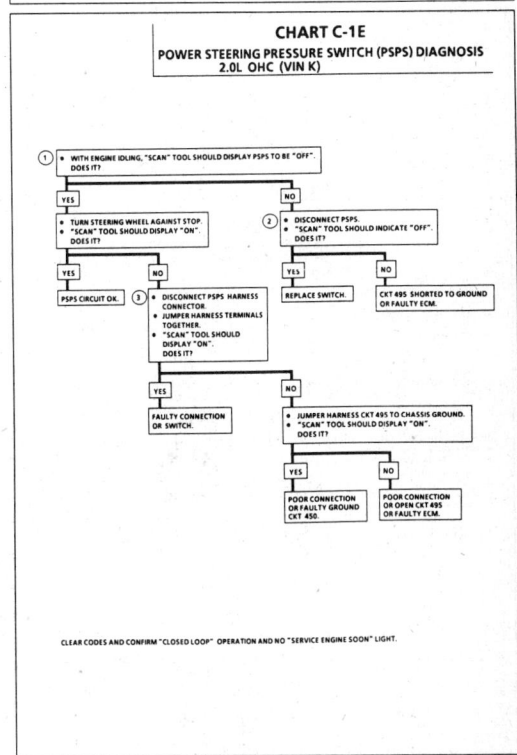

CLEAR CODES AND CONFIRM "CLOSED LOOP" OPERATION AND NO "SERVICE ENGINE SOON" LIGHT.

1988–90 2.0L (VIN K, OHC) – ALL MODELS

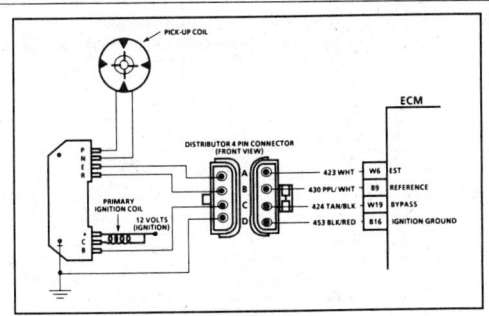

CHART C-4C
IGNITION SYSTEM CHECK
(REMOTE COIL)
2.0L OHC (VIN K)

Test Description: Numbers below refer to circled numbers on the diagnostic chart.
1. Two wires are checked to ensure that an open is not present in a spark plug wire.
1A. If spark occurs with 4 terminal distributor connector disconnected, pick-up coil output is too low for EST operation.
2. A spark indicates the problem must be the distributor cap or rotor.
3. Normally, there should be battery voltage at the "C" and "+" terminals. Low voltage would indicate an open or a high resistance circuit from the distributor to the coil or ignition switch. If "C" terminal voltage was low, but "+" terminal voltage is 10 volts or more, circuit from "C" terminal to ignition coil or ignition coil primary winding is open.
4. Checks for a shorted module or grounded circuit from the ignition coil to the module. The distributor module should be turned "OFF." Normal voltage should be about 12 volts.
If the module is turned "ON," the voltage would be low, but above 1 volt. This could cause the ignition coil to fail from excessive heat.

With an open ignition coil primary winding, a small amount of voltage will leak through the module from the "Bat." to the tach terminal.
5. Applying voltage (1.5 to 8 volts) to module terminal "P" should turn the module "ON" and the tachometer terminal voltage should drop to about 7-9 volts. This test will determine whether the module or coil is faulty or if the pick-up coil is not generating the proper signal to turn the module "ON." This test can be performed by using a DC battery with a rating of 1.5 to 8 volts. The use of the test light is mainly to allow the "P" terminal to be probed more easily.
Some digital multi-meters can also be used to trigger the module by selecting ohms, usually the diode position. In this position the meter may have a voltage across its terminals which can be used to trigger the module. The voltage in the ohms position can be checked by using a second meter or by checking the manufacture's specification of the tool being used.
6. This should turn "OFF" the module and cause a spark. If no spark occurs, the fault is most likely in the ignition coil because most module problems would have been found before this point in the procedure. A module tester (J24642) could determine which is at fault.

CHART C-4C
IGNITION SYSTEM CHECK
(REMOTE COIL)
2.0L OHC (VIN K)

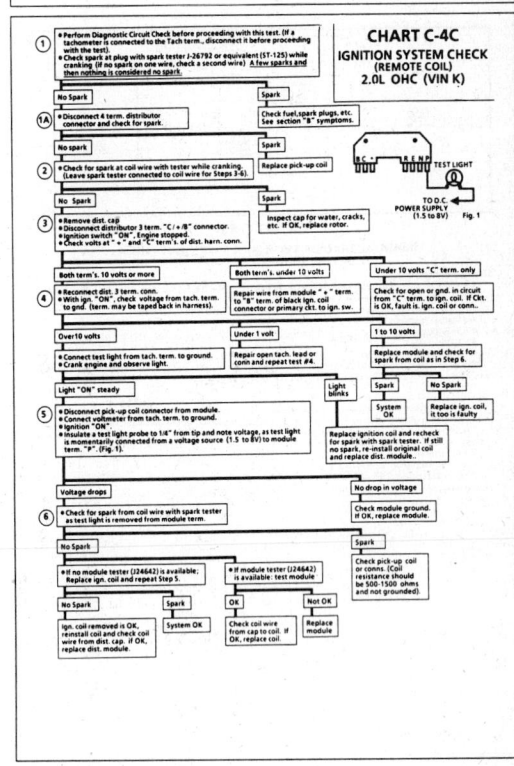

1988–90 2.0L (VIN K, OHC) – ALL MODELS

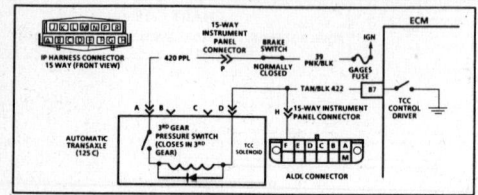

CHART C-8A
TORQUE CONVERTER CLUTCH (TCC)
(ELECTRICAL DIAGNOSIS)
2.0L OHC (VIN K)

Circuit Description:
The purpose of the automatic transaxle torque converter clutch is to eliminate the power loss of the torque converter when the vehicle is in a cruise condition. This allows the convenience of the automatic transaxle and the fuel economy of a manual transaxle.

Voltage is supplied to the TCC solenoid through the brake switch and transaxle third gear apply switch. The ECM will engage TCC by grounding CKT 422 to energize the solenoid.

The TCC will engage under the following conditions:
- Vehicle speed exceeds 30 mph (48 km/h)
- Engine coolant temperature exceeds 70°C (156°F)
- Throttle position is not changing faster than a calibrated rate (steady throttle)
- Transaxle third gear switch is closed
- Brake switch is closed

Test Description: Numbers below refer to circled numbers on the diagnostic chart.
1. Light "OFF" confirms transaxle third gear apply switch is open.
2. At 48 km/h (30 mph), the transaxle third gear switch should close. Test light will light and confirm battery supply and close brake switch.

3. Grounding the diagnostic terminal, with engine "OFF," should energize the TCC solenoid. This test checks the capability of the ECM to control the solenoid.

Diagnostic Aids:
An engine coolant thermostat that is stuck open or opens at too low a temperature may result in an inoperative TCC.

1988–90 2.0L (VIN K, OHC) – ALL MODELS

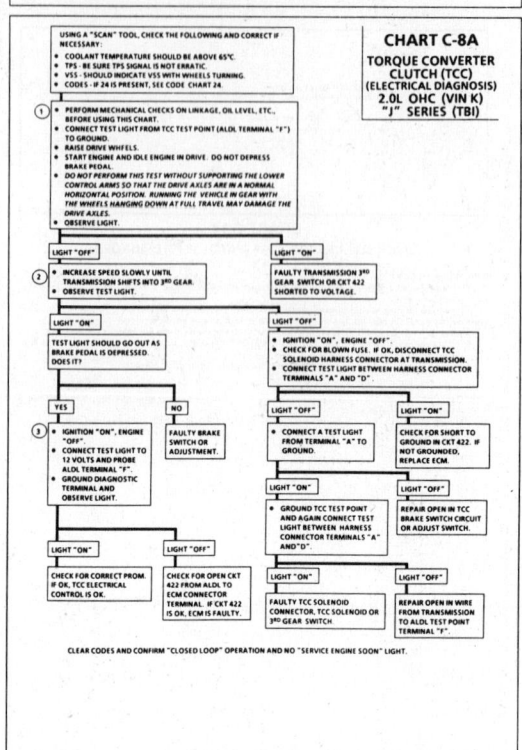

1988–90 2.0L (VIN K, OHC) – ALL MODELS

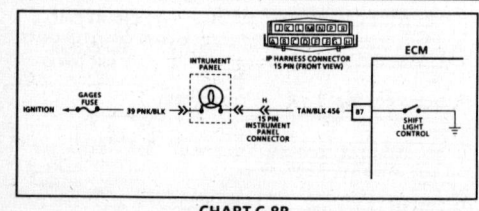

CHART C-8B
MANUAL TRANSMISSION (M/T) SHIFT LIGHT CHECK
2.0L OHC (VIN K)

Circuit Description:
The shift light indicates the best transmission shift point for maximum fuel economy. The light is controlled by the ECM and is turned "ON" by grounding CKT 456.

The ECM uses information from the following inputs to control the shift light:
- Coolant temperature
- Throttle position
- Vehicle speed
- Engine speed

The ECM uses the measured engine and vehicle speeds to calculate what gear the vehicle is in. It is this calculation which determines when the shift light should be turned "ON".

Test Description: Numbers below refer to circled numbers on the diagnostic chart.
1. This should not turn "ON" the shift light. If the light is "ON," there is a short to ground in CKT 456 or a fault in the ECM.
2. When the diagnostic terminal is grounded, the ECM should ground CKT 456, and the shift light should be "ON."

3. This checks the shift light circuit up to the ECM connector. If the shift light lights, then the ECM connector is faulty, or the ECM does not have the ability to ground the circuit.

1988–90 2.0L (VIN K, OHC) – ALL MODELS

CHART C-8B
MANUAL TRANSMISSION (M/T) SHIFT LIGHT CHECK
2.0L OHC (VIN K)

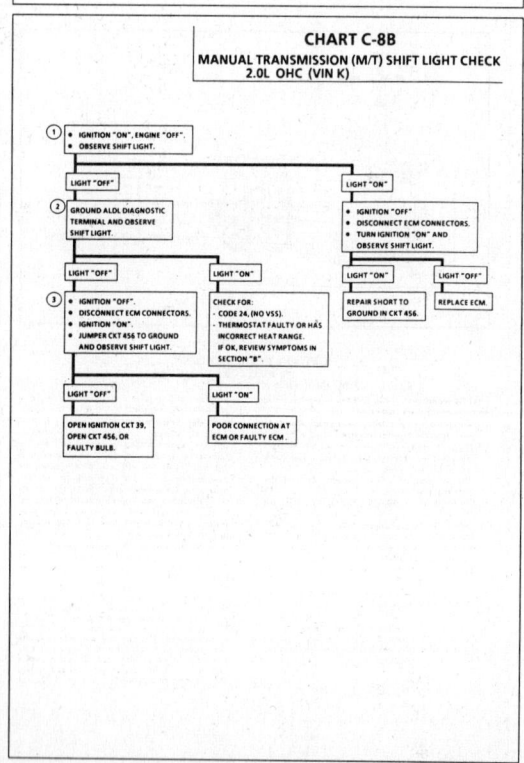

1988–90 2.0L (VIN K, OHC) – ALL MODELS

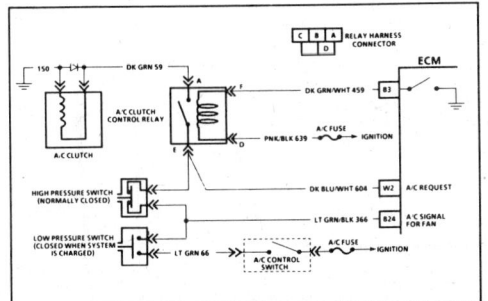

CHART C-10
A/C CLUTCH CONTROL
2.0L OHC (VIN K)

Circuit Description:

When an A/C mode is selected on the A/C control switch, ignition voltage is supplied to the compressor low pressure switch. If there is sufficient A/C refrigerant charge, the low pressure switch will be closed and complete the circuit to the closed high pressure cut-off switch and to CKTs 366 and 604. The voltage on CKT 604 to the ECM is shown by the "Scan" tool as A/C request "ON" (voltage present), "OFF" (no voltage). When a request for A/C is sensed by the ECM, the ECM will ground CKT 459 of the A/C clutch control relay, the relay contact will close, and current will flow from CKT 604 to CKT 59 and engage the A/C compressor clutch. A "Scan" tool will show the grounding of CKT 459 as A/C clutch "ON." If voltage is sensed by the ECM on CKT 366, the cooling fan will be turned "ON."

Diagnostic Aids:

Both pressure switches are located at the rear of the A/C compressor. The low pressure switch connector can be identified by a blue insert and the high pressure switch connector by a red insert.

1988–90 2.0L (VIN K, OHC) – ALL MODELS

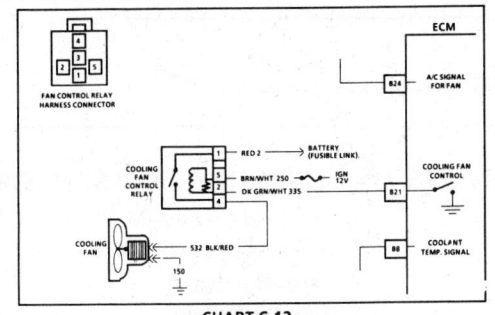

CHART C-12
ENGINE COOLING FAN
2.0L OHC (VIN K)

Circuit Description:

Battery voltage to operate the cooling fan motor is supplied to relay terminal "1." Ignition voltage to energize the relay is supplied to relay terminal "5." When the ECM grounds CKT 335, the relay is energized and the cooling fan is turned "ON." When the engine is running, the ECM will energize the cooling fan relay if a coolant temperature sensor code (14 or 15) has been set, or under the following conditions:

- A/C is "ON" and vehicle speed is less than 30 mph (48 km/h).
- Coolant temperature is greater than 108°C (230°F).

Diagnostic Aids:

If the vehicle has an overheating problem, it must be determined if the complaint was due to an actual boil over, or the coolant temperature warning light, or the temperature gage indicated over heating.

If the gage or light indicates overheating but no boilover is detected, the gage circuit should be checked. The gage accuracy can be checked by comparing the coolant sensor reading using a "Scan" tool with the gage reading.

If the engine is actually overheating and the gage indicates overheating, but the cooling fan is not turning "ON," the coolant sensor has probably shifted out of calibration and should be replaced.

1988–90 2.0L (VIN K, OHC) – ALL MODELS

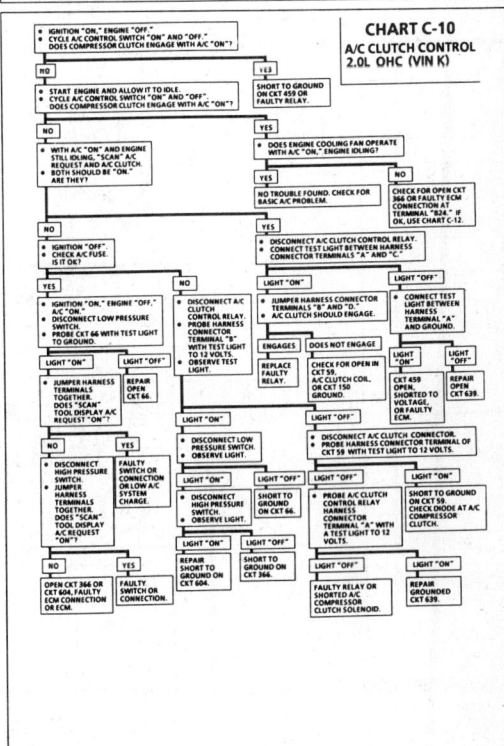

CHART C-10
A/C CLUTCH CONTROL
2.0L OHC (VIN K)

1988–90 2.0L (VIN K, OHC) – ALL MODELS

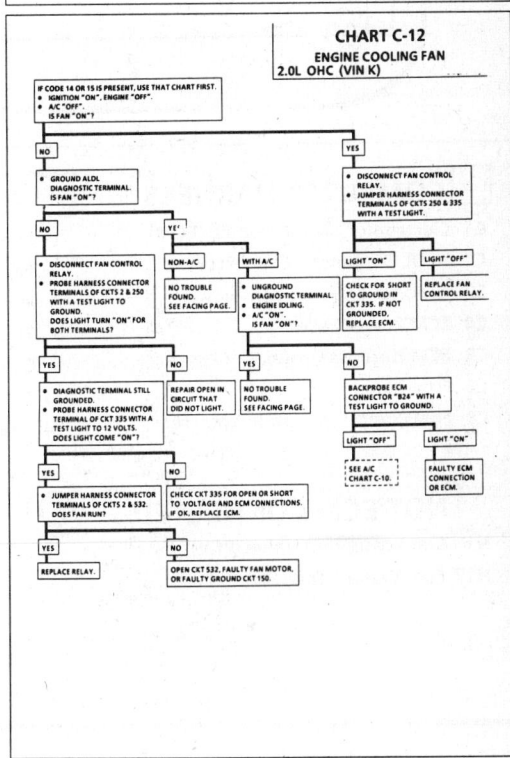

CHART C-12
ENGINE COOLING FAN
2.0L OHC (VIN K)

2.5L (VIN R) COMPONENT LOCATIONS

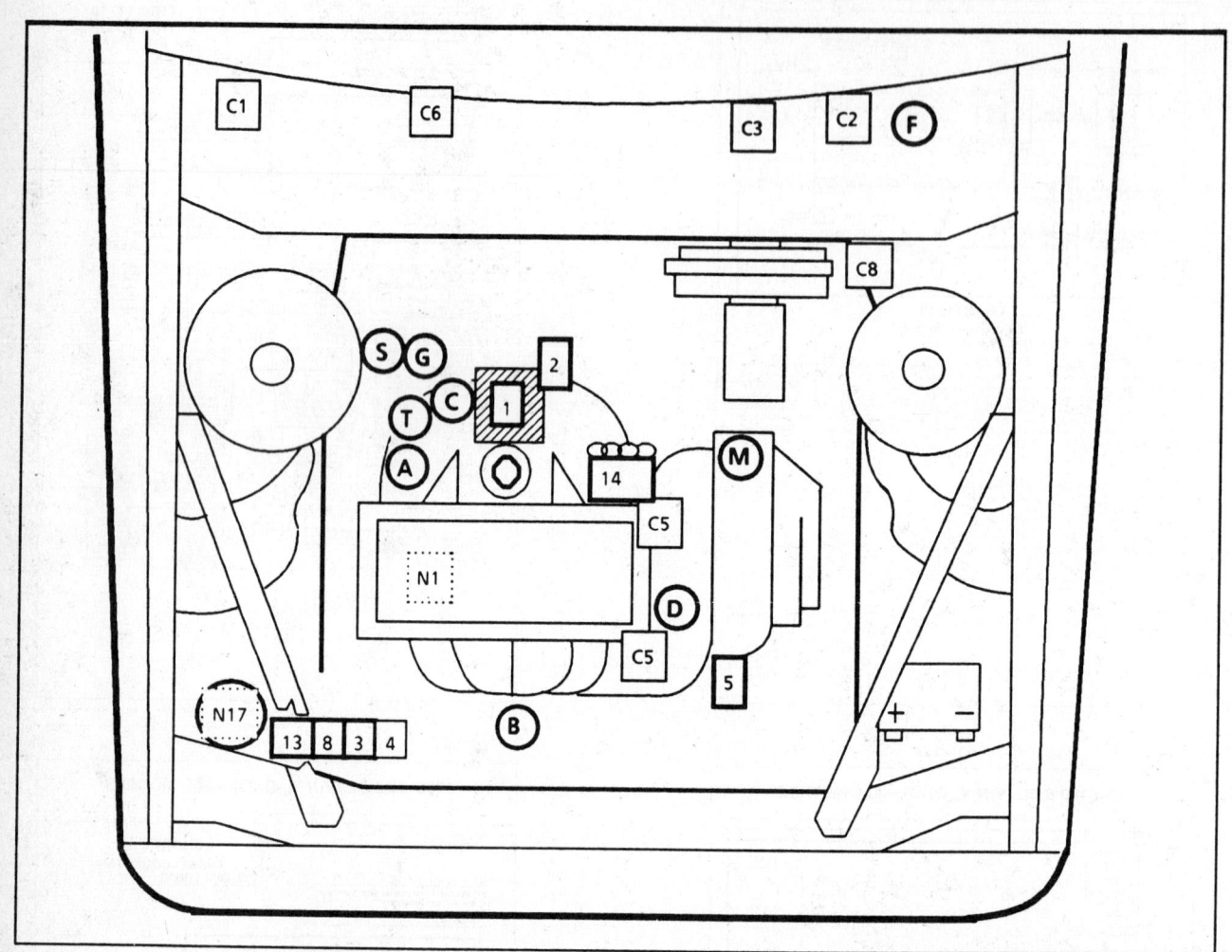

☐ COMPUTER HARNESS

- C1 Electronic Control Module (ECM)
- C2 ALDL Diagnostic Connector
- C3 "Service Engine Soon" light
- C4 ECM Power Fuse
- C5 ECM Harness Grounds
- C6 Fuse Panel
- C8 Fuel Pump Test Connector

☐ NOT ECM CONNECTED

- N1 Crankcase Vent Valve (PCV)
- N17 Fuel Vapor Canister

☐ CONTROLLED DEVICES

- 1 Fuel Injector
- 2 Idle Air Control Valve
- 3 Fuel Pump Relay
- 5 TCC Solenoid Connector
- 8 Cooling Fan Relay
- 13 A/C Compressor Relay
- 14 Direct Ignition System Assembly

⬡ Exhaust Gas Recirculation Valve

○ INFORMATION SENSORS

- A Manifold Pressure (MAP) (Mounted on Air Cleaner)
- B Exhaust Oxygen
- C Throttle Position
- D Coolant Temperature
- F Vehicle Speed Buffer Amplifier (if used)
- G Vehicle Speed PM Generator (if used)
- M P/N Switch/Neutral Start Switch
- S P/S Pressure Switch
- T Manifold Air Temperature (MAT)

2.5L (VIN R) ECM WIRING DIAGRAM (CONT.)

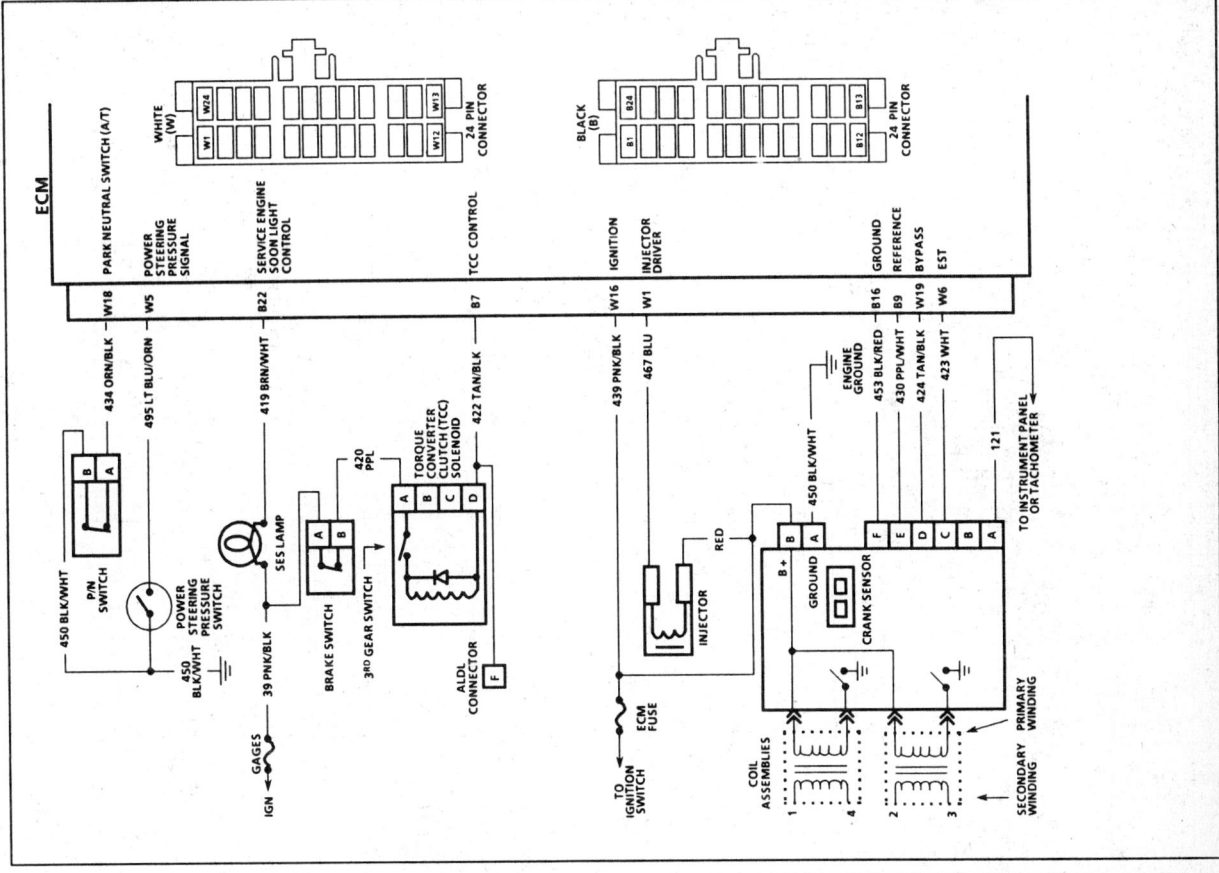

2.5L (VIN R) ECM WIRING DIAGRAM

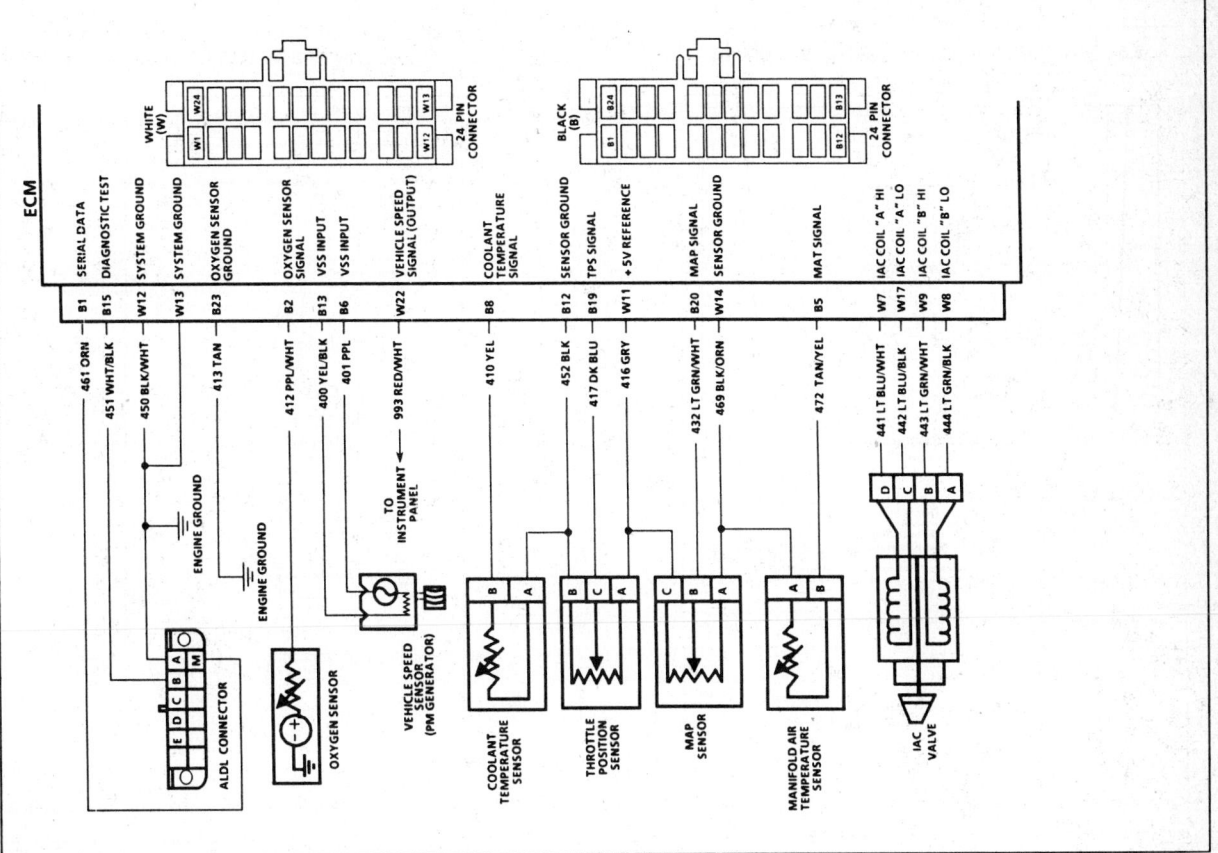

2.5L (VIN R) ECM WIRING DIAGRAM (CONT.)

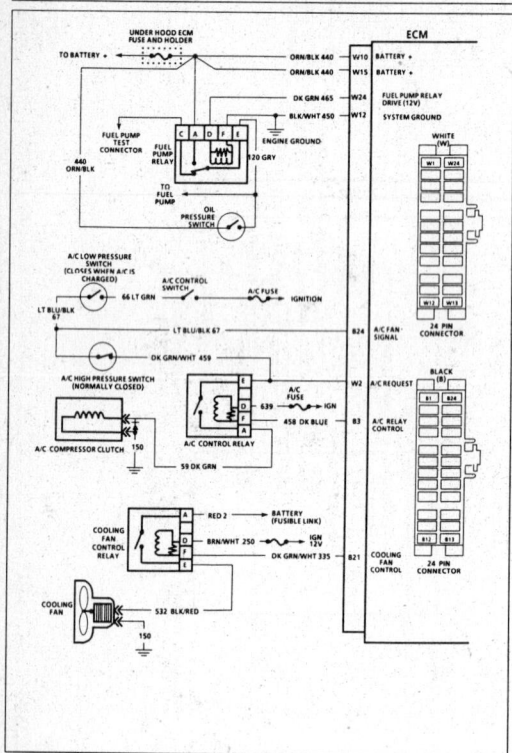

2.5L (VIN R) ECM CONNECTOR TERMINAL END VIEW

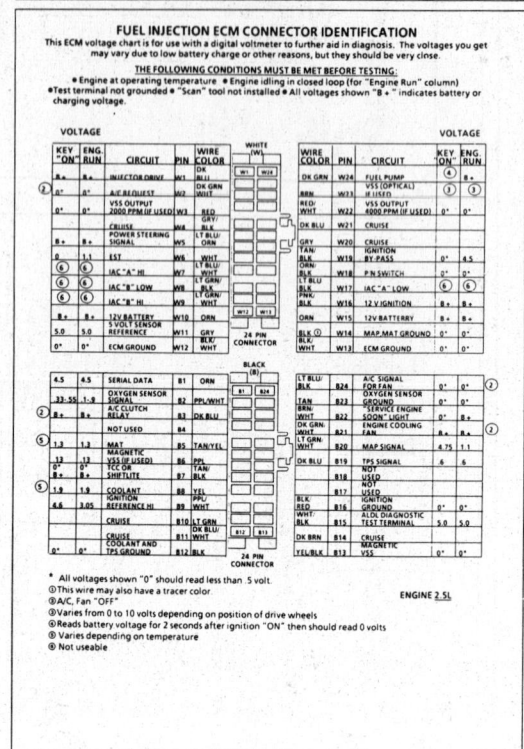

FUEL INJECTION ECM CONNECTOR IDENTIFICATION

This ECM voltage chart is for use with a digital voltmeter to further aid in diagnosis. The voltages you get may vary due to low battery charge or other reasons, but they should be very cinse.

THE FOLLOWING CONDITIONS MUST BE MET BEFORE TESTING:
- Engine at operating temperature ● Engine idling in closed loop (for "Engine Run" column)
- Test terminal not grounded ● "Scan" tool not installed ● All voltages shown "B +" indicates battery or charging voltage.

* All voltages shown "0" should read less than .5 volt.
① This wire may also have a tracer color.
② A/C Fan "OFF"
③ Varies from 0 to 10 volts depending on position of drive wheels
④ Reads battery voltage for 2 seconds after ignition "ON" then should read 0 volts
⑤ Varies depending on temperature
⑥ Not useable

ENGINE 2.5L

2.5L (VIN R) COMPONENT LOCATIONS—LUMINA

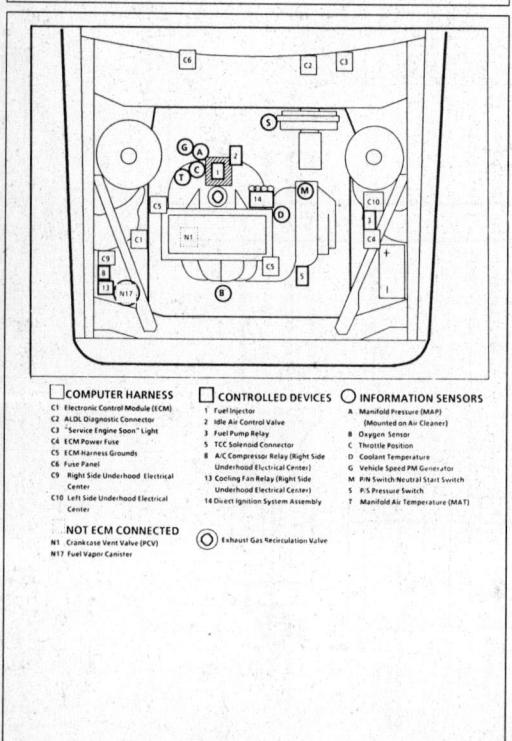

☐ COMPUTER HARNESS
C1 Electronic Control Module (ECM)
C2 ALDL Diagnostic Connector
C3 "Service Engine Soon" Light
C6 ECM Power Fuse
C5 ECM Harness Grounds
C6 Fuse Panel
C9 Right Side Underhood Electrical Center
C10 Left Side Underhood Electrical Center

NOT ECM CONNECTED
N1 Crankcase Vent Valve (PCV)
N17 Fuel Vapor Canister

◼ CONTROLLED DEVICES
1 Fuel Injector
2 Idle Air Control Valve
3 Fuel Pump Relay
5 TCC Solenoid Connector
6 A/C Compressor Relay (Right Side Underhood Electrical Center)
13 Cooling Fan Relay (Right Side Underhood Electrical Center)
14 Direct Ignition System Assembly

○ INFORMATION SENSORS
A Manifold Pressure (MAP) (Mounted on an Air Cleaner)
B Oxygen Sensor
C Throttle Position
D Coolant Temperature
G Vehicle Speed PM Generator
M P/N Switch Neutral Start Switch
S P/S Pressure Switch
T Manifold Air Temperature (MAT)

○ Exhaust Gas Recirculation Valve

2.5L (VIN R) ECM WIRING DIAGRAM—LUMINA

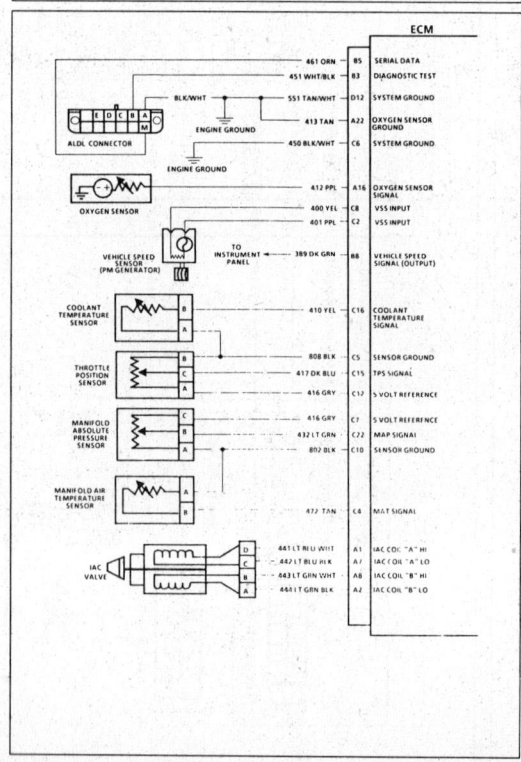

2.5L (VIN R) ECM WIRING DIAGRAM (CONT.) LUMINA

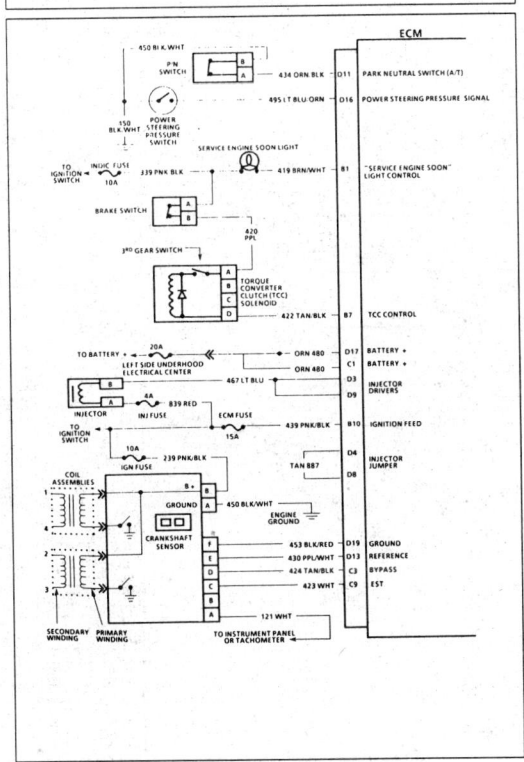

2.5L (VIN R) ECM WIRING DIAGRAM (CONT.) LUMINA

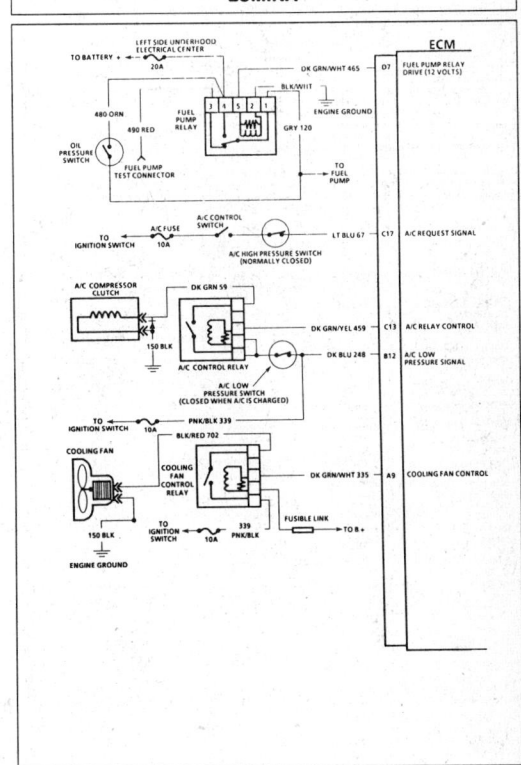

2.5L (VIN R) ECM CONNECTOR TERMINAL END VIEW—LUMINA

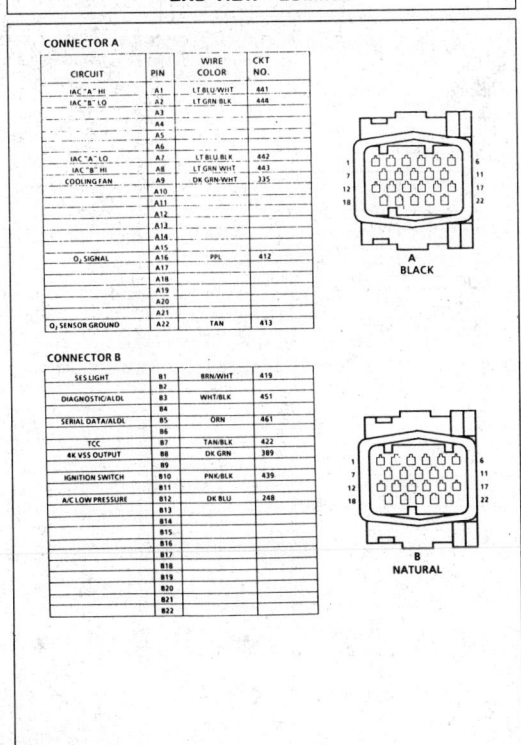

2.5L (VIN R) ECM CONNECTOR TERMINAL END VIEW (CONT.)—LUMINA

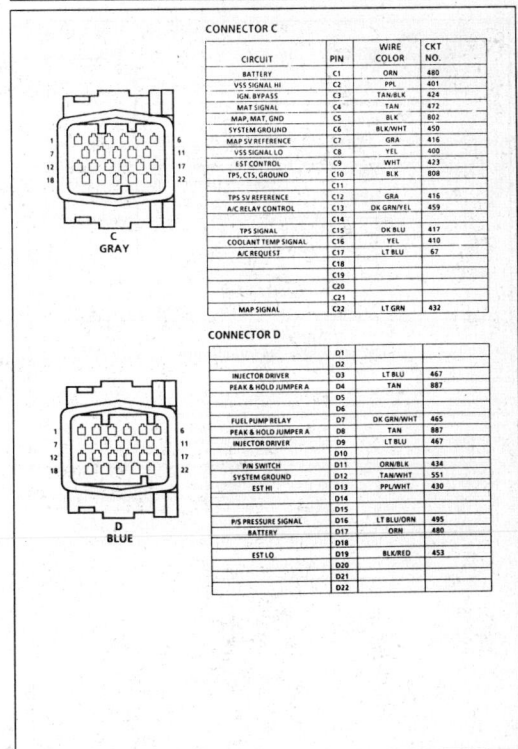

1988–90 2.5L (VIN R) – ALL MODELS EXCEPT FIERO

DIAGNOSTIC CIRCUIT CHECK

The Diagnostic Circuit Check is an organized approach to identifying a problem created by an electronic engine control system malfunction. It must be the starting point for any driveability complaint diagnosis because it directs the service technician to the next logical step in diagnosing the complaint.

The "Scan" data listed in the table may be used for comparison after completing the diagnostic circuit check and finding the on-board diagnostics functioning properly with no trouble codes displayed. The "Typical Data Values" are an average of display values recorded from normally operating vehicles and are intended to represent what a normally functioning system would typically display.

A "SCAN" TOOL THAT DISPLAYS FAULTY DATA SHOULD NOT BE USED, AND THE PROBLEM SHOULD BE REPORTED TO THE MANUFACTURER. THE USE OF A FAULTY "SCAN" TOOL CAN RESULT IN MISDIAGNOSIS AND UNNECESSARY PARTS REPLACEMENT.

Only the parameters listed below are used in this manual for diagnosis. If a "Scan" tool reads other parameters, the values are not recommended by General Motors for use in diagnosis. For more description on the values and use of the "Scan" tool to diagnosis ECM inputs, refer to the applicable component diagnosis section.

If all values are within the range illustrated, refer to "Symptoms" in Section "B".

"SCAN" TOOL DATA

Test Under Following Conditions: Idle, Upper Radiator Hose Hot, Closed Throttle, Park or Neutral, "Closed Loop", All Accessories "OFF".

"SCAN" Position	Units Displayed	Typical Data Value
Desired RPM	RPM	ECM idle command (varies with temperature)
RPM	RPM	± 50 RPM from desired rpm in drive (AUTO) ± 100 RPM from desired rpm in neutral (MANUAL)
Coolant Temperature	Degrees Celsius	85° - 105°
MAT Temperature	Degrees Celsius	10° - 90° (varies with underhood temperature and sensor location)
MAP	Volts	1 - 2 (depends on Manifold Vacuum & Barometric pressures)
BPW (base pulse width)	Milliseconds	.8 - 3.0
O₂	Volts	.1 - 1 and varies
TPS	Volts	.4 - 1.25
Throttle Angle	0 - 100%	0
IAC	Counts (steps)	1 - 50
P/N Switch	P-N and R-D-L	Park/Neutral (P/N)
INT (Integrator)	Counts	110 - 145
BLM (Block Learn Memory)	Counts	118 - 138
Open/Closed Loop	Open/Closed	"Closed Loop" (may go open with extended idle)
VSS	MPH	0
TCC	ON/OFF	"OFF"/ ("ON" with TCC commanded)
Spark Advance	Degrees	Varies
Battery	Volts	13.5 - 14.5
Fan	ON/OFF	"OFF" (below 102°C)
P/S Switch	Normal/Hi Pressure	Normal
A/C Request	Yes/No	No (Yes, with A/C requested)
A/C Clutch	ON/OFF	"OFF" ("ON", with A/C commanded "ON")

1988–90 2.5L (VIN R) – ALL MODELS EXCEPT FIERO

DIAGNOSTIC CIRCUIT CHECK
2.5L (VIN R)

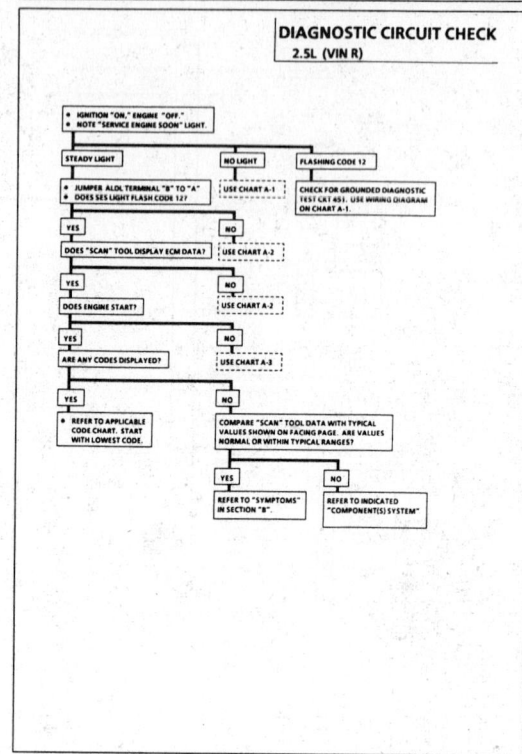

1988–90 2.5L (VIN R) – ALL MODELS EXCEPT FIERO

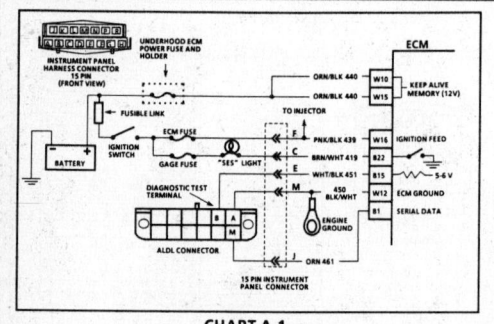

CHART A-1
NO "SERVICE ENGINE SOON" LIGHT
2.5L (VIN R)

Circuit Description:
There should always be a steady "Service Engine Soon" light when the ignition is "ON" and the engine is "OFF". Battery voltage is supplied directly to the light bulb. The Electronic Control Module (ECM) controls the light and turns it "ON" by providing a ground path through CKT 419 to the ECM.

Test Description: Numbers below refer to circled numbers on the diagnostic chart.
1. This check determines if there is a fault in the "SES" light circuit or in the ECM.
2. If CKTs 440 and 439 have voltage, the ECM grounds or the ECM is faulty.
3. Using a test light connected to 12 volts, probe each of the system ground circuits to be sure a good ground is present. See ECM terminal end view in front of this section for the ECM terminal locations of ground circuits.

Diagnostic Aids:
If engine runs OK, check the following:
- Faulty light bulb
- CKT 419 open
- Gages fuse blown. This will result in no oil or generator lights, seat belt reminder, etc.

If engine cranks, but will not run, use CHART A-3.

1988–90 2.5L (VIN R) – ALL MODELS EXCEPT FIERO

CHART A-1
NO "SERVICE ENGINE SOON" LIGHT
2.5L (VIN R)

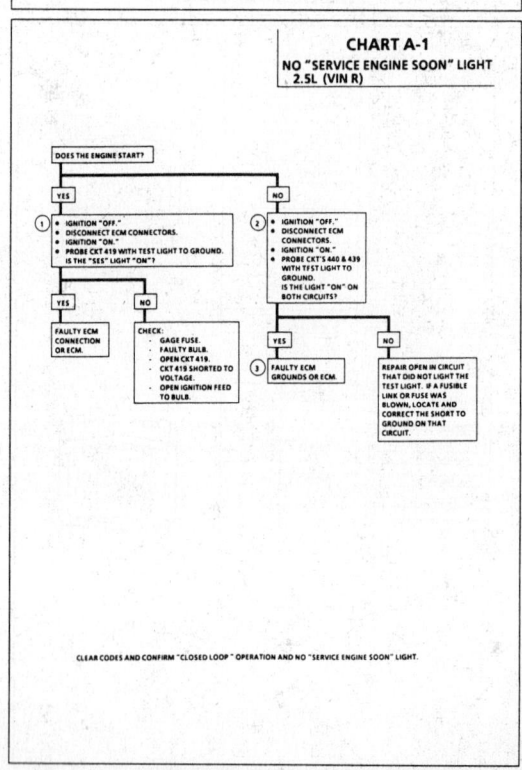

1988–90 2.5L (VIN R) – ALL MODELS EXCEPT FIERO

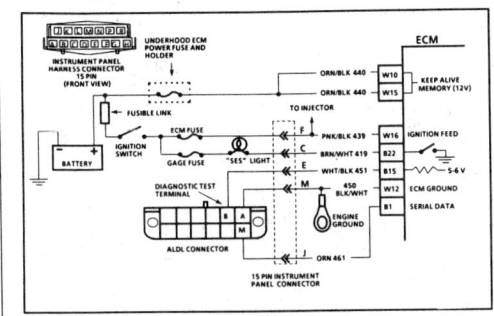

CHART A-2
NO ALDL DATA OR WON'T FLASH CODE 12
"SERVICE ENGINE SOON" LIGHT "ON" STEADY
2.5L (VIN R)

Circuit Description:
There should always be a steady "Service Engine Soon" light when the ignition is "ON" and the engine is "OFF". Battery voltage is supplied directly to the light bulb. The Electronic Control Module (ECM) controls the light and turns it "ON" by providing a ground path through CKT 419 to the ECM.

With the diagnostic terminal grounded, the light should flash a Code 12, followed by any trouble code(s) stored in memory. A steady light suggests a short to ground in the light control CKT 419, or an open in diagnostic CKT 451.

Test Description: Numbers below refer to circled numbers on the diagnostic chart.
1. If there is a problem with the ECM that causes a "Scan" tool to not read data from the ECM, then the ECM should not flash a Code 12. If Code 12 does flash, be sure that the "Scan" tool is working properly on another vehicle. If the "Scan" tool is functioning properly and CKT 461 is OK, the PROM or ECM may be at fault for the "NO ALDL" symptom.
2. If the light turns "OFF" when the ECM connector is disconnected, then CKT 419 is not shorted to ground.
3. This step will check for an open diagnostic CKT 451.
4. At this point, the "Service Engine Soon" light wiring is OK. The problem is a faulty ECM or PROM. If Code 12 does not flash, the ECM should be replaced using the original PROM. Replace the PROM only after trying an ECM, as a defective PROM is an unlikely cause of the problem.

1988–90 2.5L (VIN R) – ALL MODELS EXCEPT FIERO

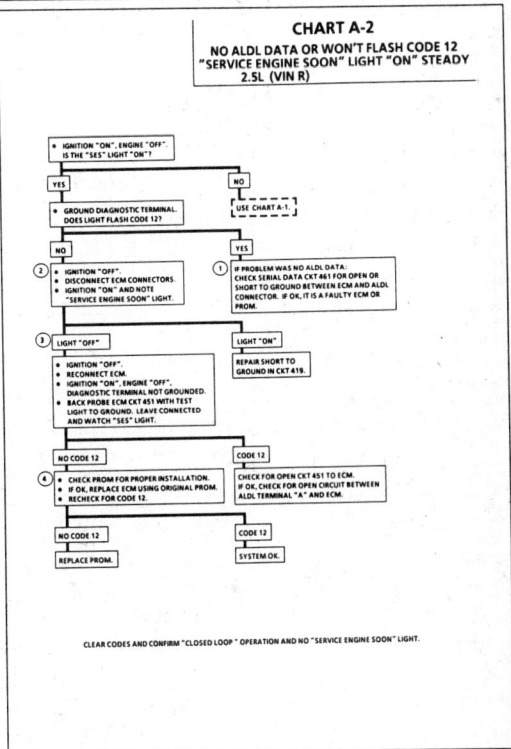

1988–90 2.5L (VIN R) – ALL MODELS EXCEPT FIERO

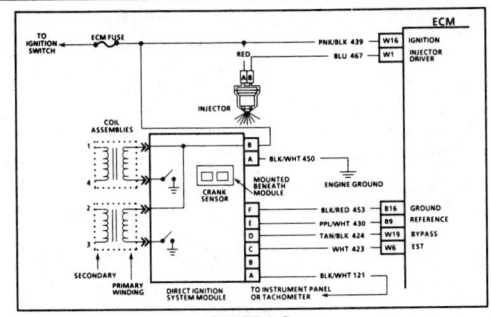

CHART A-3
(Page 1 of 3)
ENGINE CRANKS BUT WON'T RUN
2.5L (VIN R)

Circuit Description:
Before using this chart, battery condition, engine cranking speed, and fuel quantity should be checked and verified as being OK.

Test Description: Numbers below refer to numbers on the diagnostic chart.
1. A "Service Engine Soon" light "ON" is a basic test to determine if there is battery and ignition voltage at the ECM. No ALDL data may be due to an ECM problem, and CHART A-2 will diagnose the ECM. If TPS is over 2.5 volts, the engine may be in the clear flood mode, which will cause starting problems. The engine will not start without crank sensor reference pulses. The "Scan" tool should display rpm during cranking if pulses are received at the ECM.
2. Because the direct ignition system uses two plugs and wires to complete the circuit of each coil, the opposite spark plug wire must be left connected. If rpm was indicated during crank, but "No Spark" at this test indicates the ignition module is not triggering the coil.
3. While cranking the engine, there should be no fuel spray with the injector electrical connector disconnected. Replace the injector if it sprays fuel or drips.
4. The test light should flash, indicating that the ECM is controlling the injector. How bright the light flashes is not important. However, the test light should be a BT-8329 or equivalent.
5. Fuel spray from the injector indicates that fuel is available. However, the engine could be severely flooded due to too much fuel. No fuel spray from injector indicates a faulty fuel system or no ECM control of injector.

Diagnostic Aids:
- Water or foreign material can cause a no start condition during freezing weather. The engine may start after approximately 5 minutes in a heated shop. The problem may not re-occur until after an overnight park in freezing temperatures.
- An EGR valve sticking open can cause a low air/fuel ratio during cranking. Unless engine enters "Clear Flood" at the first indication of a flooding condition, it can result in a no start.
- Fuel pressure. Low fuel pressure can result in a very lean air/fuel ratio. Use CHART A-7.

1988–90 2.5L (VIN R) – ALL MODELS EXCEPT FIERO

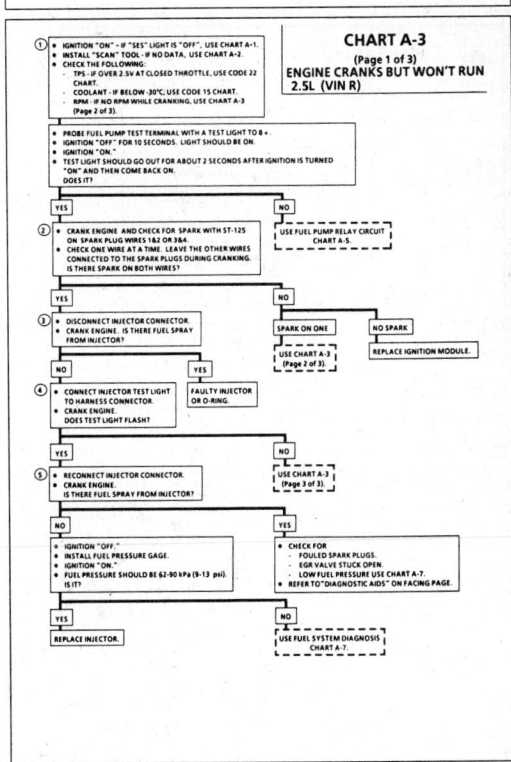

1988–90 2.5L (VIN R) – ALL MODELS EXCEPT FIERO

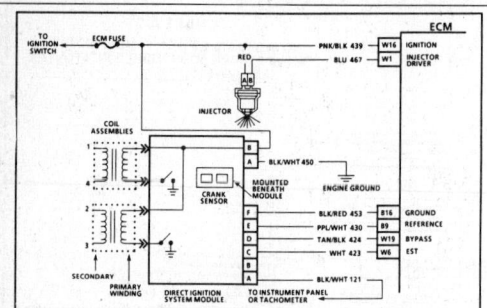

CHART A-3
(Page 2 of 3)
ENGINE CRANKS BUT WON'T RUN
2.5L (VIN R)

Circuit Description:
A magnetic crank sensor is used to determine engine crankshaft position, much the same way as the pick-up coil did in HEI type systems. The sensor is mounted in the block, near a slotted wheel on the crankshaft. The rotation of the wheel creates a flux change in the sensor, which produces a voltage signal. The DIS ignition module processes this signal and creates the reference pulses needed by the ECM to trigger the correct coil at the correct time.

If the "Scan" tool did not indicate cranking rpm, and there is no spark present at the plugs, the problem lies in the direct ignition system or the power and ground supplies to the module.

Test Description: Numbers below refer to circled numbers on the diagnostic chart.
1. The direct ignition system uses two plugs and wires to complete the circuit of each coil. The other spark plug wire in the circuit must be left connected to create a spark.
2. This test will determine if the 12 volt supply and a good ground is available at the DIS ignition module.
3. This test will determine if the ignition module is not generating the reference pulse, or if the wiring or ECM are at fault. By touching and removing a test light to 12 volts on CKT 430, a reference pulse should be generated. If rpm is indicated, the ECM and wiring are OK.
4. This test will determine if the ignition module is not triggering the problem coil, or if the tested coil is at fault. This test could also be performed by substituting a known good coil. The secondary coil winding can be checked with a DVM. There should be 5,000 to 10,000 ohms across the coil towers. There should not be any continuity from either coil tower to ground.
5. Checks for continuity of the crank sensor and connections. Also checks sensor magnetism.
6. An intermittent problem with the crank sensor could cause an intermittent "Cranks But Won't Run". Therefore, as a precaution, the crank sensor should be replaced as well.

1988–90 2.5L (VIN R) – ALL MODELS EXCEPT FIERO

CHART A-3
(Page 2 of 3)
ENGINE CRANKS BUT WON'T RUN
2.5L (VIN R)

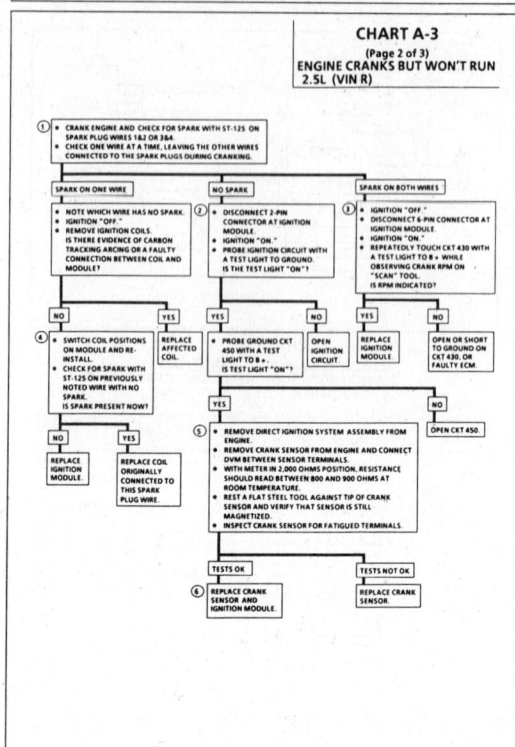

1988–90 2.5L (VIN R) – ALL MODELS EXCEPT FIERO

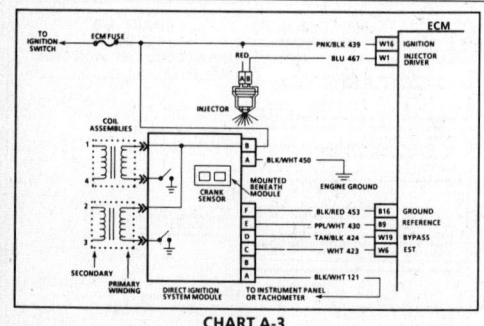

CHART A-3
(Page 3 of 3)
ENGINE CRANKS BUT WON'T RUN
2.5L (VIN R)

Circuit Description:
Ignition voltage is supplied to the fuel injector on CKT 439. The injector will be pulsed (turned "ON" and "OFF") when the ECM opens and grounds injector drive CKT 467.

Test Description: Numbers below refer to numbers on the diagnostic chart.
1. This check determines if injector connector has ignition voltage, and on only one terminal.

2. A faulty ECM may result in damage to the injector.
3. A test light connected from ECM harness terminal "W1" to ground should light due to continuity through the injector.

1988–90 2.5L (VIN R) – ALL MODELS EXCEPT FIERO

CHART A-3
(Page 3 of 3)
ENGINE CRANKS BUT WON'T RUN
2.5L (VIN R)

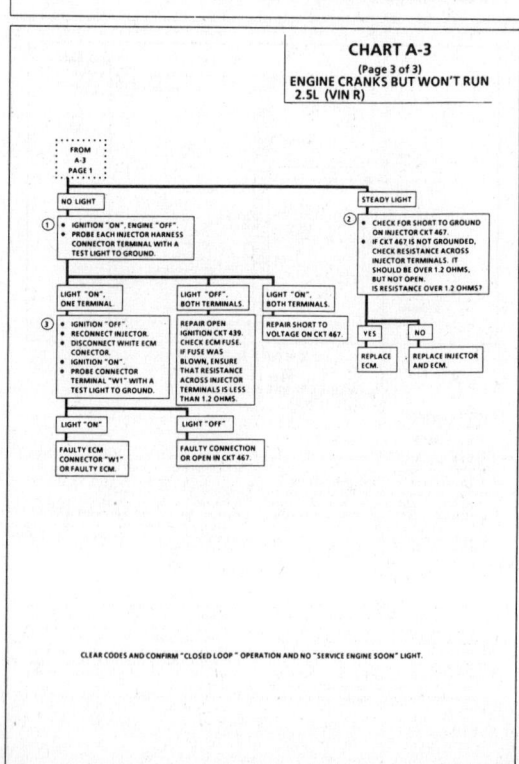

1988–90 2.5L (VIN R) – ALL MODELS EXCEPT FIERO

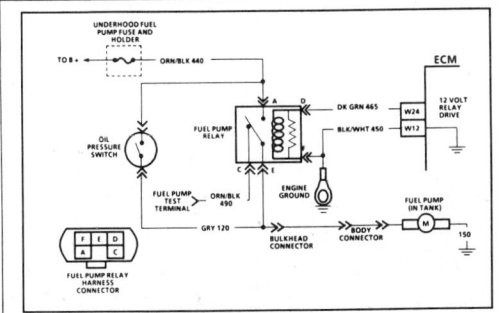

CHART A-5
FUEL PUMP RELAY CIRCUIT
2.5L (VIN R)

Circuit Description:

When the ignition switch is turned "ON", the Engine Control Module (ECM) will activate the fuel pump relay with a 12 volt signal and run the in-tank fuel pump. The fuel pump will operate as long as the engine is cranking or running and the ECM is receiving ignition reference pulses. If there are no ignition reference pulses, the ECM will no longer supply the fuel pump relay signal within 2 seconds after the ignition is turned "ON".

Should the fuel pump relay or the 12 volt relay drive from the ECM fail, the fuel pump will receive electrical current through the oil pressure switch back-up circuit.

The fuel pump "test" terminal is located in the left side of the engine compartment. When the engine is stopped, the pump can be turned "ON" by applying battery voltage to the "test" terminal.

Diagnostic Aids:

1. This check is to determine if the oil pressure switch is faulty also. Since it is a back-up component, it must be checked without the fuel pump relay connected.
2. This check will determine if the oil pressure switch is shorted internally.

An inoperative fuel pump relay can result in long cranking times. The extended crank period is caused by the time necessary for oil pressure to reach the pressure required to close the oil pressure switch and turn "ON" the fuel pump.

1988–90 2.5L (VIN R) – ALL MODELS EXCEPT FIERO

CHART A-5
FUEL PUMP RELAY CIRCUIT
2.5L (VIN R)

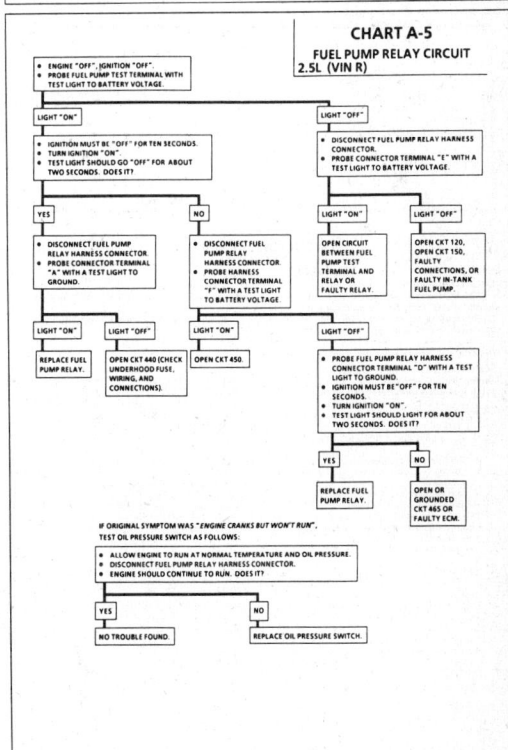

1988–90 2.5L (VIN R) – ALL MODELS EXCEPT FIERO

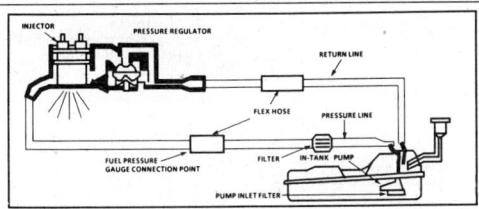

CHART A-7
FUEL SYSTEM DIAGNOSIS
2.5L (VIN R)

Circuit Description:

The fuel pump "test" terminal delivers fuel to the TBI unit where the system pressure is controlled to 62 to 90 kPa (9 to 13 psi). Excess fuel is returned to the fuel tank.

When the ignition switch is turned "ON", the Electronic Control Module (ECM) will activate the fuel pump relay with a 12 volt signal and run the in-tank fuel pump. The fuel pump will operate as long as the engine is cranking or running and the ECM is receiving ignition reference pulses. If there are no ignition reference pulses, the ECM will no longer supply the fuel pump relay signal within 2 seconds after key "ON".

Should the fuel pump relay or the 12 volt relay drive from the ECM fail, the fuel pump will receive electrical current through the oil pressure switch back-up circuit.

Test Description: Numbers below refer to circled numbers on the diagnostic chart.

1. If fuse in jumper wire blows, check CKT 120, between relay and fuel pump, for a short to ground.
2. Fuel flow at less than 62 kPa (9 psi) can cause the following:
   - System will run lean and may set Code 44. Also, hard starting and poor overall performance will result.
   - Engine surging and possible stalling during driving with satisfactory idle quality. This would be caused by a restriction in the fuel flow. As the fuel flow increases, the pressure drop across the restriction becomes great enough to starve the engine of fuel. At idle, the pressure may still be adequate. Normally, an engine with a fuel pressure less than 62 kPa (9 psi) at idle will not be driveable.

3. Gradually restricting the fuel return line allows the fuel pump to develop its maximum pressure (dead head pressure). When battery voltage is applied to the pump "test" terminal, pressure should be between 90 and 124 kPa (13 to 18 psi).
4. This test determines if the high fuel pressure is due to a restricted fuel return line or a throttle body pressure regulator problem.

Diagnostic Aids:

Improper fuel system pressure can result in one of the following symptoms:
- Cranks, but won't run
- Code 44
- Code 45
- Cuts out, may feel like ignition problem
- Poor fuel economy, loss of power
- Hesitation

CHART A-7
FUEL SYSTEM DIAGNOSIS
2.5L (VIN R)

NOTICE: *FUEL SYSTEM IS UNDER PRESSURE. TO AVOID FUEL SPILLAGE, REFER TO FIELD SERVICE PROCEDURE FOR TESTING OR MAKING REPAIRS REQUIRING DISASSEMBLY OF FUEL LINES OR FITTINGS.*

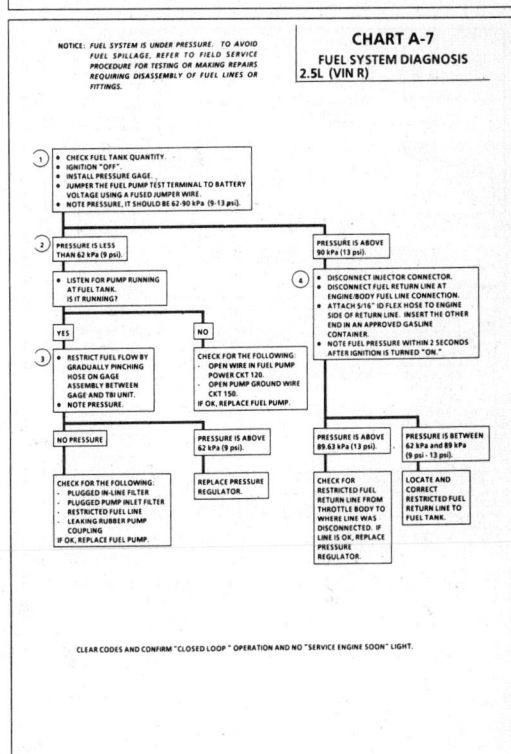

1988–90 2.5L (VIN R) – ALL MODELS EXCEPT FIERO

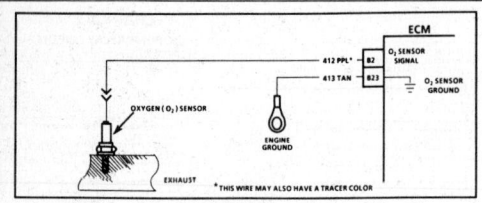

CODE 13
OXYGEN SENSOR CIRCUIT
(OPEN CIRCUIT)
2.5L (VIN R)

Circuit Description:

The ECM supplies a voltage of about .45 volt between terminals "B2" and "B23". (If measured with a 10 megohm digital voltmeter, this may read as low as .32 volt).

When the O_2 sensor reaches operating temperature, it varies this voltage from about .1 volt (exhaust is lean) to about .9 volt (exhaust is rich).

The sensor is like an open circuit and produces no voltage when it is below 360°C (600°F). An open sensor circuit, or cold sensor, causes "Open Loop" operation.

Test Description: Numbers below refer to circled numbers on the diagnostic chart.
1. Code 13 will set under the following conditions:
   - Engine at normal operating temperature.
   - At least 2 minutes have elapsed since engine start-up.
   - O_2 signal voltage is steady between .35 and .55 volt.
   - Throttle angle is above 7%.
   - All above conditions are met for about 60 seconds.

   If the conditions for a Code 13 exist, the system will not operate in "Closed Loop".
2. This test determines if the O_2 sensor is the problem or if the ECM and wiring are at fault.

3. In doing this test, use only a 10 megohm digital voltmeter. This test checks the continuity of CKTs 412 and 413. If CKT 413 is open, the ECM voltage on CKT 412 will be over .6 volt (600 mV).

Diagnostic Aids:

Normal "Scan" tool O_2 sensor voltage varies between 100mV to 999 mV (.1 and 1.0 volt) while in "Closed Loop". Code 13 sets in one minute if sensor signal voltage remains between .35 and .55 volt, but the system will go to "Open Loop" in about 15 seconds.

Verify a clean, tight ground connection for CKT 413. Open CKT(s) 412 or 413 will result in a Code 13. If Code 13 is intermittent, refer to Section "B".

1988–90 2.5L (VIN R) – ALL MODELS EXCEPT FIERO

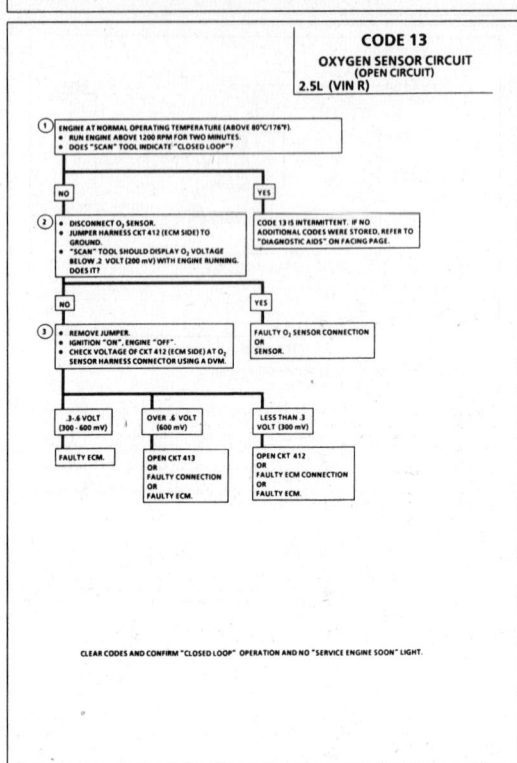

1988–90 2.5L (VIN R) – ALL MODELS EXCEPT FIERO

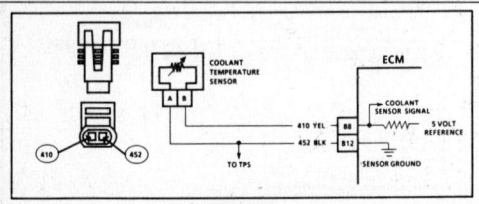

CODE 14
COOLANT TEMPERATURE SENSOR CIRCUIT
(HIGH TEMPERATURE INDICATED)
2.5L (VIN R)

Circuit Description:

The coolant temperature sensor uses a thermistor to control the signal voltage to the ECM. The ECM applies a reference voltage on CKT 410 to the sensor. When the engine is cold, the sensor (thermistor) resistance is high. The ECM will then sense a high signal voltage.

As the engine warms up, the sensor resistance decreases and the voltage drops. At normal engine operating temperature, the voltage will measure about 1.5 to 2.0 volts at ECM terminal "B8".

Coolant temperature is one of the inputs used to control the following:
- Fuel Delivery
- Electronic Spark Timing (EST)
- Cooling Fan
- Torque Converter Clutch (TCC)
- Idle Air Control (IAC)

Test Description: Numbers below refer to circled numbers on the diagnostic chart.
1. Checks to see if code was set as result of hard failure or intermittent condition.

   Code 14 will set if:
   - Signal voltage indicates a coolant temperature above 135°C (275°F) for 3 seconds.
2. This test simulates conditions for a Code 15. If the ECM recognizes the open circuit (high voltage) and displays a low temperature, the ECM and wiring are OK.

Diagnostic Aids:

A "Scan" tool reads engine temperature in degrees Celsius.

After the engine is started, the temperature should rise steadily to about 90°C, then stabilize when the thermostat opens.

If the engine has been allowed to cool to an ambient temperature (overnight), coolant temperature and Manifold Air Temperature (MAT) may be checked with a "Scan" tool and should read close to each other.

When a Code 14 is set, the ECM will turn "ON" the engine cooling fan.

A Code 14 will result if CKT 410 is shorted to ground.

If Code 14 is intermittent, refer to Section "B".

1988–90 2.5L (VIN R) – ALL MODELS EXCEPT FIERO

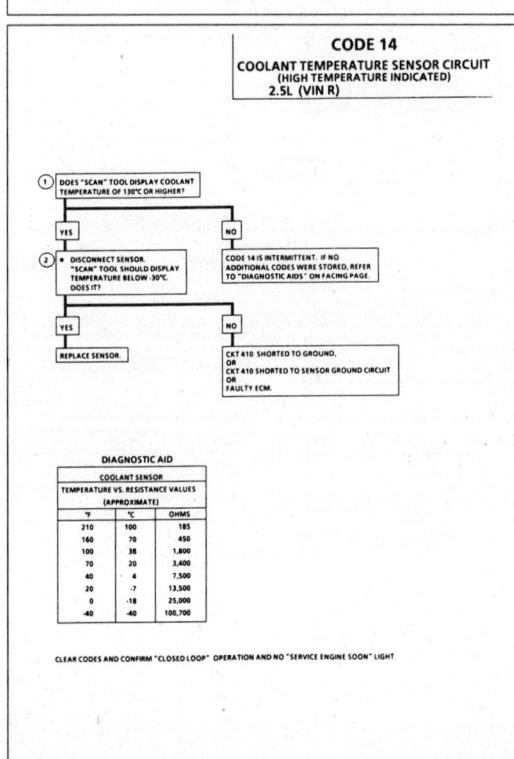

1988–90 2.5L (VIN R) – ALL MODELS EXCEPT FIERO

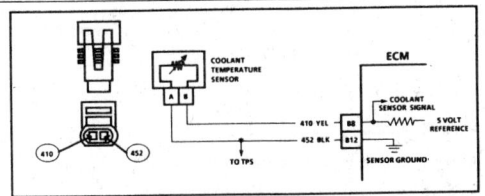

CODE 15
COOLANT TEMPERATURE SENSOR CIRCUIT
(LOW TEMPERATURE INDICATED)
2.5L (VIN R)

Circuit Description:

The coolant temperature sensor uses a thermistor to control the signal voltage to the ECM. The ECM applies a reference voltage on CKT 410 to the sensor. When the engine is cold, the sensor (thermistor) resistance is high. The ECM will then sense a high signal voltage.

As the engine warms up, the sensor resistance decreases and the voltage drops. At normal engine operating temperature, the voltage will measure about 1.5 to 2.0 volts at ECM terminal "B8".

Coolant temperature is one of the inputs used to control the following:
- Fuel Delivery
- Torque Convertor Clutch (TCC)
- Electronic Spark Timing (EST)
- Idle Air Control (IAC)
- Cooling Fan

Test Description: Numbers below refer to circled numbers on the diagnostic chart.
1. Checks to see if code was set as result of hard failure or intermittent condition.
   Code 15 will set if:
   - Signal voltage indicates a coolant temperature below -30°C (-22°F) for 60 seconds.
2. This test simulates conditions for a Code 14. If the ECM recognizes the grounded circuit (low voltage) and displays a high temperature, the ECM and wiring are OK.
3. This test will determine if there is a wiring problem or a faulty ECM. If CKT 452 is open, there may also be a Code 21 stored.

Diagnostic Aids:

A "Scan" tool reads engine temperature in degrees Celsius. After the engine is started, the temperature should rise steadily to about 90°C, then stabilize when the thermostat opens.

If the engine has been allowed to cool to an ambient temperature (overnight), coolant temperature and Manifold Air Temperature (MAT) may be checked with a "Scan" tool and should read close to each other.

When a Code 15 is set, the ECM will turn "ON" the engine cooling fan.

A Code 15 will result if CKTs 410 or 452 are open. If Code 15 is intermittent, refer to Section "B".

1988–90 2.5L (VIN R) – ALL MODELS EXCEPT FIERO

CODE 15
COOLANT TEMPERATURE SENSOR CIRCUIT
(LOW TEMPERATURE INDICATED)
2.5L (VIN R)

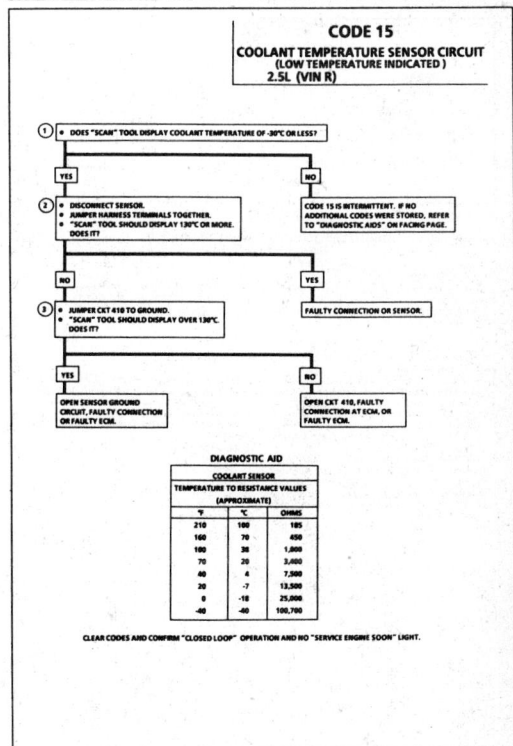

COOLANT SENSOR		
TEMPERATURE TO RESISTANCE VALUES (APPROXIMATE)		
°F	°C	OHMS
210	100	185
160	70	450
100	38	1,800
70	20	3,400
40	4	7,500
20	-7	13,500
0	-18	25,000
-40	-40	100,700

CLEAR CODES AND CONFIRM "CLOSED LOOP" OPERATION AND NO "SERVICE ENGINE SOON" LIGHT.

1988–90 2.5L (VIN R) – ALL MODELS EXCEPT FIERO

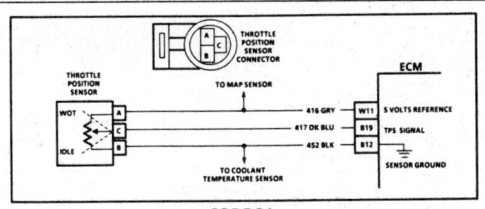

CODE 21
THROTTLE POSITION SENSOR (TPS) CIRCUIT
(SIGNAL VOLTAGE HIGH)
2.5L (VIN R)

Circuit Description:

The Throttle Position Sensor (TPS) provides a voltage signal that changes relative to the throttle valve. Signal voltage will vary from less than 1.25 volts at idle to about 5 volts at Wide Open Throttle (WOT).

The TPS signal is one of the most important inputs used by the ECM for fuel control and for many of the ECM controlled outputs.

Test Description: Numbers below refer to circled numbers on the diagnostic chart.
1. This step checks to see if Code 21 is the result of a hard failure or an intermittent condition.
   A Code 21 will set under the following conditions:
   - TPS reading above 2.5 volts
   - Engine speed less than 1800 rpm
   - MAP reading below 60 kPa
   - All of the above conditions present for at least 2 seconds.
2. This step simulates conditions for a Code 22. If the ECM recognizes the change of state, the ECM and CKTs 416 and 417 are OK.
3. This step isolates a faulty sensor, ECM, or an open CKT 452. If CKT 452 is open, there may also be a Code 15 stored.

Diagnostic Aids:

A "Scan" tool displays throttle position in volts. Closed throttle voltage should be less than 1.25 volts. TPS voltage should increase at a steady rate as throttle is moved to WOT.

A Code 21 will result if CKT 452 is open or CKT 417 is shorted to voltage. If Code 21 is intermittent, refer to Section "B".

1988–90 2.5L (VIN R) – ALL MODELS EXCEPT FIERO

CODE 21
THROTTLE POSITION SENSOR (TPS) CIRCUIT
(SIGNAL VOLTAGE HIGH)
2.5L (VIN R)

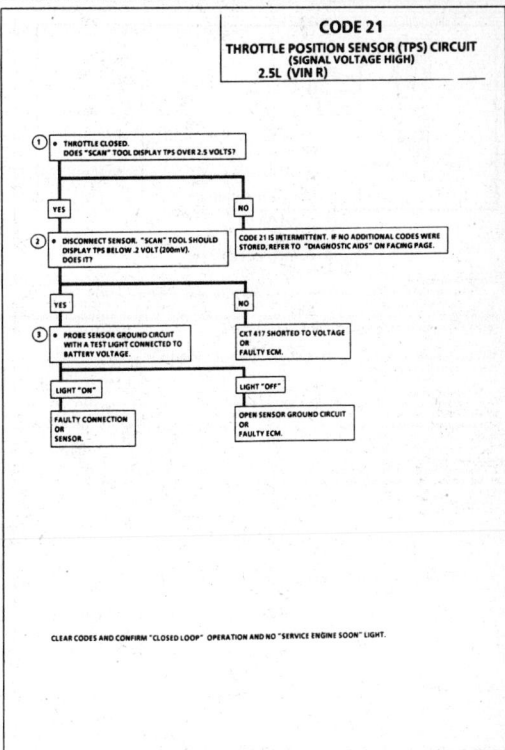

CLEAR CODES AND CONFIRM "CLOSED LOOP" OPERATION AND NO "SERVICE ENGINE SOON" LIGHT.

1988–90 2.5L (VIN R) – ALL MODELS EXCEPT FIERO

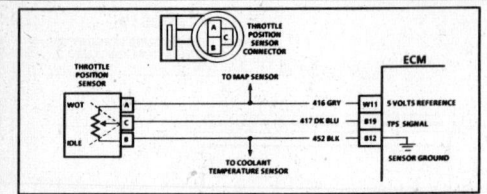

CODE 22
THROTTLE POSITION SENSOR (TPS) CIRCUIT
(SIGNAL VOLTAGE LOW)
2.5L (VIN R)

Circuit Description:
The Throttle Position Sensor (TPS) provides a voltage signal that changes relative to the throttle valve. Signal voltage will vary from less than 1.25 volts at idle to about 4.5 volts at Wide Open Throttle (WOT).
The TPS signal is one of the most important inputs used by the ECM for fuel control and for many of the ECM controlled outputs.

Test Description: Numbers below refer to circled numbers on the diagnostic chart.
1. This step checks to see if Code 22 is the result of a hard failure or an intermittent condition. A Code 22 will set when the engine is running and TPS voltage is below .2 volt (200 mV).
2. This step simulates conditions for a Code 21. If a Code 21 is set, or the "Scan" tool displays over 4 volts, the ECM and wiring are OK.
3. The "Scan" tool may not display 12 volts. The important thing is that the ECM recognizes the voltage as over 4 volts, indicating that CKT 417 and the ECM are OK.
4. If CKT 416 is shorted to ground, there may also be a stored Code 34.

Diagnostic Aids:
A "Scan" tool displays throttle position in volts. Closed throttle voltage should be less than 1.25 volts. TPS voltage should increase at a steady rate as the throttle is moved to WOT.
An open or grounded CKT 416 or 417 will result in a Code 22.
If Code 22 is intermittent, refer to Section "B".

1988–90 2.5L (VIN R) – ALL MODELS EXCEPT FIERO

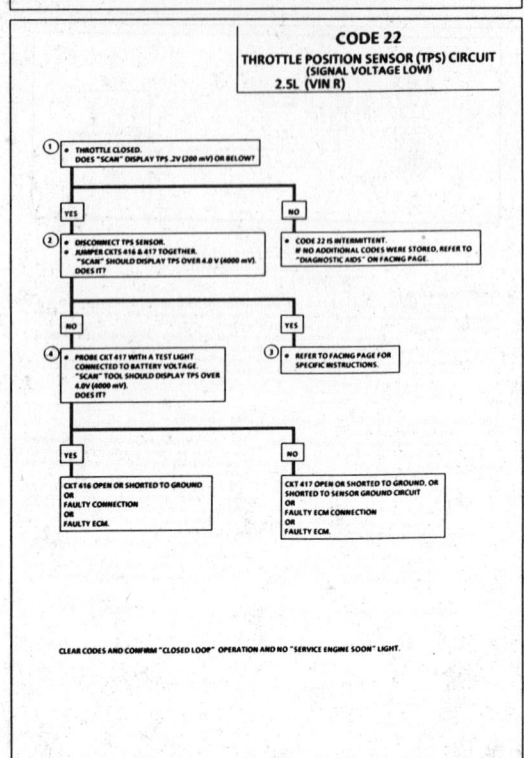

1988–90 2.5L (VIN R) – ALL MODELS EXCEPT FIERO

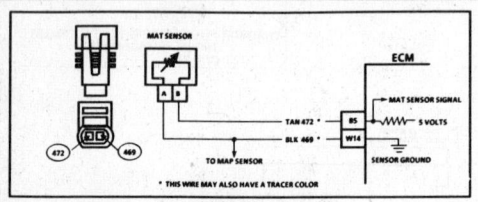

CODE 23
MANIFOLD AIR TEMPERATURE (MAT) SENSOR CIRCUIT
(LOW TEMPERATURE INDICATED)
2.5L (VIN R)

Circuit Description:
The manifold air temperature sensor uses a thermistor to control the signal voltage to the ECM. The ECM applies a reference voltage (4-6 volts) on CKT 472 to the sensor. When manifold air is cold, the sensor (thermistor) resistance is high. The ECM will then sense a high signal voltage. As the air warms, the sensor resistance becomes less and the voltage drops.

Test Description: Numbers below refer to circled numbers on the diagnostic chart.
1. This step determines if Code 23 is the result of a hard failure or an intermittent condition. Code 23 will set when signal voltage indicates a MAT temperature less than -30°C and the engine is running for longer than 58 seconds.
2. This test simulates conditions for a Code 25. If the "Scan" tool displays a high temperature, the ECM and wiring are OK.
3. This step checks continuity of CKTs 472 and 469. If CKT 469 is open there may also be a Code 33.

Diagnostic Aids:
If the engine has been allowed to cool to an ambient temperature (overnight), coolant and MAT temperatures may be checked with a "Scan" tool and should read close to each other.
A Code 23 will result if CKT 472 or 469 becomes open.
If Code 23 is intermittent, refer to Section "B".

1988–90 2.5L (VIN R) – ALL MODELS EXCEPT FIERO

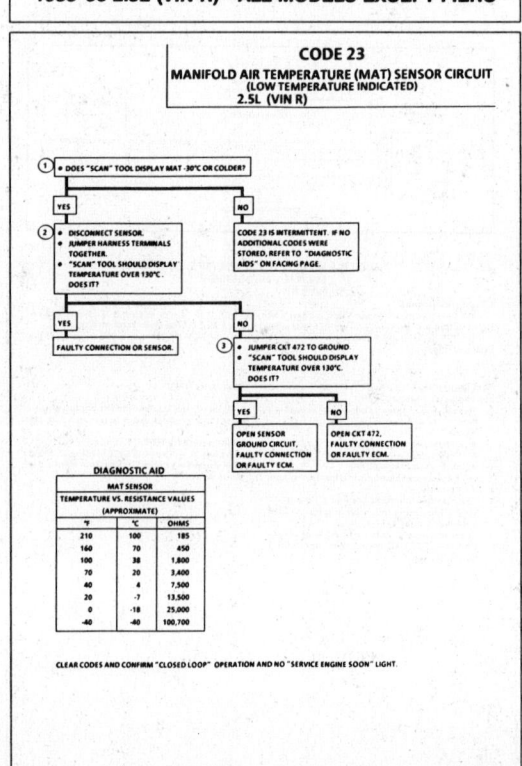

1988–90 2.5L (VIN R) – ALL MODELS EXCEPT FIERO

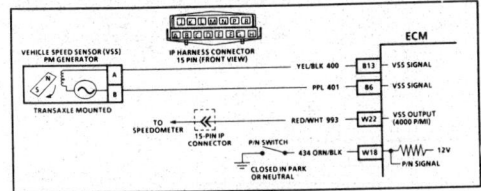

CODE 24
VEHICLE SPEED SENSOR (VSS) CIRCUIT
2.5L (VIN R)

Circuit Description:

Vehicle speed information is provided to the ECM by the vehicle speed sensor. It is a Permanent Magnet (PM) generator and is mounted in the transmission. The PM generator produces a pulsing voltage whenever vehicle speed is over about 3 mph. The A/C voltage level and the number of pulses increases with vehicle speed. The ECM converts the pulsing voltage to miles per hour. This data is used for calculations and can be displayed on a "Scan" tool.

The function of VSS buffer used in past model years has been incorporated into the ECM. The ECM supplies the necessary signal for the instrument panel (4000 pulses per mile) for operating the speedometer and the odometer.

Test Description:

Code 24 will set if vehicle speed equals 0 mph when:
* Engine speed is between 1400 and 3600 rpm.
* TPS is less than 2%.
* Low load condition (low MAP voltage, high manifold vacuum).
* Transmission not in park or neutral.
* All above conditions are met for 5 seconds.

These conditions are met during a road load deceleration.

Disregard a Code 24 that sets when the drive wheels are not turning. This can be caused by a faulty park/neutral switch circuit.

The PM generator only produces a signal if the drive wheels are turning greater than 3 mph.

Diagnostic Aids:

"Scan" tool should indicate a vehicle speed whenever the drive wheels are turning greater than 3 mph.

A problem in CKT 993 will not affect the VSS input or the readings on a "Scan" tool.

Check CKTs 400 and 401 for proper connections to be sure they are clean and tight and the harness is routed correctly. Refer to "Intermittents" in Section "B".

(A/T) - A faulty or misadjusted park/neutral switch can result in a false Code 24. Use a "Scan" tool and check for the proper signal while in a drive range. Refer to CHART C-1A for the P/N switch check.

1988–90 2.5L (VIN R) – ALL MODELS EXCEPT FIERO

CODE 24
VEHICLE SPEED SENSOR (VSS) CIRCUIT
2.5L (VIN R)

DISREGARD CODE 24 IF SET WHILE DRIVE WHEELS ARE NOT TURNING. REFER TO "DIAGNOSTIC AIDS" ON FACING PAGE.

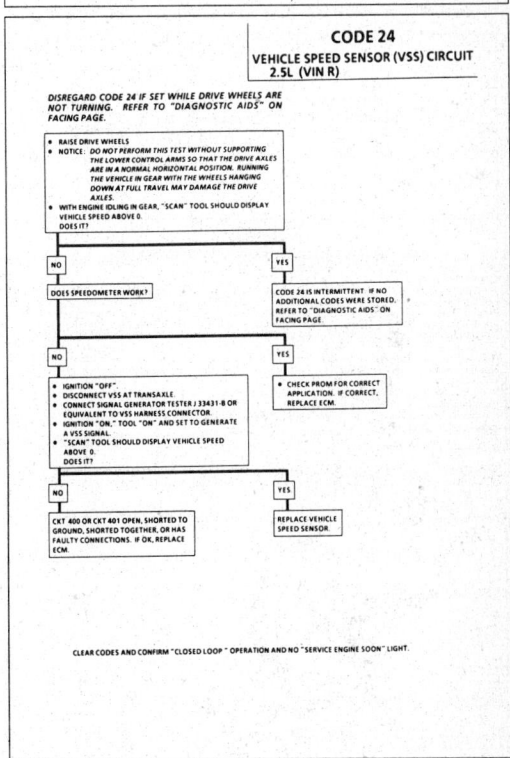

CLEAR CODES AND CONFIRM "CLOSED LOOP" OPERATION AND NO "SERVICE ENGINE SOON" LIGHT.

1988–90 2.5L (VIN R) – ALL MODELS EXCEPT FIERO

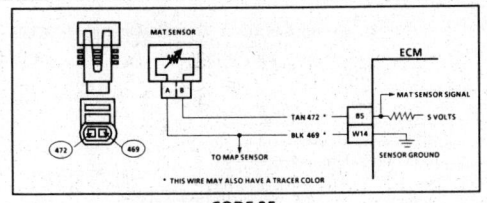

CODE 25
MANIFOLD AIR TEMPERATURE (MAT) SENSOR CIRCUIT
(HIGH TEMPERATURE INDICATED)
2.5L (VIN R)

Circuit Description:

The manifold air temperature sensor uses a thermistor to control the signal voltage to the ECM. The ECM applies a reference voltage (4-6 volts) on CKT 472 to the sensor. When manifold air is cold, the sensor (thermistor) resistance is high. Therefore, the ECM will see a high signal voltage. As the air warms, the sensor resistance becomes less and the voltage drops.

Test Description: Numbers below refer to circled numbers on the diagnostic chart.

1. This check determines if the Code 25 is the result of a hard failure or an intermittent condition. Code 25 will set if a MAT temperature greater than 135°C is detected for a time longer than 2 seconds and the ignition is "ON".

Diagnostic Aids:

If the engine has been allowed to cool to an ambient temperature (overnight), coolant and MAT temperatures may be checked with a "Scan" tool and should read close to each other.

A Code 25 will result if CKT 472 is shorted to ground.

If Code 25 is intermittent, refer to Section "B".

1988–90 2.5L (VIN R) – ALL MODELS EXCEPT FIERO

CODE 25
MANIFOLD AIR TEMPERATURE (MAT) SENSOR CIRCUIT
(HIGH TEMPERATURE INDICATED)
2.5L (VIN R)

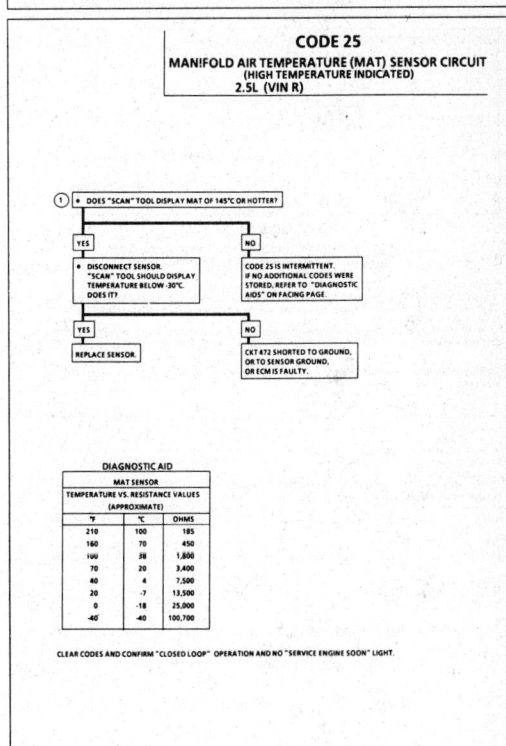

DIAGNOSTIC AID		
MAT SENSOR		
TEMPERATURE VS. RESISTANCE VALUES		
(APPROXIMATE)		
°F	°C	OHMS
210	100	185
160	70	450
100	38	1,800
70	20	3,400
40	4	7,500
20	-7	13,500
0	-18	25,000
-40	-40	100,700

CLEAR CODES AND CONFIRM "CLOSED LOOP" OPERATION AND NO "SERVICE ENGINE SOON" LIGHT.

1988–90 2.5L (VIN R) – ALL MODELS EXCEPT FIERO

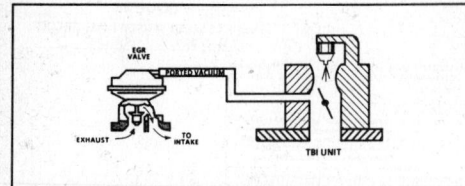

CODE 32
EXHAUST GAS RECIRCULATION (EGR) SYSTEM FAILURE
2.5L (VIN R)

Circuit Description:

A properly operating EGR system will directly affect the air/fuel requirements of the engine. Since the exhaust gas introduced into the air/fuel mixture is an inert gas (contains very little or no oxygen), less fuel is required to maintain a correct air/fuel ratio. If the EGR system were to become inoperative, the inert exhaust gas would be replaced with air and the air/fuel mixture would be leaner. The ECM would compensate for the lean condition by adding fuel, resulting in higher block learn values.

The engine control system operates between two block learn cells, a closed throttle cell, and an open throttle cell. Since EGR is not used at idle, the closed throttle cell would not be affected by EGR system operation. The open throttle cell is affected by EGR operation and, when the EGR system is operating properly, the block learn values in both cells should be close to being the same. If the EGR system was inoperative, the block learn value in the open throttle cell would change (become higher) to compensate for the resulting lean system, but the block learn value in the closed throttle cell would not change.

This change or difference in block learn values is used to monitor EGR system performance. When the change becomes too great, a Code 32 is set.

Diagnostic Aids:

The Code 32 chart is a functional check of the EGR system. If the EGR system works properly, but a Code 32 has been set, check other items that could result in high block learn values in the open throttle cell, but not in the closed throttle cell.

CHECK:

EGR Passages
  Restricted or blocked

MAP Sensor
  A MAP sensor may shift in calibration enough to affect fuel delivery. Use CHART C-1D, MAP output check.

1988–90 2.5L (VIN R) – ALL MODELS EXCEPT FIERO

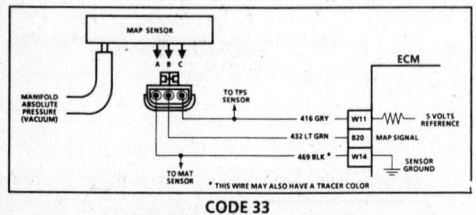

CODE 33
MANIFOLD ABSOLUTE PRESSURE (MAP) SENSOR CIRCUIT
(SIGNAL VOLTAGE HIGH - LOW VACUUM)
2.5L (VIN R)

Circuit Description:

The Manifold Absolute Pressure sensor (MAP) responds to changes in manifold pressure (vacuum). The ECM receives this information as a signal voltage that will vary from about 1 to 1.5 volts at closed throttle (idle) to 4-4.5 volts at wide open throttle (low vacuum).

If the MAP sensor fails, the ECM will substitute a fixed MAP value and use the Throttle Position Sensor (TPS) to control fuel delivery.

Test Description: Numbers below refer to the circled numbers on the diagnostic chart.

1. This step will determine if Code 33 is the result of a hard failure or an intermittent condition.
   A Code 33 will set under the following conditions:
   - MAP signal voltage is too high (low vacuum)
   - TPS less than 4%
   - These conditions exist longer than 48 seconds
2. This step simulates conditions for a Code 34. If the ECM recognizes the change, the ECM and CKTs 416 and 432 are OK. If CKT 469 is open, there may also be a stored Code 23.

Diagnostic Aids:

With the ignition "ON" and the engine stopped, the manifold pressure is equal to atmospheric pressure and the signal voltage will be high. This information is used by the ECM as an indication of vehicle altitude and is referred to as BARO. Comparison of this BARO reading with a known good vehicle with the same sensor is a good way to check accuracy of a "suspect" sensor. Readings should be within .4 volt.

A Code 33 will result if CKT 469 is open or if CKT 432 is shorted to voltage or to CKT 416.

If Code 33 is intermittent, refer to Section "B".

1988–90 2.5L (VIN R) – ALL MODELS EXCEPT FIERO

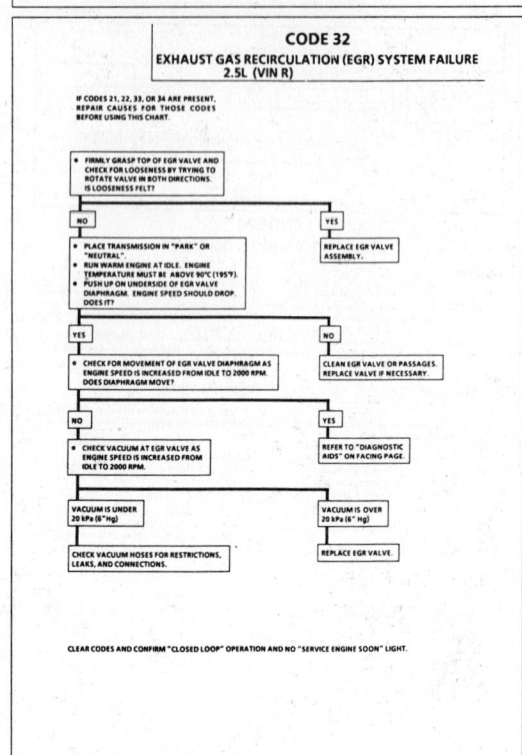

1988–90 2.5L (VIN R) – ALL MODELS EXCEPT FIERO

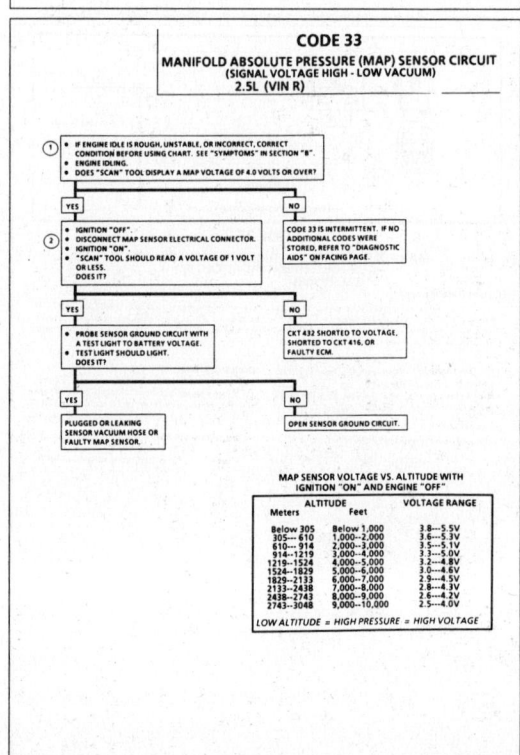

1988–90 2.5L (VIN R) – ALL MODELS EXCEPT FIERO

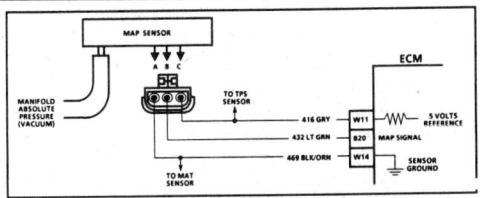

CODE 34
MANIFOLD ABSOLUTE PRESSURE (MAP) SENSOR CIRCUIT
(SIGNAL VOLTAGE LOW - HIGH VACUUM)
2.5L (VIN R)

Circuit Description:

The Manifold Absolute Pressure sensor (MAP) responds to changes in manifold pressure (vacuum). The ECM receives this information as a signal voltage that will vary from about 1 to 1.5 volts at closed throttle (idle) to 4 - 4.5 volts at wide open throttle (low vacuum).

If the MAP sensor fails, the ECM will substitute a fixed MAP value and use the Throttle Position Sensor (TPS) to control fuel delivery.

Test Description: Numbers below refer to circled numbers on the diagnostic chart.

1. This step determines if Code 34 is the result of a hard failure or an intermittent condition. A Code 34 will set when MAP signal voltage is too low and the ignition is "ON."
2. Jumpering harness terminals "B" to "C" (5 volts to signal circuit) will determine if the sensor is at fault, or if there is a problem with the ECM or wiring.
3. The "Scan" tool may not display 12 volts. The important thing is that the ECM recognizes the voltage as more than 4 volts, indicating that the ECM and CKT 432 are OK.

Diagnostic Aids:

With the ignition "ON" and the engine stopped, the manifold pressure is equal to atmospheric pressure and the signal voltage will be high. This information is used by the ECM as an indication of vehicle altitude and is referred to as BARO. Comparison of this BARO reading with a known good vehicle with the same engine is a good way to check accuracy of a "suspect" sensor. Readings should be within .4 volt.

A Code 34 will result if CKT(s) 416 or 432 are open or shorted to ground.

If CKT 416 is open or shorted to ground, there may also be a stored Code 22.

If Code 34 is intermittent, refer to Section "B".

1988–90 2.5L (VIN R) – ALL MODELS EXCEPT FIERO

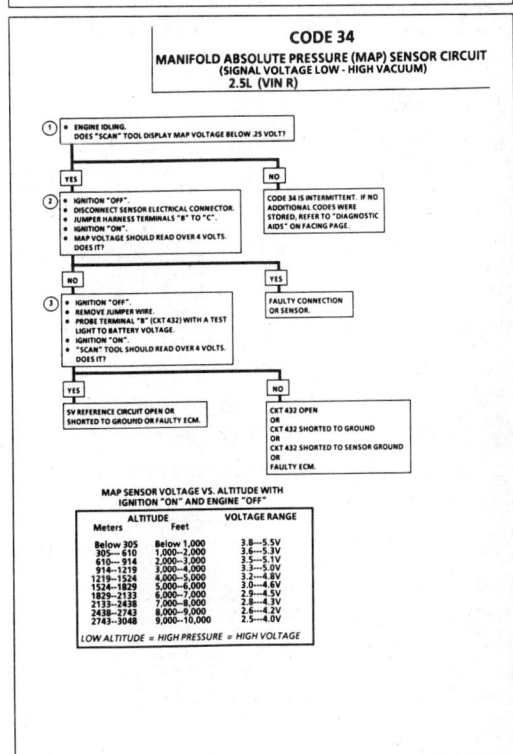

CODE 34
MANIFOLD ABSOLUTE PRESSURE (MAP) SENSOR CIRCUIT
(SIGNAL VOLTAGE LOW - HIGH VACUUM)
2.5L (VIN R)

MAP SENSOR VOLTAGE VS. ALTITUDE WITH IGNITION "ON" AND ENGINE "OFF"

ALTITUDE		VOLTAGE RANGE
Meters	Feet	
Below 305	Below 1,000	3.8---5.5V
305— 610	1,000—2,000	3.6---5.3V
610— 914	2,000—3,000	3.5---5.1V
914—1219	3,000—4,000	3.3---5.0V
1219—1524	4,000—5,000	3.2---4.8V
1524—1829	5,000—6,000	3.0---4.6V
1829—2133	6,000—7,000	2.9---4.5V
2133—2438	7,000—8,000	2.8---4.3V
2438—2743	8,000—9,000	2.6---4.2V
2743—3048	9,000—10,000	2.5---4.0V

LOW ALTITUDE = HIGH PRESSURE = HIGH VOLTAGE

1988–90 2.5L (VIN R) – ALL MODELS EXCEPT FIERO

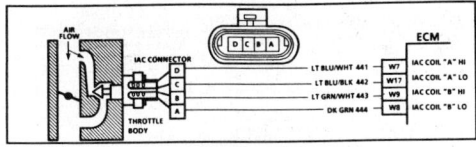

CODE 35
IDLE SPEED ERROR
2.5L (VIN R)

Circuit Description:

Code 35 will set when the closed throttle engine speed is 150 rpm above or below the desired idle speed for 20 seconds. Review general description in Section "C".

Test Description: Numbers below refer to circled numbers on the diagnostic chart.

1. Continue with test, even if will not idle. If idle is low enough, "Scan" will display 80 or more counts, or steps. If idle is high, it will display "0" counts. Occasionally an erratic or unstable idle may occur. Engine speed may vary 200 rpm or more up and down. Disconnect the IAC valve. If the condition is unchanged, the IAC valve is not at fault.
2. When the engine was stopped, the IAC valve retracted (more air) to a fixed "Park" position for increased air flow and idle speed during the next engine start. This test determines if IAC valve is functioning.
3. Be sure to disconnect the IAC valve prior to grounding the diagnostic "test" terminal. The test light will confirm the ECM signals by a steady or flashing light on all circuits.
4. There is a remote possibility that one of the circuits is shorted to voltage, which would have been indicated by a steady light. Disconnect ECM and turn the ignition "ON" and probe terminals to check for this condition.

Diagnostic Aids:

A slow, unstable idle may be caused by a system problem that cannot be overcome by the IAC valve. "Scan" tool counts will be above 60 counts if too low, and zero counts if too high.

If idle is too high, stop the engine. Turn ignition "ON" and ground diagnostic terminal. Wait 30-45 seconds for IAC valve to seat, then disconnect IAC valve. Start engine. If idle speed is above 800 rpm, locate and correct vacuum leak.

- **System too lean (High Air/Fuel Ratio)**
  The idle speed may be too high or too low. Engine speed may vary up and down and disconnecting the IAC valve does not help. Code 44 may be set. "Scan" tool and/or voltmeter will read an oxygen sensor output less than 300 mV (.3 volt). Check for low regulated fuel pressure or water in fuel. A lean exhaust, with an oxygen sensor fixed above 800 mV (.8 volt) will be a contaminated sensor, usually from silicone. This may also set a Code 45.
- **System too rich (Low Air/Fuel Ratio)**
  The idle speed will be too low. "Scan" tool counts will usually be above 80. System is obviously rich and may exhibit black smoke in exhaust. "Scan" tool and/or voltmeter will read an oxygen sensor signal fixed above 800 mV (.8 volt). Check for high fuel pressure or leaking or sticking injector.
- **Throttle Body**
  Remove IAC valve and inspect bore for foreign material or evidence of IAC valve dragging in the bore.
- **IAC Valve Connections**
  IAC valve connections should be carefully checked for proper contact.
- **PCV Valve**
  An incorrect or faulty PCV valve may result in an incorrect idle speed.
- Refer to "Rough, Unstable, Incorrect Idle or Stalling" in "Symptoms" in Section "B".

1988–90 2.5L (VIN R) – ALL MODELS EXCEPT FIERO

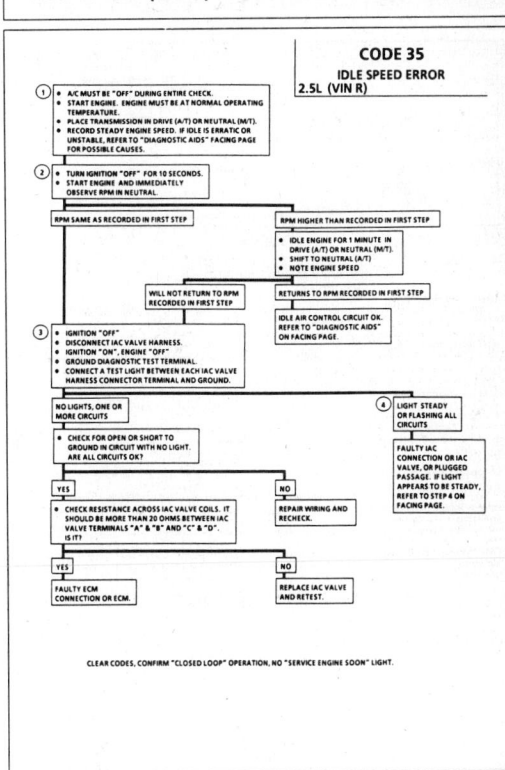

CODE 35
IDLE SPEED ERROR
2.5L (VIN R)

1988–90 2.5L (VIN R) – ALL MODELS EXCEPT FIERO

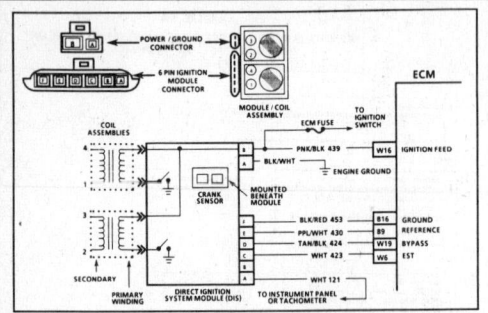

CODE 42
ELECTRONIC SPARK TIMING (EST) CIRCUIT
2.5L (VIN R)

Circuit Description:

The DIS module sends a reference signal to the ECM when the engine is cranking. While the engine speed is under 400 rpm, the DIS module controls the ignition timing. When the system is running on the ignition module (no voltage on the bypass line), the ignition module grounds the EST signal. The ECM expects to sense no voltage on the EST line during this condition. If it senses a voltage, it sets Code 42 and will not enter the EST mode.

When the engine speed exceeds 400 rpm, the ECM applies 5 volts to the bypass line to switch the timing to ECM control (EST). If the bypass line is open or grounded, once the rpm for EST control is reached, the ignition module will not switch to EST mode. This results in low EST voltage and the setting of Code 42. If the EST line is grounded, the ignition module will switch to EST, but because the line is grounded, there will be no EST signal. A Code 42 will be set.

Test Description: Numbers below refer to circled numbers on the diagnostic chart.
1. Code 42 means the ECM has sensed an open or short to ground in the EST or bypass circuits. This test confirms Code 42 and that the fault causing the code is present.
2. Checks for a normal EST ground path through the ignition module. An EST CKT 423, shorted to ground, will also read less than 500 ohms, but this will be checked later.
3. As the test light voltage contacts CKT 424, the module should switch, causing the ohmmeter to "overrange" if the meter is in the 1000-2000 ohms

position. Selecting the 10 - 20,000 ohms position will indicate a reading above 5000 ohms. The important thing is that the module "switched".
4. The module did not switch and this step checks for:
   - EST CKT 423 shorted to ground
   - Bypass CKT 424 open
   - Faulty ignition module connection or module.
5. Confirms that Code 42 is a faulty ECM and not an intermittent in CKT(s) 423 or 424.

Diagnostic Aids:

The "Scan" tool does not have any ability to help diagnose a Code 42 problem.
If Code 42 is intermittent, refer to Section "B".

1988–90 2.5L (VIN R) – ALL MODELS EXCEPT FIERO

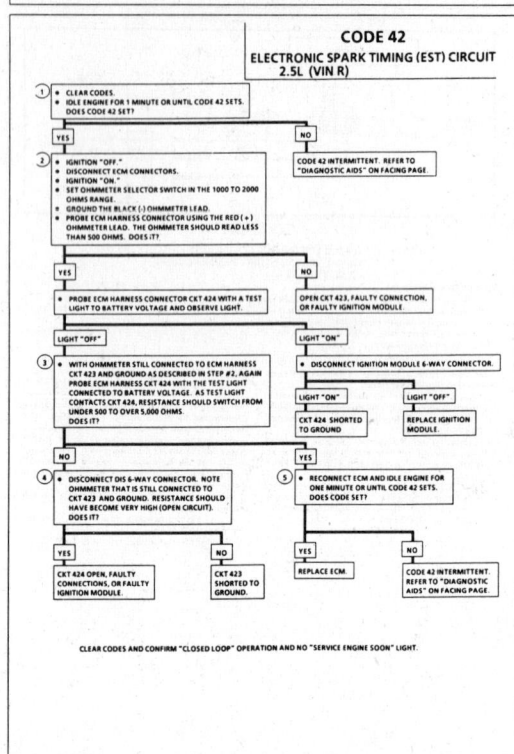

1988–90 2.5L (VIN R) – ALL MODELS EXCEPT FIERO

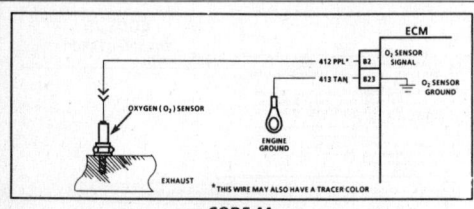

CODE 44
OXYGEN SENSOR CIRCUIT
(LEAN EXHAUST INDICATED)
2.5L (VIN R)

Circuit Description:

The ECM supplies a voltage of about .45 volt between terminals "B2" and "B23". (If measured with a 10 megohm digital voltmeter, this may read as low as .32 volt.)

When the O_2 sensor reaches operating temperature, it varies this voltage from about .1 volt (exhaust is lean) to about .9 volt (exhaust is rich).

The sensor is like an open circuit and produces no voltage when it is below 360°C (600°F). An open sensor circuit, or cold sensor, causes "Open Loop" operation.

Test Description: Numbers below refer to numbers on the diagnostic chart.
1. Code 44 is set when the O_2 sensor signal voltage on CKT 412 remains below .2 volt for 60 seconds or more and the system is operating in "Closed Loop".

Diagnostic Aids:

Using the "Scan" tool, observe the block learn value at different engine speeds. The "Scan" tool also displays the block learn cells so the block learn values can be checked in each of the cells to determine when the Code 44 may have been set. If the conditions for Code 44 exists, the block learn values will be around 150 or higher.
Check the following possible causes:

- O_2 Sensor Wire. Sensor pigtail may be mispositioned and contacting the exhaust manifold.
  Check for ground in wire between connector and sensor.
- Fuel Contamination. Water, even in small amounts, near the in-tank fuel pump inlet can be delivered to the injector. The water causes a lean exhaust and can set a Code 44.
- Fuel Pressure. System will be lean if fuel pressure is too low. It may be necessary to monitor fuel pressure while driving the car at various road speeds and/or loads to confirm. See "Fuel System Diagnosis", CHART A-7.
- Exhaust Leaks. If there is an exhaust leak, the engine can cause outside air to be pulled into the exhaust and past the sensor. Vacuum or crankcase leaks can cause a lean condition.
- If Code 44 is intermittent, refer to Section "B".

1988–90 2.5L (VIN R) – ALL MODELS EXCEPT FIERO

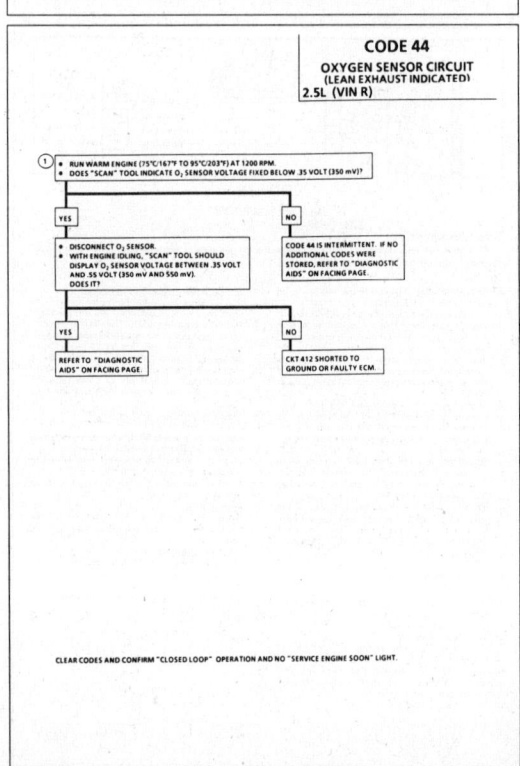

1988–90 2.5L (VIN R) – ALL MODELS EXCEPT FIERO

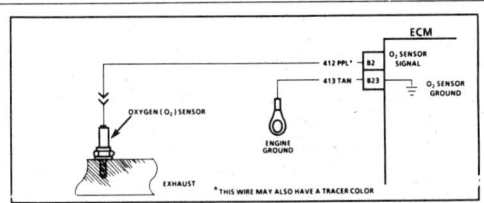

ECM

412 PPL* — B2 — O₂ SENSOR SIGNAL
413 TAN — B23 — O₂ SENSOR GROUND

OXYGEN (O₂) SENSOR

ENGINE GROUND

EXHAUST

* THIS WIRE MAY ALSO HAVE A TRACER COLOR

CODE 45
OXYGEN SENSOR CIRCUIT
(RICH EXHAUST INDICATED)
2.5L (VIN R)

Circuit Description:

The ECM supplies a voltage of about .45 volt between terminals "B2" and "B23". (If measured with a 10 megohm digital voltmeter, this may read as low as .32 volt).

When the O₂ sensor reaches operating temperature, it varies this voltage from about .1 volt (exhaust is lean) to about .9 volt (exhaust is rich).

The sensor is like an open circuit and produces no voltage when it is below 360°C (600°F). An open sensor circuit, or cold sensor, causes "Open Loop" operation.

Test Description: Numbers below refer to circled numbers on the diagnostic chart.

1. Code 45 is set when the O₂ sensor signal voltage on CKT 412 remains above .7 volt under the following conditions:
   - 30 seconds or more
   - System is operating in "Closed Loop"
   - Engine run time after start is 1 minute or more
   - Throttle angle is between 3% and 45%

Diagnostic Aids:

Code 45, or rich exhaust, is most likely caused by one of the following:

- **Fuel Pressure.** System will go rich, if pressure is too high. The ECM can compensate for some increase. However, if it gets too high, a Code 45 will be set. See "Fuel System Diagnosis", CHART A-7.
- **Leaking Injector.** See CHART A-7.
- **HEI Shielding.** An open ground CKT 453 may result in EMI, or induced electrical "noise". The ECM looks at this "noise" as reference pulse. The additional pulses result in a higher than actual engine speed signal. The ECM then delivers too much fuel causing the system to go rich. The engine tachometer will also show higher than actual engine speed, which can help in diagnosing this problem.

- **Canister Purge.** Check for fuel saturation. If full of fuel, check canister control and hoses.
- **MAP Sensor.** An output that causes the ECM to sense a higher than normal manifold pressure (low vacuum) can cause the system to go rich. Disconnecting the MAP sensor will allow the ECM to set a fixed value for the MAP sensor. Substitute a different MAP sensor if the rich condition is gone, while the sensor is disconnected.
- **TPS.** An intermittent TPS output will cause the system to operate richly due to a false indication of the engine accelerating.
- **O₂ Sensor Contamination.** Inspect oxygen sensor for silicone contamination from fuel, or use of improper RTV sealant. The sensor may have a white, powdery coating and result in a high but false signal voltage (rich exhaust indication). The ECM will then reduce the amount of fuel delivered to the engine causing a severe surge driveability problem.
- **EGR Valve.** EGR sticking open at idle is usually accompanied by a rough idle and/or stall condition.
  If Code 45 is intermittent, refer to Section "B".

1988–90 2.5L (VIN R) – ALL MODELS EXCEPT FIERO

CODE 45
OXYGEN SENSOR CIRCUIT
(RICH EXHAUST INDICATED)
2.5L (VIN R)

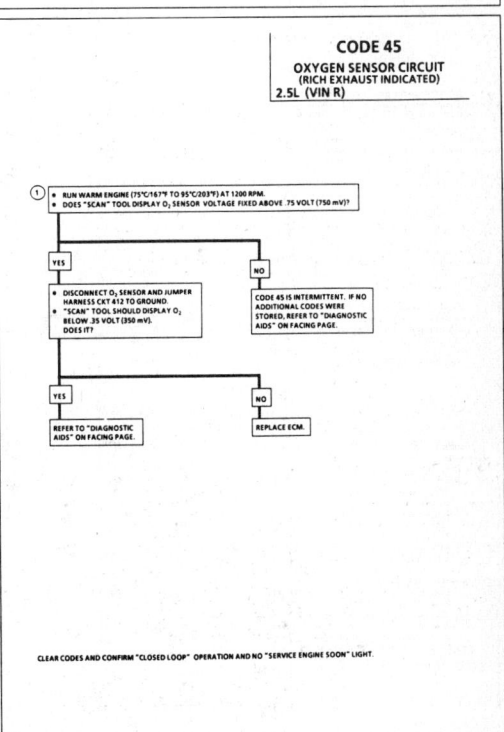

① • RUN WARM ENGINE (75°C/167°F TO 95°C/203°F) AT 1200 RPM.
   • DOES "SCAN" TOOL DISPLAY O₂ SENSOR VOLTAGE FIXED ABOVE .75 VOLT (750 mV)?

YES
- DISCONNECT O₂ SENSOR AND JUMPER HARNESS CKT 412 TO GROUND.
- "SCAN" TOOL SHOULD DISPLAY O₂ BELOW .35 VOLT (350 mV). DOES IT?

NO
CODE 45 IS INTERMITTENT. IF NO ADDITIONAL CODES WERE STORED, REFER TO "DIAGNOSTIC AIDS" ON FACING PAGE.

YES
REFER TO "DIAGNOSTIC AIDS" ON FACING PAGE.

NO
REPLACE ECM.

CLEAR CODES AND CONFIRM "CLOSED LOOP" OPERATION AND NO "SERVICE ENGINE SOON" LIGHT.

1988–90 2.5L (VIN R) – ALL MODELS EXCEPT FIERO

CODE 51
CODE 53
2.5L (VIN R)

CODE 51
PROM ERROR
(FAULTY OR INCORRECT PROM)

CHECK THAT ALL PINS ARE FULLY INSERTED IN THE SOCKET AND THAT PROM IS PROPERLY SEATED.
IF OK, REPLACE PROM, CLEAR MEMORY, AND RECHECK. IF CODE 51 REAPPEARS, REPLACE ECM.

CLEAR CODES AND CONFIRM "CLOSED LOOP" OPERATION AND NO "SERVICE ENGINE SOON" LIGHT.

CODE 53
SYSTEM OVER VOLTAGE

THIS CODE INDICATES THAT THERE IS A BASIC GENERATOR PROBLEM.
- CODE 53 WILL SET IF BATTERY VOLTAGE AT THE ECM IS GREATER THAN 16.9 VOLTS FOR AT LEAST 50 SECONDS.
- CHECK AND REPAIR CHARGING SYSTEM.

CLEAR CODES AND CONFIRM "CLOSED LOOP" OPERATION AND NO "SERVICE ENGINE SOON" LIGHT.

1988–90 2.5L (VIN R) – ALL MODELS EXCEPT FIERO

SECTION B
SYMPTOMS
TABLE OF CONTENTS

Performing Symptom Diagnosis
Intermittents
Hard Start
Rough, Unstable, or incorrect Idle, Stalling
Poor Gas Mileage
Detonation/Spark Knock
Lack of Power, Sluggish, or Spongy
Surges and/or Chuggle
Cuts Out, Misses
Hesitation, Sag, Stumble
Excessive Exhaust Emissions or Odors
Dieseling, Run-On
Backfire

PERFORMING SYMPTOM DIAGNOSIS

The DIAGNOSTIC CIRCUIT CHECK should be performed before using this section. The purpose of this section is to locate the source of a driveability or emissions problem when other diagnostic procedures cannot be used. This may be because of difficulties in locating a suspected sub-system or component.

Many driveability related problems can be eliminated by following the procedures found in Service Bulletins. These bulletins supersede this manual. Be sure to check all bulletins related to the complaint or suspected system.

If the engine cranks but will not run, use CHART A-3.

The sequence of the checks listed in this section is not intended to be followed as on a step-by-step procedure. The checks are listed such that the less difficult and time consuming operations are performed before more difficult ones.

Most of the symptom procedures call for a careful visual and physical check. *The importance of this step cannot be stressed too strongly.* It can lead to correcting a problem without further checks, and can save valuable time. This procedure includes checking the following:

- Vacuum hoses for splits, kinks, and proper connections, as shown on the underhood Emission Control Information label
- Throttle body and intake manifold for leaks
- Ignition wires for cracking, hardness, proper routing, and carbon tracking
- Wiring for proper connections, pinches, and cuts

1988–90 2.5L (VIN R) – ALL MODELS EXCEPT FIERO

INTERMITTENTS

Definition: Problem may or may not activate the "Service Engine Soon" light or store a trouble code.

DO NOT use the trouble code charts in Section "A" for intermittent problems. The fault must be present to locate the problem. If a fault is intermittent, the use of trouble code charts may result in the replacement of good parts.

- Most intermittent problems are caused by faulty electrical connections or wiring. Perform careful checks of suspected circuits for
  - Poor mating of the connector halves and terminals not fully seated in the connector body (backed out).
  - Improperly formed or damaged terminals. All connector terminals in problem circuit should be carefully reformed to increase contact tension.
  - Poor terminal to wire connection. This requires removing the terminal from the connector body to check
- If a visual and physical check does not locate the cause of the problem, the car can be driven with a voltmeter connected to a suspected circuit or a "Scan" tool may be used. An abnormal voltage reading while the problem occurs indicates that the problem may be in that circuit.
- Check for loss of trouble code memory. To check, disconnect the TPS and allow the engine to idle until the "Service Engine Soon" light turns "ON." Code 22 should be stored and kept in memory when the ignition is turned "OFF" for at least 10 seconds. If not, the ECM is faulty.
- An intermittent SES light and no trouble codes may be caused by
  - Electrical system interference caused by a defective relay, ECM driven solenoid, or switch. They can cause a sharp electrical surge. Normally, the problem will occur when the faulty component is operated.
  - Improper installation of electrical options, such as lights, 2-way radios, etc.
  - EST wires which should be routed away from spark plug wires, ignition system components, and generator. Ground wire from ECM to ignition system which may be faulty.
  - Ignition secondary wire shorted to ground.
  - "Service Engine Soon" light and diagnostic test terminal circuits intermittently shorted to ground.
  - Faulty ECM grounds.

HARD START

Definition: Engine cranks well but does not start for a long time. Engine does eventually start, but may or may not continue to run.

Perform careful visual and physical check as described at the beginning of Section "B".
Perform "Diagnostic Circuit Check."

- CHECK
  - Fuel for poor quality, "stale" fuel, and water contamination
  - Ignition wires for shorts or faulty insulation
  - Ignition coil connections
  - Fuel pump relay. Connect test light between pump test terminal and ground. Light should be on for 2 seconds following ignition "ON." If not, use CHART A-5.
  - Secondary ignition voltage output with ST-125 tester
  - Spark plugs. Look for wetness, cracks, improper gap, burned electrodes, and heavy deposits. Visually inspect ignition system for moisture, dust, cracks, burns, etc.
  - For faulty ECM and ignition grounds
  - PROM for correct application. Spray plug wires with fine water mist to check for shorts.
  - For possibility of misfiring, crossfiring, or cutting out under load or at idle. If possible, refer to the "Misfire" Chart in Section "C4."
  - For improper crank sensor resistance or faulty connections
  - EGR operation. Use CHART C-7.
  - Idle Air Control system. Use Code 35 chart.
  - Fuel system for restricted filter or improper pressure. Use CHART A-7.
  - Injector and TBI assembly for leakage. Pressurize system by energizing fuel pump through the underhood fuel pump test connector.
  - Coolant sensor for a shift in calibration. Use Code 14 or Code 15 chart.

1988–90 2.5L (VIN R) – ALL MODELS EXCEPT FIERO

- TPS for sticking or binding. TPS voltage should read less than 1.25 volts on a "Scan" tool.
- In-tank fuel pump check valve. A faulty valve would allow the fuel in the lines to drain back to the tank after the engine is stopped. To check for this condition, conduct the following test.
  1. Ignition "OFF."
  2. Disconnect fuel line at the filter.
  3. Remove the tank filler cap.
  4. Connect a radiator test pump to the line and apply 103 kPa (15 psi) pressure. If the pressure will hold for 60 seconds, the check valve is OK.
- For the possibility of an exhaust restriction or improper valve timing by performing the following test.
  1. With engine at normal operating temperature, connect a vacuum gauge to any convenient vacuum port on intake manifold.
  2. Run engine at 1000 rpm and record vacuum reading.
  3. Increase engine speed slowly to 2500 rpm. Note vacuum reading at steady 2500 rpm.
  4. If vacuum at 2500 rpm decreases more than 3" Hg from reading at 1000 rpm, the exhaust system should be inspected for restrictions.
  5. Disconnect exhaust pipe from engine and repeat Steps 3 & 4. If vacuum still drops more than 3" Hg with exhaust disconnected, check valve timing. General Information

Engine valve timing and compression.

ROUGH, UNSTABLE, OR INCORRECT IDLE, STALLING

Definition: The engine runs unevenly at idle. If severe, the car may shake. Also, the idle speed may vary (called "hunting"). Either condition may be severe enough to cause stalling. Engine idles at incorrect speed.

Perform careful visual and physical check as described at the beginning of Section "B".
Perform "Diagnostic Circuit Check."

- CHECK
  - MAP sensor. Use CHART C1-D.
  - Throttle for sticking shaft or binding linkage. This will cause a high TPS voltage (open throttle indication) and the ECM will not control idle. TPS voltage should be less than 1.25 volts with throttle closed.
  - Ignition wires for shorts or faulty insulation.
  - Ignition system for moisture, dust, cracks, burns, etc. Spray plug wires with fine water mist to check for shorts.
  - For possibility of misfiring, crossfiring, or cutting out under load or at idle. If present, refer to the "Misfire" Chart
  - Secondary ignition voltage output with ST-1 tester
  - Ignition coil connections.
  - ECM and ignition system for faulty grounds.
  - Proper operation of EST.
  - Spark plugs. Look for wetness, cracks, improper gap, burned electrodes, and heavy deposits.
  - Fuel system for restricted filter or improper pressure. Use CHART A-7.
  - Injector and TBI assembly for leakage. Pressurize system by energizing fuel pump through the underhood fuel pump test connector.
  - EGR operation. Use CHART C-7.
  - For vacuum leaks at intake manifold gasket.
  - Idle Air Control system. Use Code 35 chart.
  - Electrical system voltage. IAC valve will not move if voltage is below 9 volts or greater than 17.8 volts. Also check battery cables and ground straps for poor contact. Erratic voltage will cause the IAC valve to change its position, resulting in poor idle quality.
  - PCV valve for proper operation by placing finger over inlet hole in valve end several times. Valve should snap back. If not, replace valve. Ensure that valve is correct part. Also check PCV hose.
  - Canister purge system for proper operation. Use CHART C-3.
  - PROM for correct application.

1988–90 2.5L (VIN R) – ALL MODELS EXCEPT FIERO

- Throttle shaft or TPS for sticking or binding. TPS voltage should read less than 1.25 volts on a "Scan" tool with the throttle closed.
- MAP sensor output. Use CHART C1-D and/or check sensor by comparing it to the output ona similar vehicle if possible.
- Oxygen sensor for silicone contamination from contaminated fuel or use of improper RTV sealant. The sensor will have a white, powdery coating and will cause a high but false signal voltage (rich exhaust indication). The ECM will reduce the amount of fuel delivered to the engine, causing a severe driveability problem.
- Coolant sensor for a shift in calibration. Use Code 14 or Code 15 chart.
- A/C refrigerant pressure for high pressure. Check for overcharging or faulty pressure switch.
- P/N switch circuit on vehicle with automatic transmission. Use CHART C-1A.
- Alternator output voltage. Repair if less than 9 volts or more than 16 volts.
- Power steering. Use CHART C-1E. The ECM should compensate for power steering loads. Loss of this signal would be most noticeable when steering loads are high such as during parking.
- Engine valve timing and compression.
- For worn or incorrect basic engine parts such as cam, heads, pistons, etc. Also check for bent pushrods, worn rocker arms, and broken or weak valve springs.
- For the possibility of an exhaust restriction or improper valve timing, perform the following test.
  1. With engine at normal operating temperature, connect a vacuum gage to any convenient vacuum port on intake manifold.
  2. Run engine at 1000 rpm and record vacuum reading.
  3. Increase engine speed slowly to 2500 rpm. Note vacuum reading at steady 2500 rpm.
  4. If vacuum at 2500 rpm decreases more than 3" Hg from reading at 1000 rpm, the exhaust system should be inspected for restrictions.
  5. Disconnect exhaust pipe from engine and repeat Steps 3 & 4. If vacuum still drops more than 3" Hg with exhaust disconnected, check valve timing.

For overheating and possible causes. Look for the following.
  - Low or incorrect coolant solution. It should be a 50/50 mix of GM #1052753 anti-freeze coolant (or equivalent) and water.
  - Loose water pump belt
  - Restricted air flow to radiator, or restricted water flow through radiator
  - Faulty or incorrect thermostat
  - Inoperative electric cooling fan circuit
- If the system is running RICH, (block learn less than 118), refer to "Diagnostic Aids" on facing page of Code 45.
- If the system is running LEAN, (block learn greater than 138), refer to "Diagnostic Aids" on facing page of Code 44.

POOR GAS MILEAGE

Definition: Gas mileage, as measured by an actual road test, is noticeably lower than expected. Gas mileage is noticeably lower than it was during a previous actual road test.

Perform careful visual and physical check as described at the beginning of Section "B".
Perform "Diagnostic Circuit Check.

- CHECK
  - Proper operation of EST.
  - For possibility of misfiring, crossfiring, or cutting out under load or at idle. If present, refer to the "Misfire" Chart
  - Spark plugs. Look for wetness, cracks, improper gap, burned electrodes, and heavy deposits.
  - Spark plugs for correct heat range
  - Fuel for poor quality, "stale" fuel, and water contamination.
  - Fuel system for restricted filter or improper pressure. Use CHART A-7.
  - Injector and TBI assembly for leakage. Pressurize system by energizing fuel pump through the underhood fuel pump test connector.
  - EGR operation. Use CHART C-7.

1988–90 2.5L (VIN R) – ALL MODELS EXCEPT FIERO

- For vacuum leaks at intake manifold gasket
- Air cleaner element (filter) for dirt or plugging
- Idle Air Control system. Use Code 35 chart.
- Canister purge system for proper operation. Use CHART C-3.
- PROM for correct application
- Throttle shaft or TPS for sticking or binding. TPS voltage should read less than 1.25 volts on a "Scan" tool with the throttle closed.
- MAP sensor output. Use CHART C1-D and/or check sensor by comparing it to the output on a similar vehicle if possible.
- Oxygen sensor for silicone contamination from contaminated fuel or use of improper RTV sealant. The sensor will have a white, powdery coating and will cause a high but false signal voltage (rich exhaust indication). The ECM will reduce the amount of fuel delivered to the engine, causing a severe driveability problem.
- Coolant sensor for a shift in calibration. Use Code 14 or Code 15 chart.
- Vehicle speed sensor (VSS) input with a "Scan" tool to make sure reading of VSS matches that of vehicle speedometer.
- A/C relay operation. A/C should cut out at wide open throttle. Use CHART C-10.
- A/C refrigerant pressure for high pressure. Check for overcharging or faulty pressure switch.
- Alternator output voltage. Repair if less than 9 volts or more than 16 volts.
- Cooling fan operation. Use CHART C-12.
- Power steering. Use CHART C-1E. The ECM should compensate for power steering loads. Loss of this signal would be most noticeable when steering loads are high such as during parking.
- Transmission torque converter operation.
- Transmission for proper shift points.
- Transmission torque converter clutch operation. Use CHART C-8.
- Engine valve timing and compression.
- For worn or incorrect basic engine parts such as cam, heads, pistons, etc. Also check for bent pushrods, worn rocker arms, and broken or weak valve springs.
- For the possibility of an exhaust restriction or improper valve timing by performing the following test.
  1. With engine at normal operating temperature, connect a vacuum gauge to any convenient vacuum port on intake manifold.
  2. Run engine at 1000 rpm and record vacuum reading.
  3. Increase engine speed slowly to 2500 rpm. Note vacuum reading at steady 2500 rpm.
  4. If vacuum at 2500 rpm decreases more than 3" Hg from reading at 1000 rpm, the exhaust system should be inspected for restrictions.
  5. Disconnect exhaust pipe from engine and repeat Steps 3 & 4. If vacuum still drops more than 3" Hg with exhaust disconnected, check valve timing.
- Thermostat for incorrect heat range or being inoperative
- Check driver's driving habits and vehicle conditions which affect gas mileage.
  - Suggest driver read "Important Facts on Fuel Economy" in Owner's Manual.
  - Is A/C "ON" full time (Defroster mode "ON")?
  - Are tires at correct pressure?
  - Are excessively heavy loads being carried?
  - Is acceleration often heavy?
  - Are the wheels aligned correctly?
  - Is the speedometer calibrated correctly?
  - Are the vehicle brakes dragging?
  - Is the brake switch applying excessive force on the brake pedal?
- If the system is running RICH, (block learn less than 118), refer to "Diagnostic Aids" on facing page of Code 45.

1988–90 2.5L (VIN R) – ALL MODELS EXCEPT FIERO

DETONATION/SPARK KNOCK

Definition: A mild to severe ping, usually worse under acceleration. The engine makes sharp metallic knocks that change with throttle opening.

Perform careful visual and physical check as described at the beginning of Section "B".

Perform "Diagnostic Circuit Check."

- **CHECK**
  - Ignition wires for shorts or faulty insulation
  - For possibility of misfiring, crossfiring, or cutting out under load or at idle. If present, refer to the "Misfire" Chart
  - Spark plugs for correct heat range
  - Fuel for poor quality, "stale" fuel, and water contamination
  - Fuel system for restricted filter or improper pressure. Use CHART A-7.
  - For excessive oil entering combustion chamber. Oil will reduce the effective octane of fuel.
  - EGR operation. Use CHART C-7.
  - For vacuum leaks at intake manifold gasket
  - PCV valve for proper operation by placing finger over inlet hole in valve end several times. Valve should snap back. If not, replace valve. Ensure that valve is correct part. Also check PCV hose.
  - MAP sensor output. Use CHART C1-D and/or check sensor by comparing it to the output on a Similar vehicle, if possible.
  - Coolant sensor for a shift in calibration
  - Oxygen sensor for silicone contamination from contaminated fuel or use of improper RTV sealant. The sensor will have a white, powdery coating and will cause a high but false signal voltage (rich exhaust indication). The ECM will reduce the amount of fuel delivered to the engine, causing a severe driveability problem
  - Vehicle speed sensor (VSS) input with a "Scan" tool to make sure reading of VSS matches that of vehicle speedometer.
  - Transmission torque converter operation.
  - Transmission for proper shift points.
  - Transmission torque converter clutch operation. Use CHART C-8.
  - Vehicle brakes for dragging
  - PROM for correct application
  - For overheating and possible causes. Look for the following.
    - Low or incorrect coolant solution. It should be a 50/50 mix of GM #1052753 anti-freeze coolant (or equivalent) and water.
    - Loose water pump belt
    - Restricted air flow to radiator or restricted water flow through radiator
    - Faulty or incorrect thermostat
    - Inoperative electric cooling fan circuit
  - Engine valve timing and compression.
  - For worn or incorrect basic engine parts such as cam, heads, pistons, etc. Also check for bent pushrods, worn rocker arms, and broken or weak valve springs.
  - For the possibility of an exhaust restriction or improper valve timing by performing the following test.
    1. With engine at normal operating temperature, connect a vacuum gauge to any convenient vacuum port on intake manifold.
    2. Run engine at 1000 rpm and record vacuum reading.
    3. Increase engine speed slowly to 2500 rpm. Note vacuum reading at steady 2500 rpm.
    4. If vacuum at 2500 rpm decreases more than 3" Hg from reading at 1000 rpm, the exhaust system should be inspected for restrictions.
    5. Disconnect exhaust pipe from engine and repeat Steps 3 & 4. If vacuum still drops more than 3" Hg with exhaust disconnected, check valve timing.
- Remove internal engine carbon with top engine cleaner.
- If the system is running LEAN, (block learn greater than 138), refer to "Diagnostic Aids" on facing page of Code 44.

1988–90 2.5L (VIN R) – ALL MODELS EXCEPT FIERO

LACK OF POWER, SLUGGISH, OR SPONGY

Definition: Engine delivers less than expected power. There is little or no increase in speed when the accelerator pedal is depressed partially.

Perform careful visual and physical check as described at the beginning of Section "B".

Perform "Diagnostic Circuit Check."

- **CHECK**
  - Ignition wires for shorts or faulty insulation
  - Ignition system for moisture, dust, cracks, burns, etc. Spray plug wires with fine water mist to check for shorts.
  - For possibility of misfiring, crossfiring, or cutting out under load or at idle. If present, refer to the "Misfire" Chart
  - Secondary ignition voltage output with ST-125 tester.
  - Ignition coil connections
  - ECM and ignition system for faulty grounds
  - Proper operation of EST.
  - Spark plugs. Look for wetness, cracks, improper gap, burned electrodes, and heavy deposits.
  - Spark plugs for correct heat range
  - Fuel for poor quality, "stale" fuel, and water contamination
  - Fuei system for restricted filter or improper pressure. Use CHART A-7.
  - EGR operation. Use CHART C-7.
  - For vacuum leaks at intake manifold gasket
  - Air cleaner element (filter) for dirt or plugging
  - PROM for correct application
  - Throttle shaft or TPS for sticking or binding. TPS voltage should read less than 1.25 volts on a "Scan" tool with the throttle closed.
  - MAP sensor output. Use CHART C1-D and/or check sensor by comparing it to the output on a similar vehicle if possible.
  - Oxygen sensor for silicone contamination from contaminated fuel or use of improper RTV sealant. The sensor will have a white, powdery coating and will cause a high but false signal voltage (rich exhaust indication). The ECM will reduce the amount of fuel delivered to the engine, causing a severe driveability problem.
  - Coolant sensor for a shift in calibration. Use Code 14 or Code 15 chart.
  - Vehicle speed sensor (VSS) input with a "Scan" tool to make sure reading of VSS matches that of vehicle speedometer.
  - Engine for improper or worn camshaft.
  - A/C relay operation. A/C should cut out at wide open throttle. Use CHART C-10.
  - A/C refrigerant pressure for high pressure. Check for overcharging or faulty pressure switch.
  - Alternator output voltage. Repair if less than 9 volts or more than 16 volts.
  - Cooling fan operation. Use CHART C-12.
  - Power steering. Use CHART C-1E. The ECM should compensate for power steering loads. Loss of this signal would be most noticeable when steering loads are high such as during parking.
  - Transmission torque converter operation.
  - Transmission for proper shift points.
  - Transmission torque converter clutch operation. Use CHART C-8.
  - Vehicle brakes for dragging
  - Engine valve timing and compression.
  - For worn or incorrect basic engine parts such as cam, heads, pistons, etc. Also check for bent pushrods, worn rocker arms, and broken or weak valve springs.
  - For the possibility of an exhaust restriction or improper valve timing by performing the following test.
    1. With engine at normal operating temperature, connect a vacuum gauge to any convenient vacuum port on intake manifold.
    2. Run engine at 1000 rpm and record vacuum reading.
    3. Increase engine speed slowly to 2500 rpm. Note vacuum reading at steady 2500 rpm.
    4. If vacuum at 2500 rpm decreases more than 3" Hg from reading at 1000 rpm, the exhaust system should be inspected for restrictions.

1988–90 2.5L (VIN R) – ALL MODELS EXCEPT FIERO

5. Disconnect exhaust pipe from engine and repeat Steps 3 & 4. If vacuum still drops more than 3" Hg with exhaust disconnected, check valve timing.

- For overheating and possible causes. Look for the following:
- Low or incorrect coolant solution. It should be a 50/50 mix of GM #1052753 anti-freeze coolant (or equivalent) and water.
  - Loose water pump belt.
  - Restricted air flow to radiator, or restricted water flow through radiator.
  - Faulty or incorrect thermostat.
  - Inoperative electric cooling fan circuit. See CHART C-12.
- If the system is running RICH (block learn less than 118), refer to "Diagnostic Aids" on facing page of Code 45.
- If the system is running LEAN (block learn greater than 138), refer to "Diagnostic Aids" on facing page of Code 44.

SURGES AND/OR CHUGGLE

Definition: Engine power variation under steady throttle or cruise. Feels like the car speeds up and slows down with no change in the accelerator pedal.

Perform careful visual and physical check as described at the beginning of Section "B".

Perform "Diagnostic Circuit Check."

- **CHECK**
  - Ignition wires for shorts or faulty insulation
  - Ignition system for moisture, dust, cracks, burns, etc. Spray plug wires with fine water mist to check for shorts.
  - For possibility of misfiring, crossfiring, or cutting out under load or at idle. If present, refer to the "Misfire" Chart
  - Secondary ignition voltage output with ST-125 tester.
  - Ignition coil connections.
  - ECM and ignition system for faulty grounds.
  - Proper operation of EST.
  - Spark plugs. Look for wetness, cracks, improper gap, burned electrodes, and heavy deposits.
  - Spark plugs for correct heat range.
  - Fuel for poor quality, "stale" fuel, and water contamination.
  - Fuel system for restricted filter or improper pressure. Use CHART A-7.
  - Injector and TBI assembly for leakage. Pressurize system by energizing fuel pump through the underhood fuel pump test connector.
  - EGR operation. Use CHART C-7.
  - For vacuum leaks at intake manifold gasket.
  - Idle Air Control system. Use Code 35 chart.
  - Electrical system voltage. IAC valve will not move if voltage is below 9 volts or greater than 17.8 volts. Also check battery cables and ground straps for poor contact. Erratic voltage will cause the IAC valve to change its position, resulting in poor idle quality.
  - PCV valve for proper operation by placing finger over inlet hole in valve end several times. Valve should snap back. If not, replace valve. Ensure that valve is correct part. Also check PCV hose.
  - Canister purge system for proper operation. Use CHART C-3.
  - PROM for correct application.
  - Throttle shaft or TPS for sticking or binding. TPS voltage should read less than 1.25 volts on a "Scan" tool with the throttle closed.
  - MAP sensor output. Use CHART C1-D and/or check sensor by comparing it to the output on a similar vehicle, if possible.
  - Oxygen sensor for silicone contamination from contaminated fuel or use of improper RTV sealant. The sensor will have a white, powdery coating and will cause a high but false signal voltage (rich exhaust indication). The ECM will reduce the amount of fuel delivered to the engine, causing a severe driveability problem.
  - Coolant sensor for a shift in calibration. Use Code 14 or Code 15 chart.
  - Vehicle speed sensor (VSS) input with a "Scan" tool to make sure reading of VSS matches that of vehicle speedometer.
  - A/C relay operation. A/C should cut out at wide open throttle. Use CHART C-10.
  - P/N switch circuit on vehicle with automatic transmission. Use CHART C-1A.
  - Transmission torque converter clutch operation. Use CHART C-8.

1988–90 2.5L (VIN R) – ALL MODELS EXCEPT FIERO

- For the possibility of an exhaust restriction or improper valve timing by performing the following test:
  1. With engine at normal operating temperature, connect a vacuum gauge to any convenient vacuum port on intake manifold.
  2. Run engine at 1000 rpm and record vacuum reading.
  3. Increase engine speed slowly to 2500 rpm. Note vacuum reading at steady 2500 rpm.
  4. If vacuum at 2500 rpm decreases more than 3" Hg from reading at 1000 rpm, the exhaust system should be inspected for restrictions.
  5. Disconnect exhaust pipe from engine and repeat Steps 3 & 4. If vacuum still drops more than 3" Hg with exhaust disconnected, check valve timing.
- Engine valve timing and compression.
- For worn or incorrect basic engine parts such as cam, heads, pistons, etc. Also check for bent pushrods, worn rocker arms, and broken or weak valve springs.
- If the system is running RICH, (block learn less than 118), refer to "Diagnostic Aids" on facing page of Code 45.
- If the system is running LEAN, (block learn greater than 138), refer to "Diagnostic Aids" on facing page of Code 44.

CUTS OUT, MISSES

Definition: Steady pulsation or jerking that follows engine speed, usually more pronounced as engine load increases. The exhaust has a steady spitting sound at idle or low speed.

Perform careful visual and physical check as described at the beginning of Section "B".

Perform "Diagnostic Circuit Check."

- **CHECK**
  - Ignition wires for shorts or faulty insulation
  - Ignition system for moisture, dust, cracks, burns, etc. Spray plug wires with fine water mist to check for shorts.
  - For possibility of misfiring, crossfiring, or cutting out under load or at idle. If present, refer to the "Misfire" Chart
  - Secondary ignition voltage output with ST-125 tester
  - Ignition coil connections
  - ECM and ignition system for faulty grounds
  - Proper operation of EST.
  - Spark plugs. Look for wetness, cracks, improper gap, burned electrodes, and heavy deposits.
  - Spark plugs for correct heat range
  - For improper crank sensor resistance or faulty connections.
  - Fuel for poor quality, "stale" fuel, and water contamination
  - Fuel system for restricted filter or improper pressure. Use CHART A-7.
  - Throttle shaft or TPS for sticking or binding. TPS voltage should read less than 1.25 volts on a "Scan" tool with the throttle closed.

HESITATION, SAG, STUMBLE

Definition: Momentary lack of response as the accelerator is pushed down. Can occur at all vehicle speeds. Usually most severe when first trying to make the car move, as from a stop sign. May cause the engine to stall if severe enough.

Perform careful visual and physical check as described at the beginning of Section "B".

Perform "Diagnostic Circuit Check."

- **CHECK**
  - Ignition wires for shorts or faulty insulation
  - Ignition system for moisture, dust, cracks, burns, etc. Spray plug wires with fine water mist to check for shorts.
  - For possibility of misfiring, crossfiring, or cutting out under load or at idle. If present, refer to the "Misfire" Chart
  - Secondary ignition voltage output with ST-125 tester
  - Ignition coil connections
  - ECM and ignition system for faulty grounds
  - Proper operation of EST. See Section "C4".
  - Spark plugs. Look for wetness, cracks, improper gap, burned electrodes, and heavy deposits.
  - Spark plugs for correct heat range
  - Fuel for poor quality, "stale" fuel, and water contamination

---

1988–90 2.5L (VIN R) – ALL MODELS EXCEPT FIERO

- Fuel system for restricted filter or improper pressure. Use CHART A-7.
- EGR operation. Use CHART C-7.
- For vacuum leaks at intake manifold gasket.
- Air cleaner element (filter) for dirt or plugging
- Idle Air Control system. Use Code 35 chart.
- Check electrical system voltage. IAC valve will not move if voltage is below 9 volts or greater than 17.8 volts. Also check battery cables and ground straps for poor contact. Erratic voltage will cause the IAC valve to change its position, resulting in poor idle quality.
- PCV valve for proper operation by placing finger over inlet hole in valve and several times. Valve should snap back. If not, replace valve. Ensure that valve is correct part. Also check PCV hose.
- Canister purge system for proper operation. Use CHART C-3.
- PROM for correct application
- Throttle shaft or TPS for sticking or binding TPS voltage should read less than 1.25 volts on a "Scan" tool with the throttle closed.
- MAP sensor output. Use CHART C1-D and/or check sensor by comparing it to the output on a similar vehicle, if possible.
- Oxygen sensor for silicone contamination from contaminated fuel or use of improper RTV sealant. The sensor will have a white, powdery coating and will cause a high but false signal voltage (rich exhaust indication). The ECM will reduce the amount of fuel delivered to the engine, causing a severe driveability problem.
- Coolant sensor for a shift in calibration. Use Code 14 or Code 15 chart.
- A/C relay operation. A/C should cut out at wide open throttle. Use CHART C-10.
- A/C refrigerant pressure for high pressure. Check for overcharging or faulty pressure switch.
- P/N switch circuit on vehicle with automatic transmission. Use CHART C-1A.
- Alternator output voltage. Repair if less than 9 volts or more than 16 volts.
- Transmission torque converter operation.

- Transmission for proper shift points.

- Transmission torque converter clutch operation. Use CHART C-8.
- Vehicle brakes for dragging
- Engine valve timing and compression. See General Information - Diagnosis (Section "6")
- For the possibility of an exhaust restriction or improper valve timing by performing the following test.
  1. With engine at normal operating temperature, connect a vacuum gauge to any convenient vacuum port on intake manifold
  2. Run engine at 1000 rpm and record vacuum reading.
  3. Increase engine speed slowly to 2500 rpm. Note vacuum reading at steady 200 rpm.
  4. If vacuum at 2500 rpm decreases more than 3" Hg from reading at 1000 rpm, the exhaust system should be inspected for restrictions.
  5. Disconnect exhaust pipe from engine and repeat Steps 3 & 4. If vacuum still drops more than 3" Hg with exhaust disconnected, check valve timing.
- For worn or incorrect basic engine parts such as cam, heads, pistons, etc. Also check for bent pushrods, worn rocker arms, and broken or weak valve springs.
- For overheating and possible causes. Look for the following.
  - Low or incorrect coolant solution. It should be a 50/50 mix of GM #1052753 anti-freeze coolant (or equivalent) and water.
  - Loose water pump belt.
  - Restricted air flow to radiator, or restricted water flow through radiator
  - Faulty or incorrect thermostat
  - Inoperative electric cooling fan circuit. See CHART C-12.
- If the system is running RICH (block learn less than 118), refer to "Diagnostic Aids" on facing page of Code 45.
- If the system is running LEAN (block learn greater than 138), refer to "Diagnostic Aids" on facing page of Code 44.

1988–90 2.5L (VIN R) – ALL MODELS EXCEPT FIERO

EXCESSIVE EXHAUST EMISSIONS OR ODORS
Definition: Vehicle fails an emission test or vehicle has excessive "rotten egg" smell. (Excessive odors do not necessarily indicate excessive emissions).

Perform careful visual and physical check as described at the beginning of Section "B".
Perform "Diagnostic Circuit Check."
- **CHECK**
  - EGR valve not opening. Use CHART C-7.
  - Vacuum leaks.
  - Faulty coolant system and/or coolant fan operation. Use CHART C-12.
  - Remove carbon with top engine cleaner. Follow instructions on can.
- If the system is running RICH (block learn less than 118), refer to "Diagnostic Aids" on facing page of Code 45.
- If the system is running LEAN (block learn greater than 138), refer to "Diagnostic Aids" on facing page of Code 44.

- If emission test indicates excessive NO_x, check for items which cause car to run lean or too hot.
- If emission test indicates excessive HC and CO or exhaust has excessive odors, check for items which cause car to run RICH.
  - Incorrect fuel pressure. Use CHART A-7.
  - Fuel loading of evaporative vapor canister. Use CHART C-3.
  - PCV valve plugging, sticking, or blocked PCV hose. Check for fuel in crankcase.
  - Catalytic converter lead contamination. (Look for removal of fuel filler neck restrictor.)
  - Improper fuel cap installation.
  - Faulty spark plugs, plug wires, or ignition components.

DIESELING, RUN-ON
Definition: Engine continues to run after key is turned "OFF", but runs very roughly. (If engine runs smoothly, check ignition switch).

Perform careful visual and physical check as described at the beginning of Section "B".
Perform "Diagnostic Circuit Check."
- **CHECK**
  - Injector and TBI assembly for leakage. Pressurize system by energizing fuel pump through the fuel pump test connector.

BACKFIRE
Definition: Fuel ignites in intake manifold or in exhaust system, making a loud popping sound.

Perform careful visual and physical check as described at the beginning of Section "B".
Perform "Diagnostic Circuit Check."
- **CHECK**
  - EGR operation for valve being open all the time. Use CHART C-7.
  - Intake manifold gasket for leaks.
  - For possibility of misfiring, crossfiring, or cutting out under load or at idle. If present, refer to the "Misfire" Chart.
  - Spark plugs. Look for wetness, cracks, improper gap, burned electrodes, and heavy deposits.
  - Ignition coil connections

- Ignition system for moisture, dust, cracks, burns, etc. Spray plug wires with fine water mist to check for shorts.
- ECM and ignition system for faulty grounds
- Secondary ignition voltage output with ST-125 tester
- For vacuum leaks at intake manifold gasket
- Engine valve timing and compression.

- For worn or incorrect basic engine parts such as cam, heads, pistons, etc. Also check for bent pushrods, worn rocker arms, and broken or weak valve springs.

1988–90 2.5L (VIN R) – ALL MODELS EXCEPT FIERO

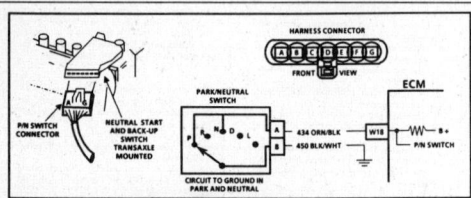

CHART C-1A
PARK/NEUTRAL SWITCH DIAGNOSIS
2.5L (VIN R)

Circuit Description:
The park/neutral switch contacts are a part of the neutral start switch and are closed to ground in park or neutral, and open in drive ranges.
The ECM supplies ignition voltage through a current limiting resistor to CKT 434 and senses a closed switch, when the voltage on CKT 434 drops to less than one volt.
The ECM uses the P/N signal as one of the inputs to control idle air control and VSS diagnostics.

Test Description: Numbers below refer to circled numbers on the diagnostic chart.
1. Checks for a closed switch to ground in park position. Different makes of "Scan" tools will read P/N differently. Refer to tool operations manual for type of display used.

2. Checks for an open switch in drive range.
3. Be sure "Scan" indicates "Drive", even while wiggling shifter to test for an intermittent or misadjusted switch in drive range.

1988–90 2.5L (VIN R) – ALL MODELS EXCEPT FIERO

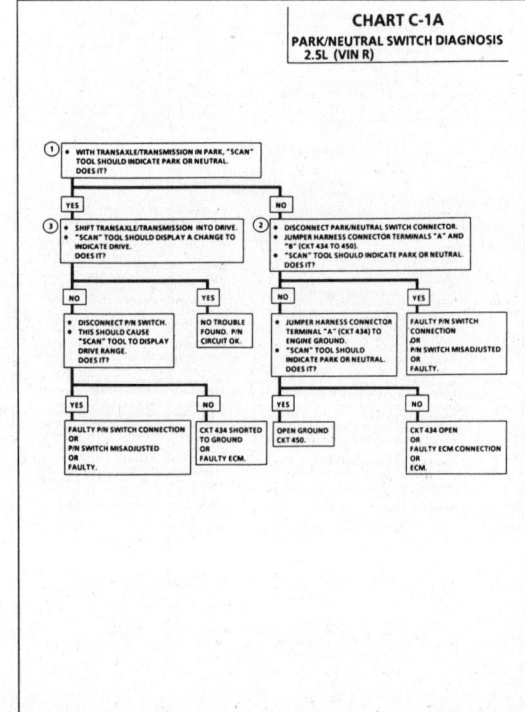

1988–90 2.5L (VIN R) – ALL MODELS EXCEPT FIERO

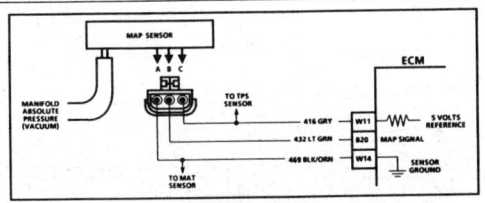

CHART C-1D
MANIFOLD ABSOLUTE PRESSURE (MAP) OUTPUT CHECK
2.5L (VIN R)

Circuit Description:
   The Manifold Absolute Pressure (MAP) sensor measures manifold pressure (vacuum) and sends that signal to the ECM. The MAP sensor is mainly used to calculate engine load, which is a fundamental input for spark advance and fuel calculations. The MAP sensor is also used to determine the barometric pressure.

Test Description: Numbers below refer to circled numbers on the diagnostic chart.
1. Checks MAP sensor output voltage to the ECM. This voltage, without engine running, represents a barometer reading to the ECM. Comparison of this BARO reading with a known good vehicle with the same sensor is a good way to check the accuracy of a "suspect" sensor. Readings should be within .4 volt.

2. Applying 34 kPa (10" Hg) vacuum to the MAP sensor should cause the voltage to be 1.2 to 2.3 volts less than the voltage at Step 1. Upon applying vacuum to the sensor, the change in voltage should be instantaneous. A slow voltage change indicates a faulty sensor.
3. Check vacuum hose to sensor for leaking or restriction. Ensure that no other vacuum devices are connected to the MAP hose.

1988–90 2.5L (VIN R) – ALL MODELS EXCEPT FIERO

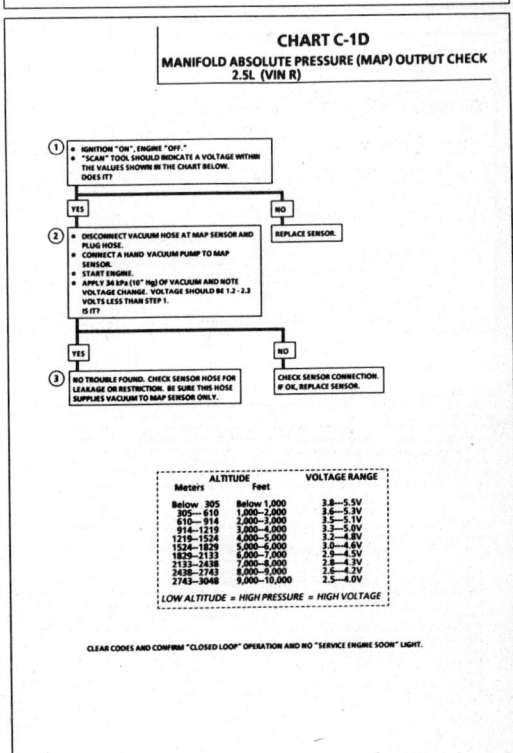

1988–90 2.5L (VIN R) – ALL MODELS EXCEPT FIERO

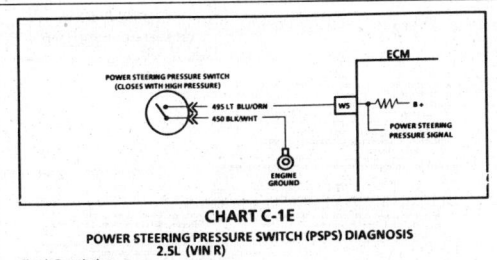

CHART C-1E
POWER STEERING PRESSURE SWITCH (PSPS) DIAGNOSIS
2.5L (VIN R)

Circuit Description:
   The power steering pressure switch is normally open to ground, and CKT 495 will be near the battery voltage.
   Turning the steering wheel increases power steering oil pressure and pump load on the engine. The pressure switch will close before the load can cause an idle problem.
   Closing the switch causes CKT 495 to read less than 1 volt and the ECM will increase the idle air rate.
 • A pressure switch that will not close, or an open CKT 495 or 450, may cause the engine to stop when power steering loads are high.
 • A switch that will not open, or a CKT 495 shorted to ground, may affect idle quality.

Test Description: Numbers below refer to circled numbers on the diagnostic chart.
1. Different makes of "Scan" tools may display the state of this switch in different ways. Refer to "Scan" tool user's manual to determine how this input is indicated.

2. Checks to determine if CKT 495 is shorted to ground.
3. This should simulate a closed switch.

1988–90 2.5L (VIN R) – ALL MODELS EXCEPT FIERO

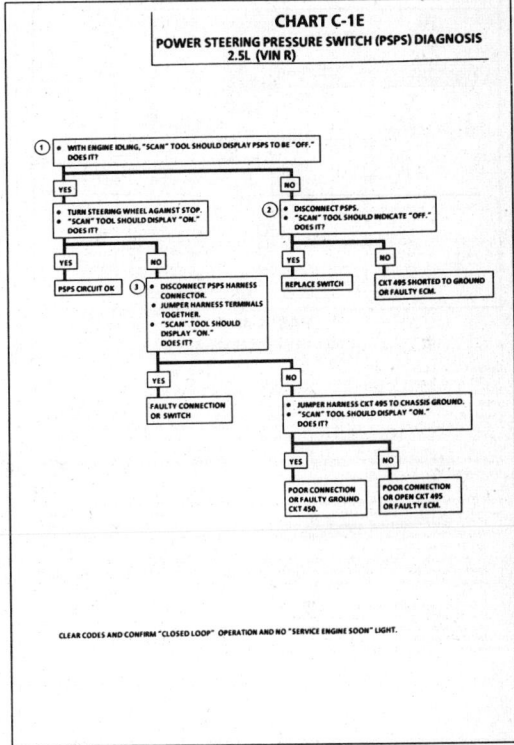

1988–90 2.5L (VIN R) – ALL MODELS EXCEPT FIERO

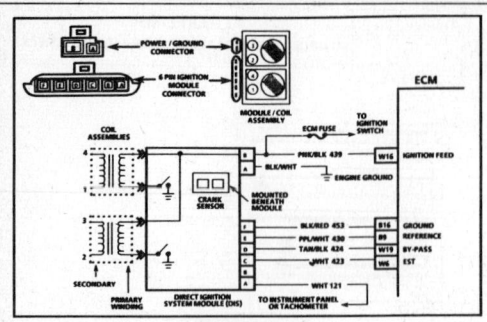

CHART C-4D-1
DIS MISFIRE AT IDLE
2.5L (VIN R)

Circuit Description:

The Direct Ignition System (DIS) uses a waste spark method of distribution. In this type of system, the ignition module triggers a dual coil, resulting in both connected spark plugs firing at the same time. One cylinder is on its compression stroke at the same time that the other is on the exhaust stroke, resulting in a lower energy requirement to fire the spark plug in the cylinder on its exhaust stroke. This leaves the remainder of the high voltage to be used to fire the spark plug which is in the cylinder on its compression stroke. On this application, the crank sensor is mounted to the bottom of the coil/module assembly and protrudes through the block to within approximately .050" of the crankshaft reluctor. Since the reluctor is a machined portion of the crankshaft and the crank sensor is mounted in a fixed position on the block, timing adjustments are not possible or necessary.

Test Description: Numbers below refer to circled numbers on the diagnostic chart.

1. If the "Misfire" complaint exists under load only, CHART C-4D-2 must be used. Engine rpm should drop approximately equally on all plug leads.
2. A spark tester such as a ST-125 must be used because it is essential to verify adequate available secondary voltage at the spark plug. (25,000 volts.)
3. If the spark jumps the test gap after grounding the opposite plug wire, it indicates excessive resistance in the plug which was bypassed. A faulty or poor connection at that plug could also result in the miss condition. Also check for carbon deposits inside the spark plug boot.

4. If carbon tracking is evident, replace coil and be sure plug wires relating to that coil are clean and tight. Excessive wire resistance or faulty connections could have caused the coil to be damaged.
5. If the no spark condition follows the suspected coil, that coil is faulty. Otherwise, the ignition module is the cause of no spark. This test could also be performed by substituting a known good coil for the one causing the no spark condition.

1988–90 2.5L (VIN R) – ALL MODELS EXCEPT FIERO

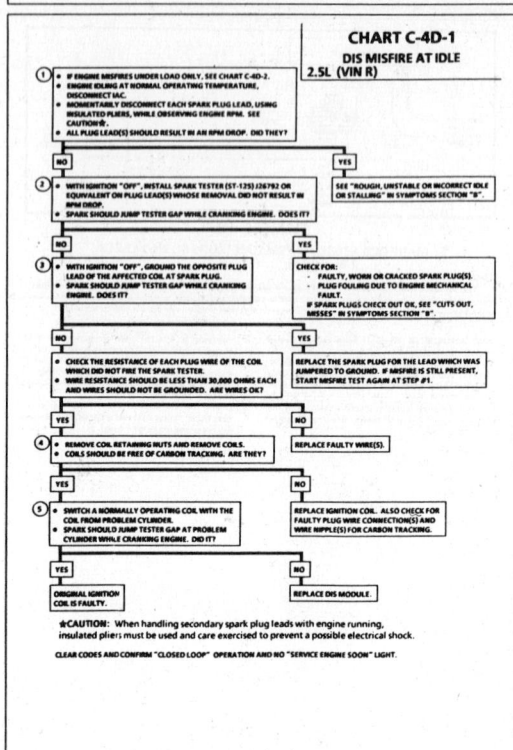

CHART C-4D-1
DIS MISFIRE AT IDLE
2.5L (VIN R)

★ **CAUTION:** When handling secondary spark plug leads with engine running, insulated pliers must be used and care exercised to prevent a possible electrical shock.

CLEAR CODES AND CONFIRM "CLOSED LOOP" OPERATION AND NO "SERVICE ENGINE SOON" LIGHT.

1988–90 2.5L (VIN R) – ALL MODELS EXCEPT FIERO

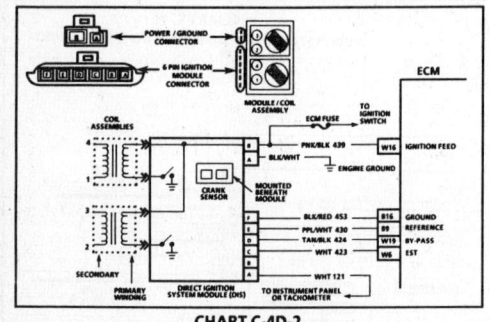

CHART C-4D-2
DIS MISFIRE UNDER LOAD
2.5L (VIN R)

Circuit Description:

The Direct Ignition System (DIS) uses a waste spark method of distribution. In this type of system, the ignition module triggers a dual coil, resulting in both connected spark plugs firing at the same time. One cylinder is on its compression stroke at the same time that the other is on the exhaust stroke, resulting in a lower energy requirement to fire the spark plug in the cylinder on its exhaust stroke. This leaves the remainder of the high voltage to be used to fire the spark plug which is in the cylinder on its compression stroke. On this application, the crank sensor is mounted to the bottom of the coil/module assembly and protrudes through the block to within approximately .050" of the crankshaft reluctor. Since the reluctor is a machined portion of the crankshaft and the crank sensor is mounted in a fixed position on the block, timing adjustments are not possible or necessary.

Test Description: Numbers below refer to circled numbers on the diagnostic chart.

1. If the "Misfire" complaint exists at idle only, CHART C-4D-1 must be used. A spark tester such as a ST-125 must be used because it is essential to verify adequate available secondary voltage at the spark plug. (25,000 volts). Spark should jump the test gap on all 4 leads. This simulates a "load" condition.
2. If the spark jumps the tester gap after grounding the opposite plug wire, it indicates excessive resistance in the plug which was bypassed. A faulty or poor connection at that plug could also result in the miss condition. Also check for carbon deposits inside the spark plug boot.

3. If carbon tracing is evident replace coil and be sure plug wires relating to that coil are clean and tight. Excessive wire resistance or faulty connections could have caused the coil to be damaged.
4. If the no spark condition follows the suspected coil, that coil is faulty. Otherwise, the ignition module is the cause of no spark. This test could also be performed by substituting a known good coil for the one causing the no spark condition.

1988–90 2.5L (VIN R) – ALL MODELS EXCEPT FIERO

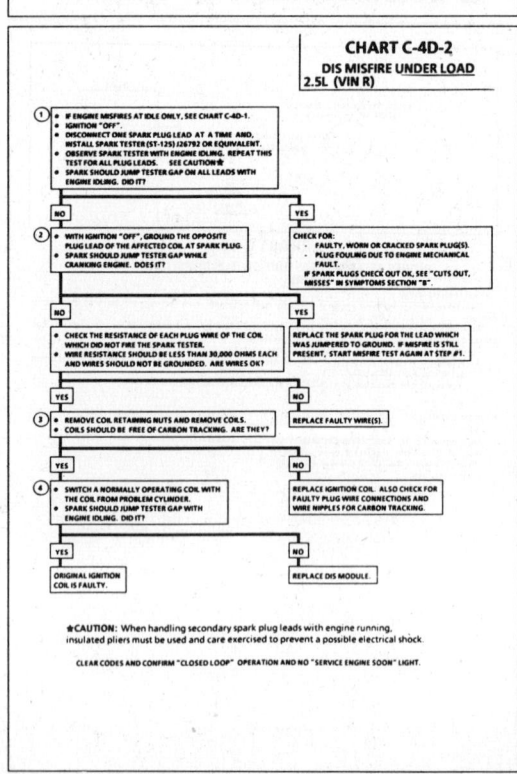

CHART C-4D-2
DIS MISFIRE UNDER LOAD
2.5L (VIN R)

★ **CAUTION:** When handling secondary spark plug leads with engine running, insulated pliers must be used and care exercised to prevent a possible electrical shock.

CLEAR CODES AND CONFIRM "CLOSED LOOP" OPERATION AND NO "SERVICE ENGINE SOON" LIGHT.

1988–90 2.5L (VIN R) – ALL MODELS EXCEPT FIERO

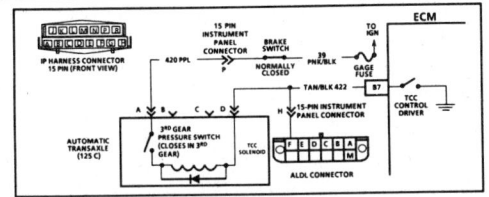

CHART C-8A
TORQUE CONVERTER CLUTCH (TCC) CIRCUIT
2.5L (VIN R)

Circuit Description:

The purpose of the automatic transaxle torque converter clutch is to eliminate the power loss of the torque converter when the vehicle is in a cruise condition. This allows the convenience of the automatic transaxle and the fuel economy of a manual transaxle.

Fused ignition voltage is supplied to the TCC solenoid through the brake switch and transaxle third gear apply switch. The ECM will engage TCC by grounding CKT 422 to energize the solenoid.

TCC will engage under the following conditions:
- Vehicle speed is above 30 mph (48 km/h).
- Engine at normal operating temperature (above 70°C, 156°F).
- Throttle position sensor output not changing, indicating a steady road speed.
- Transaxle third gear switch closed.
- Brake switch closed.

Test Description: Numbers below refer to circled numbers on the diagnostic chart.

1. Light "OFF" confirms transaxle third gear apply switch is open.
2. At 30 km (48 mph/h), the transaxle third gear switch should close. Test light will turn "ON" and confirm ignition voltage in circuit and close brake switch in closed position.
3. Grounding the diagnostic terminal with the engine "OFF" should energize the TCC solenoid. This test checks the capability of the ECM to control the solenoid.

Diagnostic Aids:

An engine coolant thermostat that is stuck open or opens at too low a temperature may result in an inoperative TCC.

1988–90 2.5L (VIN R) – ALL MODELS EXCEPT FIERO

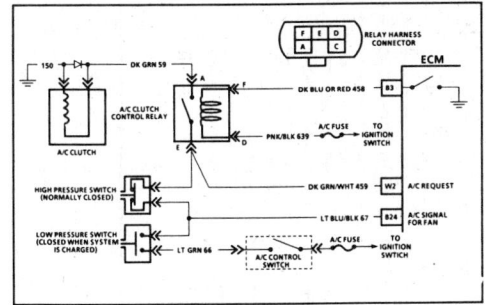

CHART C-10
A/C CLUTCH CONTROL
2.5L (VIN R)

Circuit Description:

When an A/C mode is selected on the A/C control switch, ignition voltage is supplied to the compressor low pressure switch. If there is sufficient A/C refrigerant pressure, the low pressure switch will be closed and complete the circuit to the closed high pressure cut-off switch and to CKTs 67 and 459. The voltage on CKT 459 to the ECM is shown by the "Scan" tool as A/C request "ON" (voltage present), "OFF" (no voltage). When a request for A/C is sensed by the ECM, the ECM will ground CKT 458 of the A/C clutch control relay, the relay contact will close, and current will flow from CKT 459 to CKT 59 to engage the A/C compressor clutch. A "Scan" tool will show the grounding of CKT 458 as A/C clutch "ON". If voltage is sensed by the ECM on CKT 67, the cooling fan will be turned "ON".

Test Description: Numbers below refer to circled numbers on the diagnostic chart.

1. The ECM will energize the A/C relay only when the engine is running. This test will determine if the relay or CKT 458 is faulty.
2. The low pressure and high pressure switches must be closed so that the A/C request signal (12 volts) will be present at the ECM.
3. A short to ground in any part of the A/C request or A/C clutch control circuits could be the cause of the blown fuse.
4. With the engine idling and A/C "ON," the ECM should be grounding CKT 458, causing the test light to be "ON."
5. Determines if the signal is reaching the low pressure switch on CKT 66 from the A/C control panel. The signal should be present only when the A/C mode or defrost mode has been selected.

Diagnostic Aids:

If complaint was insufficient cooling, the problem may be caused by an inoperative cooling fan or A/C pressure fan switch. The engine cooling fan should turn "ON" when A/C pressure exceeds a value to open the switch, which causes the ECM to energize the cooling fan relay. See CHART C-12 for cooling fan diagnosis.

The A/C clutch will be disengaged if a high power steering pressure signal is detected by the ECM. Refer to CHART C1-E.

1988–90 2.5L (VIN R) – ALL MODELS EXCEPT FIERO

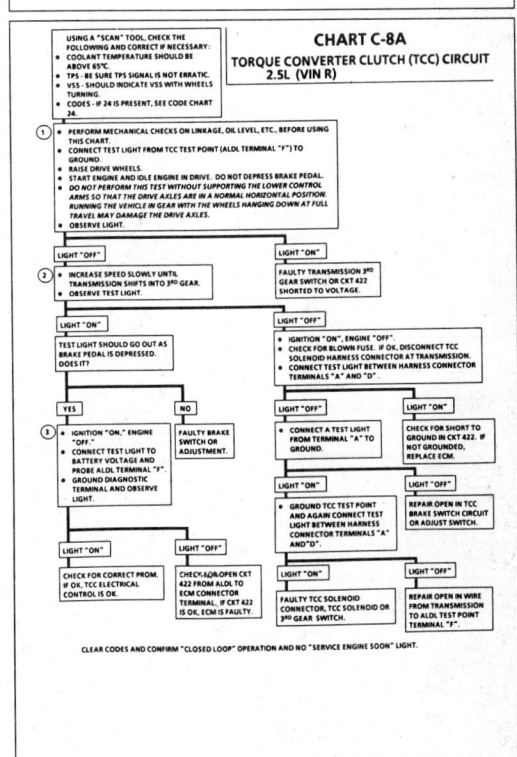

CHART C-8A
TORQUE CONVERTER CLUTCH (TCC) CIRCUIT
2.5L (VIN R)

1988–90 2.5L (VIN R) – ALL MODELS EXCEPT FIERO

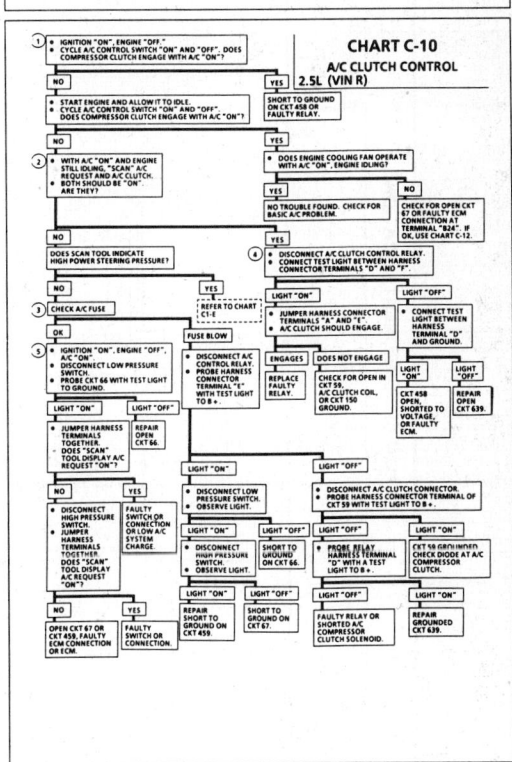

CHART C-10
A/C CLUTCH CONTROL
2.5L (VIN R)

1988–90 2.5L (VIN R) – ALL MODELS EXCEPT FIERO

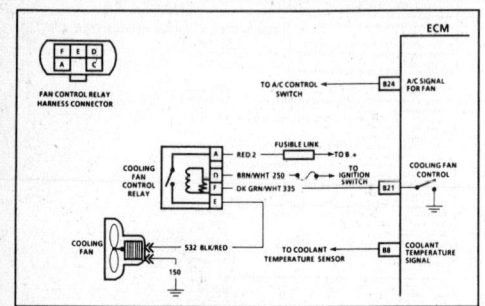

FAN CONTROL RELAY HARNESS CONNECTOR

CHART C-12
ENGINE COOLING FAN
2.5L (VIN R)

Circuit Description:

Battery voltage to operate the cooling fan motor is supplied to relay terminal "A". Ignition voltage to energize the relay is supplied to relay terminal "D". When the ECM grounds CKT 335, the relay is energized and the cooling fan is turned "ON". When the engine is running, the ECM will turn the cooling fan "ON" if a coolant temperature sensor code (14 or 15) has been set, or under the following conditions:
* A/C is "ON" and vehicle speed is less than 30 mph (48 km/h).
* Coolant temperature is greater than 108°C (230°F).

Diagnostic Aids:

If the vehicle has an overheating problem, it must be determined if the complaint was due to an actual boil over, the coolant temperature warning light, or the temperature gage indicated overheating.

If the gage or light indicates overheating, but no boilover is detected, the gage circuit should be checked.

The gage accuracy can be checked by comparing the coolant sensor reading on a "Scan" tool with the gage reading.

If the engine is actually overheating and the gage indicates overheating, but the cooling fan is not turning "ON", the coolant sensor has probably shifted out of calibration and should be replaced.

1988–90 2.5L (VIN R) – ALL MODELS EXCEPT FIERO

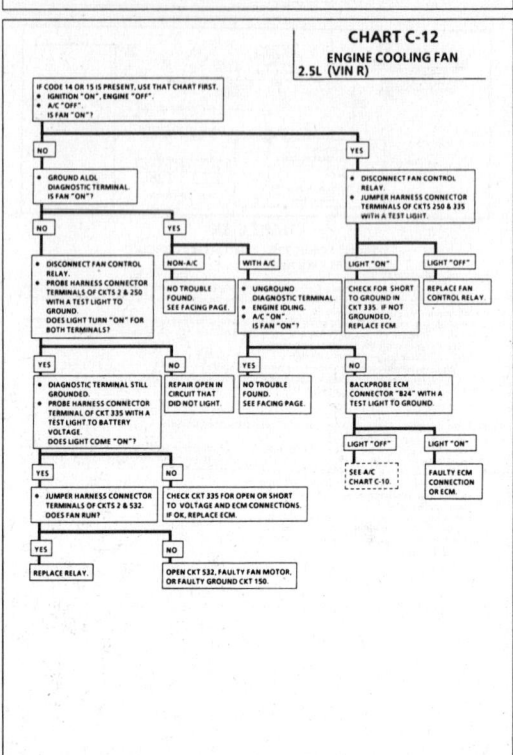

CHART C-12
ENGINE COOLING FAN
2.5L (VIN R)

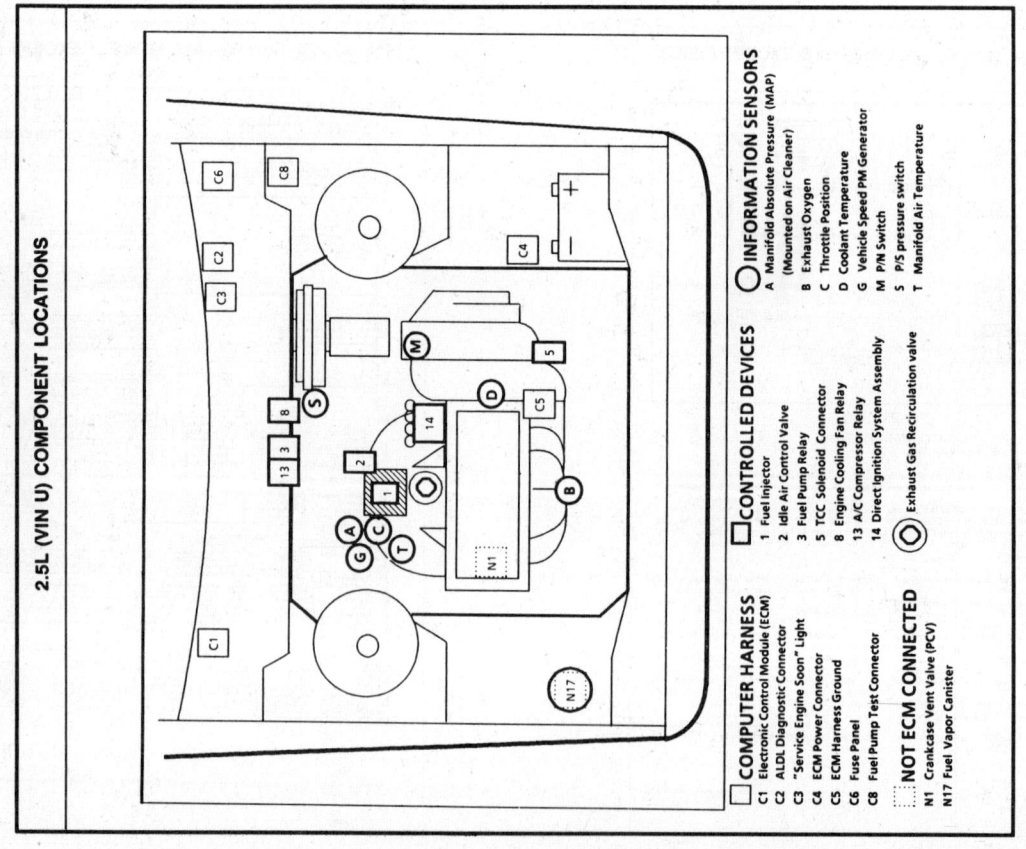

2.5L (VIN U) COMPONENT LOCATIONS

COMPUTER HARNESS
C1 Electronic Control Module (ECM)
C2 ALDL Diagnostic Connector
C3 "Service Engine Soon" Light
C4 ECM Power Connector
C5 ECM Harness Ground
C6 Fuse Panel
C8 Fuel Pump Test Connector

NOT ECM CONNECTED
N1 Crankcase Vent Valve (PCV)
N17 Fuel Vapor Canister

CONTROLLED DEVICES
1 Fuel Injector
2 Idle Air Control Valve
3 Fuel Pump Relay
8 TCC Solenoid Connector
8 Engine Cooling Fan Relay
13 A/C Compressor Relay
14 Direct Ignition System Assembly
Exhaust Gas Recirculation valve

INFORMATION SENSORS
A Manifold Absolute Pressure (MAP) (Mounted on Air Cleaner)
B Exhaust Oxygen
C Throttle Position
D Coolant Temperature
G Vehicle Speed PM Generator
M P/N Switch
S P/S pressure switch
T Manifold Air Temperature

2.5L (VIN U) ECM WIRING DIAGRAM

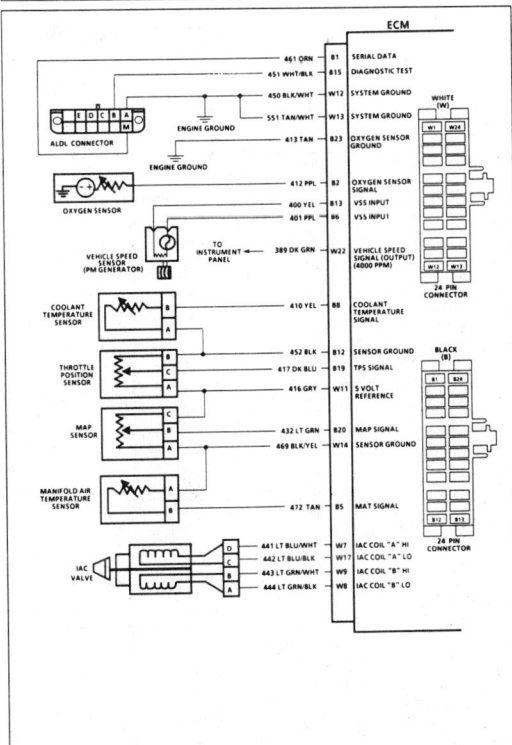

2.5L (VIN U) ECM WIRING DIAGRAM (CONT.)

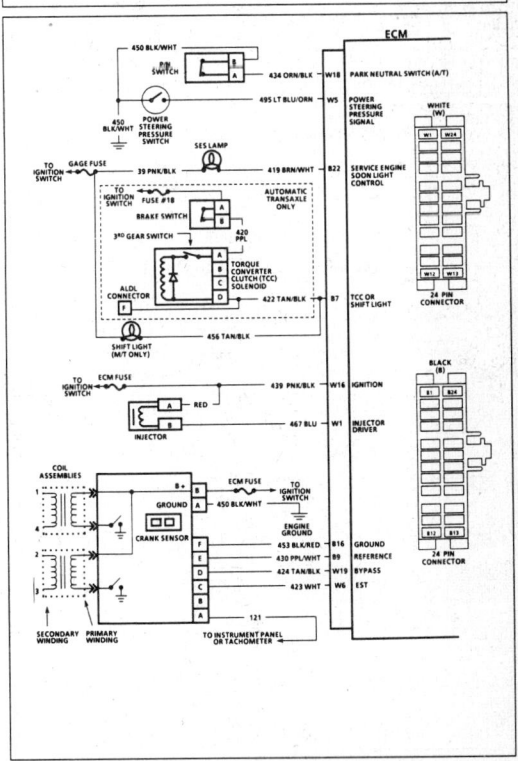

2.5L (VIN U) ECM WIRING DIAGRAM (CONT.)

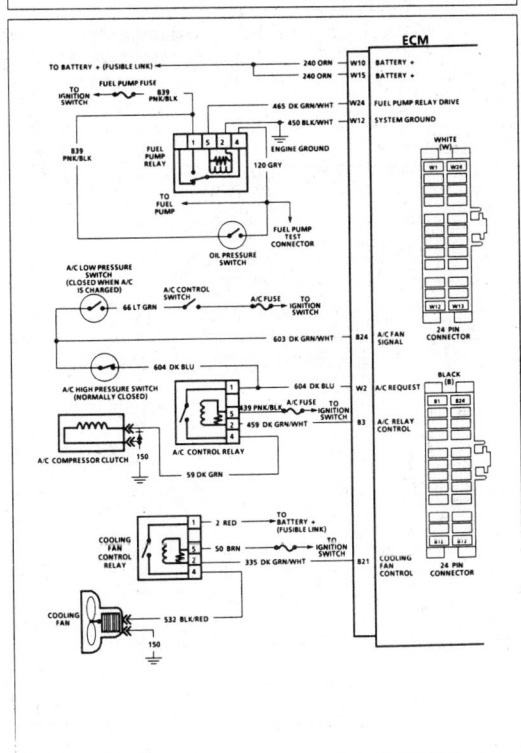

2.5L (VIN U) ECM CONNECTOR TERMINAL END VIEW

FUEL INJECTION ECM CONNECTOR IDENTIFICATION

This ECM voltage chart is for use with a digital voltmeter to further aid in diagnosis. The voltages you get may vary due to low battery charge or other reasons, but they should be very close.

THE FOLLOWING CONDITIONS MUST BE MET BEFORE TESTING:
- Engine at operating temperature • Engine idling in "Closed Loop" (for "Engine Run" column)
- Test terminal not grounded • "Scan" tool not installed • All voltages shown "B +" indicates battery or charging voltage.

VOLTAGE

KEY "ON"	ENG. RUN	CIRCUIT	PIN	WIRE COLOR
① 0*	B+	INJECTOR DRIVE	W1	DK BLU
0*	0*	A/C REQUEST	W2	DK BLU
			W3	
			W4	GRY
B+	B+	POWER STEERING SIGNAL	W5	LT BLU/ORN
0	1.1	EST	W6	WHT
④	④	IAC "A" HI	W7	LT BLU/WHT
④	④	IAC "B" HI	W9	DK GRN/WHT
B+	B+	12V BATTERY	W10	ORN
5.0	5.0	5 VOLT SENSOR REFERENCE	W11	GRY
0*	0*	ECM GROUND	W12	BLK/WHT

WIRE COLOR	PIN	CIRCUIT	KEY "ON"	ENG. RUN
DK GRN/WHT	W24	FUEL PUMP	②	B+
	W23	NOT USED		
DK GRN	W22	VSS OUTPUT 4000 PPM	0*	0*
DK BLU	W21	CRUISE		
GRY	W20	CRUISE		
TAN/ BLK	W19	IGNITION BYPASS	0*	4.5
ORN/ WHT	W18	P/N SWITCH	0*	
LT BLU/ BLK	W17	IAC "A" LOW	④	④
PNK/ BLK	W16	12V IGNITION	B+	B+
LT BLU	W15	12V BATTERY	B+	B+
BLK/YEL	W14	MAP,MAT GROUND	0*	0*
TAN/ WHT	W13	ECM GROUND	0*	0*

KEY "ON"	ENG. RUN	CIRCUIT	PIN	WIRE COLOR
4.5	4.5	SERIAL DATA	B1	ORN
		OXYGEN SENSOR SIGNAL	B2	PPL
33-55	1-.9	A/C CLUTCH RELAY	B3	DK GRN/BLK
		NOT USED	B4	
① B+	B+	A/C RELAY CONTROL	B3	
⑤ 1.3	1.3	MAT	B5	TAN
		MAGNETIC VSS (IF USED)	B6	TAN/BLK
0*	0*	TCC OR SHIFTLITE	B7	TAN/BLK
B+	B+	COOLANT	B8	PPL
③ 1.9	1.9	IGNITION REFERENCE HI	B9	PPL/WHT
4.6	3.05		B10	LT GRN
		CRUISE	B11	DK BLU/WHT
		CRUISE	B11	WHT
0*	0*	COOLANT AND TPS GROUND	B12	

WIRE COLOR	PIN	CIRCUIT	KEY "ON"	ENG. RUN
DK GRN/ WHT	B24	A/C SIGNAL FOR FAN	0*	①
DK GRN/ WHT	B23	OXYGEN SENSOR GROUND	0*	0*
BRN/ WHT	B22	SERVICE ENGINE SOON" LIGHT	0*	B+
DK GRN/ WHT	B21	ENGINE COOLING FAN	0*	①
LT GRN	B20	MAP SIGNAL	4.75	1.1
DK BLU	B19	TPS SIGNAL	.6	.6
WHT	B18			
	B17	NOT USED		
BLK/ RED	B16	ENGINE GROUND	0*	0*
WHT/ BLK	B15	ALDL DIAGNOSTIC TEST TERMINAL	5.0	5.0
DK BRN	B14	CRUISE		
YEL	B13	VSS	0*	0*

* "All voltages shown "0" should read less than .5 volt."
① A/C, Fan "OFF"
② Reads battery voltage for 2 seconds after ignition "ON" then should read 0 volts
④ Varies depending on temperature
⑤ Not useable

ENGINE 2.5L

1988–90 2.5L (VIN U) – ALL MODELS

DIAGNOSTIC CIRCUIT CHECK

The Diagnostic Circuit Check is an organized approach to identifying a problem created by an electronic engine control system malfunction. It must be the starting point for any driveability complaint diagnosis because it directs the service technician to the next logical step in diagnosing the complaint.

The "Scan" data listed in the table may be used for comparison after completing the diagnostic circuit check and finding the on-board diagnostics functioning properly and no trouble codes displayed. The "Typical Data Values" are an average of display values recorded from normally operating vehicles and are intended to represent what a normally functioning system would typically display.

A "SCAN" TOOL THAT DISPLAYS FAULTY DATA SHOULD NOT BE USED, AND THE PROBLEM SHOULD BE REPORTED TO THE MANUFACTURER. THE USE OF A FAULTY "SCAN" TOOL CAN RESULT IN MISDIAGNOSIS AND UNNECESSARY PARTS REPLACEMENT.

Only the parameters listed below are used in this manual for diagnosis. If a "Scan" tool reads other parameters, the values are not recommended by General Motors for use in diagnosis. For more description on the values and use of the "Scan" tool to diagnosis ECM inputs, refer to the applicable component diagnosis section.

If all values are within the range illustrated, refer to symptoms in Section "B".

"SCAN" TOOL DATA

Test Under Following Conditions: Idle, Upper Radiator Hose Hot, Closed Throttle, Park or Neutral, "Closed Loop", All Accessories "OFF".

"SCAN" Position	Units Displayed	Typical Data Value
Desired RPM	RPM	ECM idle command (varies with temperature)
RPM	RPM	± 50 RPM from desired rpm in drive (AUTO) ± 100 RPM from desired rpm in neutral (MANUAL)
Coolant Temperature	Degrees Celsius	85 - 105
MAT Temperature	Degrees Celsius	10 - 90 (varies with underhood temperature and sensor location)
MAP	Volts	1 - 2 (varies with manifold and barometric pressures)
BPW (base pulse width)	Milliseconds	.8 - 3.0
O₂	Volts	.1 - 1 (varies continuously)
TPS	Volts	.4 - 1.25
Throttle Angle	0 - 100%	0
IAC	Counts (steps)	1 - 50
P/N Switch	P-N and R-D-L	Park/Neutral (P/N)
INT (Integrator)	Counts	110 - 145
BLM (Block Learn Memory)	Counts	118 - 138
Open/Closed Loop	Open/Closed	"Closed Loop" (may enter "Open Loop" with extended idle)
VSS	MPH	0
TCC	ON/OFF	"OFF"
Spark Advance	Degrees	Varies
Battery	Volts	13.5 - 14.5
Fan	ON/OFF	"OFF" (coolant temperature below 102°C)
P/S Switch	Normal/Hi Pressure	Normal
A/C Request	Yes/No	No
A/C Clutch	ON/OFF	"OFF"
Shift Light (M/T)	ON/OFF	"OFF"

1988–90 2.5L (VIN U) – ALL MODELS

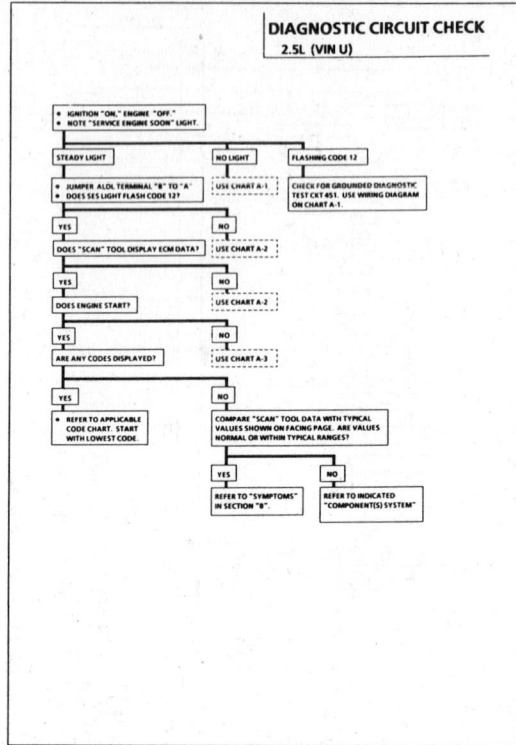

DIAGNOSTIC CIRCUIT CHECK
2.5L (VIN U)

1988–90 2.5L (VIN U) – ALL MODELS

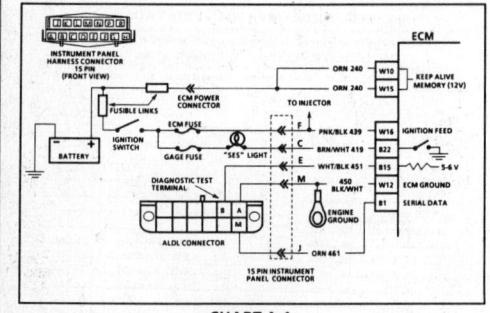

CHART A-1
NO "SERVICE ENGINE SOON" LIGHT
2.5L (VIN U)

Circuit Description:

There should always be a steady "Service Engine Soon" light when the ignition is "ON" and engine stopped. Battery voltage is supplied directly to the light bulb. The electronic control module (ECM) controls the light and turns it "ON" by providing a ground path through CKT 419 to the ECM.

Test Description: Numbers below refer to circled numbers on the diagnostic chart.
1. This check determines if there is a fault in the "SES" light circuit or in the ECM.
2. If CKTs 240 and 439 have voltage, the ECM grounds or the ECM is faulty.
3. Using a test light connected to battery voltage, probe each of the system ground circuits to be sure a good ground is present. See ECM terminal end view in front of this section for the ECM terminal locations of ground circuits.

Diagnostic Aids:

If engine runs OK, check the following:
- Faulty light bulb
- CKT 419 open
- Gage fuse blown. This will result in no oil or generator lights, seat belt reminder, etc.

If engine cranks but will not run, use CHART A-3.

1988–90 2.5L (VIN U) – ALL MODELS

CHART A-1
NO "SERVICE ENGINE SOON" LIGHT
2.5L (VIN U)

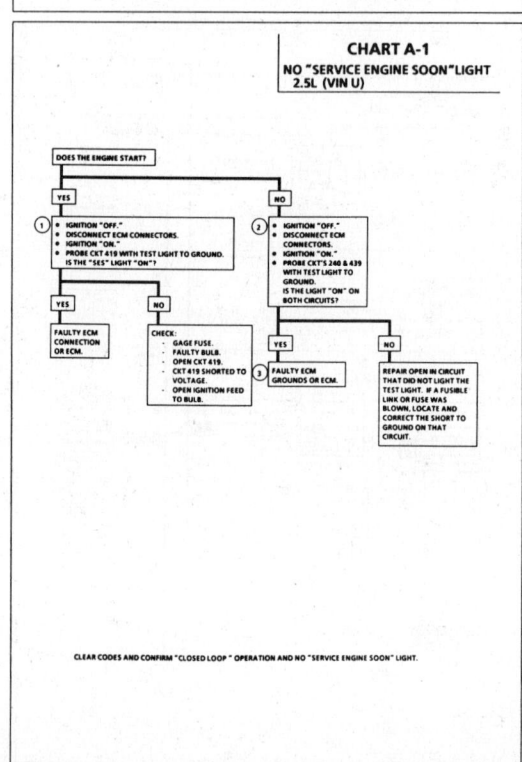

CLEAR CODES AND CONFIRM "CLOSED LOOP" OPERATION AND NO "SERVICE ENGINE SOON" LIGHT.

1988–90 2.5L (VIN U) – ALL MODELS

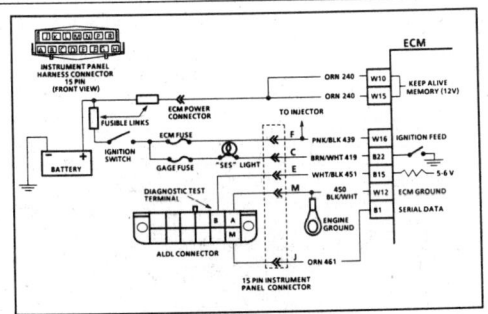

CHART A-2
NO ALDL DATA OR WON'T FLASH CODE 12
"SERVICE ENGINE SOON" LIGHT "ON" STEADY
2.5L (VIN U)

Circuit Description:
There should always be a steady "Service Engine Soon" light when the ignition is "ON" and the engine is "OFF." Battery voltage is supplied directly to the light bulb. The electronic control module (ECM) controls the light and will turn it "ON" by grounding CKT 419 at the ECM.
With the diagnostic terminal grounded, the light should flash a Code 12 followed by any trouble code(s) stored in memory. A steady light suggests a short to ground in the light control CKT 419 or an open in diagnostic CKT 451.

Test Description: Numbers below refer to circled numbers on the diagnostic chart.
1. If there is a problem with the ECM that causes a "Scan" tool to not read serial data, then the ECM should not flash a Code 12. If Code 12 does flash, be sure that the "Scan" tool is working properly on another vehicle. If the "Scan" tool is functioning properly and CKT 461 is OK, the PROM or ECM may be at fault for the No ALDL symptom.
2. If the light turns "OFF" when the ECM connector is disconnected, then CKT 419 is not shorted to ground.

3. This step will check for an open diagnostic CKT 451.
4. At this point, the "Service Engine Soon" light wiring is OK. The problem is a faulty ECM or PROM. If Code 12 does not flash, the ECM should be replaced using the original PROM. Replace the PROM only after trying an ECM, as a defective PROM is an unlikely cause of the problem.

1988–90 2.5L (VIN U) – ALL MODELS

CHART A-2
NO ALDL DATA OR WON'T FLASH CODE 12
"SERVICE ENGINE SOON" LIGHT "ON" STEADY
2.5L (VIN U)

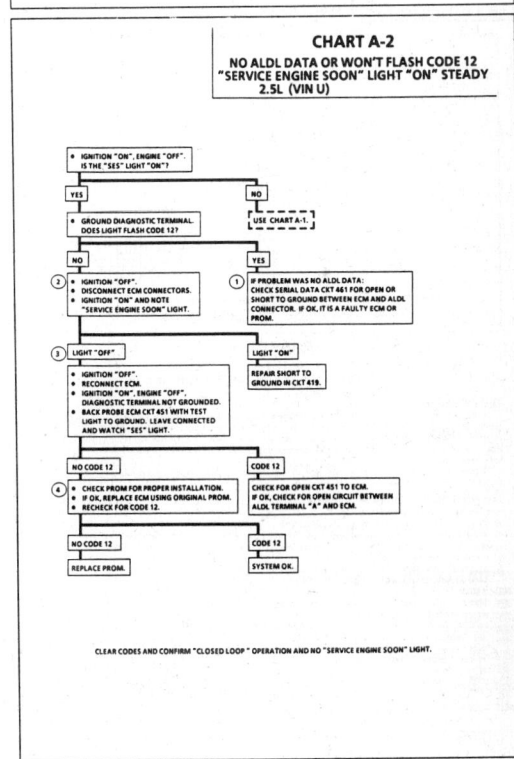

1988–90 2.5L (VIN U) – ALL MODELS

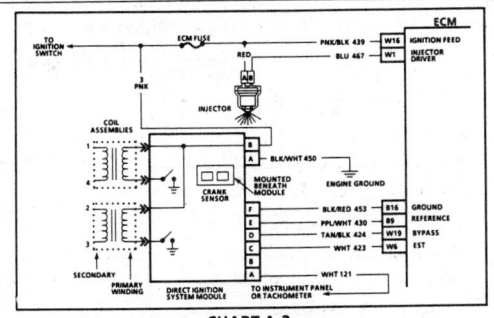

CHART A-3
(Page 1 of 3)
ENGINE CRANKS BUT WON'T RUN
2.5L (VIN U)

Circuit Description:
Before using this chart, battery condition, engine cranking speed, and fuel quantity should be checked and verified as being OK.

Test Description: Numbers below refer to circled numbers on the diagnostic chart.
1. A "Service Engine Soon" light "ON" is a basic test to determine if there is battery and ignition voltage at the ECM. No ALDL data may be due to an ECM problem, and CHART A-2 will diagnose the ECM. If TPS is over 2.5 volts, the engine may be in the clear flood mode, which will cause starting problems. The engine will not start without crank sensor reference pulses. The "Scan" tool should display rpm during cranking if pulses are received at the ECM.
2. Because the direct ignition system uses two plugs and wires to complete the circuit of each coil, the opposite spark plug wire must be left connected. If rpm was indicated during crank, the ignition module is receiving a crank signal, but "No Spark" at this test indicates the ignition module is not triggering the coil.
3. While cranking the engine, there should be no fuel spray with the injector electrical connector disconnected. Replace the injector if it sprays fuel or drips.

4. The test light should flash, indicating that the ECM is controlling the injector. How bright the light flashes is not important. However, the test light should be a BT 8329 or equivalent.
5. Fuel spray from the injector indicates that fuel is available. However, the engine could be severely flooded due to too much fuel. No fuel spray from injector indicates a faulty fuel system or no ECM control of injector.

Diagnostic Aids:
• Water or foreign material can cause a no start condition during freezing weather. The engine may start after 5 or 6 minutes in a heated shop. The problem may not re-occur until after an overnight park in freezing temperatures.
• An EGR valve sticking open can cause a low air/fuel ratio during cranking. Unless engine enters "Clear Flood" at the first indication of a flooding condition, it can result in a no start.
• Fuel pressure: Low fuel pressure can result in a very lean air/fuel ratio. Use CHART A-7.

1988–90 2.5L (VIN U) – ALL MODELS

CHART A-3
(Page 1 of 3)
ENGINE CRANKS BUT WON'T RUN
2.5L (VIN U)

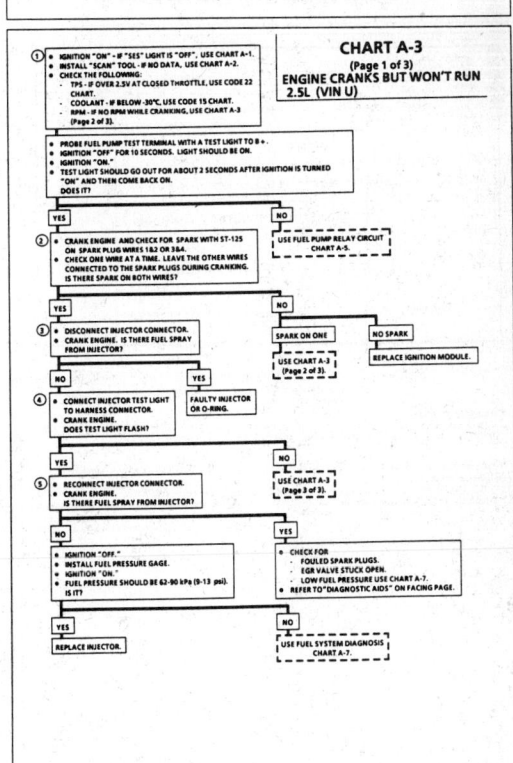

1988–90 2.5L (VIN U) – ALL MODELS

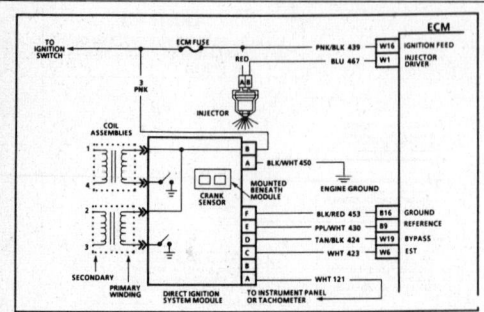

CHART A-3
(Page 2 of 3)
ENGINE CRANKS BUT WON'T RUN
2.5L (VIN U)

Circuit Description:

A magnetic crank sensor is used to determine engine crankshaft position, much the same way as the pick-up coil did in HEI type systems. The sensor is mounted in the block, near a slotted wheel on the crankshaft. The rotation of the wheel creates a flux change in the sensor, which produces a voltage signal. The DIS ignition module processes this signal and creates the reference pulses needed by the ECM to trigger the correct coil at the correct time.

If the "Scan" tool did not indicate cranking rpm, and there is no spark present at the plugs, the problem lies in the direct ignition system or the power and ground supplies to the module.

Test Description: Numbers below refer to circled numbers on the diagnostic chart.

1. The direct ignition system uses two plugs and wires to complete the circuit of each coil. The other spark plug wire in the circuit must be left connected to create a spark. The DIS ignition module.
2. This test will determine if a 12 volt supply and a good ground is available at the DIS ignition module.
3. This test will determine if the ignition module is not generating the reference pulse, or if the wiring or ECM are at fault. By touching and removing a test light to battery voltage on CKT 430, a reference pulse should be generated. If rpm is indicated, the ECM and wiring are OK.
4. This test will determine if the ignition module is not triggering the problem coil, or if the tested coil is at fault. This test could also be performed by substituting a known good coil. The secondary coil winding can be checked with a DVM. There should be 5,000 to 10,000 ohms across the coil towers. There should not be any continuity from either coil tower to ground.
5. Checks for continuity of the crank sensor and connections. Also checks sensor magnetism.

Diagnostic Aids:

An intermittent problem with crankshaft sensor could cause a "Cranks But Won't Run" condition. Therefore, the crankshaft sensor must be replaced first. If the "Cranks But Won't Run" condition persists, then replace the DIS module.

1988–90 2.5L (VIN U) – ALL MODELS

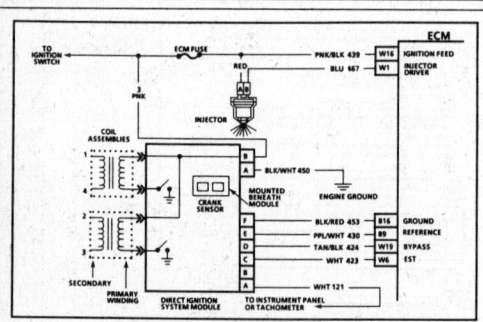

CHART A-3
(Page 3 of 3)
ENGINE CRANKS BUT WON'T RUN
2.5L (VIN U)

Circuit Description:

Ignition voltage is supplied to the fuel injector on CKT 439. The injector will be pulsed (turned "ON" and "OFF") when the ECM opens and grounds injector drive CKT 467.

Test Description: Numbers below refer to circled numbers on the diagnostic chart.

1. This check determines if injector connector has ignition voltage, and on only one terminal.
2. A faulty ECM may result in damage to the injector.
3. A test light connected from ECM harness terminal "W1" to ground should light due to continuity through the injector.

1988–90 2.5L (VIN U) – ALL MODELS

CHART A-3
(Page 2 of 3)
ENGINE CRANKS BUT WON'T RUN
2.5L (VIN U)

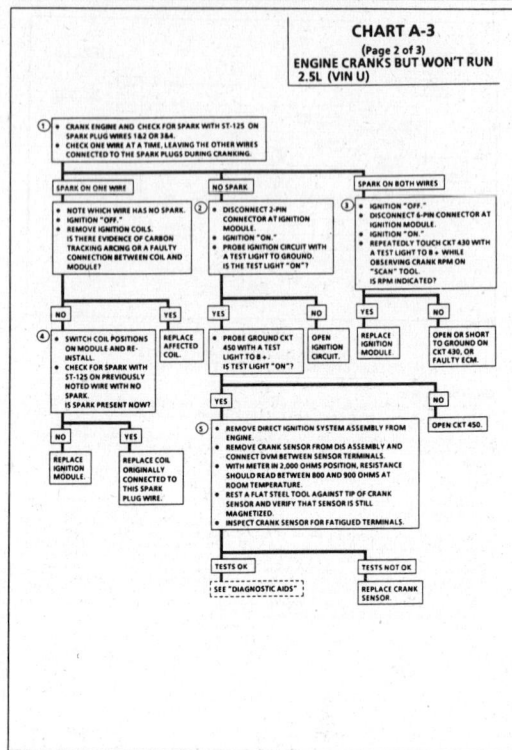

1988–90 2.5L (VIN U) – ALL MODELS

CHART A-3
(Page 3 of 3)
ENGINE CRANKS BUT WON'T RUN
2.5L (VIN U)

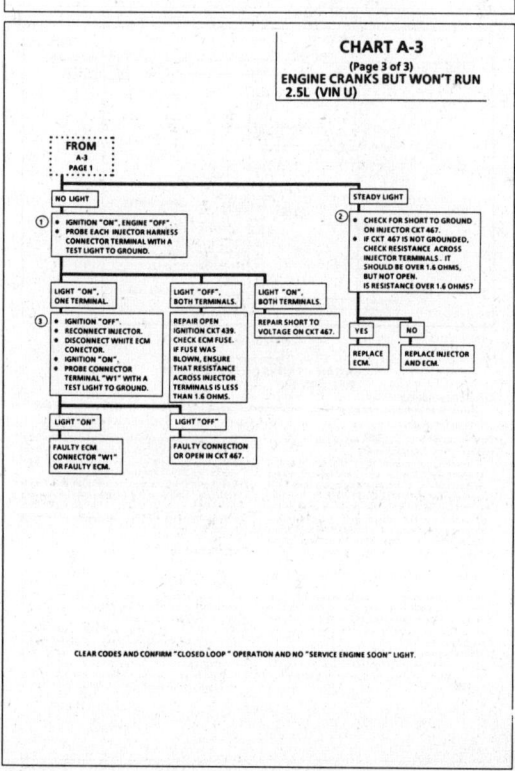

CLEAR CODES AND CONFIRM "CLOSED LOOP" OPERATION AND NO "SERVICE ENGINE SOON" LIGHT.

1988–90 2.5L (VIN U) – ALL MODELS

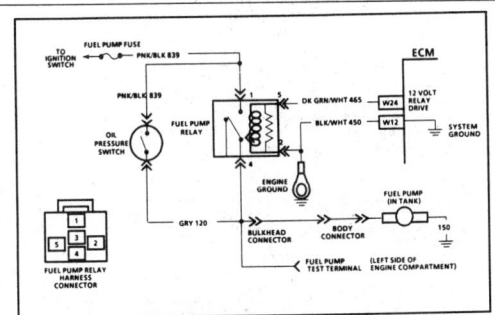

CHART A-5
FUEL PUMP RELAY CIRCUIT
2.5L (VIN U)

Circuit Description:

When the ignition switch is turned "ON," the engine control module (ECM) will activate the fuel pump relay with a 12 volt signal and run the in-tank fuel pump. The fuel pump will operate as long as the engine is cranking or running and the ECM is receiving ignition reference pulses. If there are no ignition reference pulses, the ECM will no longer supply the fuel pump relay signal within 2 seconds after the ignition is turned "ON."

Should the fuel pump relay or the 12 volt relay drive from the ECM fail, the fuel pump will receive electrical current through the oil pressure switch back-up circuit.

The fuel pump test terminal is located in the left side of the engine compartment. When the engine is stopped, the pump can be turned "ON" by applying battery voltage to the test terminal.

1. This check is to determine if the oil pressure switch is faulty also. Since it is a back-up component, it must be checked without the fuel pump relay connected.
2. This check will determine if the oil pressure switch is shorted internally.

Diagnostic Aids:

An inoperative fuel pump relay can result in long cranking times. The extended crank period is caused by the time necessary for oil pressure to reach the pressure required to close the oil pressure switch and turn "ON" the fuel pump.

1988–90 2.5L (VIN U) – ALL MODELS

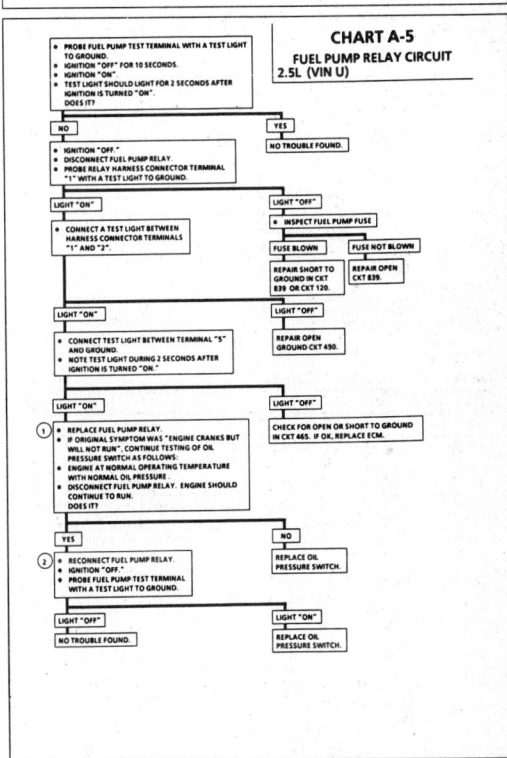

CHART A-5
FUEL PUMP RELAY CIRCUIT
2.5L (VIN U)

1988–90 2.5L (VIN U) – ALL MODELS

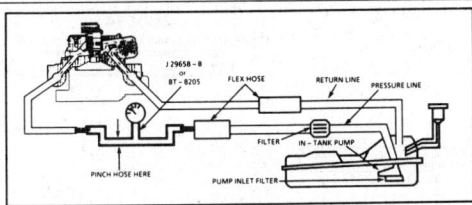

CHART A-7
FUEL SYSTEM DIAGNOSIS
2.5L (VIN U)

Circuit Description:

The fuel pump delivers fuel to the TBI unit where the system pressure is controlled to 62 to 90 kPa (9 to 13 psi). Excess fuel is returned to the fuel tank.

When the ignition switch is turned "ON," the electronic control module (ECM) will activate the fuel pump relay with a 12 volt signal and run the in-tank fuel pump. The fuel pump will operate as long as the engine is cranking or running and the ECM is receiving ignition reference pulses. If there are no ignition reference pulses, the ECM will no longer supply the fuel pump relay signal within 2 seconds after key "ON."

Should the fuel pump relay or the 12 volt relay drive from the ECM fail, the fuel pump will receive electrical current through the oil pressure switch back-up circuit.

Test Description: Numbers below refer to circled numbers on the diagnostic chart.
1. If fuse in jumper wire blows, check CKT 120, between relay and fuel pump, for a short to ground.
2. Fuel flow at less than 62 kPa (9 psi) can cause the following
   - System will run lean and may set Code 44. Also, hard starting and poor overall performance will result.
   - Engine surging and possible stalling during driving with satisfactory idle quality. This would be caused by a restriction in the fuel flow. As the fuel flow increases, the pressure drop across the restriction becomes great enough to starve the engine of fuel. At idle, the pressure may still be adequate. Normally, an engine with a fuel pressure less than 62 kPa (9 psi) at idle will not be driveable.
3. Gradually restricting the fuel return line allows the fuel pump to develop its maximum pressure (dead head pressure). When battery voltage is applied to the pump "test" terminal, pressure should be between 90 and 124 kPa (13 to 18 psi).
4. This test determines if the high fuel pressure is due to a restricted fuel return line or a throttle body pressure regulator problem.

Diagnostic Aids:

Improper fuel system pressure can result in one of the following symptoms:
- Cranks, but won't run.
- Code 44.
- Code 45.
- Cuts out, may feel like ignition problem.
- Poor fuel economy, loss of power.
- Hesitation.

1988–90 2.5L (VIN U) – ALL MODELS

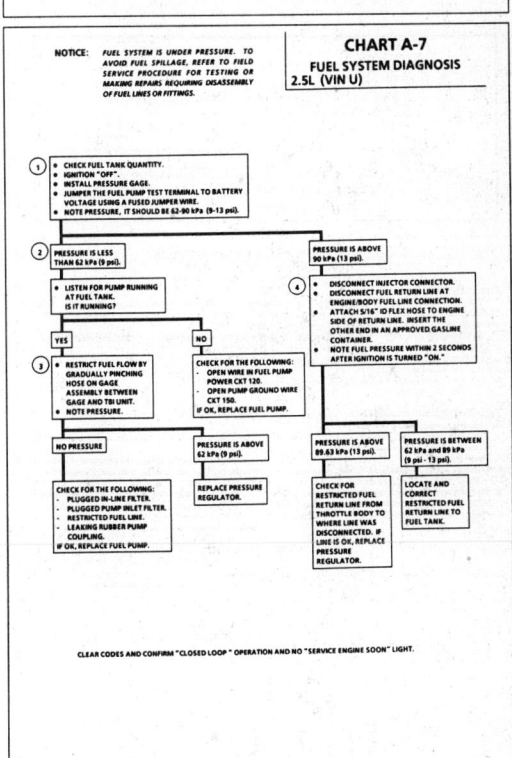

CHART A-7
FUEL SYSTEM DIAGNOSIS
2.5L (VIN U)

1988–90 2.5L (VIN U) – ALL MODELS

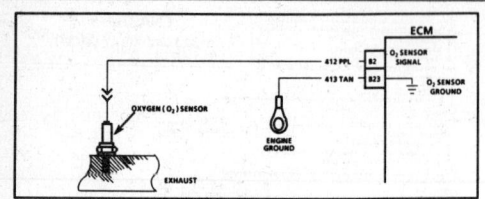

CODE 13
OXYGEN SENSOR CIRCUIT
(OPEN CIRCUIT)
2.5L (VIN U)

Circuit Description:
The ECM supplies a voltage of about .45 volt between terminals "B2" and "B23". (If measured with a 10 megohm digital voltmeter, this may read as low as .32 volt.)

When the O₂ sensor reaches operating temperature, it varies this voltage from about .1 volt (exhaust is lean) to about .9 volt (exhaust is rich).

The sensor is like an open circuit and produces no voltage when it is below 360°C (600°F). An open sensor circuit or cold sensor causes "Open Loop" operation.

Test Description: Numbers below refer to circled numbers on the diagnostic chart.
1. Code 13 will set under the following conditions:
   - Engine at normal operating temperature.
   - At least 2 minutes have elapsed since engine start-up.
   - O₂ signal voltage is steady between .35 and .55 volt.
   - Throttle angle is above 7%.
   - All above conditions are met for about 60 seconds.
   If the conditions for a Code 13 exist, the system will not operate in "Closed Loop."
2. This test determines if the O₂ sensor is the problem or if the ECM and wiring are at fault.

3. In doing this test, use only a 10 megohm digital voltmeter. This test checks the continuity of CKTs 412 and 413. If CKT 413 is open, the ECM voltage on CKT 412 will be over .6 volt (600 mV).

Diagnostic Aids:

Normal "Scan" tool O₂ sensor voltage varies between 100mV to 999 mV (.1 and 1.0 volt) while in "Closed Loop." Code 13 sets in one minute if sensor signal voltage remains between .35 and .55 volt, but the system will go to "Open Loop" in about 15 seconds.

Verify a clean, tight ground connection for CKT 413. Open CKT(s) 412 or 413 will result in a Code 13. If Code 13 is intermittent, refer to Section "B".

1988–90 2.5L (VIN U) – ALL MODELS

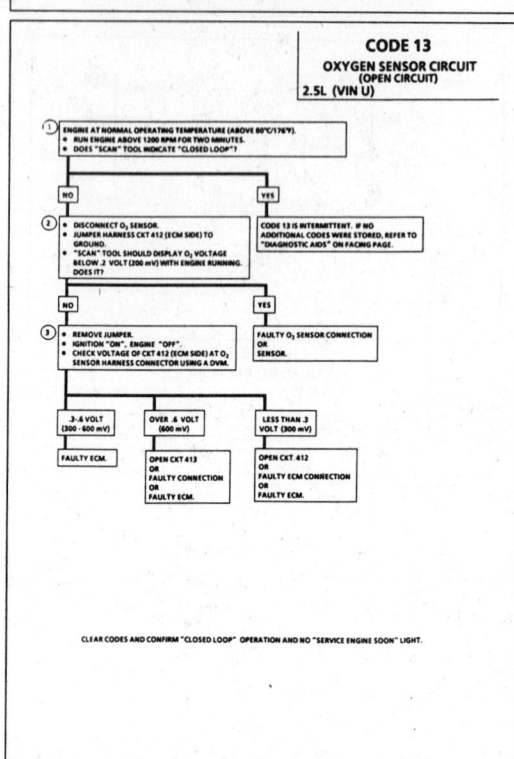

1988–90 2.5L (VIN U) – ALL MODELS

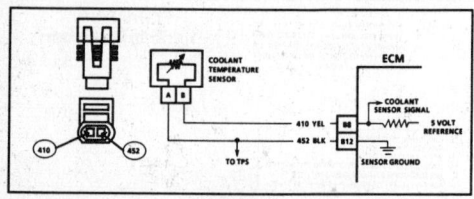

CODE 14
COOLANT TEMPERATURE SENSOR CIRCUIT
(HIGH TEMPERATURE INDICATED)
2.5L (VIN U)

Circuit Description:
The Coolant Temperature Sensor (CTS) uses a thermistor to control the signal voltage to the ECM. The ECM applies a reference voltage on CKT 410 to the sensor. When the engine is cold, the sensor (thermistor) resistance is high. The ECM will then sense a high signal voltage.

As the engine warms up, the sensor resistance decreases and the voltage drops. At normal engine operating temperature, the voltage will measure about 1.5 to 2.0 volts at ECM terminal "B8".

Coolant temperature is one of the inputs used to control the following:
- Fuel Delivery.
- Electronic Spark Timing (EST).
- Cooling Fan.
- Torque Converter Clutch (TCC).
- Idle Air Control (IAC).

Test Description: Numbers below refer to circled numbers on the diagnostic chart.
1. Checks to see if code was set as result of hard failure or intermittent condition.
   Code 14 will set if:
   - Signal voltage indicates a coolant temperature above 135°C (275°F) for 3 seconds.
2. This test simulates conditions for a Code 15. If the ECM recognizes the open circuit (high voltage) and displays a low temperature, the ECM and wiring are OK.

Diagnostic Aids:

A "Scan" tool reads engine temperature in degrees Celsius.

After the engine is started, the temperature should rise steadily to about 90°C, then stabilise when the thermostat opens.

If the engine has been allowed to cool to an ambient temperature (overnight), coolant temperature and manifold air temperature (MAT) may be checked with a "Scan" tool and should read close to each other.

When a Code 14 is set, the ECM will turn "ON" the engine cooling fan.

A Code 14 will result if CKT 410 is shorted to ground.

If Code 14 is intermittent, refer to Section "B".

1988–90 2.5L (VIN U) – ALL MODELS

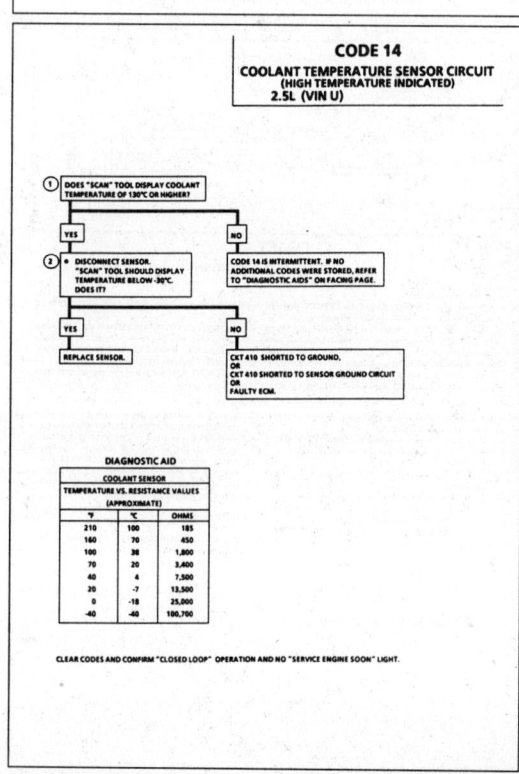

1988–90 2.5L (VIN U) – ALL MODELS

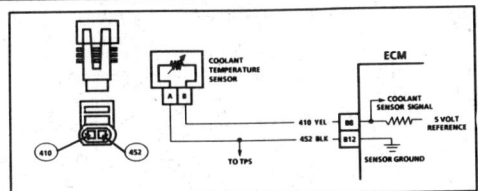

CODE 15
COOLANT TEMPERATURE SENSOR CIRCUIT
(LOW TEMPERATURE INDICATED)
2.5L (VIN U)

Circuit Description:
The Coolant Temperature Sensor (CTS) uses a thermistor to control the signal voltage to the ECM. The ECM applies a reference voltage on CKT 410 to the sensor. When the engine is cold, the sensor (thermistor) resistance is high. The ECM will then sense a high signal voltage.

As the engine warms up, the sensor resistance decreases and the voltage drops. At normal engine operating temperature, the voltage will measure about 1.5 to 2.0 volts at ECM terminal "B8".

Coolant temperature is one of the inputs used to control the following:
- Fuel Delivery.
- Electronic Spark Timing (EST).
- Cooling Fan.
- Torque Convertor Clutch (TCC).
- Idle Air Control (IAC).

Test Description: Numbers below refer to circled numbers on the diagnostic chart.
1. Checks to see if code was set as result of hard failure or intermittent condition.
   Code 15 will set if:
   - Signal voltage indicates a coolant temperature below -30°C (-22°F) for 60 seconds.
2. This test simulates conditions for a Code 14. If the ECM recognizes the grounded circuit (low voltage) and displays a high temperature, the ECM and wiring are OK.
3. This test will determine if there is a wiring problem or a faulty ECM. If CKT 452 is open, there may also be a Code 21 stored.

Diagnostic Aids:
A "Scan" tool reads engine temperature in degrees Celsius. After the engine is started, the temperature should rise steadily to about 90°C, then stabilize when the thermostat opens.

If the engine has been allowed to cool to an ambient temperature (overnight), coolant temperature and manifold air temperature (MAT) may be checked with a "Scan" tool and should read close to each other.

When a Code 15 is set, the ECM will turn "ON" the engine cooling fan.

A Code 15 will result if CKTs 410 or 452 are open. If Code 15 is intermittent, refer to Section "B".

1988–90 2.5L (VIN U) – ALL MODELS

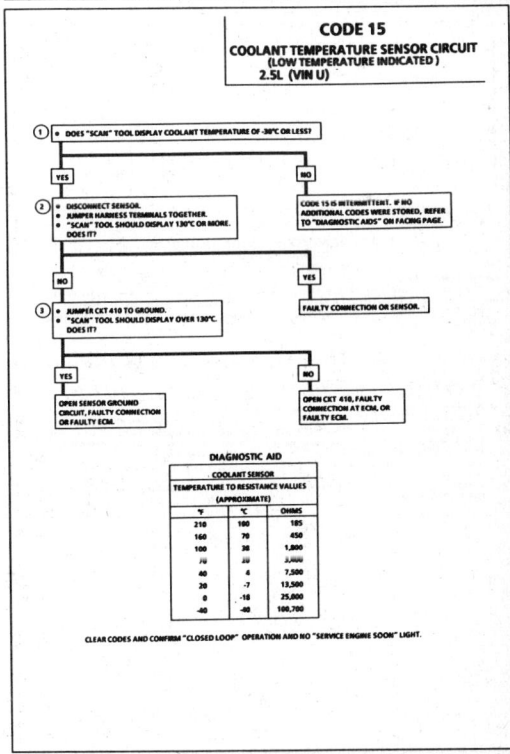

CODE 15
COOLANT TEMPERATURE SENSOR CIRCUIT
(LOW TEMPERATURE INDICATED)
2.5L (VIN U)

COOLANT SENSOR		
TEMPERATURE TO RESISTANCE VALUES (APPROXIMATE)		
°F	°C	OHMS
210	100	185
160	70	450
100	38	1,800
70	20	3,400
40	4	7,500
20	-7	13,500
0	-18	25,000
-40	-40	100,700

DIAGNOSTIC AID

CLEAR CODES AND CONFIRM "CLOSED LOOP" OPERATION AND NO "SERVICE ENGINE SOON" LIGHT.

1988–90 2.5L (VIN U) – ALL MODELS

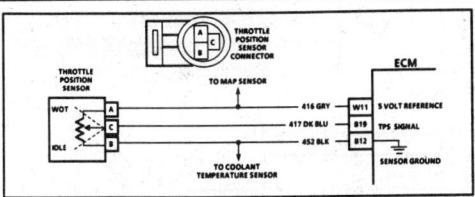

CODE 21
THROTTLE POSITION SENSOR (TPS) CIRCUIT
(SIGNAL VOLTAGE HIGH)
2.5L (VIN U)

Circuit Description:
The Throttle Position Sensor (TPS) provides a voltage signal that changes relative to the throttle valve. Signal voltage will vary from less than 1.25 volts at idle to about 5 volts at wide open throttle (WOT). The TPS signal is one of the most important inputs used by the ECM for fuel control and for many of the ECM controlled outputs.

Test Description: Numbers below refer to circled numbers on the diagnostic chart.
1. This step checks to see if Code 21 is the result of a hard failure or an intermittent condition.
   A Code 21 will set if:
   - TPS reading above 2.5 volts
   - Engine speed less than 1800 rpm
   - MAP reading below 60 kPa
   - All of the above conditions present for 2 seconds
2. This step simulates conditions for a Code 22. If the ECM recognizes the change of state, the ECM and CKTs 416 and 417 are OK.

3. This step isolates a faulty sensor ECM or an open CKT 452. If CKT 452 is open, there may also be a Code 15 stored.

Diagnostic Aids:
A "Scan" tool displays throttle position in volts. Closed throttle voltage should be less than 1.25 volts. TPS voltage should increase at a steady rate as throttle is moved to WOT.

A Code 21 will result if CKT 452 is open or CKT 417 is shorted to voltage. If Code 21 is intermittent, refer to Section "B".

1988–90 2.5L (VIN U) – ALL MODELS

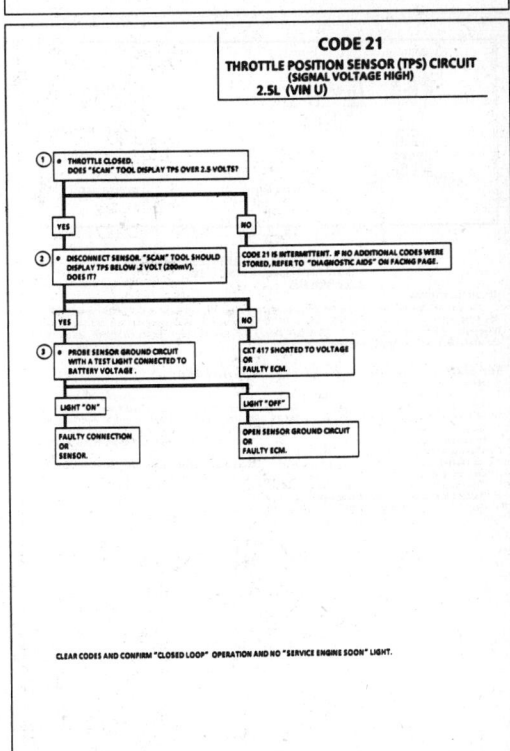

CODE 21
THROTTLE POSITION SENSOR (TPS) CIRCUIT
(SIGNAL VOLTAGE HIGH)
2.5L (VIN U)

CLEAR CODES AND CONFIRM "CLOSED LOOP" OPERATION AND NO "SERVICE ENGINE SOON" LIGHT.

1988–90 2.5L (VIN U) – ALL MODELS

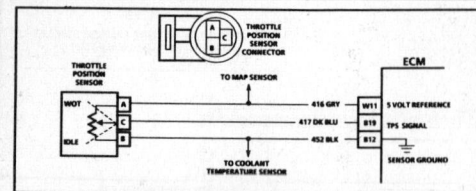

CODE 22
THROTTLE POSITION SENSOR (TPS) CIRCUIT
(SIGNAL VOLTAGE LOW)
2.5L (VIN U)

Circuit Description:
The Throttle Position Sensor (TPS) provides a voltage signal that changes, relative to the throttle valve. Signal voltage will vary from less than 1.25 volts at idle to about 4.5 volts at wide open throttle (WOT).
The TPS signal is one of the most important inputs used by the ECM for fuel control and for many of the ECM controlled outputs.

Test Description: Numbers below refer to circled numbers on the diagnostic chart.
1. This step checks to see if Code 22 is the result of a hard failure or an intermittent condition.
A Code 22 will set if the engine is running and TPS voltage is below .2 volt (200 mV).
2. This step simulates conditions for a Code 21. If a Code 21 is set or the "Scan" tool display over 4 volts, the ECM and wiring are OK.
3. The "Scan" tool may not display 12 volts. The important thing is that the ECM recognizes the voltage as over 4 volts, indicating that CKT 417 and the ECM are OK.

4. If CKT 416 is shorted to ground, there may also be a stored Code 34.

Diagnostic Aids:
A "Scan" tool displays throttle position in volts. Closed throttle voltage should be less than 1.25 volts. TPS voltage should increase at a steady rate as throttle is moved to WOT.
An open or grounded 416 or 417 will result in a Code 22.
If Code 22 is intermittent, refer to Section "B".

1988–90 2.5L (VIN U) – ALL MODELS

CODE 22
THROTTLE POSITION SENSOR (TPS) CIRCUIT
(SIGNAL VOLTAGE LOW)
2.5L (VIN U)

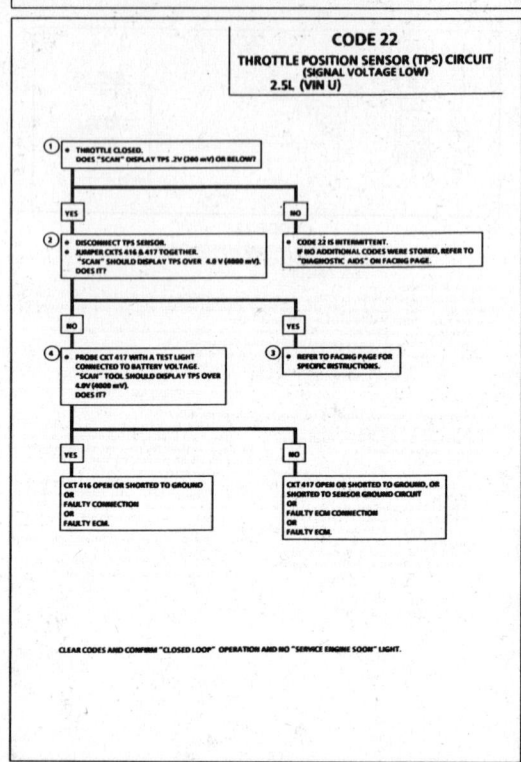

CLEAR CODES AND CONFIRM "CLOSED LOOP" OPERATION AND NO "SERVICE ENGINE SOON" LIGHT.

1988–90 2.5L (VIN U) – ALL MODELS

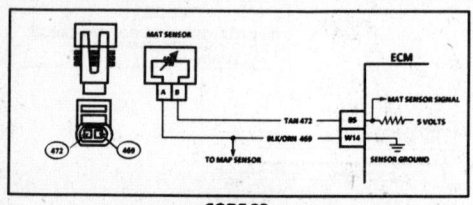

CODE 23
MANIFOLD AIR TEMPERATURE (MAT) SENSOR CIRCUIT
(LOW TEMPERATURE INDICATED)
2.5L (VIN U)

Circuit Description:
The Manifold Air Temperature (MAT) sensor uses a thermistor to control the signal voltage to the ECM. The ECM applies a reference voltage (4-6 volts) on CKT 472 to the sensor. When manifold air is cold, the sensor (thermistor) resistance is high. The ECM will then sense a high signal voltage. As the air warms, the sensor resistance becomes less and the voltage drops.

Test Description: Numbers below refer to circled numbers on the diagnostic chart.
1. This step determines if Code 23 is the result of a hard failure or an intermittent condition. Code 23 will set when signal voltage indicates a MAT temperature less than -30°C and the engine is running for longer than 58 seconds.
2. This test simulates conditions for a Code 25. If the "Scan" tool displays a high temperature, the ECM and wiring are OK.
3. This step checks continuity of CKTs 472 and 469. If CKT 469 is open there may also be a Code 33.

Diagnostic Aids:
If the engine has been allowed to cool to an ambient temperature (overnight), coolant and MAT temperatures may be checked with a "Scan" tool and should read close to each other.
A Code 23 will result if CKT 472 or 469 becomes open.
If Code 23 is intermittent, refer to Section "B".

1988–90 2.5L (VIN U) – ALL MODELS

CODE 23
MANIFOLD AIR TEMPERATURE (MAT) SENSOR CIRCUIT
(LOW TEMPERATURE INDICATED)
2.5L (VIN U)

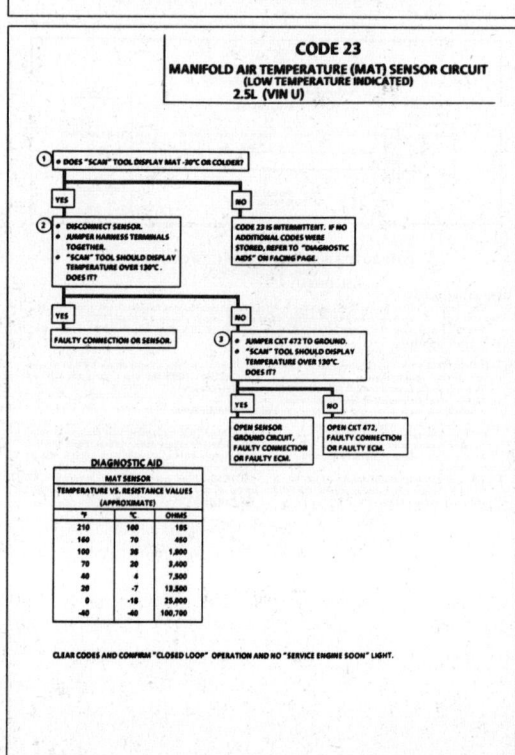

CLEAR CODES AND CONFIRM "CLOSED LOOP" OPERATION AND NO "SERVICE ENGINE SOON" LIGHT.

1988–90 2.5L (VIN U) – ALL MODELS

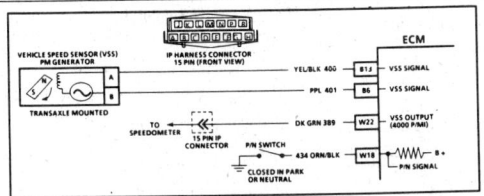

CODE 24
VEHICLE SPEED SENSOR (VSS) CIRCUIT
2.5L (VIN U)

Circuit Description:

Vehicle speed information is provided to the ECM by the Vehicle Speed Sensor (VSS). It is a permanent magnet (PM) generator and is mounted in the transmission. The PM generator produces a pulsing voltage whenever vehicle speed is over about 3 mph. The A/C voltage level and the number of pulses increases with vehicle speed. The ECM converts the pulsing voltage to miles per hour. This data is used for calculations and can be displayed on a "Scan" tool.

The function of VSS buffer used in past model years has been incorporated into the ECM. The ECM supplies the necessary signal for the instrument panel (4000 pulses per mile) for operating the speedometer and the odometer.

Test Description:

Code 24 will set if vehicle speed equals 0 mph when:
- Engine speed is between 1400 and 3600 rpm.
- TPS is less than 2%.
- Low load condition (low MAP voltage, high manifold vacuum).
- Transmission not in park or neutral.
- All above conditions are met for 5 seconds.

These conditions are met during a road load deceleration.

Disregard a Code 24 that sets when the drive wheels are not turning. This can be caused by a faulty park/neutral switch circuit.

The PM generator only produces a signal if the drive wheels are turning greater than 3 mph.

Diagnostic Aids:

"Scan" tool should indicate a vehicle speed whenever the drive wheels are turning greater than 3 mph.

A problem in CKT 993 will not affect the VSS input or the readings on a "Scan" tool.

Check CKTs 400 and 401 for proper connections to be sure they are clean and tight and the harness is routed correctly. Refer to intermittents in Section "B".

(A/T) – A faulty or misadjusted park/neutral switch can result in a false Code 24. Use a "Scan" tool and check for the proper signal while in a drive range. Refer to CHART C-1A for the P/N switch check.

1988–90 2.5L (VIN U) – ALL MODELS

DISREGARD CODE 24 IF SET WHILE DRIVE WHEELS ARE NOT TURNING. REFER TO "DIAGNOSTIC AIDS" ON FACING PAGE.

CODE 24
VEHICLE SPEED SENSOR (VSS) CIRCUIT
2.5L (VIN U)

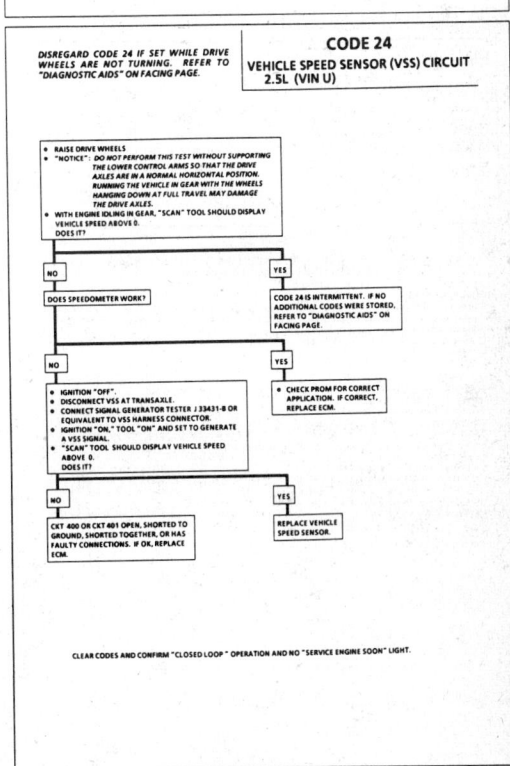

CLEAR CODES AND CONFIRM "CLOSED LOOP" OPERATION AND NO "SERVICE ENGINE SOON" LIGHT.

1988–90 2.5L (VIN U) – ALL MODELS

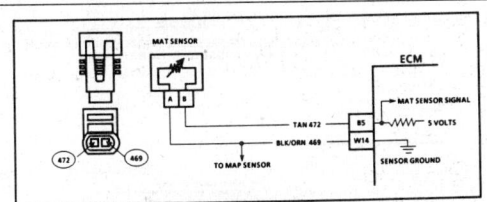

CODE 25
MANIFOLD AIR TEMPERATURE (MAT) SENSOR CIRCUIT
(HIGH TEMPERATURE INDICATED)
2.5L (VIN U)

Circuit Description:

The Manifold Air Temperature (MAT) sensor uses a thermistor to control the signal voltage to the ECM. The ECM applies a reference voltage (4-6 volts) on CKT 472 to the sensor. When manifold air is cold, the sensor (thermistor) resistance is high. Therefore, the ECM will see a high signal voltage. As the air warms, the sensor resistance becomes less and the voltage drops.

Test Description: Numbers below refer to circled numbers on the diagnostic chart.

1. This check determines if Code 25 is the result of a hard failure or an intermittent condition. Code 25 will set if a MAT temperature greater than 135°C is detected for a time longer than 2 seconds and the ignition is "ON."

Diagnostic Aids:

If the engine has been allowed to cool to an ambient temperature (overnight), coolant and MAT temperatures may be checked with a "Scan" tool and should read close to each other.

A Code 25 will result if CKT 472 is shorted to ground.

If Code 25 is intermittent, refer to Section "B".

1988–90 2.5L (VIN U) – ALL MODELS

CODE 25
MANIFOLD AIR TEMPERATURE (MAT) SENSOR CIRCUIT
(HIGH TEMPERATURE INDICATED)
2.5L (VIN U)

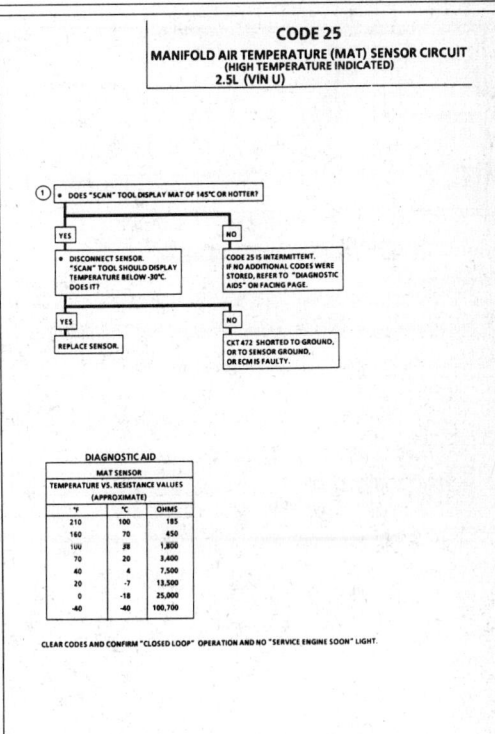

CLEAR CODES AND CONFIRM "CLOSED LOOP" OPERATION AND NO "SERVICE ENGINE SOON" LIGHT.

DIAGNOSTIC AID

MAT SENSOR

TEMPERATURE VS. RESISTANCE VALUES
(APPROXIMATE)

°F	°C	OHMS
210	100	185
160	70	450
100	38	1,800
70	20	3,400
40	4	7,500
20	-7	13,500
0	-18	25,000
-40	-40	100,700

1988–90 2.5L (VIN U) – ALL MODELS

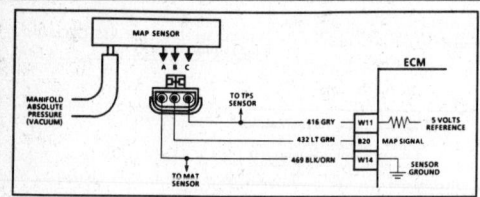

CODE 33
MANIFOLD ABSOLUTE PRESSURE (MAP) SENSOR CIRCUIT
(SIGNAL VOLTAGE HIGH - LOW VACUUM)
2.5L (VIN U)

Circuit Description:
The Manifold Absolute Pressure (MAP) sensor responds to changes in manifold pressure (vacuum). The ECM receives this information as a signal voltage that will vary from about 1 to 1.5 volts, at closed throttle idle, to 4-4.5 volts at wide open throttle (low vacuum).
If the MAP sensor fails, the ECM will substitute a fixed MAP value and use the Throttle Position Sensor (TPS) to control fuel delivery.

Test Description: Numbers below refer to circled numbers on the diagnostic chart.
1. This step will determine if Code 33 is the result of a hard failure or an intermittent condition.
   A Code 33 will be set under the following conditions:
   - MAP signal voltage is too high (low vacuum).
   - TPS less than 4%.
   - These conditions exist longer than 48 seconds.
2. This step simulates conditions for a Code 34. If the ECM recognizes the change, the ECM and CKTs 416 and 432 are OK. If CKT 469 is open, there may also be a stored Code 23.

Diagnostic Aids:
With the ignition "ON" and the engine stopped, the manifold pressure is equal to atmospheric pressure and the signal voltage will be high. This information is used by the ECM as an indication of vehicle altitude and is referred to as BARO. Comparison of this BARO reading with a known good vehicle with the same sensor is a good way to check accuracy of a "suspect" sensor. Readings should be within 4 volt.
A Code 33 will result if CKT 469 is open or if CKT 432 is shorted to voltage or to CKT 416.
If Code 33 is intermittent, refer to Section "B".

1988–90 2.5L (VIN U) – ALL MODELS

CODE 33
MANIFOLD ABSOLUTE PRESSURE (MAP) SENSOR CIRCUIT
(SIGNAL VOLTAGE HIGH - LOW VACUUM)
2.5L (VIN U)

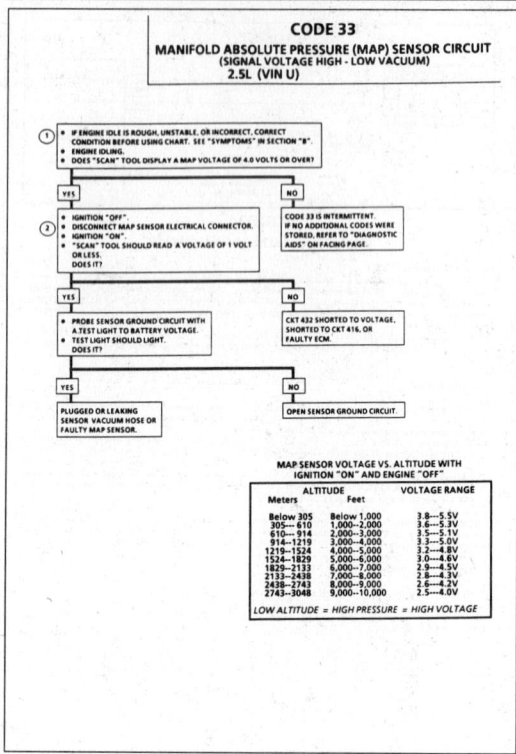

ALTITUDE		VOLTAGE RANGE
Meters	Feet	
Below 305	Below 1,000	3.8—5.5V
305— 610	1,000—2,000	3.6—5.3V
610— 914	2,000—3,000	3.5—5.1V
914—1219	3,000—4,000	3.3—5.0V
1219—1524	4,000—5,000	3.2—4.8V
1524—1829	5,000—6,000	3.0—4.6V
1829—2133	6,000—7,000	2.9—4.5V
2133—2438	7,000—8,000	2.8—4.3V
2438—2743	8,000—9,000	2.6—4.2V
2743—3048	9,000—10,000	2.5—4.0V

LOW ALTITUDE = HIGH PRESSURE = HIGH VOLTAGE

1988–90 2.5L (VIN U) – ALL MODELS

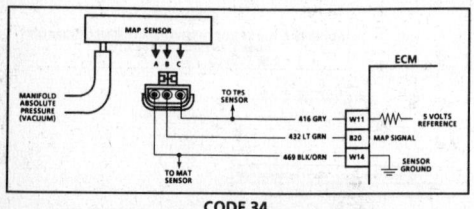

CODE 34
MANIFOLD ABSOLUTE PRESSURE (MAP) SENSOR CIRCUIT
(SIGNAL VOLTAGE LOW - HIGH VACUUM)
2.5L (VIN U)

Circuit Description:
The Manifold Absolute Pressure (MAP) sensor responds to changes in manifold pressure (vacuum). The ECM receives this information as a signal voltage that will vary from about 1 to 1.5 volts at closed throttle (idle), to 4 - 4.5 volts at wide open throttle (low vacuum).
If the MAP sensor fails, the ECM will substitute a fixed MAP value and use the Throttle Position Sensor (TPS) to control fuel delivery.

Test Description: Numbers below refer to circled numbers on the diagnostic chart.
1. This step determines if Code 34 is the result of a hard failure or an intermittent condition. A Code 34 will be set when MAP signal voltage is too low and the ignition is "ON."
2. Jumpering harness terminals "B" to "C" (5 volts to signal circuit) will determine if the sensor is at fault, or if there is a problem with the ECM or wiring.
3. The "Scan" tool may not display 12 volts. The important thing is that the ECM recognizes the voltage as more than 4 volts, indicating that the ECM and CKT 432 are OK.

Diagnostic Aids:
With the ignition "ON" and the engine stopped, the manifold pressure is equal to atmospheric pressure and the signal voltage will be high. This information is used by the ECM as an indication of vehicle altitude and is referred to as BARO. Comparison of this BARO reading with a known good vehicle with the same sensor is a good way to check accuracy of a "suspect" sensor. Readings should be within 4 volt.
A Code 34 will result if CKT(s) 416 or 432 are open or shorted to ground.
If CKT 416 is open or shorted to ground, there may also be a stored Code 22.
If Code 34 is intermittent, refer to Section "B".

1988–90 2.5L (VIN U) – ALL MODELS

CODE 34
MANIFOLD ABSOLUTE PRESSURE (MAP) SENSOR CIRCUIT
(SIGNAL VOLTAGE LOW - HIGH VACUUM)
2.5L (VIN U)

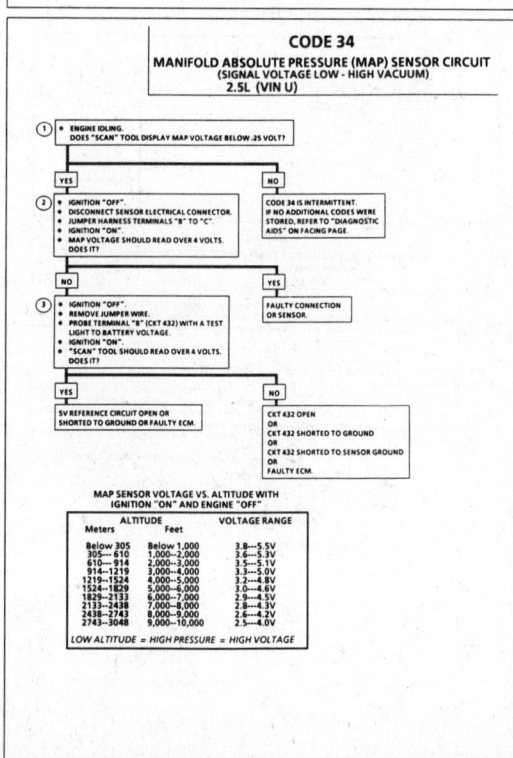

ALTITUDE		VOLTAGE RANGE
Meters	Feet	
Below 305	Below 1,000	3.8—5.5V
305— 610	1,000—2,000	3.6—5.3V
610— 914	2,000—3,000	3.5—5.1V
914—1219	3,000—4,000	3.3—5.0V
1219—1524	4,000—5,000	3.2—4.8V
1524—1829	5,000—6,000	3.0—4.6V
1829—2133	6,000—7,000	2.9—4.5V
2133—2438	7,000—8,000	2.8—4.3V
2438—2743	8,000—9,000	2.6—4.2V
2743—3048	9,000—10,000	2.5—4.0V

LOW ALTITUDE = HIGH PRESSURE = HIGH VOLTAGE

1988–90 2.5L (VIN U) – ALL MODELS

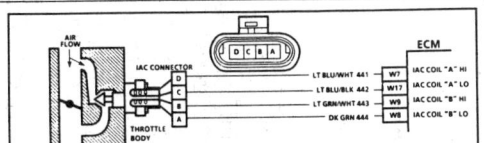

ECM	
LT BLU/WHT 441	W7 — IAC COIL "A" HI
LT BLU/BLK 442	W17 — IAC COIL "A" LO
LT GRN/WHT 443	W9 — IAC COIL "B" HI
DK GRN 444	W8 — IAC COIL "B" LO

CODE 35
IDLE SPEED ERROR
2.5L (VIN U)

Circuit Description:

Code 35 will set when the closed throttle engine speed is 150 rpm above or below the desired idle speed for 20 seconds.

Test Description: Numbers below refer to circled numbers on the diagnostic chart.

1. Continue with test, even if engine will not idle. If idle is too low, "Scan" tool will display 80 or more counts, or steps. If idle is high, it will display "0" counts. Occasionally an erratic or unstable idle may occur. Engine speed may vary 200 rpm or more up and down. Disconnect the IAC valve. If the condition is unchanged, the IAC valve is not at fault.
2. When the engine was stopped, the IAC valve retracted (more air) to a fixed "Park" position for increased air flow and idle speed during the next engine start. This test determines if IAC valve is functioning.
3. Be sure to disconnect the IAC valve prior to grounding the diagnostic test terminal. The test light will confirm the ECM signals by a steady or flashing light on all circuits.
4. There is a remote possibility that one of the circuits is shorted to voltage, which would have been indicated by a steady light. Disconnect ECM and turn the ignition "ON" and probe terminals to check for this condition.

Diagnostic Aids:

A slow, unstable idle may be caused by a system problem that cannot be overcome by the IAC valve. "Scan" tool counts will be above 60 counts if too low, and zero counts if too high.

If idle is too high, stop the engine. Turn ignition "ON" and ground diagnostic terminal. Wait 30-45 seconds for IAC valve to seat, then disconnect IAC valve. Start engine. If idle speed is above 800 rpm, locate and correct vacuum leak.

• **System too lean (High Air/Fuel Ratio)**
The idle speed may be too high or too low. Engine speed may vary up and down and disconnecting the IAC valve does not help. Code 44 may be set. "Scan" tool and/or voltmeter will read an oxygen sensor output less than 300 mV (.3 volt). Check for low regulated fuel pressure or water in fuel. A lean exhaust, with an oxygen sensor output fixed above 800 mV (.8 volt) will be a contaminated sensor, usually from silicone. This may also set a Code 45.
• **System too rich (Low Air/Fuel Ratio)**
The idle speed will be too low. "Scan" tool counts will usually be above 80. System is obviously rich and may exhibit black smoke in exhaust. "Scan" tool and/or voltmeter will read an oxygen sensor signal fixed above 800 mV (.8 volt). Check for high fuel pressure or leaking or sticking injector.
• **Throttle Body**
Remove IAC valve and inspect bore for foreign material or evidence of IAC valve dragging in the bore.
• **IAC Valve Connections**
IAC valve connections should be carefully checked for proper contact.
• **PCV Valve**
An incorrect or faulty PCV valve may result in an incorrect idle speed.
• Refer to "Rough, Unstable, Incorrect Idle or Stalling" in "Symptoms" in Section "B".

1988–90 2.5L (VIN U) – ALL MODELS

CODE 35
IDLE SPEED ERROR
2.5L (VIN U)

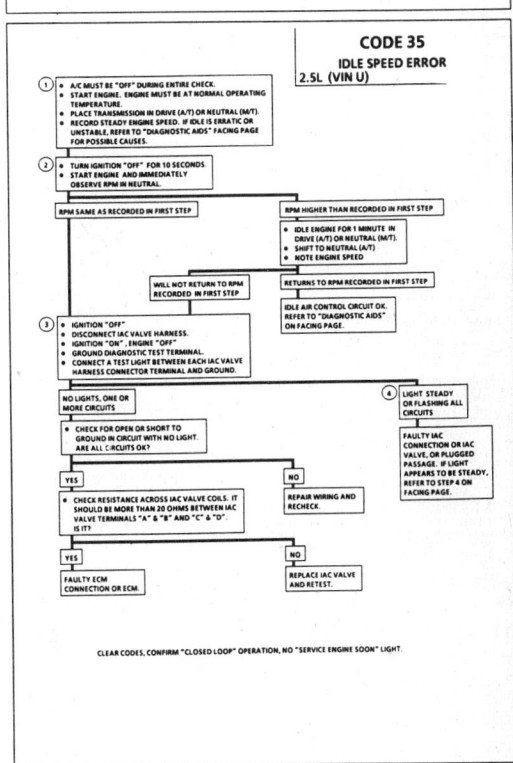

CLEAR CODES, CONFIRM "CLOSED LOOP" OPERATION, NO "SERVICE ENGINE SOON" LIGHT.

1988–90 2.5L (VIN U) – ALL MODELS

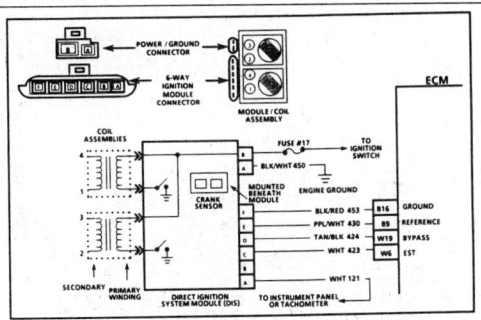

ECM	
BLK/RED 453	B16 — GROUND
PPL/WHT 430	B9 — REFERENCE
TAN/BLK 424	W19 — BYPASS
WHT 423	W6 — EST

CODE 42
ELECTRONIC SPARK TIMING (EST) CIRCUIT
2.5L (VIN U)

Circuit Description:

The DIS module sends a reference signal to the ECM when the engine is cranking. While the engine speed is under 400 rpm, the DIS module controls the ignition timing. When the system is running on the ignition module (no voltage on the bypass line), the ignition module grounds the EST signal. The ECM expects to sense no voltage on the EST line during this condition. If it senses a voltage, it sets Code 42 and will not enter the EST mode.

When the engine speed exceeds 400 rpm, the ECM applies 5 volts to the bypass line to switch the timing to ECM control (EST). If the bypass line is open or grounded, once the rpm for EST control is reached, the ignition module will not switch to EST mode. This results in low EST voltage and the setting of Code 42. If the EST line is grounded, the ignition module will switch to EST, but because the line is grounded, there will be no EST signal. A Code 42 will be set.

Test Description: Numbers below refer to circled numbers on the diagnostic chart.

1. Code 42 means the ECM has sensed an open or short to ground in the EST or bypass circuits. This test confirms Code 42 and that the fault causing the code is present.
2. Checks for a normal EST ground path through the ignition module. CKT 423, shorted to ground, will also read less than 500 ohms, but this will be checked later.
3. As the test light voltage contacts CKT 424, the module should switch, causing the ohmmeter to "overrange" if the meter is in the 1000-2000 ohms

position. Selecting the 10 - 20,000 ohms position will indicate a reading above 5000 ohms. The important thing is that the module "switched."
4. The module did not switch and this step checks for
   • EST CKT 423 shorted to ground.
   • Bypass CKT 424 open.
   • Faulty ignition module connection or module.
5. Confirms that Code 42 is a faulty ECM and not an intermittent in CKT(s) 423 or 424.

Diagnostic Aids:

The "Scan" tool does not have any ability to help diagnose a Code 42 problem.

If Code 42 is intermittent, refer to Section "B".

1988–90 2.5L (VIN U) – ALL MODELS

CODE 42
ELECTRONIC SPARK TIMING (EST) CIRCUIT
2.5L (VIN U)

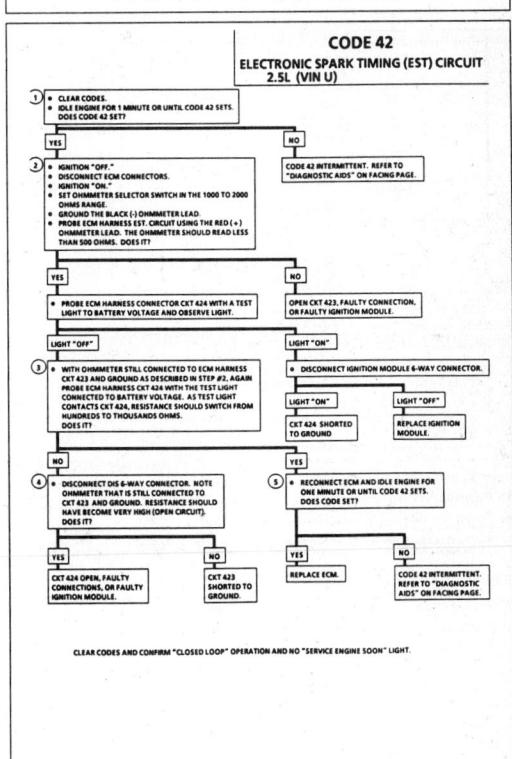

CLEAR CODES AND CONFIRM "CLOSED LOOP" OPERATION AND NO "SERVICE ENGINE SOON" LIGHT.

1988–90 2.5L (VIN U) – ALL MODELS

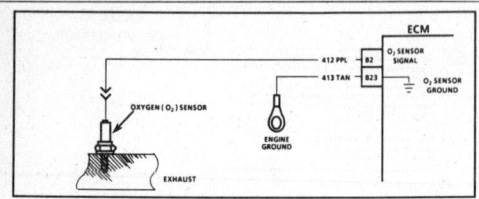

CODE 44
OXYGEN SENSOR CIRCUIT
(LEAN EXHAUST INDICATED)
2.5L (VIN U)

Circuit Description:

The ECM supplies a voltage of about .45 volt between terminals "B2" and "B23". (If measured with a 10 megohm digital voltmeter, this may read as low as .32 volt.)

When the O_2 sensor reaches operating temperature, it varies this voltage from about .1 volt (exhaust is lean) to about .9 volt (exhaust is rich).

The sensor is like an open circuit and produces no voltage when it is below 360°C (600°F). An open sensor circuit, or cold sensor, causes "Open Loop" operation.

Test Description: Numbers below refer to circled numbers on the diagnostic chart.

1. Code 44 is set when the O_2 sensor signal voltage on CKT 412 remains below .2 volt for 60 seconds or more and the system is operating in "Closed Loop".

Diagnostic Aids:

Using the "Scan" tool, observe the block learn value at different engine speeds. The "Scan" tool also displays the block learn cells so the block learn values can be checked in each of the cells to determine when the Code 44 may have been set. If the conditions for Code 44 exists, the block learn values will be around 150 or higher.

Check the following possible causes:

- O_2 Sensor Wire. Sensor pigtail may be mispositioned and contacting the exhaust manifold.
  Check for ground in wire between connector and sensor.
- Fuel Contamination. Water, even in small amounts, near the in-tank fuel pump inlet can be delivered to the injector. The water causes a lean exhaust and can set a Code 44.
- Fuel Pressure. System will be lean if fuel pressure is too low. It may be necessary to monitor fuel pressure while driving the car at various road speeds and/or loads to confirm. See "Fuel System Diagnosis," CHART A-7.
- Exhaust Leaks. If there is an exhaust leak, the engine can cause outside air to be pulled into the exhaust and past the sensor. Vacuum or crankcase leaks can cause a lean condition.
- If Code 44 is intermittent, refer to Section "B".

1988–90 2.5L (VIN U) – ALL MODELS

CODE 44
OXYGEN SENSOR CIRCUIT
(LEAN EXHAUST INDICATED)
2.5L (VIN U)

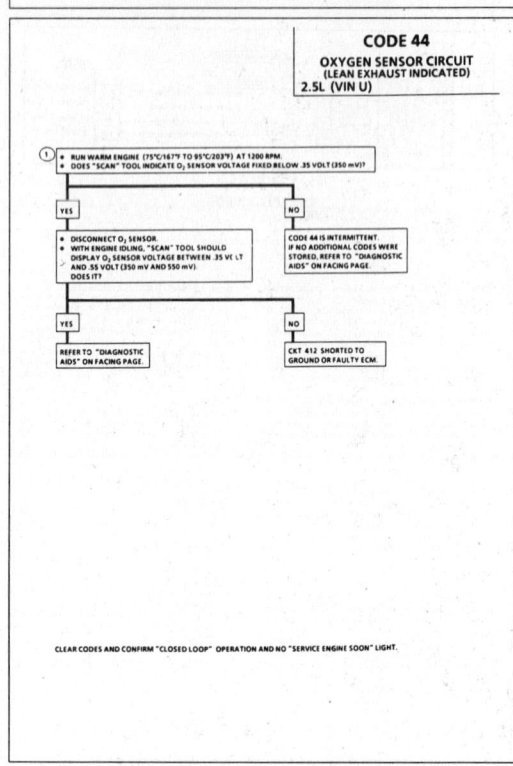

CLEAR CODES AND CONFIRM "CLOSED LOOP" OPERATION AND NO "SERVICE ENGINE SOON" LIGHT.

1988–90 2.5L (VIN U) – ALL MODELS

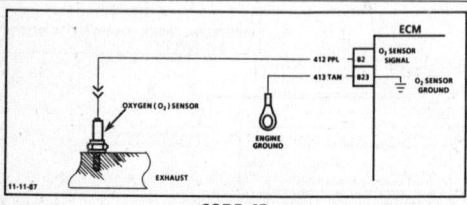

CODE 45
OXYGEN SENSOR CIRCUIT
(RICH EXHAUST INDICATED)
2.5L (VIN U)

Circuit Description:

The ECM supplies a voltage of about .45 volt between terminals "B2" and "B23". (If measured with a 10 megohm digital voltmeter, this may read as low as .32 volt).

When the O_2 sensor reaches operating temperature, it varies this voltage from about .1 volt (exhaust is lean) to about .9 volt (exhaust is rich).

The sensor is like an open circuit and produces no voltage when it is below 360°C (600°F). An open sensor circuit or cold sensor causes "Open Loop" operation.

Test Description: Numbers below refer to circled numbers on the diagnostic chart.

1. Code 45 is set when the O_2 sensor signal voltage on CKT 412 remains above .7 volt under the following conditions:
   - 30 seconds or more.
   - System is operating in "Closed Loop."
   - Engine run time after start is 1 minute or more.
   - Throttle angle is between 3% and 45%.

Diagnostic Aids:

Code 45 or rich exhaust is most likely caused by one of the following:
- Fuel Pressure. System will go rich, if pressure is too high. The ECM can compensate for some increase. However, if it gets too high, a Code 45 will be set. See "Fuel System Diagnosis," CHART A-7.
- Leaking Injector. See CHART A-7.
- HEI Shielding. An open ground CKT 453 may result in EMI, or induced electrical "noise." The ECM looks at this "noise" as reference pulses. The additional pulses result in a higher than actual engine speed signal. The ECM then delivers too much fuel causing the system to go rich. The engine tachometer will also show higher than actual engine speed, which can help in diagnosing this problem.

- Canister Purge. Check for fuel saturation. If full of fuel, check canister control and hoses. See "Canister Purge," Section "C3".
- MAP Sensor. An output that causes the ECM to sense a higher than normal manifold pressure (low vacuum) can cause the system to go rich. Disconnecting the MAP sensor will allow the ECM to set a fixed value for the MAP sensor. Substitute a different MAP sensor if the rich condition is gone, while the sensor is disconnected.
- TPS. An intermittent TPS output will cause the system to operate richly due to a false indication of the engine accelerating.
- O_2 Sensor Contamination. Inspect oxygen sensor for silicone contamination from fuel, or use of improper RTV sealant. The sensor may have a white, powdery coating and result in a high but false signal voltage (rich exhaust indication). The ECM will then reduce the amount of fuel delivered to the engine causing a severe surge driveability problem.
- EGR Valve. EGR sticking open at idle is usually accompanied by a rough idle and/or stall condition.
  If Code 45 is intermittent, refer to Section "B".

1988–90 2.5L (VIN U) – ALL MODELS

CODE 45
OXYGEN SENSOR CIRCUIT
(RICH EXHAUST INDICATED)
2.5L (VIN U)

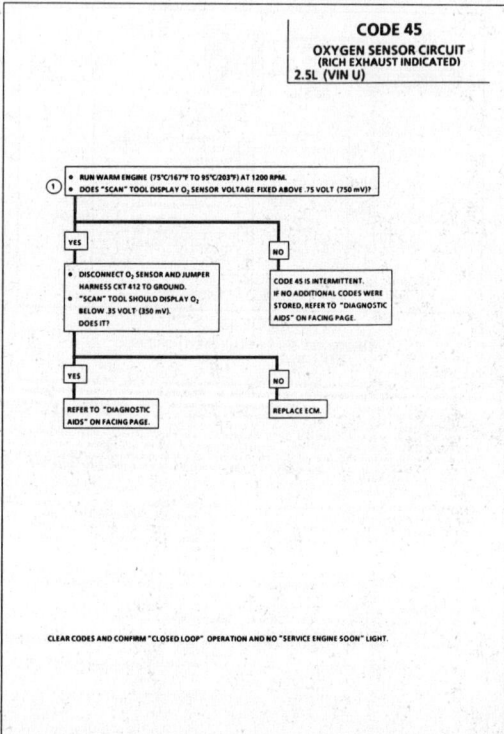

CLEAR CODES AND CONFIRM "CLOSED LOOP" OPERATION AND NO "SERVICE ENGINE SOON" LIGHT.

1988–90 2.5L (VIN U) – ALL MODELS

CODE 51
CODE 53
2.5L (VIN U)

CODE 51
PROM ERROR
(FAULTY OR INCORRECT PROM)

> CHECK THAT ALL PINS ARE FULLY INSERTED IN THE SOCKET AND THAT PROM IS PROPERLY SEATED.
> IF OK, REPLACE PROM, CLEAR MEMORY, AND RECHECK. IF CODE 51 REAPPEARS, REPLACE ECM.

CLEAR CODES AND CONFIRM "CLOSED LOOP" OPERATION AND NO "SERVICE ENGINE SOON" LIGHT.

CODE 53
SYSTEM OVERVOLTAGE

> THIS CODE INDICATES THAT THERE IS A BASIC GENERATOR PROBLEM.
> • CODE 53 WILL SET IF BATTERY VOLTAGE AT THE ECM IS GREATER THAN 16.9
>   VOLTS FOR AT LEAST 50 SECONDS.
> • CHECK AND REPAIR CHARGING SYSTEM.

CLEAR CODES AND CONFIRM "CLOSED LOOP" OPERATION AND NO "SERVICE ENGINE SOON" LIGHT.

1988–90 2.5L (VIN U) – ALL MODELS

SECTION B
SYMPTOMS
TABLE OF CONTENTS

Performing Symptom Diagnosis
Intermittents
Hard Start
Rough, Unstable, or Incorrect Idle, Stalling
Poor Gas Mileage
Detonation/Spark Knock
Lack of Power, Sluggish, or Spongy
Surges and/or Chuggle
Cuts Out, Misses
Hesitation, Sag, Stumble
Excessive Exhaust Emissions or Odors
Dieseling, Run-On
Backfire

PERFORMING SYMPTOM DIAGNOSIS

The DIAGNOSTIC CIRCUIT CHECK should be performed before using this section. The purpose of this section is to locate the source of a driveability or emissions problem when other diagnostic procedures cannot be used. This may be because of difficulties in locating a suspected sub-system or component.

Many driveability related problems can be eliminated by following the procedures found in Service Bulletins. These bulletins supersede this manual. Be sure to check all bulletins related to the complaint or suspected system.

If the engine cranks but will not run, use CHART A-3.

The sequence of the checks listed in this section is not intended to be followed as on a step-by-step procedure. The checks are listed such that the less difficult and time consuming operations are performed before more difficult ones.

Most of the symptom procedures call for a careful visual and physical check. *The importance of this step cannot be stressed too strongly.* It can lead to correcting a problem without further checks, and **can save valuable time**. This procedure includes checking the following.

- Vacuum hoses for splits, kinks, and proper connections, as shown on the underhood Emission Control Information label
- Throttle body and intake manifold for leaks
- Ignition wires for cracking, hardness, proper routing, and carbon tracking
- Wiring for proper connections, pinches, and cuts

1988–90 2.5L (VIN U) – ALL MODELS

INTERMITTENTS

Definition: Problem may or may not activate the "Service Engine Soon" light or store a trouble code.

DO NOT use the trouble code charts in Section "A" for intermittent problems. The fault must be present to locate the problem. If a fault is intermittent, the use of trouble code charts may result in the replacement of good parts.

- Most intermittent problems are caused by faulty electrical connections or wiring. Perform careful checks of suspected circuits for
  - Poor mating of the connector halves and terminals not fully seated in the connector body (backed out).
  - Improperly formed or damaged terminals. All connector terminals in problem circuit should be carefully reformed to increase contact tension.
  - Poor terminal-to wire connection. This requires removing the terminal from the connector body to check
- If a visual and physical check does not locate the cause of the problem, the car can be driven with a voltmeter connected to a suspected circuit or a "Scan" tool may be used. An abnormal voltage reading while the problem occurs indicates that the problem may be in that circuit.

- Check for loss of trouble code memory. To check, disconnect the TPS and allow the engine to idle until the "Service Engine Soon" light turns "ON." Code 22 should be stored and kept in memory when the ignition is turned "OFF" for at least 10 seconds. If not, the ECM is faulty.
- An intermittent SES light and no trouble codes may be caused by
  - Electrical system interference caused by a defective relay, ECM driven solenoid, or switch. They can cause a sharp electrical surge. Normally, the problem will occur when the faulty component is operated.
  - Improper installation of electrical options, such as lights, 2-way radios, etc.
  - EST wires which should be routed away from spark plug wires, ignition system components, and generator. Ground wire from ECM to ignition system which may be faulty.
  - Ignition secondary wire shorted to ground.
  - "Service Engine Soon" light and diagnostic test terminal circuits intermittently shorted to ground.
  - Faulty ECM grounds.

HARD START

Definition: Engine cranks well but does not start for a long time. Engine does eventually start, but may or may not continue to run.

Perform careful visual and physical check as described at the beginning of Section "B".
Perform "Diagnostic Circuit Check."
- CHECK
  - Fuel for poor quality, "stale" fuel, and water contamination.
  - Ignition wires for shorts or faulty insulation
  - Ignition coil connections
  - Fuel pump relay. Connect test light between pump test terminal and ground. Light should be on for 2 seconds following ignition "ON." If not, use CHART A-5.
  - Secondary ignition voltage output with ST-125 tester
  - Spark plugs. Look for wetness, cracks, improper gap, burned electrodes, and heavy deposits. Visually inspect ignition system for moisture, dust, cracks, burns, etc.
  - For faulty ECM and ignition grounds

- PROM for correct application.
  - Spray plug wires with fine water mist to check for shorts.
- For possibility of misfiring, crossfiring, or cutting out under load or at idle. If possible, refer to the "Misfire" Chart
- For improper crank sensor resistance or faulty connections
- EGR operation. Use CHART C-7.
- Idle Air Control system. Use Code 35 chart.
- Fuel system for restricted filter or improper pressure. Use CHART A-7.
- Injector and TBI assembly for leakage. Pressurize system by energizing fuel pump through the underhood fuel pump test connector.
- Coolant sensor for a shift in calibration. Use Code 14 or Code 15 chart.

1988–90 2.5L (VIN U) – ALL MODELS

- TPS for sticking or binding. TPS voltage should read less than 1.25 volts on a "Scan" tool.
- In-tank fuel pump check valve. A faulty valve would allow the fuel in the lines to drain back to the tank after the engine is stopped. To check for this condition, conduct the following test.
  1. Ignition "OFF."
  2. Disconnect fuel line at the filter.
  3. Remove the tank filler cap.
  4. Connect a radiator test pump to the line and apply 103 kPa (15 psi) pressure. If the pressure will hold for 60 seconds, the check valve is OK.
- For the possibility of an exhaust restriction or improper valve timing by performing the following test.
  1. With engine at normal operating temperature, connect a vacuum gauge to any convenient vacuum port on intake manifold.

  2. Run engine at 1000 rpm and record vacuum reading.
  3. Increase engine speed slowly to 2500 rpm. Note vacuum reading at steady 2500 rpm.
  4. If vacuum at 2500 rpm decreases more than 3" Hg from reading at 1000 rpm, the exhaust system should be inspected for restrictions.
  5. Disconnect exhaust pipe from engine and repeat Steps 3 & 4. If vacuum still drops more than 3" Hg with exhaust disconnected, check valve timing.
  General Information

  Engine valve timing and compression.

ROUGH, UNSTABLE, OR INCORRECT IDLE, STALLING

Definition: The engine runs unevenly at idle. If severe, the car may shake. Also, the idle speed may vary (called "hunting"). Either condition may be severe enough to cause stalling. Engine idles at incorrect speed.

Perform careful visual and physical check as described at the beginning of Section "B."
Perform "Diagnostic Circuit Check."
- CHECK
  - MAP sensor. Use CHART C1-D.
  - Throttle for sticking shaft or binding linkage. This will cause a high TPS voltage (open throttle indication) and the ECM will not control idle. TPS voltage should be less than 1.25 volts with throttle closed.
  - Ignition wires for shorts or faulty insulation
  - Ignition system for moisture, dust, cracks, burns, etc. Spray plug wires with fine water mist to check for shorts.
  - For possibility of misfiring, crossfiring, or cutting out under load or at idle. If present, refer to the "Misfire" Chart
  - Secondary ignition voltage output with ST-1 tester
  - Ignition coil connections
  - ECM and ignition system for faulty grounds.
  - Proper operation of EST.
  - Spark plugs. Look for wetness, cracks, improper gap, burned electrodes, and heavy deposits.

- Fuel system for restricted filter or improper pressure. Use CHART A-7.
- Injector and TBI assembly for leakage. Pressurize system by energizing fuel pump through the underhood fuel pump test connector.
- EGR operation. Use CHART C-7.
- For vacuum leaks at intake manifold gasket.
- Idle Air Control system. Use Code 35 chart.
- Electrical system voltage. IAC valve will not move if voltage is below 9 volts or greater than 17.8 volts. Also check battery cables and ground straps for poor contact. Erratic voltage will cause the IAC valve to change its position, resulting in poor idle quality.
- PCV valve for proper operation by placing finger over inlet hole in valve end several times. Valve should snap back. If not, replace valve. Ensure that valve is correct part. Also check PCV hose.
- Canister purge system for proper operation. Use CHART C-3.
- PROM for correct application

1988–90 2.5L (VIN U) – ALL MODELS

- Throttle shaft or TPS for sticking or binding. TPS voltage should read less than 1.25 volts on a "Scan" tool with the throttle closed.
- MAP sensor output. Use CHART C1-D and/or check sensor by comparing it to the output on a similar vehicle if possible.
- Oxygen sensor for silicone contamination from contaminated fuel or use of improper RTV sealant. The sensor will have a white, powdery coating and will cause a high but false signal voltage (rich exhaust indication). The ECM will reduce the amount of fuel delivered to the engine, causing a severe driveability problem.
- Coolant sensor for a shift in calibration. Use Code 14 or Code 15 chart.
- A/C refrigerant pressure for high pressure. Check for overcharging or faulty pressure switch.
- P/N switch circuit on vehicle with automatic transmission. Use CHART C-1A.
- Alternator output voltage. Repair if less than 9 volts or more than 16 volts.
- Power steering. Use CHART C-1E. The ECM should compensate for power steering loads. Loss of this signal would be most noticeable when steering loads are high such as during parking.
- Engine valve timing and compression.

- For worn or incorrect basic engine parts such as cam, heads, pistons, etc. Also check for bent pushrods, worn rocker arms, and broken or weak valve springs.

- For the possibility of an exhaust restriction or improper valve timing, perform the following test.
  1. With engine at normal operating temperature, connect a vacuum gage to any convenient vacuum port on intake manifold.
  2. Run engine at 1000 rpm and record vacuum reading.
  3. Increase engine speed slowly to 2500 rpm. Note vacuum reading at steady 2500 rpm.
  4. If vacuum at 2500 rpm decreases more than 3" Hg from reading at 1000 rpm, the exhaust system should be inspected for restrictions.
  5. Disconnect exhaust pipe from engine and repeat Steps 3 & 4. If vacuum still drops more than 3" Hg with exhaust disconnected, check valve timing.
- For overheating and possible causes. Look for the following.
  - Low or incorrect coolant solution. It should be a 50/50 mix of GM #1052753 anti-freeze coolant (or equivalent) and water.
  - Loose water pump belt.
  - Restricted air flow to radiator, or restricted water flow through radiator.
  - Faulty or incorrect thermostat.
  - Inoperative electric cooling fan circuit.
- If the system is running RICH, (block learn less than 118), refer to "Diagnostic Aids" on facing page of Code 45.
- If the system is running LEAN, (block learn greater than 138), refer to "Diagnostic Aids" on facing page of Code 44.

POOR GAS MILEAGE

Definition: Gas mileage, as measured by an actual road test, is noticeably lower than expected. Gas mileage is noticeably lower than it was during a previous actual road test.

Perform careful visual and physical check as described at the beginning of Section "B".
Perform "Diagnostic Circuit Check."
- **CHECK**
  - Proper operation of EST.
  - For possibility of misfiring, crossfiring, or cutting out under load or at idle. If present, refer to the "Misfire" Chart
  - Spark plugs. Look for wetness, cracks, improper gap, burned electrodes, and heavy deposits.

- Spark plugs for correct heat range
- Fuel for poor quality, "stale" fuel, and water contamination.
- Fuel system for restricted filter or improper pressure. Use CHART A-7.
- Injector and TBI assembly for leakage. Pressurize system by energizing fuel pump through the underhood fuel pump test connector.
- EGR operation. Use CHART C-7.

1988–90 2.5L (VIN U) – ALL MODELS

- For vacuum leaks at intake manifold gasket
- Air cleaner element (filter) for dirt or plugging
- Idle Air Control system. Use Code 35 chart.
- Canister purge system for proper operation. Use CHART C-3.
- PROM for correct application

- Throttle shaft or TPS for sticking or binding. TPS voltage should read less than 1.25 volts on a "Scan" tool with the throttle closed.
- MAP sensor output. Use CHART C1-D and/or check sensor by comparing it to the output on a similar vehicle if possible.
- Oxygen sensor for silicone contamination from contaminated fuel or use of improper RTV sealant. The sensor will have a white, powdery coating and will cause a high but false signal voltage (rich exhaust indication). The ECM will reduce the amount of fuel delivered to the engine, causing a severe driveability problem.
- Coolant sensor for a shift in calibration. Use Code 14 or Code 15 chart.
- Vehicle speed sensor (VSS) input with a "Scan" tool to make sure reading of VSS matches that of vehicle speedometer.
- A/C relay operation. A/C should cut out at wide open throttle. Use CHART C-10.
- A/C refrigerant pressure for high pressure. Check for overcharging or faulty pressure switch.
- Alternator output voltage. Repair if less than 9 volts or more than 16 volts.
- Cooling fan operation. Use CHART C-12.
- Power steering. Use CHART C-1E. The ECM should compensate for power steering loads. Loss of this signal would be most noticeable when steering loads are high such as during parking.
- Transmission torque converter operation.
- Transmission for proper shift points.
- Transmission torque converter clutch operation. Use CHART C-8.

- Engine valve timing and compression.

- For worn or incorrect basic engine parts such as cam, heads, pistons, etc. Also check for bent pushrods, worn rocker arms, and broken or weak valve springs.
- For the possibility of an exhaust restriction or improper valve timing by performing the following test.
  1. With engine at normal operating temperature, connect a vacuum gauge to any convenient vacuum port on intake manifold.
  2. Run engine at 1000 rpm and record vacuum reading.
  3. Increase engine speed slowly to 2500 rpm. Note vacuum reading at steady 2500 rpm.
  4. If vacuum at 2500 rpm decreases more than 3" Hg from reading at 1000 rpm, the exhaust system should be inspected for restrictions.
  5. Disconnect exhaust pipe from engine and repeat Steps 3 & 4. If vacuum still drops more than 3" Hg with exhaust disconnected, check valve timing.
- Thermostat for incorrect heat range or being inoperative.
- Check driver's driving habits and vehicle conditions which affect gas mileage.
  - Suggest driver read "Important Facts on Fuel Economy" in Owner's Manual.
  - Is A/C "ON" full time (Defroster mode "ON")?
  - Are tires at correct pressure?
  - Are excessively heavy loads being carried?
  - Is acceleration often heavy?
  - Are the wheels aligned correctly?
  - Is the speedometer calibrated correctly?
  - Are the vehicle brakes dragging?
  - Is the brake switch applying excessive force on the brake pedal?
- If the system is running RICH, (block learn less than 118), refer to "Diagnostic Aids" on facing page of Code 45.

1988–90 2.5L (VIN U) – ALL MODELS

DETONATION/SPARK KNOCK

Definition: A mild to severe ping, usually worse under acceleration. The engine makes sharp metallic knocks that change with throttle opening.

Perform careful visual and physical check as described at the beginning of Section "B".
Perform "Diagnostic Circuit Check."
- **CHECK**
  - Ignition wires for shorts or faulty insulation
  - For possibility of misfiring, crossfiring, or cutting out under load or at idle. If present, refer to the "Misfire" Chart
  - Spark plugs for correct heat range
  - Fuel for poor quality, "stale" fuel, and water contamination
  - Fuel system for restricted filter or improper pressure. Use CHART A-7.
  - For excessive oil entering combustion chamber. Oil will reduce the effective octane of fuel.
  - EGR operation. Use CHART C-7.
  - For vacuum leaks at intake manifold gasket
  - PCV valve for proper operation by placing finger over inlet hole in valve end several times. Valve should snap back. If not, replace valve. Ensure that valve is correct part. Also check PCV hose.
  - MAP sensor output. Use CHART C1-D and/or check sensor by comparing it to the output on a similar vehicle, if possible.
  - Coolant sensor for a shift in calibration
  - Oxygen sensor for silicone contamination from contaminated fuel or use of improper RTV sealant. The sensor will have a white, powdery coating and will cause a high but false signal voltage (rich exhaust indication). The ECM will reduce the amount of fuel delivered to the engine, causing a severe driveability problem
  - Vehicle speed sensor (VSS) input with a "Scan" tool to make sure reading of VSS matches that of vehicle speedometer.
- Transmission torque converter operation.
- Transmission for proper shift points.
- Transmission torque converter clutch operation. Use CHART C-8.
- Vehicle brakes for dragging

- PROM for correct application

- For overheating and possible causes. Look for the following.
  - Low or incorrect coolant solution. It should be a 50/50 mix of GM #1052753 anti-freeze coolant (or equivalent) and water.
  - Loose water pump belt
  - Restricted air flow to radiator or restricted water flow through radiator
  - Faulty or incorrect thermostat
  - Inoperative electric cooling fan circuit
- Engine valve timing and compression.

- For worn or incorrect basic engine parts such as cam, heads, pistons, etc. Also check for bent pushrods, worn rocker arms, and broken or weak valve springs.

- For the possibility of an exhaust restriction or improper valve timing by performing the following test.
  1. With engine at normal operating temperature, connect a vacuum gauge to any convenient vacuum port on intake manifold.
  2. Run engine at 1000 rpm and record vacuum reading.
  3. Increase engine speed slowly to 2500 rpm. Note vacuum reading at steady 2500 rpm.
  4. If vacuum at 2500 rpm decreases more than 3" Hg from reading at 1000 rpm, the exhaust system should be inspected for restrictions.
  5. Disconnect exhaust pipe from engine and repeat Steps 3 & 4. If vacuum still drops more than 3" Hg with exhaust disconnected, check valve timing.
- Remove internal engine carbon with top engine cleaner.
- If the system is running LEAN, (block learn greater than 138), refer to "Diagnostic Aids" on facing page of Code 44.

1988–90 2.5L (VIN U) – ALL MODELS

LACK OF POWER, SLUGGISH, OR SPONGY

Definition: Engine delivers less than expected power. There is little or no increase in speed when the accelerator pedal is depressed partially.

Perform careful visual and physical check as described at the beginning of Section "B".
Perform "Diagnostic Circuit Check."
- **CHECK**
  - Ignition wires for shorts or faulty insulation
  - Ignition system for moisture, dust, cracks, burns, etc. Spray plug wires with fine water mist to check for shorts.
  - For possibility of misfiring, crossfiring, or cutting out under load or at idle. If present, refer to the "Misfire" Chart
  - Secondary ignition voltage output with ST-125 tester.
  - Ignition coil connections
  - ECM and ignition system for faulty grounds
  - Proper operation of EST.
  - Spark plugs. Look for wetness, cracks, improper gap, burned electrodes, and heavy deposits.
  - Spark plugs for correct heat range
  - Fuel for poor quality, "stale" fuel, and water contamination
  - Fuel system for restricted filter or improper pressure. Use CHART A-7.
  - EGR operation. Use CHART C-7.
  - For vacuum leaks at intake manifold gasket
  - Air cleaner element (filter) for dirt or plugging
  - PROM for correct application

  - Throttle shaft or TPS for sticking or binding. TPS voltage should read less than 1.25 volts on a "Scan" tool with the throttle closed.
  - MAP sensor output. Use CHART C1-D and/or check sensor by comparing it to the output on a similar vehicle if possible.
  - Oxygen sensor for silicone contamination from contaminated fuel or use of improper RTV sealant. The sensor will have a white, powdery coating and will cause a high but false signal voltage (rich exhaust indication). The ECM will reduce the amount of fuel delivered to the engine, causing a severe driveability problem.
  - Coolant sensor for a shift in calibration. Use Code 14 or Code 15 chart.
  - Vehicle speed sensor (VSS) input with a "Scan" tool to make sure reading of VSS matches that of vehicle speedometer.

- Engine for improper or worn camshaft.
- A/C relay operation. A/C should cut out at wide open throttle. Use CHART C-10.
- A/C refrigerant pressure for high pressure. Check for overcharging or faulty pressure switch.
- Alternator output voltage. Repair if less than 9 volts or more than 16 volts.
- Cooling fan operation. Use CHART C-12.
- Power steering. Use CHART C-1E. The ECM should compensate for power steering loads. Loss of this signal would be most noticeable when steering loads are high such as during parking.
- Transmission torque converter operation.

- Transmission for proper shift points.

- Transmission torque converter clutch operation. Use CHART C-8.
- Vehicle brakes for dragging
- Engine valve timing and compression.

- For worn or incorrect basic engine parts such as cam, heads, pistons, etc. Also check for bent pushrods, worn rocker arms, and broken or weak valve springs.

- For the possibility of an exhaust restriction or improper valve timing by performing the following test.
  1. With engine at normal operating temperature, connect a vacuum gauge to any convenient vacuum port on intake manifold.
  2. Run engine at 1000 rpm and record vacuum reading.
  3. Increase engine speed slowly to 2500 rpm. Note vacuum reading at steady 2500 rpm.
  4. If vacuum at 2500 rpm decreases more than 3" Hg from reading at 1000 rpm, the exhaust system should be inspected for restrictions.

1988–90 2.5L (VIN U)–ALL MODELS

5. Disconnect exhaust pipe from engine and repeat Steps 3 & 4. If vacuum still drops more than 3" Hg with exhaust disconnected, check valve timing.

- For overheating and possible causes. Look for the following:
- Low or incorrect coolant solution. It should be a 50/50 mix of GM #1052753 anti-freeze coolant (or equivalent) and water.
  - Loose water pump belt.
  - Restricted air flow to radiator, or restricted water flow through radiator.
  - Faulty or incorrect thermostat.
  - Inoperative electric cooling fan circuit. See CHART C-12.
- If the system is running RICH (block learn less than 118), refer to "Diagnostic Aids" on facing page of Code 45.
- If the system is running LEAN (block learn greater than 138), refer to "Diagnostic Aids" on facing page of Code 44.

SURGES AND/OR CHUGGLE

Definition: Engine power variation under steady throttle or cruise. Feels like the car speeds up and slows down with no change in the accelerator pedal.

Perform careful visual and physical check as described at the beginning of Section "B".

Perform "Diagnostic Circuit Check."

- **CHECK**
  - Ignition wires for shorts or faulty insulation.
  - Ignition system for moisture, dust, cracks, burns, etc. Spray plug wires with fine water mist to check for shorts.
  - For possibility of misfiring, crossfiring, or cutting out under load or at idle. If present, refer to the "Misfire" Chart
  - Secondary ignition voltage output with ST-125 tester.
  - Ignition coil connections.
  - ECM and ignition system for faulty grounds.
  - Proper operation of EST.
  - Spark plugs. Look for wetness, cracks, improper gap, burned electrodes, and heavy deposits.
  - Spark plugs for correct heat range.
  - Fuel for poor quality, "stale" fuel, and water contamination.
  - Fuel system for restricted filter or improper pressure. Use CHART A-7.
  - Injector and TBI assembly for leakage. Pressurize system by energizing fuel pump through the underhood fuel pump test connector.
  - EGR operation. Use CHART C-7.
  - For vacuum leaks at intake manifold gasket.
  - Idle Air Control system. Use Code 35 chart.
  - Electrical system voltage. IAC valve will not move if voltage is below 9 volts or greater than 17.8 volts. Also check battery cables and ground straps for poor contact. Erratic voltage will cause the IAC valve to change its position, resulting in poor idle quality.
  - PCV valve for proper operation by placing finger over inlet hole in valve end several times. Valve should snap back. If not, replace valve. Ensure that valve is correct part. Also check PCV hose.
  - Canister purge system for proper operation. Use CHART C-3.
  - PROM for correct application.
  - Throttle shaft or TPS for sticking or binding. TPS voltage should read less than 1.25 volts on a "Scan" tool with the throttle closed.
  - MAP sensor output. Use CHART C1-D and/or check sensor by comparing it to the output on a similar vehicle, if possible.
  - Oxygen sensor for silicone contamination from contaminated fuel or use of improper RTV sealant. The sensor will have a white, powdery coating and will cause a high but false signal voltage (rich exhaust indication). The ECM will reduce the amount of fuel delivered to the engine, causing a severe driveability problem.
  - Coolant sensor for a shift in calibration. Use Code 14 or Code 15 chart.
  - Vehicle speed sensor (VSS) input with a "Scan" tool to make sure reading of VSS matches that of vehicle speedometer.
  - A/C relay operation. A/C should cut out at wide open throttle. Use CHART C-10.
  - P/N switch circuit on vehicle with automatic transmission. Use CHART C-1A.
  - Transmission torque converter clutch operation. Use CHART C-8.

1988–90 2.5L (VIN U)–ALL MODELS

- For the possibility of an exhaust restriction or improper valve timing by performing the following test:
  1. With engine at normal operating temperature, connect a vacuum gauge to any convenient vacuum port on intake manifold.
  2. Run engine at 1000 rpm and record vacuum reading.
  3. Increase engine speed slowly to 2500 rpm. Note vacuum reading at steady 2500 rpm.
  4. If vacuum at 2500 rpm decreases more than 3" Hg from reading at 1000 rpm, the exhaust system should be inspected for restrictions.
  5. Disconnect exhaust pipe from engine and repeat Steps 3 & 4. If vacuum still drops more than 3" Hg with exhaust disconnected, check valve timing.
- Engine valve timing and compression.
- For worn or incorrect basic engine parts such as cam, heads, pistons, etc. Also check for bent pushrods, worn rocker arms, and broken or weak valve springs.
- If the system is running RICH, (block learn less than 118), refer to "Diagnostic Aids" on facing page of Code 45.
- If the system is running LEAN, (block learn greater than 138), refer to "Diagnostic Aids" on facing page of Code 44.

CUTS OUT, MISSES

Definition: Steady pulsation or jerking that follows engine speed, usually more pronounced as engine load increases. The exhaust has a steady spitting sound at idle or low speed.

Perform careful visual and physical check as described at the beginning of Section "B".

Perform "Diagnostic Circuit Check."

- **CHECK**
  - Ignition wires for shorts or faulty insulation
  - Ignition system for moisture, dust, cracks, burns, etc. Spray plug wires with fine water mist to check for shorts.
  - For possibility of misfiring, crossfiring, or cutting out under load or at idle. If present, refer to the "Misfire" Chart
  - Secondary ignition voltage output with ST-125 tester
  - Ignition coil connections
  - ECM and ignition system for faulty grounds
  - Proper operation of EST. .
  - Spark plugs. Look for wetness, cracks, improper gap, burned electrodes, and heavy deposits.
  - Spark plugs for correct heat range
  - For improper crank sensor resistance or faulty connections
  - Fuel for poor quality, "stale" fuel, and water contamination
  - Fuel system for restricted filter or improper pressure. Use CHART A-7.
  - Throttle shaft or TPS for sticking or binding. TPS voltage should read less than 1.25 volts on a "Scan" tool with the throttle closed.

HESITATION, SAG, STUMBLE

Definition: Momentary lack of response as the accelerator is pushed down. Can occur at all vehicle speeds. Usually most severe when first trying to make the car move, as from a stop sign. May cause the engine to stall if severe enough.

Perform careful visual and physical check as described at the beginning of Section "B".

Perform "Diagnostic Circuit Check."

- **CHECK**
  - Ignition wires for shorts or faulty insulation
  - Ignition system for moisture, dust, cracks, burns, etc. Spray plug wires with fine water mist to check for shorts.
  - For possibility of misfiring, crossfiring, or cutting out under load or at idle. If present, refer to the "Misfire" Chart
  - Secondary ignition voltage output with ST-125 tester
  - Ignition coil connections
  - ECM and ignition system for faulty grounds
  - Proper operation of EST.
  - Spark plugs. Look for wetness, cracks, improper gap, burned electrodes, and heavy deposits.
  - Spark plugs for correct heat range
  - Fuel for poor quality, "stale" fuel, and water contamination

1988–90 2.5L (VIN U)–ALL MODELS

- Fuel system for restricted filter or improper pressure. Use CHART A-7.
- EGR operation. Use CHART C-7.
- For vacuum leaks at intake manifold gasket
- Air cleaner element (filter) for dirt or plugging
- Idle Air Control system. Use Code 35 chart.
- Check electrical system voltage. IAC valve will not move if voltage is below 9 volts or greater than 17.8 volts. Also check battery cables and ground straps for poor contact. Erratic voltage will cause the IAC valve to change its position, resulting in poor idle quality.
- PCV valve for proper operation by placing finger over inlet hole in valve end several times. Valve should snap back. If not, replace valve. Ensure that valve is correct part. Also check PCV hose.
- Canister purge system for proper operation. Use CHART C-3.
- PROM for correct application
- Throttle shaft or TPS for sticking or binding TPS voltage should read less than 1.25 volts on a "Scan" tool with the throttle closed.
- MAP sensor output. Use CHART C1-D and/or check sensor by comparing it to the output on a similar vehicle, if possible.
- Oxygen sensor for silicone contamination from contaminated fuel or use of improper RTV sealant. The sensor will have a white, powdery coating and will cause a high but false signal voltage (rich exhaust indication). The ECM will reduce the amount of fuel delivered to the engine, causing a severe driveability problem.
- Coolant sensor for a shift in calibration. Use Code 14 or Code 15 chart.
- A/C relay operation. A/C should cut out at wide open throttle. Use CHART C-10.
- A/C refrigerant pressure for high pressure. Check for overcharging or faulty pressure switch.
- P/N switch circuit on vehicle with automatic transmission. Use CHART C-1A.
- Alternator output voltage. Repair if less than 9 volts or more than 16 volts.
- Transmission torque converter operation.
- Transmission for proper shift points.
- Transmission torque converter clutch operation. Use CHART C-8.
- Vehicle brakes for dragging
- Engine valve timing and compression.
- For the possibility of an exhaust restriction or improper valve timing by performing the following test.
  1. With engine at normal operating temperature, connect a vacuum gauge to any convenient vacuum port on intake manifold.
  2. Run engine at 1000 rpm and record vacuum reading.
  3. Increase engine speed slowly to 2500 rpm. Note vacuum reading at steady 200 rpm.
  4. If vacuum at 2500 rpm decreases more than 3" Hg from reading at 1000 rpm, the exhaust system should be inspected for restrictions.
  5. Disconnect exhaust pipe from engine and repeat Steps 3 & 4. If vacuum still drops more than 3" Hg with exhaust disconnected, check valve timing.
- For worn or incorrect basic engine parts such as cam, heads, pistons, etc. Also check for bent pushrods, worn rocker arms, and broken or weak valve springs.
- For overheating and possible causes. Look for the following.
  - Low or incorrect coolant solution. It should be a 50/50 mix of GM #1052753 anti-freeze coolant (or equivalent) and water.
  - Loose water pump belt
  - Restricted air flow to radiator, or restricted water flow through radiator
  - Faulty or incorrect thermostat
  - Inoperative electric cooling fan circuit. See CHART C-12.
- If the system is running RICH (block learn less than 118), refer to "Diagnostic Aids" on facing page of Code 45.
- If the system is running LEAN (block learn greater than 138), refer to "Diagnostic Aids" on facing page of Code 44.

1988–90 2.5L (VIN U)–ALL MODELS

EXCESSIVE EXHAUST EMISSIONS OR ODORS

Definition: Vehicle fails an emission test or vehicle has excessive "rotten egg" smell. (Excessive odors do not necessarily indicate excessive emissions).

Perform careful visual and physical check as described at the beginning of Section "B".

Perform "Diagnostic Circuit Check."

- **CHECK**
  - EGR valve not opening. Use CHART C-7.
  - Vacuum leaks
  - Faulty coolant system and/or coolant fan operation. Use CHART C-12.
  - Remove carbon with top engine cleaner. Follow instructions on can.
- If the system is running RICH (block learn less than 118), refer to "Diagnostic Aids" on facing page of Code 45.
- If the system is running LEAN (block learn greater than 138), refer to "Diagnostic Aids" on facing page of Code 44.
- If emission test indicates excessive NO_x, check for items which cause car to run lean or too hot.
- If emission test indicates excessive HC and CO or exhaust has excessive odors, check for items which cause car to run RICH.
  - Incorrect fuel pressure. Use CHART A-7.
  - Fuel loading of evaporative vapor canister. Use CHART C-3.
  - PCV valve plugging, sticking, or blocked PCV hose. Check for fuel in crankcase.
  - Catalytic converter lead contamination (Look for removal of fuel filler neck restrictor.)
  - Improper fuel cap installation
  - Faulty spark plugs, plug wires, or ignition components.

DIESELING, RUN-ON

Definition: Engine continues to run after key is turned "OFF", but runs very roughly. (If engine runs smoothly, check ignition switch).

Perform careful visual and physical check as described at the beginning of Section "B".

Perform "Diagnostic Circuit Check."

- **CHECK**

  Injector and TBI assembly for leakage. Pressurize system by energizing fuel pump through the fuel pump test connector.

BACKFIRE

Definition: Fuel ignites in intake manifold or in exhaust system, making a loud popping sound.

Perform careful visual and physical check as described at the beginning of Section "B".

Perform "Diagnostic Circuit Check."

- **CHECK**
  - EGR operation for valve being open all the time. Use CHART C-7.
  - Intake manifold gasket for leaks
  - For possibility of misfiring, crossfiring, or cutting out under load or at idle. If present, refer to the "Misfire" Chart
  - Spark plugs. Look for wetness, cracks, improper gap, burned electrodes, and heavy deposits.
  - Ignition coil connections
  - Ignition system for moisture, dust, cracks, burns, etc. Spray plug wires with fine water mist to check for shorts.
  - ECM and ignition system for faulty grounds
  - Secondary ignition voltage output with ST-125 tester
  - For vacuum leaks at intake manifold gasket
  - Engine valve timing and compression.
  - For worn or incorrect basic engine parts such as cam, heads, pistons, etc. Also check for bent pushrods, worn rocker arms, and broken or weak valve springs.

1988–90 2.5L (VIN U) – ALL MODELS

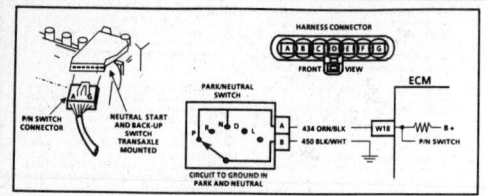

CHART C-1A
PARK/NEUTRAL SWITCH DIAGNOSIS
2.5L (VIN U)

Circuit Description:

The Park/Neutral (P/N) switch contacts are a part of the neutral start switch and are closed to ground in park or neutral, and open in drive ranges.

The ECM supplies ignition voltage through a current limiting resistor to CKT 434 and senses a closed switch, when the voltage on CKT 434 drops to less than one volt.

The ECM uses the P/N signal as one of the inputs to control idle air control and VSS diagnostics.

Test Description: Numbers below refer to circled numbers on the diagnostic chart.
1. Checks for a closed switch to ground in park position. Different makes of "Scan" tools will read P/N differently. Refer to tool operations manual for type of display used.

2. Checks for an open switch in drive range.
3. Be sure "Scan" indicates "Drive", even while wiggling shifter to test for an intermittent or misadjusted switch in drive range.

1988–90 2.5L (VIN U) – ALL MODELS

CHART C-1A
PARK/NEUTRAL SWITCH DIAGNOSIS
2.5L (VIN U)

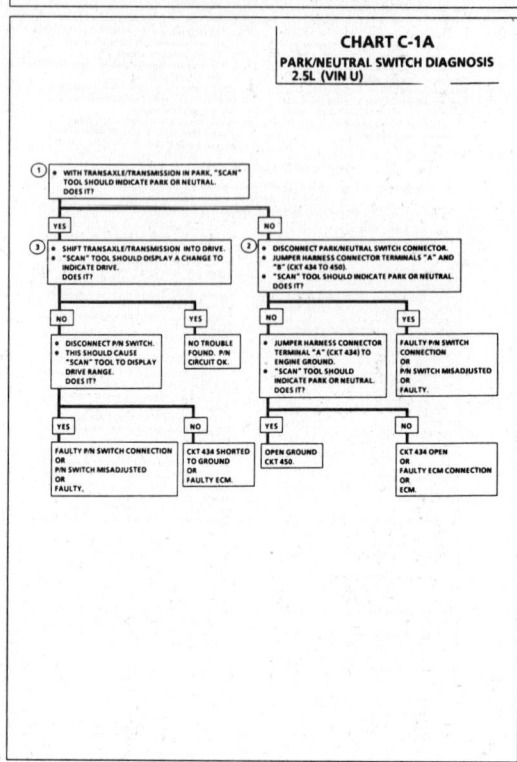

1988–90 2.5L (VIN U) – ALL MODELS

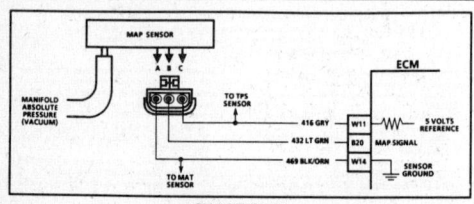

CHART C-1D
MANIFOLD ABSOLUTE PRESSURE (MAP) OUTPUT CHECK
2.5L (VIN U)

Circuit Description:

The Manifold Absolute Pressure (MAP) sensor measures manifold pressure (vacuum) and sends that signal to the ECM. The MAP sensor is mainly used to calculate engine load, which is a fundamental input for spark advance and fuel calculations. The MAP sensor is also used to determine the barometric pressure.

Test Description: Numbers below refer to circled numbers on the diagnostic chart.
1. Checks MAP sensor output voltage to the ECM. This voltage, without engine running, represents a barometer reading to the ECM. Comparison of this BARO reading with a known good vehicle with the same sensor is a good way to check the accuracy of a "suspect" sensor. Readings should be within .4 volt.

2. Applying 34 kPa (10" Hg) vacuum to the MAP sensor should cause the voltage to be 1.2 to 2.3 volts less than the voltage at Step 1. Upon applying vacuum to the sensor, the change in voltage should be instantaneous. A slow voltage change indicates a faulty sensor.
3. Check vacuum hose to sensor for leaking or restriction. Ensure that no other vacuum devices are connected to the MAP hose.

1988–90 2.5L (VIN U) – ALL MODELS

CHART C-1D
MANIFOLD ABSOLUTE PRESSURE (MAP) OUTPUT CHECK
2.5L (VIN U)

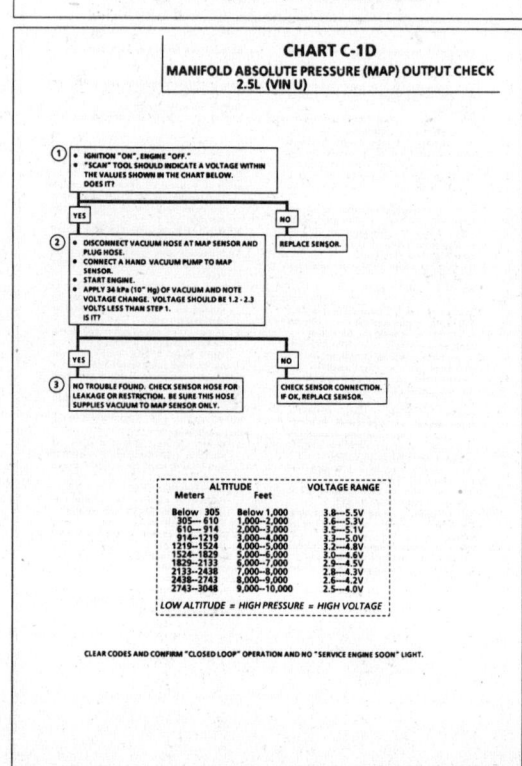

1988–90 2.5L (VIN U) – ALL MODELS

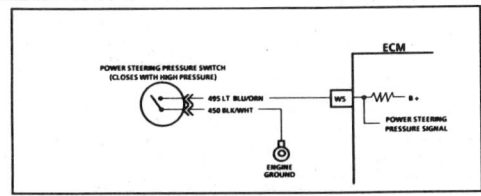

CHART C-1E
POWER STEERING PRESSURE SWITCH (PSPS) DIAGNOSIS
2.5L (VIN U)

Circuit Description:
The Power Steering Pressure Switch (PSPS) is normally open to ground, and CKT 495 will be near the battery voltage.
Turning the steering wheel increases power steering oil pressure and pump load on the engine. The pressure switch will close before the load can cause an idle problem.
Closing the switch causes CKT 495 to read less than 1 volt and the ECM will increase the idle air rate.
- A pressure switch that will not close, or an open CKT 495 or 450, may cause the engine to stop when power steering loads are high.
- A switch that will not open, or a CKT 495 shorted to ground, may affect idle quality.

Test Description: Numbers below refer to circled numbers on the diagnostic chart.
1. Different makes of "Scan" tools may display the state of this switch in different ways. Refer to "Scan" tool user's manual to determine how this input is indicated.
2. Checks to determine if CKT 495 is shorted to ground.
3. This should simulate a closed switch.

1988–90 2.5L (VIN U) – ALL MODELS

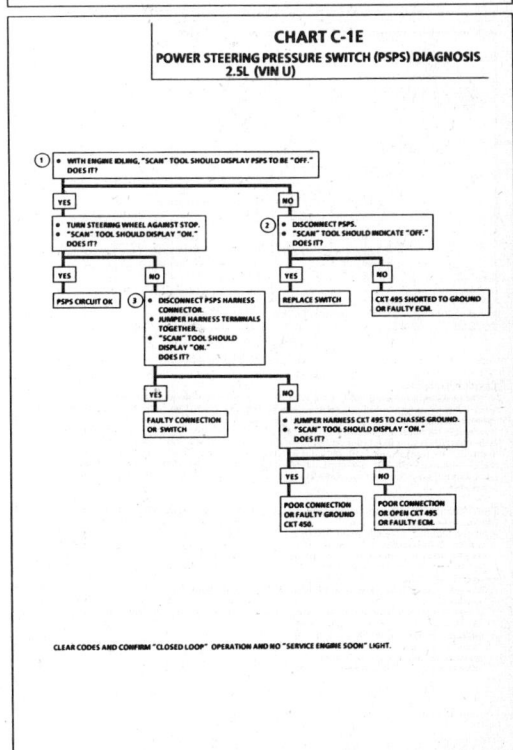

1988–90 2.5L (VIN U) – ALL MODELS

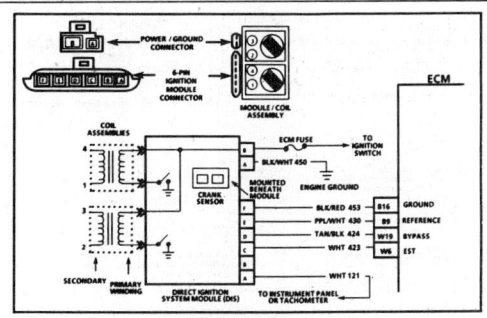

CHART C-4D-1
"DIS" MISFIRE AT IDLE
2.5L (VIN U)

Circuit Description:
The Direct Ignition System (DIS) uses a waste spark method of distribution. In this type of system, the ignition module triggers a dual coil, resulting in both contacts spark plugs firing at the same time. One cylinder is on its compression stroke at the same time that the other is on the exhaust stroke, resulting in a lower energy requirement to fire the spark plug in the cylinder on its exhaust stroke. This leaves the remainder of the high voltage to be used to fire the spark plug which is in the compression stroke. On this application, the crank sensor is mounted to the bottom of the coil/module assembly and protrudes through the block to within approximately .050" of the crankshaft reluctor. Since the reluctor is a machined portion of the crankshaft and the crank sensor is mounted in a fixed position on the block, timing adjustments are not possible or necessary.

Test Description: Numbers below refer to circled numbers on the diagnostic chart.
1. If the "Misfire" complaint exists under load only, CHART C-4D-2 must be used. Engine rpm should drop approximately equally on all plug loads.
2. A spark tester such as a ST-125 must be used because it is essential to verify adequate available secondary voltage at the spark plug. (25,000 volts.)
3. If the spark jumps the test gap after grounding the opposite plug wire, it indicates excessive resistance in the plug which was bypassed. A faulty or poor connection at that plug could also result in the miss condition. Also check for carbon deposits inside the spark plug boot.
4. If carbon tracking is evident, replace coil and be sure plug wires relating to that coil are clean and tight. Excessive wire resistance or faulty connections could have caused the coil to be damaged.
5. If the no spark condition follows the suspected coil, the ignition module is faulty. Otherwise, the ignition module is the cause of no spark. This test could also be performed by substituting a known good coil for the one causing the no spark condition.

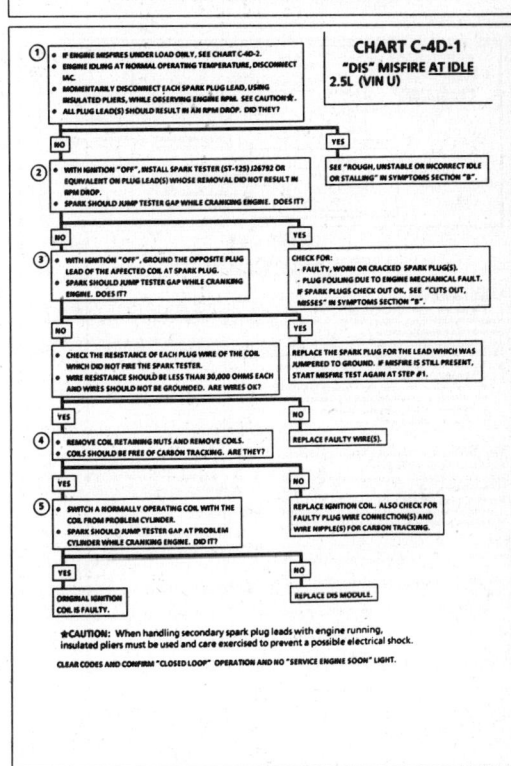

1988–90 2.5L (VIN U) – ALL MODELS

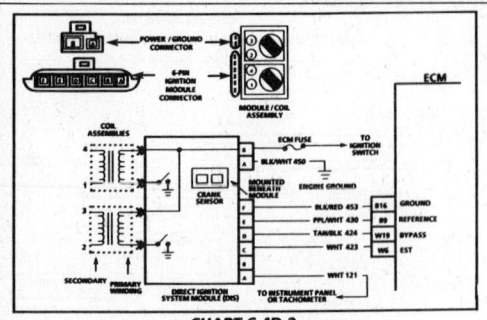

CHART C-4D-2
"DIS" MISFIRE UNDER LOAD
2.5L (VIN U)

Circuit Description:

The direct ignition system (DIS) uses a waste spark method of distribution. In this type of system, the ignition module triggers a dual coil, resulting in both connected spark plugs firing at the same time. One cylinder is on its compression stroke at the same time that the other is on the exhaust stroke, resulting in a lower energy requirement to fire the spark plug in the cylinder on its exhaust stroke. This leaves the remainder of the high voltage to be used to fire the spark plug which is in the cylinder on its compression stroke. On this application, the crank sensor is mounted to the bottom of the coil/module assembly and protrudes through the block to within approximately .050" of the crankshaft reluctor. Since the reluctor is a machined portion of the crankshaft and the crank sensor is mounted in a fixed position on the block, timing adjustments are not possible or necessary.

Test Description: Numbers below refer to circled numbers on the diagnostic chart.

1. If the "Misfire" complaint exists at idle only, CHART C-4D-1 must be used. A spark tester such as a ST-125 must be used because it is essential to verify adequate available secondary voltage at the spark plug. (25,000 volts). Spark should jump the test gap on all 4 leads. This simulates a "load" condition.

2. If the spark jumps the tester gap after grounding the opposite plug wire, it indicates excessive resistance in the plug which was bypassed. A faulty or poor connection at that plug could also result in the miss condition. Also check for carbon deposits inside the spark plug boot.

3. If carbon tracking is evident replace coil and be sure plug wires relating to that coil are clean and tight. Excessive wire resistance or faulty connections could have caused the coil to be damaged.

4. If the no spark condition follows the suspected coil, that coil is faulty. Otherwise, the ignition module is the cause of no spark. This test could also be performed by substituting a known good coil for the one causing the no spark condition.

1988–90 2.5L (VIN U) – ALL MODELS

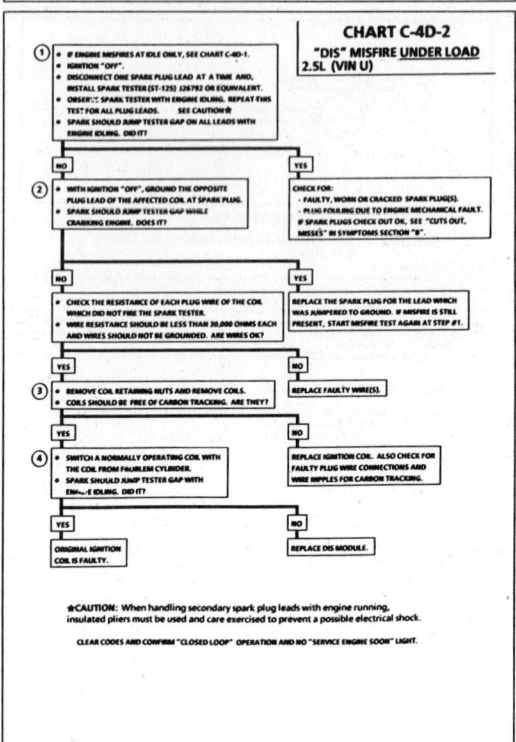

CHART C-4D-2
"DIS" MISFIRE UNDER LOAD
2.5L (VIN U)

★CAUTION: When handling secondary spark plug leads with engine running, insulated pliers must be used and care exercised to prevent a possible electrical shock.

CLEAR CODES AND CONFIRM "CLOSED LOOP" OPERATION AND NO "SERVICE ENGINE SOON" LIGHT.

1988–90 2.5L (VIN U) – ALL MODELS

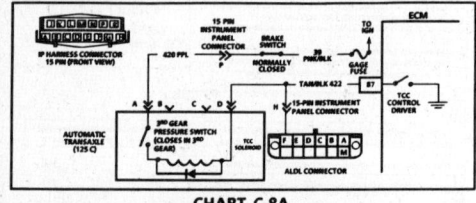

CHART C-8A
TORQUE CONVERTER CLUTCH (TCC)
(ELECTRICAL DIAGNOSIS)
2.5L (VIN U)

Circuit Description:

The purpose of the Torque Converter Clutch (TCC) is to eliminate the power loss of the torque converter when the vehicle is in a cruise condition. This allows the convenience of the automatic transaxle and the fuel economy of a manual transaxle.

Voltage is supplied to the TCC solenoid through the brake switch and transaxle third gear apply switch. The ECM will engage TCC by grounding CKT 422 to energize the solenoid.

The TCC will engage under the following conditions:
- Vehicle speed exceeds 30 mph (48 km/h)
- Engine coolant temperature exceeds 70°C (156°F)
- Throttle position is not changing faster than a calibrated rate (steady throttle)
- Transaxle third gear switch is closed
- Brake switch is closed

Test Description: Numbers below refer to circled numbers on the diagnostic chart.

1. Light "OFF" confirms transaxle third gear apply switch is open.

2. At 48 km/h (30 mph), the transaxle third gear switch should close. Test light will light and confirm battery supply and closed brake switch.

3. Grounding the diagnostic terminal, with engine "OFF," should energize the TCC solenoid. This test checks the capability of the ECM to control the solenoid.

Diagnostic Aids:

An engine coolant thermostat that is stuck open or opens at too low a temperature may result in an inoperative TCC.

1988–90 2.5L (VIN U) – ALL MODELS

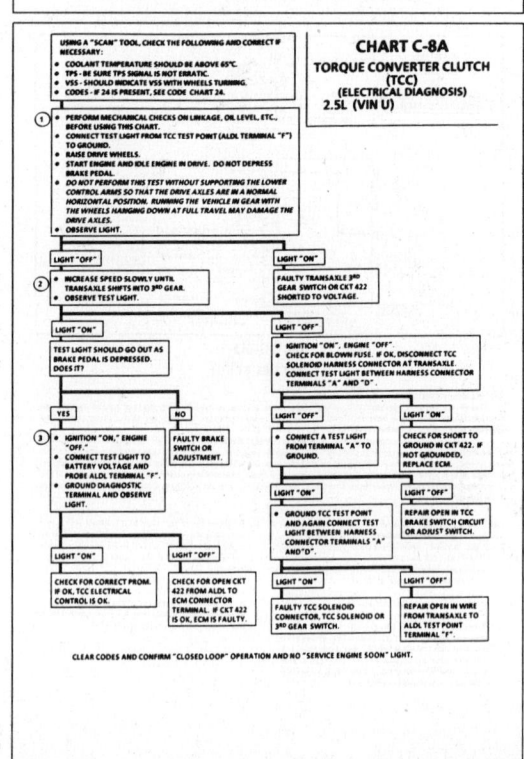

CHART C-8A
TORQUE CONVERTER CLUTCH (TCC)
(ELECTRICAL DIAGNOSIS)
2.5L (VIN U)

CLEAR CODES AND CONFIRM "CLOSED LOOP" OPERATION AND NO "SERVICE ENGINE SOON" LIGHT.

1988–90 2.5L (VIN U) – ALL MODELS

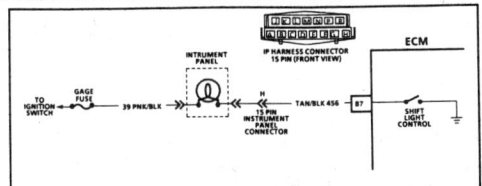

CHART C-8B
MANUAL TRANSMISSION (M/T) SHIFT LIGHT CHECK
2.5L (VIN U)

Circuit Description:
The shift light indicates the best transmission shift point for maximum fuel efficiency. The light is controlled by the ECM and is turned "ON" by grounding CKT 456.
The ECM uses information from the following inputs to control the shift light:
- Coolant temperature
- Throttle position
- Vehicle speed
- Engine speed

The ECM uses the measured engine and vehicle speeds to calculate what gear the vehicle is in. It is this calculation which determines when the shift light should be turned "ON".

Test Description: Numbers below refer to circled numbers on the diagnostic chart.
1. This should not turn "ON" the shift light. If the light is "ON," there is a short to ground in CKT 456 or a fault in the ECM.
2. When the diagnostic terminal is grounded, the ECM should ground CKT 456, and the shift light should be "ON."

3. This checks the shift light circuit up to the ECM connector. If the shift light lights, then the ECM connector is faulty or the ECM does not have the ability to ground the circuit.

1988–90 2.5L (VIN U) – ALL MODELS

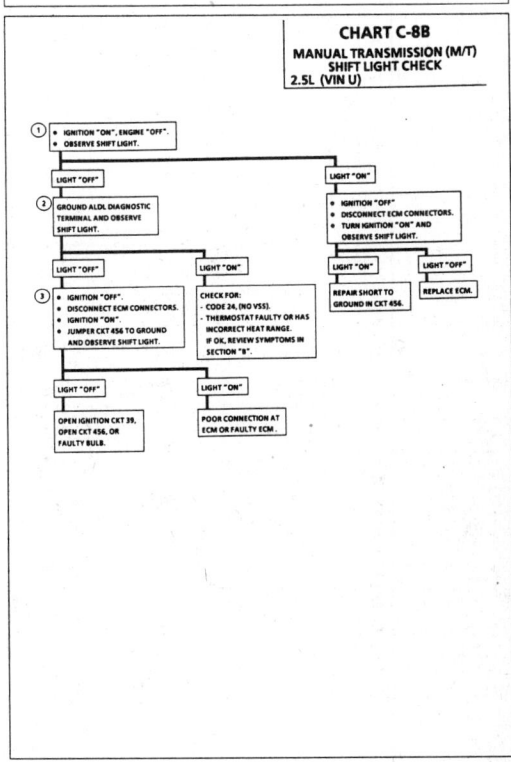

1988–90 2.5L (VIN U) – ALL MODELS

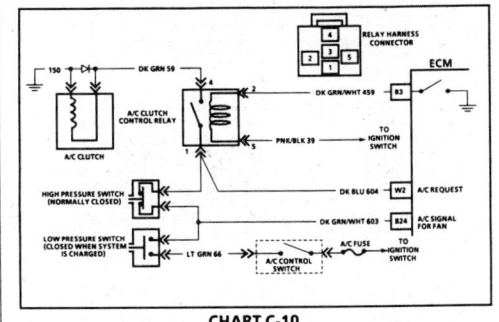

CHART C-10
A/C CLUTCH CONTROL
2.5L (VIN U)

Circuit Description:
When an A/C mode is selected on the A/C control switch, ignition voltage is supplied to the compressor low pressure switch. If there is sufficient A/C refrigerant pressure, the low pressure switch will be closed and complete the circuit to the closed high pressure cut-off switch and to CKTs 67 and 604. The voltage on CKT 604 to the ECM is shown by the "Scan" tool as A/C request "ON" (voltage present) or "OFF" (no voltage). When a request for A/C is sensed by the ECM, the ECM will ground CKT 459 of the A/C clutch control relay, the relay contact will close, and current will flow from CKT 604 to CKT 59 to engage the A/C compressor clutch. A "Scan" tool will show the grounding of CKT 459 as A/C clutch "ON." If voltage is sensed by the ECM on CKT 67, the cooling fan will be turned "ON."

Test Description: Numbers below refer to circled numbers on the diagnostic chart.
1. The ECM will energize the A/C relay only when the engine is running. This test will determine if the relay or CKT 459 is faulty.
2. The low pressure and high pressure switches must be closed so that the A/C request signal (12 volts) will be present at the ECM.
3. A short to ground in any part of the A/C request or A/C clutch control circuits could be the cause of the blown fuse.
4. With the engine idling and A/C "ON," the ECM should be grounding CKT 459, causing the test light to be "ON."
5. Determines if the signal is reaching the low pressure switch on CKT 66 from the A/C control

panel. The signal should be present only when the A/C mode or defrost mode has been selected.

Diagnostic Aids:

If complaint was insufficient cooling, the problem may be caused by an inoperative cooling fan or A/C pressure fan switch. The engine cooling fan should turn "ON" when A/C pressure exceeds a value to open the switch, which causes the ECM to energize the cooling fan relay. See CHART C-12 for cooling fan diagnosis.

The A/C clutch will be disengaged if a high power steering pressure signal is detected by the ECM. Refer to CHART C1-E.

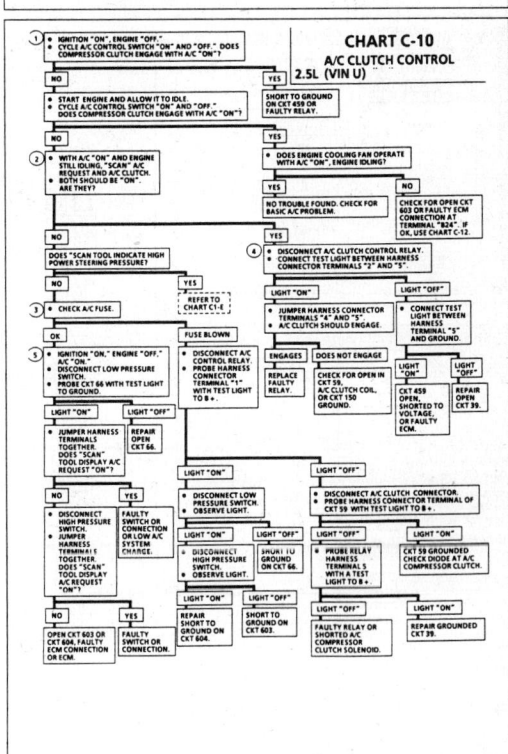

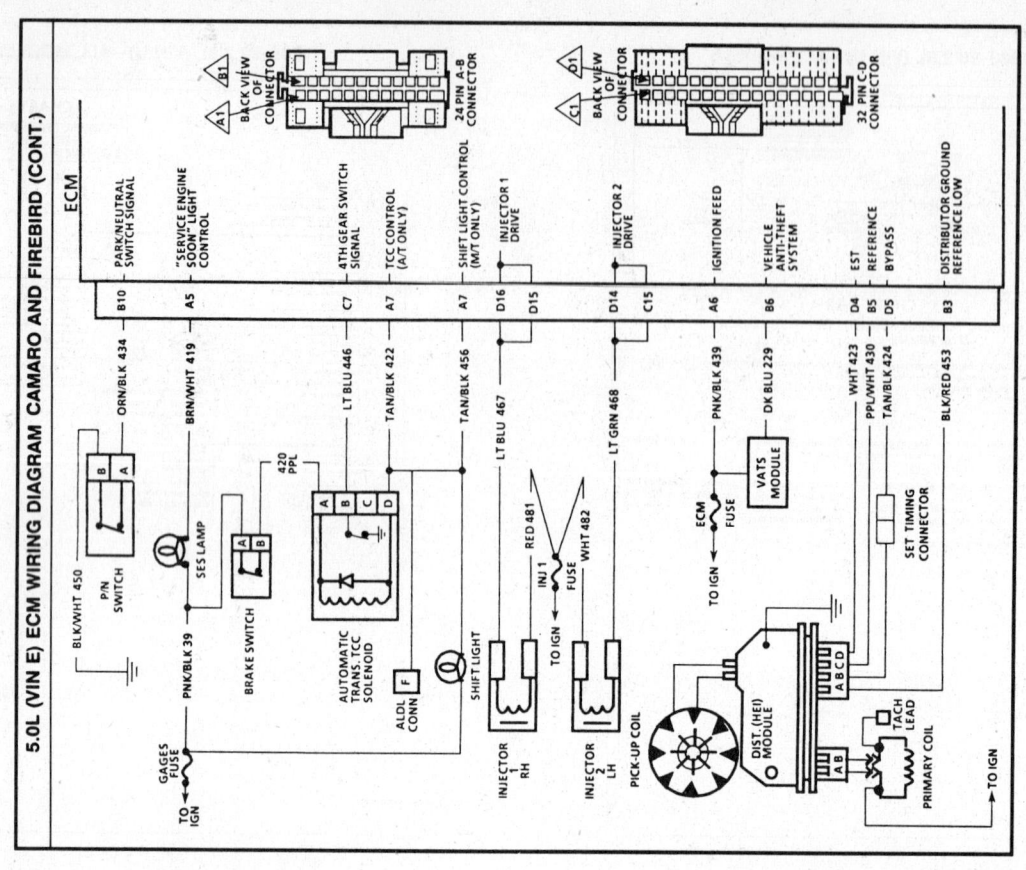

5.0L (VIN E) ECM WIRING DIAGRAM CAMARO AND FIREBIRD (CONT.)

5.0L (VIN E) ECM CONNECTOR TERMINAL END VIEW—CAMARO AND FIREBIRD

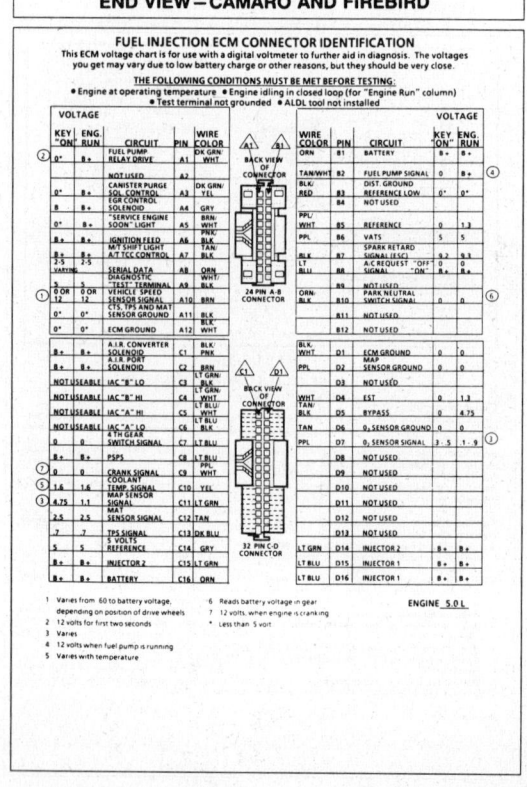

5.0L (VIN E) COMPONENT LOCATIONS
CAMARO AND FIREBIRD

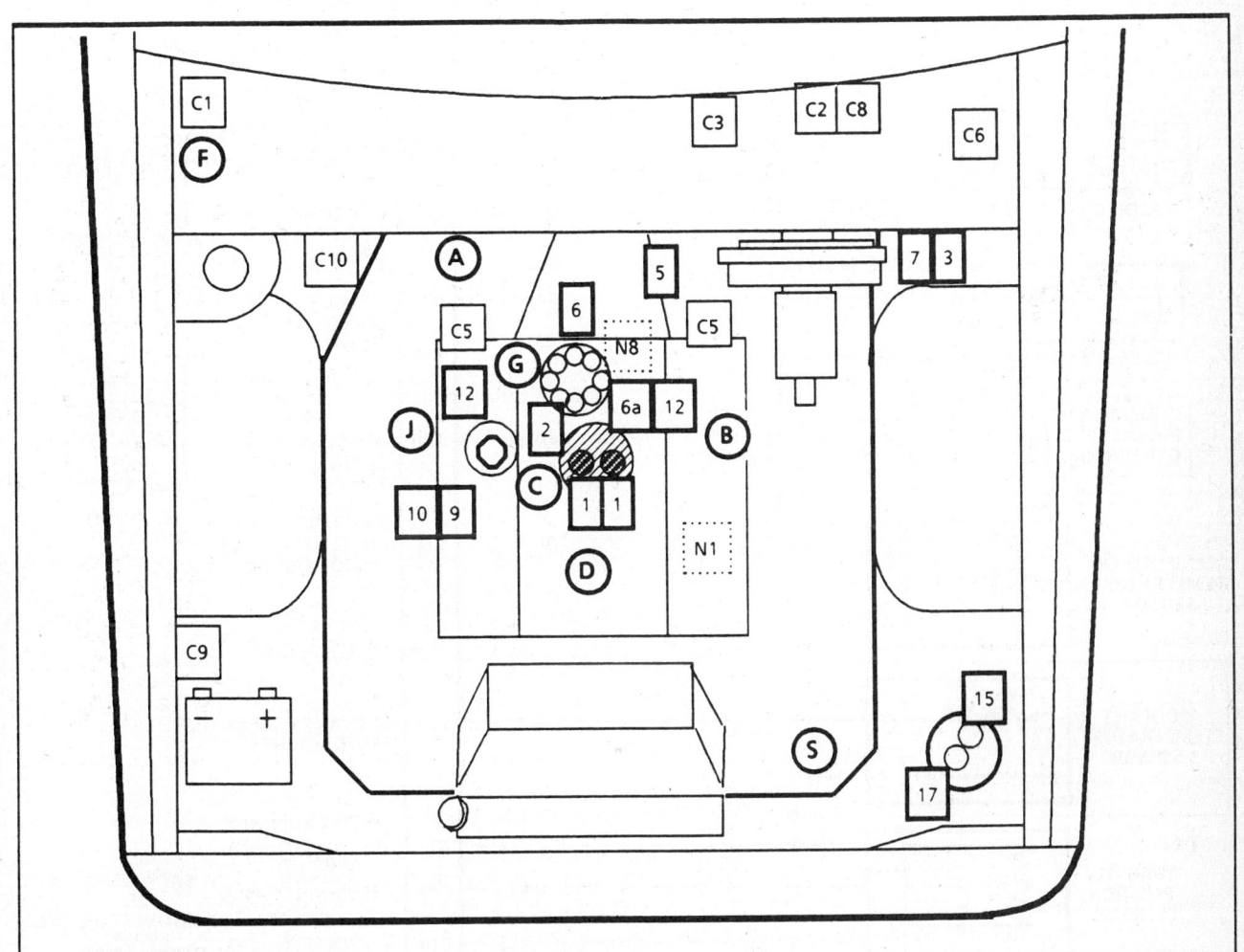

☐ COMPUTER HARNESS

C1 Electronic Control Module
C2 ALDL diagnostic connector
C3 "SERVICE ENGINE SOON" light
C5 ECM harness grounds
C6 Fuse panel
C8 Fuel pump test connector
      (Terminal "G" of ALDL Connector)
C9 Fuel pump/ECM fuse
C10 Set timing connector

☐ NOT ECM CONNECTED

N1 Crankcase vent valve (PCV)
N8 Oil pressure switch

☐ CONTROLLED DEVICES

1   Fuel injectors
2   Idle air control motor
3   Fuel pump relay
5   Torque Conv. Clutch connector
6   EST distributor
6a Remote ignition coil
7   Electronic Spark Control module
9   AIR port solenoid
10 AIR converter solenoid
12 Exh. Gas Recirc. vacuum solenoid
15 Fuel vapor canister solenoid
17 Fuel vapor canister
⬡ Exhaust Gas Recirculation valve

◯ INFORMATION SENSORS

A Manifold Absolute Pressure
B Exhaust oxygen
C Throttle position
D Coolant temperature
F Vehicle speed (buffer)
G MAT (in air cleaner)
J ESC knock
S Power Steering Pressure Switch

5.0L (VIN E) ECM WIRING DIAGRAM CAMARO AND FIREBIRD

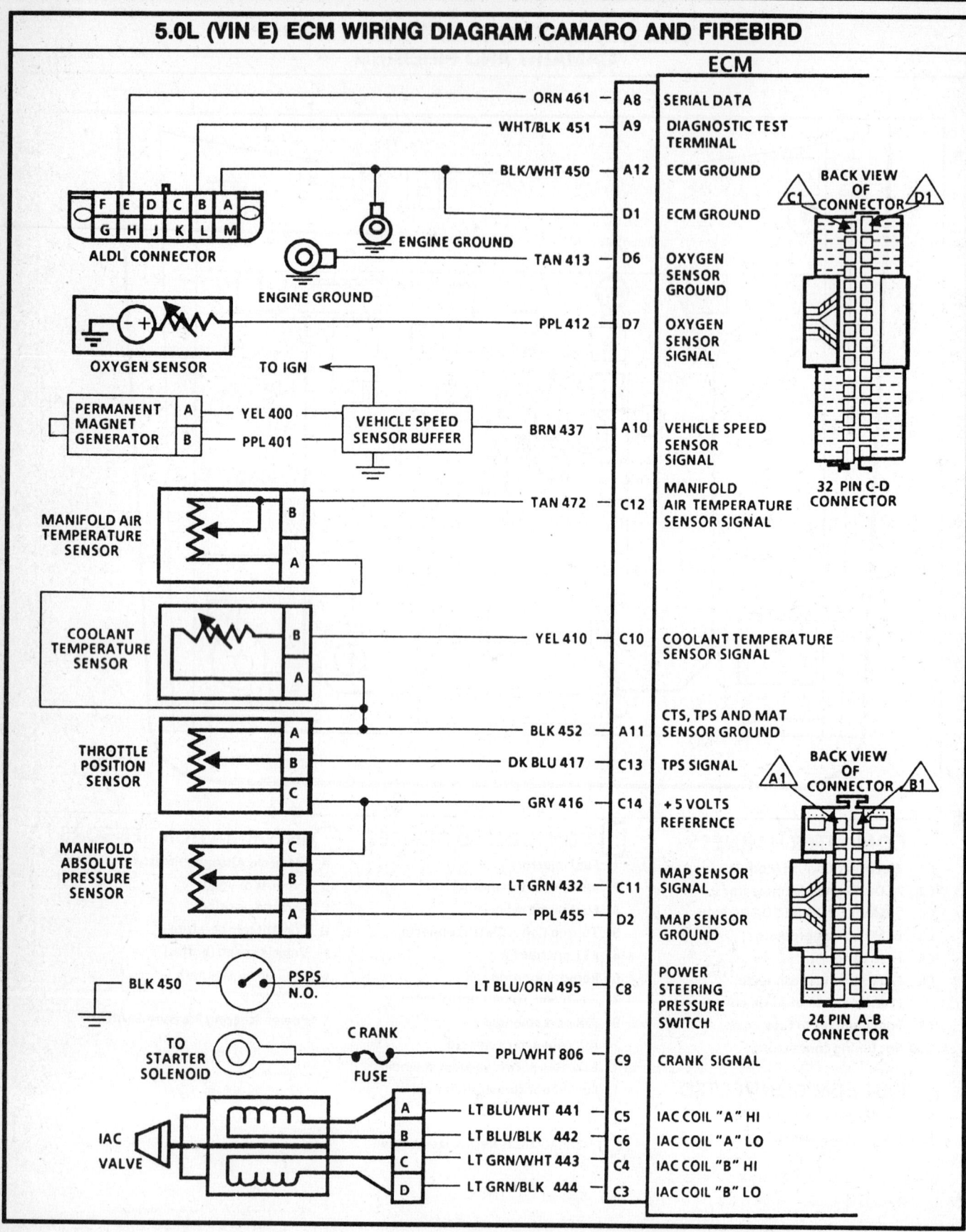

5.0L (VIN E) ECM WIRING DIAGRAM CAMARO AND FIREBIRD (CONT.)

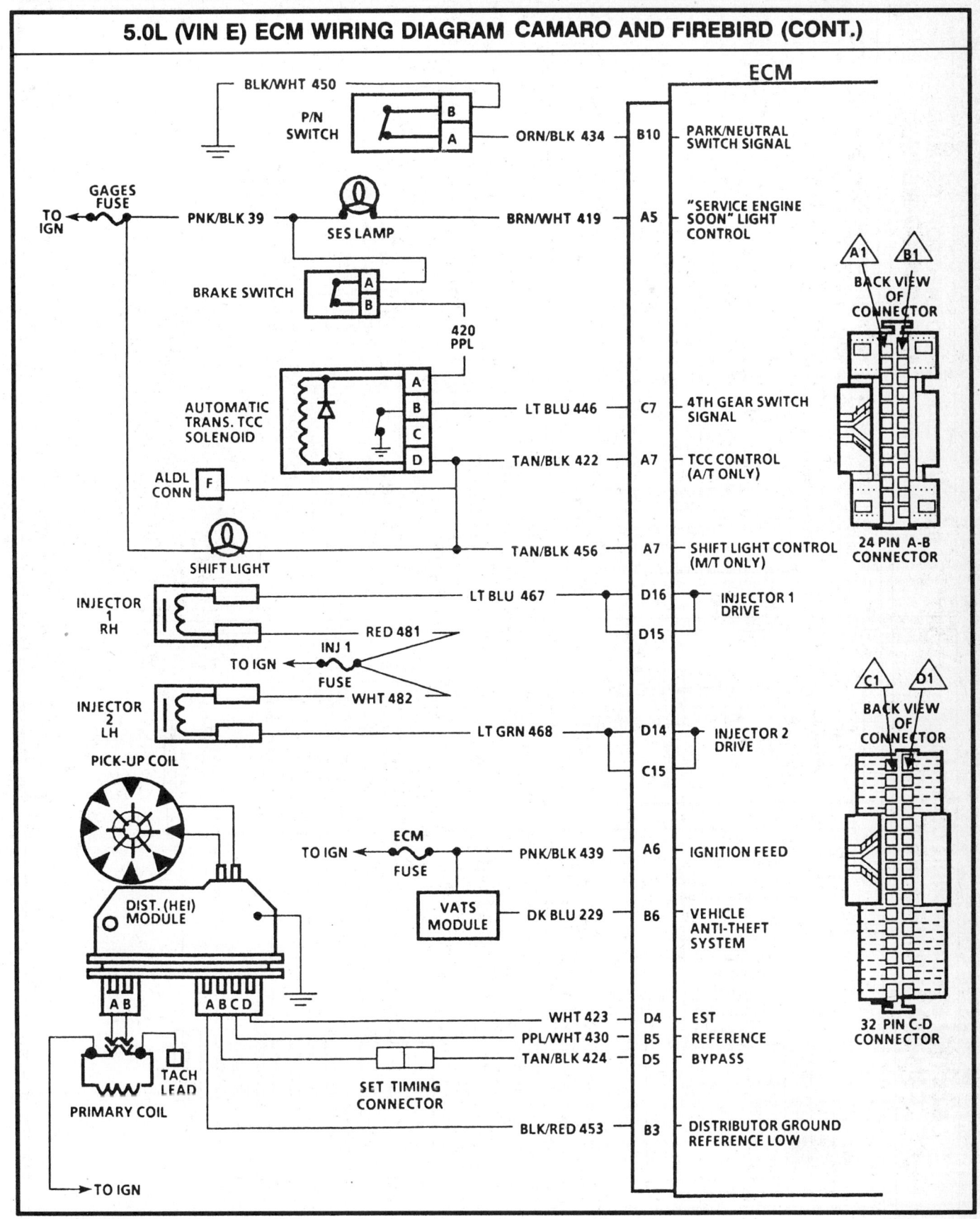

5.0L (VIN E) ECM CONNECTOR TERMINAL END VIEW—CAMARO AND FIREBIRD

FUEL INJECTION ECM CONNECTOR IDENTIFICATION

This ECM voltage chart is for use with a digital voltmeter to further aid in diagnosis. The voltages you get may vary due to low battery charge or other reasons, but they should be very close.

THE FOLLOWING CONDITIONS MUST BE MET BEFORE TESTING:
- Engine at operating temperature
- Engine idling in closed loop (for "Engine Run" column)
- Test terminal not grounded
- ALDL tool not installed

WIRE COLOR	PIN	CIRCUIT	KEY "ON" RUN	ENG. RUN
ORN	B1	BATTERY	B+	B+
TAN/WHT	B2	FUEL PUMP SIGNAL	0	B+
BLK/RED	B3	DIST. GROUND	0*	0*
	B4	REFERENCE LOW		
PPL/WHT	B5	REFERENCE	0	1.3
PPL	B6	VATS	5	5
BLK	B7	SPARK RETARD SIGNAL (ESC)	9.2	9.3
LT BLU	B8	A/C REQUEST SIGNAL	0	0
ORN/BLK	B9	PARK/NEUTRAL SWITCH SIGNAL "OFF" "ON"	B+	B+
	B10	NOT USED	0	0
	B11	NOT USED		
	B12	NOT USED		

WIRE COLOR	PIN	CIRCUIT	KEY "ON" RUN	ENG. RUN
BLK/WHT	D1	ECM GROUND	0	0
PPL	D2	MAP SENSOR GROUND	0	0
	D3	NOT USED		
WHT/BLK	D4	EST	0	1.3
TAN/BLK	D5	BYPASS	0	4.75
TAN	D6	O₂ SENSOR SIGNAL	.3 - .5	.1 - .9
PPL	D7	O₂ SENSOR GROUND		
	D8	NOT USED		
	D9	NOT USED		
	D10	NOT USED		
	D11	NOT USED		
	D12	NOT USED		
	D13	NOT USED		
LT GRN	D14	INJECTOR 2	B+	B+
LT BLU	D15	INJECTOR 1	B+	B+
LT BLU	D16	INJECTOR 1	B+	B+

ENGINE 5.0L

24 PIN A-B CONNECTOR — BACK VIEW OF CONNECTOR

32 PIN C-D CONNECTOR — BACK VIEW OF CONNECTOR

WIRE COLOR	PIN	CIRCUIT	KEY "ON" RUN	ENG. RUN
DK GRN/WHT	A1	FUEL PUMP RELAY DRIVE	0*	B+
	A2	NOT USED		
DK GRN/YEL	A3	CANISTER PURGE EGR CONTROL SOLENOID	0*	B+
GRY	A4	EGR CONTROL SOLENOID	B	B+
BRN/WHT	A5	"SERVICE ENGINE SOON" LIGHT	0*	B+
BLK/PNK	A6	IGNITION FEED	B+	B+
TAN/WHT	A7	M/T SHIFT LIGHT A/T TCC CONTROL	B+	B+
BLK	A8	SERIAL DATA	2-5	2-5
ORN	A9	"TEST TERMINAL" DIAGNOSTIC	VARYING	
WHT/BLK	A10	VEHICLE SPEED SENSOR SIGNAL	0 OR 12	0 OR 12
BRN	A11	CTS, TPS AND MAT SENSOR GROUND	0*	0*
BLK/WHT	A12	ECM GROUND	0*	0*

WIRE COLOR	PIN	CIRCUIT	KEY "ON" RUN	ENG. RUN
BLK/PNK	C1	A.I.R. CONVERTER SOLENOID	B+	B+
BRN	C2	A.I.R. PORT SOLENOID	B+	B+
LT GRN/BLK	C3	IAC "B" LO	NOT USEABLE	
LT GRN/WHT	C4	IAC "B" HI	NOT USEABLE	
LT BLU/WHT	C5	IAC "A" HI	0	0
LT BLU/BLK	C6	IAC "A" LO	0	0
LT BLU	C7	4TH GEAR SWITCH SIGNAL	B+	B+
LT BLU	C8	PSPS	0	0
WHT	C9	CRANK SIGNAL	1.6	1.6
YEL	C10	COOLANT TEMP SIGNAL	1.6	
LT GRN	C11	MAP SENSOR SIGNAL	1.1	1.1
TAN	C12	MAT SENSOR SIGNAL	2.5	2.5
DK BLU	C13	TPS SIGNAL	.7	.7
GRY	C14	5 VOLTS REFERENCE	5	5
LT GRN	C15	INJECTOR 2	B+	B+
ORN	C16	BATTERY	B+	B+

1 Varies from .60 to battery voltage, depending on position of drive wheels
3 Varies
4 12 volts for first two seconds
5 Varies with temperature
6 Reads battery voltage in gear
7 12 volts, when engine is cranking
* Less than 5 volt

5.0L (VIN E) ECM WIRING DIAGRAM CAMARO AND FIREBIRD (CONT.)

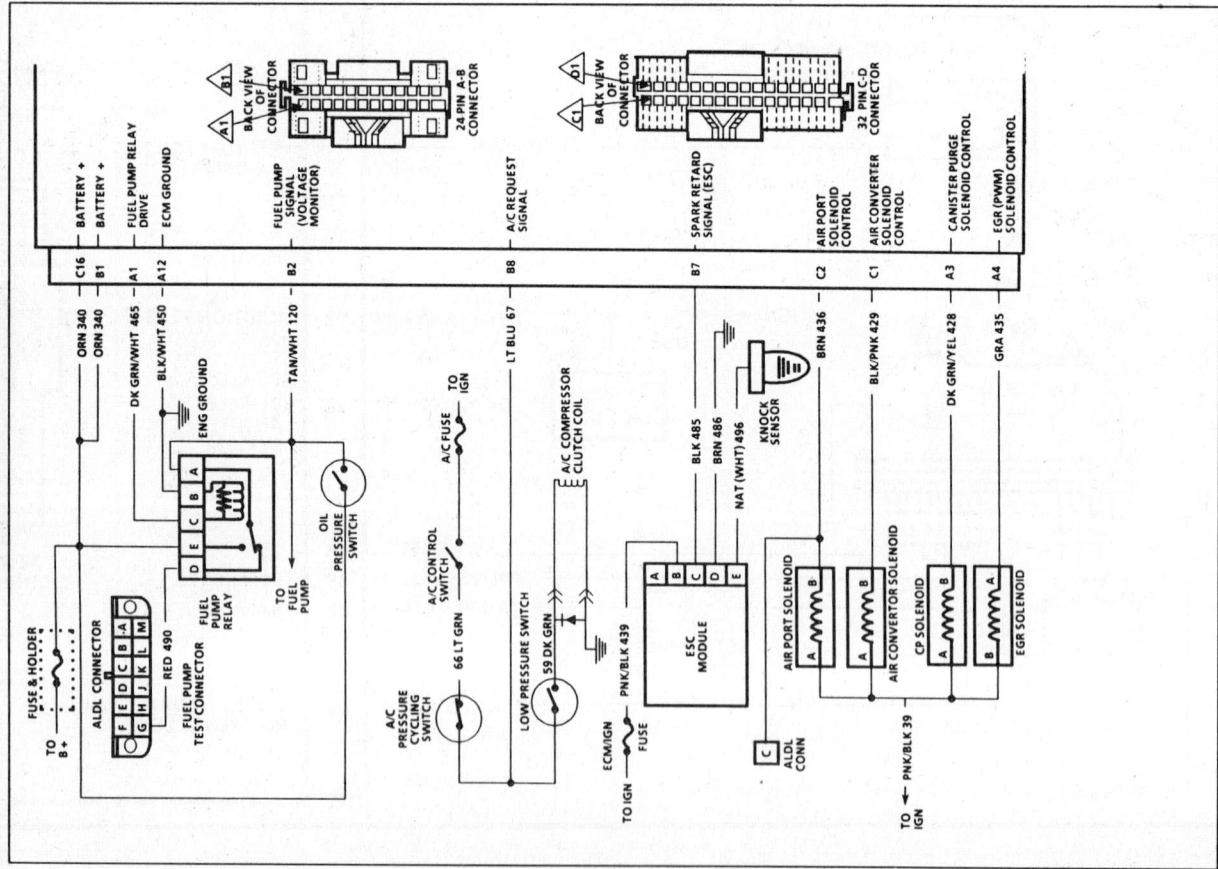

1988–90 5.0L (VIN E) – CAMARO AND FIREBIRD

DIAGNOSTIC CIRCUIT CHECK

The Diagnostic Circuit Check is an organized approach to identifying a problem created by an electronic engine control system malfunction. It must be the starting point for any driveability complaint diagnosis, because it directs the service technician to the next logical step in diagnosing the complaint.

The "Scan Data" listed in the table may be used for comparison, after completing the diagnostic circuit check and finding the on-board diagnostics functioning properly and no trouble codes displayed. The "Typical Values" are an average of display values recorded from normally operating vehicles and are intended to represent what a normally functioning system would typically display.

A "SCAN" TOOL THAT DISPLAYS FAULTY DATA SHOULD NOT BE USED, AND THE PROBLEM SHOULD BE REPORTED TO THE MANUFACTURER. THE USE OF A FAULTY "SCAN" CAN RESULT IN MISDIAGNOSIS AND UNNECESSARY PARTS REPLACEMENT.

Only the parameters listed below are used in this manual for diagnosis. If a "Scan" tool reads other parameters, the values are not recommended by General Motors for use in diagnosis. For more description on the values and use of the "Scan" to diagnose ECM inputs, refer to the applicable diagnosis section in Section "C". If all values are within the range illustrated, refer to "Symptoms" in Section "B".

"SCAN" DATA

Idle / Upper Radiator Hose Hot / Closed Throttle / Park or Neutral / Closed Loop / Acc. off

"SCAN" Position	Units Displayed	Typical Data Value
Coolant Temp	C°	85° - 105°
TPS	Volts	.4 - 1.25
MAP	Volts	1 - 2 (depends on Vac. & Baro pressure)
INT (Integrator)	Counts	Varies
BLM (Block Learn)	Counts	118 - 138
IAC	Counts (steps)	1 - 50
RPM	RPM	650 ± 100 RPM (550 ± 50 in drive)
O₂	Volts	.001 - .999 and varies
Open/Closed Loop	Open/Closed	Closed Loop (may go open with extended idle)
A/C Request	Yes/No	No (yes, with A/C requested)
P/N Switch	P/N and RDL	Park/Neutral (P/N)
TCC	On/Off	Off/ (on, with TCC commanded)
VSS	MPH	0
Battery	Volts	13.5 - 14.5
Air Divert	Normal/Divert	Normal
Air Switch	Converter/Port	Converter
Knock Signal	Yes/No	No
4th Gear	Yes/No	No (Yes, if in 4th Gear)
MAT Temp	C°	10° - 90° (Depends on air cleaner temperature).
Power Steering Pressure Switch	Normal/Hi Pressure	Normal

1988–90 5.0L (VIN E) – CAMARO AND FIREBIRD

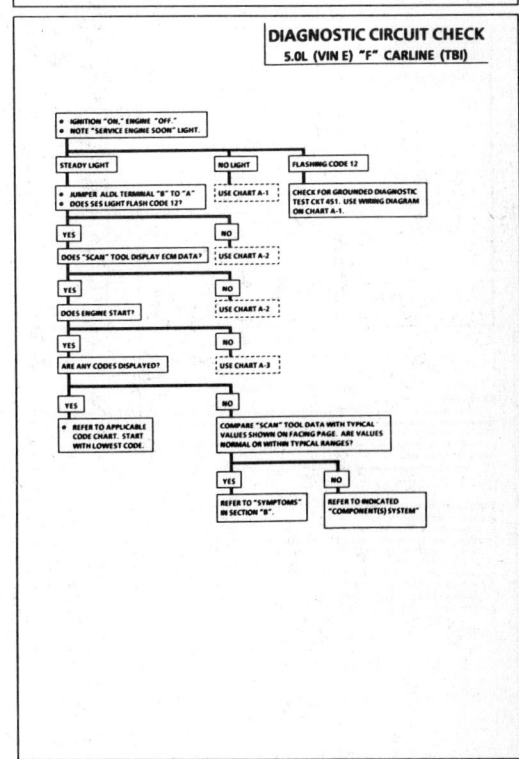

DIAGNOSTIC CIRCUIT CHECK
5.0L (VIN E) "F" CARLINE (TBI)

1988–90 5.0L (VIN E) – CAMARO AND FIREBIRD

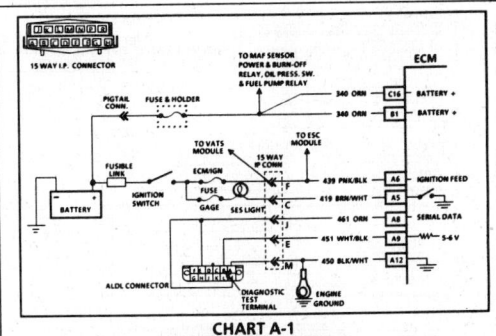

CHART A-1
NO "SERVICE ENGINE SOON" LIGHT
5.0L (VIN E) "F" CARLINE (TBI)

Circuit Description:

There should always be a steady "Service Engine Soon" light, when the ignition is "ON" and engine stopped. Battery is supplied directly to the light bulb. The electronic control module (ECM) will control the light and turn it "ON" by providing a ground path through CKT 419 to the ECM.

Test Description: Numbers below refer to circled numbers on the diagnostic chart.

1. Battery feed CKT 340 is protected by a 20 amp in-line fuse. If this fuse was blown, refer to wiring diagram on the facing page of Code 54.
2. Using a test light connected to 12 volts, probe each of the system ground circuits to be sure a good ground is present. See ECM terminal end view in front of this section for ECM pin locations of ground circuits.

Diagnostic Aids:

Engine runs ok, check:
- Faulty light bulb
- CKT 419 open
- Gage fuse blown. This will result in no oil light, generator light, seat belt reminder, etc.

Engine cranks, but will not run, check:
- Continuous battery - fuse or fusible link open.
- ECM/Ignition fuse open.
- Battery CKT 340 to ECM open
- Ignition CKT 439 to ECM open
- Poor connection to ECM

1988–90 5.0L (VIN E) – CAMARO AND FIREBIRD

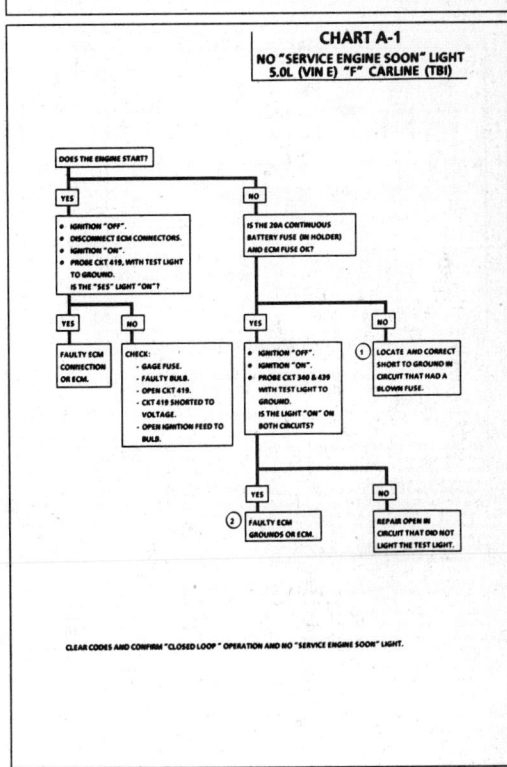

CHART A-1
NO "SERVICE ENGINE SOON" LIGHT
5.0L (VIN E) "F" CARLINE (TBI)

1988–90 5.0L (VIN E) – CAMARO AND FIREBIRD

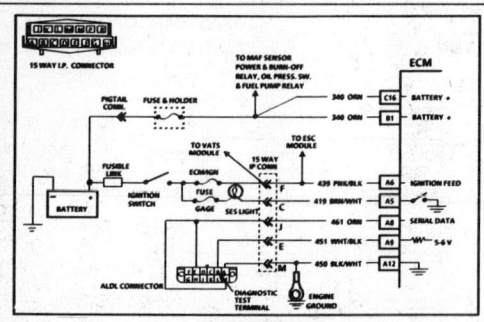

CHART A-2
NO ALDL DATA OR WILL NOT FLASH CODE 12
"SERVICE ENGINE SOON" LIGHT "ON" STEADY
5.0L (VIN E) "F" CARLINE (TBI)

Circuit Description:
There should always be a steady "Service Engine Soon" light, when the ignition is "ON" and engine stopped. Ignition voltage is supplied directly to the light bulb. The electronic control module (ECM) will turn the light "ON" by grounding CKT 419 in the ECM.

With the diagnostic "test" terminal grounded, the light should flash a Code 12, followed by any trouble code(s) stored in memory.

A steady light suggests a short to ground in the light control CKT 419, or an open in diagnostic CKT 451.

Test Description: Numbers below refer to circled numbers on the diagnostic chart.
1. If there is a problem with the ECM that causes a "Scan" tool to not read serial data, the ECM should not flash a Code 12. If Code 12 does flash, be sure that the "Scan" tool is working properly on another vehicle. If the "Scan" is functioning properly and CKT 461 is OK, the PROM or ECM may be at fault for the NO ALDL symptom.
2. If the light goes "OFF" when the ECM connector is disconnected, then CKT 419 is not shorted to ground.
3. This step will check for an open diagnostic CKT 451.
4. At this point, the "Service Engine Soon" light wiring is OK. The problem is a faulty ECM or PROM. If Code 12 does not flash, the ECM should be replaced using the original PROM. Replace the PROM only after trying an ECM, as a defective PROM is an unlikely cause of the problem.

1988–90 5.0L (VIN E) – CAMARO AND FIREBIRD

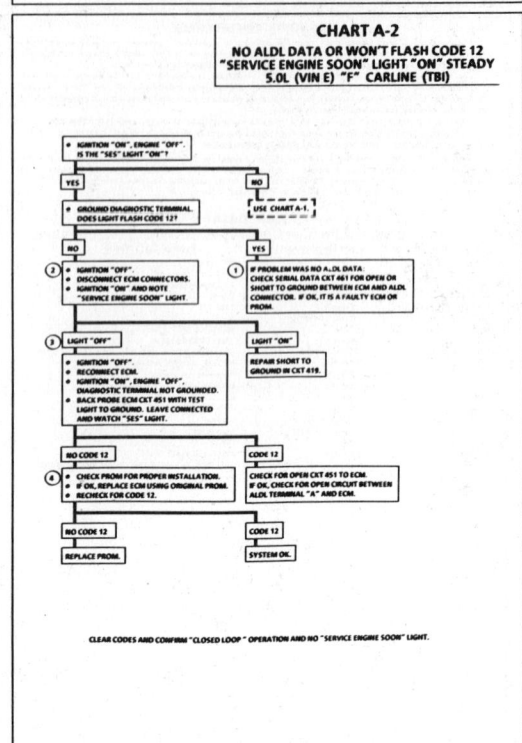

CHART A-2
NO ALDL DATA OR WON'T FLASH CODE 12
"SERVICE ENGINE SOON" LIGHT "ON" STEADY
5.0L (VIN E) "F" CARLINE (TBI)

1988–90 5.0L (VIN E) – CAMARO AND FIREBIRD

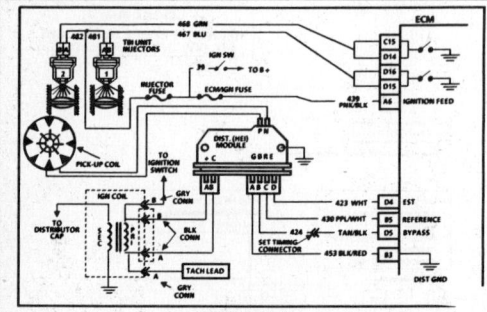

CHART A-3
(Page 1 of 2)
ENGINE CRANKS BUT WILL NOT RUN
5.0L (VIN E) "F" CARLINE (TBI)

Circuit Description:
This chart assumes that battery condition and engine cranking speed are OK, and there is adequate fuel in the tank.

Test Description: Numbers below refer to circled numbers on the diagnostic chart.
1. A "Service Engine Soon" light "ON" is a basic test to determine if there is a 12 volt supply and ignition 12 volts to ECM. No ALDL may be due to an ECM problem and CHART A-2 will diagnose the ECM. If TPS is over 2.5 volts, the engine may be in the clear flood mode, which will cause starting problems.
2. No spark may be caused by one of several components related to the Ignition/EST System. CHART C-4 will address all problems related to the causes of a no spark condition.
3. Fuel spray from the injector(s) indicates that fuel is available. However, the engine could be severely flooded due to too much fuel.
4. While cranking engine, there should be no fuel spray with injector disconnected. Replace an injector if it sprays fuel or drips like a leaking water faucet.
5. The fuel pressure will drop after the fuel pump stops running due to a controlled bleed in the fuel system.

Use of the fuel pressure gage will determine if fuel system pressure is enough for the engine to start and run. The key may have to be cycled, "ON" and "OFF," 2 or more times for accurate reading.
6. No fuel spray from injector indicates a faulty fuel system or no ECM control of injector.
7. This test will determine if the ignition module is not generating the reference pulse, or if the wiring or ECM is at fault. By touching and removing a test light to 12 volts on CKT 430, a reference pulse should be generated. If injector test light blinks, the ECM and wiring are OK.

Diagnostic Aids:
- Water or foreign material in fuel system can cause a no start during freezing weather.
- An EGR valve sticking open can cause a low air/fuel ratio during cranking
- Fuel pressure: Low fuel pressure can result in a very lean air/fuel ratio. See CHART A-7.
- A grounded CKT 423 (EST) may cause a "No-Start" or a "Start then Stall" condition.

1988–90 5.0L (VIN E) – CAMARO AND FIREBIRD

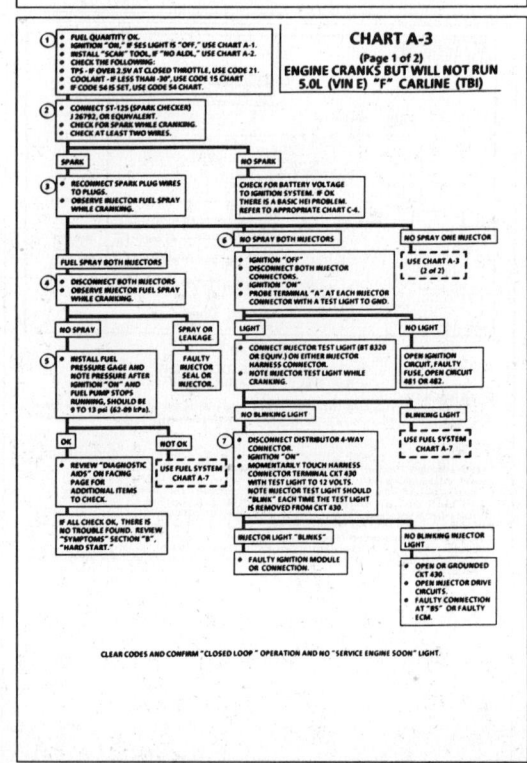

CHART A-3
(Page 1 of 2)
ENGINE CRANKS BUT WILL NOT RUN
5.0L (VIN E) "F" CARLINE (TBI)

1988–90 5.0L (VIN E) – CAMARO AND FIREBIRD

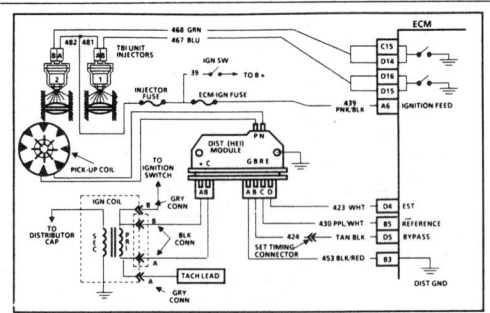

CHART A-3
(Page 2 of 2)
ENGINE CRANKS BUT WILL NOT RUN
5.0L (VIN E) "F" CARLINE (TBI)

Circuit Description:
This chart assumes that battery condition and engine cranking speed are OK, and there is adequate fuel in the tank.

Test Description: Numbers below refer to circled numbers on the diagnostic chart.
1. No fuel spray from one injector indicates a faulty fuel injector or no ECM control of injector. If the test light "blinks" while cranking, then ECM control should be considered OK. Be sure test light makes good contact between connector terminals during test. The light may be a little dim when "blinking." This is due to current draw of the test light. How bright it "blinks" is not important. However, the test light should be a BT 8320 or equivalent.

2. CKT 481 and CKT 482 supply ignition voltage to the injectors. Probe each connector terminal with a test light to ground. There should be a light on at one terminal. If the test light confirms ignition voltage at the connector, the ECM injector control CKT 467 or CKT 468 may be open. Reconnect the injector, and using a test light connected to ground, check for a light at the applicable ECM connector terminal ("D14" or "D16"). A light at this point indicates that the injector drive circuit involved is OK.
If an ECM repeat failure has occurred, the injector is shorted. Replace the injector and ECM.

1988–90 5.0L (VIN E) – CAMARO AND FIREBIRD

CHART A-3
(Page 2 of 2)
ENGINE CRANKS BUT WILL NOT RUN
5.0L (VIN E) "F" CARLINE (TBI)

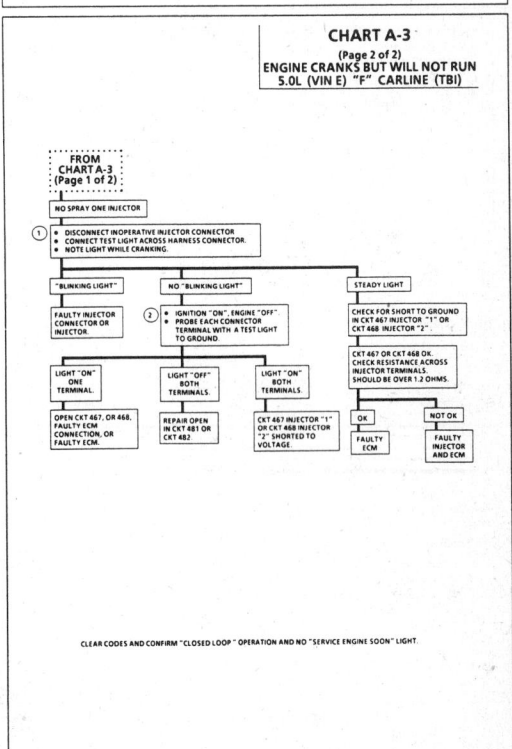

CLEAR CODES AND CONFIRM "CLOSED LOOP" OPERATION AND NO "SERVICE ENGINE SOON" LIGHT.

1988–90 5.0L (VIN E) – CAMARO AND FIREBIRD

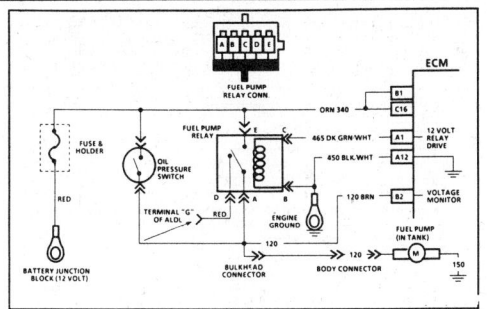

CHART A-7
(Page 1 of 2)
FUEL SYSTEM DIAGNOSIS
5.0L (VIN E) "F" CARLINE (TBI)

Circuit Description:
When the ignition switch is turned "ON," the electronic control module (ECM) will turn "ON" the in-tank fuel pump. It will remain "ON" as long as the engine is cranking or running, and the ECM is receiving ignition reference pulses.
If there are no reference pulses, the ECM will shut "OFF" the fuel pump within 2 seconds after key "ON."
The pump will deliver fuel to the TBI unit where the system pressure is controlled from 62 to 90 kPa (9 to 13 psi). Excess fuel is then returned to the fuel tank.
When the engine is stopped, the pump can be turned "ON" by applying battery voltage to terminal "G" of the ALDL connector.

Test Description: Numbers below refer to circled numbers on the diagnostic chart.
1. Fuel pressure should be noted while fuel pump is running. Fuel pressure will drop immediately after fuel pump stops running due to a controlled bleed in the fuel system.
The fuel pump test connector is terminal "G" of the ALDL connector.

Diagnostic Aids:
Improper fuel system pressure can result in one of the following symptoms.
- Cranks, but will not run
- Code 44
- Code 45
- Cuts out, may feel like ignition problem
- Poor fuel economy, loss of power
- Hesitation

1988–90 5.0L (VIN E) – CAMARO AND FIREBIRD

CHART A-7
(Page 1 of 2)
FUEL SYSTEM DIAGNOSIS
5.0L (VIN E) "F" CARLINE (TBI)

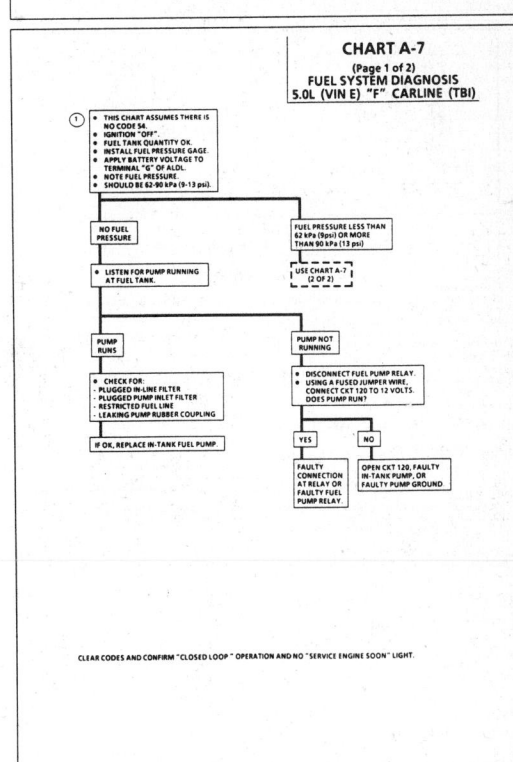

CLEAR CODES AND CONFIRM "CLOSED LOOP" OPERATION AND NO "SERVICE ENGINE SOON" LIGHT.

1988–90 5.0L (VIN E) – CAMARO AND FIREBIRD

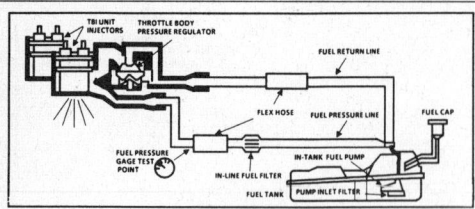

CHART A-7
(Page 2 of 2)
FUEL SYSTEM DIAGNOSIS
5.0L (VIN E) "F" CARLINE (TBI)

Test Description: Numbers below refer to circled numbers on the diagnostic chart.

1. Fuel pressure less than 62 kPa (9 psi) falls into two areas:
   - Amount of fuel to injectors OK but, pressure is too low. System will be lean and may set Code 44. Also, hard starting cold and poor overall performance.
   - Restricted flow causing pressure drop. Normally, a vehicle with a fuel pressure of less than 62 kPa (9 psi) at idle will not be driveable. However, if the pressure drop occurs only while driving, the engine will surge then stop as pressure begins to drop.

2. Restricting the fuel supply line between the gage and TBI unit and turning the fuel pump "ON" will determine if the fuel pump can supply enough fuel pressure at the injector to operate properly, above 62 kPa (9 psi).

   NOTICE: Do not restrict the fuel return line as this may damage the fuel pressure regulator.

3. This test determines if the high fuel pressure is due to a restricted fuel return line or a throttle body pressure regulator problem.

1988–90 5.0L (VIN E) – CAMARO AND FIREBIRD

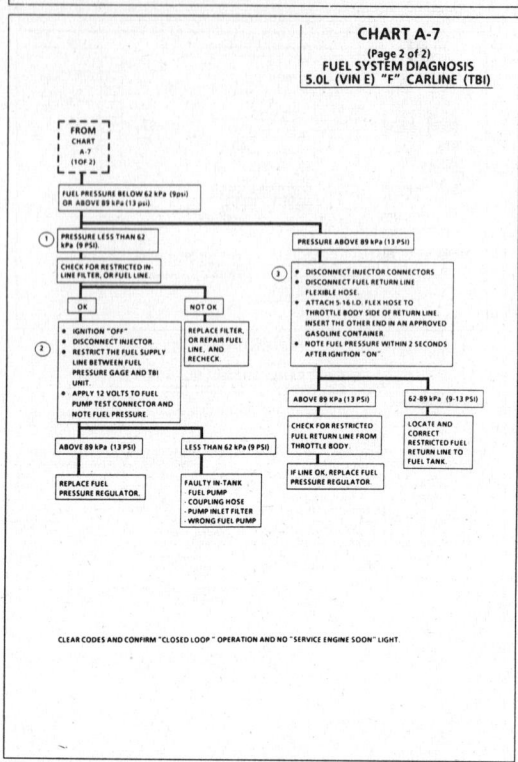

CHART A-7
(Page 2 of 2)
FUEL SYSTEM DIAGNOSIS
5.0L (VIN E) "F" CARLINE (TBI)

1988–90 5.0L (VIN E) – CAMARO AND FIREBIRD

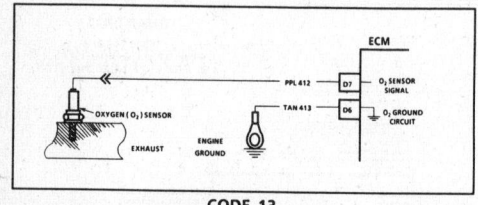

CODE 13
OXYGEN SENSOR CIRCUIT
(OPEN CIRCUIT)
5.0L (VIN E) "F" CARLINE (TBI)

Circuit Description:
The ECM supplies a voltage of about .45 volt between terminals "D7" and "D6". (If measured with a 10 megohm digital voltmeter, this may read as low as .32 volt). The O₂ sensor varies the voltage within a range of about 1 volt if the exhaust is rich, down through about .10 volt if exhaust is lean.
The sensor is like an open circuit and produces no voltage when it is below 360°C (600°F). An open sensor circuit or cold sensor causes "Open Loop" operation.

Test Description: Numbers below refer to circled numbers on the diagnostic chart.
1. Code 13 will be set.
   - Engine at normal operating temperature.
   - At least 2 minutes engine time after start.
   - O₂ signal voltage steady between .35 and .55 volts.
   - Rpm above 1600.
   - Throttle position sensor signal above 5% (about .3 volts above closed throttle voltage).
   - All conditions must be met for about 60 seconds.
   If the conditions for a Code 13 exist, the system will not go "Closed Loop."
2. This will determine if the sensor is at fault or the wiring or ECM is the cause of the Code 13.

3. To perform this test, use only a high impedence digital volt ohmmeter. This test checks the continuity of CKTs 412 and 413. If CKT 413 is open, the ECM voltage on CKT 412 will be over .6 volts (600 mV).

Diagnostic Aids:

Normal "Scan" voltage varies between 100 mV to 999 mV (.1 and 1.0 volt), while in "Closed Loop." Code 13 sets in one minute, if voltage remains between .35 and .55 volts.
Refer to "Intermittents" in Section "B".

1988–90 5.0L (VIN E) – CAMARO AND FIREBIRD

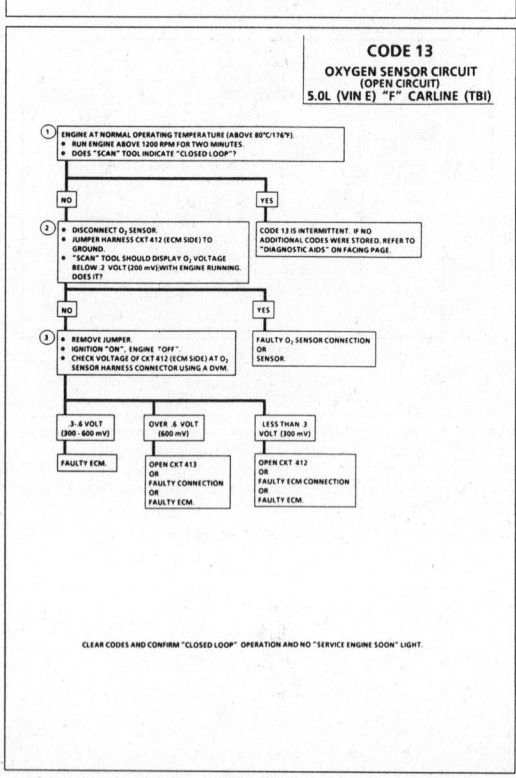

CODE 13
OXYGEN SENSOR CIRCUIT
(OPEN CIRCUIT)
5.0L (VIN E) "F" CARLINE (TBI)

1988–90 5.0L (VIN E) – CAMARO AND FIREBIRD

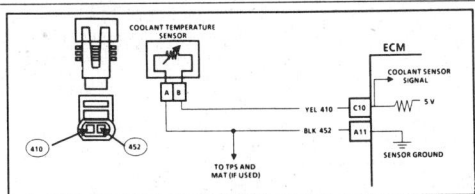

CODE 14
COOLANT TEMPERATURE SENSOR CIRCUIT
(HIGH TEMPERATURE INDICATED)
5.0L (VIN E) "F" CARLINE (TBI)

Circuit Description:

The coolant temperature sensor uses a thermistor to control the signal voltage at the ECM. The ECM applies a voltage on CKT 410 to the sensor. When the engine coolant is cold the sensor (thermistor) resistance is high, therefore the ECM will see high signal voltage.

As the engine coolant warms, the sensor resistance becomes less, and the voltage drops. At normal engine operating temperature (85 C-95 C or 185 F-203°F) the coolant sensor signal voltage will be about 1.5 to 2.0 volts.

Test Description: Numbers below refer to circled numbers on the diagnostic chart.

1. Code 14 will be set if:
   - Signal voltage indicates a coolant temperature above 135 C (275°F) for 2 seconds.
2. This test will determine if CKT 410 is shorted to ground which will cause the conditions for Code 14.

Diagnostic Aids:

Check harness routing for a potential short to ground in CKT 410.

"Scan" tool displays engine temperature in degrees centigrade. After engine is started, the temperature should rise steadily to about 90°C (194°F) then stabilize when the thermostat opens.

Refer to "Intermittents" in Section "B".

1988–90 5.0L (VIN E) – CAMARO AND FIREBIRD

CODE 14
COOLANT TEMPERATURE SENSOR CIRCUIT
(HIGH TEMPERATURE INDICATED)
5.0L (VIN E) "F" CARLINE (TBI)

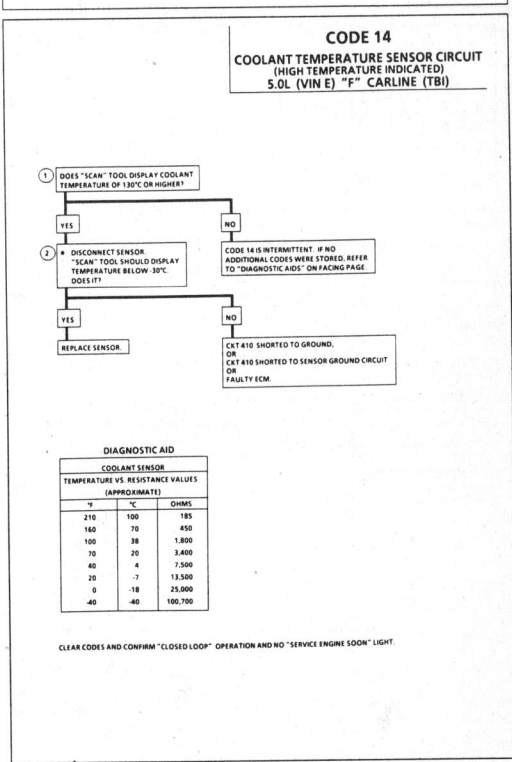

DIAGNOSTIC AID

COOLANT SENSOR
TEMPERATURE VS. RESISTANCE VALUES
(APPROXIMATE)

°F	°C	OHMS
210	100	185
160	70	450
100	38	1,800
70	20	3,400
40	4	7,500
20	-7	13,500
0	-18	25,000
-40	-40	100,700

CLEAR CODES AND CONFIRM "CLOSED LOOP" OPERATION AND NO "SERVICE ENGINE SOON" LIGHT.

1988–90 5.0L (VIN E) – CAMARO AND FIREBIRD

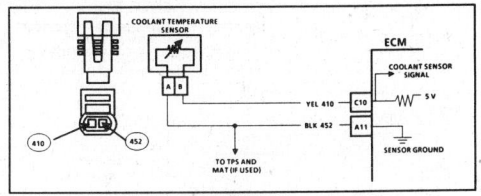

CODE 15
COOLANT TEMPERATURE SENSOR CIRCUIT
(LOW TEMPERATURE INDICATED)
5.0L (VIN E) "F" CARLINE (TBI)

Circuit Description:

The coolant temperature sensor uses a thermistor to control the signal voltage at the ECM. The ECM applies a voltage on CKT 410 to the sensor. When the engine coolant is cold, the sensor (thermistor) resistance is high, therefore, the ECM will see high signal voltage.

As the engine coolant warms, the sensor resistance becomes less, and the voltage drops. At normal engine operating temperature (85°C-95°C or 185°F-203°F) the coolant sensor signal voltage will be about 1.5 to 2.0 volts at the ECM.

Test Description: Numbers below refer to circled numbers on the diagnostic chart.

1. Code 15 will set if:
   - Engine running longer than 30 seconds.
   - Engine coolant temperature less than 30°C (-22°F), for 3 seconds.
2. This test simulates a Code 14. If the ECM recognizes the low signal voltage, (high temperature) and the "Scan" reads 130°C (266°F) or above, the ECM and wiring are OK.
3. This test will determine if CKT 410 is open. There should be 5 volts present at sensor connector if measured with a DVM.

Diagnostic Aids:

A "Scan" tool reads engine coolant temperature in degrees centigrade. After engine is started the temperature should rise steadily to about 90°C (194°F) then stabilize when the thermostat opens.

If Code 21 is also set, check CKT 452 for faulty wiring or connections. Check terminals at sensor for good contact.

Refer to "Intermittents" in Section "B".

1988–90 5.0L (VIN E) – CAMARO AND FIREBIRD

CODE 15
COOLANT TEMPERATURE SENSOR CIRCUIT
(LOW TEMPERATURE INDICATED)
5.0L (VIN E) "F" CARLINE (TBI)

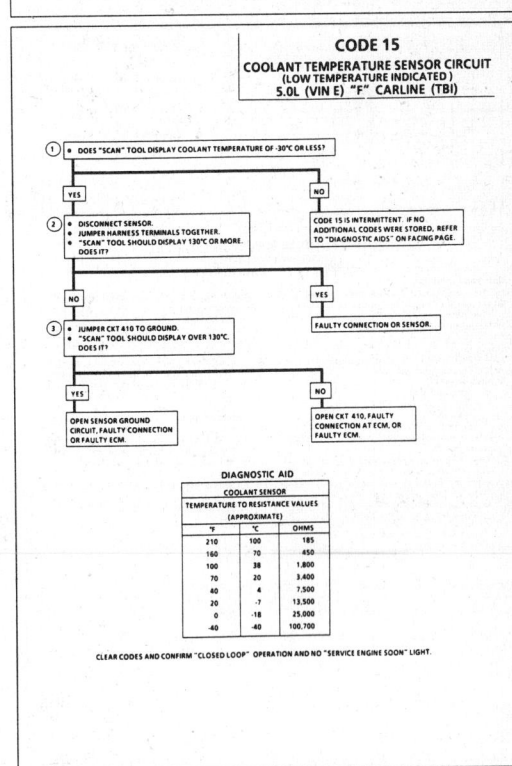

DIAGNOSTIC AID

COOLANT SENSOR
TEMPERATURE TO RESISTANCE VALUES
(APPROXIMATE)

°F	°C	OHMS
210	100	185
160	70	450
100	38	1,800
70	20	3,400
40	4	7,500
20	-7	13,500
0	-18	25,000
-40	-40	100,700

CLEAR CODES AND CONFIRM "CLOSED LOOP" OPERATION AND NO "SERVICE ENGINE SOON" LIGHT.

1988–90 5.0L (VIN E) — CAMARO AND FIREBIRD

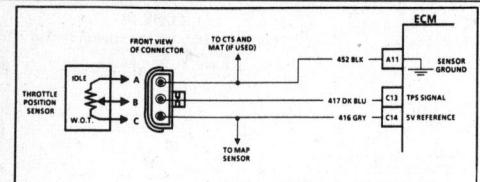

CODE 21
THROTTLE POSITION SENSOR (TPS) CIRCUIT
(SIGNAL VOLTAGE HIGH)
5.0L (VIN E) "F" CARLINE (TBI)

Circuit Description:

The throttle position sensor (TPS) provides a voltage signal that changes relative to the throttle blade. Signal voltage will vary from about .5 at idle to about 5 volts at wide open throttle.

The TPS signal is one of the most important inputs used by the ECM for fuel control and for most of the ECM controlled outputs.

Test Description: Numbers below refer to circled numbers on the diagnostic chart.
1. Code 21 will set if:
   - TPS signal voltage is greater than 2.5 volts
   - All conditions met for 8 seconds
   - MAP less than 52 kPa for greater than 15" HG
2. With the TPS sensor disconnected, the TPS voltage should go low if the ECM and wiring are OK.
3. Probing CKT 452 with a test light checks the 5 volt return circuit, because a faulty 5 volts return will cause a Code 21.

Diagnostic Aids:

A "Scan" tool reads throttle position in volts. The signal voltage when the throttle is closed should be less than 1.25 volts. With ignition "ON" or at idle, voltage should increase at a steady rate as throttle is moved toward WOT.

An open in CKT 452 will result in a Code 21. Refer to "Intermittents" in Section "B".

1988–90 5.0L (VIN E) — CAMARO AND FIREBIRD

CODE 21
THROTTLE POSITION SENSOR (TPS) CIRCUIT
(SIGNAL VOLTAGE HIGH)
5.0L (VIN E) "F" CARLINE (TBI)

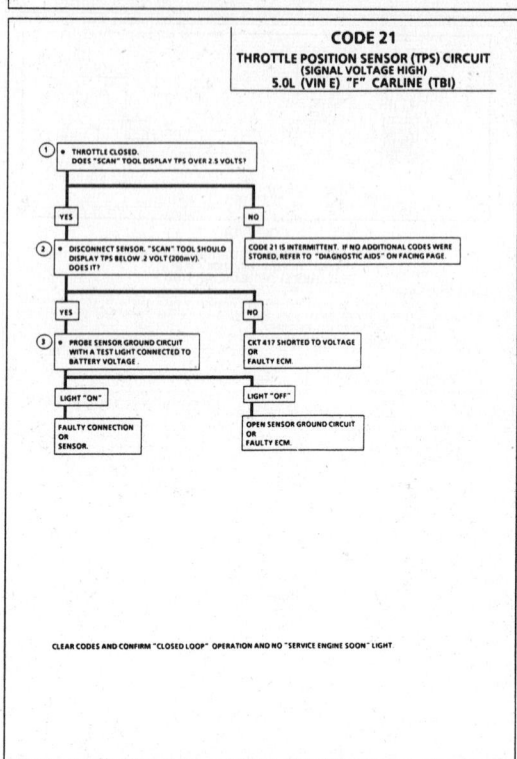

CLEAR CODES AND CONFIRM "CLOSED LOOP" OPERATION AND NO "SERVICE ENGINE SOON" LIGHT.

1988–90 5.0L (VIN E) — CAMARO AND FIREBIRD

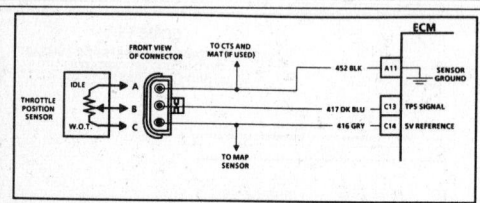

CODE 22
THROTTLE POSITION SENSOR (TPS) CIRCUIT
(SIGNAL VOLTAGE LOW)
5.0L (VIN E) "F" CARLINE (TBI)

Circuit Description:

The throttle position sensor (TPS) provides a voltage signal that changes relative to the throttle blade. Signal voltage will vary from about .5 at idle to about 5 volts at wide open throttle.

The TPS signal is one of the most important inputs used by the ECM for fuel control and for most of the ECM controlled outputs.

Test Description: Numbers below refer to circled numbers on the diagnostic chart.
1. Code 22 will set if:
   - Engine running
   - TPS signal voltage is less than about .2 volt for 3 seconds.
2. Simulates Code 21: (high voltage) If the ECM recognizes the high signal voltage the ECM and wiring are OK.
3. The TPS has an auto zeroing feature. If the voltage reading is within the range of 0.35 to 0.7 volts, the ECM will use that value as closed throttle. If the voltage reading is out of the auto zero range at closed throttle replace the TPS, refer to Section "6E2-C1" On-Car Service.

4. This simulates a high signal voltage to check for an open in CKT 417. The "Scan" tool will not read up to 12 volts, but what is important is that the ECM recognizes the signal on CKT 417.

Diagnostic Aids:

The signal voltage when the throttle is closed should be less than 1.25 volts. With ignition "ON" or at idle, voltage should increase at a steady rate as throttle is moved toward WOT.

An open or short to ground in CKT 416 or CKT 417 will result in a Code 22. Refer to "Intermittents" in Section "B".

1988–90 5.0L (VIN E) — CAMARO AND FIREBIRD

CODE 22
THROTTLE POSITION SENSOR (TPS) CIRCUIT
(SIGNAL VOLTAGE LOW)
5.0L (VIN E) "F" CARLINE (TBI)

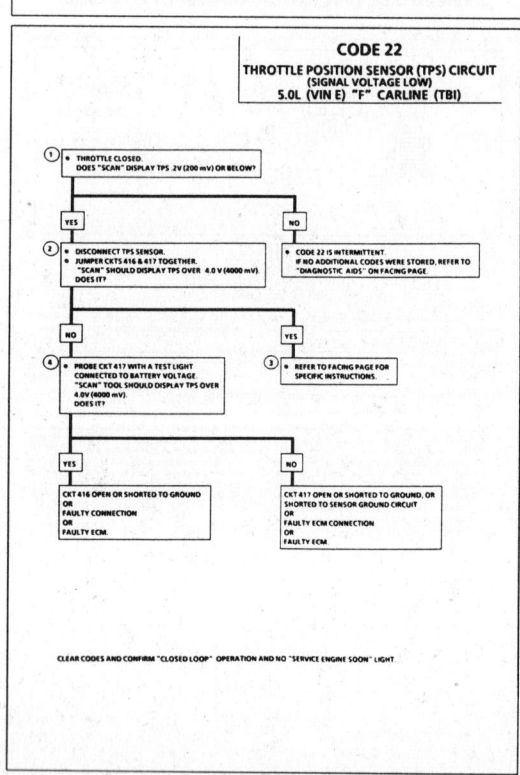

CLEAR CODES AND CONFIRM "CLOSED LOOP" OPERATION AND NO "SERVICE ENGINE SOON" LIGHT.

1988–90 5.0L (VIN E) – CAMARO AND FIREBIRD

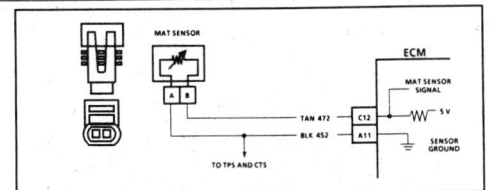

CODE 23
MANIFOLD AIR TEMPERATURE (MAT) SENSOR CIRCUIT
(LOW TEMPERATURE INDICATED)
5.0L (VIN E) "F" CARLINE (TBI)

Circuit Description:

The manifold air temperature (MAT) sensor uses a thermistor to control the signal voltage to the ECM. The ECM applies a voltage (about 5 volts) on CKT 472 to the sensor. When the manifold air is cold, the sensor (thermistor) resistance is high, therefore, the ECM will see a high signal voltage. If the manifold air is warm, the sensor (thermistor) resistance is low, therefore, the ECM will see a low voltage.

Test Description: Numbers below refer to circled numbers on the diagnostic chart.
1. Code 23 will set if:
   - A signal voltage indicates a manifold air temperature below -30°C (-22°F) for 12 seconds.
   - Time since engine start is 1 minute or longer.
2. A Code 23 will set, due to an open sensor, wire or connection. This test will determine if the wiring and ECM are OK.
3. This will determine if the MAT sensor signal CKT 472 or the MAT sensor ground CKT 452 is open.

Diagnostic Aids:

A "Scan" tool indicates the temperature of the air in the air cleaner because the MAT sensor is mounted in the air cleaner.

Carefully check harness and connections for possible open CKT 472 or CKT 452.

Refer to "Intermittents" in Section "B".

If the engine has been allowed to sit overnight, the manifold air temperature and coolant temperature values should read within a few degrees of each other. After the engine is started, the MAT will increase due to Thermac operation and underhood temperatures.

1988–90 5.0L (VIN E) – CAMARO AND FIREBIRD

CODE 23
MANIFOLD AIR TEMPERATURE (MAT) SENSOR CIRCUIT
(LOW TEMPERATURE INDICATED)
5.0L (VIN E) "F" CARLINE (TBI)

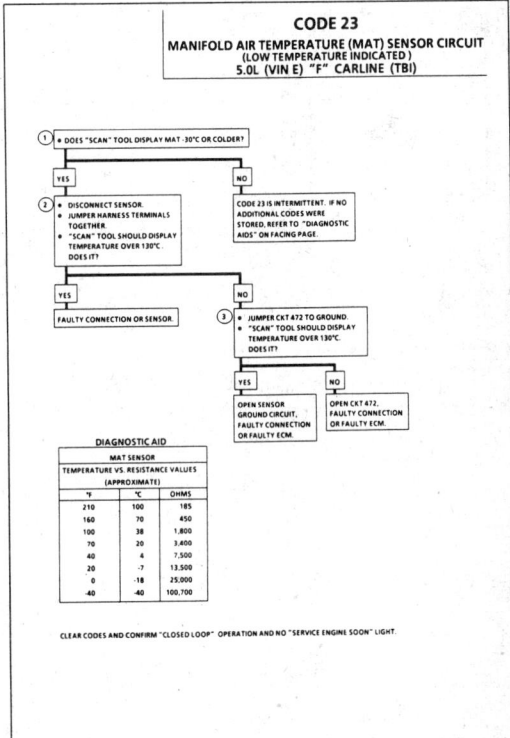

DIAGNOSTIC AID		
MAT SENSOR		
TEMPERATURE VS. RESISTANCE VALUES (APPROXIMATE)		
°F	°C	OHMS
210	100	185
160	70	450
100	38	1,800
70	20	3,400
40	4	7,500
20	-7	13,500
0	-18	25,000
-40	-40	100,700

CLEAR CODES AND CONFIRM "CLOSED LOOP" OPERATION AND NO "SERVICE ENGINE SOON" LIGHT.

1988–90 5.0L (VIN E) – CAMARO AND FIREBIRD

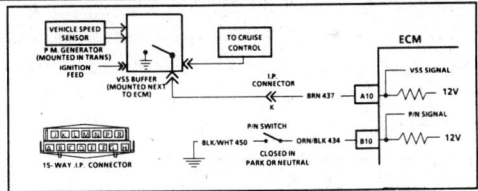

CODE 24
VEHICLE SPEED SENSOR (VSS) CIRCUIT
5.0L (VIN E) "F" CARLINE (TBI)

Circuit Description:

The ECM applies and monitors 12 volts on CKT 437. CKT 437 connects to the vehicle speed sensor which alternately grounds CKT 437 when drive wheels are turning. This pulsing action takes place about 2000 times per mile and the ECM will calculate vehicle speed based on the time between "pulses".

A "Scan" reading should closely match with speedometer reading while drive wheels turning.

Disregard a Code 24 set when drive wheels are not turning.

Test Description: Numbers below refer to circled numbers on the diagnostic chart.
1. Code 24 will set if:
   - CKT 437 voltage is constant
   - Engine speed is between 1400 and 3600 rpm
   - TPS is less than 2% throttle opening
   - Low load condition
   - Not in park or neutral
   - All conditions must be met for 4 seconds

   These conditions are met during a road load deceleration.
2. A voltage of less than 1 volt, at the 15-way I/P connector indicates that the CKT 437 wire may be shorted to ground. Disconnect CKT 437 at the vehicle speed sensor buffer.

If voltage remains less than 10 volts, then CKT 437 wire is grounded or open. If 437 is not grounded or open, check for a faulty ECM connector or ECM.

Diagnostic Aids:

If "Scan" displays vehicle speed, check park/neutral switch CHART C-1A on vehicle with automatic transmission. If switch is OK, check for intermittent connections. An open or short to ground in CKT 437 will result in a Code 24. Refer to Section "8A" for complete wiring diagram.

Refer to "Intermittents" in Section "B".

1988–90 5.0L (VIN E) – CAMARO AND FIREBIRD

CODE 24
VEHICLE SPEED SENSOR (VSS) CIRCUIT
5.0L (VIN E) "F" CARLINE (TBI)

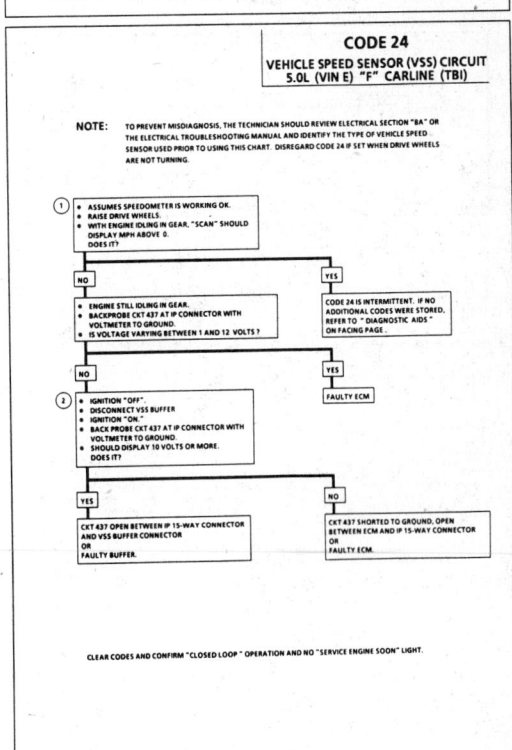

CLEAR CODES AND CONFIRM "CLOSED LOOP" OPERATION AND NO "SERVICE ENGINE SOON" LIGHT.

1988–90 5.0L (VIN E) – CAMARO AND FIREBIRD

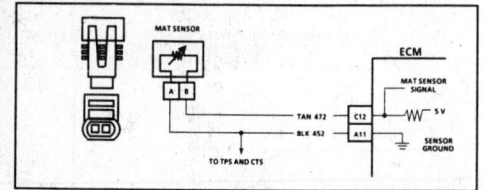

CODE 25
MANIFOLD AIR TEMPERATURE (MAT) SENSOR CIRCUIT
(HIGH TEMPERATURE INDICATED)
5.0L (VIN E) "F" CARLINE (TBI)

Circuit Description:
The manifold air temperature (MAT) sensor uses a thermistor to control the signal voltage to the ECM. The ECM applies a voltage (about 5 volts) on CKT 472 to the sensor. When manifold air is cold, the sensor (thermistor) resistance is high, therefore, the ECM will see a high signal voltage. If the manifold air is warm, the sensor (thermistor) resistance is low, therefore, the ECM will see a low signal voltage.

Test Description: Numbers below refer to circled numbers on the diagnostic chart.
1. Code 25 will set if:
   - Signal voltage indicates a manifold air temperature greater than 150°C (302°F) for 2 seconds.
   - Time since engine start is 2 minutes or longer.
   - Vehicle speed has to be greater than 5 MPH.

Diagnostic Aids:
Manifold air temperature on a "Scan" tool indicates the temperature of the air in the air cleaner, because the MAT sensor is located in the air cleaner. If the engine has been allowed to sit overnight, the manifold air temperature and coolant temperature values should read within a few degrees of each other. After the engine is started, the MAT will increase due to Thermac operation and underhood temperatures, however, MAT will rarely exceed 80°C. If a higher MAT than 80°C is noted, check for proper Thermac operation. Use Section "C14".
Check harness routing for possible short to ground in CKT 472.
Refer to "Intermittents" in Section "B".

1988–90 5.0L (VIN E) – CAMARO AND FIREBIRD

CODE 25
MANIFOLD AIR TEMPERATURE (MAT) SENSOR CIRCUIT
(HIGH TEMPERATURE INDICATED)
5.0L (VIN E) "F" CARLINE (TBI)

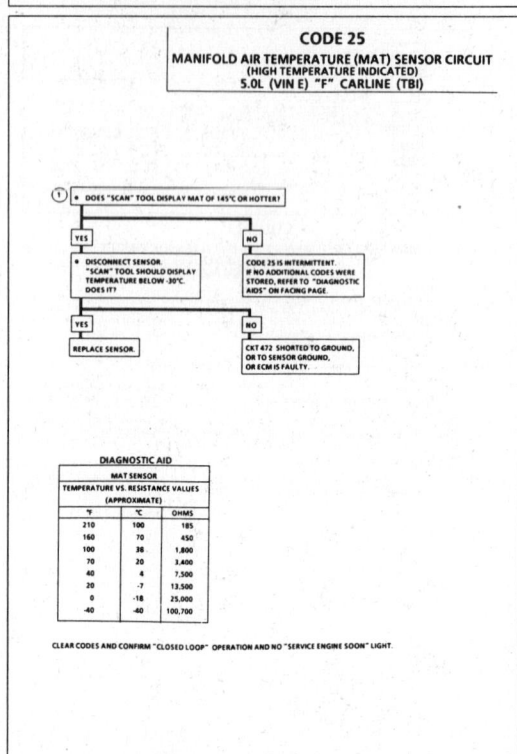

DIAGNOSTIC AID		
MAT SENSOR		
TEMPERATURE VS. RESISTANCE VALUES		
(APPROXIMATE)		
°F	°C	OHMS
210	100	185
160	70	450
100	38	1,800
70	20	3,400
40	4	7,500
20	-7	13,500
0	-18	25,000
-40	-40	100,700

CLEAR CODES AND CONFIRM "CLOSED LOOP" OPERATION AND NO "SERVICE ENGINE SOON" LIGHT.

1988–90 5.0L (VIN E) – CAMARO AND FIREBIRD

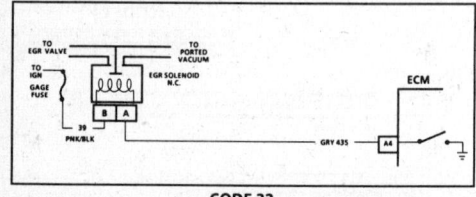

CODE 32
EXHAUST GAS RECIRCULATION (EGR) CIRCUIT
5.0L (VIN E) "F" CARLINE (TBI)

Circuit Description:
The ECM operates a solenoid to control the exhaust gas recirculation (EGR) valve. This solenoid is normally closed. By providing a ground path, the ECM energizes the solenoid which allows vacuum to pass to the EGR valve.
The ECM monitors EGR effectiveness by de-energizing the EGR control solenoid thereby shutting off vacuum to the EGR valve diaphragm. With the EGR valve closed, manifold vacuum will be greater than it was during normal EGR operation and this change will be relayed to the ECM by the MAP sensor. If the change is not within the calibrated window, a Code 32 will be set.
The ECM will check EGR operation when:
- Vehicle speed is above 50 mph
- Engine vacuum is between 40 and 51 kPa (or 12" HG and 15" HG)
- No change in throttle position while test is being run

Test Description: Numbers below refer to circled numbers on the diagnostic chart.
1. Checks for EGR solenoid stuck open.
2. Checks for EGR solenoid always being energized.
3. Grounding diagnostic "test" terminal should energize EGR solenoid and vacuum should drop.
4. Negative backpressure EGR valve should hold vacuum with engine "OFF".
5. When engine is started, exhaust backpressure should cause vacuum to bleed off and valve to fully close.

Diagnostic Aids:
Vacuum lines should be thoroughly checked for internal restrictions. The ECM uses the MAP sensor for checking EGR operation. If there is a question of MAP sensor accuracy use CHART.C-1D MAP output check in Section "C".
If no problems are found refer to "Intermittents" in Section "B".

1988–90 5.0L (VIN E) – CAMARO AND FIREBIRD

CODE 32
EXHAUST GAS RECIRCULATION (EGR) CIRCUIT
5.0L (VIN E) "F" CARLINE (TBI)

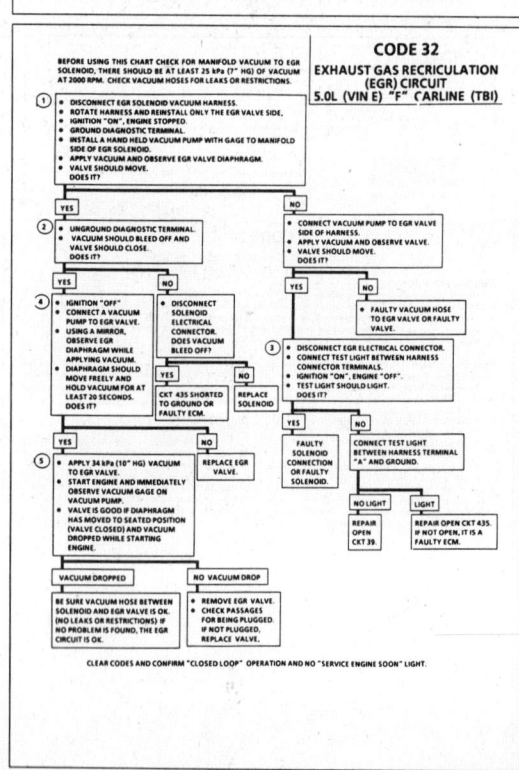

CLEAR CODES AND CONFIRM "CLOSED LOOP" OPERATION AND NO "SERVICE ENGINE SOON" LIGHT.

1988–90 5.0L (VIN E) – CAMARO AND FIREBIRD

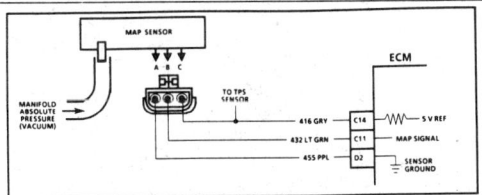

CODE 33
MANIFOLD ABSOLUTE PRESSURE (MAP) SENSOR CIRCUIT
(SIGNAL VOLTAGE HIGH - LOW VACUUM)
5.0L (VIN E) "F" CARLINE (TBI)

Circuit Description:

The manifold absolute pressure sensor (MAP) responds to changes in manifold pressure (vacuum). The ECM receives this information as a signal voltage that will vary from about 1-1.5 volts at idle to 4-4.5 volts at wide open throttle.

A "Scan" displays manifold pressure in volts. Low pressure (high vacuum) reads a low voltage while a high pressure (low vacuum) reads a high voltage.

If the MAP sensor fails the ECM will substitute a fixed MAP value and use the throttle position sensor (TPS) to control fuel delivery.

Test Description: Numbers below refer to circled numbers on the diagnostic chart.
1. Code 33 will set when:
   - Signal is too high, (kPa greater than 68 kPa or less than 9" HG), for a time greater than 5 seconds.
   - TPS less than 4%

   Engine misfire or a low unstable idle may set Code 33. Disconnect MAP sensor and system will go into backup mode. If the misfire or idle condition remains, see "Symptoms" in Section "B".
2. If the ECM recognizes the low MAP signal, the ECM and wiring are OK.

Diagnostic Aids:

If the idle is rough or unstable refer to "Symptoms" in Section "B" for items which can cause an unstable idle.

An open in CKT 455 will result in a Code 33.

With the ignition "ON" and the engine stopped, the manifold pressure is equal to atmospheric pressure and the signal voltage will be high. This information is used by the ECM as an indication of vehicle altitude and is referred to as BARO. Comparison of this BARO reading with a known good vehicle with the same sensor is a good way to check accuracy of a "suspect" sensor. Reading should be the same, ± .4 volt.

Also CHART C-1D can be used to test the MAP sensor.

Refer to "Intermittents" in Section "B".

1988–90 5.0L (VIN E) – CAMARO AND FIREBIRD

CODE 33
MANIFOLD ABSOLUTE PRESSURE (MAP) SENSOR CIRCUIT
(SIGNAL VOLTAGE HIGH - LOW VACUUM)
5.0L (VIN E) "F" CARLINE (TBI)

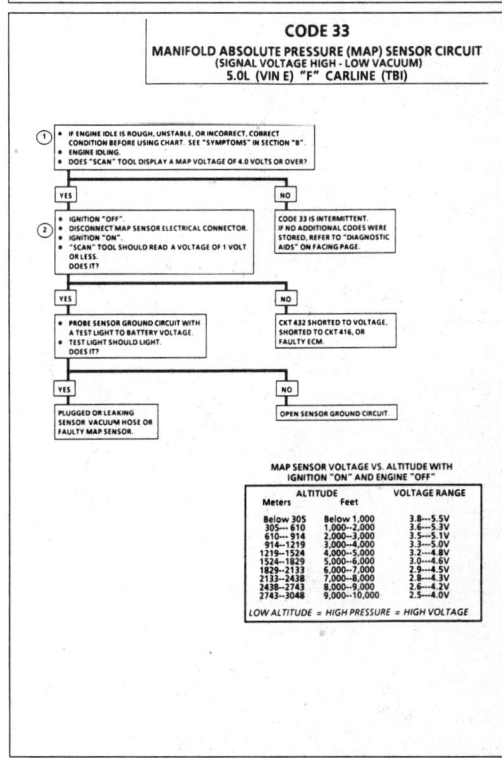

ALTITUDE		VOLTAGE RANGE
Meters	Feet	
Below 305	Below 1,000	3.8–5.5V
305–610	1,000–2,000	3.6–5.3V
610–914	2,000–3,000	3.5–5.1V
914–1219	3,000–4,000	3.3–5.0V
1219–1524	4,000–5,000	3.2–4.8V
1524–1829	5,000–6,000	3.0–4.6V
1829–2133	6,000–7,000	2.9–4.5V
2133–2438	7,000–8,000	2.8–4.3V
2438–2743	8,000–9,000	2.6–4.2V
2743–3048	9,000–10,000	2.5–4.0V

LOW ALTITUDE = HIGH PRESSURE = HIGH VOLTAGE

MAP SENSOR VOLTAGE VS. ALTITUDE WITH IGNITION "ON" AND ENGINE "OFF"

1988–90 5.0L (VIN E) – CAMARO AND FIREBIRD

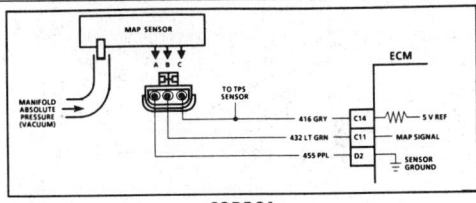

CODE 34
MANIFOLD ABSOLUTE PRESSURE (MAP) SENSOR CIRCUIT
(SIGNAL VOLTAGE LOW - HIGH VACUUM)
5.0L (VIN E) "F" CARLINE (TBI)

Circuit Description:

The manifold absolute pressure sensor (MAP) responds to changes in manifold pressure (vacuum). The ECM receives this information as a signal voltage that will vary from about 1-1.5 volts at idle to 4-4.5 volts at wide open throttle.

A "Scan" displays manifold pressure in volts. Low pressure (high vacuum) reads a low voltage while a high pressure (low vacuum) reads a high voltage.

If the MAP sensor fails the ECM will substitute a fixed MAP value and use the throttle position sensor (TPS) to control fuel delivery.

Test Description: Numbers below refer to circled numbers on the diagnostic chart.
1. Code 34 will set when:
   - Signal is too low, (less than 14 kPa or greater than 28" HG) and engine running less than 1200 rpm

   OR
   - Engine running greater than 1200 rpm
   - Throttle position greater than 21% (over 1.5 volts)
2. If the ECM recognizes the high MAP signal, the ECM and wiring are OK.
3. The "Scan" tool may not display 12 volts. The important thing is that the ECM recognizes the voltage as more than 4 volts, indicating that the ECM and CKT 432 are OK.

Diagnostic Aids:

An intermittent open in CKTs 432 or 416 will result in a Code 34.

With the ignition "ON" and engine stopped, the manifold pressure is equal to atmospheric pressure and the signal voltage will be high. This information is used by the ECM as an indication of vehicle altitude and is referred to as BARO. Comparison of this BARO reading with a known good vehicle with the same sensor is a good way to check accuracy of a "suspect" sensor. Reading should be the same, ± .4 volts.

Also CHART C-1D can be used to test the MAP sensor.

Refer to "Intermittents" in Section "B".

1988–90 5.0L (VIN E) – CAMARO AND FIREBIRD

CODE 34
MANIFOLD ABSOLUTE PRESSURE (MAP) SENSOR CIRCUIT
(SIGNAL VOLTAGE LOW - HIGH VACUUM)
5.0L (VIN E) "F" CARLINE (TBI)

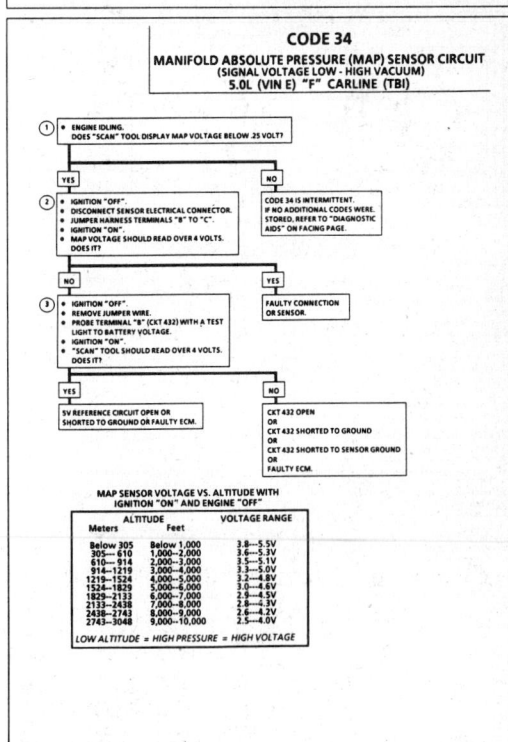

ALTITUDE		VOLTAGE RANGE
Meters	Feet	
Below 305	Below 1,000	3.8–5.5V
305–610	1,000–2,000	3.6–5.3V
610–914	2,000–3,000	3.5–5.1V
914–1219	3,000–4,000	3.3–5.0V
1219–1524	4,000–5,000	3.2–4.8V
1524–1829	5,000–6,000	3.0–4.6V
1829–2133	6,000–7,000	2.9–4.5V
2133–2438	7,000–8,000	2.8–4.3V
2438–2743	8,000–9,000	2.6–4.2V
2743–3048	9,000–10,000	2.5–4.0V

LOW ALTITUDE = HIGH PRESSURE = HIGH VOLTAGE

MAP SENSOR VOLTAGE VS. ALTITUDE WITH IGNITION "ON" AND ENGINE "OFF"

1988–90 5.0L (VIN E) – CAMARO AND FIREBIRD

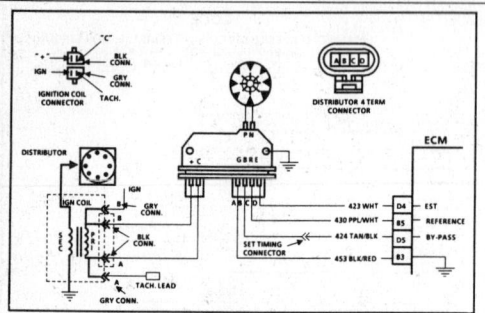

CODE 42
ELECTRONIC SPARK TIMING (EST) CIRCUIT
5.0L (VIN E) "F" CARLINE (TBI)

Circuit Description:

When the system is running on the ignition module, that is, no voltage on the bypass line, the ignition module grounds the EST signal. The ECM expects to see no voltage on the EST line during this condition. If it sees a voltage, it sets Code 42 and will not go into the EST mode.

When the rpm for EST is reached (about 400 rpm), and bypass voltage applied, the EST should no longer be grounded in the ignition module so the EST voltage should be varying.

If the bypass line is open or grounded, the ignition module will not switch to EST mode so the EST voltage will be low and Code 42 will be set.

If the EST line is grounded, the ignition module will switch to EST, but because the line is grounded there will be no EST signal. A Code 42 will be set.

Test Description: Numbers below refer to circled numbers on the diagnostic chart.

1. Code 42 means the ECM has seen an open or short to ground in the EST or bypass circuits. This test confirms that Code 42 and that the fault causing the code is present.
2. Checks for a normal EST ground path through the ignition module. An EST CKT 423 shorted to ground will also read less than 500 ohms; however, this will be checked later.
3. As the test light voltage touches CKT 424, the module should switch causing the ohmmeter to "overrange" if the meter is in the 1000-2000 ohms position. Selecting the 10-20,000 ohms position will indicate above 5000 ohms. The important thing is that the module "switched."

4. The module did not switch and this step checks for:
   - EST CKT 423 shorted to ground
   - Bypass CKT 424 open
   - Faulty ignition module connection or module
5. Confirms that Code 42 is a faulty ECM and not an intermittent in CKTs 423 or 424.

Diagnostic Aids:

If a Code 42 was stored and the customer complains of a "Hard Start", the problem is most likely a grounded EST line (CKT 423).

The "Scan" tool does not have any ability to help diagnose a Code 42 problem.

A PROM not fully seated in the ECM can result in a Code 42.

Refer to "Intermittents" in Section "B".

1988–90 5.0L (VIN E) – CAMARO AND FIREBIRD

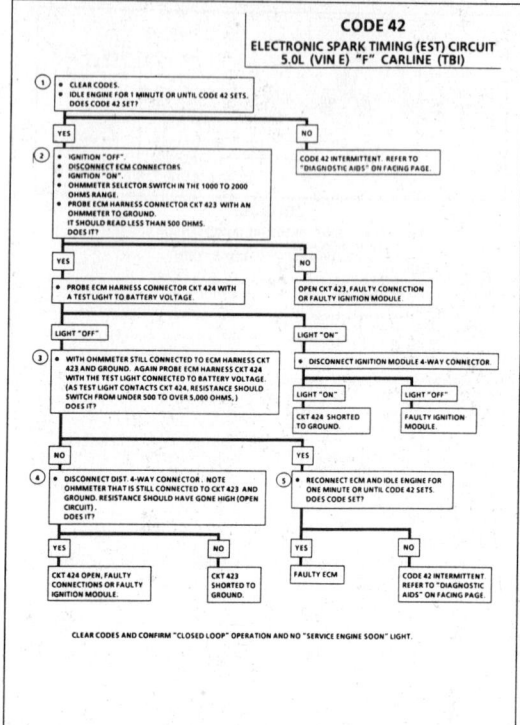

CODE 42
ELECTRONIC SPARK TIMING (EST) CIRCUIT
5.0L (VIN E) "F" CARLINE (TBI)

1988–90 5.0L (VIN E) – CAMARO AND FIREBIRD

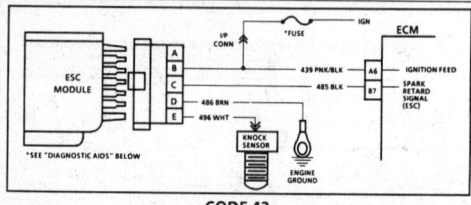

CODE 43
ELECTRONIC SPARK CONTROL (ESC) CIRCUIT
5.0L (VIN E) "F" CARLINE (TBI)

Circuit Description:

Electronic spark control is accomplished with a module that sends a voltage signal to the ECM. As the knock sensor detects engine knock, the voltage from the ESC module to the ECM drops, and this signals the ECM to retard timing. The ECM will retard the timing when knock is detected and rpm is above about 900 rpm.

Code 43 means the ECM has read low voltage on CKT 485 for longer than 5 seconds with the engine running or the system has failed the functional check.

This system performs a functional check once per start up to check the ESC system. To perform this test the ECM will advance the spark when coolant is above 95°C (194°F) and at a high load condition (near WOT). The ECM then checks the signal at "B7" to see if a knock is detected. The functional check is performed once per start up, if knock is detected when coolant is below 95°C (194°F) no check is done. The functional check will not be run. If the functional check fails, the "Service Engine Soon" light will remain "ON" until ignition is turned "OFF" or until a knock signal is detected.

Test Description: Numbers below refer to circled numbers on the diagnostic chart.

1. If the conditions for a Code 43 are present the "Scan" will always display "yes". There should not be a knock at idle unless an internal engine problem, or a system problem exists.
2. This test will determine if the system is functioning at this time. Usually a knock signal can be generated by tapping on the block close to the area of the sensor.
3. Because Code 43 sets when the signal voltage on CKT 485 remains low this test should cause the signal on CKT 485 to go high. The 12 volts signal should be seen by the ECM as "no knock" if the ECM and wiring are OK.
4. This test will determine if the knock signal is being detected on CKT 496 or if the ESC module is at fault.

5. If CKT 496 is routed to close to secondary ignition wires the ESC module may see the interference as a knock signal.
6. This checks the ground circuit to the module. An open ground will cause the voltage on CKT 485 to be about 12 volts which would cause the Code 43 functional test to fail.
7. Contacting CKT 496 with a test light to 12 volts should generate a knock signal. This will determine if the ESC module is operating correctly.

Diagnostic Aids:

* = ECM/IGN fuse

Code 43 can be caused by a faulty connection at the knock sensor at the ESC module or at the ECM. Also check CKT 485 for possible open or short to ground.

Refer to "Intermittents" in Section "B".

1988–90 5.0L (VIN E) – CAMARO AND FIREBIRD

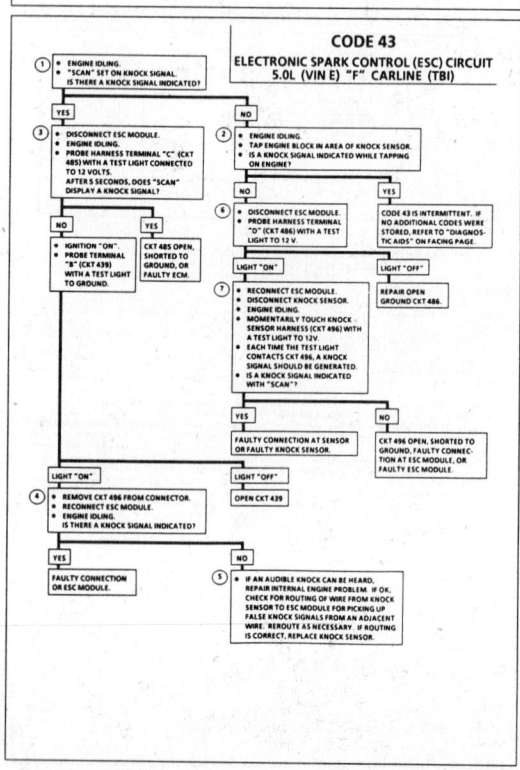

CODE 43
ELECTRONIC SPARK CONTROL (ESC) CIRCUIT
5.0L (VIN E) "F" CARLINE (TBI)

1988–90 5.0L (VIN E) – CAMARO AND FIREBIRD

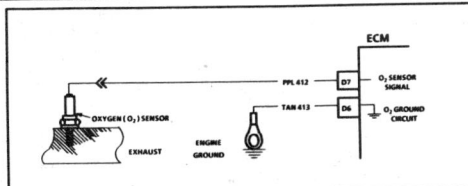

CODE 44
OXYGEN SENSOR CIRCUIT
(LEAN EXHAUST INDICATED)
5.0L (VIN E) "F" CARLINE (TBI)

Circuit Description:

The ECM supplies a voltage of about .45 volt between terminals "D6" and "D7". (If measured with a 10 megohm digital voltmeter, this may read as low as .32 volt.) The O_2 sensor varies the voltage within a range of about 1 volt if the exhaust is rich, down through about .10 volt if exhaust is lean.

The sensor is like an open circuit and produces no voltage when it is below about 360°C (600°F). An open sensor circuit or cold sensor causes "Open Loop" operation.

Test Description: Numbers below refer to circled numbers on the diagnostic chart.

1. Code 44 is set when the O_2 sensor signal voltage on CKT 412:
   - Remains below .2 volt for 50 seconds
   - And the system is operating in "Closed Loop"

Diagnostic Aids:

Using the "Scan", observe the block learn values at different rpm and air flow conditions to determine when the Code 44 may have been set. If the conditions for Code 44 exists the block learn values will be around 150.

- O_2 Sensor Wire Sensor pigtail may be mispositioned and contacting the exhaust manifold.
- Check for intermittent ground in wire between connector and sensor.

- MAP Sensor A manifold absolute pressure (MAP) sensor output that causes the ECM to sense a higher than normal vacuum will cause the system to go lean. Disconnect the MAP sensor and if the lean condition is gone, replace the sensor.
- Lean Injector(s)
- Fuel Contamination Water, even in small amounts, near the in-tank fuel pump inlet can be delivered to the injectors. The water causes a lean exhaust and can set a Code 44.
- Fuel Pressure System will be lean if pressure is too low. It may be necessary to monitor fuel pressure while driving the car at various road speeds and/or loads to confirm. See "Fuel System Diagnosis", CHART A-7.
- Exhaust Leaks If there is an exhaust leak, the engine can cause outside air to be pulled into the exhaust and past the sensor. Vacuum or crankcase leaks can cause a lean condition.
- AIR System. Be sure air is not being directed to the exhaust ports while in "Closed Loop." If the block learn value goes down while squeezing air hose to left side exhaust ports, refer to CHART C-6.
- If the above are OK, it is a faulty oxygen sensor.

1988–90 5.0L (VIN E) – CAMARO AND FIREBIRD

CODE 44
OXYGEN SENSOR CIRCUIT
(LEAN EXHAUST INDICATED)
5.0L (VIN E) "F" CARLINE (TBI)

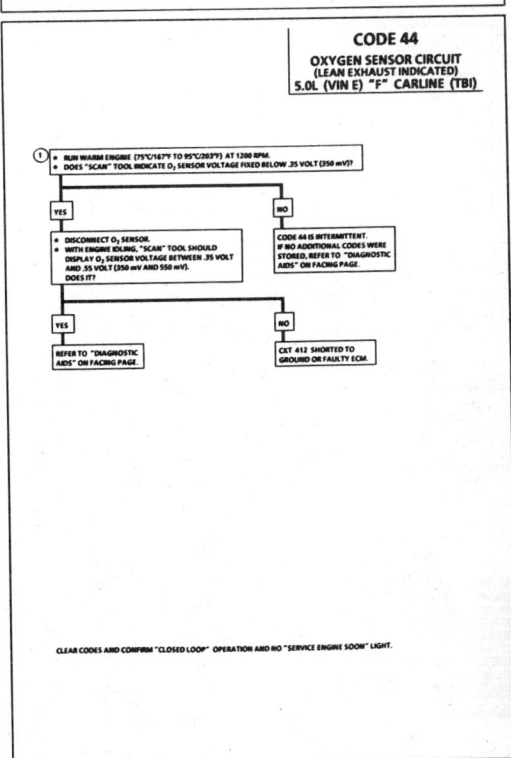

1988–90 5.0L (VIN E) – CAMARO AND FIREBIRD

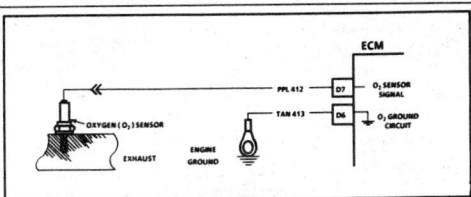

CODE 45
OXYGEN SENSOR CIRCUIT
(RICH EXHAUST INDICATED)
5.0L (VIN E) "F" CARLINE (TBI)

Circuit Description:

The ECM supplies a voltage of about .45 volt between terminals "D6" and "D7". (If measured with a 10 megohm digital voltmeter, this may read as low as .32 volt.) The O_2 sensor varies the voltage within a range of about 1 volt if the exhaust is rich, down through about .10 volt if exhaust is lean.

The sensor is like an open circuit and produces no voltage when it is below about 360°C (600°F). An open sensor circuit or cold sensor causes "Open Loop" operation.

Test Description: Numbers below refer to circled numbers on the diagnostic chart.

1. Code 45 is set when the O_2 sensor signal voltage on CKT 412:
   - Remains above .7 volt for 50 seconds; and in "Closed Loop"
   - Engine time after start is 1 minute or more
   - Throttle angle greater than 2% (about .2 volt above idle voltage) but less than 25%

Diagnostic Aids:

Using the "Scan", observe the block learn values at different rpm conditions to determine when the Code 45 may have been set. If the conditions for Code 45 exists, The block learn values will be around 115.

- Fuel Pressure System will go rich if pressure is too high. The ECM can compensate for some increase. However, if it gets too high, a Code 45 may be set. See "Fuel System Diagnosis", CHART A-7.
- Leaking injector See CHART A-7
- Check for fuel contaminated oil

- HEI Shielding An open ground CKT 453 (ignition system reference low) may result in EMI, or induced electrical "noise". The ECM looks at this "noise" as reference pulses. The additional pulses result in higher than actual engine speed signal. The ECM then delivers too much fuel, causing system to go rich. Engine tachometer will also show higher than actual engine speed, which can help in diagnosing this problem.
- Canister Purge Check for fuel saturation. If full of fuel, check canister control and hoses. See "Canister Purge", Section "C3".
- MAP Sensor An output that causes the ECM to sense a lower than normal vacuum can cause the system to go rich. Disconnecting the MAP sensor will allow the ECM to set a fixed value for the sensor. Substitute a different MAP sensor if the rich condition is gone while the sensor is disconnected.
- TPS An intermittent TPS output will cause the system to go rich, due to a false indication of the engine accelerating.

1988–90 5.0L (VIN E) – CAMARO AND FIREBIRD

CODE 45
OXYGEN SENSOR CIRCUIT
(RICH EXHAUST INDICATED)
5.0L (VIN E) "F" CARLINE (TBI)

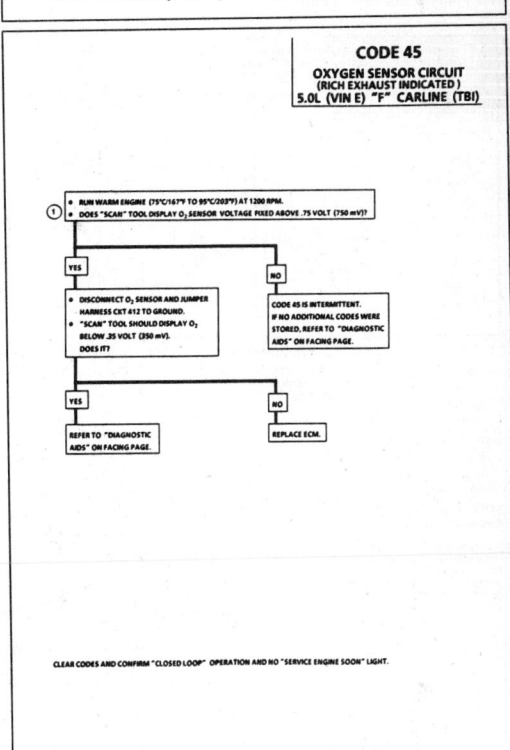

1988–90 5.0L (VIN E) – CAMARO AND FIREBIRD

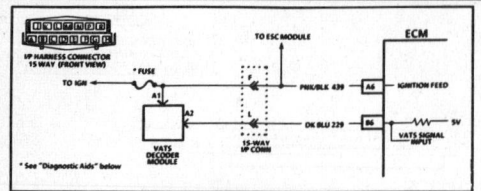

CODE 53
VEHICLE ANTI-THEFT SYSTEM (VATS) CIRCUIT
5.0L (VIN E) "F" CARLINE (TBI)

Circuit Description:

The VATS system is designed to disable vehicle operation if the incorrect key or starting procedure is used. The VATS decoder module sends a signal to the ECM if the correct key is being used. If the proper signal does not reach the ECM on CKT 229, the ECM will not pulse the injectors "ON" and thus not allow the vehicle to be started.

Code 53 will set, if the proper signal is not being received on CKT 229 ECM when the ignition is turned "ON". Code 53 does not store in the ECM memory but is only present when the conditions stated above are met.

Test Description: Numbers below refer to circled numbers on the diagnostic chart.

1. If the engine cranks, and a Code 53 is stored, it indicates that the portion of the module which generates the signal to the ECM is not operating or CKT 229 is open or shorted to ground.
   If the decoder module is found to be OK, the ECM may be at fault, but this is not a likely condition.
2. If Code 53 is stored, and the engine will not crank, it indicates that there is a VATS problem or an incorrect key or starting procedure is being used.

Diagnostic Aids:

* = ECM/IGN fuse

1988–90 5.0L (VIN E) – CAMARO AND FIREBIRD

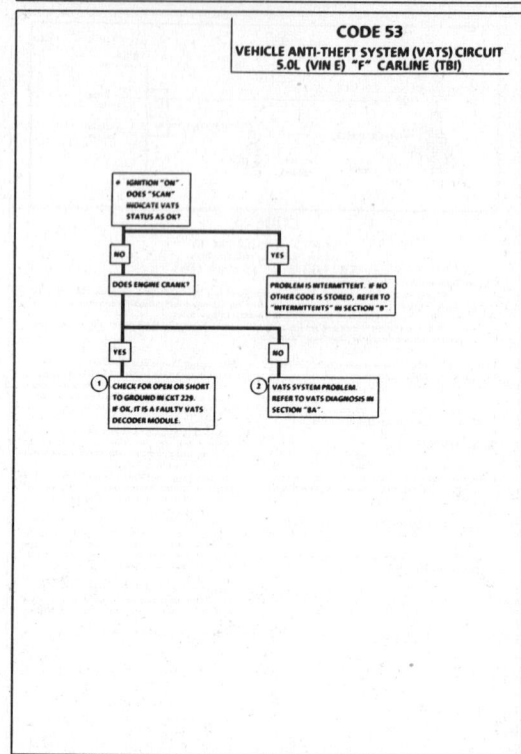

CODE 53
VEHICLE ANTI-THEFT SYSTEM (VATS) CIRCUIT
5.0L (VIN E) "F" CARLINE (TBI)

1988–90 5.0L (VIN E) – CAMARO AND FIREBIRD

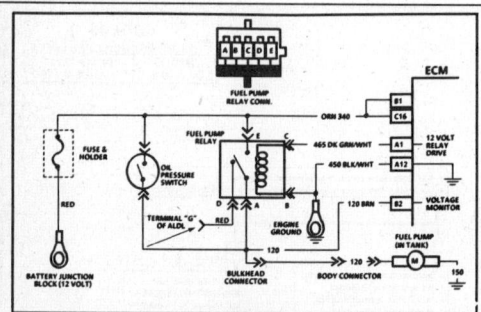

CODE 54
FUEL PUMP CIRCUIT
(LOW VOLTAGE)
5.0L (VIN E) "F" CARLINE (TBI)

Circuit Description:

When the ignition switch is turned "ON", the electronic control module (ECM) will activate the fuel pump relay and run the in-tank fuel pump. The fuel pump will operate as long as the engine is cranking or running, and the ECM is receiving ignition reference pulses.

If there are no reference pulses, the ECM will shut "OFF" the fuel pump within 2 seconds after ignition "ON", or engine stops.

Should the fuel pump relay, or the 12 volt relay drive from the ECM fail, the fuel pump will be run through an oil pressure switch back-up circuit.

Code 54 will set if the ECM does not see the 12 volts signal at terminal "B2" during the 2 seconds that the ECM is energizing the fuel pump relay.

Diagnostic Aids:

An inoperative fuel pump relay can result in long cranking times, particularly if the engine is cold or engine oil pressure is low. The extended crank period is caused by the time necessary for oil pressure to build enough to close the oil pressure switch and turn "ON" the fuel pump.

1988–90 5.0L (VIN E) – CAMARO AND FIREBIRD

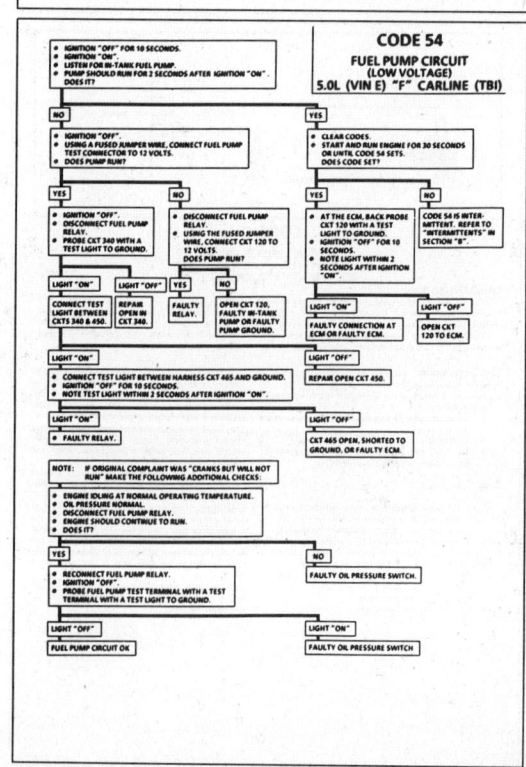

CODE 54
FUEL PUMP CIRCUIT
(LOW VOLTAGE)
5.0L (VIN E) "F" CARLINE (TBI)

1988–90 5.0L (VIN E) – CAMARO AND FIREBIRD

CODE 51
PROM ERROR
(FAULTY OR INCORRECT PROM)
5.0L (VIN E) "F" CARLINE (TBI)

CHECK THAT ALL PINS ARE FULLY INSERTED IN THE SOCKET. IF OK, REPLACE PROM, CLEAR MEMORY AND RECHECK. IF CODE 51 REAPPEARS, REPLACE ECM

CLEAR ALL CODES AND CONFIRM "CLOSED LOOP" OPERATION AND NO "SERVICE ENGINE SOON" LIGHT

CODE 52
CALPAK ERROR
(FAULTY OR INCORRECT CALPAK)

CHECK THAT ALL PINS ARE FULLY INSERTED IN THE SOCKET. IF OK, REPLACE CALPAK, CLEAR MEMORY AND RECHECK. IF CODE 52 REAPPEARS, REPLACE ECM

CLEAR ALL CODES AND CONFIRM "CLOSED LOOP" OPERATION AND NO "SERVICE ENGINE SOON" LIGHT

CODE 55
ECM ERROR

REPLACE ELECTRONIC CONTROL MODULE (ECM)

CLEAR ALL CODES AND CONFIRM "CLOSED LOOP" OPERATION AND NO "SERVICE ENGINE SOON" LIGHT

1988–90 5.0L (VIN E) – CAMARO AND FIREBIRD

SECTION B
SYMPTOMS

TABLE OF CONTENTS

Before Starting
Intermittents
Hard Start
Surges and/or Chuggle
Lack of Power, Sluggish, or Spongy
Detonation/Spark Knock
Hesitation, Sag, Stumble
Cuts Out, Misses
Poor Fuel Economy
Stalling, Rough, Unstable or Incorrect Idle
Excessive Exhaust Emissions or Odors
Dieseling, Run-On
Backfire
Restricted Exhaust System Check (All Engines)
    CHART B-1

BEFORE STARTING

Before using this section you should have performed the "Diagnostic Circuit Check."

Verify the customer complaint, and locate the correct SYMPTOM below. Check the items indicated under that symptom.

If the ENGINE CRANKS BUT WILL NOT RUN, see CHART A-3.

Several of the following symptom procedures call for a careful visual (physical) check. The importance of this step cannot be stressed too strongly -- it can lead to correcting a problem without further checks and can save valuable time.

This check should include:

- Vacuum hoses for splits, kinks, and proper connections, as shown on Vehicle Emission Control Information Label.
- Air leaks at throttle body mounting and intake manifold.
- Ignition wires for cracking, hardness, proper routing, and carbon tracking.
- Wiring for proper connections, pinches, and cuts. If wiring harness or connector repair is necessary

The following symptoms cover several engines. To determine if a particular system or component is used, refer to the ECM wiring diagrams for application.

1988–90 5.0L (VIN E) – CAMARO AND FIREBIRD

INTERMITTENTS

Problem may or may not turn "ON" the "Service Engine Soon" light, or store a code.

DO NOT use the Trouble Code Charts in Section "A" for intermittent problems. The fault must be present to locate the problem. If a fault is intermittent, use of Trouble Code Charts may result in replacement of good parts.

- Most intermittent problems are caused by faulty electrical connections or wiring. Perform careful check of suspect circuits for:

  Poor mating of the connector halves, or terminals, not fully seated in the connector body (backed out).

  Improperly formed or damaged terminals. All connector terminals in problem circuit should be carefully reformed or replaced to insure proper contact tension.

  Poor terminal to wire connection. This requires removing the terminal from the connector body to check

- If a visual (physical) check does not find the cause of the problem, the car can be driven with a voltmeter connected to a suspected circuit or a "Scan" tool may be used. An abnormal voltage on "Scan" reading, when the problem occurs, indicates the problem is in that circuit. If the wiring and connectors check OK and a Trouble Code was stored for a circuit having a

sensor, except for Codes 44 and 45, substitute a known good sensor and recheck.

- Loss of Trouble Code Memory. To check, disconnect TPS and idle engine until "Service Engine Soon" light comes "ON." Code 22 should be stored, and kept in memory, when ignition is turned "OFF" for at least 10 seconds. If not, the ECM is faulty.

- An intermittent "Service Engine Soon" light, and No Trouble Codes, may be caused by:

  Electrical system interference caused by a defective relay, ECM driven solenoid, or switch. They can cause a sharp electrical surge. Normally, the problem will occur when the faulty component is operated.

  Improper installation of electrical options, such as lights, 2-way radios, etc.

  EST wires should be routed away from spark plug wires, ignition system components, and generator. Wire for CKT 453 from ECM to ignition system should be a good ground.

  Ignition secondary shorted to ground.

  CKT 419 ("Service Engine Soon" light) or CKT 451 ("Diagnostic 'Test' Terminal") intermittently shorted to ground.

  ECM power grounds.

HARD START

Definition: Engine cranks OK, but does not start for a long time. Does eventually run, or may start but immediately dies.

- Perform careful check as described at start of Section "B".
- Make sure driver is using correct starting procedure.
- CHECK:
  - For water contaminated fuel
  - Fuel system pressure CHART A-7
  - TPS for sticking or binding should read less than 2 volt on a "Scan" tool.
  - No crank signal: see CHART C-1B
  - EGR operation. CHART C-7
  - Fuel pump relay - Connect test light between pump test terminal and ground. Light should be "ON" for 2 seconds following ignition "ON." If not, refer to CODE 54 CHART
  - For a faulty in tank fuel pump check valve, which would allow the fuel in the lines to drain back to the tank after the engine is stopped. To check for this condition:
    1. Ignition "OFF."

  2. Disconnect fuel line at the filter.
  3. Remove the tank filler cap.
  4. Connect a radiator test pump to the fuel line and apply 103 kPa (15 psi) pressure. If the pressure will hold for 60 seconds, the check valve is OK.
  - Check ignition system for:
    - Worn distributor shaft.
    - Bare and shorted wires.
    - Pickup coil resistance and connections.
    - Loose ignition coil connections.
    - Moisture in distributor cap.
    - Spark plugs; wet plugs, cracks, wear, improper gap, burned electrodes, or heavy deposits.
  - If engine starts but then, immediately stalls, disconnect the set timing connector. If engine then starts, and runs OK, replace distributor pickup coil.
  - Check CKT 423 (EST) for short to ground

1988–90 5.0L (VIN E) – CAMARO AND FIREBIRD

SURGES AND/OR CHUGGLE

Definition: Engine power variation, under steady throttle or cruise. Feels like the car speeds up and slows down, with no change in the accelerator pedal.

- Use a "Scan" tool to make sure reading of VSS matches vehicle speedometer.
- CHECK:
  - For intermittent EGR at idle. See appropriate CHART C-7.
  - Inline fuel filter for dirt or restriction.
  - Fuel pressure. See CHART A-7.
  - Generator output voltage. Repair, if less than 9, or more than 16 volts.
  - TCC Operation. See CHART C-8A.

- Inspect Oxygen sensor for silicon contamination from fuel, or use of improper RTV sealant. The sensor may have a white, powdery coating and result in a high but false signal voltage (rich exhaust indication). The ECM will then reduce the amount of fuel delivered to the engine, causing a severe driveability problem.
- Remove spark plugs. Check for cracks, wear, improper gap, burned electrodes, or heavy deposits. Also, check condition of the rest of the Ignition System.

LACK OF POWER, SLUGGISH, OR SPONGY

Definition: Engine delivers less than expected power. Little or no increase in speed, when accelerator pedal is pushed down part way.

- Compare customer's car to similar car. Make sure the customer's car has an actual problem.
- Remove air cleaner and check air filter for dirt, or for being plugged. Replace as necessary.
- If there is spray from only one injector, then, there is a malfunction in the injector assembly, or in the signal to the injector assembly. The malfunction can be isolated, by switching the injector connectors. If the problem remains with the original injector, after switching the connector, then, the injector is defective. Replace the injector. If the problem moves with the injector connector, then, the problem is an improper signal in the injector circuits, see CHART A-3.
- CHECK:
  - Ignition timing. See Vehicle Emission Control Information Label.
  - For restricted fuel filter, contaminated fuel or improper fuel pressure. See CHART A-7.
  - ECM Grounds.
  - EGR operation for being open, or partly open, all time. See CHART C-7.
  - Generator output voltage. Repair, if less than 9, or more than 16 volts.

  - Engine valve timing and compression.
  - Engine, for proper or worn camshaft.

  - Transmission torque converter operation.

  - Secondary ignition voltage, using a scope or ST-125.
  - Proper operation of EST.
- Check exhaust system for restriction. See CHART B-1:
  1. With engine at normal operating temperature, connect a vacuum gage to any convenient vacuum port on intake manifold.
  2. Run engine at 1000 rpm and record vacuum reading.
  3. Increase rpm slowly to 2500 rpm. Note vacuum reading at steady 2500 rpm.
  4. If vacuum at 2500 rpm decreases more than 3", from reading at 1000 rpm, the exhaust system should be inspected for restrictions.
  5. Disconnect exhaust pipe from engine and repeat steps 3 & 4. If vacuum still drops more than 3", with exhaust disconnected, check valve timing.

1988–90 5.0L (VIN E)—CAMARO AND FIREBIRD

DETONATION / SPARK KNOCK

Definition: A mild to severe ping, usually worse under acceleration. The engine makes sharp metallic knocks that change with throttle opening.

- **CHECK** for obvious overheating problems.
  - Low coolant.
  - Loose water pump belt.
  - Restricted air flow to radiator, or restricted water flow thru radiator.
  - Faulty or incorrect thermostat.
  - Coolant Temperature Sensor (CTS), which has shifted in value.
  - Correct coolant solution - should be a 50/50 mix of GM #1052753 anti-freeze coolant (or equivalent) and water.
- **CHECK**:
  - For poor fuel quality, proper octane rating.
  - For correct PROM.
  - Spark plugs for correct heat range.
  - Proper operation of Thermac.
  - ESC system operation. See CHARTC-5.
  - Fuel system for low pressure. See CHART A-7.
  - EGR system for not opening. See CHART C-7.
  - Ignition timing. See Vehicle Emission Control Information Label.
  - For proper transmission shift points.
  - TCC operation. See CHART C-8.
  - For incorrect basic engine parts such as cam, heads, pistons, etc.
  - Excessive oil entering combustion chamber.
- Remove carbon with top engine cleaner, follow instructions on can.
- If there is spray from only one injector, then, there is a malfunction in the injector assembly, or in the signal to the injector assembly. The malfunction can be isolated, by switching the injector connectors. If the problem remains with the original injector, after switching the connector, then, the injector is defective. Replace the injector. If the problem moves with the injector connector, then, the problem is an improper signal in the injector circuits. See CHART A-3.

HESITATION, SAG, STUMBLE

Definition: Momentary lack of response as the accelerator is pushed down. Can occur at all car speeds. Usually most severe when first trying to make the car move, as from a stop sign. May cause the engine to stall if severe enough.

- Perform careful visual (physical) check as described at start of Section "B".
- **CHECK**:
  - Fuel pressure. See CHART A-7.
  - Water contaminated fuel.
  - TPS for binding or sticking.
  - Generator output voltage. Repair if less than 9 or more than 16 volts.
  - For open Ignition System ground, CKT 453.
  - Canister purge system for proper operation.
  - EGR valve operation. See CHART C-7.
  - Ignition timing. See Vehicle Emission Control Information Label.

CUTS OUT, MISSES

Definition: Steady pulsation or jerking that follows engine speed, usually more pronounced as engine load increases. The exhaust has a steady spitting sound at idle or low speed.

- Perform careful visual (physical) check as described at start of Section "B".
- If Ignition System is suspected of causing a miss at idle or cutting, out under load:
- **Check** for missing cylinder by:
  1. Start engine, allow engine to stabilize then disconnect IAC motor. Remove one spark plug wire at a time, using insulated pliers.

CAUTION: **Do not perform this test for more than 2 minutes, as this may cause damage to the catalytic converter.**

  2. If there is an rpm drop, on all cylinders, (equal to within 50 rpm), go to STALLING, ROUGH, UNSTABLE OR INCORRECT IDLE symptom. Reconnect IAC motor with engine running, ignition "OFF."
  3. If there is no rpm drop on one or more cylinders, or excessive variation in drop, check for spark, on the suspected cylinder(s) with J 26792 (ST-125) Spark Tester or equivalent.

  If there is spark, remove spark plug(s) in these cylinders and check for:
  - Cracks
  - Wear
  - Improper Gap
  - Burned Electrodes

1988–90 5.0L (VIN E)—CAMARO AND FIREBIRD

  - Heavy Deposits
  - Perform compression check on questionable cylinder.
- Check wire resistance (should not exceed 30,000 ohms), also, check rotor and distributor cap.
- If the previous checks did not find the problem:
  - Visually inspect ignition system for moisture, dust, cracks, burns, etc. Spray plug wires with fine water mist to check for shorts.
  - Fuel System - Plugged fuel filter, water, low pressure. See CHART A-7.
  - Perform compression check.
  - Valve Timing.

  Remove rocker covers. Check for bent pushrods, worn rocker arms, broken or weak valve springs, worn camshaft lobes. Repair as necessary.
- If there is spray from only one injector, then, there is a malfunction in the injector assembly, or in the signal to the injector assembly. The malfunction can be isolated, by switching the injector connectors. If the problem remains with the original injector, after switching the connector, then, the injector is defective. Replace the injector. If the problem moves with the injector connector, then, the problem is an improper signal in the injector circuits, see CHART A-3.

POOR FUEL ECONOMY

Definition: Fuel economy, as measured by an actual road test, is noticeably lower than expected. Also, economy is noticeably lower than it was on this car at one time, as previously shown by an actual road test.

- **CHECK**:
  - Engine thermostat for faulty part (always open) or for wrong heat range.
  - Fuel Pressure. See CHART A-7.
- Check owner's driving habits.
  - Is A/C "ON" full time (Defroster mode "ON")?
  - Are tires at correct pressure?
  - Are excessively heavy loads being carried?
  - Is acceleration too much, too often?
  - Suggest driver read "Important Facts on Fuel Economy" in Owner's Manual.
- Perform "Diagnostic Circuit Check."
- Check air cleaner element (filter) for dirt or being plugged.
- Check for proper calibration of speedometer.
- Visually (physically) Check:
  - Vacuum hoses for splits, kinks, and proper connections, as shown on Vehicle Emission Control Information Label.
  - Ignition wires for cracking, hardness, and proper connections.
- Check Ignition timing. See Vehicle Emission Control Information Label.
- Remove spark plugs. Check for cracks, wear, improper gap, burned electrodes or heavy deposits. Repair or replace, as necessary.
- Check compression.
- Check TCC for proper operation. See CHART C-8. Use "Scan" tool if available.
- Check for dragging brakes.
- Suggest owner fill fuel tank and recheck fuel economy.
- Check for exhaust system restriction. See CHART B-1.

STALLING, ROUGH, UNSTABLE OR INCORRECT IDLE

Definition: The engine runs unevenly at idle. If bad enough, the car may shake. Also, the idle may vary in rpm (called "hunting"). Either condition may be severe enough to cause stalling. Engine idles at incorrect speed.

- **CHECK**:
  - P/N switch circuit. See CHART C-1A.
  - For injector(s) leaking. Check fuel pressure, see CHART A-7.
  - IAC - See CHART C-2C.
  - If a sticking throttle shaft or binding linkage causes a high TPS voltage (open throttle indication), the ECM will not control idle. Monitor TPS voltage. "Scan" and/or Voltmeter should read less than 1.2 volts with throttle closed.
  - Vacuum leaks can cause higher than normal idle.
  - EGR "ON," while idling, will cause roughness, stalling, and hard starting. See CHART C-7.
  - Battery cables and ground straps should be clean and secure. Erratic voltage will cause IAC to change its position, resulting in poor idle quality.
  - IAC valve will not move, if system voltage is below 9, or greater than 16 volts.
  - Use "Scan" tool to determine if ECM is receiving A/C request signal.

1988–90 5.0L (VIN E)—CAMARO AND FIREBIRD

- Ignition timing. See Vehicle Emission Control Information Label.
- MAP Sensor - Ignition "ON," engine stopped. Compare MAP voltage with known good vehicle. Voltage should be the same ± 400 mV (.4 volts).

  or

  Start and idle engine. Disconnect MAP sensor electrical connector. If idle improves, substitute a known good MAP sensor and recheck.
- A/C refrigerant pressure too high. Check for overcharge or faulty pressure switch.
- PCV valve for proper operation by placing finger over inlet hole in valve end several times. Valve should snap back. If not, replace valve.
- Run a cylinder compression check. See Section "6".
- Inspect Oxygen sensor for silicon contamination from fuel, or use of improper RTV sealant. The sensor will have a white, powdery coating, and will result in a high but false signal voltage (rich exhaust indication). The ECM will then reduce the amount of fuel delivered to the engine, causing a severe driveability problem.

EXCESSIVE EXHAUST EMISSIONS OR ODORS

Definition: Vehicle fails an emission test. Vehicle has excessive "rotten egg" smell. Excessive odors do not necessarily indicate excessive emissions.

- Perform "Diagnostic Circuit Check."
- IF TEST SHOWS EXCESSIVE CO AND HC, (or also has excessive odors):
  - Check items which cause car to run RICH.
  - Make sure engine is at normal operating temperature.
- **CHECK**:
  - Fuel pressure. See CHART A-7.
  - Incorrect timing. See Vehicle Emission Control Information Label.
  - Fuel Vapor Canister for fuel loading. See CHART C-3.
  - PCV valve for being plugged, stuck, or blocked PCV hose, or fuel in the crankcase.
  - Spark plugs, plug wires, and ignition components.
  - Check for lead contamination of catalytic converter (look for removal of fuel filler neck restrictor).
  - Check for properly installed fuel cap.
  - If the system is running rich, (block learn less than 118), refer to "Diagnostic Aids" on facing page of Code 45.
- IF TEST SHOWS EXCESSIVE NOx:
  - Check items which cause car to run LEAN, or too HOT.
- **CHECK**:
  - EGR valve for not opening. See CHART C-7.
  - Vacuum leaks.
  - Coolant system and coolant fan for proper operation. See CHART C-12.
  - Remove carbon with top engine cleaner, follow instructions on can.
  - Check Ignition timing for excessive base advance. See Vehicle Emission Control Information Label.
- If the system is running lean, (block learn greater than 138), refer to "Diagnostic Aids" on facing page of Code 44.

DIESELING, RUN-ON

Definition: Engine continues to run after key is turned "OFF," but runs very roughly. If engine runs smoothly, check ignition switch and adjustment.

- Check injector(s) for leaking. Apply 12 volts to fuel pump test terminal (ALDL Terminal "G") to turn "ON" fuel pump and pressurize fuel system. Visually check injector(s) and TBI assembly for fuel leakage.

BACKFIRE

Definition: Fuel ignites in intake manifold, or in exhaust system, making a loud popping noise.

- **CHECK**:
  - EGR operation for being open all the time. See CHART C-7.
  - Output voltage of ignition coil(s).
  - For crossfire between spark plugs (distributor cap, spark plug wires, and proper routing of plug wires).
  - Ignition timing - See Vehicle Emission Control Information Label.
  - For faulty spark plugs and/or plug wires or boots.
  - Faulty A.I.R. check valve.
- Perform a compression check - look for sticking or leaking valves.
  - For proper valve timing.
  - Broken or worn valve train parts.

1988–90 5.0L (VIN E)—CAMARO AND FIREBIRD

CHART B-1
RESTRICTED EXHAUST SYSTEM CHECK
ALL ENGINES

Proper diagnosis for a restricted exhaust system is essential before any components are replaced. Either of the following procedures may be used for diagnosis, depending upon engine or tool used:

CHECK AT A. I. R. PIPE:

1. Remove the rubber hose at the exhaust manifold A.I.R. pipe check valve. Remove check valve.
2. Connect a fuel pump pressure gauge to a hose and nipple from a Propane Enrichment Device (J 26911) (see illustration).
3. Insert the nipple into the exhaust manifold A.I.R. pipe.

OR

CHECK AT O_2 SENSOR:

1. Carefully remove O_2 sensor.
2. Install Borroughs exhaust backpressure tester (BT 8515 or BT 8603) or equivalent in place of O_2 sensor (see illustration).
3. After completing test described below, be sure to coat threads of O_2 sensor with anti-seize compound P/N 5613695 or equivalent prior to re-installation.

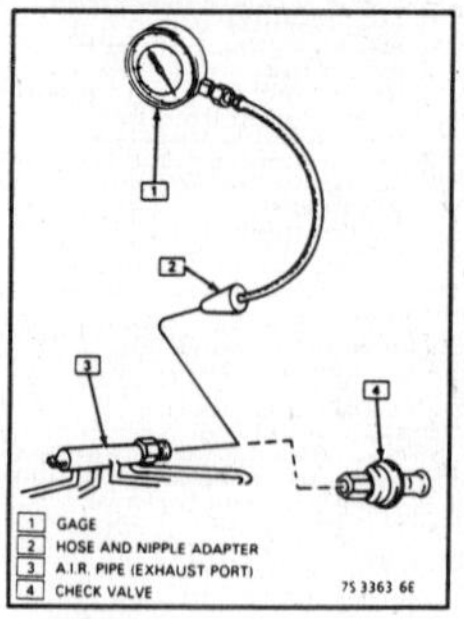
1 GAGE
2 HOSE AND NIPPLE ADAPTER
3 A.I.R. PIPE (EXHAUST PORT)
4 CHECK VALVE
7S 3363 6E

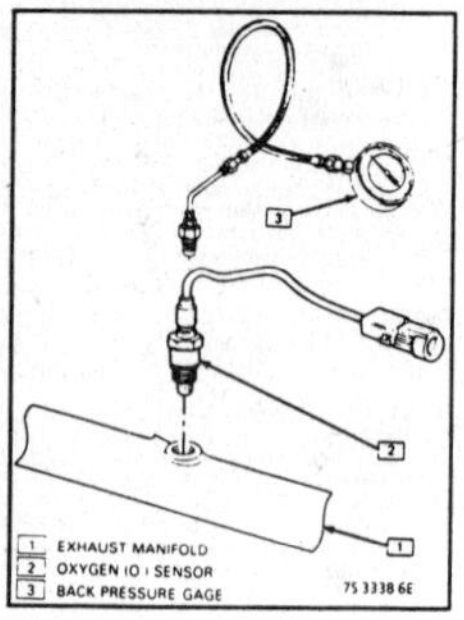
1 EXHAUST MANIFOLD
2 OXYGEN (O) SENSOR
3 BACK PRESSURE GAGE
7S 3338 6E

DIAGNOSIS:

1. With the engine idling at normal operating temperature, observe the exhaust system backpressure reading on the gage. Reading should not exceed 8.6 kPa (1.25 psi).
2. Increase engine speed to 2000 rpm and observe gage. Reading should not exceed 20.7 kPa (3 psi).
3. If the backpressure at either speed exceeds specification, a restricted exhaust system is indicated.
4. Inspect the entire exhaust system for a collapsed pipe, heat distress, or possible internal muffler failure.
5. If there are no obvious reasons for the excessive backpressure, the catalytic converter is suspected to be restricted and should be replaced using current recommended procedures.

1988–90 5.0L (VIN E) – CAMARO AND FIREBIRD

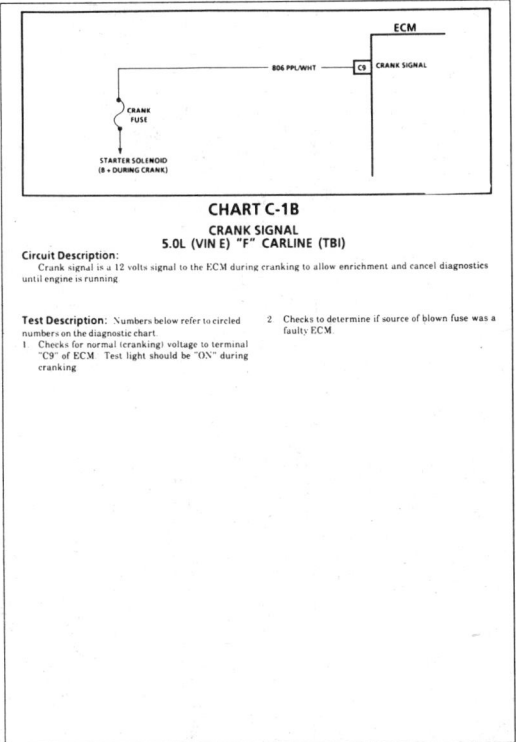

CHART C-1B
CRANK SIGNAL
5.0L (VIN E) "F" CARLINE (TBI)

Circuit Description:
Crank signal is a 12 volts signal to the ECM during cranking to allow enrichment and cancel diagnostics until engine is running.

Test Description: Numbers below refer to circled numbers on the diagnostic chart.
1. Checks for normal (cranking) voltage to terminal "C9" of ECM. Test light should be "ON" during cranking.

2. Checks to determine if source of blown fuse was a faulty ECM.

1988–90 5.0L (VIN E) – CAMARO AND FIREBIRD

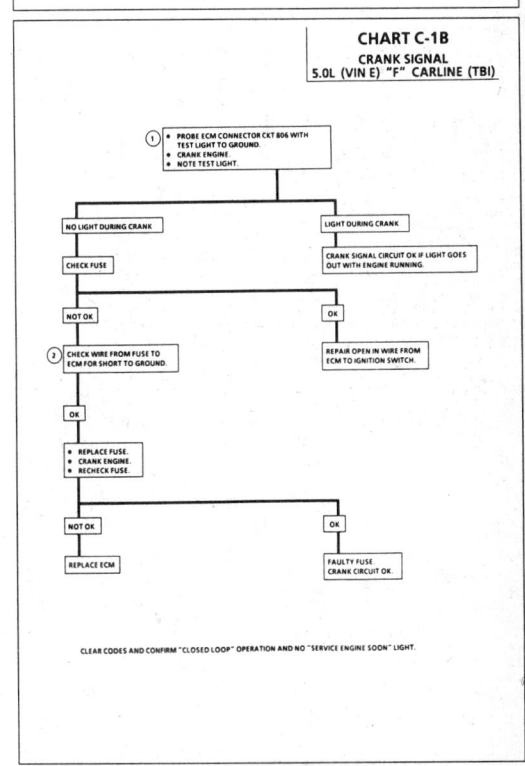

1988–90 5.0L (VIN E) – CAMARO AND FIREBIRD

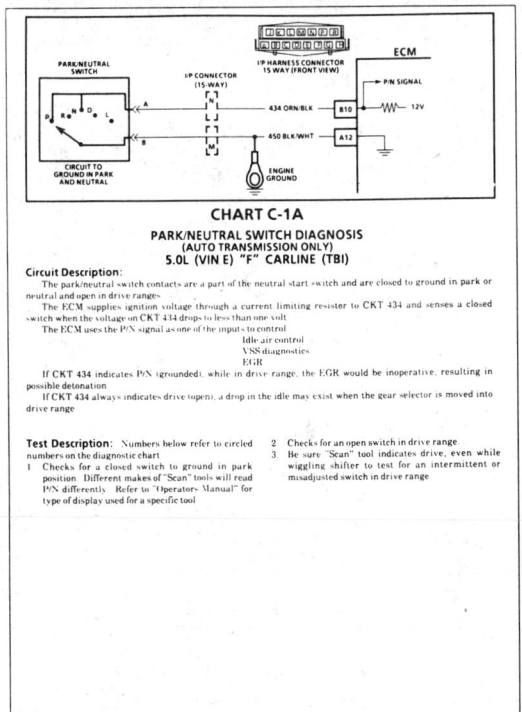

CHART C-1A
PARK/NEUTRAL SWITCH DIAGNOSIS
(AUTO TRANSMISSION ONLY)
5.0L (VIN E) "F" CARLINE (TBI)

Circuit Description:
The park/neutral switch contacts are a part of the neutral start switch and are closed to ground in park or neutral and open in drive ranges.
The ECM supplies ignition voltage through a current limiting resistor to CKT 434 and senses a closed switch when the voltage on CKT 434 drops to less than one volt.
The ECM uses the P/N signal as one of the inputs to control
    Idle air control
    VSS diagnostics
    EGR
If CKT 434 indicates P/N (grounded), while in drive range, the EGR would be inoperative, resulting in possible detonation.
If CKT 434 always indicates drive (open), a drop in the idle may exist when the gear selector is moved into drive range.

Test Description: Numbers below refer to circled numbers on the diagnostic chart.
1. Checks for a closed switch to ground in park position. Different makes of "Scan" tools will read P/N differently. Refer to "Operators Manual" for type of display used for a specific tool.

2. Checks for an open switch in drive range.
3. Be sure "Scan" tool indicates drive, even while wiggling shifter to test for an intermittent or misadjusted switch in drive range.

1988–90 5.0L (VIN E) – CAMARO AND FIREBIRD

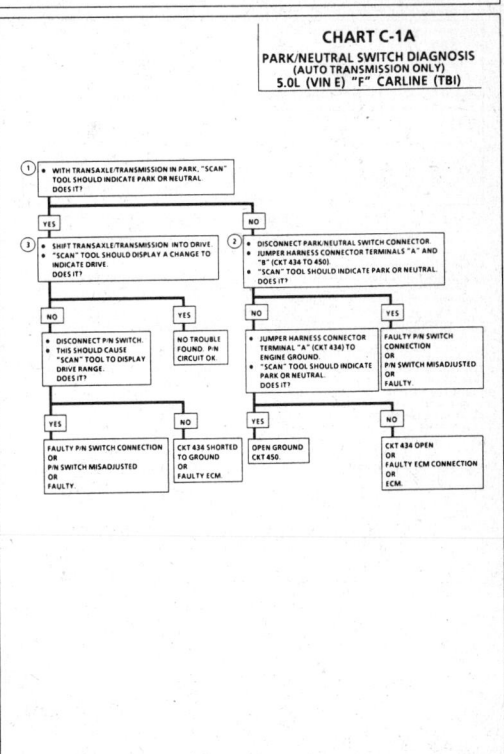

1988–90 5.0L (VIN E) – CAMARO AND FIREBIRD

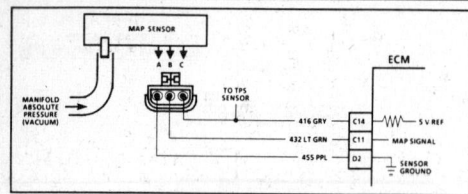

CHART C-1D
MAP OUTPUT CHECK
5.0L (VIN E) "F" CARLINE (TBI)

Circuit Description:
The Manifold Absolute Pressure (MAP) sensor measures manifold pressure (vacuum) and sends that signal to the ECM. The ECM uses this information for fuel and spark control.

Test Description: Numbers below refer to circled numbers on the diagnostic chart.
1. Checks MAP sensor output voltage to the ECM. This voltage, without engine running, represents a barometer reading to the ECM.
2. Applying 34 kPa (10" Hg) vacuum to the MAP sensor should cause the voltage to be 1.2 - 2.3 volts less than the voltage at Step 1. Upon applying vacuum to the sensor, the change in voltage should be instantaneous. A slow voltage change indicates a faulty sensor.
3. Check vacuum hose to sensor for leaking or restriction. Make sure no other vacuum devices are connected to the MAP hose.

Diagnostic Aids:

With the ignition "ON" and the engine stopped, the manifold pressure is equal to atmospheric pressure and the signal voltage will be high. This information is used by the ECM as an indication of vehicle altitude and is referred to as BARO. Comparison of this BARO reading with a known good vehicle with the same sensor is a good way to check accuracy of a "suspect" sensor. Reading should be the same, ± .4 volt.

1988–90 5.0L (VIN E) – CAMARO AND FIREBIRD

CHART C-1D
MAP OUTPUT CHECK
5.0L (VIN E) "F" CARLINE (TBI)

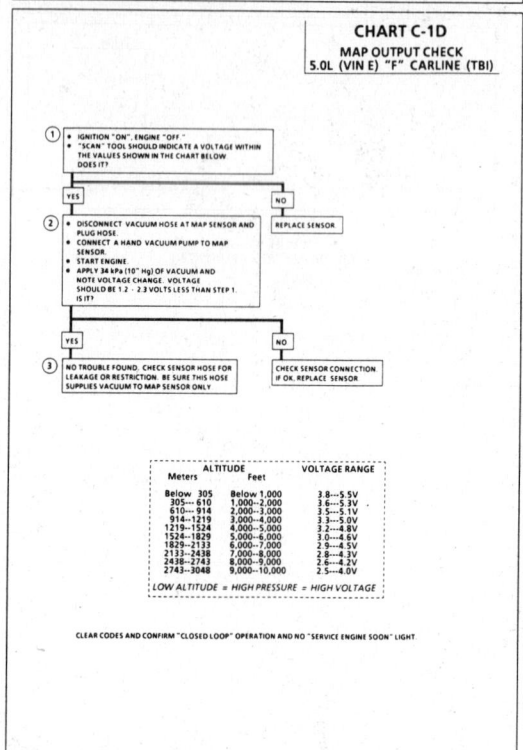

ALTITUDE		VOLTAGE RANGE
Meters	Feet	
Below 305	Below 1,000	3.8---5.5V
305--- 610	1,000---2,000	3.6---5.3V
610--- 914	2,000---3,000	3.5---5.1V
914---1219	3,000---4,000	3.3---5.0V
1219---1524	4,000---5,000	3.2---4.8V
1524---1829	5,000---6,000	3.0---4.6V
1829---2133	6,000---7,000	2.9---4.5V
2133---2438	7,000---8,000	2.8---4.3V
2438---2743	8,000---9,000	2.6---4.2V
2743---3048	9,000---10,000	2.5---4.0V

LOW ALTITUDE = HIGH PRESSURE = HIGH VOLTAGE

CLEAR CODES AND CONFIRM "CLOSED LOOP" OPERATION AND NO "SERVICE ENGINE SOON" LIGHT.

1988–90 5.0L (VIN E) – CAMARO AND FIREBIRD

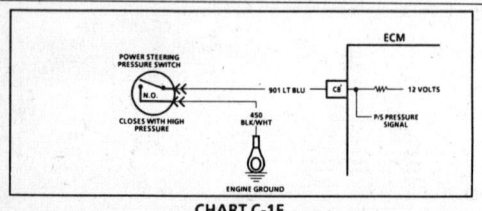

CHART C-1E
POWER STEERING PRESSURE SWITCH (PSPS) DIAGNOSIS
5.0L (VIN E) "F" CARLINE (TBI)

Circuit Description:
The power steering pressure switch is normally open to ground, and CKT 901 will be near battery voltage. Turning the steering wheel increases power steering oil pressure and applies more load on an idling engine. The pressure switch will close before the load can cause an idle problem.
Closing the switch causes CKT 901 to read less than one volt and the ECM will increase the idle air rate.
- A pressure switch that will not close, or an open CKT 901 or 450, may cause the engine to stop when power steering loads are high.
- A switch that will not open, or a CKT 901 shorted to ground, may affect idle quality.

Test Description: Numbers below refer to circled numbers on the diagnostic chart.
1. Different makes of "Scan" tools may display the state of this switch in different ways. Refer to "Scan" tool upgrading to determine how this input is indicated.
2. Checks to determine if CKT 901 is shorted to ground.
3. This should simulate a closed switch.

1988–90 5.0L (VIN E) – CAMARO AND FIREBIRD

CHART C-1E
POWER STEERING PRESSURE SWITCH (PSPS) DIAGNOSIS
5.0L (VIN E) "F" CARLINE (TBI)

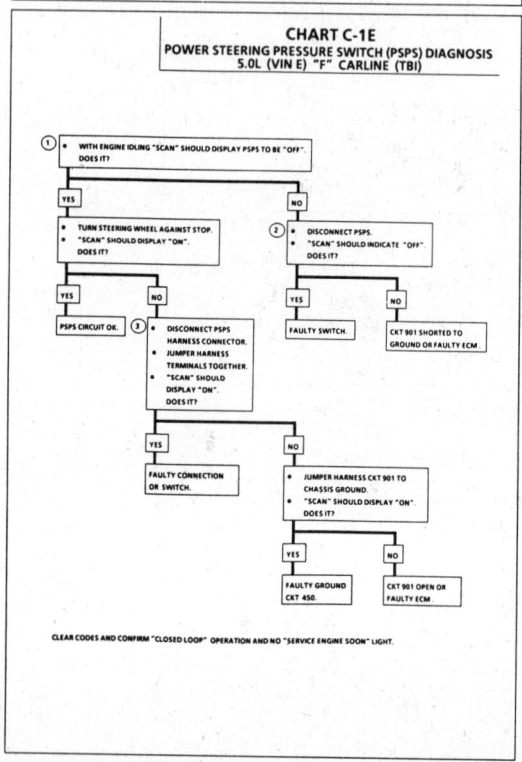

CLEAR CODES AND CONFIRM "CLOSED LOOP" OPERATION AND NO "SERVICE ENGINE SOON" LIGHT.

1988–90 5.0L (VIN E) – CAMARO AND FIREBIRD

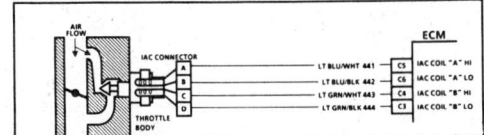

CHART C-2C
IDLE AIR CONTROL (IAC) VALVE CHECK
5.0L (VIN E) "F" CARLINE (TBI)

Circuit Description:

The ECM controls idle rpm with the IAC valve. To increase idle rpm, the ECM moves the IAC valve out, allowing more air to pass by the throttle plate. To decrease rpm, it moves the IAC valve in, reducing air flow by the throttle plate. A "Scan" tool will read the ECM commands to the IAC valve in counts. The higher the counts, the more air allowed (higher idle). The lower the counts, the less air allowed (lower idle)

Test Description: Numbers below refer to circled numbers on the diagnostic chart.

1. Continue with test, even if engine will not idle. If idle is too low, "Scan" will display 80 or more counts, or steps. If idle is high, it will display "0" counts. Occasionally, an erratic or unstable idle may occur. Engine speed may vary 200 rpm, or more, up and down. Disconnect IAC. If the condition is unchanged, the IAC is not at fault.
2. When the engine was stopped, the IAC Valve retracted (more air) to a fixed "Park" position for increased air flow and idle speed during the next engine start. A "Scan" tool will display 100 or more counts. When performing this test, immediately note rpm on start up, because, on a warm engine, the rpm will decrease rapidly.
3. Be sure to disconnect the IAC valve prior to this test. The test light will confirm the ECM signals by a steady or flashing light on all circuits.
4. There is a remote possibility that one of the CKTs is shorted to voltage, which would have been indicated by a steady light. Disconnect ECM and turn the ignition "ON" and probe terminals to check for this condition.

Diagnostic Aids:

A slow unstable idle may be caused by a system problem that cannot be overcome by the IAC. "Scan" counts will be above 60 counts, if too low, and "0" counts, if engine speed is too high.

If idle is too high, stop engine. Ignition "ON" Ground diagnostic terminal. Wait 30 seconds for IAC to seat, then, disconnect IAC. Unground diagnostic terminal and start engine. If idle speed is above 450

rpm in drive, locate and correct vacuum leak. If rpm is less than 450 rpm, adjust minimum idle speed, or correct other conditions, which may affect idle. Refer to "Rough Unstable or Incorrect Idle", in "Symptoms", Section "B"

* **System too lean (High Air/Fuel Ratio)**
  Idle speed may be too high or too low. Engine speed may vary up and down, disconnecting IAC does not help. May set Code 44.
  "Scan" and/or Voltmeter will read an oxygen sensor output less than 300 mv (.3 volt). Check for low regulated fuel pressure or water in fuel. A lean exhaust, with an oxygen sensor output fixed above 800 mv (.8 volt), will be a contaminated sensor, usually silicone. This may also set a Code 45.

* **System too rich (Low Air/Fuel Ratio)**
  Idle speed too low. "Scan" counts usually above 80. System obviously rich and may exhibit black smoke exhaust.
  "Scan" tool and/or Voltmeter will read an oxygen sensor signal fixed above 800 mv (.8 volt).
  Check
  - High fuel pressure
  - Injector leaking or sticking

* **Throttle Body.** Remove IAC and inspect bore for foreign material or evidence of IAC valve dragging the bore.
* If above are all OK, refer to "Rough, Unstable, Incorrect Idle or Stalling", in "Symptoms", Section "B"

1988–90 5.0L (VIN E) – CAMARO AND FIREBIRD

CHART C-2C
IDLE AIR CONTROL (IAC) VALVE CHECK
5.0L (VIN E) "F" CARLINE (TBI)

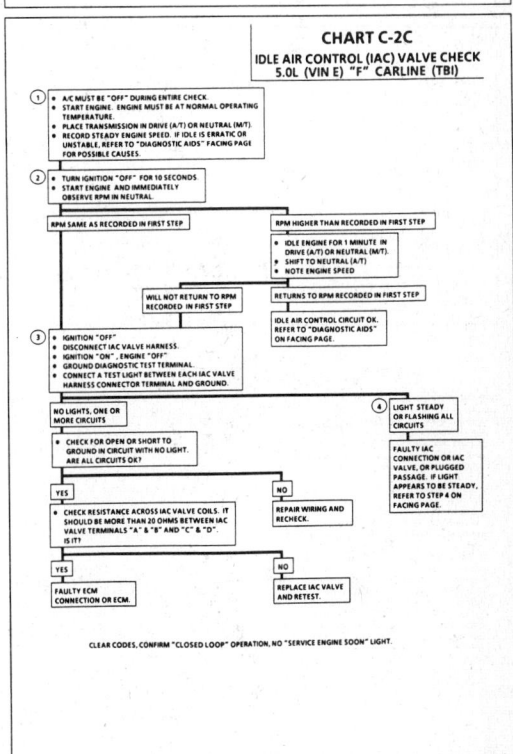

1988–90 5.0L (VIN E) – CAMARO AND FIREBIRD

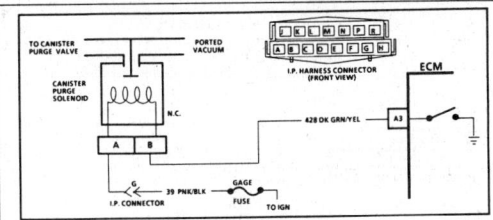

CHART C-3
CANISTER PURGE VALVE CHECK
5.0L (VIN E) "F" CARLINE (TBI)

Circuit Description:

Canister purge is controlled by a solenoid that allows ported manifold vacuum to purge the fuel vapor canister when energized. The ECM supplies a ground to energize the solenoid (purge "ON").

If the following conditions are met with the engine running, the canister purge solenoid is energized (purge "ON").

* Engine run time after start more than 1 minute.
* Coolant temperature above 80°C (176°F).
* Vehicle speed above 5 mph.
* Throttle "OFF" idle. TPS signal about .75 volt.

Test Description: Numbers below refer to circled numbers on the diagnostic chart.

1. Checks to see if the canister purge solenoid is opened or closed. The canister purge solenoid is energized in this step so it should be open.
2. Completes functional check, by grounding diagnostic "test" terminal. This should, normally, de-energize the canister purge solenoid and allow the vacuum to drop (purge "ON").
3. Checks for a complete circuit. Normally, there is battery voltage on CKT 39, and the ECM provides a ground on CKT 428. A shorted canister purge solenoid could cause an open circuit in the ECM.

Diagnostic Aids:

Normal operation of the canister purge solenoid is described as follows:

With ignition "ON", engine stopped, diagnostic "test" terminal ungrounded, the canister purge solenoid will be energized.

With ignition "ON", engine "OFF", diagnostic "test" terminal grounded, the canister purge solenoid will be de-energized.

1988–90 5.0L (VIN E) – CAMARO AND FIREBIRD

CHART C-3
CANISTER PURGE VALVE CHECK
5.0L (VIN E) "F" CARLINE (TBI)

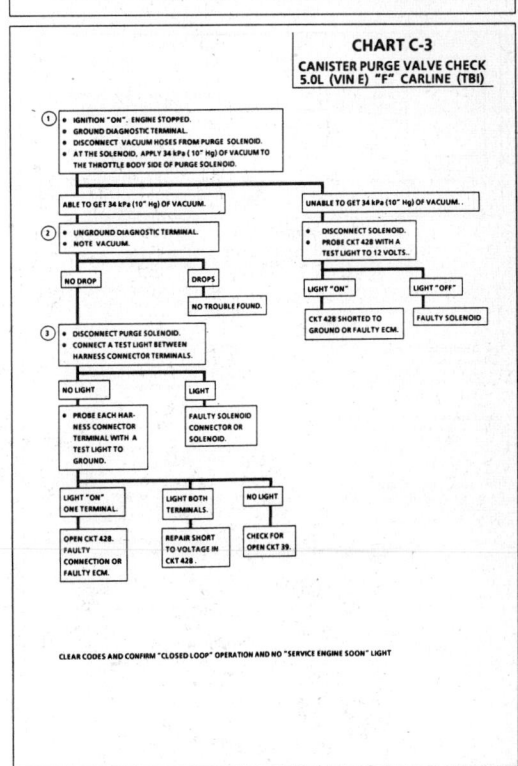

1988–90 5.0L (VIN E) – CAMARO AND FIREBIRD

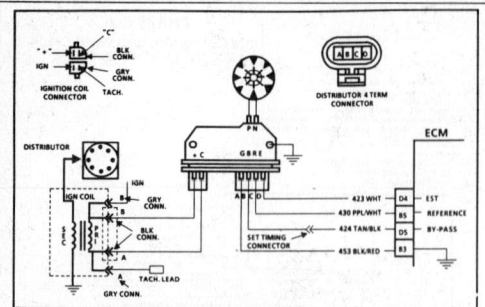

CHART C-4
IGNITION SYSTEM CHECK
(REMOTE COIL/SEALED MODULE CONNECTOR DISTRIBUTOR)
5.0L (VIN E) "F" CARLINE (TBI)

Test Description: Numbers below refer to circled numbers on the diagnostic chart.

1. Two wires are checked, to ensure that an open is not present in a spark plug wire.
1A. If spark occurs with EST connector disconnected, pick-up coil output is too low for EST operation.
2. A spark indicates the problem must be the distributor cap or rotor.
3. Normally, there should be battery voltage at the "C" and "+" terminals. Low voltage would indicate an open or a high resistance circuit from the distributor to the coil or ignition switch. If "C" terminal voltage was low, but "+" terminal voltage is 10 volts or more, circuit from "C" terminal to ignition coil or ignition coil primary winding is open.
4. Checks for a shorted module or grounded circuit from the ignition coil to the ignition module. The distributor module should be turned "OFF", so normal voltage should be about 12 volts.
   If the module is turned "ON", the voltage would be low, but about 1 volt. This could cause the ignition coil to fail from excessive heat.
   With an open ignition coil primary winding, a small amount of voltage will leak through the module from the "Batt+" to the "tach" terminal.

5. Applying a voltage (1.5 to 8 volts) to module terminal "P" should turn the module "ON" and the "tach" terminal voltage should drop to about 7-9 volts. This test will determine whether the module or coil is faulty or if the pick-up coil is not generating the proper signal to turn the module "ON". This test can be performed by using a DC battery with a rating of 1.5 to 8 volts. The use of the test light is mainly to allow the "P" terminal to be probed more easily. Some digital multi-meters can also be used to trigger the module by selecting ohms, usually the diode position. In this position the meter may have a voltage across it's terminals which can be used to trigger the module. The voltage in the ohm's position can be checked by using a second meter or by checking the manufacture's specification of the tool being used.
6. This should turn "OFF" the module and cause a spark. If no spark occurs, the fault is most likely in the ignition coil because most ignition coil problems would have been found before this point in the procedure. A module tester could determine which is at fault.

1988–90 5.0L (VIN E) – CAMARO AND FIREBIRD

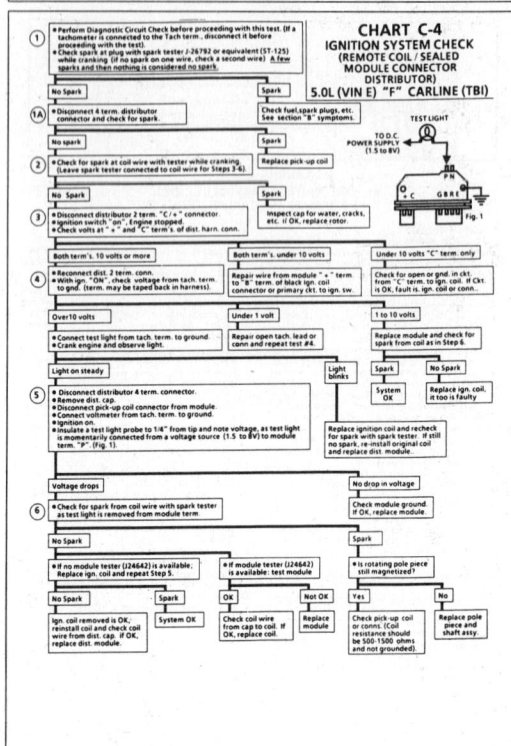

CHART C-4
IGNITION SYSTEM CHECK
(REMOTE COIL / SEALED MODULE CONNECTOR DISTRIBUTOR)
5.0L (VIN E) "F" CARLINE (TBI)

1988–90 5.0L (VIN E) – CAMARO AND FIREBIRD

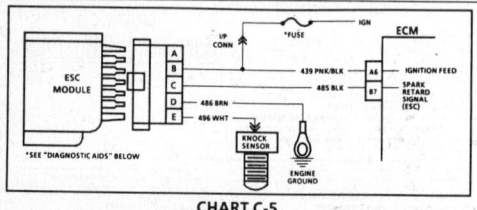

CHART C-5
ELECTRONIC SPARK CONTROL (ESC) SYSTEM CHECK
(ENGINE KNOCK, POOR PERFORMANCE, OR POOR ECONOMY)
5.0L (VIN E) "F" CARLINE (TBI)

Circuit Description:
Electronic Spark Control (ESC) is accomplished with a module that sends a voltage signal to the ECM. When the knock sensor detects engine knock, the voltage from the ESC module to the ECM is shut "OFF" and this signals the ECM to retard timing, if engine rpm is over about 900.

Test Description: Numbers below refer to circled numbers on the diagnostic chart.

1. If a Code 43 is not set, but a knock signal is indicated while running at 1500 rpm, listen for an internal engine noise. Under a no load condition, there should not be any detonation, and if knock is indicated, an internal engine problem may exist.
2. Usually a knock signal can be generated by tapping on the exhaust manifold. This test can also be performed at idle. Test number 1 was run at 1500 rpm, to determine if a constant knock signal was present, which would affect engine performance.
3. This tests whether the knock signal is due to the knock sensor, a basic engine problem, or the ESC module.
4. If the ESC module ground circuit is faulty, the ESC module will not function correctly. The test light should light indicating the ground circuit is OK.

5. Contacting CKT 496, with a test light to 12 volts, should generate a knock signal to determine whether the knock sensor is faulty, or the ESC module can't recognize a knock signal.

Diagnostic Aids:

* = ECM/IGN Fuse

"Scan" tools have two positions to diagnose the ESC system. The knock signal can be monitored to see if the knock sensor is detecting a knock condition and if the ESC module is functioning, knock signal should display "YES", whenever detonation is present. The knock retard position on the "Scan" displays the amount of spark retard the ECM is commanding. The ECM can retard the timing up to 20 degrees.
If the ESC system checks OK, but detonation is the complaint, refer to "Detonation/Spark Knock" in Section "B"

1988–90 5.0L (VIN E) – CAMARO AND FIREBIRD

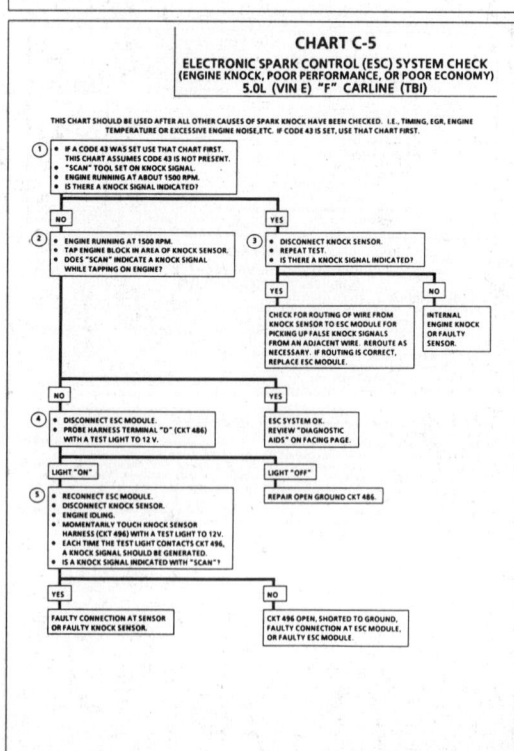

CHART C-5
ELECTRONIC SPARK CONTROL (ESC) SYSTEM CHECK
(ENGINE KNOCK, POOR PERFORMANCE, OR POOR ECONOMY)
5.0L (VIN E) "F" CARLINE (TBI)

1988–90 5.0L (VIN E) – CAMARO AND FIREBIRD

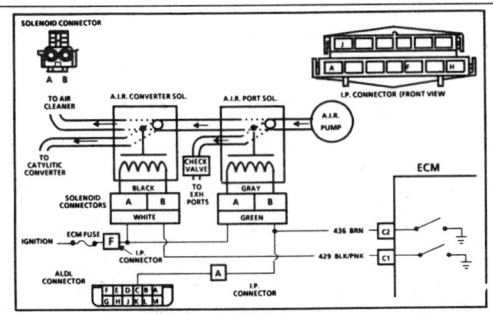

CHART C-6
AIR MANAGEMENT CHECK - PEDES VALVE
(PRESSURE OPERATED ELECTRIC DIVERT/ELECTRIC SWITCHING)
5.0L (VIN E) "F" CARLINE (TBI)

Circuit Description:

Air management is controlled by a port valve and a converter valve, each with an ECM controlled electrical solenoid. When the solenoid is grounded by the ECM, A.I.R. pump pressure will activate the valve and allow pump air to be directed as follows:

Neither solenoid grounded by the ECM - Air pump air diverted to atmosphere. Converter solenoid grounded by the ECM - Air pump air to converter.

Port solenoid grounded by the ECM - Air pump air to exhaust ports.

Test Description: Numbers below refer to circled numbers on the diagnostic chart.

1. This is a system functional check. Air is directed to ports during "Open Loop" and all engine start in "Open Loop" even on a warm engine. Since the air to the ports time is very short on some engines, prepare to observe port air prior to engine start up. On some engines, this can be done by squeezing a hose. On others, steel pipes have to be disconnected.
2. This should normally set a Code 22. When any code is set, the ECM opens the ground to the A.I.R. control valve and allows air to divert. This checks for ECM response to a fault. A ground in the A.I.R. control valve circuit to the ECM would prevent divert action.
3. This checks for a grounded circuit to the ECM. Test light "OFF" is normal and would indicate the circuit is not grounded.
4. Checks for an open in the solenoid control circuits. Grounding the diagnostic "test" terminal should ground both solenoid circuits. Normally, the test light should be "ON" which indicates the problem is not in the ECM or wiring but at the solenoid connections or A.I.R. control valve itself.
5. Checks for a grounded port (switching valve) solenoid circuit. Test light "OFF" would indicate the circuit is normal and fault is in the valve.

1988–90 5.0L (VIN E) – CAMARO AND FIREBIRD

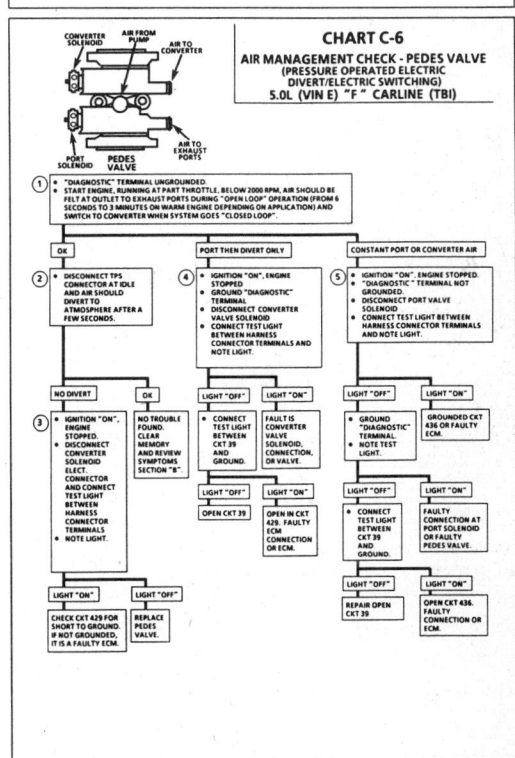

CHART C-6
AIR MANAGEMENT CHECK - PEDES VALVE
(PRESSURE OPERATED ELECTRIC DIVERT/ELECTRIC SWITCHING)
5.0L (VIN E) "F" CARLINE (TBI)

1988–90 5.0L (VIN E) – CAMARO AND FIREBIRD

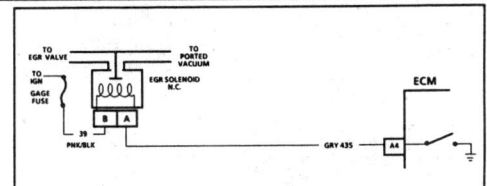

CHART C-7
EXHAUST GAS RECIRULATION (EGR) CHECK
5.0L (VIN E) "F" CARLINE (TBI)

Circuit Description:

The ECM operates a solenoid to control the Exhaust Gas Recirculation (EGR) valve. This solenoid is normally closed. By providing a ground path, the ECM energizes the solenoid which allows vacuum to pass to the EGR valve. The ECM control of the EGR is based on the following inputs:
- Engine coolant temperature above 25°C (77°F).
- TPS off idle
- MAP

If Code 24 is stored, use that chart first.

Test Description: Numbers below refer to circled numbers on the diagnostic chart.

1. By grounding the diagnostic "test" terminal, the EGR solenoid should be energized and allow vacuum to be applied to the EGR valve and the vacuum should hold.
2. When the diagnostic "test" terminal is ungrounded, the vacuum to the EGR valve should bleed off through a vent in the EGR solenoid and the valve should close. The gage may or may not bleed off but this does not indicate a problem.
3. This test will determine if the electrical control part of the system is at fault or if the connector or solenoid is at fault.
4. This system uses a negative backpressure EGR valve which should hold vacuum with engine "OFF."
5. When engine is started, exhaust backpressure should cause vacuum to bleed off and valve should fully close.

Diagnostic Aids:

Vacuum lines should be <u>thoroughly</u> checked for internal restrictions. The ECM uses the MAP sensor for checking EGR operation. If there is a question of MAP sensor accuracy, use CHART C-1D MAP output check in Section "C".

If no problems are found, refer to "Intermittents" in Section "B".

1988–90 5.0L (VIN E) – CAMARO AND FIREBIRD

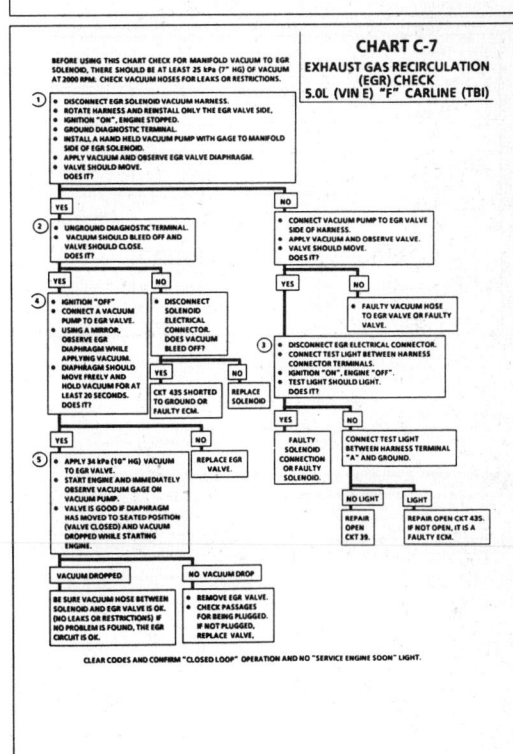

CHART C-7
EXHAUST GAS RECIRCULATION (EGR) CHECK
5.0L (VIN E) "F" CARLINE (TBI)

1988–90 5.0L (VIN E) – CAMARO AND FIREBIRD

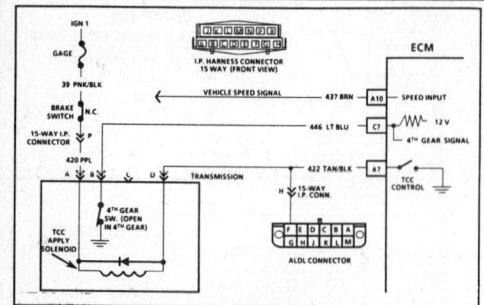

CHART C-8A
(Page 1 of 2)
TORQUE CONVERTER CLUTCH (TCC)
(ELECTRICAL DIAGNOSIS)
5.0L (VIN E) "F" CARLINE (TBI)

Circuit Description:

The purpose of the automatic transmission torque converter clutch is to eliminate the power loss of the torque converter, when the vehicle is in a cruise condition. This allows the convenience of the automatic transmission and the fuel economy of a manual transmission.

Fused battery ignition is supplied to the TCC solenoid through the brake switch, the ECM will engage TCC by grounding CKT 422 to energize the TCC solenoid.

TCC will engage when:
- Vehicle speed above 24 mph
- Engine at normal operating temperature (above 70°C, 156°F)
- Throttle position sensor output not changing, indicating a steady road speed
- Brake switch closed

Test Description: Numbers below refer to circled numbers on the diagnostic chart.
1. Confirms 12 volt supply as well as continuity of TCC circuit.
2. Grounding the diagnostic "test" terminal with engine "OFF", should energize the capability of the ECM to control the solenoid.
3. TCC solenoid coil resistance must measure more than 20 ohms. Less resistance will cause early failure of the ECM "Driver". Using an ohmmeter, check the solenoid coil resistance of all ECM

controlled solenoids and relays before installing a replacement ECM. Replace any solenoid or relay that measures less than 20 ohms.

Diagnostic Aids:

An engine coolant thermostat that is stuck open or opens at too low a temperature, may result in an inoperative TCC.

1988–90 5.0L (VIN E) – CAMARO AND FIREBIRD

CHART C-8A
(Page 1 of 2)
TORQUE CONVERTER CLUTCH (TCC)
(ELECTRICAL DIAGNOSIS)
5.0L (VIN E) "F" CARLINE (TBI)

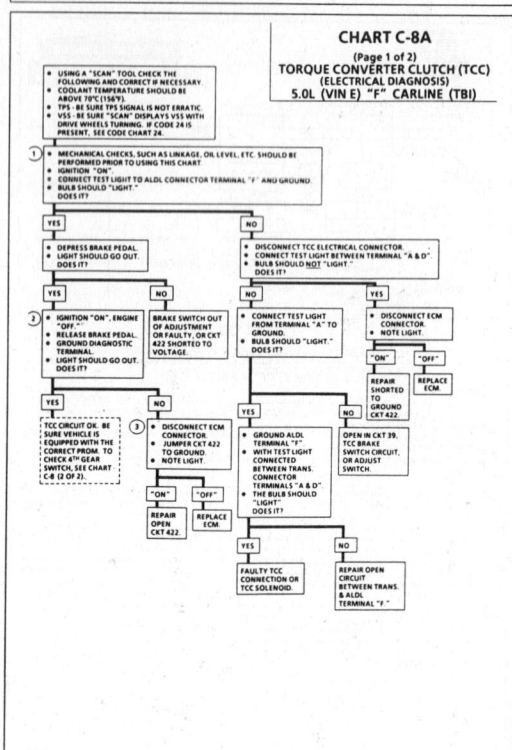

1988–90 5.0L (VIN E) – CAMARO AND FIREBIRD

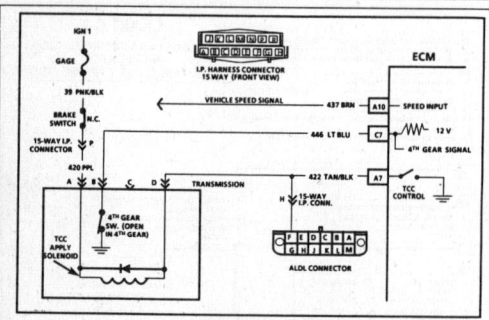

CHART C-8A
(Page 2 of 2)
700-4R TRANSMISSION
(ELECTRICAL DIAGNOSIS)
5.0L (VIN E) "F" CARLINE (TBI)

Circuit Description:

A 4th gear switch (mounted in the transmission) opens when the transmission shifts into 4th gear, and this switch is used by the ECM to modify TCC lock and unlock points, when in a 4-3 downshift maneuver.

Test Description: Numbers below refer to circled numbers on the diagnostic chart.
1. Unless the switch or CKT 446 is open the "Scan" should display "NO", indicating the transmission is not in 4th gear. The 4th gear switch should only be open while in 4th gear.
2. This step determines if the ECM and wiring are OK. Grounding CKT 446 should cause the "Scan" to display "NO", indicating the transmission is not in 4th gear.
3. Checks the operation of the 4th gear switch. When the transmission shifts into 4th gear the switch should open and the "Scan" should display "YES".
4. Disconnecting the TCC connector simulates an open switch to determine if CKT 446 is shorted to ground or the problem is in the transmission.

Diagnostic Aids:

A road test may be necessary to verify the customer complaint. If the "Scan" indicates TCC is turning "ON" and "OFF" erratically, check the state of the 4th gear switch to be sure it is not changing states under a steady throttle position. If the switch is changing states, check connections and wire routing carefully. Also if the 4th gear switch is always open the TCC may engage as soon as sufficient oil pressure is reached.

1988–90 5.0L (VIN E) – CAMARO AND FIREBIRD

CHART C-8A
(Page 2 of 2)
700-4R TRANSMISSION
(ELECTRICAL DIAGNOSIS)
5.0L (VIN E) "F" CARLINE (TBI)

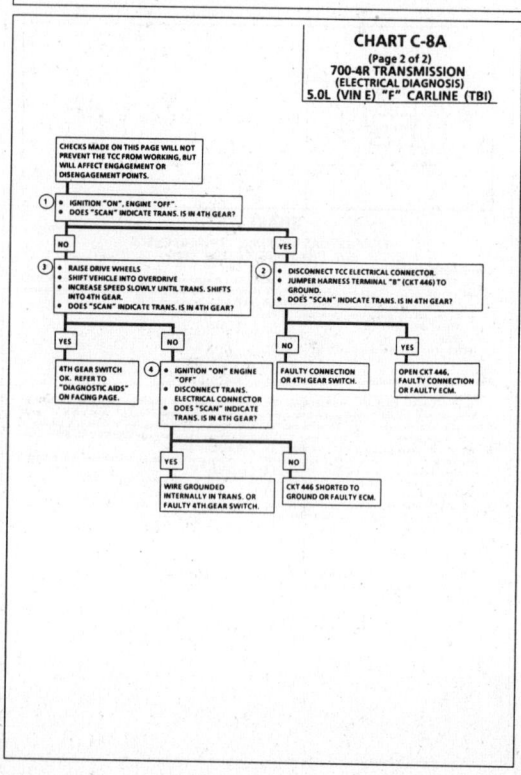

1988–90 5.0L (VIN E) – CAMARO AND FIREBIRD

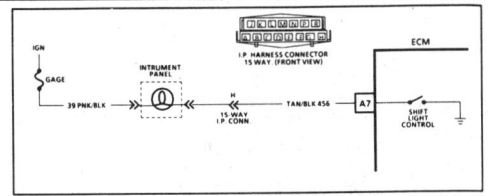

CHART C-8B
MANUAL TRANSMISSION (M/T)
SHIFT LIGHT CHECK
5.0L (VIN E) "F" CARLINE (TBI)

Circuit Description:

The shift light indicates the best transmission shift point for maximum fuel economy. The shift light is controlled by the ECM and is turned "ON" by grounding CKT 456.

The ECM uses information from the following inputs to control the shift light:
- Coolant temperature
- TPS
- VSS
- RPM

The ECM uses the measured rpm and the vehicle speed to calculate what gear the vehicle is in. It is this calculation that determines when the shift light should be turned "ON."

Test Description: Numbers below refer to circled numbers on the diagnostic chart.

1. This should not turn "ON" the shift light. If the light is "ON," there is a short to ground in CKT 456 wiring, or a fault in the ECM.
2. When the diagnostic "test" terminal is grounded, the ECM should ground CKT 456, and the shift light should come "ON."

3. This checks the shift light circuit up to the ECM connector. If the shift light illuminates, then the ECM connector is faulty, or the ECM does not have the ability to ground the circuit.

1988–90 5.0L (VIN E) – CAMARO AND FIREBIRD

CHART C-8B
MANUAL TRANSMISSION (M/T)
SHIFT LIGHT CHECK
5.0L (VIN E) "F" CARLINE (TBI)

①
- IGNITION "ON", ENGINE "OFF".
- OBSERVE SHIFT LIGHT.

LIGHT "OFF" → **②**
- GROUND ALDL DIAGNOSTIC TERMINAL AND OBSERVE SHIFT LIGHT.

LIGHT "ON" →
- IGNITION "OFF".
- DISCONNECT ECM CONNECTORS.
- TURN IGNITION "ON" AND OBSERVE SHIFT LIGHT.

LIGHT "OFF" → **③**
- IGNITION "OFF".
- DISCONNECT ECM CONNECTORS.
- IGNITION "ON".
- JUMPER CKT 456 TO GROUND AND OBSERVE SHIFT LIGHT.

LIGHT "ON" → CHECK FOR:
- CODE 24, (NO VSS).
- THERMOSTAT FAULTY OR HAS INCORRECT HEAT RANGE. IF OK, REVIEW SYMPTOMS IN SECTION "B".

LIGHT "OFF" → REPAIR SHORT TO GROUND IN CKT 456.

LIGHT "ON" → REPLACE ECM.

LIGHT "OFF" → OPEN IGNITION CKT 39, OPEN CKT 456, OR FAULTY BULB.

LIGHT "ON" → POOR CONNECTION AT ECM OR FAULTY ECM.

4.3L (VIN Z, B-CARLINE) COMPONENT LOCATIONS

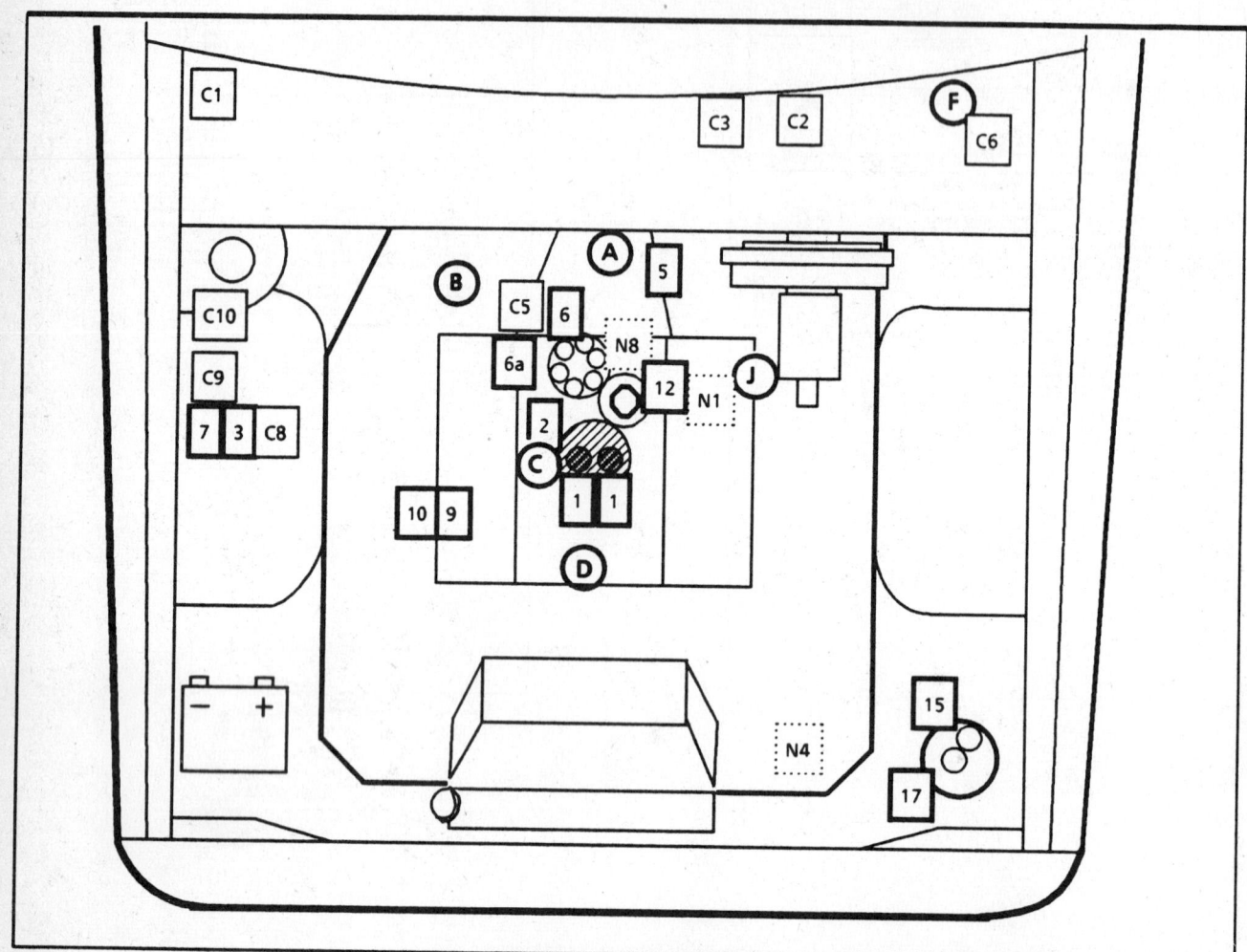

☐ **COMPUTER HARNESS**

C1	Electronic Control Module
C2	ALDL diagnostic connector
C3	"SERVICE ENGINE SOON" light
C5	ECM harness grounds
C6	Fuse panel
C8	Fuel pump test connector
C9	Fuel pump fuse & ECM power
C10	Set timing connector

☐ **NOT ECM CONNECTED**

N1	Crankcase vent valve (PCV)
N4	PS Switch
N8	Oil pressure switch

☐ **CONTROLLED DEVICES**

1	Fuel injectors
2	Idle air control motor
3	Fuel pump relay
5	Torque Conv. Clutch connector
6	EST distributor
6a	Remote ignition coil
7	Electronic Spark Control module
9	AIR port solenoid
10	AIR convertor solenoid
12	Exh. Gas Recirc. vacuum solenoid
15	Fuel vapor canister solenoid
17	Fuel vapor canister
⬡	Exhaust Gas Recirculation valve

◯ **INFORMATION SENSORS**

A	Manifold Absolute Pressure
B	Exhaust oxygen
C	Throttle position
D	Coolant temperature
F	Vehicle speed (buffer)
J	ESC knock

4.3L (VIN Z, B-CARLINE) ECM WIRING DIAGRAM

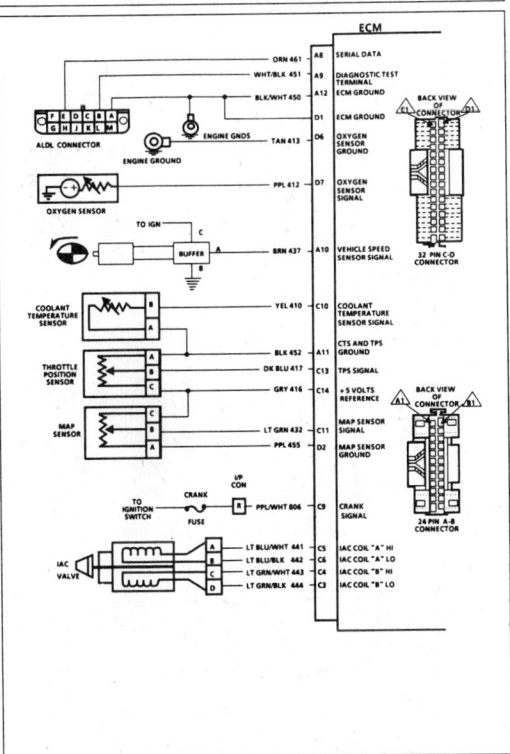

4.3L (VIN Z, B-CARLINE) ECM WIRING DIAGRAM (CONT.)

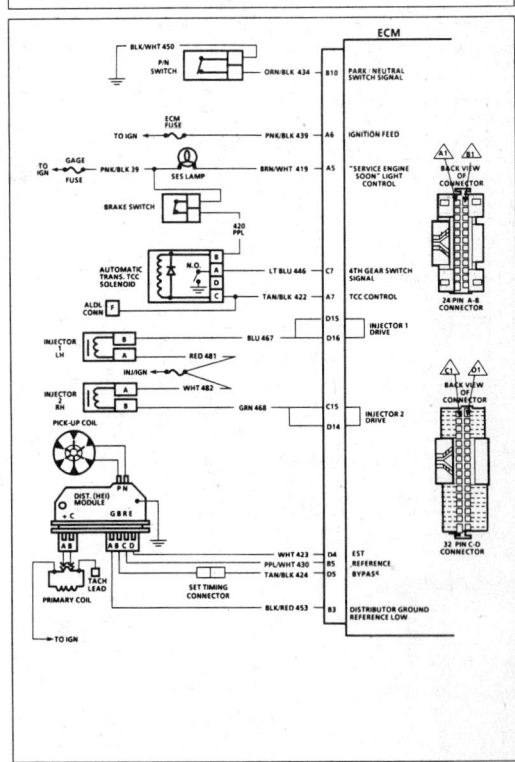

4.3L (VIN Z, B-CARLINE) ECM WIRING DIAGRAM (CONT.)

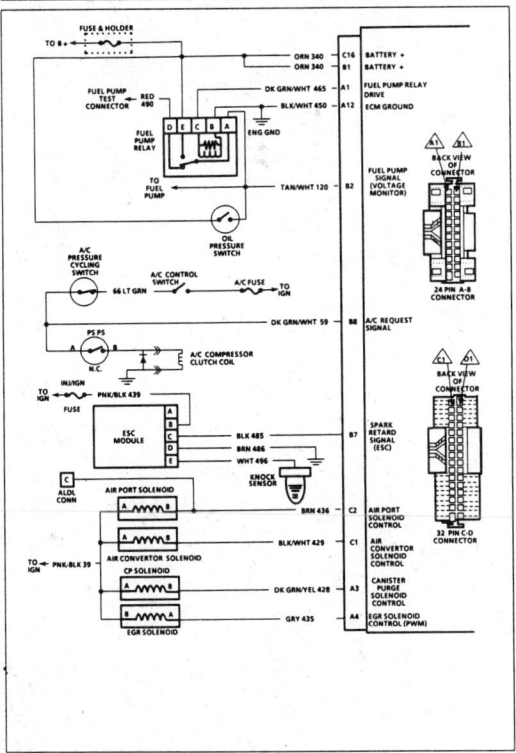

4.3L (VIN Z, B-CARLINE) ECM CONNECTOR TERMINAL END VIEW

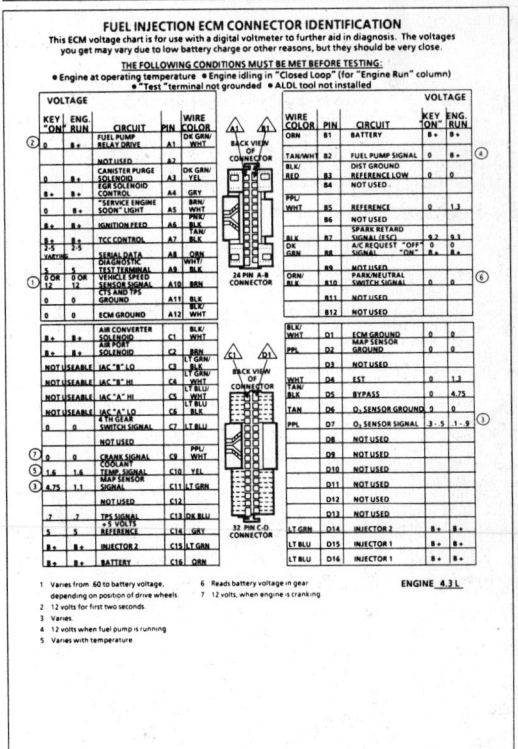

5.0L AND 5.7L (VIN E AND 7, B-CARLINE) COMPONENT LOCATIONS

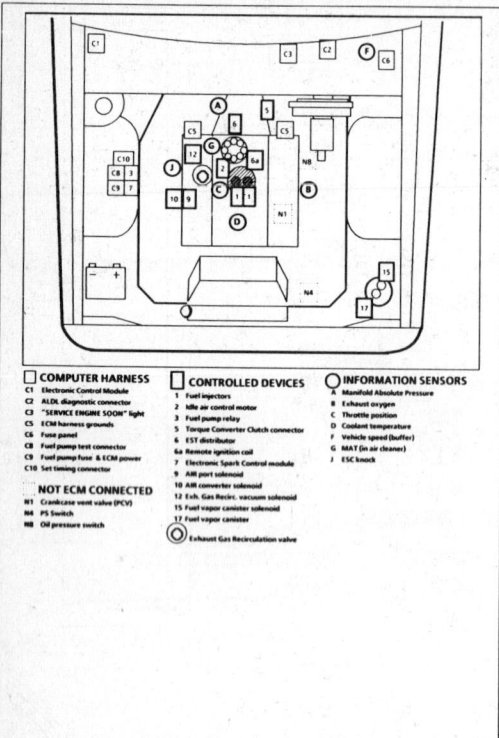

5.0L AND 5.7L (VIN E AND 7, B-CARLINE) ECM WIRING DIAGRAM

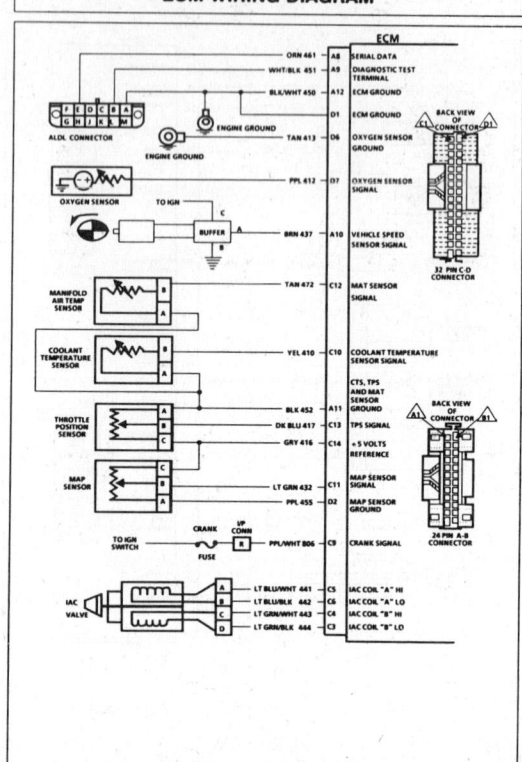

5.0L AND 5.7L (VIN E AND 7, B-CARLINE) ECM WIRING DIAGRAM (CONT.)

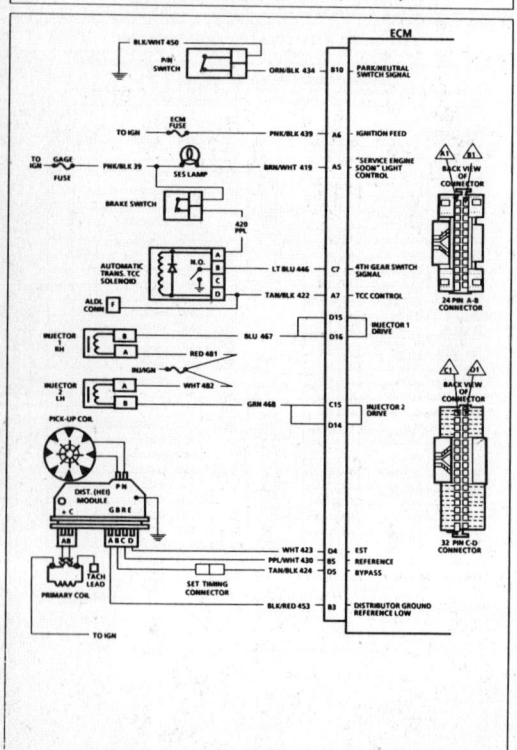

5.0L AND 5.7L (VIN E AND 7, B-CARLINE) ECM WIRING DIAGRAM (CONT.)

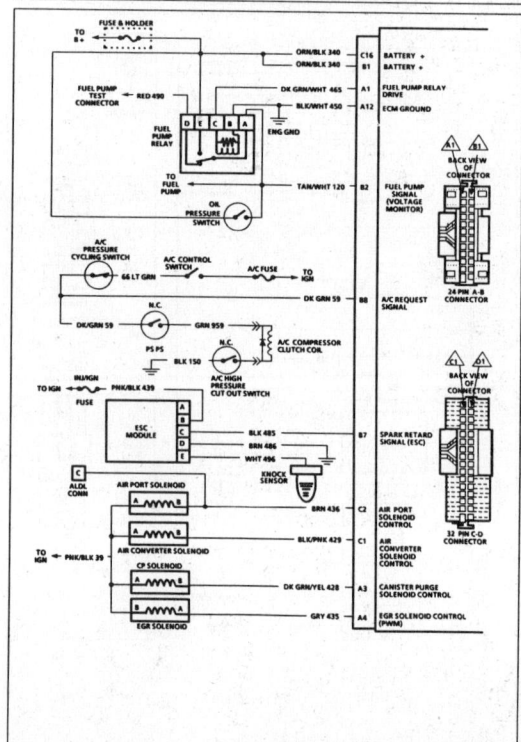

5.0L AND 5.7L (VIN E AND 7, B-CARLINE) ECM CONNECTOR TERMINAL END VIEW

FUEL INJECTION ECM CONNECTOR IDENTIFICATION

This ECM voltage chart is for use with a digital voltmeter to further aid in diagnosis. The voltages you get may vary due to low battery charge or other reasons, but they should be very close.

THE FOLLOWING CONDITIONS MUST BE MET BEFORE TESTING:
- Engine at operating temperature
- Engine idling in "Closed Loop" (for "Engine Run" column)
- "Test" terminal not grounded
- ALDL tool not installed

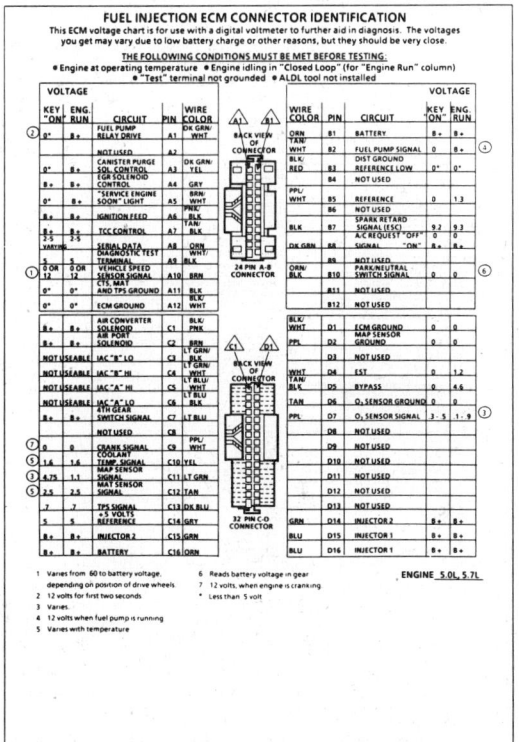

VOLTAGE KEY "ON"	ENG. RUN	CIRCUIT	PIN	WIRE COLOR
0*	B+	FUEL PUMP RELAY DRIVE	A1	DK GRN/WHT
		NOT USED	A2	
0*	B+	CANISTER PURGE SOL. CONTROL	A3	DK GRN/YEL
B+	B+	EGR SOLENOID CONTROL	A4	GRY
0*	B+	"SERVICE ENGINE SOON" LIGHT	A5	BRN/WHT
B+	B+	IGNITION FEED	A6	PNK/BLK
B+	B+	TCC CONTROL	A7	TAN/BLK
2.5 VARYING	2.5	SERIAL DATA DIAGNOSTIC TEST TERMINAL	A8	ORN
5	5		A9	BLK
0 OR 12	0 OR 12	VEHICLE SPEED SENSOR SIGNAL	A10	BRN
0*	0*	CTS, MAT AND TPS GROUND	A11	BLK
0*	0*	ECM GROUND	A12	BLK/WHT
B+	B+	AIR CONVERTER SOLENOID	C1	BLK/PNK
B+	B+	AIR PORT SOLENOID	C2	BRN
NOT USEABLE		IAC "B" LO	C3	LT GRN/BLK
NOT USEABLE		IAC "B" HI	C4	LT GRN/WHT
NOT USEABLE		IAC "A" HI	C5	LT BLU/WHT
NOT USEABLE		IAC "A" LO	C6	LT BLU/BLK
B+	B+	4TH GEAR SWITCH SIGNAL	C7	LT BLU
		NOT USED	C8	
0	0	CRANK SIGNAL	C9	PPL/WHT
1.6	1.6	COOLANT TEMP SIGNAL MAP SENSOR	C10	YEL
4.75	1.1	SIGNAL	C11	LT GRN
2.5	2.5	MAT SENSOR SIGNAL	C12	TAN
7	7	TPS SIGNAL	C13	DK BLU
5	5	+5 VOLTS REFERENCE	C14	GRY
B+	B+	INJECTOR 2	C15	GRN
B+	B+	BATTERY	C16	ORN

WIRE COLOR	PIN	CIRCUIT	VOLTAGE KEY "ON"	ENG. RUN
ORN	B1	BATTERY	B+	B+
WHT	B2	FUEL PUMP SIGNAL	0	B+
BLK/RED	B3	DIST GROUND		
	B4	REFERENCE LOW	0*	0*
PPL/WHT	B5	NOT USED		
	B6	REFERENCE	0	1.3
BLK	B7	SPARK RETARD SIGNAL (ESC)	9.2	9.3
DK GRN	B8	A-C REQUEST "OFF" SIGNAL "ON"	0	0
	B9	NOT USED	0	0
ORN/BLK	B10	PARK/NEUTRAL SWITCH SIGNAL	0	0
	B11	NOT USED		
	B12	NOT USED		
BLK/WHT	D1	ECM GROUND	0	0
PPL	D2	MAP SENSOR GROUND	0	0
	D3	NOT USED		
WHT/TAN/BLK	D4	EST	0	1.2
	D5	BYPASS	0	4.6
TAN	D6	O₂ SENSOR GROUND	0	0
PPL	D7	O₂ SENSOR SIGNAL	3-5	1-9
	D8	NOT USED		
	D9	NOT USED		
	D10	NOT USED		
	D11	NOT USED		
	D12	NOT USED		
	D13	NOT USED		
GRN	D14	INJECTOR 2	B+	B+
BLU	D15	INJECTOR 1	B+	B+
BLU	D16	INJECTOR 1	B+	B+

1. Varies from .60 to battery voltage, depending on position of drive wheels.
2. 12 volts for first two seconds.
3. Varies.
4. 12 volts when fuel pump is running.
5. Varies with temperature.
6. Reads battery voltage in gear.
7. 12 volts, when engine is cranking.
* Less than 5 volt.

ENGINE 5.0L, 5.7L

1988–90 B-CARLINE

DIAGNOSTIC CIRCUIT CHECK
4.3L, 5.0L & 5.7L "B" CARLINE (TBI)

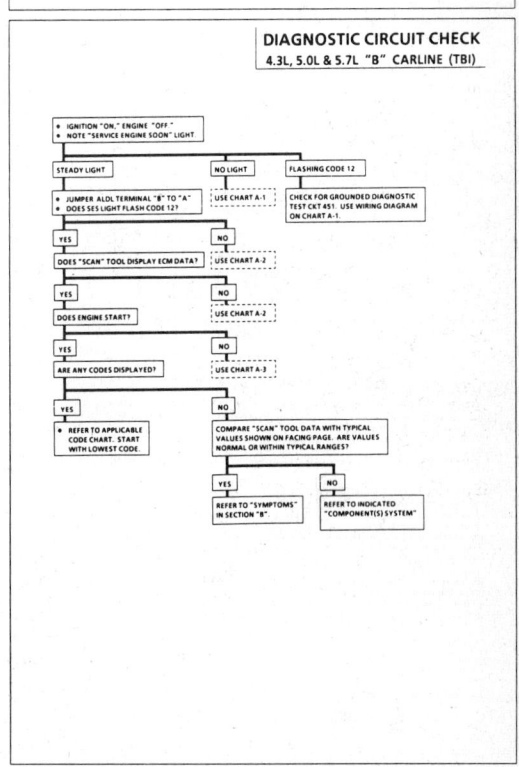

1988–90 B-CARLINE

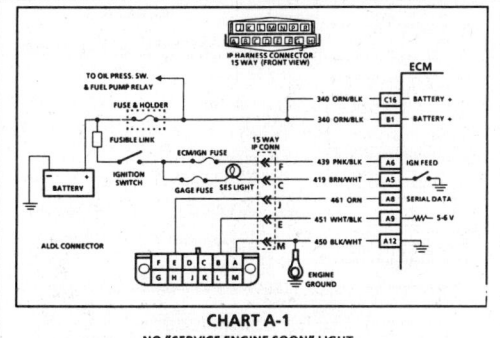

IP HARNESS CONNECTOR 15 WAY (FRONT VIEW)

		ECM
340 ORN/BLK	C16	BATTERY +
340 ORN/BLK	B1	BATTERY +
439 PNK/BLK	A6	IGN FEED
419 BRN/WHT	A5	
461 ORN	A8	SERIAL DATA
451 WHT/BLK	A9	5–6 V
450 BLK/WHT	A12	

CHART A-1
NO "SERVICE ENGINE SOON" LIGHT
4.3L, 5.0L & 5.7L "B" CARLINE (TBI)

Circuit Description:

There should always be a steady "Service Engine Soon" light, when the ignition is "ON" and engine stopped. Ignition voltage is supplied directly to the light bulb. The electronic control module (ECM) will control the light and turn it "ON" by providing a ground path through CKT 419 to the ECM.

Test Description: Numbers below refer to circled numbers on the diagnostic chart.

1. Battery feed CKT 340 is protected by a 20A in-line fuse. If this fuse was blown, refer to wiring diagram on the facing page of Code 54.
2. Using a test light connected to 12 volts, probe each of the system ground circuits to be sure a good ground is present. See ECM terminal end view in front of this section for ECM pin locations of ground circuits.

Diagnostic Aids:

Engine runs OK, check:
- Faulty light bulb
- CKT 419 open
- Gage fuse blown. This will result in no oil, or generator lights, seat belt reminder, etc.

Engine cranks, but will not run.
- Continuous battery - fuse or fusible link open
- ECM ignition fuse open
- Battery CKT 340 to ECM open
- Ignition CKT 439 to ECM open
- Poor connection to ECM

1988–90 B-CARLINE

CHART A-1
NO "SERVICE ENGINE SOON" LIGHT
4.3L, 5.0L & 5.7L "B" CARLINE (TBI)

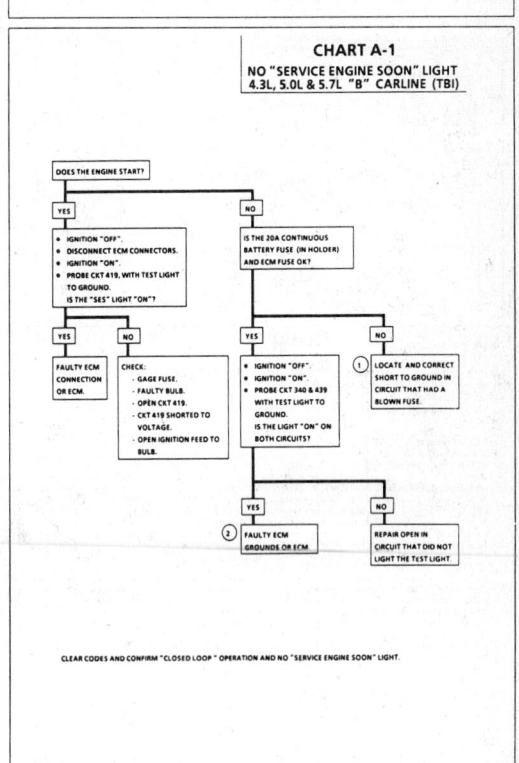

CLEAR CODES AND CONFIRM "CLOSED LOOP" OPERATION AND NO "SERVICE ENGINE SOON" LIGHT.

1988–90 B-CARLINE

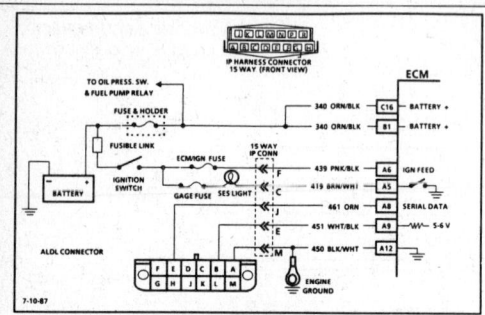

CHART A-2
NO ALDL DATA OR WON'T FLASH CODE 12
"SERVICE ENGINE SOON" LIGHT "ON" STEADY
4.3L, 5.0L & 5.7L "B" CARLINE (TBI)

Circuit Description:

There should always be a steady "Service Engine Soon" light, when the ignition is "ON" and engine stopped. Ignition voltage is supplied directly to the light bulb. The electronic control module (ECM) will turn the light "ON" by grounding CKT 419 at the ECM.

With the diagnostic terminal grounded, the light should flash a Code 12, followed by any trouble code(s) stored in memory.

A steady light suggests a short to ground in the light control CKT 419, or an open in diagnostic CKT 451.

Test Description: Numbers below refer to circled numbers on the diagnostic chart.

1. If there is a problem with the ECM that causes a "Scan" tool to not read serial data, then the ECM should not flash a Code 12. If Code 12 does flash, be sure that the "Scan" tool is working properly on another vehicle. If the "Scan" is functioning properly and CKT 461 is OK, the Mem-Cal or ECM may be at fault for the no ALDL symptom.

2. If the light goes "OFF," when the ECM connector is disconnected, then CKT 419 is not shorted to ground.

3. This step will check for an open diagnostic CKT 451.

4. At this point, the "Service Engine Soon" light wiring is OK. The problem is a faulty ECM or PROM. If Code 12 does not flash, the ECM should be replaced using the original PROM. Replace the PROM only after trying an ECM, as a defective PROM is an unlikely cause of the problem.

1988–90 B-CARLINE

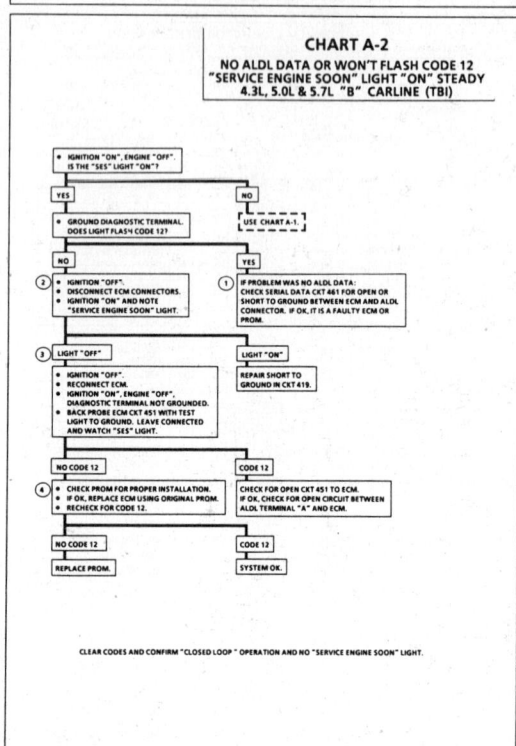

CHART A-2
NO ALDL DATA OR WON'T FLASH CODE 12
"SERVICE ENGINE SOON" LIGHT "ON" STEADY
4.3L, 5.0L & 5.7L "B" CARLINE (TBI)

CLEAR CODES AND CONFIRM "CLOSED LOOP" OPERATION AND NO "SERVICE ENGINE SOON" LIGHT.

1988–90 B-CARLINE

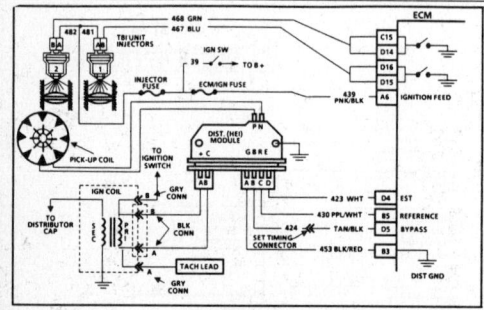

CHART A-3
(Page 1 of 2)
ENGINE CRANKS BUT WILL NOT RUN
4.3L, 5.0L & 5.7L "B" CARLINE (TBI)

Circuit Description:

This chart assumes that battery condition and engine cranking speed are OK, and there is adequate fuel in the tank.

Test Description: Numbers below refer to circled numbers on the diagnostic chart.

1. A "Service Engine Soon" light "ON" is a basic test to determine if there is a 12 volts supply and ignition 12 volts to ECM. No ALDL may be due to an ECM problem and CHART A-2 will diagnose the ECM. If TPS is over 2.5 volts, the engine may be in the clear flood mode which will cause starting problems.

2. No spark may be caused by one of several components related to the ignition system. CHART C-4 will address all problems related to the causes of a no spark condition.

3. Fuel spray from the injector(s) indicates that fuel is available. However, the engine could be severely flooded due to too much fuel.

4. While cranking engine, there should be no fuel spray with injector disconnected. Replace an injector if it sprays fuel or drips like a leaking water faucet.

5. The fuel pressure will drop after the fuel pump stops running due to a controlled bleed in the fuel system. Use of the fuel pressure gage will

determine if fuel system pressure is enough for engine to start and run.

6. No fuel spray from injector indicates a faulty fuel system or no ECM control of injector.

7. This test will determine if the ignition module is not generating the reference pulse, or if the wiring or ECM are at fault. By touching and removing a test light connected to 12 volts on CKT 430, a reference pulse should be generated. If injector test light blinks, the ECM and wiring are OK.

Diagnostic Aids:

• Water or foreign material in fuel system can cause a no start during freezing weather.
• An EGR sticking open can cause a low air/fuel ratio during cranking.
• Fuel pressure: Low fuel pressure can result in a very lean air/fuel ratio. See CHART A-7.
• A grounded CKT 423 (EST) may cause a "no-start" or a "start then stall" condition.

1988–90 B-CARLINE

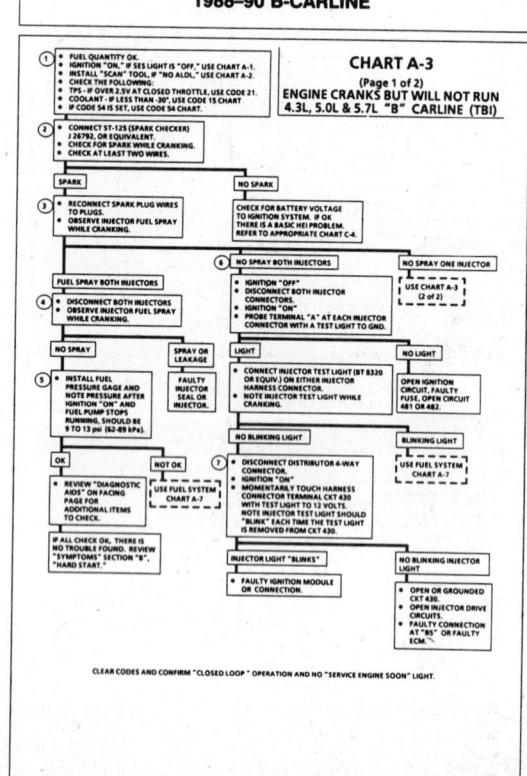

CHART A-3
(Page 1 of 2)
ENGINE CRANKS BUT WILL NOT RUN
4.3L, 5.0L & 5.7L "B" CARLINE (TBI)

CLEAR CODES AND CONFIRM "CLOSED LOOP" OPERATION AND NO "SERVICE ENGINE SOON" LIGHT.

1988–90 B-CARLINE

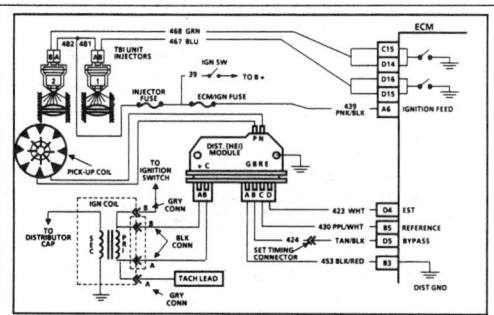

CHART A-3
(Page 2 of 2)
ENGINE CRANKS BUT WILL NOT RUN
4.3L, 5.0L & 5.7L "B" CARLINE (TBI)

Circuit Description:
This chart assumes that battery condition and engine cranking speed are OK, and there is adequate fuel in the tank.

Test Description: Numbers below refer to circled numbers on the diagnostic chart.
1. No fuel spray from one injector indicates a faulty fuel injector or no ECM control of injector. If the test light "blinks" while cranking, then ECM control should be considered OK. Be sure test light makes good contact between connector terminals during test. The light may be a little dim when "blinking." This is due to current draw of the test light. How bright it "blinks" is not important. The test light bulb should be a BT 8320 or equivalent.

2. CKTs 481 and 482 supply ignition voltage to the injectors. Probe each connector terminal with a test light to ground. There should be a light "ON" at one terminal. If the test light confirms ignition voltage at the connector, the ECM injector control CKT 467 or CKT 468 may be open. Reconnect the injector, and using a test light connected to ground, check for a light at the applicable ECM connector terminal ("D14" or "D16"). A light at this point indicates that the injector drive circuit involved is OK.
If an ECM repeat failure has occurred, the injector is shorted. Replace the injector and ECM.

1988–90 B-CARLINE

CHART A-3
(Page 2 of 2)
ENGINE CRANKS BUT WILL NOT RUN
4.3L, 5.0L & 5.7L "B" CARLINE (TBI)

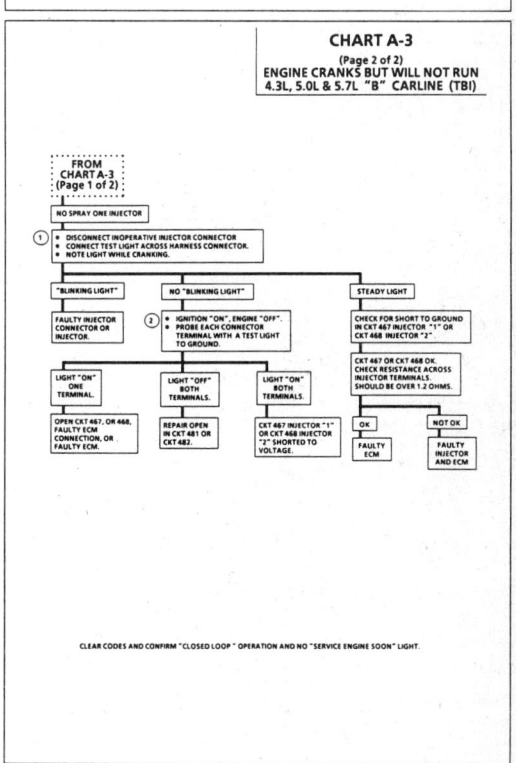

1988–90 B-CARLINE

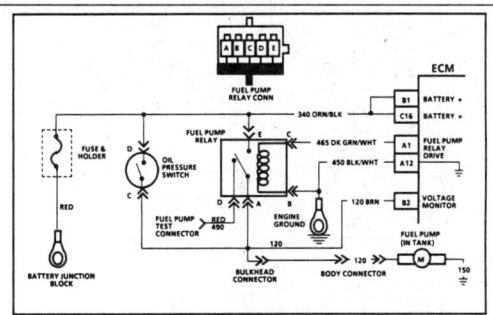

CHART A-7
(Page 1 of 2)
FUEL SYSTEM DIAGNOSIS
4.3L, 5.0L & 5.7L "B" CARLINE (TBI)

Circuit Description:
When the ignition switch is turned "ON," the electronic control module (ECM) will turn "ON" the in-tank fuel pump. It will remain "ON" as long as the engine is cranking or running, and the ECM is receiving ignition reference pulses.
If there are no reference pulses, the ECM will shut "OFF" the fuel pump within 2 seconds after key "ON."
The pump will deliver fuel to the TBI unit, where the system pressure is controlled to 62 to 90 kPa (9 to 13 psi). Excess fuel is then returned to the fuel tank.
The fuel pump test terminal is located in the passenger side of the engine compartment.

Test Description: Numbers below refer to circled numbers on the diagnostic chart.
1. Fuel pressure should be noted while fuel pump is running. Fuel pressure will drop immediately after fuel pump stops running due to a controlled bleed in the fuel system.

Diagnostic Aids:
Improper fuel system pressure can result in one of the following symptoms:
- Cranks, but won't run
- Code 44
- Code 45
- Cuts out, may feel like ignition problem
- Poor fuel economy, loss of power
- Hesitation

1988–90 B-CARLINE

CHART A-7
(Page 1 of 2)
FUEL SYSTEM DIAGNOSIS
4.3L, 5.0L & 5.7L "B" CARLINE (TBI)

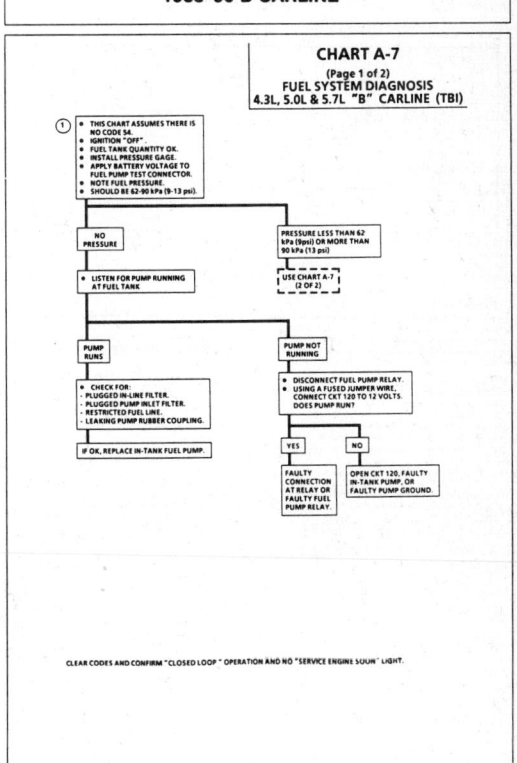

1988–90 B-CARLINE

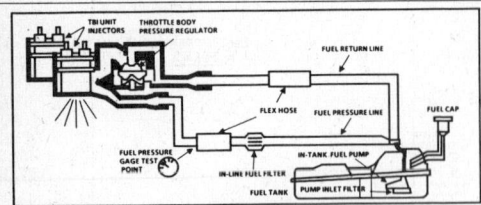

CHART A-7
(Page 2 of 2)
FUEL SYSTEM DIAGNOSIS
4.3L, 5.0L & 5.7L "B" CARLINE (TBI)

Test Description: Numbers below refer to circled numbers on the diagnostic chart.

1. Fuel Pressure less than 62 kPa (9 psi) falls into two areas:
   - Regulated pressure less than 62 kPa (9 psi). Amount of fuel to injectors OK but, pressure is too low. System will be lean and may set Code 44. Also, hard starting cold and poor overall performance.
   - Restricted flow causing pressure drop. Normally, a vehicle with a fuel pressure of less than 62 kPa (9 psi) at idle will not be driveable. However, if the pressure drop occurs only while driving, the engine will normally surge then stop as pressure begins to drop rapidly.

2. Restricting the fuel supply line between the gage and TBI unit and turning the fuel pump "ON" will determine if the fuel pump can supply enough fuel pressure at the injector to operate properly.

   NOTICE: Do not restrict the fuel return line as this may damage the fuel pressure regulator.

3. This test determines if the high fuel pressure is due to a restricted fuel return line or a throttle body pressure regulator problem.

1988–90 B-CARLINE

CHART A-7
(Page 2 of 2)
FUEL SYSTEM DIAGNOSIS
4.3L, 5.0L & 5.7L "B" CARLINE (TBI)

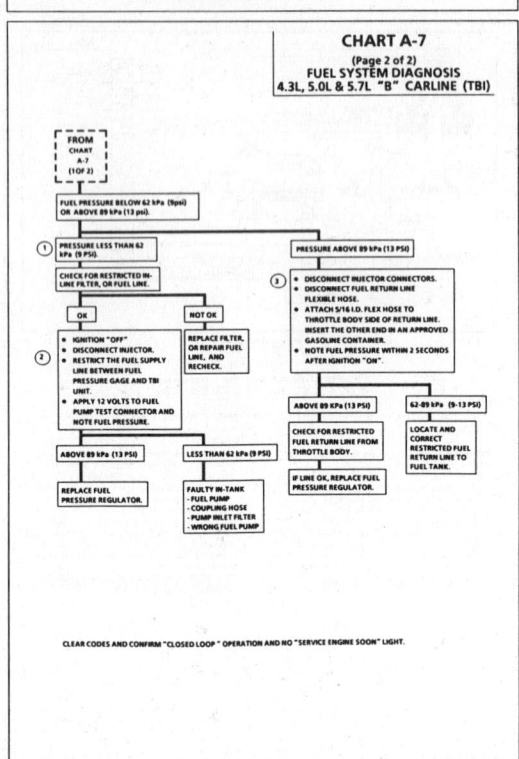

1988–90 B-CARLINE

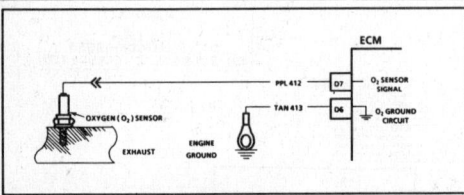

CODE 13
OXYGEN SENSOR CIRCUIT
(OPEN CIRCUIT)
4.3L, 5.0L & 5.7L "B" CARLINE (TBI)

Circuit Description:
The ECM supplies a voltage of about .45 volt between terminals "D7" and "D6". (If measured with a 10 megohm digital voltmeter, this may read as low as .32 volt). The O₂ sensor varies the voltage within a range of about 1 volt if the exhaust is rich, down through about .10 volt if exhaust is lean.
The sensor is like an open circuit and produces no voltage when it is below 360°C (600°F). An open sensor circuit or cold sensor causes "Open Loop" operation.

Test Description: Numbers below refer to circled numbers on the diagnostic chart.

1. Code 13 will set:
   - Engine at normal operating temperature
   - At least 2 minutes engine time after start
   - O₂ signal voltage steady between .35 and .55 volt.
   - rpm above 1600
   - Throttle position sensor signal above 5% (about .3 volt above closed throttle voltage).
   - All conditions must be met for about 60 seconds.
   If the conditions for a Code 13 exist, the system will not go "Closed Loop".

2. This will determine if the sensor is at fault or the wiring or ECM is the cause of the Code 13.

3. In doing this test, use only a high impedence digital volt ohmmeter. This test checks the continuity of CKT(s) 412 and 413. If CKT 413 is open, the ECM voltage on CKT 412 will be over .6 volt (600 mV).

Diagnostic Aids:
Normal "Scan" voltage varies between 100 mV (.1 and 1.0 volt), while in "Closed Loop". Code 13 sets in one minute, if voltage remains between .35 and .55 volt, but the system will go "Open Loop" in about 15 seconds.
Refer to "Intermittents" in Section "B".

1988–90 B-CARLINE

CODE 13
OXYGEN SENSOR CIRCUIT
(OPEN CIRCUIT)
4.3L, 5.0L & 5.7L "B" CARLINE (TBI)

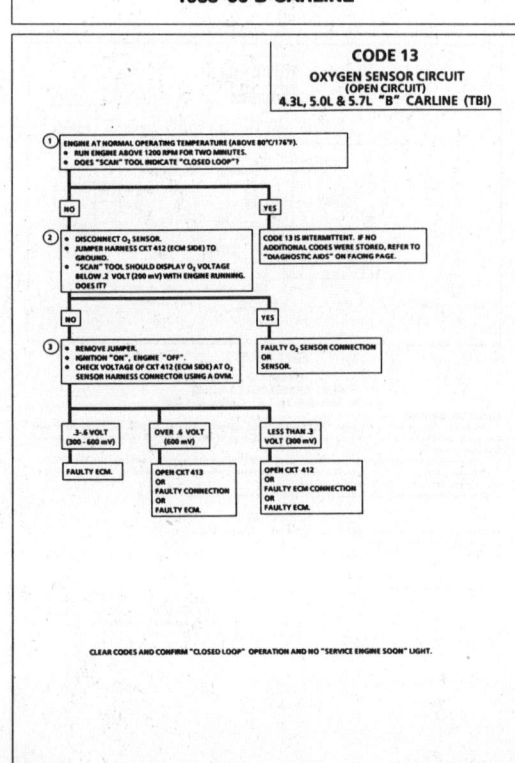

1988–90 B-CARLINE

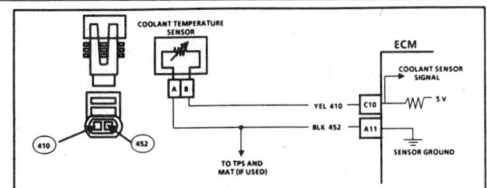

CODE 14
COOLANT TEMPERATURE SENSOR CIRCUIT
(HIGH TEMPERATURE INDICATED)
4.3L, 5.0L & 5.7L "B" CARLINE (TBI)

Circuit Description:
The coolant temperature sensor uses a thermistor to control the signal voltage to the ECM. The ECM applies a voltage on CKT 410 to the sensor. When the engine is cold the sensor (thermistor) resistance is high, therefore the ECM will see high signal voltage.
As the engine warms, the sensor resistance becomes less, and the voltage drops. At normal engine operating temperature (85°C - 95°C or 185°F - 203°F) the voltage will measure about 1.5 to 2.0 volts.

Test Description: Numbers below refer to circled numbers on the diagnostic chart.
1. Code 14 will set if:
   - Signal voltage indicates a coolant temperature above 135°C (275°F) for 2 seconds.
2. This test will determine if CKT 410 is shorted to ground which will cause the conditions for Code 14.

Diagnostic Aids:
Check harness routing for a potential short to ground in CKT 410.
"Scan" tool displays engine temperature in degrees centigrade. After engine is started, the temperature should rise steadily to about 90°C (194°F) then stabilize when thermostat opens.
Refer to "Intermittents" in Section "B".

1988–90 B-CARLINE

CODE 14
COOLANT TEMPERATURE SENSOR CIRCUIT
(HIGH TEMPERATURE INDICATED)
4.3L, 5.0L & 5.7L "B" CARLINE (TBI)

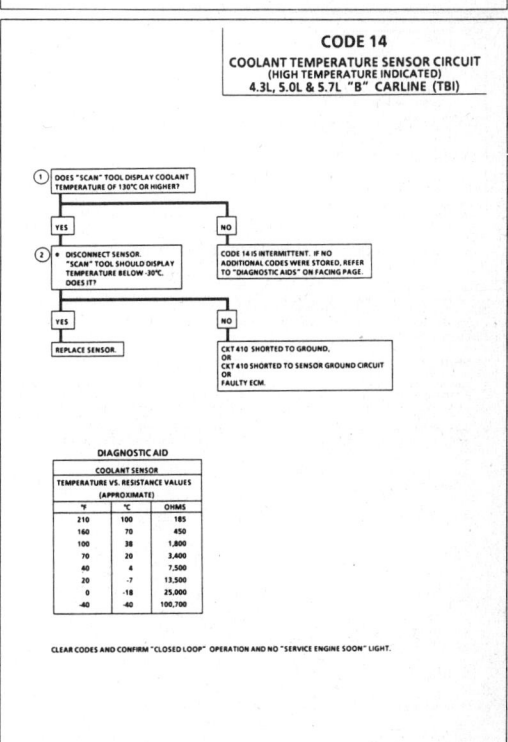

DIAGNOSTIC AID

COOLANT SENSOR		
TEMPERATURE VS. RESISTANCE VALUES		
(APPROXIMATE)		
°F	°C	OHMS
210	100	185
160	70	450
100	38	1,800
70	20	3,400
40	4	7,500
20	-7	13,500
0	-18	25,000
-40	-40	100,700

CLEAR CODES AND CONFIRM "CLOSED LOOP" OPERATION AND NO "SERVICE ENGINE SOON" LIGHT.

1988–90 B-CARLINE

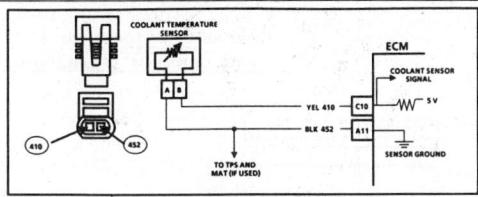

CODE 15
COOLANT TEMPERATURE SENSOR CIRCUIT
(LOW TEMPERATURE INDICATED)
4.3L, 5.0L & 5.7L "B" CARLINE (TBI)

Circuit Description:
The coolant temperature sensor uses a thermistor to control the signal voltage to the ECM. The ECM applies a voltage on CKT 410 to the sensor. When the engine is cold the sensor (thermistor) resistance is high, therefore the ECM will see high signal voltage.
As the engine warms, the sensor resistance becomes less, and the voltage drops. At normal engine operating temperature (85°C - 95°C or 185°F - 203°F) the voltage will measure about 1.5 to 2.0 volts at the ECM.

Test Description: Numbers below refer to circled numbers on the diagnostic chart.
1. Code 15 will set if:
   - Engine running longer than 30 seconds
   - Coolant temperature less than -30°C (-22°F), for 3 seconds.
2. This test simulates a Code 14. If the ECM recognizes the low signal voltage, (high temperature) and "Scan" reads 130°C (266°F) or above, the ECM and wiring are OK.
3. This test will determine if CKT 410 is open. There should be 5 volts present at sensor connector if measured with a DVM.

Diagnostic Aids:
A "Scan" tool reads engine temperature in degrees centigrade. After engine is started the temperature should rise steadily to about 90°C (194°F) then stabilize when thermostat opens.
If Code 21 is also set, check CKT 452 for faulty wiring or connections. Check terminals at sensor for good contact.
Refer to "Intermittents" in Section "B".

1988–90 B-CARLINE

CODE 15
COOLANT TEMPERATURE SENSOR CIRCUIT
(LOW TEMPERATURE INDICATED)
4.3L, 5.0L & 5.7L "B" CARLINE (TBI)

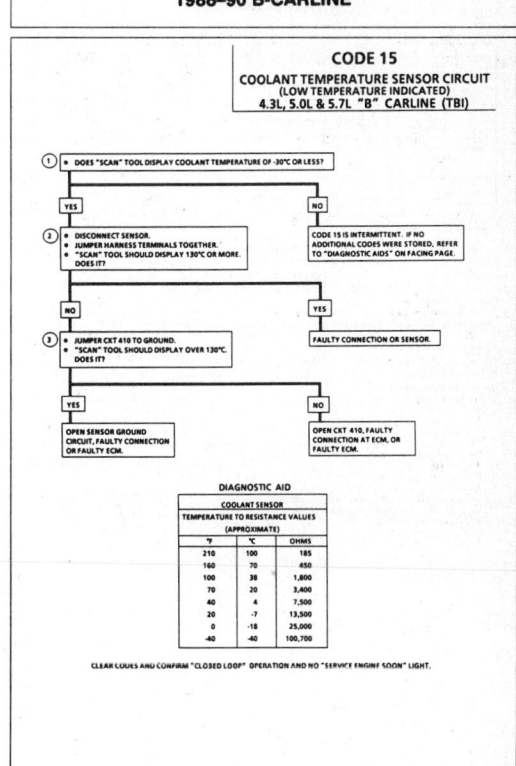

DIAGNOSTIC AID

COOLANT SENSOR		
TEMPERATURE TO RESISTANCE VALUES		
(APPROXIMATE)		
°F	°C	OHMS
210	100	185
160	70	450
100	38	1,800
70	20	3,400
40	4	7,500
20	-7	13,500
0	-18	25,000
-40	-40	100,700

CLEAR CODES AND CONFIRM "CLOSED LOOP" OPERATION AND NO "SERVICE ENGINE SOON" LIGHT.

1988–90 B-CARLINE

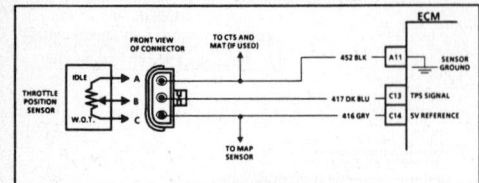

CODE 21
THROTTLE POSITION SENSOR (TPS) CIRCUIT
(SIGNAL VOLTAGE HIGH)
4.3L, 5.0L & 5.7L "B" CARLINE (TBI)

Circuit Description:
The throttle position sensor (TPS) provides a voltage signal that changes relative to the throttle blade. Signal voltage will vary from about .5 at idle to about 5 volts at wide open throttle.
The TPS signal is one of the most important inputs used by the ECM for fuel control and for most of the ECM control outputs.

Test Description: Numbers below refer to circled numbers on the diagnostic chart.
1. Code 21 will set if:
   - TPS signal voltage is greater than 2.5 volts
   - All conditions met for 8 seconds
   - MAP less than 52 kPa (15" Hg)
2. With the TPS sensor disconnected, the TPS voltage should go low if the ECM and wiring are OK.
3. Probing CKT 452 with a test light to 12 volts checks the sensor ground circuit. A faulty sensor ground will cause a Code 21.

Diagnostic Aids:
A "Scan" tool reads throttle position in volts. Should read less than 1.25 volts with throttle closed and ignition "ON" or at idle. Voltage should increase at a steady rate as throttle is moved toward WOT.
An open in CKT 452 will result in a Code 21.
Refer to "Intermittents" in Section "B".

1988–90 B-CARLINE

CODE 21
THROTTLE POSITION SENSOR (TPS) CIRCUIT
(SIGNAL VOLTAGE HIGH)
4.3L, 5.0L & 5.7L "B" CARLINE (TBI)

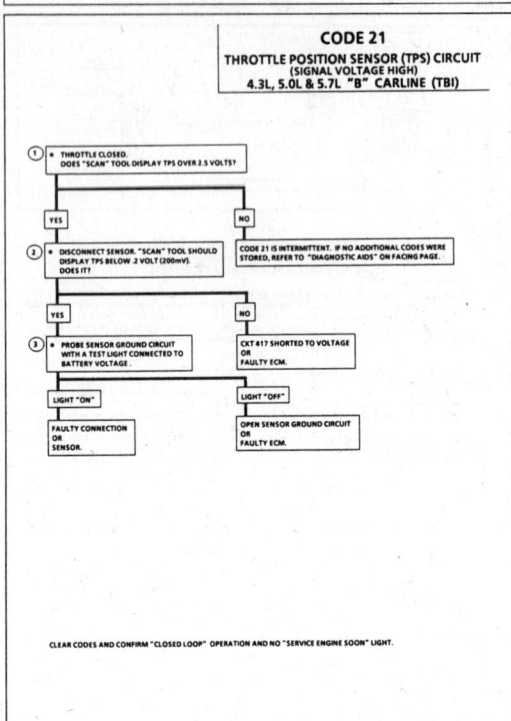

CLEAR CODES AND CONFIRM "CLOSED LOOP" OPERATION AND NO "SERVICE ENGINE SOON" LIGHT.

1988–90 B-CARLINE

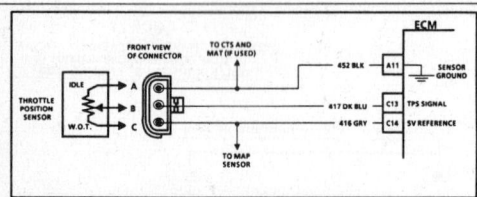

CODE 22
THROTTLE POSITION SENSOR (TPS) CIRCUIT
(SIGNAL VOLTAGE LOW)
4.3L, 5.0L & 5.7L "B" CARLINE (TBI)

Circuit Description:
The throttle position sensor (TPS) provides a voltage signal that changes relative to the throttle blade. Signal voltage will vary from about .5 at idle to about 5 volts at wide open throttle.
The TPS signal is one of the most important inputs used by the ECM for fuel control and for most of the ECM control outputs.

Test Description: Numbers below refer to circled numbers on the diagnostic chart.
1. Code 22 will set if:
   - Engine is running
   - TPS signal voltage is less than about .2 volt for 2 seconds
2. Simulates Code 21: (High voltage). If the ECM recognizes the high signal voltage then the ECM and wiring are OK.
3. Replace throttle position sensor.
4. This simulates a high signal voltage to check for an open in CKT 417. The "Scan" tool will not read up to 12 volts, but what's important is that the ECM recognizes the signal on CKT 417.

Diagnostic Aids:
A "Scan" tool reads throttle position in volts. Should read less than 1.25 volts with throttle closed and ignition "ON" or at idle. Voltage should increase at a steady rate as throttle is moved toward WOT.
An open or short to ground in CKTs 416 or 417 will result in a Code 22.
If a Code 22 is also set check CKT 416 carefully for open or short to ground.
Refer to "Intermittents" in Section "B".

1988–90 B-CARLINE

CODE 22
THROTTLE POSITION SENSOR (TPS) CIRCUIT
(SIGNAL VOLTAGE LOW)
4.3L, 5.0L & 5.7L "B" CARLINE (TBI)

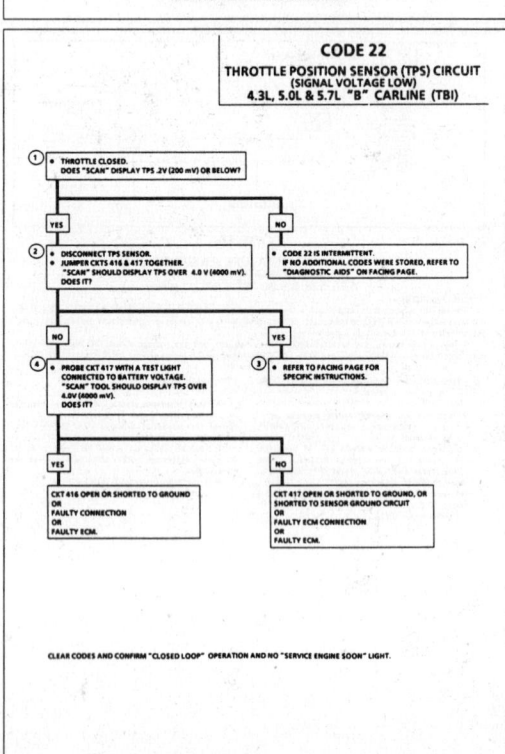

CLEAR CODES AND CONFIRM "CLOSED LOOP" OPERATION AND NO "SERVICE ENGINE SOON" LIGHT.

1988–90 B-CARLINE

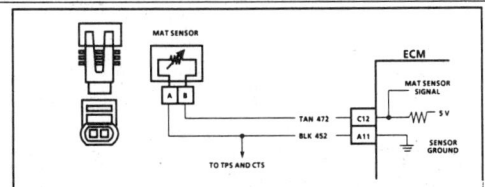

CODE 23
MANIFOLD AIR TEMPERATURE (MAT) SENSOR CIRCUIT
(LOW TEMPERATURE INDICATED)
5.0L & 5.7L "B" CARLINE (TBI)

Circuit Description:
The MAT sensor uses a thermistor to control the signal voltage to the ECM. The ECM applies a voltage (4-6 volts) on CKT 472 to the sensor. When the air is cold, the sensor (thermistor) resistance is high, therefore, the ECM will see a high signal voltage. If the air is warm, the sensor resistance is low, therefore, the ECM will see a low voltage.

Test Description: Numbers below refer to circled numbers on the diagnostic chart.
1. Code 23 will set if:
   - A signal voltage indicates a manifold air temperature below -30°C (-22°F).
   - Time since engine start is 5 minutes or longer
2. A Code 23 will set, due to an open sensor, wire or connection. This test will determine if the wiring and ECM are OK.
3. This will determine if the signal CKT 472 or the sensor ground is open.

Diagnostic Aids:
A "Scan" tool indicates the temperature of the air in the air cleaner because the MAT sensor is mounted in the air cleaner.
Carefully check harness and connections for possible open CKT 472 or 452.
Refer to "Intermittents" in Section "B".
If the engine has been allowed to sit overnight, the manifold air temperature and coolant temperature values should read within a few degrees of each other. After the engine is started, the MAT will increase due to Thermac operation and underhood temperatures.

1988–90 B-CARLINE

CODE 23
MANIFOLD AIR TEMPERATURE (MAT) SENSOR CIRCUIT
(LOW TEMPERATURE INDICATED)
5.0L & 5.7L "B" CARLINE (TBI)

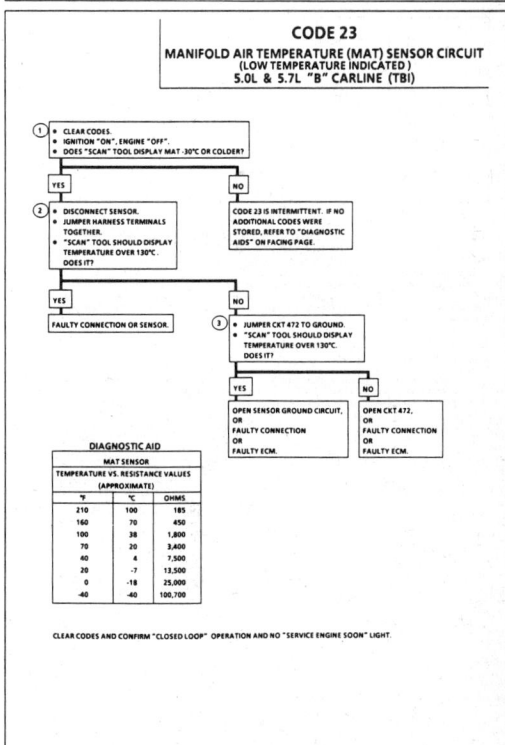

DIAGNOSTIC AID

MAT SENSOR		
TEMPERATURE VS. RESISTANCE VALUES		
(APPROXIMATE)		
°F	°C	OHMS
210	100	185
160	70	450
100	38	1,800
70	20	3,400
40	4	7,500
20	-7	13,500
0	-18	25,000
-40	-40	100,700

CLEAR CODES AND CONFIRM "CLOSED LOOP" OPERATION AND NO "SERVICE ENGINE SOON" LIGHT.

1988–90 B-CARLINE

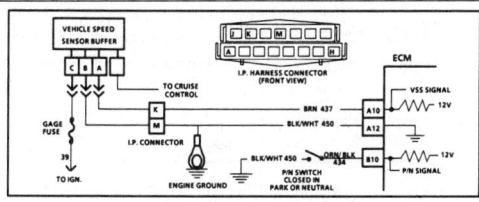

CODE 24
VEHICLE SPEED SENSOR (VSS) CIRCUIT
4.3L, 5.0L & 5.7L "B" CARLINE (TBI)

Circuit Description:
The ECM applies and monitors 12 volts on CKT 437. CKT 437 connects to the vehicle speed sensor which alternately grounds CKT 437 when drive wheels are turning. This pulsing action takes place about 2000 times per mile when the vehicle speed sensor calculate vehicle speed based on the time between "pulses."
"Scan" reading should closely match with speedometer reading with drive wheels turning.
** To prevent misdiagnosis, the technician should review "Electrical" Section "8A" or the Electrical Troubleshooting Manual and identify the type of vehicle speed sensor used prior to using this chart. Disregard a Code 24 set when drive wheels are not turning.

Test Description: Numbers below refer to circled numbers on the diagnostic chart.
1. Code 24 will set if vehicle speed equals 0 mph when:
   - Engine speed is between 1200 and 4400 rpm
   - TPS is less than 2% (closed throttle)
   - Low load condition (low MAP voltage, high manifold vacuum).
   - All conditions met for 5 seconds.
   These conditions are met during a road load deceleration. Disregard Code 24 that sets when drive wheels are not turning.
2. 8-12 volts, at the I/P connector, indicates CKT 437 is open between the I/P connector and the VSS, or there is a faulty vehicle speed sensor. A voltage of less than 1 volt, at the I/P connector, indicates that CKT 437 wire is shorted to ground. If, after disconnecting CKT 437 at the vehicle speed sensor, the voltage reads above 10 volts, the vehicle speed sensor is faulty. If voltage remains less than 8 volts, then CKT 437 wire is grounded. If 437 is not grounded, there is a faulty connection at the ECM, or a faulty ECM.

Diagnostic Aids:
"Scan" should indicate a vehicle speed whenever the drive wheels are turning.
A faulty or misadjusted park/neutral switch can result in a false Code 24. Use a "Scan" and check for proper signal while in drive. Refer to CHART C-1A for P/N switch diagnosis check.
If all OK, refer to "Intermittents" in Section "B".

1988–90 B-CARLINE

CODE 24
VEHICLE SPEED SENSOR (VSS) CIRCUIT
4.3L, 5.0L & 5.7L "B" CARLINE (TBI)

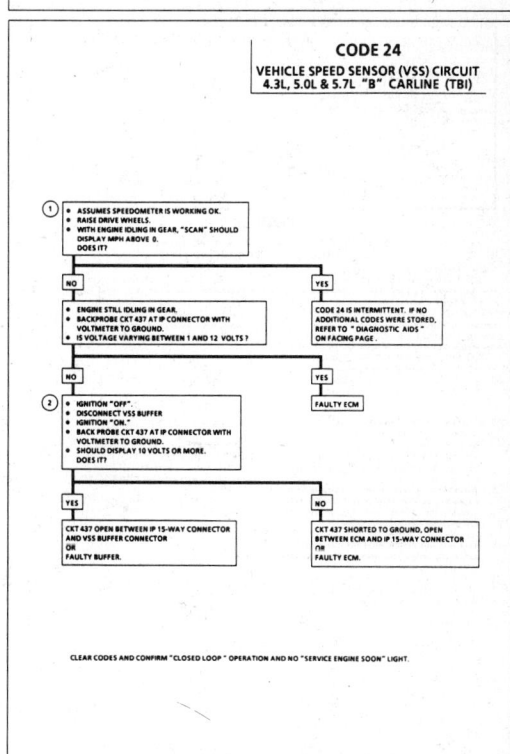

CLEAR CODES AND CONFIRM "CLOSED LOOP" OPERATION AND NO "SERVICE ENGINE SOON" LIGHT.

1988–90 B-CARLINE

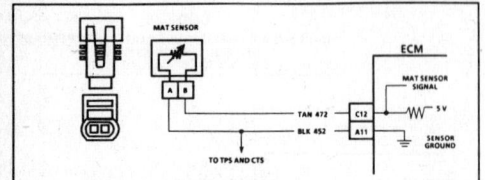

CODE 25
MANIFOLD AIR TEMPERATURE (MAT) SENSOR CIRCUIT
(HIGH TEMPERATURE INDICATED)
5.0L & 5.7L "B" CARLINE (TBI)

Circuit Description:
The manifold air temperature sensor uses a thermistor to control the signal voltage to the ECM. The ECM applies a voltage (4-6) on CKT 472 to the sensor. When manifold air is cold, the sensor (thermistor) resistance is high, therefore, the ECM will see a high signal voltage. As the air warms, the sensor resistance becomes less, and the voltage drops.

Test Description: Numbers below refer to circled numbers on the diagnostic chart.
1. Code 25 will set if:
- Signal voltage indicates a manifold air temperature greater than 150°C (302°F).
- Time since engine start is 5 minutes or longer.

Diagnostic Aids:
Manifold air temperature on a "Scan" tool indicates the temperature of the air in the air cleaner, because the MAT sensor is located in the air cleaner. If the engine has been allowed to sit overnight, the manifold air temperature and coolant temperature values should read within a few degrees of each other. After the engine is started, the MAT will increase due to Thermac operation and underhood temperatures, however, MAT will rarely exceed 80°C (176°F). If a higher MAT than 80°C (176°F) is noted, check for proper Thermac operation.
Check harness routing for possible short to ground in CKT 472.
Refer to "Intermittents" in Section "B".

1988–90 B-CARLINE

CODE 25
MANIFOLD AIR TEMPERATURE (MAT) SENSOR CIRCUIT
(HIGH TEMPERATURE INDICATED)
5.0L & 5.7L "B" CARLINE (TBI)

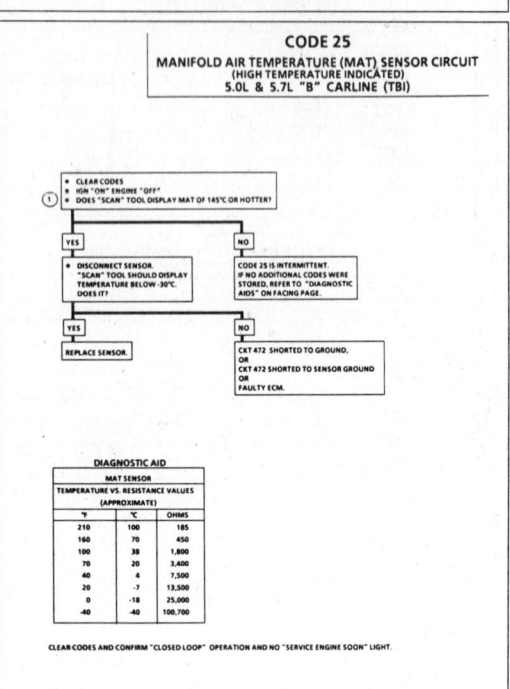

1988–90 B-CARLINE

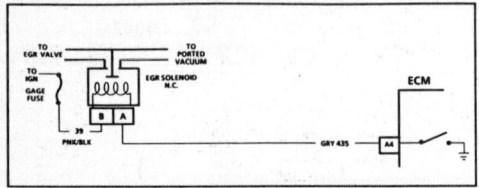

CODE 32
EXHAUST GAS RECIRCULATION (EGR) CIRCUIT
4.3L, 5.0L & 5.7L "B" CARLINE (TBI)

Circuit Description:
The ECM operates a solenoid to control the Exhaust Gas Recirculation (EGR) valve. This solenoid is normally closed. By providing a ground path, the ECM energizes the solenoid which then allows vacuum to pass to the EGR valve.
The ECM monitors EGR effectiveness by de-energizing the EGR control solenoid thereby shutting off vacuum to the EGR diaphragm. With the EGR valve closed, manifold vacuum will be greater than it was during normal EGR operation and this change will be relayed to the ECM by the MAP sensor. If the change is not within the calibrated window, a Code 32 will be set.
The ECM will check EGR operation when:
- Vehicle speed is above 50 mph.
- Manifold absolute pressure is between 35 and 55 kPa (5" and 8" Hg).
- No change in throttle position while test is being run.

Test Description: Numbers below refer to circled numbers on the diagnostic chart.
1. By grounding the diagnostic terminal, the EGR solenoid should be energized and allow vacuum to be applied to the EGR valve and the vacuum should hold.
2. When the diagnostic terminal is ungrounded, the vacuum to the EGR valve should bleed off through a vent in the solenoid and the valve should close. The gage may or may not bleed off but this does not indicate a problem.
3. This test will determine if the electrical control part of the system is at fault or if the connector or solenoid is at fault.
4. This system uses a negative backpressure valve which should hold vacuum with engine "OFF."
5. When engine is started, exhaust backpressure should cause vacuum to bleed off and valve should fully close.

Diagnostic Aids:
Vacuum lines should be thoroughly checked for internal restrictions. The ECM uses the MAP sensor for checking EGR operation. If there is a question of MAP sensor accuracy use CHART C-1D MAP output check
If no problems are found refer to "Intermittents" in Section "B".

1988–90 B-CARLINE

CODE 32
EXHAUST GAS RECIRCULATION (EGR) CIRCUIT
4.3L, 5.0L & 5.7L "B" CARLINE (TBI)

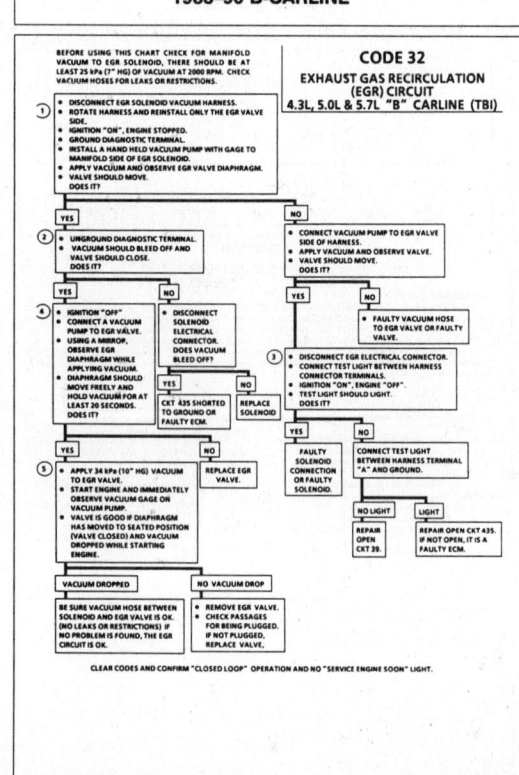

1988–90 B-CARLINE

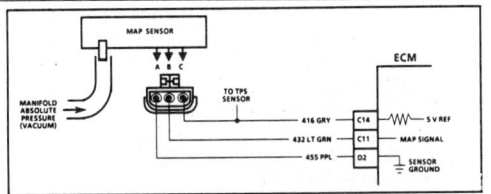

CODE 33
MANIFOLD ABSOLUTE PRESSURE (MAP) SENSOR CIRCUIT
(SIGNAL VOLTAGE HIGH - LOW VACUUM)
4.3L, 5.0L & 5.7L "B" CARLINE (TBI)

Circuit Description:
The manifold absolute pressure (MAP) sensor responds to changes in manifold pressure (vacuum). The ECM receives this information as a signal voltage that will vary from about 1-1.5 volts at idle to 4-4.5 volts at wide open throttle.

A "Scan" displays manifold pressure in volts. Low pressure (high vacuum) reads a low voltage while a high pressure (low vacuum) reads a high voltage.

If the MAP sensor fails the ECM will substitute a fixed MAP value and use the throttle position sensor (TPS) to control fuel delivery.

Test Description: Numbers below refer to circled numbers on the diagnostic chart.
1. Code 33 will set when:
   - Signal is too high, (kPa greater than 68 kPa or less than 9" Hg) for a time greater than 5 seconds.
   - TPS less than 4%
   Engine misfire or a low unstable idle may set Code 33. Disconnect MAP sensor and system will go into backup mode. If the misfire or idle condition remains, see "Symptoms" in Section "B".
2. If the ECM recognizes the low MAP signal, the ECM and wiring are OK.

Diagnostic Aids:
If idle is rough or unstable refer to "Symptoms" in Section "B" for items which can cause an unstable idle.

An open in CKT 455 or the connection will result in a Code 33.

With the ignition "ON" and the engine stopped, the manifold pressure is equal to atmospheric pressure and the signal voltage will be high. This information is used by the ECM as an indication of vehicle altitude and is referred to as BARO. Comparison of this BARO reading with a known good vehicle with the same sensor is a good way to check accuracy of a "suspect" sensor. Reading should be the same ± .4 volt.

Refer to "Intermittents" in Section "B".

1988–90 B-CARLINE

CODE 33
MANIFOLD ABSOLUTE PRESSURE (MAP) SENSOR CIRCUIT
(SIGNAL VOLTAGE HIGH - LOW VACUUM)
4.3L, 5.0L & 5.7L "B" CARLINE (TBI)

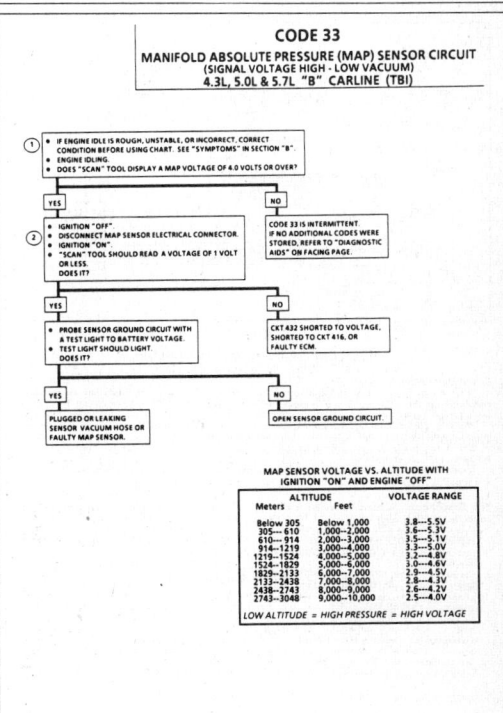

MAP SENSOR VOLTAGE VS. ALTITUDE WITH IGNITION "ON" AND ENGINE "OFF"

ALTITUDE		VOLTAGE RANGE
Meters	Feet	
Below 305	Below 1,000	3.8—5.5V
305— 610	1,000–2,000	3.6—5.3V
610— 914	2,000–3,000	3.5—5.1V
914—1219	3,000–4,000	3.3—5.0V
1219—1524	4,000–5,000	3.2—4.8V
1524—1829	5,000–6,000	3.0—4.6V
1829—2133	6,000–7,000	2.9—4.5V
2133—2438	7,000–8,000	2.8—4.3V
2438—2743	8,000–9,000	2.6—4.2V
2743—3048	9,000–10,000	2.5—4.0V

LOW ALTITUDE = HIGH PRESSURE = HIGH VOLTAGE

1988–90 B-CARLINE

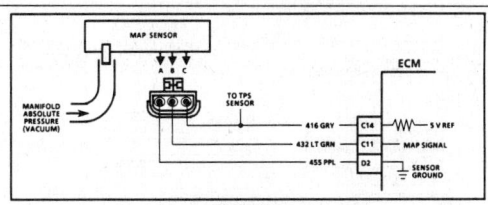

CODE 34
MANIFOLD ABSOLUTE PRESSURE (MAP) SENSOR CIRCUIT
(SIGNAL VOLTAGE LOW - HIGH VACUUM)
4.3L, 5.0L & 5.7L "B" CARLINE (TBI)

Circuit Description:
The manifold absolute pressure (MAP) sensor responds to changes in manifold pressure (vacuum). The ECM receives this information as a signal voltage that will vary from about 1-1.5 volts at idle to 4-4.5 volts at wide open throttle.

If the MAP sensor fails the ECM will substitute a fixed MAP value and use the throttle position sensor (TPS) to control fuel delivery.

Test Description: Numbers below refer to circled numbers on the diagnostic chart.
1. Code 34 will set when:
   - Signal is too low (kPa less than 14 kPa or greater than 28" Hg) and engine running less than 1200 rpm.
   OR
   - Engine running greater than 1200 rpm
   - Throttle position greater than 21% (over 1.5 volts).
2. If the ECM recognizes the high MAP signal, the ECM and wiring are OK.
3. The "Scan" tool may not display 12 volts. The important thing is that the ECM recognizes the voltage as more than 4 volts, indicating that the ECM and CKT 432 are OK.

Diagnostic Aids:
An intermittent open in CKTs 432 or 416 will result in a Code 34.

With the ignition "ON" and the engine running less, the manifold pressure is equal to atmospheric pressure and the signal voltage will be high. This information is used by the ECM as an indication of vehicle altitude and is referred to as BARO. Comparison of this BARO reading with a known good vehicle with the same sensor is a good way to check accuracy of a "suspect" sensor. Reading should be the same ± .4 volt.

Also CHART C-1D can be used to test the MAP sensor.

Refer to "Intermittents" in Section "B".

1988–90 B-CARLINE

CODE 34
MANIFOLD ABSOLUTE PRESSURE (MAP) SENSOR CIRCUIT
(SIGNAL VOLTAGE LOW - HIGH VACUUM)
4.3L, 5.0L & 5.7L "B" CARLINE (TBI)

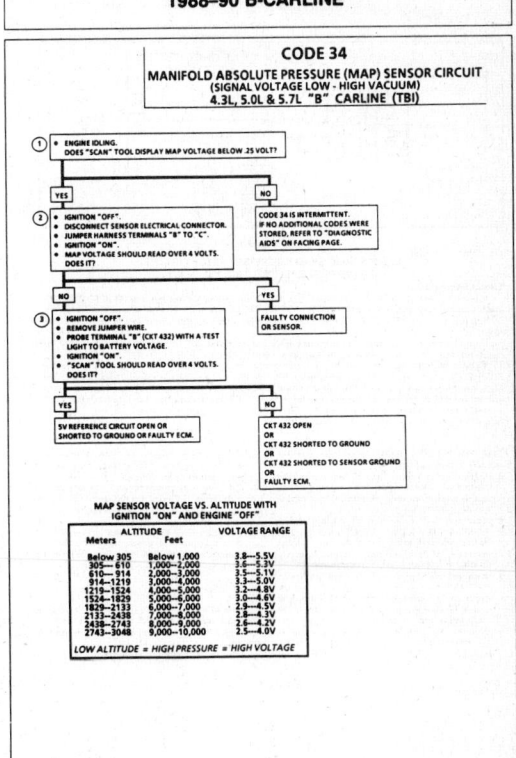

MAP SENSOR VOLTAGE VS. ALTITUDE WITH IGNITION "ON" AND ENGINE "OFF"

ALTITUDE		VOLTAGE RANGE
Meters	Feet	
Below 305	Below 1,000	3.8—5.5V
305— 610	1,000–2,000	3.6—5.3V
610— 914	2,000–3,000	3.5—5.1V
914—1219	3,000–4,000	3.3—5.0V
1219—1524	4,000–5,000	3.2—4.8V
1524—1829	5,000–6,000	3.0—4.6V
1829—2133	6,000–7,000	2.9—4.5V
2133—2438	7,000–8,000	2.8—4.3V
2438—2743	8,000–9,000	2.6—4.2V
2743—3048	9,000–10,000	2.5—4.0V

LOW ALTITUDE = HIGH PRESSURE = HIGH VOLTAGE

1988–90 B-CARLINE

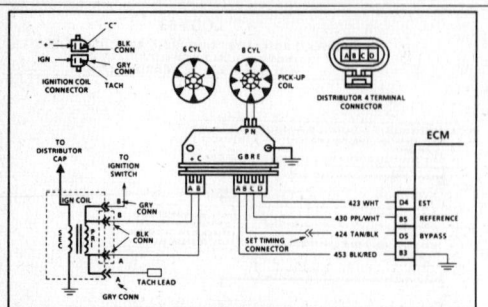

CODE 42
ELECTRONIC SPARK TIMING (EST) CIRCUIT
4.3L, 5.0L & 5.7L "B" CARLINE (TBI)

Circuit Description:

When the system is running on the ignition module, that is, no voltage on the bypass line, the ignition module grounds the EST signal. The ECM expects to see no voltage on the EST line during this condition. If it sees a voltage, it sets Code 42 and will not go into the EST mode.

When the rpm for EST is reached (about 400 rpm), and bypass voltage applied, the EST should no longer be grounded in the ignition module so the EST voltage should be varying.

If the bypass line is open or grounded, the ignition module will not switch to EST mode so the EST voltage will be low and Code 42 will be set.

If the EST line is grounded, the ignition module will switch to EST, but because the line is grounded there will be no EST signal. A Code 42 will be set.

Test Description: Numbers below refer to circled numbers on the diagnostic chart.
1. Code 42 means the ECM has seen an open or short to ground in the EST or bypass circuits. This test confirms that Code 42 and that the fault causing the code is present.
2. Checks for a normal EST ground path through the ignition module. An EST CKT 423 shorted to ground will also read less than 500 ohms; however, this will be checked later.
3. As the test light voltage touches CKT 424, the module should switch causing the ohmmeter to "overrange" if the meter is in the 1000-2000 ohms position. Selecting the 10-20,000 ohms position will indicate above 5000 ohms. The important thing is that the module "switched."

4. The module did not switch and this step checks for:
   - EST CKT 423 shorted to ground
   - Bypass CKT 424 open
   - Faulty ignition module connection or module
5. Confirms that Code 42 is a faulty ECM and not an intermittent in CKTs 423 or 424.

Diagnostic Aids:

If a Code 42 was stored and the customer complains of a "hard start," the problem is most likely a grounded EST line (CKT 423).

The "Scan" tool does not have any ability to help diagnose a Code 42 problem.

A PROM not fully seated in the ECM can result in a Code 42.

Refer to "Intermittents" in Section "B".

1988–90 B-CARLINE

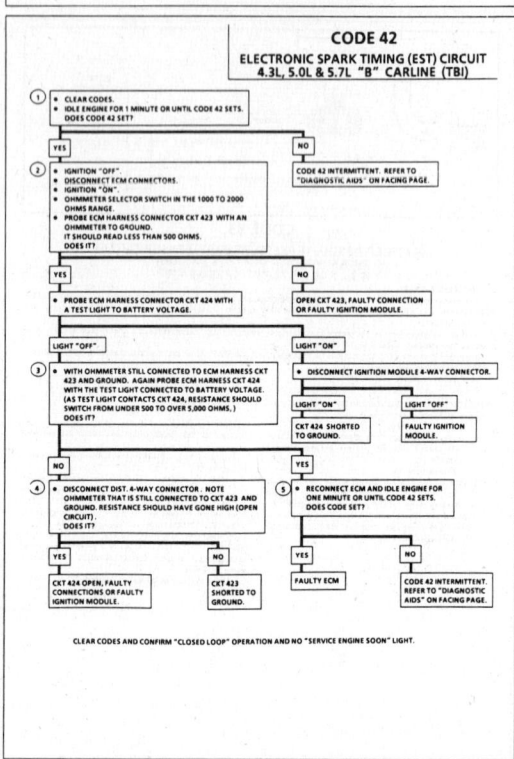

1988–90 B-CARLINE

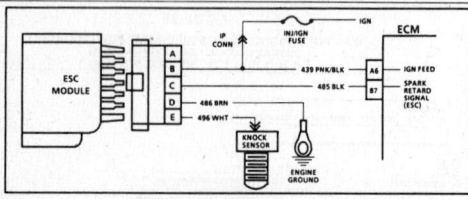

CODE 43
ELECTRONIC SPARK CONTROL (ESC) CIRCUIT
4.3L, 5.0L & 5.7L "B" CARLINE (TBI)

Circuit Description:

Electronic spark control is accomplished with a module that sends a voltage signal to the ECM. When the knock sensor detects engine knock, the voltage from the ESC module to the ECM drops, and this signals the ECM to retard the timing when knock is detected and rpm is above about 900 rpm.

Code 43 means the ECM has seen low voltage at CKT 485 terminal "B7" for longer than 5 seconds with the engine running or the system has failed the functional check.

This system performs a functional check once per start up to check the ESC system. To perform this test the ECM will advance the spark when coolant is above 95°C (194°F) and at a high load condition (near W.O.T.). The ECM then checks the signal at "B7" to see if a knock is detected. The functional check is performed once per start up and if knock is detected when coolant is below 95°C (194°F) the test has passed and the functional check will not be run. If the functional check fails, the "Service Engine Soon" light will remain "ON" until ignition is turned "OFF" or until a knock signal is detected.

Test Description: Numbers below refer to circled numbers on the diagnostic chart.
1. If the conditions for a Code 43 are present the "Scan" will always display "yes." There should not be a knock at idle unless an internal engine problem, or a system problem exists.
2. This test will determine if the system is functioning at this time. Usually a knock signal can be generated by tapping on the right exhaust manifold. If no knock signal is generated try tapping on block close to the area of the sensor.
3. Because Code 43 sets when the signal voltage on CKT 485 remains low this test should cause the signal on CKT 485 to go high. The 12 volts signal should be seen by the ECM as "no knock" if the ECM and wiring are OK.
4. This test will determine if the knock signal is being detected on CKT 496 or if the ESC module is at fault.

5. If CKT 496 is routed too close to secondary ignition wires the ESC module may see the interference as a knock signal.
6. This checks the ground circuit to the module. An open ground will cause the voltage on CKT 485 to be about 12 volts which would cause the Code 43 functional test to fail.
7. Contacting CKT 496 with a test light to 12 volts should generate a knock signal. This will determine if the ESC module is operating correctly.

Diagnostic Aids:

Code 43 can be caused by a faulty connection at the knock sensor, the ESC module or at the ECM. Also check CKT 485 for possible open or short to ground.

Refer to "Intermittents" in Section "B".

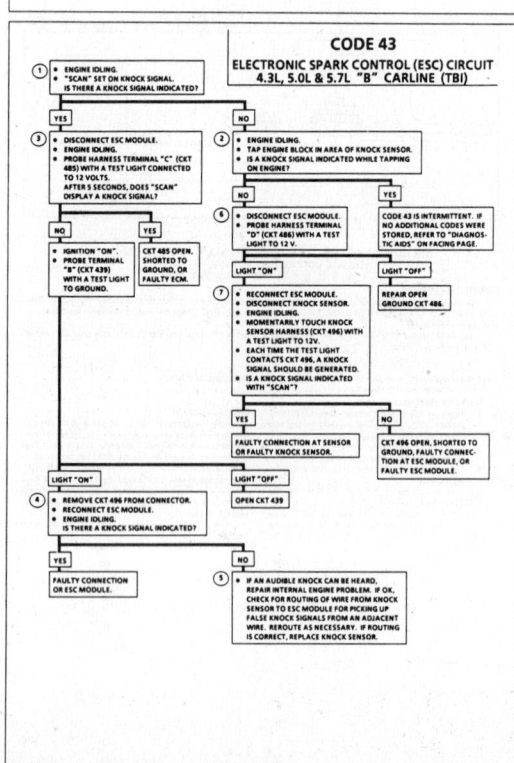

1988–90 B-CARLINE

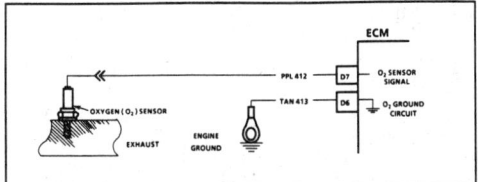

CODE 44
OXYGEN SENSOR CIRCUIT
(LEAN EXHAUST INDICATED)
4.3L, 5.0L & 5.7L "B" CARLINE (TBI)

Circuit Description:

The ECM supplies a voltage of about .45 volt between terminals "D6" and "D7". (If measured with a 10 megohm digital voltmeter, this may read as low as .32 volt). The O₂ sensor varies the voltage within a range of about 1 volt if the exhaust is rich, down through about .10 volt if exhaust is lean.

The sensor is like an open circuit and produces no voltage when it is below about 360°C (600°F). An open sensor circuit or cold sensor causes "Open Loop" operation.

Test Description: Numbers below refer to circled numbers on the diagnostic chart.
1. Code 44 is set when the O₂ sensor signal voltage on CKT 412.
   * Remains below .2 volt for 50 seconds
   * And the system is operating in "Closed Loop."

Diagnostic Aids:

Using the "Scan," observe the block learn values at different rpm and air flow conditions to determine when the Code 44 may have been set. If the conditions for Code 44 exists the block learn values will be around 150.
* O₂ Sensor Wire. Sensor pigtail may be mispositioned and contacting the exhaust manifold.
* Check for intermittent ground in wire between connector and sensor.

* MAP Sensor. A MAP sensor output that causes the ECM to sense a higher than normal vacuum will cause the system to go lean. Disconnect the MAP sensor and if the lean condition is gone, replace the sensor.
* Lean Injector(s).
* Fuel Contamination. Water, even in small amounts, near the in-tank fuel pump inlet can be delivered to the injectors. The water causes a lean exhaust and can set a Code 44.
* Fuel Pressure. System will be lean if pressure is too low. It may be necessary to monitor fuel pressure while driving the car at various road speeds and/or loads to confirm. See "Fuel System Diagnosis" CHART A-7.
* Exhaust Leaks. If there is an exhaust leak, the engine can pull outside air into the exhaust and past the sensor. Vacuum or crankcase leaks can cause a lean condition.
* If the above are OK, it is a faulty oxygen sensor.

1988–90 B-CARLINE

CODE 44
OXYGEN SENSOR CIRCUIT
(LEAN EXHAUST INDICATED)
4.3L, 5.0L & 5.7L "B" CARLINE (TBI)

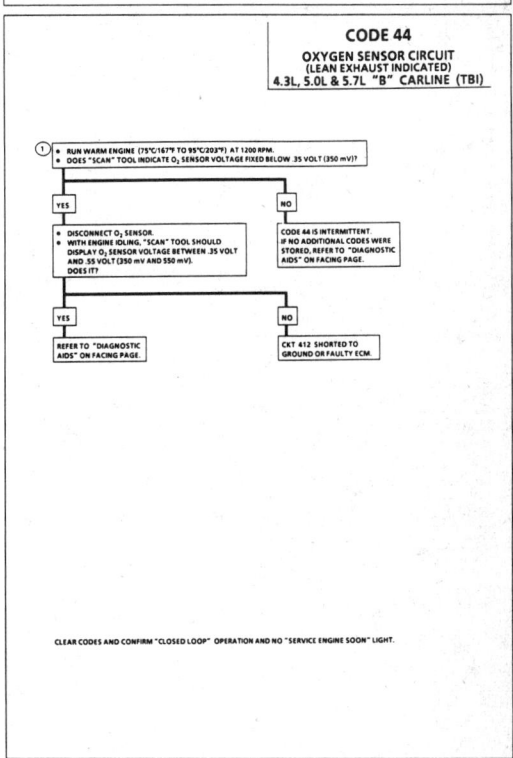

1988–90 B-CARLINE

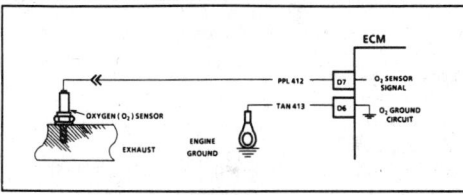

CODE 45
OXYGEN SENSOR CIRCUIT
(RICH EXHAUST INDICATED)
4.3L, 5.0L & 5.7L "B" CARLINE (TBI)

Circuit Description:

The ECM supplies a voltage of about .45 volt between terminals "D6" and "D7". (If measured with a 10 megohm digital voltmeter, this may read as low as .32 volt). The O₂ sensor varies the voltage within a range of about 1 volt if the exhaust is rich, down through about .10 volt if exhaust is lean.

The sensor is like an open circuit and produces no voltage when it is below about 360°C (600°F). An open sensor circuit or cold sensor causes "Open Loop" operation.

Test Description: Numbers below refer to circled numbers on the diagnostic chart.
1. Code 45 is set when the O₂ sensor signal voltage or CKT 412.
   * Remains above .7 volt for 30 seconds; and in "Closed Loop."
   * Engine time after start is 1 minute or more.
   * Throttle angle greater than 2% (about .2 volt above idle voltage) but less than 20%.

Diagnostic Aids:

Using the "Scan," observe the block learn values at different rpm conditions to determine when the Code 45 may have been set. If the conditions for Code 45 exists, the block learn values will be around 115.
* Fuel Pressure. System will go rich if pressure is too high. The ECM can compensate for some increase. However, if it gets too high, a Code 45 may be set. See "Fuel System Diagnosis" CHART A-7.
* Leaking injector. See CHART A-7.
* Check for fuel contaminated oil.

* HEI Shielding. An open ground CKT 453 (distributor ground reference low) may result in Electro-Magnetic Interference (EMI), or induced electrical "noise." The ECM looks at this "noise" as reference pulses. The additional pulses result in higher than actual engine speed signal. The ECM then delivers too much fuel, causing system to go rich. Engine tachometer will also show higher than actual engine speed, which can help in diagnosing this problem.
* Canister purge. Check for fuel saturation. If full of fuel, check canister control and hoses.

* MAP sensor. An output that causes the ECM to sense a lower than normal vacuum can cause the system to go rich. Disconnecting the MAP sensor will allow the ECM to set a fixed value for the sensor. Substitute a different MAP sensor if the rich condition is gone while the sensor is disconnected.
* TPS. An intermittent TPS output will cause the system to go rich, due to a false indication of the engine accelerating.

1988–90 B-CARLINE

CODE 45
OXYGEN SENSOR CIRCUIT
(RICH EXHAUST INDICATED)
4.3L, 5.0L & 5.7L "B" CARLINE (TBI)

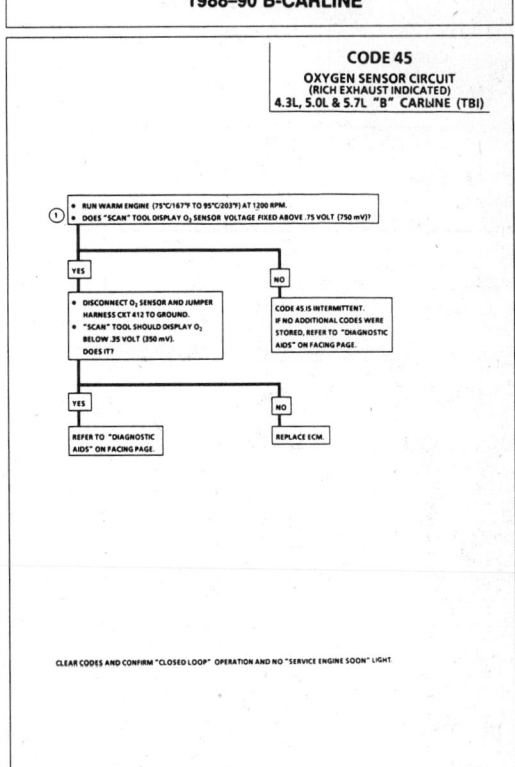

1988–90 B-CARLINE

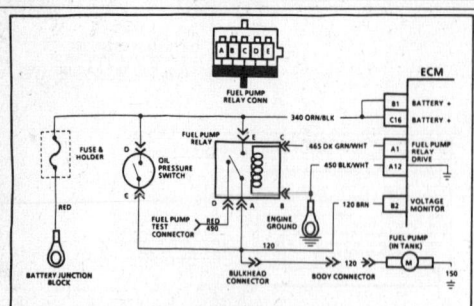

CODE 54
FUEL PUMP CIRCUIT
(LOW VOLTAGE)
4.3L, 5.0L & 5.7L "B" CARLINE (TBI)

Circuit Description:

When the ignition switch is turned "ON," the electronic control module (ECM) will activate the fuel pump relay and run the in-tank fuel pump. The fuel pump will operate as long as the engine is cranking or running, and the ECM is receiving ignition reference pulses.

If there are no reference pulses, the ECM will shut "OFF" the fuel pump within 2 seconds after key "ON." Should the fuel pump relay, or the 12 volts relay drive from the ECM fail, the fuel pump will be run through an oil pressure switch back-up circuit.

Diagnostic Aids:

An inoperative fuel pump relay can result in long cranking times, particularly if the engine is cold or engine oil pressure is low. The extended crank period is caused by the time necessary for oil pressure to build enough to close the oil pressure switch and turn "ON" the fuel pump.

1988–90 B-CARLINE

CODE 54
FUEL PUMP CIRCUIT
(LOW VOLTAGE)
4.3L, 5.0L & 5.7L "B" CARLINE (TBI)

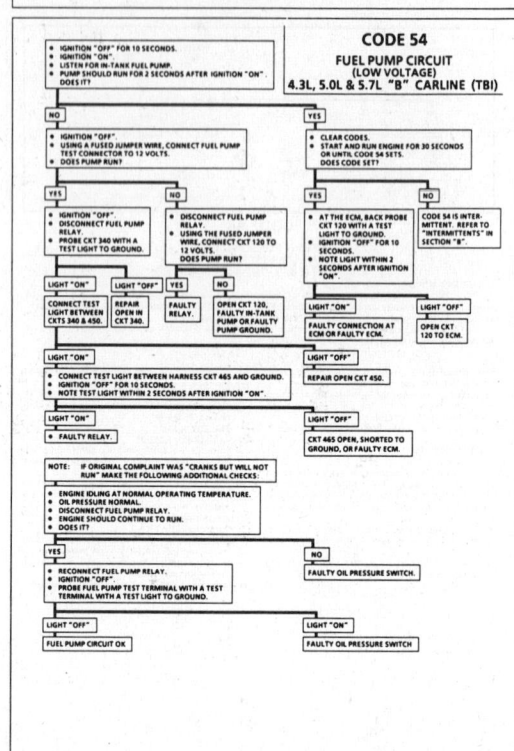

1988–90 B-CARLINE

CODE 51
PROM ERROR
(FAULTY OR INCORRECT PROM)

> CHECK THAT ALL PINS ARE FULLY INSERTED IN THE SOCKET. IF OK, REPLACE PROM, CLEAR MEMORY AND RECHECK. IF CODE 51 REAPPEARS, REPLACE ECM.

CLEAR ALL CODES AND CONFIRM "CLOSED LOOP" OPERATION AND NO "SERVICE ENGINE SOON" LIGHT

CODE 52
CALPAK ERROR
(FAULTY OR INCORRECT CALPAK)

> CHECK THAT ALL PINS ARE FULLY INSERTED IN THE SOCKET. IF OK, REPLACE CALPAK, CLEAR MEMORY AND RECHECK. IF CODE 52 REAPPEARS, REPLACE ECM.

CLEAR ALL CODES AND CONFIRM "CLOSED LOOP" OPERATION AND NO "SERVICE ENGINE SOON" LIGHT

CODE 55
ECM ERROR

> REPLACE ELECTRONIC CONTROL MODULE (ECM).

CLEAR ALL CODES AND CONFIRM "CLOSED LOOP" OPERATION AND NO "SERVICE ENGINE SOON" LIGHT

1988–90 B-CARLINE

SECTION B
SYMPTOMS

TABLE OF CONTENTS

Before Starting
Intermittents
Hard Start
Surges and/or Chuggle
Lack of Power, Sluggish, or Spongy
Detonation/Spark Knock
Hesitation, Sag, Stumble
Cuts Out, Misses
Poor Fuel Economy
Stalling, Rough, Unstable or Incorrect Idle
Excessive Exhaust Emissions or Odors
Dieseling, Run-On
Backfire.
Restricted Exhaust System Check (All Engines)
  CHART B-1

BEFORE STARTING

Before using this section you should have performed the "Diagnostic Circuit Check."

Verify the customer complaint, and locate the correct SYMPTOM below. Check the items indicated under that symptom.

If the ENGINE CRANKS BUT WILL NOT RUN, see CHART A-3.

Several of the following symptom procedures call for a careful visual (physical) check. The importance of this step cannot be stressed too strongly - it can lead to correcting a problem without further checks and can save valuable time.

This check should include:

- Vacuum hoses for splits, kinks, and proper connections, as shown on Vehicle Emission Control Information Label.
- Air leaks at throttle body mounting and intake manifold.
- Ignition wires for cracking, hardness, proper routing, and carbon tracking.
- Wiring for proper connections, pinches, and cuts.

1988–90 B-CARLINE

INTERMITTENTS

Problem may or may not turn "ON" the "Service Engine Soon" light, or store a code.

DO NOT use the Trouble Code Charts in Section "A" for intermittent problems. The fault must be present to locate the problem. If a fault is intermittent, use of Trouble Code Charts may result in replacement of good parts.

- Most intermittent problems are caused by faulty electrical connections or wiring. Perform careful check of suspect circuits for:
  - Poor mating of the connector halves, or terminals, not fully seated in the connector body (backed out).
  - Improperly formed or damaged terminals. All connector terminals in problem circuit should be carefully reformed or replaced to insure proper contact tension.
  - Poor terminal to wire connection. This requires removing the terminal from the connector body to check.
- If a visual (physical) check does not find the cause of the problem, the car can be driven with a voltmeter connected to a suspected circuit or a "Scan" tool may be used. An abnormal voltage reading, when the problem occurs, indicates the problem may be in that circuit. If the wiring and connectors check OK, and a Trouble Code was stored for a circuit having a sensor, except for Codes 44 and 45, substitute a known good sensor and recheck.
- Loss of Trouble Code Memory. To check, disconnect TPS and idle engine until "Service Engine Soon" light comes "ON." Code 22 should be stored, and kept in memory, when ignition is turned "OFF" for at least 10 seconds. If not, the ECM is faulty.
- An intermittent "Service Engine Soon" light, and No Trouble Codes, may be caused by:
  - Electrical system interference caused by a defective relay, ECM driven solenoid, or switch. They can cause a sharp electrical surge. Normally, the problem will occur when the faulty component is operated.
  - Improper installation of electrical options, such as lights, 2-way radios, etc.
  - EST wires should be routed away from spark plug wires, ignition system components, and generator. Wire for CKT 453 from ECM to ignition system should be a good ground.
  - Ignition secondary shorted to ground.
  - CKT 419 ("Service Engine Soon" light) or CKT 451 (Diagnostic Test Terminal) intermittently shorted to ground.
  - ECM power grounds.

HARD START

Definition: Engine cranks OK, but does not start for a long time. Does eventually run, or may start but immediately dies.

- **CHECK:**
  - For water contaminated fuel.
  - Fuel system pressure CHART A-7.
  - TPS for sticking or binding should read less than .7 volt on a "Scan" tool.
  - No crank signal; see CHART C-1B.
  - EGR operation; CHART C-7.
  - Fuel pump relay - Connect test light between pump test terminal and ground. Light should be "ON" for 2 seconds following ignition "ON." If not, refer to CODE 54 CHART.
  - For a faulty in-tank fuel pump check valve, which would allow the fuel in the lines to drain back to the tank after the engine is stopped. To check for this condition:
    1. Ignition "OFF."
    2. Disconnect fuel line at the filter.
    3. Remove the tank filler cap.
    4. Connect a radiator test pump to the fuel line and apply 103 kPa (15 psi) pressure. If the pressure will hold for 60 seconds, the check valve is OK.
- Check ignition system for:
  - Proper output with ST-125.
  - Worn distributor shaft.
  - Bare and shorted wires.
  - Pickup coil resistance and connections.
  - Loose ignition coil connections.
  - Moisture in distributor cap.
  - Spark plugs, wet plugs, cracks, wear, improper gap, burned electrodes, or heavy deposits.
- If engine starts but then, immediately stalls, disconnect the set timing connector. If engine then starts, and runs OK, replace distributor pickup coil.
- Check CKT 423 (EST) for short to ground.

1988–90 B-CARLINE

SURGES AND/OR CHUGGLE

Definition: Engine power variation, under steady throttle or cruise. Feels like the car speeds up and slows down, with no change in the accelerator pedal.

- Use a "Scan" tool to make sure reading of VSS matches vehicle speedometer.
- **CHECK:**
  - For intermittent EGR at idle. See appropriate CHART C-7.
  - Inline fuel filter for dirt or restriction.
  - Fuel pressure. See CHART A-7.
  - Generator output voltage. Repair, if less than 9, or more than 16 volts.
  - TCC Operation. See CHART C-8A.
- Inspect Oxygen sensor for silicon contamination from fuel, or use of improper RTV sealant. The sensor may have a white, powdery coating and result in a high but false signal voltage (rich exhaust indication). The ECM will then reduce the amount of fuel delivered to the engine, causing a severe driveability problem.
- Remove spark plugs. Check for cracks, wear, improper gap, burned electrodes, or heavy deposits. Also, check condition of the rest of the Ignition System.

LACK OF POWER, SLUGGISH, OR SPONGY

Definition: Engine delivers less than expected power. Little or no increase in speed, when accelerator pedal is pushed down part way.

- Compare customer's car to similar unit. Make sure the customer's car has an actual problem.
- Remove air cleaner and check air filter for dirt, or for being plugged. Replace as necessary.
- If there is spray from only one injector, then, there is a malfunction in the injector assembly, or in the signal to the injector assembly. The malfunction can be isolated, by switching the injector connectors. If the problem remains with the original injector, after switching the connector, then, the injector is defective. Replace the injector. If the problem moves with the injector connector, then, the problem is an improper signal in the injector circuits, see CHART A-3.
- **CHECK:**
  - Ignition timing. See Vehicle Emission Control Information Label.
  - For restricted fuel filter, contaminated fuel or improper fuel pressure. See CHART A-7.
  - ECM Grounds.
  - EGR operation for being open, or partly open, all the time. See CHART C-7.
  - Generator output voltage. Repair, if less than 9, or more than 16 volts.
  - Engine valve timing and compression.
  - Engine, for proper or worn camshaft.
  - Transmission torque converter operation.
  - Secondary ignition voltage, using a scope or ST-125.
  - Proper operation of EST.
- Check exhaust system for restriction. See CHART B-1:
  1. With engine at normal operating temperature, connect a vacuum gage to any convenient vacuum port on intake manifold.
  2. Run engine at 1000 rpm and record vacuum reading.
  3. Increase rpm slowly to 2500 rpm. Note vacuum reading at steady 2500 rpm.
  4. If vacuum at 2500 rpm decreases more than 3", from reading at 1000 rpm, the exhaust system should be inspected for restrictions.
  5. Disconnect exhaust pipe from engine and repeat steps 3 & 4. If vacuum still drops more than 3", with exhaust disconnected, check valve timing.

1988–90 B-CARLINE

DETONATION / SPARK KNOCK

Definition: A mild to severe ping, usually worse under acceleration. The engine makes sharp metallic knocks that change with throttle opening.

- **CHECK** for obvious overheating problems.
  - Low coolant.
  - Loose water pump belt.
  - Restricted air flow to radiator, or restricted water flow thru radiator.
  - Faulty or incorrect thermostat.
  - Coolant Temperature Sensor, which has shifted in value.
  - Correct coolant solution - should be a 50/50 mix of GM #1052753 anti-freeze coolant (or equivalent) and water.
- **CHECK:**
  - For poor fuel quality, proper octane rating.
  - For correct PROM.
  - Spark plugs for correct heat range.
  - Proper operation of Thermac.
  - ESC system operation. See CHARTC-5.
  - Fuel system for low pressure. See CHART A-7.
  - EGR system for not opening. See CHART C-7.
  - Ignition timing. See Vehicle Emission Control Information Label.
  - For proper transmission shift points.
  - TCC operation. See CHART C-8.
  - For incorrect basic engine parts such as cam, heads, pistons, etc.
  - Excessive oil entering combustion chamber.
- Remove carbon with top engine cleaner, follow instructions on can.
- If there is spray from only one injector, then, there is a malfunction in the injector assembly, or in the signal to the injector assembly. The malfunction can be isolated, by switching the injector connectors. If the problem remains with the original injector, after switching the connector, then, the injector is defective. Replace the injector. If the problem moves with the injector connector, then, the problem is an improper signal in the injector circuits. See CHART A-3.

HESITATION, SAG, STUMBLE

Definition: Momentary lack of response as the accelerator is pushed down. Can occur at all car speeds. Usually most severe when first trying to make the car move, as from a stop sign. May cause the engine to stall if severe enough.

- Perform careful visual (physical) check, as described at start of Section "B".
- **CHECK:**
  - Fuel pressure. See CHART A-7.
  - Water contaminated fuel.
  - TPS for binding or sticking.
  - Generator output voltage. Repair if less than 9 or more than 16 volts.
  - For open Ignition System ground, CKT 453.
  - Canister purge system for proper operation.
  - EGR valve operation. See CHART C-7.
  - Ignition timing. See Vehicle Emission Control Information Label.

CUTS OUT, MISSES

Definition: Steady pulsation or jerking that follows engine speed, usually more pronounced as engine load increases. The exhaust has a steady spitting sound at idle or low speed.

- Perform careful visual (physical) check, as described at start of Section "B".
- If Ignition System is suspected of causing a miss at idle or cutting, out under load:
- **Check** for missing cylinder by:
  1. Start engine, allow engine to stabilize then disconnect IAC motor. Remove one spark plug wire at a time, using insulated pliers.

     CAUTION: Do not perform this test for more than 2 minutes, as this may cause damage to the catalytic converter.
  2. If there is an rpm drop, on all cylinders, (equal to within 50 rpm), go to STALLING, ROUGH, UNSTABLE OR INCORRECT IDLE symptom. Reconnect IAC motor with engine running, ignition "OFF".
  3. If there is no rpm drop on one or more cylinders, or excessive variation in drop, check for spark, on the suspected cylinder(s) with J 26792 (ST-125) Spark Tester or equivalent.

     If there is spark, remove spark plug(s) in these cylinders and check for:
     - Cracks
     - Wear
     - Improper Gap
     - Burned Electrodes

1988–90 B-CARLINE

- Heavy Deposits
- Perform compression check on questionable cylinder.
- Check wire resistance (should not exceed 30,000 ohms), also, check rotor and distributor cap.
- If the previous checks did not find the problem:
  - Visually inspect ignition system for moisture, dust, cracks, burns, etc. Spray plug wires with fine water mist to check for shorts.
  - Fuel System - Plugged fuel filter, water, low pressure. See CHART A-7.
  - Perform compression check.
  - Valve Timing.
  - Remove rocker covers. Check for bent pushrods, worn rocker arms, broken or weak valve springs, worn camshaft lobes. Repair as necessary.
- If there is spray from only one injector, then, there is a malfunction in the injector assembly, or in the signal to the injector assembly. The malfunction can be isolated, by switching the injector connectors. If the problem remains with the original injector, after switching the connector, then, the injector is defective. Replace the injector. If the problem moves with the injector connector, then, the problem is an improper signal in the injector circuits, see CHART A-3.

POOR FUEL ECONOMY

Definition: Fuel economy, as measured by an actual road test, is noticeably lower than expected. Also, economy is noticeably lower than it was on this car at one time, as previously shown by an actual road test.

- **CHECK:**
  - Engine thermostat for faulty part (always open) or for wrong heat range.
  - Fuel Pressure. See CHART A-7.
- Check owner's driving habits.
  - Is A/C "ON" full time (Defroster mode "ON")?
  - Are tires at correct pressure?
  - Are excessively heavy loads being carried?
  - Is acceleration too much, too often?
  - Suggest driver read "Important Facts on Fuel Economy" in Owner's Manual.
- Perform "Diagnostic Circuit Check."
- Check air cleaner element (filter) for dirt or being plugged.
- Check for proper calibration of speedometer.
- Visually (physically) Check:
  - Vacuum hoses for splits, kinks, and proper connections, as shown on Vehicle Emission Control Information Label.
  - Ignition wires for cracking, hardness, and proper connections.
- Check Ignition timing. See Vehicle Emission Control Information Label.
- Remove spark plugs. Check for cracks, wear, improper gap, burned electrodes or heavy deposits. Repair or replace, as necessary.
- Check compression.
- Check TCC for proper operation. See CHART C-8. Use "Scan" tool if available.
- Check for dragging brakes.
- Suggest owner fill fuel tank and recheck fuel economy.
- Check for exhaust system restriction. See CHART B-1.

STALLING, ROUGH, UNSTABLE OR INCORRECT IDLE

Definition: The engine runs unevenly at idle. If bad enough, the car may shake. Also, the idle may vary in rpm (called "hunting"). Either condition may be severe enough to cause stalling. Engine idles at incorrect speed.

- **CHECK:**
  - P/N switch circuit. See CHART C-1A.
  - For injector(s) leaking. Check fuel pressure, see CHART A-7.
  - IAC - See CHART C-2C.
  - If a sticking throttle shaft or binding linkage causes a high TPS voltage (open throttle indication), the ECM will not control idle. Monitor TPS voltage. "Scan" and/or Voltmeter should read less than 1.2 volts with throttle closed.
  - Vacuum leaks can cause higher than normal idle.
  - EGR "ON," while idling, will cause roughness, stalling, and hard starting. See CHART C-7.
  - Battery cables and ground straps should be clean and secure. Erratic voltage will cause IAC to change its position, resulting in poor idle quality.
  - IAC valve will not move, if system voltage is below 9, or greater than 16 volts.
  - Use "Scan" tool to determine if ECM is receiving A/C request signal.

1988–90 B-CARLINE

- Ignition timing. See Vehicle Emission Control Information Label.
- MAP Sensor - Ignition "ON", engine stopped. Compare MAP voltage with known good vehicle. Voltage should be the same ± 400 mV (.4 volts).

or

- Start and idle engine. Disconnect MAP sensor electrical connector. If idle improves, substitute a known good MAP sensor and recheck.
- A/C refrigerant pressure too high. Check for overcharge or faulty pressure switch.

- PCV valve for proper operation by placing finger over inlet hole in valve end several times. Valve should snap back. If not, replace valve.
- Run a cylinder compression check. See Section "6".
- Inspect Oxygen sensor for silicon contamination from fuel, or use of improper RTV sealant. The sensor will have a white, powdery coating, and will result in a high but false signal voltage (rich exhaust indication). The ECM will then reduce the amount of fuel delivered to the engine, causing a severe driveability problem.

EXCESSIVE EXHAUST EMISSIONS OR ODORS

Definition: Vehicle fails an emission test. Vehicle has excessive "rotten egg" smell. Excessive odors do not necessarily indicate excessive emissions.

- Perform "Diagnostic Circuit Check."
- IF TEST SHOWS EXCESSIVE CO and HC, (or also has excessive odors):
  - Check items which cause car to run RICH.
  - Make sure engine is at normal operating temperature.
- CHECK:
  - Fuel pressure. See CHART A-7.
  - Incorrect timing. See Vehicle Emission Control Information Label.
  - Fuel Vapor Canister for fuel loading. See CHART C-3.
  - PCV valve for being plugged, stuck, or blocked PCV hose, or fuel in the crankcase.
  - Spark plugs, plug wires, and ignition components.
  - Check for lead contamination of catalytic converter (look for removal of fuel filler neck restrictor).
  - Check for properly installed fuel cap.

- If the system is running rich, (block learn less than 118), refer to "Diagnostic Aids" on facing page of Code 45.
- IF TEST SHOWS EXCESSIVE NOx:
  - Check items which cause car to run LEAN, or too HOT.
- CHECK:
  - EGR valve for not opening. See CHART C-7.
  - Vacuum leaks.
  - Coolant system and coolant fan for proper operation. See CHART C-12.
  - Remove carbon with top engine cleaner, follow instructions on can.
  - Check Ignition timing for excessive base advance. See Vehicle Emission Control Information Label.
- If the system is running lean, (block learn greater than 138), refer to "Diagnostic Aids" on facing page of Code 44.

DIESELING, RUN-ON

Definition: Engine continues to run after key is turned "OFF," but runs very roughly. If engine runs smoothly, check ignition switch and adjustment.

- Check injector(s) for leaking. Apply 12 volts to fuel pump test terminal to turn "ON" fuel pump and pressurize fuel system.
- Visually check injector(s) and TBI assembly for fuel leakage.

BACKFIRE

Definition: Fuel ignites in intake manifold, or in exhaust system, making a loud popping noise.

- CHECK:
  - EGR operation for being open all the time. See CHART C-7.
  - Output voltage of ignition coil(s).
  - For crossfire between spark plugs (distributor cap, spark plug wires, and proper routing of plug wires).

- For faulty spark plugs and/or plug wires or boots.
- Faulty A.I.R. check valve.
- Perform a compression check - look for sticking or leaking valves.
  - For proper valve timing.
  - Broken or worn valve train parts.

1988–90 B-CARLINE

CHART B-1
RESTRICTED EXHAUST SYSTEM CHECK
ALL ENGINES

Proper diagnosis for a restricted exhaust system is essential before any components are replaced. Either of the following procedures may be used for diagnosis, depending upon engine or tool used:

CHECK AT A. I. R. PIPE:

1. Remove the rubber hose at the exhaust manifold A.I.R. pipe check valve. Remove check valve.
2. Connect a fuel pump pressure gauge to a hose and nipple from a Propane Enrichment Device (J 26911) (see illustration).
3. Insert the nipple into the exhaust manifold A.I.R. pipe.

OR CHECK AT O₂ SENSOR:

1. Carefully remove O₂ sensor.
2. Install Borroughs exhaust backpressure tester (BT 8515 or BT 8603) or equivalent in place of O₂ sensor (see illustration).
3. After completing test described below, be sure to coat threads of O₂ sensor with anti-seize compound P/N 5613695 or equivalent prior to re-installation.

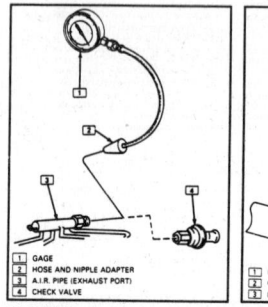
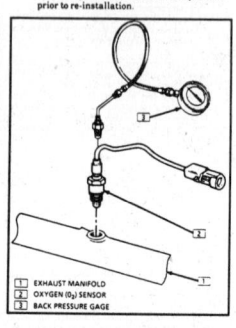

1	GAGE
2	HOSE AND NIPPLE ADAPTER
3	A.I.R. PIPE (EXHAUST PORT)
4	CHECK VALVE

1	EXHAUST MANIFOLD
2	OXYGEN (O₂) SENSOR
3	BACK PRESSURE GAGE

DIAGNOSIS:

1. With the engine idling at normal operating temperature, observe the exhaust system backpressure reading on the gage. Reading should not exceed 8.6 kPa (1.25 psi).
2. Increase engine speed to 2000 rpm and observe gage. Reading should not exceed 20.7 kPa (3 psi).
3. If the backpressure at either speed exceeds specification, a restricted exhaust system is indicated.
4. Inspect the entire exhaust system for a collapsed pipe, heat distress, or possible internal muffler failure.
5. If there are no obvious reasons for the excessive backpressure, the catalytic converter is suspected to be restricted and should be replaced using current recommended procedures.

1988–90 B-CARLINE

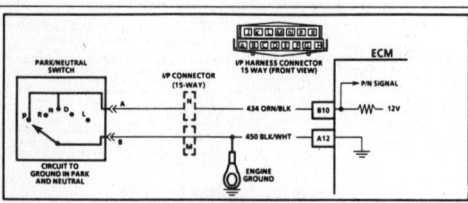

CHART C-1A
PARK/NEUTRAL SWITCH DIAGNOSIS
(AUTO TRANSMISSION ONLY)
4.3L, 5.0L & 5.7L "B" CARLINE (TBI)

Circuit Description:

The park/neutral switch contacts are a part of the neutral start switch and are closed to ground in park or neutral, and open in drive ranges.

The ECM supplies ignition voltage through a current limiting resistor to CKT 434 and senses a closed switch, when the voltage on CKT 434 drops to less than one volt.

The ECM uses the P/N signal as one of the inputs to control:

Idle air control
VSS diagnostics

Test Description: Numbers below refer to circled numbers on the diagnostic chart.

1. Checks for a closed switch to ground in park position. Different makes of "Scan" tools will read P/N differently. Refer to tool operations manual for type of display used.
2. Checks for an open switch in drive range.

3. Be sure "Scan" indicates drive, even while wiggling shifter to test for an intermittent or misadjusted switch in drive range.

1988–90 B-CARLINE

CHART C-1A
PARK/NEUTRAL SWITCH DIAGNOSIS
(AUTO TRANSMISSION ONLY)
4.3L, 5.0L & 5.7L "B" CARLINE (TBI)

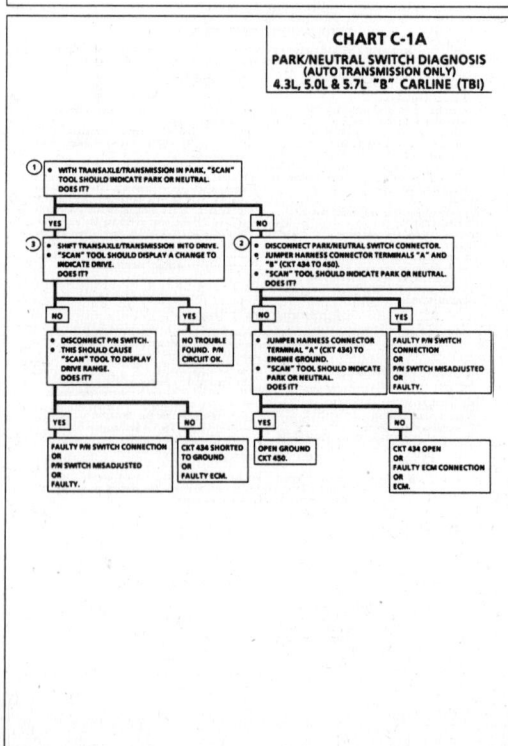

1988–90 B-CARLINE

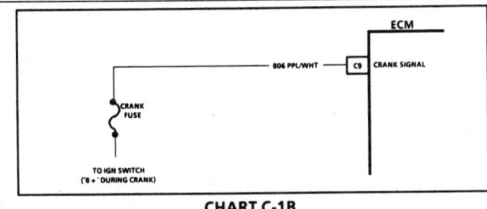

CHART C-1B
CRANK SIGNAL
4.3L, 5.0L & 5.7L "B" CARLINE (TBI)

Circuit Description:
Crank signal is a 12 volts signal to the ECM during cranking to allow enrichment and cancel diagnostics until the engine is running.

Test Description: Numbers below refer to circled numbers on the diagnostic chart.
1. Checks for normal (cranking) voltage to terminal "C9" of ECM. Test light should be "ON" during cranking.

2. Checks to determine if source of blown fuse was a faulty ECM.

1988–90 B-CARLINE

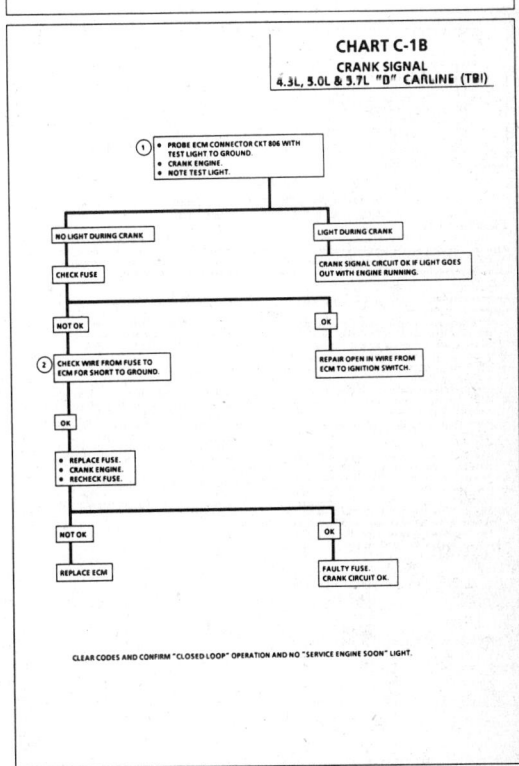

1988–90 B-CARLINE

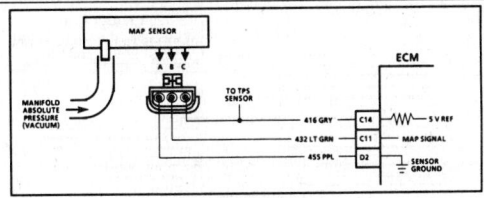

CHART C-1D
MAP OUTPUT CHECK
4.3L, 5.0L & 5.7L "B" CARLINE (TBI)

Circuit Description:
The Manifold Absolute Pressure sensor (MAP) measures manifold pressure (vacuum) and sends that signal to the ECM. The ECM uses this information for fuel and spark control.

Test Description: Numbers below refer to circled numbers on the diagnostic chart.
1. Checks MAP sensor output voltage to the ECM. This voltage, without engine running, represents a barometer reading to the ECM.
2. Applying 34 kPa (10" Hg) vacuum to the MAP sensor should cause the voltage to be 1.2 volts less than the voltage at Step 1. Upon applying vacuum to the sensor, the change in voltage should be instantaneous. A slow voltage change indicates a faulty sensor.

3. Check vacuum hose to sensor for leaking or restriction. Be sure no other vacuum devices are connected to the MAP hose.
With the ignition "ON" and the engine stopped, the manifold pressure is equal to atmospheric pressure and the signal voltage will be high. This information is used by the ECM as an indication of vehicle altitude and is referred to as BARO. Comparison of this BARO reading with a known good vehicle with the same sensor is a good way to check accuracy of a "suspect" sensor. Reading should be the same ± .4 volt.

1988–90 B-CARLINE

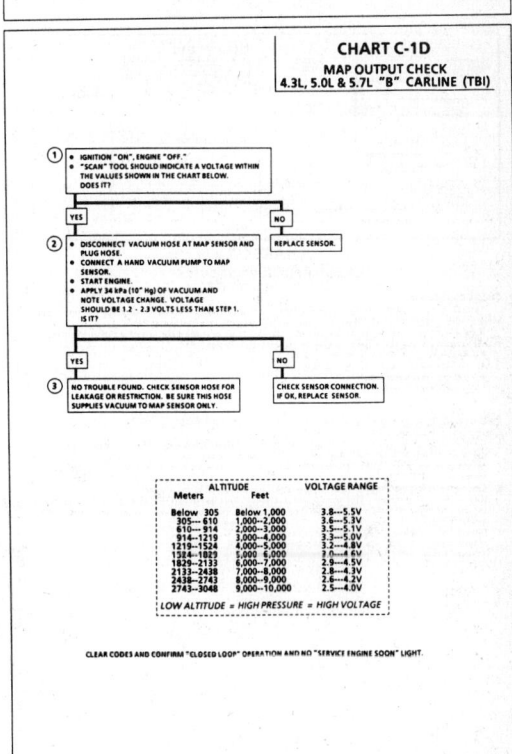

1988–90 B-CARLINE

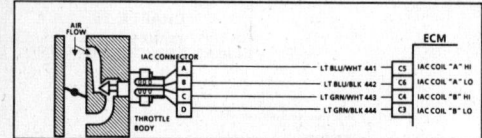

CHART C-2C
IDLE AIR CONTROL (IAC) VALVE CHECK
4.3L, 5.0L & 5.7L "B" CARLINE (TBI)

Circuit Description:
The ECM controls idle rpm with the IAC valve. To increase idle rpm, the ECM moves the IAC valve out, allowing more air to pass by the throttle plate. To decrease rpm, it moves the IAC valve in, reducing air flow by the throttle plate. A "Scan" tool will read the ECM commands to the IAC valve in counts. The higher the counts, the more air allowed (higher idle). The lower the counts, the less air allowed (lower idle).

Test Description: Numbers below refer to circled numbers on the diagnostic chart.
1. Continue with test, even if engine will not idle. If idle is to low, "Scan" will display 80 or more counts, or steps. If idle is high, it will display "0" counts. Occasionally, an erratic or unstable idle may occur. Engine speed may vary 200 rpm, or more, up and down. Disconnect IAC. If the condition is unchanged, the IAC is not at fault.
2. When the engine was stopped, the IAC valve retracted (more air) to a fixed "park" position for increased air flow and idle speed during the next engine start. A "Scan" will display 100 or more counts. When performing this test, immediately note rpm on start up, because, on a warm engine, the rpm will decrease rapidly.
3. Be sure to disconnect the IAC valve prior to this test. The test light will confirm the ECM signals by a steady or flashing light on all circuits.
4. There is a remote possibility that one of the circuits is shorted to voltage, which would have been indicated by a steady light on all circuits. Disconnect ECM and turn the ignition "ON" and probe terminals to check for this condition.

IAC VALVE RESET PROCEDURE

- Ignition "OFF" for 10 seconds
- Start and run engine for 5 seconds
- Ignition "OFF" for 10 seconds

Diagnostic Aids:

A rough or unstable idle may be caused by a system problem that cannot be overcome by the IAC. "Scan" counts will be above 60 counts, if engine speed is too low, or "0" counts, if engine speed is too high.
- System too lean (High Air/Fuel Ratio)
  Idle speed may be too high or too low. Engine speed may vary up and down, disconnecting IAC does not help.
- System too rich (Low Air/Fuel Ratio)
  Idle speed too low. "Scan" counts usually above 80. System obviously rich and may exhibit black exhaust smoke.
  Check:
  - High fuel pressure
  - Injector leaking or sticking
- If above are all OK, refer to "Rough, Unstable, Incorrect Idle or Stalling", in "Symptoms", Section "B".

1988–90 B-CARLINE

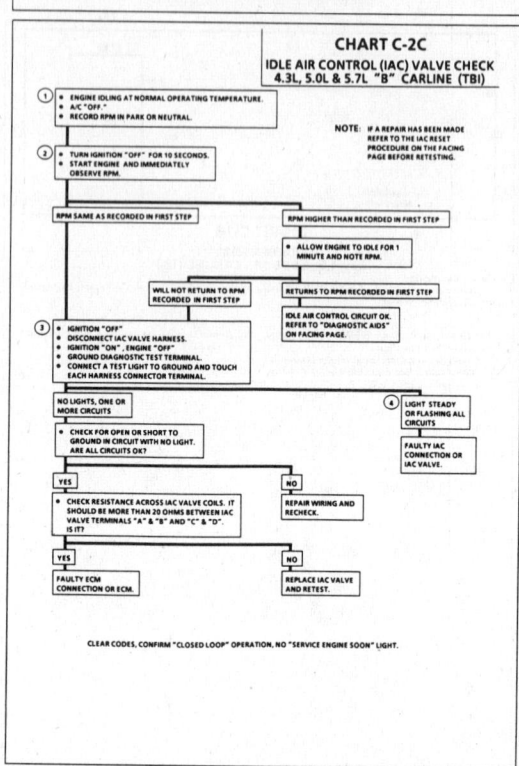

CHART C-2C
IDLE AIR CONTROL (IAC) VALVE CHECK
4.3L, 5.0L & 5.7L "B" CARLINE (TBI)

NOTE: IF A REPAIR HAS BEEN MADE REFER TO THE IAC RESET PROCEDURE ON THE FACING PAGE BEFORE RETESTING.

CLEAR CODES, CONFIRM "CLOSED LOOP" OPERATION, NO "SERVICE ENGINE SOON" LIGHT.

1988–90 B-CARLINE

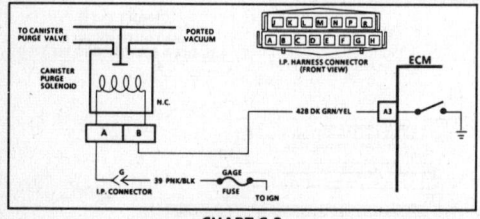

CHART C-3
CANISTER PURGE VALVE CHECK
4.3L "B" CARLINE (TBI)

Circuit Description:
Canister purge is controlled by a solenoid that allows ported manifold vacuum to purge the fuel vapor canister when energized. The ECM supplies a ground to energize the solenoid (purge "ON").
If the diagnostic "test" terminal is grounded with the engine stopped or the following conditions are met with the engine running the canister purge solenoid is energized (purge "ON").
- Engine run time after start more than 1 min.
- Coolant temperature above 80°C (176°F).
- Vehicle speed above 5 mph.
- Not at closed throttle.

Test Description: Numbers below refer to circled numbers on the diagnostic chart.
1. Checks to see if the canister purge solenoid is opened or closed. The canister purge solenoid is, normally, de-energized in this step; so it should be open.
2. Completes functional check, by grounding the diagnostic "test" terminal. This should energize the canister purge solenoid and allow the vacuum to drop (purge "ON").
3. Checks for a complete circuit. Normally, there is battery voltage on CKT 39, and the ECM provides a ground on CKT 428.

Diagnostic Aids:

Normal operation of the canister purge solenoid is described as follows:
With the ignition "ON," engine "OFF," diagnostic "test" terminal ungrounded, the canister purge solenoid will be de-energized.
With ignition "ON," engine "OFF," diagnostic "test" terminal grounded, the canister purge solenoid will be energized.

1988–90 B-CARLINE

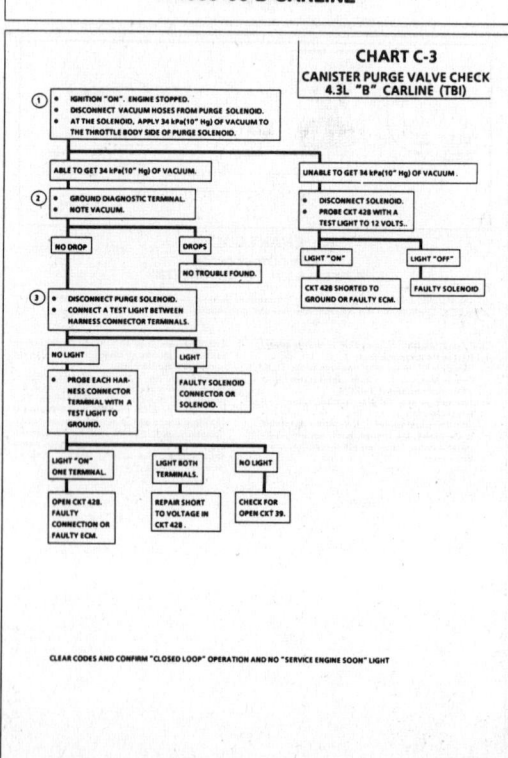

CHART C-3
CANISTER PURGE VALVE CHECK
4.3L "B" CARLINE (TBI)

CLEAR CODES AND CONFIRM "CLOSED LOOP" OPERATION AND NO "SERVICE ENGINE SOON" LIGHT

1988–90 B-CARLINE

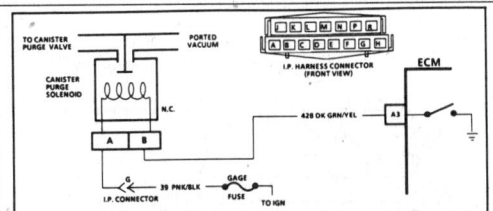

CHART C-3
CANISTER PURGE VALVE CHECK
5.0L (VIN E) & 5.7L (VIN 7) "B" CARLINE (TBI)

Circuit Description:

Canister purge is controlled by a solenoid that allows ported manifold vacuum to purge the fuel vapor canister when energized. The ECM supplies a ground to energize the solenoid (purge "ON"). If the following conditions are met with the engine running, the canister purge solenoid is energized (purge "ON").

- Engine run time after start more than 1 minute
- Coolant temperature above 80°C (176°F)
- Vehicle speed above 5 mph
- Not closed throttle

Test Description: Numbers below refer to circled numbers on the diagnostic chart.
1. Checks to see if the canister purge solenoid is opened or closed. The canister purge solenoid is energized in this step so it should be open.
2. Completes functional check, by grounding the diagnostic "test" terminal. This should, normally, de-energize the canister purge solenoid and allow the vacuum to drop (purge "ON").
3. Checks for a complete circuit. Normally, there is battery voltage on CKT 39, and the ECM provides a ground on CKT 428. A shorted canister purge solenoid could cause an open circuit in the ECM.

Diagnostic Aids:

Normal operation of the canister purge solenoid is described as follows:
With ignition "ON," engine "OFF," diagnostic "test" terminal ungrounded, the canister purge solenoid will be energized.
With ignition "ON," engine "OFF," diagnostic "test" terminal grounded, the canister purge solenoid will be de-energized.

1988–90 B-CARLINE

CHART C-3
CANISTER PURGE VALVE CHECK
5.0L (VIN E) & 5.7L (VIN 7) "B" CARLINE (TBI)

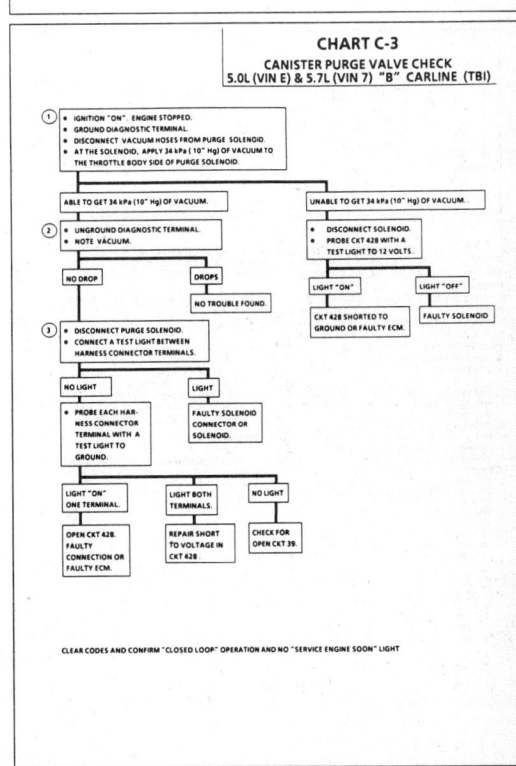

1988–90 B-CARLINE

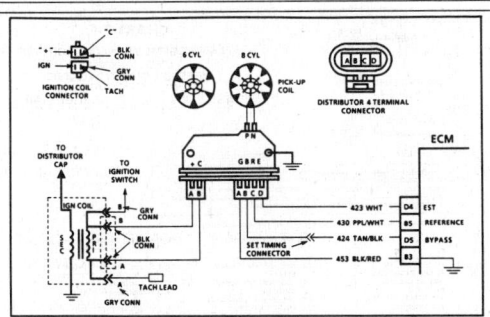

CHART C-4B
IGNITION SYSTEM CHECK
(REMOTE COIL/SEALED MODULE CONNECTOR DISTRIBUTOR)
4.3L, 5.0L & 5.7L "B" CARLINE (TBI)

Test Description: Numbers below refer to circled numbers on the diagnostic chart.
1. Two wires are checked, to ensure that an open is not present in a spark plug wire.
1A. If spark occurs with EST connector disconnected, pick-up coil output is too low for EST operation.
2. A spark indicates the problem must be the distributor cap or rotor.
3. Normally, there should be battery voltage at the "C" and "+" terminals. Low voltage would indicate an open or a high resistance circuit from the distributor to the coil or ignition switch. If "C" terminal voltage was low, but "+" terminal voltage is 10 volts or more, circuit from "C" terminal to ignition coil or ignition coil primary winding is open.
4. Checks for a shorted module or grounded circuit from the ignition coil to the module. The distributor module should be turned "OFF," so normal voltage should be about 12 volts.
   If the module is turned "ON," the voltage would be low, but above 1 volt. This could cause the ignition coil to fail from excessive heat.
   With an open ignition coil primary winding, a small amount of voltage will leak through the module from the "battery" to the "tach" terminal.
5. Applying a voltage (1.5 to 8 volts) to module terminal "P" should turn the module "ON" and the "tach" terminal voltage should drop to about 7-9 volts. This test will determine whether the module or coil is faulty or if the pick-up coil is not generating the proper signal to turn the module "ON". This test can be performed by using a DC battery with a rating of 1.5 to 8 volts. The use of the test light is mainly to allow the "P" terminal to be probed more easily. Some digital multi-meters can also be used to trigger the module by selecting ohms, usually the diode position. In this position the meter may have a voltage across it's terminals which can be used to trigger the module. The voltage in the ohm's position can be checked by using a second meter or by checking the manufacture's specification of the tool being used.
6. This should turn "OFF" the module and cause a spark. If no spark occurs, the fault is most likely in the ignition coil because most module problems would have been found before this point in the procedure. A module tester could determine which is at fault.

CHART C-4B
IGNITION SYSTEM CHECK
(REMOTE COIL/SEALED MODULE CONNECTOR DISTRIBUTOR)
4.3L, 5.0L & 5.7L "B" CARLINE (TBI)

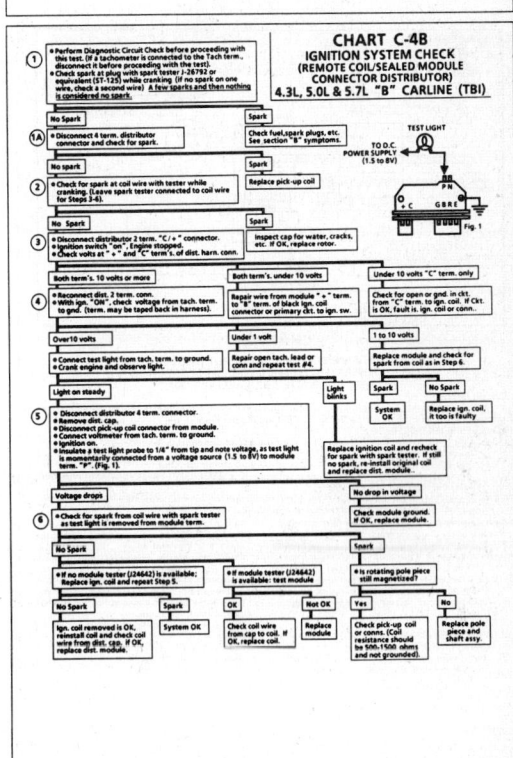

1988–90 B-CARLINE

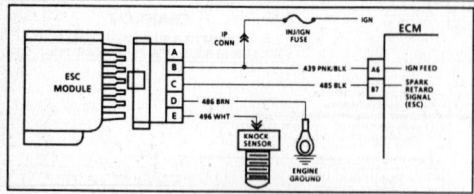

CHART C-5
ELECTRONIC SPARK CONTROL (ESC) SYSTEM CHECK
(ENGINE KNOCK, POOR PERFORMANCE, OR POOR ECONOMY)
4.3L, 5.0L & 5.7L "B" CARLINE (TBI)

Circuit Description:
Electronic spark control is accomplished with a module that sends a voltage signal to the ECM. When the knock sensor detects engine knock, the voltage from the ESC module to the ECM is shut "OFF" and this signals the ECM to retard timing, if engine rpm is over about 900.

Test Description: Numbers below refer to circled numbers on the diagnostic chart.
1. If a Code 43 is not set, but a knock signal is indicated while running at 1500 rpm, listen for an internal engine noise. Under a no load condition, there should not be any detonation, and if knock is indicated, an internal engine problem may exist.
2. Usually a knock signal can be generated by tapping on the exhaust manifold. This test can also be performed at idle. Test number 1 was run at 1500 rpm, to determine if a constant knock signal was present, which would affect engine performance.
3. This tests whether the knock signal is due to the sensor, a basic engine problem, or the ESC module.

4. If the module ground circuit is faulty, the ESC module will not function correctly. The test light should light indicating the ground circuit is OK.
5. Contacting CKT 496, with a test light to 12 volts, should generate a knock signal to determine whether the knock sensor is faulty, or the ESC module can't recognize a knock signal.

Diagnostic Aids:
If the ESC system checks OK, but detonation is the complaint, refer to "Detonation/Spark Knock" symptom in Section "B".

1988–90 B-CARLINE

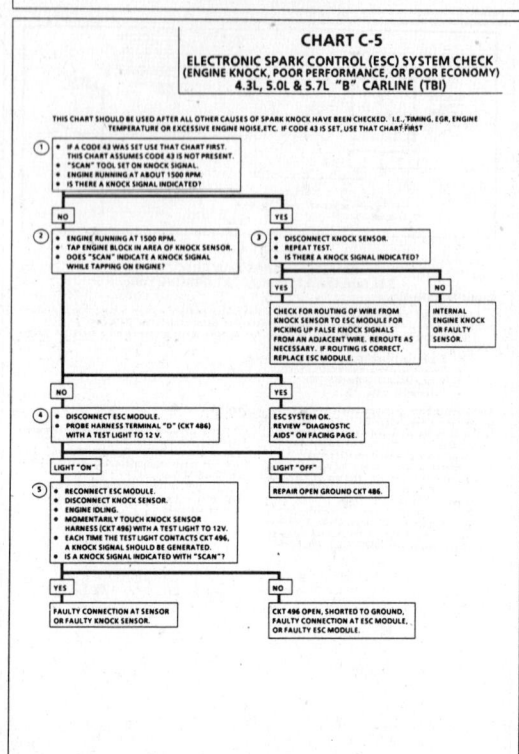

1988–90 B-CARLINE

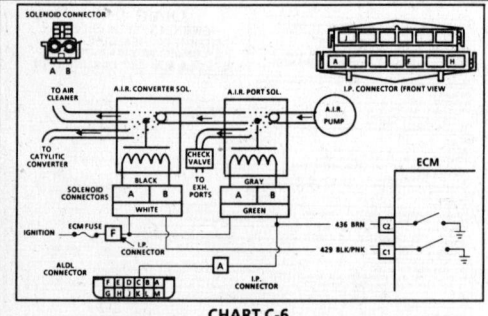

CHART C-6
AIR MANAGEMENT CHECK - CONTROL (PEDES) VALVE
(PRESSURE OPERATED ELECTRIC DIVERT/ELECTRIC SWITCHING)
4.3L, 5.0L & 5.7L "B" CARLINE (TBI)

Circuit Description:
Air management is controlled by a port valve and a converter valve, each with an ECM controlled electrical solenoid. When the solenoid is grounded by the ECM, A.I.R. pump pressure will activate the valve and allow pump air to be directed as follows:
- **Cold Mode** - The port (switch) solenoid is energized which in turn opens the port valve and allows flow to the exhaust ports.
- **Warm Mode** - The port (switch) solenoid is de-energized and the converter (divert) solenoid energized which closes the port valve and keeps the converter valve seated, thus forcing flow past the converter valve and to the catalytic converter.
- **Divert Mode** - Both solenoids are de-energized, which opens the converter valve, allowing air to take the path of least resistance, i.e., out the divert/relief tube to atmosphere.

Test Description: Numbers below refer to circled numbers on the diagnostic chart.
1. This is a system functional check. Air is directed to ports during "Open Loop" and all engines start in "Open Loop" even on a warm engine. Since the air to the ports time is very short on warm engine, prepare to observe port air lock when engine start up. On some engines, this can be done by squeezing a hose. On others, steel pipes have to be disconnected.
2. This should normally set a Code 22. When any code is set, the ECM opens the ground to the A.I.R. control valve and allows air to divert. This checks for ECM response to a fault. A ground in the A.I.R. control valve circuit to the ECM would

3. This checks for a grounded circuit to the ECM. Test light "OFF" is normal and would indicate the circuit is not grounded.
4. Checks for an open in the solenoid control circuits. Grounding the diagnostic "test" terminal should ground both solenoid circuits. Normally, the test light should be "ON" which indicates the problem is not in the ECM or wiring but at the solenoid connections or valve itself.
5. Checks for a grounded port (switching valve) solenoid circuit.
Test light "OFF" would indicate the circuit is normal and fault is in the valve.

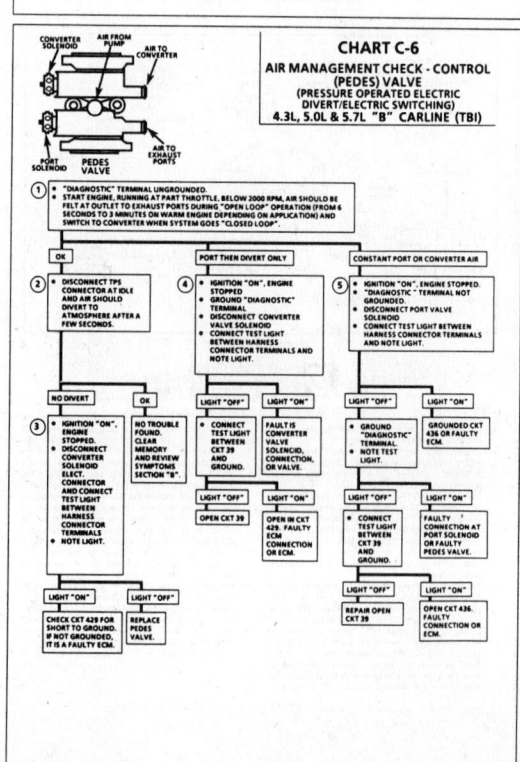

1988–90 B-CARLINE

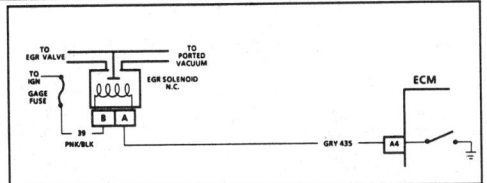

CHART C-7
EXHAUST GAS RECIRULATION (EGR) CHECK
4.3L, 5.0L & 5.7L "B" CARLINE (TBI)

Circuit Description:

The ECM operates a solenoid to control the Exhaust Gas Recirculation (EGR) valve. This solenoid is normally closed. By providing a ground path, the ECM energizes the solenoid which then allows vacuum to pass to the EGR valve. The ECM control of the EGR is based on the following inputs:
- Engine coolant temperature above 25°C (77°F).
- TPS off idle
- MAP

If Code 24 is stored, use that chart first.

Test Description: Numbers below refer to circled numbers on the diagnostic chart.
1. By grounding the diagnostic terminal, the EGR solenoid should be energized and allow vacuum to be applied to the EGR valve and the vacuum should hold.
2. When the diagnostic terminal is ungrounded, the vacuum to the EGR valve should bleed off through a vent in the EGR solenoid and the valve should close. The gage may or may not bleed off but this does not indicate a problem.
3. This test will determine if the electrical control part of the system is at fault or if the connector or solenoid is at fault.
4. This system uses a negative backpressure valve which should hold vacuum with engine "OFF."
5. When engine is started, exhaust backpressure should cause vacuum to bleed off and valve should fully close.

Diagnostic Aids:

Vacuum lines should be thoroughly checked for internal restrictions. The ECM uses the MAP sensor for checking EGR operation. If there is a question of MAP sensor accuracy use CHART C-1D MAP output check in Section "C".

If no problems are found refer to "Intermittents" in Section "B".

1988–90 B-CARLINE

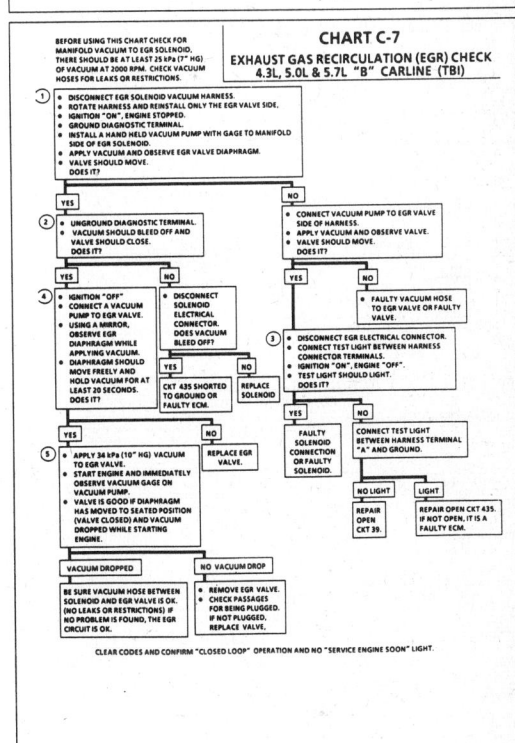

1988–90 B-CARLINE

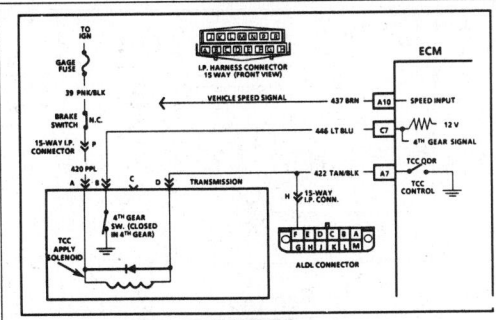

CHART C-8
(Page 1 of 2)
TORQUE CONVERTER CLUTCH (TCC) SYSTEM
(ELECTRICAL DIAGNOSIS)
4.3L, 5.0L & 5.7L "B" CARLINE (TBI)

Circuit Description:

The purpose of the automatic transmission torque converter clutch is to eliminate the power loss of the torque converter, when the vehicle is in a cruise condition. This allows the convenience of the automatic transmission and the fuel economy of a manual transmission.

Fused battery ignition is supplied to the TCC solenoid through the brake switch. The ECM will engage TCC by grounding CKT 422 to energize the TCC solenoid.

TCC will engage when:
- Vehicle speed above 24 mph
- Engine at normal operating temperature above 70°C (156°F)
- Throttle position sensor output not changing, indicating a steady road speed
- Brake switch closed

Test Description: Numbers below refer to circled numbers on the diagnostic chart.
1. Confirm 12 volts supply as well as continuity of TCC circuit.
2. Grounding the diagnostic "test" terminal with engine "OFF", should energize the capability of the ECM to control the TCC solenoid.
3. TCC solenoid coil resistance must measure more than 20 ohms. Less resistance will cause early failure of the ECM "driver". Using an ohmmeter, check the solenoid coil resistance of all ECM controlled solenoids and relays before installing a replacement ECM. Replace any solenoid or relay that measures less than 20 ohms.

Diagnostic Aids:

An engine coolant thermostat that is stuck open or opens at too low a temperature, may result in an inoperative TCC.

1988–90 B-CARLINE

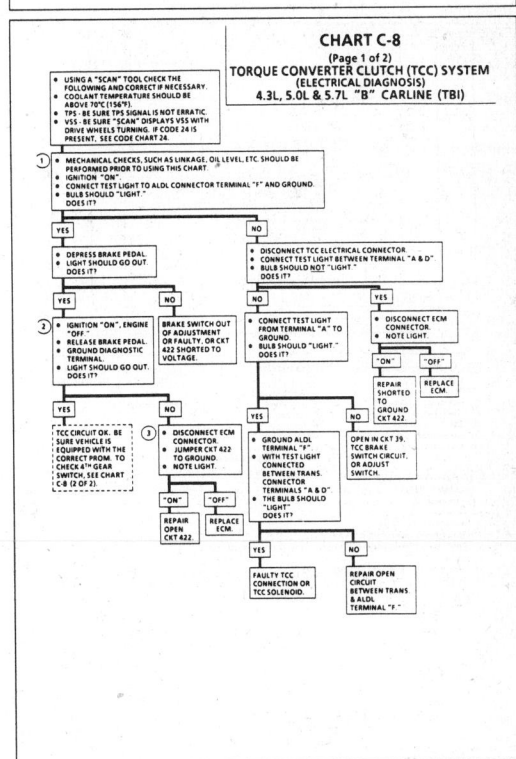

CHART C-8
(Page 1 of 2)
TORQUE CONVERTER CLUTCH (TCC) SYSTEM
(ELECTRICAL DIAGNOSIS)
4.3L, 5.0L & 5.7L "B" CARLINE (TBI)

1988–90 B-CARLINE

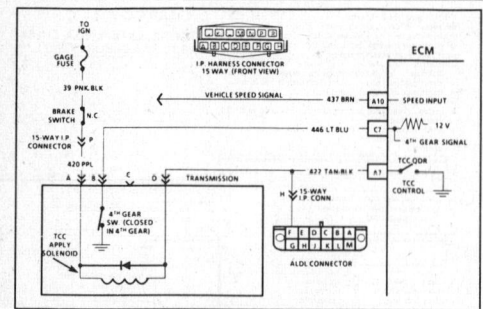

CHART C-8
(Page 2 of 2)
TORQUE CONVERTER CLUTCH (TCC) SYSTEM
(4TH GEAR SWITCH DIAGNOSIS)
4.3L, 5.0L & 5.7L "B" CARLINE (TBI)

Circuit Description:

A 4th gear switch (mounted in the transmission) closes when the transmission shifts into 4th gear, and this switch is used by the ECM to modify TCC lock and unlock points, when in a 4-3 downshift maneuver.

Test Description: Numbers below refer to circled numbers on the diagnostic chart.

1. Unless the switch or CKT 446 is open the "Scan" should display "NO", indicating the transmission is not in 4th gear. The 4th gear switch should only be closed while in 4th gear.
2. Disconnecting the TCC connector simulates an open switch to determine if CKT 446 is shorted to ground or the problem is in the transmission.
3. Checks the operation of the 4th gear switch. When the transmission shifts into 4th gear the switch should close and the "Scan" should display "YES".

4. This step determines if the ECM and wiring are OK. Grounding CKT 446 should cause the "Scan" to display "YES", indicating the transmission is in 4th gear.

Diagnostic Aids:

A road test may be necessary to verify the customer complaint. If the "Scan" indicates TCC is turning "ON" and "OFF" erratically, check the state of the 4th gear switch to be sure it is not changing states under a steady throttle position. If the switch is changing states, check connections and wire routing carefully. Also if the 4th gear switch is always open the TCC may engage as soon as sufficient oil pressure is reached.

1988–90 B-CARLINE

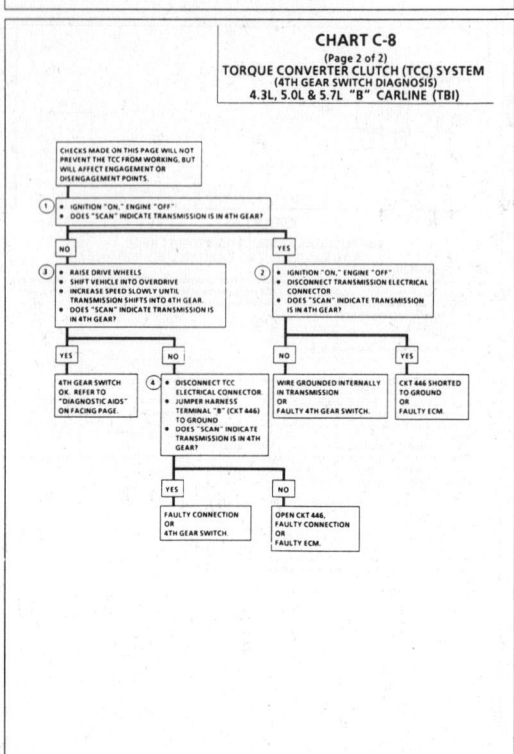

2.5L (VIN E, M-SERIES) COMPONENT LOCATIONS LIGHT TRUCKS AND VANS

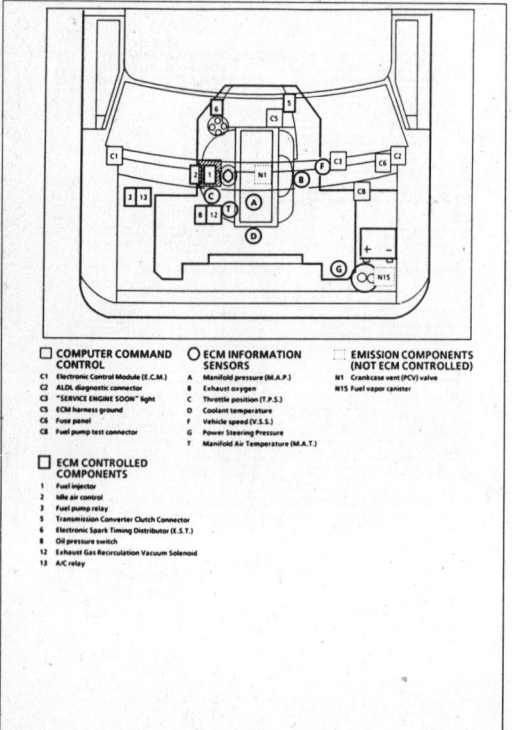

☐ **COMPUTER COMMAND CONTROL**
C1 Electronic Control Module (E.C.M.)
C2 ALDL diagnostic connector
C3 "SERVICE ENGINE SOON" light
C5 ECM harness ground
C6 Fuse panel
C8 Fuel pump test connector

☐ **ECM CONTROLLED COMPONENTS**
1 Fuel injector
2 Idle air control
3 Fuel pump relay
5 Transmission Converter Clutch Connector
6 Electronic Spark Timing Distributor (E.S.T.)
8 Oil pressure switch
12 Exhaust Gas Recirculation Vacuum Solenoid
13 A/C relay

○ **ECM INFORMATION SENSORS**
A Manifold pressure (M.A.P.)
B Exhaust oxygen
C Throttle position (T.P.S.)
D Coolant temperature
F Vehicle speed (V.S.S.)
G Power Steering Pressure
T Manifold Air Temperature (M.A.T.)

▦ **EMISSION COMPONENTS (NOT ECM CONTROLLED)**
N1 Crankcase vent (PCV) valve
N15 Fuel vapor canister

2.5L (VIN E, ST-SERIES) COMPONENT LOCATIONS LIGHT TRUCKS AND VANS

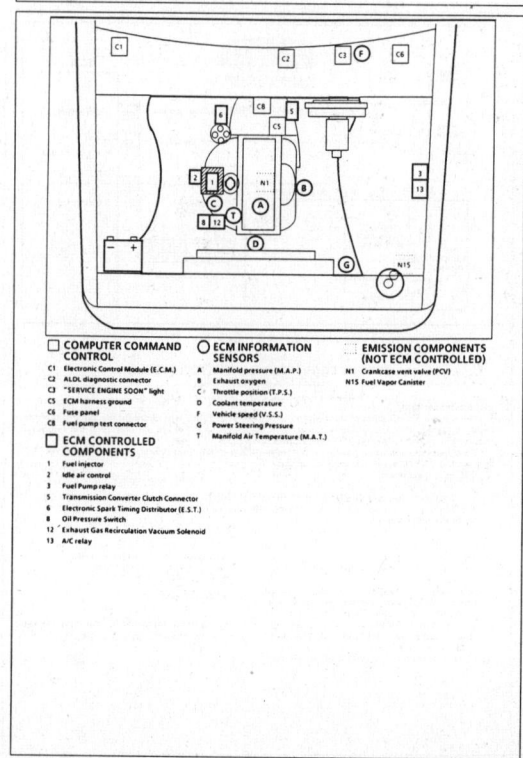

☐ **COMPUTER COMMAND CONTROL**
C1 Electronic Control Module (E.C.M.)
C2 ALDL diagnostic connector
C3 "SERVICE ENGINE SOON" light
C5 ECM harness ground
C6 Fuse panel
C8 Fuel pump test connector

☐ **ECM CONTROLLED COMPONENTS**
1 Fuel injector
2 Idle air control
3 Fuel Pump relay
5 Transmission Converter Clutch Connector
6 Electronic Spark Timing Distributor (E.S.T.)
8 Oil Pressure Switch
12 Exhaust Gas Recirculation Vacuum Solenoid
13 A/C relay

○ **ECM INFORMATION SENSORS**
A Manifold pressure (M.A.P.)
B Exhaust oxygen
C Throttle position (T.P.S.)
D Coolant temperature
F Vehicle speed (V.S.S.)
G Power Steering Pressure
T Manifold Air Temperature (M.A.T.)

▦ **EMISSION COMPONENTS (NOT ECM CONTROLLED)**
N1 Crankcase vent valve (PCV)
N15 Fuel Vapor Canister

2.5L (VIN E) ECM WIRING DIAGRAM

2.5L (VIN E) ECM WIRING DIAGRAM (CONT.)

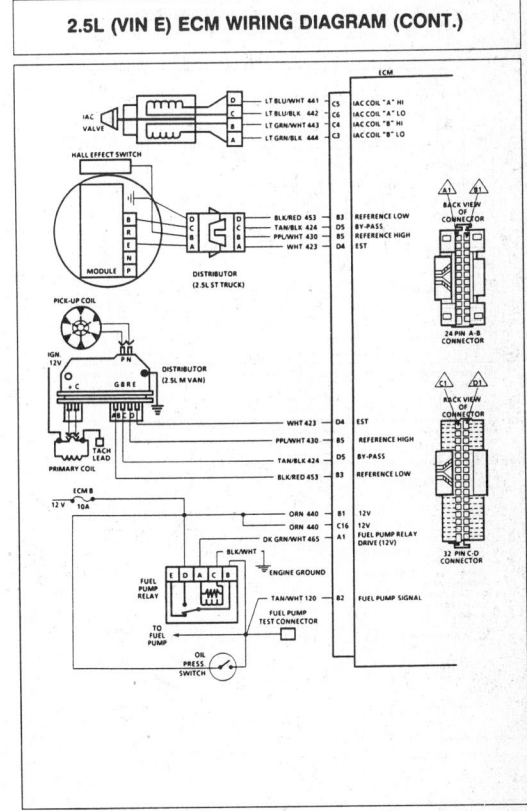

2.5L (VIN E) ECM WIRING DIAGRAM (CONT.)

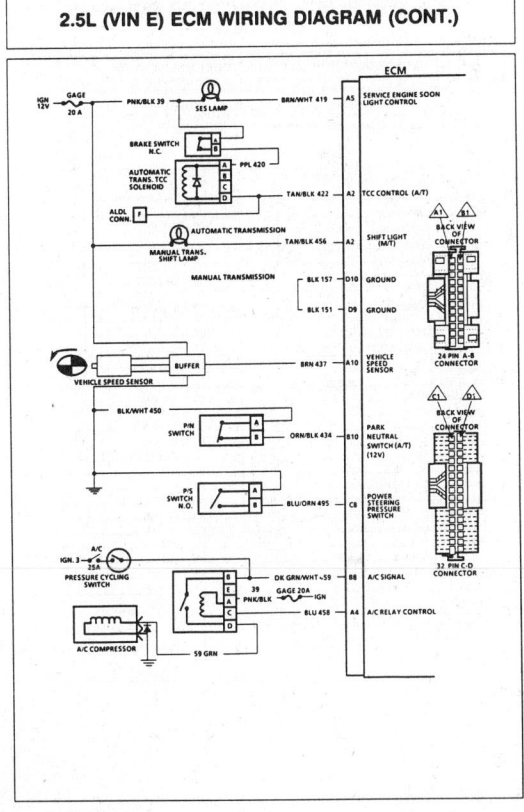

2.5L (VIN E) ECM CONNECTOR TERMINAL END VIEW

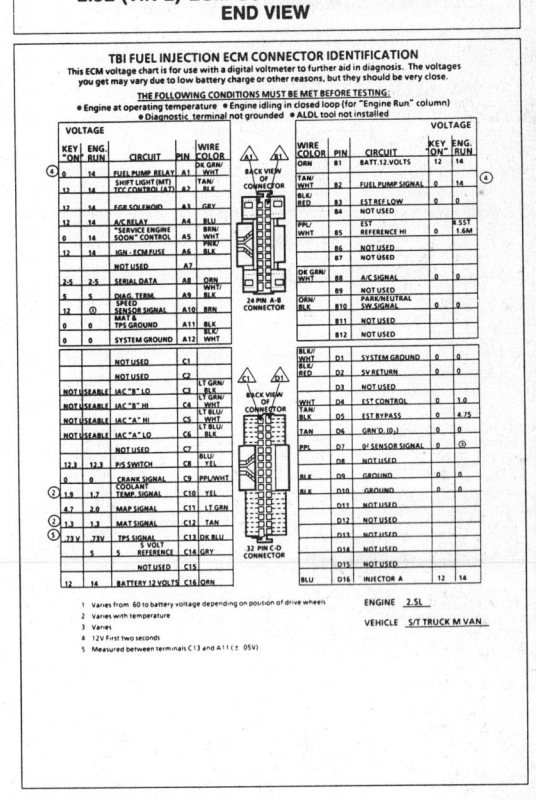

2.8L (VIN R, ST-SERIES) COMPONENT LOCATIONS — LIGHT TRUCKS AND VANS

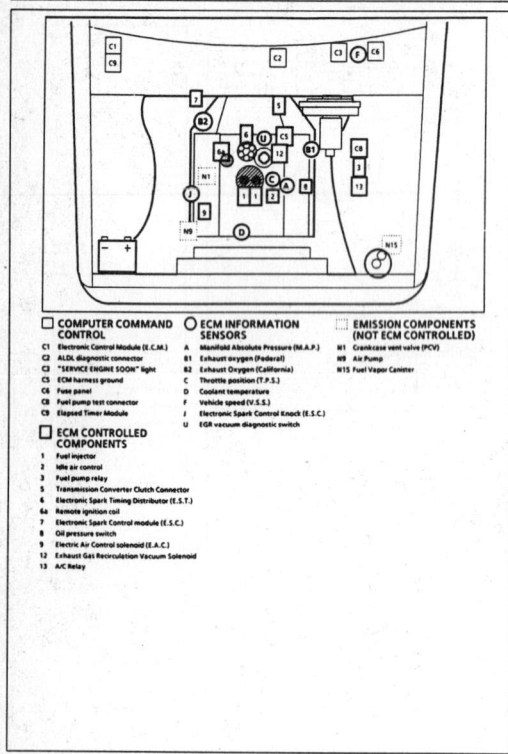

2.8L (VIN R) ECM WIRING DIAGRAM

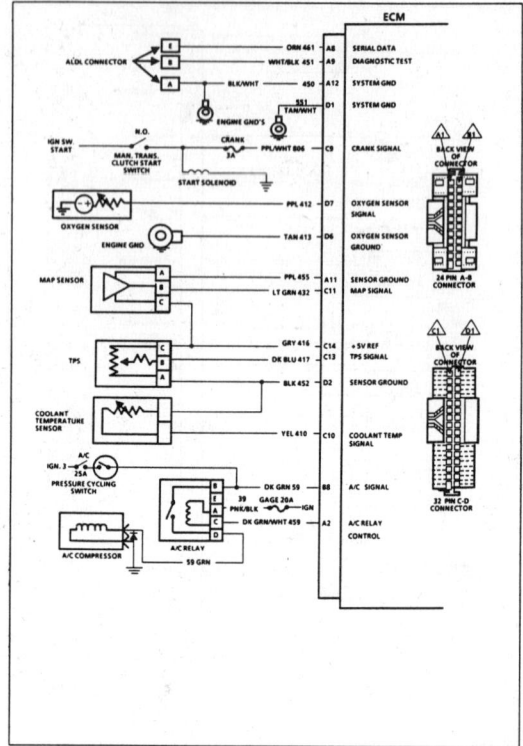

2.8L (VIN R) ECM WIRING DIAGRAM (CONT.)

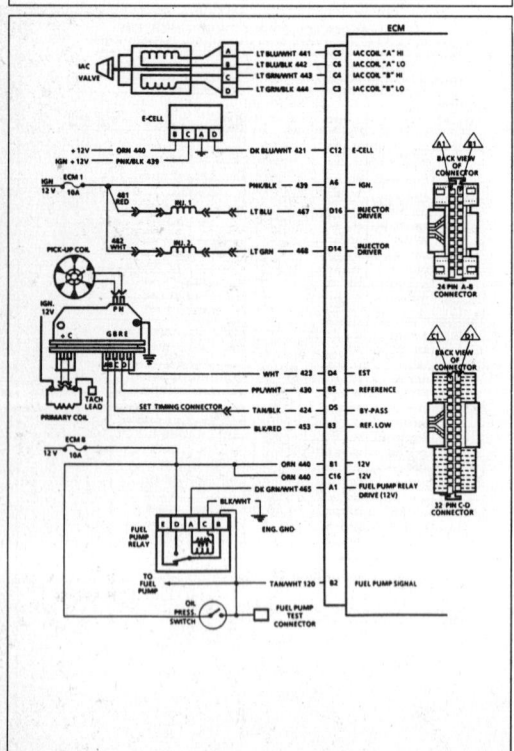

2.8L (VIN R) ECM CONNECTOR TERMINAL END VIEW

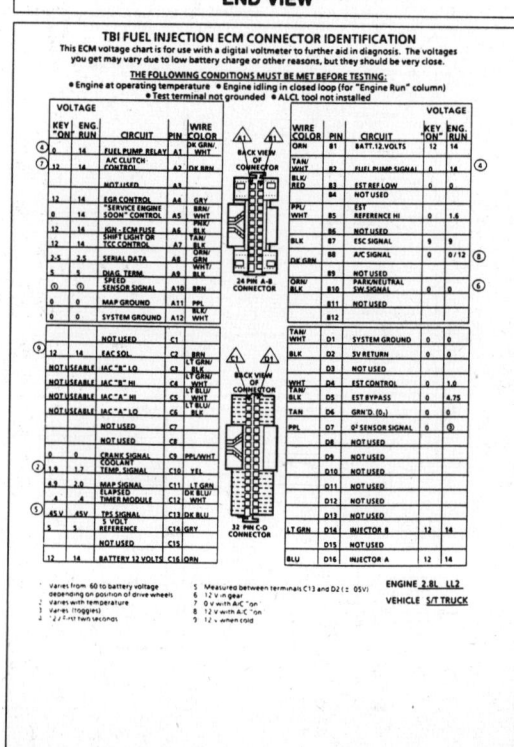

4.3L (VIN Z, ST-SERIES) COMPONENT LOCATIONS LIGHT TRUCKS AND VANS

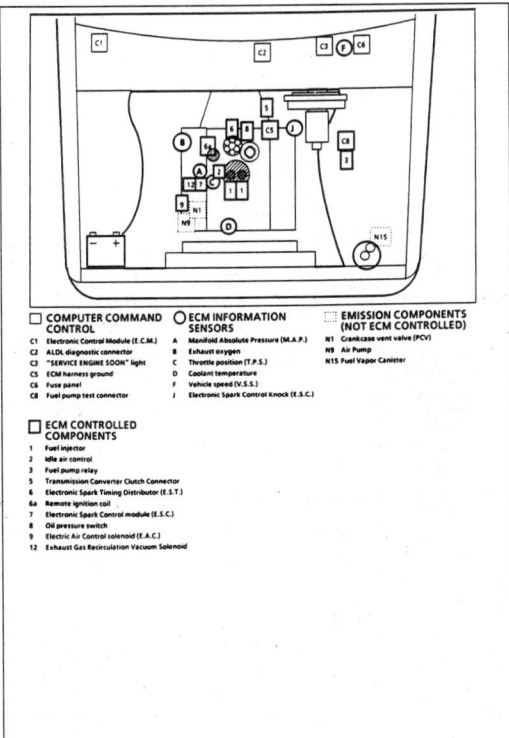

□ **COMPUTER COMMAND CONTROL**

C1   Electronic Control Module (E.C.M.)
C2   ALDL diagnostic connector
C3   "SERVICE ENGINE SOON" light
C5   ECM harness ground
C6   Fuse panel
C8   Fuel pump test connector

□ **ECM CONTROLLED COMPONENTS**

1   Fuel injector
2   Idle air control
3   Fuel pump relay
5   Transmission Converter Clutch Connector
6   Electronic Spark Timing Distributor (E.S.T.)
6a   Remote ignition coil
7   Electronic Spark Control module (E.S.C.)
8   Oil pressure switch
9   Electric Air Control solenoid (E.A.C.)
12   Exhaust Gas Recirculation Vacuum Solenoid

○ **ECM INFORMATION SENSORS**

A   Manifold Absolute Pressure (M.A.P.)
B   Exhaust oxygen
C   Throttle position (T.P.S.)
D   Coolant temperature
F   Vehicle speed (V.S.S.)
J   Electronic Spark Control Knock (E.S.C.)

▦ **EMISSION COMPONENTS (NOT ECM CONTROLLED)**

N1   Crankcase vent valve (PCV)
N9   Air Pump
N15   Fuel Vapor Canister

4.3L (VIN Z, CK-SERIES) COMPONENT LOCATIONS LIGHT TRUCKS AND VANS

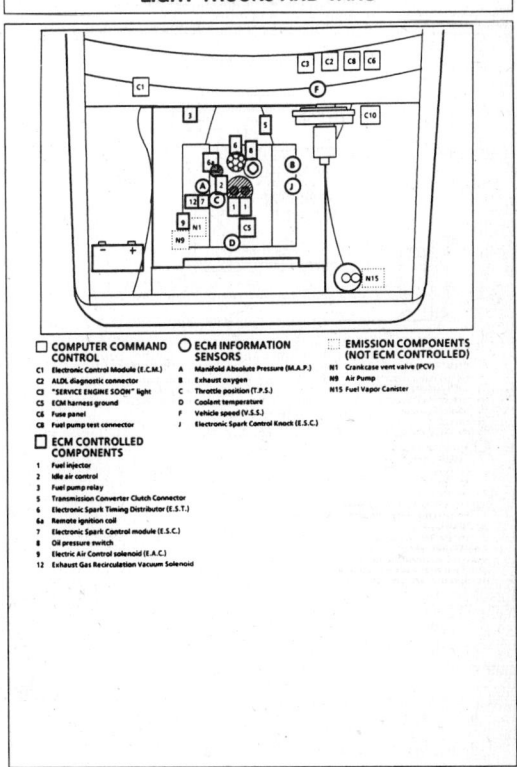

□ **COMPUTER COMMAND CONTROL**

C1   Electronic Control Module (E.C.M.)
C2   ALDL diagnostic connector
C3   "SERVICE ENGINE SOON" light
C5   ECM harness ground
C6   Fuse panel
C8   Fuel pump test connector

□ **ECM CONTROLLED COMPONENTS**

1   Fuel injector
2   Idle air control
3   Fuel pump relay
5   Transmission Converter Clutch Connector
6   Electronic Spark Timing Distributor (E.S.T.)
6a   Remote ignition coil
7   Electronic Spark Control module (E.S.C.)
8   Oil pressure switch
9   Electric Air Control solenoid (E.A.C.)
12   Exhaust Gas Recirculation Vacuum Solenoid

○ **ECM INFORMATION SENSORS**

A   Manifold Absolute Pressure (M.A.P.)
B   Exhaust oxygen
C   Throttle position (T.P.S.)
D   Coolant temperature
F   Vehicle speed (V.S.S.)
J   Electronic Spark Control Knock (E.S.C.)

▦ **EMISSION COMPONENTS (NOT ECM CONTROLLED)**

N1   Crankcase vent valve (PCV)
N9   Air Pump
N15   Fuel Vapor Canister

4.3L (VIN Z, G-SERIES) COMPONENT LOCATIONS LIGHT TRUCKS AND VANS

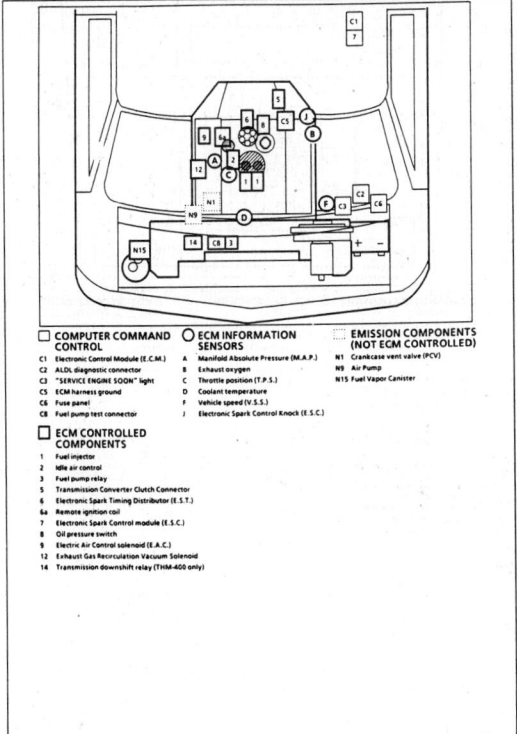

□ **COMPUTER COMMAND CONTROL**

C1   Electronic Control Module (E.C.M.)
C2   ALDL diagnostic connector
C3   "SERVICE ENGINE SOON" light
C5   ECM harness ground
C6   Fuse panel
C8   Fuel pump test connector

□ **ECM CONTROLLED COMPONENTS**

1   Fuel injector
2   Idle air control
3   Fuel pump relay
5   Transmission Converter Clutch Connector
6   Electronic Spark Timing Distributor (E.S.T.)
6a   Remote ignition coil
7   Electronic Spark Control module (E.S.C.)
8   Oil pressure switch
9   Electric Air Control solenoid (E.A.C.)
12   Exhaust Gas Recirculation Vacuum Solenoid
14   Transmission downshift relay (THM-400 only)

○ **ECM INFORMATION SENSORS**

A   Manifold Absolute Pressure (M.A.P.)
B   Exhaust oxygen
C   Throttle position (T.P.S.)
D   Coolant temperature
F   Vehicle speed (V.S.S.)
J   Electronic Spark Control Knock (E.S.C.)

▦ **EMISSION COMPONENTS (NOT ECM CONTROLLED)**

N1   Crankcase vent valve (PCV)
N9   Air Pump
N15   Fuel Vapor Canister

4.3L (VIN Z, M-SERIES) COMPONENT LOCATIONS LIGHT TRUCKS AND VANS

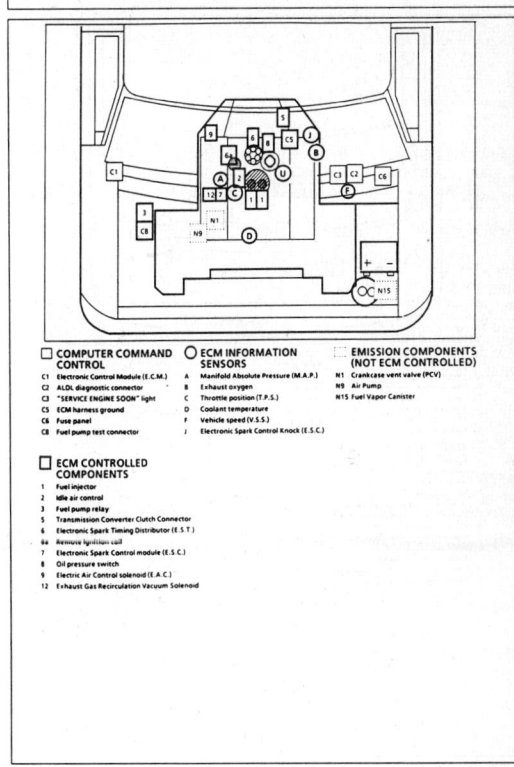

□ **COMPUTER COMMAND CONTROL**

C1   Electronic Control Module (E.C.M.)
C2   ALDL diagnostic connector
C3   "SERVICE ENGINE SOON" light
C5   ECM harness ground
C6   Fuse panel
C8   Fuel pump test connector

□ **ECM CONTROLLED COMPONENTS**

1   Fuel injector
2   Idle air control
3   Fuel pump relay
5   Transmission Converter Clutch Connector
6   Electronic Spark Timing Distributor (E.S.T.)
6a   Remote ignition coil
7   Electronic Spark Control module (E.S.C.)
8   Oil pressure switch
9   Electric Air Control solenoid (E.A.C.)
12   Exhaust Gas Recirculation Vacuum Solenoid

○ **ECM INFORMATION SENSORS**

A   Manifold Absolute Pressure (M.A.P.)
B   Exhaust oxygen
C   Throttle position (T.P.S.)
D   Coolant temperature
F   Vehicle speed (V.S.S.)
J   Electronic Spark Control Knock (E.S.C.)

▦ **EMISSION COMPONENTS (NOT ECM CONTROLLED)**

N1   Crankcase vent valve (PCV)
N9   Air Pump
N15   Fuel Vapor Canister

4.3L (VIN Z, RV-SERIES) COMPONENT LOCATIONS LIGHT TRUCKS AND VANS

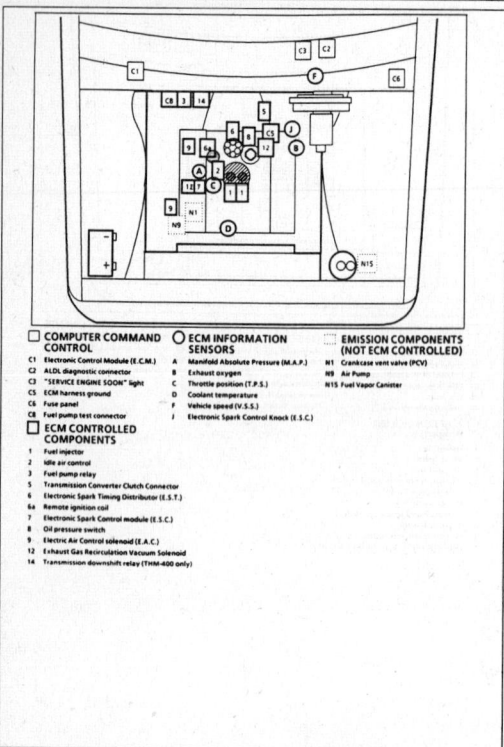

☐ COMPUTER COMMAND CONTROL

C1 Electronic Control Module (E.C.M.)
C2 ALDL diagnostic connector
C3 "SERVICE ENGINE SOON" light
C5 ECM harness ground
C6 Fuse panel
C8 Fuel pump test connector

☐ ECM CONTROLLED COMPONENTS

1   Fuel injector
2   Idle air control
3   Fuel pump relay
5   Transmission Converter Clutch Connector
6   Electronic Spark Timing Distributor (E.S.T.)
6a Remote ignition coil
7   Electronic Spark Control module (E.S.C.)
8   Oil pressure switch
9   Electric Air Control solenoid (E.A.C.)
12 Exhaust Gas Recirculation Vacuum Solenoid
14 Transmission downshift relay (THM-400 only)

○ ECM INFORMATION SENSORS

A   Manifold Absolute Pressure (M.A.P.)
B   Exhaust oxygen
C   Throttle position (T.P.S.)
D   Coolant temperature
F   Vehicle speed (V.S.S.)
J   Electronic Spark Control Knock (E.S.C.)

EMISSION COMPONENTS (NOT ECM CONTROLLED)

N1 Crankcase vent valve (PCV)
N9 Air Pump
N15 Fuel Vapor Canister

5.0L AND 5.7L (VIN H/K, CK-SERIES) COMPONENT LOCATIONS—LIGHT TRUCKS AND VANS

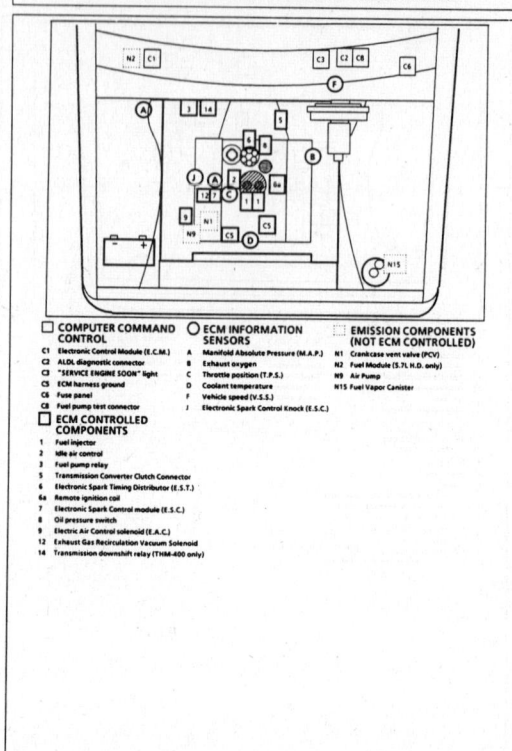

☐ COMPUTER COMMAND CONTROL

C1 Electronic Control Module (E.C.M.)
C2 ALDL diagnostic connector
C3 "SERVICE ENGINE SOON" light
C5 ECM harness ground
C6 Fuse panel
C8 Fuel pump test connector

☐ ECM CONTROLLED COMPONENTS

1   Fuel injector
2   Idle air control
3   Fuel pump relay
5   Transmission Converter Clutch Connector
6   Electronic Spark Timing Distributor (E.S.T.)
6a Remote ignition coil
7   Electronic Spark Control module (E.S.C.)
8   Oil pressure switch
9   Electric Air Control solenoid (E.A.C.)
12 Exhaust Gas Recirculation Vacuum Solenoid
14 Transmission downshift relay (THM-400 only)

○ ECM INFORMATION SENSORS

A   Manifold Absolute Pressure (M.A.P.)
B   Exhaust oxygen
C   Throttle position (T.P.S.)
D   Coolant temperature
F   Vehicle speed (V.S.S.)
J   Electronic Spark Control Knock (E.S.C.)

EMISSION COMPONENTS (NOT ECM CONTROLLED)

N1 Crankcase vent valve (PCV)
N2 Fuel Module (5.7L H.D. only)
N9 Air Pump
N15 Fuel Vapor Canister

5.0L AND 5.7L (VIN H/K, G-SERIES) COMPONENT LOCATIONS—LIGHT TRUCKS AND VANS

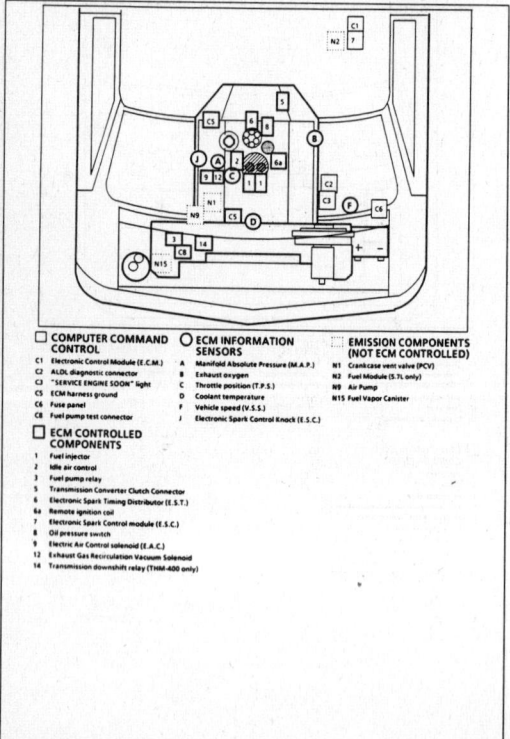

☐ COMPUTER COMMAND CONTROL

C1 Electronic Control Module (E.C.M.)
C2 ALDL diagnostic connector
C3 "SERVICE ENGINE SOON" light
C5 ECM harness ground
C6 Fuse panel
C8 Fuel pump test connector

☐ ECM CONTROLLED COMPONENTS

1   Fuel injector
2   Idle air control
3   Fuel pump relay
5   Transmission Converter Clutch Connector
6   Electronic Spark Timing Distributor (E.S.T.)
6a Remote ignition coil
7   Electronic Spark Control module (E.S.C.)
8   Oil pressure switch
9   Electric Air Control solenoid (E.A.C.)
12 Exhaust Gas Recirculation Vacuum Solenoid
14 Transmission downshift relay (THM-400 only)

○ ECM INFORMATION SENSORS

A   Manifold Absolute Pressure (M.A.P.)
B   Exhaust oxygen
C   Throttle position (T.P.S.)
D   Coolant temperature
F   Vehicle speed (V.S.S.)
J   Electronic Spark Control Knock (E.S.C.)

EMISSION COMPONENTS (NOT ECM CONTROLLED)

N1 Crankcase vent valve (PCV)
N2 Fuel Module (5.7L only)
N9 Air Pump
N15 Fuel Vapor Canister

5.0L AND 5.7L (VIN H/K, RV-SERIES) COMPONENT LOCATIONS—LIGHT TRUCKS AND VANS

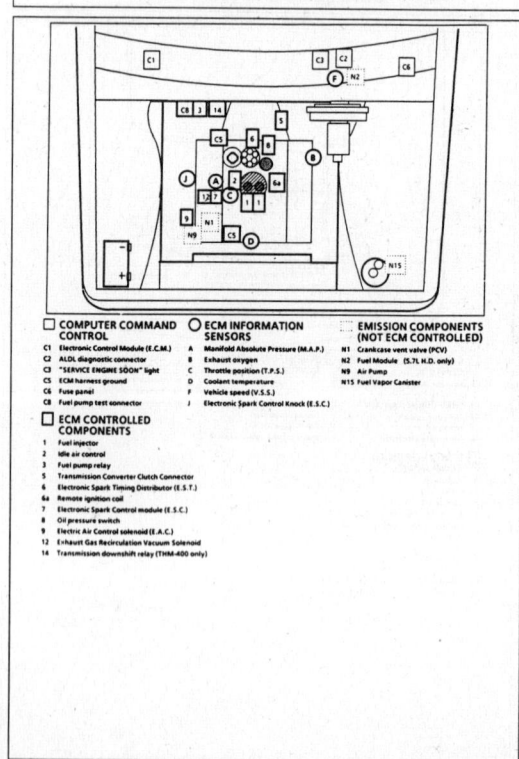

☐ COMPUTER COMMAND CONTROL

C1 Electronic Control Module (E.C.M.)
C2 ALDL diagnostic connector
C3 "SERVICE ENGINE SOON" light
C5 ECM harness ground
C6 Fuse panel
C8 Fuel pump test connector

☐ ECM CONTROLLED COMPONENTS

1   Fuel injector
2   Idle air control
3   Fuel pump relay
5   Transmission Converter Clutch Connector
6   Electronic Spark Timing Distributor (E.S.T.)
6a Remote ignition coil
7   Electronic Spark Control module (E.S.C.)
8   Oil pressure switch
9   Electric Air Control solenoid (E.A.C.)
12 Exhaust Gas Recirculation Vacuum Solenoid
14 Transmission downshift relay (THM-400 only)

○ ECM INFORMATION SENSORS

A   Manifold Absolute Pressure (M.A.P.)
B   Exhaust oxygen
C   Throttle position (T.P.S.)
D   Coolant temperature
F   Vehicle speed (V.S.S.)
J   Electronic Spark Control Knock (E.S.C.)

EMISSION COMPONENTS (NOT ECM CONTROLLED)

N1 Crankcase vent valve (PCV)
N2 Fuel Module (5.7L H.D. only)
N9 Air Pump
N15 Fuel Vapor Canister

5.7L (VIN K, P-SERIES) COMPONENT LOCATIONS LIGHT TRUCKS AND VANS

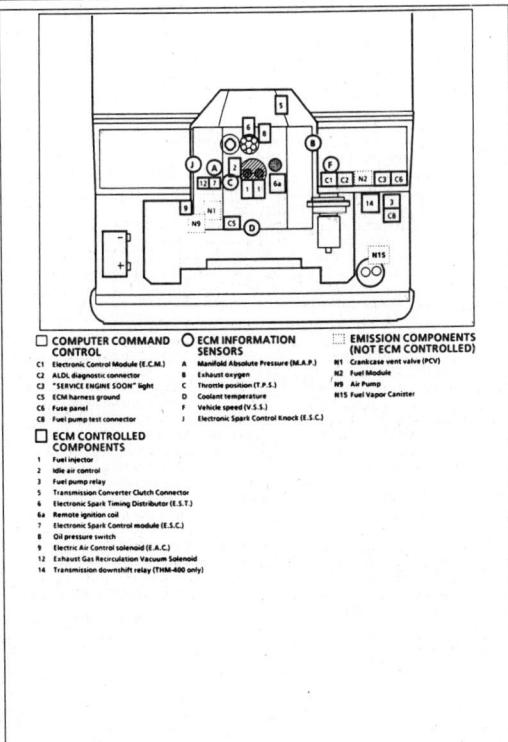

☐ **COMPUTER COMMAND CONTROL**
C1 Electronic Control Module (E.C.M.)
C2 ALDL diagnostic connector
C3 "SERVICE ENGINE SOON" light
C5 ECM harness ground
C6 Fuse panel
C8 Fuel pump test connector

◆ **ECM INFORMATION SENSORS**
A Manifold Absolute Pressure (M.A.P.)
B Exhaust oxygen
C Throttle position (T.P.S.)
D Coolant temperature
F Vehicle speed (V.S.S.)
J Electronic Spark Control Knock (E.S.C.)

▣ **EMISSION COMPONENTS (NOT ECM CONTROLLED)**
N1 Crankcase vent valve (PCV)
N2 Fuel Module
N9 Air Pump
N15 Fuel Vapor Canister

☐ **ECM CONTROLLED COMPONENTS**
1 Fuel injector
2 Idle air control
3 Fuel pump relay
5 Transmission Converter Clutch Connector
6 Electronic Spark Timing Distributor (E.S.T.)
6a Remote ignition coil
7 Electronic Spark Control module (E.S.C.)
8 Oil pressure switch
9 Electric Air Control solenoid (E.A.C.)
12 Exhaust Gas Recirculation Vacuum Solenoid
14 Transmission downshift relay (THM-400 only)

4.3L (VIN Z) AND V8 ECM WIRING DIAGRAM

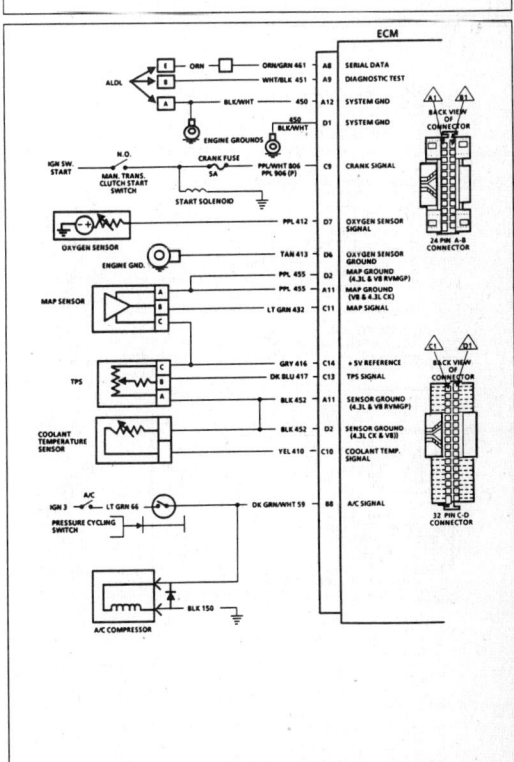

4.3L (VIN Z) AND V8 ECM WIRING DIAGRAM (CONT.)

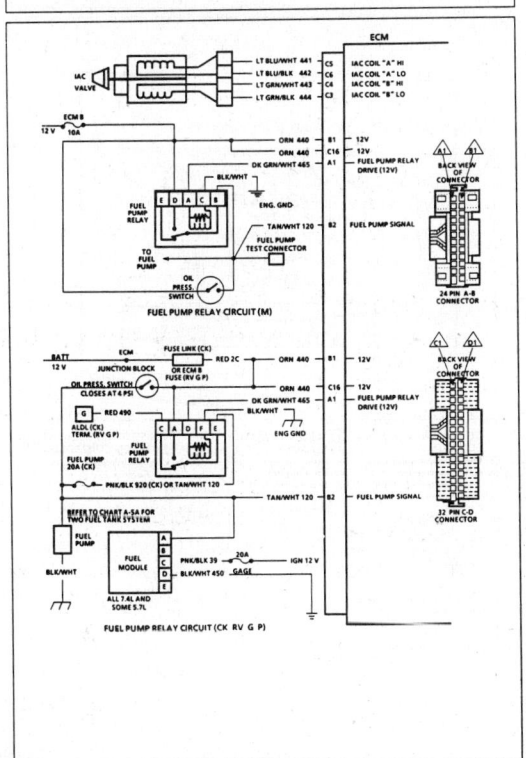

4.3L (VIN Z) AND V8 ECM WIRING DIAGRAM (CONT.)

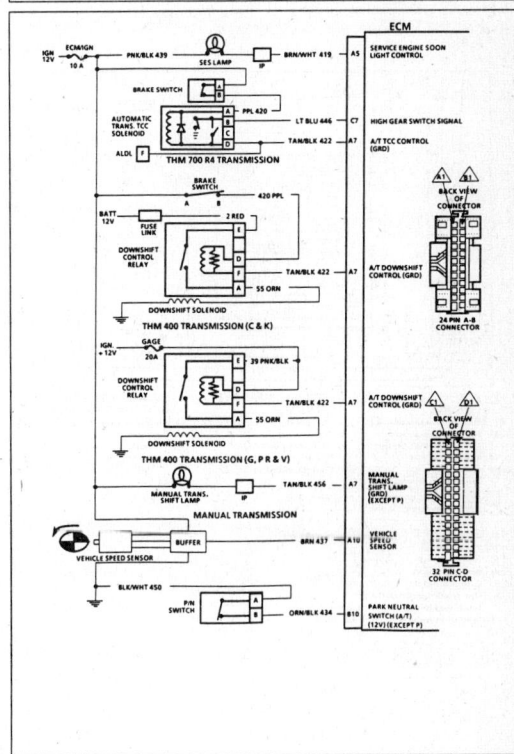

4.3L (VIN Z) AND V8 ECM WIRING DIAGRAM (CONT.)

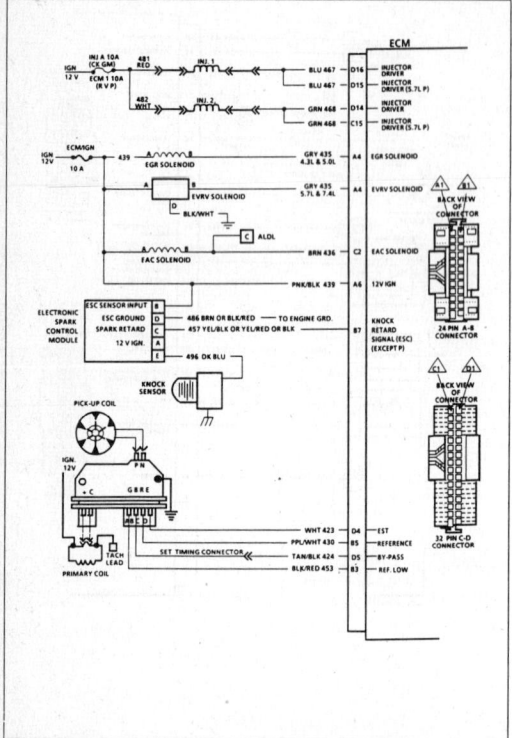

4.3L (VIN Z) AND V8 ECM CONNECTOR TERMINAL END VIEW

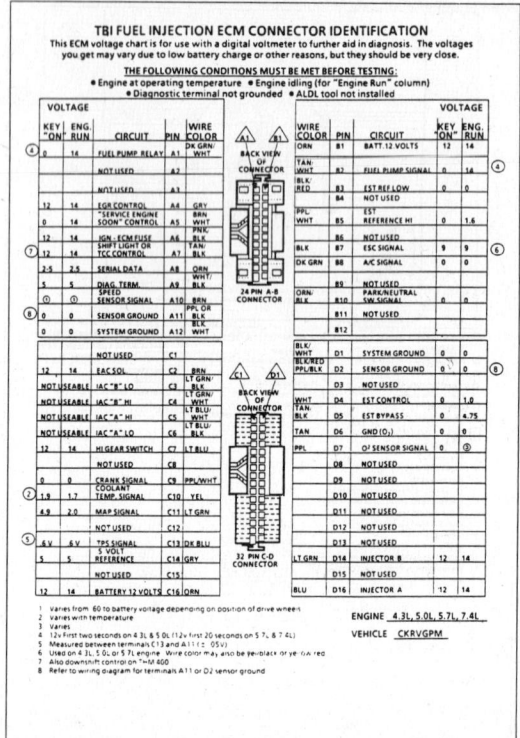

7.4L (VIN N, CK AND RV-SERIES) COMPONENT LOCATIONS—LIGHT TRUCKS AND VANS

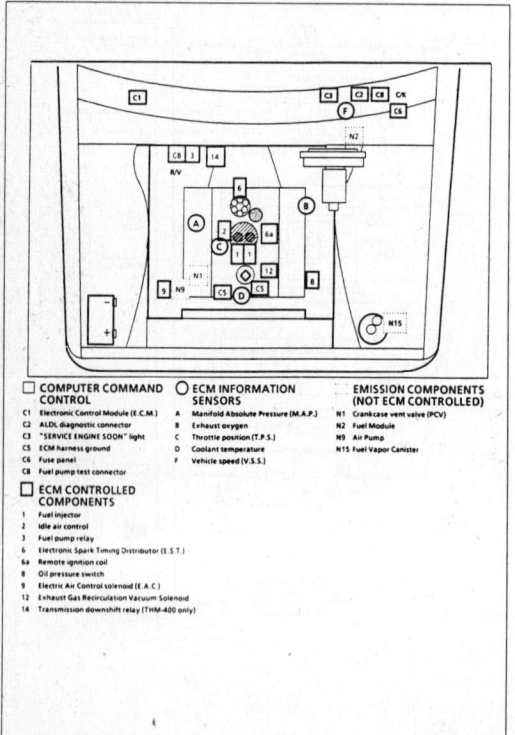

□ COMPUTER COMMAND CONTROL
C1 Electronic Control Module (E.C.M.)
C2 ALDL diagnostic connector
C3 "SERVICE ENGINE SOON" light
C5 ECM harness ground
C6 Fuse panel
C8 Fuel pump test connector

□ ECM CONTROLLED COMPONENTS
1 Fuel injector
2 Idle air control
3 Fuel pump relay
6 Electronic Spark Timing Distributor (E.S.T.)
6a Remote ignition coil
9 Oil pressure switch
9 Electric Air Control solenoid (E.A.C.)
12 Exhaust Gas Recirculation Vacuum Solenoid
14 Transmission downshift relay (THM-400 only)

○ ECM INFORMATION SENSORS
A Manifold Absolute Pressure (M.A.P.)
B Exhaust oxygen
C Throttle position (T.P.S.)
D Coolant temperature
F Vehicle speed (V.S.S.)

EMISSION COMPONENTS (NOT ECM CONTROLLED)
N1 Crankcase vent valve (PCV)
N2 Fuel Module
N9 Air Pump
N15 Fuel Vapor Canister

7.4L (VIN N, G-SERIES) COMPONENT LOCATIONS—LIGHT TRUCKS AND VANS

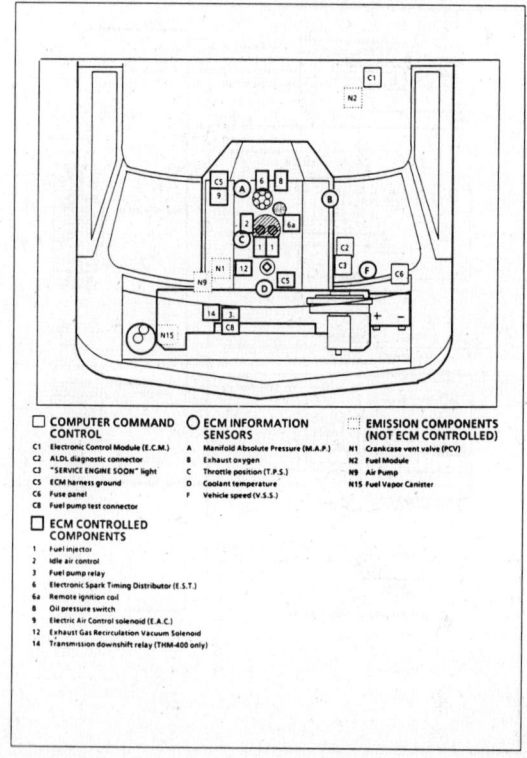

□ COMPUTER COMMAND CONTROL
C1 Electronic Control Module (E.C.M.)
C2 ALDL diagnostic connector
C3 "SERVICE ENGINE SOON" light
C5 ECM harness ground
C6 Fuse panel
C8 Fuel pump test connector

□ ECM CONTROLLED COMPONENTS
1 Fuel injector
2 Idle air control
3 Fuel pump relay
6 Electronic Spark Timing Distributor (E.S.T.)
6a Remote ignition coil
9 Oil pressure switch
9 Electric Air Control solenoid (E.A.C.)
12 Exhaust Gas Recirculation Vacuum Solenoid
14 Transmission downshift relay (THM-400 only)

○ ECM INFORMATION SENSORS
A Manifold Absolute Pressure (M.A.P.)
B Exhaust oxygen
C Throttle position (T.P.S.)
D Coolant temperature
F Vehicle speed (V.S.S.)

EMISSION COMPONENTS (NOT ECM CONTROLLED)
N1 Crankcase vent valve (PCV)
N2 Fuel Module
N9 Air Pump
N15 Fuel Vapor Canister

1988–90 LIGHT TRUCKS AND VANS

SECTION 2
DRIVEABILITY SYMPTOMS

CONTENTS

Before Starting
Poor Fuel Economy
ECM Intermittent Codes or Performance
Dieseling, Run-On
Backfire
Rough, Unstable, or Incorrect Idle, Stalling
Above Normal Emissions (Odors)
Hard Start
Surges and / or Chuggle
Lack of Power, Sluggish, or Spongy
Detonation / Spark Knock
Hesitation, Sag, Stumble
Cuts Out, Misses

BEFORE STARTING

Before using this section, you should have performed the SYSTEM CHECK.

Verify the customer complaint, and locate the correct SYMPTOM below. Check the items indicated under that symptom.

If the ENGINE CRANKS BUT WILL NOT RUN, see CHART A-3.

Refer to the related Chassis Service Manual for corrective action of driveability symptoms that are not fuel or emission related.

Careful Visual Check

Several of the following symptom procedures call for a careful visual (physical) check. This check should include:

- Vacuum hoses for splits, kinks, and proper connections, as shown on Vehicle Emission Control Information label.
- Air leaks at throttle body mounting and intake manifold.
- Ignition wires for cracking, hardness, proper routing and carbon tracking.
- Wiring for proper connections, pinches, and cuts.
  The importance of this step cannot be stressed too strongly - it can lead to correcting a problem without further checks and can save valuable time.
  The following symptoms cover several engines. To determine if a particular system or component is used, refer to the ECM wiring diagrams for application.

POOR FUEL ECONOMY

Definition: Fuel economy, as measured by an actual road test, is noticeably lower than expected. Also, economy is noticeably lower than it was on this vehicle at one time, as previously shown by an actual road test.

- **CHECK:**
- Engine thermostat for faulty part (always open) or for wrong heat range.
- Fuel pressure. See CHART A-5.
- Ignition timing. See Vehicle Emission Control Information label for procedure.
- TCC for proper operation.

1988–90 LIGHT TRUCKS AND VANS

ECM INTERMITTENT CODES OR PERFORMANCE

Problem may or may not turn "ON" the "Service Engine Soon" light, or store a code.

The ECM code charts, determine if there is a fault with a circuit, or, if there is an intermittent problem. An intermittent means that a code is stored in the ECM memory, but the circuit is OK.

- Most intermittent problems are caused by faulty electrical connections or wiring. Perform careful check of the suspected circuits for:
  - Poor mating of the connector halves, or terminals, not fully seated in the connector body (backed out).
  - Improperly formed or damaged terminals. All connector terminals, in a problem circuit, should be carefully reformed to increase contact tension.
  - Poor terminal to wire connection. This requires removing the terminal from the connector body.
- If a visual (physical) check does not find the cause of the problem, the car can be driven with a voltmeter connected to a suspected circuit. An abnormal voltage reading, when the problem occurs, indicates the problem may be in that circuit.

- Loss of code memory: To check, disconnect TPS, and idle engine until "Service Engine Soon" light comes "ON". Code 22 should be stored, and kept in memory when ignition is turned "OFF", for at least 10 seconds. If not, the ECM is faulty.
- CHECK:
- Electrical system interference caused by a defective relay, ECM driven solenoid, or switch. They can cause a sharp electrical surge. Normally, the problem will occur when the faulty component is operated.
- Improper installation of electrical options, such as lights, 2-way radios, etc.
- EST wires should be kept away from spark plug wires, distributor wires, distributor housing, coil, and generator. Wire from CKT 453 to distributor should have a good contact to ground.
- Ignition secondary shorted to ground.
- CKT's 419 and 451 intermittently shorted to ground.
- ECM power grounds.

DIESELING, RUN-ON

Definition: Engine continues to run, after key is turned "OFF", but runs very roughly. If engine runs smoothly, check ignition switch and adjustment.

- Check injector(s) for leaking. Apply 12 volts to fuel pump test terminal to turn "ON" fuel pump and prime fuel system. See CHART A-5.
- Visually check injector and TBI assembly for fuel leakage.

BACKFIRE

Definition: Fuel ignites in intake manifold, or in exhaust system, making a loud popping noise.

- CHECK:
- EGR operation, for being open all the time.
- Output voltage of ignition coil.
- For crossfire between spark plugs (distributor cap, spark plug wires, and proper routing of plug wires).
- For intermittent condition in primary ignition system.
- Engine timing - See Vehicle Emission Control Information label.
- For faulty spark plugs and/or plug wires or boots.
- For proper valve timing.
- Perform a compression check - look for sticking or leaking valves.

1988–90 LIGHT TRUCKS AND VANS

ROUGH, UNSTABLE, OR INCORRECT IDLE, STALLING

Definition: The engine runs unevenly at idle. If bad enough, the vehicle may shake. Also, the idle may vary in rpm (called "hunting"). Either condition may be severe enough to cause stalling. Engine idles at incorrect speed.

- Perform careful visual check as described at start of this section.
- **CHECK:**
- Throttle linkage for sticking or binding.
- Ignition timing. See Vehicle Emission Control Information label.
- ECM ground circuits.
- IAC system. See Code 35.
- Generator output voltage. Repair if less than 9, or more than 16, volts.
- P/N switch circuit.
- PCV valve for proper operation, by placing finger over inlet hole in valve end several times. Valve should snap back.
- Evaporative emission control system.
- A/C signal to ECM terminal "B8". "Scan" tool should indicate A/C is being requested whenever A/C is selected and the pressure cycling switch is closed.
- Controlled idle speed.

- Minimum idle air rate.
- Check A.I.R. system. There should be no A.I.R. to ports while in "Closed Loop".
- EGR valve. There should be no EGR at idle.
- Run a cylinder compression check.
- Inspect oxygen sensor for silicon contamination from fuel, or use of improper RTV sealant. The sensor will have a white, powdery coating, and will result in a high but false signal voltage (rich exhaust indication). The ECM will then reduce the amount of fuel delivered to the engine, causing a severe driveability problem.
- Check for fuel in pressure regulator hose. If present, replace regulator assembly.
- Check ignition system. wires, plugs, rotor, etc.
- Monitoring block learn will help identify the cause of the problem. If the system is running lean (block learn greater than 138), refer to "Diagnostic Aids" on facing page of Code 45. If the system is running rich (block learn less than 118), refer to "Diagnostic Aids" on facing page of Code 45.
- Stalling may be due to an incorrect idle air rate. Refer to Code 35.

EXCESSIVE EXHAUST EMISSIONS OR ODORS

Definition: Vehicle fails an emission test. May also have excessive "rotten egg" smell (hydrogen sulfide). Excessive odor does not necessarily indicate excessive emissions.

- Perform "System Check"
- IF TEST SHOWS EXCESSIVE CO AND HC, (or also has excessive odors):
- Check items which cause engine to run RICH.
  - Make sure engine is at normal operating temperature.
- **CHECK:**
- Fuel pressure. See CHART A-6.
- Incorrect timing. See Vehicle Emission Control Information label on vehicle.
- Canister for fuel loading.
- PCV valve for being plugged or stuck, or blocked PCV hose.
- Spark plugs, plug wires, and ignition components.
- Check for lead contamination of catalytic converter (look for removal of fuel filler neck restrictor).

- Check for improperly installed fuel cap.

- If the system is running rich, (block learn less than 118), refer to "Diagnostic Aids" on facing page of Code 45.
- IF TEST SHOWS EXCESSIVE NO$_x$:
- Check items which cause engine to run LEAN, or to run too hot.
- EGR valve for not opening.

Vacuum leaks
- Remove carbon with top engine cleaner. Follow instructions on can.
- Check ignition timing for excessive base advance. See Vehicle Emission Control Information label on vehicle.
- If the system is running lean, (block learn greater than 138), refer to "Diagnostic Aids" on facing page of Code 44.

1988–90 LIGHT TRUCKS AND VANS

HARD START

Definition: Engine cranks OK, but does not start for a long time. Does eventually run, or may start but immediately dies.

- **CHECK:**
- For water contaminated fuel.
- Fuel pressure.
- TPS for sticking or binding.
- EGR operation.
- Fuel pump relay (ST & M) - Connect test light between pump test terminal and ground. Light should be "ON", for 2 seconds, following ignition "ON". See CHART A-5.
  On 7.4L, and some 5.7L engines, a fuel module circuit will run the fuel pump for 20 seconds, following ignition "ON". If this circuit is not functioning, this may cause a hot, hard start condition.
- For a faulty in-tank fuel pump, check valve, which would allow the fuel in the lines to drain back to the tank after the engine is stopped. To check for this condition:
  1. Ignition "OFF".
  2. Disconnect fuel line at the filter. See Section "4".
  3. Remove the tank filler cap.
  4. Connect a radiator test pump to the line and apply 103 kPa (15 psi) pressure. If the pressure will hold for 60 seconds, the check valve is OK.

- Long cranking time but eventually runs. Refer to CHART A-3, Engine Cranks but Will Not Run.
- Check ignition system for:
  - Proper output, with J-26792 (ST-125).
  - Worn shaft.
  - Bare and shorted wires.
  - Pickup coil resistance and connections.
  - Loose ignition coil ground.
  - Moisture in distributor cap.
  - Spark plugs, wet plugs, cracks, wear, improper gap, burned electrodes, or heavy deposits.
- If engine starts, but then immediately stalls, open distributor bypass line. If engine then starts, and runs OK, replace distributor pickup coil.
- Hard start, with engine at normal operating temperature. See Crank Signal Diagnosis.

SURGES AND/OR CHUGGLE

Definition: Engine power variation under steady throttle or cruise. Feels like the vehicle speeds up and slows down with no change in the accelerator pedal.

- If a tool is available which plugs in to the ALDL connector, make sure reading of VSS matches vehicle speedometer. See Code 24.
- **CHECK:**
- For intermittent EGR at idle.
- Ignition timing. See Vehicle Emission Control Information label.
- Inline fuel filter for dirt or restriction.
- Fuel pressure. See CHART A-6.
- Generator output voltage. Repair, if less than 9 volts or more than 16 volts.
- TCC Operation.

- Inspect Oxygen sensor for silicon contamination from fuel, or use of improper RTV sealant. The sensor may have a white, powdery coating and result in a high but false signal voltage (rich exhaust indication). The ECM will then reduce the amount of fuel delivered to the engine, causing a severe driveability problem.
- Remove spark plugs. Check for cracks, wear, improper gap, burned electrodes, or heavy deposits. Also, check condition of distributor cap, rotor, and spark plug wires.

1988–90 LIGHT TRUCKS AND VANS

LACK OF POWER, SLUGGISH, OR SPONGY

Definition: Engine delivers less than expected power. Little or no increase in speed when accelerator pedal is pushed down part way.

- Compare customer's vehicle to similar unit. Make sure the customer's vehicle has an actual problem.
- Remove air cleaner and check air filter for dirt, or for being plugged. Replace as necessary.
- CHECK:
  - Ignition timing. See Vehicle Emission Control Information label.
  - For restricted fuel filter, contaminated fuel or improper fuel pressure. See CHART A-6.
  - ECM Grounds.
  - EGR operation for being open or partly open all the time.
  - Generator output voltage. Repair if less than 9 or more than 16 volts.
  - Engine valve timing and compression.
  - Engine for proper or worn camshaft.

- Check Exhaust system for restriction:
  1. With engine at normal operating temperature, connect a vacuum gage to any convenient vacuum port on intake manifold.
  2. Disconnect EGR solenoid electrical connector or connect EGR valve directly to vacuum source bypassing any switches or solenoids.
  3. Run engine at 1000 rpm and record vacuum reading.
  4. Increase rpm slowly to 2500 rpm. Note vacuum reading at steady 2500 rpm.
  5. If vacuum at 2500 rpm decreases more than 3", from reading at 1000 rpm, the exhaust system should be inspected for restrictions.
  6. Disconnect exhaust pipe from engine and repeat steps 3 & 4. If vacuum still drops, more than 3" with exhaust disconnected, check valve timing.

DETONATION / SPARK KNOCK

Definition: A mild to severe ping, usually worse under acceleration. The engine makes sharp metallic knocks that change with throttle opening.

- Check for obvious overheating problems.
  - Low coolant.
  - Loose water pump belt.
  - Restricted air flow to radiator, or restricted water flow thru radiator.
- CHECK:
  - For poor fuel quality, proper octane rating.
  - For correct PROM.
  - THERMAC for staying closed.
  - Ignition timing. See Vehicle Emission Control Information label.
  - Fuel system for low pressure. See CHART A-6

- Check EGR system for not opening.
- For proper transmission shift points.
- Check TCC operation.
- For incorrect basic engine parts such as cam, heads, pistons, etc.
- Remove carbon with top engine cleaner. Follow instructions on can.
  On vehicles with 7.4L or 5.7L (over 8500 GVW) engine and a dual catalytic converter system, check for an exhaust restriction in the rear converter area.

1988–90 LIGHT TRUCKS AND VANS

HESITATION, SAG, STUMBLE

Definition: Momentary lack of response, as the accelerator is pushed down, can occur at all vehicle speeds. Usually, most severe, when first trying to make the vehicle move, as from a stop sign. May cause the engine to stall, if severe enough.

- Perform careful visual (physical) check

- CHECK:
  - Fuel pressure. See CHART A-6
  - Water contaminated fuel.
  - Ignition timing. See Vehicle Emission Control Information label.

- Fuel pump cycle system. See CHART A-5

- TPS for binding or sticking
- Generator output voltage. Repair if less than 9 or more than 16 volts.
- For open HEI ground, CKT 453.
- Canister purge system for proper operation.

- EGR valve operation.

CUTS OUT, MISSES

Definition: Steady pulsation or jerking that follows engine speed, usually more pronounced as engine load increases. The exhaust has a steady spitting sound at idle or low speed.

- Check for missing cylinder by:
  1. Disconnecting IAC motor. Start engine. Remove one spark plug wire at a time, using insulated pliers.
  2. If there is an rpm drop on all cylinders (equal to within 50 rpm), go to "ROUGH, UNSTABLE, OR INCORRECT IDLE, STALLING" symptom. Reconnect IAC motor.
  3. If there is no rpm drop on one or more cylinders, or excessive variation in drop, check for spark on the suspected cylinder(s) with J 26792 (ST-125) Spark Gap Tool or equivalent. If no spark, see Ignition Section "6". If there is spark, remove spark plug(s) in those cylinders and check for:
     - Cracks
     - Wear
     - Improper Gap
     - Burned Electrodes
     - Heavy Deposits

- CHECK:
  - Spark plug wires by connecting ohmmeter to ends of each wire in question. If meter reads over 30,000 ohms, replace wire(s)
  - Ignition coil and secondary voltage using spark tester J-26792 (ST-125), or equivalent.
  - For restricted fuel filter. Also check fuel tank for water. See Section "4" for location.
  - For low fuel pressure. See CHART A-6 in

- Check for proper valve timing.
- Perform compression check on questionable cylinder(s) found above. If compression is low, repair as necessary.
- Visually check distributor cap and rotor for moisture, dust, cracks, burns, etc. Spray cap and plug wires with fine water mist to check for shorts.
- Remove rocker covers. Check for bent pushrods, worn rocker arms, broken valve springs, worn camshaft lobes.

1988–90 LIGHT TRUCKS AND VANS

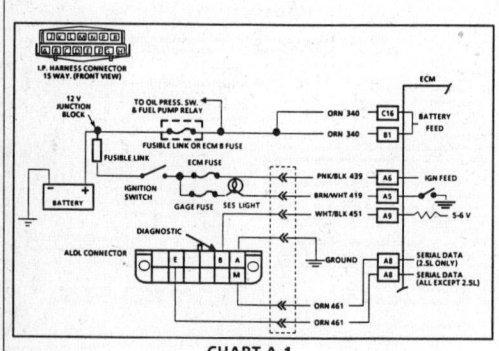

CHART A-1
NO "SERVICE ENGINE SOON" LIGHT
ALL ENGINES

Circuit Description:
There should be a steady "Service Engine Soon" light when the ignition is "ON" and engine stopped. Battery ignition voltage is supplied to the light bulb. The ECM will control the light and turn it "ON" by providing a ground path through CKT 419.

Test Description: Numbers below refer to circled numbers on the diagnostic chart.
1. If the fuse in holder is blown, refer to facing page of Code 54 for complete circuit.
2. Using a test light connected to 12 volts, probe each of the system ground circuits to be sure a good ground is present. See ECM terminal end view in front of this section for ECM pin locations of ground circuits.

Diagnostic Aids:

If the engine runs OK, check:
- Faulty light bulb.
- CKT 419 open.
- ECM terminal end.
- Gage fuse blown. This will result in no stop lights, oil or generator lights, seat belt reminder, etc.
If the engine cranks but will not run, check:
- Continuous battery-fuse or fusible link open.
- ECM ignition fuse open.
- Battery CKT 340 to ECM open.
- Ignition CKT 439 to ECM open.
- Poor connection to ECM.

1988–90 LIGHT TRUCKS AND VANS

CHART A-1
NO "SERVICE ENGINE SOON" LIGHT
ALL ENGINES

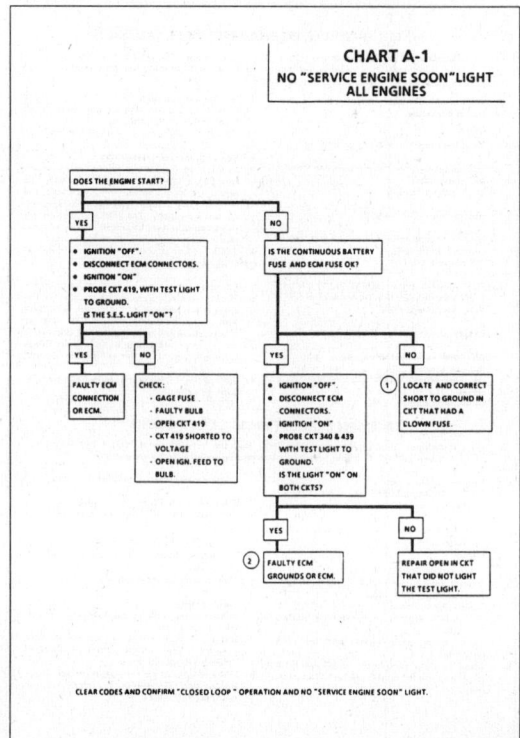

1988-90 LIGHT TRUCKS AND VANS

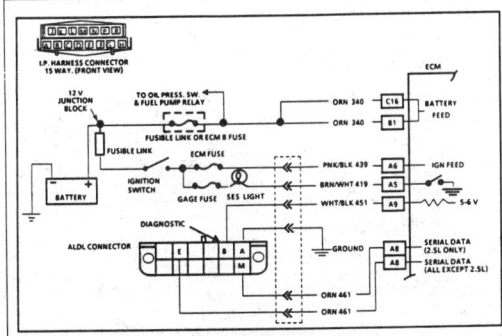

CHART A-2
NO ALDL DATA OR WON'T FLASH CODE 12
"SERVICE ENGINE SOON" LIGHT ON STEADY
ALL ENGINES

Circuit Description:
There should always be a steady "Service Engine Soon" light when the ignition is "ON" and engine stopped. Battery ignition voltage is supplied to the light bulb. The ECM will turn the light on by grounding CKT 419.

With the diagnostic terminal grounded, the light should flash a Code 12, followed by any trouble code(s) stored in memory.

A steady light suggests a short to ground in the light control CKT 419, or an open in diagnostic CKT 451.

Test Description: Numbers below refer to circled numbers on the diagnostic chart.
1. If there is a problem with the ECM that causes a "Scan" tool to not read Serial data then the ECM should not flash a Code 12. If Code 12 does flash, be sure that the "Scan" tool is working properly on another vehicle. If the "Scan" is functioning properly and CKT 461 is OK, the PROM/Mem-Cal or ECM may be at fault for the NO ALDL symptom.
2. If the light goes "OFF" when the ECM connector is disconnected, then CKT 419 is not shorted to ground.
3. This step will check for an open diagnostic CKT 451.

4. At this point the "Service Engine Soon" light wiring is OK. The problem is a faulty ECM or PROM/Mem-Cal. If Code 12 does not flash, the ECM should be replaced using the original PROM/Mem-Cal. Replace the PROM/Mem-Cal only after trying an ECM, as a defective PROM/Mem-Cal is an unlikely cause of the problem.

1988-90 LIGHT TRUCKS AND VANS

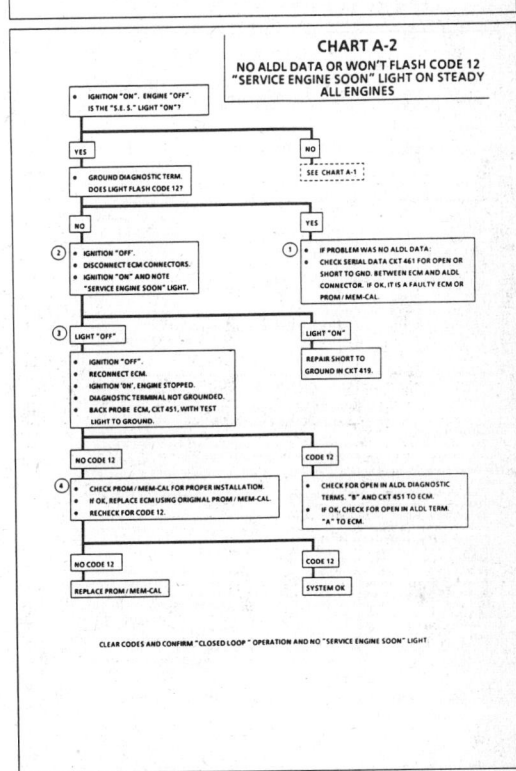

CHART A-2
NO ALDL DATA OR WON'T FLASH CODE 12
"SERVICE ENGINE SOON" LIGHT ON STEADY
ALL ENGINES

1988-90 LIGHT TRUCKS AND VANS

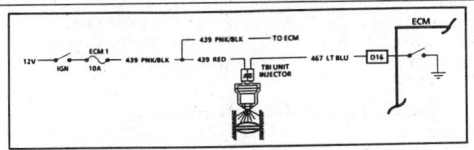

CHART A-3
ENGINE CRANKS BUT WILL NOT RUN
2.5L ENGINE

Circuit Description:
This chart assumes that battery condition and engine cranking speed are OK, and there is adequate fuel in the tank. This chart should be used on engines using the Model 700 throttle body.

Test Description: Numbers below refer to circled numbers on the diagnostic chart.
1. A "Service Engine Soon" light "ON" is a basic test to determine if there is a 12 volt supply and ignition 12 volts to ECM. No ALDL may be due to an ECM problem and CHART A-2 will diagnose the ECM. If TPS is over 2.5 volts the engine may be in the clear flood mode which will cause starting problems. If coolant sensor is below -30°C, ECM will provide fuel for this extremely cold temperature which will severely flood the engine.
2. Voltage at the spark plug is checked using spark tester tool ST125 (J26792) or equivalent. No spark indicates a basic ignition problem.
3. While cranking engine there should be no fuel spray with injector disconnected. Replace an injector if it sprays fuel or drips like a leaking water faucet.

4. Use an injector test light like J34730, BT8329A or equivalent, to test injector circuit. A blinking light indicates the ECM is controlling the injector.
5. This test will determine if there is fuel pressure at the injector and that the injector is operating.

Diagnostic Aids:
If no trouble is found in the fuel pump circuit or ignition system and the cause of a "Engine Cranks But Will Not Run" has not been found, check for:
- Fouled spark plugs
- EGR valve stuck open
- Low fuel pressure. See CHART A-6.
- Water or foreign material in the fuel system.
- A ground CKT 423 (EST) may cause a "No Start" or a "Start then Stall" condition.
- Basic engine problem.

1988-90 LIGHT TRUCKS AND VANS

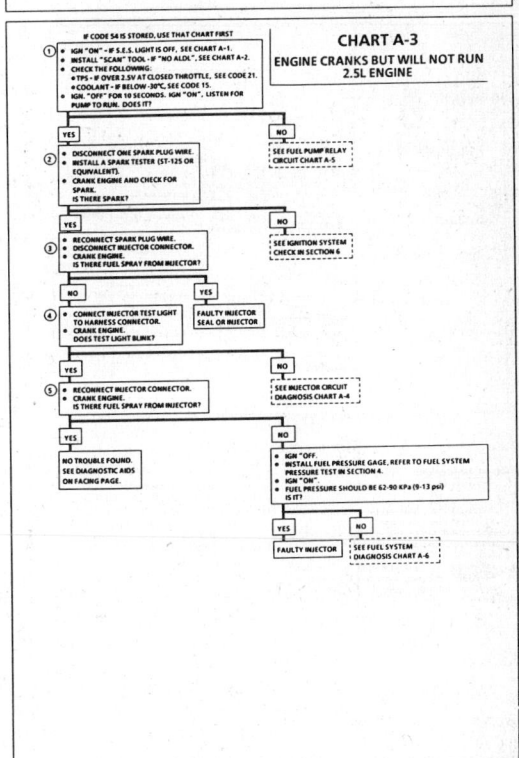

CHART A-3
ENGINE CRANKS BUT WILL NOT RUN
2.5L ENGINE

1988–90 LIGHT TRUCKS AND VANS

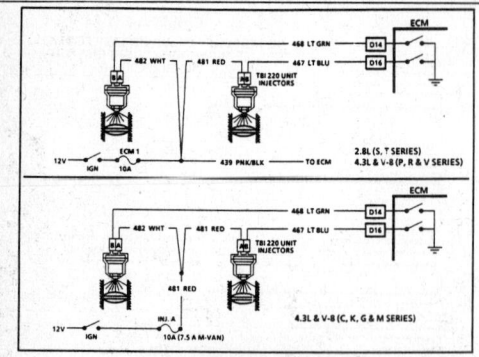

2.8L (S, T SERIES)
4.3L & V-8 (P, R & V SERIES)

4.3L & V-8 (C, K, G & M SERIES)

CHART A-3
ENGINE CRANKS BUT WILL NOT RUN
ALL ENGINES EXCEPT 2.5L

Circuit Description:

This chart assumes that battery condition and engine cranking speed are OK, and there is adequate fuel in the tank. This chart should be used on engines using the Model 220 throttle body.

Test Description: Numbers below refer to circled numbers on the diagnostic chart.

1. A "Service Engine Soon" light "ON" is a basic test to determine if there is a 12 volt supply and ignition 12 volts to ECM. No ALDL may be due to an ECM problem and CHART A-2 will diagnose the ECM. If TPS is over 2.5 volts the engine may be in the clear flood mode which will cause starting problems. If coolant sensor is below -30°C, the ECM will provide fuel for this extremely cold temperature which will severely flood the engine.

2. Voltage at the spark plug is checked using Spark Tester tool ST125 (J26792) or equivalent. No spark indicates a basic ignition problem

3. While cranking engine there should be no fuel spray with injectors disconnected. Replace an injector if it sprays fuel or drips like a leaking water faucet

4. Use an injector test light like BT8320, or equivalent, to test each injector circuit. A blinking light indicates the ECM is controlling the injectors.

5. This test will determine if there is fuel pressure at the injectors and that the injectors are operating.

Diagnostic Aids:

If no trouble is found in the fuel pump circuit or ignition system and the cause of a "Engine Cranks But Will Not Run" has not been found, check for:
- Fouled spark plugs
- EGR valve stuck open
- Low fuel pressure. See CHART A-6.
- Water or foreign material in the fuel system
- A grounded CKT 423 (EST) may cause a "No Start" or a "Start then Stall" condition.
- Basic engine problem.

1988–90 LIGHT TRUCKS AND VANS

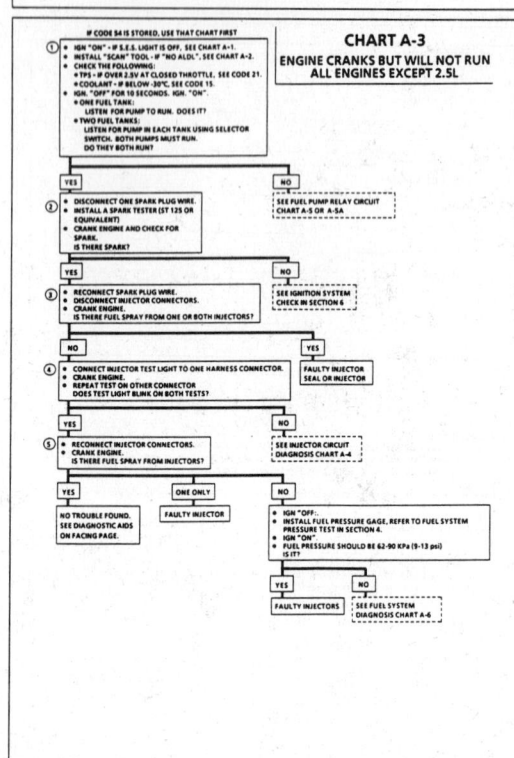

CHART A-3
ENGINE CRANKS BUT WILL NOT RUN
ALL ENGINES EXCEPT 2.5L

1988–90 LIGHT TRUCKS AND VANS

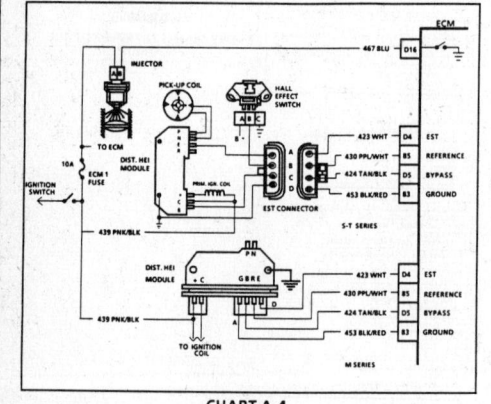

S-T SERIES

M SERIES

CHART A-4
INJECTOR CIRCUIT DIAGNOSIS
2.5L ENGINE

Circuit Description:

This chart should only be used if diagnosis in CHART A-3 indicated an injector circuit problem.

Test Description: Numbers below refer to circled numbers on the diagnostic chart.

1. This test will determine if the ignition module is generating a reference pulse, if the wiring is at fault or if the ECM is at fault. By touching and removing a test light, connected to 12 volts, to CKT 430, a reference pulse should be generated. If injector test light blinks, the ECM and wiring are OK

2. This step tests for 12 volts to the injector It will also determine if there is a short to voltage on the ECM side of the circuit.

3. This test checks for continuity to the ECM.

1988–90 LIGHT TRUCKS AND VANS

CHART A-4
INJECTOR CIRCUIT DIAGNOSIS
2.5L ENGINE

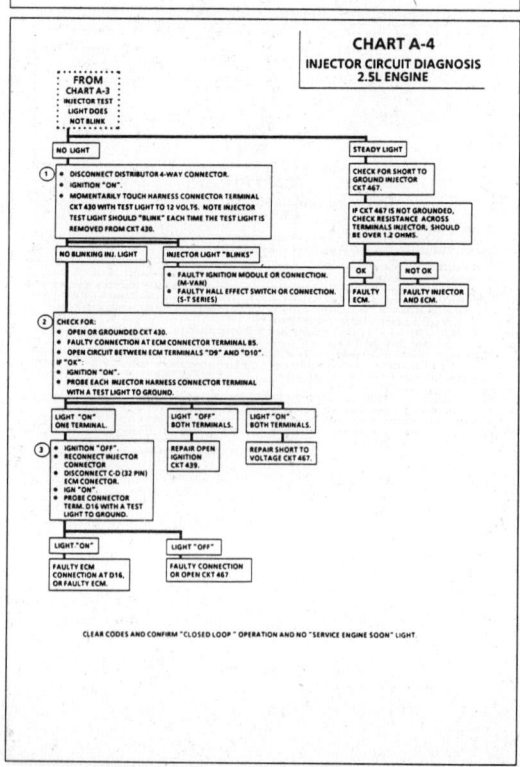

1988–90 LIGHT TRUCKS AND VANS

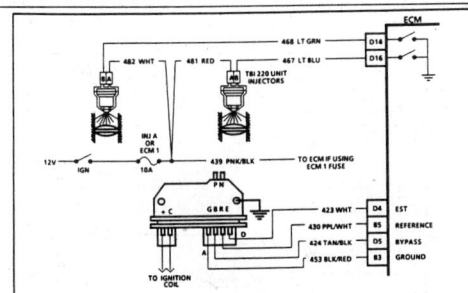

CHART A-4
INJECTOR CIRCUIT DIAGNOSIS
ALL ENGINES EXCEPT 2.5L

Circuit Description:
This chart should only be used if diagnosis in CHART A-3 indicated an injector circuit problem. If both injector circuits fail to blink when tested, diagnose one injector circuit at a time.

Test Description: Numbers below refer to circled numbers on the diagnostic chart.
1. This test will determine if the ignition module is generating a reference pulse, if the wiring is at fault or if the ECM is at fault. By touching and removing a test light, connected to 12 volts, to CKT 430, a reference pulse should be generated. If injector test light blinks, the ECM and wiring are OK.
2. This step tests for 12 volts to the injector. It will also determine if there is a short to voltage on the ECM side of the circuit.
3. This test checks for continuity to the ECM.

1988–90 LIGHT TRUCKS AND VANS

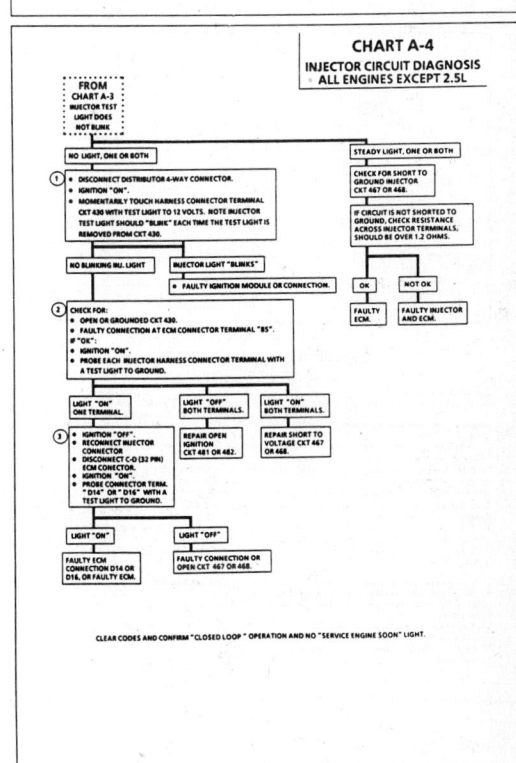

CHART A-4
INJECTOR CIRCUIT DIAGNOSIS
ALL ENGINES EXCEPT 2.5L

1988–90 LIGHT TRUCKS AND VANS

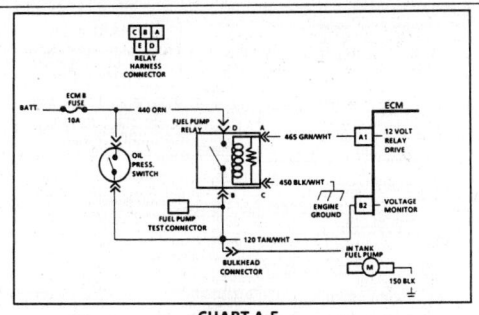

CHART A-5
FUEL PUMP RELAY CIRCUIT DIAGNOSIS
S, T & M SERIES

Circuit Description:
When the ignition switch is turned "ON", the ECM will turn "ON" the in-tank fuel pump. It will remain "ON" as long as the engine is cranking or running, and the ECM is receiving distributor reference pulses. If there are no reference pulses, the ECM will shut "OFF" the fuel pump within 2 seconds after ignition "ON" or engine stops.
The pump will deliver fuel to the TBI unit where the system pressure is controlled to about 62 to 90 kPa (9 to 13 psi). Excess fuel is then returned to the fuel tank.

Test Description: Numbers below refer to circled numbers on the diagnostic chart.
1. Turns "ON" the fuel pump if CKT 120 wiring is OK. If the pump runs, it maybe a fuel pump relay circuit problem, which the following steps will locate.
2. The next two steps check for power and ground circuits to the relay.
3. Determines if ECM can control the relay.
4. The oil pressure switch serves as a backup for the fuel pump relay to help prevent a no start situation. If the fuel pump relay was found to be inoperative, the oil pressure switch circuit should also be tested to determine why it did not operate the fuel pump.

1988–90 LIGHT TRUCKS AND VANS

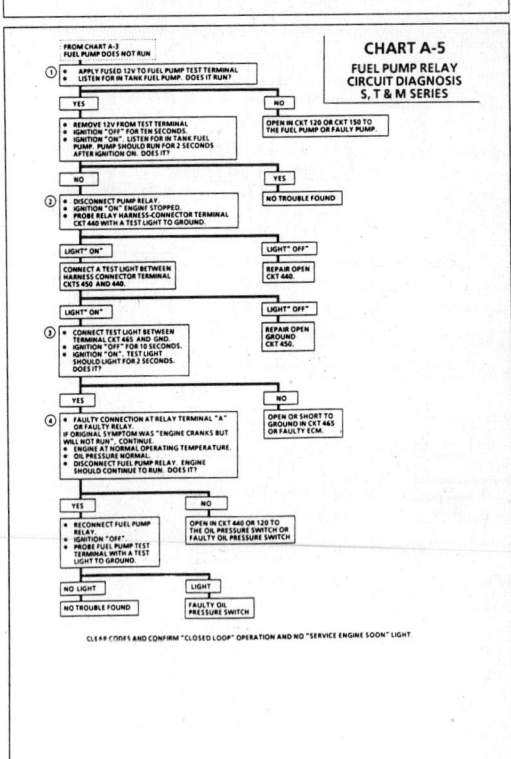

CHART A-5
FUEL PUMP RELAY CIRCUIT DIAGNOSIS
S, T & M SERIES

1988–90 LIGHT TRUCKS AND VANS

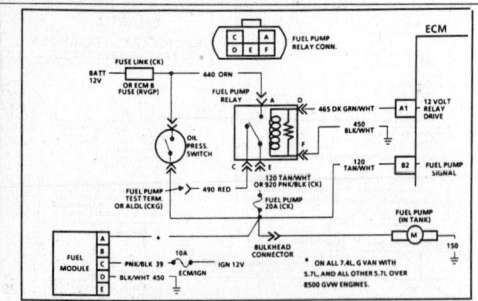

CHART A-5
FUEL PUMP RELAY CIRCUIT DIAGNOSIS
(ONE FUEL TANK)
C, K, R, V, G & P SERIES

Circuit Description:

When the ignition switch is turned "ON", the Electronic Control Module (ECM) will turn "ON" the in-tank fuel pump. It will remain "ON" as long as the engine is cranking or running, and the ECM is receiving distributor reference pulses. If there are no reference pulses, the ECM will shut "OFF" the fuel pump within 2 seconds after ignition "ON" or engine stops except when a fuel module is used.

The pump will deliver fuel to the TBI unit where the system pressure is controlled to about 62 to 90 kPa (9 to 13 psi). Excess fuel is then returned to the fuel tank.

A fuel module is used on all 7.4L, G van with 5.7L, and all other 5.7L over 8500 GVW engines to correct a hot restart (vapor lock) during a high ambient condition. It is designed to over-ride the ECM two second pump operation and will run the fuel pump for twenty seconds at initial ignition "ON".

Test Description: Numbers below refer to circled numbers on the diagnostic chart.

1. This procedure applies direct voltage to run the fuel pump. If the pump runs, it may be a fuel pump relay circuit problem which the following step will locate.
2. This step checks voltage from the battery and the ground circuit to the relay.
3. This test determines if there is voltage from the ECM, terminal A1, to terminal "D" on the relay connector.
4. This completes the fuel pump relay circuit but if this diagnosis was used because the engine would not run then oil pressure switch should also be diagnosed.

Diagnostic Aids:

- An inoperative fuel module may be the cause of a hot stall/no start condition. Check for power and ground circuit to the fuel module and a complete circuit to the pump from terminal "A". If OK, and the pump does not run for the specified 20 seconds at initial ignition "ON", replace the Fuel Module.

1988–90 LIGHT TRUCKS AND VANS

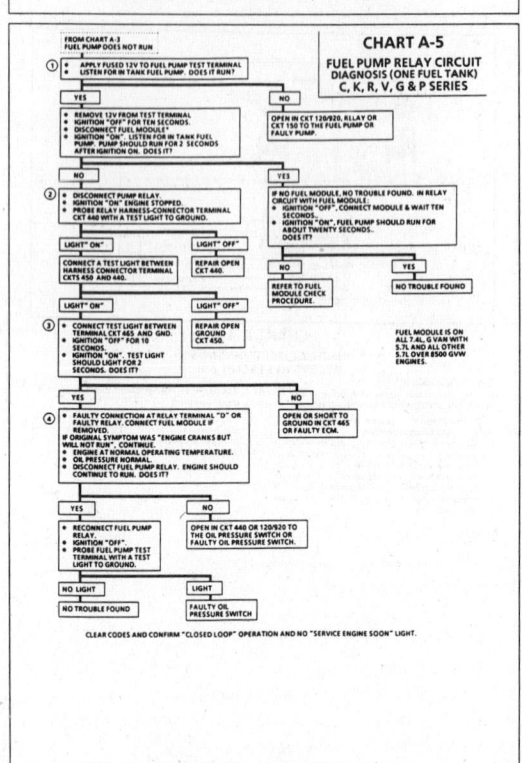

FUEL PUMP RELAY CIRCUIT DIAGNOSIS
RV-SERIES (2 FUEL TANKS)

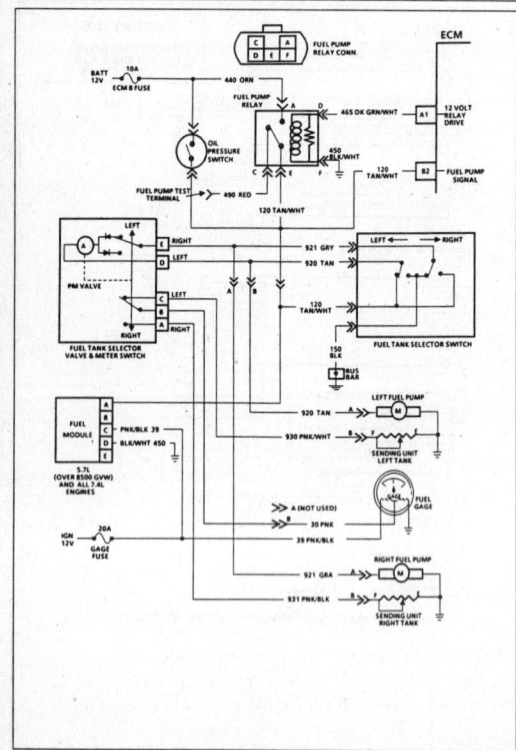

1988–90 RV-SERIES

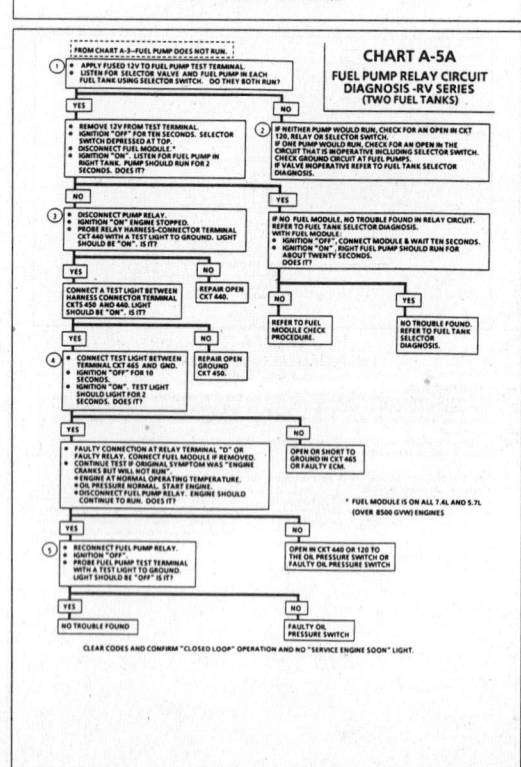

1988–90 LIGHT TRUCKS AND VANS

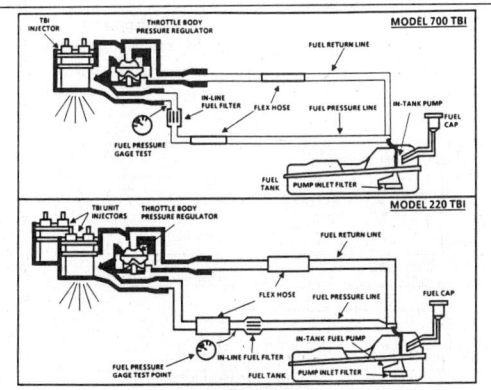

CHART A-6
FUEL SYSTEM PRESSURE TEST
ALL ENGINES

Circuit Description:
When the fuel pump is running, fuel is delivered to the injector(s) and then to the regulator where the system pressure is controlled to about 62 to 90 kPa (9 to 13 psi). Excess fuel is then returned to the fuel tank.

Test Description: Numbers below refer to circled numbers on the diagnostic chart.
1. Pressure but less than 62 kPa (9 psi) falls into two areas:
   - Regulated pressure but less than 62 kPa (9 psi) - Amount of fuel to injector OK but pressure is too low. System will be lean running and may set Code 44. Also, hard starting cold and poor overall performance.
   - Restricted flow causing pressure drop - Normally, a vehicle with a fuel pressure of less than 62 kPa (9 psi) at idle will not be driveable. However, if the pressure drop occurs only while driving, the engine will normally surge then stop as pressure begins to drop rapidly. Restricting the fuel return line allows the fuel pump to develop its maximum pressure (dead head pressure). When battery voltage is applied to the

pump test connector, pressure should be from 90 to 124 kPa (13 to 18 psi).
3. This test determines if the high fuel pressure is due to a restricted fuel return line or a throttle body pressure regulator problem.

Diagnostic Aids:
- If the vehicle is equipped with a fuel module, the module must be disconnected before performing the fuel system pressure test. Refer to Section "4"
- Fuel system is under pressure. To avoid fuel spillage, refer to procedures in Section "4" for testing or making repairs requiring disassembly of fuel lines or fittings.
- On V6 or V8 engine, the fuel pressure drops to almost zero psi after pump shuts "off".

1988–90 LIGHT TRUCKS AND VANS

CHART A-6
FUEL SYSTEM PRESSURE TEST
ALL ENGINES

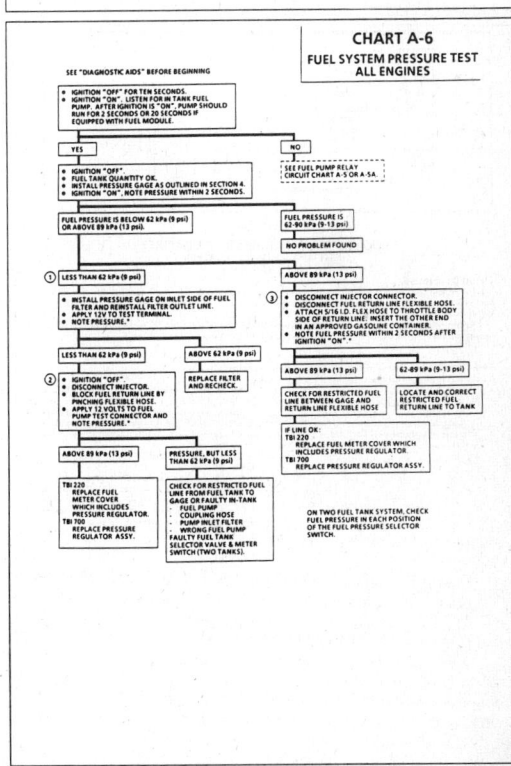

1988–90 LIGHT TRUCKS AND VANS

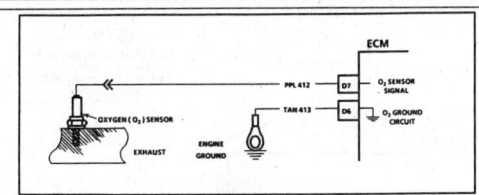

CODE 13
OXYGEN SENSOR OPEN CIRCUIT
ALL ENGINES

Circuit Description:
The ECM supplies a voltage of about .45 volt between terminals "D7" and "D6". (If measured with a 10 megohm digital voltmeter, this may read as low as .32 volts.) The Oxygen sensor varies the voltage within a range of about 1 volt if the exhaust is rich, down through about .10 volt if exhaust is lean.
The sensor is like an open circuit and produces no voltage when it is below 315° C (600°F). An open sensor circuit or cold sensor causes "Open Loop" operation.

Test Description: Numbers below refer to circled numbers on the diagnostic chart.
1. Code 13 will set if:
   - Engine at normal operating temperature
   - At least 2 minutes engine time after start.
   - Oxygen sensor signal voltage steady between .35 and .55 volts.
   - Throttle position sensor signal above 4%.
   - All conditions must be met for about 60 seconds.
   If the conditions for a Code 13 exist, the system will not go "Closed Loop".
2. This will determine if the sensor is at fault or the wiring or ECM is the cause of Code 13.
3. In doing this test, use only a high impedance digital volt ohm meter. This test checks the continuity of CKT's 412 and 413 because if CKT 413 is open the ECM voltage on CKT 412 will be over .6 volts (600 mv).

Diagnostic Aids:
Normal "Scan" voltage varies between 100 mv to 999 mv (.1 and 1.0 volt) while in closed loop. Code 13 sets in one minute if voltage remains between .35 and .55 volts, but the system will go "Open Loop" in about 15 seconds. Refer to "ECM Intermittent Codes or Performance" in Section "2".

1988–90 LIGHT TRUCKS AND VANS

CODE 13
OXYGEN SENSOR OPEN CIRCUIT
ALL ENGINES

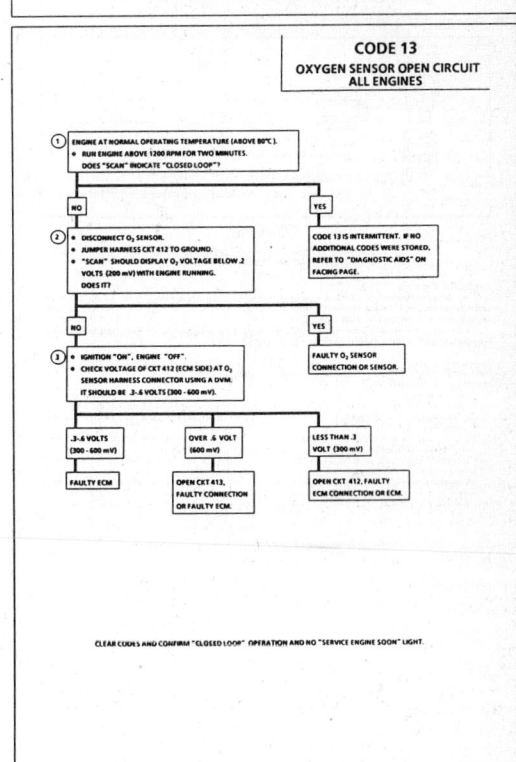

1988–90 LIGHT TRUCKS AND VANS

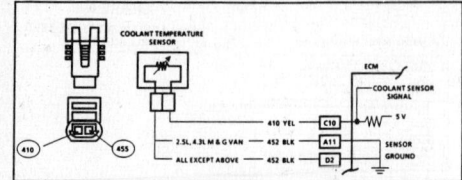

CODE 14
COOLANT TEMPERATURE SENSOR CIRCUIT
(HIGH TEMPERATURE INDICATED)
ALL ENGINES

Circuit Description:

The coolant temperature sensor is a thermistor that controls the signal voltage to the ECM. The ECM applies a voltage on CKT 410 to the sensor. When the engine is cold, the sensor (thermistor) resistance is high, therefore the ECM will see high signal voltage.

As the engine warms, the sensor resistance becomes less and the voltage drops. At normal engine operating temperature (85°C to 95°C), the voltage will measure about 1.5 to 2.0 volts.

Test Description: Numbers below refer to circled numbers on the diagnostic chart.
1. Code 14 will set if:
   - Signal voltage indicates a coolant temperature above 130°C (266°F) for 3 seconds.
2. This test will determine if CKT 410 is shorted to ground which will cause the conditions for Code 14.

Diagnostic Aids:

Check harness routing for a potential short to ground in CKT 410.

"Scan" tool displays engine temp. in degrees centigrade. After engine is started, the temperature should rise steadily to about 90°C then stabilize when thermostat opens.

See "ECM Intermittent Codes or Performance" in Section "2".

The "Temperature to Resistance Value" scale at the right may be used to test the coolant sensor at various temperature levels to evaluate the possibility of a "slewed" (mis-scaled) sensor. A "slewed" sensor could result in poor driveability complaints.

1988–90 LIGHT TRUCKS AND VANS

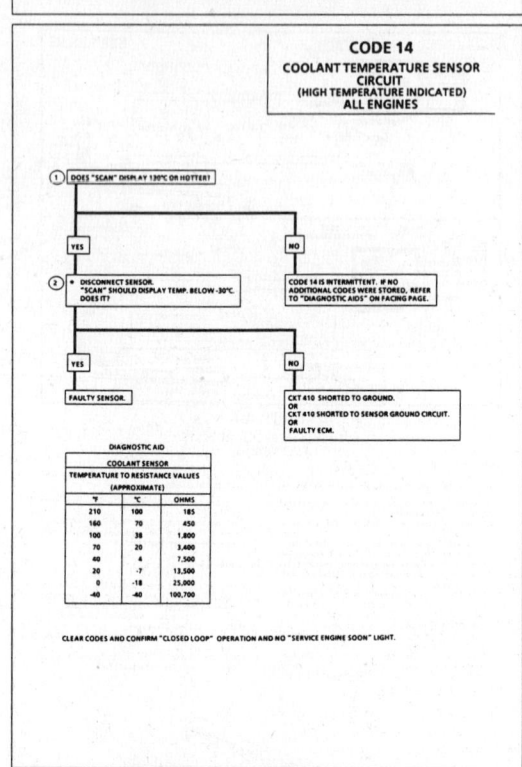

CODE 14
COOLANT TEMPERATURE SENSOR
CIRCUIT
(HIGH TEMPERATURE INDICATED)
ALL ENGINES

DIAGNOSTIC AID

COOLANT SENSOR		
TEMPERATURE TO RESISTANCE VALUES		
(APPROXIMATE)		
°F	°C	OHMS
210	100	185
160	70	450
100	38	1,800
70	20	3,400
40	4	7,500
20	-7	13,500
0	-18	25,000
-40	-40	100,700

CLEAR CODES AND CONFIRM "CLOSED LOOP" OPERATION AND NO "SERVICE ENGINE SOON" LIGHT.

1988–90 LIGHT TRUCKS AND VANS

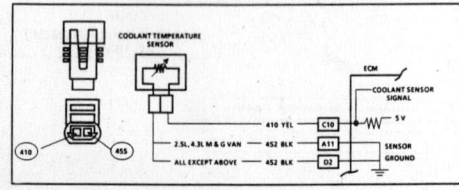

CODE 15
COOLANT TEMPERATURE SENSOR CIRCUIT
(LOW TEMPERATURE INDICATED)
ALL ENGINES

Circuit Description:

The coolant temperature sensor is a thermistor that controls the signal voltage to the ECM. The ECM applies a voltage on CKT 410 to the sensor. When the engine is cold, the sensor (thermistor) resistance is high, therefore the ECM will see high signal voltage.

As the engine warms, the sensor resistance becomes less and the voltage drops. At normal engine operating temperature (85°C to 95°C), the voltage will measure about 1.5 to 2.0 volts.

Test Description: Numbers below refer to circled numbers on the diagnostic chart.
1. Code 15 will set if:
   - Signal voltage indicates a coolant temperature less than -44°C (-47°F) for 3 seconds.
2. This test simulates a Code 14. If the ECM recognizes the low signal voltage, (high temp.) and the "Scan" reads 130°C or above, the ECM and wiring are OK.
3. This test will determine if CKT 410 is open. There should be 5 volts present at sensor connector if measured with a DVOM.

Diagnostic Aids:

A "Scan" tool reads engine temperature in degrees centigrade. After engine is started, the temperature should rise steadily to about 90°C then stabilize when thermostat opens.

A faulty connection, or an open in CKT 410 or 452 will results in a Code 15.

See "ECM Intermittent Codes on Performance" in Section "2".

The "Temperature To Resistance Value" scale at the right may be used to test the coolant sensor at various temperature levels to evaluate the possibility of a "slewed" (mis-scaled) sensor. A "slewed" sensor could result in poor driveability complaints.

1988–90 LIGHT TRUCKS AND VANS

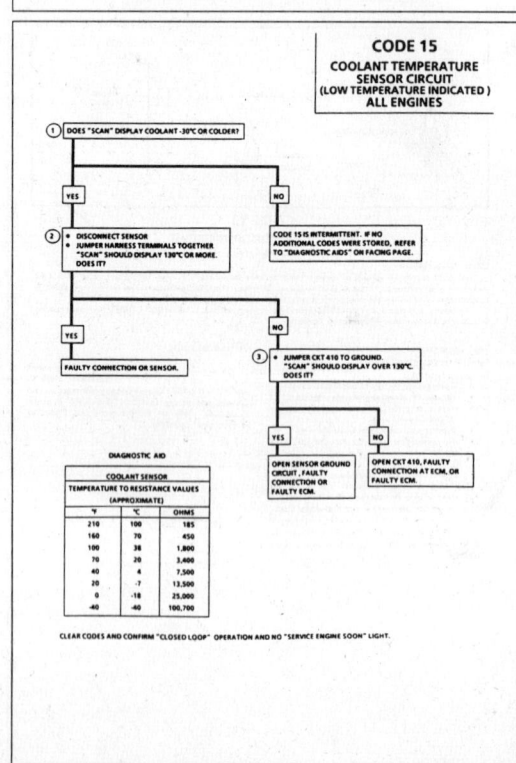

CODE 15
COOLANT TEMPERATURE
SENSOR CIRCUIT
(LOW TEMPERATURE INDICATED)
ALL ENGINES

DIAGNOSTIC AID

COOLANT SENSOR		
TEMPERATURE TO RESISTANCE VALUES		
(APPROXIMATE)		
°F	°C	OHMS
210	100	185
160	70	450
100	38	1,800
70	20	3,400
40	4	7,500
20	-7	13,500
0	-18	25,000
-40	-40	100,700

CLEAR CODES AND CONFIRM "CLOSED LOOP" OPERATION AND NO "SERVICE ENGINE SOON" LIGHT.

1988–90 LIGHT TRUCKS AND VANS

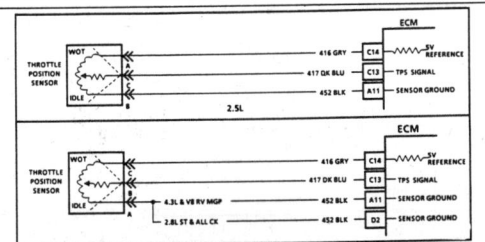

CODE 21
THROTTLE POSITION SENSOR (TPS) CIRCUIT
(SIGNAL VOLTAGE HIGH)
ALL ENGINES

Circuit Description:
The Throttle Position Sensor (TPS) provides a voltage signal that changes relative to the throttle blade angle. Signal voltage will vary from about .5 volts at idle to about 5 volts at wide open throttle.
The TPS signal is one of the most important inputs used by the ECM for fuel control and for most of the ECM control outputs.

Test Description: Numbers below refer to circled numbers on the diagnostic chart.
1. Code 21, will set if:
   * Engine running
   * TPS signal voltage is greater than about 3.5 volts
   * All conditions met for 5 seconds.
   OR
   * TPS signal voltage over 4.5 volts with ignition "ON".
   With throttle closed, the TPS should read less than .70 volts. If it doesn't, check adjustment.
2. With the TPS sensor disconnected, the TPS voltage should go low if the ECM and wiring is OK.
3. Probing CKT 452 with a test light checks the 5 volts return CKT, because a faulty 5 volts return will cause a Code 21.

Diagnostic Aids:
A "Scan" tool reads throttle position in volts. Should read about .73 volts (2.5L), .45 volts (2.8L), .60 volts (4.3L & V8) ± .75 volts with throttle closed and ignition on or at idle. Voltage should increase at a steady rate as throttle is moved toward WOT.
Also some "Scan" tools will read throttle angle. 0% = closed throttle 100% = WOT.
Refer to Section "2" for "ECM Intermittent Codes or Performance".
"Scan" TPS while depressing accelerator pedal with engine stopped and ignition "ON". Display should vary from below 2.5 volts (2500 mv) when throttle was closed, to over 4.5 volts (4500 mv) when throttle is held at wide open throttle position.

1988–90 LIGHT TRUCKS AND VANS

CODE 21
THROTTLE POSITION SENSOR (TPS) CIRCUIT
(SIGNAL VOLTAGE HIGH)
ALL ENGINES

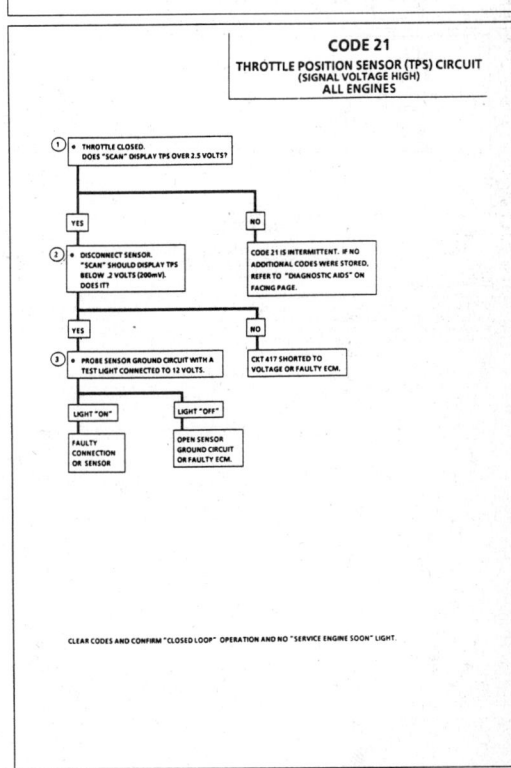

CLEAR CODES AND CONFIRM "CLOSED LOOP" OPERATION AND NO "SERVICE ENGINE SOON" LIGHT.

1988–90 LIGHT TRUCKS AND VANS

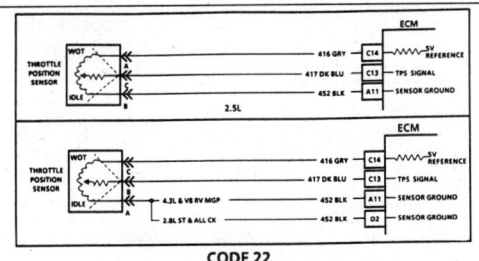

CODE 22
THROTTLE POSITION SENSOR (TPS) CIRCUIT
(SIGNAL VOLTAGE LOW)
ALL ENGINES

Circuit Description:
The Throttle Position Sensor (TPS) provides a voltage signal that changes relative to the throttle blade. Signal voltage will vary from about .5 at idle to about 5 volts at wide open throttle.
The TPS signal is one of the most important inputs used by the ECM for fuel control and for most of the ECM control outputs.

Test Description: Numbers below refer to circled numbers on the diagnostic chart.
1. Code 22, will set if:
   * Engine running
   * TPS signal voltage is less than about .2 volt for 3 seconds.
2. Simulates Code 21: (high voltage) If the ECM recognizes the high signal voltage the ECM and wiring are OK.
3. TPS - 2.8L: Refer to "Adjustable TPS Output Check" in this section. TPS - except 2.8L: Replace TPS.
4. This simulates a high signal voltage to check for an open in CKT 417.

Diagnostic Aids:
A "Scan" tool reads throttle position in volts. Should read about .73 volts (2.5L), .48 volts (2.8L), .60 volts (4.3L & V8) ± .08 volts with throttle closed and ignition on or at idle. Voltage should increase at a steady rate as throttle is moved toward WOT.
An open or short to ground in CKTs 416 or 417 will result in a Code 22.
Refer to Section "2" for "ECM Intermittent Codes or Performance".
"Scan" TPS while depressing accelerator pedal with engine stopped and ignition "ON". Display should vary from below 2.5 volts (2500 mv) when throttle was closed, to over 4.5 volts (4500 mv) when throttle is held at wide open throttle position.

1988–90 LIGHT TRUCKS AND VANS

CODE 22
THROTTLE POSITION SENSOR
(TPS) CIRCUIT
(SIGNAL VOLTAGE LOW)
ALL ENGINES

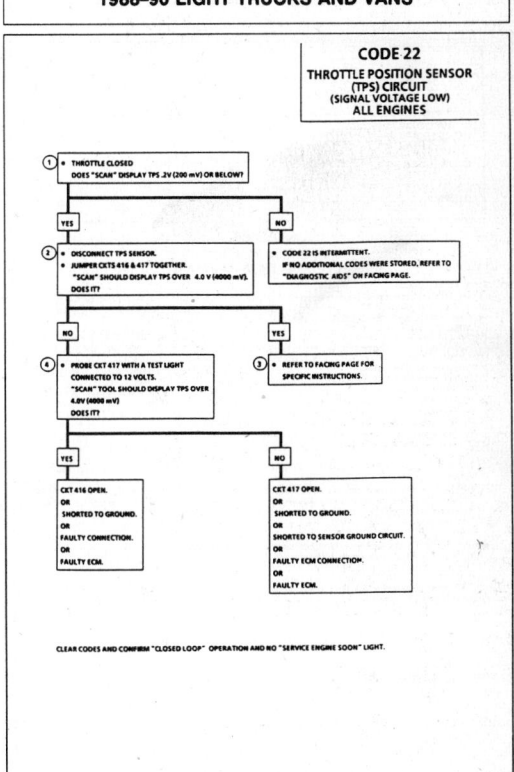

CLEAR CODES AND CONFIRM "CLOSED LOOP" OPERATION AND NO "SERVICE ENGINE SOON" LIGHT.

1988–90 LIGHT TRUCKS AND VANS

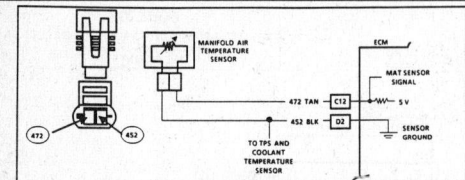

CODE 23
MANIFOLD AIR TEMPERATURE (MAT) SENSOR CIRCUIT
(LOW TEMPERATURE INDICATED)
2.5L ENGINE

Circuit Description:

The Manifold Air Temperature (MAT) Sensor is a thermistor that controls the signal voltage to the ECM. The ECM applies a voltage (4-6 volts) on CKT 472 to the sensor. When the air is cold, the sensor (thermistor) resistance is high, therefore the ECM will see a high signal voltage. If the air is warm, the sensor resistance is low therefore the ECM will see a low voltage.

Test Description: Numbers below refer to circled numbers on the diagnostic chart.
1. Code 23 will set if all conditions are met.
   - A signal voltage indicates a manifold air temperature below -30°C (-22°F) for 12 seconds.
   - Time since engine start is 1 minute or longer.
   - No VSS (vehicle not moving)
2. A Code 23 will set, due to an open sensor, wire, or connection. This test will determine if the wiring and ECM are OK.
3. This will determine if the signal CKT 472 or the 5V return CKT 452 is open.

Diagnostic Aids:

A "Scan" tool reads temperature of the air entering the engine and should read close to ambient air temperature when engine is cold, and rises as underhood temperature increases.

Carefully check harness and connections for possible open CKT 472 or 452.

Refer to Section "2" for "ECM Intermittent Codes or Performance."

The "Temperature to Resistance Value" chart at the right may be used to test the MAT sensor at various temperature levels to evaluate the possibility of a "slewed" (mis-scaled) sensor. A "slewed" sensor could result in poor driveability complaints.

1988–90 LIGHT TRUCKS AND VANS

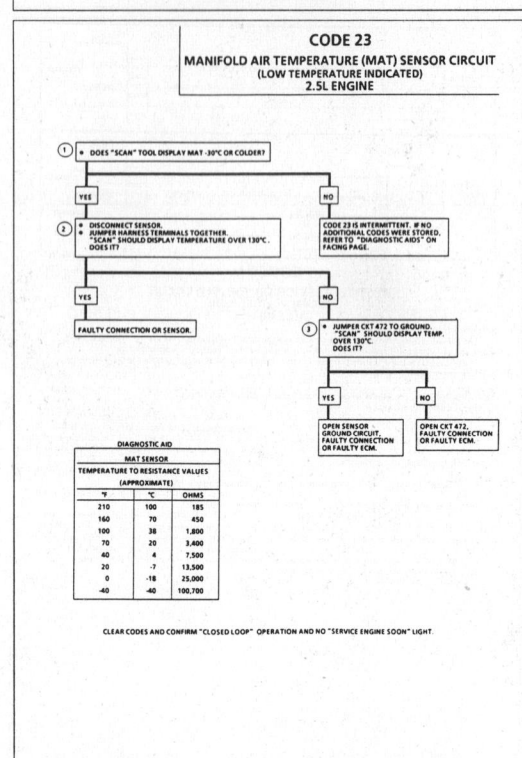

CODE 23
MANIFOLD AIR TEMPERATURE (MAT) SENSOR CIRCUIT
(LOW TEMPERATURE INDICATED)
2.5L ENGINE

DIAGNOSTIC AID
MAT SENSOR
TEMPERATURE TO RESISTANCE VALUES
(APPROXIMATE)

°F	°C	OHMS
210	100	185
160	70	450
100	38	1,800
70	20	3,400
40	4	7,500
20	-7	13,500
0	-18	25,000
-40	-40	100,700

CLEAR CODES AND CONFIRM "CLOSED LOOP" OPERATION AND NO "SERVICE ENGINE SOON" LIGHT.

1988–90 LIGHT TRUCKS AND VANS

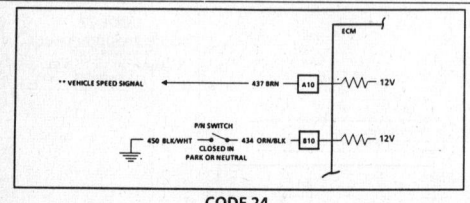

CODE 24
VSS CIRCUIT FAULT
ALL ENGINES

Circuit Description:

The ECM applies and monitors 12 volts on CKT 437. CKT 437 connects to the Vehicle Speed Sensor (VSS) which alternately grounds CKT 437 when drive wheels are turning. This pulsing action takes place about 2000 times per mile and the ECM will calculate vehicle speed based on the time between "pulses".

A "Scan" tool reading should closely match with speedometer reading with drive wheels turning.

Test Description: Numbers below refer to circled numbers on the diagnostic chart.
Code 24 will set if:
   - CKT 437 voltage is constant.
   - Engine speed is more than 200 rpm.
   - Vehicle speed signal (voltage on terminal "A9" is less than 10 mph (16 k/mh)
   - All conditions must be met for 10 seconds.
These conditions are met during a road load deceleration.
1. This test monitors the ECM voltage on CKT 437. With the wheels turning, the pulsing action will result in a varying voltage. The variation will be greater at low wheel speeds to an average of 4-6 volts at about 20 mph (32 km/h)
2. A voltage of less than 1 volt at the ECM connector indicates that the CKT 437 wire is shorted to ground. Disconnect CKT 437 at the vehicle speed sensor. If voltage now reads above 10 volts, the vehicle speed sensor is faulty. If voltage remains less than 10 volt, then CKT 437 wire is grounded. If 437 is not grounded, check for a faulty ECM connector or ECM.

3. A steady 8-12 volts at the ECM connector indicates CKT 437 is open or a faulty vehicle speed sensor.
4. This is normal voltage which indicates a possible intermittent condition.

Diagnostic Aids:

1. "Scan" reading should closely match with speedometer reading, with drive wheels turning.
2. Check park/neutral switch diagnosis chart if vehicle equipped with automatic transmission.
3. If park/neutral switch is OK, refer to "ECM Intermittent Codes or Performance" in Section "2".

1988–90 LIGHT TRUCKS AND VANS

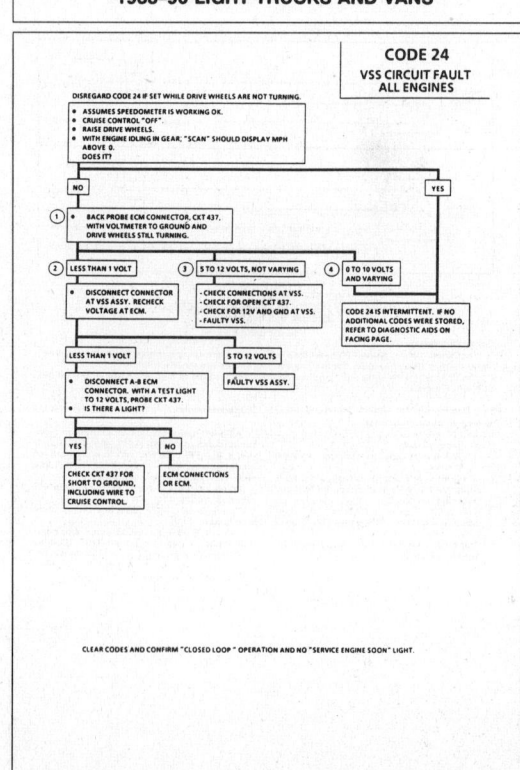

CODE 24
VSS CIRCUIT FAULT
ALL ENGINES

CLEAR CODES AND CONFIRM "CLOSED LOOP" OPERATION AND NO "SERVICE ENGINE SOON" LIGHT.

1988–90 LIGHT TRUCKS AND VANS

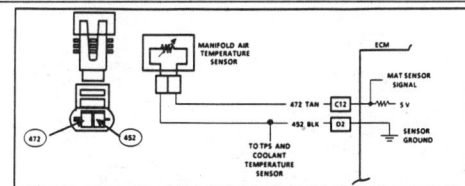

CODE 25
MANIFOLD AIR TEMPERATURE (MAT) SENSOR CIRCUIT
(HIGH TEMPERATURE INDICATED)
2.5L ENGINE

Circuit Description:

The Manifold Air Temperature (MAT) Sensor is a thermistor that controls the signal voltage to the ECM. The ECM applies a voltage (4-6 volts) on CKT 472 to the sensor. When the air is cold, the sensor (thermistor) resistance is high, therefore the ECM will see a high signal voltage. As the air warms, the sensor resistance becomes less, and the voltage drops.

Test Description: Numbers below refer to circled numbers on the diagnostic chart.

1 Code 25 will set if:
- Signal voltage indicates a manifold air temperature below 150°C (302°F) for 2 seconds.
- Time since engine start is 1 minute or longer.
- A vehicle speed is present.

Diagnostic Aids:

A "Scan" tool reads temperature of the air entering the engine and should read close to ambient air temperature when engine is cold, and rises as underhood temperature increases.

Check harness routing for possible short to ground in CKT 472.

Refer to Section "2" for "ECM Intermittent Codes or Performance."

The "Temperature to Resistance Value" scale at the right may be used to test the MAT sensor at various temperature levels to evaluate the possibility of a "slewed" (mis-scaled) sensor. A "slewed" sensor could result in poor driveability complaints.

1988–90 LIGHT TRUCKS AND VANS

CODE 25
MANIFOLD AIR TEMPERATURE (MAT) SENSOR CIRCUIT
(HIGH TEMPERATURE INDICATED)
ALL ENGINES

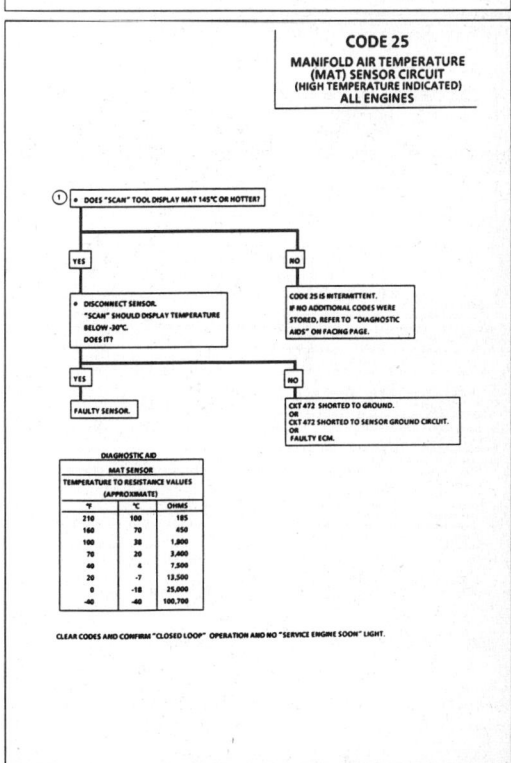

DIAGNOSTIC AID

MAT SENSOR		
TEMPERATURE TO RESISTANCE VALUES		
(APPROXIMATE)		
°F	°C	OHMS
210	100	185
160	70	450
100	38	1,800
70	20	3,400
40	4	7,500
20	-7	13,500
0	-18	25,000
-40	-40	100,700

CLEAR CODES AND CONFIRM "CLOSED LOOP" OPERATION AND NO "SERVICE ENGINE SOON" LIGHT.

1988–90 LIGHT TRUCKS AND VANS

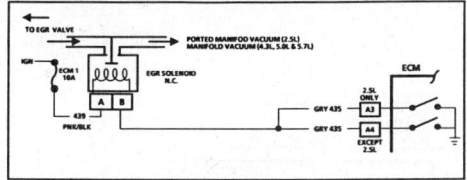

CODE 32
EXHAUST GAS RECIRCULATION (EGR) SYSTEM
2.5L & 5.0L
4.3L (EXCEPT ST)
5.7L (UNDER 8500 GW)

Circuit Description:

The ECM operates a solenoid to control the Exhaust Gas Recirculation (EGR) valve. This solenoid is normally closed. By providing a ground path, the ECM energizes the solenoid which then allows vacuum to pass to the EGR valve.

The ECM monitors EGR effectiveness by de-energizing the EGR control solenoid thereby shutting off vacuum to the EGR valve diaphragm. With the EGR valve closed, fuel integrator counts will be greater than they were during normal EGR operation. If the change is not within the calibrated window, a Code 32 will be set.

The ECM will check EGR operation when:
- Vehicle speed is above 50 mph.
- Engine vacuum is between 40 and 51 kPa.
- No change in throttle position while test is being run.

Test Description: Numbers below refer to circled numbers on the diagnostic chart.

1. By grounding the diagnostic terminal, the EGR solenoid should be energized and allow vacuum to be applied to the EGR valve and the vacuum should hold.
2. When the diagnostic terminal is ungrounded, the vacuum to the EGR valve should bleed off through a vent in the solenoid and the valve should close. The gage may or may not bleed off but this does not indicate a problem.
3. This test will determine if the electrical control part of the system is at fault or if the connector or solenoid is at fault.
4. This system uses a negative backpressure valve which should hold vacuum with engine "OFF".
5. When engine is started, exhaust backpressure should cause vacuum to bleed off and valve should fully close.

Diagnostic Aids:
- Before replacing ECM, use an ohmmeter and check the resistance of each ECM controlled relay and solenoid coil. Refer to ECM QDR Check procedure, Figure 3-18. See ECM wiring diagram for coil term. I.D. of solenoid(s) and relay(s) to be checked. Replace any solenoid where resistance measures less than 20 ohms.

1988–90 LIGHT TRUCKS AND VANS

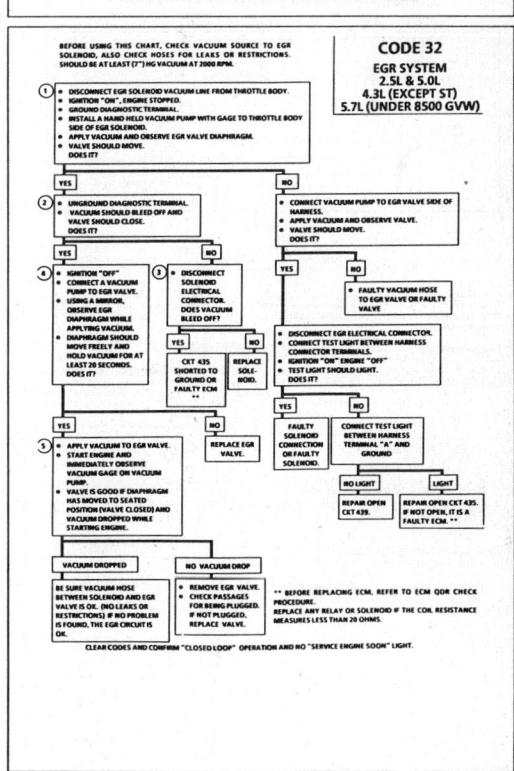

CODE 32
EGR SYSTEM
2.5L & 5.0L
4.3L (EXCEPT ST)
5.7L (UNDER 8500 GVW)

CLEAR CODES AND CONFIRM "CLOSED LOOP" OPERATION AND NO "SERVICE ENGINE SOON" LIGHT.

1988–90 LIGHT TRUCKS AND VANS

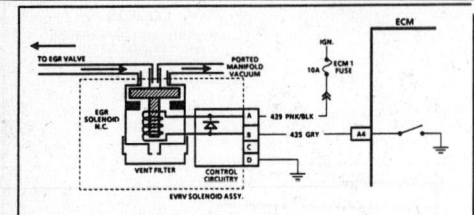

CODE 32
EGR SYSTEM
2.8L, 4.3L (ST), 7.4L & 5.7L (OVER 8500 GVW)

Circuit Description:
The ECM operates a solenoid to control the Exhaust Gas Recirculation (EGR) valve. This solenoid is normally closed. By providing a ground path, the ECM energizes the solenoid which then allows vacuum to pass to the EGR valve.

The ECM monitors EGR effectiveness by de-energizing the EGR control solenoid thereby shutting off vacuum to the EGR valve diaphragm. With the EGR valve closed, fuel integrator counts will be greater than they were during normal EGR operation. If the change is not within the calibrated window, a Code 32 will be set.

The ECM will check EGR operation when:
- Vehicle speed is above 50 mph.
- Engine vacuum is between 40 and 51 kPa.
- No change in throttle position while test is being run.

Test Description: Numbers below refer to circled numbers on the diagnostic chart.
1. With the ignition "ON," engine stopped, the solenoid should not be energized and vacuum should not pass to the EGR valve. Grounding the diagnostic terminal will energize the solenoid and allow vacuum to pass to valve.
2. Checks for plugged EGR passages. If passages are plugged, the engine may have severe detonation on acceleration.
3. The EGR solenoid will not be energized in Park or Neutral. This test will determine if the Park/Neutral switch input is being received by the ECM.

Diagnostic Aids:
- Before replacing ECM, use an ohmmeter and check the resistance of each ECM controlled relay and solenoid coil. Refer To ECM QDR check procedure, Figure 3-18.
  See ECM wiring diagram for coil terminal identification of solenoid(s) and relay(s) to be checked.
  Replace any solenoid where resistance measures less than 20 ohms.

1988–90 LIGHT TRUCKS AND VANS

IF ANY OTHER CODES ARE STORED, DIAGNOSE THEM FIRST.

CODE 32
EGR SYSTEM
2.8L, 4.3L (ST), 7.4L & 5.7L (OVER 8500 GVW)

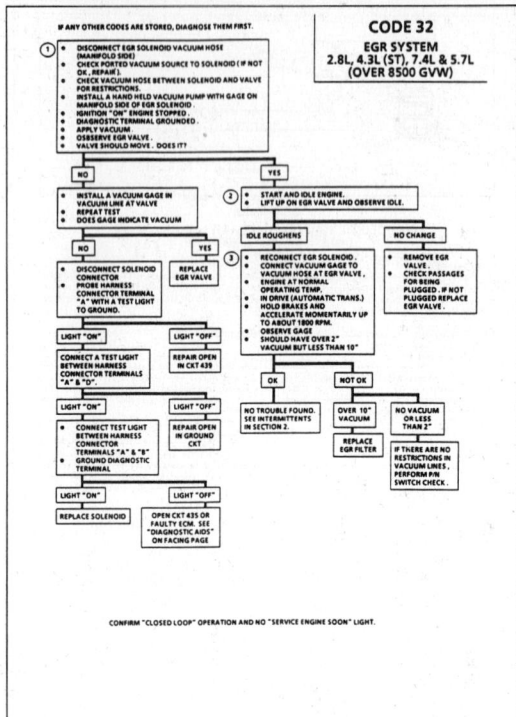

CONFIRM "CLOSED LOOP" OPERATION AND NO "SERVICE ENGINE SOON" LIGHT.

1988–90 LIGHT TRUCKS AND VANS

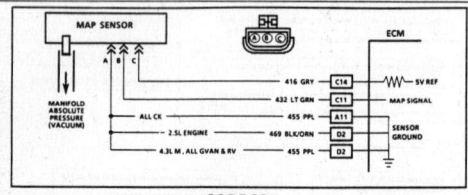

CODE 33
MAP SENSOR CIRCUIT
SIGNAL VOLTAGE HIGH
(LOW VACUUM)
ALL ENGINES

Circuit Description:
The Manifold Absolute Pressure (MAP) Sensor responds to changes in manifold pressure (vacuum). The ECM receives this information as a signal voltage that will vary from about 1 to 1.5 volts at idle to 4-4.5 volts at wide open throttle.

Test Description: Numbers below refer to circled numbers on the diagnostic chart.
1. Code 33 will set when:
   - Signal is too high for a time greater than 6 seconds.
   Engine misfire or a low unstable idle may set Code 33.
   - Engine Running:
   Manifold pressure greater than 75.3 kPa (A/C "OFF") 81.2 kPa (A/C "ON")
   Throttle angle less than 2%
   Conditions met for 2 seconds.
2. With the MAP sensor disconnect the ECM; should see a low voltage if the ECM and wiring is OK.

Diagnostic Aids:
The "Altitude To Voltage" scale at the right may be used to test the MAP sensor at a specific altitude level to evaluate the possibility of a "slewed" (mis-scaled) sensor. A "slewed" sensor could result in poor driveability complaints.

Engine misfire or a low unstable idle may set Code 33. Disconnect MAP sensor and system will go into backup mode. If the misfire or idle condition remains, see "Driveability Symptoms" in Section "2."

Refer to "ECM Intermittent Codes or Performance" in Section "2."

1988–90 LIGHT TRUCKS AND VANS

CODE 33
MAP SENSOR CIRCUIT
SIGNAL VOLTAGE HIGH
(LOW VACUUM)
ALL ENGINES

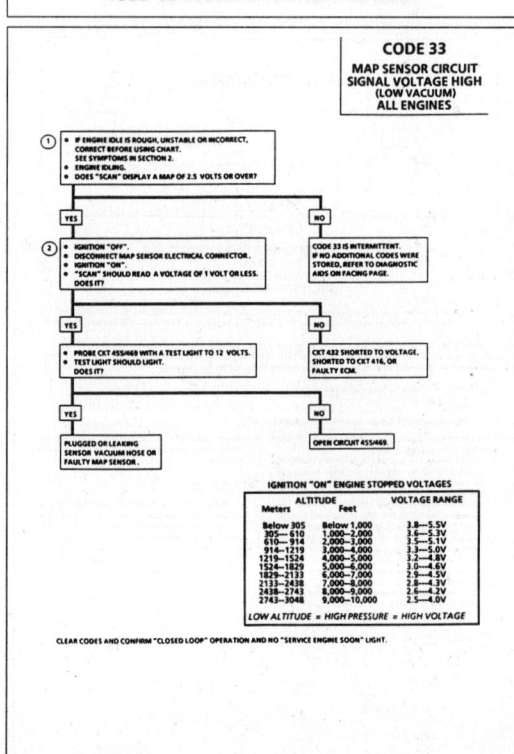

ALTITUDE		VOLTAGE RANGE
Meters	Feet	
Below 305	Below 1,000	3.8—5.5V
305— 610	1,000—2,000	3.6—5.3V
610— 914	2,000—3,000	3.5—5.1V
914—1219	3,000—4,000	3.3—5.0V
1219—1524	4,000—5,000	3.2—4.8V
1524—1829	5,000—6,000	3.0—4.6V
1829—2133	6,000—7,000	2.9—4.5V
2133—2438	7,000—8,000	2.8—4.3V
2438—2743	8,000—9,000	2.6—4.2V
2743—3048	9,000—10,000	2.5—4.0V

LOW ALTITUDE = HIGH PRESSURE = HIGH VOLTAGE

CLEAR CODES AND CONFIRM "CLOSED LOOP" OPERATION AND NO "SERVICE ENGINE SOON" LIGHT.

1988–90 LIGHT TRUCKS AND VANS

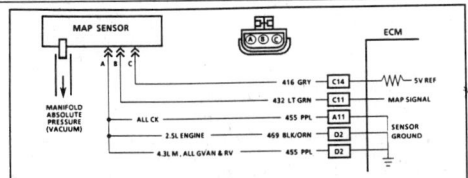

CODE 34
**MAP SENSOR CIRCUIT
SIGNAL VOLTAGE LOW
(HIGH VACUUM)
ALL ENGINES**

Circuit Description:
The Manifold Absolute Pressure (MAP) Sensor responds to changes in manifold pressure (vacuum). The ECM receives this information as a signal voltage that will vary from about 1 to 1.5 volts at idle to 4-4.5 volts at wide open throttle.
If the MAP sensor fails the ECM will substitute a fixed MAP value and use the Throttle Position Sensor (TPS) to control fuel delivery.

Test Description: Numbers below refer to circled numbers on the diagnostic chart.
1. Code 34 will set when:
   - When engine is less than 600 rpm.
   - Manifold pressure reading less than 13 kPa, conditions met for 1 second
   or
   - Engine is greater than 600 rpm.
   - Throttle angle over 20%.
   - manifold pressure less than 13 kPa conditions met for 1 second.
2. This tests to see if the sensor is at faulty for the low voltage or if there is a ECM or wiring problem.
3. This simulates a high signal voltage to check of an open in CKT 432. If the test light is bright during this test, CKT 432 is probable shorted to ground. If "Scan" reads over 4 volts at this test CKT 416 can be checked by measuring the voltage at terminal "C". (should be 5 volts)

Diagnostic Aids:
An intermittent open in CKTs 416 will result in a Code 34.
Refer to "ECM Intermittent Codes or Performance" in Section "2".
The "Altitude to Voltage" scale at the right may be used to test the MAP sensor at a specific altitude level to evaluate the possibility of a "slewed" (misscaled) sensor. A "slewed" sensor could result in poor driveability complaints.

1988–90 LIGHT TRUCKS AND VANS

CODE 34
**MAP SENSOR CIRCUIT
SIGNAL VOLTAGE LOW
(HIGH VACUUM)
ALL ENGINES**

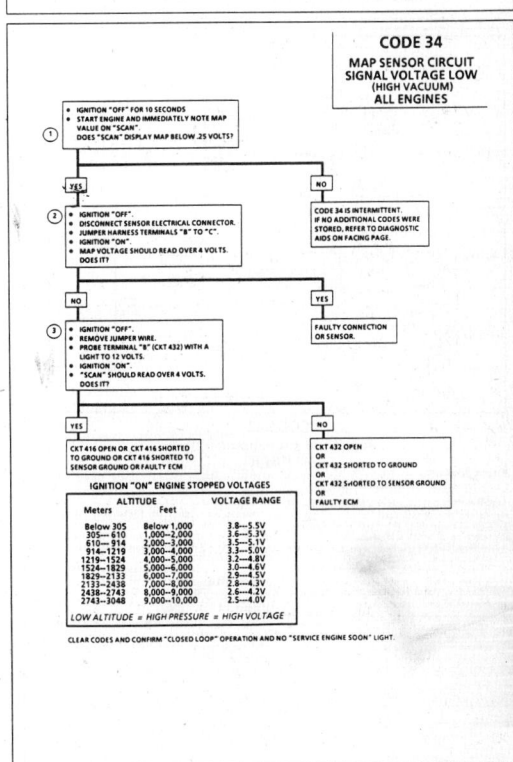

ALTITUDE		VOLTAGE RANGE
Meters	Feet	
Below 305	Below 1,000	3.8—5.5V
305—610	1,000—2,000	3.6—5.3V
610—914	2,000—3,000	3.5—5.1V
914—1219	3,000—4,000	3.3—5.0V
1219—1524	4,000—5,000	3.2—4.8V
1524—1829	5,000—6,000	3.0—4.6V
1829—2133	6,000—7,000	2.9—4.5V
2133—2438	7,000—8,000	2.8—4.3V
2438—2743	8,000—9,000	2.6—4.2V
2743—3048	9,000—10,000	2.5—4.0V

LOW ALTITUDE = HIGH PRESSURE = HIGH VOLTAGE

CLEAR CODES AND CONFIRM "CLOSED LOOP" OPERATION AND NO "SERVICE ENGINE SOON" LIGHT.

1988–90 LIGHT TRUCKS AND VANS

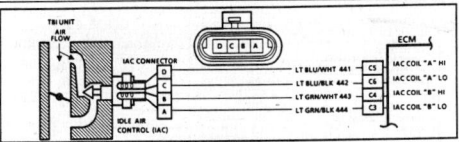

CODE 35
**IDLE AIR CONTROL (IAC) SYSTEM
2.5L ENGINE**

Circuit Description:
Code 35 will set when the closed throttle engine speed is 100 rpm above or below the correct engine idle speed for 45 seconds.

Test Description: Numbers below refer to circled numbers on the diagnostic chart.
1. Continue with test even if engine will not idle. If idle is too low, "Scan" will display 80 or more counts, or steps. If idle is high, it will display "0" counts.
Occasionally, an erratic or unstable idle may occur. Engine speed may vary 200 rpm or more up and down. Disconnect IAC. If the condition is unchanged, the IAC is not faulty, there is a system problem. Proceed to paragraph three below.
2. When the engine was stopped, the IAC valve retracted (more air) to a fixed "Park" position for increased air flow and idle speed during the next engine start. A "Scan" will display 100 or more counts.
3. He sure to disconnect IAC valve prior to this test. The test light will confirm the ECM signals by a steady or flashing light on all circuits.
4. There is a remote possibility that one of the circuits is shorted to voltage which would have been indicated by a steady light. Disconnect ECM and turn the ignition "ON" and probe terminals to check for voltage.

Diagnostic Aids:
A slow unstable idle may be caused by a system problem that cannot be overcome by the IAC. "Scan" counts will be above 60 counts if idle too low and "0" counts if too high.
If idle is too high, stop engine. Ignition "ON" Ground diagnostic terminal. Wait a few seconds for IAC to seat, then disconnect IAC. Start engine. If idle speed is above 800 ± 50 rpm, locate and correct vacuum leak.

- **System too lean.** (High air/fuel ratio)
Idle speed may be too high or too low. Engine speed may vary up and down, disconnecting IAC does not help. May set Code 44.
"Scan" and/or Voltmeter will read an oxygen sensor output less than 300 mv (.3 volts). Check for low regulated fuel pressure or water in fuel. A lean exhaust with an oxygen sensor output fixed above 800 mv (.8 volts) will be a contaminated sensor, usually silicone. This may also set a Code 45.
- **System too rich** (Low air/fuel ratio)
Idle speed too low. "Scan" counts usually above 80. System obviously rich and may exhibit black smoke exhaust.
"Scan" tool and/or voltmeter will read an oxygen sensor signal fixed above 800 mv (.8 volts). Check:
  - High fuel pressure
  - Injector leaking or sticking.
- **Throttle body.**
Remove IAC and inspect bore for foreign material or evidence of IAC valve dragging the bore.
- **A/C Compressor or relay failure**
See if A/C diagnosis circuit if shorted to ground. If the relay is faulty, an idle problem may exist.
- Refer to "Rough, Unstable, Incorrect Idle or Stalling" in "Driveability Symptoms" in Section "2".

1988–90 LIGHT TRUCKS AND VANS

CODE 35
**IDLE AIR CONTROL (IAC) SYSTEM
2.5L ENGINE**

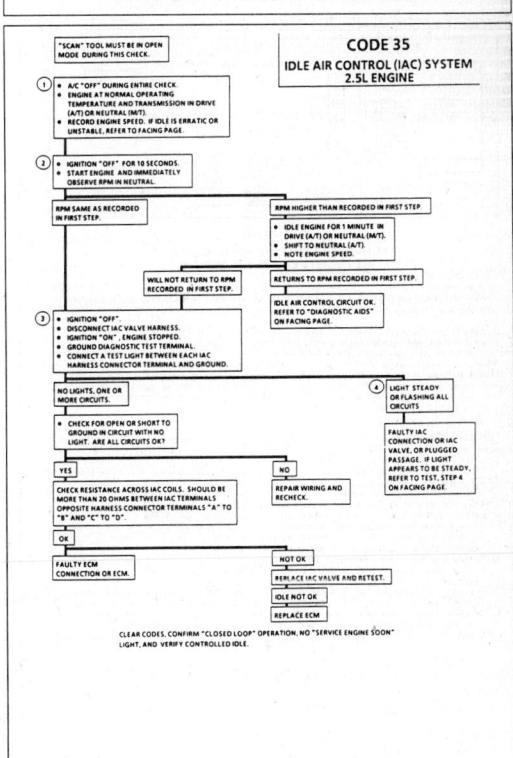

CLEAR CODES, CONFIRM "CLOSED LOOP" OPERATION, NO "SERVICE ENGINE SOON" LIGHT, AND VERIFY CONTROLLED IDLE.

1988–90 LIGHT TRUCKS AND VANS

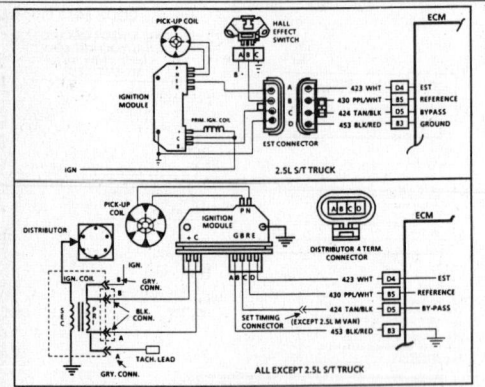

CODE 42
ELECTRONIC SPARK TIMING (EST)
ALL ENGINES

Circuit Description
Refer to page 3-9 for EST and Code 42.

Test Description: Numbers below refer to circled numbers on the diagnostic chart.
1. Code 42 means the ECM has seen an open or short to ground in the EST or bypass circuits. This test confirms Code 42 and that the fault causing the code is present.
2. Checks for a normal EST ground path through the ignition module. An EST CKT 423 shorted to ground will also read less than 500 ohms; however, this will be checked later.
3. As the test light voltage touches CKT 424, the module should switch causing the ohmmeter to "overrange" if the meter is in the 100-200 ohms position

Selecting the 10-20,000 ohms position will indicate above 5000 ohms. The important thing is that the module "switched"
4. The module did not switch and this step checks for:
   • EST CKT 423 shorted to ground.
   • Bypass CKT 424 open.
   • Faulty ignition module connection or module.
5. Confirms that Code 42 is a faulty ECM and not an intermittent in CKTs 423 or 424.

Diagnostic Aids:
The "Scan" tool does not have any ability to help diagnose a Code 42 problem.
Refer to Section "2" for "ECM Intermittent Code or Performance."

1988–90 LIGHT TRUCKS AND VANS

CODE 42
ELECTRONIC SPARK TIMING (EST)
ALL ENGINES

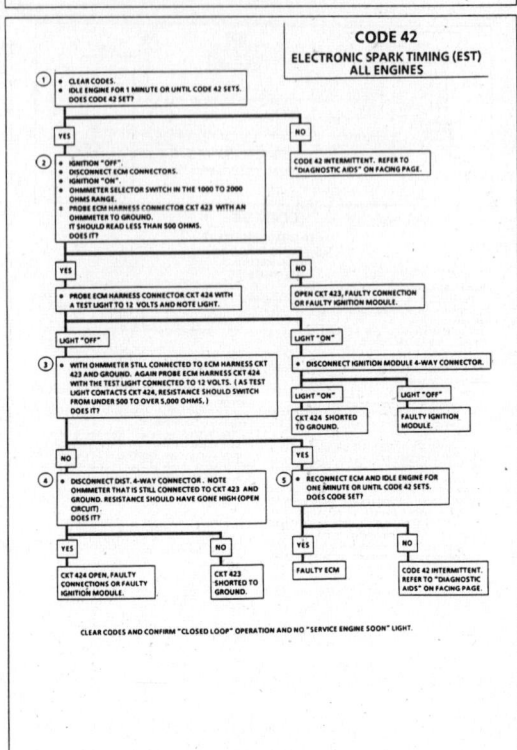

1988–90 LIGHT TRUCKS AND VANS

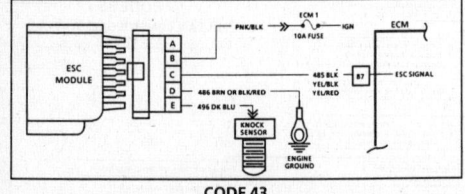

CODE 43
ELECTRONIC SPARK CONTROL (ESC) CIRCUIT
ALL ENGINES EXCEPT 2.5L AND 7.4L

Circuit Description:
Electronic spark control is accomplished with a module that sends a voltage signal to the ECM. As the knock sensor detects engine knock, the voltage from the ESC module to the ECM drops, and this signals the ECM to retard the timing. The ECM will retard the timing when knock is detected and rpm is above about 900 n.
Code 43 means the ECM has seen low voltage at CKT 485 terminal "B7" for longer then 5 seconds with the ine running or the system has failed the functional check.
This system performs a functional check per start-up to check the ESC system. To perform this test, the ECM will advance the spark when coolant is above 95°C and at a high load condition (near W.O.T.). The ECM then checks the signal at "B7" to see if a knock is detected. The functional check is performed once per start-up and if knock is detected when coolant is below 95°C (194°F), the test has passed and the functional check will not be run. If the functional check fails, the "Service Engine Soon" light will remain on until ignition is turned "OFF" or until a knock signal is detected.

Test Description: Numbers below refer to circled numbers on the diagnostic chart.
1. If the conditions for a Code 43 are present, the "Scan" will always display "yes". There should not be a knock at idle unless an internal engine problem, or a system problem exists.
2. This test will determine if the system is functioning at this time. Usually a knock signal can be generated by tapping on the right exhaust manifold. If no knock signal is generated, try tapping on block close to the area of the sensor.
3. Because Code 43 sets when the signal voltage on CKT 485 remains low, this test should cause the signal on CKT 485 to go high. The 12 volts signal should be seen by the ECM as "no knock" if the ECM and wiring are OK
4. This test will determine if the knock signal is being detected on CKT 496 or if the ESC module is at fault.

5. If CKT 496 is routed to close to secondary ignition wires, the ESC module may see the interference as a knock signal.
6. This checks the ground circuit to the module. An open ground will cause the voltage on CKT 485 to be about 12 volts which would cause the Code 43 functional test to fail.
7. Connecting CKT 496 with a test light to 12 volts should generate a knock signal. This will determine if the ESC module is operating correctly.

Diagnostic Aids:
Code 43 can be caused by a faulty connection at the knock sensor at the ESC module or at the ECM. Also check CKT 485 for possible open or short to ground.
Refer to Section "2" for "ECM Intermittent Codes or Performance"

1988–90 LIGHT TRUCKS AND VANS

CODE 43
ELECTRONIC SPARK CONTROL (ESC) CIRCUIT
ALL ENGINES EXCEPT 2.5L AND 7.4L

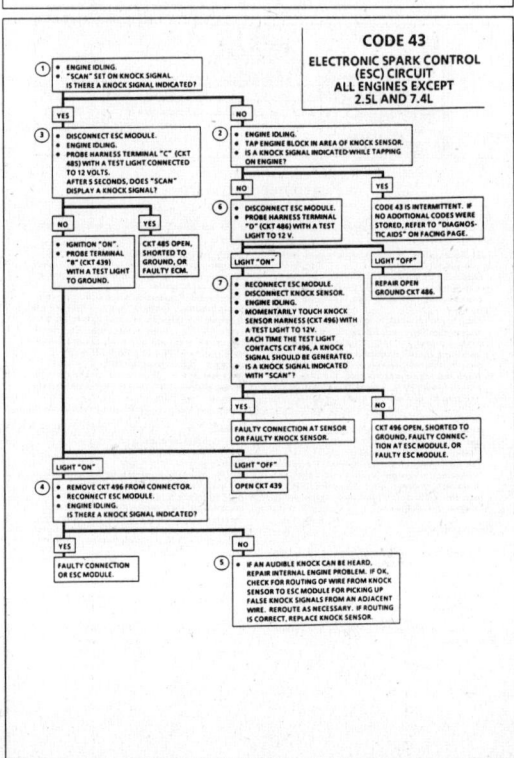

1988–90 LIGHT TRUCKS AND VANS

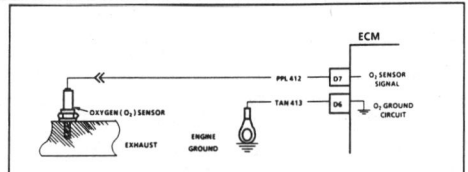

CODE 44
LEAN EXHAUST INDICATED
ALL ENGINES

Circuit Description:

The ECM supplies a voltage of about .45 volts between terminals "D6" and "D7". (If measured with a 10 meg ohm digital voltmeter, this may read as low as .32 volts). The oxygen sensor varies the voltage within a range of about 1 volt if the exhaust is rich, down through about .10 volts if exhaust is lean.

The sensor is like an open circuit and produces no voltage when it is below about 315°C (600°F). An open sensor circuit or cold sensor causes "Open Loop" operation.

Test Description: Numbers below refer to circled numbers on the diagnostic chart.
1. Code 44 is set when the oxygen sensor signal voltage on CKT 412:
   - Remains below .2 volts for 20 seconds.
   - And the system is operating in "Closed Loop".

Diagnostic Aids:

Using the "Scan", observe the block learn values at different rpm and air flow conditions. The "Scan" also displays the block learn cells, so the block learn values can be checked in each of the cells to determine when the Code 44 may have been set. If the conditions for Code 44 exists, the block learn values will be around 150.
- **Oxygen Sensor Wire.** Sensor pigtail may be mispositioned and contacting the exhaust manifold.
- Check for intermittent ground in wire between connector and sensor.

- **Fuel Contamination.** Water, even in small amounts, near the in-tank fuel pump inlet, can be delivered to the injectors. The water causes a lean exhaust and can set a Code 44.
- **Fuel Pressure.** System will be lean if pressure is too low. It may be necessary to monitor fuel pressure while driving the vehicle at various road speeds an/or loads to confirm. See Fuel System diagnosis.
- **AIR System.** Be sure air is not being directed to the exhaust ports while in "Closed Loop". If the block learn value goes down while squeezing air hose to left side of exhaust ports, refer to Section 8. If the above are OK, it is a faulty oxygen sensor.
- **CKT 413.** If CKT 413 is open, the voltage at terminal "D7" will be over one volt.
- **Sensor Harness.** Sensor pigtail may be mispositioned and contacting the exhaust manifold.
- If all check OK, the oxygen sensor is faulty.

1988–90 LIGHT TRUCKS AND VANS

CODE 44
LEAN EXHAUST INDICATED
ALL ENGINES

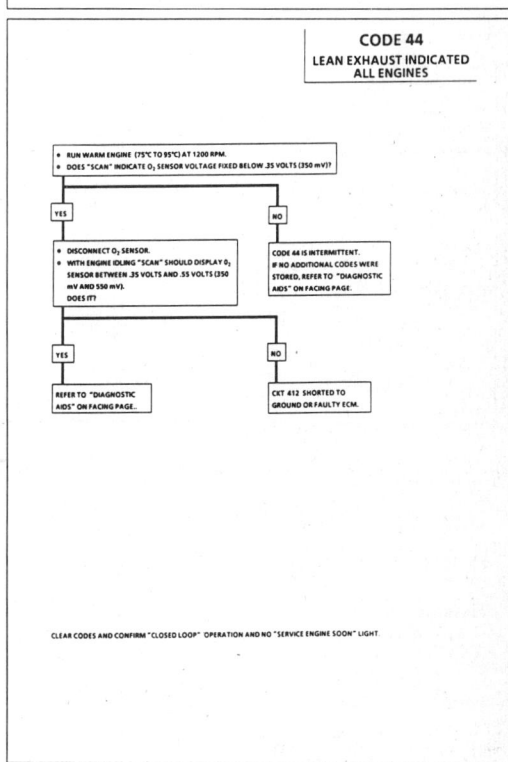

1988–90 LIGHT TRUCKS AND VANS

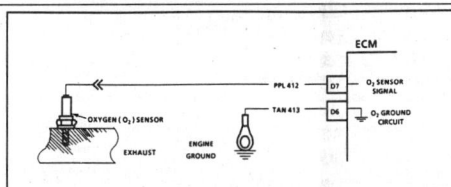

CODE 45
RICH EXHAUST INDICATED
ALL ENGINES

Circuit Description:

The ECM supplies a voltage of about .45 volts between terminals "D6" and "D7". (If measured with a 10 meg ohm digital voltmeter, this may read as low as .32 volts). The oxygen sensor varies the voltage within a range of about 1 volt if the exhaust is rich, down through about .10 volts if exhaust is lean.

The sensor is like an open circuit and produces no voltage when it is below about 315 °C (600°F). An open sensor circuit or cold sensor causes "Open Loop" operation.

Test Description: Numbers below refer to circled numbers on the diagnostic chart.
1. Code 45 is set when the oxygen sensor signal voltage or CKT 412:
   - Remains above .7 volts for 50 seconds, and in "Closed Loop".
   - Engine time after start is 1 minute or more.
   - Throttle angle greater than 2%. (about .2 volts above idle voltage)

Diagnostic Aids:

Using the "Scan", observe the block learn values at different rpm and air flow conditions. The "Scan" also displays the block cells, so the block learn values can be checked in each of the cells to determine when the Code 45 may have been set. If the conditions for Code 45 exists, the block learn values will be around 115.
- **Fuel Pressure** System will go rich if pressure is too high. The ECM can compensate for some increase. However, if it gets too high, a Code 45 may be set. See Fuel System diagnosis chart.
- **Leaking Injector.**
- Check for fuel contaminated oil.

- **HEI Shielding.** An open ground CKT 453 (ignition system reference low) may result in EMI, or induced electrical "noise". The ECM looks at this "noise" as reference pulses. The additional pulses result in a higher than actual engine speed signal. The ECM then delivers too much fuel, causing system to go rich. Engine tachometer will also show higher than actual engine speed which can help in diagnosing this problem.
- **Canister Purge.** Check for fuel saturation. If full of fuel, check canister control and hoses.
- **MAP Sensor.** An output that causes the ECM to sensor a higher than normal manifold pressure (low vacuum) can cause the system to go rich. Disconnecting the MAP sensor will allow the ECM to set a fixed value for the MAP sensor. Substitute a different MAP sensor if the rich condition is gone while the sensor is disconnected.
- **Pressure Regulator.** Check for leaking fuel pressure regulator diaphragm by checking for presence of liquid fuel in the vacuum line to the regulator.
- Check for leaking fuel pressure regulator diaphragm by checking vacuum line to regulator for fuel.
- **TPS.** An intermittent TPS output will cause the system to go rich, due to a false indication of the engine accelerating.

1988–90 LIGHT TRUCKS AND VANS

CODE 45
RICH EXHAUST INDICATED
ALL ENGINES

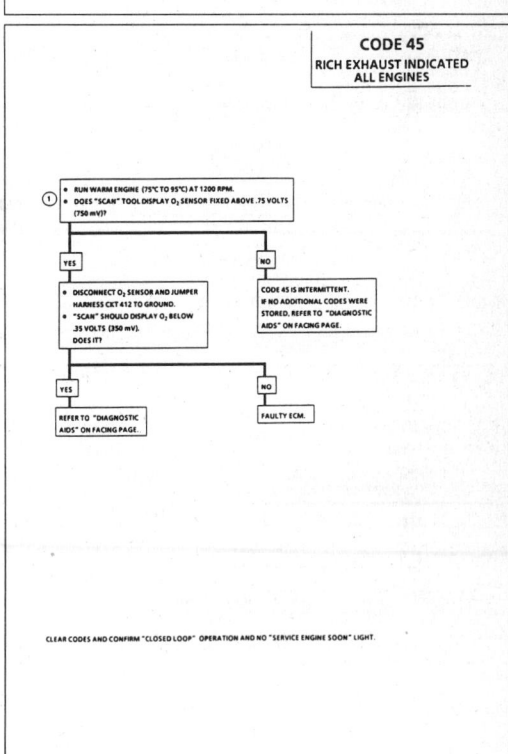

1988–90 LIGHT TRUCKS AND VANS

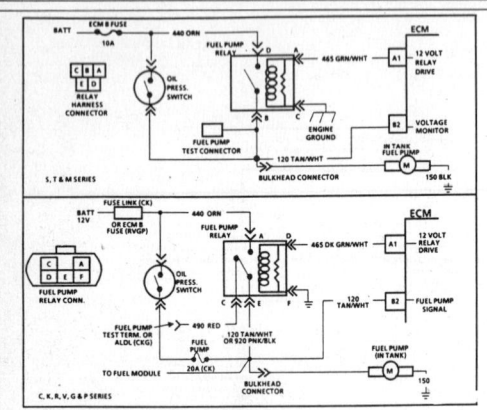

CODE 54
FUEL PUMP CIRCUIT
(LOW VOLTAGE)

Circuit Description

The status of the fuel pump CKT 120 is monitored by the ECM at terminal "B2" and is used to compensate fuel delivery based on system voltage. This signal is also used to store a code if the fuel pump relay is defective or fuel pump voltage is lost while the engine is running. There should be about 12 volts on CKT 120 for at least 2 seconds after the ignition is turned, or any time reference pulses are being received by the ECM.

Code 54 will set if the voltage at terminal "B2" is less than 2 volts for 1.5 seconds since the last reference pulse was received. This code is designed to detect a faulty relay, causing extended crank time, and the code will help the diagnosis of an engine that "Cranks But Will Not Run."

If a fault is detected during start-up, the "Service Engine Soon" light will stay "ON" until the ignition is cycled "OFF".

Diagnostic Aids:

- See "ECM Intermittent Codes or Performance" in Section "2"

1988–90 LIGHT TRUCKS AND VANS

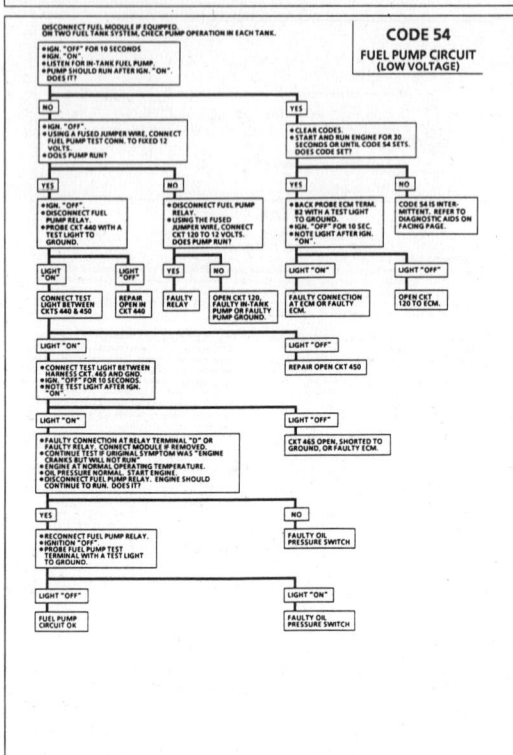

CODE 54
FUEL PUMP CIRCUIT
(LOW VOLTAGE)

1988–90 LIGHT TRUCKS AND VANS

CODE 51	CODE 51
CODE 52	FAULTY MEM-CAL
CODE 53	(2.5L ENGINE)
CODE 55	OR
	PROM PROBLEM
	(EXCEPT 2.5L ENGINE)

CHECK THAT ALL PINS ARE FULLY INSERTED IN THE SOCKET. IF OK, REPLACE PROM, CLEAR MEMORY, AND RECHECK. IF CODE 51 REAPPEARS, REPLACE ECM.

CODE 52
FUEL CALPAK MISSING
(EXCEPT 2.5L ENGINE)

CHECK FOR MISSING CALPAK AND THAT ALL PIN ARE FULLY INSERTED IN THE SOCKET. IF OK, REPLACE ECM.

CODE 53
SYSTEM OVER VOLTAGE
(2.5L ENGINE)

THIS CODE INDICATES THERE IS A BASIC GENERATOR PROBLEM.
- CODE 53 WILL SET IF VOLTAGE AT ECM TERMINAL B1 IS GREATER THAN 17.1 VOLTS FOR 2 SECONDS.
- CHECK AND REPAIR CHARGING SYSTEM.

CODE 55
ALL ENGINES
EXCEPT 2.5L ENGINE

BE SURE ECM GROUNDS ARE OK AND THAT MEM-CAL IS PROPERLY LATCHED. IF OK REPLACE ELECTRONIC CONTROL MODULE (ECM).

CLEAR CODES AND CONFIRM "CLOSED LOOP" OPERATION AND NO "SERVICE ENGINE SOON" LIGHT.

1988–90 LIGHT TRUCKS AND VANS

RESTRICTED EXHAUST SYSTEM CHECK
ALL ENGINES

Proper diagnosis for a restricted exhaust system is essential before any components are replaced. Either of the following procedures may be used for diagnosis, depending upon engine or tool used:

CHECK AT A.I.R. PIPE: **OR**   **CHECK AT O₂ SENSOR:**

CHECK AT A.I.R. PIPE:	CHECK AT O_2 SENSOR:
1. Remove the rubber hose at the exhaust manifold A.I.R. pipe check valve. Remove check valve.	1. Carefully remove O_2 sensor.
2. Connect a fuel pump pressure gauge to a hose and nipple from a Propane Enrichment Device (J26911) (see illustration).	2. Install Borroughs Exhaust Backpressure Tester (BT 8515 or BT 8603) or equivalent in place of O_2 sensor (see illustration).
3. Insert the nipple into the exhaust manifold A.I.R. pipe.	3. After completing test described below, be sure to coat threads of O_2 sensor with anti-seize compound P/N 5613695 or equivalent prior to re-installation.

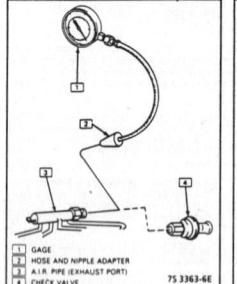

1	GAGE
2	HOSE AND NIPPLE ADAPTER
3	A.I.R. PIPE (EXHAUST PORT)
4	CHECK VALVE

7S 3363-6E

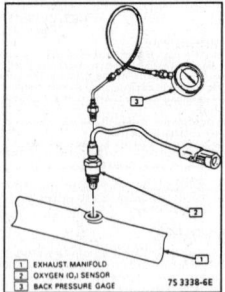

1	EXHAUST MANIFOLD
2	OXYGEN (O₂) SENSOR
3	BACK PRESSURE GAGE

7S 3338-6E

DIAGNOSIS:

1. With the engine idling at normal operating temperature, observe the exhaust system backpressure reading on the gauge. Reading should not exceed 1¼ psi (8.6 kPa).
2. Accelerate engine to 2000 rpm and observe gauge. Reading should not exceed 3 psi (20.7 kPa).
3. If the backpressure, at either rpm, exceeds specification, a restricted exhaust system is indicated.
4. Inspect the entire exhaust system for a collapsed pipe, heat distress, or possible internal muffler failure.
5. If there are no obvious reasons for the excessive backpressure, a restricted catalytic converter should be suspected and replaced using current recommended procedures.

1988–90 LIGHT TRUCKS AND VANS

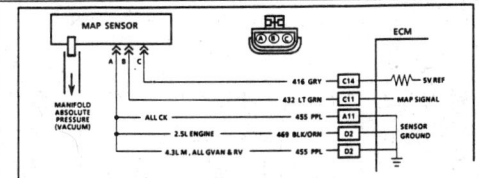

MAP OUTPUT CHECK
ALL ENGINES

Circuit Description:
The Manifold Absolute Pressure (MAP) sensor measures manifold pressure (vacuum) and sends that signal to the ECM. The MAP sensor is mainly used for fuel calculation, when the ECM is running in the throttle body backup mode. The MAP sensor is also used to determine the barometric pressure and to help calculate fuel delivery.

Test Description: Numbers below refer to circled numbers on the diagnostic chart.
1. Checks MAP sensor output voltage to the ECM. This voltage, without engine running, represents a barometer reading to the ECM.

2. Applying 34 kPa (10" hg) vacuum to the MAP sensor should cause the voltage to be 1.2 volts less than the voltage at Step 1. Upon applying vacuum to the sensor, the change in voltage should be instantaneous. A slow voltage change indicates a faulty sensor.
3. Check vacuum hose to sensor for leaking or restriction. Be sure no other vacuum devices are connected to the MAP hose.

1988–90 LIGHT TRUCKS AND VANS

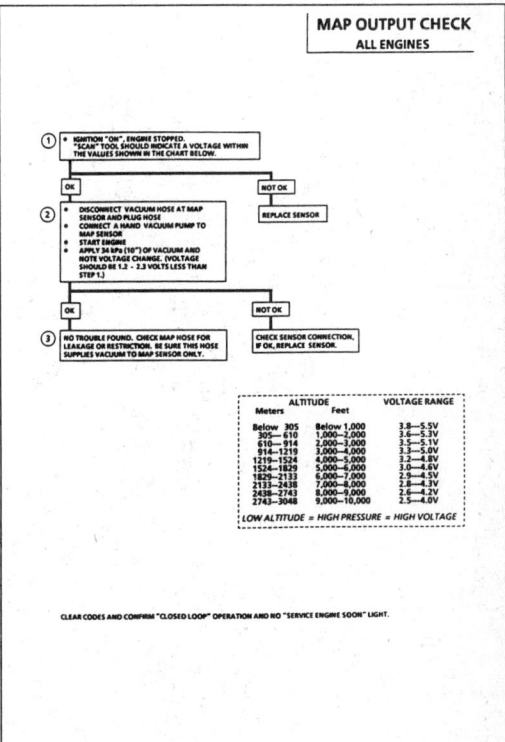

1988–90 LIGHT TRUCKS AND VANS

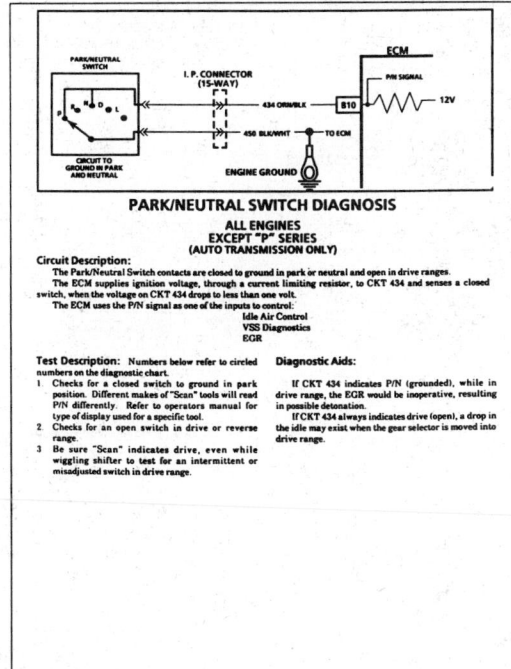

PARK/NEUTRAL SWITCH DIAGNOSIS
ALL ENGINES
EXCEPT "P" SERIES
(AUTO TRANSMISSION ONLY)

Circuit Description:
The Park/Neutral Switch contacts are closed to ground in park or neutral and open in drive ranges.
The ECM supplies ignition voltage, through a current limiting resistor, to CKT 434 and senses a closed switch, when the voltage on CKT 434 drops to less than one volt.
The ECM uses the P/N signal as one of the inputs to control:
Idle Air Control
VSS Diagnostics
EGR

Test Description: Numbers below refer to circled numbers on the diagnostic chart.
1. Checks for a closed switch to ground in park position. Different makes of "Scan" tools will read P/N differently. Refer to operators manual for type of display used for a specific tool.
2. Checks for an open switch in drive or reverse range.
3. Be sure "Scan" indicates drive, even while wiggling shifter to test for an intermittent or misadjusted switch in drive range.

Diagnostic Aids:
If CKT 434 indicates P/N (grounded), while in drive range, the EGR would be inoperative, resulting in possible detonation.
If CKT 434 always indicates drive (open), a drop in the idle may exist when the gear selector is moved into drive range.

1988–90 LIGHT TRUCKS AND VANS

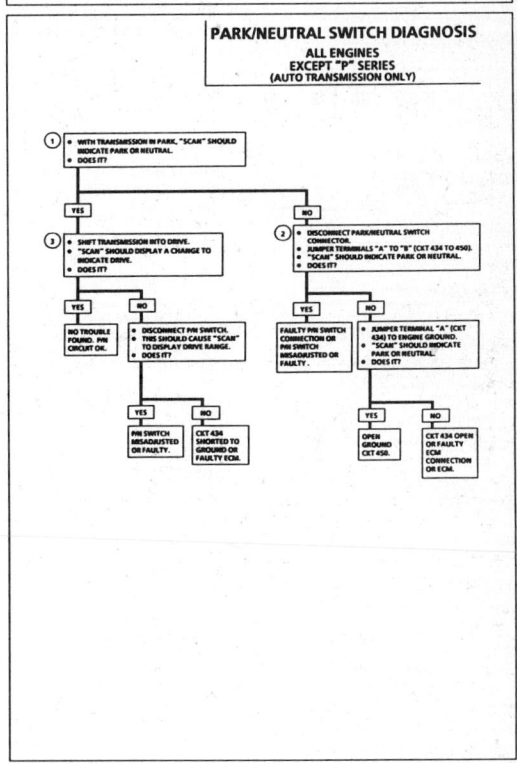

1988–90 LIGHT TRUCKS AND VANS

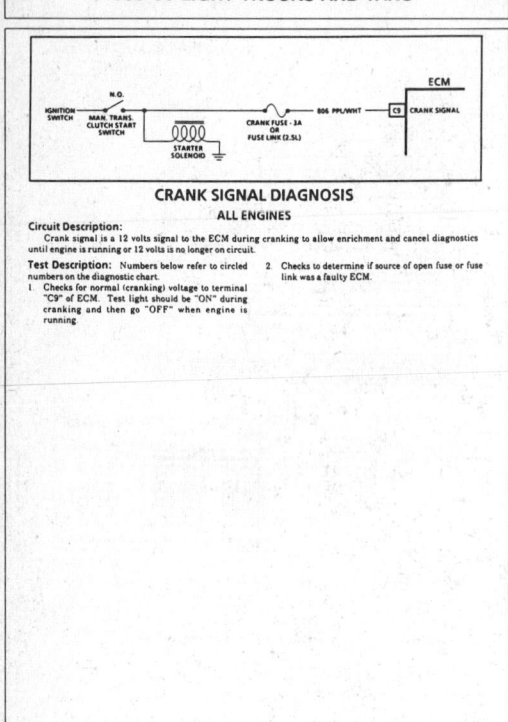

CRANK SIGNAL DIAGNOSIS
ALL ENGINES

Circuit Description:
Crank signal is a 12 volts signal to the ECM during cranking to allow enrichment and cancel diagnostics until engine is running or 12 volts is no longer on circuit.

Test Description: Numbers below refer to circled numbers on the diagnostic chart.
1. Checks for normal (cranking) voltage to terminal "C9" of ECM. Test light should be "ON" during cranking and then go "OFF" when engine is running.

2. Checks to determine if source of open fuse or fuse link was a faulty ECM.

1988–90 LIGHT TRUCKS AND VANS

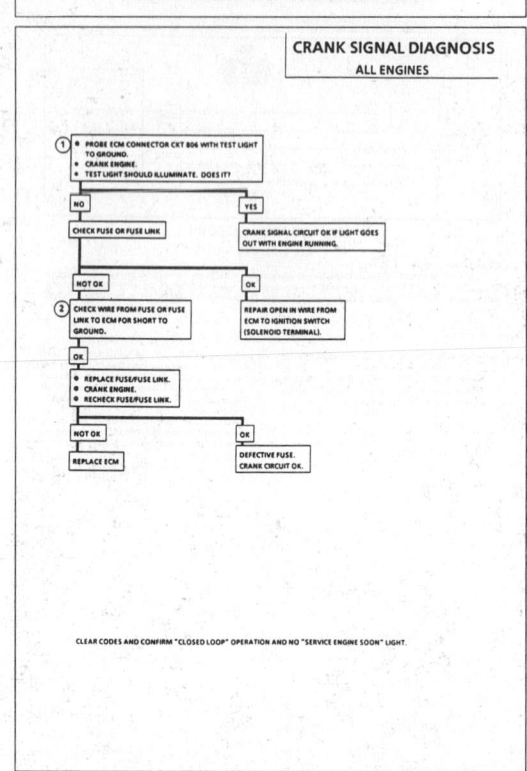

1988–90 LIGHT TRUCKS AND VANS

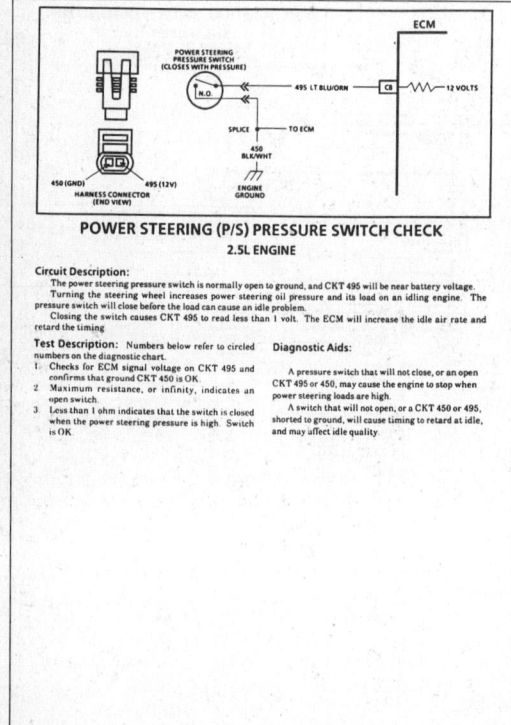

POWER STEERING (P/S) PRESSURE SWITCH CHECK
2.5L ENGINE

Circuit Description:
The power steering pressure switch is normally open to ground, and CKT 495 will be near battery voltage. Turning the steering wheel increases power steering oil pressure and its load on an idling engine. The pressure switch will close before the load can cause an idle problem. Closing the switch causes CKT 495 to read less than 1 volt. The ECM will increase the idle air rate and retard the timing.

Test Description: Numbers below refer to circled numbers on the diagnostic chart.
1. Checks for ECM signal voltage on CKT 495 and confirms that ground CKT 450 is OK.
2. Maximum resistance, or infinity, indicates an open switch.
3. Less than 1 ohm indicates that the switch is closed when the power steering pressure is high. Switch is OK.

Diagnostic Aids:
A pressure switch that will not close, or an open CKT 495 or 450, may cause the engine to stop when power steering loads are high.
A switch that will not open, or a CKT 450 or 495, shorted to ground, will cause timing to retard at idle, and may affect idle quality.

1988–90 LIGHT TRUCKS AND VANS

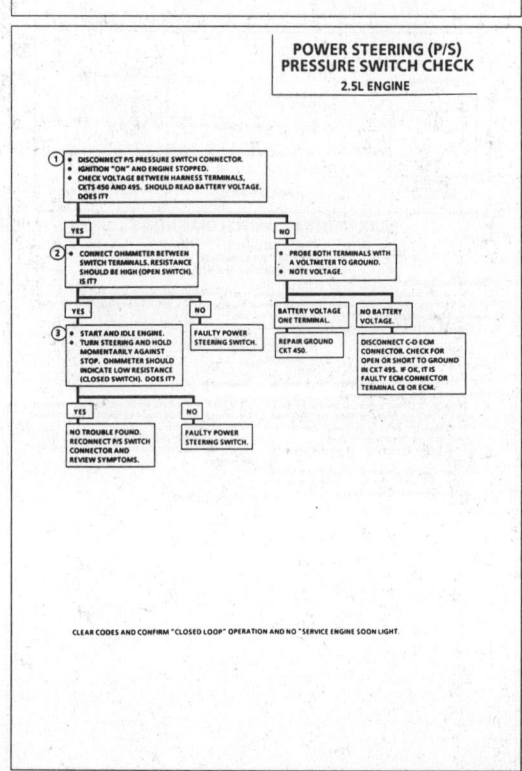

1988–90 LIGHT TRUCKS AND VANS

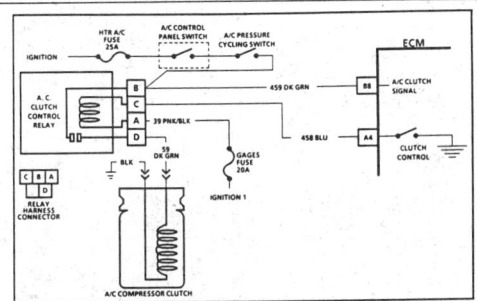

A/C CLUTCH CONTROL DIAGNOSIS
(Page 1 of 2)
2.5L ENGINE

Circuit Description:
ECM control of the A/C clutch improves idle quality and performance by:
- Delaying clutch apply until the idle air rate is increased.
- Releasing clutch when idle speed is too low.
- Releasing clutch at wide open throttle.
- Smooths cycling of the compressor by providing additional fuel at the instant clutch is applied.

Turning on air conditioning supplies CKT 459 battery voltage to the clutch control relay and terminal B8. After a time delay of about 1/2 second the ECM will ground terminal "A4," CKT 458, and close the control relay. A/C compressor clutch will engage.

Test Description: Numbers below refer to circled numbers on the diagnostic chart.
1. Checks for low refrigerant as cause for no A/C.
2. This and following tests check for faulty A/C control relay.

1988–90 LIGHT TRUCKS AND VANS

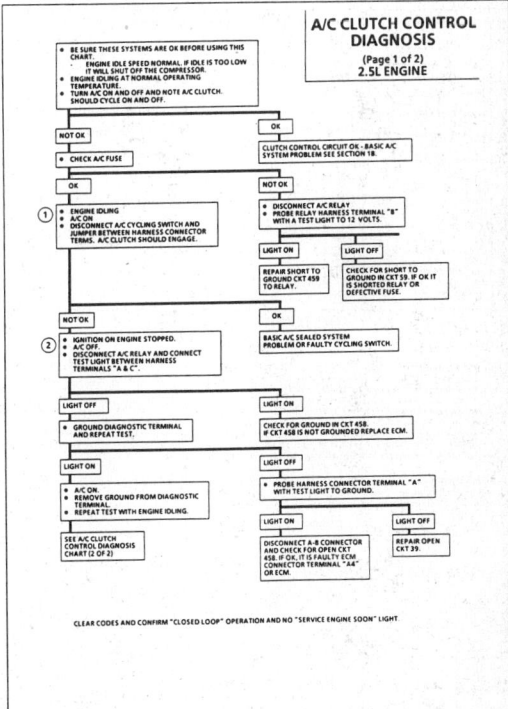

A/C CLUTCH CONTROL DIAGNOSIS
(Page 1 of 2)
2.5L ENGINE

CLEAR CODES AND CONFIRM "CLOSED LOOP" OPERATION AND NO "SERVICE ENGINE SOON" LIGHT.

1988–90 LIGHT TRUCKS AND VANS

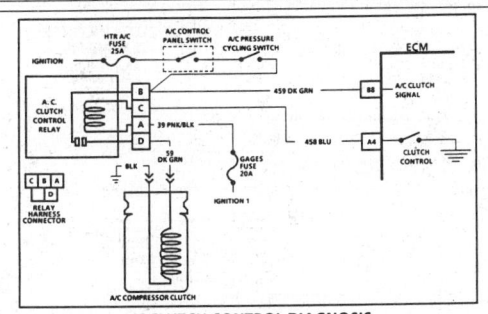

A/C CLUTCH CONTROL DIAGNOSIS
(Page 2 of 2)
2.5L ENGINE

Circuit Description:
ECM control of the A/C clutch improves idle quality and performance by:
- Delaying clutch apply until the idle air rate is increased.
- Releasing clutch when idle speed is too low.
- Releasing clutch at wide open throttle.
- Smooths cycling of the compressor by providing additional fuel at the instant clutch is applied.

Turning on air conditioning supplies CKT 459 battery voltage to the clutch control relay and terminal "B8". After a time delay of about 1/2 second, the ECM will ground terminal "A4", CKT 458, and close the control relay. A/C compressor clutch will engage.

Test Description: Numbers below refer to circled numbers on the diagnostic chart.
3. Checks for faulty cycling switch.

Solenoids and relays are turned "ON" or "OFF" by the ECM, using internal electronic switches called "drivers". Each driver is part of a group of four, called Quad-Drivers. Failure of one driver can damage any other driver in the set.

Solenoid and relay coil resistance must measure more than 20 ohms. Less resistance will cause early failure of the ECM "driver." Using an ohmmeter, check the coil resistance of the A/C relay before replacing the ECM.

Diagnostic Aids:

Before replacing ECM, use ohmmeter and check resistance of each ECM controlled relay or solenoid coil. See ECM wiring diagram for coil terminal identification for solenoid(s) and relay(s) to be checked.

Replace any relay or solenoid that measures less than 20 ohms.

1988–90 LIGHT TRUCKS AND VANS

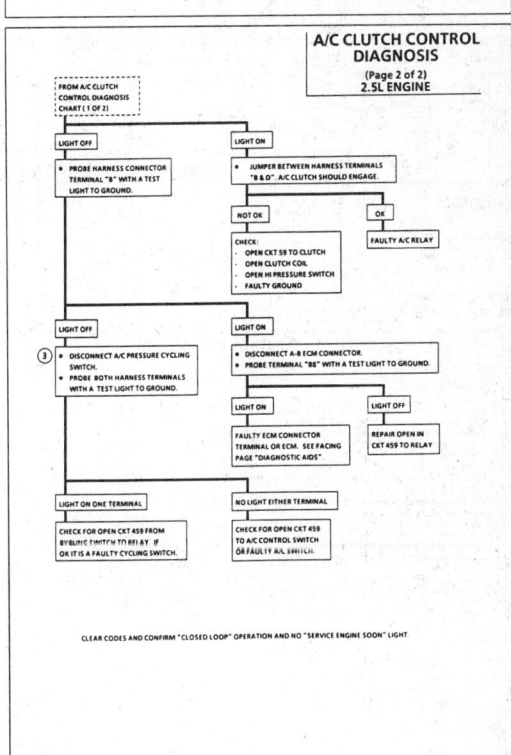

A/C CLUTCH CONTROL DIAGNOSIS
(Page 2 of 2)
2.5L ENGINE

CLEAR CODES AND CONFIRM "CLOSED LOOP" OPERATION AND NO "SERVICE ENGINE SOON" LIGHT.

1988–90 LIGHT TRUCKS AND VANS

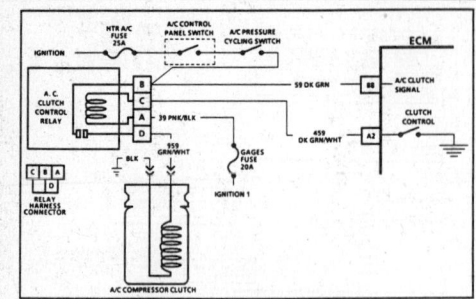

A/C CLUTCH CONTROL DIAGNOSIS
(Page 1 of 2)
2.8L ENGINE

Circuit Description:
ECM control of the A/C clutch improves idle quality and performance by:
- Delaying clutch apply until the idle air rate is increased.
- Releasing clutch when idle speed is too low.
- Releasing clutch at wide open throttle.
- Smooths cycling of the compressor by providing additional fuel at the instant clutch is applied.

Turning on air conditioning supplies CKT 59 battery voltage to the clutch control relay and terminal "B8" of the ECM connector. After a time delay of about 1/2 second the ECM will ground terminal "A2" of the ECM connector, CKT 459, and close the control relay. A/C compressor clutch will engage.

Test Description: Numbers below refer to circled numbers on the diagnostic chart.
1. Checks for low refrigerant as cause for no A/C.

2. This and following tests check for faulty A/C control relay.

1988–90 LIGHT TRUCKS AND VANS

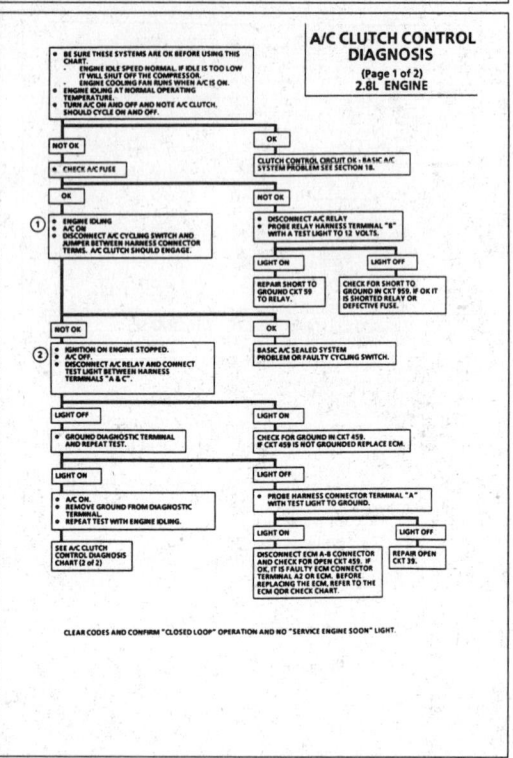

A/C CLUTCH CONTROL DIAGNOSIS
(Page 1 of 2)
2.8L ENGINE

1988–90 LIGHT TRUCKS AND VANS

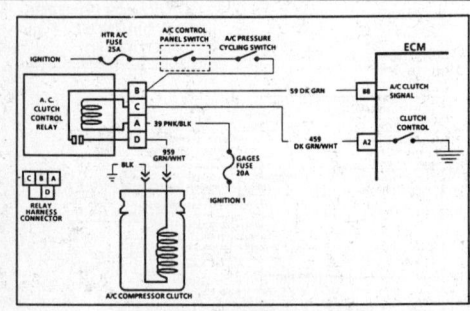

A/C CLUTCH CONTROL DIAGNOSIS
(Page 2 of 2)
2.8L ENGINE

Circuit Description:
ECM control of the A/C clutch improves idle quality and performance by;
- Delaying clutch apply until the idle air rate is increased.
- Releasing clutch when idle speed is too low.
- Releasing clutch at wide open throttle.
- Smooths cycling of the compressor by providing additional fuel at the instant clutch is applied.

Turning on air conditioning supplies CKT 59 battery voltage to the clutch control relay and terminal "B8" of the ECM connector. After a time delay of about 1/2 second the ECM will ground terminal "A2" of the ECM connector, CKT 459, and close the control relay. A/C compressor clutch will engage.

Test Description: Numbers below refer to circled numbers on the diagnostic chart.
3. Checks for faulty cycling switch.
- Solenoids and relays are turned "ON" or "OFF" by the ECM, using internal electronic switches called "drivers". Each driver is part of a group of four, called Quad-Drivers. Failure of one driver can damage any other driver in the set.
Solenoid and relay coil resistance must measure more than 20 ohms. less resistance will cause early failure of the ECM "driver". Using an ohmmeter, check the coil resistance of the A/C relay before replacing the ECM.

Diagnostic Aids:
Before replacing ECM, use ohmmeter and check resistance of each ECM controlled relay or solenoid coil. Refer to ECM QDR Check (Figure 3-18). See ECM wiring diagram for coil terminal identification for solenoid(s) and relay(s) to be checked.
Replace any relay or solenoid that measures less than 20 ohms.

1988–90 LIGHT TRUCKS AND VANS

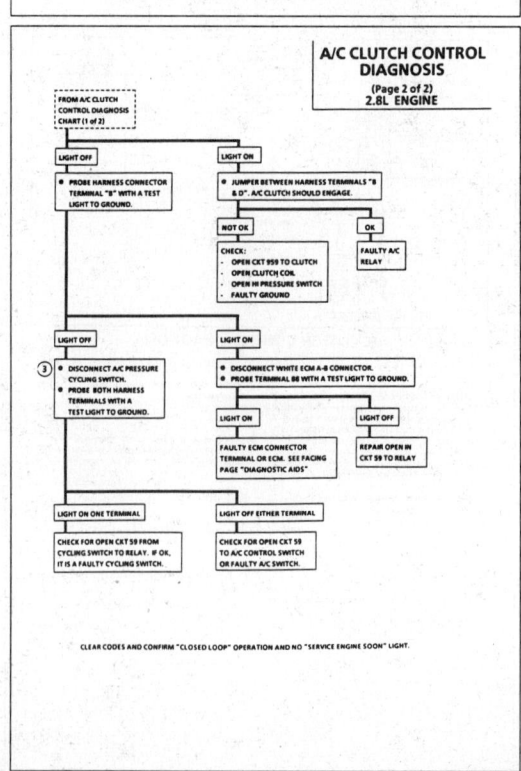

A/C CLUTCH CONTROL DIAGNOSIS
(Page 2 of 2)
2.8L ENGINE

1988–90 LIGHT TRUCKS AND VANS

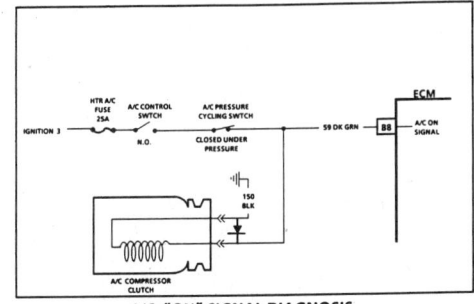

A/C "ON" SIGNAL DIAGNOSIS
4.3L AND V8 ENGINE

Circuit And Test Description:

Turning on the air conditioning supplies CKT 59 battery voltage to the A/C compressor clutch and to terminal "B8" of the ECM connector to increase idle air rate and maintain idle speed.

The ECM does not control the A/C compressor clutch, therefore, if A/C does not function, refer to the A/C section of the service manual for diagnosis of the system.

If A/C is operating properly and idle speed dips too low when the A/C compressor turns "ON" or flares too high when the A/C compressor turns "OFF," check for an open CKT 59 to the ECM. If circuits are OK, it is a faulty ECM connector terminal "B8" or ECM.

1988–90 LIGHT TRUCKS AND VANS

ECM QDR CHECK PROCEDURE
V6 OR V8 ENGINE

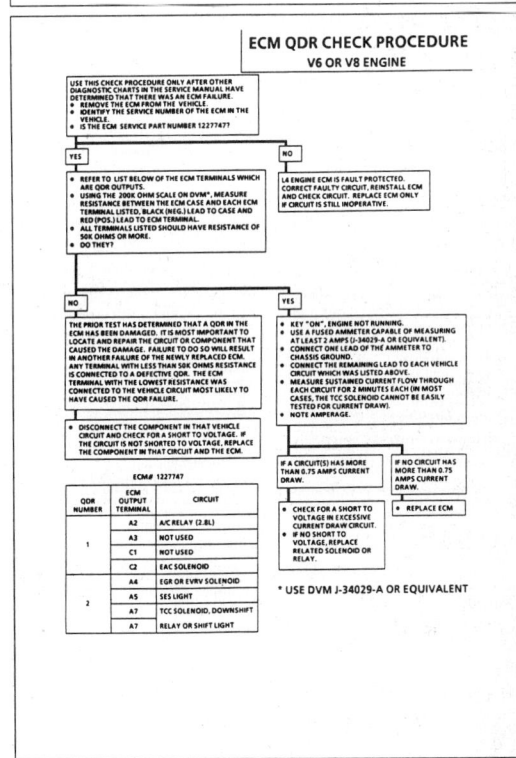

1988–90 LIGHT AND MEDIUM TRUCK AND VAN

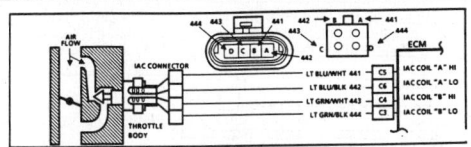

IDLE AIR CONTROL CHECK
ALL ENGINES EXCEPT 2.5L

Circuit Description:

The ECM sends voltage pulses to the IAC motor winding causing the motor shaft and valve to move "in" and "out" a given distance for each pulse (called counts) received. This movement controls air flow around the throttle plate, which, in turn, controls engine idle speed.

Test Description:

Numbers below refer to circled numbers on the diagnostic chart.
"Scan" tool must be in open mode during test.
Keep A/C "OFF" during entire check.

1. Test with engine in drive and continue with test, even if idle is erratic. If idle is too low, "Scan" will display 80, or more, counts or steps. Engine speed may vary 200 rpm, or more, up and down. Disconnect IAC. If the condition is unchanged, the IAC is not at fault. There is a system problem. Proceed to step "3" below.

2. When the engine was stopped with ignition "OFF", the IAC valve retracted (more air) to a fixed "Park" position, for increased air flow and idle speed during the next engine start. A "Scan" will display 100 or more counts.

3. Be sure to disconnect the IAC valve prior to this test. The test light will confirm the ECM signals by a steady or flashing light, on all circuits.

4. There is a remote possibility that one of the circuits is shorted to voltage, which would have been indicated by a steady light. Disconnect ECM and turn the ignition "ON" and probe terminals to check for this condition.

Diagnostic Aids:

An unstable idle may be a system problem that cannot be overcome by the IAC. "Scan" counts will be above 80 counts, if too low, and 0 counts, if too high.

- If IAC valve pintle position counts are low or zero, check with vacuum leaks at vacuum fitting, tees and hoses, the throttle body and the intake manifold. A bottomed (zero count) IAC valve pintle may result in an idle speed above specification. Refer to minimum air rate check.

- If IAC valve pintle position counts are high, look for carbon build-up in the IAC valve air inlet passage or evidence of tampering with stop screw. Also check for low engine power or excessive accessory loads.

- **System too lean (high tailpipe air/fuel ratio)** - Engine speed may vary up and down and disconnecting the IAC may not stabilize engine speed. If "Scan" and/or Voltmeter reads an oxygen sensor output less than 300 mv (.3v), check for low regulated fuel pressure or water in fuel. A code 44 (lean O_2 sensor) may be set. A lean tailpipe exhaust with an oxygen sensor output fixed above 800 mv (.8v) could be a contaminated sensor, usually Silicon. This may set a Code 45 (rich O_2 sensor) even with lean tailpipe exhaust.

- **System too rich (low tailpipe air/fuel ratio)** - System obviously rich and may exhibit black smoke exhaust. "Scan" tool and/or Voltmeter will read an oxygen sensor signal fixed above 800mv (.8v).
  Check for:
  - Injector leaking or sticking
  - High fuel pressure
  - Air leak in MAP transducer line
  - If O_2 is normal or low, inspect for air being pumped into exhaust manifold in front of O_2 sensor (or exhaust leak).

- **Throttle body** - Remove IAC and inspect bore for foreign material or evidence of IAC valve dragging the bore. (Repair as required)

- **A/C Compressor or relay failure** - See A/C diagnosis if circuit is shorted to ground. If the relay is faulty, idle problem may exist.
  Refer to Rough, Unstable, Incorrect Idle or Stalling, in Section 2.

1988–90 LIGHT AND MEDIUM TRUCK AND VAN

IDLE AIR CONTROL (IAC) SYSTEM CHECK
ALL ENGINES EXCEPT 2.5L

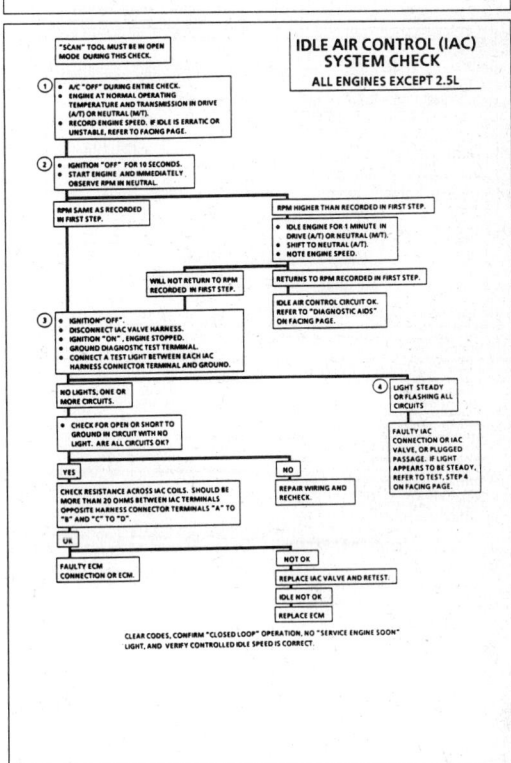

1988–90 LIGHT AND MEDIUM TRUCK AND VAN

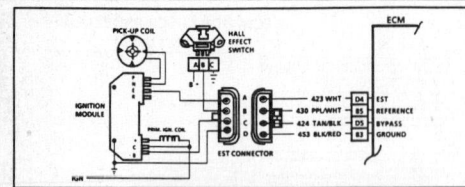

IGNITION SYSTEM CHECK
(REMOTE COIL)
2.5L S/T TRUCK

Test Description: Numbers below refer to circled numbers on the diagnostic chart.

1. Two wires are checked, to ensure that an open is not present in a spark plug wire.
1A. If spark occurs with 4 terminal distributor connector disconnected, pick-up coil output is too low for EST operation.
2. A spark indicates the problem must be the distributor cap or rotor.
3. Normally, there should be battery voltage at the "C" and "+" terminals. Low voltage would indicate an open or a high resistance circuit from the distributor to the coil or ignition switch. If "C" term. voltage was low, but "+" term. voltage is 10 volts or more, circuit from "C" term. to ignition coil or ignition coil primary winding is open.
4. Checks for a shorted module or grounded circuit from the ignition coil to the module. The dist. module should be turned "OFF", so normal voltage should be about 12 volts.
   If the module is turned "ON", the voltage would be low, but above 1 volt. This could cause the ign. coil to fail from excessive heat.
   With an open ignition coil primary winding, a small amount of voltage will leak through the module from the "Bat." to the tach terminal.

5. Applying a voltage (1.5 to 8 volts) to module terminal "P" should turn the module "ON" and the tach. term. voltage should drop to about 7-9 volts. This test will determine whether the module or coil is faulty or if the pick-up coil is not generating the proper signal to turn the module "ON". This test can be performed using a DC battery with a rating of 1.5 to 8 volts. The use of the test light is mainly to allow the "P" terminal to be probed more easily.
   Some digital multi-meters can also be used to trigger the module by selecting ohms, usually the diode position. In this position, the meter may have a voltage across its terminals which can be used to trigger the module. The voltage in the ohm's position can be checked by using a second meter or by checking the manufacture's specification of the tool being used.
6. This should turn "OFF" the module and cause a spark. If no spark occurs, the fault is most likely in the ignition coil because most module problems would have been found before this point in the procedure. A module tester (J24642) could determine which is at fault.

Diagnostic Aids:

The "Scan" tool does not have any ability to help diagnose a ignition system check.
Refer to Section "2" for "ECM Intermittent Codes or Performance".

1988–90 LIGHT AND MEDIUM TRUCK AND VAN

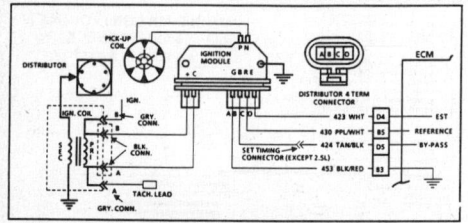

IGNITION SYSTEM CHECK
(REMOTE COIL / SEALED MODULE CONNECTOR DISTRIBUTOR)
ALL ENGINES EXCEPT 2.5L S/T TRUCK

Test Description: Numbers below refer to circled numbers on the diagnostic chart.

1. Two wires are checked, to ensure that an open is not present in a spark plug wire.
1A. If spark occurs with EST connector disconnected, pick-up coil output is too low for EST operation.
2. A spark indicates the problem must be the distributor cap or rotor.
3. Normally, there should be battery voltage at the "C" and "+" terminals. Low voltage would indicate an open or a high resistance circuit from the distributor to the coil or ignition switch. If "C" term. voltage was low, but "+" term. voltage is 10 volts or more, circuit from "C" term. to ignition coil or ignition coil primary winding is open.
4. Checks for a shorted module or grounded circuit from the ignition coil to the module. The distributor module should be turned "OFF", so normal voltage should be about 12 volts.

If the module is turned "ON", the voltage would be low, but above 1 volt. This could cause the ignition coil to fail from excessive heat.
With an open ignition coil primary winding, a small amount of voltage will leak through the module from the "Bat." to the tach terminal.
5. Checks for an open module, or circuit to it. 12 volts applied to the module "P" terminal should turn the module "ON" and the voltage should drop to about 7-9 volts.
6. This should turn "OFF" the module and cause a spark. If no spark occurs, the fault is most likely in the ignition coil because most module problems would have been found before this point in the procedure. A module tester could determine which is at fault.

Diagnostic Aids:

The "Scan" tool does not have any ability to help diagnose an ignition system check.
Refer to Section "2" for "ECM Intermittent Codes or Performance".

1988–90 LIGHT AND MEDIUM TRUCK AND VAN

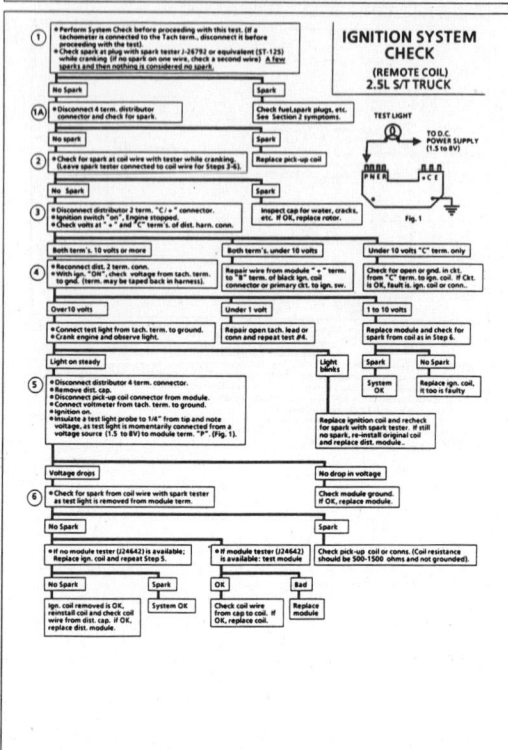

IGNITION SYSTEM CHECK
(REMOTE COIL)
2.5L S/T TRUCK

1988–90 LIGHT AND MEDIUM TRUCK AND VAN

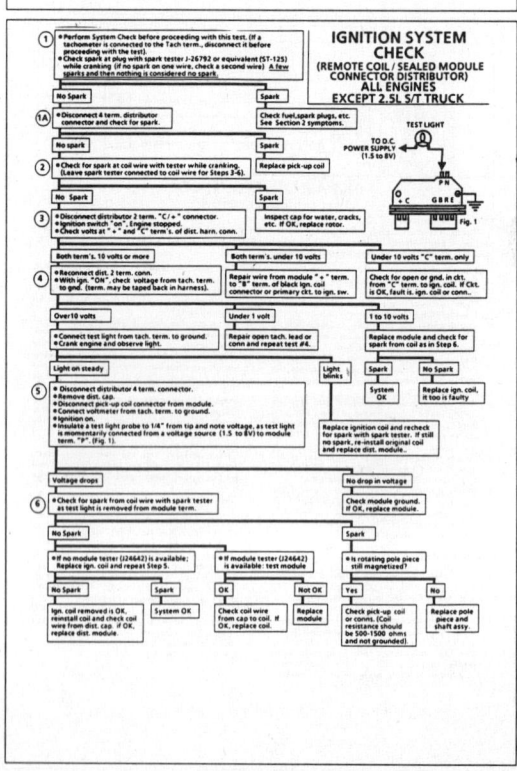

IGNITION SYSTEM CHECK
(REMOTE COIL / SEALED MODULE CONNECTOR DISTRIBUTOR)
ALL ENGINES EXCEPT 2.5L S/T TRUCK

1988–90 LIGHT AND MEDIUM TRUCK AND VAN

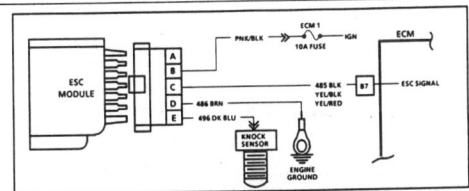

ELECTRONIC SPARK CONTROL SYSTEM CHECK
ALL ENGINES EXCEPT 2.5L AND 7.4L

Circuit Description:

Electronic spark control is accomplished with a module that sends a voltage signal to the ECM. As the knock sensor detects engine knock, the voltage from the ESC module to the ECM is shut "OFF" and this signals the ECM to retard timing, if engine rpm is over about 900.

Test Description: Numbers below refer to circled numbers on the diagnostic chart.

1. If A Code 43 is not set, but a knock signal is indicated while running at 1500 rpm, listen for an internal engine noise. Under a no load condition there should not be any detonation, and if knock is indicated, an internal engine problem may exist.
2. Usually a knock signal can be generated by tapping on the right exhaust manifold. This test can also be performed at idle. Test number 1 was run at 1500 rpm to determine if a constant knock signal was present, which would affect engine performance.
3. This tests whether the knock signal is due to the sensor, a basic engine problem, or the ESC module.
4. If the module ground circuit is faulty, the ESC module will not function correctly. The test light should light indicating the ground circuit is OK
5. Contacting CKT 496, with a test light to 12 volts, should generate a knock signal to determine whether the knock sensor is faulty, or the ESC module can't recognize a knock signal.

Diagnostic Aids:

"Scan" tools have two positions to diagnose the ESC system. The knock signal can be monitored to see if the knock sensor is detecting a knock condition and if the ESC module is functioning, knock signal should display "yes", whenever detonation is present. The knock retard position on the "Scan" displays the amount of spark retard the ECM is commanding. The ECM can retard the timing up to 20 degrees.

If the ESC system checks OK, but detonation is the complaint, refer to Detonation/Spark Knock in Section "2".

This check should be used after other causes of spark knock have been checked such as engine timing, EGR systems, engine temperature or excessive engine noise.

1988–90 LIGHT AND MEDIUM TRUCK AND VAN

ELECTRONIC SPARK CONTROL SYSTEM CHECK
ALL ENGINES EXCEPT 2.5L AND 7.4L

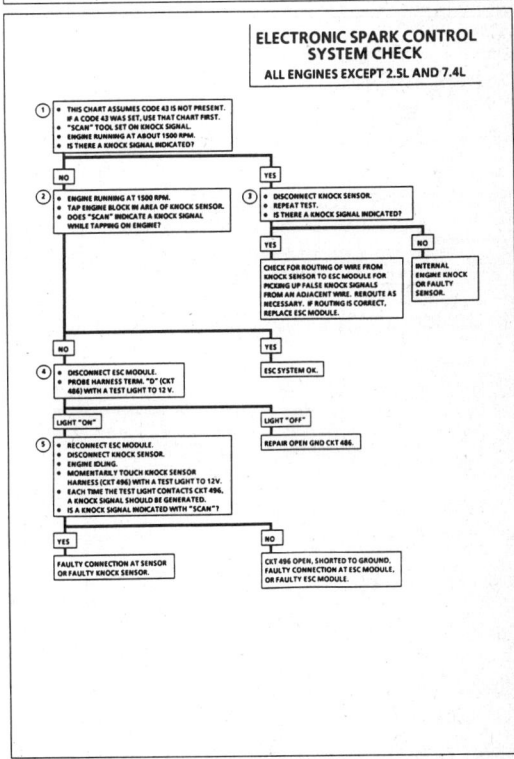

1988–90 LIGHT AND MEDIUM TRUCK AND VAN

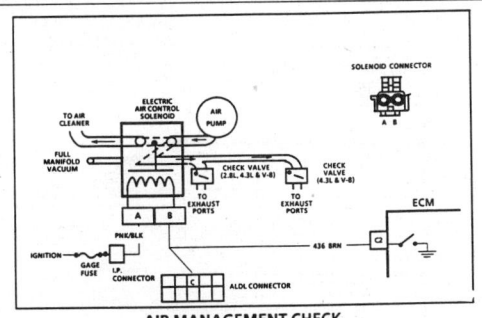

AIR MANAGEMENT CHECK
(ELECTRONIC AIR CONTROL VALVE)

Circuit Description:

An electric air control valve solenoid directs air into the exhaust ports or the air cleaner. During cold start the ECM completes the ground circuit, the EAC solenoid is energized, and air is directed to the exhaust ports. As "coolant" temperature increases, or system goes to "Closed Loop", the ECM opens the ground circuit, the EAC solenoid is de-energized, and air goes to the air cleaner. If the system is not operating properly, check manifold vacuum signal (10"Hg/34kPa) at the valve and check the electrical circuit from the solenoid to the ECM.

Test Description: Numbers below refer to circled numbers on the diagnostic chart.

1. This is a system performance test. When vehicle goes to "Closed Loop", air will switch from the ports and divert to the air cleaner.
2. Tests for a grounded electric divert circuit. Normal system light will be "OFF".
3. Checks for an open control circuit. Grounding diagnostic terminal will energize the solenoid, if ECM and circuits are normal. In this step, if test light is "ON", circuits are normal and fault is in valve connections or valve.
4. Checks for voltage from battery through a fuse to the solenoid.

1988–90 LIGHT AND MEDIUM TRUCK AND VAN

AIR MANAGEMENT CHECK
(ELECTRONIC AIR CONTROL VALVE)

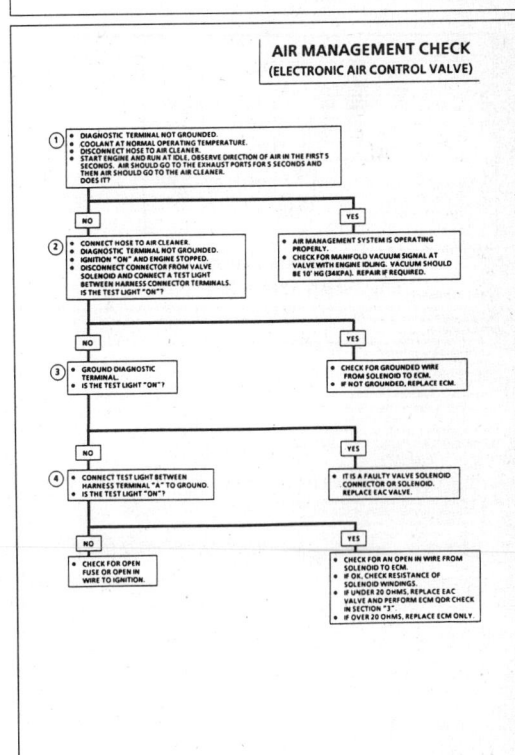

1988–90 LIGHT AND MEDIUM TRUCK AND VAN

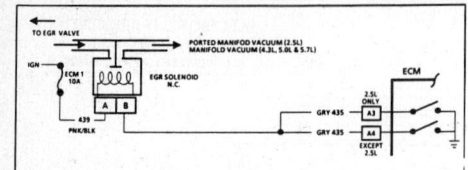

EGR SYSTEM CHECK
2.5L, 4.3L (EXCEPT ST), 5.0L & 5.7L (UNDER 8500 GVW)

Circuit Description:

The ECM operates a solenoid to control the exhaust gas recirculation (EGR) valve. This solenoid is normally closed. By providing a ground path, the ECM energizes the solenoid which then allows vacuum to pass to the EGR valve. The ECM control of the EGR is based on the following inputs:
- Engine coolant temperature - above 25°C.
- TPS - "OFF" idle
- MAP

If Code 24 is stored, use that chart first.

Code 32 will detect a faulty solenoid, vacuum supply, EGR Valve or plugged passage. This chart checks for plugged EGR passages, a sticking EGR valve, or a stuck open or inoperative solenoid.

Test Description: Numbers below refer to circled numbers on the diagnostic chart.
1. Checks for solenoid stuck open.
2. Checks for solenoid always being energized.
3. Grounding test terminal should energize solenoid and vacuum should drop.
4. Negative backpressure valve should hold vacuum with engine "OFF".
5. When engine is started, exhaust backpressure should cause vacuum to bleed off and valve to fully close.

Diagnostic Aids:
- Before replacing ECM, use an ohmmeter and check the resistance of each ECM controlled relay and solenoid coil. Refer to ECM QDR check procedure. See ECM wiring diagram for coil term. I.D. of solenoid(s) and relay(s) to be checked. Replace any solenoid where resistance measures less than 20 ohms.

1988–90 LIGHT AND MEDIUM TRUCK AND VAN

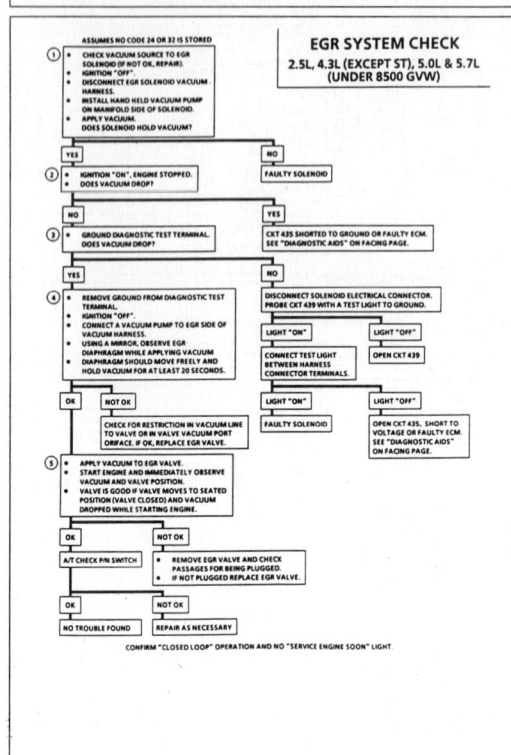

EGR SYSTEM CHECK
2.5L, 4.3L (EXCEPT ST), 5.0L & 5.7L (UNDER 8500 GVW)

1988–90 LIGHT AND MEDIUM TRUCK AND VAN

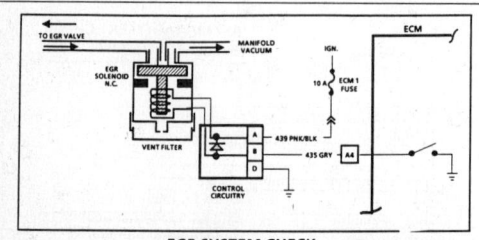

EGR SYSTEM CHECK
2.8L, 7.4L, 4.3L (ST), & 5.7L (OVER 8500 GVW)

Circuit Description:

The EGR valve is controlled by a normally closed solenoid (allows a vacuum to pass when energized). The ECM pulses the solenoid to turn "ON" and regulate the EGR. The ECM diagnoses the system using an internal EGR test procedure.

The ECM control of the EGR is based on the following inputs:
- Engine coolant temperature - above 25°C.
- TPS - "OFF" idle
- MAP

If Code 24 is stored, use that chart first.

Code 32 will detect a faulty solenoid, vacuum supply, EGR Valve or plugged passage. This chart checks for plugged EGR passages, a sticking EGR valve, or a stuck open or inoperative solenoid.

Test Description: Numbers below refer to circled numbers on the diagnostic chart.
1. With the ignition "ON", engine stopped, the solenoid should not be energized and vacuum should not pass to the EGR valve.
2. Grounding the diagnostic terminal will energize the solenoid and allow vacuum to pass to valve.
3. Checks for plugged EGR passages. If passages are plugged, the engine may have severe detonation on acceleration.
4. The EGR solenoid will not be energized in Park or Neutral This test will determine if the Park/Neutral switch input is being received by the ECM.

Diagnostic Aids:
- Before replacing ECM use ohmmeter and check resistance of each ECM controlled relay and solenoid coil. Refer to ECM QDR Check

See ECM wiring diagram for coil terminal identification of solenoid(s) and relay(s) to be checked. Replace any relay or solenoid if the coil resistance measures less than 20 ohms.

1988–90 LIGHT AND MEDIUM TRUCK AND VAN

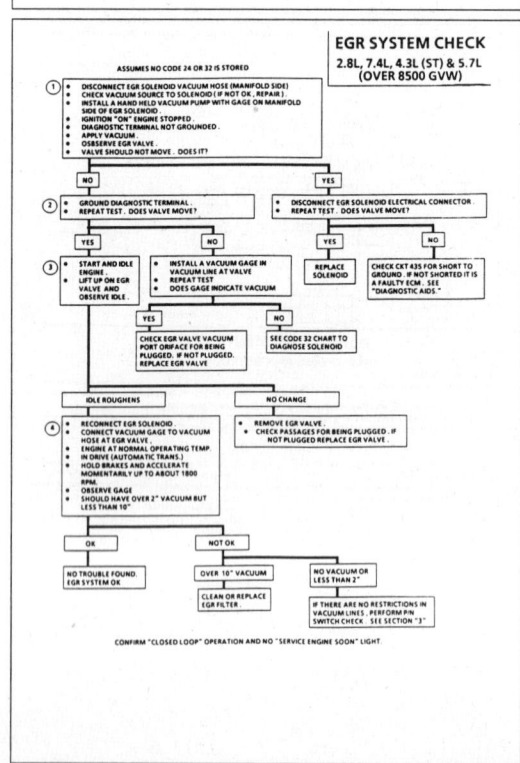

EGR SYSTEM CHECK
2.8L, 7.4L, 4.3L (ST) & 5.7L (OVER 8500 GVW)

1988–90 LIGHT AND MEDIUM TRUCK AND VAN

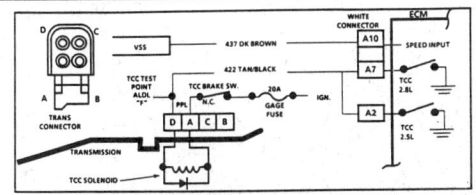

TRANSMISSION CONVERTER CLUTCH (TCC)
(ELECTRICAL DIAGNOSIS)
2.5L AND 2.8L ENGINES

Circuit Description:
The purpose of the automatic transmission converter clutch feature is to eliminate the power loss of the torque converter stage when the vehicle is in a cruise condition. This allows the convenience of the automatic transmission and the fuel economy of a manual transmission.
Fused battery ignition is supplied to the TCC solenoid through the TCC brake switch.
The ECM will engage TCC by grounding CKT 422 to energize the solenoid.
TCC will engage when:
- Vehicle speed above 24 mph (39 km/h.)
- Engine at normal operating temperature (above 65°C) (149°F).
- Throttle position sensor output not changing, indicating a steady road speed.
- Brake switch closed.
- 3rd or 4th gears.

Test Description: Numbers below refer to circled numbers on the diagnostic chart.
1. Checks continuity through brake switch and TCC solenoid.
2. Checks capability of ECM to energize solenoid. Grounding the diagnostic connector should energize the relay and cause the light to go out.
3. This test by-passes the TCC solenoid and checks for an open or short in CKT 422.

Diagnostic Aids:
Solenoid coil resistance must measure more than 20 ohms. Less resistance will cause early failure of the ECM "Driver". Refer to ECM QDR check
Using an ohm meter, check the solenoid coil resistance of all ECM controlled solenoids and relays, before installing a replacement ECM. Replace any solenoid, or relay, that measures less than 20 ohms resistance.

1988–90 LIGHT AND MEDIUM TRUCK AND VAN

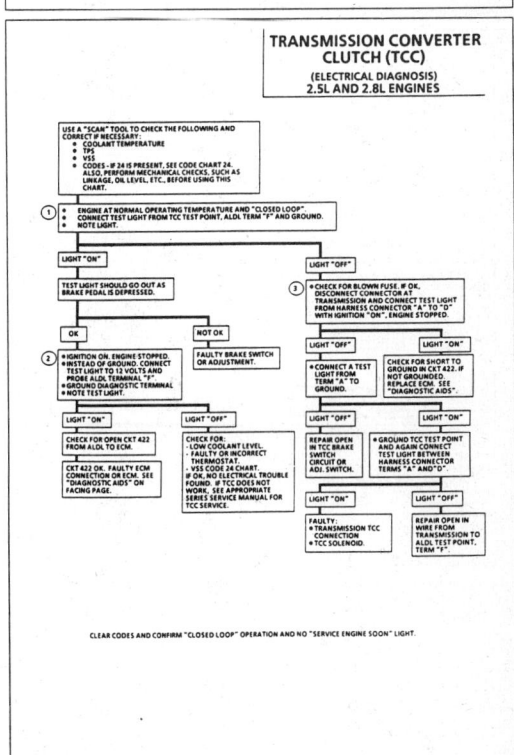

TRANSMISSION CONVERTER CLUTCH (TCC)
(ELECTRICAL DIAGNOSIS)
2.5L AND 2.8L ENGINES

1988–90 LIGHT AND MEDIUM TRUCK AND VAN

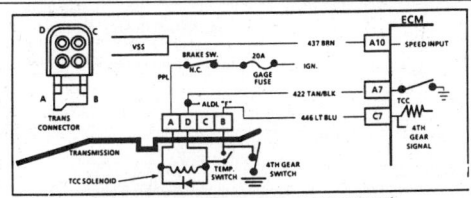

TRANSMISSION CONVERTER CLUTCH (TCC)
(ELECTRICAL DIAGNOSIS)
4.3L, 5.0L AND 5.7L (UNDER 8500 GVW)

Circuit Description:
The purpose of the automatic transmission torque converter clutch feature is to eliminate the power loss of the torque converter stage when the vehicle is in a cruise condition. This allows the convenience of the automatic transmission and the fuel economy of a manual transmission.
Fused battery ignition is supplied to the TCC solenoid through the TCC brake switch.
The ECM will engage TCC by grounding CKT 422 to energize the solenoid.
TCC will engage when:
- Vehicle speed above 30 mph (48 km/h.)
- Engine at normal operating temperature (above 65°C) (149°F).
- Throttle position sensor output not changing, indicating a steady road speed.
- Brake switch closed.
- 3rd or 4th gears.

Test Description: Numbers below refer to circled numbers on the diagnostic chart.
1. A test light on indicates battery voltage and continuity through the TCC solenoid is OK.
2. Checks for vehicle speed sensor signal to ECM using a "Scan" tool.
3. Checks for 4th gear signal to ECM. This signal will not prevent TCC engagement, but could cause a change in the engage and disengage speed points.

Diagnostic Aids:
Solenoid coil resistance must measure more than 20 ohms. Less resistance will cause early failure of the ECM "Driver". Refer to ECM QDR check
Using an ohm meter, check the solenoid coil resistance of all ECM controlled solenoids and relays before installing a replacement ECM. Replace any solenoid or relay that measures less than 20 ohms resistance.

1988–90 LIGHT AND MEDIUM TRUCK AND VAN

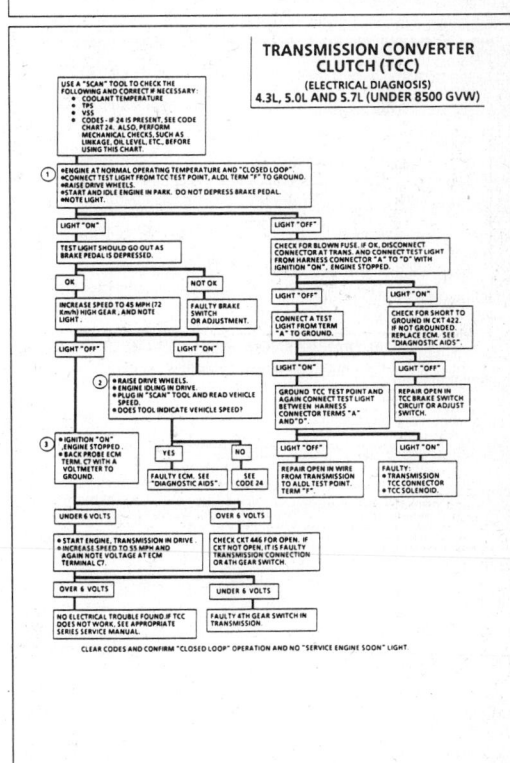

TRANSMISSION CONVERTER CLUTCH (TCC)
(ELECTRICAL DIAGNOSIS)
4.3L, 5.0L AND 5.7L (UNDER 8500 GVW)

1988–90 LIGHT AND MEDIUM TRUCK AND VAN

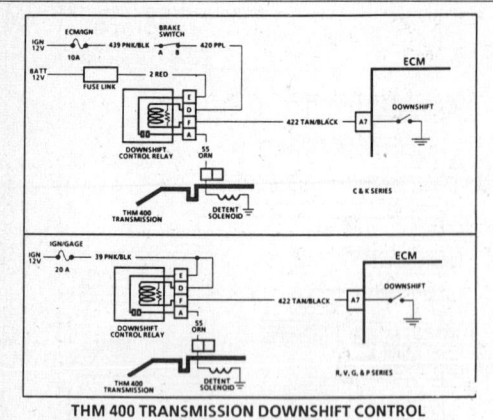

THM 400 TRANSMISSION DOWNSHIFT CONTROL
(ELECTRICAL DIAGNOSIS)

Circuit Description:
When the accelerator pedal is fully depressed, manifold vacuum in the engine drops causing the MAP sensor signal voltage to increase to approximately 4 volts. The ECM responds by grounding CKT 422 to turn "ON" the downshift control relay. The relay then sends battery voltage to the detent solenoid, which causes a forced transmission downshift.

Diagnostic Aids:
- If problem is diagnosed as being an internal transmission problem, see Section 7 of the appropriate series Service Manual as listed in the Forward.

- Relay coil resistance must measure more than 20 ohms. Less resistance will cause early failure of the ECM "Driver". Refer to ECM QDR Check.
  Using an ohm meter, check the coil resistance of all ECM controlled solenoids and relays before installing a replacement ECM. Replace any solenoid or relay that measures less than 20 ohms resistance.

1988–90 LIGHT AND MEDIUM TRUCK AND VAN

THM 400 TRANSMISSION DOWNSHIFT CONTROL
(ELECTRICAL DIAGNOSIS)

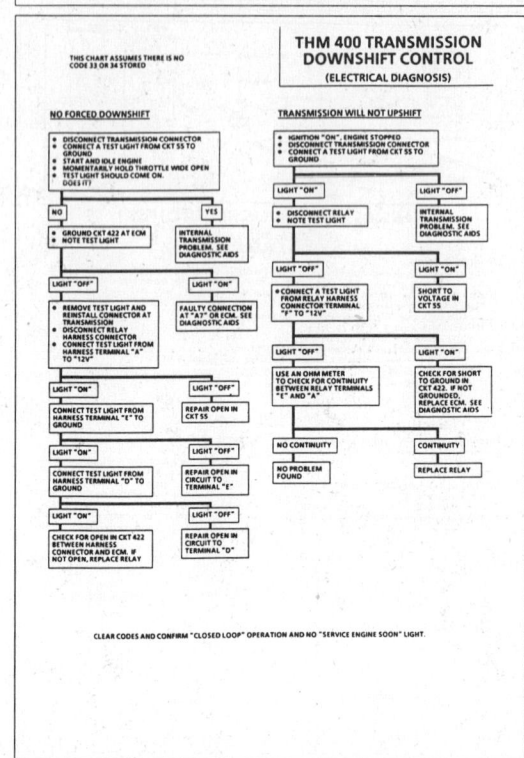

1988–90 LIGHT AND MEDIUM TRUCK AND VAN

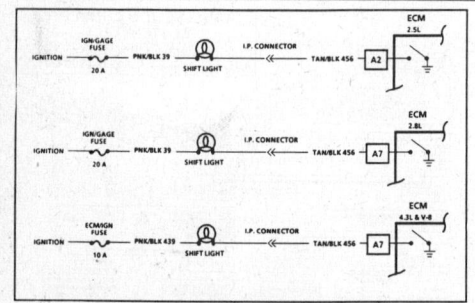

MANUAL TRANSMISSION SHIFT LIGHT CHECK
ON ALL VEHICLES BELOW 8500 GVW

Circuit Description:
The ECM uses information from the following inputs to control the shift light:
- Coolant temperature
- TPS
- VSS
- RPM

The ECM uses the measured rpm and the vehicle speed to calculate what gear the vehicle is in. It's this calculation that determines when the shift light should be turned on.

Test Description: Numbers below refer to circled numbers on the diagnostic chart.
1. This should not turn "ON" the shift light. If the light is "ON", there is a short to ground in CKT 456 wiring, or a fault in the ECM.
2. This should turn "ON" the shift light.
3. This checks for an open in the shift light circuit, or a faulty ECM.

1988–90 LIGHT AND MEDIUM TRUCK AND VAN

MANUAL TRANSMISSION SHIFT LIGHT CHECK
ON ALL VEHICLES BELOW 8500 GVW

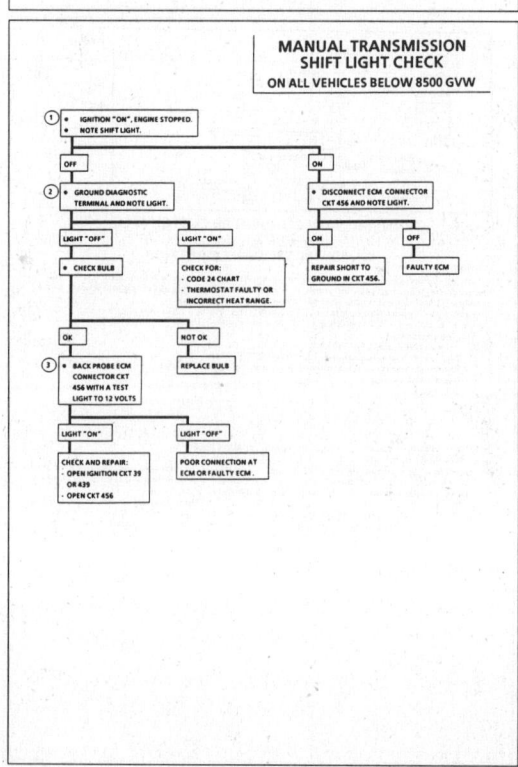

COMPONENT TESTING

Fuel System Pressure Test

A fuel system pressure test is part of several of the diagnostic charts and symptom checks. To perform this test, proceed as follows:

1. Relieve the fuel pressure from the fuel system. Turn the ignition **OFF** and remove the air cleaner assembly (if necessary).

2. Plug the Thermac vacuum port if required on the TBI unit.

3. Uncouple the fuel supply flexible hose in the engine compartment and install fuel pressure gauge J29658/BT8205 or equivalent in the pressure line or install the fuel pressure gauge into the pressure line connector located near the left engine compartment frame rail. Connection of the fuel gauge will vary accordingly to all the different engine application.

4. Be sure to tighten the fuel line to the gauge to ensure that there no leaks during testing.

5. Start the engine and observe the fuel pressure reading. The fuel pressure should be 9–13 psi (62–90 kPa).

6. Relieve the fuel pressure. Remove the fuel pressure gauge and reinstall the fuel line. Be sure to install a new O-ring on the fuel feed line.

7. Start the engine and check for fuel leaks. Stop the engine and remove the plug covering the Thermac vacuum port on the TBI unit and install the air cleaner assembly.

TPS OUTPUT CHECK TEST

WITH SCAN TOOL

The following check should be performed only when the throttle body or TPS has been replaced.

1. Use a suitable scan tool to read the TPS voltage.

2. With the ignition switch **ON** and the engine **OFF**, the TPS voltage should be less than 1.25 volts.

3. If the voltage reading is higher than specified, replace the throttle position sensor.

WITHOUT SCAN TOOL

This check should only be performed, when the throttle body or the TPS has been replaced or after the minimum idle speed has been adjusted.

1. Remove air cleaner. Disconnect the TPS harness from the TPS.

2. Using suitable jumper wires, connect a digital voltmeter J–29125–A or equivalent to the correct TPS terminals:

   a. Terminals **A** and **B** on all models except the ones listed below.

   b. Terminals **C** and **B** on the 2.0L engines using the TBI 500 model.

3. With the ignition **ON** and the engine running, The TPS voltage should be 0.450–1.25 volts at base idle to approximately 4.5 volts at wide open throttle.

4. If the reading on the TPS is out of specification, check the minimum idle speed before replacing the TPS.

5. If the voltage reading is correct, remove the voltmeter and jumper wires and reconnect the TPS connector to the sensor.

Re-install the air cleaner.

Controlled Idle Speed Check

ASTRO VAN, LIGHT TRUCKS AND VAN MODELS ONLY

Before performing the idle speed check, there should be no codes displayed, the idle air control system has been checked and the ignition timing is correct.

1. Set the parking brake and block the drive wheels.

2. Connect a suitable scan tool to the ALCL connector with the tool in open mode.

3. Start the engine and bring up to normal operating temper-

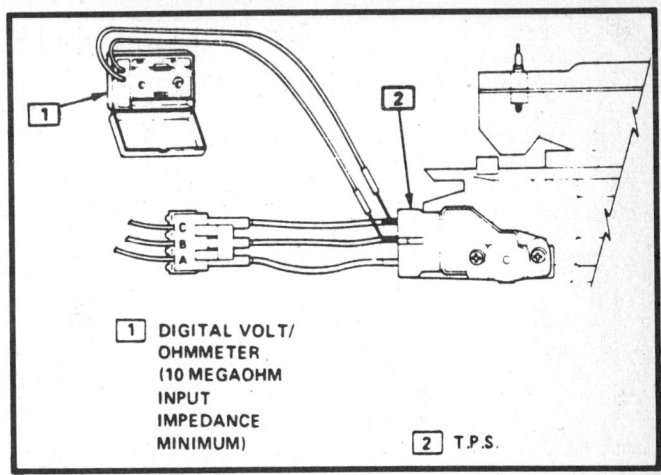

TPS voltage check

      1 DIGITAL VOLT/
          OHMMETER
          (10 MEGAOHM
          INPUT
          IMPEDANCE
          MINIMUM)   2 T.P.S.

ature. Check for the correct state of the park/neutral switch on the scan tool.

4. Check the specification in the chart provided for the controlled idle speed and the idle air control valve pintle positions (counts).

5. If within specifications, the idle is being correctly controlled by the ECM.

6. If not within specifications, check the minimum idle air rate, fuel system, idle circuit, TPS, MAP sensor, etc.

Minimum Idle Air Rate Check

ASTRO VAN, LIGHT TRUCKS AND VAN MODELS ONLY

1. Check the controlled idle speed and perform the idle air control check first.

2. Block the drive wheels and set the parking brake.

3. Start the engine and bring it up to normal operating temperature. Turn the engine to the **OFF** position.

4. Remove the air cleaner, adapter and gaskets. On the ST series vehicle, leave the thermac hose connected. Check that the throttle lever is not being bound by the throttle, T.V. or cruise control cables.

5. With the IAC valve connected, ground the diagnostic terminal (ALCL connector).

6. Turn the ignition switch to the **ON** position, do not start the engine. Wait at least 10 seconds, this will allow the pintle to extend and seat in the throttle body.

7. With the ignition switch **ON** and the engine stopped, test terminal still grounded, disconnect the IAC valve electrical connector. This disables the IAC valve seated position. Remove the ground from the diagnostic terminal.

8. Connect a suitable scan tool to the ALCL connector and place it in open mode. If a tool is not available connect a tachometer the engine.

9. Start the engine. with the transmission in **N**, allow the engine rpm to stabilize.

10. Check the rpm against the specification in the chart provided. Disregard the IAC counts on the scan tool with the IAC disconnected. If the engine has less than 500 miles on it or it is checked at a altitude above 1500 feet, the idle rpm with a seated IAC valve should be lower than the specified values.

11. If the minimum idle air rate is within specifications, no further check is required.

12. If the minimum idle air rate is not within specifications, perform the following procedures.

   a. On vehicles equipped with a tamper resistant plug covering the minimum air adjustment screw, insert a suitable tool to piece the plug and pry the plug out.

CONTROLLED IDLE SPEED

Engine	Transmission	Gear (D/N)	Idle Speed (RPM)	IAC Counts*	Open/Closed Loop**
2.5L	Man.	N	900(ST) 800(M)	5-20	CL
	Auto.	D	800(S) 650(T) 750(M)	15-40	CL
2.8L	Man.	N	800	5-20	OL
4.3L	Man.	N	500-550	2-12	CL
	Auto.	D	500-550	10-25	CL
	Auto.(1)	D	500-550	2-20	CL
5.0L	Man.	N	600	5-30	OL
	Auto.	D	500	5-30	OL
	Auto.(2)	D	550	5-30	CL
5.7L (under 8500 GVW)	Man.	N	600	5-30	OL
	Auto.	D	525	5-30	OL
5.7L (over 8500 GVW)	Man.		600	5-30	
	Man.(3)	N	600	5-30	OL
	Auto.	D	550	5-30	OL
7.4L	Man.	N	800	5-30	OL
	Auto.	D	750	5-30	OL

* Add 2 counts for engines with less than 500 miles. Add 2 counts for every 1000 ft. above sea level (4.3 L and V8).
  Add 1 count for every 1000 ft. above sea level (2.5L and 2.8 L).

** Let engine idle until proper fuel control status (open/closed loop) is reached.

(1) 4.3 ST series.

(2) 3 speed Auto in a C10 Pickup w/ Fed. emissions and no AIR system.

(3) G van or Suburban with a single catalytic converter.

MINIMUM IDLE AIR RATE

Engine	Transmission	Gear (D/N)	Engine Speed (RPM)	Open/Closed Loop*
2.5L	Man.	N	600 ± 50	CL
	Auto.	N	500 ± 50	CL
2.8L	Man.	N	700 ± 50	OL
4.3L	Man.	N	450 ± 50	CL
	Auto.	D	400 ± 50	CL
	Auto.(1)	N	475 ± 50	CL
5.0L	Man.	N	500 ± 25	OL
	Auto.	D	425 ± 25	OL
	Auto.(2)	D	425 ± 25	CL
5.7L (under 8500 GVW)	Man.	N	500 ± 25	OL
	Auto.	D	425 ± 25	OL
5.7L (over 8500 GVW)	Man.	N	550 ± 25	CL
	Auto.	D	450 ± 25	CL
7.4L	Man.	N	700 ± 25	OL
	Auto.	D	700 ± 25	OL

* Let engine idle until proper fuel control status (open/closed loop) is reached

(1) 4.3L ST series.

(2) 5.0L without AIR system.

Controlled idle sped and minimum idle air rate specifications—Astro Van, Light Trucks and Van models

b. With the engine running at normal operating temperature, adjust the stop screw to obtain the nominal rpm per specifications with seated IAC valve.

c. Turn the ignition **OFF** and reconnect the IAC valve electrical connector.

d. Disconnect scan tool or tachometer.

e. Use silicone sealant or equivalent to cover minimum air adjustment screw.

f. Install air cleaner gasket and air cleaner to engine.

Component Replacement

Due to the varied application of components, a general component replacement procedure section is outlined. The removal steps can be altered as required by the technician.

SERVICE PRECAUTIONS

When working around any part of the fuel system, take precautionary steps to prevent fire and/or explosion:

● Disconnect negative terminal from battery (except when testing with battery voltage is required).

● When ever possible, use a flashlight instead of a drop light.

● Keep all open flame and smoking material out of the area.

● Use a shop cloth or similar to catch fuel when opening a fuel system.

● Relieve fuel system pressure before servicing.

● Use eye protection.

● Always keep a dry chemical (class B) fire extinguisher near the area.

1988-90 THROTTLE BODY INJECTION SPECIFICATIONS

Model	Engine VIN	Engine	Minimum Idle rpm	TPS Voltage	Fuel Pressure (psi)
BUICK					
Century	R	2.5L	650±50	1.25 ①	9–13
Skyhawk	K	2.0L	600±50	1.25 ①	9–13
Skyhawk	1	2.0L	550±100	1.25 ①	9–13
Skylark	U	2.5L	600±50	1.25 ①	9–13
CHEVROLET					
Beretta	1	2.0L	450–650	1.25 ①	9–13
Camaro	E	5.0L	475±25	1.25 ①	9–13
Caprice	Z	4.3L	425±25	1.25 ①	9–13
Caprice	E	5.0L	475±25	1.25 ①	9–13
Cavalier	1	2.0L	500±100	1.25 ①	9–13
Celebrity	R	2.5L	600±50	1.25 ①	9–13
Corsica	1	2.0L	450–650	1.25 ①	9–13
Lumina	R	2.5L	600±50	1.25 ①	9–13
Monte Carlo	Z	4.3L	425±25	1.25 ①	9–13
Astro Van	E	2.5L	450–650	1.25 ①	9–13
Light Trucks	E	2.5L	450–650	1.25 ①	9–13
S-10 and S-15	E	2.5L	450–650	1.25 ①	9–13
Van	E	2.5L	450–650	1.25 ①	9–13
Astro Van	R	2.8L	450–650	0.48±0.06	9–13
Light Trucks	R	2.8L	450–650	0.48±0.06	9–13
S-10 and S-15	R	2.8L	450–650	0.48±0.06	9–13
Van	R	2.8L	450–650	0.48±0.06	9–13
Light Trucks	Z	4.3L	425±25	1.25 ①	9–13
S-10 and S-15	Z	4.3L	425±25	1.25 ①	9–13
Van	Z	4.3L	425±25	1.25 ①	9–13
Light Trucks	H	5.0L	400–600	1.25 ①	9–13
Van	H	5.0L	400–600	1.25 ①	9–13
Light Trucks	K	5.7L	400–600	1.25 ①	9–13
Van	K	5.7L	400–600	1.25 ①	9–13
Light Trucks	N	7.4L	400–600	1.25 ①	9–13
OLDSMOBILE					
Calais	U	2.5L	600±50	1.25 ①	9–13
Cutlass Ciera	R	2.5L	600±50	1.25 ①	9–13

1988–90 THROTTLE BODY INJECTION SPECIFICATIONS

Model	Engine VIN	Engine	Minimum Idle rpm	TPS Voltage	Fuel Pressure (psi)
Cutlass Cruiser	R	2.5L	600±50	1.25 ①	9–13
Firenza	1	2.0L	600±25	1.25 ①	9–13
Firenza	K	2.0L	600±50	1.25 ①	9–13
PONTIAC					
Fiero	R	2.5L	600±50	1.25 ①	9–13
Firebird	E	5.0L	475±25	1.25 ①	9–13
Grand Am	U	2.5L	600±50	1.25 ①	9–13
Sunbird	K	2.0L	600±50	1.25 ①	9–13
Transport	D	3.1L	700±50	1.25 ①	9–13
6000	R	2.5L	600±50	1.25 ①	9–13

NOTE: The underhood specifications sticker often reflects tune-up changes made during the production run. The sticker specifications must always be used first if they disagree with those in this chart.

① With the ignition ON and the engine OFF. The TBI unit will be at closed throttle, the TPS voltage should read less than 1.25 volts. If the reading is higher than specified, replace the TPS.

NOTE: Due to the amount of fuel pressure in the fuel lines, the fuel system should be de-pressurized before doing any work to the fuel system. To de-pressurize the fuel system, remove the fuel tank cap to relieve tank vapor pressure. Remove the fuel pump fuse and disconnect the fuel pump electrical connections at the fuel pump (if necessary) and start the vehicle. Let the vehicle run until it burns up the remaining fuel in the fuel lines. This way there will be no pressure left in the fuel system and the repair work can be performed. Some engines (like the 2.0L engine) may have a bleed valve located in the fuel pressure regulator or on the fuel line, this bleed valve may be used to bleed the fuel pressure from the system instead of using the other method. The TBI 220 model has an internal constant bleed feature and relieves the fuel pump system pressure when the engine is turned OFF. Therefore, no further pressure relief procedure is required for that model.

CLEANING AND INSPECTION

All TBI component parts, with the exception of those noted below, should be cleaned in a cold immersion cleaner such as s Carbon X (X–55) or equivalent. The throttle position sensor, idle air control valve, pressure regulator diaphragm assembly, fuel injectors or other components containing rubber should not be placed in a solvent or cleaner bath. A chemical reaction will cause these parts to swell harden or distort. Do not soak the throttle body with the above parts attached. If the throttle body assembly requires cleaning, soak time in the cleaner should be kept to a minimum. Some models have hidden throttle shaft dust seals that could lose their effectiveness by extended soaking.

1. Clean all parts thoroughly and blow dry with compressed air. Be sure that all fuel and air passages are free of dirt and burrs.

2. Inspect the mating casting surfaces for damage that could affect gasket sealing.

THREAD LOCKING COMPOUND

Service repair kits are supplied with a small vial of thread locking compound with directions fro use. If material is not available, use Loctite® 262 or GM part No. 10522624 or equivalent. In precoating the screws, do not use a higher strength locking compound than recommended, since to do so could make remov-

ing the screw extremely difficult, or result in damaging the screw head.

THROTTLE BODY

Removal and Installation
SINGLE OR DUAL INJECTOR UNIT

Due to the varied application of throttle body unit's, a general throttle body unit removal and installation procedure is outlined. The removal steps can be altered as required by the technician.

1. Depressurize the fuel system. Raise the hood, install fender covers and remove the air cleaner assembly. Disconnect the negative battery cable.

2. Disconnect the electrical connectors for the idle speed control motor, the throttle position sensor, fuel injectors, EFE and any other component necessary in order to remove the throttle body.

3. Remove the throttle return spring, cruise control, throttle linkage and downshift cable.

4. Disconnect all necessary vacuum line, the fuel inlet line, fuel return line, brake booster line, MAP sensor hose and the AIR hose. Be sure to use a back-up wrench on all metal lines.

5. Remove the PCV, EVAP and EGR hoses from the front of the throttle body.

6. Remove the 3 throttle body mounting screws and remove the throttle body and gasket.

7. Installation is the reverse order of the removal procedure. Torque the throttle body retaining screws to 15 ft. lbs.

FUEL METER COVER

Removal and Installation
EXCEPT TBI 700 UNIT

The fuel meter cover contains the pressure regulator and is only serviced as a complete preset assembly. The fuel pressure regulator is preset and plugged at the factory.

1. Depressurize the fuel system. Raise the hood, install fender covers and remove the air cleaner assembly. Disconnect the negative battery cable.

2. Disconnect electrical connector to injector by squeezing on tow tabs and pulling straight up.

3. Remove 5 screws securing fuel meter cover to fuel meter body. Notice location of 2 short screws during removal.

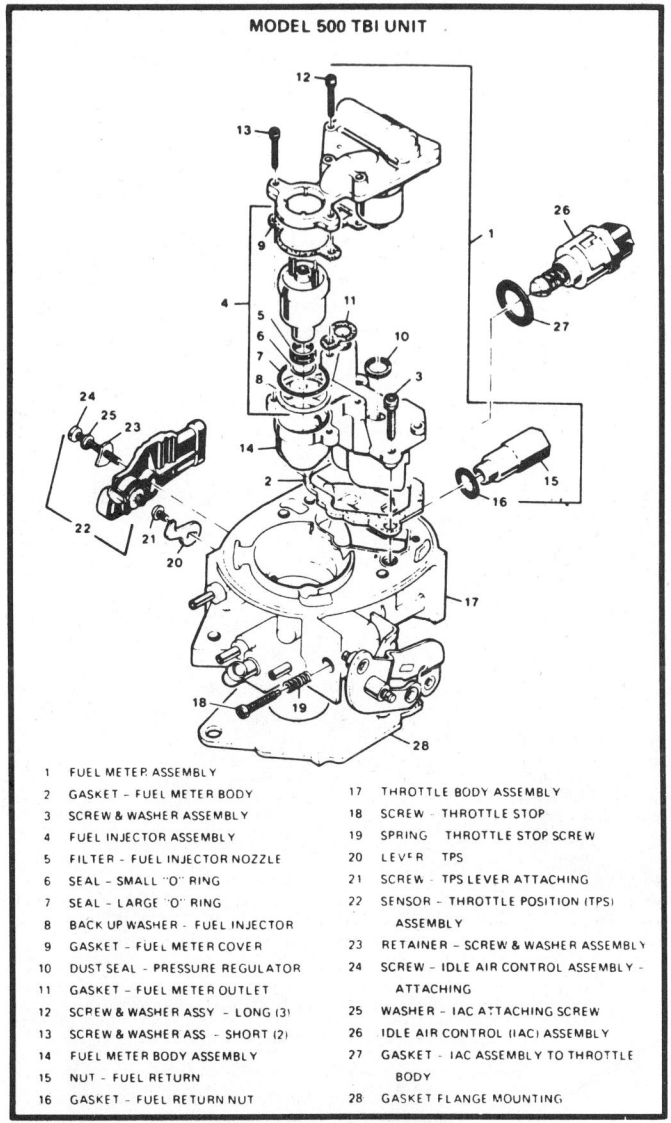

MODEL 500 TBI UNIT

1	FUEL METER ASSEMBLY	17	THROTTLE BODY ASSEMBLY
2	GASKET – FUEL METER BODY	18	SCREW – THROTTLE STOP
3	SCREW & WASHER ASSEMBLY	19	SPRING – THROTTLE STOP SCREW
4	FUEL INJECTOR ASSEMBLY	20	LEVER – TPS
5	FILTER – FUEL INJECTOR NOZZLE	21	SCREW – TPS LEVER ATTACHING
6	SEAL – SMALL "O" RING	22	SENSOR – THROTTLE POSITION (TPS) ASSEMBLY
7	SEAL – LARGE "O" RING	23	RETAINER – SCREW & WASHER ASSEMBLY
8	BACK UP WASHER – FUEL INJECTOR	24	SCREW – IDLE AIR CONTROL ASSEMBLY – ATTACHING
9	GASKET – FUEL METER COVER	25	WASHER – IAC ATTACHING SCREW
10	DUST SEAL – PRESSURE REGULATOR	26	IDLE AIR CONTROL (IAC) ASSEMBLY
11	GASKET – FUEL METER OUTLET	27	GASKET – IAC ASSEMBLY TO THROTTLE BODY
12	SCREW & WASHER ASSY – LONG (3)	28	GASKET FLANGE MOUNTING
13	SCREW & WASHER ASS – SHORT (2)		
14	FUEL METER BODY ASSEMBLY		
15	NUT – FUEL RETURN		
16	GASKET – FUEL RETURN NUT		

Exploded view of the TBI model 500 assembly

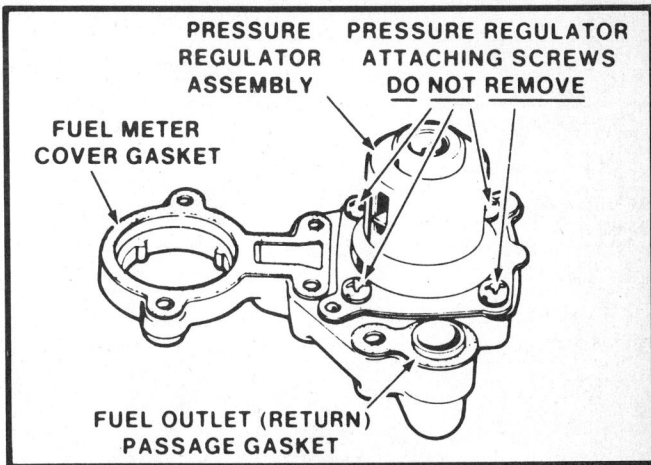

Bottom view of the fuel meter cover

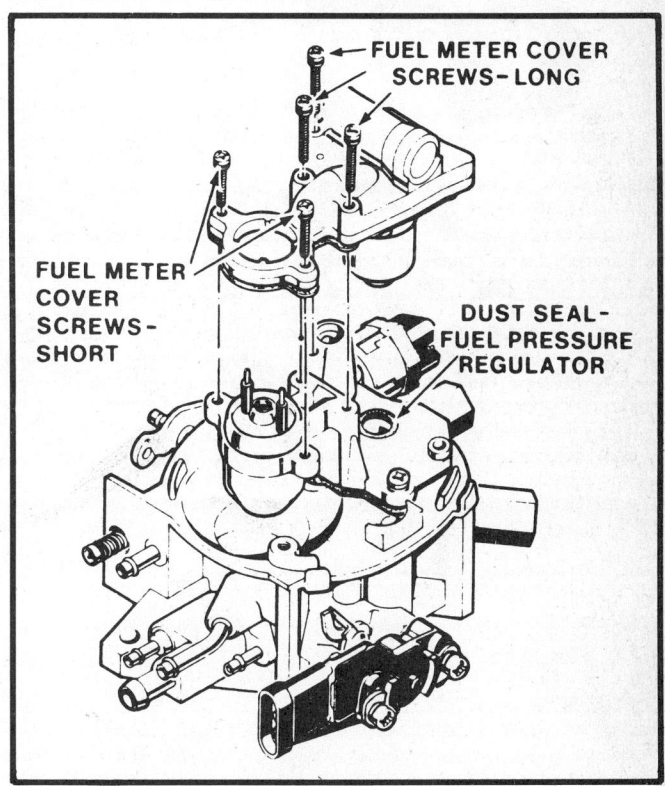

Fuel meter cover removal and installation

CAUTION

Do not remove the 4 screws securing the pressure regulator to the fuel meter cover. The fuel pressure regulator includes a large spring under heavy tension which, if accidentally released, could cause personal injury. The fuel meter cover is only serviced as a complete assembly and includes the fuel pressure regulator preset and plugged at the factor.

4. Remove the fuel meter cover assembly from the throttle body.

NOTE: Do not immerse the fuel meter cover (with pressure regulator) in any type of cleaner. Immersion of cleaner will damage the internal fuel pressure regulator diaphragms and gaskets.

5. Installation is the reverse order of the removal procedure. Be sure to use new gaskets and torque the fuel meter cover attaching screws to 28 in. lbs.

NOTE: The service kits include a small vial of thread locking compound with directions for use. If the material is not available, use part number 1052624, Loctite® 262, or equivalent. Do not use a higher strength locking compound than recommended, as this may prevent attaching screw removal or breakage of the screwhead if removal is again required.

FUEL INJECTOR

Removal and Installation
EXCEPT TBI 700 UNIT

Use care in removing the injector to prevent damage to the electrical connector pins on top of the injector, the injector fuel filter and the nozzle. The fuel injector is serviced as a complete assembly only. The fuel injector is an electrical component and should not be immersed in any type of cleaner.

1. Depressurize the fuel system. Raise the hood, install fend-

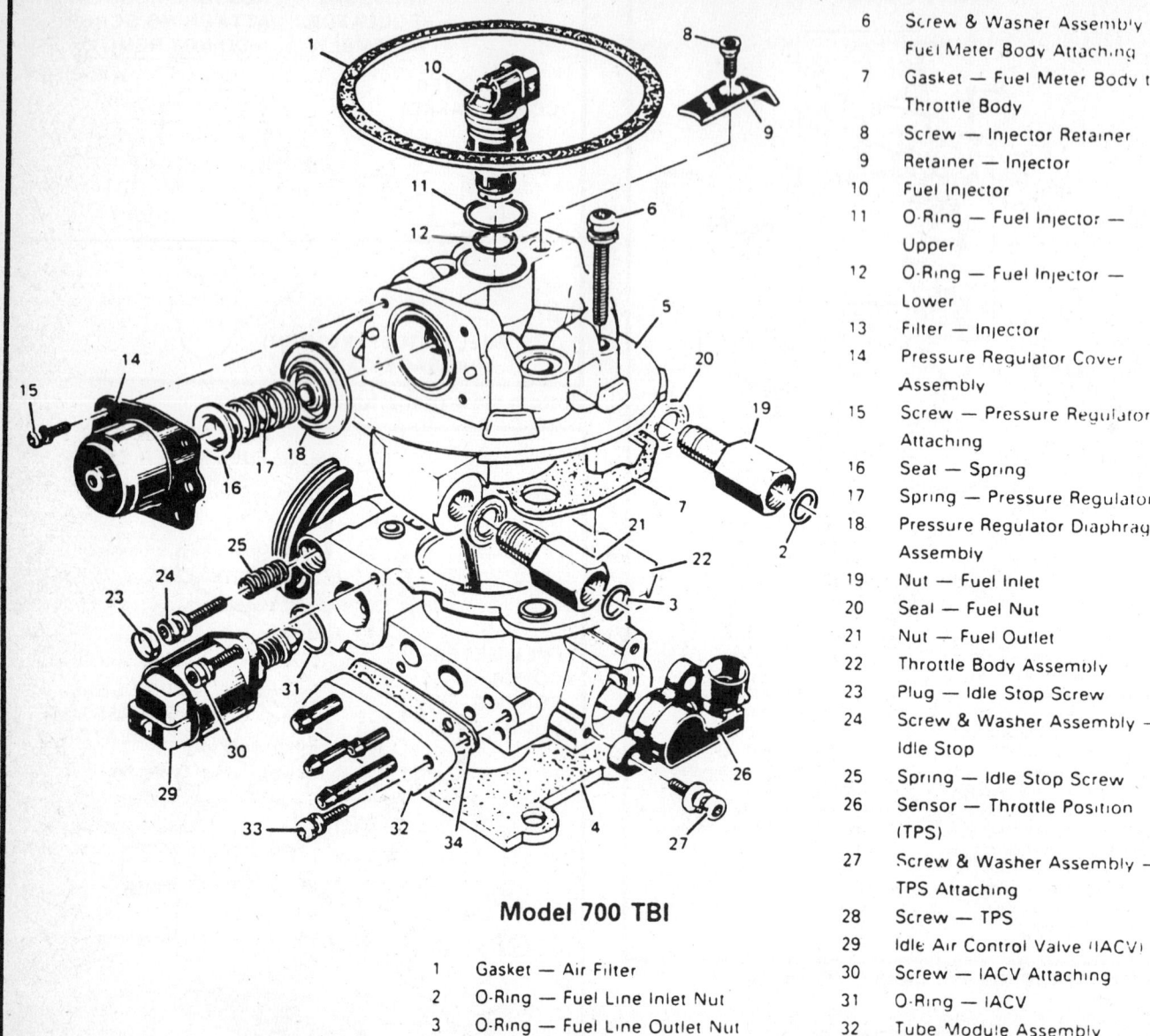

Model 700 TBI

1   Gasket — Air Filter
2   O-Ring — Fuel Line Inlet Nut
3   O-Ring — Fuel Line Outlet Nut
4   Gasket — Flange
5   Fuel Meter Assembly

6   Screw & Washer Assembly — Fuel Meter Body Attaching
7   Gasket — Fuel Meter Body to Throttle Body
8   Screw — Injector Retainer
9   Retainer — Injector
10 Fuel Injector
11 O-Ring — Fuel Injector — Upper
12 O-Ring — Fuel Injector — Lower
13 Filter — Injector
14 Pressure Regulator Cover Assembly
15 Screw — Pressure Regulator Attaching
16 Seat — Spring
17 Spring — Pressure Regulator
18 Pressure Regulator Diaphragm Assembly
19 Nut — Fuel Inlet
20 Seal — Fuel Nut
21 Nut — Fuel Outlet
22 Throttle Body Assembly
23 Plug — Idle Stop Screw
24 Screw & Washer Assembly — Idle Stop
25 Spring — Idle Stop Screw
26 Sensor — Throttle Position (TPS)
27 Screw & Washer Assembly — TPS Attaching
28 Screw — TPS
29 Idle Air Control Valve (IACV)
30 Screw — IACV Attaching
31 O-Ring — IACV
32 Tube Module Assembly
33 Screw — Manifold Attaching
34 Gasket — Tubes Manifold

Exploded view of the TBI model 700 assembly

er covers and remove the air cleaner assembly. Disconnect the negative battery cable.

2. Disconnect electrical connector to injector by squeezing on tow tabs and pulling straight up.

3. Remove the fuel meter cover assembly as previously outlined.

4. With the fuel meter cover gasket in place to prevent damage to the casting, use a suitable dowel rod and lay the dowel rod on top of the fuel meter body.

5. Insert a suitable pry tool into the small lip of the injector and pry against the dowel rod lifting the injector straight up. Tool J–26868 or equivalent can also be used.

6. Remove the injector from the fuel meter body. Remove the small O-ring at the bottom of the injector cavity. Be sure to discard both O-rings.

NOTE: The TBI unit installed on the 4.3L V6 engine in a C and K series truck have two different fuel injectors (with 2 different flow rates). Injectors having part number 5235134 (color coded orange and green) should be installed on the throttle lever side. Those with part number 5235342 (pink and brown) go on the TPS side. Be sure to replace the injector with an identical part. Injectors from other models can fit in any of the TBI units, but are calibrated for different flow rates.

7. Lubricate the new small O-ring with automatic transmission fluid. Push the new O-ring on the nozzle end of the injector, pressing the O-ring up against the injector fuel filter.

8. Install a new steel backup washer in the recess of the fuel meter body.

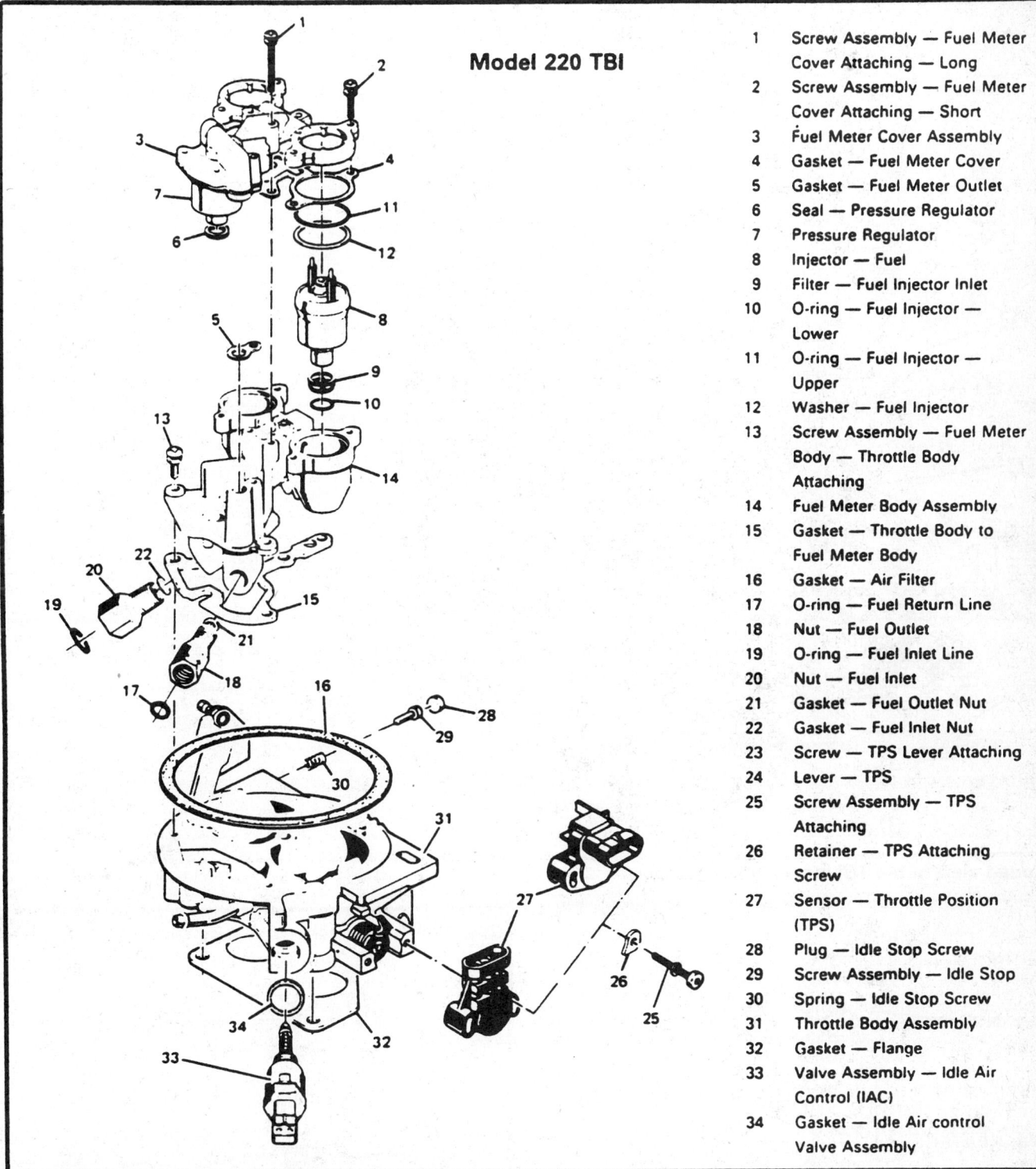

Model 220 TBI

1 Screw Assembly — Fuel Meter Cover Attaching — Long
2 Screw Assembly — Fuel Meter Cover Attaching — Short
3 Fuel Meter Cover Assembly
4 Gasket — Fuel Meter Cover
5 Gasket — Fuel Meter Outlet
6 Seal — Pressure Regulator
7 Pressure Regulator
8 Injector — Fuel
9 Filter — Fuel Injector Inlet
10 O-ring — Fuel Injector — Lower
11 O-ring — Fuel Injector — Upper
12 Washer — Fuel Injector
13 Screw Assembly — Fuel Meter Body — Throttle Body Attaching
14 Fuel Meter Body Assembly
15 Gasket — Throttle Body to Fuel Meter Body
16 Gasket — Air Filter
17 O-ring — Fuel Return Line
18 Nut — Fuel Outlet
19 O-ring — Fuel Inlet Line
20 Nut — Fuel Inlet
21 Gasket — Fuel Outlet Nut
22 Gasket — Fuel Inlet Nut
23 Screw — TPS Lever Attaching
24 Lever — TPS
25 Screw Assembly — TPS Attaching
26 Retainer — TPS Attaching Screw
27 Sensor — Throttle Position (TPS)
28 Plug — Idle Stop Screw
29 Screw Assembly — Idle Stop
30 Spring — Idle Stop Screw
31 Throttle Body Assembly
32 Gasket — Flange
33 Valve Assembly — Idle Air Control (IAC)
34 Gasket — Idle Air control Valve Assembly

Exploded view of the TBI model 220 assembly

9. Lubricate the new large O-ring with pertroleum jelly or equivalent. Install the new O-ring directly above the backup washer, pressing the O-ring down into the cavity recess. The O-ring is installed properly when it is flush with the fuel meter body casting surface.

NOTE: Do not attempt to reverse the installation of the large O-ring procedure. Install the backup washers and O-rings before the injector is located in the cavity or improper seating of the large O-ring could cause a fuel leak.

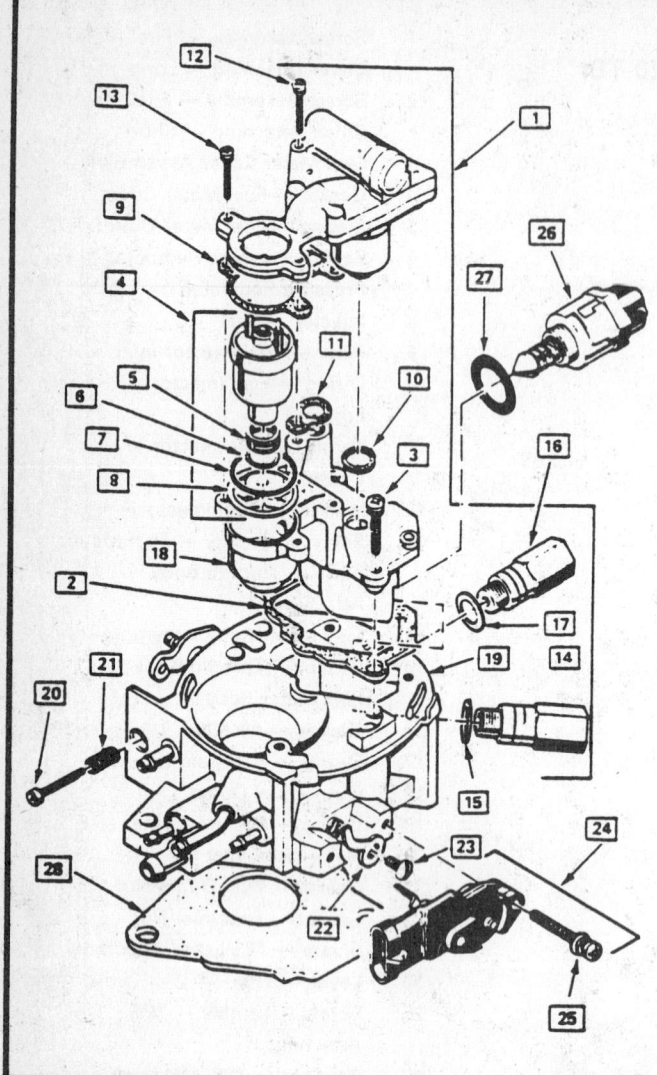

1	FUEL METER ASSEMBLY
2	GASKET – FUEL METER BODY
3	SCREW & WASHER ASSY – ATTACH. (3)
4	FUEL INJECTOR KIT
5	FILTER – FUEL INJECTOR NOZZLE
6	SEAL – SMALL "O" RING
7	SEAL – LARGE "O" RING
8	BACK-UP WASHER – FUEL INJECTOR
9	GASKET – FUEL METER COVER
10	DUST SEAL – PRESS, REGULATOR
11	GASKET – FUEL METER OUTLET
12	SCREW & WASHER ASSY – LONG (3)
13	SCREW & WASHER ASSY – SHORT (2)
14	NUT – FUEL INLET
15	GASKET – FUEL INLET NUT
16	NUT – FUEL OUTLET
17	GASKET – FUEL OUTLET NUT
18	FUEL METER BODY ASSEMBLY
19	THROTTLE BODY ASSEMBLY
20	SCREW – IDLE STOP
21	SPRING – IDLE STOP SCREW
22	LEVER – TPS
23	SCREW – TPS LEVER ATTACHING
24	SENSOR – THROTTLE POSITION KIT
25	SCREW – TPS ATTACHING (2)
26	IDLE AIR CONTROL ASSY.
27	GASKET – CONTROL ASSY. TO T.B.
28	GASKET – FLANGE MOUNTING

Exploded view of the TBI model 300 assembly

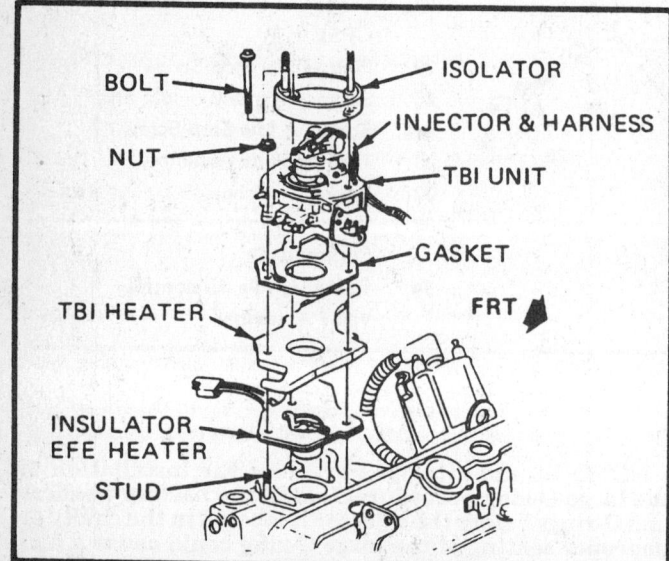

Typical TBI unit removal and installation

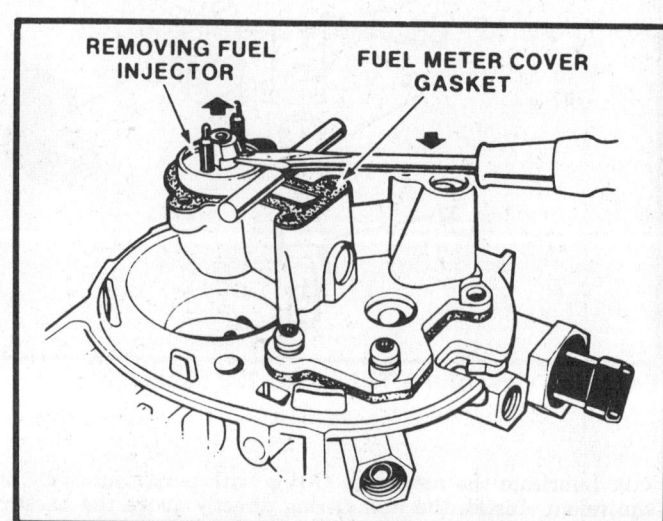

Removing the fuel injector from the throttle body assembly

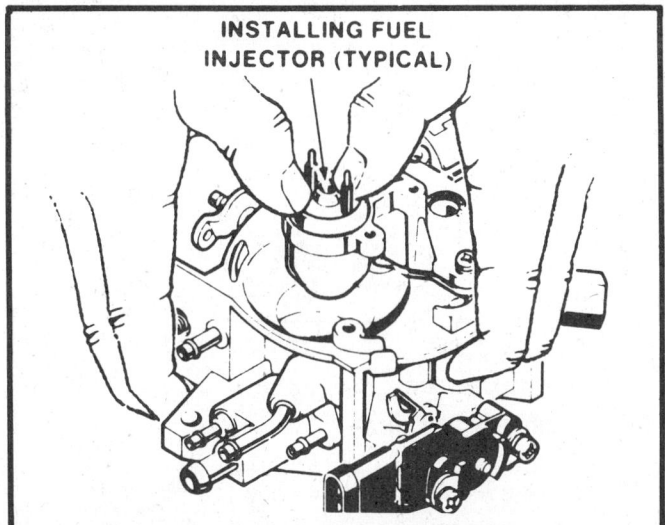

INSTALLING FUEL
INJECTOR (TYPICAL)

Installing the fuel injector into the throttle body assembly

10. Install the injector by using as pushing and twisting motion to center the nozzle O-ring in the bottom of the injector cavity and aligning the raised lug on the injector base with the notch cast into the fuel meter body.

11. Push down on the injector making sure it is fully seated in the cavity. The injector is installed correctly when the lug is seated in the notch and the electrical terminals are parallel to the throttle shaft in the throttle body.

12. Install the fuel meter cover. Install the injector electrical connector and all electrical and vacuum lines. Install the air cleaner assembly.

13. With the engine **OFF** and the ignition **ON** check for fuel leaks.

NOTE: The most widely used injector for the TBI unit is a Bosch injector, but there is also a new injector being used. This new injector will be a Multec TBI injector (a Rochester product). It is classified as a bottom feed design because the fuel enters the inlet filter near the bottom of the injector. It is designed to operate with the system fuel pressures ranging from 10-29 psi (70-200 kPa) and uses a low impedance (1.8 ohms) solenoid coil. It is used individually in the new Model 700 TBI units.

TBI 700 UNIT

Use care in removing the injector to prevent damage to the electrical connector pins on top of the injector, the injector fuel filter and the nozzle. The fuel injector is serviced as a complete assembly only. The fuel injector is an electrical component and should not be immersed in any type of cleaner.

1. Depressurize the fuel system. Raise the hood, install fender covers and remove the air cleaner assembly. Disconnect the negative battery cable.

2. Disconnect electrical connector to injector by squeezing on tow tabs and pulling straight up.

3. Remove the injector retainer screw and retainer.

4. Use a suitable dowel rod and lay the dowel rod on top of the fuel meter body.

5. Insert a suitable pry tool into the small lip of the injector and pry against the dowel rod lifting the injector straight up. Tool J–26868 or equivalent can also be used.

6. Remove the injector from the throttle body. Remove the upper and lower O-ring from the injector cavity. Be sure to discard both O-rings.

NOTE: Check the fuel injector filter for evidence of dirt and contamination. If present, check for presence of dirt in the fuel lines or fuel tank. Be sure to replace the injector with an identical part. Injectors from other models can fit in the TBI 700 unit, but are calibrated for different flow rates.

7. Lubricate the new upper and lower O-rings with automatic transmission fluid. Make sure that the upper O-ring is in the groove and the lower one is flush up against the injector fuel filter.

8. Install the injector assembly, pushing it straight into the fuel injector cavity.

NOTE: Be sure that the electrical connector end of the injector is parallel to the casting support rib and facing in the general direction of the cut-out in the fuel meter body for the wire grommet.

9. Install the injector retainer and torque the screw to 27 inch lbs. Be sure to coat the threads of the retainer screw with a suitable thread sealant.

10. With the engine **OFF** and the ignition **ON** check for fuel leaks.

FUEL METER BODY

Removal and Installation

EXCEPT TBI 700 UNIT

1. Depressurize the fuel system. Raise the hood, install fender covers and remove the air cleaner assembly. Disconnect the negative battery cable.

2. Remove the fuel meter cover assembly. Remove the fuel meter cover gasket, fuel meter outlet gasket and pressure regulator seal.

3. Remove the fuel injectors. Remove the fuel inlet and fuel outlet nuts and gaskets from the fuel meter body.

4. Remove the 3 screws and lockwashers, then remove the fuel meter body from the throttle assembly.

NOTE: Do not remove the center screw and staking at each end holding the fuel distribution skirt in the throttle body. The skirt is an integral part of the throttle body and is not serviced separately.

5. Remove the fuel meter body insulator gasket.

6. Install the new throttle body to fuel meter body gasket. Match the cut portions in the gasket with the opening in the throttle body.

7. Install the fuel meter body assembly onto the throttle body assembly.

8. Install the fuel meter body to the throttle body attaching screw assemblies, precoated with a suitable thread sealer.

9. Torque the screw assemblies to 30 inch lbs. Install the fuel inlet and outlet nuts with new gaskets to the fuel meter body assembly. Torque the inlet nut to 30 ft. lbs. (40 Nm) and the outlet nut to 21 ft. lbs (29 Nm).

10. Fuel inlet and return lines and new O-rings. Be sure to use a back-up wrench to keep the TBI nuts from turning.

11. Install the injectors with new upper and lower O-rings in the fuel meter body assembly.

12. Install the fuel meter cover gasket, fuel meter outlet gasket and pressure regulator seal.

13. Install the fuel meter cover assembly. Install the long and short fuel meter cover attaching screws assemblies, coated with a suitable thread sealer. Torque the screws to 27 inch lbs.

14. Install the electrical connectors to the fuel injectors. With the engine **OFF** and the ignition **ON** check for fuel leaks.

TBI 700 UNIT

1. Depressurize the fuel system. Raise the hood, install fend-

er covers and remove the air cleaner assembly. Disconnect the negative battery cable.

2. Remove the electrical connector from the injector. Remove the grommet with wires from the fuel meter assembly.

3. Remove the fuel inlet and outlet lines and O-rings. Be sure to use a back-up wrench to keep the TBI nuts from turning. Be sure to discard the old O-rings.

4. Remove the TBI mounting hardware. Remove the 2 fuel meter body attaching screws.

5. Remove the fuel meter assembly from the throttle body and remove the fuel meter to throttle body gasket, discard the gasket.

6. Install the new throttle body to fuel meter body gasket. Match the cut portions in the gasket with the opening in the throttle body.

7. Install the fuel meter body assembly onto the throttle body assembly.

8. Install the fuel meter body to the throttle body attaching screw assemblies, precoated with a suitable thread sealer.

9. Torque the screw assemblies to 53 inch lbs. Install the fuel inlet and outlet nuts with new gaskets to the fuel meter body assembly. Torque the inlet and outlet nut to 20 ft. lbs. (27 Nm).

10. Install the fuel inlet and return lines and new O-rings. Be sure to use a back-up wrench to keep the TBI nuts from turning.

11. Install the grommet with wires to the fuel meter assembly. Connect the electrical connector to the injector.

12. With the engine **OFF** and the ignition **ON** check for fuel leaks.

PRESSURE REGULATOR

Removal and Installation

TBI 700 UNIT ONLY

To prevent leaks, the pressure regulator diaphragm assembly must be replaced whenever the cover is removed.

1. Depressurize the fuel system. Raise the hood, install fender covers and remove the air cleaner assembly. Disconnect the negative battery cable.

2. Remove the 4 pressure regulator retaining screws, while keeping the pressure regulator compressed.

NOTE: The pressure regulator contains a large spring under heavy compression. Use care when removing the screws to prevent personal injury.

3. Check the pressure regulator seat in the fuel meter body cavity for pitting, nicks or irregularities. Use a magnifying glass if necessary. If any of the above is present, the whole fuel body casting must be replaced.

4. Install the new pressure regulator diaphragm assembly making sure it is seated in the groove in the fuel meter body.

5. Install the regulator spring seat and spring into the cover assembly.

6. Install the cover assembly over the diaphragm, while aligning the mounting holes. Be sure to use care while installing the pressure regulator to prevent misalignment of the diaphragm and possible leaks.

7. Coat the 4 regulator retaining bolts with a suitable thread sealer and torque the screws to 22 inch lbs. (2.5 Nm).

8. With the engine **OFF** and the ignition **ON** check for fuel leaks.

IDLE AIR CONTROL VALVE

Removal and Installation

1. Remove the air cleaner.

2. Disconnect the electrical connection from the idle air control assembly.

3. Using a 1¼ in. wrench, remove the idle air control assembly from the throttle body.

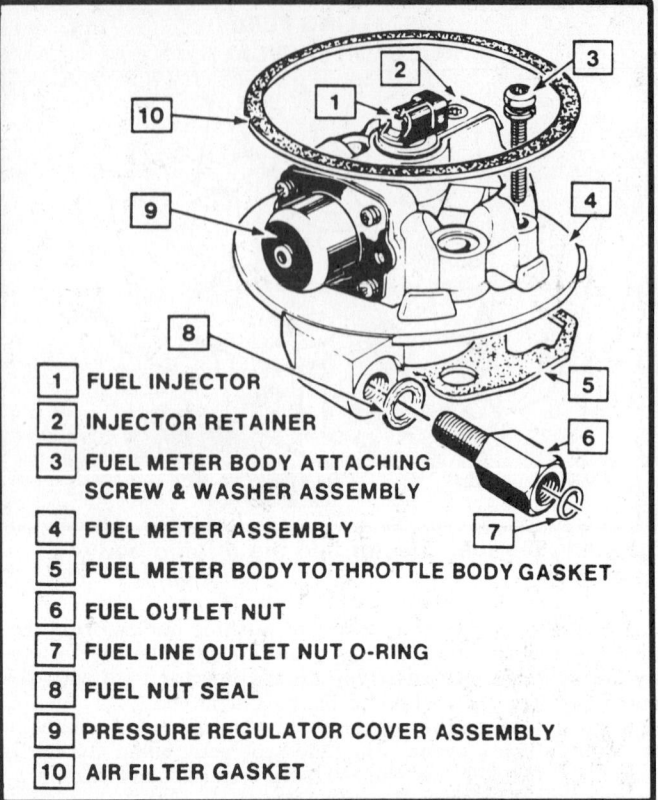

1	FUEL INJECTOR
2	INJECTOR RETAINER
3	FUEL METER BODY ATTACHING SCREW & WASHER ASSEMBLY
4	FUEL METER ASSEMBLY
5	FUEL METER BODY TO THROTTLE BODY GASKET
6	FUEL OUTLET NUT
7	FUEL LINE OUTLET NUT O-RING
8	FUEL NUT SEAL
9	PRESSURE REGULATOR COVER ASSEMBLY
10	AIR FILTER GASKET

Exploded view of the TBI 700 fuel meter assembly

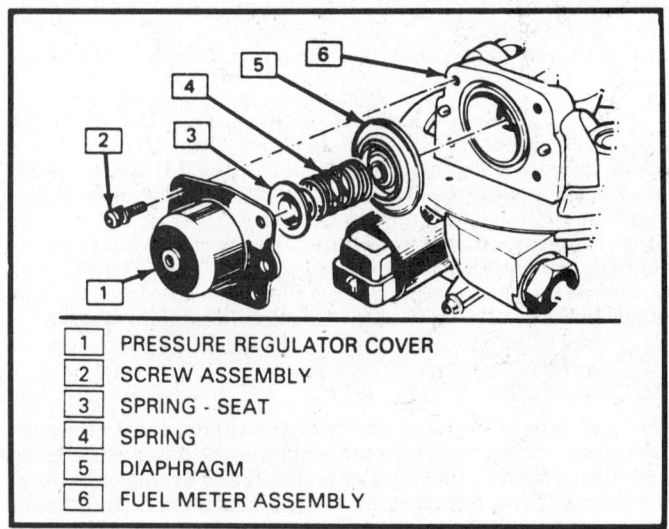

1	PRESSURE REGULATOR COVER
2	SCREW ASSEMBLY
3	SPRING - SEAT
4	SPRING
5	DIAPHRAGM
6	FUEL METER ASSEMBLY

Exploded view of the TBI 700 pressure regulator

NOTE: Before installing a new idle air control valve, measure the distance that the valve is extended. This measurement should be made from motor housing to end of the cone. The distance should be no greater than 1⅛ in. (28mm). If the cone is extended too far damage to the valve may result. The The IAC valve pintle may also be retracted by using IAC/ISC Motor Tester J–37027/BT–8256K. It is recommended not to push or pull on the IAC pintle. The force required to move the pintle mat damage the threads on the worm drive. Do not soak the IAC valve in any liquid cleaner or solvent as damage may result.

4. Be sure to identify the replacement idle air control valve as being either Type I (having a collar at the electric terminal end) or Type II (without a collar). If the measurement dimension is greater than specified, the distance must be reduced as follows:

a. Type I - Exert firm pressure on the valve to retract it. (A slight side to side movement may be helpful).

b. Type II - Compress the retaining spring from the valve while turning the valve in with a clockwise motion. Return spring to original position with the straight portion of the spring end aligned with the flat surface of the valve.

5. Install the new idle air control valve and torque the valve to 13 ft. lbs.

6. Reconnect all electrical connections. Start the engine and let it reach normal operating temperature. The ECM will reset the idle air control valve when the vehicle is driven at 30 mph or by raising the engine rpm to 2000 rpm while the diagnostic **TEST** terminal is grounded.

NOTE: Be sure to clean the IAC valve O-ring sealing surface, pintle valve seat and air passage. Use a suitable carburetor cleaner (be sure it is safe to use on systems equipped with a oxygen sensor) and a parts cleaning brush to remove the carbon deposits. Do not use a cleaner that contains methyl ethyl ketone. It is an extremely strong solvent and not necessary for this type of deposits. Shiny spots on the pintle or on the seat are normal and do not indicate a misalignment or a bent pintle shaft. If the air passage has heavy deposits, remove the throttle body for a complete cleaning. Replace the IAC O-ring with a new one.

THROTTLE POSITION SENSOR

Removal and Installation

On the Astro Van, Light Trucks and Van models equipped with the 2.8L V6 engine, the TPS is adjustable and is supplied with attaching screw retainers. On all other truck engines, the TPS is not adjustable and is not supplied with attaching screw retainers. Since these TPS configurations can be mounted interchangeable, be sure to order the correct one for your engine with the identical part number of the one being replaced.

1. Disconnect the negative battery cable. Remove the air cleaner assembly along with the necessary duct work.

2. Remove the TPS attaching screws. If the TPS is riveted to the throttle body, it will be necessary to drill out the rivets.

3. Remove the TPS from the throttle body assembly.

NOTE: The throttle position sensor is an electrical component and should not be immersed in any type of liquid solvent or cleaner, as damage may result.

4. With the throttle valve closed, install the TPS onto the throttle shaft. Rotate the TPS counterclockwise to align the mounting holes. Install the retaining screws or rivets. Torque the retaining screws to 18 inch lbs. (2.0 Nm).

5. Install the air cleaner assembly and connect the negative battery cable. Perform the TPS output check and adjust the TPS if applicable.

DIRECT IGNITION SYSTEM (DIS) ASSEMBLY

Removal and Installation

1. Disconnect the negative battery cable and remove the air cleaner assembly, if necessary. Remove and tag the DIS electrical connectors.

2. Disconnect and tag the spark plug wires, be sure to note the proper relationship of the wires to the coil.

3. Remove the DIS assembly to the block bolts. Remove the DIS assembly from the engine.

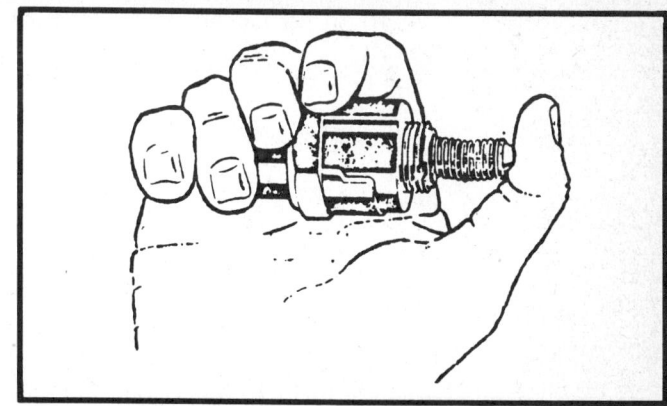

Adjusting the idle air control valve (with a collar at the electrical terminal end)

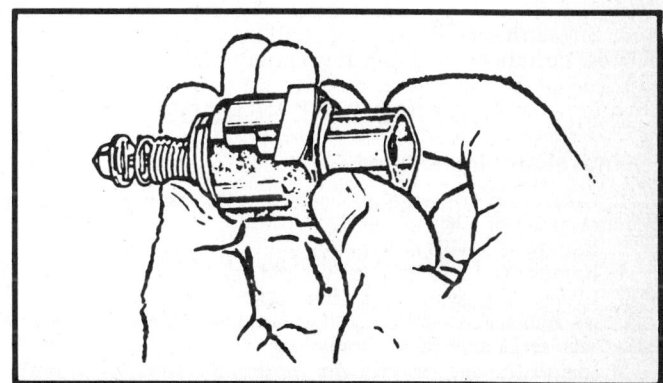

Aligning the spring under the pintle on the idle air control valve

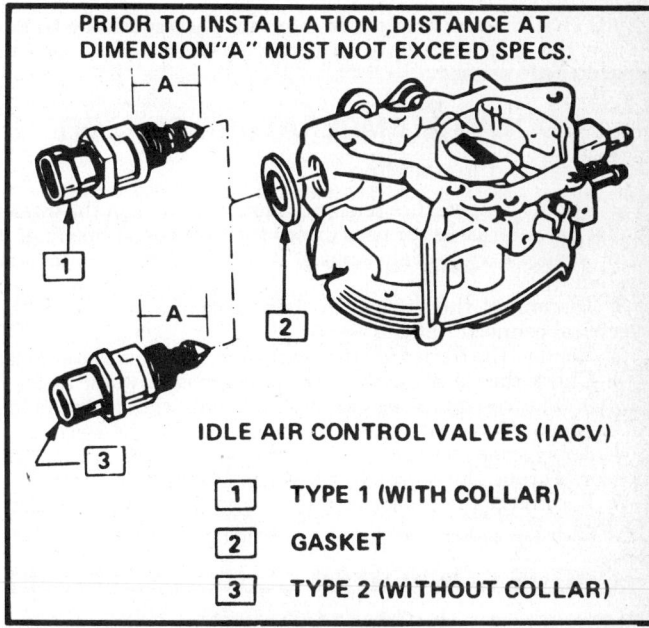

PRIOR TO INSTALLATION ,DISTANCE AT DIMENSION "A" MUST NOT EXCEED SPECS.

IDLE AIR CONTROL VALVES (IACV)

1	TYPE 1 (WITH COLLAR)
2	GASKET
3	TYPE 2 (WITHOUT COLLAR)

Typical idle air control assembly

4. Inspect the crankshaft sensor O-ring for wear, cracks or leakage. Replace as necessary (if so equipped). Lube the new O-ring with engine oil before installing.

5. Install the DIS assembly to the engine block with the retainer bolts and torque the bolts to 20 ft. lbs. (27 Nm).

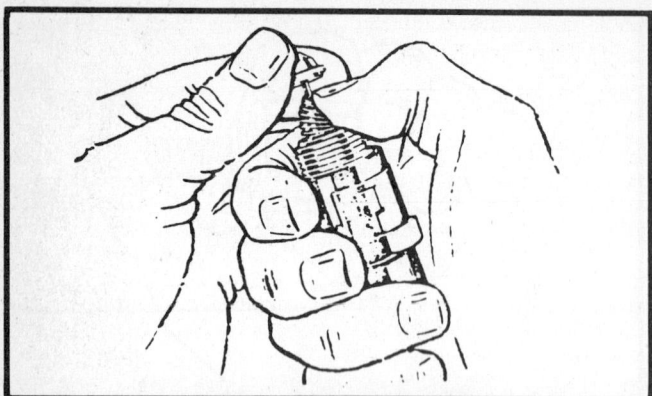

Adjusting the idle air control valve (without a collar at the electrical terminal end)

6. Install the spark plug wires to the proper coils. Install the DIS connectors and the negative battery cable.

CRANKSHAFT SENSOR

Removal and Installation

1. Disconnect the negative battery cable and remove the air cleaner assembly, if necessary.
2. Remove the DIS assembly.
3. Remove the 2 sensor screws and remove the sensor.
4. Inspect the crankshaft sensor O-ring for wear, cracks or leakage. Replace as necessary (if so equipped). Lube the new O-ring with engine oil before installing.
5. Install the sensor with the retaining screws. Torque the screws to 20 inch lbs. (2.3 Nm).
6. Install the DIS assembly and connect the negative battery cable.

NOTE:On some models it may not be necessary to remove the DIS assembly in order to gain access to the crankshaft sensor.

COOLANT TEMPERATURE SENSOR

Removal and Installation

The coolant temperature sensor is usually located on the intake manifold water jacket or near (or in) the thermostat housing. It may be necessary to drain some of the coolant from the coolant system.
1. Disconnect the negative battery cable and disconnect the electrical connector at the sensor.
2. Remove the threaded temperature sensor from the engine.
3. Check the sensor with the tip immersed in water at 50°F (15°C). The resistance across the terminals should be 4114–4743 ohms. If not within specifications, replace the sensor.
4. Apply some pipe tape to the threaded sensor and install the sensor. Torque the sensor to 22 ft. lbs. (30 Nm).
5. Reconnect the sensor connector and the negative battery cable.

TORQUE CONVERTER CLUTCH SOLENOID

Removal and Installation

1. Remove the negative battery cable. Raise and support the vehicle safely.
2. Drain the transmission fluid into a suitable drain pan. Remove the transmission pan.
3. Remove the TCC solenoid retaining screws and then remove the electrical connector, solenoid and check ball.

4. Clean and inspect all parts. Replace defective parts as necessary.
5. Install the check ball, TCC solenoid and electrical connector. Install the solenoid retaining screws and torque them to 10 ft. lbs. (14 Nm).
6. Install the transmission pan with a new gasket and torque the pan retaining bolts to 10 ft. lbs. (14 Nm).
7. Lower the vehicle and refill the transmission with the proper amount of the recommended automatic transmission fluid.

VEHICLE SPEED SENSOR

Removal and Installation

INSTRUMENT PANEL MOUNTED

1. Disconnect the negative battery cable.
2. Remove the instrument cluster from the dash board.
3. Remove the 2 screws on the back side of the instrument cluster, 1 in the optic head and 1 in the buffer amplifier.
4. Disconnect the electrical connections from the foil to the buffer amplifier.
5. Remove the vehicle speed sensor.
6. Installation is the reverse order of the removal procedures.

NOTE: If the vehicle is equipped with a digital cluster, just pull the cluster out far enough from the dash so that the optic head can be disconnected. Then remove the left side sound insulator and reach up to the instrument panel harness and disconnect the buffer amplifier connection.

TRANSMISSION MOUNTED

1. Disconnect the negative battery cable.
2. Remove the vehicle speed sensor electrical lead from the transmission.
3. Remove the VSS sensor bolt and screw retainer.
4. Remove the vehicle speed sensor assembly and O-ring.
5. Installation is the reverse order of the removal procedure. Use a new O-ring and coat with transmission fluid.

OXYGEN SENSOR

Removal and Installation

The oxygen sensor uses a permanently attached pigtail and connector. This pigtail should not be removed from the oxygen sensor. Damage or removal of the pigtail connector could effect the proper operation of the oxygen sensor. Use caution when handling the oxygen sensor. The in-line electrical connector and louvered end must be kept free of grease, dirt or other contaminants. Also avoid using cleaner solvent of any type. Do not drop or roughly handle the oxygen sensor. It is not recommended to clean or wire brush an oxygen sensor when in question, it is recommended that a new oxygen sensor be used.

NOTE: The oxygen sensor may be difficult to remove when the engine temperature is below 120°F (48°C). Excessive force may damage the threads in the exhaust manifold of exhaust pipe.

1. Start the engine and let it warm up to 120°F (48°C), stop the engine and disconnect the negative battery cable.
2. Disconnect the electrical connector from the oxygen sensor.
3. Using a special oxygen sensor socket, remove the oxygen sensor from the exhaust manifold or exhaust pipe.

NOTE: A special anti-sieze compound is used on the oxygen sensor threads. The compound consists of a liquid graphite and glass beads. The graphite will burn away, but the glass beads will remain, making the sensor

easier to remove. New or service sensors will already have the compound applied to the threads. If a oxygen sensor is removed from the engine and if for any reason it is to be reinstalled, the threads must have this anti-sieze compound applied before installation.

4. Coat the threads of the oxygen sensor with anti-sieze compound 5613695 or equivalent, if necessary.

5. Install the sensor and torque it to 30 ft. lbs (41 Nm).

6. Reconnect the electrical connector to the sensor and the negative battery cable.

MANIFOLD ABSOLUTE PRESSURE (MAP) SENSOR

Removal and Installation

1. Disconnect the negative battery cable and remove the vacuum hose from the MAP sensor.

2. Disconnect the electrical connector to the MAP sensor.

3. Remove the MAP sensor retaining screws and remove the MAP sensor.

4. Installation is the reverse order of the removal procedure.

POWER STEERING PRESSURE SWITCH

Removal and Installation

1. Disconnect the negative battery cable.

2. Disconnect the electrical connector to the switch and place a drip pan under the switch.

3. Unscrew switch and remove the switch from the power steering line.

4. Installation is the reverse order of the removal procedure. Be sure to replace the power steering fluid lost during removal of the switch.

MASS AIR FLOW (MAF) SENSOR

Removal and Installation

1. Disconnect the negative battery cable.

2. Disconnect the electrical connector to the MAF sensor.

3. Remove the air cleaner and duct assembly. Remove the sensor retaining clamps and remove the sensor.

4. Installation is the reverse order of the removal procedure.

MASS AIR TEMPERATURE (MAT) SENSOR

Removal and Installation

1. Disconnect the negative battery cable.

2. Disconnect the electrical connector to the MAT sensor.

3. Unscrew the MAT sensor from the intake manifold.

4. Installation is the reverse order of the removal procedure.

ELECTRONIC CONTROL MODULE (ECM)

To prevent possible electrostatic discharge damage to the ECM, do not touch the connector pins or soldered components on the circuit board. Service of the ECM should normally consists of either replacement of the ECM or a Mem-Cal (PROM) change.

If the diagnostic procedures call for the ECM to be replaced, the engine calibrator (Mem-Cal or PROM) and the ECM should be checked first to see if they are the correct parts. If they are, remove the Mem-Cal or PROM from the faulty ECM and install it in the new service ECM. The service ECM will not contain a new Mem-Cal or PROM.

The trouble Code 51 indicates that the Mem-Cal or PROM is installed improperly or has malfunctioned. When Code 51 is obtained, check the ECM installation for bent pins or pins not fully

seated in the socket. If it is installed correctly and the Code 51 still appears, replace the Mem-Cal or PROM.

Removal and Installation

NOTE: When replacing the production ECM with a service ECM (controller), it is important to transfer the broadcast code and production ECM number to the service ECM label. Do not record the number on the ECM cover. This will allow positive identification of the ECM parts throughout the service life of the vehicle. To prevent internal ECM damage, the ignition must be OFF when disconnecting and reconnecting the power the ECM (for example , the battery cable, ECM pigtail, ECM fuse, jumper cables, etc.).

1. Disconnect the negative battery cable.

2. Remove the glove box assembly, if necessary or remove the right hand side kick panel or what ever is necessary to gain access to the ECM.

3. Disconnect the connectors from the ECM.

4. Remove the ECM mounting hardware and remove the ECM from the passenger compartment.

5. Installation is the reverse order of the removal procedure. Be sure to remove the new ECM from its packaging and check the service number to make sure it is the same at the defective ECM.

MEM-CAL

Removal and Installation

1. Remove the ECM.

2. Remove the Mem-Cal access panel.

3. Using 2 fingers, push both retaining clips back away from the Mem-Cal. At the same time, grasp it at both ends and lift it out of the socket. Do not remove the cover of the Mem-Cal. Use of an unapproved Mem-Cal removal procedures may cause damage to the Mem-Cal or the socket.

4. Install the new Mem-Cal by pressing only the ends of the Mem-Cal. Small notches in the Mem-Cal must be aligned with the small notches in the Mem-Cal socket.

5. Press on the ends of the Mem-Cal until the retaining clips snap into the ends of the Mem-Cal. Do not press on the middle of the Mem-Cal, only on the ends.

6. Install the Mem-Cal access cover and reinstall the ECM.

FUNCTIONAL CHECK

1. Turn the ignition switch to the ON position.

2. Enter the diagnostic mode.

3. Allow a Code 12 to flash 4 times to verify that no other codes are present. This indicates that the Mem-Cal is installed properly and the ECM is working correctly.

4. If trouble Codes 41, 42, 43, 51 or 55 occurs or if the "CHECK ENGINE" light in on constantly with no codes, the Mem-Cal is not fully seated or is defective.

5. If not fully seated press down firmly on the Mem-Cal carrier If found to be defective, replace it with a new one.

NOTE: To prevent possible electrostatic discharge damage to the Mem-Cal, do not touch the component leads and do not remove the integrated circuit from the carrier.

PROM

Removal and Installation

1. Remove the ECM.

2. Remove the PROM access panel.

3. Using the rocker type PROM removal tool, engage 1 end of the PROM carrier with the hook end of the tool. Press on the

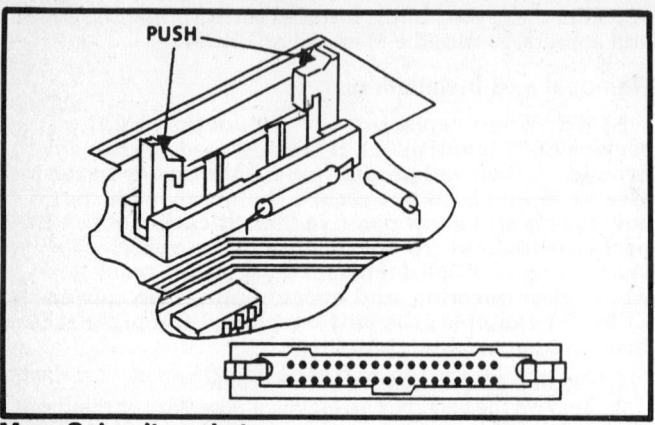

Mem-Cal unit socket

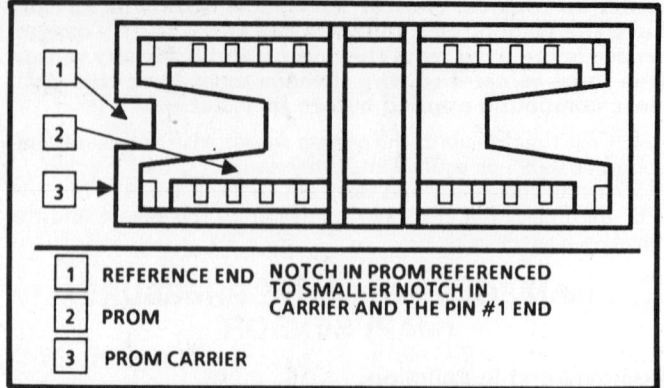

1	REFERENCE END	NOTCH IN PROM REFERENCED
2	PROM	TO SMALLER NOTCH IN CARRIER AND THE PIN #1 END
3	PROM CARRIER	

PROM located in a typical PROM carrier

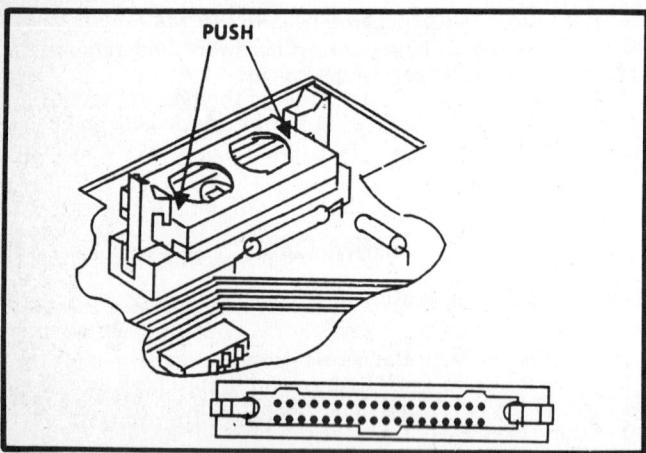

Mem-Cal unit installation

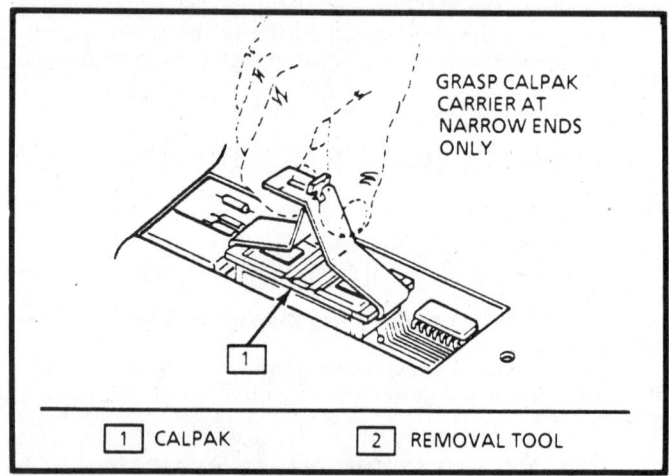

1	CALPAK	2	REMOVAL TOOL

Removing the Cal-Pak assembly

vertical bar end of the tool and rock the engaged end of the PROM carrier up as far as possible.

4. Engage the opposite end of the PROM carrier in the same manner and rock this end up as far as possible.

5. Repeat this process until the PROM carrier and PROM are free of the PROM socket. The PROM carrier with PROM in it should lift off of the PROM socket easily.

6. The PROM carrier should only be removed by using the special PROM removal tool. Other methods could cause damage to the PROM or PROM socket.

7. Before installing a new PROM, be sure the new PROM part number is the same as the old one or the as the updated number per a service bulletin.

8. Install the PROM with the small notch of the carrier aligned with the small notch in the socket. Press on the PROM carrier until the PROM is firmly seated, do not press on the PROM itself only on the PROM carrier.

9. Install the PROM access cover and reinstall the ECM.

FUNCTIONAL CHECK

1. Turn the ignition switch to the **ON** position.
2. Enter the diagnostic mode.
3. Allow a Code 12 to flash 4 times to verify that no other codes are present. This indicates that the PROM is installed properly.
4. If trouble Code 51 occurs or if the "CHECK ENGINE" light in on constantly with no codes, the PROM is not fully seated, installed backwards, has bent pins or is defective.
5. If not fully seated press down firmly on the PROM carrier.
6. If installed backwards, replace the PROM.

NOTE: Any time the PROM is installed backwards and the ignition switch is turned ON, the PROM is destroyed.

7. If the pins are bent, remove the PROM straighten the pins and reinstall the PROM. If the bent pins break or crack during straightening, discard the PROM and replace with a new PROM.

NOTE: To prevent possible electrostatic discharge damage to the PROM or Cal-Pak, do not touch the component leads and do not remove the integrated circuit from the carrier.

CAL-PAK

Removal and Installation

1. Remove the ECM.
2. Remove the Cal-Pak access panel.
3. Using the special Cal-Pak removal tool, grasp the Cal-Pak carrier at the narrow end and gently rock back and fourth while applying a firm upward force.
4. Inspect the reference end of the Cal-Pak carrier and set it aside. Do not remove the Cal-Pak from the carrier to confirm the Cal-Pak correctness.
5. Before installing a new Cal-Pak, be sure the new Cal-Pak part number is the same as the old one or the as the updated number per a service bulletin.
6. Install the Cal-Pak with the small notch of the carrier aligned with the small notch in the socket. Press on the Cal-Pak carrier until the Cal-Pak is firmly seated, do not press on the Cal-Pak itself only on the carrier.

7. Install the Cal-Pak access cover and reinstall the ECM.

Minimum Idle Speed Adjustment

The minimum idle speed should only be adjusted when installing a replacement throttle body. The idle stop screw is used to regulate the minimum idle speed of the engine. On original equipment throttle bodies, it is adjusted at the factory, then covered with a a plug to discourage unnecessary readjustment. If necessary to remove the plug, pierce the idle stop screw plug with an awl, and apply leverage to remove it.

The minimum idle speed adjustment is critical to vehicle performance and component durability. Incorrect minimum idle speed adjustment (too high) will cause the IAC valve pintle to constantly bottom out on its seat and result in early valve failure. If the minimum idle speed is adjusted to low, the vehicle may not start in cold weather or may stall during engine warm up.

BERETTA AND CORSICA WITH 2.0L ENGINE (VIN 1)

1. Block the drive wheels and apply the parking brake. Remove the air cleaner assembly and or air duct. Remove and plug any vacuum hoses on the tube manifold assembly, if so equipped. Disconnect the throttle cable.
2. Ground the diagnostic test terminal in the ALCL connector. Turn the ignition **ON** and leave the engine **OFF**. Wait at least 30 seconds, this will allow the IAC pintle to seat in the throttle body.
3. With the ignition still in the **ON** position and the engine **OFF**, with the ALCL test terminal still grounded, disconnect the IAC valve electrical connector.
4. Connect a tachometer to the engine (a scan tool can also be used) to monitor the engine speed.
5. Remove the ground from the ALCL connector test terminal.
6. Place the transmission in the **P** or **N** position. Start and run the engine until it reaches normal operating temperature. It may be necessary to depress the accelerator pedal in order to start the engine. Allow the engine idle speed to stabilize.

NOTE: The engine should be at normal operating temperature with the accessories and cooling fan off.

7. The idle speed should be set at 450–650 rpm, if not within specification, adjust as necessary.
8. Install the throttle cable, be sure that the minimum idle speed is not affected by the throttle cable. If so correct this condition.
9. Turn the ignition **OFF** and reconnect the IAC valve electrical connector. Unplug and reconnect and disconnected vacuum lines. Install the air cleaner assembly.

CAVALIER WITH 2.0L ENGINE (VIN 1), SUNBIRD WITH 2.0L ENGINE (VIN K) AND SKYHAWK WITH 2.0L ENGINE (VIN 1) AND (VIN K)

1. Block the drive wheels and apply the parking brake. Remove the air cleaner assembly and or air duct. Remove and plug any vacuum hoses on the tube manifold assembly, if so equipped.
2. Connect a suitable scan tool onto the ALCL connector. Turn the ignition switch to the **ON** position.
3. Select the **FIELD SERVICE MODE** on the scan tool , refer to the scan tool instructions.
4. This will cause the IAC valve pintle to seat in the throttle body (closing the air passage). Wait at least 45 seconds, disconnect the IAC valve connector, then exist the **FIELD SERVICE MODE**.
5. Place the transmission in the **P** or **N** position. Start and run the engine until it reaches normal operating temperature and Closed Loop operation as read on the scan tool. It may be necessary to hold the throttle open slightly to maintain an idle.
6. Select **ENGINE RPM** on the scan tool and read the engine speed.

NOTE: The engine should be at normal operating temperature and in the Closed Loop. Accessories and cooling fan should be off.

7. Make sure that the throttle and cruise control cables do not hold the throttle open. The idle speed should be set at the 550 ± 50 rpm on engines with more than 500 miles.
8. Adjust the minimum idle speed if necessary. Turn the ignition switch to the **OFF** position.
9. Connect the IAC valve electrical connector.
10. Reset IAC valve pintle position as follows:
    a. Select **ENGINE RPM** on the scan tool.
    b. Start the engine and hold the speed above 2000 rpm. Select the **FIELD SERVICE MODE** for ten seconds so as to reset the IAC valve pintle position.
    c. Exist the **FIELD SERVICE MODE** and allow the engine to return to idle.
    d. Turn the ignition switch to the **OFF** position. Restart the engine and check for proper idle operation.
11. Disconnect the scan tool and remove the blocks from the drive wheels.

FIRENZA WITH 2.0L ENGINE (VIN 1)

1. Block the drive wheels and apply the parking brake. Remove the air cleaner assembly and or air duct TBI bonnet with gasket. Remove and plug any vacuum hoses on the tube manifold assembly, if so equipped.
2. Pierce the idle stop screw plug with an awl, and apply leverage to remove it.
3. Connect a tachometer to the engine (a scan tool can also be used) to monitor the engine speed.
4. Place the transmission in the **P** or **N** position. Start and run the engine until it reaches normal operating temperature. Allow the engine idle speed to stabilize.

NOTE: The engine should be at normal operating temperature with the accessories and cooling fan off.

5. Disconnect the idle air control connector. Install IAC passage plug tool J–36377 or equivalent in the idle air passage on the TBI unit. Be sure that there are no air leaks around the passage plug tool.
6. On vehicles equipped with automatic transmission, place the selector is the **D** position before making the adjustment. Adjust the idle speed stop screw to obtain the specified idle speed (650 ± 25 rpm).
7. Place the transmission selector lever back to the **P** position, if so equipped. Turn the ignition to the **OFF** position and remove the idle air control passage plug tool and reconnect the IAC valve electrical connector.
8. Use a silicone sealant or equivalent to cover the minimum air adjustment screw.
9. Reconnect any disconnected vacuum lines. Install the air cleaner assembly, TBI bonnet and gasket.

2.5L ENGINE (VIN R)

1. Block the drive wheels and apply the parking brake. Remove the air cleaner assembly and or air duct. Remove and plug any vacuum hoses on the tube manifold assembly, if so equipped. Disconnect the throttle cable.

NOTE: If present, pierce the idle stop screw plug with an awl and apply leverage to remove it.

2. Ground the diagnostic test terminal in the ALCL connector. Turn the ignition **ON** and leave the engine **OFF**. Wait at least 30 seconds, this will allow the IAC pintle to seat in the throttle body.
3. With the ignition still in the **ON** position and the engine **OFF**, with the ALCL test terminal still grounded, disconnect the IAC valve electrical connector.
4. Connect a tachometer to the engine (a scan tool can also be used) to monitor the engine speed.

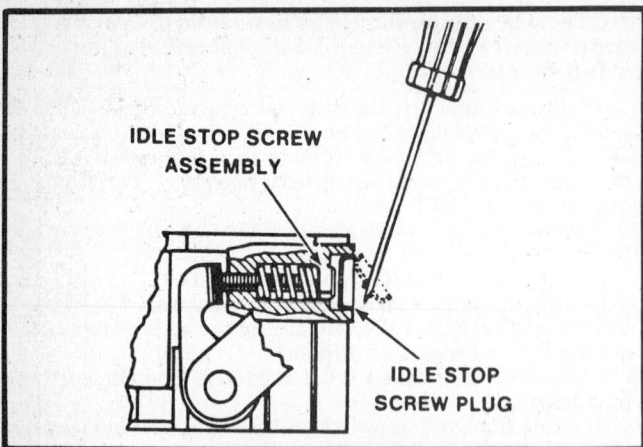

Removing the idle top screw plug

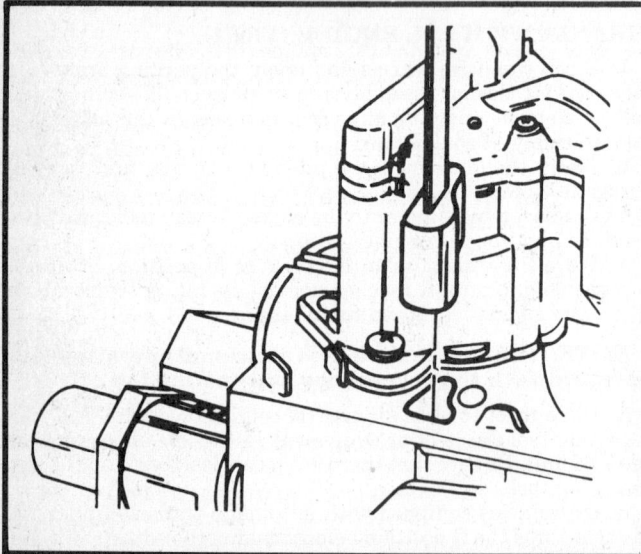

Installing special tool J–36377 or equivalent in the idle air passage of the throttle body.

5. Remove the ground from the ALCL connector test terminal.

6. Place the transmission in the **P** or **N** position. Start and run the engine until it reaches normal operating temperature. It may be necessary to depress the accelerator pedal in order to start the engine. Allow the engine idle speed to stabilize.

NOTE: The engine should be at normal operating temperature with the accessories and cooling fan off.

7. The idle speed should be set at 600 ± 50 rpm, if not within specification adjust as necessary.

8. Install the throttle cable, be sure that the minimum idle speed is not affected by the throttle cable. If so correct this condition.

9. Turn the ignition **OFF** and reconnect the IAC valve electrical connector. Unplug and reconnect and disconnected vacuum lines. Install the air cleaner assembly.

2.5L ENGINE (VIN P)

1. Block the drive wheels and apply the parking brake. Remove the air cleaner assembly and or air duct. Remove and plug any vacuum hoses on the tube manifold assembly, if so equipped. Disconnect the throttle cable.

NOTE: If present, pierce the idle stop screw plug with an awl and apply leverage to remove it.

2. Ground the diagnostic test terminal in the ALCL connector. Turn the ignition **ON** and leave the engine **OFF**. Wait at least 45 seconds, this will allow the IAC pintle to seat in the throttle body.

3. With the ignition still in the **ON** position and the engine **OFF**, with the ALCL test terminal still grounded, disconnect the IAC valve electrical connector.

4. Connect a tachometer to the engine (a scan tool can also be used) to monitor the engine speed.

5. Remove the ground from the ALCL connector test terminal.

6. Place the transmission in the **P** or **N** position. Start and run the engine until it reaches normal operating temperature. It may be necessary to depress the accelerator pedal in order to start the engine. Allow the engine idle speed to stabilize.

NOTE: The engine should be at normal operating temperature with the accessories and cooling fan off.

7. The idle speed should be set at 600 ± 50 rpm, if not within specification adjust as necessary.

8. Install the throttle cable, be sure that the minimum idle speed is not affected by the throttle cable. If so correct this condition.

9. Turn the ignition **OFF** and reconnect the IAC valve electrical connector. Use silicone sealant or equivalent to cover the minimum idle adjustment screw hole. Unplug and reconnect and disconnected vacuum lines. Install the air cleaner assembly.

CALAIS WITH 2.5L ENGINE (VIN U)

1. Block the drive wheels and apply the parking brake. Remove the air cleaner assembly and or air duct. Remove and plug any vacuum hoses on the tube manifold assembly, if so equipped.

2. Connect a suitable scan tool onto the ALCL connector. Turn the ignition switch to the **ON** position.

3. Select the **FIELD SERVICE MODE** on the scan tool , refer to the scan tool instructions.

4. This will cause the IAC valve pintle to seat in the throttle body (closing the air passage). Wait at least 45 seconds, disconnect the IAC valve connector, then exist the **FIELD SERVICE MODE**.

5. Place the transmission in the **P** or **N** position. Start and run the engine until it reaches normal operating temperature and Closed Loop operation as read on the scan tool. It may be necessary to hold the throttle open slightly to maintain an idle.

6. Select **ENGINE RPM** on the scan tool and read the engine speed.

NOTE: The engine should be at normal operating temperature and in the Closed Loop. Accessories and cooling fan should be off.

7. Make sure that the throttle and cruise control cables do not hold the throttle open. The idle speed should be set at the 600 ± 50 rpm on engines with more than 500 miles.

8. Adjust the minimum idle speed if necessary. Turn the ignition switch to the **OFF** position.

9. Connect the IAC valve electrical connector.

10. Reset IAC valve pintle position as follows:

   a. Select **ENGINE RPM** on the scan tool.

   b. Start the engine and hold the speed above 2000 rpm. Select the **FIELD SERVICE MODE** for ten seconds so as to reset the IAC valve pintle position.

   c. Exist the **FIELD SERVICE MODE** and allow the engine to return to idle.

   d. Turn the ignition switch to the **OFF** position. Restart the engine and check for proper idle operation.

11. Disconnect the scan tool and remove the blocks from the drive wheels.

GRAND AM WITH 2.5L ENGINE (VIN U)

1. Block the drive wheels and apply the parking brake. Remove the air cleaner assembly and or air duct. Remove and plug any vacuum hoses on the tube manifold assembly, if so equipped. Disconnect the throttle cable.

NOTE: If present, pierce the idle stop screw plug with an awl and apply leverage to remove it.

2. Ground the diagnostic test terminal in the ALCL connector. Turn the ignition **ON** and leave the engine **OFF**. Wait at least 45 seconds, this will allow the IAC pintle to seat in the throttle body.

3. With the ignition still in the **ON** position and the engine **OFF**, with the ALCL test terminal still grounded, disconnect the IAC valve electrical connector.

4. Connect a tachometer to the engine (a scan tool can also be used) to monitor the engine speed.

5. Remove the ground from the ALCL connector test terminal.

6. Place the transmission in the **P** or **N** position. Start and run the engine until it reaches normal operating temperature. It may be necessary to depress the accelerator pedal in order to start the engine. Allow the engine idle speed to stabilize.

NOTE: The engine should be at normal operating temperature with the accessories and cooling fan off.

7. The idle speed should be set at 600 ± 50 rpm, if not within specification adjust as necessary.

8. Install the throttle cable, be sure that the minimum idle speed is not affected by the throttle cable. If so correct this condition.

9. Turn the ignition **OFF** and reconnect the IAC valve electrical connector. Use silicone sealant or equivalent to cover the minimum idle adjustment screw hole. Unplug and reconnect and disconnected vacuum lines. Install the air cleaner assembly.

3.1L (VIN D), 4.3L (VIN Z), 5.0L (VIN E) AND 5.7L (VIN 7) ENGINES

1. Block the drive wheels and apply the parking brake. Remove the air cleaner assembly and or air duct. Remove and plug any vacuum hoses on the tube manifold assembly, if so equipped.

2. Connect a suitable scan tool onto the ALCL connector. Turn the ignition switch to the **ON** position.

3. Select the **FIELD SERVICE MODE** on the scan tool , refer to the scan tool instructions.

4. This will cause the IAC valve pintle to seat in the throttle body (closing the air passage). Wait at least 45 seconds, disconnect the IAC valve connector, then exist the **FIELD SERVICE MODE**.

5. Place the transmission in the **P** or **N** position. Start and run the engine until it reaches normal operating temperature and Closed Loop operation as read on the scan tool. It may be necessary to hold the throttle open slightly to maintain an idle.

6. Select **ENGINE RPM** on the scan tool and read the engine speed.

NOTE: The engine should be at normal operating temperature and in the Closed Loop. Accessories and cooling fan should be off.

7. Make sure that the throttle and cruise control cables do not hold the throttle open. The idle speed should be set at the the following specifications:
   a. 3.1L (VIN D) engine—700 ± 50 rpm.
   b. 4.3L (VIN Z) engine—425 ± 25 rpm.
   c. 5.0L (VIN E) engine—475 ± 25 rpm.
   d. 5.7L (VIN 7) engine—475 ± 25 rpm.

8. Adjust the minimum idle speed if necessary. Turn the ignition switch to the **OFF** position.

9. Connect the IAC valve electrical connector.

10. Reset IAC valve pintle position as follows:
    a. Select **ENGINE RPM** on the scan tool.
    b. Start the engine and hold the speed above 2000 rpm. Select the **FIELD SERVICE MODE** for 10 seconds so as to reset the IAC valve pintle position.
    c. Exist the **FIELD SERVICE MODE** and allow the engine to return to idle.
    d. Turn the ignition switch to the **OFF** position. Restart the engine and check for proper idle operation.

11. Disconnect the scan tool and remove the blocks from the drive wheels.

Controlled Idle Speed Check

ASTRO VAN, LIGHT TRUCKS AND VAN MODELS ONLY

Before performing the idle speed check, there should be no codes displayed, the idle air control system has been checked and the ignition timing is correct.

1. Set the parking brake and block the drive wheels.

2. Connect a suitable scan tool to the ALCL connector with the tool in open mode.

3. Start the engine and bring up to normal operating temperature. Check for the correct state of the park/neutral switch on the scan tool.

4. Check the specification in the chart provided for the controlled idle speed and the idle air control valve pintle positions (counts).

5. If within specifications, the idle is being correctly controlled by the ECM.

6. If not within specifications, check the minimum idle air rate, fuel system, idle circuit, Throttle Position Sensor (TPS), Manifold Absolute Pressure (MAP) sensor, etc.

Minimum Idle Air Rate Check

ASTRO VAN, LIGHT TRUCKS AND VAN MODELS ONLY

1. Check the controlled idle speed and perform the idle air control check first.

2. Block the drive wheels and set the parking brake.

3. Start the engine and bring it up to normal operating temperature. Turn the engine to the **OFF** position.

4. Remove the air cleaner, adapter and gaskets. On the ST series vehicle, leave the thermac hose connected. Check that the throttle lever is not being bound by the throttle, T.V. or cruise control cables.

5. With the IAC valve connected, ground the diagnostic terminal (ALCL connector).

6. Turn the ignition switch to the **ON** position, do not start the engine. Wait at least 10 seconds, this will allow the pintle to extend and seat in the throttle body.

7. With the ignition switch **ON** and the engine stopped, test terminal still grounded, disconnect the IAC valve electrical connector. This disables the IAC valve seated position. Remove the ground from the diagnostic terminal.

8. Connect a suitable scan tool to the ALCL connector and place it in open mode. If a tool is not available connect a tachometer the engine.

9. Start the engine. With the transmission in **N**, allow the engine rpm to stabilize.

10. Check the rpm against the specification in the chart provided. Disregard the IAC counts on the scan tool with the IAC disconnected. If the engine has less than 500 miles on it or it is checked at a altitude above 1500 feet, the idle rpm with a seated IAC valve should be lower than the specified values.

11. If the minimum idle air rate is within specifications, no further check is required.

12. If the minimum idle air rate is not within specifications, perform the following procedures.
    a. On vehicles equipped with a tamper resistant plug cover-

CONTROLLED IDLE SPEED

Engine	Transmission	Gear (D/N)	Idle Speed (RPM)	IAC Counts*	Open/Closed Loop**
2.5L	Man.	N	900(ST) 800(M)	5-20	CL
	Auto.	D	800(S) 650(T) 750(M)	15-40	CL
2.8L	Man.	N	800	5-20	OL
4.3L	Man.	N	500-550	2-12	CL
	Auto.	D	500-550	10-25	CL
	Auto.(1)	D	500-550	2-20	CL
5.0L	Man.	N	600	5-30	OL
	Auto.	D	500	5-30	OL
	Auto.(2)	D	550	5-30	CL
5.7L (under 8500 GVW)	Man.	N	600	5-30	OL
	Auto.	D	525	5-30	OL
5.7L (over 8500 GVW)	Man.		600	5-30	
	Man.(3)	N	600	5-30	OL
	Auto.	D	550	5-30	OL
7.4L	Man.	N	800	5-30	OL
	Auto.	D	750	5-30	OL

* Add 2 counts for engines with less than 500 miles. Add 2 counts for every 1000 ft. above sea level (4.3 L and V8).
  Add 1 count for every 1000 ft. above sea level (2.5L and 2.8 L).

** Let engine idle until proper fuel control status (open/closed loop) is reached.

(1) 4.3 ST series.

(2) 3 speed Auto in a C10 Pickup w/ Fed. emissions and no AIR system.

(3) G van or Suburban with a single catalytic converter.

MINIMUM IDLE AIR RATE

Engine	Transmission	Gear (D/N)	Engine Speed (RPM)	Open/Closed Loop*
2.5L	Man.	N	600 ± 50	CL
	Auto.	N	500 ± 50	CL
2.8L	Man.	N	700 ± 50	OL
4.3L	Man.	N	450 ± 50	CL
	Auto.	D	400 ± 50	CL
	Auto.(1)	N	475 ± 50	CL
5.0L	Man.	N	500 ± 25	OL
	Auto.	D	425 ± 25	OL
	Auto.(2)	D	425 ± 25	CL
5.7L (under 8500 GVW)	Man.	N	500 ± 25	OL
	Auto.	D	425 ± 25	OL
5.7L (over 8500 GVW)	Man.	N	550 ± 25	CL
	Auto.	D	450 ± 25	CL
7.4L	Man.	N	700 ± 25	OL
	Auto.	D	700 ± 25	OL

* Let engine idle until proper fuel control status (open/closed loop) is reached

(1) 4.3L ST series.

(2) 5.0L without AIR system.

Controlled idle sped and minimum idle air rate specifications—Astro Van, Light Trucks and Van models

ing the minimum air adjustment screw, the throttle body unit must be removed from the engine to remove the plug.

b. With the engine running at normal operating temperature, adjust the stop screw to obtain the nominal rpm per specifications with seated IAC valve.

c. Turn the ignition **OFF** and reconnect the IAC valve electrical connector.

d. Disconnect scan tool or tachometer.

e. Use silicone sealant or equivalent to cover minimum air adjustment screw.

f. Install air cleaner gasket and air cleaner to engine.

TPS Output Check Test

WITH SCAN TOOL

The throttle position sensor on all GM vehicles except the 2.8L (VIN R) engine are not adjustable. If the sensor is found out of specifications and the sensor is at fault it cannot be adjusted and should be replaced. The following check should be performed only when the throttle body or TPS has been replaced.

1. Use a suitable scan tool to read the TPS voltage.
2. With the ignition switch **ON** and the engine **OFF**, the TPS voltage should be less than 1.25 volts.
3. If the voltage reading is higher than specified, replace the throttle position sensor.

WITHOUT SCAN TOOL

This check should only be performed, when the throttle body or the TPS has been replaced or after the minimum idle speed has been adjusted.

1. Remove air cleaner. Disconnect the TPS harness from the TPS.
2. Using suitable jumper wires, connect a digital voltmeter J–29125–A or equivalent to the correct TPS terminals:

a. Terminals **A** and **B** on all models except the ones listed below.

b. Terminals **C** and **B** on the 2.0L engines using the TBI 500 model.

3. With the ignition **ON** and the engine running, The TPS voltage should be 0.450–1.25 volts at base idle to approximately 4.5 volts at wide open throttle.

4. If the reading on the TPS is out of specification, check the minimum idle speed before replacing the TPS.

5. If the voltage reading is correct, remove the voltmeter and jumper wires and reconnect the TPS connector to the sensor. Re-install the air cleaner.

Adjustable TPS Output Check

2.8L (VIN R) ENGINE

This check should only be performed, when the throttle body or the TPS has been replaced or after the minimum idle speed has been adjusted.

1. Remove air cleaner. Disconnect the TPS harness from the TPS.
2. Using suitable jumper wires, connect a digital voltmeter J–34029–A or equivalent from the TPS connector center terminal **B** to the outside terminal **A**. A suitable ALDL scanner can also be used to read the TPS output voltage.
3. With the ignition **ON** and the engine stopped, The TPS voltage should be less than 1.25 volts.
4. If the reading on the TPS is not within the specified range, rotate the the TPS until 0.48 ± .06 volts are obtained. If this specified voltage cannot be obtained, replace the TPS.
5. If the voltage reading is correct, remove the voltmeter and jumper wires and reconnect the TPS connector to the sensor. Re-install the air cleaner.

SEQUENTIAL FUEL INJECTION (SFI) SYSTEM

APPLICATION CHART
1988–90 Sequential Fuel Injection Vehicles

Year	Models	Series VIN	Engine	Engine VIN Code
BUICK				
1988	Century	A	3.8L	3
	Electra	C	3.8L	3
	Electra	C	3.8L	C
	LeSabre	H	3.8L	3
	LeSabre	H	3.8L	C
	Riviera	E	3.8L	C
	Reatta	E	3.8L	C
1989-90	Electra	C	3.8L	C
	LeSabre	H	3.8L	C
	Riviera	E	3.8L	C
	Reatta	E	3.8L	C

APPLICATION CHART
1988–90 Sequential Fuel Injection Vehicles

Year	Models	Series VIN	Engine	Engine VIN Code
OLDSMOBILE				
1988	Ciera	A	3.8L	3
	Delta 88	H	3.8L	3
	Delta 88	H	3.8L	C
	Ninety-Eight	C	3.8L	3
	Ninety-Eight	C	3.8L	C
	Toronado	E	3.8L	C
1989-90	Delta 88	H	3.8L	C
	Ninety-Eight	C	3.8L	3
	Toronado	E	3.8L	C
PONTIAC				
1988	Bonneville	H	3.8L	3
	Bonneville	H	3.8L	C
1989-90	Bonneville	H	3.8L	C

GENERAL INFORMATION

The Sequential Fuel Injection (SFI) system is controlled by an Electronic Control Module (ECM) which monitors engine operations and generates output signals to provide the correct air/fuel mixture, ignition timing and engine idle speed control. Input to the control unit is provided by an oxygen sensor, coolant temperature sensor, detonation sensor, hot film air mass sensor and throttle position sensor. The ECM also receives information concerning engine rpm, road speed, transmission gear position, power steering and air conditioning.

With SFI, metered fuel is timed and injected sequentially through the injectors into individual cylinder ports. Each cylinder receives 1 injection per working cycle (every 2 revolutions), just prior to the opening of the intake valve. In addition, on V6 engines, the SFI system incorporates a Computer Controlled Coil Ignition (C^3I) system that uses an electronic coil module that replaced the conventional distributor and coil used on most engines. An Electronic Spark Control (ESC) is used to adjust the spark timing. On V8 engines, the conventional High Energy Ignition (HEI) system is used.

The injection system uses solenoid-type fuel injectors, 1 at each intake port, rather than the single injector found on the earlier throttle body system. The injectors are mounted on a fuel rail and are activated by a signal from the electronic control module. The injector is a solenoid-operated valve which remains open depending on the width of the electronic pulses (length of the signal) from the ECM; the longer the open time, the more fuel is injected. In this manner, the air/fuel mixture can be precisely controlled for maximum performance with minimum emissions.

Fuel is pumped from the tank by a high pressure fuel pump, located inside the fuel tank. It is a positive displacement roller vane pump. The impeller serves as a vapor separator and precharges the high pressure assembly. A pressure regulator maintains 34–47 psi (240–315 kPa) in the fuel line to the injectors and the excess fuel is fed back to the tank.

The Mass Air Flow (MAF) sensor (V6 models) is used to measure the mass of air that is drawn into the engine cylinders. It is located just ahead of the air throttle in the intake system and consists of a heated film which measures the mass of air, rather than just the volume. A resistor is used to measure the temperature of the incoming air, the heated film and the electronic module and the MAF sensor maintains the temperature of the film at 75 degrees above ambient temperature. As the ambient (outside) air temperature rises, more energy is required to maintain

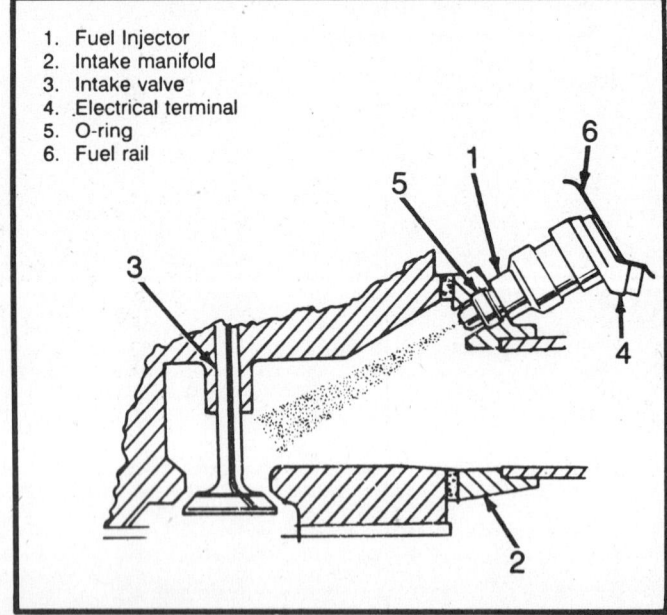

1. Fuel Injector
2. Intake manifold
3. Intake valve
4. Electrical terminal
5. O-ring
6. Fuel rail

Port fuel injection

the heated film at the higher temperature and the control unit uses this difference in required energy to calculate the mass of the incoming air. The control unit uses this information to determine the duration of fuel injection pulse, timing and EGR.

The Manifold Absolute Pressure (MAP) sensor (V8 models) measures the changes in the intake manifold pressure which result from engine load and speed changes. The pressure measured by the MAP sensor is the difference between barometric pressure (outside air) and manifold pressure (vacuum). A closed throttle engine coastdown would produce a relatively low MAP value (approximately 20–35 kPa), while wide open throttle would produce a high value (100 kPa). This high value is produced when the pressure inside the manifold is the same as outside the manifold, and 100% of outside air (or 100 kPa) is being measured. This MAP output is the opposite of what you would measure on a vacuum gauge. The use of this sensor also allows the ECM to adjust automatically for different altitude. The ECM sends a 5 volt reference signal to the MAP sensor. As the MAP changes, the electrical resistance of the sensor also changes. By monitoring the sensor output voltage the ECM can determine the manifold pressure. A higher pressure, lower vacuum (high voltage) requires more fuel, while a lower pressure, higher vacuum (low voltage) requires less fuel. The ECM uses the MAP sensor to control fuel delivery and ignition timing.

The Manifold Air Temperature (MAT) sensor (V8 models) is a thermistor mounted in the intake manifold. A thermistor is a resistor which changes resistance based on temperature. Low manifold air temperature produces a high resistance (100,000 ohms at −40°F [−40°C]), while high temperature cause low resistance (70 ohms at 266°F [130°C]). The ECM supplies a 5 volt signal to the MAT sensor through a resistor in the ECM and monitors the voltage. The voltage will be high when the manifold air is cold and low when the air is hot. By monitoring the voltage, the ECM calculates the air temperature and uses this data to help determine the fuel delivery and spark advance.

On V6 models, the throttle body incorporates an Idle Air Control (IAC) that provides for a bypass channel through which air can flow. It consists of an orifice and pintle which is controlled by the ECM through a stepper motor. The IAC provides air flow for idle and allows additional air during cold start until the engine reaches operating temperature. As the engine temperature rises, the opening through which air passes is slowly closed.

On V8 models, the Idle Speed Control (ISC) assembly consists of a motor, gear reduction section, plunger and throttle contact switch. It is mounted on the throttle body and is used to control engine idle speed. The ISC plunger acts as a movable throttle stop, which changes the primary throttle valve angle. The ECM monitors engine idle speed and causes the motor to move the ISC plunger to maintain the desired idle speed for operating condition. The position of the throttle contact switch determines whether or not the ISC controls idle speed. When the throttle lever rests against the ISC plunger, the contacts in the switch close. This causes the ECM to move the plunger to the programmed idle speed position. When the throttle lever breaks with the ISC plunger, the switch contacts open and the ECM stops sending idle speed commands. Speed is then controlled by the accelerator pedal.

The Throttle Position Sensor (TPS) provides the control unit with information on throttle position, in order to determine injector pulse width and hence correct mixture. The TPS is connected to the throttle shaft on the throttle body and consists of a potentiometer with one end connected to a 5 volt source from the ECM and the other to ground. A third wire is connected to the ECM to measure the voltage output from the TPS which changes as the throttle valve angle is changed (accelerator pedal moves). At the closed throttle position, the output is low (approximately 0.4 volts); as the throttle valve opens, the output increases to a maximum 5 volts at Wide Open Throttle (WOT). The TPS can be misadjusted open, shorted, or loose and if it is out of adjustment, the idle quality of WOT performance may be

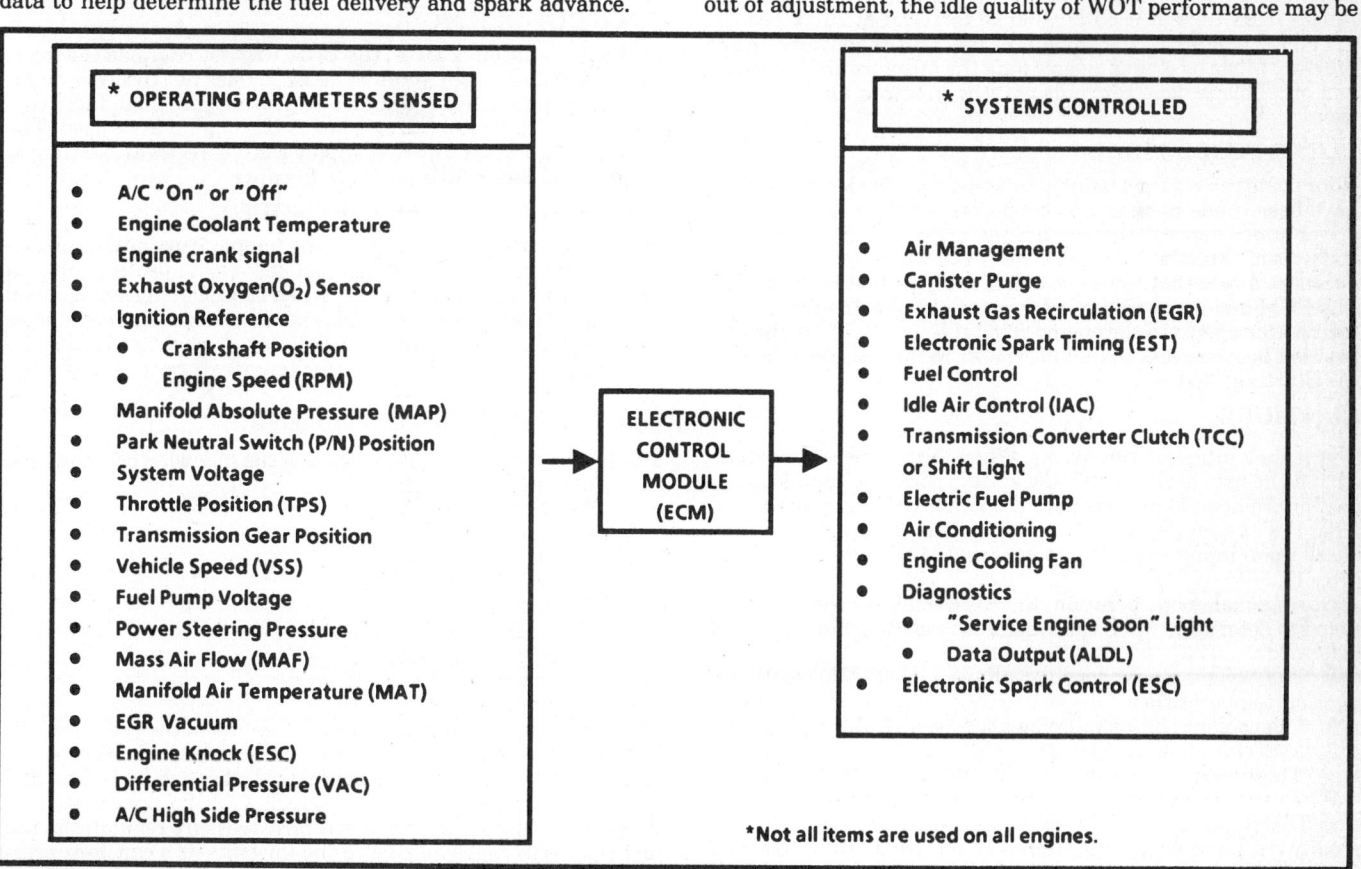

* OPERATING PARAMETERS SENSED	* SYSTEMS CONTROLLED
• A/C "On" or "Off"	
• Engine Coolant Temperature	
• Engine crank signal	• Air Management
• Exhaust Oxygen(O₂) Sensor	• Canister Purge
• Ignition Reference	• Exhaust Gas Recirculation (EGR)
• Crankshaft Position	• Electronic Spark Timing (EST)
• Engine Speed (RPM)	• Fuel Control
• Manifold Absolute Pressure (MAP)	• Idle Air Control (IAC)
• Park Neutral Switch (P/N) Position	• Transmission Converter Clutch (TCC) or Shift Light
• System Voltage	• Electric Fuel Pump
• Throttle Position (TPS)	• Air Conditioning
• Transmission Gear Position	• Engine Cooling Fan
• Vehicle Speed (VSS)	• Diagnostics
• Fuel Pump Voltage	• "Service Engine Soon" Light
• Power Steering Pressure	• Data Output (ALDL)
• Mass Air Flow (MAF)	• Electronic Spark Control (ESC)
• Manifold Air Temperature (MAT)	
• EGR Vacuum	
• Engine Knock (ESC)	
• Differential Pressure (VAC)	
• A/C High Side Pressure	

ELECTRONIC CONTROL MODULE (ECM)

*Not all items are used on all engines.

ECM inputs and outputs—3800 (VIN C) shown, 3.8L similar

poor. A loose TPS can cause intermittent bursts of fuel from the injectors and an unstable idle because the ECM thinks the throttle is moving. This should cause a trouble code to be set. Once a trouble code is set, the ECM will use a preset value for TPS and some vehicle performance may return. A small amount of engine coolant is routed through the throttle assembly to prevent freezing inside the throttle bore during cold operation.

FUEL CONTROL SYSTEM

The basic function of the fuel control system is to control the fuel delivery to the engine. The fuel is delivered to the engine by individual fuel injectors mounted on the intake manifold near each cylinder.

The main control sensor is the oxygen sensor(s) which is located in the exhaust manifold. The oxygen sensor(s) tells the ECM how much oxygen is in the exhaust gas and the ECM changes the air/fuel ratio to the engine by controlling the fuel injectors. The best mixture to minimize exhaust emissions is 14.7:1 which allows the catalytic converter to operate the most efficiently. Because of the constant measuring and adjusting of the air/fuel ratio, the fuel injection system is called a "Closed Loop" system.

Modes Of Operation

The ECM looks at voltage from several sensors to determine how much fuel to give the engine. The fuel is delivered under one of several conditions, called "Modes".

STARTING MODE

When the engine is first turned **ON**, the ECM will turn on the fuel pump relay for 2 seconds and the the fuel pump will build up pressure. The ECM then checks the coolant temperature sensor, throttle position sensor and crank sensor (V6), then the ECM determines the proper air/fuel ratio for starting. This ranges from 1.5:1 at −33°F (−36°C) to 14.7:1 at 201°F (94°C).

The ECM controls the amount of fuel that is delivered in the Starting Mode by changing how long the injectors are turned on and off. This is done by "pulsing" the injectors for very short times.

CLEAR FLOOD MODE

If for some reason the engine should become flooded, provisions have been made to clear this condition. To clear the flood, the driver must depress the accelerator pedal enough to open to wide-open throttle position. The ECM then issues injector pulses at a rate that would be equal to an air/fuel ratio of 20:1. The ECM maintains this injector rate as long as the throttle remains wide open and the engine rpm is below 600. If the throttle position becomes less than 80%, the ECM then would return to the Starting Mode.

RUN MODE

There are 2 different run modes. When the engine is first started and the rpm is above 400, the system goes into open loop operation. In open loop operation, the ECM will ignore the signal from the oxygen (O₂) sensor and calculate the injector on-time based upon inputs from the coolant and MAP sensors (or MAF sensor).

During open loop operation, the ECM analyzes the following items to determine when the system is ready to go to the closed loop mode.

1. The oxygen sensor varying voltage output. (This is dependent on temperature).
2. The coolant sensor must be above specified temperature.
3. A specific amount of time must elapse after starting the engine. These values are stored in the PROM or MEM-CAL.

When these conditions have been met, the system goes into closed loop operation In closed loop operation, the ECM will modify the pulse width (injector on-time) based upon the signal from the oxygen sensor. The ECM will decrease the on-time if the air/fuel ratio is too rich, and will increase the on-time if the air/fuel ratio is too lean.

ACCELERATION MODE

When the engine is required to accelerate, the opening of the throttle valve(s) causes a rapid increase in Manifold Absolute Pressure (MAP). This rapid increase in MAP causes fuel to condense on the manifold walls. The ECM senses this increase in throttle angle and MAP, and supplies additional fuel for a short period of time. This prevents the engine from stumbling due to too lean a mixture.

DECELERATION MODE

Upon deceleration, a leaner fuel mixture is required to reduce emission of hydrocarbons (HC) and carbon monoxide (CO). To adjust the injection on-time, the ECM uses the decrease in MAP and the decrease in throttle position to calculate a decrease in pulse width. To maintain an idle fuel ratio of 14.7:1, fuel output is momentarily reduced. This is done because of the fuel remaining in the intake manifold. The ECM can cut off the fuel completely for short periods of time.

BATTERY VOLTAGE CORRECTION MODE

The purpose of battery voltage correction is to compensate for variations in battery voltage to fuel pump and injector response. The ECM modifies the pulse width by a correction factor in the PROM or MEM-CAL. When battery voltage decreases, pulse width increases.

Battery voltage correction takes place in all operating modes. When battery voltage is low, the spark delivered by the distributor may be low. To correct this low battery voltage problem, the ECM can do any or all of the following:

a. Increase injector pulse width (increase fuel).
b. Increase idle rpm.
c. Increase ignition dwell time.

FUEL CUT-OFF MODE

When the ignition is **OFF**, the ECM will not energize the injector. Fuel will also be cut off if the ECM does not receive a reference pulse from the distributor. To prevent dieseling, fuel delivery is completely stopped as soon as the engine is stopped. The ECM will not allow any fuel supply until it receives distributor reference pulses which prevents flooding.

CONVERTER PROTECTION MODE

In this mode the ECM estimates the temperature of the catalytic converter and then modifies fuel delivery to protect the converter from high temperatures. When the ECM has determined that the converter may overheat, it will cause open loop operation and will enrichen the fuel delivery. A slightly richer mixture will then cause the converter temperature to be reduced.

Fuel Control System Components

The fuel control system is made up of the following components:

1. Fuel supply system
2. Throttle body assembly
3. Fuel injectors
4. Fuel rail
5. Fuel pressure regulator
6. Idle control
   a. Idle Air Control (IAC) — V6
   b. Idle Speed Control (ISC) — V8
7. Fuel pump
8. Fuel pump relay

The fuel control system starts with the fuel in the fuel tank. An electric fuel pump, located in the fuel tank with the fuel gauge sending unit, pumps fuel to the fuel rail through an in-line fuel filter. The pump is designed to provide fuel at a pressure above the pressure needed by the injectors. A pressure regulator in the fuel rail keeps fuel available to the injectors at a constant pressure. Unused fuel is returned to the fuel tank by a separate line.

The injectors are controlled by the ECM. They deliver fuel in one of several modes as previously described. In order to properly control the fuel supply, the fuel pump is operated by the ECM through the fuel pump relay and oil pressure switch.

THROTTLE BODY UNIT

The throttle body unit has a throttle valve to control the amount of air delivered to the engine. The TPS and IAC valve (or ISC on V8 engine) are also mounted onto the throttle body. The throttle body contains vacuum ports located at, above or below the throttle valve. These vacuum ports generate the vacuum signals needed by various components.

On some models, the engine coolant is directed through the coolant cavity at the bottom of the throttle body to warm the throttle valve and prevent icing.

FUEL RAIL

The fuel rail is mounted on top of the engine. It distributes fuel to the individual injectors. Fuel is delivered to the input end of the fuel rail by the fuel lines, goes through the rail, then to the fuel pressure regulator. The regulator keeps the fuel pressure to the injectors at a constant pressure. The remaining fuel is then returned to the fuel tank.

FUEL INJECTOR

The fuel injector is a solenoid operated device controlled by the ECM. The ECM turns on the solenoid, which opens the a valve which allows fuel delivery. The fuel, under pressure, is injected in a cone like spray pattern at the opening of the intake valve. The fuel, which is not used by the injectors, passes through the pressure regulator before returning to the fuel tank.

An injector that is partly open, will cause loss of fuel pressure after the engine is shut down, so long crank time would be noticed on some engines. Also dieseling could occur because some fuel could be delivered after the ignition is turned **OFF**.

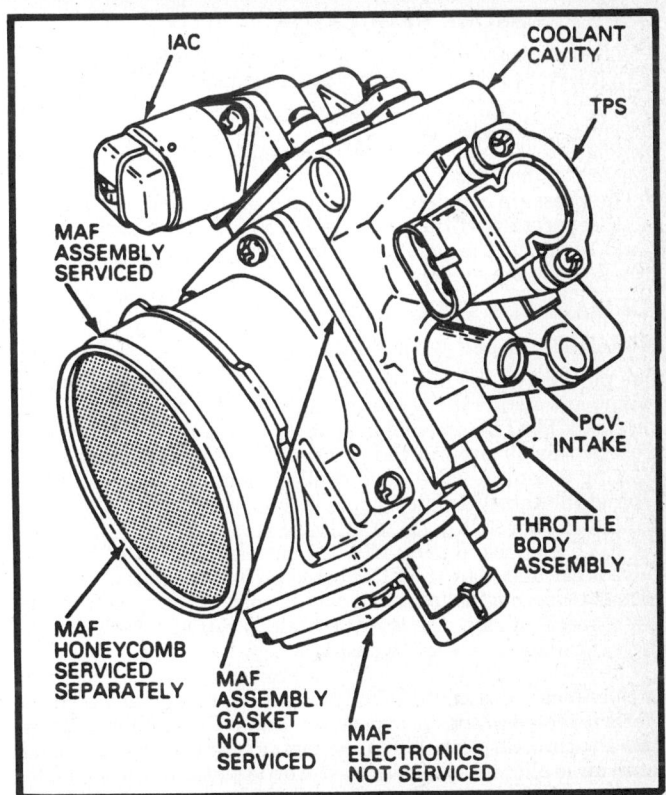

Throttle body assembly—3800 (VIN C) engine

1. Flange gasket
2. Throttle body assembly
3. Idle stop screw plug
4. Idle stop screw assembly
5. Idle stop screw assembly spring
6. Throttle position sensor (TPS)
7. TPS attaching screw assembly
8. TPS attaching screw retainer
9. TPS lever
10. Coolant cavity cover
11. Coolant cover attaching screw assembly
12. Coolant cover to throttle body O-ring
13. Idle air/vacuum signal housing assembly
14. Idle air/vacuum signal assembly screw assembly
15. Idle air/vacuum signal assembly long screw assembly
16. Idle air/vacuum signal assembly gasket
17. Idle Air Control (IAC) valve assembly
18. IAC valve assembly gasket

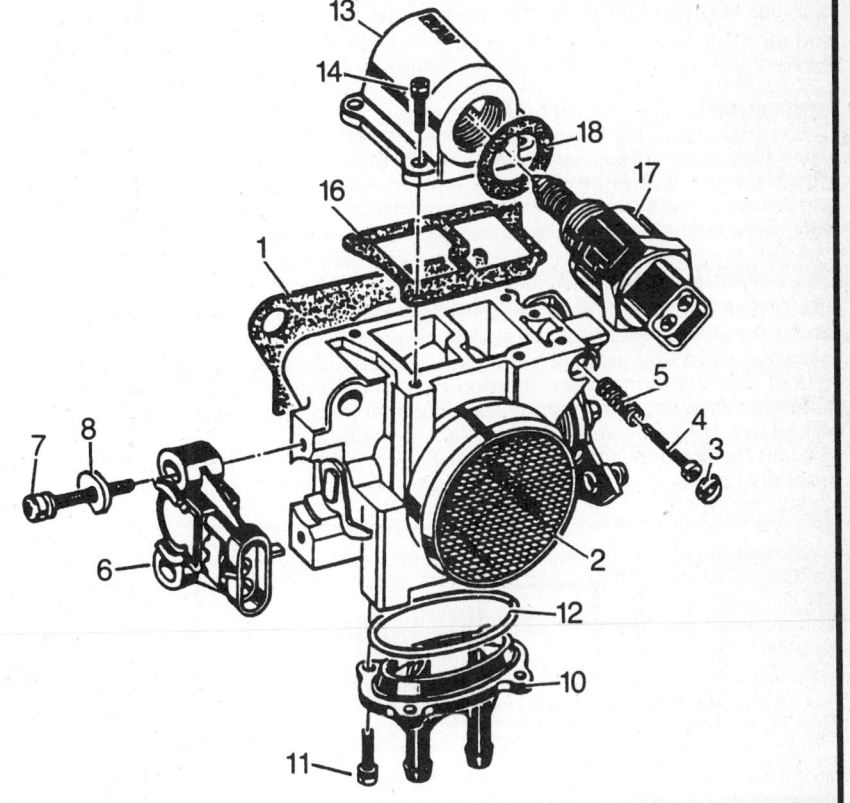

Throttle body assembly—3.8L (VIN 3) engine

FUEL PRESSURE REGULATOR

The fuel pressure regulator is a diaphragm operated relief valve with injector pressure on one side and manifold pressure on the other. The function of the regulator is to maintain a constant pressure at the injector at all times. The pressure regulator also compensates for engine load, by increasing fuel pressure when it sees low engine vacuum.

The pressure regulator is mounted on the fuel rail and serviced separately. If the pressure is too low, poor performance could result. If the pressure is too high, excessive odor and a trouble code may result.

IDLE AIR CONTROL (IAC)

Models Equipped with V6 Engine

The purpose of the Idle Air Control (IAC) system is to control engine idle speeds while preventing stalls due to changes in engine load. The IAC assembly, mounted on the throttle body, controls bypass air around the throttle plate. By extending or retracting a conical valve, a controlled amount of air can move around the throttle plate. If rpm is too low, more air is diverted around the throttle plate to increase rpm.

On V6 engines, during idle, the proper position of the IAC valve is calculated by the ECM based on battery voltage, coolant temperature, engine load, and engine rpm. If the rpm drops below a specified rate, the throttle plate is closed. The ECM will then calculate a new valve position.

The IAC motor has 255 different positions or steps. The zero, or reference position, is the fully extended position at which the pintle is seated in the air bypass seat and no air is allowed to bypass the throttle plate. When the motor is fully retracted, maximum air is allowed to bypass the throttle plate. When the motor is fully retracted, maximum air is allowed to bypass the throttle plate.

The ECM always monitors how many steps it has extended or retracted the pintle from the zero or reference position; thus, it always calculates the exact position of the motor. Once the engine has started and the vehicle has reached approximately 40 mph, the ECM will extend the motor 255 steps from whatever position it is in. This will bottom out the pintle against the seat. The ECM will call this position "0" and thus keep its zero reference updated.

The IAC only affects the engine's idle characteristics. If it is stuck fully open, idle speed is too high (too much air enters the throttle bore) If it is stuck closed, idle speed is too low (not enough air entering). If it is stuck somewhere in the middle, idle may be rough, and the engine won't respond to load changes.

FUEL PUMP RELAY CIRCUIT

The fuel pump relay is usually located on the left front inner fender (or shock tower), on the engine side of the firewall (center cowl) or in a fuse panel. The fuel pump electrical system consists of the fuel pump relay, ignition circuit and the ECM circuits are protected by a 10 amp fuse. The fuel pump relay contact switch is in the normally open (N.O.) position.

When the ignition is turned **ON**, the ECM will supply voltage to the fuel pump relay coil, closing the open contact switch. The

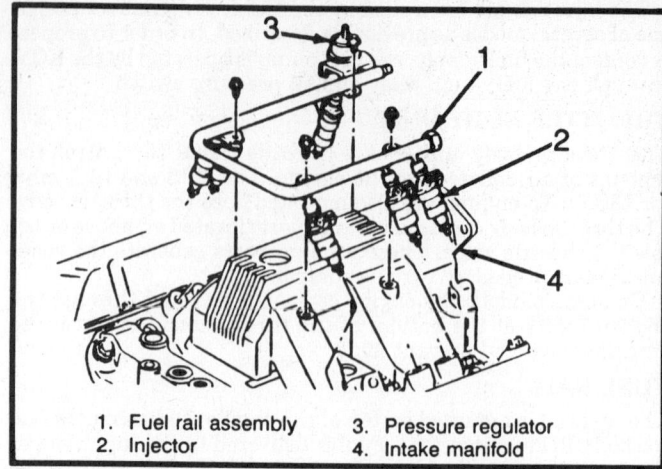

1. Fuel rail assembly   3. Pressure regulator
2. Injector   4. Intake manifold

Fuel rail and injector assembly—V6 engine

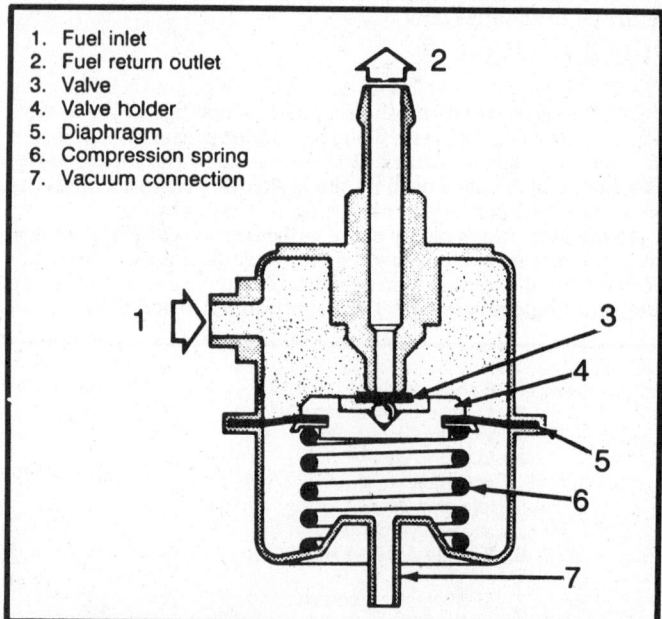

1. Fuel inlet
2. Fuel return outlet
3. Valve
4. Valve holder
5. Diaphragm
6. Compression spring
7. Vacuum connection

Fuel pressure regulator

ignition circuit 10 amp fuse can now supply ignition voltage to the circuit which feeds the relay contact switch. With the relay contacts closed, ignition voltage is supplied to the fuel pump. If the ECM does not receive a cranking signal within 2 seconds, the ECM will de-energized the fuel pump relay. The ECM will continue to supply voltage to the relay coil circuit as long as the ECM receives the rpm reference pulses from the HEI or C^3I module.

The fuel pump control circuit also includes an engine oil pressure switch with a set of normally open contacts. The switch closes at approximately 4 lbs. of oil pressure and provides a secondary battery feed path to the fuel pump. If the relay fails, the pump will continue to run using the battery feed supplied by the closed oil pressure switch. A failed fuel pump relay will result in extended engine crank times in order to build up enough oil pressure to close the switch and turn on the fuel pump.

ELECTRONIC CONTROL MODULE (ECM)

The Electronic Control Module (ECM), usually located under the instrument panel, is the control center of the fuel injection system. It constantly looks at information from various sensors, and controls the system that affect vehicle performance.

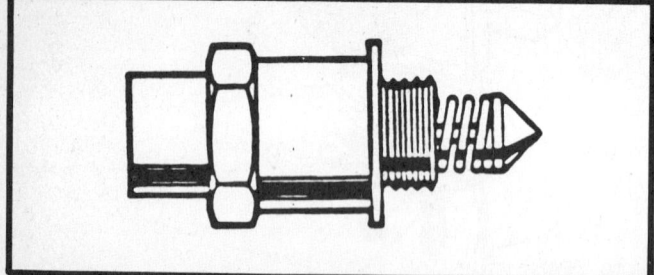

Idle Air Control (IAC)—typical

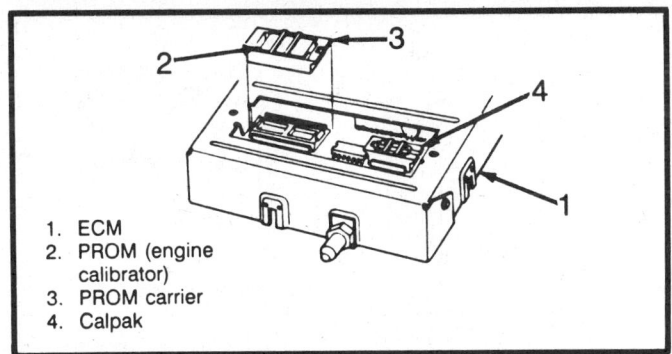

ECM PROM and CALPAK locations — 3.8L (VIN 3) engine

1. ECM
2. PROM (engine calibrator)
3. PROM carrier
4. Calpak

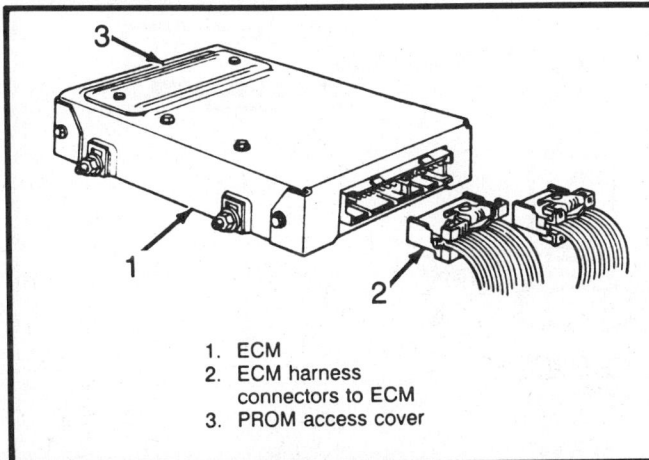

Electronic Control Module (ECM) — 3.8L (VIN 3) engine

1. ECM
2. ECM harness connectors to ECM
3. PROM access cover

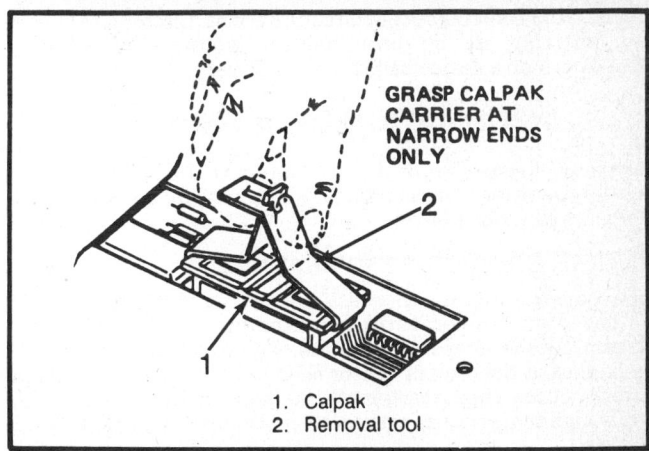

Calpak removal — 3.8L (VIN 3) engine

GRASP CALPAK CARRIER AT NARROW ENDS ONLY

1. Calpak
2. Removal tool

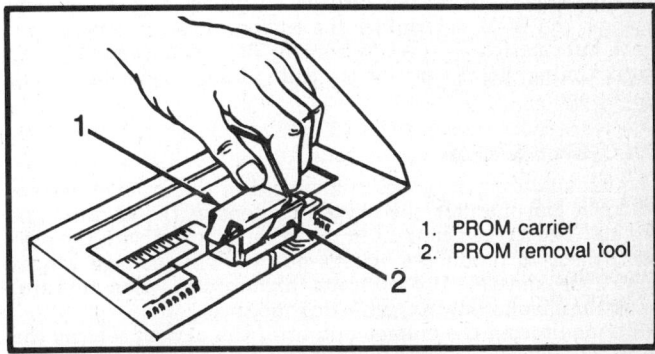

PROM removal — 3.8L (VIN 3) engine

1. PROM carrier
2. PROM removal tool

The ECM also performs the diagnostic function of the system. It can recognize operational problems, alert the driver through the "Service Engine Soon" light and store a code or codes which identify the problem areas to aid the technician in making repairs.

CALPAK

A device called a CALPAK is used to allow fuel delivery if other parts of the ECM are damaged. It has an access door in the ECM, and removal and replacement procedures are the same as with the prom. If the CALPAK is missing, a code 52 will be set.

PROM

To allow one model of the ECM to be used for many different vehicles, a device called a Calibrator (or PROM) is used. The PROM is located inside the ECM and has information on the vehicle's weight, engine, transmission, axle ratio and other components.

While one ECM part number can be used by many different vehicles, a PROM is very specific and must be used for the right vehicle. For this reason, it is very important to check the latest parts book and or service bulletin information for the correct PROM part number when replacing the PROM.

An ECM used for service (called a controller) comes without a PROM. The PROM from the old ECM must be carefully removed and installed in the new ECM.

MEM-CAL

This assembly contains the functions of the PROM, CALPAK and the ESC module used on other General Motors vehicles. Like the PROM, it contains the calibrations needed for a specific

vehicle as well as the back-up fuel control circuitry required if the rest of the ECM becomes damaged or faulty.

ECM Function

The ECM supplies either 5 or 12 volts to power various sensors and or switches. This is done through resistances in the ECM which are so high in value that a test light will not light when connected to the circuit. In some cases, even an ordinary shop voltmeter will not give an accurate reading because its resistance is too low. Therefore, a 10 megohm input impedance digital voltmeter is required to assure accurate voltage readings.

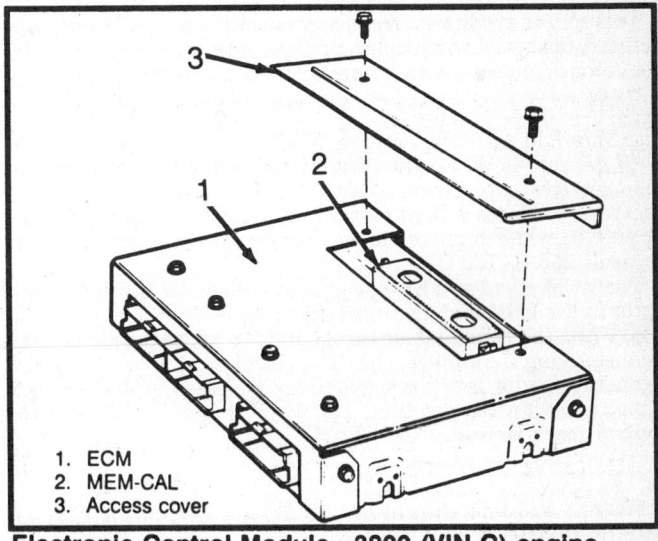

Electronic Control Module — 3800 (VIN C) engine

1. ECM
2. MEM-CAL
3. Access cover

The ECM controls output circuits such as injectors, IAC, coolant fan relay, etc. by controlling the ground circuit through transistors or a device called a quad driver.

INFORMATION DATA SENSORS

A variety of sensors provide information to the ECM regarding engine operating characteristics. These sensors and their functions are described below.

ENGINE COOLANT TEMPERATURE

The coolant sensor is a thermistor (a resistor which changes value based on temperature) mounted on the engine coolant stream. As the temperature of the engine coolant changes, the resistance of the coolant sensor changes. Low coolant temperature produces a high resistance (100,000 ohms at -40°F [-40°C]), while high temperature causes low resistance (70 ohms at 266°F [130°C]).

The ECM supplies a 5 volt signal to the coolant sensor and measures the voltage that returns. By measuring the voltage change, the ECM determines the engine coolant temperature. This information is used to control fuel management, IAC, spark timing, EGR, canister purge and other engine operating conditions.

OXYGEN SENSOR

The exhaust oxygen sensor is mounted in the exhaust system where it can monitor the oxygen content of the exhaust gas stream. The oxygen content in the exhaust reacts with the oxygen sensor to produce a voltage output. This voltage ranges from approximately 100 millivolts (high oxygen — lean mixture) to 900 millivolts (low oxygen — rich mixture).

By monitoring the voltage output of the oxygen sensor, the ECM will determine what fuel mixture command to give to the injector (lean mixture — low voltage — rich command, rich mixture — high voltage lean command).

Remember that oxygen sensor indicates to the ECM what is happening in the exhaust. It does not cause things to happen. It is a type of gauge: high oxygen content = lean mixture; low oxygen content = rich mixture. The ECM adjust fuel to keep the system working.

MASS AIR FLOW SENSOR (MAF)

The Mass Air Flow (MAF) sensor measures the amount of air which passes through it. The ECM uses this information to determine the operating condition of the engine, to control fuel delivery. A large quantity of air indicates acceleration, while a small quantity indicates deceleration or idle.

This sensor produces a frequency output between 32 and 150 hertz. A Scan tool will display air flow in terms of grams of air per second (gm/sec), with a range from 3 gm/sec to 150 gm/sec.

MANIFOLD AIR TEMPERATURE (MAT) SENSOR

The Manifold Air Temperature (MAT) sensor (is a part of the MAF sensor) is a resistor (thermistor), which changes value based on the temperature of air entering the engine. Low temperature produces a high resistance (100,000 ohms at −40°F [−40°C]), while high temperature causes low resistance (70 ohms at 266°F [130°C]).

The ECM supplies a 5 volt signal to the sensor through a resistor in the ECM and measures the voltage. The voltage will be high when the incoming air is cold and low when the air id hot. By measuring the voltage, the ECM calculates the incoming air temperature and uses this signal to compensate the MAF sensor signal based on temperature. The MAT sensor is also used to control spark timing.

VEHICLE SPEED SENSOR (VSS)

NOTE: Vehicle should not be driven without a VSS as idle quality may be affected.

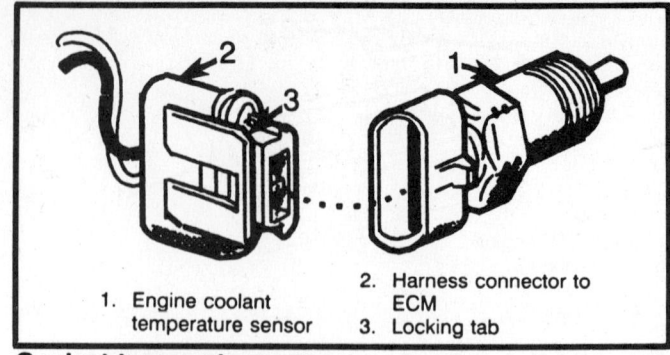

1. Engine coolant temperature sensor
2. Harness connector to ECM
3. Locking tab

Coolant temperature sensor

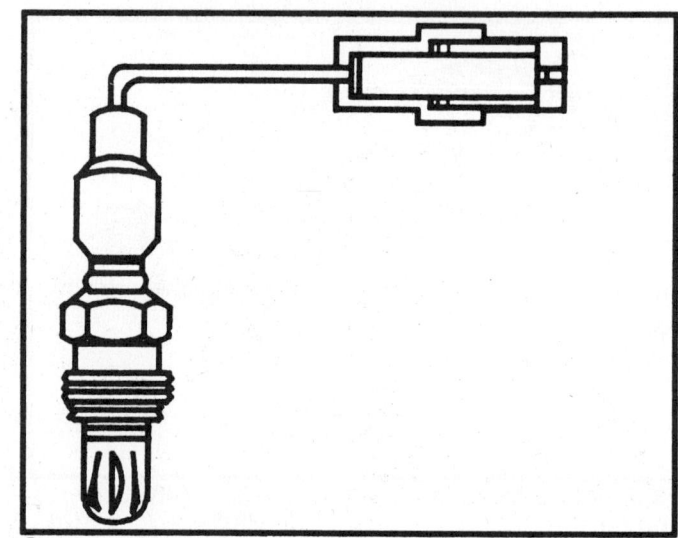

Oxygen (O_2) sensor

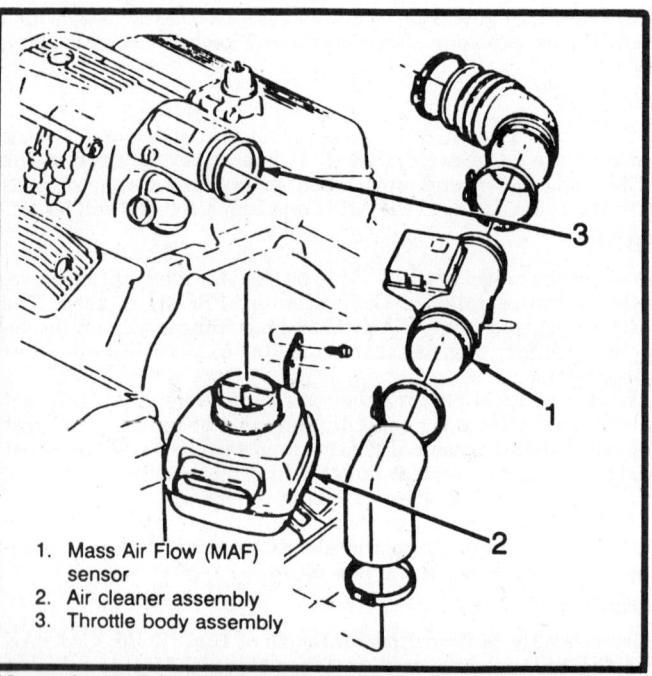

1. Mass Air Flow (MAF) sensor
2. Air cleaner assembly
3. Throttle body assembly

Mass Air Flow (MAF) sensor — 3.8L (VIN 3) engine

The Vehicle Speed Sensor (VSS) is mounted behind the speedometer in the instrument cluster. It provides electrical pulses to the ECM from the speedometer head. The pulses indicate the road speed. The ECM uses this information to operate the IAC, canister purge, and TCC.

Some vehicles use a permanent magnet (PM) generator to provide the VSS signal. The PM generator is located in the transmission and replaces the speedometer cable. The signal from the PM generator drives a stepper motor which drives the odometer.

THROTTLE POSITION SENSOR (TPS)

The Throttle Position Sensor (TPS) is connected to the throttle shaft and is controlled by the throttle mechanism. A 5 volt reference signal is sent to the TPS from the ECM. As the throttle valve angle is changed (accelerator pedal moved), the resistance of the TPS also changes. At a closed throttle position, the resistance of the TPS is high, so the output voltage to the ECM will be low (approximately 0.5 volt). As the throttle plate opens, the resistance decreases so that, at wide open throttle, the output voltage should be approximately 5 volts.

By monitoring the output voltage from the TPS, the ECM can determine fuel delivery based on throttle valve angle (driver demand). The TPS can either be misadjusted, shorted, open or loose. Misadjustment might result in poor idle or poor wide-open throttle performance. An open TPS signals the ECM that the throttle is always closed, resulting in poor performance.

This usually sets a Code 22 or E022. A shorted TPS gives the ECM a constant wide open throttle signal and should set a Code 21 or E021. A loose TPS indicates to the ECM that the throttle is moving. This causes intermittent bursts of fuel from the injector and an unstable idle.

PARK/NEUTRAL SWITCH

NOTE: Vehicle should not be driven with the park/neutral switch disconnected as idle quality may be affected in PARK or NEUTRAL.

This switch indicates to the ECM when the transmission is in **P** or **N**.
A/C Compressor Clutch Engagement. This signal indicates to the ECM that the A/C compressor clutch is engaged.

AIR CONDITIONER REQUEST SIGNAL

This signal tells the ECM that the A/C selector switch is turned **ON** and that the A/C pressure switches are closed. The ECM uses this to adjust the idle speed before turning on the A/C clutch. If this signal is not available to the ECM, the A/C compressor will be inoperative.

POWER STEERING PRESSURE SWITCH

This switch tells the ECM that the vehicle is in a parking maneuver. The ECM uses this information to compensate for additional engine load by moving the idle air control valve. The ECM will also turn off the A/C clutch when high pressure is detected.

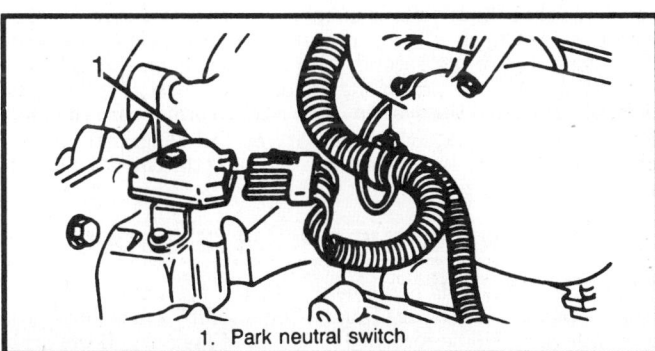

1. Park neutral switch

Park/neutral switch

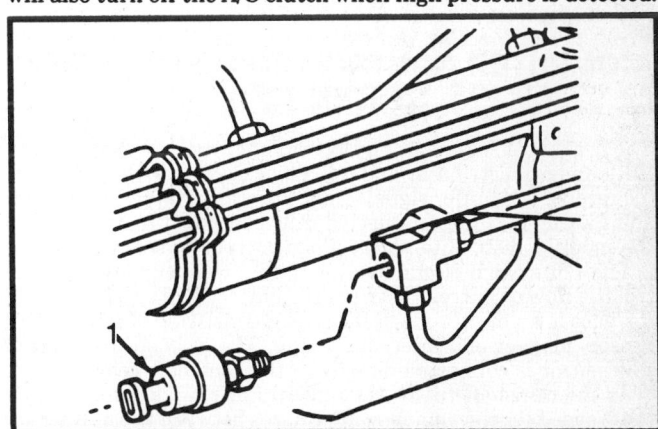

Power steering pressure switch—V6 engine

CRANKSHAFT SENSOR

The crankshaft sensor provides a signal, through the ignition module, which the ECM uses as a reference to calculate rpm and crankshaft position.

CAMSHAFT SENSOR

The cam sensor sends a signal to the ECM, which uses it as a "sync pulse" to trigger the injectors in proper sequence.

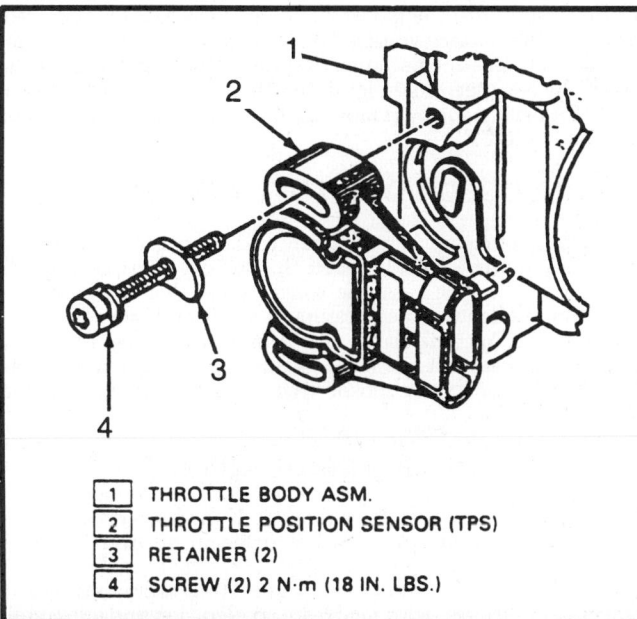

1	THROTTLE BODY ASM.
2	THROTTLE POSITION SENSOR (TPS)
3	RETAINER (2)
4	SCREW (2) 2 N·m (18 IN. LBS.)

Throttle position sensor (TPS) assembly

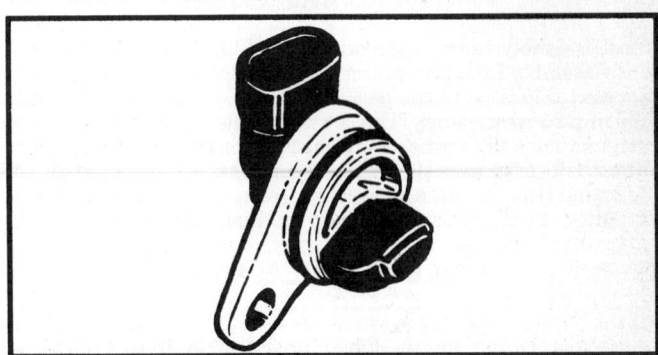

Camshaft position sensor—typical

Crankshaft sensor

DETONATION (KNOCK) SENSOR

This sensor is a piezoelectric sensor located near the back of the engine (transmission end). It generates electrical impulses which are directly proportional to the frequency of the knock which is detected. A buffer then sorts these signals and eliminates all except for those frequency range of detonation. This information is passed to the ESC module and then to the ECM, so that the ignition timing advance can be retarded until the detonation stops.

COMPUTER CONTROLLED COIL IGNITION SYSTEM (C³I)

The heart of this system is a electronic coil module that replaces the standard distributor and coil. Logic circuits within the module receive and buffer signals from the crankshaft and camshaft and by way of 3 interconnected coils contained in the cover of the module, distribute high voltage current to spark plugs.

The C³I system eliminates the need for a distributor to control the flow or current between the battery and spark plugs. In its place is an electro-magnetic sensor consisting of a hall effect sensor, magnet and interruptor ring. The gear on the shaft of this sensor is connected directly to the camshaft gear.

As the camshaft turns, the interruptor ring, moving at camshaft speed (½ the engine rpm), rotates between the hall sensor and the magnet to produce a signal which is fed to the electronic coil module. This signal provides the exact position of the valves as they open and close.

Another sensor is mounted to the crankshaft. This sensor also consists of a hall sensor, magnet and interrupt ring. As with the cam sensor, the crankshaft causes the interruptor ring to rotate between the hall sensor and magnet to produce a signal which is fed to the electronic module. This signal gives the top center position of each piston.

ALDL CONNECTOR

The Assembly Line Diagnostic Link (ALDL), or also known as the Assembly Line Communication Link (ALCL), is a diagnostic connector located in the passenger compartment usually under the instrument panel. The assembly plant were the vehicles originate use the connector to check the engine for proper operation before it leaves the plant. Terminal **B** is the diagnostic test terminal (lead) and it can be connected to terminal **A**, or ground, to enter the Diagnostic Mode or the Field Service Mode.

FIELD SERVICE MODE

If the "Test" terminal is grounded with the engine running, the system will enter the the Field Service mode. In this mode, the "Check Engine Light" or "Service Engine Soon Light" will

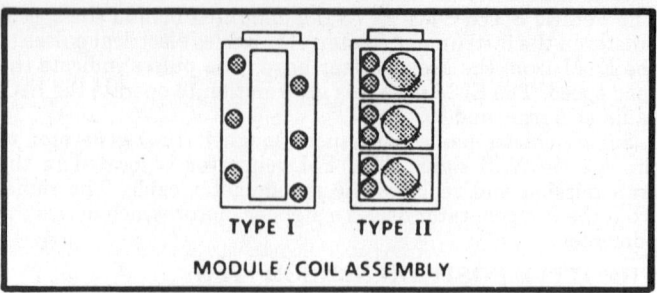

Computer Controlled Coil Ignition (C³I) system module identification

A. Ground
B. Diagnostic terminal
C. A.I.R. (if used)
D. Service engine soon light (if used)
E. Serial data
F. TCC (if used)
G. Fuel Pump (if used)
M. Serial Data (if used)

Assembly Line Data Link ALDL terminal identification

show whether the system is in open loop or closed loop. In open loop the light flashes 2½ times per second. In closed loop the light flashes once per second. Also in closed loop, the light will stay out most of the time if the system is too lean. It will stay on most of the time if the system is too rich. In either case the Field Service mode check, which is part of the Diagnostic circuit check, will lead the technician into choosing the correct diagnostic chart to refer to.

CLEARING THE TROUBLE CODES

When the ECM finds a problem with the system, the "Check Engine Light" or "Service Engine Soon Light" will come on and a trouble code will be recorded in the ECM memory. If the problem is intermittent, the light will go out after 10 seconds, when the fault goes away. However the trouble code will stay in the ECM memory until the battery voltage to the ECM is removed. Removing the battery voltage for 10 seconds will clear all trouble codes. Do this by disconnecting the ECM harness from the positive battery terminal pigtail for 10 seconds with the key in the **OFF** position, or by removing the ECM fuse for 10 seconds with the key **OFF**.

CLOSED LOOP FUEL CONTROL

The purpose of closed loop fuel control is to precisely maintain an air/fuel mixture 14.7:1. When the air/fuel mixture is maintained at 14.7:1, the catalytic converter is able to operate at maximum efficiency which results in lower emission levels.

Since the ECM controls the air/fuel mixture, it needs to check its output and correct the fuel mixture for deviations from the ideal ratio. The oxygen sensor feeds this output information back to the ECM.

ECM LEARNING ABILITY

The ECM has a "learning" capability. If the battery is disconnected the "learning" process has to begin all over again. A change may be noted in the vehicle's performance. To "teach" the ECM, insure the vehicle is at operating temperature and drive at part throttle, with moderate acceleration and idle conditions, until performance returns.

SERVICE PRECAUTION

To reduce the risk of fire and personal injury, it is necessary to relieve the fuel system pressure before servicing fuel system components.

After relieving the system pressure, a small amount of fuel may be relieved when servicing fuel lines or connections. In order to reduce the chance of personal injury, cover the fuel line fittings with a shop towel before disconnecting, to catch any fuel that may leak out. Place the towel in an approved container when disconnect is completed.

Diagnosis and Testing

NOTE: For Self-Diagnostic System and Accessing Trouble Code Memory, see Section 3, "Self-Diagnostic Systems".

Relieving Fuel System Pressure

————————— CAUTION —————————

To reduce the risk of fire and personal injury, it is necessary to relieve the fuel system pressure before servicing fuel system components. After relieving the system pressure, a small amount of fuel may be released when servicing fuel lines or connections. In order to reduce the chance of personal injury, cover the fuel line fittings with a shop towel before disconnecting, to catch any fuel that may leak out. Place the towel in an approved container when disconnect is completed.

1. Disconnect the negative battery terminal to avoid possible fuel discharge if an accidental attempt is made to start the engine.
2. Loosen the fuel filler cap to relieve tank vapor pressure.
3. Connect the gauge J–34730–1 or equivalent, to the fuel pressure connection. Wrap a shop towel around fitting while connecting the gauge to avoid spillage.
4. Install bleed hose into an approved container and open valve to bleed system pressure. Fuel connections are now safe for servicing.
5. Drain any fuel remaining in the gauge into an approved container.

Engine Performance Diagnosis

CHECK ENGINE LIGHT OR SERVICE ENGINE SOON LIGHT

The "Check Engine" or "Service Engine Soon" light on the instrument panel is used as a warning lamp to tell the driver that a problem has occurred in the electronic engine control system. When the self-diagnosis mode is activated by grounding the test terminal of the diagnostic connector, the light will flash stored trouble codes to help isolate system problems. The Electronic Control Module (ECM) has a memory that knows what certain engine sensors should be under certain conditions. If a sensor reading is not what the ECM thinks it should be, the control unit will illuminate the light and store a trouble code in its memory. The trouble code indicates what circuit the problem is in, each circuit consisting of a sensor, the wiring harness and connectors to it and the ECM.

The Assembly Line Diagnostic Link (ALDL) (described earlier in this section) can be used as a diagnostic aid. By connecting terminals **A** and **B** together with a jumper wire, the diagnostic mode is activated and the control unit will begin to flash trouble codes using the "Check Engine" or "Service Engine Soon" light.

When the test terminal is grounded with the key **ON** and the engine stopped, the ECM will display code 12 to show that the system is working. Code 12 will be displayed by the ECM 3 times, then start to display any stored trouble codes. If no trouble codes are stored, the ECM will continue to display code 12 until the test terminal is disconnected. Each trouble code will be flashed 3 times, then code 12 sill display again. The ECM will also energize all controlled relays and solenoids when in the diagnostic mode to check function. When the test terminal is grounded with the engine running, it will cause the ECM to enter the Field Service Mode. In this mode, the service engine soon light will indicate whether the system is in Open or Closed Loop operation. In open loop, the light will flash 2½ times per second; in closed loop, the light will flash once per second. In closed loop, the light will stay out most of the time if the system is too lean and will stay on most of the time if the system is too rich.

NOTE: The vehicle may be driven in the Field Service mode and system evaluated at any steady road speed. This mode us useful in diagnosing driveability problems where the system is rich or lean too long.

Trouble codes should be cleared after service is completed. To clear the trouble code memory, disconnect the battery for at least 10 seconds. This may be accomplished by disconnecting the ECM harness from the positive battery pigtail or by removing the ECM fuse.

————————— CAUTION —————————

The ignition switch must be OFF when disconnecting or reconnecting power to the ECM. The vehicle should be driven after the ECM memory is cleared to allow the system to readjust itself. The vehicle should be driven at part throttle under moderate acceleration with the engine at normal operating temperature. A change is performance should be noted initially, but normal performance should return quickly.

ELECTRONIC FUEL INJECTION ALDL TESTER INFORMATION

An ALDL display unit (ALDL tester, scanner, monitor, etc), allows a technician to read the engine control system information from the ALDL connector under the instrument panel. It can provide information faster than a digital voltmeter or ohmmeter can. The scan tool does not diagnose the exact location of the problem. The tool supplies information about the ECM (and BCM if equipped), the information that it is receiving and the commands that it is sending plus special information such as integrator and block learn. To use an ALDL display tool you should understand throughly how an engine control system operates.

An ALDL scanner or monitor puts a fuel injection system into a special test mode. This mode commands an idle speed of 1000 rpm. The idle quality cannot be evaluated with a tester plugged in. Also the test mode commands a fixed spark with no advance. On vehicles with Electronic Spark Control (ESC), there will be a fixed spark, but it will be advanced. On vehicles with ESC, there might be a serious spark knock, this spark knock could be bad enough so as not being able to road test the vehicle in the ALDL test mode. Be sure to check the tool manufacturer for instructions on special test modes which should overcome these limitations.

When a tester is used with a fuel injected engine, it by-passes the timer that keeps the system in OPEN loop for a certain period of time. When all **CLOSED** loop conditions are met, the engine will go into **CLOSED** loop as soon as the vehicle is started. This means that the air management system will not function properly and air may go directly to the converter as soon as the engine is started.

These tools cannot diagnose everything. They do not tell the technician where a problem is located in a circuit. The diagnostic charts to pinpoint the problems must still be used. These tester's do not let a technician know if a solenoid or relay has been turned on. They only tell the technician the ECM command. To find out if a solenoid has been turned on, check it with a suitable test light or digital voltmeter, or see if vacuum through the solenoid changes.

NOTE: Some vehicles will use more sensors than others. A complete general diagnostic section is outlined.

PORT FUEL INJECTION—INJECTOR BALANCE TEST
Before performing this test, the items listed below must be done.

- Check spark plugs and wires.
- Check compression.
- Check fuel injection harness for being open or shorted.

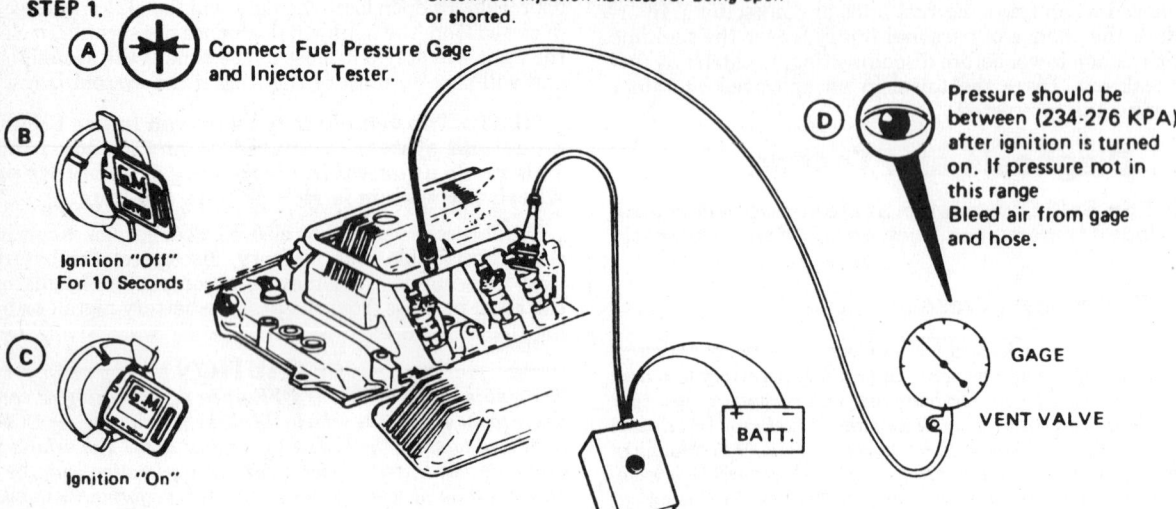

STEP 1.

(A) Connect Fuel Pressure Gage and Injector Tester.

(B) Ignition "Off" For 10 Seconds

(C) Ignition "On"

(D) Pressure should be between (234-276 KPA) after ignition is turned on. If pressure not in this range Bleed air from gage and hose.

GAGE

VENT VALVE

BATT.

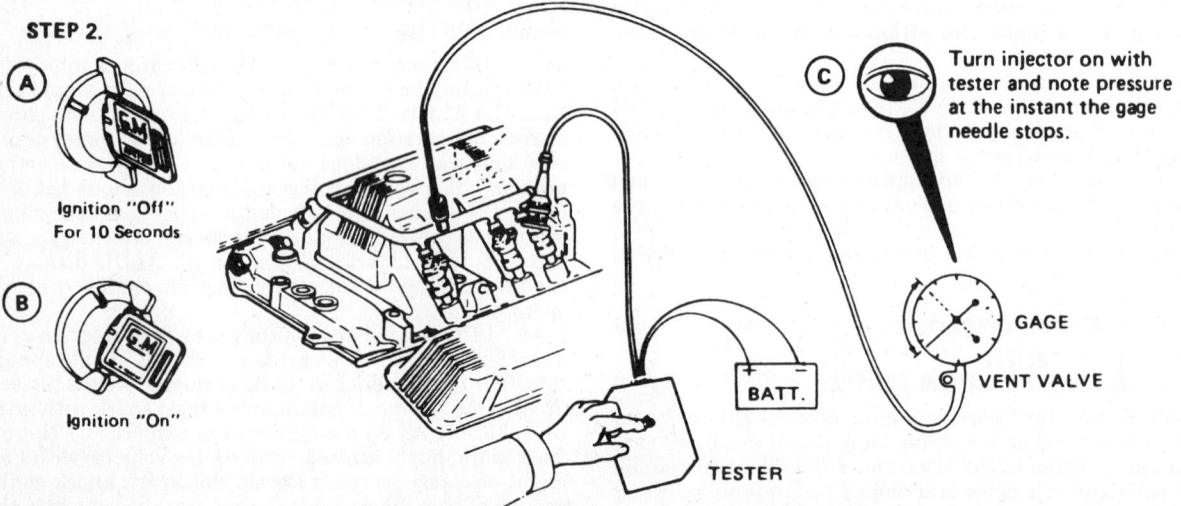

STEP 2.

(A) Ignition "Off" For 10 Seconds

(B) Ignition "On"

(C) Turn injector on with tester and note pressure at the instant the gage needle stops.

GAGE

VENT VALVE

BATT.

TESTER

STEP 3.

Repeat test as in step 2 on all injectors and record pressure drop on each.

Retest injectors that appear faulty. Replace any injectors that have a 10 KPA difference either (more or less) in pressure.

— EXAMPLE —

| CYL 1 | CYL 2 | CYL 3 | CYL 4 | CYL 5 | CYL 6 |

10 KPA LESS — FAULTY (LESS)   10 KPA MORE — FAULTY (MORE)

Fuel injector balance test

The steps and procedures can be altered as required (if necessary) by the technician according to the specific model being diagnosed and the sensors it is equipped with. The wiring diagrams and schematics may not coincide with every GM vehicle in use. If this situation should arise, use the wiring diagram or schematic as a general guide.

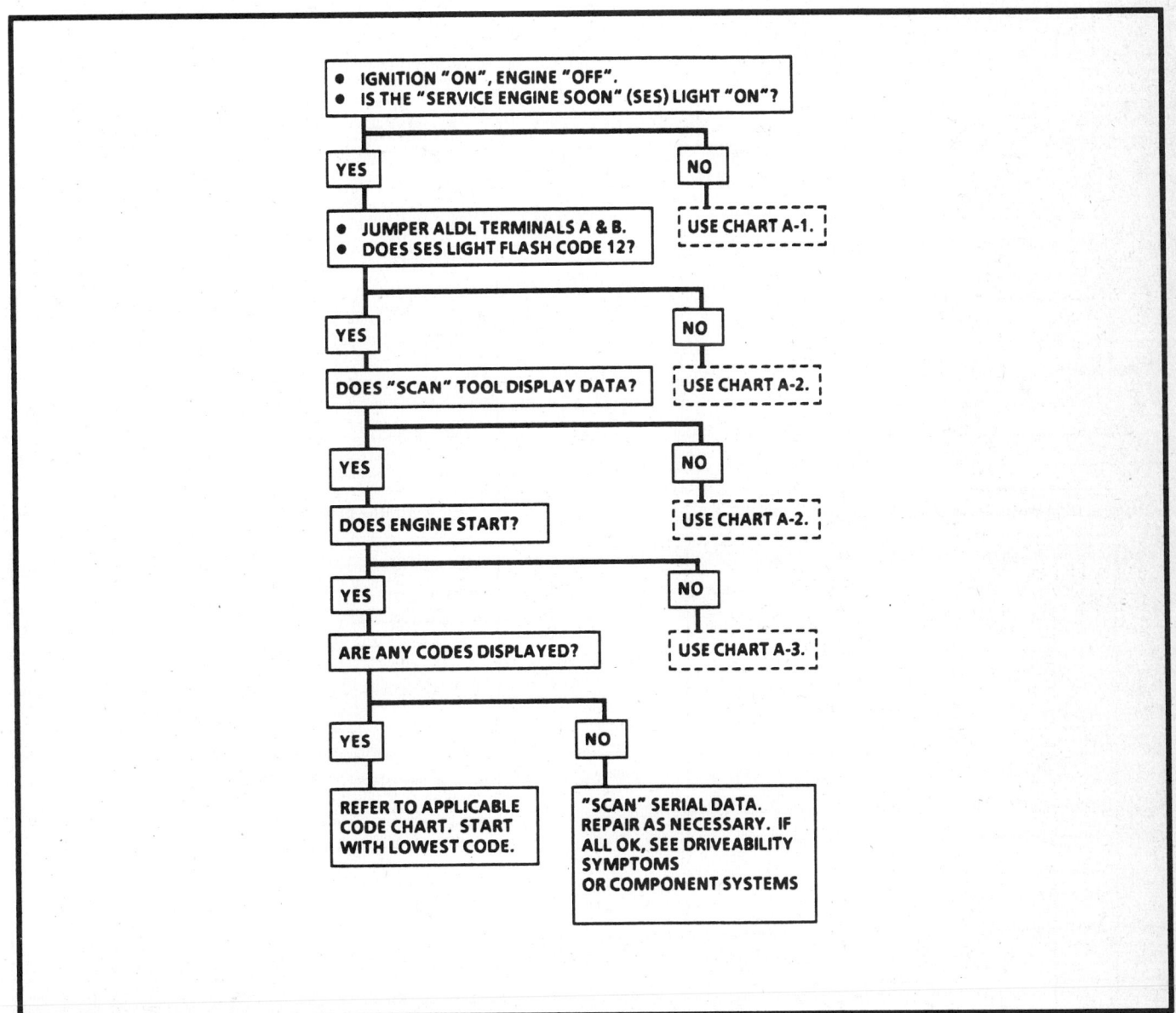

Diagnostic circuit check—all models except Toronado, Riviera, Reatta

**1988–90 VEHICLES EQUIPPED WITH
3800 (VIN C) ENGINE EXCEPT E-CAR**

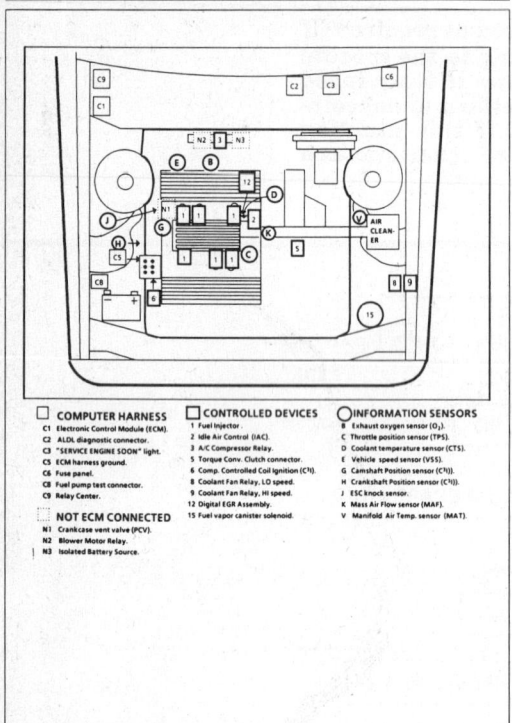

ECM WIRING DIAGRAM — 3800 (VIN C) ENGINE

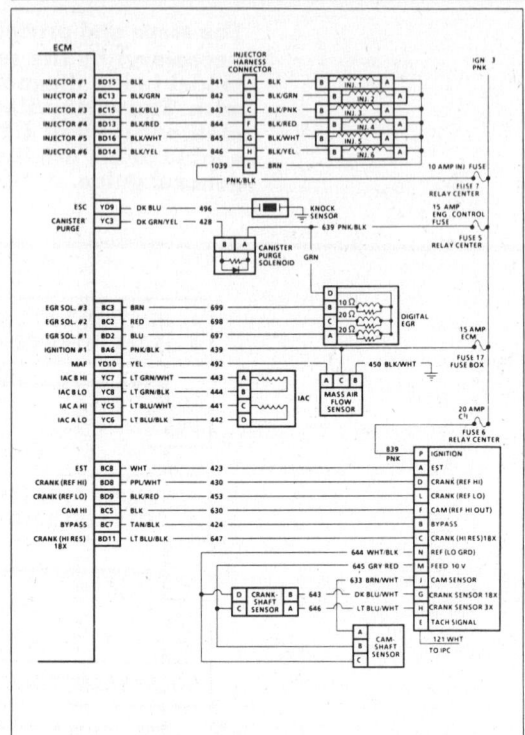

ECM WIRING DIAGRAM — 3800 (VIN C) ENGINE

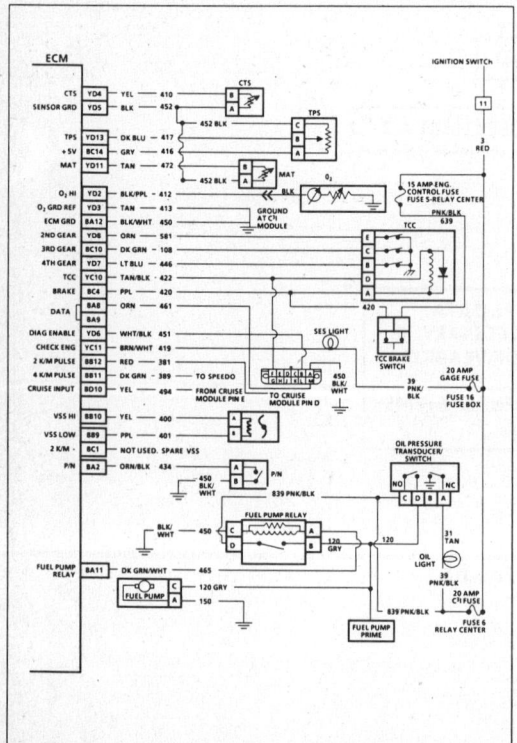

ECM WIRING DIAGRAM — 3800 (VIN C) ENGINE

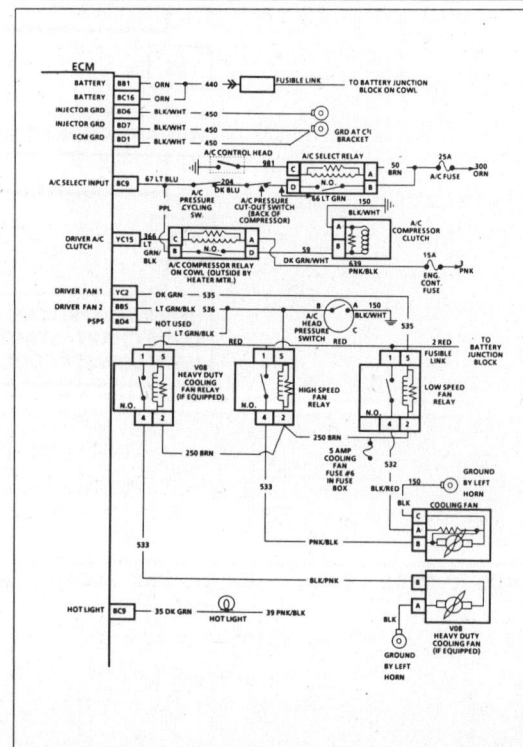

ECM CONNECTORS – 3800 (VIN C) ENGINE

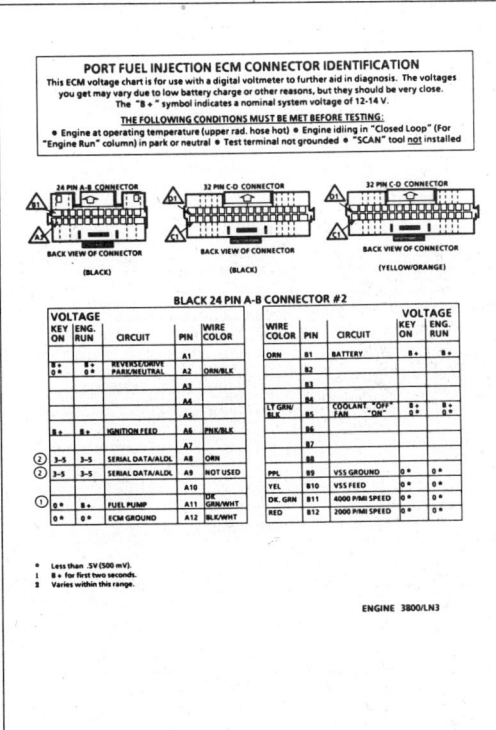

PORT FUEL INJECTION ECM CONNECTOR IDENTIFICATION

This ECM voltage chart is for use with a digital voltmeter to further aid in diagnosis. The voltages you get may vary due to low battery charge or other reasons, but they should be very close. The "B +" symbol indicates a nominal system voltage of 12-14 V.

THE FOLLOWING CONDITIONS MUST BE MET BEFORE TESTING:
- Engine at operating temperature (upper rad. hose hot)
- Engine idling in "Closed Loop" (For "Engine Run" column) in park or neutral
- Test terminal not grounded
- "SCAN" tool **not** installed

24 PIN A-B CONNECTOR — BACK VIEW OF CONNECTOR (BLACK)

32 PIN C-D CONNECTOR — BACK VIEW OF CONNECTOR (BLACK)

32 PIN C-D CONNECTOR — BACK VIEW OF CONNECTOR (YELLOW/ORANGE)

BLACK 24 PIN A-B CONNECTOR #2

VOLTAGE KEY ON	ENG. RUN	CIRCUIT	PIN	WIRE COLOR		WIRE COLOR	PIN	CIRCUIT	VOLTAGE KEY ON	ENG. RUN
			A1			ORN	B1	BATTERY	B +	B +
B +, 0*	B +, 0*	REVERSE/DRIVE PARK/NEUTRAL	A2	ORN/BLK			B2			
			A3				B3			
			A4				B4			
			A5			LT GRN/BLK	B5	COOLANT "OFF" FAN "ON"	B +, 0*	B +, 0*
B +	B +	IGNITION FEED	A6	PNK/BLK			B6			
			A7				B7			
3-5	3-5	SERIAL DATA/ALDL	A8	ORN		PPL	B9	VSS GROUND	0*	0*
3-5	3-5	SERIAL DATA/ALDL	A9	NOT USED		YEL	B10	VSS FEED	0*	0*
			A10			DK.GRN	B11	4000 P/MI SPEED	0*	0*
0*	B +	FUEL PUMP	A11	DK GRN/WHT		RED	B12	2000 P/MI SPEED	0*	0*
0*	0*	ECM GROUND	A12	BLK/WHT						

- Less than .5V (500 mV).
1 B + for first two seconds.
2 Varies within this range.

ENGINE 3800/LN3

ECM CONNECTORS – 3800 (VIN C) ENGINE

BLACK 32 PIN C-D CONNECTOR #1

VOLTAGE KEY ON	ENG. RUN	CIRCUIT	PIN	WIRE COLOR		WIRE COLOR	PIN	CIRCUIT	VOLTAGE KEY ON	ENG. RUN
0*	0*	2K MI VSS SPARE	C1	N A		BLK.WHT	D1	ECM GROUND	0*	0*
B +	B +	EGR #3	C2	RED		BLU	D2	EGR #1	B +	B +
B +	B +	EGR #2	C3	BRN			D3			
B +	B +	BRAKE SW	C4	PPL		LT BLU-ORN	D4	PSPS (NOT USED)	B +	B +
**	**	CAM HI	C5	BLK			D5			
B +	B +	2K MI VSS SPARE	C6			BLK.WHT	D6	INJ. GRND	0*	0*
0*	4.70	IGNITION BYPASS	C7	TAN.BLK		BLK.WHT	D7	INJ. GRND	0*	0*
0*	0 B +	IGNITION EST	C8	WHT		PPL.WHT	D8	IGN. REFERENCE	0*	6 V
OF-0 On-B +	OF-0 On-B +	A.C SELECT	C9	LT BLU		BLK.RED	D9	IGN GROUND	0*	0*
0*	0*	3RD GEAR INPUT	C10	DK GRN		YEL	D10	CRUISE	B +	B +
			C11			LT BLU/BLK	D11	HI.RES EST	0*	6 V
0*	0*	2 K, M CRUISE	C12	DK GRN			D12			
B +	B +	INJ. #2	C13	BLK/LTGRN		BLK.RED	D13	INJ. #4	B +	B +
5V	5V	+5V REF.	C14	GRY		BLK.YEL	D14	INJ. #6	B +	B +
B +	B +	INJ. #3	C15	BLK/LT BLU		BLK	D15			
B +	B +	BATTERY	C16	ORN		BLK.WHT	D16	INJ. #5	B +	B +

YELLOW/ORANGE 32 PIN C-D CONNECTOR #3

VOLTAGE KEY ON	ENG. RUN	CIRCUIT	PIN	WIRE COLOR		WIRE COLOR	PIN	CIRCUIT	VOLTAGE KEY ON	ENG. RUN
			C1				D1			
B +	B +	COOLANT FAN	C2	DK GRN		BLK.PPL	D2	O² HI	35-.45	1-.9
B +	B +	CANISTER PURGE	C3	DKGRN-YEL		TAN	D3	O² LO	0*	0*
			C4			YEL	D4	COOLANT TEMP	2.06	2.34
0*	0*	IAC-A-HI	C5	LT BLU/WHT		BLK	D5	GRND	0*	0*
B +	B +	IAC-A-LO	C6	LT BLU/BLK		WHT.BLK	D6	DIAG.ALDL	5V	5V
0*	0*	IAC-B-HI	C7	LT GRN/WHT		LT BLU	D7	4TH GEAR	0*	0*
B +	B +	IAC-B-LO	C8	LT GRN-BLK		DK GRN	D8	2ND GEAR	0*	0*
0*	0*	HOT IDLE	C9	DK GRN		DK BLU	D9	KNOCK SIGNAL	2.3	2.3
B +	B +	TCC	C10	TAN-BLK		YEL	D10	MAF SENSOR	0*	0*
B +	B +	SES LIGHT	C11	BRN-WHT		TAN	D11	MAT SENSOR	2.06	2.34
			C12				D12			
			C13			DK BLU	D13	TPS SIGNAL	38-.42	38-.42
			C14				D14			
0*	0*	ON "OFF"	C15	LT GRN-BLK			D15			
B +	B +	A.C RELAY	C16				D15			

** VARIES AROUND 10 VOLTS
- Less than .5V (500 mV).
2 Varies with temperature.
1 Varies within this range.

ENGINE - 3800/LN3

1988-90 VEHICLES EQUIPPED WITH 3800 (VIN C) ENGINE EXCEPT E-CAR

DIAGNOSTIC CIRCUIT CHECK

The Diagnostic Circuit Check must be the starting point for any driveability complaint diagnosis.

The diagnostic circuit check is an organized approach to identifying a problem created by an electronic engine control system malfunction because it directs the service technician to the next logical step in diagnosing the complaint.

If after completing the diagnostic circuit check and finding the on-board diagnostics functioning properly and no trouble codes displayed, a comparison of "Typical Scan Values", for the appropriate engine, may be used for comparison. The "Typical Values" are an average of display values recorded from normally operating vehicles and are intended to represent what a normally functioning system would display.

A "SCAN" TOOL THAT DISPLAYS FAULTY DATA SHOULD NOT BE USED, AND THE PROBLEM SHOULD BE REPORTED TO THE MANUFACTURER. THE USE OF A FAULTY "SCAN" CAN RESULT IN MISDIAGNOSIS AND UNNECESSARY PARTS REPLACEMENT.

TYPICAL "Scan" DATA VALUES
3800 VIN C

Idle / Upper Radiator Hose Hot / Closed Throttle / Park or Neutral / Closed Loop / Acc. off

"SCAN" Position	Units Displayed	Typical Data Value
Engine Speed	RPM	650 - 750
Coolant Temp.	C°	85° - 105°
MAT Mani Air Temp.	C°	Varies with Air Temperature
Air Flow	Gm/Sec	4 - 7
Oxygen Sensor	Millivolts	(.1 - .9)
Throt Position	Volts	.38 - .42
Idle Air Control	Counts (steps)	10 - 30
Park/Neutral	P/N and RDL	P-N
Fuel Integ.	Counts	118 - 138
Block Learn	Counts	118 - 138
Closed Loop Flag.	Open/Closed	Closed Loop (may go open with extended idle)
Vehicle Speed	MPH	0 (Zero)
Torque Conv. Cl.	On/Off	Off
Spark Advance	Degrees	16°
Knock Signal	No/Yes	No
2nd Gear	On/Off	Off
3rd Gear	On/Off	Off
4th Gear	On/Off	Off
A/C Request	Yes/No	No
PROM I.D.	Numbers	Internal I.D. Only
Exhaust Recirc.	%	0 (Zero)
LV8	0	70 - 80
Battery Voltage	Volts	13.8 Volts (varies)
Brake Switch	On/Off	Off (ON with brake pedal depressed)
Inj. Pulse width	Millivolts	4.0 - 5.0 (Varies)
QDM 1	High/Low	Low
QDM 2	High/Low	Low (High if A/C head pressure's high)
QDM 3	High/Low	Low (High with brake pedal depressed)
QDM 4	High/Low	Low

1988-90 VEHICLES EQUIPPED WITH 3800 (VIN C) ENGINE EXCEPT E-CAR

DIAGNOSTIC CIRCUIT CHECK
3800 (VIN C) (PORT)

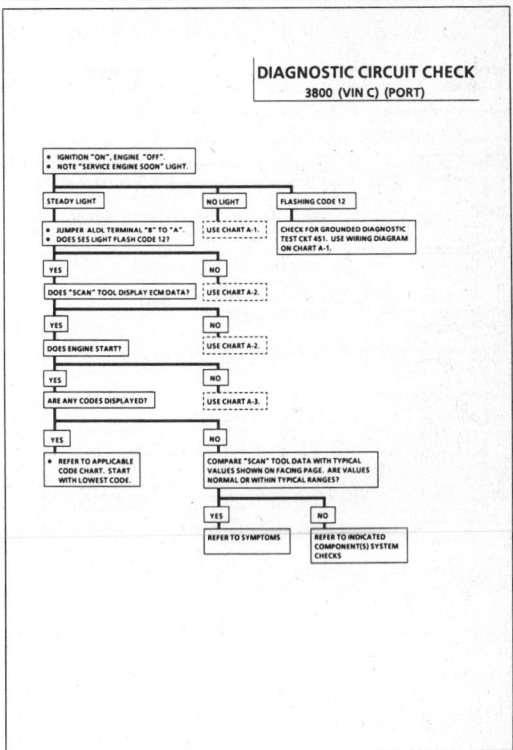

1988–90 VEHICLES EQUIPPED WITH 3800 (VIN C) ENGINE EXCEPT E-CAR

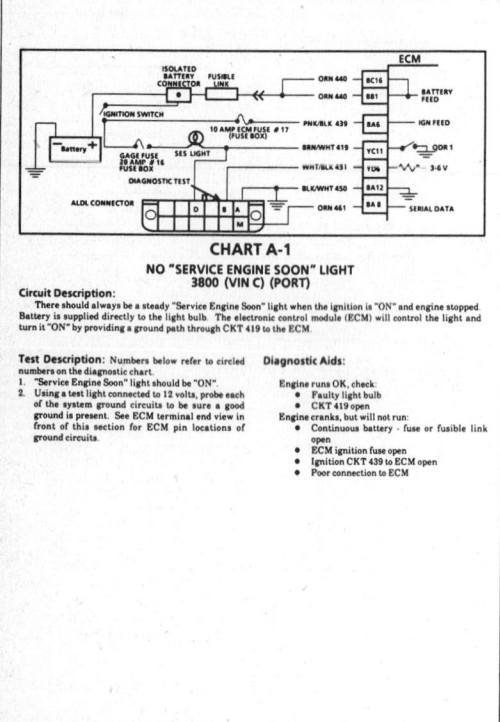

CHART A-1
NO "SERVICE ENGINE SOON" LIGHT
3800 (VIN C) (PORT)

Circuit Description:
There should always be a steady "Service Engine Soon" light when the ignition is "ON" and engine stopped. Battery is supplied directly to the light bulb. The electronic control module (ECM) will control the light and turn it "ON" by providing a ground path through CKT 419 to the ECM.

Test Description: Numbers below refer to circled numbers on the diagnostic chart.
1. "Service Engine Soon" light should be "ON".
2. Using a test light connected to 12 volts, probe each of the system ground circuits to be sure a good ground is present. See ECM terminal end view in front of this section for ECM pin locations of ground circuits.

Diagnostic Aids:
Engine runs OK, check:
- Faulty light bulb
- CKT 419 open
Engine cranks, but will not run:
- Continuous battery - fuse or fusible link open
- ECM ignition fuse open
- Ignition CKT 439 to ECM open
- Poor connection to ECM

1988–90 VEHICLES EQUIPPED WITH 3800 (VIN C) ENGINE EXCEPT E-CAR

CHART A-1
NO "SERVICE ENGINE SOON" LIGHT
3800 (VIN C) (PORT)

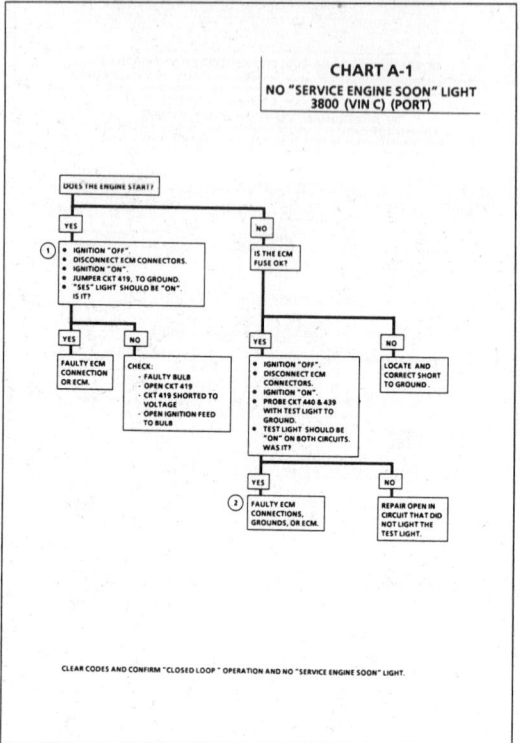

CLEAR CODES AND CONFIRM "CLOSED LOOP" OPERATION AND NO "SERVICE ENGINE SOON" LIGHT.

1988–90 VEHICLES EQUIPPED WITH 3800 (VIN C) ENGINE EXCEPT E-CAR

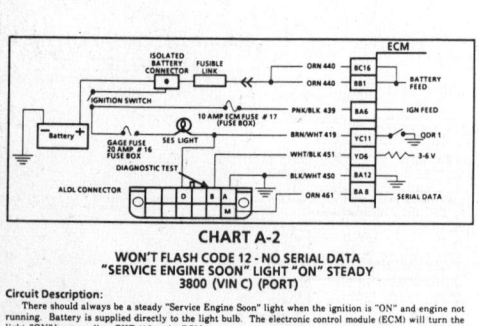

CHART A-2
WON'T FLASH CODE 12 - NO SERIAL DATA
"SERVICE ENGINE SOON" LIGHT "ON" STEADY
3800 (VIN C) (PORT)

Circuit Description:
There should always be a steady "Service Engine Soon" light when the ignition is "ON" and engine not running. Battery is supplied directly to the light bulb. The electronic control module (ECM) will turn the light "ON" by grounding CKT 419 at the ECM.
With the diagnostic terminal grounded, the light should flash a Code 12, followed by any trouble code(s) stored in memory.
A steady light suggests a short to ground in the light control CKT 419 or an open in diagnostic CKT 451.

Test Description: Numbers below refer to circled numbers on the diagnostic chart.
1. If the light goes "OFF" when the ECM connector is disconnected, then CKT 419 is not shorted to ground.
2. If there is a problem with the ECM that causes a "Scan" tool to not read serial data, then the ECM should not flash a Code 12. If Code 12 does flash, be sure that the "Scan" tool is working properly on another vehicle.

If the "Scan" is functioning properly and CKT 461 is OK, the Mem-Cal or ECM may be at fault for the NO ALDL symptom.
3. This step will check for an open diagnostic CKT 451.
4. At this point, the "Service Engine Soon" light wiring is OK. The problem is a faulty ECM or Mem-Cal. If Code 12 does not flash, the ECM should be replaced using the original Mem-Cal. Replace the Mem-Cal only after trying an ECM, as a defective Mem-Cal is an unlikely cause of the problem.

1988–90 VEHICLES EQUIPPED WITH 3800 (VIN C) ENGINE EXCEPT E-CAR

CHART A-2
WON'T FLASH CODE 12 - NO SERIAL DATA
"SERVICE ENGINE SOON" LIGHT "ON" STEADY
3800 (VIN C) (PORT)

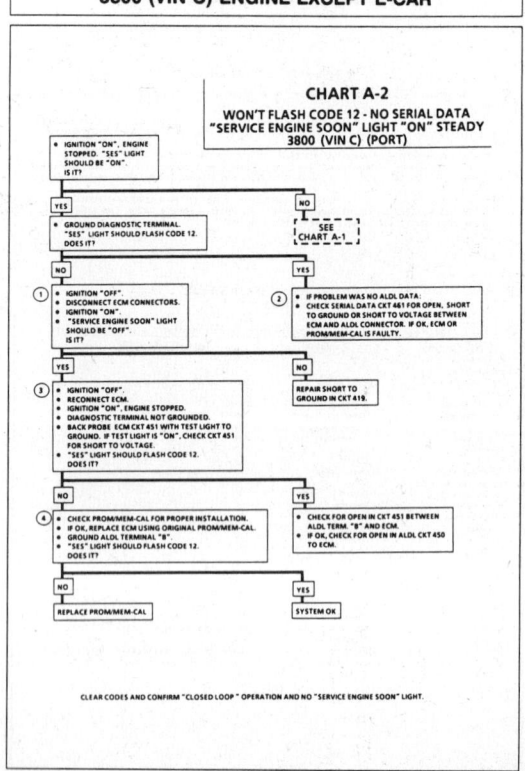

CLEAR CODES AND CONFIRM "CLOSED LOOP" OPERATION AND NO "SERVICE ENGINE SOON" LIGHT.

1988–90 VEHICLES EQUIPPED WITH 3800 (VIN C) ENGINE EXCEPT E-CAR

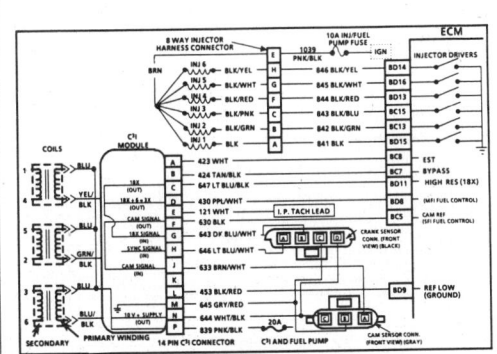

CHART A-3
(Page 1 of 4)
ENGINE CRANKS BUT WON'T RUN
3800 (VIN C) (PORT)

Circuit Description:
The C³I uses a waste spark method of spark distribution. In this type of ignition system, the C³I module triggers the correct coil pair, based on both signals from the crankshaft sensor, resulting in both spark plugs firing at the same time. One cylinder is on the compression cycle, while the other one is on the exhaust cycle, resulting in a lower energy requirement to fire the spark plug on the exhaust cycle. The remaining high voltage is used to fire the spark plug on the compression cycle.

Sequential fuel injection utilizes 6 separate injector driver circuits to activate the 6 fuel injectors. While cranking, the ECM activates all 6 injectors simultaneously (all at one time). After a calibrated engine rpm is reached, and a good cam signal has been received by the ECM on CKT 630, the injection mode of operation is changed to sequential (timed separately).

The sync signal is used only by the C³I module. It is used for spark synchronization at startup only (not passed to the ECM).

Test Description: Numbers below refer to circled numbers on the diagnostic chart.
1. This step verifies that "SES" light operation, TPS, and coolant sensor signals are normal. A blinking injector test light verifies that the ECM is monitoring the C³I reference signal and attempting to activate the injectors.
2. Both the cam and crank sensors have been verified as functioning properly, as is evidenced by the blinking injector test light. A fuel press test, at this point, will separate the diagnostic path into either a fuel related fault or ignition system malfunction.
3. The 8 terminal injector harness connector must be disconnected to avoid flooding of the engine and fouling of the spark plugs. By testing for spark on plug leads 1, 3 and 5, each ignition coil's ability to produce at least 25,000 volts is verified.
4. By testing the problem coil's control circuit with a test light, a determination can be made as to the problem being faulty or the module's internal driver for that coil being the source of the complaint.

1988–90 VEHICLES EQUIPPED WITH 3800 (VIN C) ENGINE EXCEPT E-CAR

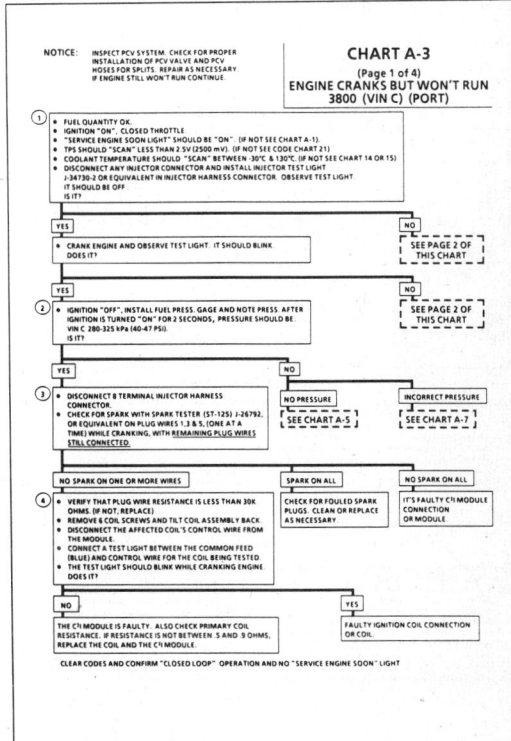

NOTICE: INSPECT PCV SYSTEM. CHECK FOR PROPER INSTALLATION OF PCV VALVE AND PCV HOSES FOR SPLITS. REPAIR AS NECESSARY. IF ENGINE STILL WON'T RUN CONTINUE.

CHART A-3
(Page 1 of 4)
ENGINE CRANKS BUT WON'T RUN
3800 (VIN C) (PORT)

1988–90 VEHICLES EQUIPPED WITH 3800 (VIN C) ENGINE EXCEPT E-CAR

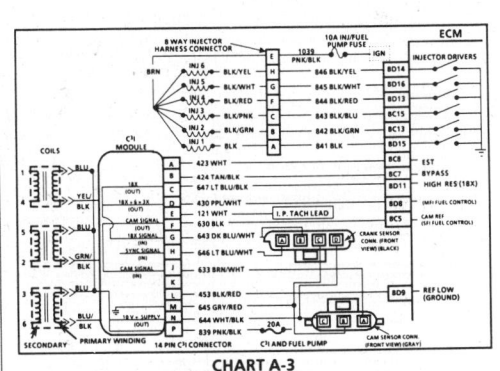

CHART A-3
(Page 2 of 4)
ENGINE CRANKS BUT WON'T RUN
3800 (VIN C) (PORT)

Circuit Description:
The C³I uses a waste spark method of spark distribution. In this type of ignition system, the C³I module triggers the correct coil pair, based on both signals from the crankshaft sensor, resulting in both spark plugs firing at the same time. One cylinder is on the compression cycle, while the other one is on the exhaust cycle, resulting in a lower energy requirement to fire the spark plug on the exhaust cycle. The remaining high voltage is used to fire the spark plug on the compression cycle.

Sequential fuel injection utilizes 6 separate injector driver circuits to activate the 6 fuel injectors. While cranking, the ECM activates all 6 injectors simultaneously (all at one time). After a calibrated engine rpm is reached, and a good cam signal has been received by the ECM on CKT 630, the injection mode of operation is changed to sequential (timed separately).

The sync signal is used only by the C³I module. It is used for spark synchronization at startup only (not passed to the ECM).

Test Description: Numbers below refer to circled numbers on the diagnostic chart.
5. Tests for battery voltage on CKT 1039. If voltage was present, the "light off" test result was caused by no activation pulse reaching the injector connector from the ECM.
6. If the fuse was blown, check CKT 839 including fuel pump relay, fuel pump, and wiring to determine the cause for the high current flow.

1988–90 VEHICLES EQUIPPED WITH 3800 (VIN C) ENGINE EXCEPT E-CAR

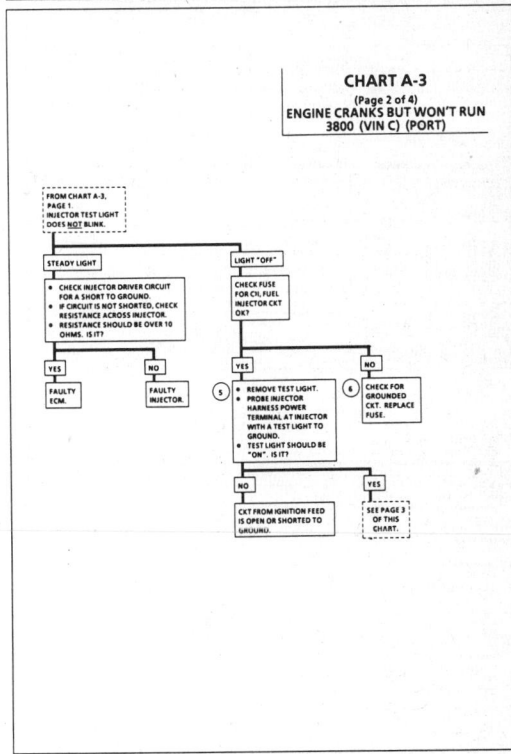

CHART A-3
(Page 2 of 4)
ENGINE CRANKS BUT WON'T RUN
3800 (VIN C) (PORT)

1988–90 VEHICLES EQUIPPED WITH 3800 (VIN C) ENGINE EXCEPT E-CAR

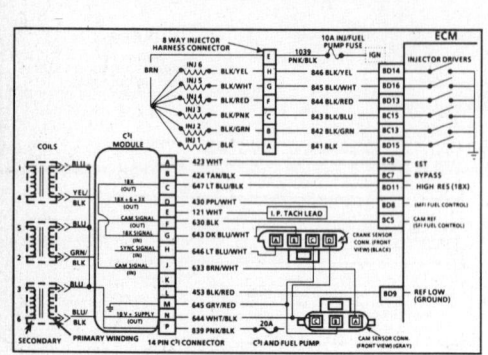

CHART A-3
(Page 3 of 4)
ENGINE CRANKS BUT WON'T RUN
3800 (VIN C) (PORT)

Circuit Description:

The C³I uses a waste spark method of spark distribution. In this type of ignition system, the C³I module triggers the correct coil pair, based on both signals from the crankshaft sensor, resulting in both spark plugs firing at the same time. One cylinder is on the compression cycle, while the other one is on the exhaust cycle, resulting in a lower energy requirement to fire the spark plug on the exhaust cycle. The remaining high voltage is used to fire the spark plug on the compression cycle.

Sequential fuel injection utilizes 6 separate injector driver circuits to activate the 6 fuel injectors. While cranking, the ECM activates all 6 injectors simultaneously (all at one time). After a calibrated engine rpm is reached, and a good cam signal has been received by the ECM on CKT 630, the injection mode of operation is changed to sequential (timed separately).

The sync signal is used only by the C³I module. It is used for spark synchronization at startup only (not passed to the ECM).

Test Description: Numbers below refer to circled numbers on the diagnostic chart.

7. The test light to 12 volts simulates a reference signal to the ECM which will result in an injector test light blink with each touch of the test light to terminal "D". You will hear various injectors click. It may take up to three touches to get an injector test light flash. CKT 430, the ECM, and the injector driver circuits are all OK.
8. If the crank sensor signal circuit terminal "A" is momentarily jumpered to the ground circuit terminal "C" and the engine is cranked, without

turning the ignition switch "OFF", the response should be an injector test light blink. This is a result of the artificial "Sync Signal" being transmitted to the C³I module which allows generation of the 3x reference signal to the C³I terminal "BD8" and the ECM activating the injector driver circuit.

9. Verifies a proper sync signal circuit voltage of 9 to 12 volts and a good ground from the C³I module to terminal "C" of the sensor connector.
10. Determines if reason for incorrect voltage reading was due to a fault in CKT 646, an open in CKT 645, or a faulty C³I module.

1988–90 VEHICLES EQUIPPED WITH 3800 (VIN C) ENGINE EXCEPT E-CAR

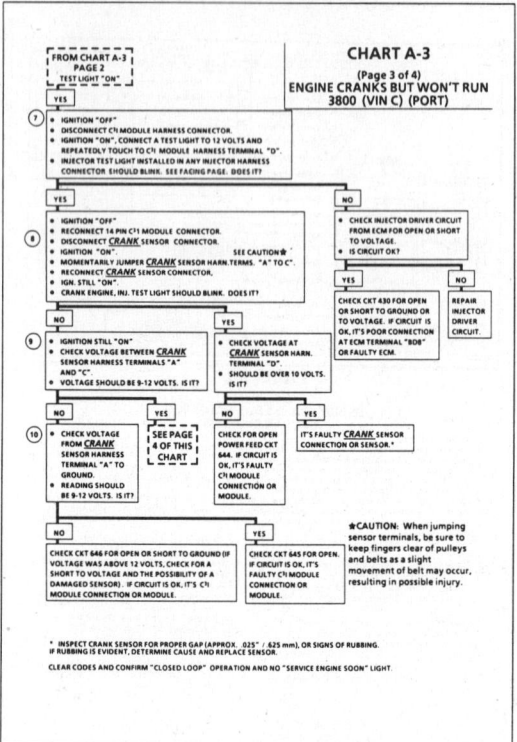

CHART A-3
(Page 3 of 4)
ENGINE CRANKS BUT WON'T RUN
3800 (VIN C) (PORT)

★CAUTION: When jumping sensor terminals, be sure to keep fingers clear of pulleys and belts as a slight movement of belt may occur, resulting in possible injury.

* INSPECT CRANK SENSOR FOR PROPER GAP (APPROX. .025" / .625 mm). IF RUBBING IS EVIDENT, DETERMINE CAUSE AND REPLACE SENSOR.

CLEAR CODES AND CONFIRM "CLOSED LOOP" OPERATION AND NO "SERVICE ENGINE SOON" LIGHT.

1988–90 VEHICLES EQUIPPED WITH 3800 (VIN C) ENGINE EXCEPT E-CAR

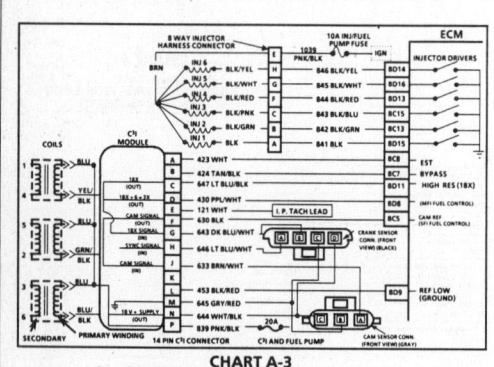

CHART A-3
(Page 4 of 4)
ENGINE CRANKS BUT WON'T RUN
3800 (VIN C) (PORT)

Circuit Description:

The C³I uses a waste spark method of spark distribution. In this type of ignition system, the C³I module triggers the correct coil pair, based on both signals from the crankshaft sensor, resulting in both spark plugs firing at the same time. One cylinder is on the compression cycle, while the other one is on the exhaust cycle, resulting in a lower energy requirement to fire the spark plug on the exhaust cycle. The remaining high voltage is used to fire the spark plug on the compression cycle.

Sequential fuel injection utilizes 6 separate injector driver circuits to activate the 6 fuel injectors. While cranking, the ECM activates all 6 injectors simultaneously (all at one time). After a calibrated engine rpm is reached, and a good cam signal has been received by the ECM on CKT 630, the injection mode of operation is changed to sequential (timed separately).

The sync signal is used only by the C³I module. It is used for spark synchronization at startup only (not passed to the ECM).

Test Description: Numbers below refer to circled numbers on the diagnostic chart.

11. Jumping the crank sensor harness terminals "A" and "C" together simulates a sync signal to the C³I module. By jumping the crank sensor harness terminals "B" and "C" together, a crank signal is simulated. This signal will then cause the ECM to 1) Energize the fuel pump relay for 2 seconds, (you can hear it clicking "ON" and "OFF"), 2) You should also be able to hear several, if not all, injectors activate.

12. Verifies a proper crank signal circuit voltage of 9 to 12 volts and a good ground from the C³I module to terminal "C" of the sensor connector.
13. Determines if reason for incorrect voltage reading was due to a fault in CKT 643, an open in CKT 645, or a faulty C³I module.

1988–90 VEHICLES EQUIPPED WITH 3800 (VIN C) ENGINE EXCEPT E-CAR

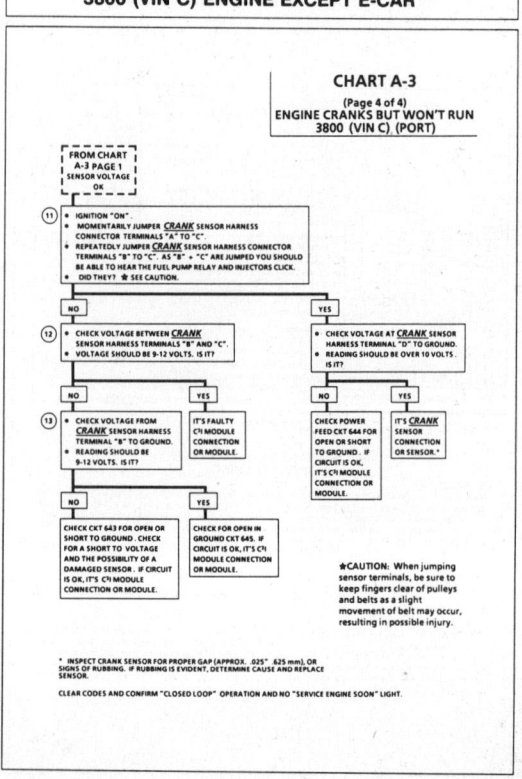

CHART A-3
(Page 4 of 4)
ENGINE CRANKS BUT WON'T RUN
3800 (VIN C) (PORT)

★CAUTION: When jumping sensor terminals, be sure to keep fingers clear of pulleys and belts as a slight movement of belt may occur, resulting in possible injury.

* INSPECT CRANK SENSOR FOR PROPER GAP (APPROX. .025" / .625 mm). SIGNS OF RUBBING. IF RUBBING IS EVIDENT, DETERMINE CAUSE AND REPLACE SENSOR.

CLEAR CODES AND CONFIRM "CLOSED LOOP" OPERATION AND NO "SERVICE ENGINE SOON" LIGHT.

1988–90 VEHICLES EQUIPPED WITH 3800 (VIN C) ENGINE EXCEPT E-CAR

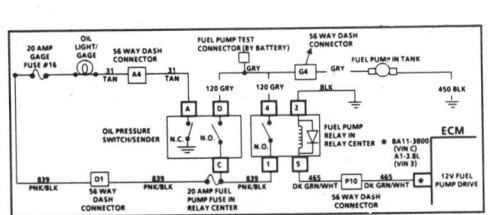

CHART A-5
(Page 1 of 2)
FUEL SYSTEM ELECTRICAL TEST
(STANDARD CLUSTER)
3800 (VIN C) (PORT)

Circuit Description:
    When the ignition switch is turned "ON", the electronic control module (ECM) will energize the fuel pump relay which completes the circuit to the in-tank fuel pump. It will remain "ON" as long as the engine is cranking or running and the ECM is receiving C3 reference pulses. If there are no reference pulses, the ECM will de-energize the fuel pump relay within 2 seconds after key "ON" or the engine is stopped.
    The fuel pump will deliver fuel to the fuel rail and injectors, then to the pressure regulator where the system pressure is controlled. Excess fuel pressure is bypassed back to the fuel tank. When the engine is stopped, the pump can be turned "ON" by applying battery voltage to the test terminal located in the engine compartment.
    Improper fuel system pressure may contribute to one or all of the following symptoms:
- Cranks but won't run
- Code 44 or 45
- Cuts out, may feel like ignition problem
- Hesitation, loss of power or poor fuel economy

Test Description: Numbers below refer to circled numbers on the diagnostic chart.
1. If the fuse is blown, a short to ground in CKTs 120, 839, or the fuel pump itself is the cause.
2. This step determines if the fuel pump circuit is being controlled by the ECM. The ECM should energize the fuel pump relay and turn the fuel pump "ON". If the engine is not cranking or running, the ECM should de-energize the relay and/or fuel pump within 2 seconds after the ignition is turned "ON".

3. Applying B+ to the pump prime connector turns "ON" the fuel pump. This validates CKT 120 wiring. If the pump runs, it is a basic fuel delivery problem.
4. This test will determine if a short to ground on CKT 120 caused the fuse to blow. To prevent a mis-diagnosis, be sure the fuel pump is disconnected before the test.
5. Checks for a short to ground in the fuel pump relay harness CKT 839.

1988–90 VEHICLES EQUIPPED WITH 3800 (VIN C) ENGINE EXCEPT E-CAR

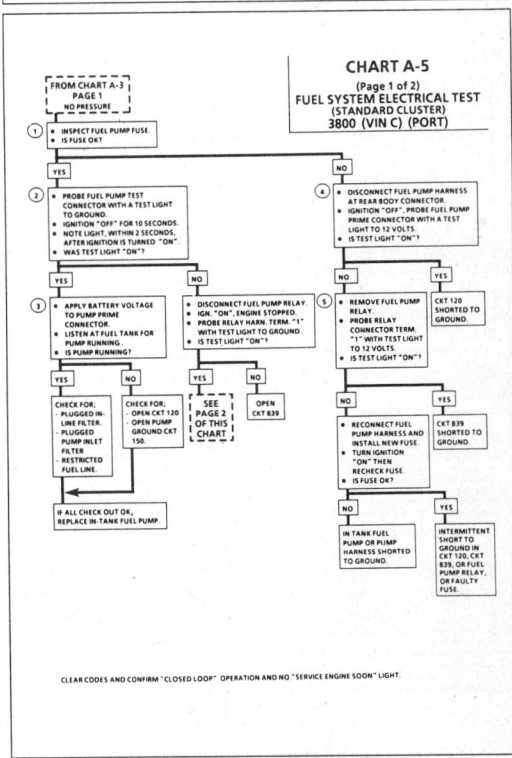

CHART A-5
(Page 1 of 2)
FUEL SYSTEM ELECTRICAL TEST
(STANDARD CLUSTER)
3800 (VIN C) (PORT)

CLEAR CODES AND CONFIRM "CLOSED LOOP" OPERATION AND NO "SERVICE ENGINE SOON" LIGHT.

1988–90 VEHICLES EQUIPPED WITH 3800 (VIN C) ENGINE EXCEPT E-CAR

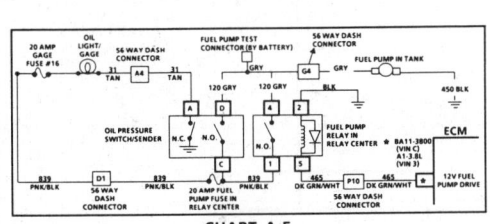

CHART A-5
(Page 2 of 2)
FUEL SYSTEM ELECTRICAL TEST
(STANDARD CLUSTER)
3800 (VIN C) (PORT)

Circuit Description:
    When the ignition switch is turned "ON", the electronic control module (ECM) will energize the fuel pump relay which completes the circuit to the in-tank fuel pump. It will remain "ON" as long as the engine is cranking or running and the ECM is receiving C3 reference pulses. If there are no reference pulses, the ECM will de-energize the fuel pump relay within 2 seconds after key "ON" or the engine is stopped.

Test Description: Numbers below refer to circled numbers on the diagnostic chart.
6. Checks for open in the fuel pump relay ground CKT 450.
7. Determines if the ECM is in control of the fuel pump through CKT 465 (terminal "A").
8. The fuel pump control circuit includes an engine oil pressure switch with a separate set of normally open contacts. The switch closes at about (4 lbs) 28 kPa of oil pressure and provides a second battery feed path to the fuel pump. If the relay fails, the pump will run using the battery feed supplied by the closed oil pressure switch.

This step checks the oil pressure switch to be sure it provides battery feed to the fuel pump should the pump relay fail. A failed pump relay will result in extended engine crank time because of the time required to build enough oil pressure to close the oil pressure switch and turn "ON" the fuel pump. There may be instances when the relay has failed but the engine will not crank fast enough to build enough oil pressure to close the switch. This is a faulty oil pressure switch can result in "Engine Cranks But Won't Run".

1988–90 VEHICLES EQUIPPED WITH 3800 (VIN C) ENGINE EXCEPT E-CAR

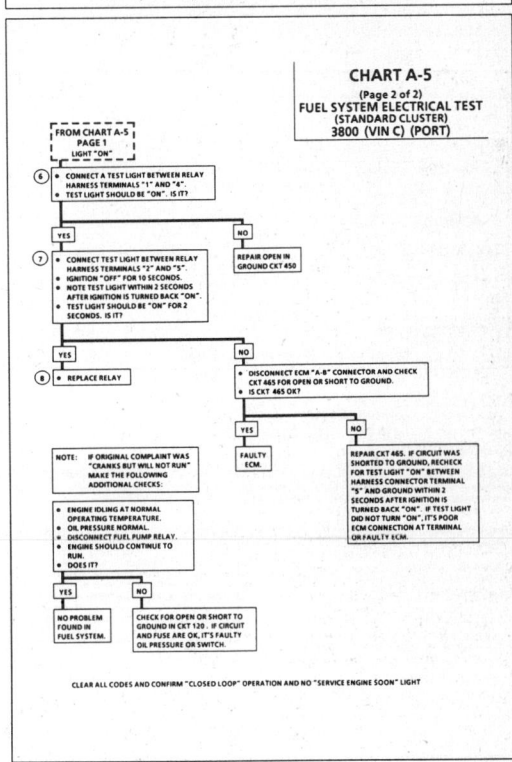

CHART A-5
(Page 2 of 2)
FUEL SYSTEM ELECTRICAL TEST
(STANDARD CLUSTER)
3800 (VIN C) (PORT)

CLEAR ALL CODES AND CONFIRM "CLOSED LOOP" OPERATION AND NO "SERVICE ENGINE SOON" LIGHT

1988–90 VEHICLES EQUIPPED WITH 3800 (VIN C) ENGINE EXCEPT E-CAR

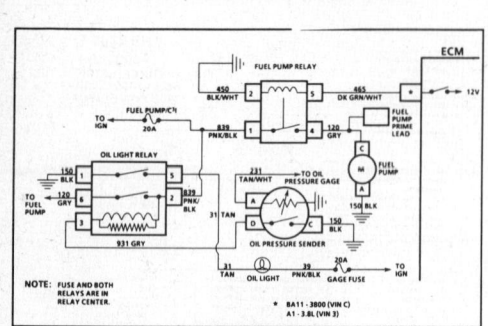

CHART A-5
(Page 1 of 2)
FUEL SYSTEM ELECTRICAL TEST
(U21 CLUSTER ONLY)
3800 (VIN C) (PORT)

Circuit Description:

When the ignition switch is turned "ON", the electronic control module (ECM) will energize the fuel pump relay which completes the circuit to the in-tank fuel pump. It will remain "ON" as long as the engine is cranking or running and the ECM is receiving C³I reference pulses. If there are no reference pulses, the ECM will de-energize the fuel pump relay within 2 seconds after key "ON" or the engine is stopped.

The fuel pump will deliver fuel to the fuel rail and injectors, then to the pressure regulator, where the system pressure is controlled. Excess fuel pressure is bypassed back to the fuel tank. When the engine is stopped, the pump can be turned "ON" by applying battery voltage to the test terminal in the engine compartment.

Improper fuel system pressure may contribute to one or all of the following symptoms:
- Cranks but won't run
- Code 44 or 45
- Cuts out, may feel like ignition problem
- Hesitation, loss of power or poor fuel economy

Test Description: Numbers below refer to circled numbers on the diagnostic chart.

1. If the fuse is blown, a short to ground in CKTs 120, 839, or the fuel pump itself is the cause.
2. Determines if the fuel pump circuit is being controlled by the fuel pump relay. The ECM should energize the fuel pump relay. If the engine is not cranking or running, the ECM should de-energize the relay within 2 seconds after the ignition is turned "ON"

3. Turns "ON" the fuel pump if CKT 120 wiring is OK. If the pump runs, it is a basic fuel delivery problem.
4. This test will determine if a short to ground on CKT 120 caused the fuse to blow. To prevent a mis-diagnosis, be sure the fuel pump is disconnected before the test.
5. Checks for a short to ground in the fuel pump relay harness CKT 839.

1988–90 VEHICLES EQUIPPED WITH 3800 (VIN C) ENGINE EXCEPT E-CAR

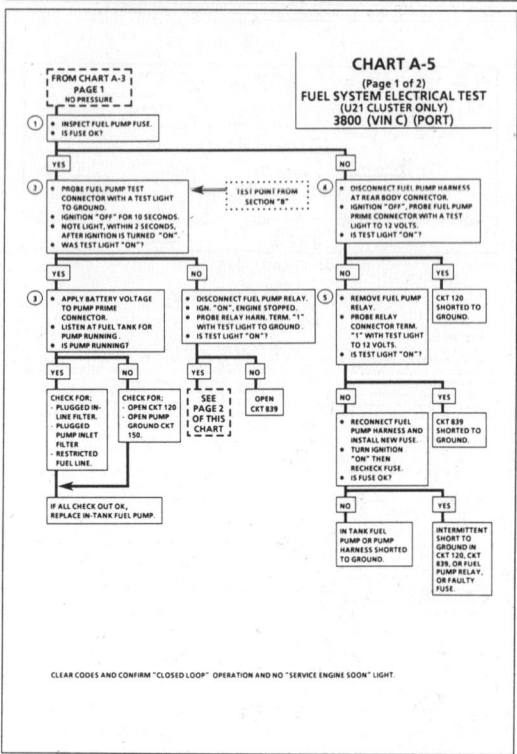

CLEAR CODES AND CONFIRM "CLOSED LOOP" OPERATION AND NO "SERVICE ENGINE SOON" LIGHT.

1988–90 VEHICLES EQUIPPED WITH 3800 (VIN C) ENGINE EXCEPT E-CAR

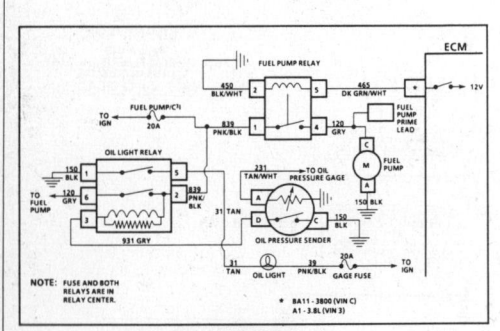

CHART A-5
(Page 2 of 2)
FUEL SYSTEM ELECTRICAL TEST
(U21 CLUSTER ONLY)
3800 (VIN C) (PORT)

Circuit Description:

When the ignition switch is turned "ON", the electronic control module (ECM) will energize the fuel pump relay which completes the circuit to the in-tank fuel pump. It will remain "ON" as long as the engine is cranking or running and the ECM is receiving C³I reference pulses. If there are no reference pulses, the ECM will de-energize the fuel pump relay within 2 seconds after key "ON" or the engine is stopped.

Test Description: Numbers below refer to circled numbers on the diagnostic chart.

6. Checks for open in the fuel pump relay ground, CKT 450.
7. Determines if the ECM is in control of the fuel pump relay through CKT 465 (terminal "A").
8. The fuel pump control circuit includes an engine oil pressure switch with a separate set of normally open contacts. The switch closes at about (4 lbs) 28 kPa of oil pressure and provides a second battery feed path to the fuel pump. If the relay fails, the pump will run using the battery feed supplied by the closed oil pressure switch.

This step checks the oil pressure switch to be sure it provides battery feed so the pump relay fail. A failed pump will result in extended engine crank time because of the time required to build enough oil pressure to close the oil pressure switch and turn "ON" the fuel pump. There may be instances when the relay has failed but the engine will not crank fast enough to build enough oil pressure to close the switch. This or a faulty oil pressure switch can result in "Engine Cranks But Will Not Run".

1988–90 VEHICLES EQUIPPED WITH 3800 (VIN C) ENGINE EXCEPT E-CAR

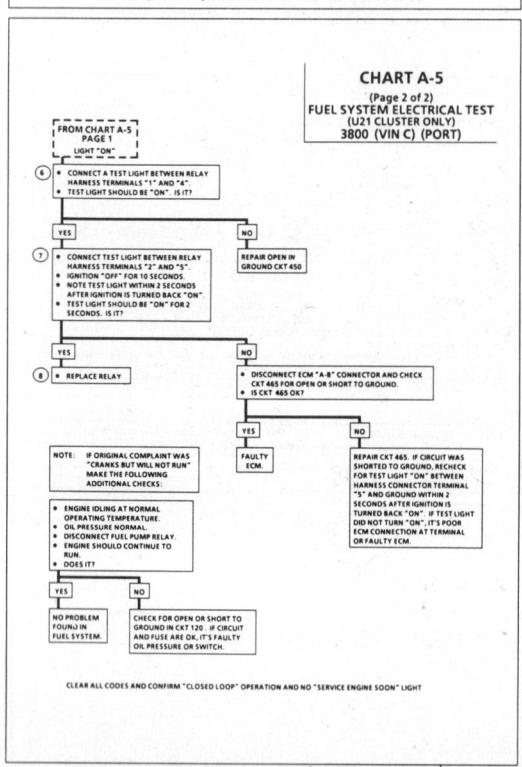

CLEAR ALL CODES AND CONFIRM "CLOSED LOOP" OPERATION AND NO "SERVICE ENGINE SOON" LIGHT

1988–90 VEHICLES EQUIPPED WITH 3800 (VIN C) ENGINE EXCEPT E-CAR

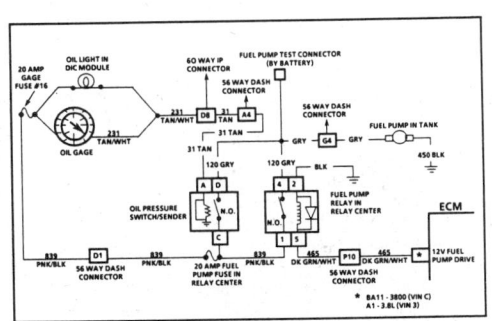

CHART A-5
(Page 1 of 2)
FUEL SYSTEM ELECTRICAL TEST
(U21 WITH UW1)
3800 (VIN C) (PORT)

Circuit Description:
When the ignition switch is turned "ON", the electronic control module (ECM) will energize the fuel pump relay which completes the circuit to the in-tank fuel pump. It will remain "ON" as long as the engine is cranking or running and the ECM is receiving C3I reference pulses. If there are no reference pulses, the ECM will de-energize the fuel pump relay within 2 seconds after key "ON" or the engine is stopped.
The fuel pump will deliver fuel to the fuel rail and injectors then to the pressure regulator where the system pressure is controlled. Excess fuel pressure is bypassed back to the fuel tank. When the engine is stopped, the pump can be turned "ON" by applying battery voltage to the test terminal located in the engine compartment.
Improper fuel system pressure may contribute to one or all of the following symptoms:
 • Cranks but won't run
 • Code 44 or 45
 • Cuts out, may feel like ignition problem
 • Hesitation, loss of power or poor fuel economy

Test Description: Numbers below refer to circled numbers on the diagnostic chart.
1. If the fuse is blown, a short to ground in CKTs 120, 839 or the fuel pump itself is the cause.
2. Determines if the fuel pump circuit is being controlled by the ECM. The ECM should energize the fuel pump relay. The engine is not cranking or running so the ECM should de-energize the relay within 2 seconds after ignition is turned "ON".

3. Turns "ON" the fuel pump if CKT 120 wiring is OK. If the pump runs, it is a basic fuel delivery problem.
4. This test will determine if a short to ground on CKT 120 caused the fuse to blow. To prevent a mis-diagnosis, be sure the fuel pump is disconnected before the test.
5. Checks for a short to ground in the fuel pump relay harness CKT 839.

1988–90 VEHICLES EQUIPPED WITH 3800 (VIN C) ENGINE EXCEPT E-CAR

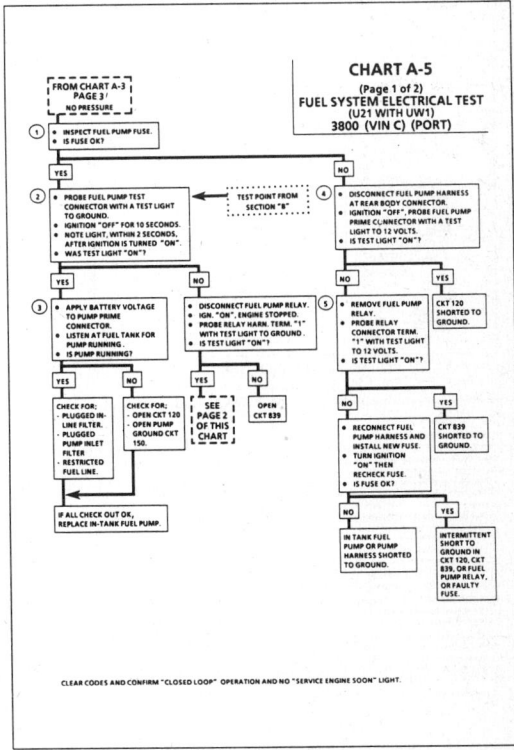

1988–90 VEHICLES EQUIPPED WITH 3800 (VIN C) ENGINE EXCEPT E-CAR

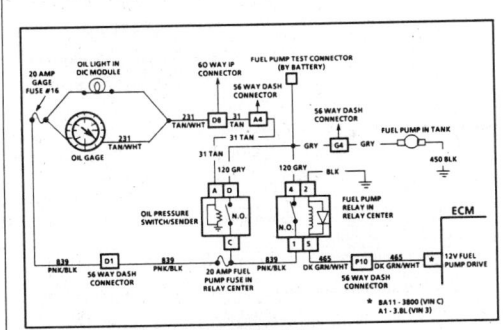

CHART A-5
(Page 2 of 2)
FUEL SYSTEM ELECTRICAL TEST
(U21 WITH UW1)
3800 (VIN C) (PORT)

Circuit Description:
When the ignition switch is turned "ON", the electronic control module (ECM) will energize the fuel pump relay, which completes the circuit to the in-tank fuel pump. It will remain "ON" as long as the engine is cranking or running and the ECM is receiving C3I reference pulses. If there are no reference pulses, the ECM will de-energize the fuel pump relay in 2 seconds after key "ON" or the engine is stopped.

Test Description: Numbers below refer to circled numbers on the diagnostic chart.
6. Checks for open in the fuel pump relay ground, CKT 450.
7. Determines if the ECM is in control of the fuel pump relay through CKT 465 (terminal "A").
8. The fuel pump control circuit includes an engine oil pressure switch with a separate set of normally open contacts. The switch closes at about (4 lbs) 28 kPa of oil pressure and provides a second battery feed path to the fuel pump. If the relay fails, the pump will run using the battery feed supplied by the closed oil pressure switch.

This step checks the oil pressure switch to be sure it provides battery feed to the fuel pump should the pump relay fail. A failed pump relay will result in extended engine crank time because of the time required to build enough oil pressure to close the oil pressure switch and turn "ON" the fuel pump. There may be instances when the relay has failed but the engine will not crank fast enough to build enough oil pressure to close the switch. This or a faulty oil pressure switch can result in "Engine Cranks But Won't Run".

1988–90 VEHICLES EQUIPPED WITH 3800 (VIN C) ENGINE EXCEPT E-CAR

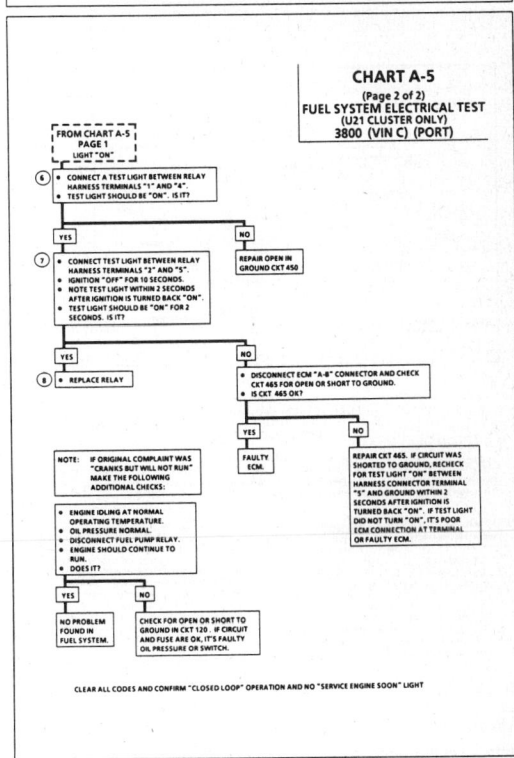

1988–90 VEHICLES EQUIPPED WITH 3800 (VIN C) ENGINE EXCEPT E-CAR

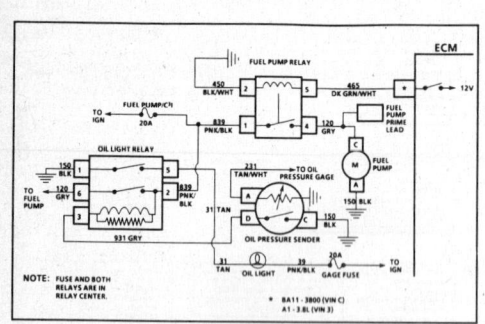

CHART A-5
(Page 1 of 2)
FUEL SYSTEM ELECTRICAL TEST
(U52 CLUSTER ONLY)
3800 (VIN C) (PORT)

Circuit Description:
When the ignition switch is turned "ON," the electronic control module (ECM) will energize the fuel pump relay which completes the circuit to the in-tank fuel pump. It will remain "ON" as long as the engine is cranking or running and the ECM is receiving C3I reference pulses. If there are no reference pulses, the ECM will de-energize the fuel pump relay within 2 seconds after key "ON" or the engine is stopped.

The fuel pump will deliver fuel to the fuel rail and injectors, then to the pressure regulator, where the system pressure is controlled. Excess fuel pressure is bypassed back to the fuel tank. When the engine is stopped, the pump can be turned "ON" by applying battery voltage to the test terminal in the engine compartment.

Improper fuel system pressure may contribute to one or all of the following symptoms:
- Cranks but won't run.
- Code 44 or 45.
- Cuts out, may feel like ignition problem.
- Hesitation, loss of power or poor fuel economy.

Test Description: Numbers below refer to circled numbers on the diagnostic chart.
1. If the fuse is blown, a short to ground in CKTs 120, 839, or the fuel pump itself is the cause.
2. Determines if the fuel pump circuit is being controlled by the ECM. The ECM should energize the fuel pump relay. If the engine is not cranking or running, the ECM should de-energize the relay within 2 seconds after the ignition is turned "ON."
3. Turns "ON" the fuel pump if CKT 120 wiring is OK. If the pump runs, it is a basic fuel delivery problem.
4. This test will determine if a short to ground on CKT 120 caused the fuse to blow. To prevent a mis-diagnosis, be sure the fuel pump is disconnected before the test.
5. Checks for a short to ground in the fuel pump relay harness CKT 839.

1988–90 VEHICLES EQUIPPED WITH 3800 (VIN C) ENGINE EXCEPT E-CAR

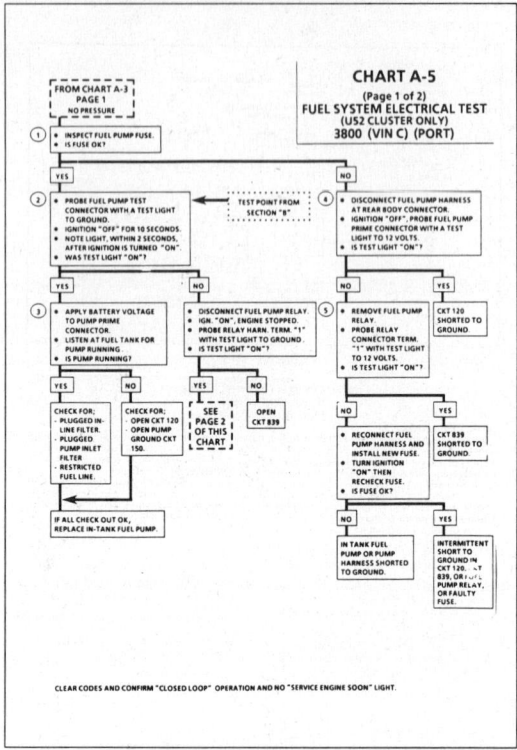

CLEAR CODES AND CONFIRM "CLOSED LOOP" OPERATION AND NO "SERVICE ENGINE SOON" LIGHT.

1988–90 VEHICLES EQUIPPED WITH 3800 (VIN C) ENGINE EXCEPT E-CAR

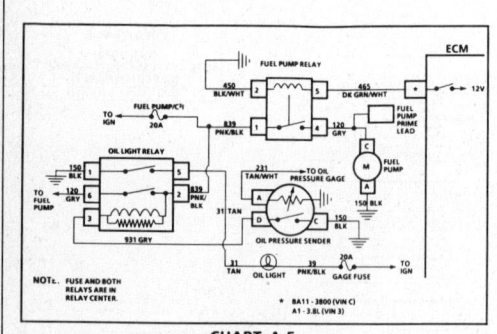

CHART A-5
(Page 2 of 2)
FUEL SYSTEM ELECTRICAL TEST
(U52 CLUSTER ONLY)
3800 (VIN C) (PORT)

Circuit Description:
When the ignition switch is turned "ON," the electronic control module (ECM) will energize the fuel pump relay which completes the circuit to the in-tank fuel pump. It will remain "ON" as long as the engine is cranking or running and the ECM is receiving C3I reference pulses. If there are no reference pulses, the ECM will de-energize the fuel pump relay within 2 seconds after key "ON" or the engine is stopped.

Test Description: Numbers below refer to circled numbers on the diagnostic chart.
6. Checks for open in the fuel pump relay ground. CKT 450.
7. Determines if the ECM is in control of the fuel pump relay through CKT 465 (terminal "A").
8. The fuel pump control circuit includes an engine oil pressure switch with a separate set of normally open contacts. The switch closes when 14 lbs/28 kPa of oil pressure and provides a second battery feed path to the fuel pump. If the relay fails, the pump will run using the battery feed supplied by the closed oil pressure switch.

This step checks the oil pressure switch to be sure it provides battery feed to the fuel pump relay should the pump relay fail. A failed oil pressure relay will result in extended engine crank time because of the time required to build enough oil pressure to close the oil pressure switch and turn "O" the fuel pump. There may be instances when the relay has failed but the engine will not crank fast enough to build enough oil pressure to close the switch. This or a faulty oil pressure switch can result in "Engine Cranks But Will Not Run."

1988–90 VEHICLES EQUIPPED WITH 3800 (VIN C) ENGINE EXCEPT E-CAR

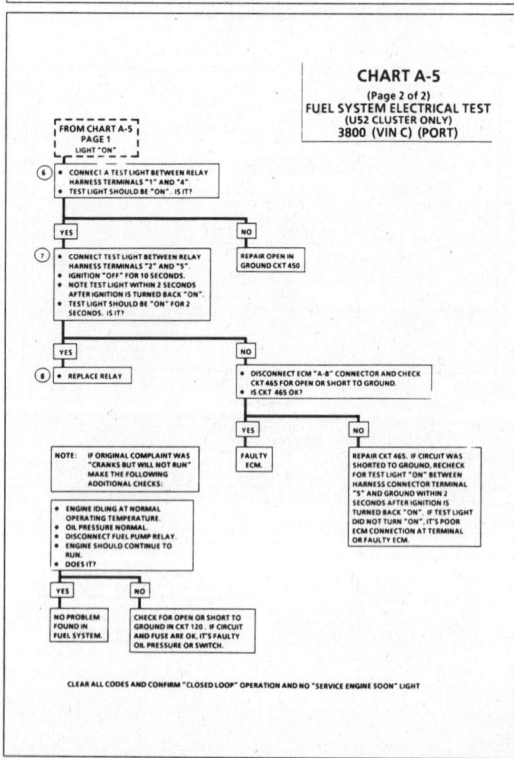

CHART A-5
(Page 2 of 2)
FUEL SYSTEM ELECTRICAL TEST
(U52 CLUSTER ONLY)
3800 (VIN C) (PORT)

CLEAR ALL CODES AND CONFIRM "CLOSED LOOP" OPERATION AND NO "SERVICE ENGINE SOON" LIGHT

1988–90 VEHICLES EQUIPPED WITH 3800 (VIN C) ENGINE EXCEPT E-CAR

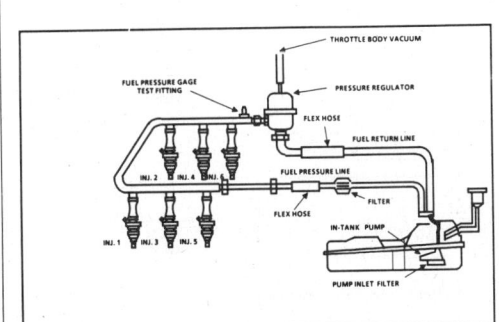

CHART A-7
(Page 1 of 2)
FUEL SYSTEM PRESSURE TEST
3800 (VIN C) (PORT)

Circuit Description:
The fuel pump will deliver fuel to the fuel rail and injectors, then to the pressure regulator, where the system pressure is controlled. Excess fuel pressure is bypassed back to the fuel tank. When the engine is stopped, the pump can be turned "ON" by applying battery voltage to the test terminal located in the engine compartment.

Improper fuel system pressure may contribute to one or all of the following symptoms:
- Cranks but won't run
- Code 44 or 45

Test Description: Numbers below refer to circled numbers on the diagnostic chart.

CAUTION: To reduce the risk of fire and personal injury, it is necessary to relieve the fuel system pressure before servicing fuel system components. To do this:

- Disconnect the fuel tank harness connector
- Crank engine - engine will start and run until fuel supply remaining in fuel pipes is consumed
- Engage starter for 3.0 seconds to assure relief of any remaining pressure
1. - Install pressure gage J-34730-1 to fuel pressure tap
   - Connect the fuel tank harness connector
   - Start engine. With ignition "ON", pump pressure is controlled by spring pressure and throttle body vacuum within the pressure regulator assembly.
   - Ignition "OFF" for 10 seconds. Pressure should not leak down after the fuel pump is shut "OFF".
2. When the engine is idling, the throttle body vacuum is high and is applied to the fuel regulator diaphragm. This will offset the spring and result in a lower fuel pressure.
3. The application of 12-14 inches of vacuum to the pressure regulator should result in a fuel pressure less than step 1.
4. Pressure that leaks down may be caused by one of the following:
   - In-tank fuel pump check valve not holding
   - Pump coupling hose leaking
   - Fuel pressure regulator valve leaking
   - Injector sticking open

1988–90 VEHICLES EQUIPPED WITH 3800 (VIN C) ENGINE EXCEPT E-CAR

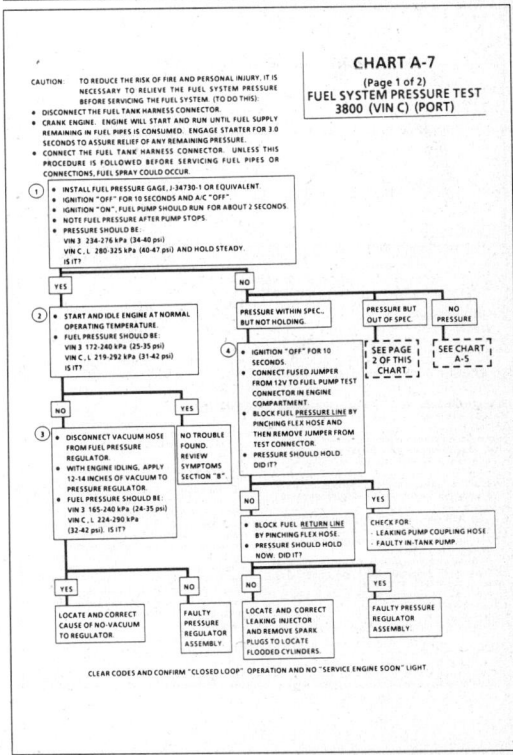

CHART A-7
(Page 1 of 2)
FUEL SYSTEM PRESSURE TEST
3800 (VIN C) (PORT)

CLEAR CODES AND CONFIRM "CLOSED LOOP" OPERATION AND NO "SERVICE ENGINE SOON" LIGHT

1988–90 VEHICLES EQUIPPED WITH 3800 (VIN C) ENGINE EXCEPT E-CAR

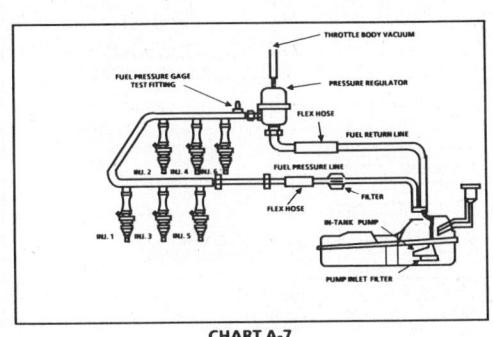

CHART A-7
(Page 2 of 2)
FUEL SYSTEM PRESSURE TEST
3800 (VIN C) (PORT)

Circuit Description:
The fuel pump will deliver fuel to the fuel rail and injectors, then to the pressure regulator, where the system pressure is controlled. Excess fuel pressure is bypassed back to the fuel tank. When the engine is stopped, the pump can be turned "ON" by applying battery voltage to the test terminal located in the engine compartment.

Improper fuel system pressure may contribute to one or all of the following symptoms:
- Cranks but won't run
- Code 44 or 45
- Cuts out, may feel like ignition problem
- Hesitation, loss of power, or poor fuel economy

Test Description: Numbers below refer to circled numbers on the diagnostic chart.
5. Pressure but less than specifications falls into two areas:
   - Regulated pressure but less than specifications: the amount of fuel to injectors is OK, but pressure is too low. The system will be lean running and may set Code 44. Also, possible hard starting cold and overall poor performance.
   - Restricted flow causing pressure drop. Normally, a vehicle with a fuel pressure of less than 165 kPa (24 psi) at idle will not be driveable.

However, if the pressure drop occurs only while driving, the engine will normally surge then stop as pressure begins to drop rapidly.
6. Restricting the fuel return line allows the fuel pump to develop its maximum pressure (dead head pressure). When battery voltage is applied to the pump test terminal, pressure should be above 517 kPa (75 psi).
7. This test determines if the high fuel pressure is due to a restricted fuel return line or a pressure regulator problem.

1988–90 VEHICLES EQUIPPED WITH 3800 (VIN C) ENGINE EXCEPT E-CAR

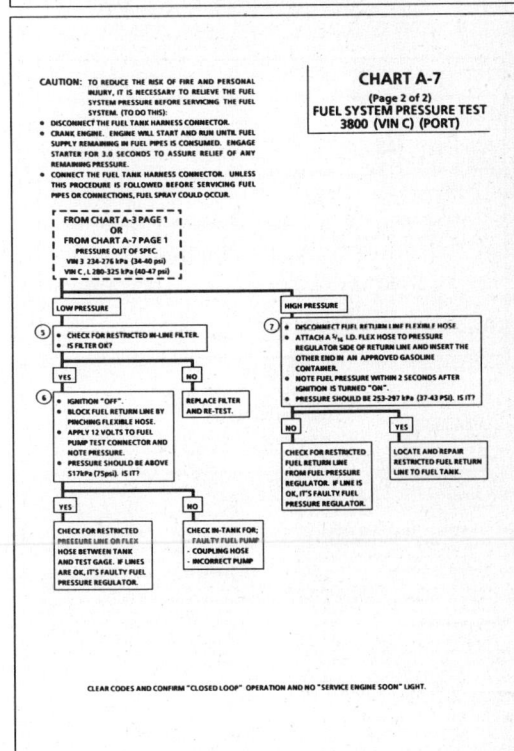

CHART A-7
(Page 2 of 2)
FUEL SYSTEM PRESSURE TEST
3800 (VIN C) (PORT)

CLEAR CODES AND CONFIRM "CLOSED LOOP" OPERATION AND NO "SERVICE ENGINE SOON" LIGHT

1988–90 VEHICLES EQUIPPED WITH 3800 (VIN C) ENGINE EXCEPT E-CAR

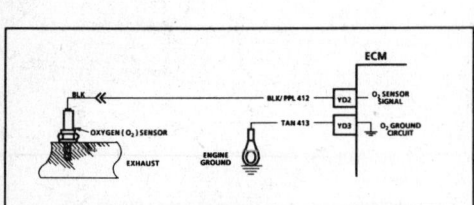

CODE 13
OXYGEN SENSOR CIRCUIT
(OPEN CIRCUIT)
3800 (VIN C) (PORT)

Circuit Description:
The ECM supplies a voltage of about .45 volt between terminals "YD2" and "YD3". (If measured with a 10 megohm digital voltmeter, this may read as low as .32 volt.) The O₂ sensor varies the voltage within a range of about 1 volt if the exhaust is rich, down through about .10 volt if exhaust is lean.
The sensor is like an open circuit and produces no voltage when it is below 360° C (600°F). An open oxygen sensor circuit or cold oxygen sensor causes "Open Loop" operation.

Test Description: Numbers below refer to circled numbers on the diagnostic chart.
1. Code 13 will set if:
   - Engine at normal operating temperature
   - Engine run time more than 40 seconds
   - O₂ signal voltage is steady between .35 and .55 volt
   - Throttle position sensor signal above .55 volt
   - All conditions must be met for about 30 seconds.
   If the conditions for a Code 13 exists, the system will not go to "Closed Loop".
2. This will determine if the sensor or the wiring is the cause of the Code 13.
3. In doing this test use only a high impedence digital volt ohmmeter. This test checks the continuity of CKTs 412 and 413. If CKT 413 is open the ECM voltage on CKT 412 will be over .6 volt (600 mV).

Diagnostic Aids:
An intermittent may be caused by a poor connection, rubbed through wire insulation, or a wire broken inside the insulation.
Check For:
- **Poor Connection or Damaged Harness** Inspect harness connectors for backed out terminals, improper mating, broken locks, improperly formed or damaged terminals, poor terminal to wire connection, and damaged harness.
- **Intermittent Test** If connections and harness check OK, "Scan" O₂ sensor voltage while moving related connectors and wiring harness, with warm engine running at part throttle in "Closed Loop". If the failure is induced, the "O₂ sensor" reading will change from its normal fluctuating voltage (above 600 mV and below 300 mV) to a fixed value around 450 mV. This may help to isolate the location of the malfunction.

1988–90 VEHICLES EQUIPPED WITH 3800 (VIN C) ENGINE EXCEPT E-CAR

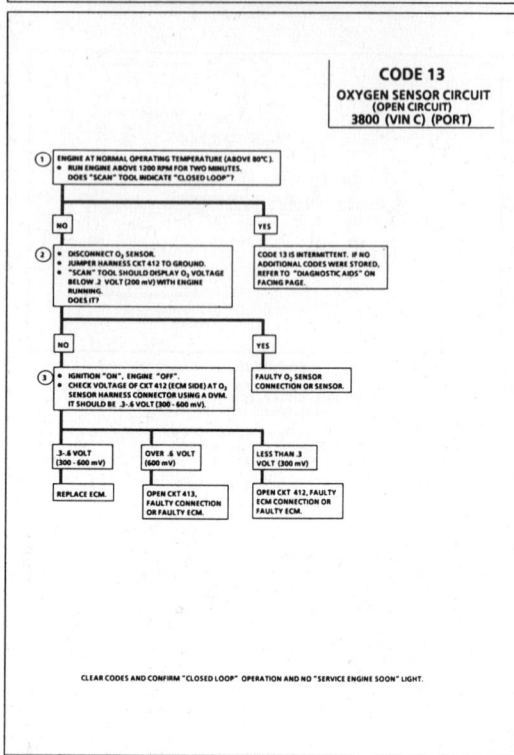

1988–90 VEHICLES EQUIPPED WITH 3800 (VIN C) ENGINE EXCEPT E-CAR

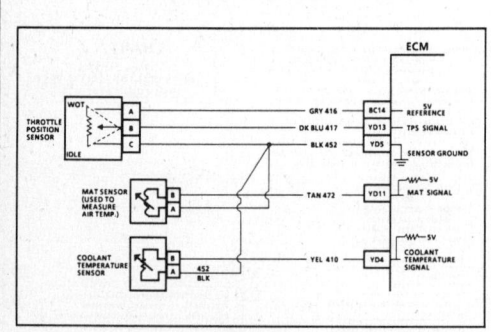
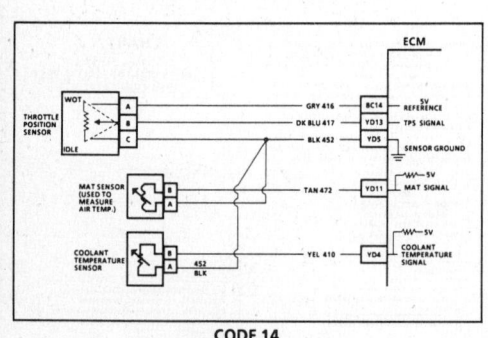

CODE 14
COOLANT TEMPERATURE SENSOR CIRCUIT
(HIGH TEMPERATURE INDICATED)
3800 (VIN C) (PORT)

Circuit Description:
The coolant temperature sensor uses a thermistor to control the signal voltage to the ECM. The ECM applies a voltage on CKT 410 to the sensor. When the engine is cold the sensor (thermistor) resistance is high, therefore, the ECM will see high signal voltage.
As the engine warms, the sensor resistance becomes less, and the voltage drops. At normal engine operating temperature (85°C to 95°C), the voltage will measure about 1.5 to 2.0 volts.

Test Description: Numbers below refer to circled numbers on the diagnostic chart.
1. Code 14 will set if:
   - Signal voltage indicates a coolant temperature above 140°C (284°F) for 4 seconds.
2. This test will determine if CKT 410 is shorted to ground which will cause the conditions for Code 14.
   If Code 14 is set, the ECM will use a default coolant temperature value of 48.5°C, 119°F, for fuel control.

Diagnostic Aids:
"Scan" tool displays engine temperature in degrees centigrade. After engine is started, the temperature should rise steadily to about 90°C then stabilize when thermostat opens.
An intermittent may be caused by a poor connection, rubbed through wire insulation, or a wire broken inside the insulation.

Check For:
- **Poor Connection or Damaged Harness** Inspect ECM harness connectors for backed out terminal "YD4", improper mating, broken locks, improperly formed or damaged terminals, poor terminal to wire connection, and damaged harness.
- **Intermittent Test** If connections and harness check OK, "Scan" coolant temperature while moving related connectors and wiring harness. If the failure is induced, the "coolant temperature" display will change. This may help to isolate the location of the malfunction.
- **Shifted Sensor** The "Temperature To Resistance Value" scale may be used to test the coolant sensor at various temperature levels to evaluate the possibility of a "shifted" (mis-scaled) sensor, which may result in driveability complaints.

1988–90 VEHICLES EQUIPPED WITH 3800 (VIN C) ENGINE EXCEPT E-CAR

CODE 14
COOLANT TEMPERATURE SENSOR CIRCUIT
(HIGH TEMPERATURE INDICATED)
3800 (VIN C) (PORT)

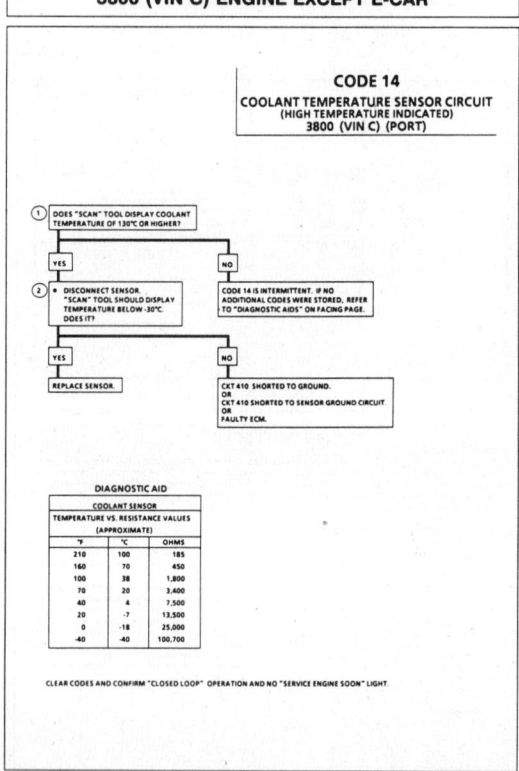

DIAGNOSTIC AID

COOLANT SENSOR TEMPERATURE VS. RESISTANCE VALUES (APPROXIMATE)		
°F	°C	OHMS
210	100	185
160	70	450
100	38	1,800
70	20	3,400
40	4	7,500
20	-7	13,500
0	-18	25,000
-40	-40	100,700

CLEAR CODES AND CONFIRM "CLOSED LOOP" OPERATION AND NO "SERVICE ENGINE SOON" LIGHT.

1988–90 VEHICLES EQUIPPED WITH 3800 (VIN C) ENGINE EXCEPT E-CAR

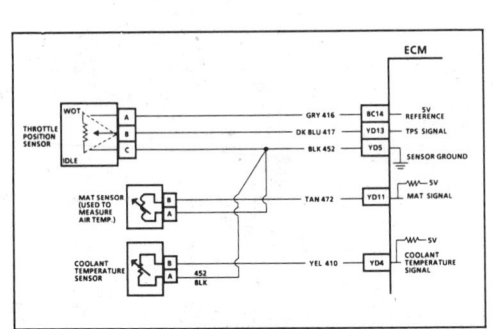

CODE 15
COOLANT TEMPERATURE SENSOR CIRCUIT
(LOW TEMPERATURE INDICATED)
3800 (VIN C) (PORT)

Circuit Description:

The coolant temperature sensor uses a thermistor to control the signal voltage to the ECM. The ECM applies a voltage on CKT 410 to the sensor. When the engine is cold, the sensor (thermistor) resistance is high, therefore, the ECM will see high signal voltage.

As the engine warms, the sensor resistance becomes less, and the voltage drops. At normal engine operating temperature (85°C to 95°C) the voltage will measure about 1.5 to 2.0 volts at the ECM.

Test Description: Numbers below refer to circled numbers on the diagnostic chart.

1. Code 15 will set if:
   - Engine is running
   - Signal voltage indicates a coolant temperature less than -40°C (-40°F) for at least 4 seconds.
2. This test simulates a Code 14. If the ECM recognizes the low signal voltage, (high temperature) and the "Scan" reads 130°C, the ECM and wiring are OK.
3. This test will determine if CKT 410 is open. There should be 5 volts present at sensor connector if measured with a DVM.

Note: If Code 15 is set, the ECM will use 48.5°C (119°F) for fuel control.

Diagnostic Aids:

A "Scan" tool reads engine temperature in degrees centigrade. After engine is started the temperature should rise steadily to about 90°C then stabilize when thermostat opens.

An intermittent may be caused by a poor connection, rubbed through wire insulation or a wire broken inside the insulation.

Check For:
- **Poor Connection or Damaged Harness** Inspect ECM harness connectors for backed out terminal "YD4", improper mating, broken locks, improperly formed or damaged terminals, poor terminal to wire connection and damaged harness.
- **Intermittent Test** If connections and harness check OK, "Scan" coolant temperature while moving related connectors and wiring harness. If the failure is induced, the display will change. This may help to isolate the location of the malfunction.
- **Shifted Sensor** The "Temperature To Resistance Value" scale may be used to test the coolant sensor at various temperature levels to evaluate the possibility of a "shifted" (mis-scaled) sensor which may result in driveability complaints.

A faulty connection, or an open in CKTs 410 or 452 will result in a Code 15. If Code 23 is also set, check CKT 452 for faulty wiring or connections. Check terminals at sensor for good contact.

1988–90 VEHICLES EQUIPPED WITH 3800 (VIN C) ENGINE EXCEPT E-CAR

CODE 15
COOLANT TEMPERATURE SENSOR CIRCUIT
(LOW TEMPERATURE INDICATED)
3800 (VIN C) (PORT)

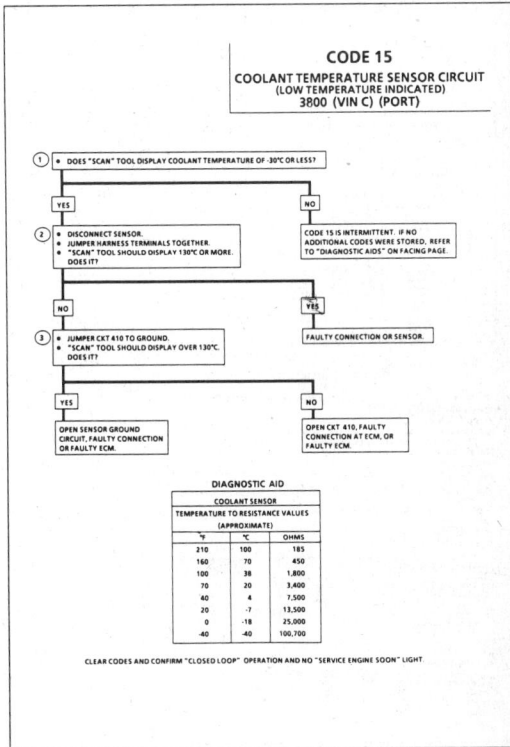

DIAGNOSTIC AID		
COOLANT SENSOR		
TEMPERATURE TO RESISTANCE VALUES (APPROXIMATE)		
°F	°C	OHMS
210	100	185
160	70	450
100	38	1,800
70	20	3,400
40	4	7,500
20	-7	13,500
0	-18	25,000
-40	-40	100,700

CLEAR CODES AND CONFIRM "CLOSED LOOP" OPERATION AND NO "SERVICE ENGINE SOON" LIGHT.

1988–90 VEHICLES EQUIPPED WITH 3800 (VIN C) ENGINE EXCEPT E-CAR

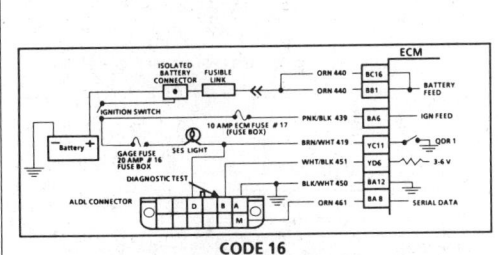

CODE 16
SYSTEM VOLTAGE HIGH
3800 (VIN C) (PORT)

Circuit Description:

The ECM monitors battery or system voltage on CKT 440 to terminals "BB1" and "BC16". If the ECM detects voltage above 16 volts for more than 10 seconds, it will turn the SES light "ON" and set Code 16 in memory.

Test Description: Numbers below refer to circled numbers on the diagnostic chart.

1. Test generator output as outlined in "6D3" to determine proper operation of the voltage regulator. Run engine at moderate speed and measure voltage across the battery. If over 16 volts, repair generator as outlined in "6D3".

Diagnostic Aids:

An intermittent may be caused by a poor connection, rubbed through insulation, a wire broken inside the insulation or poor ECM grounds.

Check For:
- **Poor Connection or Damaged Harness** Inspect ECM harness connectors for backed out terminal "BC16" or "BB1," improper mating, broken locks, improperly formed or damaged terminals, poor terminal to wire connection and damaged harness.
- **Intermittent Test** If connections and harness checks OK, monitor battery voltage display while moving related connectors. If the failure is induced, the battery voltage will abruptly change. This may help to isolate the location of the malfunction. An engine stall while manipulating the harness indicates that the ECM has lost voltage at terminal "BC16" or "BB1." Check for loose connectors in CKT 440.

Note: Charging battery with a battery charger and starting the engine may set a Code 16.

1988–90 VEHICLES EQUIPPED WITH 3800 (VIN C) ENGINE EXCEPT E-CAR

CODE 16
SYSTEM VOLTAGE HIGH
3800 (VIN C) (PORT)

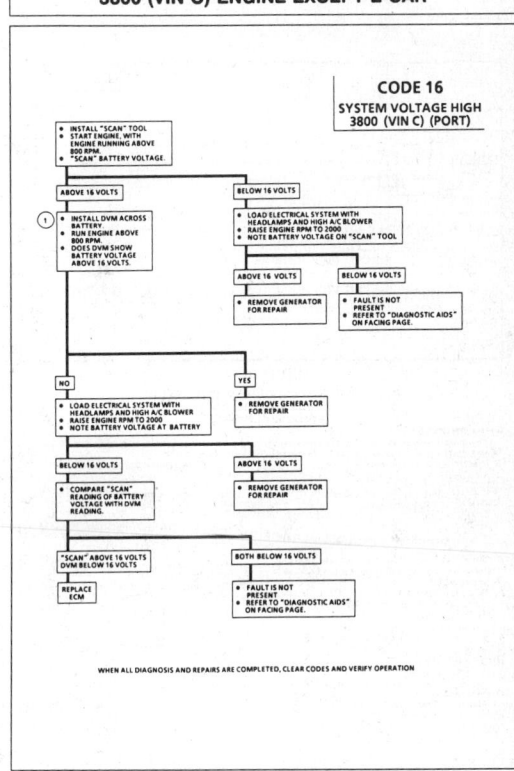

WHEN ALL DIAGNOSIS AND REPAIRS ARE COMPLETED, CLEAR CODES AND VERIFY OPERATION

1988–90 VEHICLES EQUIPPED WITH 3800 (VIN C) ENGINE EXCEPT E-CAR

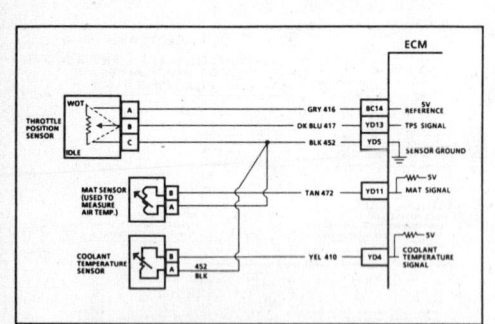

CODE 21
THROTTLE POSITION SENSOR (TPS) CIRCUIT
(SIGNAL VOLTAGE HIGH)
3800 (VIN C) (PORT)

Circuit Description:
The throttle position sensor (TPS) provides a voltage signal that changes relative to throttle blade angle. Signal voltage will vary from about .4 at idle to about 5 volts at wide open throttle.
The TPS signal is one of the most important inputs used by the ECM for fuel control and for most of the ECM control outputs.

Test Description: Numbers below refer to circled numbers on the diagnostic chart.
1. Code 21 will set if:
   - TPS voltage is greater than 4.9 volts at any time.
   - Engine is running and air flow is less than 15 gm/sec
   - TPS signal voltage is greater than 2.5 volts
   - Code 34 not present
   - All conditions met for 5 seconds
   With closed throttle, ignition "ON", or at idle, voltage at YD13 should be .38-.42 volt.
2. When the TPS sensor is disconnected, the TPS voltage will go low and a Code 22 will set. Therefore, the ECM and wiring are OK.
3. Probing CKT 452 with a test light checks the sensor ground CKT. A faulty sensor ground circuit will cause a Code 21.
Note: If a Code 21 is set, the ECM will use a defaulted value for TPS of about .5 volt.

Diagnostic Aids:
A "Scan" tool reads throttle position in volts. With closed throttle, ignition "ON" or at idle, voltage should be .38-.42 volt.

Also some "Scan" tools will read throttle angle 0% = closed throttle 100% = WOT.
An open in CKT 452 will result in a Code 21. Check For:
- Poor Connection or Damaged Harness Inspect ECM harness connectors for backed out terminal "YD13", improper mating, broken locks, improperly formed or damaged terminals, poor terminal to wire connection, and damaged harness.
- Intermittent Test If connections and harness check OK, monitor TPS voltage while moving related connectors and wiring harness. If the failure is induced, the display will change. This may help to isolate the location of the malfunction.
- TPS Scaling Observe TPS voltage display while depressing accelerator pedal with engine stopped and ignition "ON". Display should vary from closed throttle TPS voltage when throttle was closed, to over (4.5 volts) 4500 mV when throttle is held at wide open throttle position.

1988–90 VEHICLES EQUIPPED WITH 3800 (VIN C) ENGINE EXCEPT E-CAR

CODE 21
THROTTLE POSITION SENSOR (TPS) CIRCUIT
(SIGNAL VOLTAGE HIGH)
3800 (VIN C) (PORT)

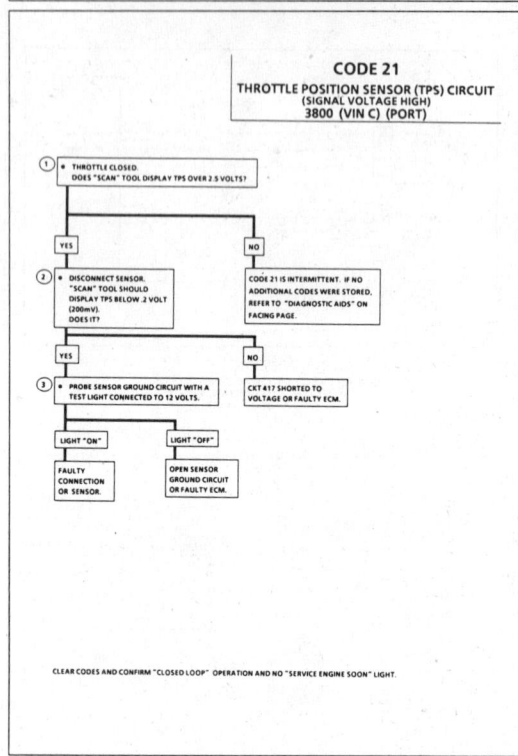

CLEAR CODES AND CONFIRM "CLOSED LOOP" OPERATION AND NO "SERVICE ENGINE SOON" LIGHT.

1988–90 VEHICLES EQUIPPED WITH 3800 (VIN C) ENGINE EXCEPT E-CAR

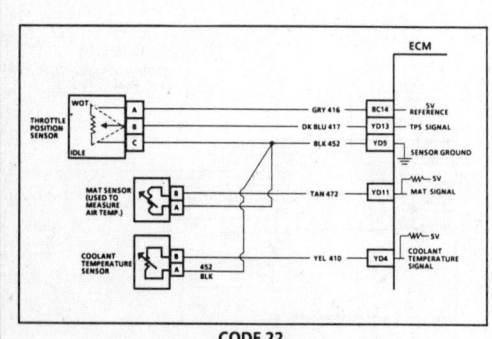

CODE 22
THROTTLE POSITION SENSOR (TPS) CIRCUIT
(SIGNAL VOLTAGE LOW)
3800 (VIN C) (PORT)

Circuit Description:
The throttle position sensor (TPS) provides a voltage signal that changes relative to throttle blade angle. Signal voltage will vary from about .4 at idle to about 5 volts at wide open throttle.
The TPS signal is one of the most important inputs used by the ECM for fuel control and for most of the ECM control outputs.

Test Description: Numbers below refer to circled numbers on the diagnostic chart.
1. Code 22 will set if:
   - The ignition key is "ON"
   - TPS signal voltage is less than .1 volt for 4 seconds
2. Simulates Code 21: (high voltage) If ECM recognizes the high signal voltage the ECM and wiring are OK.
3. With closed throttle, ignition "ON" or at idle, voltage at "YD13" should be .38-.42 volt. If not, check adjustment.
4. Simulates a high signal voltage. Checks CKT 417 for an open.

Diagnostic Aids:
A "Scan" tool reads throttle position in volts. Voltage should increase at a steady rate as throttle is moved toward WOT.

Also some "Scan" tools will read throttle angle 0% = closed throttle 100% = WOT.

An open or short to ground in CKTs 416 or 417 will result in a Code 22.
Check For:
- Poor Connection or Damaged Harness Inspect ECM harness connectors for backed out terminal "YD13", improper mating, broken locks, improperly formed or damaged terminals, and damaged harness.
- Intermittent Test If connections and harness check OK, monitor TPS voltage display while moving related connectors and wiring harness. If the failure is induced, the display will change. This may help to isolate the location of the malfunction.
- TPS Scaling Observe TPS voltage display while depressing accelerator pedal with engine stopped and ignition "ON". Display should vary from closed throttle TPS voltage when throttle was closed, to over 4.5 volts (4500 mV) when throttle is held at wide open throttle position.

1988–90 VEHICLES EQUIPPED WITH 3800 (VIN C) ENGINE EXCEPT E-CAR

CODE 22
THROTTLE POSITION SENSOR (TPS) CIRCUIT
(SIGNAL VOLTAGE LOW)
3800 (VIN C) (PORT)

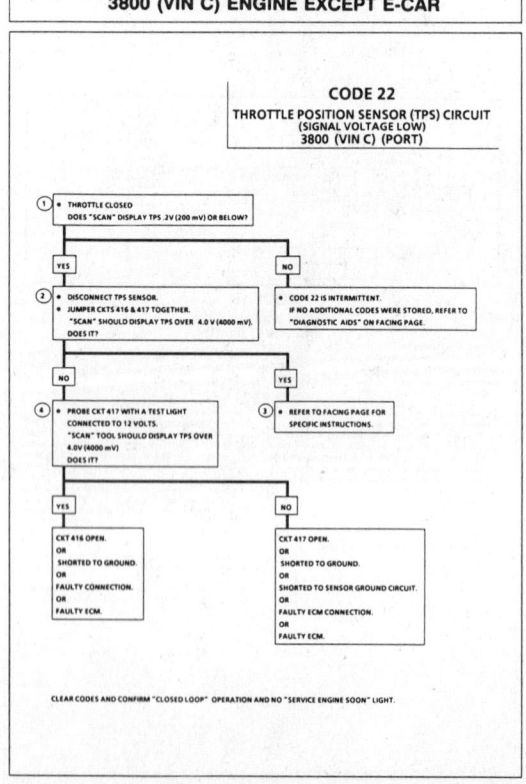

CLEAR CODES AND CONFIRM "CLOSED LOOP" OPERATION AND NO "SERVICE ENGINE SOON" LIGHT.

1988–90 VEHICLES EQUIPPED WITH 3800 (VIN C) ENGINE EXCEPT E-CAR

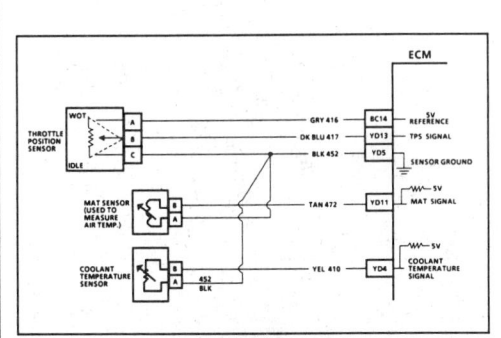

CODE 23
MANIFOLD AIR TEMPERATURE (MAT) SENSOR CIRCUIT
(LOW TEMPERATURE INDICATED)
3800 (VIN C) (PORT)

Circuit Description:

The MAT sensor is a thermistor. The ECM applies a voltage (about 5 volts) on CKT 472 to the sensor. When the air is cold, the sensor (thermistor) resistance is high, therefore, the ECM will measure a high signal voltage. If the air is warm, the sensor resistance is low, therefore, the ECM will measure a low voltage.

Test Description: Numbers below refer to circled numbers on the diagnostic chart.

Code 23 will set if:
- A signal voltage indicates a manifold air temperature below −40°C (-40°F) for 4 seconds

Due to the conditions necessary to set a Code 23, the "Service Engine Soon" light will only stay "ON" while the fault is present.

1. A "Scan" tool may not be used to diagnose this fault, due to the ECM transmitting "default" (substitute) values when the fault is present. A Code 23 will set, due to an open sensor, wire, or connection. This test determines if the wiring and ECM are OK.
2. If the resistance is greater than 25,000 ohms, inspect the air cleaner assembly for the presence of ice. If OK, replace the sensor.

1988–90 VEHICLES EQUIPPED WITH 3800 (VIN C) ENGINE EXCEPT E-CAR

CODE 23
MANIFOLD AIR TEMPERATURE (MAT) SENSOR CIRCUIT
(LOW TEMPERATURE INDICATED)
3800 (VIN C) (PORT)

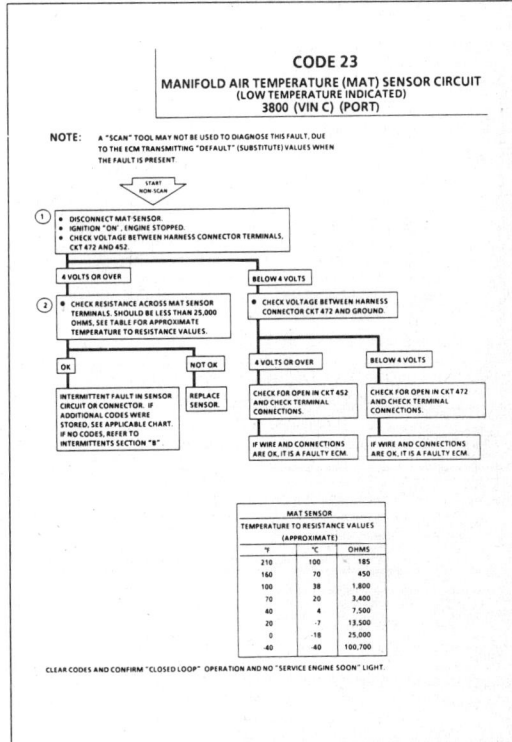

MAT SENSOR		
TEMPERATURE TO RESISTANCE VALUES		
(APPROXIMATE)		
°F	°C	OHMS
210	100	185
160	70	450
100	38	1,800
70	20	3,400
40	4	7,500
20	-7	13,500
0	-18	25,000
-40	-40	100,700

CLEAR CODES AND CONFIRM "CLOSED LOOP" OPERATION AND NO "SERVICE ENGINE SOON" LIGHT.

1988–90 VEHICLES EQUIPPED WITH 3800 (VIN C) ENGINE EXCEPT E-CAR

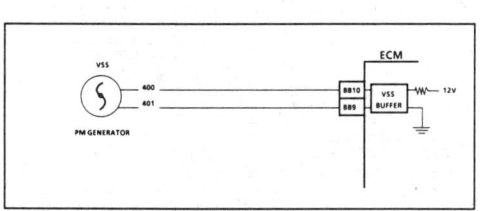

CODE 24
VEHICLE SPEED SENSOR (VSS) CIRCUIT
3800 (VIN C) (PORT)

Circuit Description:

Vehicle speed information is provided to the ECM by the vehicle speed sensor, a permanent magnet (PM) generator mounted in the transmission. The PM generator produces a pulsing voltage whenever vehicle speed is over about 3 mph. The A/C voltage level and the number of pulses increases with vehicle speed. The ECM converts the pulsing voltage to mph, and the mph can be displayed with a "Scan" tool.

The function of the VSS buffer, used in past model years, has been incorporated into the ECM. The ECM supplies the necessary signal for the instrument panel (4004 pulses per mile) for operating the speedometer and the odometer.

Test Description: Numbers below refer to circled numbers on the diagnostic chart.

1. Code 24 will set if vehicle speed signal equals less than 3 mph when:
   - Engine is running
   - No Code 29 or 31
   - When the vehicle is in 4th gear
   - All conditions met for 40 seconds

   The PM generator only produces a signal if drive wheels are turning greater than 3 mph.
2. Before replacing the ECM, check the Mem-Cal for correct application.

Diagnostic Aids:

"Scan" should indicate a vehicle speed whenever the drive wheels are turning greater than 3 mph.

Check CKT 400 and 401 for proper connections to be sure they are clean and tight and the harness is routed correctly. Refer to "Intermittents" in Section "B".

1988–90 VEHICLES EQUIPPED WITH 3800 (VIN C) ENGINE EXCEPT E-CAR

CODE 24
VEHICLE SPEED SENSOR (VSS) CIRCUIT
3800 (VIN C) (PORT)

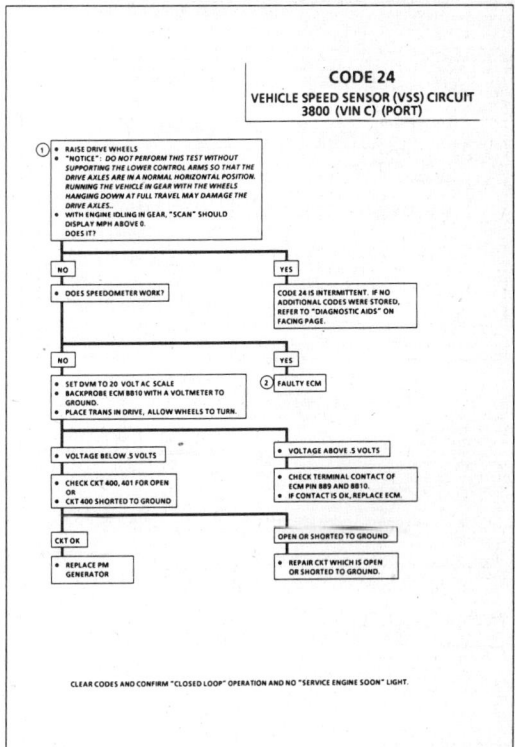

CLEAR CODES AND CONFIRM "CLOSED LOOP" OPERATION AND NO "SERVICE ENGINE SOON" LIGHT.

1988–90 VEHICLES EQUIPPED WITH 3800 (VIN C) ENGINE EXCEPT E-CAR

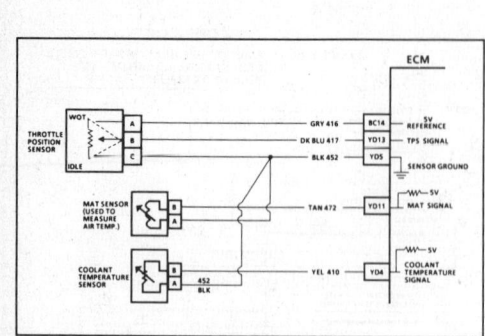

CODE 25
MANIFOLD AIR TEMPERATURE (MAT) SENSOR CIRCUIT
(HIGH TEMPERATURE INDICATED)
3800 (VIN C) (PORT)

Circuit Description:
The MAT sensor is a thermistor. The ECM applies a voltage (about 5 volts) on CKT 472 to the sensor. When air is cold, the sensor (thermistor) resistance is high, therefore, the ECM will measure a high signal voltage. If the air is warm, the sensor resistance is low, and the ECM will measure a low voltage.

Test Description: Numbers below refer to circled numbers on the diagnostic chart.
Code 25 will set if:
- Signal voltage indicates a manifold air temperature greater than 135°C (275°F).
- Vehicle speed is greater than 35 mph.
- Both of the above requirements are met for at least 16 seconds.

Due to the conditions necessary to set a Code 25, the "Service Engine Soon" light will only stay "ON" while the fault is present.

1. A "Scan" tool may not be used to diagnose this fault due to the ECM transmitting "default" (substitute) values when the fault is present. If voltage is above 4 volts, the ECM and wiring are OK.
2. If the resistance is less than 185 ohms, replace the sensor.

1988–90 VEHICLES EQUIPPED WITH 3800 (VIN C) ENGINE EXCEPT E-CAR

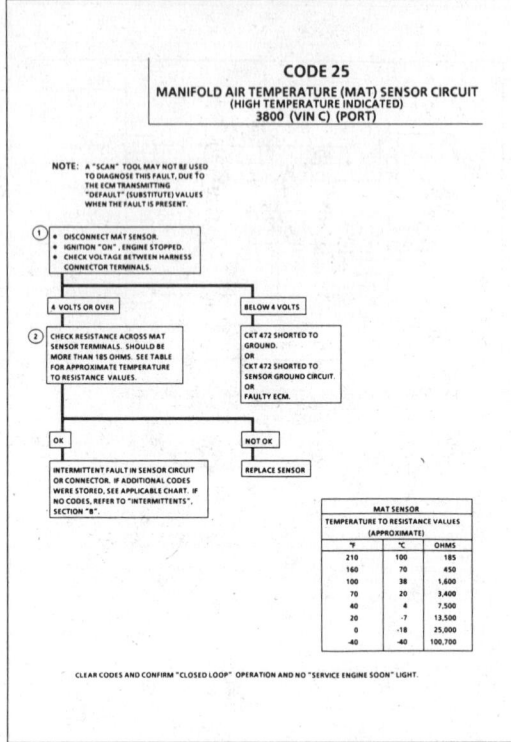

CODE 25
MANIFOLD AIR TEMPERATURE (MAT) SENSOR CIRCUIT
(HIGH TEMPERATURE INDICATED)
3800 (VIN C) (PORT)

MAT SENSOR
TEMPERATURE TO RESISTANCE VALUES (APPROXIMATE)

°F	°C	OHMS
210	100	185
160	70	450
100	38	1,600
70	20	3,400
40	4	7,500
20	-7	13,500
0	-18	25,000
-40	-40	100,700

CLEAR CODES AND CONFIRM "CLOSED LOOP" OPERATION AND NO "SERVICE ENGINE SOON" LIGHT.

1988–90 VEHICLES EQUIPPED WITH 3800 (VIN C) ENGINE EXCEPT E-CAR

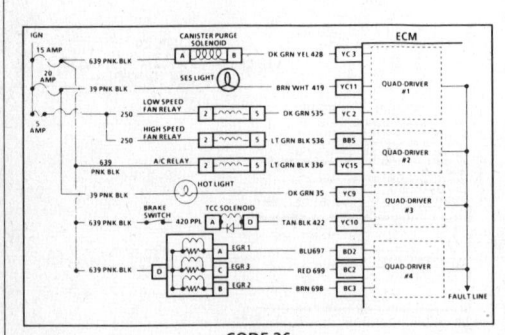

CODE 26
(Page 1 of 3)
QUAD-DRIVER (QDM) CIRCUIT
3800 (VIN C) (PORT)

Circuit Description:
The ECM is used to control several components such as these illustrated above. The ECM controls these devices through the use of a quad driver module (QDM). When the ECM is commanding a component "ON", the voltage potential of the output circuit will be "low" (near 0 volts). When the ECM is commanding the output circuit to a component "OFF", the voltage potential of the circuit will be "high" (near battery voltage). The primary function of the QDM is to supply the ground for the component being controlled.

Each QDM has a fault line which is monitored by the ECM. The fault line signal is available on the data stream for "Scan" tool test equipment. The ECM will compare the voltage at the QDM based on accepted values of the fault line. If the QDM fault detection circuit senses a voltage other than the accepted value, the fault line will go from a "low" signal to a "high" signal on the data stream and a Code 26 will set if applicable.

Some QDM circuits will switch from "low" to "high" normally. Examples: QDM 2 - If A/C pressure switch closes and turns "ON" high speed coolant fan or QDM3 - If the brake is depressed. These conditions are normal and the code 26 is set. These are accepted conditions. A fault on QDM #2 will not set a Code 26, diagnosis may be done by making sure the A/C is turned "OFF".

Test Description: Numbers below refer to the diagnostic chart.
1. The ECM does not know which controlled circuit caused the Code 26 so this chart will go through each of the circuits to determine which is at fault.
This test checks the "Service Engine Soon" light driver and the "Service Engine Soon" light circuit.

2. QDM symptoms
TCC Inoperative - Code 39
EGR Inoperative - Codes 63, 64, 65
Hot Light - "ON" all the time/"OFF" during bulb check.
Coolant fan on low speed all the time or won't come "ON" at all.
Poor driveability due to 100% canister purge.

1988–90 VEHICLES EQUIPPED WITH 3800 (VIN C) ENGINE EXCEPT E-CAR

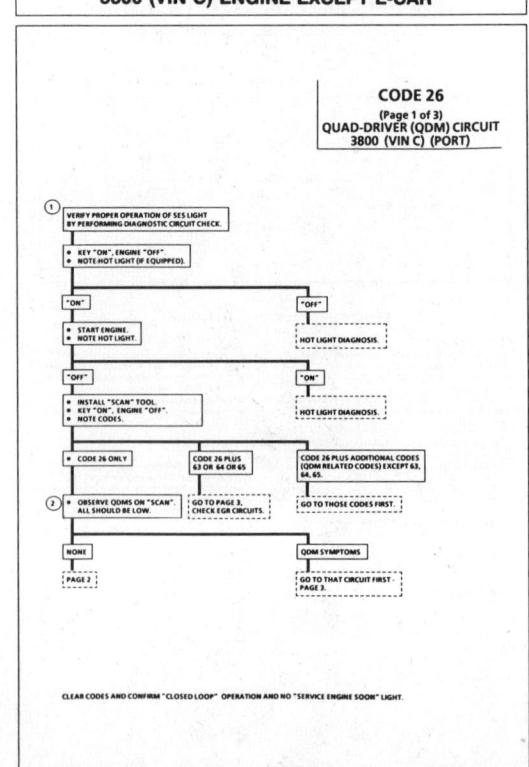

CODE 26
(Page 1 of 3)
QUAD-DRIVER (QDM) CIRCUIT
3800 (VIN C) (PORT)

CLEAR CODES AND CONFIRM "CLOSED LOOP" OPERATION AND NO "SERVICE ENGINE SOON" LIGHT.

1988–90 VEHICLES EQUIPPED WITH 3800 (VIN C) ENGINE EXCEPT E-CAR

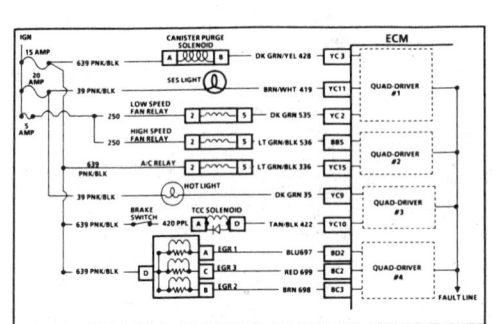

CODE 26
(Page 2 of 3)
QUAD-DRIVER (QDM) CIRCUIT
3800 (VIN C) (PORT)

Circuit Description:

The ECM is used to control several components such as those illustrated above. The ECM controls these devices through the use of a quad driver module (QDM). When the ECM is commanding a component "ON", the voltage potential of the output circuit will be "low" (near 0 volts). When the ECM is commanding the output circuit to a component "OFF", the voltage potential of the circuit will be "high" (near battery voltage). The primary function of the QDM is to supply the ground for the component being controlled.

Each QDM has a fault line which is monitored by the ECM. The fault line signal is available on the data stream for "Scan" tool test equipment. The ECM will compare the voltage at the QDM based on accepted values of the fault line. If the QDM fault detection circuit senses a voltage other than the accepted value, the fault line will go from a "low" signal on the data stream to a "high" signal and a Code 26 will be set if applicable.

Some QDM circuits will switch from "low" to "high" normally. Examples: QDM 2 - If A/C pressure switch closes and turns "ON" high speed coolant fan or QDM3 - If the brake is depressed. These conditions are normal and no code 26 is set. These are accepted conditions. A fault on QDM #2 will not set a Code 26, diagnosis may be done by making sure the A/C is turned "OFF".

Test Description: Numbers below refer to circled numbers on the diagnostic chart.

3. This test will determine which circuit is out of specifications. All circuits EXCEPT "YC11", the "hot light" should be B + when key is "ON", engine not running. The diagnostic test terminal is not grounded.

Diagnostic Aids:

Monitor the voltage at each terminal while moving related harness connectors, including ECM harness. If the failure is induced, the voltage will change. This may help locate the intermittent. Check for bent pins at ECM and ECM connector terminals. If code re-occurs with no apparent connector problem, replace ECM.

1988–90 VEHICLES EQUIPPED WITH 3800 (VIN C) ENGINE EXCEPT E-CAR

CODE 26
(Page 2 of 3)
QUAD-DRIVER (QDM) CIRCUIT
3800 (VIN C) (PORT)

CIRCUIT NOT ISOLATED BY PRIOR STEPS (VOLTAGE TEST)

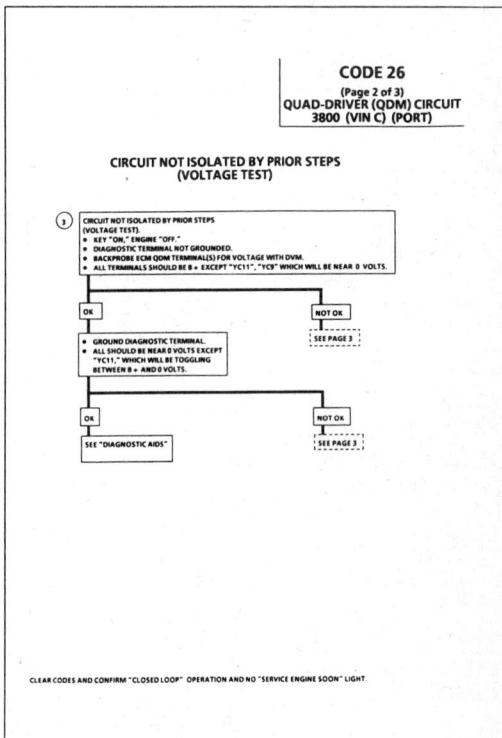

CLEAR CODES AND CONFIRM "CLOSED LOOP" OPERATION AND NO "SERVICE ENGINE SOON" LIGHT.

1988–90 VEHICLES EQUIPPED WITH 3800 (VIN C) ENGINE EXCEPT E-CAR

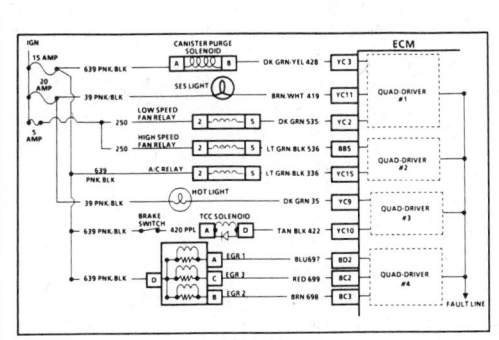

CODE 26
(Page 3 of 3)
QUAD-DRIVER (QDM) CIRCUIT
3800 (VIN C) (PORT)

Circuit Description:

The ECM is used to control several components such as those illustrated above. The ECM controls these devices through the use of a quad driver module (QDM). When the ECM is commanding a component "ON", the voltage potential of the output circuit will be "low" (near 0 volts). When the ECM is commanding the output circuit to a component "OFF", the voltage potential of the circuit will be "high" (near battery voltage). The primary function of the QDM is to supply the ground for the component being controlled.

Each QDM has a fault line which is monitored by the ECM. The fault line signal is available on the data stream for "Scan" tool test equipment. The ECM will compare the voltage at the QDM based on accepted values of the fault line. If the QDM fault detection circuit senses a voltage other than the accepted value, the fault line will go from a "low" signal on the data stream to a "high" signal and a Code 26 will be set if applicable.

Some QDM circuits will switch from "low" to "high" normally. Examples: QDM 2 - If A/C pressure switch closes and turns "ON" high speed coolant fan or QDM3 - If the brake is depressed. These conditions are normal and no code 26 is set. These are accepted conditions. A fault on QDM #2 will not set a Code 26, diagnosis may be done by making sure the A/C is turned "OFF".

Test Description: Numbers below refer to circled numbers on the diagnostic chart.

4. This test will determine if the problem is the circuit or the component. As the factory installed ECM is protected with an internal fuse, it is highly unlikely that the ECM needs to be replaced.

1988–90 VEHICLES EQUIPPED WITH 3800 (VIN C) ENGINE EXCEPT E-CAR

CODE 26
(Page 3 of 3)
QUAD-DRIVER (QDM) CIRCUIT
3800 (VIN C) (PORT)

CIRCUIT ISOLATED FROM PRIOR CHARTS

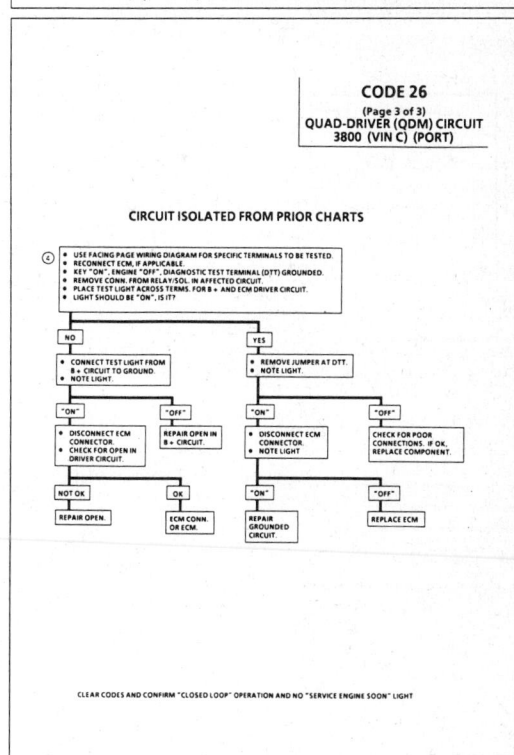

CLEAR CODES AND CONFIRM "CLOSED LOOP" OPERATION AND NO "SERVICE ENGINE SOON" LIGHT.

1988–90 VEHICLES EQUIPPED WITH 3800 (VIN C) ENGINE EXCEPT E-CAR

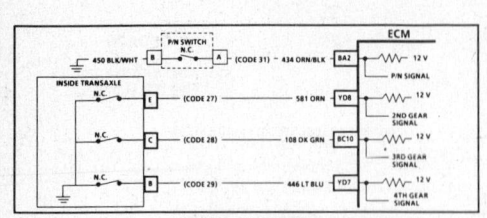

CODE 27, 28, 29
GEAR SWITCH DIAGNOSIS
3800 (VIN C) (PORT)

Circuit Description:
The gear switches are located inside the transaxle. They are pressure operated switches, normally closed. The ECM supplies 12 volts through each selected circuit to the switch. In any condition other than the specific gear application, the signal line monitors low voltage, low potential. As road speed increases, hydraulic pressure applies the specific gear clutches and the gear switch opens. At this time, the ECM monitors a high, 12 volts potential, and interprets this to indicate that gear is applied. The ECM uses the gear signals to control fuel delivery (and TCC).

Test Description: Numbers below refer to circled numbers on the diagnostic chart.
Code 27 will set if:
- No Code 29 present
- CKT 581 indicates ground or closed switch for 10 seconds when vehicle is in 4th gear operation.
- CKT 581 indicates drive (drive) when the engine is first started.
Code 28 will set if:
- CKT 108 indicates ground or closed switch for 10 seconds when vehicle is in 4th gear operation.
- CKT 108 indicates an open (drive) when the engine is first started.
Code 29 will set if:
- CKT 446 indicates ground or closed switch for 10 seconds when vehicle is in 4th gear operation.
- CKT 446 indicates an open (drive) when the engine is first started.
1. Must use a DVM. A test light will not light due to the very low current being supplied by the ECM.
2. Checks to see if CKT is grounded through the switch.
3. Checks for a good, properly operating switch and checks CKT within transaxle for an improper ground.

Diagnostic Aids:
An intermittent may be caused by a poor connection, mis-routed harness, rubbed through wire insulation, or a wire broken inside the insulation.
Check For:
- <u>Poor Connection</u> at ECM pins. Inspect harness connectors for backed out terminals, improper mating, broken locks, improperly formed or damaged terminals, and poor terminal to wire connection.
- <u>Mis-routed Harness</u> Inspect wiring harness to insure that it is not too close to high voltage wires, such as spark plug leads.
- <u>Damaged Harness</u> Inspect harness for damage. If harness appears OK, "Scan" while moving related connectors and wiring harness. A change in display would indicate the intermittent fault location.

1988–90 VEHICLES EQUIPPED WITH 3800 (VIN C) ENGINE EXCEPT E-CAR

CODE 27, 28, 29
GEAR SWITCH DIAGNOSIS
3800 (VIN C) (PORT)

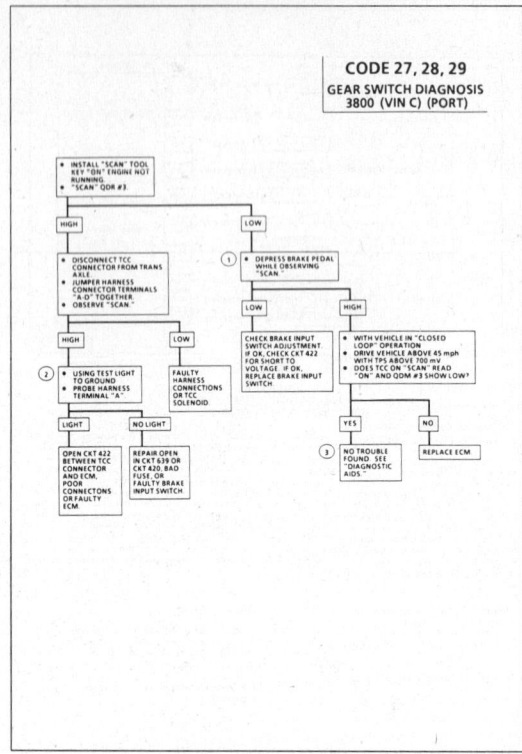

1988–90 VEHICLES EQUIPPED WITH 3800 (VIN C) ENGINE EXCEPT E-CAR

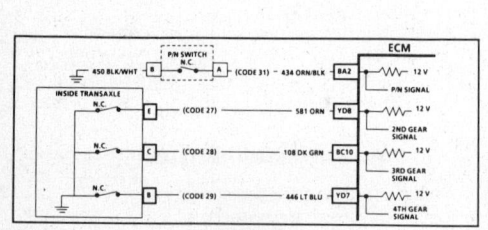

CODE 31
PARK/NEUTRAL SWITCH CIRCUIT
3800 (VIN C) (PORT)

Circuit Description:
The park/neutral switch contacts are a part of the neutral start switch and are closed to ground in park or neutral and open in drive ranges.
The ECM supplies ignition voltage through a current limiting resistor to CKT 434 and senses a closed switch when the voltage on CKT 434 drops to less than one volt.
The ECM uses the P/N signal as one of the inputs to control:
Idle Speed (IAC)
Exhaust Gas Recirculation (EGR)
and Vehicle Speed Sensor Diagnostics (VSS)
Code 31 Symptoms:
- CKT 434 indicates P/N (grounded) while in drive range, 4th gear TCC engaged, EGR would be inoperative, resulting in possible detonation.
- CKT 434 indicates drive (open) at start up, a dip in the idle may exist when the gear selector is moved into drive range.
- Transaxle 4th gear switch has an intermittent open, the ECM thinks vehicle is in 4th gear and will set a Code 31.

Test Description: Numbers below refer to circled numbers on the diagnostic chart.
1. Checks for a closed switch to ground in park position. Different makes of "Scan" tools will read P/N differently. Refer to "Tool Operator's" manual for type of display used for a specific tool.
Code 31 will set if:
- CKT 434 indicates an open for 4 consecutive starts.
- CKT 434 indicates a grounded CKT when transmission is in 4th gear, TCC engaged and no Code 29 for 10 seconds.

2. Checks for an open switch in drive range.
3. Be sure "Scan" indicates drive, even while wiggling shifter, to test for an intermittent or misadjusted switch in drive or overdrive range.

1988–90 VEHICLES EQUIPPED WITH 3800 (VIN C) ENGINE EXCEPT E-CAR

CODE 31
PARK/NEUTRAL SWITCH CIRCUIT
3800 (VIN C) (PORT)

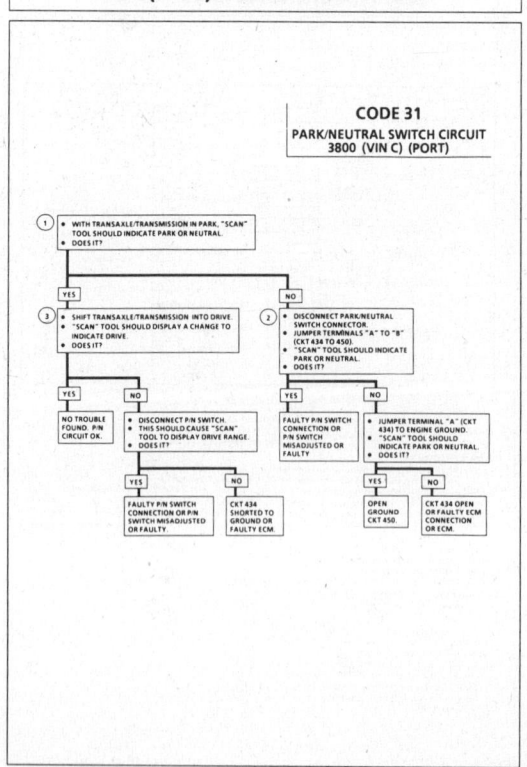

1988–90 VEHICLES EQUIPPED WITH 3800 (VIN C) ENGINE EXCEPT E-CAR

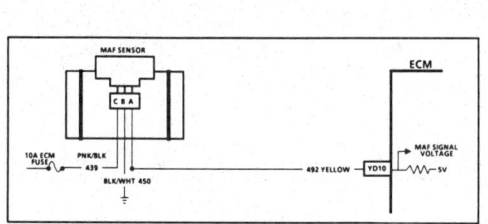

CODE 34
MASS AIR FLOW (MAF) SENSOR CIRCUIT
(GM/SEC LOW)
3800 (VIN C) (PORT)

Circuit Description:

The mass air flow (MAF) sensor measures the flow of air which passes through it in a given time. The ECM uses this information to monitor the operating condition of the engine for fuel delivery calculations. A large quantity of air movement indicates acceleration, while a small quantity indicates deceleration or idle.

The MAF sensor produces a frequency signal which cannot be easily measured. The sensor can be diagnosed using the procedures on this chart.

Code 34 will set when of the following conditions exists:
- Engine running
- No MAF sensor signal for 1.5 mS
- Above conditions for over 4 seconds

Test Description: Numbers below refer to circled numbers on the diagnostic chart.
1. This step checks to see if ECM recognizes a problem.
2. A voltage reading at sensor harness connector terminal "A" of less than 4 or over 6 volts indicates a fault in CKT 492 or poor connection.
3. Verifies that both ignition voltage and a good ground circuit are available.

Diagnostic Aids:

An intermittent may be caused by a poor connection, mis-routed harness, rubbed through wire insulation, or a wire broken inside the insulation.

Check For:
- Poor connection at ECM pin "YD10". Inspect harness connectors for backed out terminals, improper mating, broken locks, improperly formed or damaged terminals, and poor terminal to wire connection.
- Mis-routed Harness Inspect MAF sensor harness to insure that it is not too close to high voltage wires, such as spark plug leads.
- Damaged Harness Inspect harness for damage. If harness appears OK, "Scan" while moving related connectors and wiring harness. A change in display would indicate the intermittent fault location.

1988–90 VEHICLES EQUIPPED WITH 3800 (VIN C) ENGINE EXCEPT E-CAR

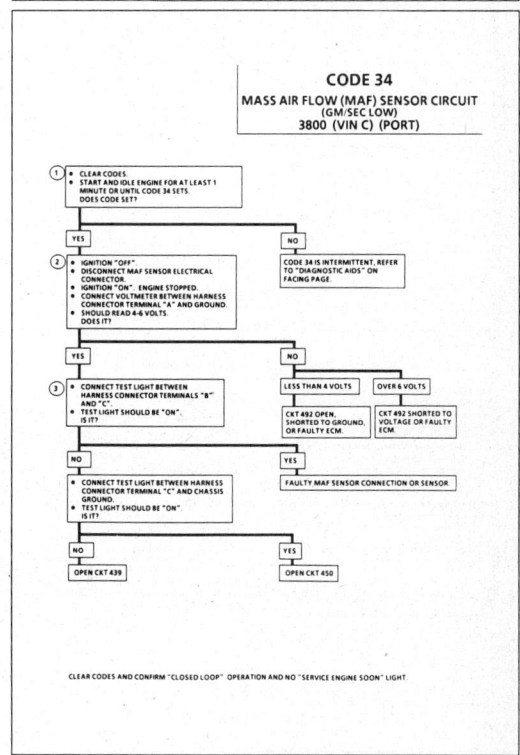

CODE 34
MASS AIR FLOW (MAF) SENSOR CIRCUIT
(GM/SEC LOW)
3800 (VIN C) (PORT)

CLEAR CODES AND CONFIRM "CLOSED LOOP" OPERATION AND NO "SERVICE ENGINE SOON" LIGHT.

1988–90 VEHICLES EQUIPPED WITH 3800 (VIN C) ENGINE EXCEPT E-CAR

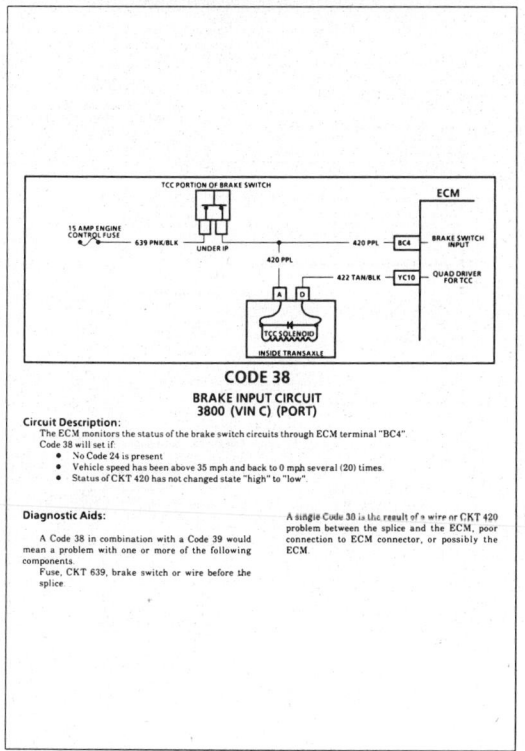

CODE 38
BRAKE INPUT CIRCUIT
3800 (VIN C) (PORT)

Circuit Description:

The ECM monitors the status of the brake switch circuits through ECM terminal "BC4".

Code 38 will set if:
- No Code 24 is present
- Vehicle speed has been above 35 mph and back to 0 mph several (20) times.
- Status of CKT 420 has not changed state "high" to "low".

Diagnostic Aids:

A Code 38 in combination with a Code 39 would mean a problem with one or more of the following components.

Fuse, CKT 639, brake switch or wire before the splice.

A single Code 38 is the result of a wire or CKT 420 problem between the splice and the ECM, poor connection to ECM connector, or possibly the ECM.

1988–90 VEHICLES EQUIPPED WITH 3800 (VIN C) ENGINE EXCEPT E-CAR

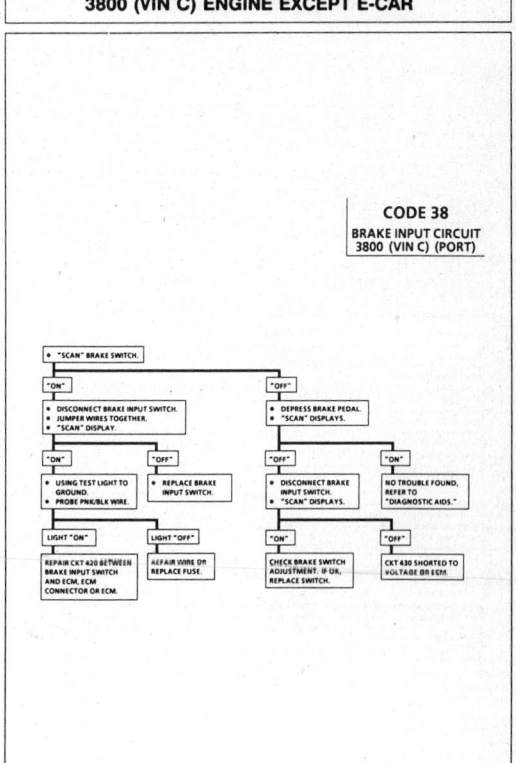

CODE 38
BRAKE INPUT CIRCUIT
3800 (VIN C) (PORT)

1988–90 VEHICLES EQUIPPED WITH 3800 (VIN C) ENGINE EXCEPT E-CAR

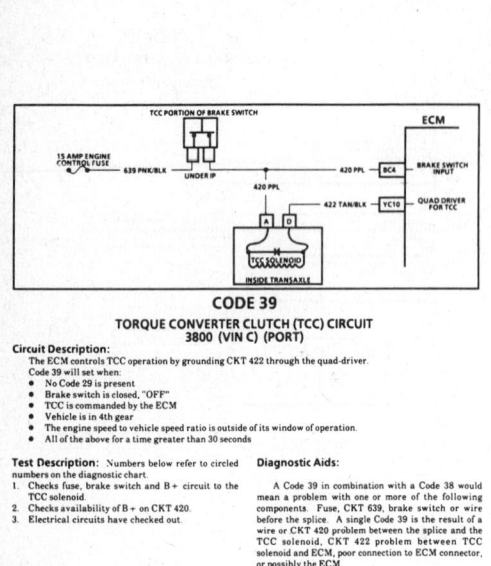

CODE 39
TORQUE CONVERTER CLUTCH (TCC) CIRCUIT
3800 (VIN C) (PORT)

Circuit Description:
The ECM controls TCC operation by grounding CKT 422 through the quad-driver.
Code 39 will be set when:
- No Code 29 is present
- Brake switch is closed, "OFF"
- TCC is commanded by the ECM
- Vehicle is in 4th gear
- The engine speed to vehicle speed ratio is outside of its window of operation
- All of the above for a time greater than 30 seconds

Test Description: Numbers below refer to circled numbers on the diagnostic chart.
1. Checks fuse, brake switch and B+ circuit to the TCC solenoid.
2. Checks availability of B+ on CKT 420.
3. Electrical circuits have checked out.

Diagnostic Aids:
A Code 39 in combination with a Code 38 would mean a problem with one or more of the following components. Fuse, CKT 639, brake switch or wire before the splice. A single Code 39 is the result of a wire or CKT 420 problem between the splice and the TCC solenoid, CKT 422 problem between TCC solenoid and ECM, poor connection to ECM connector, or possibly the ECM.

1988–90 VEHICLES EQUIPPED WITH 3800 (VIN C) ENGINE EXCEPT E-CAR

CODE 39
TORQUE CONVERTER CLUTCH (TCC) CIRCUIT
3800 (VIN C) (PORT)

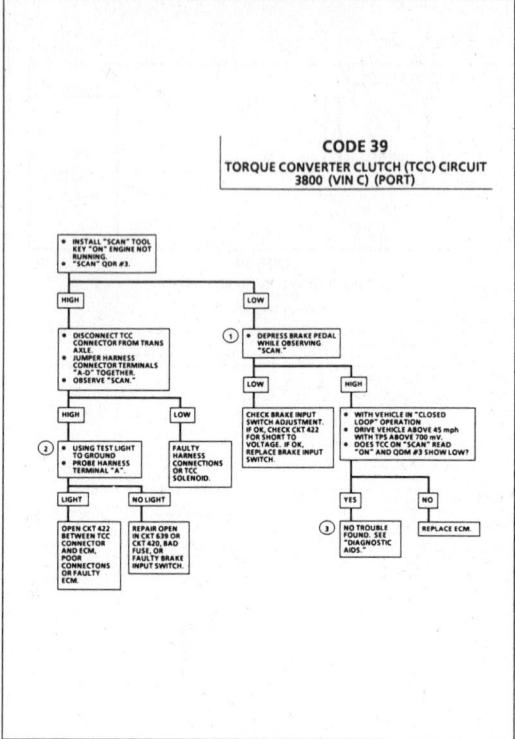

1988–90 VEHICLES EQUIPPED WITH 3800 (VIN C) ENGINE EXCEPT E-CAR

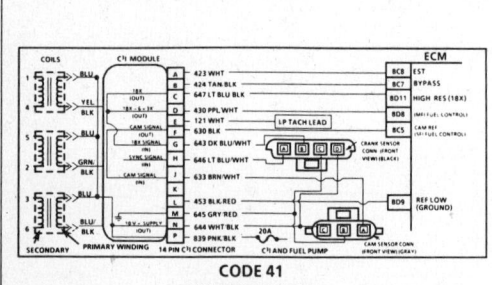

CODE 41
CAM SENSOR CIRCUIT
3800 (VIN C) (PORT)

Circuit Description:
The 3800 engine uses the simultaneous mode of fuel injection during start-up. As engine speed attains the 400 rpm level and a cam signal has been received by the ECM from the C³I module, the fuel injection switches modes to sequential injection. This is accomplished by use of a cam interrupter magnet and a cam sensor "hall effect" switch. The cam sensor sends a signal (sync-pulse) to the ignition module when cylinder #1 is 25° after top dead center on the compression stroke. This signal is used to start sequential fuel injection with the proper cylinder. If the cam signal is lost to the ECM, the engines fuel delivery will switch back to the simultaneous mode of operations. THE ENGINE WILL CONTINUE TO RUN. IT WILL RESTART AFTER SHUT DOWN.
Code 41 is set when the following conditions are met.
- Engine is running   • Cam sensor signal not received by ECM for last 2 seconds

Test Description: Numbers below refer to circled numbers on the diagnostic chart.
1. Checks to see if ECM recognizes a problem and sets Code 41.
2. This step verifies proper operation of CKTs 633, 644, and 645.
3. Step validates the integrity of CKT 630 from C³I module to ECM.
4. If the camshaft gear magnet is interfacing with the cam sensor the voltage reading will be zero, bumping engine will cause the condition to go away.
5. If the voltage reading of "BC5" is constantly varying and connection to terminal "BC5" are good, the ECM is faulty.

Diagnostic Aids:
An intermittent may be caused by a poor connection, rubbed through wire insulation or a wire broken inside the insulation.
Check For:
- Poor Connection or Damaged Harness Inspect ECM harness connectors for backed out terminal "BC5", improper mating, broken locks, improperly formed or damaged terminals, poor terminal to wire connection and damaged harness
- Intermittent Test If connections and harness check OK, monitor a digital voltmeter connected from ECM terminal "BC5" while moving related connectors and wiring harness. If the failure is induced, the voltage reading will change. This may help to isolate the location of the malfunction.

1988–90 VEHICLES EQUIPPED WITH 3800 (VIN C) ENGINE EXCEPT E-CAR

CODE 41
CAM SENSOR CIRCUIT
3800 (VIN C) (PORT)

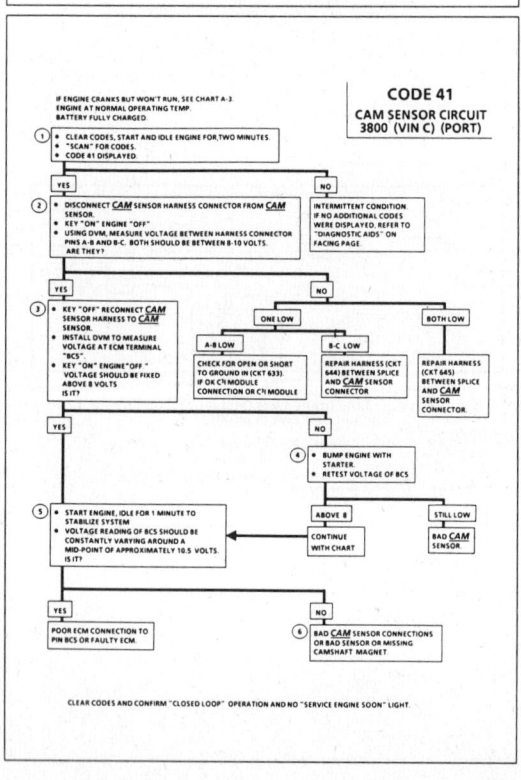

1988–90 VEHICLES EQUIPPED WITH 3800 (VIN C) ENGINE EXCEPT E-CAR

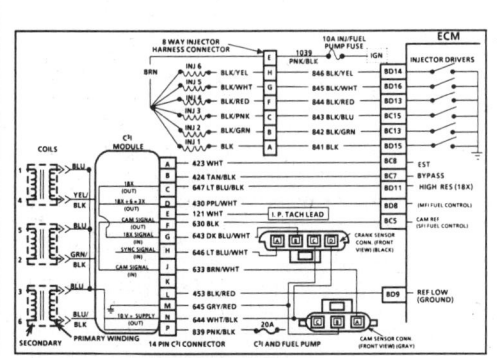

CODE 42
ELECTRONIC SPARK TIMING (EST) CIRCUIT
3800 (VIN C) (PORT)

Circuit Description:
The C³I module sends a reference signal to the ECM when the engine is cranking. Under 400 rpm, the C³I module controls ignition timing. When the engine speed exceeds 400 rpm, the ECM sends a 5 volts signal on the bypass CKT 424 to switch timing to ECM control. An open or ground in the EST circuit will stall the engine and set a Code 42. The engine can be re-started but will run on module timing.

To set a Code 42 the following conditions must be met:
- Engine speed greater than 600 rpm with no EST pulse for 200 mS (open or grounded CKT 423), or
- ECM commanding bypass mode (open or grounded CKT 424).

Test Description: Numbers below refer to circled numbers on the diagnostic chart.
1. Checks to see if ECM recognizes a problem. If it does not set Code 42, it is an intermittent problem and could be due to a loose connection.
2. With the ECM disconnected, the ohmmeter should be reading less than 200 ohms, which is the normal resistance of the EST circuit. A higher resistance would indicate a fault in CKT 423, a poor C³I module connection, or a faulty C³I module.
3. If test light was "ON" when connected from 12 volts to ECM harness terminal "BC7", either CKT 424 is shorted to ground or the C³I module is faulty.
4. Checks to see if C³I module switches when the bypass circuit is energized by 12 volts through the test light. If the C³I module actually switches, the ohmmeter reading should shift to over 8,000 ohms.
5. Disconnecting the ignition module should make the ohmmeter read as if it were monitoring an open circuit (infinite reading). Otherwise, CKT 423 is shorted to ground.

Diagnostic Aids:
An intermittent may be caused by a poor connection, rubbed through wire insulation, or a wire broken inside the insulation. Check For:
- **Poor Connection or Damaged Harness** Inspect ECM harness connectors for backed out terminals "BC7" or "BC8", improper mating, broken locks, improperly formed or damaged terminals, poor terminal to wire connection, and damaged harness.
- **Intermittent Test** If connections and harness check OK, a digital voltmeter connected from affected terminal to ground while moving related connectors and wiring harness. If the failure is induced, the voltage reading will change.

1988–90 VEHICLES EQUIPPED WITH 3800 (VIN C) ENGINE EXCEPT E-CAR

CODE 42
ELECTRONIC SPARK TIMING (EST) CIRCUIT
3800 (VIN C) (PORT)

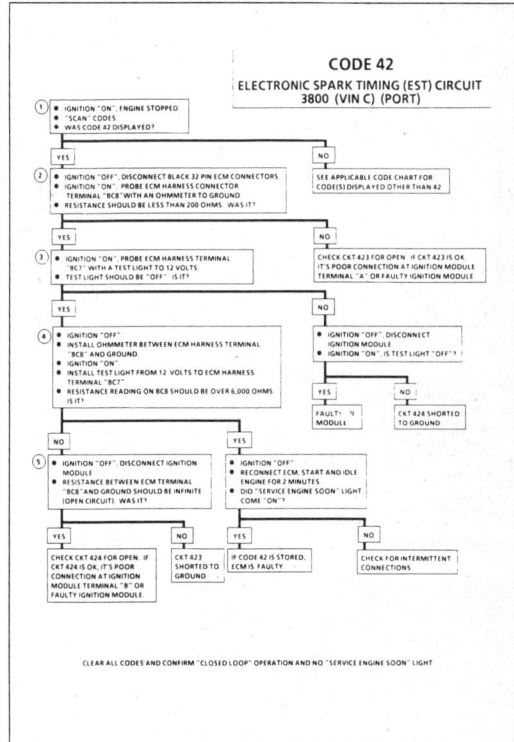

CLEAR ALL CODES AND CONFIRM "CLOSED LOOP" OPERATION AND NO "SERVICE ENGINE SOON" LIGHT

1988–90 VEHICLES EQUIPPED WITH 3800 (VIN C) ENGINE EXCEPT E-CAR

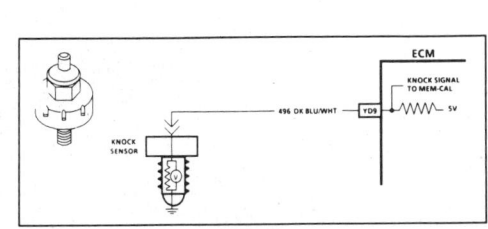

CODE 43
ELECTRONIC SPARK CONTROL (ESC) CIRCUIT
3800 (VIN C) (PORT)

Circuit Description:
The knock sensor is used to detect engine detonation and the ECM will retard the electronic spark timing based on the signal being received. The circuitry within the knock sensor causes the ECM's supplied 5 volts signal to be pulled down, so that under a no knock condition CKT 496 would measure about 2.5 volts. The knock sensor produces an AC signal which rides on the 2.5 volts DC voltage. The amplitude and signal frequency is dependent upon the knock level.

If CKT 496 becomes open or shorted to ground, the voltage will either go above 3.5 volts or below 1.5 volts. If either of these conditions are met for 20 seconds, a Code 43 will be stored.

Test Description: Numbers below refer to circled numbers on the diagnostic chart.
1. Code 43 will set if:
- Coolant temperature is over 90°C.
- MAT temperature is over 0°C.
- High engine load based on LV8 and rpm
- Voltage on CKT 496 goes above 3.5 volts or below 1.5 volts
- All conditions present for 20 seconds

If a Code 43 is detected, the ECM will retard spark timing by 10 degrees.

If an audible knock is heard from the engine, repair the internal engine problem, normally no knock should be detected at idle.

2. If tapping on the engine lift hook does not produce a knock signal, try tapping engine closer to sensor before proceeding.
3. The ECM has a 5 volts pull-up resistor which should be present at the knock sensor terminal.
4. This test determines if the knock sensor is faulty or if the ESC portion of the Mem-Cal is faulty.

Diagnostic Aids:
Check CKT 496 for a potential open or short to ground. Also check for proper installation of mem-cal.

1988–90 VEHICLES EQUIPPED WITH 3800 (VIN C) ENGINE EXCEPT E-CAR

CODE 43
ELECTRONIC SPARK CONTROL (ESC) CIRCUIT
3800 (VIN C) (PORT)

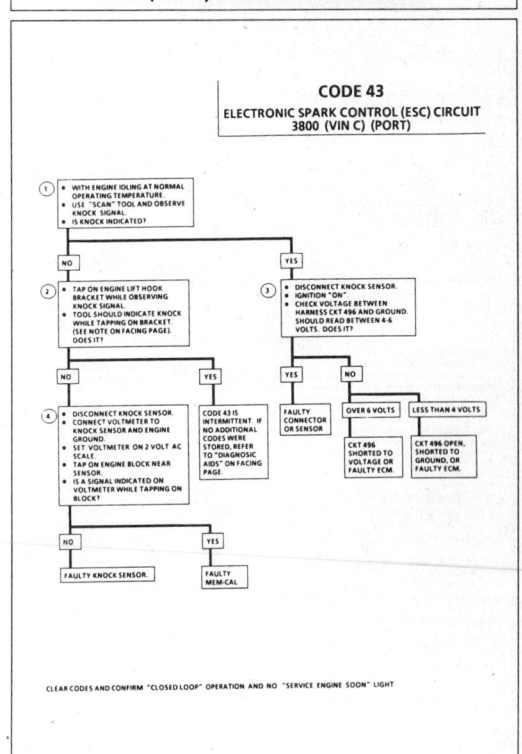

CLEAR CODES AND CONFIRM "CLOSED LOOP" OPERATION AND NO "SERVICE ENGINE SOON" LIGHT

1988-90 VEHICLES EQUIPPED WITH 3800 (VIN C) ENGINE EXCEPT E-CAR

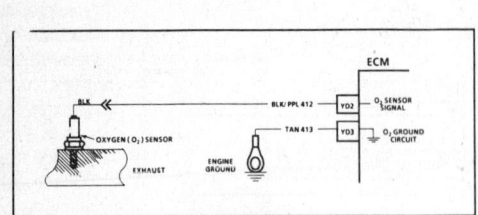

CODE 44
OXYGEN SENSOR CIRCUIT
(LEAN EXHAUST INDICATED)
3800 (VIN C) (PORT)

Circuit Description:

The ECM supplies a voltage of about .45 volt (450 mV) between terminals "YD2" and "YD3". (If measured with a 10 megohm digital voltmeter, this may read as low as .32 volt.) The O_2 sensor varies the voltage within a range of about 1 volt, (1000 mV) if the exhaust is rich, down through about .10 volt (100 mV) if exhaust is lean.

The sensor is like an open circuit and produces no voltage when it is below about 360°C (600°F). An open sensor circuit or cold sensor causes "Open Loop" operation.

Code 44 is set when the O_2 sensor signal voltage on CKT 412
- Remains below .25 volt for up to 4.5 minutes.
- The system is operating in "Closed Loop."

Test Description: Numbers below refer to circled numbers on the diagnostic chart.
1. Running the engine at 1000 rpm keeps the O_2 sensor hot, so an accurate display voltage is maintained.
   Opening the O_2 sensor wire should result in a voltage display of between 350 and 550 mV. If the display is still fixed below 350 mV, the fault is a short to ground in CKT 412 or the ECM is faulty.

Diagnostic Aids:

Using the "Scan", observe the block learn values at different rpm and air flow conditions. The "Scan" also displays the block cells, so the block learn values can be checked in each of the cells to determine when the Code 44 may have been set. If the conditions for Code 44 exists, the block learn values will be around 150.
- O_2 Sensor Wire Sensor pigtail may be mispositioned and contacting the exhaust manifold.
- Check for intermittent ground in wire between connector and sensor.

- MAF Sensor A mass air flow (MAF) sensor output that causes the ECM to sense a lower than normal air flow will cause the system to go lean. Disconnect the MAF sensor and if the lean condition is gone, replace the MAF sensor.
- Lean Injector(s) Perform injector balance test CHART C-2A.
- Fuel Contamination Water, even in small amounts, near the in-tank fuel pump inlet can be delivered to the injectors. The water causes a lean exhaust and can set a Code 44.
- Fuel Pressure System will be lean if pressure is too low. It may be necessary to monitor fuel pressure while driving the car at various road speeds and/or loads to confirm. See "Fuel System Diagnosis", CHART A-7.
- Exhaust Leaks If there is an exhaust leak, the engine can cause outside air to be pulled into the exhaust and past the sensor. Vacuum or crankcase leaks can cause a lean condition.
- If the above are OK, it is a faulty oxygen sensor.

CODE 44
OXYGEN SENSOR CIRCUIT
(LEAN EXHAUST INDICATED)
3800 (VIN C) (PORT)

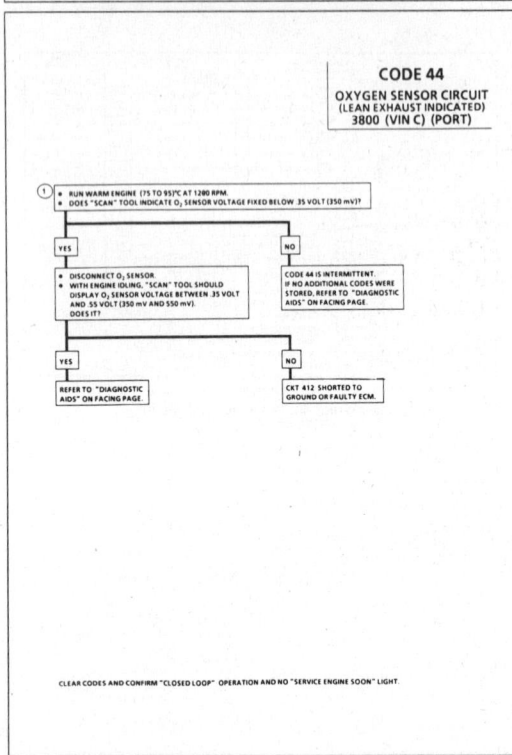

CLEAR CODES AND CONFIRM "CLOSED LOOP" OPERATION AND NO "SERVICE ENGINE SOON" LIGHT.

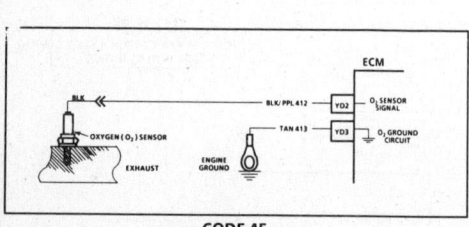

CODE 45
OXYGEN SENSOR CIRCUIT
(RICH EXHAUST INDICATED)
3800 (VIN C) (PORT)

Circuit Description:

The ECM supplies a voltage of about .45 volt (450 mV) between terminals "YD2" and "YD3". (If measured with a 10 megohm digital voltmeter, this may read as low as .32 volt.) The O_2 sensor varies the voltage within a range of about 1 volt (1000 mV) if the exhaust is rich, down through about .10 volt (100 mV) if exhaust is lean.

The sensor is like an open circuit and produces no voltage when it is below about 360°C (600°F). An open sensor circuit or cold sensor causes "Open Loop" operation.

Test Description: Numbers below refer to circled numbers on the diagnostic chart.
1. Code 45 is set when the O_2 sensor signal voltage or CKT 412:
   - Remains above .75 volt for 2 minutes and in "Closed Loop."
   - Throttle angle between .7 and 1.4 volts
   - Engine time after start is 1 minute or more
   - No Code 21 or Code 22

Diagnostic Aids:

Using the "Scan", observe the block learn values at different rpm and air flow conditions. The "Scan" also displays the block cells, so the block learn values can be checked in each of the cells to determine when the Code 45 may have been set. If the conditions for Code 45 exists, the block learn values will be around 115.
- Fuel Pressure System will go rich if pressure is too high. The ECM can compensate for some increase. However, if it gets too high, a Code 45 may be set. See "Fuel System Diagnosis", CHART A-7.

- Rich Injector Perform injector balance test CHART C-2A.
- Leaking Injector See CHART A-7.
- Check for fuel contaminated oil.
- Canister Purge Check for fuel saturation. If full of fuel, check canister control and hoses. See "Canister Purge", Section "C3".
- MAF Sensor An output that causes the ECM to sense a higher than normal airflow can cause the system to go rich. Disconnecting the MAF sensor will allow the ECM to set a fixed value for the sensor. Substitute a different MAF sensor if the rich condition is gone while the sensor is disconnected.
- Check for leaking fuel pressure regulator diaphragm by checking vacuum line to regulator for fuel.
- TPS An intermittent TPS output will cause the system to go rich due to a false indication of the engine accelerating.
- EGR An EGR staying open (especially at idle) will cause the O_2 sensor to indicate a rich exhaust and this could result in a Code 45.

CODE 45
OXYGEN SENSOR CIRCUIT
(RICH EXHAUST INDICATED)
3800 (VIN C) (PORT)

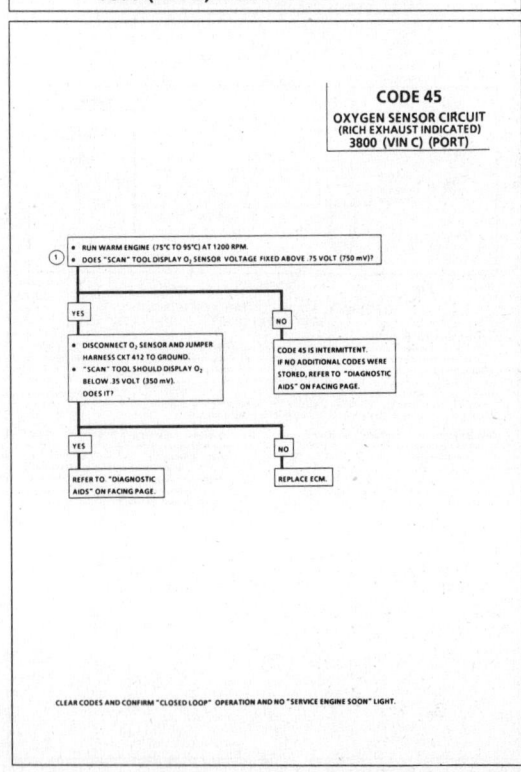

CLEAR CODES AND CONFIRM "CLOSED LOOP" OPERATION AND NO "SERVICE ENGINE SOON" LIGHT.

1988–90 VEHICLES EQUIPPED WITH 3800 (VIN C) ENGINE EXCEPT E-CAR

CODE 48

MISFIRE DIAGNOSIS
3800 (VIN C) (PORT)

If multiple codes are set, go to the lowest code first.
Repairing for a Code 13, 44, or 45 may correct Code 48.

Test Description:
Code 48 will set if the following:
- TPS is between .48 and 1.02 volts.
- Rpm is between 1250 and 2100.
- Mph is between 50 and 60.
- O_2 cross counts greater than 21.
- All of the above for 3 seconds.

O_2 Sensor Test:
Code 48 could be set if the O_2 sensor is degraded and cannot travel over the full rich to lean voltage range. This narrowed range could allow O_2 cross counts to be above the value necessary to set the code.

```
┌─────────────────────────────────────┐
│ • WITH "SCAN" TOOL INSTALLED, VERIFY │
│   ENGINE IS AT NORMAL OPERATING      │
│   TEMPERATURE AND IN "CLOSED LOOP".  │
└─────────────────────────────────────┘
              │
┌─────────────────────────────────────┐
│ • ENGINE IDLING IN PARK.             │
│ • SELECT O2 SENSOR POSITION ON "SCAN"│
│ • RAPIDLY FLASH THE THROTTLE FROM    │
│   IDLE TO NEAR WIDE OPEN THROTTLE    │
│   AND BACK WHILE OBSERVING O2 VOLTAGE│
│ • REPEAT IF NECESSARY TO CONFIRM     │
│   VOLTAGE RANGE, AND "CLOSED LOOP".  │
└─────────────────────────────────────┘
         │                    │
┌──────────────────┐  ┌──────────────────┐
│ VOLTAGE EXCEEDS  │  │ VOLTAGE REMAINS  │
│ 250-750 mV RANGE.│  │ WITHIN 250-750   │
│                  │  │ mV RANGE         │
└──────────────────┘  └──────────────────┘
         │                    │
┌──────────────────┐  ┌──────────────────┐
│ O2 SENSOR OK,    │  │ REPLACE O2 SENSOR│
│ SEE "DIAGNOSTIC  │  │                  │
│ AIDS".           │  │                  │
└──────────────────┘  └──────────────────┘
```

Diagnostic Aids:

1. **Ignition system checks:**
   Remove each spark plug and inspect (fouled, cracked, worn)
   Fouled -- check ignition wires (hi resistance, damage, poor connections, grounds)
   check coil and module operation
   check basic engine problem (see 3 below)
   Cracked or worn -- replace as necessary

2. **Fuel system checks:**
   Restricted fuel system (injectors, fuel pump, lines, and filter)
   Injectors -- perform injector balance test
   verify each injector circuit with tool J-34730-2 or equivalent
   Fuel Pump --verify proper fuel pressure and fuel quality
   Lines and Filter --verify no restrictions in lines or filter

3. **Basic Engine Checks:**
   Unless spark plug(s) inspection identifies a specific cylinder(s), road test vehicle under test conditions to reverify Code 48 prior to engine disassembly.
   Basic engine (valves, compression, camshaft, lifters)
   Compression -- check rings, pistons, valves
   Valves -- check for burned, weak springs, broken parts, worn or loose guide
   Camshaft -- check for worn or broken
   Lifters -- check for worn, broken
   For additional items see Section "B" under "Rough Unstable Idle", "Hard Start", or "Hesitation, Sag, or Stumble".

1988–90 VEHICLES EQUIPPED WITH 3800 (VIN C) ENGINE EXCEPT E-CAR

CODE 51

MEM-CAL ERROR
(FAULTY OR INCORRECT MEM-CAL)
3800 (VIN C) (PORT)

```
┌──────────────────────────────────────────────────┐
│ CHECK THAT ALL PINS ARE FULLY INSERTED IN THE     │
│ SOCKET. IF OK, REPLACE PROM, CLEAR MEMORY AND     │
│ RECHECK. IF CODE 51 REAPPEARS, REPLACE ECM.       │
└──────────────────────────────────────────────────┘
```

1988–90 VEHICLES EQUIPPED WITH 3800 (VIN C) ENGINE EXCEPT E-CAR

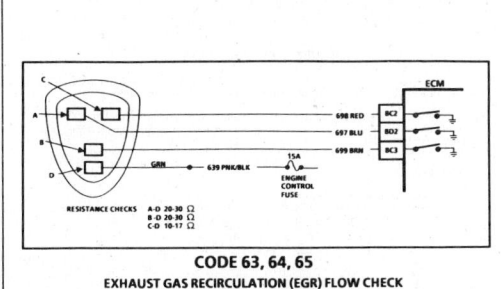

CODE 63, 64, 65

EXHAUST GAS RECIRCULATION (EGR) FLOW CHECK
3800 (VIN C) (PORT)

Circuit Description:
Codes 63, 64, and 65 are EGR flow test failures. The ECM, on a closed throttle coast down, will cycle the solenoids "ON" and "OFF" individually and look for a resulting change in engine rpm and O_2 sensor activity.

Test Description: Numbers below refer to circled numbers on the diagnostic chart.
If a code is set, inspect EGR for damage. Disconnect the 4 wire connector at EGR.
Install a fused jumper from the battery to terminal "D" of the EGR.
Start and idle engine, and with a jumper, ground terminals "A", "B", & "C" one at a time. You should be able to discern a change in engine rpm as the terminal is grounded.
Terminal "A" should result in a small change and "C" in a large change in rpm.

Diagnostic Aids:
An intermittent may be caused by a poor connection, rubbed-through wire insulation, or a wire broken inside the insulation. Check For:
- **Poor Connection or Damaged Harness** Inspect ECM harness connectors for backed out terminals "BC2", "BC3", and "BD2", improper mating, broken locks, improperly formed or damaged terminals, poor terminal to wire connection, and damaged harness.
- **Intermittent Test** If connections and harness check OK, connect a digital voltmeter from effected terminals to ground while moving related connectors and wiring harness. If the failure is induced, the voltage reading will change.

1988–90 VEHICLES EQUIPPED WITH 3800 (VIN C) ENGINE EXCEPT E-CAR

CODE 63, 64, 65

EXHAUST GAS RECIRCULATION (EGR) FLOW CHECK
3800 (VIN C) (PORT)

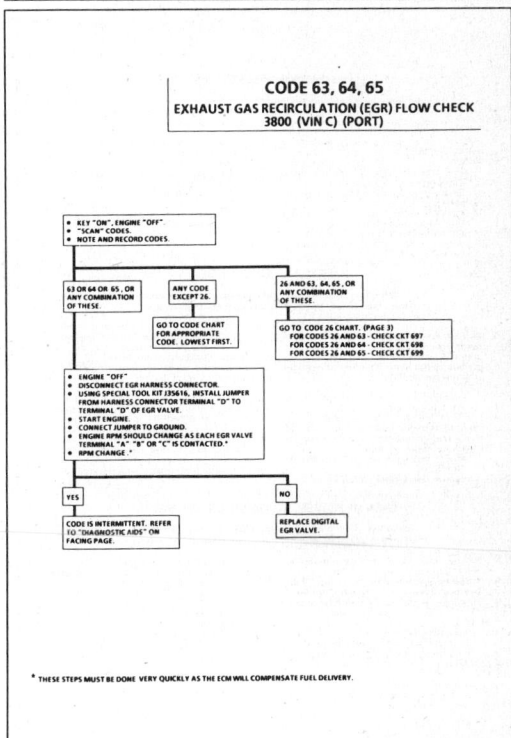

* THESE STEPS MUST BE DONE VERY QUICKLY AS THE ECM WILL COMPENSATE FUEL DELIVERY.

1988–90 VEHICLES EQUIPPED WITH 3800 (VIN C) ENGINE EXCEPT E-CAR

SECTION B

SYMPTOMS

Before Starting
Intermittents
Hard Start
Hesitation, Sag, Stumble
Surges and/or Chuggle
Lack of Power, Sluggish, or Spongy
Detonation/Spark Knock
Cuts Out, Misses
Backfire
Poor Fuel Economy
Dieseling, Run-On
Rough, Unstable, or Incorrect Idle, Stalling
Excessive Exhaust Emissions Or Odors
Chart B-1 - Restricted Exhaust System Check

BEFORE STARTING

Before using this section you should have performed the DIAGNOSTIC CIRCUIT CHECK and found out that:
1. The ECM and "Service Engine Soon" light are operating.
2. There are no trouble codes stored, or there is a trouble code but no "Service Engine Soon" light.
3. The fuel control system is operating OK (by performing field service mode check).
   Verify the customer complaint, and locate the correct SYMPTOM. Check the items indicated under that symptom.
   If the ENGINE CRANKS BUT WILL NOT RUN, see CHART A-3.

Several of the symptom procedures below call for a careful visual/physical check. This check should include:
- ECM B+ wires for being clean and tight at starter and/or junction block.
- ECM grounds for being clean and tight
- Vacuum hoses for splits, kinks, and proper connections, as shown on emission control information label.
- Air leaks at throttle body mounting and intake manifold.
- Air leaks between MAF sensor and throttle body.
- Ignition wires for cracking, hardness, proper routing, and carbon tracking.
- Wiring for proper connections, pinches, and cuts.
The importance of this step cannot be stressed too strongly - it can lead to correcting a problem without further checks and can save valuable time.

1988–90 VEHICLES EQUIPPED WITH 3800 (VIN C) ENGINE EXCEPT E-CAR

INTERMITTENTS

Problem may or may not turn "ON" the "Service Engine Soon" light or store a code.

DO NOT use the trouble code charts in Section "A" for intermittent problems. The fault must be present to locate the problem. If a fault is intermittent, use of trouble code charts may result in replacement of good parts.
- Most intermittent problems are caused by faulty electrical connections or wiring. Perform a careful check as described at start of this section. Check for:
- Poor mating of the connector halves, or terminals not fully seated in the connector body (backed out).
- Improperly formed or damaged terminals. All connector terminals in problem circuit should be carefully reformed to increase contact tension.
- Poor terminal to wire connection. This requires removing the terminal from the connector body to check.
- If a visual/physical check does not find the cause of the problem, the car may be driven with a voltmeter connected to a suspect circuit. An abnormal voltage reading, when the problem occurs, indicates the problem may be in that circuit. If the wiring and connectors check OK and a trouble code was stored for a circuit having a sensor, except for Codes 43, 44, and 45, substitute a known good sensor and recheck.

An intermittent "Service Engine Soon" light with no stored code may be caused by
- Ignition coil shorted to ground, arcing at spark plug wires or plugs.
- "Service Engine Soon" light wire to ECM shorted to ground. (CKT 419).
- Diagnostic "Test" terminal wire to ECM, shorted to ground. (CKT 451)
- ECM power grounds. See ECM wiring diagrams.
- Loss of trouble code memory. To check, disconnect TPS and idle engine until "Service Engine Soon" light comes "ON." Code 22 should be stored, and kept in memory when ignition is turned "OFF". If not, the ECM is faulty.
- Check for an electrical system interference caused by a defective relay, ECM driven solenoid, or switch. They can cause a sharp electrical surge. Normally, the problem will occur when the faulty component is operated.
- Check for improper installation of electrical options, such as lights, 2-way radios, etc.
- EST wires should be kept away from spark plug wires, coil and generator.
- Check for open diode across A/C compressor clutch, and for other open diodes (see wiring diagrams).

HARD START

Definition: Engine cranks OK, but does not start for a long time. Does eventually run, or may start but immediately dies.

- Perform careful check as described at start of Section "B".
- Make sure driver is using correct starting procedure.
- CHECK:
  - TPS for sticking or binding or a high TPS voltage with the throttle closed.
  - High resistance in coolant sensor circuit or sensor itself. See CODE 15 CHART or with a "Scan" tool compare coolant temperature with ambient temperature on a cold engine.
  - Fuel pressure CHART A-7.
  - Water contaminated fuel.
  - EGR operation. Be sure valves seat properly and are not staying open. See CHART C-7.
  - Fuel pump relay - See CHART A-7.

Ignition system - Check for:
  Proper output from ST-125.
  Bare and shorted wires.
  Loose C3I module ground, mounting screws.
- A faulty in-tank fuel pump check valve will allow the fuel in the lines to drain back to the tank after the engine is stopped. To check for this condition:
  Perform fuel system diagnosis, CHART A-7.
- Remove spark plugs. Check for wet plugs, cracks, wear, improper gap, burned electrodes, or heavy deposits. Repair or replace as necessary.

1988–90 VEHICLES EQUIPPED WITH 3800 (VIN C) ENGINE EXCEPT E-CAR

HESITATION, SAG, STUMBLE

Definition: Momentary lack of response as the accelerator is pushed down. Can occur at all car speeds. Usually most severe when first trying to make the car move, as from a stop sign. May cause the engine to stall if severe enough.

- Perform careful visual check as described at start of Section "B".
- CHECK:
  - Fuel pressure. See CHART A-7. Also Check for water contaminated fuel.
  - Spark plugs for being fouled or faulty wiring.
  - PROM number. Also check service bulletins for latest PROM.
  - TPS for binding or sticking. Voltage should increase at a steady rate as throttle is moved toward WOT

- Generator output voltage. Repair if less than 9 or more than 16 volts.
- C3I ground, CKT 453.
- Canister purge system for proper operation. See CHART C-3.
- EGR - See CHART C-7.
- Engine Thermostat - functioning correctly and proper heat range.
- Perform injector balance test CHART C-2A.

SURGES AND/OR CHUGGLE

Definition: Engine power variation under steady throttle or cruise. Feels like the car speeds up and slows down with no change in the accelerator pedal.

- Be sure driver understands transmission converter clutch and A/C compressor operation in owner's manual.
- Perform careful visual inspection as described at start of Section "B".
- CHECK:
  - Generator output voltage. Repair if less than 9 or more than 16 volts.
  - If a "Scan" tool is available which plugs in to the ALDL connector, make sure reading of VSS matches vehicle speedometer. See introduction explaining "Scan" tool positions.
  - EGR - There should be no EGR at idle. See CHART C-7.
  - Vacuum lines for kinks or leaks.

- In-line fuel filter. Replace if dirty or plugged.
- Fuel pressure while condition exists. See CHART A-7.
- Inspect oxygen sensor for silicone contamination, from fuel or use of improper RTV sealant. The sensor may have a white powdery coating. This will result in a high but false signal voltage (rich exhaust indication). The ECM will then reduce the amount of fuel delivered to the engine, causing a severe driveability problem.
- Remove spark plugs. Check for cracks, wear, improper gap, burned electrodes, or heavy deposits.
- Check spark plug leads.

LACK OF POWER, SLUGGISH, OR SPONGY

Definition: Engine delivers less than expected power. Little or no increase in speed when accelerator pedal is pushed down part way.

- Perform careful visual check as described at start of Section "B".
- Compare customer's car to similar unit. Make sure the customer's car has an actual problem.
- Remove air cleaner and check air filter for dirt, or for being plugged. Replace as necessary.
- CHECK:
  - Restricted fuel filter, contaminated fuel or improper fuel pressure. See CHART A-7.
  - ECM power grounds -See wiring diagrams.
  - EGR operation for being open or partly open all the time - CHART C-7.

- Exhaust system for possible restriction.
  - Inspect exhaust system for damaged or collapsed pipes.
  - Inspect muffler for heat distress or possible internal failure.
- Generator output voltage. Repair if less than 9 or more than 16 volts.
- Engine valve timing and compression.
- Engine for proper or worn camshaft.

- Secondary voltage using a shop ocilliscope or a spark tester J-26792 (ST-125) or equivalent.

1988–90 VEHICLES EQUIPPED WITH 3800 (VIN C) ENGINE EXCEPT E-CAR

DETONATION /SPARK KNOCK

Definition: A mild to severe ping, usually worse under acceleration. The engine makes sharp metallic knocks that change with throttle opening. Sounds like popcorn popping.

- Check for obvious overheating problems:
  - Low/weak coolant mixture.
  - Inoperative thermostat.
  - Restricted air/water flow through radiator.
  - Inoperative electric coolant fan circuit. See CHART C-12.
- CHECK:
  - EGR system for not opening - CHART C-7.
  - ESC system for no retard - see CHART C-5.
  - TCC operation - see CHART C-8.

- Fuel system pressure. See CHART A-7.
- Remove carbon with top engine cleaner. Follow instructions on can.
  - Check for leaking valve oil seals.
  - Check for poor fuel quality, proper octane rating.
  - Check for correct PROM.
- Check for incorrect basic engine parts such as cam, heads, pistons, etc.

CUTS OUT, MISSES

Definition: Steady pulsation or jerking that follows engine speed, usually more pronounced as engine load increases. The exhaust has a steady spitting sound at idle or low speed.

- Perform careful visual (physical) check as described at start of Section "B".
- If engine "Misses at Idle", see CHART C-4F-1.
- If engine "Misses Under Load", see Chart C-4F-2.
- If above checks did not discover cause of problem, check the following:
  - Disconnect all injector harness connectors. Connect J-34730-2 injector test light or equivalent 6 volt test light between the harness terminals of each injector connector and note light while cranking. If test light fails to blink at any connector, it is a faulty injector drive harness, connector, or terminal.
  - Fuel pump - Plugged fuel filter, water, low pressure. See CHART A-7.

- Perform the injector balance test. See CHART C-2A.
- On C3I systems, a misfire may be caused by a misaligned crank sensor and wear on rotating interrupter. Inspect for proper clearance at each vane, using tool J-31179 or equivalent and procedure detailed in Section Sensor should be replaced if it shows evidence of having been rubbed by the interrupter.
- Perform compression check.
- Remove rocker covers. Check for bent pushrods, worn rocker arms, broken valve springs, worn camshaft lobes. Repair as necessary.
- Valve timing.

1988–90 VEHICLES EQUIPPED WITH 3800 (VIN C) ENGINE EXCEPT E-CAR

BACKFIRE

Definition: Fuel ignites in intake manifold, or in exhaust system, making a loud popping noise.

- **CHECK:**
  - Compression - Look for sticking or leaking valves.
  - EGR operation for being open all the time. See CHART C-7.
  - EGR gasket for faulty or loose fit.
  - Valve timing.
  - Output voltage of ignition coil using a shop ociliscope or spark tester J-26792 (ST-125) or equivalent.
- Spark plugs for crossfire. Also inspect spark plug wires, and proper routing of plug wires.
- Ignition system for intermittent condition.
- Engine timing - see emission control information label.

POOR FUEL ECONOMY

Definition: Fuel economy, as measured by an actual road test, is noticeably lower than expected. Also, economy is noticeably lower than it was on this car at one time, as previously shown by an actual road test.

- **CHECK:**
  - Engine thermostat for faulty part (always open) or for wrong heat range.
  - Fuel pressure. See CHART A-5. Check owner's driving habits.
    - Is A/C "ON" full time (defroster mode "ON")?
    - Are tires at correct pressure?
    - Are excessively heavy loads being carried?
    - Is acceleration too much, too often?
    - Suggest driver read "Important Facts on Fuel Economy" in owner's manual.
  - Perform "Diagnostic Circuit Check".
- Check air cleaner element (filter) for dirt or being plugged.
- Check for proper calibration of speedometer.
- Visually (physically) Check:
  - Vacuum hoses for splits, kinks and proper connections as shown on Vehicle Emissions Control Information label.
  - Ignition wires for cracking, hardness, and proper connections.
- Remove spark plugs. Check for cracks, wear, improper gap, burned electrodes or heavy deposits. Repair or replace as necessary.
- Check compression.
- Check TCC for proper operation. See CHART C-8, use "Scan" tool if available.
- Check for dragging brakes.
- Suggest owner fill fuel tank and recheck fuel economy.
- Check for exhaust system restriction.

DIESELING, RUN-ON

Definition: Engine continues to run after key is turned "OFF," but runs very roughly. If engine runs smoothly, check ignition switch and adjustment.

- Check injectors for leaking. See CHART A-7.

1988–90 VEHICLES EQUIPPED WITH 3800 (VIN C) ENGINE EXCEPT E-CAR

ROUGH, UNSTABLE, OR INCORRECT IDLE, STALLING

Definition: The engine runs unevenly at idle. If bad enough, the car may shake. Also, the idle may vary in rpm (called "hunting"). Either condition may be bad enough to cause stalling. Engine idles at incorrect speed.

- Perform careful visual check as described at start of Section "B".
- Clean injectors.
- **CHECK:**
  - Throttle linkage for sticking or binding.
  - TPS for sticking or binding, be sure output is stable at idle and adjustment specification is correct.
  - IAC system.
  - Generator output voltage. Repair if less than 9 or more than 16 volts.
  - P/N switch circuit. Code 31, 3800 (VIN C), or use "Scan" tool, and be sure tool indicates vehicle is in drive with gear selector in drive.
  - Injector balance. See CHART C-2A.
  - PCV valve for proper operation by placing finger over inlet hole in valve end several times. Valve should snap back. If not, replace valve. See Section "C13".
  - Evaporative emission control system. CHART C-3.
  - ECM ground circuits.
  - EGR valve: There should be no EGR at idle.
- Monitoring block learn values may help identify the cause of the problem. If the system is running lean (block learn greater than 138) refer to "Diagnostic Aids" on facing page of Code 44. If the system is running rich (block learn values less than 118) refer to "Diagnostic Aids" on facing page of Code 45.
- Run a cylinder compression check.
- Check for fuel in pressure regulator hose. If present, replace regulator assembly.
- Check ignition system, wires and plugs.
- Check for loose or damaged gaskets MAF between sensor and throttle body intake.
- Disconnect MAF sensor and if condition is corrected, replace sensor. "Scan" tool should read about 4-8 grams per second at idle.
- If problem exists with A/C "ON", check A/C system operation CHART C-10.

EXCESSIVE EXHAUST EMISSIONS OR ODORS

Definition: Vehicle fails an emission test. Vehicle has excessive "rotten egg" smell. Excessive odors do not necessarily indicate excessive emissions.

- Perform "Diagnostic Circuit Check."
- IF TEST SHOWS EXCESSIVE CO AND HC, (or also has excessive odors):
  - Check items which cause car to run RICH.
    - Make sure engine is at normal operating temperature.
  - **CHECK:**
    - Fuel pressure. See CHART A-7.
    - Canister for fuel loading. See CHART C-3.
    - Injector balance. See CHART C-2A.
    - PCV valve for being plugged, stuck, or blocked PCV hose or fuel in the crankcase.
    - Spark plugs, plug wires, and ignition components.
    - Check for lead contamination of catalytic converter (look for removal of fuel filler neck restrictor).
    - Check for properly installed fuel cap.
- If the system is running rich (block learn less than 118), refer to "Diagnostic Aids" on facing page of Code 45.
- IF TEST SHOWS EXCESSIVE NOx:
  - Check items which cause car to run LEAN, or to run too hot.
    - EGR valve for not opening. See CHART C-7.
    - Vacuum leaks.
    - Coolant system and coolant fan for proper operation. See CHART C-12.
    - Remove carbon with top engine cleaner. Follow instructions on can.
- If the system is running lean (block learn greater than 138), refer to "Diagnostic Aids" on facing page of Code 44.

1988–90 VEHICLES EQUIPPED WITH 3800 (VIN C) ENGINE EXCEPT E-CAR

CHART C-2A

INJECTOR BALANCE TEST

The injector balance tester is a tool used to turn the injector on for a precise amount of time, thus spraying a measured amount of fuel into the manifold. This causes a drop in fuel rail pressure that we can record and compare between each injector. All injectors should have the same amount of pressure drop (± 10 kpa). Any injector with a pressure drop that is 10 kpa (or more) greater than the average drop of the other injectors should be considered faulty and replaced. Any injector with a pressure drop of less than 10 kpa from the average drop of the other injectors should be cleaned and retested.

STEP 1

Engine "cool down" period (10 minutes) is necessary to avoid irregular readings due to "Hot Soak" fuel boiling. Disconnect harness connectors at all injectors, and connect injector tester J-34730-3, or equivalent, to one injector. With ignition "OFF" connect fuel gauge J347301 or equivalent to fuel pressure tap. Wrap a shop towel around fitting while connecting gage to avoid fuel spillage. At this point, insert clear tubing attached to vent valve into a suitable container and bleed air from gauge and hose to insure accurate gauge operation, by applying B + to fuel pump test terminal. Repeat this step until all air is bled from gauge.

STEP 2

Apply B + to fuel pump test terminal until fuel pressure reaches its maximum, about 3 seconds. Record this initial pressure reading. Energize tester one time and note pressure at its lowest point (Disregard any slight pressure increase after drop hits low point.) By subtracting this second pressure reading from the initial pressure, we have the actual amount of injector pressure drop. See note below.

STEP 3

Repeat step 2 on each injector and compare the amount of drop. Usually, good injectors will have virtually the same drop. Clean and retest any injector that has a pressure difference of 10kPa, either more or less than the average of the other injectors on the engine. Replace any injector that also fails the retest. If the pressure drop of all injectors is within 10kPa of this average, the injectors appear to be flowing properly. Reconnect them and review "Symptoms", Section "B".

NOTE: *The entire test should not be repeated more than once without running the engine to prevent flooding. (This includes any retest on faulty injectors).*

1988–90 VEHICLES EQUIPPED WITH 3800 (VIN C) ENGINE EXCEPT E-CAR

NOTE: If injectors are suspected of being dirty, they should be cleaned using an approved tool and procedure prior to performing this test. The fuel pressure test in Chart A-7, should be completed prior to this test.

CHART C-2A

INJECTOR BALANCE TEST 3800 (VIN C) (PORT)

Step 1. If engine is at operation temperature start at number 1, otherwise you can start at number 2.
1. Engine "cool down" period (10 minutes) is necessary to avoid irregular readings due to "Hot Soak" Fuel Boiling.
2. Disconnect harness connectors at all injectors and connect injector tester J-34730-3, or equivalent, to one injector.
3. With ignition "OFF" connect fuel gauge J-347301 or equivalent to Fuel Pressure Tap. NOTE: Wrap a shop towel around fitting while connecting gauge to avoid fuel spray or spillage.
4. Apply B + to fuel pump test terminal and install bleed off to fuel pressure gauge.

Step 2. Run test:
1. Apply B + to Fuel Pump Test Terminal. Allow about 3 seconds for fuel pressure to reach its maximum.
2. Record gauge pressure. (Pressure must hold steady, if not see the Fuel System diagnosis, Chart A-7.)
3. Turn injector on, by depressing button on injector tester, and note pressure at the instant the gauge needle stops.

Step 3. Repeat step 2 on all injectors and record pressure drop on each. Clean and retest injectors that appear faulty (Any injectors that have a 10 kPa difference, either more or less, in pressure from the average). If no problem is found, review Symptoms Section B.

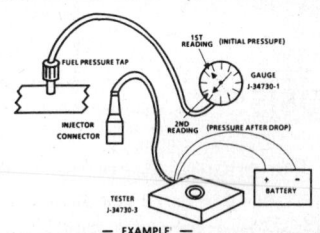

— EXAMPLE —

CYLINDER	1	2	3	4	5	6
1ST READING	225	225	225	225	225	225
2ND READING	100	100	100	90	100	115
AMOUNT OF DROP	125	125	125	135	125	110
	OK	OK	OK	FAULTY, RICH (TOO MUCH) (FUEL DROP)	OK	FAULTY, LEAN (TOO LITTLE) (FUEL DROP)

CHART B-1

RESTRICTED EXHAUST SYSTEM CHECK

Proper diagnosis for a Restricted Exhaust System is essential before any components are replaced. The following procedure may be used for diagnosis:

CHECK AT O_2 SENSOR:

1. Carefully remove O_2 sensor.
2. Install exhaust backpressure tester (BT 8515 or BT 8603) or equivalent in place of O_2 sensor (see illustration).
3. After completing test described below, be sure to coat threads of O_2 sensor with anti-seize compound P/N 5613695 or equivalent prior to re-installation.

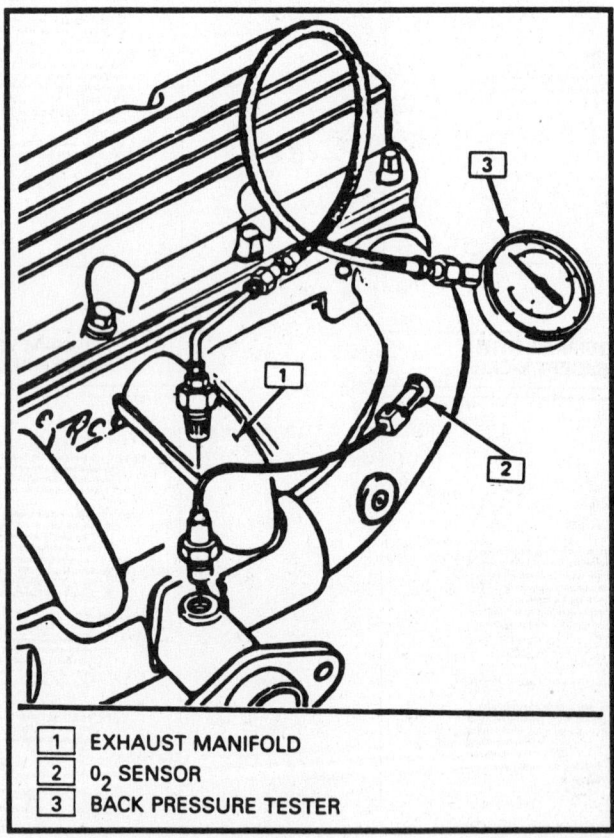

1	EXHAUST MANIFOLD
2	O_2 SENSOR
3	BACK PRESSURE TESTER

DIAGNOSIS:

1. With the engine at normal operating temperature and running at 2500 rpm, observe the exhaust system backpressure reading on the gauge.
2. If the backpressure exceeds 1 1/4 psi (8.62 kPa), a restricted exhaust system is indicated.
3. Inspect the entire exhaust system for a collapsed pipe, heat distress, or possible internal muffler failure.
4. If there are no obvious reasons for the excessive backpressure, a restricted catalytic converter should be suspected, and replaced using current recommended procedures.

1988–90 VEHICLES EQUIPPED WITH 3800 (VIN C) ENGINE EXCEPT E-CAR

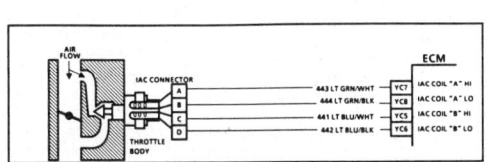

CHART C-2B
IDLE AIR CONTROL (IAC) VALVE CHECK
3800 (VIN C) (PORT)

Circuit Description:
The ECM controls idle rpm with the IAC valve. To increase idle rpm, the ECM retracts the IAC pintle, allowing more air to bypass the throttle plate. To decrease rpm, it extends the IAC pintle valve, reducing air flow through the IAC valve port in the throttle body. A "Scan" tool will read the ECM commands to the IAC valve in counts. The higher the counts, the more air allowed (higher idle). The lower the counts, the less air allowed (lower idle).

Test Description: Numbers below refer to circled numbers on the diagnostic chart.
1. Continue with test, even if engine will not idle. If idle is too low, "Scan" will display 80 or more counts, or steps. If idle is high, it will display "0" counts. Occasionally, an erratic or unstable idle may occur. Engine speed may vary 200 rpm or more up and down. Disconnect IAC. If the condition is unchanged, the IAC is not at fault.
2. When the engine was stopped, the IAC pintle retracted (more air) to a fixed "Park" position for increased air flow and idle speed during the next engine start. A "Scan" will display 100 or more counts.
3. Be sure to disconnect the IAC valve prior to this test. The test light will confirm the ECM signals by a steady or flashing light on all circuits.
4. There is a remote possibility that one of the circuits is shorted to voltage, which would have been indicated by a steady light. Disconnect ECM and turn the ignition "ON" and probe terminals to check for this condition.

Diagnostic Aids:
A slow, unstable idle may be caused by a system problem that cannot be overcome by the IAC. "Scan" counts will be above 60 counts, if idle speed is too low. If idle speed is too high, IAC counts will be "0".

- System lean (High Air/Fuel Ratio)
Idle speed may be too high or too low. Engine speed may vary up and down. May set Code 44.
"Scan" and/or voltmeter will read an oxygen sensor output less than 300 mV (.3volt). Check for low regulated fuel pressure or water in fuel. A lean exhaust, with an oxygen sensor output fixed above 800 mV (.8 volt), will be a contaminated sensor, usually silicone. This may also set Code 45.
- System rich (Low Air/Fuel Ratio)
Idle speed too low. "Scan" counts usually above 80. System obviously rich and may exhibit black smoke exhaust.
"Scan" tool and/or voltmeter will read an oxygen sensor signal fixed above 800 mV (.8 volt).
Check:
   - High fuel pressure
   - Injector leaking or sticking
- Throttle Body - Remove IAC and inspect bore for foreign material or evidence of IAC pintle dragging the bore.
- Refer to "Rough, Unstable, Incorrect Idle or Stalling" in "Symptoms" in Section "B".

1988–90 VEHICLES EQUIPPED WITH 3800 (VIN C) ENGINE EXCEPT E-CAR

CHART C-2B
IDLE AIR CONTROL (IAC) VALVE CHECK
3800 (VIN C) (PORT)

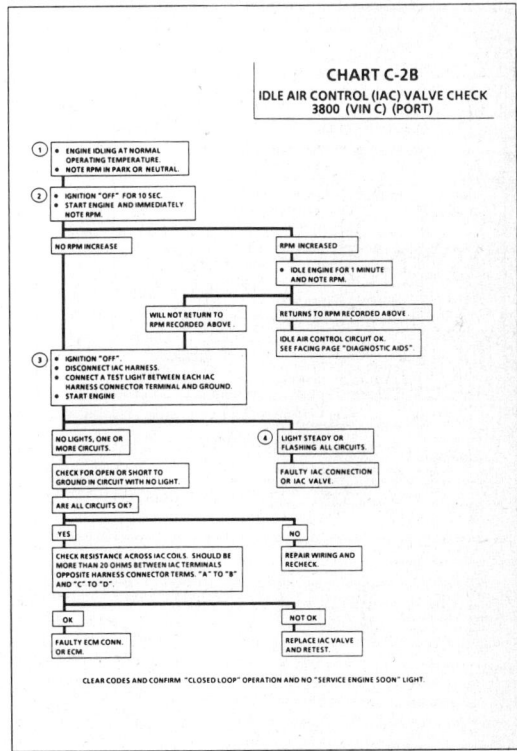

CLEAR CODES AND CONFIRM "CLOSED LOOP" OPERATION AND NO "SERVICE ENGINE SOON" LIGHT.

1988–90 VEHICLES EQUIPPED WITH 3800 (VIN C) ENGINE EXCEPT E-CAR

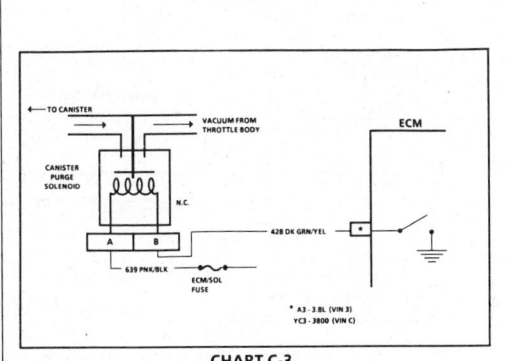

CHART C-3
CANISTER PURGE VALVE CHECK
3800 (VIN C) (PORT)

Circuit Description:
Canister purge is controlled by a solenoid that allows vacuum to purge the canister when energized. The ECM supplies a ground to energize the solenoid (purge "ON").
If the diagnostic test terminal is grounded with the engine stopped or the following is met with the engine running the purge solenoid is energized (purge "ON").
- Engine run time after start more than 1 minute.
- Coolant temperature above 80°C (176°F).
- Vehicle speed above 5 mph (8 km/h).
- Throttle "OFF" idle. TPS signal above .75 volt.

Test Description: Numbers below refer to circled numbers on the diagnostic chart.
1. Checks to see if the solenoid is opened or closed. The solenoid is normally de-energized in this step, so it should be closed.
2. Completes functional check by grounding test terminal. This should normally energize the solenoid and allow the vacuum to drop (purge "ON").
3. Checks for open or shorted solenoid circuit.

1988–90 VEHICLES EQUIPPED WITH 3800 (VIN C) ENGINE EXCEPT E-CAR

CHART C-3
CANISTER PURGE VALVE CHECK
3800 (VIN C) (PORT)

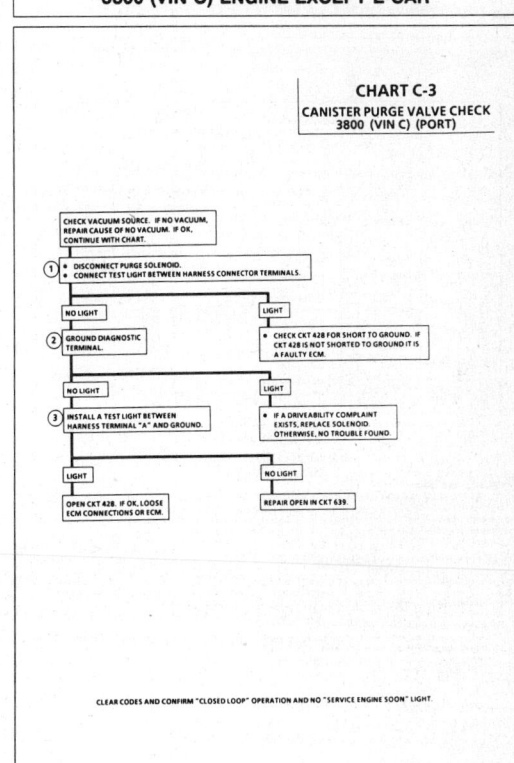

CLEAR CODES AND CONFIRM "CLOSED LOOP" OPERATION AND NO "SERVICE ENGINE SOON" LIGHT.

1988–90 VEHICLES EQUIPPED WITH 3800 (VIN C) ENGINE EXCEPT E-CAR

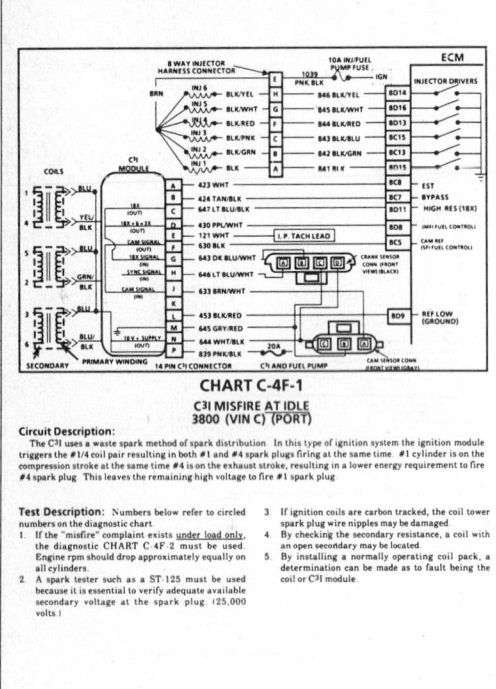

CHART C-4F-1

C3I MISFIRE AT IDLE
3800 (VIN C) (PORT)

Circuit Description:
The C3I uses a waste spark method of spark distribution. In this type of ignition system the ignition module triggers the #1/4 coil pair resulting in both #1 and #4 spark plugs firing at the same time. #1 cylinder is on the compression stroke at the same time #4 is on the exhaust stroke, resulting in a lower energy requirement to fire #4 spark plug. This leaves the remaining high voltage to fire #1 spark plug.

Test Description: Numbers below refer to circled numbers on the diagnostic chart.
1. If the "misfire" complaint exists under load only, the diagnostic CHART C-4F-2 must be used. Engine rpm should drop approximately equally on all cylinders.
2. A spark tester such as a ST-125 must be used because it is essential to verify adequate available secondary voltage at the spark plug. (25,000 volts.)
3. If ignition coils are carbon tracked, the coil tower spark plug wire nipples may be damaged.
4. By checking the secondary resistance, a coil with an open secondary may be located.
5. By installing a normally operating coil pack, a determination can be made as to fault being the coil or C3I module.

1988–90 VEHICLES EQUIPPED WITH 3800 (VIN C) ENGINE EXCEPT E-CAR

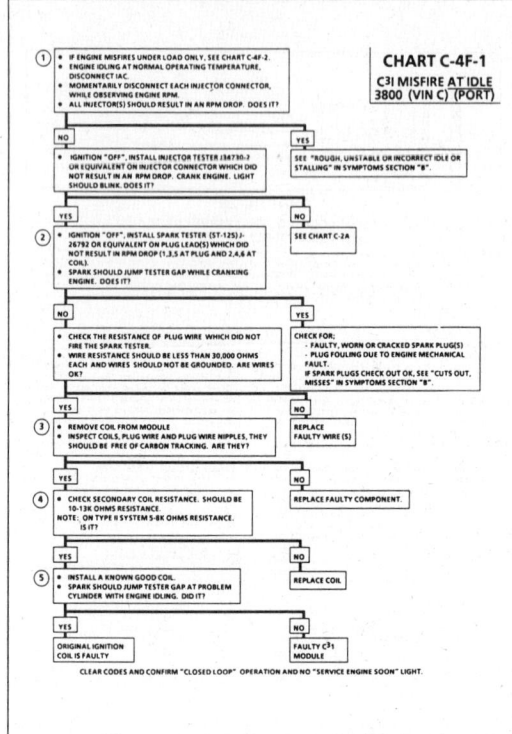

CHART C-4F-1

C3I MISFIRE AT IDLE
3800 (VIN C) (PORT)

1988–90 VEHICLES EQUIPPED WITH 3800 (VIN C) ENGINE EXCEPT E-CAR

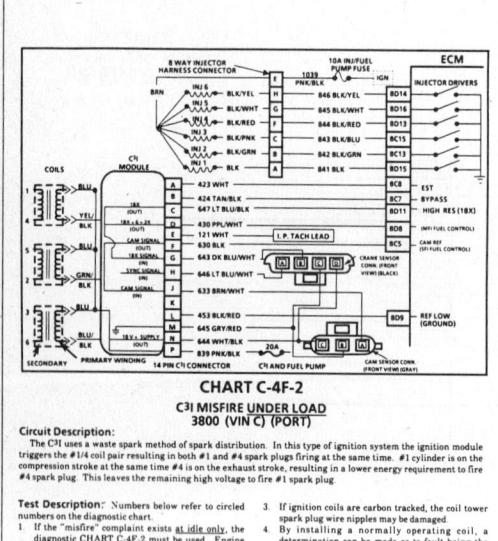

CHART C-4F-2

C3I MISFIRE UNDER LOAD
3800 (VIN C) (PORT)

Circuit Description:
The C3I uses a waste spark method of spark distribution. In this type of ignition system the ignition module triggers the #1/4 coil pair resulting in both #1 and #4 spark plugs firing at the same time. #1 cylinder is on the compression stroke at the same time #4 is on the exhaust stroke, resulting in a lower energy requirement to fire #4 spark plug. This leaves the remaining high voltage to fire #1 spark plug.

Test Description: Numbers below refer to circled numbers on the diagnostic chart.
1. If the "misfire" complaint exists at idle only, the diagnostic CHART C-4F-2 must be used. Engine rpm should drop approximately equally on all cylinders.
2. A spark tester such as a ST-125 must be used because it is essential to verify adequate available secondary voltage at the spark plug. (25,000 volts.) Spark should jump the tester gap on all 6 leads. This simulates a "load" condition.
3. If ignition coils are carbon tracked, the coil tower spark plug wire nipples may be damaged.
4. By installing a normally operating coil, a determination can be made as to fault being the coil or C3I module.

1988–90 VEHICLES EQUIPPED WITH 3800 (VIN C) ENGINE EXCEPT E-CAR

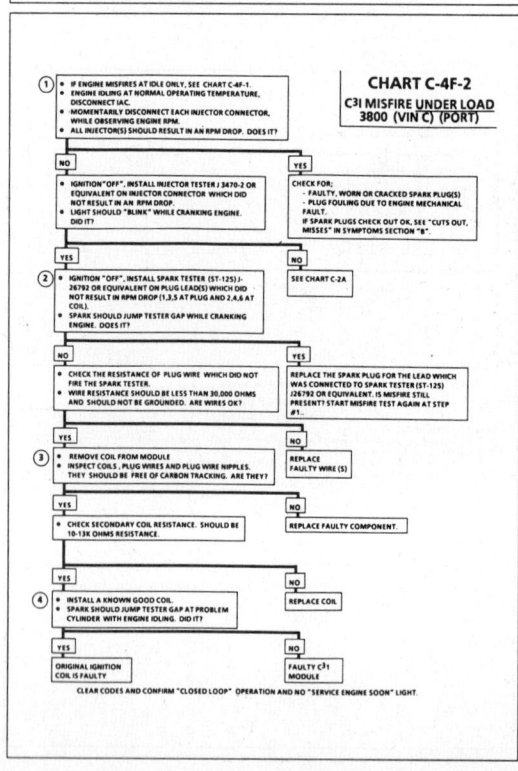

CHART C-4F-2

C3I MISFIRE UNDER LOAD
3800 (VIN C) (PORT)

1988–90 VEHICLES EQUIPPED WITH 3800 (VIN C) ENGINE EXCEPT E-CAR

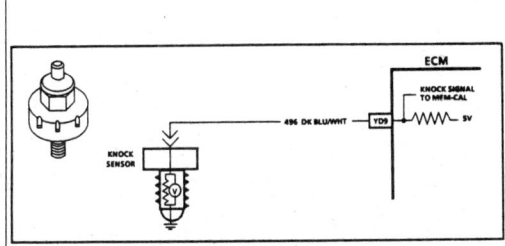

CHART C-5
ELECTRONIC SPARK CONTROL (ESC) SYSTEM CHECK
3800 (VIN C) (PORT)

Circuit Description:
The knock sensor is used to detect engine detonation and the ECM will retard the electronic spark timing based on the signal being received. The circuitry within the knock sensor causes the ECM's supplied 5 volt signal to be pulled down so that under a no knock condition, CKT 496 would measure about 2.5 volts. The knock sensor produces an A/C signal which rides on the 2.5 volts DC voltage. The amplitude and frequency are dependent upon the knock level.
The Mem-Cal used with this engine contains the functions which were part of the remotely mounted ESC modules used on other GM vehicles. The ESC portion of the Mem-Cal then sends a signal to other parts of the ECM which adjusts the spark timing to retard the spark and reduce the detonation.

Test Description: Numbers below refer to circled numbers on the diagnostic chart.
1. With engine idling, there should not be a knock signal present at the ECM because detonation is not likely under a no load condition.
2. Tapping on the engine lift hook should simulate a knock signal to determine if the sensor is capable of detecting detonation. If no knock is detected, try tapping on engine block closer to sensor before replacing sensor.
3. If the engine has an internal problem which is creating a knock, the knock sensor may be responding to the internal failure.

4. This test determines if the knock sensor is faulty or if the ESC portion of the Mem-Cal is faulty. If it is determined that the Mem-Cal is faulty, be sure that it is properly installed and latched into place. If not properly installed, repair and retest.

Diagnostic Aids:

While observing knock signal on the "Scan," there should be an indication that knock is present when detonation can be heard. Detonation is most likely to occur under high engine load conditions.

1988–90 VEHICLES EQUIPPED WITH 3800 (VIN C) ENGINE EXCEPT E-CAR

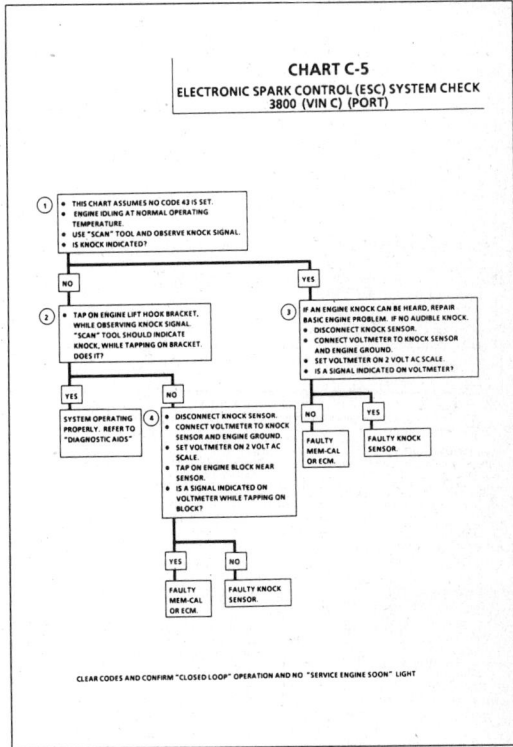

1988–90 VEHICLES EQUIPPED WITH 3800 (VIN C) ENGINE EXCEPT E-CAR

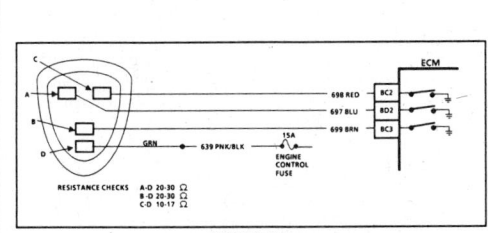

CHART C-7
EXHAUST GAS RECIRCULATION (EGR) FLOW CHECK
3800 (VIN C) (PORT)

Circuit Description:
The digital (EGR) valve is designed to accurately supply EGR to an engine independent of intake manifold vacuum. The valve controls EGR flow from the exhaust to the intake manifold through three orifices which increment in size to produce seven combinations. When a solenoid is energized, the armature with attached shaft and swivel pintle is lifted opening the orifice.
The flow accuracy is dependent on metering orifice size only, which results in improved control

Test Description: Numbers below refer to circled numbers on the diagnostic chart.
If a code is set, inspect EGR for damage. Disconnect the 4 wire connector at EGR.
Install a fused jumper from the battery to terminal "D" of the EGR.

Start and idle engine and with a jumper, ground terminals "A," "B" and "C" one at a time. You should be able to discern a change in engine rpm as the terminal is grounded.
Terminal "A" should result in a small change and "C" in a large change in rpm. The engine may stall when "C" is grounded.

1988–90 VEHICLES EQUIPPED WITH 3800 (VIN C) ENGINE EXCEPT E-CAR

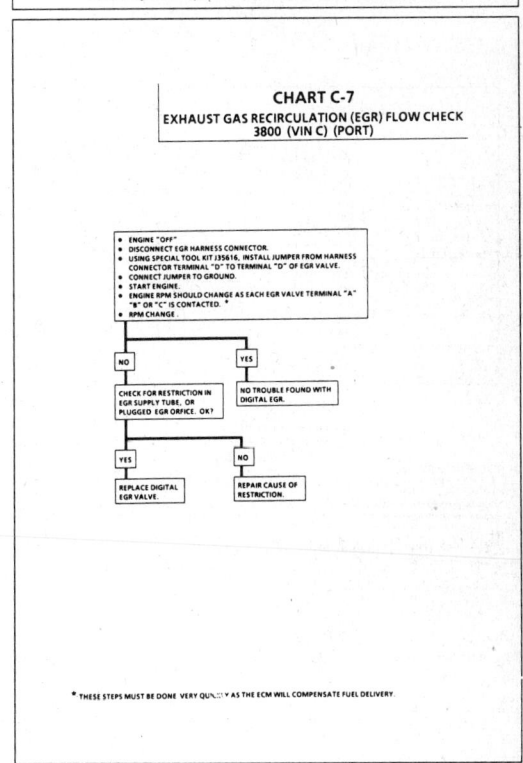

1988-90 VEHICLES EQUIPPED WITH 3800 (VIN C) ENGINE EXCEPT E-CAR

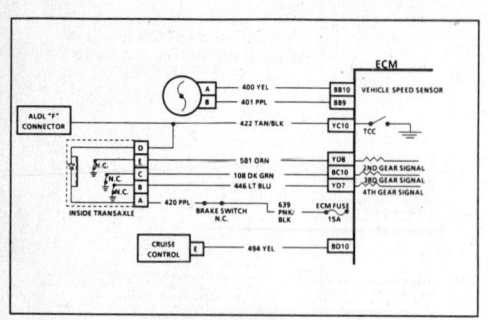

CHART C-8A
(Page 1 of 2)
TORQUE CONVERTER CLUTCH (TCC)
(ELECTRICAL DIAGNOSIS)
3800 (VIN C) (PORT)

Circuit Description:

The purpose of the torque converter clutch feature is to eliminate the power loss of the torque converter when the vehicle is in a cruise condition. This allows the convenience of the automatic and the fuel economy of a manual transaxle. The heart of the system is a solenoid located inside the transaxle which is controlled by the ECM.

When the solenoid coil is activated ("ON,") the torque converter clutch is applied which results in straight through mechanical coupling from the engine to the wheels. When the transaxle (TCC) solenoid is deactivated, the torque converter clutch is released which allows the torque converter to operate in the conventional manner (fluidic coupling between engine and transaxle).

TCC will engage when.
- Engine warmed up.
- Vehicle speed above a calibrated value (about 28 mph 45 km/h).
- Throttle position sensor output not changing, indicating a steady road speed.
- Brake switch closed.

Test Description: Numbers below refer to circled numbers on the diagnostic chart.
1. This test checks the continuity of the TCC circuit from the fuse to the ALDL connector.
2. When the brake pedal is released, the light should come back "ON" and then go "OFF" when the diagnostic terminal is grounded. This tests CKT 422 and the TCC driver in the ECM.

Diagnostic Aids:

The "Scan" tool only indicates when the ECM has turned "ON" the TCC driver and this does not confirm that the TCC has engaged. To determine if TCC is functioning properly, road test the vehicle. Engine rpm should decrease when the "Scan" indicates the TCC driver has turned "ON."

1988-90 VEHICLES EQUIPPED WITH 3800 (VIN C) ENGINE EXCEPT E-CAR

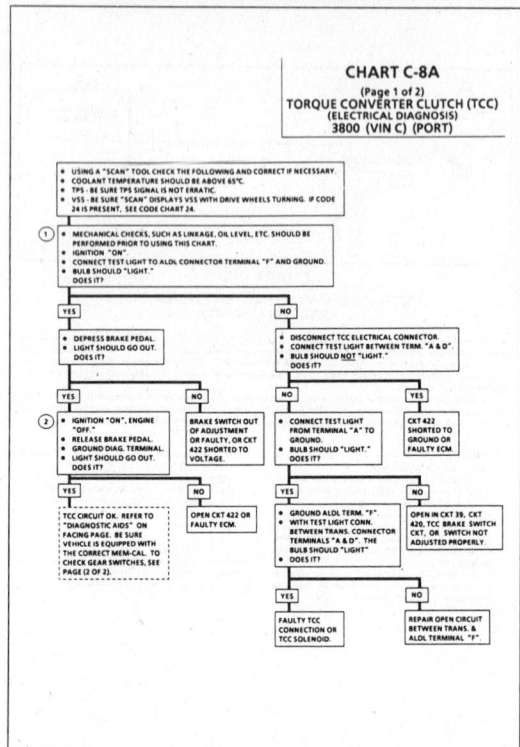

CHART C-8A
(Page 1 of 2)
TORQUE CONVERTER CLUTCH (TCC)
(ELECTRICAL DIAGNOSIS)
3800 (VIN C) (PORT)

1988-90 VEHICLES EQUIPPED WITH 3800 (VIN C) ENGINE EXCEPT E-CAR

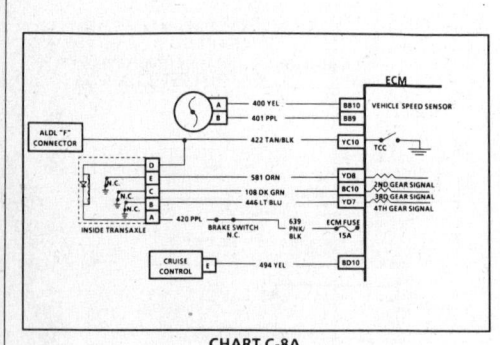

CHART C-8A
(Page 2 of 2)
TORQUE CONVERTER CLUTCH (TCC)
(ELECTRICAL DIAGNOSIS)
3800 (VIN C) (PORT)

Circuit Description:

Each gear switch opens when the appropriate clutch is applied. All gear switches are open in fourth gear.

Test Description: Numbers below refer to circled numbers on the diagnostic chart.
1. Some "Scan" tools display the state of these switches in different ways. Be familiar with the type of tool being used. All switches should be in the closed state during this test, the tool should read the same for 2nd, 3rd or 4th gear switches.
2. Determines whether the switch or signal circuit is open. The circuit can be checked for an open by measuring the voltage (with a voltmeter) at the TCC connector. Should be about 12 volts.

3. Because the switch(s) should be grounded in this step, disconnecting the TCC connector should cause the "Scan" switch state to change.
4. The switch state should change when the vehicle shifts into 3rd gear

Diagnostic Aids:

If vehicle is road tested because of a TCC related problem, be sure the switch states do not change while in 4th gear because the TCC will disengage. If switches change state, carefully check wire routing and connections.

1988-90 VEHICLES EQUIPPED WITH 3800 (VIN C) ENGINE EXCEPT E-CAR

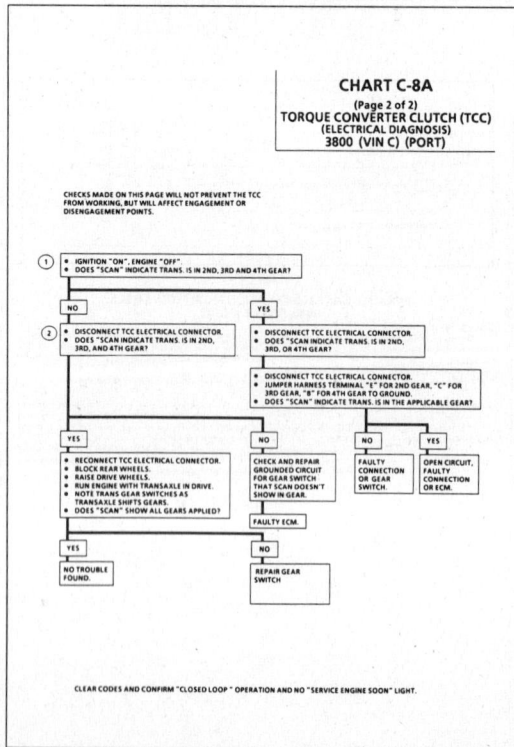

CHART C-8A
(Page 2 of 2)
TORQUE CONVERTER CLUTCH (TCC)
(ELECTRICAL DIAGNOSIS)
3800 (VIN C) (PORT)

1988–90 VEHICLES EQUIPPED WITH 3800 (VIN C) ENGINE EXCEPT E-CAR

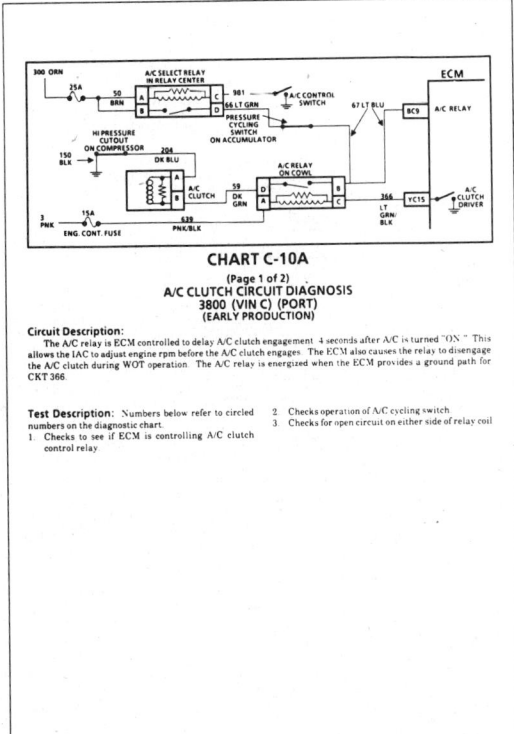

CHART C-10A
(Page 1 of 2)
A/C CLUTCH CIRCUIT DIAGNOSIS
3800 (VIN C) (PORT)
(EARLY PRODUCTION)

Circuit Description:
The A/C relay is ECM controlled to delay A/C clutch engagement 4 seconds after A/C is turned "ON." This allows the IAC to adjust engine rpm before the A/C clutch engages. The ECM also causes the relay to disengage the A/C clutch during WOT operation. The A/C relay is energized when the ECM provides a ground path for CKT 366.

Test Description: Numbers below refer to circled numbers on the diagnostic chart.
1. Checks to see if ECM is controlling A/C clutch control relay.
2. Checks operation of A/C cycling switch.
3. Checks for open circuit on either side of relay coil.

1988–90 VEHICLES EQUIPPED WITH 3800 (VIN C) ENGINE EXCEPT E-CAR

CHART C-10A
(Page 1 of 2)
A/C CLUTCH CIRCUIT DIAGNOSIS
3800 (VIN C) (PORT)
(EARLY PRODUCTION)

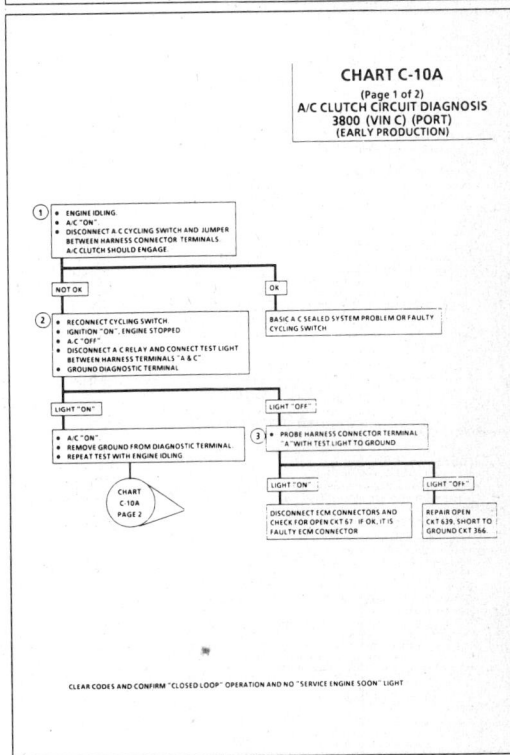

1988–90 VEHICLES EQUIPPED WITH 3800 (VIN C) ENGINE EXCEPT E-CAR

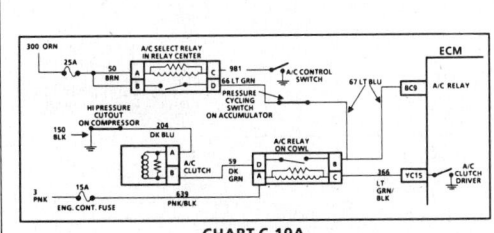

CHART C-10A
(Page 2 of 2)
A/C CLUTCH CIRCUIT DIAGNOSIS
3800 (VIN C) (PORT)
(EARLY PRODUCTION)

Circuit Description:
The A/C relay is ECM controlled to delay A/C clutch engagement 4 seconds after A/C is turned "ON." This allows the IAC to adjust engine rpm before the A/C clutch engages. The ECM also causes the relay to disengage the A/C clutch during WOT operation. The A/C relay is energized when the ECM provides a ground path for CKT 366.

Test Description: Numbers below refer to circled numbers on the diagnostic chart.
1. Checks for battery voltage to relay through CKT 67.
2. Substitutes for relay to determine if problem is in relay or in CKT 59, A/C clutch coil, high pressure, switch, or ground.
3. Checks for open in CKT 67 between cycling switch and A/C fuse, or open CKT 67 to relay.
4. Checks to see that A/C "ON" signal is getting to ECM through CKT 67. A test light "OFF" at this time indicates CKT 67 is open between the cycling switch and the ECM.

1988–90 VEHICLES EQUIPPED WITH 3800 (VIN C) ENGINE EXCEPT E-CAR

CHART C-10A
(Page 2 of 2)
A/C CLUTCH CIRCUIT DIAGNOSIS
3800 (VIN C) (PORT)
(EARLY PRODUCTION)

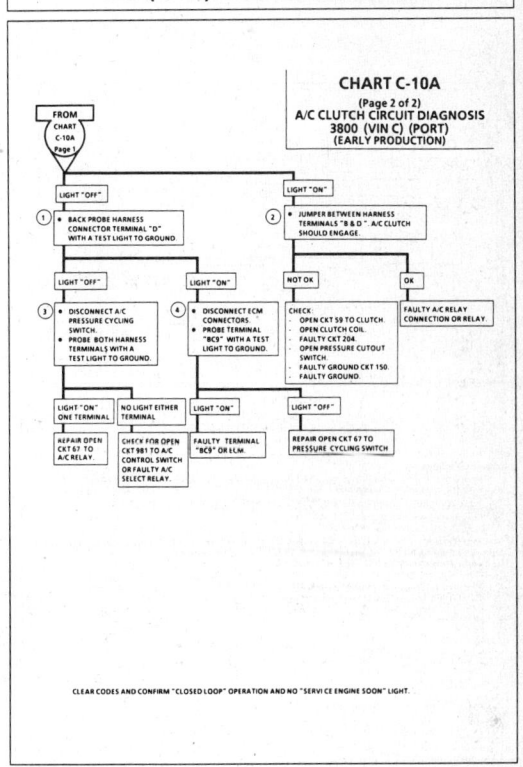

1988–90 VEHICLES EQUIPPED WITH 3800 (VIN C) ENGINE EXCEPT E-CAR

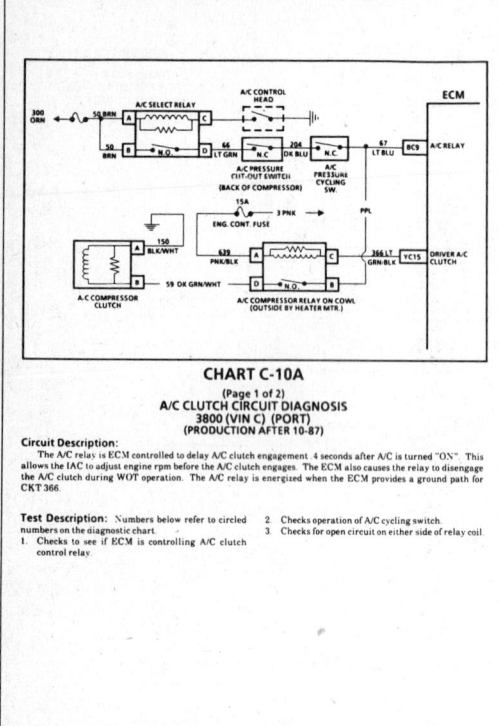

CHART C-10A
(Page 1 of 2)
A/C CLUTCH CIRCUIT DIAGNOSIS
3800 (VIN C) (PORT)
(PRODUCTION AFTER 10-87)

Circuit Description:
The A/C relay is ECM controlled to delay A/C clutch engagement 4 seconds after A/C is turned "ON". This allows the IAC to adjust engine rpm before the A/C clutch engages. The ECM also causes the relay to disengage the A/C clutch during WOT operation. The A/C relay is energized when the ECM provides a ground path for CKT 366.

Test Description: Numbers below refer to circled numbers on the diagnostic chart.
1. Checks to see if ECM is controlling A/C clutch control relay.
2. Checks operation of A/C cycling switch.
3. Checks for open circuit on either side of relay coil.

1988–90 VEHICLES EQUIPPED WITH 3800 (VIN C) ENGINE EXCEPT E-CAR

CHART C-10A
(Page 1 of 2)
A/C CLUTCH CIRCUIT DIAGNOSIS
3800 (VIN C) (PORT)
(PRODUCTION AFTER 10-87)

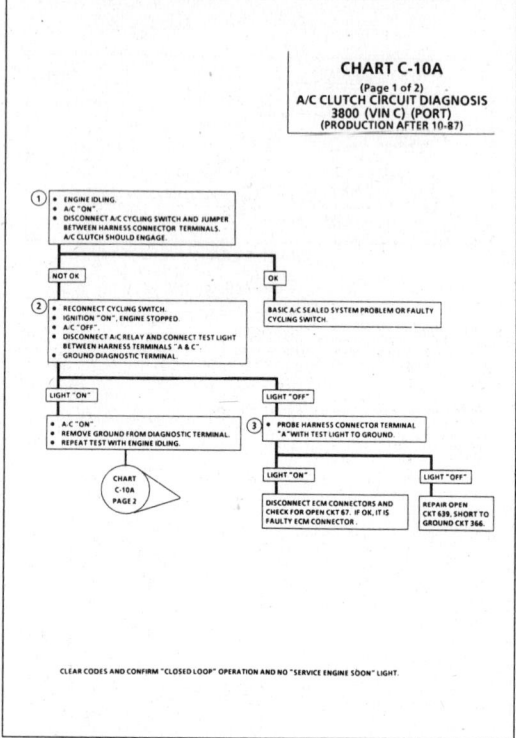

CLEAR CODES AND CONFIRM "CLOSED LOOP" OPERATION AND NO "SERVICE ENGINE SOON" LIGHT.

1988–90 VEHICLES EQUIPPED WITH 3800 (VIN C) ENGINE EXCEPT E-CAR

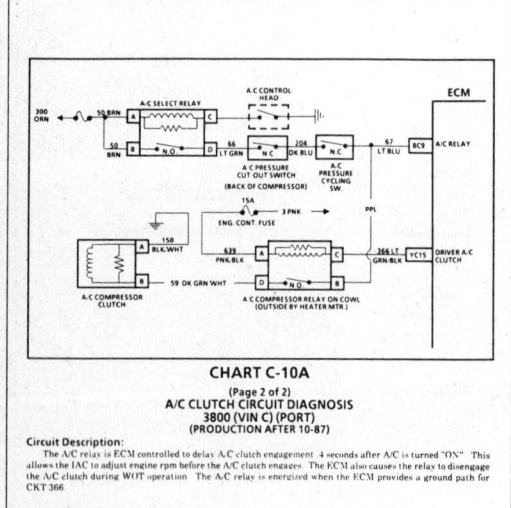

CHART C-10A
(Page 2 of 2)
A/C CLUTCH CIRCUIT DIAGNOSIS
3800 (VIN C) (PORT)
(PRODUCTION AFTER 10-87)

Circuit Description:
The A/C relay is ECM controlled to delay A/C clutch engagement 4 seconds after A/C is turned "ON". This allows the IAC to adjust engine rpm before the A/C clutch engages. The ECM also causes the relay to disengage the A/C clutch during WOT operation. The A/C relay is energized when the ECM provides a ground path for CKT 366.

Test Description: Numbers below refer to the diagnostic chart.
1. Checks for battery voltage to relay through CKT 67.
2. Substitutes for relay to determine if problem is in relay or in CKT 59, A/C clutch coil, high pressure, switch, or ground.
3. Checks for open in CKT 67 between cycling switch and A/C fuse, or open CKT 67 to relay.
4. Checks to see that A/C "ON" signal is getting to ECM through CKT 67. A test light "OFF" at this time indicates CKT 67 is open between the cycling switch and the ECM.

1988–90 VEHICLES EQUIPPED WITH 3800 (VIN C) ENGINE EXCEPT E-CAR

CHART C-10A
(Page 2 of 2)
A/C CLUTCH CIRCUIT DIAGNOSIS
3800 (VIN C) (PORT)
(PRODUCTION AFTER 10-87)

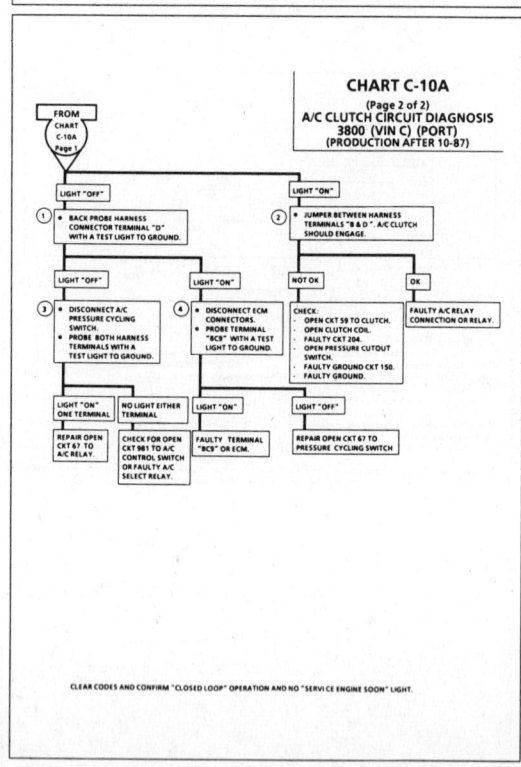

CLEAR CODES AND CONFIRM "CLOSED LOOP" OPERATION AND NO "SERVICE ENGINE SOON" LIGHT.

1988–90 VEHICLES EQUIPPED WITH 3800 (VIN C) ENGINE EXCEPT E-CAR

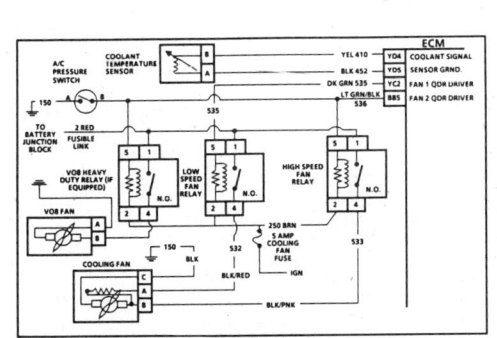

CHART C-12A
COOLANT FAN CHECK
3800 (VIN C) (PORT)

Circuit Description:
Power for the fan motor comes through the fusible link to terminal "1" on all relays. The relays are energized when current flows to ground through the ECM (quad-drivers).
Low Speed Relay - The ECM energizes the relay through terminal "YC2" when the coolant temperature reaches 101°C (208°F).
High Speed Relay - The high speed relay is energized by the ECM or the A/C pressure switch. If the A/C refrigerant pressure reaches 275 psi (1896 kPa) or the coolant temperature reaches 108°C (226°F), the high speed fan relay is energized.

Test Description: Numbers below refer to circled numbers on the diagnostic chart.
1. Grounding the diagnostic test terminal should cause the ECM to ground CKT 535 and the fan should run in low speed.
2. Grounding the coolant temperature switch harness terminal will check CKT 536 and will also check the high speed fan control relay.
3. Separates and checks ECM driver circuit and relay to fan circuit for open circuit or faulty relay.
4. This step checks to see if the coolant temperature switch is grounding and is grounded when the light comes "ON." The switch should close at 108°C (226°F).
5. This will check the A/C pressure switch and related wiring from the switch to the fan control relay. If poor A/C performance is noted, the A/C pressure switch should be checked by a qualified A/C repair person. The low speed fan should come "ON," if high pressure exceeds 260 psi (1782 kPa).

1988–90 VEHICLES EQUIPPED WITH 3800 (VIN C) ENGINE EXCEPT E-CAR

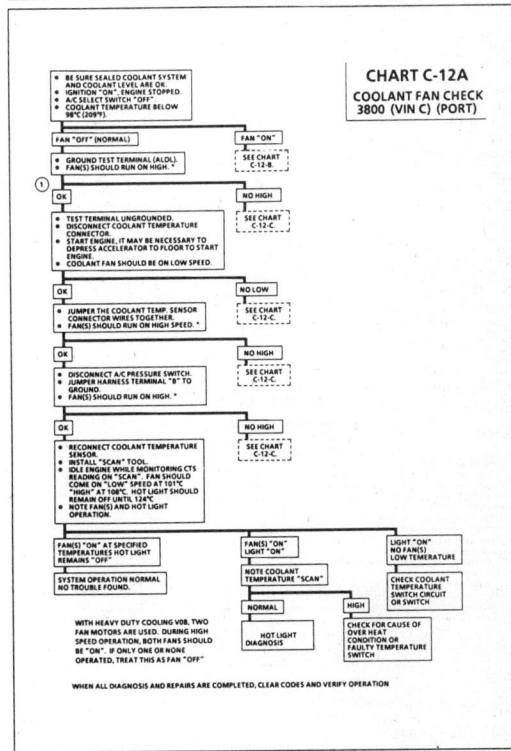

CHART C-12A
COOLANT FAN CHECK
3800 (VIN C) (PORT)

1988–90 VEHICLES EQUIPPED WITH 3800 (VIN C) ENGINE EXCEPT E-CAR

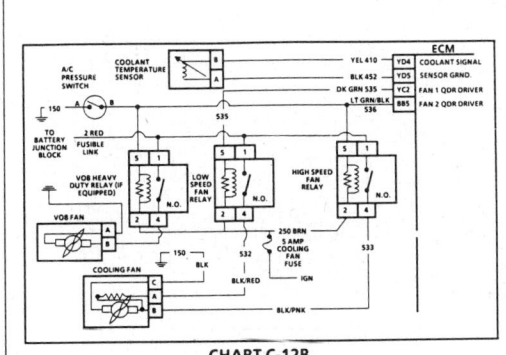

CHART C-12B
FAN "ON" AT ALL TIMES
COOLANT FAN CHECK
3800 (VIN C) (PORT)

Circuit Description:
Power for the fan motor comes through the fusible link to terminal "1" on all relays. The relays are energized when current flows to ground through the ECM (quad drivers).
Low Speed Relay - The ECM energizes the relay through terminal "YC2" when the coolant temperature reaches 101°C (208°F).
High Speed Relay - The high speed relay is energized by the ECM or the A/C pressure switch. If the A/C refrigerant pressure reaches 275 psi (1896 kPa) or the coolant temperature reaches 108°C (226°F), the high speed fan relay is energized.

Test Description: Numbers below refer to circled numbers on the diagnostic chart.
1. Checks to see if CKT 535 is shorted to ground, which would keep the relay grounded at all times.
2. Checks to see if CKT 536 is shorted to ground. A light indicates the wire is shorted to ground the following steps will isolate the short.
3. If the test light is "OFF," after disconnecting, the ECM is shorted internally. Before replacing the ECM, be sure and check the resistance value of low speed side of the fan control relay. Replace, if resistance is less than 20 ohms. Also, be sure the CKT 535 is not shorted to B + and check the resistance of the canister purge solenoid and replace if under 20 ohms.

1988–90 VEHICLES EQUIPPED WITH 3800 (VIN C) ENGINE EXCEPT E-CAR

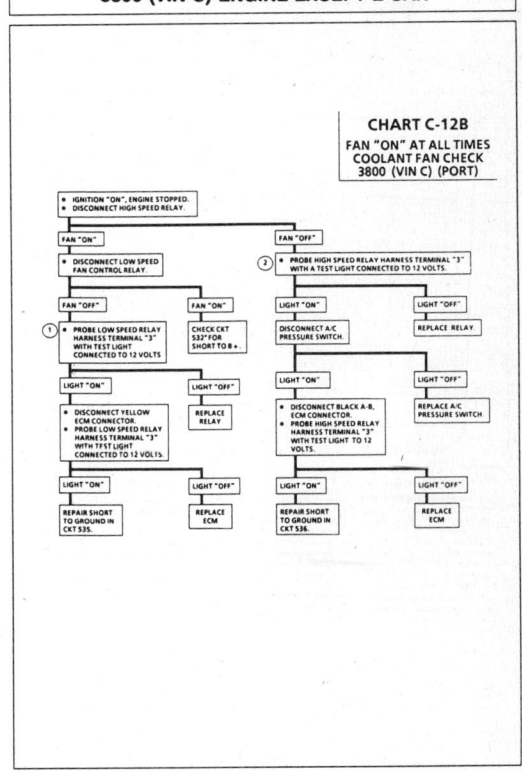

CHART C-12B
FAN "ON" AT ALL TIMES
COOLANT FAN CHECK
3800 (VIN C) (PORT)

1988–90 VEHICLES EQUIPPED WITH 3800 (VIN C) ENGINE EXCEPT E-CAR

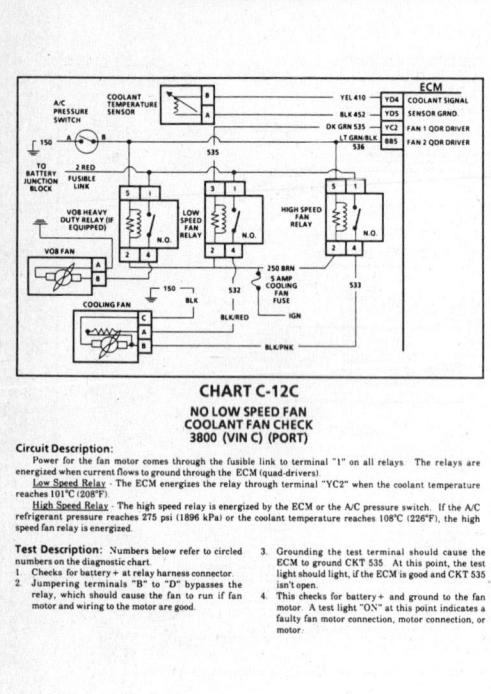

CHART C-12C
NO LOW SPEED FAN
COOLANT FAN CHECK
3800 (VIN C) (PORT)

Circuit Description:
Power for the fan motor comes through the fusible link to terminal "1" on all relays. The relays are energized when current flows to ground through the ECM (quad-drivers).

Low Speed Relay - The ECM energizes the relay through terminal "YC2" when the coolant temperature reaches 101°C (208°F).

High Speed Relay - The high speed relay is energized by the ECM or the A/C pressure switch. If the A/C refrigerant pressure reaches 275 psi (1896 kPa) or the coolant temperature reaches 108°C (226°F), the high speed fan relay is energized.

Test Description: Numbers below refer to circled numbers on the diagnostic chart.
1. Checks for battery + at relay harness connector.
2. Jumpering terminals "B" to "D" bypasses the relay, which should cause the fan to run if fan motor and wiring to the motor are good.
3. Grounding the test terminal should cause the ECM to ground CKT 535. At this point, the test light should light, if the ECM is good and CKT 535 isn't open.
4. This checks for battery + and ground to the fan motor. A test light "ON" at this point indicates a faulty fan motor connection, motor connection, or motor.

1988–90 VEHICLES EQUIPPED WITH 3800 (VIN C) ENGINE EXCEPT E-CAR

CHART C-12C
NO LOW SPEED FAN
COOLANT FAN CHECK
3800 (VIN C) (PORT)

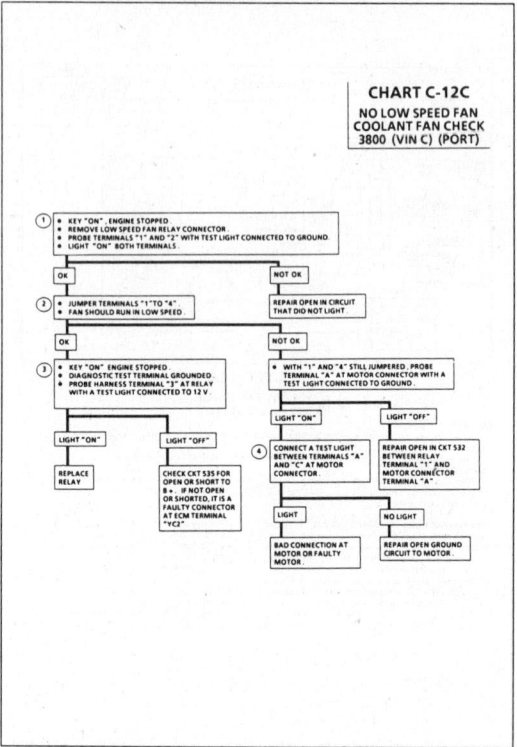

1988 A-CAR EQUIPPED WITH 3.8L (VIN 3) ENGINE

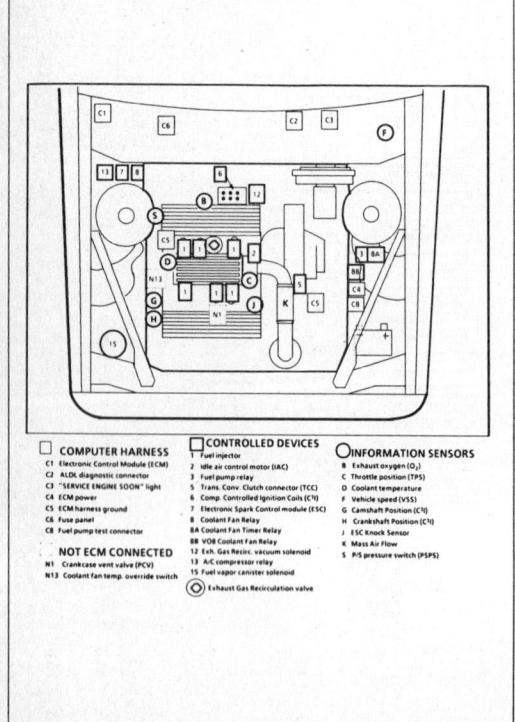

☐ **COMPUTER HARNESS**
C1 Electronic Control Module (ECM)
C2 ALDL diagnostic connector
C3 "SERVICE ENGINE SOON" light
C4 ECM power
C5 ECM harness ground
C6 Fuse panel
C8 Fuse pump test connector

NOT ECM CONNECTED
N1 Crankcase vent valve (PCV)
N13 Coolant fan temp. override switch

☐ **CONTROLLED DEVICES**
1 Fuel injector
2 Idle air control motor (IAC)
3 Fuel pump relay
5 Trans. Conv. Clutch connector (TCC)
6 Comp. Controlled Ignition Coils (C³I)
7 Electronic Spark Control module (ESC)
8 Coolant Fan Relay
8A Coolant Fan Timer Relay
8B VO8 Coolant Fan Relay
12 Exh. Gas Recirc. vacuum solenoid
13 A/C compressor relay
15 Fuel vapor canister solenoid

○ **INFORMATION SENSORS**
B Exhaust oxygen (O₂)
5 Throttle position (TPS)
D Coolant temperature
F Vehicle speed (VSS)
G Camshaft Position (C³I)
H Crankshaft Position (C³I)
J ESC Knock Sensor
K Mass Air Flow
5 P/S pressure switch (PSPS)

⊙ Exhaust Gas Recirculation valve

1988 C AND H-CAR EQUIPPED WITH 3.8L (VIN 3) ENGINE

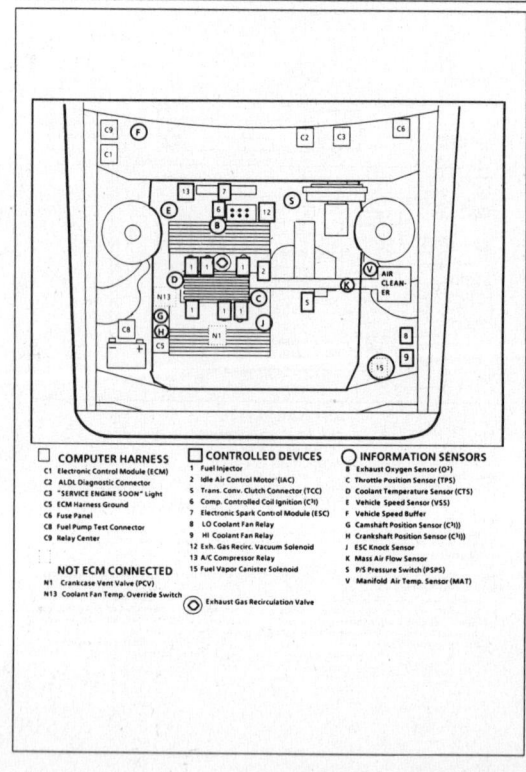

☐ **COMPUTER HARNESS**
C1 Electronic Control Module (ECM)
C2 ALDL Diagnostic Connector
C3 "SERVICE ENGINE SOON" Light
C5 ECM Harness Ground
C6 Fuse Panel
C8 Fuel Pump Test Connector
C9 Relay Center

NOT ECM CONNECTED
N1 Crankcase Vent Valve (PCV)
N13 Coolant Fan Temp. Override Switch

☐ **CONTROLLED DEVICES**
1 Fuel Injector
2 Idle Air Control Motor (IAC)
3 Fuel Pump Relay
5 Trans. Conv. Clutch Connector (TCC)
6 Comp. Controlled Coil Ignition (C³I)
7 Electronic Spark Control Module (ESC)
8 LO Coolant Fan Relay
9 HI Coolant Fan Relay
12 Exh. Gas Recirc. Vacuum Solenoid
15 Fuel Vapor Canister Solenoid

○ **INFORMATION SENSORS**
B Exhaust Oxygen Sensor (O²)
C Throttle Position Sensor (TPS)
D Coolant Temperature Sensor (CTS)
E Vehicle Speed Sensor (VSS)
F Vehicle Speed Buffer
G Camshaft Position Sensor (C³I)
H Crankshaft Position Sensor (C³I)
J ESC Knock Sensor
K Mass Air Flow Sensor
5 P/S Pressure Switch (PSPS)
V Manifold Air Temp. Sensor (MAT)

⊙ Exhaust Gas Recirculation Valve

ECM WIRING DIAGRAM — A-CAR EQUIPPED WITH 3.8L (VIN 3) ENGINE

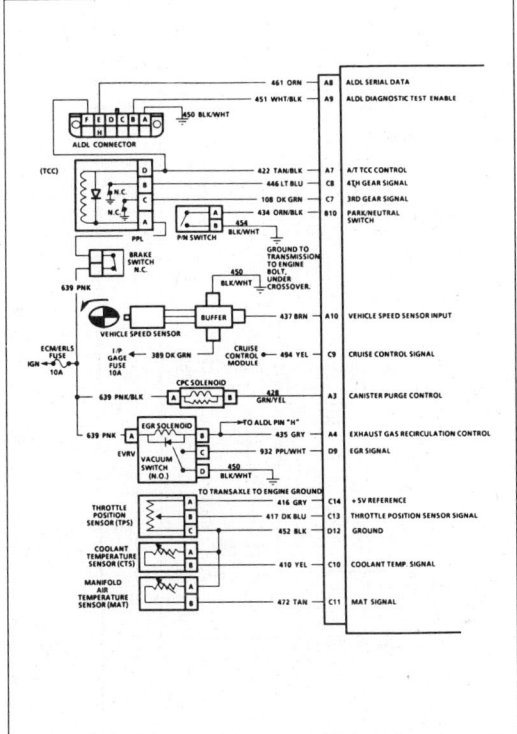

ECM WIRING DIAGRAM — A-CAR EQUIPPED WITH 3.8L (VIN 3) ENGINE

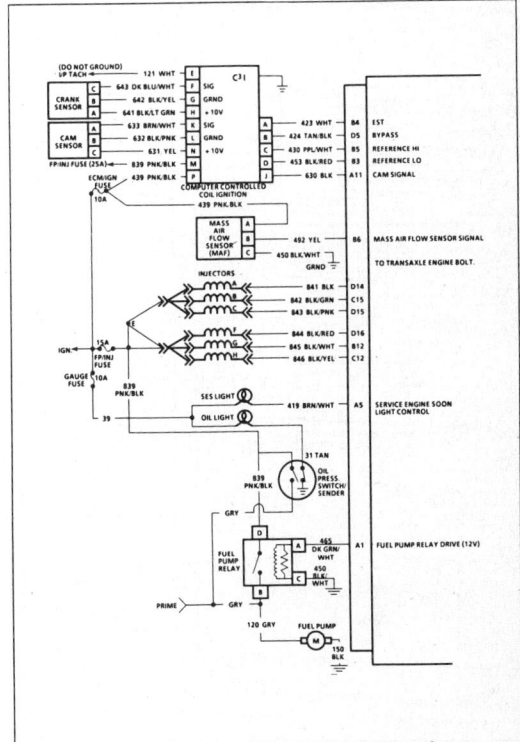

ECM WIRING DIAGRAM — A-CAR EQUIPPED WITH 3.8L (VIN 3) ENGINE

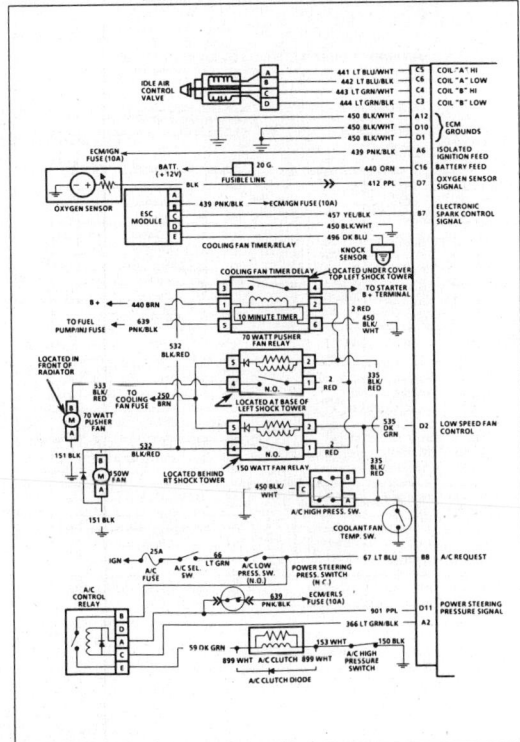

ECM WIRING DIAGRAM — C AND H-CAR EQUIPPED WITH 3.8L (VIN 3) ENGINE

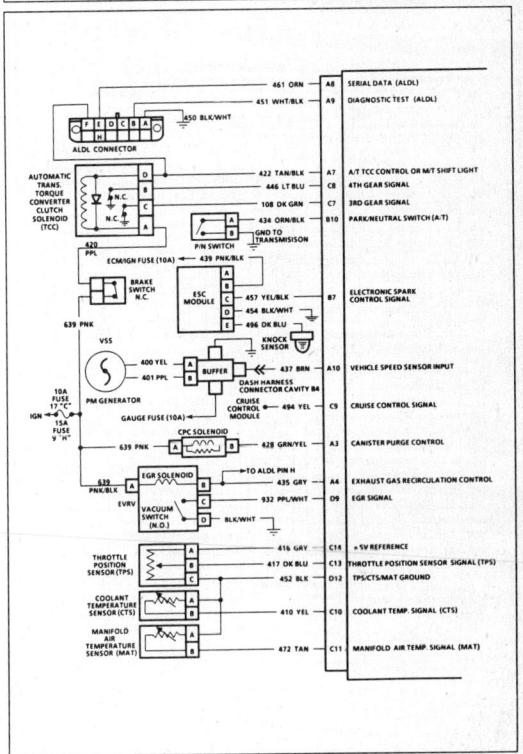

ECM WIRING DIAGRAM—C AND H-CAR EQUIPPED WITH 3.8L (VIN 3) ENGINE

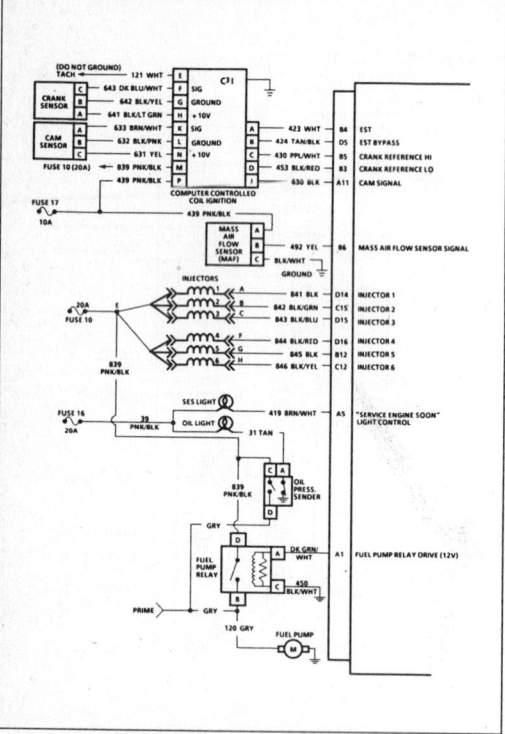

ECM WIRING DIAGRAM—C AND H-CAR EQUIPPED WITH 3.8L (VIN 3) ENGINE

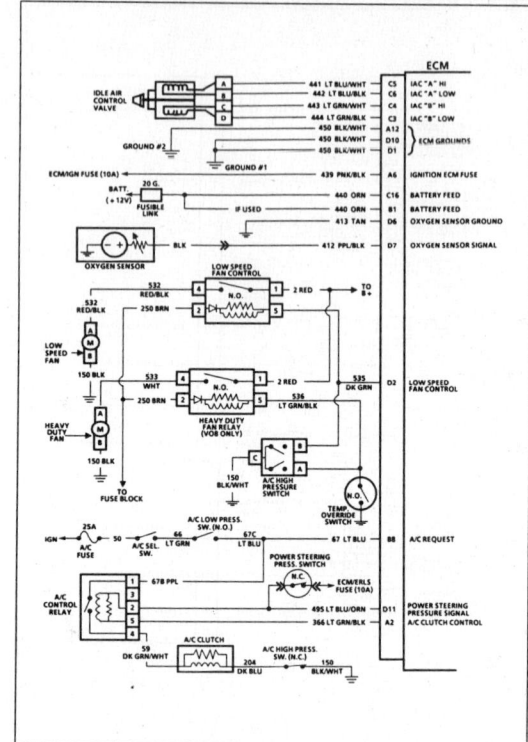

ECM CONNECTORS—A-CAR EQUIPPED WITH 3.8L (VIN 3) ENGINE

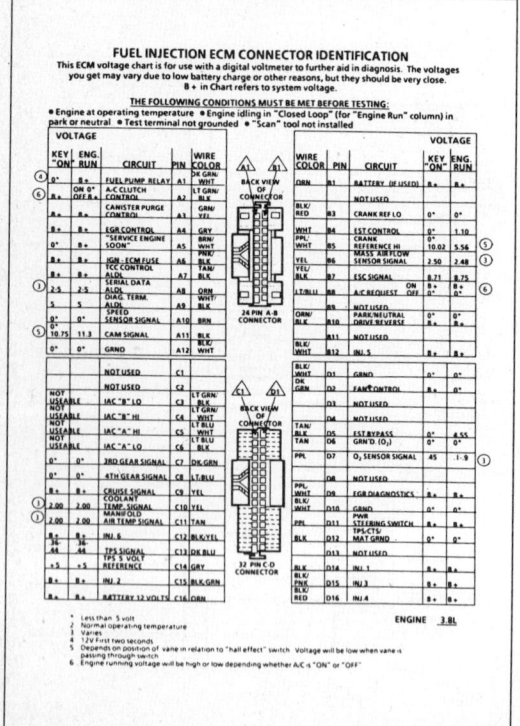

ECM CONNECTORS—C AND H-CAR EQUIPPED WITH 3.8L (VIN 3) ENGINE

1988 VEHICLES EQUIPPED WITH 3.8L (VIN 3) ENGINE

DIAGNOSTIC CIRCUIT CHECK

The diagnostic circuit check must be the starting point for any driveability complaint diagnosis.

The diagnostic circuit check is an organized approach to identifying a problem created by an electronic engine control system malfunction because it directs the service technician to the next logical step in diagnosing the complaint.

If after completing the diagnostic circuit check and finding the on-board diagnostics functioning properly and no trouble codes displayed, a comparison of "Typical Scan Values", for the appropriate engine, may be used for comparison. The "Typical Values" are an average of display values recorded from normally operating vehicles and are intended to represent what a normally functioning system would display. **A "SCAN" TOOL THAT DISPLAYS FAULTY DATA SHOULD NOT BE USED, AND THE PROBLEM SHOULD BE REPORTED TO THE MANUFACTURER. THE USE OF A FAULTY "SCAN" CAN RESULT IN MISDIAGNOSIS AND UNNECESSARY PARTS REPLACEMENT.**

TYPICAL "Scan" DATA VALUES
3.8L (VIN 3)

Idle / Upper Radiator Hose Hot / Closed Throttle / Park or Neutral / Closed Loop / Acc. off

"SCAN" Position	Units Displayed	Typical Data Value
Engine Speed	RPM	Varies
Coolant Temp.	C°	85° - 105°
MAT Mani Air Temp.	C°	Varies with Air Temperature
Air Flow	Gm/Sec	4 - 7
Oxygen Sensor	Millivolts	1 - 9
Throt. Position	Vol	36 - 44
Idle Air Control	Counts (steps)	10 - 50
Park/Neutral	P/N and RDL	P-N
Fuel Integ.	Counts	118 - 138
Block Learn	Counts	118 - 138
Closed Loop Flag.	Open/Closed	Closed Loop (may go open with extended idle)
Vehicle Speed	MPH	0 (Zero)
Torque Conv. Cl.	On/Off	Off
EGR Diag. Switch	On/Off	Off
Spark Advance	Degrees	Varies
Knock Signal	No/Yes	No
Power Steering	Normal	Normal
3rd Gear (440-T4)	On/Off	Off
4th Gear (440-T4)	On/Off	Off
A/C Request	Yes/No	No
PROM I.D.	Numbers	Internal I.D. Only
EGR Duty Cycle	%	0 - 50
LV8	0	30 - 40

NOTE: IF ALL VALUES ARE WITHIN THE RANGE ILLUSTRATED, REFER TO SYMPTOMS IN SECTION "B".

1988 VEHICLES EQUIPPED WITH 3.8L (VIN 3) ENGINE

DIAGNOSTIC CIRCUIT CHECK
3.8L (VIN 3) (PORT)

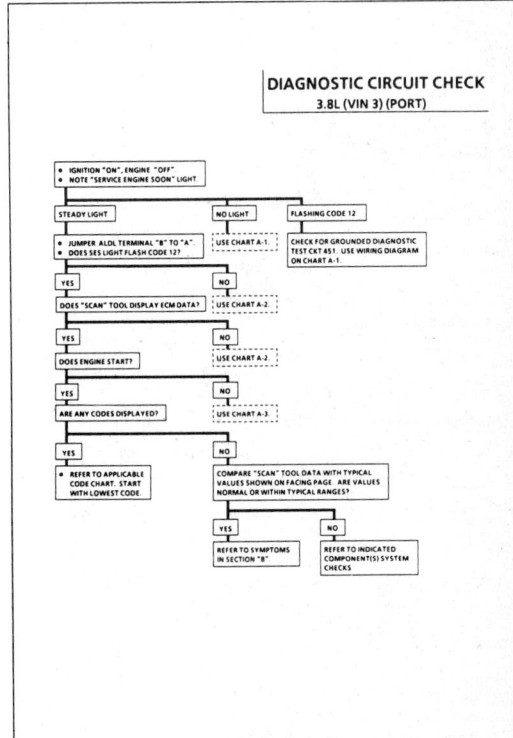

1988 VEHICLES EQUIPPED WITH 3.8L (VIN 3) ENGINE

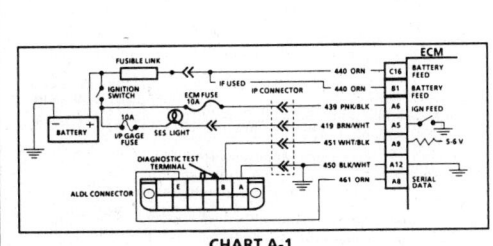

CHART A-1
NO "SERVICE ENGINE SOON" LIGHT
3.8L (VIN 3) (PORT)

Circuit Description:
There should always be a steady "Service Engine Soon" light when the ignition is "ON" and engine not running. Battery is supplied directly to the light bulb. The electronic control module (ECM) will control the light and turn it "ON" by providing a ground path through CKT 419 to the ECM.

Test Description: Numbers below refer to circled numbers on the diagnostic chart.
1. "Service Engine Soon" light should be "ON" as the test light provides the ground.
2. Using a test light connected to 12 volts, probe each of the system ground circuits to be sure a good ground is present. See ECM terminal end view in front of this section for ECM pin locations of ground circuits.

Diagnostic Aids:
Engine runs OK, check.
- Faulty light bulb
- CKT 419 open
Engine cranks, but will not run.
- Continuous battery - fuse or fusible link open
- ECM ignition fuse open
- Ignition CKT 439 to ECM open
- Poor connection to ECM

1988 VEHICLES EQUIPPED WITH 3.8L (VIN 3) ENGINE

CHART A-1
NO "SERVICE ENGINE SOON" LIGHT
3.8L (VIN 3) (PORT)

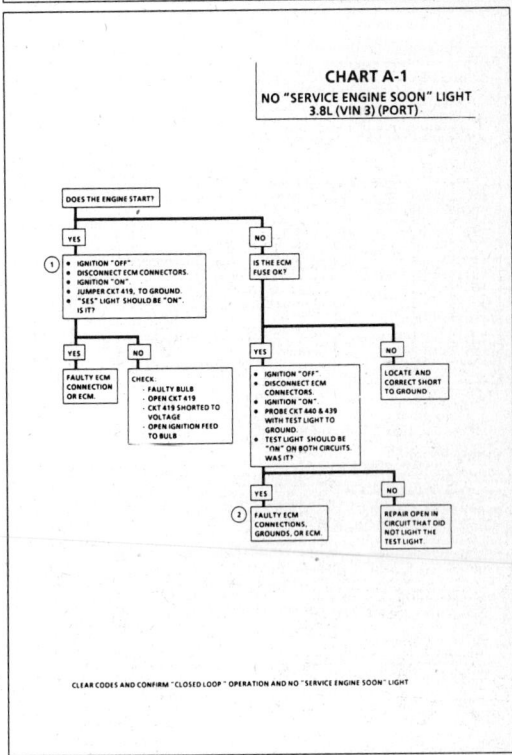

CLEAR CODES AND CONFIRM "CLOSED LOOP" OPERATION AND NO "SERVICE ENGINE SOON" LIGHT

1988 VEHICLES EQUIPPED WITH 3.8L (VIN 3) ENGINE

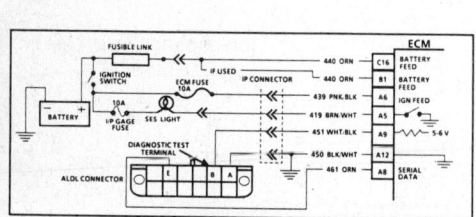

CHART A-2
WON'T FLASH CODE 12
"SERVICE ENGINE SOON" LIGHT "ON" STEADY
3.8L (VIN 3) (PORT)

Circuit Description:
There should always be a steady "Service Engine Soon" light when the ignition is "ON" and engine not running. Battery is supplied directly to the light bulb. The electronic control module (ECM) will turn the light "ON" by grounding CKT 419 at the ECM.

With the diagnostic terminal grounded, the light should flash a Code 12, followed by any trouble code(s) stored in memory.

A steady light suggests a short to ground in the light control CKT 419 or an open in diagnostic CKT 451.

Test Description: Numbers below refer to circled numbers on the diagnostic chart.
1. If the light goes "OFF" when the ECM connector is disconnected, then CKT 419 is not shorted to ground.
2. If there is a problem with the ECM that causes a "Scan" tool to not read serial data, then the ECM should not flash a Code 12. If Code 12 does flash, be sure that the "Scan" tool is working properly on another vehicle.

3. If the "Scan" is functioning properly and CKT 451 is OK, the PROM or ECM may be at fault for the NO ALDL symptom.
4. This step will check for an open diagnostic CKT 451.
5. At this point, the "Service Engine Soon" light wiring is OK. The problem is a faulty ECM or PROM. If Code 12 does not flash, the ECM should be replaced using the original PROM. Replace the PROM only after trying an ECM, as a defective PROM is an unlikely cause of the problem.

1988 VEHICLES EQUIPPED WITH 3.8L (VIN 3) ENGINE

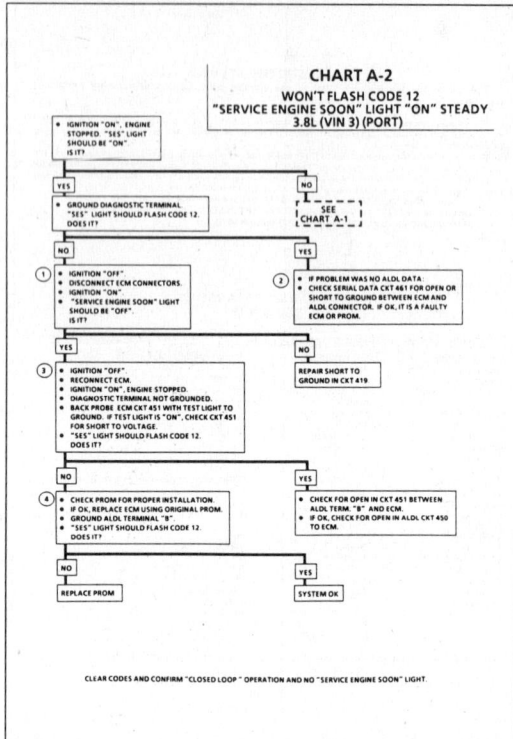

1988 VEHICLES EQUIPPED WITH 3.8L (VIN 3) ENGINE

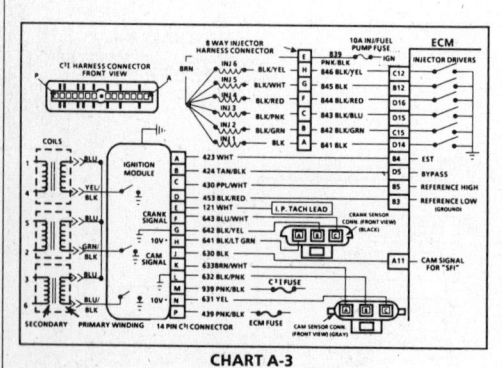

CHART A-3
(Page 1 of 3)
ENGINE CRANKS BUT WON'T RUN "TYPE I C³I"
3.8L (VIN 3) (PORT)

Circuit Description:
The C³I uses a waste spark method of spark distribution. In this type of ignition system, the ignition module triggers the #1/4 coil pair resulting in both #1 and #4 spark plugs firing at the same time. #1 cylinder is on the compression stroke at the same time #4 is on the exhaust stroke, resulting in a lower energy requirement to fire #4 spark plug. This leaves the remaining high voltage to fire #1 spark plug.

The sequential fuel injection type of fuel delivery system utilizes 6 separate injector driver circuits to activate the 6 fuel injectors. While cranking, the ECM activates all 6 injectors simultaneously (all at one time). After a calibrated engine rpm is reached and a good Cam signal has been received by the ECM, the injection mode of operation is changed to sequential (timed sequence).

Test Description: Numbers below refer to circled numbers on the diagnostic chart.
1. Identification of "TYPE I" or "TYPE II" ignition system is very important, because "TYPE I" diagnostics will not work on "TYPE II" systems. Identification can be made by comparing the position of the coil towers with the drawing at the top of the chart. This step verifies "SES" light operation, TPS, and coolant sensor signals are normal. A blinking injector test light verifies that the ECM is monitoring the C³I reference signal and attempting to activate the injectors.
2. Both the Cam and Crank Sensors have been verified as functioning properly, as is evidenced by the blinking injector test light. A fuel pressure test at this point will separate the diagnostic path into either a fuel related fault or ignition system malfunction.

3. The 8 terminal injector harness connector must be disconnected to avoid flooding of the engine and fouling of the spark plugs. By testing for spark on plug leads 1, 3, and 5, each ignition coil's ability to produce at least 25,000 volts is verified.
4. By testing the problem coil's control circuit with a test light, a determination can be made as to the problem coil being faulty or the module's internal driver for that coil being the source of the complaint.
5. An injector with a resistance value of less than 10 ohms (shorted) must be replaced.
6. Tests for battery voltage on CKT 839. If voltage was present, the light "OFF" test result was caused by no activation pulse reaching the injector connector from the ECM.

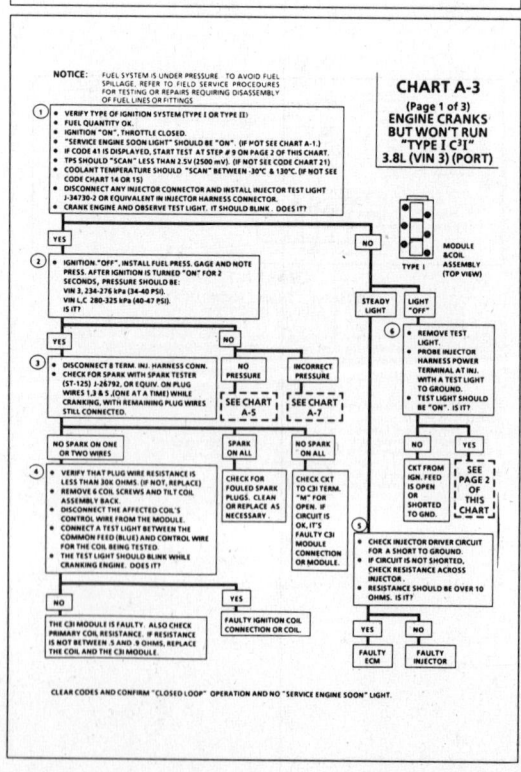

1988 VEHICLES EQUIPPED WITH 3.8L (VIN 3) ENGINE

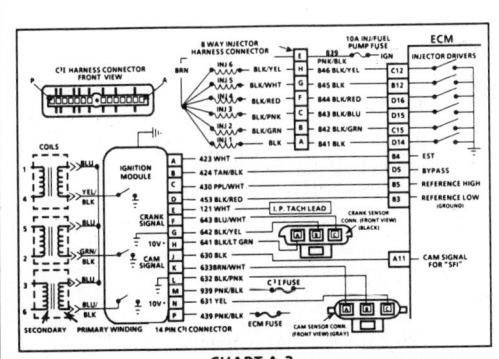

CHART A-3
(Page 2 of 3)
ENGINE CRANKS BUT WON'T RUN "TYPE I C³I"
3.8L (VIN 3) (PORT)

Circuit Description:
For timing of spark plug firing, a Cam sensor "hall effect" switch is used. The Cam sensor sends a signal (sync-pulse) to the ignition module when cylinder #1 is 25° after top dead center on the compression stroke. This signal is used to start the correct coil firing sequence and to enable sequential fuel injection. The engine will continue to run if the Cam signal to the ignition module (CKT 633) is lost while running, however, will not restart after shut down and a Code 41 will be stored.

The Crank sensor sends a signal to the ignition module and then to the ECM for reference rpm and crankshaft position. There are three windows in a disc (interruptor), which is mounted to the harmonic balancer. As these windows pass by the sensor the next coil is triggered.

Test Description: Numbers below refer to circled numbers on the diagnostic chart.
7. Verifies ignition feed voltage at terminal "P" of the C³I module. Less than battery voltage would be an indication of a CKT 439 fault.
8. The test light to 12volts simulates an injector signal to the ECM which will result in an injector test light blink. This test validates CKT 430, the ECM and the injector driver circuits.
9. If the Cam Sensor signal circuit terminal "A" is momentarily jumpered to the ground circuit terminal "B" and the engine is cranked, without turning the ignition switch "OFF", the response should be an injector test light blink. This is a

result of the artificial "cam signal" being transmitted through the C³I module to ECM terminal "A11" and the ECM activating the injector driver circuit.
10. Verifies a proper Cam signal circuit voltage of 6 to 9 volts and a good ground from the C³I module to terminal "B" of the sensor connector.
11. Determines if reason for incorrect voltage reading was due to a fault in CKT 633, an open in CKT 632, or a faulty C³I module.

1988 VEHICLES EQUIPPED WITH 3.8L (VIN 3) ENGINE

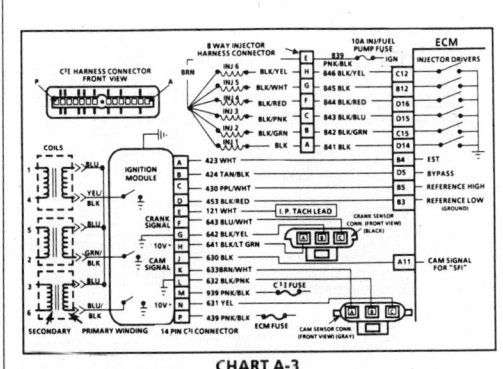

CHART A-3
(Page 3 of 3)
ENGINE CRANKS BUT WON'T RUN "TYPE I C³I"
3.8L (VIN 3) (PORT)

Circuit Description:
For timing of spark plug firing, a Cam sensor "hall effect" switch is used. The Cam sensor sends a signal (sync-pulse) to the ignition module when cylinder #1 is 25° after top dead center on the compression stroke. This signal is used to start the correct coil firing sequence and to enable sequential fuel injection. The engine will continue to run if the Cam signal to the ignition module (CKT 633) is lost while running, however, will not restart after shut down and a Code 41 will be stored.

The Crank sensor sends a signal to the ignition module and then to the ECM for reference rpm and crankshaft position. There are three windows in a disc (interruptor) which is mounted to the harmonic balancer. As these windows pass by the sensor the next coil is triggered.

Test Description: Numbers below refer to circled numbers on the diagnostic chart.
12. Jumping the Cam Sensor harness terminals "A" and "B" together simulates a Cam signal to the C³I module. Then, by repeatedly jumping the Crank sensor harness terminals "B" and "C" together, a Crank signal is simulated which should result in the injector test light blinking.

13. Verifies a proper Crank signal circuit voltage of 6 to 9 volts and a good ground from the C³I module to terminal "B" of the sensor connector.
14. Determines if reason for incorrect voltage reading was due to a fault in CKT 643, an open in CKT 642, or a faulty C³I module.

1988 VEHICLES EQUIPPED WITH 3.8L (VIN 3) ENGINE

CHART A-3
(Page 2 of 3)
ENGINE CRANKS BUT WON'T RUN "TYPE I C³I"
3.8L (VIN 3) (PORT)

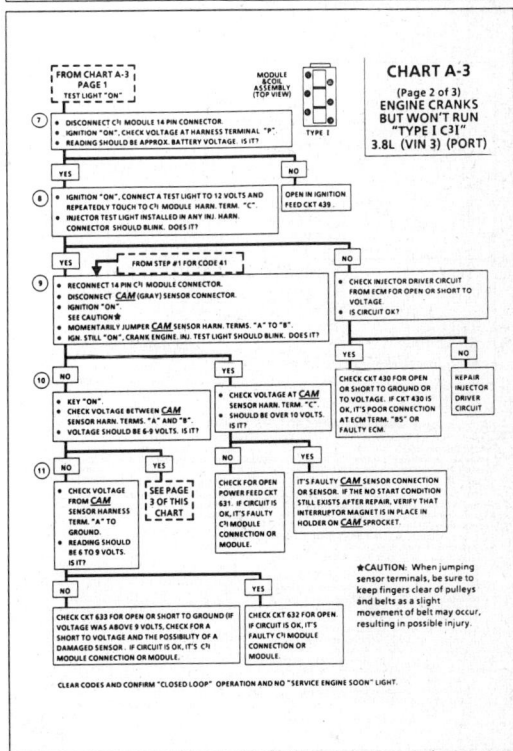

★CAUTION: When jumping sensor terminals, be sure to keep fingers clear of pulleys and belts as a slight movement of belt may occur, resulting in possible injury.

CLEAR CODES AND CONFIRM "CLOSED LOOP" OPERATION AND NO "SERVICE ENGINE SOON" LIGHT.

1988 VEHICLES EQUIPPED WITH 3.8L (VIN 3) ENGINE

CHART A-3
(Page 3 of 3)
ENGINE CRANKS BUT WON'T RUN "TYPE I C³I"
3.8L (VIN 3) (PORT)

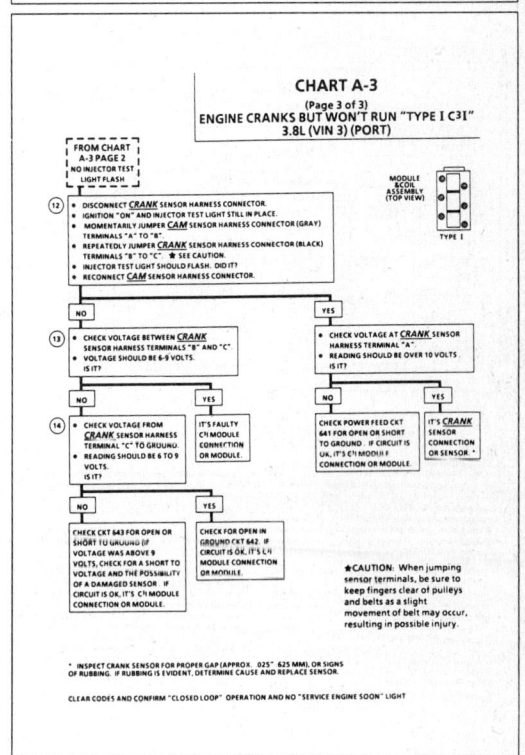

★CAUTION: When jumping sensor terminals, be sure to keep fingers clear of pulleys and belts as a slight movement of belt may occur, resulting in possible injury.

* INSPECT CRANK SENSOR FOR PROPER GAP (APPROX. .025" .625 MM) OR SIGNS OF RUBBING. IF RUBBING IS EVIDENT, DETERMINE CAUSE AND REPLACE SENSOR.

CLEAR CODES AND CONFIRM "CLOSED LOOP" OPERATION AND NO "SERVICE ENGINE SOON" LIGHT

1988 VEHICLES EQUIPPED WITH 3.8L (VIN 3) ENGINE

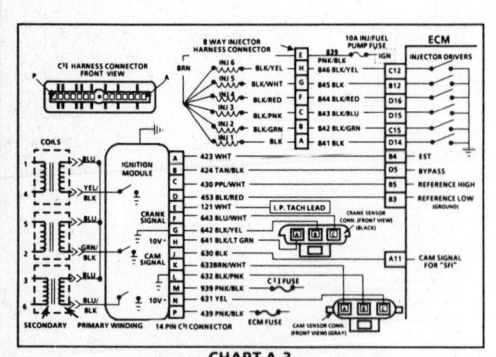

CHART A-3
(Page 1 of 3)
ENGINE CRANKS BUT WON'T RUN "TYPE II C³I"
3.8L (VIN 3) (PORT)

Circuit Description:
The C³I uses a waste spark method of spark distribution. In this type of ignition system the ignition module triggers the #1/4 coil pair resulting in both #1 and #4 spark plugs firing at the same time. #1 cylinder is on the compression stroke at the same time #4 is on the exhaust stroke, resulting in a lower energy requirement to fire the #4 spark plug. This leaves the remaining high voltage to fire #1 spark plug.

The sequential fuel injection type of fuel delivery system utilizes 6 separate injector driver circuits to activate the 6 fuel injectors. While cranking, the ECM activates all 6 injectors simultaneously (all at one time). After a calibrated engine rpm is reached, and a good cam signal has been received by the ECM, the injection mode of operation is changed to sequential (timed seperately).

Test Description: Numbers below refer to circled numbers on the diagnostic chart.

1. Identification of "TYPE I" or "TYPE II" ignition system is very important, because "TYPE II" diagnostics will not work on "TYPE I" systems. Identification can be made by comparing the position of the coil towers with the drawing at the top of the chart. This step verifies "SES" light, TPS, and coolant temperature signals are normal. A blinking injector test light verifies that the ECM is monitoring the C³I reference signal and attempting to activate the injectors.

2. The blinking injector test light verifies a working Cam Sensor. A fuel pressure test, at this point, will seperate the diagnostic path into either a fuel related fault or ignition system malfunction.

3. The 8 terminal injector harness connector must be disconnected to avoid flooding of the engine and fouling of the spark plugs. By testing for spark on plug leads 1, 3 and 5, each ignition coil's ability to produce at least 25,000 volts is verified. An erratic spark may be an indication of an open secondary winding in one of the ignition coils. This can result in internal arcing which may seek ground through the C³I module resulting in a stall or "Cranks But Won't Run" condition.

4. By switching the problem coil with a working one, a determination can be made as to the problem coil being faulty or the module's internal driver for that coil being the source of the complaint.

5. An injector with a resistance value of less than 10 ohms (shorted) must be replaced.

6. Tests for battery voltage on CKT 839. If voltage was present, the light "OFF" resulted from the injector not receiving a pulse from the ECM.

1988 VEHICLES EQUIPPED WITH 3.8L (VIN 3) ENGINE

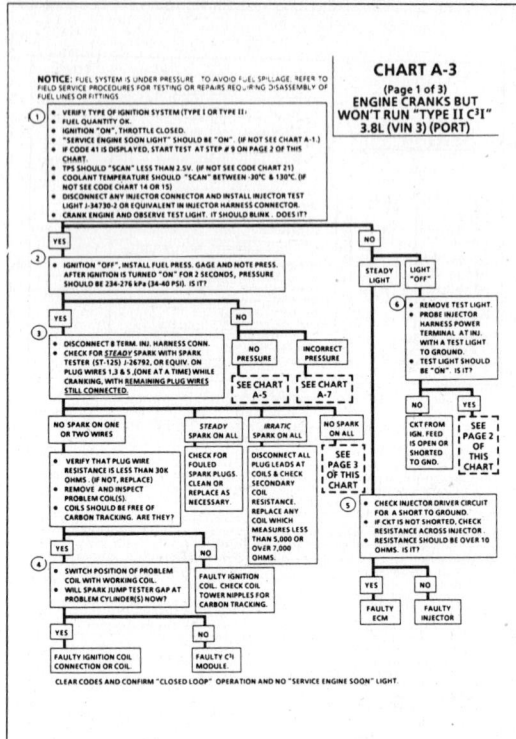

NOTICE: FUEL SYSTEM IS UNDER PRESSURE. TO AVOID FUEL SPILLAGE, REFER TO FIELD SERVICE PROCEDURES FOR TESTING OR REPAIRS REQUIRING DISASSEMBLY OF FUEL LINES OR FITTINGS.

CHART A-3
(Page 1 of 3)
ENGINE CRANKS BUT WON'T RUN "TYPE II C³I"
3.8L (VIN 3) (PORT)

CLEAR CODES AND CONFIRM "CLOSED LOOP" OPERATION AND NO "SERVICE ENGINE SOON" LIGHT.

1988 VEHICLES EQUIPPED WITH 3.8L (VIN 3) ENGINE

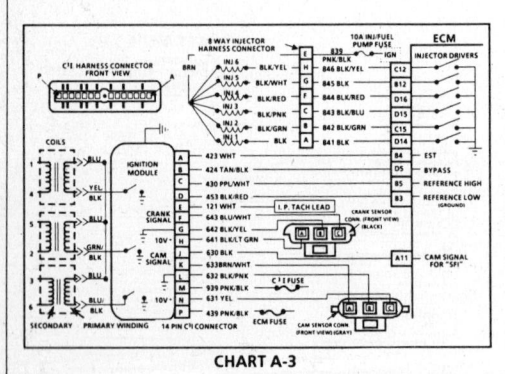

CHART A-3
(Page 2 of 3)
ENGINE CRANKS BUT WON'T RUN "TYPE II C³I"
3.8L (VIN 3) (PORT)

Circuit Description:
For timing of spark plug firing, a Cam sensor "hall effect" switch is used. The Cam sensor sends a signal (sync-pulse) to the ignition module when cylinder #1 is 25° after top dead center on the compression stroke. This signal is used to start the correct sequence of coil firing and to enable sequential fuel injection. The engine will continue to run if the Cam signal, CKT 633 to the ignition module, is lost while running. The engine will not restart after shut down and a Code 41 will be stored.

The Crank sensor sends a signal to the ignition module and then to the ECM for reference rpm and crankshaft position. There are three windows in a disc (interruptor) which is mounted to the harmonic balancer. As these windows pass by the sensor the next coil is triggered.

Test Description: Numbers below refer to circled numbers on the diagnostic chart.

7. Verifies ignition feed voltage at terminal "M" of the C³I module. Less than battery voltage would be an indication of a CKT 839 fault.

8. The test light to 12 volts simulates a reference signal to the ECM which will result in an injector test light blink, this validates CKT 430, the ECM, and the injector driver circuit are all OK.

9. If the Cam Sensor signal circuit terminal "A" is jumped to the ground circuit terminal "B", the response should be an injector test light blink.

This is a result of this artificial "Cam Signal" being transmitted through the C³I module to ECM terminal "A11" and the ECM activating the injector driver circuit.

10. Verifies a proper Cam Signal circuit voltage of 6 to 9 volts and a good ground from the C³I module to terminal "B" of the sensor connector.

11. Determines if incorrect voltage reading was due to a fault in CKT 633, an open in CKT 632, or a faulty C³I module.

1988 VEHICLES EQUIPPED WITH 3.8L (VIN 3) ENGINE

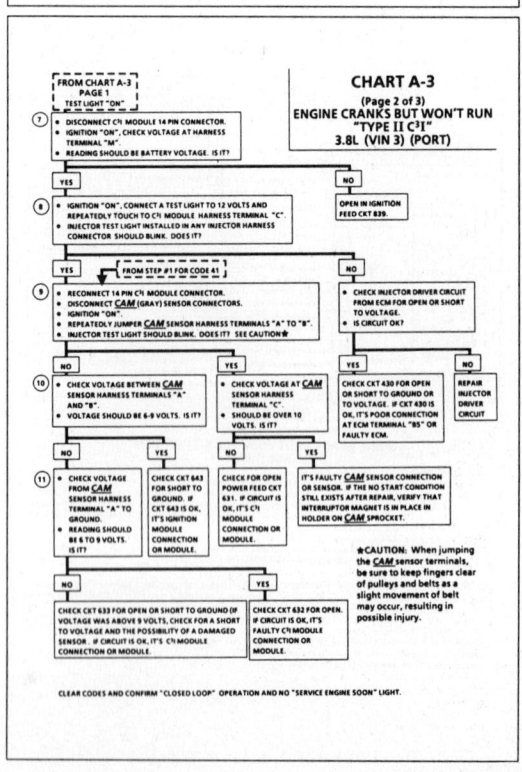

CHART A-3
(Page 2 of 3)
ENGINE CRANKS BUT WON'T RUN
"TYPE II C³I"
3.8L (VIN 3) (PORT)

★CAUTION: When jumping the CAM sensor terminals, be sure to keep fingers clear of pulleys and belts as a slight movement of belt may occur, resulting in possible injury.

CLEAR CODES AND CONFIRM "CLOSED LOOP" OPERATION AND NO "SERVICE ENGINE SOON" LIGHT.

1988 VEHICLES EQUIPPED WITH 3.8L (VIN 3) ENGINE

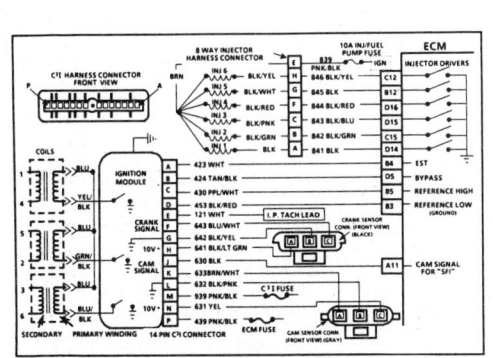

CHART A-3
(Page 3 of 3)
ENGINE CRANKS BUT WON'T RUN "TYPE II C³I"
3.8L (VIN 3) (PORT)

Circuit Description:

For timing of spark plug firing, a Cam sensor "hall effect" switch is used. The Cam sensor sends a signal (sync-pulse) to the ignition module when cylinder #1 is 25° after top dead center on the compression stroke. This signal is used to start the correct sequence of coil firing and to enable sequential fuel injection. The engine will continue to run if the Cam signal to the ignition module (CKT 633) is lost while running, however, will not restart after shut down and a Code 41 will be stored.

The Crank sensor sends a signal to the ignition module and then to the ECM for reference rpm and crankshaft position. There are three windows in a disc (interruptor) which is mounted to the harmonic balancer. As these windows pass by the sensor the next coil is triggered.

Test Description: Numbers below refer to circled numbers on the diagnostic chart.

12. Jumping the Crank Sensor harness terminals "B" and "C" together simulates a Crank signal to the C³I module. Then, by repeatedly jumping the Cam sensor harness terminals "A" and "B" together a cam signal is simulated which should result in a spark.

13. Verifies a proper Crank signal circuit voltage of 6 to 9 volts and a good ground from the C³I module to terminal "B" at the sensor connector.

14. Determines if reason for incorrect voltage reading was due to a fault in CKT 643, an open in CKT 642, or a faulty C³I module.

1988 VEHICLES EQUIPPED WITH 3.8L (VIN 3) ENGINE

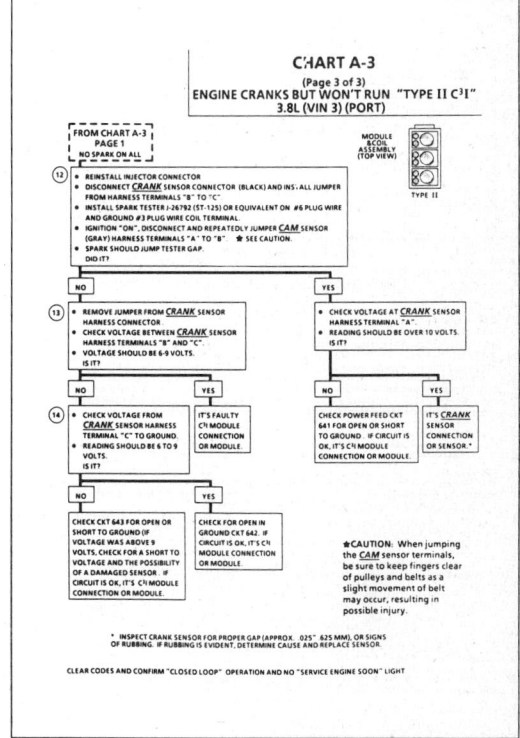

CHART A-3
(Page 3 of 3)
ENGINE CRANKS BUT WON'T RUN "TYPE II C³I"
3.8L (VIN 3) (PORT)

★CAUTION: When jumping the Cam sensor terminals, be sure to keep fingers clear of pulleys and belts as a slight movement of belt may occur, resulting in possible injury.

* INSPECT CRANK SENSOR FOR PROPER GAP (APPROX. .025" 625 MM), OR SIGNS OF RUBBING. IF RUBBING IS EVIDENT, DETERMINE CAUSE AND REPLACE SENSOR.

CLEAR CODES AND CONFIRM "CLOSED LOOP" OPERATION AND NO "SERVICE ENGINE SOON" LIGHT.

1988 A-CAR EQUIPPED WITH 3.8L (VIN 3) ENGINE

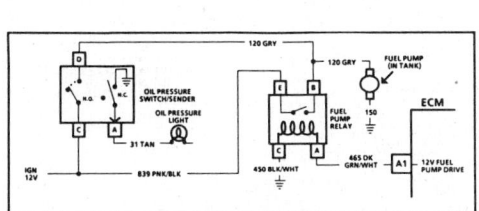

CHART A-5
(Page 1 of 2)
FUEL SYSTEM ELECTRICAL TEST
3.8L (VIN 3) "A" SERIES (PORT)

Circuit Description:

When the ignition switch is turned "ON", the electronic control module (ECM) will energize the fuel pump relay which completes the circuit to the in-tank fuel pump. It will remain "ON" as long as the engine is cranking or running and the ECM is receiving C³I reference pulses. If there are no reference pulses, the ECM will de-energize the fuel pump relay within 2 seconds after key "ON" or the engine is stopped.

The fuel pump will deliver fuel to the fuel rail and injectors then to the pressure regulator where the system pressure is controlled. Excess fuel pressure is bypassed back to the fuel tank. When the engine is stopped the pump can be turned "ON" by applying battery voltage to the test terminal located in the engine compartment.

Improper fuel system pressure may contribute to one or all of the following symptoms:
* Cranks but won't run.
* Code 44 or 45.
* Cuts out, may feel like ignition problem.
* Hesitation, loss of power or poor fuel economy.

Test Description: Numbers below refer to circled numbers on the diagnostic chart.

1. If the fuse is blown, a short to ground in CKTs 120, 939 or the fuel pump itself is the cause.

2. Determines if the fuel pump circuit is being controlled by the ECM. The ECM should energize the fuel pump relay. The engine is not cranking or running so the ECM should de-energize the relay within 2 seconds after ignition is turned "ON".

3. Turns "ON" the fuel pump if CKT 120 wiring is OK. If the pump runs, it is a basic fuel delivery problem.

4. This test will determine if a short to ground on CKT 120 caused the fuse to blow. To prevent a mis-diagnosis, be sure the fuel pump is disconnected before the test.

5. Checks for a short to ground in the fuel pump relay harness CKT 839.

1988 A-CAR EQUIPPED WITH 3.8L (VIN 3) ENGINE

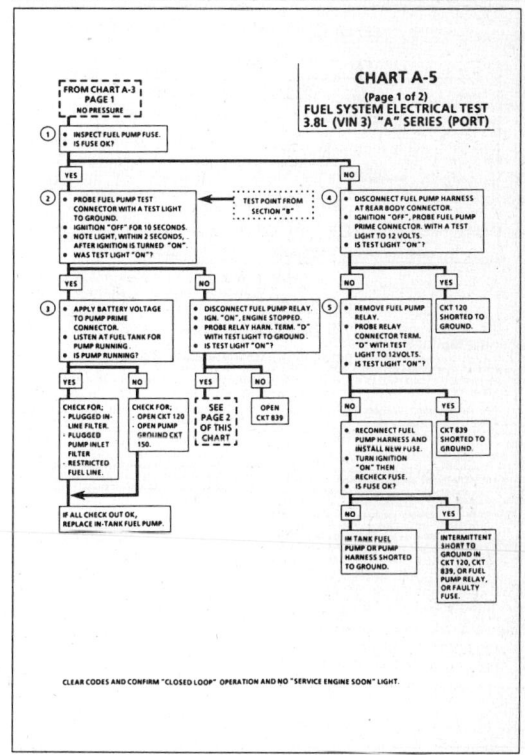

CHART A-5
(Page 1 of 2)
FUEL SYSTEM ELECTRICAL TEST
3.8L (VIN 3) "A" SERIES (PORT)

CLEAR CODES AND CONFIRM "CLOSED LOOP" OPERATION AND NO "SERVICE ENGINE SOON" LIGHT.

1988 VEHICLES EQUIPPED WITH 3.8L (VIN 3) ENGINE

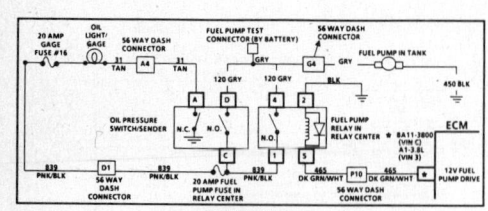

CHART A-5
(Page 1 of 2)
FUEL SYSTEM ELECTRICAL TEST
(STANDARD CLUSTER)
3.8L (VIN 3) (PORT)

Circuit Description:
When the ignition switch is turned "ON", the electronic control module (ECM) will energize the fuel pump relay which completes the circuit to the in-tank fuel pump. It will remain "ON" as long as the engine is cranking or running and the ECM is receiving C³I reference pulses. If there are no reference pulses, the ECM will de-energize the fuel pump relay within 2 seconds after key "ON" or the engine is stopped.

The fuel pump will deliver fuel to the fuel rail and injectors then to the pressure regulator where the system pressure is controlled. Excess fuel pressure is bypassed back to the fuel tank. When the engine is stopped, the pump can be turned "ON" by applying battery voltage to the test terminal located in the engine compartment.

Improper fuel system pressure may contribute to one or all of the following symptoms
- Cranks but won't run
- Code 44 or 45
- Cuts out, may feel like ignition problem
- Hesitation, loss of power or poor fuel economy

Test Description: Numbers below refer to circled numbers on the diagnostic chart.
1. If the fuse is blown, a short to ground in CKTs 120, 839 or the fuel pump itself is the cause.
2. Determines if the fuel pump circuit is being controlled by the ECM. The ECM should energize the fuel pump relay. The engine is not cranking or running so the ECM should de-energize the relay within 2 seconds after ignition is turned "ON".

3. Turns "ON" the fuel pump if CKT 120 wiring is OK. If the pump runs, it is a basic fuel delivery problem.
4. This test will determine if a short to ground on CKT 120 caused the fuse to blow. To prevent a mis-diagnosis, be sure the fuel pump is disconnected before the test.
5. Checks for a short to ground in the fuel pump relay harness CKT 839.

1988 VEHICLES EQUIPPED WITH 3.8L (VIN 3) ENGINE

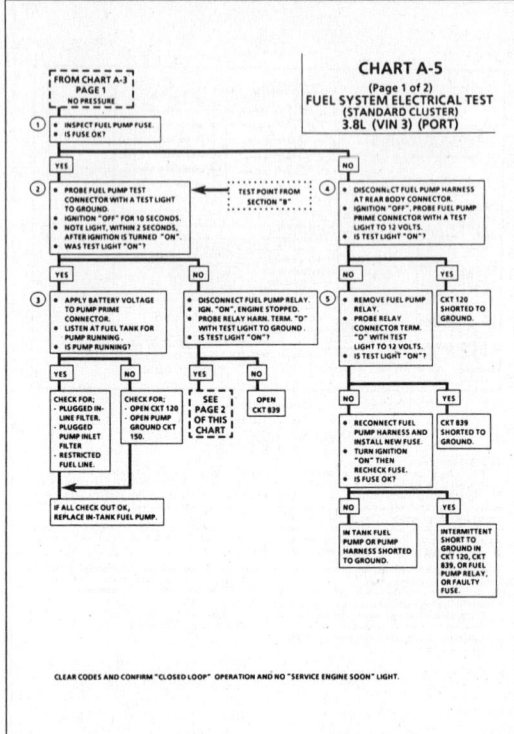

CHART A-5
(Page 1 of 2)
FUEL SYSTEM ELECTRICAL TEST
(STANDARD CLUSTER)
3.8L (VIN 3) (PORT)

CLEAR CODES AND CONFIRM "CLOSED LOOP" OPERATION AND NO "SERVICE ENGINE SOON" LIGHT.

1988 VEHICLES EQUIPPED WITH 3.8L (VIN 3) ENGINE

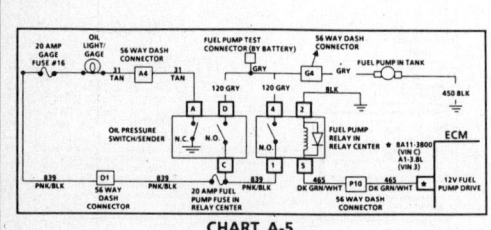

CHART A-5
(Page 2 of 2)
FUEL SYSTEM ELECTRICAL TEST
(STANDARD CLUSTER)
3.8L (VIN 3) (PORT)

Circuit Description:
When the ignition switch is turned "ON", the electronic control module (ECM) will energize the fuel pump relay which completes the circuit to the in-tank fuel pump. It will remain "ON" as long as the engine is cranking or running, and the ECM is receiving "C³I" reference pulses. If there are no reference pulses, the ECM will de-energize the fuel pump relay within 2 seconds after key "ON", or the engine is stopped.

Test Description: Numbers below refer to circled numbers on the diagnostic chart.
6. Checks for open in the fuel pump relay ground, CKT 450.
7. Determines if the ECM is in control of the fuel pump relay through CKT 465 (terminal "A").
8. The fuel pump control circuit includes an engine oil pressure switch with a separate set of normally open contacts. The switch closes at about (4 lbs) 28 kPa of oil pressure and provides a second battery feed path to the fuel pump. If the relay fails, the pump will continue to run using the battery feed supplied by the closed oil pressure switch. This step checks the oil pressure switch to be sure it provides battery feed to the fuel pump should the pump relay fail.

A failed pump relay will result in extended engine crank time because of the time required to build enough oil pressure to close the oil pressure switch and turn "ON" the fuel pump. There may be instances when the relay has failed but the engine will not crank fast enough to build enough oil pressure to close the switch. This or a faulty oil pressure switch can result in "Engine Cranks But Will Not Run."

1988 VEHICLES EQUIPPED WITH 3.8L (VIN 3) ENGINE

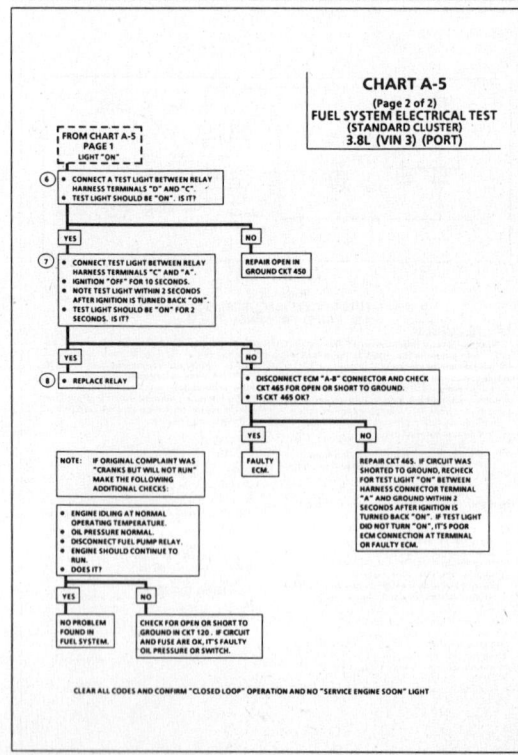

CHART A-5
(Page 2 of 2)
FUEL SYSTEM ELECTRICAL TEST
(STANDARD CLUSTER)
3.8L (VIN 3) (PORT)

CLEAR ALL CODES AND CONFIRM "CLOSED LOOP" OPERATION AND NO "SERVICE ENGINE SOON" LIGHT

1988 VEHICLES EQUIPPED WITH 3.8L (VIN 3) ENGINE

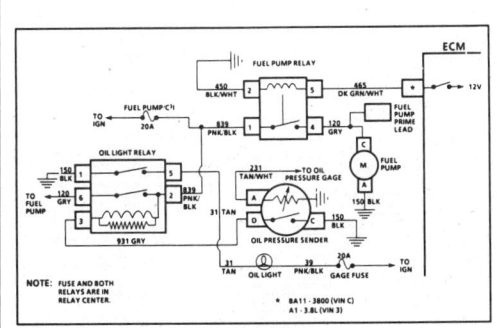

CHART A-5
(Page 1 of 2)
FUEL SYSTEM ELECTRICAL TEST
(U21 CLUSTER)
3.8L (VIN 3) (PORT)

NOTE: FUSE AND BOTH RELAYS ARE IN RELAY CENTER.

* BA11 - 3800 (VIN C)
A1 - 3.8L (VIN 3)

Circuit Description:
When the ignition switch is turned "ON", the electronic control module (ECM) will energize the fuel pump relay which completes the circuit to the in-tank fuel pump. It will remain on as long as the engine is cranking or running, and the ECM is receiving "C3I" reference pulses. If there are no reference pulses, the ECM will de-energize the fuel pump relay within 2 seconds after key "ON", or the engine is stopped.

The fuel pump will deliver fuel to the fuel rail and injectors, then to the pressure regulator, where the system pressure is controlled. Excess fuel pressure is bypassed back to the fuel tank. When the engine is stopped, the pump can be turned "ON" by applying battery voltage to the test terminal in the engine compartment.

Improper fuel system pressure may contribute to one or all of the following symptoms:
* Cranks but won't run
* Code 44 or 45
* Cuts out, may feel like ignition problem
* Hesitation, loss of power or poor fuel economy

Test Description: Numbers below refer to circled numbers on the diagnostic chart.
1. If the fuse is blown, a short to ground in CKTs 120, 839, or the fuel pump itself is the cause.
2. Determines if the fuel pump circuit is being controlled by the ECM. The ECM should energize the fuel pump relay. The engine is not cranking or running so the ECM should de-energize the relay within 2 seconds after ignition is turned "ON".

3. Turns "ON" the fuel pump if CKT 120 wiring is OK. If the pump runs, it is a basic fuel delivery problem.
4. This test will determine if a short to ground on CKT 120 caused the fuse to blow. To prevent a mis-diagnosis, be sure the fuel pump is disconnected before the test.
5. Checks for a short to ground in the fuel pump relay harness CKT 839.

1988 VEHICLES EQUIPPED WITH 3.8L (VIN 3) ENGINE

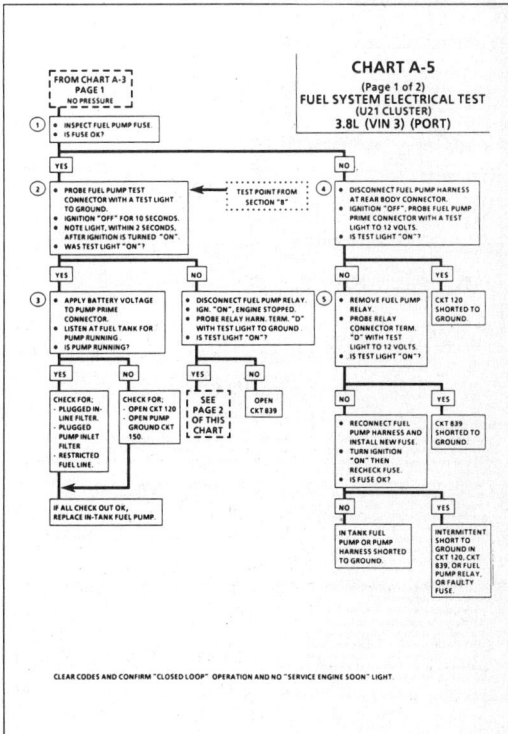

CHART A-5
(Page 1 of 2)
FUEL SYSTEM ELECTRICAL TEST
(U21 CLUSTER)
3.8L (VIN 3) (PORT)

CLEAR CODES AND CONFIRM "CLOSED LOOP" OPERATION AND NO "SERVICE ENGINE SOON" LIGHT.

1988 VEHICLES EQUIPPED WITH 3.8L (VIN 3) ENGINE

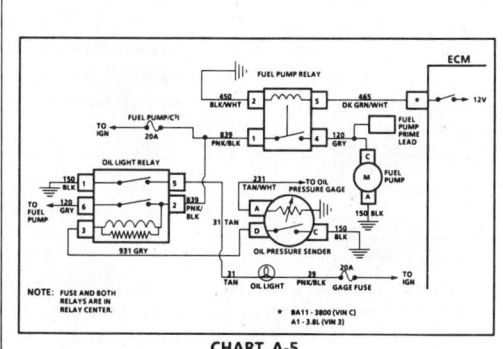

CHART A-5
(Page 2 of 2)
FUEL SYSTEM ELECTRICAL TEST
(U21 CLUSTER)
3.8L (VIN 3) (PORT)

NOTE: FUSE AND BOTH RELAYS ARE IN RELAY CENTER.

* BA11 - 3800 (VIN C)
A1 - 3.8L (VIN 3)

Circuit Description:
When the ignition switch is turned "ON", the electronic control module (ECM) will energize the fuel pump relay which completes the circuit to the in-tank fuel pump. It will remain on as long as the engine is cranking or running and the ECM is receiving "C3I" reference pulses. If there are no reference pulses, the ECM will de-energize the fuel pump relay within 2 seconds after key "ON", or the engine is stopped.

Test Description: Numbers below refer to circled numbers on the diagnostic chart.
6. Checks for open in the fuel pump relay ground, CKT 450.
7. Determines if the ECM is in control of the fuel pump relay through CKT 465 (terminal "A").
8. The fuel pump control circuit includes an engine oil pressure switch with a separate set of normally open contacts. The switch closes at about (4 lbs) 28 kPa of oil pressure and provides a second battery feed path to the fuel pump. If the relay fails, the pump will continue to run using the battery feed supplied by the closed oil pressure switch. This step checks the oil pressure switch to be sure it provides battery feed to the fuel pump should the pump relay fail.

A failed pump relay will result in extended engine crank time, because of the time required to build enough oil pressure to close the oil pressure switch and turn "ON" the fuel pump. There may be instances when the relay has failed but the engine will not crank fast enough to build enough oil pressure to close the switch. This or a faulty oil pressure switch can result in "Engine Cranks But Will Not Run".

1988 VEHICLES EQUIPPED WITH 3.8L (VIN 3) ENGINE

CHART A-5
(Page 2 of 2)
FUEL SYSTEM ELECTRICAL TEST
(U21 CLUSTER)
3.8L (VIN 3) (PORT)

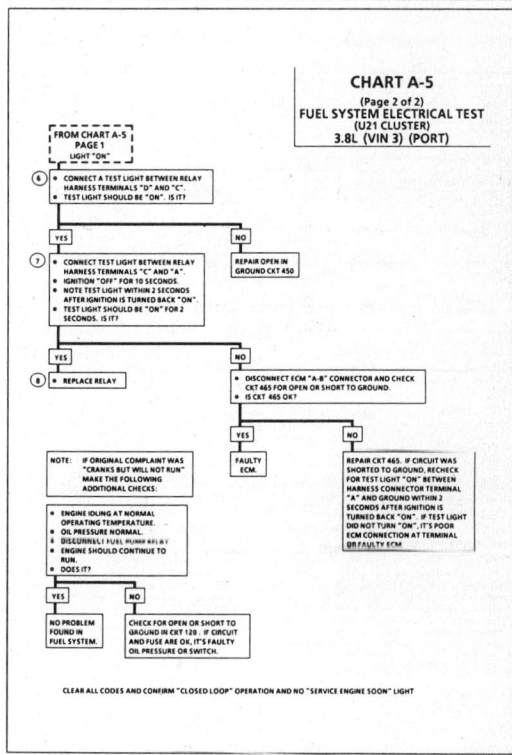

CLEAR ALL CODES AND CONFIRM "CLOSED LOOP" OPERATION AND NO "SERVICE ENGINE SOON" LIGHT

1988 VEHICLES EQUIPPED WITH 3.8L (VIN 3) ENGINE

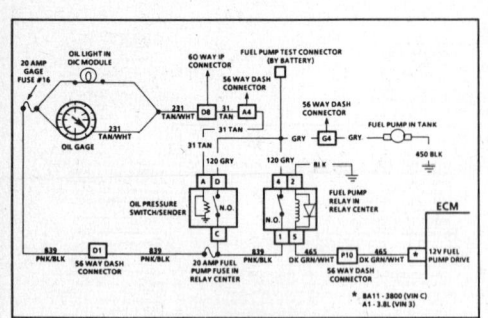

CHART A-5
(Page 1 of 2)
FUEL SYSTEM ELECTRICAL TEST
(U21 WITH UW1)
3.8L (VIN 3) (PORT)

Circuit Description:
When the ignition switch is turned "ON", the electronic control module (ECM) will energize the fuel pump relay which completes the circuit to the in-tank fuel pump. It will remain "ON" as long as the engine is cranking or running and the ECM is receiving C3I reference pulses. If there are no reference pulses, the ECM will de-energize the fuel pump relay within 2 seconds after key "ON" or the engine is stopped.

The fuel pump will deliver fuel to the fuel rail and injectors then to the pressure regulator where the system pressure is controlled. Excess fuel pressure is bypassed back to the fuel tank. When the engine is stopped, the pump can be turned "ON" by applying battery voltage to the test terminal located in the engine compartment.

Improper fuel system pressure may contribute to one or all of the following symptoms:
- Cranks but won't run
- Code 44 or 45
- Cuts out, may feel like ignition problem
- Hesitation, loss of power or poor fuel economy

Test Description: Numbers below refer to circled numbers on the diagnostic chart.
1. If the fuse is blown, a short to ground in CKTs 120, 839 or the fuel pump itself is the cause.
2. Determines if the fuel pump circuit is being controlled by the ECM. The ECM should energize the fuel pump relay. The engine is not cranking or running so the ECM should de-energize the relay within 2 seconds after ignition is turned "ON".

3. Turns "ON" the fuel pump if CKT 120 wiring is OK. If the pump runs, it is a basic fuel delivery problem.
4. This test will determine if a short to ground on CKT 120 caused the fuse to blow. To prevent a mis-diagnosis, be sure the fuel pump is disconnected before the test.
5. Checks for a short to ground in the fuel pump relay harness CKT 839.

1988 VEHICLES EQUIPPED WITH 3.8L (VIN 3) ENGINE

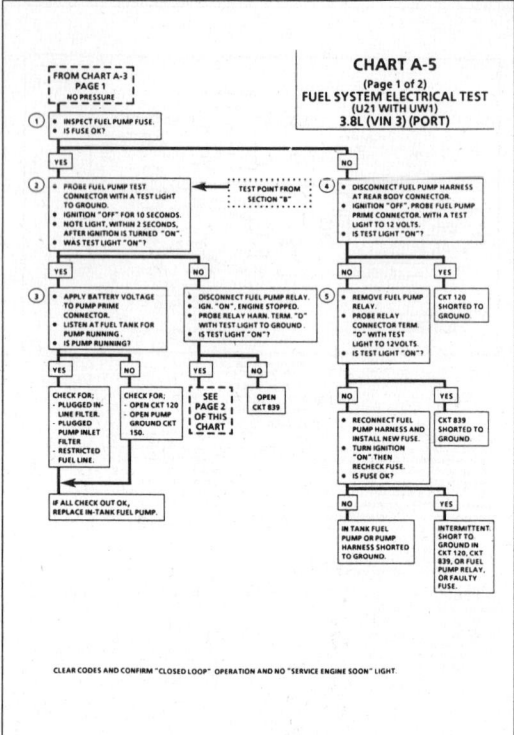

CLEAR CODES AND CONFIRM "CLOSED LOOP" OPERATION AND NO "SERVICE ENGINE SOON" LIGHT.

1988 VEHICLES EQUIPPED WITH 3.8L (VIN 3) ENGINE

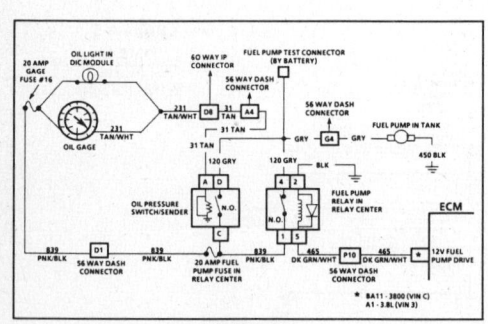

CHART A-5
(Page 2 of 2)
FUEL SYSTEM ELECTRICAL TEST
(U21 WITH UW1)
3.8L (VIN 3) (PORT)

Circuit Description:
When the ignition switch is turned "ON", the electronic control module (ECM) will energize the fuel pump relay, which completes the circuit to the in-tank fuel pump. It will remain "ON" as long as the engine is cranking or running and the ECM is receiving C3I reference pulses. If there are no reference pulses, the ECM will de-energize the fuel pump relay within 2 seconds after key "ON" or the engine is stopped.

Test Description: Numbers below refer to circled numbers on the diagnostic chart.
6. Checks for open in the fuel pump relay ground, CKT 450.
7. Determines if the ECM is in control of the fuel pump relay through CKT 465 (terminal "A").
8. The fuel pump control circuit includes an engine oil pressure switch with a separate set of normally open contacts. The switch closes at about 4 lbs/28 kPa of oil pressure and provides a second battery feed path to the fuel pump. If the relay fails, the pump will run using the battery feed supplied by the closed oil pressure switch.

This step checks the oil pressure switch to be sure it provides battery feed to the fuel pump relay should the pump relay fail. A failed pump relay will result in extended engine crank time because of the time required to build enough oil pressure to close the oil pressure switch and turn "ON" the fuel pump. There may be instances when the relay has failed but the engine will not crank fast enough to build enough oil pressure to close the switch. This or a faulty oil pressure switch can result in "Engine Cranks But Won't Run".

1988 VEHICLES EQUIPPED WITH 3.8L (VIN 3) ENGINE

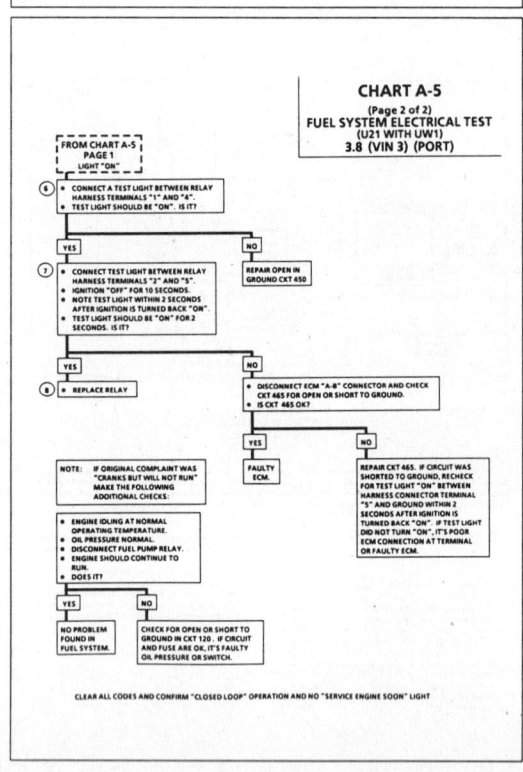

CLEAR ALL CODES AND CONFIRM "CLOSED LOOP" OPERATION AND NO "SERVICE ENGINE SOON" LIGHT

1988 VEHICLES EQUIPPED WITH 3.8L (VIN 3) ENGINE

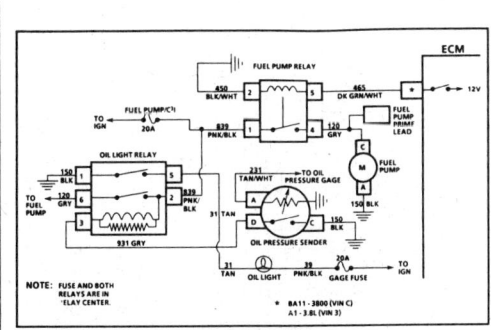

CHART A-5
(Page 1 of 2)
FUEL SYSTEM ELECTRICAL TEST
(U52 CLUSTER)
3.8L (VIN 3) (PORT)

Circuit Description:
When the ignition switch is turned "ON", the electronic control module (ECM) will energize the fuel pump relay which completes the circuit to the in-tank fuel pump. It will remain on as long as the engine is cranking or running, and the ECM is receiving "C3I" reference pulses. If there are no reference pulses, the ECM will de-energize the fuel pump relay within 2 seconds after key "ON", or the engine is stopped.

The fuel pump will deliver fuel to the fuel rail and injectors, then to the pressure regulator, where the system pressure is controlled. Excess fuel pressure is bypassed back to the fuel tank. When the engine is stopped, the pump can be turned "ON" by applying battery voltage to the test terminal in the engine compartment.

Improper fuel system pressure may contribute to one or all of the following symptoms
- Cranks but won't run
- Code 44 or 45
- Cuts out, may feel like ignition problem
- Hesitation, loss of power or poor fuel economy

Test Description: Numbers below refer to circled numbers on the diagnostic chart.
1. If the fuse is blown, a short to ground in CKTs 120, 839, or the fuel pump itself is the cause.
2. Determines if the fuel pump circuit is being controlled by the ECM. The ECM should energize the fuel pump relay. The engine is not cranking or running so the ECM should de-energize the relay within 2 seconds after ignition is turned "ON".

3. Turns "ON" the fuel pump if CKT 120 wiring is OK. If the pump runs, it is a basic fuel delivery problem
4. This test will determine if a short to ground on CKT 120 caused the fuse to blow. To prevent a mis-diagnosis, be sure the fuel pump is disconnected before the test.
5. Checks for a short to ground in the fuel pump relay harness CKT 839.

1988 VEHICLES EQUIPPED WITH 3.8L (VIN 3) ENGINE

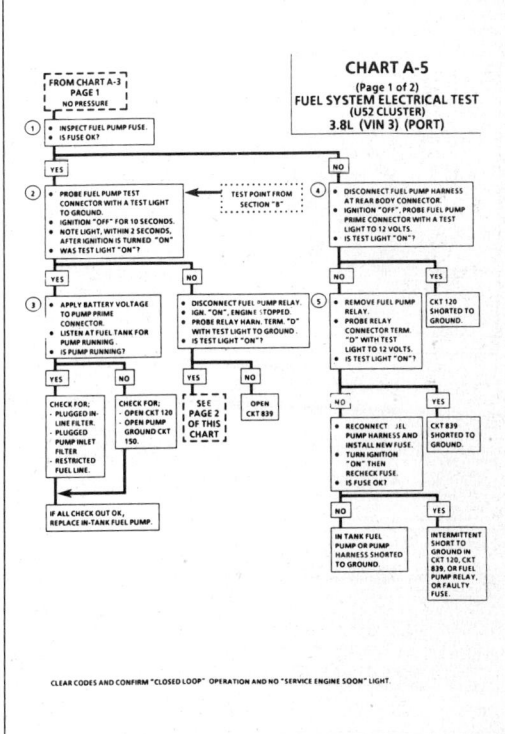

CLEAR CODES AND CONFIRM "CLOSED LOOP" OPERATION AND NO "SERVICE ENGINE SOON" LIGHT.

1988 VEHICLES EQUIPPED WITH 3.8L (VIN 3) ENGINE

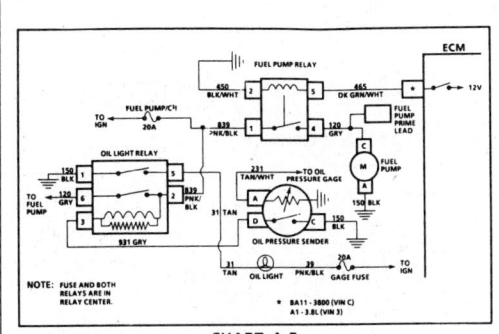

CHART A-5
(Page 2 of 2)
FUEL SYSTEM ELECTRICAL TEST
(U52 CLUSTER)
3.8L (VIN 3) (PORT)

Circuit Description:
When the ignition switch is turned "ON", the electronic control module (ECM) will energize the fuel pump relay which completes the circuit to the in-tank fuel pump. It will remain on as long as the engine is cranking or running and the ECM is receiving "C3I" reference pulses. If there are no reference pulses, the ECM will de-energize the fuel pump relay within 2 seconds after key "ON", or the engine is stopped.

Test Description: Numbers below refer to circled numbers on the diagnostic chart.
6. Checks for open in the fuel pump relay ground, CKT 450.
7. Determines if the ECM is in control of the fuel pump relay through CKT 465 (terminal "A").
8. The fuel pump control circuit includes an engine oil pressure switch with a separate set of normally open contacts. The switch closes at about 4 lbs/28 kPa of oil pressure and provides a second battery feed path to the fuel pump. If the relay fails, the pump will continue to run using the battery feed supplied by the closed oil pressure switch. This step checks the oil pressure switch to be sure it provides battery feed to the fuel pump should the pump relay fail.

A failed pump relay will result in extended engine crank time, because of the time required to build enough oil pressure to close the oil pressure switch and turn "ON" the fuel pump. There may be instances when the relay has failed but the engine will not crank fast enough to build enough oil pressure to close the switch. This or a faulty oil pressure switch can result in "Engine Cranks But Will Not Run"

1988 VEHICLES EQUIPPED WITH 3.8L (VIN 3) ENGINE

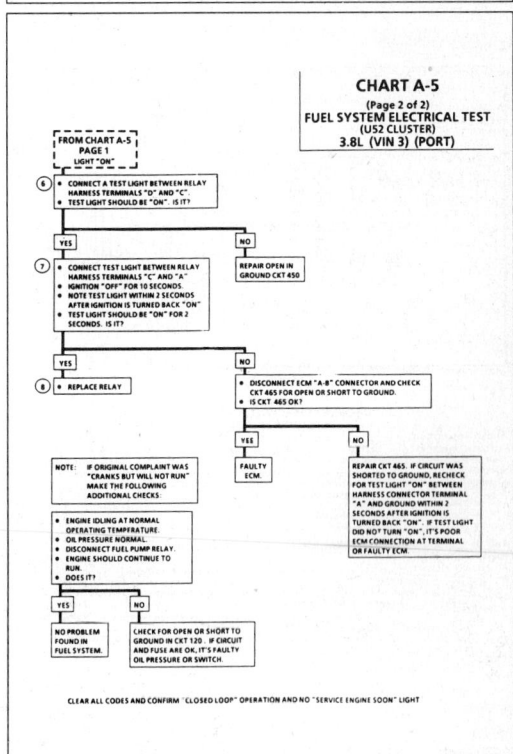

CLEAR ALL CODES AND CONFIRM "CLOSED LOOP" OPERATION AND NO "SERVICE ENGINE SOON" LIGHT

1988 VEHICLES EQUIPPED WITH 3.8L (VIN 3) ENGINE

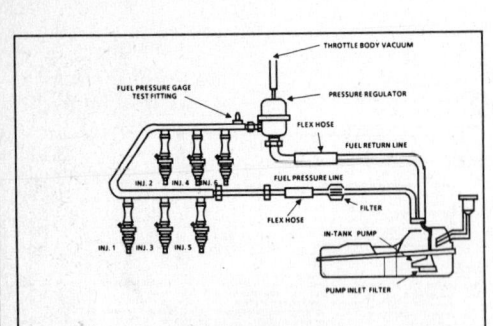

CHART A-7
(Page 1 of 2)
FUEL SYSTEM PRESSURE TEST
3.8L (VIN 3) (PORT)

Circuit Description:
The fuel pump will deliver fuel to the fuel rail and injectors, then to the pressure regulator, where the system pressure is controlled. Excess fuel pressure is bypassed back to the fuel tank. When the engine is stopped, the pump can be turned "ON" by applying battery voltage to the test terminal located in the engine compartment.

Improper fuel system pressure may contribute to one or all of the following symptoms:
- Cranks but won't run
- Cuts out, may feel like ignition problem
- Code 44 or 45
- Hesitation, loss of power, or poor fuel economy

Test Description: Numbers below refer to circled numbers on the diagnostic chart.

CAUTION: To reduce the risk of fire and personal injury, it is necessary to relieve the fuel pressure before servicing fuel system components. To do this:
- Disconnect the fuel tank harness connector.
- Crank engine - engine will start and run until fuel supply remaining in fuel pipes is consumed.
- Engage starter for 3.0 seconds to assure relief of any remaining pressure.
1. • Install pressure gage J-34730-1 to fuel pressure tap.
   • Connect the fuel tank harness conector.
   • Start engine. With ignition "ON", pump pressure is controlled by spring pressure and throttle body vacuum within the pressure regulator assembly.
   • Ignition "OFF". Pressure should not leak down after the fuel pump is shut "OFF".
2. When the engine is idling, the throttle body vacuum is high and is applied to the fuel regulator diaphragm. This will offset the spring and result in a lower fuel pressure.
3. The application of 12-14 inches of vacuum to the fuel pressure regulator should result in a fuel pressure less than step 1.
4. Pressure that leaks down may be caused by one of the following:
   In-tank fuel pump check valve not holding
   Pump coupling hose is sking
   Fuel pressure regulator valve leaking
   Injector sticking open

1988 VEHICLES EQUIPPED WITH 3.8L (VIN 3) ENGINE

CAUTION: TO REDUCE THE RISK OF FIRE AND PERSONAL INJURY, IT IS NECESSARY TO RELIEVE THE FUEL SYSTEM PRESSURE BEFORE SERVICING THE FUEL SYSTEM. (TO DO THIS):
- DISCONNECT THE FUEL TANK HARNESS CONNECTOR.
- CRANK ENGINE. ENGINE WILL START AND RUN UNTIL FUEL SUPPLY REMAINING IN FUEL PIPES IS CONSUMED. ENGAGE STARTER FOR 3.0 SECONDS TO ASSURE RELIEF OF ANY REMAINING PRESSURE.
- CONNECT THE FUEL TANK HARNESS CONNECTOR. UNLESS THIS PROCEDURE IS FOLLOWED BEFORE SERVICING FUEL PIPES OR CONNECTIONS, FUEL SPRAY COULD OCCUR.

CHART A-7
(Page 1 of 2)
FUEL SYSTEM PRESSURE TEST
3.8L (VIN 3) (PORT)

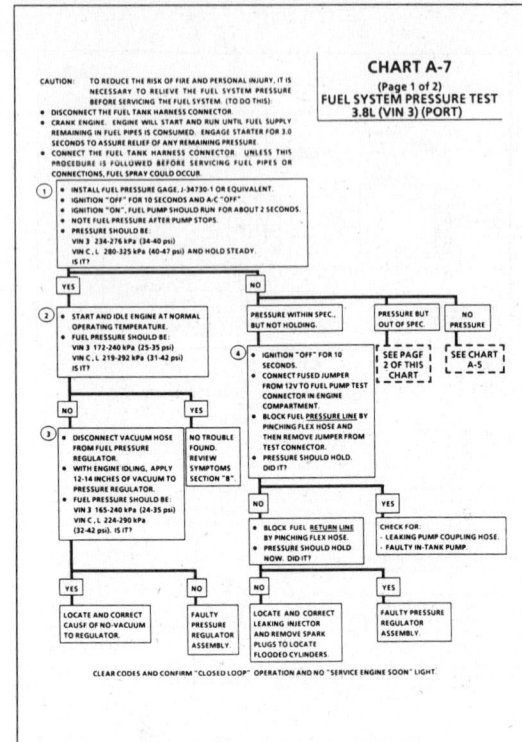

CLEAR CODES AND CONFIRM "CLOSED LOOP" OPERATION AND NO "SERVICE ENGINE SOON" LIGHT.

1988 VEHICLES EQUIPPED WITH 3.8L (VIN 3) ENGINE

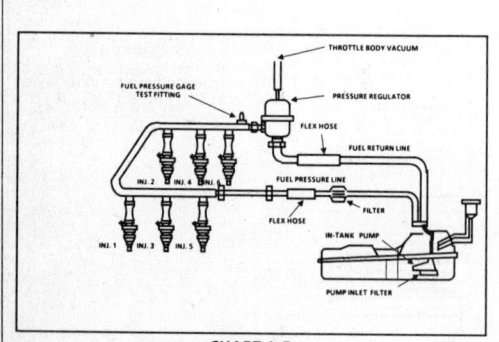

CHART A-7
(Page 2 of 2)
FUEL SYSTEM PRESSURE TEST
3.8L (VIN 3) (PORT)

Circuit Description:
The fuel pump will deliver fuel to the fuel rail and injectors, then to the pressure regulator, where the system pressure is controlled. Excess fuel pressure is bypassed back to the fuel tank. When the engine is stopped, the pump can be turned "ON" by applying battery voltage to the test terminal located in the engine compartment.

Improper fuel system pressure may contribute to one or all of the following symptoms:
- Cranks but won't run
- Code 44 or 45
- Cuts out, may feel like ignition problem
- Hesitation, loss of power, or poor fuel economy

Test Description: Numbers below refer to circled numbers on the diagnostic chart.
5. Pressure but less than specifications falls into two areas:
   - Regulated pressure but less than specifications. The amount of fuel to injectors is OK but pressure is too low. The system will be lean running and may set Code 44. Also, possible hard starting cold and overall poor performance.
   - Restricted flow causing pressure drop. Normally, a vehicle with a fuel pressure of less than 165 kPa (24 psi) at idle will not be driveable.

However, if the pressure drop occurs only while driving, the engine will normally surge then stop as pressure begins to drop rapidly.
6. Restricting the fuel return line allows the fuel pump to develop its maximum pressure (dead head pressure). When battery voltage is applied to the pump test terminal, pressure should be above 517 kPa (75 psi).
7. This test determines if the high fuel pressure is due to a restricted fuel return line or a pressure regulator problem.

1988 VEHICLES EQUIPPED WITH 3.8L (VIN 3) ENGINE

CAUTION: TO REDUCE THE RISK OF FIRE AND PERSONAL INJURY, IT IS NECESSARY TO RELIEVE THE FUEL SYSTEM PRESSURE BEFORE SERVICING THE FUEL SYSTEM. (TO DO THIS):
- DISCONNECT THE FUEL TANK HARNESS CONNECTOR.
- CRANK ENGINE. ENGINE WILL START AND RUN UNTIL FUEL SUPPLY REMAINING IN FUEL PIPES IS CONSUMED. ENGAGE STARTER FOR 3.0 SECONDS TO ASSURE RELIEF OF ANY REMAINING PRESSURE.
- CONNECT THE FUEL TANK HARNESS CONNECTOR. UNLESS THIS PROCEDURE IS FOLLOWED BEFORE SERVICING FUEL PIPES OR CONNECTIONS, FUEL SPRAY COULD OCCUR.

CHART A-7
(Page 2 of 2)
FUEL SYSTEM PRESSURE TEST
3.8L (VIN 3) (PORT)

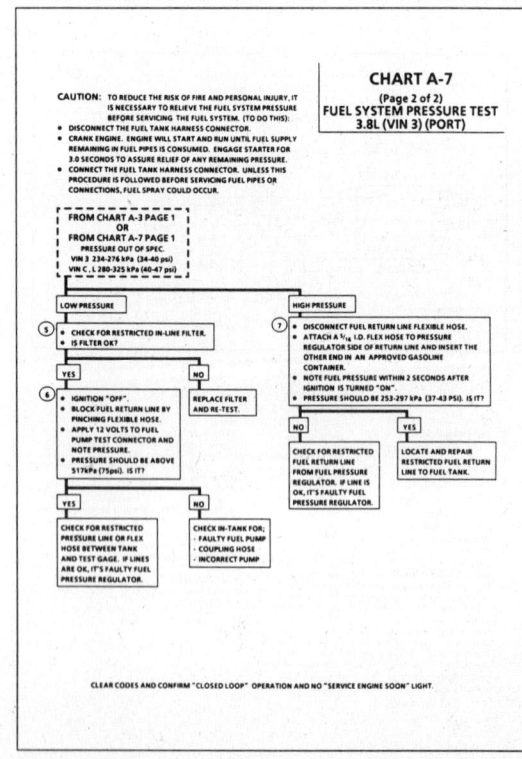

CLEAR CODES AND CONFIRM "CLOSED LOOP" OPERATION AND NO "SERVICE ENGINE SOON" LIGHT.

1988 VEHICLES EQUIPPED WITH 3.8L (VIN 3) ENGINE

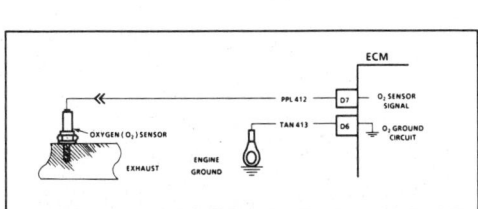

CODE 13
OXYGEN SENSOR CIRCUIT
(OPEN CIRCUIT)
3.8L (VIN 3) (PORT)

Circuit Description:

The ECM supplies a voltage of about .45 volts between terminals "D6" and "D7". (If measured with a 10 megohm digital voltmeter, this may read as low as .32 volts.) The O_2 sensor varies the voltage within a range of about 1 volts if the exhaust is rich, down through about .10 volts if exhaust is lean.

The sensor is like an open circuit and produces no voltage when it is below 360° C (600°F). An open sensor circuit or cold sensor causes "Open Loop" operation.

Test Description: Numbers below refer to circled numbers on the diagnostic chart.

1. Code 13 will set if
   - Engine at normal operating temperature
   - Up to 2 minutes engine time after start
   - O_2 signal voltage steady between .35 and .55 volts
   - Throttle position sensor signal above .55 volts
   - All conditions must be met for about 60 seconds

   If the conditions for a Code 13 exist, the system will not go "Closed Loop".

2. This will determine if the sensor or the wiring is the cause of the Code 13.

3. In doing this test, use only a high impedence digital volt ohmmeter. This test only checks the continuity of CKTs 412 and 413 because if CKT 413 is open, the ECM voltage on CKT 412 will be over .6 volts (600 mV).

Diagnostic Aids:

An intermittent may be caused by a poor connection, rubbed through wire insulation, or a wire broken inside the insulation.
Check For
- <u>Poor Connection or Damaged Harness</u> Inspect ECM harness connectors for backed out terminals "D7" or "D6", improper mating, broken locks, improperly formed or damaged terminals, poor terminal to wire connection, and damaged harness.
- <u>Intermittent Test</u> If connections and harness check OK, "Scan" O_2 sensor voltage while moving related connectors and wiring harness, with warm engine running at part throttle in "Closed Loop". If the failure is induced, the "O_2 sensor voltage" reading will change from its normal fluctuating voltage (above 600 mV and below 300 mV) to a fixed value around 450 mV. This may help to isolate the location of the malfunction.

1988 VEHICLES EQUIPPED WITH 3.8L (VIN 3) ENGINE

CODE 13
OXYGEN SENSOR CIRCUIT
(OPEN CIRCUIT)
3.8L (VIN 3) (PORT)

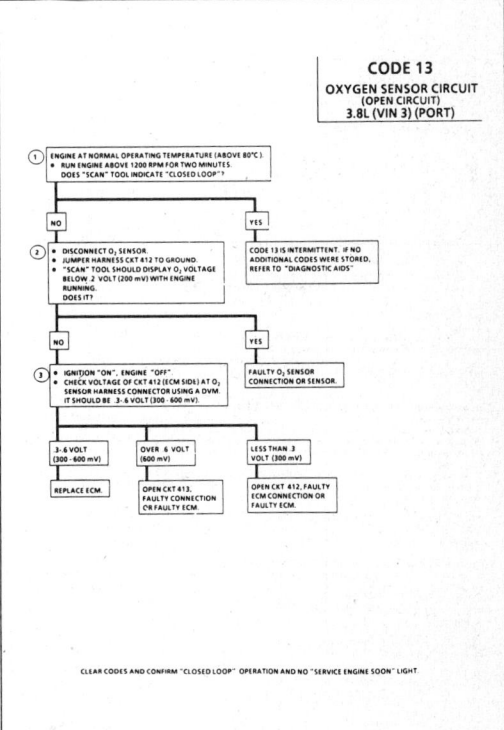

CLEAR CODES AND CONFIRM "CLOSED LOOP" OPERATION AND NO "SERVICE ENGINE SOON" LIGHT.

1988 VEHICLES EQUIPPED WITH 3.8L (VIN 3) ENGINE

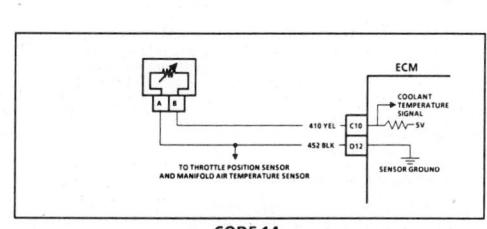

CODE 14
COOLANT TEMPERATURE SENSOR CIRCUIT
(HIGH TEMPERATURE INDICATED)
3.8L (VIN 3) (PORT)

Circuit Description:

The coolant temperature sensor uses a thermistor to control the signal voltage to the ECM. The ECM applies a voltage on CKT 410 to the sensor. When the engine is cold the sensor (thermistor) resistance is high, therefore the ECM will see high signal voltage.

As the engine warms, the sensor resistance becomes less, and the voltage drops. At normal engine operating temperature (85°C to 95°C) the voltage will measure about 1.5 to 2.0 volts.

Test Description: Numbers below refer to circled numbers on the diagnostic chart.

1. Code 14 will set if
   - Signal voltage indicates a coolant temperature above 135°C (275°F) for calibrated time.

2. This test will determine if CKT 410 is shorted to ground which will cause the conditions for Code 14.

Diagnostic Aids:

"Scan" tool displays engine temperature in degrees centigrade. After engine is started, the temperature should rise steadily to about 90°C then stabilize when thermostat opens.

An intermittent may be caused by a poor connection, rubbed through wire insulation, or a wire broken inside the insulation.

Check For:
- <u>Poor Connection or Damaged Harness</u> Inspect ECM harness connectors for backed out terminals "C10" or "D12", improper mating, broken locks, improperly formed or damaged terminals, poor terminal to wire connection, and damaged harness.
- <u>Intermittent Test</u> If connections and harness check OK, "Scan" coolant temperature while moving related connectors and wiring harness. If the failure is induced, the "coolant temperature" display will change. This may help to isolate the location of the malfunction.
- <u>Shifted Sensor</u> The "Temperature To Resistance Value" scale may be used to test the coolant sensor at various temperature levels to evaluate the possibility of a "shifted" (mis-scaled) sensor which may result in driveability complaints.

1988 VEHICLES EQUIPPED WITH 3.8L (VIN 3) ENGINE

CODE 14
COOLANT TEMPERATURE SENSOR CIRCUIT
(HIGH TEMPERATURE INDICATED)
3.8L (VIN 3) (PORT)

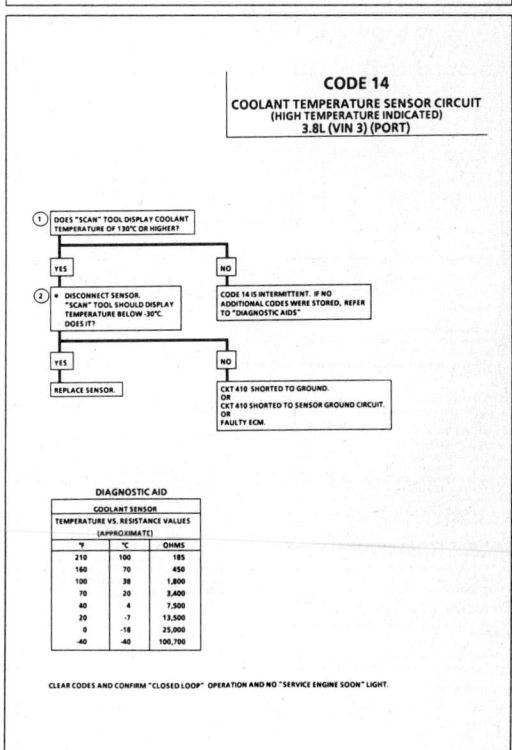

DIAGNOSTIC AID

COOLANT SENSOR TEMPERATURE VS. RESISTANCE VALUES (APPROXIMATE)		
°F	°C	OHMS
210	100	185
160	70	450
100	38	1,800
70	20	3,400
40	4	7,500
20	-7	13,500
0	-18	25,000
-40	-40	100,700

CLEAR CODES AND CONFIRM "CLOSED LOOP" OPERATION AND NO "SERVICE ENGINE SOON" LIGHT.

1988 VEHICLES EQUIPPED WITH 3.8L (VIN 3) ENGINE

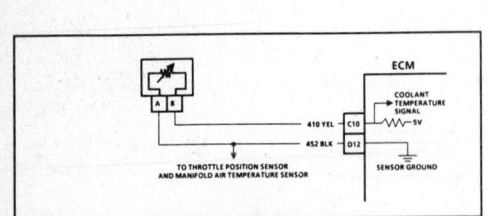

CODE 15
COOLANT TEMPERATURE SENSOR CIRCUIT
(LOW TEMPERATURE INDICATED)
3.8L (VIN 3) (PORT)

Circuit Description:

The coolant temperature sensor uses a thermistor to control the signal voltage to the ECM. The ECM applies a voltage on CKT 410 to the sensor. When the engine is cold the sensor (thermistor) resistance is high, therefore the ECM will see high signal voltage.

As the engine warms, the sensor resistance becomes less, and the voltage drops. At normal engine operating temperature (85°C to 95°C) the voltage will measure about 1.5 to 2.0 volts at the ECM.

Test Description: Numbers below refer to circled numbers on the diagnostic chart.

1. Code 15 will set if
   - Signal voltage indicates a coolant temperature less than -44°C (-47°F) for at least 3 seconds.
2. This test simulates a Code 14. If the ECM recognizes the low signal voltage (high temperature) and the "Scan" reads 130°C, the ECM and wiring are OK.
3. This test will determine if CKT 410 is open. There should be 5 volts present at sensor connector if measured with a DVM.

Diagnostic Aids:

A "Scan" tool reads engine temperature in degrees centigrade. After engine is started, the temperature should rise steadily to about 90°C then stabilize when thermostat opens.

An intermittent may be caused by a poor connection, rubbed through wire insulation, or a wire broken inside the insulation.

Check For:
- Poor Connection or Damaged Harness Inspect ECM harness connectors for backed out terminals "C10" or "D12", improper mating, broken locks, improperly formed or damaged terminals, poor terminal to wire connection, and damaged harness.
- Intermittent Test. If connections and harness check OK, "Scan" coolant temperature while moving related connectors and wiring harness. If the failure is induced, the display will change. This may help to isolate the location of the malfunction.
- Shifted Sensor. The "Temperature To Resistance Value" scale may be used to test the coolant sensor at various temperature levels to evaluate the possibility of a "shifted" (mis-scaled) sensor which may result in driveability complaints.

A faulty connection or an open in CKTs 410 or 452 will result in a Code 15.

If Code 23 or 63 is also set, check CKT 452 for faulty wiring or connections. Check terminals at sensor for good contact.

1988 VEHICLES EQUIPPED WITH 3.8L (VIN 3) ENGINE

CODE 15
COOLANT TEMPERATURE SENSOR CIRCUIT
(LOW TEMPERATURE INDICATED)
3.8L (VIN 3) (PORT)

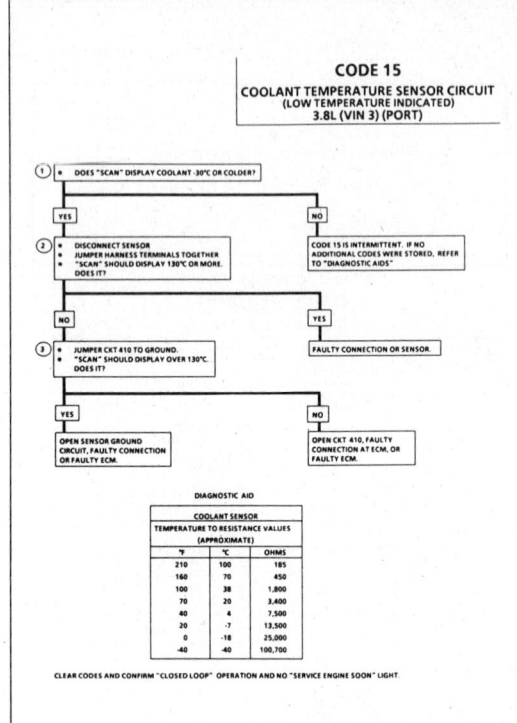

COOLANT SENSOR		
TEMPERATURE TO RESISTANCE VALUES		
(APPROXIMATE)		
°F	°C	OHMS
210	100	185
160	70	450
100	38	1,800
70	20	3,400
40	4	7,500
20	-7	13,500
0	-18	25,000
-40	-40	100,700

CLEAR CODES AND CONFIRM "CLOSED LOOP" OPERATION AND NO "SERVICE ENGINE SOON" LIGHT.

1988 VEHICLES EQUIPPED WITH 3.8L (VIN 3) ENGINE

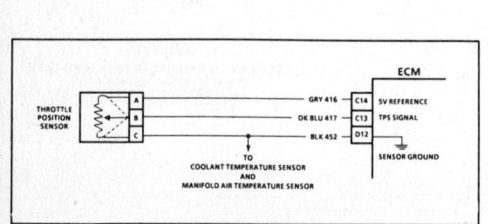

CODE 21
THROTTLE POSITION SENSOR (TPS) CIRCUIT
(SIGNAL VOLTAGE HIGH)
3.8L (VIN 3) (PORT)

Circuit Description:

The throttle position sensor (TPS) provides a voltage signal that changes relative to throttle blade angle. Signal voltage will vary from about .4 at idle to about 5 volts at wide open throttle.

The TPS signal is one of the most important inputs used by the ECM for fuel control and for most of the ECM control outputs.

Test Description: Numbers below refer to circled numbers on the diagnostic chart.

1. Code 21 will set if
   - Engine is running
   - TPS signal voltage is greater than 2.5 volts
   - Code 33 or 34 not present at first start up
   - All conditions met for 5 seconds
   With closed throttle, ignition "ON" or at idle, voltage should be .36- .44 volts.
2. With the TPS sensor disconnected, the TPS voltage should go low if the ECM and wiring is OK.
3. Probing CKT 452 with a test light checks the sensor ground CKT. A faulty sensor ground circuit will cause a Code 21.

Diagnostic Aids:

A "Scan" tool reads throttle position in volts. With closed throttle, ignition "ON" or at idle, voltage should be 3.8L .36-.44 volts.

Also some "Scan" tools will read throttle angle 0% = closed throttle, 100% = WOT.

An open in CKT 452 will result in a Code 21. Refer to "Intermittents" in Section "B".

Check For:
- Poor Connection or Damaged Harness Inspect ECM harness connectors for backed out terminals "C13" or "D12", improper mating, broken locks, improperly formed or damaged terminals, poor terminal to wire connection, and damaged harness.
- Intermittent Test If connections and harness check OK, monitor TPS voltage display while moving related connectors and wiring harness. If the failure is induced, the display will change. This may help to isolate the location of the malfunction.
- TPS Scaling Observe TPS voltage display while depressing accelerator pedal with engine stopped and ignition "ON" Display should vary from closed throttle TPS voltage when throttle was closed, to over (4.5 volts) 4500 mV when throttle is held at wide open throttle position.

1988 VEHICLES EQUIPPED WITH 3.8L (VIN 3) ENGINE

CODE 21
THROTTLE POSITION SENSOR (TPS) CIRCUIT
(SIGNAL VOLTAGE HIGH)
3.8L (VIN 3) (PORT)

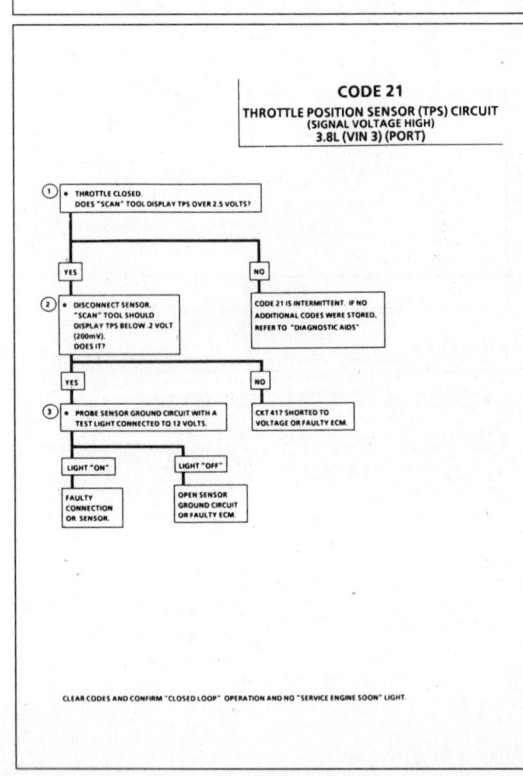

CLEAR CODES AND CONFIRM "CLOSED LOOP" OPERATION AND NO "SERVICE ENGINE SOON" LIGHT.

1988 VEHICLES EQUIPPED WITH 3.8L (VIN 3) ENGINE

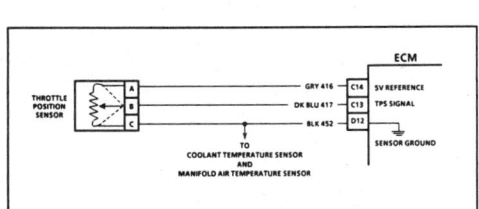

CODE 22
THROTTLE POSITION SENSOR (TPS) CIRCUIT
(SIGNAL VOLTAGE LOW)
3.8L (VIN 3) (PORT)

Circuit Description:
The throttle position sensor (TPS) provides a voltage signal that changes relative to throttle blade angle. Signal voltage will vary from about .4 at idle to about 5 volts at wide open throttle.
The TPS signal is one of the most important inputs used by the ECM for fuel control and for most of the ECM control outputs.

Test Description: Numbers below refer to circled numbers on the diagnostic chart.
1. Code 22 will set if
   - Engine is running
   - TPS signal voltage is less than .2 volts for 3 seconds
2. Simulates Code 21: (high voltage) If ECM recognizes the high signal voltage the ECM and wiring are OK.
3. With closed throttle, ignition "ON" or at idle, voltage should be .36-.44 volts. If not, check adjustment.
4. Simulates a high signal voltage. Checks CKT 417 for an open.

Diagnostic Aids:

A "Scan" tool reads throttle position in volts. Voltage should increase at a steady rate as throttle is moved toward WOT.
Also some "Scan" tools will read throttle angle 0% = closed throttle 100% = WOT.

An open or short to ground in CKTs 416 or 417 will result in a Code 22.
Check For
- **Poor Connection or Damaged Harness** Inspect ECM harness connectors for backed out terminals "C13" or "D12", improper mating, broken locks, improperly formed or damaged terminals, poor terminal to wire connection and damaged harness.
- **Intermittent Test** If connections and harness check OK, monitor TPS voltage display while moving related connectors and wiring harness. If the failure is induced, the display will change. This may help to isolate the location of the malfunction.
- **TPS Scaling** Observe TPS voltage display while depressing accelerator pedal with engine stopped and ignition "ON". Display should vary from closed throttle TPS voltage when throttle was closed, to over 4.5 volts (4500 mV) when throttle is held at wide open throttle position.

1988 VEHICLES EQUIPPED WITH 3.8L (VIN 3) ENGINE

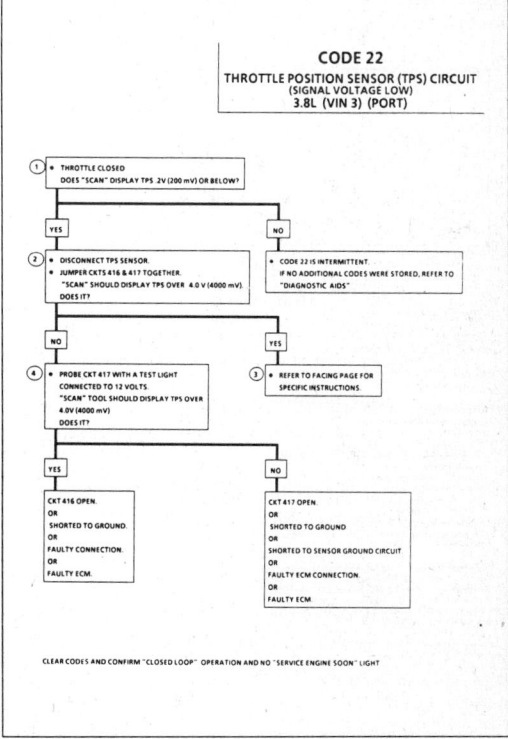

1988 VEHICLES EQUIPPED WITH 3.8L (VIN 3) ENGINE

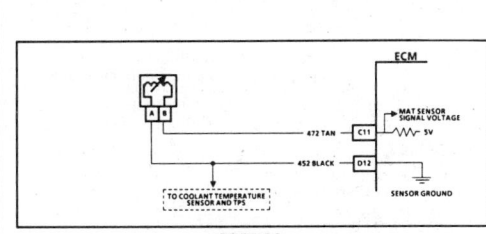

CODE 23
MANIFOLD AIR TEMPERATURE (MAT) SENSOR CIRCUIT
(LOW TEMPERATURE INDICATED)
3.8L (VIN 3) (PORT)

Circuit Description:
The MAT sensor uses a thermistor to control the signal voltage to the ECM. The ECM applies a voltage (about 5 volts) on CKT 472 to the sensor. When the air is cold, the sensor (thermistor) resistance is high, therefore, the ECM will see a high signal voltage. If the air is warm, the sensor resistance is low, therefore, the ECM will see a low voltage.

Test Description: Numbers below refer to circled numbers on the diagnostic chart.
Code 23 will set if:
- A signal voltage indicates a manifold air temperature below −40°C (−40°F) for 4 seconds
Due to the conditions necessary to set a Code 23, the "Service Engine Soon" light will only stay "ON" while the fault is present.

1. A "Scan" tool may not be used to diagnose this fault, due to the ECM transmitting "default" (substitute) values, when the fault is present. A Code 23 is set, due to an open sensor, wire, or connection. This test determines if the wiring and ECM are OK.
2. If the resistance is greater than 25,000 ohms, inspect air cleaner assembly for the presence of ice. If OK, replace the sensor.

1988 VEHICLES EQUIPPED WITH 3.8L (VIN 3) ENGINE

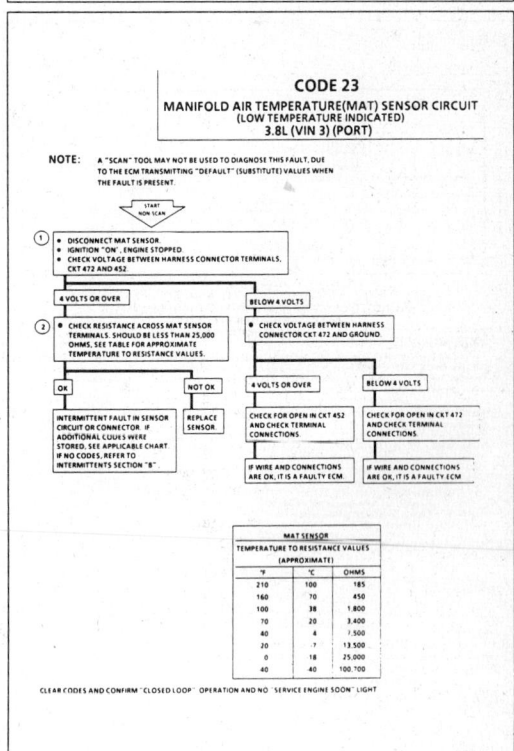

1988 VEHICLES EQUIPPED WITH 3.8L (VIN 3) ENGINE

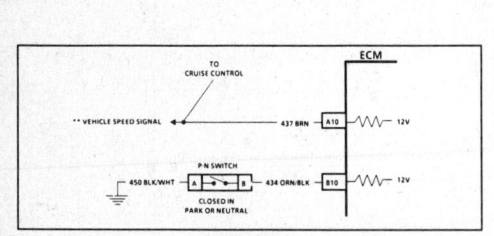

CODE 24
VEHICLE SPEED SENSOR (VSS) CIRCUIT
3.8L (VIN 3) (PORT)

Circuit Description:
The ECM applies and monitors 12 volts on CKT 437. CKT 437 connects to the vehicle speed buffer, which alternately grounds CKT 437 when drive wheels are turning. This pulsing action takes place about 2000 times per mile and the ECM will calculate vehicle speed based on the time between "pulses"
"Scan" reading should closely match with speedometer reading with drive wheels turning

Test Description: Numbers below refer to circled numbers on the diagnostic chart.
1. Code 24 will set if vehicle speed signal equals 0 mph when:
   - Engine speed is between 1400 and 3600 rpm
   - TPS is less than 2%
   - Low load condition (low air flow)
   - Not in park or neutral
   - All conditions met for 5 seconds.
2. 8-12 volts at the IP connector indicates CKT 437 is open between the IP connector and the VSS or there is a faulty vehicle speed sensor. A voltage of less than 1 volts at the IP connector indicates that CKT 437 wire is shorted to ground. If after disconnecting CKT 437 at the vehicle speed sensor the voltage reads above 10 volts, the vehicle speed sensor is faulty. If voltage remains less than 8 volts, then CKT 437 wire is grounded. If CKT 437 is not grounded, there is a faulty connection at the ECM or a faulty ECM.

Diagnostic Aids:
"Scan" should indicate a vehicle speed whenever the drive wheels are turning greater than 3 mph. If speedometer works OK, and "Scan" reads zero mph, a shorted cruise module could cause a Code 24 to set.

A faulty or misadjusted park/neutral switch can result in a false Code 24. Use a "Scan" and check for proper signal while in drive overdrive. Refer to CHART C-1A for P/N switch diagnosis check.
If all OK, refer to "Intermittents" in Section "B"

1988 VEHICLES EQUIPPED WITH 3.8L (VIN 3) ENGINE

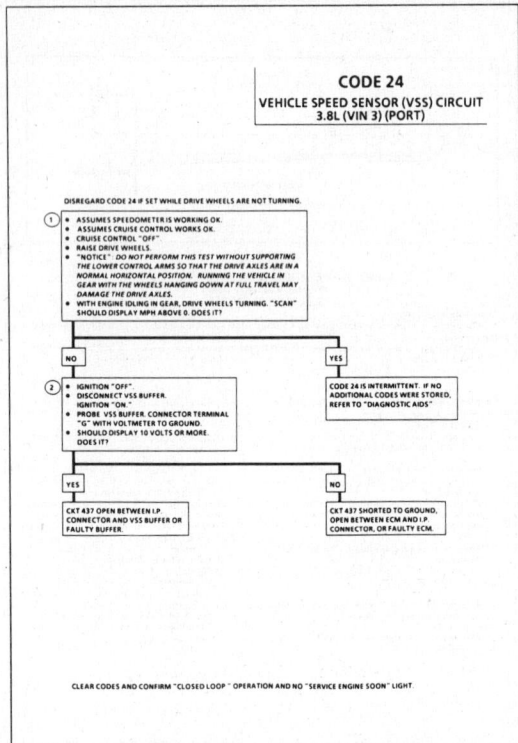

1988 VEHICLES EQUIPPED WITH 3.8L (VIN 3) ENGINE

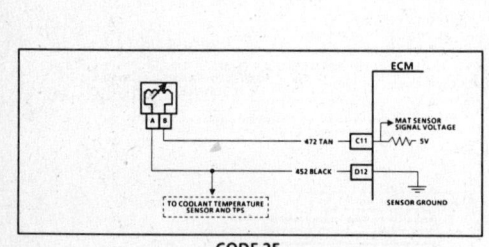

CODE 25
MANIFOLD AIR TEMPERATURE (MAT) SENSOR CIRCUIT
(HIGH TEMPERATURE INDICATED)
3.8L (VIN 3) (PORT)

Circuit Description:
The MAT sensor uses a thermistor to control the signal voltage to the ECM. The ECM applies a voltage of 4-6 volts on CKT 472 to the sensor. When manifold air is cold, the sensor (thermistor) resistance is high, therefore, the ECM will see a high signal voltage. If the air warms, the sensor resistance becomes less, and the voltage drops.

Test Description: Numbers below refer to circled numbers on the diagnostic chart.
Code 25 will set if:
- Signal voltage indicates a manifold air temperature greater than 135°C (275°F)
- A vehicle speed is present
- Both of the above requirements are met for at least 30 seconds
Due to the conditions necessary to set a Code 25, the "Service Engine Soon" light will only stay "ON" while the fault is present.

1. A "Scan" tool may not be used to diagnose this fault due to the ECM transmitting "default" (substitute) values when the fault is present. If voltage is above 4 volts, the ECM and wiring are OK.
2. If the resistance is less than 100 ohms, replace the sensor.

1988 VEHICLES EQUIPPED WITH 3.8L (VIN 3) ENGINE

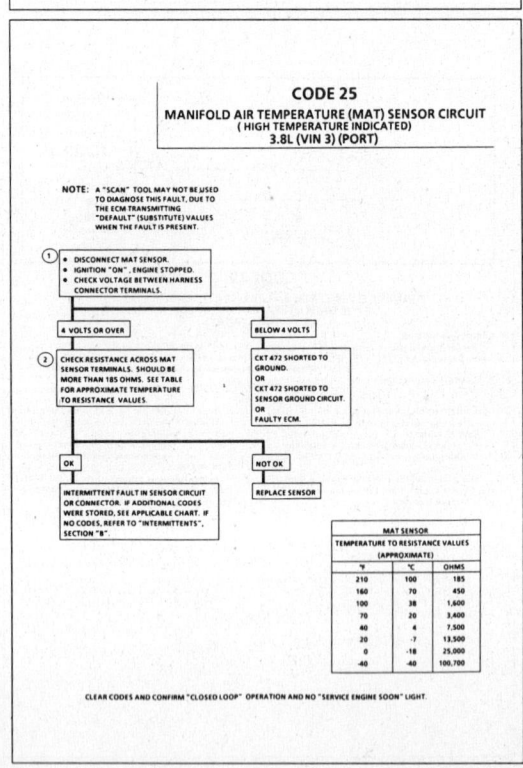

1988 VEHICLES EQUIPPED WITH 3.8L (VIN 3) ENGINE

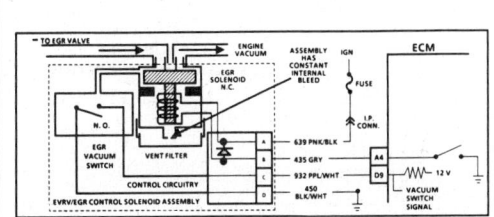

CODE 32
EXHAUST GAS RECIRCULATION (EGR) CIRCUIT
3.8L (VIN 3) (PORT)

Circuit Description:

The EGR valve is opened by engine vacuum. In order to control and monitor EGR application an electronic vacuum regulator valve is used (EVRV). The EVRV is composed of two devices: 1. EGR solenoid, normally closed (vacuum blocked). 2. EGR vacuum switch, normally open, (no current flow).

EGR vacuum control is accomplished by the ECM grounding CKT 435. This energizes the EGR solenoid. This is done thousands of times a second. By varing the length of "ON" time, as compared to "OFF" time, pulse width modulation (PWM), the ECM controls the vacuum source to the EGR valve.

EGR is monitored by the ECM through the EGR vacuum switch. The EGR vacuum switch, a normally open electrical switch, has an orfice built in which restricts the vacuum signal to the EGR vacuum switch. When sufficent vacuum reaches the EGR vacuum switch to close the electrical switch, there should also be sufficient vacuum to open the EGR valve.

Code 32 will be set if the vacuum switch closes at idle, or if it does not close under load (less than WOT).
- Engine running
- Code 33 or 34 not present
- Coolant temperature above 42.5°C (108°F)
- LV8 reading less than 144 counts
- No vacuum to EGR (switch open)
- Conditions exist over 5 seconds

Test Description: Numbers below refer to circled numbers on the diagnostic chart.
1. "Scan" displays the condition of the EGR diagnostic switch. In park or neutral, the display should read "NO" or "OFF" (open switch).
2. Under moderate engine load, the display will switch from "NO" to "YES" or "OFF" to "ON".
3. Checks the integrity of the 12 volt feed and ground circuits. If these circuits check OK, the fault is elsewhere in the EVRV/EGR control circuit.
4. A test light connected between terminals "A" and "B" will verify the integrity of the ECM wiring and check for proper ECM operation.
5. If "YES" was displayed at idle, disconnect the EVRV harness. If display remains "YES", the fault is either a short to ground in CKT 357 or the ECM.
6. If the EGR display switches from "YES" to "NO" when the EVRV is disconnected, the fault is either in the EVRV/EGR solenoid, CKT 435 or the ECM.

Probing at terminal "B" will further isolate the fault. If the test light is "ON", disconnect ECM A-B connector before checking CKT 435 for a short to ground since the short could be inside the ECM

Diagnostic Aids:

An intermittent may be caused by a poor connection, rubbed through wire insulation, or a wire broken inside the insulation. Check For:
- Poor Connection or Damaged Harness Inspect ECM harness connectors for backed out terminal "D9", improper mating, broken locks, improperly formed or damaged terminals, poor terminal to wire connection, and damaged harness.
- Intermittent Test If connections and harness checks out OK, "Scan" EVRV switch, while moving related connectors and wiring harness. If the failure is induced, the "EVRV Switch" display will change. This may help to isolate the location of the malfunction.

1988 VEHICLES EQUIPPED WITH 3.8L (VIN 3) ENGINE

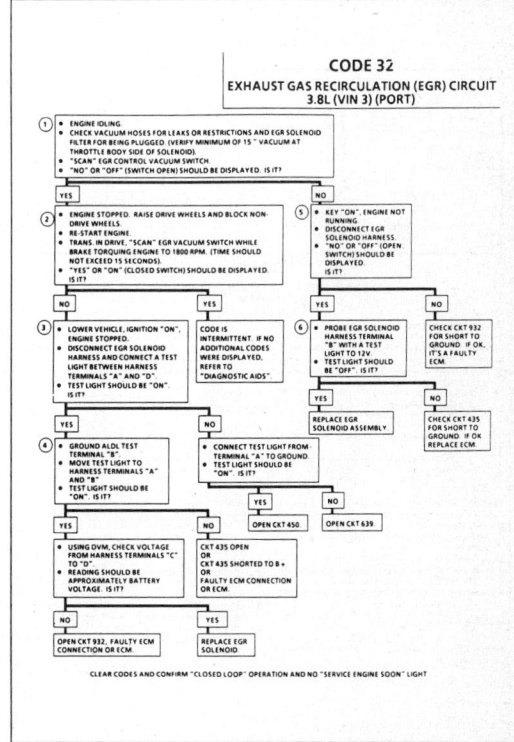

CODE 32
EXHAUST GAS RECIRCULATION (EGR) CIRCUIT
3.8L (VIN 3) (PORT)

CLEAR CODES AND CONFIRM "CLOSED LOOP" OPERATION AND NO "SERVICE ENGINE SOON" LIGHT

1988 VEHICLES EQUIPPED WITH 3.8L (VIN 3) ENGINE

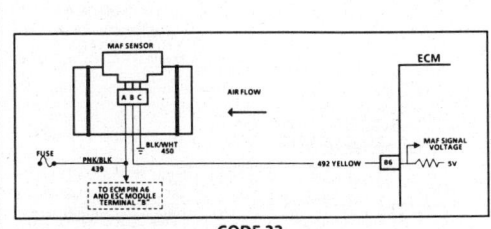

CODE 33
MASS AIR FLOW (MAF) SENSOR CIRCUIT
(GM/SEC HIGH)
3.8L (VIN 3) (PORT)

Circuit Description:

The MAF sensor measures the flow of air entering the engine. The sensor produces a frequency output between 32 and 150 herts (3 gm/sec to 150 gm/sec.). A large quantity (high frequency) indicates acceleration, and a small quantity (low frequency) indicates deceleration or idle. This information is used by the ECM for fuel control and is converted by a "Scan" tool to read out the air flow in grams per second. A normal reading is about 4-7 grams per second at idle and increases with rpm.

Test Description: Numbers below refer to circled numbers on the diagnostic chart.
1. Code 33 will set if
   - Ignition "ON" and air flow exceeds 20 gm/sec
   OR
   - Engine is running less than 800 rpm
   - TPS is 10% or less
   - Air flow greater than 150 grams per second (high frequency)
   - All of the above are met for 5 seconds or more

Diagnostic Aids:

The "Scan" tool is not of much use in diagnosing this code because when the code sets, gm/sec will display a default value. However, it may be useful in comparing the signal of a problem vehicle with that of a known good running one.

Check For:
- Poor Connection or Damaged Harness Inspect ECM harness connectors for backed out terminal "B6", improper mating, broken locks, improperly formed or damaged terminals, poor terminal to wire conection, and damaged harness.
- Intermittent Test If connections and harness check OK, "Scan" MAF while moving related connectors and wiring harness. If the failure is induced, the MAF display will change. This may help to isolate the location of the malfunction.
- Mis-routed Harness Inspect MAF sensor harness to insure that it is not too close to high voltage wires, such as spark plug leads.

1988 VEHICLES EQUIPPED WITH 3.8L (VIN 3) ENGINE

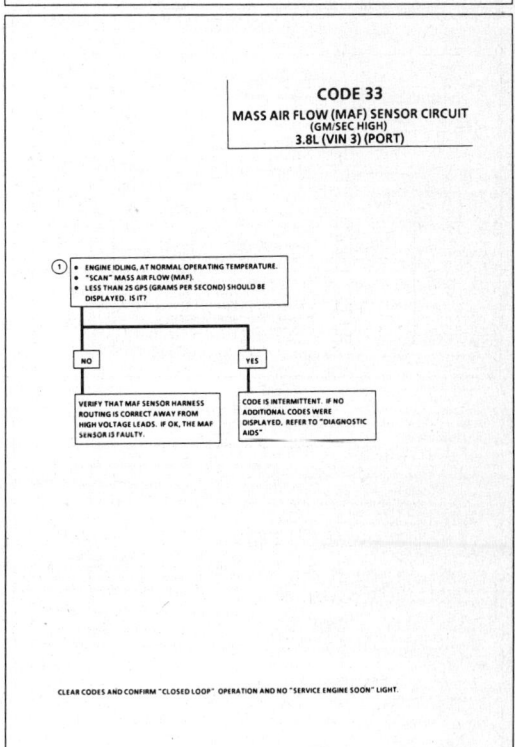

CODE 33
MASS AIR FLOW (MAF) SENSOR CIRCUIT
(GM/SEC HIGH)
3.8L (VIN 3) (PORT)

CLEAR CODES AND CONFIRM "CLOSED LOOP" OPERATION AND NO "SERVICE ENGINE SOON" LIGHT.

1988 VEHICLES EQUIPPED WITH 3.8L (VIN 3) ENGINE

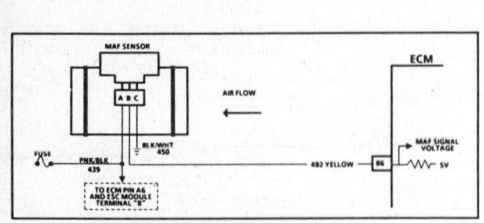

CODE 34
MASS AIR FLOW (MAF) SENSOR CIRCUIT
(GM/SEC LOW)
3.8L (VIN 3) (PORT)

Circuit Description:

The mass air flow (MAF) sensor measures the flow of air, which passes through it in a given time. The ECM uses this information to monitor the operating condition of the engine in calculating fuel delivery. A large quantity of air movement indicates acceleration, while a small quantity indicates deceleration or idle.

The MAF sensor produces a frequency signal which cannot be easily measured. The sensor can be diagnosed using the procedures on this chart.

Code 34 will set when either of the following sets of conditions exists:

- Engine running
- No MAF sensor signal for 250 mS

OR

- Engine running over 1400 rpm
- TPS over 2.5V
- Air flow less than 10 grams per second (low frequency)
- Above conditions for over 10 seconds

Test Description: Numbers below refer to circled numbers on the diagnostic chart.
1. This step checks for a loose or damaged air duct, which could set Code 34, and also checks to see if ECM recognizes a problem.
2. A voltage reading at sensor harness connector terminal "B" of less than 4 or over 6 volts indicates a fault in CKT 492 or poor connection
3. Verifies that both ignition voltage and a good ground circuit are available.

Diagnostic Aids:

An intermittent may be caused by a poor connection, mis-routed harness, rubbed through wire insulation, or a wire broken inside the insulation.

Check For:
- **Poor Connection - at ECM pin "B-6"** Inspect harness connectors for backed out terminals, improper mating, broken locks, improperly formed or damaged terminals, and poor terminal to wire connection.
- **Mis-routed Harness** Inspect MAF sensor harness to insure that it is not too close to high voltage wires, such as spark plug leads.
- **Damaged Harness** Inspect harness for damage. If harness appears OK, "Scan" while moving related connectors and wiring harness. A change in display would indicate the intermittent fault location.

1988 VEHICLES EQUIPPED WITH 3.8L (VIN 3) ENGINE

CODE 34
MASS AIR FLOW (MAF) SENSOR CIRCUIT
(GM/SEC LOW)
3.8L (VIN 3) (PORT)

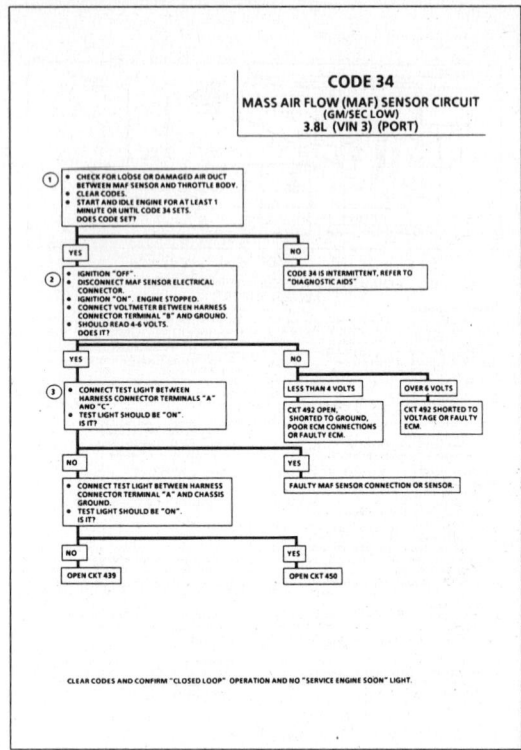

1988 VEHICLES EQUIPPED WITH 3.8L (VIN 3) ENGINE

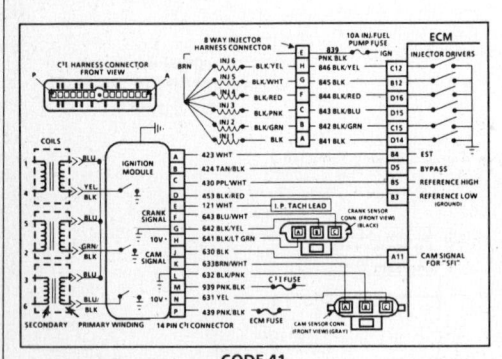

CODE 41
CAM SENSOR CIRCUIT
3.8L (VIN 3) (PORT)

Circuit Description:

For timing of spark plug firing, a cam sensor (hall effect) switch is used. The cam sensor sends a signal (sync-pulse) to the ignition module, when cylinder #1 is 25° after top dead center on the compression stroke. This signal is used to start the correct coil firing sequence, and to enable sequential fuel injection. The engine will continue to run if the cam signal is lost while running, however, will not restart after shut down. If the failure is in the cam signal output portion of the C³I module (terminal "J") or the "SFI" cam signal CKT 630 to ECM terminal "A11", the ECM will switch to run in the simultaneous fuel injection mode and continue to run. The engine can be re-started but will continue to run in the simultaneous mode as long as the fault is present. In either failure mode a Code 41 will be stored.

Code 41 is set when the following conditions are met:
- Engine is running
- Cam sensor signal not received by ECM in last 1 second interval

Test Description: Numbers below refer to circled numbers on the diagnostic chart.
1. Checks to see if ECM recognizes a problem, and sets Code 41. If the engine cranks but will not start and a Code 41 was displayed, the fault is in the ignition system portion of the cam sensor circuit and should be diagnosed using CHART A-3 "Cranks But Will Not Run"
2. The voltage to ECM terminal "A11" is supplied by the ignition module. If voltage reading is below 6 volts, the fault is in CKT 630, a poor connection at the C³I module.

Diagnostic Aids:

An intermittent may be caused by a poor connection, rubbed through wire insulation, or a wire broken inside the insulation.

Check For:
- **Poor Connection or Damaged Harness** Inspect ECM harness connectors for backed out terminal "A11", improper mating, broken locks, improperly formed or damaged terminals, poor terminal to wire connection, and damaged harness.
- **Intermittent Test** If connections and harness check OK, monitor a digital voltmeter connected from ECM terminal "A11" to ground while moving related connectors and wiring harness. If the failure is induced, the voltage reading will change. This may help to isolate the location of the malfunction.

1988 VEHICLES EQUIPPED WITH 3.8L (VIN 3) ENGINE

CODE 41
CAM SENSOR CIRCUIT
3.8L (VIN 3) (PORT)

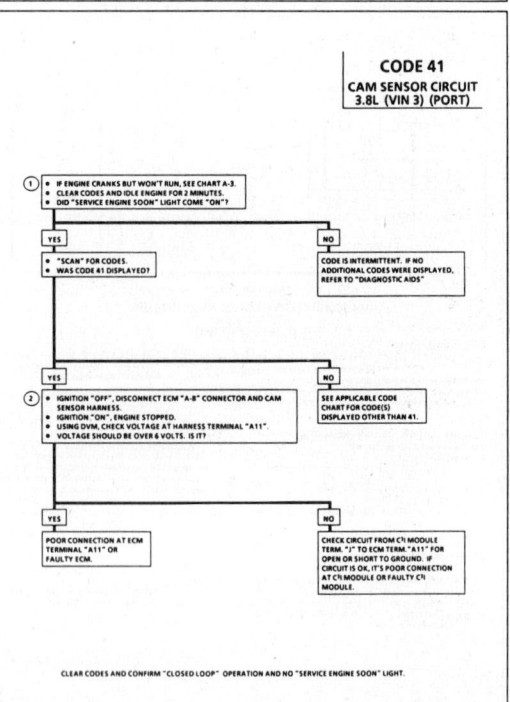

1988 VEHICLES EQUIPPED WITH 3.8L (VIN 3) ENGINE

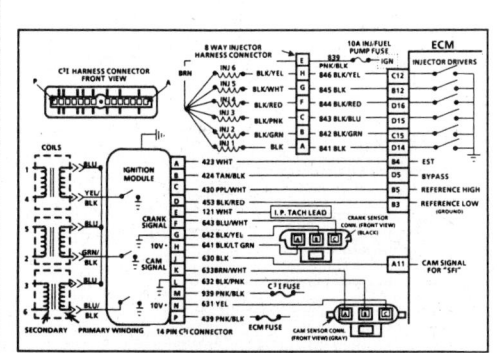

CODE 42
ELECTRONIC SPARK TIMING (EST) CIRCUIT
3.8L (VIN 3) (PORT)

Circuit Description:

The C³I module sends a reference signal to the ECM when the engine is cranking. While the engine is under 400 rpm, the C³I module controls the ignition timing. When the engine speed exceeds 400 rpm, the ECM sends a 5 volts signal on the bypass CKT 424 to switch the timing to ECM control through the EST CKT 423. An open or ground in the EST circuit will stall the engine and set a Code 42. The engine can be re-started but will run on module timing.

To set a Code 42 the following conditions must be met.
- Engine speed greater than 600 rpm with no EST pulse for 200 mS (open or grounded CKT 423), or
- ECM commanding bypass mode (open or grounded CKT 424)

Test Description: Numbers below refer to circled numbers on the diagnostic chart.
1. Checks to see if ECM recognizes a problem. If it does not set Code 42 at this point, it is an intermittent problem and could be due to a loose connection.
2. With the ECM disconnected, the ohmmeter should be reading less than 200 ohms, which is the normal resistance of the EST circuit through the C³I module. A higher resistance would indicate a fault in CKT 423, a poor C³I module connection, or a faulty C³I module.
3. If test light was "ON", connected from 12 volts to ECM harness terminal "D5", either CKT 423 is shorted to ground, or the C³I module is faulty.
4. Checks to see if C³I module switches when the bypass circuit is energized by 12 volts through the test light. If the C³I module actually switches, the ohmmeter reading should shift to over 8,000 ohms.
5. Disconnecting the ignition module should make the ohmmeter read as if it were monitoring an open circuit (infinite reading). Otherwise, CKT 423 is shorted to ground.

Diagnostic Aids:
An intermittent may be caused by a poor connection, rubbed through wire insulation, or a wire broken inside the insulation. Check For:
- **Poor Connection or Damaged Harness** Inspect ECM harness connectors for backed out terminals "B4" or "D5", improper mating, broken locks, improperly formed or damaged terminals, poor terminal to wire connection, and damaged harness.
- **Intermittent Test** If connections and harness check OK, a digital voltmeter voltage reading, when connected from the affected terminal to ground, may change when the affected wires and/or connectors are moved and the failure is induced.

1988 VEHICLES EQUIPPED WITH 3.8L (VIN 3) ENGINE

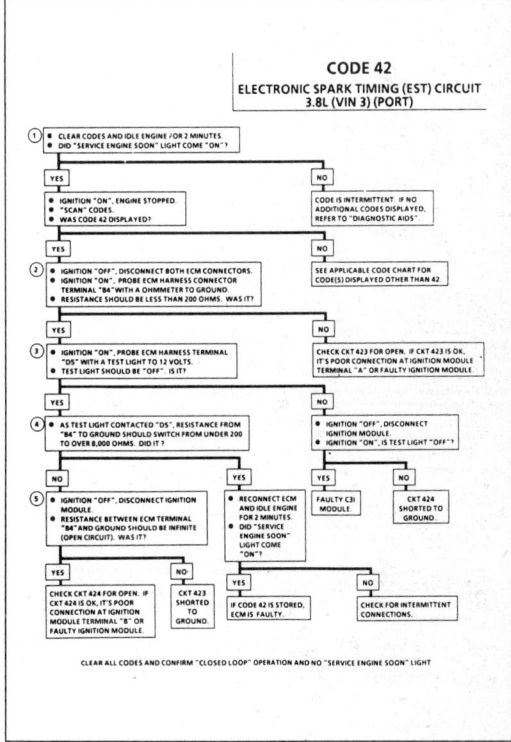

1988 VEHICLES EQUIPPED WITH 3.8L (VIN 3) ENGINE

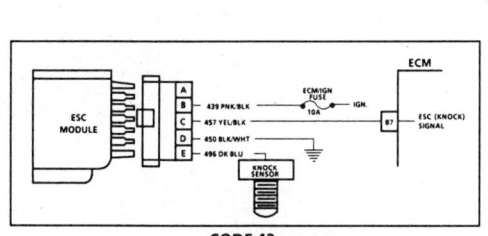

CODE 43
ELECTRONIC SPARK CONTROL (ESC) CIRCUIT
3.8L (VIN 3) (PORT)

Circuit Description:

The ESC system is comprised of a knock sensor and an ESC module.

The ESC module sends a voltage signal (8 to 10 volts) to the ECM.

When the sensor detects detonation, the module turns "OFF" the circuit to the ECM, and the voltage at ECM terminal "B7" drops to 0 volts. The ECM then retards EST as much as 20° in one (1) degree increments to reduce detonation. This happens fast and frequently enough that if looking at this signal with a DVM, you will not see 0 volts, but an average voltage somewhat less than what is normal with no detonation.

A loss of the knock sensor signal or a loss of ground at the ESC module would cause the signal at the ECM to remain high. The ECM would control ignition timing (EST) as if no detonation were occurring. The EST would not be retarded, and detonation could become severe enough under heavy engine load conditions to result in pre-ignition and potential engine damage.

Loss of the ESC signal to the ECM would cause the ECM to constantly retard the EST to its max retard of 20° from the spark table. This could result in sluggish performance and cause a Code 43 to set.

Code 43 will be set when:
- Engine Running
- ESC input signal has been low more than 2.2 seconds

Test Description: Numbers below refer to circled numbers on the diagnostic chart.
1. If the ECM data (knock signal) display is fluctuating widely, the ECM is monitoring a low voltage signal on CKT 457 at ECM terminal "B7".
2. Probing ESC harness terminal "C" with a test light connected to 12 volts should result in the "OLD PA3" (knock signal) display holding a steady reading due to over 8 volts having been applied to ECM terminal "B7" through CKT 457.
3. If over 6 volts is measured at ECM terminal "B7", CKT 457 is OK and the fault is due to a poor connection at the ECM or the ECM is faulty.

Diagnostic Aids:
An intermittent may be caused by a poor connection, rubbed through wire insulation or a wire broken inside the insulation.
Check For:
- **Poor Connection or Damaged Harness** Inspect ECM harness connectors for backed out terminal "B7", improper mating, broken locks, improperly formed or damaged terminals, poor terminal to wire connection, and damaged harness.
- **Intermittent Test** If connections and harness check OK, "Scan" knock signal (OLD PA3) while moving related connectors and wiring harness. If the failure is induced, the display will change. This may help to isolate the location of the malfunction.

1988 VEHICLES EQUIPPED WITH 3.8L (VIN 3) ENGINE

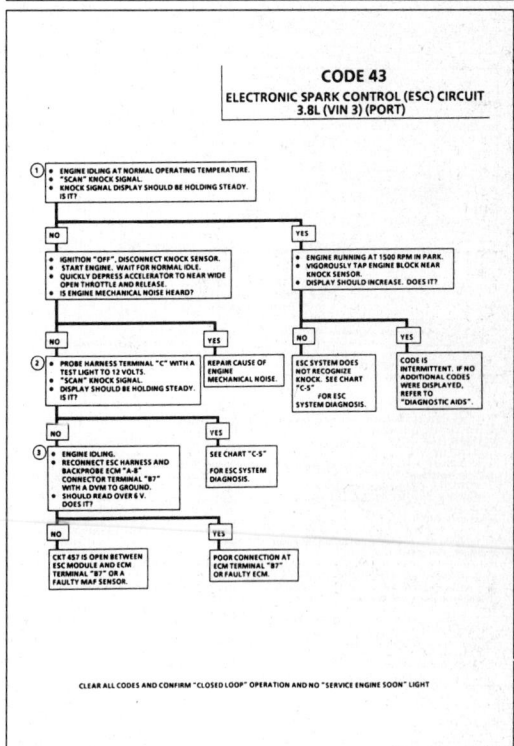

1988 VEHICLES EQUIPPED WITH 3.8L (VIN 3) ENGINE

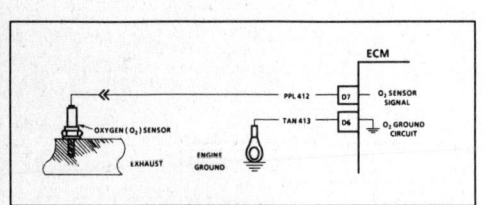

CODE 44
OXYGEN SENSOR CIRCUIT
(LEAN EXHAUST INDICATED)
3.8L (VIN 3) (PORT)

Circuit Description:

The ECM supplies a voltage of about .45 volt (450 mV) between terminals "D6" and "D7". (If measured with a 10 megohm digital voltmeter, this may read as low as .32 volt). The O_2 sensor varies the voltage within a range of about 1 volt (1000 mV) if the exhaust is rich, down through about .10 volt (100 mV) if exhaust is lean.

The sensor is like an open circuit and produces no voltage when it is below about 360°C (600°F). An open sensor circuit or cold sensor causes "Open Loop" operation.

Test Description: Numbers below refer to circled numbers on the diagnostic chart.
1. Code 44 is set, when the O_2 sensor signal voltage on CKT 412
   - Remains below .2 volt for 60 seconds or more
   - And the system is operating in "Closed Loop"

Diagnostic Aids:

Using the "Scan", observe the block learn values at different rpm and air flow conditions. The "Scan" also displays the block cells, so the block learn values can be checked in each of the cells to determine when the Code 44 exists, the block learn values will be around 150.
- O_2 Sensor Wire Sensor pigtail may be mispositioned and contacting the exhaust manifold.
- Check for intermittent ground in wire between connector and sensor.

- MAF Sensor. A mass air flow (MAF) sensor output that causes the ECM to sense a lower than normal air flow will cause the system to go lean. Disconnect the MAF sensor and if the lean condition is gone, replace the MAF sensor.
- Lean Injector(s) Perform injector balance test CHART C-2A.
- Fuel Contamination Water, even in small amounts, near the in-tank fuel pump inlet can be delivered to the injectors. The water causes a lean exhaust and can set a Code 44.
- Fuel Pressure System will be lean if pressure is too low. It may be necessary to monitor fuel pressure while driving the car at various road speeds and/or loads to confirm. See "Fuel System Diagnosis", CHART A-7.
- Exhaust Leaks If there is an exhaust leak, the engine can cause outside air to be pulled into the exhaust and past the sensor. Vacuum or crankcase leaks can cause a lean condition.
- If the above are OK, it is a faulty oxygen sensor.

1988 VEHICLES EQUIPPED WITH 3.8L (VIN 3) ENGINE

CODE 44
OXYGEN SENSOR CIRCUIT
(LEAN EXHAUST INDICATED)
3.8L (VIN 3) (PORT)

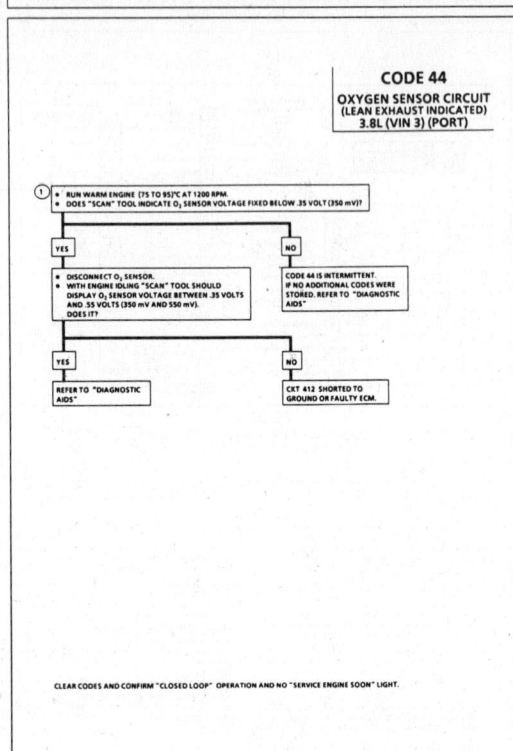

CLEAR CODES AND CONFIRM "CLOSED LOOP" OPERATION AND NO "SERVICE ENGINE SOON" LIGHT.

1988 VEHICLES EQUIPPED WITH 3.8L (VIN 3) ENGINE

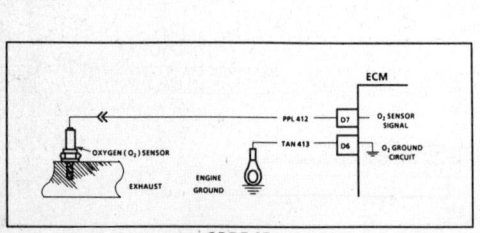

CODE 45
OXYGEN SENSOR CIRCUIT
(RICH EXHAUST INDICATED)
3.8L (VIN 3) (PORT)

Circuit Description:

The ECM supplies a voltage of about .45volt (450 mV) between terminals "D6" and "D7". (If measured with a 10 megohm digital voltmeter, this may read as low as .32 volt.) The O_2 sensor varies the voltage within a range of about 1volt (1000 mV) if the exhaust is rich, down through about .10 volt (100 mV) if exhaust is lean.

The sensor is like an open circuit and produces no voltage when it is below about 360°C (600°F). An open sensor circuit or cold sensor causes "Open Loop" operation.

Test Description: Numbers below refer to circled numbers on the diagnostic chart.
1. Code 45 is set, when the O_2 sensor signal voltage or CKT 412:
   - Remains above .7 volt for 30 seconds and in "Closed Loop"
   - Throttle angle between 3% and 45%
   - Engine time after start is 1 minute or more

Diagnostic Aids:

Using the "Scan", observe the block learn values at different rpm and air flow conditions. The "Scan" also displays the block cells, so the block learn values can be checked in each of the cells to determine when the Code 45 may have been set. If the conditions for Code 45 exists, the block learn values will be around 115.
- Fuel Pressure System will go rich if pressure is too high. The ECM can compensate for some increase. However, if it gets too high, a Code 45 may be set. See "Fuel System Diagnosis", CHART A-7.

- Rich Injector Perform injector balance test CHART C-2A.
- Leaking Injector See CHART A-7.
- Check for fuel contaminated oil.
- Canister Purge Check for fuel saturation. If full of fuel, check canister control and hoses. See "Canister Purge".
- MAF Sensor An output that causes the ECM to sense a higher than normal airflow can cause the system to go rich. Disconnecting the MAF sensor will allow the ECM to set a fixed value for the sensor. Substitute a different MAF sensor if the rich condition is gone while the sensor is disconnected or replace MAF sensor.
- Check for leaking fuel pressure regulator diaphragm by checking vacuum line to regulator for fuel.
- TPS An intermittent TPS output will cause the system to go rich due to a false indication of the engine accelerating.
- EGR An EGR staying open (especially at idle) will cause the O_2 sensor to indicate a rich exhaust and this could result in a Code 45.

1988 VEHICLES EQUIPPED WITH 3.8L (VIN 3) ENGINE

CODE 45
OXYGEN SENSOR CIRCUIT
(RICH EXHAUST INDICATED)
3.8L (VIN 3) (PORT)

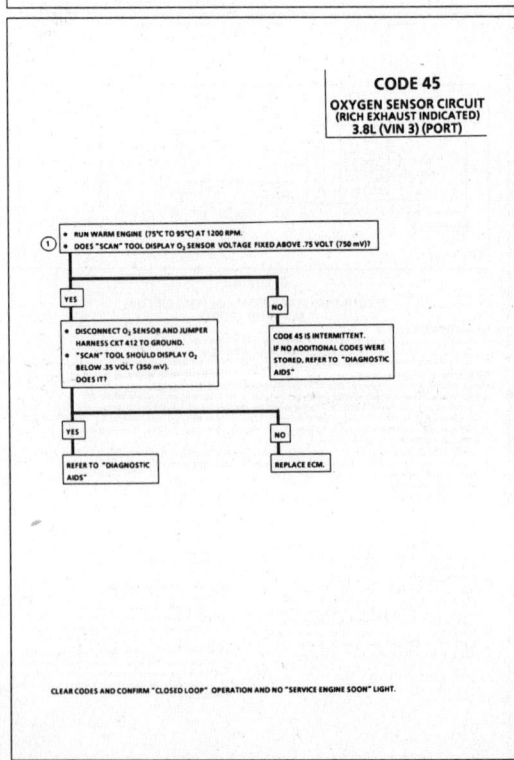

CLEAR CODES AND CONFIRM "CLOSED LOOP" OPERATION AND NO "SERVICE ENGINE SOON" LIGHT.

1988 VEHICLES EQUIPPED WITH 3.8L (VIN 3) ENGINE

CODE 51
PROM ERROR
(FAULTY OR INCORRECT PROM)

CHECK THAT ALL PINS ARE FULLY INSERTED IN THE SOCKET. IF OK, REPLACE PROM, CLEAR MEMORY AND RECHECK. IF CODE 51 REAPPEARS, REPLACE ECM.

CLEAR ALL CODES AND CONFIRM "CLOSED LOOP" OPERATION AND NO "SERVICE ENGINE SOON" LIGHT

CODE 52
CALPAK ERROR
(FAULTY OR INCORRECT CALPAK)

CHECK THAT ALL PINS ARE FULLY INSERTED IN THE SOCKET. IF OK, REPLACE CALPAK, CLEAR MEMORY AND RECHECK. IF CODE 52 REAPPEARS, REPLACE ECM.

CLEAR ALL CODES AND CONFIRM "CLOSED LOOP" OPERATION AND NO "SERVICE ENGINE SOON" LIGHT

CODE 55
ECM ERROR

REPLACE ELECTRONIC CONTROL MODULE (ECM).

CLEAR ALL CODES AND CONFIRM "CLOSED LOOP" OPERATION AND NO "SERVICE ENGINE SOON" LIGHT

1988 VEHICLES EQUIPPED WITH 3.8L (VIN 3) ENGINE

SECTION B
SYMPTOMS

Before Starting
Intermittents
Hard Start
Hesitation, Sag, Stumble
Surges and/or Chuggle
Lack of Power, Sluggish, or Spongy
Detonation/Spark Knock
Cuts Out, Misses
Backfire
Poor Fuel Economy
Dieseling, Run On
Rough, Unstable, or Incorrect Idle, Stalling
Excessive Exhaust Emissions Or Odors
Chart B-1 - Restricted Exhaust System Check

BEFORE STARTING

Before using this section you should have performed the DIAGNOSTIC CIRCUIT CHECK and found out that:
1. The ECM and "Service Engine Soon" light are operating.
2. There are no trouble codes stored, or there is a trouble code but no "Service Engine Soon" light.
3. The fuel control system is operating OK (by performing field service mode check).
Verify the customer complaint, and locate the correct SYMPTOM. Check the items indicated under that symptom.
If the ENGINE CRANKS BUT WILL NOT RUN, see CHART A-3.

Several of the symptom procedures below call for a careful visual/physical check. This check should include:
• ECM grounds for being clean and tight.
• Vacuum hoses for splits, kinks, and proper connections, as shown on emission control information label.
• Air leaks at throttle body mounting and intake manifold.
• Air leaks between MAF sensor and throttle body.
• Ignition wires for cracking, hardness, proper routing, and carbon tracking.
• Wiring for proper connections, pinches, and cuts.
The importance of this step cannot be stressed too strongly - it can lead to correcting a problem without further checks and can save valuable time.

1988 VEHICLES EQUIPPED WITH 3.8L (VIN 3) ENGINE

INTERMITTENTS

Problem may or may not turn "ON" the "Service Engine Soon" light, or store a code.

DO NOT use the trouble code charts in Section "A" for intermittent problems. The fault must be present to locate the problem. If a fault is intermittent, use of trouble code charts may result in replacement of good parts.
• Most intermittent problems are caused by faulty electrical connections or wiring. Perform a careful check as described at start of this section. Check for:
  • Poor mating of the connector halves, or terminals not fully seated in the connector body (backed out).
  • Improperly formed or damaged terminals. All connector terminals in problem circuit should be carefully reformed to increase contact tension.
  • Poor terminal to wire connection. This requires removing the terminal from the connector body to check. See introduction to Section "6E".
• If a visual/physical check does not find the cause of the problem, the car may be driven with a voltmeter connected to a suspected circuit. An abnormal voltage reading, when the problem occurs, indicates the problem may be in that circuit. If the wiring and connectors check OK and a trouble code was stored for a circuit having a sensor, except for Codes 43, 44, and 45, substitute a known good sensor and recheck.

An intermittent "Service Engine Soon" light with no stored code may be caused by:
• Ignition coil shorted to ground, arcing at spark plug wires or plugs.
• "Service Engine Soon" light wire to ECM shorted to ground. (CKT 419).
• Diagnostic "Test" terminal wire to ECM, shorted to ground. (CKT 451)
• ECM power grounds. See ECM wiring diagrams.
• Loss of trouble code memory. To check, disconnect TPS and idle engine until "Service Engine Soon" light comes "ON". Code 22 should be stored, and kept in memory when ignition is turned "OFF". If not, the ECM is faulty.
• Check for an electrical system interference caused by a defective relay, ECM driven solenoid, or switch. They can cause a sharp electrical surge. Normally, the problem will occur when the faulty component is operated.
• Check for improper installation of electrical options, such as lights, 2-way radios, etc.
• EST wires should be kept away from spark plug wires, coil and generator.
• Check for open diode across A/C compressor clutch, and for other open diodes (see wiring diagrams).

HARD START

Definition: Engine cranks OK, but does not start for a long time. Does eventually run, or may start but immediately dies.

• Perform careful check as described at start of Section "B".
• Make sure driver is using correct starting procedure.
• CHECK:
  - TPS for sticking or binding or a high TPS voltage with the throttle closed.
  - High resistance in coolant sensor circuit or sensor itself. See CODE 15 CHART or with a "Scan" tool compare coolant temperature with ambient temperature on a cold engine.
  - Fuel pressure CHART A-7.
  - Water contaminated fuel.
  - EGR operation. Be sure valve seats properly and is not staying open. See CHART C-7.
  - Fuel pump relay - See CHART A-7.

- Ignition system - Check for:
  Proper output with ST-125.
  Bare and shorted wires.
  Loose C3I module ground, mounting screws.
• A faulty in-tank fuel pump check valve will allow the fuel in the lines to drain back to the tank after the engine is stopped. To check for this condition:
  Perform fuel system diagnosis. CHART A-7.
• Remove spark plugs. Check for wet plugs, cracks, wear, improper gap, burned electrodes, or heavy deposits. Repair or replace as necessary.

1988 VEHICLES EQUIPPED WITH 3.8L (VIN 3) ENGINE

HESITATION, SAG, STUMBLE

Definition: Momentary lack of response as the accelerator is pushed down. Can occur at all car speeds. Usually most severe when first trying to make the car move, as from a stop sign. May cause the engine to stall if severe enough.

• Perform careful visual check as described at start of Section "B".
• CHECK:
  - Fuel pressure. See CHART A-7. Also Check for water contaminated fuel.
  - Spark plugs for being fouled or faulty wiring.
  - PROM number. Also check service bulletins for latest PROM.
  - TPS for binding or sticking. Voltage should increase at a steady rate as throttle is moved toward WOT.

- Generator output voltage. Repair if less than 9 or more than 16 volts.
- C3I ground, CKT 453.
- Canister purge system for proper operation. See CHART C-3.
- EGR - See CHART C-7.
- Engine Thermostat - functioning correctly and proper heat range.
• Perform injector balance test CHART C-2A.

SURGES AND/OR CHUGGLE

Definition: Engine power variation under steady throttle or cruise. Feels like the car speeds up and slows down with no change in the accelerator pedal.

• Be sure driver understands transmission converter clutch and A/C compressor operation in owner's manual.
• Perform careful visual inspection as described at start of Section "B".
• CHECK:
  - Generator output voltage. Repair if less than 9 or more than 16 volts.
■ • If a "Scan" tool is available which plugs in to the ALDL connector, make sure reading of VSS matches vehicle speedometer. See introduction explaining "Scan" tool positions.
  - EGR - There should be no EGR at idle. See CHART C-7.
  - EGR solenoid vent.
  - Vacuum lines for kinks or leaks.

- In-line fuel filter. Replace if dirty or plugged.
- Fuel pressure while condition exists. See CHART A-7.
• Inspect oxygen sensor for silicone contamination, from fuel or use of improper RTV sealant. The sensor may have a white powdery coating. This will result in a high but false signal voltage (rich exhaust indication). The ECM will then reduce the amount of fuel delivered to the engine, causing a severe driveability problem.
• Remove spark plugs. Check for cracks, wear, improper gap, burned electrodes, or heavy deposits.
• Check spark plug leads.

LACK OF POWER, SLUGGISH, OR SPONGY

Definition: Engine delivers less than expected power. Little or no increase in speed when accelerator pedal is pushed down part way.

• Perform careful visual check as described at start of Section "B".
• Compare customer's car to similar unit. Make sure the customer's car has an actual problem.
• Remove air cleaner and check air filter for dirt, or for being plugged. Replace as necessary.
• CHECK:
  - Restricted fuel filter, contaminated fuel or improper fuel pressure. See CHART A-7.
  - ECM power grounds - See wiring diagrams.
  - EGR operation for being open or partly open all the time - CHART C-7.

- Exhaust system for possible restriction:
  - Inspect exhaust system for damaged or collapsed pipes.
  - Inspect muffler for heat distress or possible internal failure.
- Generator output voltage. Repair if less than 9 or more than 16 volts.
- Engine valve timing and compression.
- Engine for proper or worn camshaft.
- Secondary voltage using a shop ociliiscope or a spark tester J-26792 (ST-125) or equivalent.

1988 VEHICLES EQUIPPED WITH 3.8L (VIN 3) ENGINE

DETONATION /SPARK KNOCK

Definition: A mild to severe ping, usually worse under acceleration. The engine makes sharp metallic knocks that change with throttle opening. Sounds like popcorn popping.

- Check for obvious overheating problems.
  - Low/weak coolant mixture.
  - Inoperative thermostat.
  - Restricted air/water flow through radiator.
  - Inoperative electric coolant fan circuit. See CHART C-12.
- CHECK:
  - EGR system for not opening - CHART C-7.
  - ESC system for no retard - see CHART C-5.
  - TCC operation - see CHART C-8.

- Fuel system pressure. See CHART A-7.
- Remove carbon with top engine cleaner. Follow instructions on can.
  - Check for leaking valve oil seals.
  - Check for poor fuel quality, proper octane rating.
  - Check for correct PROM.

- Check for incorrect basic engine parts such as cam, heads, pistons, etc.

CUTS OUT, MISSES

Definition: Steady pulsation or jerking that follows engine speed, usually more pronounced as engine load increases. The exhaust has a steady spitting sound at idle or low speed.

- Perform careful visual (physical) check as described at start of Section "B".
- Misses at idle:
  1. With engine idling, disconnect IAC motor. Remove one spark plug wire at a time using insulated pliers.
  2. If there is an rpm drop on all cylinders, go to ROUGH, UNSTABLE, OR INCORRECT IDLE, OR STALLING symptom. Reconnect IAC motor.
  3. If there is no rpm drop on one or more cylinders, or excessive variation in drop, check for spark on the suspected cylinder(s) with J26792 (ST-125) spark tester or equivalent.
  If spark exists:
  Inspect plugs for:
  - Cracks, wear, improper gap.
  - Burned electrodes, or heavy deposits.
  - Perform compression check on questionable cylinder.
  If no spark on suspected cylinders:
  Ground the opposite plug wire of the effected coil pair. If tester now sparks replace the spark plug of the wire which was jumpered to ground. If tester still does not spark, refer to appropriate ignition chart.
- Cuts out under load:
  Test for spark on each wire with an J26792 spark tester or equivalent. If there is spark on all cylinders, remove plugs and check for:
  - Cracks, wear, improper gap.
  - Burned electrodes, or heavy deposits.

If no spark at plug(s):
Ground the opposite plug wire of the effected coil pair. If tester now sparks replace the spark plug of the wire which was jumpered to ground. If tester still does not spark, refer to appropriate ignition chart.
- If above checks did not correct problem, check the following:
  - Disconnect all injector harness connectors. Connect J-34730-2 injector test light or equivalent 6 volt test light between the harness terminals of each injector connector and note light while cranking. If test light fails to blink at any connector, it is a faulty injector drive circuit harness, connector, or terminal.
  - Perform the injector balance test. See CHART C-2A.
  - Fuel system - Plugged fuel filter, water, low pressure. See CHART A-7.
  - Valve timing.
  - Remove rocker covers. Check for bent pushrods, worn rocker arms, broken valve springs, worn camshaft lobes. Repair as necessary.
  - Perform compression check.
  - On C³I systems, a misfire may be caused by a misaligned crank sensor or bent vane on rotating disc. Inspect for proper clearance at each vane, approx. .025" (.635 mm). Sensor should be replaced if it shows evidence of rubbing.

1988 VEHICLES EQUIPPED WITH 3.8L (VIN 3) ENGINE

BACKFIRE

Definition: Fuel ignites in intake manifold, or in exhaust system, making a loud popping noise.

- CHECK:
  - Compression - Look for sticking or leaking valves.
  - EGR operation for being open all the time. See CHART C-7.
  - EGR gasket for faulty or loose fit.
  - Valve timing.
  - Output voltage of ignition coil using a shop ocilliscope or spark tester J-26792 (ST-125) or equivalent.

- Spark plugs for crossfire. Also inspect spark plug wires, and proper routing of plug wires.
  Ignition system for intermittent condition.

POOR FUEL ECONOMY

Definition: Fuel economy, as measured by an actual road test, is noticeably lower than expected. Also, economy is noticeably lower than it was on this car at one time, as previously shown by an actual road test.

- CHECK:
  - Engine thermostat for faulty part (always open) or for wrong heat range.
  - Fuel pressure. See CHART A-5.
- Check owner's driving habits.
  - Is A/C "ON" full time (Defroster mode "ON")?
  - Are tires at correct pressure?
  - Are excessively heavy loads being carried?
  - Is acceleration too much, too often?
  - Suggest driver read "Important Facts on Fuel Economy" in owner's manual.
- Perform "Diagnostic Circuit Check."
- Check air cleaner element (filter) for dirt or being plugged.
- Check for proper calibration of speedometer.

- Visually (physically) Check:
  - Vacuum hoses for splits, kinks and proper connections as shown on Vehicle Emissions Control Information label.
  - Ignition wires for cracking, hardness, and proper connections.
- Remove spark plugs. Check for cracks, wear, improper gap, burned electrodes or heavy deposits. Repair or replace as necessary.
- Check compression.
- Check TCC for proper operation. See CHART C-8, use "Scan" tool if available.
- Check for dragging brakes.
- Suggest owner fill fuel tank and recheck fuel economy.
- Check for exhaust system restriction.

DIESELING, RUN-ON

Definition: Engine continues to run after key is turned "OFF", but runs very roughly. If engine runs smoothly, check ignition switch and adjustment.

- Check injectors for leaking. See CHART A-7.

1988 VEHICLES EQUIPPED WITH 3.8L (VIN 3) ENGINE

ROUGH, UNSTABLE, OR INCORRECT IDLE, STALLING

Definition: The engine runs unevenly at idle. If bad enough, the car may shake. Also, the idle may vary in rpm (called "hunting"). Either condition may be bad enough to cause stalling. Engine idles at incorrect speed.

- Perform careful visual check as described at start of Section "B".
- Clean injectors.
- CHECK:
  - Throttle linkage for sticking or binding.
  - TPS for sticking or binding, be sure output is stable at idle and adjustment specification is correct.
  - IAC system.
  - Generator output voltage. Repair if less than 9 or more than 16 volts.
  - P/N switch circuit. See CHART C-1A 3.8L (VIN 3) Code 31, 3800 (VIN C), or use "Scan" tool, and be sure tool indicates vehicle is in drive with gear selector in drive.
  - Injector balance. See CHART C-2A.
  - PCV valve for proper operation by placing finger over inlet hole in valve and several times. Valve should snap back. If not, replace valve.
  - Evaporative emission control system. CHART C-3.
  - Power steering pressure switch circuit (3.8L VIN 3 only). The state of the switch should only change when wheels are turned up against the stops. See CHART C-1B.

- Minimum idle speed. Incorrect minimum idle speed may be caused by foreign material accumulation in the throttle bore, on the throttle valve or on the throttle shaft.
  - ECM ground circuits.
  - EGR valve: There should be no EGR at idle.
- Monitoring block learn values may help identify the cause of the problem. If the system is running lean (block learn greater than 138) refer to "Diagnostic Aids" on facing page of Code 44. If the system is running rich (block learn values less than 118) refer to "Diagnostic Aids" on facing page of Code 45.
- Run a cylinder compression check.
- Check for fuel in pressure regulator hose. If present, replace regulator assembly.
- Check ignition system; wires and plugs.
- Check for loose or damaged MAF duct between sensor and throttle body.
- Disconnect MAF sensor and if condition is corrected, replace sensor. "Scan" tool should read about 4-8 grams per second at idle.
- If problem exists with A/C "ON", check A/C system operation CHART C-10.

EXCESSIVE EXHAUST EMISSIONS OR ODORS

Definition: Vehicle fails an emission test. Vehicle has excessive "rotten egg" smell. Excessive odors do not necessarily indicate excessive emissions.

- Perform "Diagnostic Circuit Check."
- IF TEST SHOWS EXCESSIVE CO AND HC, (or also has excessive odors):
  - Check items which cause car to run RICH.
    - Make sure engine is at normal operating temperature.
  - CHECK:
    - Fuel pressure. See CHART A-7.
    - Canister for fuel loading. See CHART C-3.
    - Injector balance. See CHART C-2A.
    - PCV valve for being plugged, stuck, or blocked PCV hose or fuel in the crankcase.
    - Spark plugs, plug wires, and ignition components.
    - Check for lead contamination of catalytic converter (look for removal of fuel filler neck restrictor).
    - Check for properly installed fuel cap.

- If the system is running rich, (block learn less than 118) refer to "Diagnostic Aids" on facing page of Code 45.
- IF TEST SHOWS EXCESSIVE NOx:
  - Check items which cause car to run LEAN, or to run too hot.
    - EGR valve for not opening. See CHART C-7.
    - Vacuum leaks.
    - Coolant system and coolant fan for proper operation. See CHART C-12.
    - Remove carbon with top engine cleaner. Follow instructions on can.
- If the system is running lean (block learn greater than 138), refer to "Diagnostic Aids" on facing page of Code 44.

1988 VEHICLES EQUIPPED WITH 3.8L (VIN 3) ENGINE

CHART B-1
RESTRICTED EXHAUST SYSTEM CHECK

Proper diagnosis for a Restricted Exhaust System is essential before any components are replaced. The following procedure may be used for diagnosis:

CHECK AT O₂ SENSOR:

1. Carefully remove O₂ sensor.
2. Install exhaust backpressure tester (BT 8515 or BT 8603) or equivalent in place of O₂ sensor (see illustration).
3. After completing test described below, be sure to coat threads of O₂ sensor with anti-seize compound P/N S613695 or equivalent prior to re-installation.

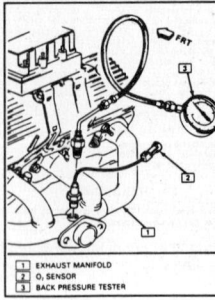

1	EXHAUST MANIFOLD
2	O₂ SENSOR
3	BACK PRESSURE TESTER

DIAGNOSIS:

1. With the engine at normal operating temperature and running at 2500 rpm, observe the exhaust system backpressure reading on the gauge.
2. If the backpressure exceeds 1 1/4 psi (8.62 kPa), a restricted exhaust system is indicated.
3. Inspect the entire exhaust system for a collapsed pipe, heat distress, or possible internal muffler failure.
4. If there are no obvious reasons for the excessive backpressure, a restricted catalytic converter should be suspected, and replaced using current recommended procedures.

1988 VEHICLES EQUIPPED WITH 3.8L (VIN 3) ENGINE

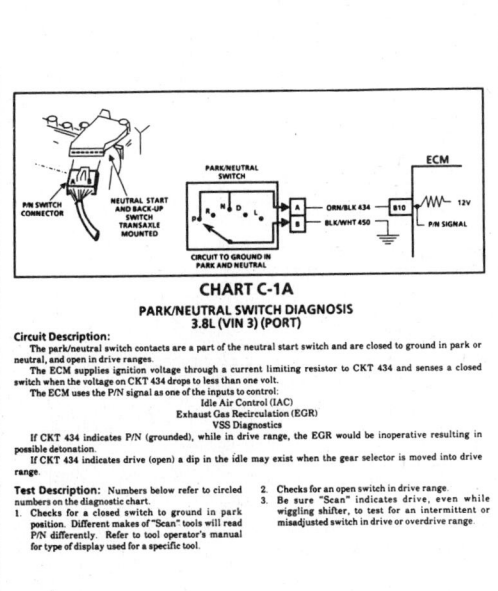

CHART C-1A
PARK/NEUTRAL SWITCH DIAGNOSIS
3.8L (VIN 3) (PORT)

Circuit Description:
The park/neutral switch contacts are a part of the neutral start switch and are closed to ground in park or neutral, and open in drive ranges.

The ECM supplies ignition voltage through a current limiting resistor to CKT 434 and senses a closed switch when the voltage on CKT 434 drops to less than one volt.

The ECM uses the P/N signal as one of the inputs to control:
Idle Air Control (IAC)
Exhaust Gas Recirculation (EGR)
VSS Diagnostics

If CKT 434 indicates P/N (grounded), while in drive range, the EGR would be inoperative resulting in possible detonation.

If CKT 434 indicates drive (open) a dip in the idle may exist when the gear selector is moved into drive range.

Test Description: Numbers below refer to circled numbers on the diagnostic chart.
1. Checks for a closed switch to ground in park position. Different makes of "Scan" tools will read P/N differently. Refer to tool operator's manual for type of display used for a specific tool.

2. Checks for an open switch in drive range.
3. Be sure "Scan" indicates drive, even while wiggling shifter, to test for an intermittent or misadjusted switch in drive or overdrive range.

1988 VEHICLES EQUIPPED WITH 3.8L (VIN 3) ENGINE

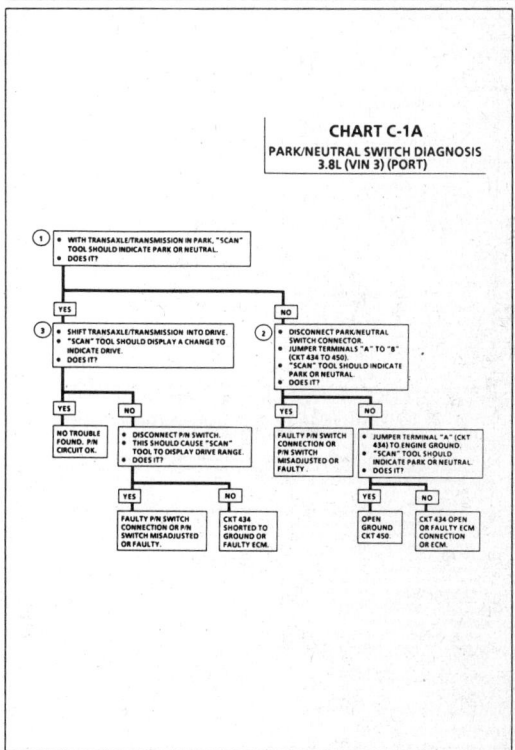

1988 A-CAR EQUIPPED WITH 3.8L (VIN 3) ENGINE

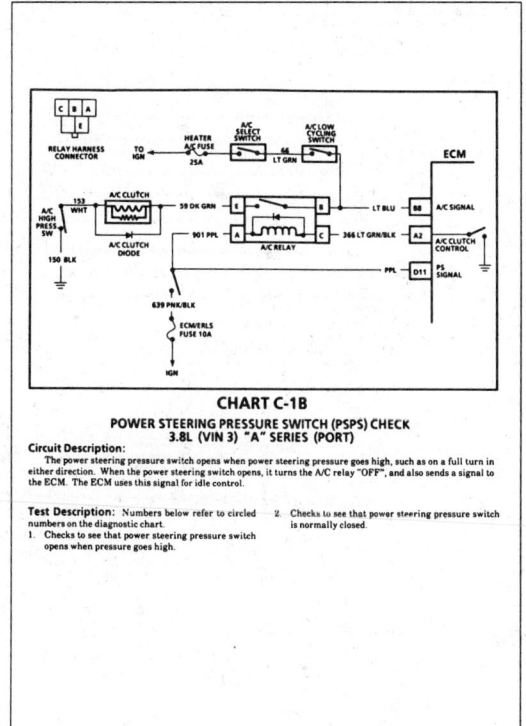

CHART C-1B
POWER STEERING PRESSURE SWITCH (PSPS) CHECK
3.8L (VIN 3) "A" SERIES (PORT)

Circuit Description:
The power steering pressure switch opens when power steering pressure goes high, such as on a full turn in either direction. When the power steering switch opens, it turns the A/C relay "OFF", and also sends a signal to the ECM. The ECM uses this signal for idle control.

Test Description: Numbers below refer to circled numbers on the diagnostic chart.
1. Checks to see that power steering pressure switch opens when pressure goes high.

2. Checks to see that power steering pressure switch is normally closed.

1988 A-CAR EQUIPPED WITH 3.8L (VIN 3) ENGINE

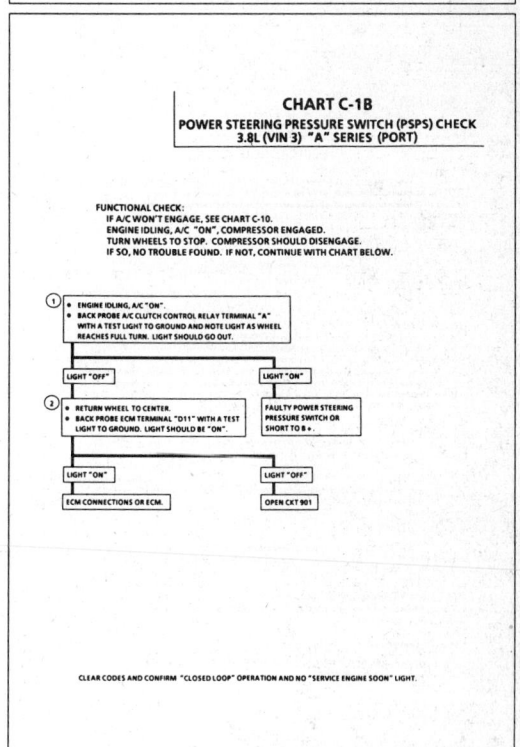

1988 C AND H-CAR EQUIPPED WITH 3.8L (VIN 3) ENGINE

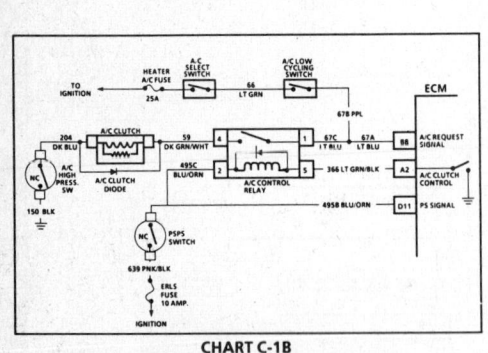

CHART C-1B
POWER STEERING PRESSURE SWITCH (PSPS) CHECK
3.8L (VIN 3) (PORT)

Circuit Description:
The power steering pressure switch opens when power steering pressure goes high, such as on a full turn in either direction. When the power steering switch opens, it turns the A/C relay "OFF", and also sends a signal to the ECM. The ECM uses this signal for idle control.

Test Description: Numbers below refer to circled numbers on the diagnostic chart:
1. Checks to see that power steering pressure switch opens when pressure goes high.
2. Checks to see that power steering pressure switch is normally closed.

1988 C AND H-CAR EQUIPPED WITH 3.8L (VIN 3) ENGINE

CHART C-1B
POWER STEERING PRESSURE SWITCH (PSPS) CHECK
3.8L (VIN 3) (PORT)

FUNCTIONAL CHECK:
IF A/C WON'T ENGAGE SEE CHART C-10.
ENGINE IDLING, A/C "ON" COMPRESSOR ENGAGED.
TURN WHEELS TO STOP. COMPRESSOR SHOULD DISENGAGE.
IF SO, NO TROUBLE FOUND. IF NOT, CONTINUE WITH CHART BELOW.

① • ENGINE IDLING, A/C "ON".
   • BACK PROBE A/C CONTROL RELAY TERMINAL "2" WITH A TEST LIGHT TO GROUND AND NOTE LIGHT AS WHEEL REACHES FULL TURN.

| LIGHT "OFF" | LIGHT "ON" |

② • RETURN WHEEL TO CENTER.
   • BACK PROBE ECM TERMINAL "D11" WITH A TEST LIGHT TO GROUND.

FAULTY POWER STEERING PRESSURE SWITCH OR SHORT TO B+.

| LIGHT "ON" | LIGHT "OFF" |

ECM CONNECTIONS OR ECM.

OPEN CKT 495.

CLEAR CODES AND CONFIRM "CLOSED LOOP" OPERATION AND NO "SERVICE ENGINE SOON" LIGHT.

1988 VEHICLES EQUIPPED WITH 3.8L (VIN 3) ENGINE

CHART C-2A

INJECTOR BALANCE TEST

The injector balance tester is a tool used to turn the injector on for a precise amount of time, thus spraying a measured amount of fuel into the manifold. This causes a drop in fuel rail pressure that we can record and compare between each injector. All injectors should have the same amount of pressure drop (± 10 kpa). Any injector with a pressure drop that is 10 kpa (or more) greater than the average drop of the other injectors should be considered faulty and replaced. Any injector with a pressure drop of less than 10 kpa from the average drop of the other injectors should be cleaned and retested.

STEP 1
Engine "cool down" period (10 minutes) is necessary to avoid irregular readings due to "Hot Soak" fuel boiling.
Disconnect harness connectors at all injectors, and connect injector tester J-34730-3, or equivalent, to one injector. With ignition "OFF" connect fuel gauge J347301 or equivalent to fuel pressure tap. Wrap a shop towel around fitting while connecting gage to avoid fuel spillage. At this point, insert clear tubing attached to vent valve into a suitable container and bleed air from gauge and hose to insure accurate gauge operation, by applying B+ to fuel pump test terminal. Repeat this step until all air is bled from gauge

STEP 2
apply B+ to fuel pump test terminal until fuel pressure reaches its maximum, about 3 seconds. Record this initial pressure reading. Energize tester one time and note pressure at its lowest point (Disregard any slight pressure increase after drop hits low point.). By subtracting this second pressure reading from the initial pressure, we have the actual amount of injector pressure drop. See note below.

STEP 3
Repeat step 2 on each injector and compare the amount of drop. Usually, good injectors will have virtually the same drop. Clean and retest any injector that has a pressure difference of 10kPa, either more or less than the average of the other injectors on the engine. Replace any injector that also fails the retest. If the pressure drop of all injectors is within 10kPa of this average, the injectors appear to be flowing properly. Reconnect them and review Symptoms, Section B.

NOTE: The entire test should not be repeated more than once without running the engine to prevent flooding. (This includes any retest on faulty injectors).

1988 VEHICLES EQUIPPED WITH 3.8L (VIN 3) ENGINE

NOTE: If injectors are suspected of being dirty, they should be cleaned using an approved tool and procedure prior to performing this test. The fuel pressure test in Chart A-7, should be completed prior to this test.

CHART C-2A
INJECTOR BALANCE TEST
3.8L (VIN 3) (PORT)

Step 1. If engine is at operation temperature start at number 1, otherwise you can start at number 2.
1. Engine "cool down" period (10 minutes) is necessary to avoid Irregular Readings due to "Hot Soak" Fuel Boiling.
2. Disconnect harness connectors at all injectors and connect injector tester J-34730-3, or equivalent, to one injector.
3. With ignition "OFF" connect fuel gauge J-347301 or equivalent to Fuel Pressure Tap. NOTE: Wrap a shop towel around fitting while connecting gauge to avoid fuel spray or spillage.
4. Apply B+ to fuel pump test terminal and bleed off to fuel pressure test gauge.

Step 2. Run test:
1. Apply B+ to Fuel Pump Test Terminal. Allow about 3 seconds for fuel pressure to reach its maximum.
2. Record gauge pressure. (Pressure must hold steady, if not see the Fuel System diagnosis, Chart A-7).
3. Turn injector on, by depressing button on injector tester, and note pressure at the instant the gauge needle stops.

Step 3.
1. Repeat step 2 on all injectors and record pressure drop on each. Clean and retest injectors that appear faulty (Any injectors that have a 10 kPa difference, either more or less, in pressure from the average). If no problem is found, review Symptoms Section B.

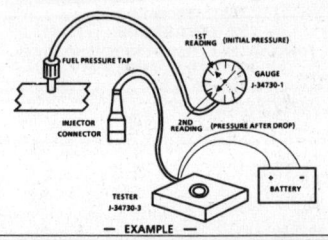

— EXAMPLE —

CYLINDER	1	2	3	4	5	6
1ST READING	225	225	225	225	225	225
2ND READING	100	100	100	90	100	115
AMOUNT OF DROP	125	125	125	135	125	110
	OK	OK	OK	FAULTY, RICH (TOO MUCH) (FUEL DROP)	OK	FAULTY, LEAN (TOO LITTLE) (FUEL DROP)

1988 VEHICLES EQUIPPED WITH 3.8L (VIN 3) ENGINE

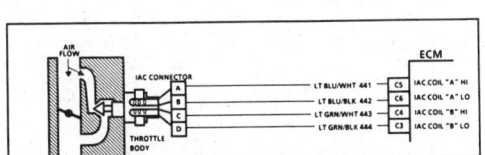

LT BLU/WHT 441	IAC COIL "A" HI	C5
LT BLU/BLK 442	IAC COIL "A" LO	C6
LT GRN/WHT 443	IAC COIL "B" HI	C4
LT GRN/BLK 444	IAC COIL "B" LO	C3

CHART C-2B
IDLE AIR CONTROL (IAC) VALVE CHECK
3.8L (VIN 3) (PORT)

Circuit Description:

The ECM controls idle rpm with the IAC valve. To increase idle rpm, the ECM retracts the IAC pintle valve out of its seat, allowing more air to pass by the throttle plate. To decrease rpm, it extends the IAC pintle valve in to its seat, reducing air flow by the throttle plate. A "Scan" tool will read the ECM commands to the IAC valve in counts. The higher the counts, the more air allowed (higher idle). The lower the counts, the less air allowed (lower idle).

Test Description: Numbers below refer to circled numbers on the diagnostic chart.

1. Continue with test, even if engine will not idle. If idle is too low, "Scan" will display 80 or more counts or steps. If idle is high, it will display "0" counts. Occasionally an erratic or unstable idle may occur. Engine speed may vary 200 rpm or more up and down. Disconnect IAC. If the condition is unchanged, the IAC is not at fault.
2. When the engine was stopped, the IAC valve retracted (more air) to a fixed "Park" position for increased air flow and idle speed during the next engine start. A "Scan" will display 100 or more counts.
3. Be sure to disconnect the IAC valve prior to this test. The test light will confirm the ECM signals by a steady or flashing light on all circuits.
4. There is a remote possibility that one of the circuits is shorted to voltage which would have been indicated by a steady light. Disconnect ECM and turn the ignition "ON" and probe terminals to check for this condition.

Diagnostic Aids:

A slow unstable idle may be caused by a system problem that cannot be overcome by the IAC. "Scan" counts will be above 60 counts if too low and "0" counts if too high.

If idle is too high, stop engine. Ignition"ON". Ground diagnostic terminal. Wait 30 seconds for IAC to seat, then disconnect IAC. Unground diagnostic terminal and start engine. If idle speed is above 800 ± 50 rpm, locate and correct vacuum leak.

- **System too lean (High Air/Fuel Ratio)**
  Idle speed may be too high or too low. Engine speed may vary up and down, disconnecting IAC does not help. May set Code 44. "Scan" and/or voltmeter will read an oxygen sensor output less than 300 mV (.3V). Check for low regulated fuel pressure or water in fuel. A lean exhaust, with an oxygen sensor output fixed above 800 mV (.8V) will be a contaminated sensor, usually silicone. This may also set a Code 45.
- **System too rich (Low Air/Fuel Ratio)**
  Idle speed too low. "Scan" counts usually above 80. System obviously rich and may exhibit black smoke exhaust. "Scan" tool and/or voltmeter will read an oxygen sensor signal fixed above 800 mV (.8V). Check:
  - High fuel pressure
  - Injector leaking or sticking
- **Throttle Body**
  Remove IAC and inspect bore for foreign material or evidence of IAC valve dragging the bore.
- Refer to "Rough, Unstable, Incorrect Idle or Stalling" in symptoms in Section "B".

1988 VEHICLES EQUIPPED WITH 3.8L (VIN 3) ENGINE

CHART C-2B
IDLE AIR CONTROL (IAC) VALVE CHECK
3.8L (VIN 3) (PORT)

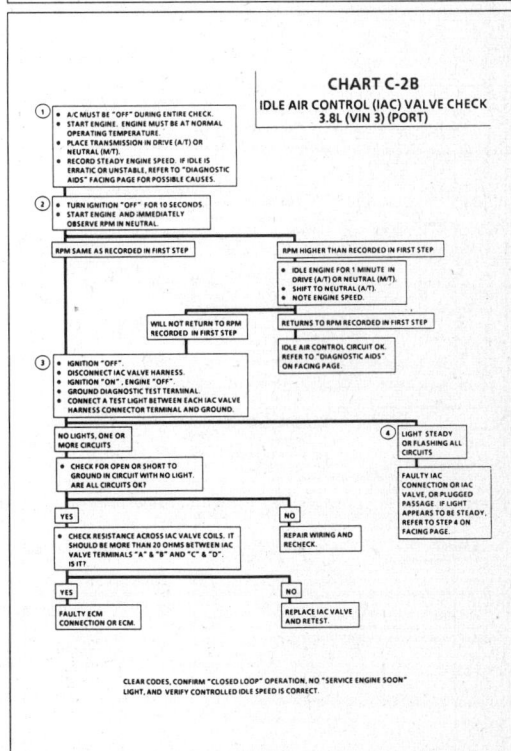

CLEAR CODES, CONFIRM "CLOSED LOOP" OPERATION, NO "SERVICE ENGINE SOON" LIGHT, AND VERIFY CONTROLLED IDLE SPEED IS CORRECT.

1988 VEHICLES EQUIPPED WITH 3.8L (VIN 3) ENGINE

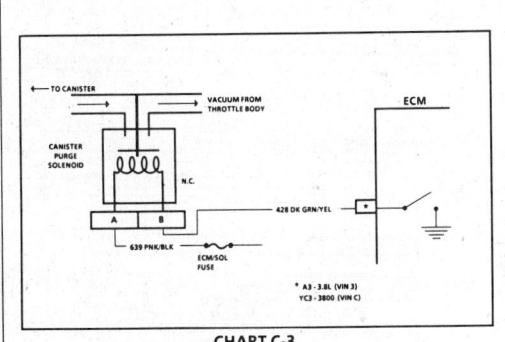

CHART C-3
CANISTER PURGE VALVE CHECK
3.8L (VIN 3) (PORT)

Circuit Description:

Canister purge is controlled by a solenoid that allows vacuum to purge the canister when energized. The ECM supplies a ground to energize the solenoid (purge "ON").

If the diagnostic test terminal is grounded with the engine stopped or the following is met with the engine running the purge solenoid is energized (purge "ON").
- Engine run time after start more than 1 minute.
- Coolant temperature above 80°C (176°F).
- Vehicle speed above 5 mph (8 km/h).
- Throttle "OFF" idle. TPS signal above .75V.

Test Description: Numbers below refer to circled numbers on the diagnostic chart.

1. Checks to see if the ECM or circuit is supplying the ground to energize the solenoid. The solenoid is normally de-energized in this step so the light should be "OFF."

2. Completes functional check by grounding test terminal. This should normally energize the solenoid and allow the vacuum to drop (purge "ON").

3. Checks for open or shorted solenoid circuit.

1988 VEHICLES EQUIPPED WITH 3.8L (VIN 3) ENGINE

CHART C-3
CANISTER PURGE VALVE CHECK
3.8L (VIN 3) (PORT)

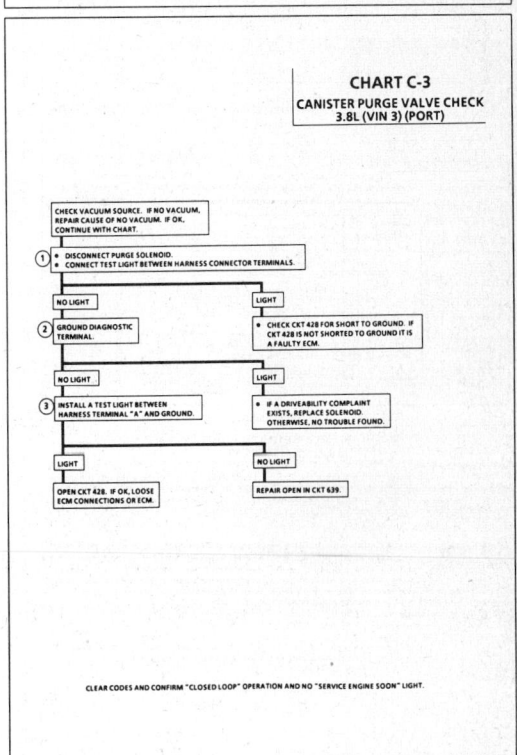

CLEAR CODES AND CONFIRM "CLOSED LOOP" OPERATION AND NO "SERVICE ENGINE SOON" LIGHT.

1988 VEHICLES EQUIPPED WITH 3.8L (VIN 3) ENGINE

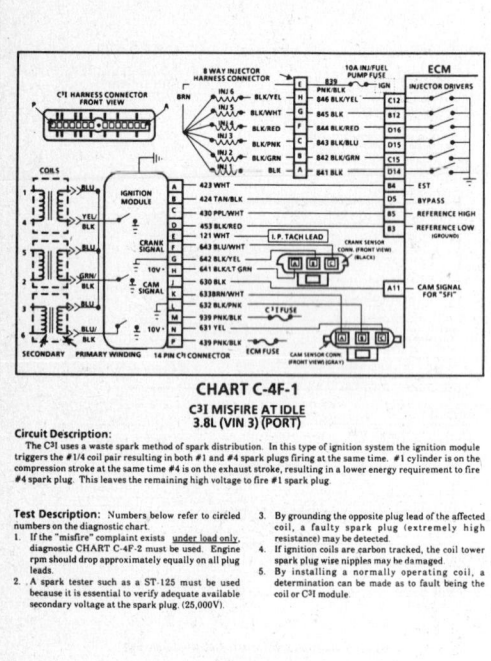

CHART C-4F-1
C³I MISFIRE AT IDLE
3.8L (VIN 3) (PORT)

Circuit Description:
The C³I uses a waste spark method of spark distribution. In this type of ignition system the ignition module triggers the #1/4 coil pair resulting in both #1 and #4 spark plugs firing at the same time. #1 cylinder is on the compression stroke at the same time #4 is on the exhaust stroke, resulting in a lower energy requirement to fire #4 spark plug. This leaves the remaining high voltage to fire #1 spark plug.

Test Description: Numbers below refer to circled numbers on the diagnostic chart.
1. If the "misfire" complaint exists <u>under load only</u>, diagnostic CHART C-4F-2 must be used. Engine rpm should drop approximately equally on all plug leads.
2. A spark tester such as a ST-125 must be used because it is essential to verify adequate available secondary voltage at the spark plug. (25,000V).
3. By grounding the opposite plug lead of the affected coil, a faulty spark plug (extremely high resistance) may be detected.
4. If ignition coils are carbon tracked, the coil tower spark plug wire nipples may be damaged.
5. By installing a normally operating coil, a determination can be made as to fault being the coil or C³I module.

1988 VEHICLES EQUIPPED WITH 3.8L (VIN 3) ENGINE

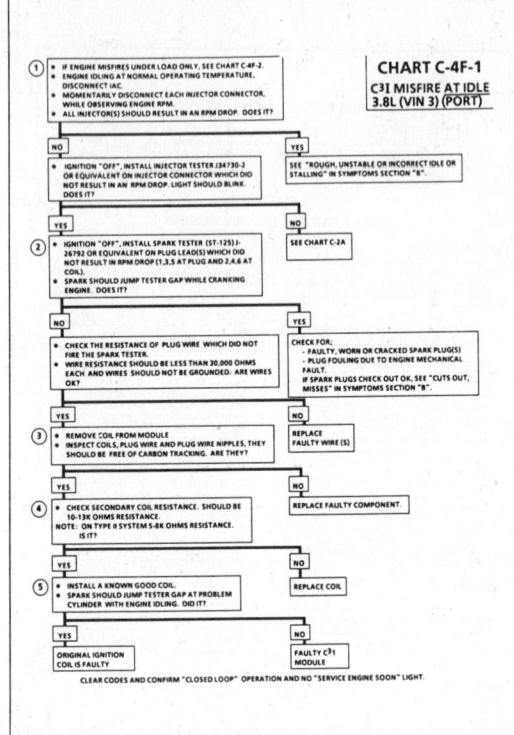

CHART C-4F-1
C³I MISFIRE AT IDLE
3.8L (VIN 3) (PORT)

1988 VEHICLES EQUIPPED WITH 3.8L (VIN 3) ENGINE

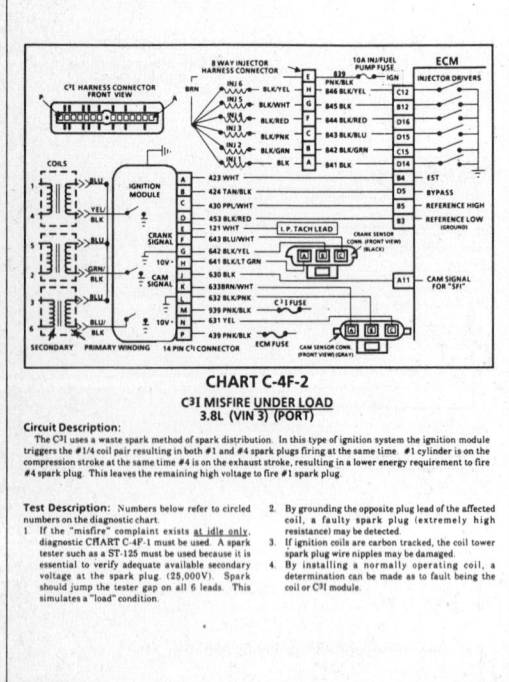

CHART C-4F-2
C³I MISFIRE UNDER LOAD
3.8L (VIN 3) (PORT)

Circuit Description:
The C³I uses a waste spark method of spark distribution. In this type of ignition system the ignition module triggers the #1/4 coil pair resulting in both #1 and #4 spark plugs firing at the same time. #1 cylinder is on the compression stroke at the same time #4 is on the exhaust stroke, resulting in a lower energy requirement to fire #4 spark plug. This leaves the remaining high voltage to fire #1 spark plug.

Test Description: Numbers below refer to circled numbers on the diagnostic chart.
1. If the "misfire" complaint exists <u>at idle only</u>, diagnostic CHART C-4F-1 must be used. A spark tester such as a ST-125 must be used because it is essential to verify adequate available secondary voltage at the spark plug. (25,000V). Spark should jump the tester gap on all 6 leads. This simulates a "load" condition.
2. By grounding the opposite plug lead of the affected coil, a faulty spark plug (extremely high resistance) may be detected.
3. If ignition coils are carbon tracked, the coil tower spark plug wire nipples may be damaged.
4. By installing a normally operating coil, a determination can be made as to fault being the coil or C³I module.

1988 VEHICLES EQUIPPED WITH 3.8L (VIN 3) ENGINE

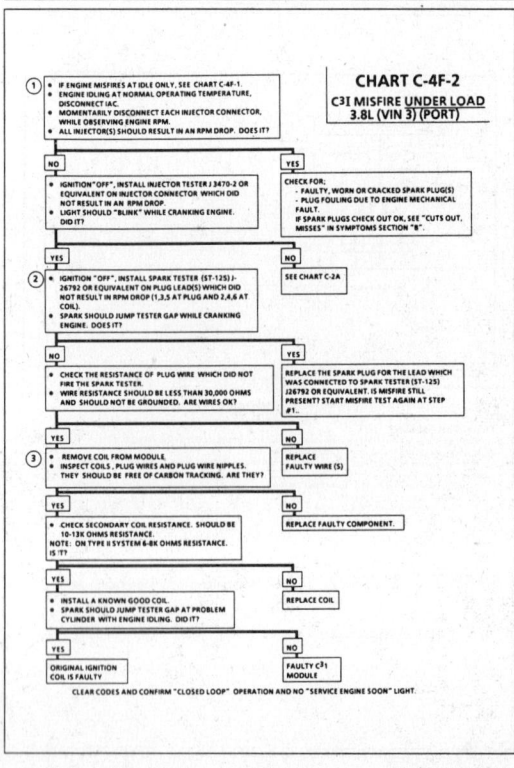

CHART C-4F-2
C³I MISFIRE UNDER LOAD
3.8L (VIN 3) (PORT)

1988 VEHICLES EQUIPPED WITH 3.8L (VIN 3) ENGINE

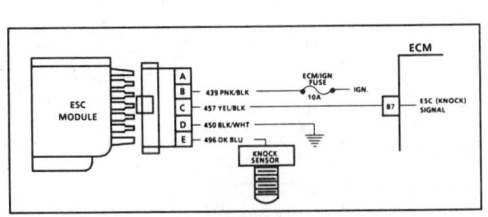

CHART C-5
ELECTRONIC SPARK CONTROL (ESC) CIRCUIT
3.8L (VIN 3) (PORT)

Circuit Description:
The ESC system is comprised of a knock sensor and an ESC module.

As long as the ESC module is sending a voltage signal (8 to 10V) to the ECM (no detonation detected by the ESC sensor) the ECM provides normal spark advance.

When the sensor detects detonation, the module turns "OFF" the circuit to the ECM and the voltage at ECM terminal "B7" drops to 0V. The ECM then retards EST as much as 20° to reduce detonation. This happens fast and frequently enough that if looking at this signal with a DVM, you won't see 0V, but an average voltage somewhat less than what is normal with no detonation.

A loss of the knock sensor signal or a loss of ground at ESC module would cause the signal at the ECM to remain high. This condition would result in the ECM controlling EST as if no detonation were occurring. The EST would not be retarded, and detonation could become severe enough under heavy engine load conditions to result in pre-ignition and potential engine damage.

Loss of the ESC signal to the ECM would cause the ECM to constantly retard EST. This could result in sluggish performance and cause a Code 43 to set.

Test Description: Numbers below refer to circled numbers on the diagnostic chart.
1. Tests ESC system's ability to detect detonation and retard the ignition timing.
2. By disconnecting the ESC module, the ECM monitors a low voltage at terminal "B7" and should retard the ignition timing.
3. After approximately 4 seconds, the "Service Engine Soon" light will come "ON" and Code 43 will be stored.

4. Checks for proper voltage output (measured on A/C scale) of knock sensor. Low or no voltage would indicate an open circuit to terminal "E" or faulty sensor.
5. Checks to see if constant retard is due to a faulty knock sensor or module or if a false voltage signal is being transmitted on the wire from the knock sensor by induction from an adjacent wire, such as a spark plug wire, ignition wire, etc. Reroute wires as necessary.

1988 VEHICLES EQUIPPED WITH 3.8L (VIN 3) ENGINE

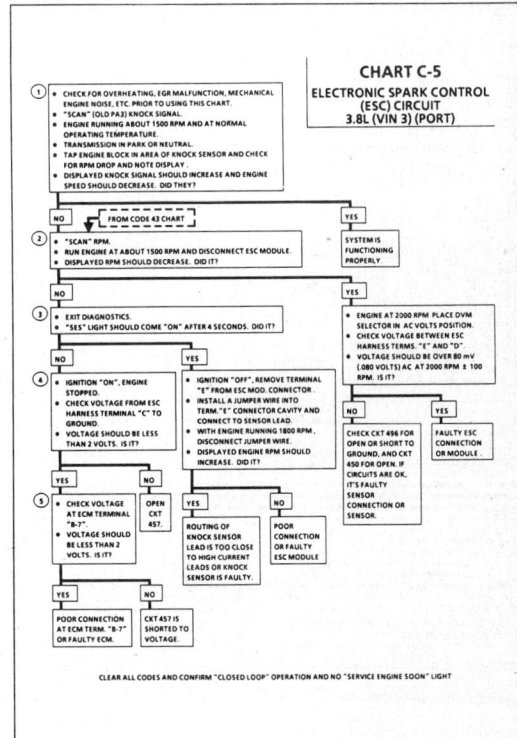

1988 VEHICLES EQUIPPED WITH 3.8L (VIN 3) ENGINE

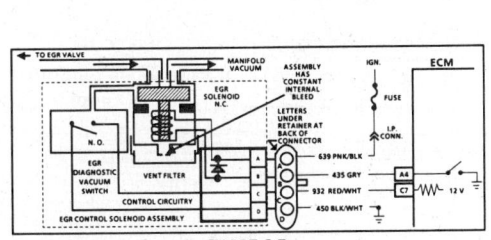

CHART C-7
EXHAUST GAS RECIRCULATION (EGR) CHECK
3.8L (VIN 3) (PORT)

Circuit Description:
The EGR valve is opened by manifold vacuum to let exhaust gas flow into the intake manifold. The exhaust gas then moves with the air/fuel mixture into the combustion chamber. If too much exhaust gas enters, combustion will not occur. For this reason, very little exhaust gas is allowed to pass through the valve, especially at idle. The EGR valve is usually open under the following conditions:
- Warm engine operation
- Above idle speed

The amount of exhaust gas recirculated is controlled by variations in vacuum and the EGR vacuum control solenoid.

Test Description: Numbers below refer to circled numbers on the diagnostic chart.
1. Checks for a sticking EGR valve. If sticking, remove and examine valve to determine whether it can be cleaned, or must replaced. A sticking EGR valve will most likely cause a rough idle condition.
2. Checks for plugged EGR passages. If passages are plugged, the engine may have severe detonation on acceleration.

1988 VEHICLES EQUIPPED WITH 3.8L (VIN 3) ENGINE

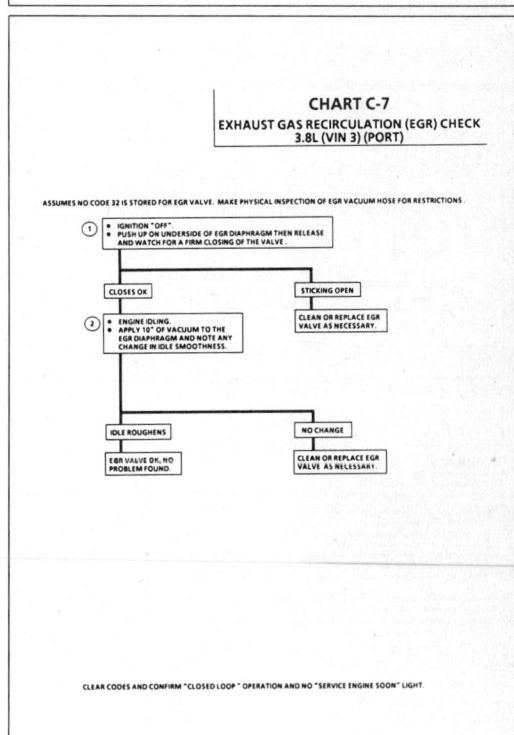

1988 VEHICLES EQUIPPED WITH 3.8L (VIN 3) ENGINE

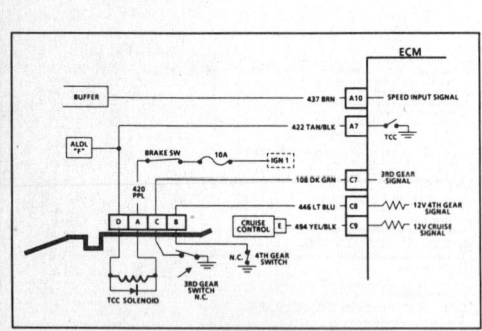

CHART C-8A
(Page 1 of 2)
TORQUE CONVERTER CLUTCH (TCC)
(ELECTRICAL DIAGNOSIS)
3.8L (VIN 3) (PORT)

Circuit Description:

The purpose of the torque converter clutch feature is to eliminate the power loss of the torque converter when the vehicle is in a cruise condition. This allows the convenience of the automatic and the fuel economy of a manual transaxle. The heart of the system is a solenoid located inside the transaxle which is controlled by the ECM.

When the solenoid coil is activated ("ON"), the torque converter clutch is applied which results in straight through mechanical coupling from the engine to the wheels. When the transmission solenoid is deactivated, the torque converter clutch is released which allows the torque converter to operate in the conventional manner (fluidic coupling between engine and transmission).

TCC will engage when:
* Engine warmed up
* Vehicle speed above a calibrated value (about 28 mph 45 km/h)
* Throttle position sensor output not changing, indicating a steady road speed.
* Brake switch closed.

NOTE: A vehicle with a 3.8L engine, 440-T4 transaxle and factory installed cruise control. Engagement of the cruise control system will result in loss of TCC in 3rd gear. However, 4th gear TCC operation will be maintained.

Test Description: Numbers below refer to circled numbers on the diagnostic chart.
1. This test checks the continuity of the TCC circuit from the fuse to the ALDL connector.
2. When the brake pedal is released, the light should come back "ON" and then go "OFF", when the diagnostic terminal is grounded. This tests CKT 422 and the TCC driver in the ECM.

Diagnostic Aids:

The "Scan" tool only indicates when the ECM has turned "ON" the TCC driver and this does not confirm that the TCC has engaged. To determine if TCC is functioning properly, road test the vehicle. Engine rpm should decrease when the "Scan" indicates the TCC driver has turned "ON".

1988 VEHICLES EQUIPPED WITH 3.8L (VIN 3) ENGINE

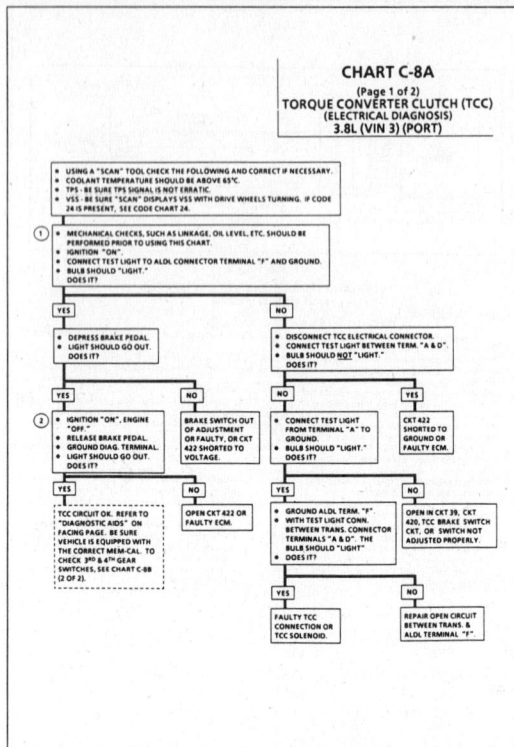

CHART C-8A
(Page 1 of 2)
TORQUE CONVERTER CLUTCH (TCC)
(ELECTRICAL DIAGNOSIS)
3.8L (VIN 3) (PORT)

1988 VEHICLES EQUIPPED WITH 3.8L (VIN 3) ENGINE

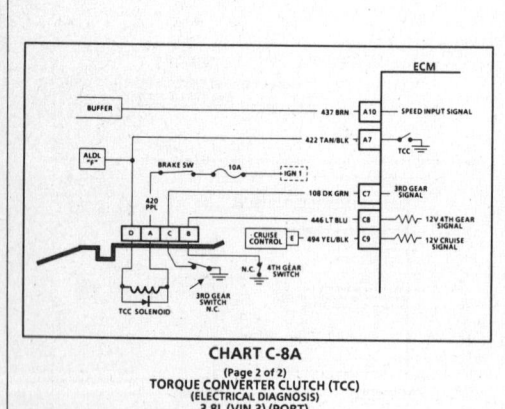

CHART C-8A
(Page 2 of 2)
TORQUE CONVERTER CLUTCH (TCC)
(ELECTRICAL DIAGNOSIS)
3.8L (VIN 3) (PORT)

Circuit Description:

The 3rd gear switch in this vehicle is open in 3rd and 4th gear. The fourth gear switch is open in fourth gear.

Test Description: Numbers below refer to circled numbers on the diagnostic chart.
1. Some "Scan" tools display the state of these switches in different ways. Be familiar with the type of tool being used. Since both switches should be in the closed state during this test, the tool should read the same for either the 3rd or 4th gear switch.
2. Determines whether the switch or signal circuit is open. The circuit can be checked for an open by measuring the voltage (with a voltmeter) at the TCC connector. Should be about 12V.

3. Because the switch(s) should be grounded in this step, disconnecting the TCC connector should cause the "Scan" switch state to change.
4. The switch state should change when the vehicle shifts into 3rd gear.

Diagnostic Aids:

If vehicle is road tested because of a TCC related problem, be sure the switch states do not change while in 4th gear because the TCC will disengage. If switches change state, carefully check wire routing and connections.

1988 VEHICLES EQUIPPED WITH 3.8L (VIN 3) ENGINE

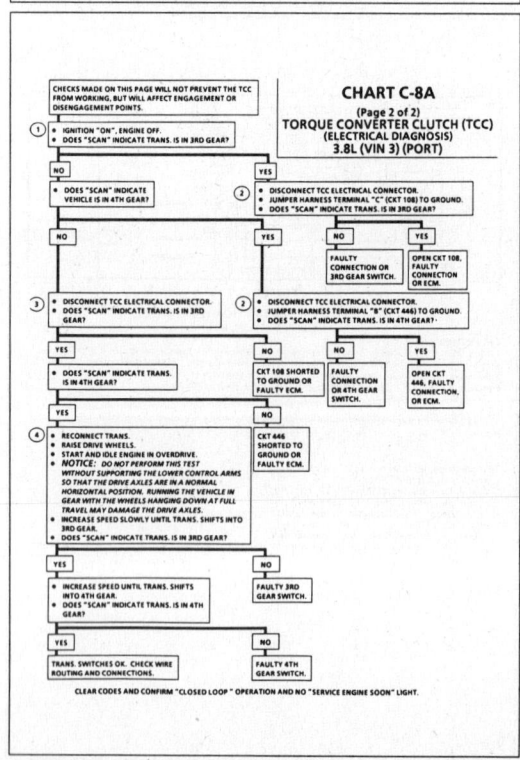

CHART C-8A
(Page 2 of 2)
TORQUE CONVERTER CLUTCH (TCC)
(ELECTRICAL DIAGNOSIS)
3.8L (VIN 3) (PORT)

1988 A-CAR EQUIPPED WITH 3.8L (VIN 3) ENGINE

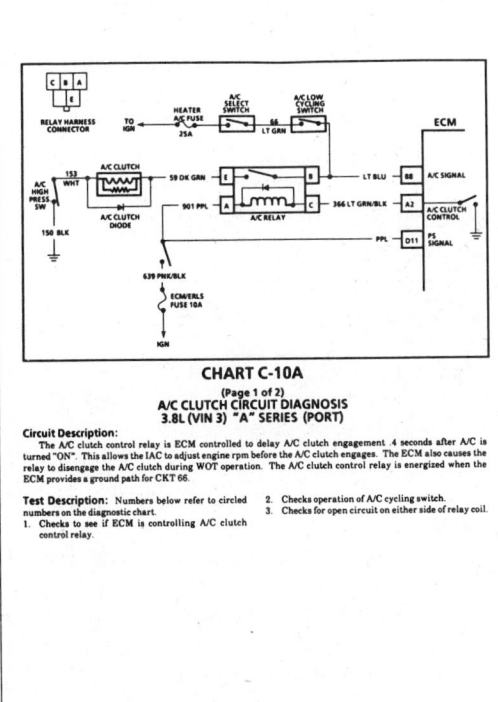

CHART C-10A
(Page 1 of 2)
A/C CLUTCH CIRCUIT DIAGNOSIS
3.8L (VIN 3) "A" SERIES (PORT)

Circuit Description:
The A/C clutch control relay is ECM controlled to delay A/C clutch engagement .4 seconds after A/C is turned "ON". This allows the IAC to adjust engine rpm before the A/C clutch engages. The ECM also causes the relay to disengage the A/C clutch during WOT operation. The A/C clutch control relay is energized when the ECM provides a ground path for CKT 66.

Test Description: Numbers below refer to circled numbers on the diagnostic chart.
1. Checks to see if ECM is controlling A/C clutch control relay.
2. Checks operation of A/C cycling switch.
3. Checks for open circuit on either side of relay coil.

1988 A-CAR EQUIPPED WITH 3.8L (VIN 3) ENGINE

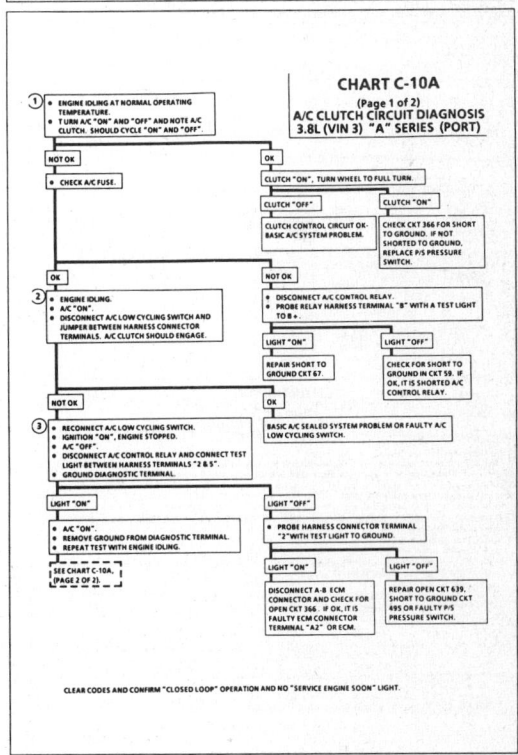

1988 C AND H-CAR EQUIPPED WITH 3.8L (VIN 3) ENGINE

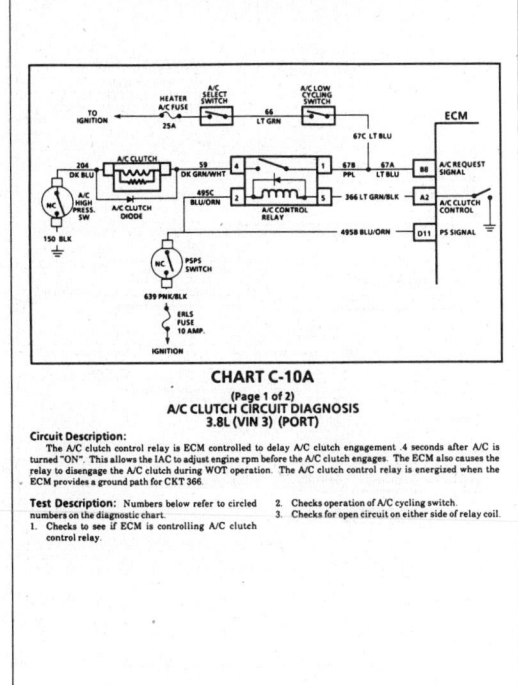

CHART C-10A
(Page 1 of 2)
A/C CLUTCH CIRCUIT DIAGNOSIS
3.8L (VIN 3) (PORT)

Circuit Description:
The A/C clutch control relay is ECM controlled to delay A/C clutch engagement .4 seconds after A/C is turned "ON". This allows the IAC to adjust engine rpm before the A/C clutch engages. The ECM also causes the relay to disengage the A/C clutch during WOT operation. The A/C clutch control relay is energized when the ECM provides a ground path for CKT 366.

Test Description: Numbers below refer to circled numbers on the diagnostic chart.
1. Checks to see if ECM is controlling A/C clutch control relay.
2. Checks operation of A/C cycling switch.
3. Checks for open circuit on either side of relay coil.

1988 C AND H-CAR EQUIPPED WITH 3.8L (VIN 3) ENGINE

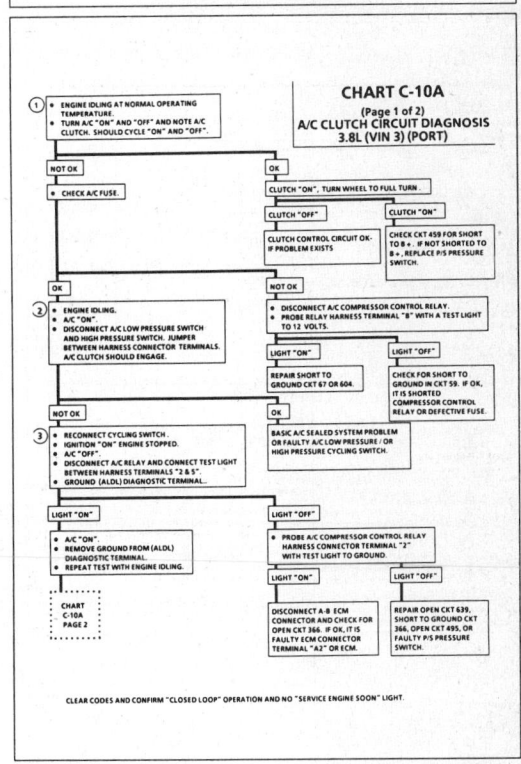

1988 C AND H-CAR EQUIPPED WITH 3.8L (VIN 3) ENGINE

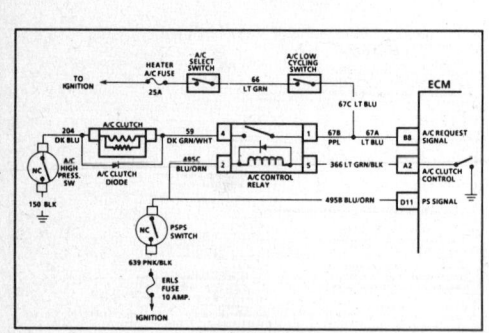

CHART C-10A
(Page 2 of 2)
A/C CLUTCH CIRCUIT DIAGNOSIS
3.8L (VIN 3) (PORT)

Circuit Description:
The A/C clutch control relay is ECM controlled to delay A/C clutch engagement .4 seconds after A/C is turned "ON". This allows the IAC to adjust engine rpm before the A/C clutch engages. The ECM also causes the relay to disengage the A/C clutch during WOT operation. The A/C clutch control relay is energized when the ECM provides a ground path for CKT 366.

Test Description: Numbers below refer to circled numbers on the diagnostic chart.
1. Checks for battery voltage to relay through CKT 67.
2. Substitutes for relay to determine if problem is in relay or in CKT 59, A/C clutch coil, high pressure, switch, or ground.

3. Checks for open in CKT 67 between cycling switch and A/C fuse, or open CKT 67 to relay.
4. Checks to see that A/C "ON" signal is getting to ECM through CKT 67. A test light "OFF" at this time indicates CKT 67 is open between the cycling switch and the ECM.

1988 C AND H-CAR EQUIPPED WITH 3.8L (VIN 3) ENGINE

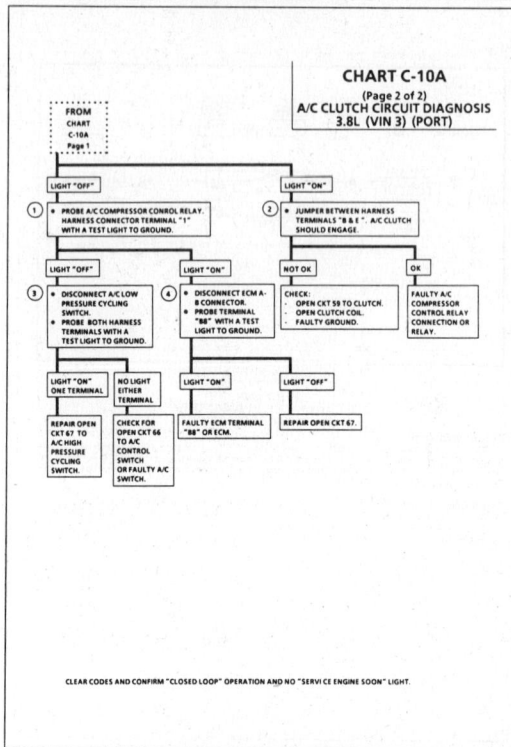

CLEAR CODES AND CONFIRM "CLOSED LOOP" OPERATION AND NO "SERVICE ENGINE SOON" LIGHT.

1988 A-CAR EQUIPPED WITH 3.8L (VIN 3) ENGINE

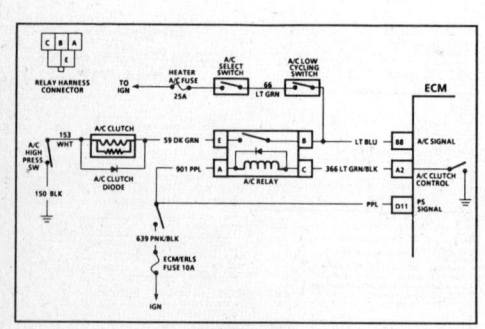

CHART C-10B
(Page 2 of 2)
A/C CLUTCH CIRCUIT DIAGNOSIS
3.8L (VIN 3) "A" SERIES (PORT)

Circuit Description:
The A/C clutch control relay is ECM controlled to delay A/C clutch engagement .4 seconds after A/C is turned "ON". This allows the IAC to adjust engine rpm before the A/C clutch engages. The ECM also causes the relay to disengage the A/C clutch during WOT operation. The A/C clutch control relay is energized when the ECM provides a ground path for CKT 66.

Test Description: Numbers below refer to circled numbers on the diagnostic chart.
1. Checks for battery voltage to relay through CKT 67.
2. Substitutes for relay to determine if problem is in relay or in CKT 59, A/C clutch coil, high pressure, switch, or ground.

3. Checks for open in CKT 67 between cycling switch and A/C fuse, or open CKT 67 to relay.
4. Checks to see that A/C "ON" signal is getting to ECM through CKT 67. A test light "OFF" at this time indicates CKT 67 is open between the cycling switch and the ECM.

1988 A-CAR EQUIPPED WITH 3.8L (VIN 3) ENGINE

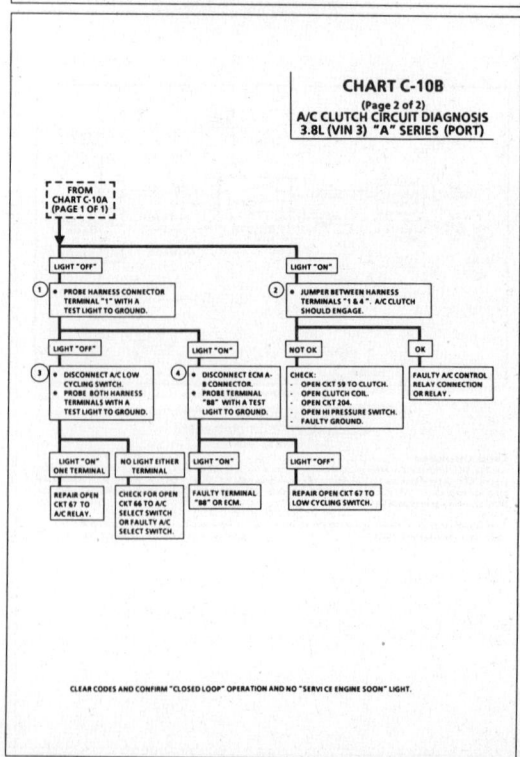

CHART C-10B
(Page 2 of 2)
A/C CLUTCH CIRCUIT DIAGNOSIS
3.8L (VIN 3) "A" SERIES (PORT)

CLEAR CODES AND CONFIRM "CLOSED LOOP" OPERATION AND NO "SERVICE ENGINE SOON" LIGHT.

1988 C AND H-CAR EQUIPPED WITH 3.8L (VIN 3) ENGINE

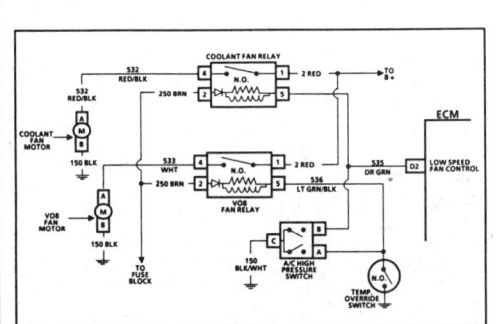

CHART C-12A
COOLANT FAN CHECK
3.8L (VIN 3) (PORT)

Circuit Description:
Power for the coolant fan motor comes through the fusible link to terminal "1" on all relays. The relays are energized when current flows to ground through the activation of the A/C High Pressure Switch, Temp. override switch and/or the ECM.

Coolant Fan Relay - The coolant fan relay is energized by the ECM or the A/C high pressure switch. The ECM energizes the relay through terminal "D2" when the coolant temperature reaches 98°C (208°F). The coolant fan relay is also energized through the A/C high pressure switch, when refrigerant pressure reaches 150 psi (1034 kPa).

V08 Coolant Fan - The optional V08 fan relay is energized by the A/C high pressure switch and/or coolant temp. override switch.

Test Description: Numbers below refer to circled numbers on the diagnostic chart.
1. Grounding the diagnostic test terminal should cause the ECM to ground CKT 535 and the coolant fan should run.
2. Grounding the temp. override switch harness terminal will check CKT 536 and will also check the V08 fan control circuit.
3. Separates and checks relay driver circuit and relay to fan circuit for open circuit or faulty relay.
4. This step checks to see if the temp. override switch is grounding and is grounded when the light comes "ON". The switch should close at 108°C (226°F).
5. The following steps will check the high pressure switches and related wiring from the switch to the fan control relay. If poor A/C performance is noted, the A/C pressure switches should be checked by a qualified A/C repair person. The V08 fan motor should come "ON" if A/C system pressure exceeds 260 psi (1782 kPa).

1988 C AND H-CAR EQUIPPED WITH 3.8L (VIN 3) ENGINE

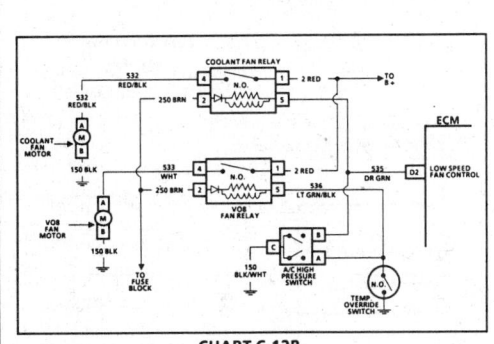

CHART C-12B
COOLANT FAN "ON" AT ALL TIMES
3.8L (VIN 3) (PORT)

Circuit Description:
Power for the coolant fan motor comes through the fusible link to terminal "1" on all relays. The relays are energized when current flows to ground through the activation of the A/C high pressure switch, temp. override switch and/or the ECM.

Coolant Fan Relay - The coolant fan relay is energized by the ECM or the A/C high pressure switch. The ECM energizes the relay through terminal "D2" when the coolant temperature reaches 98°C (208°F). The coolant fan relay is also energized through the A/C high pressure switch, when refrigerant pressure reaches 150 psi (1034 kPa).

V08 Coolant Fan - The optional V08 fan relay is energized by the A/C high pressure switch and/or coolant temp. override switch.

Test Description: Numbers below refer to circled numbers on the diagnostic chart.
1. Checks to see if CKT 535 is shorted to ground which would keep the coolant fan relay grounded at all times.
2. Checks to see if CKT 536 is shorted to ground. A light indicates the wire is shorted to ground.
3. If the test light is "OFF" after disconnecting, the ECM is shorted internally.

1988 C AND H-CAR EQUIPPED WITH 3.8L (VIN 3) ENGINE

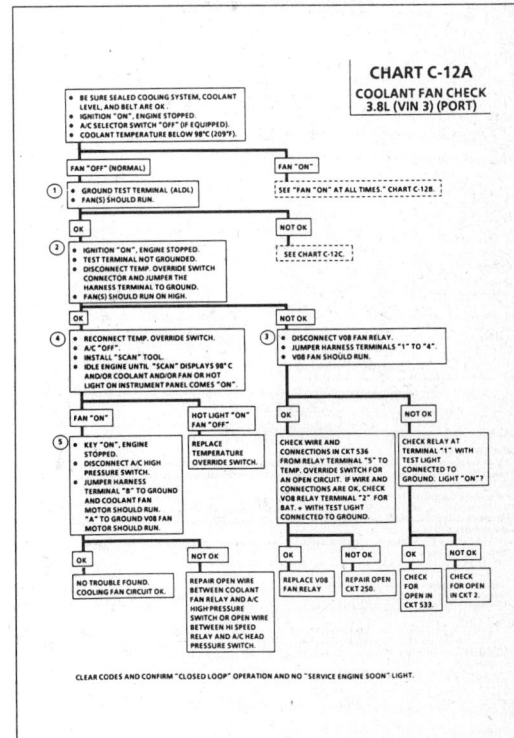

CHART C-12A
COOLANT FAN CHECK
3.8L (VIN 3) (PORT)

1988 C AND H-CAR EQUIPPED WITH 3.8L (VIN 3) ENGINE

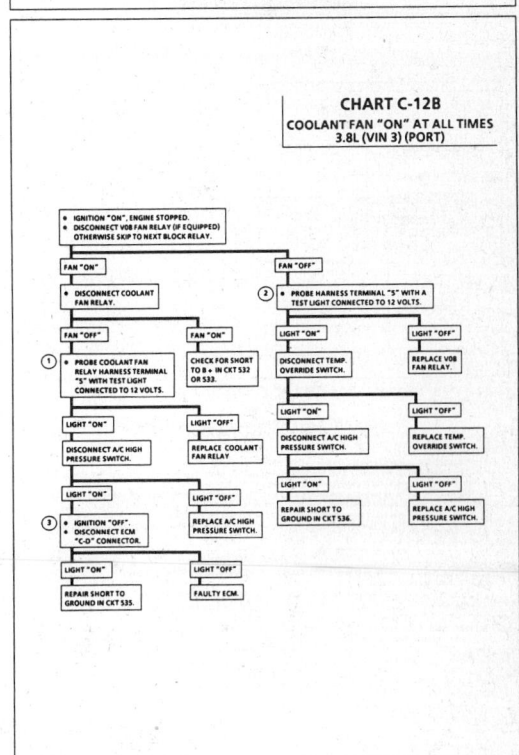

CHART C-12B
COOLANT FAN "ON" AT ALL TIMES
3.8L (VIN 3) (PORT)

1988 C AND H-CAR EQUIPPED WITH 3.8L (VIN 3) ENGINE

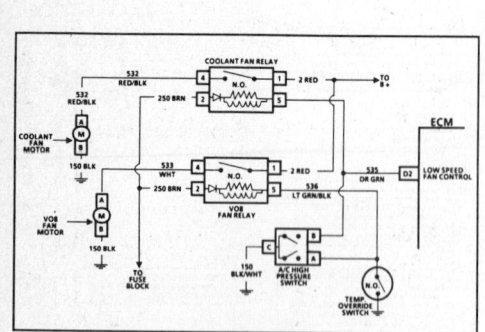

CHART C-12C
NO COOLANT FAN
3.8L (VIN 3) (PORT)

Circuit Description:
Power for the coolant fan motor comes through the fusible link to terminal "1" on all relays. The relays are energized when current flows to ground through the activation of the A/C high pressure switch, temp. override switch and/or the ECM.

Coolant Fan Relay - The coolant fan relay is energized by the ECM or the A/C high pressure switch. The ECM energizes the relay through terminal "D2" when the coolant temperature reaches 98°C (208°F). The coolant fan relay is also energized through the A/C high pressure switch, when refrigerant pressure reaches 150 psi (1034 kPa).

VO8 Coolant Fan - The optional VO8 fan relay is energized by the A/C high pressure switch and/or coolant temp. override switch.

Test Description: Numbers below refer to circled numbers on the diagnostic chart.
1. Checks for B+ at coolant fan relay harness connector.
2. Jumpering terminals "1" to "4" bypasses the relay, which should cause the coolant fan motor to run if coolant fan motor and wiring to the motor are good.
3. Grounding the test terminal should cause the ECM to ground CKT 535. At this point, the test light should light if the ECM is good and CKT 535 isn't open.
4. This checks for B+ and ground to the fan motor. A test light "ON" at this point indicates a faulty coolant fan motor connection, or coolant fan motor.

1988 C AND H-CAR EQUIPPED WITH 3.8L (VIN 3) ENGINE

CHART C-12C
NO COOLANT FAN
3.8L (VIN 3) (PORT)

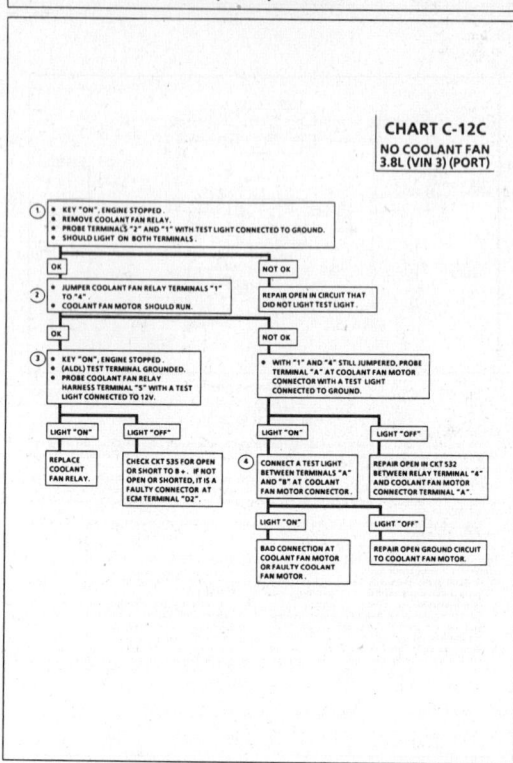

1988 A-CAR EQUIPPED WITH 3.8L (VIN 3) ENGINE

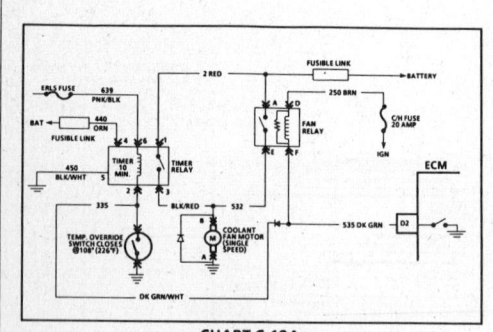

CHART C-12A
(Page 1 of 3)
COOLANT FAN CHECK (WITHOUT A/C)
3.8L (VIN 3) "A" SERIES (PORT)

Circuit Description:
The coolant fan motor is energized through a fan and/or timer relay. Power for the coolant fan motor comes through the fusible link to terminal "1" of the timer relay and "A" of the fan relay. The relays are energized when current flows to ground through the activation of the temp. override switch or ECM.

Fan Relay - The fan relay is energized by the ECM or temp. override switch. The ECM energizes the relay through terminal "D2" when the coolant temperature reaches 98°C (208°F) and vehicle speed is below 45 mph.

Timer Relay - The timer relay is energized by the temp. override switch. If the coolant temperature is 108°C (226°F) or higher when the ignition switch is shut "OFF", the timer relay is energized for 10 minutes or until the coolant temperature is lowered below 108°C (226°F).

Test Description: Numbers below refer to circled numbers on the diagnostic chart.
1. Checks to see if coolant fan runs at all times with ignition "ON".
2. Grounding the test terminal in the ALDL connector should cause the ECM to ground CKT 535. At this point, the coolant fan motor should run.
3. Grounding the temp. override switch harness to ground should cause the fan relay to turn "ON" the coolant fan motor. If the coolant fan motor runs, the coolant fan motor, fan relay and CKT 335 to the temp. override switch are good.
4. Checks to see if temp. override switch operates correctly. Test light should come "ON" at about 108°C (226°F) and the "hot light" on the instrument panel should come "ON" above 116°C (241°F).
5. Checks to see if timer relay works. With the temp. override switch harness grounded and ignition "OFF", coolant fan motor should run until ground jumper is removed.

1988 A-CAR EQUIPPED WITH 3.8L (VIN 3) ENGINE

CHART C-12A
(Page 1 of 3)
COOLANT FAN CHECK (WITHOUT A/C)
3.8L (VIN 3) "A" SERIES (PORT)

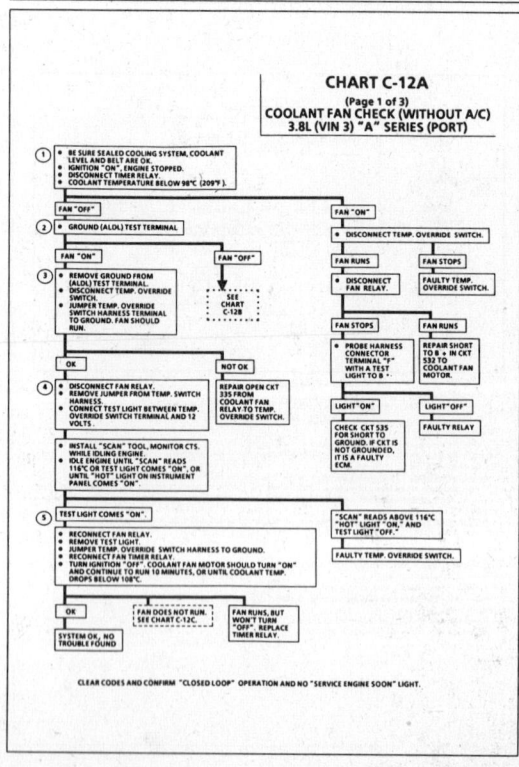

1988 A-CAR EQUIPPED WITH 3.8L (VIN 3) ENGINE

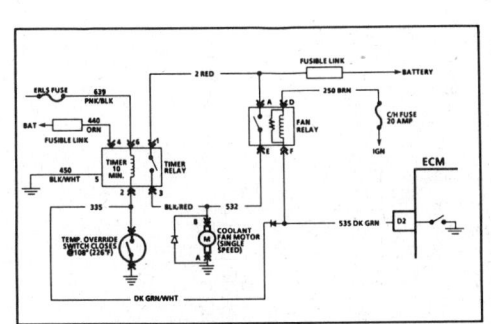

CHART C-12B
(Page 2 of 3)
COOLANT FAN CHECK (WITHOUT A/C)
3.8L (VIN 3) "A" SERIES (PORT)

Circuit Description:
The coolant fan motor is energized through a fan and/or timer relay. Power for the coolant fan motor comes through the fusible link to terminal "1" of the timer relay and "A" of the fan relay. The relays are energized when current flows to ground through the activation of the temp. override switch or ECM.

Fan Relay - The fan relay is energized by the ECM or temp. override switch. The ECM energizes the relay through terminal "D2" when the coolant temperature reaches 98°C (208°F) and vehicle speed is below 45 mph.

Timer Relay - The timer relay is energized by the temp. override switch. If the coolant temperature is 108°C (226°F) or higher when the ignition switch is shut "OFF", the timer relay is energized for 10 minutes or until the coolant temperature is lowered below 108°C (226°F).

Test Description: Numbers below refer to circled numbers on the diagnostic chart.
1. This check bypasses the fan relay and applies B+ directly to the coolant fan motor through CKT 532. At this point, the coolant fan motor should run.
2. Checks to see if problem is the fan relay, wiring or ECM.
3. Checks for presence of B+ to relay connector terminal "5".
4. Checks for presence of B+ to relay connector terminal "A".
5. Checks for an open ground or coolant fan motor circuit. Then checks for open in CKT 532 between fan relay and coolant fan motor or faulty coolant fan motor.

1988 A-CAR EQUIPPED WITH 3.8L (VIN 3) ENGINE

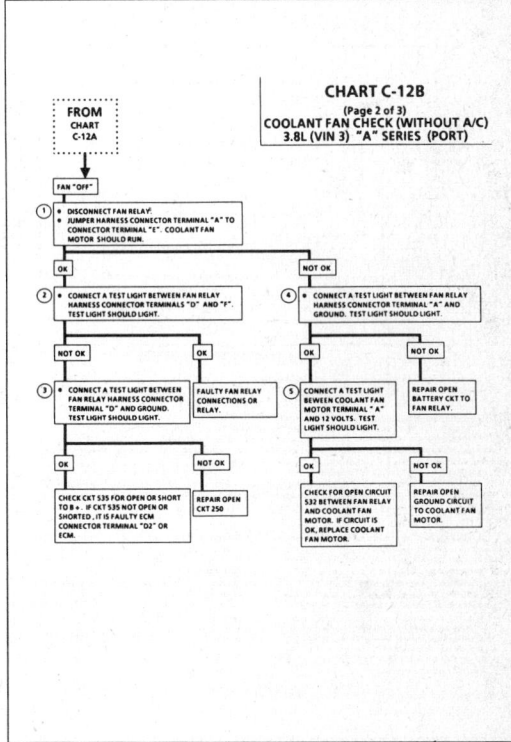

1988 A-CAR EQUIPPED WITH 3.8L (VIN 3) ENGINE

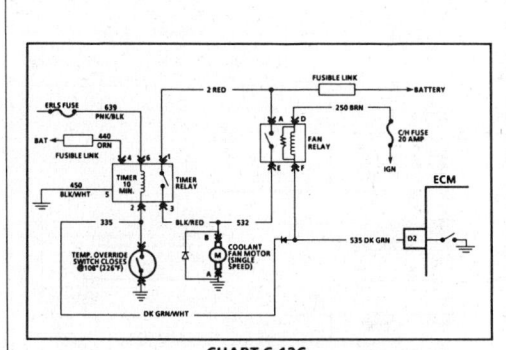

CHART C-12C
(Page 3 of 3)
COOLANT FAN CHECK (WITHOUT A/C)
3.8L (VIN 3) "A" SERIES (PORT)

Circuit Description:
The coolant fan motor is energized through a fan and/or timer relay. Power for the coolant fan motor comes through the fusible link to terminal "1" of the timer relay and "A" of the fan relay. The relays are energized when current flows to ground through the activation of the temp. override switch or ECM.

Fan Relay - The fan relay is energized by the ECM or temp. override switch. The ECM energizes the relay through terminal "D2" when the coolant temperature reaches 98°C (208°F) and vehicle speed is below 45 mph.

Timer Relay - The timer relay is energized by the temp. override switch. If the coolant temperature is 108°C (226°F) or higher when the ignition switch is shut "OFF", the timer relay is energized for 10 minutes or until the coolant temperature is lowered below 108°C (226°F).

Test Description: Numbers below refer to circled numbers on the diagnostic chart.
1. Checks B+ at terminals "1" and "4" of the timer relay.
2. Checks to be sure CKT 639 does not have voltage when the ignition switch is "OFF". If CKT 639 has voltage with the ignition "OFF", the timer relay will not turn the coolant fan motor "ON".
3. Checks to be sure CKT 450 is a good ground.

1988 A-CAR EQUIPPED WITH 3.8L (VIN 3) ENGINE

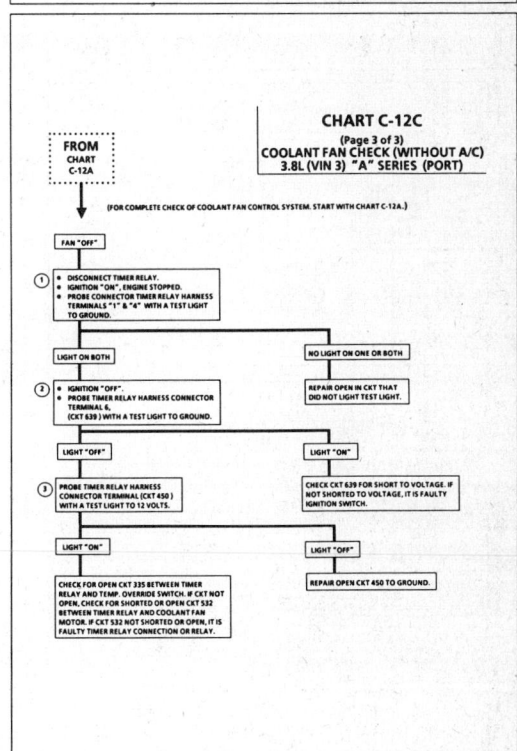

1988 A-CAR EQUIPPED WITH 3.8L (VIN 3) ENGINE

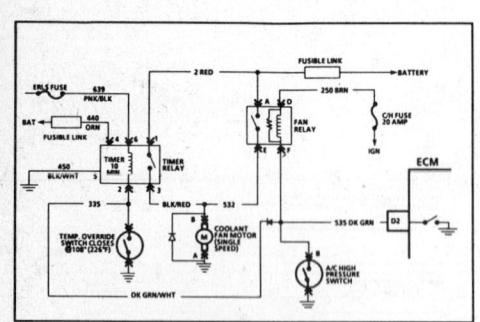

CHART C-12D
(Page 1 of 3)
COOLANT FAN CHECK (WITH A/C-LESS V08)
3.8L (VIN 3) "A" SERIES (PORT)

Circuit Description:
The coolant fan motor can be energized through two (2) relays; a fan relay, or timer relay.
Power for the coolant fan motor comes through the fusible link to both relays. The relays are energized when current flows to ground, through the activation of the A/C high pressure switch, temp. override switch, and/or the ECM.
Fan Relay - The fan relay is energized by the ECM, A/C high pressure switch and/or the temp. override switch. The ECM energizes the fan relay through terminal "D2", when coolant temperature reaches 98°C (208°F). The A/C high pressure switch energizes the fan relay when refrigerant pressure reaches 300 psi (2068 kPa).

Test Description: Numbers below refer to circled numbers on the diagnostic chart.
1. Checks to see if coolant fan motor runs at all times with ignition "ON".
2. Checks the test terminal in the ALDL connector, should cause the ECM to ground CKT 535. At this point, the coolant fan motor should run.
3. Grounding the temp. override switch harness to ground should cause the fan relay to turn "ON" the coolant fan motor. If the coolant fan motor runs, the fan relay and CKT 335 to the temp. override switch are good.
4. Grounding the A/C high pressure switch terminal should cause the fan relay to close, turning "ON" the coolant fan motor, proving the fan relay and circuit are OK.
5. Checks to see if the temp. override switch operates correctly. Test light should come "ON" at about 108°C (226°F), and the "hot light", on the instrument panel should come "ON" above 116°C (241°F).
6. Checks to see if the fan timer relay works. With temp. override switch harness grounded and ignition "OFF", fan should run until ground jumper is removed.

1988 A-CAR EQUIPPED WITH 3.8L (VIN 3) ENGINE

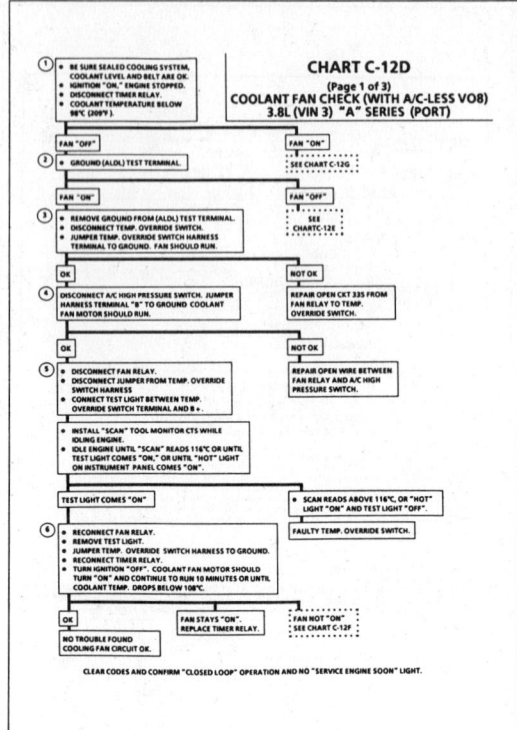

1988 A-CAR EQUIPPED WITH 3.8L (VIN 3) ENGINE

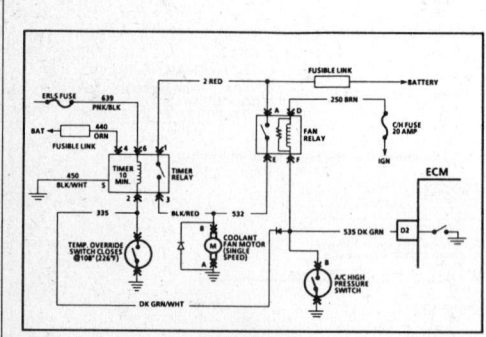

CHART C-12E
(Page 2 of 3)
COOLANT FAN CHECK (WITH A/C-LESS V08)
3.8L (VIN 3) "A" SERIES (PORT)

Circuit Description:
Power for the coolant fan motor comes through the fusible link to both relays. The relays are energized when current flows to ground, through the activation of the A/C high pressure switch, temp. override switch, and/or the ECM.
Fan Relay - The fan relay is energized by the ECM, A/C high pressure switch and/or the temp. override switch. The ECM energizes the fan relay through terminal "D2", when coolant temperature reaches 98°C (208°F). The A/C high pressure switch energizes the fan relay when refrigerant pressure reaches 300 psi (2068 kPa).

Test Description: Numbers below refer to circled numbers on the diagnostic chart.
1. This check bypasses the fan relay and applies B+ directly to the coolant fan motor through CKT 532. At this point, the coolant fan motor should run.
2. Checks to see if problem is the fan relay, wiring, or ECM.
3. Checks for presence of B+ to relay connector terminal "5".
4. Checks for presence of B+ to relay connector terminal "A".
5. Checks for an open ground or coolant fan motor circuit, then checks for an open in CKT 532, between fan relay and coolant fan motor or faulty coolant fan motor.

1988 A-CAR EQUIPPED WITH 3.8L (VIN 3) ENGINE

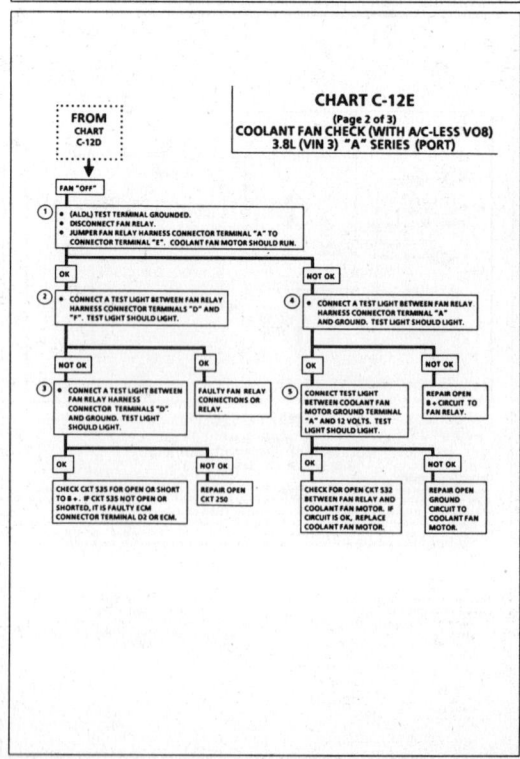

1988 A-CAR EQUIPPED WITH 3.8L (VIN 3) ENGINE

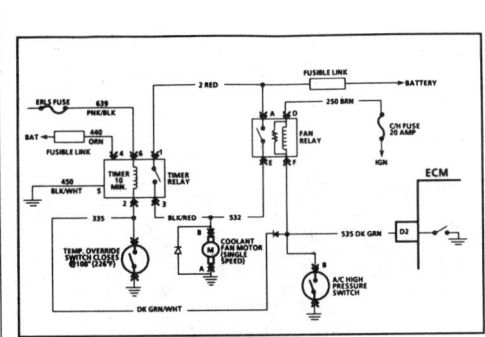

CHART C-12F
(Page 3 of 3)
COOLANT FAN CHECK (WITH A/C-LESS VO8)
3.8L (VIN 3) "A" SERIES (PORT)

Circuit Description:
The coolant fan motor can be energized through two (2) relays; a fan relay, or timer relay.
Power for the coolant fan motor comes through the fusible link to both relays. The relays are energized when current flows to ground, through the activation of the A/C high pressure switch, temp. override switch, and/or the ECM.
Fan Relay - The fan relay is energized by the ECM, A/C high pressure switch and/or the temp. override switch. The ECM energizes the fan relay through terminal "D2", when coolant temperature reaches 98°C (208°F). The A/C high pressure switch energizes the fan relay when refrigerant pressure reaches 300 psi (2068 kPa).

Test Description: Numbers below refer to circled numbers on the diagnostic chart.
1. Checks for B+ at terminals "1" and "4" to the timer relay.

2. Checks to be sure CKT 639 does not have voltage when the ignition switch is "OFF". If CKT 639 has voltage, with the ignition "OFF", the timer relay will not turn the fan "ON".
3. Checks to be sure CKT 450 is a good ground.

1988 A-CAR EQUIPPED WITH 3.8L (VIN 3) ENGINE

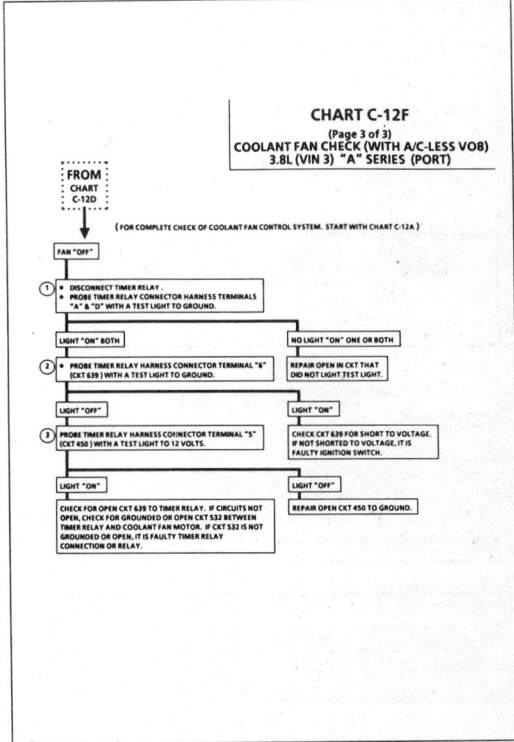

CHART C-12F
(Page 3 of 3)
COOLANT FAN CHECK (WITH A/C-LESS VO8)
3.8L (VIN 3) "A" SERIES (PORT)

1988 A-CAR EQUIPPED WITH 3.8L (VIN 3) ENGINE

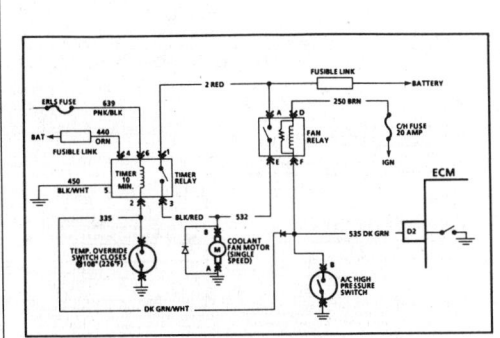

CHART C-12G
(Page 3 of 3)
COOLANT FAN "ON" AT ALL TIMES (WITH A/C-LESS VO8)
3.8L (VIN 3) "A" SERIES (PORT)

Circuit Description:
The coolant fan motor can be energized through two (2) relays; a fan relay, or timer relay.
Power for the coolant fan motor comes through the fusible link to both relays. The relays are energized when current flows to ground, through the activation of the A/C high pressure switch, temp. override switch, and/or the ECM.
Fan Relay - The fan relay is energized by the ECM, A/C high pressure switch and/or the temp. override switch. The ECM energizes the fan relay through terminal "D2", when coolant temperature reaches 98°C (208°F). The A/C high pressure switch energizes the fan relay when refrigerant pressure reaches 300 psi (2068 kPa).

Test Description: Numbers below refer to circled numbers on the diagnostic chart.
1. This step will separate the problem as either the timer relay or the fan relay.
2. Checks for short to voltage in CKT 532.
3. Checks to see if CKT 535 is shorted to ground, which would keep the relay grounded at all times.

4. Checks to see if CKT 335 is shorted to ground. A light indicates the wire is shorted to ground and following the steps will isolate the short.
5. If the test light is "OFF" after disconnecting, the ECM is shorted internally.

1988 A-CAR EQUIPPED WITH 3.8L (VIN 3) ENGINE

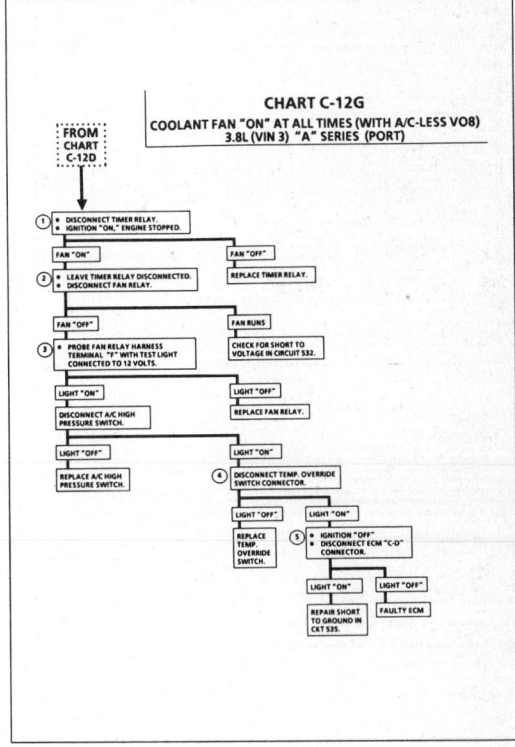

CHART C-12G
COOLANT FAN "ON" AT ALL TIMES (WITH A/C-LESS VO8)
3.8L (VIN 3) "A" SERIES (PORT)

1988 A-CAR EQUIPPED WITH 3.8L (VIN 3) ENGINE

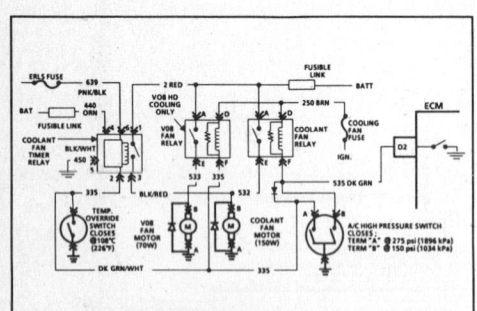

CHART C-12H
(Page 1 of 3)
COOLANT FAN CHECK (WITH A/C & VO8 HD COOLING)
3.8L (VIN 3) "A" SERIES (PORT)

Circuit Description:
On VO8 (heavy duty cooling) systems, the standard coolant fan motor and the VO8 fan motor are energized through a coolant fan relay and a VO8 fan relay and/or a coolant fan timer relay. Power for the coolant fan motors come through the fusible link to all relays. The relays are energized when current flows to ground through the activation of the A/C high pressure switch, temp. override switch and/or the ECM.
Coolant Fan Relay - The coolant fan relay is energized by the temp. override switch, A/C high pressure switch and/or the ECM. The ECM energizes the coolant fan relay through terminal "D2" when the coolant temperature reaches 98°C (208°F), and vehicle speed is below 45 mph. The relay is also energized through terminal "B" of the A/C high pressure switch (150 psi) and/or the temp. override switch (108°C).
VO8 Fan Relay - The pusher fan is installed as part of the heavy duty cooling package (VO8) and is turned "ON" when A/C system pressure reaches 275 psi (1896 kPa) or coolant temperature reaches 108°C (226°F).
Coolant Fan Relay - The coolant fan relay is energized by the temp. override switch. If the coolant temperature is 108°C (226°F) or higher when the ignition switch is turned "OFF", the timer relay is energized for 10 minutes or until the coolant temperature is lowered below 108°C (226°F). The standard coolant fan is the only fan that will run with the ignition switch "OFF".

Test Description: Numbers below refer to circled numbers on the diagnostic chart.
1. Checks to see if fans run at all times with ignition "ON".
2. Grounding the test terminal, in the ALDL connector, should cause the ECM to ground CKT 535. At this point, the standard coolant fan should run.
3. Grounding the temp. override switch harness to ground should cause the relays to turn "ON" both fans. If the fans run, the fan control relays and CKT 335 to the temp. override switch are good.
4. Grounding each pressure switch terminal should cause relays to close, turning "ON" fans, proving relays and circuits are OK.
5. Checks to see if temp. override switch operates correctly. Test light should come "ON" at about 108°C (226°F), and the "hot light" on the instrument panel should come "ON" above 116°C (241°F).
6. Checks to see if coolant fan relay works. With temp. override switch harness grounded, the coolant fan motor should run until ground jumper is removed.

1988 A-CAR EQUIPPED WITH 3.8L (VIN 3) ENGINE

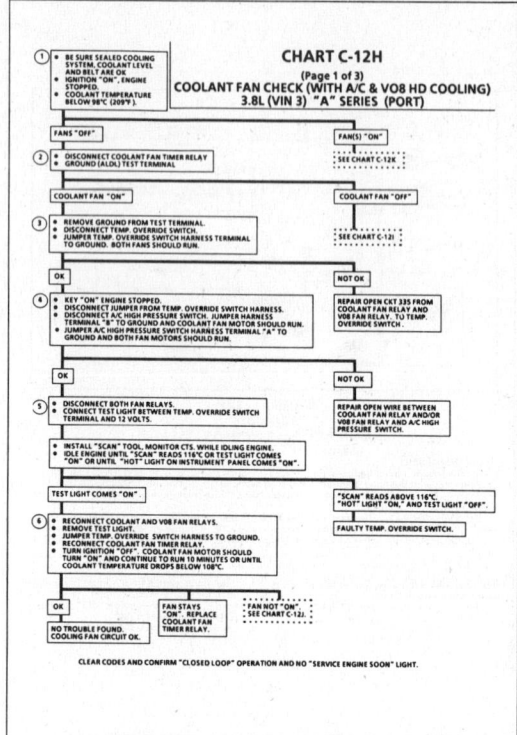

1988 A-CAR EQUIPPED WITH 3.8L (VIN 3) ENGINE

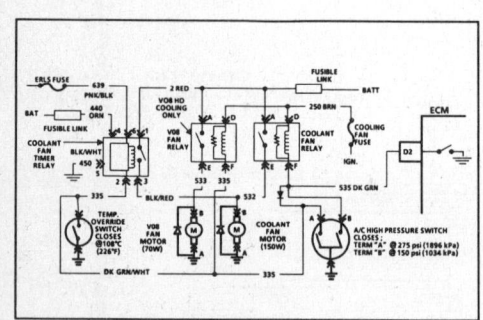

CHART C-12I
(Page 2 of 3)
COOLANT FAN CHECK (WITH A/C & VO8 HD COOLING)
3.8L (VIN 3) "A" SERIES (PORT)

Circuit Description:
On VO8 (heavy duty cooling) systems, the standard coolant fan motor and the VO8 fan motor are energized through a coolant fan relay and a VO8 fan relay and/or a coolant fan timer relay. Power for the coolant fan motors come through the fusible link to all relays. The relays are energized when current flows to ground through the activation of the A/C high pressure switch, temp. override switch and/or the ECM.
Coolant Fan Relay - The coolant fan relay is energized by the temp. override switch, A/C high pressure switch and/or the ECM. The ECM energizes the coolant fan relay through terminal "D2" when the coolant temperature reaches 98°C (208°F), and vehicle speed is below 45 mph. The relay is also energized through terminal "B" of the A/C high pressure switch (150 psi) and/or the temp. override switch (108°C).
VO8 Fan Relay - The pusher fan is installed as part of the heavy duty cooling package (VO8) and is turned "ON" when A/C system pressure reaches 275 psi (1896 kPa) or coolant temperature reaches 108°C (226°F).
Coolant Fan Relay - The coolant fan relay is energized by the temp. override switch. If the coolant temperature is 108°C (226°F) or higher when the ignition switch is turned "OFF", the timer relay is energized for 10 minutes or until the coolant temperature is lowered below 108°C (226°F). The standard coolant fan is the only fan that will run with the ignition switch "OFF".

Test Description: Numbers below refer to circled numbers on the diagnostic chart.
1. This check bypasses the coolant fan relay and applies B+ to the coolant fan motor through CKT 532. At this point, the fan should run.
2. Checks to see if problem is the coolant fan relay, wiring, or ECM.
3. Checks for presence of ignition voltage to relay connector terminal "5".
4. Checks for presence of B+ to coolant fan relay connector terminal "A".
5. Checks for an open ground or fan motor circuit, then checks for an open in CKT 532 between coolant fan relay and motor, or faulty motor.

1988 A-CAR EQUIPPED WITH 3.8L (VIN 3) ENGINE

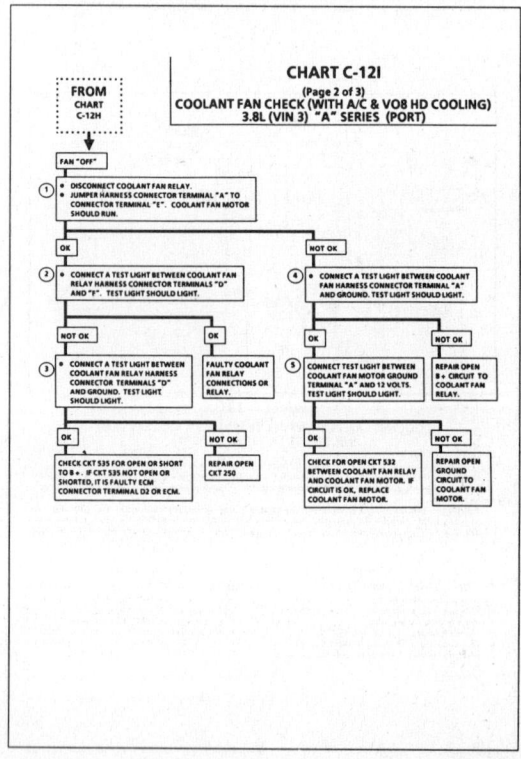

1988 A-CAR EQUIPPED WITH 3.8L (VIN 3) ENGINE

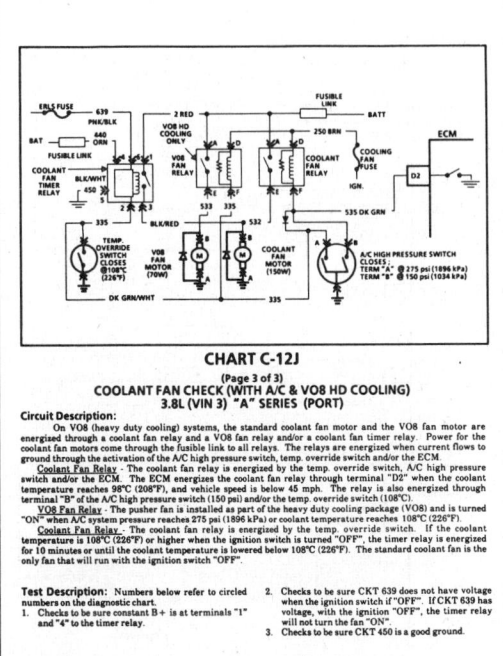

CHART C-12J
(Page 3 of 3)
COOLANT FAN CHECK (WITH A/C & VO8 HD COOLING)
3.8L (VIN 3) "A" SERIES (PORT)

Circuit Description:
On VO8 (heavy duty cooling) systems, the standard coolant fan motor and the VO8 fan motor are energized through a coolant fan relay and a VO8 fan relay and/or a coolant fan timer relay. Power for the coolant fan motors come through the fusible link to all relays. The relays are energized when current flows to ground through the activation of the A/C high pressure switch, temp. override switch and/or the ECM.

Coolant Fan Relay - The coolant fan relay is energized by the temp. override switch, A/C high pressure switch and/or the ECM. The ECM energizes the coolant fan relay through terminal "D2" when the coolant temperature reaches 98°C (208°F), and vehicle speed is below 45 mph. The relay is also energized through terminal "B" of the A/C high pressure switch (150 psi) and/or the temp. override switch (108°C).

VO8 Fan Relay - The pusher fan is installed as part of the heavy duty cooling package (VO8) and is turned "ON" when A/C system pressure reaches 275 psi (1896 kPa) or coolant temperature reaches 108°C (226°F).

Coolant Fan Relay - The coolant fan relay is energized by the temp. override switch. If the coolant temperature is 108°C (226°F) or higher when the ignition switch is turned "OFF", the timer relay is energized for 10 minutes or until the coolant temperature is lowered below 108°C (226°F). The standard coolant fan is the only fan that will run with the ignition switch "OFF".

Test Description: Numbers below refer to circled numbers on the diagnostic chart.
1. Checks to be sure constant B+ is at terminals "1" and "4" to the timer relay.
2. Checks to be sure CKT 639 does not have voltage when the ignition switch is if "OFF". If CKT 639 has voltage, with the ignition switch "OFF", the timer relay will not turn the fan "ON".
3. Checks to be sure CKT 450 is a good ground.

1988 A-CAR EQUIPPED WITH 3.8L (VIN 3) ENGINE

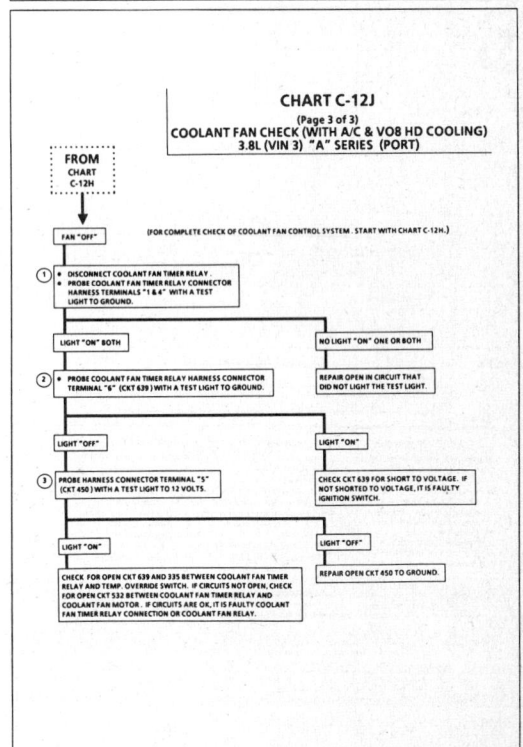

1988 A-CAR EQUIPPED WITH 3.8L (VIN 3) ENGINE

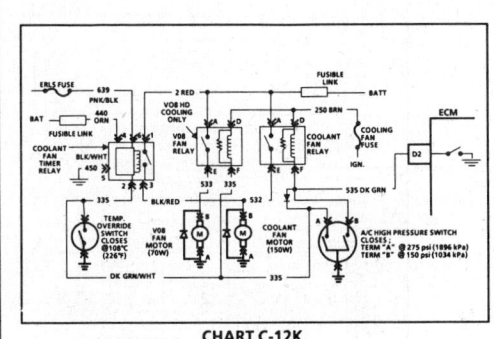

CHART C-12K
FAN(S) "ON" AT ALL TIMES (WITH A/C & VO8 HD COOLING)
3.8L (VIN 3) "A" SERIES (PORT)

Circuit Description:
On VO8 (heavy duty cooling) systems, the standard coolant fan motor and the VO8 fan motor are energized through a coolant fan relay and a VO8 fan relay and/or a coolant fan timer relay. Power for the coolant fan motors come through the fusible link to all relays. The relays are energized when current flows to ground through the activation of the A/C high pressure switch, temp. override switch and/or the ECM.

Coolant Fan Relay - The coolant fan relay is energized by the temp. override switch, A/C high pressure switch and/or the ECM. The ECM energizes the coolant fan relay through terminal "D2" when the coolant temperature reaches 98°C (208°F), and vehicle speed is below 45 mph. The relay is also energized through terminal "B" of the A/C high pressure switch (150 psi) and/or the temp. override switch (108°C).

VO8 Fan Relay - The pusher fan is installed as part of the heavy duty cooling package (VO8) and is turned "ON" when A/C system pressure reaches 275 psi (1896 kPa) or coolant temperature reaches 108°C (226°F).

Coolant Fan Relay - The coolant fan relay is energized by the temp. override switch. If the coolant temperature is 108°C (226°F) or higher when the ignition switch is turned "OFF", the timer relay is energized for 10 minutes or until the coolant temperature is lowered below 108°C (226°F). The standard coolant fan is the only fan that will run with the ignition switch "OFF".

Test Description: Numbers below refer to circled numbers on the diagnostic chart.
1. This step will separate the problem between the coolant fan timer relay or the coolant fan relay.
2. Checks for short to voltage in CKT 532 or 533.
3. Checks to see if CKT 335 or 535 is shorted to ground, which would keep the coolant fan relay grounded at all times.
4. Checks to see if CKT 335 is shorted to ground. A light indicates the wire is shorted to ground and the following steps will isolate the short.
5. If the test light is "OFF" after disconnecting, the ECM is shorted internally.

1988 A-CAR EQUIPPED WITH 3.8L (VIN 3) ENGINE

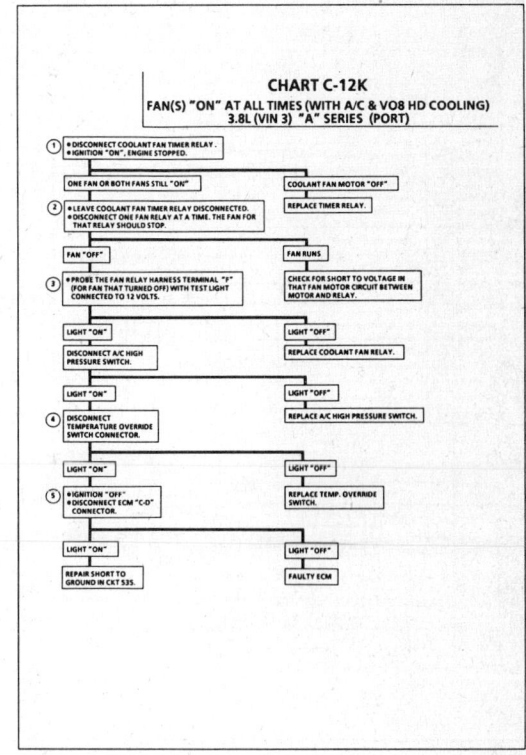

1988–90 E-CAR EQUIPPED WITH 3800 (VIN C) ENGINE

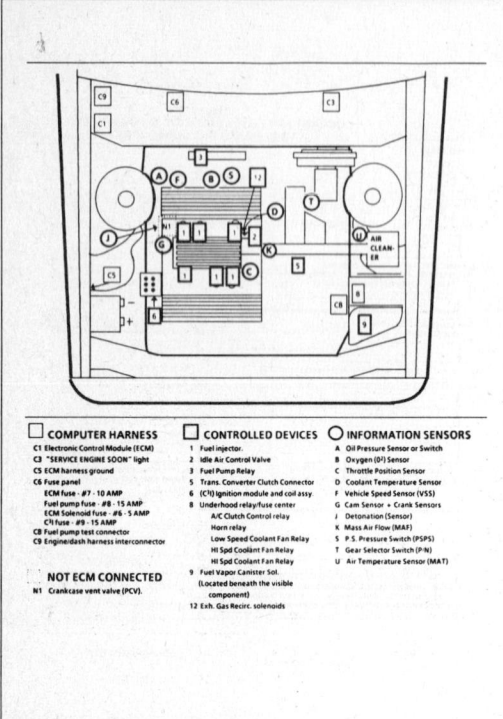

☐ **COMPUTER HARNESS**
C1 Electronic Control Module (ECM)
C3 "SERVICE ENGINE SOON" light
C5 ECM harness ground
C6 Fuse panel
  ECM fuse - #7 - 10 AMP
  Fuel pump fuse - #8 - 15 AMP
  ECM Solenoid fuse - #6 - 5 AMP
  C7 fuse - #9 - 15 AMP
C8 Fuel pump test connector
C9 Engine/dash harness interconnector

NOT ECM CONNECTED
N1 Crankcase vent valve (PCV).

☐ **CONTROLLED DEVICES**
1 Fuel injector.
2 Idle Air Control Valve
3 Fuel Pump Relay
4 Trans. Converter Clutch Connector
5 (C¹⁰)Ignition module and coil assy.
8 Underhood relay/fuse center
  A/C Clutch Control relay
  Horn relay
  Low Speed Coolant Fan Relay
  HI Spd Coolant Fan Relay
  HI Spd Coolant Fan Relay
9 Fuel Vapor Canister Sol.
  (Located beneath the visible component)
12 Exh. Gas Recirc. solenoids

◯ **INFORMATION SENSORS**
A Oil Pressure Sensor or Switch
B Oxygen (0²) Sensor
C Throttle Position Sensor
D Coolant Temperature Sensor
F Vehicle Speed Sensor (VSS)
G Cam Sensor + Crank Sensors
J Detonation (Sensor)
K Mass Air Flow (MAF)
S P.S. Pressure Switch (PSPS)
T Gear Selector Switch (P/N)
U Air Temperature Sensor (MAT)

1988–90 E-CAR EQUIPPED WITH 3800 (VIN C) ENGINE

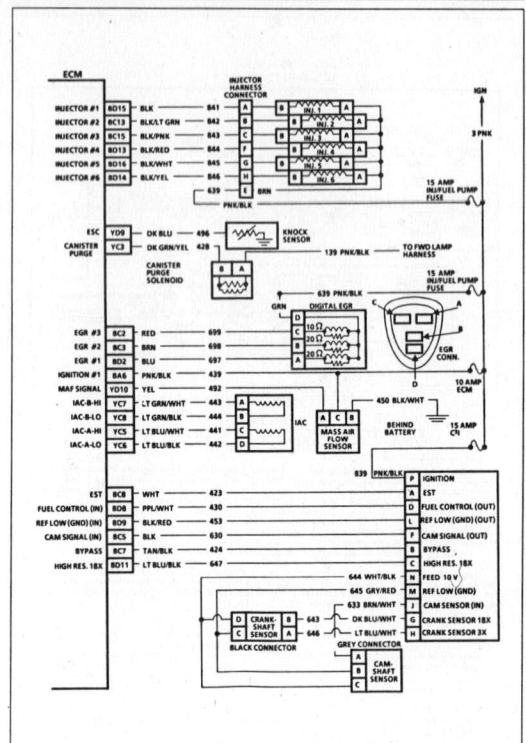

1988–90 E-CAR EQUIPPED WITH 3800 (VIN C) ENGINE

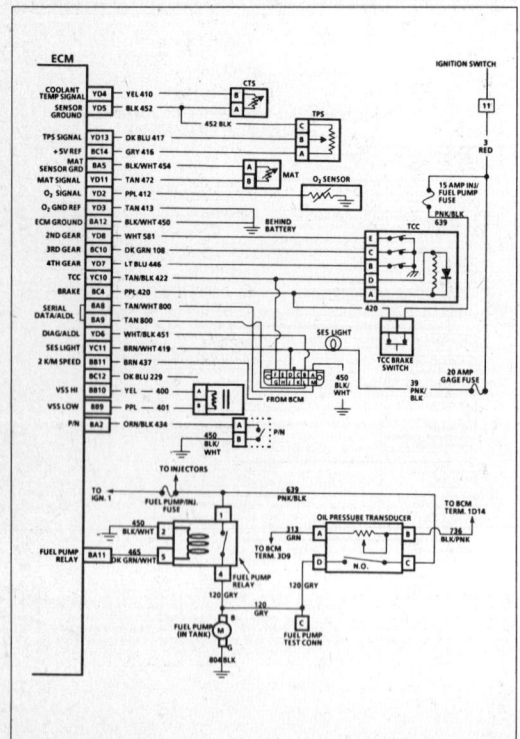

1988–90 E-CAR EQUIPPED WITH 3800 (VIN C) ENGINE

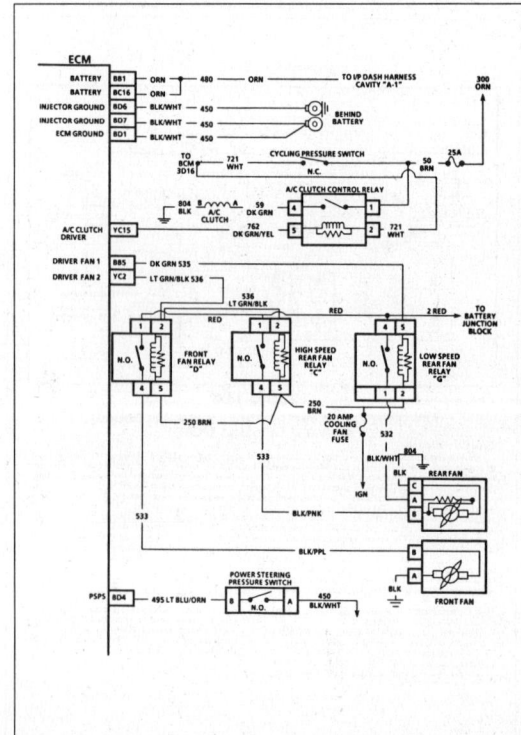

1988–90 E-CAR EQUIPPED WITH 3800 (VIN C) ENGINE

PORT FUEL INJECTION ECM CONNECTOR IDENTIFICATION
This ECM voltage chart is for use with a digital voltmeter to further aid in diagnosis. The voltages you get may vary due to low battery charge or other reasons, but they should be very close. The "B +" symbol indicates a nominal system voltage of 12-14 V.

THE FOLLOWING CONDITIONS MUST BE MET BEFORE TESTING:
- Engine at operating temperature (upper rad. hose hot) • Engine idling in "Closed Loop" (For "Engine Run" column) in park or neutral • Test terminal not grounded • "SCAN" tool not installed

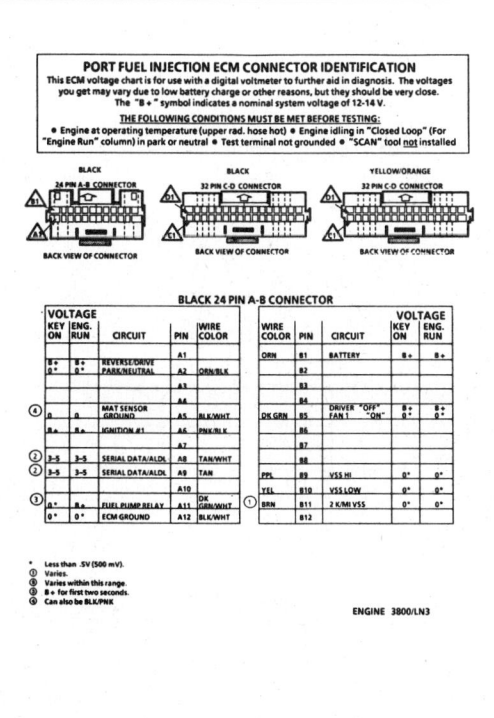

BACK VIEW OF CONNECTOR (24 PIN A-B CONNECTOR) — BACK VIEW OF CONNECTOR (32 PIN C-D CONNECTOR) — BACK VIEW OF CONNECTOR (32 PIN C-D CONNECTOR)

BLACK 24 PIN A-B CONNECTOR

VOLTAGE KEY ON	ENG. RUN	CIRCUIT	PIN	WIRE COLOR					WIRE COLOR	PIN	CIRCUIT	VOLTAGE KEY ON	ENG. RUN
			A1						ORN	B1	BATTERY	B+	B+
B+ 0*	B+ 0*	REVERSE/DRIVE PARK/NEUTRAL	A2	ORN/BLK						B2			
			A3							B3			
			A4							B4			
④		MAT SENSOR GROUND	A5	BLK/WHT					DK GRN	B5	DRIVER "OFF" FAN 1 "ON"	B+ 0*	B+ 0*
		IGNITION #1	A6	PNK/BLK						B6			
			A7							B7			
② 3-5	3-5	SERIAL DATA/ALDL	A8	TAN/WHT						B8			
② 3-5	3-5	SERIAL DATA/ALDL	A9	TAN					PPL	B9	VSS HI	0*	0*
			A10						YEL	B10	VSS LOW	0*	0*
③ B+	B+	FUEL PUMP RELAY	A11	DK GRN/WHT					BRN	B11	2 K/MI VSS	0*	0*
0*	0*	ECM GROUND	A12	BLK/WHT						B12			

ENGINE 3800/LN3

* Less than .5V (500 mV).
① Varies.
② Varies within this range.
③ B + for first two seconds.
④ Can also be BLK/PNK.

1988–90 E-CAR EQUIPPED WITH 3800 (VIN C) ENGINE

BLACK 32 PIN C-D CONNECTOR

VOLTAGE KEY ON	ENG. RUN	CIRCUIT	PIN	WIRE COLOR			WIRE COLOR	PIN	CIRCUIT	VOLTAGE KEY ON	ENG. RUN
① 0*	0*	2K/MI VSS SPARE	C1	N/A			BLK/WHT	D1	ECM GROUND	0*	0*
B+	B+	EGR #3	C2	RED			BLU	D2	EGR #1	B+	B+
B+	B+	EGR #2	C3	BRN				D3			
B+	B+	BRAKE INPUT	C4	PPL			LT BLU/ORN	D4	PSPS	B+	B+
**	**	CAM SIGNAL	C5	BLK				D5			
			C6				BLK/WHT	D6	INJECTOR GROUND	0*	0*
0*	4.70	BYPASS	C7	TAN/BLK			BLK/WHT	D7	INJECTOR GROUND	0*	0*
0*	0-8+	EST	C8	WHT			PPL/WHT	D8	FUEL CONTROL REFERENCE LOW	0*	6 V
			C9				BLK/RED	D9	(GRN/WHT)	0*	6 V
0*	0*	3RD GEAR	C10	DK GRN				D10			
			C11				LT BLU/BLK	D11	HIGH RES. (18 x)	0*	6 V
			C12					D12			
B+	B+	INJECTOR #2	C13	BLK/LTGRN			BLK/RED	D13	INJECTOR #4	B+	B+
5V	5V	+5V REFERENCE	C14	GRY			BLK/YEL	D14	INJECTOR #6	B+	B+
B+	B+	INJECTOR #3	C15	BLK/PNK			BLK	D15	INJECTOR #1	B+	B+
B+	B+	BATTERY	C16	ORN			BLK/WHT	D16	INJECTOR #5	B+	B+

YELLOW/ORANGE 32 PIN C-D CONNECTOR

VOLTAGE KEY ON	ENG. RUN	CIRCUIT	PIN	WIRE COLOR			WIRE COLOR	PIN	CIRCUIT	VOLTAGE KEY ON	ENG. RUN
			C1					D1			
0*	0*	"ON" DRIVER "OFF" FAN 2	C2	DK GRN/BLK			PPL	D2	O₂ SIGNAL	.35 - 45	1-9 ②
B+	B+	CANISTER PURGE	C3	DKGRN/YEL			TAN	D3	O₂ GROUND	0*	0*
			C4				YEL	D4	COOLANT TEMP SIGNAL	2.04	2.18 ①
0*	0*	IAC-A-HI	C5	LT BLU/WHT			BLK	D5	SENSOR GROUND	0*	0*
0*	0*	IAC-A-LO	C6	LT BLU/BLK			WHT/BLK	D6	DIAG/ALDL	5V	5V
0*	0*	IAC-B-HI	C7	LT GRN/WHT			LT BLU	D7	4TH GEAR	0*	0*
0*	0*	IAC-B-LO	C8	LT GRN/BLK			WHT	D8	2ND GEAR	0*	0*
			C9				DK BLU	D9	ESC SIGNAL	2.3	2.3
B+	B+	TCC	C10	TAN/BLK			YEL	D10	MAF SIGNAL	0*	2.3 ①
0*	B+	SES LIGHT	C11	BRN/WHT			TAN	D11	MAT SIGNAL	1.8	1.8 ①
			C12					D12			
			C13				DK BLU	D13	TPS SIGNAL	.33-46	.33-46 ②
			C14					D14			
0*	0*	"ON" A/C CLUTCH "OFF" DRIVER	C15	LT GRN/YEL				D15			
			C16					D16			

** VARIES AROUND 10 VOLTS.
* Less than .5V (500 mV)   ③ Varies within this range
① Varies   ④ Varies with temperature.

ENGINE - 3800/LN3

1988–90 E-CAR EQUIPPED WITH 3800 (VIN C) ENGINE

DIAGNOSTIC CIRCUIT CHECK
(Using a "SCAN" Tool)

The Diagnostic Circuit Check must be the starting point for any driveability complaint diagnosis. Before using this you should perform a careful visual/physical check of the ECM and engine grounds for being clean and tight.

The diagnostic circuit check is an organized approach to identifying a problem created by an electronic engine control system malfunction because it directs the service technician to the next logical step in diagnosing the complaint.

If after completing the diagnostic circuit check and finding the on-board diagnostics functioning properly and no trouble codes displayed, a comparison of "Typical Scan Values," for the appropriate engine, may be used for comparison. The "Typical Values" are an average of display values recorded from normally operating vehicles and are intended to represent what a normally functioning system would display.

A "SCAN" TOOL THAT DISPLAYS FAULTY DATA SHOULD NOT BE USED, AND THE PROBLEM SHOULD BE REPORTED TO THE MANUFACTURER. THE USE OF A FAULTY "SCAN" CAN RESULT IN MISDIAGNOSIS AND UNNECESSARY PARTS REPLACEMENT.

Only the parameters listed below are used in this manual for diagnosis. If a "Scan" reads other parameters, the values are not recommended by General Motors for use in diagnosis. For more description on the values and use of the "Scan" to diagnose ECM inputs, refer to the applicable diagnosis.   If all values are within the range illustrated, refer to "Symptoms" in Section "B".

TYPICAL "SCAN" DATA VALUES
3800 VIN C

Idle / Upper Radiator Hose Hot / Closed Throttle / Park or Neutral / Closed Loop / Acc. off

"SCAN" Position	Units Displayed	Typical Data Value
Engine Speed	RPM	650 - 750
Coolant Temp.	C°	85° - 105°
Mani Air Temp.	C°	Varies with Air Temperature
Throt. Position	Volts	.33- .46
LV8	0	70 - 80
Oxygen Sensor	Millivolts	(100-900)
Inj Pulse width	Milliseconds	4.0 - 5.0 (Varies)
Spark Advance	Degrees	20°
Mass Air Flow	Gm/Sec	4 - 7
Fuel Integ.	Counts	118 - 138 *
Block Learn	Counts	118 - 138 *
Open/Closed Loop	Open/Closed	Closed Loop (may go open with extended idle)
Knock Signal	No/Yes	No
Idle Air Control	Counts (steps)	10 - 30
Park/Neutral	P/N and RDL	P-N
Vehicle Speed	MPH, KPH	0 (Zero)
Torque Conv Cl	On/Off	Off
Battery Voltage	Volts	13.8 Volts (varies)
Exhaust Recirc	%	0 (Zero)
A/C Request	Yes/No	No
Brake Switch	No/Yes	No (Yes with brake pedal depressed)
Power Steering	High Press/Normal	Normal
QDM 1	High/Low	Low
QDM 2	High/Low	Low (High if A/C head pressure's high)
QDM 3	High/Low	Low
QDM 4	High/Low	Low (High with brake pedal depressed)
2nd Gear	On/Off	Off
3rd Gear	On/Off	Off
4th Gear	On/Off	Off
PROM I.D.	Numbers	Internal I.D. Only

*A poor ECM ground at the CII mounting bracket to engine, could cause fuel integ and block learn to read around 150, make a careful physical inspection of this critical connection.

NOTE: IF ALL VALUES ARE WITHIN THE RANGE ILLUSTRATED, REFER TO "SYMPTOMS" IN SECTION "B".

1988–90 E-CAR EQUIPPED WITH 3800 (VIN C) ENGINE

Additional "Scan" Values
3800 (VIN C)

Idle / Upper Radiator Hose Hot / Closed Throttle / Park or Neutral / "Closed Loop" / Acc. "OFF"

"SCAN" Position	Units Displayed	Typical Data Value
Desired Idle	RPM	Varies
O₂ Cross Counts	Counts	Varies
Air Fuel Ratio	Air Fuel Ratio	14.7 varies
Rich/Lean Flag	Rich/Lean	Constantly changing
Block Learn Cell	Cell #	No (Except in high areas)
Knock Retard	Degrees	0°
Purge Duty Cycle	%	Varies
IAC Learned	Yes/No	No
Air Fuel Learned	Yes/No	No
A/C Clutch	Yes/No	No (Yes with A/C "ON")
Cruise Engaged	Yes/No	No
Fan 1	ON/OFF	OFF
Fan 2	ON/OFF	OFF

DESIRED IDLE - SCAN TOOL RANGE 0-3175
The idle speed that is commanded by the ECM. The ECM will compensate for various engine loads to keep the engine at the desired rpm.
O₂ CROSS COUNTS - SCAN TOOL RANGE 0-255
The number of times the oxygen sensor voltage crosses over the rich/lean threshold since the last data transmission to the scan tool.
AIR FUEL RATIO - SCAN TOOL RANGE 0.00 - 99.99
The reading reflects the commanded value. This should be at or near 14.7. A lower number indicates a richer commanded air fuel mixture while a higher number indicates a leaner mixture.
RICH/LEAN FLAG - SCAN TOOL STATES RICH/LEAN
Refers to the oxygen sensor voltage corresponding to the intake air/fuel ratio of 14.7 or 450 mV. An oxygen sensor reading above the threshold means the intake mixture is rich, and a sensor reading below 450 mV indicates a lean fuel mixture.
ALTITUDE - SCAN TOOL STATES YES/NO
This is an internal ECM parameter. The ECM compares engine load (from the MAF sensor) with the throttle position. If a greater than expected value for the TPS is read for a given engine load, the ECM assumes it is at high altitude.
BLOCK LEARN CELL - SCAN TOOL RANGE 0-15
Block learn is dependent upon engine speed and either manifold air pressure or mass air flow readings. A plot of rpm vs MAP or MAF is broken into 16 cells. BLM cell indicates which cell is currently active

KNOCK RETARD - SCAN TOOL RANGE 0-90
This is the amount of degrees of advance pulled out of the spark advance in response to ESC counts.
PURGE DUTY CYCLE - SCAN TOOL RANGE 0-100%
A proportional signal used to control the canister purge function. 0% implies the CCP value is closed while 100% implies that the value is fully open.
IAC LEARNED - SCAN TOOL STATES YES/NO
Each time the car is driven, the IAC system learns how many counts is required to produce a given desired idle. After the engine is started and accelerates, then drops to idle, the IAC system learns the proper number of counts for the desired idle.
AIR FUEL LEARNED - SCAN TOOL STATES YES/NO
When the air fuel system is learning (YES) the block learn is responding to the fuel integrator. If the air fuel learn reads (NO) then block learn will not respond to change in fuel integrator. Normally the fuel system will start learning as soon as the system goes into closed loop.
A/C CLUTCH - SCAN TOOL STATES ON/OFF
A feedback signal to the ECM that lets the ECM know if the A/C clutch is on or off.
CRUISE ENGAGED - SCAN TOOL STATES ON/OFF
The scan tool will read on if the car is operating in cruise control.
FAN 1, FAN 2 SCAN TOOL STATES ON/OFF
The scan tool will read "ON" for #1 fan when the ECM completes the ground circuit for the #1 fan relay coil. For the #2 fan it will read "ON" when the ECM or the A/C head pressure switch completes the ground circuit for the relay coil.

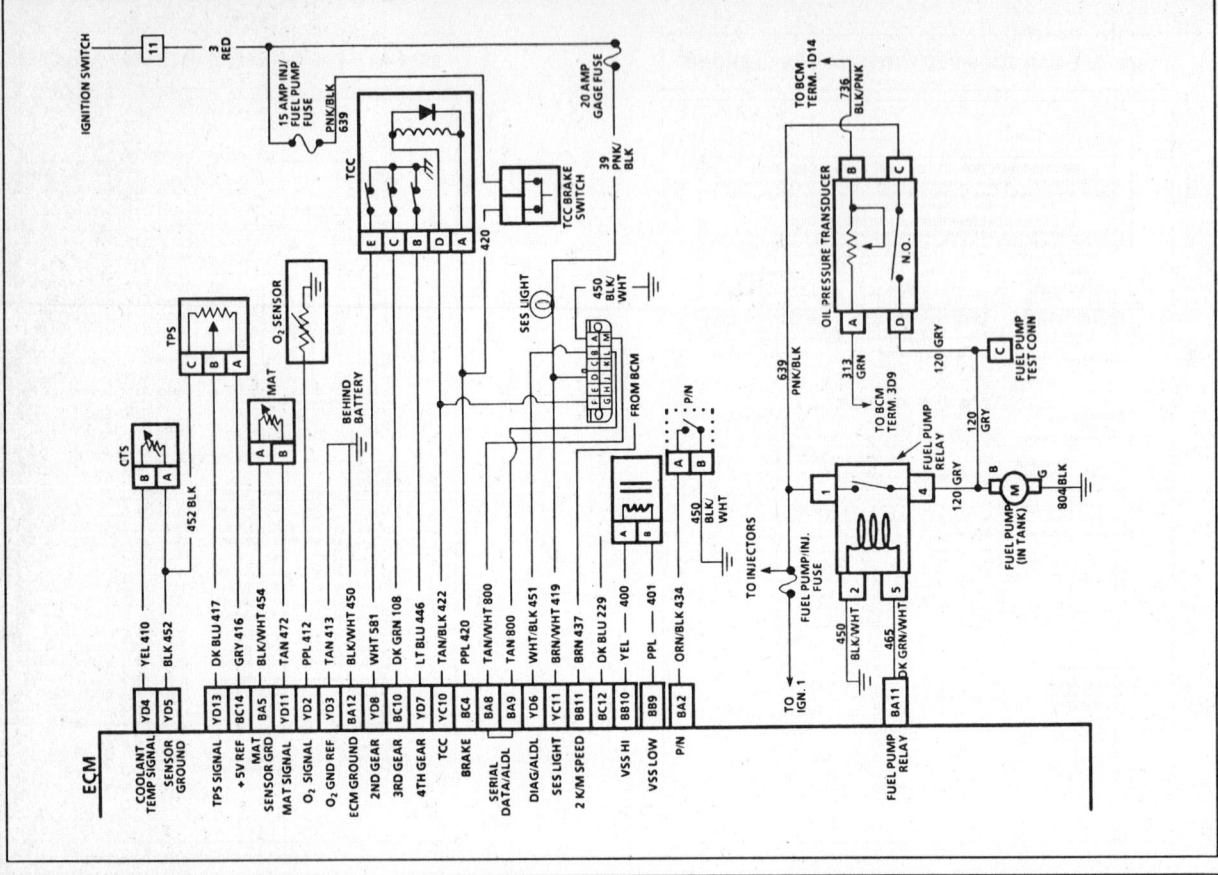

ECM WIRING DIAGRAM—3800 (VIN C) ENGINE

ECM WIRING DIAGRAM—3800 (VIN C) ENGINE

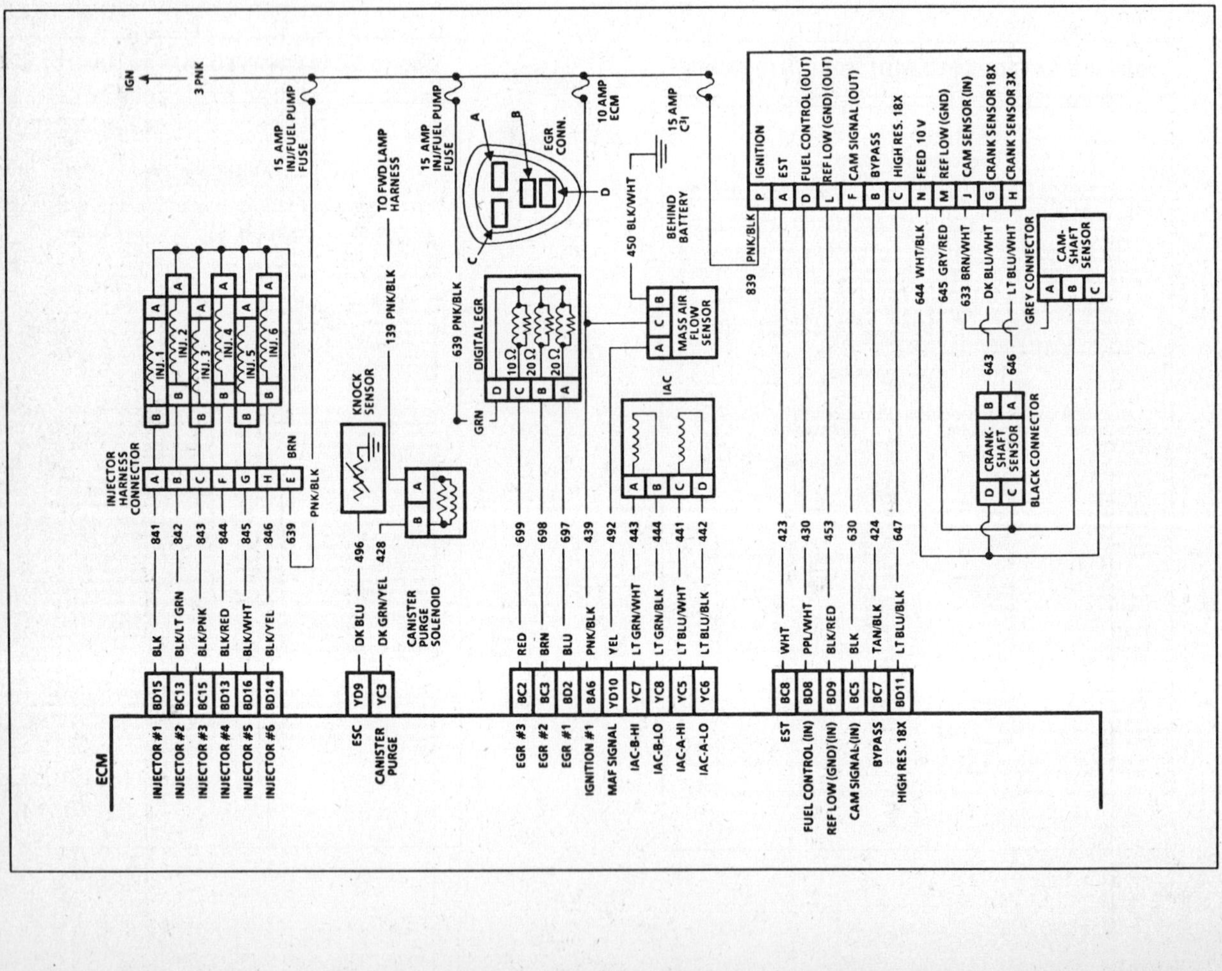

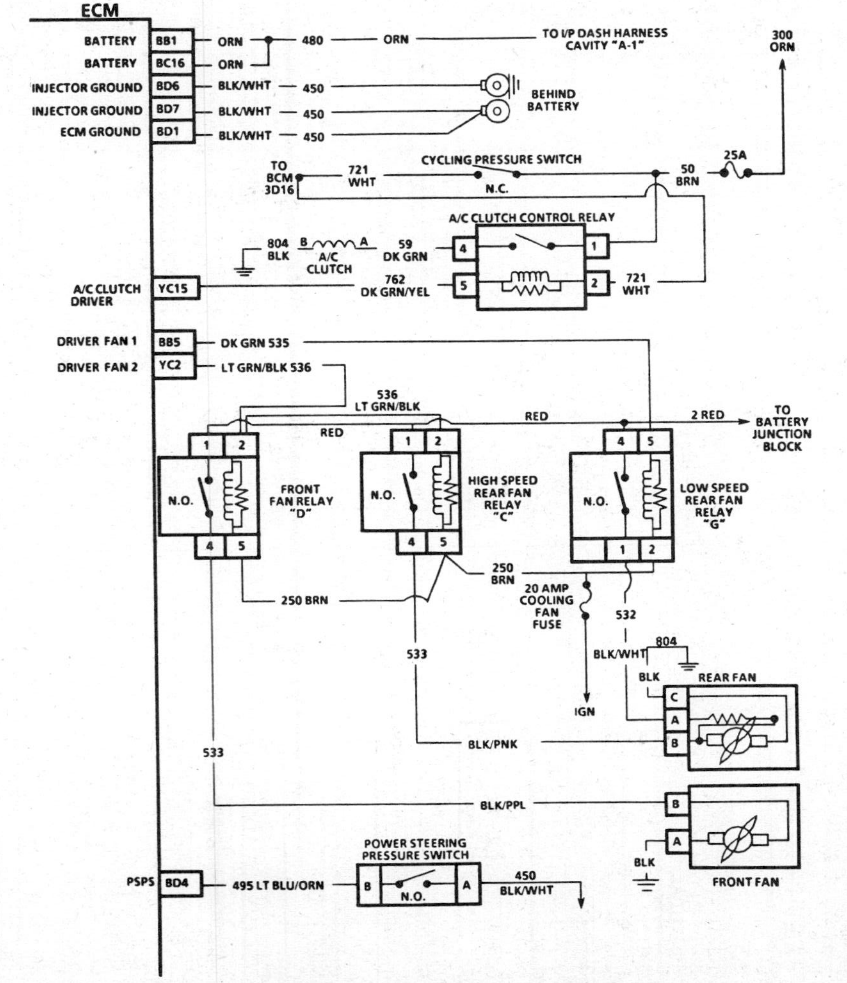

ECM CONNECTORS—3800 (VIN C) ENGINE

PORT FUEL INJECTION ECM CONNECTOR IDENTIFICATION

This ECM voltage chart is for use with a digital voltmeter to further aid in diagnosis. The voltages you get may vary due to low battery charge or other reasons, but they should be very close. The "B+" symbol indicates a nominal system voltage of 12-14 V.

THE FOLLOWING CONDITIONS MUST BE MET BEFORE TESTING:
• Engine at operating temperature (upper rad. hose hot) • Engine idling in "Closed Loop" (For "Engine Run" column) in park or neutral • Test terminal not grounded • "SCAN" tool not installed

YELLOW/ORANGE
32 PIN C-D CONNECTOR
BACK VIEW OF CONNECTOR
D1 C1

BLACK
32 PIN C-D CONNECTOR
BACK VIEW OF CONNECTOR
D1 C1

BLACK
24 PIN A-B CONNECTOR
BACK VIEW OF CONNECTOR
B1 A1

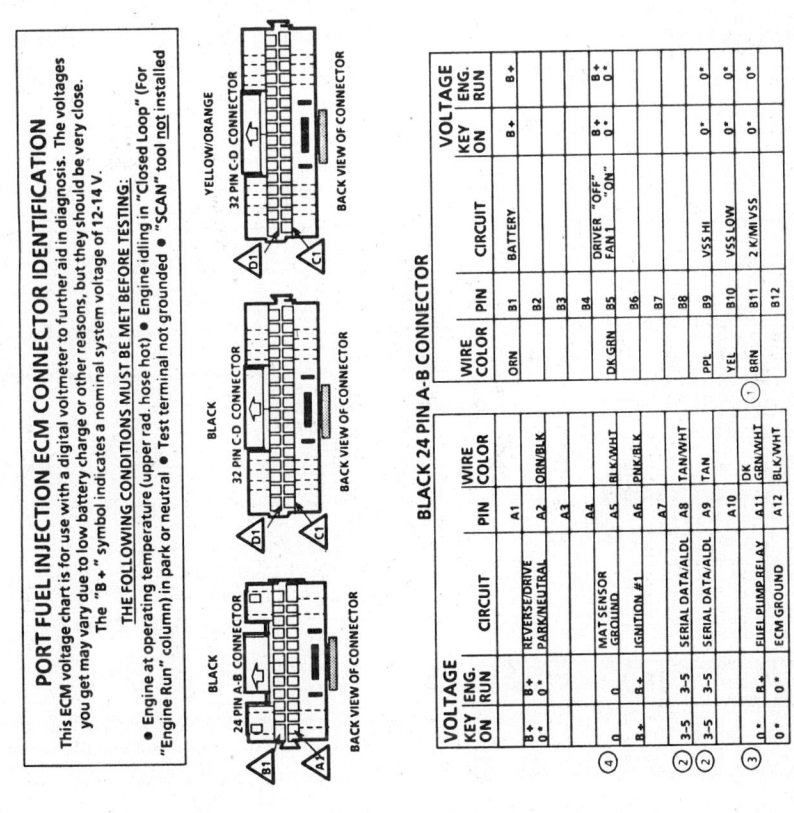

BLACK 24 PIN A-B CONNECTOR

VOLTAGE KEY ON	ENG RUN	CIRCUIT	PIN	WIRE COLOR
			A1	
B+*	0*	REVERSE/DRIVE PARK/NEUTRAL	A2	ORN/BLK
			A3	
			A4	
0 ④	0	MAT SENSOR GROUND	A5	BLK/WHT
B+	B+	IGNITION #1	A6	PNK/BLK
			A7	
3-5 ②	3-5	SERIAL DATA/ALDL	A8	TAN/WHT
3-5 ②	3-5	SERIAL DATA/ALDL	A9	TAN
			A10	
0* ③	B+	FUEL PUMP RELAY	A11	DK GRN/WHT
0*	0*	ECM GROUND	A12	BLK/WHT

WIRE COLOR	PIN	CIRCUIT	VOLTAGE KEY ON	ENG RUN
ORN	B1	BATTERY	B+	B+
	B2			
	B3			
	B4			
DK GRN	B5	DRIVER FAN 1 (DRIVER "OFF" / "ON")	B+ / 0*	B+ / 0*
	B6			
	B7			
PPL	B9	VSS HI	0*	0*
YEL	B10	VSS LOW	0*	0*
BRN	B11	2 K/MI VSS ①	0*	0*
	B12			

* Less than .5V (500 mV).
① Varies.
② Varies within this range.
③ B+ for first two seconds.
④ Can also be BLK/PNK

ENGINE 3800/LN3

ECM WIRING DIAGRAM—3800 (VIN C) ENGINE

ECM

Terminal	Circuit
BB1	BATTERY — ORN 300 ORN (TO U/P DASH HARNESS CAVITY "A-1")
BC16	BATTERY — ORN 480
BD6	INJECTOR GROUND — BLK/WHT 450 (BEHIND BATTERY)
BD7	INJECTOR GROUND — BLK/WHT 450
BD1	ECM GROUND — BLK/WHT 450 (TO BCM 3D16)
YC15	A/C CLUTCH DRIVER — DK GRN/YEL 762
BB5	DRIVER FAN 1 — DK GRN 535
YC2	DRIVER FAN 2 — LT GRN/BLK 536
BD4	PSPS — LT BLU/ORN 495 / BLK/WHT 450

CYCLING PRESSURE SWITCH — 721 WHT — N.C.
A/C CLUTCH CONTROL RELAY — 50 BRN — 25A
A/C CLUTCH — 804 BLK / 59 DK GRN
LOW SPEED REAR FAN RELAY "G" — N.O.
HIGH SPEED REAR FAN RELAY "C" — N.O.
FRONT FAN RELAY "D" — N.O.
536 LT GRN/BLK — 250 BRN — RED — 2 RED (TO BATTERY JUNCTION BLOCK)
20 AMP COOLING FAN FUSE
532 BLK/WHT — 533 — 250 BRN — IGN
REAR FAN — BLK 804 — BLK/PNK
FRONT FAN — BLK/PPL — BLK
POWER STEERING PRESSURE SWITCH — 450 BLK/WHT — N.O.

ECM CONNECTORS — 3800 (VIN C) ENGINE

BLACK 32 PIN C-D CONNECTOR

VOLTAGE KEY ON	ENG. RUN	CIRCUIT	PIN	WIRE COLOR
0*	0*	2K/MI VSS SPARE	C1	N/A
B+	B+	EGR #3	C2	RED
B+	B+	EGR #2	C3	BRN
B+	B+	BRAKE INPUT	C4	PPL
**	**	CAM SIGNAL	C5	BLK
			C6	
0*	4.70	BYPASS	C7	TAN-BLK
0*	0-8+	EST	C8	WHT
			C9	
0*	0*	3RD GEAR	C10	DK GRN
			C11	
			C12	
B+	B+	INJECTOR #2	C13	BLK/LTGRN
5V	5V	+5V REFERENCE	C14	GRY
B+	B+	INJECTOR #3	C15	BLK/PNK
B+	B+	BATTERY	C16	ORN

WIRE COLOR	PIN	CIRCUIT	VOLTAGE KEY ON	ENG. RUN
BLK/WHT	D1	ECM GROUND	0*	0*
BLU	D2	EGR #1	B+	B+
	D3			
LT BLU/ORN	D4	PSPS	B+	B+
	D5			
BLK/WHT	D6	INJECTOR GROUND	0*	0*
BLK/WHT	D7	INJECTOR GROUND	0*	0*
PPL/WHT	D8	FUEL CONTROL REFERENCE LOW	0*	6V
BLK/RED	D9	INJECTOR (GROUND)	0*	0*
	D10			
LT BLU/BLK	D11	HIGH RES. (18 x)	0*	6V
	D12			
BLK/RED	D13	INJECTOR #4	B+	B+
BLK/YEL	D14	INJECTOR #6	B+	B+
BLK	D15	INJECTOR #1	B+	B+
BLK/WHT	D16	INJECTOR #5	B+	B+

YELLOW/ORANGE 32 PIN C-D CONNECTOR

VOLTAGE KEY ON	ENG. RUN	CIRCUIT	PIN	WIRE COLOR
			C1	
0*	0*	"ON" DRIVER	C2	DK GRN/BLK
B+	B+	"OFF" FAN 2	C3	DKGRN/YEL
B+	B+	CANISTER PURGE	C4	
0*	0*	IAC-A-HI	C5	LT BLU/WHT
0*	0*	IAC-A-LO	C6	LT BLU/BLK
0*	0*	IAC-B-HI	C7	LT GRN/WHT
0*	0*	IAC-B-LO	C8	LT GRN/BLK
			C9	
0*	0*	TCC	C10	TAN/BLK
0*	B+	SES LIGHT	C11	BRN/WHT
			C12	
			C13	
			C14	
0*	0*	"ON" A/C CLUTCH	C15	LT GRN/YEL
B+	B+	"OFF" DRIVER	C16	

WIRE COLOR	PIN	CIRCUIT	VOLTAGE KEY ON	ENG. RUN
	D1			
PPL	D2	O₂ SIGNAL	.35 - .45	1 - 9
TAN	D3	O₂ GROUND	0*	0*
YEL	D4	COOLANT TEMP SIGNAL	2.04	2.18
BLK	D5	COOLANT TEMP SENSOR GROUND	0*	0*
WHT/BLK	D6	DIAG/ALDL	5V	5V
LT BLU	D7	4TH GEAR	0*	0*
WHT	D8	2ND GEAR	0*	0*
DK BLU	D9	ESC SIGNAL	2.3	2.3
YEL	D10	MAF SIGNAL	0*	2.3
TAN	D11	MAT SIGNAL	1.8	1.8
	D12			
DK BLU	D13	TPS SIGNAL	.33 - .46	.33 - .46
	D14			
	D15			
	D16			

** — VARIES AROUND 10 VOLTS.
* — Less than .5V (500 mV)
* — Varies
① — Varies within this range
② — Varies with temperature

ENGINE - 3800/LN3

1988–90 E-CAR EQUIPPED WITH 3800 (VIN C) ENGINE

DIAGNOSTIC CIRCUIT CHECK
(Using a "SCAN" Tool)

The Diagnostic Circuit Check must be the starting point for any driveability complaint diagnosis. Before using this you should perform a careful visual/physical check of the ECM and engine grounds for being clean and tight.

The diagnostic circuit check is an organized approach to identifying a problem created by an electronic engine control system malfunction because it directs the service technician to the next logical step in diagnosing the complaint.

After completing the diagnostic circuit check and finding the on-board diagnostics functioning properly and no trouble codes displayed, a comparison of "Typical Values," for the appropriate engine, may be used for comparison. The "Typical Values" are an average of display values recorded from normally operating vehicles and are intended to represent what a normally functioning system would display.

A "SCAN" TOOL THAT DISPLAYS FAULTY DATA SHOULD NOT BE USED, AND THE PROBLEM SHOULD BE REPORTED TO THE MANUFACTURER. THE USE OF A FAULTY "SCAN" CAN RESULT IN MISDIAGNOSIS AND UNNECESSARY PARTS REPLACEMENT.

Only those parameters listed below are used in this manual for diagnosis. If a "Scan" reads other parameters, the values are not recommended by General Motors for use in diagnosis. For more description on the values and use of the "Scan" to diagnose ECM inputs, refer to "Symptoms" in Section "B". If all values are within the range illustrated, refer to the applicable chart.

TYPICAL "Scan" DATA VALUES
3800 VIN C

Idle / Upper Radiator Hose Hot / Closed Throttle / Park or Neutral / Closed Loop / Acc. off

"SCAN" Position	Units Displayed	Typical Data Value
Engine Speed	RPM	650 - 750
Coolant Temp.	C°	85° - 105°
Mani Air Temp.	C°	Varies with Air Temperature
Throt. Position	Volts	.33 - .46
LVB	0	70 - 80
Oxygen Sensor	Millivolts	(100-900)
Inj Pulse width	Milliseconds	4.0 - 5.0 (Varies)
Spark Advance	Degrees	20°
Mass Air Flow	Gm/Sec	4 - 7
Fuel Integ.	Counts	118 - 138 *
Block Learn	Counts	118 - 138 *
Open/Closed Loop	Open/Closed	Closed Loop (may go open with extended idle)
Knock Signal	No/Yes	No
Idle Air Control	Counts (steps)	10 - 30
Park/Neutral	P/N and RDL	P-N
Vehicle Speed	MPH, KPH	0 (Zero)
Torque Conv. Cl	On/Off	Off
Battery Voltage	Volts	13.8 Volts (varies)
Exhaust Recirc	%	0 (Zero)
A/C Request	Yes/No	No
Brake Switch	No/Yes	No (Yes with brake pedal depressed)
Power Steering	High Press/Normal	Normal
QDM 1	High/Low	Low
QDM 2	High/Low	Low (High if A/C head pressure's high)
QDM 3	High/Low	Low (High with brake pedal depressed)
QDM 4	High/Low	Low
2nd Gear	On/Off	Off
3rd Gear	On/Off	Off
4th Gear	On/Off	Off
PROM I.D.	Numbers	Internal I.D. Only

*A poor ECM ground at the CH mounting bracket to engine, could cause fuel integ and block learn to read around 150, make a careful physical inspection of this critical connection.

NOTE: IF ALL VALUES ARE WITHIN THE RANGE ILLUSTRATED, REFER TO "SYMPTOMS" IN SECTION "B".

1988–90 E-CAR EQUIPPED WITH 3800 (VIN C) ENGINE

DIAGNOSTIC CIRCUIT CHECK
3800 (VIN C) "E" CARLINE (PORT)

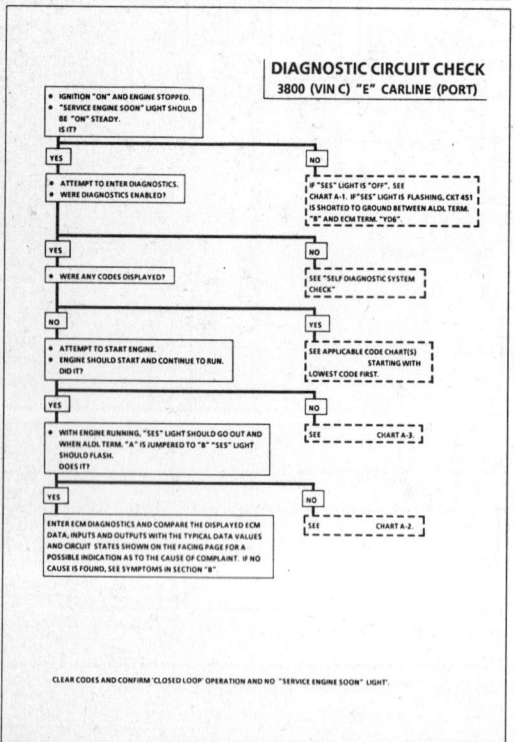

- IGNITION "ON" AND ENGINE STOPPED.
- "SERVICE ENGINE SOON" LIGHT SHOULD BE "ON" STEADY.
IS IT?

→ NO: IF "SES" LIGHT IS "OFF," SEE CHART A-1. IF "SES" LIGHT IS FLASHING, CKT 451 IS SHORTED TO GROUND BETWEEN ALDL TERM. "B" AND ECM TERM. "D6".

YES

- ATTEMPT TO ENTER DIAGNOSTICS.
- WERE DIAGNOSTICS ENABLED?

→ NO: SEE "SELF DIAGNOSTIC SYSTEM CHECK".

YES

- WERE ANY CODES DISPLAYED?

→ YES: SEE APPLICABLE CODE CHART(S) STARTING WITH LOWEST CODE FIRST.

NO

- ATTEMPT TO START ENGINE.
- ENGINE SHOULD START AND CONTINUE TO RUN.
DID IT?

→ NO: SEE CHART A-3.

YES

- WITH ENGINE RUNNING, "SES" LIGHT SHOULD GO OUT AND WHEN ALDL TERM. "A" IS JUMPERED TO "B" "SES" LIGHT SHOULD FLASH. DOES IT?

→ NO: SEE CHART A-2.

YES

ENTER ECM DIAGNOSTICS AND COMPARE THE DISPLAYED ECM DATA, INPUTS AND OUTPUTS WITH THE TYPICAL DATA VALUES AND CIRCUIT STATES ON THE FACING PAGE FOR A POSSIBLE INDICATION AS TO THE CAUSE OF COMPLAINT. IF NO CAUSE IS FOUND, SEE SYMPTOMS IN SECTION "B"

CLEAR CODES AND CONFIRM "CLOSED LOOP" OPERATION AND NO "SERVICE ENGINE SOON" LIGHT.

1988–90 E-CAR EQUIPPED WITH 3800 (VIN C) ENGINE

Additional "Scan" Values
3800 (VIN C)

Idle / Upper Radiator Hose Hot / Closed Throttle / Park or Neutral / "Closed Loop" / Acc. "OFF"

"SCAN" Position	Units Displayed	Typical Data Value
Desired Idle	RPM	Varies
O₂ Cross Counts	Counts	Varies
Air Fuel Ratio	Air Fuel Ratio	14.7 varies
Rich/Lean Flag	Rich/Lean	Constantly changing
Altitude	Yes/No	No (Except in high areas)
Block Learn Cell	Cell #	0
Knock Retard	Degrees	0°
Purge Duty Cycle	%	Varies
IAC Learned	Yes/No	No
Air Fuel Learned	Yes/No	No
A/C Clutch	Yes/No	No (Yes with A/C "ON")
Cruise Engaged	Yes/No	No
Fan 1	ON/OFF	OFF
Fan 2	ON/OFF	OFF

DESIRED IDLE - SCAN TOOL RANGE 0-3175
The idle speed that is commanded by the ECM. The ECM will compensate for various engine loads to keep the engine at the desired rpm.

O₂ CROSS COUNTS - SCAN TOOL RANGE 0-255
The number of times the oxygen sensor voltage crosses over the rich/lean threshold since the last data transmission to the scan tool

AIR FUEL RATIO - SCAN TOOL RANGE 0.00 - 99.99
The reading reflects the commanded value. This should be at or near 14.7. A lower number indicates a richer commanded air/fuel mixture while a higher number indicates a leaner mixture.

RICH/LEAN FLAG - SCAN TOOL STATES RICH/LEAN
Refers to the oxygen sensor voltage corresponding to the intake air/fuel ratio of 14.7 or 450 mV. An oxygen sensor reading above the threshold means the intake mixture is rich, and a sensor reading below 450 mV indicates a lean fuel mixture.

ALTITUDE - SCAN TOOL STATES YES/NO
This is an internal ECM parameter. The ECM compares engine load (from the MAF sensor) with the throttle position. If a greater than expected value for the TPS is read for a given engine load, the ECM assumes it is at high altitude.

BLOCK LEARN CELL - SCAN TOOL RANGE 0-15
Block learn is dependent upon engine speed and either manifold air pressure or mass air flow readings. A plot of rpm vs MAP or MAF is broken into 16 cells. BLM cell indicates which cell is currently active.

KNOCK RETARD - SCAN TOOL RANGE 0-90
This is the amount of degrees of advance pulled out of the spark advance in response to ESC counts.

PURGE DUTY CYCLE - SCAN TOOL RANGE 0-100%
A proportional signal used to control the canister purge function. 0% implies the CCP value is closed while 100% implies that the value is fully open

IAC LEARNED - SCAN TOOL STATES YES/NO
Each time the car is driven, the IAC system learns how many counts is required to produce a given desired idle. After the engine is started and accelerates, then drops to idle, the IAC system learns the proper number of counts for the desired idle.

AIR FUEL LEARNED - SCAN TOOL STATES YES/NO
When the air fuel system is learning (YES) the block learn is responding to the fuel integrator. If the air fuel learn reads (NO) then block learn will not respond to changes in fuel integrator. Normally the fuel system will start learning as soon as the system goes into closed loop.

A/C CLUTCH - SCAN TOOL STATES ON/OFF
A feedback signal to the ECM that lets the ECM know if the A/C clutch is on or off.

CRUISE ENGAGED - SCAN TOOL STATES ON/OFF
The scan tool will read on if the car is operating in cruise control.

FAN 1, FAN 2 - SCAN TOOL STATES ON/OFF
The scan tool will read "ON" for #1 fan when the ECM completes the ground circuit for the #1 fan relay coil. For the #2 fan it will read "ON" when the ECM or the A/C head pressure switch completes the ground circuit for the relay coil

ALL ECM DIAGNOSTIC CODE CHARTS ARE IN THIS SECTION

ECM DIAGNOSTIC CODES

When the ECM diagnostics are activated, the display will show all codes stored in the ECM and BCM. (See Diagnostic Codes Chart below for the code list and the comments relative to each code.) The code comments can be a great aid in helping to diagnose an emission or driveability problem.

Clear all codes before making a repair (ECM and BCM). Some ECM related problems will set a BCM code as well. After repair has been made, again clear all codes that may have been set during testing and repair. Confirm "Closed Loop" operation and make sure that there is no "Service Engine Soon" Light.

Clearing trouble codes is covered in a three step procedure.

See this Section for the wiring diagram, Diagnostic Circuit Check, and complete explanation of each code.

ECM DIAGNOSTIC CODES

CODE	DESCRIPTION	COMMENTS
E013	Open O_2 Sensor Circuit (*Canister Purge)	A
E014	Coolant Sensor Circuit - High Temperature	A – B
E015	Coolant Sensor Circuit - Low Temperature	A – B
E016	System Voltage Out Of Range (*All Solenoids)	A
E021	TPS Circuit Failure – Signal Voltage High (*TCC)	A
E022	TPS Circuit Failure – Signal Voltage Low (*TCC)	A
E023	Manifold Air Temperature - Low Temperature Indicated	A
E024	Vehicle Speed Sensor Circuit Failure (*TCC)	A
E025	Manifold Air Temperature - High Temperature Indicated	A
E026	Quad Driver (QDM) Error	A
E027	Second Gear Switch Circuit	
E028	Third Gear Switch Circuit	
E029	Fourth Gear Switch Circuit Open	
E031	Park/Neutral Switch Circuit	
E034	Mass Air Flow Sensor Frequency Low	A
E038	Brake Switch Circuit	
E039	Torque Convertor Clutch (TCC) Circuit	
E041	CAM Sensor Circuit - C^3I Module to ECM	A
E042	Electronic Spark Timing Or Bypass Circuit Failure	A – D
E043	Electronic Spark Control System Failure	A
E044	Lean Exhaust Indication	A – C
E045	Rich Exhaust Indication	A – C
E046	Power Steering Pressure Switch Circuit Open - (*A/C Clutch and Cruise)	
E047	ECM–BCM Data - (*A/C Clutch and Cruise)	
E048	Misfire	A - D - E
E063	EGR - Flow Problem - Small	A
E064	EGR - Flow Problem - Medium	A
E065	EGR - Flow Problem - Large	A

DIAGNOSTIC CODE COMMENTS

"A"	"Service Engine Soon" Message Displayed
"B"	Forces Cooling Fans "ON"
"C"	Forces Open Loop Operation
"D"	Causes System to Operate in Bypass Spark Mode
"E"	Causes System to Operate in Back up Fuel Mode

* Functions are disengaged while specified malfunctions remain current.

1988–90 E-CAR EQUIPPED WITH 3800 (VIN C) ENGINE

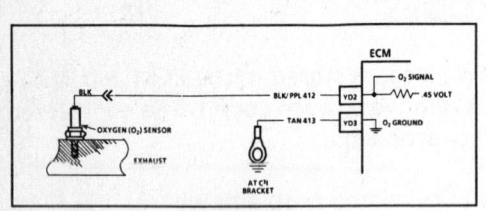

CODE E013
OXYGEN SENSOR CIRCUIT
(OPEN CIRCUIT)
3800 (VIN C) "E" CARLINE (PORT)

Circuit Description:
The ECM supplies a voltage of about .45 volt between terminals "YD2" and "YD3". The O₂ sensor varies the voltage within a range of about 1 volt if the exhaust is rich, down through about .10 volt if exhaust is lean.
The sensor is like an open circuit and produces no voltage when it is below 360°C (600°F). An open oxygen sensor circuit or cold oxygen sensor causes "Open Loop" operation.

Test Description: Numbers below refer to circled numbers on the diagnostic chart.
1. Code E013 will set:
   - No Codes E021 or E022
   - Engine at normal operating temperature
   - Engine run time more than 40 seconds
   - O₂ signal voltage ED07 steady between 0.35 and 0.55 volt
   - Throttle position sensor signal above .55 volt
   - All conditions must be met for about 30 seconds.
   If the conditions for a Code E013 exists, the system will not go to "Closed Loop."
2. This will determine if the sensor or the wiring is the cause of the Code E013.
3. In doing this test use only a high impedance digital volt ohmmeter. This test checks the continuity of CKTs 412 and 413. If CKT 413 is open, the ECM voltage on CKT 412 will be over .6 volt (600 mV).

Diagnostic Aids:
An intermittent may be caused by a poor connection, rubbed through wire insulation, or a wire broken inside the insulation.
Check For:
- Poor Connection or Damaged Harness Inspect harness connectors for backed out terminals, improper mating, broken locks, improperly formed or damaged terminals, poor terminal to wire connection, and damaged harness.
- Intermittent Test If connections and harness check OK, observe ED07 display, O₂ sensor voltage while moving related connectors and wiring harness with warm engine running at part throttle in "Closed Loop." If the failure is induced, the "O₂ sensor voltage" reading will change from its normal fluctuating voltage (above 0.60 volt and below 0.30 volt) to a fixed value around 0.45 volt. This may help to isolate the location of the malfunction.

1988–90 E-CAR EQUIPPED WITH 3800 (VIN C) ENGINE

CODE E013
OXYGEN SENSOR CIRCUIT
(OPEN CIRCUIT)
3800 (VIN C) "E" CARLINE (PORT)

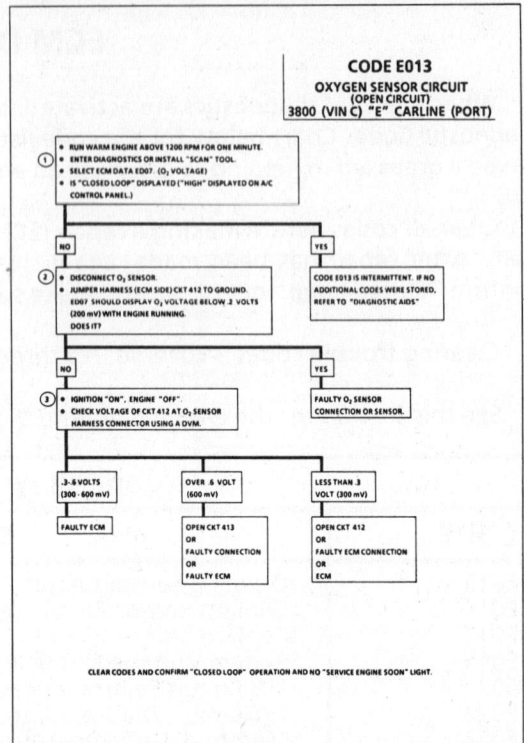

1988–90 E-CAR EQUIPPED WITH 3800 (VIN C) ENGINE

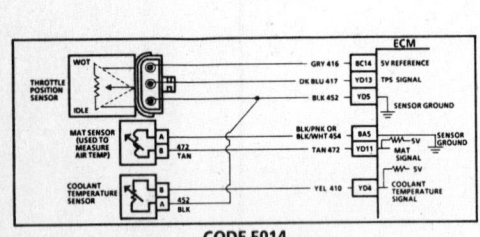

CODE E014
COOLANT TEMPERATURE SENSOR CIRCUIT
(HIGH TEMPERATURE INDICATED)
3800 (VIN C) "E" CARLINE (PORT)

Circuit Description:
The coolant temperature sensor uses a thermistor to control the signal voltage to the ECM. The ECM applies a voltage on CKT 410 to the sensor. When the engine is cold, the sensor (thermistor) resistance is high, therefore, the ECM will see high signal voltage.
As the engine warms, the sensor resistance becomes less, and the voltage drops. At normal engine operating temperature (85°C to 95°C), the voltage will measure about 1.5 to 2.0 volts.

Test Description: Numbers below refer to circled numbers on the diagnostic chart.
1. Code E014 will set if:
   - Engine run time is greater than 10 seconds.
   - Signal voltage indicates a coolant temperature above 130°C (282°F) for 4 seconds.
2. This test will determine if CKT 410 is shorted to ground which will cause the conditions for Code E014.
   If Code E014 is set, the ECM will use 48.5°C, 119°F, for fuel control.

Diagnostic Aids:
On-board diagnostics ED04 displays engine temp in degrees centigrade. After engine is started, the temperature should rise steadily to about 90°C then stabilize when thermostat opens.
An intermittent may be caused by a poor connection, rubbed through wire insulation, or a wire broken inside the insulation.

Check For:
- Poor Connection or Damaged Harness Inspect ECM harness connectors for backed out terminal "YD4," improper mating, broken locks, improperly formed or damaged terminals, poor terminal to wire connection, and damaged harness.
- Intermittent Test If connections and harness check OK, observe ED04, coolant temperature while moving related connectors and wiring harness. If the failure is induced, the "coolant temperature" display will change. This may help to isolate the location of the malfunction.
- Shifted Sensor The 'Temperature To Resistance Value" scale may be used to test the coolant sensor at various temperature levels to evaluate the possibility of a "shifted" (mis-scaled) sensor, which may result in driveability complaints.

1988–90 E-CAR EQUIPPED WITH 3800 (VIN C) ENGINE

CODE E014
COOLANT TEMPERATURE SENSOR CIRCUIT
(HIGH TEMPERATURE INDICATED)
3800 (VIN C) "E" CARLINE (PORT)

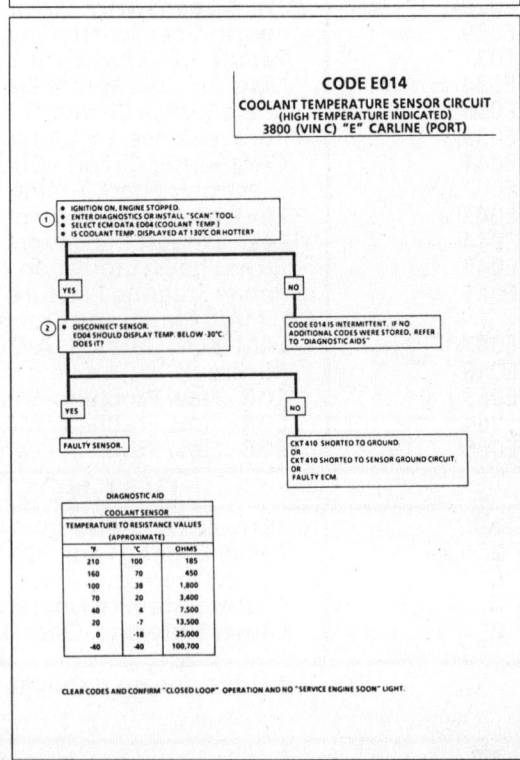

DIAGNOSTIC AID

COOLANT SENSOR		
TEMPERATURE TO RESISTANCE VALUES (APPROXIMATE)		
°F	°C	OHMS
210	100	185
160	70	450
100	38	1,800
70	20	3,400
40	4	7,500
20	-7	13,500
0	-18	25,000
-40	-40	100,700

1988–90 E-CAR EQUIPPED WITH 3800 (VIN C) ENGINE

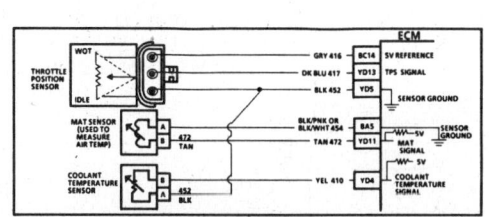

CODE E015
COOLANT TEMPERATURE SENSOR CIRCUIT
(LOW TEMPERATURE INDICATED)
3800 (VIN C) "E" CARLINE (PORT)

Circuit Description:

The coolant temperature sensor uses a thermistor to control the signal voltage to the ECM. The ECM applies a voltage on CKT 410 to the sensor. When the engine is cold, the sensor (thermistor) resistance is high, therefore, the ECM will see high signal voltage.

As the engine warms, the sensor resistance becomes less, and the voltage drops. At normal engine operating temperature (85°C to 95°C) the voltage will measure about 1.5 to 2.0 volts at the ECM.

Test Description: Numbers below refer to circled numbers on the diagnostic chart.

1. Code E015 will set if:
   - Signal voltage indicates a coolant temperature less than -40°C (-40°F) for at least 2 seconds.
2. This test simulates a Code E014. If the ECM recognizes the low signal voltage, (high temperature) and the CTS display reads 130°C, the ECM and wiring are OK.
3. This test will determine if CKT 410 is open. There should be 5 volts present at sensor connector if measured with a DVM.

Note: If Code E015 is set, the ECM will use 48.5°C (119°F) for fuel control.

Diagnostic Aids:

The on-board diagnostics reads engine temperature in degrees centigrade. After engine is started, the temperature should rise steadily to about 90°C then stabilize when thermostat opens.

An intermittent may be caused by a poor connection, rubbed through wire insulation or a wire broken inside the insulation.

Check For:
- **Poor Connection or Damaged Harness** Inspect ECM harness connectors for backed out terminal "YD4", improper mating, broken locks, improperly formed or damaged terminals, poor terminal to wire connection and damaged harness.
- **Intermittent Test** If connections and harness check OK, observe ED04, coolant temperature while moving related connectors and wiring harness. If the failure is induced, the display will change. This may help to isolate the location of the malfunction.
- **Shifted Sensor** The "Temperature To Resistance Value" scale may be used to test the coolant sensor at various temperature levels to evaluate the possibility of a "shifted" (mis-scaled) sensor which may result in driveability complaints.

A faulty connection, or an open in CKTs 410 or 452 will result in a Code E015.

If Code E021 is also set, check CKT 452 for faulty wiring or connections. Check terminals at sensor for good contact. Refer to "Intermittents" in Section "B".

1988–90 E-CAR EQUIPPED WITH 3800 (VIN C) ENGINE

CODE E015
COOLANT TEMPERATURE SENSOR CIRCUIT
(LOW TEMPERATURE INDICATED)
3800 (VIN C) "E" CARLINE (PORT)

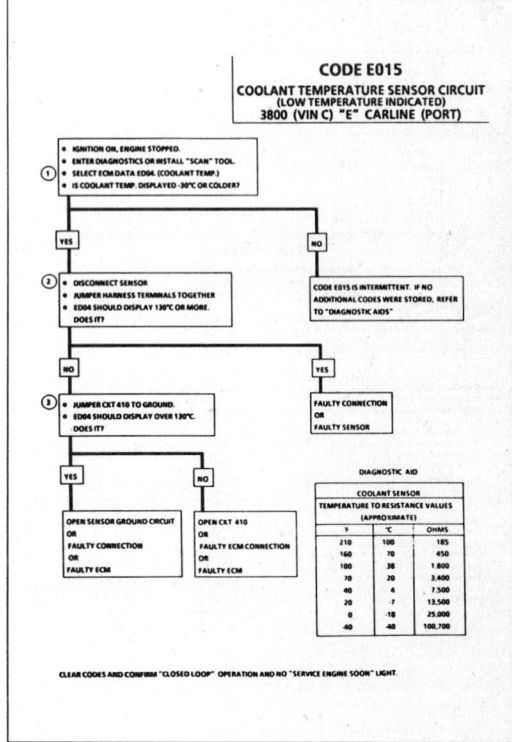

1988–90 E-CAR EQUIPPED WITH 3800 (VIN C) ENGINE

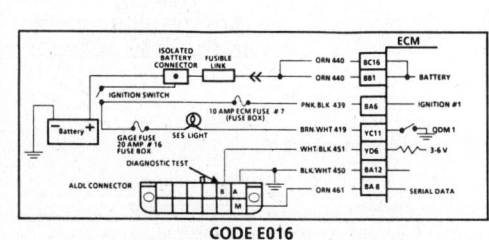

CODE E016
BATTERY VOLTAGE TOO HIGH
3800 (VIN C) "E" CARLINE (PORT)

Circuit Description:

The ECM monitors battery or system voltage on CKT 480 to terminals "BB1" and "BC16". If the ECM detects voltage above 16 volts, it will turn the "SES" light "ON" and set Code E016 in memory.

Test Description: Numbers below refer to circled numbers on the diagnostic chart.

1. Test generator output to determine proper operation of the voltage regulator. Run engine at moderate speed and measure voltage across the battery. If over 16 volts, repair generator.

Diagnostic Aids:

An intermittent may be caused by a poor connection, rubbed through insulation, a wire broken inside the insulation or poor ECM grounds.

Check For:
- **Poor Connection or Damaged Harness** Inspect ECM harness connectors for backed out terminal "BC16" or "BB1", improper mating, broken locks, improperly formed or damaged terminals, poor terminal to wire connection and damaged harness.
- **Intermittent Test** If connections and harness checks OK, observe ED10 battery voltage while moving related connectors. If the failure is induced, the battery voltage will abruptly change. This may help to isolate the location of the malfunction. An engine stall while manipulating the harness indicates that the ECM has lost voltage at terminal "BC16" or "BB1". Check for loose connectors in CKT 480.

Note: Charging battery with a battery charger and starting the engine may set a Code E016.

1988–90 E-CAR EQUIPPED WITH 3800 (VIN C) ENGINE

CODE E016
BATTERY VOLTAGE TOO HIGH
3800 (VIN C) "E" CARLINE (PORT)

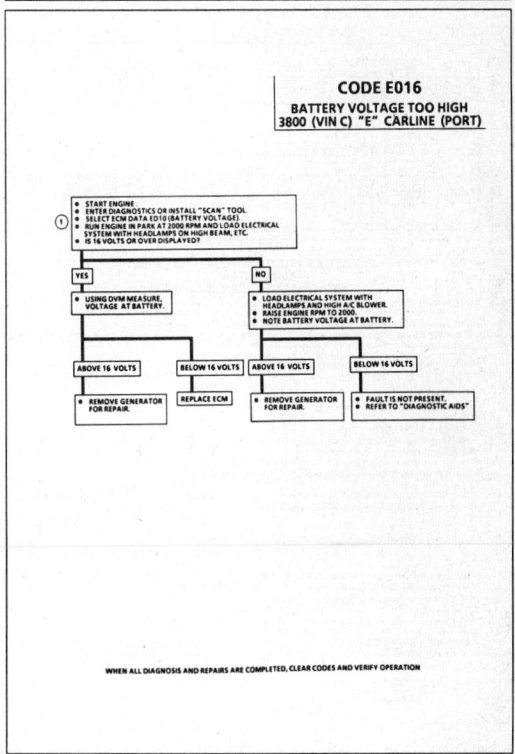

1988–90 E-CAR EQUIPPED WITH 3800 (VIN C) ENGINE

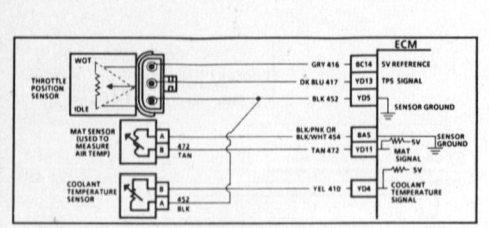

CODE E021
THROTTLE POSITION SENSOR (TPS) CIRCUIT
(SIGNAL VOLTAGE HIGH)
3800 (VIN C) "E" CARLINE (PORT)

Circuit Description:

The Throttle Position Sensor (TPS) provides a voltage signal that changes relative to throttle blade angle. Signal voltage will vary from about .4 at idle to about 5 volts at wide open throttle.

The TPS signal is one of the most important inputs used by the ECM for fuel control and for most of the ECM control outputs.

Test Description: Numbers below refer to circled numbers on the diagnostic chart.

1. Code E021 will set if:
   - Engine is running and air flow is less than 15 gm/sec.
   - TPS signal voltage is greater than 2.5 volts.
   - Code E033 or E034 not present at first start up.
   - All conditions met for 5 seconds.

   With closed throttle, ignition "ON," or at idle, voltage at "YD13" should be .34-.46 volt. If not, check adjustment.

2. When the TPS sensor is disconnected, the TPS voltage will go low and a Code E022 will set. Therefore, the ECM and wiring are OK.

3. Probing CKT 452 with a test light checks the sensor ground CKT. A faulty sensor ground circuit will cause a Code E021.

Diagnostic Aids:

The on board diagnostics ED01 reads throttle position in volts. With closed throttle, ignition "ON" or at idle, voltage should be .34-.46 volt.

An open in CKT 452 will result in a Code E021. Refer to "Intermittents" in Section "B".
Check For:
- **Poor Connection or Damaged Harness** Inspect ECM harness connectors for backed out terminal "YD13," improper mating, broken locks, improperly formed or damaged terminals, poor terminal to wire connection, and damaged harness.
- **Intermittent Test** If connections and harness check OK, observe ED01, TPS voltage while moving related connectors and wiring harness. If the failure is induced, the display will change. This may help to isolate the location of the malfunction.
- **TPS Scaling** Observe ED01 TPS voltage while depressing accelerator pedal with engine stopped and ignition "ON." Display should vary from closed throttle TPS voltage when throttle was closed, to over 4.0 volts (4000 mV) when throttle is held at wide open throttle position.

1988–90 E-CAR EQUIPPED WITH 3800 (VIN C) ENGINE

CODE E021
THROTTLE POSITION SENSOR (TPS) CIRCUIT
(SIGNAL VOLTAGE HIGH)
3800 (VIN C) "E" CARLINE (PORT)

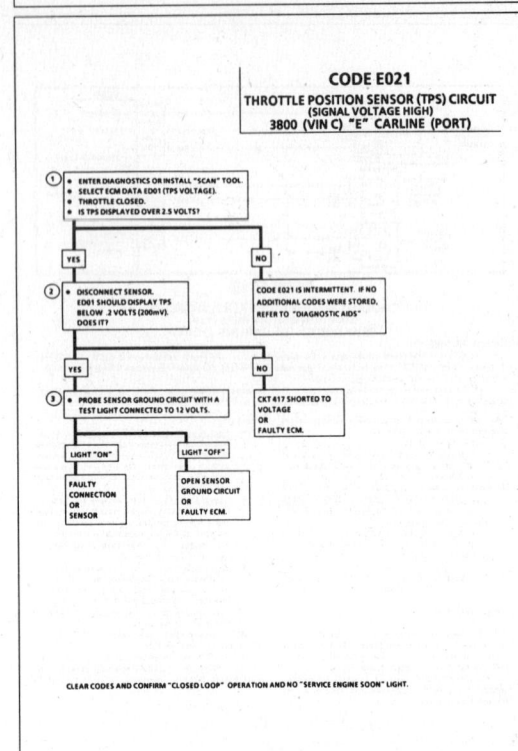

CLEAR CODES AND CONFIRM "CLOSED LOOP" OPERATION AND NO "SERVICE ENGINE SOON" LIGHT.

1988–90 E-CAR EQUIPPED WITH 3800 (VIN C) ENGINE

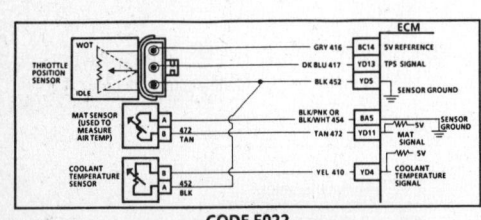

CODE E022
THROTTLE POSITION SENSOR (TPS) CIRCUIT
(SIGNAL VOLTAGE LOW)
3800 (VIN C) "E" CARLINE (PORT)

Circuit Description:

The Throttle Position Sensor (TPS) provides a voltage signal that changes relative to throttle blade angle. Signal voltage will vary from about .4 at idle to about 5 volts at wide open throttle.

The TPS signal is one of the most important inputs used by the ECM for fuel control and for most of the ECM control outputs.

Test Description: Numbers below refer to circled numbers on the diagnostic chart.

1. Code E022 will set if:
   - Engine is running.
   - TPS signal voltage is less than 2 volt for 4 seconds.

2. Simulates Code E021 (high voltage) If ECM recognizes the high signal voltage, the ECM and wiring are OK.

3. With closed throttle, ignition "ON," or at idle, voltage at "YD13" should be .34-.46 volt. If not, check adjustment.

4. Simulates a high signal voltage. Checks CKT 417 for an open.

Diagnostic Aids:

The on-board diagnostics ED01 reads throttle position in volts. Voltage should increase at a steady rate as throttle is moved toward WOT.

If CKT 417 is open or grounded when the vehicle engine is started, a high idle may result. ED01 or a "Scan" tool may read the following with CKT 417 open or grounded.

TPS = .04 volt or less
RPM > 1000 in park
Desired idle = 725
IAC counts = 40
SES light "ON" Code 22 set.

If the intermittent is repaired without cycling the key "OFF," ED01 or the "Scan" may read the following.

TPS = .42 volt
RPM > 1500 in park
Desired idle = 725
IAC counts > 55
SES light "OFF"

An open or short to ground in CKTs 416 or 417 will result in a Code E022.
Check For:
- **Poor Connection or Damaged Harness** Inspect ECM harness connectors for backed out terminal "YD13," improper mating, broken locks, improperly formed or damaged terminals, poor terminal to wire connection, and damaged harness.
- **Intermittent Test** If connections and harness check OK, observe ED01 TPS voltage while moving related connectors and wiring harness. If the failure is induced, the display will change. This may help to isolate the location of the malfunction.
- **TPS Scaling** Observe ED01 TPS voltage while depressing accelerator pedal with engine stopped and ignition "ON." Display should vary from closed throttle TPS voltage when throttle was closed, to over 4.0 volts (4000 mV) when throttle is held at wide open throttle position.

1988–90 E-CAR EQUIPPED WITH 3800 (VIN C) ENGINE

CODE E022
THROTTLE POSITION SENSOR (TPS) CIRCUIT
(SIGNAL VOLTAGE LOW)
3800 (VIN C) "E" CARLINE (PORT)

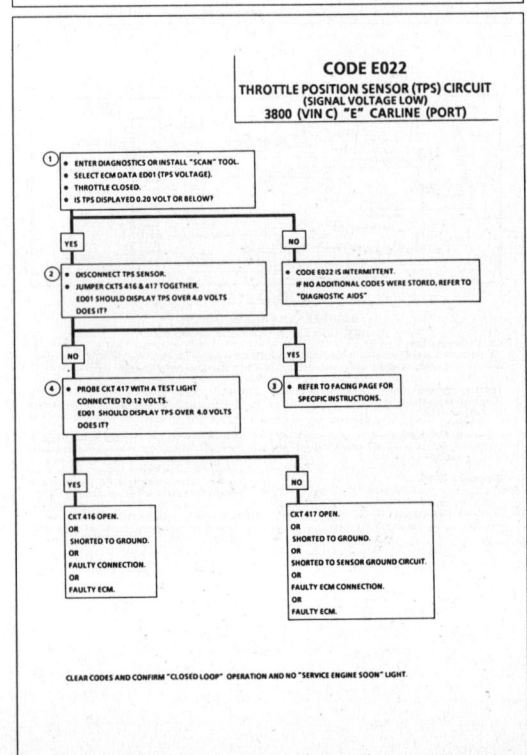

CLEAR CODES AND CONFIRM "CLOSED LOOP" OPERATION AND NO "SERVICE ENGINE SOON" LIGHT.

1988–90 E-CAR EQUIPPED WITH 3800 (VIN C) ENGINE

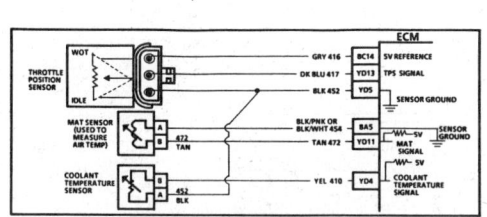

CODE E023
MANIFOLD AIR TEMPERATURE (MAT) SENSOR CIRCUIT
(LOW TEMPERATURE INDICATED)
3800 (VIN C) "E" CARLINE (PORT)

Circuit Description:
The Manifold Air Temperature (MAT) sensor is a thermistor. The ECM applies a voltage (about 5 volts) on CKT 472 to the sensor. When the air is cold, the sensor (thermistor) resistance is high, therefore, the ECM will measure a high signal voltage on "YD11". If the air is warm, the sensor resistance is low, therefore, the ECM will measure a low voltage on "YD11".

Test Description: Numbers below refer to circled numbers on the diagnostic chart.
Code E023 will set if:
- A signal voltage indicates a manifold air temperature below -40°C (-40°F) for 4 seconds.
Due to the conditions necessary to set a Code E023, the "Service Engine Soon" light will only stay "ON" while the fault is present.
1. The on-board diagnostics display (ED23) may not be used to diagnose this fault. A Code E023 will set, due to an open sensor, wire, or connection. This test determines if the wiring and ECM are OK.
2. If the resistance is greater than 25,000 ohms, inspect the air cleaner assembly for the presence of ice. If OK, replace the sensor.

Diagnostic Aids:

An intermittent may be caused by a poor connection, rubbed through wire insulation or a wire broken inside the insulation.
Check For:
- Poor Connection or Damaged Harness Inspect ECM harness connectors for backed out terminals "BA5", "YD11", improper mating, broken locks, improperly formed or damaged terminals, poor terminal to wire connection, and damaged harness.
- Intermittent Test If connections and harness check OK, observe ED23 MAT display while moving related connectors and wiring harness with warm engine running. If the failure is induced, the MAT display may change to a -40° temperature reading. This may help to isolate the location of the malfunction.

1988–90 E-CAR EQUIPPED WITH 3800 (VIN C) ENGINE

CODE E023
MANIFOLD AIR TEMPERATURE (MAT) SENSOR CIRCUIT
(LOW TEMPERATURE INDICATED)
3800 (VIN C) "E" CARLINE (PORT)

NOTE: ON BOARD DIAGNOSTICS MAY NOT BE USED TO DIAGNOSE THIS FAULT.

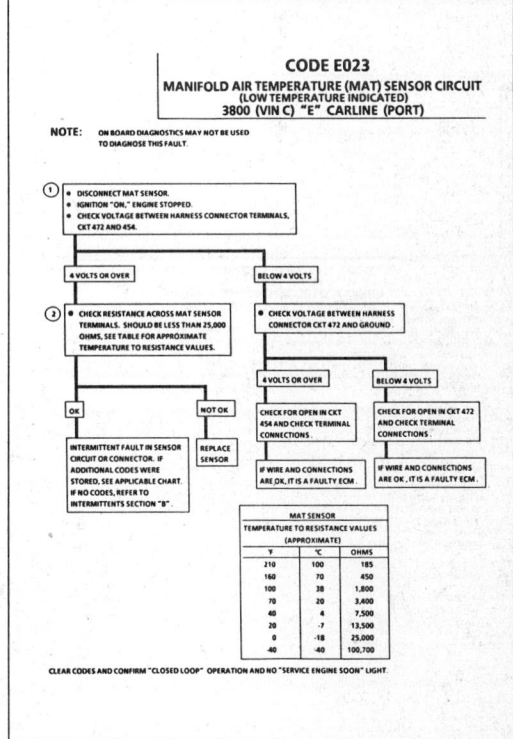

MAT SENSOR		
TEMPERATURE TO RESISTANCE VALUES		
(APPROXIMATE)		
°F	°C	OHMS
210	100	185
160	70	450
100	38	1,800
70	20	3,400
40	4	7,500
20	-7	13,500
0	-18	25,000
-40	-40	100,700

CLEAR CODES AND CONFIRM "CLOSED LOOP" OPERATION AND NO "SERVICE ENGINE SOON" LIGHT.

1988–90 E-CAR EQUIPPED WITH 3800 (VIN C) ENGINE

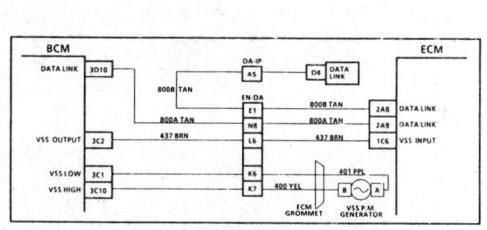

CODE E024
VEHICLE SPEED SENSOR (VSS) CIRCUIT
3800 (VIN C) "E" SERIES (PORT)

Circuit Description:
The VSS system incorporates three major components, the vehicle speed generator, the BCM, and the ECM. The vehicle speed sensor is a permanent magnet generator assembly attached to the transaxle. As the vehicle moves, the generator creates a "sine wave" electrical pulse which is monitored by the BCM. In the BCM, this is amplified and "cleaned up" via the same process used in a buffer. Part of this "cleaning up" involves changing the "sine wave" signal to a "square wave" or "ON/OFF" type signal. By determining the length of time between the "ON" and "OFF" portions of this signal, the BCM can interpret the vehicle speed. It then transmits this data via CKT 437 to the ECM.
Because of the circuitry involved, it is highly unlikely that a Code E024 alone will be caused by a faulty VSS. However, a combination of Codes E024 and B124 could indicate this. In the case of a Code B124 accompanying a Code E024, see the appropriate charts in the BCM section first.
To set a Code E024, the following conditions must exist:
- Engine running between 1500 and 4000 rpm.
- Codes E033 or E034 not present.
- LV8 reading (Diagnostics Data ED23) Between 40 and 150 counts.
- Vehicle speed less than 3 mph.
- Gear selector not in park or neutral.
- Above conditions for 40 seconds.

Test Description: Numbers below refer to circled numbers on the diagnostic chart.
1. If a Code B124 is present also, refer to the accompanying charts for diagnosis of B124 before attempting to diagnose E024.
2. If ECM data ED12 (Vehicle Speed) displays "0" mph, the ECM is not receiving a VSS input from the BCM.
3. A "0" mph display at this point, indicates a fault which should have set a Code B124. Refer to that chart for the remainder of diagnosis. A display greater than "0" indicates a faulty in CKT 437, the connections at the ECM or BCM, or a faulty ECM.

1988–90 E-CAR EQUIPPED WITH 3800 (VIN C) ENGINE

CODE E024
VEHICLE SPEED SENSOR (VSS) CIRCUIT
3800 (VIN C) "E" SERIES (PORT)

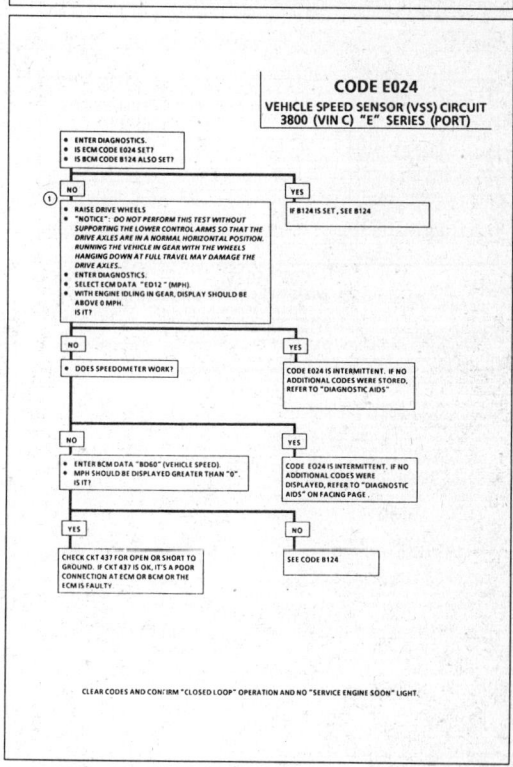

CLEAR CODES AND CONFIRM "CLOSED LOOP" OPERATION AND NO "SERVICE ENGINE SOON" LIGHT.

1988–90 E-CAR EQUIPPED WITH 3800 (VIN C) ENGINE

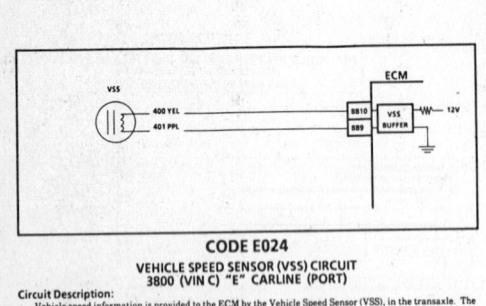

CODE E024
VEHICLE SPEED SENSOR (VSS) CIRCUIT
3800 (VIN C) "E" CARLINE (PORT)

Circuit Description:
Vehicle speed information is provided to the ECM by the Vehicle Speed Sensor (VSS), in the transaxle. The A/C voltage level and the number of pulses increases with vehicle speed. The ECM converts the pulsing voltage to mph, and the mph can be displayed with a "Scan" tool.
The function of the VSS buffer, used in past model years, has been incorporated into the ECM. The ECM supplies the necessary signal for the instrument panel (4004 pulses per mile) for operating the speedometer and the odometer.

Test Description: Numbers below refer to circled numbers on the diagnostic chart.
1. Code E024 will set if vehicle speed signal equals less than 3 mph when:
   - Engine is running
   - No Code E029 or E031
   - When the vehicle is in 4th gear
   - All conditions met for 40 seconds
   The PM generator only produces a signal if drive wheels are turning greater then 3 mph.
2. Before replacing the ECM, check the Mem-Cal for correct application.

Diagnostic Aids:
The "On-Board Diagnostic Display" should indicate a vehicle speed whenever the drive wheels are turning greater than 3 mph.
Check CKT 400 and 401 for proper connections to be sure they are clean and tight and the harness is routed correctly. Refer to "Intermittents" in Section "B".

1988–90 E-CAR EQUIPPED WITH 3800 (VIN C) ENGINE

CODE E024
VEHICLE SPEED SENSOR (VSS) CIRCUIT
3800 (VIN C) "E" CARLINE (PORT)

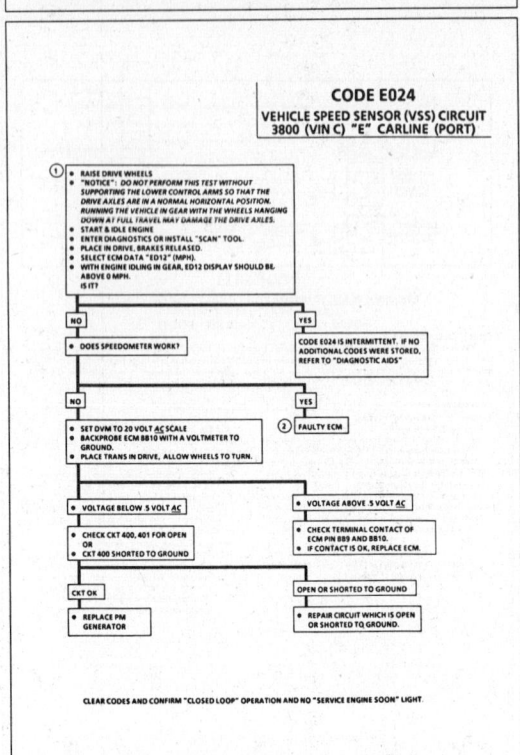

CLEAR CODES AND CONFIRM "CLOSED LOOP" OPERATION AND NO "SERVICE ENGINE SOON" LIGHT.

1988–90 E-CAR EQUIPPED WITH 3800 (VIN C) ENGINE

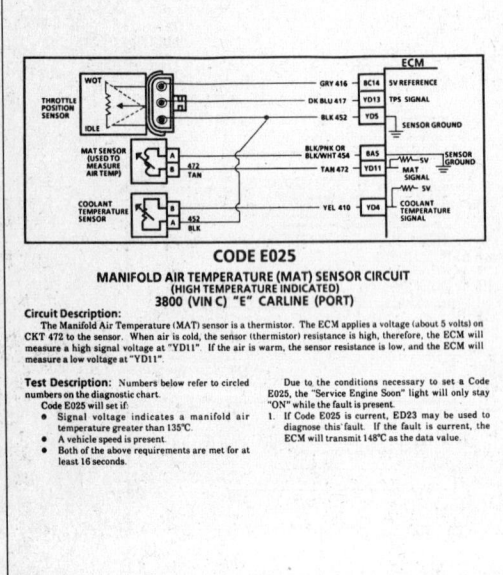

CODE E025
MANIFOLD AIR TEMPERATURE (MAT) SENSOR CIRCUIT
(HIGH TEMPERATURE INDICATED)
3800 (VIN C) "E" CARLINE (PORT)

Circuit Description:
The Manifold Air Temperature (MAT) sensor is a thermistor. The ECM applies a voltage (about 5 volts) on CKT 472 to the sensor. When air is cold, the sensor (thermistor) resistance is high, therefore, the ECM will measure a high signal voltage at "YD11". If the air is warm, the sensor resistance is low, and the ECM will measure a low voltage at "YD11".

Test Description: Numbers below refer to circled numbers on the diagnostic chart.
Code E025 will set if:
- Signal voltage indicates a manifold air temperature greater than 135°C.
- A vehicle speed is present.
- Both of the above requirements are met for at least 16 seconds.

Due to the conditions necessary to set a Code E025, the "Service Engine Soon" light will only stay "ON" while the fault is present.
1. If Code E025 is current, ED23 may be used to diagnose this fault. If the fault is current, the ECM will transmit 148°C as the data value.

1988–90 E-CAR EQUIPPED WITH 3800 (VIN C) ENGINE

CODE E025
MANIFOLD AIR TEMPERATURE (MAT) SENSOR CIRCUIT
(HIGH TEMPERATURE INDICATED)
3800 (VIN C) "E" CARLINE (PORT)

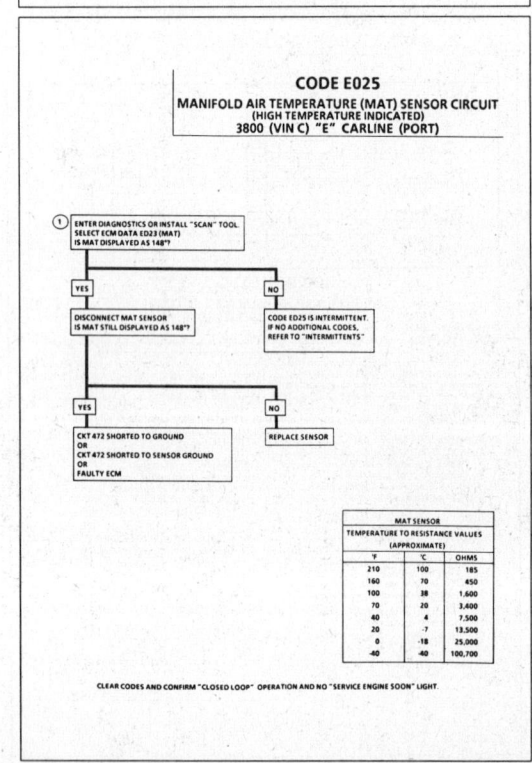

MAT SENSOR		
TEMPERATURE TO RESISTANCE VALUES (APPROXIMATE)		
°F	°C	OHMS
210	100	185
160	70	450
100	38	1,600
70	20	3,400
40	4	7,500
20	-7	13,500
0	-18	25,000
-40	-40	100,700

CLEAR CODES AND CONFIRM "CLOSED LOOP" OPERATION AND NO "SERVICE ENGINE SOON" LIGHT.

1988–90 E-CAR EQUIPPED WITH 3800 (VIN C) ENGINE

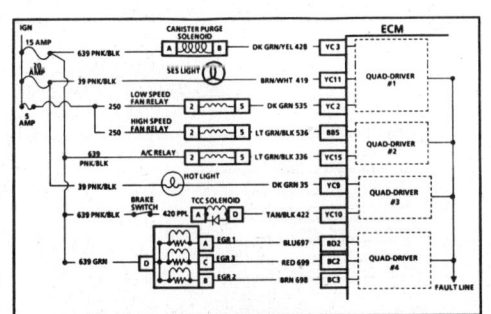

CODE E026
(Page 1 of 3)
QUAD-DRIVER (QDM) CIRCUIT
3800 (VIN C) "E" CARLINE (PORT)

Circuit Description:
The ECM is used to control several components such as those illustrated above. The ECM controls these devices through the use of a Quad-Driver Module (QDM). When the ECM is commanding a component "ON," the voltage potential of the output circuit will be "low" (near 0 volt). When the ECM is commanding the output circuit to a component "OFF," the voltage potential of the circuit will be "high" (near battery voltage). The primary function of the QDM is to supply the ground for the component being controlled. Each QDM has a fault line which is monitored by the ECM. The fault line signal is available on the data stream for "Scan" tool test equipment.

The ECM will compare the voltage at the QDM based on accepted values of the fault line. If the QDM fault detection circuit senses a voltage other than the accepted value, the fault line will go from a "low" signal on the data stream to a "high" signal and a Code E026 will set if applicable.

NOTICE: Some QDM circuits will switch from "low" to "high" normally. Examples: QDM 2 - If A/C pressure switch closes and turns "ON" high speed coolant fan QDM 3 - If the brake is depressed. These conditions are normal and no Code E026 is set. These are accepted conditions. A fault on QDM 2 will not set a Code E026, diagnosis may be done by making sure the A/C is turned "OFF."

NOTICE: Some "Scan" tools will cause a false Code E026 to be set if the engine is running and the service brake is depressed for over 30 seconds.

Test Description: Numbers below refer to circled numbers on the diagnostic chart.
1. The ECM does not know which controlled circuit caused the Code E026 so this chart will go through each of the circuits to determine which is at fault. This test checks the "Service Engine Soon" light driver and the "Service Engine Soon" light circuit.

2. QDM Symptoms:
   - TCC - Inoperative - Code E039
   - EGR - Inoperative - Codes E063, E064, E065
   - Coolant fan on low speed all the time or won't come "ON" at all
   - Poor driveability due to 100% canister purge

1988–90 E-CAR EQUIPPED WITH 3800 (VIN C) ENGINE

CODE E026
(Page 1 of 3)
QUAD-DRIVER (QDM) CIRCUIT
3800 (VIN C) "E" CARLINE (PORT)

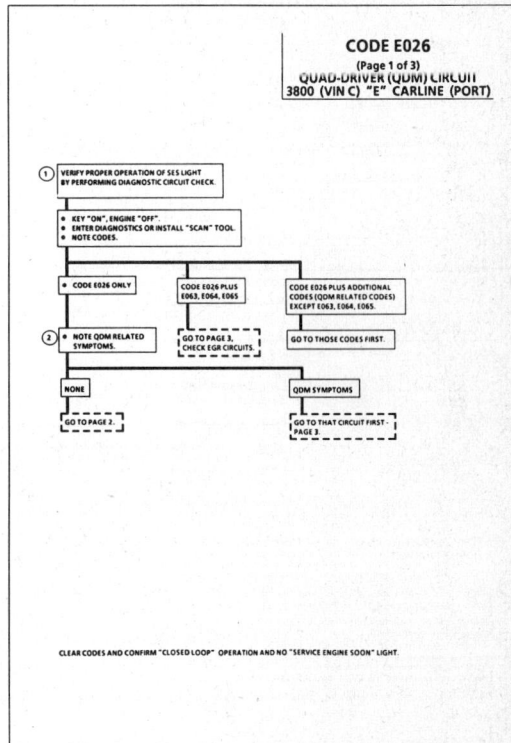

1988–90 E-CAR EQUIPPED WITH 3800 (VIN C) ENGINE

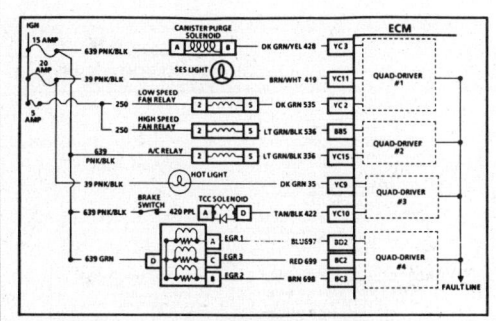

CODE E026
(Page 2 of 3)
QUAD-DRIVER (QDM) CIRCUIT
3800 (VIN C) "E" CARLINE (PORT)

Circuit Description:
The ECM is used to control several components such as those illustrated above. The ECM controls these devices through the use of a Quad-Driver Module (QDM). When the ECM is commanding a component "ON," the voltage potential of the output circuit will be "low" (near 0 volt). When the ECM is commanding the output circuit to a component "OFF," the voltage potential of the circuit will be "high" (near battery voltage). The primary function of the QDM is to supply the ground for the component being controlled. Each QDM has a fault line which is monitored by the ECM. The fault line signal is available on the data stream for "Scan" tool test equipment.

The ECM will compare the voltage at the QDM based on accepted values of the fault line. If the QDM fault detection circuit senses a voltage other than the accepted value, the fault line will go from a "low" signal on the data stream to a "high" signal and a Code E026 will set if applicable.

NOTICE: Some QDM circuits will switch from "low" to "high" normally. Examples: QDM 2 - If A/C pressure switch closes and turns "ON" high speed coolant fan QDM 3 - If the brake is depressed. These conditions are normal and no Code E026 is set. These are accepted conditions. A fault on QDM 2 will not set a Code E026, diagnosis may be done by making sure the A/C is turned "OFF."

NOTICE: Some "Scan" tools will cause a Code E026 to be set if, the engine is running and the service brake is depressed for over 30 seconds.

Test Description: Numbers below refer to circled numbers on the diagnostic chart.
3. This test will determine which circuit is out of specifications. All circuits EXCEPT "YC11," the "SES" light and "YC9", the "hot light" should be B+ when key is "ON", engine not running. The diagnostic "test" terminal is not grounded.

Diagnostic Aids:
An intermittent may be caused by a poor connection, rubbed through wire insulation or a wire broken inside the insulation. Check for:
- Poor Connection or Damaged Harness - Inspect ECM harness connectors for backed out terminals, improper mating, broken locks, improperly formed or damaged terminals, poor terminal to wire connection, and damaged harness.

1988–90 E-CAR EQUIPPED WITH 3800 (VIN C) ENGINE

CODE E026
(Page 2 of 3)
QUAD-DRIVER (QDM) CIRCUIT
3800 (VIN C) "E" CARLINE (PORT)

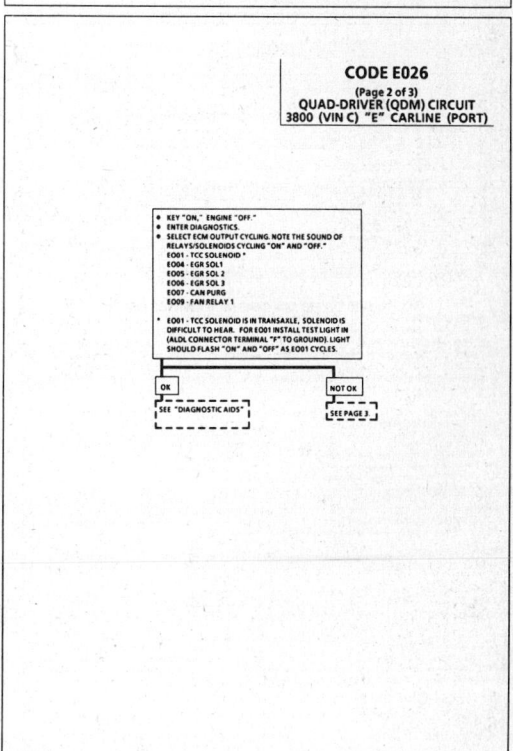

1988–90 E-CAR EQUIPPED WITH 3800 (VIN C) ENGINE

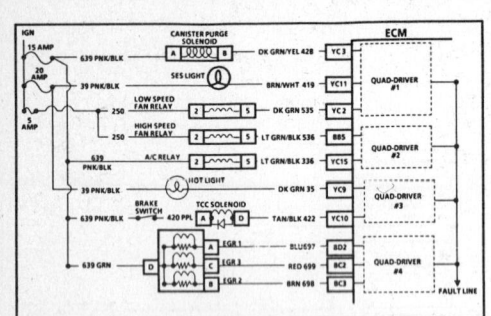

CODE E026
(Page 3 of 3)
QUAD-DRIVER (QDM) CIRCUIT
3800 (VIN C) "E" CARLINE (PORT)

Circuit Description:

The ECM is used to control several components such as those illustrated above. The ECM controls these devices through the use of a Quad-Driver Module (QDM). When the ECM is commanding a component "ON," the voltage potential of the output circuit will be "low" (near 0 volt). When the ECM is commanding the output circuit to a component "OFF," the voltage potential of the circuit will be "high" (near battery voltage). The primary function of the QDM is to supply the ground for the component being controlled. Each QDM has a fault line which is monitored by the ECM. The fault line signal is available on the data stream for "Scan" tool test equipment.

The ECM will compare the voltage at the QDM based on accepted values of the fault line. If the QDM fault detection circuit senses a voltage other than the accepted value, the fault line will go from a "low" signal on the data stream to a "high" signal and a Code E026 will set if applicable.

NOTICE: Some QDM circuits will switch from "low" to "high" normally. Examples: QDM 2 - If A/C pressure switch closes and turns "ON" high speed coolant fan QDM 3 - If the brake is depressed. These conditions are normal and no Code E026 is set. These are accepted conditions. A fault on QDM 2 will not set a Code E026, diagnosis may be done by making sure the A/C is turned "OFF."

NOTICE: Some "Scan" tools will cause a false Code E026 to be set if, the engine is running and the service brake is depressed for over 30 seconds.

Test Description: Numbers below refer to circled numbers on the diagnostic chart.
4. This test will determine if the problem is the circuit or the component. As the factory installed ECM is protected with an internal fuse, it is highly unlikely that the ECM needs to be replaced.

1988–90 E-CAR EQUIPPED WITH 3800 (VIN C) ENGINE

CODE E026
(Page 3 of 3)
QUAD-DRIVER (QDM) CIRCUIT
3800 (VIN C) "E" CARLINE (PORT)

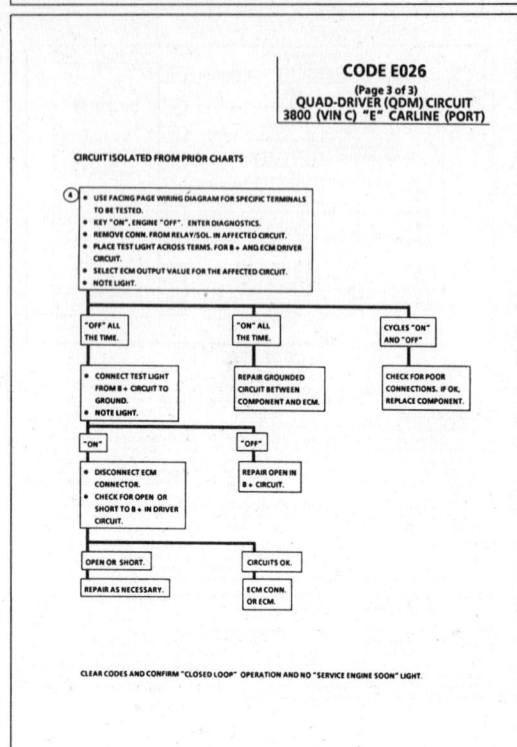

1988–90 E-CAR EQUIPPED WITH 3800 (VIN C) ENGINE

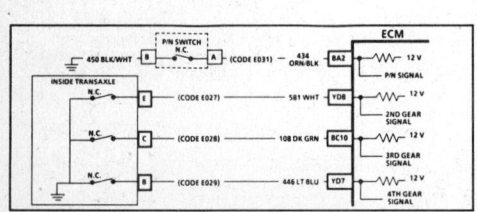

CODES E027, E028, E029
GEAR SWITCH DIAGNOSIS
3800 (VIN C) "E" CARLINE (PORT)

Circuit Description:

The gear switches are located inside the transaxle. They are pressure operated switches, normally closed. The ECM supplies 12 volts through each selected circuit to the switch. In any condition other than the specific gear application, the signal line monitors low voltage, low potential. As road speed increases, hydraulic pressure applies the specific gear clutches, and the gear switch opens. At this time, the ECM monitors a high 12 volts potential and interprets this to indicate that gear is applied. The ECM uses the gear signals to control fuel delivery (and TCC).

Test Description: Numbers below refer to circled numbers on the diagnostic chart.
1. Code E027 will set if:
- No Code 29 present.
- CKT 581 indicates ground or closed switch when vehicle is in 4th gear operation.
- CKT 581 indicates an open (drive) when the engine is first started.
- CKT 446 is open when the engine is first started.
Code E028 will set if:
- CKT 108 indicates ground or closed switch when vehicle is in 4th gear operation.
- CKT 108 indicates an open (drive) when the engine is first started.
- CKT 446 is open when the engine is first started.
Code E029 will set if:
- CKT 446 indicates ground or closed switch when vehicle is in 4th gear operation.
- CKT 446 indicates an open (drive) when the engine is first started.

2. Checks to see if circuit is grounded through the switch.
3. Checks for a good, properly operating switch and checks circuit within transaxle for an improper ground

Diagnostic Aids:

An intermittent may be caused by a poor connection, mis-routed harness, rubbed through wire insulation, or a wire broken inside the insulation.
Check For:
- Poor Connection At ECM Pins: Inspect harness connectors for backed out terminals, improper mating, broken locks, improperly formed or damaged terminals, and poor terminal to wire connection.
- Mis-routed Harness: Inspect wiring harness to insure that it is not too close to high voltage wires, such as spark plug leads.
- Damaged Harness: Inspect harness for damage. If harness appears OK, observe E179, E180 & E182 while moving related connectors and wiring harness. A change in display would indicate the intermittent fault location.

Note: Step 1 must use a DVM. A test light will not light due to the very low current being supplied by the ECM.

1988–90 E-CAR EQUIPPED WITH 3800 (VIN C) ENGINE

CODES E027, E028, E029
GEAR SWITCH DIAGNOSIS
3800 (VIN C) "E" CARLINE (PORT)

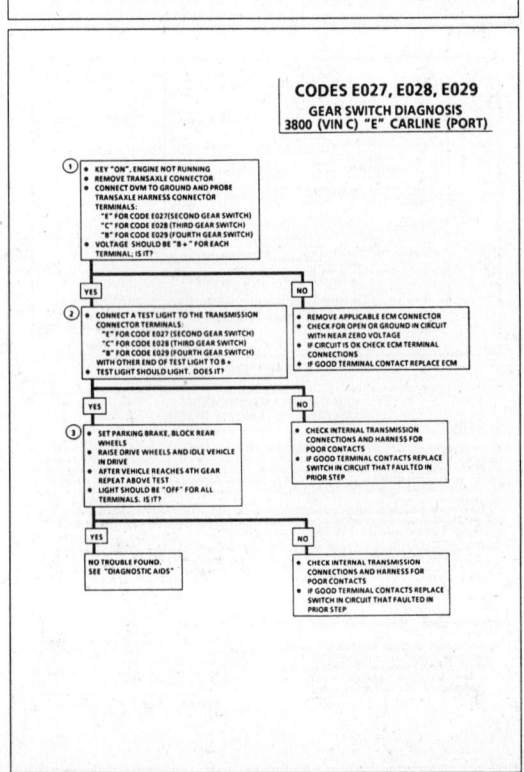

1988–90 E-CAR EQUIPPED WITH 3800 (VIN C) ENGINE

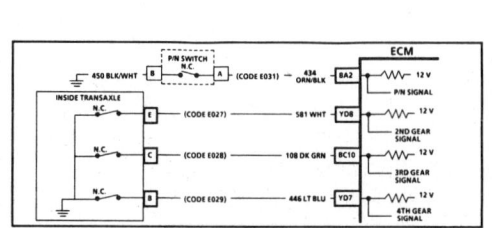

CODE E031
PARK/NEUTRAL SWITCH CIRCUIT
3800 (VIN C) "E" CARLINE (PORT)

Circuit Description:
The park/neutral switch contacts are a part of the neutral start switch and are closed to ground in park or neutral and open in drive ranges.
The ECM supplies ignition voltage through a current limiting resistor to CKT 434 and senses a closed switch when the voltage on CKT 434 drops to less than one volt.
The ECM uses the P/N signal as one of the inputs to control:
- Idle Speed (IAC)
- Exhaust Gas Recirculation (EGR)
- Spark timing
- Fuel control

Code E031 will set:
- If CKT 434 indicates P/N (grounded) while in drive range, 4th gear TCC engaged, EGR would be inoperative, resulting in possible detonation.
- If CKT 434 indicates drive (open) at start up, a dip in the idle may exist when the gear selector is moved into drive range.
- If transaxle 4th gear switch has an intermittent open, the ECM thinks vehicle is in 4th gear and may set a Code E031.

Test Description: Numbers below refer to circled numbers on the diagnostic chart.
1. Checks for a closed switch to ground in park position.
2. Checks for an open switch in drive range.
3. Be sure on-board diagnostics E174 indicates drive ("HI"), even while wiggling shifter, to test for an intermittent or misadjusted switch in drive or overdrive range.

1988–90 E-CAR EQUIPPED WITH 3800 (VIN C) ENGINE

CODE E031
PARK/NEUTRAL SWITCH CIRCUIT
3800 (VIN C) "E" CARLINE (PORT)

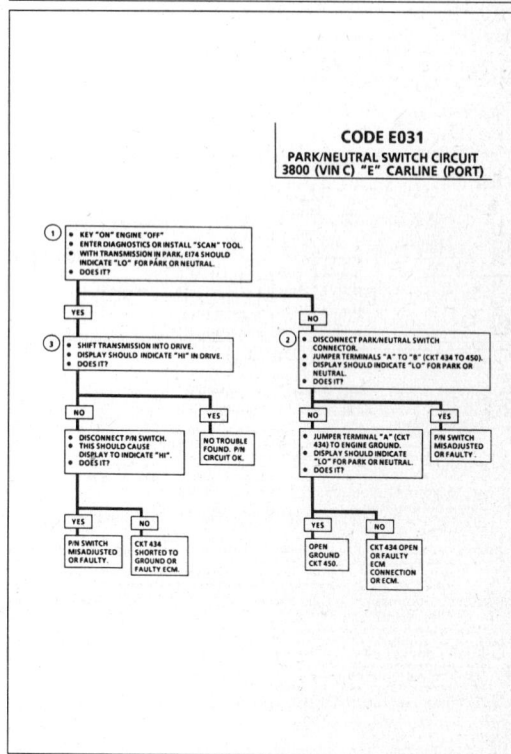

1988–90 E-CAR EQUIPPED WITH 3800 (VIN C) ENGINE

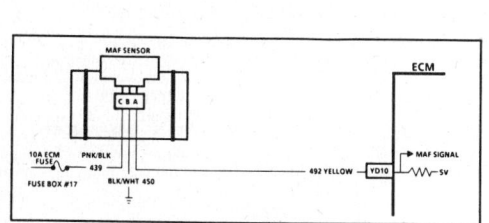

CODE E034
MASS AIR FLOW (MAF) SENSOR CIRCUIT
3800 (VIN C) "E" CARLINE (PORT)

Circuit Description:
The Mass Air Flow (MAF) sensor measures the flow of air which passes through it in a given time. The ECM uses this information to monitor the operating condition of the engine for fuel delivery calculations. A large quantity of air movement indicates acceleration, while a small quantity indicates deceleration or idle.
The MAF sensor produces a frequency signal, which cannot be easily measured. The sensor can be diagnosed using the procedures on this chart.
Code E034 will set, when either of the following sets of conditions exists:
- Engine running
- No MAF sensor signal for 250 mS
- Above conditions for over 4 seconds

Test Description: Numbers below refer to circled numbers on the diagnostic chart.
1. This step checks to see if ECM recognizes a problem.
2. A voltage reading at sensor harness connector terminal "A" of less than 4 or over 6 volts indicates a fault in CKT 492 or poor connection.
3. Verifies that both ignition voltage and a good ground circuit are available.

Diagnostic Aids:
An intermittent may be caused by a poor connection, mis-routed harness, rubbed through wire insulation, or a wire broken inside the insulation. Check For:
- Poor Connection At ECM Pin "YD10". Inspect harness connectors for backed out terminals, improper mating, broken locks, improperly formed or damaged terminals, and poor terminal to wire connection.
- Mis-routed Harness Inspect MAF sensor harness to insure that it is not too close to high voltage wires, such as spark plug leads.
- Damaged Harness Inspect harness for damage. If harness appears OK, observe ED21 while moving related connectors and wiring harness. A change in display would indicate the intermittent fault location.
- Plugged Air Intake Filter. A WOT acceleration from a stop should cause the MAF reading (ED21) to range from about 4.7 at idle to 100 or greater at the time of the 1-2 shift. If not, check for restriction.

1988–90 E-CAR EQUIPPED WITH 3800 (VIN C) ENGINE

CODE E034
MASS AIR FLOW (MAF) SENSOR CIRCUIT
3800 (VIN C) "E" CARLINE (PORT)

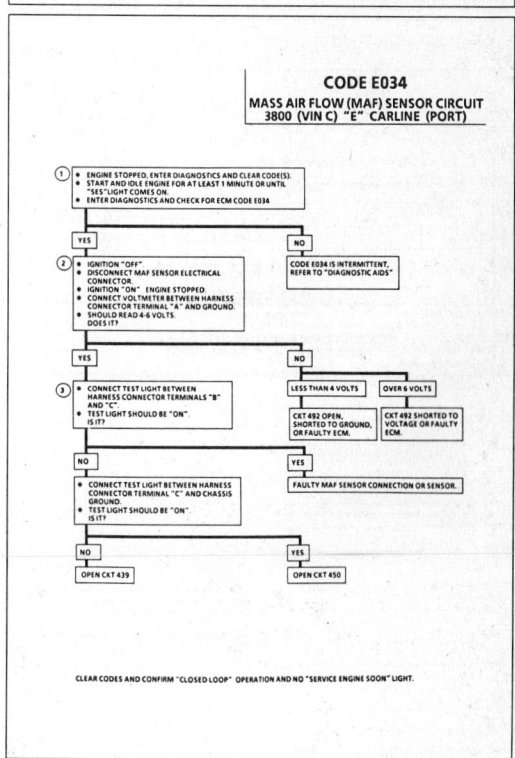

1988–90 E-CAR EQUIPPED WITH 3800 (VIN C) ENGINE

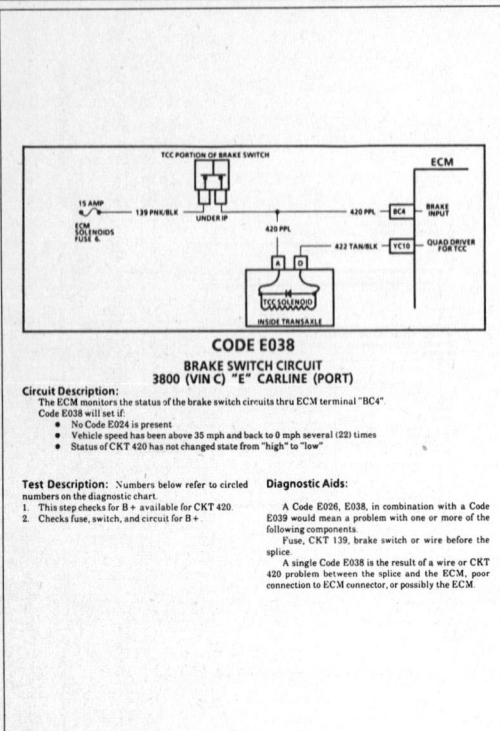

CODE E038
BRAKE SWITCH CIRCUIT
3800 (VIN C) "E" CARLINE (PORT)

Circuit Description:
The ECM monitors the status of the brake switch circuits thru ECM terminal "BC4".
Code E038 will be set if:
- No Code E024 is present
- Vehicle speed has been above 35 mph and back to 0 mph several (22) times
- Status of CKT 420 has not changed state from "high" to "low"

Test Description: Numbers below refer to circled numbers on the diagnostic chart.
1. This step checks for B+ available for CKT 420.
2. Checks fuse, switch, and circuit for B+.

Diagnostic Aids:
A Code E026, E038, in combination with a Code E039 would mean a problem with one or more of the following components.
Fuse, CKT 139, brake switch or wire before the splice.
A single Code E038 is the result of a wire or CKT 420 problem between the splice and the ECM, poor connection to ECM connector, or possibly the ECM.

1988–90 E-CAR EQUIPPED WITH 3800 (VIN C) ENGINE

CODE E038
BRAKE SWITCH CIRCUIT
3800 (VIN C) "E" CARLINE (PORT)

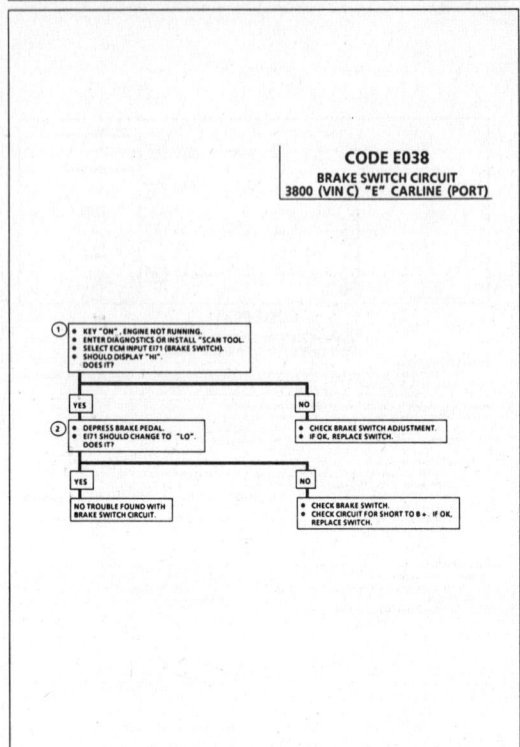

1988–90 E-CAR EQUIPPED WITH 3800 (VIN C) ENGINE

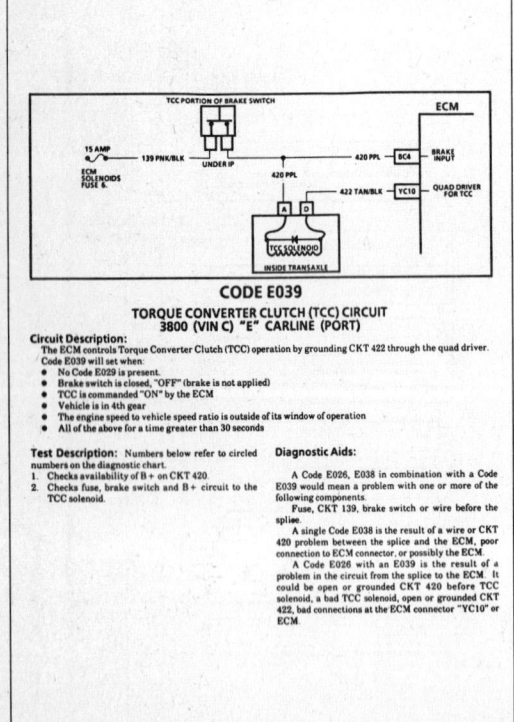

CODE E039
TORQUE CONVERTER CLUTCH (TCC) CIRCUIT
3800 (VIN C) "E" CARLINE (PORT)

Circuit Description:
The ECM controls Torque Converter Clutch (TCC) operation by grounding CKT 422 through the quad driver.
Code E039 will be set when:
- No Code E029 is present.
- Brake switch is closed, "OFF" (brake is not applied)
- TCC is commanded "ON" by the ECM
- Vehicle is in 4th gear
- The engine speed to vehicle speed ratio is outside of its window of operation
- All of the above for a time greater than 30 seconds

Test Description: Numbers below refer to circled numbers on the diagnostic chart.
1. Checks availability of B+ on CKT 420.
2. Checks fuse, brake switch and B+ circuit to the TCC solenoid.

Diagnostic Aids:
A Code E026, E038 in combination with a Code E039 would mean a problem with one or more of the following components.
Fuse, CKT 139, brake switch or wire before the splice.
A single Code E038 is the result of a wire or CKT 420 problem between the splice and the ECM, poor connection to ECM connector, or possibly the ECM.
A Code E026 with an E039 is the result of a problem in the circuit from the splice to the ECM. It could be open or grounded CKT 420 before TCC solenoid, a bad TCC solenoid, open or grounded CKT 422, bad connections at the ECM connector "YC10" or ECM.

1988–90 E-CAR EQUIPPED WITH 3800 (VIN C) ENGINE

CODE E039
TORQUE CONVERTER CLUTCH (TCC) CIRCUIT
3800 (VIN C) "E" CARLINE (PORT)

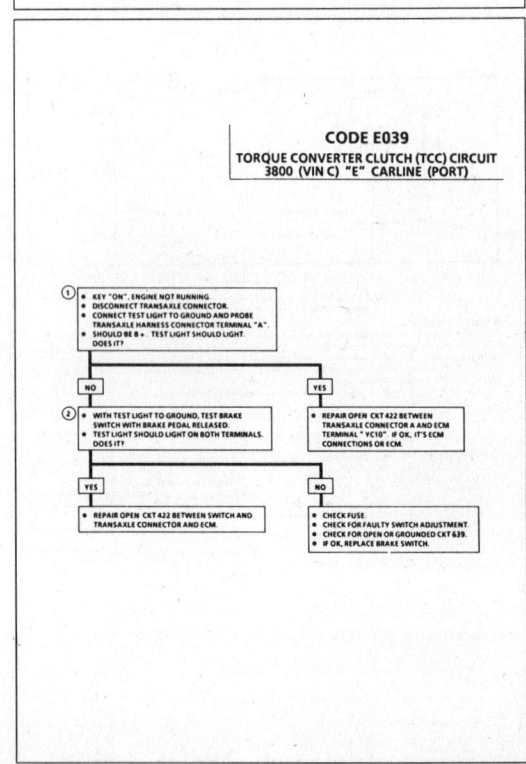

1988–90 E-CAR EQUIPPED WITH 3800 (VIN C) ENGINE

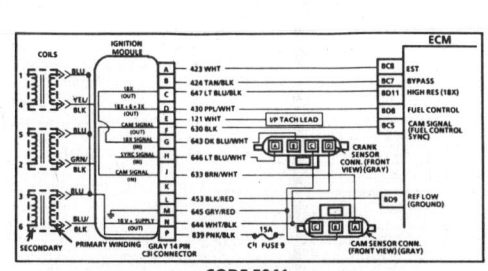

CODE E041
CAM SENSOR CIRCUIT
3800 (VIN C) "E" CARLINE (PORT)

Circuit Description:

During cranking, the ignition module monitors the dual crank sensor sync signal. The sync signal is used to determine the correct cylinder pair to spark first. After the sync signal has been processed by the ignition module it sends a fuel control reference pulse to the ECM. When the ECM receives this pulse it will command all six injectors to open for one (priming) shot of fuel in all cylinders. After the priming, the injectors are left off for the next six fuel control reference pulses (two crankshaft revolutions from the ignition module. This allows each cylinder a chance to use the fuel from the (priming) shot. During this waiting period, a cam pulse will have been received by the ECM. Now the ECM begins to operate the injectors sequentially, based on true camshaft position. However, if the cam signal is not present at start-up a Code 41 will set and the ECM will start sequential fuel delivery in any old random pattern. A 1 in 6 chance that fuel delivery is correct.

Code E041 is set when the following conditions are met:
- Engine is running.   • Cam sensor signal not received by ECM for last 2 seconds.

Test Description: Numbers below refer to circled numbers on the diagnostic chart.
1. Checks to see if ECM recognizes a problem and sets a Code E041.
2. This step verifies proper operation of CKTs 633, 644, and 645.
3. Step validates the integrity of CKT 630 from the ignition module to ECM.
4. If the camshaft gear magnet is interfacing with the cam sensor the voltage reading will be zero, bumping engine will cause the condition to go away.
5. If the voltage reading of "BC5" is constantly varying and connection to terminal "BC5" are good, the ECM is faulty.

Diagnostic Aids:

An intermittent may be caused by a poor connection, rubbed through wire insulation or a wire broken inside the insulation.
Check For:
- Poor Connection or Damaged Harness Inspect ECM harness connectors for backed out terminal "BC5", improper mating, broken locks, improperly formed or damaged terminals, poor terminal to wire connection and damaged harness.
- Intermittent Test If connections and harness check OK, monitor a digital voltmeter connected from ECM terminal "BC5" to ground while moving related connectors and wiring harness. If the failure is induced, the voltage reading will change. This may help to isolate the location of the malfunction.

1988–90 E-CAR EQUIPPED WITH 3800 (VIN C) ENGINE

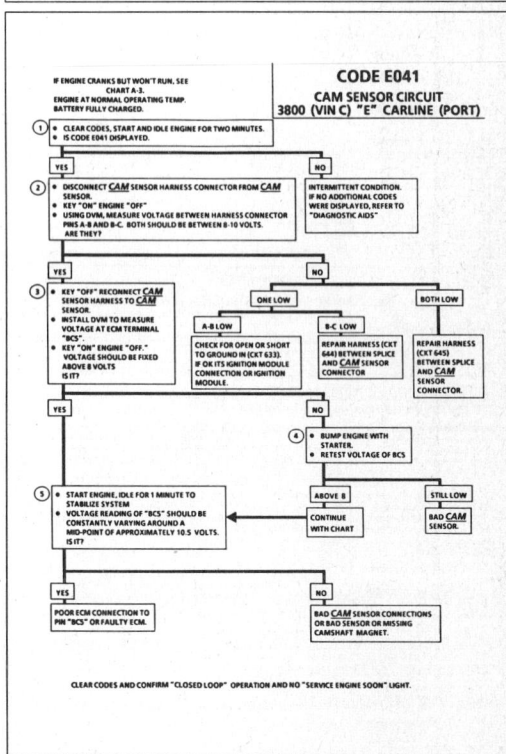

1988–90 E-CAR EQUIPPED WITH 3800 (VIN C) ENGINE

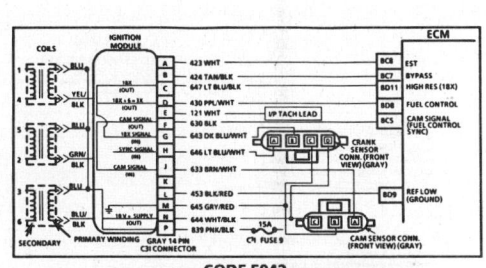

CODE E042
ELECTRONIC SPARK TIMING (EST) CIRCUIT
3800 (VIN C) "E" CARLINE (PORT)

Circuit Description:

The ignition module sends a reference signal to the ECM when the engine is cranking. Under 400 rpm, the ignition module controls ignition timing. When the engine speed exceeds 400 rpm, the ECM sends a 5 volts signal on the bypass CKT 424 to switch timing to ECM control. An open or ground in the EST or bypass circuit will stall the engine but can be re-started but will run on module timing.

To set a Code E042 the following conditions must be met:
- Engine speed greater than 600 rpm with no EST pulse for 200 mS (open or grounded CKT 423), or
- ECM commanding bypass mode (open or grounded CKT 424)

Test Description: Numbers below refer to circled numbers on the diagnostic chart.
1. Checks to see if ECM recognizes a problem. If it does not set Code E042, it is an intermittent problem and could be due to a loose connection.
2. With the ECM disconnected, the ohmmeter should be reading less than 200 ohms, which is the normal resistance of the EST circuit through the ignition module. A higher resistance would indicate a fault in CKT 423, a poor ignition module connection, or a faulty ignition module.
3. If test light was "ON" when connected from 12 volts to ECM harness terminal "BC7", either CKT 424 is shorted to ground or the ignition module is faulty.
4. Checks to see if ignition module switches when the bypass circuit is energized by 12 volts through the test light. If the ignition module actually switches, the ohmmeter reading should shift to over 6,000 ohms.
5. Disconnecting the ignition module should make the ohmmeter read as if it were monitoring an open circuit (infinite reading). Otherwise, CKT 423 is shorted to ground.

Diagnostic Aids:

An intermittent may be caused by a poor connection, rubbed through wire insulation, or a wire broken inside the insulation. Check For:
- Poor Connection or Damaged Harness Inspect ECM harness connectors for backed out terminals "BC7" or "BC8", improper mating, broken locks, improperly formed or damaged terminals, poor terminal to wire connection, and damaged harness.
- Intermittent Test If connections and harness check OK, a digital voltmeter connected to affected terminal to ground while moving related connectors and wiring harness. If the failure is induced, the voltage reading will change.

1988–90 E-CAR EQUIPPED WITH 3800 (VIN C) ENGINE

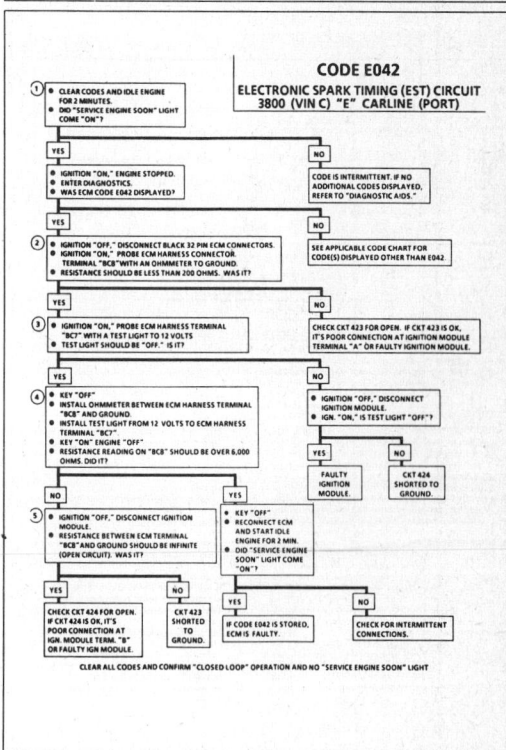

1988–90 E-CAR EQUIPPED WITH 3800 (VIN C) ENGINE

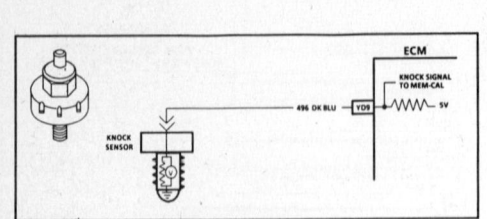

ECM
KNOCK SIGNAL TO MEM-CAL
496 DK BLU — YD9 — 5V

KNOCK SENSOR

CODE E043
ELECTRONIC SPARK CONTROL (ESC) CIRCUIT
3800 (VIN C) "E" CARLINE (PORT)

Circuit Description:
The knock sensor is used to detect engine detonation, and the ECM will retard the electronic spark timing based on the signal being received. The circuitry within the knock sensor causes the ECM 5 volts to be pulled down so that under a no knock condition CKT 496 would measure about 2.5 volts. The knock sensor produces an AC signal which rides on the 2.5 volts DC voltage. The amplitude and signal frequency is dependent upon the knock level.
If CKT 496 becomes open or shorted to ground, the voltage will either go above 3.5 volts or below 1.5 volts. If either of these conditions are met about 5 seconds, a Code E043 will be stored.

Test Description: Numbers below refer to circled numbers on the diagnostic chart.
1. Code E043 will set if:
   • Voltage on CKT 496 goes above 3.5 volts or below 1.5 volts
   • All conditions present for 5 seconds
   If an audible knock is heard from the engine, repair the internal engine problem, normally no knock should be detected at idle
2. If tapping on the engine lift hook does not produce a knock signal, try tapping engine closer to sensor before proceeding

3. The ECM has a 5 volts pull-up resistor which should be present at the knock sensor terminal.
4. This test determines if the knock sensor is faulty or if the ESC portion of the Mem-Cal is faulty.

Diagnostic Aids:
Check CKT 496 for a potential open or short to ground. Also check for proper installation of Mem-Cal.

1988–90 E-CAR EQUIPPED WITH 3800 (VIN C) ENGINE

CODE E043
ELECTRONIC SPARK CONTROL (ESC) CIRCUIT
3800 (VIN C) "E" CARLINE (PORT)

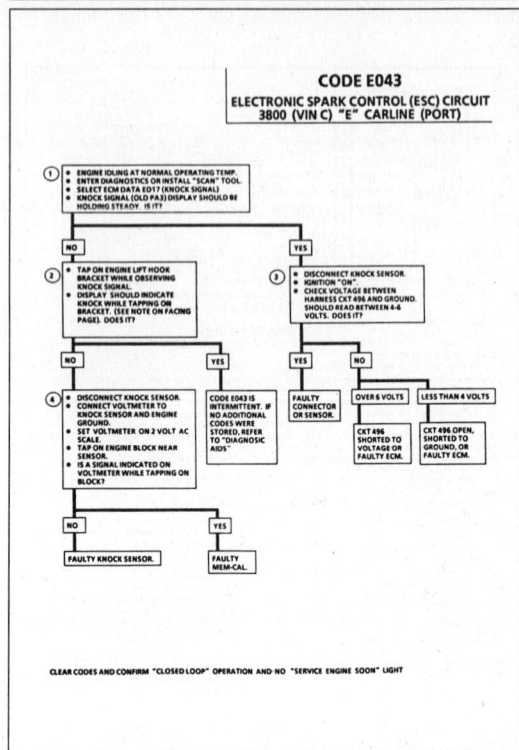

CLEAR CODES AND CONFIRM "CLOSED LOOP" OPERATION AND NO "SERVICE ENGINE SOON" LIGHT

1988–90 E-CAR EQUIPPED WITH 3800 (VIN C) ENGINE

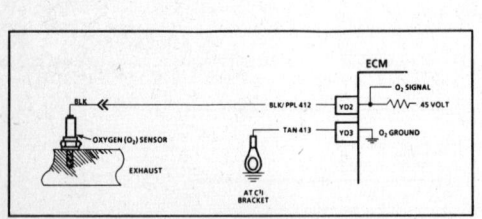

ECM
BLK/PPL 412 — YD2 — O₂ SIGNAL — 45 VOLT
OXYGEN (O₂) SENSOR
EXHAUST
TAN 413 — YD3 — O₂ GROUND
AT CH BRACKET

CODE E044
OXYGEN SENSOR CIRCUIT
(LEAN EXHAUST INDICATED)
3800 (VIN C) "E" CARLINE (PORT)

Circuit Description:
The ECM supplies a voltage of about .45 volt (450 mV) between terminals "YD2" and "YD3". (If measured with a 10 megohm digital voltmeter, this may read as low as .32 volt.) The O₂ sensor varies the voltage within a range of about 1 volt, (1000 mV) if the exhaust is rich, down through about .10 volt (100 mV) if the exhaust is lean.
The sensor is like an open circuit and produces no voltage when it is below about 360°C (600°F). An open sensor circuit or cold sensor causes "Open Loop" operation.
Code E044 is set when the O₂ sensor signal voltage on CKT 412:
   • Remains below .2 volt for up to 4.5 minutes
   • The system is operating in "Closed Loop"

Test Description: Numbers below refer to circled numbers on the diagnostic chart.
1. Running the engine at 1000 rpm keeps the O₂ sensor hot, so an accurate display voltage is maintained.
2. Opening the O₂ sensor wire should result in a voltage display of between 350 and 550 mV. If the display is still fixed below 350 mV, the fault is a short to ground in CKT 412 or the ECM is faulty.

Diagnostic Aids:
Using the on-board diagnostics ED20, observe the block learn values at different rpm and air flow conditions. ED20 displays the block cells, so the block learn values can be checked in each of the cells to determine when the Code E044 may have been set. If the conditions for Code E044 exists, the block learn values will be around 150.
   • **O₂ Sensor Wire** Sensor pigtail may be mispositioned and contacting the exhaust manifold.
   • Check for intermittent ground in wire between connector and sensor.

   • **MAF Sensor** A Mass Air Flow (MAF) sensor output that causes the ECM to sense a lower than normal air flow will cause the system to go lean. Disconnect the MAF sensor and if the lean condition is gone, replace the MAF sensor.
   • **Lean Injector(s)** Perform injector balance test CHART C-2A.
   • **Fuel Contamination** Water, even in small amounts, near the in-tank fuel pump inlet can be delivered to the injectors. The water causes a lean exhaust and can set a Code E044.
   • **Fuel Pressure** System will be lean if fuel pressure is too low. It may be necessary to monitor fuel pressure while driving the car at various road speeds and/or loads to confirm. See "Fuel System Diagnosis" CHART A-7.
   • **Exhaust Leaks** If there is an exhaust leak, the engine can cause outside air to be pulled into the exhaust and past the sensor.
   • Vacuum or crankcase leaks can cause a lean condition and/or possibly a high idle
   • If the above are OK, it is a faulty oxygen sensor.

1988–90 E-CAR EQUIPPED WITH 3800 (VIN C) ENGINE

CODE E044
OXYGEN SENSOR CIRCUIT
(LEAN EXHAUST INDICATED)
3800 (VIN C) "E" CARLINE (PORT)

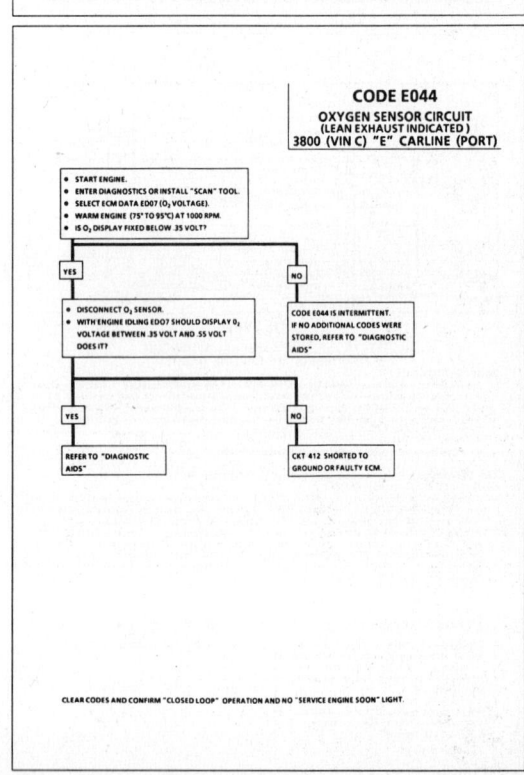

CLEAR CODES AND CONFIRM "CLOSED LOOP" OPERATION AND NO "SERVICE ENGINE SOON" LIGHT

1988–90 E-CAR EQUIPPED WITH 3800 (VIN C) ENGINE

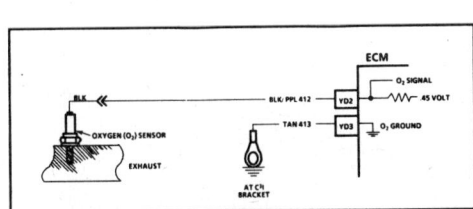

CODE E045
OXYGEN SENSOR CIRCUIT
(RICH EXHAUST INDICATED)
3800 (VIN C) "E" CARLINE (PORT)

Circuit Description:

The ECM supplies a voltage of about .45 volt (450 mV) between terminals "YD2" and "YD3". (If measured with a 10 megohm digital voltmeter, this may read as low as .32 volt.) The O_2 sensor varies the voltage within a range of about 1 volt (1000 mV) if the exhaust is rich, down through about .10 volt (100 mV) if exhaust is lean.

The sensor is like an open circuit and produces no voltage when it is below about 360°C (600°F). An open sensor circuit or cold sensor causes "Open Loop" operation.

Code E045 is set when the O_2 sensor signal voltage or CKT 412:
- Remains above .7 volt for 30 seconds and in "Closed Loop"
- Throttle angle between 0.54 and 2.08 volts
- Engine time after start is 1 minute or more

Diagnostic Aids:

Using the on-board diagnostics ED20, observe the block learn values at different rpm and air flow conditions. ED20 displays the block cells, so the block learn values can be checked in each of the cells to determine when the Code E045 may have been set. If the conditions for Code E045 exists, the block learn values will be around 115.

- **Fuel Pressure** System will go rich if pressure is too high. The ECM can compensate for some increase. However, if it gets too high, a Code E045 may be set. See "Fuel System Diagnosis" CHART A-7.
- **Rich Injector** Perform injector balance test CHART C-2A.
- **Leaking Injector** See CHART A-7.
- Check for fuel contaminated oil.

- **Canister Purge** Check for fuel saturation. If full of fuel, check canister control and hoses. See "Canister Purge".
- **MAF Sensor** An output that causes the ECM to sense a higher than normal airflow can cause the system to go rich. Disconnecting the MAF sensor will allow the ECM to set a fixed value for the sensor. Substitute a different MAF sensor if the rich condition is gone while the sensor is disconnected.
- Check for leaking fuel pressure regulator diaphragm by checking vacuum line to regulator for fuel.
- **TPS** An intermittent TPS output will cause the system to go rich, due to a false indication of the engine accelerating.
- **EGR** An EGR staying open (especially at idle) will cause the O_2 sensor to indicate a rich exhaust, and this could result in a Code E045.

1988–90 E-CAR EQUIPPED WITH 3800 (VIN C) ENGINE

CODE E045
OXYGEN SENSOR CIRCUIT
(RICH EXHAUST INDICATED)
3800 (VIN C) "E" CARLINE (PORT)

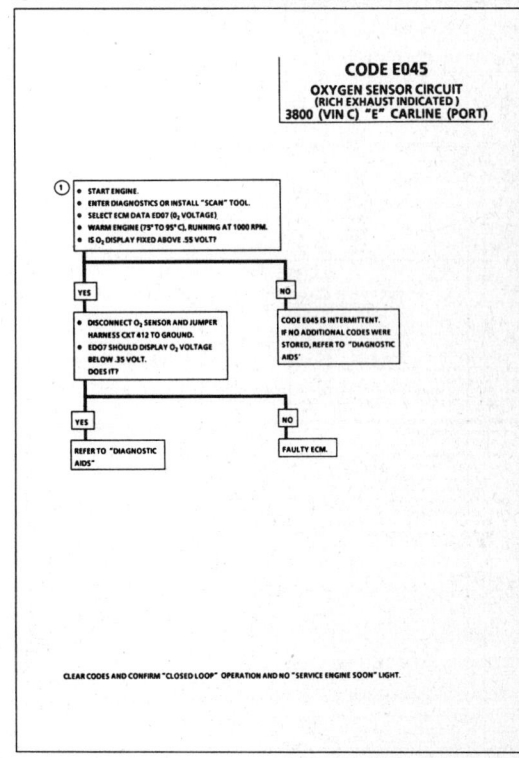

1988–90 E-CAR EQUIPPED WITH 3800 (VIN C) ENGINE

CODE E046

**POWER STEERING PRESSURE SWITCH CIRCUIT
3800 (VIN C) "E" CARLINE (PORT)**

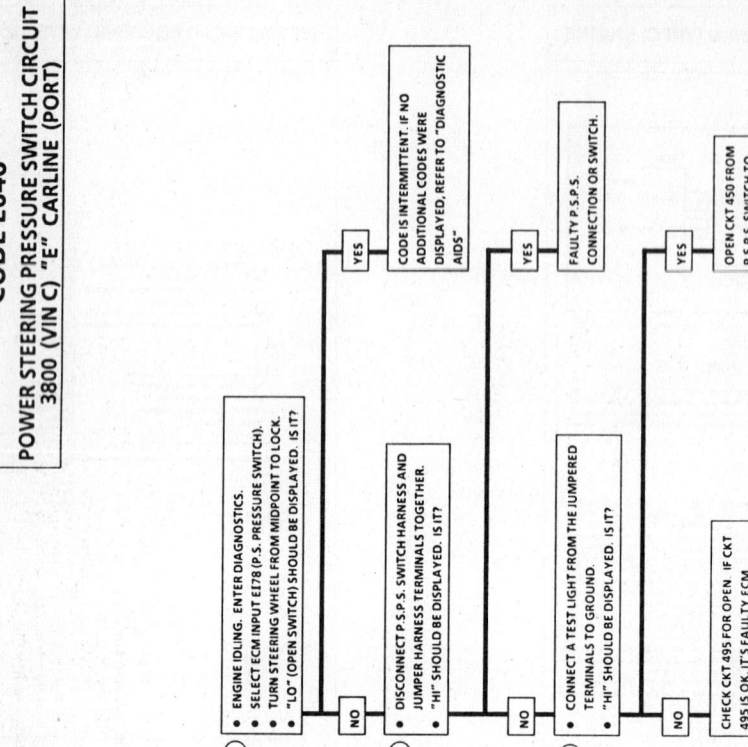

1988–90 E-CAR EQUIPPED WITH 3800 (VIN C) ENGINE

CODE E046

**POWER STEERING PRESSURE SWITCH CIRCUIT
3800 (VIN C) "E" CARLINE (PORT)**

Circuit Description:

The power steering pressure switch is incorporated as an ECM discrete input signal representing the parasitic load placed on the engine during high power steering demand periods, such as parking. This load may cause the engine to stall under high power steering pump pressure conditions, therefore, the ECM compensates by automatically increasing the engine idle speed, via IAC whenever the switch closes and a low voltage is monitored at ECM terminal "BD4." This low voltage indicates a high pressure within the power steering system resulting from high demand. The ECM also opens the ground for the A/C compressor relay so the A/C clutch is disengaged to further reduce engine load.

To set a Code E046, the following conditions must be met:
• Closed power steering switch. E178 indicates "LO" (low voltage potential).
• Vehicle speed is greater than 40 mph.
• Both conditions existing for a time greater than 25 seconds.

Test Description: Numbers below refer to circled numbers on the diagnostic chart.

1. Tests the power steering pressure switch for proper operation by using the "self-diagnostics" system. "LO" should be displayed when high power steering pressure loads are created.

2. This step determines whether the fault is in the switch or the circuit. A jumpered switch connector should normally indicate "HI" if the circuit is complete. If so, the power steering pressure switch or connection is faulty.

3. If the display switches to "HI" when CKT 495 is grounded, the fault is an open in CKT 450.

Diagnostic Aids:

An intermittent may be caused by a poor connection, rubbed through wire insulation or a wire broken inside the insulation.
Check For:
• **Poor Connection or Damaged Harness** Inspect ECM harness connectors for backed out terminal "BD4," improper mating, broken locks, improperly formed or damaged terminals, poor terminal to wire connection, and damaged harness.
• **Intermittent Test** If connections and harness check OK, monitor E178 (power steering pressure switch) display while moving related connectors and wiring harness. If the failure is induced, the power steering pressure switch display will abruptly change. This may help to isolate the location of the malfunction.

1988–90 E-CAR EQUIPPED WITH 3800 (VIN C) ENGINE

CODE E047
ECM-BCM DATA
3800 (VIN C) "E" CARLINE (PORT)

CHECK FOR MOMENTARY LOSS OF BCM'S 7 VOLTS (CKT 807), IGNITION #3 (CKT 750), CPS WAKE-UP (CKT 555) AND PROM. REFERENCE "B" CHARTS

IF CODE E047 IS SET ALONG WITH ANY OF THE FOLLOWING CODES SET AS "HISTORY", B334, B335, B336, B337, B552 AND B553, SEE CHART ON "MULTIPLE INTERMITTENT CODES".

1988–90 E-CAR EQUIPPED WITH 3800 (VIN C) ENGINE

CODE E048
MISFIRE DIAGNOSIS
3800 (VIN C) "E" CARLINE (PORT)

If multiple codes are set, go to the lowest code first.
Repairing for a Code E013, E044 OR E045 may correct Code E048

Test Description:

Code 48 will set if the following:
- TPS is between .48 and 1.30 volts.
- Rpm is between 1300 and 2100.
- Mph is between 50 and 60.
- O_2 cross counts greater than 21.
- All of the above for 30 seconds.

O_2 Sensor Test:

Code 48 could be set if the O_2 sensor is degraded and cannot travel over the full rich to lean voltage range. This narrowed range could allow O_2 cross counts to be above the value necessary to set the code.

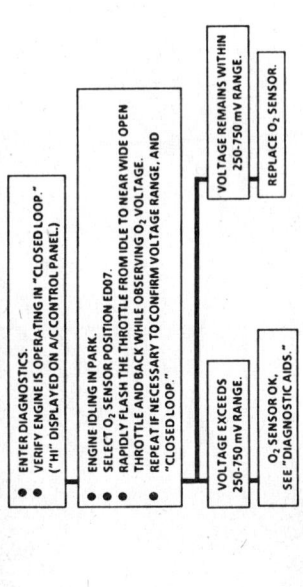

- ENTER DIAGNOSTICS.
- VERIFY ENGINE IS OPERATING IN "CLOSED LOOP." ("HI" DISPLAYED ON A/C CONTROL PANEL.)

- ENGINE IDLING IN PARK.
- SELECT O_2 SENSOR POSITION ED07.
- RAPIDLY FLASH THE THROTTLE FROM IDLE TO NEAR WIDE OPEN THROTTLE AND BACK WHILE OBSERVING O_2 VOLTAGE.
- REPEAT IF NECESSARY TO CONFIRM VOLTAGE RANGE, AND
- "CLOSED LOOP."

VOLTAGE EXCEEDS 250-750 mV RANGE. → O_2 SENSOR OK, SEE "DIAGNOSTIC AIDS."

VOLTAGE REMAINS WITHIN 250-750 mV RANGE. → REPLACE O_2 SENSOR.

Diagnostic Aids:

1. Ignition system checks:
Remove each spark plug and inspect (fouled, cracked, worn)
Fouled -- check ignition wires (hi resistance, damage, poor connections, grounds)
check coil and module operation
check basic engine problem (see 3 below)
Cracked or worn -- replace as necessary

2. Fuel system checks:
Restricted fuel system (injectors, fuel pump, lines, and filter)
Injectors -- perform injector balance test
verify each injector circuit with tool J34730-2 or equivalent
Fuel Pump --verify proper fuel pressure and fuel quality
Lines and Filter --verify no restrictions in lines or filter

3. Basic Engine Checks:
Unless spark plug(s) inspection identifies a specific cylinder(s), road test vehicle under test conditions to reverify Code E048 prior to engine disassembly.
Basic engine (valves, compression, camshaft, lifters)
Compression -- check rings, pistons, valves
Valves --check for burned, weak springs, broken parts, worn or loose guide
Camshaft --check for worn or broken
Lifters --check for worn, broken
For additional items see Section "B" under "Rough Unstable Idle,"
"Hard Start," or "Hesitation, Sag, or Stumble."

1988–90 E-CAR EQUIPPED WITH 3800 (VIN C) ENGINE

ACCESS COVER

MEM-CAL

ECM TOP COVER

1988–90 E-CAR EQUIPPED WITH 3800 (VIN C) ENGINE

CODE E051
MEM-CAL ERROR
(FAULTY OR INCORRECT MEM-CAL)
3800 (VIN C) "E" CARLINE (PORT)

CHECK THAT ALL PINS ARE FULLY INSERTED IN THE SOCKET. IF OK, REPLACE MEM-CAL, CLEAR MEMORY AND RECHECK. IF CODE E051 REAPPEARS, REPLACE ECM.

NOTICE: To prevent possible Electrostatic Discharge damage:
• Do Not touch the ECM connector pins or soldered components on the ECM circuit board.
• When handling the MEM-CAL, Do Not touch the component leads, and Do Not remove integrated circuit from carrier.

1988–90 E-CAR EQUIPPED WITH 3800 (VIN C) ENGINE

CODES E063, E064, E065
EXHAUST GAS RECIRCULATION (EGR) CIRCUIT 3800 (VIN C) "E" CARLINE (PORT)

- KEY "ON", ENGINE "OFF".
- ENTER DIAGNOSTICS.
- NOTE AND RECORD CODES.

CODE E063, E064, E065, OR ANY COMBINATION OF THESE.

ANY CODE EXCEPT E026.

GO TO CODE CHART FOR APPROPRIATE CODE. LOWEST FIRST.

CODE E026 AND E063, E064, E065, OR ANY COMBINATION OF THESE.

GO TO CODE E026 CHART.
FOR CODES E026 AND E063 - CHECK CKT 697
FOR CODES E026 AND E064 - CHECK CKT 698
FOR CODES E026 AND E065 - CHECK CKT 699

(1)
- KEY "ON", ENGINE RUNNING
- ENTER DIAGNOSTICS MODE.
- ENTER ECM OVERRIDES
  ES02 (SMALL EGR)
  ES03 (MEDIUM EGR)
  ES04 (LARGE EGR)
- ENGINE RPM SHOULD CHANGE AS EACH SOLENOID IS CYCLED "ON" OR "OFF".

DID RPM CHANGE ON EACH?

YES

NO

CODE IS INTERMITTENT. REFER TO "DIAGNOSTIC AIDS".

REPLACE DIGITAL EGR VALVE.

* THESE STEPS MUST BE DONE VERY QUICKLY AS THE ECM WILL COMPENSATE FUEL DELIVERY.

1988–90 E-CAR EQUIPPED WITH 3800 (VIN C) ENGINE

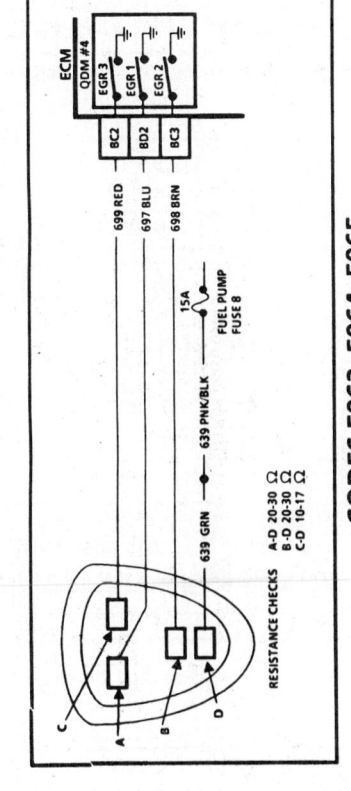

ECM
QDM #4
EGR 3
EGR 1
EGR 2

BC2
BD2
BC3

699 RED
697 BLU
698 BRN

639 GRN ——— 639 PNK/BLK

15A
FUEL PUMP
FUSE 8

RESISTANCE CHECKS
A-D 20-30
B-D 20-30
C-D 10-17

CODES E063, E064, E065
EXHAUST GAS RECIRCULATION (EGR) CIRCUIT 3800 (VIN C) "E" CARLINE (PORT)

Circuit Description:
Codes E063, E064 and E065 are EGR flow test failures. The ECM, on a closed throttle coast down, will cycle the solenoids "ON" and "OFF" individually and look for a resulting change in engine rpm and O_2 sensor activity.

Test Description: Numbers below refer to circled numbers on the diagnostic chart.
1. If a code is set, inspect EGR for damage. See note below.
Start and idle engine, enter ECM overrides ES02, ES03, ES04. You should be able to discern a change in engine rpm as the ECM turns "ON" the EGR. ES02 should result in a small change and ES04 in a large change in rpm. The engine may stall when ES04 is turned "ON."

Note: If the digital EGR valve shows signs of excessive heat, possibly slightly melted, check the exhaust system for blockage (possibly a plugged converter) using procedure found on CHART B-1. If the exhaust system is restricted, repair the cause, one of which might be an injector, which is open due to one of the following: a. stuck
b. grounded driver circuit
c. possibly a faulty ECM

If this condition is found, the oil should be checked for possible fuel contamination.

Diagnostic Aids:

An intermittent may be caused by a poor connection, rubbed-through wire insulation, or a wire broken inside the insulation. Check For:

- **Poor Connection or Damaged Harness** Inspect ECM harness connectors for backed out terminals "BD2", "BC2", and "BC3", improper mating, broken locks, improperly formed or damaged terminals, poor terminal to wire connection, and damaged harness.
- **Intermittent Test** If connections and harness check OK, connect a digital voltmeter from effected terminals to ground while moving related connectors and wiring harness. If the failure is induced, the voltage reading will change.

BCM WIRING DIAGRAM – 1988–90 RIVIERA AND REATTA

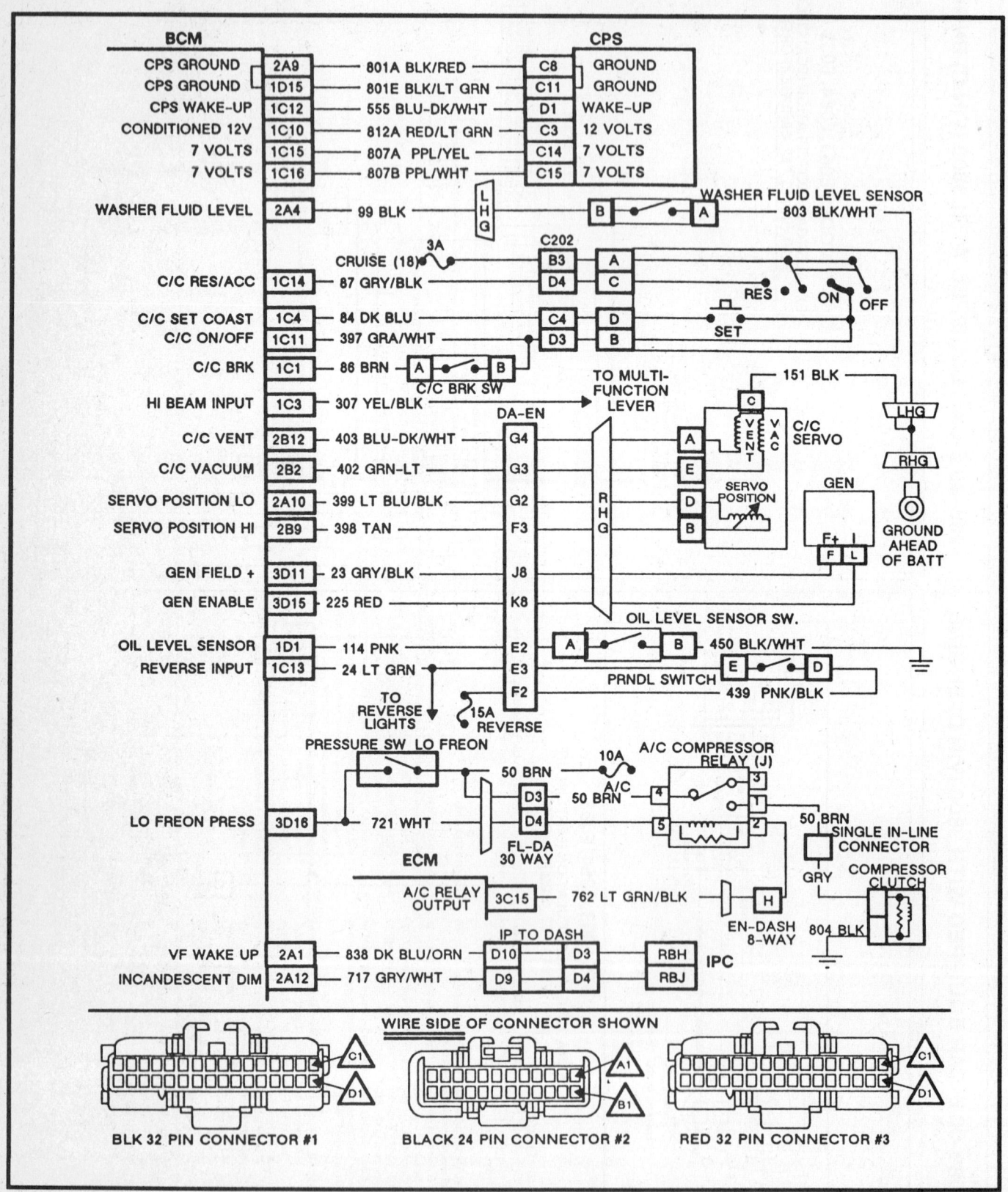

BCM WIRING DIAGRAM – 1988–90 RIVIERA AND REATTA

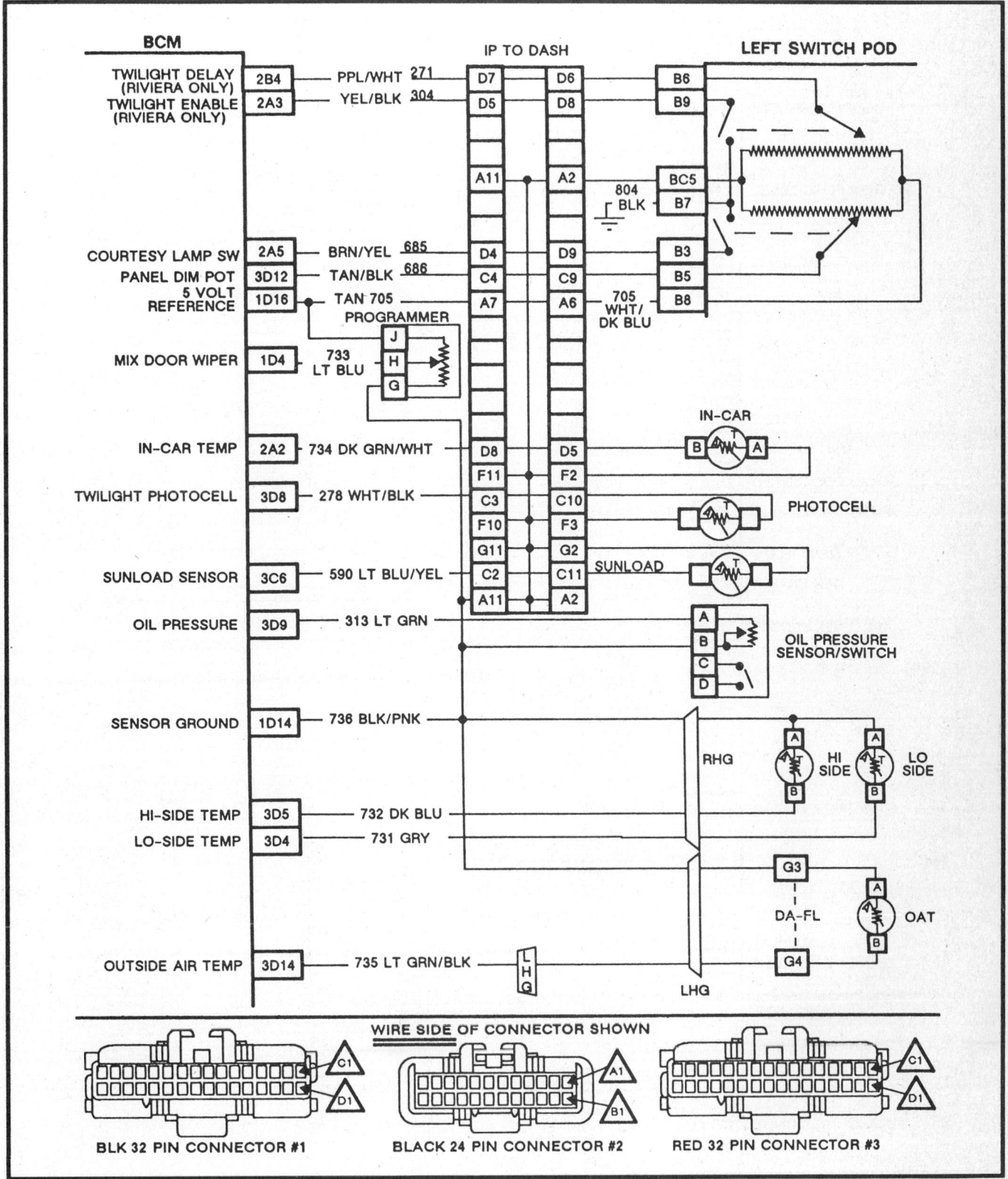

**BCM WIRING DIAGRAM
1988–90 RIVIERA AND REATTA**

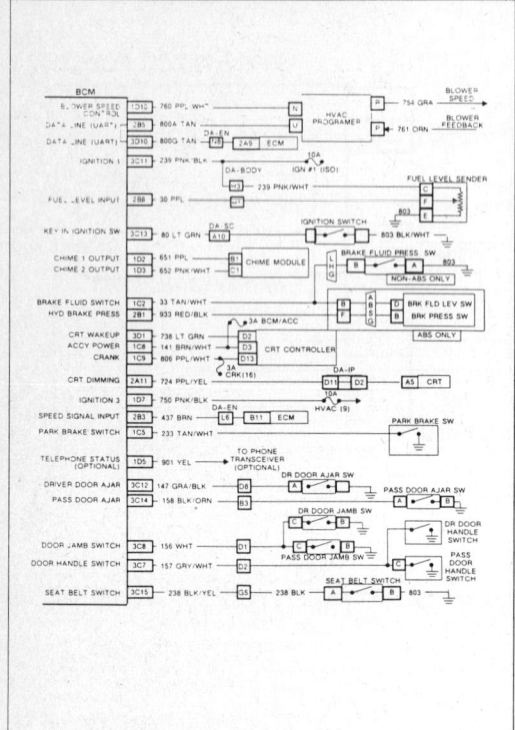

BCM WIRING DIAGRAM – 1988–90 RIVIERA

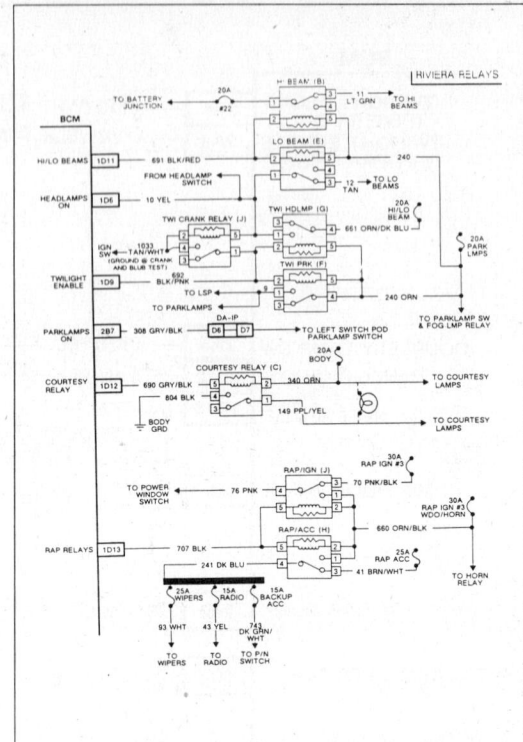

BCM WIRING DIAGRAM – 1988–90 REATTA

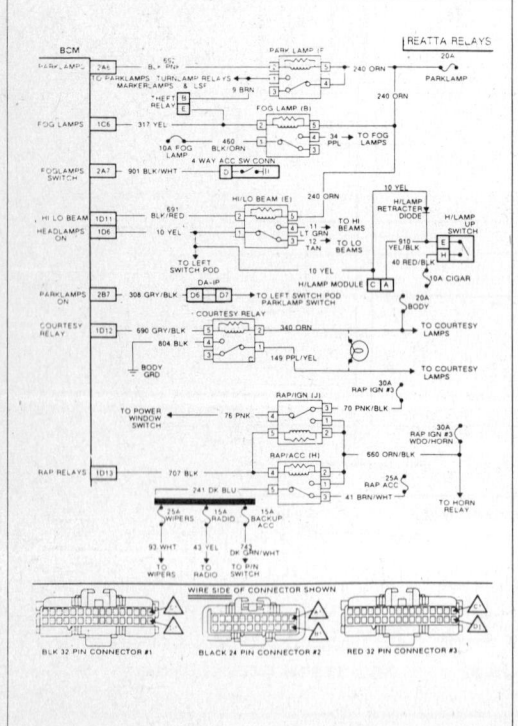

BCM CONNECTOR VIEW – 1988–90 REATTA

BCM CONNECTOR VIEW – 1988–90 RIVIERA

CKT DESC	/ COLOR	CKT NO	WIRE SIDE		CKT NO	/ COLOR	CKT DESC
SPARE					738	LT GRN	CRTC WAKE-UP
SPARE							SPARE
SPARE							SPARE
SPARE					731	GRY	A/C LO SIDE TEMP
SPARE					732	DK BLU	A/C HI SIDE TEMP
SUNLOAD SENSOR	LT BLU/YEL	590					SPARE
DOOR HANDLE SWITCH	GRY/WHT	157			276	WHT/BLK	TWI PHOTOCELL
DOOR JAMB SWITCH	WHT	156			313	LT GRN	OIL PRESSURE
					800G	TAN	SERIAL DATA
					23	GRY/BLK	REGULATOR SIG
FUEL LEV REF – IGN #1	PNK/BLK	239			686	TAN/BLK	DIM CONTROL
DRIVER DOOR AJAR	GRY/BLK	147					SPARE
KEY IN IGNITION	LT GRN	80			735	LT GRN/BLK	OUTSIDE TEMP
PASS. DOOR AJAR	BLK/ORN	158			225	RED	GEN SIGNAL
SEAT BELT SW	BLK/YEL	238			721	WHT	LO REF PRESSURE
SPARE							
VF WAKE-UP	DK BLU/ORN	838			933	RED/BLK	LO BRAKE BOOSTER PRES
IN CAR TEMP	DK GRN/WHT	734			402	LT GRN	CRUISE VAC ON
TWILIGHT ENABLE	YEL/BLK	304			437	BRN	VEHICLE SPEED FROM ECM
WASHER FLUID LEVEL	BLK	99			271	PPL/WHT	TWILIGHT DELAY
COURTESY LT SW	BRN/YEL	685			800A	TAN	SERIAL DATA
SPARE					308	GRY/BLK	PARK LAMP ENABLE
SPARE					30	PPL	FUEL LEVEL WIPER
CPS GROUND	BLK/RED	801			398	TAN	CRUISE SERVO HI
CRUISE SERVO LO	LT BLU/BLK	399					SPARE
CRT DIMMING	PPL/YEL	724					SPARE
INCANDESCENT DIM	GRY/WHT	717			403	DK BLU/WHT	CRUISE VENT ON
CRUISE BRAKE SW	BRN	86			114	PNK	OIL LEVEL SENSOR
BRAKE FLD LEV SW	TAN/WHT	33			651	PPL	CHIME 1
HI BEAM SW	YEL/BLK	307			652	PNK/WHT	CHIME 2
C.C. SET/COAST SW	DK BLU	84			733	LT BLU	MIX DR. WIPER
PARK BRAKE SW	TAN/WHT	233			901	YEL	TELEPHONE STATUS
SPARE					10	YEL	HEADLAMPS ON
SPARE					750	PNK/BLK	IGNITION #3
ACCESSORY POWER	BRN/WHT	141					SPARE
CRANK INPUT	PPL/WHT	806			692	BLK/PNK	TWILIGHT RELAYS
CPS 12 VOLT	RED/LT GRN	812			760	PPL/WHT	BLOWER CONTROL
CRUISE ON/OFF	GRY/BLK	397			691	BLK/RED	HI/LO BEAM RELAY
CPS WAKE-UP	DK BLU/WHT	555			690	GRY/BLK	COURTESY LT RELAY
REVERSE GEAR	LT GRN	24			707	BLK	RAP RELAYS
C.C. RESUME/ACCEL	GRY/BLK	87			736	BLK/PNK	5V RETURN (GROUND)
CPS 7 VOLT	PPL/YEL	807A			801	BLK/LT GRN	CPS GROUND
CPS 7 VOLT	PPL/WHT	807B			705	TAN	5V REFERENCE

WIRE SIDE
RIVIERA

1988–90 RIVIERA AND REATTA

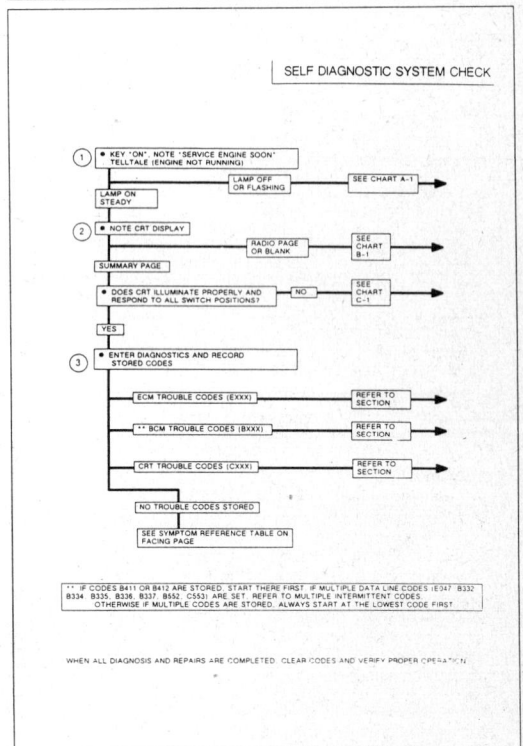

SELF DIAGNOSTIC SYSTEM CHECK

1988–90 RIVIERA AND REATTA

CHART B-1

RADIO PAGE / BLANK CRT DISPLAY

Circuit Description:

From the system check, we have determined that the radio page or a blank screen is the default at key on. In this state diagnostic mode cannot be entered. The system indicates a problem with either the serial data line linking the CRTC to the BCM (Radio page display) or a CRT/CRTC problem (blank screen).

NOTE: When viewing the CRT screen to note what page is showing or if it not lighting up, always touch the Radio border switch to make sure the screen is not blank because of the display blanking feature used to save battery power in retained cessory power (RAP), or accessory mode. All questions concerning the state of the different displays are the state they end up in after "WOW", display on entry routine is finished, if activated.

Test Description: Numbers below refer to circled numbers on the diagnostic chart.

1) Noting if the CRT screen is blank or showing some function of the Radio Category will break the problem in half. Since the Radio functions do not rely upon the rest of the computer system to operate, a completely blank screen indicates a problem with the CRT-CRTC's display capability.

2) If the "Electrical Problem" telltale is "ON" when some function of the Radio page is displayed on the CRT, the integrity of the data CKT needs to be verified.

3) Odometer display in the IPC, when the "Electrical Problem" telltale is "OFF", will verify data communication between the IPC and the BCM if actual miles are displayed. This confirms data communications just as no "Electrical Problem" telltale does. A second verification is necessary as the whole system may be down and the IPC would not be able to display the "Electrical Problem" telltale indicating loss of data.

4) Since the BCM is the device that turns the display "ON" (Wake-Up), operating the courtesy lights

will verify if the reason the system is not working is because the BCM is not working and cannot wake-up the other devices. The only micro-controlled device required to operate the courtesy lights is the BCM; turning "ON" the courtesy lights from the panel lamp switch verifies BCM operation.

5) If the CRT screen is blank and the IPC's "Electrical Problem" telltale is "ON", the problem must be in that the CRTC is not getting 12 volts from the CPS. The Cps is sending 12 volts out as the "Service Engine Soon" telltale worked. The problem is also not the BCM as the radio did not work either, and the radio functions do not require the rest of the computer system to be working. If the IPC's "Electrical Problem" telltale is "OFF", the rest of the system is working and the problem must be within the CRT-CRTC. The problem cannot be data as radio functions would have worked, nor could it be loss of 12 volts as loss of 12 volts to the CRTC would prevent data communication to the CRTC and the IPC would turn on the "Electrical Problem" telltale.

1988–90 RIVIERA AND REATTA

CHART B-1
CANNOT ENTER DIAGNOSTICS

NOTE: ALL QUESTIONS ABOUT THE STATUS OF DISPLAYS (IPC TELLTALES AND CRT) ARE THEIR STATE AFTER THE "WOW" DISPLAY ON ENTRY ROUTINE FINISHES.

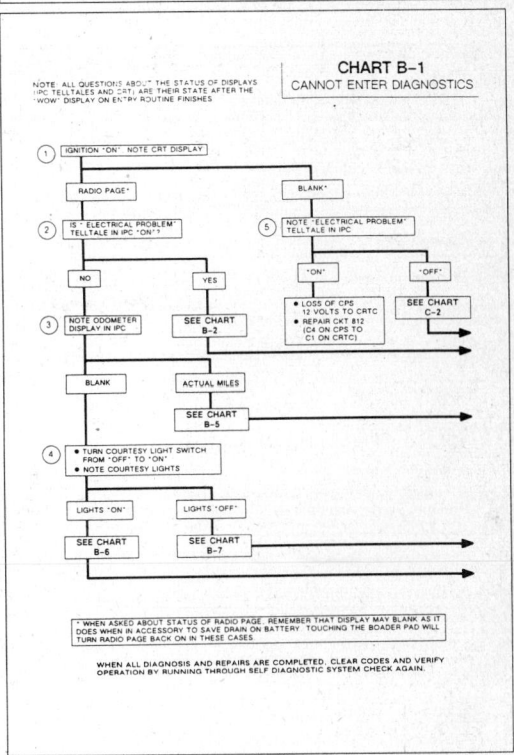

* WHEN ASKED ABOUT STATUS OF RADIO PAGE, REMEMBER THAT DISPLAY MAY BLANK AS IT DOES WHEN IN ACCESSORY TO SAVE DRAIN ON BATTERY. TOUCHING THE BOADER PAD WILL TURN RADIO PAGE BACK ON IN THESE CASES.

WHEN ALL DIAGNOSIS AND REPAIRS ARE COMPLETED, CLEAR CODES AND VERIFY OPERATION BY RUNNING THROUGH SELF DIAGNOSTIC SYSTEM CHECK AGAIN.

1988–90 RIVIERA AND REATTA

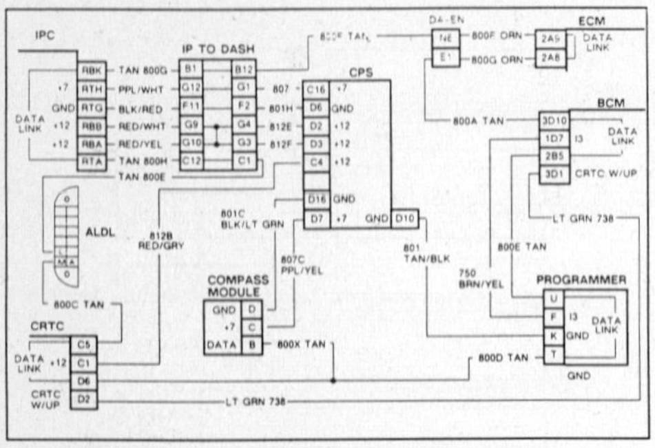

CHART B–2

DATA PROBLEM

Test Description: Numbers below refer to circled numbers on the diagnosuc chart.

1) The "Electrical Problem" telltale in the IPC will come "ON" if the IPC or CRTC has lost data. Since diagnostics cannot yet be accessed to isolate the fault, noting the odometer display will verify if communication is lost or, if the IPC is receiving data from the BCM. If the IPC is receiving data from the BCM the odometer will display actual miles, not "ERROR".

2) Because the IPC and BCM can exchange data, the data line cannot be shorted. Since the IPC odometer displays actual miles the reason the "Electrical Trouble" telltale is "ON" is because the CRTC, not the IPC, has lost data. In order for this to occur, CKT 800F must be OK and there must be two opens in the data line, one on either side of the CRTC isolating it from the BCM, or the CRTC is at fault. To verify the integrity of the data line between the ALDL and the IPC, remove the ALDL cover and jumper the data line to ground. If the data line is OK, the odometer will display "Error" as jumpering the data line to ground will stop all communcations.

3) If the odometer did not display "Error" there is an open in the data line between the ALDL and the CRTC along with another open between the CRTC and the BCM, or the CRTC is bad. To determine if the CRTC is bad or if there are two opens, reconnect the ALDL cover and ground the data line at the CRTC. IF the data line is open, the IPC will continue to display actual miles.

1988–90 RIVIERA AND REATTA

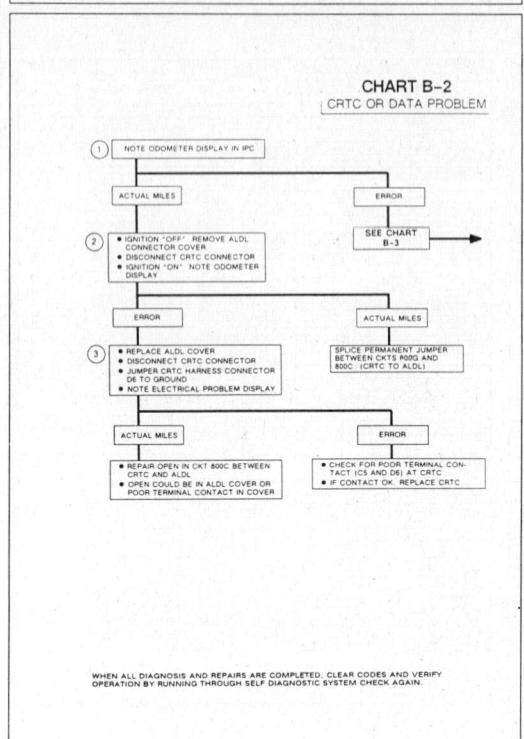

CHART B–2
CRTC OR DATA PROBLEM

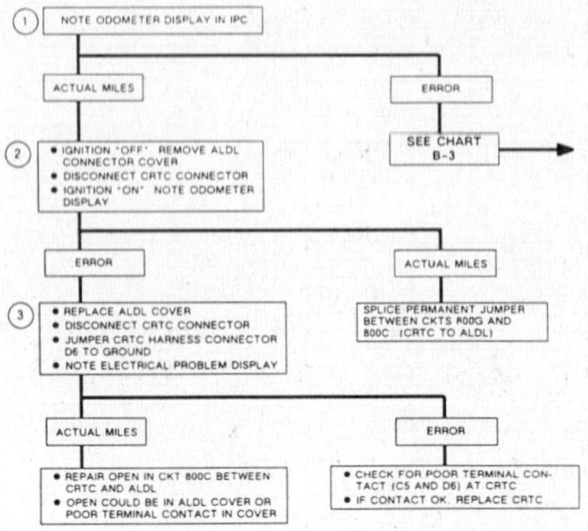

WHEN ALL DIAGNOSIS AND REPAIRS ARE COMPLETED, CLEAR CODES AND VERIFY OPERATION BY RUNNING THROUGH SELF DIAGNOSTIC SYSTEM CHECK AGAIN.

1988–90 RIVIERA AND REATTA

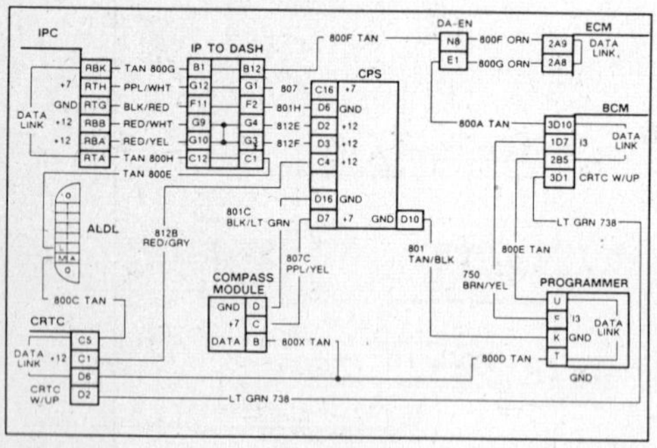

CHART B–3

NO COMMUNICATION

Circuit Description:

Since both the CRT and the IPC are unable to communicate with the BCM, the data line must be shorted to ground, the data line has two opens, or the BCM is not communicating.

Test Description: Numbers below refer to circled numbers on the diagnostic chart.

1) If the problem is a short in the data line (loop), breaking the loop apart, trying to get a short out of the loop, will be indicated by communication being restored to those devices left on the loop with the BCM. By removing the ALDL cover and disconnecting the programmer, we can eliminate the CRTC, programmer, and CKTs 800C and 800B from the loop. If the "Electrical Problem" telltale goes OFF, the problem was in what was disconnected from the loop.

2) If the CRT turns ON to the summary page after the ECM was disconnected and the programmer was reconnected, the problem is in the IPC, ECM, [illegible] data lines between the ECM and the IPC, [illegible] IPC and the ALDL.

3) [illegible]

4) If the summary page does not appear after the ECM was reconnected and the IPC was disconnected, the fault imust be in the data line between the IPC and the ECM. If the summary page did not appear, the problem is in the IPC or the data line between the IPC and the ALDL as this is the only thing not in the loop.

5) If the CRT stays on the radio page after the ECM was disconnected and the programmer was reconnected, the problem must be the BCM or the data line on either side of the BCM. Measuring voltage at the ALDL will determine if the problem is due to an open or a short.

6) Measuring voltage at the other pin of the ALDL with BCM connector #3 disconnected will isolate where the short is.

7) Jumpering the ALDL to the BCM will determine if [illegible]

1988–90 RIVIERA AND REATTA

CHART B–3
NO COMMUNICATION

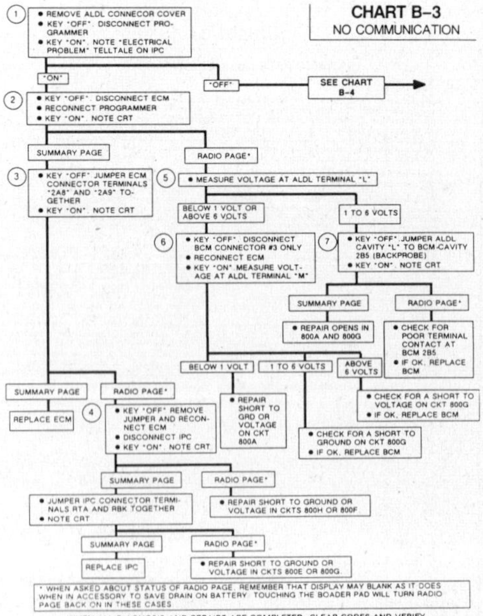

* WHEN ASKED ABOUT STATUS OF RADIO PAGE, REMEMBER THAT DISPLAY MAY BLANK AS IT DOES WHEN IN ACCESSORY TO SAVE DRAIN ON BATTERY. TOUCHING THE BOADER PAD WILL TURN RADIO PAGE BACK ON IN THESE CASES

WHEN ALL DIAGNOSIS AND REPAIRS ARE COMPLETED, CLEAR CODES AND VERIFY OPERATION BY RUNNING THROUGH SELF DIAGNOSTIC SYSTEM CHECK AGAIN.

1988–90 RIVIERA AND REATTA

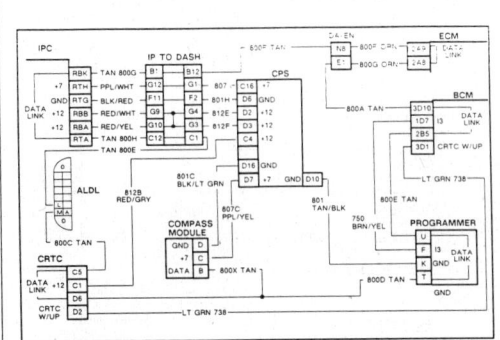

CHART B-4

NO COMMUNICATION CONT'D

Circuit Description:

Since the "Electrical Problem" telltale on the IPC goes out when the ALDL cover and programmer are disconnected, the fault must be in the section of the data loop that has been isolated from the BCM.

Test Description: Numbers below refer to circled numbers on the diagnostic chart.

1) If the "Electrical Problem" telltale on the IPC stays out when the ALDL cover is replaced, the problem must be the programmer, as it is the only thing not connected to the data loop. An open ground CKT, at the programmer, will cause the data link to be inoperative.

2) If the "Electrical Problem" telltale is on the IPC stays on, even after the CRTC has been disconnected and the ALDL cover is replaced, the problem is in the data circuit between the CRTC and the ALDL conector.

3) If the "Electrical Problem" telltale is on the IPC goes OFF after the CRTC was disconnected and stays off when the programmer is reconnected, the CRTC must be bad as it is the only thing in the data loop.

CHART B-5

CRTC PROBLEM

Circuit Description:

Since all we can get on the CRT is a Radio Function page and we know the system is communicating as there is no "Electrical Problem" telltale and the odometer is displaying actual mileage, there must be a problem with the CRTC's ignition-on wake-up signal.

Test Description: Numbers below refer to circled numbers on the diagnostic chart.

1) By grounding the circuit at the BCM we are doing what the BCM should be doing. If the CRT now works, the BCM connector or BCM is faulty.

2) If the CRT still does not work, grounding the circuit at the CRTC will check the integrity of CKT 738. If the CRT still does not work, the CRTC connector or CRTC is faulty.

1988–90 RIVIERA AND REATTA

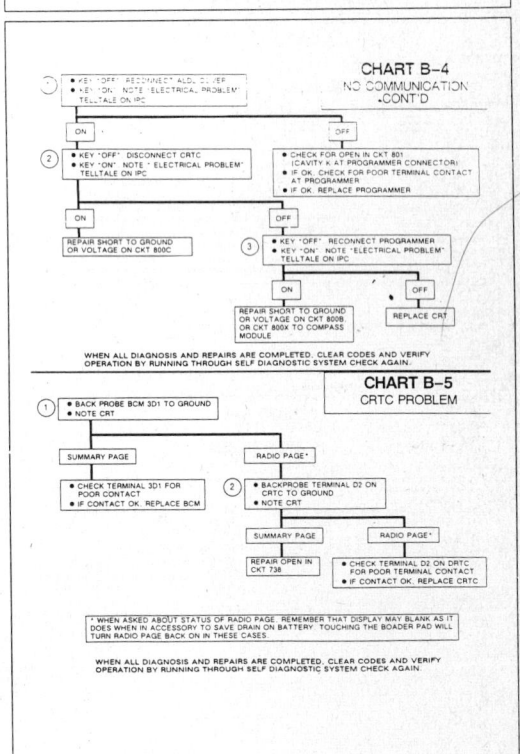

CHART B-4
NO COMMUNICATION
.CONT'D

WHEN ALL DIAGNOSIS AND REPAIRS ARE COMPLETED. CLEAR CODES AND VERIFY OPERATION BY RUNNING THROUGH SELF DIAGNOSTIC SYSTEM CHECK AGAIN.

CHART B-5
CRTC PROBLEM

* WHEN ASKED ABOUT STATUS OF RADIO PAGE. REMEMBER THAT DISPLAY MAY BLANK AS IT DOES WHEN IN ACCESSORY TO SAVE DRAIN ON BATTERY. TOUCHING THE BOADER PAD WILL TURN RADIO PAGE BACK ON IN THESE CASES.

WHEN ALL DIAGNOSIS AND REPAIRS ARE COMPLETED. CLEAR CODES AND VERIFY OPERATION BY RUNNING THROUGH SELF DIAGNOSTIC SYSTEM CHECK AGAIN.

1988–90 RIVIERA AND REATTA

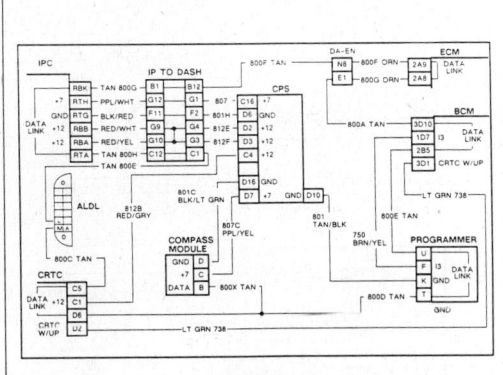

CHART B-6

NO COMMUNICATION TO IPC WITH ONLY RADIO FUNCTION

Circuit Description:

Because the CRT can only display Radio functions and the IPC cannot communicate with BCM as the odometer is blank, the whole system is down, or the IPC has lost its logic ground which stops all communication. The BCM wakes-up ("turns-on") the displays, however, we know the problem is not a totally dead BCM as it turned the courtesy lights on. This could mean the BCM itself is not waking-up when the ignition is on.

Test Description: Numbers below refer to circled numbers on the diagnostic chart.

1) Since the BCM uses the igition 3 circuit to know when its time to turn on when the ignition is on, first check the 10 Amp fuse.

2) If the fuse is blown, use another fuse to determine if the short is intermittent before isolating the fault.

3) If the fuse was not blown, check voltage at the BCM with a test light. If the test light does not light, the circuit is open.

4) If the test light lights, remove the IPC to determine if the fault is loss of IPC logic ground, or faulty BCM connector, or BCM. If the problem was loss of logic ground to the IPC, when the IPC is removed, the CRT will work.

5) Verifying the integrity of the ground circuit at the IPC will determine if the circuit is open, or if the problem is a faulty IPC connector or IPC.

1988–90 RIVIERA AND REATTA

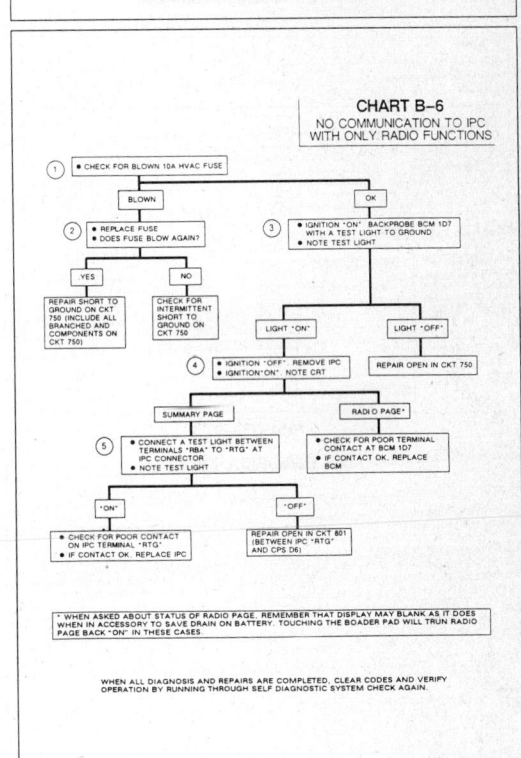

CHART B-6
NO COMMUNICATION TO IPC
WITH ONLY RADIO FUNCTIONS

* WHEN ASKED ABOUT STATUS OF RADIO PAGE. REMEMBER THAT DISPLAY MAY BLANK AS IT DOES WHEN IN ACCESSORY TO SAVE DRAIN ON BATTERY. TOUCHING THE BOADER PAD WILL TRUN RADIO PAGE BACK "ON" IN THESE CASES.

WHEN ALL DIAGNOSIS AND REPAIRS ARE COMPLETED. CLEAR CODES AND VERIFY OPERATION BY RUNNING THROUGH SELF DIAGNOSTIC SYSTEM CHECK AGAIN.

1988–90 RIVIERA AND REATTA

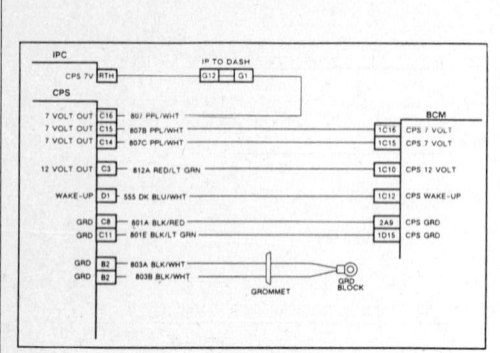

CHART B-7

BCM PROBLEM

Circuit Description:

Since the BCM did not turn the courtesy lights on from the panel lamp switch, the BCM is not working, or the whole system lost its logic voltage 7 volt. The CPS must have and be providing power and ground as the CRT can display radio functions.

Test Description: Numbers below refer to circled numbers on the diagnostic chart.

1) If the problem is that the BCM lost its ground the rest of the system would go to default modes and the climate control fan would be on HI speed. When the rest of the system is in default mode we know the BCM has 12 volts because it needs this much to establish defaults.

2) Verifying the integrity of the BCM grounds at the BCM will determine if the problem is wiring (including CPS and CPS connector) or the BCM connector or the BCM.

3) If the rest of the system did not default, verifing the integrity of the 12 volt circuit to the BCM, at

the BCM, will determine if the fault is in the 12 volt circuit or the CPS not providing its logic voltage (7 volts).

4) If the 12 volt circuit to the BCM checked OK, verifying the integrity of the 7 volt logic circuits to the BCM, at the BCM, will determine if the problem is the CPS not providing the logic 7 volts or if the BCM connector, the BCM PROM, or the BCM is at fault.

5) If 7 volts was not present at the BCM, measuring these circuits again at the CPS will determine if the CPS is not supplying 7 volts or if these circuits are open.

1988–90 RIVIERA AND REATTA

CHART B-7
BCM PROBLEM

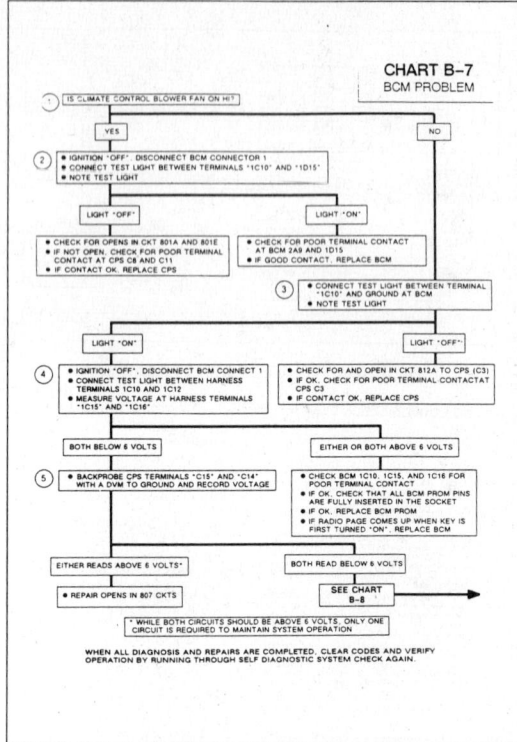

WHEN ALL DIAGNOSIS AND REPAIRS ARE COMPLETED, CLEAR CODES AND VERIFY OPERATION BY RUNNING THROUGH SELF DIAGNOSTIC SYSTEM CHECK AGAIN.

1988–90 RIVIERA AND REATTA

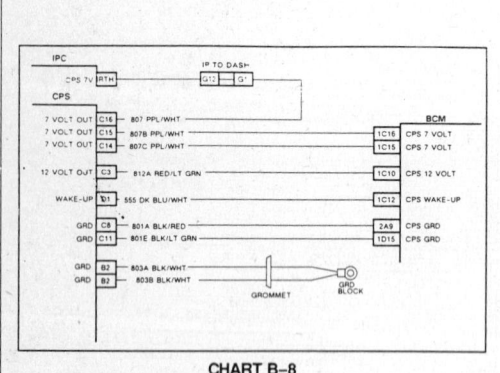

CHART B-8

LOSS OF LOGIC – 7 VOLTS

Circuit Description:

Since the logic 7 volts is not coming out of the CPS the BCM must not be waking it up or there is a circuit failure with the CPS's 7 volt logic circuit including the CPS.

Test Description: Numbers below refer to circled numbers on the diagnostic chart.

1) Since we have established that the CPS is not providing 7 volts out and that the BCM has 12 volts, jumpering 12 volts to the wake-up ("turn-on") circuit at the CPS will verify the integrity of the wake-up circuit that is normally provided by the BCM.

2) If 7 volts still does not leave the CPS when the wake-up circuit is jumpered to 12 volts.

sequentially checking resistance to ground on each of the CPS's 7 volt circuits will verify the integrity of each branch of the circuit and its components. If circuits check OK, the CPS is faults. This check also includes a short to ground on the 12 volt wake-up signal to the CPS from the BCM. When a 7 volt line shorts to ground, there is no fuse to blow as the CPS will internally sense the short and automatically cut power out.

1988–90 RIVIERA AND REATTA

CHART B-8
LOSS OF LOGIC – 7 VOLTS

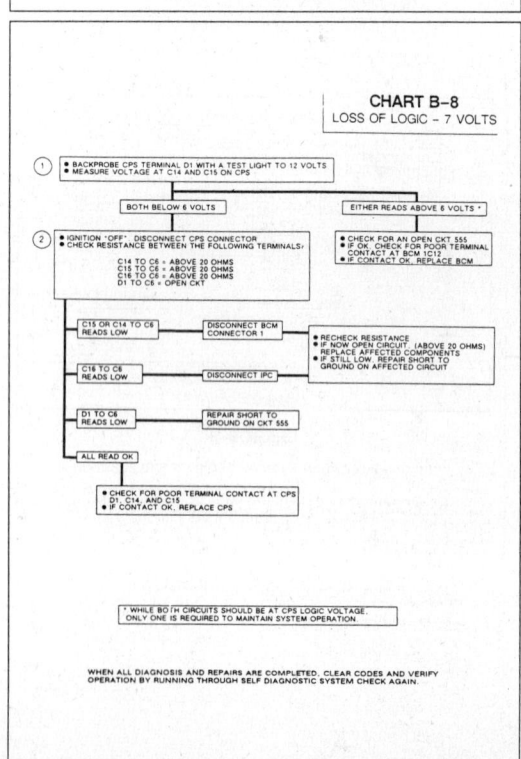

WHEN ALL DIAGNOSIS AND REPAIRS ARE COMPLETED, CLEAR CODES AND VERIFY OPERATION BY RUNNING THROUGH SELF DIAGNOSTIC SYSTEM CHECK AGAIN.

1988–90 RIVIERA AND REATTA

CHART C-1
CRT PROBLEM

Circuit Description:
Following the "Self-Diagnostic System Check" first verifies the integrity of the circuits to the CRTC. Because the CRTC, which drives the CRT, is known to be in communication with the rest of the system, the problem must be with the CRT, CRTC, or the circuitry between the two. At this point, the problem is with the display of the picture on the CRT, or its ability to interpret switch positions touched on the screen. Chart C-1 breaks the problem down into symptoms so the cause of the problem can be identified quickly.

Test Description: Numbers below refer to circled numbers on the diagnostic chart.

1) Following CHART C-3 will isolate problems that can cause the "BEEPER" to not work or sound all the time, or the switches on the face of the screen, including the border switches, to not be recognized. (No switch can be recognized).

2) Following CHART C-4 will isolate problems that can cause some but not all of the switches to work. (At least one works).

3) Following CHART C-5 will isolate problems that can cause the CRT's picture to roll.

4) If the CRT has a border switch LED problem, the CRT must be replaced, unless all LEDs do not function in which case the "BEEPER" and/or switches will also not work if the cause is a circuit problem. It is unlikely that all of the LEDs would burn out all at once.

1988–90 RIVIERA AND REATTA

CHART C-1
CRT PROBLEM

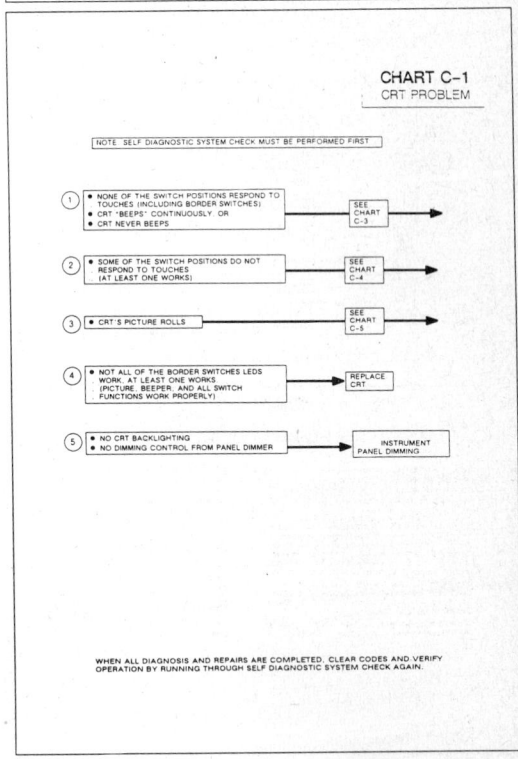

WHEN ALL DIAGNOSIS AND REPAIRS ARE COMPLETED, CLEAR CODES AND VERIFY OPERATION BY RUNNING THROUGH SELF DIAGNOSTIC SYSTEM CHECK AGAIN.

1988–90 RIVIERA AND REATTA

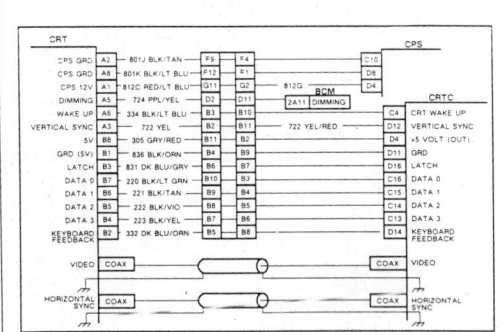

CHART C-2
NO PICTURE

Circuit Description:
Since the problem has been narrowed down to the CRT screen not being able to light, when using tester J-34914 it is not necessary to go through the whole test sequence, only see if the tester can illuminate the screen.

Test Description: Numbers below refer to circled numbers on the diagnostic chart.

1) First noting if any of the border LEDs work will verify the integrity of the 12 volts from the CPS, as this circuit is used to operate the CRT and feed the LEDs.

2) Disconnecting the CRT from the vehicle and connecting the tester to the CRT will verify the integrity of the CRT. The only thing to look for here is the ability of the CRT to light up. If it cannot light up with the tester connected, the CRT connector or CRT is at fault. Make sure the CRT tester's fuse is not blown and that the tester is working properly.

3) If no LEDs light, and the CRT did light with the tester, verify the integrity of the 12 volt circuit from the CPS to the CRT using the CRT's ground. The problem must be power, ground, or wake-up ("turn-on").

4) If test light does not not light, use a known good ground to determine if the problem is the 12 volt circuit or the CRT ground.

5) If the test light lights when checking from the CRT 12 volt to CRT ground circuits, connecting the tester to the harness end at the CRTC will determine if the wake-up ("turn-on") circuit from the CRTC to the CRT (including CRT terminal contact) is at fault or if the CRTC connector or CRT is at fault.

6) We know that the power, ground, and wake-up signals are present at at least one of the LED lights, and that the CRT is OK as the tester was able to light the screen. Connecting the tester to the harness at the CRTC will determine if the CRTC COAX connector or CRT is at fault or if the the video circuit is open or shorted, or if the dimming circuit is shorted to ground, or poor terminal contact at the CRT. Because the tester provides the dimming signal to the CRT but returns control of the dimming signal to the BCM when the tester is connected to the harness at the CRTC, a shorted to ground dimming circuit will cause a blank screen.

1988–90 RIVIERA AND REATTA

CHART C-2
NO PICTURE

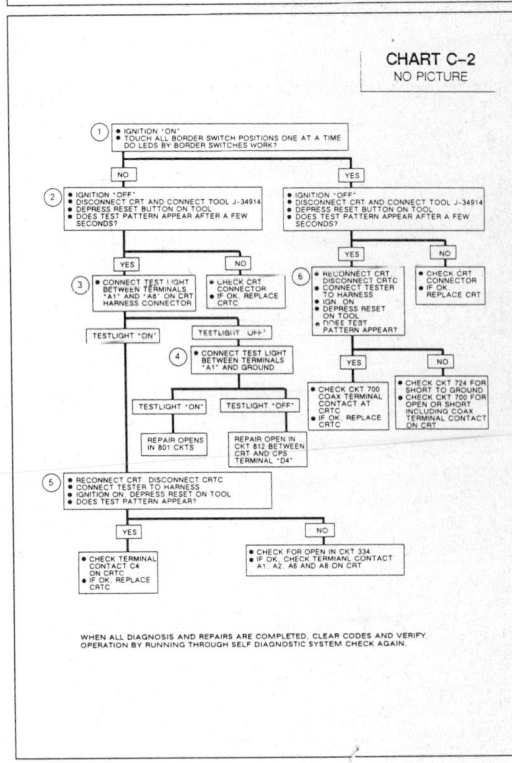

WHEN ALL DIAGNOSIS AND REPAIRS ARE COMPLETED, CLEAR CODES AND VERIFY OPERATION BY RUNNING THROUGH SELF DIAGNOSTIC SYSTEM CHECK AGAIN.

1988–90 RIVIERA AND REATTA

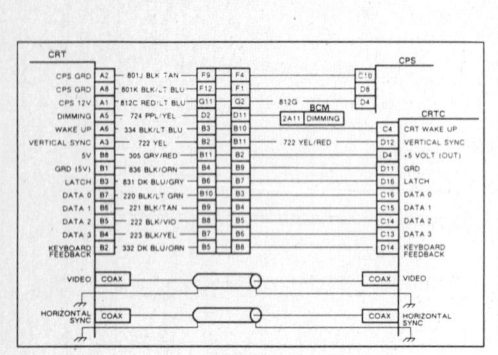

CHART C–3

NO SWITCHES RESPOND OR BEEPER PROBLEM

Circuit Description:
Because we have a picture on the screen, we know the CRT has power, ground, wake-up, video, vertical sync, and horizontal sync signals.

Test Description: Numbers below refer to circled numbers on the diagnostic chart.

1) The summary page will appear first when the key is turned "on". Since diagnostics cannot be accessed, pressing the "Metric" switch on the IPC will verify the BCM's ability to listen to or hear the messages from the other devices on the data line. Viewing the temperature display on the CRT will indicate if the BCM is OK, if the display change form °F to °C.

2) Connecting the tester to the CRT, with the CRT

removed, will verify the integrity of the CRT. If the red light on the tester never lights, or if the CRT beeps continuously, or never beeps, replace the CRT.

3) When the tester was connected to only the CRT, the CRT was OK if the tester lit up the CRT, followed by three green lights on the tester, then the CRT dimmed and brightened followed by a 4th green light and a double "Beep". By moving the tester to the harness at the CRTC and noting tester lights and beeps, the problem can be narrowed down to a faulty circuit or CRTC.

1988–90 RIVIERA AND REATTA

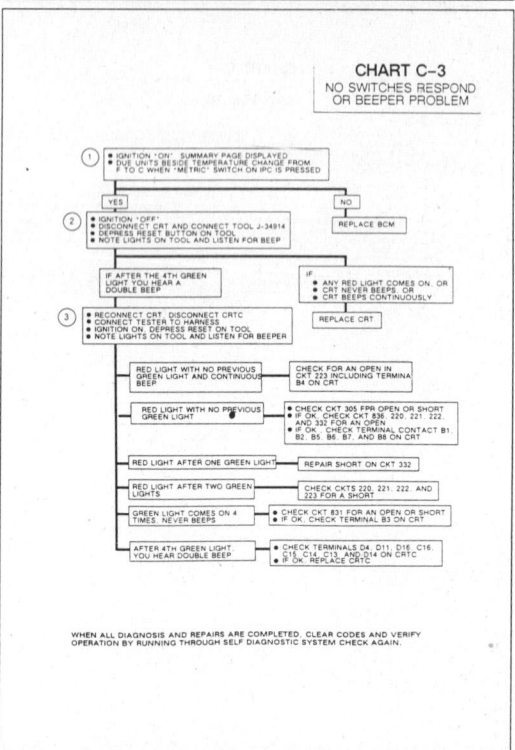

CHART C–3
NO SWITCHES RESPOND OR BEEPER PROBLEM

WHEN ALL DIAGNOSIS AND REPAIRS ARE COMPLETED, CLEAR CODES AND VERIFY OPERATION BY RUNNING THROUGH SELF DIAGNOSTIC SYSTEM CHECK AGAIN.

1988–90 RIVIERA AND REATTA

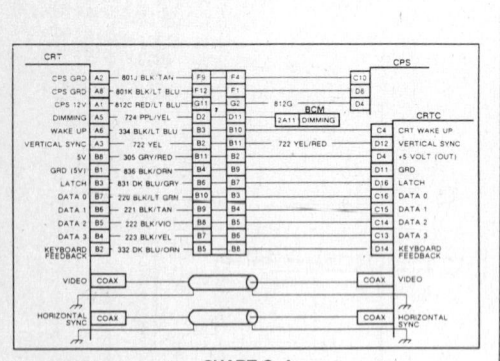

CHART C–4

SOME SWITCHES DO NOT RESPOND

Circuit Description:
If the problem is that some, or only one particular switch pad does not work, chances are the problem is in the CRT.

Test Description: Numbers below refer to circled numbers on the diagnostic chart.

1) Connect tester to CRT only, letting it run its automatic check, to allow individual switch checks to be performed. If, during the automatic check, the tester ever lights its red light or does not beep after the 4th green light, replace the CRT.

2) After the CRT has successfully passed the automatic test, indicated by a double beep after the 4th green light, the individual switch tests can be performed. This is done by touching the border switches or outlined box(s) on the screen, representing all switch combinations, and listening

for beeps. If you hear a beep, the switch works and is being recognized. If it does not beep, the switch is not touching. Be sure you are not touching two switches at once, and touching only the switch that you want. Touching two switches at once will stop the "Beep". This can be useful in verifying that you are on the right switch. When you touch other switches around the one you initially touched and are still touching, the beep stops if the switches are working properly. If the tester beeps with every switch touched, but when the CRT is in the car a switch is not recognized, replace the CRTC. Make sure some switches work, if all switches do not work, you should be on Chart C–3.

1988–90 RIVIERA AND REATTA

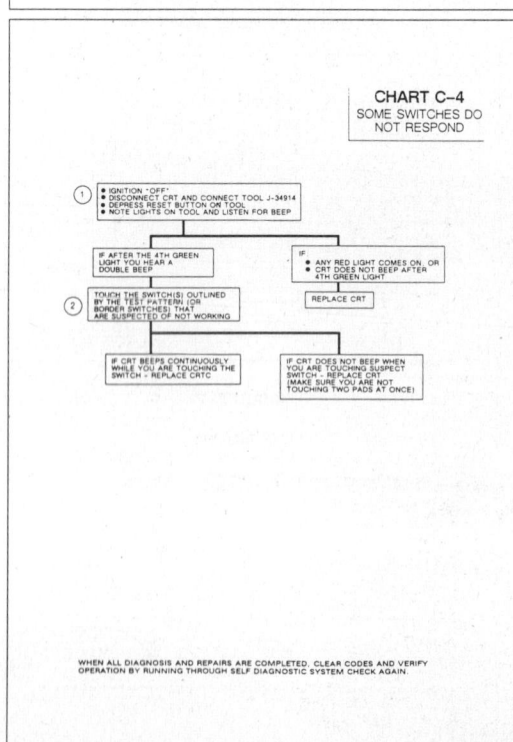

CHART C–4
SOME SWITCHES DO NOT RESPOND

WHEN ALL DIAGNOSIS AND REPAIRS ARE COMPLETED, CLEAR CODES AND VERIFY OPERATION BY RUNNING THROUGH SELF DIAGNOSTIC SYSTEM CHECK AGAIN.

1988–90 RIVIERA AND REATTA

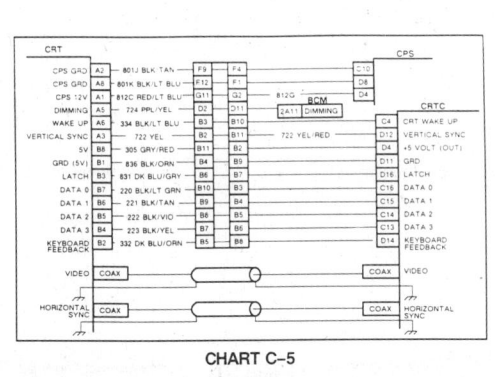

CHART C-5
PICTURE ROLLS

Circuit Description:

This chart is used if the only problem is the picture rolling interested in noting if the picture rolls with the tester connected.

Test Description: Numbers below refer to circled numbers on the diagnostic chart.

1) Connecting the tester to the CRT only will verify the integrity of the CRT.
2) Connecting the tester to the harness at the CRTC will verify the integrity of the vertical and

When using the J-34914 tester here we are only horizontal sync circuits. If the picture does not roll, the CRTC connectors or CRTC are at fault.

3) Noting which way the picture rolls will determine which circuit is at fault. The horizontal sync circuit problem will appear as a snowy picture and the picture roll will be more difficult to see.

1988–90 RIVIERA AND REATTA

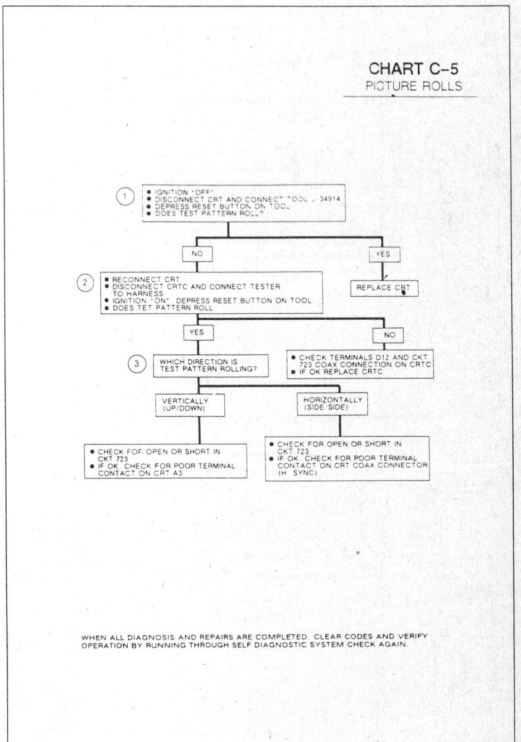

BCM WIRING DIAGRAM — 1988–90 TORONADO

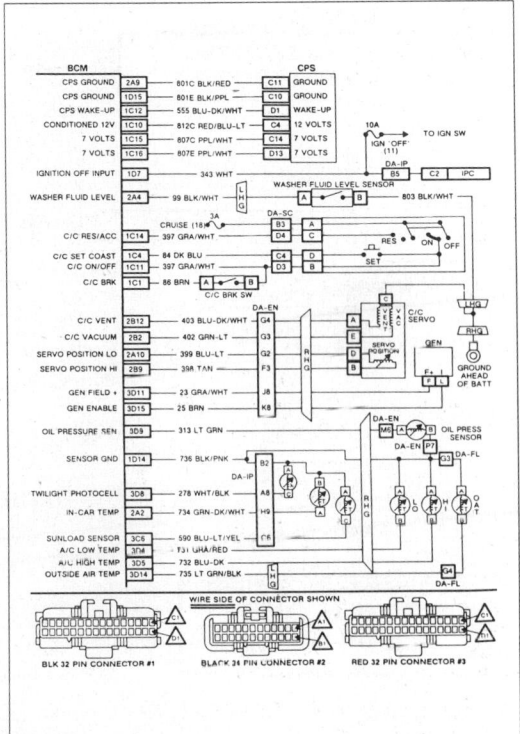

BCM WIRING DIAGRAM — 1988–90 TORONADO

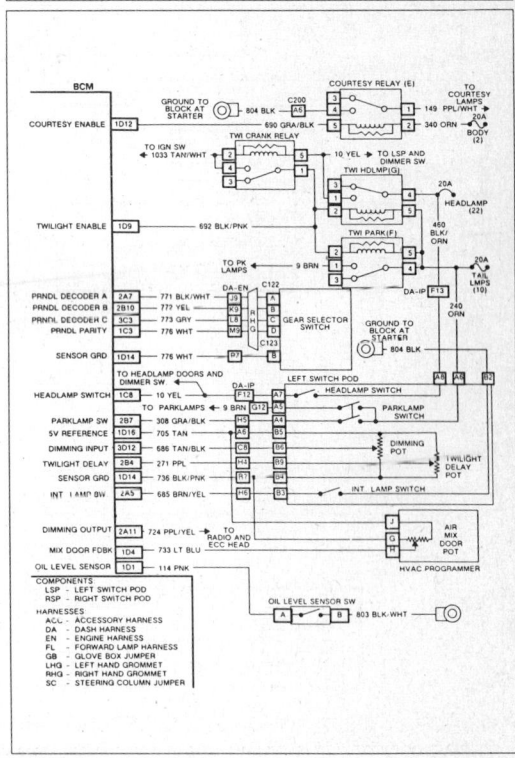

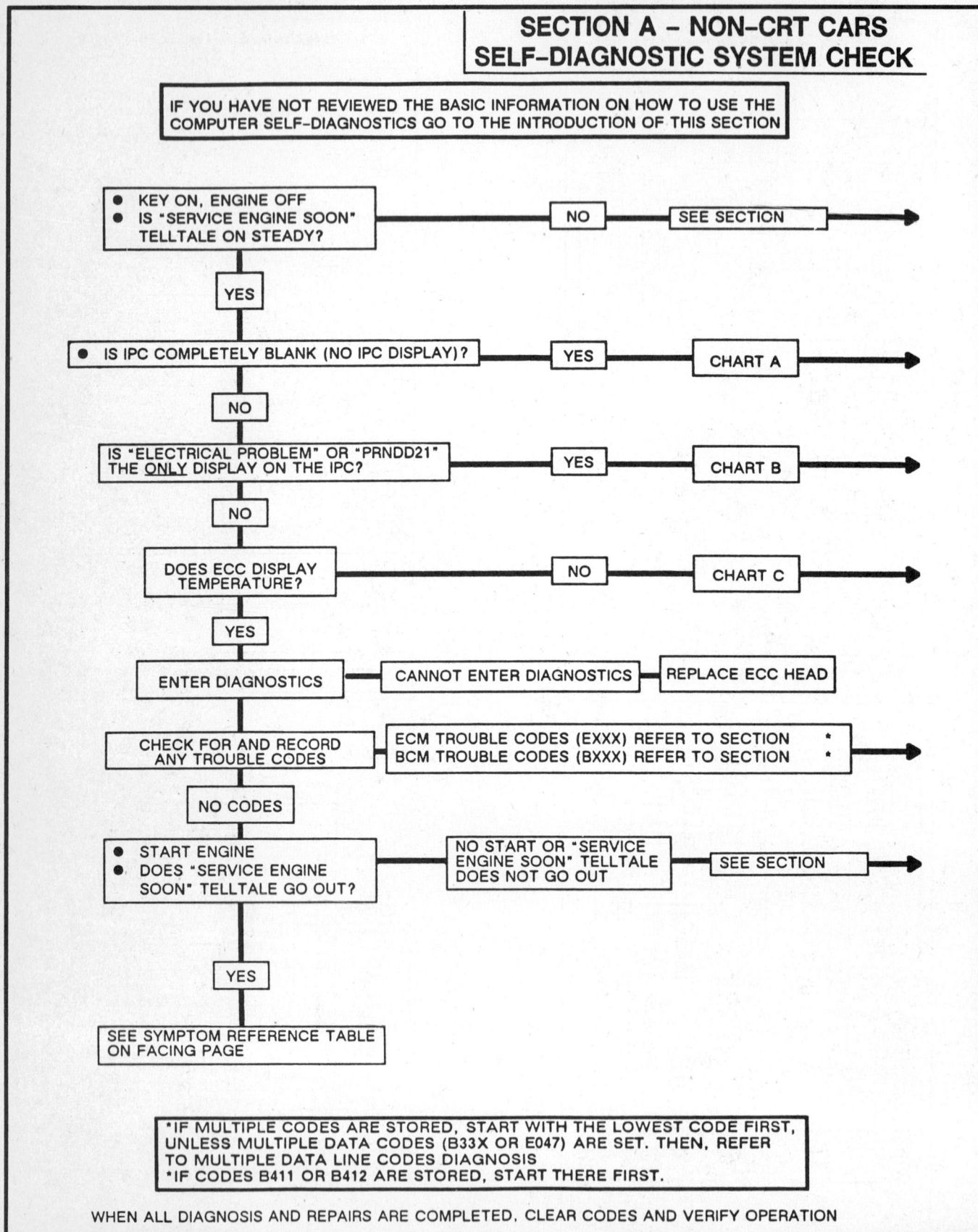

SECTION A – NON–CRT CARS
SELF–DIAGNOSTIC SYSTEM CHECK

IF YOU HAVE NOT REVIEWED THE BASIC INFORMATION ON HOW TO USE THE
COMPUTER SELF-DIAGNOSTICS GO TO THE INTRODUCTION OF THIS SECTION

- KEY ON, ENGINE OFF
- IS "SERVICE ENGINE SOON"
  TELLTALE ON STEADY? → NO → SEE SECTION

YES

- IS IPC COMPLETELY BLANK (NO IPC DISPLAY)? → YES → CHART A

NO

IS "ELECTRICAL PROBLEM" OR "PRNDD21"
THE ONLY DISPLAY ON THE IPC? → YES → CHART B

NO

DOES ECC DISPLAY
TEMPERATURE? → NO → CHART C

YES

ENTER DIAGNOSTICS ← CANNOT ENTER DIAGNOSTICS → REPLACE ECC HEAD

CHECK FOR AND RECORD
ANY TROUBLE CODES → ECM TROUBLE CODES (EXXX) REFER TO SECTION   *
                     BCM TROUBLE CODES (BXXX) REFER TO SECTION   *

NO CODES

- START ENGINE
- DOES "SERVICE ENGINE
  SOON" TELLTALE GO OUT? → NO START OR "SERVICE
                            ENGINE SOON" TELLTALE
                            DOES NOT GO OUT → SEE SECTION

YES

SEE SYMPTOM REFERENCE TABLE
ON FACING PAGE

*IF MULTIPLE CODES ARE STORED, START WITH THE LOWEST CODE FIRST,
UNLESS MULTIPLE DATA CODES (B33X OR E047) ARE SET. THEN, REFER
TO MULTIPLE DATA LINE CODES DIAGNOSIS
*IF CODES B411 OR B412 ARE STORED, START THERE FIRST.

WHEN ALL DIAGNOSIS AND REPAIRS ARE COMPLETED, CLEAR CODES AND VERIFY OPERATION

BCM WIRING DIAGRAM—1988–90 TORONADO

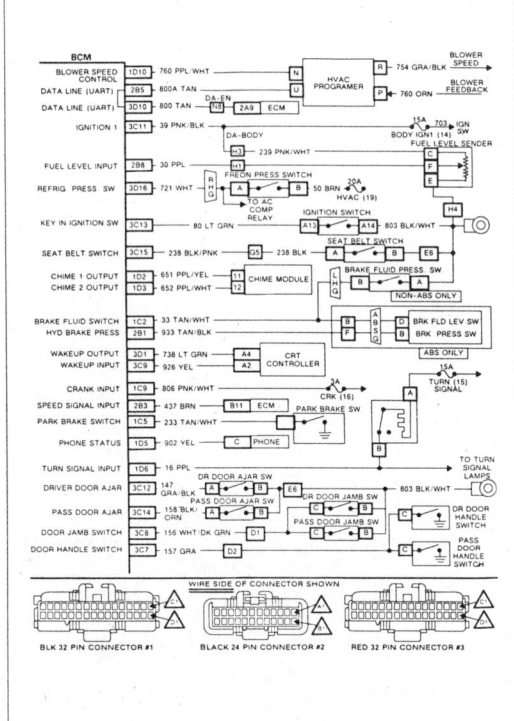

BCM CONNECTOR VIEW—1988–90 TORONADO

1988–90 TORONADO

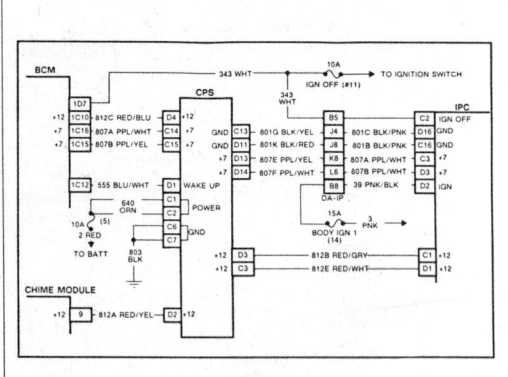

CHART A

LOSS OF IPC DISPLAY

Circuit Description:

A loss of IPC display and messages can result from the following:
- Loss of +12 volt Power to the IPC
- Loss of ground to the IPC
- Loss of +7 volts to the IPC
- Ground in +7 volt CKT 807 anywhere along its length
- Internal CPS, BCM, or IPC fault.

Test Description: Numbers below refer to circled numbers on the diagnostic chart.

1. Checks to see if the BCM is awake. If not, the cause of the blank display is power and ground related.
2. Checks CPS power and ground sources.
3. Checks to see which the CPS has lost, power or ground.
4. Checks for ignition source and ground to the IPC. Normally with the key "ON" battery voltage should be present at C2.
5. Fuse #5 feeds power to the CPS.
6. Checks to see if the fuse #5 failure is due to a short to ground on any branch of circuit 812.
7. Checks to see if the short to ground is inside a component or in the circuit itself.
8. Checks to see which the IPC lost, ignition power or ground.
9. Checks whether the loss of ignition power is due to an open or short to ground.

1988–90 TORONADO

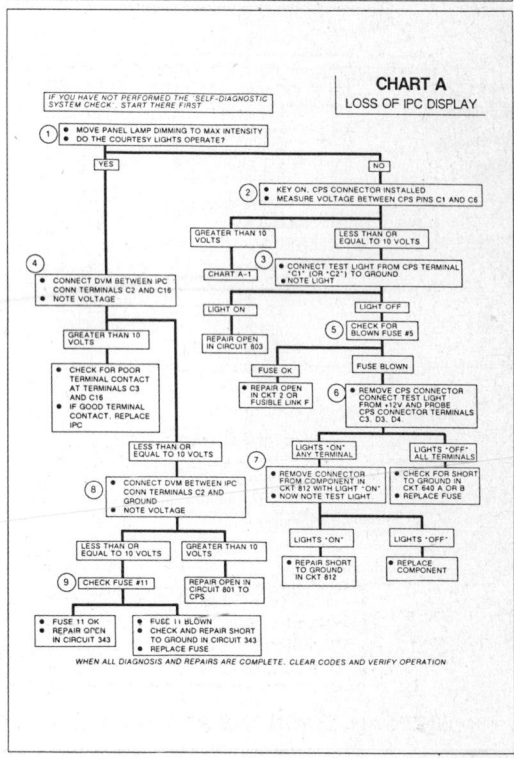

CHART A
LOSS OF IPC DISPLAY

1988–90 TORONADO

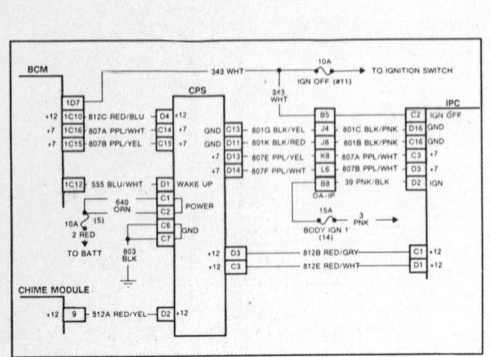

CHART A–1

CENTRAL POWER SUPPLY (CPS) CHECK

Circuit Description:

The Central Power Supply (CPS) provides a filtered 12 volts, 7 volts, and ground for most of the major computers in the vehicle. The BCM is continuously supplied 12 volts from the CPS. When a "wake-up" signal such as opening a door or turning "ON" the ignition is received at the BCM, a 12 volt signal is sent from the BCM to the CPS to turn "ON" the 7 volt power supply. The 7 volt powers the various computers that control the serial data communications.

The BCM needs constant 12 volt supply from the CPS plus a ground to function. The BCM must supply the wake-up signal to activate the 7V supply in the CPS.

Test Description: Numbers below refer to circled numbers on the diagnostic chart.

1. Check for wake-up signal back to the CPS. If test

light is "ON", the wake-up signal is occurring and the fault is in the 7 volt circuit. No light indicates that the wake-up is not occurring.
2. Checks for source of short by testing all 7V circuits and components on those circuits for a short to ground. If the 7V circuits are OK, then fault is in the IPC or BCM.
3. Checks whether the BCM or IPC are faulty.
4. Checks the 12V power circuit from the CPS to the BCM.
5. Checks at the BCM for the 12V into and wake-up out of the BCM.
6. Checks for source of ignition 3 at the BCM. Ignition 3 is a source of wake-up signal
7. Checks to see if the BCM is receiving a crank signal. If there is 12V on pin 3C11 (the crank circuit), the BCM goes to sleep and brings the CPS down too.

1988–90 TORONADO

CHART A–1
CPS CHECK

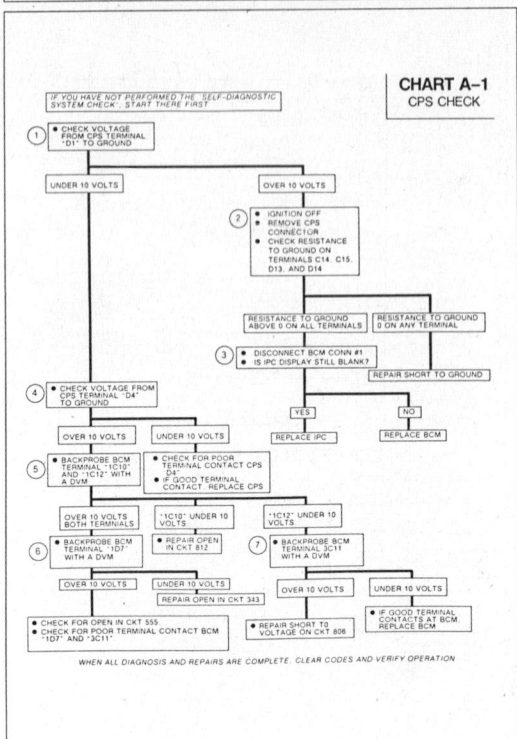

1988–90 TORONADO

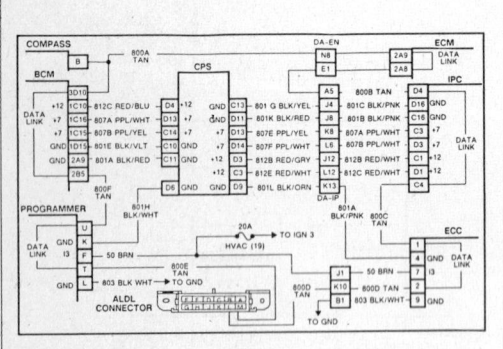

CHART B

LOSS OF SERIAL DATA

Circuit Description:

The IPC indicates a loss of serial communication by displaying only the message "Electrical Problem". All other segments are blank. If the display is fully illuminated and an "Electrical Problem" message is also displayed, enter diagnostics and proceed to the appropriate code chart. Do not use CHART A.

Loss of serial data communications can occur for the following reasons:

- Short to ground anywhere in the 800 CKT.
- Short to voltage anywhere in the 800 CKT.
- Two opens in the 800 CKT.

- Loss of 7 volt supply which powers serial communication.
- Internal BCM or IPC fault.

Test Description: Numbers below refer to circled numbers on the diagnostic chart.

1. Checks to see if the ECC head has lost data communication along with the IPC.
2. Since the IPC has lost data line communications, creating an open in the data line at the ALDL connector should remove data communications from the ECC head as well. If not, the IPC is receiving data but is unable to communicate due to a poor connection or an internal problem.

1988–90 TORONADO

CHART B
LOSS OF SERIAL DATA

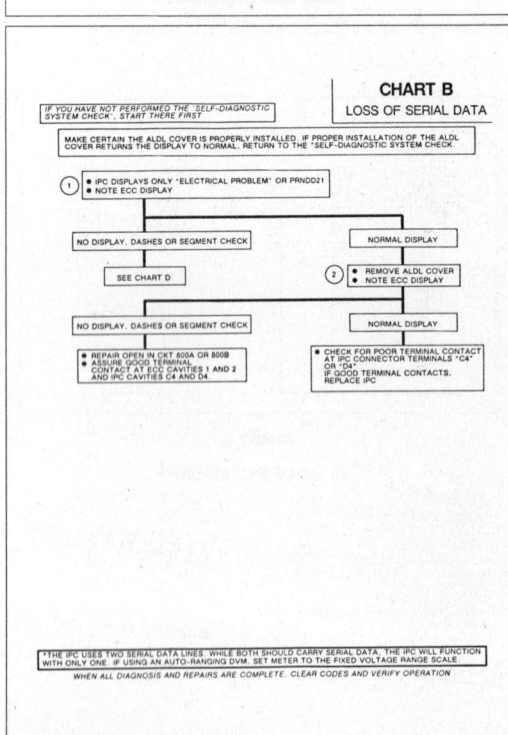

1988–90 TORONADO

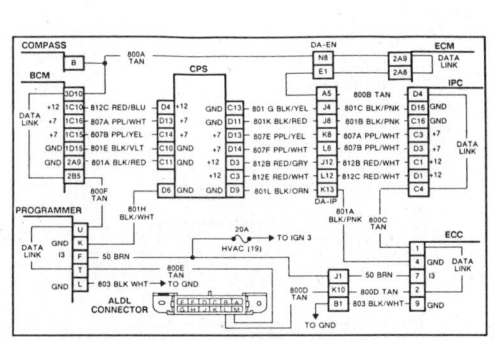

CHART C
LOSS OF ECC COMMUNICATIONS

Circuit Description:

When the ECC communications are lost service diagnostics usually cannot be entered but a quick scan of the panel will usually help isolate the source of the fault. The ECC panel is powered by ignition. If the power source is lost the secondary display will usually display a segment check. Loss of serial data will result in three dashes being shown on the panel VF display. This indicates power and ground are OK. If the panel displays normal outside or interior set temperature, power, ground, and serial communications are being received thus panel replacement is indicated.

Test Description: Numbers below refer to circled numbers on the diagnostic chart.

1. This step checks for power and ground to the panel. Test light should be "ON" for both power to ground checks.

2. Step checks for serial data voltage. Normal voltage should be above 1 volt and varying. A steady voltage indicates a second open exists, or the ALDL cover is off. If a second open is suspected, see CHART B-5.

1988–90 TORONADO

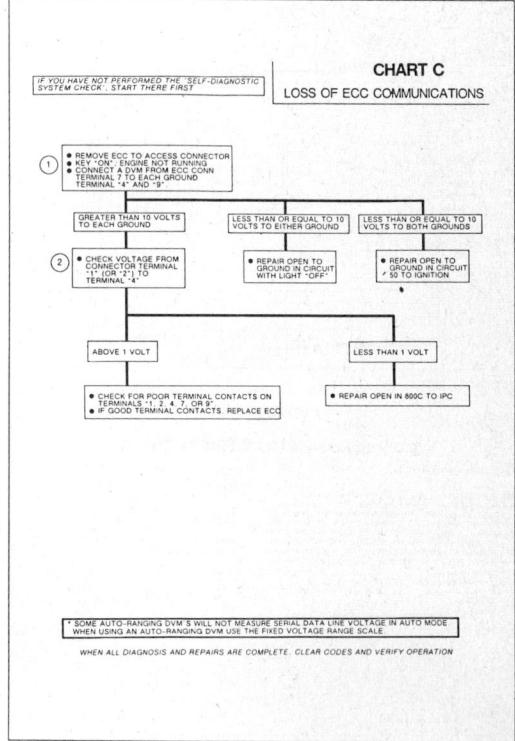

1988–90 TORONADO

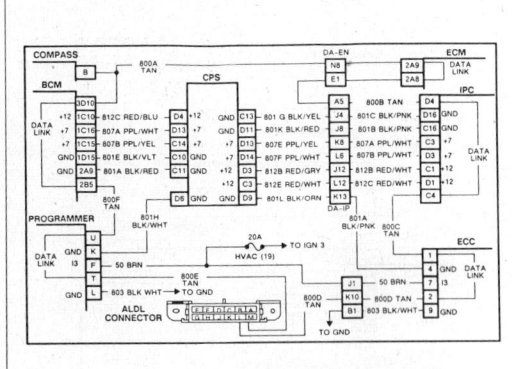

CHART D
LOSS OF COMMUNICATIONS

Circuit Description:

Test Description: Numbers below refer to circled numbers on the diagnostic chart.

1. Checks continuity of the data line (circuit 800). Since the data line is a loop, resistance around it should be close to 0 ohms.

2. Checks for a short to ground on the data line.

3. Checks voltage on the data line under normal operating conditions. Normal data line voltage is between 1 and 5 volts and continually fluctuating. A fixed voltage or a voltage above 5 volts indicates a problem on the data line which may cause the entire communications network to be inoperative.

1988–90 TORONADO

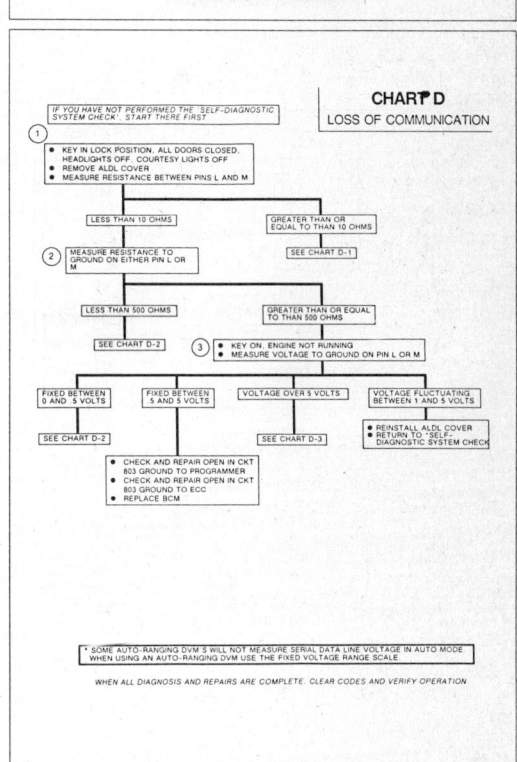

1988–90 TORONADO

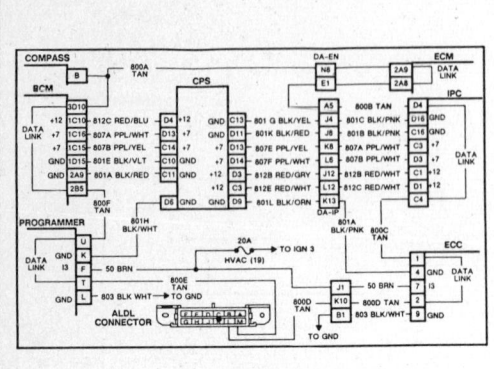

CHART D–1

SERIAL DATA CIRCUIT OPEN

Circuit Description:

The serial data line is redundant at most of the devices that it communicates with. While only one line need be connected to each device for that system to function, when making repairs both lines should be made operational. If when the ALDL cover is removed the system shuts down, a single open existed before, and a second open is being created by removing the ALDL

cover.

Test Description: Numbers below refer to circled numbers on the diagnostic chart.

1. This test and all subsequent tests are designed to subdivide the data line. This is necessary in order to pinpoint the location of an open in the data line. This is not the only way to pinpoint an open, but, in general, will most quickly lead to the open.

1988–90 TORONADO

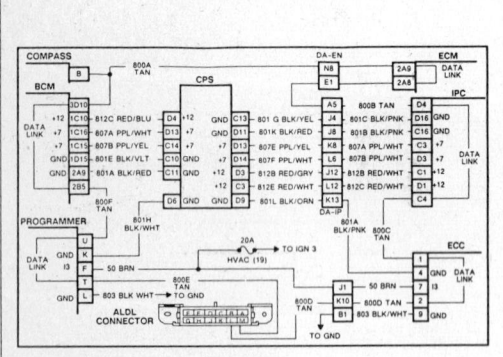

CHART D–2

GROUNDED SERIAL DATA CIRCUIT

Circuit Description:

With a grounded serial data circuit the BCM will not be able to communicate with any other devices in the system. Voltage measured anywhere along the circuit will be zero. The display on the IPC will be "Electrical Problem" and the ECC will display dashes.

Test Description: Numbers below refer to circled numbers on the diagnostic chart.

1. This step splits the system in half to help isolate

the source of the grounded serial data line. If taking the ECM out of the system allows both halves of serial data line to rise above 1 volt the source of the short has been located. The remaining steps follow the same pattern in that the components are removed from the serial data line until the normal voltage range is again measured. When the section of the circuit is isolated a simple check of the wire for a short to ground differentiates between a grounded wire or a ground in the device remote.

1988–90 TORONADO

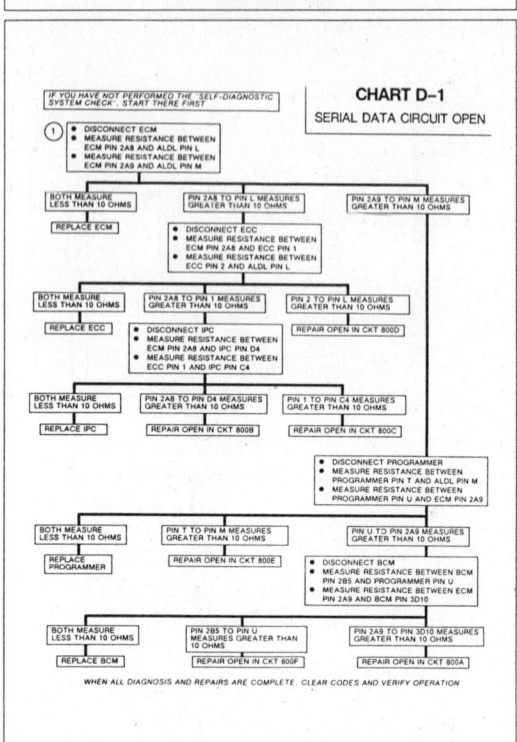

1988–90 TORONADO

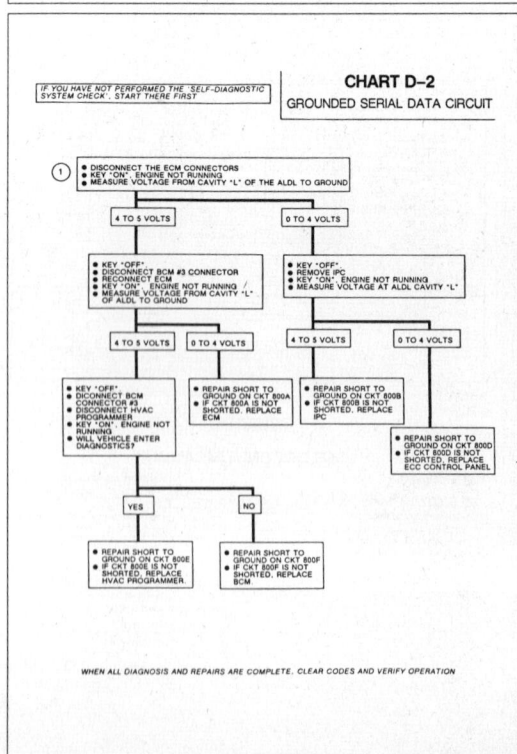

1988–90 TORONADO

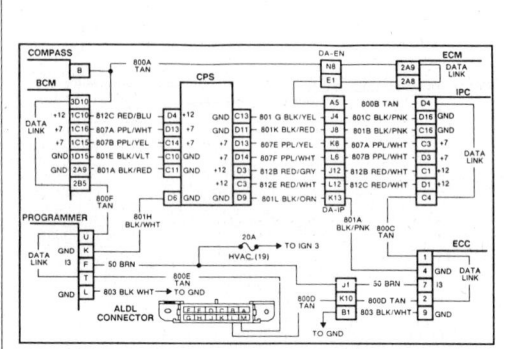

CHART D–3

SERIAL DATA CIRCUIT SHORTED TO VOLTAGE

Circuit Description:

With a shorted serial data circuit the BCM will not be able to communicate with the other devices in the system. Voltage measured anywhere along the circuit will be above 5 volts. The only display on the IPC is "Electrical Problem".

Test Description: Numbers below refer to circled numbers on the diagnostic chart.

1. This step splits the system in half to help isolate

the source of the short in the serial data line. If taking the ECM out of the system allows both halves of the serial data line to fall below 5 volts, the source of the short was in the ECM. The remaining steps follow the same pattern in that the components are removed from the serial data line until the normal voltage range is again measured.

1988–90 TORONADO

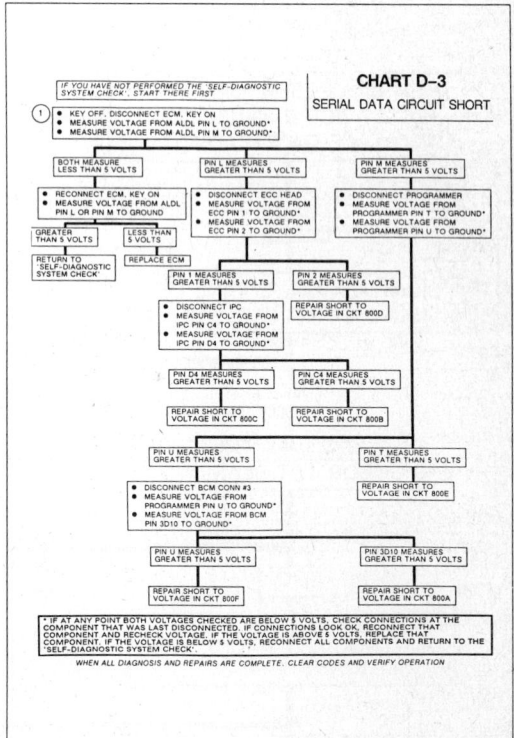

CHART D–3
SERIAL DATA CIRCUIT SHORT

1988–90 TORONADO

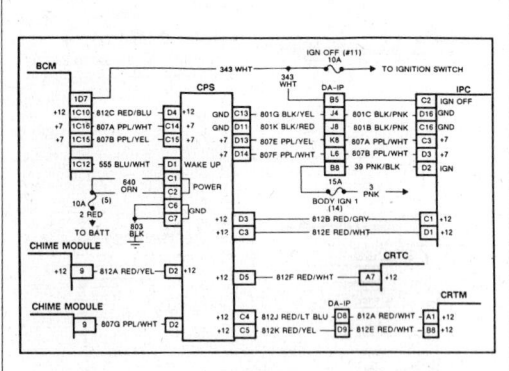

CHART A

LOSS OF IPC DISPLAY

Circuit Description:

A loss of IPC display and messages can result from the following:

- Loss of +12 volt Power to the IPC
- Loss of ground to the IPC
- Loss of +7 volts to the IPC
- Ground in +7 volt CKT 807 anywhere along its length
- Internal BCM or IPC fault.

Test Description: Numbers below refer to circled numbers on the diagnostic chart.

1. Checks to see if the BCM is awake. If not, the cause of the blank display is power and ground related.

2. Checks CPS power and ground sources.
3. Checks to see which the CPS has lost, power or ground.
4. Checks for ignition source and ground to the IPC. Normally with the key "ON" battery voltage should be present at C2.
5. Fuse #5 feeds power to the CPS.
6. Checks to see if the fuse #5 failure is due to a short to ground on any branch of circuit 812.
7. Checks to see if the short to ground is inside a component or in the circuit itself.
8. Checks to see which the IPC lost, ignition power or ground.
9. Checks whether the loss of ignition power is due to an open or short to ground.

1988–90 TORONADO

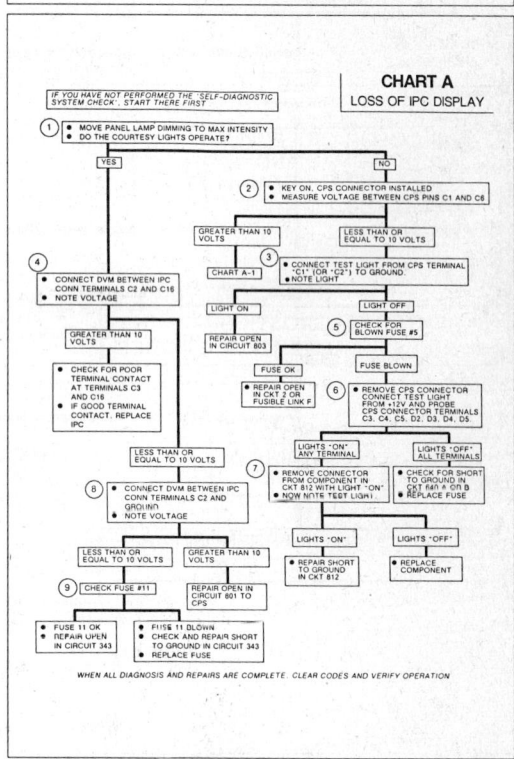

CHART A
LOSS OF IPC DISPLAY

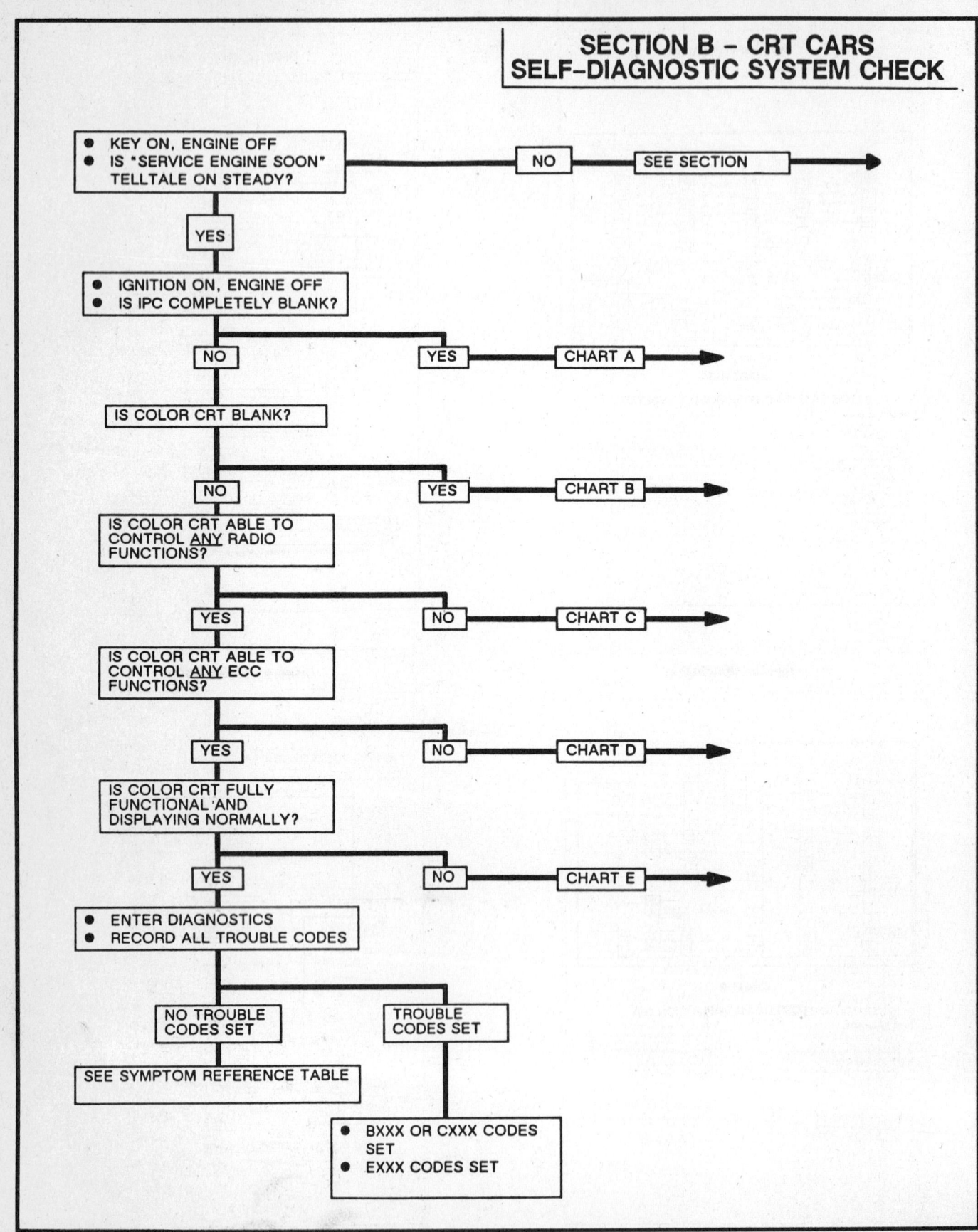

SECTION B – CRT CARS
SELF–DIAGNOSTIC SYSTEM CHECK

- KEY ON, ENGINE OFF
- IS "SERVICE ENGINE SOON" TELLTALE ON STEADY? — NO → SEE SECTION

YES

- IGNITION ON, ENGINE OFF
- IS IPC COMPLETELY BLANK?

NO / YES → CHART A

IS COLOR CRT BLANK?

NO / YES → CHART B

IS COLOR CRT ABLE TO CONTROL ANY RADIO FUNCTIONS?

YES / NO → CHART C

IS COLOR CRT ABLE TO CONTROL ANY ECC FUNCTIONS?

YES / NO → CHART D

IS COLOR CRT FULLY FUNCTIONAL AND DISPLAYING NORMALLY?

YES / NO → CHART E

- ENTER DIAGNOSTICS
- RECORD ALL TROUBLE CODES

NO TROUBLE CODES SET / TROUBLE CODES SET

SEE SYMPTOM REFERENCE TABLE

- BXXX OR CXXX CODES SET
- EXXX CODES SET

1988–90 TORONADO

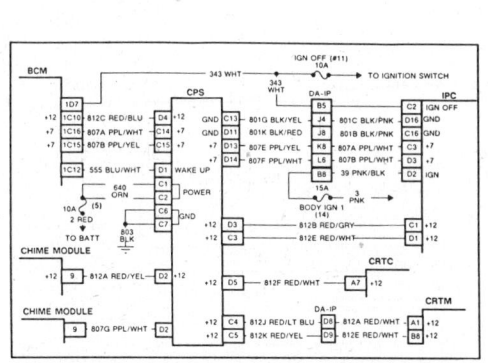

CHART A-1
CENTRAL POWER SUPPLY (CPS) CHECK

Circuit Description:

The Central Power Supply (CPS) provides a filtered 12 volts, 7 volts, and ground for most of the major computers in the system. The BCM is continuously supplied 12 volts from the CPS. When a "wake-up" signal such as opening a door or turning "ON" the ignition is received at the BCM, a 12 volt signal is sent from the BCM to the CPS to turn "ON" the 7 volt power supply. The 7 volt powers the various computers that control the serial data communications.

The BCM needs constant 12 volt supply from the CPS plus a ground to function. The BCM must supply the wake-up signal to activate the 7 volt supply in the CPS.

Test Description: Numbers below refer to circled numbers on the diagnostic chart.

1 Check for wake-up signal back to the CPS. If test

light is "ON", the wake-up signal is occurring and the fault is in the 7 volt circuit. No light indicates that the wake-up is not occurring.

2. Checks for source of short by testing all 7V circuits and components on those circuits for a short to ground. If the 7V circuits are OK, then fault is in the IPC or BCM.

3. Checks whether the BCM or IPC are faulty.

4. Checks the 12V power circuit from the CPS to the BCM.

5. Checks at the BCM for the 12V into and wake-up out of the BCM.

6. Checks for source of ignition 3 at the BCM. Ignition 3 is a source of wake-up signal.

7. Checks to see if the BCM is receiving a crank signal. If there is 12V on pin 3C11 (the crank circuit), the BCM goes to sleep and brings the CPS down too.

1988–90 TORONADO

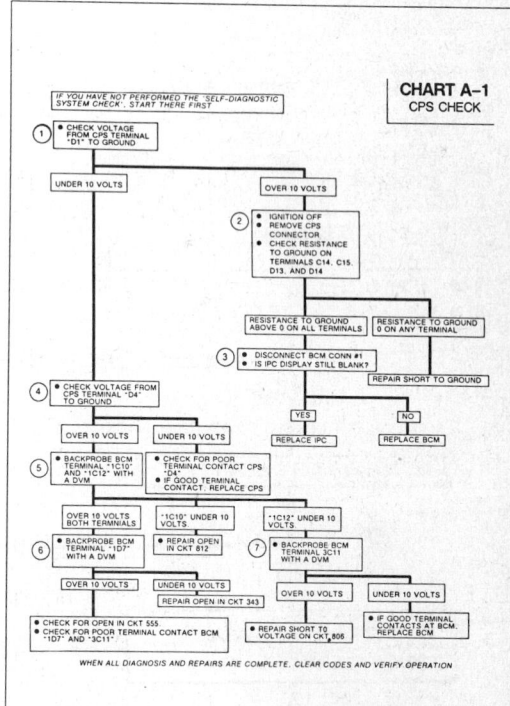

CHART A-1
CPS CHECK

1988–90 TORONADO

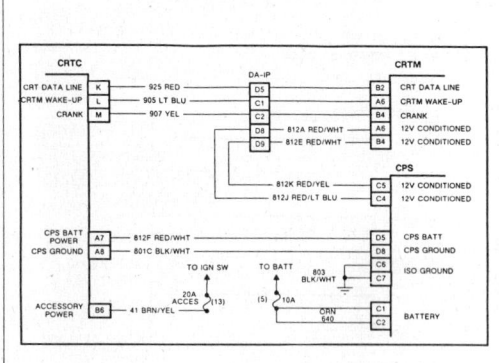

CHART B
COLOR CRT BLANK WITH IGNITION ON

Circuit Description:

Test Description: Numbers below refer to circled numbers on the diagnostic chart.
1. Checks power supplies to the CPS and CRTC.

2. Determines whether the CRT/CRTC system is awake and operating.
3. Checks voltage on the CRTC to CRT data line. Normal voltage is between 1.5 and 4.5 volts and fluctuates some.
4. Checks whether the CRT or CRTC are causing the low voltage reading on the data line.

1988–90 TORONADO

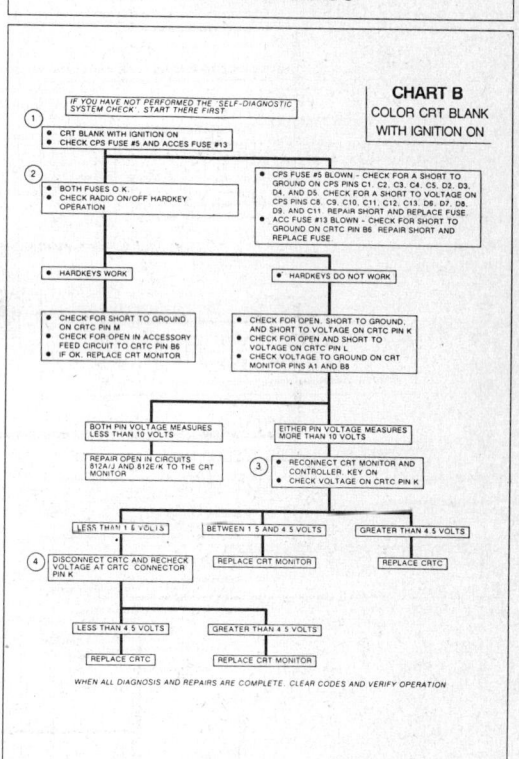

CHART B
COLOR CRT BLANK WITH IGNITION ON

1988–90 TORONADO

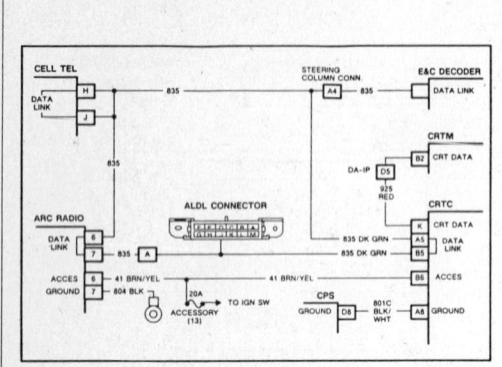

CHART C
COLOR CRT UNABLE TO CONTROL RADIO

Circuit Description:

Test Description: Numbers below refer to circled numbers on the diagnostic chart.

1. Be sure and check hardkeys for all subsystems, ECC, Radio and BCM. This will determine whether the problem is in the E&C communications line or in the CRT system itself.
2. The E&C communication line differs from the 800 circuit data line in that it is not always active. Only

when a radio button is activated or when the display is updated does the line voltage fluctuate. This allows us to check and see if the CRTC can and is communicating when a radio button is pressed. This test may be used with other components on the line as well (i.e the steering wheel controls and the cellular telephone).

3. Checks to see if the data line shorted to ground. Normal voltage is 12 volts and will drop 1 or 2 volts when a component is communicating on it.
4. Checks if the CRTC has the power and ground it uses to drive the data line communication.
5. Checks if the CRTC is shorting the E&C bus to ground.

1988–90 TORONADO

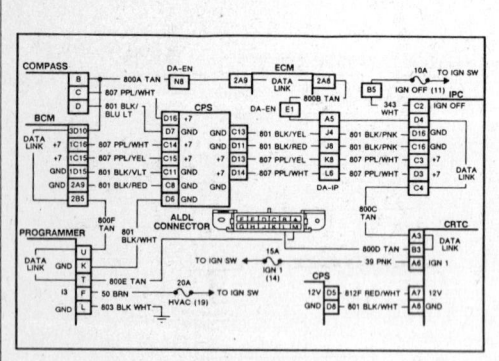

CHART D
COLOR CRT UNABLE TO CONTROL ECC

Circuit Description:

The IPC indicates a loss of serial data communication by displaying only the message "Electrical Problem". All other segments are blank. If the display is fully illuminated and an "Electrical Problem" message is also displayed, enter diagnostics and proceed to the appropriate code chart. Do not use CHART A.

Loss of serial data communications can occur for the following reasons:

- Short to ground anywhere in the 800 CKT.
- Short to voltage anywhere in the 800 CKT.
- Double open in the 800 CKT.
- Loss of 7 volt supply which powers serial communication.
- Internal BCM or IPC fault.

Test Description: Numbers below refer to circled numbers on the diagnostic chart.

1. If the status page is accessible, the CRTC is receiving ignition power and should be capable of communicating with the BCM.
2. Checks to see if the CRTC has lost communication with the BCM.
3. Checks to see if the IPC has lost communication with the BCM as well.
4. Checks continuity of the data line (circuit 800). Since the data line is a loop, resistance around it should be close to 0 ohms.
5. Checks for a short to ground on the data line.
6. Checks voltage on the data line under normal operating conditions. Normal data line voltage is between 1 and 5 volts and continually fluctuating. A fixed voltage or a voltage above 5 volts indicates a problem on the data line which may cause the entire communications network to be inoperative.

1988–90 TORONADO

CHART C
COLOR CRT UNABLE TO
CONTROL RADIO

IF YOU HAVE NOT PERFORMED THE 'SELF-DIAGNOSTIC SYSTEM CHECK', START THERE FIRST

① ARE ALL CRT HARDKEYS INOPERATIVE?

NO — YES

YES → REPLACE CRT MONITOR

② ★ IF CAR IS EQUIPPED WITH STEERING WHEEL CONTROLS AND STEERING WHEEL CONTROLS ARE FUNCTIONAL, CHECK CONNECTION AT CRTC. IF OK, REPLACE CRT CONTROLLER
- REMOVE ALDL CONNECTOR COVER
- SET DVOM ON 2 VOLT AC SCALE
- MEASURE VOLTAGE BETWEEN ALDL TERMINALS J AND A WHILE ACTUATING CRT RADIO BUTTONS
- DURING EACH BUTTON ACTUATION, A MOMENTARY VOLTAGE READING OF APPROXIMATELY 1 VOLT SHOULD REGISTER

- MOMENTARY VOLTAGE READING IS REGISTERED
- REFER TO RADIO DIAGNOSIS

③ • MOMENTARY VOLTAGE READING IS NOT REGISTERED
- CHECK DC VOLTAGE READING ON ALDL TERMINALS J AND A

10 VOLTS OR GREATER | LESS THAN 10 VOLTS

④ • DISCONNECT CRTC
- MEASURE VOLTAGE BETWEEN CRTC PINS B6 AND A8

⑤ • CHECK FOR SHORT TO GROUND ON CKT 835
- IF NO SHORT, DISCONNECT CRTC AND RECHECK VOLTAGE BETWEEN ALDL TERMINALS J AND A

10 VOLTS OR GREATER | LESS THAN 10 VOLTS

- CHECK FOR SHORT TO VOLTAGE ON CKT 835
- IF NO SHORT, REPLACE COLOR CRT CONTROLLER

REPAIR OPEN IN CIRCUIT 801C

10 VOLTS OR GREATER | LESS THAN 10 VOLTS

REPLACE CRTC | REFER TO RADIO DIAGNOSIS

WHEN ALL DIAGNOSIS AND REPAIRS ARE COMPLETE, CLEAR CODES AND VERIFY OPERATION

1988–90 TORONADO

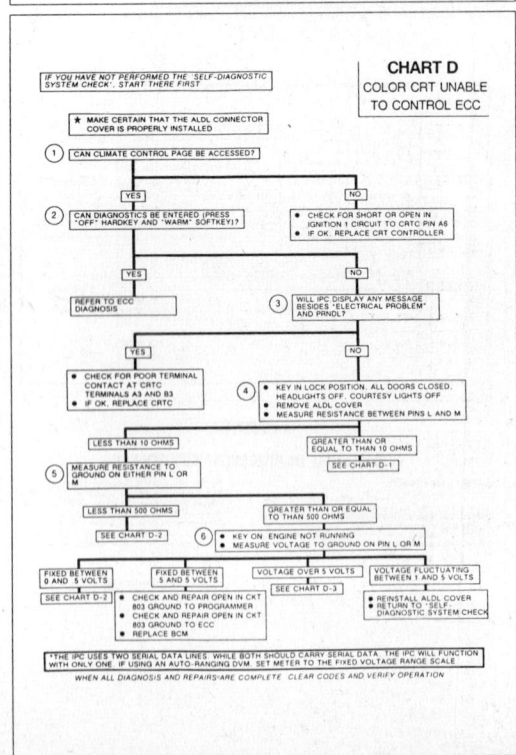

CHART D
COLOR CRT UNABLE
TO CONTROL ECC

WHEN ALL DIAGNOSIS AND REPAIRS ARE COMPLETE, CLEAR CODES AND VERIFY OPERATION

1988–90 TORONADO

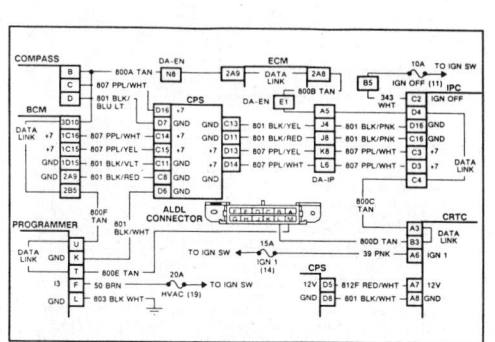

CHART D-1

SERIAL DATA CIRCUIT OPEN

Circuit Description:

The serial data line is redundant at most of the devices that it communicates with. While only one line need be connected to each device for that system to function, when making repairs both lines should be made operational. If when the ALDL cover is removed the system shuts down, a single open existed before, and a second open is being created by removing the ALDL

cover.

Test Description: Numbers below refer to circled numbers on the diagnostic chart.

1. This test and all subsequent tests are designed to subdivide the data line. This is necessary in order to pinpoint the location of an open in the data line. This is not the only way to pinpoint an open, but, in general, will most quickly lead to the open.

1988–90 TORONADO

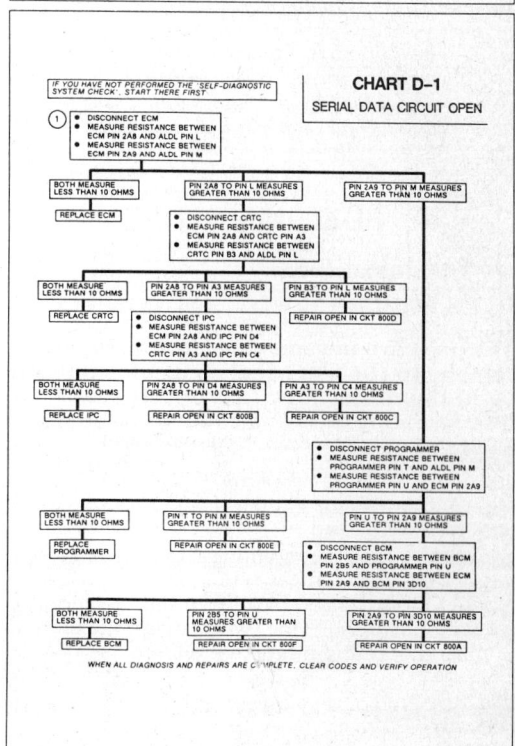

1988–90 TORONADO

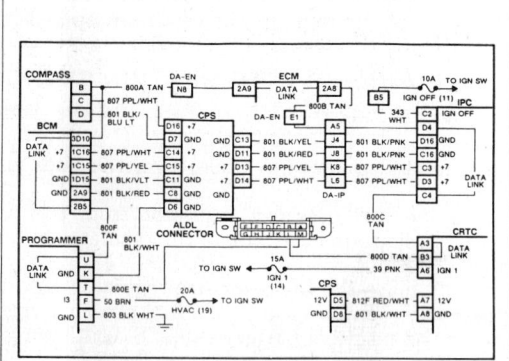

CHART D-2

GROUNDED SERIAL DATA CIRCUIT

Circuit Description:

With a grounded serial data circuit the BCM will not be able to communicate with any other devices in the system. Voltage measured anywhere along the circuit will be zero. The display on the IPC will be "Electrical Problem" and the CRT will display normally, but HVAC controls will not function.

Test Description: Numbers below refer to circled numbers on the diagnostic chart.

1. This step splits the system in half to help isolate

the source of the grounded serial data line. If taking the ECM out of the system allows both halves of serial data line to rise above 1 volt the source of the short has been located.

The remaining steps follow the same pattern in that the components are removed from the serial data line until the normal voltage range is again measured. When the section of the circuit is isolated a simple check of the wire for a short to ground differentiates between a grounded wire or a ground in the device remote.

1988–90 TORONADO

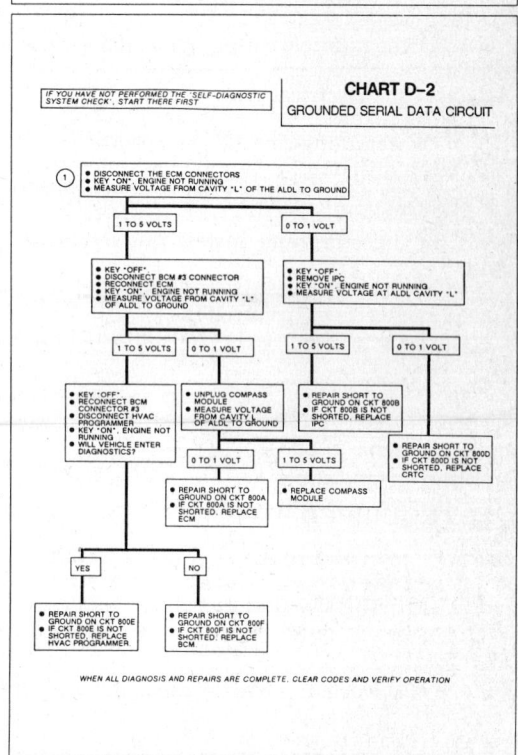

Component Replacement

--- CAUTION ---

To reduce the risk of fire and personal injury, it is necessary to relieve the fuel system pressure before servicing fuel system components. After relieving the system pressure, a small amount of fuel may be released when servicing fuel lines or connections. In order to reduce the chance of personal injury, cover the fuel line fittings with a shop towel before disconnecting, to catch any fuel that may leak out. Place the towel in an approved container when disconnect is completed.

FUEL INJECTORS

Removal and Installation

VEHICLES EQUIPPED WITH 3.8L (VIN 3) ENGINE

NOTE: Use care in removing the fuel injectors to prevent damage to the electrical connector pins on the injector and the nozzle. The fuel injector is serviced as a complete assembly only and should not be immersed in any kind of cleaner.

1. Relieve fuel system pressure.
2. Remove the injector electrical connections.
3. Remove the fuel rail.
4. Remove the injector retaining clip (if used). Separate the injector from the fuel rail.
5. Installation is the reverse of removal. Replace the O-rings with new and lubricate O-rings with engine oil before installing injectors into fuel rail and intake manifold.

VEHICLES EQUIPPED WITH 3800 (VIN C) ENGINE

NOTE: Use care in removing the injectors to prevent damage to the electrical connector pins on the injector and the nozzle. The fuel injector is serviced as a complete assembly only. The fuel injector is an electrical component and should not be immersed in any type of cleaner.

1. Disconnect the negative battery cable. Relieve the fuel system pressure.
2. Disconnect the electrical connections.
3. Remove the fuel rail.
4. Remove the retaining clips (if equipped).
5. Remove the injectors.
6. To install, use new O-rings. Coat the O-rings with engine oil.
7. Install the injectors and install the retaining clips (if equipped).
8. Install the fuel rail.
9. Connect the electrical connections.
10. Connect the negative battery cable. Start the engine and check for leaks.

FUEL RAIL

Removal and Installation

VEHICLES EQUIPPED WITH 3800 (VIN C) ENGINE

NOTE: When servicing the fuel rail assembly, precautions must be taken to prevent dirt and other contaminants from entering the fuel passages. It is recommended that fittings be capped and holes plugged during servicing.

1. Disconnect the negative battery cable. Relieve the fuel system pressure.
2. Disconnect the electrical connections.
3. Remove the fuel rail.

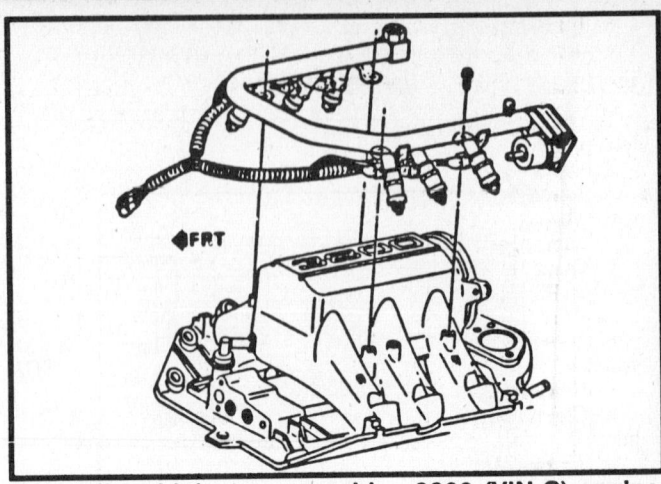

Fuel rail and injector assembly—3800 (VIN C) engine shown

4. Before disassembly, the fuel rail assembly may be cleaned with a spray type engine cleaner, such as AC Delco® X–30A or equivalent, following the package instructions. The fuel rail should not be immersed in liquid solvent.
5. To install, use new O-rings. Coat the O-rings with engine oil.
6. Install the fuel rail.
7. Connect the electrical connections.
8. Connect the negative battery cable. Start the engine and check for leaks.

FUEL PRESSURE REGULATOR

Removal and Installation

VEHICLES EQUIPPED WITH 3.8L (VIN3) ENGINE

1. Relieve fuel system pressure.
2. Remove pressure regulator from fuel rail. Place a rag around the base of the regulator to catch any spilled fuel.
3. Installation is the reverse of removal.

VEHICLES EQUIPPED WITH 3800 (VIN C) ENGINE

1. Disconnect the negative battery cable. Relieve the fuel system pressure.
2. Disconnect the vacuum hose from the fuel pressure regulator.
3. Remove the fuel pressure regulator from the fuel rail. Place a shop towel around the base of the regulator to catch any spilled fuel.
4. To install, place the fuel pressure regulator onto the fuel rail. Connect the vacuum hose to the fuel pressure regulator.
5. Connect the negative battery cable. Start the engine and check for leaks.

THROTTLE BODY ASSEMBLY

Removal and Installation

VEHICLES EQUIPPED WITH 3.8L (VIN 3) ENGINE

1. Disconnect the negative battery cable.
2. Remove the air inlet duct. The idle air control valve and throttle position sensor connectors.
3. Remove and mark all necessary vacuum lines. Remove and plug the 2 coolant hoses.
4. Remove the throttle, T.V. and cruise control cables.
5. Remove the throttle body retaining bolts and then remove the throttle body assembly. Discard the flange gasket.

6. Installation is the reverse order of the removal procedure. Torque the retaining bolts to 11 ft. lbs (15 Nm).

MODELS EQUIPPED WITH 3800 (VIN C) ENGINE

1. Disconnect the negative battery cable. Relieve the fuel system pressure.
2. Drain the radiator coolant. Remove the air inlet duct.
3. Remove the throttle cable, T.V. cable and the cruise control cable.
4. Disconnect the 2 vacuum hoses.
5. Disconnect the Idle Air Control (IAC), Throttle Position Sensor (TPS) and the Mass Air Flow (MAF) sensor electrical connectors.
6. Remove the 2 bolts holding the throttle body to the EGR adapter.
7. Remove the throttle assembly.
8. Clean the gasket surface on the EGR adapter and throttle body.
9. To install, use a new flange gasket.
10. Install the throttle body assembly.
11. Install the throttle body retaining bolts and torque the bolts to 11 ft. lbs. (15 Nm).
12. Connect the throttle cable, T.V. cable and the cruise control cable.
13. Connect the 2 vacuum lines to the throttle body.
14. Connect the IAC, TPS and MAF sensor electrical connectors.
15. Install the air inlet duct.
16. Refill the radiator to replace lost coolant.
17. Connect the negative battery cable. Start the engine and check for leaks.

IDLE AIR CONTROL VALVE

Removal and Installation

VEHICLES EQUIPPED WITH 3.8L (VIN 3) ENGINE

1. Remove electrical connector from idle air control valve.
2. Remove the idle air control valve using a suitable wrench.
3. Installation is the reverse of removal. Before installing the idle air control valve, measure the distance that the valve is extended. Measurement should be made from the motor housing to the end of the cone. The distance should not exceed 1⅛ in., or damage to the valve may occur when installed. Use a new gasket and turn the ignition **ON** then **OFF** again to allow the ECM to reset the idle air control valve.

NOTE: Identify replacement IAC valve as being either Type 1 (with collar at electric terminal end) or Type 2 (without collar). If measuring distance is greater than specified above, proceed as follows:

a. Type 1—Press on valve firmly to retract it.
b. Type 2—Compress retaining spring from valve while turning valve in with a clockwise motion. Return spring to original position with straight portion of spring end aligned with flat surface of valve.

NOTE: The ECM will reset the IAC whenever the ignition switch is turned ON and then OFF.

VEHICLES EQUIPPED WITH 3800 (VIN C) ENGINE

1. Disconnect the negative battery cable.
2. Disconnect the connector from the IAC valve.
3. Remove the IAC attaching screws and remove the IAC.

NOTE: Before installing the new IAC valve, measure the distance that the valve is extended. Measurement should be made from the motor housing to the end of the cone. Distance should be no greater than 1⅛ in. (28mm). If the cone is extended too far, adjustment is required or damage may occur to the valve when installed.

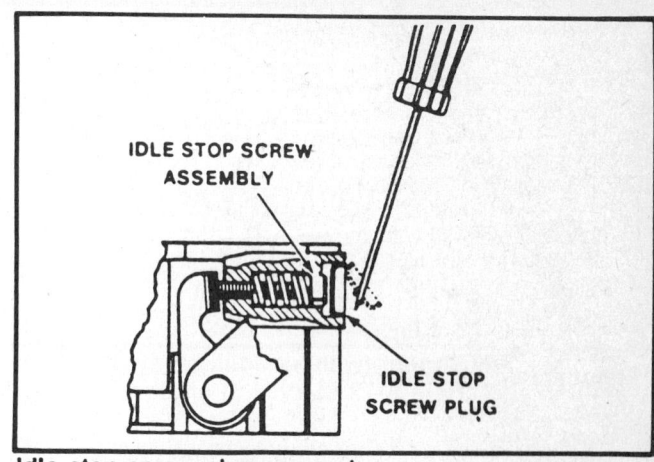

Idle stop screw plug removal

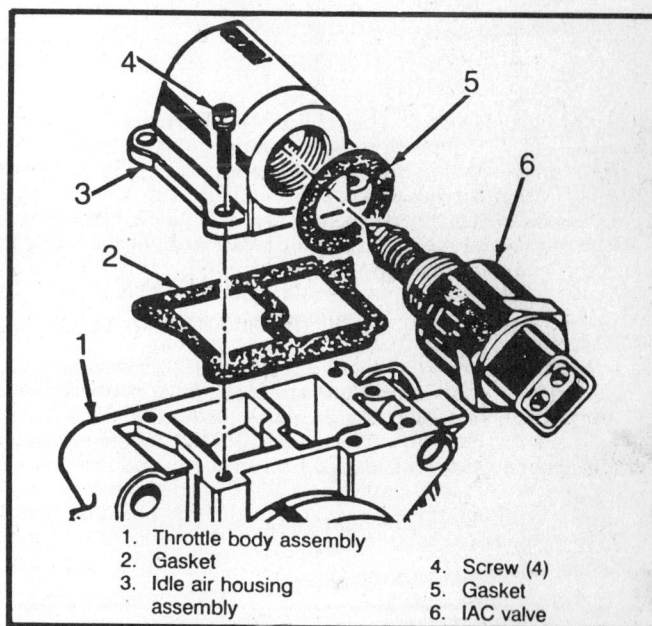

1. Throttle body assembly
2. Gasket
3. Idle air housing assembly
4. Screw (4)
5. Gasket
6. IAC valve

Idle Air Control (IAC) and housing assembly—3.8L (VIN 3)

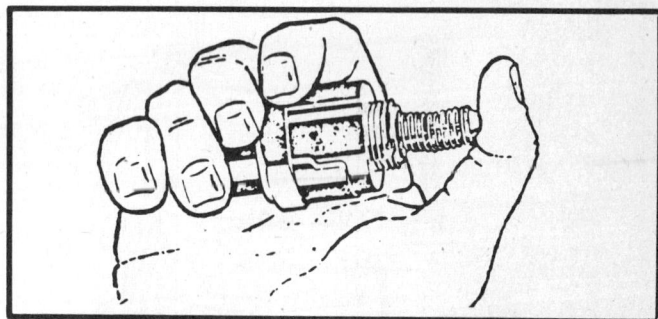

IAC valve adjustment (with collar at electric terminal end) Type 1—3.8L (VIN 3) and 3800 (VIN C) engine

4. To install, use a new O-ring on the IAC and install the IAC.
5. Install the IAC attaching screws and tighten to 13 ft. lb (18 Nm).
6. Connect the electrical connector to the IAC valve.
7. Start the engine and allow the engine to reach operating temperature.

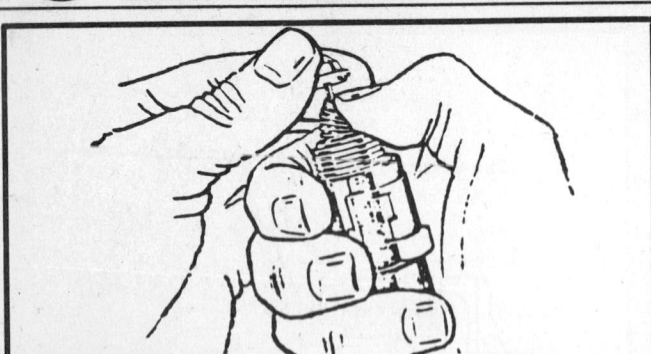

IAC valve adjustment (without collar at electric terminal end) Type 2 – 3.8L (VIN 3) engine

8. The ECM will reset the IAC, whenever the ignition switch is turned **ON** and then **OFF**.

Minimum Idle Speed Adjustment

VEHICLES EQUIPPED WITH 3.8L (VIN 3) OR 3800 (VIN C) ENGINE

The throttle stop screw that is used to adjust the idle speed of the vehicle, is adjusted to specifications at the factory. The throttle stop screw is then covered with a steel plug to prevent the unnecessary readjustment in the field. If it is necessary to gain access to the throttle stop screw, the following procedure will allow access to the throttle stop screw without removing the throttle body unit from the manifold.

1. Apply the parking brake and block the drive wheels. Remove the plug from the idle stop screw by piercing it first with a suitable tool, then applying leverage to the tool to lift the plug out.

2. Leave the Idle Air Control (IAC) valve connected and ground the diagnostic terminal (ALDL connector).

3. Turn the ignition switch to the **ON** position, do not start the engine. Wait for at least 30 seconds (this allows the IAC valve pintle to extend and seat in the throttle body).

4. With the ignition switch still in the **ON** position, disconnect IAC electrical connector.

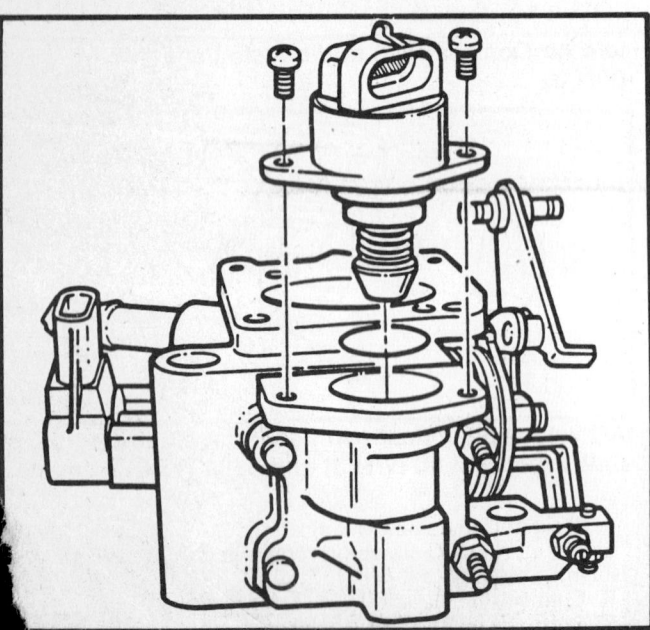

C valve servicing – 3800 (VIN C) engine

5. Remove the ground from the diagnostic terminal and start the engine. Let the engine reach normal operating temperature.

6. With the engine in the drive position adjust the idle stop screw to obtain the correct specifications.

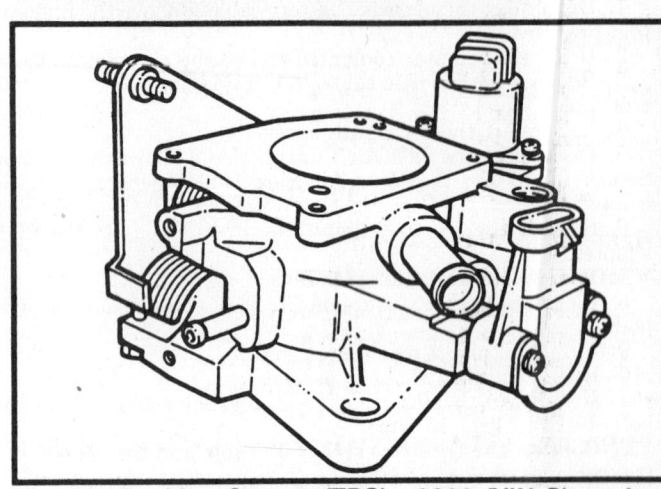

Throttle Position Sensor (TPS) – 3800 (VIN C) engine

THROTTLE POSITION SENSOR

Removal and Installation

VEHICLES EQUIPPED WITH 3.8L (VIN 3) ENGINE

1. Remove the electrical connector.
2. Remove the 2 attaching screws, lockwashers and retainers.
3. Remove the TPS.
4. Installation is the reverse of removal. Apply Loctite® or equivalent to the threads of the attaching screws. DO NOT TIGHTEN THE ATTACHING SCREWS UNTIL THE TPS HAS BEEN ADJUSTED.

VEHICLES EQUIPPED WITH 3800 (VIN C) ENGINE

1. Disconnect the negative battery cable.
2. Disconnect the electrical connector from the TPS.
3. Disconnect the PCV vent hose.
4. Remove the TPS attaching screws.
5. Remove the TPS.
6. To install, with the throttle valve in the normal closed idle position, install the TPS on the throttle body assembly, making sure the TPS pickup lever is located above the tang on the throttle actuator lever.
7. Install the TPS retainer and attaching screws using a thread locking compound on the screws. Loctite® 262 or equivalent should be used. Do not tighten the attaching screw until the TPS is adjusted.

Adjustment

VEHICLES EQUIPPED WITH 3.8L (VIN 3) ENGINE

1. With the TPS attaching screws loose, install 3 jumper wires between TPS and harness connector.

2. With the ignition switch **ON**, use a digital voltmeter connected to terminals **B** and **C** and adjust TPS to obtain 0.40 ± 0.05 volts.

3. Tighten the attaching screws, then recheck reading to insure the adjustment has not changed.

NOTE: If the TPS is being adjusted ONLY, remove screws, add thread locking compound (Loctite® or equivalent), then reinstall the screws.

4. With the ignition switch **OFF**, remove the jumper wires and connect the harness to the TPS.

VEHICLES EQUIPPED WITH 3800 (VIN C) ENGINE

1. Install 3 jumper wires between the TPS and the harness connector or use a Scan tool.
2. With the ignition switch **ON**, use a digital voltmeter connected to terminals **B** and **C** and adjust the TPS to obtain 0.33–0.46 volts.
3. Tighten the screws to 18 inch lbs. (2 Nm), then recheck the reading to insure the adjustment has not changed.
4. With the ignition switch **OFF**, remove the jumper wires and connect the harness to the TPS.

MASS AIR FLOW (MAF) SENSOR

Removal and Installation

VEHICLES EQUIPPED WITH 3.8L (VIN 3) ENGINE

1. Disconnect the negative battery cable.
2. Disconnect the electrical connector from the MAF sensor.
3. Loosen the clamps on the fresh air ducts.
4. Remove the fresh air ducts and remove the MAF sensor.
5. To install, reverse the removal procedure.

VEHICLES EQUIPPED WITH 3800 (VIN C) ENGINE

1. Disconnect the negative battery cable.
2. Loosen the clamps on the ends of the fresh air duct.
3. Remove the fresh air duct.
4. Disconnect the electrical connector from the MAF sensor.
5. Remove the 4 MAF sensor attaching screws to the throttle body.

NOTE: The MAF sensor is the electronic portion (black) plus the aluminum housing to which it is attached. (They are a matched set.)

6. Remove the MAF sensor.

NOTE: Do not attempt to remove the gasket. It is not removable. It is not serviced separately.

7. To install, position the MAF sensor onto the throttle body.
8. Install the 4 attaching screws holding the MAF sensor to the throttle body.
9. Connect the electrical connector to the MAF sensor.
10. Install the fresh air duct and clamps.
11. Connect the negative battery cable.

CRANKSHAFT SENSOR

Removal and Installation

VEHICLES EQUIPPED WITH 3.8L (VIN 3) ENGINE

1. Disconnect crank sensor harness connector.
2. Using a 28mm socket and pull handle, rotate the harmonic balancer until any window in the interrupter is aligned with the crank sensor.
3. Loosen the pinch bolt on the sensor pedestal until the sensor is free to slide in the pedestal.
4. Remove the pedestal to engine mounting bolts.
5. While manipulating the sensor within the pedestal, carefully remove the sensor and pedestal as a unit.
6. To install, loosen the pinch bolt on the new sensor pedestal until the sensor is free to slide the pedestal.
7. Verify that the window in the interrupter is still properly positioned and install sensor and pedestal as a unit while making sure that the interrupter ring is aligned within the proper slot.
8. Install pedestal to engine mounting bolts and torque to 22 ft. lbs. (30 Nm).

VEHICLES EQUIPPED WITH 3800 (VIN C) ENGINE

1. Disconnect the negative battery cable.
2. Disconnect the serpentine belt from the crankshaft pulley.

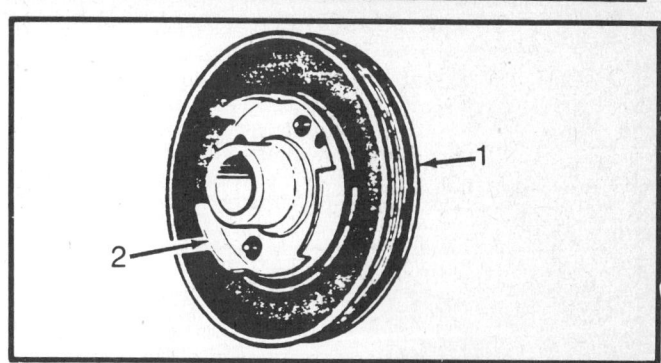

Harmonic balancer – 3.8L (VIN 3) engine

3. Raise the vehicle and support safely.
4. Remove the right front wheel assembly.
5. Remove the right inner fender access cover.
6. Remove the crankshaft harmonic balancer retaining bolt using a 28mm socket.
7. Remove the crankshaft harmonic balancer.
8. Disconnect the sensor electrical connector.
9. Remove the sensor and pedestal from the block face.
10. Remove the sensor from the pedestal.
11. To install, loosely install the crankshaft sensor on the pedestal.
12. Position the sensor with the pedestal attached on special tool J–37089 or equivalent.
13. Position the special tool on the crankshaft.
14. Install the bolts to hold the pedestal to the block and torque to 14–28 ft. lbs. (20–40 Nm).
15. Torque the pedestal pinch bolt to 30–35 inch lbs. (3–4 Nm).
16. Remove the special tool J–37089 or equivalent.
17. Place the special tool J–37089 or equivalent, on the harmonic balancer and turn. If any vane of the harmonic balancer touches the tool, replace the balancer assembly.
18. Install the balancer on the crankshaft.
19. Install the crankshaft bolt and torque to 200–239 ft. lbs. (270–325 Nm).
20. Install the inner fender shield.
21. Install the wheel assembly and lower the vehicle.
22. Install the serpentine belt.
23. Connect the negative battery cable.

Adjustment

VEHICLES EQUIPPED WITH 3.8L (VIN 3) ENGINE

1. Using a 28mm socket and pull handle, rotate the harmonic balancer until the interrupter ring(s) fills the sensor slot(s) and

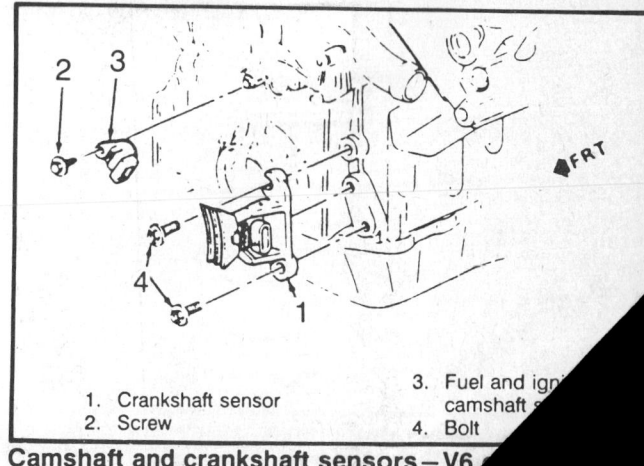

1. Crankshaft sensor
2. Screw
3. Fuel and ign[...]
   camshaft s[...]
4. Bolt

Camshaft and crankshaft sensors – V6 [...]

edge of interrupter window is aligned with edge of the deflector on the pedestal.

2. Insert adjustment tool (J–36179 or equivalent) into the gap between sensor and interrupter on each side of interrupter ring. If gauge will not slide past sensor on either side of interrupter ring, the sensor is out of adjustment or interrupter ring is bent. This clearance should be checked at 3 positions around the outer interrupter ring, approximately 120 degrees apart.

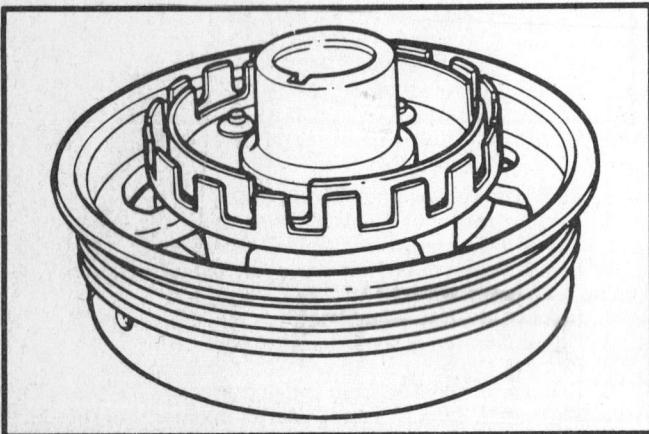

Harmonic balancer—3800 (VIN C) engine

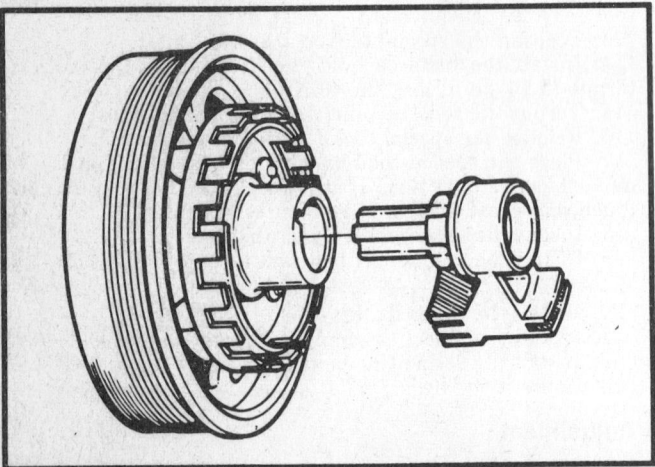

Tool J–37089 installation—3800 (VIN C) engine

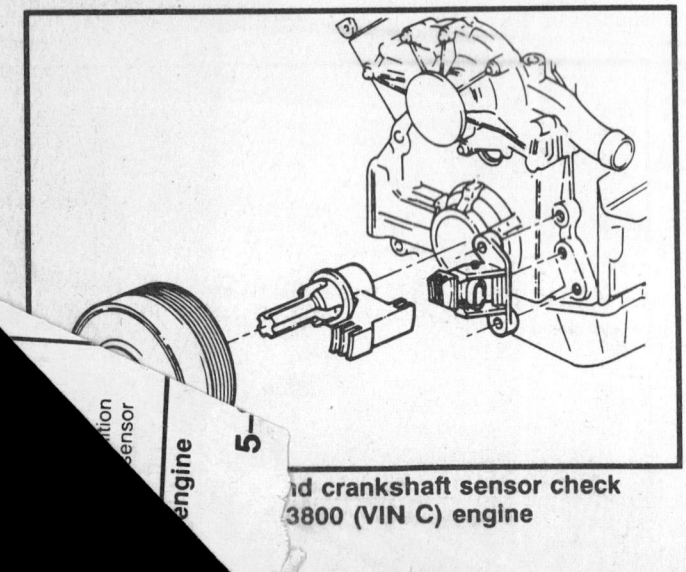

...d crankshaft sensor check
3800 (VIN C) engine

NOTE: If found out of adjustment, the sensor should be removed and inspected for potential damage.

3. Loosen the pinch bolt on sensor pedestal and insert adjustment tool (J–36179 or equivalent) into the gap between sensor and interrupter on each side of interrupter ring.

4. Slide the sensor into contact against gauge and interrupter ring.

5. Torque sensor retaining pinch bolt to 30 inch lbs. (3.4 Nm) while maintaining light pressure on sensor against gauge and interrupter ring. This clearance should be checked again, at 3 positions around the interrupter ring, approximately 120 degrees apart. If interrupter ring contacts sensor at any point during harmonic balancer rotation, the interrupter ring has excessive runout and must be replaced.

CAMSHAFT POSITION SENSOR

Removal and Installation

1. Disconnect negative battery cable.
2. Remove attaching screw securing sensor.
3. Disconnect 3 terminal sensor connector and remove sensor.
4. Installation is the reverse of removal.

OXYGEN SENSOR

Removal and Installation

NOTE: The oxygen sensor uses a permanently attached pigtail and connector. This pigtail should not be removed from the oxygen sensor. Damage or removal of the pigtail or connector could affect proper operation of the oxygen sensor.

The oxygen sensor is installed in the exhaust manifold and is removed in the same manner as a spark plug. The sensor may be difficult to remove when the engine temperature is below 120°F (48°C) (so it may be a good idea to warm the engine up for approximately 2 minutes before removing the sensor) and excessive force may damage threads in the exhaust manifold or exhaust pipe. Exercise care when handling the oxygen sensor; the electrical connector and louvered end must be kept free of grease, dirt, or other contaminants. Avoid using cleaning solvents of any kind and don't drop or roughly handle the sensor. A special anti-seize compound is used on the oxygen sensor threads when installing and care should be used NOT to get compound on the sensor itself. Disconnect the negative battery cable when servicing the oxygen sensor and torque to 30 ft. lbs. (41 Nm) when installing.

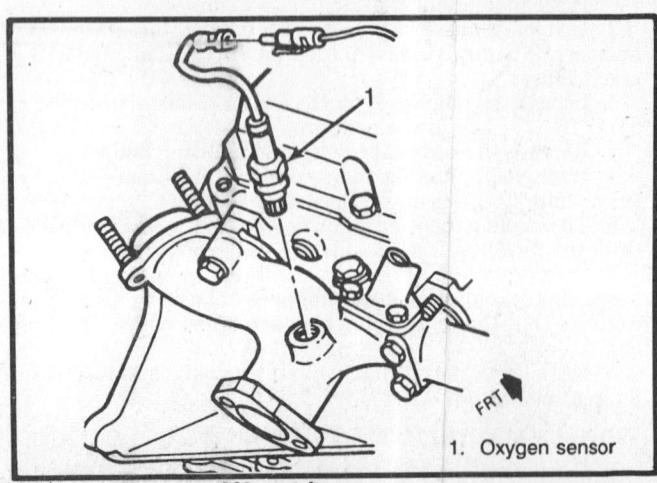

1. Oxygen sensor

Oxygen sensor—V6 engine